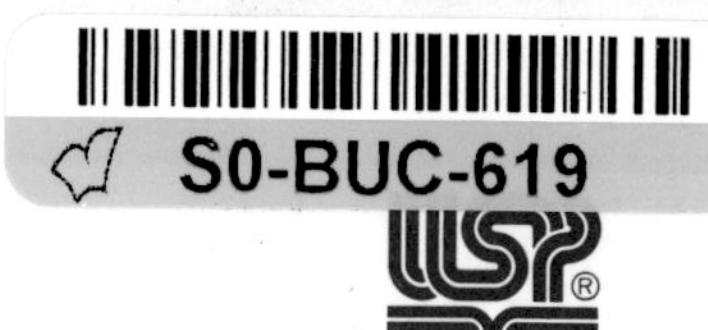

TABLE OF METRIC–APOTHECARY
Approximate Dose Equivalents

When prepared dosage forms such as tablets, capsules, etc., are prescribed in the metric system, the pharmacist may dispense the corresponding *approximate* equivalent in the apothecary system, and vice versa, as indicated in this table.

However, to calculate quantities required in pharmaceutical formulas, use the *exact* equivalents (see page 1881). For prescription compounding, use the exact equivalents rounded to three significant figures. Where expressed in the metric and apothecary systems, statements of quantity or strength in the labeling of drug products shall utilize the *exact* equivalents. (Also see *Volumetric Apparatus* ⟨31⟩, *Weights and Balances* ⟨41⟩, and *Prescription Balances and Volumetric Apparatus* ⟨1176⟩.)

NOTE—A milliliter (mL) is the *approximate* equivalent of a cubic centimeter (cc).

LIQUID MEASURE

Metric		*Apothecary*
1000	mL	1 quart
750	mL	1½ pints
500	mL	1 pint
250	mL	8 fluid ounces
200	mL	7 fluid ounces
100	mL	3½ fluid ounces
50	mL	1¾ fluid ounces
30	mL	1 fluid ounce
15	mL	4 fluid drams
10	mL	2½ fluid drams
8	mL	2 fluid drams
5	mL	1¼ fluid drams
4	mL	1 fluid dram
3	mL	45 minims
2	mL	30 minims
1	mL	15 minims
0.75	mL	12 minims
0.6	mL	10 minims
0.5	mL	8 minims
0.3	mL	5 minims
0.25	mL	4 minims
0.2	mL	3 minims
0.1	mL	1½ minims
0.06	mL	1 minim
0.05	mL	¾ minim
0.03	mL	½ minim

WEIGHT

Metric		*Apothecary*
30	g	1 ounce
15	g	4 drams
10	g	2½ drams
7.5	g	2 drams
6	g	90 grains
5	g	75 grains
4	g	60 grains (1 dram)
3	g	45 grains
2	g	30 grains (½ dram)
1.5	g	22 grains
1	g	15 grains
750	mg	12 grains
600	mg	10 grains
500	mg	7½ grains
400	mg	6 grains
300	mg	5 grains
250	mg	4 grains
200	mg	3 grains
150	mg	2½ grains
125	mg	2 grains
100	mg	1½ grains
75	mg	1¼ grains
60	mg	1 grain
50	mg	¾ grain
40	mg	⅔ grain
30	mg	½ grain
25	mg	⅜ grain
20	mg	⅓ grain
15	mg	¼ grain
12	mg	⅕ grain
10	mg	⅙ grain
8	mg	⅛ grain
6	mg	1/10 grain
5	mg	1/12 grain
4	mg	1/15 grain
3	mg	1/20 grain
2	mg	1/30 grain
1.5	mg	1/40 grain
1.2	mg	1/50 grain
1	mg	1/60 grain
800	µg	1/80 grain
600	µg	1/100 grain
500	µg	1/120 grain
400	µg	1/150 grain
300	µg	1/200 grain
250	µg	1/250 grain
200	µg	1/300 grain
150	µg	1/400 grain
120	µg	1/500 grain
100	µg	1/600 grain

1990

USP XXII
NF XVII

The United States Pharmacopeia
The National Formulary

By authority of the United States Pharmacopeial Convention, Inc., meeting at Washington, D.C., March 22–24, 1985. Prepared by the Committee of Revision and published by the Board of Trustees

Official from January 1, 1990

UNITED STATES PHARMACOPEIAL CONVENTION, INC.
12601 Twinbrook Parkway, Rockville, MD 20852

NOTICE AND WARNING

Concerning U. S. Patent or Trademark Rights

The inclusion in the Pharmacopeia or in the National Formulary of a monograph on any drug in respect to which patent or trademark rights may exist shall not be deemed, and is not intended as, a grant of, or authority to exercise, any right or privilege protected by such patent or trademark. All such rights and privileges are vested in the patent or trademark owner, and no other person may exercise the same without express permission, authority, or license secured from such patent or trademark owner.

Concerning Use of USP or NF Text

Attention is called to the fact that USP and NF text is fully copyrighted. Authors and others wishing to use portions of the text should request permission to do so from the Secretary of the USPC Board of Trustees.

12601 Twinbrook Parkway, Rockville, MD 20852

Library of Congress Catalog Card Number 83-640088
ISSN 0195-7996
ISBN 0-913595-37-3 (cloth)
0-913595-38-1 (leather)

Printed by Mack Printing Company, Easton, PA 18042

USP XXII

THE UNITED STATES PHARMACOPEIA

Official from January 1, 1990

TWENTY-SECOND REVISION

12601 Twinbrook Parkway, Rockville, MD 20852

Contents

USP XXII

People

UNITED STATES PHARMACOPEIAL CONVENTION 1985–1990

Officers

Arthur Hull Hayes, Jr., M.D.
President
Hawthorne, NY

Joseph M. Benforado, M.D.
Vice-President
Madison, WI

Paul F. Parker, D.Sc.
Treasurer
Lexington, KY

Frederick E. Shideman, M.D., Ph.D.[6]
Past President
Minneapolis, MN

William M. Heller, Ph.D.
Secretary
Rockville, MD

Board of Trustees

John T. Fay, Jr., Ph.D.[1]
Chairman
Orange, CA

Joseph M. Benforado, M.D.
ex officio
Madison, WI

John V. Bergen, Ph.D.[1]
Villanova, PA

Joseph P. Buckley, Ph.D.[2]
Houston, TX

Estelle G. Cohen, M.A.[3]
Baltimore, MD

J. Richard Crout, M.D.[4]
Rockville, MD

James T. Doluisio, Ph.D.[2]
Austin, TX

Arthur Hull Hayes, Jr., M.D.
ex officio
Hawthorne, NY

Leo E. Hollister, M.D.[4]
Houston, TX

Paul F. Parker, D.Sc.
ex officio
Lexington, KY

Frederick E. Shideman, M.D., Ph.D.[6]
(1985–1988)
ex officio
Minneapolis, MN

General Committee of Revision

William M. Heller, Ph.D., *Executive Director, USPC*, 12601 Twinbrook Parkway, Rockville, MD 20852
Chairman, ex officio

Lee T. Grady, Ph.D.[5]
Director, Drug Standards Division

Keith W. Johnson[5]
Director, Research and Development, Drug Information Division

Faye G. Abdellah, R.N., Ed.D., Sc.D.
Rockville, MD

Jerry R. Allison, Ph.D.
Syracuse, NY

Thomas J. Ambrosio, Ph.D.
Somerville, NJ

Mary A. Amini, Ph.D.
(1985–1988)
Cranbury, NJ

William F. Appel, D.Sc.
St. Paul, MN

John W. ApSimon, Ph.D.
(1988–)
Ottawa, Ontario, Canada

Norman W. Atwater, Ph.D.
Princeton, NJ

Gilbert S. Banker, Ph.D.
Minneapolis, MN

Donald R. Bennett, M.D., Ph.D.
Chicago, IL

Martin I. Blake, Ph.D.
Skokie, IL

Judy P. Boehlert, Ph.D.
(1988–)
Nutley, NJ

Saul Boyarsky, M.D., J.D.
St. Louis, MO

James C. Boylan, Ph.D.
Abbott Park, IL

Lynn R. Brady, Ph.D.
Seattle, WA

---

[1] At large.
[2] Representing pharmacy.
[3] Public member.
[4] Representing medicine.
[5] 12601 Twinbrook Parkway, Rockville, MD 20852.
[6] Deceased.

Edwin D. Bransome, Jr., M.D.
Augusta, GA

William H. Briner, Capt., B.S.
Durham, NC

Herbert S. Carlin, D.Sc.
Chappaqua, NY

Lester Chafetz, Ph.D.
Kansas City, MO

Virginia C. Chamberlain
Silver Spring, MD

Zak T. Chowhan, Ph.D.
Palo Alto, CA

Sebastian G. Ciancio, D.D.S.
Buffalo, NY

Jordan L. Cohen, Ph.D.
Lexington, KY

Murray S. Cooper, Ph.D.
Islamarada, FL

Lloyd E. Davis, D.V.M., Ph.D.
Urbana, IL

Clyde R. Erskine, B.S., M.B.A.
Philadelphia, PA

James D. Finkelstein, M.D.
Washington, DC

Harry W. Fischer, M.D.
Rochester, NY

Edward A. Fitzgerald, Ph.D.
Bethesda, MD

Douglas R. Flanagan, Ph.D.
Iowa City, IA

John P. Fletcher, M.S.
South Charleston, WV

Klaus G. Florey, Ph.D.
New Brunswick, NJ

William O. Foye, Ph.D.
Boston, MA

Edward D. Frohlich, M.D.
New Orleans, LA

Salvatore A. Fusari, Ph.D.
St. Clair Shores, MI

Joseph F. Gallelli, Ph.D.
Bethesda, MD

Burton J. Goldstein, M.D.
Miami, FL

Alan Gray, Ph.D.
Wyncote, PA

Marvin F. Grostic, Ph.D.
Kalamazoo, MI

Michael J. Groves, Ph.D.
Chicago, IL

J. Keith Guillory, Ph.D.
Iowa City, IA

John J. Halki, M.D., Ph.D.
Dayton, OH

Samir A. Hanna, Ph.D.
Syracuse, NY

Stanley L. Hem, Ph.D.
West Lafayette, IN

David W. Hughes, Ph.D.
Ottawa, Ontario, Canada

Rodney D. Ice, Ph.D.
Quapaw, OK

Norman C. Jamieson, Ph.D.
(*1988–*)
St. Louis, MO

Raymond N. Johnson, Ph.D.
Rouses Point, NY

Herbert E. Kaufman, M.D.
New Orleans, LA

Donald Kaye, M.D.
Philadelphia, PA

William J. Keller, Ph.D.
(*1985–1988*)
Monroe, LA

B. J. Kennedy, M.D.
Minneapolis, MN

Jay S. Keystone, M.D.
Toronto, Ontario, Canada

Boen T. Kho, Ph.D.
Plattsburgh, NY

Lewis J. Leeson, Ph.D.
Summit, NJ

Herbert Letterman, Ph.D.
(*1988–*)
Somerville, NJ

Robert D. Lindeman, M.D.
Albuquerque, NM

Charles H. Lochmuller, Ph.D
Durham, NC

Dean H. Lockwood, M.D.
Rochester, NY

Jennifer M. H. Loggie, M.D.
Cincinnati, OH

Edward G. Lovering, Ph.D.
Ottawa, Ontario, Canada

John R. Markus, B.S.
Rockville, MD

Robert A. Mathews, Ph.D.
(*1985–1986*)
Washington, DC

Warren A. McAllister, Ph.D.
Greenville, NC

Thomas Medwick, Ph.D.
Piscataway, NJ

Robert F. Morrissey, Ph.D.
Somerville, NJ

Terry E. Munson, B.S.
Fairfax, VA

Harold R. Nace, Ph.D.
Barrington, RI

Wendel L. Nelson, Ph.D.
Seattle, WA

Sharon C. Northup, Ph.D.
(*1986–*)
Round Lake, IL

Stanley P. Owen, Ph.D.
Federal Way, WA

James W. Pate, M.D.
Memphis, TN

Garnet E. Peck, Ph.D.
West Lafayette, IN

Robert V. Petersen, Ph.D.
Salt Lake City, UT

James R. Rankin, B.S.
Santa Fe, NM

Christopher T. Rhodes, Ph.D.
Kingston, RI

Jay Roberts, Ph.D.
Philadelphia, PA

Joseph R. Robinson, Ph.D.
Madison, WI

Theodore J. Roseman, Ph.D.
Morton Grove, IL

Leonard P. Rybak, M.D., Ph.D.
Springfield, IL

Andrew J. Schmitz, Jr., M.S.
New York, NY

Stephen G. Schulman, Ph.D.
Gainesville, FL

Ralph F. Shangraw, Ph.D.
Baltimore, MD

Albert L. Sheffer, M.D.
Boston, MA

Eli Shefter, Ph.D.
South San Francisco, CA

Eric B. Sheinin, Ph.D.
Rockville, MD

Jane C. Sheridan, Ph.D.
(*1985–1988*)
Nutley, NJ

Edward B. Silberstein, M.D.
Cincinnati, OH

Joseph E. Sinsheimer, Ph.D.
Ann Arbor, MI

Marilyn Dix Smith, Ph.D.
Stony Point, NY

Robert V. Smith, Ph.D.
Pullman, WA

E. John Staba, Ph.D.
Minneapolis, MN

Robert S. Stern, M.D.
Boston, MA

James T. Stewart, Ph.D.
Athens, GA

Murray M. Tuckerman, Ph.D.
Philadelphia, PA

Samuel M. Tuthill, Ph.D.
(*1985–1988*)
St. Louis, MO

Gerald W. Wallace, Ph.D.
Indianapolis, IN

Walter L. Way, M.D.
San Francisco, CA

Robert G. Wolfangel, Ph.D.
St. Louis, MO

Lynn C. Yeoman, Ph.D.
Houston, TX

George Zografi, Ph.D.
Madison, WI

Executive Committee of Revision

WILLIAM M. HELLER, PH.D., *Chairman of the Executive Committee of Revision*

The members of the General Committee of Revision who served annual terms on the Executive Committee of Revision during the present revision cycle are designated below.

(1984–1985)	(1985–1986)	(1986–1987)	(1987–1988)	(1988–1989)
Martin I. Blake	Martin I. Blake	Martin I. Blake	Martin I. Blake	Martin I. Blake
Herbert S. Carlin	Herbert S. Carlin	William H. Briner	William H. Briner	James R. Rankin
Klaus G. Florey	Marvin F. Grostic	Marvin F. Grostic	Marvin F. Grostic	Robert S. Stern
Jennifer M.H. Loggie	James R. Rankin	James R. Rankin	James R. Rankin	Murray M. Tuckerman
James R. Rankin	Murray M. Tuckerman	Murray M. Tuckerman	Murray M. Tuckerman	Walter L. Way
Jane C. Sheridan	Walter L. Way	Walker L. Way	Walter L. Way	George Zografi

USP Drug Nomenclature Committee for 1985–1990

WILLIAM M. HELLER, PH.D., *Chairman*
Donald R. Bennett, M.D., Ph.D.; Herbert S. Carlin, D.Sc.; Lester Chafetz, Ph.D.; Lloyd E. Davis, D.V.M., Ph.D.; John J. Halki, M.D., Ph.D.; David W. Hughes, Ph.D.; Andrew J. Schmitz, Jr., M.S.; Ralph F. Shangraw, Ph.D.; Gerald W. Wallace, Ph.D.

Drug Standards Division Executive Committee and Subcommittees for 1985–1990

WILLIAM M. HELLER, PH.D., *Chairman of the Drug Standards Division Executive Committee*

The Chairmen of the Subcommittees and, ex officio, the Executive Director, constitute the Drug Standards Division Executive Committee.

ANT. Antibiotics

KLAUS G. FLOREY, PH.D., *Chairman*
Lynn R. Brady, Ph.D.; Salvatore A. Fusari, Ph.D.; Samir A. Hanna, Ph.D.; David W. Hughes, Ph.D.; Marilyn Dix Smith, Ph.D.

B&M. Biochemistry and Microbiology

ANDREW J. SCHMITZ, JR., M.S., *Chairman*
Edwin D. Bransome, Jr., M.D.; Virginia C. Chamberlain; Murray S. Cooper, Ph.D.; Edward A. Fitzgerald, Ph.D.; Alan Gray, Ph.D.; Robert F. Morrissey, Ph.D.; Terry E. Munson, B.S.; Lynn C. Yeoman, Ph.D.

CH1. Steroids

LESTER CHAFETZ, PH.D., *Chairman*
William O. Foye, Ph.D.; Charles H. Lochmuller, Ph.D.; Harold R. Nace, Ph.D.; Marilyn Dix Smith, Ph.D.

CH2. Inorganics

JAMES T. STEWART, PH.D., *Chairman*
Stanley L. Hem, Ph.D.; Norman C. Jamieson, Ph.D. (1988–); John R. Markus, B.S.; Warren A. McAllister, Ph.D.; Eric B. Sheinin, Ph.D.; Samuel M. Tuthill, Ph.D. (1985–1988)

CH3. Chemistry

NORMAN W. ATWATER, PH.D., *Chairman*
Douglas R. Flanagan, Ph.D.; Raymond N. Johnson, Ph.D.; Wendel L. Nelson, Ph.D.; Robert V. Smith, Ph.D.

CH4. Natural Products

GERALD W. WALLACE, PH.D., *Chairman*
John W. ApSimon, Ph.D. (1988–); Lynn R. Brady, Ph.D.; William J. Keller, Ph.D. (1985–1988); Edward G. Lovering, Ph.D.; Joseph E. Sinsheimer, Ph.D.; E. John Staba, Ph.D.

CH5. Chemistry

J. KEITH GUILLORY, PH.D., *Chairman*
Jerry R. Allison, Ph.D.; Judy P. Boehlert, Ph.D. (1988–); Raymond N. Johnson, Ph.D. (1985–1988); Boen T. Kho, Ph.D.; Stephen G. Schulman, Ph.D.

C&M. Containers and Materials

THOMAS MEDWICK, PH.D., *Chairman*
Thomas J. Ambrosio, Ph.D.; Mary A. Amini, Ph.D. (1985–1988); Herbert S. Carlin, D.Sc.; Clyde R. Erskine, B.S., M.B.A.; Herbert Letterman, Ph.D. (1988–)

GCR. General Chapters and Reagents

DAVID W. HUGHES, PH.D., *Chairman (1988–)*
SAMUEL M. TUTHILL, PH.D., *Chairman (1985–1988)*
Martin I. Blake, Ph.D.; Marvin F. Grostic, Ph.D.; Norman C. Jamieson, Ph.D. (1988–); Robert A. Mathews, Ph.D. (1985–1986); Murray M. Tuckerman, Ph.D.

HHC. Home Health Care

WILLIAM F. APPEL, D.SC., *Chairman*
James C. Boylan, Ph.D.; Herbert S. Carlin, D.Sc.; Michael J. Groves, Ph.D.; Robert D. Lindeman, M.D.

IVT. In-vitro Toxicity

ROBERT V. PETERSEN, PH.D., *Chairman*
Lloyd E. Davis, D.V.M., Ph.D. (1985–1987); David W. Hughes, Ph.D.; Robert A. Mathews, Ph.D. (1985–1986); Sharon C. Northup, Ph.D. (1986–); E. John Staba, Ph.D.; Lynn C. Yeoman, Ph.D.

M&S. Medical and Surgical Products

STANLEY P. OWEN, PH.D., *Chairman*
Thomas J. Ambrosio, Ph.D.; Robert F. Morrissey, Ph.D.; Eli Shefter, Ph.D.; Robert G. Wolfangel, Ph.D.

PAR. Parenteral Products

JOSEPH F. GALLELLI, PH.D., *Chairman*
James C. Boylan, Ph.D.; Herbert S. Carlin, D.Sc.; Michael J. Groves, Ph.D.; Terry E. Munson, B.S.; Theodore J. Roseman, Ph.D.

PH1. Dosage Forms and Excipients

RALPH F. SHANGRAW, PH.D., *Chairman*
Gilbert S. Banker, Ph.D.; Zak T. Chowhan, Ph.D.; John P. Fletcher, M.S.; Garnet E. Peck, Ph.D.; George Zografi, Ph.D.

PH2. Biopharmaceutics

JORDAN L. COHEN, PH.D., *Chairman*
Lewis J. Leeson, Ph.D.; Christopher T. Rhodes, Ph.D.; Joseph R. Robinson, Ph.D.; Theodore J. Roseman, Ph.D.; Eli Shefter, Ph.D.

PUR. Chemical Purity

MARVIN F. GROSTIC, *Chairman (1988–)*
JANE C. SHERIDAN, PH.D., *Chairman (1985–1988)*
John P. Fletcher, M.S.; David W. Hughes, Ph.D. (1985–1988); Edward G. Lovering, Ph.D. (1988–); Sharon C. Northup, Ph.D. (1986–)

RAD. Radiopharmaceuticals

ROBERT G. WOLFANGEL, PH.D., *Chairman*
William H. Briner, Capt., B.S.; Rodney D. Ice, Ph.D.; Edward B. Silberstein, M.D.

USP Reference Standards Committee

LEE T. GRADY, PH.D., *Chairman, ex officio*
Norman W. Atwater, Ph.D.; Lester Chafetz, Ph.D.; Klaus G. Florey, Ph.D.; Joseph F. Gallelli, Ph.D.; J. Keith Guillory, Ph.D.; Andrew J. Schmitz, Jr.; James T. Stewart, Ph.D.; Samuel M. Tuthill, Ph.D., (1985–1988); Gerald W. Wallace, Ph.D.

USP/FDA Antibiotic Monograph Subcommittee for 1985–1990

WILLIAM W. WRIGHT, PH.D., *Chairman*

Representing FDA, Rockville, MD:
Patricia N. Cushing (1985–1986); Peter Dionne; Katharine A. Freeman; Bernard Goldstein (1985–1986); Joseph H. Graham, Ph.D.; Richard Norton; Robert A. Rippere; Lola G. Wayland

Representing USP, Rockville, MD:
Roger Dabbah, Ph.D.; Lee T. Grady, Ph.D.; Aubrey S. Outschoorn, Lic. Med. Surg., Ph.D.

Drug Standards Division Advisory Panels for 1985–1990

Panel on Biological Tests and Assays

TERRY E. MUNSON, B.S., *Chairman*, Fairfax, VA; Robert L. Amos, Indianapolis, IN; Suggy Chrai, New Brunswick, NJ; Janet C. Curry, New York, NY; Herbert N. Prince, Ph.D., Fairfield, NJ

Panel on Biotechnology-derived Products

ANDREW J. SCHMITZ, JR., M.S., *Chairman*, New York, NY; Robert L. Garnick, Ph.D., South San Francisco, CA; John B. Landis, Ph.D., Kalamazoo, MI; Frederick C. Pearson, Ph.D., Boulder, CO; Stanley Stein, Piscataway, NJ; Kathryn Zoon, Ph.D., Bethesda, MD

Panel on Bulk Packaging

GARNET E. PECK, PH.D., *Chairman*, West Lafayette, IN; Terry Benney, Ph.D., Philadelphia, PA; Gregory Haines, Union, NJ; Gordon E. Mallett, Ph.D., Indianapolis, IN

Panel on Home Health Care

JOHN LEVCHUK, PH.D., *Chairman*, Rockville, MD; Patricia DeLaPointe, Hibbing, MN; Paulette Egging, Portland, OR; Kenneth M. Hale, Columbus, OH; Lawrence A. Robinson, Memphis, TN

Panel on In-vitro Pregnancy Tests

ROBERT F. MORRISSEY, PH.D., *Chairman*, Somerville, NJ; Kaiser Aziz, Ph.D., Silver Spring, MD; Irwin Hindberg, Ph.D., Ottawa, Ontario, Canada; Leonard L. Kaplan, Ph.D., Raritan, NJ; Norman P. Kubasik, Ph.D., Rochester, NY

Panel on In-vitro Toxicity Testing

ROBERT V. PETERSEN, PH.D., *Chairman*, Salt Lake City, UT; Scott A. Burton, Ph.D., St. Paul, MN; Thomas F. Genova, Ph.D., Somerville, NJ; Alan Goldberg, Ph.D., Baltimore, MD; Richard D. Henry, Philadelphia, PA; Barbara H. Keech, Vienna, VA; Tibor Matula, Ph.D., Ottawa, Ontario, Canada; Jerry Nelson, Ph.D., Salt Lake City, UT; Daniel L. Prince, Ph.D., Fairfield, NJ; Thomas D. Sabourin, Ph.D., Columbus, OH; Adelbert L. Stagg, Ph.D., Research Triangle Park, NC; Dennis M. Stark, D.V.M., Ph.D., New York, NY; Richard F. Wallin, D.V.M., Ph.D., Northwood, OH

Panel on Moisture Specifications

GEORGE ZOGRAFI, PH.D., *Chairman*, Madison, WI; R. Gary Hollenbeck, Ph.D., Baltimore, MD; Sharon Laughlin, Ph.D., Groton, CT; Mikal J. Pikal, Ph.D., Indianapolis, IN; Joseph B. Schwartz, Ph.D., Philadelphia, PA; Lynn Van Campen, Ph.D., Ridgefield, CT

Panel on Radiopharmaceuticals

WILLIAM H. BRINER, CAPT., B.S., *Chairman*, Durham, NC; Jacqueline M. Calhoun, Gaithersburg, MD; Paul Early, Cleveland, OH; Ronald D. Finn, Ph.D., Bethesda, MD; Carol S. Marcus, Ph.D., M.D., Torrance, CA

Panel on Sterility and Microbial Attributes

MURRAY S. COOPER, PH.D., *Chairman*, Islamarada, FL; Frank W. Adair, Ph.D., Summit, NJ; William C. Alegnani, Ph.D., Rochester, MI; R. Michael Enzinger, Ph.D., Kalamazoo, MI; Henry Jarocha, R.Ph., Rochester, NY; Eugene A. Timm, Ph.D., Rochester, MI; C. Searle Wadley, North Chicago, IL

Panel on Sterilization Indicators

VIRGINIA C. CHAMBERLAIN, *Chairman*, Silver Spring, MD; Henry L. Avallone, North Brunswick, NJ; David Bekus, Somerville, NJ; Robert Berube, Ph.D., St. Paul, MN; Frank B. Engley, Jr., Ph.D., Columbia, MO; Gary S. Graham, Ph.D., Erie, PA; Lois A. Jones, Research Triangle Park, NC; Karl Kereluk, Ph.D.,* Rouses Point, NY; Ruth B. Kundsin, Sc.D., Boston, MA; Patrick McCormick, Ph.D., Rochester, NY; Gregg A. Mosley, Belgrade, MT; Theron E. Odlaug, Ph.D., Elkhart, IN; Gordon S. Oxborrow, Minneapolis, MN

Drug Information Division Executive Committee for 1985–1990

WILLIAM M. HELLER, PH.D., *Chairman of the Drug Information Division Executive Committee*
Faye G. Abdellah, R.N., Ed.D., Sc.D., William F. Appel, D.Sc., James C. Boylan, Ph.D., Herbert S. Carlin, D.Sc., Sebastian G. Ciancio, D.D.S., Lloyd E. Davis, D.V.M., Ph.D., Edward D. Frohlich, M.D., Jay S. Keystone, M.D., Jennifer M. H. Loggie, M.D., James R. Rankin, B.S., Albert L. Sheffer, M.D., Robert V. Smith, Ph.D.

Drug Information Division Advisory Panels for 1985–1990

The Chairman of each Panel is a member of the General Committee of Revision.

Panel on Anesthesiology

WALTER L. WAY, M.D., *Chairman*, San Francisco, CA; Frederic Berry, M.D., Charlottesville, VA; Roy Cronnelly, M.D., Ph.D., Somerset, CA; Dennis Mangano, M.D., Ph.D., San Francisco, CA; W. Jerry Merrell, M.D., Gainesville, FL; Carl Rosow, M.D., Ph.D., Boston, MA; Bradley Smith, M.D., Nashville, TN; Paul White, M.D., Ph.D., St. Louis, MO

Panel on Cardiovascular and Renal Drugs

EDWARD D. FROHLICH, M.D., *Chairman*, New Orleans, LA; Emmanuel L. Bravo, M.D., Cleveland, OH; James E. Doherty, M.D., Little Rock, AR; Garabed Eknoyan, M.D., Houston, TX; Ruth Eshleman, Ph.D., West Kingston, RI; Edward Genton, M.D., New Orleans, LA; Thomas M. Glenn, Ph.D., South San Francisco, CA; Norman K. Hollenberg, M.D., Ph.D., Boston, MA; John L. Juergens, M.D., Rochester, MN; Michael Lesch, M.D., Chicago, IL; Benjamin F. McGraw, Pharm.D., Kansas City, MO; Patrick A. McKee, M.D., Oklahoma City, OK; Bernard L. Mirkin, Ph.D., M.D., Minneapolis, MN; Burton E. Sobel, M.D., St. Louis, MO; W. David Watkins, M.D., Ph.D., Durham, NC

Panel on Clinical Immunology

ALBERT L. SHEFFER, M.D., *Chairman*, Boston, MA; John Baum, M.D., Rochester, NY; Jonathan S. Coblyn, M.D., Brookline, MA; Elliot F. Ellis, M.D., Buffalo, NY; Patricia A. Fraser, M.D., Boston, MA; Thomas Gilman, Pharm.D., Los Angeles, CA;

*Deceased.

Stephen R. Kaplan, M.D., Providence, RI; Sandra Koehler, Milwaukee, WI; Floyd Malveaux, M.D., Ph.D., Washington, DC; Edward B. Nelson, M.D., Ph.D., Houston, TX; Robert E. Reisman, M.D., Buffalo, NY; Daniel J. Stechschulte, M.D., Kansas City, KS; Martin D. Valentine, M.D., Baltimore, MD

Panel on Consumer Interest

JAMES RANKIN, B.S., *Chairman*, Santa Fe, NM; Judith Brown, B.A., Washington, DC; Jose Camacho, Austin, TX; Margaret A. Charters, Ph.D., Syracuse, NY; Jennifer Cross, San Francisco, CA; Gabriel Daniel, Washington, DC; John Forbes, Rio Piedras, PR; Jerome Halperin, Edison, NJ; Anita M. LeValdo, Window Rock, AZ; Janice Lieberman, Buffalo, NY; Esther Peterson, Washington, DC; Ruth Richards, M.A., M.P.H., Los Angeles, CA; T. Donald Rucker, Ph.D., Chicago, IL; Gordon Schiff, M.D., Chicago, IL

Panel on Dentistry

SEBASTIAN G. CIANCIO, D.D.S., *Chairman*, Buffalo, NY; Donald F. Adams, D.D.S., Portland, OR; Karen Baker, M.S., Iowa City, IA; Priscilla C. Bourgault, Ph.D., Maywood, IL; Frederick A. Curro, D.M.D., Ph.D., Jersey City, NJ; Phyllis Eliasberg, Boston, MA; Tommy W. Gage, D.D.S., Ph.D., Dallas, TX; Stephen F. Goodman, D.D.S., New York, NY; Zack Kasloff, D.D.S., Winnipeg, Manitoba, Canada; Joseph Margarone, D.D.S., Buffalo, NY; Michael Newman, D.D.S., Los Angeles, CA; James W. Smudski, D.M.D., Ph.D., Pittsburgh, PA; Clarence L. Trummel, D.D.S., Ph.D., Farmington, CT; Raymond P. White, Jr., D.D.S., Ph.D., Chapel Hill, NC

Panel on Dermatology

ROBERT S. STERN, M.D., *Chairman*, Boston, MA; Richard D. Baughman, M.D., Hanover, NH; Michael Bigby, M.D., Brookline, MA; Henry Jolly, M.D., New Orleans, LA; W. Stuart Maddin, M.D., F.R.C.P., Vancouver, British Columbia, Canada; Milton Orkin, M.D., Minneapolis, MN; Edgar Benton Smith, M.D., Galveston, TX; John Strauss, M.D., Iowa City, IA; Dennis West, M.S., Chicago, IL; Gail Zimmerman, B.A., Portland, OR

Panel on Diagnostic Agents—Nonradioactive

HARRY W. FISCHER, M.D., *Chairman*, Rochester, NY; James A. Nelson, M.D., Seattle, WA; Robert L. Siegle, M.D., San Antonio, TX; Jovitas Skucas, M.D., Rochester, NY; William M. Thompson, M.D., Minneapolis, MN; Gerald L. Wolf, Ph.D., M.D., Pittsburgh, PA

Panel on Endocrinology

DEAN H. LOCKWOOD, M.D., *Chairman*, Rochester, NY; Louis V. Avioli, M.D., St. Louis, MO; Edwin D. Bransome, Jr., M.D., Augusta, GA; P. Reed Larsen, M.D., Boston, MA; Marvin E. Levin, M.D., St. Louis, MO; Marvin M. Lipman, M.D., Mt. Vernon, NY; Walter J. Meyer, III., M.D., Galveston, TX; Rita Nemchik, R.N., M.S., Philadelphia, PA; Maria New, M.D., New York, NY; John A. Owen, M.D., Charlottesville, VA; Robert W. Rebar, M.D., Cincinnati, OH; Thomas H. Wiser, Pharm.D., Buies Creek, NC

Panel on Family Practice

DONALD R. BENNETT, M.D., PH.D., *Chairman*, Chicago, IL; Paul C. Brucker, M.D., Philadelphia, PA; Allan H. Bruckheim, M.D., Harrison, NY; Robert E. Davis, Pharm.D., Englewood, CO; Robert Guthrie, M.D., Columbus, OH; Marlene Haffner, M.D., Rockville, MD; Edward L. Langston, M.D., R.Ph. Flora, IN; Jack M. Rosenberg, Pharm.D., Ph.D., Brooklyn, NY; John Thornburg, D.O., Ph.D., East Lansing, MI

Panel on Gastroenterology

JAMES D. FINKELSTEIN, M.D., *Chairman*, Washington, DC; Rosemary R. Berardi, Pharm.D., Ann Arbor, MI; James J. Cerda, M.D., Gainesville, FL; Gerald Friedman, M.D., New York, NY; Donald J. Glotzer, M.D., Boston, MA; Louis Y. Korman, M.D., Washington, DC; Suzanne Rosenthal, New York, NY

Panel on Geriatrics

JAY ROBERTS, PH.D., *Chairman*, Philadelphia, PA; Susan D. Abbott, M.P.A., Washington, DC; Benjamin Calesnick, M.D., Philadelphia, PA; Philip P. Gerbino, Pharm.D., Philadelphia, PA; David J. Greenblatt, M.D., Boston, MA; Peter P. Lamy, Ph.D., Baltimore, MD; Joseph V. Levy, Ph.D., San Francisco, CA; John O. Lindower, M.D., Dayton, OH; Marcus M. Reidenberg, M.D., New York, NY; Daphne A. Roe, M.D., Ithaca, NY; John W. Rowe, M.D., Boston, MA; Alexander M. M. Shepherd, M.D., Ph.D., San Antonio, TX; William Simonson, Pharm.D., Portland, OR; Betsy Todd, R.N., Bronx, NY; Elliot S. Vesell, M.D., Hershey, PA; Robert E. Vestal, M.D., Boise, ID

Panel on Hematologic and Neoplastic Disease

B. J. KENNEDY, M.D., *Chairman*, Minneapolis, MN; Laurence H. Baker, D.O., Detroit, MI; Barbara D. Blumberg, Philadelphia, PA; Helene G. Brown, B.S., Los Angeles, CA; M. Robert Cooper, M.D., Winston-Salem, NC; William Dana, Pharm.D., Houston, TX; William E. Evans, Pharm. D., Memphis, TN; John Lazo, M.D., Pittsburgh, PA; Barbara R. Medvec, R.N., B.S.N., Ann Arbor, MI; Grace Powers Monaco, Washington, DC; John Penner, M.D., East Lansing, MI; Samuel G. Taylor, IV, M.D., Chicago, IL; John Yarbro, M.D., Ph.D., Columbia, MO

Panel on Infectious Disease Therapy

DONALD KAYE, M.D., *Chairman*, Philadelphia, PA; C. Glenn Cobbs, M.D., Birmingham, AL; John J. Dennehy, M.D., Danville, PA; Earl H. Freimer, M.D., Toledo, OH; Susan M. Garabedian-Ruffalo, Pharm.D., Glendale, CA; Calvin M. Kunin, M.D., Columbus, OH; John D. Nelson, M.D., Dallas, TX; Reuben Ramphal, M.D., Gainesville, FL; Harold C. Standiford, M.D., Baltimore, MD; Roy T. Steigbigel, M.D., Stony Brook, NY; Paul F. Wehrle, M.D., Los Angeles, CA

Panel on Neurological and Psychiatric Disease

BURTON J. GOLDSTEIN, M.D., *Chairman*, F.A.C.P., Miami, FL; Alex A. Cardoni, M.S., Hartford, CT; James C. Cloyd, Pharm.D., Minneapolis, MN; N. Michael Davis, M.S., Miami, FL; Richard Dorsey, M.D., Newport Beach, CA; Larry Ereshefsky, Pharm.D., San Antonio, TX; W. Edwin Fann, M.D., Houston, TX; Kathleen M. Foley, M.D., New York, NY; Tracy R. Gordy, M.D., Austin, TX; I.K. Ho, Ph.D., Jackson, MS; Chung Y. Hsu, M.D., Ph.D., Charleston, SC; J. Kiffin Penry, M.D., Winston-Salem, NC; Michael A. Taylor, M.D., North Chicago, IL; Sister Ann Walton, St. Paul, MN; Michael Weintraub, M.D., Rochester, NY; Stanley van den Noort, M.D., Irvine, CA

Panel on Nursing Practice

FAYE ABDELLAH, R.N., ED.D., SC.D., *Chairman*, Rockville, MD; Col. Naldean Borg, Tacoma, WA; Mecca S. Cranley, R.N., Ph.D., Madison, WI; Barbara A. Durand, R.N.C., Ed.D., F.A.A.N., Chicago, IL; Hector Hugo Gonzales, R.N., Ph.D., San Antonio, TX; Laurie Martin Gunter, Ph.D., R.N., F.A.A.N., Seattle, WA; Gloria S. Hope, R.N., Ph.D., Washington, DC; Jean Marshall, B.A., R.N., Lakewood, NJ; Ida M. Martinson, R.N., Ph.D., San Francisco, CA; Loretta Nowakowski, R.N., Ph.D., New Castle, PA; Carol P. Patton, R.N., Ph.D., J.D., Ann Arbor, MI; Sharon S. Rising, R.N., C.N.M., Cheshire, CT; Franklin A. Shaffer, R.N., Ed.D., New York, NY; Margaret D. Sovie, R.N., Ph.D., Philadelpha, PA; Una Beth Westfall, R.N., Seattle, WA

Panel on Nutrition and Electrolytes

ROBERT D. LINDEMAN, M.D., *Chairman*, Albuquerque, NM; William O. Berndt, Ph.D., Omaha, NE; Steven B. Heymsfield, M.D., New York, NY; Bonnie Liebman, M.S., Washington, DC; Timothy Lipman, M.D., Washington, DC; Sudesh K. Mahajan, M.D., Allen Park, MI; David J. Martin, Pharm.D., Atlanta, GA; Robert M. Russell, M.D., Boston, MA; Harold H. Sandstead, M.D., Galveston, TX; William J. Stone, M.D., Nashville, TN; Carlos A. Vaamonde, M.D., Miami, FL; Stanley Wallach, M.D., Bay Pines, FL; Robert Whang, M.D., Oklahoma City, OK

Panel on Obstetrics and Gynecology

JOHN J. HALKI, M.D., PH.D., *Chairman*, Dayton, OH; Rudi Ansbacher, M.D., Ann Arbor, MI; Alvin L. Brekken, M.D., Dallas, TX; John G. Daley, M.D., Wilmington, NC; James W. Daly, M.D., Columbia, MO; Dale R. Dunnihoo, M.D., Ph.D., Shreveport, LA; Marilynn C. Frederiksen, M.D., Chicago, IL; R. Don Gambrell, Jr., M.D., Augusta, GA; Larry C. Gilstrap, M.D., Dallas, TX; Douglas D. Glover, M.D., Morgantown, WV; Barbara A. Hayes, M.A., New Rochelle, NY; Art Jacknowitz, M.Sc., Pharm. D., Morgantown, WV; Russell K. Laros, Jr., M.D., San Francisco, CA; William A. Nahhas, M.D., Dayton, OH; Robert C. Park, M.D., Washington, DC; Robert J. Sokol, M.D., Detroit, MI

Panel on Ophthalmology

HERBERT E. KAUFMAN, M.D., *Chairman*, New Orleans, LA; Steven R. Abel, Pharm.D., Indianapolis, IN; Jules Baum, M.D., Boston, MA; Bernard Becker, M.D., St. Louis, MO; Robert P. Burns, M.D., Columbia, MO; Howard M. Leibowitz, M.D., Boston, MA; Henry S. Metz, M.D., Rochester, NY; Joel S. Mindel, M.D., Ph.D., New York, NY; Steven Podos, M.D., New York, NY

Panel on Otorhinolaryngology

LEONARD P. RYBAK, M.D., PH.D., *Chairman*, Springfield, IL; Robert E. Brummett, Ph.D., Portland, OR; David N. F. Fairbanks, M.D., Washington, DC; Linda J. Gardiner, M.D., New Haven, CT; David Hilding, M.D., Price, UT; Richard L. Mabry, M.D., Dallas, TX; William J. Richtsmeier, M.D., Ph.D., Baltimore, MD; Richard W. Waguespack, M.D., Birmingham, AL

Panel on Parasitic Disease Therapy

JAY S. KEYSTONE, M.D., *Chairman*, Toronto, Ontario, Canada; Michele Barry, M.D., New Haven, CT; Frank J. Bia, M.D., Guilford, CT; Robert Goldsmith, M.D., San Francisco, CA; Timothy Johnston, M.D., Pharm.D., Merced, CA; Elaine C. Jong, M.D., Seattle, WA; William Kammerer, M.D., Hershey, PA; Donald J. Krogstad, M.D., St. Louis, MO; John D. MacLean, M.D., Montreal, Quebec, Canada; Edward K. Markell, Ph.D., M.D., Berkeley, CA; J. Joseph Marr, M.D., Denver, CO; Theodore E. Nash, M.D., Bethesda, MD; Franklin A. Neva, M.D., Bethesda, MD; Murray Wittner, M.D., Bronx, NY

Panel on Pediatrics

JENNIFER M. H. LOGGIE, M.D., *Chairman*, Cincinnati, OH; Jacob Aranda, M.D., Montreal, Quebec, Canada; H. Verdain Barnes, M.D., Dayton, OH; Cheston M. Berlin, Jr., M.D., Hershey, PA; Anne Harrison Clark, Washington, DC; W. Manford Gooch, III, M.D., Salt Lake City, UT; Ralph E. Kauffman, M.D., Detroit, MI; Joan M. Korth-Bradley, Pharm.D., Edmonton, Alberta, Canada; Robert H. Levin, Pharm.D., San Francisco, CA; Paul A. Palmisano, M.D., Birmingham, AL; Albert W. Pruitt, M.D., Augusta, GA; Philip D. Walson, M.D., Columbus, OH; Sumner J. Yaffe, M.D., Bethesda, MD

Panel on Pharmacy Practice

WILLIAM F. APPEL, D.SC., *Chairman*, St. Paul, MN; Henry Cade, B.S., Chicago, IL; Herbert S. Carlin, D.Sc., Chappaqua, NY; Olya Duzey, M.S., Reed City, MI; Frances Hall Grogan, B.S., Wickliffe, KY; Ned Heltzer, M.S., Philadelphia, PA; James E. Hosch, B.S., Kansas City, KS; Patricia A. Kramer, B.S., Bismarck, ND; Shirley P. McKee, B.S., Houston, TX; Thomas P. Reinders, Pharm.D., Richmond, VA; Lorie G. Rice, B.A., M.P.H., Sacramento, CA; Al Sebok, B.S., Twinsburg, OH; Stephen M. Sleight, M.S., Bay Pines, FL; William E. Smith, Pharm.D., M.P.H., Long Beach, CA; Thomas C. Snader, Pharm.D., Furlong, PA; J. Richard Wuest, Pharm.D., Cincinnati, OH

Panel on Radiopharmaceuticals

EDWARD B. SILBERSTEIN, M.D., *Chairman*, Cincinnati, OH; Neil M. Abel, M.B.A., M.S., Rockville, MD; William H. Briner, Capt., B.S., Durham, NC; Henry Chilton, Pharm.D., Winston-Salem, NC; Jan M. Ellerhorst-Ryan, R.N., M.S.N., C.S., Cincinnati, OH; Richard Holmes, M.D., Columbia, MO; William Kaplan, M.D., Boston, MA; David L. Laven, C.R.Ph., F.A.S.C.P., Bay Pines, FL; Merle K. Loken, M.D., Ph.D., Minneapolis, MN; Norman L. McElroy, M.A., Washington, DC; William B. Nelp, M.D., Seattle, WA; Buck A. Rhodes, Ph.D., Albuquerque, NM; Barry A. Siegel, M.D., St. Louis, MO; Guy Simmons, Ph.D., Lexington, KY; Dennis P. Swanson, R.Ph., M.S., Pittsburgh, PA; David A. Weber, Ph.D., Upton, NY; Henry N. Wellman, M.D., Indianapolis, IN

Panel on Surgical Drugs and Devices

JAMES W. PATE, M.D., *Chairman*, Memphis, TN; C. Andrew Bassett, M.D., New York, NY; Terry Baumann, Pharm.D., Detroit, MI; Peter J. Fabri, M.D., Tampa, FL; Susan Bartlett Foote, J.D., Berkeley, CA; Jack Hirsh, M.D., Hamilton, Ontario, Canada; Larry R. Pilot, Esq., Washington, DC; Lary A. Robinson, M.D., Omaha, NE; H. Harlan Stone, M.D., Cleveland, OH; Clark Watts, M.D., Columbia, MO

Panel on Urology

SAUL BOYARSKY, M.D., J.D., *Chairman*, St. Louis, MO; John Belis, M.D., Hershey, PA; Michael Boileau, M.D., Bend, OR; Culley C. Carson, M.D., Durham, NC; Warren Heston, Ph.D., New York, NY; Mark V. Jarowenko, M.D., Hershey, PA, Marguerite Lippert, M.D., Charlottesville, VA; Penelope A. Longhurst, Ph.D., Philadelphia, PA; Michael G. Mawhinney, Ph.D., Morgantown, WV; Harris Nagler, M.D., New York, NY; Randall H. Rowland, M.D., Ph.D., Indianapolis, IN; J. Patrick Spirnak, M.D., Cleveland, OH; William F. Tarry, M.D., Morgantown, WV; Alan J. Wein, M.D., Philadelphia, PA; Robert Weiss, M.D., New Haven, CT

Panel on Veterinary Medicine

LLOYD E. DAVIS, D.V.M., PH.D., *Chairman*, Urbana, IL; Arthur L. Aronson, D.V.M., Ph.D., Raleigh, NC; Nicholas H. Booth, D.V.M., Ph.D., Jacksonville, FL; Gordon L. Coppoc, D.V.M., Ph.D., West Lafayette, IN; Sidney A. Ewing, D.V.M., Ph.D., Stillwater, OK; Stuart D. Forney, M.S., Fort Collins, CO; Diane K. Gerken, D.V.M., Ph.D., Columbus, OH; William G. Huber, D.V.M., Blacksburg, VA; William L. Jenkins, D.V.M., Ph.D., Baton Rouge, LA; Robert W. Phillips, D.V.M., Ph.D., Fort Collins, CO; Thomas E. Powers, D.V.M., Ph.D., Columbus, OH; Charles R. Short, D.V.M., Ph.D., Baton Rouge, LA; Richard H. Teske, D.V.M., Rockville, MD; Jeffrey R. Wilcke, D.V.M., M.S., Blacksburg, VA

Assistants During 1985–1990

Participants in the Revision and Reference Standards Programs during the Period 1985–1990 Who Are Not Otherwise Mentioned

Thérèse Agami, Englewood Cliffs, NJ
Abu S. Alam, Melrose Park, IL
Richard Albert, Washington, DC
Thomas G. Alexander, Washington, DC
Robert J. Allara, Piscataway, NJ
Gaylord D. Anthony, Morris Plains, NJ
Bruce Aronson, Springfield Gardens, NY
Nana K. Athanikar, Irvine, CA
L. Larry Augsburger, Baltimore, MD
Philippe Baetz, LePlessis-Robinson, France
Margaret E. Baier, Chicago, IL
Grant Barlow*, Rochester, NY
Irving D. Becker, Allentown, PA
David L. Benjamin, Rochester, NY
Gregory Bennett, East Hanover, NJ
Harriet Benson, Palo Alto, CA
Walter R. Benson, Washington, DC
Bruce G. Berger, Nutley, NJ
Ira R. Berry, Elizabeth, NJ
William F. Beyer, Kalamazoo, MI
Charles H. Bibart, Kalamazoo, MI
William Biles, Central Islip, NY
Margaret L. Blackmon, Research Triangle Park, NC
Jerome I. Bodin, Cranbury, NJ
Ben Borsje, Weesp, The Netherlands
Charles E. Boufford, Wyandotte, MI
Paul A. Bouis, Phillipsburg, NJ
George A. Boulet, North Chicago, IL
Edward S. Brewton, Columbus, OH
Frederick R. Brofazi, Summit, NJ
James F. Buckley, Rahway, NJ
V. Bühler, Ludwigshaven, Fed. Rep. of Germany
Walter C. Cantrell, Syracuse, NY
Robert Y. Cardner, Wilmington, DE
Frederick J. Carleton, Philadelphia, PA
Tony R. Carrie, Evansville, IN
Armand R. Casola, Rockville, MD
John A. Caughlan, St. Louis, MO
Gretchen W. Cecchini, Rochester, NY
Wei-Wei Chang, Morris Plains, NJ
Kenneth G. Chapman, Groton, CT
Herman N. Clark, Centersville, GA
Howard E. Cmejla, New York, NY
Standiford H. Cox, Indianapolis, IN
Erwin Coyne, Chicago, IL
E. Craenhals, Brussels, Belgium
Donald R. Daoust Cranbury, NJ
Ronald J. Darnowski, Rahway, NJ
Nicholas J. DeAngelis, Kenilworth, NJ
Ellen J. deVries, Weesp, The Netherlands
James G. Dickinson, Morgantown, WV
Allan H. Doane, Morris Plains, NJ
Durward F. Dodgen, Washington, DC
Rubin Drucker, Flushing, NY
Robert J. Duane, Randolph, MA
Thomas D. Duffy, Edinburgh, Scotland
Robert Egli, Schaffhausen, Switzerland
James T. Elfstrum, Westport, CT
Robert W. Elkas, Pearl River, NY
Jawed Fareed, Chicago, IL
John S. Finlayson, Bethesda, MD
Valerie A. Flourney, Washington, DC
Robert G. Flynn, St. Louis, MO
Arnold Goldfarb, Melville, NY
Mason P. Goldman, Rockville, MD
Ralph Gomez, Nutley, NJ
Donald E. Grant, West Haven, CT
William Grosse, III, West Point, PA
William F. Haddad, Carmel, NY
Jerome A. Halperin, Summit, NJ
Yusuf K. Hamied, Bombay, India
Henry F. Hammer, Amelia Island, FL
Robert W. Hanson, Rahway, NJ
William T. Hensler, Bedford, MA
Atilla A. Hincal, Ankara, Turkey
H. Donald Hochstein, Rockville, MD
Norman E. Holden, Upton, L.I., NY
Keith D. Holmes, Research Triangle Park, NC
Rosa Holmes, East Hanover, NJ
J. Hoogmartens, Leuvein, Belgium
Seymour Hyden, Springfield Gardens, NY
Harold Jacobson, New Brunswick, NJ
Jay J. Jadeja, Bronx, NY
Jack P. Jeffries, Greenford, Middlesex, England
Erik H. Jensen, Kalamazoo, MI
C. A. Johnson*, London, England
Lawrence Jones, St. Louis, MO
Jose C. Joseph, North Chicago, IL
William Juhl, St. Louis, MO
Yvonne Juhl, St. Louis, MO
Gary R. Kamzar, Wilmington, DE
Arnold Kaufman, Palo Alto, CA
Ernest L. Kelly, Fort Washington, PA
John Kerins, Paramus, NJ
Martha B. Killelea, North Billerica, MA
G. Kinoshita, Osaka, Japan
Ross Kirchoffer, St. Louis, MO
Richard E. Kolinski, St. Louis, MO
Ronald G. Komejan, Zeeland, MI
Michael S. Korczynski, North Chicago, IL
James B. Kottemann, Washington, DC
Paul Kucera, Pearl River, NY
Gene Kulyk, Chicago, IL
Leonard Laerten, Flushing, NY
J. B. Laidler, Amersham, England
M. Stephen Lajoie, Princeton, NJ
A. Lasis, Scarborough, ONT, Canada
Ronald J. Lauback, Syracuse, NY
Edgar A. Lazo-Wasem, Sunnyvale, CA
Wayne S. Lehrman, Stafford, VA
Sally J. Lenhart, Bedford, OH
Henry H. Lerner, New Brunswick, NJ
Klaus Lingner, Basle, Switzerland
Julio Litovsky, Kenilworth, NJ
Barbara Lobberding, Broomfield, CO
Carmen N. Mangieri, South Orange, NJ
Peter E. Manni, Philadelphia, PA
Arnold D. Marcus, Hillside, NJ
Michel Margosis, Washington, DC
Vincent A. Marino, Broomfield, CO
J. C. Marques, Lisbon, Portugal
Michael Martin-Smith, Ware, England
Dave Martinez, Bellefonte, PA
George L. Mattok, Ottawa, ONT, Canada
Ian J. McGilveray, Ottawa, ONT, Canada
Eugene J. McGonigle, Kenilworth, NJ
Theodore H. Meltzer, Arlington, VA
Joy E. Merritt, Columbus, OH
Fred H. Meyer, St. Louis, MO
Michel S. Michailidis, Rahway, NJ
Ralph F. Michielli, Rahway, NJ
Lloyd C. Miller, Escondido, CA
David J. Miner, Indianapolis, IN
Barbara A. Morin, Tewksbury, MA
Andrew D. Morris, Indianapolis, IN
Lewis L. Moyer, Jr., West Point, PA
James W. Munson, Crawley, England
Genisio Murano, Bethesda, MD
Martin C. Musolf, Hemlock, MI
Anna Napiorkowski, Milwaukee, WI
V. G. Nayak, Bombay, India
Randy Nelson, Lincoln, NB
Sidney D. Nelson, Seattle, WA
Joanne Nikitakis, Washington, DC
John P. Obedier, North Chicago, IL
Maureen O'Conner, Palo Alto, CA
Bengt Ohrner, Solna, Sweden
Jeffery L. Otto, Broomfield, CO
T. L. Overman, London, England
John M. Padfield, Ware, England
Fernand J. Pellerin, Paris, France
Solomon C. Pflag, Philadelphia, PA
Barry Poretskin, Eatontown, NJ
Norman J. Pound, Ottawa, ONT, Canada
Stanley Rasberry, Gaithersburg, MD
Suresh C. Rastogi, Rockville, MD
Michael Raya, Springfield Gardens, NY
Van D. Reif, Philadelphia, PA
Peter H. Rheinstein, Rockville, MD
William G. Richardson, Indianapolis, IN
Steven Richheimer, Denver, CO
Joachim Richter, East Berlin, DDR
Alan R. Rogers, London, England
Gerhard Rotzler, Basel, Switzerland
Robert J. Saccaro, Newark, NJ
E. G. Bud Schmidt, St. Paul, MN
Lorne A. Schnell, Kankakee, IL
Peter J. Schorn, Strasbourg, France
Alan J. Sheppard, Washington, DC
Arvin P. Shroff, Rockville, MD
Frank Simione, Rockville, MD
Don Slee, Arlington, VA
Edward Smith, Washington, DC
Sidney J. Smith, Markham, ONT, Canada
Antoinette Spears, Libertyville, IL
Jean-Marc Spieser, Strasbourg, France
Louis B. Stadler, Cincinnati, OH
James R. Stoker, New York, NY
Elmer H. Stotz, Rochester, NY
Robert I. Stearns, St. Louis, MO
F. R. Stravs, Dallas, TX
James E. Swon, Edison, NJ
John A. Syverson, New York, NY
Robert D. Tackitt, Philadelphia, PA
Tsutomu Takahashi, Osaka, Japan
Anthony J. Taraszka, Kalamazoo, MI
Nancy Thrutchley, Kansas City, MO
Kinji Toyoda, Tokyo, Japan
Roger B. Trigg, London, England
Ron Turton, Whitby, ONT, Canada
Norlin Tymes, Baltimore, MD
H. W. Unterman, Netanya, Israel
Mark D. Van Arendonk, Kalamazoo, MI
A. van den Hoek, Straatweg, The Netherlands
Hubert Vanderhaeghe, Leuven, Belgium
C. van der Vlies, Delft, The Netherlands
Luciano Virgili, New Brunswick, NJ
Ruth Wantz, Milwaukee, WI
Fred M. Warnaar, Amersfoort, The Netherlands
Melvin H. Weinswig, Madison, WI
Elliott T. Weisman, Wilmington, DE
Charles D. Wentling, Indianapolis, IN
Monika Westermark, Stockholm, Sweden
Thomas X. White, Washington, DC
H. Willemsen, Straatweg, The Netherlands
Charles M. Wilson, Cincinnati, OH
D. S. Wooliscroft, Nottingham, England
John E. Zarembo, Stratham, NH

* Deceased.

Members of the United States Pharmacopeial Convention and the Institutions and Organizations Represented as of November 1, 1988

Current Officers and Board of Trustees

President: Arthur Hull Hayes, Jr., M.D.,* President and Chief Executive Officer, EM Industries, Inc., 5 Skyline Drive, Hawthorne, NY 10532

Vice President: Joseph M. Benforado, M.D.,* 730 Seneca Place, Madison, WI 53711

Past President: Frederick E. Shideman, M.D., Ph.D.,**

Treasurer: Paul F. Parker, D.Sc.,* Pharmacy Consultant, Paul Parker, Inc., 917 Celia Lane, Lexington, KY 40504

Representing Medicine: J. Richard Crout, M.D.,* Boehringer Mannheim Corporation, 1301 Piccard Drive, Rockville, MD 20850
Leo E. Hollister, M.D., Harris County Psychiatric Center, P.O. Box 20249, Houston, TX 77225

Representing Pharmacy: Joseph P. Buckley, Ph.D.,* Texas Medical Center, Rm. 431, University of Houston, 1441 Moursund, Houston, TX 77030

James T. Doluisio, Ph.D.,* College of Pharmacy, University of Texas at Austin, Austin, TX 78712

Public Member: Estelle G. Cohen, M.A.* 5813 Greenspring Avenue, Baltimore, MD 21209

At Large: John V. Bergen, Ph.D.,* National Committee for Clinical Laboratory Standards, 771 E. Lancaster Avenue, Villanova, PA 19085
John T. Fay, Jr., Ph.D.,* Bergen Brunswig Corporation, 4000 Metropolitan Drive, Orange, CA 92668

Executive Director and Secretary: William M. Heller, Ph.D.,* 12601 Twinbrook Parkway, Rockville, MD 20852

United States Government Services

Department of the Army: Col. Frank A. Cammarata,* 5111 Leesburg Pike, Falls Church, VA 22041

Department of Health & Human Services: Richard R. Ashbaugh,* 5600 Fishers Lane, Rm. 11-03, Rockville, MD 20857

Food and Drug Administration: Peter H. Rheinstein, M.D., J.D.,* Director, Office of Drug Standards, Center for Drug Evaluation and Research, 5600 Fishers Lane, HFD-200, Rockville, MD 20857

National Institute of Standards & Technology: Stanley D. Rasberry,* Chemistry, B311, National Institute of Standards & Technology, Gaithersburg, MD 20899

Office of the Chief of Naval Operations, U.S. Navy: Commander Ronnie E. Whiten, MSC, USN, Defense Medical Standardization Board, Fort Detrick, Frederick, MD 21701

Office of the Surgeon General, U.S. Air Force: Lt. Col. John M. Hammond,* USAF/SCB, Bolling AFB, DC 20332

United States Public Health Service: ASG Richard M. Church, 5600 Fishers Lane, Rockville, MD 20857

U.S. Office of Consumer Affairs: Robert F. Steeves, J.D.,* Dep. Spec. Adv. to the President for Consumer Affairs, 1725 I Street, N.W., Suite 1003, Washington, DC 20201

Veterans Administration: Stephen M. Sleight, M.Sc.,* Pharmacy Service (119), Veterans Administration Medical Center, Bay Pines, FL 33504

National Organizations

American Association of Pharmaceutical Scientists: Ralph F. Shangraw, Ph.D., University of Maryland, School of Pharmacy, 20 N. Pine Street, Baltimore, MD 21201

American Chemical Society: Samuel M. Tuthill, Ph.D.,* P.O. Box 5439, St. Louis, MO 63147

American Dental Association: Edgar W. Mitchell, Ph.D.,* American Dental Association, 211 E. Chicago Avenue, Chicago, IL 60611

American Hospital Association: William R. Reid, Community Hospital of Roanoke Valley, 101 Elm Avenue, S.E., Box 12946, Roanoke, VA 24029

American Medical Association: John C. Ballin, Ph.D.,* American Medical Association, 535 N. Dearborn Street, Chicago, IL 60610

American Nurses' Association, Inc.: Jean Marshall, B.A., R.N.,* Employee Relations Coordinator, Paul Kimball Hospital, 600 River Avenue, Lakewood, NJ 08701

American Pharmaceutical Association: John F. Schlegel, Pharm.D.,* President, American Pharmaceutical Association, 2215 Constitution Avenue, N.W., Washington, DC 20037

American Society for Clinical Pharmacology & Therapeutics: William B. Abrams, M.D.,* Merck Sharp & Dohme Research Laboratories, West Point, PA 19486

American Society of Hospital Pharmacists: R. David Anderson,* 6 Pelham Greene West, Waynesboro, VA 22980

American Society for Pharmacology & Experimental Therapeutics: Marilyn E. Hess, Ph.D.,* School of Medicine, University of Pennsylvania, Philadelphia, PA 19104

American Society for Quality Control: Theodore C. Fleming,* 7125 Monterrey Drive, Fort Worth, TX 76112

American Veterinary Medical Association: L. Meyer Jones, D.V.M., Ph.D.,* 1225 St. Andrews Drive, Pinehurst, NC 28374

Association of Food and Drug Officials: David R. Work, J.D.,* Executive Director, North Carolina State Board of Pharmacy, P. O. Box H, Carrboro, NC 27510

Association of Official Analytical Chemists: James B. Kottemann,* Food and Drug Administration, 200 C Street, S.W., HFN-004, Washington, DC 20204

Chemical Manufacturers Association: Andrew J. Schmitz, Jr.,* Pfizer, Inc., 235 East 42nd Street, New York, NY 10017

Cosmetic, Toiletry & Fragrance Association, Inc.: G. N. McEwen, Jr., Ph.D., Cosmetic, Toiletry & Fragrance Association, Inc., 1110 Vermont Avenue, N.W., Suite 800, Washington, DC 20005

Health Industry Manufacturers Association: James F. Jorkasky, Director,* Manufacturing & Quality Programs, Health Industry Manufacturers Association, 1030 15th Street, N.W., Washington, DC 20005

National Association of Boards of Pharmacy: Carmen A. Catizone, M.S., R.Ph., Executive Director, National Association of Boards of Pharmacy, O'Hare Corporate Center, 1300 Higgins Road, Suite 103, Park Ridge, IL 60068

National Association of Chain Drug Stores, Inc.: Donald Bell,* Thrift Drug Company, 615 Alpha Drive, Pittsburgh, PA 15230

National Association of Retail Druggists: William N. Tindall, Ph.D.,* 205 Daingerfield Road, Alexandria, VA 22314

National Wholesale Druggists' Association: Ronald J. Streck, Vice President,* Government Affairs, National Wholesale Druggists' Association, 105 Oronoco Street, P. O. Box 238, Alexandria, VA 22313

Parenteral Drug Association, Inc.: Sol Motola, Ph.D.,* Bausch & Lomb, Personal Products Division, 1400 North Goodman Street, Rochester, NY 14692

Pharmaceutical Manufacturers Association: John Jennings, M.D.,* Pharmaceutical Manufacturers Association, 1100 Fifteenth Street, N.W., Washington, DC 20005

The Proprietary Association: R. William Soller, Ph.D.,* Vice President, Scientific Affairs, The Proprietary Association, 1150 Connecticut Avenue, N.W. Washington, DC 20036

Other Organizations and Institutions

ALABAMA

University of Alabama, School of Medicine: Robert B. Diasio, M.D.,* School of Medicine, University of Alabama, Birmingham, AL 35294

University of South Alabama, College of Medicine: Samuel J. Strada, Ph.D., Prof. and Chairman, Dept. of Pharmacology, Univ. of South Alabama, College of Medicine, 3190 MSB, Mobile, AL 36688

Auburn University, School of Pharmacy: Kenneth N. Barker, Ph.D.,* Department of Pharmacy Care Systems, School of Pharmacy, Auburn University, Auburn, AL 36849

Samford University, School of Pharmacy: Stanley V. Susina, Ph.D.,* School of Pharmacy, Samford University, 800 Lakeshore Drive, Birmingham, AL 35229

*Present at the 1985 Quinquennial Meeting.
**Deceased.

Medical Association of the State of Alabama: Paul A. Palmisano, M.D.,* Professor of Pediatrics, University of Alabama in Birmingham, University Station, Birmingham, AL 35294

ALASKA

Alaska Pharmaceutical Association: Jacqueline L. Warren, 11731 Trails End Road, Anchorage, AK 99516

ARIZONA

University of Arizona, College of Medicine: Kenneth A. Conrad, M.D., Department of Internal Medicine, College of Medicine, University of Arizona, 1501 North Campbell Avenue, Tucson, AZ 85724

University of Arizona, College of Pharmacy: Samuel H. Yalkowsky, Ph.D.,* College of Pharmacy, University of Arizona, Tucson, AZ 85721

Arizona Pharmacy Association: Mr. Michael J. Henry, 19322 E. Calle de Flors, Queen Creek, AZ 85242

ARKANSAS

University of Arkansas, College of Medicine: James E. Doherty, III, M.D., VA Hospital, 4300 West 7th Street, Little Rock, AR 72205

University of Arkansas for Medical Sciences, College of Pharmacy: James R. McCowan, Ph.D., College of Pharmacy, University of Arkansas for Medical Sciences, 4301 W. Markham Street, Slot 522, Little Rock, AR 72205

Arkansas Pharmacists Association: Marcus W. Jordin, Ph.D.,* 309 Brookside Drive, Little Rock, AR 72205

CALIFORNIA

Loma Linda University, School of Medicine: Ralph E. Cutler, M.D.,* Department of Medicine, Loma Linda School of Medicine, Anderson and Barton, Loma Linda, CA 92354

Stanford University, School of Medicine: Phyllis Gardner, M.D., School of Medicine, Stanford University, CV-291, Stanford, CA 94305

University of California, Davis, School of Medicine: Larry Stark, Ph.D.,* Department of Pharmacology, School of Medicine, University of California, Davis, CA 95616

University of California, Los Angeles, School of Medicine: Don H. Catlin,* Department of Pharmacology, School of Medicine, CHS, University of California, Los Angeles, CA 90024

University of California, San Francisco, School of Medicine: Walter L. Way, M.D.,* Department of Anesthesia, Rm. S-436, University of California, San Francisco, CA 94143

University of Southern California, School of Medicine: Samuel P. Bessman, M.D., MUDD 414, 1333 San Pablo Street, Los Angeles, CA 90033

University of California, San Francisco, School of Pharmacy: Jack Cooper, Ph.G.,* School of Pharmacy, University of California, San Francisco, CA 94143

University of Southern California, School of Pharmacy: Robert T. Koda, Pharm.D., Ph.D.,* School of Pharmacy, University of Southern California, 1985 Zonal Avenue, Los Angeles, CA 90033

University of the Pacific, School of Pharmacy: Alice Jean Matuszak, Ph.D., 751 Brookside Road, Stockton, CA 95207

California Pharmacists Association: Max Stollman,* 8314 Wilshire Boulevard, Beverly Hills, CA 90211

COLORADO

University of Colorado, School of Medicine: Antonia Vernadakis, Ph.D., Prof., Depts. of Psych. & Pharm., School of Medicine, University of Colorado, 4200 E. 9th Avenue, Box C263, Denver, CO 80262

University of Colorado, School of Pharmacy: Duane C. Bloedow, Ph.D.,* 3630 Silver Plume Lane, Boulder, CO 80303

Colorado Medical Society: Franklin Lee Bowling, M.D., 1001 E. Oxford Lane, Englewood, CO 80110

Colorado Pharmacal Association: Thomas G. Arthur R. Ph.,* 9852 Corsair Drive, Conifer, CO 80433

CONNECTICUT

University of Connecticut, School of Medicine: Paul F. Davern,* Director of Pharmacy, University of Connecticut Health Center, 261 Farmington Avenue, Farmington, CT 06032

University of Connecticut, School of Pharmacy: Max W. Miller, Ph.D.,* School of Pharmacy, University of Connecticut, Storrs, CT 06268

Connecticut Pharmaceutical Association: Henry A. Palmer, Ph.D.,* 26 Timber Drive, Storrs, CT 06268

Connecticut State Medical Society: James E. O'Brien, M.D., 31 Surrey Drive, Wethersfield, CT 06109

DELAWARE

Delaware Pharmaceutical Society: Charles J. O'Connor, 2100 West 17th Street, Wilmington, DE 19806

Medical Society of Delaware: Jeffry I. Komins, M.D.,* 2323 Pennsylvania Avenue, Wilmington, DE 19806

DISTRICT OF COLUMBIA

Georgetown University, School of Medicine: Arthur Raines, Ph.D., School of Medicine, Georgetown University, 3900 Reservoir Road, N.W., Washington, DC 20007

Howard University, College of Medicine: Robert E. Taylor, M.D., Ph.D., Department of Medicine, Howard University Hospital, 2041 Georgia Avenue, N.W., Washington, DC 20060

Howard University, College of Pharmacy & Pharmacal Sciences: Wendell T. Hill, Jr., Pharm.D.,* 2300 Fourth Street, N.W., Washington, DC 20059

Medical Society of the District of Columbia: Michael D. Abramowitz, M.D.,* 111 Michigan Avenue, N.W., Washington, DC 20010

FLORIDA

Florida A & M University, College of Pharmacy and Pharmaceutical Sciences: Henry Lewis, III, Pharm.D.,* College of Pharmacy and Pharmaceutical Sciences, Florida A & M University, Tallahassee, FL 32307

University of Florida, College of Medicine: Thomas F. Muther, Ph.D.,* Department of Pharmacology & Therapeutics, College of Medicine, University of Florida, Box J-267, J. Hillis Miller Health Center, Gainesville, FL 32610

University of Florida, College of Pharmacy: Michael A. Schwartz, Ph.D.,* Box J-4, J. Hillis Miller Health Center, Gainesville, FL 32610

University of South Florida, College of Medicine: Joseph J. Krzanowski, Jr., Ph.D.,* 12901 N. 30th Street, Box 9, Tampa, FL 33612

Florida Pharmacy Association: George Browning,* 8552 Sylvan Drive, Melbourne, FL 32901

GEORGIA

Medical College of Georgia, School of Medicine: Merle W. Riley, Ph.D.,* Department of Pharmacology & Toxicology, Medical College of Georgia, 1120 Fifteenth Street, Augusta, GA 30912

Mercer University, Southern School of Pharmacy: A. Vincent Lopez, Ph.D.,* Southern School of Pharmacy, Mercer University, 345 Boulevard, N.E., Atlanta, GA 30312

Morehouse School of Medicine: Ralph W. Trottier, Ph.D.,* Morehouse School of Medicine, 720 Westview Drive, S.W., Atlanta, GA 30310

University of Georgia, College of Pharmacy: Howard C. Ansel, Ph.D.,* Dean, College of Pharmacy, University of Georgia, Athens, GA 30602

Georgia Pharmaceutical Association: Charles L. Braucher, Ph.D., College of Pharmacy, University of Georgia, Athens, GA 30602

Medical Association of Georgia: E. D. Bransome, Jr., M.D.,* Professor of Medicine, Medical College of Georgia, Augusta, GA 30912

IDAHO

Idaho State University, College of Pharmacy: Eugene I. Isaacson, Ph.D.,* 1619 East Terry, Pocatello, ID 83201

ILLINOIS

Chicago Medical School/University of Health Sciences: Seymour Ehrenpreis, Ph.D.,* Department of Pharmacology, University of Health Sciences/The Chicago Medical School, 3333 N. Green Bay Road, N. Chicago, IL 60064

Loyola University of Chicago Stritch School of Medicine: John E. Nelson, M.D., Assist. Professor of Medicine, Dept. of Medicine, 2160 S. First Avenue, Maywood, IL 60153

Southern Illinois University, School of Medicine: Ronald A. Browning, Ph.D.,* Department of Medical Physiology and Pharmacology, Southern Illinois University School of Medicine, Carbondale, IL 62901

University of Illinois, College of Medicine: Marten M. Kernis, Ph.D., College of Medicine, University of Illinois, 1853 West Polk Street, Chicago, IL 60612

University of Illinois, College of Pharmacy: Martin I. Blake, Ph.D.,* 9023 Kenton Avenue, Skokie, IL 60076

Illinois Pharmacists Association: Ronald W. Gottrich,* 1817 Clearview Dr., Springfield, IL 62704

Illinois State Medical Society: Vincent A. Costanzo, Jr., M.D.,* 18304 Maple, Lansing, IL 60438

*Present at the 1985 Quinquennial Meeting.

INDIANA

Butler University, College of Pharmacy: Margaret A. Shaw, Ph.D.,* College of Pharmacy, Butler University, 4600 Sunset Avenue, Indianapolis, IN 46208

Purdue University, School of Pharmacy and Pharmacal Sciences: Adelbert M. Knevel, Ph.D.,* School of Pharmacy & Pharmacal Sciences, Purdue University, West Lafayette, IN 47907

Indiana State Medical Association: Edward Langston, M.D., 203 North Division, Flora, IN 46924

IOWA

University of Iowa, College of Medicine: John E. Kasik, M.D., Ph.D., Internal Medicine 1A02 VA, University of Iowa, Iowa City, IA 52242

Drake University, College of Pharmacy: Wendell Southard, Ph.D.,* College of Pharmacy, Drake University, Des Moines, IA 50311

University of Iowa, College of Pharmacy: Robert A. Wiley, Ph.D.,* Dean, College of Pharmacy, University of Iowa, Iowa City, IA 52242

Iowa Pharmacists Association: Robert Osterhaus,* 124 S. Main, Maquoketa, IA 52060

KANSAS

University of Kansas Medical Center, School of Medicine: Edward J. Walaszek, Ph.D.,* School of Medicine, University of Kansas Medical Center, 39th and Rainbow Boulevard, Kansas City, KS 66103

University of Kansas, School of Pharmacy: Siegfried Lindenbaum, Ph.D.,* 1025 Holiday Drive, Lawrence, KS 66044

Kansas Pharmacists Association: John Owen,* 1202 Eastmoor, Wichita, KS 67207

KENTUCKY

University of Kentucky, College of Medicine: Edgar T. Iwamoto, Ph.D., College of Medicine, University of Kentucky, Pharmacology, 800 Rose Street, Lexington, KY 40536

University of Kentucky, College of Pharmacy: Patrick P. DeLuca, Ph.D.,* 3292 Nantucket Drive, Lexington, KY 40502

University of Louisville, School of Medicine: Peter P. Rowell, Ph.D.,* Department of Pharmacology & Toxicology, School of Medicine, University of Louisville, Louisville, KY 40292

Kentucky Medical Association: Ellsworth C. Seeley, M.D.,* 820 South Limestone, Annex 4, Lexington, KY 40536

Kentucky Pharmacists Association: Chester L. Parker, Pharm.D.,* 1816 Darien Drive, Lexington, KY 40504

LOUISIANA

Northeast Louisiana University, School of Pharmacy: Robert D. Kee, Ph.D.,* Turtledove Drive, Monroe, LA 71203

Tulane University, School of Medicine: Floyd R. Domer, Ph.D.,* Department of Pharmacology, School of Medicine, Tulane University, 1430 Tulane Avenue, New Orleans, LA 70112

Xavier University of Louisiana, College of Pharmacy: Josephine Daigle, Ph.D.,* 7325 Palmetto Street, New Orleans, LA 70125

Louisiana Pharmacists Association: William G. Day,* 13114 Country Manor, Baton Rouge, LA 70816

Louisiana State Medical Society: John Adriani, M.D.,* 67 N. Park Place, New Orleans, LA 70124

MARYLAND

Johns Hopkins University, School of Medicine: E. Robert Feroli, Pharm.D.,* Johns Hopkins Hospital, 600 N. Wolfe Street (526 Osler), Baltimore, MD 21205

Uniformed Services University of the Health Sciences, F. Edward Hebert School of Medicine: Jeffrey D. Lazar, M.D., Department of Pharmacology, USUHS, 4301 Jones Bridge Road, Bethesda, MD 20814

University of Maryland, School of Medicine: James I. Hudson, M.D.,* Dean's Office, School of Medicine, University of Maryland, 655 W. Baltimore Street, Baltimore, MD 21201

University of Maryland, School of Pharmacy: Larry L. Augsburger, Ph.D.,* 20 North Pine Street, Baltimore, MD 21201

Maryland Pharmacists Association: Paul Freiman,* 3 Pipe Hill Court, Baltimore, MD 21209

Medical and Chirurgical Faculty of the State of Maryland: Frederick H. Wilheim, M.D., 5807 Annapolis Road, Hyattsville, MD 20784

MASSACHUSETTS

Boston University, School of Medicine: Edward W. Pelikan, M.D.,* Department of Pharmacology, Boston University School of Medicine, 80 East Concord Street, Boston, MA 02118

Harvard Medical School: Peter Goldman, M.D., Harvard Medical School, 25 Shattuck Street, Boston, MA 02115

Massachusetts College of Pharmacy & Allied Health Sciences: William O. Foye, Ph.D.,* 179 Longwood Avenue, Boston, MA 02115

Northeastern University, College of Pharmacy and Allied Health Professions: Larry N. Swanson, Pharm.D.,* 124 Cobble Hill Road, Warwick, RI 02886

University of Massachusetts Medical School: Brian Johnson, M.D.,* University of Massachusetts Medical Center, 55 Lake Avenue, North, Worcester, MA 01605

Massachusetts Medical Society: Edward J. Khantzian, M.D., Cambridge Hospital, 1493 Cambridge Street, Cambridge, MA 02139

Massachusetts State Pharmaceutical Association: Bertram A. Nicholas, M.Sc.,* 179 Longwood Avenue, Boston, MA 02115

MICHIGAN

Ferris State College, School of Pharmacy: Gerald W. A. Slywka, Ph.D.,* 7630 Crestview Drive, Reed City, MI 49677

University of Michigan, College of Pharmacy: Ara G. Paul, Ph.D., Dean, College of Pharmacy, University of Michigan, Ann Arbor, MI 48109-1065

Wayne State University, College of Pharmacy and Allied Health Professions: Janardan Nagwekar, Ph.D., College of Pharmacy and Allied Health Professions, Shapero Hall, Room 511, Wayne State University, Detroit, MI 48202

Wayne State University, School of Medicine: Ralph E. Kauffman, M.D.,* 3901 Beaubien, Detroit, MI 48201

Michigan Pharmacists Association: Salvador Pancorbo, Ph.D., Pharm.D.,* College of Pharmacy & Allied Health Professions, Wayne State University, Detroit, MI 48202

MINNESOTA

University of Minnesota, College of Pharmacy: Edward G. Rippie, Ph.D.,* 2 N. Mallard Road, North Oaks, MN 55127

University of Minnesota Medical School, Minneapolis: Jack W. Miller, Ph.D.,* University of Minnesota Medical School, 3-260 Millard Hall, 435 Delaware Street, S.E., Minneapolis, MN 55455

Minnesota State Pharmaceutical Association: Arnold D. Delger,* 1533 Grantham Street, St. Paul, MN 55108

MISSISSIPPI

University of Mississippi, School of Medicine: Richard L. Klein, Ph.D.,* Department of Pharmacology & Toxicology, University of Mississippi Medical Center, Jackson, MS 39216-4504

University of Mississippi, School of Pharmacy: Robert W. Cleary, Ph.D.,* Department of Pharmaceutics, University of Mississippi, University, MS 38677

Mississippi Pharmacists Association: Phylliss M. Moret R.Ph.,* Mississippi Pharmacists Association, 341 Edgewood Terrace Drive, Jackson, MS 39206

MISSOURI

St. Louis College of Pharmacy: John W. Zuzack, Ph.D.,* St. Louis College of Pharmacy, 4588 Parkview Place, St. Louis, MO 63110

St. Louis University, School of Medicine: Alvin H. Gold, Ph.D.,* 1402 S. Grand Boulevard, St. Louis, MO 63104

University of Missouri, Columbia, School of Medicine: John W. Yarbro, M.D., Ph.D.,* N408 Health Sciences Center, Columbia, MO 65212

University of Missouri, Kansas City, School of Medicine: David Rush, Pharm.D.,* Truman Medical Center-East, Little Blue & Lee's Summit Roads, Kansas City, MO 64139

University of Missouri, Kansas City, School of Pharmacy: Wayne M. Brown, Ph.D.,* School of Pharmacy, University of Missouri-Kansas City, 5005 Rockhill Road, Kansas City, MO 64110-2499

Washington University, School of Medicine: H. Mitchell Perry, Jr., M.D.,* School of Medicine, Washington University, Box 8048, 660 S. Euclid Avenue, St. Louis, MO 63110

Missouri Pharmaceutical Association: James R. Boyd, P.D.,* 215 Shirley Ridge Drive, St. Charles, MO 63303

MONTANA

University of Montana, School of Pharmacy & Allied Health Professions: Donald H. Canham, Ph.D., School of Pharmacy, University of Montana, Missoula, MT 59812

Montana State Pharmaceutical Association: Robert H. Likewise, 4376 Head Drive, Helena, MT 59601

*Present at the 1985 Quinquennial Meeting.

NEBRASKA

Creighton University, School of Medicine: Michael C. Makoid, Ph.D.,* School of Medicine, Creighton University, 2500 California Street, Omaha, NE 68178

Creighton University, School of Pharmacy and Allied Health Professions: James M. Crampton, Ph.D., School of Pharmacy and Allied Health Professions, Creighton University, California and 24th Streets, Omaha, NE 68178

University of Nebraska, College of Medicine: Manuchair Ebadi, Ph.D., Department of Pharmacology, University of Nebraska, College of Medicine, 42nd Street and Dewey Avenue, Omaha, NE 68105

University of Nebraska, College of Pharmacy: Clarence T. Ueda, Pharm.D., Ph.D., Dean, College of Pharmacy, University of Nebraska, 42nd and Dewey Avenues, Omaha, NE 68105

Nebraska Pharmacists Association: Rex C. Higley, R.P.,* 3110 South 42nd, Lincoln, NE 68506

NEVADA

University of Nevada, Reno, School of Medicine: Iain L. O. Buxton, D.Ph.,* Department of Pharmacology, School of Medicine, University of Nevada, Reno, NV 89557

Nevada State Medical Association: Lawrence P. Matheis, 3660 Baker Lane #101, Reno, NV 89509

NEW HAMPSHIRE

Dartmouth Medical School: James J. Kresel, Ph.D.,* Dartmouth-Hitchcock Medical Center, Hanover, NH 03756

New Hampshire Pharmaceutical Association: William J. Lancaster,* 4 Woodmore Drive, Hanover, NH 03755

NEW JERSEY

University of Medicine and Dentistry of New Jersey, New Jersey Medical School: Sheldon B. Gertner, Ph.D.,* New Jersey Medical School, UMDNJ, 100 Bergen Street, Newark, NJ 07103

Rutgers, The State University of New Jersey, College of Pharmacy: Thomas Medwick, Ph.D.,* College of Pharmacy, Rutgers, The State University of New Jersey, P. O. Box 789, Piscataway, NJ 08854

Medical Society of New Jersey: Joseph N. Micale, M.D., 914-85th Street, North Bergen, NJ 07047

New Jersey Pharmaceutical Association: Stephen J. Csubak, Ph.D.,* 4 Decision Way East, Washington Crossing, PA 18977

NEW MEXICO

University of New Mexico, College of Pharmacy: William M. Hadley, Ph.D., Dean, College of Pharmacy, University of New Mexico, Albuquerque, NM 87131

New Mexico Pharmaceutical Association: Hugh Kabat, Ph.D.,* College of Pharmacy, University of New Mexico, Albuquerque, NM 87131

NEW YORK

Albert Einstein College of Medicine of Yeshiva University: Dr. Walter G. Levine,* Albert Einstein College of Medicine, 1300 Morris Park Avenue, New York, NY 10461

City University of New York, Mt. Sinai School of Medicine: Joel S. Mindel, M.D., Ph.D.,* Department of Ophthalmology, Annenberg Bldg. 22-14, Mt. Sinai School of Medicine, 1 Gustave L. Levy Place, New York, NY 10029

Columbia University College of Physicians and Surgeons: Dr. Norman Kahn,* Department of Pharmacology, Columbia University, 630 West 168th Street, New York, NY 10032

Cornell University Medical College: W. Y. Chan, Ph.D.,* Cornell University Medical College, New York, NY 10021

Long Island University, Arnold and Marie Schwartz College of Pharmacy and Health Sciences: John J. Sciarra, Ph.D.,* 8 Allen Drive, Locust Valley, NY 11560

New York Medical College: Mario A. Inchiosa, Jr., Ph.D.,* Department of Pharmacology, New York Medical College, Valhalla, NY 10595

State University of New York, Buffalo, School of Medicine: Robert J. McIsaac, Ph.D.,* Department of Pharmacology and Therapeutics, School of Medicine, 127 Farber Hall, SUNY at Buffalo, Buffalo, NY 14214

State University of New York, Buffalo, School of Pharmacy: Walter D. Conway, Ph.D., School of Pharmacy, SUNY at Buffalo, H565 Hochstetter Hall, Buffalo, NY 14260

State University of New York, Stony Brook, School of Medicine: Dr. Arthur P. Grollman, Dept. of Pharmacological Science, School of Medicine, SUNY at Stony Brook, BHS T-8 140, Stony Brook, NY 11794

St. John's University, College of Pharmacy and Allied Health Professions: Andrew J. Bartilucci, Ph.D.,* College of Pharmacy and Allied Health Professions, St. John's University, Grand Central and Utopia Parkways, Jamaica, NY 11439

Union University, Albany College of Pharmacy: Barry S. Reiss, Ph.D.,* Albany College of Pharmacy, Union University, 106 New Scotland Avenue, Albany, NY 12208

University of Rochester, School of Medicine and Dentistry: Michael Weintraub, M.D., Ph.D,* University of Rochester Medical Center, Box 644, Rochester, NY 14642

Medical Society of the State of New York: Richard S. Blum, M.D., 25 Spruce Dr., East Hills, NY 11576

Pharmaceutical Society of the State of New York: Walter Singer, R.Ph., Ph.D., Pharmaceutical Society of the State of New York, Pine West Plaza IV, Washington Avenue Extension, Albany, NY 12205

NORTH CAROLINA

Bowman Gray School of Medicine, Wake Forest University: Jack W. Strandhoy, Ph.D.,* Department of Physiology/Pharmacology, Bowman Gray School of Medicine, Wake Forest University, 300 S. Hawthorne Road, Winston-Salem, NC 27103

Duke University, School of Medicine: William J. Murray, M.D., Ph.D.,* Box 3061, Duke University Medical Center, Durham, NC 27710

East Carolina University, School of Medicine: Wallace R. Wooles, Ph.D. Chairman, Dept. of Pharmacology, School of Medicine, East Carolina University, Greenville, NC 27834

University of North Carolina, Chapel Hill, School of Medicine: Tai-Chan Peng, M.D.,* Department of Pharmacology, School of Medicine, Faculty Laboratory Office Building 231H, University of North Carolina, Chapel Hill, NC 27514

University of North Carolina, Chapel Hill, School of Pharmacy: Richard J. Kowalsky, Pharm.D.,* School of Pharmacy, 24 Beard Hall 200H, University of North Carolina, Chapel Hill, NC 27514

North Carolina Pharmaceutical Association: George H. Cocolas, Ph.D.,* Beard Hall 200H, Chapel Hill, NC 27514

NORTH DAKOTA

North Dakota State University, College of Pharmacy: William M. Henderson, Ph.D.,* College of Pharmacy, North Dakota State University, Fargo, ND 58105

North Dakota Pharmaceutical Association: William H. Shelver, Ph.D., College of Pharmacy, North Dakota State University, Fargo, ND 58105

OHIO

Case Western Reserve University, School of Medicine: Kenneth A. Scott, Ph.D.,* School of Medicine, Case Western Reserve University, 2119 Abington Road, Cleveland, OH 44106

Medical College of Ohio at Toledo: Robert D. Wilkerson, Ph.D.,* Department of Pharmacology, Medical College of Ohio, 3000 Arlington Avenue, Toledo, OH 43699

Northeastern Ohio University, College of Medicine: Ralph E. Berggren, M.D., Vice Provost for Academic Affairs, College of Medicine, Northeastern Ohio University, Rootstown, OH 44272

Ohio Northern University, College of Pharmacy and Allied Health Sciences: Dr. Ajaz S. Hussain, College of Pharmacy and Allied Health Sciences, Ohio Northern University, Ada, OH 45810

Ohio State University, College of Medicine: Gopi A. Tejwani, Ph.D., Department of Pharmacology, College of Medicine, Ohio State University, 5086 Graves Hall, 333 West 10th Avenue, Columbus, OH 43210-1239

Ohio State University, College of Pharmacy: Michael C. Gerald, Ph.D.,* College of Pharmacy, Ohio State University, 500 West 12th Avenue, Columbus, OH 43210

University of Cincinnati, College of Medicine: Leonard T. Sigell, Ph.D.,* College of Medicine, University of Cincinnati, Rm. 7701, 231 Bethesda Avenue, Mail Location No. 144, Cincinnati, OH 45267-0144

University of Cincinnati, College of Pharmacy: Henry S. I. Tan, Ph.D.,* College of Pharmacy, University of Cincinnati, 136 Health Professions Building, 3223 Eden Avenue, Mail Location No. 4, Cincinnati, OH 45267

University of Toledo, College of Pharmacy: Norman F. Billups, Ph.D., Dean, College of Pharmacy, University of Toledo, Toledo, OH 43606

Wright State University, School of Medicine: John O. Lindower, M.D., Ph.D.,* 3301 Stonebridge Road, Kettering, OH 45419

*Present at the 1985 Quinquennial Meeting.

Ohio State Medical Association: Ray W. Gifford, Jr., M.D.,* 3479 Glen Allen Drive, Cleveland, OH 44121

Ohio State Pharmaceutical Association: J. Richard Wuest, Pharm.D.,* 2720 Topichills Drive, Cincinnati, OH 45211

OKLAHOMA

Oral Roberts University, School of Medicine: Jimmie L. Valentine, Ph.D.,* 8181 South Lewis, Tulsa, OK 74137

Southwestern Oklahoma State University, School of Pharmacy: William G. Waggoner, Ph.D.,* School of Pharmacy, Southwestern Oklahoma State University, 100 Campus Drive, Weatherford, OK 73096

University of Oklahoma, College of Pharmacy: Loyd V. Allen, Jr., Ph.D.,* College of Pharmacy, University of Oklahoma, 1110 N. Stonewall, P. O. Box 26901, Oklahoma City, OK 73190

Oklahoma Pharmaceutical Association: Carl D. Lyons,* Skyline Terrace Nursing Center, 6202 E. 61st, Tulsa, OK 74136

Oklahoma State Medical Association: Clinton Nicholas Corder, M.D., Ph.D.,* 1000 N. Lee, P. O. Box 205, Oklahoma City, OK 73101

OREGON

Oregon Health Sciences University, School of Medicine: Hall Downes, M.D., Ph.D., Pharmacology (SM) L221, Oregon Health Sciences University, Portland, OR 97201

Oregon State University, College of Pharmacy: Freya F. Hermann, College of Pharmacy, Oregon State University, Corvallis, OR 97331

Oregon Medical Association: Richard E. Lahti, M.D.,* 2350 S.W. Multnomah Boulevard, Portland, OR 97219

Oregon State Pharmacists Association: Mrs. Hallie L. Lahti,* 1601 S.E. Oak Shore Lane, Milwaukie, OR 97222

PENNSYLVANIA

Duquesne University, School of Pharmacy: Lawrence H. Block, Ph.D., School of Pharmacy, Mellon Hall of Science, Room 441, Duquesne University, Pittsburgh, PA 15282

Hahnemann University, School of Medicine: Vincent J. Zarro, M.D., Hahnemann University M.S., 431 Broad & Vine Streets, Philadelphia, PA 19102

Medical College of Pennsylvania: Athole G. McNeil Jacobi, M.D., F.F.A.R.C.S.,* Medical College of Pennsylvania, 3300 Henry Avenue, Philadelphia, PA 19129

Pennsylvania State University, College of Medicine: John D. Connor, Ph.D.,* Milton S. Hershey Medical Center, Pennsylvania State University, P. O. Box 850, Hershey, PA 17033

Philadelphia College of Pharmacy and Science: Alfonso R. Gennaro, Ph.D.,* Philadelphia College of Pharmacy and Science, 43rd Street and Kingsessing Mall, Philadelphia, PA 19104

Temple University, School of Medicine: Charles A. Papacostas, Ph.D.,* Professor Emeritus, Temple University School of Medicine, 260 North Bent Street, Wyncote, PA 19095

Temple University, School of Pharmacy: Murray Tuckerman, Ph.D.,* School of Pharmacy, Temple University, 3307 N. Broad Street, Philadelphia, PA 19140

Thomas Jefferson University, Jefferson Medical College: C. Paul Bianchi, Ph.D.,* Jefferson Medical College, Thomas Jefferson University, 1020 Locust Street, Philadelphia, PA 19107

University of Pennsylvania, School of Medicine: George B. Koelle, M.D., Ph.D., Department of Pharmacology, School of Medicine, University of Pennsylvania, 36th and Hamilton Walk, Philadelphia, PA 19104-6084

University of Pittsburgh, School of Medicine: Robert H. McDonald, Jr., M.D.,* School of Medicine, University of Pittsburgh, 448 Scaife Hall, 3550 Terrace Street, Pittsburgh, PA 15261

University of Pittsburgh School of Pharmacy: Randy P. Juhl, Ph.D., Dean, School of Pharmacy, University of Pittsburgh, 1103 Salk Hall, Pittsburgh, PA 15261

Pennsylvania Medical Society: Benjamin Calesnick, M.D.,* 646 W. Springfield Road, Springfield, PA 19064

Pennsylvania Pharmaceutical Association: Joseph A. Mosso R.Ph.,* 319 Carolyn Avenue, Latrobe, PA 15650

PUERTO RICO

Ponce School of Medicine: Dr. Arthur L. Hupka, Dept. of Pharmacology, Ponce School of Medicine, P.O. Box 7004, Ponce, PR 00732

Universidad Central del Caribe School of Medicine: Jesús Santos-Martínez, Ph.D., Department of Physiology, School of Medicine, Universidad Central del Caribe, P.O. Box 935, Cayey, PR 00634

University of Puerto Rico, College of Pharmacy: Andrés Malavé, Ph.D., Dean, College of Pharmacy, University of Puerto Rico, G.P.O. Box 5067, San Juan, PR 00936-5067

RHODE ISLAND

Brown University Program in Medicine: Michael C. Wiemann, M.D.,* Roger Williams General Hospital, 825 Chalkstone Avenue, Providence, RI 02908

University of Rhode Island, College of Pharmacy: Christopher Rhodes, Ph.D.,* Department of Pharmaceutics, University of Rhode Island, Kingston, RI 02881

SOUTH CAROLINA

Medical University of South Carolina, College of Medicine: James F. Cooper, Pharm.D., 1056 Fort Sumter Dr., Charleston, SC 29412

Medical University of South Carolina, College of Pharmacy: Paul J. Niebergall, Ph.D.,* Department of Pharmaceutical Sciences, Medical University of South Carolina, 171 Ashley Avenue, Charleston, SC 29425

University of South Carolina, College of Pharmacy: Robert L. Beamer, Ph.D.,* College of Pharmacy, University of South Carolina, Columbia, SC 29208

SOUTH DAKOTA

South Dakota State University, College of Pharmacy: Gary S. Chappell, Ph.D., College of Pharmacy, South Dakota State University, Box 2201, Brookings, SD 57007

TENNESSEE

East Tennessee State University, Quillen-Dishner College of Medicine: Ernest A. Daigneault, Ph.D.,* 104 Hillside Road, Johnson City, TN 37601

Meharry Medical College, School of Medicine: Dolores C. Shockley, Ph.D.,* Meharry Medical College, 1005 D. B. Todd Boulevard, Nashville, TN 37208

University of Tennessee, College of Medicine: Murray Heimberg, M.D., Ph.D.,* University of Tennessee Center for Health Sciences, 874 Union Avenue, Room 100, Memphis, TN 38163

University of Tennessee, College of Pharmacy: Michael R. Ryan, Ph.D., Dean, College of Pharmacy, University of Tennessee, Center for the Health Sciences, 874 Union Avenue, Room 109, Memphis, TN 38163

TEXAS

Texas Southern University, College of Pharmacy and Health Sciences: Mary Ann Galley, Pharm.D.,* College of Pharmacy and Health Sciences, Texas Southern University, 3100 Cleburne Street, Houston, TX 77004

Texas Tech University, School of Medicine: Thomas W. Hale, Ph.D.,* Texas Tech University Health Science Center, Regional Academic Health Center at Amarillo, 1400 Wallace Boulevard, Amarillo, TX 79106

University of Houston, College of Pharmacy: Joseph P. Buckley, Ph.D.,* Texas Medical Center, Rm. 431, University of Houston, 1441 Moursund, Houston, TX 77030

University of Texas, Austin, College of Pharmacy: James T. Doluisio, Ph.D.,* College of Pharmacy, University of Texas at Austin, Austin, TX 78712

University of Texas, Medical School, Galveston: Wayne R. Snodgrass, M.D., Ph.D., Professor of Pharmacology & Pediatrics, University of Texas Medical Branch, Galveston, TX 77550

University of Texas Medical School, Houston: Larry K. Pickering, M.D.,* Medical School at Houston, University of Texas, 6431 Fannin, P. O. Box 20708, Houston, TX 77030

University of Texas Medical School, San Antonio: Arthur H. Briggs, M.D.,* University of Texas Health Science Center at San Antonio, 7703 Floyd Curl Drive, San Antonio, TX 78284

Texas Medical Association: Robert H. Barr, M.D., P.O. Box 25249, Houston, TX 77005

Texas Pharmaceutical Association: Shirley McKee,* 11213 Shannon Hills, Houston, TX 77099

UTAH

University of Utah, College of Pharmacy: Harold H. Wolf, Ph.D., Dean, College of Pharmacy, University of Utah, 201 Skaggs Hall, Salt Lake City, UT 84109

University of Utah, School of Medicine: Douglas E. Rollins, M.D., Ph.D.,* Department of Pharmacology, School of Medicine, University of Utah, 50 N. Medical Drive, Salt Lake City, UT 84132

Utah Pharmaceutical Association: Robert V. Petersen, Ph.D.,* College of Pharmacy, University of Utah, Salt Lake City, UT 84112

*Present at the 1985 Quinquennial Meeting.

VERMONT

University of Vermont, College of Medicine: John J. McCormack, Ph.D., Dept. of Pharmacology, College of Medicine, Given Medical Building, University of Vermont, Burlington, VT 05405

VIRGINIA

Eastern Virginia Medical School: Desmond R. H. Gourley, Ph.D.,* Department of Pharmacology, Eastern Virginia Medical School, 700 Olney Road, Norfolk, VA 23501

Medical College of Virginia/Virginia Commonwealth University, School of Medicine: Albert J. Wasserman, M.D., Box 565, MCV Station, Richmond, VA 23298

Medical College of Virginia/Virginia Commonwealth University, School of Pharmacy: William H. Barr, Pharm.D., Ph.D.,* MCV School of Pharmacy, Virginia Commonwealth University, MCV Station Box 581, Richmond, VA 23298-0001

University of Virginia, School of Medicine: Peyton E. Weary, M.D.,* Chairman, Department of Dermatology, School of Medicine, University of Virginia, Box 134-Medical Center, Charlottesville, VA 22908

Medical Society of Virginia: William J. Hagood, Jr., M.D.,* P. O. Box 158, Clover, VA 24534

Virginia Pharmaceutical Association: Elmer R. Deffenbaugh, Jr.,* 1407 Cummings Drive, Richmond, VA 23220

WASHINGTON

University of Washington, School of Medicine: David W. Johnson, M.D., Director, Area Health Education Ctr., WAMI Regional Programs, School of Medicine, SC-64, University of Washington, Seattle, WA 98195

University of Washington, School of Pharmacy: Lynn R. Brady, Ph.D.,* Asst. Dean, School of Pharmacy, T-341 Health Sciences, SC-69, University of Washington, Seattle, WA 98195

Washington State University, College of Pharmacy: William E. Johnson, Ph.D.,* College of Pharmacy, Washington State University, Pullman, WA 99164-6510

Washington State Pharmacists Association: Danial Baker, College of Pharmacy, WSU-Spokane, W. 601 First Avenue, Spokane, WA 99204

WEST VIRGINIA

Marshall University, School of Medicine: John L. Szarek, Ph.D., Department of Pharmacology, Marshall University School of Medicine, Huntington, WV 25704-2901

West Virginia University Medical Center, School of Pharmacy: Sidney A. Rosenbluth, Ph.D.,* Dean, School of Pharmacy, West Virginia University Medical Center, Morgantown, WV 26506

West Virginia Pharmacists Association: Art Jacknowitz, Pharm.D.,* 329 Wagner Road, Morgantown, WV 26505

WISCONSIN

Medical College of Wisconsin: Richard I. H. Wang, M.D., Ph.D.,* VA Medical Center, 5000 W. National Avenue, Wood, WI 53193

University of Wisconsin, Madison, School of Pharmacy: Chester A. Bond, Pharm.D., School of Pharmacy, University of Wisconsin, Madison, 425 N. Charter Street, Madison, WI 53706

University of Wisconsin Medical School, Madison: Joseph M. Benforado, M.D.,* 730 Seneca Place, Madison, WI 53711

Wisconsin Pharmacists Association: Dennis Dziczkowski,* 11330 W. Woodside Drive, Hales Corners, WI 53130

WYOMING

University of Wyoming, School of Pharmacy: Kenneth F. Nelson, Box 3375, Laramie, WY 82071

Wyoming Pharmaceutical Association: Linda G. Sutherland School of Pharmacy, University of Wyoming, Laramie, WY 82071

MEMBERS-AT-LARGE

Norman W. Atwater, Ph.D.,* Box 85, Zion Rd., Hopewell, NJ 08525

Herbert S. Carlin, D.Sc.,* 62 Hilltop Drive, Chappaqua, NY 10514

Lester Chafetz, Ph.D.,* School of Pharmacy, University of Missouri, Kansas City, MO 64110

J. Richard Crout, M.D.,* Boehringer Mannheim Corporation, 1301 Piccard Drive, Rockville, MD 20850

Lloyd E. Davis, D.V.M., Ph.D.,* Professor of Clinical Pharmacology, University of Illinois at Urbana-Champaign, College of Veterinary Medicine, 1102 W. Hazelwood Drive, Urbana, IL 61801

Leroy Fevang, Executive Director, Canadian Pharmaceutical Association, 101 - 1815 Alta Vista, Ottawa, Ontario K1G 3Y6, Canada

Fred T. Mahaffey, Pharm.D., 116 North Hamlin, Park Ridge, IL 60068

Stuart L. Nightingale, M.D.,* Associate Commissioner for Health Affairs, Department of Health & Human Services, Public Health Service, Food and Drug Administration, 5600 Fishers Lane, Rockville, MD 20857

Daniel A. Nona, Ph.D.,* Executive Director, The American Council on Pharmaceutical Education, 311 West Superior Street, Chicago, IL 60610

Mark Novitch, M.D.,* The Upjohn Company, Kalamazoo, MI 49001

Donald O. Schiffman, Ph.D.,* Secretary, United States Adopted Names Council, 535 North Dearborn Street, Chicago, IL 60610

Albert L. Sheffer, M.D.,* 110 Francis Street, Boston, MA 02215

Eugene A. Timm, Ph.D.,* Immuno-U.S., Inc., 1200 Parkdale Road, Rochester, MI 48063

Carl E. Trinca, Ph.D.,* Executive Director, American Association of Colleges of Pharmacy, 1426 Prince Street, Alexandria, VA 22314

Lawrence C. Weaver, Ph.D.,* Vice President, Professional Relations, Pharmaceutical Manufacturers Association, 1100 Fifteenth Street, N.W., Washington, DC 20005

Members-at-Large (Representing Other Countries That Provide Legal Status To USP or NF)

Salman A. Alfarsi, Ph.D., General Director, General Laboratories and Blood Bank, Saudi Arabia

Thomas D. Arias, Ph.D.,* Departamento del Investigación y Docenia, Universidad de Panamá, Apartado Postal 10767, Estafeta Universitaria, Panamá, Republica de Panamá

Denys Cook, Ph.D.,* Director, Drug Research Laboratories, Health Protection Branch, Tunney's Pasture, Ottawa, Ontario, K1A 0L2, Canada

Quintin L. Kintanar, M.D., Ph.D., Deputy Director General, National Science & Technology Authority, Bicutan, Taguig, Philippines

Kun Ho Yong, Director, Department of Safety Research, NIH, Republic of Korea, 5 Nokbun-Dong, Eunpyung-Ku, Seoul, 122, Korea

Members-at-Large (Public)

Estelle G. Cohen, M.A.* 5813 Greenspring Avenue, Baltimore, MD 21209

Alexander Grant,* Associate Commissioner for Consumer Affairs, Department of Health & Human Services, Public Health Service, Food and Drug Administration, Parklawn Bldg., Room 16-85, 5600 Fishers Lane, Rockville, MD 20857

Paul G. Rogers,* Chairman of the Board, National Council on Patient Information & Education, 666 11th Street, N.W., Washington, DC 20006

Frances M. West,* Secretary, State of Delaware, Department of Community Affairs, 156 South State Street, P. O. Box 1401, Dover, DE 19901

Ex-Officio Members

William H. Barr,* Pharm.D., Ph.D.,* School of Pharmacy, Medical College of Virginia, Health Sciences Div. of VCU, Richmond, VA 23219

Peter Goldman, M.D., Food and Drug Laboratory, 665 Huntington Avenue, Boston, MA 02115

Felix B. Gorrell,* 5035 35th Road N., Arlington, VA 22207

William J. Kinnard, Jr., Ph.D.,* Dean, School of Pharmacy, University of Maryland, 20 North Pine Street, Baltimore, MD 21201

Irwin Lerner, President and Chief Executive Officer, Hoffmann-La-Roche, Inc., Nutley, NJ 07110

John A. Owen, Jr., M.D.,* Box 242, University of Virginia Hospital, Charlottesville, VA 22901

Harry C. Shirkey, M.D.,* 1216 Paxton Avenue, Cincinnati, OH 45208

*Present at the 1985 Quinquennial Meeting.

Committee Chairmen

Credentials Committee: Peyton E. Weary, M.D.,* Chairman, Dept. of Dermatology, School of Medicine, University of Virginia, Box 134 Medical Center, Charlottsville, VA 22908

Nominating Committee for the General Committee of Revision: Klaus G. Florey, Ph.D.,* Squibb Institute for Medical Research, P. O. Box 191, New Brunswick, NJ 08903

Nominating Committee for Officers and Trustees: Ray W. Gifford, M.D.,* 3479 Glen Allen Drive, Cleveland Heights, OH 44121

Resolutions Committee: Arthur Hull Hayes, Jr., M.D.,* President and Chief Executive Officer, EM Industries, Inc., 5 Skyline Drive, Hawthorne, NY 10532

General Committee of Revision: William M. Heller, Ph.D.,* 12601 Twinbrook Parkway, Rockville, MD 20852

Honorary Members

George F. Archambault, Pharm.D., J.D.,* 5916 Melvern Drive, Bethesda, MD 20817

Lloyd C. Miller, Ph.D.,* 1625 Skyhawk Road, Escondido, CA 92025

Harry C. Shirkey, M.D.,* 1216 Paxton Avenue, Cincinnati, OH 45208

*Present at the 1985 Quinquennial Meeting.

Articles of Incorporation

The United States Pharmacopoeial Convention of May 1900, directed the Board of Trustees to take out articles of incorporation for the Convention under the laws of the District of Columbia. While the Convention had appointed certain of its officers to sign the articles, some changes were necessary because the laws aforesaid require that a majority of the Incorporators be residents of the District of Columbia.

These preliminaries having been arranged, the following certificate of incorporation was drawn up, signed, and recorded, finally, on the eleventh day of July, 1900:

Certificate of Incorporation

This is to certify that we, whose names are hereunto subscribed, citizens of the United States, of full age, and a majority citizens of the District of Columbia, do associate ourselves together pursuant to the provisions of sections 545–552 inclusive of the Revised Statutes of the United States relating to the District of Columbia and of the Act of Congress to amend the same, approved the twenty-third day of April, 1884, under the corporate name of The United States Pharmacopoeial Convention.

This Association is organized for a period of nine hundred and ninety-nine years.

The particular objects and business of this Association are the encouragement and promotion of the science and art of medicine and pharmacy by selecting by research and experiment and other proper methods and by naming such materials as may be properly used as medicines and drugs with formulas for their preparation; by establishing one uniform standard and guide for the use of those engaged in the practice of medicine and pharmacy in the United States whereby the identity, strength, and purity of all such medicines and drugs may be accurately determined, and for other like and similar purposes; and by printing and distributing at suitable intervals such formulas and the results of such and similar selections, names and determinations among the members of this Association, pharmacists, and physicians generally in the United States and others interested in pharmacy and medicine.

The management and control of the affairs, funds, and property of this Association for the first year of its existence shall be vested in a Board of Trustees consisting of the seven following persons:

ALBERT E. EBERT.
SAMUEL A. D. SHEPPARD.
WILLIAM S. THOMPSON.
CHARLES E. DOHME.
GEORGE W. SLOAN.
HORATIO C. WOOD.
CHARLES RICE.

In testimony whereof we have hereunto set our hands and affixed our seals this seventh day of July, 1900.

WILLIAM S. THOMPSON. [SEAL]
G. LLOYD MAGRUDER. [SEAL]
JOHN T. WINTER. [SEAL]
THOMAS C. SMITH. [SEAL]
MURRAY G. MOTTER. [SEAL]
WILLIAM M. MEW. [SEAL]
FRANK M. CRISWELL. [SEAL]

Constitution and Bylaws

THE UNITED STATES PHARMACOPEIAL CONVENTION

Under which USP XXII and NF XVII were prepared

As revised by the Convention at its Quinquennial meeting, March 22–24, 1985

CONSTITUTION

Article I—Name and Objects

Section 1. The corporate name of this corporation shall be "The United States Pharmacopoeial Convention," hereinafter referred to in this Constitution as the Pharmacopeial Convention. The alternative spelling, "Pharmacopeial," also shall be considered as official.

Section 2. Its objects shall be those declared in its Certificate of Incorporation, and include (a) the revision and publication of the Pharmacopeia of the United States of America and the National Formulary, also referred to as The United States Pharmacopeia and as the USP and as the NF, respectively, and of the supplements thereto; and (b) the publication regularly or at suitable times of other information of related scientific purpose. The term authorized shall be applied to such information when it has been prepared in accordance with the rules and procedures adopted by the Committee of Revision, or otherwise by direction of the Board of Trustees. Approval of the Board of Trustees shall be required for publication in either case.

Section 3. The Pharmacopeia of the United States of America and the National Formulary, together with their supplements, are published by and issued under the authority of the Pharmacopeial Convention; they are prepared and regularly revised entirely or in part by a committee or committees of representative medical and pharmaceutical experts and other appropriately qualified individuals; their primary purpose is to provide authoritative standards and specifications for materials and substances and their preparations that are used in the practice of the healing arts; they establish titles, definitions, descriptions, and standards for identity, quality, strength, purity, packaging and labeling, and also, where practicable, bioavailability, stability, procedures for proper handling and storage, and methods for their examination and formulas for their manufacture or preparation.

Article II—Membership

Section 1. The members of the Pharmacopeial Convention shall consist of: accredited delegates representing the following educational institutions, professional and scientific organizations, and designated divisions of the Federal Government, hereafter designated as eligible organizations, and others as herein provided.

In this Constitution and Bylaws "state" or "United States" includes each of the several states, the District of Columbia, the Commonwealth of Puerto Rico, and all territorial possessions of the United States of America.

(*a*) Colleges and schools of medicine in the United States, accredited by the Liaison Committee on Medical Education sponsored by the Association of American Medical Colleges and by the Council on Medical Education of the American Medical Association; colleges and schools of pharmacy in the United States with programs accredited by the American Council on Pharmaceutical Education; state medical associations or state medical societies, which are constituents of the American Medical Association; and state pharmaceutical associations or state pharmaceutical societies, which are entitled to voting delegates in the House of Delegates of the American Pharmaceutical Association.

(*b*) The American Chemical Society, the American Dental Association, the American Hospital Association, the American Medical Association, the American Nurses Association, the American Pharmaceutical Association, the American Society for Clinical Pharmacology and Therapeutics, the American Society of Hospital Pharmacists, the American Society for Pharmacology and Experimental Therapeutics, the American Society for Quality Control, the American Veterinary Medical Association, the Association of Food and Drug Officials, the Association of Official Analytical Chemists, the Chemical Manufacturers Association, the Cosmetic, Toiletry, and Fragrance Association, the Health Industry Manufacturers Association, The National Association of Boards of Pharmacy, the National Association of Chain Drug Stores, the National Association of Retail Druggists, the National Wholesale Druggists' Association, the Parenteral Drug Association, the Pharmaceutical Manufacturers Association, and The Proprietary Association; or the successors thereto.

(*c*) The Bureau of Medicine and Surgery, United States Navy; the Department of Medicine and Surgery, Veterans Administration; the Food and Drug Administration; the National Bureau of Standards; the office of the Surgeon General, United States Air Force; the office of the Surgeon General, United States Army; the office of the Surgeon General, United States Public Health Service; the office of the Secretary of the Department of Health and Human Services; and the Office of the President of the United States to represent Consumer Affairs; or the successors thereto.

(*d*) An individual who has been appointed to be a delegate by one of not more than five organizations that have been designated as eligible organizations by the Board of Trustees. Such designation by the Board of Trustees shall be effective until the next stated meeting.

(*e*) The Officers, the Board of Trustees, and the Chairmen of Committees directed to report at a stated meeting, serving by authority of the preceding stated Pharmacopeial Convention or the Constitution and Bylaws, also shall be members, ex officio, of the Pharmacopeial Convention.

(*f*) Not more than twenty members-at-large, selected by the Board of Trustees for their special competence in relation to the administrative or scientific needs of the Pharmacopeia, also shall be members of the Pharmacopeial Convention until the next stated meeting.

(*g*) Not more than five members-at-large selected by the Board of Trustees to serve as public members also shall be members of the Pharmacopeial Convention until the next stated meeting.

(*h*) Not more than five members-at-large selected by the Board of Trustees to represent nations, other than the United States, that give official recognition to the standards of the USP or the NF also shall be members of the Pharmacopeial Convention until the next stated meeting.

(*i*) Nominees for elective offices, selected in accordance with Chapter IX, Sections 2 and 3, of the Bylaws, shall be members of the Pharmacopeial Convention for the stated meeting when their names are presented as nominees in accordance with Chapter IV, Section 13 (*h*), of the Bylaws, whether present or not.

Section 2. Each eligible organization entitled to representation in the Pharmacopeial Convention, as provided for in Article II, Section 1, of this Constitution shall elect or appoint in such manner as it may determine, one delegate and one alternate delegate. One of these shall be accredited and seated as a member of the Pharmacopeial Convention, as provided in the Bylaws.

Section 3. Each delegate and alternate delegate, at the time of his or her election or appointment, and each member of the Convention, during his or her membership, except those named under Article II, Sections 1 (*f*) and (*g*), shall be an officer or a member of the academic staff of the educational institution, an officer, active member, or employee of the organization, or an employee of the division of the Federal Government, that he or she represents.

Section 4. Honorary Members. By unanimous vote the Board of Trustees may elect honorary members to the Convention in recognition of their distinguished contribution to drug standardization, the sciences of medicine and pharmacy, and the public health. Each honorary member shall have the rights, privileges, and duties prescribed by the Constitution and Bylaws and defined by the Board of Trustees except the right to vote, to hold office, or to otherwise determine the policy and administration of the Convention.

Article III—Officers

Section 1. The officers of the Pharmacopeial Convention shall be a President and a Vice-President, each of whom, at the time of his or her election, shall be a member of the Pharmacopeial Convention; a Past President; a Treasurer; and a Secretary. The President, the Vice-President, and the Treasurer shall be elected by ballot at each stated meeting. The Secretary of the Board of Trustees shall act as Secretary of the Pharmacopeial Convention.

Section 2. Vacancies in these offices shall be filled as provided in the Bylaws.

Article IV—Board of Trustees

The members of the Pharmacopeial Convention shall elect a Board of Trustees from the members of the Convention at each stated meeting, as provided in the Bylaws.

Article V—The General Committee of Revision

The members of the Pharmacopeial Convention shall elect a General Committee of Revision at each stated meeting, as provided in the Bylaws. The members of the Committee shall act as the Committee until its successor committee shall have been elected and shall have assumed the duties of the Committee.

Article VI—Meetings

Section 1. The stated meetings of the Pharmacopeial Convention shall be held in the United States at five-year intervals, beginning in 1980. The Board of Trustees is authorized to determine the day within the year and the place of the meeting.

Section 2. The President shall call a special meeting upon the written request of not fewer than fifty members of the Pharmacopeial Convention.

Article VII—Quorum

At any session of a meeting of the Pharmacopeial Convention, one hundred members who are eligible to vote shall constitute a quorum.

Article VIII—Amendments

Section 1. Every proposition to amend this Constitution, except as hereinafter provided, shall be submitted to the Board of Trustees in writing, subscribed to by at least five members of the Pharmacopeial Convention, not later than one hundred and twenty days prior to a stated or a special meeting and, after review by the Board, shall be submitted, together with the Board's recommendation, not later than sixty days prior to a stated or a special meeting of the Pharmacopeial Convention, to the medical and pharmaceutical press, to those authorized to send delegates to the Pharmacopeial Convention, and to those persons who are authorized to attend the stated or a special meeting. Except as provided in Section 4, at the next ensuing meeting of the Pharmacopeial Convention, after the expiration of the said sixty days, upon receiving the affirmative votes of not less than three-fourths of the members present and voting, the proposition shall become a part of this Constitution.

Section 2. An amendment also may be proposed in writing by ten or more members at any session of a stated meeting except the final session. Such amendment shall be referred to the Board of Trustees, which shall report upon it at the next succeeding session when, upon receiving the affirmative votes of not less than seven-eighths of the members present and voting, it shall become a part of this Constitution.

Section 3. The Board of Trustees may at its discretion mail to the members of the Convention a proposition to amend or alter this Constitution. A period of sixty days shall be allowed for the submission of comments on the proposition. A summary of the comments shall accompany the mail ballot by which the vote of the member may be recorded and returned within ninety days of the date appearing on the ballot. Upon receiving the affirmative vote of seven-eighths of the members voting, with a total of not less than one hundred members voting, the proposition shall become a part of this Constitution.

BYLAWS

Chapter I—The President

Section 1. The President shall preside at all meetings of The United States Pharmacopeial Convention, hereinafter referred to in these Bylaws as the Pharmacopeial Convention. In the absence of the President or in the event of his or her inability to preside, the Vice-President, or in the absence or inability of both, a President pro tempore, to be appointed by the Board of Trustees, shall conduct the meeting.

Section 2. The presiding officer shall present an address at the stated meeting.

Section 3. The presiding officer shall have the right to call a member to the chair in order that he or she may take the floor in debate. The presiding officer shall not vote except in case of a tie, when he or she shall cast the deciding ballot.

Section 4. Not later than six months before the stated meeting of the Pharmacopeial Convention, the President shall appoint, subject to the advice and consent of the Board of Trustees, five members of the Convention as a Nominating Committee for Officers and Trustees, two of whom shall be members of the Board of Trustees. Of the members appointed from the Board of Trustees, one shall be representative of the pharmaceutical sciences and one shall be representative of the medical sciences.

Section 5. The President shall appoint all committees not otherwise provided for in these Bylaws.

Section 6. Not later than one year prior to the stated meeting of the Pharmacopeial Convention, the President, through the Secretary, shall mail a notice of the meeting and a request to prepare for the authorization or reauthorization of a delegate and an alternate delegate to each of those organizations eligible for representation under the Constitution. He or she shall mail a second notice of the meeting and the request for the authorization or reauthorization of a delegate and an alternate delegate not later than six months prior to the stated meeting. He or she shall also request medical, pharmaceutical, and other appropriate journals to publish the announcement of the said meeting. At suitable intervals the President, through the Secretary, may request by mail from an eligible organization, authorization or reauthorization of a delegate and alternate delegate.

Section 7. The President shall be a member, ex officio, of the Board of Trustees, and of all committees except the Committee of Revision.

Section 8. The President shall act as President until his or her successor shall have been elected and installed.

Section 9. In the event of a vacancy in the office of President, the Vice-President shall become President, or in the event the Vice-President is unable to serve, the Board of Trustees shall elect a President.

Chapter II—The Vice-President

Section 1. The Vice-President shall preside in the absence of the President or in the event of his or her inability to serve.

Section 2. The Vice-President shall become President in case of a vacancy in the office of President.

Section 3. The Vice-President shall be a member, ex officio, of the Board of Trustees.

Section 4. The Vice-President shall act as Vice-President until his or her successor shall have been elected and installed.

Section 5. A vacancy in the office of Vice-President shall be filled by the Board of Trustees.

Chapter III—The Past President

Section 1. The Past President shall be the immediate past president.

Section 2. The Past President shall be a member, ex officio, of the Board of Trustees.

Section 3. In the event the immediate past president is unable to serve as Past President, the office shall remain vacant.

Chapter IV—The Executive Director and Secretary

Section 1. The Executive Director shall be elected by the Board of Trustees as provided in Chapter VI, Section 18, of these Bylaws.

Section 2. The Executive Director shall be a member, ex officio, and Chairman of the General Committee of Revision and of the Executive Committee of Revision.

Section 3. The Executive Director shall organize the General Committee of Revision in accordance with Chapter VII of these bylaws. The Executive Director shall appoint all other committees required for the work of revision.

Section 4. The Executive Director shall have charge of the work of Revision, and shall cause to be prepared the final manuscript of The United States Pharmacopeia, the National Formulary, and other authorized publications, and any supplements thereto.

Section 5. The Executive Director, the Executive Committee of Revision, and the Division Executive Committees shall consult on matters of general policy concerning the planning and executing of the revision of the Pharmacopeia, the National Formulary, and other authorized publications, and of supplements thereto.

Section 6. Whenever the occasion requires, the Executive Director shall certify in writing to the Board of Trustees that a new United States Pharmacopeia and new National Formulary, or a new supplement, has been completed, that each section of it has been approved by the Subcommittee to which its revision or preparation was assigned, and that the text in its entirety has been approved by the Division Executive Committee and the Executive Committee of Revision. The Executive Director shall also certify in writing to the Board of Trustees that an authorized publication or its supplement has been completed in accordance with existing rules and procedures.

Section 7. The Executive Director shall present an annual report to the Board of Trustees and also shall present a report of the General Committee of Revision at the stated meeting of the Pharmacopeial Convention.

Section 8. The Executive Director shall fill vacancies in the General Committee of Revision with the advice and consent of the General Committee of Revision and of the Board of Trustees.

Section 9. The Executive Director, with the advice and consent of the Board of Trustees, may appoint a committee composed of individuals from other nations who by law recognize or utilize the standards of the United States Pharmacopeia or the National Formulary to provide advice to the Committee of Revision.

Section 10. The Executive Director shall act as Executive Director until his or her successor shall have been elected and shall have assumed the duties of the office.

Section 11. In case of a vacancy in the position of Executive Director, the Chairman of the Board of Trustees shall appoint an Executive Director pro tempore, who shall serve until an Executive Director shall have been elected by the Board of Trustees.

Section 12. The Executive Director shall serve as the Secretary of the Board of Trustees and as Secretary of the Pharmacopeial Convention.

Section 13. Acting as the Secretary, the Executive Director shall:

(*a*) Keep the minutes of each meeting of the Pharmacopeial Convention and receive all reports, essays, and papers presented to the said meeting;

(*b*) Compile abstracts of the proceedings of the stated and special meetings and submit them to the Board of Trustees and, upon its order, arrange for their publication;

(*c*) At each stated meeting of the Pharmacopeial Convention, present a summary of the minutes of the preceding stated meeting;

(*d*) Present a report at the stated meeting of the Pharmacopeial Convention;

(*e*) Present to the Pharmacopeial Convention papers handed in for such purpose by the President;

(*f*) Record the ayes and nays when required;

(*g*) Notify the chairman and other members of each committee of their appointment, giving each member the name of the chairman of that committee and the names of the members and stating the business upon which the committee is to act;

(*h*) Immediately before the opening of the stated meeting of the Pharmacopeial Convention, post, in a conspicuous place in the meeting hall, a roll containing the names of the Officers of the Pharmacopeial Convention, the members of the Board of Trustees, the members of the General Committee of Revision, the Chairmen of Committees directed to report at the stated meeting, the persons selected by the Board of Trustees as members-at-large, the organizations designated as eligible organizations by the Board of Trustees, the persons selected by the Nominating Committee for Officers and Trustees as nominees for elected office, the delegates whose credentials have been approved by the Credentials Committee, and the order of business;

(*i*) During the stated meeting of the Pharmacopeial Convention, provide for each member two official ballots. One bailot shall contain the names of those nominated for President, for Vice-President, and for Treasurer of the Pharmacopeial Convention, and for the seven elective memberships on the Board of Trustees. A second ballot shall contain the names of those nominated for the General Committee of Revision, classified according to the outline of specific classifications of service prepared by the Executive Committee of Revision as provided in Chapter VII, Section 13, of these Bylaws, and designate the number to be elected in each classification, as provided in said outline;

(*j*) Deliver all reports, papers, and other documents of the Pharmacopeial Convention to the Board of Trustees;

(*k*) Issue such notices, announcements, and requests as provided for in Chapter I, Section 6, of these Bylaws, together with credential forms which shall be mailed with the second notice, and also mail the notices of special meetings as provided in Chapter XIII, Section 1, of these Bylaws;

(*l*) Submit to the Nominating Committee for the General Committee of Revision, not later than ninety days prior to the stated meeting of the Pharmacopeial Convention, all recommendations that have been submitted for membership on the General Committee of Revision;

(*m*) Acknowledge in writing the receipt of all resignations of membership and report them to the Board of Trustees;

(*n*) Serve as Secretary of all committees appointed by the President.

Section 14. In the absence of the Executive Director or in the event of his or her inability to serve as Secretary, the Chairman of the Board of Trustees shall appoint a Secretary pro tempore.

Section 15. The Executive Director shall act as Secretary until his or her successor shall have assumed the duties of the office.

Chapter V—The Treasurer

Section 1. The Treasurer shall receive all moneys coming from any source to the Pharmacopeial Convention and shall pay out moneys only on the authorization of and in accordance with the procedures adopted by the Board of Trustees.

Section 2. The Treasurer shall present annually to the Board of Trustees a complete financial statement and shall present at the stated meeting of the Pharmacopeial Convention a complete financial statement for the period since the preceding stated meeting.

The above statements of the Treasurer shall be audited by a certified public accountant.

Section 3. The Treasurer shall be a member, ex officio, of the Board of Trustees.

Section 4. During any temporary disability or absence of the Treasurer, an Acting Treasurer may be appointed by the Board of Trustees to serve in place of the Treasurer.

Section 5. The Treasurer shall act as Treasurer until his or her successor shall have been elected and installed.

Section 6. A vacancy in the office of Treasurer shall be filled by the Board of Trustees.

Chapter VI—The Board of Trustees

Section 1. The Board of Trustees shall consist of seven members of the Pharmacopeial Convention elected by plurality vote at the stated meeting; and the President, the Past President, the Vice-President, and the Treasurer, ex officio.

Of the seven elected members of the Board of Trustees, at least two shall be representative of the pharmaceutical sciences, and at least two shall be representative of the medical sciences. At least one of the elected members of the Board of Trustees shall be a public member.

The members of the Board of Trustees shall not be members of the General Committee of Revision.

No person shall be eligible to be an elected member of the Board of Trustees who, at the adjournment of the stated meeting, shall have served two successive terms as an elected member of the Board of Trustees.

Section 2. The officers of the Board of Trustees shall consist of a Chairman who shall be elected by a majority vote of the Board, and a Secretary. The Chairman of the Board of Trustees shall serve as a member, ex officio, of the Credentials Committee. The Secretary of the Board of Trustees need not be a member of the Board of Trustees or a member of the Pharmacopeial Convention. The Secretary of the Board of Trustees shall serve as Secretary of the Convention.

Section 3. The Board of Trustees shall have the right to transact business by correspondence. A motion by mail shall require no seconder.

Section 4. The Board of Trustees shall meet annually at such time and place as it shall determine. Special meetings of the Board of Trustees shall be called upon the written request of at least four members. The Chairman shall have the right to call a special meeting whenever he shall deem it necessary. Six members of the Board of Trustees shall constitute a quorum.

Section 5. Members of the Board of Trustees may participate in a special meeting of such Board by means of a telephone conference or similar communications equipment by means of which all persons participating in the meeting can hear each other at the same time, and participation by such means shall constitute presence in person at a meeting.

Section 6. No member of the Board of Trustees shall receive compensation for his services. Members of the Board of Trustees shall be reimbursed from the funds of the Pharmacopeial Convention for traveling and other necessary expenses that may be incurred by them in the performance of their duties.

Section 7. The Board of Trustees shall manage and control the affairs, funds, and property of the Pharmacopeial Convention, except as is herein otherwise provided.

The Board of Trustees may invest the funds of the Pharmacopeial Convention, authorize the payment of all moneys due for services performed, transact business involving financial or other matters that may be for the best interests of the Pharmacopeial Convention, and perform such other duties as the Pharmacopeial Convention may direct. Any and all contracts or agreements authorized by the Pharmacopeial Convention or by the Board of Trustees shall be executed on behalf of the Pharmacopeial Convention by the Chairman and the Secretary of the Board of Trustees, or by such other person or persons as the Board may designate.

Section 8. The Board of Trustees shall authorize a general revision of the United States Pharmacopeia, the National Formulary, or other authorized publication at suitable intervals and such supplements during the intervening periods as may be deemed necessary.

Upon receipt of written certification from the Executive Director that a new United States Pharmacopeia or National Formulary or a new supplement has been completed, that each section of it has been approved by the Subcommittee to which its revision or preparation had been assigned, and that it has been approved in its entirety by the Division Executive Committee and the Executive Committee of Revision, the Board of Trustees shall then act upon releasing the text and upon designating the date when it is to become official, said date to be reasonably distant from the date of its release. Upon the receipt of written certification from the Executive Director that an authorized publication or supplement has been completed in accordance with the rules and procedures adopted under Section 9, the Board of Trustees shall then act upon releasing the text.

Section 9. The Board of Trustees shall review all rules and procedures made or adopted by the General Committee of Revision and the Executive Committee of Revision and shall determine whether the rules and procedures conflict with the Constitution and Bylaws and whether they are sufficient to ensure the accuracy and adequacy of the text of the Pharmacopeia and the National Formulary and other authorized publications. The Board shall determine whether such rules and procedures shall be sufficient to provide adequate notice and opportunity for comment and full and impartial consideration of all proposed changes in, and additions to, the text. Where such conflict or insufficiency is found, the Board shall propose such amendments as may be indicated and, after having given the General Committee of Revision and the Executive Committee of Revision an opportunity to be heard, shall revise or modify the rules and procedures to bring about conformity with the Constitution and Bylaws and to ensure their sufficiency.

Section 10. The Board of Trustees shall be informed of the decisions of the Executive Committee of Revision on appeals for reconsideration, revision, or abrogation of standards adopted and decisions made by a Subcommittee of the Committee of Revision, Division Executive Committee, or other Committee of Revision body and, upon consideration of the facts, may request the entire Committee of Revision to review the decision.

Section 11. The advice and consent of the Board of Trustees shall be required for filling vacancies on the General Committee of Revision.

Section 12. Upon recommendation of the Executive Committee of Revision, the Board of Trustees may remove any member of the General Committee of Revision for cause.

Section 13. The Board of Trustees shall issue an annual financial report of the Pharmacopeial Convention audited by a certified public accountant.

Section 14. The Board of Trustees shall arrange for the bonding of Officers and employees to the extent necessary to protect the interests of the Pharmacopeial Convention.

Section 15. The Board of Trustees shall maintain a headquarters for conducting the business of the Pharmacopeial Convention.

Section 16. In the absence of or in the event of the inability of the President and Vice-President to preside at a meeting of the Pharmacopeial Convention, the Board of Trustees shall appoint a President pro tempore to conduct the meeting.

Section 17. During any temporary inability or absence of the Treasurer, an Acting Treasurer may be appointed by the Board of Trustees.

Section 18. At least sixty days prior to each stated meeting of the Pharmacopeial Convention, or at any time in the event of a vacancy in the position, the Board of Trustees shall elect an Executive Director. The Executive Director shall serve at the will of the Board of Trustees. In case of a vacancy in the position of Executive Director, the Chairman of the Board of Trustees shall appoint an Executive Director pro tempore, who shall serve until the Board of Trustees shall have elected an Executive Director. The Board of Trustees may request the advice of the Executive Committee of Revision before electing an Executive Director.

Section 19. The Board of Trustees shall provide for such assistants to the Executive Director as may be deemed necessary.

Section 20. The Board of Trustees shall select four members of the Pharmacopeial Convention, at least two of whom are representative of the medical and pharmaceutical sciences, who are neither members of the General Committee of Revision or of the Board of Trustees, nor salaried employees or Officers of the Pharmacopeial Convention, to serve on the Nominating Committee for the General Committee of Revision.

Section 21. The Board of Trustees may select not more than twenty persons for their special competence in relation to the administrative or scientific needs of the Pharmacopeia, to be members-at-large of the current or ensuing Pharmacopeial Convention.

Section 22. The Board of Trustees may select not more than five persons to be members-at-large of the current or ensuing Pharmacopeial Convention to serve as public members.

Section 23. The Board of Trustees may select not more than five persons to serve as members-at-large of the current or ensuing Pharmacopeial Convention to represent nations, other than the United States, that give official recognition to the standards of the USP or the NF.

Section 24. The Board of Trustees may designate not more than five organizations as eligible organizations until the next stated meeting and request the appointment of one delegate by each, who, upon approval of the Credentials Committee, shall be seated as a member of the Pharmacopeial Convention.

Section 25. The Board of Trustees may invite the national professional medical and pharmaceutical associations of other nations to send representatives as observers to a stated meeting of the Pharmacopeial Convention.

Section 26. The Chairman of the Board of Trustees shall present a report for the Board at each stated meeting of the Pharmacopeial Convention.

Section 27. The Board of Trustees shall receive from the Secretary abstracts of the stated and special meetings of the Pharmacopeial Convention and shall arrange for their publication.

Section 28. The Board of Trustees shall receive and review any proposition to amend the Constitution and Bylaws and shall recommend such action as it deems advisable as provided in the Constitution and Bylaws.

Section 29. The Board of Trustees shall fill vacancies for unexpired terms of office of the seven elected members of the Board of Trustees, each of whom, at the time of his or her election, shall be a member of the Pharmacopeial Convention; the President, in case the Vice-President is unable to serve; the Vice-President; the Treasurer; and the Executive Director and Secretary; as herein provided.

Section 30. Any officer or member of the Board of Trustees may be removed for cause by the Board of Trustees by an affirmative vote of not less than two-thirds of the members of the Board, other than himself or herself.

Section 31. The Board of Trustees shall approve the outline of specific classifications of service prepared by the Executive Committee of Revision and deemed necessary for the work of the Committee of Revision in preparing the next revisions of the United States Pharmacopeia, the National Formulary, and other authorized publications, and the number of persons to be elected to each classification.

Section 32. Between stated meetings of the Pharmacopeial Convention, the Board of Trustees may, at its discretion, mail to the members of the Convention a proposal to be acted upon by mail vote. On any such mail vote, not less than one hundred members shall cast a ballot to constitute a valid action and a majority of those voting shall determine the action.

Chapter VII—The General and Executive Committees of Revision

Section 1. The General Committee of Revision shall be composed of appropriately qualified individuals, including public members, together with the Executive Director, ex officio. The General Committee of Revision shall be elected at the stated meeting by a plurality of the votes cast by the members of the Pharmacopeial Convention as provided in these Bylaws. The Executive Director shall be Chairman of the General Committee of Revision.

Section 2. The General Committee of Revision shall be organized by the Executive Director with the approval of the Board of Trustees into appropriate divisions; each of these may be under the supervision of a Director, who shall act under the authority of the Executive Director. Each Division may be organized further by the Executive Director and may be organized into Subcommittees. The rosters of the Subcommittees shall be subject to the advice and consent of the Committee of Revision members in that Division. The appointment of the Chairmen of the Subcommittees by the Executive Director shall be subject to the approval of the respective Subcommittees.

No person who, during his or her service on the Committee, has a direct or indirect financial interest that conflicts or appears to conflict with his or her duties and responsibilities, may participate in the selection for admission to the United States Pharmacopeia or the National Formulary of therapeutic articles in which he or she has any financial interest.

Section 3. During the period between stated meetings additional Committee members may be added to the Committee of Revision if the Executive Committee of Revision and the Board of Trustees determine that such a need exists. Members of the Convention shall be requested to submit the names of persons whose qualifications entitle them to consideration as nominees for membership on the General Committee of Revision. The Executive Committee of Revision, insofar as possible, shall nominate twice the number of qualified persons, from among whom the Committee of Revision shall elect the additional Committee members who shall be added to the Committee roster until the next stated meeting.

Section 4. Each Division shall have a Division Executive Committee. Where a Division is organized into Subcommittees, the Chairmen of the Subcommittees and the Executive Director, ex officio, shall constitute the Division Executive Committee.

Otherwise, a Division Executive Committee shall be appointed by the Executive Director, subject to the advice and consent of the members of the Committee of Revision in that Division.

The Executive Committee of Revision shall be appointed by the Executive Director subject to the advice and consent of the Division Executive Committees. The Executive Director shall be Chairman of the Executive Committee of Revision.

Section 5. The General Committee of Revision and the Executive Committee of Revision shall make such rules and adopt such procedures, not in conflict with the Constitution and Bylaws, as are sufficient to ensure the accuracy and adequacy of the text of the Pharmacopeia and the National Formulary and other authorized publications, and to provide for adequate notice and opportunity for comment and full and impartial consideration of all proposed changes in, and additions to, the text, as may be necessary to the proper discharge of their respective functions. The rules and procedures shall be made available to the Members of the Pharmacopeial Convention for comment. The Committee shall submit the rules and procedures so made or adopted for review by the Board and shall comment upon any proposals made by the latter pursuant to such review.

Section 6. Each Subcommittee shall undertake the revision and preparation of assigned portions of each new Pharmacopeia and National Formulary, and other authorized publications, having regard for the principles of revision laid down by the Pharmacopeial Convention at the last meeting. Upon completion of its assignment, each Subcommittee shall review, with a view to approval, the final text of those portions for which it has been responsible.

Section 7. The General Committee of Revision, through the Executive Director, shall present a report at the stated meeting of the Pharmacopeial Convention.

Section 8. The members of the Division Executive Committees shall be members, ex officio, of the Nominating Committee for the General Committee of Revision.

Section 9. A vacancy in the General Committee of Revision shall be filled by the Executive Director with the advice and consent of the General Committee of Revision and of the Board of Trustees.

Section 10. The Executive Committee of Revision, the Division Executive Committees, and the Executive Director shall consult on matters of general policy concerning the planning and executing of the revision of the Pharmacopeia and the National Formulary and other authorized publications and of supplements thereto. The Executive Committee of Revision and the Division Executive Committees shall review, with a view to approval, the proposals of all Subcommittees and other committees for monographs on articles newly admitted and for revisions of, additions to, and deletions from existing text. The respective Division Executive Committee shall authorize that appropriate publicity be given to all such proposals prior to their final approval and shall consider suggestions for changes arising from the publicity thus given.

Section 11. The Executive Committee of Revision shall receive and rule upon all appeals for reconsideration, revision or abrogation of standards adopted and decisions made by a USP Subcommittee, Division Executive Committee, or other Committee of Revision body. Each decision of the Executive Committee of Revision on such appeals shall be referred to the Board of Trustees which, upon consideration of the facts, may request the entire Committee of Revision to review the decision. After such review, which shall be given as promptly as practicable, the Committee of Revision may uphold or reverse the Executive Committee's decision by a majority vote of the members voting.

Section 12. The Executive Committee of Revision shall approve the completed text before a new United States Pharmacopeia or a new National Formulary or a new supplement is released for publication.

Section 13. At least one year before the stated meeting of the Pharmacopeial Convention, the Executive Committee of Revision shall prepare and publish an outline of specific classifications of service deemed necessary to the efficient prosecution of the work of the next revisions of The United States Pharmacopeia and the National Formulary and other authorized publications and a statement of the number of persons to be elected in each classification. Upon approval of the Board of Trustees, this outline shall be followed in submitting to the Nominating Committee for the General Committee of Revision the names of persons to be considered for membership on the General Committee of Revision that is to be elected at the forthcoming stated meeting of the Pharmacopeial Convention.

Section 14. The Executive Committee of Revision may recommend to the Board of Trustees the removal, for cause, of any member of the General Committee of Revision.

Chapter VIII—The Committee on Resolutions

Section 1. The Committee on Resolutions, hereinafter referred to in this Chapter as the Committee, shall consist of five members of the Pharmacopeial Convention.

Prior to the stated meeting of the Pharmacopeial Convention, the President shall appoint five members of the Pharmacopeial Convention to serve as members of the Committee, one of whom shall be designated as Chairman, and such additional associates, who need not be members of the Convention, as may be required to aid the Committee.

Section 2. The Committee shall be responsible for reviewing the recommendations made in the addresses and reports of the Officers, Committees, and Board of Trustees; reviewing resolutions that have been submitted to it in written form at least thirty days prior to a stated meeting by eligible organizations and prior to the first session by individual members of the Pharmacopeial Convention, conferring with the parties concerned whenever necessary about the intent or any other aspect of the resolution; for drafting statements in resolution form that shall reflect the official policy of the Pharmacopeial Convention; and for presenting all resolutions to the Convention for consideration, unless later introduction of a resolution by a member is permitted by two-thirds vote of the members present and voting.

Section 3. The Committee shall make an initial report in at least one session prior to the Final Session of the meeting; the final report of the Committee and the vote on the resolutions submitted by the Committee shall be conducted in the Final Session.

Chapter IX—The Nominating Committee for Officers and Trustees

Section 1. The Nominating Committee for Officers and Trustees, hereinafter referred to in this Chapter as the Committee, shall consist of five members of the Pharmacopeial Convention, one of whom shall be designated as Chairman, appointed by the President with the advice and consent of the Board of Trustees, not later than six months before the stated meeting.

Section 2. The Committee shall nominate from the membership of the Pharmacopeial Convention or the General Committee of Revision two members for the office of President, two members for the office of Vice-President, and one or more persons, who need not be members of the Convention, for the office of Treasurer.

Section 3. The Committee shall nominate from the membership of the Pharmacopeial Convention or the General Committee of Revision fourteen members for the elected office of Trustee, divided into four classes: four shall represent pharmaceutical sciences; four shall represent medical sciences; four shall be without restriction concerning their affiliations; and two shall represent the public.

Section 4. The report of the Committee shall be submitted to the Secretary of the Pharmacopeial Convention not later than sixty days before the stated meeting; not later than forty-five days before the stated meeting, the Secretary shall notify the nominees of their selection by the Committee and the date of the stated meeting.

Section 5. The Committee shall present its report to the Convention, at which time additional nominations of Convention or Committee of Revision members may be made from the floor for Officers and the Board of Trustees, which shall be seconded by at least four members of the Pharmacopeial Convention and submitted in writing at once to the Secretary. A nomination for Trustee shall indicate in which of the four classes the nomination is being made.

Chapter X—The Nominating Committee for the General Committee of Revision

Section 1. The Nominating Committee for the General Committee of Revision, hereinafter referred to in this Chapter as the Committee, shall consist of the members of the Division Executive Committees, ex officio, together with four members of the Pharmacopeial Convention at least two of whom are representative of the medical and pharmaceutical sciences, selected by the Board of Trustees, as provided in Chapter VI, Section 20, of these Bylaws. The President of the Pharmacopeial Convention and the Executive Director shall be members, ex officio, of the Committee.

The Committee shall be organized at least one year prior to the stated meeting of the Pharmacopeial Convention and shall elect its own Chairman. The Secretary of the Convention shall serve as Secretary of the Committee.

Section 2. Not later than seven months prior to the stated meeting of the Pharmacopeial Convention, the Secretary of the Committee shall issue to all who are entitled to representation in the Pharmacopeial Convention, as provided in Article II, Section 1, of the Constitution, and to the members of the General Committee of Revision, a request that they submit the names of persons whose qualifications entitle them to consideration as nominees for membership on the General Committee of Revision. Each name submitted shall be accompanied by a statement on an official form supplied by the Committee indicating affiliations, academic and professional backgrounds, and the qualifications for the specific classification or classifications of membership on the General Committee of Revision as set forth in the outline referred to in Chapter VII, Section 13, of these Bylaws. All recommendations shall be mailed to the Secretary of the Pharmacopeial Convention not later than one hundred and twenty days prior to the stated meeting.

All recommendations received by the Secretary shall be submitted to the Committee not later than ninety days prior to the stated meeting and shall be given primary consideration by the Committee in selecting nominees for membership on the General Committee of Revision.

Section 3. At the stated meeting of the Pharmacopeial Convention, the Committee shall submit, through the Secretary, to each member, a list of names of persons whom it deems qualified to serve on the General Committee of Revision, together with a statement covering the classification or classifications of the work of revision for which each nominee is deemed best qualified, and the affiliations, academic and professional backgrounds and specific qualifications of each nominee.

The Committee shall nominate, insofar as possible, twice the number of qualified persons to be elected to meet each need shown in the outline referred to in Chapter VII, Section 13, of these Bylaws.

Additional nominations may be made from the floor for membership on the General Committee of Revision and shall be seconded on the floor by at least four members of the Pharmacopeial Convention. Each such nomination shall be submitted on the official form and presented at once to the Secretary of the Pharmacopeial Convention.

Chapter XI—The Credentials Committee

Section 1. The Credentials Committee shall consist of five members and shall be appointed by the President, who shall also appoint the Chairman, not later than sixty days after the stated meeting of the Pharmacopeial Convention. The Chairman of the Board of Trustees, and the President, and the Secretary of the Pharmacopeial Convention shall be members, ex officio, of the Credentials Committee. Members of the Credentials Committee shall retain their Pharmacopeial Convention membership until their successors are appointed.

Section 2. The Credentials Committee shall examine carefully the credentials of all delegates or alternate delegates in the absence of delegates. Credentials issued in blank, with the names of the delegates and alternate delegates to be inserted subsequently by other than the regularly constituted officers of the appointing educational institution, organization, or designated division of the Federal Government shall not be accepted. Credential forms presented for the first time at a meeting shall be independently verified before acceptance. Immediately prior to the opening of the stated meeting of the Pharmacopeial Convention, the Committee shall furnish to the Secretary of the Pharmacopeial Convention a list of delegates whose credentials have been approved by the Committee and a list of alternate delegates whom it has approved in lieu of delegates who are absent.

Section 3. The Credentials Committee shall make a report to the Pharmacopeial Convention concerning all credentials that have not been approved. A majority vote of the members of the Pharmacopeial Convention shall be required to overrule a decision of the Credentials Committee.

Chapter XII—Members

Section 1. Each delegate shall present his credentials to the Credentials Committee and, upon report by that Committee that the credentials are approved, shall be accredited and be seated as a member. If a delegate is absent from the stated meeting the alternate delegate shall present credentials as Alternate Delegate to the Credentials Committee and, upon report that the credentials are approved, shall be accredited and be seated as the member for that stated meeting.

Section 2. The Officers of the Pharmacopeial Convention, the Board of Trustees, the Chairmen of Committees directed to report at the stated meeting, serving by authority of the preceding stated Pharmacopeial Convention or the Constitution and Bylaws shall be members, ex officio, of the Pharmacopeial Convention. The persons selected by the Board of Trustees as members-at-large shall be members of the Convention.

Section 3. Nominees selected by the Nominating Committee for Officers and Trustees for the Board of Trustees, the office of President and the office of Vice-President shall be members of the Pharmacopeial Convention for the stated meeting when their names are presented as nominees in accordance with Chapter IV, Section 13 (*h*), of the Bylaws, whether present or not.

Section 4. A member of the Pharmacopeial Convention shall continue to be a member until the next stated meeting of the Convention unless he or she dies, resigns, or fails to meet the requirements as provided in Article II, Section 3, of the Constitution or until his or her successor has been authorized by the

Eligible Organization and his or her successor's credentials approved by the Credentials Committee.

Section 5. A member shall represent not more than one educational institution, organization, or designated division of the Federal Government.

Section 6. A member may be expelled for improper conduct or for violation of the Constitution and Bylaws, by a vote of not less than two-thirds of the members present and voting.

Section 7. Resignations of membership in the Pharmacopeial Convention shall be submitted in writing to the Secretary of the Convention. All resignations shall be acknowledged in writing by the Secretary and shall be reported to the Board of Trustees.

Chapter XIII—Meetings

Section 1. Meetings of the Pharmacopeial Convention shall be held as provided in Article VI of the Constitution.

The call for a stated meeting of the Pharmacopeial Convention shall be issued as provided in Chapter I, Section 6, of these Bylaws.

The call for a special meeting of the Pharmacopeial Convention shall state the purpose for which the meeting is called, and shall be mailed by the Secretary to all members and to all those entitled to representation under the Constitution, not later than sixty days prior to the meeting. The business of a special meeting shall be limited to the purpose for which the meeting has been called.

Section 2. Each stated or special meeting of the Pharmacopeial Convention shall consist of an opening and a final session, and such number of intervening sessions as the Convention may direct in order to conduct the business of the Convention. At least one hour shall elapse between any two sessions, and no session shall be opened at a time earlier than that announced at the session immediately preceding.

Any session may be recessed by a majority vote and shall be reconvened at the time specified in the recess motion.

Section 3. The Secretary, with the advice of the President, shall prepare the order of business for each stated meeting and shall distribute it to the Members, Delegates, and Alternate Delegates at least thirty days prior to the meeting. That order of business shall govern unless otherwise directed by the affirmative votes of not less than two-thirds of the members present and voting.

On all questions of procedure, the latest edition of Robert's Rules of Order, Newly Revised, shall prevail, unless otherwise specified herein.

Chapter XIV—The Corporate Seal

The seal of the Corporation shall contain the words: "The United States Pharmacopeial Convention. Corporate—1900—seal, D.C."

Chapter XV—Amendments

Section 1. Every proposition to amend these Bylaws, except as hereinafter provided, shall be submitted in accordance with the procedures to amend the Constitution as set forth in Article VIII, Sections 1 and 2, of the Constitution, and shall, under either procedure, upon receiving the affirmative votes of not less than three-fourths of the members present and voting, become a part of these Bylaws.

Section 2. The Board of Trustees may at its discretion mail to the members of the Pharmacopeial Convention a proposition to amend or alter these Bylaws. A period of sixty days shall be allowed for the submission of comments on the proposition. A summary of the comments shall accompany the mail ballot by which the vote of the member may be recorded and returned within ninety days of the date printed on the ballot. Upon receiving the affirmative vote of seven-eighths of the members voting, with a total of not less than one hundred members voting, it shall become a part of these Bylaws.

Chapter XVI—Dissolution

In the event of dissolution of the Corporation, the Board of Trustees shall after paying or making provision for the payment of all of the liabilities of the Corporation, dispose of all of the assets exclusively for the purposes of the Corporation in such manner as the Board of Trustees shall determine to such organization or organizations, organized and operated exclusively for charitable, educational or scientific purposes as at the time shall qualify as an exempt organization or organizations under Section 501(c) (3) of the Internal Revenue Code of 1954 (or the corresponding provision of any future United States Internal Revenue Law).

Rules and Procedures

Rules and Procedures of the General and Executive Committees of Revision for 1985–1990

The following rules and procedures of the General and Executive Committees of Revision for 1985–1990 were adopted in accordance with the USPC Bylaws. These rules and procedures have been set up within the principles set forth in the USPC Constitution and Bylaws, for the purposes of ensuring the accuracy and adequacy of the text of the USP and the NF and other authorized publications, and of providing for adequate notice and opportunity for comment and full and impartial consideration of all proposed revisions.

Each new General and Executive Committee of Revision adopts at the start of the quinquennium its rules and procedures, which are then made available to the members of the Pharmacopeial Convention for comment. The rules and procedures then are reviewed by the USPC Board of Trustees and, if adjudged to be not in conflict with the Constitution and Bylaws, they become duly adopted for the quinquennium.

RULES AND PROCEDURES of the GENERAL AND EXECUTIVE COMMITTEES OF REVISION for 1985–1990

I. The Executive Director

Section 1. The Executive Director is Chairman of the Executive Committee and the General Committee of Revision and exercises the authority vested by and responsibilities as provided in the Bylaws of the Pharmacopeial Convention and shall operate and be governed by these rules.

Section 2. As provided in the Bylaws, that Division of the General Committee of Revision that deals with the standardization of drugs and other articles in the United States Pharmacopeia (USP) and the National Formulary (NF) shall be under the supervision of the Director of the Drug Standards Division, who shall act under the authority of and in accordance with the directions of the Executive Director.

Section 3. As provided in the Bylaws, that Division of the General Committee of Revision that deals with the development of drug use information in *USP Dispensing Information* (*USP DI*) and its related publications and products shall be under the supervision of a Director, who shall act under the authority of and in accordance with the directions of the Executive Director.

II. The General Committee of Revision

Section 1. The General Committee of Revision shall function to determine the content of the USP, the NF, the USP DI, and other authorized publications with respect to: drug substances, dosage forms, medical devices, diagnostic agents, and other articles to be included; use information; general notices and requirements and standards and specifications for the individual articles admitted; assays and tests adequate for showing compliance with the adopted standards and specifications; and sections on general information, reagents, and reference tables.

Section 2. To achieve the objectives in foregoing Section 1, the Committee of Revision shall:

(a) Draft and adopt for USP and NF monographs standards and specifications for identity, strength, quality, purity, stability and bioavailability, tests and methods of assay, definitions, descriptions, formulas for manufacturing and preparation, or other parameters necessary to prescribe an identity for the articles, and specifications for packaging and labeling, with a view to helping to ensure that the article concerned shall be suitable for its intended use; and, consistent with the requirements of the monographs, draft and adopt the text of the General Notices and other sections of the official compendia.

(b) Provide in USP DI monographs, where practicable, information on the proper uses and precautions in using the articles; information on side effects; dosing information; patient consultation guidelines; and labeling, packaging, and storage, and other appropriate information related to the dispensing and use of the articles, including information in lay language.

Section 3. Upon approval by the Executive Committee, and the appropriate Division Executive Committee to whom the task is to be assigned, the Executive Director may assign extra compendial tasks related to the purposes of the USP Convention.

Section 4. The General Committee of Revision, through its chairman, the Executive Director, shall present a report at the stated meetings of the Pharmacopeial Convention.

Section 5. The General Committee of Revision shall advise and consent with respect to the filling of any vacancy on the Committee by the Executive Director.

Section 6. From nominees selected by the Executive Committee, the General Committee of Revision shall elect additional Committee members, during the revision period.

Section 7. The General Committee of Revision shall comprise two Divisions as provided in Chapter VII, Section 2, of the Bylaws. The Drug Standards Division (DSD) shall be responsible for revision of the United States Pharmacopeia and the National Formulary, and related publications; the Drug Information Division (DID), for USP Dispensing Information and related publications.

Section 8. The Drug Standards Division (DSD) shall be organized into not fewer than nine Subcommittees of which the membership shall be as determined by the Executive Director with the advice and consent of the members of the Committee of Revision in that Division.

Section 9. The Drug Information Division (DID) shall be organized into Expert Advisory Panels. Each Expert Advisory Panel shall be chaired or co-chaired by a member of the Committee of Revision appointed by the Executive Director. Members of each advisory panel shall be appointed by the Executive Director with the advice and consent of the Chairman of the respective Panel.

Section 10. Each DSD Subcommittee or DID Panel shall work within the areas determined by the Executive Director.

Section 11. Each DSD Subcommittee shall prepare assigned portions of each new compendium or authorized publication, taking into account any principles of revision laid down by the Pharmacopeial Convention. Each Subcommittee shall review, with a view to approval, the final text of those portions for which it has been responsible.

Section 12. Each DID Expert Advisory Panel shall review the information or text appearing in *USP DI Review* for accuracy and appropriateness relative to its specialty area and/or role in the health-care delivery system.

In the event that general agreement or consensus on a particular point is not evident as determined by the panel chairman in consultation with the Division Director, the matter shall be referred to the DID Executive Committee for final determination.

Section 13. A Subcommittee or Panel may meet upon the call of its chairman. Where an expenditure of USP funds is involved in connection with such a meeting, prior authorization shall be obtained from USP headquarters.

III. The Executive Committee of Revision and Division Executive Committees

Section 1. The Executive Committee of Revision shall be appointed annually by the Executive Director and shall be composed of six members of the General Committee of Revision—three to represent the Drug Standards Division and three to represent the Drug Information Division. Such appointments shall be made subject to the advice and consent of the respective Division Executive Committee concerning the three nominees representing that Division.

The Drug Standards Division Executive Committee shall be composed of the Chairmen of the DSD Subcommittees and the Executive Director, ex officio. The Drug Information Division Executive Committee shall be composed of Revision Committee members appointed by the Executive Director, subject to the advice and consent of all of the members of the Committee of Revision in the Drug Information Division, plus the Executive Director, ex officio.

Section 2. The Division Executive Committees and the Executive Committee of Revision shall carry out such assignments as have been directed to them by the Pharmacopeial Convention, the USPC Board of Trustees, or the General Committee of Revision.

Section 3. The Executive Committee and the Division Executive Committee concerned shall consult with the Executive Director or his designate on general policy concerning the planning and executing of the revision of the respective compendia and related products.

Section 4. The respective Division Executive Committee shall authorize appropriate public notice for all proposals as described in Rule IV, Section 6, following, prior to their final approval.

Section 5. The Drug Information Division Executive Committee shall determine the parameters of content, format and style of the information to be contained in USP DI and related publications, and shall approve the assessment of consensus achieved to allow publication of USP DI and related publications. The Drug Standards Division Executive Committee shall review, with a view to approval, the proposals of the Division Subcommittees for admission or omission of monographs and for changes in and additions to or deletions from any portion of USP or NF text.

Section 6. The Executive Director, subject to the advice and consent of the Division Executive Committees, shall appoint a Drug Nomenclature Committee.

The Editor of USAN/USP DDN shall serve as secretary to the Committee and act under the authority of the Executive Director.

Section 7. The Drug Nomenclature Committee shall be responsible for promoting uniformity and consistency among official titles in the USP and the NF and the USP DI, coining suitable titles where such are needed because of the lack of public (nonproprietary) names or to improve the names already in use. The titles shall be in harmony with convenience in prescribing and with accepted tenets of general usage and shall be simple, useful, and clearly distinguishing and differentiating.

Section 8. Proposed changes by USPC staff of titles previously appearing in *USAN and the USP Dictionary of Drug Names* (USAN/USP DDN) shall be published in *Pharmacopeial Forum* and *USP DI Review* for comment.

Section 9. The Drug Nomenclature Committee shall determine the parameters of content, format, and style of the information contained in USAN/USP DDN and shall receive and rule upon all questions and appeals for reconsideration of text prepared by USPC staff and appearing in USAN/USP DDN.

Appeals from nomenclature decisions of the Drug Nomenclature Committee shall be referred to the Executive Committee of Revision in accordance with Rule IV, Section 9.

Section 10. The Drug Standards Division Executive Committee shall be responsible for ascertaining the initial and continued suitability of materials to be utilized for reference materials as provided for in the United States Pharmacopeia and the National Formulary, or otherwise deemed necessary. All reference materials shall be handled, tested, and approved prior to release in accordance with the procedures adopted by the Division Executive Committee.

Appeals from decisions of the Division Executive Committee shall be referred to the Executive Committee in accordance with Rule IV, Section 9.

Section 11. Not less than one year before the stated meeting of the Pharmacopeial Convention, the Executive Committee shall prepare, with the assistance of the Division Executive Committees, and publish an outline of specific classifications of service deemed necessary to the efficient prosecution of the work of the next revisions of the USP, the NF, the USP DI, and other authorized publications. The outline shall be accompanied by a statement of the number of persons to be elected in each classification. This outline shall be followed by the Nominating Committee in submitting the names of persons to be considered for membership on the General Committee of Revision that is to be elected at the forthcoming stated meeting of the Pharmacopeial Convention.

Section 12. The Division Executive Committee members shall serve on the Nominating Committee to select the nominees for the General Committee of Revision for the forthcoming revision period.

Section 13. The Executive Committee may recommend to the Board of Trustees the removal, for violation of these rules or otherwise for cause, of any member from the General Committee of Revision.

Section 14. The Executive Committee shall make determinations in conjunction with the Board of Trustees with respect to a Division Executive Committee's claim of need for additional members to be added to the Committee prior to the next stated meeting of the Pharmacopeial Convention.

Section 15. After consultation with the Division Executive Committee, the Executive Committee shall nominate, insofar as possible, twice the number of qualified persons for the allotted expansion positions, from among whom the General Committee of Revision shall elect the additional members to the Committee.

IV. Procedures

Section 1. A written communication directed to the USPC that calls into question a standard, procedure, or other written text proposed in *Pharmacopeial Forum* or *USP DI Review*, or appearing in the USP, the NF, the USP DI, or other authorized publication, or that criticizes a lack thereof, and that requests consideration of a change with respect to the matter called into question and that involves a question of medical or scientific judgment and that is accompanied by reasonable grounds and, where necessary, supporting data, shall be forwarded to the appropriate committee, subcommittee, or panel for consideration. Since changes in compendial standards or analytical procedures may be adopted prior to or without governmental approval, a request for consideration of a modification or change may be refused unless accompanied by data deemed sufficient to justify modification of an approved New Drug Application.

Section 2. Internal communications to a committee, subcommittee, or panel shall issue generally from USPC headquarters in the form of Bulletins or Memoranda, the text of which generally shall be supplied by the chairman of the group, or by other member or members with the approval of the chairman, or by the Executive Director or his designate.

Section 3. Public communications to a committee, subcommittee, or panel, as a whole, and to the General Committee of Revision as a whole, shall issue from USPC headquarters in the form of *Pharmacopeial Forum* or *USP DI Review*, the text of which generally shall be supplied by the chairman of the group, or by other member or members with the approval of the chairman. Signed explanations of the text may be issued by the Executive Director or his designate without specific approval of the chairman of the group.

Section 4. A written communication that requests consideration of a change in a standard, procedure, or other written text may be published in *Pharmacopeial Forum* or *USP DI Review* and/or sent to known interested parties for comment.

Section 5. Committee and subcommittee members and panelists shall forward copies of all pertinent correspondence to the group chairman and to the Division Director.

Section 6. In order to provide for adequate notice and opportunity for comment, all proposed additions, deletions, or modifications of standards, procedures, or other written text shall be published in *Pharmacopeial Forum* or *USP DI Review*, except in instances where the Executive Director determines for good cause that it is impracticable, unnecessary, or contrary to the public interest.

All such published proposals shall be accompanied by a date determined by the Division Executive Committee by which comments must be received for consideration. Such date shall be not less than 30 days from the date of publication, unless it is determined by the Executive Director for good cause that such period should be shortened or eliminated.

Section 7. Any proposal published in *Pharmacopeial Forum* or *USP DI Review* for comment that does not evoke significant adverse comment or objection, or that as a result of the comments received does not result in material modification, need not be reprinted in final format prior to publication in the USP, the NF, the USP DI, or other authorized publication.

Section 8. Any proposal published in *Pharmacopeial Forum* or *USP DI Review* that does evoke significant adverse comment or objection shall be reprinted in final format prior to publication in the USP, the NF, the USP DI, or other authorized publication. Except for good cause to the contrary, such publication shall be accompanied by a summary or abstract of each type of comment received and may also be accompanied by a statement of the Subcommittee, the Division Executive Committee, or the Executive Director. Where a proposal has been materially modified, it shall be reprinted in *Pharmacopeial Forum* or *USP DI Review* in accordance with Section 6 of this Rule unless otherwise determined by the Executive Director for good cause.

Section 9. A matter may properly be appealed only prior to its appearance in the United States Pharmacopeia, the National Formulary, or the USP Dispensing Information or, in the case of a reference material, prior to its issuance. Once publication or issuance has occurred, a request for consideration of a revision is the appropriate procedure. Failure of the Committee of Revision to take action on a request for consideration of a revision may properly be appealed at any time.

The Executive Committee of Revision shall receive and rule upon all appeals for reconsideration, revision or abrogation of standards adopted or decisions made by a committee, subcommittee, division executive committee, or panel in those cases where the appellant has specifically requested a hearing by the Executive Committee. Such appeal shall be submitted to the Executive Director and shall be accompanied by the evidence in support of appellant's request.

A decision of the Executive Committee on such appeals shall be referred to the Board of Trustees which, upon consideration of the facts, may request the entire Committee of Revision to review the decision. After such review, which shall be given as promptly as practicable, the Committee of Revision may uphold or reverse the Executive Committee's decision by a majority of the members voting.

Section 10. An appeal of a decision that originated in the Executive Committee shall be referred to the Board of Trustees which, upon consideration of the facts, may request the Division Executive Committees to review the decision. After such review, which shall be given as promptly as practicable, the members of the Division Executive Committees may uphold or reverse the Executive Committee's decision by a majority of the members voting.

Section 11. The Drug Standards Division Executive Committee and the Executive Committee of Revision shall approve the text of the USP and NF, including Supplements and Interim Revision Announcements and other revisions thereof, prior to publication.

Except in case of emergency, approval of the text by the Division Executive Committee and the Executive Committee of Revision shall not be simultaneous and the Executive Committee shall be given an additional ten days in which to cast its ballots.

Section 12. The Executive Committee shall have the authority to delay or postpone the effective or official date of any requirement or textual material appearing in the USP, the NF, the USP DI, or other authorized text.

A request for consideration of postponement shall be accompanied by a statement of the grounds upon which the postponement is requested and appropriate supporting data. A request for consideration of postponement received within 30 days of the effective or official date may be refused by the Executive Director if, solely within his discretion, he considers the request untimely or lacking adequate supporting data. Such decision shall be unappealable. Where postponement is not granted, the matter will be considered as a request for consideration of revision.

Section 13. Where a new Reference Standard is required in a new monograph or in a revision of an existing monograph but is not yet available, the official date of any portion of the monograph utilizing such Reference Standard shall be postponed until it is available. During the period of such unavailability, the Executive Director shall publish in the official Reference Standards Catalog the name of such Reference Standard and indicate that it is not yet available. Subsequently, the availability of the Reference Standard shall be announced in the official Reference Standards catalog, and the effective date of the postponed portion(s) of the monograph shall be announced in the next official USP-NF Supplement or Interim Revision Announcement.

Section 14. General decisions may be reached by the Executive Committee, General Committee of Revision, Subcommittee, Division Executive Committee, or Panel through a simple majority vote of those voting on a ballot issued by mail (or by telephone, at the discretion of its chairman); or, upon presentation of specific, positive proposals in *Pharmacopeial Forum* or *USP DI Review* or in the Circulars or Bulletins accompanied by an announcement that approval shall be assumed in the absence of word to the contrary within a stated, reasonable period of time (i.e., silence is assent).

Section 15. The results of voting shall be reported in subsequent numbers of *PF* or *USP DI Review* or in Circulars, Letters, Communiques, or Bulletins, and such report may include the substance of comments received. The showing of substantial opposition by a minority may be the basis for a subsequent reconsideration of the issue involved, following circulation of the adverse comments.

Section 16. Any organizational group of the Committee of Revision may hold a special meeting by means of a conference telephone call or similar communications equipment by means of which all persons participating in the meeting shall hear one another in the same period of time, and participation by such means shall constitute presence in person at a meeting.

V. Special Panels

Section 1. Ad hoc panels shall be appointed by the Executive Director. These advisory panels may be set up where special advice is needed, or where the potential need appears to extend indefinitely into the future. The advisory panels shall serve in a fact-finding, advisory capacity. Special ad hoc advisory panels shall report to the Executive Director or his designate. Recommendations of these advisory panels shall be referred to the appropriate committees, subcommittees, or panels, division executive committees, or Executive Committee for consideration.

Section 2. Ad hoc advisory panels may be set up for specific studies, frequently of the sort involving organized collaborative effort among laboratories, with a view to being discharged when the work has been completed and reported. An ad hoc advisory panel may be established also for a specific, continuing purpose.

Section 3. Joint panels may be established in cooperation with other organizations having interests in common with those of a Division of the Committee of Revision.

Section 4. The Executive Director, with the advice and consent of the Board of Trustees, may appoint a committee composed of individuals from other nations who by law recognize or utilize the standards of the United States Pharmacopeia or the National Formulary, or the USP DI data base, to provide information and advice to the Committee of Revision.

VI. Standards of Conduct

Section 1. A Committee of Revision member shall conduct himself according to ethical behavior of the highest order.

Section 2. A Committee of Revision member shall not vote on the approval of any item or proposal in which he has a conflict of interest, or the appearance of a conflict of interest.

Section 3. A Committee of Revision member shall not be assigned the sole responsibility to work on a monograph in which he has a conflict of interest; he may, however, provide scientific information to the Committee member who is responsible for the monograph.

Section 4. A Committee of Revision member who, during his membership, represents anyone else before the compendia in a matter in which he has at any time participated personally and substantially in the course of his Committee membership shall no longer personally and substantially participate in the matter on behalf of the compendia.

Section 5. A Committee of Revision member shall not use information not generally available to those outside the Committee of Revision for the special benefit of a business or other entity through which he has a financial interest. Information intentionally not publicly available shall be held confidential by the Committee member and shall not be divulged to his employer or client. In case of doubt whether information is generally available to the public, the member should confer with the Executive Director or his designate.

Section 6. A Committee of Revision member shall not use his membership in any way that is, or appears to be, motivated by private gain for himself or other persons, particularly those with whom he has family, business, or financial ties.

VII. Statement of Financial Interests Required

Section 1. Each Committee of Revision member shall submit to the Executive Director a statement of all other employment and financial interests that relate either directly or indirectly to his duties and responsibilities, which have not been disclosed previously to the Convention Members in the report of the Nominating Committee for the Committee of Revision or are not evident by identity of his employer.

Section 2. Each Committee of Revision member shall submit such statement as prescribed in Section 1 not later than August 1, 1985, and shall keep it current throughout the revision period by the submission of supplementary statements as required.

Section 3. The Executive Director shall be responsible for reviewing the statements of employment and financial interests. Where no conflict of interest is disclosed by the review of the statements, no further action is necessary. Where a question of conflict of interest arises, the Executive Director shall work with the Committee member to resolve the matter or shall refer the question to the Executive Committee for further consideration and advice.

Section 4. Except as specified in the foregoing Sections, such statements shall be regarded as wholly confidential.

VIII. General

(1) On all questions of procedure, the latest edition of Robert's Rules of Order shall prevail, unless otherwise specified herein.

(2) A copy of the rules of the General and Executive Committees of Revision shall be made available upon written request.

(3) The Committee of Revision shall serve and shall act as the Committee until its successor committee shall have been elected and installed.

(4) The following publications, or their successors, shall be deemed authorized:

(a) USP Dispensing Information and related publications incorporating portions of its text.

(b) USAN and the USP Dictionary of Drug Names.

(5) The use of terminology in these rules and procedures and in communications of the Committee indicating a particular gender shall be interpreted to apply to both genders.

IX. Amendments

These rules and procedures may be amended during the revision period in accordance with Chapter VII, Section 5, of the Bylaws.

USPC Communications Policy

The United States Pharmacopeia and the National Formulary are recognized as official compendia and are referenced in various statutes as a basis for determining the strength, quality, purity, packaging, and labeling of drugs and related articles. As such, the texts may have significant legal consequences. The Pharmacopeial Convention attempts to publish the compendia in plain, explicit language in accordance with Pharmacopeial editorial principles; nevertheless, questions sometimes arise regarding their interpretation.

It is important to understand that it is the text in the USP-NF that is so recognized, not the comments, explanations, interpretations, or pronouncements of the Pharmacopeial Convention that may appear outside of those two compendia. Since the compendia have such legal consequences, the language must stand on its own. The Convention cannot give an official "ex post facto" interpretation to one party, thereby placing other parties without that interpretation at a possible disadvantage. The requirements must be uniformly and equally available to all parties. Any interpretation provided by USP staff, an individual Revision Committee member or Advisory Panel member, etc., regarding USP-NF requirements or language is to be understood to be a personal one reflecting only the opinion of the individual providing the interpretation and not necessarily an official position of the Pharmacopeial Convention.

The conformance of an individual product to compendial requirements is properly a matter of compliance, not of standards-setting. It is the USPC policy that the Convention does not provide an interpretation or opinion as to whether a particular article does or does not comply with compendial requirements. Any such opinion or interpretation by a member of the General Committee of Revision, an Advisory Panel, the Board of Trustees, or the USPC staff is to be interpreted as a personal opinion and is not to be construed as a USPC interpretation.

Formal statements reflecting the policy, opinion or position of the Pharmacopeial Convention may emanate only in the following manner:

(1) By approval of the Executive Committee of Revision on all scientific matters.

(2) By approval of the USPC Board of Trustees in all matters relating to the affairs, funds, or property of the USPC.

(3) By resolution adopted by the Pharmacopeial Convention at a stated meeting.

Adopted by the USPC Board of Trustees,
January 1984
WILLIAM M. HELLER, *Secretary*

USPC Document Disclosure Policy

I. Listed below are documents available for public viewing at USP headquarters during regular office hours, 8:30 a.m. to 5:00 p.m. [NOTE—Permission to utilize the library shall be obtained at least 24 hours in advance from the Manager of USP Purchasing and Facilities, telephone (301) 881-0666, but is freely given. Permission is required because of space limitations.]

(A) All volumes and periodicals contained in the library.

(B) The official compendia, Supplements, Addenda, and Interim Revision Announcements.

(C) USP Dispensing Information and USP DI Update, and their extracts and spin-offs.

(D) *USAN and the USP Dictionary of Drug Names.*

(E) *Pharmacopeial Forum* and *USP DI Review.*

(F) Other volumes published by the USPC.

(G) USP Material Safety Data Sheets. Information required to be maintained by the Department of Labor, Occupational Safety and Health Administration or other government law or regulation regarding substances on the premises—available only on a need-to-know basis.

(H) Pharmacy Board Bulletins.

(I) USP Reference Standards Catalog.

(J) Proceedings of USP Convention Meetings and USPC current membership list.

(K) List of current members of the General Committee of Revision and its Advisory Panels.

(L) Rules and Procedures of the General and Executive Committees of Revision.

(M) List of names of Staff Members.

(N) Weekly Meeting Schedule.

(O) The USP Standard, News Releases, and other USP-prepared, mass-produced materials mailed to third parties.

(P) USPC Member Memoranda.

II. Copies of the following USP documents are available only on written specific request. Except as indicated in Part IV, copies of the following documents are ordinarily available on request.

(A) Correspondence between USPC personnel and third parties.

(B) Memoranda of telephone conversations between USPC personnel and third parties.

(C) Memoranda of meetings between USPC personnel and third parties.

(D) Copies of requests for documents.

(E) Copies of speeches or public presentations by USPC personnel.

(F) Copies of USP Convention committee reports:

(1) Report of the Nominating Committee for the General Committee of Revision.

(2) Report of the Constitution and Bylaws Committee.

(3) Report of the Nominating Committee for the Board of Trustees.

(G) The auditors' financial statements for the USP Convention.

(H) Edited text prepared from USP Drug Product Problem Reports after deletion of names and any information that would identify the person submitting the information or third parties involved with the report (e.g., a patient, physician, hospital, etc.)

III. The following information is not disclosable or available for public inspection:

(A) Internal or interoffice communications, including

(1) USP Committee of Revision Circulars, Bulletins, Ballots, Memoranda, Communiques, Letters, and marginal comments and communications on pages of *USP DI Review* and *PF;*

(2) USPC Board of Trustees Letters and Memoranda;

(3) Reports of the USP Drug Research and Testing Laboratory.

(B) Except as indicated in II (G), documents containing financial information.

(C) Scientific data on Reference Standards unless in the judgment of the Secretary of the Convention special scientific reason exists.

(D) Documents containing data about USPC personnel whether or not originating in the USPC.

(E) USP Committee of Revision conflict of interest statements.

(F) Documents proposing candidates or nominees for election or appointment to USPC committees or advisory panels.

(G) Documents revealing: the location of USP Reference Standards; identity of suppliers of the materials or purchasers of the materials; amounts of individual reference materials purchased or distributed; inventory control procedures; identity of information sources for USP Material Safety Data Sheets.

(H) Documents containing any trade or commercial secrets or security arrangements, except that documents containing trade secrets or confidential commercial secrets submitted to USPC by a third party that ordinarily would be contained in a New Drug Application or Supplement thereto may be disclosed to the Food and Drug Administration upon its request in its review of any revision or proposed revision of the United States Pharmacopeia or the National Formulary.

Documents containing any trade secrets or confidential commercial secrets shall be specifically designated as such when submitted to USPC. Only such documents so designated will be handled in accordance with this provision.

(I) Laboratory and Medical Devices Product Problem Reports, and communications relating thereto.

(J) USP Drug Product Problem Reports, except if in the judgment of the Secretary such disclosure is in furtherance of the objectives of the reporting program.

(K) Documents containing the terms of business arrangements or agreements with third parties.

(L) All communications between USPC personnel that are in written form or subsequently reduced to writing.

(M) The home addresses and home telephone numbers of USPC personnel.

(N) Specific assignments of USP Committeemen without the permission of the Committeeman.

IV. (A) The term USPC personnel as used in these rules includes members of the USPC Board of Trustees, the USP Committee of Revision and Advisory Panels, individuals assigned text-review responsibilities, and staff. It is USPC policy that USPC staff not appear as expert witnesses in any case regarding or relating to drugs.

(B) The USPC will make the fullest disclosure of records possible consistent with the rights of individuals to privacy, the property rights of persons in trade secrets and confidential commercial or financial information, and the need of the Convention to promote frank internal deliberations and to pursue its activities without disruption.

(C) If a record contains both disclosable and nondisclosable information, the nondisclosable information will be deleted and the remaining record will be disclosed unless the two are so inextricably intertwined that it is not feasible to separate them or unless release of the disclosable information would compromise or impinge upon the nondisclosable portion of the record.

(D) Requests for records should be made to:

Secretary
United States Pharmacopeial Convention, Inc.
12601 Twinbrook Parkway
Rockville, Maryland 20852

who shall be responsible for making the decision as to the disclosure of the records.

(E) Documents submitted to the USPC specifically designated as confidential will be available only to USPC personnel and will be distributed on a need-to-know basis to USP Committeemen and advisory panelists as determined by the Executive Director.

(F) Names of USPC personnel ordinarily will not be deleted from disclosable records except where it is deemed necessary for the protection of the individual.

(G) Because of staff and financial limitations, it is the policy of the USPC not to conduct searches of its files in response to requests for general information regarding the effects of drugs, unless it is determined by the Secretary that the request is directed to the content of a USPC publication.

V. The following fees shall be imposed for disclosure of any record pursuant to these rules.

(A) Copying: 25¢ per copy of each page.

(B) Clerical searchers: $10.00 for each quarter-hour spent by clerical personnel searching for and producing a requested record, including time spent copying any such records.

(C) Nonclerical searchers: $25.00 for each quarter-hour spent by scientific or professional personnel searching for a requested record, including time spent reviewing and preparing and copying any such records.

(D) Postage, insurance, and special fees will be charged on an actual cost basis.

(E) For computerized records, add to the above (1) through (4) the sum of the actual costs of the computer time involved and the supplies or materials necessary to produce the requested records.

(F) In lieu of fees imposed by (V) (A) through (E) above, the Secretary may establish uniform fees for any particular type of document (e.g., material safety data sheets, edited text from USP Drug Product Problem Reports, etc.).

(G) The Secretary may waive payment of any fee.

(H) For requests of records involving a fee of more than $50.00, the records shall be forwarded as soon as possible after receipt of prepayment from the person requesting the records.

(I) The Secretary shall have ultimate authority to refuse to grant a request for disclosure. In situations where the confidentiality of the data or information is uncertain and there is a request for public disclosure, the USPC will consult with the person who has submitted the data or information or who would be affected by disclosure.

(J) The Secretary may, in determining whether to grant a request for disclosure, take into account the loss that would result from diverting staff personnel from their other responsibilities.

(K) Only copies of existing records may be made available. All existing records are subject to routine destruction.

Proceedings

Abstract of the Proceedings of the United States Pharmacopeial Convention of 1985

The United States Pharmacopeial Convention held its third quinquennial meeting in Washington, D.C., on March 22–24, 1985, President Frederick E. Shideman, M.D., Ph.D.,* presiding.

A survey of the makeup of the Convention with regard to the attendance at the meeting from various groups named in the Constitution was reported by Joseph M. Benforado, M.D., Chairman of the Credentials Committee. The total number of voting members present was 227.

A total of 183 delegates and 14 alternate delegates were in attendance. Only 25 organizations that had appointed delegates were not represented. Seven of the eight federal agencies entitled to representation were present, as were 21 of the national organizations, 13 of the state medical associations, 66 of the schools of medicine, 28 of the state pharmaceutical associations, and 61 of the schools of pharmacy.

Fred T. Mahaffey, Pharm. D., of the newly admitted National Association of Boards of Pharmacy, presented credentials which were accepted.

Those present also included 16 members-at-large selected by the Board of Trustees for their expertise in Pharmacopeial affairs, three public members-at-large, two foreign members-at-large, six committee chairmen, and nine officers and members of the Board of Trustees.

The first half-day of the meeting was devoted largely to a scientific program, with a symposium on inactive ingredients in pharmaceutical dosage forms and the Pharmacopeia, and a symposium on home parenterals and the Pharmacopeia, each symposium being followed by an audience participation session. The symposium on inactive ingredients was addressed by Jack Cooper, University of California School of Pharmacy, from the standpoint of the necessity for inactive ingredients; Albert L. Sheffer, M.D., Chairman of the Advisory Panel on Allergy, Immunology, and Connective Tissue Disease of the USP Committee of Revision, reviewed clinical problems associated with inactive ingredients. The symposium on home parenterals was addressed from the point of view of two pharmaceutical manufacturers that had established community pharmacies: James C. Boylan, Ph.D., of Abbott Laboratories and Donald S. Ebersman, Ph.D., of Travenol Laboratories; from the point of view of a pharmacist providing home parenterals by William F. Appel, a pharmacist on the USP Committee of Revision; and from the point of view of a state board of pharmacy by Robert E. Snyder of the Maryland Board of Pharmacy.

On March 22 and March 23, luncheon addresses were presented by, respectively, C. A. Johnson,* Secretary and Scientific Director of the British Pharmacopoeia Commission, and Paul G. Rogers, Chairman of the Board of the National Council on Patient Information and Education (NCPIE).

The Convention was addressed also by Frank E. Young, M.D., Ph.D., Commissioner of Food and Drugs; Joseph D. Williams, Chairman of the Board, Pharmaceutical Manufacturers Association; James A. Main, J.D., Chairman of the Board, American Pharmaceutical Association; and Joseph F. Boyle, M.D., President, American Medical Association. President Shideman, Treasurer Parker, and Chairman Kinnard reported to the Convention on their stewardship of USPC affairs during the years 1980 to 1985. As Executive Director, Chairman of the Committee of Revision, Secretary of the USPC Board of Trustees, and Secretary of the Convention for the five years, William M. Heller, Ph.D., reported on his responsibilities in each of those four capacities.

The Committee on Constitution and Bylaws, under the chairmanship of James T. Doluisio, Ph.D., held the view that revisions adopted by the delegates at the previous three quinquennial meetings had worked well to serve the expanding activities of the Convention, so that the revisions proposed by this Committee were relatively few. Most of the changes adopted by the delegates at the 1985 quinquennial meeting serve to clarify or describe more accurately the current operations.

The Constitution was revised to recognize the National Association of Boards of Pharmacy (NABP) as an organization eligible to appoint a delegate to the Convention. A revision of the chapter on the Board of Trustees in the Bylaws provided that, between stated meetings of the Convention, the members may vote by mail when deemed necessary or desirable by the USPC Board of Trustees; not fewer than 100 members shall cast such mail ballot to constitute a valid action, and a majority of those voting shall determine the action. The provisions whereby the Board of Trustees appoints members-at-large were clarified to state that the respective numbers of such members are the maximums that may be appointed at any one point in time, and that such appointments may be for either the current or the ensuing Convention. A revision of the appeals procedure further ensured due process by allowing the USPC Board of Trustees to refer a decision of any Committee of Revision subcommittee, committee, or other body that is under appeal to the entire Committee of Revision for review of such decision. (Formerly, the Executive Committee of the Committee of Revision was the primary appeals body of the Committee of Revision.) A revision of the chapter on General and Executive Committees of Revision enabled the USP Drug Information Division to have representation from all major health-care specialties by means of advisory panels as distinct from the subcommittee system used by the USP Drug Standards Division. Thus, in the structure of the Committee of Revision, the members of the Drug Information Division Executive Committee are appointed with the approval of the Revision Committee members in that Division, which is not bound to the subcommittee system. The members of the Drug Standards Division Executive Committee are subcommittee members appointed with the approval of the Revision Committee members in that Division, which also is not bound by the subcommittee system, but traditionally utilizes it. This amendment of the structure of the Committee of Revision allows the Executive Director to organize the Divisions on the basis of the nature of the tasks to be performed by the respective Divisions.

As Chairman of the Committee on Resolutions, Arthur Hull Hayes, Jr., M.D., reminded the members that the resolutions, although representing the thinking of the Convention, constituted only recommendations to the incoming Board of Trustees and General Committee of Revision. The following resolutions were adopted, some after extensive discussion and amendments of the original submission.

*Deceased.

RESOLUTION NO. 1—*Rx to OTC Switch*

WHEREAS the USP Committee of Revision has prepared patient language information for practically all drugs, and

WHEREAS the information is processed through an elaborate national advisory panel system that incorporates all of the medical specialties, the several health professions involved in drug therapy, and consumers also, and

WHEREAS the process is open also to comment by manufacturers and regulators and other interested parties, and

WHEREAS the information is already in widespread use, either in its entirety or in abstracts, by national, state and local associations, and individual health-care givers, therefore

Be it resolved that the Committee of Revision explore with the Food and Drug Administration the mechanisms by which the USP might assist in selecting drugs appropriate for switch from Rx to OTC status and in formulating labeling for them.

RESOLUTION NO. 2—*Supplemental Practitioner and Patient Information for OTC Drugs*

WHEREAS the USP Committee of Revision has prepared practitioner and patient language information for practically all drugs, including OTC drugs, and

WHEREAS the information is processed through an elaborate national advisory panel system that incorporates all of the medical specialties, the several health professions involved in drug therapy, and consumers also, and

WHEREAS the process is open also to comment by manufacturers and regulators and other interested parties, and

WHEREAS the information is already in widespread use, either in its entirety or in abstracts, by national, state, and local associations, and individual health-care givers, therefore

Be it resolved that for drugs sold without prescription, the Committee of Revision consider developing information for practitioners and patient langauge information that may be used by a health practitioner as part of his or her professional practice.

RESOLUTION NO. 3—*Pharmaceutic Ingredients in Dosage-form Labeling*

WHEREAS the USP Committee of Revision requires the disclosure of all ingredients, including those considered to be pharmacologically inactive, in Injections, ophthalmic preparations, and topical preparations, and

WHEREAS it would add to safety in the use of pharmaceutical preparations to identify, on the label or in the labeling, all ingredients in all dosage forms, and

WHEREAS the USP commends the efforts of the Proprietary Association and the Pharmaceutical Manufacturers Association and their members to list inactive ingredients on their drug product labels, therefore

Be it resolved that the Committee of Revision consider requiring the declaration of all ingredients on the label or in the labeling of all dosage forms included in the USP, with regard for the programs of the Proprietary Association and the Pharmaceutical Manufacturers Association.

RESOLUTION NO. 4—*Compounding Information in the Pharmacopeia*

WHEREAS the United States Pharmacopeial Convention has received a request from the American Society of Hospital Pharmacists to include in the United States Pharmacopeia certain types of information, and

WHEREAS the American Society of Hospital Pharmacists has indicated these data are very important in pharmacy practice, therefore

Be it resolved that the Committee of Revision examine the desirability and feasibility of developing, with a view to inclusion in the United States Pharmacopeia, the following types of information:

(1) the short-term stability of drugs when dissolved in common diluents and stored in common standardized containers and/or delivery systems at room, refrigerator, and freezer temperatures;

(2) pKa and minimum solubility of drugs in common diluents; and

(3) pH, osmolality, and osmolarity of reconstituted injectables and liquid dosage forms.

RESOLUTION NO. 5—*Standards for Repackaged and Compounded Parenterals*

WHEREAS the USP Committee of Revision provides standards for articles and tests, for compliance with those standards, and

WHEREAS those standards are enforceable at the practitioner level as well as at the manufacturer level, and

WHEREAS it is increasingly the practice not to wait until the time of administration to repackage parenterals into injection devices or to mix one injectable with another, and

WHEREAS the timc between preparation and administration, and the facilities and equipment reasonably available in professional practice, do not generally permit the utilization of techniques, practices, and procedures applicable to large manufacturing operations, therefore

Be it resolved that the Committee of Revision be charged with the responsibility for providing standards and test methods; specifications for packaging, labeling, and storage; guidelines for appropriate documentation; and, where necessary, procedures for compounding parenteral preparations.

RESOLUTION NO. 6—*Standards for Non-USA Articles*

WHEREAS USP Reference Standards and compendial standards are of increasing usefulness internationally, therefore

Be it resolved that the Committee of Revision study the possibility of developing a separate section or other mechanism in the compendia to encompass articles not available in the United States but which need standards, particularly in other countries that utilize the USP.

RESOLUTION NO. 7—*Standards for Medical Devices and Diagnostic Agents*

WHEREAS the USP Committee of Revision has established a comprehensive national advisory panel system that incorporates all of the health professions involved in drug therapy, as well as consumers, and

WHEREAS the process is also open to comment by drug manufacturers and regulators and other interested parties, and

WHEREAS the Food and Drug Administration has identified over a thousand medical devices and diagnostic agents that are in need of standards, therefore

Be it resolved that the Committee of Revision study the feasibility of identifying through its advisory panels those medical devices for which the USP could appropriately set standards, and explore with the FDA, methods for implementing such standards.

RESOLUTION NO. 8—*Chemical Safety Information*

WHEREAS the safety of chemicals is and will be of increasing concern to health professionals and the public, and

WHEREAS the USP has developed information regarding the physical, chemical, and toxicological characteristics of chemicals used as drug reference substances, therefore

Be it resolved that the Committee of Revision and the Board of Trustees study the possibility of publishing information on the physical, chemical, and toxicological properties of chemicals used as drug reference substances, to assist health professionals.

RESOLUTION NO. 9—*Increase Use of USP DI*

WHEREAS the USP Committee of Revision has developed an outstanding and comprehensive source of information on drugs in the form of *USP DI*, the use of which by the health-care professions would benefit the public, therefore

Be it resolved that the Board of Trustees continue and expand its efforts to increase the awareness and use of *USP DI* systems by all health-care professionals.

The report of the Nominating Committee for Officers and the Board of Trustees was presented by Ray W. Gifford, M.D., and that of the Nominating Committee for the General Committee of Revision was presented by Klaus G. Florey, Ph.D. No nominations were made from the floor. By ballot vote, the Convention elected the Officers, Board of Trustees, and General Committee of Revision, as shown beginning on page vi.

An Honorary Membership in the Convention was presented to Harry C. Shirkey, M.D., D. Sc., and Honor Citations were presented to Kurt L. Loening, Ph.D., Fred A. Morecombe,* B.S., and Mary C. Griffiths, B.A., M.S.L.S., by William J. Kinnard, Jr., Ph.D., Chairman of the USPC Board of Trustees, on behalf of the Board.

Presentation of Honorary Membership

Harry C. Shirkey, M.D., D.Sc.

Introduction by Dr. Kinnard: Beginning in 1953, Dr. Harry C. Shirkey served both USP and NF in separate capacities, and then, after the consolidation of the official compendia in 1975, served them both together—a circumstance that exemplifies the very kind of unifying influence his work and talents have exerted on the two professions (pharmacy and medicine) that claim his allegiance. Over the past thirty-two years, as a member of the USP Committee of Revision from 1953 to 1980, a member of the NF Board from 1960 to 1975, and Vice-President of this Convention from 1980 to 1985, Dr. Shirkey has been a loyal and diligent advocate of the highest standards for medicine, pharmacy, and the public health.

CITATION: Awarded by the Board of Trustees assembled at Washington, D.C., in grateful recognition of your distinguished contribution to drug standardization, the sciences of medicine and pharmacy, and the public health during the past thirty-two years; as a dedicated and constructive member of the USP Committee of Revision (1953-1980), member of the NF Board (1960-1975), and Vice-President of this Convention and member of the Board of Trustees (1980-1985), outstanding pediatrician, pharmacist, educator, lecturer, author, and editor.

Presentation of Honor Citations

Kurt L. Loening, Ph.D.

Introduction by Dr. Kinnard: Dr. Kurt L. Loening has been with Chemical Abstracts Service of Columbus, Ohio, since 1951, and holds the position of Director of Nomenclature. Although he has not served on the Committee of Revision, he has earned a place of special esteem as one of those truly exceptional volunteer experts whose numbers are few but whose years of dedicated contributions are many. An outstanding chemist, author, and specialist in chemical nomenclature, and a meticulous scholar, he has consistently brought the USP, NF, and USAN chemical structures and names and related information up to date in such manner as to impart notable prestige to these aspects of the Pharmacopeial program, over the past twenty-one years.

CITATION: Awarded by the Board of Trustees assembled at Washington, D.C., in grateful recognition of your sustained interest in and meritorious contributions to the USP-NF revision program and all editions of *USAN and the USP Dictionary of Drug Names* since 1964, with respect to the structure and systematic nomenclature of chemical compounds. The magnitude of your volunteer efforts in these professional and scientific endeavors during the past twenty-one years is matched only by that of your loyal and selfless sharing of your expert knowledge and judgment.

Fred A. Morecombe

Introduction by Dr. Kinnard: Fred A. Morecombe was elected to the Committee of Revision in 1975 and again in 1980, but his outstanding contributions as a volunteer expert had begun as early as 1951. From the outset, he was an eager, outspoken, hard-working proponent of quality standards for our nation's drugs. His highly creditable, dedicated, thorough, prompt work for 35 years in behalf of USP and NF has earned him a permanent and unique place in the Pharmacopeial history and heritage.

CITATION: Awarded by the Board of Trustees assembled at Washington, D. C., in grateful recognition of your sustained interest in and meritorious contributions to the USP-NF revision program, with respect to reagent specifications, general test procedures, and development and validation of standards for drugs, during the past 35 years. As a member of the USP headquarters staff (1964–1968) and as a member of the Committee of Revision (1975–1985), you added a further full measure of diligent and distinguished service to an already established record of some two decades of voluntary contributions to the work of USP and NF revision. In devoting this much of your life to these compendial endeavors, you indeed gave new life to the Pharmacopeial tradition of quality drug standardization and public service.

Mary C. Griffiths

Introduction by Dr. Kinnard: Mary C. Griffiths joined the USP staff in New York City in 1951, and Dr. Lloyd C. Miller, then Director of Revision, soon had her in charge of "getting the book out." Now, 34 years later she is still in charge of getting "the book" out—the USP-NF and all other USP books and publications. Her title, Director of Publication Services, carries with it a heavy responsibility as she coordinates a staff of 13 and oversees a half million dollars worth of computer systems, using her skills of management and her qualities of efficiency and dedication to accomplish the task of getting the USP publications not only "out" but "out on time." Her attention to detail and her high regard for "accuracy" have resulted in USP publications of which we can all be proud.

CITATION: Awarded by the Board of Trustees assembled at Washington, D. C., in grateful recognition of your consummate interest in and meritorious contributions to the USP-NF revision program for over three decades. The results of your efforts have contributed to the outstanding reputation enjoyed by these texts for editorial accuracy and dependability. Your dedication to perfection is sustained by a tenaciousness of work habits and an unmitigated loyalty to USP tradition that is an inspiration to both the Committee of Revision and your fellow staff members.

Adjournment

Following installation of the newly elected officers and Board of Trustees, and the Executive Director, and the announcement of the results of the Committee of Revision election, the Convention was adjourned subject to the call of the President.

*Deceased.

History of the Pharmacopeia of the United States

In January 1817,[1] Dr. Lyman Spalding, of New York City, submitted to the Medical Society of the County of New York a plan to create a National Pharmacopoeia.[2] He proposed dividing the United States as then known into four districts—Northern, Middle, Southern, and Western. The "Western" District embraced all states west of Pennsylvania and the Southern District all states south of the District of Columbia.

The plan provided for calling a Convention in each of these districts, to be composed of delegates from all medical societies and schools situated within them. Each District Convention was to draft a Pharmacopeia and appoint delegates to a General Convention, to be held in Washington, D.C. At this General Convention the four District Pharmacopeias were to be compiled into a single National Pharmacopeia.

Doctor Spalding's plan was approved by the committee to which it was referred, and subsequently, through the agency of the Medical Society of the State of New York, it went into effect. This society issued circulars requesting the cooperation of the several incorporated State Medical Societies and such medical bodies as constituted a faculty in any incorporated university or college in the United States. Where there was as yet no incorporated medical society, college, or school, voluntary associations of physicians and surgeons were invited to assist in the undertaking.

The U.S. Pharmacopeial Convention—This general plan succeeded, and the first United States Pharmacopeial Convention assembled in Washington, D.C., on January 1, 1820. Samuel L. Mitchill, M.D., was elected President, and Thomas T. Hewson, M.D., was elected Secretary.

Draft pharmacopeias were submitted to the Convention only by the Northern and Middle Districts. These were reviewed and consolidated and, after adoption, were referred to a Publication Committee, of which Dr. Lyman Spalding was Chairman. The first U.S. Pharmacopeia was published on December 15, 1820, in both Latin and English. Within its 272 pages were listed 217 drugs considered worthy of recognition.

Before adjourning, the first Convention adopted a Constitution and Bylaws, with provisions for subsequent meetings of the Convention and a revised Pharmacopeia every 10 years. In 1900, the Pharmacopoeial Convention was incorporated in the District of Columbia (see page xx). Altogether, Convention meetings have been held in Washington, D.C., including a special meeting (as distinct from a regularly scheduled meeting) held on April 14, 1973.

At the 1940 meeting, the Convention directed that the Pharmacopeia be revised every 5 years. Authority for the issue of interim supplements whenever necessary to maintain satisfactory standards had been granted 40 years earlier. The 1940 Convention also arranged for revising the Constitution and Bylaws, a step that required calling an interim session in 1942. Some of the revised Bylaws took effect at once, one important result of which was to entrust the Board of Trustees with the responsibility of naming the Director of Pharmacopeial Revision. Also, under the new Bylaws, a Nominating Committee was to select nominees for the Committee of Revision prior to each decennial meeting.

The last decennial meeting was held in 1970, and at that meeting the Convention, by constitutional amendments, placed all Pharmacopeial activities—not just the revision of the Pharmacopeia itself—on a 5-year cycle rather than on a decennial basis; created the new post of Executive Director; provided for expanded informational services in connection with USP drugs; and generally paved the way for more continuous representation and communication between the Convention officials and the Convention membership. Henceforth, stated meetings of the Convention were to be held every 5 years; there would be a newly elected Committee of Revision every 5 years; and the Convention membership was to be kept continually apprised of Pharmacopeial activities so that it would be familiar with pertinent issues in advance of the quinquennial meeting.

The observance of the sesquicentennial anniversary of the founding of the United States Pharmacopeia was marked by two events[3] held in Washington during 1970. The first event was a ceremony conducted at the United States Capitol on January 2, 1970, in the very room where USP was founded on January 1, 1820—then the new Senate chamber but now known as the Old Senate Chamber. The second event was an all-day program held on April 9, 1970, in connection with the 1970 decennial meeting of the Convention.

At the first quinquennial meeting, held in 1975, resolutions were adopted that led to immediately significant changes in the direction of the revision program and the overall activities of the USP Convention. One resolution urged the Committee of Revision to provide standards and tests for unit-dose packages and dispensing packages of drugs and to establish a classification system whereby practitioners may readily determine the suitability of such containers in dispensing practice. Other resolutions advocated the formation of a Resources Development Committee to solicit voluntary contributions toward support of the USP Drug Research and Testing Laboratory, and the extension of standards-setting by USP-NF criteria to all drugs having proven efficiency as therapeutic agents. The last-mentioned objective coincides

[1] Much of the USP history appears in the prefaces of the individual editions of the USP. USP XIII, pages xvii to xli, gives a rather detailed résumé of the history up to the time of its publication.

[2] With the Fourteenth Revision, the spelling "Pharmacopoeia" was changed to "Pharmacopeia," and although the corporate title has not been changed formally, the use of the diphthong is being discontinued generally.

[3] A record of these commemorative events is included in *150 Years of USP—Abstract of Proceedings, Decennial Convention, April 8–10, 1970; USP Sesquicentennial Observances, January 2 and April 9, 1970*, published in 1972.

with the sense of another resolution adopted, advocating that USPC "directly assist and benefit practitioners of medicine and pharmacy in their use of drugs," as a result of which Subcommittee 1 (Scope) and Subcommittee 2 (Posology and Related Information) served in 1975–1980 primarily in the preparation of the new annual volume, *USP Dispensing Information* (USP DI). Information under the headings, *Usual pediatric dose* and *Dispensing information*, had been introduced into the therapeutic information sections of the monographs of USP XIX. USP DI, now a separate publication and not part of the Pharmacopeia, greatly expands upon those earlier beginnings of providing information for the pharmacist and for the physician and other practitioners and now includes information for the patient as well.

Membership in the USP Convention and Revision Committee—The first United States Pharmacopeial Convention and the first Revision Committee were composed exclusively of physicians. In 1830, and again in 1840, prominent pharmacists were invited to assist in the new revision. In recognition of their contributions, pharmacists were given full membership in the Convention of 1850 and have participated regularly ever since. By 1870, the Pharmacopeia was almost entirely in the hands of pharmacists, and vigorous efforts were required to revive interest in it among physicians. The present Constitution provides a greater opportunity for physicians than pharmacists to become members; in practice, representation is about equal.

Beginning about 1935, formally organized advisory groups were appointed to assist the Committee of Revision both on matters of policy and on technical problems. These groups have been called upon to an increasing extent during the past 50 years. Such expert panels are particularly valuable in assessing the current status and value of drug information and drug production and control procedures. These advisory panels make it possible for the Pharmacopeia to reflect more promptly the advances in analytical methods and for USP DI to provide complete and current drug use information.

Policy on Scope—The first USP Convention chose as a primary objective of the Pharmacopeia the compilation of a select list of medicines of each decennial period, including the most efficient forms for their application.

The preface of the Pharmacopeia of 1820 reads in part:

"It is the object of a Pharmacopeia to select from among substances which possess medicinal power, those, the utility of which is most fully established and best understood; and to form from them preparations and compositions, in which their powers may be exerted to the greatest advantage. It should likewise distinguish those articles by convenient and definite names, such as may prevent trouble or uncertainty in the intercourse of physicians and apothecaries.

"The value of a Pharmacopeia depends upon the fidelity with which it conforms to the best state of medical knowledge of the day. Its usefulness depends upon the sanction it receives from the medical community and the public; and the extent to which it governs the language and practice of those for whose use it is intended."

Through the years this policy on USP admissions was maintained with increasing collaboration of all branches of medicine, until 1975. Despite steadfastness in admitting only drugs of established merit, the number of USP drugs rose steadily with each succeeding revision. Beginning in 1975, with the purpose of establishing standards for all efficacious drugs, the consolidated USP-NF revision program provided a substantial number of new monographs. USP XXII–NF XVII includes some 1077 more official monographs than were included in the individual compendia published at the end of 1974.

With each passing decade, more and more drugs are subject to patent rights. It was USP XII (1942) that first stated that "medicinal substances and preparations" could be recognized in the Pharmacopeia, "even when trademarked or manufactured under a process or product patent . . . . the question of therapeutic value only being considered."

Evolution of the Pharmacopeia—During the 170 years since the appearance of the first Pharmacopeia in the United States, the art and science of both medicine and pharmacy have changed remarkably. Prior to 1820 the drugs in the physician's armamentarium had been much the same for centuries. The Pharmacopeia of 1820 reflected the fact that the apothecary of that day was able personally to collect many of the vegetable drugs required and to make from them the simple mixtures and preparations required by the physician. Also, many physicians of that time prepared much of the medicine they administered. With the passage of time, these functions were transferred to others, until by 1880 chemical and pharmaceutical manufacturing had become well established in the United States and the pharmacist was relying heavily upon commercial sources of supply.

USP VI, published during the 1880–1890 decade, included for the first time tests and assays to set the standards of strength and purity for drugs. The emphasis on objective standards increased immensely in 1906, under the new Pure Food and Drugs Act, and has continued to do so with each successive revision of the Pharmacopeia.

USP Dispensing Information (**USP DI**)—USP VIII initiated the practice of including recommended doses, and subsequent revisions extended the amount of information useful to pharmacists and physicians. With the discontinuance of the limited Scope policy, the advisory panel system that had selected drugs for USP recognition turned its attention instead to developing a consensus on the totality of relevant information about each drug that is needed in the post-prescribing situations, i.e., by the pharmacist in dispensing it, the nurse in administering it, and physicians other than the prescriber who may be serving the patient—and even lay-language information for the patient. Thus, USP XX and NF XV were published in 1980, containing only the tests and standards, while the consensus drug-use information was published separately under the name USP Dispensing Information (USP DI). The latter was intended to be published annually, with bimonthly updates. Beginning with the 1983 edition, separate volumes were published: USP DI Volume I, *Drug Information for the Health Care Provider*, and USP DI Volume II, *Advice for the Patient*. In 1989, a third volume was added, encompassing the Food and Drug Administration's "Orange Book" and entitled "Approved Drug Products and Legal Requirements."

Legislative Recognition—The authority of the United States Pharmacopeia was recognized as early as 1848, when the Drugs and Medicine Act made reference to USP and several European pharmacopeias for standards of strength and purity of imported drugs. The definition of adulteration whereby USP monograph standards were recognized was also embodied in several state statutes in the late nineteenth century. Thus, in 1906, when the first Pure Food and Drugs Act was adopted by Congress, recognition of the expertise of the compilers of the USP and the NF was already a fact of life. This law designated the USP and the NF as the source of the standards of strength, quality, and purity of medicinal products recognized therein when sold in interstate commerce for medicinal use. Both the Congress and the states have continued and extended the recognition of the USP and the NF as "official compendia."

In the 1980's legal recognition of various types has been accorded USP DI, most often in connection with pharmacists' counseling of patients.

The USP Spanish Edition—Between 1908 and 1958, eight Spanish editions of the USP were published to meet the needs of Spanish-speaking countries, especially Puerto Rico, the Philippine Islands, and Cuba. USP XV was the latest revision to appear in Spanish. The USP has been adopted officially by several Central and South American countries, as well as by various nations in other parts of the world.

The Pharmacopeia Building—For more than a century, the USP headquarters office was located where the Chairman of the Committee of Revision attended his regular professional duties. The 1940 Convention, however, authorized the establishment of a permanent headquarters. Thus, in 1949, a building at 46 Park Avenue, in New York City, was purchased largely with contributed funds raised through the efforts of E. Fullerton Cook, then Chairman of the Committee of Revision.

Dedicated in 1950, the Pharmacopeia Building served increasingly as a focal point in the development of standards for drugs of all kinds. It served as the distributing center for all USP Reference Standards, which in themselves added to the prestige of the United States Pharmacopeia in all quarters of the world. In 1968, the headquarters office was moved to temporary quarters in Bethesda, Maryland, and the building in New York was sold. In 1971, a headquarters building in Rockville, Maryland, was purchased, and most of the staff was moved into the newly refurbished and refurnished facility by the end of that year. The Reference Standards unit was transferred to the new quarters by the end of 1972. In 1989, a new main building will be completed, attached to the original, making a total of 95,000 square feet of office and laboratory space.

The Changing Character of Drug Standards—The technology of drug standardization underwent a remarkable transition during the 1950–60 decade. Chemical analysis attained new levels of precision and sensitivity, albeit at a decided loss of simplicity as instrumental methods replaced classical methods of analysis. Thus, vastly improved techniques of isolation, coupled with refinements in quantitative instrumental measurement, provided means of analyzing drug principles for the first time or served to replace more costly and time-consuming biologic assays. During that decade, the need first arose for standardizing radioactive drugs, and the use of radioactive tracers was introduced into drug analysis. This entirely new departure in pharmaceutical chemistry was promptly recognized in the Pharmacopeia, in keeping with its policy of maintaining standards, tests, and assays on a current basis.

From 1960 on, still further advances in the techniques of drug standardization continued to revolutionize the various emphases on and approaches to USP tests and standards. Most notable have been content uniformity of dosage units and dissolution tests of oral solid dosage forms.

Unification of the Official Compendia—On July 5, 1974, the Executive Directors of the United States Pharmacopeial Convention and the American Pharmaceutical Association, respectively, announced jointly the unification of the two official drug compendia, the United States Pharmacopeia and the National Formulary, through acquisition of the NF by the USP Convention, Inc. The agreement was reached after 4 years of merger discussions, and completion of the details involved constitutional revisions by both the APhA and the Pharmacopeial Convention. Unification of the USP and NF is in keeping with the intent of resolutions adopted by the Convention in 1970 and at the 1973 special meeting. Pursuant to the acquisition, by the USP Convention, of the National Formulary on January 2, 1975, responsibility for keeping both official compendia current was consolidated. Thus, all Supplements and Interim Revision Announcements issued thenceforth served to bring both up to date. Beginning in 1980, a change in the scope of each compendium was introduced to obviate relabeling of drug products with the publication of each new edition. All drug substances and drug products were covered in USP, rather than limiting it only to the "best" drugs. NF, on the other hand, was devoted exclusively to pharmaceutic ingredients.

International Drug Standards—Since 1902, when the first of two international conferences on drug standards was held at Brussels, Belgium, the United States Pharmacopeia has contributed actively to the establishment of international drug standards. Since 1937, the Chairman of the USP Committee of Revision or, as in recent years, his designated director of drug standards revision, has served on a committee of experts that was charged first with "the unification of pharmacopeias" (under the Health Organization of the League of Nations) and, later, under the Interim Commission and the World Health Organization, with preparing the Pharmacopoea Internationalis (PhI). The PhI may be used in whole or in part by any country in establishing national drug specifications. Two editions of the PhI have appeared, both of which drew heavily upon the United States Pharmacopeia and information made available through the USP Committee of Revision. The second edition was entitled, "Specifications for the Quality Control of Pharmaceutical Preparations." A third edition is being published in a number of volumes over a period of years, again under the title, "The International Pharmacopoeia." In more recent years, the European Pharmacopoeia was established, under the direction of the Council of Europe (Partial Agreement) in accordance with the Convention on the Elaboration of a European Pharmacopoeia. The first volume of the European Pharmacopoeia was published in 1969. A Second Edition is being published as a series of volumes. The United States Pharmacopeia has maintained contact with and exchanged information with the European Pharmacopoeia since its inception. Similar relations have been maintained with a number of national pharmacopeias, in particular the British Pharmacopoeia, under a policy of collaborating, insofar as possible, in improving drug standards wherever such an interest exists.

Preface to USP XXII

The 5 years that have elapsed since publication of USP XXI–NF XVI have seen undiminished progress in setting public standards for compendial articles. This is the third Revision that consolidates USP and NF into a single volume. Consolidation of these two historic works and the energetic revision programs of the past two quinquennia account for the heft of this volume. A total of 420 new monographs were adopted in the 1985–1990 revision period. Of that total, 389 are USP and 31 are NF monographs. A sum of 173 USP XXI–NF XVI monographs have been deleted. Due to the changes made, this volume exhibits 254 more monographs than did the former volume and includes 11 new general chapters. The monographs added were the result of a system of priorities based both on the medical merit and the extent of use of any given article consistent with the resolutions of the past several meetings of the Pharmacopeial Convention. Improvements in the public standards for existing monographs and for general chapters were numerous; about 3900 individual revisions were processed during the past 5 years. Because it is not possible to portray the breadth and depth of all of those improved or added standards in this preface, highlights have been selected that address major trends or issues.

The 1985–1990 General Committee of Revision

The General Committee of Revision (see page vi) is responsible for ensuring the "accuracy and adequacy" of the *United States Pharmacopeia* and the *National Formulary*, the compendia of legally recognized standards for drug quality, purity, strength, packaging, and labeling, and for the consensus of *USP Dispensing Information.* The Committee is the decision-making body on all USP-NF and USP DI revisions and additions made possible by advances in science, therapy, and technology.

The 1985-1990 General Committee of Revision, 101 in all, comprised 26 members qualified in the medical sciences, 1 member representative of consumer interest, and 73 members qualified in the pharmaceutical and allied sciences, together with the Executive Director, who is Chairman of the Committee (see page vi). The 1985-1990 revision period marked an expansion in the number of Canadians serving on the Committee of Revision, although communication with Canadian scientists has been a regular feature of compendial programs for decades.

The effects of major changes in the Committee organization that were made in 1980 have continued. Subcommittees are smaller and therefore the work load has been lessened for the Subcommittee chairmen. And Subcommittees have more across-the-board members so that monographs receive attention from several points of view. For example, a monograph on an injection can be viewed by separate experts with regard to problems of chemical identity and assay as well as those problems inherent in parenteral products.

The 17 Subcommittees for the 1985-1990 revision period were supplemented by the appointment of 10 Advisory Panels, each of which reported to a specific Subcommittee (see page viii).

Changing Times—Continuous expansion and improvement are evident in the pharmaceutical world in any period as long as a quinqennium. The last 5 years were punctuated by major developments in science and in laws and regulations. Two such historic events involved (1) the growth of research and development of biotechnology-derived products and (2) changes in production and distribution of generic drugs and bulk pharmaceutical chemicals.

Explosive growth occurred in research and development of products from biotechnology. Since a number of these products are entering into general use, the Committee has initiated the groundwork for a broad Pharmacopeial response, the fruits of which will be published in Supplements to this volume during the 1990–1995 cycle. Public standards thus will be kept in equilibrium with the state of pharmaceutical science and industry.

Generic products have long been a feature, albeit a limited one, of North American pharmaceuticals. The fundamental limits on growth have been legal and political, not scientific. The 1984 Drug Price Competition and Patent Term Restoration Act (Waxman-Hatch Act) redefined those limits in its attempt to facilitate the marketing of generic products and to stimulate future research leading to new drugs and to new uses for existing drugs. Substantial impact on compendia resulted. Changes in the production and distribution of both drug products and bulk pharmaceutical chemicals on an international scale presented challenges to standards setting. Commentary on revisions proposed in the journal *Pharmacopeial Forum* (*PF*) sharpened and grew. Reference Standard usage patterns changed. The General Committees of Revision responded to this act by making numerous changes in USP-NF policies, in general tests and assays, and in hundreds of individual monographs.

Revision Features of USP XXII

Animal Tests—The Committee of Revision operates under a policy favoring the reduction, refinement, or replacement of whole animal procedures. New general chapters redefine the USP pattern of use of in-vivo and in-vitro methods: *Biological Reactivity Tests, In-vivo* ⟨88⟩ and *Biological Reactivity Tests, In-vitro* ⟨87⟩, where a cell culture method using mouse fibroblasts appears. Other in-vitro methods are being pursued. Consistent with the negligible scientific yield of safety tests in mice, these were viewed as an unnecessary expense and were deleted from antibiotic monographs. Chromatographic science has afforded acceptable alternatives in some other monographs.

Antibiotics—Both government and industry have recognized that quality standards and analytical methods for antibiotics have not kept pace with improvements in standards for other classes of drugs. As a consequence of the batch certification authority granted in 1962 to the federal Food and Drug Administration by amendment of the Act (section 507), certification methods in use by the FDA became the de-facto basis for quality standards. The Pharmacopeia only provided outline monographs for antibiotics. Because the USP revision process offers an active forum for scientific dialogue as well as a responsive mechanism for analytical improvements, and because the FDA reevaluated the continued utility of certification, the agency decided to discontinue that large program. Officials of the agency then indicated a desire that the USP fully resume its role of standards setting with respect to antibiotics.

The Committee of Revision accordingly took all of the necessary steps to structure the USP XXI antibiotic monographs to avoid referencing the Code of Federal Regulations (CFR) provisions. Extensive revisions of the 257 USP XX antibiotic monographs brought them into line with compendial style and included details of analytical procedures and citations to general chapters. Five new general chapters were prepared, and several others were revised to accommodate the attributes unique to antibiotics. With the 140 new monographs that were prepared, USP XXI included about 400 antibiotic monographs. During the 1985-1990 revision cycle, the Committee of Revision concentrated on improving the tests and assays in the USP XXI antibiotic monographs. As examples, over 100 new identification tests were added to antibiotic dosage-form monographs, and selective liquid chromatographic assays replaced several dozen microbial, colorimetric, spectrophotometric, and titrimetric procedures. About 80 new antibiotic monographs are included in USP XXII. Technical coordination and cooperation between USP and FDA in this broad revision effort was achieved by the Technical Subcommittee of the Joint USP-FDA Committee on Antibiotic Quality.

Drug Release and Bioavailability—In 1970, *Dissolution* tests were introduced into six monographs in the *United States Pharmacopeia* and into six monographs in the *National Formulary* as an initial response to concerns about the physiological availability of drug products. The subsequent 5-year period was devoted more to developing and publicizing the usefulness of the tests and to gaining experience with them in manufacturing and governmental laboratories. During the revision period leading to USP XX, a policy favored the inclusion of a *Dissolution* test in essentially all Tablet and Capsule monographs, resulting in about 60 monographs with dissolution requirements. A more forceful exercise of this policy led to about 400 dissolution requirements in USP XXI. Focus during the 1985–1990 quinquennium was possible, therefore, on the more challenging applications of drug-release testing; there are 23 drug-release requirements in this volume in addition to the 481 dissolution requirements.

Concern continued during this revision period for the bioavailability and physiological availability of formulated articles. Of equal significance was the continuing recognition of the value of dissolution testing as a tool for quality control. Thus, equivalence in dissolution behavior is sought in light of both bioavailability and quality-control considerations.

Experience has demonstrated that where a medically significant difference in bioavailability has been found among supposedly identical articles, a dissolution test has been efficacious in discriminating among these articles. Because the USP sets forth attributes of an acceptable article, such a discriminating test is satisfactory because the dissolution standard can exclude definitively any unacceptable article. Therefore, no compendial requirements for animal or human tests of bioavailability are necessary. The practical problem has been the obverse. Dissolution tests are so discriminating of formulation factors that may only sometimes affect bioavailability that it is not uncommon for a clinically acceptable article to perform poorly in a typical dissolution test. In such cases, the Committee of Revision has been mindful of including as many acceptable articles as possible, without setting forth dissolution specifications so generous as to raise reasonable scientific concern for bioinequivalence.

Medically significant cases of bioinequivalence rest mainly on just four causal factors: particle size of an active ingredient; magnesium stearate in excess as a lubricant-glidant; coatings, especially shellac; and inadequate disintegrant. Every one of these factors is reactive to dissolution testing. There is no known medically significant bioinequivalence problem with articles where 75 percent of an article is dissolved in water at 37° in 45 minutes using either official apparatus at the usual speed. The assistance of acid sometimes is justified. A majority of monographs have such requirements.

MODIFIED-RELEASE DOSAGE FORMS—Modern pharmaceutical technology makes possible the design of dosage forms that modify the timing and the rate of release of the drug substance from the drug product. The dosage form becomes more than a simple vehicle for drug storage, portability, and administration. And physiological factors are of inescapable significance.

For compendial purposes, a modified-release dosage form was defined in the previous revision cycle as one for which the drug-release characteristics of time course and/or location are chosen to accomplish therapeutic or convenience objectives not offered by conventional dosage forms such as solutions, ointments, or promptly dissolving dosage forms. For the present, only two types of modified-release dosage forms are recognized and defined: extended-release and delayed-release. An extended-release dosage form is defined as one that allows at least a twofold reduction in dosing frequency as compared to that drug presented as a conventional dosage form (e.g., as a solution or a prompt drug-releasing, conventional solid dosage form). A delayed-release dosage form is defined as one that releases a drug (or drugs) at a time other than promptly after administration. Enteric-coated articles are delayed-release dosage forms.

USP XXII contains 18 monographs for delayed-release articles and 5 for extended-release articles. Extended-release articles exist for which there are a number of successful formulations, each based on a different method for achieving modified release (for Case Three in the USP implementation policy, see *Theophylline Extended-release Capsules*). In each case, tailor-made selections of test apparatus, media, procedures, and test time can be made, provided that the product labeling is unambiguous and is consistent with the formulation-specific standard that applies to that article. Just as the labeling of strengths is necessary, information on the biological profiles of the concentrations of the active ingredients released by the various formulations is required on labels for practitioners. Addition of such labeling requirements represents a new step in standardizing these products and a significant improvement in public standards.

Transdermal drug therapy research has resulted in the manufacture of articles now in general use. Accordingly, the Subcommittee on Biopharmaceutics has reviewed Nitroglycerin Patches and adopted specific test apparatus and procedures so that each distinctive formulation will be subject to appropriate requirements.

Excipients and Formulas—Wherever possible, most of the long-standing, prescription-like composition formulas were removed from USP XXII. As a consequence, monographs for most flavors and fragrances in the *National Formulary* were deleted. The NF has also been expanded, with polymeric materials leading the way, so that about 360 excipients are described in either USP XXII or NF XVII. A new chapter inserted in USP XXII, *Water-Solid Interactions in Pharmaceutical Systems* ⟨1241⟩, deals with the importance of the exposure of solid pharmaceutical materials to water during processing and storage.

During the last revision cycle the American Pharmaceutical Association and The Pharmaceutical Society of Great Britain published an unofficial volume describing attributes of many excipients entitled the *Handbook of Pharmaceutical Excipients.* This handbook is important to the compendial revision of NF just as *Analytical Profiles of Drug Substances* has been useful to the Subcommittees in developing and improving analytical methods.

Impurities—A consequence of the primacy of separation science in pharmaceutical analysis throughout the past two quinquennia was the adoption of hundreds of chromatographic tests and assays. As the resolution of an article into its components became easier, more attention could be paid to the now apparent minor components. There was, consequently, a rethinking of Pharmacopeial policies and selection of tests and assays. Besides the hundreds of revisions of older monographs and the establishment of new monographs, there were revisions in the *General Notices* statements on these subjects and additions of two new general chapters, *Ordinary Impurities* ⟨466⟩ and *Impurities in Official Articles* ⟨1086⟩. Organic volatile impurities are the subject of an especially intensive effort at this time.

Radiopharmaceuticals—In USP XXII, the Becquerel unit was adopted to eventually replace the long-familiar Curie as a matter of international agreement. Both units are to be labeled for the present. Only the Becquerel is expected to appear in USP XXIII.

Diagnostic use of positron emitters is of recent application. This new development led to adoption in this volume of a monograph for Fludeoxyglucose F 18 Injection.

Scanning Electron Microscopy—Consistent with the view that the Pharmacopeia needs to be in equilibrium with the state of pharmaceutical analysis and to reflect any important component thereof, a general chapter, *Scanning Electron Microscopy* (SEM) ⟨1181⟩, has been added. SEM is a powerful tool that allows direct examination of the form and surfaces of solids of microscopic specimens and, through fluorescence, allows interpretation of the elemental composition of the components of a specimen. SEM is applicable to excipients, isolated particulate matter, and packaging materials. So far, no requirements for SEM have been inserted in any USP articles.

Reagent Standards—Success in conducting Pharmacopeial tests and assays is dependent on the use of reagents of the highest quality. Efforts have been made to ascertain where special precautions are essential and to provide suitable specifications. The development and revision of reagent specifications for USP XXII–NF XVII has been an assignment of the USP Subcommittee on General Chapters and Reagents (GCR). This has changed the stature of the reagents section by placing responsibility for the reagent specifications within the USP General Committee of Revision, along with the continuing assistance of volunteer experts in the chemical industry.

Reference is made in the section on Reagents to the specifications prepared and published under the aegis of the American Chemical Society because these specifications are being followed closely by reagent producers in the United States. Reliance is placed on the ACS specifications because generally for reagents covered in the current edition of ACS *Reagent Chemicals*, the corresponding USP entries specify the use of the ACS reagent grades of the substances.

Stability—Monographs do not include specific sections that deal directly with this major aspect of drug product quality, primarily because stability is a concern of the total monograph in (1) general requirements for expiration dates; (2) monograph packaging and storage requirements; (3) standards for packaging materials and containers; (4) standards for repackaging and storage; (5) limit tests for decomposition products; and (6) frequent selections of stability-indicating assays. The informational general chapter, *Stability Considerations in Dispensing Practice* ⟨1191⟩, discusses drug product stability.

System Suitability Tests—These tests are based on the concept that instruments, reagents, packings, conditions, procedural details, detectors, electronic accessories, and even analysts constitute a single system that is amenable to an overall test of system function. Reliable chromatographic performance, for example, may require specifications for resolution, column efficiency, peak tailing, precision of replicates, or extremes of conditions. Such tests obviate the necessity to specify a multitude of instrumental settings, model numbers, names of manufacturers, packings and lot numbers, and other physical-chemical and engineering characteristics as well as the distribution of official lots of chromatographic packings. System suitability tests were introduced in USP XIX and have come into general application.

Test Results, Statistics, and Standards—Interpretation of results from official tests and assays requires an understanding of the nature and style of compendial standards, in addition to an understanding of the scientific and mathematical aspects of laboratory analysis.

Confusion of compendial standards with release tests and with statistical sampling plans occasionally occurs. Compendial standards define what an acceptable article is and give test procedures that demonstrate that the article is in compliance. These standards apply at any time in the life of the article from production to consumption. The manufacturer's release specifications, and compliance with good manufacturing practices generally, are developed and followed to assure that the article will indeed comply with compendial standards until its expiration date, when stored as directed. Thus, when tested from the viewpoint of commercial or regulatory compliance, any specimen tested as directed in the monograph for that article shall comply (see *Test and Assays* under *General Notices*).

Tests and assays in this Pharmacopeia prescribe operation on a single specimen, that is, the singlet determination, which is the minimum sample on which the attributes of a compendial article should be measured. Some tests, such as those for *Dissolution* and *Uniformity of dosage units*, require multiple-dosage units in conjunction with a decision scheme. These tests, albeit using a number of dosage units, are in fact the singlet determinations of those particular attributes of the specimen. These procedures should not be confused with statistical sampling plans. The compendial procedures demonstrate compliance of the attributes of an article with compendial standards for a specimen (of one or more dosage units) that is subjected to analysis. Repeats, replicates, or extrapolations of results to larger populations are neither specified nor proscribed by the compendia; such decisions are dependent on the objectives of the testing. Commercial or regulatory compliance testing, or manufacturer's release testing, may or may not require examination of additional specimens, in accordance with predetermined guidelines or sampling strategies.

Format and Style—To expedite compilation and printing of the book, cross-references throughout the monographs, general chapters, and reagent specifications are made without the use of page numbers. Cross-references to other monographs cite the other monograph title in *italics;* cross-references to chapters cite the italicized chapter title and/or the chapter number (in angle brackets; see *Guide to General Chapters*, page 1468).

As a part of cooperation with the U.S. Adopted Names (USAN) program, the word order for the names of most organic compounds follows the USAN principle that the pharmacologically active portion is named first.

Decisions affecting the alphabetic order of the monographs must always be arbitrary and, to some extent, in conflict with the well-established rules of indexing. For example, a monograph on the salt form of a drug substance precedes the monographs on the dosage forms bearing only the active moiety in their titles, which alphabetically would appear first. As another example, the group of monographs for the products containing insulin are placed together, since they are all closely interrelated with respect to both content and use. Also, monographs on radioactive pharmaceuticals are placed in alphabetic sequence according to the isotope. (For example, Sodium Iodide I 123 Capsules appears in the sequence governed by the word "iodine.") Cross-references are provided as a guide to the location, and the index lists the monographs individually as well as in a grouping under Radioactive Pharmaceuticals.

A running head at the top of each page denotes, in italics, the section of the monographs or general chapters in which a page occurs. And in alphabetical order in boldface type, the keyword for the first monograph (or the complete or partial title of the first chapter) on a left-hand page is at the top left. The keyword for the last monograph (or the complete or partial title of the last chapter) on a right-hand page is at the top right.

The general chapters pertaining to Pharmacopeial requirements are grouped under the main heading, *General Tests and Assays*. General chapters that "contain no standards, tests, or assays, nor other mandatory specifications, with respect to any Pharmacopeial article," are grouped under the main heading *General Information.*

The *Combined Index to USP XXII and NF XVII* includes display groups of similar titles, e.g., *Acids*. Its comprehensive coverage is intended to make the index not only more useful, but also more used.

Many U.S. and Canadian brand names of USP articles are given in the companion volumes of *USP Dispensing Information* (USP DI Volume I, *Drug Information for the Health Care Professional; USP DI Volume II, Advice for the Patient;* and USP DI Volume III, *Approved Drug Products and Legal Requirements*).

The USAN Program—A cooperative effort inaugurated in 1961 between the American Medical Association and the United States Pharmacopeial Convention, Inc., flourished from the outset. The organizing agencies were joined, in January 1964, by the American Pharmaceutical Association, then the publisher of the *National Formulary*, to form the United States Adopted Names (USAN) Council. In 1967, the Council invited the federal Food and Drug Administration to join them so that the Council's work could be coordinated with the requirements of the federal Food, Drug, and Cosmetic Act. The

Council consists of five persons who understand the needs and problems of naming drugs. The Council's output is incorporated, along with other names for drugs (including public, proprietary, chemical, and code-designated names) in an annual book, *USAN and the USP Dictionary of Drug Names*, published by the U.S. Pharmacopeial Convention, Inc.

Chemical Names, CAS Registry Numbers, and Graphic Formulas—The chemical subtitles given in the monographs are Index names used by the Chemical Abstracts Service (CAS) of the American Chemical Society. They are provided only in monographs in which the titles specify substances that are distinctly definable chemical entities.

The first subtitle is the inverted form of the systematic chemical name developed by the CAS, in general accordance with the rules established over the years by the International Union of Pure and Applied Chemistry (IUPAC) and the International Union of Biochemistry (IUB), and is employed in the current issues of *Chemical Abstracts* (*CA*). The second name, given in uninverted form, is of a systematic type formerly used in *CA*; it is identical with, or closely resembles, the chemical name sanctioned and employed by the IUPAC and by the World Health Organization (WHO).

While the IUPAC names make generous use of nonsystematic and semisystematic (often referred to as trivial) names and qualifying terms, all of which impede electromechanical manipulation, the CAS names are fully systematic for most substances and are amenable to search and retrieval. The two subtitles referred to in the foregoing paragraphs are frequently identical, and a *CA* synonym is occasionally supplied as a third subtitle.

Monographs carrying chemical subtitles generally carry also CAS Registry Numbers. These italicized, bracketed numbers function independently of nomenclature as invariant numerical designators of unique unambiguous chemical substances in the CAS registry and thus find wide, convenient use.

Consonant with the employment of *CA* nomenclature, and also in the interest of uniformity of style, the orientation of ring systems and the depiction of stereoisomeric features in graphic formulas are generally consistent with CAS practices. The assurance of conformance with these CAS principles has been largely due to the untiring efforts and many contributions by Kurt L. Loening, Ph.D., of Columbus, Ohio, who for many years has been a USP consultant on chemical names, graphic formulas, and related information.

Drug Nomenclature—A USP Drug Nomenclature Committee was appointed in 1986 and subsequently met on several occasions to provide recommendations for the Division Executive Committees. The Committee's work does not overlap that of the USAN Council. Rather, it is complementary and is concerned with standardization of compendial titles, particularly dosage-form titles and combination drug products.

Pharmacopeial Forum—The bimonthly *Pharmacopeial Forum* (*PF*), the journal of drug standards development and official compendia revision, presents proposed USP and NF texts for public review and comment. *PF* enables reviewers to see at a glance both the text that is proposed for deletion and the text that is proposed for addition or modification.

PF is offered by subscription to all interested parties. Thus, the public and those concerned with drug standards, as well as the Committee of Revision and its Advisory Panels, have access to the latest proposed revisions in the official standards and tests and can readily transmit their comments, suggestions, and data to USP headquarters for consideration by the Revision Committee. A cumulative index for the revision cycle appears in each issue of *PF* for maximum convenience to the user.

In addition to presenting proposals recommended by the Committee of Revision in a section entitled *In-process Revision*, *PF* contains several other sections. A *Headquarters Column* gives information on publication deadlines, news or statements from the U.S. Pharmacopeial Convention, Inc., and the Committee of Revision, summaries of issues discussed by the Drug Standards Division Executive Committee, and various tabulations or lists that aid one in keeping track of the multifaceted revision program. Reports or statements of authoritative committees, scientific articles relevant to compendial issues, general commentaries by interested parties, and collations of comments received in response to policy initiatives are published in the *Stimuli to the Revision Process* section.

Reference Standards—The catalog listing the collection of official USP and NF Reference Standards is kept up-to-date through bimonthly publication in *PF*. The listing identifies new items, replacement lots, those lots of a single item that are simultaneously official, those lots deleted from official status, and a projection of items eventually to be adopted. Purchase order information is included. The names of distributors who can facilitate international availability of these items are suggested.

Conferences—Consistent with the USP policy of emphasis on public participation in standards setting, conferences are held to allow interactive examination of selected topics. Most revision work is handled through publication of proposals in *Pharmacopeial Forum*, by correspondence, through Subcommittee meetings, by telecommunications, and by attendance of staff and Subcommittee members at regular meetings of scientific and professional societies. There arises from time to time a cluster of related topics that calls for conferences as yet another information-gathering mechanism, one better suited to complex issues of interest to diverse parties.

In addition to the subject-oriented conferences listed below, two other conference-like meetings were developed late in the quinquennium. Starting with the Eighth Supplement, a date was announced for an Open Conference on the contents of the Supplement. Following the publication of each new issue of *Pharmacopeial Forum* since Jan.-Feb. 1988, an Open House has been held to discuss the contents of *PF*. Both of these new meetings have been conducted on an experimental basis to determine their worthiness in the standards-setting process.

The Summary Proceedings of the following USP Conferences that took place in this quinquennium have been published by USPC:[1]

Drug Substance Purity, October 1986, Alexandria, Virginia.

Alternative Methods for Toxicity Testing—the Invitro Option, January 1988, Marco Island, Florida.

Current Revision Issues, April 1988, Alexandria, Virginia.

Current Revision Issues, May 1989, Colonial Williamsburg, Williamsburg, Virginia.

Biotechnology-derived Products, September 1989, Corpus Christi, Texas.

Legal Status of the Official Compendia—References to the *United States Pharmacopeia* and the *National Formulary* occur in numerous statutes and regulating articles used in medical practice; the most significant one is recognition of the Official Compendia in the federal Food, Drug, and Cosmetic Act. These statutes usually empower the governmental agency to enforce the law using certain defined aspects of the compendia. Most commonly recognized are USP and NF standards for determining the identity, strength, quality, and purity of the articles and specifications for packaging and labeling.

In addition, many statutes and regulations governing the practice of pharmacy reference USP and NF requirements for packaging and storage of drugs. During the 1985–1990 revision period, newly developed USP requirements for customized patient medication packages (see *Containers* ⟨661⟩) were incorporated into several state laws.

Aside from statutory recognition, USP and NF standards serve an important role in ordinary commercial transactions. Thus, although the compendia are not published in response to any statutes, statutory and commercial use places upon the Committee of Revision implied strictures with respect to the need for clarity of presentation and for reliability of USP and NF standards and procedures. This pertains both to national and to international commerce.

Foreign Use of the USP—USP standards have been developed and improved for more than a century and a half, primarily for pharmaceutical items used in the United States. Interest in the USP outside of the United States has always existed, and during the past revision period that interest increased notably. More than 40 countries have declared the USP to be their legal compendium, or have required its use along with other compendia or sources of drug or device standards. Representatives from five of these countries (Canada, Panama, Philippines, Saudi Arabia, and South Korea) were invited to serve as members of the Pharmacopeial Convention for the 1985-1990 revision period. Visitors from abroad are invited to USP headquarters for exchanges of information and discussions on how USP standards could become more extensively available. Several members of the Committee of Revision and two staff members are members of WHO expert advisory panels: one on the International Pharmacopoeia and Pharmaceutical Preparations, and the other on Biological Standardization. Staff and Committee of Revision members have attended meetings of expert committees and study groups of international organizations concerned with drug standards and supply, have participated in international meetings of nongovernmental organizations on a variety of compendia-related topics, have acted as consultants for such bodies, and have visited standard-setting agencies in the Americas and on other continents.

USP Drug Research and Testing Laboratory—The Laboratory contributes significantly to the revision of compendial tests and standards and to the establishment of USP Reference Standards. Continuing expansion of laboratory and reference standard needs was the impetus for construction of a new USP headquarters building in 1988-89.

Dispensing Information—Categories and indications for the use of drugs, routes of administration and dosage schedules, sizes and strengths of commercially available products, and other notes to the pharmacist and suggestions for advising the patient were removed from the USP and the NF in 1980. All of this information is published separately in greatly expanded form in the authorized publication, *USP Dispensing Information*—Volume I, *Drug Information for the Health Care Professional* and Volume II, *Advice for the Patient*. Volume III, *Approved Drug Products and Legal Requirements*, a composite of legal and federal subjects, was first published in 1989. As an "authorized" publication, USP DI is prepared in accordance with the rules and procedures adopted for it by the Committee of Revision. USP DI is not limited to those drugs that are included in the USP or the NF, nor vice versa. But USP DI and USP-NF share the goal of including all drugs in use in the United States of America.

Participants

Food and Drug Administration—Close working relations with the FDA continued as a hallmark of the revision program, both with individual scientists and with FDA laboratories and other components. Formal liaison efforts were conducted primarily through the Compendial Operations Branch of the FDA. During this past revision cycle, the Branch operated under the leadership of Charma A. Konner (1985-1987) and Joel S. Davis (1987 to present). Mr. Davis's staff consists of Keith E. Egli, Linda J. McGee, Meade D. North, Robert A. Rippere, Victor J. Warner, Jr., and Nathan Weston. This Branch is responsible for obtaining and coordinating agency comments on proposals appearing in *Pharmacopeial Forum*, coordinating the Compendial Monograph Evaluation and Development (CMED) project, and serving as an official point of contact for the compendia. As a result of the activities of the Compendial Operations Branch of the FDA, there have been many suggestions for improvement as well as a greater degree of consistency between FDA and compendial requirements.

The Center for Drug Analysis in St. Louis, Missouri, under the direction of Dr. Thomas P. Layloff,

[1]Available from USPC Order Processing Department, 12601 Twinbrook Parkway, Rockville, MD 20852. Toll-free telephone number: 1-800-227-USPC.

is a constant participant in the revision process. Not only did this laboratory continue from the past cycle to do extensive development and review of tests and assays, but during this cycle it also became the primary governmental participant in the ongoing evaluation of established and proposed new USP Reference Standards.

PMA Biological and Quality Control Sections—These Sections of the Pharmaceutical Manufacturers Association (PMA) have long histories of interaction with the USP and the NF revision programs. Subject- and task-oriented PMA committees, which are organized by the Sections, concern themselves with subjects that are of direct interest to members of the Committee of Revision. Subjects are often assigned to these committees in response to initiatives from the USP-NF revision process. The Sections are able to do broad-based surveys of members and operate multilaboratory collaborative studies. However, as in the case of interactions between the Committee of Revision and government-sponsored programs relevant to standards setting, not all of these efforts have led to adoption in form or substance by the Committee of Revision. Nevertheless, a series of important contributions have been made to each recent revision cycle. Notable contributions during the 1985-1990 cycle include collaborative studies on, and preparation of, dissolution calibrators; improved test methods for particulate matter in small-volume injections; development of a consensus approach to the validation of compendial tests and assays that has been adopted herein as a general chapter, *Validation of Compendial Methods* ⟨1225⟩; and development of consensus statements on a rational approach to setting limits on impurities in bulk pharmaceutical chemicals that, along with definitions of impurities established through the Committee of Revision, have been formed into a new general chapter, *Impurities in Official Articles* ⟨1086⟩.

Other Associations—Guidelines for voluntary labeling of inactive ingredients of preparations were prepared by the PMA, the Proprietary Association, and the Generic Pharmaceutical Industry Association. In response to a resolution adopted by the Pharmacopeial Convention in 1985, these guidelines were adapted as an informational chapter, *Labeling of Inactive Ingredients* ⟨1091⟩.

Survey data on particulate contamination of large-volume injections was supplied by the Health Industry Manufacturers Association.

USP Drug Product Problem Reporting Program—The former Drug Product Problem Reporting Program, operated under contract between USPC and the Food and Drug Administration, was terminated in January 1988 and replaced with the USP Drug Product Problem Reporting (USP DPPR) Program. This program contributes significantly to the revision of compendial standards through the sharing of information reported by health professionals and by industry responses to those reported problems.

USP Fellowships—To stimulate training and research pertinent to drug standards, USPC has offered a number of fellowships to doctoral candidates. Each applicant must be sponsored by a faculty member who is a member of the current Committee of Revision or a member of an Advisory Panel.

The purposes of the USP Fellowships are to promote postgraduate research in areas related to compendial standards, thereby encouraging the education and training of Fellows in the sciences applicable to drug standardization. Recognition is also given to university faculty members elected to the USP Committee of Revision or serving on an Advisory Panel. During the 1985–1989 period, awards were made to 24 individuals at 13 different institutions.

Credits—As with any group effort, the weight of responsibility for successfully producing the Pharmacopeia falls more heavily on some than on others. The Revision Committee as a whole (see page vi) served as the elected agent of the USP Convention in producing this Pharmacopeia. But credit is due also to the people who were not on the Revision Committee who contributed out of a sense of public service and thereby greatly enhanced the Committee's effectiveness (see page xii).

Individual contributions of some members of the Revision Committee merit special mention, particularly the chairmen of the 17 USP Subcommittees (see page viii) who shared their management talents as well as their scientific knowledge. Special recognition should be given to Norman W. Atwater, Lester Chafetz, Jordan L. Cohen, Klaus G. Florey, Andrew J. Schmitz, Jr., and Ralph F. Shangraw, who worked strenuously to respond to major challenges in public standards.

The 1985–1990 General Committee of Revision has been broadly productive. The majority of the members completed their assignments creditably and showed commendable initiative and imagination in pursuing problems that arose. The progress made in public standards during the past 5 years is due to many large and small steps. Thanks and appreciation are also due to all participants who responded to lesser demands made by the 1985–1990 cycle.

Retirements—Notable individuals retired during or near the end of this quinquennium. William M. Heller, Ph.D., Executive Director, leaves after 22 years on the USP staff, following 10 years as an elected member of the Committee of Revision. Mary C. Griffiths, Director of Publication Services, retired after 36 years on the USP staff. Aubrey S. Outschoorn, L.M.S., Ph.D., and Edgar E. Theimer retired from the staff of the Drug Standards Division after 11 and 10 years, respectively. Dr. Outschoorn was appointed Scientist Emeritus. Jane C. Sheridan, Ph.D., and Samuel M. Tuthill, Ph.D., left the DSD Executive Committee. Those who will carry on look forward to advice and counsel from these highly regarded people.

HEADQUARTERS STAFF

An active and effective Committee of Revision requires a responsive and supportive staff. Staff members who served during the 1985-1990 revision period included:

Drug Standards Division—LEE T. GRADY, PH.D., *Director;* CHARLES H. BARNSTEIN, PH.D., *Assistant Director, Drug Standards Revision;* J. JOSEPH BELSON; ROGER

DABBAH, PH.D.; BARBARA B. HUBERT; FATIMA N. JOHNSON, PH.D.; AUBREY S. OUTSCHOORN, LIC. MED. SURG., PH.D. (ret.); W. LARRY PAUL, PH.D.; SHARON L. RYAN; V. SRINIVASAN, PH.D.; EDGAR E. THEIMER, M.S. (ret.); WILLIAM W. WRIGHT, PH.D.

DSD TECHNICAL SERVICES—ROBERT H. KING, *Assistant Director;* THERESA H. LEE, *Information Specialist*

USP REFERENCE STANDARDS—JOAN M. STAPLETON, *Supervisor;* JAZMIN L. ARISTORENAS; JONAIDA J. BANAS; CHITRA M. BHAT; SUSAN D. ZAWACKI

USP HAZARD COMMUNICATIONS PROGRAM—ROBERT C. MOORE, *Coordinator;* DANIEL D. BAKER; KIMBERLY MARTINELLI

USP DRUG RESEARCH AND TESTING LABORATORY—RICHARD F. LINDAUER, PH.D., *Director;* L. VALENTIN FEYNS, PH.D., *Supervisor;* VIVIAN A. GRAY, *Supervisor;* ANNE M. KELLEY, *Supervisor;* PATRICIA A. WHITE, PH.D., *Supervisor;* BETTY L. GOLDMAN, *Coordinator;* ROY E. ADKINS, JR.; WILLIAM E. BROWN; BEN C. CHOW, PH.D.; MILAGROS DE VERA; MARIA J. GLASGOW; G. STEVE HARTONG; DANIEL L. HAZELRIGG; G. SCOTT KOBLER; BENJAMIN LAY, M.S.; SUE C. LEE, PH.D.; ANTHONY J. MANNA; TIMOTHY J. O'HANLON; RICHARD P. PRICE; ARTHUR N. STROHL, PH.D.; KAILAS D. THAKKER, PH.D.; SARAH L. VINCENT

SUPPORTING STAFF—ANN K. FERGUSON, *Administrative Supervisor;* ANJU K. MALHOTRA, *Staff Assistant;* ANNE M. BEARY; MILDRED R. GARRISON; FRANCES RAMPP

Drug Information Division—KEITH W. JOHNSON, *Director, Research and Development;* GEORGIE M. CATHEY, *Assistant Director;* DIANA M. BLAIS; SANDRA LEE BOYER; NANCY LEE DASHIELL; WANDA J. JANICKI; ESTHER H. KLEIN; ANGELA MENDEZ MAYO; ERIC M. ORR; CAROL A. PAMER; JAIME RAMIREZ; DOROTHY RAYMOND; CATHERINE E. SZABO; MARTHA A.D. WOLLAM; GORDON K. WURSTER

Publication Services—PATRICIA H. MORGENSTERN, *Director;* SANDRA F. BOYNTON, *Managing Editor;* JESUSA D. CORDOVA, ELLEN E. ELOVITZ, CAROLYN A. FLEEGER, SUZANNE C. LASSANDRO, MELISSA M. SMITH

TYPESETTING—JEAN E. DALE, *Typesetting Systems Coordinator;* LORI E. FOSTER, *Supervisor;* DEBORAH R. CONNELLY, LAUREN TAYLOR DAVIS, MYRA M. SPENCER, RUTH E. THOMPSON

GRAPHICS—GAIL M. ORING, *Supervisor;* ELIZABETH BOZIEVICH, STAMATIA CALOMERIS

WORD PROCESSING—BARBARA A. BOWMAN, *Supervisor;* PAMELA J. HOES, MICHELINE TRANQUILLE

USP Drug Product Problem Reporting Program and Medical Device and Laboratory Product Problem Reporting Program—DIANE M. MCGINNIS, *Coordinator;* SHAWN C. BECKER, *Promotion Assistant;* ANN CORKEN, *DPPR Program Assistant;* ANNE PAULA THOMPSON, *PRP Program Assistant;* ELIDA AMEZQUITA; JEAN CANADA; SUSAN CLAGETT; ANN DALAGER; BANDANA DAS; FELISA ENRIQUEZ; RETHA FORD; DOROTHY KOLSKY; LAURA MILLER, R.N.; SHANTY RAMKISSOON; LATA RAO; SUSMITA SAMANTA, M.D.; CATHY R. WOOD; MARY SUSAN ZMUDA, R.N.

Administrative—WENDY M. AUSTIN, *Marketing Coordinator;* CATHY M. FERRERE, *Director of Marketing;* ALICE E. KIMBALL, *Director of Professional Affairs;* J. ROBERT STRANG, *Director of Operations;* JOSEPH G. VALENTINO, *Executive Associate;* RUSSELL L. WILLIAMS, *Controller*

WILLIAM M. HELLER, PH.D.
Executive Director, USPC, and
Chairman, USP Committee of Revision

Admissions

Articles Admitted to USP XXI[1] and NF XVI[2] by Supplement

First Supplement (January 1, 1985)

Acetohydroxamic Acid[1]
Acetohydroxamic Acid Tablets[1]
Adenine[1]
Alprostadil[1]
Alprostadil Injection[1]
Sterile Amoxicillin[1]
Sterile Amoxicillin for Suspension[1]
Amoxicillin Oral Suspension[1]
Ampicillin Boluses[1]
Ampicillin Soluble Powder[1]
Sterile Ampicillin Suspension[1]
Anticoagulant Citrate Phosphate Dextrose Adenine Solution[1]
Soluble Bacitracin Methylene-Disalicylate[1]
Bacitracin Methylene Disalicylate Soluble Powder[1]
Bacitracin Zinc Soluble Powder[1]
Baclofen[1]
Baclofen Tablets[1]
Butalbital and Aspirin Tablets[1]
Cellulose Sodium Phosphate[1]
Cimetidine[1]
Cimetidine Tablets[1]
Citric Acid, Magnesium Oxide, and Sodium Carbonate Irrigation[1]
Cloxacillin Benzathine[1]
Cloxacillin Benzathine Intramammary Infusion[1]
Sterile Cloxacillin Benzathine[1]
Cloxacillin Sodium Intramammary Infusion[1]
Sterile Cloxacillin Sodium[1]
Cromolyn Sodium Inhalation[1]
Cromolyn Sodium Nasal Solution[1]
Cyclomethicone[2]
Dihydrostreptomycin Sulfate[1]
Dihydrostreptomycin Sulfate Boluses[1]
Dihydrostreptomycin Sulfate Injection[1]
Sterile Dihydrostreptomycin Sulfate[1]
Diphenhydramine Citrate[1]
Dipivefrin Hydrochloride[1]
Dipivefrin Hydrochloride Ophthalmic Solution[1]
Guanabenz Acetate[1]
Guanabenz Acetate Tablets[1]
Halcinonide[1]
Halcinonide Cream[1]
Halcinonide Ointment[1]
Halcinonide Topical Solution[1]
Hetacillin Potassium Intramammary Infusion[1]
Hetacillin Potassium Oral Suspension[1]
Hetacillin Potassium Tablets[1]
Hydroxypropyl Methylcellulose[1]
Hydroxypropyl Methylcellulose Phthalate[2]
Ketoconazole Tablets[1]
Levonorgestrel[1]
Levonorgestrel and Ethinyl Estradiol Tablets[1]
Metaproterenol Sulfate[1]
Metaproterenol Sulfate Inhalation Aerosol[1]
Metaproterenol Sulfate Inhalation Solution[1]
Metaproterenol Sulfate Tablets[1]
Miconazole Nitrate Vaginal Suppositories[1]
Neomycin Sulfate and Hydrocortisone Acetate Lotion[1]
Nitroglycerin Injection[1]
Oxygen 93 Percent[1]
Penicillin G Procaine Intramammary Infusion[1]
Penicillin G Procaine and Dihydrostreptomycin Sulfate Intramammary Infusion[1]
Penicillin G Procaine and Novobiocin Sodium Intramammary Infusion[1]
Penicillin V Benzathine Oral Suspension[1]
Yellow Phenolphthalein[1]
Potassium Chloride, Potassium Bicarbonate, and Potassium Citrate Effervescent Tablets for Oral Solution[1]
Prednisone Syrup[1]
Sodium Stearyl Fumarate[2]
Sterile Spectinomycin Hydrochloride for Suspension[1]
Technetium Tc 99m Disofenin Injection[1]
Triethyl Citrate[2]
Vincristine Sulfate Injection[1]

Second Supplement (July 1, 1985)

Acetaminophen and Diphenhydramine Citrate Tablets[1]
Alclometasone Dipropionate[1]
Alclometasone Dipropionate Cream[1]
Alclometasone Dipropionate Ointment[1]
Alprazolam[1]
Alprazolam Tablets[1]
Ammonium Alum[1]
Potassium Alum[1]
Aspirin, Alumina, and Magnesia Tablets[1]
Purified Bentonite[2]
Calcium Carbonate Oral Suspension[1]
Carisoprodol[1]
Carisoprodol and Aspirin Tablets[1]
Carisoprodol, Aspirin, and Codeine Phosphate Tablets[1]
Chlortetracycline Hydrochloride Tablets[1]
Clotrimazole Lotion[1]
Dextrin[2]
Dinoprost Tromethamine[1]
Dinoprost Tromethamine Injection[1]
Dyphylline Elixir[1]
Furazolidone Oral Suspension[1]
Furazolidone Tablets[1]
Hexylene Glycol[2]
Hydrocortisone and Acetic Acid Otic Solution[1]
Hydrocoritsone Butyrate[1]
Lidocaine and Epinephrine Injection[1]
Magnesium Oxide Capsules[1]
Magnesium Oxide Tablets[1]
Metaproterenol Sulfate Syrup[1]
Mineral Oil Enema[1]
Minoxidil[1]
Minoxidil Tablets[1]
Nadolol[1]
Nadolol Tablets[1]
Pentazocine and Naloxone Hydrochlorides Tablets[1]
Piroxicam[1]
Piroxicam Capsules[1]
Propylene Glycol Alginate[2]
Stannous Fluoride Gel[1]
Tetracycline Boluses[1]
Tetracycline Hydrochloride and Novobiocin Sodium Tablets[1]
Tetracycline Hydrochloride, Novobiocin Sodium, and Prednisolone Tablets[1]
Tetracycline Phosphate Complex and Novobiocin Sodium Capsules[1]
Tioconazole[1]
Valproic Acid[1]
Valproic Acid Capsules[1]
Valproic Acid Syrup[1]
Zinc Gluconate[1]

Third Supplement (January 1, 1986)

Acetaminophen Oral Solution[1]
Amoxicillin Intramammary Infusion[1]
Aspirin Delayed-release Capsules[1]
Aspirin Delayed-release Tablets[1]
Aspirin and Codeine Phosphate Tablets[1]
Aspirin, Codeine Phosphate, Alumina, and Magnesia Tablets[1]
Bupivacaine in Dextrose Injection[1]
Ceftizoxime Sodium[1]
Ceftizoxime Sodium Injection[1]
Chlorocresol[2]
Chlortetracycline Bisulfate[1]
Chlortetracycline and Sulfamethazine Bisulfates Soluble Powder[1]
Chlortetracycline Hydrochloride Soluble Powder[1]
Cromolyn Sodium Ophthalmic Solution[1]
Diflorasone Diacetate[1]
Diflorasone Diacetate Cream[1]
Diflorasone Diacetate Ointment[1]
Dihydroxyaluminum Aminoacetate Capsules[1]
Estradiol Vaginal Cream[1]
Magnesium Trisilicate Tablets[1]
Meprednisone[1]
Myristyl Alcohol[2]
Nalorphine Hydrochloride[1]
Nalorphine Hydrochloride Injection[1]
Neomycin Sulfate and Flurandrenolide Lotion[1]
Neomycin and Polymyxin B Sulfates Cream[1]
Octyldodecanol[2]
Polyethylene Excipient[2]
Potassium Benzoate[2]
Prednisone Oral Solution[1]

Psyllium Hydrophilic Mucilloid for Oral Suspension[1]
Racepinephrine[1]
Racepinephrine Inhalation Solution[1]
Racepinephrine Hydrochloride[1]
Tetracycline Hydrochloride Soluble Powder[1]
Theophylline and Guaifenesin Oral Solution[1]
Triazolam[1]
Triazolam Tablets[1]

Fourth Supplement (November 15, 1986)

Acetaminophen and Codeine Phosphate Elixir[1]
Alumina and Magnesium Trisilicate Tablets[1]
Dried Aluminium Hydroxide Gel Capsules[1]
Sterile Amdinocillin[1]
Amoxicillin and Clavulanate Potassium for Oral Suspension[1]
Amoxicillin and Clavulanate Potassium Tablets[1]
Anthralin Cream[1]
Antimony Sodium Tartrate[1]
Buffered Aspirin Tablets[1]
Biotin[1]
Sterile Cefonicid Sodium[1]
Ceforanide for Injection[1]
Sterile Ceforanide[1]
Clavulanate Potassium[1]
Sterile Clavulanate Potassium[1]
Clotrimazole and Betamethasone Dipropionate Cream[1]
Colestipol Hydrochloride[1]
Colestipol Hydrochloride for Oral Suspension[1]
Desoximetasone Ointment[1]
Erythromycin and Benzoyl Peroxide Topical Gel[1]
Etidronate Disodium Tablets[1]
Hard Fat[2]
Indomethacin Suppositories[1]
Methoxsalen Capsules[1]
Methyldopa and Chlorothiazide Tablets[1]
Neomycin and Polymyxin B Sulfates and Hydrocortisone Acetate Cream[1]
Nystatin and Triamcinolone Acetonide Cream[1]
Nystatin and Triamcinolone Acetonide Ointment[1]
Oxycodone and Acetaminophen Capsules[1]
Oxycodone and Acetaminophen Tablets[1]
Oxycodone Hydrochloride[1]
Sterile Penicillin G Procaine and Dihydrostreptomycin Sulfate Suspension[1]
Sterile Penicillin G Procaine, Dihydrostreptomycin Sulfate, and Prednisolone Suspension[1]
Penicillin G Procaine, Neomycin and Polymyxin B Sulfates, and Hydrocortisone Acetate Topical Suspension[1]
Perphenazine and Amitriptyline Hydrochloride Tablets[1]
Polymyxin B Sulfate and Bacitracin Zinc Topical Aerosol[1]
Polymyxin B Sulfate and Bacitracin Zinc Topical Powder[1]
Potassium Gluconate and Potassium Citrate Oral Solution[1]
Potassium Gluconate, Potassium Citrate, and Ammonium Chloride Oral Solution[1]
Praziquantel[1]
Prilocaine and Epinephrine Injection[1]
Reserpine and Chlorothiazide Tablets[1]
Sodium Gluconate[1]
Sugar Spheres[2]
Sterile Ticarcillin Disodium and Clavulanate Potassium[1]
Tocainide Hydrochloride[1]

Fifth Supplement (May 15, 1987)

Compressed Air[1]
Alumina and Magnesium Carbonate Oral Suspension[1]
Betamethasone Sodium Phosphate Injection[1]
Calcium Citrate[1]
Carisoprodol Tablets[1]
Cefotaxime Sodium[1]
Cefotaxime Sodium Injection[1]
Cefoxitin Sodium[1]
Cefoxitin Sodium Injection[1]
Sterile Ceftriaxone Sodium[1]
Sterile Cefuroxime Sodium[1]
Cephalothin Sodium[1]
Disopyramide Phosphate Extended-release Capsules[1]
Doxycycline Hyclate Delayed-release Capsules[1]
Erythromycin Delayed-release Tablets[1]
Fluocinonide Topical Solution[1]
Gentistic Acid Ethanolamide[2]
Glyceryl Behenate[2]
Guanadrel Sulfate[1]
Guanadrel Sulfate Tablets[1]
Iodohippurate Sodium I 123 Injection[1]
Isosorbide Dinitrate Extended-release Capsules[1]
Isosorbide Dinitrate Tablets[1]
Isosorbide Dinitrate Chewable Tablets[1]
Isosorbide Dinitrate Extended-release Tablets[1]
Isosorbide Dinitrate Sublingual Tablets[1]
Methylparaben Sodium[2]
Metoprolol Tartrate Injection[1]
Nonoxynol 9[1]
Oxybutynin Chloride[1]
Oxybutynin Chloride Syrup[1]
Oxybutynin Chloride Tablets[1]
Oxyquinoline Sulfate[2]
Sterile Penicillin G Procaine, Dihydrostreptomycin Sulfate, Chlorpheniramine Maleate, and Dexamethasone Suspension[1]
Pindolol[1]
Pindolol Tablets[1]
Polyethylene Oxide[2]
Potassium Gluconate and Potassium Chloride Oral Solution[1]
Potassium Gluconate and Potassium Chloride for Oral Solution[1]
Probucol[1]
Propylparaben Sodium[2]
Quinidine Sulfate Extended-release Tablets[1]
Sodium Chloride Ophthalmic Ointment[1]
Invert Sugar Injection[1]
Sulfamethoxazole and Trimethoprim Concentrate for Injection[1]
Tocainide Hydrochoride Tablets[1]
Trihexyphenidyl Hydrochloride Extended-release Capsules[1]
Vancomycin Hydrochloride Capsules[1]

Sixth Supplement (November 15, 1987)

Acetaminophen, Aspirin, and Caffeine Capsules[1]
Acetaminophen, Aspirin, and Caffeine Tablets[1]
Alumina, Magnesium Carbonate, and Magnesium Oxide Tablets[1]
Basic Aluminum Carbonate Gel[1]
Dried Basic Aluminum Carbonate Gel Capsules[1]
Dried Basic Aluminum Carbonate Gel Tablets[1]
Azatadine Maleate Tablets[1]
Bacitracin and Polymyxin B Sulfate Topical Aerosol[1]
Bumetanide[1]
Bumetanide Tablets[1]
Chloroxylenol[1]
Copper Gluconate[1]
Cyclosporine[1]
Cyclosporine Concentrate for Injection[1]
Cyclosporine Oral Solution[1]
Sterile Erythromycin Lactobionate[1]
Halazepam[1]
Halazepam Tablets[1]
Isotretinoin[1]
Lorazepam[1]
Lorazepam Injection[1]
Lorazepam Tablets[1]
Magnesium Hydroxide Paste[1]
Malathion[1]
Malathion Lotion[1]
Mexiletine Hydrochloride[1]
Mexiletine Hydrochloride Capsules[1]
Topical Light Mineral Oil[1]
Neomycin and Polymyxin B Sulfates, Bacitracin Zinc, and Lidocaine Ointment[1]
Neomycin Sulfate and Triamcinolone Acetonide Cream[1]
Nitrofurantoin Capsules[1]
Phenylpropanolamine Hydrochloride Extended-release Capsules[1]
Potassium and Sodium Bicarbonates and Citric Acid Effervescent Tablets for Oral Solution[1]
Sodium Polystyrene Sulfonate Suspension[1]
Tocopherols Excipient[2]
Zein[2]

Seventh Supplement (May 15, 1988)

Albuterol[1]
Albuterol Sulfate[1]
Alumina, Magnesia, and Simethicone Oral Suspension[1]
Alumina, Magnesia, and Simethicone Tablets[1]
Antipyrine, Benzocaine, and Phenylephrine Hydrochloride Otic Solution[1]
Aztreonam for Injection[1]
Sterile Aztreonam[1]
Calcium Glubionate Syrup[1]
Calcium Lactobionate[1]
Captopril[1]
Carbamide Peroxide[1]
Cisplatin[1]
Cisplatin for Injection[1]
Erythromycin Pledgets[1]
Hydrocortisone Butyrate Cream[1]
Indomethacin Extended-release Capsules[1]
Iopamidol[1]
Iopamidol Injection[1]

Lidocaine Hydrochloride Oral Topical Solution[1]
Magaldrate and Simethicone Oral Suspension[1]
Magaldrate and Simethicone Tablets[1]
Methylphenidate Hydrochloride Extended-release Tablets[1]
Metoclopramide Injection[1]
Metoclopramide Tablets[1]
Metoclopramide Hydrochloride[1]
Metoprolol Tartrate and Hydrochlorothiazide Tablets[1]
Potassium Metabisulfite[2]
Safflower Oil[1]
Salicylamide[1]
Selenious Acid[1]
Selenious Acid Injection[1]
Sulfacetamide Sodium and Prednisolone Acetate Ophthalmic Ointment[1]
Tetracaine Hydrochloride in Dextrose Injection[1]

Eighth Supplement (November 15, 1988)

Alumina and Magnesium Carbonate Tablets[1]
Aminophylline Oral Solution[1]
Bismuth Subgallate[1]
Bumetanide Injection[1]
Carbomer 910[2]
Carbomer 934[2]
Carbomer 940[2]
Carbomer 941[2]
Carbomer 1342[2]
Cefoperazone Sodium[1]
Cefoperazone Sodium Injection[1]
Sterile Cefotetan Disodium[1]
Ceftazidime for Injection[1]
Sterile Ceftazidime[1]
Dihydroergotamine Mesylate, Heparin Sodium, and Lidocaine Hydrochloride Injection[1]
Ethylcellulose Aqueous Dispersion[2]
Flurbiprofen Sodium[1]
Flurbiprofen Sodium Ophthalmic Solution[1]
Labetalol Hydrochloride[1]
Labetalol Hydrochloride Injection[1]
Labetalol Hydrochloride Tablets[1]
Levobunolol Hydrochloride[1]
Levobunolol Hydrochloride Ophthalmic Solution[1]
Levocarnitine[1]
Levocarnitine Oral Solution[1]
Manganese Gluconate[1]
Oxprenolol Hydrochloride[1]
Oxycodone Hydrochloride Oral Solution[1]
Pimozide[1]
Pimozide Tablets[1]
Potassium Chloride in Dextrose and Sodium Chloride Injection[1]
Potassium Chloride in Sodium Chloride Injection[1]
Praziquantel Tablets[1]
Prednisolone Acetate Ophthalmic Suspension[1]
Procainamide Hydrochloride Tablets[1]
Procainamide Hydrochloride Extended-release Tablets[1]
Promethazine Hydrochloride Suppositories[1]
Propranolol Hydrochloride and Hydrochlorothiazide Tablets[1]
Oral Rehydration Salts[1]
Tioconazole Cream[1]
Xenon Xe 127[1]

New Admissions to the Official Compendia

New Admissions to USP XXII

Acetaminophen and Codeine Phosphate Capsules
Acetaminophen and Codeine Phosphate Tablets
Sterile Ampicillin Sodium and Sulbactam Sodium
Aspirin Effervescent Tablets for Oral Solution
Aspirin Extended-release Tablets
Aspirin, Codeine Phosphate, and Caffeine Capsules
Aspirin, Codeine Phosphate, and Caffeine Tablets
Activated Attapulgite
Colloidal Activated Attapulgite
Brompheniramine Maleate and Pseudoephedrine Sulfate Syrup
Butoconazole Nitrate
Butoconazole Nitrate Cream
Calcium Acetate
Chlordiazepoxide Hydrochloride and Clidinium Bromide Capsules
Chlorpheniramine Maleate Extended-release Capsules
Chlorpromazine Hydrochloride Oral Concentrate
Chlorazepate Dipotassium
Diltiazem Hydrochloride
Dilitiazem Tablets
Dyphylline and Guaifenesin Elixir
Dyphylline and Guaifenesin Tablets
Enalapril Maleate
Enalapril Maleate Tablets
Ethotoin
Ethotoin Tablets
Famotidine
Famotidine Tablets
Fludeoxyglucose F 18 Injection
Hydroxypropyl Cellulose Ocular System
Indium In 111 Oxyquinoline Solution
Metoclopramide Oral Solution
Norfloxacin
Norfloxacin Tablets
Nystatin Lozenges
Oxyprenolol Hydrochloride
Oxyprenolol Tablets
Oxycodone Tablets
Oxycodone and Aspirin Tablets
Oxymetazoline Hydrochloride Ophthalmic Solution
Phenylpropanolamine Hydrochloride Extended-release Tablets
Prednisolone Syrup
Propranolol Hydrochloride Extended-release Capsules
Propranolol Hydrochloride and Hydrochlorothiazide Extended-release Capsules
Ranitidine Hydrochloride
Ranitidine Injection
Ranitidine Tablets
Spironolactone and Hydrochlorothiazide Tablets
Sterile Sulbactam Sodium
Sulfacetamide Sodium and Prednisolone Acetate Ophthalmic Suspension
Theophylline Extended-release Capsules
Theophylline in Dextrose Injection
Triamterene and Hydrochlorothiazide Extended Capsules
Triamterene and Hydrochlorothiazide Tablets
Trientine Hydrochloride
Trientine Hydrochloride Capsules
Verapamil Hydrochloride
Verapamil Injection
Verapamil Tablets

New Admission to NF XVII

Ferric Oxide

Changes in Official Titles

USP XXII Title	USP XXI Title
Alcohol in Dextrose Injection	Alcohol and Dextrose Injection
Ammonium Chloride Delayed-release Tablets	Ammonium Chloride Tablets
Inulin in Sodium Chloride Injection	Inulin and Sodium Chloride Injection
Mannitol in Sodium Chloride Injection	Mannitol and Sodium Chloride Injection
Neomycin and Polymyxin B Sulfates, Bacitracin, and Lidocaine Ointment	Neomycin and Polymyxin B Sulfates, Bacitracin Zinc, and Lidocaine Ointment
Oxtriphylline Delayed-release Tablets	Oxtriphylline Tablets

NOTE—No changes in official titles were newly introduced in NF XVII.

Official Titles Changed by Supplement

Parenthetic notation following the title indicates the Supplement (S) in which the title change became official in USP XXI or in NF XVI.

Current Title	Former Title
Erythromycin Delayed-release Capsules (5S)[1]	Erythromycin Capsules
Fluorescein Injection (2S)[1]	Fluorescein Sodium Injection
Monobenzone Cream (2S)[1]	Monobenzone Ointment
Silicon Dioxide (3S)[2]	Silica Gel
Triple Sulfa Vaginal Cream (8S)[1]	Sulfathiazole, Sulfacetamide, and Sulfabenzamide Vaginal Cream
Triple Sulfa Vaginal Tablets (8S)[1]	Sulfathiazole, Sulfacetamide, and Sulfabenzamide Vaginal Tablets

[1] USP XXI.
[2] NF XVI.

Articles Included in USP XXI but Not Included in USP XXII or in NF XVII

Acetaminophen Elixir
Alphaprodine Hydrochloride
Alphaprodine Hydrochloride Injection
Aminosalicylate Calcium
Aminosalicylate Calcium Capsules
Aminosalicylate Calcium Tablets
Amobarbital Elixir

Bephenium Hydroxynaphthoate
Bephenium Hydroxynaphthoate for Oral Suspension
Betamethasone Valerate Topical Aerosol

Carbachol
Carbachol Intraocular Solution
Carbachol Ophthalmic Solution
Carbarsone
Carbarsone Capsules
Chlorcyclizine Hydrochloride
Chlorotrianisene
Chlorotrianisene Capsules
Chlorphenoxamine Hydrochloride
Chlorphenoxamine Hydrochloride Tablets
Chlorzoxazone and Acetaminophen Tablets
Cobalamin Concentrate
Cold Cream
Cyclomethycaine Sulfate
Cyclomethycaine Sulfate Cream
Cyclomethycaine Sulfate Jelly
Cyclomethycaine Sulfate Ointment
Cyclomethycaine Sulfate Suppositories
Cyclopentamine Hydrochloride
Cyclopentamine Hydrochloride Nasal Solution
Cyclothiazide
Cyclothiazide Tablets

Danthron
Danthron Tablets
Decavitamin Capsules
Decavitamin Tablets
Dehydrocholate Sodium Injection
Deslanoside
Deslanoside Injection
Dextrothyroxine Sodium
Dextrothyroxine Sodium Tablets
Dichlorphenamide
Dichlorphenamide Tablets
Dicumarol Capsules
Diethylstilbestrol Suppositories
Diphenylpyraline Hydrochloride
Diphenylpyraline Hydrochloride Tablets

Floxuridine
Sterile Floxuridine
Flurothyl

Glycine Irrigation
Glycobiarsol
Glycobiarsol Tablets
Gold Sodium Thiomalate
Gold Sodium Thiomalate Injection
Guanethidine Sulfate
Guanethidine Sulfate Tablets

Hexafluorenium Bromide
Hexafluorenium Bromide Injection
Hexavitamin Capsules
Hexavitamin Tablets
Hexobarbital
Hexobarbital Tablets
Hexylresorcinol
Hexylresorcinol Pills

Isobucaine Hydrochloride
Isobucaine Hydrochloride and Epinephrine Injection

Levallorphan Tartrate
Levallorphan Tartrate Injection
Sulfurated Lime Topical Solution
Lymphogranuloma Venereum Antigen

Mecamylamine Hydrochloride
Mecamylamine Hydrochloride Tablets
Meningococcal Polysaccharide Vaccine, Groups A and C Combined
Mepenzolate Bromide
Mepenzolate Bromide Syrup
Mepenzolate Bromide Tablets
Metaraminol Bitartrate
Metaraminol Bitartrate Injection
Methaqualone
Methaqualone Tablets
Methiodal Sodium
Methiodal Sodium Injection
Methoxyphenamine Hydrochloride
Methscopolamine Bromide Injection
Methylthiouracil

Nandrolone Phenpropionate
Nandrolone Phenpropionate Injection
Neomycin and Polymyxin B Sulfates and Bacitracin Zinc Topical Aerosol
Neomycin and Polymyxin B Sulfates and Bacitracin Zinc Topical Powder
Neostigmine Bromide
Neostigmine Bromide Tablets
Neostigmine Methylsulfate
Neostigmine Methylsulfate Injection
Nitromersol Tincture

Sterile Paraldehyde
Parathyroid Injection
Pentobarbital Sodium Elixir
Phenacaine Hydrochloride
Phenobarbital Sodium Tablets
Phenolsulfonphthalein
Phenolsulfonphthalein Injection
Phenoxybenzamine Hydrochloride
Phenoxybenzamine Hydrochloride Capsules
Sterile Phenytoin Sodium
Pine Tar
Potassium Chloride Elixir
Potassium Permanganate Tablets for Topical Solution
Procarbazine Hydrochloride
Procarbazine Hydrochloride Capsules
Pyroxylin
Pyrrobutamine Phosphate

Racemethionine
Racemethionine Capsules
Racemethionine Tablets
Rosin

Sodium Iodide Solution
Effervescent Sodium Phosphate

Tetrachloroethylene
Tetrachloroethylene Capsules
Tetracycline and Amphotericin B Capsules
Tetracycline and Amphotericin B Oral Suspension
Thiethylperazine Malate
Thiethylperazine Malate Injection
Tranylcypromine Sulfate
Tranylcypromine Sulfate Tablets
Trichloroacetic Acid

Vinyl Ether

Warfarin Potassium
Warfarin Potassium Tablets

Zinc-Eugenol Cement
Zinc Gelatin
Zinc Gelatin Impregnated Gauze
Zomepirac Sodium
Zomepirac Sodium Tablets

Articles Included in NF XVI but Not Included in NF XVII or in USP XXII

Acacia Syrup
Diluted Acetic Acid
Anise Oil

Caraway
Caraway Oil
Cardamom Oil
Cardamom Seed
Compound Cardamom Tincture
Cherry Juice
Cherry Syrup
Cinnamon
Cinnamon Oil
Clove Oil
Cocoa

Cocoa Syrup
Coriander Oil

Eriodictyon
Eriodictyon Fluidextract
Aromatic Eriodictyon Syrup
Eucalyptus Oil

Fennel Oil
Red Ferric Oxide
Yellow Ferric Oxide

Glycyrrhiza
Pure Glycyrrhiza Extract
Glycyrrhiza Fluidextract

Lavender Oil
Lemon Oil

Nutmeg Oil

Orange Flower Water
Orange Oil
Sweet Orange Peel Tincture
Compound Orange Spirit
Orange Syrup

Pine Needle Oil
Polypropylene Glycol

Spearmint
Spearmint Oil

General Notices and Requirements

Applying to Standards, Tests, Assays, and Other Specifications of the United States Pharmacopeia

Guide to GENERAL NOTICES AND REQUIREMENTS

The General Notices and Requirements (hereinafter referred to as the "General Notices") provide in summary form the basic guidelines for the interpretation and application of the standards, tests, assays, and other specifications of The United States Pharmacopeia and obviate the need to repeat throughout the book those requirements that are pertinent in numerous instances.

Where exceptions to the General Notices are made, the wording in the individual monograph or general test chapter takes precedence and specifically indicates the directions or the intent. To emphasize that such exceptions do exist, the General Notices employ where indicated a qualifying expression such as "unless otherwise specified." Thus, it is understood that the specific wording of standards, tests, assays, and other specifications is binding wherever deviations from the General Notices exist. By the same token, where no language is given specifically to the contrary, the General Notices apply.

TITLE

The full title of this book, including its supplements, is The Pharmacopeia of the United States of America, Twenty-second Revision. This title may be abbreviated to United States Pharmacopeia, Twenty-second Revision, or to USP XXII. The United States Pharmacopeia, Twenty-second Revision, supersedes all earlier revisions. Where the term USP is used, without further qualification, during the period in which this Pharmacopeia is official, it refers only to USP XXII and any supplement(s) thereto.

"OFFICIAL" AND "OFFICIAL ARTICLES"

The word "official," as used in this Pharmacopeia or with reference hereto, is synonymous with "Pharmacopeial," with "USP," and with "compendial."

The designation USP in conjunction with the official title on the label of an article is a reminder that the article purports to comply with USP standards; such specific designation on the label does not constitute a representation, endorsement, or incorporation by the manufacturer's labeling of the informational material contained in the USP monograph, nor does it constitute assurance by USP that the article is known to comply with USP standards. The standards apply equally to articles bearing the official titles or names derived by transposition of the definitive words of official titles or transposition in the order of the names of two or more active ingredients in official titles, whether or not the added designation "USP" is used. Names considered to be synonyms of the official titles may not be used for official titles.

Where an article differs from the standards of strength, quality, and purity, as determined by the application of the assays and tests, set forth for it in the Pharmacopeia, its difference shall be plainly stated on its label. Where an article fails to comply in identity with the identity prescribed in the USP, or contains an added substance that interferes with the prescribed assays and tests, such article shall be designated by a name that is clearly distinguishing and differentiating from any name recognized in the Pharmacopeia.

Articles listed herein are official and the standards set forth in the monographs apply to them only when the articles are intended or labeled for use as drugs or as medical devices and when bought, sold, or dispensed for these purposes or when labeled as conforming to this Pharmacopeia.

An article is deemed to be recognized in this Pharmacopeia when a monograph for the article is published in it, including its supplements, addenda, or other interim revisions, and an official date is generally or specifically assigned to it.

The following terminology is used for distinguishing the articles for which monographs are provided: an *official substance* is an active drug entity or a pharmaceutic ingredient (see also NF XVII) or a component of a finished device for which the monograph title includes no indication of the nature of the finished form; an *official preparation* is a *drug product* or a *finished device*. It is the finished or partially finished (e.g., as in the case of a sterile solid to be constituted into a solution for administration) preparation or product of one or more official substances formulated for use on or for the patient; an *article* is an item for which a monograph is provided, whether an official substance or an official preparation.

ATOMIC WEIGHTS AND CHEMICAL FORMULAS

The atomic weights used in computing molecular weights and the factors in the assays and elsewhere are those recommended in 1981 by the IUPAC Commission on Atomic Weights.

Chemical formulas, other than those in the definitions, tests, and assays, are given for purposes of information and calculation. The format within a given monograph is such that after the official title, the primarily informational portions of the text appear first, followed by the text comprising requirements, the latter section of the monograph being introduced by a boldface double-arrow symbol **»**. (Graphic formulas and chemical nomenclature provided as information in the individual monographs are discussed in the *Preface.*)

ABBREVIATIONS

The expression FDA refers to the U. S. Food and Drug Administration; NIST refers to National Institute of Standards and Technology (formerly NBS, the National Bureau of Standards). The expression FCC refers to the current edition of the Food Chemicals Codex, including its supplements. The term PhI refers to the International Pharmacopoeia, published as a recommendation on international standards of strength, quality, and purity for drugs by the World Health Organization (WHO). The expressions AAMI, ACS, ANSI, AOAC, ASTM, and ATCC refer, respectively, to the Association for the Advancement of Medical Instrumentation, the American Chemical Society, the American National Standards Institute, the Association of Official Analytical

Chemists, the American Society for Testing and Materials, and the American Type Culture Collection.

The term RS refers to Reference Standard as stated under *Reference Standards* in the General Notices.

The terms CS and TS refer to Colorimetric Solution and Test Solution, respectively (see under *Reagents, Indicators, and Solutions*). The term VS refers to Volumetric Solution as stated under *Solutions* in the General Notices.

The term *PF* refers to *Pharmacopeial Forum*, the journal of drug standards development and official compendia revision.

Abbreviated Statements in Monographs—Incomplete sentences are employed in various portions of the monographs for directness and brevity. Where the limit tests are so abbreviated, it is to be understood that the chapter numbers (shown in angle brackets) designate the respective procedures to be followed, and that the values specified after the colon are the required limits.

SIGNIFICANT FIGURES AND TOLERANCES

Where limits are expressed numerically herein, the upper and lower limits of a range are inclusive so that the range consists of the two values themselves and all intermediate values, but no values outside the limits. The limits expressed in monograph definitions and tests, regardless of whether the values are expressed as percentages or as absolute numbers, are considered significant to the last digit shown.

Equivalence Statements in Titrimetric Procedures—The directions for titrimetric procedures conclude with a statement of the weight of the analyte that is equivalent to each mL of the standardized titrant. In such an equivalence statement, it is to be understood that the number of significant figures in the concentration of the titrant corresponds to the number of significant figures in the weight of the analyte. Blank corrections are to be made for all titrimetric assays where appropriate (see *Titrimetry* ⟨541⟩).

Tolerances—The limits specified in the monographs for Pharmacopeial articles are established with a view to the use of these articles as drugs, except where the monograph indicates otherwise. The use of the molecular formula for the active ingredient(s) named in defining the required strength of a Pharmacopeial article is intended to designate the chemical entity or entities having absolute (100 percent) purity.

A dosage form shall be formulated with the intent to provide 100 percent of the quantity of each ingredient declared on the label. Where the content of an ingredient is known to decrease with time, an amount in excess of that declared on the label may be introduced into the dosage form at the time of manufacture to assure compliance with the content requirements of the monograph throughout the expiration period. The tolerances and limits stated in the definitions in the monographs for Pharmacopeial articles allow for such overages, and for analytical error, for unavoidable variations in manufacturing and compounding, and for deterioration to an extent considered insignificant under practical conditions.

The specified tolerances are based upon such attributes of quality as might be expected to characterize an article produced from suitable raw materials under recognized principles of good manufacturing practice.

The existence of compendial limits or tolerances does not constitute a basis for a claim that an official substance that more nearly approaches 100 percent purity "exceeds" the Pharmacopeial quality. Similarly, the fact that an article has been prepared to closer tolerances than those specified in the monograph does not constitute a basis for a claim that the article "exceeds" the Pharmacopeial requirements.

ALCOHOL

All statements of percentages of alcohol, such as under the heading, *Alcohol content*, refer to percentage, by volume, of C_2H_5OH at 15.56°. Where reference is made to "C_2H_5OH," the chemical entity possessing absolute (100 percent) strength is intended.

Alcohol—Where "alcohol" is called for in formulas, tests, and assays, the monograph article *Alcohol* is to be used.

Dehydrated Alcohol—Where "dehydrated alcohol" (absolute alcohol) is called for in tests and assays, the monograph article *Dehydrated Alcohol* is to be used.

Denatured Alcohol—Specially denatured alcohol formulas are available for use in accordance with federal statutes and regulations of the Internal Revenue Service. A suitable formula of specially denatured alcohol may be substituted for Alcohol in the manufacture of Pharmacopeial preparations intended for internal or topical use, provided that the denaturant is volatile and does not remain in the finished product. A finished product that is intended for topical application to the skin may contain specially denatured alcohol, provided that the denaturant is either a normal ingredient or a permissible added substance; in either case the denaturant must be identified on the label of the topical preparation. Where a process is given in the individual monograph, the preparation so made must be identical with that prepared by the given process.

PHARMACOPEIAL FORUM

Pharmacopeial Forum (*PF*) is the USP journal of drug standards development and official compendia revision. *Pharmacopeial Forum* is the working document of the USP Committee of Revision, the major portion of which presents proposals for revising standards of USP and NF. In addition to this *In-process Revision*, each number of *PF* includes the current *Reference Standards Catalog* and, under *Stimuli to the Revision Process*, contributed articles that present reports and discussions of subjects relating directly to compendial standards.

REAGENT STANDARDS

The proper conduct of the Pharmacopeial tests and assays and the reliability of the results depend, in part, upon the quality of the reagents used in the performance of the procedures. Unless otherwise specified, reagents are to be used that conform to the standards set forth in the current edition of *Reagent Chemicals* published by the American Chemical Society. Where such ACS reagent standards are not available or where for various reasons the required purity differs, compendial specifications for reagents of acceptable quality are provided. (See *Reagents, Indicators, and Solutions.*) Listing of these reagents, including the indicators and solutions employed as reagents, in no way implies that they have therapeutic utility; furthermore, any reference to USP in their labeling shall include also the term "reagent" or "reagent grade."

REFERENCE STANDARDS

USP Reference Standards are authentic specimens that have been approved by the USP Reference Standards Committee as suitable for use as comparison standards in compendial tests and assays. (See *USP Reference Standards* ⟨11⟩.) Currently official lots of USP Reference Standards are published in *Pharmacopeial Forum.*

Where first referred to in a monograph, the name of a USP Reference Standard is generally spelled out in full. However, where a USP Reference Standard is referred to thereafter in an assay or a test in this compendium, the words "Reference Standard" are abbreviated to "RS."

Where a test or an assay calls for the use of a compendial article rather than for a USP Reference Standard as a material standard of reference, a substance meeting all of the compendial monograph requirements for that article is to be used.

UNITS OF POTENCY

For those products for which it is necessary to express the potency in terms of units by reference to a suitable working standard (usually a USP Reference Standard), the individual monographs refer to USP Units of activity. Unless otherwise indicated, USP Units are equivalent to the corresponding international units, where such exist or formerly existed, or to the units of activity established by the Food and Drug Administration in the case of antibiotics and biological products.

INGREDIENTS AND PROCESSES

Official preparations are prepared from ingredients that meet the requirements of the compendial monographs for those individual ingredients for which monographs are provided (see also NF XVII). Water used as an ingredient of official preparations meets the requirements for *Purified Water*, for *Water for Injection*, or for one of the sterile forms of water covered by a monograph in this Pharmacopeia.

Potable water meeting the requirements for drinking water as set forth in the regulations of the federal Environmental Protection Agency may be used in the preparation of official substances.

Official substances are prepared according to recognized principles of good manufacturing practice and from ingredients complying with specifications designed to assure that the resultant substances meet the requirements of the compendial monographs (see also *Foreign Substances and Impurities* under *Tests and Assays*).

Preparations for which a complete composition is given in this Pharmacopeia, unless specifically exempted herein or in the individual monograph, are to contain only the ingredients named in the formulas. However, there may be deviation from the specified processes or methods of compounding, though not from the ingredients or proportions thereof, provided the finished preparation conforms to the relevant standards laid down herein and to preparations produced by following the specified process.

Where a monograph on a preparation calls for an ingredient in an amount expressed on the dried basis, the ingredient need not be dried prior to use if due allowance is made for the water or other volatile substances present in the quantity taken.

Unless specifically exempted elsewhere in this Pharmacopeia, the identity, strength, quality, and purity of an official article are determined by the definition, physical properties, tests, assays, and other specifications relating to the article, whether incorporated in the monograph itself, in the General Notices, or in the section, *General Chapters.*

Added Substances—An official substance, as distinguished from an official preparation, contains no added substances except where specifically permitted in the individual monograph. Where such addition is permitted, the label indicates the name(s) and amount(s) of any added substance(s).

Unless otherwise specified in the individual monograph, or elsewhere in the General Notices, suitable substances such as bases, carriers, coatings, colors, flavors, preservatives, stabilizers, and vehicles may be added to an official preparation to enhance its stability, usefulness, or elegance or to facilitate its preparation. Such substances are regarded as unsuitable and are prohibited unless (a) they are harmless in the amounts used, (b) they do not exceed the minimum quantity required to provide their intended effect, (c) their presence does not impair the bioavailability or the therapeutic efficacy or safety of the official preparation, and (d) they do not interfere with the assays and tests prescribed for determining compliance with the Pharmacopeial standards.

The air in a container of an official preparation may be evacuated or be replaced by carbon dioxide, helium, nitrogen, or other suitable gases, which fact shall be declared on the label unless otherwise specified in the individual monograph.

Colors—Added substances employed solely to impart color may be incorporated into official preparations, except those intended for parenteral or ophthalmic use, in accordance with the regulations pertaining to the use of colors issued by the Food and Drug Administration, provided such added substances are otherwise appropriate in all respects. (See also *Added Substances* under *Injections* ⟨1⟩.)

Ointments and Suppositories—In the preparation of ointments and suppositories, the proportions of the substances constituting the base may be varied to maintain a suitable consistency under different climatic conditions, provided the concentrations of active ingredients are not varied.

TESTS AND ASSAYS

Apparatus—A specification for a definite size or type of container or apparatus in a test or assay is given solely as a recommendation. Where volumetric flasks or other exact measuring, weighing, or sorting devices are specified, this or other equipment of at least equivalent accuracy shall be employed. (See also *Volumetric Apparatus* ⟨31⟩.) Where low-actinic or light-resistant containers are specified, clear containers that have been rendered opaque by application of a suitable coating or wrapping may be used.

Where an instrument for physical measurement, such as a spectrophotometer, is specified in a test or assay by its distinctive name, another instrument of equivalent or greater sensitivity and accuracy may be used. In order to obtain solutions having concentrations that are adaptable to the working range of the instrument being used, solutions of proportionately higher or lower concentrations may be prepared according to the solvents and proportions thereof that are specified for the procedure.

Where a particular brand or source of a material, instrument, or piece of equipment, or the name and address of a manufacturer or distributor, is mentioned (ordinarily in a footnote), this identification is furnished solely for informational purposes as a matter of convenience, without implication of approval, endorsement, or certification. Items capable of equal or better performance may be used if these characteristics have been validated.

Where the use of a centrifuge is specified, the directions are predicated upon the use of apparatus having an effective radius of about 20 cm (8 inches) and driven at a speed sufficient to clarify the supernatant layer within 15 minutes.

Unless otherwise specified, for chromatographic tubes and columns the diameter specified refers to internal diameter (ID); for other types of tubes and tubing the diameter specified refers to outside diameter (OD).

Steam Bath—Where the use of a steam bath is directed, exposure to actively flowing steam or to another form of regulated heat, corresponding in temperature to that of flowing steam, may be used.

Water Bath—Where the use of a water bath is directed without qualification with respect to temperature, a bath of vigorously boiling water is intended.

Foreign Substances and Impurities—Tests for the presence of foreign substances and impurities are provided to limit such substances to amounts that are unobjectionable under conditions in which the article is customarily employed (see also *Impurities in Official Articles* ⟨1086⟩).

While one of the primary objectives of the Pharmacopeia is to assure the user of official articles of their identity, strength, quality, and purity, it is manifestly impossible to include in each monograph a test for every impurity, contaminant, or adulterant that might be present, including microbial contamination. These may arise from a change in the source of material or from a change in the processing, or may be introduced from extraneous sources. Tests suitable for detecting such occurrences, the presence of which is inconsistent with good pharmaceutical practice, should be employed in addition to the tests provided in the individual monograph.

Procedures—Assay and test procedures are provided for determining compliance with the Pharmacopeial standards of identity, strength, quality, and purity.

In performing the assay or test procedures in this Pharmacopeia, it is expected that safe laboratory practices will be followed. This includes the utilization of precautionary measures, protective equipment, and work practices consistent with the chemicals and procedures utilized. Prior to undertaking any assay or procedure described in this Pharmacopeia, the individual should be aware of the hazards associated with the chemicals and the procedures and means of protecting against them. This Pharmacopeia is not designed to describe such hazards or protective measures.

Every compendial article in commerce shall be so constituted that when examined in accordance with these assay and test procedures, it meets all of the requirements in the monograph defining it. However, it is not to be inferred that application of every analytical procedure in the monograph to samples from every production batch is necessarily a prerequisite for assuring compliance with Pharmacopeial standards before the batch is released for distribution. Data derived from manufacturing process validation studies and from in-process controls sometimes may provide greater assurance that a batch meets a particular monograph requirement than analytical data derived from an examination of finished units drawn from that batch. On the basis of such assurances, the analytical procedures in the monograph may be omitted by the manufacturer in judging compliance of the batch with the Pharmacopeial standards.

Automated procedures employing the same basic chemistry as those assay and test procedures given in the monograph are recognized as being equivalent in their suitability for determining compliance. Conversely, where an automated procedure is given in the monograph, manual procedures employing the same basic chemistry are recognized as being equivalent in their suitability for determining compliance. Compliance may be determined also by the use of alternative methods, chosen for advantages in accuracy, sensitivity, precision, selectivity, or adaptability to automation or computerized data reduction or in other special circumstances. However, Pharmacopeial standards and procedures are interrelated; therefore, where a difference appears or in the event of dispute, only the result obtained by the procedure given in this Pharmacopeia is conclusive.

In the performance of assay or test procedures, not less than the specified number of dosage units should be taken for analysis. Proportionately larger or smaller quantities than the specified weights and volumes of assay or test substances and Reference Standards may be taken, provided the measurement is made with at least equivalent accuracy and provided that any subsequent steps, such as dilutions, are adjusted accordingly to yield concentrations equivalent to those specified and are made in such manner as to provide at least equivalent accuracy.

Where it is directed in an assay or a test that a certain quantity of substance or a counted number of dosage units is to be examined, the specified quantity or number is a minimal figure chosen only for convenience of analytical manipulation; it is not intended to restrict the total quantity of substance or number of units that may be subjected to the assay or test or that should be tested in accordance with good manufacturing practices.

Where it is directed in the assay of Tablets to "weigh and finely powder not less than" a given number, usually 20, of the Tablets, it is intended that a counted number of Tablets shall be weighed and reduced to a powder. The portion of the powdered tablets taken for assay is representative of the whole Tablets and is, in turn, weighed accurately. The result of the assay is then related to the amount of active ingredient per Tablet by multiplying this result by the average Tablet weight and dividing by the weight of the portion taken for the assay.

Similarly, where it is directed in the assay of Capsules to remove, as completely as possible, the contents of not less than a given number, usually 20, of the Capsules, it is intended that a counted number of Capsules should be carefully opened and the contents quantitatively removed, combined, mixed, and weighed accurately. The portion of mixed Capsules contents taken for the assay is representative of the contents of the Capsules and is, in turn, weighed accurately. The result of the assay is then related to the amount of active ingredient per Capsule by multiplying this result by the average weight of Capsule content and dividing by the weight of the portion taken for the assay.

Where the definition in a monograph states the tolerances as being "calculated on the dried [or anhydrous or ignited] basis," the directions for drying or igniting the sample prior to assaying are generally omitted from the *Assay* procedure. Assay and test procedures may be performed on the undried or unignited substance and the results calculated on the dried, anhydrous, or ignited basis, provided a test for *Loss on drying*, or *Water*, or *Loss on ignition*, respectively, is given in the monograph. Where the presence of moisture or other volatile material may interfere with the procedure, previous drying of the substance is specified in the individual monograph and is obligatory.

Throughout a monograph that includes a test for *Loss on drying* or *Water*, the expression "previously dried" without qualification signifies that the substance is to be dried as directed under *Loss on drying* or *Water* (gravimetric determination).

Unless otherwise directed in the test or assay in the individual monograph or in a general chapter, USP Reference Standards are to be dried before use, or used without prior drying, specifically in accordance with the instructions given in the section, *Reference Standard(s)* and on the label of the Reference Standard. Where the label instructions differ in detail from those in the monograph section, the label text is determinative.

In stating the appropriate quantities to be taken for assays and tests, the use of the word "about" indicates a quantity within 10 percent of the specified weight or volume. However, the weight or volume taken is accurately determined and the calculated result is based upon the exact amount taken. The same tolerance applies to specified dimensions.

Where the use of a pipet is directed for measuring a specimen or an aliquot in conducting a test or an assay, the pipet conforms to the standards set forth under *Volumetric Apparatus* ⟨31⟩, and is to be used in such manner that the error does not exceed the limit stated for a pipet of its size. Where a pipet is specified, a suitable buret, conforming to the standards set forth under *Volumetric Apparatus* ⟨31⟩, may be substituted. Where a "to contain" pipet is specified, a suitable volumetric flask may be substituted.

Expressions such as "25.0 mL" and "25.0 mg," used with respect to volumetric or gravimetric measurements, indicate that the quantity is to be "accurately measured" or "accurately weighed" within the limits stated under *Volumetric Apparatus* ⟨31⟩ or under *Weights and Balances* ⟨41⟩.

The term "transfer" is used generally to specify a quantitative manipulation.

The term "concomitantly," used in such expressions as "concomitantly determine" or "concomitantly measured," in directions for assays and tests, is intended to denote that the determinations or measurements are to be performed in immediate succession. See also *Use of Reference Standards* under *Spectrophotometry and Light-scattering* ⟨851⟩.

Blank Determination—Where it is directed that "any necessary correction" be made by a blank determination, the determination is to be conducted using the same quantities of the same reagents treated in the same manner as the solution or mixture containing the portion of the substance under assay or test, but with the substance itself omitted.

Desiccator—The expression "in a desiccator" specifies the use of a tightly closed container of suitable size and design that maintains an atmosphere of low moisture content by means of silica gel or other suitable desiccant.

A "vacuum desiccator" is one that maintains the low-moisture atmosphere at a reduced pressure of not more than 20 mm of mercury or at the pressure designated in the individual monograph.

Dilution—Where it is directed that a solution be diluted "quantitatively and stepwise," an accurately measured portion is to be diluted by adding water or other solvent, in the proportion indicated, in one or more steps. The choice of apparatus to be used should take into account the relatively larger errors generally associated with using small-volume volumetric apparatus (see *Volumetric Apparatus* ⟨31⟩).

Drying to Constant Weight—The specification "dried to constant weight" means that the drying shall be continued until two consecutive weighings do not differ by more than 0.50 mg per g of substance taken, the second weighing following an additional hour of drying.

Filtration—Where it is directed to "filter," without further qualification, the intent is that the liquid be filtered through suitable filter paper or equivalent device until the filtrate is clear.

Identification Tests—The Pharmacopeial tests headed *Identification* are provided as an aid in verifying the identity of articles as they are purported to be, such as those taken from labeled containers. Such tests, however specific, are not necessarily sufficient to establish proof of identity; but failure of an article taken from a labeled container to meet the requirements of a prescribed identification test indicates that the article may be mislabeled. Other tests and specifications in the monograph often contribute to establishing or confirming the identity of the article under examination.

Ignition to Constant Weight—The specification "ignite to constant weight" means that the ignition shall be continued, at 800 ± 25° unless otherwise indicated, until two consecutive weighings do not differ by more than 0.50 mg per g of substance taken, the second weighing following an additional 15-minute ignition period.

Indicators—Where the use of a test solution ("TS") as an indicator is specified in a test or an assay, approximately 0.2 mL, or 3 drops, of the solution shall be added, unless otherwise directed.

Logarithms—Logarithms used in the assays are to the base 10.

Negligible—This term indicates a quantity not exceeding 0.50 mg.

Odor—Terms such as "odorless," "practically odorless," "a faint characteristic odor," or variations thereof, apply to examination, after exposure to the air for 15 minutes, of either a freshly opened package of the article (for packages containing not more than 25 g) or (for larger packages) of a portion of about 25 g of the article that has been removed from its package to an open evaporating dish of about 100-mL capacity. An odor designation is descriptive only and is not to be regarded as a standard of purity for a particular lot of an article.

Pressure Measurements—The term "mm of mercury" used with respect to measurements of blood pressure, pressure within an apparatus, or atmospheric pressure refers to the use of a suitable manometer or barometer calibrated in terms of the pressure exerted by a column of mercury of the stated height.

Solutions—Unless otherwise specified in the individual monograph, all solutions called for in tests and assays are prepared with *Purified Water.*

An expression such as "(1 in 10)" means that 1 part *by volume* of a liquid is to be diluted with, or 1 part *by weight* of a solid is to be dissolved in, sufficient of the diluent or solvent to make the volume of the finished solution 10 parts *by volume.*

An expression such as "(20:5:2)" means that the respective numbers of parts, by volume, of the designated liquids are to be mixed, unless otherwise indicated.

The notation "(VS)" after a specified volumetric solution indicates that such solution is standardized in accordance with directions given in the individual monograph or under *Volumetric Solutions* in the section, *Reagents, Indicators, and Solutions*, and is thus differentiated from solutions of approximate normality or molarity.

Where a standardized solution of a specific concentration is called for in a test or an assay, a solution of other normality or molarity may be used, provided allowance is made for the difference in concentration and provided the error of measurement is not increased thereby.

Specific Gravity—Unless otherwise stated, the specific gravity basis is 25°/25°, i.e., the ratio of the weight of a substance in air at 25° to the weight of an equal volume of water at the same temperature.

Temperatures—Unless otherwise specified, all temperatures in this Pharmacopeia are expressed in centigrade (Celsius) degrees, and all measurements are made at 25°. Where "controlled room temperature" is specified, a temperature range between 15° and 30° is intended.

Time Limit—In the conduct of tests and assays, 5 minutes shall be allowed for the reaction to take place unless otherwise specified.

Vacuum—The term "in vacuum" denotes exposure to a pressure of less than 20 mm of mercury unless otherwise indicated.

Where drying in vacuum over a desiccant is directed in the individual monograph, a vacuum desiccator or a vacuum drying pistol, or other suitable vacuum drying apparatus, is to be used.

Water—Where water is called for in tests and assays, *Purified Water* is to be used. For special kinds of water such as "carbon dioxide-free water," see the introduction to the section, *Reagents, Indicators, and Solutions.*

Water and *Loss on Drying*—Where the water of hydration or adsorbed water of a Pharmacopeial article is determined by the titrimetric method, the test is generally given under the heading *Water.* Where the determination is made by drying under specified conditions, the test is generally given under the heading *Loss on drying.* However, *Loss on drying* is most often given as the heading where the loss in weight is known to represent residual volatile constituents including organic solvents as well as water.

Description—Information on the "description" pertaining to an article, which is relatively general in nature, is provided in the reference table, *Description and Relative Solubility of USP and NF Articles*, in this Pharmacopeia for those who use, prepare, and dispense drugs and/or related articles, solely to indicate properties of an article complying with monograph standards. The properties are not in themselves standards or tests for purity even though they may indirectly assist in the preliminary evaluation of an article.

Solubility—The statements concerning solubilities given in the reference table, *Description and Relative Solubility of USP and NF Articles*, for Pharmacopeial articles are not standards or tests for purity but are provided primarily as information for those who use, prepare, and dispense drugs and/or related articles. Only where a quantitative solubility test is given, and is designated as such, is it a test for purity.

The approximate solubilities of Pharmacopeial substances are indicated by the descriptive terms in the accompanying table.

Descriptive Term	Parts of Solvent Required for 1 Part of Solute
Very soluble	Less than 1
Freely soluble	From 1 to 10
Soluble	From 10 to 30
Sparingly soluble	From 30 to 100
Slightly soluble	From 100 to 1000
Very slightly soluble	From 1000 to 10,000
Practically insoluble, or Insoluble	10,000 and over

Soluble Pharmacopeial articles, when brought into solution, may show traces of physical impurities, such as minute fragments of filter paper, fibers, and other particulate matter, unless limited or excluded by definite tests or other specifications in the individual monographs.

PRESERVATION, PACKAGING, STORAGE, AND LABELING

Containers—The *container* is that which holds the article and is or may be in direct contact with the article. The *immediate container* is that which is in direct contact with the article at all times. The *closure* is a part of the container.

Prior to its being filled, the container should be clean. Special precautions and cleaning procedures may be necessary to ensure that each container is clean and that extraneous matter is not introduced into or onto the article.

The container does not interact physically or chemically with the article placed in it so as to alter the strength, quality, or purity of the article beyond the official requirements.

The Pharmacopeial requirements for the use of specified containers apply also to articles as packaged by the pharmacist or other dispenser, unless otherwise indicated in the individual monograph.

Tamper-resistant Packaging—The container or individual carton of a sterile article intended for ophthalmic or otic use, except where extemporaneously compounded for immediate dispensing on prescription, shall be so sealed that the contents cannot be used without obvious destruction of the seal.

Articles intended for sale without prescription are also required to comply with the tamper-resistant packaging and labeling requirements of the Food and Drug Administration where applicable.

Preferably, the immediate container and/or the outer container or protective packaging utilized by a manufacturer or distributor for all dosage forms that are not specifically exempt is designed so as to show evidence of any tampering with the contents.

Light-resistant Container (see *Light Transmission* under *Containers* ⟨661⟩)—A light-resistant container protects the contents from the effects of light by virtue of the specific properties of the material of which it is composed, including any coating applied to it. Alternatively, a clear and colorless or a translucent container may be made light-resistant by means of an opaque covering, in which case the label of the container bears a statement that the opaque covering is needed until the contents are to be used or administered. Where it is directed to "protect from light" in an individual monograph, preservation in a light-resistant container is intended.

Where an article is required to be packaged in a light-resistant container, and if the container is made light-resistant by means of an opaque covering, a single-use, unit-dose container or mnemonic pack for dispensing may not be removed from the outer opaque covering prior to dispensing.

Well-closed Container—A well-closed container protects the contents from extraneous solids and from loss of the article under the ordinary or customary conditions of handling, shipment, storage, and distribution.

Tight Container—A tight container protects the contents from contamination by extraneous liquids, solids, or vapors, from loss of the article, and from efflorescence, deliquescence, or evaporation under the ordinary or customary conditions of handling, shipment, storage, and distribution, and is capable of tight re-closure. Where a tight container is specified, it may be replaced by a hermetic container for a single dose of an article.

A gas cylinder is a metallic tight container designed to hold a gas under pressure. As a safety measure, for carbon dioxide, cyclopropane, helium, nitrous oxide, and oxygen, the Pin-index Safety System of matched fittings is recommended for cylinders of Size E or smaller.

NOTE—Where packaging and storage in a *tight container* or a *well-closed container* is specified in the individual monograph, the container utilized for an article when dispensed on prescription meets the requirements under *Containers—Permeation* ⟨671⟩.

Hermetic Container—A hermetic container is impervious to air or any other gas under the ordinary or customary conditions of handling, shipment, storage, and distribution.

Single-unit Container—A single-unit container is one that is designed to hold a quantity of drug product intended for administration as a single dose or a single finished device intended for use promptly after the container is opened. Preferably, the immediate container and/or the outer container or protective packaging shall be so designed as to show evidence of any tampering with the contents. Each single-unit container shall be labeled to indicate the identity, quantity and/or strength, name of the manufacturer, lot number, and expiration date of the article.

Single-dose Container (see also *Containers for Injections* under *Injections* ⟨1⟩)—A single-dose container is a single-unit container for articles intended for parenteral administration only. A single-dose container is labeled as such. Examples of single-dose containers include pre-filled syringes, cartridges, fusion-sealed containers, and closure-sealed containers when so labeled.

Unit-dose Container—A unit-dose container is a single-unit container for articles intended for administration by other than the parenteral route as a single dose, direct from the container.

Multiple-unit Container—A multiple-unit container is a container that permits withdrawal of successive portions of the contents without changing the strength, quality, or purity of the remaining portion.

Multiple-dose Container (see also *Containers for Injections* under *Injections* ⟨1⟩)—A multiple-dose container is a multiple-unit container for articles intended for parenteral administration only.

Storage Temperature—Specific directions are stated in some monographs with respect to the temperatures at which Pharmacopeial articles shall be stored, where it is considered that storage at a lower or a higher temperature may produce undesirable results. Such directions apply except where the label on an article states a different storage temperature on the basis of stability studies of that particular formulation. The conditions are defined by the following terms.

Cold—Any temperature not exceeding 8° (46°F). A *refrigerator* is a cold place in which the temperature is maintained thermostatically between 2° and 8° (36° and 46°F). A *freezer* is a cold place in which the temperature is maintained thermostatically between −20° and −10° (−4° and 14°F).

Cool—Any temperature between 8° and 15° (46° and 59°F). An article for which storage in a cool place is directed may, alternatively, be stored in a *refrigerator*, unless otherwise specified in the individual monograph.

Room Temperature—The temperature prevailing in a working area. *Controlled room temperature* is a temperature maintained thermostatically between 15° and 30° (59° and 86°F).

Warm—Any temperature between 30° and 40° (86° and 104°F).

Excessive Heat—Any temperature above 40° (104°F).

Protection from Freezing—Where, in addition to the risk of breakage of the container, freezing subjects an article to loss of strength or potency, or to destructive alteration of its characteristics, the container label bears an appropriate instruction to protect the article from freezing.

Storage in Bulk—Articles in bulk packages are exempt from the storage requirements when the contained articles are intended for manufacture or for subsequent repackaging for the dispenser or distributor.

Storage under Non-specific Conditions—Where no specific storage directions or limitations are provided in the individual monograph, it is to be understood that the storage conditions include protection from moisture, freezing, and excessive heat.

Labeling—The term "labeling" designates all labels and other written, printed, or graphic matter upon an immediate container of an article or upon, or in, any package or wrapper in which it is enclosed, except any outer shipping container. The term "label" designates that part of the labeling upon the immediate container.

A shipping container, unless such container is also essentially the immediate container or the outside of the consumer package, is exempt from the labeling requirements of this Pharmacopeia.

Articles in this Pharmacopeia are subject to compliance with such labeling requirements as may be promulgated by governmental bodies in addition to the Pharmacopeial requirements set forth for the articles.

Amount of Ingredient per Dosage Unit—The strength of a drug product is preferably expressed in terms of µg or mg of the therapeutically active moiety. The full name of the chemical compound is also used in the content declaration.

Pharmacopeial articles in capsule, tablet, or other unit dosage form shall be labeled to express the quantity of each active ingredient contained in each such unit. Pharmacopeial drug products not in unit dosage form shall be labeled to express the quantity of each active ingredient in each mL or in each g, or to express the percentage of each such ingredient (see *Percentage Measurements*), except that oral liquids or solids intended to be constituted to yield oral liquids may, alternatively, be labeled in terms of each 5-mL portion of the liquid or resulting liquid.

In order to help minimize the possibility of errors in the dispensing and administration of drugs, the quantity of active ingredient when expressed in whole numbers shall be shown without a decimal point that is followed by a terminal zero (e.g., express as 4 mg [not 4.0 mg]).

Labeling Vitamin-containing Products—The vitamin content of Pharmacopeial preparations shall be stated on the label in metric units per dosage unit. The amounts of vitamins A, D, and E may be stated also in USP Units. Quantities of vitamin A declared in metric units refer to the equivalent amounts of retinol (vitamin A alcohol).

Labeling Parenteral and Topical Preparations—The label of a preparation intended for parenteral or topical use states the names of all added substances (see *Added Substances* in these General Notices, and see *Labeling* under *Injections* ⟨1⟩), and, in the case of parenteral preparations, also their amounts or proportions, except that for substances added for adjustment of pH or to achieve isotonicity, the label may indicate only their presence and the reason for their addition.

Labeling Electrolytes—The concentration and dosage of electrolytes for replacement therapy (e.g., sodium chloride or potassium chloride) shall be stated on the label in milliequivalents (mEq). The label of the product shall indicate also the quantity of ingredient(s) in terms of weight or percentage concentration.

Special Capsules and Tablets—The label of any form of Capsule or Tablet intended for administration other than by swallowing intact bears a prominent indication of the manner in which it is to be used.

Expiration Date—The label of an official drug product shall bear an expiration date. The monographs for some preparations specify the expiration date that shall appear on the label. In the absence of a specific requirement in the individual monograph for a drug product, the label shall bear an expiration date assigned for the particular formulation and package of the article, with the following exception: the label need not show an expiration date in the case of a drug product packaged in a container that is intended for sale without prescription and the labeling of which states no dosage limitations, and which is stable for not less than 3 years when stored under the prescribed conditions.

Where an official article is required to bear an expiration date, such article shall be dispensed solely in, or from, a container labeled with an expiration date, and the date on which the article is dispensed shall be within the labeled expiry period. The expiration date identifies the time during which the article may be expected to meet the requirements of the Pharmacopeial monograph provided it is kept under the prescribed storage conditions. The expiration date limits the time during which the article may be dispensed or used. Where an expiration date is stated only in terms of the month and the year, it is a representation that the intended expiration date is the last day of the stated month.

For articles requiring constitution prior to use, a suitable beyond-use date for the constituted product shall be identified in the labeling.

In determining an appropriate period of time during which a prescription drug may be retained by a patient after its dispensing, the dispenser shall take into account, in addition to any other relevant factors, the nature of the drug; the container in which it was packaged by the manufacturer and the expiration date thereon; the characteristics of the patient's container, if the article is repackaged for dispensing; the expected storage conditions to which the article may be exposed; and the expected length of time of the course of therapy. Unless otherwise required, the dispenser may, on taking into account the foregoing, place on the label of a multiple-unit container a suitable beyond-use date to limit the patient's use of the article. Unless otherwise specified in the individual monograph, such beyond-use date shall be not later than (a) the expiration date on the manufacturer's container, or (b) one year from the date the drug is dispensed, whichever is earlier.

VEGETABLE AND ANIMAL DRUGS

The requirements for vegetable and animal drugs apply to the articles as they enter commerce; however, lots of such drugs intended solely for the manufacture or isolation of volatile oils, alkaloids, glycosides, or other active principles may depart from such requirements.

Statements of the distinctive microscopic structural elements in powdered drugs of animal or vegetable origin may be included in the individual monograph as a means of determining identity, quality, or purity.

Foreign Matter—Vegetable and animal drugs are to be free from pathogenic organisms (see *Microbiological Attributes of Non-sterile Pharmaceutical Products* ⟨1111⟩), and are to be as free as reasonably practicable from microorganisms, insects, and other animal contamination, including animal excreta. They shall show no abnormal discoloration, abnormal odor, sliminess, or other evidence of deterioration.

The amount of foreign inorganic matter in vegetable or animal drugs, estimated as *Acid-insoluble ash*, shall not exceed 2 percent of the weight of the drug, unless otherwise specified in the individual monograph.

Before vegetable drugs are ground or powdered, stones, dust, lumps of soil, and other foreign inorganic matter are removed by mechanical or other suitable means.

In commerce it is seldom possible to obtain vegetable drugs that are without some adherent or admixed, innocuous, foreign matter, which usually is not detrimental. No poisonous, dangerous, or otherwise noxious foreign matter or residues may be present. Foreign matter includes any part of the plant not specified as constituting the drug.

Preservation—Vegetable or animal substances may be protected from insect infestation or microbiological contamination by means of suitable agents or processes that leave no harmful residues.

WEIGHTS AND MEASURES

The metric system of weights and measures is used in this Pharmacopeia. The metric and other units, and the abbreviations commonly employed, are as follows:

Ci = curie
mCi = millicurie
µCi = microcurie
Eq = gram-equivalent weight (equivalent)
mEq = milliequivalent

nCi = nanocurie
Mrad = megarad
m = meter
dm = decimeter
cm = centimeter
mm = millimeter
μm = micrometer (0.001 mm)
nm = nanometer*
kg = kilogram
g = gram
mg = milligram
μg; mcg = microgram†
ng = nanogram
pg = picogram
dL = deciliter
L = liter
mL = milliliter; ‡
μL = microliter
mol = gram-molecular weight (mole)
mmol = millimole
Osmol = osmole
mOsmol = milliosmole
Hz = hertz
kHz = kilohertz
MHz = megahertz
MeV = million electron volts
keV = kilo-electron volt
mV = millivolt
psi = pounds per square inch
Pa = pascal
kPa = kilopascal
g = gravity (in centrifugation)

The International System of Units (SI) is also used in all radiopharmaceutical monographs. The abbreviations commonly employed are as follows.

Bq = becquerel
kBq = kilobecquerel
MBq = megabecquerel
GBq = gigabecquerel
Gy = gray
mGy = milligray

* Formerly the abbreviation mμ (for millimicron) was used.

† Formerly the abbreviation mcg was used in the Pharmacopeial monographs; however, the symbol μg now is more widely accepted and thus is used in this Pharmacopeia. The term "gamma," symbolized by γ, is frequently used for microgram in biochemical literature.

NOTE—The abbreviation mcg is still commonly employed to denote microgram(s) in labeling and in prescription writing. Therefore, for purposes of labeling, "mcg" may be used to denote microgram(s).

‡ One milliliter (mL) is used herein as the equivalent of 1 cubic centimeter (cc).

CONCENTRATIONS

Molal, molar, and normal solution concentrations are indicated throughout this Pharmacopeia for most chemical assay and test procedures (see also *Volumetric Solutions* in the section, *Reagents, Indicators, and Solutions*). Molality is designated by the symbol *m* preceded by a number that is the number of moles of the designated solute contained in one kilogram of the designated solvent. Molarity is designated by the symbol *M* preceded by a number that is the number of moles of the designated solute contained in an amount of the designated solvent that is sufficient to prepare one liter of solution. Normality is designated by the symbol *N* preceded by a number that is the number of equivalents of the designated solute contained in an amount of the designated solvent that is sufficient to prepare one liter of solution.

Percentage Measurements—Percentage concentrations are expressed as follows:

Percent weight in weight—(w/w) expresses the number of g of a constituent in 100 g of solution or mixture.

Percent weight in volume—(w/v) expresses the number of g of a constituent in 100 mL of solution, and is used regardless of whether water or another liquid is the solvent.

Percent volume in volume—(v/v) expresses the number of mL of a constituent in 100 mL of solution.

The term *percent* used without qualification means, for mixtures of solids and semisolids, percent weight in weight; for solutions or suspensions of solids in liquids, percent weight in volume; for solutions of liquids in liquids, percent volume in volume; and for solutions of gases in liquids, percent weight in volume. For example, a 1 percent solution is prepared by dissolving 1 g of a solid or semisolid, or 1 mL of a liquid, in sufficient solvent to make 100 mL of the solution.

In the dispensing of prescription medications, slight changes in volume owing to variations in room temperatures may be disregarded.

Official Monographs for USP XXII

Absorbable Dusting Powder—*see* Dusting Powder, Absorbable

Absorbable Gelatin Film—*see* Gelatin Film, Absorbable

Absorbable Gelatin Sponge—*see* Gelatin Sponge, Absorbable

Absorbable Surgical Suture—*see* Suture, Absorbable Surgical

Absorbent Gauze—*see* Gauze, Absorbent

Acacia—*see* Acacia NF

Acetaminophen

HO–C₆H₄–NHCOCH₃

$C_8H_9NO_2$ 151.16

Acetamide, *N*-(4-hydroxyphenyl)-.

4′-Hydroxyacetanilide   [*103-90-2*].

» Acetaminophen contains not less than 98.0 percent and not more than 101.0 percent of $C_8H_9NO_2$, calculated on the anhydrous basis.

Packaging and storage—Preserve in tight, light-resistant containers.

Reference standard—*USP Acetaminophen Reference Standard*—Dry over silica gel for 18 hours before using.

Identification—

A: The infrared absorption spectrum of a potassium bromide dispersion of it, previously dried over a suitable desiccant, exhibits maxima only at the same wavelengths as that of a similar preparation of USP Acetaminophen RS.

B: The ultraviolet absorption spectrum of a 1 in 200,000 solution of it in a 1 in 100 solution of 0.1 *N* hydrochloric acid in methanol exhibits maxima and minima at the same wavelengths as that of a similar solution of USP Acetaminophen RS, concomitantly measured.

C: To 10 mL of a 1 in 100 solution add 1 drop of ferric chloride TS: a violet-blue color is produced.

Melting range ⟨741⟩: between 168° and 172°.

Water, *Method I* ⟨921⟩: not more than 0.5%.

Residue on ignition ⟨281⟩: not more than 0.1%.

Chloride ⟨221⟩—Shake 1.0 g with 25 mL of water, filter, and add 1 mL of 2 *N* nitric acid and 1 mL of silver nitrate TS: the filtrate shows no more chloride than corresponds to 0.20 mL of 0.020 *N* hydrochloric acid (0.014%).

Sulfate ⟨221⟩—Shake 1.0 g with 25 mL of water, filter, add 2 mL of 1 *N* acetic acid, then add 2 mL of barium chloride TS: the mixture shows no more sulfate than corresponds to 0.20 mL of 0.020 *N* sulfuric acid (0.02%).

Sulfide—Place about 2.5 g in a 50-mL beaker. Add 5 mL of alcohol and 1 mL of 3 *N* hydrochloric acid. Moisten a piece of lead acetate test paper with water, and fix to the underside of a watch glass. Cover the beaker with the watch glass so that part of the lead acetate paper hangs down near the pouring spout of the beaker. Heat the contents of the beaker on a hot plate just to boiling: no coloration or spotting of the test paper occurs.

Heavy metals, *Method II* ⟨231⟩: 0.001%.

Readily carbonizable substances ⟨271⟩—Dissolve 0.50 g in 5 mL of sulfuric acid TS: the solution has no more color than Matching Fluid A.

Free *p*-aminophenol—Transfer 5.0 g to a 100-mL volumetric flask, and dissolve in about 75 mL of a mixture of equal volumes of methanol and water. Add 5.0 mL of alkaline nitroferricyanide solution (prepared by dissolving 1 g of sodium nitroferricyanide and 1 g of anhydrous sodium carbonate in 100 mL of water), dilute with a mixture of equal volumes of methanol and water to volume, mix, and allow to stand for 30 minutes. Concomitantly determine the absorbances of this solution and of a freshly prepared solution of *p*-aminophenol, similarly prepared at a concentration of 2.5 µg per mL, using the same quantities of the same reagents, in 1-cm cells, at the maximum at about 710 nm, with a suitable spectrophotometer, using 5.0 mL of alkaline nitroferricyanide solution diluted with a mixture of equal volumes of methanol and water to 100 mL as the blank: the absorbance of the test solution does not exceed that of the standard solution, corresponding to not more than 0.005% of *p*-aminophenol.

***p*-Chloroacetanilide**—Transfer 1.0 g to a glass-stoppered, 15-mL centrifuge tube, add 5.0 mL of ether, shake by mechanical means for 30 minutes, and centrifuge at 1000 rpm for 15 minutes or until a clean separation is obtained. On a suitable thin-layer chromatographic plate (see *Chromatography* ⟨621⟩), coated with a 0.25-mm layer of chromatographic silica gel mixture, apply 200 µL of the supernatant liquid, in 40-µL portions, to obtain a single spot not more than 10 mm in diameter. Similarly apply 40 µL of a Standard solution in ether containing 10 µg of *p*-chloroacetanilide per mL, and allow the spots to dry. Develop the chromatogram, in an unsaturated chamber, with a solvent system consisting of a mixture of solvent hexane and acetone (75:25), until the solvent front has moved three-fourths of the length of the plate. Remove the plate from the developing chamber, mark the solvent front, and allow the solvent to evaporate. Locate the spots in the chromatogram by examination under short-wavelength ultraviolet light: any spot obtained from the solution under test, at an R_f value corresponding to the principal spot from the Standard solution, is not greater in size or intensity than the principal spot obtained from the Standard solution, corresponding to not more than 0.001% of *p*-chloroacetanilide.

Assay—Dissolve about 120 mg of Acetaminophen, accurately weighed, in 10 mL of methanol in a 500-mL volumetric flask, dilute with water to volume, and mix. Transfer 5.0 mL of this solution to a 100-mL volumetric flask, dilute with water to volume, and mix. Concomitantly determine the absorbances of this solution and of a Standard solution of USP Acetaminophen RS, in the same medium, at a concentration of about 12 µg per mL in 1-cm cells, at the wavelength of maximum absorbance at about 244 nm, with a suitable spectrophotometer, using water as the blank. Calculate the quantity, in mg, of $C_8H_9NO_2$ in the Acetaminophen taken by the formula:

$$10C(A_U/A_S),$$

in which C is the concentration, in µg per mL, of USP Acetaminophen RS in the Standard solution, and A_U and A_S are the absorbances of the solution of Acetaminophen and the Standard solution, respectively.

Acetaminophen Capsules

» Acetaminophen Capsules contain not less than 90.0 percent and not more than 110.0 percent of the labeled amount of $C_8H_9NO_2$.

Packaging and storage—Preserve in tight containers.

Reference standard—*USP Acetaminophen Reference Standard*—Dry over silica gel for 18 hours before using.

Identification—

A: The ultraviolet absorption spectrum of the solution of the Capsules prepared for the measurement of absorbance in the *Assay* exhibits maxima and minima at the same wavelengths as that of a similar solution of USP Acetaminophen RS, concomitantly measured.

B: Triturate an amount of the contents of the Capsules, equivalent to about 1 g of acetaminophen, with 30 mL of warm alcohol, cool, and filter. To 3 mL of the filtrate add 10 mL of water and 1 drop of ferric chloride TS, and mix: a violet-blue color is produced.

Dissolution ⟨711⟩—

Medium: water; 900 mL.

Apparatus 2: 50 rpm.

Time: 45 minutes.

Procedure—Determine the amount of $C_8H_9NO_2$ dissolved from ultraviolet absorbances at the wavelength of maximum absorbance at about 249 nm of filtered portions of the solution under test, suitably diluted with *Dissolution Medium*, if necessary, in comparison with a Standard solution having a known concentration of USP Acetaminophen RS in the same medium.

Tolerances—Not less than 75% (Q) of the labeled amount of $C_8H_9NO_2$ is dissolved in 45 minutes.

Uniformity of dosage units ⟨905⟩: meet the requirements.

Assay—

Standard preparation and *Chromatographic column*—Prepare as directed in the *Assay* under *Acetaminophen Elixir*.

Assay preparation—Weigh the contents of not fewer than 20 Acetaminophen Capsules. Mix the contents, and transfer an accurately weighed portion of the powder, equivalent to about 250 mg of acetaminophen, to a 250-mL volumetric flask, add 2 mL of 1 *N* sodium hydroxide, dilute with water to volume, mix, and filter, discarding the first 20 mL of the filtrate. Proceed as directed for *Assay preparation* in the *Assay* under *Acetaminophen Elixir*, beginning with "Transfer 2.0 mL of this solution to a 100-mL beaker."

Procedure—Proceed as directed for *Procedure* in the *Assay* under *Acetaminophen Elixir*. Calculate the quantity, in mg, of $C_8H_9NO_2$ in the portion of Capsules taken by the formula:

$$31.25C(A_U/A_S),$$

in which C is the concentration, in µg per mL, of USP Acetaminophen RS in the *Standard preparation*, and A_U and A_S are the absorbances of the *Assay preparation* and the *Standard preparation*, respectively.

Acetaminophen Capsules, Chlorzoxazone and—*see* Chlorzoxazone and Acetaminophen Capsules

Acetaminophen Capsules, Oxycodone and—*see* Oxycodone and Acetaminophen Capsules

Acetaminophen Oral Solution

» Acetaminophen Oral Solution contains not less than 90.0 percent and not more than 110.0 percent of the labeled amount of $C_8H_9NO_2$.

Packaging and storage—Preserve in tight containers.

Reference standard—*USP Acetaminophen Reference Standard*—Dry over silica gel for 18 hours before using.

Identification—The ultraviolet absorption spectrum of a portion of the *Assay preparation* employed for measurement of absorbance in the *Assay* exhibits maxima and minima at the same wavelengths as that of a similar solution of USP Acetaminophen RS, concomitantly measured.

pH ⟨791⟩: between 3.8 and 6.1.

Alcohol content (if present), *Method II* ⟨611⟩: between 90.0% and 115.0% of the labeled amount of C_2H_5OH, determined by the gas-liquid chromatographic procedure, acetone being used as the internal standard.

Assay—

Standard preparation—Transfer about 80 mg of USP Acetaminophen RS, accurately weighed, to a 100-mL volumetric flask, add methanol to volume, and mix. Transfer 10.0 mL of this solution to a second 100-mL volumetric flask, dilute with methanol to volume, and mix. Transfer 10.0 mL of the resulting solution to a 100-mL volumetric flask, add 1 mL of 0.1 *N* hydrochloric acid, then add methanol to volume, and mix.

Chromatographic column—Pack a pledget of fine glass wool in the base of a chromatographic tube (25-mm × 250-mm tube to which is fused a 5-cm length of 7-mm tubing) with the aid of a tamping rod about 45 cm in length and having a disk with a diameter about 1 mm less than that of the tube. To 2 g of purified siliceous earth in a 100-mL beaker add 2.0 mL of a solution containing 1.0 g of sodium bicarbonate and 4.5 g of sodium carbonate in each 100 mL, and mix until a fluffy mixture is obtained. Transfer the mixture to the chromatographic tube, and tamp gently to compress the material to a uniform mass.

Assay preparation—Transfer an accurately measured volume of Acetaminophen Oral Solution, equivalent to about 250 mg of acetaminophen, to a 250-mL volumetric flask, add 2 mL of 1 *N* sodium hydroxide, dilute with water to volume, and mix. Transfer 2.0 mL of this solution to a 100-mL beaker, add 1 drop of hydrochloric acid, swirl to mix, then add 3.0 g of purified siliceous earth. Mix, and transfer to the chromatographic column. Scrub the beaker with 1 g of purified siliceous earth mixed with 2 drops of water, transfer the washings to the column, and tamp gently. Place a pledget of fine glass wool on top of the column. Wash the column with 100 mL of water-saturated chloroform, and discard the eluate. Elute the acetaminophen with 150 mL of water-saturated ether, collecting the eluate in a 400-mL beaker. Evaporate the ether on a steam bath with the aid of a current of air just to dryness. [NOTE—Avoid prolonged drying, to prevent loss of acetaminophen.] Without delay, dissolve the residue in a solvent mixture consisting of 1 mL of dilute hydrochloric acid (1 in 100) per 100 mL of methanol, and transfer to a 50-mL volumetric flask. Rinse the beaker with the solvent mixture, adding the rinsings to the flask, dilute with the solvent mixture to volume, and mix. Transfer 10.0 mL of this solution to a second 50-mL volumetric flask, dilute with the solvent mixture to volume, and mix.

Procedure—Concomitantly determine the absorbances of the *Standard preparation* and the *Assay preparation* in 1-cm cells at the wavelength of maximum absorbance at about 249 nm, with a suitable spectrophotometer, using a solvent mixture consisting of 1 mL of 0.1 *N* hydrochloric acid per 100 mL of methanol as the blank. Calculate the quantity, in mg, of $C_8H_9NO_2$ in each mL of the Oral Solution taken by the formula:

$$31.25(C/V)(A_U/A_S),$$

in which C is the concentration, in µg per mL, of USP Acetaminophen RS in the *Standard preparation*, V is the volume, in mL, of Oral Solution taken, and A_U and A_S are the absorbances of the *Assay preparation* and the *Standard preparation*, respectively.

Acetaminophen for Effervescent Oral Solution

» Acetaminophen for Effervescent Oral Solution contains, in each 100 g, not less than 5.63 g and not more than 6.88 g of $C_8H_9NO_2$.

Packaging and storage—Preserve in tight containers.

Reference standard—*USP Acetaminophen Reference Standard*—Dry over silica gel for 18 hours before using.

Identification—

A: A 10-g portion dissolves, with effervescence, in 200 mL of water.

B: The ultraviolet absorption spectrum of the solution prepared for the measurement of absorbance in the *Assay* exhibits maxima and minima at the same wavelengths as that of a similar solution of USP Acetaminophen RS, concomitantly measured.

Minimum fill ⟨755⟩—FOR SOLID PACKAGED IN MULTIPLE-UNIT CONTAINERS: meets the requirements.

Uniformity of dosage units ⟨905⟩—FOR SOLID PACKAGED IN SINGLE-UNIT CONTAINERS: meets the requirements.

Assay—Dissolve about 10 g of Acetaminophen for Effervescent Oral Solution, accurately weighed, in about 200 mL of 0.1 *N* hydrochloric acid in a 500-mL volumetric flask, using gentle heat if necessary, until effervescence subsides, then dilute with 0.1 *N* hydrochloric acid to volume, and mix. Transfer 10.0 mL of this solution to a 100-mL volumetric flask, dilute with 0.1 *N* hydrochloric acid to volume, and mix. Transfer 10.0 mL of the resulting solution to a 100-mL volumetric flask, dilute with 0.1 *N* hydrochloric acid to volume, and mix. Concomitantly determine the absorbances of this solution and of a Standard solution of USP Acetaminophen RS, in the same medium, at a concentration of about 12 µg per mL in 1-cm cells, at the wavelength of maximum absorbance at about 243 nm, with a suitable spectrophotometer, using water as the blank. Calculate the quantity, in mg, of $C_8H_9NO_2$ in the portion of Acetaminophen for Effervescent Oral Solution taken by the formula:

$$50C(A_U/A_S),$$

in which C is the concentration, in µg per mL, of USP Acetaminophen RS in the Standard solution, and A_U and A_S are the absorbances of the solution of Acetaminophen for Effervescent Oral Solution and the Standard solution, respectively.

Acetaminophen Oral Suspension

» Acetaminophen Oral Suspension is a suspension of Acetaminophen in a suitable aqueous vehicle. It contains not less than 90.0 percent and not more than 110.0 percent of the labeled amount of $C_8H_9NO_2$.

Packaging and storage—Preserve in tight containers.

Reference standard—*USP Acetaminophen Reference Standard*—Dry over silica gel for 18 hours before using.

Identification—Transfer a volume of Oral Suspension, equivalent to about 240 mg of acetaminophen, to a separator, add 50 mL of ethyl acetate, and shake well. Filter the ethyl acetate extract through a funnel containing glass wool and about 10 g of anhydrous sodium sulfate. Collect the filtrate in a beaker, and evaporate on a steam bath to dryness. Dry the residue in vacuum over silica gel: the crystals so obtained respond to *Identification test A* under *Acetaminophen.*

pH ⟨791⟩: between 5.4 and 6.9.

Assay—

Standard preparation—Dissolve an accurately weighed quantity of USP Acetaminophen RS in water, and dilute quantitatively and stepwise, if necessary, with water to obtain a solution having a known concentration of about 100 µg per mL.

Assay preparation—Transfer an accurately measured volume of Oral Suspension, previously well-shaken, equivalent to about 100 mg of acetaminophen, to a 100-mL volumetric flask, add 60 mL of water, insert the stopper, and shake by mechanical means for 30 minutes. Dilute with water to volume, and mix. Transfer 10.0 mL of this solution to a 100-mL volumetric flask, dilute with water to volume, and mix.

Procedure—Transfer 10.0 mL each of the *Standard preparation*, the *Assay preparation*, and water to provide a blank, to separate 50-mL volumetric flasks, and treat each as follows: Add 2.0 mL of 6 *N* hydrochloric acid, and mix. Add 5.0 mL of sodium nitrite solution (1 in 10), mix, and allow to stand for 15 minutes. Add 5.0 mL of ammonium sulfamate solution (15 in 100), and swirl gently, allowing the solution to cool to room temperature. Add 15.0 mL of 2.5 *N* sodium hydroxide, allow to cool to room temperature, dilute with water to volume, and mix. Concomitantly determine the absorbances of the solutions obtained from the *Standard preparation* and the *Assay preparation* in 1-cm cells relative to the blank at the wavelength of maximum absorbance at about 430 nm, with a suitable spectrophotometer. Calculate the quantity, in mg, of $C_8H_9NO_2$ in each mL of the Oral Suspension taken by the formula:

$$(C/V)(A_U/A_S),$$

in which C is the concentration, in µg per mL, of USP Acetaminophen RS in the *Standard preparation*, V is the volume, in mL, of Oral Suspension taken, and A_U and A_S are the absorbances of the solutions from the *Assay preparation* and the *Standard preparation*, respectively.

Acetaminophen Tablets

» Acetaminophen Tablets contain not less than 90.0 percent and not more than 110.0 percent of the labeled amount of $C_8H_9NO_2$.

Packaging and storage—Preserve in tight containers.

Reference standard—*USP Acetaminophen Reference Standard*—Dry over silica gel for 18 hours before using.

Identification—

A: The ultraviolet absorption spectrum of the solution of the Tablets prepared for measurement of absorbance in the *Assay* exhibits maxima and minima at the same wavelengths as that of a similar solution of USP Acetaminophen RS, concomitantly measured.

B: Triturate an amount of powdered Tablets, equivalent to about 1 g of acetaminophen, with 30 mL of warm alcohol, cool, and filter. To 3 mL of the filtrate add 10 mL of water and 1 drop of ferric chloride TS, and mix: a violet-blue color is produced.

Dissolution ⟨711⟩—

Medium: pH 5.8 phosphate buffer (see *Buffer Solutions* in the section, *Reagents, Indicators, and Solutions*); 900 mL.

Apparatus 2: 50 rpm.

Time: 30 minutes.

Procedure—Determine the amount of $C_8H_9NO_2$ dissolved from ultraviolet absorbances at the wavelength of maximum absorbance at about 243 nm of filtered portions of the solution under test, suitably diluted with *Dissolution Medium*, if necessary, in comparison with a Standard solution having a known concentration of USP Acetaminophen RS in the same medium.

Tolerances—Not less than 80% (*Q*) of the labeled amount of $C_8H_9NO_2$ is dissolved in 30 minutes.

Uniformity of dosage units ⟨905⟩: meet the requirements.

Assay—

Standard preparation and *Chromatographic column*—Prepare as directed in the *Assay* under *Acetaminophen Elixir*.

Assay preparation—Weigh and finely powder not less than 20 Acetaminophen Tablets. Transfer an accurately weighed portion of the powder, equivalent to about 250 mg of acetaminophen, to a 250-mL volumetric flask, add 2 mL of 1 *N* sodium hydroxide, dilute with water to volume, mix, and filter, discarding the first 20 mL of the filtrate. Proceed as directed for *Assay preparation* in the *Assay* under *Acetaminophen Elixir*, beginning with "Transfer 2.0 mL of this solution to a 100-mL beaker."

Procedure—Proceed as directed for *Procedure* in the *Assay* under *Acetaminophen Elixir*. Calculate the quantity, in mg, of $C_8H_9NO_2$ in the portion of Tablets taken by the formula:

$$31.25C(A_U/A_S),$$

in which *C* is the concentration, in µg per mL, of USP Acetaminophen RS in the *Standard preparation*, and A_U and A_S are the absorbances of the *Assay preparation* and the *Standard preparation*, respectively.

Acetaminophen Tablets, Oxycodone and—*see* Oxycodone and Acetaminophen Tablets

Acetaminophen Tablets, Propoxyphene Hydrochloride and—*see* Propoxyphene Hydrochloride and Acetaminophen Tablets

Acetaminophen Tablets, Propoxyphene Napsylate and—*see* Propoxyphene Napsylate and Acetaminophen Tablets

Acetaminophen and Aspirin Tablets

» Acetaminophen and Aspirin Tablets contain not less than 90.0 percent and not more than 110.0 percent of the labeled amounts of acetaminophen ($C_8H_9NO_2$) and aspirin ($C_9H_8O_4$).

Packaging and storage—Preserve in tight containers.

Reference standards—*USP Acetaminophen Reference Standard*—Dry over silica gel for 18 hours before using. *USP Aspirin Reference Standard*—Dry over silica gel for 5 hours before using. *USP Salicylic Acid Reference Standard*—Dry over silica gel for 3 hours before using.

Identification—The retention times of the major peaks in the chromatogram of the *Assay preparation* correspond to those of the *Standard preparation*, relative to the internal standard, as obtained in the *Assay*.

Dissolution ⟨711⟩—

Medium: water; 900 mL.

Apparatus 2: 50 rpm.

Time: 45 minutes.

Mobile phase—Prepare as directed under *Assay*.

Solvent mixture—Prepare as directed under *Assay*.

Internal standard solution—Prepare a solution of benzoic acid in methanol having a concentration of about 1 mg per mL.

Standard preparation I—Dissolve an accurately weighed quantity of USP Salicylic Acid RS in the *Solvent mixture* to obtain a solution having a known concentration of about 70 µg per mL. Combine 4.0 mL of this solution and 1.0 mL of the *Internal standard solution*, and mix.

Standard preparation II—Dissolve accurately weighed quantities of USP Acetaminophen RS and USP Aspirin RS in the *Solvent mixture* to obtain a solution having known concentrations of about 360 µg of acetaminophen and about 360 µg of aspirin per mL. Combine 4.0 mL of this solution and 1.0 mL of the *Internal standard solution*, and mix.

Test preparation—Combine 4.0 mL of a filtered portion of the solution under test and 1.0 mL of the *Internal standard solution*, and mix.

Chromatographic system—Proceed as directed under *Assay*.

Procedure—Separately inject equal volumes (about 20 µL) of the two *Standard preparations* and the *Test preparation* into the chromatograph, record the chromatograms, and measure the responses for the major peaks. The relative retention times are about 0.3 for acetaminophen, 0.4 for salicylic acid, 0.6 for aspirin, and 1.0 for benzoic acid. Determine the amount of acetaminophen ($C_8H_9NO_2$) dissolved by the formula:

$$90(C/W)(R_U/R_S),$$

in which *C* is the concentration, in µg per mL, of USP Acetaminophen RS in *Standard preparation II*, R_U and R_S are the relative peak response ratios obtained from the *Test preparation* and *Standard preparation II*, respectively, and *W* is the labeled amount, in mg, of acetaminophen. Determine the amount of aspirin ($C_9H_8O_4$) dissolved by the formula:

$$[[90C_1(R_{U_1}/R_S)] + [90C_2(R_{U_2}/R_{S_2})(1.3044)]]/W,$$

in which C_1 and C_2 are the concentrations, in µg per mL, of USP Aspirin RS in *Standard preparation II* and USP Salicylic Acid RS in *Standard preparation I*, respectively, R_{U_1} and R_{S_1} are the relative peak response ratios for the aspirin peak and the internal standard peak obtained from the *Test preparation* and *Standard preparation II*, respectively, R_{U_2} and R_{S_2} are the relative peak response ratios for the salicylic acid peak and the internal standard peak obtained from the *Test preparation* and *Standard preparation I*, respectively, and *W* is the labeled amount, in mg, of aspirin.

Tolerances—Not less than 75% (*Q*) of the labeled amount of $C_8H_9NO_2$ and not less than 75% (*Q*) of the labeled amount of $C_9H_8O_4$ are dissolved in 45 minutes.

Uniformity of dosage units ⟨905⟩: meet the requirements for *Content Uniformity* with respect to acetaminophen and to aspirin.

Salicylic acid—

Solvent mixture, Mobile phase, Internal standard solution, and *Chromatographic system*—Prepare as directed in the *Assay*.

Procedure—Dissolve a suitable quantity of USP Salicylic Acid RS, accurately weighed, in *Solvent mixture* to obtain a solution having a known concentration of about 1.0 mg per mL. Transfer 1.0-mL, 5.0-mL, and 10.0-mL portions, respectively, of this solution to separate 100-mL volumetric flasks, add 10.0 mL of *Internal standard solution* to each flask, dilute with *Solvent mixture* to volume, and mix. Chromatograph these three Standard solutions as directed in the *Assay*. Plot the ratios of the peak responses for salicylic acid and benzoic acid for each of the Standard solutions versus concentrations, in mg per mL, of salicylic acid, and draw the straight line best fitting the three plotted points. From the graph so obtained, and from the ratio of the peak responses for salicylic acid and benzoic acid in the chromatogram of the *Assay preparation* as obtained in the *Assay*, determine the concentration, in mg per mL, of salicylic acid ($C_7H_6O_3$) in the *Assay preparation*, and calculate the percentage of salicylic acid in relation to the concentration of aspirin in the *Assay preparation*, as determined in the *Assay*. Not more than 3.0% is found.

Assay—[NOTE—Use clean, dry glassware. Inject the *Standard preparation* and the *Assay preparation* promptly after preparation.]

Solvent mixture—Prepare a mixture of chloroform, methanol, and glacial acetic acid (78:20:2).

Mobile phase—Transfer 225 mg of tetramethylammonium hydroxide pentahydrate to a 1000-mL flask, and add 750 mL of water, 125 mL of methanol, 125 mL of acetonitrile, and 1.0 mL of glacial acetic acid. Stir for 3 minutes, filter through a membrane filter (0.5-µm or finer porosity), and degas.

Internal standard solution—Dissolve benzoic acid in *Solvent mixture* to obtain a solution having a concentration of about 20 mg per mL.

Standard preparation—Transfer about 325 mg of USP Acetaminophen RS and about 325 mg of USP Aspirin RS, each accurately weighed, to a 100-mL volumetric flask, add 10.0 mL of *Internal standard solution*, dilute with *Solvent mixture* to volume, and mix.

Assay preparation—Weigh and finely powder not less than 20 Acetaminophen and Aspirin Tablets. Transfer an accurately weighed portion of the powder, equivalent to about 325 mg of acetaminophen, to a 100-mL volumetric flask, add 10.0 mL of *Internal standard solution* and about 50 mL of *Solvent mixture*, and sonicate for about 3 minutes. Dilute with *Solvent mixture* to volume, and mix. Filter a portion of this solution through a 2.5-μm or finer porosity filter, and use the filtrate as the *Assay preparation*.

Chromatographic system—The liquid chromatograph is equipped with a 280-nm detector and a 3.9-mm × 30-cm column that contains packing L1. The flow rate is about 2 mL per minute. Chromatograph four replicate injections of the *Standard preparation*, and record the peak responses as directed under *Procedure:* the relative standard deviation for either analyte is not more than 3.0%.

Procedure—Separately inject equal volumes (about 5 μL) of the *Standard preparation* and the *Assay preparation* into the chromatograph, record the chromatograms, and measure the responses for the major peaks. The retention times are about 2, 3, 5, and 8 minutes for acetaminophen, salicylic acid (if present), aspirin, and benzoic acid, respectively. Calculate the quantity, in mg, of acetaminophen ($C_8H_9NO_2$) in the portion of Tablets taken by the formula:

$$100C(R_U/R_S),$$

in which C is the concentration, in mg per mL, of USP Acetaminophen RS in the *Standard preparation*, and R_U and R_S are the ratios of the peak responses of acetaminophen and benzoic acid obtained from the *Assay preparation* and the *Standard preparation*, respectively. Calculate the quantity, in mg, of aspirin ($C_9H_8O_4$) in the portion of Tablets taken by the same formula, except to read "USP Aspirin RS" where "USP Acetaminophen RS" is specified, and "aspirin" where "acetaminophen" is specified.

Acetaminophen, Aspirin, and Caffeine Capsules

» Acetaminophen, Aspirin, and Caffeine Capsules contain not less than 90.0 percent and not more than 110.0 percent of the labeled amounts of acetaminophen ($C_8H_9NO_2$), aspirin ($C_9H_8O_4$), and caffeine ($C_8H_{10}N_4O_2$).

Packaging and storage—Preserve in tight containers.

Reference standards—*USP Acetaminophen Reference Standard*—Dry over silica gel for 18 hours before using. *USP Aspirin Reference Standard*—Dry over silica gel for 5 hours before using. *USP Caffeine Reference Standard*—Dry at 80° for 4 hours before using. *USP Salicylic Acid Reference Standard*—Dry over silica gel for 3 hours before using.

Identification—The retention times of the major peaks in the chromatogram of the *Assay preparation* correspond to those of the *Standard preparation*, relative to the internal standard, obtained as directed in the *Assay*.

Dissolution ⟨711⟩—

Medium: water; 900 mL.

Apparatus 1: 100 rpm.

Time: 45 minutes.

Mobile phase, Internal standard solution, Solvent mixture, Standard stock solution, and *Chromatographic system*—Proceed as directed in the *Assay*.

Standard preparation—Transfer 20.0 mL of *Standard stock solution*, 3.0 mL of *Internal standard solution*, and 20 mL of water to a 50-mL volumetric flask, mix, and allow to stand for about 30 seconds. Dilute with *Solvent mixture* to volume, and mix. Use within 8 hours.

Test preparation—Transfer 20.0 mL of a filtered portion of the solution under test to a 50-mL volumetric flask, add 3.0 mL of *Internal standard solution* and 20 mL of *Solvent mixture*, mix, and allow to stand for 30 seconds. Dilute with *Solvent mixture* to volume, and mix.

Procedure—Proceed as directed for *Procedure* in the *Assay*. Calculate the quantities, in mg, of acetaminophen ($C_8H_9NO_2$), aspirin ($C_9H_8O_4$), and caffeine ($C_8H_{10}N_4O_2$) dissolved by the formula:

$$2250C(R_U/R_S),$$

in which C is the concentration, in mg per mL, of the appropriate USP Reference Standard in the *Standard preparation*, and R_U and R_S are the ratios of the peak responses of the corresponding analyte and internal standard peaks of the solution under test and the *Standard preparation*, respectively.

Tolerances—Not less than 75% (*Q*) of the labeled amounts of $C_8H_9NO_2$, $C_9H_8O_4$, and $C_8H_{10}N_4O_2$ are dissolved in 45 minutes.

Uniformity of dosage units ⟨905⟩: meet the requirements for *Content Uniformity* with respect to acetaminophen, aspirin, and caffeine.

Salicylic acid—

Mobile phase and *Solvent mixture*—Prepare as directed in the *Assay*.

Standard preparation—Dissolve an accurately weighed quantity of USP Salicylic Acid RS in *Solvent mixture* to obtain a solution having a known concentration of about 1 mg per mL. Transfer 2.0 mL of the resulting solution to a 100-mL volumetric flask, dilute with *Solvent mixture* to volume, and mix. This solution contains about 0.02 mg of USP Salicylic Acid RS per mL.

Test preparation—Weigh the contents of not less than 20 Capsules, and calculate the average weight per Capsule. Mix the combined contents of the Capsules, and transfer an accurately weighed portion, equivalent to about 250 mg of aspirin, to a 100-mL volumetric flask. Add about 75 mL of *Solvent mixture*, and shake by mechanical means for 30 minutes. Dilute with *Solvent mixture* to volume, and mix.

Chromatographic system—Proceed as directed for *Chromatographic system* in the *Assay*, except to use a 302-nm detector. Chromatograph the *Standard preparation*, and record the responses as directed under *Procedure:* the tailing factor is not more than 1.6, and the relative standard deviation for replicate injections is not more than 3.0%.

Procedure—Separately inject equal volumes (about 10 μL) of the *Standard preparation* and the *Test preparation* into the chromatograph, record the chromatograms, and measure the responses for the salicylic acid peaks. Calculate the percentage of salicylic acid in the portion of Capsules taken by the formula:

$$100(C/a)(r_U/r_S),$$

in which C is the concentration, in mg per mL, of USP Salicylic Acid RS in the *Standard preparation*, a is the quantity, in mg, of aspirin in the portion of Capsules taken, based on the labeled amount, and r_U and r_S are the salicylic acid peak responses of the *Test preparation* and the *Standard preparation*, respectively: not more than 3.0% is found.

Assay—

Mobile phase—Prepare a suitable mixture of water, methanol, and glacial acetic acid (69:28:3). Make adjustments if necessary (see *System Suitability* under *Chromatography* ⟨621⟩).

Internal standard solution—Prepare a solution of benzoic acid in methanol containing about 6 mg per mL.

Solvent mixture—Prepare a mixture of methanol and glacial acetic acid (95:5).

Standard stock solution—Dissolve accurately weighed quantities of USP Acetaminophen RS, USP Aspirin RS, and USP Caffeine RS in *Solvent mixture* to obtain a solution having known concentrations of about 0.25 mg of USP Acetaminophen RS per mL, 0.25*J* mg of USP Aspirin RS per mL, and 0.25*J′* mg of USP Caffeine RS per mL, *J* being the ratio of the labeled amount, in mg, of aspirin to the labeled amount, in mg, of acetaminophen per Capsule, and J' being the ratio of the labeled amount, in mg, of caffeine to the labeled amount, in mg, of acetaminophen per Capsule.

Standard preparation—Transfer 20.0 mL of *Standard stock solution* and 3.0 mL of *Internal standard solution* to a 50-mL volumetric flask, dilute with *Solvent mixture* to volume, and mix. This solution contains about 0.1 mg of USP Acetaminophen RS, $0.1J$ mg of USP Aspirin RS, and $0.1J'$ mg of USP Caffeine RS per mL.

Assay preparation—Weigh the contents of not less than 20 Capsules, and calculate the average weight per Capsule. Mix the combined contents of the Capsules, and transfer an accurately weighed portion, equivalent to about 250 mg of acetaminophen, to a 100-mL volumetric flask. Add about 75 mL of *Solvent mixture*, and shake by mechanical means for 30 minutes. Dilute with *Solvent mixture* to volume, and mix. Transfer 2.0 mL of this solution and 3.0 mL of *Internal standard solution* to a 50-mL volumetric flask, dilute with *Solvent mixture* to volume, and mix.

Chromatographic system—The liquid chromatograph is equipped with a 275-nm detector and a 4.6-mm × 10-cm column that contains 5-μm packing L1, and is maintained at 45 ± 1°. The flow rate is about 2 mL per minute. Chromatograph the *Standard preparation*, and record the responses as directed under *Procedure:* the tailing factor for each analyte peak is not more than 1.2, the resolution, R, between any of the analyte and internal standard peaks is not less than 1.4, and the relative standard deviation for replicate injections is not more than 2.0%.

Procedure—Separately inject equal volumes (about 10 μL) of the *Standard preparation* and the *Assay preparation* into the chromatograph, record the chromatograms, and measure the responses for the major peaks. The relative retention times are about 0.3 for acetaminophen, 0.5 for caffeine, 0.8 for aspirin, 1.0 for benzoic acid, and 1.2 for salicylic acid. Calculate the quantities, in mg, of acetaminophen ($C_8H_9NO_2$), aspirin ($C_9H_8O_4$), and caffeine ($C_8H_{10}N_4O_2$) in the portion of Capsules taken by the formula:

$$2500C(R_U/R_S),$$

in which C is the concentration, in mg per mL, of the appropriate USP Reference Standard in the *Standard preparation*, and R_U and R_S are the ratios of the peak responses of the corresponding analyte and internal standard peaks of the *Assay preparation* and the *Standard preparation*, respectively.

Acetaminophen, Aspirin, and Caffeine Tablets

» Acetaminophen, Aspirin, and Caffeine Tablets contain not less than 90.0 percent and not more than 110.0 percent of the labeled amounts of acetaminophen ($C_8H_9NO_2$), aspirin ($C_9H_8O_4$), and caffeine ($C_8H_{10}N_4O_2$).

Packaging and storage—Preserve in well-closed containers.

Reference standards—*USP Acetaminophen Reference Standard*—Dry over silica gel for 18 hours before using. *USP Aspirin Reference Standard*—Dry over silica gel for 5 hours before using. *USP Caffeine Reference Standard*—Dry at 80° for 4 hours before using. *USP Salicylic Acid Reference Standard*—Dry over silica gel for 3 hours before using.

Identification—The retention times of the major peaks in the chromatogram of the *Assay preparation* correspond to those of the *Standard preparation*, relative to the internal standard, obtained as directed in the *Assay*.

Dissolution ⟨711⟩—

Medium: water; 900 mL.

Apparatus 2: 100 rpm.

Time: 60 minutes.

Mobile phase, *Internal standard solution*, *Solvent mixture*, *Standard stock solution*, and *Chromatographic system*—Proceed as directed in the *Assay*.

Standard preparation—Transfer 20.0 mL of *Standard stock solution*, 3.0 mL of *Internal standard solution*, and 20 mL of water to a 50-mL volumetric flask, mix, and allow to stand for about 30 seconds. Dilute with *Solvent mixture* to volume, and mix. Use within 8 hours.

Test preparation—Transfer 20.0 mL of a filtered portion of the solution under test to a 50-mL volumetric flask, add 3.0 mL of *Internal standard solution* and 20 mL of *Solvent mixture*, mix, and allow to stand for 30 seconds. Dilute with *Solvent mixture* to volume, and mix.

Procedure—Proceed as directed for *Procedure* in the *Assay*. Calculate the quantities, in mg, of acetaminophen ($C_8H_9NO_2$), aspirin ($C_9H_8O_4$), and caffeine ($C_8H_{10}N_4O_2$) dissolved by the formula:

$$2250C(R_U/R_S),$$

in which C is the concentration, in mg per mL, of the appropriate USP Reference Standard in the *Standard preparation*, and R_U and R_S are the ratios of the peak responses of the corresponding analyte and internal standard peaks of the solution under test and the *Standard preparation*, respectively.

Tolerances—Not less than 75% (Q) of the labeled amounts of $C_8H_9NO_2$, $C_9H_8O_4$, and $C_8H_{10}N_4O_2$ are dissolved in 60 minutes.

Uniformity of dosage units ⟨905⟩: meet the requirements for *Content Uniformity* with respect to acetaminophen, aspirin, and caffeine.

Salicylic acid—

Mobile phase and *Solvent mixture*—Prepare as directed in the *Assay*.

Standard preparation—Dissolve an accurately weighed quantity of USP Salicylic Acid RS in *Solvent mixture* to obtain a solution having a known concentration of about 1 mg per mL. Transfer 2.0 mL of the resulting solution to a 100-mL volumetric flask, dilute with *Solvent mixture* to volume, and mix. This solution contains about 0.02 mg of USP Salicylic Acid RS per mL.

Test preparation—Weigh and finely powder not less than 20 Tablets. Transfer an accurately weighed portion of the powder, equivalent to about 250 mg of aspirin, to a 100-mL volumetric flask. Add about 75 mL of *Solvent mixture*, and shake by mechanical means for 30 minutes. Dilute with *Solvent mixture* to volume, and mix.

Chromatographic system—Proceed as directed for *Chromatographic system* in the *Assay*, except to use a 302-nm detector. Chromatograph the *Standard preparation*, and record the responses as directed under *Procedure:* the tailing factor is not more than 1.6, and the relative standard deviation for replicate injections is not more than 3.0%.

Procedure—Separately inject equal volumes (about 10 μL) of the *Standard preparation* and the *Test preparation* into the chromatograph, record the chromatograms, and measure the responses for the salicylic acid peaks. Calculate the percentage of salicylic acid in the portion of Tablets taken by the formula:

$$100(C/a)(r_U/r_S),$$

in which C is the concentration, in mg per mL, of USP Salicylic Acid RS in the *Standard preparation*, a is the quantity, in mg, of aspirin in the portion of Tablets taken, based on the labeled amount, and r_U and r_S are the salicylic acid peak responses of the *Test preparation* and the *Standard preparation*, respectively: not more than 3.0% is found.

Assay—

Mobile phase—Prepare a suitable mixture of water, methanol, and glacial acetic acid (69:28:3). Make adjustments if necessary (see *System Suitability* under *Chromatography* ⟨621⟩).

Internal standard solution—Prepare a solution of benzoic acid in methanol containing about 6 mg per mL.

Solvent mixture—Prepare a mixture of methanol and glacial acetic acid (95:5).

Standard stock solution—Dissolve accurately weighed quantities of USP Acetaminophen RS, USP Aspirin RS, and USP Caffeine RS in *Solvent mixture* to obtain a solution having known concentrations of about 0.25 mg of USP Acetaminophen RS per mL, $0.25J$ mg of USP Aspirin RS per mL, and $0.25J'$ mg of USP Caffeine RS per mL, J being the ratio of the labeled amount, in mg, of aspirin to the labeled amount, in mg, of acetaminophen per Capsule, and J' being the ratio of the labeled amount, in mg, of caffeine to the labeled amount, in mg, of acetaminophen per Capsule.

Standard preparation—Transfer 20.0 mL of *Standard stock solution* and 3.0 mL of *Internal standard solution* to a 50-mL volumetric flask, dilute with *Solvent mixture* to volume, and mix. This solution contains about 0.1 mg of USP Acetaminophen RS, $0.1J$ mg of USP Aspirin RS, and $0.1J'$ mg of USP Caffeine RS per mL.

Assay preparation—Weigh and finely powder not less than 20 Acetaminophen, Aspirin, and Caffeine Tablets. Transfer an accurately weighed portion of the powder, equivalent to about 250 mg of acetaminophen, to a 100-mL volumetric flask. Add about 75 mL of *Solvent mixture*, and shake by mechanical means for 30 minutes. Dilute with *Solvent mixture* to volume, and mix. Transfer 2.0 mL of this solution and 3.0 mL of *Internal standard solution* to a 50-mL volumetric flask, dilute with *Solvent mixture* to volume, and mix.

Chromatographic system—The liquid chromatograph is equipped with a 275-nm detector and a 4.6-mm × 10-cm column that contains 5-µm packing L1, and is maintained at 45 ± 1°. The flow rate is about 2 mL per minute. Chromatograph the *Standard preparation*, and record the responses as directed under *Procedure:* the tailing factor for each analyte peak is not more than 1.2, the resolution, *R*, between any of the analyte and internal standard peaks is not less than 1.4, and the relative standard deviation for replicate injections is not more than 2.0%.

Procedure—Separately inject equal volumes (about 10 µL) of the *Standard preparation* and the *Assay preparation* into the chromatograph, record the chromatograms, and measure the responses for the major peaks. The relative retention times are about 0.3 for acetaminophen, 0.5 for caffeine, 0.8 for aspirin, 1.0 for benzoic acid, and 1.2 for salicylic acid. Calculate the quantities, in mg, of acetaminophen ($C_8H_9NO_2$), aspirin ($C_9H_8O_4$), and caffeine ($C_8H_{10}N_4O_2$) in the portion of Tablets taken by the formula:

$$2500C(R_U/R_S),$$

in which *C* is the concentration, in mg per mL, of the appropriate USP Reference Standard in the *Standard preparation*, and R_U and R_S are the ratios of the peak responses of the corresponding analyte and internal standard peaks of the *Assay preparation* and the *Standard preparation*, respectively.

Acetaminophen and Codeine Phosphate Capsules

» Acetaminophen and Codeine Phosphate Capsules contain not less than 90.0 percent and not more than 110.0 percent of the labeled amounts of acetaminophen ($C_8H_9NO_2$) and codeine phosphate ($C_{18}H_{21}NO_3 . H_3PO_4 . \frac{1}{2}H_2O$).

Packaging and storage—Preserve in tight, light-resistant containers.

Reference standards—*USP Acetaminophen Reference Standard*—Dry over silica gel for 18 hours before using. *USP Codeine Phosphate Reference Standard*—This is the hemihydrate form of codeine phosphate. Dry at 105° for 18 hours before using.

Identification—

A: The retention times of the major peaks in the chromatogram of the *Assay preparation* correspond to those of the *Standard preparation* obtained as directed in the *Assay*.

B: Transfer a portion of Capsule contents, equivalent to about 12 mg of codeine phosphate, to a separator, add 5 mL of water, 1 mL of ammonium hydroxide, and 5 mL of methylene chloride, shake for 1 minute, and allow the layers to separate. Use the clear, lower layer as the test solution. Prepare a Standard solution of USP Acetaminophen RS and USP Codeine Phosphate RS in methanol containing 12 mg of each per mL. On a suitable thin-layer chromatographic plate (see *Chromatography* ⟨621⟩), coated with a 0.25-mm layer of chromatographic silica gel mixture, apply 10 µL of each solution. Allow the spots to dry, and develop the chromatogram in a solvent system consisting of a mixture of methanol and ammonium hydroxide (49 : 1) until the solvent front has moved about three-fourths of the length of the plate. Remove the plate from the developing chamber, mark the solvent front, and allow the solvent to evaporate. Locate the spots on the plate by examination under short-wavelength ultraviolet light: the R_f values of the two principal spots obtained from the test solution correspond to those obtained from the Standard solution.

Dissolution ⟨711⟩—

Medium: 0.1 *N* hydrochloric acid; 900 mL.

Apparatus 2: 50 rpm.

Time: 30 minutes.

Procedure—Determine the amounts of acetaminophen ($C_8H_9NO_2$) and codeine phosphate hemihydrate ($C_{18}H_{21}NO_3 . H_3PO_4 . \frac{1}{2}H_2O$) dissolved, employing the procedure set forth in the *Assay*, except to use 0.1 *N* hydrochloric acid to prepare the *Codeine phosphate standard stock solution*, and to make any other necessary volumetric adjustments.

Tolerances—Not less than 75% (*Q*) of the labeled amounts of $C_8H_9NO_2$ and $C_{18}H_{21}NO_3 . H_3PO_4 . \frac{1}{2}H_2O$ are dissolved in 30 minutes.

Uniformity of dosage units ⟨905⟩: meet the requirements.

Assay—

Mobile phase—Dissolve 4.44 g of docusate sodium in 1000 mL of a mixture of methanol, water, tetrahydrofuran, and phosphoric acid (600:360:40:1) with stirring, and filter through a membrane filter of 0.45-µm or finer porosity. Make adjustments if necessary (see *System Suitability* under *Chromatography* ⟨621⟩).

Solvent mixture—Mix methanol and water (60:40).

Codeine phosphate standard stock solution—Dissolve an accurately weighed quantity of USP Codeine Phosphate RS in water to obtain a solution having a known concentration of about 0.33 mg per mL.

Standard preparation—To a 100-mL volumetric flask transfer about 66 mg of USP Acetaminophen RS and $200J$ mL of *Codeine phosphate standard stock solution*, *J* being the ratio of the labeled amount, in mg, of codeine phosphate hemihydrate to that of acetaminophen, dilute with *Solvent mixture* to volume, and mix. This solution contains about 0.66 mg of acetaminophen and $0.66J$ mg of codeine phosphate hemihydrate per mL. Prepare this solution fresh daily.

Assay preparation—Remove, as completely as possible, the contents of not less than 20 Acetaminophen and Codeine Phosphate Capsules, weigh, and mix. Transfer an accurately weighed portion of the combined contents, equivalent to about 550 mg of acetaminophen, to a 250-mL volumetric flask. Add about 150 mL of *Solvent mixture*, and shake by mechanical means for 10 minutes. Dilute with *Solvent mixture* to volume, and mix. Transfer 15.0 mL of the resulting solution to a 50-mL volumetric flask, dilute with *Solvent mixture* to volume, and mix. Filter a portion of this solution through a membrane filter of 0.45-µm or finer porosity, discarding the first few mL of the filtrate, and use the filtrate as the *Assay preparation*.

Chromatographic system (see *Chromatography* ⟨621⟩)—The liquid chromatograph is equipped with a 280-nm detector and a 3.9-mm × 30-cm column that contains packing L1. The flow rate is about 1.5 mL per minute. Chromatograph the appropriate *Standard preparation*, and record the peak responses as directed under *Procedure:* the resolution, *R*, between the acetaminophen and codeine phosphate peaks is not less than 3.0, and the relative standard deviation of the peak responses is not more than 2.0% and 3.0%, respectively.

Procedure—Separately inject equal volumes (about 15 µL) of the *Standard preparation* and the *Assay preparation* into the chromatograph, record the chromatograms, and measure the responses for the major peaks. The relative retention times are about 0.3 for acetaminophen, and 1.0 for codeine phosphate. Calculate the quantity, in mg, of acetaminophen ($C_8H_9NO_2$) in the portion of Capsules taken by the formula:

$$(2500C/3)(r_U/r_S),$$

in which *C* is the concentration, in mg per mL, of USP Acetaminophen RS in the *Standard preparation*, and r_U and r_S are the peak responses of acetaminophen obtained from the *Assay preparation* and the *Standard preparation*, respectively. Calculate the quantity, in mg, of codeine phosphate hemihydrate ($C_{18}H_{21}NO_3 . H_3PO_4 . \frac{1}{2}H_2O$) in the portion of Capsules taken by the formula:

$$(406.37/397.36)2500C/3(r_U/r_S),$$

in which 406.37 and 397.36 are the molecular weights of codeine phosphate hemihydrate and anhydrous codeine phosphate, respectively, C is the concentration, in mg per mL, of USP Codeine Phosphate RS in the *Standard preparation*, and r_U and r_S are the peak responses of codeine phosphate obtained from the *Assay preparation* and the *Standard preparation*, respectively.

Acetaminophen and Codeine Phosphate Elixir

» Acetaminophen and Codeine Phosphate Elixir contains not less than 90.0 percent and not more than 110.0 percent of the labeled amounts of acetaminophen ($C_8H_9NO_2$) and codeine phosphate ($C_{18}H_{21}NO_3 . H_3PO_4 . \frac{1}{2}H_2O$).

Packaging and storage—Preserve in tight, light-resistant containers.

Reference standards—*USP Acetaminophen Reference Standard*—Dry over silica gel for 18 hours before using. *USP Codeine Phosphate Reference Standard*—This is the hemihydrate form of codeine phosphate. Dry at 105° for 18 hours before using.

Identification—

A: The retention times of the major peaks in the chromatograms of the *Assay preparations* correspond to those of the *Standard preparations* obtained as directed in the *Assay for acetaminophen* and the *Assay for codeine phosphate*, respectively.

B: Transfer a volume of Elixir, equivalent to about 12 mg of codeine phosphate, to a separator, add 1 mL of ammonium hydroxide and 5 mL of methylene chloride, shake for 1 minute, and allow the layers to separate. Use the clear, lower layer as the test solution. Prepare a Standard solution of USP Acetaminophen RS and USP Codeine Phosphate RS in methanol containing 12 mg of each per mL. On a suitable thin-layer chromatographic plate (see *Chromatography* ⟨621⟩), coated with a 0.25-mm layer of chromatographic silica gel mixture, apply 10 µL of each solution. Allow the spots to dry, and develop the chromatogram in a solvent system consisting of a mixture of methanol and ammonium hydroxide (49:1) until the solvent front has moved about three-fourths of the length of the plate. Remove the plate from the developing chamber, mark the solvent front, and allow the solvent to evaporate. Locate the spots on the plate by examination under short-wavelength ultraviolet light: the R_f values of the two principal spots obtained from the test solution correspond to those obtained from the Standard solution.

pH ⟨791⟩: between 4.0 and 6.1.

Alcohol content ⟨611⟩: between 6.0% and 8.5% of C_2H_5OH, determined by the gas-liquid chromatographic procedure, acetone being used as the internal standard.

Assay for acetaminophen—

Mobile phase—Prepare a suitable mixture of water and methanol (7:3), and degas. Make adjustments if necessary (see *System Suitability* under *Chromatography* ⟨621⟩).

Standard preparation—Dissolve an accurately weighed quantity of USP Acetaminophen RS in *Mobile phase* to obtain a solution having a known concentration of about 0.48 mg per mL.

Assay preparation—Transfer an accurately measured volume of Acetaminophen and Codeine Phosphate Elixir, equivalent to about 120 mg of acetaminophen, to a 250-mL volumetric flask, dilute with *Mobile phase* to volume, and mix.

Chromatographic system (see *Chromatography* ⟨621⟩)—The liquid chromatograph is equipped with a 280-nm detector and a 3.9-mm × 30-cm column that contains packing L1. The flow rate is about 2.0 mL per minute. Chromatograph the *Standard preparation*, and record the peak responses as directed under *Procedure:* the column efficiency determined from the analyte peak is not less than 1000 theoretical plates, the tailing factor for the analyte peak is not more than 1.5, and the relative standard deviation for replicate injections is not more than 2.0%.

Procedure—Separately inject equal volumes (about 10 µL) of the *Standard preparation* and the *Assay preparation* into the chromatograph, record the chromatograms, and measure the responses for the acetaminophen peaks. Calculate the quantity, in mg, of acetaminophen ($C_8H_9NO_2$) in each mL of the Elixir taken by the formula:

$$250(C/V)(r_U/r_S),$$

in which C is the concentration, in mg per mL, of USP Acetaminophen RS in the *Standard preparation*, V is the volume, in mL, of Elixir taken, and r_U and r_S are the peak responses of acetaminophen obtained from the *Assay preparation* and the *Standard preparation*, respectively.

Assay for codeine phosphate—

Mobile phase—Dissolve 4.44 g of docusate sodium in 1000 mL of a mixture of methanol, water, tetrahydrofuran, and phosphoric acid (600:360:40:1) with stirring, and filter through a membrane filter of 0.45-µm or finer porosity. Make adjustments if necessary (see *System Suitability* under *Chromatography* ⟨621⟩).

Solvent mixture—Mix water and methanol (7:3).

Standard preparation—Dissolve an accurately weighed quantity of USP Codeine Phosphate RS in *Solvent mixture* to obtain a solution having a known concentration of about 0.12 mg per mL.

Assay preparation—Transfer an accurately measured volume of Acetaminophen and Codeine Phosphate Elixir, equivalent to about 12 mg of codeine phosphate hemihydrate, to a 100-mL volumetric flask, dilute with *Solvent mixture* to volume, and mix.

Chromatographic system (see *Chromatography* ⟨621⟩)—The liquid chromatograph is equipped with a 280-nm detector and a 3.9-mm × 30-cm column that contains packing L1. The flow rate is about 1.5 mL per minute. Chromatograph the *Standard preparation*, and record the peak responses as directed under *Procedure:* the column efficiency determined from the analyte peak is not less than 1500 theoretical plates, the tailing factor for the analyte peak is not more than 2.0, and the relative standard deviation for replicate injections is not more than 3.0%.

Procedure—Separately inject equal volumes (about 10 µL) of the *Standard preparation* and the *Assay preparation* into the chromatograph, record the chromatograms, and measure the responses for the codeine peaks. Calculate the quantity, in mg, of codeine phosphate hemihydrate ($C_{18}H_{21}NO_3 . H_3PO_4 . \frac{1}{2}H_2O$) in each mL of the Elixir taken by the formula:

$$(406.37/397.36)(100C/V)(r_U/r_S),$$

in which 406.37 and 397.36 are the molecular weights of codeine phosphate hemihydrate and anhydrous codeine phosphate, respectively, C is the concentration, in mg per mL, of USP Codeine Phosphate RS in the *Standard preparation*, V is the volume, in mL, of Elixir taken, and r_U and r_S are the peak responses of codeine phosphate obtained from the *Assay preparation* and the *Standard preparation*, respectively.

Acetaminophen and Codeine Phosphate Tablets

» Acetaminophen and Codeine Phosphate Tablets contain not less than 90.0 percent and not more than 110.0 percent of the labeled amounts of acetaminophen ($C_8H_9NO_2$) and codeine phosphate ($C_{18}H_{21}NO_3 . H_3PO_4 . \frac{1}{2}H_2O$).

Packaging and storage—Preserve in tight, light-resistant containers.

Reference standards—*USP Acetaminophen Reference Standard*—Dry over silica gel for 18 hours before using. *USP Codeine Phosphate Reference Standard*—This is the hemihydrate form of codeine phosphate. Dry at 105° for 18 hours before using.

Identification—

A: The retention times of the major peaks in the chromatogram of the *Assay preparation* correspond to those of the *Standard preparation*, obtained as directed in the *Assay*.

B: A quantity of finely powdered Tablets, equivalent to about 12 mg of codeine phosphate, responds to *Identification test B* under *Acetaminophen and Codeine Phosphate Capsules*.

Dissolution ⟨711⟩—

Medium: 0.1 *N* hydrochloric acid; 900 mL.

Apparatus 2: 50 rpm.

Time: 45 minutes.

Procedure—Determine the amounts of acetaminophen ($C_8H_9NO_2$) and codeine phosphate hemihydrate ($C_{18}H_{21}NO_3 . H_3PO_4 . \frac{1}{2}H_2O$) dissolved, employing the procedure set forth in the *Assay*, except to use 0.1 *N* hydrochloric acid to prepare the *Codeine phosphate standard stock solution* and to make any other necessary volumetric adjustments.

Tolerances—Not less than 75% (*Q*) of the labeled amounts of $C_8H_9NO_2$ and $C_{18}H_{21}NO_3 . H_3PO_4 . \frac{1}{2}H_2O$ are dissolved in 45 minutes.

Uniformity of dosage units ⟨905⟩: meet the requirements.

Assay—

Mobile phase, Solvent mixture, Codeine phosphate standard stock solution, Standard preparation, and *Chromatographic system*—Prepare as directed in the *Assay* under *Acetaminophen and Codeine Phosphate Capsules*.

Assay preparation—Weigh and finely powder not less than 20 Acetaminophen and Codeine Phosphate Tablets. Transfer an accurately weighed portion of the powder, equivalent to about 550 mg of acetaminophen, to a 250-mL volumetric flask. Add about 150 mL of *Solvent mixture* and shake by mechanical means for 10 minutes. Dilute with *Solvent mixture* to volume, and mix. Transfer 15.0 mL of the resulting solution to a 50-mL volumetric flask, dilute with *Solvent mixture* to volume, and mix. Filter a portion of this solution through a membrane filter of 0.45-µm or finer porosity, discarding the first few mL of the filtrate, and use the filtrate as the *Assay preparation*.

Procedure—Proceed as directed in the *Assay* under *Acetaminophen and Codeine Phosphate Capsules*. Calculate the quantity, in mg, of acetaminophen ($C_8H_9NO_2$) in the portion of Tablets taken by the formula:

$$2500C/3(r_U/r_S),$$

in which *C* is the concentration, in mg per mL, of USP Acetaminophen RS in the *Standard preparation*, and r_U and r_S are the peak responses of acetaminophen obtained from the *Assay preparation* and the *Standard preparation*, respectively. Calculate the quantity, in mg, of codeine phosphate hemihydrate ($C_{18}H_{21}NO_3 . H_3PO_4 . \frac{1}{2}H_2O$), in the portion of Tablets taken by the formula:

$$(406.37/397.36)2500C/3(r_U/r_S),$$

in which 406.37 and 397.36 are the molecular weights of codeine phosphate hemihydrate and anhydrous codeine phosphate, respectively, *C* is the concentration, in mg per mL, of USP Codeine Phosphate RS in the *Standard preparation*, and r_U and r_S are peak responses of codeine phosphate obtained from the *Assay preparation* and the *Standard preparation*, respectively.

Acetaminophen and Diphenhydramine Citrate Tablets

» Acetaminophen and Diphenhydramine Citrate Tablets contain not less than 90.0 percent and not more than 110.0 percent of the labeled amounts of acetaminophen ($C_8H_9NO_2$) and diphenhydramine citrate ($C_{17}H_{21}NO . C_6H_8O_7$).

Packaging and storage—Preserve in tight containers.

Reference standards—*USP Acetaminophen Reference Standard*—Dry over silica gel for 18 hours before using. *USP Diphenhydramine Citrate Reference Standard*—Dry at 105° for 3 hours before using.

Identification—The retention times of the major peaks in the chromatograms of the *Assay preparations*, obtained as directed in the *Assay for acetaminophen* and in the *Assay for diphenhydramine citrate*, relative to the retention times of the respective internal standards, correspond to those of the respective *Standard preparations*.

Dissolution ⟨711⟩—

Medium: water; 900 mL.

Apparatus 2: 50 rpm.

Time: 45 minutes.

Procedure—Determine the amount of acetaminophen ($C_8H_9NO_2$) and of diphenhydramine citrate ($C_{17}H_{21}NO . C_6H_8O_7$) dissolved, employing the procedures set forth in the *Assay for acetaminophen* and the *Assay for diphenhydramine citrate*, respectively, making any necessary volumetric adjustments.

Tolerances—Not less than 75% (*Q*) of the labeled amounts of $C_8H_9NO_2$ and $C_{17}H_{21}NO . C_6H_8O_7$ are dissolved in 45 minutes.

Uniformity of dosage units ⟨905⟩: meet the requirements for *Content Uniformity* with respect to acetaminophen and to diphenhydramine citrate.

Assay for acetaminophen—

Mobile phase—Prepare a suitable degassed and filtered mixture of water and methanol (60:40), making adjustments if necessary (see *System Suitability* under *Chromatography* ⟨621⟩).

Internal standard solution—Prepare a solution of guaifenesin in a mixture of water and methanol (4:1) to obtain a solution containing 8.0 mg per mL.

Standard preparation—Transfer about 50 mg of USP Acetaminophen RS, accurately weighed, to a 100-mL volumetric flask. Dissolve in 2.5 mL of methanol, dilute with water to volume, and mix. Transfer 2.0 mL of this solution to a 50-mL volumetric flask, add 5.0 mL of *Internal standard solution*, dilute with *Mobile phase* to volume, and mix.

Assay preparation—Weigh and finely powder not less than 20 Acetaminophen and Diphenhydramine Citrate Tablets. Transfer an accurately weighed portion of the powder, equivalent to about 500 mg of acetaminophen, to a 100-mL volumetric flask, add 25 mL of methanol, and shake by mechanical means for 10 minutes. Dilute with water to volume, and mix. Transfer 10.0 mL of this solution to a 100-mL volumetric flask, dilute with water to volume, and mix. Transfer 2.0 mL of this solution to a 50-mL volumetric flask, add 5.0 mL of *Internal standard solution*, dilute with *Mobile phase* to volume, and mix.

Chromatographic system (see *Chromatography* ⟨621⟩)—The liquid chromatograph is equipped with a 254-nm detector and a 4.6-mm × 15-cm column that contains 5-µm packing L7. The flow rate is about 1 mL per minute. The column temperature is maintained at about 35 ± 0.5°. Chromatograph the *Standard preparation*, and record the responses as directed under *Procedure:* the column efficiency as determined from the analyte peak is not less than 1000 theoretical plates, the tailing factor for the analyte peak is not more than 1, the resolution, *R*, between the analyte and internal standard peaks is not less than 6.0, and the relative standard deviation of the peak response ratios for replicate injections is not more than 2.5%.

Procedure—Separately inject equal volumes (about 10 µL) of the *Standard preparation* and the *Assay preparation* into the chromatograph, record the chromatograms, and measure the responses for the major peaks. The relative retention times are about 0.5 for acetaminophen and 1.0 for guaifenesin. Calculate the quantity, in mg, of acetaminophen ($C_8H_9NO_2$) in the portion of Tablets taken by the formula:

$$10W_S(R_U/R_S),$$

in which W_S is the weight, in mg, of USP Acetaminophen RS taken, and R_U and R_S are the ratios of the peak response of acetaminophen to that of the internal standard obtained from the *Assay preparation* and the *Standard preparation*, respectively.

Assay for diphenhydramine citrate—

Mobile phase—Prepare a suitable degassed and filtered mixture of methanol, water, and glacial acetic acid (61:38:1) containing 1.0813 g of sodium 1-octanesulfonate in each 1000 mL of solution, making adjustments if necessary (see *System Suitability* under *Chromatography* ⟨621⟩).

Solvent mixture—Prepare a mixture of methanol and water (1:1).

Internal standard solution—Prepare a solution of xylometazoline hydrochloride in water having a concentration of about 8 mg per mL.

Standard preparation—Transfer about 38 mg of USP Diphenhydramine Citrate RS, accurately weighed, to a 100-mL volumetric flask containing 500 mg of acetaminophen. Add 5.0 mL of *Internal standard solution* and about 50 mL of *Solvent mixture*, and mix until solution is complete. Dilute with *Solvent mixture* to volume, and mix.

Assay preparation—Weigh and finely powder not less than 20 Acetaminophen and Diphenhydramine Citrate Tablets. Transfer an accurately weighed portion of the powder, equivalent to about 38 mg of diphenhydramine citrate, to a 100-mL volumetric flask, add about 65 mL of *Solvent mixture*, and shake by mechanical means for about 15 minutes. Add 5.0 mL of *Internal standard solution*, dilute with *Solvent mixture* to volume, and mix.

Chromatographic system (see *Chromatography* ⟨621⟩)—The liquid chromatograph is equipped with a 265-nm detector and a 3.9-mm × 30-cm column that contains packing L1. The flow rate is about 1.5 mL per minute. The column temperature is maintained at about 35 ± 0.5°. Chromatograph the *Standard preparation*, and record the responses as directed under *Procedure:* the column efficiency as determined from the analyte peak is not less than 1000 theoretical plates, the tailing factor for the analyte peak is not more than 1.7, the resolution, *R*, between the analyte and internal standard peaks is not less than 2.5, and the relative standard deviation of the peak response ratios for replicate injections is not more than 2.0%.

Procedure—Separately inject equal volumes (about 50 µL) of the *Standard preparation* and the *Assay preparation* into the chromatograph, record the chromatograms, and measure the peak responses for the diphenhydramine citrate and xylometazoline hydrochloride peaks. The relative retention times are about 0.3 for acetaminophen, 0.7 for diphenhydramine citrate, and 1.0 for xylometazoline hydrochloride, respectively. Calculate the quantity, in mg, of diphenhydramine citrate ($C_{17}H_{21}NO.C_6H_8O_7$) in the portion of Tablets taken by the formula:

$$W_S(R_U/R_S),$$

in which W_S is the weight, in mg, of USP Diphenhydramine Citrate RS taken, and R_U and R_S are the ratios of the peak response of diphenhydramine citrate to that of the internal standard obtained from the *Assay preparation* and the *Standard preparation*, respectively.

Acetazolamide

H_2NO_2S–[1,3,4-thiadiazole ring]–$NHCOCH_3$

$C_4H_6N_4O_3S_2$ 222.24

Acetamide, *N*-[5-(aminosulfonyl)-1,3,4-thiadiazol-2-yl]-.

N-(5-Sulfamoyl-1,3,4-thiadiazol-2-yl)acetamide [*59-66-5*].

» Acetazolamide contains not less than 98.0 percent and not more than 102.0 percent of $C_4H_6N_4O_3S_2$, calculated on the anhydrous basis.

Packaging and storage—Preserve in well-closed containers.

Reference standard—*USP Acetazolamide Reference Standard*—Dry at 105° for 4 hours before using.

Identification—

A: The infrared absorption spectrum of a potassium bromide dispersion of it exhibits maxima only at the same wavelengths as that of a similar preparation of USP Acetazolamide RS. If a difference appears, dissolve portions of both the test specimen and the Reference Standard in methanol, evaporate the solutions to dryness, and repeat the test on the residues.

B: Dissolve about 100 mg in 5 mL of 1 *N* sodium hydroxide. Add 5 mL of a solution made by dissolving 100 mg of hydroxylamine hydrochloride and 80 mg of cupric sulfate in 10 mL of water. Mix, and heat the resulting pale yellow solution on a steam bath for 5 minutes: a clear, bright yellow solution is produced. No heavy precipitate or dark brown color results after the mixing or heating.

Water, *Method I* ⟨921⟩: not more than 0.5%.

Residue on ignition ⟨281⟩: not more than 0.1%.

Chloride ⟨221⟩—Digest 1.5 g with 75 mL of water at about 70° for 5 minutes. Cool to room temperature, and filter: a 25-mL portion of the filtrate shows no more chloride than corresponds to 0.10 mL of 0.020 *N* hydrochloric acid (0.014%).

Sulfate ⟨221⟩—A 25-mL portion of the filtrate prepared in the test for *Chloride* shows no more sulfate than corresponds to 0.20 mL of 0.020 *N* sulfuric acid (0.04%).

Selenium ⟨291⟩: 0.003%, a 200-mg specimen being used.

Heavy metals, *Method II* ⟨231⟩: 0.002%.

Silver-reducing substances—Thoroughly wet 5.0 g with alcohol. Add 125 mL of water, 10 mL of nitric acid, and 5.0 mL of 0.1 *N* silver nitrate VS. Stir with a mechanical stirrer for 30 minutes. Filter, add 5 mL of ferric ammonium sulfate TS to the filtrate, and titrate with 0.1 *N* ammonium thiocyanate VS to a reddish brown end-point: not less than 4.8 mL of 0.1 *N* ammonium thiocyanate is required.

Ordinary impurities ⟨466⟩—

Test solution: a mixture of acetone and methanol (1:1).

Standard solution: a mixture of acetone and methanol (1:1).

Eluant: a mixture of *n*-propyl alcohol and 1 *N* ammonium hydroxide (88:12).

Visualization: 1.

Assay—Dissolve about 200 mg of Acetazolamide, accurately weighed, in a small volume of pyridine in a 10-mL volumetric flask, add the solvent to volume, and mix. Similarly, dissolve an accurately weighed quantity of USP Acetazolamide RS in pyridine to obtain a Standard solution having a known concentration of about 20 mg per mL. Concomitantly determine the absorbances of both solutions in 0.1-mm cells at the wavelength of maximum absorbance at about 7.38 µm, with a suitable infrared spectrophotometer, using pyridine as the blank. Calculate the quantity, in mg, of $C_4H_6N_4O_3S_2$ in the portion of Acetazolamide taken by the formula:

$$10C(A_U/A_S),$$

in which *C* is the concentration, in mg per mL, of USP Acetazolamide RS in the Standard solution, and A_U and A_S are the absorbances of the solution of Acetazolamide and the Standard solution, respectively.

Acetazolamide Tablets

» Acetazolamide Tablets contain not less than 95.0 percent and not more than 105.0 percent of the labeled amount of $C_4H_6N_4O_3S_2$.

Packaging and storage—Preserve in well-closed containers.

Reference standard—*USP Acetazolamide Reference Standard*—Dry at 105° for 4 hours before using.

Identification—Extract a quantity of finely powdered Tablets, equivalent to about 500 mg of acetazolamide, with 50 mL of acetone. Filter, and add sufficient solvent hexane to the filtrate to cause formation of a heavy, white precipitate. Collect the precipitate on a medium-porosity, sintered-glass funnel, and dry with suction: the acetazolamide so obtained responds to the *Identification tests* under *Acetazolamide.*

Dissolution ⟨711⟩—

Medium: 0.1 *N* hydrochloric acid; 900 mL.

Apparatus 1: 100 rpm.

Time: 60 minutes.

Procedure—Determine the amount of $C_4H_6N_4O_3S_2$ dissolved from ultraviolet absorbances at the wavelength of maximum absorbance at about 265 nm of filtered portions of the solution under test, suitably diluted with *Dissolution Medium*, if necessary, in comparison with a Standard solution having a known concentration of USP Acetazolamide RS in the same medium.

Tolerances—Not less than 80% (*Q*) of the labeled amount of $C_4H_6N_4O_3S_2$ is dissolved in 60 minutes.

Uniformity of dosage units ⟨905⟩: meet the requirements.

Assay—

Mobile phase—Dissolve 4.1 g of anydrous sodium acetate in 950 mL of water, add 20 mL of methanol and 30 mL of acetonitrile, and mix. Adjust with glacial acetic acid to a pH of 4.0 ± 0.05. Filter and degas the solution. Make adjustments if necessary (see *System Suitability* under *Chromatography* ⟨621⟩).

Standard acetazolamide stock solution—Transfer about 25 mg of USP Acetazolamide RS, accurately weighed, to a 25-mL volumetric flask, add 2.5 mL of 0.5 *N* sodium hydroxide, and mix to dissolve. Dilute with water to volume, and mix.

Internal standard solution—Transfer about 100 mg of sulfadiazine to a 100-mL volumetric flask, add 10 mL of 0.5 *N* sodium hydroxide, and mix to dissolve. Dilute with water to volume, and mix.

Standard preparation—Transfer 10.0 mL of *Standard acetazolamide stock solution* and 10.0 mL of *Internal standard solution* to a 100-mL volumetric flask, add 10 mL of 0.5 *N* sodium hydroxide, dilute with water to volume, and mix to obtain a solution having a known concentration of about 0.1 mg of USP Acetazolamide RS per mL.

Assay preparation—Weigh and finely powder not less than 20 Acetazolamide Tablets. Transfer an accurately weighed portion of the powder, equivalent to about 100 mg of acetazolamide, to a 100-mL volumetric flask, add 10 mL of 0.5 *N* sodium hydroxide, and sonicate for 5 minutes. Cool to room temperature, dilute with water to volume, and mix. Filter a portion of this solution, discarding the first 20 mL of the filtrate. Transfer 10.0 mL of the clear filtrate to a 100-mL volumetric flask, add 10.0 mL of *Internal standard solution* and 10 mL of 0.5 *N* sodium hydroxide, dilute with water to volume, and mix.

Chromatographic system (see *Chromatography* ⟨621⟩)—The liquid chromatograph is equipped with a 254-nm detector and a 4.6-mm × 25-cm column that contains packing L1. The flow rate is about 2 mL per minute. Chromatograph the *Standard preparation*, and record the peak responses as directed under *Procedure:* the resolution, *R*, between the analyte and internal standard peaks is not less than 2.0, and the relative standard deviation of the ratios of the analyte peak response to the internal standard peak response for replicate injections is not more than 1.0%.

Procedure—Separately inject equal volumes (about 20 µL) of the *Standard preparation* and the *Assay preparation* into the chromatograph, record the chromatograms, and measure the responses for the major peaks. The relative retention times are about 0.7 for acetazolamide and 1.0 for sulfadiazine. Calculate the quantity, in mg, of $C_4H_6N_4O_3S_2$ in the portion of Tablets taken by the formula:

$$1000C(R_U/R_S),$$

in which *C* is the concentration, in mg per mL, of USP Acetazolamide RS in the *Standard preparation*, and R_U and R_S are the peak response ratios of the analyte peak to the internal standard peak obtained from the *Assay preparation* and the *Standard preparation*, respectively.

Sterile Acetazolamide Sodium

$C_4H_5N_4NaO_3S_2$ 244.22

Acetamide, *N*-[5-(aminosulfonyl)-1,3,4-thiadiazol-2-yl]-, monosodium salt.

N-(5-Sulfamoyl-1,3,4-thiadiazol-2-yl)acetamide monosodium salt [*1424-27-7*].

» Sterile Acetazolamide Sodium is prepared from Acetazolamide with the aid of Sodium Hydroxide. It is suitable for parenteral use. The contents of each container, when constituted as directed in the labeling, yields a solution containing not less than 95.0 percent and not more than 110.0 percent of the labeled amount of acetazolamide ($C_4H_6N_4O_3S_2$).

Packaging and storage—Preserve in *Containers for Sterile Solids* as described under *Injections* ⟨1⟩, preferably of Type III glass.

Reference standard—*USP Acetazolamide Reference Standard*—Dry at 105° for 4 hours before using.

Completeness of solution ⟨641⟩—A 1.0-g portion dissolves in 10 mL of carbon dioxide–free water to yield a clear solution.

Constituted solution—At the time of use, the constituted solution prepared from Sterile Acetazolamide Sodium meets the requirements for *Constituted Solutions* under *Injections* ⟨1⟩.

Identification—

A: Dissolve about 500 mg in 5 mL of water, add 2 drops of hydrochloric acid, and allow the mixture to stand for about 15 minutes. Filter through a fine sintered-glass funnel, wash with several small portions of water, and dry in vacuum over silica gel for 3 hours: the crystals so obtained respond to the *Identification tests* under *Acetazolamide*.

B: It responds to the tests for *Sodium* ⟨191⟩.

pH ⟨791⟩: between 9.0 and 10.0, in a freshly prepared solution (1 in 10).

Other requirements—It meets the requirements for *Sterility Tests* ⟨71⟩, *Uniformity of Dosage Units* ⟨905⟩, and *Labeling* under *Injections* ⟨1⟩.

Assay—Dissolve the contents of 1 container of Sterile Acetazolamide Sodium in an accurately measured volume of water corresponding to the volume of solvent specified in the labeling. Dilute a portion of the solution quantitatively and stepwise with water to obtain a solution having a concentration of about 500 µg of acetazolamide per mL. Pipet 5 mL of the solution into a 250-mL volumetric flask, add 25 mL of 1 *N* hydrochloric acid, then add water to volume, and mix. Dissolve an accurately weighed quantity of USP Acetazolamide RS in sodium hydroxide solution (1 in 100) to obtain a Standard solution having a known concentration of about 100 µg per mL. Dilute 10.0 mL of this solution with 0.1 *N* hydrochloric acid to 100 mL. Concomitantly determine the absorbances of both solutions at the wavelength of maximum absorbance at about 265 nm, with a suitable spectrophotometer, using 0.1 *N* hydrochloric acid as the blank. Calculate the quantity, in µg, of $C_4H_6N_4O_3S_2$ in the 5.0-mL portion of the solution of Sterile Acetazolamide Sodium taken by the formula:

$$25C(A_U/A_S),$$

in which *C* is the concentration, in µg per mL, of USP Acetazolamide RS in the Standard solution, and A_U and A_S are the absorbances of the solution from the Sterile Acetazolamide Sodium and the Standard solution, respectively.

Acetic Acid—*see* Acetic Acid NF

Acetic Acid Otic Solution, Hydrocortisone and—*see* Hydrocortisone and Acetic Acid Otic Solution

Glacial Acetic Acid

CH_3COOH

$C_2H_4O_2$ 60.05

Acetic acid.

Acetic acid [*64-19-7*].

» Glacial Acetic Acid contains not less than 99.5 percent and not more than 100.5 percent, by weight, of $C_2H_4O_2$.

Packaging and storage—Preserve in tight containers.

Identification—A mixture of 1 volume of it with 2 volumes of water responds to the tests for *Acetate* ⟨191⟩.

Congealing temperature ⟨651⟩: not lower than 15.6°.

Nonvolatile residue—Evaporate 20 mL in a tared dish, and dry at 105° for 1 hour: the weight of the residue does not exceed 1.0 mg.

Chloride—Dilute 1.0 mL with 20 mL of water, and add 5 drops of silver nitrate TS: no opalescence is produced.

Sulfate—Dilute 1.0 mL with 10 mL of water, and add 1 mL of barium chloride TS: no turbidity is produced.

Heavy metals ⟨231⟩—To the residue obtained in the test for *Nonvolatile residue* add 8 mL of 0.1 *N* hydrochloric acid, warm gently until solution is complete, dilute with water to 100 mL, and use 20 mL of the solution: the limit is 5 ppm.

Readily oxidizable substances—Dilute 2.0 mL in a glass-stoppered vessel with 10 mL of water, and add 0.10 mL of 0.10 *N* potassium permanganate: the pink color is not changed to brown within 2 hours.

Assay—Measure about 2 mL of Glacial Acetic Acid into a glass-stoppered flask, previously tared while containing about 20 mL of water, and weigh again to obtain the weight of the substance under assay. Add 20 mL of water, then add phenolphthalein TS, and titrate with 1 *N* sodium hydroxide VS. Each mL of 1 *N* sodium hydroxide is equivalent to 60.05 mg of $C_2H_4O_2$.

Acetic Acid Irrigation

» Acetic Acid Irrigation is a sterile solution of Glacial Acetic Acid in Water for Injection. It contains, in each 100 mL, not less than 237.5 mg and not more than 262.5 mg of $C_2H_4O_2$.

Packaging and storage—Preserve in single-dose containers, preferably of Type I or Type II glass. It may be packaged in suitable plastic containers.

Identification—Evaporate 100 mL to about 10 mL: the resulting solution responds to the tests for *Acetate* ⟨191⟩.

Pyrogen—After neutralization with sodium hydroxide TS to a pH between 6.0 and 7.0, it meets the requirements of the *Pyrogen Test* ⟨151⟩.

pH ⟨791⟩: between 2.8 and 3.4.

Other requirements—It meets the requirements under *Injections* ⟨1⟩, except that the container in which it is packaged may be designed to empty rapidly and may exceed 1000 mL in capacity.

Assay—Pipet 50 mL of Acetic Acid Irrigation into a 150-mL conical flask, add 2 drops of phenolphthalein TS, and titrate with 0.1 *N* sodium hydroxide VS. Each mL of 0.1 *N* sodium hydroxide is equivalent to 6.005 mg of $C_2H_4O_2$.

Acetic Acid Otic Solution

» Acetic Acid Otic Solution is a solution of Glacial Acetic Acid in a suitable nonaqueous solvent. It contains not less than 85.0 percent and not more than 130.0 percent of the labeled amount of $C_2H_4O_2$.

Packaging and storage—Preserve in tight containers.

Identification—

A: Dilute 5 mL with 10 mL of water, and adjust with 1 *N* sodium hydroxide to a pH of about 7. Add ferric chloride TS: a deep red color is produced, and it is destroyed by the addition of hydrochloric acid.

B: Warm it with sulfuric acid and alcohol: ethyl acetate, recognizable by its characteristic odor, is evolved.

pH ⟨791⟩: between 2.0 and 4.0, when diluted with an equal volume of water.

Assay—Transfer an accurately known quantity of Acetic Acid Otic Solution, containing about 100 mg of glacial acetic acid, to a 250-mL conical flask, and add 5 mL of saturated sodium chloride solution and about 40 mL of water. Add 3 drops of phenolphthalein TS, and titrate with 0.1 *N* sodium hydroxide VS to a faint pink end-point. Each mL of 0.1 *N* sodium hydroxide is equivalent to 6.005 mg of $C_2H_4O_2$.

Acetohexamide

CH_3CO–(benzene ring)–$SO_2NHCONH$–(cyclohexane ring)

$C_{15}H_{20}N_2O_4S$ 324.39

Benzenesulfonamide, 4-acetyl-*N*-[[cyclohexylamino]carbonyl]-.

1-[(*p*-Acetylphenyl)sulfonyl]-3-cyclohexylurea [*968-81-0*].

» Acetohexamide contains not less than 97.0 percent and not more than 101.0 percent of $C_{15}H_{20}N_2O_4S$, calculated on the dried basis.

Packaging and storage—Preserve in well-closed containers.

Reference standard—*USP Acetohexamide Reference Standard*—Dry at 105° for 3 hours before using.

Identification—

A: The infrared absorption spectrum of a potassium bromide dispersion of it exhibits maxima only at the same wavelengths as that of a similar preparation of USP Acetohexamide RS.

B: The ultraviolet absorption spectrum of a 1 in 100,000 solution of it in 0.01 *N* sodium hydroxide exhibits maxima and minima at the same wavelengths as that of a similar solution of USP Acetohexamide RS, and the respective absorptivities, calculated on the dried basis, at the wavelength of maximum absorbance at about 247 nm do not differ by more than 3.0%.

Melting range ⟨741⟩: between 182.5° and 187°.

Loss on drying ⟨731⟩—Dry it at 105° for 3 hours: it loses not more than 1.0% of its weight.

Selenium ⟨291⟩: 0.003%, a 200-mg specimen mixed with 200 mg of magnesium oxide being used.

Residue on ignition ⟨281⟩: not more than 0.1%.

Heavy metals, *Method II* ⟨231⟩: 0.002%.

Assay—Dissolve about 300 mg of Acetohexamide, accurately weighed, in 40 mL of dimethylformamide, add 5 drops of thymol blue TS, and titrate, using a magnetic stirrer, with 0.1 *N* sodium methoxide VS to a blue end-point. Perform a blank determination, and make any necessary correction. Each mL of 0.1 *N* sodium methoxide is equivalent to 32.44 mg of $C_{15}H_{20}N_2O_4S$.

Acetohexamide Tablets

» Acetohexamide Tablets contain not less than 93.0 percent and not more than 107.0 percent of the labeled amount of $C_{15}H_{20}N_2O_4S$.

Packaging and storage—Preserve in well-closed containers.

Reference standard—*USP Acetohexamide Reference Standard*—Dry at 105° for 3 hours before using.

Identification—Evaporate on a steam bath to dryness a 20-mL portion of the diluted chloroform solution prepared as directed in the *Assay:* the residue meets the requirements of *Identification test A* under *Acetohexamide.*

Dissolution ⟨711⟩—

Medium: pH 7.6 phosphate buffer (see *pH* ⟨791⟩); 900 mL.

Apparatus 1: 100 rpm.

Time: 60 minutes.

Procedure—Determine the amount of $C_{15}H_{20}N_2O_4S$ dissolved from ultraviolet absorbances at the wavelength of maximum absorbance at about 245 nm of filtered portions of the solution under test, suitably diluted with *Dissolution Medium,* if necessary, using *Dissolution Medium* as the blank, in comparison with a Standard solution having a known concentration of USP Acetohexamide RS in the same medium.

Tolerances—Not less than 50% (*Q*) of the labeled amount of $C_{15}H_{20}N_2O_4S$ is dissolved in 60 minutes.

Uniformity of dosage units ⟨905⟩: meet the requirements.

Assay—Weigh and finely powder not less than 20 Acetohexamide Tablets. Transfer an accurately weighed portion of the powder, equivalent to about 500 mg of acetohexamide, to a 100-mL volumetric flask, add 60 mL of 0.1 *N* sodium hydroxide, and shake for 30 minutes. Dilute with water to volume, mix, and filter, discarding the first 20 mL of the filtrate. Transfer 20.0 mL of the subsequent filtrate to a 125-mL separator, add 2 mL of 3 *N* hydrochloric acid, and extract with four 40-mL portions of chloroform, filtering each portion through chloroform-washed paper into a 200-mL volumetric flask. Dilute with chloroform to volume, and mix. Transfer 20.0 mL of this solution to a suitable beaker, and evaporate on a steam bath to dryness. Transfer the residue, with the aid of 0.1 *N* sodium hydroxide, to a 100-mL volumetric flask, add 0.1 *N* sodium hydroxide to volume, and mix. Transfer 10.0 mL of this solution to a third 100-mL volumetric flask, dilute with water to volume, and mix. Concomitantly determine the absorbances of the solution from the Tablets and a Standard solution prepared from USP Acetohexamide RS, in the same medium, at a concentration of about 10 µg per mL, in 1-cm cells, at the wavelength of maximum absorbance at about 247 nm, with a suitable spectrophotometer, using 0.01 *N* sodium hydroxide as the blank. Calculate the quantity, in mg, of $C_{15}H_{20}N_2O_4S$ in the portion of Tablets taken by the formula:

$$50C(A_U/A_S),$$

in which C is the concentration, in µg per mL, of USP Acetohexamide RS in the Standard solution, and A_U and A_S are the absorbances of the solution from the Tablets and the Standard solution, respectively.

Acetohydroxamic Acid

$$CH_3-C(=O)-N(OH)H$$

$C_2H_5NO_2$ 75.07
N-Acetyl hydroxyacetamide.
Acetohydroxamic acid [*546-88-3*].

» Acetohydroxamic Acid, dried over phosphorus pentoxide for 16 hours, contains not less than 98.0 percent and not more than 101.0 percent of $C_2H_5NO_2$.

Packaging and storage—Preserve in tight containers, and store in a cool, dry place.

Reference standard—*USP Acetohydroxamic Acid Reference Standard*—Dry over phosphorus pentoxide for 16 hours before using.

Completeness of solution ⟨641⟩—A 1.0-g portion dissolves in 10 mL of water to yield a clear solution.

Color of solution—Dissolve 1.0 g in 5 mL of water: the absorbance, determined in a 1-cm cell in the wavelength range between 400 nm and 750 nm in a suitable spectrophotometer, water being used as the blank, is not greater than 0.050.

Identification—

A: The infrared absorption spectrum of a potassium bromide dispersion of it, previously dried, exhibits maxima only at the same wavelengths as that of a similar preparation of USP Acetohydroxamic Acid RS.

B: To 10 mL of a solution (1 in 50) add 2 drops of potassium permanganate TS: the pink color of the permanganate disappears.

Loss on drying ⟨731⟩—Dry it over phosphorus pentoxide for 16 hours: it loses not more than 1.0% of its weight.

Residue on ignition ⟨281⟩: not more than 0.1%.

Heavy metals, *Method I* ⟨231⟩—Dissolve 1 g in 23 mL of water, and add 2 mL of 1 *N* acetic acid: the limit is 0.002%.

Hydroxylamine—

pH 7.4 buffer—Dissolve 1.36 g of monobasic potassium phosphate in about 950 mL of water, adjust by the dropwise addition of 1 *N* potassium hydroxide to a pH of 7.4, dilute with water to 1000 mL, and mix.

Pyridoxal 5-phosphate solution—Dissolve 20 mg of pyridoxal 5-phosphate monohydrate in 50 mL of *pH 7.4 buffer* in a low-actinic flask. Prepare fresh before use.

Standard preparation—Dissolve an accurately weighed quantity of hydroxylamine hydrochloride in 0.035 *N* potassium hydroxide, and dilute quantitatively and stepwise with the same solvent to obtain a solution having a known concentration of about 20 µg per mL.

Test preparation—Transfer about 1 g of Acetohydroxamic Acid, previously dried and accurately weighed, to a 100-mL volumetric flask, add 4.0 mL of 0.035 *N* potassium hydroxide, and swirl to dissolve.

Procedure—Transfer 2.0 mL of the *Standard preparation* to a 100-mL volumetric flask, add 2.0 mL of water, and mix. To this flask, and to the flask containing the *Test preparation*, add 4.0 mL of *Pyridoxal 5-phosphate solution*, and mix. After 8 minutes, accurately timed, dilute the contents of both flasks with *pH 7.4 buffer* to volume, and mix. Without delay, concomitantly determine the fluorescence intensities of the solutions from the *Standard preparation* and the *Test preparation* in a fluorometer at an excitation wavelength of 350 nm and an emission wavelength of 450 nm, setting the instrument to zero with a blank solution prepared by transferring 4.0 mL of water and 4.0 mL of *Pyridoxal 5-phosphate solution* to a 100-mL volumetric flask, diluting with *pH 7.4 buffer* to volume, and mixing. Calculate the percentage of hydroxylamine by the formula:

$$(33.03/69.50)(0.0002C/W)(I_U/I_S),$$

in which 33.03 and 69.50 are the molecular weights of hydroxylamine and hydroxylamine hydrochloride, respectively, C is the concentration, in µg per mL, of hydroxylamine hydrochloride in the *Standard preparation*, W is the weight, in g, of Acetohydroxamic Acid taken, and I_U and I_S are the fluorescence values of the solutions from the *Test preparation* and the *Standard preparation*, respectively. [NOTE—If the fluorescence intensity of the solution from the *Test preparation* is off-scale, repeat the test using a smaller specimen.] The limit is 0.5%.

Assay—

Ferric chloride solution—Dissolve 4 g of ferric chloride in about 50 mL of 0.1 *N* hydrochloric acid, dilute with the same solvent to make 200 mL, and mix.

Standard preparation—Dissolve an accurately weighed quantity of USP Acetohydroxamic Acid RS in 0.1 *N* hydrochloric acid to obtain a solution having a known concentration of about 500 µg per mL.

Assay preparation—Transfer about 250 mg of Acetohydroxamic Acid, previously dried and accurately weighed, to a 500-mL volumetric flask, dilute with 0.1 *N* hydrochloric acid to volume, and mix.

Procedure—Transfer 10.0 mL each of the *Standard preparation*, the *Assay preparation*, and 0.1 *N* hydrochloric acid to provide the blank, to separate 100-mL volumetric flasks. To each flask add about 50 mL of 0.1 *N* hydrochloric acid and 10.0 mL of *Ferric chloride solution*, mix, dilute with 0.1 *N* hydrochloric acid to volume, and mix. Without delay, concomitantly determine the absorbances of the solutions from the *Standard preparation* and the *Assay preparation* at the wavelength of maximum absorbance at about 502 nm, with a suitable spectrophotometer, using the blank to set the instrument. Calculate the quantity, in mg, of $C_2H_5NO_2$ in the portion of Acetohydroxamic Acid taken by the formula:

$$0.5C(A_U/A_S),$$

in which C is the concentration, in µg per mL, of USP Acetohydroxamic Acid RS in the *Standard preparation*, and A_U and A_S are the absorbances of the *Assay preparation* and the *Standard preparation*, respectively.

Acetohydroxamic Acid Tablets

» Acetohydroxamic Acid Tablets contain not less than 90.0 percent and not more than 110.0 percent of the labeled amount of $C_2H_5NO_2$.

Packaging and storage—Preserve in tight containers.

Reference standard—*USP Acetohydroxamic Acid Reference Standard*—Dry over phosphorus pentoxide for 16 hours before using.

Identification—Tablets produce a purple color when mixed with an acidic solution of ferric chloride.

Dissolution ⟨711⟩—

Medium: water; 900 mL.

Apparatus 1: 100 rpm.

Time: 30 minutes.

Procedure—Determine the amount of $C_2H_5NO_2$ dissolved, employing the procedure set forth in the *Assay*, making any necessary volumetric adjustments.

Tolerances—Not less than 85% (Q) of the labeled amount of $C_2H_5NO_2$ is dissolved in 30 minutes.

Uniformity of dosage units ⟨905⟩: meet the requirements.

Hydroxylamine—

pH 7.4 buffer, Pyridoxal 5-phosphate solution, and *Standard preparation*—Prepare as directed in the test for *Hydroxylamine* under *Acetohydroxamic Acid.*

Test preparation—Weigh and finely powder not less than 20 Tablets. Transfer an accurately weighed portion of the powder, equivalent to about 250 mg of acetohydroxamic acid, to a glass-stoppered, 50-mL centrifuge tube, add 25.0 mL of water, shake for about 2 minutes, and centrifuge. Use the clear solution so obtained as the *Test preparation.*

Procedure—Proceed as directed for *Procedure* in the test for *Hydroxylamine* under *Acetohydroxamic Acid.* Calculate the percentage of hydroxylamine in the Tablets by the formula:

$$(33.03/69.50)(0.05)(C/a)(I_U/I_S),$$

in which a is the quantity, in mg, of acetohydroxamic acid in the portion of Tablets taken, based on the labeled amount, and the other terms are as defined therein. The limit is 0.5%.

Assay—

Ferric chloride solution and *Standard preparation*—Prepare as directed in the *Assay* under *Acetohydroxamic Acid.*

Assay preparation—Weigh and finely powder not less than 20 Acetohydroxamic Acid Tablets. Transfer an accurately weighed portion of the powder, equivalent to about 500 mg of acetohydroxamic acid, to a 1000-mL volumetric flask, add about 500 mL of 0.1 N hydrochloric acid, and shake for 1 minute. Dilute with 0.1 N hydrochloric acid to volume, and mix. Filter, discarding the first 40 mL of the filtrate. Use the clear filtrate as the *Assay preparation.*

Procedure—Proceed as directed for *Procedure* in the *Assay* under *Acetohydroxamic Acid.* Calculate the quantity, in mg, of $C_2H_5NO_2$ in the portion of Tablets taken by the formula:

$$C(A_U/A_S).$$

Acetone—*see* Acetone NF

Acetophenazine Maleate

$CH_2CH_2CH_2$—N(piperazine)N—CH_2CH_2OH; phenothiazine with $COCH_3$ · 2 HC—COOH / HC—COOH

$C_{23}H_{29}N_3O_2S.2C_4H_4O_4$ 643.71

Ethanone, 1-[10-[3-[4-(2-hydroxyethyl)-1-piperazinyl]propyl]-10*H*-phenothiazin-2-yl]-, (*Z*) 2-butenedioate (1:2) (salt).

10-[3-[4-(2-Hydroxyethyl)-1-piperazinyl]propyl]phenothiazin-2-yl methyl ketone maleate (1:2) (salt) [*5714-00-1*].

» Acetophenazine Maleate, dried at 65° for 4 hours, contains not less than 97.0 percent and not more than 103.0 percent of $C_{23}H_{29}N_3O_2S.2C_4H_4O_4$.

Packaging and storage—Preserve in tight, light-resistant containers.

Reference standard—*USP Acetophenazine Maleate Reference Standard*—Dry at 65° for 4 hours before using.

NOTE—Throughout the following procedures, protect test or assay specimens, the Reference Standard, and solutions containing them, by conducting the procedures without delay, under subdued light, or using low-actinic glassware.

Identification—

A: The infrared absorption spectrum of a mineral oil dispersion of it, previously dried, exhibits maxima only at the same wavelengths as that of a similar preparation of USP Acetophenazine Maleate RS.

B: The ultraviolet absorption spectrum of a 1 in 100,000 solution in methanol exhibits maxima and minima at the same wavelengths as that of a similar solution of USP Acetophenazine Maleate RS, concomitantly measured, and the respective absorptivities, calculated on the dried basis, at the wavelength of maximum absorbance at about 243 nm do not differ by more than 3.0%.

C: Prepare a solution in methanol containing 1 mg per mL. On a suitable thin-layer chromatographic plate (see *Chromatography* ⟨621⟩), coated with a 0.25-mm layer of chromatographic silica gel mixture, apply 10 µL of this solution and 10 µL of a solution of USP Acetophenazine Maleate RS in methanol containing 1 mg per mL. Allow the spots to dry, and develop the chromatogram in a solvent system consisting of a mixture of acetone and ammonium hydroxide (95:5) until the solvent front has moved about three-fourths of the length of the plate. Remove the plate from the developing chamber, mark the solvent front, and allow the solvent to evaporate. Locate the spots using long-wavelength ultraviolet light: the R_f value of the principal spot obtained from the test solution corresponds to that obtained from the Standard solution.

Loss on drying ⟨731⟩—Dry it at 65° for 4 hours: it loses not more than 0.5% of its weight.

Residue on ignition ⟨281⟩: not more than 0.1%.

Ordinary impurities ⟨466⟩—

Test solution: methanol.

Standard solution: methanol.

Application volume: 40 µL.

Eluant: a mixture of toluene, chloroform, methanol, and ammonium hydroxide (40:10:10:1).

Visualization: 1.

Assay—Dissolve about 500 mg of Acetophenazine Maleate, previously dried and accurately weighed, in 50 mL of glacial acetic acid, warming slightly to effect solution. Cool to room temperature, add 10 mL of acetic anhydride, and allow to stand for 5 minutes. Add 1 drop of crystal violet TS, and titrate with 0.1 N perchloric acid VS to a green-yellow end-point. Perform a blank determination, and make any necessary correction. Each mL of 0.1 N perchloric acid is equivalent to 32.19 mg of $C_{23}H_{29}N_3O_2S.2C_4H_4O_4$.

Acetophenazine Maleate Tablets

» Acetophenazine Maleate Tablets contain not less than 90.0 percent and not more than 110.0 percent of the labeled amount of $C_{23}H_{29}N_3O_2S.2C_4H_4O_4$.

Packaging and storage—Preserve in tight, light-resistant containers.

Reference standard—*USP Acetophenazine Maleate Reference Standard*—Dry at 65° for 4 hours before using.

NOTE—Throughout the following procedures, protect test or assay specimens, the Reference Standard, and solutions containing them, by conducting the procedures without delay, under subdued light, or using low-actinic glassware.

Identification—Heat a quantity of finely powdered Tablets, equivalent to about 20 mg of acetophenazine maleate, with 20 mL of methanol on a steam bath for 5 minutes, with occasional swirling. Cool to room temperature, add methanol to restore to original volume, shake vigorously for 2 minutes, and filter. A 10-µL portion of the filtrate responds to *Identification test C* under *Acetophenazine Maleate*.

Dissolution ⟨711⟩—

Medium: 0.1 *N* hydrochloric acid; 900 mL.

Apparatus 2: 50 rpm.

Time: 45 minutes.

Procedure—Determine the amount of $C_{23}H_{29}N_3O_2S.2C_4H_4O_4$ dissolved from ultraviolet absorbances at the wavelength of maximum absorbance at about 278 nm of filtered portions of the solution under test, suitably diluted with *Dissolution Medium*, if necessary, in comparison with a Standard solution having a known concentration of USP Acetophenazine Maleate RS in the same medium.

Tolerances—Not less than 75% (*Q*) of the labeled amount of $C_{23}H_{29}N_3O_2S.2C_4H_4O_4$ is dissolved in 45 minutes.

Uniformity of dosage units ⟨905⟩: meet the requirements.

Assay—

Standard preparation—Transfer about 20 mg of USP Acetophenazine Maleate RS, accurately weighed, to a separator, and proceed as directed under *Assay preparation*, beginning with "add 50 mL of sodium chloride solution (1 in 5)." The concentration of USP Acetophenazine Maleate RS in the *Standard preparation* is about 10 µg per mL.

Assay preparation—Weigh and finely powder not less than 20 Acetophenazine Maleate Tablets. Transfer an accurately weighed portion of the powder, equivalent to about 20 mg of acetophenazine maleate, to a separator, add 50 mL of sodium chloride solution (1 in 5) and sufficient 2.5 *N* sodium hydroxide to adjust to a pH of 11, and allow to cool to room temperature. Extract with four 40-mL portions of solvent hexane. Extract the combined solvent hexane solution with three 50-mL portions of 0.1 *N* hydrochloric acid, combining the acid extracts in a 200-mL volumetric flask. Dilute with 0.1 *N* hydrochloric acid to volume, and mix. Transfer 10.0 mL of this solution to a 100-mL volumetric flask, dilute with 0.1 *N* hydrochloric acid to volume, and mix.

Procedure—Concomitantly determine the absorbances of the *Assay preparation* and the *Standard preparation* in 1-cm cells at the wavelength of maximum absorbance at about 278 nm, with a suitable spectrophotometer, using 0.1 *N* hydrochloric acid as the blank. Calculate the quantity, in mg, of $C_{23}H_{29}N_3O_2S.$-$2C_4H_4O_4$ in the portion of Tablets taken by the formula:

$$2C(A_U / A_S),$$

in which *C* is the concentration, in µg per mL, of USP Acetophenazine Maleate RS in the *Standard preparation*, and A_U and A_S are the absorbances of the *Assay preparation* and the *Standard preparation*, respectively.

Acetylcholine Chloride

$$CH_3\overset{O}{\overset{\|}{C}}O(CH_2)_2N^+(CH_3)_3 \quad Cl^-$$

$C_7H_{16}ClNO_2$ 181.66

Ethanaminium, 2-(acetyloxy)-*N*,*N*,*N*-trimethyl-, chloride.

Choline acetate (ester) chloride.

(2-Hydroxyethyl)trimethylammonium chloride, acetate [*60-31-1*].

» Acetylcholine Chloride contains not less than 98.0 percent and not more than 102.0 percent of C_7H_{16}-$ClNO_2$, calculated on the dried basis.

Packaging and storage—Preserve in tight containers.

Reference standard—*USP Acetylcholine Chloride Reference Standard*—Dry at 105° for 3 hours before using.

Identification—

A: The infrared absorption spectrum of a potassium bromide dispersion of it, previously dried, exhibits maxima only at the same wavelengths as that of a similar preparation of USP Acetylcholine Chloride RS.

B: To 5 mL of a solution (1 in 10) add 5 mL of silver nitrate TS: a white, curdy precipitate, which is soluble in ammonium hydroxide but insoluble in nitric acid, is formed.

Melting range, *Class I* ⟨741⟩: between 149° and 152°.

Acidity—Dissolve 100 mg in 10 mL of recently boiled water, and add at once 1 drop of bromothymol blue TS: not more than 0.50 mL of 0.010 *N* sodium hydroxide is required in order to produce a color change.

Loss on drying ⟨731⟩—Dry it at 105° for 3 hours: it loses not more than 1.0% of its weight.

Residue on ignition ⟨281⟩: not more than 0.2%.

Chloride content—Transfer about 280 mg, accurately weighed, to a porcelain casserole, and add 140 mL of water and 1 mL of dichlorofluorescein TS. Mix, and titrate with 0.1 *N* silver nitrate VS until the silver chloride flocculates and the mixture acquires a faint pink color. Each mL of 0.1 *N* silver nitrate is equivalent to 3.545 mg of Cl. Not less than 19.3% and not more than 19.8% of Cl, calculated on the dried basis, is found.

Assay—Weigh accurately about 400 mg of Acetylcholine Chloride, and dissolve in 15 mL of water in a glass-stoppered conical flask. Add 40.0 mL of 0.1 *N* sodium hydroxide VS, and heat on a steam bath for 30 minutes. Insert the stopper, allow to cool, add phenolphthalein TS, and titrate the excess alkali with 0.1 *N* sulfuric acid VS. Perform a blank determination (see *Residual Titrations* under *Titrimetry* ⟨541⟩). Each mL of 0.1 *N* sodium hydroxide is equivalent to 18.17 mg of $C_7H_{16}ClNO_2$.

Acetylcholine Chloride for Ophthalmic Solution

» Acetylcholine Chloride for Ophthalmic Solution is a sterile mixture of Acetylcholine Chloride with Mannitol or other suitable diluent, prepared by freeze-drying. Each container contains not less than 90.0 percent and not more than 115.0 percent of the labeled amount of $C_7H_{16}ClNO_2$.

Packaging and storage—Preserve in *Containers for Sterile Solids* as described under *Injections* ⟨1⟩.

Reference standard—*USP Acetylcholine Chloride Reference Standard*—Dry at 105° for 3 hours before using.

Constituted solution—At the time of use, the constituted solution prepared from Acetylcholine Chloride for Ophthalmic Solution meets the requirements for *Constituted Solutions* under *Injections* ⟨1⟩.

Identification—

A: On a line 2 cm from the bottom edge of a thin-layer chromatographic plate coated with a 0.25-mm layer of aluminum oxide apply 2-µL portions of a test solution, containing about 10 mg of acetylcholine chloride per mL, and of a Standard solution, containing about the same concentration of USP Acetylcholine Chloride RS. Develop the chromatogram, without delay, in a vapor-saturated chamber, using a solvent system consisting of the upper layer obtained by mixing water, butyl alcohol, and glacial acetic acid (100:80:20) and allowing to separate completely. Allow the solvent front to move about 10 cm beyond the initial spotting line, remove the plate, and dry it with the aid of a current of warm air. Immediately spray it with a solution freshly prepared by dissolving 250 mg of cobaltous chloride in 50 mL of water and diluting with alcohol to 100 mL. Dry the plate as before, and immediately spray it with a solution prepared by dissolving 1.0 g of potassium ferrocyanide in 100 mL of water and diluting with 50 mL of alcohol. Dry the plate as before: the

R_f value and color of the principal spot obtained from the test solution correspond to those obtained from the Standard solution.

B: Dissolve a portion, equivalent to about 20 mg of acetylcholine chloride, in about 2 mL of water, add 1 drop of nitric acid and 1 mL of silver nitrate TS: a curdy, white precipitate, soluble in an excess of 6 *N* ammonium hydroxide, is formed.

Acidity—It meets the requirements of the test for *Acidity* under *Acetylcholine Chloride*, an amount equivalent to 100 mg of acetylcholine chloride being used.

Water, *Method I* ⟨921⟩—Perform the titration in the original container, observing precautions against contact with water or moist atmosphere. Adjust the concentration of the reagent so that the titration volume approaches but does not exceed the capacity of the container. Titrate to an amber color that persists for 15 seconds after mixing. Not more than 1.0% of water is found.

Other requirements—It meets the requirements under *Sterility Tests* ⟨71⟩ and *Uniformity of Dosage Units* ⟨905⟩.

Assay—

Mobile phase—Add 1.03 g of sodium 1-heptanesulfonate to a mixture of 900 mL of water and 10 mL of methanol. Mix, then add sufficient glacial acetic acid and ammonium hydroxide, if necessary, to adjust the solution to a pH of 4.0. Add 50 mL of acetonitrile, then add water to make 1000 mL, and mix. Slight variation of the amount of acetonitrile may be required to improve resolution or adjust retention time. Degas the solution.

Standard preparation—Dissolve an accurately weighed quantity of USP Acetylcholine Chloride RS in *Mobile phase*, and dilute quantitatively and stepwise with *Mobile phase* to obtain a solution having a known concentration about equal to that of the acetylcholine chloride in the *Assay preparation*.

Assay preparation—Transfer the contents of 1 container of Acetylcholine Chloride for Ophthalmic Solution to a 10-mL volumetric flask with the aid of *Mobile phase*, add *Mobile phase* to volume, and mix.

Chromatographic system—Use a liquid chromatograph fitted with a 30-cm × 3.9-mm stainless steel column packed with packing L1, and a refractive index detector. The flow rate is about 2 mL per minute. Chromatograph replicate 50-µL injections of the *Standard preparation*, and record the peak response: the relative standard deviation is not more than 3.5%. Chromatograph a solution containing about 0.2% each of acetylcholine chloride and choline chloride: the resolution is not less than 2.0.

Procedure—Separately inject equal volumes (about 50 µL) of the *Standard preparation* and the *Assay preparation* into the chromatograph, record the chromatograms, and measure the responses. Calculate the quantity, in mg, of $C_7H_{16}ClNO_2$ in the container taken by the formula:

$$10C(r_U/r_S),$$

in which C is the concentration, in mg per mL, of USP Acetylcholine Chloride RS in the *Standard preparation*, and r_U and r_S are the peak responses obtained from the *Assay preparation* and the *Standard preparation*, respectively.

Acetylcysteine

```
        NHCOCH3
           |
HSCH2----C----COOH
           |
           H
```

$C_5H_9NO_3S$ 163.19

L-Cysteine, *N*-acetyl-.

N-Acetyl-L-cysteine [*616-91-1*].

» Acetylcysteine contains not less than 98.0 percent and not more than 102.0 percent of $C_5H_9NO_3S$, calculated on the dried basis.

Packaging and storage—Preserve in tight containers.

Reference standard—*USP Acetylcysteine Reference Standard*—Dry at a pressure of about 50 mm of mercury at 70° for 4 hours before using.

Identification—The infrared absorption spectrum of a potassium bromide dispersion of it, previously dried, exhibits maxima only at the same wavelengths as that of a similar preparation of USP Acetylcysteine RS.

Melting range, *Class I* ⟨741⟩: between 104° and 110°.

Specific rotation ⟨781⟩—In a 25-mL volumetric flask mix 1.25 g with 1 mL of disodium ethylenediaminetetraacetate solution (1 in 100), add 7.5 mL of sodium hydroxide solution (1 in 25), and mix to dissolve. Dilute to volume with pH 7.0 buffer (prepared by mixing 29.5 mL of 1 *N* sodium hydroxide, 50 mL of 1 *M* monobasic potassium phosphate, and sufficient water to make 100 mL and, using a pH meter, adjusting to a pH of 7.0 ± 0.1 by adding, as necessary, more of either solution): the specific rotation, calculated on the dried basis, compared with a blank prepared with the same amounts of the same reagents, is between +21° and +27°.

pH ⟨791⟩: between 2.0 and 2.8, in a solution (1 in 100).

Loss on drying ⟨731⟩—Dry it at a pressure of about 50 mm of mercury at 70° for 4 hours: it loses not more than 1.0% of its weight.

Residue on ignition ⟨281⟩—Transfer to a tared fused silica dish about 2 g, weigh accurately, heat on a hot plate until thoroughly charred, cool, add 1 mL of sulfuric acid, and heat gently until fuming ceases. Ignite at 600° until the carbon is consumed. Not more than 0.5% is found.

Heavy metals, *Method II* ⟨231⟩—In a dropwise manner [*Caution—Exercise care, since explosion may occur*], wet the test specimen with 2 mL of nitric acid, and proceed as directed for the *Test preparation:* the limit is 0.001%.

Assay—

Mobile phase—Dissolve 6.8 g of monobasic potassium phosphate in 1000 mL of water, filter through a membrane filter (0.45-µm porosity), and degas.

Internal standard solution—Dissolve about 1 g of DL-phenylalanine in 200 mL of freshly prepared sodium bisulfite solution (1 in 2000).

Standard preparation—Dissolve an accurately weighed quantity of USP Acetylcysteine RS in sodium bisulfite solution (1 in 2000) to obtain a solution having a known concentration of about 10 mg per mL. Pipet 10.0 mL of this solution and 10.0 mL of *Internal standard solution* into a 200-mL volumetric flask, dilute with sodium bisulfite solution (1 in 2000) to volume, and mix to obtain a *Standard preparation* having a known concentration of about 0.5 mg per mL.

Assay preparation—Transfer about 1000 mg of Acetylcysteine, accurately weighed, to a 100-mL volumetric flask. Dissolve in sodium bisulfite solution (1 in 2000), dilute with the same solvent to volume, and mix. Pipet 10.0 mL of this solution and 10.0 mL of *Internal standard solution* into a 200-mL volumetric flask, dilute with sodium bisulfite solution (1 in 2000) to volume, and mix.

Chromatographic system (see *Chromatography* ⟨621⟩)—The liquid chromatograph is equipped with a 214-nm detector and a 3.9-mm × 30-cm column that contains packing L1. The flow rate is about 1.5 mL per minute. Chromatograph the *Standard preparation*, and record the peak responses as directed under *Procedure:* the relative standard deviation for replicate injections is not more than 2.0% and the resolution, R, between acetylcysteine and DL-phenylalanine is not less than 6.

Procedure—Separately inject equal volumes (about 5 µL) of the *Standard preparation* and the *Assay preparation* into the chromatograph, record the chromatograms, and measure the responses for the major peaks. The relative retention times are about 0.4 for acetylcysteine and 1.0 for DL-phenylalanine. Calculate the quantity, in mg, of $C_5H_9NO_3S$ in the Acetylcysteine taken by the formula:

$$2000C(R_U/R_S),$$

in which C is the concentration, in mg per mL, of USP Acetylcysteine RS in the *Standard preparation*, and R_U and R_S are the ratios of the peak response of acetylcysteine to that of DL-phenylalanine obtained from the *Assay preparation* and the *Standard preparation*, respectively.

Acetylcysteine Solution

» Acetylcysteine Solution is a sterile solution of Acetylcysteine in water, prepared with the aid of Sodium Hydroxide. It contains not less than 90.0 percent and not more than 110.0 percent of the labeled amount of $C_5H_9NO_3S$.

Packaging and storage—Preserve in single-dose or in multiple-dose containers, preferably of Type I glass, tightly closed with a glass or polyethylene or polytef-coated elastomeric closure.

Reference standard—*USP Acetylcysteine Reference Standard*—Dry at a pressure of about 50 mm of mercury at 70° for 4 hours before using.

Identification—Place 2 mL in a 10-mL beaker, and adjust to a pH of about 3 (pH indicator paper) using 3 *N* hydrochloric acid. Add 500 mg to 1 g of finely powdered sodium chloride, in two portions of about 200 mg each initially, and then in smaller portions (about 25 mg), stirring after each addition, until a precipitate is formed. Allow to stand at room temperature for 15 minutes, and collect the residue by suction filtration: the acetylcysteine so obtained, after being dried as directed in the test for *Loss on drying* under *Acetylcysteine*, responds to the *Identification test* under *Acetylcysteine*.

Sterility—It meets the requirements under *Sterility Tests* ⟨71⟩.

pH ⟨791⟩: between 6.0 and 7.5.

Assay—

Mobile phase, Internal standard solution, and *Standard preparation*—Prepare as directed in the *Assay* under *Acetylcysteine*.

Assay preparation—Pipet a volume of Acetylcysteine Solution, equivalent to about 1000 mg of acetylcysteine, into a 100-mL volumetric flask, dilute with sodium bisulfite solution (1 in 2000) to volume, and mix. Pipet 10.0 mL of this solution and 10.0 mL of *Internal standard solution* into a 200-mL volumetric flask, dilute with sodium bisulfite solution (1 in 2000) to volume, and mix.

Chromatographic system—Proceed as directed for *Chromatographic system* in the *Assay* under *Acetylcysteine*.

Procedure—Proceed as directed for *Procedure* in the *Assay* under *Acetylcysteine*. Calculate the quantity, in mg, of $C_5H_9NO_3S$ in each mL of the Solution taken by the formula:

$$2000(C/V)(R_U/R_S),$$

in which *C* is the concentration, in mg per mL, of USP Acetylcysteine RS in the *Standard preparation*, *V* is the volume, in mL, of Solution taken, and R_U and R_S are the ratios of the peak response of acetylcysteine to that of DL-phenylalanine obtained from the *Assay preparation* and the *Standard preparation*, respectively.

Acetylcysteine and Isoproterenol Hydrochloride Inhalation Solution

» Acetylcysteine and Isoproterenol Hydrochloride Inhalation Solution is a sterile solution of Acetylcysteine and Isoproterenol Hydrochloride in water. It contains not less than 90.0 percent and not more than 110.0 percent of the labeled amount of acetylcysteine ($C_5H_9NO_3S$), and not less than 90.0 percent and not more than 115.0 percent of the labeled amount of isoproterenol hydrochloride ($C_{11}H_{17}NO_3.HCl$).

Packaging and storage—Preserve in single-dose or in multiple-dose containers, preferably of Type I glass, tightly closed with a glass or polyethylene closure.

Reference standards—*USP Acetylcysteine Reference Standard*—Dry at a pressure of about 50 mm of mercury at 70° for 4 hours before using. *USP Isoproterenol Hydrochloride Reference Standard*—Dry in vacuum over phosphorus pentoxide for 4 hours before using.

Identification—

A: Place 2 mL in a 10-mL beaker, and adjust with 3 *N* hydrochloric acid to a pH of about 3 (pH indicator paper). Add 500 mg to 1 g of finely powdered sodium chloride, in two portions of about 200 mg each initially, and then in smaller portions (about 25 mg), stirring after each addition, until a precipitate is formed. Allow to stand at room temperature for 15 minutes, and collect the residue by suction filtration: the acetylcysteine so obtained, after being dried as directed in the test for *Loss on drying* under *Acetylcysteine*, responds to the *Identification test* under *Acetylcysteine*.

B: *Ferro-citrate solution* and *Buffer solution*—Prepare as directed under *Epinephrine Assay* ⟨391⟩.

Procedure—Place a volume of Inhalation Solution, equivalent to about 0.26 mg of isoproterenol hydrochloride, in a test tube with 3 mL of 0.1 *M* mercuric chloride, and mix. Add 100 µL of *Ferro-citrate solution* and 1.0 mL of *Buffer solution*, and mix: the presence of isoproterenol hydrochloride is confirmed by the development of a purple color.

Sterility—It meets the requirements under *Sterility Tests* ⟨71⟩.

pH ⟨791⟩: between 6.0 and 7.0.

Assay for acetylcysteine—

Mobile phase, Internal standard solution, and *Standard preparation*—Prepare as directed in the *Assay* under *Acetylcysteine*.

Assay preparation—Pipet a volume of Acetylcysteine and Isoproterenol Hydrochloride Inhalation Solution, equivalent to about 1000 mg of acetylcysteine, into a 100-mL volumetric flask, dilute with sodium bisulfite solution (1 in 2000) to volume, and mix. Pipet 10 mL of this solution and 10 mL of *Internal standard solution* into a 200-mL volumetric flask, dilute with sodium bisulfite solution (1 in 2000) to volume, and mix.

Chromatographic system—Proceed as directed for *Chromatographic system* in the *Assay* under *Acetylcysteine*.

Procedure—Proceed as directed for *Procedure* in the *Assay* under *Acetylcysteine*. Calculate the quantity, in mg, of $C_5H_9NO_3S$ in each mL of the Inhalation Solution taken by the formula:

$$2000(C/V)(R_U/R_S),$$

in which *C* is the concentration, in mg per mL, of USP Acetylcysteine RS in the *Standard preparation*, *V* is the volume, in mL, of Inhalation Solution taken, and R_U and R_S are the ratios of the peak response of acetylcysteine to that of DL-phenylalanine obtained from the *Assay preparation* and the *Standard preparation*, respectively.

Assay for isoproterenol hydrochoride—

Mobile phase—Dissolve 13.6 g of monobasic potassium phosphate in 1000 mL of water, and filter through a membrane filter (0.45-µm porosity). Add 20.0 mL of methanol, mix, and degas.

Internal standard solution—Place about 150 mg of acetaminophen in a 500-mL volumetric flask, add 5 mL of glacial acetic acid, dilute with water to volume, and mix.

Standard preparation—Transfer about 30 mg of USP Isoproterenol Hydrochloride RS, accurately weighed, to a 200-mL volumetric flask. Add 200 mg of sodium bisulfite, dilute with water to volume, and mix. Pipet 10 mL of this solution and 10 mL of *Internal standard solution* into a 25-mL volumetric flask, add dilute glacial acetic acid (1 in 100) to volume, and mix to obtain a *Standard preparation* having a known isoproterenol hydrochloride concentration of about 0.06 mg per mL.

Assay preparation—Transfer an accurately measured volume of Inhalation Solution, equivalent to about 1.5 mg of isoproterenol hydrochloride, and 10 mL of *Internal standard solution* to a 25-mL volumetric flask, add dilute glacial acetic acid (1 in 100) to volume, and mix.

Chromatographic system (see *Chromatography* ⟨621⟩)—The liquid chromatograph is equipped with a 280-nm detector and a 3.9-mm × 40-cm column that contains packing L1. The flow rate is about 2 mL per minute. Chromatograph the *Standard preparation*, and record the peak responses as directed under *Procedure*: the relative standard deviation for replicate injections is not more than 2.0% and the resolution, *R*, between isoproterenol hydrochloride and acetaminophen is not less than 6.

Procedure—Separately inject equal volumes (about 25 µL) of the *Standard preparation* and the *Assay preparation* into the

chromatograph, record the chromatograms, and measure the responses for the major peaks. The relative retention times are about 0.5 for isoproterenol hydrochloride and 1.0 for acetaminophen. Calculate the quantity, in mg, of $C_{11}H_{17}NO_3.HCl$ in each mL of the Inhalation Solution taken by the formula:

$$(25C/V)(R_U/R_S),$$

in which C is the concentration, in mg per mL, of USP Isoproterenol Hydrochloride RS in the *Standard preparation*, V is the volume, in mL, of Inhalation Solution taken, and R_U and R_S are the ratios of the peak responses of isoproterenol hydrochloride to those of acetaminophen obtained from the *Assay preparation* and the *Standard preparation*, respectively.

Acids—*see complete list in index*

Acrisorcin

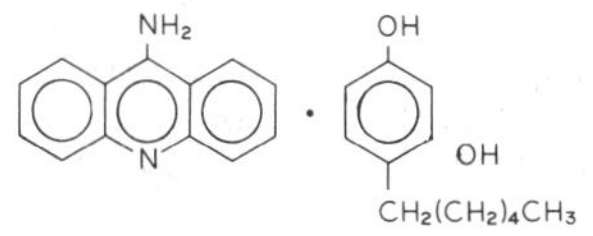

$C_{12}H_{18}O_2.C_{13}H_{10}N_2$ 388.51
1,3-Benzenediol, 4-hexyl-, compd. with 9-acridinamine (1:1).
4-Hexylresorcinol compound with 9-aminoacridine (1:1) [*7527-91-5*].

» Acrisorcin contains not less than 97.0 percent and not more than 101.0 percent of $C_{12}H_{18}O_2.C_{13}H_{10}N_2$, calculated on the dried basis.

Packaging and storage—Preserve in well-closed containers.

Reference standard—*USP Acrisorcin Reference Standard*—Dry at 105° for 3 hours before using.

Identification—

A: The infrared absorption spectrum of a mineral oil dispersion of it, previously dried, exhibits maxima only at the same wavelengths as that of a similar preparation of USP Acrisorcin RS.

B: The ultraviolet absorption spectrum of a 1 in 400,000 solution of it in methanol exhibits maxima and minima at the same wavelengths as that of a similar preparation of USP Acrisorcin RS, concomitantly measured, and the respective absorptivities, calculated on the dried basis, at the wavelength of maximum absorbance at about 261 nm do not differ by more than 3.0%.

Loss on drying ⟨731⟩—Dry it at 105° for 3 hours: it loses not more than 1.0% of its weight.

Residue on ignition ⟨281⟩: not more than 0.2%.

Assay—Weigh accurately about 750 mg of Acrisorcin, and dissolve in about 25 mL of glacial acetic acid. Add 5 mL of acetic anhydride, and stir for about 5 minutes. Titrate with 0.1 *N* perchloric acid VS, determining the end-point potentiometrically, using a calomel-glass electrode system. Perform a blank determination, and make any necessary correction. Each mL of 0.1 *N* perchloric acid is equivalent to 38.85 mg of $C_{12}H_{18}O_2.C_{13}H_{10}N_2$.

Acrisorcin Cream

» Acrisorcin Cream is Acrisorcin in a suitable water-miscible base. It contains not less than 90.0 percent and not more than 110.0 percent of the labeled amount of $C_{12}H_{18}O_2.C_{13}H_{10}N_2$.

Packaging and storage—Preserve in collapsible tubes or in tight containers.

Reference standard—*USP Acrisorcin Reference Standard*—Dry at 105° for 3 hours before using.

Identification—The ultraviolet absorption spectrum of the *Assay preparation*, prepared as directed in the *Assay*, exhibits maxima at 380 ± 2, 400 ± 2, and 425 ± 2 nm.

Minimum fill ⟨755⟩: meets the requirements.

Assay—

Standard preparation—Dissolve a suitable quantity of USP Acrisorcin RS, accurately weighed, in 0.3 *N* hydrochloric acid to obtain a solution having a known concentration of about 20 µg per mL.

Assay preparation—Transfer an accurately weighed portion of Acrisorcin Cream, equivalent to about 2 mg of acrisorcin, to a separator, add 50 mL of chloroform, and extract with four 20-mL portions of 0.3 *N* hydrochloric acid. Transfer the combined extracts to a 100-mL volumetric flask, dilute with 0.3 *N* hydrochloric acid to volume, and mix. (Retain a portion of this solution for the *Identification test*.)

Procedure—Concomitantly determine the absorbances of the solutions in 1-cm cells at the wavelength of maximum absorption at about 400 nm, with a suitable spectrophotometer, using 0.3 *N* hydrochloric acid as the blank. Calculate the quantity, in mg, of $C_{12}H_{18}O_2.C_{13}H_{10}N_2$ in the portion of Cream taken by the formula:

$$0.1C(A_U/A_S),$$

in which C is the concentration, in µg per mL, of USP Acrisorcin RS in the *Standard preparation*, and A_U and A_S are the absorbances of the *Assay preparation* and the *Standard preparation*, respectively.

Activated Charcoal—*see* Charcoal, Activated

Adenine

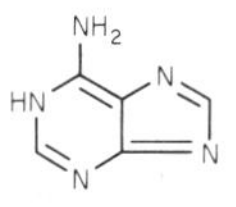

$C_5H_5N_5$ 135.13
1*H*-Purin-6-amine.
1,6-Dihydro-6-iminopurine [*73-24-5*].

» Adenine contains not less than 98.0 percent and not more than 102.0 percent of $C_5H_5N_5$, calculated on the dried basis.

Packaging and storage—Preserve in well-closed containers.

Reference standard—*USP Adenine Reference Standard*—Dry at 110° for 4 hours before using.

Identification—The infrared absorption spectrum of a potassium bromide dispersion of it, previously dried, exhibits maxima only at the same wavelengths as that of a similar preparation of USP Adenine RS.

Loss on drying ⟨731⟩—Dry it at 110° for 4 hours: it loses not more than 1.0% of its weight.

Residue on ignition ⟨281⟩: not more than 0.1%.

Heavy metals, *Method II* ⟨231⟩: 0.001%.

Organic impurities—

pH 7.0 phosphate buffer—Dissolve 4.54 g of monobasic potassium phosphate in water to make 500 mL of solution. Dissolve 4.73 g of anhydrous dibasic sodium phosphate in water to make 500 mL of solution. Mix 38.9 mL of the monobasic potassium phosphate solution with 61.1 mL of the dibasic sodium phosphate solution. Adjust, if necessary, by the dropwise addition of the dibasic sodium phosphate solution to a pH of 7.0.

Standard preparations—Dissolve a suitable quantity of USP Adenine RS, accurately weighed, in hot water, cool, and dilute quantitatively with water to obtain a solution having a known concentration of about 0.19 mg per mL. Pipet 5-mL portions into three 100-mL volumetric flasks, dilute with 0.10 *N* hydrochloric acid, 0.010 *N* sodium hydroxide, and *pH 7.0 phosphate buffer*, respectively, to volume, and mix.

Test preparations—Proceed as directed for *Standard preparations*, using the specimen in place of the Reference Standard.

Procedure—Obtain the ultraviolet absorption spectra of the *Test preparations* and the *Standard preparations* concomitantly, in 1-cm cells, covering the range between 220 nm and 320 nm, using water as the blank. The respective absorptivities, calculated on the dried basis, at the wavelengths of maximum absorbance, for each pair of corresponding solutions do not differ by more than 2.0%.

Nitrogen content, *Method II* ⟨461⟩—Not less than 50.2% and not more than 53.4% is found, calculated on the dried basis.

Assay—

0.1 N Perchloric acid—Standardize 0.1 *N* perchloric acid VS as follows: Transfer about 300 mg of potassium biphthalate, accurately weighed, to a 150-mL beaker, and dissolve in 80 mL of a mixture of 100 mL of glacial acetic acid and 300 mL of acetic anhydride, by stirring. Titrate with the perchloric acid solution, determining the end-point potentiometrically. Each 20.42 mg of potassium biphthalate is equivalent to 1 mL of 0.1 *N* perchloric acid.

Procedure—Transfer about 200 mg of Adenine, accurately weighed, to a 150-mL beaker, and dissolve in 80 mL of a mixture of 100 mL of glacial acetic acid and 300 mL of acetic anhydride, by stirring. Titrate with *0.1 N Perchloric acid*, determining the end-point potentiometrically. Perform a blank determination, and make any necessary correction. Each mL of *0.1 N Perchloric acid* is equivalent to 13.514 mg of $C_5H_5N_5$.

Adhesive Bandage—*see* Bandage, Adhesive

Adhesive Tape—*see* Tape, Adhesive

Adsorbed for Adult Use, Tetanus and Diphtheria Toxoids—*see* Tetanus and Diphtheria Toxoids Adsorbed for Adult Use

Adsorbed Diphtheria Toxoid—*see* Diphtheria Toxoid Adsorbed

Adsorbed Diphtheria and Tetanus Toxoids—*see* Diphtheria and Tetanus Toxoids Adsorbed

Adsorbed Diphtheria and Tetanus Toxoids and Pertussis Vaccine—*see* Diphtheria and Tetanus Toxoids and Pertussis Vaccine Adsorbed

Adsorbed Pertussis Vaccine—*see* Pertussis Vaccine Adsorbed

Adsorbed Tetanus Toxoid—*see* Tetanus Toxoid Adsorbed

Aerosols—*see complete list in index*

Agar—*see* Agar NF

Compressed Air

» Compressed Air is a natural or synthetic mixture of gases consisting largely of nitrogen and oxygen. It contains not less than 19.5 percent and not more than 23.5 percent, by volume, of O_2.

Packaging and storage—Preserve in cylinders or in a low pressure collecting tank. Containers used for Compressed Air are not to be treated with any toxic, sleep-inducing, or narcosis-producing compounds, and are not to be treated with any compound that would be irritating to the respiratory tract when the Compressed Air is used.

Notes—Reduce the container pressure by means of a regulator. Measure the gases with a gas volume meter downstream from the detector tube in order to minimize contamination or change of the specimens.

The various detector tubes called for in the respective tests are listed under *Reagents* in the section, *Reagents, Indicators, and Solutions.*

Labeling—Where it is piped directly from the collecting tank to the point of use, label each outlet "Compressed Air."

Water and oil—Support 1 container in an inverted position (with the valve at the bottom) for 5 minutes. Cautiously open the valve slightly, maintaining the container in an inverted position. Vent the gas with a barely audible flow against a stainless steel mirror for a few seconds: no liquid is discernible on the mirror.

Odor—Carefully open the container valve to produce a moderate flow of gas. Do not direct the gas stream toward the face, but deflect a portion of the stream toward the nose: no appreciable odor is discernible.

Carbon dioxide—Pass 1050 ± 50 mL through a carbon dioxide detector tube at the rate specified for the tube: the indicator change corresponds to not more than 0.05%.

Carbon monoxide—Pass 1050 ± 50 mL through a carbon monoxide detector tube at the rate specified for the tube: the indicator change corresponds to not more than 0.001%.

Nitric oxide and nitrogen dioxide—Pass 550 ± 50 mL through a nitric oxide–nitrogen dioxide detector tube at the rate specified for the tube: the indicator change corresponds to not more than 2.5 ppm.

Sulfur dioxide—Pass 1050 ± 50 mL through a sulfur dioxide detector tube at the rate specified for the tube: the indicator change corresponds to not more than 5 ppm.

Assay—Determine the oxygen concentration of Compressed Air using an electrochemical cell analyzer (see *Oxygen Determination* ⟨468⟩), readable to 0.1% of oxygen and calibrated with ambient air to an accuracy of ±0.2% of oxygen.

Alanine

$$CH_3-\overset{\displaystyle H}{\underset{\displaystyle NH_2}{C}}-COOH$$

$C_3H_7NO_2$ 89.09
L-Alanine.
L-Alanine [*56-41-7*].

» Alanine contains not less than 98.5 percent and not more than 101.5 percent of $C_3H_7NO_2$, as L-alanine, calculated on the dried basis.

Packaging and storage—Preserve in well-closed containers.

Reference standard—*USP L-Alanine Reference Standard*—Dry at 105° for 3 hours before using.

Identification—The infrared absorption spectrum of a potassium bromide dispersion of it, previously dried, exhibits maxima only at the same wavelengths as that of a similar preparation of USP L-Alanine RS.

Specific rotation ⟨781⟩: between +13.7° and +15.1°, calculated on the dried basis, determined in a solution in 6 *N* hydrochloric acid containing 10 g in each 100 mL.

pH ⟨791⟩: between 5.5 and 7.0 in a solution (1 in 20).

Loss on drying ⟨731⟩—Dry it at 105° for 3 hours: it loses not more than 0.2% of its weight.

Residue on ignition ⟨281⟩: not more than 0.15%.

Chloride ⟨221⟩—A 1.0-g portion shows no more chloride than corresponds to 0.70 mL of 0.020 *N* hydrochloric acid (0.05%).

Sulfate ⟨221⟩—A 1.0-g portion shows no more sulfate than corresponds to 0.30 mL of 0.020 *N* sulfuric acid (0.03%).

Arsenic ⟨211⟩: 1.5 ppm.

Iron ⟨241⟩: 0.003%.

Heavy metals, *Method I* ⟨231⟩: 0.0015%.

Assay—Transfer about 80 mg of Alanine, accurately weighed, to a 125-mL flask, dissolve in a mixture of 3 mL of formic acid and 50 mL of glacial acetic acid, and titrate with 0.1 *N* perchloric acid VS, determining the end-point potentiometrically. Perform a blank determination, and make any necessary correction. Each mL of 0.1 *N* perchloric acid is equivalent to 8.909 mg of $C_3H_7NO_2$.

Albumin Aggregated Injection, Iodinated I 131—*see under "iodine"*

Albumin Aggregated Injection, Technetium Tc 99m—*see under "technetium"*

Albumin Human

» Albumin Human conforms to the regulations of the federal Food and Drug Administration concerning biologics (640.80 to 640.86) (see *Biologics* ⟨1041⟩). It is a sterile, non-pyrogenic preparation of serum albumin obtained by fractionating material (source blood, plasma, serum, or placentas) from healthy human donors, the source material being tested for the absence of hepatitis B surface antigen. It is made by a process that yields a product that is safe for intravenous use. Not less than 96 percent of its total protein is albumin. It is a solution containing, in each 100 mL, either 25 g of serum albumin osmotically equivalent to 500 mL of normal human plasma, or 20 g equivalent to 400 mL, or 5 g equivalent to 100 mL, or 4 g equivalent to 80 mL thereof, and contains not less than 93.75 percent and not more than 106.25 percent of the labeled amount in the case of the solution containing 4 g in each 100 mL, and not less than 94.0 percent and not more than 106.0 percent of the labeled amount in the other cases. It contains no added antimicrobial agent, but may contain sodium acetyltryptophanate with or without sodium caprylate as a stabilizing agent. It has a sodium content of not less than 130 mEq per liter and not more than 160 mEq per liter. It has a heme content such that the absorbance of a solution, diluted to contain 1 percent of protein, in a 1-cm holding cell, measured at a wavelength of 403 nm, is not more than 0.25. It meets the requirements of the tests for heat stability and for pH.

Packaging and storage—Preserve at the temperature indicated on the label.

Expiration date—The expiration date is not later than 5 years after issue from manufacturer's cold storage (5°, 3 years) if labeling recommends storage between 2° and 10°; not later than 3 years after issue from manufacturer's cold storage (5°, 3 years) if labeling recommends storage at temperatures not higher than 37°; and not later than 10 years after date of manufacture if in a hermetically sealed metal container and labeling recommends storage between 2° and 10°.

Labeling—Label it to state that it is not to be used if it is turbid and that it is to be used within 4 hours after the container is entered. Label it also to state the osmotic equivalent in terms of plasma, the sodium content, and the type of source material (venous plasma, placental plasma, or both) from which it was prepared. Label it also to indicate that additional fluids are needed when the 20-g-per-100-mL or 25-g-per-100-mL product is administered to a markedly dehydrated patient.

Albumin Injection, Iodinated I 125—*see under "iodine"*

Albumin Injection, Iodinated I 131—*see under "iodine"*

Albumin Injection, Technetium Tc 99m—*see* Technetium Tc 99m Albumin Injection

Albuterol

HOCH₂ / HO— benzene ring —CHCH₂NHC(CH₃)₃ / OH

$C_{13}H_{21}NO_3$ 239.31

1,3-Benzenedimethanol, α^1-[[(1,1-dimethylethyl)amino]methyl]-4-hydroxy-.

α^1-[(*tert*-Butylamino)methyl]-4-hydroxy-*m*-xylene-α,α'-diol [*18559-94-9*].

» Albuterol contains not less than 98.5 percent and not more than 101.0 percent of $C_{13}H_{21}NO_3$, calculated on the anhydrous basis.

Packaging and storage—Preserve in well-closed, light-resistant containers.

Reference standard—*USP Albuterol Reference Standard*—Do not dry before using.

Identification—

A: The infrared absorption spectrum of a potassium bromide dispersion of it exhibits maxima only at the same wavelengths as that of a similar preparation of USP Albuterol RS.

B: The ultraviolet absorption spectrum of a 1 in 12,500 solution of it in 0.1 *N* hydrochloric acid exhibits maxima and minima at the same wavelengths as that of a similar solution of USP Albuterol RS, concomitantly measured.

Water, *Method I* ⟨921⟩: not more than 0.5%.

Residue on ignition ⟨281⟩: not more than 0.1%.

Chromatographic purity—

Standard preparation—Dissolve an accurately weighed quantity of USP Albuterol RS in methanol to obtain a solution having a known concentration of 0.10 mg per mL.

Test preparation—Dissolve an accurately weighed quantity of Albuterol in methanol to obtain a solution having a concentration of 20 mg per mL.

Procedure—Apply 10-µL aliquots of the *Test preparation* and the *Standard preparation* to separate points on a thin-layer chromatographic plate (see *Chromatography* ⟨621⟩) coated with a 0.25-mm layer of chromatographic silica gel. Place the plate in a saturated chromatographic chamber, and develop the chromatogram in a solvent system consisting of a mixture of methyl isobutyl ketone, isopropyl alcohol, ethyl acetate, water, and ammonium hydroxide (50:45:35:18:3) until the solvent front has moved three-fourths of the length of the plate. Remove the plate from the developing chamber, air-dry, and expose it to iodine vapor: any spot, other than the principal spot, obtained from the *Test preparation* is not greater in size and intensity than the spot produced by the *Standard preparation* (0.5%), and the sum of the impurities is not greater than 2.0%.

Assay—Dissolve about 400 mg of Albuterol, accurately weighed, in 50 mL of glacial acetic acid, add 2 drops of crystal violet TS, and titrate with 0.1 *N* perchloric acid VS. Perform a blank determination, and make any necessary correction. Each mL of 0.1 *N* perchloric acid is equivalent to 23.93 mg of $C_{13}H_{21}NO_3$.

Albuterol Sulfate

$(C_{13}H_{21}NO_3)_2 \cdot H_2SO_4$ 576.70

1,3-Benzenedimethanol, α^1-[[(1,1-dimethylethyl)amino]-methyl]-4-hydroxy-, sulfate (2:1) (salt).

α^1-[(*tert*-Butylamino)methyl]-4-hydroxy-*m*-xylene-α,α'-diol sulfate (2:1) (salt) [*51022-70-9*].

» Albuterol Sulfate contains not less than 98.5 percent and not more than 101.0 percent of $(C_{13}H_{21}NO_3)_2 \cdot H_2SO_4$, calculated on the anhydrous basis.

Packaging and storage—Preserve in well-closed, light-resistant containers.

Reference standard—*USP Albuterol Sulfate Reference Standard*—Do not dry before using.

Identification—

A: The infrared absorption spectrum of a potassium bromide dispersion of it exhibits maxima only at the same wavelengths as that of a similar preparation of USP Albuterol Sulfate RS.

B: The ultraviolet absorption spectrum of a 1 in 12,500 solution of it in 0.1 *N* hydrochloric acid exhibits maxima and minima at the same wavelengths as that of a similar solution of USP Albuterol Sulfate RS, concomitantly measured.

C: Shake a quantity of it, equivalent to 4 mg of albuterol, with 10 mL of water, and filter: the filtrate so obtained responds to the tests for *Sulfate* ⟨191⟩.

Water, *Method I* ⟨921⟩: not more than 0.5%.

Residue on ignition ⟨281⟩: not more than 0.1%.

Chromatographic purity—It meets the requirements of the test for *Chromatographic purity* under *Albuterol*, except to read Albuterol Sulfate in place of Albuterol.

Assay—Dissolve about 900 mg of Albuterol Sulfate, accurately weighed, in 50 mL of glacial acetic acid, add 2 drops of oracet blue B TS, and titrate with 0.1 *N* perchloric acid VS. Perform a blank determination, and make any necessary correction. Each 1 mL of 0.1 *N* perchloric acid is equivalent to 57.67 mg of $(C_{13}H_{21}NO_3)_2 \cdot H_2SO_4$.

Alclometasone Dipropionate

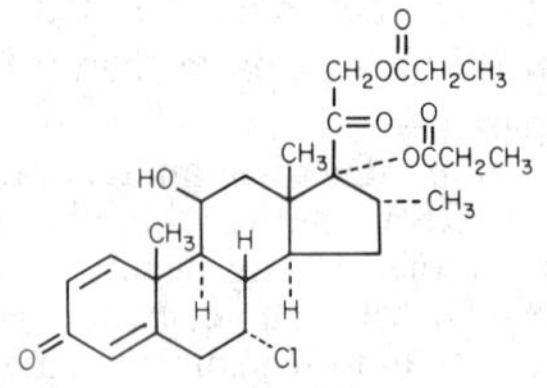

$C_{28}H_{37}ClO_7$ 521.05

Pregna-1,4-diene-3,20-dione, 7-chloro-11-hydroxy-16-methyl-17,21-bis(1-oxopropoxy)-, (7α,11β,16α)-.

7α-Chloro-11β,17,21-trihydroxy-16α-methylpregna-1,4-diene-3,20-dione 17,21-dipropionate [*66734-13-2*].

» Alclometasone Dipropionate contains not less than 97.0 percent and not more than 102.0 percent of $C_{28}H_{37}ClO_7$, calculated on the dried basis.

Packaging and storage—Preserve in tight containers.

Reference standards—*USP Alclometasone Dipropionate Reference Standard*—Do not dry before using.

Identification—

A: The infrared absorption spectrum of a mineral oil dispersion of it exhibits maxima only at the same wavelengths as that of a similar preparation of USP Alclometasone Dipropionate RS.

B: The retention time of the major peak in the chromatogram of the *Assay preparation* corresponds to that of the *Standard preparation*, both relative to the internal standard, as obtained in the *Assay*.

Specific rotation ⟨781⟩: not less than +21° and not more than +25°, calculated on the dried basis, determined in a solution in dioxane containing 30 mg per mL.

Loss on drying ⟨731⟩—Dry it in vacuum at a pressure not exceeding 5 mm of mercury at 105° for 3 hours: it loses not more than 0.5% of its weight.

Residue on ignition ⟨281⟩: not more than 0.1%.

Heavy metals, *Method II* ⟨231⟩: 0.003%.

Chromatographic purity—

Methanol-dichloromethane solution—Dilute 250 mL of methanol with dichloromethane to 500 mL, and mix.

Standard stock solution—Transfer 25 ± 1 mg of USP Alclometasone Dipropionate RS to a 100-mL volumetric flask, add *Methanol-dichloromethane solution* to volume, and mix to obtain a solution having a known concentration of 250 ± 10 µg per mL.

Standard preparations A, B, C, D, E, and F—Dilute accurately measured volumes of *Standard stock solution* quantitatively with *Methanol-dichloromethane solution* designated below (as parts by volume of *Standard stock solution* in total parts by volume of the finished *Standard preparation*) to obtain *Standard preparations*, designated below by letter, having the following concentrations and percentage assignments:

A—(4 in 5); 200 µg per mL (2.0%).
B—(3 in 5); 150 µg per mL (1.5%).
C—(2 in 5); 100 µg per mL (1.0%).
D—(3 in 10); 75 µg per mL (0.75%).
E—(2 in 10); 50 µg per mL (0.50%).
F—(1 in 10); 25 µg per mL (0.25%).

Test preparation—Transfer 50 ± 2 mg of Alclometasone Dipropionate to a 5-mL volumetric flask, add *Methanol-dichloromethane solution* to volume, and mix.

Procedure—On a suitable thin-layer chromatographic plate (see *Chromatography* ⟨621⟩), coated with a 0.25-mm layer of chromatographic silica gel mixture, apply separately 25 µL of the *Test preparation* and 25 µL of each *Standard preparation.* Position the plate in a saturated, unlined chromatographic chamber, and develop the chromatograms in a solvent system consisting of a mixture of chloroform and acetone (7:1) until the solvent front has moved about three-fourths of the length of the plate. Remove the plate from the developing chamber, mark the solvent front, and allow the solvent to evaporate. Observe the plate under short-wavelength ultraviolet light, and compare the intensities of any secondary spots observed in the chromatogram of the *Test preparation* with those of the principal spots in the chromatograms of the *Standard preparations:* the sum of the intensities of secondary spots obtained from the *Test preparation* corresponds to not more than 3.0% of related compounds.

Assay—

0.05 M Monobasic potassium phosphate—Transfer 3.40 g of monobasic potassium phosphate to a 500-mL volumetric flask, add water to volume, and mix.

Mobile phase—Prepare a filtered and degassed mixture of methanol and *0.05 M Monobasic potassium phosphate* (2:1). Make adjustments if necessary (see *System Suitability* under *Chromatography* ⟨621⟩).

Internal standard solution—Transfer about 50 mg of betamethasone dipropionate to a 25-mL volumetric flask, add methanol to volume, and mix.

Standard preparation—Transfer about 30 mg of USP Alclometasone Dipropionate RS, accurately weighed, to a 25-mL volumetric flask, add methanol to volume, and mix. Transfer 4.0 mL of this solution to a second 25-mL volumetric flask, add 4.0 mL of *Internal standard solution*, dilute with methanol to volume, and mix to obtain a *Standard preparation* having a known concentration of about 0.2 mg of USP Alclometasone Dipropionate RS per mL.

Assay preparation—Transfer about 30 mg of Alclometasone Dipropionate, accurately weighed, to a 25-mL volumetric flask, add methanol to volume, and mix. Transfer 4.0 mL of this solution to a second 25-mL volumetric flask, add 4.0 mL of *Internal standard solution*, dilute with methanol to volume, and mix.

Chromatographic system (see *Chromatography* ⟨621⟩)—The liquid chromatograph is equipped with a 254-nm detector and a 4-mm × 30-cm column that contains packing L1. The flow rate is about 1.2 mL per minute. Chromatograph the *Standard preparation*, and record the peak responses as directed under *Procedure:* the resolution, *R*, between the analyte and internal standard peaks is not less than 3.0, and the relative standard deviation for replicate injections is not more than 2%.

Procedure—Separately inject equal volumes (about 10 µL) of the *Standard preparation* and the *Assay preparation* into the chromatograph, record the chromatograms, and measure the responses for the major peaks. The relative retention times are about 0.7 for alclometasone dipropionate and 1.0 for betamethasone dipropionate. Calculate the quantity, in mg, of $C_{28}H_{37}ClO_7$ in the portion of Alclometasone Dipropionate taken by the formula:

$$150C(R_U/R_S),$$

in which C is the concentration, in mg per mL, of USP Alclometasone Dipropionate RS in the *Standard preparation*, and R_U and R_S are the peak height ratios obtained from the *Assay preparation* and the *Standard preparation*, respectively.

Alclometasone Dipropionate Cream

» Alclometasone Dipropionate Cream contains not less than 90.0 percent and not more than 110.0 percent of the labeled amount of $C_{28}H_{37}ClO_7$ in a suitable cream base.

Packaging and storage—Preserve in collapsible tubes or in tight containers.

Reference standard—*USP Alclometasone Dipropionate Reference Standard*—Do not dry before using.

Identification—

A: The retention time of the major peak in the chromatogram of the *Assay preparation* corresponds to that of the *Standard preparation*, both relative to the internal standard, as obtained in the *Assay*.

B: Place a quantity of Cream, equivalent to about 1.25 mg of alclometasone dipropionate, in a 50-mL centrifuge tube, and add 15 mL of methanol. Insert a stopper securely into the tube, and place the tube in a water bath maintained at 60° until the semisolid components melt. Remove the tube from the bath, shake vigorously until the specimen components resolidify, and place the tube in an ice-methanol bath for 15 minutes. Remove the tube from the bath, and centrifuge at 2500 rpm for 5 minutes. Transfer the clear supernatant liquid to a vial, and allow this test solution to equilibrate to room temperature. On a suitable thin-layer chromatographic plate (see *Chromatography* ⟨621⟩), coated with a 0.25-mm layer of chromatographic silica gel mixture, apply separately 20 µL of the test solution and 20 µL of a Standard solution of USP Alclometasone Dipropionate RS in methanol containing about 0.08 mg per mL, and dry the applications with the aid of a stream of nitrogen. Position the plate in a saturated, unlined chromatographic chamber, and develop the chromatograms in a solvent system consisting of a mixture of chloroform and acetone (7:1) until the solvent front has moved about three-fourths of the length of the plate. Remove the plate from the developing chamber, mark the solvent front, and allow the solvent to evaporate. Observe the plate under short-wavelength ultraviolet light: the R_f value of the principal spot obtained from the test solution corresponds to that obtained from the Standard solution.

Microbial limits—It meets the requirements of the tests for absence of *Staphylococcus aureus* and *Pseudomonas aeruginosa* under *Microbial Limit Tests* ⟨61⟩.

Minimum fill ⟨755⟩: meets the requirements.

Assay—

0.05 M Monobasic potassium phosphate—Transfer 3.40 g of monobasic potassium phosphate to a 500-mL volumetric flask, add water to volume, and mix.

Mobile phase—Prepare a filtered and degassed mixture of methanol and *0.05 M Monobasic potassium phosphate* (2:1). Make adjustments if necessary (see *System Suitability* under *Chromatography* ⟨621⟩).

Internal standard solution—Transfer about 20 mg of betamethasone dipropionate to a 50-mL volumetric flask, add methanol to volume, and mix.

Standard preparation—Transfer about 25 mg of USP Alclometasone Dipropionate RS, accurately weighed, to a 100-mL volumetric flask, add methanol to volume, and mix. Transfer 5.0 mL of this solution to a small stoppered flask, add 5.0 mL of methanol, add 5.0 mL of *Internal standard solution*, and mix to obtain a *Standard preparation* having a known concentration of about 0.08 mg of USP Alclometasone Dipropionate RS per mL.

Assay preparation—Transfer an accurately weighed quantity of Alclometasone Dipropionate Cream, equivalent to about 1.25 mg of alclometasone dipropionate, to a 50-mL centrifuge tube, add 5.0 mL of *Internal standard solution*, and add 10.0 mL of methanol. Insert a stopper securely into the tube, and place it in a water bath maintained at 60° until the semisolid components melt. Remove the tube from the bath, shake vigorously until the specimen components resolidify, and return the tube to the 60° water bath until the semisolid components melt. Remove the tube from the bath, shake vigorously until the specimen components resolidify, and place the tube in an ice-methanol bath for 15 minutes. Remove the tube from the bath, and centrifuge at 2500 rpm for 5 minutes. Transfer the clear supernatant liquid to a small stoppered flask, and allow this *Assay preparation* to equilibrate to room temperature.

Chromatographic system (see *Chromatography* ⟨621⟩)—The liquid chromatograph is equipped with a 254-nm detector and a 4-mm × 30-cm column that contains packing L1. The flow rate is about 1.2 mL per minute. Chromatograph the *Standard preparation*, and record the peak responses as directed under *Procedure:* the resolution, R, between the analyte and internal standard peaks is not less than 3.0, and the relative standard deviation for replicate injections is not more than 2%.

Procedure—Separately inject equal volumes (about 20 µL) of the *Standard preparation* and the *Assay preparation* into the chromatograph, record the chromatograms, and measure the responses for the major peaks. The relative retention times are about 0.7 for alclometasone dipropionate and 1.0 for betamethasone dipropionate. Calculate the quantity, in mg, of $C_{28}H_{37}ClO_7$ in the portion of Cream taken by the formula:

$$15C(R_U/R_S),$$

in which C is the concentration, in mg per mL, of USP Alclometasone Dipropionate RS in the *Standard preparation*, and R_U and R_S are the peak height ratios obtained from the *Assay preparation* and the *Standard preparation*, respectively.

Alclometasone Dipropionate Ointment

» Alclometasone Dipropionate Ointment contains not less than 90.0 percent and not more than 110.0 percent of the labeled amount of $C_{28}H_{37}ClO_7$, in a suitable ointment base.

Packaging and storage—Preserve in collapsible tubes or in tight containers.

Reference standard—*USP Alclometasone Dipropionate Reference Standard*—Do not dry before using.

Identification—

A: The retention time of the major peak in the chromatogram of the *Assay preparation* corresponds to that of the *Standard preparation*, both relative to the internal standard, as obtained in the *Assay*.

B: Place a quantity of Ointment, equivalent to about 1.25 mg of alclometasone dipropionate, in a 50-mL centrifuge tube, add 10 mL of 2,2,4-trimethylpentane, insert a stopper securely into the tube, and disperse the specimen using a vortexing mixer. Add 5.0 mL of a solution of methanol in water (45 in 50), insert the stopper securely, shake vigorously for 2 minutes, and centrifuge at 2500 rpm for 3 minutes. Remove the lower, aqueous alcohol phase, and transfer this test solution to a stoppered vial. On a suitable thin-layer chromatographic plate (see *Chromatography* ⟨621⟩), coated with a 0.25-mm layer of chromatographic silica gel mixture, apply separately 20 µL of the test solution and 20 µL of a Standard solution of USP Alclometasone Dipropionate RS in methanol containing about 0.25 mg per mL, and dry the applications with the aid of a stream of nitrogen. Position the plate in a saturated, unlined chromatographic chamber, and develop the chromatograms in a solvent system consisting of a mixture of chloroform and acetone (7:1) until the solvent front has moved about three-fourths of the length of the plate. Remove the plate from the developing chamber, mark the solvent front, and allow the solvent to evaporate. Observe the plate under short-wavelength ultraviolet light: the R_f value of the principal spot obtained from the test solution corresponds to that obtained from the Standard solution.

Microbial limits—It meets the requirements of the tests for absence of *Staphylococcus aureus* and *Pseudomonas aeruginosa* under *Microbial Limit Tests* ⟨61⟩.

Minimum fill ⟨755⟩: meets the requirements.

Assay—

Methanol-water solution—Dilute 450 mL of methanol with water to 500 mL, and mix.

0.05 M Monobasic potassium phosphate—Transfer 3.40 g of monobasic potassium phosphate to a 500-mL volumetric flask, add water to volume, and mix.

Mobile phase—Prepare a filtered and degassed mixture of methanol and *0.05 M Monobasic potassium phosphate* (2:1). Make adjustments if necessary (see *System Suitability* under *Chromatography* ⟨621⟩).

Internal standard solution—Transfer about 30 mg of betamethasone dipropionate to a 200-mL volumetric flask, add *Methanol-water solution* to volume, and mix.

Standard preparation—Transfer about 20 mg of USP Alclometasone Dipropionate RS, accurately weighed, to a 200-mL volumetric flask, add *Methanol-water solution* to volume, and mix. Transfer 5.0 mL of this solution to a small stoppered flask, add 5.0 mL of *Internal standard solution*, and mix to obtain a *Standard preparation* having a known concentration of about 0.05 mg of USP Alclometasone Dipropionate RS per mL.

Assay preparation—Transfer an accurately weighed quantity of Alclometasone Dipropionate Ointment, equivalent to about 0.5 mg of alclometasone dipropionate, to a 50-mL centrifuge tube, add 10 mL of 2,2,4-trimethylpentane, insert a stopper securely into the tube, and disperse the specimen using a vortexing mixer. Add 5.0 mL of *Internal standard solution* and 5.0 mL of *Methanol-water solution*, insert the stopper securely, shake vigorously for 2 minutes, and centrifuge at 2500 rpm for 3 minutes. Remove the lower, aqueous alcohol phase, and transfer this *Assay preparation* to a stoppered vial.

Chromatographic system (see *Chromatography* ⟨621⟩)—The liquid chromatograph is equipped with a 254-nm detector and a 4-mm × 30-cm column that contains packing L1. The flow rate is about 1.2 mL per minute. Chromatograph the *Standard preparation*, and record the peak responses as directed under *Procedure:* the resolution, *R*, between the analyte and internal standard peaks is not less than 3.0, and the relative standard deviation for replicate injections is not more than 2%.

Procedure—Separately inject equal volumes (about 20 µL) of the *Standard preparation* and the *Assay preparation* into the chromatograph, record the chromatograms, and measure the responses for the major peaks. The relative retention times are about 0.7 for alclometasone dipropionate and 1.0 for betamethasone dipropionate. Calculate the quantity, in mg, of $C_{28}H_{37}ClO_7$ in the portion of Ointment taken by the formula:

$$10C(R_U/R_S),$$

in which *C* is the concentration, in mg per mL, of USP Alclometasone Dipropionate RS in the *Standard preparation*, and R_U and R_S are the peak height ratios obtained from the *Assay preparation* and the *Standard preparation*, respectively.

Alcohol

C_2H_5OH

C_2H_6O 46.07

Ethanol.

Ethyl alcohol [*64-17-5*].

» Alcohol contains not less than 92.3 percent and not more than 93.8 percent, by weight, corresponding to not less than 94.9 percent and not more than 96.0 percent, by volume, at 15.56°, of C_2H_5OH.

Packaging and storage—Preserve in tight containers, remote from fire.

Identification—

A: Mix 5 drops in a small beaker with 1 mL of potassium permanganate solution (1 in 100) and 5 drops of 2 *N* sulfuric acid, and cover the beaker immediately with a filter paper moistened with a solution recently prepared by dissolving 0.1 g of sodium nitroferricyanide and 0.25 g of piperazine in 5 mL of water: an intense blue color is produced on the filter paper, the color becoming paler after a few minutes.

B: To 5 mL of a solution (1 in 10) add 1 mL of 1.0 *N* sodium hydroxide, then slowly (over a period of 3 minutes) add 2 mL of 0.1 *N* iodine: the odor of iodoform develops, and a yellow precipitate is formed within 30 minutes.

Specific gravity ⟨841⟩: between 0.812 and 0.816 at 15.56°, indicating between 92.3% and 93.8%, by weight, or between 94.9% and 96.0%, by volume, of C_2H_5OH.

Acidity—To 50 mL, in a glass-stoppered flask, add 50 mL of recently boiled water. Add phenolphthalein TS, and titrate with 0.020 *N* sodium hydroxide to a pink color that persists for 30 seconds: not more than 0.90 mL of 0.020 *N* sodium hydroxide is required for neutralization.

Nonvolatile residue—Evaporate 40 mL in a tared dish on a water bath, and dry at 105° for 1 hour: the weight of the residue does not exceed 1 mg.

Water-insoluble substances—Dilute it with an equal volume of water: the mixture is clear and remains clear for 30 minutes after cooling to 10°.

Aldehydes and other foreign organic substances—Place 20 mL in a glass-stoppered cylinder that has been thoroughly cleaned with hydrochloric acid, then rinsed with water and finally with the Alcohol to be tested. Cool the contents to approximately 15°, and add, by means of a carefully cleaned pipet, 0.10 mL of 0.10 *N* potassium permanganate, noting accurately the time of addition. Mix at once by inverting the stoppered cylinder, and allow it to stand at 15° for 5 minutes: the pink color does not entirely disappear.

Amyl alcohol and nonvolatile, carbonizable substances, etc.—Allow 25 mL to evaporate spontaneously from a porcelain dish, carefully protected from dust, until the surface of the dish is barely moist: no red or brown color is produced immediately upon the addition of a few drops of sulfuric acid.

Fusel oil constituents—Wet a piece of clean, odorless, absorbent paper with a mixture of 10 mL of Alcohol, 5 mL of water, and 1 mL of glycerin, and allow to evaporate spontaneously: no foreign odor is perceptible after the last traces of Alcohol have left the paper.

Acetone and isopropyl alcohol—To 1.0 mL add 1.0 mL of water, 1.0 mL of a saturated solution of dibasic sodium phosphate, and 3.0 mL of a saturated solution of potassium permanganate. Warm the mixture to 45° to 50°, and allow to stand until the permanganate color is discharged. Add 3.0 mL of 2.5 *N* sodium hydroxide, and filter, without washing, through a sintered-glass filter. Prepare a control by mixing 1.0 mL of the saturated solution of dibasic sodium phosphate, 3.0 mL of 2.5 *N* sodium hydroxide, 8 µg of acetone, and 5.0 mL of water. To each solution add 1 mL of furfural solution (1 in 100), allow to stand for 10 minutes, then to 1.0 mL of each solution add 3 mL of hydrochloric acid: any pink color produced in the test solution is not more intense than that in the control.

Methanol—To 1 drop add 1 drop of water, 1 drop of dilute phosphoric acid (1 in 20), and 1 drop of potassium permanganate solution (1 in 20). Mix, allow to stand for 1 minute, and add sodium bisulfite solution (1 in 20), dropwise, until the permanganate color is discharged. If a brown color remains, add 1 drop of the dilute phosphoric acid. To the colorless solution add 5 mL of freshly prepared chromotropic acid TS, and heat on a water bath at 60° for 10 minutes: no violet color appears.

Dehydrated Alcohol

C_2H_5OH

C_2H_6O 46.07
Ethanol.
Ethyl alcohol [*64-17-5*].

» Dehydrated Alcohol contains not less than 99.2 percent, by weight, corresponding to not less than 99.5 percent, by volume, at 15.56°, of C_2H_5OH.

Packaging and storage—Preserve in tight containers, remote from fire.

Identification—

A: Mix 5 drops in a small beaker with 1 mL of potassium permanganate solution (1 in 100) and 5 drops of 2 *N* sulfuric acid, and cover the beaker immediately with a filter paper moistened with a solution recently prepared by dissolving 0.1 g of sodium nitroferricyanide and 0.25 g of piperazine in 5 mL of water: an intense blue color is produced on the filter paper, the color becoming paler after a few minutes.

B: To 5 mL of a solution (1 in 10) add 1 mL of 1.0 *N* sodium hydroxide, then slowly (over a period of 3 minutes) add 2 mL of 0.1 *N* iodine: the odor of iodoform develops, and a yellow precipitate is formed within 30 minutes.

Specific gravity ⟨841⟩: not more than 0.7964 at 15.56°, indicating not less than 99.2% of C_2H_5OH by weight.

Acidity—To 50 mL, in a glass-stoppered flask, add 50 mL of recently boiled water. Add phenolphthalein TS, and titrate with 0.020 *N* sodium hydroxide to a pink color that persists for 30 seconds: not more than 0.90 mL of 0.020 *N* sodium hydroxide is required for neutralization.

Nonvolatile residue—Evaporate 40 mL in a tared dish on a water bath, and dry at 105° for 1 hour: the weight of the residue does not exceed 1 mg.

Water-insoluble substances—Dilute it with an equal volume of water: the mixture is clear and remains clear for 30 minutes after cooling to 10°.

Aldehydes and other foreign organic substances—Place 20 mL in a glass-stoppered cylinder that has been thoroughly cleaned with hydrochloric acid, then rinsed with water and finally with the Dehydrated Alcohol to be tested. Cool the contents to approximately 15°, and add, by means of a carefully cleaned pipet, 0.10 mL of 0.10 *N* potassium permanganate, noting accurately the time of addition. Mix at once by inverting the stoppered cylinder, and allow it to stand at 15° for 5 minutes: the pink color does not entirely disappear.

Amyl alcohol and nonvolatile, carbonizable substances—Allow 25 mL to evaporate spontaneously from a porcelain dish, carefully protected from dust, until the surface of the dish is barely moist: no red or brown color is produced immediately upon the addition of a few drops of sulfuric acid.

Ultraviolet absorbance—Record the ultraviolet absorption spectrum between 350 nm and 220 nm in a 1-cm cell, with water in a matched cell in the reference beam: the absorbance is not more than 0.30 at 220 nm, 0.18 at 230 nm, 0.08 at 240 nm, and 0.02 at 270 nm to 350 nm, and the curve drawn through these points is smooth.

Fusel oil constituents—Wet a piece of clean, odorless, absorbent paper with a mixture of 10 mL of Dehydrated Alcohol, 5 mL of water, and 1 mL of glycerin, and allow to evaporate spontaneously: no foreign odor is perceptible after the last traces of Dehydrated Alcohol have left the paper.

Acetone and isopropyl alcohol—To 1.0 mL add 1 mL of water, 1 mL of a saturated solution of dibasic sodium phosphate, and 3 mL of a saturated solution of potassium permanganate. Warm the mixture to 45° to 50°, and allow to stand until the permanganate color is discharged. Add 3 mL of 2.5 *N* sodium hydroxide, and filter, without washing, through a sintered-glass filter. Prepare a control containing 1 mL of the saturated solution of dibasic sodium phosphate, 3 mL of 2.5 *N* sodium hydroxide, and 8 µg of acetone in 9 mL. To each solution add 1 mL of furfural solution (1 in 100), and allow to stand for 10 minutes, then to 1.0 mL of each solution add 3 mL of hydrochloric acid: any pink color produced in the test solution is not more intense than that in the control.

Methanol—To 1 drop add 1 drop of water, 1 drop of dilute phosphoric acid (1 in 20), and 1 drop of potassium permanganate solution (1 in 20). Mix, allow to stand for 1 minute, and add sodium bisulfite solution (1 in 20), dropwise, until the permanganate color is discharged. If a brown color remains, add 1 drop of the same dilute phosphoric acid. To the colorless solution add 5 mL of freshly prepared chromotropic acid TS, and heat on a water bath at 60° for 10 minutes: no violet color appears.

Dehydrated Alcohol Injection

» Dehydrated Alcohol Injection is Dehydrated Alcohol suitable for parenteral use.

Packaging and storage—Preserve in single-dose containers, preferably of Type I glass. The container may contain an inert gas in the headspace.

Specific gravity ⟨841⟩: not more than 0.8035 at 15.56°, indicating not less than 96.8%, by weight, of C_2H_5OH.

Acidity—To 50 mL, in a glass-stoppered flask, add 50 mL of recently boiled water. Add phenolphthalein TS, and titrate with 0.020 *N* sodium hydroxide to a pink color that persists for 30 seconds: not more than 10.0 mL of 0.020 *N* sodium hydroxide is required for neutralization.

Other requirements—It meets the requirements for *Identification, Nonvolatile residue, Water-insoluble substances, Aldehydes and other foreign organic substances, Amyl alcohol and nonvolatile, carbonizable substances, Ultraviolet absorbance, Fusel oil constituents, Acetone and isopropyl alcohol*, and *Methanol* under *Dehydrated Alcohol*, and meets the requirements under *Injections* ⟨1⟩.

Rubbing Alcohol

» Rubbing Alcohol and all preparations under the classification of Rubbing Alcohols are manufactured in accordance with the requirements of the U.S. Treasury Department, Bureau of Alcohol, Tobacco, and Firearms, Formula 23-H (8 parts by volume of acetone, 1.5 parts by volume of methyl isobutyl ketone, and 100 parts by volume of ethyl alcohol) being used. It contains not less than 68.5 percent and not more than 71.5 percent by volume of dehydrated alcohol, the remainder consisting of water and the denaturants, with or without color additives, and perfume oils. Rubbing Alcohol contains, in each 100 mL, not less than 355 mg of sucrose octaacetate or not less than 1.40 mg of denatonium benzoate. The preparation may be colored with one or more color additives, listed by the FDA for use in drugs. A suitable stabilizer may be added. Rubbing Alcohol complies with the requirements of the Bureau of Alcohol, Tobacco, and Firearms of the U.S. Treasury Department.

NOTE—Rubbing Alcohol is packaged, labeled, and sold in accordance with the regulations issued by the U.S. Treasury Department, Bureau of Alcohol, Tobacco, and Firearms.

Packaging and storage—Preserve in tight containers, remote from fire.

Labeling—Label it to indicate that it is flammable.

Reference standard—*USP Denatonium Benzoate Reference Standard*—Dry at 105° for 2 hours before using.

Specific gravity ⟨841⟩: between 0.8691 and 0.8771 at 15.56° (the U.S. Government standard temperature for alcohol determination), for Rubbing Alcohol manufactured with specially denatured alcohol Formula 23-H.

Nonvolatile residue—

Where the denaturant is sucrose octaacetate—Evaporate 25.0 mL of Rubbing Alcohol in a suitable tared dish on a steam bath, and dry the residue at 105° for 1 hour: the weight of the residue is not less than 89 mg. (Retain the residue for the *Assay for sucrose octaacetate.*)

Where the denaturant is denatonium benzoate—Evaporate 200.0 mL of Rubbing Alcohol, transferred in convenient portions, in a suitable tared dish on a steam bath, and dry the residue at 105° for 1 hour: the weight of the residue is not less than 2.8 mg. (Retain the residue for the *Assay for denatonium benzoate.*)

Methanol—Dilute 0.50 mL of it with water to 1.0 mL. To 0.50 mL of the resulting solution add 1 drop of dilute phosphoric acid (1 in 20) and 1 drop of potassium permanganate solution (1 in 20). Mix, allow to stand for 1 minute, and add sodium bisulfite solution (1 in 20), dropwise, until the permanganate color is discharged. If a brown color remains, add 1 drop of dilute phosphoric acid (1 in 20). To the colorless solution add 5 mL of freshly prepared chromotropic acid TS, and heat in a water bath at 60° for 10 minutes: no violet color appears.

Assay for denatonium benzoate—

Buffer solution—Dissolve 9.23 g of anhydrous dibasic sodium phosphate in 800 mL of water, adjust with saturated citric acid solution to a pH of 4 ± 0.1, dilute with water to 1000 mL, and mix.

Standard preparation—Dissolve about 25 mg of USP Denatonium Benzoate RS, accurately weighed, in water to make 500 mL, and mix.

Assay preparation—Dissolve the residue obtained in the test for *Nonvolatile residue* in 50.0 mL of water, and transfer to a suitable flask.

Procedure—Treat the *Standard preparation*, the *Assay preparation*, and the blank similarly and concomitantly. Transfer 10.0 mL each of the *Standard preparation*, the *Assay preparation*, and the *Buffer solution* to individual 250-mL separators, and add to each 40 mL of *Buffer solution*, 10 mL of a 1 in 1000 solution of bromophenol blue in chloroform, and 60 mL of chloroform. Shake the separators vigorously for 2 minutes, allow to stand for 15 minutes, then withdraw the chloroform layers through chloroform-washed cotton into 100-mL volumetric flasks. Repeat the extraction with 20 mL of chloroform, adding the filtered chloroform extracts to the respective volumetric flasks, dilute with chloroform to volume, and mix. Without delay, concomitantly determine the absorbances of the solutions in 1-cm cells at the wavelength of maximum absorbance at about 410 nm, with a suitable spectrophotometer, using the blank to set the instrument. Calculate the quantity, in mg, of denatonium benzoate ($C_{28}H_{34}N_2O_3$) in 100 mL of Rubbing Alcohol by the formula:

$$0.025C(A_U / A_S),$$

in which C is the concentration, in µg per mL, of USP Denatonium Benzoate RS in the *Standard preparation*, and A_U and A_S are the absorbances of the solutions from the *Assay preparation* and the *Standard preparation*, respectively.

Assay for sucrose octaacetate—Using about 50 mL of 70% alcohol, transfer the residue obtained in the test for *Nonvolatile residue* to a 500-mL conical flask. Neutralize the solution with 0.1 *N* sodium hydroxide VS, using phenolphthalein TS as the indicator. Add 25.0 mL of 0.1 *N* sodium hydroxide, attach an air condenser to the flask, and reflux on a steam bath for 1 hour. Remove from the steam bath, cool quickly, and titrate the excess alkali with 0.1 *N* sulfuric acid VS, using phenolphthalein TS as the indicator. Perform a blank determination (see *Residual Titrations* under *Titrimetry* ⟨541⟩). Each mL of 0.1 *N* sodium hydroxide is equivalent to 8.483 mg of sucrose octaacetate ($C_{28}H_{38}O_{19}$).

Alcohol in Dextrose Injection

» Alcohol in Dextrose Injection is a sterile solution of Alcohol and Dextrose in Water for Injection. It contains not less than 90.0 percent and not more than 110.0 percent of the labeled amount of alcohol (C_2H_5OH), and not less than 95.0 percent and not more than 105.0 percent of the labeled amount of dextrose ($C_6H_{12}O_6 . H_2O$).

Packaging and storage—Preserve in single-dose containers, preferably of Type I or Type II glass.

Labeling—The label states the total osmolarity of the solution expressed in mOsmol per liter.

Identification—It responds to the *Identification tests* under *Dextrose* and under *Dehydrated Alcohol.*

Pyrogen—It meets the requirements of the *Pyrogen Test* ⟨151⟩. [NOTE—Dilute, with Water for Injection, Injection containing more than 5% of dextrose to give a concentration of 5% of dextrose.]

pH ⟨791⟩: between 3.5 and 6.5, determined on a portion to which 0.30 mL of saturated potassium chloride solution has been added for each 100 mL and which previously has been diluted with water, if necessary, to a concentration of not more than 5% of dextrose.

Heavy metals ⟨231⟩—Transfer a volume of Injection, equivalent to 4.0 g of dextrose, to a vessel, and adjust the volume to 25 mL by evaporation or by addition of water, as necessary: the limit is 0.0005C%, in which C is the labeled amount, in g, of $C_6H_{12}O_6 . H_2O$ per mL of Injection.

5-Hydroxymethylfurfural and related substances—Dilute an accurately measured volume of Injection, equivalent to 1.0 g of $C_6H_{12}O_6 . H_2O$, with water to 500.0 mL. Determine the absorbance of this solution in a 1-cm cell at 284 nm, with a suitable spectrophotometer, using water as the blank: the absorbance is not more than 0.25.

Other requirements—It meets the requirements under *Injections* ⟨1⟩.

Assay for alcohol—Determine by the *Distillation Method* ⟨611⟩, using a 50.0-mL portion of Alcohol in Dextrose Injection.

Assay for dextrose—Transfer an accurately measured volume of Alcohol in Dextrose Injection, containing from 2 g to 5 g of dextrose, to a 100-mL volumetric flask. Add 0.2 mL of 6 *N* ammonium hydroxide, dilute with water to volume, and mix. Determine the angular rotation in a 200-mm tube at 25° (see *Optical Rotation* ⟨781⟩). The observed rotation in degrees, multiplied by 1.0425, represents the weight, in g, of $C_6H_{12}O_6 . H_2O$ in the volume of Injection taken.

Alcohols—*see complete list in index*
Alginic Acid—*see* Alginic Acid NF

Allopurinol

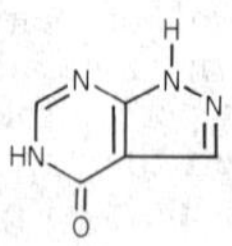

$C_5H_4N_4O$ 136.11
4*H*-Pyrazolo[3,4-*d*]pyrimidin-4-one, 1,5-dihydro-.
1,5-Dihydro-4*H*-pyrazolo[3,4-*d*]pyrimidin-4-one.
1*H*-Pyrazolo[3,4-*d*]pyrimidin-4-ol [*315-30-0*].

» Allopurinol contains not less than 98.0 percent and not more than 101.0 percent of $C_5H_4N_4O$, calculated on the dried basis.

Packaging and storage—Preserve in well-closed containers.

Reference standards—*USP Allopurinol Reference Standard*—Dry in vacuum at 105° for 5 hours before using. *USP 3-Amino-4-carboxamidopyrazole Hemisulfate Reference Standard*—Dry in vacuum at 105° for 3 hours before using.

Identification—The infrared spectrum of a potassium bromide dispersion of it exhibits maxima only at the same wavelengths as that of a similar preparation of USP Allopurinol RS.

Loss on drying ⟨731⟩—Dry it in vacuum at 105° for 5 hours: it loses not more than 0.5% of its weight.

Chromatographic impurities—

Standard preparation—Dissolve a suitable quantity of USP 3-Amino-4-carboxamidopyrazole Hemisulfate RS in 6 *N* ammonium hydroxide to obtain a solution having a known concentration of 50 µg per mL.

Test preparation—Dissolve 250 mg of Allopurinol in a mixture of 9 volumes of 6 *N* ammonium hydroxide and 1 volume of 1 *N* sodium hydroxide to make 10.0 mL, and mix.

Procedure—Apply 10 µL each of the *Standard preparation* and *Test preparation* on a thin-layer chromatographic plate coated with a 0.16-mm layer of chromatographic cellulose containing a fluorescent indicator. Develop the chromatogram in a chromatographic chamber containing a solvent prepared by shaking 200 mL of *n*-butyl alcohol and 200 mL of 6 *N* ammonium hydroxide, discarding the lower layer, and adding 20 mL of *n*-butyl alcohol to the top layer. When the solvent front is 1 cm from the top of the plate, remove the plate, air-dry, and examine under ultraviolet light: the intensity of any secondary spot produced by the *Test preparation* is not greater than that of the spot from the *Standard preparation* (0.2%).

Assay—Dissolve about 100 mg of Allopurinol, accurately weighed, in 30 mL of dimethylformamide, warming if necessary. Titrate with 0.1 *N* tetrabutylammonium hydroxide VS, determining the end-point potentiometrically, using a calomel-glass electrode system and taking the necessary precautions to prevent the absorption of atmospheric carbon dioxide. Perform a blank determination, and make any necessary correction. Each mL of 0.1 *N* tetrabutylammonium hydroxide is equivalent to 13.61 mg of $C_5H_4N_4O$.

Allopurinol Tablets

» Allopurinol Tablets contain not less than 93.0 percent and not more than 107.0 percent of the labeled amount of $C_5H_4N_4O$.

Packaging and storage—Preserve in well-closed containers.

Reference standard—*USP Allopurinol Reference Standard*—Dry in vacuum at 105° for 5 hours before using.

Identification—Extract a quantity of finely powdered Tablets, equivalent to about 50 mg of allopurinol, by trituration with 10 mL of 0.1 *N* sodium hydroxide. Filter, acidify the filtrate with 1 *N* acetic acid, collect the precipitated allopurinol (allow 10 to 15 minutes for sufficient precipitation to occur), wash the precipitate with 3 mL of dehydrated alcohol, in portions, and finally wash with 4 mL of anhydrous ethyl ether. Allow to dry in air for 15 minutes, then dry at 105° for 3 hours: the residue so obtained responds to *Identification test A* under *Allopurinol.*

Dissolution ⟨711⟩—

Medium: 0.1 *N* hydrochloric acid; 900 mL.

Apparatus 2: 75 rpm.

Time: 45 minutes.

Procedure—Determine the amount of $C_5H_4N_4O$ dissolved from ultraviolet absorbances at the wavelength of maximum absorbance at about 250 nm of filtered portions of the solution under test, suitably diluted with 0.1 *N* hydrochloric acid, in comparison with a Standard solution having a known concentration of USP Allopurinol RS in the same medium.

Tolerances—Not less than 75% (*Q*) of the labeled amount of $C_5H_4N_4O$ is dissolved in 45 minutes.

Uniformity of dosage units ⟨905⟩: meet the requirements.

Assay—[NOTE—Do not allow the *Mobile phase* to remain in the column overnight. After performing the procedure, flush the system with water for not less than 20 minutes, and then flush with methanol for 20 minutes.]

Mobile phase—Prepare a filtered and degassed, aqueous, 0.05 *M* solution of monobasic ammonium phosphate.

Internal standard solution—On the day of use, dissolve about 50 mg of hypoxanthine in 10 mL of 0.1 *N* sodium hydroxide, shake by mechanical means until dissolved (about 10 minutes), dilute with water to 50 mL, and mix.

Standard preparation—On the day of use, transfer about 50 mg of USP Allopurinol RS, accurately weighed, to a 50-mL volumetric flask, add 10 mL of 0.1 *N* sodium hydroxide, shake by mechanical means for 10 minutes, dilute with water to volume, and mix. Transfer 4.0 mL of this solution and 2.0 mL of *Internal standard solution* to a 200-mL volumetric flask, dilute with *Mobile phase* to volume, and mix.

Assay preparation—Weigh and finely powder not less than 20 Allopurinol Tablets. Transfer an accurately weighed portion of the powder, equivalent to about 50 mg of allopurinol, to a 50-mL volumetric flask, add 10 mL of 0.1 *N* sodium hydroxide, shake by mechanical means for 10 minutes, add water to volume, and mix. [NOTE—From this point, conduct the remainder of the *Assay* without delay.] Filter, rejecting the first 10 mL of the filtrate. Transfer 4.0 mL of the filtrate and 2.0 mL of *Internal standard solution* to a 200-mL volumetric flask, dilute with *Mobile phase* to volume, and mix.

Chromatographic system (see *Chromatography* ⟨621⟩)—The liquid chromatograph is equipped with a 254-nm detector and a 4-mm × 30-cm column that contains packing L1. The flow rate is about 1.5 mL per minute. Chromatograph the *Standard preparation*, and record the peak responses as directed under *Procedure:* the resolution, *R*, between the analyte and internal standard peaks is not less than 5, and the relative standard deviation for replicate injections is not more than 3.0%.

Procedure—Separately inject equal volumes (about 15 µL) of the *Standard preparation* and the *Assay preparation* into the chromatograph, record the chromatograms, and measure the responses for the major peaks. The relative retention times are about 0.6 for hypoxanthine and 1.0 for allopurinol. Calculate the quantity, in mg, of $C_5H_4N_4O$ in the portion of Tablets taken by the formula:

$$2.5C(R_U/R_S),$$

in which *C* is the concentration, in µg per mL, of USP Allopurinol RS in the *Standard preparation*, and R_U and R_S are the peak response ratios obtained from the *Assay preparation* and the *Standard preparation*, respectively.

Almond Oil—*see* Almond Oil NF

Aloe

» Aloe is the dried latex of the leaves of *Aloe barbadensis* Miller (*Aloe vera* Linné), known in commerce as Curaçao Aloe, or of *Aloe ferox* Miller and hybrids of this species with *Aloe africana* Miller and *Aloe spicata* Baker, known in commerce as Cape Aloe (Fam. Liliaceae).

Aloe yields not less than 50.0 percent of water-soluble extractive.

Botanic characteristics—

Curaçao Aloe—Brownish black, opaque masses. Its fractured surface is uneven, waxy, and somewhat resinous.

Cape Aloe—Dusky to dark brown irregular masses, the surfaces of which are often covered with a yellowish powder. Its fracture is smooth and glassy.

Powdered Aloe—Yellow, yellowish brown to olive-brown in color. When mounted in a bland expressed oil, it appears as greenish yellow to reddish brown angular or irregular fragments, the hues of which depend to some extent upon the thickness of the fragments.

Identification—

A: Powdered Aloe dissolves in nitric acid with effervescence, forming a reddish brown to brown or green solution.

B: Intimately mix in a flask or bottle 1 g of finely powdered Aloe with 25 mL of cold water, shake the mixture occasionally during 2 hours, transfer to a filter, and wash the filter and residue with sufficient cold water to make the filtrate measure 100 mL: the color of the filtrate, viewed in the bulb of a 100-mL volumetric flask, is dark orange with Curaçao Aloe, and greenish yellow with Cape Aloe. The filtrate darkens on standing.

C: To 5 mL of the filtrate obtained in *Identification test B* add 2 mL of nitric acid: the mixture exhibits a reddish orange color with Curaçao Aloe, and a reddish brown color which changes rapidly to green with Cape Aloe.

D: Mix 10 mL of the filtrate obtained in *Identification test B* with 2 mL of ammonium hydroxide: the mixture exhibits an amber color with Cape Aloe, and a dark amber color with Curaçao Aloe.

Water, *Method III* ⟨921⟩: not more than 12.0%, determined by drying at 105° for 5 hours. For Aloe that is not powdered, crush it in a mortar until it passes through a No. 40 sieve, and mix the ground material before weighing the sample.

Total ash ⟨561⟩: not more than 4.0%.

Alcohol-insoluble substances—Add about 1 g of powdered Aloe, accurately weighed, to 50 mL of alcohol in a flask. Heat the mixture to boiling, and maintain at incipient boiling for 15 minutes, replacing any loss by evaporation. Remove from the heat, and shake the mixture at intervals during 1 hour, filter through a small dried and tared filter paper or a dried and tared filtering crucible, and wash the residue on the filter with alcohol until the last washing is colorless. Dry the residue at 105° to constant weight: the weight of the residue does not exceed 10.0% of the weight of Aloe taken.

Assay—Macerate about 2 g of Aloe, accurately weighed, in about 70 mL of water in a suitable flask. Shake the mixture during 8 hours at 30-minute intervals, and allow it to stand for 16 hours without shaking. Filter, and wash the flask and residue with small portions of water, passing the washings through the filter, until the filtrate measures 100.0 mL. Evaporate a 50-mL aliquot of the filtrate in a tared dish on a steam bath to dryness, and dry at 110° to constant weight. The weight of water-soluble extractive so obtained is not less than 50.0% of the weight of Aloe taken.

Alprazolam

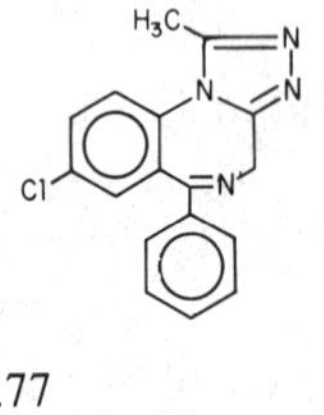

$C_{17}H_{13}ClN_4$ 308.77

4*H*-[1,2,4]Triazolo[4,3-*α*][1,4]benzodiazepine, 8-chloro-1-methyl-6-phenyl-.

8-Chloro-1-methyl-6-phenyl-4*H*-*s*-triazolo[4,3-*α*][1,4]benzodiazepine [*28981-97-7*].

» Alprazolam contains not less than 98.0 percent and not more than 102.0 percent of $C_{17}H_{13}ClN_4$.

Caution—Care should be taken to prevent inhaling particles of Alprazolam and exposing the skin to it.

Packaging and storage—Preserve in well-closed containers.

Reference standard—*USP Alprazolam Reference Standard*—Do not dry before using.

Identification—

A: The infrared absorption spectrum of a mineral oil dispersion of it exhibits maxima only at the same wavelengths as that of a similar preparation of USP Alprazolam RS, except that in the region of 880 to 890 cm^{-1} the maxima will vary with the ratios of alprazolam polymorphs.

B: Dissolve a suitable quantity in alcohol to obtain a solution having a concentration of 4 µg per mL: the ultraviolet absorption spectrum of this solution exhibits maxima and minima at the same wavelengths as that of a similar solution of USP Alprazolam RS, concomitantly measured, and the respective absorptivities, calculated on the dried basis, at the wavelength of maximum absorbance at about 220 nm, do not differ by more than 3.0%.

Loss on drying ⟨731⟩—Dry it at a pressure of not more than 5 mm of mercury at 60° for 16 hours: it loses not more than 0.5% of its weight.

Residue on ignition ⟨281⟩: not more than 0.5%.

Heavy metals, *Method II* ⟨231⟩: 0.002%.

Chromatographic purity—

METHOD A—

Test solution—Prepare a solution of Alprazolam in chloroform containing about 2 mg per mL.

Chromatographic system (see *Chromatography* ⟨621⟩)—The gas chromatograph is equipped with a flame-ionization detector, and contains a 3-mm × 120-cm glass column packed with 3 percent phase G6 on support S1AB. The column and injector port are maintained at a temperature of about 240°. The detector is maintained at a temperature about 20° to 50° above the column temperature. The carrier gas is helium.

Procedure—[NOTE—Allow about three times the elution time of the major component between successive injections.] Chromatograph about 4 µL of the *Test solution*, and record the chromatograms. Calculate the total impurities, in percentage, from the equation:

$$(100)(r_A + r_B + \ldots r_I)/(r_A + r_B + \ldots r_I + r),$$

where r_A, r_B, ... r_I are the responses for each peak other than the alprazolam peak present in the *Test solution*, and r is the response of the alprazolam peak in the *Test solution*. The total amount of impurities detected is not more than 1.0%.

METHOD B—

Standard solutions—Prepare a 4.0 mg per mL solution of USP Alprazolam RS in chloroform. Separately dilute 1, 3, and 5 mL of this solution to 100 mL with chloroform to obtain 0.1%, 0.3%, and 0.5% *Standard solutions*, respectively.

Test solution—Prepare a solution in chloroform containing 40 mg per mL.

Procedure—Separately apply 10 µL each of the *Test solution* and each *Standard solution* to a thin-layer chromatographic plate (see *Chromatography* ⟨621⟩) coated with a 0.50-mm layer of chromatographic silica gel mixture, and allow the spots to dry. Develop the chromatogram in a solvent system consisting of a mixture of chloroform, acetone, ethyl acetate, and methanol (50:50:50:5). Allow the solvent front to move about three-fourths of the length of the plate, remove the plate, and allow to dry. Repeat the development process a second time. Examine the plate under short-wavelength ultraviolet light and estimate the amount of any spots, other than the principal spot, in the chromatogram of the *Test solution:* no individual spot is greater in size or intensity than the spot produced by the 0.3% *Standard solution*, and the sum of any such spots detected is not greater than 1%.

Assay—

Internal standard solution—Prepare a solution of triazolam in acetonitrile having a concentration of about 0.3 mg per mL.

Standard preparation—Transfer about 2.5 mg of USP Alprazolam RS, accurately weighed, to a 100-mL volumetric flask, dissolve in 10.0 mL of *Internal standard solution*, dilute with acetonitrile to volume, and mix.

Assay preparation—Prepare as directed under *Standard preparation* using Alprazolam instead of the Reference Standard.

Mobile phase—Prepare a degassed and filtered solution containing acetonitrile, chloroform, 1-butanol, water, and glacial acetic acid (850:80:50:20:0.5). Make adjustments if necessary (see *System Suitability* under *Chromatography* ⟨621⟩).

Chromatographic system (see *Chromatography* ⟨621⟩)—The liquid chromatograph is equipped with a 254-nm detector and a 4.6-mm × 30-cm column that contains packing L3. The flow rate is about 2 mL per minute. Chromatograph the *Standard preparation*, and record the peak responses as directed under *Procedure:* the resolution, *R*, between the internal standard and alprazolam is not less than 2.0, and the relative standard deviation for replicate injections is not more than 2.0%.

Procedure—Separately inject equal volumes of the *Standard preparation* and the *Assay preparation* into the chromatograph, record the chromatograms, and measure the responses for the major peaks. The relative retention times are about 0.6 for the internal standard and 1.0 for alprazolam. Calculate the quantity, in mg, of $C_{17}H_{13}ClN_4$ in the portion of Alprazolam taken by the formula:

$$100C(R_U/R_S),$$

in which C is the concentration, in mg per mL, of USP Alprazolam RS in the *Standard preparation*, and R_U and R_S are the peak response ratios obtained from the *Assay preparation* and the *Standard preparation*, respectively.

Alprazolam Tablets

» Alprazolam Tablets contain not less than 90.0 percent and not more than 110.0 percent of the labeled amount of $C_{17}H_{13}ClN_4$.

Packaging and storage—Preserve in tight, light-resistant containers.

Reference standard—*USP Alprazolam Reference Standard*—Do not dry before using.

Identification—Dissolve an amount of finely powdered Tablets, equivalent to about 15 mg of alprazolam, in 10 mL of sodium carbonate solution (1 in 100). Add 15 mL of chloroform, and shake vigorously for 30 minutes. Centrifuge, withdraw the aqueous layer, and transfer the chloroform to a clean container. Add about 200 mg of potassium bromide. Evaporate the chloroform from this mixture to dryness, and dry the dispersion in vacuum at 60° for 24 hours. Grind this dispersion into a fine powder. Prepare a suitable pellet for testing by placing about 100 mg of dried potassium bromide into a die. Sprinkle about 20 mg of the finely ground, alprazolam–potassium bromide dispersion onto the dried potassium bromide layer, and cover with another specimen of about 100 mg of dried potassium bromide: the infrared absorption spectrum of the potassium bromide dispersion so obtained exhibits maxima characteristic of alprazolam, as compared to that of a similar preparation of USP Alprazolam RS, at the following wavenumbers: at 1609, 1578, 1566, 1539, 1530, 1487, 1445, 1428, 1379, 1337, and 1320 wavenumbers in the region of 1650 to 1300 cm^{-1}; at 970, 932, 891, 826, 797, 779, 746, 696, 669, 658, and 640 wavenumbers in the region of 975 to 600 cm^{-1}.

Dissolution ⟨711⟩—

Stock buffer solution—Dissolve 160 g of monobasic potassium phosphate and 40 g of dibasic potassium phosphate in water, and dilute with water to obtain 2.0 liters of solution. Add, with mixing, phosphoric acid or potassium hydroxide solution (45 in 100) as necessary to adjust the solution such that, when this *Stock buffer solution* is diluted 1 in 10 with water, the resulting solution has a pH of 6.0 ± 0.1.

Working buffer solution—Prepare a 1 in 10 dilution of *Stock buffer solution* in water to obtain a *Working buffer solution* having a pH of 6.0 ± 0.1.

Medium: Working buffer solution; 500 mL.

Apparatus 1: 100 rpm.

Time: 30 minutes.

Procedure—

Stock standard solution—Prepare a solution in methanol of USP Alprazolam RS having a known concentration of about 0.05 mg per mL.

Working standard solution—Add 50 mL of *Stock buffer solution* and 250 mL of water to a 500-mL volumetric flask. Add to the flask 5.0 mL of *Stock standard solution* for every 0.25 mg of alprazolam contained in the Tablet being assayed. Dilute with water to volume, and mix.

Mobile phase—Prepare a degassed and filtered solution of *Working buffer solution*, acetonitrile, and tetrahydrofuran (60:35:5). Make adjustments if necessary (see *System Suitability* under *Chromatography* ⟨621⟩).

Chromatographic system (see *Chromatography* ⟨621⟩)—The liquid chromatograph is equipped with a 254-nm detector and a 4.6-mm × 10-cm analytical column that contains packing L7. The flow rate is about 1 mL per minute. Chromatograph the *Standard preparation*, and record the peak responses as directed under *Procedure:* the number of theoretical plates is not less than 500, and the relative standard deviation for replicate injections is not more than 3.0%.

Procedure—Separately inject equal volumes of a filtered portion of the solution under test and the *Working standard solution* into the chromatograph, record the chromatograms, and measure the responses for the major peaks. Calculate the quantity of $C_{17}H_{13}ClN_4$ dissolved based on the peak responses obtained from the solution under test and the *Working standard solution*.

Tolerances—Not less than 80% (*Q*) of the labeled amount of $C_{17}H_{13}ClN_4$ is dissolved in 30 minutes.

Uniformity of dosage units ⟨905⟩: meet the requirements for *Content Uniformity*.

Procedure for content uniformity—

Mobile phase—Prepare as directed in the *Assay* under *Alprazolam*.

Internal standard solution—Prepare a solution of triazolam in acetonitrile having a concentration of about 0.032 mg per mL.

Test preparation—Transfer 1 Tablet to a container. Add about 0.4 mL of water directly onto the Tablet, allow the Tablet to stand for 2 minutes, and then swirl the container to disperse the Tablet. For every 0.25 mg of alprazolam contained in the Tablet, add 10.0 mL of *Internal standard solution* to the container. Shake, and centrifuge if necessary.

Standard preparation—Prepare a solution in *Internal standard solution* of USP Alprazolam RS having a known concentration of about 0.025 mg per mL.

Chromatographic system and *Procedure*—Proceed as directed in the *Assay* under *Alprazolam*. Calculate the quantity, in mg, of $C_{17}H_{13}ClN_4$ in the Tablet by the formula:

$$CV(R_U/R_S),$$

where V is the volume, in mL, of *Internal standard solution* in the *Test preparation*, and the other terms are as defined in the *Procedure* for *Alprazolam*.

Water, *Method I* ⟨921⟩: not more than 8.0%.

Assay—

Mobile phase, Internal standard solution, and *Standard preparation*—Prepare as directed in the *Assay* under *Alprazolam*.

Assay preparation—Place 6 Tablets in a container. For Tablets containing 1 mg or less of alprazolam, add 2 mL of water. For Tablets containing more than 1 mg of alprazolam, add 8 mL of water. Swirl to disperse the Tablets. Add a known, accurately measured volume of *Internal standard solution* such that the ratio of the volume, in mL, of *Internal standard solution* to the weight, in mg, of alprazolam is within a range of 3 to 4.5. Shake for 10 minutes, and centrifuge if necessary. Quantitatively dilute a volume of the resultant solution with acetonitrile to ten times its volume, and mix.

Chromatographic system and *Procedure*—Proceed as directed in the *Assay* under *Alprazolam*. Calculate the quantity, in mg, of $C_{17}H_{13}ClN_4$ in the Tablets taken by the formula:

$$10CV(R_U/R_S),$$

in which V is the volume, in mL, of *Internal standard solution* added, and the other terms are as defined therein.

Alprostadil

$C_{20}H_{34}O_5$ 354.49

Prost-13-en-1-oic acid, 11,15-dihydroxy-9-oxo-, (11α,13*E*,15*S*)-.

(1*R*,2*R*,3*R*)-3-Hydroxy-2-[(*E*)-(3*S*)-3-hydroxy-1-octenyl]-5-oxo-cyclopentane heptanoic acid [*745-65-3*].

» Alprostadil contains not less than 95.0 percent and not more than 105.0 percent of $C_{20}H_{34}O_5$, calculated on the anhydrous basis.

Caution—Great care should be taken to prevent inhaling particles of Alprostadil and exposing the skin to it.

Packaging and storage—Preserve in tight containers, in a refrigerator.

Reference standards—*USP Alprostadil Reference Standard*—Store in a freezer. Do not dry before using. *USP Prostaglandin A_1 Reference Standard*—Store in a freezer. Do not dry before using.

Identification—The infrared absorption spectrum of a mineral oil dispersion of it exhibits maxima only at the same wavelengths as that of a similar preparation of USP Alprostadil RS.

Water, *Method I* ⟨921⟩: not more than 0.5%, using 0.5 g.

Residue on ignition ⟨281⟩: not more than 0.5%, using 0.3 g.

Prostaglandin A_1, prostaglandin B_1, and 13,14-dihydroprostaglandin E_1—

Internal standard solution—Prepare a solution in methylene chloride containing about 0.02 mg of methyltestosterone per mL.

Mobile phase—Add 1.0 mL of water to 7.5 mL of *tert*-amyl alcohol. Add this solution to 1000 mL of methylene chloride, mix, and filter before use.

Standard preparation—Dissolve an accurately weighed quantity of USP Prostaglandin A_1 RS in methylene chloride, and dilute quantitatively with methylene chloride to obtain a *Standard stock solution* having a known concentration of about 0.1 mg per mL. Evaporate a 2.0-mL aliquot to dryness using nitrogen. Add 200 µL of a freshly prepared 1 in 50 solution of α-bromo-2-acetonaphthone in acetonitrile. Swirl to wash down the sides of the vial. Add 100 µL of a freshly prepared 1 in 100 solution of diisopropylethylamine in acetonitrile, swirl again, and allow to stand at room temperature for not less than 90 minutes. Evaporate with the aid of a stream of nitrogen to dryness, add 20.0 mL of *Internal standard solution*, and mix.

Test preparation—Accurately weigh about 2 mg of Alprostadil, dissolve in about 2 mL of dehydrated alcohol, and *gently* evaporate to dryness using nitrogen. Proceed as directed for *Standard preparation* beginning with "Add 200 µL of a freshly prepared 1 in 50 solution. . ." Dissolve the residue in 2.0 mL of *Internal standard solution.*

Chromatographic system (see *Chromatography* ⟨621⟩)—The liquid chromatograph is equipped with a 254-nm detector and a 4-mm × 30-cm column that contains packing L3. The flow rate is about 2 mL per minute. Chromatograph the *Standard preparation*, and record the peak responses as directed under *Procedure:* the resolution, *R*, between the USP Prostaglandin A_1 RS and internal standard peaks is not less than 4.0, and the relative standard deviation for replicate injections is not more than 2.5%.

Procedure—[NOTE—Where peak responses are indicated, use peak heights.] Separately inject equal volumes (about 25 µL) of the *Standard preparation* and the *Test preparation* into the chromatograph, record the chromatograms, and measure the responses for the peaks. The relative retention times are about 1.3 and 1.4 for prostaglandin A_1 and B_1, respectively, and 1.0 for the internal standard. Prostaglandin B_1 co-elutes with 13,14-dihydroprostaglandin E_1. Calculate the quantity, in mg, of prostaglandin A_1 by the formula:

$$0.2C_S(R_U/R_S),$$

in which C_S is the concentration of USP Prostaglandin A_1 RS, in mg per mL in the *Standard stock solution*, and R_U and R_S are the ratios of the peak responses for prostaglandin A_1 and internal standard peaks obtained from the *Test preparation* and the *Standard preparation*, respectively. Not more than 1.5% of prostaglandin A_1 is found.

Calculate the quantity, in mg, of prostaglandin B_1 and 13,14-dihydroprostaglandin E_1, as a summation, from the formula:

$$[0.2C_S(R_U/R_S)]/F,$$

where *F* is a response factor defined as the ratio of retention time of prostaglandin A_1 to prostaglandin B_1. The other terms are as previously defined. A total content of not more than 0.5% is found.

Foreign prostaglandins—

Mobile phase—Prepare a solution containing methylene chloride, 1,3-butanediol, and water (995:5:0.5).

Test preparation—Dissolve about 2 mg of Alprostadil, accurately weighed, in 2 mL of dehydrated alcohol, and gently evaporate to dryness using nitrogen.

Chromatographic system—The liquid chromatograph is capable of providing column pressure up to about 2000 psi and is equipped with a suitable recorder, a detector for monitoring ultraviolet light at 254 nm and a 4-mm × 30-cm stainless steel column containing packing L3.

Procedure—Add 200 µL of a 1 in 50 solution of α-bromo-2′-acetonaphthone in acetonitrile to the *Test preparation* washing the sides of the container, and swirl. Add 100 µL of a 1 in 100 solution of diisopropylethylamine in acetonitrile in a similar manner. Heat the container at 45° for about 60 minutes and evaporate to dryness using nitrogen. Add 10 mL of methylene chloride to dissolve the residue. Inject approximately 10 µL of the resulting solution. Adjust the attenuation for adequate recorder detection. [NOTE—Allow each chromatogram to run not less than 30 minutes before another injection.] Neglect peaks eluting in the first 8 minutes after injection (prostaglandin A_1 and prostaglandin B_1). Normalize areas for the alprostadil peak recorder attenuation. Calculate the percentage of each foreign prostaglandin by the formula:

$$(A_i \times 100)/(A_a + A_b + A_c \ldots + A_A),$$

in which A_i is the area of each individual peak, A_a, A_b, etc., are the areas of the foreign prostaglandin peaks, and A_A is the area of the alprostadil peak. Not more than 3.0% total is found and no single foreign prostaglandin is greater than 2.0%.

Assay—

Mobile phase—Proceed as directed for *Mobile phase* under *Foreign prostaglandins.*

Internal standard solution—Prepare a solution in methylene chloride containing about 0.4 mg of methylprednisolone per mL. Filter if necessary.

Standard preparation—Dissolve an accurately weighed quantity of USP Alprostadil RS in dehydrated alcohol, and dilute quantitatively with dehydrated alcohol to obtain a *Standard stock preparation* having a known concentration of about 0.5 mg per mL. *Gently* evaporate a 2.0-mL aliquot to dryness using nitrogen. Proceed as directed under *Assay preparation* beginning with "Add 200 µL of a 1 in 50."

Assay preparation—Accurately weigh about 5 mg of Alprostadil and dissolve in 10.0 mL of dehydrated alcohol. *Gently* evaporate a 2.0-mL aliquot to dryness using nitrogen. Add 200 µL of a 1 in 50 freshly prepared solution of α-bromo-2′-acetonaphthone in acetonitrile, washing down the inside of the container, and swirl. Add 100 µL of a freshly prepared 1 in 100 solution of diisopropylethylamine in acetonitrile to the container in a similar manner. Cap and heat the container at 45° for one hour, swirling occasionally. If the entire sample does not dissolve using the derivatization mixture, the specimen should be discarded. Evaporate the solution using nitrogen, add 10.0 mL of *Internal standard solution* to the container, and mix. Use sonication if undissolved drug is observed. If incomplete dissolution is still observed, discard the specimen.

Chromatographic system—Proceed as directed in *Chromatographic system* under *Prostaglandin A_1, prostaglandin B_1, and 13,14-dihydroprostaglandin E_1.* The resolution, *R*, is not less than 4.0 between the peaks for alprostadil and the internal standard and the relative standard deviation for replicate injections of the *Standard preparation* is not greater than 2.5%. The approximate relative retention times of alprostadil and methylprednisolone are 0.6 and 1.0, respectively.

Procedure—[NOTE—Where peak responses are indicated, use peak heights.] Inject equal volumes of the derivatized *Standard preparation* and derivatized *Assay preparation* into the liquid chromatograph. Record the chromatograms and measure the peak responses at equivalent relative retention times. Calculate the quantity, in mg, of $C_{20}H_{34}O_5$ in the portion of Alprostadil taken by the formula:

$$10C(R_U/R_S),$$

in which *C* is the concentration, in mg per mL, of USP Alprostadil RS in the *Standard stock preparation*, and R_U and R_S are the ratios of the peak responses for the alprostadil and internal standard peaks obtained from the *Assay preparation* and the *Standard preparation*, respectively.

Alprostadil Injection

» Alprostadil Injection is a sterile solution of Alprostadil in Dehydrated Alcohol. It contains not less than 90.0 percent and not more than 115.0 percent of the labeled amount of alprostadil ($C_{20}H_{34}O_5$).

Packaging and storage—Preserve in single-dose containers, preferably of Type I glass. Store in a refrigerator.

Reference standard—*USP Alprostadil Reference Standard*—Store in a freezer. Do not dry beforc using.

Identification—Dry an amount of Injection, equivalent to about 2 mg of alprostadil, on about 500 mg of spectroscopic grade potassium bromide at about 40° to 50° under vacuum. Prepare a pellet from this mixture: the infrared absorption of it exhibits maxima only at the same wavelengths as that of a similar preparation of USP Alprostadil RS, prepared from an ethanolic solution.

Bacterial endotoxins—When tested as directed under *Bacterial Endotoxins Test* ⟨85⟩, the USP Endotoxin RS being used, it contains not more than 5 USP Endotoxin units per 100 µg of alprostadil.

Sterility—It meets the requirements under *Sterility Tests* ⟨71⟩, when tested as directed in the section, *Test Procedures Using Membrane Filtration.*

Water, *Method I* ⟨921⟩: not more than 0.4%.

Other requirements—It meets the requirements under *Injections* ⟨1⟩.

Assay—

Internal standard preparation, Standard preparation, and *Chromatographic system*—Proceed as directed in the *Assay* under *Alprostadil.*

Mobile phase—Prepare as directed for *Mobile phase* in the test for *Foreign prostaglandins* under *Alprostadil.*

Assay preparation—Pool the contents of several containers of Alprostadil Injection and *gently* evaporate an accurately measured volume, equivalent to about 1 mg of alprostadil, to dryness using nitrogen. Proceed as directed for *Assay preparation* in the *Assay* under *Alprostadil* beginning with "Add 200 µL of . . . ."

Procedure—Proceed as directed for *Procedure* in the *Assay* under *Alprostadil.* Calculate the quantity, in mg, of $C_{20}H_{34}O_5$ in the volume of Injection taken by the formula:

$$C(R_U/R_S),$$

in which the terms are as previously defined.

Ammonium Alum

$AlNH_4(SO_4)_2 \cdot 12H_2O$ 453.32

Sulfuric acid, aluminum ammonium salt (2:1:1), dodecahydrate.

Aluminum ammonium sulfate (1:1:2) dodecahydrate [*7784-26-1*].

Anhydrous 237.14 [*7784-25-0*].

» Ammonium Alum contains not less than 99.0 percent and not more than 100.5 percent of $AlNH_4(SO_4)_2$, calculated on the dried basis.

Identification—

A: Add 1 *N* sodium hydroxide dropwise to a solution of it (1 in 20): a precipitate is formed, and it dissolves in an excess of the reagent with the evolution of ammonia, recognizable by its odor and by its alkaline effect upon moistened red litmus paper exposed to the vapor.

B: A solution (1 in 20) responds to the tests for *Aluminum* ⟨191⟩ and for *Sulfate* ⟨191⟩.

Loss on drying ⟨731⟩—Transfer 2.0 g, in a tared porcelain crucible, to a muffle furnace at 200°. Increase the temperature to 300°, and dry at 300° to constant weight. Cool in a desiccator, and weigh: it loses between 45.0% and 48.0% of its weight.

Alkalies and alkaline earths—Completely precipitate the aluminum from a boiling solution of 1 g of it in 100 mL of water by the addition of sufficient 6 *N* ammonium hydroxide to render the solution distinctly alkaline to methyl red TS, and filter. Evaporate the filtrate to dryness, and ignite: the weight of the residue does not exceed 5 mg (0.5%).

Arsenic, *Method I* ⟨211⟩: 3 ppm.

Heavy metals, *Method I* ⟨231⟩—Dissolve 1 g in water to make 20 mL, and add 5 mL of 0.1 *N* hydrochloric acid. Evaporate the solution in a porcelain evaporating dish to dryness. Treat the residue with 20 mL of water, and add 50 mg of hydroxylamine hydrochloride. Heat the solution on a steam bath for 10 minutes, cool, dilute with water to 25 mL, and proceed as directed, except to add 50 mg of hydroxylamine hydrochloride to the *Standard Preparation:* the limit is 0.002%.

Iron—Add 5 drops of potassium ferrocyanide TS to 20 mL of a solution (1 in 150): no blue color is produced immediately.

Assay—

Disodium ethylenediaminetetraacetate titrant—Dissolve 18.6 g of disodium ethylenediaminetetraacetate in water to make 1000 mL, and standardize the solution as follows: Weigh accurately about 2 g of aluminum wire, transfer to a 1000-mL volumetric flask, and add 50 mL of a mixture of hydrochloric acid and water (1:1). Swirl the flask to ensure contact of the aluminum and the acid, and allow the reaction to proceed until all of the aluminum has dissolved. Dilute with water to volume, and mix. Pipet 10 mL of this solution into a 250-mL beaker, add, in the order named and with continuous stirring 25.0 mL of *Disodium ethylenediaminetetraacetate titrant* and 20 mL of acetic acid–ammonium acetate buffer TS, and boil gently for 5 minutes. Cool, and add 50 mL of alcohol, and 2 mL of dithizone TS. Titrate with 0.05 *M* zinc sulfate VS to a bright rose-pink color. Perform a blank determination, substituting 10 mL of water for the aluminum solution, and make any necessary correction. Calculate the molarity of the solution by the formula:

$$W/26.98V,$$

in which *W* is the weight, in mg, of aluminum in the portion of solution taken, and *V* is the volume, in mL, of *Disodium ethylenediaminetetraacetate titrant* consumed.

Procedure—Transfer about 800 mg of Ammonium Alum, accurately weighed, to a 400-mL beaker, moisten with 1 mL of glacial acetic acid, and add 50 mL of water, 50.0 mL of *Disodium ethylenediaminetetraacetate titrant*, and 20 mL of acetic acid–ammonium acetate buffer TS. Warm on a steam bath until solution is complete, and boil gently for 5 minutes. Cool, add 50 mL of alcohol and 2 mL of dithizone TS, and titrate with 0.05 *M* zinc sulfate VS to a bright rose-pink color. Perform a blank determination, and make any necessary correction. Each mL of 0.05 *M Disodium ethylenediaminetetraacetate titrant* is equivalent to 11.86 mg of $AlNH_4(SO_4)_2$.

Potassium Alum

$AlK(SO_4)_2 \cdot 12H_2O$ 474.38

Sulfuric acid, aluminum potassium salt (2:1:1), dodecahydrate.

Aluminum potassium sulfate (1:1:2) dodecahydrate [*7784-24-9*].

Anhydrous 258.19 [*10043-67-1*].

» Potassium Alum contains not less than 99.0 percent and not more than 100.5 percent of $AlK(SO_4)_2$, calculated on the dried basis.

Packaging and storage—Preserve in well-closed containers.

Identification—

A: Add 1 *N* sodium hydroxide dropwise to a solution of it (1 in 20): a precipitate is formed which dissolves in an excess of the reagent. Ammonia is not evolved (*distinction from Ammonium Alum*).

B: Hold it in a nonluminous flame: a violet color is imparted to the flame.

C: Add 10 mL of sodium bitartrate TS to 5 mL of a saturated solution of it: a white, crystalline precipitate is formed within 30 minutes.

D: A solution (1 in 20) responds to the tests for *Aluminum* ⟨191⟩ and for *Sulfate* ⟨191⟩.

Loss on drying ⟨731⟩—Transfer 2.0 g, in a tared porcelain crucible, to a muffle furnace at 200°. Increase the temperature to 400°, and dry at 400° to constant weight. Cool in a desiccator, and weigh: it loses between 43.0% and 46.0% of its weight.

Arsenic, *Method I* ⟨211⟩: 3 ppm.

Heavy metals, *Method I* ⟨231⟩—Dissolve 1 g in water to make 20 mL, and add 5 mL of 0.1 *N* hydrochloric acid. Evaporate the solution in a porcelain evaporating dish to dryness. Treat the residue with 20 mL of water, and add 50 mg of hydroxylamine hydrochloride. Heat the solution on a steam bath for 10 minutes, cool, dilute with water to 25 mL, and proceed as directed, except to add 50 mg of hydroxylamine hydrochloride to the *Standard Preparation:* the limit is 0.002%.

Iron—Add 5 drops of potassium ferrocyanide TS to 20 mL of a solution (1 in 150): no blue color is produced immediately.

Assay—

Disodium ethylenediaminetetraacetate titrant—Prepare and standardize as directed in the *Assay* under *Ammonium Alum.*

Procedure—Transfer about 800 mg of Potassium Alum, accurately weighed, to a 400-mL beaker, moisten with 1 mL of glacial acetic acid, and add 50 mL of water, 50.0 mL of *Disodium ethylenediaminetetraacetic titrant*, and 20 mL of acetic acid–ammonium acetate buffer TS. Warm on a steam bath until solution is complete, and boil gently for 5 minutes. Cool, add 50 mL of alcohol and 2 mL of dithizone TS, and titrate 0.05 *M* zinc sulfate VS to a bright rose-pink color. Perform a blank determination, and make any necessary correction. Each mL of 0.05 *M Disodium ethylenediaminetetraacetate titrant* is equivalent to 12.91 mg of $AlK(SO_4)_2$.

Alumina Oral Suspension, Magnesia and—*see* Magnesia and Alumina Oral Suspension

Alumina Tablets, Magnesia and—*see* Magnesia and Alumina Tablets

Alumina and Magnesia Oral Suspension

» Alumina and Magnesia Oral Suspension is a mixture containing aluminum hydroxide [$Al(OH)_3$] and Magnesium Hydroxide [$Mg(OH)_2$], with $Al(OH)_3$ predominating. It contains the equivalent of not less than 90.0 percent and not more than 110.0 percent of the labeled amounts of aluminum hydroxide [$Al(OH)_3$] and magnesium hydroxide [$Mg(OH)_2$]. It may contain a flavoring agent, and may contain suitable antimicrobial agents.

Packaging and storage—Preserve in tight containers, and avoid freezing.

Labeling—Oral Suspension may be labeled to state the aluminum hydroxide content in terms of the equivalent amount of dried aluminum hydroxide gel, on the basis that each mg of dried gel is equivalent to 0.765 mg of $Al(OH)_3$.

Identification—

A: To a solution of 5 g in 10 mL of 3 *N* hydrochloric acid add 5 drops of methyl red TS, heat to boiling, add 6 *N* ammonium hydroxide until the color of the solution changes to deep yellow, then continue boiling for 2 minutes, and filter: the filtrate responds to the tests for *Magnesium* ⟨191⟩.

B: Wash the precipitate obtained in *Identification test A* with hot ammonium chloride solution (1 in 50), and dissolve the precipitate in hydrochloric acid: the solution responds to the tests for *Aluminum* ⟨191⟩.

Microbial limits ⟨61⟩—Its total aerobic microbial count does not exceed 100 per mL, and it meets the requirements of the test for absence of *Escherichia coli.*

Acid-neutralizing capacity ⟨301⟩—The acid consumed by the minimum single dose recommended in the labeling is not less than 5 mEq, and not less than the number of mEq calculated by the formula:

$$0.55(0.0385A) + 0.8(0.0343M),$$

in which 0.0385 and 0.0343 are the theoretical acid-neutralizing capacities, in mEq, of $Al(OH)_3$ and $Mg(OH)_2$, respectively, and A and M are the quantities, in mg, of $Al(OH)_3$ and $Mg(OH)_2$ in the specimen tested, based on the labeled quantities.

pH ⟨791⟩: between 7.3 and 7.9.

Chloride ⟨221⟩—Dissolve 5.0 g in the minimum volume of nitric acid required to achieve complete solution, add 1 mL of acid in excess, then add water to make 100 mL, and filter: a 10-mL portion of the filtrate shows no more chloride than corresponds to 1.0 mL of 0.020 *N* hydrochloric acid (0.14%).

Sulfate ⟨221⟩—Dissolve 5.0 g in 5 mL of 3 *N* hydrochloric acid, with gentle heating. Cool, add water to make 250 mL, mix, and filter: a 20-mL portion of the filtrate shows no more sulfate than corresponds to 0.40 mL of 0.020 *N* sulfuric acid (0.1%).

Other requirements—It meets the requirements of the tests for *Arsenic* and *Heavy metals* under *Aluminum Hydroxide Gel.*

Assay for aluminum hydroxide—

Disodium ethylenediaminetetraacetate titrant—Prepare and standardize as directed in the *Assay* under *Ammonium Alum.*

Assay preparation—Transfer an accurately measured quantity of Alumina and Magnesia Oral Suspension, previously well shaken in its original container, equivalent to about 1200 mg of aluminum hydroxide, to a suitable beaker. Add 20 mL of water, stir, and slowly add 10 mL of hydrochloric acid. Heat gently, if necessary, to aid solution, cool, and filter into a 200-mL volumetric flask. Wash the filter with water into the flask, add water to volume, and mix.

Procedure—Pipet 10 mL of *Assay preparation* into a 250-mL beaker, add 20 mL of water, then add, in the order named and with continuous stirring, 25.0 mL of *Disodium ethylenediaminetetraacetate titrant* and 20 mL of acetic acid–ammonium acetate buffer TS, and heat near the boiling point for 5 minutes. Cool, add 50 mL of alcohol and 2 mL of dithizone TS, and mix. Titrate with 0.05 *M* zinc sulfate VS until the color changes from green-violet to rose-pink. Perform a blank determination, substituting 10 mL of water for the *Assay preparation*, and make any necessary correction. Each mL of 0.05 *M Disodium ethylenediaminetetraacetate titrant* consumed is equivalent to 3.900 mg of $Al(OH)_3$.

Assay for magnesium hydroxide—

Assay preparation—Prepare as directed in the *Assay for aluminum hydroxide.*

Procedure—Pipet a volume of *Assay preparation*, equivalent to about 40 mg of magnesium hydroxide, into a 400-mL beaker, add 200 mL of water and 20 mL of triethanolamine, and stir. Add 10 mL of ammonia–ammonium chloride buffer TS and 3 drops of an eriochrome black indicator solution prepared by dissolving 200 mg of eriochrome black T in a mixture of 15 mL of triethanolamine and 5 mL of dehydrated alcohol, and mix. Cool the solution to between 3° and 4° by immersion of the beaker in an ice bath, then remove, and titrate with 0.05 *M* disodium ethylenediaminetetraacetate VS to a blue end-point. Perform a blank determination, substituting 10 mL of water for the *Assay preparation*, and make any necessary correction. Each mL of 0.05 *M* disodium ethylenediaminetetraacetate consumed is equivalent to 2.916 mg of $Mg(OH)_2$.

Alumina and Magnesia Tablets

» Alumina and Magnesia Tablets contain not less than 90.0 percent and not more than 110.0 percent of the labeled amounts of aluminum hydroxide [$Al(OH)_3$] and magnesium hydroxide [$Mg(OH)_2$].

Packaging and storage—Preserve in well-closed containers.

Labeling—Tablets prepared with the use of Dried Aluminum Hydroxide Gel may be labeled to state the aluminum hydroxide content in terms of the equivalent amount of dried aluminum hydroxide gel, on the basis that each mg of dried gel is equivalent to 0.765 mg of $Al(OH)_3$.

Identification—

A: To a 0.7-g portion of finely powdered Tablets add 10 mL of 3 *N* hydrochloric acid and 5 drops of methyl red TS, heat to boiling, and add 6 *N* ammonium hydroxide until the color of the solution changes to deep yellow. Continue boiling for 2 minutes, and filter: the filtrate responds to the tests for *Magnesium* ⟨191⟩.

B: Wash the precipitate obtained in *Identification test A* with a hot solution of ammonium chloride (1 in 50), and dissolve the precipitate in hydrochloric acid: the solution responds to the tests for *Aluminum* ⟨191⟩.

Disintegration ⟨701⟩: 10 minutes, simulated gastric fluid TS being substituted for water in the test.

Uniformity of dosage units ⟨905⟩: meet the requirements for *Weight Variation* with respect to alumina and to magnesia.

Acid-neutralizing capacity ⟨301⟩—The acid consumed by the minimum single dose recommended in the labeling is not less than 5 mEq, and not less than the number of mEq calculated by the formula:

$$0.55(0.0385A) + 0.8(0.0343M),$$

in which 0.0385 and 0.0343 are the theoretical acid-neutralizing capacities, in mEq, of $Al(OH)_3$ and $Mg(OH)_2$, respectively, and *A* and *M* are the quantities, in mg, of $Al(OH)_3$ and $Mg(OH)_2$ in the specimen tested, based on the labeled quantities.

Assay for aluminum hydroxide—

Disodium ethylenediaminetetraacetate titrant—Prepare and standardize as directed in the *Assay* under *Ammonium Alum*.

Assay preparation—Weigh and finely powder not less than 20 Alumina and Magnesia Tablets. Transfer an accurately weighed portion of the powder, equivalent to about 1200 mg of aluminum hydroxide, to a 150-mL beaker, add 20 mL of water, stir, and slowly add 30 mL of 3 *N* hydrochloric acid. Proceed as directed for *Assay preparation* in the *Assay for aluminum hydroxide* under *Alumina and Magnesia Oral Suspension*, beginning with "Heat gently, if necessary."

Procedure—Proceed as directed for *Procedure* in the *Assay for aluminum hydroxide* under *Alumina and Magnesia Oral Suspension*. Each mL of 0.05 *M Disodium ethylenediaminetetraacetate titrant* is equivalent to 3.900 mg of $Al(OH)_3$.

Assay for magnesium hydroxide—

Assay preparation—Prepare as directed in the *Assay for aluminum hydroxide*.

Procedure—Proceed as directed for *Procedure* in the *Assay for magnesium hydroxide* under *Alumina and Magnesia Oral Suspension*.

Alumina, and Magnesia Tablets, Aspirin,—*see* Aspirin, Alumina, and Magnesia Tablets

Alumina, and Magnesia Tablets, Aspirin, Codeine Phosphate,—*see* Aspirin, Codeine Phosphate, Alumina, and Magnesia Tablets

Alumina, Magnesia, and Calcium Carbonate Oral Suspension

» Alumina, Magnesia, and Calcium Carbonate Oral Suspension contains not less than 90.0 percent and not more than 110.0 percent of the labeled amounts of aluminum hydroxide [$Al(OH)_3$], magnesium hydroxide [$Mg(OH)_2$], and calcium carbonate ($CaCO_3$).

Packaging and storage—Preserve in tight containers, and avoid freezing.

Labeling—Oral Suspension may be labeled to state the aluminum hydroxide content in terms of the equivalent amount of dried aluminum hydroxide gel, on the basis that each mg of dried gel is equivalent to 0.765 mg of $Al(OH)_3$.

Identification—

A: To 5 g of Oral Suspension add 25 mL of 2 *N* sulfuric acid, stir, and allow to stand for 5 minutes. Add 25 mL of alcohol, stir, and place in an ice bath for 30 minutes. Filter while cold, retaining the filtrate for *Identification tests B* and *C*. Wash the precipitate with 50 mL of 0.75 *N* sulfuric acid, and discard the washings: the precipitate so obtained, dissolved in 3 *N* hydrochloric acid and filtered, responds to the tests for *Calcium* ⟨191⟩.

B: To the filtrate obtained in *Identification test A* add 5 drops of methyl red TS, and heat to boiling. Add 6 *N* ammonium hydroxide until the color of the solution changes to deep yellow, continue boiling for 2 minutes, and filter through hardened filter paper. (Retain the filtrate for *Identification test C*.) Wash the precipitate with 350 mL of a hot ammonium chloride solution (1 in 50), discarding the washings: the precipitate so obtained, dissolved in 3 *N* hydrochloric acid, responds to the tests for *Aluminum* ⟨191⟩.

C: The filtrate obtained in *Identification test B* responds to the tests for *Magnesium* ⟨191⟩.

Microbial limits—Its total aerobic microbial count does not exceed 100 per mL, and it meets the requirements of the test for absence of *Escherichia coli* and *Pseudomonas aeruginosa* under *Microbial Limit Tests* ⟨61⟩.

pH ⟨791⟩: between 7.5 and 8.5.

Chloride ⟨221⟩—Dissolve 5.0 g in 3 mL of nitric acid, add water to make 100 mL, and filter: a 10.0-mL portion of the filtrate shows no more chloride than corresponds to 1.0 mL of 0.020 *N* hydrochloric acid (0.14%).

Sulfate ⟨221⟩—Dissolve 5.0 g in 7 mL of 3 *N* hydrochloric acid, with gentle heating. Cool, add water to make 250 mL, mix, and filter: a 20.0-mL portion of the filtrate shows no more sulfate than corresponds to 0.40 mL of 0.020 *N* sulfuric acid (0.1%).

Other requirements—It meets the requirements of the tests for *Arsenic* and *Heavy metals* under *Aluminum Hydroxide Gel*.

Assay for aluminum hydroxide—

Disodium ethylenediaminetetraacetate titrant—Prepare and standardize as directed in the *Assay* under *Alum*.

Assay preparation—Transfer an amount of Alumina, Magnesia, and Calcium Carbonate Oral Suspension, previously well shaken in its original container, equivalent to about 600 mg of aluminum hydroxide, to a tared beaker, and weigh accurately. Add 20 mL of water, stir, and slowly add 40 mL of 3 *N* hydrochloric acid. Heat gently, if necessary, to aid solution, cool, and transfer to a 200-mL volumetric flask. Wash the beaker with water, adding the washings to the flask, add water to volume, and mix.

Procedure—Pipet 10 mL of *Assay preparation* into a 250-mL beaker, add 20 mL of water, then add, in the order named and with continuous stirring, 25.0 mL of 0.05 *M Disodium ethylenediaminetetraacetate titrant* and 20 mL of acetic acid–ammonium acetate buffer TS, and heat the solution near the boiling temperature for 5 minutes. Cool, add 50 mL of alcohol and 2 mL of dithizone TS, and mix. Titrate with 0.05 *M* zinc sulfate VS until the color changes from green-violet to rose-pink. Perform a blank determination, substituting 10 mL of water for the *Assay preparation*, and make any necessary correction. Each mL of 0.05 *M Disodium ethylenediaminetetraacetate titrant* consumed is equivalent to 3.900 mg of $Al(OH)_3$.

Assay for magnesium hydroxide—

Assay preparation—Prepare as directed in the *Assay* for *aluminum hydroxide.*

Procedure—Pipet a volume of *Assay preparation*, equivalent to about 40 mg of magnesium hydroxide, into a 400-mL beaker, add 200 mL of water and 20 mL of trolamine, and mix. Add 50 mL of ammonia–ammonium chloride buffer TS and 2 drops of eriochrome black indicator solution (prepared by dissolving 200 mg of eriochrome black T in a mixture of 15 mL of trolamine and 5 mL of dehydrated alcohol, and mixing). Cool the solution to between 3° and 4° by immersing the beaker in an ice bath, and titrate with 0.05 *M* disodium ethylenediaminetetraacetate VS until the color changes to pure blue. Perform a blank determination, substituting 10 mL of water for the *Assay preparation*, and make any necessary correction. From the volume of 0.05 *M* sodium ethylenediaminetetraacetate consumed, subtract the volume of 0.05 *M* disodium ethylenediaminetetraacetate consumed in the *Assay for calcium carbonate.* Each mL of 0.05 *M* sodium ethylenediaminetetraacetate is equivalent to 2.916 mg of $Mg(OH)_2$.

Assay for calcium carbonate—

Assay preparation—Prepare as directed in the *Assay for aluminum hydroxide.*

Procedure—Pipet a volume of *Assay preparation*, equivalent to about 50 mg of calcium carbonate, into a 400-mL beaker, and add 200 mL of water, 5 mL of sodium hydroxide solution (1 in 2), and 250 mg of hydroxy naphthol blue indicator. Stir with a magnetic stirrer, and titrate immediately with 0.05 *M* disodium ethylenediaminetetraacetate VS until the solution is distinctly blue. Each mL of 0.05 *M* disodium ethylenediaminetetraacetate is equivalent to 5.004 mg of $CaCO_3$.

Alumina, Magnesia, and Calcium Carbonate Tablets

» Alumina, Magnesia, and Calcium Carbonate Tablets contain not less than 90.0 percent and not more than 110.0 percent of the labeled amounts of aluminum hydroxide [$Al(OH)_3$], magnesium hydroxide [$Mg(OH)_2$], and calcium carbonate ($CaCO_3$).

Packaging and storage—Preserve in well-closed containers.

Labeling—Label the Tablets to indicate that they are to be chewed before being swallowed. Tablets prepared with the use of Dried Aluminum Hydroxide Gel may be labeled to state the aluminum hydroxide content in terms of the equivalent amount of dried aluminum hydroxide gel, on the basis that each mg of dried gel is equivalent to 0.765 mg of $Al(OH)_3$.

Identification—To a 3-g portion of finely powdered Tablets add 25 mL of water and 25 mL of 2 *N* sulfuric acid, stir, and heat on a steam bath for 10 minutes. Cool, add 50 mL of alcohol, and stir: the mixture so obtained responds to *Identification tests A, B,* and *C* under *Alumina, Magnesia, and Calcium Carbonate Oral Suspension*, beginning under *Identification test A* with "place in an ice bath for 30 minutes."

Disintegration ⟨701⟩: 45 minutes.

Uniformity of dosage units ⟨905⟩: meet the requirements for *Weight Variation* with respect to alumina, to magnesia, and to calcium carbonate.

Assay for aluminum hydroxide—

Assay preparation—Weigh and finely powder not less than 20 Tablets. Transfer an accurately weighed portion of the powder, equivalent to about 600 mg of aluminum hydroxide, to a beaker, add 20 mL of water, and slowly add 40 mL of 3 *N* hydrochloric acid, with mixing. Heat the mixture to boiling, cool, and filter into a 200-mL volumetric flask. Wash the beaker with water, adding the washings to the filter. Add water to volume, and mix.

Procedure—Proceed as directed for *Procedure* in the *Assay for aluminum hydroxide* under *Alumina, Magnesia, and Calcium Carbonate Oral Suspension.* Each mL of 0.05 *M Disodium ethylenediaminetetraacetate titrant* is equivalent to 3.900 mg of $Al(OH)_3$.

Assay for magnesium hydroxide—

Assay preparation—Prepare as directed in the *Assay for aluminum hydroxide.*

Procedure—Proceed as directed for *Procedure* in the *Assay for magnesium hydroxide* under *Alumina, Magnesia, and Calcium Carbonate Oral Suspension.* Each mL of 0.05 *M* disodium ethylenediaminetetraacetate is equivalent to 2.916 mg of $Mg(OH)_2$.

Assay for calcium carbonate—

Assay preparation—Prepare as directed in the *Assay for aluminum hydroxide.*

Procedure—Proceed as directed for *Procedure* in the *Assay for calcium carbonate* under *Alumina, Magnesia, and Calcium Carbonate Oral Suspension.* Each mL of 0.05 *M* disodium ethylenediaminetetraacetate is equivalent to 5.004 mg of $CaCO_3$.

Alumina, Magnesia, and Simethicone Oral Suspension

» Alumina, Magnesia, and Simethicone Oral Suspension contains the equivalent of not less than 90.0 percent and not more than 115.0 percent of the labeled amounts of aluminum hydroxide [$Al(OH)_3$] and magnesium hydroxide [$Mg(OH)_2$], and an amount of polydimethylsiloxane [$-(CH_3)_2SiO-]_n$ that is not less than 85.0 percent and not more than 115.0 percent of the labeled amount of simethicone.

Packaging and storage—Preserve in tight containers, and avoid freezing.

Labeling—Oral Suspension may be labeled to state the aluminum hydroxide content in terms of the equivalent amount of dried aluminum hydroxide gel, on the basis that each mg of dried gel is equivalent to 0.765 mg of $Al(OH)_3$. Label it to state the sodium content if it is greater than 1 mg per mL.

Reference standard—*USP Polydimethylsiloxane Reference Standard*—Keep container tightly closed. Do not dry before using.

Identification—

A: The infrared absorption spectrum, determined in a 0.5-mm cell, of the solution employed for measurement of the absorbance of polydimethylsiloxane, prepared as directed in the *Assay for polydimethylsiloxane*, exhibits maxima only at the same wavelengths as that of the solution employed for measurement of the absorbance of USP Polydimethylsiloxane RS, prepared as directed in the *Assay for polydimethylsiloxane.*

B: To a solution of 5 g in 10 mL of 3 *N* hydrochloric acid add 5 drops of methyl red TS, heat to boiling, add 6 *N* ammonium hydroxide until the color of the solution just changes to deep yellow, then continue boiling for 2 minutes, and filter: the filtrate so obtained responds to the tests for *Magnesium* ⟨191⟩.

C: Wash the precipitate obtained in *Identification test B* with hot ammonium chloride solution (1 in 50), and dissolve the precipitate in hydrochloric acid. Divide this solution into two portions: the dropwise addition of 6 *N* ammonium hydroxide to one portion yields a gelatinous white precipitate, which does not dissolve in an excess of 6 *N* ammonium hydroxide. The dropwise addition of 1 *N* sodium hydroxide to the other portion yields a gelatinous white precipitate, which dissolves in an excess of 1 *N* sodium hydroxide, leaving some turbidity.

Microbial limits ⟨61⟩—Its total aerobic microbial count does not exceed 100 per mL, and it meets the requirements of the test for absence of *Escherichia coli.*

Acid-neutralizing capacity ⟨301⟩—The acid consumed by the minimum single dose recommended in the labeling is not less than 5 mEq, and not less than the number of mEq calculated by the formula:

$$0.55(0.0385A) + 0.8(0.0343M),$$

in which 0.0385 and 0.0343 are the theoretical acid-neutralizing capacities, in mEq, of $Al(OH)_3$ and $Mg(OH)_2$, respectively, and

A and *M* are the quantities, in mg, of $Al(OH)_3$ and $Mg(OH)_2$ in the specimen tested, based on the labeled quantities.

pH ⟨791⟩: between 7.0 and 8.6.

Defoaming activity—

Foaming solution—Dissolve 500 µg of FD&C Blue No. 1 and 1 g of octoxynol 9 in 100 mL of 0.1 *N* hydrochloric acid.

Procedure—[NOTE—For each test, employ a clean, unused, 250-mL glass jar.] Transfer a volume of Oral Suspension, equivalent to 20 mg of simethicone, to a clean, unused, cylindrical 250-mL glass jar, fitted with a 50-mm cap, containing 100 mL of *Foaming solution* that has been warmed to 37°. Proceed as directed for *Procedure* in the test for *Defoaming activity* under *Simethicone*, beginning with "Cap the jar." The defoaming activity time does not exceed 45 seconds.

Sodium—

Potassium chloride solution—Prepare a solution of potassium chloride in water containing 38 mg per mL.

Sodium chloride stock solution—Dissolve a suitable quantity of sodium chloride, previously dried at 105° for 2 hours and accurately weighed, in water, and dilute quantitatively and stepwise with water to obtain a solution containing 25.42 µg per mL (10 µg of sodium per mL).

Standard preparations—On the day of use, transfer 4.0 mL of 1 *N* hydrochloric acid and 10.0 mL of *Potassium chloride solution* to each of two 100-mL volumetric flasks. To the respective flasks add 5.0 mL and 10.0 mL of *Sodium chloride stock solution.* Dilute with water to volume, and mix. These solutions contain about 0.5 µg and 1.0 µg of sodium per mL, respectively.

Test preparation—Transfer 5.0 mL of Alumina, Magnesia, and Simethicone Oral Suspension, previously well-shaken in its original container, to a 100-mL volumetric flask, add 50 mL of 1 *N* hydrochloric acid, boil for 15 minutes, cool to room temperature, dilute with water to volume, and mix. Filter, discarding the first few mL of the filtrate. Transfer 5.0 mL of the filtrate to a 100-mL volumetric flask containing 10.0 mL of *Potassium chloride solution*, dilute with water to volume, and mix.

Procedure—Concomitantly determine the absorbances of the *Standard preparations* and the *Test preparation* at the sodium emission line at 589.0 nm with a suitable atomic absorption spectrophotometer (see *Spectrophotometry and Light-scattering* ⟨851⟩) equipped with a sodium hollow-cathode lamp and an air-acetylene flame, using as a blank a solution prepared by pipeting 4 mL of 1 *N* hydrochloric acid and 10.0 mL of *Potassium chloride solution* into a 100-mL volumetric flask, diluting with water to volume, and mixing. Plot the absorbances of the *Standard preparations* versus concentrations, in µg per mL, of sodium and draw a straight line between the plotted points. From the graph so obtained, determine the concentration, *C*, in µg per mL, of sodium in the *Test preparation.* Calculate the quantity, in mg, of sodium in each mL of Oral Suspension by the formula:

$$0.4C.$$

Assay for aluminum hydroxide—

Disodium ethylenediaminetetraacetate titrant—Prepare and standardize as directed in the *Assay* under *Ammonium Alum.*

Assay preparation—Transfer an accurately measured volume of Alumina, Magnesia, and Simethicone Oral Suspension, previously well-shaken in its original container, equivalent to about 800 mg of aluminum hydroxide, to a suitable beaker. Add 20 mL of water, stir, and slowly add 10 mL of hydrochloric acid. Heat gently, if necessary, to aid solution, cool, and filter into a 200-mL volumetric flask. Wash the filter with water into the flask, add water to volume, and mix.

Procedure—Pipet 10 mL of *Assay preparation* into a 250-mL beaker, add 20 mL of water, then add, in the order named and with continuous stirring, 25.0 mL of *Disodium ethylenediaminetetraacetate titrant* and 20 mL of acetic acid–ammonium acetate buffer TS, and heat near the boiling temperature for 5 minutes. Cool, add 50 mL of alcohol and 2 mL of dithizone TS, and mix. Titrate with 0.05 *M* zinc sulfate VS until the color changes from green-violet to rose-pink. Perform a blank determination, substituting 10 mL of water for the *Assay preparation*, and making any necessary correction. Each mL of 0.05 *M Disodium ethylenediaminetetraacetate titrant* consumed is equivalent to 3.900 mg of $Al(OH)_3$.

Assay for magnesium hydroxide—

Assay preparation—Prepare as directed in the *Assay for aluminum hydroxide.*

Procedure—Pipet a volume of *Assay preparation*, equivalent to about 40 mg of magnesium hydroxide, into a 400-mL beaker, add 200 mL of water and 20 mL of triethanolamine, and stir. Add 10 mL of ammonia–ammonium chloride buffer TS and 3 drops of an eriochrome black indicator solution prepared by dissolving 200 mg of eriochrome black T in a mixture of 15 mL of triethanolamine and 5 mL of dehydrated alcohol, and mix. Cool the solution to between 3° and 4° by immersion of the beaker in an ice bath, then remove, and titrate with 0.05 *M* disodium ethylenediaminetetraacetate VS to a blue end-point. Perform a blank determination, substituting water for the *Assay preparation*, and make any necessary correction. Each mL of 0.05 *M* disodium ethylenediaminetetraacetate consumed is equivalent to 2.916 mg of $Mg(OH)_2$.

Assay for polydimethylsiloxane—Transfer an accurately measured volume of Alumina, Magnesia, and Simethicone Oral Suspension, equivalent to about 50 mg of simethicone, to a suitable round, narrow-mouth, screw-capped, 120-mL bottle, add 40 mL of 0.1 *N* sodium hydroxide, and swirl to disperse. Add 25.0 mL of carbon tetrachloride, close the bottle securely with a cap having an inert liner, and shake for 15 minutes, accurately timed, on a reciprocating shaker (e.g., about 200 oscillations per minute and a stroke of 38 ± 2 mm). Transfer the mixture to a 125-mL separator. Remove about 5 mL of the lower, organic layer to a screw-capped, 15-mL test tube containing 0.5 g of anhydrous sodium sulfate. Close the tube with a screw-cap having an inert liner, agitate vigorously, and centrifuge the mixture until a clear supernatant liquid (*Assay preparation*) is obtained. Prepare a *Standard preparation* similarly, except to dissolve about 50 mg of USP Polydimethylsiloxane RS, accurately weighed, in 25.0 mL of carbon tetrachloride, add 40 mL of 0.1 *N* sodium hydroxide, and add a volume of water equal to that of the specimen of Oral Suspension taken. Prepare a blank by mixing 10 mL of carbon tetrachloride with 0.5 g of anhydrous sodium sulfate and centrifuging to obtain a clear supernatant liquid. Concomitantly determine the absorbances of the solutions in 0.5-mm cells at the wavelength of maximum absorbance at about 7.9 µm, with a suitable infrared spectrophotometer, using the blank to set the instrument. Calculate the quantity, in mg, of $[-(CH_3)_2SiO-]_n$ in each mL of the Oral Suspension taken by the formula:

$$(W/V)(A_U/A_S),$$

in which *W* is the weight, in mg, of USP Polydimethylsiloxane RS used in preparing the *Standard preparation*, *V* is the volume, in mL, of Oral Suspension taken, and A_U and A_S are the absorbances of the *Assay preparation* and the *Standard preparation*, respectively.

Alumina, Magnesia, and Simethicone Tablets

» Alumina, Magnesia, and Simethicone Tablets contain the equivalent of not less than 90.0 percent and not more than 115.0 percent of the labeled amounts of aluminum hydroxide $[Al(OH)_3]$ and magnesium hydroxide $[Mg(OH)_2]$, and an amount of polydimethylsiloxane $[-(CH_3)_2SiO-]_n$ that is not less than 85.0 percent and not more than 115.0 percent of the labeled amount of simethicone.

Packaging and storage—Preserve in well-closed containers.

Labeling—Label Tablets to indicate that they are to be chewed before being swallowed. Label Tablets to state the sodium content if it is greater than 5 mg per Tablet. Tablets may be labeled to state the aluminum hydroxide content in terms of the equivalent amount of dried aluminum hydroxide gel, on the basis that each mg of dried gel is equivalent to 0.765 mg of $Al(OH)_3$.

Reference standard—*USP Polydimethylsiloxane Reference Standard*—Keep container tightly closed. Do not dry before using.

Identification—

A: The infrared absorption spectrum, determined in a 0.5-mm cell, of the solution employed for measurement of the absorbance of polydimethylsiloxane, prepared as directed in the *Assay for polydimethylsiloxane,* exhibits maxima only at the same wavelengths as that of the solution employed for measurement of the absorbance of USP Polydimethylsiloxane RS, prepared as directed in the *Assay for polydimethylsiloxane.*

B: To a portion of finely powdered Tablets, equivalent to about 600 mg of magnesium hydroxide, add 25 mL of 3 *N* hydrochloric acid and 25 mL of water, and mix. Boil gently for 2 minutes. Allow to cool, and filter. Add 5 drops of methyl red TS, heat to boiling, and add 6 *N* ammonium hydroxide until the color of the solution just turns to deep yellow. Continue boiling for 2 minutes, and filter: the filtrate so obtained responds to the tests for *Magnesium* ⟨191⟩.

C: Wash the precipitate obtained in *Identification test B* with a hot solution of ammonium chloride (1 in 50), and dissolve the precipitate in hydrochloric acid: the solution so obtained responds to *Identification test C* under *Alumina, Magnesia, and Simethicone Oral Suspension.*

Uniformity of dosage units ⟨905⟩: meet the requirements for *Weight Variation* with respect to aluminum hydroxide and to magnesium hydroxide.

Acid-neutralizing capacity ⟨301⟩—The acid consumed by the minimum single dose recommended in the labeling is not less than 5 mEq, and not less than the number of mEq calculated by the formula:

$$0.55(0.0385A) + 0.8(0.0343M),$$

in which 0.0385 and 0.0343 are the theoretical acid-neutralizing capacities, in mEq, of $Al(OH)_3$ and $Mg(OH)_2$, respectively, and A and M are the quantities, in mg, of $Al(OH)_3$ and $Mg(OH)_2$ in the specimen tested, based on the labeled quantities.

Defoaming activity—

Foaming solution—Dissolve 500 µg of FD&C Blue No. 1 and 1 g of octoxynol 9 in 100 mL of 0.1 *N* hydrochloric acid.

Procedure—[NOTE—For each test, employ a clean, unused, 250-mL glass jar.] Transfer a quantity of finely powdered Tablets, passed completely through an 80-mesh sieve, equivalent to 20 mg of simethicone, to a clean, unused, cylindrical 250-mL glass jar, fitted with a 50-mm cap, containing 100 mL of *Foaming solution* that has been warmed to 37°. Proceed as directed for *Procedure* in the test for *Defoaming activity* under *Simethicone,* beginning with "Cap the jar." The defoaming activity time does not exceed 45 seconds.

Sodium—

Potassium chloride solution, Sodium chloride stock solution, and *Standard preparations*—Prepare as directed in the test for *Sodium content* under *Alumina, Magnesia, and Simethicone Oral Suspension.*

Test preparation—Weigh and finely powder not less than 20 Tablets. Transfer an accurately weighed portion of the powder, equivalent to the average weight of 1 Tablet, to a 100-mL volumetric flask. Add 50 mL of 1 *N* hydrochloric acid, boil for 15 minutes, cool to room temperature, dilute with water to volume, and mix. Filter, discarding the first few mL of the filtrate. Transfer 5.0 mL of the filtrate to a 100-mL volumetric flask containing 10.0 mL of *Potassium chloride solution,* dilute with water to volume, and mix.

Procedure—Proceed as directed in the test for *Sodium* under *Alumina, Magnesia, and Simethicone Oral Suspension.* Calculate the quantity, in mg, of sodium per Tablet by the formula:

$$2C.$$

Assay for aluminum hydroxide—

Disodium ethylenediaminetetraacetate titrant—Prepare and standardize as directed in the *Assay* under *Ammonium Alum.*

Assay preparation—Weigh and finely powder not less than 20 Alumina, Magnesia, and Simethicone Tablets. Transfer an accurately weighed portion of the powder, equivalent to about 800 mg of aluminum hydroxide, to a 150-mL beaker, add 20 mL of water, stir, and slowly add 30 mL of 3 *N* hydrochloric acid. Heat gently, if necessary, to aid solution, cool to room temperature, and filter into a 200-mL volumetric flask. Wash the filter with water into the flask, add water to volume, and mix.

Procedure—Proceed as directed for *Procedure* in the *Assay for aluminum hydroxide* under *Alumina, Magnesia, and Simethicone Oral Suspension.*

Assay for magnesium hydroxide—

Assay preparation—Prepare as directed in the *Assay for aluminum hydroxide.*

Procedure—Proceed as directed for *Procedure* in the *Assay for magnesium hydroxide* under *Alumina, Magnesia, and Simethicone Oral Suspension.*

Assay for polydimethylsiloxane—Weigh and finely powder not less than 20 Alumina, Magnesia, and Simethicone Tablets. Transfer an accurately weighed portion of the powder, equivalent to about 33 mg of simethicone, to a suitable round, narrow-mouth, screw-capped, 120-mL bottle, add 20.0 mL of carbon tetrachloride, and swirl to disperse. Add 35 mL of 5 *N* hydrochloric acid, close the bottle securely with a cap having an inert liner, and shake for 30 minutes, accurately timed, on a reciprocating shaker (e.g., about 200 oscillations per minute and a stroke of 38 ± 2 mm). Transfer the mixture to a 125-mL separator, and allow to separate. Remove the lower, organic layer to a screw-capped, centrifuge tube containing about 2 g of anhydrous sodium sulfate. Close the tube with a screw-cap having an inert liner, agitate vigorously, and centrifuge the mixture until a clear supernatant liquid (*Assay preparation*) is obtained. Similarly prepare a *Standard preparation,* using about 33 mg of USP Polydimethylsiloxane RS, accurately weighed. Prepare a blank by mixing 10 mL of carbon tetrachloride with about 1 g of anhydrous sodium sulfate and centrifuging to obtain a clear supernatant liquid. Concomitantly determine the absorbances of the solutions in 0.5-mm cells at the wavelength of maximum absorbance at about 7.9 µm, with a suitable infrared spectrophotometer, using the blank to set the instrument. Calculate the quantity, in mg, of $[-(CH_3)_2SiO-]_n$ in the portion of Tablets taken by the formula:

$$(W)(A_U/A_S),$$

in which W is the weight, in mg, of USP Polydimethylsiloxane RS used to prepare the *Standard preparation,* and A_U and A_S are the absorbances of the *Assay preparation* and the *Standard preparation,* respectively.

Alumina and Magnesium Carbonate Oral Suspension

» Alumina and Magnesium Carbonate Oral Suspension contains the equivalent of not less than 90.0 percent and not more than 110.0 percent of the labeled amounts of aluminum hydroxide [$Al(OH)_3$] and magnesium carbonate ($MgCO_3$).

Packaging and storage—Preserve in tight containers, and avoid freezing.

Identification—

A: Place about 1 g in a flask equipped with a stopper and glass tubing, the tip of which is immersed in calcium hydroxide TS in a test tube. Add 5 mL of 3 *N* hydrochloric acid to the flask, and immediately insert the stopper: gas evolves in the flask and a precipitate is formed in the test tube.

B: To a solution of 5 g in 10 mL of 3 *N* hydrochloric acid add 5 drops of methyl red TS, heat to boiling, add 6 *N* ammonium hydroxide until the color of the solution changes to deep yellow, then continue boiling for 2 minutes, and filter: the filtrate responds to the tests for *Magnesium* ⟨191⟩.

C: Wash the precipitate obtained in *Identification test B* with a hot solution of ammonium chloride (1 in 50), and dissolve the precipitate in hydrochloric acid: the solution responds to the tests for *Aluminum* ⟨191⟩.

Microbial limits ⟨61⟩—Its total aerobic microbial count does not exceed 100 per mL, and it meets the requirements of the test for absence of *Escherichia coli*, *Salmonella* species, *Staphylococcus aureus*, and *Pseudomonas aeruginosa*.

pH ⟨791⟩: between 7.5 and 9.5.

Assay for aluminum hydroxide—

Potassium chloride solution—Prepare a solution containing 4.5 g of potassium chloride in each 100 mL.

Aluminum stock solution—Transfer 1.000 g of aluminum wire to a 1000-mL volumetric flask, and add 50 mL of 6 *N* hydrochloric acid. Swirl to ensure contact of the aluminum and the acid, and allow the reaction to proceed until all of the aluminum has dissolved. Dilute with water to volume, and mix.

Standard preparations—To separate 100-mL volumetric flasks, each containing 10 mL of *Potassium chloride solution*, transfer 9.0 mL, 10.0 mL, and 11.0 mL, respectively, of *Aluminum stock solution*, dilute with water to volume, and mix. These *Standard preparations* contain 90.0 µg, 100.0 µg, and 110.0 µg of aluminum per mL, respectively.

Assay preparation—Transfer an accurately measured quantity of Alumina and Magnesium Carbonate Oral Suspension, previously well shaken in its original container, equivalent to about 75 mg of aluminum hydroxide, to a suitable beaker. Add 25 mL of 6 *N* hydrochloric acid, and heat on a steam bath for 30 minutes, with occasional swirling. Cool, and transfer with the aid of water to a 250-mL volumetric flask containing 25 mL of *Potassium chloride solution*. Dilute with water to volume, mix, and filter.

Procedure—Concomitantly determine the absorbances of the *Standard preparations* and the *Assay preparation* at the aluminum emission line at 309.3 nm, with a suitable atomic absorption spectrophotometer (see *Spectrophotometry and Light-scattering* ⟨851⟩) equipped with an aluminum hollow-cathode lamp and a nitrous oxide–acetylene flame, using water as the blank. Calculate the quantity, in mg, of $Al(OH)_3$ in the portion of Oral Suspension taken by the formula:

$$(78.00/26.98)(0.25)(A_U/R_S),$$

in which 78.00 is the molecular weight of aluminum hydroxide, 26.98 is the atomic weight of aluminum, A_U is the absorbance of the *Assay preparation*, and R_S is the average of the ratios of the absorbances of the *Standard preparations* to their respective concentrations, in µg of aluminum per mL.

Assay for magnesium carbonate—

Lanthanum chloride solution—Prepare a solution of lanthanum chloride in water containing 5 mg per mL.

Magnesium stock solution—Transfer 1.000 g of magnesium metal to a 1000-mL volumetric flask containing 50 mL of water, and slowly add 10 mL of hydrochloric acid. Dilute with water to volume, and mix.

Standard preparations—To separate 100-mL volumetric flasks, each containing 10 mL of *Lanthanum chloride solution*, transfer 1.70 mL and 1.80 mL, respectively, of *Magnesium stock solution*, dilute with water to volume, and mix. These *Standard preparations* contain 1.7 µg and 1.8 µg of magnesium per mL, respectively.

Assay preparation—Dilute an accurately measured volume of the *Assay preparation* prepared as directed in the *Assay for aluminum hydroxide* quantitatively with water to obtain a solution having a concentration of about 6 µg of magnesium carbonate per mL.

Procedure—Concomitantly determine the absorbances of the *Standard preparations* and the *Assay preparation* at the magnesium emission line at 285.2 nm, with a suitable atomic absorption spectrophotometer (see *Spectrophotometry and Light-scattering* ⟨851⟩) equipped with a magnesium hollow-cathode lamp and an air-acetylene flame, using water as the blank. Calculate the quantity, in mg, of $MgCO_3$ in the portion of Oral Suspension taken by the formula:

$$(84.31/24.31)(L/D)(A_U/R_S),$$

in which 84.31 is the molecular weight of magnesium carbonate, 24.31 is the atomic weight of magnesium, L is the labeled quantity, in mg, of magnesium carbonate in the portion of Oral Suspension taken, D is the concentration, in µg of magnesium carbonate per mL, of the *Assay preparation*, based on the labeled amount of magnesium carbonate in the portion of Oral Suspension taken and the extent of dilution, A_U is the absorbance of the *Assay preparation*, and R_S is the average of the ratios of the absorbances of the *Standard preparations* to their respective concentrations, in µg of magnesium per mL.

Alumina and Magnesium Carbonate Tablets

» Alumina and Magnesium Carbonate Tablets contain the equivalent of not less than 90.0 percent and not more than 110.0 percent of the labeled amounts of aluminum hydroxide [$Al(OH)_3$] and magnesium carbonate ($MgCO_3$).

Packaging and storage—Preserve in tight containers.

Identification—

A: Place about 1 g of finely powdered Tablets in a flask equipped with a stopper and glass tubing, the tip of which is immersed in calcium hydroxide TS in a test tube. Add 5 mL of 3 *N* hydrochloric acid to the flask, and immediately insert the stopper: gas evolves in the flask and a precipitate is formed in the test tube.

B: To a 7-g portion of finely powdered Tablets add 10 mL of 3 *N* hydrochloric acid and 5 drops of methyl red TS, heat to boiling, and add 6 *N* ammonium hydroxide until the color of the solution changes to deep yellow. Continue boiling for 2 minutes, and filter: the filtrate responds to the tests for *Magnesium* ⟨191⟩.

C: Wash the precipitate obtained in *Identification test B* with a hot solution of ammonium chloride (1 in 50), and dissolve the precipitate in hydrochloric acid: the solution responds to the tests for *Aluminum* ⟨191⟩.

Disintegration ⟨701⟩: 10 minutes, simulated gastric fluid TS being substituted for water in the test.

Uniformity of dosage units ⟨905⟩: meet the requirements for *Weight Variation* with respect to aluminum hydroxide and to magnesium carbonate.

Acid-neutralizing capacity ⟨301⟩: not less than 5 mEq of acid is consumed by the minimum single dose recommended in the labeling.

Assay for aluminum hydroxide—

Potassium chloride solution—Prepare a solution containing 38.1 g of potassium chloride in each 1000 mL.

Digestion fluid—Mix 5 mL of hydrochloric acid, 10 mL of nitric acid, and 10 mL of water, and use promptly.

Aluminum stock solution—Transfer 1.000 g of aluminum metal to a 1000-mL volumetric flask, and add 50 mL of 6 *N* hydrochloric acid. Swirl to ensure contact of the aluminum and the acid, and allow the reaction to proceed until all of the aluminum has dissolved. Dilute with water to volume, and mix.

Standard preparations—To separate 100-mL volumetric flasks transfer 3.0 mL, 4.0 mL, and 5.0 mL of *Aluminum stock solution*, respectively. To each flask add 10 mL of *Potassium chloride solution* and 7.5 mL of *Digestion fluid*, dilute with water to volume, and mix. These *Standard preparations* contain 30 µg, 40 µg, and 50 µg of aluminum per mL, respectively.

Assay preparation—Weigh and finely powder not less than 20 Alumina and Magnesium Carbonate Tablets. Transfer an accurately weighed portion of the powder, equivalent to about 30 mg of aluminum hydroxide, to a 100-mL volumetric flask, add 25-mL of *Digestion fluid*, and heat on a steam bath for 30 minutes or on a hot plate until the volume is reduced by about one-half. Cool, dilute with water to volume, and mix. Filter, discarding the first 20 mL of the filtrate. Transfer 15.0 mL of the filtrate to a 50-mL volumetric flask, add 5.0 mL of *Potassium chloride solution*, dilute with water to volume, and mix. [NOTE—Reserve a portion of the filtrate for use in the *Assay for magnesium carbonate*.]

Procedure—Concomitantly determine the absorbances of the *Standard preparations* and the *Assay preparation* at the aluminum emission line at 309.3 nm, with a suitable atomic absorption spectrophotometer (see *Spectrophotometry and Light-scattering* ⟨851⟩) equipped with an aluminum hollow-cathode lamp and a nitrous oxide–acetylene flame, using water as the blank. Plot the absorbances of the *Standard preparations* versus concentration, in μg per mL, of aluminum, and draw the straight line best fitting the three plotted points. From the graph so obtained, determine the concentration, C, in μg per mL, of aluminum in each mL of the *Assay preparation*. Calculate the quantity, in mg, of aluminum hydroxide [$Al(OH)_3$] in the portion of Tablets taken by the formula:

$$(78.00/26.98)(C/3),$$

in which 78.00 is the molecular weight of aluminum hydroxide, and 26.98 is the atomic weight of aluminum.

Assay for magnesium carbonate—

Lanthanum chloride solution—Transfer 17.6 g of lanthanum chloride to a 1000-mL volumetric flask, add 500 mL of water, and carefully add 50 mL of hydrochloric acid. Mix, and allow to cool. Dilute with water to volume, and mix.

Digestion fluid—Mix 5 mL of hydrochloric acid, 10 mL of nitric acid, and 10 mL of water, and use promptly.

Magnesium stock solution—Transfer 1.000 g of magnesium metal to a 1000-mL volumetric flask containing 50 mL of water, and slowly add 10 mL of hydrochloric acid. Dilute with water to volume, and mix. Transfer 5.0 mL of this solution to a 500-mL volumetric flask, dilute with water to volume, and mix.

Standard preparations—To separate 100-mL volumetric flasks transfer 4.0 mL, 6.0 mL, and 8.0 mL of *Magnesium stock solution*, respectively. To each flask add 0.5 mL of *Digestion fluid* and 10 mL of *Lanthanum chloride solution*, dilute with water to volume, and mix. These *Standard preparations* contain 0.40 μg, 0.60 μg, and 0.80 μg of magnesium per mL, respectively.

Assay preparation—Transfer an accurately measured volume of the filtrate used to prepare the *Assay preparation* in the *Assay for aluminum hydroxide*, equivalent to about 0.4 mg of magnesium carbonate, to a 200-mL volumetric flask, add 20 mL of *Lanthanum chloride solution*, dilute with water to volume, and mix.

Procedure—Concomitantly determine the absorbances of the *Standard preparations* and the *Assay preparation* at the magnesium emission line at 285.2 nm, with a suitable atomic absorption spectrophotometer (see *Spectrophotometry and Light-scattering* ⟨851⟩) equipped with a magnesium hollow-cathode lamp and an air-acetylene flame, using water as the blank. Plot the absorbances of the *Standard preparations* versus concentration, in μg per mL, of magnesium, and draw the straight line best fitting the three plotted points. From the graph so obtained, determine the concentration, C, in μg per mL, of magnesium in each mL of the *Assay preparation*. Calculate the quantity, in mg, of magnesium carbonate ($MgCO_3$) in the portion of Tablets taken by the formula:

$$(84.31/24.31)(20C/V),$$

in which 84.31 is the molecular weight of magnesium carbonate, 24.31 is the atomic weight of magnesium, and V is the volume taken, in mL, of the *Assay preparation* prepared as directed in the *Assay for aluminum hydroxide*.

Alumina, Magnesium Carbonate, and Magnesium Oxide Tablets

» Alumina, Magnesium Carbonate, and Magnesium Oxide Tablets contain the equivalent of not less than 90.0 percent and not more than 110.0 percent of the labeled amounts of aluminum hydroxide [$Al(OH)_3$] and magnesium carbonate ($MgCO_3$), and not less than 85.0 percent and not more than 115.0 percent of the labeled amount of magnesium oxide (MgO).

Packaging and storage—Preserve in tight containers.

Identification—

A: Place about 3 g of finely powdered Tablets in a flask equipped with a stopper and glass tubing, the tip of which is immersed in calcium hydroxide TS in a test tube. Add 5 mL of 3 N hydrochloric acid to the flask, and immediately insert the stopper: gas evolves in the flask and a precipitate is formed in the test tube.

B: To the solution in the flask obtained in *Identification test A* add 5 drops of methyl red TS, and heat to boiling. Add 6 N ammonium hydroxide until the color of the solution changes to deep yellow, continue boiling for 2 minutes, and filter through hardened filter paper. (Retain the filtrate for *Identification test C*.) Wash the precipitate with 350 mL of a hot ammonium chloride solution (1 in 50), discarding the washings: the precipitate so obtained, dissolved in 3 N hydrochloric acid, responds to the tests for *Aluminum* ⟨191⟩.

C: The filtrate obtained in *Identification test B* responds to the tests for *Magnesium* ⟨191⟩.

Disintegration ⟨701⟩: 10 minutes, simulated gastric fluid TS being substituted for water in the test.

Uniformity of dosage units ⟨905⟩: meet the requirements for *Weight Variation* with respect to alumina, to magnesium carbonate, and to magnesium oxide.

Acid-neutralizing capacity ⟨301⟩: not less than 5 mEq of acid is consumed by the minimum single dose recommended in the labeling.

Assay for aluminum hydroxide—

Disodium ethylenediaminetetraacetate titrant—Prepare and standardize as directed in the *Assay* under *Ammonium Alum*.

Assay preparation—Weigh and finely powder not less than 20 Alumina, Magnesium Carbonate, and Magnesium Oxide Tablets. Transfer an accurately weighed portion of the powder, equivalent to about 1200 mg of aluminum hydroxide, to a 150-mL beaker, add 20 mL of water, stir, and slowly add 30 mL 3 N hydrochloric acid. Heat gently, if necessary, to aid solution, cool, and filter into a 200-mL volumetric flask. Wash the filter with water into the flask, add water to volume, and mix.

Procedure—Pipet 10 mL of the *Assay preparation* into a 250-mL beaker, add 20 mL of water, then add, in the order named and with continuous stirring, 25.0 mL of *Disodium ethylenediaminetetraacetate titrant* and 20 mL of acetic acid–ammonium acetate buffer TS, and heat near the boiling point for 5 minutes. Cool, add 50 mL of alcohol and 2 mL of dithizone TS, and mix. Titrate with 0.05 M zinc sulfate VS until the color changes from green-violet to rose-pink. Perform a blank determination, substituting 10 mL of water for the *Assay preparation*, and make any necessary correction. Each mL of 0.05 *M Disodium ethylenediaminetetraacetate titrant* consumed is equivalent to 3.900 mg of $Al(OH)_3$.

Assay for magnesium carbonate—Weigh and finely powder not less than 20 Alumina, Magnesium Carbonate, and Magnesium Oxide Tablets. Transfer an accurately weighed portion of the powder, equivalent to about 750 mg of magnesium carbonate, to a 250-mL conical flask fitted with a two-hole stopper. Fill the lower transverse section of a U-shaped drying tube of about 15-mm internal diameter and 15-cm height with loosely packed glass wool. Place in one arm of the tube about 5 g of anhydrous calcium chloride, and accurately weigh the tube and the contents. Into the other arm of the tube place 9.5 g to 10.5 g of soda lime, and again weigh accurately. Insert stoppers in the open arms of the U-tube, and connect the side tube of the arm filled with soda lime to a calcium chloride drying tube, which in turn is connected to one of the holes in the stopper of the 250-mL conical flask. Attach a dropping funnel to the other hole in the stopper of the 250-mL conical flask. Add 100 mL of water and 10 mL of a mixture of hydrochloric acid and nitric acid (4 : 1) to the 250-mL conical flask through the dropping funnel, and close the dropping funnel. Heat the 250-mL conical flask at 95° for 1 hour, and allow the evolved carbon dioxide to pass through the U-tube. Replace the dropping funnel with a source of carbon dioxide–free air, and pass the carbon dioxide–free air through the apparatus at a rate of about 75 mL per minute for 30 minutes. Disconnect the U-tube, cool to room temperature, remove the stoppers, and weigh. The increase in weight corresponds to the quantity of carbon dioxide evolved. Calculate the quantity, in mg, of magnesium carbonate in each Tablet taken by the formula:

$$(84.31/44.01)(I)(W_A/W_P),$$

in which 84.31 and 44.01 are the molecular weights of magnesium carbonate and carbon dioxide, respectively, I is the quantity, in mg, of carbon dioxide evolved from the portion of Tablets taken,

W_A is the average weight, in g, of 1 Tablet, and W_P is the weight, in g, of the portion of Tablets taken.

Assay for magnesium oxide—Weigh and finely powder not less than 20 Alumina, Magnesium Carbonate, and Magnesium Oxide Tablets. Transfer an accurately weighed portion of the powder, equivalent to about 1000 mg of magnesium carbonate and magnesium oxide combined, to a beaker, add 20 mL of water, and slowly add 40 mL of 3 *N* hydrochloric acid, with mixing. Heat the mixture to boiling, cool, and filter into a 200-mL volumetric flask. Wash the beaker with water, adding the washings to the filter. Add water to volume, and mix. Transfer 20.0 mL of this solution to a 400-mL beaker, add 180 mL of water and 20 mL of triethanolamine, and stir. Add 10 mL of ammonia–ammonium chloride buffer TS and 3 drops of an eriochrome black indicator solution prepared by dissolving 200 mg of eriochrome black T in a mixture of 15 mL of triethanolamine and 5 mL of dehydrated alcohol, and mix. Cool the solution to between 3° and 4° by immersion of the beaker in an ice bath, then remove, and titrate with 0.05 *M* disodium ethylenediaminetetraacetate VS to a blue end-point. Perform a blank determination, substituting 20 mL of water for the assay solution, and make any necessary correction. Each mL of 0.05 *M* disodium ethylenediaminetetraacetate consumed is equivalent to 1.216 mg of Mg. Calculate the quantity, in mg, of magnesium equivalent in each Tablet taken by the formula:

$$10T(W_A/W_P),$$

in which *T* is the magnesium equivalent obtained in the titration, W_A is the average weight, in g, of 1 Tablet, and W_P is the weight, in g, of the portion of Tablets taken.

Calculate the quantity, in mg, of magnesium oxide in each Tablet taken by the formula:

$$(40.30/24.31)(A - 0.2883B),$$

in which 40.30 and 24.31 are the molecular weight of magnesium oxide and the atomic weight of magnesium, respectively, *A* is the quantity, in mg, of magnesium equivalent in each Tablet, and *B* is the quantity, in mg, of magnesium carbonate in each Tablet, as determined in the *Assay for magnesium carbonate.*

Alumina and Magnesium Trisilicate Oral Suspension

» Alumina and Magnesium Trisilicate Oral Suspension contains the equivalent of not less than 90.0 percent and not more than 110.0 percent of the labeled amounts of aluminum hydroxide [$Al(OH)_3$] and magnesium trisilicate ($Mg_2Si_3O_8$).

Packaging and storage—Preserve in tight containers.

Identification—

A: To a mixture of 5 mL in 10 mL of 3 *N* hydrochloric acid add 5 drops of methyl red TS, heat to boiling, add 6 *N* ammonium hydroxide until the color of the solution changes to deep yellow, then continue boiling for 2 minutes, and filter: the filtrate responds to the tests for *Magnesium* ⟨191⟩.

B: Wash the solids on the filter obtained in *Identification test A* with hot ammonium chloride solution (1 in 50), add 10 mL of 3 *N* hydrochloric acid, and filter: the filtrate responds to the tests for *Aluminum* ⟨191⟩.

C: Transfer the filter paper and contents from *Identification test B* to a small platinum dish, ignite, cool in a desiccator, and weigh. Moisten the residue with water and add 6 mL of hydrofluoric acid. Evaporate to dryness, ignite for 5 minutes, cool in a desiccator, and weigh: a loss of more than 10% in relation to the weight of the residue from the initial ignition indicates SiO_2.

Acid-neutralizing capacity ⟨301⟩: not less than 5 mEq of acid is consumed by the minimum single dose recommended in the labeling.

pH ⟨791⟩: between 7.5 and 8.5.

Assay for aluminum hydroxide—

Disodium ethylenediaminetetraacetate titrant—Prepare and standardize as directed in the *Assay* under *Ammonium Alum.*

Assay preparation—Transfer about 10 g of well-shaken Alumina and Magnesium Trisilicate Oral Suspension to a tared beaker, and weigh accurately. Add 50 mL of water and 10 mL of hydrochloric acid, and digest on a steam bath for 1 hour. Cool, and filter into a 200-mL volumetric flask, washing the filter with water into the flask. Dilute with water to volume, and mix.

Procedure—Pipet 20 mL of *Assay preparation* into a 250-mL beaker, add 20 mL of water, then add, in the order named and with continuous stirring, 25.0 mL of *Disodium ethylenediaminetetraacetate titrant* and 20 mL of acetic acid–ammonium acetate buffer TS, and heat near the boiling point for 5 minutes. Cool, add 50 mL of alcohol and 2 mL of dithizone TS, and mix. Titrate with 0.05 *M* zinc sulfate VS until the color changes from green-violet to rose-pink. Perform a blank determination, substituting 20 mL of water for the *Assay preparation*, and make any necessary correction. Each mL of 0.05 *M Disodium ethylenediaminetetraacetate titrant* consumed is equivalent to 3.900 mg of $Al(OH)_3$.

Assay for magnesium trisilicate—

Assay preparation—Prepare as directed in the *Assay for aluminum hydroxide.*

Procedure—Pipet 20 mL of *Assay preparation* into a 400-mL beaker, add 180 mL of water and 20 mL of triethanolamine, and stir. Add 10 mL of ammonia–ammonium chloride buffer TS and 3 drops of an eriochrome black indicator solution prepared by dissolving 200 mg of eriochrome black T in a mixture of 15 mL of triethanolamine and 5 mL of dehydrated alcohol, and mix. Cool the solution to between 3° and 4° by immersion of the beaker in an ice bath, then remove and titrate with 0.05 *M* disodium ethylenediaminetetraacetate VS to a blue end-point. Perform a blank determination, substituting 20 mL of water for the *Assay preparation*, and make any necessary correction. Each mL of 0.05 *M* disodium ethylenediaminetetraacetate consumed is equivalent to 6.521 mg of $Mg_2Si_3O_8$.

Alumina and Magnesium Trisilicate Tablets

» Alumina and Magnesium Trisilicate Tablets contain not less than 90.0 percent and not more than 110.0 percent of the labeled amounts of aluminum hydroxide [$Al(OH)_3$] and magnesium trisilicate ($Mg_2Si_3O_8$).

Packaging and storage—Preserve in well-closed containers.

Labeling—Tablets prepared with the use of Dried Aluminum Hydroxide Gel may be labeled to state the aluminum hydroxide content in terms of the equivalent amount of dried aluminum hydroxide gel, on the basis that each mg of dried gel is equivalent to 0.765 mg of $Al(OH)_3$.

Identification—One powdered Tablet responds to the *Identification tests* under *Alumina and Magnesium Trisilicate Oral Suspension.*

Disintegration ⟨701⟩: 10 minutes, simulated gastric fluid TS being substituted for water in the test.

Uniformity of dosage units ⟨905⟩: meet the requirements for *Weight Variation* with respect to aluminum hydroxide and to magnesium trisilicate.

Acid-neutralizing capacity ⟨301⟩—Not less than 5 mEq of acid is consumed by the minimum single dose recommended in the labeling.

Assay for aluminum hydroxide—

Disodium ethylenediaminetetraacetate titrant—Prepare and standardize as directed in the *Assay* under *Ammonium Alum.*

Assay preparation—Weigh and finely powder not less than 20 Alumina and Magnesium Trisilicate Tablets. Transfer an accurately weighed portion of the powder, equivalent to about 600 mg of aluminum hydroxide, to a beaker, add 20 mL of water, stir, and slowly add 40 mL of 3 *N* hydrochloric acid. Heat gently, if necessary, to aid solution, cool, and transfer to a 200-mL volumetric flask. Wash the beaker with water, adding the washings to the flask, add water to volume, and mix.

Procedure—Pipet 10 mL of *Assay preparation* into a 250-mL beaker, add 20 mL of water, then add, in the order named and with continuous stirring, 25.0 mL of 0.05 *M Disodium ethylenediaminetetraacetate titrant* and 20 mL of acetic acid–ammonium acetate buffer TS, and heat the solution near the boiling temperature for 5 minutes. Cool, add 50 mL of alcohol and 2 mL of dithizone TS, and mix. Titrate with 0.05 *M* zinc sulfate VS until the color changes from green-violet to rose-pink. Perform a blank determination, substituting 10 mL of water for the *Assay preparation*, and make any necessary correction. Each mL of 0.05 *M Disodium ethylenediaminetetraacetate titrant* consumed is equivalent to 3.900 mg of $Al(OH)_3$.

Assay for magnesium trisilicate—

Assay preparation—Prepare as directed in the *Assay for aluminum hydroxide*.

Procedure—Pipet a volume of *Assay preparation*, equivalent to about 100 mg of magnesium trisilicate, into a 400-mL beaker, add 200 mL of water and 20 mL of triethanolamine, and mix. Add 50 mL of ammonia–ammonium chloride buffer TS and 2 drops of eriochrome black indicator solution (prepared by dissolving 200 mg of eriochrome black T in a mixture of 15 mL of triethanolamine and 5 mL of dehydrated alcohol, and mixing). Cool the solution to between 3° and 4° by immersing the beaker in an ice bath, and titrate with 0.05 *M* disodium ethylenediaminetetraacetate VS until the color changes to pure blue. Perform a blank determination, substituting 10 mL of water for the *Assay preparation*, and make any necessary correction. Each mL of 0.05 *M* disodium ethylenediaminetetraacetate consumed is equivalent to 6.521 mg of $Mg_2Si_3O_8$.

Aluminum Acetate Topical Solution

$Al(OCOCH_3)_3$

$C_6H_9AlO_6$ 204.12

Acetic acid, aluminum salt.

Aluminum acetate [*139-12-8*].

» Aluminum Acetate Topical Solution yields, from each 100 mL, not less than 1.20 g and not more than 1.45 g of aluminum oxide (Al_2O_3), and not less than 4.24 g and not more than 5.12 g of acetic acid ($C_2H_4O_2$), corresponding to not less than 4.8 g and not more than 5.8 g of aluminum acetate ($C_6H_9AlO_6$). Aluminum Acetate Topical Solution may be stabilized by the addition of not more than 0.6 percent of Boric Acid.

Aluminum Subacetate Topical Solution	545 mL
Glacial Acetic Acid	15 mL
Purified Water, a sufficient quantity, to make	1000 mL

Add the Glacial Acetic Acid to the Aluminum Subacetate Topical Solution and sufficient water to make 1000 mL. Mix, and filter, if necessary.

Note—Dispense only clear Aluminum Acetate Topical Solution.

Packaging and storage—Preserve in tight containers.

Identification—It responds to the tests for *Aluminum* ⟨191⟩ and for *Acetate* ⟨191⟩.

pH ⟨791⟩: between 3.6 and 4.4.

Limit of boric acid—Pipet 25 mL into 75 mL of water in a conical flask. Add 3 mL of phenolphthalein TS, then add 0.5 *N* sodium hydroxide VS from a buret until a faint pink color is obtained. Heat to boiling, and again neutralize. Add 150 mL of glycerin to the neutralized solution, and titrate with 0.5 *N* sodium hydroxide VS. Perform a blank determination in a similar manner. From the volume of 0.5 *N* sodium hydroxide VS used after the addition of the glycerin, subtract the volume used in the blank. Each mL of 0.5 *N* sodium hydroxide is equivalent to 30.92 mg of H_3BO_3.

Heavy metals ⟨231⟩—Dilute 2 mL of it with water to 25 mL: the limit is 0.001%.

Assay for aluminum oxide—

Disodium ethylenediaminetetraacetate titrant—Prepare and standardize as directed in the *Assay* under *Ammonium Alum*.

Procedure—Pipet 25 mL of Aluminum Acetate Topical Solution into a 250-mL volumetric flask, add 5 mL of hydrochloric acid, dilute with water to volume, and mix. Pipet 25 mL of this solution into a 250-mL beaker, and add, in the order named and with continuous stirring, 25.0 mL of *Disodium ethylenediaminetetraacetate titrant* and 20 mL of acetic acid–ammonium acetate buffer TS, then heat the solution near the boiling point for 5 minutes. Cool, and add 50 mL of alcohol and 2 mL of dithizone TS. Titrate the solution with 0.05 *M* zinc sulfate VS to a bright rose-pink color. Perform a blank determination, substituting water for the sample, and make any necessary correction. Each mL of 0.05 *M Disodium ethylenediaminetetraacetate titrant* is equivalent to 2.549 mg of Al_2O_3.

Assay for acetic acid—Pipet 20 mL of Aluminum Acetate Topical Solution into a Kjeldahl flask containing a mixture of 20 mL of phosphoric acid and 150 mL of water. Connect the flask to a condenser, the delivery tube from which dips beneath the surface of 50.0 mL of 0.5 *N* sodium hydroxide VS contained in a receiving flask. Distil about 160 mL, then remove the delivery tube from below the surface of the liquid, allow the distilling flask to cool, add 50 mL of water, and distil an additional 40 to 45 mL into the receiving flask. Add phenolphthalein TS to the distillate, and titrate the excess 0.5 *N* sodium hydroxide VS with 0.5 *N* sulfuric acid VS. Each mL of 0.5 *N* sodium hydroxide is equivalent to 30.03 mg of $C_2H_4O_2$.

Basic Aluminum Carbonate Gel

» Basic Aluminum Carbonate Gel contains the equivalent of not less than 90.0 percent and not more than 110.0 percent of the labeled amount of aluminum hydroxide [$Al(OH)_3$].

Packaging and storage—Preserve in tight containers, and avoid freezing.

Identification—

A: Place about 1 g in a flask equipped with a stopper and glass tubing, the tip of which is immersed in calcium hydroxide TS in a test tube. Add 5 mL of 3 *N* hydrochloric acid to the flask, and immediately insert the stopper: gas evolves in the flask and a precipitate is formed in the test tube.

B: The solution remaining in the flask responds to the tests for *Aluminum* ⟨191⟩.

Microbial limits ⟨61⟩—Its total aerobic microbial count does not exceed 100 per mL, and it meets the requirements of the test for the absence of *Escherichia coli*.

pH ⟨791⟩: between 5.5 and 8.0.

Chloride—Transfer a quantity of the Gel, equivalent to about 7.6 g of $Al(OH)_3$, to a porcelain dish. Add 0.1 mL of potassium chromate TS and 25 mL of water. Stir, and add 0.10 *N* silver nitrate until a faint, persistent pink color is obtained: not more than 8.0 mL of 0.10 *N* silver nitrate is required (0.37%).

Sulfate ⟨221⟩—Add 5.0 mL of 3 *N* hydrochloric acid to a quantity of the Gel, equivalent to about 3.8 g of $Al(OH)_3$, and heat to dissolve the specimen under test. Cool, dilute with water to 250 mL, and filter if necessary: a 20-mL portion of the filtrate shows no more sulfate than corresponds to 0.20 mL of 0.020 *N* sulfuric acid (0.06%).

Arsenic, *Method I* ⟨211⟩—Prepare a *Standard Preparation* as directed in the test for *Arsenic* ⟨211⟩, except to prepare it to contain 5 μg of arsenic instead of 3 μg. Prepare the *Test Preparation* as follows: Dissolve a quantity of the Gel, equivalent to about 6.3 g of $Al(OH)_3$, in 20 mL of 7 *N* sulfuric acid. The limit is 0.8 ppm.

Heavy metals ⟨231⟩—Dissolve a quantity of the Gel, equivalent to about 3.0 g of $Al(OH)_3$, in 10 mL of 3 *N* hydrochloric acid with the aid of heat, filter, if necessary, and dilute with water to 25 mL: the limit is 6.7 ppm.

Assay—

Disodium ethylenediaminetetraacetate titrant—Prepare and standardize as directed in the *Assay* under *Ammonium Alum.*

Procedure—Transfer an accurately weighed quantity of Basic Aluminum Carbonate Gel, equivalent to about 1.2 g of $Al(OH)_3$, to a beaker, add 15 mL of hydrochloric acid, and heat gently until solution is complete. Cool, transfer to a 500-mL volumetric flask, dilute with water to volume, and mix. Pipet 20 mL of this solution into a 250-mL beaker, and add, in the order named and with continuous stirring, 25.0 mL of *Disodium ethylenediaminetetraacetate titrant* and 20 mL of acetic acid–ammonium acetate buffer TS, then heat the solution near the boiling point for 5 minutes. Cool, and add 50 mL of alcohol and 2 mL of dithizone TS. Titrate the solution with 0.05 *M* zinc sulfate VS until the color changes from green-violet to rose-pink. Perform a blank determination, substituting 20 mL of water for the sample, and make any necessary correction. Each mL of 0.05 *M Disodium ethylenediaminetetraacetate titrant* consumed is equivalent to 3.900 mg of $Al(OH)_3$.

Dried Basic Aluminum Carbonate Gel Capsules

» Dried Basic Aluminum Carbonate Gel Capsules contain the equivalent of not less than 90.0 percent and not more than 110.0 percent of the labeled amount of aluminum hydroxide [$Al(OH)_3$].

Packaging and storage—Preserve in well-closed containers.

Labeling—Capsules may be labeled to state the aluminum hydroxide content in terms of the equivalent amount of dried aluminum hydroxide gel, on the basis that each mg of dried aluminum hydroxide gel is equivalent to 0.765 mg of $Al(OH)_3$.

Identification—

A: Place a portion of Capsule contents, equivalent to about 500 mg of aluminum hydroxide, in a flask equipped with a stopper and glass tubing, the tip of which is immersed in calcium hydroxide TS in a test tube. Add 10 mL of 3 *N* hydrochloric acid to the flask, and immediately insert the stopper: gas evolves in the flask and a precipitate is formed in the test tube.

B: The solution remaining in the flask responds to the tests for *Aluminum* ⟨191⟩.

Disintegration ⟨701⟩: 10 minutes, simulated gastric fluid TS being substituted for water in the test.

Uniformity of dosage units ⟨905⟩: meet the requirements.

Acid-neutralizing capacity ⟨301⟩: not less than 5 mEq of acid is consumed by the minimum single dose recommended in the labeling, and not less than 85.0% of the expected mEq value, calculated from the results of the *Assay*, is obtained. Each mg of $Al(OH)_3$ has an expected acid-neutralizing capacity value of 0.0385 mEq.

Assay—

Disodium ethylenediaminetetraacetate titrant—Prepare and standardize as directed in the *Assay* under *Ammonium Alum.*

Procedure—Weigh accurately the contents of not less than 20 Dried Basic Aluminum Carbonate Gel Capsules, and mix. Transfer an accurately weighed portion of the powder, equivalent to about 1.2 g of aluminum hydroxide, to a beaker, add 15 mL of hydrochloric acid, and heat until dissolved. Dilute with water to about 100 mL, mix, and filter quantitatively into a 500-mL volumetric flask, washing the filter with water. Proceed as directed in the *Assay* under *Basic Aluminum Carbonate Gel*, beginning with "dilute with water to volume." Each mL of 0.05 *M Disodium ethylenediaminetetraacetate titrant* is equivalent to 3.900 mg of $Al(OH)_3$.

Dried Basic Aluminum Carbonate Gel Tablets

» Dried Basic Aluminum Carbonate Gel Tablets contain the equivalent of not less than 90.0 percent and not more than 110.0 percent of the labeled amount of aluminum hydroxide [$Al(OH)_3$].

Packaging and storage—Preserve in well-closed containers.

Labeling—Tablets may be labeled to state the aluminum hydroxide content in terms of the equivalent amount of dried aluminum hydroxide gel, on the basis that each mg of dried aluminum hydroxide gel is equivalent to 0.765 mg of $Al(OH)_3$.

Identification—

A: Place a quantity of finely ground Tablets, equivalent to about 500 mg of aluminum hydroxide, in a flask equipped with a stopper and glass tubing, the tip of which is immersed in calcium hydroxide TS in a test tube. Add 5 mL of 3 *N* hydrochloric acid to the flask, and immediately insert the stopper: gas evolves in the flask and a precipitate is formed in the test tube.

B: The solution remaining in the flask responds to the tests for *Aluminum* ⟨191⟩.

Disintegration ⟨701⟩: 10 minutes, simulated gastric fluid TS being substituted for water in the test.

Uniformity of dosage units ⟨905⟩: meet the requirements.

Acid-neutralizing capacity ⟨301⟩: not less than 5 mEq of acid is consumed by the minimum single dose recommended in the labeling, and not less than 85.0% of the expected mEq value, calculated from the results of the *Assay*, is obtained. Each mg of $Al(OH)_3$ has an expected acid-neutralizing capacity value of 0.0385 mEq.

Assay—

Disodium ethylenediaminetetraacetate titrant—Prepare and standardize as directed in the *Assay* under *Ammonium Alum.*

Procedure—Weigh and finely powder not less than 20 Dried Basic Aluminum Carbonate Gel Tablets. Transfer an accurately weighed portion of the powder, equivalent to about 1.2 g of aluminum hydroxide, to a beaker, add 15 mL of hydrochloric acid, and heat until dissolved. Dilute with water to about 100 mL, mix, and filter quantitatively into a 500-mL volumetric flask, washing the filter with water. Proceed as directed in the *Assay* under *Basic Aluminum Carbonate Gel*, beginning with "dilute with water to volume." Each mL of 0.05 *M Disodium ethylenediaminetetraacetate titrant* is equivalent to 3.900 mg of $Al(OH)_3$.

Aluminum Chloride

$AlCl_3 \cdot 6H_2O$ 241.43
Aluminum chloride, hexahydrate.
Aluminum chloride hexahydrate [*7784-13-6*].
Anhydrous 133.34 [*7446-70-0*].

» Aluminum Chloride contains not less than 95.0 percent and not more than 102.0 percent of $AlCl_3$, calculated on the anhydrous basis.

Packaging and storage—Preserve in tight containers.

Identification—A solution (1 in 10) responds to the tests for *Aluminum* and for *Chloride* ⟨191⟩.

Water, *Method I* ⟨921⟩: between 42.0% and 48.0%.

Sulfate—The addition of 0.2 mL of barium chloride TS to 10 mL of a solution (1 in 100) produces no turbidity within 1 minute.

Alkalies and alkaline earths—To a boiling solution of 1.0 g in 150 mL of water add a few drops of methyl red TS, then add 6

N ammonium hydroxide until the color of the solution just changes to a distinct yellow. Add hot water to restore the volume to 150 mL, and filter while hot. Evaporate 75 mL of the filtrate to dryness, and ignite to constant weight: the weight of the residue does not exceed 2.5 mg (0.5%).

Arsenic, *Method I* ⟨211⟩: 8 ppm.

Heavy metals, *Method I* ⟨231⟩—Dissolve 1 g in 1 mL of 1 *N* acetic acid and sufficient water to make 25 mL: the limit is 0.002%.

Iron ⟨241⟩—Dissolve 1.0 g in 45 mL of water, and add 2 mL of hydrochloric acid: the limit is 0.001%.

Assay—

Disodium ethylenediaminetetraacetate titrant—Prepare and standardize as directed in the *Assay* under *Alum*.

Procedure—Transfer to a 250-mL volumetric flask about 5 g of Aluminum Chloride, accurately weighed, dissolve in water, dilute with water to volume, and mix. Transfer 10.0 mL of the solution to a 250-mL beaker, and add, in the order named and with continuous stirring, 25.0 mL of *Disodium ethylenediaminetetraacetate titrant*, and 20 mL of acetic acid–ammonium acetate buffer TS, and boil gently for 5 minutes. Cool, and add 50 mL of alcohol and 2 mL of dithizone TS. Titrate with 0.05 *M* zinc sulfate VS to a bright rose-pink color. Perform a blank determination, substituting 10 mL of water for the assay preparation, and make any necessary correction. Each mL of 0.05 *M Disodium ethylenediaminetetraacetate titrant* is equivalent to 6.667 mg of $AlCl_3$.

Aluminum Hydroxide Gel

$Al(OH)_3$ 78.00
Aluminum hydroxide.
Aluminum hydroxide [*21645-51-2*].

» Aluminum Hydroxide Gel is a suspension, each 100 g of which contains the equivalent of not less than 5.5 g and not more than 6.7 g of aluminum hydroxide [$Al(OH)_3$], in the form of amorphous aluminum hydroxide in which there is a partial substitution of carbonate for hydroxide. It may contain Peppermint Oil, Glycerin, Sorbitol, Sucrose, Saccharin, or other suitable flavors, and it may contain suitable antimicrobial agents.

Packaging and storage—Preserve in tight containers, and avoid freezing.

Identification—A solution in hydrochloric acid responds to the tests for *Aluminum* ⟨191⟩.

Microbial limits ⟨61⟩—Its total aerobic microbial count does not exceed 100 per mL, and it meets the requirements of the test for the absence of *Escherichia coli*.

Acid-neutralizing capacity ⟨301⟩: not less than 65.0% of the expected mEq value, calculated from the results of the *Assay*, is obtained. Each mg of $Al(OH)_3$ has an expected acid-neutralizing capacity value of 0.0385 mEq.

pH ⟨791⟩: between 5.5 and 8.0, determined potentiometrically.

Chloride—Transfer 10 g to a porcelain dish. Add 0.1 mL of potassium chromate TS and 25 mL of water. Stir, and add 0.10 *N* silver nitrate until a faint, persistent pink color is obtained: not more than 8.0 mL of 0.10 *N* silver nitrate is required (0.28%).

Sulfate ⟨221⟩—Add 5.0 mL of 3 *N* hydrochloric acid to 5 g of it, and heat to dissolve the specimen under test. Cool, dilute with water to 250 mL, and filter if necessary: a 20-mL portion of the filtrate shows no more sulfate than corresponds to 0.20 mL of 0.020 *N* sulfuric acid (0.05%).

Arsenic, *Method I* ⟨211⟩—Prepare a *Standard Preparation* as directed in the test for *Arsenic* ⟨211⟩, except to prepare it to contain 5 µg of arsenic instead of 3 µg. Prepare the *Test Preparation* as follows: Dissolve 8.3 g of Aluminum Hydroxide Gel in 20 mL of 7 *N* sulfuric acid. The limit is 0.6 ppm.

Heavy metals ⟨231⟩—Dissolve 4.0 g in 10 mL of 3 *N* hydrochloric acid with the aid of heat, filter, if necessary, and dilute with water to 25 mL: the limit is 5 ppm.

Assay—

Disodium ethylenediaminetetraacetate titrant—Prepare and standardize as directed in the *Assay* under *Ammonium Alum*.

Procedure—Transfer about 25 g, accurately weighed, of Aluminum Hydroxide Gel to a beaker, add 15 mL of hydrochloric acid, and heat gently until solution is complete. Cool, transfer to a 500-mL volumetric flask, dilute with water to volume, and mix. Pipet 20 mL of this solution into a 250-mL beaker, and add, in the order named and with continuous stirring, 25.0 mL of *Disodium ethylenediaminetetraacetate titrant* and 20 mL of acetic acid–ammonium acetate buffer TS, then heat the solution near the boiling point for 5 minutes. Cool, and add 50 mL of alcohol and 2 mL of dithizone TS. Titrate the solution with 0.05 *M* zinc sulfate VS until the color changes from green-violet to rose-pink. Perform a blank determination, substituting 20 mL of water for the sample, and make any necessary correction. Each mL of 0.05 *M Disodium ethylenediaminetetraacetate titrant* consumed is equivalent to 3.900 mg of $Al(OH)_3$.

Dried Aluminum Hydroxide Gel

$Al(OH)_3$ 78.00
Aluminum hydroxide [*21645-51-2*].

» Dried Aluminum Hydroxide Gel is an amorphous form of aluminum hydroxide in which there is a partial substitution of carbonate for hydroxide. It contains the equivalent of not less than 76.5 percent of $Al(OH)_3$, and it may contain varying quantities of basic aluminum carbonate and bicarbonate.

Packaging and storage—Preserve in tight containers.

Labeling—Where the quantity of dried aluminum hydroxide gel equivalent is stated in the labeling of any preparation, this shall be understood to be on the basis that each mg of dried gel is equivalent to 0.765 mg of $Al(OH)_3$.

Reference standard—*USP Dried Aluminum Hydroxide Gel Reference Standard*—Do not dry before using.

Identification—

A: The infrared absorption spectrum of a potassium bromide dispersion of it exhibits maxima only at the same wavelengths as that of a similar preparation of USP Dried Aluminum Hydroxide Gel RS.

B: Dissolve 500 mg in 10 mL of 3 *N* hydrochloric acid, with gentle warming: the solution responds to the tests for *Aluminum* ⟨191⟩.

Acid-neutralizing capacity ⟨301⟩: not less than 25.0 mEq per g, 400 mg being tested as directed for *Powders* under *Test Preparation*.

pH ⟨791⟩: not higher than 10.0, in an aqueous dispersion (1 in 25).

Chloride ⟨221⟩—Dissolve 1.0 g in 30 mL of 2 *N* nitric acid, heat to boiling, add water to make 100 mL, and filter: a 5.0-mL portion of the filtrate, diluted with an equal volume of water, shows no more chloride than corresponds to 0.60 mL of 0.020 *N* hydrochloric acid (0.85%).

Sulfate ⟨221⟩—Dissolve 330 mg in 15 mL of 3 *N* hydrochloric acid, heat to boiling, add water to make 250 mL, and filter: a 25-mL portion of the filtrate shows no more sulfate than corresponds to 0.20 mL of 0.020 *N* sulfuric acid (0.6%).

Arsenic, *Method I* ⟨211⟩—Dissolve 1.5 g in 80 mL of 7 *N* sulfuric acid, and dilute with water to 220 mL: 55 mL of the resulting solution meets the requirements of the test, the addition of 20 mL of 7 *N* sulfuric acid specified under *Procedure* being omitted. The limit is 8 ppm.

Heavy metals ⟨231⟩—Dissolve 330 mg in 10 mL of 3 *N* hydrochloric acid with the aid of heat, filter if necessary, and dilute with water to 25 mL: the limit is 0.006%.

Assay—

Disodium ethylenediaminetetraacetate titrant—Prepare and standardize as directed in the *Assay* under *Ammonium Alum.*

Procedure—Weigh accurately about 2 g of Dried Aluminum Hydroxide Gel, and dissolve in 15 mL of hydrochloric acid, with the aid of heat. Cool, transfer to a 500-mL volumetric flask, dilute with water to volume, and mix. Pipet 20 mL of this solution into a 250-mL beaker, and add, in the order named and with continuous stirring, 25.0 mL of *Disodium ethylenediaminetetraacetate titrant* and 20 mL of acetic acid–ammonium acetate buffer TS, then heat the solution near the boiling point for 5 minutes. Cool, and add 50 mL of alcohol and 2 mL of dithizone TS. Titrate the solution with 0.05 *M* zinc sulfate VS to a bright rose-pink color. Perform a blank determination, substituting 20 mL of water for the sample solution, and make any necessary correction. Each mL of 0.05 *M Disodium ethylenediaminetetraacetate titrant* is equivalent to 3.900 mg of $Al(OH)_3$.

Dried Aluminum Hydroxide Gel Capsules

» Dried Aluminum Hydroxide Gel Capsules contain not less than 90.0 percent and not more than 110.0 percent of the labeled amount of aluminum hydroxide [$Al(OH)_3$].

Packaging and storage—Preserve in well-closed containers.

Labeling—Capsules may be labeled to state the aluminum hydroxide content in terms of the equivalent amount of dried aluminum hydroxide gel, on the basis that each mg of dried gel is equivalent to 0.765 mg of $Al(OH)_3$.

Identification—Digest a portion of Capsule contents, equivalent to about 500 mg of aluminum hydroxide, with 10 mL of 3 *N* hydrochloric acid with gentle warming, and filter: the filtrate so obtained responds to the tests for *Aluminum* ⟨191⟩.

Disintegration ⟨701⟩: 10 minutes, simulated gastric fluid TS being substituted for water in the test.

Uniformity of dosage units ⟨905⟩: meet the requirements.

Acid-neutralizing capacity ⟨301⟩: not less than 5 mEq of acid is consumed by the minimum single dose recommended in the labeling, and not less than 55.0% of the expected mEq value, calculated from the labeled quantity of $Al(OH)_3$, is obtained. Each mg of $Al(OH)_3$ has an expected acid-neutralizing capacity value of 0.0385 mEq.

Assay—

Disodium ethylenediaminetetraacetate titrant—Prepare and standardize as directed in the *Assay* under *Ammonium Alum.*

Procedure—Weigh accurately the contents of not less than 20 Dried Aluminum Hydroxide Gel Capsules, and mix. Transfer an accurately weighed portion of the powder, equivalent to about 1.2 g of aluminum hydroxide, to a beaker, add 15 mL of hydrochloric acid, and heat until dissolved. Dilute with water to about 100 mL, mix, and filter quantitatively into a 500-mL volumetric flask, washing the filter with water. Proceed as directed in the *Assay* under *Dried Aluminum Hydroxide Gel*, beginning with "dilute with water to volume." Each mL of 0.05 *M Disodium ethylenediaminetetraacetate titrant* is equivalent to 3.900 mg of $Al(OH)_3$.

Dried Aluminum Hydroxide Gel Tablets

» Dried Aluminum Hydroxide Gel Tablets contain not less than 90.0 percent and not more than 110.0 percent of the labeled amount of aluminum hydroxide [$Al(OH)_3$].

Packaging and storage—Preserve in well-closed containers.

Labeling—Tablets may be labeled to state the aluminum hydroxide content in terms of the equivalent amount of dried aluminum hydroxide gel, on the basis that each mg of dried gel is equivalent to 0.765 mg of $Al(OH)_3$.

Identification—Digest a portion of finely powdered Tablets, equivalent to about 500 mg of dried aluminum hydroxide gel, with 10 mL of 3 *N* hydrochloric acid with gentle warming, and filter: the filtrate responds to the tests for *Aluminum* ⟨191⟩.

Disintegration ⟨701⟩: 10 minutes, simulated gastric fluid TS being substituted for water in the test.

Uniformity of dosage units ⟨905⟩: meet the requirements for *Weight Variation.*

Acid-neutralizing capacity ⟨301⟩: not less than 5 mEq of acid is consumed by the minimum single dose recommended in the labeling, and not less than 55.0% of the expected mEq value, calculated from the labeled quantity of $Al(OH)_3$, is obtained. Each mg of $Al(OH)_3$ has an expected acid-neutralizing capacity value of 0.0385 mEq.

Assay—

Disodium ethylenediaminetetraacetate titrant—Prepare and standardize as directed in the *Assay* under *Ammonium Alum.*

Procedure—Weigh and finely powder not less than 20 Dried Aluminum Hydroxide Gel Tablets. Weigh accurately a portion of the powder, equivalent to about 1.2 g of aluminum hydroxide, add 15 mL of hydrochloric acid, and heat until dissolved. Dilute with water to about 100 mL, mix, and filter quantitatively into a 500-mL volumetric flask, washing the filter with water. Proceed as directed in the *Assay* under *Dried Aluminum Hydroxide Gel*, beginning with "dilute with water to volume." Each mL of 0.05 *M Disodium ethylenediaminetetraacetate titrant* is equivalent to 3.900 mg of $Al(OH)_3$.

Aluminum Monostearate—*see* Aluminum Monostearate NF

Aluminum Phosphate Gel

Phosphoric acid, aluminum salt (1:1).
Aluminum phosphate (1:1) [*7784-30-7*].

» Aluminum Phosphate Gel is a water suspension containing not less than 4.0 percent and not more than 5.0 percent (w/w) of aluminum phosphate ($AlPO_4$). It may contain sodium benzoate, benzoic acid, or other suitable agent, in an amount not exceeding 0.5 percent, as a preservative.

Packaging and storage—Preserve in tight containers.

Identification—

A: A solution of it in hydrochloric acid responds to the tests for *Aluminum* ⟨191⟩.

B: A solution of it in 2 *N* nitric acid responds to the tests for *Phosphate* ⟨191⟩.

pH ⟨791⟩: between 6.0 and 7.2.

Chloride ⟨221⟩—Transfer 25 g to a beaker with the aid of about 50 mL of water, add 5 mL of nitric acid, mix, then add, with stirring, 30.0 mL of 0.1 *N* silver nitrate VS. Warm on a steam bath for 30 minutes, filter, and wash the precipitate with water acidified with nitric acid. To the filtrate add ferric ammonium sulfate TS, and titrate the excess silver nitrate with 0.1 *N* ammonium thiocyanate VS. Each mL of 0.1 *N* silver nitrate is equivalent to 3.545 mg of Cl. Not more than 0.16% of Cl is found.

Soluble phosphate—Filter 20 g, and wash the residue with 30 mL of water. Add to the filtrate 2 mL of nitric acid, heat to 60°, and add 20 mL of ammonium molybdate TS. Heat at 50° for 30 minutes, filter, wash the precipitate with dilute nitric acid (1 in 36), then wash with potassium nitrate solution (1 in 100) until the last portion of the filtrate is not acid to litmus paper. Dissolve the precipitate in 50.0 mL of 0.5 *N* sodium hydroxide VS, add phenolphthalein TS, and titrate the excess alkali with 0.5 *N* hydrochloric acid VS. Each mL of 0.5 *N* sodium hydroxide is equivalent to 2.065 mg of PO_4. The soluble phosphate, calculated as PO_4, does not exceed 0.30%.

Sulfate ⟨221⟩—Add 10 mL of 3 *N* hydrochloric acid to 10 g of Gel, and heat to boiling. Cool, dilute with water to 250 mL, and filter, if necessary. A 10-mL portion of the solution shows no more sulfate than corresponds to 0.20 mL of 0.020 *N* sulfuric acid (0.05%).

Arsenic, *Method I* ⟨211⟩—Prepare the *Test Preparation* by dissolving 5.0 g of Gel in the smallest necessary volume of 3 *N* hydrochloric acid. The limit is 0.6 ppm.

Heavy metals, *Method I* ⟨231⟩—Dissolve 8.0 g in 5 mL of 3 *N* hydrochloric acid, warming if necessary, dilute with water to 25 mL, and adjust with 6 *N* ammonium hydroxide to a pH between 1.9 and 2.1. Transfer to a 50-mL color-comparison tube, dilute with water to 40 mL, and mix to obtain the *Test Preparation.* Similarly prepare a *Standard Preparation*, using 4.0 mL of *Standard Lead Solution:* the limit is 5 ppm.

Assay—To about 20 g of Aluminum Phosphate Gel, accurately weighed, in a 100-mL volumetric flask, add nitric acid to effect solution, dilute with water to volume, and mix. Transfer 10.0 mL of this solution to a 400-mL beaker, dilute with water to 100 mL, heat to 60°, add an excess of ammonium molybdate TS, and maintain at 50° for 30 minutes. Filter, and wash the precipitate with dilute nitric acid (1 in 36), then with potassium nitrate solution (1 in 100) until the last portion of the filtrate is not acid to litmus paper. Dissolve the precipitate in 50.0 mL of 0.5 *N* sodium hydroxide VS, add phenolphthalein TS, and titrate the excess sodium hydroxide with 0.5 *N* sulfuric acid VS. Each mL of 0.5 *N* sodium hydroxide is equivalent to 2.651 mg of $AlPO_4$.

Aluminum Stearate Suspension, Sterile Penicillin G Procaine with—*see* Penicillin G Procaine with Aluminum Stearate Suspension, Sterile

Aluminum Subacetate Topical Solution

$Al(OH)(OCOCH_3)_2$

$C_4H_7AlO_5$ 162.08
Aluminum, bis(acetato-*O*)hydroxy-.
Bis(acetato)hydroxyaluminum.
Basic aluminum acetate [*142-03-0; 8000-61-1*].

» Aluminum Subacetate Topical Solution yields, from each 100 mL, not less than 2.30 g and not more than 2.60 g of aluminum oxide (Al_2O_3), and not less than 5.43 g and not more than 6.13 g of acetic acid ($C_2H_4O_2$). It may be stabilized by the addition of not more than 0.9 percent of boric acid.

Aluminum Subacetate Topical Solution may be prepared as follows.

Aluminum Sulfate	145 g
Acetic Acid	160 mL
Precipitated Calcium Carbonate	70 g
Purified Water, a sufficient quantity, to make	1000 mL

Dissolve the Aluminum Sulfate in 600 mL of cold water, filter the solution, and add the Precipitated Calcium Carbonate gradually, in several portions, with constant stirring. Then slowly add the Acetic Acid, mix, and set the mixture aside for 24 hours. Filter the product with the aid of vacuum if necessary, returning the first portion of the filtrate to the funnel. Wash the magma on the filter with small portions of cold water, until the total filtrate measures 1000 mL.

Packaging and storage—Preserve in tight containers.

Identification—It responds to the tests for *Aluminum* ⟨191⟩ and for *Acetate* ⟨191⟩.

pH ⟨791⟩: between 3.8 and 4.6.

Limit of boric acid—Proceed as directed in the test for *Limit of boric acid* under *Aluminum Acetate Topical Solution.*

Assay for aluminum oxide—

Disodium ethylenediaminetetraacetate titrant—Prepare and standardize as directed in the *Assay* under *Ammonium Alum.*

Procedure—Pipet 20 mL of Aluminum Subacetate Topical Solution into a 250-mL volumetric flask, add 5 mL of hydrochloric acid, dilute with water to volume, and mix. Pipet 25 mL of this dilution into a 250-mL beaker, and proceed as directed for *Procedure* in the *Assay for aluminum oxide* under *Aluminum Acetate Topical Solution*, beginning with "add, in the order named." Each mL of 0.05 *M Disodium ethylenediaminetetraacetate titrant* is equivalent to 2.549 mg of Al_2O_3.

Assay for acetic acid—Proceed as directed in the *Assay for acetic acid* under *Aluminum Acetate Topical Solution.*

Aluminum Sulfate

$Al_2(SO_4)_3 \cdot xH_2O$ (anhydrous) 342.14
Sulfuric acid, aluminum salt (3:2), hydrate.
Aluminum sulfate (2:3) hydrate [*17927-65-0*].
Anhydrous 342.14 [*10043-01-3*].

» Aluminum Sulfate contains not less than 54.0 percent and not more than 59.0 percent of $Al_2(SO_4)_3$. It contains a varying amount of water of crystallization.

Packaging and storage—Preserve in well-closed containers.

Identification—A solution (1 in 10) responds to the tests for *Aluminum* and for *Sulfate* ⟨191⟩.

pH ⟨791⟩: not less than 2.9, in a solution (1 in 20).

Water, *Method I* ⟨921⟩: not less than 41.0% and not more than 46.0%.

Alkalies and alkaline earths—To a boiling solution of 1.0 g in 150 mL of water add a few drops of methyl red TS and then add 6 *N* ammonium hydroxide until the color of the solution just changes to a distinct yellow. Add hot water to restore the volume to 150 mL, and filter while hot. Evaporate 75 mL of the filtrate to dryness, and ignite to constant weight: not more than 2 mg of residue remains (0.4%).

Ammonium salts—Heat 1 g with 10 mL of 1 *N* sodium hydroxide on a steam bath for 1 minute: the odor of ammonia is not perceptible.

Arsenic, *Method I* ⟨211⟩: 3 ppm.

Heavy metals ⟨231⟩—Dissolve 0.50 g in 1 mL of 1 *N* acetic acid, and dilute with water to 25 mL. The limit is 0.004%.

Iron—To 20 mL of a solution (1 in 150) add 0.3 mL of potassium ferrocyanide TS: no blue color is produced immediately.

Assay—

Disodium ethylenediaminetetraacetate titrant—Prepare and standardize as directed in the *Assay* under *Ammonium Alum.*

Procedure—Transfer about 7.5 g of Aluminum Sulfate, accurately weighed, to a 250-mL volumetric flask, and dissolve in water. Dilute with water to volume, mix, and pipet 10 mL of the solution into a 250-mL beaker. Proceed as directed in the *Assay for aluminum oxide* under *Aluminum Acetate Topical Solution*, beginning with "add, in the order named." Each mL of 0.05 *M Disodium ethylenediaminetetraacetate titrant* is equivalent to 8.554 mg of $Al_2(SO_4)_3$.

Amantadine Hydrochloride

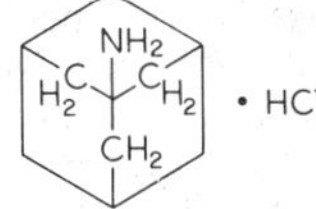

$C_{10}H_{17}N.HCl$ 187.71
Tricyclo[3.3.1.1^{3,7}]decan-1-amine, hydrochloride.
1-Adamantanamine hydrochloride [*665-66-7*].

» Amantadine Hydrochloride contains not less than 98.5 percent and not more than 101.5 percent of $C_{10}H_{17}N.HCl$.

Packaging and storage—Preserve in well-closed containers.

Reference standard—*USP Amantadine Hydrochloride Reference Standard*—Do not dry before using.

Clarity and color of solution—Dissolve 2 g in 10 mL of water: the solution is clear and nearly colorless.

Identification—Dissolve about 50 mg in 10 mL of 0.1 *N* hydrochloric acid, and filter. Transfer the filtrate to a suitable separator, add 1 mL of 5 *N* sodium hydroxide, and extract with 5 mL of methylene chloride. Filter the extract through anhydrous sodium sulfate, and rinse the anhydrous sodium sulfate with 2 mL of methylene chloride: the infrared absorption spectrum, determined in a 1-mm cell, of the solution so obtained exhibits maxima only at the same wavelengths as that of a similar preparation of USP Amantadine Hydrochloride RS.

pH ⟨791⟩: between 3.0 and 5.5, in a solution (1 in 5).

Heavy metals, *Method I* ⟨231⟩—Use 1 mL of 1 *N* acetic acid in the *Test preparation:* the limit is 0.001%.

Assay—Dissolve about 120 mg of Amantadine Hydrochloride, accurately weighed, in a mixture of 30 mL of glacial acetic acid and 10 mL of mercuric acetate TS, and titrate with 0.1 *N* perchloric acid VS, determining the end-point potentiometrically, using suitable electrodes. Perform a blank determination, and make any necessary correction. Each mL of 0.1 *N* perchloric acid is equivalent to 18.77 mg of $C_{10}H_{17}N.HCl$.

Amantadine Hydrochloride Capsules

» Amantadine Hydrochloride Capsules contain not less than 95.0 percent and not more than 105.0 percent of the labeled amount of $C_{10}H_{17}N.HCl$.

Packaging and storage—Preserve in tight containers.

Reference standard—*USP Amantadine Hydrochloride Reference Standard*—Do not dry before using.

Identification—Place the contents of Capsules, equivalent to about 200 mg of amantadine hydrochloride, in a vessel, dissolve in 0.1 *N* hydrochloric acid, and filter. Transfer the filtrate to a separator, add 1 mL of 5 *N* sodium hydroxide, and extract with 5 mL of methylene chloride. Filter the extract through anhydrous sodium sulfate, and rinse the anhydrous sodium sulfate with 2 mL of methylene chloride: the infrared absorption spectrum, determined in a 1-mm cell, of the solution so obtained exhibits maxima only at the same wavelengths as that of a similarly prepared solution of USP Amantadine Hydrochloride RS.

Dissolution ⟨711⟩—

Medium: water; 900 mL.

Apparatus 1: 100 rpm.

Time: 45 minutes.

Bromocresol purple solution—Dissolve 0.5 g of bromocresol purple sodium salt in 500 mL of 0.33 *N* acetic acid (19 mL of glacial acetic acid per liter). Filter through a 1-μm cellulose acetate membrane filter, if necessary, to clarify.

Standard solution—Dissolve an accurately weighed quantity of USP Amantadine Hydrochloride RS in water to obtain a solution having a known concentration of about 0.11 mg per mL.

Procedure—Separately pipet 1.5 mL of the *Standard solution*, 1.5 mL of a filtered portion of the solution under test, and 1.5 mL of water to provide a blank, respectively, into three glass-stoppered, 50-mL centrifuge tubes. Add 4.0 mL of *Bromocresol purple solution* and 15.0 mL of chloroform to each tube, shake vigorously for 20 seconds, and centrifuge. Concomitantly determine the absorbances of the chloroform extracts obtained from the *Standard solution* and the solution under test against the chloroform extract from the blank at the wavelength of maximum absorbance at about 408 nm, and calculate the amount of $C_{10}H_{17}N.HCl$ dissolved.

Tolerances—Not less than 75% (*Q*) of the labeled amount of $C_{10}H_{17}N.HCl$ is dissolved in 45 minutes.

Uniformity of dosage units ⟨905⟩: meet the requirements.

Assay—Dissolve 10 Amantadine Hydrochloride Capsules in 20 mL of water by warming on a steam bath, with shaking. Cool, add 20 mL of 5 *N* sodium hydroxide and 100.0 mL of hexane, insert the stopper, and shake gently for 15 minutes. Centrifuge to separate the layers, and transfer 10.0 mL of the supernatant hexane layer to a conical flask. Add 30 mL of glacial acetic acid, and titrate with 0.1 *N* perchloric acid VS, determining the end-point potentiometrically, using suitable electrodes. Perform a blank determination, and make any necessary correction. Each mL of 0.1 *N* perchloric acid is equivalent to 18.77 mg of $C_{10}H_{17}N.HCl$.

Amantadine Hydrochloride Syrup

» Amantadine Hydrochloride Syrup contains not less than 95.0 percent and not more than 105.0 percent of the labeled amount of $C_{10}H_{17}N.HCl$.

Packaging and storage—Preserve in tight containers.

Reference standard—*USP Amantadine Hydrochloride Reference Standard*—Do not dry before using.

Identification—Place a volume of Syrup, equivalent to about 200 mg of amantadine hydrochloride, in a vessel, dissolve in 0.1 *N* hydrochloric acid, and filter. Transfer the filtrate to a separator, add 10 mL of 0.5 *N* sodium hydroxide, and extract with 5 mL of methylene chloride. Filter the extract through anhydrous sodium sulfate, and rinse the anhydrous sodium sulfate with 2 mL of methylene chloride: the infrared absorption spectrum, determined in a 1-mm cell, of the solution so obtained exhibits maxima only at the same wavelengths as that of a similarly prepared solution of USP Amantadine Hydrochloride RS.

Assay—Transfer an accurately measured volume of Amantadine Hydrochloride Syrup, equivalent to about 200 mg of amantadine hydrochloride, to a separator, add 5 mL of 5 *N* sodium hydroxide and 20.0 mL of hexane, shake vigorously for 15 minutes, and allow the layers to separate. Transfer 10.0 mL of the hexane layer to a beaker containing 30 mL of glacial acetic acid, and titrate with 0.1 *N* perchloric acid VS, determining the end-point potentiometrically, using suitable electrodes. Perform a blank determination, and make any necessary correction. Each mL of 0.1 *N* perchloric acid is equivalent to 18.77 mg of $C_{10}H_{17}N.HCl$.

Amcinonide

$C_{28}H_{35}FO_7$ 502.58
Pregna-1,4-diene-3,20-dione, 21-(acetyloxy)-16,17-[cyclopentylidenebis(oxy)]-9-fluoro-11-hydroxy-, (11β,16α)-.
9-Fluoro-11β,16α,17,21-tetrahydroxypregna-1,4-diene-3,20-dione cyclic 16,17-acetal with cyclopentanone, 21-acetate [*51022-69-6*].

» Amcinonide contains not less than 97.0 percent and not more than 102.0 percent of $C_{28}H_{35}FO_7$, calculated on the dried basis.

Packaging and storage—Preserve in well-closed containers.

Reference standard—*USP Amcinonide Reference Standard*—Dry at 105° for 4 hours before using.

Identification—

A: The infrared absorption spectrum of a potassium bromide dispersion of it, previously dried, exhibits maxima only at the same wavelengths as that of a similar preparation of USP Amcinonide RS.

B: The ultraviolet absorption spectrum of a 1 in 25,000 solution in methanol exhibits maxima and minima at the same wavelengths as that of a similar preparation of USP Amcinonide RS, concomitantly measured, and the respective absorptivities, calculated on the dried basis, at the wavelength of maximum absorbance at about 238 nm, do not differ by more than 3.0%.

Specific rotation ⟨781⟩: between +89.4° and +94.0°, calculated on the dried basis, determined in a solution in chloroform containing 100 mg in each 10 mL.

Loss on drying ⟨731⟩—Dry it at 105° for 4 hours: it loses not more than 1.0% of its weight.

Heavy metals, *Method II* ⟨231⟩: 0.002%.

Assay—

Mobile phase—Prepare a solution of isopropyl alcohol in methylene chloride (up to 4 in 100) such that the retention time of amcinonide is between 3.5 and 4.0 minutes.

Standard preparation—Dissolve an accurately weighed portion of USP Amcinonide RS in *Mobile phase* to obtain a solution having a known concentration of about 0.4 mg per mL.

Assay preparation—Using about 20 mg of Amcinonide, accurately weighed, prepare as directed under *Standard preparation.*

Chromatographic system (see *Chromatography* ⟨621⟩)—The liquid chromatograph is equipped with a 254-nm detector and a 2.1-mm × 60-cm column that contains packing L4. The flow rate is about 1 mL per minute. Adjust the operating parameters such that the peak obtained from the *Standard preparation* is about 0.6 full scale.

System suitability preparation—Prepare a solution in *Mobile phase* containing about 0.1 mg of USP Amcinonide RS per mL and about 0.2 mg of triamcinolone diacetate per mL.

System suitability test—Inject 20 µL of the *System suitability preparation* into the chromatograph: the resolution factor between triamcinolone diacetate and amcinonide is not less than 1.90, and the tailing factor for the amcinonide peak is not more than 1.35.

Procedure—Introduce equal volumes (about 20 µL) of the *Assay preparation* and the *Standard preparation* into the chromatograph by means of a sampling valve, and record the chromatogram. Measure the peak responses at corresponding retention times, obtained from the *Assay preparation* and the *Standard preparation,* and calculate the quantity, in mg, of $C_{28}H_{35}FO_7$ in the portion of Amcinonide taken by the formula:

$$50C(r_U/r_S),$$

in which C is the concentration, in mg per mL, of USP Amcinonide RS in the *Standard preparation,* and r_U and r_S are the peak responses obtained from the *Assay preparation* and the *Standard preparation,* respectively.

Amcinonide Cream

» Amcinonide Cream is Amcinonide in a suitable cream base. It contains not less than 90.0 percent and not more than 115.0 percent of the labeled amount of $C_{28}H_{35}FO_7$.

Packaging and storage—Preserve in tight containers.

Reference standard—*USP Amcinonide Reference Standard*—Dry at 105° for 4 hours before using.

Identification—Place 2 g of the Cream in a 150-mL beaker, add 50 mL of chloroform and 15 g of anhydrous sodium sulfate, and stir with a glass rod to dissolve the specimen. Filter the solution, and clarify the filtrate, if necessary, by the further addition of anhydrous sodium sulfate and a second filtration. Evaporate the filtrate to dryness, and dissolve the residue in chloroform to obtain a solution containing about 100 µg of amcinonide per mL. Apply 25 µL of this solution, and 25 µL of a solution of USP Amcinonide RS in chloroform containing 100 µg per mL, on a line parallel to and about 3 cm from the bottom edge of a thin-layer chromatographic plate (see *Chromatography* ⟨621⟩) coated with a 0.25-mm layer of chromatographic silica gel. Place the plate in a developing chamber containing and equilibrated with ether. Develop the chromatogram until the solvent front has moved about 12 cm above the line of application. Remove the plate, allow the solvent to evaporate, and locate the spots on the plate by viewing under short-wavelength ultraviolet light: the intensity and the R_f value of the principal spot obtained from the solution under test are similar to those of the spot obtained from the Standard solution.

Microbial limits—It meets the requirements of the tests for absence of *Staphylococcus aureus* and *Pseudomonas aeruginosa* under *Microbial Limit Tests* ⟨61⟩.

Minimum fill ⟨755⟩: meets the requirements.

pH ⟨791⟩: between 3.5 and 5.2.

Assay—

Mobile phase, Chromatographic system, System suitability preparation, and *System suitability test*—Proceed as directed in the *Assay* under *Amcinonide.*

Standard preparation—Dissolve an accurately weighed portion of USP Amcinonide RS in *Mobile phase* to obtain a solution having a known concentration of about 0.1 mg per mL.

Assay preparation—Transfer an accurately weighed quantity of Amcinonide Cream, equivalent to about 1 mg of amcinonide, to a separator. Add 30 mL of a mixture of chloroform and cyclohexane (1:1), 9 mL of water, and 1 mL of 1 *N* sulfuric acid. Shake vigorously for 2 to 3 minutes, and allow the phases to separate. Filter the chloroform-cyclohexane phase through anhydrous sodium sulfate, previously soaked with a mixture of chloroform and cyclohexane (1:1), supported in a filter funnel, into a glass chromatographic column measuring about 1.3 cm × 45 cm, fitted with a stopcock and glass-wool plug, and containing 8 g of 60- to 100-mesh chromatographic magnesium silicate. Allow the chloroform-cyclohexane to pass through the column. Repeat the extraction with three 30-mL portions of the mixture of chloroform and cyclohexane (1:1), passing the extracts through the column, and discarding the eluates. Elute the amcinonide from the column with two 40-mL portions of a mixture of 1 volume of water, 10 volumes of methanol, and 89 volumes of acetone, collecting the eluate in a suitable receiving vessel. Evaporate the eluate at a temperature below 50°, with the aid of a gentle current of filtered air, to dryness. Dissolve the residue in a small amount of *Mobile phase,* and transfer to a 10-mL volumetric flask with the aid of small portions of *Mobile phase.* Dilute with *Mobile phase* to volume, and mix.

Procedure—Proceed as directed for *Procedure* in the *Assay* under *Amcinonide.* Calculate the quantity, in mg, of $C_{28}H_{35}FO_7$ in the portion of Cream taken by the formula:

$$10C(r_U/r_S),$$

in which C is the concentration, in mg per mL, of USP Amcinonide RS in the *Standard preparation,* and r_U and r_S are the peak responses obtained from the *Assay preparation* and the *Standard preparation,* respectively.

Amcinonide Ointment

» Amcinonide Ointment is Amcinonide in a suitable ointment base. It contains not less than 90.0 percent and not more than 115.0 percent of the labeled amount of $C_{28}H_{35}FO_7$.

Packaging and storage—Preserve in tight containers.

Reference standard—*USP Amcinonide Reference Standard*—Dry at 105° for 4 hours before using.

Identification—It responds to the *Identification test* under *Amcinonide Cream.*

Microbial limits—It meets the requirements of the tests for absence of *Staphylococcus aureus* and *Pseudomonas aeruginosa* under *Microbial Limit Tests* ⟨61⟩.

Minimum fill ⟨755⟩: meets the requirements.

Assay—

Mobile phase—Prepare a mixture of methanol and water (80:20), and filter through a 0.45-µm membrane filter before use.

Internal standard solution—Dissolve dibutyl phthalate in methanol to obtain a solution having a concentration of about 500 µg per mL.

Standard preparation—Dissolve an accurately weighed portion of USP Amcinonide RS in *Mobile phase* to obtain a solution having a known concentration of about 0.1 mg per mL. Mix 5.0 mL of this solution and 5.0 mL of *Internal standard solution* in a suitable glass-stoppered vessel to obtain a *Standard preparation* having a known concentration of about 0.05 mg per mL.

Assay preparation—Transfer an accurately weighed quantity of Amcinonide Ointment, equivalent to about 0.5 mg of amcinonide, to a glass centrifuge tube. Add 5.0 mL of *Internal standard solution,* insert the stopper loosely, and heat in a 50° water bath for 3 minutes until the ointment forms small beadlets when shaken. Insert the stopper securely in the tube, and mix for 4 minutes, using a test-tube mixer, while the ointment is still molten, then allow to cool to room temperature. Add 5.0 mL of *Mobile phase*, mix, and filter through paper.

Chromatographic system (see *Chromatography* ⟨621⟩)—The liquid chromatograph is equipped with a 254-nm detector and a 4.6-mm × 25-cm column that contains packing L1. The flow rate is about 1 mL per minute. Adjust the operating parameters such that the amcinonide peak obtained from the *Standard preparation* is about 0.5 full scale. Typical retention times are about 7 minutes for amcinonide and about 9 minutes for the internal standard. Inject three replicate 20-µL portions of the *Standard preparation* into the chromatograph, and record the peak heights as directed under *Procedure:* the resolution factor between amcinonide and dibutyl phthalate is not less than 2.0, and the relative standard deviation is not more than 3.0%.

Procedure—Separately inject equal volumes (about 20 µL) of the *Standard preparation* and the *Assay preparation* into the chromatograph. Measure the peak heights of the amcinonide and the internal standard peaks at the same retention times obtained from the *Assay preparation* and the *Standard preparation.* Calculate the quantity, in mg, of $C_{28}H_{35}FO_7$ in the portion of Ointment taken by the formula:

$$10C(R_U/R_S),$$

in which *C* is the concentration, in mg per mL, of USP Amcinonide RS in the *Standard preparation*, and R_U and R_S are the ratios of the peak heights of amcinonide to internal standard obtained from the *Assay preparation* and the *Standard preparation*, respectively.

Sterile Amdinocillin

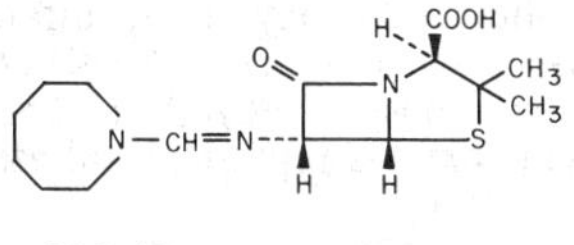

$C_{15}H_{23}N_3O_3S$ 325.43

4-Thia-1-azabicyclo[3.2.0]heptane-2-carboxylic acid, 6-[[(hexahydro-1*H*-azepin-1-yl)methylene]amino]-3,3-dimethyl-7-oxo-, [2*S*-(2α,5α,6β)]-.

(2*S*,5*R*,6*R*)-6-[[(Hexahydro-1*H*-azepin-1-yl)-methylene]-amino]-3,3-dimethyl-7-oxo-4-thia-1-azabicyclo[3.2.0]-heptane-2-carboxylic acid [*32887-01-7*].

» Sterile Amdinocillin is amdinocillin suitable for parenteral use. It contains not less than 950 µg and not more than 1,050 µg of $C_{15}H_{23}N_3O_3S$ per mg, calculated on the anhydrous basis. In addition, where packaged for dispensing, it contains not less than 90.0 percent and not more than 120.0 percent of the labeled amount of $C_{15}H_{23}N_3O_3S$.

Packaging and storage—Preserve in *Containers for Sterile Solids* as described under *Injections* ⟨1⟩.

Reference standard—*USP Amdinocillin Reference Standard*—Do not dry before using.

Constituted solution—At the time of use, the constituted solution prepared from Sterile Amdinocillin meets the requirements for *Constituted Solutions* under *Injections* ⟨1⟩.

Identification—

A: The infrared absorption spectrum of a potassium bromide dispersion of it exhibits maxima only at the same wavelengths as that of a similar preparation of USP Amdinocillin RS.

B: The chromatogram of the *Assay preparation* obtained as directed in the *Assay* exhibits a major peak for amdinocillin the retention time of which corresponds to that exhibited in the chromatogram of the *Standard preparation* obtained as directed in the *Assay.*

Crystallinity ⟨695⟩: meets the requirements.

Pyrogen—It meets the requirements of the *Pyrogen Test* ⟨151⟩, the test dose being 1 mL per kg of a solution prepared by diluting Sterile Amdinocillin with pyrogen-free saline TS to a concentration of 40 mg of amdinocillin per mL.

Sterility—It meets the requirements under *Sterility Tests* ⟨71⟩, when tested as directed in the section, *Test Procedures Using Membrane Filtration.*

pH ⟨791⟩: between 4.0 and 6.2, in a solution (1 in 10).

Water, *Method I* ⟨921⟩: not more than 0.5%.

Particulate matter ⟨788⟩: meets the requirements under *Small-volume Injections.*

Hexamethyleneimine—

Solvent mixture—Mix 80 volumes of chloroform, 20 volumes of alcohol, and 30 volumes of glacial acetic acid.

Chromatographic plate—Use a suitable thin-layer chromatographic plate (see *Chromatography* ⟨621⟩) coated with a 0.25-mm layer of chromatographic silica gel. Dry the plate at 100° for 2 hours before using.

Spray reagent—Dissolve 850 mg of bismuth nitrate in a mixture of 10 mL of glacial acetic acid and 40 mL of water. In a separate container, dissolve 20 g of potassium iodide in 50 mL of water. Mix the two solutions, and add dilute sulfuric acid (1 in 10) to make 500 mL. Add 7.5 ± 2.5 g of iodine, and mix until solution is complete.

Standard preparations—[NOTE—Use low-actinic glassware.] Dissolve an accurately weighed quantity of hexamethyleneimine in water to obtain a solution having a concentration of 0.80 mg per mL (*Standard preparation 1*). Dilute 5.0 mL of this solution with 5.0 mL of water (*Standard preparation 2*).

Test preparation—Dissolve 80 mg of Sterile Amdinocillin in 2.0 mL of water. [NOTE—Use low-actinic glassware, and apply the *Test preparation* on the chromatographic plate immediately after preparation.]

Procedure—Apply 5 µL each of the *Standard preparations* and the *Test preparation* to the thin-layer chromatographic plate. [NOTE—Apply the *Test preparation* last.] Dry immediately under a stream of nitrogen, and immediately chromatograph. Develop the chromatogram in an unlined, unsaturated chamber containing freshly prepared *Solvent mixture* until the solvent front has moved three-fourths of the length of the plate. Remove the plate from the developing chamber, mark the solvent front, and air-dry the plate thoroughly with the aid of a current of air. Spray with *Spray reagent:* any spot from the *Test preparation* at an R_f value corresponding to that of the principal spots obtained from the *Standard preparations* is not greater in size or intensity than the spot obtained from *Standard preparation 2* (1.0%). Where the Sterile Amdinocillin is packaged for dispensing, any such spot from the *Test preparation* is not greater in size or intensity than the spot obtained from *Standard preparation 1* (2.0%).

Other requirements—It meets the requirements for *Uniformity of Dosage Units* ⟨905⟩ and for *Labeling* under *Injections* ⟨1⟩.

Assay—

Mobile phase—Dissolve 1.36 g of monobasic potassium phosphate in water to obtain 1000 mL of solution, and adjust with phosphoric acid or 10 *N* sodium hydroxide to a pH of 5.0 ± 0.1. Mix 85 volumes of this solution with 15 volumes of acetonitrile. Make adjustments if necessary (see *System Suitability* under *Chromatography* ⟨621⟩).

Standard preparation—Dissolve a suitable quantity of USP Amdinocillin RS, accurately weighed, in water to obtain a solution having a known concentration of about 100 µg of amdinocillin ($C_{15}H_{23}N_3O_3S$) per mL.

Assay preparation 1—Using a suitable quantity of Sterile Amdinocillin, accurately weighed, proceed as directed under *Standard preparation.*

Assay preparation 2 (where it is represented as being in a single-dose container)—Constitute Sterile Amdinocillin in a volume of water, accurately measured, corresponding to the volume of solvent specified in the labeling. Withdraw all of the withdrawable contents, using a suitable hypodermic needle and syringe, and dilute quantitatively with water to obtain a solution containing about 100 µg of amdinocillin per mL.

Assay preparation 3 (where the label states the quantity of amdinocillin in a given volume of constituted solution)—Constitute Sterile Amdinocillin in a volume of water, accurately measured, corresponding to the volume of solvent specified in the labeling. Dilute an accurately measured volume of the constituted solution quantitatively with water to obtain a solution containing about 100 µg of amdinocillin per mL.

Chromatographic system (see *Chromatography* ⟨621⟩)—The liquid chromatograph is equipped with a 220-nm detector and a 4.6-mm × 25-cm column that contains 5-µm packing L1. The flow rate is about 1 mL per minute. Chromatograph the *Standard preparation*, and record the peak responses as directed under *Procedure:* the column efficiency determined from the analyte peak is not less than 1500 theoretical plates, the tailing factor for the analyte peak is not more than 2.5, the resolution, *R*, between the analyte peak and the impurity peak closest to it is not less than 2.5, and the relative standard deviation for replicate injections is not more than 2.0%.

Procedure—Separately inject equal volumes (about 20 µL) of the *Standard preparation* and the *Assay preparation* into the chromatograph, record the chromatograms, and measure the responses for the major peaks. Calculate the quantity, in µg, of amdinocillin per mg of Sterile Amdinocillin taken by the formula:

$$1000(C/M)(r_U/r_S),$$

in which *C* is the concentration, in mg of amdinocillin ($C_{15}H_{23}N_3O_3S$) per mL, of the *Standard preparation, M* is the concentration, in mg per mL, of the *Assay preparation* based on the weight of Sterile Amdinocillin taken and the extent of dilution, and r_U and r_S are the peak responses obtained from the *Assay preparation* and the *Standard preparation*, respectively. Calculate the quantity, in mg of amdinocillin ($C_{15}H_{23}N_3O_3S$), withdrawn from the container, or in the portion of constituted solution taken by the formula:

$$(L/D)(C)(r_U/r_S),$$

in which *L* is the labeled quantity, in mg of amdinocillin ($C_{15}H_{23}N_3O_3S$), in the container, or in the volume of constituted solution taken, and *D* is the concentration, in mg of amdinocillin ($C_{15}H_{23}N_3O_3S$) per mL, of *Assay preparation 2* or *Assay preparation 3*, based on the labeled quantity in the container or in the portion of constituted solution taken, respectively, and the extent of dilution.

Amikacin

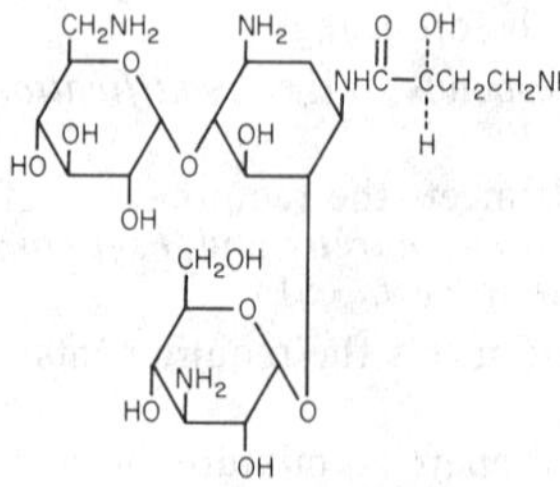

$C_{22}H_{43}N_5O_{13}$ 585.61

D-Streptamine, *O*-3-amino-3-deoxy-α-D-glucopyranosyl-(1→6)-*O*-[6-amino-6-deoxy-α-D-glucopyranosyl(1→4)]-*N*[1]-(4-amino-2-hydroxy-1-oxobutyl)-2-deoxy-, (*S*)-.

O-3-Amino-3-deoxy-α-D-glucopyranosyl(1→4)-*O*-[6-amino-6-deoxy-α-D-glucopyranosyl(1→6)]-*N*[3]-(4-amino-L-2-hydroxybutyryl)-2-deoxy-L-streptamine [*37517-28-5*].

» Amikacin has a potency of not less than 900 µg of $C_{22}H_{43}N_5O_{13}$ per mg, calculated on the anhydrous basis.

Packaging and storage—Preserve in tight containers.

Reference standard—*USP Amikacin Reference Standard*—Do not dry before using.

Identification—Prepare a solution of it in water containing 6 mg per mL. On a suitable thin-layer chromatographic plate (see *Chromatography* ⟨621⟩), coated with a 0.25-mm layer of chromatographic silica gel mixture, apply 3 µL of this test solution, 3 µL of a Standard solution of USP Amikacin RS containing 6 mg per mL, and 3 µL of a mixture of equal volumes of the two solutions. Place the plate in a suitable chromatographic chamber, and develop the chromatogram by continuous flow with a mixture of methanol, ammonium hydroxide, and chloroform (60:30:25) for 5.5 hours. Remove the plate from the chamber, allow the solvent to evaporate, and heat the plate at 110° for 15 minutes. Immediately locate the spots on the plate by spraying with a 1 in 100 solution of ninhydrin in a mixture of butanol and pyridine (100:1): amikacin appears as a pink spot, and the spots obtained from the test solution and from the mixture of test solution and Standard solution, respectively, correspond in distance from the origin to that obtained from the Standard solution.

Specific rotation ⟨781⟩: between +97° and +105°, calculated on the anhydrous basis, determined in a solution containing 20 mg per mL.

Crystallinity ⟨695⟩: meets the requirements.

pH ⟨791⟩: between 9.5 and 11.5, in a solution containing 10 mg per mL.

Water, *Method I* ⟨921⟩: not more than 8.5%.

Residue on ignition ⟨281⟩: not more than 1.0%, the charred residue being moistened with 2 mL of nitric acid and 5 drops of sulfuric acid.

Assay—Proceed with Amikacin as directed under *Antibiotics—Microbial Assays* ⟨81⟩.

Amikacin Sulfate Injection

$C_{22}H_{43}N_5O_{13} \cdot 2H_2SO_4$ 781.75

D-Streptamine, *O*-3-amino-3-deoxy-α-D-glucopyranosyl(1→6)-*O*-[6-amino-6-deoxy-α-D-glucopyranosyl(1→4)]-*N*[1]-(4-amino-2-hydroxy-1-oxobutyl)-2-deoxy-, (*S*)-, sulfate (1:2) (salt).

O-3-Amino-3-deoxy-α-D-glucopyranosyl(1→4)-*O*-[6-amino-6-deoxy-α-D-glucopyranosyl(1→6)]-*N*[3]-(4-amino-L-2-hydroxybutyryl)-2-deoxy-L-streptamine sulfate (1:2) [*39831-55-5*].

» Amikacin Sulfate Injection is a sterile solution of Amikacin in Water for Injection prepared with the aid of Sulfuric Acid and containing suitable buffers and preservatives. It contains not less than 90.0 percent and not more than 120.0 percent of the labeled amount of amikacin ($C_{22}H_{43}N_5O_{13}$).

Packaging and storage—Preserve in single-dose or in multiple-dose containers, preferably of Type I or Type III glass.

Reference standard—*USP Amikacin Reference Standard*—Do not dry before using.

Identification—Dilute it with water to obtain a solution containing 6 mg per mL: the resulting solution responds to the *Identification test* under *Amikacin.*

Pyrogen—When diluted, if necessary, with Sterile Water for Injection to a concentration of 25 mg of amikacin per mL, it meets the requirements of the *Pyrogen Test* ⟨151⟩, the test dose being 1.0 mL per kg.

pH ⟨791⟩: between 3.5 and 5.5.

Particulate matter ⟨788⟩: meets the requirements under *Small-volume Injections.*

Other requirements—It meets the requirements under *Injections* ⟨1⟩.

Assay—Proceed with Amikacin Sulfate Injection as directed under *Antibiotics—Microbial Assays* ⟨81⟩, using an accurately measured volume of Injection diluted quantitatively with water to obtain a *Test Dilution* having a concentration assumed to be equal to the median dose level of the Standard.

Amiloride Hydrochloride

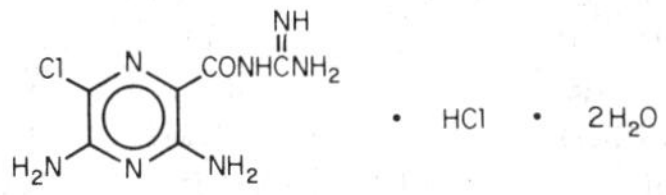

$C_6H_8ClN_7O.HCl.2H_2O$ 302.12

Pyrazinecarboxamide, 3,5-diamino-*N*-(aminoiminomethyl)-6-chloro-, monohydrochloride dihydrate.

N-Amidino-3,5-diamino-6-chloropyrazinecarboxamide monohydrochloride dihydrate [*17440-83-4*].

» Amiloride Hydrochloride contains not less than 98.0 percent and not more than 101.0 percent of $C_6H_8ClN_7O.HCl$, calculated on the anhydrous basis.

Packaging and storage—Preserve in well-closed containers.

Reference standard—*USP Amiloride Hydrochloride Reference Standard*—Do not dry; determine the *Water* content by *Method I* before using.

Identification—

A: The infrared absorption spectrum of a mineral oil dispersion of it exhibits maxima only at the same wavelengths as that of a similar, undried, preparation of USP Amiloride Hydrochloride RS.

B: Prepare an aqueous solution having a known concentration of about 600 µg of amiloride hydrochloride per mL, and dilute quantitatively and stepwise with 0.1 *N* hydrochloric acid to obtain a solution having a concentration of about 9.6 µg per mL: the ultraviolet absorption spectrum of the solution so prepared exhibits maxima and minima at the same wavelengths as that of a similar solution of USP Amiloride Hydrochloride RS, concomitantly measured.

C: It responds to the tests for *Chloride* ⟨191⟩.

Acidity—Dissolve 1.0 g in 100 mL of a mixture of methanol and water (1:1), and titrate with 0.10 *N* sodium hydroxide, determining the end-point potentiometrically: not more than 0.30 mL is required (0.1% as HCl).

Water, *Method I* ⟨921⟩: between 11.0% and 13.0%.

Residue on ignition ⟨281⟩: not more than 0.1%.

Heavy metals, *Method II* ⟨231⟩: 0.002%.

Chromatographic purity—

Standard preparations—Prepare a series of solutions, *A, B, C, D, E,* and *F,* of USP Amiloride Hydrochloride RS in a mixture of methanol and chloroform (4:1) having concentrations of 4000, 40, 20, 8, 4, and 2 µg per mL, respectively.

Test preparation—Prepare a solution of Amiloride Hydrochloride in a mixture of methanol and chloroform (4:1) having a concentration of 4 mg per mL.

Procedure—On a suitable thin-layer chromatographic plate (see *Chromatography* ⟨621⟩), coated with a 0.25-mm layer of chromatographic silica gel and previously washed with methanol, separately apply 5-µL portions of *Standard preparations A, B, C, D, E,* and *F* and the *Test preparation.* Dry the spots with a stream of nitrogen, and develop the chromatograms in a solvent system consisting of tetrahydrofuran and 3 *N* ammonium hydroxide (15:2) until the solvent front has moved about three-fourths of the length of the plate. Remove the plate from the developing chamber, mark the solvent front, allow to air-dry, and examine the plate under long-wavelength ultraviolet light: the R_f value of the principal spot obtained from the *Test preparation* corresponds to that obtained from *Standard preparation A.* Estimate the levels of any additional spots observed in the chromatogram of the *Test preparation* by comparison with the principal spots in the chromatograms of *Standard preparations B, C, D, E,* and *F:* the sum of the intensities of any additional spots observed is not greater than that of the principal spot obtained from *Standard preparation B* (equivalent to 1%).

Assay—Dissolve about 450 mg of Amiloride Hydrochloride, accurately weighed, in 100 mL of glacial acetic acid, add 10 mL of mercuric acetate TS and 15 mL of dioxane, and mix. Add crystal violet TS, and titrate with 0.1 *N* perchloric acid VS to a blue end-point. Perform a blank determination, and make any necessary correction. Each mL of 0.1 *N* perchloric acid is equivalent to 26.61 mg of $C_6H_8ClN_7O.HCl$.

Amiloride Hydrochloride Tablets

» Amiloride Hydrochloride Tablets contain not less than 90.0 percent and not more than 110.0 percent of the labeled amount of $C_6H_8ClN_7O.HCl$.

Packaging and storage—Preserve in well-closed containers.

Reference standard—*USP Amiloride Hydrochloride Reference Standard*—Do not dry; determine the *Water* content by *Method I* before using.

Identification—

A: The retention time of the major peak in the chromatogram obtained from the *Assay preparation* corresponds to that obtained from the *Standard preparation.*

B: Transfer an amount of finely ground Tablets, equivalent to 5 mg of amiloride hydrochloride, to a 25-mL volumetric flask, add methanol to volume, mix, and filter. Separately apply 10 µL each of the filtrate and a Standard solution of USP Amiloride Hydrochloride RS in methanol, containing 0.2 mg per mL, to a thin-layer chromatographic plate (see *Chromatography* ⟨621⟩) coated with a 0.25-mm layer of chromatographic silica gel mixture. Develop the chromatogram in a solvent system consisting of a mixture of tetrahydrofuran and 3 *N* ammonium hydroxide (22:3) until the solvent front has moved about three-fourths of the length of the plate. Remove the plate from the developing chamber, air-dry, and examine under short-wavelength ultraviolet light: the R_f value of the principal spot in the chromatogram of the test solution corresponds to that obtained from the Standard solution.

Dissolution ⟨711⟩—

Medium: 0.1 *N* hydrochloric acid; 900 mL.

Apparatus 2: 50 rpm.

Time: 30 minutes.

Procedure—Determine the amount of $C_6H_8ClN_7O.HCl$ dissolved from ultraviolet absorbances at the wavelength of maximum absorbance at about 363 nm of filtered portions of the solution under test, suitably diluted with 0.1 *N* hydrochloric acid, in comparison with a Standard solution having a known concentration of USP Amiloride Hydrochloride RS in the same medium. An amount of methanol not to exceed 2% of the total volume of the Standard solution may be used to dissolve the amiloride hydrochloride.

Tolerances—Not less than 80% (*Q*) of the labeled amount of $C_6H_8ClN_7O.HCl$ is dissolved in 30 minutes.

Uniformity of dosage units ⟨905⟩: meet the requirements.

Procedure for content uniformity—Transfer 1 finely powdered Tablet to a 100-mL volumetric flask, add 60 mL of 0.1 *N* hydrochloric acid, and shake by mechanical means for 30 minutes. Dilute with 0.1 *N* hydrochloric acid to volume, mix, and centrifuge a portion of the mixture. Dilute an accurately measured portion of the clear supernatant liquid quantitatively to obtain a solution containing about 10 µg of amiloride hydrochloride per mL. Concomitantly determine the absorbances of this solution and a Standard solution of USP Amiloride Hydrochloride RS in the same medium having a known concentration of about 10 µg per mL, at the wavelength of maximum absorbance at about 363 nm, with a suitable spectrophotometer, using 0.1 *N* hydrochloric acid as the blank. Calculate the quantity, in mg, of $C_6H_8ClN_7O.HCl$ in the Tablet by the formula:

$$(TC/D)(A_U/A_S),$$

in which *T* is the labeled quantity, in mg, of amiloride hydrochloride in the Tablet, *C* is the concentration, in µg per mL, of USP Amiloride Hydrochloride RS, on the anhydrous basis, in the Standard solution, *D* is the concentration, in µg per mL, of amiloride hydrochloride in the solution from the Tablet, based upon the labeled quantity per Tablet and the extent of dilution, and A_U and A_S are the absorbances of the solution from the Tablets and the Standard solution, respectively.

Assay—

Buffer solution—Dissolve 136 g of monobasic potassium phosphate in 800 mL of water, and adjust by the addition of phosphoric acid, with mixing, to a pH of 3.0. Dilute with water to 1000 mL, and mix.

Mobile phase—Prepare a suitable degassed mixture of water, methanol, and *Buffer solution* (71:25:4).

Standard preparation—Dissolve a suitable quantity of USP Amiloride Hydrochloride RS in methanol to obtain a solution having a known concentration of about 1.0 mg of amiloride hydrochloride per mL. Transfer 5.0 mL of the solution to a 50-mL volumetric flask, and add 10.0 mL of methanol and 2.0 mL of 0.1 *N* hydrochloric acid. Dilute with water to volume, and mix. The concentration of USP Amiloride Hydrochloride RS in the *Standard preparation* is about 0.1 mg per mL.

Assay preparation—Weigh and finely powder not less than 20 Amiloride Hydrochloride Tablets. Transfer an accurately weighed portion of the powder, equivalent to about 5 mg of amiloride hydrochloride, to a 50-mL volumetric flask containing 15.0 mL of methanol and 2.0 mL of 0.1 *N* hydrochloric acid. Sonicate for 10 minutes, dilute with water to volume, sonicate for an additional 10 minutes, mix, and filter.

Chromatographic system (see *Chromatography* ⟨621⟩)—The liquid chromatograph is equipped with a 286-nm detector and a 3.9-mm × 30-cm column that contains packing L1. The flow rate is about 1 mL per minute. Chromatograph five replicate injections of the *Standard preparation*, and record the peak responses as directed under *Procedure:* the relative standard deviation is not more than 2.0%, and the tailing factor of the major peak is not more than 2.0.

Procedure—Separately inject equal volumes (about 10 µL) of the *Standard preparation* and the *Assay preparation* into the chromatograph, record the chromatograms, and measure the responses for the major peaks. Calculate the quantity, in mg, of $C_6H_8ClN_7O.HCl$ in the portion of Tablets taken by the formula:

$$50C(r_U/r_S),$$

in which *C* is the concentration, in mg per mL, of USP Amiloride Hydrochloride RS, on the anhydrous basis, in the *Standard preparation*, and r_U and r_S are the peak responses obtained from the *Assay preparation* and the *Standard preparation*, respectively.

Amiloride Hydrochloride and Hydrochlorothiazide Tablets

» Amiloride Hydrochloride and Hydrochlorothiazide Tablets contain not less than 90.0 percent and not more than 110.0 percent of the labeled amounts of amiloride hydrochloride ($C_6H_8ClN_7O.HCl$) and hydrochlorothiazide ($C_7H_8ClN_3O_4S_2$).

Packaging and storage—Preserve in well-closed containers.

Reference standards—*USP Amiloride Hydrochloride Reference Standard*—Do not dry; determine the *Water* content by *Method I* before using. *USP Hydrochlorothiazide Reference Standard*—Dry at 105° for 1 hour before using.

Identification—

A: The retention times of the major peaks in the chromatogram obtained from the *Assay preparation* correspond to those obtained from the *Standard preparation.*

B: Transfer an amount of finely ground Tablets, equivalent to 5 mg of amiloride hydrochloride, to a 25-mL volumetric flask, add methanol to volume, mix, and filter. Separately apply 10 µL each of the filtrate, a Standard solution of USP Amiloride Hydrochloride RS in methanol containing 0.2 mg per mL, and a Standard solution of USP Hydrochlorothiazide RS in methanol containing 2 mg per mL to a thin-layer chromatographic plate (see *Chromatography* ⟨621⟩) coated with a 0.25-mm layer of chromatographic silica gel mixture. Develop the chromatogram in a solvent system consisting of a mixture of tetrahydrofuran and 3 *N* ammonium hydroxide (22:3) until the solvent front has moved about three-fourths of the length of the plate. Remove the plate from the developing chamber, air-dry, and examine under short-wavelength ultraviolet light: the R_f values of the amiloride hydrochloride and hydrochlorothiazide spots obtained from the test solution correspond to those obtained from the corresponding Standard solutions.

Dissolution ⟨711⟩—

Medium: 0.1 *N* hydrochloric acid; 900 mL.

Apparatus 2: 50 rpm.

Time: 30 minutes.

Procedure—Determine the amount of amiloride hydrochloride ($C_6H_8ClN_7O.HCl$) and hydrochlorothiazide ($C_7H_8ClN_3O_4S_2$) dissolved from ultraviolet absorbances at the wavelengths of maximum absorbance at about 363 nm for amiloride hydrochloride and 270 nm for hydrochlorothiazide (corrected for interference from amiloride hydrochloride on the basis of the absorbances of amiloride hydrochloride at 270 nm and 363 nm) of filtered portions of the solution under test, suitably diluted with 0.1 *N* hydrochloric acid, in comparison with Standard solutions having known concentrations of USP Amiloride Hydrochloride RS and USP Hydrochlorothiazide RS in the same medium. An amount of methanol not to exceed 2% of the total volume of the Standard solution may be used to dissolve the amiloride hydrochloride.

Tolerances—Not less than 80% (*Q*) of the labeled amount of $C_6H_8ClN_7O.HCl$ and 75% (*Q*) of the labeled amount of $C_7H_8ClN_3O_4S_2$ are dissolved in 30 minutes.

Uniformity of dosage units ⟨905⟩: meet the requirements for *Content uniformity* with respect to amiloride hydrochloride and to hydrochlorothiazide.

Assay—

Buffer solution and *Mobile phase*—Prepare as directed in the *Assay* under *Amiloride Hydrochloride Tablets.*

Standard preparation—Dissolve a suitable quantity of USP Amiloride Hydrochloride RS in methanol to obtain a solution having a known concentration of about 1.0 mg of amiloride hydrochloride per mL. Transfer 10.0 mL of this solution to a 100-mL volumetric flask containing 100 mg of USP Hydrochlorothiazide RS, accurately weighed, and 20.0 mL of methanol. Add 4.0 mL of 1 *N* hydrochloric acid, dilute with water to volume, and mix. The concentrations of USP Amiloride Hydrochloride RS and USP Hydrochlorothiazide RS in the *Standard preparation* are about 0.1 mg per mL and 1 mg per mL, respectively.

Assay preparation—Weigh and finely powder not less than 20 Amiloride Hydrochloride and Hydrochlorothiazide Tablets. Transfer an accurately weighed portion of the powder, equivalent to about 5 mg of amiloride hydrochloride, to a 50-mL volumetric flask. Add 15.0 mL of methanol, and 2.0 mL of 1 *N* hydrochloric acid. Sonicate for 10 minutes, dilute with water to volume, sonicate for an additional 10 minutes, mix, and filter.

Chromatographic system (see *Chromatography* ⟨621⟩)—The liquid chromatograph is equipped with a 286-nm detector and a 3.9-mm × 30-cm column that contains packing L1. The flow rate is about 1 mL per minute. Chromatograph five replicate injections of the *Standard preparation*, and record the peak responses as directed under *Procedure:* the relative standard deviation is not more than 2.0%, and the resolution factor between hydrochlorothiazide and amiloride hydrochloride is not less than 2.0.

Procedure—Separately inject equal volumes (about 10 µL) of the *Standard preparation* and the *Assay preparation* into the chromatograph, record the chromatograms, and measure the responses for the major peaks. The relative retention times are about 0.7 for hydrochlorothiazide and 1.0 for amiloride hydrochloride. Calculate the quantity, in mg, of amiloride hydrochloride ($C_6H_8ClN_7O.HCl$) in the portion of Tablets taken by the formula:

$$50C(r_U/r_S),$$

in which *C* is the concentration, in mg per mL, of USP Amiloride Hydrochloride RS, on the anhydrous basis, in the *Standard preparation*, and r_U and r_S are the peak responses for amiloride hydrochloride obtained from the *Assay preparation* and the *Standard preparation*, respectively. Similarly calculate the quantity, in mg, of hydrochlorothiazide ($C_7H_8ClN_3O_4S_2$) in the portion of Tablets taken by the formula:

$$50C(r_U/r_S),$$

in which *C* is the concentration, in mg per mL, of USP Hydrochlorothiazide RS in the *Standard preparation*, and r_U and r_S are the peak responses of hydrochlorothiazide obtained from the *Assay preparation* and the *Standard preparation*, respectively.

Aminoacetic Acid—*see* Glycine

Aminoacetic Acid Irrigation—*see* Glycine Irrigation

Aminobenzoate Potassium

» Aminobenzoate Potassium contains not less than 98.5 percent and not more than 101.0 percent of $C_7H_6KNO_2$, calculated on the dried basis.

Packaging and storage—Preserve in well-closed containers.

Reference standard—*USP Aminobenzoate Potassium Reference Standard*—Dry at 105° for 2 hours before using.

Identification—

A: The ultraviolet absorption spectrum of a 1 in 100,000 solution in 0.001 *N* sodium hydroxide exhibits maxima and minima only at the same wavelengths as that of a similar solution of USP Aminobenzoate Potassium RS, concomitantly measured.

B: Dissolve about 400 mg in 10 mL of water, add 1 mL of 3 *N* hydrochloric acid, filter, and wash the precipitate with two 5-mL portions of cold water. Recrystallize from alcohol the precipitate so obtained, and dry at 110° for 1 hour: the *p*-aminobenzoic acid so obtained melts between 186° and 189°.

C: A solution (1 in 100) responds to the flame test for *Potassium* ⟨191⟩.

pH ⟨791⟩: between 8.0 and 9.0, in a solution (1 in 20).

Loss on drying ⟨731⟩—Dry it at 105° for 2 hours: it loses not more than 1.0% of its weight.

Chloride ⟨221⟩—A 1.4-g portion shows no more chloride than corresponds to 0.4 mL of 0.020 *N* hydrochloric acid (0.02%).

Sulfate ⟨221⟩—A 1.4-g portion shows no more sulfate than corresponds to 0.3 mL of 0.020 *N* sulfuric acid (0.02%).

Heavy metals, *Method II* ⟨231⟩: 0.002%.

Volatile diazotizable substances—

Standard preparation—Dissolve 10 mg of *p*-toluidine in 5 mL of methanol in a 100-mL volumetric flask, add water to volume, and mix. Transfer 1 mL to a 100-mL volumetric flask, dilute with water to volume, and mix.

Test preparation—Transfer 5.0 g of Aminobenzoate Potassium to a suitable flask, and add a volume of 1.25 *N* sodium hydroxide that is just sufficient to dissolve the test specimen and to render the solution just alkaline to phenolphthalein TS. Dilute with water to 50 mL, and steam-distil the solution, collecting about 95 mL of the distillate in a 100-mL volumetric flask. Add water to volume, and mix.

Procedure—Transfer 20.0-mL portions of the *Standard preparation* and the *Test preparation* to separate 100-mL beakers, and transfer 20.0 mL of water to a third 100-mL beaker to provide the blank. Treat each as follows: Add 5.0 mL of 1 *N* hydrochloric acid, and cool in an ice bath. Add 2.0 mL of 0.1 *M* sodium nitrite dropwise, with stirring, allow to stand for 5 minutes for the diazotization reaction to be complete, add quickly to 10.0 mL of a cold solution of guaiacol (freshly prepared by dissolving 0.20 g of guaiacol in 100 mL of 1 *N* sodium hydroxide), mix, and allow to stand for 30 minutes. Concomitantly determine the absorbances of the solutions at the wavelength of maximum absorbance at about 405 nm, with a suitable spectrophotometer, using the blank to set the instrument: the absorbance of the solution obtained from the *Test preparation* does not exceed that of the solution obtained from the *Standard preparation*, corresponding to not more than 0.002% of volatile diazotizable substances, as *p*-toluidine.

Assay—Transfer about 500 mg of Aminobenzoate Potassium, accurately weighed, to a suitable vessel, add 25 mL of water and 25 mL of 3 *N* hydrochloric acid, mix, and cool in an ice bath. Titrate with 0.1 *M* sodium nitrite VS, determining the end-point potentiometrically, using a calomel-platinum electrode system. Each mL of 0.1 *M* sodium nitrite is equivalent to 17.52 mg of $C_7H_6KNO_2$.

Aminobenzoate Potassium Capsules

» Aminobenzoate Potassium Capsules contain not less than 90.0 percent and not more than 110.0 percent of the labeled amount of $C_7H_6KNO_2$.

Packaging and storage—Preserve in well-closed containers.

Reference standard—*USP Aminobenzoate Potassium Reference Standard*—Dry at 105° for 2 hours before using.

Identification—Dissolve about 1 g of the Capsule contents in 25 mL of water, add 5 mL of 3 *N* hydrochloric acid, and wash the precipitate with two 5-mL portions of cold water. Recrystallize from alcohol the precipitate so obtained, and dry at 110° for 1 hour: the *p*-aminobenzoic acid so obtained melts between 186° and 189°.

Dissolution ⟨711⟩—

Medium: water; 900 mL.

Apparatus 1: 100 rpm.

Time: 45 minutes.

Procedure—Determine the amount of $C_7H_6KNO_2$ dissolved from ultraviolet absorbances at the wavelength of maximum absorbance at about 270 nm of filtered portions of the solution under test, suitably diluted with water, in comparison with a Standard solution having a known concentration of USP Aminobenzoate Potassium RS in the same medium.

Tolerances—Not less than 75% (*Q*) of the labeled amount of $C_7H_6KNO_2$ is dissolved in 45 minutes.

Uniformity of dosage units ⟨905⟩: meet the requirements.

Assay—

Standard preparation—Prepare a solution of USP Aminobenzoate Potassium RS having a known concentration of about 5 µg per mL.

Assay preparation—Remove as completely as possible, and combine, the contents of not less than 20 Aminobenzoate Potassium Capsules. Transfer an accurately weighed portion of the combined contents, equivalent to about 100 mg of aminobenzoate potassium, to a 200-mL volumetric flask, add 150 mL of water, shake by mechanical means for 30 minutes, dilute with water to volume, mix, and filter. Pipet 2 mL of the filtrate into a 200-mL volumetric flask, dilute with water to volume, and mix.

Procedure—Concomitantly determine the absorbances of the *Standard preparation* and the *Assay preparation* at the wavelength of maximum absorbance at about 270 nm, with a suitable spectrophotometer, using water as the blank. Calculate the quantity, in mg, of $C_7H_6KNO_2$ in the portion of Capsule contents taken by the formula:

$$20C(A_U/A_S),$$

in which C is the concentration, in µg per mL, of USP Aminobenzoate Potassium RS in the *Standard preparation*, and A_U and A_S are the absorbances of the *Assay preparation* and the *Standard preparation*, respectively.

Aminobenzoate Potassium for Oral Solution

» Aminobenzoate Potassium for Oral Solution contains not less than 90.0 percent and not more than 110.0 percent of the labeled amount of $C_7H_6KNO_2$.

Packaging and storage—Preserve in tight containers.

Reference standard—*USP Aminobenzoate Potassium Reference Standard*—Dry at 105° for 2 hours before using.

Identification—

A: The ultraviolet absorption spectrum of a 1 in 20,000 solution exhibits maxima and minima only at the same wavelengths as that of a similar solution of USP Aminobenzoate Potassium RS, concomitantly measured.

B: Dissolve about 400 mg in 10 mL of water, add 1 mL of 3 *N* hydrochloric acid, filter, and wash the precipitate with two 5-mL portions of cold water. Recrystallize from alcohol the precipitate so obtained, and dry at 110° for 1 hour: the *p*-aminobenzoic acid so obtained melts between 186° and 189°.

pH ⟨791⟩: between 7.0 and 9.0, in a solution (1 in 10).

Minimum fill ⟨755⟩—

FOR SOLID PACKAGED IN MULTIPLE-UNIT CONTAINERS: meets the requirements.

Uniformity of dosage units ⟨905⟩—FOR SOLID PACKAGED IN SINGLE-UNIT CONTAINERS: meets the requirements.

Assay—Transfer about 100 mg of Aminobenzoate Potassium for Oral Solution, accurately weighed, to a suitable vessel, add 5 mL of hydrochloric acid and 50 mL of water, mix, cool to 15°, and add 25 g of crushed ice. Titrate with 0.1 *M* sodium nitrite VS, determining the end-point potentiometrically, using a calomel-platinum electrode system. Each mL of 0.1 *M* sodium nitrite is equivalent to 17.52 mg of $C_7H_6KNO_2$.

Aminobenzoate Potassium Tablets

» Aminobenzoate Potassium Tablets contain not less than 90.0 percent and not more than 110.0 percent of the labeled amount of $C_7H_6KNO_2$.

Packaging and storage—Preserve in well-closed containers.

Reference standard—*USP Aminobenzoate Potassium Reference Standard*—Dry at 105° for 2 hours before using.

Identification—Proceed as directed for *Aminobenzoate Potassium Capsules*, using 1 g of finely powdered Tablets.

Dissolution ⟨711⟩—

Medium: water; 900 mL.
Apparatus 1: 100 rpm.
Time: 45 minutes.

Procedure—Determine the amount of $C_7H_6KNO_2$ dissolved from ultraviolet absorbances at the wavelength of maximum absorbance at about 270 nm of filtered portions of the solution under test, suitably diluted with water, in comparison with a Standard solution having a known concentration of USP Aminobenzoate Potassium RS in the same medium.

Tolerances—Not less than 75% (*Q*) of the labeled amount of $C_7H_6KNO_2$ is dissolved in 45 minutes.

Uniformity of dosage units ⟨905⟩: meet the requirements.

Assay—

Standard preparation—Prepare a solution of USP Aminobenzoate Potassium RS having a known concentration of about 5 µg per mL.

Assay preparation and *Procedure*—Weigh and finely powder not less than 20 Aminobenzoate Potassium Tablets. Using a portion of the powdered Tablets, equivalent to about 100 mg of aminobenzoate potassium, proceed as directed in the *Assay* under *Aminobenzoate Potassium Capsules*.

Aminobenzoate Sodium

» Aminobenzoate Sodium contains not less than 98.5 percent and not more than 101.0 percent of $C_7H_6NNaO_2$, calculated on the dried basis.

Packaging and storage—Preserve in well-closed containers.

Reference standard—*USP Aminobenzoate Sodium Reference Standard*—Dry at 105° for 2 hours before using.

Identification—

A: The ultraviolet absorption spectrum of a 1 in 100,000 solution in 0.001 *N* sodium hydroxide exhibits maxima and minima only at the same wavelengths as that of a similar solution of USP Aminobenzoate Sodium RS, concomitantly measured.

B: Dissolve about 400 mg in 10 mL of water, add 1 mL of 3 *N* hydrochloric acid, filter, and wash the precipitate with two 5-mL portions of cold water. Recrystallize from alcohol the precipitate so obtained, and dry at 110° for 1 hour: the *p*-aminobenzoic acid so obtained melts between 186° and 189°.

C: A solution (1 in 100) responds to the flame test for *Sodium* ⟨191⟩.

pH ⟨791⟩: between 8.0 and 9.0, in a solution (1 in 20).

Loss on drying ⟨731⟩—Dry it at 105° for 2 hours: it loses not more than 1.0% of its weight.

Chloride ⟨221⟩—A 1.4-g portion shows no more chloride than corresponds to 0.4 mL of 0.020 *N* hydrochloric acid (0.02%).

Sulfate ⟨221⟩—A 1.4-g portion shows no more sulfate than corresponds to 0.3 mL of 0.020 *N* sulfuric acid (0.02%).

Heavy metals, *Method II* ⟨231⟩: 0.002%.

Volatile diazotizable substances—

Standard preparation—Dissolve 10 mg of *p*-toluidine in 5 mL of methanol in a 100-mL volumetric flask, add water to volume, and mix. Transfer 1 mL to a 100-mL volumetric flask, dilute with water to volume, and mix.

Test preparation—Transfer 5.0 g of Aminobenzoate Sodium to a suitable flask, and add a volume of 1.25 *N* sodium hydroxide that is just sufficient to dissolve the test specimen and to render the solution just alkaline to phenolphthalein TS. Dilute with water to 50 mL, and steam-distil the solution, collecting about 95 mL of the distillate in a 100-mL volumetric flask. Add water to volume, and mix.

Procedure—Transfer 20.0-mL portions of the *Standard preparation* and the *Test preparation* to separate 100-mL beakers, and transfer 20.0 mL of water to a third 100-mL beaker to provide the blank. Treat each as follows: Add 5.0 mL of 1 *N* hydrochloric acid, and cool in an ice bath. Add 2.0 mL of 0.1 *M* sodium nitrite dropwise, with stirring, allow to stand for 5 minutes for the diazotization reaction to be complete, add quickly to 10.0 mL of a cold solution of guaiacol (freshly prepared by dissolving 0.20 g of guaiacol in 100 mL of 1 *N* sodium hydroxide), mix, and allow to stand for 30 minutes. Concomitantly determine the absorbances of the solutions at the wavelength of maximum absorbance at about 405 nm, with a suitable spectrophotometer, using the blank to set the instrument: the absorbance of the solution obtained from the *Test preparation* does not exceed that of the solution obtained from the *Standard preparation*, corresponding to not more than 0.002% of volatile diazotizable substances, as *p*-toluidine.

Assay—Transfer about 500 mg of Aminobenzoate Sodium, accurately weighed, to a suitable vessel, add 25 mL of water and 25 mL of 3 *N* hydrochloric acid, mix, and cool in an ice bath. Titrate with 0.1 *M* sodium nitrite VS, determining the end-point potentiometrically, using a calomel-platinum electrode system. Each mL of 0.1 *M* sodium nitrite is equivalent to 15.91 mg of $C_7H_6NNaO_2$.

Aminobenzoic Acid

$H_2N-C_6H_4-COOH$

$C_7H_7NO_2$ 137.14
Benzoic acid, 4-amino.
p-Aminobenzoic acid [*150-13-0*].

» Aminobenzoic Acid contains not less than 98.5 percent and not more than 101.5 percent of $C_7H_7NO_2$, calculated on the dried basis.

Packaging and storage—Preserve in tight, light-resistant containers.

Reference standard—*USP Aminobenzoic Acid Reference Standard*—Dry at 105° for 2 hours before using.

Identification—

A: The infrared absorption spectrum of a potassium bromide dispersion of it exhibits maxima only at the same wavelengths as that of a similar preparation of USP Aminobenzoic Acid RS.

B: The ultraviolet absorption spectrum of a 1 in 200,000 solution in 0.001 *N* sodium hydroxide exhibits maxima and minima at the same wavelengths as that of a similar solution of USP Aminobenzoic Acid RS, concomitantly measured.

Melting range ⟨741⟩: between 186° and 189°.

Loss on drying ⟨731⟩—Dry it at 105° for 2 hours: it loses not more than 0.2% of its weight.

Residue on ignition ⟨281⟩: not more than 0.1%.

Heavy metals, *Method II* ⟨231⟩: 0.002%.

Volatile diazotizable substances—

Standard preparation—Dissolve 10 mg of *p*-toluidine in 5 mL of methanol in a 100-mL volumetric flask, add water to volume, and mix. Transfer 1 mL to a 100-mL volumetric flask, dilute with water to volume, and mix.

Test preparation—Transfer 5.0 g of Aminobenzoic Acid to a suitable flask, and add a volume of 1.25 *N* sodium hydroxide that is just sufficient to dissolve the test specimen and to render the solution just alkaline to phenolphthalein TS. Dilute with water to 50 mL, and steam-distil the solution, collecting about 95 mL of the distillate in a 100-mL volumetric flask. Add water to volume, and mix.

Procedure—Transfer 20.0-mL portions of the *Standard preparation* and the *Test preparation* to separate 100-mL beakers, and transfer 20.0 mL of water to a third 100-mL beaker to provide the blank. Treat each as follows: Add 5.0 mL of 1 *N* hydrochloric acid, and cool in an ice bath. Add 2.0 mL of 0.1 *M* sodium nitrite dropwise, with stirring, allow to stand for 5 minutes in order for the diazotization reaction to be complete, add quickly to 10.0 mL of a cold solution of guaiacol (freshly prepared by dissolving 0.20 g of guaiacol in 100 mL of 1 *N* sodium hydroxide), mix, and allow to stand for 30 minutes. Concomitantly determine the absorbances of the solutions in 1-cm cells at the wavelength of maximum absorbance at about 405 nm, with a suitable spectrophotometer, using the blank to set the instrument: the absorbance of the solution obtained from the *Test preparation* does not exceed that of the solution obtained from the *Standard preparation*, corresponding to not more than 0.002% of volatile diazotizable substances, as *p*-toluidine.

Ordinary impurities ⟨466⟩—

Test solution: alcohol.

Standard solution: alcohol.

Eluant: a mixture of toluene, ethyl acetate, and alcohol (60:20:20), in a nonequilibrated chamber.

Visualization: 1.

Assay—Weigh accurately about 250 mg of Aminobenzoic Acid, and proceed as directed under *Nitrite Titration* ⟨451⟩. Each mL of 0.1 *M* sodium nitrite is equivalent to 13.71 mg of $C_7H_7NO_2$.

Aminobenzoic Acid Gel

» Aminobenzoic Acid Gel contains not less than 90.0 percent and not more than 110.0 percent of the labeled amount of $C_7H_7NO_2$.

Packaging and storage—Preserve in tight, light-resistant containers.

Reference standard—*USP Aminobenzoic Acid Reference Standard*—Dry at 105° for 2 hours before using.

Identification—

A: The retention time of the major peak in the chromatogram of the *Assay preparation* is the same as that of the *Standard preparation* obtained as directed in the *Assay*.

B: Dissolve a portion of Gel in alcohol, and dilute with alcohol to obtain a solution having a concentration of about 5 µg of aminobenzoic acid per mL: the ultraviolet absorption spectrum of the solution so obtained exhibits maxima and minima at the same wavelengths as that of a solution containing 5 µg of USP Aminobenzoic Acid RS per mL of alcohol, concomitantly measured.

Minimum fill ⟨755⟩: meets the requirements.

pH ⟨791⟩: between 4.0 and 6.0.

Alcohol content, *Method II* ⟨611⟩: between 42.3% and 54.0% (w/w) of C_2H_5OH.

Assay—

Mobile phase—Mix 300 mL of methanol and 10 mL of glacial acetic acid with 690 mL of water. Allow the mixture to cool, and filter, if necessary, through a suitable microporous membrane filter. Degas the solution.

Internal standard solution—Dissolve salicylic acid in methanol, by sonicating, to obtain a solution having a concentration of about 7 mg per mL.

Standard preparation—Dissolve, by sonicating, an accurately weighed quantity of USP Aminobenzoic Acid RS in methanol, dilute quantitatively with methanol to obtain a solution having a known concentration of about 0.42 mg per mL, and mix. Pipet 5 mL of this solution and 5 mL of the *Internal standard solution* into a 50-mL volumetric flask, dilute with methanol to volume, and mix. Filter through 0.6-µm filter paper. Throughout the preparation, protect against actinic light.

Assay preparation—Transfer an accurately weighed quantity of Aminobenzoic Acid Gel, equivalent to about 4.2 mg of aminobenzoic acid, to a 100-mL volumetric flask, and add 10.0 mL of *Internal standard preparation* and about 50 mL of methanol. Shake or sonicate, as necessary, dilute with methanol to volume, and mix. Filter, if necessary, through filter paper (Whatman No. 41 or equivalent). Filter through 0.6-µm filter paper. Throughout this preparation, protect against actinic light.

Chromatographic system (see *Chromatography* ⟨621⟩)—The liquid chromatograph is equipped with a 280-nm detector and a 3.9-mm × 30-cm column that contains packing L11. The flow rate is about 1.0 mL per minute. Chromatograph replicate 15-µL injections of *Standard preparation* until the response ratio variability is within 1.0% of average. The resolution factor is not less than 3.0 between aminobenzoic acid and salicylic acid.

Procedure—Separately inject equal volumes (about 15 µL) of the *Standard preparation* and the *Assay preparation* into the chromatograph, record the chromatograms, and measure the responses for the major peaks. The retention time of salicylic acid is about 3.0 relative to that of aminobenzoic acid as 1.0. Calculate the quantity, in mg, of $C_7H_7NO_2$ in the portion of Gel taken by the formula:

$$100C(R_U/R_S),$$

in which C is the concentration, in mg per mL, of USP Aminobenzoic Acid RS in the *Standard preparation*, and R_U and R_S are the ratios of the peak responses of the aminobenzoic acid peak to the salicylic acid peak obtained from the *Assay preparation* and the *Standard preparation*, respectively.

Aminobenzoic Acid Topical Solution

» Aminobenzoic Acid Topical Solution contains, in each mL, not less than 45 mg and not more than 55 mg of $C_7H_7NO_2$.

Packaging and storage—Preserve in tight, light-resistant containers.

Identification—

A: To 1 mL of Topical Solution add 1 mL of 1 *N* sodium hydroxide, and add, in the order named, 0.5 mL of potassium iodide TS, 0.5 mL of 3 *N* hydrochloric acid, and 0.5 mL of sodium hypochlorite TS: a brown precipitate is formed.

B: To 1 mL of Topical Solution add 2 mL of 3 *N* hydrochloric acid, and cool to about 10°. Add 1 mL of sodium nitrite solution (1 in 100), then add a solution prepared by mixing 50 mg of 2-naphthol with 3 mL of sodium hydroxide solution (1 in 10): a red color is produced.

Specific gravity ⟨841⟩: not less than 0.895 and not more than 0.905.

Alcohol content ⟨611⟩: between 65% and 75% of C_2H_5OH.

Assay—Transfer 5 mL of Aminobenzoic Acid Topical Solution, accurately measured, to a suitable open vessel, evaporate on a steam bath to dryness, and proceed as directed under *Nitrite Titration* ⟨451⟩, beginning with "Add 20 mL of hydrochloric acid." Each mL of 0.1 *M* sodium nitrite is equivalent to 13.71 mg of $C_7H_7NO_2$.

Aminocaproic Acid

$NH_2CH_2(CH_2)_3CH_2COOH$

$C_6H_{13}NO_2$ 131.17
Hexanoic acid, 6-amino-.
6-Aminohexanoic acid [*60-32-2*].

» Aminocaproic Acid contains not less than 98.5 percent and not more than 100.5 percent of $C_6H_{13}NO_2$, calculated on the anhydrous basis.

Packaging and storage—Preserve in tight containers.

Reference standard—*USP Aminocaproic Acid Reference Standard*—Dry at 105° for 30 minutes before using.

Identification—The infrared absorption spectrum of a potassium bromide dispersion of it, previously dried at 105° for 30 minutes, exhibits maxima only at the same wavelengths as that of a similar preparation of USP Aminocaproic Acid RS.

Water, *Method I* ⟨921⟩: not more than 0.5%.

Residue on ignition ⟨281⟩: not more than 0.1%.

Heavy metals, *Method II* ⟨231⟩: 0.002%.

Assay—

Mobile phase—Transfer 11 g of 1-pentanesulfonic acid sodium salt and 40 g of anhydrous sodium sulfate to a 2-liter volumetric flask, and dissolve in about 500 mL of water. Add 20 mL of 1 *N* sulfuric acid and 30 mL of acetonitrile, dilute with water to volume, and mix. Filter, and degas. Make adjustments if necessary (see *System Suitability* under *Chromatography* ⟨621⟩).

Standard preparation—Dissolve an accurately weighed quantity of USP Aminocaproic Acid RS in *Mobile phase* to obtain a solution having a known concentration of about 2.5 mg per mL.

Resolution solution—Mix 20 µL of benzyl alcohol with 100 mL of water. Dilute 1.0 mL of this solution with the *Standard preparation* to 10 mL.

Assay preparation—Transfer about 1.25 g of Aminocaproic Acid, accurately weighed, to a 50-mL volumetric flask, dilute with water to volume, and mix. Transfer 5.0 mL of the resulting solution to a 50-mL volumetric flask, dilute with *Mobile phase* to volume, and mix.

Chromatographic system (see *Chromatography* ⟨621⟩)—The liquid chromatograph is equipped with a 210-nm detector and a 30-cm × 4-mm column that contains packing L1. The flow rate is about 2 mL per minute. Chromatograph the *Resolution solution* as directed under *Procedure:* the resolution, *R*, between the benzyl alcohol and aminocaproic acid peaks is not less than 7.0. The aminocaproic acid peak elutes prior to the benzyl alcohol peak. Chromatograph the *Standard preparation*, and record the peak responses as directed under *Procedure:* the relative standard deviation for replicate injections is not more than 1.0%.

Procedure—Separately inject equal volumes (about 50 µL) of the *Standard preparation* and the *Assay preparation* into the chromatograph, record the chromatograms, and measure the areas for the major peaks. Caprolactam, if present, has a retention time about 2.8 times that of aminocaproic acid. Calculate the quantity, in mg, of $C_6H_{13}NO_2$ in the portion of Aminocaproic Acid taken by the formula:

$$500C(r_U/r_S),$$

in which C is the concentration, in mg per mL, of USP Aminocaproic Acid RS in the *Standard preparation*, and r_U and r_S are the aminocaproic acid peak areas obtained from the *Assay preparation* and the *Standard preparation*, respectively.

Aminocaproic Acid Injection

» Aminocaproic Acid Injection is a sterile solution of Aminocaproic Acid in Water for Injection. It contains not less than 95.0 percent and not more than 107.5 percent of the labeled amount of $C_6H_{13}NO_2$.

Packaging and storage—Preserve in single-dose or in multiple-dose containers, preferably of Type I glass.

Reference standard—*USP Aminocaproic Acid Reference Standard*—Dry at 105° for 30 minutes before using.

Identification—Mix 2 mL of Injection, added dropwise, with 100 mL of acetone, rapidly stirring the mixture with a glass rod to induce crystallization. Allow the mixture to stand for 15 minutes, and filter through a medium-porosity, sintered-glass filter. Wash the crystals with 25 mL of acetone, apply vacuum to remove the solvent, dry at 105° for 30 minutes, and cool: the residue so obtained responds to the *Identification test* under *Aminocaproic Acid*.

Pyrogen—Aminocaproic Acid Injection, diluted, if necessary, with water for injection to a concentration of 250 mg per mL, meets the requirements of the *Pyrogen Test* ⟨151⟩, the test dose being 1 mL per kg.

pH ⟨791⟩: between 6.0 and 7.6.

Other requirements—It meets the requirements under *Injections* ⟨1⟩.

Assay—

Mobile phase, Standard preparation, Resolution solution, and *Chromatographic system*—Proceed as directed in the *Assay* under *Aminocaproic Acid*.

Assay preparation—Transfer an accurately measured volume of Aminocaproic Acid Injection, equivalent to about 1.25 g of aminocaproic acid, to a 50-mL volumetric flask, dilute with water to volume, and mix. Transfer 5.0 mL of the resulting solution to a 50-mL volumetric flask, dilute with *Mobile phase* to volume, and mix.

Procedure—Separately inject equal volumes (about 50 µL) of the *Standard preparation* and the *Assay preparation* into the chromatograph, record the chromatograms, and measure the areas for the major peaks. Calculate the quantity, in mg, of $C_6H_{13}NO_2$ in each mL of the Injection taken by the formula:

$$500(C/V)(r_U/r_S),$$

in which C is the concentration, in mg per mL, of USP Aminocaproic Acid RS in the *Standard preparation*, V is the volume, in mL, of Injection taken, and r_U and r_S are the aminocaproic acid peak areas obtained from the *Assay preparation* and the *Standard preparation*, respectively.

Aminocaproic Acid Syrup

» Aminocaproic Acid Syrup contains not less than 95.0 percent and not more than 115.0 percent of the labeled amount of $C_6H_{13}NO_2$.

Packaging and storage—Preserve in tight containers.

Reference standard—*USP Aminocaproic Acid Reference Standard*—Dry at 105° for 30 minutes before using.

Identification—Mix about 1 g of ion-exchange resin (strongly acidic styrene-divinylbenzene cation-exchange resin) with about 10 mL of 1 *N* hydrochloric acid in a 100-mL beaker. Decant and discard the hydrochloric acid, and wash the resin with five 10-mL portions of water, decanting and discarding the liquid following each washing. Place the washed resin in a glass-stoppered, 125-mL conical flask, and add a volume of Syrup, equivalent to about 250 mg of aminocaproic acid, and 10 mL of water. Insert the stopper in the flask, and shake by mechanical means for 30 minutes. Transfer the resin slurry to a medium-porosity, sintered-glass funnel, wash with about 100 mL of water, apply suction to filter, and discard the washing. Place a 100-mL beaker under the stem of the funnel, add 10 mL of 1 *N* hydrochloric acid to the resin, stir for 4 to 5 minutes, and filter by applying suction. Evaporate the filtrate on a steam bath to dryness, dry at 105° for 1 hour, and cool: the residue so obtained responds to the *Identification test* under *Aminocaproic Acid*.

pH ⟨791⟩: between 6.1 and 6.6.

Assay—Transfer an accurately measured volume of Aminocaproic Acid Syrup, equivalent to about 250 mg of aminocaproic acid, to a 250-mL beaker, add 80 mL of glacial acetic acid, and mix. Add 10 drops of a 1 in 500 solution of crystal violet in chlorobenzene, and titrate with 0.1 *N* perchloric acid in dioxane VS to a blue end-point. Perform a blank determination, and make any necessary correction. Each mL of 0.1 *N* perchloric acid is equivalent to 13.12 mg of $C_6H_{13}NO_2$.

Aminocaproic Acid Tablets

» Aminocaproic Acid Tablets contain not less than 95.0 percent and not more than 105.0 percent of the labeled amount of $C_6H_{13}NO_2$.

Packaging and storage—Preserve in tight containers.

Reference standard—*USP Aminocaproic Acid Reference Standard*—Dry at 105° for 30 minutes before using.

Identification—Triturate 2 Tablets with 10 mL of water, and filter into 100 mL of acetone. Swirl the mixture, and allow to stand for 15 minutes to complete crystallization. Filter through a medium-porosity, sintered-glass filter, and wash the crystals with 25 mL of acetone. Apply vacuum to remove the solvent, then dry at 105° for 30 minutes, and cool: the residue so obtained responds to the *Identification test* under *Aminocaproic Acid*.

Dissolution ⟨711⟩—

Medium: water; 900 mL.

Apparatus 1: 100 rpm.

Time: 45 minutes.

pH 9.5 borate buffer—Dissolve 6.185 g of boric acid and 7.930 g of potassium chloride in about 1000 mL of water, add 60 mL of 1.0 *N* sodium hydroxide, and mix. Dilute with water to 2000 mL, mix, and add 1.0 *N* sodium hydroxide, if necessary, to adjust to a pH of 9.5 ± 0.1.

Standard preparation—Dissolve an accurately weighed quantity of USP Aminocaproic Acid RS in water, and dilute quantitatively with water to obtain a solution having a known concentration of about 0.5 mg per mL.

Procedure—Into three separate 50-mL volumetric flasks pipet (a) 1 mL of a filtered portion of the solution under test, (b) 1 mL of the *Standard preparation*, and (c) 1 mL of water to provide a blank. Add 20.0 mL of *pH 9.5 borate buffer* and 3.0 mL of freshly prepared β-naphthoquinone-4-sodium sulfonate solution (1 in 500) to each, swirl to mix, and place the 3 flasks in a water bath maintained at a temperature of 65 ± 5° for 45 minutes. Cool, dilute each with water to volume, and mix. Determine the amount of $C_6H_{13}NO_2$ dissolved from absorbances, at the wavelength of maximum absorbance at about 460 nm, obtained from the test solution in comparison with those obtained from the Standard solution, using the blank to set the instrument.

Tolerances—Not less than 75% (*Q*) of the labeled amount of $C_6H_{13}NO_2$ is dissolved in 45 minutes.

Uniformity of dosage units ⟨905⟩: meet the requirements.

Assay—Weigh and finely powder not less than 20 Aminocaproic Acid Tablets. Transfer an accurately weighed portion of the powder, equivalent to about 500 mg of aminocaproic acid, to a beaker, add about 100 mL of glacial acetic acid, heat gently to effect solution, and cool. Add 10 drops of a 1 in 500 solution of crystal violet in chlorobenzene, and titrate with 0.1 *N* perchloric acid in dioxane VS to a blue end-point. Perform a blank determination, and make any necessary correction. Each mL of 0.1 *N* perchloric acid is equivalent to 13.12 mg of $C_6H_{13}NO_2$.

Aminoglutethimide

$C_{13}H_{16}N_2O_2$ 232.28

2,6-Piperidinedione, 3-(4-aminophenyl)-3-ethyl-.

2-(*p*-Aminophenyl)-2-ethylglutarimide [*125-84-8*].

» Aminoglutethimide contains not less than 98.0 percent and not more than 102.0 percent of $C_{13}H_{16}N_2O_2$, calculated on the dried basis.

Packaging and storage—Preserve in well-closed containers.

Reference standards—*USP Aminoglutethimide Reference Standard*—Dry at 105° to constant weight before using. *USP m-Aminoglutethimide Reference Standard*—Do not dry before using.

Identification—

A: The infrared absorption spectrum of a mineral oil dispersion of it, previously dried, exhibits maxima only at the same wavelengths as that of a similar preparation of USP Aminoglutethimide RS.

B: The ultraviolet absorption spectrum of a 1 in 100,000 solution in methanol exhibits maxima and minima at the same wavelengths as that of a similar solution of USP Aminoglutethimide RS, concomitantly measured, and the respective absorptivities, calculated on the dried basis, at the wavelength of maximum absorbance at about 242 nm do not differ by more than 2.0%.

pH ⟨791⟩: between 6.2 and 7.3, in a 1 in 1000 solution in dilute methanol (1 in 20).

Loss on drying ⟨731⟩—Dry it at 105° to constant weight: it loses not more than 0.5% of its weight.

Residue on ignition ⟨281⟩: not more than 0.1%.

Heavy metals, *Method II* ⟨231⟩: 0.001%.

Sulfate—Add 1.0 mL of 3 *N* hydrochloric acid to 100 mL of a 1 in 1000 solution in dilute methanol (1 in 20), add 2.0 mL of barium chloride TS, and mix: no turbidity is produced.

***m*-Aminoglutethimide and related compounds**—

Aminoglutethimide standard solution—Dissolve 20 mg of USP Aminoglutethimide RS, accurately weighed, in 1.0 mL of methanol.

m-Aminoglutethimide standard stock solution—Dissolve 10 mg of USP *m*-Aminoglutethimide RS in 10.0 mL of methanol.

Standard dilution I—Dilute 1.0 mL of *m-Aminoglutethimide standard stock solution* with methanol to 10.0 mL (equivalent to 0.5%).

Standard dilution II—Dilute 2.0 mL of *m-Aminoglutethimide standard stock solution* with methanol to 10.0 mL (equivalent to 1.0%).

Standard dilution III—Dilute 4.0 mL of *m-Aminoglutethimide standard stock solution* with methanol to 10.0 mL (equivalent to 2.0%).

Test solution—Dissolve 20 mg of Aminoglutethimide, accurately weighed, in 1.0 mL of methanol.

Detecting reagent—

SOLUTION A—Dissolve 2.5 g of 4,4′-tetramethyldiaminodiphenylmethane in 10 mL of glacial acetic acid. Add 50 mL of water, mix, and filter.

SOLUTION B—Dissolve 5.0 g of potassium iodide in 100 mL of water.

SOLUTION C—Dissolve 0.3 g ninhydrin in 90 mL of water, add 10 mL of glacial acetic acid, and mix.

Combine *Solutions A* and *B*, add 1.5 mL of *Solution C*, and mix.

Procedure—On a suitable thin-layer chromatographic plate (see *Chromatography* ⟨621⟩), coated with a 0.25-mm layer of chromatographic silica gel mixture, previously dried at 100° for 30 minutes, apply separately 5 µL each of the *Test solution*, *Aminoglutethimide standard solution*, and the three *Standard dilutions*. Develop the chromatogram in a solvent system consisting of a mixture of chloroform and ethyl acetate (70:30) until the solvent front has moved about three-fourths of the length of the plate. Remove the plate, and allow the solvent to evaporate. Repeat the development process two times, allowing the solvent to evaporate between successive developments. When the third development is complete, mark the solvent front, and allow the plate to air-dry for not less than 1 hour. Saturate a developing chamber with chlorine vapor by placing a beaker containing 0.5 g of potassium permanganate in the chamber, and adding 5 mL of 6 *N* hydrochloric acid to the beaker. Expose the plate to chlorine vapor for 5 minutes, allow to dry in an exhaust hood for 1 hour, spray with *Detecting reagent*, and compare the chromatograms: the R_f value, color, and intensity of the principal spot obtained from the *Test solution* correspond to those obtained from the *Standard solution*.

m-Aminoglutethimide—The intensity of any spot due to *m*-aminoglutethimide in the chromatogram from the *Test preparation* is not greater than that of the principal spot from *Standard dilution III*, corresponding to not more than 2.0% of *m*-aminoglutethimide.

Related compounds—The sum of the intensities of any secondary spots other than the spot due to *m*-aminoglutethimide in the chromatogram from the *Test preparation* is not greater than the intensity of the principal spot from *Standard dilution II*.

Assay—Dissolve about 225 mg of Aminoglutethimide, accurately weighed, in 30 mL of glacial acetic acid. Titrate with 0.1 *N* perchloric acid VS, determining the end-point potentiometrically, using a glass electrode and a calomel electrode containing saturated lithium chloride in glacial acetic acid (see *Titrimetry* ⟨541⟩). Perform a blank determination, and make any necessary correction. Each mL of 0.1 *N* perchloric acid is equivalent to 23.23 mg of $C_{13}H_{16}N_2O_2$.

Aminoglutethimide Tablets

» Aminoglutethimide Tablets contain not less than 90.0 percent and not more than 110.0 percent of the labeled amount of $C_{13}H_{16}N_2O_2$.

Packaging and storage—Preserve in well-closed containers.

Reference standard—*USP Aminoglutethimide Reference Standard*—Dry at 105° to constant weight before using.

Identification—Transfer 500 mg of finely powdered Tablets to a suitable container, add 25 mL of acetone, mix, and filter. Evaporate the filtrate at room temperature to dryness, and dry the residue in vacuum over silica gel for 2 hours: the infrared absorption spectrum of a mineral oil dispersion of the residue so obtained exhibits maxima only at the same wavelengths as that of a similar preparation of USP Aminoglutethimide RS.

Dissolution ⟨711⟩—

Medium: dilute hydrochloric acid (7 in 1000); 1000 mL.

Apparatus 1: 100 rpm.

Time: 30 minutes.

Procedure—Determine the amount of $C_{13}H_{16}N_2O_2$ dissolved from ultraviolet absorbances at the wavelength of maximum absorbance at about 237 nm of filtered portions of the solution under test, suitably diluted with pH 7.5 phosphate buffer, in comparison with a Standard solution having a known concentration of USP Aminoglutethimide RS in the same medium.

Tolerances—Not less than 70% (*Q*) of the labeled amount of $C_{13}H_{16}N_2O_2$ is dissolved in 30 minutes.

Uniformity of dosage units ⟨905⟩: meet the requirements.

Assay—

Standard preparation—Dissolve an accurately weighed quantity of USP Aminoglutethimide RS in methanol, and dilute quantitatively and stepwise with methanol to obtain a solution having a known concentration of about 10 µg per mL.

Assay preparation—Weigh and finely powder not less than 20 Aminoglutethimide Tablets. Transfer an accurately weighed portion of the powder, equivalent to about 250 mg of aminoglutethimide, to a 100-mL volumetric flask. Add about 75 mL of methanol, shake by mechanical means for 10 minutes, dilute with methanol to volume, mix, and filter. Dilute with methanol quantitatively and stepwise to obtain a final solution containing about 10 µg per mL.

Procedure—Concomitantly determine the absorbances of the solutions in 1-cm cells at the wavelength of maximum absorbance at about 242 nm, with a suitable spectrophotometer, using methanol as a blank. Calculate the quantity, in mg, of $C_{13}H_{16}N_2O_2$ in the portion of Tablets taken by the formula:

$$25C(A_U/A_S),$$

in which *C* is the concentration, in µg per mL, of USP Aminoglutethimide RS in the *Standard preparation*, and A_U and A_S are the absorbances of the solutions from the *Assay preparation* and the *Standard preparation*, respectively.

Aminohippurate Sodium Injection

H_2N–C_6H_4–$CONHCH_2COONa$

$C_9H_9N_2NaO_3$ 216.17

Glycine, *N*-(4-aminobenzoyl)-, monosodium salt.

Monosodium *p*-aminohippurate [*94-16-6*].

» Aminohippurate Sodium Injection is a sterile solution of Aminohippuric Acid in Water for Injection prepared with the aid of Sodium Hydroxide. It contains not less than 95.0 percent and not more than 105.0 percent of the labeled amount of $C_9H_9N_2NaO_3$.

Packaging and storage—Preserve in single-dose or in multiple-dose containers, preferably of Type I glass.

Identification—

A: A volume of Injection, equivalent to 100 mg of aminohippuric acid, diluted to 50 mL and acidified with hydrochloric acid, responds to *Identification test B* under *Aminohippuric Acid*.

B: Transfer a volume of Injection, equivalent to about 200 mg of aminohippurate sodium, to a test tube, and add, in the order named, 2 mL of potassium iodide TS, 10 mL of water, and 5 mL of sodium hypochlorite TS: a red color is produced.

C: It responds to the flame test for *Sodium* ⟨191⟩.

pH ⟨791⟩: between 6.7 and 7.6.

Other requirements—It meets the requirements under *Injections* ⟨1⟩.

Assay—Transfer to a 200-mL volumetric flask an accurately measured volume of Aminohippurate Sodium Injection, equivalent to about 1 g of aminohippurate sodium, and dilute with water to volume. Transfer 50.0 mL of the solution to a suitable container, add 5 mL of hydrochloric acid, and proceed as directed under *Nitrite Titration* ⟨451⟩, beginning with "cool to 15°." Each mL of 0.1 *M* sodium nitrite is equivalent to 21.62 mg of $C_9H_9N_2NaO_3$.

Aminohippuric Acid

$C_9H_{10}N_2O_3$ 194.19
Glycine, *N*-(4-aminobenzoyl)-.
p-Aminohippuric acid [*61-78-9*].

» Aminohippuric Acid contains not less than 98.0 percent and not more than 100.5 percent of $C_9H_{10}N_2O_3$, calculated on the dried basis.

Packaging and storage—Preserve in tight, light-resistant containers.

Reference standard—*USP Aminohippuric Acid Reference Standard*—Do not dry before using.

Identification—

A: The infrared absorption spectrum of a potassium bromide dispersion of it exhibits maxima only at the same wavelengths as that of a similar preparation of USP Aminohippuric Acid RS.

B: Dissolve about 10 mg in 5 mL of water, and add 0.5 mL of 3 *N* hydrochloric acid, 0.5 mL of sodium nitrite solution (1 in 10), and a solution of 0.20 g of 2-naphthol in 10 mL of 6 *N* ammonium hydroxide: a red color is produced.

Loss on drying ⟨731⟩—Dry it at 105° for 2 hours: it loses not more than 0.25% of its weight.

Residue on ignition ⟨281⟩: not more than 0.25%.

Heavy metals, *Method II* ⟨231⟩: 0.001%.

Assay—Transfer about 150 mg, accurately weighed, of Aminohippuric Acid to a beaker or a casserole. Add 5 mL of hydrochloric acid and 50 mL of water, and proceed as directed under *Nitrite Titration* ⟨451⟩, beginning with "stir until dissolved." Each mL of 0.1 *M* sodium nitrite is equivalent to 19.42 mg of $C_9H_{10}N_2O_3$.

Aminophylline

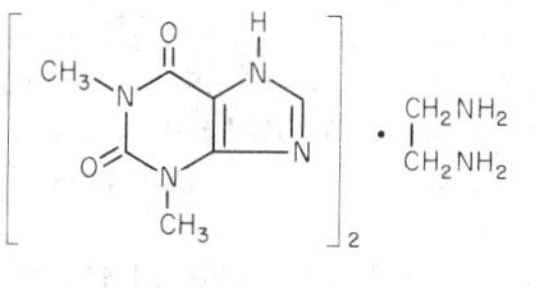

$C_{16}H_{24}N_{10}O_4$ (anhydrous) 420.43
1*H*-Purine-2,6-dione, 3,7-dihydro-1,3-dimethyl-, compd. with 1,2-ethanediamine (2:1).
Theophylline compound with ethylenediamine (2:1) [*317-34-0*].
Dihydrate 456.46 [*49746-06-7*].

» Aminophylline is anhydrous or contains not more than two molecules of water of hydration. It contains not less than 84.0 percent and not more than 87.4 percent of anhydrous theophylline ($C_7H_8N_4O_2$), calculated on the anhydrous basis.

Packaging and storage—Preserve in tight containers.

Labeling—Label it to indicate whether it is anhydrous or hydrous, and also to state the content of anhydrous theophylline.

Identification—

A: Dissolve about 500 mg in 20 mL of water, add, with constant stirring, 1 mL of 3 *N* hydrochloric acid, filter (retain the filtrate), wash the precipitate with small portions of cold water, and dry at 105° for 1 hour: the precipitate of theophylline so obtained melts between 270° and 274°.

B: To about 10 mg of the dried precipitate obtained in *Identification test A*, contained in a porcelain dish, add 1 mL of hydrochloric acid and 100 mg of potassium chlorate, evaporate on a steam bath to dryness, and invert the dish over a vessel containing a few drops of 6 *N* ammonium hydroxide: the residue acquires a purple color, which is destroyed by solutions of fixed alkalies.

C: To the filtrate obtained in *Identification test A* add 0.5 mL of benzenesulfonyl chloride and 5 mL of 1 *N* sodium hydroxide to render alkaline, shake by mechanical means for 10 minutes, add 5 mL of 3 *N* hydrochloric acid to acidify, chill, collect the precipitated disulfonamide of ethylenediamine, wash well with water, recrystallize from water, and dry at 105° for 1 hour: the dried precipitate melts between 164° and 171°.

Water, *Method I* ⟨921⟩: not more than 0.75% (anhydrous form) and not more than 7.9% (hydrous form), determined on 1.5 g of it, a mixture of 25 mL of chloroform and 25 mL of methanol being used in place of the methanol solvent.

Residue on ignition ⟨281⟩: not more than 0.15%.

Ethylenediamine content—Dissolve about 500 mg of Aminophylline, accurately weighed, in 30 mL of water, add methyl orange TS, and titrate with 0.1 *N* hydrochloric acid VS. Each mL of 0.1 *N* hydrochloric acid is equivalent to 3.005 mg of $C_2H_8N_2$. The content of ethylenediamine ($C_2H_8N_2$) is between 157 mg and 175 mg per g of $C_7H_8N_4O_2$ found in the *Assay*.

Assay—Place about 250 mg of Aminophylline, accurately weighed, in a 250-mL conical flask, add 50 mL of water and 8 mL of 6 *N* ammonium hydroxide, and gently warm the mixture on a steam bath until complete solution is effected. Add 20.0 mL of 0.1 *N* silver nitrate VS, mix, heat to boiling, and boil for 15 minutes. Cool to between 5° and 10° for 20 minutes, then filter, preferably through a filtering crucible under reduced pressure, and wash the precipitate with three 10-mL portions of water. Acidify the combined filtrate and washings with nitric acid, and add an excess of 3 mL of the acid. Cool, add 2 mL of ferric ammonium sulfate TS, and titrate the excess silver nitrate with 0.1 *N* ammonium thiocyanate VS. Each mL of 0.1 *N* silver nitrate is equivalent to 18.02 mg of $C_7H_8N_4O_2$.

Aminophylline Enema

» Aminophylline Enema is an aqueous solution of Aminophylline, prepared with the aid of Ethylenediamine. It contains an amount of aminophylline equivalent to not less than 90.0 percent and not more than 110.0 percent of the labeled amount of anhydrous theophylline ($C_7H_8N_4O_2$).

Aminophylline Enema may contain an excess of ethylenediamine, but no other substance may be added for the purpose of pH adjustment.

Packaging and storage—Preserve in single-dose or in multiple-dose containers.

Labeling—Label the Enema to state the content of anhydrous theophylline.

Reference standard—*USP Theophylline Reference Standard*—Dry at 105° for 4 hours before using.

Identification—Dilute a volume of Enema, equivalent to about 500 mg of aminophylline, with water to about 20 mL, and add, with constant stirring, 1 mL of 3 *N* hydrochloric acid or sufficient to precipitate the theophylline completely. Filter, wash with small portions of cold water until free from chloride, and dry at 105°

for 4 hours. Use the precipitate and the filtrate for the following tests.

A: The infrared spectrum of a potassium bromide dispersion of the theophylline obtained as directed above exhibits maxima only at the same wavelengths as that of a similar preparation of USP Theophylline RS.

B: The filtrate obtained as directed above responds to *Identification test C* under *Aminophylline.*

pH ⟨791⟩: between 9.0 and 9.5.

Ethylenediamine content—Measure accurately a volume of Enema, equivalent to about 500 mg of aminophylline, and dilute with water, if necessary, to make about 30 mL. Add methyl orange TS, and titrate with 0.1 *N* hydrochloric acid VS. Each mL of 0.1 *N* hydrochloric acid is equivalent to 3.005 mg of $C_2H_8N_2$. The Enema contains between 218 mg and 267 mg of ethylenediamine ($C_2H_8N_2$) per g of $C_7H_8N_4O_2$ found in the *Assay.*

Assay—

Standard preparation—Dissolve an accurately weighed quantity of USP Theophylline RS in dilute hydrochloric acid (1 in 100), and dilute quantitatively and stepwise with the same solvent to obtain a solution having a known concentration of about 8 µg per mL.

Assay preparation—Pipet an accurately measured volume of Aminophylline Enema, equivalent to about 500 mg of aminophylline, into a 500-mL volumetric flask, dilute with water to volume, and mix. Pipet 5 mL of this solution into a second 500-mL volumetric flask, add 50 mL of dilute hydrochloric acid (1 in 10), dilute with water to volume, and mix.

Procedure—Concomitantly determine the absorbances of the *Standard preparation* and the *Assay preparation* in 1-cm cells at the wavelength of maximum absorbance at about 270 nm, with a suitable spectrophotometer, using dilute hydrochloric acid (1 in 100) as the blank. Calculate the quantity, in mg, of $C_7H_8N_4O_2$ in each mL of the Enema taken by the formula:

$$50(C/V)(A_U/A_S),$$

in which *C* is the concentration, in µg per mL, of USP Theophylline RS in the *Standard preparation*, *V* is the volume, in mL, of Enema taken, and A_U and A_S are the absorbances of the *Assay preparation* and the *Standard preparation*, respectively.

Aminophylline Injection

» Aminophylline Injection is a sterile solution of Aminophylline in Water for Injection, or is a sterile solution of Theophylline in Water for Injection prepared with the aid of Ethylenediamine. It contains, in each mL, an amount of aminophylline equivalent to not less than 93.0 percent and not more than 107.0 percent of the labeled amount of anhydrous theophylline ($C_7H_8N_4O_2$).

Aminophylline Injection may contain an excess of Ethylenediamine, but no other substance may be added for the purpose of pH adjustment.

Note—Do not use the Injection if crystals have separated.

Packaging and storage—Preserve in single-dose containers from which carbon dioxide has been excluded, preferably of Type I glass, protected from light.

Labeling—Label the Injection to state the content of anhydrous theophylline.

Identification—Dilute a volume of Injection, equivalent to about 500 mg of aminophylline, with water to about 20 mL, and add, with constant stirring, 1 mL of 3 *N* hydrochloric acid or enough to precipitate the theophylline completely. Filter: the filtrate responds to *Identification test C* under *Aminophylline.*

Wash the precipitate of theophylline with a small portion of cold water, and dry at 105° for 1 hour: the theophylline so obtained melts between 270° and 274°, and responds to *Identification test B* under *Aminophylline.*

pH ⟨791⟩: between 8.6 and 9.0.

Particulate matter ⟨788⟩: meets the requirements under *Small-volume Injections.*

Other requirements—It meets the requirements under *Injections* ⟨1⟩.

Ethylenediamine content—Accurately measure a volume of Injection, equivalent to about 500 mg of aminophylline, and dilute with water, if necessary, to make about 30 mL. Add methyl orange TS, and titrate with 0.1 *N* hydrochloric acid VS. Each mL of 0.1 *N* hydrochloric acid is equivalent to 3.005 mg of $C_2H_8N_2$. The Injection contains between 166 mg and 192 mg of ethylenediamine ($C_2H_8N_2$) per g of $C_7H_8N_4O_2$ found in the *Assay.*

Assay—Transfer to a 250-mL conical flask an accurately measured volume of Aminophylline Injection, equivalent to about 250 mg of aminophylline, and dilute with water to about 40 mL. Add 8 mL of 6 *N* ammonium hydroxide, and proceed as directed in the *Assay* under *Aminophylline*, beginning with "Add 20.0 mL of 0.1 *N* silver nitrate VS." Each mL of 0.1 *N* silver nitrate is equivalent to 18.02 mg of $C_7H_8N_4O_2$.

Aminophylline Oral Solution

» Aminophylline Oral Solution is an aqueous solution of Aminophylline, prepared with the aid of Ethylenediamine. It contains an amount of aminophylline equivalent to not less than 90.0 percent and not more than 110.0 percent of the labeled amount of anhydrous theophylline ($C_7H_8N_4O_2$).

Aminophylline Oral Solution may contain an excess of ethylenediamine, but no other substance may be added for the purpose of pH adjustment.

Packaging and storage—Preserve in tight containers.

Labeling—Label the Oral Solution to state the content of anhydrous theophylline.

Identification—

A: Place a volume of Oral Solution, equivalent to about 500 mg of aminophylline, in a suitable container, and add, with constant stirring, 1 mL of 3 *N* hydrochloric acid or an amount sufficient to precipitate the theophylline completely. Filter (retain the filtrate), wash the precipitate with small portions of cold water until free from chloride, and dry at 105° for 1 hour: the theophylline so obtained melts between 270° and 274°.

B: The filtrate from *Identification test A* responds to *Identification test C* under *Aminophylline.*

pH ⟨791⟩: between 8.7 and 9.3.

Ethylenediamine content—Accurately measure a volume of Oral Solution, equivalent to about 500 mg of aminophylline, and dilute with water, if necessary, to make about 30 mL. Add methyl orange TS, and titrate with 0.1 *N* hydrochloric acid VS. Each mL of 0.1 *N* hydrochloric acid is equivalent to 3.005 mg of $C_2H_8N_2$. The Oral Solution contains between 176 mg and 221 mg of ethylenediamine ($C_2H_8N_2$) per g of $C_7H_8N_4O_2$ found in the *Assay.*

Assay—Transfer to a 250-mL conical flask an accurately measured volume of Aminophylline Oral Solution, equivalent to about 315 mg of aminophylline, and dilute with water to about 40 mL. Add 8 mL of 6 *N* ammonium hydroxide, and proceed as directed in the *Assay* under *Aminophylline*, beginning with "Add 20.0 mL of 0.1 *N* silver nitrate VS." Each mL of 0.1 *N* silver nitrate is equivalent to 18.02 mg of $C_7H_8N_4O_2$.

Aminophylline Suppositories

» Aminophylline Suppositories contain an amount of aminophylline equivalent to not less than 90.0 percent and not more than 110.0 percent of the labeled amount of anhydrous theophylline ($C_7H_8N_4O_2$).

Packaging and storage—Preserve in well-closed containers, in a cold place.

Labeling—Label the Suppositories to state the content of anhydrous theophylline.

Identification—

A: Evaporate to about one-half its volume on a steam bath a portion, equivalent to about 500 mg of aminophylline, of the water solution prepared in the *Assay*. Adjust with 1 *N* sodium hydroxide to a pH of 7.0, chill, and filter the crystals of theophylline. Save the filtrate, free from the washings, for use in *Identification test B:* the crystals, after being washed with small portions of ice-cold water and dried at 105° for 1 hour, melt between 270° and 274°, and respond to *Identification test B* under *Aminophylline.*

B: The filtrate from *Identification test A* responds to *Identification test C* under *Aminophylline.*

Ethylenediamine content—Weigh accurately a portion of the stirred, congealed mass of the Suppositories used for the *Assay*, equivalent to about 500 mg of aminophylline, and place in a 500-mL conical flask. Add 150 mL of a mixture of equal volumes of alcohol and ether, and warm gently under reflux for 30 minutes, with occasional swirling. Cool to room temperature, and titrate with 0.1 *N* hydrochloric acid VS, using a glass–modified calomel electrode system (replace the saturated potassium chloride solution of the calomel electrode with methanol saturated with lithium chloride). Each mL of 0.1 *N* hydrochloric acid is equivalent to 3.005 mg of $C_2H_8N_2$. The Suppositories contain between 152 mg and 190 mg of ethylenediamine ($C_2H_8N_2$) per g of $C_7H_8N_4O_2$ found in the *Assay*.

Assay—Tare a small dish and a glass rod, place in the dish not less than 5 Aminophylline Suppositories, and heat on a steam bath until melted. Mix the melt by stirring it with the rod, cool while stirring, and weigh. Weigh accurately a portion of the mass, equivalent to about 1 g of aminophylline, place it in a beaker, add 60 mL of hot water and 3 mL of nitric acid, and heat on a steam bath for 15 minutes with frequent stirring. Cool, transfer to a separator with the aid of 40 mL of ether, shake well, and allow to separate, using a few mL of alcohol, if necessary, to bring about separation of any emulsion that has formed. Draw the water layer into a 100-mL volumetric flask, wash the ether with two 15-mL portions of water, adding the washings to the volumetric flask, dilute with water to volume, and mix. Transfer an accurately measured portion of the solution, equivalent to about 250 mg of aminophylline, to a 250-mL conical flask, add 10 mL of 6 *N* ammonium hydroxide and about 20 mL of 0.1 *N* silver nitrate VS, and heat on a steam bath for 15 minutes. Cool to between 5° and 10° for 20 minutes, then filter, preferably through a filtering crucible of fine porosity under reduced pressure, and wash the precipitate with small portions of water until the last washing gives not more than a faint opalescence with hydrochloric acid. Dissolve the precipitate by pouring over it small volumes of warm 2 *N* nitric acid, receiving the solution in a conical flask. Wash the filtering crucible a few times with warm water acidified with nitric acid, receiving the washings in the same flask. Cool, add 2 mL of ferric ammonium sulfate TS, and titrate with 0.1 *N* ammonium thiocyanate VS. Each mL of 0.1 *N* ammonium thiocyanate is equivalent to 18.02 mg of $C_7H_8N_4O_2$.

Aminophylline Tablets

» Aminophylline Tablets contain an amount of aminophylline equivalent to not less than 93.0 percent and not more than 107.0 percent of the labeled amount of anhydrous theophylline ($C_7H_8N_4O_2$).

NOTE—The ammoniacal odor present in the vapor space above Aminophylline Tablets is often quite strong, especially when bottles having suitably tight closures are newly opened. This is due to ethylenediamine vapor pressure build-up, a natural condition in the case of aminophylline.

Packaging and storage—Preserve in tight containers.

Labeling—Label the Tablets to state the content of anhydrous theophylline.

Reference standard—*USP Theophylline Reference Standard*—Dry at 105° for 4 hours before using.

Identification—

A: Macerate a quantity of Tablets, equivalent to about 500 mg of aminophylline, with 25 mL of water, and filter: the filtrate is alkaline to litmus. To the filtrate add 1 mL of 3 *N* hydrochloric acid, stir, and chill, if necessary, to precipitate the theophylline. Filter, and retain the filtrate, free from washings. Wash the crystals on the filter with small quantities of ice-cold water, and dry at 105° for 1 hour: the theophylline so obtained responds to *Identification test B* under *Aminophylline*, and when recrystallized from water and dried at 105° for 1 hour, melts between 270° and 274°.

B: The filtrate obtained in *Identification test A* responds to *Identification test C* under *Aminophylline.*

Disintegration ⟨701⟩ FOR ENTERIC-COATED TABLETS: 30 minutes, determined as directed under *Enteric-coated Tablets.*

Dissolution ⟨711⟩—

FOR UNCOATED OR PLAIN COATED TABLETS—

Medium: water; 900 mL.

Apparatus 2: 50 rpm.

Time: 45 minutes.

Procedure—Determine the amount of anhydrous theophylline ($C_7H_8N_4O_2$) dissolved from ultraviolet absorbances at the wavelength of maximum absorbance at about 269 nm of filtered portions of the solution under test, suitably diluted with water, if necessary, in comparison with a Standard solution having a known concentration of USP Theophylline RS in the same medium.

Tolerances—Not less than 75% (*Q*) of the labeled amount of $C_7H_8N_4O_2$ is dissolved in 45 minutes.

Uniformity of dosage units ⟨905⟩: meet the requirements.

Procedure for content uniformity—Place 1 Tablet in a 250-mL volumetric flask, add about 200 mL of water, and shake by mechanical means until disintegration is complete. Add water to volume, and mix. Filter a portion of the mixture, discarding the first 20 mL of the filtrate. Concomitantly determine the absorbances of this solution, quantitatively diluted, if necessary, and a Standard solution of USP Theophylline RS having a known concentration of about 10 µg per mL, in 1-cm cells at the wavelength of maximum absorbance at about 269 nm, with a suitable spectrophotometer, using water as the blank. Calculate the quantity, in mg, of anhydrous theophylline ($C_7H_8N_4O_2$) in the Tablet by the formula:

$$(TC/D)(A_U/A_S),$$

in which T is the labeled quantity, in mg, of anhydrous theophylline in the Tablet, D is the concentration, in µg per mL, of theophylline in the solution from the Tablet, based on the labeled quantity per Tablet and the extent of dilution, C is the concentration, in µg per mL, of USP Theophylline RS in the Standard solution, and A_U and A_S are the absorbances of the solution from the Tablet and the Standard solution, respectively.

Ethylenediamine content—Weigh accurately a portion of the powdered Tablets prepared in the *Assay*, equivalent to about 350 mg of aminophylline, transfer to a 100-mL conical flask, add 20 mL of water, and digest at 50°, with frequent shaking, for 30 minutes. Cool, filter into a 250-mL conical flask, and wash with water until the last washing is neutral to litmus. To the combined filtrate and washings add methyl orange TS, and titrate with 0.1 *N* hydrochloric acid VS. Each mL of 0.1 *N* hydrochloric acid is equivalent to 3.005 mg of $C_2H_8N_2$. The Tablets contain between 152 mg and 178 mg of ethylenediamine ($C_2H_8N_2$) per g of $C_7H_8N_4O_2$ found in the *Assay.*

Assay—Weigh and finely powder not less than 20 Aminophylline Tablets. Transfer a portion of the powder, equivalent to about 2 g of aminophylline, to a 200-mL volumetric flask with the aid of a mixture of 50 mL of water and 15 mL of 6 *N* ammonium hydroxide, and allow to stand for 30 minutes with frequent shaking, warming to about 50° if necessary, to effect solution of the aminophylline. Cool the mixture to room temperature if it has been warmed, add water to volume, and mix. Centrifuge about 50 mL of the mixture, and pipet a portion of the clear supernatant liquid, equivalent to about 250 mg of aminophylline, into a 250-mL conical flask, and dilute with water, if necessary, to make about 40 mL. Add 8 mL of 6 *N* ammonium hydroxide, and proceed as directed in the *Assay* under *Aminophylline*, beginning with "Add 20.0 mL of 0.1 *N* silver nitrate VS." Each mL of 0.1 *N* silver nitrate is equivalent to 18.02 mg of $C_7H_8N_4O_2$.

Aminosalicylate Sodium

$C_7H_6NNaO_3.2H_2O$ 211.15

Benzoic acid, 4-amino-2-hydroxy-, monosodium salt, dihydrate.

Monosodium 4-aminosalicylate dihydrate [*6018-19-5*].

Anhydrous 175.12 [*133-10-8*].

» Aminosalicylate Sodium contains not less than 98.0 percent and not more than 101.0 percent of $C_7H_6NNaO_3$, calculated on the anhydrous basis.

Note—Prepare solutions of Aminosalicylate Sodium within 24 hours of administration. Under no circumstances use a solution if its color is darker than that of a freshly prepared solution.

Packaging and storage—Preserve in tight, light-resistant containers, protected from excessive heat.

Reference standards—*USP Aminosalicylic Acid Reference Standard*—Dry in vacuum at 50° for 1 hour before using. *USP m-Aminophenol Reference Standard*—Do not dry before using.

Clarity and color of solution—One g dissolves in 10 mL of water to give a clear solution that has not more than a faint yellow color. One g dissolves in a freshly prepared mixture of 5 mL of nitric acid and 45 mL of water to give a clear solution that has not more than a slight color.

Identification—

A: Dissolve 250 mg in 3 mL of 1 *N* sodium hydroxide, transfer to a 500-mL volumetric flask, dilute with water to volume, and mix. Transfer a 5-mL aliquot to a 250-mL volumetric flask containing 12.5 mL of pH 7 phosphate buffer (see under *Solutions* in the section, *Reagents, Indicators, and Solutions*), dilute with water to volume, and mix. This solution, when compared in a suitable spectrophotometer against a blank of the same buffer in the same concentration, exhibits absorbance maxima at 265 ± 2 nm and 299 ± 2 nm, and the ratio A_{265}/A_{299} is between 1.50 and 1.56.

B: Place about 1 g in a small, round-bottom flask, and add 10 mL of acetic anhydride. Heat the flask on a steam bath for 30 minutes, add 40 mL of water, mix, filter, cool, and allow to stand until the diacetyl derivative has crystallized. Collect the precipitate on a filter, wash well with water, and dry at 105° for 1 hour: the diacetyl derivative so obtained melts between 191° and 197°.

C: Dissolve 50 mg in 5 mL of water, add 1 mL of 3 *N* hydrochloric acid, and filter if necessary. To the filtrate add 1 drop of ferric chloride TS: a violet color is produced.

D: A solution of it responds to the tests for *Sodium* ⟨191⟩.

pH ⟨791⟩: between 6.5 and 8.5, in a solution (1 in 50).

Water, *Method I* ⟨921⟩: between 16.0% and 18.0%.

Chloride ⟨221⟩—Dissolve 0.50 g in a mixture of 5 mL of nitric acid and 15 mL of water: the solution shows no more chloride than corresponds to 0.30 mL of 0.020 *N* hydrochloric acid (0.042%).

Heavy metals, *Method II* ⟨231⟩: 0.003%.

***m*-Aminophenol**—

Mobile phase and *Internal standard solution*—Prepare as directed in the *Assay* under *Aminosalicylic Acid.*

Standard preparation and *Chromatographic system*—Prepare as directed in the test for *m-Aminophenol* under *Aminosalicylic Acid.*

Test preparation—Use the *Assay preparation,* prepared as directed in the *Assay.*

Procedure—Separately inject equal volumes (about 20 µL) of the *Standard preparation* and the *Test preparation* into the chromatograph, record the chromatograms, and measure the responses for the major peaks. The relative retention times are about 0.66 for sulfanilamide and 1.0 for *m*-aminophenol. Calculate the percentage of *m*-aminophenol, in relation to the quantity of aminosalicylate sodium in the portion of Aminosalicylate Sodium taken by the formula:

$$10(C/W)(R_U/R_S),$$

in which *C* is the concentration, in µg per mL, of USP *m*-Aminophenol RS in the *Standard preparation, W* is the quantity of aminosalicylate sodium, in mg, in the portion of Aminosalicylate Sodium taken, as determined in the *Assay*, and R_U and R_S are the ratios of the response of the *m*-aminophenol peak to the response of the sulfanilamide peak obtained from the *Test preparation* and the *Standard preparation,* respectively: not more than 0.25% of *m*-aminophenol is found.

Hydrogen sulfide, sulfur dioxide, and amyl alcohol—Dissolve about 500 mg in 5 mL of 1 *N* sodium hydroxide, add 6 mL of 3 *N* hydrochloric acid, and stir vigorously: no odor of hydrogen sulfide or sulfur dioxide is perceptible, and not more than a faint odor of amyl alcohols is perceptible. A piece of moistened lead acetate test paper held over the mixture does not become discolored.

Assay—

Mobile phase, Internal standard solution, Standard preparation, and *Chromatographic system*—Prepare as directed in the *Assay* under *Aminosalicylic Acid.*

Assay preparation—Transfer about 69 mg of Aminosalicylate Sodium, accurately weighed, to a 100-mL low-actinic volumetric flask, add 50 mL of *Mobile phase,* and swirl to dissolve. Add 10.0 mL of *Internal standard solution,* dilute with *Mobile phase* to volume, and mix.

Procedure—Proceed as directed for *Procedure* in the *Assay* under *Aminosalicylic Acid.* Calculate the quantity of $C_7H_6NNaO_3$, in mg, in the Aminosalicylate Sodium taken by the formula:

$$(175.12/153.14)(100C)(R_U/R_S),$$

in which 175.12 and 153.14 are the molecular weights of anhydrous aminosalicylate sodium and aminosalicylic acid, respectively, *C* is the concentration, in mg per mL, of USP Aminosalicylic Acid RS in the *Standard preparation,* and R_U and R_S are the ratios of the response of the aminosalicylic acid peak to the response of the acetaminophen peak obtained from the *Assay preparation* and the *Standard preparation,* respectively.

Aminosalicylate Sodium Tablets

» Aminosalicylate Sodium Tablets contain not less than 95.0 percent and not more than 105.0 percent of the labeled amount of $C_7H_6NNaO_3.2H_2O$.

Packaging and storage—Preserve in tight, light-resistant containers, protected from excessive heat.

Reference standards—*USP Aminosalicylic Acid Reference Standard*—Dry in vacuum at 50° for 1 hour before using. *USP m-Aminophenol Reference Standard*—Do not dry before using.

Identification—Digest a quantity of powdered Tablets, equivalent to about 3 g of aminosalicylate sodium, with 40 mL of water, and filter. Add to the filtrate 15 mL of 1 *N* acetic acid, and allow to stand until precipitation has occurred. Collect the precipitate on a filter, wash well with water, and dry at 105° for 30 minutes: the residue responds to the following tests.

A: Place about 1 g in a small, round-bottom flask, and add 10 mL of acetic anhydride. Heat the flask on a steam bath for 30 minutes, add 40 mL of water, mix, filter, cool, and allow to stand until the diacetyl derivative has crystallized. Collect the precipitate on a filter, wash well with water, and dry at 105° for 1 hour: the diacetyl derivative so obtained melts between 191° and 197°.

B: Shake 0.1 g with 10 mL of water, and filter. To 5 mL of the filtrate add 1 drop of ferric chloride TS: a violet color is produced.

Dissolution ⟨711⟩—

Medium: water; 900 mL.

Apparatus 1: 100 rpm.

Time: 45 minutes.

Procedure—Determine the amount of $C_7H_6NNaO_3 . 2H_2O$ dissolved, employing the procedure set forth in the *Assay*, making any necessary modifications.

Tolerances—Not less than 75% (*Q*) of the labeled amount of $C_7H_6NNaO_3 . 2H_2O$ is dissolved in 45 minutes.

Uniformity of dosage units ⟨905⟩: meet the requirements.

***m*-Aminophenol**—

Mobile phase and *Internal standard solution*—Prepare as directed in the *Assay* under *Aminosalicylic Acid.*

Standard preparation and *Chromatographic system*—Prepare as directed in the test for *m-Aminophenol* under *Aminosalicylic Acid.*

Test preparation—Use the *Assay preparation*, prepared as directed in the *Assay.*

Procedure—Proceed as directed for *Procedure* in the test for *m-Aminophenol* under *Aminosalicylic Acid.* Calculate the percentage of *m*-aminophenol, in relation to the quantity of aminosalicylate sodium in the portion of Tablets taken by the formula:

$$100(C/W)(R_U/R_S),$$

in which *C* is the concentration, in µg per mL, of USP *m*-Aminophenol RS in the *Standard preparation, W* is the quantity of aminosalicylate sodium, in mg, in the portion of Tablets taken, as determined in the *Assay*, and R_U and R_S are the ratios of the response of the *m*-aminophenol peak to the response of the sulfanilamide peak obtained from the *Test preparation* and the *Standard preparation*, respectively: not more than 1.0% of *m*-aminophenol is found.

Assay—

Mobile phase, Internal standard solution, Standard preparation, and *Chromatographic system*—Prepare as directed in the *Assay* under *Aminosalicylic Acid.*

Assay preparation—Weigh and finely powder not less than 20 Aminosalicylate Sodium Tablets. Transfer an accurately weighed portion of the powder, equivalent to about 690 mg of aminosalicylate sodium, to a 100-mL low-actinic volumetric flask. Add 50 mL of *Mobile phase*, and shake for about 5 minutes. Dilute with *Mobile phase* to volume, and mix. Filter, and transfer 10.0 mL of the clear filtrate to a low-actinic, 100-mL volumetric flask containing 10.0 mL of *Internal standard solution*, dilute with *Mobile phase* to volume, and mix.

Procedure—Proceed as directed for *Procedure* in the *Assay* under *Aminosalicylic Acid.* Calculate the quantity, in mg, of $C_7H_6NNaO_3 . 2H_2O$ in the portion of Tablets taken by the formula:

$$(211.15/153.14)(1000C)(R_U/R_S),$$

in which 211.15 and 153.14 are the molecular weights of aminosalicylate sodium dihydrate and aminosalicylic acid, respectively, *C* is the concentration, in mg per mL, of USP Aminosalicylic Acid RS in the *Standard preparation*, and R_U and R_S are the ratios of the response of the aminosalicylic acid peak to the response of the acetaminophen peak obtained from the *Assay preparation* and the *Standard preparation*, respectively.

Aminosalicylic Acid

$C_7H_7NO_3$ 153.14

Benzoic acid, 4-amino-2-hydroxy-.
4-Aminosalicylic acid [*65-49-6*].

» Aminosalicylic Acid contains not less than 98.5 percent and not more than 100.5 percent of $C_7H_7NO_3$, calculated on the anhydrous basis.

NOTE—Under no circumstances use a solution prepared from Aminosalicylic Acid if its color is darker than that of a freshly prepared solution.

Packaging and storage—Preserve in tight, light-resistant containers, at a temperature not exceeding 30°.

Reference standards—*USP Aminosalicylic Acid Reference Standard*—Dry in vacuum at 50° for 1 hour before using. *USP m-Aminophenol Reference Standard*—Do not dry before using.

Clarity and color of solution—One g dissolves in 10 mL of sodium bicarbonate solution (1 in 15) to form a clear solution that has not more than a faint yellow color. One g dissolves in a freshly prepared mixture of 5 mL of nitric acid and 45 mL of water to form a clear solution that has not more than a slight color.

Identification—

A: Dissolve 0.25 g in 3 mL of 1 *N* sodium hydroxide, transfer to a 500-mL volumetric flask, dilute with water to volume, and mix. Transfer a 5-mL aliquot to a 250-mL volumetric flask containing 12.5 mL of pH 7 phosphate buffer (see *Buffer Solutions* in the section, *Reagents, Indicators, and Solutions*), dilute with water to volume, and mix. This solution, when compared in a suitable spectrophotometer against a blank of the same buffer in the same concentration, exhibits absorbance maxima at 265 ± 2 and 299 ± 2 nm, and the ratio A_{265}/A_{299} is between 1.50 and 1.56.

B: Place about 1 g in a small, round-bottom flask, and add 10 mL of acetic anhydride. Heat the flask on a steam bath for 30 minutes, add 40 mL of water, mix, filter, cool, and allow to stand until the diacetyl derivative has crystallized. Collect the precipitate on a filter, wash well with water, and dry at 105° for 1 hour: the diacetyl derivative so obtained melts between 191° and 197°.

C: Shake 0.1 g with 10 mL of water, and filter. To 5 mL of the filtrate add 1 drop of ferric chloride TS: a violet color is produced.

pH ⟨791⟩: between 3.0 and 3.7, in a saturated solution.

Water, *Method I* ⟨921⟩: not more than 0.5%.

Residue on ignition ⟨281⟩: not more than 0.2%.

Chloride ⟨221⟩—Dissolve 0.50 g in a mixture of 5 mL of nitric acid and 15 mL of water: the solution shows no more chloride than corresponds to 0.30 mL of 0.020 *N* hydrochloric acid (0.042%).

Heavy metals, *Method II* ⟨231⟩: 0.003%.

***m*-Aminophenol**—

Mobile phase—Prepare as directed in the *Assay.*

Internal standard solution—Prepare a solution of sulfanilamide in *Mobile phase* having a concentration of about 5 µg per mL.

Standard preparation—Dissolve an accurately weighed quantity of USP *m*-Aminophenol RS in *Mobile phase* to obtain a solution having a known concentration of about 12 µg per mL. Transfer 10.0 mL of this solution and 10.0 mL of *Internal standard solution* to a 100-mL low-actinic volumetric flask, dilute with *Mobile phase* to volume, and mix.

Test preparation—Transfer about 50 mg of Aminosalicylic Acid, accurately weighed, to a 100-mL low-actinic volumetric flask, add 50 mL of *Mobile phase*, and swirl to dissolve. Add 10.0 mL of *Internal standard solution*, dilute with *Mobile phase* to volume, and mix.

Chromatographic system (see *Chromatography* ⟨621⟩)—The liquid chromatograph is equipped with a 280-nm detector and a 4.6-mm × 25-cm column that contains 10-µm packing L1. The flow rate is about 1.5 mL per minute. Chromatograph the *Standard preparation*, and record the peak responses as directed under *Procedure:* the resolution, *R*, between the *m*-aminophenol and sulfanilamide peaks is not less than 2.5, and the relative standard deviation for replicate injections is not more than 7%.

Procedure—[NOTE—After use, wash the column for 30 minutes with a filtered and degassed mixture of methanol, water, and phosphoric acid (77:23:0.6), and then wash for 30 minutes with a filtered and degassed mixture of methanol and water (50:50).] Separately inject equal volumes (about 20 µL) of the *Standard preparation* and the *Test preparation* into the chromatograph, record the chromatograms, and measure the responses for the major peaks. The relative retention times are about 0.66 for sulfanilamide and 1.0 for *m*-aminophenol. Calculate the percentage of *m*-aminophenol, in relation to the quantity of aminosalicylic acid in the portion of Aminosalicylic Acid taken by the formula:

$$10(C/W)(R_U/R_S),$$

in which C is the concentration, in µg per mL, of USP *m*-Aminophenol RS in the *Standard preparation*, W is the quantity of aminosalicylic acid, in mg, in the portion of Aminosalicylic Acid taken, as determined in the *Assay*, and R_U and R_S are the ratios of the response of the *m*-aminophenol peak to the response of the sulfanilamide peak obtained from the *Test preparation* and the *Standard preparation*, respectively: not more than 0.25% of *m*-aminophenol is found.

Hydrogen sulfide, sulfur dioxide, and amyl alcohol—Dissolve about 500 mg in 5 mL of 1 *N* sodium hydroxide, add 6 mL of 3 *N* hydrochloric acid, and stir vigorously: no odor of hydrogen sulfide or sulfur dioxide is perceptible, and not more than a faint odor of amyl alcohols is perceptible. A piece of moistened lead acetate test paper held over the mixture does not become discolored.

Assay—

Mobile phase—Prepare a mixture of 425 mL of 0.05 *M* dibasic sodium phosphate, 425 mL of 0.05 *M* monobasic sodium phosphate, and 150 mL of methanol containing 1.9 g of tetrabutylammonium hydroxide. Filter, and degas. Make adjustments if necessary (see *System Suitability* under *Chromatography* ⟨621⟩).

Internal standard solution—Prepare a solution of acetaminophen in *Mobile phase* having a concentration of about 5 mg per mL.

Standard preparation—Transfer about 50 mg of USP Aminosalicylic Acid RS, accurately weighed, to a 100-mL low-actinic volumetric flask, add 50 mL of *Mobile phase*, and swirl to dissolve. Add 10.0 mL of *Internal standard solution*, dilute with *Mobile phase* to volume, and mix.

Assay preparation—Prepare as directed for *Standard preparation*, except to use Aminosalicylic Acid instead of USP Aminosalicylic Acid RS.

Chromatographic system (see *Chromatography* ⟨621⟩)—The chromatograph is equipped with a 254-nm detector and a 4.6-mm × 25-cm column that contains packing L1. The flow rate is about 1.5 mL per minute. Chromatograph replicate injections of the *Standard preparation*, and record the peak responses as directed under *Procedure:* the relative standard deviation of the ratios of the response of the aminosalicylic acid peak to the response of the acetaminophen peak is not more than 1.0%, and the resolution, R, between aminosalicylic acid and acetaminophen is not less than 1.7.

Procedure—[NOTE—After use, wash the column for 30 minutes with a filtered and degassed mixture of methanol, water, and phosphoric acid (77:23:0.6), and then wash for 30 minutes with a filtered and degassed mixture of methanol and water (50:50).] Separately inject equal volumes (about 20 µL) of the *Standard preparation* and the *Assay preparation* into the chromatograph, record the chromatograms, and measure the responses for the major peaks: the relative retention times are about 0.83 for acetaminophen and 1.0 for aminosalicylic acid. Calculate the quantity, in mg, of $C_7H_7NO_3$ in the Aminosalicylic Acid taken by the formula:

$$100C(R_U/R_S),$$

in which C is the concentration, in mg per mL, of USP Aminosalicylic Acid RS in the *Standard preparation*, and R_U and R_S are the ratios of the response of the aminosalicylic acid peak to the response of the acetaminophen peak obtained from the *Assay preparation* and the *Standard preparation*, respectively.

Aminosalicylic Acid Tablets

» Aminosalicylic Acid Tablets contain not less than 95.0 percent and not more than 105.0 percent of the labeled amount of $C_7H_7NO_3$.

Packaging and storage—Preserve in tight, light-resistant containers, at a temperature not exceeding 30°.

Reference standards—*USP Aminosalicylic Acid Reference Standard*—Dry in vacuum at 50° for 1 hour before using. *USP m-Aminophenol Reference Standard*—Do not dry before using.

Identification—Macerate a portion of powdered Tablets, equivalent to about 2 g of aminosalicylic acid, with 50 mL of a mixture of 1 volume of acetone and 2 volumes of chloroform, and filter. Evaporate the filtrate with the aid of a current of warm air to dryness: the residue so obtained responds to *Identification tests B* and *C* under *Aminosalicylic Acid*.

Dissolution ⟨711⟩—

Medium: pH 7.5 phosphate buffer (see *Buffer Solutions* in the section, *Reagents, Indicators, and Solutions*); 900 mL.

Apparatus 1: 100 rpm.

Time: 45 minutes.

Procedure—Determine the amount of $C_7H_7NO_3$ dissolved, employing the procedure set forth in the *Assay*, making any necessary modifications.

Tolerances—Not less than 75% (Q) of the labeled amount of $C_7H_7NO_3$ is dissolved in 45 minutes.

Uniformity of dosage units ⟨905⟩: meet the requirements.

***m*-Aminophenol**—

Mobile phase and *Internal standard solution*—Prepare as directed in the *Assay* under *Aminosalicylic Acid*.

Standard preparation and *Chromatographic system*—Prepare as directed in the test for *m-Aminophenol* under *Aminosalicylic Acid*.

Test preparation—Use the *Assay preparation*, prepared as directed in the *Assay*.

Procedure—Proceed as directed for *Procedure* in the test for *m-Aminophenol* under *Aminosalicylic Acid*. Calculate the percentage of *m*-aminophenol, in relation to the quantity of aminosalicylic acid in the portion of Tablets taken by the formula:

$$100(C/W)(R_U/R_S),$$

in which C is the concentration, in µg per mL, of USP *m*-Aminophenol RS in the *Standard preparation*, W is the quantity of aminosalicylic acid, in mg, in the portion of Tablets taken, as determined in the *Assay*, and R_U and R_S are the ratios of the response of the *m*-aminophenol peak to the response of the sulfanilamide peak obtained from the *Test preparation* and the *Standard preparation*, respectively: not more than 1.0% of *m*-aminophenol is found.

Assay—

Mobile phase, Internal standard solution, Standard preparation, and *Chromatographic system*—Prepare as directed in the *Assay* under *Aminosalicylic Acid*.

Assay preparation—Weigh and finely powder not less than 20 Aminosalicylic Acid Tablets. Transfer an accurately weighed portion of the powder, equivalent to about 500 mg of aminosalicylic acid, to a 100-mL low-actinic volumetric flask. Add 50 mL of *Mobile phase*, and shake for about 5 minutes. Dilute with *Mobile phase* to volume, and mix. Filter, and transfer 10.0 mL of the clear filtrate to a 100-mL low-actinic volumetric flask containing 10.0 mL of *Internal standard solution*, dilute with *Mobile phase* to volume, and mix.

Procedure—Proceed as directed for *Procedure* in the *Assay* under *Aminosalicylic Acid*. Calculate the quantity, in mg, of $C_7H_7NO_3$ in the portion of Tablets taken by the formula:

$$1000C(R_U/R_S),$$

in which C is the concentration, in mg per mL, of USP Aminosalicylic Acid RS in the *Standard preparation*, and R_U and R_S are the ratios of the response of the aminosalicylic acid peak to the response of the acetaminophen peak obtained from the *Assay preparation* and the *Standard preparation*, respectively.

Amitriptyline Hydrochloride

CHCH$_2$CH$_2$N(CH$_3$)$_2$ · HCl

$C_{20}H_{23}N.HCl$ 313.87

1-Propanamine, 3-(10,11-dihydro-5*H*-dibenzo[*a,d*]cyclohepten-5-ylidene)-*N,N*-dimethyl-, hydrochloride.

10,11-Dihydro-*N,N*-dimethyl-5*H*-dibenzo[*a,d*]cycloheptene-$\Delta^{5,\gamma}$-propylamine hydrochloride [*549-18-8*].

» Amitriptyline Hydrochloride contains not less than 99.0 percent and not more than 100.5 percent of $C_{20}H_{23}N.HCl$, calculated on the dried basis.

Packaging and storage—Preserve in well-closed containers.

Reference standard—*USP Amitriptyline Hydrochloride Reference Standard*—Dry at a pressure not exceeding 5 mm of mercury at 60° to constant weight before using.

Identification—

A: The infrared absorption spectrum of a potassium bromide dispersion of it, previously dried, exhibits maxima only at the same wavelengths as that of a similar preparation of USP Amitriptyline Hydrochloride RS.

B: The ultraviolet absorption spectrum of a 1 in 100,000 solution in methanol exhibits a maximum at the same wavelength as that of a similar solution of USP Amitriptyline Hydrochloride RS, concomitantly measured, and the respective absorptivities, calculated on the dried basis, at the wavelength of maximum absorbance at about 239 nm do not differ by more than 3.0%.

C: It responds to the tests for *Chloride* ⟨191⟩.

Melting range ⟨741⟩: between 195° and 199°.

pH ⟨791⟩: between 5.0 and 6.0, in a solution (1 in 100).

Loss on drying ⟨731⟩—Dry it at a pressure not exceeding 5 mm of mercury at 60° to constant weight: it loses not more than 0.5% of its weight.

Residue on ignition ⟨281⟩: not more than 0.1%.

Heavy metals, *Method II* ⟨231⟩: 0.001%.

Chromatographic purity—

Standard preparations—Dissolve USP Amitriptyline Hydrochloride RS in methanol, and mix to obtain a solution having a known concentration of 0.8 mg per mL. Dilute this solution quantitatively with methanol to obtain *Standard preparations*, designated below by letter, having the following compositions:

Dilution	Concentration (μg RS per mL)	Percentage (%, for comparison with test specimen)
A (1 in 2)	400	1.0
B (1 in 4)	200	0.5
C (1 in 5)	160	0.4
D (1 in 10)	80	0.2
E (1 in 20)	40	0.1

Test preparation—Dissolve an accurately weighed quantity of Amitriptyline Hydrochloride in methanol to obtain a solution containing 40 mg per mL.

Procedure—On a suitable thin-layer chromatographic plate (see *Chromatography* ⟨621⟩), coated with a 0.25-mm layer of chromatographic silica gel mixture, apply separately 10 μL of the *Test preparation* and 10 μL of each *Standard preparation.* Allow the applications to dry, position the plate in a chromatographic chamber, and develop the chromatograms in a solvent system consisting of a mixture of chloroform, methanol, and ammonium hydroxide (135:15:1) until the solvent front has moved about three-fourths of the length of the plate. Remove the plate from the developing chamber, mark the solvent front, and allow the solvent to evaporate. Examine the plate under short-wavelength ultraviolet light. Compare the intensities of any secondary spots observed in the chromatogram of the *Test preparation* with those of the principal spots in the chromatograms of the *Standard preparations.* [NOTE—Disregard any spots observed at the origins of the chromatograms]. No secondary spot from the chromatogram of the *Test preparation* is larger or more intense than the principal spot obtained from *Standard preparation B* (0.5%), and the sum of the intensities of all secondary spots obtained from the *Test preparation* corresponds to not more than 1.0%. Disregard any spot in the chromatogram of the *Test preparation* that is smaller or less intense than the principal spot obtained from *Standard preparation E* (0.1%).

Assay—Dissolve about 1 g of Amitriptyline Hydrochloride, accurately weighed, in 30 mL of glacial acetic acid, warming slightly if necessary to effect solution. Cool, add 10 mL of mercuric acetate TS, then add crystal violet TS, and titrate with 0.1 *N* perchloric acid VS to a green end-point. Perform a blank determination, and make any necessary correction. Each mL of 0.1 *N* perchloric acid is equivalent to 31.39 mg of $C_{20}H_{23}N.HCl$.

Amitriptyline Hydrochloride Injection

» Amitriptyline Hydrochloride Injection is a sterile solution of Amitriptyline Hydrochloride in Water for Injection. It contains not less than 90.0 percent and not more than 110.0 percent of the labeled amount of $C_{20}H_{23}N.HCl$.

Packaging and storage—Preserve in single-dose or in multiple-dose containers, preferably of Type I glass.

Reference standard—*USP Amitriptyline Hydrochloride Reference Standard*—Dry at a pressure not exceeding 5 mm of mercury at 60° to constant weight before using.

Identification—

A: Pipet 1 mL of Injection into a 125-mL separator containing 10 mL of water and 1 mL of 1 *N* sodium hydroxide, mix, extract with two 10-mL portions of methylene chloride, and evaporate the extracts on a steam bath just to dryness. Dissolve the residue in methanol, add 1 mL of 1.2 *N* hydrochloric acid, then add methanol to make 100 mL. Dilute 10 mL of this solution with methanol to 100 mL: the ultraviolet absorption spectrum of this solution exhibits a maximum at the same wavelength as that of a similar solution of USP Amitriptyline Hydrochloride RS, concomitantly measured.

B: The retention time of the major peak in the chromatogram of the *Assay preparation* corresponds to that in the chromatogram of the *Standard preparation,* as obtained in the *Assay.*

C: Pipet a volume of Injection, equivalent to 50 mg of amitriptyline hydrochloride, into a separator containing 25 mL of water. Proceed as directed in the test for *Identification—Organic Nitrogenous Bases* ⟨181⟩, beginning with "In a second separator," and using water in place of 0.01 *N* hydrochloric acid in the Reference Standard solution: the solution of the test specimen so obtained meets the requirements of the test for *Identification—Organic Nitrogenous Bases* ⟨181⟩.

Pyrogen—Amitriptyline Hydrochloride Injection, diluted with Sodium Chloride Injection containing 0.9 percent of NaCl to a concentration of 2.5 mg of amitriptyline hydrochloride per mL, meets the requirements of the *Pyrogen Test* ⟨151⟩, the test dose being 1 mL per kg.

pH ⟨791⟩: between 4.0 and 6.0.

Other requirements—It meets the requirements under *Injections* ⟨1⟩.

Assay—

Phosphate buffer—Dissolve 11.04 g of monobasic sodium phosphate in 900 mL of water, adjust with phosphoric acid to a pH of 2.5 ± 0.5, dilute with water to make 1000 mL, and mix.

Mobile phase—Prepare a filtered and degassed mixture of *Phosphate buffer* and acetonitrile (58:42). Make adjustments if necessary (see *System Suitability* under *Chromatography* ⟨621⟩).

Standard preparation—Dissolve an accurately weighed quantity of USP Amitriptyline Hydrochloride RS in water, and dilute quantitatively with water to obtain a solution having a known concentration of about 0.2 mg per mL.

Assay preparation—Transfer an accurately measured volume of Injection, equivalent to about 50 mg of amitriptyline hydrochloride, to a 250-mL volumetric flask, dilute with water to volume, and mix.

Chromatographic system (see *Chromatography* ⟨621⟩)—The liquid chromatograph is equipped with a 254-nm detector and a 4-mm × 30-cm column that contains packing L1. The flow rate is about 2 mL per minute. Chromatograph the *Standard preparation*, and record the peak responses as directed under *Procedure:* the tailing factor is not more than 2.0, the column efficiency is not less than 800 theoretical plates, and the relative standard deviation for replicate injections is not more than 2.0%.

Procedure—Separately inject equal volumes (about 20 µL) of the *Standard preparation* and the *Assay preparation* into the chromatograph, record the chromatograms, and measure the responses for the major peaks. Calculate the quantity, in mg, of $C_{20}H_{23}N.HCl$ in each mL of the Injection taken by the formula:

$$250(C/V)(r_U/r_S),$$

in which C is the concentration, in mg per mL, of USP Amitriptyline Hydrochloride RS in the *Standard preparation*, V is the volume, in mL, of Injection taken, and r_U and r_S are the peak responses obtained from the *Assay preparation* and the *Standard preparation*, respectively.

Amitriptyline Hydrochloride Tablets

» Amitriptyline Hydrochloride Tablets contain not less than 90.0 percent and not more than 110.0 percent of the labeled amount of $C_{20}H_{23}N.HCl$.

Packaging and storage—Preserve in well-closed containers.

Reference standard—*USP Amitriptyline Hydrochloride Reference Standard*—Dry at a pressure not exceeding 5 mm of mercury at 60° to constant weight before using.

Identification—

A: Transfer a quantity of finely powdered Tablets, equivalent to about 10 mg of amitriptyline hydrochloride, to a 100-mL volumetric flask, add 50 mL of methanol, shake well, then add methanol to volume. Filter a portion of this solution, and dilute 10 mL of the filtrate with methanol to 100 mL: the ultraviolet absorption spectrum of this solution exhibits a maximum at the same wavelength as that of a similar solution of USP Amitriptyline Hydrochloride RS, concomitantly measured.

B: The retention time of the major peak in the chromatogram of the *Assay preparation* corresponds to that in the chromatogram of the *Standard preparation*, as obtained in the *Assay*.

Dissolution ⟨711⟩—

Medium: 0.1 *N* hydrochloric acid; 900 mL.

Apparatus 1: 100 rpm.

Time: 45 minutes.

Procedure—Determine the amount of $C_{20}H_{23}N.HCl$ dissolved from ultraviolet absorbances at the wavelength of maximum absorbance at about 239 nm of filtered portions of the solution under test, suitably diluted with *Dissolution Medium*, if necessary, in comparison with a Standard solution having a known concentration of USP Amitriptyline Hydrochloride RS in the same medium.

Tolerances—Not less than 75% (*Q*) of the labeled amount of $C_{20}H_{23}N.HCl$ is dissolved in 45 minutes.

Uniformity of dosage units ⟨905⟩: meet the requirements.

Assay—

Phosphate buffer, Mobile phase, and *Chromatographic system*—Prepare as directed in the *Assay* under *Amitriptyline Hydrochloride Injection.*

Standard preparation—Dissolve an accurately weighed quantity of USP Amitriptyline Hydrochloride RS in *Mobile phase*, and dilute quantitatively with *Mobile phase* to obtain a solution having a known concentration of about 0.2 mg per mL.

Assay preparation—Transfer 20 Amitriptyline Hydrochloride Tablets to a 500-mL volumetric flask, add 250 mL of *Mobile phase*, and shake the mixture for 1 hour or until the tablets have disintegrated. Add *Mobile phase* to volume, mix, and filter. Dilute an accurately measured volume (V_F mL) of the clear filtrate quantitatively with *Mobile phase* to obtain a solution (V_A mL) containing about 0.2 mg of amitriptyline hydrochloride per mL.

Procedure—Separately inject equal volumes (about 20 µL) of the *Standard preparation* and the *Assay preparation* into the chromatograph, record the chromatograms, and measure the responses for the major peaks. Calculate the quantity, in mg, of $C_{20}H_{23}N.HCl$ in each Tablet taken by the formula:

$$500(C/20)(V_A/V_F)(r_U/r_S),$$

in which C is the concentration, in mg per mL, of USP Amitriptyline Hydrochloride RS in the *Standard preparation*, and r_U and r_S are the peak responses obtained from the *Assay preparation* and the *Standard preparation*, respectively.

Amitriptyline Hydrochloride Tablets, Chlordiazepoxide and—*see* Chlordiazepoxide and Amitriptyline Hydrochloride Tablets

Amitriptyline Hydrochloride Tablets, Perphenazine and—*see* Perphenazine and Amitriptyline Hydrochloride Tablets

Ammonia Solution, Strong—*see* Ammonia Solution, Strong NF

Aromatic Ammonia Spirit

» Aromatic Ammonia Spirit is a hydroalcoholic solution that contains, in each 100 mL, not less than 1.7 g and not more than 2.1 g of total NH_3, and Ammonium Carbonate corresponding to not less than 3.5 g and not more than 4.5 g of $(NH_4)_2CO_3$.

Packaging and storage—Preserve in tight, light-resistant containers, at a temperature not exceeding 30°.

Alcohol content, *Method I* ⟨611⟩: between 62.0% and 68.0% of C_2H_5OH.

Assay for total NH_3—Transfer 10.0 mL to a 250-mL conical flask containing about 50 mL of water. Add 30.0 mL of 0.5 *N* sulfuric acid VS, and boil until the solution becomes clear. Cool, add methyl red TS, and titrate the excess acid with 0.5 *N* sodium hydroxide VS. Perform a blank determination (see *Residual Titrations* under *Titrimetry* ⟨541⟩). Each mL of 0.5 *N* sulfuric acid is equivalent to 8.515 mg of NH_3.

Assay for ammonium carbonate—Transfer 10.0 mL to a flask of about 300-mL capacity. Add 30 mL of 0.5 *N* sodium hydroxide, and boil the mixture, replacing the water lost by evaporation, until the vapors no longer turn moistened red litmus paper blue. Cool, dilute with 100 mL of cold, carbon dioxide–free water, add about 6 drops of phenolphthalein TS, then add just enough 0.5 *N* sulfuric acid VS to discharge the color of the phenolphthalein. Add methyl orange TS, and titrate with 0.5 *N* sulfuric acid VS. Perform a blank determination (see *Residual Titrations* under *Titrimetry* ⟨541⟩). Each mL of 0.5 *N* sulfuric acid consumed in the titration with methyl orange TS is equivalent to 48.04 mg of $(NH_4)_2CO_3$.

Ammoniated Mercury—*see* Mercury, Ammoniated

Ammoniated Mercury Ointment—*see* Mercury Ointment, Ammoniated

Ammoniated Mercury Ophthalmic Ointment—*see* Mercury Ophthalmic Ointment, Ammoniated

Ammonium Carbonate—*see* Ammonium Carbonate NF

Ammonium Chloride

NH_4Cl 53.49
Ammonium chloride.
Ammonium chloride [*12125-02-9*].

» Ammonium Chloride contains not less than 99.5 percent and not more than 100.5 percent of NH_4Cl, calculated on the dried basis.

Packaging and storage—Preserve in tight containers.

Identification—A solution (1 in 10) responds to the tests for *Ammonium* ⟨191⟩ and for *Chloride* ⟨191⟩.

pH ⟨791⟩: between 4.6 and 6.0, in a solution (1 in 20).

Loss on drying ⟨731⟩—Dry it over silica gel for 4 hours: it loses not more than 0.5% of its weight.

Residue on ignition ⟨281⟩—Add 1 mL of sulfuric acid to about 2 g, accurately weighed, and heat the mixture gently until volatilization is complete: the residue is white, and when ignited, not more than 0.1% of nonvolatile substance remains.

Thiocyanate—Acidify 10 mL of a solution (1 in 10) with hydrochloric acid, and add a few drops of ferric chloride TS: no orange-red color is produced.

Heavy metals, *Method I* ⟨231⟩: 0.001%.

Assay—Weigh accurately about 100 mg of Ammonium Chloride, and dissolve in about 100 mL of water in a porcelain casserole. Add 1 mL of dichlorofluorescein TS, mix, and titrate with 0.1 *N* silver nitrate VS until the silver chloride flocculates and the mixture turns a faint pink color. Each mL of 0.1 *N* silver nitrate is equivalent to 5.349 mg of NH_4Cl.

Ammonium Chloride Injection

» Ammonium Chloride Injection is a sterile solution of Ammonium Chloride in Water for Injection. It contains not less than 95.0 percent and not more than 105.0 percent of the labeled amount of NH_4Cl. Hydrochloric acid may be added to adjust the pH.

Packaging and storage—Preserve in single-dose or in multiple-dose containers, preferably of Type I or Type II glass.

Labeling—The label states the content of ammonium chloride in terms of weight and of milliequivalents in a given volume.
The label states also the total osmolar concentration in mOsmol per liter or per mL. The label states that the Injection is not for direct injection but is to be diluted with Sodium Chloride Injection to the appropriate strength before use.

Identification—It responds to the tests for *Ammonium* ⟨191⟩ and for *Chloride* ⟨191⟩.

Pyrogen—It meets the requirements of the *Pyrogen Test* ⟨151⟩, the test dose being 5 mL per kg of a solution containing not more than 10 mg of ammonium chloride in each mL, injected over a period of 1 minute.

pH ⟨791⟩: between 4.0 and 6.0, in a concentration of not more than 100 mg of ammonium chloride per mL.

Particulate matter ⟨788⟩: meets the requirements under *Small-volume Injections*.

Chloride content—Transfer an accurately measured volume of Injection, evaporated, if necessary, equivalent to about 2 g of ammonium chloride, to a 200-mL volumetric flask, dilute with water to volume, and mix. Pipet 10 mL of this solution into a porcelain casserole, dilute with water to about 100 mL, and add 1 mL of dichlorofluorescein TS. Mix, and titrate with 0.1 *N* silver nitrate VS until the silver chloride flocculates and the mixture turns a faint pink color. Each mL of 0.1 *N* silver nitrate is equivalent to 3.545 mg of Cl. The content of Cl is between 63.0% and 70.3% of the labeled amount of ammonium chloride.

Other requirements—It meets the requirements under *Injections* ⟨1⟩.

Assay—Transfer an accurately measured volume of Ammonium Chloride Injection, equivalent to about 200 mg of ammonium chloride, to a 500-mL Kjeldahl flask, dilute with water to 200 mL, mix, and add 50 mL of sodium hydroxide solution (2 in 5). Immediately connect the flask by means of a distillation trap to a well-cooled condenser, the delivery tube of which dips into 40 mL of boric acid solution (1 in 25) contained in a suitable receiver. Heat to boiling, and distil about 200 mL. Cool the liquid in the receiver, if necessary, then add methyl red TS, and titrate with 0.1 *N* sulfuric acid VS. Perform a blank determination, and make any necessary correction. Each mL of 0.1 *N* sulfuric acid is equivalent to 5.349 mg of NH_4Cl.

Ammonium Chloride Oral Solution, Potassium Gluconate, Potassium Citrate, and—*see* Potassium Gluconate, Potassium Citrate, and Ammonium Chloride Oral Solution

Ammonium Chloride Delayed-release Tablets

» Ammonium Chloride Delayed-release Tablets contain not less than 94.0 percent and not more than 106.0 percent of the labeled amount of NH_4Cl. Ammonium Chloride Delayed-release Tablets are enteric-coated.

Packaging and storage—Preserve in tight containers.

Identification—A filtered solution of finely powdered Tablets, equivalent to ammonium chloride solution (1 in 10), responds to the tests for *Ammonium* ⟨191⟩ and for *Chloride* ⟨191⟩.

Disintegration ⟨701⟩: 2 hours, determined as directed for *Enteric-coated Tablets*.

Thiocyanate—Powder and dissolve in water a sufficient number of Tablets to make about 25 mL of ammonium chloride solution (1 in 10), and filter. Acidify 10 mL of the solution with hydrochloric acid, and add a few drops of ferric chloride TS: no reddish orange color is produced.

Assay—Weigh and finely powder not less than 20 Ammonium Chloride Delayed-release Tablets. Transfer an accurately weighed portion of the powder, equivalent to about 200 mg of ammonium chloride, to a 500-mL Kjeldahl flask, and add 200 mL of water and 50 mL of sodium hydroxide solution (2 in 5). Immediately connect the flask by means of a distillation trap to a well-cooled condenser, the delivery tube of which dips into 40 mL of boric acid solution (1 in 25) contained in a suitable receiver. Heat to boiling, and distil about 200 mL. Cool the liquid in the receiver, if necessary, then add methyl red TS, and titrate with 0.1 *N* sulfuric acid VS. Perform a blank determination, and make any necessary correction. Each mL of 0.1 *N* sulfuric acid is equivalent to 5.349 mg of NH_4Cl.

Ammonium Molybdate

$(NH_4)_6Mo_7O_{24}.4H_2O$ 1235.86
Molybdate ($Mo_7O_{24}^{6-}$), hexaammonium, tetrahydrate.
Hexaammonium molybdate tetrahydrate
[*12054-85-2*].

» Ammonium Molybdate contains not less than 99.3 percent and not more than 101.8 percent of $(NH_4)_6Mo_7O_{24}.4H_2O$.

Packaging and storage—Preserve in tight containers.

Identification—Dissolve 0.1 g in 5 mL of water, and add 1 mL of dibasic sodium phosphate TS: a yellow precipitate is formed, and it is soluble in an excess of 6 *N* ammonium hydroxide.

Insoluble substances—Dissolve 20 g in 200 mL of water in a beaker, heat to boiling, cover, and heat on a steam bath for 1 hour. Filter the hot solution through a tared filtering crucible, wash the insoluble residue with hot water, and dry at 105° for 2 hours: the weight of the residue so obtained does not exceed 1 mg (0.005%).

Chloride ⟨221⟩—A 0.5-g portion shows no more chloride than 0.30 mL of 0.001 *N* hydrochloric acid (0.002%).

Nitrate—Dissolve 1 g in 10 mL of water containing 5 mg of sodium chloride, and add 0.10 mL of a 1 in 1000 solution of indigo carmine in 3.6 *N* sulfuric acid: the blue color is not completely discharged in 5 minutes.

Sulfate ⟨221⟩—A 0.25-g portion shows no more sulfate than corresponds to 1.0 mL of 0.001 *N* sulfuric acid (0.02%).

Arsenate, phosphate, and silicate—Dissolve 2.5 g in 70 mL of water in a container other than one of glass. Adjust with 1.2 *N* hydrochloric acid to a pH of between 3 and 4, transfer to a glass container, add 2 mL of bromine TS, and adjust with 1.2 *N* hydrochloric acid to a pH of 1.8 ± 0.1. Heat almost to boiling, and cool to room temperature. Dilute with water to 90 mL, add 10 mL of hydrochloric acid, and transfer to a separator. Add 1 mL of butyl alcohol and 30 mL of 4-methyl-2-pentanone, shake vigorously, and allow the phases to separate. Discard the aqueous phase, and wash the ketone phase with three successive 10-mL portions of 1.2 *N* hydrochloric acid, discarding the washings. To the washed ketone phase add 10 mL of 1.2 *N* hydrochloric acid to which has just been added 0.2 mL of a freshly prepared 1 in 50 solution of stannous chloride in hydrochloric acid. Similarly treat a control solution prepared by dissolving 0.5 g of specimen in 70 mL of water and adding 2 mL of sodium silicate solution (1 in 20,000): any blue color in the test solution does not exceed that in the control solution.

Phosphate—Dissolve 20 g in 100 mL of 3 *N* ammonium hydroxide, add 3.5 mL of ferric nitrate solution (1 in 10), and allow to stand for about 15 minutes. Warm gently to coagulate the precipitate, and filter. Wash the filter several times with 1.5 *N* ammonium hydroxide. Then wash the filter with 60 mL of warm 4 *N* nitric acid to dissolve the residue on the filter, collecting the filtrate in a glass-stoppered, 250-mL conical flask. Add 13 mL of ammonium hydroxide, warm to 40°, add 50 mL of ammonium molybdate TS, shake for 5 minutes, and allow to stand at 40° for 2 hours. Similarly treat 100 mL of a Standard solution prepared by dissolving 143.3 mg of dried monobasic potassium phosphate in water to make 1000 mL, and then diluting 1.0 mL of this solution with 3 *N* ammonium hydroxide to 100 mL: any yellow precipitate formed from the test solution does not exceed that obtained from the Standard solution (5 ppm).

Magnesium and alkali salts—Dissolve 5.0 g in 50 mL of water, and filter. To the filtrate add 0.5 g of sodium carbonate and 25 mL of 2.5 *N* sodium hydroxide. Boil the solution gently for 5 minutes, cool, and filter through an ignited and tared filter. Wash the filter with 1 *N* ammonium hydroxide. Ignite the filter at 800 ± 25° for 30 minutes: the weight of the residue so obtained does not exceed 1 mg (0.02%).

Heavy metals—Dissolve 2.0 g in about 20 mL of water, add 10 mL of 2.5 *N* sodium hydroxide and 2 mL of ammonium hydroxide, and dilute with water to 40 mL. To 10 mL of this stock solution add 1.0 mL of *Standard Lead Solution*, prepared as directed under *Heavy Metals* ⟨231⟩, and dilute with water to 40 mL (*Control solution*). Dilute the remaining 30-mL portion of the stock solution with water to 40 mL (*Test solution*). To both solutions add 10 mL of freshly prepared hydrogen sulfide TS: any color in the *Test solution* does not exceed that in the *Control solution* (0.001%).

Assay—Dissolve about 1 g of Ammonium Molybdate, accurately weighed, in a mixture of 10 mL of water and 1 mL of ammonium hydroxide in a 250-mL volumetric flask, dilute with water to volume, and mix. Filter the solution, and transfer 50.0 mL of the filtrate to a 600-mL beaker. Add 250 mL of water, 20 g of ammonium chloride, 15 mL of hydrochloric acid, and 0.15 mL of methyl orange TS, heat nearly to boiling, and add 18 mL of lead acetate TS. To the hot solution add, slowly and with constant stirring, a saturated solution of ammonium acetate until the color turns yellow, and then add 15 mL of lead acetate TS. Cover the beaker, and heat just below the boiling temperature until the precipitate has settled. Filter through a tared, porous porcelain crucible, wash with seven or eight successive portions of a mixture of water, saturated ammonium acetate solution, and nitric acid (890:100:10), and finally wash with three successive portions of hot water. Ignite to constant weight at 560° to 625°. Each mg of the lead molybdate so obtained is equivalent to 0.4809 mg of $(NH_4)_6Mo_7O_{24} \cdot 4H_2O$.

Ammonium Molybdate Injection

» Ammonium Molybdate Injection is a sterile solution of Ammonium Molybdate in Water for Injection. It contains not less than 85.0 percent and not more than 115.0 percent of the labeled amount of molybdenum (Mo).

Packaging and storage—Preserve in single-dose or in multiple-dose containers, preferably of Type I or Type II glass.

Labeling—Label the Injection to indicate that it is to be diluted to the appropriate strength with Sterile Water for Injection or other suitable fluid prior to administration.

Identification—

A: The *Assay preparation*, prepared as directed in the *Assay*, exhibits an absorption maximum at about 313 nm when tested as directed for *Procedure* in the *Assay*.

B: Add 0.3 mL of alkaline mercuric–potassium iodide TS to 5 mL of Injection: a reddish brown color develops.

C: Evaporate 50 mL of Injection on a steam bath to a volume of about 1 mL, and add 1 mL of monobasic potassium phosphate solution (1 in 10): a yellow precipitate is formed, and it dissolves in an excess of 6 *N* ammonium hydroxide.

Pyrogen—It meets the requirements of the *Pyrogen Test* ⟨151⟩, the test dose being 10 mL of Injection per kg.

pH ⟨791⟩: between 3.0 and 6.0.

Particulate matter ⟨788⟩: meets the requirements under *Small-volume Injections*.

Other requirements—It meets the requirements under *Injections* ⟨1⟩.

Assay—

Ammonium hydroxide diluent—Dilute 40 mL of ammonium hydroxide with water to 1000 mL. Store in a plastic bottle.

Sodium sulfate solution—Dissolve 1 g of sodium sulfate in water to make 100 mL.

Molybdenum stock solution—Transfer about 1.84 g of previously assayed Ammonium Molybdate, accurately weighed, to a 1000-mL volumetric flask, dilute with *Ammonium hydroxide diluent* to volume, and mix. This solution contains the equivalent of 1000 µg of molybdenum per mL.

Standard preparations—Transfer 0, 1.0, 2.0, 3.0, and 4.0 mL, respectively, of *Molybdenum stock solution* to separate 100-mL volumetric flasks, and to the respective flasks add 5.0, 4.0, 3.0, 2.0, and 1.0 mL of *Ammonium hydroxide diluent*. To each flask add 10 mL of *Sodium sulfate solution*, dilute with water to volume, and mix. These *Standard preparations* contain, respectively, 0, 10, 20, 30, and 40 µg of molybdenum per mL.

Assay preparation—Transfer an accurately measured volume of Ammonium Molybdate Injection, equivalent to about 500 µg of molybdenum, to a 25-mL volumetric flask, add 1.25 mL of *Ammonium hydroxide diluent* and 2.5 mL of *Sodium sulfate solution*, dilute with water to volume, and mix.

Procedure—Concomitantly determine the absorbances of the *Standard preparations* and the *Assay preparation* at the molybdenum emission line of 313.3 nm with a suitable atomic absorption spectrophotometer (see *Spectrophotometry and Light-scattering* ⟨851⟩) equipped with a molybdenum hollow-cathode lamp and a nitrous oxide–acetylene reducing flame, using water as the blank. Plot the absorbances of the *Standard preparations* versus concentration, in µg per mL, of molybdenum, and draw the straight line best fitting the five plotted points. From the graph so obtained, determine the concentration, in µg per mL, of molybdenum in the *Assay preparation.* Calculate the quantity, in µg, of molybdenum (Mo) in each mL of the Injection taken by the formula:

$$25C/V,$$

in which C is the concentration, in µg per mL, of molybdenum in the *Assay preparation*, and V is the volume, in mL, of Injection taken.

Ammonium Phosphate—*see* Ammonium Phosphate NF

Amobarbital

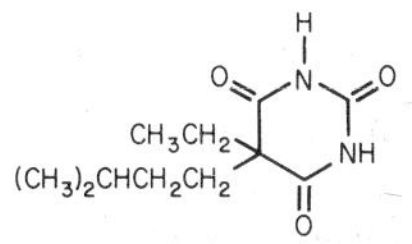

$C_{11}H_{18}N_2O_3$ 226.27
2,4,6(1*H*,3*H*,5*H*)-Pyrimidinetrione, 5-ethyl-5-(3-methylbutyl)-.
5-Ethyl-5-isopentylbarbituric acid [*57-43-2*].

» Amobarbital contains not less than 98.5 percent and not more than 101.0 percent of $C_{11}H_{18}N_2O_3$, calculated on the dried basis.

Packaging and storage—Preserve in well-closed containers.

Reference standard—*USP Amobarbital Reference Standard*—Dry at 105° for 4 hours before using.

Identification—

A: The infrared absorption spectrum of a potassium bromide dispersion of it exhibits maxima only at the same wavelengths as that of a similar preparation of USP Amobarbital RS.

B: The ultraviolet absorption spectrum of a 1 in 100,000 solution of it in a 1 in 20 solution of alcohol in pH 9.6 alkaline borate buffer (see under *Solutions* in the section, *Reagents, Indicators, and Solutions*) exhibits maxima and minima at the same wavelengths as that of a similar solution of USP Amobarbital RS, concomitantly measured, and the respective absorptivities, calculated on the dried basis, at the wavelength of maximum absorbance at about 239 nm do not differ by more than 3.0%.

C: Boil 0.2 g with 10 mL of 1 *N* sodium hydroxide: ammonia is evolved.

Melting range, *Class Ia* ⟨741⟩: between 156° and 161°, but the range between beginning and end of melting does not exceed 3°.

Loss on drying ⟨731⟩—Dry it at 105° for 4 hours: it loses not more than 1.0% of its weight.

Residue on ignition ⟨281⟩: not more than 0.1%.

Assay—Dissolve about 450 mg of Amobarbital, accurately weighed, in 60 mL of dimethylformamide in a 125-mL conical flask. Add 4 drops of thymol blue TS, and titrate with 0.1 *N* sodium methoxide VS, using a magnetic stirrer and taking precautions against the absorption of atmospheric carbon dioxide. Perform a blank determination, and make any necessary correction. Each mL of 0.1 *N* sodium methoxide is equivalent to 22.63 mg of $C_{11}H_{18}N_2O_3$.

Amobarbital Tablets

» Amobarbital Tablets contain not less than 90.0 percent and not more than 110.0 percent of the labeled amount of $C_{11}H_{18}N_2O_3$.

Packaging and storage—Preserve in well-closed containers.

Reference standard—*USP Amobarbital Reference Standard*—Dry at 105° for 4 hours before using.

Identification—Triturate a quantity of finely powdered Tablets, equivalent to about 500 mg of amobarbital, with 10 mL of solvent hexane, and decant the liquid as rapidly as possible. Treat the residue again with 5 mL of solvent hexane, decant, and evaporate any hexane remaining in the residue on a steam bath. Dissolve the residue by warming with 10 mL of sodium carbonate TS, and filter. Add hydrochloric acid, dropwise, to the filtrate until no more precipitate is formed. Collect the precipitate on a filter, wash it well with small portions of cold water until the last washing is free from chloride, and dry at 105° for 4 hours: the residue so obtained responds to *Identification test A* under *Amobarbital.*

Dissolution ⟨711⟩—

Medium: pH 7.6 phosphate buffer (see *pH* ⟨791⟩); 500 mL.

Apparatus 1: 100 rpm.

Time: 30 minutes.

Procedure—Determine the amount of $C_{11}H_{18}N_2O_3$ dissolved from ultraviolet absorbances at the wavelength of maximum absorbance at about 239 nm of filtered portions of solution under test, suitably diluted with *Dissolution Medium,* if necessary, using *Dissolution Medium* as the blank, in comparison with a Standard solution having a known concentration of USP Amobarbital RS in the same medium.

Tolerances—Not less than 70% (Q) of the labeled amount of $C_{11}H_{18}N_2O_3$ is dissolved in 30 minutes.

Uniformity of dosage units ⟨905⟩: meet the requirements.

Procedure for content uniformity—Transfer 1 finely powdered Tablet to a 200-mL volumetric flask with the aid of first, 10 mL of alcohol, and then about 150 mL of pH 9.6 alkaline borate buffer (see *Buffer Solutions* in the section, *Reagents, Indicators, and Solutions*). Shake the mixture vigorously, dilute with pH 9.6 alkaline borate buffer to volume, mix, and filter. Dilute the filtrate quantitatively with a 1 in 20 solution of alcohol in pH 9.6 alkaline borate buffer to obtain a Test solution having a concentration of about 10 µg per mL. Concomitantly determine the absorbances of this solution and of a Standard solution of USP Amobarbital RS in the same medium, having a known concentration of about 10 µg per mL, in 1-cm cells, at the wavelength of maximum absorbance at about 239 nm, with a suitable spectrophotometer, using a 1 in 20 solution of alcohol in pH 9.6 alkaline borate buffer as the blank. Calculate the quantity, in mg, of $C_{11}H_{18}N_2O_3$ in the Tablet by the formula:

$$(TC/D)(A_U/A_S),$$

in which T is the labeled quantity, in mg, of amobarbital in the Tablet, C is the concentration, in µg per mL, of USP Amobarbital RS in the Standard solution, D is the concentration, in µg per mL, of amobarbital in the Test solution, on the basis of the labeled quantity per Tablet and the extent of dilution, and A_U and A_S are the absorbances of the Test solution and the Standard solution, respectively.

Assay—

Internal standard—Barbital.

Internal standard solution—Dissolve an accurately weighed quantity of barbital in chloroform, and dilute quantitatively with chloroform to obtain a solution having a known concentration of about 0.6 mg per mL.

Standard preparation—Dissolve accurately weighed quantities of USP Amobarbital RS and barbital in chloroform, and dilute quantitatively with chloroform to obtain a solution that contains, in each mL, known amounts of about 1 mg of USP Amobarbital RS and about 0.6 mg of barbital.

Assay preparation—Weigh and finely powder not less than 20 Amobarbital Tablets. Transfer an accurately weighed portion of the powder, equivalent to about 50 mg of amobarbital, to a separator, add 15 mL of water and 1 mL of hydrochloric acid, and extract with five 25-mL portions of chloroform. Filter the extracts through about 15 g of anhydrous sodium sulfate that is supported on a funnel by a small pledget of glass wool. Collect the combined filtrate in a 200-mL volumetric flask, wash the sodium sulfate with 25 mL of chloroform, collecting the washing with the filtrate, dilute with chloroform to volume, and mix. Combine 4.0 mL of this solution with 1.0 mL of *Internal standard solution* in a suitable container, and reduce the volume to about

1 mL by evaporation, with the aid of a stream of dry nitrogen, at room temperature.

Chromatographic system and *System suitability*—Proceed as directed for *Chromatographic System* and *System Suitability* under *Barbiturate Assay* ⟨361⟩, the resolution, *R*, between amobarbital and barbital being not less than 2.5. [NOTE—Relative retention times are, approximately, 0.5 for barbital and 1.0 for amobarbital.]

Procedure—Proceed as directed for *Procedure* under *Barbiturate Assay* ⟨361⟩. Calculate the quantity, in mg, of $C_{11}H_{18}N_2O_3$ in the portion of Tablets taken by the formula:

$$50(R_U)(Q_S)(C_i)/(R_S).$$

Amobarbital Sodium

$C_{11}H_{17}N_2NaO_3$ 248.26

2,4,6(1*H*,3*H*,5*H*)-Pyrimidinetrione, 5-ethyl-5-(3-methylbutyl)-, monosodium salt.

Sodium 5-ethyl-5-isopentylbarbiturate [*64-43-7*].

» Amobarbital Sodium contains not less than 98.5 percent and not more than 100.5 percent of $C_{11}H_{17}N_2NaO_3$, calculated on the dried basis.

Packaging and storage—Preserve in tight containers.

Reference standard—*USP Amobarbital Reference Standard*—Dry at 105° for 4 hours before using.

Completeness of solution ⟨641⟩—Mix 1.0 g with 10 mL of carbon dioxide–free water: after 1 minute, the solution is clear and free from undissolved solid.

Identification—

A: The infrared absorption spectrum of a potassium bromide dispersion of the residue of amobarbital obtained in the *Assay* exhibits maxima only at the same wavelengths as that of a similar preparation of USP Amobarbital RS.

B: Ignite about 200 mg: the residue effervesces with acid and responds to the tests for *Sodium* ⟨191⟩.

pH ⟨791⟩: between 9.6 and 10.4, in the solution prepared for the test for *Completeness of solution*.

Loss on drying ⟨731⟩—Dry about 1 g, accurately weighed, at 105° for 4 hours: it loses not more than 2.0% of its weight.

Heavy metals, *Method II* ⟨231⟩: 0.003%.

Assay—Dissolve about 500 mg of Amobarbital Sodium, accurately weighed, in about 15 mL of water in a separator. To the solution add 2 mL of hydrochloric acid, shake well, and completely extract the liberated amobarbital with 25-mL portions of chloroform. Test for completeness of extraction by extracting with an additional 10-mL portion of chloroform and evaporating the solvent: not more than 0.5 mg of residue remains. Filter the combined extract through a glass filter funnel into a tared beaker, and wash the separator and the filter with several small portions of chloroform. Evaporate the combined filtrate and washings on a steam bath with the aid of a current of air, dry the residue at 105° for 30 minutes, cool, and weigh. The weight of the residue, multiplied by 1.097, represents the weight of $C_{11}H_{17}N_2NaO_3$.

Amobarbital Sodium Capsules

» Amobarbital Sodium Capsules contain not less than 90.0 percent and not more than 110.0 percent of the labeled amount of $C_{11}H_{17}N_2NaO_3$.

Packaging and storage—Preserve in tight containers.

Reference standard—*USP Amobarbital Reference Standard*—Dry at 105° for 4 hours before using.

Identification—Transfer a quantity of the Capsule contents, equivalent to about 200 mg of amobarbital sodium, to a separator, and dissolve in 25 mL of water. Add 2 mL of hydrochloric acid and 50 mL of chloroform, and shake. Filter the chloroform extract through a funnel containing a pledget of glass wool and about 2 g of anhydrous sodium sulfate, collecting the filtrate in a beaker, and evaporate on a steam bath to dryness: the residue so obtained responds to *Identification test A* under *Amobarbital Sodium*.

Dissolution ⟨711⟩—

Medium: water; 500 mL.

Apparatus 1: 100 rpm.

Time: 60 minutes.

Procedure—Determine the amount of $C_{11}H_{17}N_2NaO_3$ dissolved from ultraviolet absorbances at the wavelength of maximum absorbance at about 239 nm of filtered portions of the solution under test mixed with sufficient sodium hydroxide to provide a concentration of 0.1 *N* sodium hydroxide, and, suitably diluted with *Dissolution Medium*, if necessary, in comparison with a Standard solution having a known concentration of USP Amobarbital RS in 0.1 *N* sodium hydroxide.

Tolerances—Not less than 75% (*Q*) of the labeled amount of $C_{11}H_{17}N_2NaO_3$ is dissolved in 60 minutes.

Uniformity of dosage units ⟨905⟩: meet the requirements.

Assay—Weigh accurately not less than 20 Amobarbital Sodium Capsules. Empty most of the contents of the Capsules into a weighing bottle. Place the nearly emptied Capsules in a beaker, and wash the remaining contents from them by means of small portions of alcohol, followed by small portions of ether. Remove the ether by evaporation, weigh the empty capsules, and calculate the net weight of the Capsule contents. Transfer a quantity of the contents, accurately weighed and equivalent to about 200 mg of amobarbital sodium, to a separator with the aid of about 25 mL of water. When the specimen has dissolved, saturate the solution with sodium chloride. To the solution add 2 mL of hydrochloric acid, shake well, and extract the liberated amobarbital with five 25-mL portions of chloroform. Filter the combined chloroform extracts through a funnel containing a pledget of glass wool and about 2 g of anhydrous sodium sulfate. Collect the filtrate in a beaker, and wash the separator and the filter with several small portions of chloroform. Evaporate the solvent on a steam bath with the aid of a current of air to dryness. Dissolve the amobarbital in the beaker in a mixture of 5 mL of water and 2.5 mL of ammonium hydroxide. Transfer the dissolved amobarbital, with the aid of several small portions of water, to a 500-mL volumetric flask, add water to volume, and mix. Pipet into separate, 200-mL volumetric flasks 5 mL each of this solution and a solution of USP Amobarbital RS, prepared with the aid of dilute ammonium hydroxide (1 in 200) to yield a Standard solution having a known concentration of about 400 µg per mL. To each flask add 0.25 mL of ammonium hydroxide, then add water to volume, and mix. Concomitantly determine the absorbances of both solutions in 1-cm cells at the wavelength of maximum absorbance at about 239 nm, with a suitable spectrophotometer, using water as the blank. Calculate the quantity, in mg, of $C_{11}H_{17}N_2NaO_3$ in the portion of Capsules taken by the formula:

$$0.5(1.097)C(A_U/A_S),$$

in which 1.097 is the ratio of the molecular weight of amobarbital sodium to that of amobarbital, *C* is the concentration, in µg per mL, of USP Amobarbital RS in the Standard solution, and A_U and A_S are the absorbances of the solution from the Capsule contents and the Standard solution, respectively.

Amobarbital Sodium Capsules, Secobarbital Sodium and—*see* Secobarbital Sodium and Amobarbital Sodium Capsules

Sterile Amobarbital Sodium

» Sterile Amobarbital Sodium is Amobarbital Sodium suitable for parenteral use.

Packaging and storage—Preserve in *Containers for Sterile Solids* as described under *Injections* ⟨1⟩.

Reference standard—*USP Amobarbital Reference Standard*—Dry at 105° for 4 hours before using.

Constituted solution—At the time of use, the constituted solution prepared from Sterile Amobarbital Sodium meets the requirements for *Constituted Solutions* under *Injections* ⟨1⟩.

Loss on drying ⟨731⟩—Dry about 1 g, accurately weighed, at 105° for 4 hours: it loses not more than 1.0% of its weight.

Other requirements—It conforms to the definition, responds to the *Identification tests*, and meets the requirements for *Completeness of solution*, *pH*, *Heavy metals*, and *Assay* under *Amobarbital Sodium*. It meets also the requirements for *Sterility Tests* ⟨71⟩, *Uniformity of Dosage Units* ⟨905⟩, and *Labeling* under *Injections* ⟨1⟩.

Amodiaquine

$C_{20}H_{22}ClN_3O$ 355.87

Phenol, 4-[(7-chloro-4-quinolinyl)amino]-2-[(diethylamino)-methyl]-.

4-[(7-Chloro-4-quinolyl)amino]-α-(diethylamino)-*o*-cresol [*86-42-0*].

» Amodiaquine contains not less than 97.0 percent and not more than 103.0 percent of $C_{20}H_{22}ClN_3O$, calculated on the anhydrous basis.

Packaging and storage—Preserve in tight containers.

Reference standard—*USP Amodiaquine Hydrochloride Reference Standard*—Do not dry; determine the water content titrimetrically at the time of use.

Identification—

A: Dissolve 20 mg of USP Amodiaquine Hydrochloride RS in 10 mL of water in a separator, add 1 mL of ammonium hydroxide, and extract by shaking with 25 mL of chloroform. Draw off and evaporate the chloroform extract, and dry the residue at 105° for 2 hours: the infrared absorption spectrum of a potassium bromide dispersion of Amodiaquine exhibits maxima only at the same wavelengths as that of a similar preparation of the amodiaquine isolated from the Reference Standard.

B: The ultraviolet absorption spectrum of a 1 in 100,000 solution in 0.1 *N* hydrochloric acid exhibits maxima and minima at the same wavelengths as that of a similar solution of USP Amodiaquine Hydrochloride RS.

Water, *Method I* ⟨921⟩: not more than 0.5%.

Residue on ignition ⟨281⟩: not more than 0.2%.

Chromatographic purity—

Standard preparations—To 20 mg of USP Amodiaquine Hydrochloride RS in a glass-stoppered test tube add 1.0 mL of chloroform (saturated with ammonium hydroxide), and shake vigorously for 2 minutes. Allow the solids to settle, and decant the liquid into a second test tube (*Solution A*). Prepare a second solution by diluting 1.0 volume of *Solution A* with chloroform (saturated with ammonium hydroxide) to obtain 200 volumes of solution (*Solution B*).

Test preparation—Dissolve 150 mg of Amodiaquine in 10 mL of chloroform (saturated with ammonium hydroxide).

Procedure—On a suitable thin-layer chromatographic plate (see *Chromatography* ⟨621⟩), coated with a 0.25-mm layer of chromatographic silica gel mixture, apply 10-μL portions of *Solution A*, *Solution B*, and the *Test preparation*. Allow the spots to dry, and develop the chromatogram in a solvent system consisting of a mixture of chloroform (saturated with ammonium hydroxide) and dehydrated alcohol (9:1) until the solvent front has moved about three-fourths of the length of the plate. Remove the plate from the developing chamber, mark the solvent front, allow the solvent to evaporate, and examine the plate under short-wavelength ultraviolet light: the chromatograms show principal spots at about the same R_f value, and no secondary spot, if present in the chromatogram from the *Test preparation*, is more intense than the principal spot obtained from *Solution B*.

Assay—Transfer about 300 mg of Amodiaquine, accurately weighed, to a 200-mL volumetric flask, add 0.1 *N* hydrochloric acid to volume, and mix. Pipet 10.0 mL of the resulting solution into a 1000-mL volumetric flask, add 0.1 *N* hydrochloric acid to volume, and mix. Concomitantly determine the absorbances of this solution and a Standard solution of USP Amodiaquine Hydrochloride RS in the same medium having a known concentration of about 15 μg per mL, in 1-cm cells at the wavelength of maximum absorbance at about 342 nm, with a suitable spectrophotometer, using 0.1 *N* hydrochloric acid as the blank. Calculate the quantity, in mg, of $C_{20}H_{22}ClN_3O$ in the portion of Amodiaquine taken by the formula:

$$(355.87/428.79)(20C)(A_U/A_S),$$

in which 355.87 and 428.79 are the molecular weights of amodiaquine and anhydrous aminodiaquine hydrochloride, respectively, C is the concentration, in μg per mL, calculated on the anhydrous basis, of USP Amodiaquine Hydrochloride RS in the Standard solution, and A_U and A_S are the absorbances of the solution of Amodiaquine Hydrochloride and the Standard solution, respectively.

Amodiaquine Hydrochloride

$C_{20}H_{22}ClN_3O.2HCl.2H_2O$ 464.82

Phenol, 4-[(7-chloro-4-quinolinyl)amino]-2-[(diethylamino)-methyl]-, dihydrochloride, dihydrate.

4-[(7-Chloro-4-quinolyl)amino]-α-(diethylamino)-*o*-cresol dihydrochloride dihydrate [*6398-98-7*].

Anhydrous 428.79 [*69-44-3*].

» Amodiaquine Hydrochloride contains not less than 97.0 percent and not more than 103.0 percent of $C_{20}H_{22}ClN_3O.2HCl$, calculated on the anhydrous basis.

Packaging and storage—Preserve in tight containers.

Reference standard—*USP Amodiaquine Hydrochloride Reference Standard*—Do not dry; determine the water content titrimetrically at the time of use.

Completeness of solution ⟨641⟩—A solution of 200 mg in 10 mL of water is clear.

Identification—

A: Dissolve 20 mg in 10 mL of water in a separator, add 1 mL of ammonium hydroxide, and extract by shaking with 25 mL of chloroform. Draw off and evaporate the chloroform extract, and dry the residue at 105° for 2 hours: the infrared absorption spectrum of a potassium bromide dispersion of the amodiaquine so obtained exhibits maxima only at the same wavelengths as that of a similar preparation of USP Amodiaquine Hydrochloride RS.

B: The ultraviolet absorption spectrum of a 1 in 100,000 solution of it in dilute hydrochloric acid (1 in 100) exhibits maxima and minima at the same wavelengths as that of a similar preparation of USP Amodiaquine Hydrochloride RS.

C: A solution of it responds to the tests for *Chloride* ⟨191⟩.

Water, *Method I* ⟨921⟩: not less than 7.0% and not more than 9.0%.

Residue on ignition ⟨281⟩: not more than 0.2%.

Chromatographic purity—

Standard preparations—To 20 mg of USP Amodiaquine Hydrochloride RS in a glass-stoppered test tube add 1.0 mL of chloroform (saturated with ammonium hydroxide), and shake vigorously for 2 minutes. Allow the solids to settle, and decant the liquid into a second test tube (*Solution A*). Prepare a second solution by diluting 1.0 volume of *Solution A* with sufficient chloroform (saturated with ammonium hydroxide) to obtain 200 volumes of solution (*Solution B*).

Test preparation—To 200 mg of Amodiaquine Hydrochloride in a glass-stoppered test tube add 10 mL of chloroform (saturated with ammonium hydroxide), and shake vigorously for 2 minutes. Allow the solids to settle, and decant the liquid into a second test tube.

Procedure—On a suitable thin-layer chromatographic plate (see *Chromatography* ⟨621⟩), coated with a 0.25-mm layer of chromatographic silica gel mixture, apply 10 µL portions of *Solution A*, *Solution B*, and the *Test preparation*. Allow the spots to dry, and develop the chromatogram in a solvent system consisting of a mixture of chloroform (saturated with ammonium hydroxide) and dehydrated alcohol (9:1) until the solvent front has moved about three-fourths of the length of the plate. Remove the plate from the developing chamber, mark the solvent front, allow the solvent to evaporate, and examine the plate under short-wavelength ultraviolet light: the chromatograms show principal spots at about the same R_f value, and no secondary spot, if present in the chromatogram from the *Test preparation*, is more intense than the principal spot obtained from *Solution B*.

Assay—Transfer about 300 mg of Amodiaquine Hydrochloride, accurately weighed, to a 200-mL volumetric flask, add dilute hydrochloric acid (1 in 100) to volume, and mix. Pipet 10.0 mL of the solution into a 1000-mL volumetric flask, add dilute hydrochloric acid (1 in 100) to volume, and mix. Concomitantly determine the absorbances of this solution and a solution of undried USP Amodiaquine Hydrochloride RS in the same medium having a known concentration of about 15 µg per mL, in 1-cm cells at the wavelength of maximum absorbance at about 342 nm, with a suitable spectrophotometer, using dilute hydrochloric acid (1 in 100) as the blank. Calculate the quantity, in mg, of $C_{20}H_{22}ClN_3O.2HCl$ in the portion of Amodiaquine Hydrochloride taken by the formula:

$$20C(A_U/A_S),$$

in which C is the concentration, in µg per mL, calculated on the anhydrous basis, of USP Amodiaquine RS in the Standard solution, and A_U and A_S are the absorbances of the solution of Amodiaquine Hydrochloride and the Standard solution, respectively.

Amodiaquine Hydrochloride Tablets

» Amodiaquine Hydrochloride Tablets contain an amount of amodiaquine hydrochloride ($C_{20}H_{22}ClN_3O.2HCl.2H_2O$) equivalent to not less than 93.0 percent and not more than 107.0 percent of the labeled amount of amodiaquine ($C_{20}H_{22}ClN_3O$).

Packaging and storage—Preserve in tight containers.

Reference standard—*USP Amodiaquine Hydrochloride Reference Standard*—Do not dry; determine the water content titrimetrically at the time of use.

Identification—

A: Powder 1 or more Tablets, and transfer a portion of the powder, equivalent to about 50 mg of amodiaquine, to a 125-mL separator. Add 20 mL of water, and shake for 1 minute. Add 25 mL of chloroform and 1 mL of ammonium hydroxide, shake for 2 minutes, and when settled, filter the chloroform extract through cotton that previously has been rinsed with chloroform, collecting the extract in a vessel suitable for evaporation. Evaporate the chloroform, and dry the residue at 105° for 1 hour: the infrared absorption spectrum of a potassium bromide dispersion of the amodiaquine so obtained exhibits maxima only at the same wavelengths as that of a similar preparation of USP Amodiaquine Hydrochloride RS.

B: The ultraviolet absorption spectrum of the solution prepared from the powdered Tablets as directed in the *Assay* exhibits maxima and minima at the same wavelengths as that of a 1 in 67,000 solution of USP Amodiaquine Hydrochloride RS in dilute hydrochloric acid (1 in 100).

Dissolution ⟨711⟩—

Medium: water; 900 mL.

Apparatus 2: 50 rpm.

Time: 30 minutes.

Procedure—Determine the amount of $C_{20}H_{22}ClN_3O.2HCl.2H_2O$ dissolved from ultraviolet absorbances at the wavelength of maximum absorbance at about 342 nm of filtered portions of the solution under test, suitably diluted with water, if necessary, in comparison with a Standard solution having a known concentration of USP Amodiaquine Hydrochloride RS in the same medium.

Tolerances—An amount of amodiaquine hydrochloride ($C_{20}H_{22}ClN_3O.2HCl.2H_2O$) equivalent to not less than 75% (*Q*) of the labeled amount of amodiaquine ($C_{20}H_{22}ClN_3O$) is dissolved in 30 minutes.

Uniformity of dosage units ⟨905⟩: meet the requirements.

Assay—Weigh and finely powder not less than 20 Amodiaquine Hydrochloride Tablets. Transfer an accurately weighed portion of the powder, equivalent to about 300 mg of amodiaquine, to a 250-mL beaker, add about 100 mL of dilute hydrochloric acid (1 in 100), and heat on a steam bath for about 15 minutes with occasional stirring. Cool the mixture to room temperature, transfer to a 200-mL volumetric flask, add dilute hydrochloric acid (1 in 100) to volume, and mix. Pipet 10 mL of the clear supernatant liquid into a 125-mL separator. Add about 10 mL of dilute hydrochloric acid (1 in 100), and wash with 20 mL of chloroform, discarding the washing. Add 4.5 mL of 1 *N* sodium hydroxide, and extract with four 25-mL portions of chloroform. Extract the combined chloroform solutions with three 50-mL portions of dilute hydrochloric acid (1 in 100). Combine the acid extracts in a 200-mL volumetric flask, add dilute hydrochloric acid (1 in 100) to volume, and mix. Pipet 20 mL of this solution into a 100-mL volumetric flask, add dilute hydrochloric acid (1 in 100) to volume, and mix. Concomitantly determine the absorbances of this solution and a solution of undried USP Amodiaquine Hydrochloride RS, in the same medium having a known concentration of about 15 µg per mL in 1-cm cells at the wavelength of maximum absorbance at about 342 nm, with a suitable spectrophotometer, using dilute hydrochloric acid (1 in 100) as the blank. Calculate the quantity, in mg, of $C_{20}H_{22}ClN_3O.2HCl.2H_2O$ in the portion of Tablets taken by the formula:

$$21.68C(A_U/A_S),$$

in which C is the concentration, in µg per mL, calculated on the anhydrous basis, of the USP Amodiaquine Hydrochloride RS in the Standard solution, and A_U and A_S are the absorbances of the solution from the Tablets and the Standard solution, respectively. The quantity determined, multiplied by 0.7656, gives the equivalent quantity of $C_{20}H_{22}ClN_3O$.

Amoxicillin

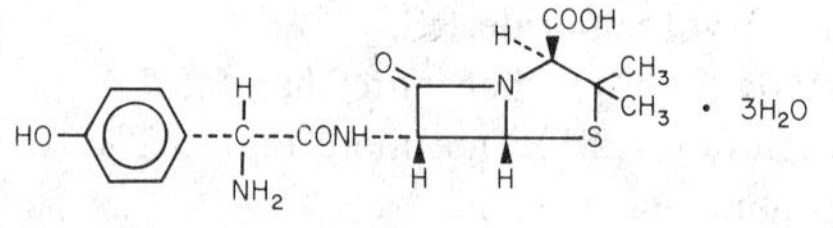

$C_{16}H_{19}N_3O_5S.3H_2O$ 419.45

4-Thia-1-azabicyclo[3.2.0]heptane-2-carboxylic acid, 6-[[amino(4-hydroxyphenyl)acetyl]amino]-3,3-dimethyl-7-oxo-, trihydrate [2*S*-[2α,5α,6β(*S**)]]-.

(2*S*,5*R*,6*R*)-6-[(*R*)-(−)-2-Amino-2-(*p*-hydroxyphenyl)acetamido]-3,3-dimethyl-7-oxo-4-thia-1-azabicyclo[3.2.0]heptane-2-carboxylic acid trihydrate [*61336-70-7*].

Anhydrous 365.40 [*26787-78-0*].

» Amoxicillin contains not less than 90.0 percent of $C_{16}H_{19}N_3O_5S$, calculated on the anhydrous basis. It has a potency equivalent to not less than 900 µg and not more than 1050 µg of $C_{16}H_{19}N_3O_5S$ per mg, calculated on the anhydrous basis.

Packaging and storage—Preserve in tight containers, at controlled room temperature.

Labeling—Label it to indicate that it is to be used in the manufacture of nonparenteral drugs only.

Reference standard—*USP Amoxicillin Reference Standard*—Do not dry before using.

Identification—The infrared absorption spectrum of a potassium bromide dispersion of it exhibits maxima only at the same wavelengths as that of a similar preparation of USP Amoxicillin RS.

Crystallinity ⟨695⟩: meets the requirements.

pH ⟨791⟩: between 3.5 and 6.0, in a solution containing 2 mg per mL.

Water, *Method I* ⟨921⟩: between 11.5% and 14.5%.

Dimethylaniline—

Internal standard solution—Prepare a solution of naphthalene in cyclohexane containing about 50 µg per mL.

Standard preparation—Transfer 50.0 mg of *N,N*-dimethylaniline to a 50-mL volumetric flask, add 25 mL of 1 *N* hydrochloric acid, swirl to dissolve, dilute with water to volume, and mix. Transfer 5.0 mL of the resulting solution to a 250-mL volumetric flask, dilute with water to volume, and mix. To a suitable centrifuge tube add 1.0 mL of this solution, 5.0 mL of 1 *N* sodium hydroxide, and 1.0 mL of *Internal standard solution*, shake vigorously for 1 minute, and centrifuge. Use the clear supernatant liquid as the *Standard preparation.*

Test preparation—Transfer 1.0 g of Amoxicillin to a suitable centrifuge tube, add 5 mL of 1 *N* sodium hydroxide, swirl to dissolve the specimen, add 1.0 mL of *Internal standard solution*, shake vigorously for 1 minute, and centrifuge. Use the clear supernatant liquid as the *Test preparation.*

Chromatographic system (see *Chromatography* ⟨621⟩)—The gas chromatograph is equipped with a flame-ionization detector and a 2-mm × 2-m column packed with 3 percent liquid phase G3 on silanized packing S1A and is maintained at 120°. Nitrogen is used as the carrier gas, flowing at the rate of about 30 mL per minute.

Procedure—[NOTE—Use peak areas where peak responses are indicated.] Inject equal volumes (about 20 µL) of the *Standard preparation* and the *Test preparation* into the chromatograph, record the chromatograms, and measure the responses for the major peaks. The ratio of the response of any dimethylaniline peak to the response of the naphthalene peak obtained from the *Test preparation* is not greater than that obtained from the *Standard preparation* (0.002%).

Concordance—Calculate the *Acid titration–base titration concordance* by the formula:

$$A - B,$$

the *Acid titration–potency concordance* by the formula:

$$A - (P_A/10),$$

and the *Base titration–potency concordance* by the formula:

$$B - (P_A/10),$$

in which A and B are the *Content of amoxicillin*, in percent, as determined by the *Titration of acidic function* and the *Titration of basic function*, respectively, in the test for *Content of amoxicillin*, and P_A is the potency, in µg per mg, as determined in the *Assay:* none of the concordance values exceed 6.0%.

Content of amoxicillin—

Titration of acidic function—Dissolve about 100 mg of Amoxicillin, accurately weighed, in 20 mL of dimethylsulfoxide and 30 mL of methanol (add methanol after the specimen under test has dissolved in dimethylsulfoxide). Stir by mechanical means, and titrate with 0.02 *N* lithium methoxide in methanol VS, determining the end-point potentiometrically using a glass electrode and a calomel electrode containing potassium chloride–saturated methanol (see *Titrimetry* ⟨541⟩). Perform a blank determination, and make any necessary correction. Each mL of 0.02 *N* lithium methoxide is equivalent to 7.309 mg of $C_{16}H_{19}N_3O_5S$.

Titration of basic function—Dissolve about 100 mg of Amoxicillin, accurately weighed, in 50 mL of glacial acetic acid. Titrate with 0.02 *N* perchloric acid VS, determining the end-point potentiometrically using a glass electrode and a calomel electrode containing potassium chloride–saturated methanol (see *Titrimetry* ⟨541⟩). Perform a blank determination, and make any necessary correction. Each mL of 0.02 *N* perchloric acid is equivalent to 7.309 mg of $C_{16}H_{19}N_3O_5S$.

Assay—

Standard preparation—Prepare as directed for *Standard Preparation* under *Iodometric Assay—Antibiotics* ⟨425⟩, using USP Amoxicillin RS.

Assay preparation—Proceed with Amoxicillin as directed for *Assay Preparation* under *Iodometric Assay—Antibiotics* ⟨425⟩.

Procedure—Proceed as directed for *Procedure* under *Iodometric Assay—Antibiotics* ⟨425⟩. Calculate the potency, in µg of $C_{16}H_{19}N_3O_5S$ per mg, of the Amoxicillin taken by the formula:

$$(F)(B - I)/2D,$$

in which D is the concentration, in mg per mL, of the *Assay preparation*, based on the weight of Amoxicillin taken and the extent of dilution.

Amoxicillin Capsules

» Amoxicillin Capsules contain the equivalent of not less than 90.0 percent and not more than 120.0 percent of the labeled amount of $C_{16}H_{19}N_3O_5S$.

Packaging and storage—Preserve in tight containers, at controlled room temperature.

Reference standard—*USP Amoxicillin Reference Standard*—Do not dry before using.

Identification—Prepare a test solution containing the equivalent of 4 mg of amoxicillin per mL by adding 0.1 *N* hydrochloric acid to powder from Amoxicillin Capsules. Prepare a Standard solution of USP Amoxicillin RS in 0.1 *N* hydrochloric acid containing 4 mg per mL. Use within 10 minutes after preparation. Apply separately 5 µL of each solution on a thin-layer chromatographic plate coated with a 0.25-mm layer of chromatographic silica gel mixture (see *Chromatography* ⟨621⟩). Place the plate in a suitable chromatographic chamber, and develop the chromatogram in a solvent system consisting of a mixture of methanol, chloroform, water, and pyridine (90:80:30:10). When the solvent front has moved about three-fourths of the length of the plate, remove the plate from the chamber, and dry with warm air for 10 minutes. Locate the spots on the plate by spraying lightly with a solution of ninhydrin in alcohol containing 3 mg per mL, and dry at 110° for 15 minutes: the R_f value of the principal spot obtained from the test solution corresponds to that obtained from the Standard solution.

Dissolution ⟨711⟩—

Medium: water; 900 mL.

Apparatus 1: 100 rpm.

Time: 90 minutes.

Procedure—Determine the amount of $C_{16}H_{19}N_3O_5S.3H_2O$ dissolved from ultraviolet absorbances at the wavelength of maximum absorbance at about 272 nm of filtered portions of the solution under test, suitably diluted with water, if necessary, in comparison with a Standard solution having a known concentration of USP Amoxicillin RS in the same medium.

Tolerances—Not less than 80% (Q) of the labeled amount of $C_{16}H_{19}N_3O_5S.3H_2O$ is dissolved in 90 minutes.

Uniformity of dosage units ⟨905⟩: meet the requirements.

Water, *Method I* ⟨921⟩: not more than 14.5%.

Assay—

Standard preparation—Prepare as directed for *Standard preparation* under *Iodometric Assay—Antibiotics* ⟨425⟩, using USP Amoxicillin RS.

Assay preparation—Place not less than 5 Amoxicillin Capsules in a high-speed glass blender jar containing an accurately measured volume of water, and blend for 4 ± 1 minutes. Dilute an accurately measured volume of this stock solution quantitatively and stepwise with water to obtain an *Assay preparation* containing about 1 mg of amoxicillin per mL.

Procedure—Proceed with Amoxicillin Capsules as directed for *Procedure* under *Iodometric Assay—Antibiotics* ⟨425⟩. Calculate the quantity, in mg, of $C_{16}H_{19}N_3O_5S$ in each Capsule taken by the formula:

$$(T/D)(F/2000)(B - I),$$

in which T is the labeled quantity, in mg, of amoxicillin in each Capsule, and D is the concentration, in mg per mL, of amoxicillin in the *Assay preparation* on the basis of the labeled quantity in each Capsule and the extent of dilution.

Amoxicillin Intramammary Infusion

» Amoxicillin Intramammary Infusion is a suspension of Amoxicillin in a suitable vegetable oil vehicle. It contains not less than 90.0 percent and not more than 120.0 percent of the labeled amount of $C_{16}H_{19}N_3O_5S$. It contains a suitable dispersing agent and preservative.

Packaging and storage—Preserve in well-closed disposable syringes.

Labeling—Label it to indicate that it is intended for veterinary use only.

Reference standard—*USP Amoxicillin Reference Standard*—Do not dry before using.

Identification—Transfer a quantity of Intramammary Infusion, equivalent to about 60 mg of amoxicillin, to a 50-mL centrifuge tube, add 25 mL of toluene, mix, and centrifuge. Decant and discard the toluene. Wash the residue with four 25-mL portions of toluene, sonicating for about 30 seconds after each addition of toluene. Dry the residue in vacuum over silica gel. Add 15 mL of 0.1 *N* hydrochloric acid to the residue, and mix: the solution so obtained responds to the *Identification test* under *Amoxicillin Capsules*.

Water, *Method I* ⟨921⟩: not more than 1.0%, 20 mL of a mixture of carbon tetrachloride, chloroform, and methanol (2:2:1) being used in place of methanol in the titration vessel.

Assay—Proceed as directed for ampicillin under *Antibiotics—Microbial Assays* ⟨81⟩, except to read "Amoxicillin" where "Ampicillin" is specified and to use USP Amoxicillin RS. Expel the contents of 1 syringe of Amoxicillin Intramammary Infusion into a high-speed glass blender jar containing 499.0 mL of *Buffer No. 3* and 1.0 mL of polysorbate 80, and blend for 3 to 5 minutes. Allow to stand for about 10 minutes, and dilute an accurately measured volume of the aqueous phase quantitatively and stepwise with *Buffer No. 3* to obtain a *Test Dilution* having a concentration assumed to be equal to the median dose level of the Standard.

Sterile Amoxicillin

» Sterile Amoxicillin is Amoxicillin suitable for parenteral use. It has a potency equivalent to not less than 900 μg and not more than 1050 μg of $C_{16}H_{19}N_3O_5S$ per mg, calculated on the anhydrous basis.

Packaging and storage—Preserve in *Containers for Sterile Solids* as described under *Injections* ⟨1⟩.

Labeling—Label it to indicate that it is intended for veterinary use only.

Reference standard—*USP Amoxicillin Reference Standard*—Do not dry before using.

Pyrogen—It meets the requirements of the *Pyrogen Test* ⟨151⟩, the test dose being 1.0 mL per kg of a solution in 0.05 *N* sodium hydroxide containing 20 mg of amoxicillin per mL.

Sterility—It meets the requirements under *Sterility Tests* ⟨71⟩, when tested as directed in the section, *Test Procedures for Direct Transfer to Test Media*, except to use Fluid Thioglycollate Medium containing polysorbate 80 solution (1 in 200) and an amount of sterile penicillinase sufficient to inactivate the amoxicillin in each tube, to use Soybean-Casein Digest Medium containing polysorbate 80 solution (1 in 200) and an amount of sterile penicillinase sufficient to inactivate the amoxicillin in each tube, and to shake the tubes once daily.

Other requirements—It responds to the *Identification test* and meets the requirements for *Crystallinity, pH, Water, Dimethylaniline, Concordance, Content of amoxicillin,* and *Assay* under *Amoxicillin.*

Sterile Amoxicillin for Suspension

» Sterile Amoxicillin for Suspension is a sterile mixture of Amoxicillin and one or more suitable buffers, preservatives, stabilizers, and suspending agents. It contains not less than 90.0 percent and not more than 120.0 percent of the labeled amount of $C_{16}H_{19}N_3O_5S$.

Packaging and storage—Preserve in *Containers for Sterile Solids* as described under *Injections* ⟨1⟩.

Labeling—Label it to indicate that it is for veterinary use only.

Reference standard—*USP Amoxicillin Reference Standard*—Do not dry before using.

Identification—Prepare a test solution containing the equivalent of 4 mg of amoxicillin per mL by adding 0.1 *N* hydrochloric acid to Sterile Amoxicillin for Suspension. Allow the solution to stand for 5 minutes before use: the solution responds to the *Identification test* under *Amoxicillin Capsules*.

Pyrogen—It meets the requirements of the *Pyrogen Test* ⟨151⟩, the test dose being 1.0 mL per kg of a solution in 0.05 *N* sodium hydroxide containing 20 mg of amoxicillin per mL.

Sterility—It meets the requirements under *Sterility Tests* ⟨71⟩, when tested as directed in the section, *Test Procedures for Direct Transfer to Test Media*, except to use Fluid Thioglycollate Medium containing polysorbate 80 solution (1 in 200) and an amount of sterile penicillinase sufficient to inactivate the amoxicillin in each tube, to use Soybean-Casein Digest Medium containing polysorbate 80 solution (1 in 200) and an amount of sterile penicillinase sufficient to inactivate the amoxicillin in each tube, and to shake the tubes once daily.

pH ⟨791⟩: between 5.0 and 7.0, in the suspension constituted as directed in the labeling.

Water, *Method I* ⟨921⟩: between 11.0% and 14.0%.

Assay—

Standard preparation—Using USP Amoxicillin RS, prepare as directed for *Standard Preparation* under *Iodometric Assay—Antibiotics* ⟨425⟩.

Assay preparation 1 (where it is represented as being in a single-dose container)—Constitute Sterile Amoxicillin for Suspension as directed in the labeling. Withdraw all of the withdrawable contents, using a hypodermic needle and syringe, and dilute quantitatively with water to obtain a solution containing about 1 mg of amoxicillin per mL. Pipet 2 mL of this solution into each of two glass-stoppered, 125-mL conical flasks.

Assay preparation 2 (where the label states the quantity of amoxicillin in a given volume of constituted suspension)—Constitute Sterile Amoxicillin for Suspension as directed in the labeling. Dilute an accurately measured volume of the constituted suspension quantitatively with water to obtain a solution containing about 1 mg of amoxicillin per mL. Pipet 2 mL of this solution into each of two glass-stoppered, 125-mL conical flasks.

Procedure—Proceed as directed for *Procedure* under *Iodometric Assay—Antibiotics* ⟨425⟩. Calculate the quantity, in mg, of amoxicillin in the container, or in the portion of constituted suspension taken by the formula:

$$(L/D)(F/2000)(B - I),$$

in which L is the labeled quantity, in mg, of amoxicillin in the container, or in the volume of constituted suspension taken, and D is the concentration, in mg of amoxicillin per mL, of *Assay preparation 1* or of *Assay preparation 2* on the basis of the labeled quantity in the container or in the portion of constituted suspension taken, respectively, and the extent of dilution.

Amoxicillin Oral Suspension

» Amoxicillin Oral Suspension is a suspension of Amoxicillin in Soybean Oil. It contains not less than 90.0 percent and not more than 120.0 percent of the labeled amount of $C_{16}H_{19}N_3O_5S$.

Packaging and storage—Preserve in multiple-dose containers equipped with a suitable dosing pump.

Labeling—Label it to indicate that it is for veterinary use only.

Reference standard—*USP Amoxicillin Reference Standard*—Do not dry before using.

Identification—Shake a portion of Oral Suspension with a mixture of acetone and 0.1 *N* hydrochloric acid (4:1) to obtain a solution containing about 1 mg of amoxicillin per mL. The solution responds to the *Identification test* under *Amoxicillin Capsules*.

Water, *Method I* ⟨921⟩: not more than 2.0%, 20 mL of a mixture of carbon tetrachloride, chloroform, and methanol (2:2:1) being used in place of methanol in the titration vessel.

Assay—

Standard preparation—Prepare as directed for *Standard Preparation* under *Iodometric Assay—Antibiotics* ⟨425⟩, using USP Amoxicillin RS.

Assay preparation—Using the dosing pump, deliver a number of doses of Amoxicillin Oral Suspension, equivalent to about 250 mg of amoxicillin, to a separator containing 100 mL of hexanes, and shake vigorously. Add 140 mL of water, and shake for 5 minutes. Allow the layers to separate, and drain the lower, aqueous layer into a 250-mL volumetric flask. Repeat the extraction with two 50-mL portions of water. Combine the aqueous extracts in the volumetric flask, dilute with water to volume, and mix.

Procedure—Proceed with Amoxicillin Oral Suspension as directed for *Procedure* under *Iodometric Assay—Antibiotics* ⟨425⟩. Calculate the quantity, in mg, of $C_{16}H_{19}N_3O_5S$ in each dose of Oral Suspension taken by the formula:

$$(250/N)(F/2000)(B - I),$$

in which *N* is the number of doses taken, and the other terms are as defined therein.

Amoxicillin for Oral Suspension

» Amoxicillin for Oral Suspension contains the equivalent of not less than 90.0 percent and not more than 120.0 percent of the labeled amount of $C_{16}H_{19}N_3O_5S$. It contains one or more suitable buffers, colors, flavors, preservatives, stabilizers, sweeteners, and suspending agents.

Packaging and storage—Preserve in tight containers, at controlled room temperature.

Reference standard—*USP Amoxicillin Reference Standard*—Do not dry before using.

Identification—Prepare a solution containing the equivalent of 4 mg of amoxicillin by adding 0.1 *N* hydrochloric acid to a portion of Amoxicillin for Oral Suspension. Allow the solution to stand for 5 minutes before use: the solution responds to the *Identification test* under *Amoxicillin Capsules*.

Uniformity of dosage units ⟨905⟩—

FOR SOLID PACKAGED IN SINGLE-UNIT CONTAINERS: meets the requirements.

pH ⟨791⟩: between 5.0 and 7.5, in the suspension constituted as directed in the labeling.

Water, *Method I* ⟨921⟩: not more than 3.0%.

Assay—

Standard preparation—Prepare as directed for *Standard preparation* under *Iodometric Assay—Antibiotics* ⟨425⟩, using USP Amoxicillin RS.

Assay preparation—Dilute an accurately measured volume of Amoxicillin for Oral Suspension, constituted as directed in the labeling, freshly mixed and free from air bubbles, quantitatively and stepwise in *Buffer No. 1* (see *Media and Diluents* under *Antibiotics—Microbial Assays* ⟨81⟩) to obtain a solution containing about 1 mg of amoxicillin per mL.

Procedure—Proceed as directed for *Procedure* under *Iodometric Assay—Antibiotics* ⟨425⟩. Calculate the quantity, in mg, of $C_{16}H_{19}N_3O_5S$ in each mL of the constituted Oral Suspension taken by the formula:

$$(T/D)(F/2000)(B - I),$$

in which *T* is the labeled quantity, in mg per mL, of amoxicillin in the constituted Amoxicillin for Oral Suspension, and *D* is the concentration, in mg per mL, of amoxicillin in the *Assay preparation* on the basis of the labeled quantity in the constituted Amoxicillin for Oral Suspension and the extent of dilution.

Amoxicillin Tablets

» Amoxicillin Tablets contain not less than 90.0 percent and not more than 120.0 percent of the labeled amount of $C_{16}H_{19}N_3O_5S$.

Packaging and storage—Preserve in tight containers, at controlled room temperature.

Labeling—Label chewable Tablets to indicate that they are to be chewed before swallowing. Label film-coated Tablets or Tablets having a diameter of greater than 15 mm to indicate that they are intended for veterinary use only.

Reference standard—*USP Amoxicillin Reference Standard*—Do not dry before using.

Identification—To a portion of powdered Tablets add 0.1 *N* hydrochloric acid to obtain a test solution containing 4 mg per mL. Prepare a Standard solution of USP Amoxicillin RS in 0.1 *N* hydrochloric acid containing 4 mg per mL. Use within 10 minutes after preparation. Apply separately 5 µL of each solution on a thin-layer chromatographic plate coated with a 0.25-mm layer of chromatographic silica gel mixture (see *Chromatography* ⟨621⟩). Place the plate in a suitable chromatographic chamber, and develop the chromatogram in a solvent system consisting of a mixture of methanol, chloroform, water, and pyridine (90:80:30:10). When the solvent front has moved about three-fourths of the length of the plate, remove the plate from the chamber, and dry with the aid of a current of warm air for 10 minutes. Locate the spots on the plate by spraying lightly with a solution of ninhydrin in alcohol containing 3 mg per mL, and dry at 110° for 15 minutes: the R_f value of the principal spot obtained from the test solution corresponds to that obtained from the Standard solution.

Disintegration ⟨701⟩: 30 minutes, simulated gastric fluid TS being used as the test medium in place of water (chewable Tablets and Tablets having a diameter of greater than 15 mm are exempt from this requirement).

Water, *Method I* ⟨921⟩: not more than 6.0%, or not more than 7.0% in the case of film-coated Tablets, or not more than 7.5% in the case of Tablets having a diameter greater than 15 mm.

Assay—

Standard preparation—Prepare as directed for *Standard preparation* under *Iodometric Assay—Antibiotics* ⟨425⟩, using USP Amoxicillin RS.

Assay preparation—Place not less than 5 Amoxicillin Tablets in a high-speed glass blender jar containing an accurately measured volume of *Buffer No. 1*, and blend for 4 ± 1 minutes. Dilute an accurately measured volume of this stock solution quantitatively with *Buffer No. 1* to obtain an *Assay preparation* containing about 1 mg of amoxicillin per mL.

Procedure—Proceed as directed for *Procedure* under *Iodometric Assay—Antibiotics* ⟨425⟩, except to add 0.1 mL of 1.2

N hydrochloric acid immediately after the 0.1 *N* iodine VS in performing the *Blank determination*. Calculate the quantity, in mg, of $C_{16}H_{19}N_3O_5S$ in each Tablet taken by the formula:

$$(T/D)(F/2000)(B - I),$$

in which *T* is the labeled quantity, in mg, of amoxicillin in each Tablet, and *D* is the concentration, in mg per mL, of amoxicillin in the *Assay preparation* on the basis of the labeled quantity in each Tablet and the extent of dilution.

Amoxicillin and Clavulanate Potassium for Oral Suspension

» Amoxicillin and Clavulanate Potassium for Oral Suspension contains the equivalent of not less than 90.0 percent and not more than 120.0 percent of the labeled amount of amoxicillin ($C_{16}H_{19}N_3O_5S$) and the equivalent of not less than 90.0 percent and not more than 125.0 percent of the labeled amount of clavulanic acid ($C_8H_9NO_5$). It contains one or more suitable buffers, colors, flavors, preservatives, stabilizers, sweeteners, and suspending agents.

Packaging and storage—Preserve in tight containers, at controlled room temperature.

Reference standards—*USP Amoxicillin Reference Standard*—Do not dry before using. *USP Clavulanate Lithium Reference Standard*—Keep container tightly closed and protected from light. Do not dry before using.

Identification—The retention times of the major peaks in the chromatogram of the *Assay preparation* correspond to those of the *Standard preparation*, as obtained in the *Assay*.

pH ⟨791⟩: between 4.8 and 6.6, in the suspension constituted as directed in the labeling.

Water, *Method I* ⟨921⟩: not more than 7.5%, where the label indicates that after constitution as directed, the suspension contains 25 mg of amoxicillin per mL, or not more than 8.5%, where the label indicates that after constitution as directed, the suspension contains 50 mg of amoxicillin per mL.

Assay—

pH 4.4 sodium phosphate buffer—Dissolve 7.8 g of monobasic sodium phosphate in 900 mL of water, adjust with phosphoric acid or 10 *N* sodium hydroxide to a pH of 4.4 ± 0.1, dilute with water to make 1000 mL, and mix.

Mobile phase—Prepare a suitable mixture of *pH 4.4 sodium phosphate buffer* and methanol (95:5), and filter through a membrane filter of 0.5-μm or finer porosity. Make adjustments if necessary (see *System Suitability* under *Chromatography* ⟨621⟩).

Standard preparation—Dissolve accurately weighed quantities of USP Amoxicillin RS and USP Clavulanate Lithium RS in water to obtain a solution having known concentrations of about 0.5 mg per mL and 0.2 mg per mL, respectively.

Assay preparation—Dilute an accurately measured volume of the Amoxicillin and Clavulanic Acid for Oral Suspension, constituted as directed in the labeling, quantitatively with water to obtain a solution containing about 0.5 mg of amoxicillin per mL. Stir by mechanical means for 10 minutes, and filter. Use the filtrate as the *Assay preparation* within 1 hour of the dilution of the suspension.

Chromatographic system (see *Chromatography* ⟨621⟩)—The liquid chromatograph is equipped with a 220-nm detector and a 4-mm × 30-cm column that contains 3- to 10-μm packing L1. The flow rate is about 2 mL per minute. Chromatograph the *Standard preparation*, and record the peak responses as directed under *Procedure:* the column efficiency determined from the analyte peak is not less than 550 theoretical plates, the tailing factor for the analyte peak is not more than 1.5, and the relative standard deviation for replicate injections is not more than 2.0%. Chromatograph the *Standard preparation*, and record the peak responses as directed under *Procedure:* the resolution, *R*, between the amoxicillin and clavulanic acid peaks is not less than 3.5.

Procedure—Separately inject equal volumes (about 20 μL) of the *Standard preparation* and the *Assay preparation* into the chromatograph, record the chromatograms, and measure the responses for the major peaks. The relative retention times are about 0.5 for clavulanic acid and 1.0 for amoxicillin. Calculate the quantity, in mg, of amoxicillin ($C_{16}H_{19}N_3O_5S$) in each mL of the constituted Oral Suspension taken by the formula:

$$(CP)(L/1000D)(r_U/r_S),$$

in which *C* is the concentration, in mg per mL, of USP Amoxicillin RS in the *Standard preparation*, *P* is the potency, in μg of amoxicillin per mg, of USP Amoxicillin RS, *L* is the labeled quantity, in mg per mL, of amoxicillin in the constituted Amoxicillin and Clavulanate Potassium for Oral Suspension, *D* is the concentration, in mg per mL, of amoxicillin in the *Assay preparation* on the basis of the labeled quantity of amoxicillin in the constituted Amoxicillin and Clavulanate Potassium for Oral Suspension, the volume of constituted Oral Suspension taken, and the extent of dilution, and r_U and r_S are the amoxicillin peak responses obtained from the *Assay preparation* and the *Standard preparation*, respectively. Calculate the quantity, in mg, of clavulanic acid ($C_8H_9NO_5$) in each mL of the constituted Amoxicillin and Clavulanate Potassium for Oral Suspension taken by the formula:

$$(L/D)(CP/1000)(r_U/r_S),$$

in which *L* is the labeled quantity, in mg per mL, of clavulanic acid in the constituted Amoxicillin and Clavulanate Potassium for Oral Suspension, *D* is the concentration, in mg per mL, of clavulanic acid in the *Assay preparation* on the basis of the labeled quantity of clavulanic acid in the constituted Oral Suspension, the volume of constituted Oral Suspension taken, and the extent of dilution, *C* is the concentration, in mg per mL, of USP Clavulanate Lithium RS in the *Standard preparation*, *P* is the designated potency, in μg of clavulanic acid per mg, of the USP Clavulanate Lithium RS, and r_U and r_S are the clavulanic acid peak responses obtained from the *Assay preparation* and the *Standard preparation*, respectively.

Amoxicillin and Clavulanate Potassium Tablets

» Amoxicillin and Clavulanate Potassium Tablets contain the equivalent of not less than 90.0 percent and not more than 120.0 percent of the labeled amounts of amoxicillin ($C_{16}H_{19}N_3O_5S$) and clavulanic acid ($C_8H_9NO_5$).

Packaging and storage—Preserve in tight containers.

Labeling—Label chewable Tablets to include the word "chewable" in juxtaposition to the official name. The labeling indicates that chewable Tablets may be chewed before being swallowed or may be swallowed whole. Tablets intended for veterinary use only are so labeled.

Reference standards—*USP Amoxicillin Reference Standard*—Do not dry before using. *USP Clavulanate Lithium Reference Standard*—Keep container tightly closed and protected from light. Do not dry before using.

Identification—The retention times of the major peaks in the chromatogram of the *Assay preparation* correspond to those of the *Standard preparation*, as obtained in the *Assay*.

Disintegration ⟨701⟩: For Tablets labeled for veterinary use only, 30 minutes, simulated gastric fluid TS being substituted for water in the test.

Dissolution ⟨711⟩—[NOTE—Tablets labeled for veterinary use only are exempt from this requirement.]

Medium: water; 900 mL.
Apparatus 2: 75 rpm.
Time: 30 minutes.

Procedure—Determine the amount of amoxicillin ($C_{16}H_{19}N_3O_5S$) dissolved, employing the procedure set forth in the *Assay*, making any necessary volumetric adjustments.

Tolerances—Not less than 85% (*Q*) of the labeled amount of $C_{16}H_{19}N_3O_5S$ is dissolved in 30 minutes.

Uniformity of dosage units ⟨905⟩: meet the requirements for *Weight Variation* with respect to amoxicillin and for *Content Uniformity* with respect to clavulanic acid.

Water, *Method I* ⟨921⟩: not more than 6.0% where the Tablets are labeled as being chewable; not more than 7.0% where the labeled amount of amoxicillin in each Tablet is 250 mg or less; not more than 10.0% where the labeled amount of amoxicillin in each Tablet is greater than 250 mg.

Assay—

pH 4.4 sodium phosphate buffer, Mobile phase, Standard preparation, and *Chromatographic system*—Prepare as directed in the *Assay* under *Amoxicillin and Clavulanate Potassium for Oral Suspension.*

Assay preparation—Dissolve not less than 10 Amoxicillin and Clavulanate Potassium Tablets, accurately counted, in water with the aid of mechanical stirring, transfer to a suitable volumetric flask, dilute with water to volume, and mix. Filter a portion of this solution, discarding the first 10 mL of the filtrate. Dilute an accurately measured volume of the filtrate quantitatively and stepwise with water to obtain a solution containing about 0.5 mg of amoxicillin per mL. Use this *Assay preparation* within 1 hour.

Procedure—Proceed as directed for *Procedure* in the *Assay* under *Amoxicillin and Clavulanate Potassium for Oral Suspension.* Calculate the quantity, in mg, of amoxicillin ($C_{16}H_{19}N_3O_5S$) in each Tablet taken by the formula:

$$(L/D)(CP/1000)(r_U/r_S),$$

in which *L* is the labeled quantity, in mg, of amoxicillin in each Tablet, *D* is the concentration, in mg per mL, of amoxicillin in the *Assay preparation* on the basis of the number of Tablets taken, the labeled quantity of amoxicillin in each Tablet, and the extent of dilution, *C* is the concentration, in mg per mL, of USP Amoxicillin RS in the *Standard preparation, P* is the potency, in µg of amoxicillin per mg, of USP Amoxicillin RS, and r_U and r_S are the amoxicillin peak responses obtained from the *Assay preparation* and the *Standard preparation*, respectively. Calculate the quantity, in mg, of clavulanic acid ($C_8H_9NO_5$) in each Tablet taken by the formula:

$$(L/D)(CP/1000)(r_U/r_S),$$

in which *L* is the labeled quantity, in mg, of clavulanic acid in each Tablet, *D* is the concentration, in mg per mL, of clavulanic acid in the *Assay preparation*, on the basis of the number of Tablets taken, the labeled quantity of clavulanic acid in each Tablet, and the extent of dilution, *C* is the concentration, in mg per mL, of USP Clavulanate Lithium RS in the *Standard preparation, P* is the designated potency, in µg of clavulanic acid per mg, of USP Clavulanate Lithium RS, and r_U and r_S are the clavulanic acid peak responses obtained from the *Assay preparation* and the *Standard preparation*, respectively.

Amphetamine Sulfate

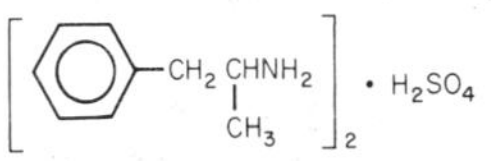

$(C_9H_{13}N)_2 \cdot H_2SO_4$ 368.49

Benzeneethanamine, α-methyl-, sulfate (2:1), (±)-.

(±)-α-Methylphenethylamine sulfate (2:1) [*60-13-9*].

» Amphetamine Sulfate, dried at 105° for 2 hours, contains not less than 98.0 percent and not more than 100.5 percent of $(C_9H_{13}N)_2 \cdot H_2SO_4$.

Packaging and storage—Preserve in well-closed containers.

Reference standard—*USP Dextroamphetamine Sulfate Reference Standard*—Dry at 105° for 2 hours before using.

Identification—

A: Dissolve about 100 mg in 5 mL of water, add 5 mL of 1 *N* sodium hydroxide, cool to about 10°, add 1 mL of a mixture of 1 volume of benzoyl chloride and 2 volumes of absolute ether, insert the stopper, and shake well for 3 minutes. Filter the precipitate, wash with about 10 mL of cold water, and recrystallize from diluted alcohol: the crystals of the benzoyl derivative of amphetamine so obtained, after drying at 80° for 2 hours, melt between 131° and 135°, the procedure for *Class I* being used (see *Melting Range or Temperature* ⟨741⟩).

B: A solution (1 in 10) responds to the tests for *Sulfate* ⟨191⟩.

Loss on drying ⟨731⟩—Dry it at 105° for 2 hours: it loses not more than 1.0% of its weight.

Residue on ignition ⟨281⟩: not more than 0.2%.

Dextroamphetamine—A solution (1 in 50) is optically inactive.

Ordinary impurities ⟨466⟩—

Test solution: methanol.

Standard solution: methanol.

Eluant: a mixture of methanol and ammonium hydroxide (50:1).

Visualization: 1.

Assay—

Standard preparation—Prepare as directed under *Amphetamine Assay* ⟨331⟩.

Assay preparation—Dissolve about 125 mg of Amphetamine Sulfate, previously dried and accurately weighed, in 25 mL of hydrochloric acid solution (1 in 100) in a 50-mL volumetric flask, dilute with the solvent to volume, and mix. Pipet 2.0 mL of the solution into a 100-mL beaker containing 3 g of purified siliceous earth, and mix until a fluffy mixture is obtained.

Procedure—Proceed as directed under *Amphetamine Assay* ⟨331⟩. Calculate the quantity, in mg, of $(C_9H_{13}N)_2 \cdot H_2SO_4$ in the portion of Amphetamine Sulfate taken by the formula:

$$0.25C[(A_{U257} - A_{U280})/(A_{S257} - A_{S280})],$$

in which *C* is the concentration, in µg per mL, of USP Dextroamphetamine Sulfate RS in the *Standard preparation.*

Amphetamine Sulfate Tablets

» Amphetamine Sulfate Tablets contain not less than 93.0 percent and not more than 107.0 percent of the labeled amount of $(C_9H_{13}N)_2 \cdot H_2SO_4$.

Packaging and storage—Preserve in well-closed containers.

Reference standard—*USP Dextroamphetamine Sulfate Reference Standard*—Dry at 105° for 2 hours before using.

Identification—Macerate a quantity of powdered Tablets, equivalent to about 50 mg of amphetamine sulfate, with 10 mL of water for 30 minutes, and filter into a small flask. To the filtrate add 3 mL of 1 *N* sodium hydroxide. Cool to about 10° to 15°, add 1 mL of a mixture of 1 volume of benzoyl chloride and 2 volumes of absolute ether, insert the stopper, and shake well for 3 minutes. Filter the precipitate, wash with about 15 mL of cold water, and recrystallize twice from diluted alcohol: the crystals of the benzoyl derivative of amphetamine so obtained, after drying at 80° for 2 hours, melt between 131° and 135°, the procedure for *Class I* being used (see *Melting Range or Temperature* ⟨741⟩).

Dissolution ⟨711⟩—

Medium: water; 500 mL.

Apparatus 1: 100 rpm.

Time: 45 minutes.

Mobile phase—Dissolve 1.1 g of sodium 1-heptanesulfonate in 575 mL of water. Add 25 mL of dilute glacial acetic acid (14 in 100) and 400 mL of methanol. Adjust by the dropwise addition of glacial acetic acid to a pH of 3.3 ± 0.1, if necessary, filter, and degas the solution. Make adjustments if necessary (see *System Suitability* under *Chromatography* ⟨621⟩).

Chromatographic system (see *Chromatography* ⟨621⟩)—The liquid chromatograph is equipped with a 254-nm detector and a 3.9-mm × 30-cm column that contains packing L1. The flow rate is about 1 mL per minute. Chromatograph replicate injections of the Standard solution, and record the peak responses as directed under *Procedure:* the relative standard deviation is not more than 2.0%.

Procedure—Inject a volume (about 500 µL) of a filtered portion of the solution under test into the chromatograph, record the chromatogram, and measure the response for the major peak. Calculate the quantity of $(C_9H_{13}N)_2 . H_2SO_4$ dissolved in comparison with a Standard solution having a known concentration of USP Dextroamphetamine Sulfate RS in the same medium and similarly chromatographed.

Tolerances—Not less than 75% (*Q*) of the labeled amount of $(C_9H_{13}N)_2 . H_2SO_4$ is dissolved in 45 minutes.

Uniformity of dosage units ⟨905⟩: meet the requirements.

Assay—

Standard preparation—Prepare as directed under *Amphetamine Assay* ⟨331⟩.

Assay preparation—Weigh and finely powder not less than 20 Amphetamine Sulfate Tablets. Transfer an accurately weighed portion of the powder, equivalent to about 5 mg of amphetamine sulfate, to a 100-mL beaker, add 2 mL of hydrochloric acid solution (1 in 100), swirl gently to wet the powder thoroughly, warm on a steam bath for about 1 minute, with occasional gentle swirling, and cool. Add 3 g of purified siliceous earth, and mix until a fluffy mixture is obtained.

Procedure—Proceed as directed under *Amphetamine Assay* ⟨331⟩. Calculate the quantity, in mg, of $(C_9H_{13}N)_2 . H_2SO_4$ in the portion of Tablets taken by the formula:

$$0.01C[(A_{U257} - A_{U280})/(A_{S257} - A_{S280})],$$

in which *C* is the concentration, in µg per mL, of USP Dextroamphetamine Sulfate RS in the *Standard preparation.*

Amphotericin B

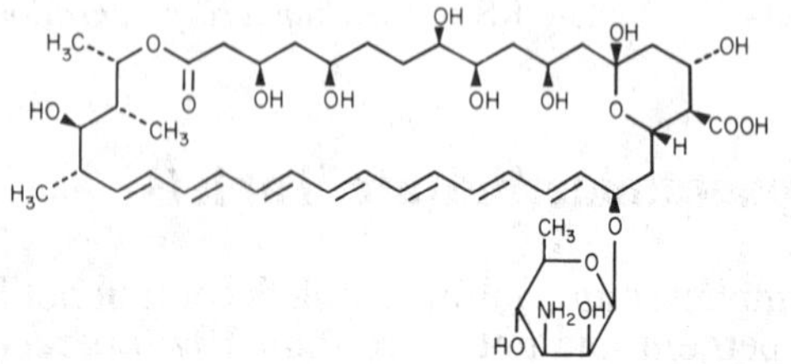

$C_{47}H_{73}NO_{17}$ 924.09

Amphotericin B.

Amphotericin B.

[1*R*-(1*R**,3*S**,5*R**,6*R**,9*R**,11*R**,15*S**,16*R**,17*R**,18*S**,-19*E*,21*E*,23*E*,25*E*,27*E*,29*E*,31*E*,33*R**,35*S**,36*R**,-37*S**)]-33-[(3-Amino-3,6-dideoxy-β-D-mannopyranosyl)-oxy]-1,3,5,6,9,11,17,37-octahydroxy-15,16,18-trimethyl-13-oxo-14,39-dioxabicyclo[33.3.1]nonatriaconta-19,21,23,25,-27,29,31-heptaene-36-carboxylic acid [*1397-89-3*].

» Amphotericin B has a potency of not less than 750 µg of $C_{47}H_{73}NO_{17}$ per mg, calculated on the dried basis.

Packaging and storage—Preserve in tight, light-resistant containers, in a cold place.

Labeling—Label it to state whether it is intended for use in preparing dermatological and oral dosage forms or parenteral dosage forms.

Reference standards—*USP Amphotericin B Reference Standard*—Dry in vacuum at a pressure not exceeding 5 mm of mercury at 60° for 3 hours before using. *USP Nystatin Reference Standard*—Dry in vacuum at a pressure not exceeding 5 mm of mercury at 40° for 2 hours before using.

Identification—The ultraviolet absorption spectrum of the *Test preparation,* prepared as directed under *Amphotericin A,* exhibits maxima in the range of 240 nm to 320 nm at the same wavelengths as that of the *Amphotericin B standard preparation* prepared under *Amphotericin A,* except that an extra peak may occur at about 304 nm in the spectrum of the *Test preparation.* The ultraviolet and visible absorption spectra of a solution obtained by diluting the *Test preparation* with 9 volumes of methanol exhibits maxima in the range of 320 nm to 400 nm at the same wavelengths as that of a similar preparation of the *Amphotericin B standard preparation.*

Loss on drying ⟨731⟩—Dry about 100 mg, accurately weighed, in a capillary-stoppered bottle in vacuum at a pressure not exceeding 5 mm of mercury at 60° for 3 hours: it loses not more than 5.0% of its weight.

Residue on ignition ⟨281⟩: not more than 0.5%, the charred residue being moistened with 2 mL of nitric acid and 5 drops of sulfuric acid. [NOTE—Amphotericin B intended for use in preparing dermatological creams, lotions, and ointments, and oral suspensions and capsules, yields not more than 3.0%.]

Amphotericin A—

Test preparation—Dissolve about 50 mg of Amphotericin B, accurately weighed, in 10.0 mL of dimethyl sulfoxide in a 50-mL volumetric flask, dilute with methanol to volume, and mix. Transfer 4.0 mL of this solution to a 50-mL volumetric flask, dilute with methanol to volume, and mix.

Nystatin standard preparation—Dissolve about 20 mg of USP Nystatin RS, accurately weighed, in 40.0 mL of dimethyl sulfoxide in a 200-mL volumetric flask, dilute with methanol to volume, and mix. Transfer 4.0 mL of this solution to a 50-mL volumetric flask, dilute with methanol to volume, and mix.

Amphotericin B standard preparation—Dissolve about 50 mg of Amphotericin B RS, accurately weighed, in 10.0 mL of dimethyl sulfoxide in a 50-mL volumetric flask, dilute with methanol to volume, and mix. Transfer 4.0 mL of this solution to a 50-mL volumetric flask, dilute with methanol to volume, and mix. Prepare this solution fresh daily.

Procedure—Concomitantly determine the absorbances of the *Standard preparations* and the *Test preparation* in 1-cm cells at 304 nm and at 282 nm, with a suitable spectrophotometer, using a 1 in 62.5 solution of dimethyl sulfoxide in methanol as the blank. Calculate the percentage of amphotericin A by the formula:

$$25W_N[(A_{B_{282}} \times A_{U_{304}}) - (A_{B_{304}} \times A_{U_{282}})]/[(A_{B_{282}} \times A_{N_{304}}) - (A_{B_{304}} \times A_{N_{282}})]W_U,$$

in which W_N is the weight, in mg, of USP Nystatin RS taken, $A_{B_{282}}$ and $A_{B_{304}}$ are the absorbances of the *Amphotericin B standard preparation* at 282 nm and 304 nm, respectively, $A_{N_{282}}$ and $A_{N_{304}}$ are the absorbances of the *Nystatin standard preparation* at 282 nm and 304 nm, respectively, $A_{U_{282}}$ and $A_{U_{304}}$ are the absorbances of the *Test preparation* at 282 nm and 304 nm, respectively, and W_U is the weight, in mg, of the Amphotericin B taken: not more than 5%, calculated on the dried basis, is found. [NOTE—Amphotericin B intended for use in preparing dermatological creams, lotions, and ointments, and oral suspensions and capsules, contains not more than 15% of amphotericin A, calculated on the dried basis.]

Assay—Proceed with Amphotericin B as directed under *Antibiotics—Microbial Assays* ⟨81⟩.

Amphotericin B Cream

» Amphotericin B Cream contains not less than 90.0 percent and not more than 125.0 percent of the labeled amount of amphotericin B.

Packaging and storage—Preserve in collapsible tubes, or in other well-closed containers.

Reference standard—*USP Amphotericin B Reference Standard*—Dry in vacuum at a pressure not exceeding 5 mm of mercury at 60° for 3 hours before using.

Minimum fill ⟨755⟩: meets the requirements.

Assay—Proceed as directed for amphotericin B under *Antibiotics—Microbial Assays* ⟨81⟩, blending a suitable accurately weighed portion of Amphotericin B Cream in a high-speed blender with a sufficient accurately measured volume of dimethyl sulfoxide to give a convenient concentration. Dilute an accurately measured volume of this solution quantitatively with dimethyl sulfoxide to obtain a stock solution having a concentration of about 20 µg of amphotericin B per mL. Dilute an accurately measured volume of this stock solution quantitatively with *Buffer No. 10* to obtain a *Test Dilution* having a concentration assumed to be equal to the median dose level of the Standard.

Amphotericin B for Injection

» Amphotericin B for Injection is a sterile complex of amphotericin B and deoxycholate sodium and one or more suitable buffers. It contains not less than 90.0 percent and not more than 120.0 percent of the labeled amount of $C_{47}H_{73}NO_{17}$.

Packaging and storage—Preserve in *Containers for Sterile Solids* as described under *Injections* ⟨1⟩, in a refrigerator and protected from light.

Labeling—Label it to indicate that it is intended for use by intravenous infusion to hospitalized patients only, and that the solution should be protected from light during administration.

Reference standard—*USP Amphotericin B Reference Standard*—Dry in vacuum at a pressure not exceeding 5 mm of mercury at 60° for 3 hours before using.

Pyrogen—It meets the requirements of the *Pyrogen Test* ⟨151⟩, the test dose being 0.5 mL per kg of a solution in Sterile Water for Injection containing 2.0 mg of amphotericin B per mL, except that under *Test Interpretation and Continuation* temperatures of 0.6° are replaced by 1.1°, temperatures of 1.4° by 3.0°, and the temperature of 3.7° by 8.0°.

Sterility—It meets the requirements under *Sterility Tests* ⟨71⟩, when tested as directed in the section, *Test Procedures Using Membrane Filtration*, 50 mg from each container being tested.

pH ⟨791⟩: between 7.2 and 8.0, in an aqueous solution containing 10 mg of amphotericin B per mL.

Loss on drying ⟨731⟩—Dry about 100 mg in a capillary-stoppered bottle in vacuum at a pressure not exceeding 5 mm of mercury at 60° for 3 hours: it loses not more than 8.0% of its weight.

Other requirements—It meets the requirements for *Uniformity of Dosage Units* ⟨905⟩ and *Labeling* under *Injections* ⟨1⟩.

Assay—

Assay preparation 1 (where it is packaged as a single-dose container)—Constitute Amphotericin B for Injection as directed in the labeling. Withdraw all of the withdrawable contents, using a suitable hypodermic needle and syringe, and dilute quantitatively and stepwise with dimethyl sulfoxide to obtain a solution containing about 20 µg of amphotericin B per mL.

Assay preparation 2 (where the labeling states the quantity of amphotericin B in a given volume of constituted solution)—Constitute Amphotericin B for Injection as directed in the labeling. Withdraw an accurately measured volume of the resultant solution, using a suitable hypodermic needle and syringe, and dilute quantitatively and stepwise with dimethyl sulfoxide to obtain a solution containing about 20 µg of amphotericin B per mL.

Procedure—Proceed as directed for amphotericin B under *Antibiotics—Microbial Assays* ⟨81⟩, using an accurately measured volume of *Assay preparation* diluted quantitatively and stepwise with *Buffer No. 10* to obtain a *Test Dilution* having a concentration assumed to be equal to the median dose level of the Standard.

Amphotericin B Lotion

» Amphotericin B Lotion contains not less than 90.0 percent and not more than 125.0 percent of the labeled amount of amphotericin B.

Packaging and storage—Preserve in well-closed containers.

Reference standard—*USP Amphotericin B Reference Standard*—Dry in vacuum at a pressure not exceeding 5 mm of mercury at 60° for 3 hours before using.

Minimum fill ⟨755⟩: meets the requirements.

pH ⟨791⟩: between 5.0 and 7.0.

Assay—Proceed as directed for amphotericin B under *Antibiotics—Microbial Assays* ⟨81⟩, dissolving a suitable accurately measured volume of Amphotericin B Lotion quantitatively in sufficient dimethyl sulfoxide to give a convenient concentration. Dilute an accurately measured volume of this solution quantitatively with dimethyl sulfoxide to obtain a stock solution having a concentration of about 20 µg of amphotericin B per mL. Dilute an accurately measured volume of this stock solution quantitatively with *Buffer No. 10* to obtain a *Test Dilution* having a concentration assumed to be equal to the median dose level of the Standard.

Amphotericin B Ointment

» Amphotericin B Ointment is Amphotericin B in a suitable ointment base. It contains not less than 90.0 percent and not more than 125.0 percent of the labeled amount of amphotericin B.

Packaging and storage—Preserve in collapsible tubes, or in other well-closed containers.

Reference standard—*USP Amphotericin B Reference Standard*—Dry in vacuum at a pressure not exceeding 5 mm of mercury at 60° for 3 hours before using.

Minimum fill ⟨755⟩: meets the requirements.

Water, *Method I* ⟨921⟩: not more than 1.0%, 20 mL of a mixture of carbon tetrachloride, chloroform, and methanol (2:2:1) being used in place of methanol in the titration vessel.

Assay—Proceed as directed for amphotericin B under *Antibiotics—Microbial Assays* ⟨81⟩, using an accurately weighed portion of Amphotericin B Ointment, equivalent to about 30 mg of amphotericin B, mixed with 10.0 mL of ether in a suitable glass-stoppered conical flask and allowed to stand, with intermittent shaking, for 1 hour. Add 20.0 mL of dimethyl sulfoxide and shake by mechanical means for 10 minutes. Dilute quantitatively and stepwise with dimethyl sulfoxide to a concentration of approximately 20 µg per mL. Dilute an accurately measured volume of this stock solution quantitatively with *Buffer No. 10* to obtain a *Test Dilution* having a concentration assumed to be equal to the median dose level of the Standard.

Ampicillin

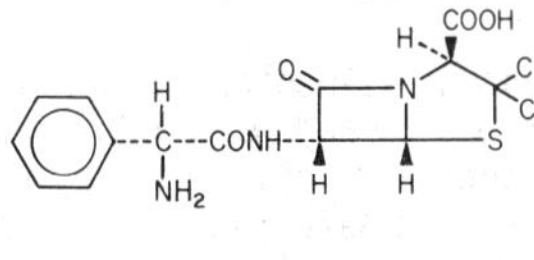

$C_{16}H_{19}N_3O_4S$ (anhydrous) 349.40

4-Thia-1-azabicyclo[3.2.0]heptane-2-carboxylic acid, 6-[(aminophenylacetyl)amino]-3,3-dimethyl-7-oxo-, [2*S*-[2α,5α,6β(*S**)]]-.

(2*S*,5*R*,6*R*)-6-[(*R*)-2-Amino-2-phenylacetamido]-3,3-dimethyl-7-oxo-4-thia-1-azabicyclo[3.2.0]heptane-2-carboxylic acid [*69-53-4*].

Trihydrate 403.45 [*7177-48-2*].

» Ampicillin is anhydrous or contains three molecules of water of hydration. It contains not less than 900 µg and not more than 1050 µg of $C_{16}H_{19}N_3O_4S$ per mg, calculated on the anhydrous basis.

Packaging and storage—Preserve in tight containers.

Labeling—Label it to indicate whether it is anhydrous or is the trihydrate. Where the quantity of ampicillin is indicated in the labeling of any preparation containing Ampicillin, this shall be understood to be in terms of anhydrous ampicillin ($C_{16}H_{19}N_3O_4S$).

Reference standard—*USP Ampicillin Reference Standard*—Do not dry before using.

Identification—The infrared absorption spectrum of a potassium bromide dispersion of it exhibits maxima only at the same wavelengths as that of a similar preparation of USP Ampicillin RS.

Crystallinity ⟨695⟩: meets the requirements.

pH ⟨791⟩: between 3.5 and 6.0, in a solution containing 10 mg per mL.

Loss on drying ⟨731⟩—Dry about 100 mg, accurately weighed, in vacuum at a pressure not exceeding 5 mm of mercury at 60° for 3 hours: anhydrous ampicillin loses not more than 2.0% of its weight; ampicillin trihydrate loses between 12.0% and 15.0% of its weight.

Dimethylaniline—

Internal standard solution—Prepare a solution of naphthalene in cyclohexane containing about 50 µg per mL.

Standard preparation—Transfer 50.0 mg of *N,N*-dimethylaniline to a 50-mL volumetric flask, add 25 mL of 1 *N* hydrochloric acid, swirl to dissolve, dilute with water to volume, and mix. Transfer 5.0 mL of the resulting solution to a 250-mL volumetric flask, dilute with water to volume, and mix. To a suitable centrifuge tube add 1.0 mL of this solution, 5.0 mL of 1 *N* sodium hydroxide, and 1.0 mL of *Internal standard solution*, shake vigorously for 1 minute, and centrifuge. Use the clear supernatant liquid as the *Standard preparation*.

Test preparation—Transfer 1.0 g of Ampicillin to a suitable centrifuge tube, add 5 mL of 1 *N* sodium hydroxide, swirl to dissolve the specimen, add 1.0 mL of *Internal standard solution*, shake vigorously for 1 minute, and centrifuge. Use the clear supernatant liquid as the *Test preparation*.

Chromatographic system (see *Chromatography* ⟨621⟩)—The gas chromatograph is equipped with a flame-ionization detector and a 2-mm × 2-m column packed with 3 percent liquid phase G3 on silanized packing S1A and is maintained at 120°. Nitrogen is used as the carrier gas, flowing at the rate of about 30 mL per minute.

Procedure—[NOTE—Use peak areas where peak responses are indicated.]—Inject equal volumes (about 20 µL) of the *Standard preparation* and the *Test preparation* into the chromatograph, record the chromatograms, and measure the responses for the major peaks. The ratio of the response of any dimethylaniline peak to the response of the naphthalene peak obtained from the *Test preparation* is not greater than that obtained from the *Standard preparation* (0.002%).

Assay—

Mobile phase—Prepare a suitable filtered and degassed mixture of water, acetonitrile, 1 *M* monobasic potassium phosphate, and 1 *N* acetic acid (909:80:10:1). Make adjustments if necessary (see *System Suitability* under *Chromatography* ⟨621⟩).

Diluent—Mix 10 mL of 1 *M* monobasic potassium phosphate and 1 mL of 1 *N* acetic acid, dilute with water to 1000 mL, and mix.

Standard preparation—Dissolve a suitable quantity of USP Ampicillin RS, accurately weighed, in *Diluent* to obtain a solution having a known concentration of about 1 mg per mL, using shaking and sonication, if necessary, to achieve complete dissolution. Use this solution promptly after preparation.

Assay preparation—Transfer an accurately weighed quantity of Ampicillin, equivalent to about 100 mg of anhydrous ampicillin, to a 100-mL volumetric flask, add about 75 mL of *Diluent*, shake and sonicate, if necessary, to achieve complete dissolution, dilute with *Diluent* to volume, and mix. Use this solution promptly after preparation.

Resolution solution—Dissolve caffeine in *Standard preparation* to obtain a solution containing about 0.12 mg per mL.

Chromatographic system (see *Chromatography* ⟨621⟩)—The liquid chromatograph is equipped with a 254-nm detector, a 4-mm × 5-cm pre-column containing 5- to 10-µm packing L1, and a 4-mm × 30-cm analytical column containing 5- to 10-µm packing L1. The flow rate is about 2 mL per minute. Chromatograph the *Resolution solution*, and record the peak responses as directed under *Procedure:* the resolution, *R*, between the caffeine and the ampicillin peaks is not less than 2.0. The relative retention times are about 0.5 for ampicillin and 1.0 for caffeine. Chromatograph the *Standard preparation*, and record the peak responses as directed under *Procedure:* the capacity factor, *k′*, is not more than 2.5, the tailing factor is not more than 1.4, and the relative standard deviation for replicate injections is not more than 2.0%.

Procedure—Separately inject equal volumes (about 20 µL) of the *Standard preparation* and the *Assay preparation* into the chromatograph, record the chromatograms, and measure the responses for the major peaks. Calculate the quantity, in µg, of ampicillin ($C_{13}H_{19}NO_4S$) in each mg of the Ampicillin taken by the formula:

$$100(CP/W)(r_U/r_S),$$

in which *C* is the concentration, in mg per mL, of USP Ampicillin RS in the *Standard preparation*, *P* is the potency, in µg of ampicillin per mg, of USP Ampicillin RS, *W* is the weight, in mg, of Ampicillin taken, and r_U and r_S are the peak responses obtained from the *Assay preparation* and the *Standard preparation*, respectively.

Ampicillin Boluses

» Ampicillin Boluses contain an amount of ampicillin (as the trihydrate) equivalent to not less than 90.0 percent and not more than 120.0 percent of the labeled amount of $C_{16}H_{19}N_3O_4S$.

Packaging and storage—Preserve in tight containers.

Labeling—Label Boluses to indicate that they are for veterinary use only.

Reference standard—*USP Ampicillin Reference Standard*—Do not dry before using.

Identification—Powder 1 or more Boluses, and prepare a solution containing the equivalent of 10 mg of ampicillin per mL in a mixture of acetone and 0.1 *N* hydrochloric acid (4:1): the resulting solution responds to the *Identification test* under *Ampicillin Capsules*.

Uniformity of dosage units ⟨905⟩: meet the requirements.

Loss on drying ⟨731⟩: not more than 5.0%.

Assay—

Standard preparation—Prepare as directed for *Standard Preparation* under *Iodometric Assay—Antibiotics* ⟨425⟩, using USP Ampicillin RS.

Assay preparation—Place not less than 5 Ampicillin Boluses in a high-speed glass blender jar containing an accurately measured volume of water, and blend for 4 ± 1 minutes. Dilute an accurately measured volume of this stock solution quantitatively and stepwise with water to obtain an *Assay preparation* containing about 1.25 mg of ampicillin per mL.

Procedure—Proceed as directed for *Procedure* under *Iodometric Assay—Antibiotics* ⟨425⟩. Calculate the quantity, in mg, of $C_{16}H_{19}N_3O_4S$ in each Bolus taken by the formula:

$$(T/D)(F/2000)(B - I),$$

in which *T* is the labeled quantity, in mg, of ampicillin in each Bolus, and *D* is the concentration, in mg per mL, of ampicillin in the *Assay preparation* on the basis of the labeled quantity in each Bolus and the extent of dilution.

Ampicillin Capsules

» Ampicillin Capsules contain an amount of ampicillin (anhydrous or as the trihydrate) equivalent to not less than 90.0 percent and not more than 120.0 percent of the labeled amount of $C_{16}H_{19}N_3O_4S$.

Packaging and storage—Preserve in tight containers.

Labeling—Label Capsules to indicate whether the ampicillin therein is in the anhydrous form or is the trihydrate.

Reference standard—*USP Ampicillin Reference Standard*—Do not dry before using.

Identification—Prepare a solution containing about 5 mg of ampicillin per mL, using powder from Ampicillin Capsules, in a mixture of acetone and 0.1 *N* hydrochloric acid (4:1). Prepare a Standard solution of USP Ampicillin RS to contain 5 mg per mL in the same solvent mixture. Apply separately 2 µL of each solution on a thin-layer chromatographic plate (see *Chromatography* ⟨621⟩) coated with a 0.25-mm layer of chromatographic silica gel mixture. Place the plate in a suitable chromatographic chamber, and develop the chromatogram in a solvent system consisting of a mixture of acetone, water, toluene, and glacial acetic acid (650:100:100:25). When the solvent front has moved about three-fourths of the length of the plate, remove the plate from the chamber, mark the solvent front, and allow to air-dry. Locate the spots on the plate by spraying lightly with a solution of ninhydrin in alcohol containing 3 mg per mL, and dry at 90° for 15 minutes: the R_f value of the principal spot obtained from the solution under test corresponds to that obtained from the Standard solution.

Dissolution ⟨711⟩—

Medium: water; 900 mL.

Apparatus 1: 100 rpm.

Time: 45 minutes.

Procedure—Determine the amount of $C_{16}H_{19}N_3O_4S$ by a suitable validated spectrophotometric analysis of a filtered portion of the solution under test, suitably diluted with *Dissolution Medium*, if necessary, in comparison with a Standard solution having a known concentration of USP Ampicillin RS in the same medium.

Tolerances—Not less than 75% (*Q*) of the labeled amount of $C_{16}H_{19}N_3O_4S$ is dissolved in 45 minutes.

Uniformity of dosage units ⟨905⟩: meet the requirements.

Loss on drying ⟨731⟩—Dry about 100 mg, accurately weighed, of the powder obtained from 4 Capsules, in vacuum at a pressure not exceeding 5 mm of mercury at 60° for 3 hours: powder from Capsules containing anhydrous ampicillin loses not more than 4.0% of its weight; powder from Capsules containing ampicillin trihydrate loses between 10.0% and 15.0% of its weight.

Assay—

Standard preparation—Prepare as directed for *Standard Preparation* under *Iodometric Assay—Antibiotics* ⟨425⟩, using USP Ampicillin RS.

Assay preparation—Place not less than 5 Ampicillin Capsules in a high-speed glass blender jar containing an accurately measured volume of water, and blend for 4 ± 1 minutes. Dilute an accurately measured volume of this stock solution quantitatively and stepwise with water to obtain an *Assay preparation* containing about 1.25 mg of ampicillin per mL.

Procedure—Proceed as directed for *Procedure* under *Iodometric Assay—Antibiotics* ⟨425⟩. Calculate the quantity, in mg, of $C_{16}H_{19}N_3O_4S$ in each Capsule taken by the formula:

$$(T/D)(F/2000)(B - I),$$

in which *T* is the labeled quantity, in mg, of ampicillin in each Capsule, and *D* is the concentration, in mg per mL, of ampicillin in the *Assay preparation* on the basis of the labeled quantity in each Capsule and the extent of dilution.

Ampicillin Soluble Powder

» Ampicillin Soluble Powder is a dry mixture of ampicillin (as the trihydrate) and one or more suitable diluents and stabilizing agents. It contains not less than 90.0 percent and not more than 120.0 percent of the labeled amount of $C_{16}H_{19}N_3O_4S$.

Packaging and storage—Preserve in tight containers.

Labeling—Label it to indicate that it is for veterinary use only.

Reference standard—*USP Ampicillin Reference Standard*—Do not dry before using.

Identification—Dissolve a quantity in a mixture of acetone and 0.1 *N* hydrochloric acid (4:1) to obtain a solution containing 10 mg of ampicillin per mL: the resulting solution responds to the *Identification test* under *Ampicillin Capsules*.

pH ⟨791⟩: between 3.5 and 6.0, in an aqueous solution containing the equivalent of 20 mg of ampicillin per mL.

Water, *Method I* ⟨921⟩: not more than 5.0%.

Assay—

Standard preparation—Prepare as directed for *Standard Preparation* under *Iodometric Assay—Antibiotics* ⟨425⟩, using USP Ampicillin RS.

Assay preparation—Transfer an accurately weighed quantity of Ampicillin Soluble Powder, equivalent to about 125 mg of ampicillin, to a 100-mL volumetric flask, dissolve in water, dilute with water to volume, and mix.

Procedure—Proceed as directed for *Procedure* under *Iodometric Assay—Antibiotics* ⟨425⟩. Calculate the quantity, in mg, of $C_{16}H_{19}N_3O_4S$ in the portion of Soluble Powder taken by the formula:

$$(F/20)(B - I).$$

Sterile Ampicillin

» Sterile Ampicillin is the trihydrate form of Ampicillin suitable for parenteral use. It contains not less than 900 µg and not more than 1050 µg of $C_{16}H_{19}N_3O_4S$ per mg, calculated on the dried basis.

Packaging and storage—Preserve in *Containers for Sterile Solids* as described under *Injections* ⟨1⟩.

Labeling—Label it to indicate that it is the trihydrate. Where the quantity of ampicillin is indicated in the labeling of any preparation containing Sterile Ampicillin, this shall be understood to be in terms of anhydrous ampicillin ($C_{16}H_{19}N_3O_4S$).

Reference standard—*USP Ampicillin Reference Standard*—Do not dry before using.

Pyrogen—It meets the requirements of the *Pyrogen Test* ⟨151⟩, the test dose being 1.0 mL per kg of a solution in 0.05 *N* sodium hydroxide containing 20 mg of ampicillin per mL.

Sterility—It meets the requirements under *Sterility Tests* ⟨71⟩, when tested as directed in the section, *Test Procedures Using Membrane Filtration*, except to dissolve 6 g in 800 mL of *Fluid D* containing sufficient sterile penicillinase to inactivate the ampicillin and swirl the vessel until solution is complete before filtering.

Loss on drying ⟨731⟩—Dry about 100 mg, accurately weighed, in vacuum at a pressure not exceeding 5 mm of mercury at 60° for 3 hours: it loses between 12.0% and 15.0% of its weight.

Other requirements—It responds to the *Identification test* and meets the requirements of the tests for *pH*, *Dimethylaniline*, and *Crystallinity* under *Ampicillin.*

Assay—Proceed with Sterile Ampicillin as directed in the *Assay* under *Ampicillin.*

Sterile Ampicillin Suspension

» Sterile Ampicillin Suspension is a sterile suspension of Ampicillin in a suitable oil vehicle. It contains an amount of ampicillin (as the trihydrate) equivalent to not less than 90.0 percent and not more than 120.0 percent of the labeled amount of $C_{16}H_{19}N_3O_4S$.

Packaging and storage—Preserve in single-dose or in multiple-dose containers, preferably of Type I giass.

Labeling—Label it to indicate that it is for veterinary use only.

Reference standard—*USP Ampicillin Reference Standard*—Do not dry before using.

Identification—Dissolve a quantity in a mixture of acetone and 0.1 *N* hydrochloric acid (4:1) to obtain a solution containing 5 mg of ampicillin per mL: the resulting solution responds to the *Identification test* under *Ampicillin Capsules.*

Sterility—It meets the requirements under *Sterility Tests* ⟨71⟩, when tested as directed in the section, *Test Procedures Using Membrane Filtration*, except to transfer the contents of each of two groups of 10 containers to separate 400-mL portions of *Fluid D* containing sufficient penicillinase to solubilize the ampicillin in the suspension. If the resultant liquid cannot be filtered, perform the test as directed in the section, *Test Procedures for Direct Transfer to Test Media*, except to use Fluid Thioglycollate Medium containing polysorbate 80 solution (1 in 200) and an amount of sterile penicillinase sufficient to inactivate the ampicillin in each tube, to use Soybean-Casein Digest Medium containing polysorbate 80 solution (1 in 200) and an amount of sterile penicillinase sufficient to inactivate the ampicillin in each tube, and to shake the tubes once daily.

Water, *Method I* ⟨921⟩: not more than 4.0%, 20 mL of a mixture of carbon tetrachloride, chloroform, and methanol (2:2:1) being used in place of methanol in the titration vessel.

Assay—

Standard preparation—Prepare as directed for *Standard Preparation* under *Iodometric Assay—Antibiotics* ⟨425⟩, using USP Ampicillin RS.

Assay preparation—Transfer an accurately measured volume of Sterile Ampicillin Suspension, equivalent to about 625 mg of ampicillin, to a separator containing 100 mL of ether. Extract with two 150-mL portions of 0.2 *N* hydrochloric acid, collecting the extracts in a 500-mL volumetric flask. Dilute with water to volume, and mix.

Procedure—Proceed as directed for *Procedure* under *Iodometric Assay—Antibiotics* ⟨425⟩. Calculate the quantity, in mg, of $C_{16}H_{19}N_3O_4S$ in each mL of the Suspension taken by the formula:

$$(F/4V)(B - I),$$

in which V is the volume, in mL, of Suspension taken.

Sterile Ampicillin for Suspension

» Sterile Ampicillin for Suspension is a dry mixture of ampicillin trihydrate and one or more suitable buffers, preservatives, stabilizers, and suspending agents. It contains the equivalent of not less than 90.0 percent and not more than 120.0 percent of the labeled amount of $C_{16}H_{19}N_3O_4S$.

Packaging and storage—Preserve in *Containers for Sterile Solids* as described under *Injections* ⟨1⟩.

Reference standard—*USP Ampicillin Reference Standard*—Do not dry before using.

Identification—Dissolve a quantity in a mixture of acetone and 0.1 *N* hydrochloric acid (4:1) to obtain a solution containing 5 mg of ampicillin per mL: the resulting solution responds to the *Identification test* under *Ampicillin Capsules.*

Pyrogen—It meets the requirements of the *Pyrogen Test* ⟨151⟩, the test dose being 1 mL per kg of a solution in 0.05 *N* sodium hydroxide containing 20 mg of ampicillin per mL.

pH ⟨791⟩: between 5.0 and 7.0, in the suspension constituted as directed in the labeling.

Loss on drying ⟨731⟩—Dry about 100 mg, accurately weighed, in vacuum at a pressure not exceeding 5 mm of mercury at 60° for 3 hours: it loses between 11.4% and 14.0% of its weight.

Other requirements—It meets the requirements for *Sterility* under *Sterile Ampicillin Suspension, Uniformity of Dosage Units* ⟨905⟩, and *Labeling* under *Injections* ⟨1⟩.

Assay—

Standard preparation—Prepare as directed for *Standard Preparation* under *Iodometric Assay—Antibiotics* ⟨425⟩, using USP Ampicillin RS.

Assay preparation—Dilute an accurately measured volume of solution of Sterile Ampicillin for Suspension, constituted as directed in the labeling, quantitatively and stepwise with water to obtain a solution containing about 1.25 mg of ampicillin per mL.

Procedure—Proceed as directed for *Procedure* under *Iodometric Assay—Antibiotics* ⟨425⟩. Calculate the quantity, in mg, of $C_{16}H_{19}N_3O_4S$ in each mL of the constituted solution of Sterile Ampicillin for Suspension taken by the formula:

$$(T/D)(F/2000)(B - I),$$

in which T is the labeled quantity, in mg per mL, of ampicillin in the constituted solution of Sterile Ampicillin for Suspension, and D is the concentration, in mg per mL, of ampicillin in the *Assay preparation* on the basis of the labeled quantity in the constituted solution of Sterile Ampicillin for Suspension and the extent of dilution.

Ampicillin for Oral Suspension

» Ampicillin for Oral Suspension contains an amount of ampicillin (anhydrous or as the trihydrate) equivalent to not less than 90.0 percent and not more than 120.0 percent of the labeled amount of $C_{16}H_{19}N_3O_4S$, when constituted as directed. It contains one or more suitable buffers, colors, flavors, preservatives, and sweetening ingredients.

Packaging and storage—Preserve in tight containers.

Labeling—Label it to indicate whether the ampicillin therein is in the anhydrous form or is the trihydrate.

Reference standard—*USP Ampicillin Reference Standard*—Do not dry before using.

Identification—Dissolve a quantity in a mixture of acetone and 0.1 *N* hydrochloric acid (4:1) to obtain a solution containing 5 mg of ampicillin per mL: the resulting solution responds to the *Identification test* under *Ampicillin Capsules.*

Uniformity of dosage units ⟨905⟩—

FOR SOLID PACKAGED IN SINGLE-UNIT CONTAINERS: meets the requirements.

pH ⟨791⟩: between 5.0 and 7.5, in the suspension constituted as directed in the labeling.

Water, *Method I* ⟨921⟩: not more than 2.5%, or not more than 5.0% if it contains ampicillin trihydrate and contains the equivalent of 100 mg of ampicillin per mL when constituted as directed in the labeling.

Assay—

Standard preparation—Prepare as directed for *Standard Preparation* under *Iodometric Assay—Antibiotics* ⟨425⟩, using USP Ampicillin RS.

Assay preparation—Dilute an accurately measured volume of Ampicillin for Oral Suspension, constituted as directed in the labeling, freshly mixed and free from air bubbles, quantitatively and stepwise with water to obtain a solution containing about 1.25 mg of ampicillin per mL.

Procedure—Proceed as directed for *Procedure* under *Iodometric Assay—Antibiotics* ⟨425⟩. Calculate the quantity, in mg, of $C_{16}H_{19}N_3O_4S$ in each mL of the constituted suspension prepared from Ampicillin for Oral Suspension taken by the formula:

$$(T/D)(F/2000)(B - I),$$

in which T is the labeled quantity, in mg per mL, of ampicillin in the constituted suspension, and D is the concentration, in mg per mL, of ampicillin in the *Assay preparation* on the basis of the labeled quantity in the constituted suspension and the extent of dilution.

Ampicillin Tablets

» Ampicillin Tablets contain an amount of Ampicillin (anhydrous form or trihydrate form) equivalent to not less than 90.0 percent and not more than 120.0 percent of the labeled amount of $C_{16}H_{19}N_3O_4S$.

Packaging and storage—Preserve in tight containers.

Labeling—Label Tablets to indicate whether the ampicillin therein is in the anhydrous form or is the trihydrate. Label chewable Tablets to indicate that they are to be chewed before being swallowed. Tablets intended for veterinary use only are so labeled.

Reference standard—*USP Ampicillin Reference Standard*—Do not dry before using.

Identification—Powder 1 or more Tablets, and prepare a solution containing 5 mg of ampicillin per mL in a mixture of acetone and 0.1 *N* hydrochloric acid (4:1): the resulting solution responds to the *Identification test* under *Ampicillin Capsules.*

Dissolution ⟨711⟩—

Medium: water; 900 mL.

Apparatus 1: 100 rpm.

Time: 45 minutes.

Procedure—Determine the amount of $C_{16}H_{19}N_3O_4S$ by a suitable validated spectrophotometric analysis of a filtered portion of the solution under test, suitably diluted with *Dissolution Medium,* if necessary, in comparison with a Standard solution having a known concentration of USP Ampicillin RS in the same medium.

Tolerances—Not less than 75% (*Q*) of the labeled amount of $C_{16}H_{19}N_3O_4S$ is dissolved in 45 minutes.

Uniformity of dosage units ⟨905⟩: meet the requirements.

Loss on drying ⟨731⟩—Dry about 100 mg of powdered Tablets in vacuum at a pressure not exceeding 5 mm of mercury at 60° for 3 hours: where the Tablets contain anhydrous ampicillin, powder from nonchewable Tablets loses not more than 4.0% of its weight; powder from chewable Tablets loses not more than 3.0% of its weight. Where the Tablets contain ampicillin as the trihydrate, powder from Tablets for veterinary use loses not more than 13.0% of its weight.

Water, *Method I* ⟨921⟩ (where chewable Tablets contain ampicillin trihydrate): not more than 5.0%; (where nonchewable Tablets contain ampicillin trihydrate): between 9.5% and 12.0%.

Assay—

Standard preparation—Prepare as directed for *Standard Preparation* under *Iodometric Assay—Antibiotics* ⟨425⟩, using USP Ampicillin RS.

Assay preparation—Place not less than 5 Ampicillin Tablets in a high-speed glass blender jar containing an accurately measured volume of water, and blend for 4 ± 1 minutes. Dilute an accurately measured volume of this stock solution with water to obtain an *Assay preparation* containing about 1.25 mg of ampicillin per mL.

Procedure—Proceed as directed for *Procedure* under *Iodometric Assay—Antibiotics* ⟨425⟩. Calculate the quantity, in mg, of $C_{16}H_{19}N_3O_4S$ in each Tablet taken by the formula:

$$(T/D)(F/2000)(B - I),$$

in which *T* is the labeled quantity, in mg, of ampicillin in each Tablet, and *D* is the concentration, in mg per mL, of ampicillin in the *Assay preparation* on the basis of the labeled quantity in each Tablet and the extent of dilution.

Ampicillin and Probenecid Capsules

» Ampicillin and Probenecid Capsules contain an amount of ampicillin (as the trihydrate) equivalent to not less than 90.0 percent and not more than 120.0 percent of the labeled amount of ampicillin ($C_{16}H_{19}N_3O_4S$) and not less than 90.0 percent and not more than 110.0 percent of the labeled amount of probenecid ($C_{13}H_{19}NO_4S$).

Packaging and storage—Preserve in tight containers.

Reference standards—*USP Ampicillin Reference Standard*—Do not dry before using. *USP Probenecid Reference Standard*—Dry at 105° for 4 hours before using.

Loss on drying ⟨731⟩—Dry about 100 mg, accurately weighed, of the powder obtained from 4 Capsules in vacuum at a pressure not exceeding 5 mm of mercury at 60° for 3 hours: it loses between 8.5% and 13.0% of its weight.

Uniformity of dosage units ⟨905⟩: meet the requirements.

Assay for ampicillin—

Standard preparation—Prepare as directed for *Standard Preparation* under *Iodometric Assay—Antibiotics* ⟨425⟩, using USP Ampicillin RS.

Assay preparation—Place not less than 5 Ampicillin and Probenecid Capsules in a high-speed glass blender jar containing an accurately measured volume of water, and blend for about 10 minutes. Dilute an accurately measured volume of this stock solution quantitatively and stepwise with water to obtain an *Assay preparation* containing about 1.25 mg of ampicillin per mL.

Procedure—Proceed as directed for *Procedure* under *Iodometric Assay—Antibiotics* ⟨425⟩. Calculate the quantity, in mg, of $C_{16}H_{19}N_3O_4S$ in each Capsule taken by the formula:

$$(L/D)(F/2000)(B - I),$$

in which *L* is the labeled quantity, in mg, of ampicillin in each Capsule, and *D* is the concentration, in mg per mL, of ampicillin in the *Assay preparation* on the basis of the labeled quantity per Capsule and the extent of dilution.

Assay for probenecid—

Standard preparation—Dissolve an accurately weighed portion of USP Probenecid RS in sodium carbonate solution (1 in 100) to obtain a solution having a known concentration of about 1 mg per mL.

Assay preparation—Place not less than 5 Ampicillin and Probenecid Capsules in a high-speed glass blender jar containing an accurately measured volume of sodium carbonate solution (1 in 100) sufficient to provide a solution containing about 1 mg of probenecid per mL, blend for about 10 minutes, and filter.

Procedure—Transfer 2.0 mL of the clear *Assay preparation* to a 125-mL separator, and add 8.0 mL of 1.0 *N* hydrochloric acid. Extract this solution with four 20-mL portions of chloroform, filtering each extract through a glass wool pledget and 6 g of chloroform-washed anhydrous sodium sulfate into a 100-mL volumetric flask. Wash the pledget and the sodium sulfate with chloroform, collecting the washings in the 100-mL volumetric flask, dilute with chloroform to volume, and mix. Treat 2.0 mL of the *Standard preparation* in the same manner. Concomitantly determine the absorbances of the solutions from the *Assay preparation* and the *Standard preparation* at the wavelength of maximum absorbance at about 257 nm, with a suitable spectrophotometer, using chloroform washed with sodium carbonate solution (1 in 100) as the blank. Calculate the quantity, in mg, of $C_{13}H_{19}NO_4S$ in each Capsule taken by the formula:

$$C(L/D)(A_U/A_S),$$

in which *C* is the concentration, in mg per mL, of USP Probenecid RS in the *Standard preparation,* *L* is the labeled quantity, in mg, of probenecid in each Capsule, *D* is the concentration, in mg per mL, of probenecid in the *Assay preparation* on the basis of the labeled quantity per Capsule and the extent of dilution, and A_U and A_S are the absorbances of the solutions from the *Assay preparation* and the *Standard preparation,* respectively.

Ampicillin and Probenecid for Oral Suspension

» Ampicillin and Probenecid for Oral Suspension contains an amount of ampicillin (as the trihydrate) equivalent to not less than 90.0 percent and not more than 120.0 percent of the labeled amount of ampicillin ($C_{16}H_{19}N_3O_4S$) and not less than 90.0 percent and not more than 110.0 percent of the labeled amount of probenecid ($C_{13}H_{19}NO_4S$). It contains one or more suitable colors, flavors, and suspending agents.

Packaging and storage—Preserve in tight, unit-dose containers.

Reference standards—*USP Ampicillin Reference Standard*—Do not dry before using. *USP Probenecid Reference Standard*—Dry at 105° for 4 hours before using.

Uniformity of dosage units ⟨905⟩—

FOR SOLID PACKAGED IN SINGLE-UNIT CONTAINERS: meets the requirements for *Content Uniformity* with respect to ampicillin and probenecid.

pH ⟨791⟩: between 5.0 and 7.5, in the suspension constituted as directed in the labeling.

Water, *Method I* ⟨921⟩: not more than 5.0%.

Assay for ampicillin—

Standard preparation—Prepare as directed for *Standard Preparation* under *Iodometric Assay—Antibiotics* ⟨425⟩, using USP Ampicillin RS.

Assay preparation—Constitute Ampicillin and Probenecid for Oral Suspension as directed in the labeling, and mix. Transfer the resulting suspension to a high-speed glass blender jar containing sufficient water to make 500.0 mL, and blend for about 10 minutes. Dilute an accurately measured volume of this stock solution quantitatively with water to obtain an *Assay preparation* containing about 1.25 mg of ampicillin per mL.

Procedure—Proceed as directed for *Procedure* under *Iodometric Assay—Antibiotics* ⟨425⟩. Calculate the quantity, in mg, of $C_{16}H_{19}N_3O_4S$ in the Ampicillin and Probenecid for Oral Suspension taken by the formula:

$$(L/D)(F/2000)(B - I),$$

in which L is the labeled quantity, in mg, of ampicillin in the Ampicillin and Probenecid for Oral Suspension, and D is the concentration, in mg per mL, of ampicillin in the *Assay preparation* on the basis of the labeled quantity in the Ampicillin and Probenecid for Oral Suspension and the extent of dilution.

Assay for probenecid—

Standard preparation—Dissolve an accurately weighed portion of USP Probenecid RS in sodium carbonate solution (1 in 100) to obtain a solution having a known concentration of about 1 mg per mL.

Assay preparation—Constitute Ampicillin and Probenecid for Oral Suspension as directed in the labeling, and mix. Dilute the resulting suspension quantitatively with sodium carbonate solution (1 in 100) to obtain a solution containing about 1 mg of probenecid per mL, mix, and filter.

Procedure—Transfer 2.0 mL of the clear *Assay preparation* to a 125-mL separator, and add 8.0 mL of 1.0 *N* hydrochloric acid. Extract this solution with four 20-mL portions of chloroform, filtering each extract through a glass wool pledget and 6 g of chloroform-washed anhydrous sodium sulfate into a 100-mL volumetric flask. Wash the pledget and the sodium sulfate with chloroform, collecting the washings in the 100-mL volumetric flask, dilute with chloroform to volume, and mix. Treat 2.0 mL of the *Standard preparation* in the same manner. Concomitantly determine the absorbances of the solutions from the *Assay preparation* and the *Standard preparation* at the wavelength of maximum absorbance at about 257 nm, with a suitable spectrophotometer, using chloroform washed with sodium carbonate solution (1 in 100) as the blank. Calculate the quantity, in mg, of $C_{13}H_{19}NO_4S$ in the Ampicillin and Probenecid for Oral Suspension taken by the formula:

$$C(L/D)(A_U/A_S),$$

in which C is the concentration, in mg per mL, of USP Probenecid RS in the *Standard preparation*, L is the labeled quantity, in mg, of probenecid in the Ampicillin and Probenecid for Oral Suspension, D is the concentration, in mg per mL, of probenecid in the *Assay preparation* on the basis of the labeled quantity in the Ampicillin and Probenecid for Oral Suspension and the extent of dilution, and A_U and A_S are the absorbances of the solutions from the *Assay preparation* and the *Standard preparation*, respectively.

Sterile Ampicillin Sodium

$C_{16}H_{18}N_3NaO_4S$ 371.39

4-Thia-1-azabicyclo[3.2.0]heptane-2-carboxylic acid, 6-[(aminophenylacetyl)amino]-3,3-dimethyl-7-oxo-, monosodium salt, [2*S*-[2α,5α,6β(*S**)]]-.

Monosodium D-(−)-6-(2-amino-2-phenylacetamido)-3,3-dimethyl-7-oxo-4-thia-1-azabicyclo[3.2.0]heptane-2-carboxylate [*69-52-3*].

» Sterile Ampicillin Sodium has a potency equivalent to not less than 845 µg and not more than 988 µg of ampicillin ($C_{16}H_{19}N_3O_4S$) per mg, calculated on the anhydrous basis. In addition, where packaged for dispensing, it contains the equivalent of not less than 90.0 percent and not more than 115.0 percent of the labeled amount of ampicillin.

Packaging and storage—Preserve in *Containers for Sterile Solids* as described under *Injections* ⟨1⟩. Protect the constituted solution from freezing.

Reference standards—*USP Ampicillin Reference Standard*—Do not dry before using. *USP Ampicillin Sodium Reference Standard*—Do not dry before using.

Constituted solution—At the time of use, the constituted solution prepared from Sterile Ampicillin Sodium meets the requirements for *Constituted Solutions* under *Injections* ⟨1⟩.

Identification—The infrared absorption spectrum of a mineral oil dispersion of it exhibits maxima only at the same wavelengths as that of a similar preparation of USP Ampicillin Sodium RS.

Crystallinity ⟨695⟩: meets the requirements.

Pyrogen—It meets the requirements of the *Pyrogen Test* ⟨151⟩, the test dose being 1 mL per kg of a solution in Sterile Water for Injection containing 20 mg of ampicillin per mL.

pH ⟨791⟩: between 8.0 and 10, in a solution containing 10.0 mg of ampicillin per mL.

Water, *Method I* ⟨921⟩: not more than 2.0%.

Particulate matter ⟨788⟩: meets the requirements under *Small-volume Injections.*

Dimethylaniline—

Internal standard solution—Dissolve 75 mg of *N,N*-diethylaniline in 25 mL of 1 *N* hydrochloric acid, and dilute quantitatively and stepwise with water to obtain a solution containing about 30 µg per mL.

Standard preparation—Transfer 50.0 mg of *N,N*-dimethylaniline to a 50-mL volumetric flask, add 25 mL of 1 *N* hydrochloric acid, swirl to dissolve, dilute with water to volume, and mix. Transfer 2.0 mL of the resulting solution to a 100-mL volumetric flask, dilute with water to volume, and mix. To a suitable centrifuge tube add 1.0 mL of this solution, 1.0 mL of 1.25 *N* sodium hydroxide, 1.0 mL of *Internal standard solution*, and 1.0 mL of cyclohexane, shake vigorously for 1 minute, and centrifuge. Use the clear supernatant liquid as the *Standard preparation.*

Test preparation—Transfer 1.0 g of Sterile Ampicillin Sodium to a suitable centrifuge tube, add 2 mL of 1.25 *N* sodium hydroxide, swirl to dissolve the specimen, add 1.0 mL of *Internal standard solution* and 1.0 mL of cyclohexane, shake vigorously for 1 minute, and centrifuge. Use the clear supernatant liquid as the *Test preparation.*

Chromatographic system (see *Chromatography* ⟨621⟩)—The gas chromatograph is equipped with a flame-ionization detector and a 4-mm × 1.5-m column packed with 3 percent liquid phase G26 on silanized packing S1A and is maintained at 80°. Nitrogen is used as the carrier gas, flowing at the rate of about 30 mL per minute.

Procedure—[NOTE—Use peak areas where peak responses are indicated.] Inject equal volumes (about 20 µL) of the *Standard preparation* and the *Test preparation* into the chromatograph, record the chromatograms, and measure the responses for the major peaks. The ratio of the response of any dimethylaniline peak to the response of the *N,N*-diethylaniline peak obtained from the *Test preparation* is not greater than that obtained from the *Standard preparation* (0.002%).

Other requirements—It meets the requirements for *Sterility Tests* ⟨71⟩, *Uniformity of Dosage Units* ⟨905⟩, and *Labeling* under *Injections* ⟨1⟩.

Assay—

Mobile phase, Diluent, Standard preparation, Resolution solution, and *Chromatographic system*—Prepare as directed in the *Assay* under *Ampicillin.*

Assay preparation 1—[NOTE—Ampicillin sodium is hygroscopic. Minimize exposure to the atmosphere, and weigh promptly.] Transfer an accurately weighed quantity of Sterile Ampicillin Sodium, equivalent to about 100 mg of anhydrous ampicillin, to a 100-mL volumetric flask, dissolve in *Diluent*, dilute with *Diluent* to volume, and mix. Use this solution promptly after preparation.

Assay preparation 2 (where it is packaged for dispensing and is represented as being in a single-dose container)—Constitute Sterile Ampicillin Sodium in an accurately measured volume of *Diluent*, corresponding to the volume of solvent specified in the labeling. Withdraw all of the withdrawable contents, using a suitable hypodermic needle and syringe, and dilute quantitatively with *Diluent* to obtain a solution containing about 1 mg of ampicillin per mL. Use this solution promptly after preparation.

Assay preparation 3 (where the label states the quantity of ampicillin in a given volume of constituted solution)—Constitute 1 container of Sterile Ampicillin Sodium in a volume of *Diluent*, accurately measured, corresponding to the volume of solvent specified in the labeling. Dilute an accurately measured portion of the constituted solution quantitatively with *Diluent* to obtain a solution having a concentration of about 1 mg of ampicillin per mL. Use this solution promptly after preparation.

Procedure—Proceed as directed for *Procedure* in the *Assay* under *Ampicillin.* Calculate the quantity, in µg, of ampicillin ($C_{16}H_{19}N_3O_4S$) in each mg of Sterile Ampicillin Sodium taken by the formula:

$$100(CP/W)(r_U/r_S),$$

in which W is the weight, in mg, of Sterile Ampicillin Sodium taken to prepare *Assay preparation 1*, and the other terms are as defined in the *Assay* under *Ampicillin.* Calculate the quantity, in mg, of ampicillin ($C_{16}H_{19}N_3O_4S$) in the container, and in the volume of constituted solution taken, by the formula:

$$(L/D)(CP/1000)(r_U/r_S),$$

in which L is the labeled quantity, in mg, of ampicillin ($C_{16}H_{19}N_3O_4S$) in the container, or in the volume of constituted solution taken, and D is the concentration, in mg of ampicillin ($C_{16}H_{19}N_3O_4S$) per mL, of *Assay preparation 2* or *Assay preparation 3*, based on the labeled quantity in the container or in the portion of constituted solution taken, respectively, and the extent of dilution, and the other terms are as previously defined.

Sterile Ampicillin Sodium and Sulbactam Sodium

» Sterile Ampicillin Sodium and Sulbactam Sodium is a sterile, dry mixture of Sterile Ampicillin Sodium and Sterile Sulbactam Sodium. It contains the equivalent of not less than 90.0 percent and not more than 115.0 percent of the labeled amounts of ampicillin ($C_{16}H_{19}N_3O_4S$) and sulbactam ($C_8H_{11}NO_5S$), the labeled amounts representing proportions of ampicillin to sulbactam of 2:1. It contains not less than 563 µg of ampicillin and 280 µg of sulbactam per mg, calculated on the anhydrous basis.

Packaging and storage—Preserve in *Containers for Sterile Solids* as described under *Injections* ⟨1⟩.

Reference standards—*USP Ampicillin Reference Standard*—Do not dry before using. *USP Sulbactam Reference Standard*—Do not dry before using.

Constituted solution—At the time of use, the constituted solution prepared from Sterile Ampicillin and Sulbactam Sodium meets the requirements for *Constituted Solutions* under *Injections* ⟨1⟩.

Identification—The retention times of the major peaks in the chromatogram of the *Assay preparation* correspond to those of the *Standard preparation*, as obtained in the *Assay.*

Pyrogen—It meets the requirements of the *Pyrogen Test* ⟨151⟩, the test dose being 1.0 mL per kg of a solution prepared by diluting Sterile Ampicillin Sodium and Sulbactam Sodium with Sterile Water for Injection to a concentration of 40 mg of ampicillin and 20 mg of sulbactam per mL.

Sterility—It meets the requirements under *Sterility Tests* ⟨71⟩, when tested as directed in the section, *Test Procedures Using Membrane Filtration.*

pH ⟨791⟩: between 8.0 and 10.0, in a solution containing 10 mg of ampicillin and 5 mg of sulbactam per mL.

Water, *Method I* ⟨921⟩: not more than 2.0%.

Particulate matter ⟨788⟩: meets the requirements under *Small-volume Injections.*

Other requirements—It meets the requirements for *Uniformity of Dosage Units* ⟨905⟩ and *Labeling* under *Injections* ⟨1⟩.

Assay—

0.005 M Tetrabutylammonium hydroxide—Dilute 6.6 mL of a 40% solution of tetrabutylammonium hydroxide with water to obtain 1800 mL of solution. Adjust with 1 *M* phosphoric acid to a pH of 5.0 ± 0.1, dilute with water to 2000 mL, and mix.

Mobile phase—Prepare a suitable filtered and degassed mixture of *0.005 M Tetrabutylammonium hydroxide* and acetonitrile (1650:350). Make adjustments if necessary (see *System Suitability* under *Chromatography* ⟨621⟩).

Standard preparation—Dissolve accurately weighed quantities of USP Ampicillin RS and USP Sulbactam RS quantitatively in *Mobile phase* to obtain a solution having known concentrations equivalent to about 0.6 mg of ampicillin and 0.3 mg of sulbactam per mL.

Resolution solution—Prepare a solution of USP Sulbactam RS in 0.01 *N* sodium hydroxide containing 0.3 mg per mL, and allow to stand for 30 minutes. Adjust with phosphoric acid to a pH of 5.0 ± 0.1. Transfer 5 mL of the solution to a 25-mL volumetric flask, add 4.25 mL of acetonitrile, dilute with *0.005 M Tetrabutylammonium hydroxide* to volume, and mix. Transfer 1 mL of this solution to a second 25-mL volumetric flask, add 15 mg of USP Ampicillin RS, dilute with *Mobile phase* to volume, and mix.

Assay preparation 1—Mix the contents of a container of Sterile Ampicillin Sodium and Sulbactam Sodium. Dissolve an accurately weighed portion of the powder quantitatively in *Mobile phase* to obtain a solution having a concentration of about 1 mg of the powder per mL.

Assay preparation 2 (where it is represented as being in a single-dose container)—Constitute a container of Sterile Ampicillin Sodium and Sulbactam Sodium with a volume of water, accurately measured, corresponding to the volume of solvent specified in the labeling. Withdraw the total withdrawable contents from the container, using a suitable hypodermic needle and syringe, and dilute quantitatively and stepwise, if necessary, with *Mobile phase* to obtain a solution containing about 0.6 mg of ampicillin and 0.3 mg of sulbactam per mL.

Assay preparation 3 (where the label states the quantities of ampicillin and sulbactam in a given volume of constituted solution)—Constitute a container of Sterile Ampicillin Sodium and Sulbactam Sodium with a volume of water, accurately measured, corresponding to the volume of solvent specified in the labeling. Dilute an accurately measured volume of the constituted solution quantitatively and stepwise, if necessary, with *Mobile phase* to obtain a solution containing about 0.6 mg of ampicillin and 0.3 mg of sulbactam per mL.

Chromatographic system (see *Chromatography* ⟨621⟩)—The liquid chromatograph is equipped with a 230-nm detector and a 4-mm × 30-cm column containing packing L1. The flow rate is about 2 mL per minute. Chromatograph the *Resolution solution*, and record the responses as directed under *Procedure:* the relative retention times are about 0.7 for ampicillin and 1.0 for the sulbactam alkaline degradation product, and the resolution, *R*, between ampicillin and the sulbactam alkaline degradation product peaks is not less than 4.0. Chromatograph the *Standard preparation*, and record the responses as directed under *Procedure:* the column efficiency determined from the sulbactam peak is not less than 3500 theoretical plates, the tailing factor is not more than 1.5, and the relative standard deviation for replicate injections is not more than 2.0%.

Procedure—[NOTE—Use peak areas where peak responses are specified.] Separately inject equal volumes (about 10 µL) of the *Standard preparation* and the *Assay preparation* into the chromatograph, record the chromatograms, and measure the responses for the major peaks. The relative retention times are about 0.35 for ampicillin and 1.0 for sulbactam. Calculate the quantities, in µg, of ampicillin ($C_{16}H_{19}N_3O_4S$) and of sulbactam ($C_8H_{11}NO_5S$), respectively, in the portion of Sterile Ampicillin Sodium and Sulbactam Sodium taken by the same formula:

$$(C_SP/C_U)(r_U/r_S),$$

in which C_S is the concentration, in mg per mL, of the appropriate USP Reference Standard in the *Standard preparation*, *P* is the assigned content, in µg per mg, of the appropriate USP Reference Standard, C_U is the concentration, in mg per mL, of Sterile Ampicillin Sodium and Sulbactam Sodium in *Assay preparation 1*, based on the weight, in mg, of powder removed from the container and the extent of dilution, and r_U and r_S are responses of the appropriate analyte peaks obtained from *Assay preparation 1* and the *Standard preparation*, respectively. Calculate the quantities of ampicillin ($C_{16}H_{19}N_3O_4S$) and of sulbactam ($C_8H_{11}NO_5S$), respectively, withdrawn from the container, or in the volume of constituted solution taken by the same formula:

$$(L/D)(C_SP)(r_U/r_S),$$

in which *L* is the labeled quantity, in mg, of ampicillin or sulbactam, as appropriate, in the container or in the volume of constituted solution taken, *D* is the concentration, in mg per mL, of ampicillin or sulbactam in *Assay preparation 2* or *Assay preparation 3*, on the basis of the labeled quantity, in mg, of ampicillin or sulbactam, as appropriate, in the container and the extent of dilution, r_U and r_S are the responses of the appropriate analyte peaks obtained from *Assay preparation 2* or *Assay preparation 3* and the *Standard preparation*, respectively, and the other terms are as defined above.

Amprolium

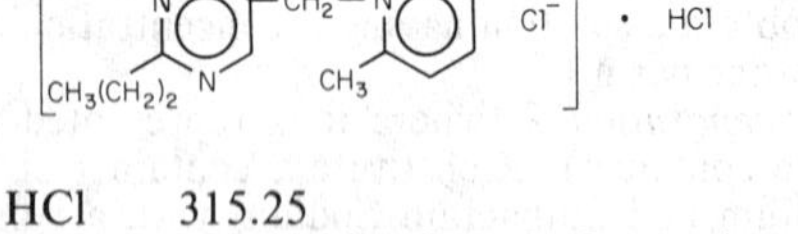

$C_{14}H_{19}ClN_4 . HCl$ 315.25

1-[(4-Amino-2-propyl-5-pyrimidinyl)methyl]-2-methylpyridinium chloride monohydrochloride.

1-[(4-Amino-2-propyl-5-pyrimidinyl)methyl]-2-picolinium chloride monohydrochloride [*121-25-5*].

» Amprolium contains not less than 97.0 percent and not more than 101.0 percent of amprolium ($C_{14}H_{19}ClN_4 . HCl$), calculated on the dried basis.

Packaging and storage—Preserve in well-closed containers.

Labeling—Label it to indicate that it is for veterinary use only.

Reference standard—*USP Amprolium Reference Standard*—Dry at a pressure not exceeding 5 mm of mercury at 100° for 3 hours before using.

Identification—

A: The infrared absorption spectrum of a mineral oil dispersion of it, previously dried, exhibits maxima only at the same wavelengths as that of a similar preparation of USP Amprolium RS.

B: The ultraviolet absorption spectrum of a 1 in 100,000 solution in 0.1 *N* hydrochloric acid exhibits maxima and minima at the same wavelengths as that of a similar solution of USP Amprolium RS, concomitantly measured, and the respective absorptivities, calculated on the dried basis, at the wavelength of maximum absorbance at about 246 nm do not differ by more than 3.0%.

Loss on drying ⟨731⟩—Dry it at a pressure not exceeding 5 mm of mercury at 100° for 3 hours: it loses not more than 1.0% of its weight.

Assay—

Potassium ferricyanide solution—Dissolve 200 mg of potassium ferricyanide in 100 mL of water.

Potassium cyanide solution—Dissolve 1.0 g of potassium cyanide in 100 mL of water.

Alcoholic potassium hydroxide—Dilute 15.0 mL of 0.2 *N* potassium hydroxide with methanol to 200 mL.

2,7-Naphthalenediol solution—Dissolve 25 mg of 2,7-naphthalenediol in 1000 mL of methanol.

Color developing solution—[NOTE—Use within 75 minutes of preparation]—Mix 90 volumes of *2,7-Naphthalenediol solution* with 5 volumes of *Potassium ferricyanide solution*, add 5 volumes of *Potassium cyanide solution*, mix, and allow to stand for 30 to 35 minutes. Add 100 volumes of *Alcoholic potassium hydroxide,* and mix.

Standard preparation—Dissolve an accurately weighed quantity of USP Amprolium RS in a mixture of methanol and water (2:1) to obtain a solution having a known concentration of about 25 µg per mL.

Assay preparation—Transfer about 100 mg of Amprolium, accurately weighed, to a 100-mL volumetric flask, dilute with a mixture of methanol and water (2:1) to volume, and mix. Transfer 5.0 mL of the resulting solution to a 200-mL volumetric flask, dilute with the same solvent to volume, and mix.

Procedure—Transfer 4.0 mL each of the *Standard preparation*, the *Assay preparation*, and the mixture of methanol and water (2:1) to provide a reagent blank, to separate tubes. Add 10.0 mL of *Color developing solution* to each tube, insert the stopper, mix, and allow to stand for 20 to 45 minutes. Concomitantly determine the absorbances of the solutions from the *Standard preparation* and the *Assay preparation* at the wavelength of maximum absorbance at about 520 nm, with a suitable spectrophotometer, using the reagent blank to set the instrument. Calculate the quantity, in mg, of $C_{14}H_{19}ClN_4 . HCl$ in the portion of Amprolium taken by the formula:

$$4C(A_U/A_S),$$

in which *C* is the concentration, in µg per mL, of USP Amprolium RS in the *Standard preparation*, and A_U and A_S are the absorbances of the solutions from the *Assay preparation* and the *Standard preparation*, respectively.

Amprolium Soluble Powder

» Amprolium Soluble Powder contains not less than 90.0 percent and not more than 110.0 percent of the labeled amount of $C_{14}H_{19}ClN_4 . HCl$.

Packaging and storage—Preserve in tight containers.

Labeling—Label it to indicate that it is for veterinary use only.

Reference standard—*USP Amprolium Reference Standard*—Dry at a pressure not exceeding 5 mm of mercury at 100° for 3 hours before using.

Identification—Dissolve a suitable portion of Soluble Powder in 0.1 *N* hydrochloric acid to obtain a solution having a concentration of about 0.01 mg of amprolium per mL, and filter: the ultraviolet absorption spectrum of the solution so obtained exhibits maxima and minima at the same wavelengths as that of a similar preparation of USP Amprolium RS, concomitantly measured.

Assay—

Potassium ferricyanide solution, Potassium cyanide solution, Alcoholic potassium hydroxide, 2,7-Naphthalenediol solution, Color developing solution, and *Standard preparation*—Prepare as directed in the *Assay* under *Amprolium.*

Assay preparation—Transfer an accurately weighed portion of Amprolium Soluble Powder, equivalent to about 50 mg of amprolium, to a glass-stoppered, 250-mL flask, add 100.0 mL of a mixture of methanol and water (2:1), shake for 20 minutes, and filter, discarding the first 15 mL of the filtrate. Transfer 5.0 mL of the subsequent filtrate to a 100-mL volumetric flask, dilute with a mixture of methanol and water (2:1) to volume, and mix.

Procedure—Proceed as directed for *Procedure* in the *Assay* under *Amprolium.* Calculate the quantity, in mg, of $C_{14}H_{19}ClN_4 . HCl$ in the portion of Soluble Powder taken by the formula:

$$2C(A_U / A_S).$$

Amprolium Oral Solution

» Amprolium Oral Solution contains not less than 93.0 percent and not more than 107.0 percent of the labeled amount of $C_{14}H_{19}ClN_4 . HCl$.

Packaging and storage—Preserve in tight containers.

Labeling—Label it to indicate that it is for veterinary use only.

Reference standard—*USP Amprolium Reference Standard*—Dry at a pressure not exceeding 5 mm of mercury at 100° for 3 hours before using.

Identification—Dilute a suitable volume of Oral Solution with 0.1 *N* hydrochloric acid to obtain a solution having a concentration of about 0.01 mg of amprolium per mL: the ultraviolet absorption spectrum of the solution so obtained exhibits maxima and minima at the same wavelengths as that of a similar solution of USP Amprolium RS, concomitantly measured.

pH ⟨791⟩: between 2.5 and 3.0.

Assay—

Potassium ferricyanide solution, Potassium cyanide solution, Alcoholic potassium hydroxide, 2,7-Naphthalenediol solution, Color developing solution, and *Standard preparation*—Prepare as directed in the *Assay* under *Amprolium.*

Assay preparation—Transfer an accurately measured volume of Amprolium Oral Solution, equivalent to about 1.9 g of amprolium, to a 1000-mL volumetric flask, dilute with water to volume, and mix. Transfer 25.0 mL of this solution to a 100-mL volumetric flask, dilute with water to volume, and mix. Transfer 5.0 mL of the resulting solution to a 100-mL volumetric flask, dilute with a mixture of methanol and water (2:1) to volume, and mix.

Procedure—Proceed as directed for *Procedure* in the *Assay* under *Amprolium.* Calculate the quantity, in g, of $C_{14}H_{19}ClN_4 . HCl$ in each mL of the Oral Solution taken by the formula:

$$(0.08C / V)(A_U / A_S),$$

in which *V* is the volume, in mL, of Oral Solution taken, and the other terms are as defined therein.

Amyl Nitrite

$C_5H_{11}NO_2$ 117.15

Mixture of nitrous acid, 2-methylbutyl ester, and nitrous acid, 3-methylbutyl ester [*8017-89-8; 110-46-3*].

» Amyl Nitrite is a mixture of the nitrite esters of 3-methyl-1-butanol and 2-methyl-1-butanol. It contains not less than 85.0 percent and not more than 103.0 percent of $C_5H_{11}NO_2$.

Caution—Amyl Nitrite is very flammable. Do not use where it may be ignited.

Packaging and storage—Preserve in tight containers, and store in a cool place, protected from light.

Reference standard—*USP Benzyl Benzoate Reference Standard*—After opening the ampul, store in a tightly closed container, protected from light.

Identification—

A: The NMR spectrum recorded as directed in the *Assay* exhibits, among other peaks, a doublet with a band centered at about 1 ppm and a multiplet with a band centered at about 4.8 ppm representing methyl protons and methylene protons alpha to the nitrite group, respectively, both relative to the tetramethylsilane singlet at 0 ppm.

B: To a few drops of it add a mixture of 1 mL of ferrous sulfate TS and 5 mL of 3 *N* hydrochloric acid: a greenish brown color is produced.

Specific gravity ⟨841⟩: between 0.870 and 0.876.

Acidity—To 0.30 mL in a glass-stoppered cylinder add a mixture of 0.60 mL of 0.1 *N* sodium hydroxide, 10 mL of water, and 1 drop of phenolphthalein TS, and invert the cylinder three times: the red tint of the water layer is still perceptible.

Nonvolatile residue—Allow 10 mL to evaporate at room temperature in a tared evaporating dish, *in a well-ventilated hood,* and dry the residue at 105° for 1 hour: the weight of the residue does not exceed 2 mg (0.02%).

Total nitrites—Inject a portion of Amyl Nitrite of suitable volume, but not more than 2 µL, into a suitable gas chromatograph (see *Chromatography* ⟨621⟩) equipped with a thermal conductivity detector. Under typical conditions, the instrument contains a 2-m × 3-mm column packed with a methyl polysiloxane oil, 25% by weight on suitable calcined diatomite, the column is maintained at about 80°, the injection port and detector block are maintained about 10° above the temperature of the column, and helium is used as a carrier gas at a flow rate of about 60 mL per minute. From the area under the curve, calculate the percentage (a/a) of total nitrites, represented by the area under the main peak of the chromatogram, in the Amyl Nitrite taken: not less than 97.0% is found.

Assay—

Solvent—Carbon tetrachloride.

Internal standard—USP Benzyl Benzoate RS.

Procedure—Transfer 4 mEq to 5 mEq of *Internal standard,* accurately weighed, to a semimicro sampling tube, add 2 to 3 mL of carbon tetrachloride, apply a sampling valve and septum,* thereby sealing the tube, and determine the weight of the sealed assembly. Open the valve, introduce about 500 µL of Amyl Nitrite with a syringe, close the valve, and determine the weight of the sealed assembly when it has attained constant weight. Shake the sampling tube and valve assembly, and transfer about 500 µL of the solution to a precision NMR tube as directed for *Absolute Method of Quantitation* under *Nuclear Magnetic Resonance* ⟨761⟩. With no spinning, or with the spinning adjusted so that the spinning side bands of neither the substance under assay nor the *Internal standard* interfere with the regions to be integrated, record as A_S the average area of the *Internal standard* singlet appearing at about 5.3 ppm, representing the methylene protons of benzyl benzoate, and record as A_U the average area of the multiplet with a band center at about 4.8 ppm, representing the alpha methylene protons of amyl nitrite, with reference to the tetramethylsilane singlet at 0 ppm. Calculate the quantity of $C_5H_{11}NO_2$ in the Amyl Nitrite taken, using 58.57 as the equivalent weight of amyl nitrite (EW_U) and 106.12 as that of benzyl benzoate (EW_S).

Amyl Nitrite Inhalant

» Amyl Nitrite Inhalant contains a mixture of the nitrite esters of 3-methyl-1-butanol and 2-methyl-1-

* Suitable sampling tubes, sampling valves, and septums are available, respectively, as catalog Nos. K-749000, K-749100, and K-749102 (50 septums) or K-749101 (100 septums), from Kontes Glass Company, Vineland, NJ 08360.

butanol. It contains not less than 80.0 percent and not more than 105.0 percent of $C_5H_{11}NO_2$. It contains a suitable stabilizer.

Caution—Amyl Nitrite Inhalant is very flammable. Do not use where it may be ignited.

Packaging and storage—Preserve in tight, unit-dose glass containers, wrapped loosely in gauze or other suitable material, and store in a cool place, protected from light.

Reference standard—*USP Benzyl Benzoate Reference Standard*—After opening the ampul, store in a tightly closed container, protected from light.

Specific gravity ⟨841⟩: between 0.870 and 0.880.

Total nitrites—Remove the gauze or other covering, place the glass container of Inhalant upright in a dry ice–acetone slurry, and cool for 10 minutes. Dry the container of Inhalant, place it in a pointed glass tube, and break the container with a glass rod. Proceed as directed for *Total nitrites* under *Amyl Nitrite:* not less than 95.0% is found.

Other requirements—It responds to the *Identification tests* and meets the requirements of the test for *Acidity* under *Amyl Nitrite.*

Assay—

Solvent—Carbon tetrachloride.

Internal standard—USP Benzyl Benzoate RS.

Procedure—Remove the gauze or other covering from 1 or more Amyl Nitrite Inhalant ampuls containing a total of 300 µL to 400 µL of amyl nitrite. Weigh accurately the clean and dry intact glass ampul(s), and place the weighed specimen in a freezer for not less than 15 minutes. Transfer the chilled specimen to a glass-stoppered, 25-mL conical flask containing a solution of 4 to 5 mEq of *Internal standard*, accurately weighed, in 1 to 2 mL of carbon tetrachloride. Break the ampul(s) with a glass rod, and rinse any sample or glass fragments adhering to the glass rod with 1 mL of carbon tetrachloride into the main assay solution. Insert the stopper in the flask immediately, mix, and proceed as directed for *Absolute Method of Quantitation* under *Nuclear Magnetic Resonance* ⟨761⟩, beginning with "When dissolution has been completed." With no spinning, or with the spinning adjusted so that the spinning side bands of neither the substance under assay nor the *Internal standard* interfere with the regions to be integrated, record as A_S the average area of the *Internal standard* singlet appearing at about 5.3 ppm, representing the methylene protons of benzyl benzoate, and record as A_U the average area of the multiplet with a band center at about 4.8 ppm, representing the alpha methylene protons of amyl nitrite, with reference to the tetramethylsilane singlet at 0 ppm. Calculate the quantity of $C_5H_{11}NO_2$ in the Inhalant taken, using 58.57 as the equivalent weight of amyl nitrite (EW_U) and 106.12 as that of benzyl benzoate (EW_S). Rinse the flask containing the assay preparation with three 5-mL portions of ether, decanting each rinsing carefully to avoid loss of glass fragments, and evaporate any remaining ether with the aid of a current of dry air. Transfer the dry glass fragments to a tared watch glass, weigh, and subtract the weight of the glass fragments from that of the intact ampul(s) to obtain the weight of the Inhalant taken.

Amylene Hydrate—*see* Amylene Hydrate NF
Anethole—*see* Anethole NF
Anhydrous Lanolin—*see* Lanolin, Anhydrous

Anileridine

$C_{22}H_{28}N_2O_2$ 352.48

4-Piperidinecarboxylic acid, 1-[2-(4-aminophenyl)ethyl]-4-phenyl-, ethyl ester.

Ethyl 1-(*p*-aminophenethyl)-4-phenylisonipecotate [*144-14-9*].

» Anileridine contains not less than 98.5 percent and not more than 101.0 percent of $C_{22}H_{28}N_2O_2$, calculated on the anhydrous basis.

Packaging and storage—Preserve in tight, light-resistant containers.

Identification—

A: Dissolve 40 mg in 2.3 mL of 0.1 *N* hydrochloric acid in a 100-mL volumetric flask, dilute with water to volume, and mix (*Stock solution*). Transfer 4.0 mL of this solution to a 100-mL volumetric flask, and add 25 mL of pH 7.0 *Buffer solution.* (Prepare the *Buffer solution* by dissolving 22.73 g of anhydrous dibasic sodium phosphate and 14.52 g of monobasic potassium phosphate in water to make 1000 mL. Dilute 25 mL of the buffer with water to 100 mL: the pH, determined potentiometrically, is 7.0 ± 0.05.) Dilute with water to volume, and mix (*Solution A*). Transfer 20.0 mL of the *Stock solution* to a 100-mL volumetric flask, add 25 mL of the pH 7.0 *Buffer solution*, dilute with water to volume, and mix (*Solution B*): the ultraviolet absorption spectrum of *Solution A* exhibits a maximum at 234 ± 1 nm, and the ultraviolet absorption spectrum of *Solution B* exhibits a maximum at 285 ± 2 nm. The ratio $5A_{234}/A_{285}$ is about 8.8.

B: To 5 mL of a 1 in 5000 solution in 0.1 *N* hydrochloric acid add 2 mL of a 1 in 100 solution of *p*-dimethylaminobenzaldehyde in alcohol: a yellow color develops immediately.

Water, *Method I* ⟨921⟩: not more than 1.0%.

Residue on ignition ⟨281⟩: not more than 0.1%.

Chloride ⟨221⟩—Dissolve 180 mg in a mixture of 1 mL of nitric acid and 40 mL of water: the solution shows no more chloride than corresponds to 0.10 mL of 0.020 *N* hydrochloric acid (0.040%).

Assay—Dissolve about 350 mg of Anileridine, accurately weighed, in 50 mL of glacial acetic acid, add 1 drop of crystal violet TS, and titrate with 0.1 *N* perchloric acid VS to a blue-green endpoint. Perform a blank determination, and make any necessary correction. Each mL of 0.1 *N* perchloric acid is equivalent to 17.62 mg of $C_{22}H_{28}N_2O_2$.

Anileridine Injection

» Anileridine Injection is a sterile solution of Anileridine in Water for Injection, prepared with the aid of Phosphoric Acid. It contains not less than 90.0 percent and not more than 115.0 percent of the labeled amount of $C_{22}H_{28}N_2O_2$, as the phosphate.

Packaging and storage—Preserve in single-dose or in multiple-dose containers, preferably of Type I glass, protected from light.

Reference standard—*USP Anileridine Hydrochloride Reference Standard*—Dry at a pressure below 5 mm of mercury at 100° for 2 hours before using.

Identification—

A: Dilute a volume of Injection, equivalent to about 1.25 mg of anileridine, with water to 5 mL, and add 2 mL of a 1 in 100 solution of *p*-dimethylaminobenzaldehyde in alcohol: a yellow color develops immediately.

B: A volume of Injection, diluted with water to a concentration of about 25 mg of anileridine in 1000 mL, exhibits absorbance maxima at 234 ± 1 and 285 ± 2 nm.

pH ⟨791⟩: between 4.5 and 5.0.

Other requirements—It meets the requirements under *Injections* ⟨1⟩.

Assay—

Standard preparation—[NOTE—Prepare on the day of the assay.] Dissolve an accurately weighed quantity of USP Anileridine Hydrochloride RS in 0.1 *N* hydrochloric acid, and dilute quantitatively with the same solvent to obtain a solution having a known concentration of about 250 µg per mL. (Each mg of anileridine hydrochloride is equivalent to 0.8286 mg of anileridine.)

Assay preparation—Transfer to a 500-mL volumetric flask an accurately measured volume of Anileridine Injection, equivalent to about 100 mg of anileridine, dilute with 0.1 *N* hydrochloric acid to volume, and mix.

Procedure—Transfer 5.0 mL each of the *Standard preparation*, the *Assay preparation*, and 0.1 *N* hydrochloric acid to provide the blank to separate 200-mL volumetric flasks. To each flask add 25 mL of water, 5 mL of 1 *N* hydrochloric acid, and 5 mL of sodium nitrite solution (1 in 1000), and mix. Allow to stand for 2 minutes, then add to each flask 5 mL of ammonium sulfamate solution (1 in 200), and mix. Allow to stand for 3 minutes, then add 5 mL of *N*-(1-naphthyl)ethylenediamine dihydrochloride solution (1 in 1000), and mix. Allow to stand for 1 hour, dilute with water to volume, and mix. Determine the absorbances of the solutions in 1-cm cells at the wavelength of maximum absorbance at about 560 nm, with a suitable spectrophotometer, using the reagent blank to set the instrument. Calculate the quantity, in mg, of $C_{22}H_{28}N_2O_2$ in each mL of the Injection taken by the formula:

$$(352.48/425.40)(0.5C/V)(A_U/A_S),$$

in which 352.48 and 425.40 are the molecular weights of anileridine and anileridine hydrochloride, respectively, *C* is the concentration, in µg per mL, of USP Anileridine Hydrochloride RS in the *Standard preparation*, *V* is the volume, in mL, of Injection taken, and A_U and A_S are the absorbances of the solutions from the *Assay preparation* and the *Standard preparation*, respectively.

Anileridine Hydrochloride

$C_{22}H_{28}N_2O_2 \cdot 2HCl$ 425.40

4-Piperidinecarboxylic acid, 1-[2-(4-aminophenyl)ethyl]-4-phenyl-, ethyl ester, dihydrochloride.

Ethyl 1-(*p*-aminophenethyl)-4-phenylisonipecotate dihydrochloride [*126-12-5*].

» Anileridine Hydrochloride contains not less than 96.0 percent and not more than 102.0 percent of $C_{22}H_{28}N_2O_2 \cdot 2HCl$, calculated on the dried basis.

Packaging and storage—Preserve in tight, light-resistant containers.

Reference standard—*USP Anileridine Hydrochloride Reference Standard*—Dry at a pressure below 5 mm of mercury at 100° for 2 hours before using.

Identification—

A: The infrared absorption spectrum of a potassium bromide dispersion of it, previously dried, exhibits maxima only at the same wavelengths as that of a similar preparation of USP Anileridine Hydrochloride RS.

B: Dissolve 50 mg in water in a 100-mL volumetric flask, dilute with water to volume, and mix (*Stock solution*). Transfer 4.0 mL of this solution to a 100-mL volumetric flask, and add 25 mL of pH 7.0 *Buffer solution.* (Prepare the *Buffer solution* by dissolving 22.73 g of anhydrous dibasic sodium phosphate and 14.52 g of monobasic potassium phosphate in water to make 1000.0 mL. Dilute 25 mL of the buffer with water to 100 mL: the pH, determined potentiometrically, is 7.0 ± 0.05.) Dilute with water to volume, and mix (*Solution A*). Transfer 20.0 mL of the *Stock solution* to a 100-mL volumetric flask, add 25 mL of the pH 7.0 *Buffer solution*, dilute with water to volume, and mix (*Solution B*): the ultraviolet absorption spectrum of *Solution A* exhibits a maximum at 234 ± 1 nm, and the ultraviolet absorption spectrum of *Solution B* exhibits a maximum at 285 ± 2 nm.

C: To 5 mL of a solution (1 in 5000) add 2 mL of a 1 in 100 solution of *p*-dimethylaminobenzaldehyde in alcohol: a yellow color develops immediately.

D: A solution (1 in 100) responds to the tests for *Chloride* ⟨191⟩.

pH ⟨791⟩: between 2.5 and 3.0, in a solution (1 in 20).

Loss on drying ⟨731⟩—Dry it at a pressure below 5 mm of mercury at 100° for 2 hours: it loses not more than 1.0% of its weight.

Residue on ignition ⟨281⟩: not more than 0.1%.

Chloride content—Dissolve about 200 mg, accurately weighed, in 50 mL of water in a glass-stoppered flask. Add 25.0 mL of 0.1 *N* silver nitrate VS, then add 5 mL of 2 *N* nitric acid and 5 mL of nitrobenzene, shake vigorously, add 2 mL of ferric ammonium sulfate TS, and titrate the excess silver nitrate with 0.1 *N* ammonium thiocyanate VS. Each mL of 0.1 *N* silver nitrate is equivalent to 3.545 mg of Cl: the content of Cl is between 16.0% and 17.2%.

Assay—Dissolve about 200 mg of Anileridine Hydrochloride, accurately weighed, in 10 mL of glacial acetic acid by heating on a steam bath. Cool immediately in a cold water bath, add 5 mL of mercuric acetate TS, 20 mL of acetone, and 0.5 mL of indicator solution (70 mg of α-naphtholbenzein, 10 mg of crystal violet, and 40 mg of quinaldine red in 100 mL of glacial acetic acid), and titrate with 0.1 *N* perchloric acid VS to a gray-green end-point. Perform a blank determination, and make any necessary correction. Each mL of 0.1 *N* perchloric acid is equivalent to 21.27 mg of $C_{22}H_{28}N_2O_2 \cdot 2HCl$.

Anileridine Hydrochloride Tablets

» Anileridine Hydrochloride Tablets contain an amount of anileridine hydrochloride ($C_{22}H_{28}N_2O_2 \cdot 2HCl$) equivalent to not less than 95.0 percent and not more than 105.0 percent of the labeled amount of anileridine ($C_{22}H_{28}N_2O_2$).

Packaging and storage—Preserve in tight, light-resistant containers.

Reference standard—*USP Anileridine Hydrochloride Reference Standard*—Dry at a pressure below 5 mm of mercury at 100° for 2 hours before using.

Identification—

A: Place a quantity of finely powdered Tablets, equivalent to about 50 mg of anileridine, in a 250-mL volumetric flask, add 100 mL of water, and heat on a steam bath. Cool, dilute with water to volume, mix, and filter: five mL of the filtrate responds to *Identification test C* under *Anileridine Hydrochloride.*

B: Transfer to a 100-mL volumetric flask a quantity of finely powdered Tablets, equivalent to about 50 mg of anileridine. Add about 30 mL of water, and heat on a steam bath. Cool, dilute with water to volume, mix, and filter (*Stock solution*): the filtrate responds to *Identification test B* under *Anileridine Hydrochloride*, beginning with "Transfer 4.0 mL of this solution."

Dissolution ⟨711⟩—

Medium: 0.1 *N* hydrochloric acid; 900 mL.

Apparatus 1: 100 rpm.

Time: 45 minutes.

Procedure—Determine the amount of anileridine ($C_{22}H_{28}N_2O_2$) dissolved, employing the procedure set forth in the *Assay*, making any necessary modifications.

Tolerances—Not less than 65% (*Q*) of the labeled amount of $C_{22}H_{28}N_2O_2$ is dissolved in 45 minutes.

Uniformity of dosage units ⟨905⟩: meet the requirements.

Assay—

Standard preparation—Prepare as directed in the *Assay* under *Anileridine Injection.*

Assay preparation—Weigh and finely powder not less than 20 Anileridine Hydrochloride Tablets. Transfer an accurately weighed portion of the powder, equivalent to about 50 mg of anileridine, to a 250-mL volumetric flask. Add 25 mL of 1 *N* hydrochloric acid and 100 mL of water, and heat on a water bath. Cool, dilute with water to volume, and mix. Filter the solution, discarding the first 25 mL of the filtrate.

Procedure—Proceed as directed for *Procedure* in the *Assay* under *Anileridine Injection.* Calculate the quantity, in mg, of $C_{22}H_{28}N_2O_2$ in the portion of Tablets taken by the formula:

$$(352.48/425.40)(0.25C)(A_U/A_S),$$

in which 352.48 and 425.40 are the molecular weights of anileridine and anileridine hydrochloride, respectively, C is the concentration, in µg per mL, of USP Anileridine Hydrochloride RS in the *Standard preparation*, and A_U and A_S are the absorbances of the solutions from the *Assay preparation* and the *Standard preparation*, respectively.

Antazoline Phosphate

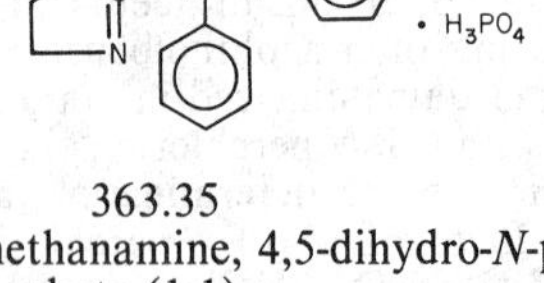

$C_{17}H_{19}N_3 . H_3PO_4$ 363.35

1*H*-Imidazole-2-methanamine, 4,5-dihydro-*N*-phenyl-*N*-(phenylmethyl)-, phosphate (1:1).

2-[(*N*-Benzylanilino)methyl]-2-imidazoline phosphate (1:1) [*154-68-7*].

» Antazoline Phosphate contains not less than 98.0 percent and not more than 101.0 percent of $C_{17}H_{19}N_3 . H_3PO_4$, calculated on the dried basis.

Packaging and storage—Preserve in tight containers.

Reference standard—*USP Antazoline Phosphate Reference Standard*—Dry at 105° for 4 hours before using.

Identification—

A: The infrared absorption spectrum of a mineral oil dispersion of it, previously dried, exhibits maxima only at the same wavelengths as that of a similar preparation of USP Antazoline Phosphate RS.

B: The R_f value of the principal spot in the chromatogram of the *Identification preparation* corresponds to that of *Standard preparation A*, as obtained in the test for *Chromatographic purity*.

Melting range, *Class Ia* ⟨741⟩: between 194° and 198°, with decomposition.

pH ⟨791⟩: between 4.0 and 5.0, in a solution (1 in 50).

Loss on drying ⟨731⟩—Dry it at 105° for 4 hours: it loses not more than 0.5% of its weight.

Chromatographic purity—

Standard preparations—Dissolve USP Antazoline Phosphate RS in methanol, and mix to obtain a solution having a known concentration of 0.10 mg per mL. Dilute quantitatively with methanol to obtain 5 *Standard preparations* having the following compositions:

Standard preparation	Dilution	Concentration (µg RS per mL)	Percentage (%, for comparison with test specimen)
A	(1 in 2)	50	0.5
B	(2 in 5)	40	0.4
C	(3 in 10)	30	0.3
D	(1 in 5)	20	0.2
E	(1 in 10)	10	0.1

Test preparation—Dissolve an accurately weighed quantity of Antazoline Phosphate in methanol to obtain a solution containing 10 mg per mL.

Identification preparation—Dilute a portion of the *Test preparation* quantitatively with methanol to obtain a solution containing 50 µg per mL.

Procedure—On a suitable thin-layer chromatographic plate (see *Chromatography* ⟨621⟩), coated with a 0.25-mm layer of chromatographic silica gel mixture, apply separately 10 µL of the *Test preparation*, 10 µL of the *Identification preparation*, and 10 µL of each *Standard preparation.* Position the plate in a chromatographic chamber, and develop the chromatograms in a solvent system consisting of a mixture of ethyl acetate, methanol, and diethylamine (17:2:1) until the solvent front has moved about three-fourths of the length of the plate. Remove the plate from the developing chamber, mark the solvent front, and allow the solvent to evaporate. Examine the plate under short-wavelength ultraviolet light, and compare the intensities of any secondary spots observed in the chromatogram of the *Test preparation* with those of the principal spots in the chromatograms of the *Standard preparations.* [NOTE—Disregard any spots observed at the origins of the chromatograms.] No secondary spot from the chromatogram of the *Test preparation* is larger or more intense than the principal spot obtained from *Standard preparation A* (0.5%), and the sum of the intensities of all secondary spots obtained from the *Test preparation* corresponds to not more than 1.0%.

Assay—Dissolve about 750 mg of Antazoline Phosphate, accurately weighed, in 50 mL of glacial acetic acid, and titrate with 0.1 *N* perchloric acid VS, determining the end-point potentiometrically, using a glass electrode and a calomel electrode containing a saturated solution of lithium chloride in glacial acetic acid (see *Titrimetry* ⟨541⟩). Perform a blank determination, and make any necessary correction. Each mL of 0.1 *N* perchloric acid is equivalent to 36.34 mg of $C_{17}H_{19}N_3 . H_3PO_4$.

Anthralin

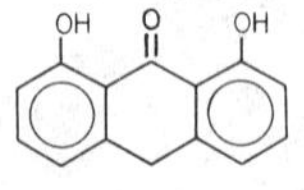

$C_{14}H_{10}O_3$ 226.23

9(10*H*)-Anthracenone, 1,8-dihydroxy-.

1,8-Dihydroxy-9-anthrone [*1143-38-0*].

» Anthralin contains not less than 97.0 percent and not more than 102.0 percent of $C_{14}H_{10}O_3$, calculated on the dried basis.

Packaging and storage—Preserve in tight containers in a cool place. Protect from light.

Reference standard—*USP Anthralin Reference Standard*—Dry over silica gel for 4 hours before using.

Identification—

A: The infrared absorption spectrum of a potassium bromide dispersion of it, previously dried over silica gel for 4 hours, exhibits maxima only at the same wavelengths as that of a similar preparation of USP Anthralin RS.

B: The ultraviolet absorption spectrum of a 1 in 100,000 solution in chloroform exhibits maxima and minima at the same wavelengths as that of a similar solution of USP Anthralin RS, concomitantly measured.

Melting range, *Class I* ⟨741⟩: between 178° and 181°.

Acidity or alkalinity—Suspend it in water, and filter: the filtrate is neutral to litmus.

Loss on drying ⟨731⟩—Dry it over silica gel for 4 hours: it loses not more than 0.5% of its weight.

Residue on ignition ⟨281⟩: not more than 0.1%.

Chloride—Add 1 g of it to 15 mL of water, mix, and filter. Acidify 5 mL of the filtrate with nitric acid, and add a few drops of silver nitrate TS: no more opalescence is produced immediately than is present in a 5-mL portion of the filtrate to which nothing has been added.

Sulfate—To 5 mL of the untreated filtrate obtained in the test for *Chloride* add 3 drops of 3 *N* hydrochloric acid and 5 drops of barium chloride TS: no more turbidity is produced than is present in a 5-mL portion of the filtrate to which nothing has been added.

Assay—[NOTE—Use low-actinic glassware.]

Internal standard solution—Dissolve a sufficient quantity of *o*-nitroaniline in a small quantity of dichloromethane, and dilute with *n*-hexane to obtain a solution having a concentration of about 500 µg per mL.

Mobile phase—Prepare a degassed mixture of *n*-hexane, dichloromethane, and glacial acetic acid (82:12:6). Make adjustments if necessary (see *System Suitability* under *Chromatography* ⟨621⟩).

System suitability preparation—Prepare a solution containing 0.1 mg of USP Anthralin RS and 0.2 mg of danthron per mL in dichloromethane. Transfer 5 mL of this solution to a 25-mL volumetric flask. Add 5.0 mL of *n*-hexane, dilute with *Mobile phase* to volume, and mix.

Solvent blank solution—Mix *Mobile phase*, *n*-hexane, and dichloromethane (3:1:1).

Standard preparation—Dissolve an accurately weighed quantity of USP Anthralin RS in dichloromethane to obtain a Standard solution having a known concentration of about 250 µg per mL. Pipet 5 mL of this solution and 5 mL of *Internal standard solution* into a 25-mL volumetric flask, dilute with *Mobile phase* to volume, and mix.

Assay preparation—Weigh accurately about 250 mg of Anthralin into a tared 100-mL flask, dilute with dichloromethane to volume after dissolution of the solid, and mix. Pipet 10 mL of the solution into a 100-mL volumetric flask, dilute with dichloromethane to volume, and mix. Pipet 5 mL of this solution and 5 mL of *Internal standard solution* into a 25-mL volumetric flask, dilute with *Mobile phase* to volume, and mix.

Chromatographic system (see *Chromatography* ⟨621⟩)—The liquid chromatograph is equipped with a 354-nm detector and a 4.6-mm × 25-cm column that contains packing L3. The flow rate is about 2 mL per minute. Chromatograph replicate injections of the *Standard preparation*, and record the peak responses as directed under *Procedure:* the relative standard deviation of the ratio of the peak responses is not more than 2.0%. Chromatograph the *System suitability preparation:* the resolution, *R*, is not less than 1.3, and the tailing factor, *T*, is not more than 1.5. Chromatograph the *Solvent blank solution:* no effect on the baseline is discernible at the retention time of anthralin.

Procedure—[NOTE—Where peak responses are indicated, use peak areas.] Separately inject equal volumes (about 10 µL) of the *Standard preparation* and the *Assay preparation* into the chromatograph, record the chromatograms, and measure the responses for the major peaks. Relative retention times are 1.0 for anthralin, about 1.2 for danthron (if present), about 1.7 for dianthrone (if present), and about 2.3 for *o*-nitroaniline. Calculate the quantity, in mg, of $C_{14}H_{10}O_3$ in the portion of anthralin taken by the formula:

$$C(R_U/R_S),$$

in which C is the concentration, in µg per mL, of USP Anthralin RS in the Standard solution, and R_U and R_S are the response ratios of the anthralin peak to the *o*-nitroaniline peak obtained from the *Assay preparation* and the *Standard preparation*, respectively.

Anthralin Cream

» Anthralin Cream is Anthralin in an aqueous (oil-in-water) or oily (water-in-oil) cream vehicle. Anthralin Cream labeled to contain more than 0.1 percent of anthralin contains not less than 90.0 percent and not more than 115.0 percent of the labeled amount of $C_{14}H_{10}O_3$, and Cream labeled to contain 0.1 percent or less of anthralin contains not less than 90.0 percent and not more than 130.0 percent of the labeled amount of $C_{14}H_{10}O_3$.

Packaging and storage—Preserve in tight containers, in a cool place. Protect from light.

Reference standard—*USP Anthralin Reference Standard*—Dry over silica gel for 4 hours before using.

Labeling—Label it to indicate whether the cream vehicle is aqueous or oily.

Assay—[NOTE—Use low-actinic glassware.]

Internal standard solution, Mobile phase, System suitability preparation, Solvent blank solution, and *Chromatographic system*—Proceed as directed in the *Assay* under *Anthralin.*

Standard preparation—Dissolve an accurately weighed quantity of USP Anthralin RS in dichloromethane to obtain a Standard solution having a known concentration of about 0.25 mg per mL. Pipet 2 mL of this solution and 2 mL of *Internal standard solution* into a 25-mL volumetric flask, dilute with *Mobile phase* to volume, and mix.

Assay preparation—Weigh accurately approximately 5 g of Anthralin Cream into a tared 100-mL beaker. Add about 20 mL of dichloromethane and 10 mL of glacial acetic acid, and stir to disperse the Cream. Transfer the contents of the beaker to a Whatman No. 4 filter paper with the aid of dichloromethane, and filter into a 100-mL volumetric flask. Thoroughly wash the precipitate with dichloromethane, and allow the washings to drain into the flask. Dilute with dichloromethane to volume, and mix. Pipet a volume of this solution, equivalent to about 0.5 mg of anthralin, and 2 mL of *Internal standard solution* into a 25-mL volumetric flask, dilute with *Mobile phase* to volume, and mix.

Procedure—Proceed as directed for the *Assay* under *Anthralin.* Calculate the quantity, in mg, of $C_{14}H_{10}O_3$ in the portion of Cream taken by the formula:

$$(200C/V)(R_U/R_S),$$

in which C is the concentration, in mg per mL, of USP Anthralin RS in the Standard solution, V is the volume, in mL, of the filtrate taken for the *Assay preparation*, and R_U and R_S are the response ratios of the anthralin peak to the *o*-nitroaniline peak obtained from the *Assay preparation* and the *Standard preparation*, respectively.

Anthralin Ointment

» Anthralin Ointment is Anthralin in a petrolatum or other oleaginous vehicle. Anthralin Ointment labeled to contain more than 0.1 percent of anthralin contains not less than 90.0 percent and not more than 115.0 percent of the labeled amount of $C_{14}H_{10}O_3$, and Ointment labeled to contain 0.1 percent or less of anthralin contains not less than 90.0 percent and not more than 130.0 percent of the labeled amount of $C_{14}H_{10}O_3$.

Packaging and storage—Preserve in tight containers, in a cool place. Protect from light.

Reference standard—*USP Anthralin Reference Standard*—Dry over silica gel for 4 hours before using.

Assay—Proceed with Anthralin Ointment as directed in the *Assay* under *Anthralin Cream.*

Anti-A Blood Grouping Serum—*see* Blood Grouping Serum, Anti-A

Anti-B Blood Grouping Serum—*see* Blood Grouping Serum, Anti-B

Anticoagulant Citrate Dextrose Solution

» Anticoagulant Citrate Dextrose Solution is a sterile solution of Citric Acid, Sodium Citrate, and Dextrose in Water for Injection. It contains in each 1000 mL:

	Solution A	Solution B
Total Citrate, expressed as citric acid, anhydrous ($C_6H_8O_7$)		
not less than	20.59 g	12.37 g
not more than	22.75 g	13.67 g
Dextrose ($C_6H_{12}O_6 \cdot H_2O$)		
not less than	23.28 g	13.96 g
not more than	25.73 g	15.44 g
Sodium (Na)		
not less than	4.90 g	2.94 g
not more than	5.42 g	3.25 g

It contains no antimicrobial agents.

Prepare Anticoagulant Citrate Dextrose Solution as follows.

	Solution A	Solution B
Citric Acid (anhydrous)	7.3 g	4.4 g
Sodium Citrate (dihydrate)	22.0 g	13.2 g
Dextrose (monohydrate) ...	24.5 g	14.7 g
Water for Injection, a sufficient quantity, to make	1000 mL	1000 mL

Dissolve the ingredients, and mix. Filter the solution until clear, place immediately in suitable containers, and sterilize.

If desired, 8 g and 4.8 g of monohydrated citric acid may be used instead of the indicated, respective amounts of anhydrous citric acid; 19.3 g and 11.6 g of anhydrous sodium citrate may be used instead of the indicated, respective amounts of dihydrated sodium citrate; and 22.3 g and 13.4 g of anhydrous dextrose may be used instead of the indicated, respective amounts of monohydrated dextrose.

Packaging and storage—Preserve in single-dose containers, of colorless, transparent, Type I or Type II glass, or of a suitable plastic material (see *Transfusion and Infusion Assemblies* ⟨161⟩).

Labeling—Label it to indicate the number of mL of Solution required per 100 mL of whole blood or the number of mL of Solution required per volume of whole blood to be collected.

Identification—It responds to the *Identification test* under *Dextrose*, and, when concentrated to one-half its volume, responds to the tests for *Citrate* ⟨191⟩ and for *Sodium* ⟨191⟩.

Pyrogen—When diluted with pyrogen-free saline TS to contain approximately 0.5% of dihydrated sodium citrate, it meets the requirements of the *Pyrogen Test* ⟨151⟩.

pH ⟨791⟩: between 4.5 and 5.5.

Chloride ⟨221⟩—A 10-mL portion shows no more chloride than corresponds to 0.50 mL of 0.020 *N* hydrochloric acid (0.0035%).

Other requirements—It meets the requirements under *Injections* ⟨1⟩.

Assay for total citrate—Proceed as directed in the *Assay for total citrate* under *Anticoagulant Citrate Phosphate Dextrose Solution*. Report total citrate per liter of sample (as citric acid, anhydrous).

Assay for sodium—Proceed as directed in the *Assay for sodium* under *Anticoagulant Citrate Phosphate Dextrose Adenine Solution*.

Assay for dextrose—Determine the angular rotation of Anticoagulant Citrate Dextrose Solution in a 200-mm tube, using sodium light at 25°. The observed rotation in degrees, multiplied by 1.0425, represents the weight of $C_6H_{12}O_6 \cdot H_2O$ in 100 mL of the solution.

Anticoagulant Citrate Phosphate Dextrose Solution

» Anticoagulant Citrate Phosphate Dextrose Solution is a sterile solution of Citric Acid, Sodium Citrate, Monobasic Sodium Phosphate, and Dextrose in Water for Injection. It contains, in each 1000 mL, not less than 2.11 g and not more than 2.33 g of monobasic sodium phosphate ($NaH_2PO_4 \cdot H_2O$); not less than 24.22 g and not more than 26.78 g of dextrose ($C_6H_{12}O_6 \cdot H_2O$); not less than 19.16 g and not more than 21.18 g of total citrate, expressed as citric acid, anhydrous ($C_6H_8O_7$); and not less than 6.21 g and not more than 6.86 g of Sodium (Na). It contains no antimicrobial agents.

Prepare Anticoagulant Citrate Phosphate Dextrose Solution as follows.

Citric Acid (anhydrous)	2.99 g
Sodium Citrate (dihydrate)	26.3 g
Monobasic Sodium Phosphate (monohydrate; $NaH_2PO_4 \cdot H_2O$)	2.22 g
Dextrose (monohydrate)	25.5 g
Water for Injection, a sufficient quantity, to make	1000 mL

Dissolve the ingredients, and mix. Filter the solution until clear, place immediately in suitable containers, and sterilize.

If desired, 3.27 g of monohydrated citric acid may be used instead of the indicated amount of anhydrous citric acid; 23.06 g of anhydrous sodium citrate may be used instead of the indicated amount of dihydrated sodium citrate; 1.93 g of anhydrous monobasic sodium phosphate may be used instead of the indicated amount of monohydrated monobasic sodium phosphate; and 23.2 g of anhydrous dextrose may be used instead of the indicated amount of monohydrated dextrose.

Packaging and storage—Preserve in single-dose containers, of colorless, transparent, Type I or Type II glass, or of a suitable plastic material (see *Transfusion and Infusion Assemblies* ⟨161⟩).

Labeling—Label it to indicate the number of mL of Solution required per 100 mL of whole blood or the number of mL of Solution required per volume of whole blood to be collected.

Identification—It responds to the *Identification test* under *Dextrose*, and to the tests for *Phosphate* ⟨191⟩, and, when concentrated to one-half its volume, responds to the tests for *Citrate* ⟨191⟩ and for *Sodium* ⟨191⟩.

Pyrogen—When diluted with pyrogen-free saline TS to contain approximately 0.5% of dihydrated sodium citrate, it meets the requirements of the *Pyrogen Test* ⟨151⟩.

pH ⟨791⟩: between 5.0 and 6.0.

Chloride ⟨221⟩—A 10-mL portion shows no more chloride than corresponds to 0.50 mL of 0.020 *N* hydrochloric acid (0.0035%).

Other requirements—It meets the requirements under *Injections* ⟨1⟩.

Assay for total citrate—

Standard preparation—Dissolve a suitable quantity of citric acid, previously dried at 90° for 3 hours and accurately weighed, in water to obtain a solution having a known concentration of about 1.0 mg of anhydrous citric acid per mL.

Assay preparation—Pipet 10 mL of Anticoagulant Citrate Phosphate Dextrose Solution into a 100-mL volumetric flask, dilute with water to volume, and mix. Pipet 5 mL of this solution into another 100-mL volumetric flask, dilute with water to volume, and mix.

Standard curve—Pipet aliquots of 8, 9, 10, 11, and 12 mL, respectively, of *Standard preparation* into separate 100-mL volumetric flasks, dilute with water to volume, and mix. Continue as directed for *Procedure*. Plot the resultant absorbances versus the respective µg-per-mL concentrations of the standard solutions.

Procedure—Pipet 1 mL of *Assay preparation* into a suitable test tube. To a second test tube add 1.0 mL of water to serve as a reference blank. To each tube add 1.3 mL of pyridine, and mix by swirling. To one tube at a time add 5.7 mL of acetic anhydride, and mix, using a rotary vortex stirrer. Immediately place in a water bath maintained at 31 ± 1.0°, and allow the color to develop for 33 ± 1 minutes. Determine the absorbance against the reference blank in 2.5-cm cells at 425 nm, taking care to measure the absorbance of each solution at the same elapsed time from mixing. Calculate the total citrate content, in mg per mL, of the Anticoagulant Citrate Phosphate Dextrose Solution taken by the formula:

$$0.2C,$$

in which *C* is the concentration, in µg per mL, of anhydrous citric acid read from the *Standard curve*.

Assay for total phosphate—[expressed as monobasic sodium phosphate, monohydrate ($NaH_2PO_4 . H_2O$)]—

1,2,4-Aminonaphtholsulfonic acid solution—Dissolve 0.5 g of 1,2,4-aminonaphtholsulfonic acid in about 150 mL of sodium bisulfite solution (3 in 20) in a 200-mL volumetric flask, warming if necessary. Add 5 mL of sodium sulfite solution (1 in 5), mix, and dilute with sodium bisulfite solution (3 in 20) to volume.

Standard preparation—Dissolve about 0.44 g of monobasic potassium phosphate, accurately weighed, in water to make 1000 mL, and mix. Pipet 25 mL of this solution into a 100-mL volumetric flask, and dilute with water to volume so as to obtain a solution having a known concentration of about 0.11 mg per mL of monobasic potassium phosphate.

Assay preparation—Dilute 5.0 mL of Anticoagulant Citrate Phosphate Dextrose Solution with water to 100 mL.

Procedure—Pipet 5 mL of *Standard preparation*, 5 mL of *Assay preparation*, and 5 mL of water, to provide a control, into separate 25-mL volumetric flasks. Treat the contents of each flask as follows: Add 10.0 mL of 1 *N* sulfuric acid, and mix. Add 2.0 mL of ammonium molybdate solution (1 in 40), and mix. Add 1.0 mL of *1,2,4-Aminonaphtholsulfonic acid solution*, dilute with water to volume, again mix, and allow to stand for 10 minutes at 20° to 25°. Determine the absorbances of the solutions from the *Standard preparation* and the *Assay preparation* against the reference solution in 1-cm cells at 660 nm, with a suitable spectrophotometer and taking care to measure the absorbance of each solution at the same elapsed time from mixing. Calculate the quantity, in mg, of $NaH_2PO_4 . H_2O$ in each mL of the Solution taken by the formula:

$$20.28C(A_U / A_S),$$

in which *C* is the concentration, in mg per mL, of KH_2PO_4 in the *Standard preparation*, and A_U and A_S are the absorbances of the solutions from the *Assay preparation* and the *Standard preparation*, respectively.

Assay for dextrose—Tare a clean, medium-porosity filtering crucible containing several carborundum boiling chips or glass beads. Pipet 50 mL of freshly mixed alkaline cupric tartrate TS into a 400-mL beaker. Add the boiling chips or glass beads from the tared crucible, 45 mL of water, and 5.0 mL of Anticoagulant Citrate Phosphate Dextrose Solution to the beaker. Heat the beaker and contents over a burner that has been adjusted to cause boiling of the solution to start in 3.5 to 4 minutes. Boil the solution for 2 minutes, accurately timed, and filter immediately through the tared crucible, taking care to transfer all of the boiling chips or glass beads to the crucible. Wash the precipitate with hot water and 10 mL of alcohol. Dry the crucible and contents at 110° to constant weight. Perform a blank determination, and correct the weight of the precipitate from the sample for any precipitate obtained in the blank. Each mg of cuprous oxide precipitate obtained from the substance under assay is equivalent to 0.496 mg of $C_6H_{12}O_6 . H_2O$.

Assay for sodium—Proceed as directed in the *Assay for sodium* under *Anticoagulant Citrate Phosphate Dextrose Adenine Solution*.

Anticoagulant Citrate Phosphate Dextrose Adenine Solution

» Anticoagulant Citrate Phosphate Dextrose Adenine Solution is a sterile solution of Citric Acid, Sodium Citrate, Monobasic Sodium Phosphate, Dextrose, and Adenine in Water for Injection. It contains, in each 1000 mL, not less than 2.11 g and not more than 2.33 g of monobasic sodium phosphate ($NaH_2PO_4 . H_2O$); not less than 30.30 g and not more than 33.50 g of dextrose ($C_6H_{12}O_6 . H_2O$); not less than 19.16 g and not more than 21.18 g of total citrate, expressed as citric acid, anhydrous ($C_6H_8O_7$); not less than 6.21 g and not more than 6.86 g of sodium (Na); and not less than 0.247 g and not more than 0.303 g of adenine ($C_5H_5N_5$). It contains no antimicrobial agents.

Prepare Anticoagulant Citrate Phosphate Dextrose Adenine Solution as follows.

Citric Acid (anhydrous)	2.99	g
Sodium Citrate (dihydrate)	26.3	g
Monobasic Sodium Phosphate (monohydrate; $NaH_2PO_4 . H_2O$)	2.22	g
Dextrose (monohydrate)	31.9	g
Adenine ($C_5H_5N_5$)	0.275	g
Water for Injection, a sufficient quantity, to make	1000	mL

Dissolve the ingredients, and mix. Filter the solution until clear, place immediately in suitable containers, and sterilize.

If desired, 3.27 g of monohydrated citric acid may be used instead of the indicated amount of anhydrous citric acid; 23.06 g of anhydrous sodium citrate may be used instead of the indicated amount of dihydrated sodium citrate; 1.93 g of anhydrous monobasic sodium phosphate may be used instead of the indicated amount of monohydrated monobasic sodium phosphate; and 29.0 g of anhydrous dextrose may be used instead of the indicated amount of monohydrated dextrose.

Packaging and storage—Preserve in single-dose containers, of colorless, transparent, Type I or Type II glass, or of a suitable plastic material (see *Transfusion and Infusion Assemblies* ⟨161⟩).

Labeling—Label it to indicate the number of mL of solution required per 100 mL of whole blood or the number of mL of solution required per volume of whole blood to be collected.

Reference standard—*USP Adenine Reference Standard*—Dry at 110° for 4 hours before using.

Pyrogen—When diluted with pyrogen-free saline TS to contain approximately 0.5% of dihydrated sodium citrate, it meets the requirements of the *Pyrogen Test* ⟨151⟩.

pH ⟨791⟩: between 5.0 and 6.0.

Chloride ⟨221⟩—A 10-mL portion shows no more chloride than corresponds to 0.50 mL of 0.020 *N* hydrochloric acid (0.0035%).

Other requirements—It meets the requirements under *Injections* ⟨1⟩.

Assay for total citrate—

Standard preparations—Dissolve a suitable quantity of citric acid, previously dried at 90° for 3 hours and accurately weighed, in water to obtain a solution having a known concentration of about 1.0 mg of anhydrous citric acid per mL. Further pipet quantities of 8, 9, 10, 11, and 12 mL of this stock standard into separate 100-mL volumetric flasks, dilute with water to volume, and mix.

Assay preparation—Pipet 5 mL of Anticoagulant Citrate Phosphate Dextrose Adenine Solution into a 1000-mL volumetric flask, dilute with water to volume, and mix.

Procedure—Pipet 1 mL of the *Assay preparation*, each *Standard preparation*, and water into separate test tubes. To each tube add 1.3 mL of pyridine, and mix by swirling. To one tube at a time add 5.7 mL of acetic anhydride, and mix, using a rotary vortex stirrer. Immediately place in a water bath maintained at 31 ± 1.0°, and allow the color to develop for 33 ± 1 minutes. Determine the absorbance against the reference blank in 1-cm cells at 425 nm, taking care to measure the absorbance of each solution at the same elapsed time from mixing. Calculate the total citrate content, in mg per mL, of the Anticoagulant Citrate Phosphate Dextrose Adenine Solution by the formula:

$$0.2C,$$

in which *C* is the concentration, in μg per mL, of anhydrous citric acid read from the Standard curve.

Assay for total phosphate—[expressed as monobasic sodium phosphate monohydrate ($NaH_2PO_4 . H_2O$)]—

Acid solution—Dilute 75 mL of sulfuric acid to 200 mL with water.

Ammonium molybdate solution—Dissolve 25 g of ammonium molybdate in 300 mL of water, and add 200 mL of *Acid solution*.

Standard preparation—Dissolve about 0.44 g of monobasic potassium phosphate, previously dried at 105° for 4 hours and accurately weighed, in water to make 1000 mL, and mix. Pipet 10 mL of this solution into a 100-mL volumetric flask, dilute with water to volume, and mix to obtain a solution having a known concentration of about 0.01 mg of phosphorus (P) per mL.

Assay preparation—Pipet 10 mL of Anticoagulant Citrate Phosphate Dextrose Adenine Solution into a 500-mL volumetric flask, dilute with water to volume, and mix.

Procedure—Pipet 5-, 10-, and 15-mL portions of the *Standard preparation* and 10 mL of the *Assay preparation* into separate 50-mL volumetric flasks. Treat the contents of each flask as follows: Add 2.5 mL of *Ammonium molybdate solution*, and mix. Add rapidly and in order, 2.5 mL of hydroquinone solution (1 in 200) and 2.5 mL of sodium sulfite solution (1 in 10), both prepared fresh daily. Dilute with water to volume, mix, and allow to stand at room temperature for 30 ± 5 minutes. Determine the absorbances against water in 1-cm cells at 660 nm with a suitable spectrophotometer. Plot the absorbances against the mg of phosphorus in the portions of *Standard preparation* taken. Calculate the quantity, in mg, of $NaH_2PO_4 . H_2O$ in each mL of the Anticoagulant Citrate Phosphate Dextrose Adenine Solution taken by the formula:

$$(137.99/30.9738)(5W),$$

in which 137.99 is the molecular weight of monobasic sodium phosphate monohydrate, 30.9738 is the atomic weight of phosphorus, and *W* is the weight, in mg, of P in the 10 mL of *Assay preparation* taken, read from the Standard curve.

Assay for sodium—

Lithium diluent solution—Transfer 1.04 g of lithium nitrate to a 1000-mL volumetric flask, add a suitable nonionic surfactant, then add water to volume, and mix. This solution contains 15 mEq of lithium per liter.

Standard preparation—Transfer 8.18 g of sodium chloride, previously dried at 105° for 2 hours and accurately weighed, to a 1000-mL volumetric flask, dilute with water to volume, and mix. This solution contains 140 mEq of sodium per liter. Transfer 50 μL of this solution to a 10-mL volumetric flask, dilute with *Lithium diluent solution* to volume, and mix.

Assay preparation—Pipet 25 mL of Anticoagulant Citrate Phosphate Dextrose Adenine Solution into a 50-mL volumetric flask, dilute with water to volume, and mix. Transfer 50 μL of this solution to a 10-mL volumetric flask, dilute with *Lithium diluent solution* to volume, and mix.

Procedure—Using a suitable flame photometer, adjusted to read zero with *Lithium diluent solution*, concomitantly determine the sodium flame emission readings for the *Standard preparation* and the *Assay preparation* at the wavelength of maximum emission at about 589 nm. Calculate the quantity, in g, of Na in 1000 mL of Anticoagulant Citrate Phosphate Dextrose Adenine Solution taken by the formula:

$$2(8.18)(22.99/58.44)(R_U/R_S),$$

in which 8.18 is the weight, in g, of sodium chloride taken to make the *Standard preparation*, 22.99 is the atomic weight of sodium, 58.44 is the molecular weight of sodium chloride, and R_U and R_S are the sodium emission readings obtained from the *Assay preparation* and the *Standard preparation*, respectively.

Assay for dextrose—Tare a clean, medium-porosity filtering crucible containing several carborundum boiling chips or glass beads. Pipet 50 mL of freshly mixed alkaline cupric tartrate TS into a 400-mL beaker. Add the boiling chips or glass beads from the tared crucible, 45 mL of water, and 5.0 mL of Anticoagulant Citrate Phosphate Dextrose Adenine Solution to the beaker. Heat the beaker and contents over a burner that has been adjusted to cause boiling of the solution to start in 3.5 to 4 minutes. Boil the solution for 2 minutes, accurately timed, and filter immediately through the tared crucible, taking care to transfer all of the boiling chips or glass beads to the crucible. Wash the precipitate with hot water and 10 mL of alcohol. Dry the crucible and contents at 110° to constant weight. Perform a blank determination, and make any necessary correction. Each mg of cuprous oxide precipitate obtained is equivalent to 0.496 mg of $C_6H_{12}O_6 . H_2O$.

Assay for adenine—

Mobile phase—Dissolve 3.45 g of ammonium dihydrogen phosphate in 950 mL of water in a 1000-mL volumetric flask, add 10 mL of glacial acetic acid, dilute with water to volume, mix, filter through a membrane filter (1-μm or finer porosity), and degas.

Standard preparations—Dissolve accurately weighed quantities of USP Adenine RS in dilute hydrochloric acid (1 in 120) in three separate volumetric flasks, dilute with the dilute hydrochloric acid solution to volume, and mix to obtain Standard preparations having known concentrations of about 0.25, 0.275, and 0.30 mg of adenine per mL, respectively. Protect from light.

Chromatographic system (see *Chromatography* ⟨621⟩)—The liquid chromatograph is equipped with a 254-nm detector and a 4-mm × 30-cm stainless steel column that contains packing L9. The flow rate is about 2.0 mL per minute. Prepare a solution containing USP Adenine RS and purine, each at about 0.275 mg per mL, in dilute hydrochloric acid (1 in 120), and chromatograph not less than four injections (about 20 μL) of this solution: the relative standard deviation of the peak response of adenine is not more than 2.5%, the relative standard deviation of the retention time of adenine is not more than 2.0%, and the resolution of adenine and purine is not less than 3.0.

Procedure—Separately inject equal volumes (about 20 μL) of the Anticoagulant Citrate Phosphate Dextrose Adenine Solution and the *Standard preparations*, record the chromatograms, and measure the responses for the major peaks. Plot the responses against the concentrations, in mg of USP Adenine RS per mL of the *Standard preparations*. Calculate the quantity, in mg, of $C_5H_5N_5$ in each mL of the Anticoagulant Citrate Phosphate Dextrose Adenine Solution taken as the value read directly from the Standard curve corresponding to the response obtained from the portion of the Anticoagulant Citrate Phosphate Dextrose Adenine Solution chromatographed.

Anticoagulant Heparin Solution

» Anticoagulant Heparin Solution is a sterile solution of Heparin Sodium in Sodium Chloride Injection. Its potency is not less than 90.0 percent and not more than 110.0 percent of the potency stated on the label in terms of USP Heparin Units. It contains not less than 0.85 percent and not more than 0.95 percent of sodium chloride (NaCl). It may be buffered. It contains no antimicrobial agents.

Heparin Sodium	75,000 Units
Sodium Chloride Injection, a sufficient quantity, to make ...	1,000 mL

Add the Heparin Sodium, in solid form or in solution, to the Sodium Chloride Injection, mix, filter if necessary, and sterilize.

Packaging and storage—Preserve in single-dose containers, of colorless, transparent, Type I or Type II glass, or of a suitable plastic material (see *Transfusion and Infusion Assemblies* ⟨161⟩).

Labeling—Label it in terms of USP Heparin Units, and to indicate the number of mL of Solution required per 100 mL of whole blood.

Reference standard—*USP Heparin Sodium Reference Standard*—Store in a cool place.

Pyrogen—It meets the requirements of the *Pyrogen Test* ⟨151⟩, the test dose being 2 mL per kg.

pH ⟨791⟩: between 5.0 and 7.5.

Other requirements—It meets the requirements under *Injections* ⟨1⟩.

Assay for heparin sodium—Proceed as directed in the *Assay* under *Heparin Sodium Injection*, substituting Anticoagulant Heparin Solution for the Injection.

Assay for sodium chloride—Pipet 10 mL of Anticoagulant Heparin Solution into a suitable container, add 2 mL of potassium chromate TS, and titrate with 0.1 *N* silver nitrate VS. Each mL of 0.1 *N* silver nitrate is equivalent to 5.844 mg of NaCl.

Anticoagulant Sodium Citrate Solution

» Anticoagulant Sodium Citrate Solution is a sterile solution of Sodium Citrate in Water for Injection. It contains, in each 100 mL, not less than 3.80 g and not more than 4.20 g of $C_6H_5Na_3O_7 . 2H_2O$. It contains no antimicrobial agents.

Sodium Citrate (dihydrate)	40 g
Water for Injection, a sufficient quantity, to make	1000 mL

NOTE—Anhydrous sodium citrate (35.1 g) may be used instead of the dihydrate.

Dissolve the Sodium Citrate in sufficient Water for Injection to make 1000 mL, and filter until clear. Place the solution in suitable containers, and sterilize.

Packaging and storage—Preserve in single-dose containers, preferably of Type I or Type II glass.

Identification—When evaporated to a concentration of 1 in 20, it responds to the tests for *Sodium* ⟨191⟩, and for *Citrate* ⟨191⟩.

pH ⟨791⟩: between 6.4 and 7.5.

Pyrogen—When diluted with Sodium Chloride Injection to contain approximately 0.5% of sodium citrate, it meets the requirements of the *Pyrogen Test* ⟨151⟩.

Other requirements—It meets the requirements under *Injections* ⟨1⟩.

Assay—Transfer 10.0 mL of Anticoagulant Sodium Citrate Solution to a 250-mL beaker, and evaporate to dryness. Add 100 mL of glacial acetic acid, stir until completely dissolved, and titrate with 0.1 *N* perchloric acid VS, determining the end-point potentiometrically using a calomel-glass electrode system. Perform a blank determination, and make any necessary correction. Each mL of 0.1 *N* perchloric acid is equivalent to 9.803 mg of $C_6H_5Na_3O_7 . 2H_2O$.

Anti-D, Anti-C, Anti-E, Anti-c, Anti-e Blood Grouping Serums—*see* Blood Grouping Serums Anti-D, Anti-C, Anti-E, Anti-c, Anti-e

Antigen, Mumps Skin Test—*see* Mumps Skin Test Antigen

Antihemophilic Factor

» Antihemophilic Factor conforms to the regulations of the federal Food and Drug Administration concerning biologics (see *Biologics* ⟨1041⟩). It is a sterile, freeze-dried powder containing the Factor VIII fraction prepared from units of human venous plasma that have been tested for the absence of hepatitis B surface antigen, obtained from whole-blood donors and pooled. It may contain Heparin Sodium or Sodium Citrate. It meets the requirements of the test for potency, by comparison with the U.S. Standard Antihemophilic Factor (Factor VIII) or with a working reference that has been calibrated with it, in containing not less than 80 percent and not more than 120 percent of the potency stated on the label, the stated potency being not less than 100 Antihemophilic Factor Units per g of protein. It meets the requirements of the test for pyrogen, the test dose being 10 Antihemophilic Factor Units per kg.

Packaging and storage—Preserve in hermetic containers, in a refrigerator, unless otherwise indicated.

Expiration date—The expiration date is not later than 2 years from date of manufacture, within which time it may be stored at room temperature and used within 6 months of the time of such storage.

Labeling—Label it to state that it is to be used within 4 hours after constitution, that it is for intravenous administration, and that a filter is to be used in the administration equipment.

Cryoprecipitated Antihemophilic Factor

» Cryoprecipitated Antihemophilic Factor conforms to the regulations of the federal Food and Drug Administration concerning biologics (640.50 to 640.57) (see *Biologics* ⟨1041⟩). It is a sterile, frozen concentrate of human antihemophilic factor prepared from the Factor VIII–rich cryoprotein fraction of human venous plasma obtained from suitable whole-blood donors from a single unit of plasma derived from whole blood or by plasmapheresis, collected and processed in a closed system. It contains no preservative. It meets the requirements of the test for potency by comparison with the U.S. Standard Antihemophilic Factor (Factor VIII) or with a working reference that has been calibrated with it, in having an average potency of not less than 80 Antihemophilic Factor Units per container, made at intervals of not more than 1 month during the dating period.

Packaging and storage—Preserve in hermetic containers at a temperature of −18° or lower.

Expiration date—The expiration date is not later than 1 year from the date of collection of source material.

Labeling—Label it to indicate the ABO blood group designation and the identification number of the donor from whom the source material was obtained. Label it also with the type and result of a serologic test for syphilis, or to indicate that it was non-reactive in such test; with the type and result of a test for hepatitis B surface antigen, or to indicate that it was non-reactive in such test; with a warning not to use it if there is evidence of breakage or thawing; with instructions to thaw it before use to a temperature between 20° and 37°, after which it is to be stored at room temperature and used as soon as possible but within 6 hours after thawing; to state that it is to be used within 4 hours after the container is entered; and to state that it is for intravenous administration, and that a filter is to be used in the administration equipment.

Anti–Human Globulin Serum—*see* Globulin Serum, Anti–Human

Antimony Potassium Tartrate

$C_8H_4K_2O_{12}Sb_2 . 3H_2O$ 667.85
Antimonate(2-), bis[μ-[2,3-dihydroxybutanedioato(4-)-$O^1,O^2:O^3,O^4$]]-di-, dipotassium, trihydrate, stereoisomer.
Dipotassium bis[μ-tartrato(4-)]diantimonate(2-) trihydrate [*28300-74-5*].
Anhydrous 613.81 [*11071-15-1*].

» Antimony Potassium Tartrate contains not less than 99.0 percent and not more than 103.0 percent of $C_8H_4K_2O_{12}Sb_2 . 3H_2O$.

Packaging and storage—Preserve in well-closed containers.

Completeness of solution ⟨641⟩: meets the requirements, a 750-mg specimen being used.

Identification—

A: When heated to redness, it chars, emits an odor resembling that of burning sugar, and leaves a blackened residue. This residue has an alkaline reaction, and when a small fragment of it is held in a nonluminous flame, the flame is tinted violet.

B: In a solution (1 in 20), acidified with hydrochloric acid, hydrogen sulfide TS produces an orange-red precipitate, which is soluble in ammonium sulfide TS and in 1 *N* sodium hydroxide.

C: It responds to the test for *Tartrate* ⟨191⟩.

Lead ⟨251⟩: not more than 0.002%.

Acidity or alkalinity—Dissolve 1.0 g in 50 mL of carbon dioxide–free water, and titrate with 0.010 *N* hydrochloric acid or 0.010 *N* sodium hydroxide to a pH of 4.5: not more than 2.0 mL is required.

Loss on drying ⟨731⟩—Dry it at 105° to constant weight: it loses not more than 2.7% of its weight.

Arsenic—Dissolve 100 mg in 5 mL of hydrochloric acid. Add 10 mL of a recently prepared solution of 20 g of stannous chloride in 30 mL of hydrochloric acid. Mix, transfer to a color-comparison tube, and allow to stand for 30 minutes. Viewed downward over a white surface, the color of the solution appears no deeper than that of a blank to which has been added 15 µg of arsenic (0.015%).

Assay—Dissolve about 500 mg of Antimony Potassium Tartrate, accurately weighed, in 50 mL of water, add 5 g of potassium sodium tartrate, 2 g of sodium borate, and 3 mL of starch TS, and immediately titrate with 0.1 *N* iodine VS to the production of a persistent blue color. Each mL of 0.1 *N* iodine is equivalent to 16.70 mg of $C_8H_4K_2O_{12}Sb_2 . 3H_2O$.

Antimony Sodium Tartrate

$C_8H_4Na_2O_{12}Sb_2$ 581.59
Antimonate(2-), bis[μ-[2,3-dihydroxybutanedioato(4-)-$O^1,O^2:O^3,O^4$]]di-, disodium, stereoisomer.
Disodium bis[μ-[L-(+)-tartrato(4-)]]diantimonate(2-) [*34521-09-0*].

» Antimony Sodium Tartrate contains not less than 98.0 percent and not more than 101.0 percent of $C_8H_4Na_2O_{12}Sb_2$, calculated on the dried basis.

Packaging and storage—Preserve in well-closed containers.

Identification—It responds to the tests for *Antimony* ⟨191⟩, for *Sodium* ⟨191⟩, and for *Tartrate* ⟨191⟩.

Acidity or alkalinity—Dissolve 1.0 g in 50 mL of carbon dioxide–free water, and titrate with 0.010 *N* hydrochloric acid or 0.010 *N* sodium hydroxide to a pH of 4.5: not more than 2.0 mL is required.

Loss on drying ⟨731⟩—Dry it at 105° to constant weight: it loses not more than 6.0% of its weight.

Arsenic, *Method II* ⟨211⟩: 8 ppm.

Lead ⟨251⟩: not more than 0.002%.

Assay—Dissolve about 500 mg of Antimony Sodium Tartrate, accurately weighed, in 50 mL of water, add 5 g of potassium sodium tartrate, 2 g of sodium borate, and 3 mL of starch TS, and immediately titrate with 0.1 *N* iodine VS to the production of a persistent blue color. Each mL of 0.1 *N* iodine is equivalent to 14.54 mg of $C_8H_4Na_2O_{12}Sb_2$.

Antipyrine

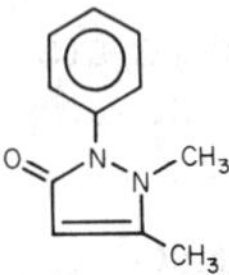

$C_{11}H_{12}N_2O$ 188.23
1,2-Dihydro-1,5-dimethyl-2-phenyl-3*H*-pyrazol-3-one.
2,3-Dimethyl-1-phenyl-3-pyrazolin-5-one [*60-80-0*].

» Antipyrine contains not less than 99.0 percent and not more than 100.5 percent of $C_{11}H_{12}N_2O$, calculated on the dried basis.

Packaging and storage—Preserve in tight containers.

Reference standard—*USP Antipyrine Reference Standard*—Dry at 60° for 2 hours before using.

Completeness and color of solution—It is completely soluble in its own weight of cold water, the solution being colorless or not more than slightly yellow when viewed transversely in a tube having a diameter of about 20 mm.

Identification—

A: The infrared absorption spectrum of a potassium bromide dispersion of it exhibits maxima only at the same wavelengths as that of a similar preparation of USP Antipyrine RS.

B: The ultraviolet absorption spectrum of a 1 in 50,000 solution in methanol exhibits maxima and minima at the same wavelengths as that of a similar solution of USP Antipyrine RS, concomitantly measured, and the respective absorptivities, calculated on the dried basis, at the point of maximum absorbance at about 266 nm do not differ by more than 3.0%.

C: Add tannic acid TS to a solution of it: a white precipitate is formed.

Melting range ⟨741⟩: between 110° and 112.5°.

Loss on drying ⟨731⟩—Dry it at 60° for 2 hours: it loses not more than 1.0% of its weight.

Residue on ignition ⟨281⟩: not more than 0.15%.

Heavy metals ⟨231⟩—Dissolve 1 g in 2 mL of 1 *N* acetic acid, and add water to make 25 mL: the limit is 0.002%.

Ordinary impurities ⟨466⟩—

Test solution: chloroform.

Standard solution: chloroform.

Eluant: a mixture of chloroform, acetone, butyl alcohol, and formic acid (60:15:15:15).

Visualization: 1.

Assay—Transfer about 100 mg of Antipyrine, accurately weighed, to a 250-mL iodine flask, and dissolve in 25 mL of water. Add 2 g of sodium acetate and 20.0 mL of 0.1 *N* iodine VS, mix, and allow to stand in a cool, dark place for 20 minutes. Add 25 mL of alcohol to dissolve the precipitate, and titrate the excess iodine with 0.1 *N* sodium thiosulfate VS, using starch TS as the indicator. Each mL of 0.1 *N* iodine is equivalent to 9.412 mg of $C_{11}H_{12}N_2O$.

Antipyrine and Benzocaine Otic Solution

» Antipyrine and Benzocaine Otic Solution is a solution of Antipyrine and Benzocaine in Glycerin. It contains not less than 90.0 percent and not more than 110.0 percent of the labeled amounts of antipyrine ($C_{11}H_{12}N_2O$) and benzocaine ($C_9H_{11}NO_2$). [NOTE—In the preparation of this Otic Solution, use Glycerin that has a low water content, in order that the Otic Solution may comply with the *Water* limit. This may be ensured by using Glycerin having a specific gravity of not less than 1.2607, corresponding to a concentration of 99.5%.]

Packaging and storage—Preserve in tight, light-resistant containers.

Reference standards—*USP Antipyrine Reference Standard*—Dry at 60° for 2 hours before using. *USP Benzocaine Reference Standard*—Dry over phosphorus pentoxide for 3 hours before using.

Identification—

A: Transfer 5 mL to a separator containing 25 mL of water, and extract the solution with two 25-mL portions of a mixture of equal volumes of ether and solvent hexane. Combine the extracts, and retain the water solution for *Identification test B.* Extract the ether-hexane solution with 50 mL of water, and discard the water layer. Evaporate the ether-hexane solution to dryness, dry the residue in vacuum at 40° to 50° for 1 hour, and dissolve the residue in 1 mL of chloroform: the infrared absorption spectrum of this solution exhibits maxima at the same wavelengths as that of a similar solution of USP Benzocaine RS, concomitantly measured.

B: Add 5 mL of 1 *N* sodium hydroxide solution to the water solution retained from *Identification test A*, and extract with two 25-mL portions of chloroform. Evaporate the combined extracts to dryness, dry the residue in vacuum at 40° to 50° for 1 hour, and dissolve the residue in 3 mL of chloroform: the infrared absorption spectrum of this solution exhibits maxima at the same wavelengths as that of a similar solution of USP Antipyrine RS, concomitantly measured.

C: The retention times of the major peaks in the chromatogram of the *Assay preparation* correspond to those of the *Standard preparation*, as obtained in the *Assay*.

Water ⟨921⟩—Determine the water content by the titrimetric method: not more than 1.0% is found.

Assay—

Ammonium acetate solution—Dissolve 7.7 g of ammonium acetate in water, dilute with water to 1000 mL, and mix.

Mobile phase—Prepare a filtered and degassed mixture of *Ammonium acetate solution* and acetonitrile (3:1). Make adjustments if necessary (see *System Suitability* under *Chromatography* ⟨621⟩).

Standard preparation—Transfer about 15 mg of USP Benzocaine RS, accurately weighed, to a 100-mL volumetric flask, add 15*J* mg of USP Antipyrine RS, accurately weighed, *J* being the ratio of the labeled amount, in mg, of antipyrine to the labeled amount, in mg, of benzocaine per mL of Antipyrine and Benzocaine Otic Solution. Add 50 mL of methanol, dissolve in methanol, dilute with methanol to volume, and mix. Transfer 10.0 mL of this solution to a 100-mL volumetric flask, dilute with *Mobile phase* to volume, and mix.

Assay preparation—Transfer an accurately measured volume of Antipyrine and Benzocaine Otic Solution, equivalent to about 15 mg of benzocaine, to a 100-mL volumetric flask. Dissolve in 50 mL of methanol, dilute with methanol to volume, and mix. Transfer 10.0 mL of this solution to a 100-mL volumetric flask, dilute with *Mobile phase* to volume, and mix.

Chromatographic system (see *Chromatography* ⟨621⟩)—The liquid chromatograph is equipped with a 280-nm detector and a 4-mm × 15-cm column that contains 5-μm packing L15. The flow rate is about 1.0 mL per minute. Chromatograph the *Standard preparation*, and record the peak responses as directed under *Procedure:* the tailing factor for the benzocaine peak is not more than 2.5, the column efficiency for the benzocaine peaks is not less than 1500 theoretical plates, and the relative standard deviation for replicate injections is not more than 2%.

Procedure—Separately inject equal volumes (about 10 μL) of the *Standard preparation* and the *Assay preparation* into the chromatograph, record the chromatograms, and measure the responses for the major peaks. The relative retention times are about 0.35 for antipyrine, and 1.0 for benzocaine. Calculate the quantity, in mg, of benzocaine ($C_9H_{11}NO_2$) in each mL of the Otic Solution taken by the formula:

$$1000(C/V)(r_U/r_S),$$

in which *C* is the concentration, in mg per mL, of USP Benzocaine RS in the *Standard preparation*, *V* is the volume, in mL, of Otic Solution taken, and r_U and r_S are the peak responses due to benzocaine in the *Assay preparation* and the *Standard preparation*, respectively. Calculate the quantity, in mg, of antipyrine ($C_{11}H_{12}N_2O$) in each mL of the Otic Solution taken by the same formula, changing the terms to refer to antipyrine instead of benzocaine.

Antipyrine, Benzocaine, and Phenylephrine Hydrochloride Otic Solution

» Antipyrine, Benzocaine, and Phenylephrine Hydrochloride Otic Solution is a solution of Antipyrine, Benzocaine, and Phenylephrine Hydrochloride in a suitable nonaqueous solvent. It contains not less than 90.0 percent and not more than 110.0 percent of the labeled amounts of antipyrine ($C_{11}H_{12}N_2O$), benzocaine ($C_9H_{11}NO_2$), and phenylephrine hydrochloride ($C_9H_{13}NO_2 . HCl$).

Packaging and storage—Preserve in tight, light-resistant containers.

Reference standards—*USP Antipyrine Reference Standard*—Dry at 60° for 2 hours before using. *USP Benzocaine Reference Standard*—Dry over phosphorus pentoxide for 3 hours before using. *USP Phenylephrine Hydrochloride Reference Standard*—Dry at 105° for 2 hours before using.

Identification—The retention times of the major peaks in the chromatograms of the *Assay preparations* correspond to those in the chromatogram of the *Standard preparation*, obtained as directed in the *Assay*.

Assay—

Mobile phase—Mix 480 mL of acetonitrile, 3520 mL of a 0.005 *M* solution of sodium 1-heptanesulfonate in water, and 4 mL of phosphoric acid.

Standard preparation—Accurately weigh about 25 mg of USP Antipyrine RS, about 25 mg of USP Benzocaine RS, and about 25 mg of USP Phenylephrine Hydrochloride RS into a 250-mL volumetric flask. Add 5 mL of a 0.5 mg-per-mL solution of para-aminobenzoic acid in *Mobile phase*. Add 150 mL of *Mobile phase*, and mix to effect solution, sonicating if necessary. Dilute with *Mobile phase* to volume, and mix.

Assay preparation A—Transfer an accurately measured volume of Antipyrine, Benzocaine, and Phenylephrine Hydrochloride Otic Solution, equivalent to about 100 mg of antipyrine, to a 50-mL volumetric flask, dilute with *Mobile phase* to volume, and mix. Pipet 5 mL of this solution into a 100-mL volumetric flask, dilute with *Mobile phase* to volume, and mix.

Assay preparation B—Transfer an accurately measured volume of Antipyrine, Benzocaine, and Phenylephrine Hydrochloride Otic Solution, equivalent to about 100 mg of benzocaine, to a 50-mL volumetric flask, dilute with *Mobile phase* to volume, and mix. Pipet 5 mL of this solution into a 100-mL volumetric flask, dilute with *Mobile phase* to volume, and mix.

Assay preparation P—Transfer an accurately measured volume of Antipyrine, Benzocaine, and Phenylephrine Hydrochloride Otic Solution, equivalent to about 5 mg of phenylephrine hydrochloride, to a 50-mL volumetric flask, dilute with *Mobile phase* to volume, and mix.

Chromatographic system (see *Chromatography* ⟨621⟩)—The liquid chromatograph is equipped with a 272-nm detector and a 4.6-mm × 30-cm column that contains packing L11. The flow rate is about 1.5 mL per minute. Chromatograph the *Standard preparation*, and record the peak responses as directed under *Procedure:* the resolution, *R*, between the phenylephrine and para-aminobenzoic acid peaks is not less than 1.5, and the relative standard deviation for replicate injections is not more than 3.0%.

Procedure—Separately inject equal volumes (about 20 or 25 µL) of the *Standard preparation* and each of the *Assay preparations* into the chromatograph, record the chromatograms, and measure the responses for the major peaks. The relative retention times are about 0.19 for para-aminobenzoic acid, 0.26 for phenylephrine, 0.64 for antipyrine, and 1.0 for benzocaine. Calculate the quantity, in mg, of $C_{11}H_{12}N_2O$ in each mL of the Otic Solution taken by the formula:

$$(C/V)(r_U/r_S),$$

in which *C* is the concentration, in µg per mL of USP Antipyrine RS in the *Standard preparation*, *V* is the volume, in mL, of Otic Solution taken, and r_U and r_S are the antipyrine peak responses obtained from *Assay preparation A* and the *Standard preparation*, respectively. Calculate the quantity, in mg, of $C_9H_{11}NO_2$ in each mL of the Otic Solution taken by the formula:

$$(C/V)(r_U/r_S),$$

in which *C* is the concentration, in µg per mL, of USP Benzocaine RS in the *Standard preparation*, *V* is the volume, in mL, of Otic Solution taken, and r_U and r_S are the benzocaine peak responses obtained from *Assay preparation B* and the *Standard preparation*, respectively. Calculate the quantity, in µg of $C_9H_{13}NO_2 \cdot HCl$ in each mL of the Otic Solution taken by the formula:

$$50(C/V)(r_U/r_S),$$

in which *C* is the concentration, in µg per mL, of USP Phenylephrine Hydrochloride RS in the *Standard preparation*, *V* is the volume, in mL, of Otic Solution taken, and r_U and r_S are the phenylephrine peak responses obtained from *Assay preparation P* and the *Standard preparation*, respectively.

Antirabies Serum

» Antirabies Serum conforms to the regulations of the federal Food and Drug Administration concerning biologics (see *Biologics* ⟨1041⟩). It is a sterile, non-pyrogenic solution containing antiviral substances obtained from the blood serum or plasma of a healthy animal, usually the horse, that has been immunized against rabies by means of vaccine. Its potency is determined in mice using the Serum Neutralization Test (SNT) or, in cell culture using the Rapid Fluorescent Focus Inhibition Test (RFFIT) in comparison with the U. S. Standard Rabies Immune Globulin. The CVS strain of rabies virus (mouse-adapted or cell culture adapted) is used as the challenge strain. It contains a suitable antimicrobial agent.

Packaging and storage—Preserve at a temperature between 2° and 8°.

Expiration date—The expiration date is not later than 2 years after date of issue from manufacturer's cold storage (5°, 1 year; or 0°, 2 years).

Labeling—Label it to indicate the species of animal in which it was prepared.

Antitoxin, Botulism—*see* Botulism Antitoxin
Antitoxin, Diphtheria—*see* Diphtheria Antitoxin
Antitoxin, Tetanus—*see* Tetanus Antitoxin

Antivenin (Crotalidae) Polyvalent

» Antivenin (Crotalidae) Polyvalent conforms to the regulations of the federal Food and Drug Administration concerning biologics (see *Biologics* ⟨1041⟩). It is a sterile, non-pyrogenic preparation derived by drying a frozen solution of specific venom-neutralizing globulins obtained from the serum of healthy horses immunized against venoms of four species of pit vipers, *Crotalus atrox*, *Crotalus adamanteus*, *Crotalus durissus terrificus*, and *Bothrops atrox* (Fam. Crotalidae). It is standardized by biological assay on mice, in terms of one dose of antivenin neutralizing the venoms in not less than the number of mouse LD_{50} stated, of *Crotalus atrox* (Western diamondback), 180; *Crotalus durissus terrificus* (South American rattlesnake), 1320; and *Bothrops atrox* (South American fer de lance), 780. It may contain a suitable preservative. When constituted as specified in the labeling, it is opalescent and contains not more than 20.0 percent of solids, determined by drying 1 mL at 105° to constant weight (±1 mg). It meets the requirements for general safety (see *Safety Tests—General* under *Biological Reactivity Tests, In-vivo* ⟨88⟩).

Packaging and storage—Preserve in single-dose containers, and avoid exposure to excessive heat.

Expiration date—The expiration date for Antivenin containing a 10% excess of potency is not more than 5 years after date of issue from manufacturer's cold storage (5°, 1 year; or 0°, 2 years).

Labeling—Label it to indicate the species of snakes against which the Antivenin is to be used, and to state that it was prepared from horse serum.

Antivenin (Latrodectus mactans)

» Antivenin (Latrodectus mactans) conforms to the regulations of the federal Food and Drug Administration concerning biologics (see *Biologics* ⟨1041⟩). It is the sterile, non-pyrogenic preparation derived by drying a frozen solution of specific venom-neutralizing globulins obtained from the serum of healthy horses immunized against venom of black widow spiders (*Latrodectus mactans*). It is standardized by biological assay on mice, in terms of one dose of antivenin neutralizing the venom of *Latrodectus mactans* in not less than 6000 mouse LD_{50}. Thimerosal 1:10,000 is added as a preservative. When constituted as specified in the labeling, it is opalescent and contains not more than 20.0 percent of solids.

Packaging and storage—Preserve in single-dose containers, and avoid exposure to excessive heat.

Expiration date—The expiration date for Antivenin containing a 10% excess of potency is not more than 5 years after date of issue from manufacturer's cold storage (5°, 1 year; 0°, 2 years).

Labeling—Label it to indicate the species of spider against which the Antivenin is to be used, that it is not intended to protect against bites from other spider species, and to state that it was prepared in the horse.

Antivenin (Micrurus Fulvius)

» Antivenin (Micrurus Fulvius) conforms to the regulations of the federal Food and Drug Administration concerning biologics (see *Biologics* ⟨1041⟩). It is the sterile, non-pyrogenic preparation derived by drying a frozen solution of specific venom-neutralizing globulins obtained from the serum of healthy horses immunized against venom of the Eastern Coral snake (*Micrurus fulvius*). It is standardized by biological assay on mice, in terms of one dose of antivenin neutralizing the venom of *Micrurus fulvius* in not less than 250 mouse LD_{50}. It may contain a suitable preservative. When constituted as specified in the labeling, it is opalescent and contains not more than 20.0 percent of solids, determined by drying 1 mL at 105° to constant weight (±1 mg). It meets the requirements for general safety (see *Safety Tests—General* under *Biological Reactivity Tests, In-vivo* ⟨88⟩).

Packaging and storage—Preserve in single-dose containers, and avoid exposure to excessive heat.

Expiration date—The expiration date for Antivenin containing a 10% excess of potency is not more than 5 years after date of issue from manufacturer's cold storage (5°, 1 year; or 0°, 2 years).

Labeling—Label it to indicate the species of snake against which the Antivenin is to be used, and to state that it was prepared in the horse.

Apomorphine Hydrochloride

HO HO N—CH3 H • HCl • ½H2O

$C_{17}H_{17}NO_2.HCl.½H_2O$ 312.80
4*H*-Dibenzo[*de,g*]quinoline-10,11-diol, 5,6,6a,7-tetrahydro-6-methyl-, hydrochloride, hemihydrate, (*R*)-.
6a*β*-Aporphine-10,11-diol hydrochloride hemihydrate [*41372-20-7*].
Anhydrous 303.79 [*314-19-2*].

» Apomorphine Hydrochloride contains not less than 98.5 percent and not more than 100.5 percent of $C_{17}H_{17}NO_2.HCl$, calculated on the dried basis.

Packaging and storage—Preserve in small, tight, light-resistant containers. Containers from which Apomorphine Hydrochloride is to be taken for immediate use in compounding prescriptions contain not more than 350 mg.

Reference standard—*USP Apomorphine Hydrochloride Reference Standard*—Dry at 105° for 2 hours before using.

Color of solution—Place 100 mg in a suitable test tube, add 10 mL of cold, oxygen-free water, and agitate gently until dissolved: the color of the resulting solution, observed promptly after the Apomorphine Hydrochloride has dissolved, is not more intense than that of a color standard prepared as follows: Dissolve 5 mg of Apomorphine Hydrochloride in 100.0 mL of water. Transfer 1.0 mL of this solution to a test tube of the same size as that used for the test solution, dilute with 6 mL of water, add 1 mL of sodium bicarbonate solution (1 in 20), and then add 0.50 mL of iodine TS. Allow to stand for 30 seconds, add 0.60 mL of sodium thiosulfate solution (1 in 40), and dilute with water to 10 mL.

Identification—

A: The infrared absorption spectrum of a potassium bromide dispersion of it, previously dried, exhibits maxima only at the same wavelengths as that of a similar preparation of USP Apomorphine Hydrochloride RS.

B: To 5 mL of a solution (1 in 100) add a slight excess of sodium bicarbonate solution (1 in 20): a white or greenish white precipitate is formed. Add 3 drops of iodine TS, and shake vigorously: an emerald-green color is produced. Add 5 mL of ether and, after vigorous shaking, allow the layers to separate: the ether is colored deep ruby-red while the water layer retains its green color.

C: Dissolve it in nitric acid: a dark purple solution is produced.

D: To a solution of it add silver nitrate TS: a white precipitate, which is insoluble in nitric acid, is formed. This precipitate soon darkens by reduction to metallic silver, the reduction being accelerated by the addition of 6 *N* ammonium hydroxide.

Specific rotation ⟨781⟩: between −60.5° and −63.0°, calculated on the dried basis, determined in a solution in dimethylsulfoxide containing the equivalent of 150 mg of apomorphine hydrochloride in each 10 mL.

Loss on drying ⟨731⟩—Dry it at 105° for 2 hours: it loses between 2.0% and 3.5% of its weight.

Residue on ignition ⟨281⟩: not more than 0.1%.

Decomposition products—Shake 100 mg with 5 mL of ether: the latter acquires not more than a pale reddish color.

Ordinary impurities ⟨466⟩—

Test solution: methanol.

Standard solution: methanol.

Eluant: a mixture of methanol and 0.1 *N* hydrochloric acid (9:1).

Visualization: 1.

Assay—Dissolve about 300 mg of Apomorphine Hydrochloride, accurately weighed, in 100 mL of glacial acetic acid with the aid of heat provided by a steam bath. Add 0.1 mL of acetic anhydride to the hot solution, and stir for 5 minutes. Cool to room temperature, add 5 mL of mercuric acetate TS and 0.25 mL of crystal violet TS, and titrate with 0.1 *N* perchloric acid VS to a blue end-point. Perform a blank determination, and make any necessary correction. Each mL of 0.1 *N* perchloric acid is equivalent to 30.38 mg of $C_{17}H_{17}NO_2.HCl$.

Apomorphine Hydrochloride Tablets

» Apomorphine Hydrochloride Tablets contain not less than 90.0 percent and not more than 110.0 percent of the labeled amount of $C_{17}H_{17}NO_2.HCl.½H_2O$.

Packaging and storage—Preserve in tight, light-resistant containers.

Reference standard—*USP Apomorphine Hydrochloride Reference Standard*—Dry at 105° for 2 hours before using.

Color of solution—Dissolve a quantity of powdered Tablets, equivalent to 5 mg of apomorphine hydrochloride, in water to make 100.0 mL. Transfer 1.0 mL of the solution to a test tube, dilute with 6 mL of water, and, if necessary, filter through a small pledget of cotton. Add 1 mL of sodium bicarbonate solution (1 in 20), then add 0.50 mL of iodine TS. Allow to stand for 30 seconds, then add 0.60 mL of sodium thiosulfate solution (1 in 40), and dilute with water to 10 mL. This solution represents the color standard.

Place a quantity of powdered Tablets, equivalent to about 50 mg of apomorphine hydrochloride, in a test tube of suitably small size, add 10.0 mL of cold, oxygen-free water, insert the stopper in the test tube, and agitate gently until no more dissolves; if necessary, filter immediately through a small pledget of cotton. The color of the solution, observed promptly after preparation, is not more intense than that of the color standard. Use closely matched test tubes for the comparison.

Identification—To 5 mL of a filtered solution of Tablets, containing about 10 mg of apomorphine hydrochloride, add a slight excess of sodium bicarbonate solution (1 in 20): a white or greenish white precipitate is formed. Add 3 drops of iodine TS, and shake vigorously: an emerald-green color is produced. Add 5 mL of ether, and, after vigorous shaking, allow the layers to separate: the ether is colored deep ruby-red while the water layer retains its green color.

Disintegration ⟨701⟩: 15 minutes.

Uniformity of dosage units ⟨905⟩: meet the requirements.

Procedure for content uniformity—Place 1 Tablet in a 500-mL volumetric flask containing 100 mL of 0.1 *N* hydrochloric acid, and shake for 15 minutes. Dilute with 0.1 *N* hydrochloric acid to volume, mix, and filter, discarding the first 20 mL of filtrate. Dilute a portion of the subsequent filtrate quantitatively and stepwise, if necessary, with 0.1 *N* hydrochloric acid to provide a solution containing approximately 12 µg of apomorphine hydrochloride per mL. Concomitantly determine the absorbances of this solution and of a solution of USP Apomorphine Hydrochloride RS in the same medium having a known concentration of about 12 µg of anhydrous apomorphine hydrochloride per mL, in 1-cm cells at the wavelength of maximum absorbance at about 273 nm, with a suitable spectrophotometer, using 0.1 *N* hydrochloric acid as the blank. Calculate the quantity, in mg, of $C_{17}H_{17}NO_2.HCl.½H_2O$ in the Tablet by the formula:

$$(312.80/303.79)(TC/D)(A_U/A_S),$$

in which 312.80 and 303.79 are the molecular weights of apomorphine hydrochloride hemihydrate and anhydrous apomorphine hydrochloride, respectively, *T* is the labeled quantity, in mg, of apomorphine hydrochloride in the Tablet, *C* is the concentration, in µg per mL, of anhydrous apomorphine hydrochloride in the Standard solution, *D* is the concentration, in µg per mL, of apomorphine hydrochloride in the solution from the Tablet, based upon the labeled quantity per Tablet and the extent of dilution, and A_U and A_S are the absorbances of the solution from the Tablet and the Standard solution, respectively.

Assay—Weigh and finely powder not less than 20 Apomorphine Hydrochloride Tablets. Dissolve an accurately weighed portion of the powder, equivalent to about 50 mg of apomorphine hydrochloride, in 25 mL of water in a separator, add 500 mg of sodium bicarbonate, and completely extract with successive small portions of ether. Combine the ether extracts in a separator, and wash them with three 5-mL portions of water. Shake the combined water washings with 10 mL of ether, and add this ether to the combined ether extracts. Extract the ether solutions with 20.0 mL of 0.02 *N* sulfuric acid VS, and wash with three 5-mL portions of water. Combine the acid extract and washings in a beaker, and warm on a steam bath to expel any residual ether. Cool, add methyl red TS, and titrate the excess acid with 0.02 *N* sodium hydroxide VS (see *Residual Titrations* ⟨541⟩). Each mL of 0.02 *N* sulfuric acid is equivalent to 6.256 mg of $C_{17}H_{17}NO_2.HCl.½H_2O$.

Arginine

$$H_2N\overset{NH}{\overset{\|}{C}}NHCH_2CH_2CH_2-\overset{H}{\underset{NH_2}{C}}-COOH$$

$C_6H_{14}N_4O_2$ 174.20

L-Arginine.

L-Arginine [*74-79-3*].

» Arginine contains not less than 98.5 percent and not more than 101.5 percent of $C_6H_{14}N_4O_2$, as L-arginine, calculated on the dried basis.

Packaging and storage—Preserve in well-closed containers.

Reference standard—*USP L-Arginine Reference Standard*—Dry at 105° for 3 hours before using.

Identification—The infrared absorption spectrum of a potassium bromide dispersion of it, previously dried, exhibits maxima only at the same wavelengths as that of a similar preparation of USP L-Arginine RS.

Specific rotation ⟨781⟩: between +26.2° and +27.6°, calculated on the dried basis, determined in a solution in 6 *N* hydrochloric acid containing 800 mg in each 10 mL.

Loss on drying ⟨731⟩—Dry it at 105° for 3 hours: it loses not more than 0.5% of its weight.

Residue on ignition ⟨281⟩: not more than 0.3%.

Chloride ⟨221⟩—A 1.0-g portion shows no more chloride than corresponds to 0.70 mL of 0.020 *N* hydrochloric acid (0.05%).

Sulfate ⟨221⟩—A 1.0-g portion shows no more sulfate than corresponds to 0.30 mL of 0.020 *N* sulfuric acid (0.03%).

Arsenic ⟨211⟩: 1.5 ppm.

Iron ⟨241⟩: 0.003%.

Heavy metals, *Method I* ⟨231⟩: 0.0015%.

Assay—Transfer about 80 mg of Arginine, accurately weighed, to a 125-mL flask, dissolve in a mixture of 3 mL of formic acid and 50 mL of glacial acetic acid, and titrate with 0.1 *N* perchloric acid VS, determining the end-point potentiometrically. Perform a blank determination, and make any necessary correction. Each mL of 0.1 *N* perchloric acid is equivalent to 8.710 mg of $C_6H_{14}N_4O_2$.

Arginine Hydrochloride

$C_6H_{14}N_4O_2.HCl$ 210.66

L-Arginine monohydrochloride.

L-(+)-Arginine monohydrochloride [*1119-34-2*].

» Arginine Hydrochloride contains not less than 98.5 percent and not more than 101.5 percent of $C_6H_{14}N_4O_2.HCl$, calculated on the dried basis.

Packaging and storage—Preserve in well-closed containers.

Reference standard—*USP Arginine Hydrochloride Reference Standard*—Dry at 105° for 2 hours before using.

Identification—

A: The infrared absorption spectrum of a potassium bromide dispersion of it exhibits maxima only at the same wavelengths as that of a similar preparation of USP Arginine Hydrochloride RS.

B: A solution (1 in 10) responds to the tests for *Chloride* ⟨191⟩.

Specific rotation ⟨781⟩: not less than +21.4° and not more than +23.6°, calculated on the dried basis, determined at 20° in a solution in 6 *N* hydrochloric acid containing 800 mg in each 10 mL.

Loss on drying ⟨731⟩—Dry it at 105° for 2 hours: it loses not more than 0.2% of its weight.

Residue on ignition ⟨281⟩: not more than 0.1%.

Sulfate ⟨221⟩—A 1.6-g portion shows no more sulfate than corresponds to 0.50 mL of 0.020 *N* sulfuric acid (0.03%).

Arsenic, *Method II* ⟨211⟩: 1.5 ppm.

Chloride content—Transfer about 350 mg, accurately weighed, to a porcelain casserole, and add 140 mL of water and 1 mL of dichlorofluorescein TS. Mix, and titrate with 0.1 *N* silver nitrate VS until the silver chloride flocculates and the mixture acquires a faint pink color. Each mL of 0.1 *N* silver nitrate is equivalent to 3.545 mg of chloride: between 16.5% and 17.1% is found.

Heavy metals ⟨231⟩—Dissolve 1.0 g in 20 mL of water, add 2 mL of 1 *N* acetic acid, dilute with water to 25 mL, and proceed as directed for *Method I* (0.002%).

Assay—Dissolve about 100 mg of Arginine Hydrochloride, accurately weighed, in 3 mL of 98 percent formic acid and 50 mL of glacial acetic acid. Add 6 mL of mercuric acetate TS, and titrate with 0.1 *N* perchloric acid VS, determining the end-point potentiometrically. Perform a blank determination, and make any necessary correction. Each mL of 0.1 *N* perchloric acid is equivalent to 10.53 mg of $C_6H_{14}N_4O_2 \cdot HCl$.

Arginine Hydrochloride Injection

» Arginine Hydrochloride Injection is a sterile solution of Arginine Hydrochloride in Water for Injection. It contains not less than 9.5 percent and not more than 10.5 percent of $C_6H_{14}N_4O_2 \cdot HCl$. It contains no antimicrobial agents.

NOTE—The chloride ion content of Arginine Hydrochloride Injection is approximately 475 mEq per liter.

Packaging and storage—Preserve in single-dose containers, preferably of Type II glass.

Reference standard—*USP Arginine Hydrochloride Reference Standard*—Dry at 105° for 2 hours before using.

Labeling—The label states the total osmolar concentration in mOsmol per liter. Where the contents are less than 100 mL, or where the label states that the Injection is not for direct injection but is to be diluted before use, the label alternatively may state the total osmolar concentration in mOsmol per mL.

Identification—

A: To 1 mL of Injection add 2 mL of a solution of 0.02% 8-hydroxyquinoline in 3 *N* sodium hydroxide, and add 1 mL of 0.1% *N*-bromosuccinimide solution: a red color is produced.

B: It responds to the tests for *Chloride* ⟨191⟩.

Pyrogen—It meets the requirements of the *Pyrogen Test* ⟨151⟩.

pH ⟨791⟩: between 5.0 and 6.5.

Other requirements—It meets the requirements under *Injections* ⟨1⟩.

Assay—

Color reagent—Dissolve 28.0 g of potassium hydroxide and 2.0 g of potassium sodium tartrate in 100 mL of water. Cool, and add, in the order named, 100 mg of 2,4-dichloro-1-naphthol, 180 mL of alcohol, and 20.0 mL of 0.475% sodium hypochlorite solution. Mix by swirling, and allow to stand at room temperature for 1 hour before using. This *Color reagent* may be stored in a glass-stoppered bottle, in a refrigerator, for 2 months.

Standard preparation—Dissolve an accurately weighed quantity of USP Arginine Hydrochloride RS in water, and dilute quantitatively and stepwise with water to obtain a solution having a known concentration of about 40 µg per mL.

Assay preparation—Pipet into a 100-mL volumetric flask a volume of Arginine Hydrochloride Injection, equivalent to 200 mg of arginine hydrochloride, add water to volume, and mix. Pipet 5 mL of this solution into a 250-mL volumetric flask, add water to volume, and mix.

Procedure—Transfer 2.0-mL portions of the *Assay preparation* and the *Standard preparation*, respectively, to separate flasks, and treat each as follows: Add 2.0 mL of potassium iodide solution (3 in 1000), mix, and allow to stand for 15 minutes. Add 6.0 mL of *Color reagent*, mix, and allow to stand for 15 minutes. Add 2.0 mL of sodium hypochlorite solution (19 in 10,000), mix, and allow to stand for 15 minutes. Concomitantly determine the absorbances of both solutions in 1-cm cells at the wavelength of maximum absorbance at about 520 nm, with a suitable spectrophotometer, using water as the blank. Calculate the quantity, in mg, of $C_6H_{14}N_4O_2 \cdot HCl$ in each mL of the Injection taken by the formula:

$$5(C/V)(A_U/A_S),$$

in which *C* is the concentration, in µg per mL, of USP Arginine Hydrochloride RS in the *Standard preparation*, *V* is the volume, in mL, of Injection taken, and A_U and A_S are the absorbances of the solutions from the *Assay preparation* and the *Standard preparation*, respectively.

Aromatic Ammonia Spirit—*see* Ammonia Spirit, Aromatic

Aromatic Cascara Fluidextract—*see* Cascara Fluidextract, Aromatic

Aromatic Castor Oil—*see* Castor Oil, Aromatic

Aromatic Elixir—*see* Aromatic Elixir NF

Ascorbic Acid

$C_6H_8O_6$ 176.13

L-Ascorbic acid.

L-Ascorbic acid [*50-81-7*].

» Ascorbic Acid contains not less than 99.0 percent and not more than 100.5 percent of $C_6H_8O_6$.

Packaging and storage—Preserve in tight, light-resistant containers.

Reference standard—*USP Ascorbic Acid Reference Standard.*

Identification—

A: The infrared absorption spectrum of a potassium bromide dispersion of it exhibits maxima only at the same wavelengths as that of a similar preparation of USP Ascorbic Acid RS.

B: A solution (1 in 50) reduces alkaline cupric tartrate TS slowly at room temperature but more readily upon heating.

Specific rotation ⟨781⟩: between +20.5° and +21.5°, determined in a solution in carbon dioxide–free water containing 1 g in each 10 mL, the optical rotation being measured immediately following the preparation of the solution.

Residue on ignition ⟨281⟩: not more than 0.1%.

Heavy metals ⟨231⟩—Dissolve 1 g in 25 mL of water: the limit is 0.002%.

Assay—Dissolve about 400 mg of Ascorbic Acid, accurately weighed, in a mixture of 100 mL of water and 25 mL of 2 *N* sulfuric acid. Add 3 mL of starch TS, and titrate at once with 0.1 *N* iodine VS. Each mL of 0.1 *N* iodine is equivalent to 8.806 mg of $C_6H_8O_6$.

Ascorbic Acid Injection

» Ascorbic Acid Injection is a sterile solution, in Water for Injection, of Ascorbic Acid prepared with the aid of Sodium Hydroxide, Sodium Carbonate, or Sodium Bicarbonate. It contains not less than 90.0 percent and not more than 110.0 percent of the labeled amount of $C_6H_8O_6$.

Packaging and storage—Preserve in light-resistant, single-dose containers, preferably of Type I or Type II glass.

Labeling—In addition to meeting the requirements for *Labeling* under *Injections* ⟨1⟩, fused-seal containers of the Injection in concentrations of 250 mg per mL and greater are labeled to indicate that since pressure may develop on long storage, precautions should be taken to wrap the container in a protective covering while it is being opened.

Identification—

A: To a volume of Injection, equivalent to 40 mg of ascorbic acid, add 4 mL of 0.1 *N* hydrochloric acid, then add 4 drops of methylene blue TS, and warm to 40°: the deep blue color becomes appreciably lighter or is completely discharged within 3 minutes.

B: To a volume of Injection, equivalent to 15 mg of ascorbic acid, add 15 mL of a solution of trichloroacetic acid (1 in 20), add about 200 mg of activated charcoal, shake the mixture vigorously for 1 minute, and filter through a small fluted filter, returning the filtrate, if necessary, until clear. To 5 mL of the filtrate add 1 drop of pyrrole, and agitate gently until dissolved, then heat in a bath at 50°: a blue color develops.

C: It responds to the flame test for *Sodium* ⟨191⟩.

pH ⟨791⟩: between 5.5 and 7.0.

Oxalate—Dilute a volume of Injection, equivalent to 50 mg of ascorbic acid, with water to 5 mL. Add 0.2 mL of acetic acid and 0.5 mL of calcium chloride TS: no turbidity is produced in 1 minute.

Other requirements—It meets the requirements under *Injections* ⟨1⟩.

Assay—Transfer to a 100-mL volumetric flask an accurately measured volume of Ascorbic Acid Injection, equivalent to about 50 mg of ascorbic acid and previously diluted with water if necessary. Add 20 mL of metaphosphoric-acetic acids TS, dilute with water to volume, and mix.

Accurately measure a volume of the dilution, equivalent to about 2 mg of ascorbic acid, into a 50-mL conical flask, add 5 mL of metaphosphoric-acetic acids TS, and titrate with standard dichlorophenol-indophenol solution until a rose-pink color persists for at least 5 seconds. Correct for the volume of the dichlorophenol-indophenol solution consumed by a mixture of 5.5 mL of metaphosphoric-acetic acids TS and 15 mL of water.

From the ascorbic acid equivalent of the standard dichlorophenol-indophenol solution calculate the ascorbic acid content in each mL of the Injection.

Ascorbic Acid Oral Solution

» Ascorbic Acid Oral Solution is a solution of Ascorbic Acid in a hydroxylic organic solvent or an aqueous mixture thereof. It contains not less than 90.0 percent and not more than 110.0 percent of the labeled amount of $C_6H_8O_6$.

Packaging and storage—Preserve in tight, light-resistant containers.

Labeling—Label Oral Solution that contains alcohol to state the alcohol content.

Identification—

A: To a volume of Oral Solution equivalent to 40 mg of ascorbic acid add 4 mL of 0.1 *N* hydrochloric acid, then add 4 drops of methylene blue TS, and warm to 40°: the deep blue color becomes appreciably lighter or is completely discharged within 3 minutes.

B: To a volume of Oral Solution equivalent to 20 mg of ascorbic acid add 15 mL of trichloroacetic acid solution (1 in 20), then add about 200 mg of activated charcoal, shake the mixture vigorously for 1 minute, and filter through a small fluted filter, returning the filtrate, if necessary, until clear. To 5 mL of the filtrate add 1 drop of pyrrole, agitate gently until dissolved, then heat in a bath at 50°: a blue color develops.

Alcohol content (if present), *Method I* ⟨611⟩: between 90.0% and 110.0% of the labeled amount of C_2H_5OH.

Assay—Transfer to a 100-mL volumetric flask an accurately measured volume of Ascorbic Acid Oral Solution, equivalent to about 50 mg of ascorbic acid, previously diluted with water if necessary, and proceed as directed in the *Assay* under *Ascorbic Acid Injection*, beginning with "Add 20 mL of metaphosphoric-acetic acids TS."

Ascorbic Acid Tablets

» Ascorbic Acid Tablets contain not less than 90.0 percent and not more than 110.0 percent of the labeled amount of $C_6H_8O_6$.

Packaging and storage—Preserve in tight, light-resistant containers.

Identification—Triturate a quantity of finely powdered Tablets with sufficient diluted alcohol to make approximately the equivalent of a 1 in 50 solution of ascorbic acid, filter, and proceed with the following tests:

A: A portion of the filtrate responds to *Identification test B* under *Ascorbic Acid*.

B: To 2 mL of the filtrate add 4 drops of methylene blue TS, and warm to 40°: the deep blue color becomes appreciably lighter or is completely discharged within 3 minutes.

C: To 1 mL of the filtrate add 15 mL of a solution of trichloroacetic acid (1 in 20), add about 200 mg of activated charcoal, shake the mixture vigorously for 1 minute, and filter through a small fluted filter, returning the filtrate, if necessary, until clear. To 5 mL of the filtrate add 1 drop of pyrrole, and agitate gently until dissolved, then heat in a bath at 50°: a blue color develops.

Disintegration ⟨701⟩: 30 minutes.

Uniformity of dosage units ⟨905⟩: meet the requirements.

Assay—Transfer not less than 20 Ascorbic Acid Tablets to a 1000-mL volumetric flask containing 250 mL of metaphosphoric-acetic acids TS. Insert the stopper in the flask, and shake by mechanical means for 30 minutes or until the tablets have disintegrated completely. Dilute with water to volume, and mix. Transfer a portion of the solution to a centrifuge tube, and centrifuge until a clear supernatant solution is obtained. Quantitatively dilute the clear supernatant solution with water, if necessary, to obtain a solution containing about 500 µg of ascorbic acid per mL. Pipet 4 mL of the solution, equivalent to about 2 mg of ascorbic acid, into a 50-mL conical flask, and proceed as directed in the *Assay* under *Ascorbic Acid Injection*, beginning with "add 5 mL of metaphosphoric-acetic acids TS."

From the ascorbic acid equivalent of the standard dichlorophenol-indophenol solution calculate the content of ascorbic acid in each Tablet.

Ascorbyl Palmitate—*see* Ascorbyl Palmitate NF
Aspartame—*see* Aspartame NF

Aspirin

$C_9H_8O_4$ 180.16
Benzoic acid, 2-(acetyloxy)-.
Salicylic acid acetate [*50-78-2*].

» Aspirin contains not less than 99.5 percent and not more than 100.5 percent of $C_9H_8O_4$, calculated on the dried basis.

Packaging and storage—Preserve in tight containers.

Reference standard—*USP Aspirin Reference Standard*—Keep container tightly closed. Dry over silica gel for 5 hours before using.

Identification—

A: Heat it with water for several minutes, cool, and add 1 or 2 drops of ferric chloride TS: a violet-red color is produced.

B: The infrared absorption spectrum of a potassium bromide dispersion of it exhibits maxima only at the same wavelengths as that of a similar preparation of USP Aspirin RS.

Loss on drying ⟨731⟩—Dry it over silica gel for 5 hours: it loses not more than 0.5% of its weight.

Residue on ignition ⟨281⟩: not more than 0.05%.

Chloride ⟨221⟩—Boil 1.5 g with 75 mL of water for 5 minutes, cool, add sufficient water to restore the original volume, and filter. A 25-mL portion of the filtrate shows no more chloride than corresponds to 0.10 mL of 0.020 *N* hydrochloric acid (0.014%).

Sulfate—Dissolve 6.0 g in 37 mL of acetone, and add 3 mL of water. Titrate potentiometrically with 0.02 *M* lead perchlorate, prepared by dissolving 9.20 g of lead perchlorate in water to make 1000 mL of solution, using a pH meter capable of a minimum reproducibility of ±0.1 mV (see *pH* ⟨791⟩) equipped with an electrode system consisting of a lead-specific electrode and a silver–silver chloride reference glass-sleeved electrode containing a 1 in 44 solution of tetraethylammonium perchlorate in glacial acetic acid (see *Titrimetry* ⟨541⟩): not more than 1.25 mL of 0.02 *M* lead perchlorate is consumed (0.04%). [NOTE—After use, rinse the lead-specific electrode with water, drain the reference electrode, flush with water, rinse with methanol, and allow to dry.]

Non-aspirin salicylates—Dissolve 2.5 g in sufficient alcohol to make 25.0 mL. To each of two matched color-comparison tubes add 48 mL of water and 1 mL of a freshly prepared, diluted ferric ammonium sulfate solution (prepared by adding 1 mL of 1 *N* hydrochloric acid to 2 mL of ferric ammonium sulfate TS and diluting with water to 100 mL). Into one tube pipet 1 mL of a standard solution of salicylic acid in water, containing 0.10 mg of salicylic acid per mL. Into the second tube pipet 1 mL of the 1 in 10 solution of Aspirin. Mix the contents of each tube: after 30 seconds, the color in the second tube is not more intense than that in the tube containing the salicylic acid (0.1%).

Heavy metals—Dissolve 2 g in 25 mL of acetone, and add 1 mL of water and 10 mL of hydrogen sulfide TS: any color produced is not darker than that of a control made with 25 mL of acetone, 2 mL of standard lead solution (see *Heavy Metals* ⟨231⟩), and 10 mL of hydrogen sulfide TS (0.001%).

Readily carbonizable substances ⟨271⟩—Dissolve 500 mg in 5 mL of sulfuric acid TS: the solution has no more color than Matching Fluid Q.

Substances insoluble in sodium carbonate TS—A solution of 500 mg in 10 mL of warm sodium carbonate TS is clear.

Assay—Place about 1.5 g of Aspirin, accurately weighed, in a flask, add 50.0 mL of 0.5 *N* sodium hydroxide VS, and boil the mixture gently for 10 minutes. Add phenolphthalein TS, and titrate the excess sodium hydroxide with 0.5 *N* sulfuric acid VS. Perform a blank determination (see *Residual Titrations* under *Titrimetry* ⟨541⟩). Each mL of 0.5 *N* sodium hydroxide is equivalent to 45.04 mg of $C_9H_8O_4$.

Aspirin, and Caffeine Capsules, Acetaminophen,—*see* Acetaminophen, Aspirin, and Caffeine Capsules
Aspirin, and Caffeine Capsules, Propoxyphene Hydrochloride,—*see* Propoxyphene Hydrochloride, Aspirin, and Caffeine Capsules
Aspirin, and Caffeine Tablets, Acetaminophen,—*see* Acetaminophen, Aspirin, and Caffeine Tablets
Aspirin, and Codeine Phosphate Tablets, Carisoprodol,—*see* Carisoprodol, Aspirin, and Codeine Phosphate Tablets

Aspirin Capsules

» Aspirin Capsules contain not less than 93.0 percent and not more than 107.0 percent of the labeled amount of $C_9H_8O_4$.

NOTE—Capsules that are enteric-coated or the contents of which are enteric-coated meet the requirements for *Aspirin Delayed-release Capsules.*

Packaging and storage—Preserve in tight containers.

Reference standard—*USP Aspirin Reference Standard*—Dry over silica gel for 5 hours before using.

Identification—

A: Heat about 100 mg of the Capsule contents with 10 mL of water for several minutes, cool, and add 1 drop of ferric chloride TS: a violet-red color is produced.

B: Shake a quantity of the contents of Aspirin Capsules, equivalent to about 500 mg of aspirin, with 10 mL of chloroform for several minutes. Centrifuge the mixture. Pour off the clear supernatant liquid and evaporate it to dryness: the residue responds to *Identification test B* under *Aspirin.*

Dissolution ⟨711⟩—

Medium: 0.05 *M* acetate buffer, prepared by mixing 2.99 g of sodium acetate trihydrate and 1.66 mL of glacial acetic acid with water to obtain 1000 mL of solution having a pH of 4.50 ± 0.05; 500 mL.

Apparatus 1: 100 rpm.

Time: 30 minutes.

Procedure—Determine the amount of $C_9H_8O_4$ dissolved from ultraviolet absorbances at the wavelength of the isosbestic point of aspirin and salicylic acid at 265 ± 2 nm of filtered portions of the solution under test, suitably diluted with *Dissolution Medium,* if necessary, in comparison with a Standard solution having a known concentration of USP Aspirin RS in the same medium. [NOTE—Prepare the Standard solution at the time of use. An amount of alcohol not to exceed 1% of the total volume of the Standard solution may be used to bring the Reference Standard into solution prior to dilution with *Dissolution Medium.*]

Tolerances—Not less than 80% (*Q*) of the labeled amount of $C_9H_8O_4$ is dissolved in 30 minutes.

Uniformity of dosage units ⟨905⟩: meet the requirements.

Free salicylic acid—

Ferric chloride–urea reagent—Dissolve by swirling, without the aid of heat, 60 g of urea in a mixture of 8 mL of ferric chloride solution (6 in 10) and 42 mL of 0.05 *N* hydrochloric acid. Adjust the resulting solution, if necessary, with 6 *N* hydrochloric acid to a pH of 3.2.

Standard preparation—Transfer 75.0 mg of salicylic acid, previously dried over silica gel for 3 hours and accurately weighed, to a 100-mL volumetric flask, add chloroform to volume, and mix. Transfer 10.0 mL of this solution to a second 100-mL volumetric flask, dilute with chloroform to volume, and mix. Transfer 10.0 mL of this last solution to a 50-mL volumetric flask containing 10 mL of methanol, 2 drops of hydrochloric acid, and 10 mL of a 1 in 10 solution of glacial acetic acid in ether, dilute with chloroform to volume, and mix.

Chromatographic column—Proceed as directed under *Column Partition Chromatography* ⟨621⟩, packing a chromatographic tube with two segments of packing material. The lower segment is a mixture of 1 g of *Solid Support* and 0.5 mL of 5 *M* phosphoric acid, and the upper segment is a mixture of 3 g of *Solid Support* and 2 mL of freshly prepared *Ferric chloride–urea reagent*.

Test preparation—Weigh accurately a portion of the contents of Aspirin Capsules, as determined by the *Assay*, equivalent to 100 mg of aspirin, mix with 10 mL of chloroform by stirring for 3 minutes, and then transfer to the chromatographic column with the aid of a few mL of chloroform. Pass 50 mL of chloroform through the column, rinse the tip of the chromatographic tube with chloroform, and discard the eluate. Prepare as a receiver a 50-mL volumetric flask containing 10 mL of methanol and 2 drops of hydrochloric acid, and elute any salicylic acid from the column by passing 10 mL of a 1 in 10 solution of glacial acetic acid in ether that has been recently saturated with water, followed by 30 mL of chloroform. Dilute the eluate with chloroform to volume, and mix.

Procedure—Concomitantly determine the absorbances of the solutions in 1-cm cells at the wavelength of maximum absorbance at about 306 nm, with a suitable spectrophotometer, using as the blank a solvent mixture of the same composition as that used for the *Standard preparation:* the absorbance of the *Test preparation* does not exceed that of the *Standard preparation* (0.75%, calculated on the labeled aspirin content).

Assay—[NOTE—In this assay use chloroform recently saturated with water.]

Standard preparation—Transfer about 50 mg of USP Aspirin RS, accurately weighed, to a 50-mL volumetric flask, add 0.5 mL of glacial acetic acid, add chloroform to volume, and mix. Transfer 5.0 mL of this solution to a 100-mL volumetric flask, dilute with a 1 in 100 solution of glacial acetic acid in chloroform to volume, and mix. The concentration of USP Aspirin RS is about 50 µg per mL.

Chromatographic column—Proceed as directed under *Column Partition Chromatography* (see *Chromatography* ⟨621⟩), packing a chromatographic tube with a mixture of 3 g of *Solid Support* and 2 mL of freshly prepared sodium bicarbonate solution (1 in 12).

Assay preparation—Remove, as completely as possible, the contents of not less than 20 Aspirin Capsules, and weigh accurately. Mix the combined contents, and transfer an accurately weighed quantity of the powder, equivalent to about 50 mg of aspirin, to a 50-mL volumetric flask containing 1 mL of a 1 in 50 solution of hydrochloric acid in methanol, add chloroform to volume, and mix. Transfer 5.0 mL of this solution to the column, wash with 5 mL and then with 25 mL of chloroform, and discard the washings. Elute into a 100-mL volumetric flask with about 10 mL of a 1 in 10 solution of glacial acetic acid in chloroform and then with about 85 mL of a 1 in 100 solution of glacial acetic acid in chloroform, dilute with the latter solvent to volume, and mix.

Procedure—Without delay, concomitantly determine the absorbances of the solutions in 1-cm cells at the wavelength of maximum absorbance at about 280 nm, with a suitable spectrophotometer, using chloroform as the blank. Calculate the quantity, in mg, of $C_9H_8O_4$ in the portion of Capsules taken by the formula:

$$C(A_U/A_S),$$

in which *C* is the concentration, in µg per mL, of USP Aspirin RS in the *Standard preparation*, and A_U and A_S are the absorbances of the *Assay preparation* and the *Standard preparation*, respectively.

Aspirin Delayed-release Capsules

» Aspirin Delayed-release Capsules contain not less than 93.0 percent and not more than 107.0 percent of the labeled amount of $C_9H_8O_4$.

Packaging and storage—Preserve in tight containers.

Labeling—The label indicates that Aspirin Delayed-release Capsules or the contents thereof are enteric-coated.

Reference standards—*USP Aspirin Reference Standard*—Dry over silica gel for 5 hours before using. *USP Salicylic Acid Reference Standard*—Dry over silica gel for 3 hours before using.

Identification—

A: Heat about 100 mg of the Capsule contents with 10 mL of water for several minutes, cool, and add 1 drop of ferric chloride TS: a violet-red color is produced.

B: Shake a quantity of the contents of Capsules, equivalent to about 500 mg of aspirin, with 10 mL of chloroform for several minutes. Centrifuge the mixture. Pour off the clear supernatant liquid and evaporate it to dryness: the residue responds to *Identification test B* under *Aspirin*.

Drug release, *Method A* ⟨724⟩—

Apparatus 1: 100 rpm.

Time: 90 minutes, for *Buffer stage*.

Diluent—Prepare a mixture of 0.1 *N* hydrochloric acid and 0.20 *M* tribasic sodium phosphate (3:1), and adjust, if necessary, with 2 *N* hydrochloric acid or 2 *N* sodium hydroxide to a pH of 6.8 ± 0.05.

Procedure—Determine the amount of $C_9H_8O_4$ dissolved by determining ultraviolet absorbances at the wavelength of the isosbestic point of aspirin and salicylic acid (about 280 nm in the *Acid stage*, and about 265 nm in the *Buffer stage*) using a filtered portion of the solution under test, diluted, if necessary, with 0.1 *N* hydrochloric acid (analyzing the *Acid stage*) and with *Diluent* (analyzing the *Buffer stage*), in comparison with a Standard solution having a known concentration of USP Aspirin RS in the same medium.

Uniformity of dosage units ⟨905⟩: meet the requirements.

Free salicylic acid—

Mobile phase and *Diluting solution*—Prepare as directed in the *Assay*.

Standard solution—Dissolve an accurately weighed quantity of USP Salicylic Acid RS in the *Standard preparation* prepared as directed in the *Assay*, to obtain a solution having a known concentration of about 0.015 mg of salicylic acid per mL.

Test preparation—Use the *Stock solution* prepared as directed for *Assay preparation* in the *Assay*.

Chromatographic system—Use the *Chromatographic system* described in the *Assay*. Chromatograph the *Standard solution*, and record the peak responses as directed under *Procedure* in the *Assay:* the relative standard deviation of the salicylic acid peak responses is not more than 4.0%. In a suitable chromatogram, the resolution, *R*, between salicylic acid and aspirin is not less than 2.0.

Procedure—Proceed as directed for *Procedure* in the *Assay*. The relative retention times are about 0.7 for salicylic acid and 1.0 for aspirin. Calculate the percentage of salicylic acid ($C_7H_6O_3$) in the portion of Capsules taken by the formula:

$$2000(C/Q_A)(r_U/r_S),$$

in which *C* is the concentration, in mg per mL, of USP Salicylic Acid RS in the *Standard solution*, Q_A is the quantity, in mg, of aspirin ($C_9H_8O_4$) in the portion of Capsules taken, as determined in the *Assay*, and r_U and r_S are the peak responses of the salicylic acid peaks obtained from the *Test preparation* and the *Standard solution*, respectively: not more than 3.0% is found.

Assay—

Mobile phase—Dissolve 2 g of sodium 1-heptanesulfonate in a mixture of 850 mL of water and 150 mL of acetonitrile, and adjust with glacial acetic acid to a pH of 3.4.

Diluting solution—Prepare a mixture of acetonitrile and formic acid (99:1).

Standard preparation—Dissolve an accurately weighed quantity of USP Aspirin RS in *Diluting solution* to obtain a solution having a known concentration of about 0.5 mg per mL.

Assay preparation—Remove, as completely as possible, the contents of not less than 20 Aspirin Delayed-release Capsules, and weigh accurately. Mix the combined contents, and transfer an accurately weighed quantity of the powder, equivalent to about 100 mg of aspirin, to a suitable container. Add 20.0 mL of *Diluting solution* and about 10 glass beads. Shake vigorously for about 10 minutes, and centrifuge (*Stock solution*). Quantitatively dilute an accurately measured volume of the *Stock solution* with 9 volumes of *Diluting solution* (*Assay preparation*). Retain the remaining portion of *Stock solution* for the test for *Salicylic acid*.

Chromatographic system (see *Chromatography* ⟨621⟩)—The liquid chromatograph is equipped with a 280-nm detector and a 4.0-mm × 30-cm column containing packing L1. The flow rate is about 2 mL per minute. Chromatograph the *Standard preparation*, and record the peak responses as directed under *Procedure:* the relative standard deviation is not more than 2.0%. In a suitable chromatogram, the tailing factor is not greater than 2.0.

Procedure—Separately inject equal volumes (about 10 µL) of the *Standard preparation* and the *Assay preparation* into the chromatograph, record the chromatograms, and measure the responses for the major peaks. Calculate the quantity, in mg, of aspirin ($C_9H_8O_4$) in the portion of Capsules taken by the formula:

$$200C(r_U / r_S),$$

in which C is the concentration, in mg per mL, of USP Aspirin RS in the *Standard preparation*, and r_U and r_S are the peak responses of the aspirin peaks obtained from the *Assay preparation* and the *Standard preparation*, respectively.

Aspirin Suppositories

» Aspirin Suppositories contain not less than 90.0 percent and not more than 110.0 percent of the labeled amount of $C_9H_8O_4$.

Packaging and storage—Preserve in well-closed containers, in a cool place.

Reference standard—*USP Aspirin Reference Standard*—Dry over silica gel for 5 hours before using.

Identification—Transfer a portion of the melted Suppositories obtained in the *Assay*, equivalent to about 1 g of aspirin, to a 125-mL conical flask. Add 20 mL of alcohol, and warm until completely disintegrated. Cool in an ice bath for 5 minutes, filter, and evaporate the filtrate to dryness: the residue responds to *Identification tests A* and *B* under *Aspirin.*

Non-aspirin salicylates—

Ferric chloride–urea reagent—To a mixture of 8 mL of ferric chloride solution (6 in 10) and 42 mL of 0.05 *N* hydrochloric acid add 60 g of urea. Dissolve the urea by swirling and without the aid of heat, and adjust the resulting solution, if necessary, by the addition of 6 *N* hydrochloric acid to a pH of 3.2. Prepare on the day of use.

Procedure—Insert a small pledget of glass wool above the stem constriction of a 20- × 2.5-cm chromatographic tube, and uniformly pack with a mixture of about 1 g of chromatographic siliceous earth and 0.5 mL of 5 *M* phosphoric acid. Directly above this layer, pack a similar mixture of about 3 g of chromatographic siliceous earth and 2 mL of *Ferric chloride–urea reagent*. Transfer to a small beaker an accurately weighed portion of the cooled mass from the previously melted Suppositories obtained in the *Assay*, equivalent to 50 mg of aspirin, add 10 mL of chloroform, warm slightly, and stir until dissolved. With the aid of 5 mL of chloroform, transfer the mixture to the chromatographic adsorption column. Pass 50 mL of chloroform in several portions through the column, rinse the tip of the chromatographic tube with chloroform, and discard the eluate. If the purple zone reaches the bottom of the tube, discard the column, and repeat the test with a smaller quantity of melted Suppositories.

Elute the adsorbed salicylic acid into a 100-mL volumetric flask containing 20 mL of methanol and 0.2 mL of hydrochloric acid by passing two 10-mL portions of a 1 in 10 solution of glacial acetic acid in water-saturated ether, and then 30 mL of chloroform, through the column, and dilute the eluate with chloroform to volume. Dissolve a suitable, accurately weighed quantity of salicylic acid in chloroform to obtain a Standard solution containing 150 µg of salicylic acid per mL. Pipet 5 mL of this solution into a 50-mL volumetric flask containing 10 mL of methanol, 0.1 mL of hydrochloric acid, and 10 mL of a 1 in 10 solution of glacial acetic acid in ether. Add chloroform to volume, and mix. Concomitantly determine the absorbances of both solutions in 1-cm cells at the wavelength of maximum absorbance at about 306 nm, using as the blank a solvent mixture of the same composition as that of the Standard solution: the absorbance of the solution from the Suppositories does not exceed that of the Standard solution (3.0%).

Assay—[NOTE—In this assay, use chloroform that recently has been saturated with water.]

Chromatographic column—Uniformly pack a chromatographic tube, as described in the test for *Non-aspirin salicylates*, with a mixture of about 3 g of chromatographic siliceous earth and 2 mL of sodium bicarbonate solution (1 in 12) prepared on the day of use.

Standard preparation—Transfer about 50 mg of USP Aspirin RS, accurately weighed, to a 50-mL volumetric flask, add 0.5 mL of glacial acetic acid, and add chloroform to volume. Transfer 5.0 mL of this solution to a 100-mL volumetric flask, add a 1 in 100 solution of glacial acetic acid in chloroform to volume, and mix.

Assay preparation—Tare a small dish and glass rod, place in the dish not less than 5 Aspirin Suppositories, heat gently on a steam bath until melted, then stir, cool while stirring, and weigh. Transfer an accurately weighed portion of the mass, equivalent to about 50 mg of aspirin, to a 50-mL volumetric flask containing 1 mL of a 1 in 50 solution of hydrochloric acid in methanol, add 40 mL of chloroform, mix, and add chloroform to volume.

Procedure—Pipet 5 mL of the *Assay preparation* into the column, wash with 5 mL and then with 25 mL of chloroform, and discard the washings. Without delay, elute into a 100-mL volumetric flask with about 10 mL of a 1 in 10 solution of glacial acetic acid in chloroform, and then with about 85 mL of a 1 in 100 solution of glacial acetic acid in chloroform, dilute with the latter solvent to volume, and mix. Without delay, concomitantly determine the absorbances of the eluted *Assay preparation* and the *Standard preparation* in 1-cm cells at the wavelength of maximum absorbance at about 280 nm, with a suitable spectrophotometer, using chloroform as the blank. Calculate the quantity, in mg, of $C_9H_8O_4$ in the portion of Suppositories taken by the formula:

$$C(A_U / A_S),$$

in which C is the concentration, in µg per mL, of USP Aspirin RS in the *Standard preparation*, and A_U and A_S are the absorbances of the *Assay preparation* and the *Standard preparation*, respectively.

Aspirin Tablets

» Aspirin Tablets contain not less than 90.0 percent and not more than 110.0 percent of the labeled amount of $C_9H_8O_4$. Tablets of larger than 81-mg size contain no sweeteners or other flavors.

NOTE—Tablets that are enteric-coated meet the requirements for *Aspirin Delayed-release Tablets.*

Packaging and storage—Preserve in tight containers. Preserve Tablets of 81-mg size or smaller in containers holding not more than 36 Tablets each.

Reference standards—*USP Aspirin Reference Standard*—Dry over silica gel for 5 hours before using. *USP Salicylic Acid Reference Standard*—Dry over silica gel for 3 hours before using.

Identification—

A: Crush 1 Tablet, boil it with 50 mL of water for 5 minutes, cool, and add 1 or 2 drops of ferric chloride TS: a violet-red color is produced.

B: Shake a quantity of finely powdered Tablets, equivalent to about 500 mg of aspirin, with 10 mL of chloroform for several minutes. Centrifuge the mixture. Pour off the clear supernatant liquid, and evaporate it to dryness: the residue so obtained responds to *Identification test B* under *Aspirin*.

Dissolution ⟨711⟩—

Medium: 0.05 *M* acetate buffer, prepared by mixing 2.99 g of sodium acetate trihydrate and 1.66 mL of glacial acetic acid with water to obtain 1000 mL of solution having a pH of 4.50 ± 0.05; 500 mL.

Apparatus 1: 50 rpm.

Time: 30 minutes.

Procedure—Determine the amount of $C_9H_8O_4$ dissolved from ultraviolet absorbances at the wavelength of the isosbestic point of aspirin and salicylic acid at 265 ± 2 nm of filtered portions of the solution under test, suitably diluted with *Dissolution Medium*, if necessary, in comparison with a Standard solution having a known concentration of USP Aspirin RS in the same medium. [NOTE—Prepare the Standard solution at the time of use. An amount of alcohol not to exceed 1% of the total volume of the Standard solution may be used to bring the Reference Standard into solution prior to dilution with *Dissolution Medium*.]

Tolerances—Not less than 80% (*Q*) of the labeled amount of $C_9H_8O_4$ is dissolved in 30 minutes.

Uniformity of dosage units ⟨905⟩: meet the requirements.

Free salicylic acid—

Mobile phase and *Diluting solution*—Prepare as directed in the *Assay*.

Standard solution—Dissolve an accurately weighed quantity of USP Salicylic Acid RS in the *Standard preparation* prepared as directed in the *Assay*, to obtain a solution having a known concentration of about 0.015 mg of salicylic acid per mL.

Test preparation—Use the *Stock solution* prepared as directed for *Assay preparation* in the *Assay*.

Chromatographic system—Use the *Chromatographic system* described in the *Assay*. Chromatograph the *Standard solution*, and record the peak responses as directed under *Procedure* in the *Assay*. The relative standard deviation of the salicylic acid peak responses is not more than 4.0%. In a suitable chromatogram, the resolution, *R*, between salicylic acid and aspirin is not less than 2.0.

Procedure—Proceed as directed for *Procedure* in the *Assay*. The relative retention times are about 0.7 for salicylic acid and 1.0 for aspirin. Calculate the percentage of salicylic acid ($C_7H_6O_3$) in the portion of Tablets taken by the formula:

$$2000(C/Q_A)(r_U/r_S),$$

in which *C* is the concentration, in mg per mL, of USP Salicylic Acid RS in the *Standard solution*, Q_A is the quantity, in mg, of aspirin ($C_9H_8O_4$) in the portion of Tablets taken, as determined in *Assay*, and r_U and r_S are the peak responses of the salicylic acid peaks obtained from the *Test preparation* and the *Standard solution*, respectively: not more than 0.3% is found. In the case of Tablets that are coated, not more than 3.0% is found.

Assay—

Mobile phase—Dissolve 2 g of sodium 1-heptanesulfonate in a mixture of 850 mL of water and 150 mL of acetonitrile, and adjust with glacial acetic acid to a pH of 3.4.

Diluting solution—Prepare a mixture of acetonitrile, and formic acid (99:1).

Standard preparation—Dissolve an accurately weighed quantity of USP Aspirin RS in *Diluting solution* to obtain a solution having a known concentration of about 0.5 mg per mL.

Assay preparation—Weigh and finely powder not less than 20 Aspirin Tablets. Transfer an accurately weighed quantity of the powder, equivalent to about 100 mg of aspirin, to a suitable container. Add 20.0 mL of *Diluting solution* and about 10 beads. Shake vigorously for about 10 minutes, and centrifuge (*Stock solution*). Quantitatively dilute an accurately measured volume of the *Stock solution* with 9 volumes of *Diluting solution* (*Assay preparation*). Retain the remaining portion of *Stock solution* for the test for *Salicylic acid*.

Chromatographic system (see *Chromatography* ⟨621⟩)—The liquid chromatograph is equipped with a 280-nm detector and a 4.0-mm × 30-cm column containing packing L1. The flow rate is about 2 mL per minute. Chromatograph the *Standard preparation*, and record the peak responses as directed under *Procedure:* the relative standard deviation is not more than 2.0%. In a suitable chromatogram, the tailing factor is not greater than 2.0.

Procedure—Separately inject equal volumes (about 10 µL) of the *Standard preparation* and the *Assay preparation* into the chromatograph, record the chromatograms, and measure the responses for the major peaks. Calculate the quantity, in mg, of aspirin ($C_9H_8O_4$) in the portion of Tablets taken by the formula:

$$200C(r_U/r_S),$$

in which *C* is the concentration, in mg per mL, of USP Aspirin RS in the *Standard preparation*, and r_U and r_S are the peak responses of the aspirin peaks obtained from the *Assay preparation* and the *Standard preparation*, respectively.

Buffered Aspirin Tablets

» Buffered Aspirin Tablets contain Aspirin and suitable buffering agents. Tablets contain not less than 90.0 percent and not more than 110.0 percent of the labeled amount of $C_9H_8O_4$.

Packaging and storage—Preserve in tight containers.

Reference standards—*USP Aspirin Reference Standard*—Dry over silica gel for 5 hours before using. *USP Salicylic Acid Reference Standard*—Dry over silica gel for 3 hours before using.

Identification—

A: Crush 1 Tablet, boil it with 50 mL of water for 5 minutes, cool, and add 1 or 2 drops of ferric chloride TS: a violet-red color is produced.

B: Shake a quantity of finely powdered Tablets, equivalent to about 500 mg of aspirin, with 10 mL of chloroform for several minutes. Centrifuge the mixture. Pour off the clear supernatant liquid, and evaporate it to dryness: the residue so obtained responds to *Identification test B* under *Aspirin*.

Dissolution ⟨711⟩—

Medium: 0.05 *M* acetate buffer, prepared by mixing 2.99 g of sodium acetate trihydrate and 1.66 mL of glacial acetic acid with water to obtain 1000 mL of solution having a pH of 4.50 ± 0.05; 500 mL.

Apparatus 2: 75 rpm.

Time: 30 minutes.

Procedure—Determine the amount of aspirin ($C_9H_8O_4$) dissolved from ultraviolet absorbances at the wavelength of the isosbestic point of aspirin and salicylic acid at 265 ± 2 nm of filtered portions of the solution under test, suitably diluted with *Dissolution Medium*, if necessary, in comparison with a Standard solution having a known concentration of USP Aspirin RS in the same medium. [NOTE—Prepare the Standard solution at the time of use. An amount of methanol not to exceed 1% of the total volume of the Standard solution may be used to bring the Reference Standard into solution prior to dilution with *Dissolution Medium*.]

Tolerances—Not less than 80% (*Q*) of the labeled amount of $C_9H_8O_4$ is dissolved in 30 minutes.

Uniformity of dosage units ⟨905⟩: meet the requirements.

Acid-neutralizing capacity ⟨301⟩: not less than 1.9 mEq of acid is consumed for each 325 mg of aspirin in the Tablets.

Free salicylic acid—

Mobile phase and *Diluting solution*—Prepare as directed in the *Assay*.

Standard solution—Dissolve an accurately weighed quantity of USP Salicylic Acid RS in the *Standard preparation* prepared as directed in the *Assay*, to obtain a solution having a known concentration of about 0.015 mg of salicylic acid per mL.

Test preparation—Use the *Stock solution* prepared as directed for *Assay preparation* in the *Assay*.

Chromatographic system—Use the *Chromatographic system* described in the *Assay*. Chromatograph the *Standard solution*, and record the peak responses as directed under *Procedure* in the *Assay*. The relative standard deviation of the salicylic acid peak responses is not more than 4.0%. In a suitable chromatogram, the resolution, *R*, between salicylic acid and aspirin is not less than 2.0.

Procedure—Proceed as directed for *Procedure* in the *Assay*. The relative retention times are about 0.7 for salicylic acid and 1.0 for aspirin. Calculate the percentage of salicylic acid ($C_7H_6O_3$) in the portion of Tablets taken by the formula:

$$2000(C/Q_A)(r_U/r_S),$$

in which C is the concentration, in mg per mL, of USP Salicylic Acid RS in the *Standard solution*, Q_A is the quantity, in mg, of aspirin ($C_9H_8O_4$) in the portion of Tablets taken, as determined in the *Assay*, and r_U and r_S are the peak responses of the salicylic acid peaks obtained from the *Test preparation* and the *Standard solution*, respectively: not more than 3.0% is found.

Assay—

Mobile phase—Dissolve 2 g of sodium 1-heptanesulfonate in a mixture of 850 mL of water and 150 mL of acetonitrile, and adjust with glacial acetic acid to a pH of 3.4.

Diluting solution—Prepare a mixture of acetonitrile and formic acid (99:1).

Standard preparation—Dissolve an accurately weighed quantity of USP Aspirin RS in *Diluting solution* to obtain a solution having a known concentration of about 0.5 mg per mL.

Assay preparation—Weigh and finely powder not less than 20 Buffered Aspirin Tablets. Transfer an accurately weighed quantity of the powder, equivalent to about 100 mg of aspirin, to a suitable container. Add 20.0 mL of *Diluting solution* and about 10 glass beads. Shake vigorously for about 10 minutes, and centrifuge (*Stock solution*). Quantitatively dilute an accurately measured volume of the *Stock solution* with 9 volumes of *Diluting solution* (*Assay preparation*). Retain the remaining portion of *Stock solution* for the test for *Salicylic acid*.

Chromatographic system (see *Chromatography* ⟨621⟩)—The liquid chromatograph is equipped with a 280-nm detector and a 4.0-mm × 30-cm column containing packing L1. The flow rate is about 2 mL per minute. Chromatograph the *Standard preparation*, and record the peak responses as directed under *Procedure*: the relative standard deviation is not more than 2.0%. In a suitable chromatogram, the tailing factor is not greater than 2.0.

Procedure—Separately inject equal volumes (about 10 µL) of the *Standard preparation* and the *Assay preparation* into the chromatograph, record the chromatograms, and measure the responses for the major peaks. Calculate the quantity, in mg, of aspirin ($C_9H_8O_4$) in the portion of Tablets taken by the formula:

$$200C(r_U/r_S),$$

in which C is the concentration, in mg per mL, of USP Aspirin RS in the *Standard preparation*, and r_U and r_S are the peak responses of the aspirin peaks obtained from the *Assay preparation* and the *Standard preparation*, respectively.

Aspirin Tablets, Acetaminophen and—*see* Acetaminophen and Aspirin Tablets

Aspirin Tablets, Butalbital and—*see* Butalbital and Aspirin Tablets

Aspirin Tablets, Carisoprodol and—*see* Carisoprodol and Aspirin Tablets

Aspirin Tablets, Pentazocine Hydrochloride and—*see* Pentazocine Hydrochloride and Aspirin Tablets

Aspirin Tablets, Propoxyphene Napsylate and—*see* Propoxyphene Napsylate and Aspirin Tablets

Aspirin Delayed-release Tablets

» Aspirin Delayed-release Tablets contain not less than 95.0 percent and not more than 105.0 percent of the labeled amount of $C_9H_8O_4$.

Packaging and storage—Preserve in tight containers.

Labeling—The label indicates that Aspirin Delayed-release Tablets are enteric-coated.

Reference standards—*USP Aspirin Reference Standard*—Dry over silica gel for 5 hours before using. *USP Salicylic Acid Reference Standard*—Dry over silica gel for 3 hours before using.

Identification—

A: Crush 1 Tablet, boil it with 50 mL of water for 5 minutes, cool, and add 1 or 2 drops of ferric chloride TS: a violet-red color is produced.

B: Shake a quantity of finely powdered Tablets, equivalent to about 500 mg of aspirin, with 10 mL of chloroform for several minutes. Centrifuge the mixture. Pour off the clear supernatant liquid, and evaporate it to dryness: the residue so obtained responds to *Identification test B* under *Aspirin*.

Drug release, *Method B* ⟨724⟩—

Apparatus 1: 100 rpm.

Time: 90 minutes, for *Buffer stage*.

Diluent—Prepare a mixture of 0.1 *N* hydrochloric acid and 0.20 *M* tribasic sodium phosphate (3:1), and adjust, if necessary, with 2 *N* hydrochloric acid or 2 *N* sodium hydroxide to a pH of 6.8 ± 0.05.

Procedure—Determine the amount of $C_9H_8O_4$ dissolved by determining ultraviolet absorbances at the wavelength of the isosbestic point of aspirin and salicylic acid (about 280 nm in the *Acid stage*, and about 265 nm in the *Buffer stage*) using a filtered portion of the solution under test, diluted, if necessary, with 0.1 *N* hydrochloric acid (analyzing the *Acid stage*) and with *Diluent* (analyzing the *Buffer stage*), in comparison with a Standard solution having a known concentration of USP Aspirin RS in the same medium.

Uniformity of dosage units ⟨905⟩: meet the requirements.

Free salicylic acid—

Mobile phase and *Diluting solution*—Prepare as directed in the *Assay*.

Standard solution—Dissolve an accurately weighed quantity of USP Salicylic Acid RS in the *Standard preparation* prepared as directed in the *Assay*, to obtain a solution having a known concentration of about 0.015 mg of salicylic acid per mL.

Test preparation—Use the *Stock solution* prepared as directed for *Assay preparation* in the *Assay*.

Chromatographic system—Use the *Chromatographic system* described in the *Assay*. Chromatograph the *Standard solution*, and record the peak responses as directed under *Procedure* in the *Assay*: the relative standard deviation of the salicylic acid peak responses is not more than 4.0%. In a suitable chromatogram, the resolution, *R*, between salicylic acid and aspirin is not less than 2.0.

Procedure—Proceed as directed for *Procedure* in the *Assay*. The relative retention times are about 0.7 for salicylic acid and 1.0 for aspirin. Calculate the percentage of salicylic acid ($C_7H_6O_3$) in the portion of Tablets taken by the formula:

$$2000(C/Q_A)(r_U/r_S),$$

in which C is the concentration, in mg per mL, of USP Salicylic Acid RS in the *Standard solution*, Q_A is the quantity, in mg, of aspirin ($C_9H_8O_4$) in the portion of Tablets taken, as determined in the *Assay*, and r_U and r_S are the peak responses of the salicylic acid peaks obtained from the *Test preparation* and the *Standard solution*, respectively: not more than 3.0% is found.

Assay—

Mobile phase—Dissolve 2 g of sodium 1-heptanesulfonate in a mixture of 850 mL of water and 150 mL of acetonitrile, and adjust with glacial acetic acid to a pH of 3.4.

Diluting solution—Prepare a mixture of acetonitrile and formic acid (99:1).

Standard preparation—Dissolve an accurately weighed quantity of USP Aspirin RS in *Diluting solution* to obtain a solution having a known concentration of about 0.5 mg per mL.

Assay preparation—Weigh and finely powder not less than 20 Aspirin Delayed-release Tablets. Transfer an accurately weighed quantity of the powder, equivalent to about 100 mg of aspirin, to a suitable container. Add 20.0 mL of *Diluting solution* and about 10 glass beads. Shake vigorously for about 10 minutes, and centrifuge (*Stock solution*). Quantitatively dilute an accurately measured volume of the *Stock solution* with 9 volumes of *Diluting solution* (*Assay preparation*). Retain the remaining portion of *Stock solution* for the test for *Salicylic acid.*

Chromatographic system (see *Chromatography* ⟨621⟩)—The liquid chromatograph is equipped with a 280-nm detector and a 4.0-mm × 30-cm column containing packing L1. The flow rate is about 2 mL per minute. Chromatograph the *Standard preparation*, and record the peak responses as directed under *Procedure:* the relative standard deviation is not more than 2.0%. In a suitable chromatogram, the tailing factor is not greater than 2.0.

Procedure—Separately inject equal volumes (about 10 µL) of the *Standard preparation* and the *Assay preparation* into the chromatograph, record the chromatograms, and measure the responses for the major peaks. Calculate the quantity, in mg, of aspirin ($C_9H_8O_4$) in the portion of Tablets taken by the formula:

$$200C(r_U / r_S),$$

in which C is the concentration, in mg per mL, of USP Aspirin RS in the *Standard preparation*, and r_U and r_S are the peak responses of the aspirin peaks obtained from the *Assay preparation* and the *Standard preparation*, respectively.

Aspirin Effervescent Tablets for Oral Solution

» Aspirin Effervescent Tablets for Oral Solution contain Aspirin and an effervescent mixture of a suitable organic acid and an alkali metal bicarbonate and/or carbonate. Tablets contain not less than 90.0 percent and not more than 110.0 percent of the labeled amount of $C_9H_8O_4$.

Packaging and storage—Preserve in tight containers.

Reference standards—*USP Aspirin Reference Standard*—Dry over silica gel for 5 hours before using. *USP Salicylic Acid Reference Standard*—Dry over silica gel for 3 hours before using.

Identification—

A: Dissolve 1 Tablet in about 50 mL of 1 *N* hydrochloric acid, boil for about 5 minutes, and allow to cool. To 2 mL of the resulting solution add 2 or 3 drops of ferric chloride TS: a violet-red color is produced.

B: Add about one-half a Tablet to 50 mL of water in a flask, and immediately stopper with a stopper fitted with tubing so that the evolved gas passes through calcium hydroxide TS: a white precipitate forms.

Solution time—Two Tablets dissolve completely in 180 mL of water at 17.5 ± 2.5° within 5 minutes.

Uniformity of dosage units ⟨905⟩: meet the requirements.

Acid-neutralizing capacity ⟨301⟩: not less than 5.0 mEq of acid is consumed by 1 Tablet.

Free salicylate—Proceed as directed for *Free salicylic acid* under *Buffered Aspirin Tablets:* not more than 8.0% is found.

Assay—Proceed as directed in the *Assay* under *Buffered Aspirin Tablets.*

Aspirin Extended-release Tablets

» Aspirin Extended-release Tablets contain not less than 95.0 percent and not more than 105.0 percent of the labeled amount of $C_9H_8O_4$.

Packaging and storage—Preserve in tight containers.

Labeling—The labeling indicates steady-state blood level profiles and confidence intervals and states the in-vitro *Drug release* test conditions of medium, apparatus and speed, times and tolerances, as directed under *Drug release.*

Reference standards—*USP Aspirin Reference Standard*—Dry over silica gel for 5 hours before using. *USP Salicylic Acid Reference Standard*—Dry over silica gel for 3 hours before using.

Identification—

A: Crush 1 Tablet, boil it with 50 mL of water for 5 minutes, cool, and add 1 or 2 drops of ferric chloride TS: a violet-red color is produced.

B: Shake a quantity of finely powdered Tablets, equivalent to about 500 mg of aspirin, with 10 mL of chloroform for several minutes. Centrifuge the mixture. Pour off the clear supernatant liquid, and evaporate it to dryness: the residue so obtained responds to *Identification test B* under *Aspirin.*

Drug release ⟨724⟩—

Medium, Times, and *Tolerances:* as specified in the *Labeling;* use *Acceptance Table 1.*

Apparatus 1 or *2:* as specified in the *Labeling* operated at the speed in the *Labeling.*

Procedure—Determine the amount of $C_9H_8O_4$ dissolved from ultraviolet absorbances at the isosbestic point at about 280 nm, using filtered portions of the solution under test, suitably diluted with 0.1 *N* hydrochloric acid, if necessary, in comparison with a Standard solution having a known concentration of USP Aspirin RS in the same medium.

Uniformity of dosage units ⟨905⟩: meet the requirements.

Free salicylic acid—

Mobile phase and *Diluting solution*—Prepare as directed in the *Assay.*

Standard solution—Dissolve an accurately weighed quantity of USP Salicylic Acid RS in the *Standard preparation* prepared as directed in the *Assay*, to obtain a solution having a known concentration of about 0.015 mg of salicylic acid per mL.

Test preparation—Use the *Stock solution* prepared as directed for *Assay preparation* in the *Assay.*

Chromatographic system—Use the *Chromatographic system* described in the *Assay.* Chromatograph the *Standard solution*, and record the peak responses as directed under *Procedure* in the *Assay:* the relative standard deviation of the salicylic acid peak responses is not more than 4.0%. In a suitable chromatogram, the resolution, *R*, between salicylic acid and aspirin is not less than 2.0.

Procedure—Proceed as directed for *Procedure* in the *Assay.* The relative retention times are about 0.7 for salicylic acid and 1.0 for aspirin. Calculate the percentage of salicylic acid ($C_7H_6O_3$) in the portion of Tablets taken by the formula:

$$2000(C / Q_A)(r_U / r_S),$$

in which C is the concentration, in mg per mL, of USP Salicylic Acid RS in the *Standard solution*, Q_A is the quantity, in mg, of aspirin ($C_9H_8O_4$) in the portion of Tablets taken, as determined in the *Assay*, and r_U and r_S are the peak responses of the salicylic acid peaks obtained from the *Test preparation* and the *Standard solution*, respectively: not more than 3.0% is found.

Assay—

Mobile phase—Dissolve 2 g of sodium 1-heptanesulfonate in a mixture of 850 mL of water and 150 mL of acetonitrile, and adjust with glacial acetic acid to a pH of 3.4.

Diluting solution—Prepare a mixture of acetonitrile and formic acid (99:1).

Standard preparation—Dissolve an accurately weighed quantity of USP Aspirin RS in *Diluting solution* to obtain a solution having a known concentration of about 0.5 mg per mL.

Assay preparation—Weigh and finely powder not less than 20 Aspirin Extended-release Tablets. Transfer an accurately weighed quantity of the powder, equivalent to about 100 mg of aspirin, to a suitable container. Add 20.0 mL of *Diluting solution* and about 10 glass beads. Shake vigorously for about 10 minutes, and centrifuge (*Stock solution*). Quantitatively dilute an accurately measured volume of the *Stock solution* with 9 volumes of *Diluting solution* (*Assay preparation*). Retain the remaining portion of *Stock solution* for the test for *Salicylic acid.*

Chromatographic system (see *Chromatography* ⟨621⟩)—The liquid chromatograph is equipped with a 280-nm detector and a 4.0-mm × 30-cm column containing packing L1. The flow rate is about 2 mL per minute. Chromatograph the *Standard preparation,* and record the peak responses as directed under *Procedure:* the relative standard deviation is not more than 2.0%. In a suitable chromatogram, the tailing factor is not greater than 2.0.

Procedure—Separately inject equal volumes (about 10 µL) of the *Standard preparation* and the *Assay preparation* into the chromatograph, record the chromatograms, and measure the responses for the major peaks. Calculate the quantity, in mg, of aspirin ($C_9H_8O_4$) in the portion of Tablets taken by the formula:

$$200C(r_U/r_S),$$

in which C is the concentration, in mg per mL, of USP Aspirin RS in the *Standard preparation,* and r_U and r_S are the peak responses of the aspirin peaks obtained from the *Assay preparation* and the *Standard preparation,* respectively.

Aspirin, Alumina, and Magnesia Tablets

» Aspirin, Alumina, and Magnesia Tablets contain not less than 90.0 percent and not more than 110.0 percent of the labeled amount of aspirin ($C_9H_8O_4$), the equivalent of not less than 90.0 percent and not more than 110.0 percent of the labeled amount of aluminum hydroxide [$Al(OH)_3$], and not less than 90.0 percent and not more than 110.0 percent of the labeled amount of magnesium hydroxide [$Mg(OH)_2$].

Packaging and storage—Preserve in tight containers.

Reference standards—*USP Aspirin Reference Standard*—Dry over silica gel for 5 hours before using. *USP Salicylic Acid Reference Standard*—Dry over silica gel for 3 hours before using.

Identification—

A: The chromatogram of the *Assay preparation* obtained as directed in the *Assay for aspirin and limit of free salicylic acid* exhibits a major peak for aspirin, the retention time of which corresponds with that exhibited in the chromatogram of the *Standard preparation* obtained as directed in the *Assay for aspirin and limit of free salicylic acid.*

B: To a 0.7-g portion of finely powdered Tablets add 20 mL of 3 *N* hydrochloric acid and 5 drops of methyl red TS, heat to boiling, and add 6 *N* ammonium hydroxide until the color of the solution changes to deep yellow. Continue boiling for 2 minutes, and filter: the filtrate so obtained responds to the tests for *Magnesium* ⟨191⟩.

C: Wash the precipitate obtained in *Identification test B* with a hot solution of ammonium chloride (1 in 50), and dissolve the precipitate in hydrochloric acid: the solution so obtained responds to the tests for *Aluminum* ⟨191⟩.

Dissolution ⟨711⟩—

Medium: 0.05 *M* acetate buffer, prepared by mixing 2.99 g of sodium acetate (trihydrate) and 1.66 mL of glacial acetic acid with water to obtain 1000 mL of solution having a pH of 4.50 ± 0.05; 900 mL.

Apparatus 2: 75 rpm.

Time: 45 minutes.

Procedure—Determine the amount of aspirin ($C_9H_8O_4$) dissolved from ultraviolet absorbances at the wavelength of the isosbestic point of aspirin and salicylic acid at 265 ± 2 nm of filtered portions of the solution under test, suitably diluted with *Dissolution Medium,* if necessary, in comparison with a Standard solution having a known concentration of USP Aspirin RS in the same medium. [NOTE—Prepare the Standard solution at the time of use. An amount of methanol not to exceed 1% of the total volume of the Standard solution may be used to bring the Reference Standard into solution prior to dilution with *Dissolution Medium.*]

Tolerances—Not less than 75% (Q) of the labeled amount of $C_9H_8O_4$ is dissolved in 45 minutes.

Uniformity of dosage units ⟨905⟩: meet the requirements for *Weight Variation* with respect to aluminum hydroxide and to magnesium hydroxide, and for *Content Uniformity* with respect to aspirin.

Acid-neutralizing capacity ⟨301⟩: not less than 1.9 mEq of acid is consumed for each 325 mg of aspirin in the Tablets.

Assay for aspirin and limit of free salicylic acid—

Mobile phase—Dissolve 225 mg of tetramethylammonium hydroxide pentahydrate and 200 mg of sodium 1-octanesulfonate in 700 mL of water. Add 150 mL of methanol, 150 mL of acetonitrile, and 1.0 mL of glacial acetic acid, and stir. [NOTE—The composition of the *Mobile phase* may be adjusted if necessary (see *System Suitability* under *Chromatography* ⟨621⟩)].

Solvent mixture—To 2 g of anhydrous citric acid add 990 mL of acetonitrile, 990 mL of chloroform, and 20 mL of formic acid, and stir for about 30 minutes. Allow to settle, and decant the clear solution into a suitable container. Use the clear solution as the *Solvent mixture.*

Internal standard solution—Dissolve phenacetin in *Solvent mixture* to obtain a solution having a concentration of about 2 mg per mL.

Salicylic acid stock standard solution—Dissolve a suitable quantity of USP Salicylic Acid RS in *Solvent mixture* to obtain a solution having a known concentration of about 1 mg per mL.

Standard preparation—Transfer about 325 mg of USP Aspirin RS, accurately weighed, to a 50-mL volumetric flask. Add 10.0 mL of *Salicylic acid stock standard solution* and 5.0 mL of *Internal standard solution,* dilute with *Solvent mixture* to volume, and mix.

Assay preparation—Weigh and finely powder not less than 20 Aspirin, Alumina, and Magnesia Tablets. Immediately transfer an accurately weighed portion of the powder, equivalent to about 325 mg of aspirin, to a screw-capped, 120-mL bottle, add 5.0 mL of *Internal standard solution* and 45.0 mL of *Solvent mixture,* cap the bottle, mix, and sonicate for 2 to 5 minutes. Centrifuge, and use a portion of the resultant clear solution as the *Assay preparation.*

Chromatographic system (see *Chromatography* ⟨621⟩)—The liquid chromatograph is equipped with a 280-nm detector and a 4-mm × 30-cm column that contains 10-µm packing L1. The flow rate is about 2 mL per minute. Chromatograph the *Standard preparation,* and record the peak responses as directed under *Procedure:* the resolution, R, between the salicylic acid, aspirin, and internal standard peaks is not less than 2.0, the tailing factor for any of these peaks is not more than 2.0, and the relative standard deviation for replicate injections is not more than 3.0%.

Procedure—Separately inject equal volumes (about 5 µL) of the *Standard preparation* and the *Assay preparation* into the chromatograph, record the chromatograms, and measure the responses for the major peaks. The relative retention times are about 0.3 for salicylic acid, 0.6 for aspirin, and 1.0 for phenacetin. [NOTE—Record each chromatogram until the chloroform peak appears at a relative retention time of about 1.8.] Calculate the quantity, in mg, of aspirin ($C_9H_8O_4$) in the portion of Tablets taken by the formula:

$$50C(R_U/R_S),$$

in which C is the concentration, in mg per mL, of USP Aspirin RS in the *Standard preparation,* and R_U and R_S are the ratios of the peak responses of aspirin and phenacetin obtained from the *Assay preparation* and the *Standard preparation,* respectively. Calculate the percentage of free salicylic acid in the Tablets taken by the formula:

$$5000(C/a)(R_U/R_S),$$

in which C is the concentration, in mg per mL, of USP Salicylic Acid RS in the *Standard preparation*, a is the quantity, in mg, of aspirin in the portion of Tablets taken, based on the labeled amount, and R_U and R_S are the ratios of the peak responses of salicylic acid and phenacetin obtained from the *Assay preparation* and the *Standard preparation*, respectively: not more than 3.0% is found.

Assay for aluminum hydroxide—

Disodium ethylenediaminetetraacetate titrant—Prepare and standardize as directed in the *Assay* under *Ammonium Alum*.

Assay preparation—Weigh and finely powder not less than 20 Aspirin, Alumina, and Magnesia Tablets. Transfer an accurately weighed portion of the powder, equivalent to about 250 mg of aluminum hydroxide, to a 150-mL beaker, add 20 mL of water, stir, and slowly add 30 mL of 3 *N* hydrochloric acid. Heat gently, if necessary, to aid solution, cool, and transfer to a 200-mL volumetric flask. Wash the beaker with water, adding the washings to the flask, add water to volume, and mix.

Procedure—Pipet 50 mL of *Assay preparation* into a 250-mL beaker, then add, in the order named and with continuous stirring, 25.0 mL of 0.05 *M Disodium ethylenediaminetetraacetate titrant* and 20 mL of acetic acid–ammonium acetate buffer TS, and heat the solution near the boiling temperature for 5 minutes. Cool, add 50 mL of alcohol and 2 mL of dithizone TS, and mix. Titrate with 0.05 *M* zinc sulfate VS until the color changes from green-violet to rose-pink. Perform a blank determination, substituting 50 mL of water for the *Assay preparation*, and make any necessary corrections. Each mL of 0.05 *M Disodium ethylenediaminetetraacetate titrant* consumed is equivalent to 3.900 mg of $Al(OH)_3$.

Assay for magnesium hydroxide—

Assay preparation—Prepare as directed in the *Assay for aluminum hydroxide*.

Procedure—Pipet a volume of *Assay preparation*, equivalent to about 80 mg of magnesium hydroxide, into a 400-mL beaker, add 200 mL of water and 20 mL of triethanolamine, and mix. Add 50 mL of ammonia–ammonium chloride buffer TS and 2 drops of eriochrome black indicator solution (prepared by dissolving 200 mg of eriochrome black T in a mixture of 15 mL of triethanolamine and 5 mL of dehydrated alcohol, and mixing). Cool the solution to between 3° and 4° by immersion of the beaker in an ice bath, then remove, and titrate with 0.05 *M* disodium ethylenediaminetetraacetate VS until the color changes to pure blue. Perform a blank determination, substituting for the *Assay preparation*, a volume of water equal to the volume of *Assay preparation* used, and make any necessary corrections. Each mL of 0.05 *M* sodium ethylenediaminetetraacetate is equivalent to 2.916 mg of $Mg(OH)_2$.

Aspirin and Codeine Phosphate Tablets

» Aspirin and Codeine Phosphate Tablets contain not less than 90.0 percent and not more than 110.0 percent of the labeled amounts of aspirin ($C_9H_8O_4$) and codeine phosphate hemihydrate ($C_{18}H_{21}NO_3 \cdot H_3PO_4 \cdot \frac{1}{2}H_2O$).

Packaging and storage—Preserve in well-closed, light-resistant containers.

Reference standards—*USP Aspirin Reference Standard*—Dry over silica gel for 5 hours before using. *USP Codeine Phosphate Reference Standard*—This is the hemihydrate form of codeine phosphate. Dry at 105° for 18 hours before using. *USP Salicylic Acid Reference Standard*—Dry over silica gel for 3 hours before using.

Identification—Dissolve a suitable quantity of USP Aspirin RS in the *Solvent mixture* prepared as directed under *Assay for aspirin and codeine phosphate and limit of free salicylic acid* to obtain a *Standard aspirin solution* containing about 3.3 mg per mL. Dissolve a suitable quantity of USP Codeine Phosphate RS in the *Solvent mixture* to obtain a *Standard codeine phosphate solution* containing about 1 mg per mL. Chromatograph these solutions as directed for *Procedure* in the *Assay for aspirin and codeine phosphate and limit of free salicylic acid*. The retention times of the major peaks in the chromatogram of the *Assay preparation* obtained as directed in the *Assay for aspirin and codeine phosphate and limit of free salicylic acid* correspond to those of the *Standard aspirin solution* and the *Standard codeine phosphate solution*, respectively.

Dissolution ⟨711⟩—

Medium: 0.05 *M* acetate buffer, prepared by mixing 2.99 g of sodium acetate trihydrate and 1.66 mL of glacial acetic acid with water to obtain 1000 mL of solution having a pH of 4.50 ± 0.05; 900 mL.

Apparatus 2: 75 rpm.

Time: 30 minutes.

Mobile phase, Solvent mixture, and *Aspirin and codeine phosphate standard preparation*—Prepare as directed in the *Assay for aspirin and codeine phosphate and limit of free salicylic acid*.

Internal standard solution—Dissolve phenacetin in methanol to obtain a solution having a concentration of about 0.07 mg per mL.

Standard solution A—Prepare a solution of USP Aspirin RS in *Solvent mixture* having an accurately known concentration of about 0.36 mg per mL.

Standard solution B—Transfer about 25 mg of USP Codeine Phosphate RS and 50 mg of USP Salicylic Acid RS, each accurately weighed, to a 100-mL volumetric flask, add 5.0 mL of methanol, and mix. Add *Dissolution Medium* to volume, and mix. Pipet 10 mL of the resulting solution into a 100-mL volumetric flask, add *Dissolution Medium* to volume, and mix.

Standard preparations A and B—Pipet 10 mL of *Standard solution A* and 10 mL of *Standard solution B* into separate containers, add 3.0 mL of the *Internal standard solution* to each container, and mix.

Test preparation—Withdraw a portion of the solution under test and filter, discarding the few mL of the filtrate. Pipet 10 mL of the filtrate and 3.0 mL of the *Internal standard solution* into a suitable container, and mix.

Chromatographic system—Proceed as directed for *Chromatographic system* in the *Assay for aspirin and codeine phosphate and limit of free salicylic acid*, except to use only the *aspirin and codeine phosphate preparation* for evaluation of the suitability of the system.

Procedure—Proceed as directed in the *Assay for aspirin and codeine phosphate and limit of free salicylic acid*, except to inject about 50 µL of the *Standard preparations* and *Test preparation*. The relative retention times are 0.3 for salicylic acid, 0.6 for aspirin, 0.8 for codeine phosphate, and 1.0 for phenacetin. Calculate the amount of codeine phosphate dissolved by comparison of the relative peak response ratios for the codeine phosphate peaks, obtained from *Standard preparation B* and the *Test solution*. Calculate the percentage of aspirin dissolved by the formula:

$$[0.9C(R_U/R_S) + 0.9C'(R'_U/R'_S)(180.16/138.12)]/3.25,$$

in which C is the concentration, in µg per mL, of USP Aspirin RS in *Standard solution A*, and R_U and R_S are the peak response ratios for the aspirin component obtained from the *Test solution* and *Standard preparation A*, respectively; and C' is the concentration, in µg per mL, of USP Salicylic Acid RS in *Standard solution B*, R'_U and R'_S are the peak response ratios for the salicylic acid component obtained from the *Test solution* and *Standard preparation B*, respectively, and 180.16 and 138.12 are the molecular weights of aspirin and salicylic acid, respectively.

Tolerances—Not less than 75% (*Q*) of the labeled amount of aspirin ($C_9H_8O_4$) and codeine phosphate hemihydrate ($C_{18}H_{21}NO_3 \cdot H_3PO_4 \cdot \frac{1}{2}H_2O$) are dissolved in 30 minutes.

Uniformity of dosage units ⟨905⟩: meet the requirements for *Content Uniformity* with respect to aspirin and codeine phosphate.

Assay for aspirin and codeine phosphate and limit of free salicylic acid—

Mobile phase—Dissolve 225 mg of tetramethylammonium hydroxide pentahydrate and 200 mg of sodium 1-octanesulfonate in 700 mL of water. Add 150 mL of methanol, 150 mL of acetonitrile, and 1.0 mL of glacial acetic acid, and stir. Filter through a membrane filter, and degas. [NOTE—The amounts of sodium 1-octanesulfonate, methanol, and acetonitrile may be varied to obtain acceptable chromatography.]

Solvent mixture—To 15 g of anhydrous citric acid add 200 mL of methanol and 20 mL of glacial acetic acid, dilute with chloroform to 1000 mL, and mix until the citric acid is dissolved.

Internal standard solution—Dissolve phenacetin in *Solvent mixture* to obtain a solution having a concentration of about 2 mg per mL.

Salicylic acid stock standard solution—Dissolve an accurately weighed quantity of USP Salicylic Acid RS in *Solvent mixture*, and dilute quantitatively with *Solvent mixture* to obtain a solution having a known concentration of about 1 mg per mL.

Salicylic acid standard preparation—Transfer 5.0 mL of *Salicylic acid stock standard solution* to a 50-mL volumetric flask, add 5.0 mL of *Internal standard solution*, dilute with *Solvent mixture* to volume, and mix.

Aspirin and codeine phosphate standard preparation—Transfer about 325 mg of USP Aspirin RS, accurately weighed, and 325*J* mg of USP Codeine Phosphate RS, accurately weighed, to a 50-mL volumetric flask, *J* being the ratio of the labeled amount, in mg, of codeine phosphate to the labeled amount, in mg, of aspirin per Tablet. Add 5.0 mL of *Salicylic acid stock standard solution* and 5.0 mL of *Internal standard solution*, dilute with *Solvent mixture* to volume, and mix.

Assay preparation—Weigh and finely powder not less than 20 Aspirin and Codeine Phosphate Tablets. Transfer an accurately weighed portion of the powder, equivalent to about 325 mg of aspirin, to a screw-capped, 120-mL bottle, add 5.0 mL of *Internal standard solution* and 45.0 mL of *Solvent mixture*, mix, and sonicate for 2 to 5 minutes. Centrifuge, and use a portion of the resultant clear solution as the *Assay preparation*. Use on the day prepared.

Chromatographic system (see *Chromatography* ⟨621⟩)—The liquid chromatograph is equipped with a 280-nm detector and a 3.9-mm × 30-cm column that contains 10-µm packing L1. The flow rate is about 2 mL per minute. Chromatograph replicate injections of the *Salicylic acid standard preparation* and the *Aspirin and codeine phosphate standard preparation*, and record the peak responses as directed under *Procedure:* the tailing factor for each analyte peak is not more than 2.0, the resolution, *R*, between salicylic acid and aspirin, between aspirin and codeine, and between codeine and phenacetin is not less than 2.0, and the relative standard deviation of the ratios of the peak responses of salicylic acid, aspirin, and codeine to the peak response of phenacetin is not more than 3.0%.

Procedure—Separately inject equal volumes (about 5 µL) of the *Salicylic acid standard preparation*, the *Aspirin and codeine phosphate standard preparation*, and the *Assay preparation* into the chromatograph, record the chromatograms, and measure the responses for the major peaks. The relative retention times for salicylic acid, aspirin, codeine, and phenacetin are about 0.3, 0.5, 0.8, and 1.0, respectively. Calculate the quantity, in mg, of aspirin ($C_9H_8O_4$) in the portion of powdered Tablets taken by the formula:

$$50C(R_U/R_S),$$

in which *C* is the concentration, in mg per mL, of USP Aspirin RS in the *Aspirin and codeine phosphate standard preparation*, and R_U and R_S are the ratios of the peak responses of aspirin and phenacetin obtained from the *Assay preparation* and the *Aspirin and codeine phosphate standard preparation*, respectively. Calculate the quantity, in mg, of codeine phosphate hemihydrate ($C_{18}H_{21}NO_3 . H_3PO_4 . ½H_2O$) in the portion of Tablets taken by the formula:

$$(406.37/397.36)(50C)(R_U/R_S),$$

in which 406.37 and 397.36 are the molecular weights of codeine phosphate hemihydrate and anhydrous codeine phosphate, respectively, *C* is the concentration, in mg per mL, of USP Codeine Phosphate RS in the *Aspirin and codeine phosphate standard preparation*, and R_U and R_S are the ratios of the peak responses of codeine phosphate and phenacetin obtained from the *Assay preparation* and the *Aspirin and codeine phosphate standard preparation*, respectively. Calculate the percentage of free salicylic acid in the Tablets taken by the formula:

$$5{,}000(C/a)(R_U/R_S),$$

in which *C* is the concentration, in mg per mL, of USP Salicylic Acid RS in the *Salicylic acid standard preparation*, *a* is the quantity, in mg, of aspirin in the portion of powdered Tablets taken, based on the labeled amount, and R_U and R_S are the ratios of the peak responses of salicylic acid and phenacetin obtained from the *Assay preparation* and the *Salicylic acid standard preparation*, respectively: not more than 3.0% is found.

Aspirin, Codeine Phosphate, Alumina, and Magnesia Tablets

» Aspirin, Codeine Phosphate, Alumina, and Magnesia Tablets contain not less than 90.0 percent and not more than 110.0 percent of the labeled amounts of aspirin ($C_9H_8O_4$), codeine phosphate hemihydrate ($C_{18}H_{21}NO_3 . H_3PO_4 . ½H_2O$), aluminum hydroxide [$Al(OH)_3$], and magnesium hydroxide [$Mg(OH)_2$].

Packaging and storage—Preserve in well-closed, light-resistant containers.

Reference standards—*USP Aspirin Reference Standard*—Dry over silica gel for 5 hours before using. *USP Codeine Phosphate Reference Standard*—This is the hemihydrate form of codeine phosphate. Dry at 105° for 18 hours before using. *USP Salicylic Acid Reference Standard*—Dry over silica gel for 3 hours before using.

Identification—

A: Tablets respond to the *Identification test* under *Aspirin and Codeine Phosphate Tablets*.

B: Tablets respond to the *Identification tests* under *Alumina and Magnesia Tablets*.

Dissolution ⟨711⟩—

Medium: 0.05 *M* acetate buffer, prepared by mixing 2.99 g of sodium acetate trihydrate and 1.66 mL of glacial acetic acid with water to obtain 1000 mL of solution having a pH of 4.50 ± 0.05; 900 mL.

Apparatus 2: 75 rpm.

Time: 30 minutes.

Mobile phase, Internal standard solution, Solvent mixture, Aspirin and codeine phosphate standard preparation, Standard solution A, Standard solution B, Standard preparations A and B, Test preparation, Chromatographic system, and *Procedure*—Proceed as directed in the test for *Dissolution* under *Aspirin and Codeine Phosphate Tablets*.

Tolerances—Not less than 75% (*Q*) of the labeled amounts of aspirin ($C_9H_8O_4$) and codeine phosphate hemihydrate ($C_{18}H_{21}NO_3.H_3PO_4 . ½H_2O$) are dissolved in 30 minutes.

Uniformity of dosage units ⟨905⟩: meet the requirements for *Content Uniformity* with respect to aspirin and codeine phosphate and for *Weight Variation* with respect to aluminum hydroxide and magnesium hydroxide.

Acid-neutralizing capacity ⟨301⟩: not less than 1.9 mEq per Tablet.

Assay for aspirin and codeine phosphate and limit of free salicylic acid—

Mobile phase, Solvent mixture, Salicylic acid stock standard solution, Salicylic acid standard preparation, Aspirin and codeine phosphate standard preparation, and *Chromatographic system*—Prepare as directed in the *Assay for aspirin and codeine phosphate and limit of free salicylic acid* under *Aspirin and Codeine Phosphate Tablets*.

Assay preparation—Weigh and finely powder not less than 20 Aspirin, Codeine Phosphate, Alumina, and Magnesia Tablets. Transfer an accurately weighed portion of the powder, equivalent to about 325 mg of aspirin, to a screw-capped, 120-mL bottle, add 5.0 mL of *Internal standard solution* and 45.0 mL of *Solvent mixture*, mix, and sonicate for 2 to 5 minutes. Centrifuge, and use a portion of the resultant clear solution as the *Assay preparation*.

Procedure—Separately inject equal volumes (about 5 µL) of the *Salicylic acid standard preparation*, the *Aspirin and codeine phosphate standard preparation*, and the *Assay preparation* into the chromatograph, record the chromatograms, and measure the responses for the major peaks. The relative retention times for salicylic acid, aspirin, codeine, and phenacetin are about 0.3, 0.5, 0.8, and 1.0, respectively. Calculate the quantity, in mg, of aspirin ($C_9H_8O_4$) in the portion of powdered Tablets taken by the formula:

$$50C(R_U/R_S),$$

in which C is the concentration, in mg per mL, of USP Aspirin RS in the *Aspirin and codeine phosphate standard preparation*, and R_U and R_S are the ratios of the peak responses of aspirin and phenacetin obtained from the *Assay preparation* and the *Aspirin and codeine phosphate standard preparation*, respectively. Calculate the quantity, in mg, of codeine phosphate hemihydrate ($C_{18}H_{21}NO_3 . H_3PO_4 . \frac{1}{2}H_2O$), in the portion of powdered Tablets taken by the formula:

$$(406.37/397.36)(50C)(R_U/R_S),$$

in which 406.37 and 397.36 are the molecular weights of codeine phosphate hemihydrate and anhydrous codeine phosphate, respectively, C is the concentration, in mg per mL, of USP Codeine Phosphate RS in the *Aspirin and codeine phosphate standard preparation*, and R_U and R_S are the ratios of the peak responses of codeine phosphate and phenacetin obtained from the *Assay preparation* and the *Aspirin and codeine phosphate standard preparation*, respectively. Calculate the percentage of free salicylic acid in the Tablets taken by the formula:

$$5{,}000(C/a)(R_U/R_S),$$

in which C is the concentration, in mg per mL, of USP Salicylic Acid RS in the *Salicylic acid standard preparation*, a is the quantity, in mg, of aspirin in the portion of Tablets taken, determined as directed above, and R_U and R_S are the ratios of the peak responses of salicylic acid and phenacetin obtained from the *Assay preparation* and the *Salicylic acid standard preparation*, respectively: not more than 3.0% is found.

Assay for aluminum hydroxide—

Disodium ethylenediaminetetraacetate titrant—Prepare and standardize as directed in the *Assay* under *Ammonium Alum*.

Assay preparation—Weigh and finely powder not less than 20 Aspirin, Codeine Phosphate, Alumina, and Magnesia Tablets. Transfer an accurately weighed portion of the powder, equivalent to about 600 mg of aluminum hydroxide, to a 150-mL beaker, add 20 mL of water, stir, and slowly add 30 mL of 3 *N* hydrochloric acid. Heat gently, if necessary, to aid solution, cool, and filter into a 200-mL volumetric flask. Wash the filter with water into the flask, add water to volume, and mix.

Procedure—Pipet 10 mL of *Assay preparation* into a 250-mL beaker, add 20 mL of water, then add, in the order named and with continuous stirring, 25.0 mL of *Disodium ethylenediaminetetraacetate titrant* and 20 mL of acetic acid–ammonium acetate buffer TS. Add 50 mL of alcohol and 2 mL of dithizone TS, and mix. Titrate with 0.05 *M* zinc sulfate VS until the color changes from green-violet to rose-pink. Perform a blank determination, substituting 10 mL of water for the *Assay preparation*, and make any necessary correction. Each mL of 0.05 *M Disodium ethylenediaminetetraacetate titrant* is equivalent to 3.900 mg of $Al(OH)_3$.

Assay for magnesium hydroxide—

Assay preparation—Prepare as directed in the *Assay for aluminum oxide*.

Procedure—Pipet a volume of *Assay preparation*, equivalent to about 40 mg of magnesium hydroxide, into a 400-mL beaker, add 200 mL of water and 20 mL of triethanolamine, and stir. Add 10 mL of ammonia–ammonium chloride buffer TS and 3 drops of an eriochrome black indicator solution prepared by dissolving 200 mg of eriochrome black T in a mixture of 15 mL of triethanolamine and 5 mL of dehydrated alcohol, and mix. Cool the solution to between 3° and 4° by immersion of the beaker in an ice bath, then remove, and titrate with 0.05 *M* disodium ethylenediaminetetraacetate VS to a blue end-point. Perform a blank determination, substituting 10 mL of water for the *Assay preparation*, and make any necessary correction. Each mL of 0.05 *M* disodium ethylenediaminetetraacetate consumed is equivalent to 2.916 mg of $Mg(OH)_2$.

Aspirin, Codeine Phosphate, and Caffeine Capsules

» Aspirin, Codeine Phosphate, and Caffeine Capsules contain not less than 90.0 percent and not more than 110.0 percent of the labeled amounts of aspirin ($C_9H_8O_4$), codeine phosphate hemihydrate ($C_{18}H_{21}NO_3 . H_3PO_4 . \frac{1}{2}H_2O$), and caffeine ($C_8H_{10}N_4O_2$).

Packaging and storage—Preserve in well-closed, light-resistant containers.

Reference standards—*USP Aspirin Reference Standard*—Dry over silica gel for 5 hours before using. *USP Codeine Phosphate Reference Standard*—This is the hemihydrate form of codeine phosphate. Dry at 105° for 18 hours before using. *USP Caffeine Reference Standard*—Dry at 80° for 4 hours before using. *USP Salicylic Acid Reference Standard*—Dry over silica gel for 3 hours before using.

Identification—The retention times of the major peaks in the chromatogram of the *Assay preparation* correspond to those of the *Standard preparation*, obtained as directed in the *Assay*.

Uniformity of dosage units ⟨905⟩: meet the requirements.

Salicylic acid—

Mobile phase—Dissolve 2 g of sodium 1-heptanesulfonate in a mixture of 825 mL of water and 175 mL of acetonitrile, and adjust with glacial acetic acid to a pH of 3.4. Make adjustments if necessary (see *System Suitability* under *Chromatography* ⟨621⟩).

Solvent mixture—Prepare a mixture of acetonitrile, alcohol-free chloroform, and formic acid (99:99:2).

Standard preparation—Dissolve an accurately weighed quantity of USP Salicylic Acid RS in *Solvent mixture* to obtain a solution having a known concentration of about 0.008 mg per mL.

Test preparation—Weigh the contents of not less than 20 Capsules, and calculate the average weight per Capsule. Mix the combined contents of the Capsules, and transfer an accurately weighed portion, equivalent to about 380 mg of aspirin, to a 100-mL volumetric flask. Add about 20 mL of *Solvent mixture*, and sonicate for about 15 minutes. Dilute with *Solvent mixture* to volume, and mix. Centrifuge a portion of this mixture, and use the clear supernatant liquid as the *Test preparation*.

Chromatographic system (see *Chromatography* ⟨621⟩)—The liquid chromatograph is equipped with a 299-nm detector and a 3.9-mm × 30-cm column that contains packing L1. The flow rate is about 2 mL per minute. Chromatograph the *Standard preparation*, and record the responses as directed under *Procedure:* the relative standard deviation for replicate injections is not more than 4.0%.

Procedure—Separately inject equal volumes (about 20 µL) of the *Test preparation* and the *Standard preparation* into the chromatograph, record the chromatograms, and measure the responses for the salicylic acid peaks. Calculate the percentage of salicylic acid in the portion of Capsules taken by the formula:

$$100(C/a)(r_U/r_S),$$

in which C is the concentration, in mg per mL, of USP Salicylic Acid RS in the *Standard preparation*, a is the quantity, in mg, of aspirin in the portion of Capsules taken, as determined in the *Assay*, and r_U and r_S are the salicylic acid peak responses obtained from the *Test preparation* and the *Standard preparation*, respectively: not more than 3.0% is found.

Assay—

Mobile phase—Prepare a suitable mixture of monobasic potassium phosphate solution (0.9 g in 1000 mL) and methanol (1:1), and filter through a suitable filter of 0.5-µm or finer porosity. Adjust, if necessary, with 0.1 *N* sodium hydroxide or 1 *M* phosphoric acid to a pH of 3.0. Make adjustments if necessary (see *System Suitability* under *Chromatography* ⟨621⟩).

Standard preparation—Dissolve accurately weighed quantities of USP Aspirin RS, USP Codeine Phosphate RS, and USP Caffeine RS in methanol to obtain a solution having known concentrations of about 0.38 mg of USP Aspirin RS per mL, 0.38J mg of USP Codeine Phosphate RS per mL, and 0.38J' mg of USP Caffeine RS per mL, J being the ratio of the labeled amount, in mg, of codeine phosphate to the labeled amount, in mg, of aspirin per Capsule, and J' being the ratio of the labeled amount, in mg, of caffeine to the labeled amount, in mg, of aspirin per Capsule. Transfer 25.0 mL of this solution to a 50-mL volumetric flask, dilute with water to volume, and mix. This solution contains about 0.19 mg of USP Aspirin RS per mL, 0.19J mg of USP Codeine Phosphate RS per mL, and 0.19J' mg of USP Caffeine RS per mL.

Assay preparation—Weigh the contents of not less than 20 Aspirin, Codeine Phosphate, and Caffeine Capsules, and calculate the average weight per Capsule. Mix the combined contents of the Capsules, and transfer an accurately weighed portion, equivalent to about 380 mg of aspirin, to a 1000-mL volumetric flask. Add about 100 mL of methanol, and sonicate for about 15 minutes. Dilute with methanol to volume, and mix. Filter a portion of this solution through a suitable filter of 0.5-µm or finer porosity, discarding the first few milliliters of the filtrate. Transfer 25.0 mL of the clear filtrate to a 50-mL volumetric flask, dilute with water to volume, and mix.

Chromatographic system (see *Chromatography* ⟨621⟩)—The liquid chromatograph is equipped with a 216-nm detector and a 4-mm × 12.5-cm column that contains packing L16. The flow rate is about 0.5 mL per minute. Chromatograph the *Standard preparation*, and record the responses as directed under *Procedure:* the resolution, R, between the codeine and caffeine peaks and between the caffeine and aspirin peaks is not less than 1.5, the tailing factor is not more than 2.0 for the aspirin peak and not more than 2.5 for the codeine and caffeine peaks, and the relative standard deviation for replicate injections is not more than 2.0%.

Procedure—Separately inject equal volumes (about 10 µL) of the *Standard preparation* and the *Assay preparation* into the chromatograph, record the chromatograms, and measure the responses for the major peaks. The relative retention times for codeine, caffeine, and aspirin are about 0.4, 0.75, and 1.0, respectively. Calculate the quantities, in mg, of aspirin ($C_9H_8O_4$) and caffeine ($C_8H_{10}N_4O_2$) in the portion of Capsules taken by the same formula:

$$2000C(r_U/r_S),$$

in which C is the concentration, in mg per mL, of the appropriate USP Reference Standard in the *Standard preparation*, and r_U and r_S are the responses of the corresponding analyte peaks of the *Assay preparation* and the *Standard preparation*, respectively. Calculate the quantity, in mg, of codeine phosphate hemihydrate ($C_{18}H_{21}NO_3 . H_3PO_4 . \frac{1}{2}H_2O$) in the portion of Capsules taken by the formula:

$$(406.37/397.36)(2000C)(r_U/r_S),$$

in which 406.37 and 397.36 are the molecular weights of codeine phosphate hemihydrate and anhydrous codeine phosphate, respectively, C is the concentration, in mg per mL, of USP Codeine Phosphate RS in the *Standard preparation*, and r_U and r_S are the responses of the codeine peaks obtained from the *Assay preparation* and the *Standard preparation*, respectively.

Aspirin, Codeine Phosphate, and Caffeine Tablets

» Aspirin, Codeine Phosphate, and Caffeine Tablets contain not less than 90.0 percent and not more than 110.0 percent of the labeled amounts of aspirin ($C_9H_8O_4$), codeine phosphate hemihydrate ($C_{18}H_{21}NO_3 . H_3PO_4 . \frac{1}{2}H_2O$), and caffeine ($C_8H_{10}N_4O_2$).

Packaging and storage—Preserve in well-closed, light-resistant containers.

Reference standards—*USP Aspirin Reference Standard*—Dry over silica gel for 5 hours before using. *USP Codeine Phosphate Reference Standard*—This is the hemihydrate form of codeine phosphate. Dry at 105° for 18 hours before using. *USP Caffeine Reference Standard*—Dry at 80° for 4 hours before using. *USP Salicylic Acid Reference Standard*—Dry over silica gel for 3 hours before using.

Identification—The retention times of the major peaks in the chromatogram of the *Assay preparation* correspond to those of the *Standard preparation*, obtained as directed in the *Assay*.

Uniformity of dosage units ⟨905⟩: meet the requirements.

Salicylic acid—

Mobile phase—Dissolve 2 g of sodium 1-heptanesulfonate in a mixture of 825 mL of water and 175 mL of acetonitrile, and adjust with glacial acetic acid to a pH of 3.4. Make adjustments if necessary (see *System Suitability* under *Chromatography* ⟨621⟩).

Solvent mixture—Prepare a mixture of acetonitrile, alcohol-free chloroform, and formic acid (99:99:2).

Standard preparation—Dissolve an accurately weighed quantity of USP Salicylic Acid RS in *Solvent mixture* to obtain a solution having a known concentration of about 0.008 mg per mL.

Test preparation—Weigh and finely powder not less than 20 Tablets. Transfer an accurately weighed portion of the powder, equivalent to about 380 mg of aspirin, to a 100-mL volumetric flask. Add about 20 mL of *Solvent mixture*, and sonicate for about 15 minutes. Dilute with *Solvent mixture* to volume, and mix. Centrifuge a portion of this mixture, and use the clear supernatant liquid as the *Test preparation*.

Chromatographic system (see *Chromatography* ⟨621⟩)—The liquid chromatograph is equipped with a 299-nm detector and a 3.9-mm × 30-cm column that contains packing L1. The flow rate is about 2 mL per minute. Chromatograph the *Standard preparation*, and record the responses as directed under *Procedure:* the relative standard deviation for replicate injections is not more than 4.0%.

Procedure—Separately inject equal volumes (about 20 µL) of the *Test preparation* and the *Standard preparation* into the chromatograph, record the chromatograms, and measure the responses for the salicylic acid peaks. Calculate the percentage of salicylic acid in the portion of Tablets taken by the formula:

$$100(C/a)(r_U/r_S),$$

in which C is the concentration, in mg per mL, of USP Salicylic Acid RS in the *Standard preparation*, a is the quantity, in mg, of aspirin in the portion of Tablets taken, as determined in the *Assay*, and r_U and r_S are the salicylic acid peak responses obtained from the *Test preparation* and the *Standard preparation*, respectively: not more than 3.0% is found.

Assay—

Mobile phase—Prepare a suitable mixture of monobasic potassium phosphate solution (0.9 g in 1000 mL) and methanol (1:1), and filter through a suitable filter of 0.5-µm or finer porosity. Adjust, if necessary, with 0.1 N sodium hydroxide or 1 M phosphoric acid to a pH of 3.0. Make adjustments if necessary (see *System Suitability* under *Chromatography* ⟨621⟩).

Standard preparation—Dissolve accurately weighed quantities of USP Aspirin RS, USP Codeine Phosphate RS, and USP Caffeine RS in methanol to obtain a solution having known concentrations of about 0.38 mg of USP Aspirin RS per mL, 0.38J mg of USP Codeine Phosphate RS per mL, and 0.38J' mg of USP Caffeine RS per mL, J being the ratio of the labeled amount, in mg, of codeine phosphate to the labeled amount, in mg, of aspirin per Capsule, and J' being the ratio of the labeled amount, in mg, of caffeine to the labeled amount, in mg, of aspirin per Capsule. Transfer 25.0 mL of this solution to a 50-mL volumetric flask, dilute with water to volume, and mix. This solution contains about 0.19 mg of USP Aspirin RS per mL, 0.19J mg of USP Codeine Phosphate RS per mL, and 0.19J' mg of USP Caffeine RS per mL.

Assay preparation—Weigh and finely powder not less than 20 Aspirin, Codeine Phosphate, and Caffeine Tablets. Transfer an accurately weighed portion of the powder, equivalent to about

380 mg of aspirin, to a 1000-mL volumetric flask. Add about 100 mL of methanol, and sonicate for about 15 minutes. Dilute with methanol to volume, and mix. Filter a portion of this solution through a suitable filter of 0.5-μm or finer porosity, discarding the first few milliliters of the filtrate. Transfer 25.0 mL of the clear filtrate to a 50-mL volumetric flask, dilute with water to volume, and mix.

Chromatographic system (see *Chromatography* ⟨621⟩)—The liquid chromatograph is equipped with a 216-nm detector and a 4-mm × 12.5-cm column that contains packing L16. The flow rate is about 0.5 mL per minute. Chromatograph the *Standard preparation*, and record the responses as directed under *Procedure:* the resolution, *R*, between the codeine and caffeine peaks and between the caffeine and aspirin peaks is not less than 1.5, the tailing factor is not more than 2.0 for the aspirin peak and not more than 2.5 for the codeine and caffeine peaks, and the relative standard deviation for replicate injections is not more than 2.0%.

Procedure—Separately inject equal volumes (about 10 μL) of the *Standard preparation* and the *Assay preparation* into the chromatograph, record the chromatograms, and measure the responses for the major peaks. The relative retention times for codeine, caffeine, and aspirin are about 0.4, 0.75, and 1.0, respectively. Calculate the quantities, in mg, of aspirin ($C_9H_8O_4$) and caffeine ($C_8H_{10}N_4O_2$) in the portion of Tablets taken by the same formula:

$$2000C(r_U/r_S),$$

in which *C* is the concentration, in mg per mL, of the appropriate USP Reference Standard in the *Standard preparation*, and r_U and r_S are the responses of the corresponding analyte peaks of the *Assay preparation* and the *Standard preparation*, respectively. Calculate the quantity, in mg, of codeine phosphate hemihydrate ($C_{18}H_{21}NO_3 . H_3PO_4 . \frac{1}{2}H_2O$) in the portion of Tablets taken by the formula:

$$(406.37/397.36)(2000C)(r_U/r_S),$$

in which 406.37 and 397.36 are the molecular weights of codeine phosphate hemihydrate and anhydrous codeine phosphate, respectively, *C* is the concentration, in mg per mL, of the USP Codeine Phosphate RS in the *Standard preparation*, and r_U and r_S are the responses of the codeine peaks obtained from the *Assay preparation* and the *Standard preparation*, respectively.

Atropine

$C_{17}H_{23}NO_3$ 289.37

Benzeneacetic acid, α-(hydroxymethyl)-8-methyl-8-azabicyclo[3.2.1]oct-3-yl ester, *endo*-(±)-.

1αH,5αH-Tropan-3α-ol (±)-tropate (ester) [*51-55-8*].

» Atropine contains not less than 99.0 percent and not more than 100.5 percent of $C_{17}H_{23}NO_3$, calculated on the anhydrous basis.

Caution—Handle Atropine with exceptional care, since it is highly potent.

Packaging and storage—Preserve in tight, light-resistant containers.

Reference standard—*USP Atropine Sulfate Reference Standard—Caution—Avoid contact.* Dry at 120° for 4 hours before using.

Identification—

A: Dissolve 30 mg of Atropine and 36 mg of USP Atropine Sulfate RS in individual 60-mL separators with the aid of 5-mL portions of water. To each separator add 1.5 mL of 1 *N* sodium hydroxide solution and 10 mL of chloroform. Shake for 1 minute, allow the layers to separate, and filter the chloroform extracts through separate filters of about 2 g of anhydrous granular sodium sulfate supported on pledgets of glass wool. Extract each aqueous layer with two additional 10-mL portions of chloroform, filtering and combining with the respective main extracts. Evaporate the chloroform solutions under reduced pressure to dryness, and dissolve each residue in 10 mL of carbon disulfide. The infrared absorption spectrum, determined in a 1-mm cell, of the solution obtained from the test specimen exhibits maxima only at the same wavelengths as that of the solution obtained from the Reference Standard.

B: To a 1 in 50 solution in 3 *N* hydrochloric acid add gold chloride TS: a lusterless precipitate is formed (*distinction from hyoscyamine, which, similarly treated, yields a lustrous precipitate*).

Melting range ⟨741⟩: between 114° and 118°.

Optical rotation ⟨781⟩—Dissolve 1 g, previously dried at 105° for 1 hour, in sufficient 50% alcohol (w/w) to obtain a volume of 20 mL at 25°: the angular rotation of this solution, a 200-mm tube being used, is between −0.70° and +0.05° (limit of hyoscyamine).

Water, *Method I* ⟨921⟩: not more than 0.2%.

Residue on ignition ⟨281⟩: not more than 0.1%.

Readily carbonizable substances ⟨271⟩—Dissolve 200 mg in 5 mL of 2 *N* sulfuric acid: the solution has no more color than Matching Fluid A, and the solution is colored no more than light yellow upon the addition of 0.2 mL of nitric acid.

Foreign alkaloids and other impurities—Prepare a solution of Atropine in methanol containing 20 mg per mL, and, by quantitative dilution of a portion of this solution with methanol, prepare a second solution of Atropine containing 1 mg per mL. On a suitable thin-layer chromatographic plate (see *Chromatography* ⟨621⟩), coated with a 0.5-mm layer of chromatographic silica gel, spot 25 μL of the first (20 mg per mL) Atropine solution, 1 μL of the second (1 mg per mL) Atropine solution, and 5 μL of a methanol solution of USP Atropine Sulfate RS containing 24 mg per mL. Allow the spots to dry, and develop the chromatogram in a solvent system consisting of a mixture of chloroform, acetone, and diethylamine (5:4:1) until the solvent front has moved about three-fourths of the length of the plate. Remove the plate from the developing chamber, mark the solvent front, and allow the solvent to evaporate. Locate the spots on the plate by spraying with potassium iodoplatinate TS: the R_f value of the principal spot obtained from each test solution corresponds to that obtained from the Reference Standard solution; no secondary spot obtained from the first Atropine solution exhibits intensity equal to or greater than the principal spot obtained from the second Atropine solution (0.2%).

Assay—Dissolve about 400 mg of Atropine, accurately weighed, in 50 mL of glacial acetic acid, add 1 drop of crystal violet TS, and titrate with 0.1 *N* perchloric acid VS to a green end-point. Perform a blank determination, and make any necessary correction. Each mL of 0.1 *N* perchloric acid is equivalent to 28.94 mg of $C_{17}H_{23}NO_3$.

Atropine Sulfate

$(C_{17}H_{23}NO_3)_2 . H_2SO_4 . H_2O$ 694.84

Benzeneacetic acid, α-(hydroxymethyl)-, 8-methyl-8-azabicyclo[3.2.1]oct-3-yl ester, *endo*-(±)-, sulfate (2:1) (salt), monohydrate.

1αH,5αH-Tropan-3-α-ol (±)-tropate (ester), sulfate (2:1) (salt) monohydrate [*5908-99-6*].

Anhydrous 676.82 [*55-48-1*].

» Atropine Sulfate contains not less than 98.5 percent and not more than 101.0 percent of $(C_{17}H_{23}NO_3)_2 . H_2SO_4$, calculated on the anhydrous basis.

Caution—Handle Atropine Sulfate with exceptional care, since it is highly potent.

Packaging and storage—Preserve in tight containers.

Reference standard—*USP Atropine Sulfate Reference Standard—Caution—Avoid contact.* Dry at 120° for 4 hours before using.

Identification—

A: It meets the requirements under *Identification—Organic Nitrogenous Bases* ⟨181⟩.

B: A solution (1 in 20) responds to the tests for *Sulfate* ⟨191⟩.

Melting temperature, *Class Ia* ⟨741⟩: not lower than 187°, determined after drying at 120° for 4 hours. NOTE—Since anhydrous Atropine Sulfate is hygroscopic, determine its melting temperature promptly on a specimen placed in the capillary tube immediately after drying.

Optical rotation—Dissolve 1 g, accurately weighed, in water to make a volume of 20 mL at 25°. Determine the angular rotation of this solution in a suitable polarimeter tube at 25° (see *Optical Rotation* ⟨781⟩). The observed rotation, in degrees, multiplied by 200, and divided by the length, in mm, of the polarimeter tube used, is between −0.60° and +0.05° (limit of hyoscyamine).

Acidity—Dissolve 1.0 g in 20 mL of water, add 1 drop of methyl red TS, and titrate with 0.020 *N* sodium hydroxide: not more than 0.30 mL is required to produce a yellow color.

Water, *Method I* ⟨921⟩: not more than 4.0%.

Residue on ignition ⟨281⟩: not more than 0.2%.

Other alkaloids—Dissolve 150 mg in 10 mL of water. To 5 mL of the solution add a few drops of platinic chloride TS: no precipitate is formed. To the remaining 5 mL of the solution add 2 mL of 6 *N* ammonium hydroxide, and shake vigorously: a slight opalescence may develop but no turbidity is produced.

Assay—Dissolve about 1 g of Atropine Sulfate, accurately weighed, in 50 mL of glacial acetic acid, and titrate with 0.1 *N* perchloric acid VS, determining the end-point potentiometrically. Perform a blank determination, and make any necessary correction. Each mL of 0.1 *N* perchloric acid is equivalent to 67.68 mg of $(C_{17}H_{23}NO_3)_2 . H_2SO_4$.

Atropine Sulfate Injection

» Atropine Sulfate Injection is a sterile solution of Atropine Sulfate in Water for Injection. It contains not less than 93.0 percent and not more than 107.0 percent of the labeled amount of $(C_{17}H_{23}NO_3)_2 . H_2SO_4 . H_2O$.

Packaging and storage—Preserve in single-dose or in multiple-dose containers, preferably of Type I glass.

Reference standard—*USP Atropine Sulfate Reference Standard*—[*Caution—Avoid contact.*] Dry at 120° for 4 hours before using.

Identification (see *Thin-layer Chromatographic Identification Test* ⟨201⟩)—

Adsorbant: chromatographic silica gel.

Developing solvent: mixture of chloroform and diethylamine (9:1).

Test preparation—Use undiluted. Apply 15 µL.

Detection reagent: potassium iodoplatinate TS.

Procedure—Proceed as directed for *Procedure* under *Thin-layer Chromatographic Identification Test* ⟨201⟩, the spots on the plate being located by spraying with *Detection reagent.*

pH ⟨791⟩: between 3.0 and 6.5.

Other requirements—It meets the requirements under *Injections* ⟨1⟩.

Assay—[NOTE—Where peak responses are indicated, use peak areas.]

Acetate buffer—Prepare a solution in water containing in each liter 0.05 mol of sodium acetate and 2.9 mL of glacial acetic acid.

Mobile phase—Transfer 5.1 g of tetrabutylammonium hydrogen sulfate to a 1-liter volumetric flask, add 50 mL of acetonitrile, and dilute with *Acetate buffer* to volume. Adjust with 5 *N* sodium hydroxide to a pH of 5.5 ± 0.1.

Standard preparation—Dissolve an accurately weighed quantity of USP Atropine Sulfate RS in water, and dilute quantitatively with water to obtain a solution having a known concentration of about 80 µg per mL.

Assay preparation—Transfer an accurately measured volume of Atropine Sulfate Injection, equivalent to about 2 mg of atropine sulfate, to a 25-mL volumetric flask, dilute with water to volume, and mix.

Resolution solution—Prepare a solution in water containing about 80 µg of *p*-hydroxybenzoic acid per mL. Dilute one volume of this solution with four volumes of *Standard preparation.*

Chromatographic system (see *Chromatography* ⟨621⟩)—The liquid chromatograph is equipped with a 254-nm detector and 30-cm × 3.9-mm column that contains packing L1. The flow rate is about 2 mL per minute. Chromatograph the *Standard preparation,* and record the peak responses as directed under *Procedure:* the relative standard deviation for replicate injections is not more than 1.5%. In a similar manner, chromatograph the *Resolution solution:* the retention time of *p*-hydroxybenzoic acid is about 1.6 relative to that of atropine, and the resolution, *R,* between the *p*-hydroxybenzoic acid and atropine peaks is not less than 2.2.

Procedure—Separately inject equal volumes (about 100 µL) of the *Standard preparation* and the *Assay preparation* into the chromatograph, record the chromatograms, and measure the responses for the major peaks. Calculate the quantity, in mg, of $(C_{17}H_{23}NO_3)_2 . H_2SO_4 . H_2O$ in each mL of the Injection taken by the formula:

$$(694.84/676.82)(25C/V)(r_U/r_S),$$

in which 694.84 and 676.82 are the molecular weights of atropine sulfate monohydrate and anhydrous atropine sulfate, respectively, C is the concentration, in mg per mL, of USP Atropine Sulfate RS in the *Standard preparation,* V is the volume, in mL, of Atropine Sulfate Injection taken, and r_U and r_S are the peak responses obtained from the *Assay preparation* and the *Standard preparation,* respectively.

Atropine Sulfate Ophthalmic Ointment

» Atropine Sulfate Ophthalmic Ointment is Atropine Sulfate in a suitable ophthalmic ointment base. It contains not less than 90.0 percent and not more than 110.0 percent of the labeled amount of $(C_{17}H_{23}NO_3)_2 . H_2SO_4 . H_2O$. It is sterile.

Packaging and storage—Preserve in collapsible ophthalmic ointment tubes.

Reference standard—*USP Atropine Sulfate Reference Standard*—Keep container tightly closed and protected from light. *Caution—Avoid contact.* Dry at 120° for 4 hours before using.

Identification—

A: Transfer a portion of Ophthalmic Ointment, equivalent to about 50 mg of atropine sulfate, to a suitable separator, and dissolve in 25 mL of ether. Add 25 mL of 0.01 *N* hydrochloric acid, shake vigorously, allow the layers to separate, and discard the organic phase. Heat the aqueous phase gently on a steam bath while passing nitrogen through the solution, to expel any residual ether. Proceed as directed under *Identification—Organic Nitrogenous Bases* ⟨181⟩, beginning with "In a second separator dissolve 50 mg."

B: Transfer about 5 g of Ophthalmic Ointment to a separator, dissolve in 50 mL of ether, and extract with 20 mL of water: the extracted solution so obtained responds to the tests for *Sulfate* ⟨191⟩.

Sterility—It meets the requirements under *Sterility Tests* ⟨71⟩.

Metal particles—It meets the requirements of the test for *Metal Particles in Ophthalmic Ointments* ⟨751⟩.

Assay—Proceed with Atropine Sulfate Ophthalmic Ointment as directed in the *Assay* under *Atropine Sulfate Ophthalmic Solution,* but in preparing the *Assay solution,* weigh accurately a portion of Ophthalmic Ointment equivalent to about 10 mg of atropine sulfate into a separator containing 50 mL of ether, shake to dissolve, extract with three 25-mL portions of 0.2 *N* sulfuric acid, collect the acid extracts in a 100-mL volumetric flask, dilute with 0.2 *N* sulfuric acid to volume, and mix. Calculate the quantity, in mg, of $(C_{17}H_{23}NO_3)_2 . H_2SO_4 . H_2O$ in the portion of Ophthalmic Ointment taken by the formula given.

Atropine Sulfate Ophthalmic Solution

» Atropine Sulfate Ophthalmic Solution is a sterile, aqueous solution of Atropine Sulfate. It contains not less than 93.0 percent and not more than 107.0 percent of the labeled amount of $(C_{17}H_{23}NO_3)_2 . H_2SO_4 . H_2O$. It may contain suitable stabilizers and antimicrobial agents.

Packaging and storage—Preserve in tight containers.

Reference standard—*USP Atropine Sulfate Reference Standard—Caution—Avoid contact.* Dry at 120° for 4 hours before using.

Identification—After evaporation to dryness or appropriate adjustment of concentration, it responds to the *Identification tests* under *Atropine Sulfate.*

Sterility—It meets the requirements under *Sterility Tests* ⟨71⟩.

pH ⟨791⟩: between 3.5 and 6.0.

Assay—

pH 9.0 buffer—Dissolve 34.8 g of dibasic potassium phosphate in 900 mL of water, and adjust to a pH of 9.0, determined electrometrically, by the addition of 3 *N* hydrochloric acid or 1 *N* sodium hydroxide, as necessary, with mixing.

Internal standard solution—Dissolve about 25 mg of homatropine hydrobromide in water, contained in a 50-mL volumetric flask, add water to volume, and mix. Prepare fresh daily.

Standard solution—Dissolve about 10 mg of USP Atropine Sulfate RS, accurately weighed, in water, contained in a 100-mL volumetric flask, add water to volume, and mix. Prepare fresh daily.

Assay solution—Transfer an accurately measured volume of Ophthalmic Solution, equivalent to about 10 mg of Atropine Sulfate, to a 100-mL volumetric flask. Dilute with water to volume, and mix.

Assay preparation and *Standard preparation*—Pipet 10 mL of *Assay solution* and 10 mL of *Standard solution*, respectively, into two different separators, and treat them identically as follows: Add 2.0 mL of *Internal standard solution* and 5.0 mL of *pH 9.0 buffer*, and adjust the solution in the separator with 1 *N* sodium hydroxide to a pH of 9.0. Extract with two 10-mL portions of methylene chloride, filter the methylene chloride extracts through 1 g of anhydrous sodium sulfate supported by a small cotton plug in a funnel into a 50-mL beaker, and evaporate under nitrogen to near-dryness. Dissolve the residue in 2.0 mL of methylene chloride.

Chromatographic system—The gas chromatograph contains a 1.8-m × 2-mm glass column packed with 3 percent phase G3 on support S1AB, conditioned as directed (see *Chromatography* ⟨621⟩). Maintain the column at 225°, and use nitrogen as the carrier gas at a flow rate of 25 mL per minute.

System suitability—Chromatograph six to ten injections of the *Standard preparation*, and record peak areas as directed under *Procedure*. The analytical system is suitable for conducting this assay if the relative standard deviation for the ratio of the peak areas does not exceed 2.0%; the resolution, *R*, is not less than 4.0; and the tailing factor does not exceed 2.0.

Procedure—Inject 1-µL portions of the *Assay preparation* and the *Standard preparation* successively into the gas chromatograph. Measure the areas under the peaks for atropine sulfate and homatropine hydrobromide in each chromatogram. Calculate the ratio, R_U, of the area of the atropine sulfate peak to the area of the internal standard peak in the chromatogram from the *Assay preparation*, and similarly calculate the ratio, R_S, in the chromatogram from the *Standard preparation.* Calculate the quantity, in mg, of $(C_{17}H_{23}NO_3)_2 . H_2SO_4 . H_2O$ in each mL of the Ophthalmic Solution taken by the formula:

$$(694.84/676.82)(W/V)(R_U/R_S),$$

in which 694.84 and 676.82 are the molecular weights of atropine sulfate monohydrate and anhydrous atropine sulfate, respectively, *W* is the weight, in mg, of USP Atropine Sulfate RS in the *Standard solution*, and *V* is the volume, in mL, of Ophthalmic Solution taken.

Atropine Sulfate Oral Solution, Diphenoxylate Hydrochloride and—*see* Diphenoxylate Hydrochloride and Atropine Sulfate Oral Solution

Atropine Sulfate Tablets

» Atropine Sulfate Tablets contain not less than 90.0 percent and not more than 110.0 percent of the labeled amount of $(C_{17}H_{23}NO_3)_2 . H_2SO_4 . H_2O$.

Packaging and storage—Preserve in well-closed containers.

Reference standard—*USP Atropine Sulfate Reference Standard—Caution—Avoid contact.* Dry at 120° for 4 hours before using.

Identification—

A: Triturate a quantity of Tablets, equivalent to about 5 mg of atropine sulfate, with 10 mL of water for a few minutes, and filter into a small separator. Render the solution alkaline with 6 *N* ammonium hydroxide, and extract with 50 mL of chloroform. Filter the chloroform layer, and evaporate to dryness. The residue so obtained meets the requirements under *Identification—Organic Nitrogenous Bases* ⟨181⟩.

B: A filtered solution of Tablets responds to the tests for *Sulfate* ⟨191⟩.

Disintegration ⟨701⟩: 15 minutes.

Uniformity of dosage units ⟨905⟩: meet the requirements.

Assay—

pH 9.0 buffer, Internal standard solution, Standard solution, Chromatographic system, and *System suitability*—Proceed as directed in the *Assay* under *Atropine Sulfate Ophthalmic Solution.*

Assay solution—Weigh and finely powder not less than 20 Atropine Sulfate Tablets. Transfer an accurately weighed portion of the powder, equivalent to about 1.0 mg of atropine sulfate, to a separator containing 5 mL of *pH 9.0 buffer*, and add, by pipet, 2.0 mL of *Internal standard solution.* Adjust with 1 *N* sodium hydroxide to a pH of 9.0, extract with two 10-mL portions of methylene chloride, filter the methylene chloride extracts through 1 g of anhydrous sodium sulfate supported by a small cotton plug in a funnel into a 50-mL beaker, and evaporate under nitrogen to dryness. Dissolve the residue in 2.0 mL of methylene chloride.

Assay preparation and *Standard preparation*, and *Procedure*—Proceed as directed in the *Assay* under *Atropine Sulfate Ophthalmic Solution.* Calculate the quantity, in mg, of $(C_{17}H_{23}NO_3)_2 . H_2SO_4 . H_2O$ in the portion of Tablets taken by the formula:

$$(694.84/676.82)(W/10)(R_U/R_S),$$

in which 694.84 and 676.82 are the molecular weights of atropine sulfate monohydrate and anhydrous atropine sulfate, respectively, and *W* is the weight, in mg, of USP Atropine Sulfate RS in the *Standard solution.*

Atropine Sulfate Tablets, Diphenoxylate Hydrochloride and—*see* Diphenoxylate Hydrochloride and Atropine Sulfate Tablets

Activated Attapulgite

» Activated Attapulgite is a highly heat-treated, processed, native magnesium aluminum silicate.

Packaging and storage—Preserve in well-closed containers.

Identification—Activated Attapulgite responds to the *Identification* test for *Colloidal Activated Attapulgite*, the characteristic peak, however, being much less intense.

Loss on drying ⟨731⟩—Dry it at 105° to constant weight: it loses not more than 4.0% of its weight.

Volatile matter—When ignited at 600° for 1 hour, it loses between 3.0% and 7.5% of its weight on the dried basis.

Loss on ignition—When ignited at 1000° for 1 hour, it loses between 4.0% and 12.0% of its weight.

Acid-soluble matter—Boil 2.0 g with 100 mL of 0.2 *N* hydrochloric acid for 5 minutes, cool, and filter. Evaporate 50 mL of the filtrate so obtained to dryness, and ignite the residue at 600°: not more than 0.25 g is found (25%).

Powder fineness ⟨811⟩—Proceed as directed in the test for *Powder fineness* under *Colloidal Activated Attapulgite*. The dry weight of the residue is not more than 0.10% of the weight of the specimen taken.

Other requirements—It meets the requirements of the tests for *Microbial limit, pH, Carbonate, Arsenic and Lead*, and *Adsorptive capacity*, under *Colloidal Activated Attapulgite*.

Colloidal Activated Attapulgite

» Colloidal Activated Attapulgite is a purified native magnesium aluminum silicate.

Packaging and storage—Preserve in well-closed containers.

Identification—Add 2 g in small portions to 100 mL of water, with vigorous agitation. Allow to stand for at least 12 hours to ensure complete hydration. Place 2 mL of the resulting mixture on a suitable glass slide, and allow to air-dry at room temperature to produce a uniform film. Place the slide in a vacuum desiccator over a free surface of ethylene glycol. Evacuate the desiccator, and close the stopcock so that the ethylene glycol saturates the desiccator chamber. Allow to stand for 12 hours. Record the X-ray diffraction pattern (see *X-ray Diffraction* ⟨941⟩), and calculate the *d* values: several peaks are observed; the characteristic peak corresponds to a *d* value between 10.3 and 10.7 Angstrom units.

Microbial limit—It meets the requirements of the test for absence of *Escherichia coli* under *Microbial Limit Tests* ⟨61⟩.

pH ⟨791⟩—Disperse 1.0 g in 10 mL of carbon dioxide–free water, and mix: the pH of the mixed dispersion so obtained is between 7.0 and 9.5.

Loss on drying ⟨731⟩—Dry it at 105° to constant weight: it loses between 5.0% and 17.0% of its weight.

Volatile matter—When ignited at 600° for 1 hour, it loses between 7.5% and 12.5% of its weight on the dried basis.

Loss on ignition—When ignited at 1000° for 1 hour, it loses between 17.0% and 27.0% of its weight.

Acid-soluble matter—Boil 2.0 g with 100 mL of 0.2 *N* hydrochloric acid for 5 minutes, cool, and filter. Evaporate 50 mL of the filtrate so obtained to dryness, and ignite the residue at 600°: not more than 0.15 g is found (15%).

Carbonate—Mix 1.0 g with 15 mL of 0.5 *N* sulfuric acid: no effervescence occurs.

Arsenic and Lead—To 5.0 g add 50 mL of 1 *N* nitric acid, and boil for 30 minutes, adding 1 *N* nitric acid at times to maintain the volume. Filter into a 100-mL volumetric flask, wash the filter with water, and dilute the combined filtrate and washings with water to volume.

Arsenic—Determine the arsenic in the solution by atomic absorption spectrometry (see *Spectrophotometry and Light-scattering* ⟨851⟩), using a graphite furnace to volatilize the arsenic, as directed by the manufacturer of the instrument used, and measuring the absorbance at 189.0 nm against a standard: not more than 2 ppm is found.

Lead—Determine the lead in the solution by atomic absorption spectrometry (see *Spectrophotometry and Light-scattering* ⟨851⟩), using a graphite furnace to volatilize the lead, as directed by the manufacturer of the instrument used, and measuring the absorbance at 283.3 nm against a standard: not more than 0.001% is found.

Powder fineness ⟨811⟩—Add 50 g to 450 mL of water containing 5 g of sodium pyrophosphate, and stir for 10 minutes. Pour the resulting dispersion slowly through a No. 325 standard sieve, and carefully wash the residue until clean. Dry the residue at 105° to constant weight: the dry weight of the residue so obtained is not more than 0.30% of the weight of specimen taken.

Adsorptive capacity—To 10 mL of a 1 in 10 suspension of the specimen in water add 80 mL of methylene blue solution (1 in 1000), and shake. Add 10 mL of barium chloride solution (1 in 50), and shake. Allow to stand for 15 minutes. Transfer 40 mL of the supernatant liquid to a 50-mL centrifuge tube, and centrifuge. To 5 mL of the clear supernatant liquid add 495 mL of water, and mix: the color of the solution so obtained is not deeper than that of a solution containing 0.15 µg of methylene blue per mL.

Aurothioglucose

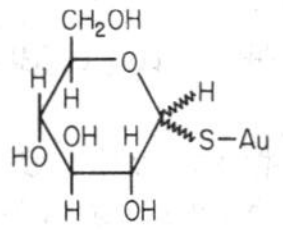

$C_6H_{11}AuO_5S$ 392.18

Gold, (1-thio-D-glucopyranosato)-.
(1-Thio-D-glucopyranosato)gold [*12192-57-3*].

» Aurothioglucose contains not less than 95.0 percent and not more than 105.0 percent of $C_6H_{11}AuO_5S$, calculated on the dried basis. It is stabilized by the addition of a small amount of Sodium Acetate.

Packaging and storage—Preserve in tight, light-resistant containers.

Reference standard—*USP Aurothioglucose Reference Standard*—Dry over phosphorus pentoxide for 24 hours before using.

Identification—

A: Dissolve a suitable quantity in water to obtain a solution containing 4 mg per mL. On a suitable thin-layer chromatographic glass microfilament sheet (see *Chromatography* ⟨621⟩), impregnated with silicic acid and a suitable fluorescing substance, apply 10 µL of this solution and 10 µL of an aqueous Standard solution of USP Aurothioglucose RS containing 4 mg per mL. Allow the spots to dry, and develop the chromatogram in a solvent system consisting of a mixture of *n*-propyl alcohol, water, and ethyl acetate (3:3:1) until the solvent front has moved about three-fourths of the length of the plate. Remove the sheet from the developing chamber, mark the solvent front, and allow the solvent to evaporate. Locate the spots on the plate by examination under short-wavelength ultraviolet light: the R_f value of the principal spot obtained from the solution under test corresponds to that obtained from the Standard solution.

B: To a portion of the filtrate obtained in the *Assay* add barium chloride TS: a heavy, white precipitate is formed.

Specific rotation ⟨781⟩: between +65° and +75°, calculated on the dried basis, determined in a solution containing 100 mg in each 10 mL.

Loss on drying ⟨731⟩—Dry it over phosphorus pentoxide for 24 hours: it loses not more than 1.0% of its weight.

Assay—Weigh accurately about 1 g of Aurothioglucose, and dissolve in 100 mL of water in a 300-mL Kjeldahl flask. Slowly add 10 mL of nitric acid, and when the reaction has subsided, boil the mixture for 5 minutes. Filter, wash well the separated gold with hot water, dry, and ignite to constant weight. The weight of the gold so obtained, multiplied by 1.991, represents the weight of $C_6H_{11}AuO_5S$ in the portion of Aurothioglucose taken.

Sterile Aurothioglucose Suspension

» Sterile Aurothioglucose Suspension is a sterile suspension of Aurothioglucose in a suitable vegetable oil. It contains not less than 90.0 percent and not more than 110.0 percent of the labeled amount of $C_6H_{11}AuO_5S$. It may contain suitable thickening agents.

Packaging and storage—Preserve in single-dose or multiple-dose containers, preferably of Type I glass. Protect from light.

Reference standard—*USP Aurothioglucose Reference Standard*—Dry over phosphorus pentoxide for 24 hours before using.

Identification—Transfer a volume of Suspension, equivalent to about 200 mg of aurothioglucose, to a centrifuge separator containing 20 mL of ethyl acetate and 50 mL of water. Shake the mixture thoroughly, and centrifuge until the liquid phases have been clearly separated. Withdraw the lower, aqueous phase, and filter, discarding the first 10 mL of the filtrate. Collect the filtrate in a glass-stoppered vessel, and proceed as directed in *Identification test A* under *Aurothioglucose*, beginning with "apply 10 µL of this solution."

Other requirements—It meets the requirements under *Injections* ⟨1⟩.

Assay—Transfer with a pipet, calibrated to contain rather than to deliver, an accurately measured volume of Sterile Aurothioglucose Suspension, equivalent to about 200 mg of aurothioglucose, to a beaker containing 400 mL of acetone. Wash the pipet into the beaker with a small quantity of acetone, mix, allow the solids to settle, and decant the supernatant liquid through a filter. Wash the solids with another 400-mL portion of acetone, and repeat the decantation. Transfer the solids to the filter with the aid of acetone, then transfer the filter and its contents to a short-necked, 300-mL Kjeldahl flask, add 5 mL of water, and proceed as directed in the *Assay* under *Gold Sodium Thiomalate*, beginning with "add 20 mL of nitric acid." The weight of gold so obtained, multiplied by 1.991, represents the weight of $C_6H_{11}AuO_5S$ in the portion of Suspension taken.

Azatadine Maleate

$C_{20}H_{22}N_2 \cdot 2C_4H_4O_4$ 522.55

5*H*-Benzo[5,6]cyclohepta[1,2-*b*]pyridine, 6,11-dihydro-11-(1-methyl-4-piperidinylidene)-, (*Z*)-2-butenedioate (1:2).

6,11-Dihydro-11-(1-methyl-4-piperidylidene)-5*H*-benzo[5,6]-cyclohepta[1,2-*b*]pyridine maleate (1:2) [*3978-86-7*].

» Azatadine Maleate contains not less than 98.0 percent and not more than 102.0 percent of $C_{20}H_{22}N_2 \cdot 2C_4H_4O_4$, calculated on the dried basis.

Packaging and storage—Preserve in well-closed containers.

Reference standard—*USP Azatadine Maleate Reference Standard*—Dry in vacuum at 60° for 3 hours before using.

Identification—

A: The infrared absorption spectrum of a mineral oil dispersion of it, previously dried, exhibits maxima only at the same wavelengths as that of a similar preparation of USP Azatadine Maleate RS.

B: The ultraviolet absorption spectrum of a 1 in 25,000 solution in 0.25 *N* hydrochloric acid in methanol exhibits maxima and minima at the same wavelengths as that of a similar solution of USP Azatadine Maleate RS, concomitantly measured.

Loss on drying ⟨731⟩—Dry it in vacuum at 60° for 3 hours: it loses not more than 1.0% of its weight.

Residue on ignition ⟨281⟩: not more than 0.1%.

Chromatographic purity—Dissolve an accurately weighed quantity in a solvent mixture consisting of toluene and methanol (1:1), to obtain a solution having a known concentration of about 7 mg of the test specimen per mL. Similarly prepare a Standard solution of USP Azatadine Maleate RS in the same medium having a known concentration of about 7 mg per mL. On a suitable thin-layer chromatographic plate (see *Chromatography* ⟨621⟩), coated with a 0.25-mm layer of chromatographic silica gel mixture, apply 100-µL portions of the test solution and the Standard solution. Allow the spots to dry, and develop the chromatogram in a solvent system consisting of a mixture of toluene, isopropyl alcohol, and diethylamine (10:10:1), until the solvent front has moved about three-fourths of the length of the plate. Remove the plate from the developing chamber, mark the solvent front, and allow the solvent to evaporate. Locate the spots on the plate by visualization under short-wavelength ultraviolet light. Separately transfer the silica gel mixture containing the principal spot from each track to suitable stoppered centrifuge tubes. [NOTE—Take care to separate the principal spots from any adjacent spots.] Similarly transfer an equal amount of silica gel from a blank section of the plate to a separate, suitable stoppered centrifuge tube. To each of the three tubes add 15.0 mL of a solvent mixture consisting of methanol and 0.5 *N* hydrochloric acid (4:1), shake vigorously for about 15 minutes, and centrifuge. Concomitantly determine the absorbances of the supernatant test solution and the Standard solution in 1-cm cells at the wavelength of maximum absorbance at about 284 nm, with a suitable spectrophotometer, using the solution obtained from the blank section of the plate as the blank. Calculate the chromatographic purity by the formula:

$$100(A_U/A_S)(C_S/C_U),$$

in which A_U and A_S are the absorbances of the test solution and the Standard solution, respectively, and C_S and C_U are the concentrations, in mg per mL, of the Standard solution and the test solution, respectively. The chromatographic purity is not less than 98.0%.

Assay—Dissolve about 650 mg of Azatadine Maleate, accurately weighed, in 50 mL of glacial acetic acid, add 2 drops of crystal violet TS, and titrate with 0.1 *N* perchloric acid VS. Perform a blank determination, and make any necessary correction. Each mL of 0.1 *N* perchloric acid is equivalent to 26.13 mg of $C_{20}H_{22}N_2 \cdot 2C_4H_4O_4$.

Azatadine Maleate Tablets

» Azatadine Maleate Tablets contain not less than 90.0 percent and not more than 110.0 percent of the labeled amount of $C_{20}H_{22}N_2 \cdot 2C_4H_4O_4$.

Packaging and storage—Preserve in well-closed containers.

Reference standard—*USP Azatadine Maleate Reference Standard*—Dry in vacuum at 60° for 3 hours before using.

Identification—Transfer 15.0 mL of the *Standard preparation* and 15.0 mL of the *Assay preparation*, respectively, prepared as directed in the *Assay*, to separate 50-mL centrifuge tubes fitted with glass stoppers. To each centrifuge tube add 10.0 mL of 1.0 *N* sodium hydroxide and 20 mL of solvent hexane, insert the stoppers, rotate the centrifuge tubes for about 15 minutes, and centrifuge. Transfer the solvent hexane extracts (upper phase) from each centrifuge tube to separate 50-mL conical flasks fitted with glass stoppers. Evaporate the solvent hexane extracts on a steam bath under a stream of nitrogen to dryness, pipet 1 mL of solvent hexane into each flask, insert the stoppers, and mix by use of a vortex mixer (or equivalent) until the residues have dissolved. Use these solutions as the Standard solution and the Test solution, respectively. On a suitable thin-layer chromatographic plate (see *Chromatography* ⟨621⟩), coated with a 0.25-mm layer of chromatographic silica gel mixture, apply separately 100 µL each of the Test solution and the Standard solution. Allow the spots to dry, and develop the chromatogram in a solvent system consisting of toluene, isopropyl alcohol, and diethylamine (10:10:1) until the solvent front has moved about three-fourths of the length of the plate. Remove the plate from the developing chamber, mark the solvent front, and allow the plate to air-dry. Examine the plate under short-wavelength ultraviolet light: the R_f value and intensity of the principal spot in the chromatogram of the Test solution correspond to those obtained from the chromatogram of the Standard solution.

Dissolution ⟨711⟩—

Medium: 0.1 *N* hydrochloric acid; 500 mL.

Apparatus 2: 50 rpm.

Time: 30 minutes.

Procedure—Determine the amount of $C_{20}H_{22}N_2 . 2C_4H_4O_4$ dissolved from ultraviolet absorbances at the wavelength of maximum absorbance at about 283 nm, using filtered portions of the solution under test, diluted with 0.1 *N* hydrochloric acid, if necessary, in comparison with a Standard solution having a known concentration of USP Azatadine Maleate RS in the same medium.

Tolerances—Not less than 80% (*Q*) of the labeled amount of $C_{20}H_{22}N_2 . 2C_4H_4O_4$ is dissolved in 30 minutes.

Uniformity of dosage units ⟨905⟩: meet the requirements.

Assay—

Standard preparation—Dissolve an accurately weighed quantity of USP Azatadine Maleate RS in 0.1 *N* hydrochloric acid, and dilute quantitatively, and stepwise if necessary, with 0.1 *N* hydrochloric acid to obtain a solution having a known concentration of about 0.06 mg per mL.

Assay preparation—Weigh and finely powder not less than 20 Azatadine Maleate Tablets. Transfer an accurately weighed portion of the powder, equivalent to about 1.5 mg of azatadine maleate, to a 50-mL flask fitted with a glass stopper. Add 25.0 mL of 0.1 *N* hydrochloric acid, insert the stopper, and shake the mixture by mechanical means for about 30 minutes. Filter the mixture into a suitable glass-stoppered vessel, discarding the first 5 mL of the filtrate.

Procedure—Separately transfer 15.0 mL of the *Standard preparation*, 15.0 mL of the *Assay preparation*, and 15.0 mL of 0.1 *N* hydrochloric acid to provide the reagent blank to three 50-mL centrifuge tubes fitted with glass stoppers. To each centrifuge tube add 10.0 mL of 1.0 *N* sodium hydroxide and 20 mL of solvent hexane, insert the stoppers, rotate the centrifuge tubes for about 15 minutes, and centrifuge until the supernatant liquids (solvent hexane phase) are clear. With the aid of separate syringes, transfer the supernatant liquids to separate 50-mL centrifuge tubes fitted with glass stoppers. Rinse each syringe with 10 mL of solvent hexane, and add the rinse to the aqueous phase from which the respective supernatant liquid was removed. Insert the stoppers, rotate each tube for about 10 minutes, and centrifuge. Transfer each supernatant liquid to the respective supernatant liquid previously collected. Pipet 15 mL of 0.1 *N* hydrochloric acid into each centrifuge tube containing the combined supernatant liquids, insert the stoppers, rotate each tube for about 15 minutes, and centrifuge. Remove and discard the supernatant liquids. Concomitantly determine the absorbances of the solutions in 1-cm cells at the wavelength of maximum absorbance at about 283 nm, with a suitable spectrophotometer zeroed with 0.1 *N* hydrochloric acid, using the prepared reagent blank. Calculate the quantity, in mg, of $C_{20}H_{22}N_2 . 2C_4H_4O_4$ in the portion of Tablets taken by the formula:

$$25C(A_U / A_S),$$

in which *C* is the concentration, in mg per mL, of USP Azatadine Maleate RS in the *Standard preparation*, and A_U and A_S are the absorbances of the solutions from the *Assay preparation* and the *Standard preparation*, respectively.

Azathioprine

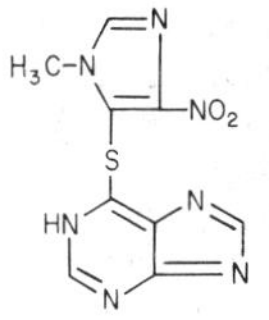

$C_9H_7N_7O_2S$ 277.26

1*H*-Purine, 6-[(1-methyl-4-nitro-1*H*-imidazol-5-yl)thio]-.

6-[(1-Methyl-4-nitroimidazol-5-yl)thio]purine [*446-86-6*].

» Azathioprine contains not less than 98.0 percent and not more than 101.5 percent of $C_9H_7N_7O_2S$, calculated on the dried basis.

Packaging and storage—Preserve in tight, light-resistant containers.

Reference standards—*USP Azathioprine Reference Standard*—Dry in vacuum at 105° for 5 hours before using. *USP Mercaptopurine Reference Standard*—Do not dry; determine the *Water* content by *Method I* ⟨921⟩ before using.

Identification—

A: The infrared absorption spectrum of a potassium bromide dispersion of it exhibits maxima only at the same wavelengths as that of a similar preparation of USP Azathioprine RS.

B: The principal spot in the test preparation chromatogram in the test for *Limit of mercaptopurine* shows the same R_f value as that obtained with the solution of USP Azathioprine RS.

Acidity or alkalinity—Shake 2.0 g with 100 mL of water for 15 minutes, and filter: 20.0 mL of the filtrate requires for neutralization not more than 0.10 mL of 0.020 *N* hydrochloric acid or not more than 0.10 mL of 0.020 *N* sodium hydroxide, methyl red TS being used as the indicator.

Loss on drying ⟨731⟩—Dry it in vacuum at 105° for 5 hours: it loses not more than 1.0% of its weight.

Residue on ignition ⟨281⟩: not more than 0.1%.

Limit of mercaptopurine—Prepare three solutions in 6 *N* ammonium hydroxide containing, respectively, 20 mg of Azathioprine per mL, 20 mg of USP Azathioprine RS per mL, and 200 µg of USP Mercaptopurine RS, on the anhydrous basis, per mL. Apply 5-µL volumes of the solutions at points about 2 cm from the bottom edge of a thin-layer chromatographic plate (see *Chromatography* ⟨621⟩) coated with a 0.25-mm layer of microcrystalline cellulose. Allow the spots to dry, and develop the chromatogram in a suitable chamber, using butyl alcohol, previously saturated with 6 *N* ammonium hydroxide, as the solvent, until the solvent front has moved about 15 cm from the point of application. Remove the plate, air-dry, and locate the spots by viewing under short- and long-wavelength ultraviolet light: any spot in the chromatogram from Azathioprine, other than the principal spot, is not more intense than the spot in the chromatogram obtained with USP Mercaptopurine RS (1.0%).

Assay—Dissolve about 300 mg of Azathioprine, accurately weighed, in 80 mL of dimethylformamide. Add 5 drops of a 1 in 100 solution of thymol blue in dimethylformamide, and titrate with 0.1 *N* tetrabutylammonium hydroxide VS, using a magnetic stirrer, and taking precautions to prevent absorption of atmospheric carbon dioxide. Perform a blank determination, and make any necessary correction. Each mL of 0.1 *N* tetrabutylammonium hydroxide is equivalent to 27.73 mg of $C_9H_7N_7O_2S$.

Azathioprine Tablets

» Azathioprine Tablets contain not less than 93.0 percent and not more than 107.0 percent of the labeled amount of $C_9H_7N_7O_2S$.

Packaging and storage—Protect from light.

Reference standard—*USP Azathioprine Reference Standard*—Dry in vacuum at 105° for 5 hours before using.

Identification—Tablets meet the requirement of the *Thin-layer Chromatographic Identification Test* ⟨201⟩, the test solution being the filtrate obtained by shaking a quantity of powdered Tablets, equivalent to about 200 mg of azathioprine, with 10 mL of 6 *N* ammonium hydroxide, the Standard solution being a solution of USP Azathioprine RS in 6 *N* ammonium hydroxide containing 20 mg per mL, and 5-µL portions of each solution being spotted on a thin-layer chromatographic plate coated with a 0.25-mm layer of microcrystalline cellulose, and butyl alcohol, previously saturated with 6 *N* ammonium hydroxide, being used for developing.

Dissolution ⟨711⟩—

Medium: water; 900 mL.

Apparatus 2: 50 rpm.

Time: 45 minutes.

Procedure—Determine the amount of $C_9H_7N_7O_2S$ dissolved from ultraviolet absorbances at the wavelength of maximum absorbance at about 280 nm of filtered portions of the solution under test, suitably diluted with *Dissolution Medium*, if necessary, in comparison with a Standard solution having a known concentration of USP Azathioprine RS in the same medium.

Tolerances—Not less than 65% (*Q*) of the labeled amount of $C_9H_7N_7O_2S$ is dissolved in 45 minutes.

Uniformity of dosage units ⟨905⟩: meet the requirements.

Assay—

Mobile phase—Dissolve 1.1 g of sodium 1-heptanesulfonate in 700 mL of water, add 300 mL of methanol, and mix. Adjust the solution with 1 *N* hydrochloric acid to a pH of 3.5. Filter the solution through a 0.8-µm solvent-resistant membrane, and degas, making adjustments if necessary (see *System Suitability* under *Chromatography* ⟨621⟩).

Standard preparation—Transfer about 25 mg of USP Azathioprine RS, accurately weighed, to a 50-mL volumetric flask. Add about 15 mL of methanol and 0.5 mL of ammonium hydroxide to the flask, swirl, and sonicate for 2 minutes. Dilute with methanol to volume, and mix. Transfer 10.0 mL of this solution to a 50-mL volumetric flask, dilute with water to volume, and mix.

Assay preparation—Weigh and finely powder not less than 20 Azathioprine Tablets. Accurately weigh a portion of the powder, equivalent to about 50 mg of azathioprine, and transfer to a 100-mL volumetric flask. Add 25 mL of methanol and 1.0 mL of ammonium hydroxide to the flask, swirl, and sonicate for 2 minutes. Dilute with methanol to volume, and mix. Allow the excipients to settle, transfer 10.0 mL of the supernatant liquid to a 50-mL volumetric flask, dilute with water to volume, and mix.

Chromatographic system (see *Chromatography* ⟨621⟩)—The liquid chromatograph is equipped with a 254-nm detector and a 4-mm × 30-cm column that contains packing L1. The flow rate is about 2.0 mL per minute. Chromatograph the *Standard preparation*, and record the peak responses as directed under *Procedure:* the column efficiency is not less than 800 theoretical plates, the tailing factor for the azathioprine peak is not more than 1.5, and the relative standard deviation for replicate injections is not more than 2.0%.

Procedure—Separately inject equal volumes (about 10 µL) of the *Standard preparation* and the *Assay preparation* into the chromatograph, record the chromatograms, and measure the responses for the major peaks. Calculate the quantity, in mg, of $C_9H_7N_7O_2S$ in the portion of Tablets taken by the formula:

$$500(C)(r_U / r_S),$$

in which *C* is the concentration, in mg per mL, of USP Azathioprine RS in the *Standard preparation*, and r_U and r_S are the peak responses for azathioprine obtained from the *Assay preparation* and the *Standard preparation*, respectively.

Azathioprine Sodium for Injection

» Azathioprine Sodium for Injection is a sterile solid prepared by the freeze-drying of an aqueous solution of Azathioprine and Sodium Hydroxide. It contains not less than 93.0 percent and not more than 107.0 percent of the labeled amount of $C_9H_7N_7O_2S$.

Packaging and storage—Preserve in *Containers for Sterile Solids* as described under *Injections* ⟨1⟩, at controlled room temperature.

Reference standards—*USP Azathioprine Reference Standard*—Dry in vacuum at 105° for 5 hours before using. *USP Mercaptopurine Reference Standard*—Do not dry; determine the *Water* content by *Method I* ⟨921⟩ before using.

Completeness of solution ⟨641⟩—The contents of 1 container are soluble in 10 mL of water, to give a clear, bright yellow solution, essentially free from foreign matter.

Identification—The principal spot in the chromatogram of the specimen under examination obtained in the test for *Limit of mercaptopurine* shows the same R_f value as that obtained with the solution of USP Azathioprine RS.

pH ⟨791⟩: between 9.8 and 11.0, the contents of 1 container being dissolved in 10 mL of water.

Water, *Method I* ⟨921⟩: not more than 5.0%.

Limit of mercaptopurine—Prepare solutions in dimethylformamide containing, respectively, 10 mg of Azathioprine Sodium for Injection per mL, 10 mg of USP Azathioprine RS per mL, and 100 µg of USP Mercaptopurine RS per mL. Apply 15 µL of the USP Mercaptopurine RS solution and 5-µL portions of the other two solutions at points about 2 cm from the bottom edge of a thin-layer chromatographic plate (see *Chromatography* ⟨621⟩) coated with a 250-µm layer of microcrystalline cellulose. Allow the spots to dry, and develop the chromatogram in a suitable chamber, using butyl alcohol, previously saturated with 5 *N* ammonium hydroxide, as the solvent, until the solvent front has moved about 15 cm from the point of application. Remove the plate, air-dry, and locate the spots by viewing under short- and long-wavelength ultraviolet light: any spot in the chromatogram from azathioprine, other than the principal spot, is not more intense than the spot in the chromatogram obtained with USP Mercaptopurine RS (3.0%).

Other requirements—It meets the requirements under *Injections* ⟨1⟩ and *Uniformity of Dosage Units* ⟨905⟩.

Assay—

Standard preparation—Transfer 25 mg of USP Azathioprine RS, accurately weighed, to a 50-mL volumetric flask, dissolve in 2.5 mL of 0.1 *N* sodium hydroxide, dilute with water to volume, and mix. Pipet 10.0 mL of this solution into a 50-mL volumetric flask, dilute with 0.1 *N* sulfuric acid to volume, and mix.

Assay preparation—Transfer the contents of 1 vial of Azathioprine Sodium for Injection with the aid of water to a 100-mL volumetric flask, dilute with water to volume, and mix. Pipet 10 mL of this solution into another 100-mL volumetric flask, dilute with 0.1 *N* sulfuric acid to volume, and mix.

Procedure—Transfer 20 mL each of the *Standard preparation* and the *Assay preparation*, separately, to polarographic cells, and deaerate for 10 minutes with nitrogen that previously has been saturated with 0.1 *N* sulfuric acid. Blanket the solution with saturated nitrogen, insert the dropping mercury electrode of a suitable polarograph, and record the polarogram from −0.60 volt to −1.00 volt, using a saturated calomel electrode as the reference electrode. Determine the height of the diffusion current as the difference between the residual current and diffusion current plateau. Calculate the quantity, in mg, of $C_9H_7N_7O_2S$ in the volume of solution from the vial used for the *Assay preparation* by the formula:

$$0.1C(i_d)_U / (i_d)_S,$$

in which *C* is the concentration, in µg per mL, of USP Azathioprine RS in the *Standard preparation*, and $(i_d)_U$ and $(i_d)_S$ are the diffusion currents of the *Assay preparation* and the *Standard preparation*, respectively.

Azeotropic Isopropyl Alcohol—*see* Isopropyl Alcohol, Azeotropic

Sterile Azlocillin Sodium

$C_{20}H_{22}N_5NaO_6S$ 483.47

4-Thia-1-azabicyclo[3.2.0]heptane-2-carboxylic acid, 3,3-dimethyl-7-oxo-6-[[[[(2-oxo-1-imidazolidinyl)carbonyl]amino]phenylacetyl]amino]-, monosodium salt, [2*S*-[2α,5α,6β(*S**)]]-.

Sodium (2*S*,5*R*,6*R*)-3,3-dimethyl-7-oxo-6-[(*R*)-2-(2-oxo-1-imidazolidinecarboxamido)-2-phenylacetamido]-4-thia-1-azabicyclo[3.2.0]heptane-2-carboxylate [*37091-65-9*].

» Sterile Azlocillin Sodium is azlocillin sodium suitable for parenteral use. It has a potency equivalent to not less than 859 μg and not more than 1000 μg of azlocillin ($C_{20}H_{23}N_5O_6S$) per mg, calculated on the anhydrous basis. In addition, where packaged for dispensing, it contains the equivalent of not less than 90.0 percent and not more than 115.0 percent of the labeled amount of azlocillin ($C_{20}H_{23}N_5O_6S$).

Packaging and storage—Preserve in *Containers for Sterile Solids* as described under *Injections* ⟨1⟩.

Reference standard—*USP Azlocillin Sodium Reference Standard*—Do not dry before using.

Constituted solution—Use 10 mL of the constituted solution prepared from Sterile Azlocillin Sodium as the *Test solution.* Prepare a *Comparison solution* by mixing 0.25 mL of 0.00002 *M* sodium chloride, 3.75 mL of water, 5.0 mL of 2.5 *N* nitric acid, and 1.0 mL of 0.1 *N* silver nitrate, and use this solution within 5 minutes. Transfer the *Test solution* and the *Comparison solution* to separate flat-bottom tubes having the same nominal diameter (about 16 mm). View the *Test solution* and the *Comparison solution* vertically against a dull black background in diffused light: the *Test solution* is clear or not more opalescent than the *Comparison solution.*

Identification—Prepare a test solution containing the equivalent of 20 mg of azlocillin per mL. Prepare a Standard solution of USP Azlocillin Sodium RS containing the equivalent of 20 mg of azlocillin per mL. Apply separately 10 μL of each solution on a thin-layer chromatographic plate coated with a 0.25-mm layer of chromatographic silica gel mixture (see *Chromatography* ⟨621⟩). Place the plate in a suitable chromatographic chamber, and develop the chromatogram with the upper phase of a mixture prepared by shaking together 50 mL of normal butyl acetate, 9 mL of butyl alcohol, 25 mL of glacial acetic acid, and 15 mL of a buffer containing 8 mg of monobasic potassium phosphate and 2.2 mg of dibasic sodium phosphate heptahydrate per mL. When the solvent front has moved about three-fourths of the length of the plate, remove the plate from the chamber, and allow to dry. Locate the spots on the plate by exposing it to iodine vapors in a closed chamber for about 10 minutes: the R_f value of the principal spot obtained from the test solution corresponds to that obtained from the Standard solution.

Specific rotation ⟨781⟩: between +170° and +200°, determined in a solution containing 10 mg per mL.

Pyrogen—It meets the requirements of the *Pyrogen Test* ⟨151⟩, the test dose being 1.0 mL per kg of a solution prepared by diluting Sterile Azlocillin Sodium with Sterile Water for Injection to a concentration of 100 mg of azlocillin per mL.

Sterility—It meets the requirements under *Sterility Tests* ⟨71⟩, when tested as directed in the section, *Test Procedures Using Membrane Filtration.*

pH ⟨791⟩: between 6.0 and 8.0, in a solution containing the equivalent of 100 mg of azlocillin per mL.

Water, *Method Ib* ⟨921⟩: not more than 2.5%. [NOTE—Weigh and transfer the specimen in an environment of low humidity to minimize absorption of atmospheric water.]

Particulate matter ⟨788⟩: meets the requirements under *Small-volume Injections.*

Other requirements—It meets the requirements for *Uniformity of Dosage Units* ⟨905⟩ and for *Labeling* under *Injections* ⟨1⟩.

Assay—

Buffer solution—Dissolve 200 g of tris(hydroxymethyl)aminomethane in water to make 1000 mL. Filter before using.

Standard preparation—Dissolve a suitable quantity of USP Azlocillin Sodium RS, accurately weighed, in water to obtain a solution having a known concentration of about 1 mg of azlocillin per mL.

Assay preparation 1—Using a suitable quantity of Sterile Azlocillin Sodium, accurately weighed, proceed as directed under *Standard preparation.*

Assay preparation 2 (where it is represented as being in a single-dose container)—Constitute Sterile Azlocillin Sodium in a volume of water, accurately measured, corresponding to the volume of solvent specified in the labeling. Withdraw all of the withdrawable contents, using a suitable hypodermic needle and syringe, and dilute quantitatively with water to obtain a solution containing about 1 mg of azlocillin per mL.

Assay preparation 3 (where the label states the quantity of azlocillin in a given volume of constituted solution)—Constitute Sterile Azlocillin Sodium in a volume of water, accurately measured, corresponding to the volume of solvent specified in the labeling. Dilute an accurately measured portion of the constituted solution quantitatively with water to obtain a solution containing about 1 mg of azlocillin per mL.

Procedure—Proceed as directed for *Procedure* in the section, *Antibiotics—Hydroxylamine Assay*, under *Automated Methods of Analysis* ⟨16⟩, except to use *Buffer solution* instead of *Acetate buffer* in line 3 of the sample manifold and in the blank manifold to use a 0.03% solution of polyoxyethylene (23) lauryl ether in line 1 instead of *Hydroxylamine hydrochloride solution, Buffer solution* in line 3 instead of 3.3 *N* sulfuric acid, and 3.3 *N* sulfuric acid in line 5 instead of *Acetate buffer.* Calculate the potency, in μg of azlocillin ($C_{20}H_{23}N_5O_6S$) per mg, of the Sterile Azlocillin Sodium taken by the formula:

$$(CP/W)(A_U/A_S),$$

in which *W* is the weight, in mg, of the Sterile Azlocillin Sodium taken in each mL of *Assay preparation 1*, and the other terms are as defined therein. Calculate the quantity, in mg, of azlocillin in the container, or in the portion of constituted solution taken, by the formula:

$$(L/D)(CP/1000)(A_U/A_S),$$

in which *L* is the labeled quantity, in mg, of azlocillin in the container, or in the volume of constituted solution taken, and *D* is the concentration, in mg per mL, of azlocillin in *Assay preparation 2* or in *Assay preparation 3*, on the basis of the labeled quantity in the container, or in the portion of constituted solution taken, respectively, and the extent of dilution.

Aztreonam for Injection

» Aztreonam for Injection is a dry mixture of Sterile Aztreonam and Arginine. It contains not less than 90.0 percent and not more than 105.0 percent of aztreonam ($C_{13}H_{17}N_5O_8S_2$), calculated on the anhydrous and arginine-free basis. Each container contains not less than 90.0 percent and not more than 120.0 percent of the labeled amount of aztreonam ($C_{13}H_{17}N_5O_8S_2$).

Packaging and storage—Preserve in *Containers for Sterile Solids* as described under *Injections* ⟨1⟩.

Reference standards—*USP* L-*Arginine Reference Standard*—Dry at 105° for 3 hours before using. *USP Aztreonam Reference Standard*—Do not dry before using. *USP Open Ring Aztreonam Reference Standard*—Do not dry before using.

Constituted solution—At the time of use, the constituted solution prepared from Aztreonam for Injection meets the requirements for *Constituted Solutions* under *Injections* ⟨1⟩.

Identification—The retention times of the major peaks in the chromatogram of the *Assay preparation* correspond to those of the *Standard preparation*, as obtained in the *Assay*.

Pyrogen—It meets the requirements of the *Pyrogen Test* ⟨151⟩, the test dose being 1.0 mL per kg of a solution prepared by diluting Aztreonam for Injection with Sterile Water for Injection to a concentration of 50 mg of aztreonam per mL.

Sterility—It meets the requirements under *Sterility Tests* ⟨71⟩, when tested as directed in the section, *Test Procedures Using Membrane Filtration*.

pH ⟨791⟩: between 4.5 and 7.5, in a solution containing 100 mg of aztreonam per mL.

Water, *Method I* ⟨921⟩: not more than 2.0%.

Particulate matter ⟨788⟩: meets the requirements under *Small-volume Injections*.

Other requirements—It meets the requirements for *Uniformity of Dosage Units* ⟨905⟩ and *Labeling* under *Injections* ⟨1⟩.

Assay and content of arginine—

Mobile phase—Dissolve 1.15 g of monobasic ammonium phosphate in about 800 mL of water. Adjust with phosphoric acid to a pH of 2.0 ± 0.1, dilute with water to 1000 mL, and mix. Prepare a suitable mixture of this solution and acetonitrile (250:750). Make adjustments if necessary (see *System Suitability* under *Chromatography* ⟨621⟩).

Standard preparation—Dissolve accurately weighed quantities of USP Aztreonam RS and USP L-Arginine RS quantitatively in *Mobile phase* to obtain a solution containing known concentrations of about 0.2 mg of each per mL.

Resolution solution—Prepare a solution in *Mobile phase* containing in each mL about 0.2 mg each of USP Aztreonam RS and USP Open Ring Aztreonam RS.

Assay preparation 1—Weigh accurately 1 container of Aztreonam for Injection. Transfer the contents of the container to a 100-mL volumetric flask. Weigh the empty container, and calculate the weight, in mg, of Aztreonam for Injection removed. Dissolve the powder in the volumetric flask in *Mobile phase*, dilute with *Mobile phase* to volume, and mix. Dilute an accurately measured volume of this solution quantitatively with *Mobile phase* to obtain a solution having a concentration of about 0.2 mg of aztreonam per mL.

Assay preparation 2—Constitute 1 container of Aztreonam for Injection with a volume of water, accurately measured, corresponding to the volume of solvent specified in the labeling, except where the capacity of the container is 100 mL or greater to constitute with 10 mL of water. Withdraw the total withdrawable contents of the container, and dilute quantitatively and stepwise, if necessary, with *Mobile phase* to obtain a solution containing about 0.2 mg of aztreonam per mL.

Chromatographic system (see *Chromatography* ⟨621⟩)—The liquid chromatograph is equipped with a 206-nm detector, a 4.6-mm × 50-cm saturator pre-column containing packing L4, and a 4-mm × 25-cm analytical column containing packing L20. The flow rate is about 1 mL per minute. Chromatograph the *Resolution solution*, and record the responses as directed under *Procedure:* the column efficiency as determined from the aztreonam peak is not less than 1000 theoretical plates, the resolution, *R*, between aztreonam and open ring aztreonam is not less than 2.0, the tailing factor for the aztreonam peak is not more than 2.0, and the relative standard deviation for replicate injections is not more than 2.0%. The relative retention times are about 0.8 for aztreonam and 1.0 for open ring aztreonam.

Procedure—Separately inject equal volumes (about 20 µL) of the *Standard preparation* and the *Assay preparation* into the chromatograph, record the chromatograms, and measure the responses for the major peaks. The relative retention times are about 0.3 for aztreonam and 1.0 for arginine. Calculate the percentage of aztreonam ($C_{13}H_{17}N_5O_8S_2$) in the Aztreonam for Injection taken by the formula:

$$0.1(C_SP_S/C_U)(r_U/r_S),$$

in which C_S is the concentration, in mg per mL, of USP Aztreonam RS in the *Standard preparation*, P_S is the assigned purity, in µg per mg, of the USP Aztreonam RS, C_U is the concentration, in mg per mL, of Aztreonam for Injection in *Assay preparation 1*, based on the weight, in mg, of Aztreonam for Injection removed from the container and the extent of dilution, and r_U and r_S are the aztreonam peak responses obtained from *Assay preparation 1* and the *Standard preparation*, respectively. Calculate the percentage of arginine ($C_6H_{14}N_4O_2$) in the Aztreonam for Injection taken by the formula:

$$100(C_S/C_U)(r_U/r_S),$$

in which C_S is the concentration, in mg per mL, of USP L-Arginine RS in the *Standard preparation*, C_U is the concentration, in mg per mL, of Aztreonam for Injection in *Assay preparation 1*, based on the weight, in mg, of Aztreonam for Injection removed from the container and the extent of dilution, and r_U and r_S are the arginine peak responses obtained from *Assay preparation 1* and the *Standard preparation*, respectively. Calculate the quantity, in mg, of aztreonam ($C_{13}H_{17}N_5O_8S_2$) in the container of Aztreonam for Injection used to prepare *Assay preparation 2* taken by the formula:

$$(C_SP_SL/1000C_U)(r_U/r_S),$$

in which C_S is the concentration, in mg per mL, of USP Aztreonam RS in the *Standard preparation*, P_S is the assigned purity, in µg per mg, of USP Aztreonam RS, *L* is the labeled quantity, in mg, of aztreonam in the container of Aztreonam for Injection, C_U is the concentration, in mg per mL, of aztreonam in *Assay preparation 2*, on the basis of the labeled quantity, in mg, of aztreonam in the container and the extent of dilution, and r_U and r_S are the aztreonam peak responses obtained from *Assay preparation 1* and the *Standard preparation*, respectively.

Sterile Aztreonam

» Sterile Aztreonam is aztreonam suitable for parenteral use. It contains not less than 90.0 percent of aztreonam ($C_{13}H_{17}N_5O_8S_2$).

Packaging and storage—Preserve in *Containers for Sterile Solids* as described under *Injections* ⟨1⟩.

Reference standards—*USP Aztreonam Reference Standard*—Do not dry before using. *USP Aztreonam E-Isomer Reference Standard*—Do not dry before using.

Identification—The infrared absorption spectrum of a dispersion of 0.5 mL of a methanol solution of it containing 3 mg per mL in 200 mg of potassium bromide exhibits maxima only at the same wavelengths as that of a similar preparation of USP Aztreonam RS.

Pyrogen—It meets the requirements of the *Pyrogen Test* ⟨151⟩, the test dose being 1.0 mL per kg of a solution prepared by diluting Sterile Aztreonam with Sterile Water for Injection containing 39 mg of pyrogen-free arginine per mL to a concentration of 50 mg of aztreonam per mL.

Sterility—It meets the requirements under *Sterility Tests* ⟨71⟩, when tested as directed in the section, *Test Procedures Using Membrane Filtration, Fluid A* to each 1000 mL of which has been added 23.4 g of sterile arginine being used.

Water, *Method I* ⟨921⟩: not more than 2.0%.

Residue on ignition ⟨281⟩: not more than 0.1%, the charred residue being moistened with 2 mL of nitric acid and 5 drops of sulfuric acid.

Heavy metals, *Method II* ⟨241⟩: 0.003%.

Assay—

Mobile phase—Dissolve 6.8 g of monobasic potassium phosphate in water to make 1000 mL, and adjust with 1 *M* phosphoric acid to a pH of 3.0 ± 0.1. Prepare a suitable mixture of this solution and methanol (4:1). Make adjustments if necessary (see *System Suitability* under *Chromatography* ⟨621⟩).

Standard preparation—Dissolve an accurately weighed quantity of USP Aztreonam RS quantitatively in *Mobile phase* to obtain a solution having a known concentration of about 1 mg per mL.

Resolution solution—Prepare a solution in *Mobile phase* containing in each mL about 0.2 mg each of USP Aztreonam RS and USP Aztreonam E-Isomer RS.

Assay preparation—Transfer about 25 mg of Sterile Aztreonam, accurately weighed, to a 25-mL volumetric flask, dissolve in *Mobile phase*, dilute with *Mobile phase* to volume, and mix.

Chromatographic system (see *Chromatography* ⟨621⟩)—The liquid chromatograph is equipped with a 270-nm detector, a 2-mm × 10-cm pre-column containing packing L2, and a 4.6-mm × 30-cm analytical column containing packing L1. The flow rate is about 1.5 mL per minute. Chromatograph the *Resolution solution*, and record the responses as directed under *Procedure:* the column efficiency as determined from the aztreonam peak is not less than 1000 theoretical plates, the resolution, *R*, between aztreonam and aztreonam E-isomer is not less than 2.0, the tailing factor for the aztreonam peak is not more than 2.0, and the relative standard deviation for replicate injections is not more than 2.0%.

Procedure—Separately inject equal volumes (about 20 µL) of the *Standard preparation* and the *Assay preparation* into the chromatograph, record the chromatograms, and measure the responses for the major peaks. The relative retention times are about 0.6 for aztreonam and 1.0 for aztreonam E-isomer. Calculate the percentage of aztreonam ($C_{13}H_{17}N_5O_8S_2$) in the Sterile Aztreonam taken by the formula:

$$2.5(C_SP_S/W)(r_U/r_S),$$

in which C_S is the concentration, in mg per mL, of USP Aztreonam RS in the *Standard preparation*, P_S is the assigned purity, in µg per mg, of the USP Aztreonam RS, W is the weight, in mg, of Sterile Aztreonam taken to prepare the *Assay preparation*, and r_U and r_S are the aztreonam peak responses obtained from the *Assay preparation* and the *Standard preparation*, respectively.

Bacampicillin Hydrochloride

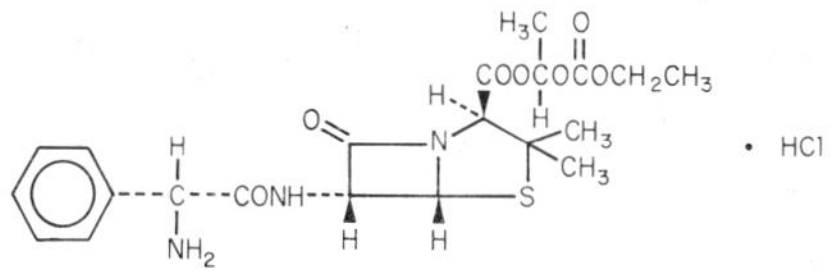

$C_{21}H_{27}N_3O_7S.HCl$ 501.98

4-Thia-1-azabicyclo[3.2.0]heptane-2-carboxylic acid, 6-[(aminophenylacetyl)amino]-3,3-dimethyl-7-oxo-, 1-[(ethoxycarbonyl)oxy]ethyl ester, monohydrochloride, [2S-[2α,5α,6β(*S**)]]-.

(2*S*,5*R*,6*R*)-6-[(*R*)-(2-Amino-2-phenylacetamido)]-3,3-dimethyl-7-oxo-4-thia-1-azabicyclo[3.2.0]heptane-2-carboxylic acid ester with ethyl 1-hydroxyethyl carbonate, monohydrochloride [*37661-08-8*].

» Bacampicillin Hydrochloride has a potency of not less than 623 µg and not more than 727 µg of ampicillin ($C_{16}H_{19}N_3O_4S$) per mg.

Packaging and storage—Preserve in tight containers.

Reference standards—*USP Ampicillin Reference Standard*—Do not dry before using. *USP Bacampicillin Hydrochloride Reference Standard*—Do not dry before using.

Identification—Prepare a test solution of it in alcohol containing 2 mg per mL. Prepare a Standard solution of USP Bacampicillin Hydrochloride RS in alcohol containing 2 mg per mL. Apply two 5-µL portions of the test solution 4.0 cm apart on a thin-layer chromatographic plate coated with a 0.25-mm layer of chromatographic silica gel mixture (see *Chromatography* ⟨621⟩). After the spots dry, apply two 5-µL portions of the Standard solution, one midway between the test solution spots and the other on one of the test solution spots. Allow the spots to dry, place the plate in a suitable chromatographic chamber, and develop the chromatogram in a solvent system consisting of a mixture of methylene chloride, chloroform, and alcohol (10:1:1). When the solvent front has moved about three-fourths of the length of the plate, remove the plate from the chamber, and allow to dry. Spray the plate with a spray reagent containing 1 g of ninhydrin and 1 mL of pyridine in each 100 mL of solution in butyl alcohol, and heat at 100° for 10 minutes: bacampicillin appears as a purple spot, and the R_f values of the spots from the test solution and from the combined test solution and Standard solution, respectively, correspond to the R_f value of the spot obtained from the Standard solution.

pH ⟨791⟩: between 3.0 and 4.5, in a solution containing 20 mg per mL.

Water, *Method I* ⟨921⟩: not more than 1.0%.

Assay—

Standard preparation—Prepare as directed for *Standard preparation* under *Iodometric Assay—Antibiotics* ⟨425⟩, using USP Ampicillin RS.

Assay preparation—Proceed with Bacampicillin Hydrochloride as directed for *Assay preparation* under *Iodometric Assay—Antibiotics* ⟨425⟩, using the *Solvent* and *Final Concentration* specified for ampicillin.

Procedure—Proceed as directed for *Procedure* under *Iodometric Assay—Antibiotics* ⟨425⟩. Calculate the potency, in µg of ampicillin ($C_{16}H_{19}N_3O_4S$) per mg, of the Bacampicillin Hydrochloride taken by the formula:

$$(F)(B - I)/(2D),$$

in which D is the concentration, in mg per mL, of the *Assay preparation*, based on the weight of Bacampicillin Hydrochloride taken and the extent of dilution.

Bacampicillin Hydrochloride for Oral Suspension

» Bacampicillin Hydrochloride for Oral Suspension contains an amount of Bacampicillin Hydrochloride equivalent to not less than 90.0 percent and not more than 125.0 percent of the labeled amount of ampicillin ($C_{16}H_{19}N_3O_4S$) when constituted as directed. It contains one or more suitable buffers, colors, flavors, suspending agents, and sweetening ingredients.

Packaging and storage—Preserve in tight containers.

Reference standards—*USP Ampicillin Reference Standard*—Do not dry before using. *USP Bacampicillin Hydrochloride Reference Standard*—Do not dry before using.

Identification—Constitute Bacampicillin Hydrochloride for Oral Suspension as directed in the labeling. Transfer a portion of the resulting suspension, equivalent to about 140 mg of ampicillin, to a 100-mL volumetric flask, add 70 mL of alcohol, shake by mechanical means for 30 minutes, dilute with alcohol to volume, and mix: the solution so obtained responds to the *Identification test* under *Bacampicillin Hydrochloride*.

Uniformity of dosage units ⟨905⟩—

FOR SOLID PACKAGED IN SINGLE-UNIT CONTAINERS: meets the requirements.

pH ⟨791⟩: between 6.5 and 8.0, in the suspension constituted as directed in the labeling.

Loss on drying ⟨731⟩—Dry about 100 mg, accurately weighed, in a capillary-stoppered bottle in vacuum at 60° for 3 hours: it loses not more than 2.0% of its weight.

Assay—

Standard preparation—Using USP Ampicillin RS, prepare as directed for *Standard preparation* under *Iodometric Assay—Antibiotics* ⟨425⟩.

Assay preparation—Transfer an accurately measured volume of Bacampicillin Hydrochloride for Oral Suspension, constituted as directed in the labeling and free from bubbles, equivalent to about 87.5 mg of ampicillin, to a 250-mL volumetric flask. Add 200 mL of a solvent mixture consisting of alcohol and 0.1 *M* phosphoric acid (4:1). Shake by mechanical means for 30 minutes, dilute with the same solvent mixture to volume, and mix. Centrifuge a portion of the resulting suspension. Pipet 4.0 mL of the clear solution so obtained into each of two glass-stoppered, 125-mL conical flasks.

Procedure—Proceed as directed for *Procedure* under *Iodometric Assay—Antibiotics* ⟨425⟩. Calculate the quantity, in mg, of $C_{16}H_{19}N_3O_4S$ in each mL of the constituted Bacampicillin Hydrochloride for Oral Suspension taken by the formula:

$$(0.0625F)(B - I)/V,$$

in which V is the volume, in mL, of constituted Bacampicillin Hydrochloride for Oral Suspension taken, and the other terms are as defined therein.

Bacampicillin Hydrochloride Tablets

» Bacampicillin Hydrochloride Tablets contain the equivalent of not less than 90.0 percent and not more than 125.0 percent of the labeled amount of ampicillin ($C_{16}H_{19}N_3O_4S$).

Packaging and storage—Preserve in tight containers.

Reference standards—*USP Ampicillin Reference Standard*—Do not dry before using. *USP Bacampicillin Hydrochloride Reference Standard*—Do not dry before using.

Identification—To a portion of powdered Tablets add alcohol to obtain a solution containing the equivalent of 2 mg of ampicillin per mL: the solution so obtained responds to the *Identification test* under *Bacampicillin Hydrochloride.*

Dissolution ⟨711⟩—

Medium: water; 900 mL.

Apparatus 2: 75 rpm.

Time: 30 minutes.

Standard preparation—Dissolve an accurately weighed quantity of USP Ampicillin RS in water to obtain a solution having a known concentration of about 0.3 mg per mL.

Procedure—Determine the amount of ampicillin ($C_{16}H_{19}N_3O_4S$) dissolved as directed for *Procedure* in the section, *Antibiotics—Hydroxylamine Assay* under *Automated Methods of Analysis* ⟨16⟩.

Tolerances—Not less than 85% (*Q*) of the labeled amount of $C_{16}H_{19}N_3O_4S$ is dissolved in 30 minutes.

Uniformity of dosage units ⟨905⟩: meet the requirements.

Water, *Method I* ⟨921⟩: not more than 2.5%.

Assay—

Standard preparation—Prepare as directed for *Standard preparation* under *Iodometric Assay—Antibiotics* ⟨425⟩, using USP Ampicillin RS.

Assay preparation—Dissolve not less than 5 Bacampicillin Hydrochloride Tablets in an accurately measured volume of water, and dilute an accurately measured volume of this stock solution quantitatively and stepwise with water to obtain an *Assay preparation* containing the equivalent of about 1.25 mg of ampicillin per mL.

Procedure—Proceed as directed for *Procedure* under *Iodometric Assay—Antibiotics* ⟨425⟩. Calculate the quantity, in mg, of ampicillin ($C_{16}H_{19}N_3O_4S$) in each Tablet taken by the formula:

$$(T/D)(F/2000)(B - I),$$

in which T is the labeled quantity, in mg, of ampicillin in each Tablet, and D is the concentration, in mg per mL, of ampicillin in the *Assay preparation*, on the basis of the labeled quantity in each Tablet and the extent of dilution.

Bacitracin

Bacitracin.
Bacitracin [*1405-87-4*].

» Bacitracin is a polypeptide produced by the growth of an organism of the *licheniformis* group of *Bacillus subtilis* (Fam. Bacillaceae). It has a potency of not less than 40 Bacitracin Units per mg.

Packaging and storage—Preserve in tight containers, and store in a cool place.

Labeling—Where it is packaged for prescription compounding, label it to indicate that it is not sterile and that the potency cannot be assured for longer than 60 days after opening, and to state the number of Bacitracin Units per milligram.

Reference standard—*USP Bacitracin Zinc Reference Standard*—Dry in vacuum at a pressure not exceeding 5 mm of mercury at 60° for 3 hours before using.

Identification—On a suitable thin-layer chromatographic plate (see *Chromatography* ⟨621⟩), coated with a 0.25-mm layer of chromatographic silica gel mixture, apply 1 µL each of a solution of Bacitracin containing 6.0 mg per mL in edetate disodium solution (1 in 100) and a similar solution of USP Bacitracin Zinc RS. Allow the spots to dry, and develop the chromatogram in a solvent system consisting of a mixture of butyl alcohol, glacial acetic acid, water, pyridine, and alcohol (60:15:10:6:5), equilibrated in the chamber for 30 minutes, until the solvent front has moved about three-fourths of the length of the plate. Remove the plate from the developing chamber, mark the solvent front, and allow the solvent to evaporate. Locate the spots by spraying the plate lightly with a 1 in 100 solution of triketohydrindene hydrate in a mixture of butyl alcohol and pyridine (99:1). Heat the plate at about 110° for about 5 minutes: the R_f value of the principal spot obtained from the solution under test corresponds to that obtained from the Standard solution.

pH ⟨791⟩: between 5.5 and 7.5, in a solution containing 10,000 Bacitracin Units per mL.

Loss on drying ⟨731⟩—Dry about 100 mg in a capillary-stoppered bottle in vacuum at a pressure not exceeding 5 mm of mercury at 60° for 3 hours: it loses not more than 5.0% of its weight.

Assay—Proceed with Bacitracin as directed under *Antibiotics—Microbial Assays* ⟨81⟩.

Bacitracin Ointment

» Bacitracin Ointment is Bacitracin in an anhydrous ointment base. It contains not less than 90.0 percent and not more than 140.0 percent of the labeled amount of bacitracin. It may contain a suitable anesthetic.

Packaging and storage—Preserve in well-closed containers containing not more than 60 g, unless labeled solely for hospital use, preferably at controlled room temperature.

Reference standard—*USP Bacitracin Zinc Reference Standard*—Dry in vacuum at a pressure not exceeding 5 mm of mercury at 60° for 3 hours before using.

Identification—Shake a quantity of Ointment, equivalent to about 2500 USP Bacitracin Units, with 20 mL of chloroform, add 5 mL of 0.1 *N* hydrochloric acid, shake vigorously, centrifuge, and use the clear supernatant liquid as the test solution. On a suitable thin-layer chromatographic plate coated with a 0.25-mm layer of chromatographic silica gel (see *Chromatography* ⟨621⟩) apply separately 10 µL of the test solution and 10 µL of a Standard bacitracin solution of USP Bacitracin Zinc RS in 0.1 *N* hydrochloric acid containing 500 USP Bacitracin Units per mL. Place the plate in a suitable chromatographic chamber, and develop the chromatogram in a solvent system consisting of a mixture of isopropyl alcohol, water, and ammonium hydroxide (24:17:3) until the solvent front has moved about three-fourths of the length of the plate. Remove the plate from the chamber, and dry at 105° for 5 minutes. Spray the plate with a 1 in 200 solution of ninhydrin in butyl alcohol, and heat the plate at 105° for 15 minutes: the R_f value of the principal spot in the chromatogram obtained from the test solution corresponds to that of the principal spot in the chromatograms obtained from the Standard solution.

Minimum fill ⟨755⟩: meets the requirements.

Water, *Method I* ⟨921⟩: not more than 0.5%, 20 mL of a mixture of carbon tetrachloride, chloroform, and methanol (2:2:1) being used in place of methanol in the titration vessel.

Assay—Proceed as directed under *Antibiotics—Microbial Assays* ⟨81⟩, using an accurately weighed portion of Bacitracin Ointment shaken with about 50 mL of ether in a separator, and extracted with four 20-mL portions of *Buffer No. 1*. Combine the buffer extracts, and dilute with *Buffer No. 1* to an appropriate volume to obtain a stock solution. Add sufficient 0.01 *N* hydrochloric acid to an accurately measured portion of the stock solution so that the amount of hydrochloric acid in the *Test Dilution* will be the same as in the median dose level of the Standard, and dilute quantitatively with *Buffer No. 1* to obtain a *Test Dilution* having a bacitracin concentration assumed to be equal to the median dose level of the Standard.

Bacitracin Ointment, Neomycin and Polymyxin B Sulfates and—*see* Neomycin and Polymyxin B Sulfates and Bacitracin Ointment

Bacitracin Ointment, Neomycin Sulfate and—*see* Neomycin Sulfate and Bacitracin Ointment

Bacitracin Ophthalmic Ointment

» Bacitracin Ophthalmic Ointment is a sterile preparation of Bacitracin in an anhydrous ointment base. It contains not less than 90.0 percent and not more than 140.0 percent of the labeled amount of bacitracin.

Packaging and storage—Preserve in collapsible ophthalmic ointment tubes.

Reference standard—*USP Bacitracin Zinc Reference Standard*—Dry in vacuum at a pressure not exceeding 5 mm of mercury at 60° for 3 hours before using.

Identification—It responds to the *Identification test* under *Bacitracin Ointment.*

Sterility—It meets the requirements under *Sterility Tests* ⟨71⟩.

Water, *Method I* ⟨921⟩: not more than 0.5%, 20 mL of a mixture of carbon tetrachloride, chloroform, and methanol (2:2:1) being used in place of methanol in the titration vessel.

Metal particles—It meets the requirements of the test for *Metal Particles in Ophthalmic Ointments* ⟨751⟩.

Assay—Proceed with Bacitracin Ophthalmic Ointment as directed in the *Assay* under *Bacitracin Ointment.*

Bacitracin Ophthalmic Ointment, Neomycin and Polymyxin B Sulfates and—*see* Neomycin and Polymyxin B Sulfates and Bacitracin Ophthalmic Ointment

Sterile Bacitracin

» Sterile Bacitracin has a potency of not less than 50 Bacitracin Units per mg. In addition, where packaged for dispensing, it contains not less than 90.0 percent and not more than 115.0 percent of the labeled amount of bacitracin.

Packaging and storage—Preserve in *Containers for Sterile Solids* as described under *Injections* ⟨1⟩, and store in a cool place.

Reference standard—*USP Bacitracin Zinc Reference Standard*—Dry in vacuum at a pressure not exceeding 5 mm of mercury at 60° for 3 hours before using.

Constituted solution—At the time of use, the constituted solution prepared from Sterile Bacitracin meets the requirements for *Constituted Solutions* under *Injections* ⟨1⟩.

Pyrogen—It meets the requirements of the *Pyrogen Test* ⟨151⟩, the test dose being 1.0 mL per kg of a solution in pyrogen-free saline TS containing 300 Bacitracin Units per mL.

Sterility—It meets the requirements under *Sterility Tests* ⟨71⟩, when tested as directed in the section, *Test Procedures Using Membrane Filtration.*

Residue on ignition ⟨281⟩: not more than 3.0%, the charred residue being moistened with 2 mL of nitric acid and 5 drops of sulfuric acid.

Heavy metals, *Method II* ⟨231⟩: not more than 0.003%.

Other requirements—It responds to the *Identification test* and meets the requirements of the tests for *pH*, and *Loss on drying* under *Bacitracin.* Where packaged for dispensing, it meets the requirements under *Injections* ⟨1⟩ and *Uniformity of Dosage Units* ⟨905⟩. Where intended for use in preparing sterile ophthalmic dosage forms, it is exempt from the requirements for *Pyrogen, Residue on ignition,* and *Heavy metals.*

Assay—

Assay preparation 1—Using a suitable quantity of Sterile Bacitracin, accurately weighed, prepare as directed for the Bacitracin Zinc RS under *Antibiotics—Microbial Assays* ⟨81⟩.

Assay preparation 2 (where it is packaged for dispensing)—Constitute 1 container of Sterile Bacitracin as directed in the labeling. Using a suitable hypodermic needle and syringe, withdraw the contents of the container, and dilute quantitatively with *Buffer No. 1* to obtain a solution containing about 100 Bacitracin Units per mL.

Assay preparation 3 (where the label states the number of Bacitracin Units in a given volume of constituted solution)—Constitute 1 container of Sterile Bacitracin as directed in the labeling. Dilute an accurately measured volume of the constituted solution quantitatively with *Buffer No. 1* to obtain a solution containing about 100 Bacitracin Units per mL.

Procedure—Proceed as directed under *Antibiotics—Microbial Assays* ⟨81⟩, using an accurately measured volume of *Assay preparation.* Add sufficient 0.01 *N* hydrochloric acid to the *Assay preparation* so that the amount of hydrochloric acid in the *Test Dilution* will be the same as in the median dose level of the Standard, and dilute quantitatively with *Buffer No. 1* to obtain a *Test Dilution* having a bacitracin concentration assumed to be equal to the median dose level of the Standard.

Bacitracin and Polymyxin B Sulfate Topical Aerosol

» Bacitracin and Polymyxin B Sulfate Topical Aerosol is a suspension of Bacitracin and Polymyxin B Sulfate in a suitable vehicle, packaged in a pressurized container with a suitable inert propellant. It contains not less than 90.0 percent and not more than 130.0 percent of the labeled amounts of bacitracin and polymyxin B.

Packaging and storage—Preserve in pressurized containers, and avoid exposure to excessive heat.

Reference standards—*USP Polymyxin B Sulfate Reference Standard*—Dry in vacuum at a pressure not exceeding 5 mm of mercury at 60° for 3 hours before using. *USP Bacitracin Zinc Reference Standard*—Dry in vacuum at a pressure not exceeding 5 mm of mercury at 60° for 3 hours before using.

Note—Prepare the specimen for the following tests and assays as follows: Maintain the container in the inverted position throughout this procedure. Store the container in a freezer at −70° for 16 to 24 hours. Remove the container from the freezer, promptly puncture the container, and allow the propellant to volatilize. Open the container, and mix the contents.

Identification—A portion of the contents of 1 container, prepared as directed above, responds to the *Identification test* under *Bacitracin Zinc and Polymyxin B Sulfate Ointment.*

Water, *Method I* ⟨921⟩: not more than 0.5%, an accurately weighed portion of the contents of 1 container, prepared as directed above, being used, and 20 mL of a mixture of carbon tetrachloride, chloroform, and methanol (2:2:1) being used in place of methanol in the titration vessel.

Assay for bacitracin—Proceed as directed for bacitracin under *Antibiotics—Microbial Assays* ⟨81⟩, using an accurately weighed portion of the contents of 1 container, prepared as directed above, equivalent to about 500 USP Bacitracin Units. Transfer to a suitable separator containing about 50 mL of ether, and extract with three 25-mL portions of *Buffer No. 1.* Combine the buffer extracts in a 100-mL volumetric flask, dilute with *Buffer No. 1* to volume, and mix. Add sufficient 0.01 *N* hydrochloric acid to an accurately measured volume of this solution so that the amount of hydrochloric acid in the *Test Dilution* will be the same as in the median dose level of the Standard, and dilute quantitatively with *Buffer No. 1* to obtain a *Test Dilution* having a bacitracin concentration assumed to be equal to the median dose level of the Standard.

Assay for polymyxin B—Proceed as directed for polymyxin B under *Antibiotics—Microbial Assays* ⟨81⟩. Transfer an accurately weighed portion of the contents of 1 container, prepared as directed above, equivalent to about 5000 USP Polymyxin B Units, to a suitable separator containing about 50 mL of ether, and extract with three 25-mL portions of *Buffer No. 6.* Combine the buffer extracts in a 100-mL volumetric flask, dilute with *Buffer No. 6* to volume, and mix. Dilute an accurately measured volume of this solution quantitatively and stepwise with *Buffer No. 6* to obtain a *Test Dilution* having a concentration of polymyxin B assumed to be equal to the median dose level of the Standard.

Bacitracin, and Hydrocortisone Acetate Ointment, Neomycin and Polymyxin B Sulfates,—*see* Neomycin and Polymyxin B Sulfates, Bacitracin, and Hydrocortisone Acetate Ointment

Bacitracin, and Hydrocortisone Acetate Ophthalmic Ointment, Neomycin and Polymyxin B Sulfates,—*see* Neomycin and Polymyxin B Sulfates, Bacitracin, and Hydrocortisone Acetate Ophthalmic Ointment

Soluble Bacitracin Methylene Disalicylate

» Soluble Bacitracin Methylene Disalicylate is a mixture of bacitracin methylene disalicylate and Sodium Bicarbonate. It has a potency of not less than 8 Bacitracin Units per mg, calculated on the dried basis.

Packaging and storage—Preserve in well-closed containers.

Labeling—Label it to indicate that it is for veterinary use only.

Reference standard—*USP Bacitracin Zinc Reference Standard*—Dry in vacuum at a pressure not exceeding 5 mm of mercury at 60° for 3 hours before using.

Loss on drying ⟨731⟩—Dry about 100 mg, accurately weighed, in a capillary-stoppered bottle in vacuum at a pressure not exceeding 5 mm of mercury at 60° for 3 hours: it loses not more than 8.5% of its weight.

pH ⟨791⟩: between 8.0 and 9.5, in a solution containing 25 mg of specimen per mL.

Assay—Transfer an accurately weighed quantity of Soluble Bacitracin Methylene Disalicylate to a high-speed glass blender jar. Add 99.0 mL of sodium bicarbonate solution (1 in 50) and 1.0 mL of polysorbate 80, and blend for about 3 minutes. Proceed as directed for bacitracin under *Antibiotics—Microbial Assays* ⟨81⟩, adding a sufficient volume of 0.01 *N* hydrochloric acid to an accurately measured volume of this solution so that the amount of hydrochloric acid in the *Test Dilution* will be the same as in the median dose level of the Standard, and diluting quantitatively with *Buffer No. 1* to obtain a *Test Dilution* having a concentration of bacitracin assumed to be equal to the median dose level of the Standard.

Bacitracin Methylene Disalicylate Soluble Powder

» Bacitracin Methylene Disalicylate Soluble Powder contains not less than 90.0 percent and not more than 120.0 percent of the labeled amount of bacitracin.

Packaging and storage—Preserve in tight containers.

Labeling—Label it to indicate that it is for veterinary use only. Label it to state the content of bacitracin in terms of grams per pound, each gram of bacitracin being equivalent to 42,000 Bacitracin Units.

Reference standard—*USP Bacitracin Zinc Reference Standard*—Dry in vacuum at a pressure not exceeding 5 mm of mercury at 60° for 3 hours before using.

Loss on drying ⟨731⟩—Dry about 100 mg, accurately weighed, in a capillary-stoppered bottle in vacuum at a pressure not exceeding 5 mm of mercury at 60° for 3 hours: it loses not more than 8.5% of its weight.

pH ⟨791⟩: between 8.0 and 9.5, in a solution containing 50 mg of specimen per mL.

Assay—Proceed with Bacitracin Methylene Disalicylate Soluble Powder as directed in the *Assay* under *Soluble Bacitracin Methylene Disalicylate.*

Bacitracin Zinc

Bacitracins, zinc complex.
Bacitracins zinc complex [*1405-89-6*].

» Bacitracin Zinc is the zinc salt of a kind of bacitracin or a mixture of two or more such salts. It has a potency of not less than 40 Bacitracin Units per mg. It contains not less than 2.0 percent and not more than 10.0 percent of zinc (Zn), calculated on the dried basis.

Packaging and storage—Preserve in tight containers, and store in a cool place.

Labeling—Label it to indicate that it is to be used in the manufacture of nonparenteral drugs only. Where it is packaged for prescription compounding, label it to indicate that it is not sterile and that the potency cannot be assured for longer than 60 days after opening, and to state the number of Bacitracin Units per milligram.

Reference standard—*USP Bacitracin Zinc Reference Standard*—Dry in vacuum at a pressure not exceeding 5 mm of mercury at 60° for 3 hours before using.

Identification—It responds to the *Identification test* under *Bacitracin.*

pH ⟨791⟩: between 6.0 and 7.5, in a (saturated) solution containing approximately 100 mg per mL.

Loss on drying ⟨731⟩—Dry about 100 mg in a capillary-stoppered bottle in vacuum at 60° for 3 hours: it loses not more than 5.0% of its weight.

Zinc content—[NOTE—The *Standard preparations* and the *Test preparation* may be diluted quantitatively with 0.001 *N* hydrochloric acid, if necessary, to obtain solutions of suitable concentrations, adaptable to the linear or working range of the instrument.]

Standard preparations—Transfer 3.11 g of zinc oxide, accurately weighed, to a 250-mL volumetric flask, add 80 mL of 1 *N* hydrochloric acid, warm to dissolve, cool, dilute with water to volume, and mix. This solution contains 10 mg of zinc per mL. Further dilute this solution with 0.001 *N* hydrochloric acid to obtain *Standard preparations* containing 0.5, 1.5, and 2.5 µg of zinc per mL, respectively.

Test preparation—Transfer about 200 mg of Bacitracin Zinc, accurately weighed, to a 100-mL volumetric flask. Dissolve in 0.01 *N* hydrochloric acid, dilute with the same solvent to volume, and mix. Pipet 2 mL of this solution into a 200-mL volumetric flask, dilute with 0.001 *N* hydrochloric acid to volume, and mix.

Procedure—Concomitantly determine the absorbances of the *Standard preparations* and the *Test preparation* at the zinc resonance line of 213.8 nm, with a suitable atomic absorption spectrophotometer (see *Spectrophotometry and Light-scattering* ⟨851⟩), equipped with a zinc hollow-cathode lamp and an air-acetylene flame, using 0.001 *N* hydrochloric acid as the blank. Plot the absorbances of the *Standard preparations* versus concentration, in µg per mL, of zinc, and draw the straight line best fitting the three plotted points. From the graph so obtained, determine the concentration, in µg per mL, of zinc in the *Test preparation.* Calculate the content of zinc, in percent, in the portion of Bacitracin Zinc taken by the formula:

$$1000C/W,$$

in which C is the concentration in µg per mL, of zinc in the *Test preparation*, and W is the weight, in mg, of the portion of Bacitracin Zinc taken.

Assay—Proceed with Bacitracin Zinc as directed under *Antibiotics—Microbial Assays* ⟨81⟩.

Bacitracin Zinc Topical Aerosol, Polymyxin B Sulfate and—*see* Polymyxin B Sulfate and Bacitracin Zinc Topical Aerosol

Bacitracin Zinc Ointment

» Bacitracin Zinc Ointment is Bacitracin Zinc in an anhydrous ointment base. It contains not less than 90.0 percent and not more than 140.0 percent of the labeled amount of bacitracin.

Packaging and storage—Preserve in well-closed containers containing not more than 60 g, unless labeled solely for hospital use, preferably at controlled room temperature.

Reference standard—*USP Bacitracin Zinc Reference Standard*—Dry in vacuum at a pressure not exceeding 5 mm of mercury at 60° for 3 hours before using.

Identification—It responds to the *Identification test* under *Bacitracin Ointment.*

Minimum fill ⟨755⟩: meets the requirements.

Water, *Method I* ⟨921⟩: not more than 0.5%, 20 mL of a mixture of carbon tetrachloride, chloroform, and methanol (2:2:1) being used in place of methanol in the titration vessel.

Assay—Proceed as directed under *Antibiotics—Microbial Assays* ⟨81⟩, using an accurately weighed portion of Bacitracin Zinc Ointment shaken with about 50 mL of ether in a separator, and extracted with four 20-mL portions of 0.01 *N* hydrochloric acid. Combine the acid extracts, and dilute with 0.01 *N* hydrochloric acid to an appropriate volume to obtain a stock solution. Dilute this stock solution quantitatively and stepwise with *Buffer No. 1* to obtain a *Test Dilution* having a concentration assumed to be equal to the median dose level of the Standard, adding additional hydrochloric acid to each test dilution of the Standard to obtain the same concentration of hydrochloric acid as in the *Test Dilution.*

Bacitracin Zinc Ointment, Neomycin and Polymyxin B Sulfates and—*see* Neomycin and Polymyxin B Sulfates and Bacitracin Zinc Ointment

Bacitracin Zinc Ointment, Neomycin Sulfate and—*see* Neomycin Sulfate and Bacitracin Zinc Ointment

Bacitracin Zinc Ophthalmic Ointment, Neomycin and Polymyxin B Sulfates and—*see* Neomycin and Polymyxin B Sulfates and Bacitracin Zinc Ophthalmic Ointment

Bacitracin Zinc Soluble Powder

» Bacitracin Zinc Soluble Powder is a mixture of bacitracin zinc and zinc proteinates. It contains not less than 90.0 percent and not more than 120.0 percent of the labeled amount of bacitracin.

Packaging and storage—Preserve in tight containers.

Labeling—Label it to indicate that it is for veterinary use only. Label it to state the content of bacitracin in terms of grams per pound, each gram of bacitracin being equivalent to 42,000 Bacitracin Units.

Reference standard—*USP Bacitracin Zinc Reference Standard*—Dry in vacuum at a pressure not exceeding 5 mm of mercury at 60° for 3 hours before using.

Loss on drying ⟨731⟩—Dry about 100 mg, accurately weighed, in a capillary-stoppered bottle in vacuum at a pressure not exceeding 5 mm of mercury at 60° for 3 hours: it loses not more than 5.0% of its weight.

Zinc content—Using Bacitracin Zinc Soluble Powder, proceed as directed for *Zinc content* under *Bacitracin Zinc.* Calculate the zinc content, in g, in relation to each 42,000 Bacitracin Units in the specimen by the formula:

$$280{,}000C/WA,$$

in which A is the bacitracin content of the specimen, in Bacitracin Units per g, and the other terms are as defined therein: it contains not more than 2.0 g of Zn for each 42,000 Bacitracin Units.

Assay—Dissolve an accurately weighed quantity of Bacitracin Zinc Soluble Powder quantitatively in 0.01 *N* hydrochloric acid to obtain a stock solution containing about 100 Bacitracin Units per mL. Proceed as directed under *Antibiotics—Microbial Assays* ⟨81⟩, using an accurately measured volume of this stock solution diluted quantitatively and stepwise with *Buffer No. 1* to obtain a *Test Dilution* having a concentration assumed to be equal to the median dose level of the Standard. In preparing each test dilution of the Standard, add additional hydrochloric acid to each to obtain the same concentration of hydrochloric acid as in the *Test Dilution.*

Bacitracin Zinc Topical Powder, Polymyxin B Sulfate and—*see* Polymyxin B Sulfate and Bacitracin Zinc Topical Powder

Sterile Bacitracin Zinc

» Sterile Bacitracin Zinc is Bacitracin Zinc suitable for use in the manufacture of sterile topical dosage forms. It has a potency of not less than 40 Bacitracin Units per mg.

Packaging and storage—Preserve in *Containers for Sterile Solids* as described under *Injections* ⟨1⟩, and store in a cool place.

Labeling—Label it to indicate that it is to be used in the manufacture of topical drugs only.

Reference standard—*USP Bacitracin Zinc Reference Standard*—Dry in vacuum at a pressure not exceeding 5 mm of mercury at 60° for 3 hours before using.

Sterility—It meets the requirements under *Sterility Tests* ⟨71⟩, when tested as directed in the section, *Test Procedures Using Membrane Filtration*, except to use *Fluid A* to each liter of which has been added 20 g of edetate disodium.

Other requirements—It responds to the *Identification test* and meets the requirements for *pH, Loss on drying, Zinc content*, and *Assay* under *Bacitracin Zinc.*

Bacitracin Zinc, and Hydrocortisone Ointment, Neomycin and Polymyxin B Sulfates—*see* Neomycin and Polymyxin B Sulfates, Bacitracin Zinc, and Hydrocortisone Ointment

Bacitracin Zinc, and Hydrocortisone Ophthalmic Ointment, Neomycin and Polymyxin B Sulfates,—*see* Neomycin and Polymyxin B Sulfates, Bacitracin Zinc, and Hydrocortisone Ophthalmic Ointment

Bacitracin Zinc, and Hydrocortisone Acetate Ophthalmic Ointment, Neomycin and Polymyxin B Sulfates,—*see* Neomycin and Polymyxin B Sulfates, Bacitracin Zinc, and Hydrocortisone Acetate and Ophthalmic Ointment

Bacitracin Zinc, and Lidocaine Ointment, Neomycin and Polymyxin B Sulfates,—*see* Neomycin and Polymyxin B Sulfates, Bacitracin Zinc, and Lidocaine Ointment

Bacitracin Zinc and Polymyxin B Sulfate Ointment

» Bacitracin Zinc and Polymyxin B Sulfate Ointment contains the equivalent of not less than 90.0 percent and not more than 130.0 percent of the labeled amounts of bacitracin and polymyxin B.

Packaging and storage—Preserve in well-closed, light-resistant containers.

Reference standards—*USP Bacitracin Zinc Reference Standard*—Dry in vacuum at a pressure not exceeding 5 mm of mercury at 60° for 3 hours before using. *USP Polymyxin B Sulfate Reference Standard*—Dry in vacuum at a pressure not exceeding 5 mm of mercury at 60° for 3 hours before using.

Identification—Shake a quantity of Ointment, equivalent to about 2500 USP Bacitracin Units, with 20 mL of chloroform, add 5 mL of 0.1 *N* hydrochloric acid, shake vigorously, centrifuge, and use the clear supernatant liquid as the test solution. On a suitable thin-layer chromatographic plate coated with a 0.25-mm layer of chromatographic silica gel (see *Chromatography* ⟨621⟩) apply separately 10 µL of the test solution, 10 µL of a Standard bacitracin solution of USP Bacitracin Zinc RS in 0.1 *N* hydrochloric acid containing 500 USP Bacitracin Units per mL, and 10 µL of a Standard polymyxin B solution of USP Polymyxin B Sulfate RS in 0.1 *N* hydrochloric acid containing 500*J* USP Polymyxin B Units per mL, *J* being the ratio of the labeled amount of USP Polymyxin B Units to the labeled amount of USP Bacitracin Units in each g of Ointment. Place the plate in a suitable chromatographic chamber, and develop the chromatogram in a solvent system consisting of a mixture of isopropyl alcohol, water, and ammonium hydroxide (24:17:3) until the solvent front has moved about three-fourths of the length of the plate. Remove the plate from the chamber, and dry at 105° for 5 minutes. Spray the plate with a 1 in 200 solution of ninhydrin in butyl alcohol, and heat the plate at 105° for 15 minutes: the R_f values of the two principal spots in the chromatogram obtained from the test solution correspond to those of the single principal spots in the chromatograms obtained from the Standard bacitracin solution and the Standard polymyxin solution, respectively.

Minimum fill ⟨755⟩: meets the requirements.

Water, *Method I* ⟨921⟩: not more than 0.5%, 20 mL of a mixture of carbon tetrachloride, chloroform, and methanol (2:2:1) being used in place of methanol in the titration vessel.

Assay for bacitracin—Proceed with Bacitracin Zinc and Polymyxin B Sulfate Ointment as directed in the *Assay* under *Bacitracin Zinc Ointment.*

Assay for polymyxin B—Proceed as directed for polymyxin B under *Antibiotics—Microbial Assays* ⟨81⟩, using an accurately weighed portion of Bacitracin Zinc and Polymyxin B Sulfate Ointment shaken with about 50 mL of ether in a separator, and extracted with four 20-mL portions of *Buffer No. 6.* Combine the aqueous extracts, and dilute with *Buffer No. 6* to an appropriate volume to obtain a stock solution. Dilute this stock solution quantitatively and stepwise with *Buffer No. 6* to obtain a *Test Dilution* having a concentration assumed to be equal to the median dose level of the Standard.

Bacitracin Zinc and Polymyxin B Sulfate Ophthalmic Ointment

» Bacitracin Zinc and Polymyxin B Sulfate Ophthalmic Ointment contains the equivalent of not less than 90.0 percent and not more than 130.0 percent of the labeled amounts of bacitracin and polymyxin B.

Packaging and storage—Preserve in collapsible ophthalmic ointment tubes.

Reference standards—*USP Bacitracin Zinc Reference Standard*—Dry in vacuum at a pressure not exceeding 5 mm of mercury at 60° for 3 hours before using. *USP Polymyxin B Sulfate Reference Standard*—Dry in vacuum at a pressure not exceeding 5 mm of mercury at 60° for 3 hours before using.

Identification—It responds to the *Identification test* under *Bacitracin Zinc and Polymyxin B Sulfate Ointment.*

Sterility—It meets the requirements under *Sterility Tests* ⟨71⟩, when tested as directed in the section, *Test Procedures Using Membrane Filtration.*

Minimum fill ⟨755⟩: meets the requirements.

Water, *Method I* ⟨921⟩: not more than 0.5%, 20 mL of a mixture of carbon tetrachloride, chloroform, and methanol (2:2:1) being used in place of methanol in the titration vessel.

Metal particles—It meets the requirements of the test for *Metal Particles in Ophthalmic Ointments* ⟨751⟩.

Assay for bacitracin—Proceed with Bacitracin Zinc and Polymyxin B Sulfate Ophthalmic Ointment as directed in the *Assay* under *Bacitracin Zinc Ointment.*

Assay for polymyxin B—Proceed as directed for polymyxin B under *Antibiotics—Microbial Assays* ⟨81⟩, using an accurately weighed portion of Bacitracin Zinc and Polymyxin B Sulfate Ophthalmic Ointment shaken with about 50 mL of ether in a separator and extracted with four 20-mL portions of *Buffer No. 6*. Combine the aqueous extracts, and dilute with *Buffer No. 6* to an appropriate volume to obtain a stock solution. Dilute this stock solution quantitatively and stepwise with *Buffer No. 6* to obtain a *Test Dilution* having a concentration assumed to be equal to the median dose level of the Standard.

Baclofen

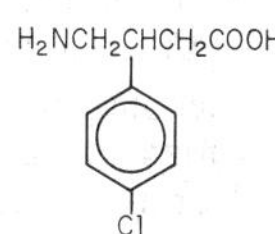

$C_{10}H_{12}ClNO_2$ 213.66
Butanoic acid, 4-amino-3-(4-chlorophenyl)-.
β-(Aminomethyl)-*p*-chlorohydrocinnamic acid [*1134-47-0*].

» Baclofen contains not less than 99.0 percent and not more than 101.0 percent of $C_{10}H_{12}ClNO_2$, calculated on the anhydrous basis.

Packaging and storage—Preserve in tight containers.

Reference standards—*USP Baclofen Reference Standard*—Do not dry; determine the *Water* content by *Method I* before using. *USP 4-(4-Chlorophenyl)-2-pyrrolidinone Reference Standard*—Do not dry. Use as is.

Identification—The infrared absorption spectrum of a mineral oil dispersion of it exhibits maxima only at the same wavelengths as that of a similar preparation of USP Baclofen RS.

Water, *Method I* ⟨921⟩: not more than 3.0%.

Residue on ignition ⟨281⟩: not more than 0.3%.

Heavy metals, *Method II* ⟨231⟩: 0.001%.

Limit of 4-(4-chlorophenyl)-2-pyrrolidinone—

Detection reagent—Dissolve 200 mg of *o*-tolidine in 2.0 mL of glacial acetic acid with the aid of a hot water bath. Dilute with water to 100.0 mL, and filter. Mix one volume of this solution with an equal volume of potassium iodide solution (0.83 in 100).

Procedure—Prepare a *Standard solution* of USP 4-(4-Chlorophenyl)-2-pyrrolidinone RS in a mixture of alcohol and glacial acetic acid (4:1), to contain 0.1 mg per mL. Prepare a *Test solution* containing 10.0 mg of Baclofen per mL in a mixture of alcohol and glacial acetic acid (4:1). Apply 10 µL of the *Standard solution* and 10 µL of the *Test solution* on a suitable thin-layer chromatographic plate coated with a 0.25-mm layer of chromatographic silica gel. Place the plate in a suitable chromatographic chamber (see *Chromatography* ⟨621⟩), and develop the chromatogram in a solvent system consisting of a mixture of butyl alcohol, glacial acetic acid, and water (4:1:1) until the solvent front has moved about three-fourths of the length of the plate. Remove the plate from the chamber, and dry with the aid of a current of warm air.

Transfer the dry plate to another chromatographic chamber containing a beaker with 1 g of potassium permanganate. Add 10 mL of dilute hydrochloric acid (4 in 10) to the beaker, cover the chamber, and allow the chamber to become saturated with chlorine gas. Expose the plate to the chlorine gas for 8 minutes. Remove the plate and expose to the air for 2 minutes, then spray with freshly prepared *Detection reagent:* the intensity of the secondary spot from the *Test solution* is not greater than that of the principal spot from the *Standard solution* (1.0%).

Assay—Dissolve about 40 mg of Baclofen, accurately weighed, in 10 mL of glacial acetic acid TS, and titrate with 0.1 *N* perchloric acid VS, determining the end-point potentiometrically, using a glass electrode and a calomel electrode containing a saturated solution of lithium chloride in glacial acetic acid (see *Titrimetry* ⟨541⟩). Perform a blank determination, and make any necessary correction. Each mL of 0.1 *N* perchloric acid is equivalent to 21.37 mg of $C_{10}H_{12}ClNO_2$.

Baclofen Tablets

» Baclofen Tablets contain not less than 90.0 percent and not more than 110.0 percent of the labeled amount of $C_{10}H_{12}ClNO_2$.

Packaging and storage—Preserve in well-closed containers.

Reference standards—*USP Baclofen Reference Standard*—Do not dry; determine the *Water* content by *Method I* before using. *USP 4-(4-Chlorophenyl)-2-pyrrolidinone Reference Standard*—Do not dry. Use as is.

Identification—Transfer a portion of powdered Tablets, equivalent to about 50 mg of baclofen, to a glass-stoppered, 40-mL centrifuge tube. Add 10.0 mL of a mixture of dehydrated alcohol and glacial acetic acid (4:1), shake by mechanical means for 30 minutes, and centrifuge. Apply 20 µL of this solution and 20 µL of a *Standard solution* containing 5 mg of USP Baclofen RS per mL in a mixture of dehydrated alcohol and glacial acetic acid (4:1) on a suitable thin-layer chromatographic plate coated with a 0.25-mm layer of chromatographic silica gel. Place the plate in a suitable chromatographic chamber (see *Chromatography* ⟨621⟩) containing a solvent system consisting of a mixture of butyl alcohol, glacial acetic acid, and water (4:1:1), and develop the chromatogram until the solvent front has moved about three-fourths of the length of the plate. Remove the plate from the chamber, and dry in a current of warm air. Spray with a detecting reagent consisting of 0.4 g of ninhydrin in 95 mL of butyl alcohol and 5 mL of dilute glacial acetic acid (1 in 10), until the plate is slightly wet. Place the plate in an oven maintained at 100° for 10 minutes: the R_f value of the principal orange-red spot obtained from the solution from the Tablets corresponds to that obtained from the *Standard solution.*

Dissolution ⟨711⟩—

Medium: 0.1 *N* hydrochloric acid; 500 mL for Tablets containing 10 mg or less of drug and 1000 mL for Tablets containing more than 10 mg of drug.

Apparatus 2: 50 rpm.

Time: 30 minutes.

Mobile phase—Prepare a suitable degassed and filtered mixture of 0.3 *N* acetic acid, methanol, and 0.36 *M* sodium 1-pentanesulfonate (550:440:20). Make adjustments if necessary (see *System Suitability* under *Chromatography* ⟨621⟩).

Chromatographic system (see *Chromatography* ⟨621⟩)—The liquid chromatograph is equipped with a 254-nm detector and a 3.9-mm × 30-cm column that contains packing L1. The flow rate is about 0.6 mL per minute. Chromatograph replicate injections of the *Standard solution*, and record the peak responses as directed under *Procedure:* the relative standard deviation is not more than 2.0%.

Procedure—Inject an accurately measured volume (about 190 µL) of a filtered portion of the solution under test into the chromatograph by means of a microsyringe or a sampling valve, record the chromatogram, and measure the response for the major peak. Calculate the quantity of $C_{10}H_{12}ClNO_2$ dissolved in comparison with a Standard solution having a known concentration of USP Baclofen RS in the same medium and similarly chromatographed.

Tolerances—Not less than 75% (*Q*) of the labeled amount of $C_{10}H_{12}ClNO_2$ is dissolved in 30 minutes.

Limit of 4-(4-chlorophenyl)-2-pyrrolidinone—

Diluting solution—Transfer 100 mL of 1 *M* nitric acid to a 1000-mL volumetric flask, add 300 mL of methanol, dilute with water to volume, and mix. If the mixture is cloudy, filter it through a fine-porosity, sintered-glass funnel prior to use.

Mobile phase—Transfer 1.36 g of tetramethylammonium nitrate to a 1000-mL volumetric flask. Pipet 20.0 mL of 1 *M* nitric acid into the flask, dilute with water to volume, mix, filter, and degas. Make adjustments if necessary (see *System Suitability* under *Chromatography* ⟨621⟩).

Standard preparation—Dissolve accurately weighed quantities of USP 4-(4-Chlorophenyl)-2-pyrrolidinone RS and USP

Baclofen RS in *Diluting solution*, and dilute quantitatively with *Diluting solution* to obtain a *Standard preparation* having known concentrations of 0.06 mg per mL and 1.2 mg per mL, respectively.

Test preparation—Transfer a portion of accurately weighed powdered Tablets, equivalent to about 30 mg of baclofen, to a 25-mL volumetric flask, and pipet 10 mL of *Diluting solution* into the flask. Sonicate, then shake by mechanical means for 30 minutes. Filter through glass wool, pipet 2 mL of the filtrate into a glass-stoppered, 10-mL centrifuge tube, and centrifuge.

Chromatographic system (see *Chromatography* ⟨621⟩)—The liquid chromatograph is equipped with a 254-nm detector and a 2.1-mm × 1-m column that contains packing L6. The flow rate is about 0.7 mL per minute. Chromatograph the *Standard preparation*, and record the peak responses as directed under *Procedure:* the resolution, *R*, between the baclofen and 4-(4-chlorophenyl)-2-pyrrolidinone peaks is not less than 1.5, and the relative standard deviation for replicate injections is not more than 1.5.

Procedure—Separately inject equal volumes (about 10 µL) of the *Standard preparation* and the *Assay preparation* into the chromatograph, record the chromatograms, and measure the responses for the major peaks. The relative retention times are about 0.5 for 4-(4-chlorophenyl)-2-pyrrolidinone and 1.0 for baclofen. Calculate the quantity, in mg, of 4-(4-chlorophenyl)-2-pyrrolidinone in the portion of Tablets taken by the formula:

$$10C(r_U/r_S),$$

in which *C* is the concentration, in mg per mL, of USP 4-(4-Chlorophenyl)-2-pyrrolidinone RS in the *Standard preparation*, and r_U and r_S are the peak responses of 4-(4-chlorophenyl)-2-pyrrolidinone obtained from the *Test preparation* and the *Standard preparation*, respectively. Not more than 5.0% is present.

Uniformity of dosage units ⟨905⟩: meet the requirements.

Procedure for content uniformity—

Standard solutions—Using accurately weighed quantities of USP Baclofen RS, prepare *Standard solutions*, S_1, S_2, and S_3, with concentrations of 0.17, 0.20, and 0.23 mg per mL (anhydrous basis) in 90% aqueous methanol (prepared by diluting 900 mL of methanol with water to 1000 mL).

Test solution—Place 1 Tablet in a suitable volumetric flask, and add 40 mL of 90% aqueous methanol. Insert the stopper in the flask, shake by mechanical means for 30 minutes, dilute with 90% aqueous methanol to volume to obtain a final concentration of 0.2 mg of baclofen (labeled amount) per mL, and filter.

Procedure—Pipet 10 mL of the *Test solution*, 10 mL of the *Standard solutions* S_1, S_2, S_3, and 10 mL of 90% aqueous methanol to provide the blank, into individual glass-stoppered, 40-mL centrifuge tubes. Add 2.0 mL of 10% glacial acetic acid in methanol to each tube, insert the stopper, and mix. Add 5.0 mL of a solution of salicylaldehyde in methanol (1 in 10) to each tube, insert the stopper, and mix. Wait 45 minutes, and concomitantly determine the absorbances of the solutions in a 1-cm cell at the wavelength of maximum absorbance at about 400 nm, with a suitable spectrophotometer, using the blank to set the instrument. Prepare a standard curve by plotting the concentration, in mg per mL, versus the absorbance values of the three *Standard solutions* S_1, S_2, S_3. Using this Standard curve, determine the concentration, C_U, in mg per mL, of the *Test solution*. Calculate the quantity, in mg, of $C_{10}H_{12}ClNO_2$ in the Tablet by the formula:

$$C_U \times D = \text{mg baclofen per Tablet},$$

in which *D* is the dilution factor of the *Test solution*.

Assay—Weigh and finely powder not less than 20 Baclofen Tablets. Using a portion of the powder, equivalent to about 30 mg of baclofen, proceed as directed for *Procedure for content uniformity* under *Uniformity of dosage units*.

Bacteriostatic Sodium Chloride Injection—*see* Sodium Chloride Injection, Bacteriostatic

Bacteriostatic Water for Injection—*see* Water for Injection, Bacteriostatic

Adhesive Bandage

» Adhesive Bandage consists of a compress of four layers of Type I Absorbent Gauze, or other suitable material, affixed to a film or fabric coated with a pressure-sensitive adhesive substance. It is sterile. The compress may contain a suitable antimicrobial agent and may contain one or more suitable colors. The adhesive surface is protected by a suitable removable covering.

Packaging and storage—Package Adhesive Bandage that does not exceed 15 cm (6 inches) in width individually in such manner that sterility is maintained until the individual package is opened. Package individual packages in a second protective container.

Labeling—The label of the second protective container bears a statement that the contents may not be sterile if the individual package has been damaged or previously opened, and it bears the names of any added antimicrobial agents. Each individual package is labeled to indicate the dimensions of the compress and the name of the manufacturer, packer, or distributor, and each protective container indicates also the address of the manufacturer, packer, or distributor.

Sterility—It meets the requirements under *Sterility Tests* ⟨71⟩.

Gauze Bandage

» Gauze Bandage is Type I Absorbent Gauze. Its length is not less than 98.0 percent of that declared on the label, and its average width is not more than 1.6 mm less than the declared width. It contains no dye or other additives.

Packaging and storage—Gauze Bandage that has been rendered sterile is so packaged that the sterility of the contents of the package is maintained until the package is opened for use.

Labeling—The width and length of the Bandage and the number of pieces contained, and the name of the manufacturer, packer, or distributor, are stated on the package. The designation "non-sterilized" or "not sterilized" appears prominently on the package unless the Gauze Bandage has been rendered sterile, in which case it may be labeled to indicate that it is sterile and that the contents may not be sterile if the package bears evidence of damage or if the package has been previously opened.

Note—Before determining the thread count, dimensions, and weight, hold the Bandage, unrolled, for not less than 4 hours in a standard atmosphere of 65 ± 2% relative humidity at 21 ± 1.1° C (70 ± 2° F).

Thread count—Count the number of warp and filling threads of it in areas of 1.27 cm (½ inch) square at 5 points evenly spread along the center line of the Bandage, no point being within 30.5 cm (12 inches) of either end of the Bandage, and calculate the average number of threads per 2.54 cm (1 inch) in each direction. A variation of not more than 3 threads per inch is allowed in either warp or filling, provided that the combined variations do not exceed 5 threads per square inch.

Width—Measure its width at each of the 5 points selected for the determination of the thread count: the average of 5 measurements is not more than 1.6 mm (1/16 inch) less than the labeled width of the Bandage.

Length—Measure the length of the unrolled Gauze Bandage, smoothed without tension, along the center line of the Gauze Bandage: the length is not less than 98.0% of the labeled length of the Bandage.

Weight—Weigh the entire Bandage: the calculated weight in g per 0.894 square meter (1 linear yard *Type I gauze*), using the measurements obtained as described in the two paragraphs just preceding, is not less than 39.2 g.

Absorbency—Hold a rolled Gauze Bandage horizontal to and almost in contact with the surface of water at 25°, and allow it to drop lightly upon the water: complete submersion takes place in not more than 30 seconds.

Sterility—Gauze Bandage that has been rendered sterile meets the requirements under *Sterility Tests* ⟨71⟩.

Other requirements—It meets the requirements of the tests for *Ignited residue, Acid or alkali, and Dextrin or starch, in water extract, Residue on ignition, Fatty matter*, and *Alcohol-soluble dyes* under *Absorbent Gauze.*

Barium Hydroxide Lime

» Barium Hydroxide Lime is a mixture of barium hydroxide octahydrate and Calcium Hydroxide. It may contain also Potassium Hydroxide and may contain an indicator that is inert toward anesthetic gases such as Ether, Cyclopropane, and Nitrous Oxide, and that changes color when the Barium Hydroxide Lime no longer can absorb carbon dioxide.

Caution—Since Barium Hydroxide Lime contains a soluble form of barium, it is toxic if swallowed.

Packaging and storage—Preserve in tight containers.

Labeling—If an indicator has been added, the name and color change of such indicator are stated on the container label. The container label indicates also the mesh size in terms of standard-mesh sieve sizes (see *Powder Fineness* ⟨811⟩).

Identification—

A: Place a granule of it on a piece of moistened red litmus paper: the paper turns blue immediately.

B: A 1 in 10 solution of it in 6 *N* acetic acid responds to the tests for *Barium* ⟨191⟩ and for *Calcium* ⟨191⟩, and it may respond also to the flame test for *Potassium* ⟨191⟩.

Size of granules—Screen 100 g for 5 minutes as directed under *Powder Fineness—Method for Determining Uniformity of Fineness* ⟨811⟩, using a mechanical shaker. It passes completely through a No. 2 standard-mesh sieve, and not more than 2.0% passes through a No. 40 standard-mesh sieve. Not more than 7.0% is retained on the coarse-mesh sieve, and not more than 15.0% passes through the fine-mesh sieve designated on the label.

Loss on drying ⟨731⟩—Weigh accurately, in a tared weighing bottle, about 10 g, and dry at 105° for 2 hours: it loses between 11.0% and 16.0% of its weight.

Hardness—Screen 200 g on a mechanical sieve shaker (see *Powder Fineness* ⟨811⟩) having a frequency of oscillation of 285 ± 3 cycles per minute, for 3 minutes, to remove granules coarser than 4-mesh and finer than 8-mesh. Weigh 50 g of the granules retained on the screen, and place them in a hardness pan of the following description: the hardness pan has a diameter of 200 mm and a concave brass bottom, and the bottom of the pan is 7.9 mm thick at the circumference and 3.2 mm thick at the center and has an inside spherical radius of curvature of 109 cm. Add 15 steel balls of 7.9-mm diameter, and shake on a mechanical sieve shaker for 30 minutes. Remove the steel balls, brush the contents of the hardness pan onto a sieve of the fine-mesh size designated on the label, shake for 3 minutes on the mechanical sieve shaker, and weigh: the percentage of Barium Hydroxide Lime retained on the screen is not less than 75.0, and represents the hardness.

Carbon dioxide absorbency—Fill the lower transverse section of a U-shaped drying tube of about 15-mm internal diameter and 15-cm height with loosely packed glass wool. Place in one arm of the tube about 5 g of anhydrous calcium chloride, and accurately weigh the tube and the contents. Into the other arm of the tube place 9.5 g to 10.5 g of Barium Hydroxide Lime, and again weigh accurately. Insert stoppers in the open arms of the U-tube, and connect the side tube of the arm filled with Barium Hydroxide Lime to a calcium chloride drying tube, which in turn is connected to a suitable source of supply of carbon dioxide. Pass the carbon dioxide through the U-tube at a rate of 75 mL per minute for 20 minutes, accurately timed. Disconnect the U-tube, cool to room temperature, remove the stoppers, and weigh: the increase in weight is not less than 19.0% of the weight of Barium Hydroxide Lime used for the test.

Barium Sulfate

$BaSO_4$ 233.39
Sulfuric acid, barium salt (1:1).
Barium sulfate (1:1) [*7727-43-7*].

» Barium Sulfate contains not less than 97.5 percent and not more than 100.5 percent of $BaSO_4$.

Packaging and storage—Preserve in well-closed containers.

Identification—

A: Mix 0.5 g with 2 g each of anhydrous sodium carbonate and anhydrous potassium carbonate, heat the mixture in a crucible until fusion is complete, treat the resulting fused mass with hot water, and filter: the filtrate, acidified with hydrochloric acid, responds to the tests for *Sulfate* ⟨191⟩.

B: Dissolve a portion of the well-washed residue from *Identification test A* in 6 *N* acetic acid: the solution responds to the tests for *Barium* ⟨191⟩.

Bulkiness—Place 5.0 g, previously sifted through a No. 60 standard sieve, in a dry, graduated, glass-stoppered cylinder having the 50-mL graduation 11 cm to 14 cm from the bottom. Dilute with water to make the mixture measure 50 mL. Shake the mixture briskly for 1 minute, accurately timed, and set it aside for sedimentation: the Barium Sulfate does not settle below the 11-mL graduation within 15 minutes.

Acidity or alkalinity—Digest 1 g with 20 mL of water for 5 minutes: the water remains neutral to litmus.

Sulfide—To 10 g in a 500-mL conical flask add 100 mL of 0.3 *N* hydrochloric acid. Cover the mouth of the flask with a circle of filter paper that has been moistened at the area over the mouth of the flask with 0.15 mL of lead acetate TS, the paper being held in place with a string tied around the neck of the flask. Boil the mixture gently for 10 minutes, taking care to avoid spattering the paper: any darkening of the paper is not greater than that produced by a similarly treated control consisting of 100 mL of 0.3 *N* hydrochloric acid containing 5 µg of sulfide (S) (0.5 ppm).

Acid-soluble substances—Cool the mixture obtained in the test for *Sulfide*, add water to restore approximately the original volume, and filter it through paper that previously has been washed with a mixture of 10 mL of 3 *N* hydrochloric acid and 90 mL of water, returning the first portions, if necessary, to obtain a clear filtrate. Evaporate 50 mL of the filtrate on a steam bath to dryness, and add 2 drops of hydrochloric acid and 10 mL of hot water. Filter again through acid-washed paper, prepared as directed above, wash the filter with 10 mL of hot water, and evaporate the combined filtrate and washings in a tared dish on a steam bath to dryness: the residue, when dried at 105° for 1 hour, weighs not more than 15 mg (0.3%).

Soluble barium salts—Treat the residue obtained in the test for *Acid-soluble substances* with 10 mL of water, filter the solution through a filter previously washed with 100 mL of 0.3 *N* hydrochloric acid, and add 0.5 mL of 2 *N* sulfuric acid: any turbidity formed within 30 minutes is not greater than that produced in a similarly treated control consisting of 10 mL of water containing 50 µg of Ba (0.001%) and 0.5 mL of 2 *N* sulfuric acid.

Arsenic, *Method I* ⟨211⟩—To 3.75 g in a 150-mL beaker add 40 mL of water and 10 mL of nitric acid, cover, and digest on a steam bath for 1 hour, with frequent stirring. Filter, wash the insoluble residue with 20 mL of 7 *N* sulfuric acid, in small portions, and then with 10 mL of water, receiving the filtrate and washings in an arsine generator flask. Evaporate just to the production of white fumes, cool, wash down the sides of the flask, and again evaporate to white fumes. Repeat the washing and evaporation. Cool, and cautiously dilute with water to 55 mL: the resulting solution meets the requirements of the test, the addition of 20 mL of 7 *N* sulfuric acid specified under *Procedure* being omitted. The limit is 0.8 ppm.

Heavy metals ⟨231⟩—Boil 4.0 g with a mixture of 2 mL of glacial acetic acid and 48 mL of water for 10 minutes. Dilute with water to 50 mL, filter, and use 25 mL of the filtrate: the limit is 0.001%.

Assay—Weigh accurately not less than 0.58 g and not more than 0.62 g of Barium Sulfate in a tared platinum crucible, add 10 g of anhydrous sodium carbonate, and mix by rotating the crucible. Fuse over a blast burner until a clear melt is obtained, and heat for an additional 30 minutes. Cool, place the crucible in a 400-mL beaker, add 250 mL of water, stir with a glass rod, and heat to dislodge the melt. Remove the crucible from the beaker, and wash well with water, collecting the washings in the beaker. Rinse the inside of the crucible with 2 mL of 6 *N* acetic acid and then with water, again collecting the washings in the beaker, and continue heating and stirring until the melt is disintegrated. Cool the beaker in an ice bath until the precipitate settles, decant the clear liquid through filter paper (Whatman No. 40 or equivalent), taking care to transfer as little precipitate as possible to the paper. Wash twice by decantation as follows: Wash down the inside of the beaker with about 10 mL of cold sodium carbonate solution (1 in 50), swirl the contents of the beaker, allow the precipitate to settle, and decant the supernatant liquid through the same filter paper as before, transferring as little precipitate as possible. Place the beaker containing the bulk of the barium carbonate precipitate under the funnel, wash the filter paper with five 1-mL portions of 3 *N* hydrochloric acid, and wash the paper well with water. [NOTE—the solution may be slightly hazy.] Add 100 mL of water, 5.0 mL of hydrochloric acid, 10.0 mL of ammonium acetate solution (2 in 5), 25 mL of potassium dichromate solution (1 in 10), and 10.0 g of urea. Cover the beaker with a watch glass, and digest at 80° to 85° for not less than 16 hours. Filter while hot through a tared, fine-porosity, sintered-glass crucible, transferring all of the precipitate with the aid of a rubber-tipped stirring rod. Wash the precipitate with potassium dichromate solution (1 in 200), and finally with about 20 mL of water. Dry at 105° for 2 hours, cool, and weigh: the weight of the barium chromate so obtained, multiplied by 0.9213, represents the weight of $BaSO_4$.

Barium Sulfate for Suspension

» Barium Sulfate for Suspension is a dry mixture of Barium Sulfate and one or more suitable dispersing and/or suspending agents. It contains not less than 90.0 percent of $BaSO_4$. It may contain one or more suitable colors, flavors, fluidizing agents, and preservatives.

Packaging and storage—Preserve in well-closed containers.

Identification—Ignite a 1-g portion to constant weight: the ignited residue responds to *Identification tests A* and *B* under *Barium Sulfate.*

pH ⟨791⟩: between 4.0 and 10.0, in a 60% (w/w) aqueous suspension.

Loss on drying ⟨731⟩—Dry it at 105° for 4 hours: it loses not more than 1.0% of its weight.

Heavy metals, *Method III* ⟨231⟩—Boil 2.0 g with a mixture of 1 mL of glacial acetic acid and 24 mL of water for 10 minutes. Filter while hot through a membrane filter of 0.45-μm pore size, and wash with a small amount of hot water, collecting the washing with the filtrate. Add half of a mixture of 8 mL of sulfuric acid and 10 mL of nitric acid. Warm gently, and proceed as directed under *Test Preparation—If the substance is a solid,* beginning with "add additional portions of the same acid mixture." The limit is 0.001%.

Other requirements—It meets the requirements of the tests for *Sulfide* and *Arsenic* under *Barium Sulfate.*

Assay—Weigh accurately a quantity of Barium Sulfate for Suspension, equivalent to about 0.60 g of $BaSO_4$, in a tared platinum crucible, and ignite over a low flame until any organic matter is thoroughly carbonized. Cool, cautiously add 0.5 mL of nitric acid and 0.5 mL of sulfuric acid, and continue the ignition over a low flame until the residue becomes gray in color, then ignite over the full heat of a blast burner. Allow the contents of the crucible to cool to room temperature.

NOTE 1—*If the specimen contains a silicate, such as bentonite, proceed as follows:* Add 10 mL of water and 1 mL of sulfuric acid to the residue in the crucible, mix, and add 10 mL of hydrofluoric acid. Heat gently over a low flame until fumes of sulfur trioxide appear. Add 5 mL more of hydrofluoric acid, heat again over a low flame to the appearance of dense fumes, and continue heating until the sulfuric acid has been completely volatilized. Allow the contents of the crucible to cool.

NOTE 2—*If the specimen does not contain a silicate, omit the treatment of the specimen with hydrofluoric and sulfuric acids.*

Add to the treated or untreated specimen in the platinum crucible 10 g of anhydrous sodium carbonate, fuse over a blast burner until a clear melt is obtained, and heat for an additional 30 minutes. Proceed as directed in the *Assay* under *Barium Sulfate,* beginning with "Cool, place the crucible in a 400-mL beaker."

Basic Aluminum Carbonate Gel—*see* Aluminum Carbonate Gel, Basic

Basic Fuchsin—*see* Fuchsin, Basic

BCG Vaccine

» BCG Vaccine conforms to the regulations of the federal Food and Drug Administration concerning biologics (see *Biologics* ⟨1041⟩). It is a dried, living culture of the bacillus Calmette-Guérin strain of *Mycobacterium tuberculosis* var. *bovis,* grown in a suitable medium from a seed strain of known history that has been maintained to preserve its capacity for conferring immunity. It contains an amount of viable bacteria such that inoculation, in the recommended dose, of tuberculin-negative persons results in an acceptable tuberculin conversion rate. It is free from other organisms, and contains a suitable stabilizer. It contains no antimicrobial agent.

NOTE—Use the Vaccine immediately after its constitution, and discard any unused portion after 2 hours.

Packaging and storage—Preserve in hermetic containers, preferably of Type I glass, at a temperature between 2° and 8°.

Expiration date—The expiration date is not later than 6 months after date of issue, or not later than 1 year after date of issue if stored at a temperature below 5°.

Beclomethasone Dipropionate

$C_{28}H_{37}ClO_7$ 521.05

Pregna-1,4-diene-3,20-dione, 9-chloro-11-hydroxy-16-methyl-17,21-bis(1-oxopropoxy)-, (11β,16β)-.

9-Chloro-11β,17,21-trihydroxy-16β-methylpregna-1,4-diene-3,20-dione 17,21-dipropionate [*5534-09-8*].

Monohydrate 539.07.

» Beclomethasone Dipropionate is anhydrous or contains one molecule of water of hydration. It contains not less than 97.0 percent and not more than

103.0 percent of $C_{28}H_{37}ClO_7$, calculated on the dried basis.

Packaging and storage—Preserve in well-closed containers.

Reference standards—*USP Beclomethasone Dipropionate Reference Standard*—Dry at 105° for 3 hours before using. *USP Testosterone Propionate Reference Standard*—Dry in vacuum over silica gel for 4 hours before using.

Identification—The infrared absorption spectrum of a mineral oil dispersion of it, previously dried, exhibits maxima only at the same wavelengths as that of a similar preparation of USP Beclomethasone Dipropionate RS.

Specific rotation ⟨781⟩: between +88° and +94°, calculated on the dried basis, determined in a solution in dioxane containing 100 mg in each 10 mL.

Loss on drying ⟨731⟩—Dry it at 105° for 3 hours: the anhydrous form loses not more than 0.5% of its weight; the monohydrate form loses between 2.8% and 3.8% of its weight.

Residue on ignition ⟨281⟩: not more than 0.1%.

Assay—

Mobile solvent—Prepare a suitable degassed solution of 3 volumes of acetonitrile in 2 volumes of water, such that the retention time of beclomethasone dipropionate is approximately 6 minutes and that of testosterone propionate is approximately 10 minutes.

Internal standard solution—Dissolve a suitable quantity of USP Testosterone Propionate RS, accurately weighed, in methanol to obtain a solution having a concentration of about 1.2 mg per mL.

Standard preparation—Dissolve a suitable quantity of USP Beclomethasone Dipropionate RS, accurately weighed, in methanol to obtain a solution having a concentration of about 1.4 mg per mL. Transfer 4.0 mL of this solution to a suitable vial, and add 4.0 mL of *Internal standard solution*, to obtain a solution having a known concentration of about 0.7 mg per mL with respect to the Reference Standard and 0.6 mg per mL with respect to the internal standard.

Assay preparation—Weigh accurately about 70 mg of Beclomethasone Dipropionate, transfer to a 50-mL volumetric flask, dilute with methanol to volume, and mix. Transfer 4.0 mL of this solution to a suitable vial, and add 4.0 mL of *Internal standard solution*.

Procedure—Introduce equal volumes (between 5 µL and 25 µL) of the *Assay preparation* and the *Standard preparation* into a high-performance liquid chromatograph (see *Chromatography* ⟨621⟩) operated at room temperature, by means of a suitable microsyringe or sampling valve, adjusting the specimen size and other operating parameters such that the peak obtained with the internal standard in the *Standard preparation* is about 0.6 to 0.9 full-scale. Typically, the apparatus is fitted with a 4-mm × 30-cm column packed with packing L1 and is equipped with an ultraviolet detector capable of monitoring absorption at 254 nm, a suitable recorder, and a pump capable of operating at a column pressure of up to 3500 psi. In a suitable chromatogram, the coefficient of variation for five replicate injections of the *Standard preparation* is not more than 3.0%. Calculate the quantity, in mg, of $C_{28}H_{37}ClO_7$ in the portion of Beclomethasone Dipropionate taken by the formula:

$$100C(R_U/R_S),$$

in which C is the concentration, in mg per mL, of USP Beclomethasone Dipropionate RS in the *Standard preparation*, and R_U and R_S are the peak height ratios of beclomethasone dipropionate to the internal standard, at equivalent retention times, obtained from the *Assay preparation* and the *Standard preparation*, respectively.

Belladonna Extract

» Belladonna Extract contains, in each 100 g, not less than 1.15 g and not more than 1.35 g of the alkaloids of belladonna leaf.

Pilular Belladonna Extract

Prepare the extract by percolating 1000 g of Belladonna Leaf, using a mixture of 3 volumes of alcohol and 1 volume of water as the menstruum. Macerate the drug for 16 hours, and then percolate it at a moderate rate. Evaporate the percolate under reduced pressure and at a temperature not exceeding 60° to a pilular consistency, and adjust the remaining extract, after assaying, by dilution with liquid glucose so that the finished Extract will contain, in each 100 g, 1.25 g of the alkaloids of belladonna leaf.

Powdered Belladonna Extract

Prepare the extract by percolating 1000 g of Belladonna Leaf, using alcohol as the menstruum. Macerate the drug for 16 hours, and then percolate it slowly. Evaporate the percolate under reduced pressure and at a temperature not exceeding 60° to a soft extract, add 50 g of dry starch, and continue the evaporation, at the same temperature, until the product is dry. Powder the residue. The extract may be deprived of its fat by treating either the soft extract first obtained, or the dry and powdered extract, as directed under *Extracts* (see *Pharmaceutical Dosage Forms* ⟨1151⟩). Assay the powdered residue, and add sufficient starch, previously dried at 100°, to obtain a finished Extract containing 1.25 g of the alkaloids of belladonna leaf in each 100 g. Mix the powders, and pass the Extract through a fine sieve.

Packaging and storage—Preserve in tight containers, at a temperature not exceeding 30°.

Reference standards—*USP Atropine Sulfate Reference Standard—Caution—Avoid contact.* Dry at 120° for 4 hours before using. *USP Homatropine Hydrobromide Reference Standard*—Dry at 105° for 2 hours before using. *USP Scopolamine Hydrobromide Reference Standard*—Dry at 105° for 3 hours before using.

Assay—

pH 9.5 phosphate buffer—Dissolve 34.8 g of dibasic potassium phosphate in 900 mL of water, and adjust to a pH of 9.5, determined electrometrically, by the addition of 3 *N* hydrochloric acid or sodium hydroxide, with mixing.

Internal standard solution—Dissolve about 40 mg of USP Homatropine Hydrobromide RS, accurately weighed, in about 25 mL of dilute sulfuric acid (1 in 350) in a 50-mL volumetric flask, add the same dilute acid to volume, and mix. Prepare fresh on the day of use.

Standard preparation—Dissolve about 10 mg of USP Scopolamine Hydrobromide RS, accurately weighed, in about 5 mL of dilute sulfuric acid (1 in 350) in a 10-mL volumetric flask, add the same dilute acid to volume, and mix (*Solution A*). Dissolve about 20 mg of USP Atropine Sulfate RS, accurately weighed, in about 25 mL of dilute sulfuric acid (1 in 350) in a 50-mL volumetric flask, add 2.0 mL of *Solution A*, and mix. Add dilute sulfuric acid (1 in 350) to volume, and mix. Prepare fresh on the day of use.

Extraction blank—Place about 10 mL of dilute sulfuric acid (1 in 350) in a 60-mL separator. Proceed as directed under *Assay preparation*, beginning with "then add 15 mL of chloroform." The blank chromatogram contains no significant interferences at the locus of atropine, scopolamine, or homatropine.

Assay preparation—Weigh accurately about 0.5 g of Belladonna Extract, transfer to a 125-mL conical flask, and add 40 mL of dilute sulfuric acid (1 in 350). Heat to a temperature not above 45°, and stir to hasten solution. Filter the solution through filter paper into a 100-mL volumetric flask. Wash the flask and the filter with two 20-mL portions of warmed dilute sulfuric acid

(1 in 350), and collect the washings in the 100-mL volumetric flask. Add dilute sulfuric acid (1 in 350) to volume, and mix.

Pipet 10 mL of this solution into a 60-mL separator. To the separator add 1.0 mL of *Internal standard solution*, then add 15 mL of chloroform, shake vigorously, allow the layers to separate, and discard the chloroform layer. (If emulsions are formed, a *mixed solvent* consisting of chloroform and isopropyl alcohol (10:3) may be substituted for chloroform throughout the extraction procedure.) Add another 15 mL of chloroform, and extract again, discarding the chloroform phase. Add 15 mL of *pH 9.5 phosphate buffer* and sufficient 1 *N* sodium hydroxide to yield a final pH between 9.0 and 9.5. Add 15 mL of chloroform, shake vigorously, and allow the layers to separate. Filter the organic phase through 10 g of anhydrous sodium sulfate (see *Suitability for alkaloid assays* under *Sodium Sulfate, Anhydrous*, in the section, *Reagents, Indicators, and Solutions*), previously washed with chloroform and supported in a funnel with a small pledget of glass wool, into a suitable container. Extract again with two 15-mL portions of chloroform, again collecting the clarified organic phase. Wash the sodium sulfate and the tip of the funnel with 5 mL of chloroform. Evaporate the combined organic phases under reduced pressure, at a temperature below 45°, add 1 mL of chloroform, and mix to dissolve the alkaloids, taking care to wet the sides of the container.

Standard curve—Prepare three *Standard solutions* as follows: Pipet into three separate 60-mL separators 1.0-, 2.0-, and 3.0-mL portions, respectively, of *Standard preparation*, and add 9.0, 8.0, and 7.0 mL, respectively, of dilute sulfuric acid (1 in 350). Proceed as directed under *Assay preparation*, beginning with "add 1.0 mL of *Internal standard solution*."

Chromatographic system—Under typical condition, the instrument contains a 1.2-m × 4-mm glass column packed with 3 percent G3 on S1AB. The column may be cured and conditioned as specified under *Gas Chromatography* (see *Chromatography* ⟨621⟩). The column is maintained at a temperature of about 215°, and the injection port and detector block at about 240°, and dry helium is used as a carrier gas at a flow rate of about 65 mL per minute.

System suitability—Chromatograph six to ten injections of the solution, and record peak areas as directed under *Procedure*. The analytical system is suitable for conducting this assay if the relative standard deviation for the ratio, R_A, calculated by the formula:

$$100 \times (\text{standard deviation/mean ratio}),$$

does not exceed 2.0%; the resolution, *R*, between a_H and a_A is not less than 3; and the tailing factor (the sum of the distances from peak center to the leading edge and to the tailing edge divided by twice the distance from peak center to the leading edge), measured at 5% of the peak height of a_A, does not exceed 2.0.

Procedure—Inject a portion (about 5 µL) of each *Standard solution* into a suitable gas chromatograph equipped with a flame-ionization detector. Measure the areas, a_A, a_H, and a_S, of the atropine, homatropine, and scopolamine peaks, respectively, in each chromatogram, and calculate the ratios A_A and A_S by the formulas:

$$a_A/a_H \text{ and } a_S/a_H.$$

Plot the *Standard curves* of the values of R_A and R_S against the amounts, in mg, of atropine and scopolamine in the solutions. (The ratio of the molecular weight of atropine to that of anhydrous atropine sulfate is 0.8551, and the ratio of the molecular weight of scopolamine to that of anhydrous scopolamine hydrobromide is 0.7894.) Inject a portion of the *Assay preparation* into the chromatograph, obtain the chromatogram area ratios, measure the peak areas, and calculate the area ratios, as with the *Standard solutions*. Record from the *Standard curve* the quantities, in mg, of atropine and scopolamine in the volume of specimen taken. Add the quantity, in mg, of atropine and scopolamine, and multiply by 10 to obtain the weight, in mg, of alkaloids in the portion of Belladonna Extract taken.

Belladonna Extract Tablets

» Belladonna Extract Tablets contain not less than 90.0 percent and not more than 110.0 percent of the labeled amount of the alkaloids of belladonna leaf.

Packaging and storage—Preserve in tight, light-resistant containers.

Reference standards—*USP Atropine Sulfate Reference Standard*—Keep container tightly closed and protected from light. [*Caution—Avoid contact.*] Dry at 100° to constant weight before using. *USP Homatropine Hydrobromide Reference Standard*—Dry at 105° for 2 hours before using. *USP Scopolamine Hydrobromide Reference Standard*—Store tightly closed, protected from light. Dry at 105° for 3 hours before using.

Identification—Macerate a quantity of powdered Tablets, equivalent to about 5 mg of the alkaloids of belladonna extract, with 20 mL of water, and transfer to a separator. Render the solution alkaline with 6 *N* ammonium hydroxide, and extract the alkaloids with 50 mL of chloroform. Filter the chloroform layer, divide it into two equal portions, and evaporate to dryness: the residue responds to the following tests.

A: To one portion of the dry residue add 2 drops of nitric acid, evaporate on a steam bath to dryness, and add a few drops of alcoholic potassium hydroxide TS: a violet color is produced.

B: Dissolve the other portion of the residue in 1 mL of dilute hydrochloric acid (1 in 120), and add gold chloride TS, dropwise with shaking, until a definite precipitate separates. Slowly heat until the precipitate dissolves, and allow the solution to cool: a lusterless precipitate is produced.

Disintegration ⟨701⟩: 30 minutes.

Uniformity of dosage units ⟨905⟩: meet the requirements.

Assay—

Internal standard solution, Standard preparation, Extraction blank, Standard curve, Chromatographic system, and *System suitability*—Proceed as directed in the *Assay* under *Belladonna Extract.*

Assay preparation—Weigh and finely powder not less than 20 Belladonna Extract Tablets. Transfer an accurately weighed portion of the powder, equivalent to about 600 µg of atropine and scopolamine, to a 60-mL separator, add 10.0 mL of dilute sulfuric acid (1 in 350), and sonicate to dissolve as much as possible of the specimen. Proceed as directed for *Assay preparation* in the *Assay* under *Belladonna Leaf*, beginning with "add 1.0 mL of *Internal standard solution.*"

Procedure—Proceed as directed for *Procedure* in the *Assay* under *Belladonna Extract*. Record from the *Standard curves* the quantities, in mg, of atropine and scopolamine in the weight of specimen taken.

Belladonna Leaf

» Belladonna Leaf consists of the dried leaf and flowering or fruiting top of *Atropa belladonna* Linné or of its variety *acuminata* Royle ex Lindley (Fam. Solanaceae). Belladonna Leaf yields not less than 0.35 percent of the alkaloids of belladonna leaf.

Packaging and storage—Preserve in well-closed containers and avoid long exposure to direct sunlight. Preserve powdered Belladonna Leaf in light-resistant containers.

Reference standards—*USP Atropine Sulfate Reference Standard*—Keep container tightly closed and protected from light. [*Caution—Avoid contact.*] Dry at 100° to constant weight before using. *USP Homatropine Hydrobromide Reference Standard*—Dry at 105° for 2 hours before using. *USP Scopolamine Hydrobromide Reference Standard*—Store tightly closed, protected from light. Dry at 105° for 3 hours before using.

Botanic characteristics—

Belladonna Leaf—Usually partly matted together, crumpled or broken leaves, together with some smaller stems and a number of flowers and fruits. The leaves are thin and brittle, mostly light green to moderate olive-green. The lamina is mostly from 5 to 25 cm in length and from 4 to 12 cm in width and possesses an ovate-lanceolate to broadly ovate outline, an acute to acuminate apex, an entire margin, an acute to somewhat decurrent base and slightly hairy surface, the hairs being more abundant along the veins; when broken transversely, it shows numerous light-colored dots (crystal cells) visible with a lens. The petiole is slender and usually up to 4 cm in length. The flowers possess a campanulate corolla with 5 small, reflexed lobes, purplish to yellowish purple, becoming faded to brown or dusky yellow or yellow, a green, 5-lobed calyx, 5 epipetalous stamens, and a superior, bilocular ovary with numerous ovules. The fruit is subglobular, dark yellow to yellowish brown to dusky red or black, up to about 12 mm in width and sometimes subtended by the persistent calyx and containing numerous flattened, somewhat reniform seeds, the latter up to about 2 mm in width. The stems are more or less flattened and hollow and finely hairy when young.

Histology—Leaf: The epidermis of the lamina possesses wavy anticlinal walls and a distinctly striated cuticle. Stomata are more numerous in the lower epidermis and are surrounded by 3 or 4 neighboring cells, one of which is smaller than the others. The nonglandular hairs are uniseriate and up to 6-celled. Short club-shaped glandular hairs with a 1-celled stalk and multicellular head and long glandular hairs with a uniseriate stalk and unicellular head occur on both epidermises. The mesophyll consists of a single layer of palisade parenchyma beneath which occurs spongy parenchyma, the latter with scattered cells filled with microcrystals. The midrib contains an arc of bicollateral bundles, collenchyma beneath upper epidermis, and scattered parenchyma cells with microcrystals. *Stem:* The stem shows an epidermis with striated cuticle and few hairs, a distinct endodermis, small strands of long, thin-walled, slightly lignified pericyclic fibers, and a circle of bicollateral bundles. The parenchyma of the cortex and pith is interspersed with crystal cells. *Flower:* The calyx possesses numerous glandular hairs with uniseriate stalks and 1- to 3-celled glandular heads. The corolla shows a papillose inner epidermis and an outer epidermis with glandular hairs similar to those of the calyx. The pollen grains, when mounted in chloral hydrate solution, are subspherical, about 40 µm in diameter, tricolpate, having 3 germinal furrows and rows of pits between the ridges on the exine. *Fruit:* The epicarp exhibits polygonal epidermal cells with a striated cuticle and stomata. The mesocarp consists of large pulp cells some of which contain rosette aggregate crystals of calcium oxalate. *Seed:* The seed is characterized by an epidermis of large, wavy-walled cells with prominent ridges over the anticlinal walls.

Powdered Belladonna Leaf—Light olive-brown to moderate olive-green in color. The following are among the elements of identification: the separate microcrystals, the dark gray crystal cells, the cuticular striping of the epidermal cells, the vessels with ellipsoidal bordered pits, the fibers of the stem, and occasional hairs and pollen grains. Rosette aggregates of calcium oxalate and fragments of the seed occur when the drug contains belladonna fruits. Examine Belladonna Leaf for hairs having a papillose cuticle and for raphides of calcium oxalate: their presence indicates adulteration.

Acid-insoluble ash ⟨561⟩: not more than 3.0%.

Belladonna stems—The proportion of belladonna stems over 10 mm in diameter does not exceed 3.0%.

Assay—

Internal standard solution, Standard preparation, Extraction blank, Standard curve, Chromatographic system, and *System suitability*—Proceed as directed in the *Assay* under *Belladonna Extract.*

Assay preparation—Moisten 10 g, previously reduced to a moderately coarse powder and accurately weighed, with a mixture of 8 mL of ammonium hydroxide, 10 mL of alcohol, and 20 mL of ether, and extract the alkaloids by either of the methods given in the following two paragraphs. If necessary, reduce the volume of the extract to 100 mL by evaporation on a steam bath.

I—Place the moistened drug in a continuous-extraction thimble, and allow maceration to proceed overnight, then extract with ether for 3 hours, or longer if necessary to effect complete extraction.

II—Place the moistened drug in a small percolator, and allow maceration to proceed overnight. Percolate slowly with a mixture of 3 volumes of ether and 1 volume of chloroform. Continue the percolation until the residue from 3 to 4 mL of percolate last passed, when dissolved in dilute sulfuric acid (1 in 70) and treated with mercuric iodide TS, shows not more than a faint turbidity.

Transfer the extract to a separator with the aid of ether. Extract with five 15-mL portions of dilute sulfuric acid (1 in 70), filtering each portion drawn off into a 100-mL volumetric flask. Wash the filter with dilute sulfuric acid (1 in 70), and collect the washings in the flask. Add dilute sulfuric acid (1 in 70) to volume, and mix. Dilute 20.0 mL of the resulting solution with the same dilute acid to 100.0 mL.

Pipet 10 mL of this solution into a 60-mL separator. To the separator add 1.0 mL of *Internal standard solution,* then add 15 mL of chloroform, shake vigorously, allow the layers to separate, and discard the chloroform layer. (If emulsions are formed, a *mixed solvent* consisting of chloroform–isopropyl alcohol (10:3) may be substituted for chloroform throughout the extraction procedure.) Add another 15 mL of chloroform, and extract again, discarding the chloroform phase. Add 15 mL of pH 9.5 phosphate buffer and sufficient 1 *N* sodium hydroxide to yield a final pH between 9.0 and 9.5. Add 15 mL of chloroform, shake vigorously, and allow the layers to separate. Filter the organic phase through 10 g of anhydrous sodium sulfate (see *Suitability for alkaloid assays* under *Sodium Sulfate, Anhydrous,* in the section, *Reagent Specifications*), previously washed with chloroform and supported in a funnel with a small pledget of glass wool, into a suitable container. Extract again with two 15-mL portions of chloroform, again collecting the clarified organic phase. Wash the sodium sulfate and the tip of the funnel with 5 mL of chloroform. Evaporate the combined organic phases under reduced pressure, at a temperature below 45°, add 1 mL of chloroform, and mix to dissolve the alkaloids, taking care to wet the sides of the container.

Procedure—Record from the *Standard curves* the quantities, in mg, of atropine and scopolamine in the weight of the specimen taken. Proceed as directed for *Procedure* in the *Assay* under *Belladonna Extract,* through the next-to-the-last sentence. Add the quantity, in mg, of atropine and scopolamine, and multiply by 50 to obtain the weight, in mg, of alkaloids in the portion of Belladonna Leaf taken.

Belladonna Tincture

» Belladonna Tincture yields, from each 100 mL, not less than 27 mg and not more than 33 mg of the alkaloids of belladonna leaf.

Belladonna Leaf, in moderately coarse powder	100 g
To make about	1000 mL

Prepare a tincture by Process P as modified for assayed *Tinctures* (see *Pharmaceutical Dosage Forms* ⟨1151⟩), using a mixture of 3 volumes of alcohol and 1 volume of water as the menstruum. Finally adjust the Tincture to contain, in each 100 mL, 30 mg of the alkaloids of belladonna leaf.

Packaging and storage—Preserve in tight, light-resistant containers, and avoid exposure to direct sunlight and to excessive heat.

Reference standards—*USP Atropine Sulfate Reference Standard*—Keep container tightly closed and protected from light. [*Caution—Avoid contact.*] Dry at 100° to constant weight before using. *USP Homatropine Hydrobromide Reference Standard*—Dry at 105° for 2 hours before using. *USP Scopolamine Hydrobromide Reference Standard*—Store tightly closed, protected from light. Dry at 105° for 3 hours before using.

Alcohol content, *Method II* ⟨611⟩: between 65.0% and 70.0% of C_2H_5OH, determined by the gas-liquid chromatographic procedure, acetone being used as the internal standard.

Assay—Proceed with Belladonna Tincture as directed in the *Assay* under *Belladonna Leaf*, but in preparing the *Assay preparation*, pipet 2 mL of Belladonna Tincture (in place of "10 mL of this solution") into a 60-mL separator containing 10 mL of dilute sulfuric acid (1 in 350). Record from the *Standard curve* the quantities, in mg, of atropine and scopolamine in the specimen. Add the quantity, in mg, of atropine and scopolamine, and multiply by 50 to obtain the weight, in mg, of alkaloids per 100 mL.

Bendroflumethiazide

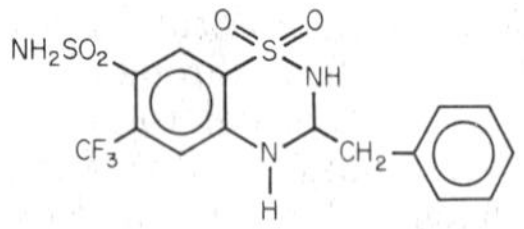

$C_{15}H_{14}F_3N_3O_4S_2$ 421.41

2*H*-1,2,4-Benzothiadiazine-7-sulfonamide, 3,4-dihydro-3-(phenylmethyl)-6-(trifluoromethyl)-, 1,1-dioxide.

3-Benzyl-3,4-dihydro-6-(trifluoromethyl)-2*H*-1,2,4-benzothiadiazine-7-sulfonamide 1,1-dioxide [*73-48-3*].

» Bendroflumethiazide contains not less than 98.0 percent and not more than 102.0 percent of $C_{15}H_{14}F_3N_3O_4S_2$, calculated on the anhydrous basis.

Packaging and storage—Preserve in tight containers.

Reference standards—*USP Bendroflumethiazide Reference Standard*—Dry over silica gel for 4 hours before using. *USP 2,4-Disulfamyl-5-trifluoromethylaniline Reference Standard*—Keep container tightly closed and protected from light. Dry in vacuum over silica gel for 4 hours before using.

Identification—

A: The infrared absorption spectrum of a potassium bromide dispersion of it, previously dried over silica gel for 4 hours, exhibits maxima only at the same wavelengths as that of a similar preparation of USP Bendroflumethiazide RS.

B: The ultraviolet absorption spectrum of a 1 in 100,000 solution in methanol exhibits maxima and minima at the same wavelengths as that of a similar solution of USP Bendroflumethiazide RS, concomitantly measured, and the respective absorptivities, calculated on the anhydrous basis, at the wavelength of maximum absorbance at about 271 nm do not differ by more than 4.0%.

C: Mix 5 mL of dilute hydrochloric acid (1 in 2) with 20 mg of Bendroflumethiazide, boil gently for 1 minute, and cool in an ice bath. Add in succession 0.5 mL of sodium nitrite solution (1 in 1000), 0.5 mL of ammonium sulfamate solution (1 in 200), and 0.5 mL of *N*-(1-naphthyl)ethylenediamine dihydrochloride solution (1 in 1000): a deep red color is produced.

Water, *Method I* ⟨921⟩: not more than 0.5%.

Residue on ignition ⟨281⟩: not more than 0.2%.

Heavy metals, *Method II* ⟨231⟩: 0.002%.

Selenium ⟨291⟩—The absorbance from the *Test Solution* prepared with 100 mg of Bendroflumethiazide and 100 mg of magnesium oxide, is not greater than one-half that from the *Standard Solution* (0.003%).

Limit of 2,4-disulfamyl-5-trifluoromethylaniline—[NOTE—Use low-actinic glassware for the *Test preparation* and the *Standard preparation.*]

Mobile phase—Dissolve 5.62 g of sodium chloride and 1.97 g of anhydrous sodium acetate in 1000 mL of water in a 2-liter volumetric flask. Add 4.0 mL of glacial acetic acid and 800 mL of methanol, dilute with water to volume, mix, filter, and degas.

Standard preparation—Dissolve an accurately weighed quantity of USP 2,4-Disulfamyl-5-trifluoromethylaniline RS in methanol to obtain a solution having a known concentration of about 0.75 µg per mL.

Test preparation—Transfer about 25 mg of Bendroflumethiazide, accurately weighed, to a 100-mL volumetric flask, add methanol to volume, and mix. Transfer 10.0 mL of this solution to a 50-mL volumetric flask, dilute with methanol to volume, and mix.

Chromatographic system (see *Chromatography* ⟨621⟩)—The liquid chromatograph is equipped with a 270-nm detector and a 4.6-mm × 30-cm column that contains packing L11 maintained at a temperature of 35 ± 5°. The flow rate is about 1.5 mL per minute. Chromatograph five replicate injections of the *Standard preparation*, and record the peak responses as directed under *Procedure:* the relative standard deviation is not more than 3.0%, and the resolution factor between the methanol and 2,4-disulfamyl-5-trifluoromethylaniline is not less than 1.4.

Procedure—Separately inject equal volumes (about 20 µL) of the *Standard preparation* and the *Test preparation* into the chromatograph, record the chromatograms, and measure the peak response of the 2,4-disulfamyl-5-trifluoromethylaniline. Calculate the percentage of 2,4-disulfamyl-5-trifluoromethylaniline in the Bendroflumethiazide taken by the formula:

$$100(C_S/C_U)(r_U/r_S),$$

in which C_S is the concentration, in µg per mL, of USP 2,4-Disulfamyl-5-trifluoromethylaniline RS in the *Standard preparation*, C_U is the concentration, in µg per mL, of bendroflumethiazide in the *Test preparation*, and r_U and r_S are the peak responses obtained from the *Test preparation* and the *Standard preparation*, respectively. Not more than 1.5% of 2,4-disulfamyl-5-trifluoromethylaniline is found.

Assay—Dissolve about 190 mg of Bendroflumethiazide, accurately weighed, in 80 mL of pyridine in a 250-mL, tall-form beaker in a well-ventilated hood. Add 3 drops of a saturated solution of azo violet in methanol, cover the beaker, and gently bubble nitrogen through the solution for 5 minutes, being careful to avoid any contact between the solution and the cover. Raise the nitrogen delivery tube above the solution surface and, maintaining a gentle flushing with nitrogen and stirring with a magnetic or mechanical stirring device, add 0.1 *N* sodium methoxide VS from a 10-mL buret inserted through an opening in the cover. Titrate to a blue end-point, approaching the end-point at a rate of 1 or 2 drops per second. Perform a blank determination, and make any necessary correction. Each mL of 0.1 *N* sodium methoxide is equivalent to 21.07 mg of $C_{15}H_{14}F_3N_3O_4S_2$.

Bendroflumethiazide Tablets

» Bendroflumethiazide Tablets contain not less than 90.0 percent and not more than 110.0 percent of the labeled amount of $C_{15}H_{14}F_3N_3O_4S_2$.

Packaging and storage—Preserve in tight containers.

Reference standard—*USP Bendroflumethiazide Reference Standard*—Dry over silica gel for 4 hours before using.

Identification—The retention time of the major peak in the chromatogram of the *Assay preparation* corresponds to that of the *Standard preparation* obtained as directed in the *Assay.*

Dissolution ⟨711⟩—[NOTE—Protect solutions from light throughout this test.]

Medium: 0.1 *N* hydrochloric acid; 900 mL.

Apparatus 2: 50 rpm.

Time: 45 minutes.

Procedure—Determine the amount of $C_{15}H_{14}F_3N_3O_4S_2$ dissolved from ultraviolet absorbances at the wavelength of maximum absorbance at about 271 nm of filtered portions of the solution under test, suitably diluted with water, if necessary, in comparison with a Standard solution having a known concentration of USP Bendroflumethiazide RS in the same medium.

Tolerances—Not less than 75% (*Q*) of the labeled amount of $C_{15}H_{14}F_3N_3O_4S_2$ is dissolved in 45 minutes.

Uniformity of dosage units ⟨905⟩: meet the requirements.

Assay—[NOTE—Use low-actinic glassware for the *Assay preparation* and the *Standard preparation.*]

Mobile phase—Dissolve 5.62 g of sodium chloride and 1.97 g of anhydrous sodium acetate in 1000 mL of water in a 2-liter volumetric flask. Add 4.0 mL of glacial acetic acid and 800 mL of methanol, dilute with water to volume, mix, filter, and degas.

Standard preparation—Dissolve an accurately weighed quantity of USP Bendroflumethiazide RS in methanol, and dilute quantitatively and stepwise with methanol to obtain a solution having a known concentration of about 50 µg per mL.

Assay preparation—Weigh and finely powder not less than 20 Bendroflumethiazide Tablets. Transfer an accurately weighed portion of the powder, equivalent to about 5 mg of bendroflumethiazide, to a 100-mL volumetric flask, add about 70 mL of methanol, and sonicate for 15 minutes, with occasional shaking. Dilute with methanol to volume, mix, and centrifuge a portion of the solution for 15 minutes.

Chromatographic system (see *Chromatography* ⟨621⟩)—The liquid chromatograph is equipped with a 270-nm detector and a 4.6-mm × 30-cm column that contains packing L11 maintained at a temperature of 35 ± 5°. The flow rate is about 1.5 mL per minute. Chromatograph five replicate injections of the *Standard preparation*, and record the peak responses as directed under *Procedure:* the relative standard deviation is not more than 3.0%, and the tailing factor is not more than 2.0.

Procedure—Separately inject equal volumes (about 20 µL) of the *Standard preparation* and the *Assay preparation* into the chromatograph, record the chromatograms, and measure the response of the major peak. Calculate the quantity, in mg, of $C_{15}H_{14}F_3N_3O_4S_2$ in the portion of Tablets taken by the formula:

$$0.1C(r_U/r_S),$$

in which C is the concentration, in µg per mL, of USP Bendroflumethiazide RS in the *Standard preparation*, and r_U and r_S are the peak responses obtained from the *Assay preparation* and the *Standard preparation*, respectively.

Benoxinate Hydrochloride

$COOCH_2CH_2N(C_2H_5)_2$ · $O(CH_2)_3CH_3$ · NH_2 · HCl

$C_{17}H_{28}N_2O_3 \cdot HCl$ 344.88

Benzoic acid, 4-amino-3-butoxy-, 2-(diethylamino)ethyl ester, monohydrochloride.

2-(Diethylamino)ethyl 4-amino-3-butoxybenzoate monohydrochloride [*5987-82-6*].

» Benoxinate Hydrochloride contains not less than 98.5 percent and not more than 101.5 percent of $C_{17}H_{28}N_2O_3 \cdot HCl$, calculated on the dried basis.

Packaging and storage—Preserve in well-closed containers.

Reference standard—*USP Benoxinate Hydrochloride Reference Standard*—Dry at 105° for 3 hours before using.

Identification—

A: The infrared absorption spectrum of a potassium bromide dispersion of it, previously dried, exhibits maxima only at the same wavelengths as that of a similar preparation of USP Benoxinate Hydrochloride RS.

B: The ultraviolet absorption spectrum of a solution (1 in 67,000) exhibits maxima and minima at the same wavelengths as that of a similar solution of USP Benoxinate Hydrochloride RS, concomitantly measured.

C: A solution (1 in 100) responds to the tests for *Chloride* ⟨191⟩.

Loss on drying ⟨731⟩—Dry it at 105° for 3 hours: it loses not more than 1.0% of its weight.

Residue on ignition ⟨281⟩: not more than 0.2%.

Ordinary impurities ⟨466⟩—

Test solution: methanol.

Standard solution: methanol.

Eluant: a mixture of chloroform, cyclohexane, and diethylamine (5:4:1).

Visualization: 12.

Assay—Dissolve about 250 mg of Benoxinate Hydrochloride, accurately weighed, in a mixture of 20 mL of glacial acetic acid and 20 mL of acetic anhydride, and titrate with 0.1 *N* perchloric acid VS, determining the end-point potentiometrically. Perform a blank determination, and make any necessary correction. Each mL of 0.1 *N* perchloric acid is equivalent to 34.49 mg of $C_{17}H_{28}N_2O_3 \cdot HCl$.

Benoxinate Hydrochloride Ophthalmic Solution

» Benoxinate Hydrochloride Ophthalmic Solution is a sterile solution of Benoxinate Hydrochloride in water. It contains not less than 95.0 percent and not more than 105.0 percent of the labeled amount of $C_{17}H_{28}N_2O_3 \cdot HCl$.

Packaging and storage—Preserve in tight containers.

Reference standard—*USP Benoxinate Hydrochloride Reference Standard*—Dry at 105° for 3 hours before using.

Identification—Dilute a volume of Solution, equivalent to about 50 mg of benoxinate hydrochloride, with 0.01 *N* hydrochloric acid to 25 mL, and proceed as directed under *Identification—Organic Nitrogenous Bases* ⟨181⟩, beginning with "Transfer the liquid to a separator": the solution meets the requirements of the test.

Sterility—It meets the requirements under *Sterility Tests* ⟨71⟩.

pH ⟨791⟩: between 3.0 and 6.0.

Assay—

Standard preparation—Dissolve an accurately weighed quantity of USP Benoxinate Hydrochloride RS in 0.1 *N* hydrochloric acid to obtain a solution having a known concentration of about 400 µg per mL.

Assay preparation—Transfer a volume of Benoxinate Hydrochloride Ophthalmic Solution, equivalent to about 20 mg of benoxinate hydrochloride, to a separator containing 15 mL of water, add 1 mL of ammonium hydroxide, and extract with five 20-mL portions of ether. Wash the combined ether extracts with 10 mL of water, extract the water washing with 10 mL of ether, and add this ether extract to the main extract. Extract the ether solution with three 5-mL portions of 0.1 *N* hydrochloric acid, collect the acid extracts in a 50-mL volumetric flask, dilute with 0.1 *N* hydrochloric acid to volume, and mix.

Procedure—Transfer 5.0 mL each of the *Standard preparation*, the *Assay preparation*, and 0.1 *N* hydrochloric acid to provide a blank, to separate 200-mL volumetric flasks. Dilute the contents of each flask with water to volume, and mix. Concomitantly determine the absorbances of the solutions in 1-cm cells at the wavelength of maximum absorbance at about 308 nm, with a suitable spectrophotometer, using the blank to set the instrument. Calculate the quantity, in mg, of $C_{17}H_{28}N_2O_3 \cdot HCl$ in each mL of the Ophthalmic Solution taken by the formula:

$$(0.05C/V)(A_U/A_S),$$

in which C is the concentration, in µg per mL, of USP Benoxinate Hydrochloride RS in the *Standard preparation*, V is the volume, in mL, of Ophthalmic Solution taken, and A_U and A_S are the absorbances of the solutions from the *Assay preparation* and the *Standard preparation*, respectively.

Bentonite—*see* Bentonite NF

Bentonite Magma—*see* Bentonite Magma NF

Bentonite, Purified—*see* Bentonite, Purified NF

Benzaldehyde—*see* Benzaldehyde NF

Benzaldehyde Elixir, Compound—*see* Benzaldehyde Elixir, Compound NF

Benzalkonium Chloride—*see* Benzalkonium Chloride NF

Benzalkonium Chloride Solution—*see* Benzalkonium Chloride Solution NF

Benzethonium Chloride

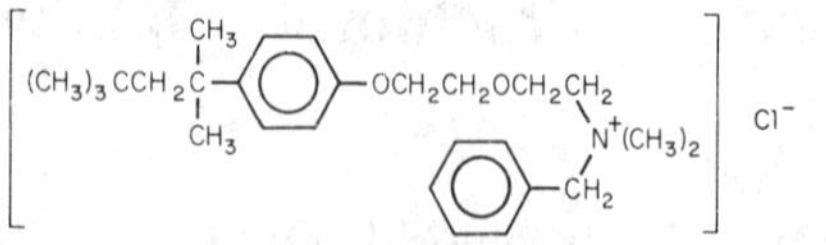

$C_{27}H_{42}ClNO_2$ 448.09

Benzenemethanaminium, *N,N*-dimethyl-*N*-[2-[2-[4-(1,1,3,3-tetramethylbutyl)phenoxy]ethoxy]ethyl]-, chloride.

Benzyldimethyl[2-[2-[*p*-(1,1,3,3-tetramethylbutyl)-phenoxy]ethoxy]ethyl]ammonium chloride [*121-54-0*].

» Benzethonium Chloride contains not less than 97.0 percent and not more than 103.0 percent of $C_{27}H_{42}ClNO_2$, calculated on the dried basis.

Packaging and storage—Preserve in tight, light-resistant containers.

Identification—

A: To 1 mL of a solution (1 in 100) add 2 mL of alcohol, 0.5 mL of 2 *N* nitric acid, and 1 mL of silver nitrate TS: a white precipitate, which is insoluble in 2 *N* nitric acid but soluble in 6 *N* ammonium hydroxide, is formed.

B: A solution (1 in 100) forms precipitates with 2 *N* nitric acid and with mercuric chloride TS, both of which dissolve upon the addition of alcohol.

C: Dissolve 0.1 g in 1 mL of sulfuric acid, add 0.1 g of potassium nitrate, and heat on a steam bath for 3 minutes. Cautiously dilute the solution with water to 10 mL, add 0.5 g of granulated zinc, and warm the mixture for 10 minutes. Cool, add 0.2 g of sodium nitrite to 1 mL of the clear liquid, and add this mixture to 20 mg of naphthol potassium disulfonate in 1 mL of ammonium hydroxide: the solution turns orange-red, and a brown precipitate may be formed.

Melting range ⟨741⟩: between 158° and 163°, the specimen having been dried previously.

Loss on drying ⟨731⟩—Dry it at 105° for 4 hours: it loses not more than 5.0% of its weight.

Residue on ignition ⟨281⟩: not more than 0.1%.

Ammonium compounds—To 5 mL of a solution (1 in 50) add 3 mL of 1 *N* sodium hydroxide, and heat to boiling: the odor of ammonia is not perceptible.

Assay—Dissolve about 0.3 g of Benzethonium Chloride, accurately weighed, in 75 mL of water contained in a glass-stoppered, 250-mL flask. Add 0.4 mL of bromophenol blue solution (1 in 2000), 10 mL of chloroform, and 1 mL of 1 *N* sodium hydroxide. Titrate with 0.02 *M* sodium tetraphenylboron VS until the blue color disappears from the chloroform layer. Add the last portions of the sodium tetraphenylboron solution dropwise, agitating vigorously after each addition. Each mL of 0.02 *M* sodium tetraphenylboron is equivalent to 8.962 mg of $C_{27}H_{42}ClNO_2$.

Benzethonium Chloride Topical Solution

» Benzethonium Chloride Topical Solution contains not less than 95.0 percent and not more than 105.0 percent of the labeled amount of $C_{27}H_{42}ClNO_2$.

Packaging and storage—Preserve in tight, light-resistant containers.

Identification—The residue obtained by evaporating, on a steam bath, a volume of Topical Solution, equivalent to about 200 mg of benzethonium chloride, responds to the *Identification tests* under *Benzethonium Chloride*.

Oxidizing substances—To 5 mL add 0.5 mL of potassium iodide TS and a few drops of 3 *N* hydrochloric acid: the solution does not acquire a yellow color.

Nitrites—To 1 drop of Topical Solution on a spot plate add 1 drop each of glacial acetic acid, sulfanilic acid in acetic acid (1 in 100), and α-naphthylamine–acetic acid solution (prepared by boiling 30 mg of α-naphthylamine in 70 mL of water, decanting the colorless solution from the blue-violet residue, and mixing with 30 mL of glacial acetic acid): no red color develops in the resulting solution within 10 minutes.

Assay—Transfer a volume of Benzethonium Chloride Topical Solution, equivalent to about 200 mg of benzethonium chloride, to a glass-stoppered flask, and proceed as directed in the *Assay* under *Benzethonium Chloride*, beginning with "Add 0.4 mL of bromophenol blue solution (1 in 2000)."

Benzethonium Chloride Tincture

» Benzethonium Chloride Tincture contains, in each 100 mL, not less than 190 mg and not more than 210 mg of $C_{27}H_{42}ClNO_2$.

Benzethonium Chloride	2 g
Alcohol	685 mL
Acetone	100 mL
Purified Water, a sufficient quantity, to make	1000 mL

Dissolve the Benzethonium Chloride in a mixture of the Alcohol and the Acetone. Add sufficient Purified Water to make 1000 mL.

NOTE—Benzethonium Chloride Tincture may be colored by the addition of any suitable color or combination of colors certified by the FDA for use in drugs.

Packaging and storage—Preserve in tight, light-resistant containers.

Identification—The residue obtained by evaporating 50 mL on a steam bath responds to *Identification tests A* and *B* under *Benzethonium Chloride*.

Specific gravity ⟨841⟩: between 0.868 and 0.876.

Alcohol and acetone content—To a 100-mL volumetric flask transfer 20.0 mL of Benzethonium Chloride Tincture and 5.0 mL of methanol as the internal standard, dilute with water to volume, and mix. Similarly prepare four 100-mL standard solutions in water, each containing 5.0 mL of methanol as the internal standard, and individually containing, respectively, 11.0 mL of dehydrated alcohol, 14.0 mL of dehydrated alcohol, 1.7 mL of acetone, and 2.2 mL of acetone. Inject 0.8 µL of the solution containing the substance under test into a suitable gas chromatograph equipped with a flame-ionization detector, and record the chromatogram. Similarly and successively record the chromatograms for 0.8-µL injected volumes of the four standard solutions. Under typical conditions, the instrument contains a 120-cm × 4-mm column packed with a suitable type of support, such as 80- to 100-mesh S3; the column is maintained at about 120°; the injection port and detector block are maintained at about 240°; and dry helium is used as the carrier gas at a flow rate of about 90 mL per minute. From the respective chromatograms obtained as described previously, calculate the ratios of peak areas for alcohol to internal standard and for acetone to internal standard.

Calculate the percentage of alcohol and of acetone in the Tincture by the formula:

$$[A(Y - Z) + B(Z - X)]/(Y - X),$$

in which *A* and *B* are the percentage of alcohol, or of acetone, in the lower and higher standards, respectively; *X*, *Y*, and *Z* are the ratios of the alcohol peak areas, or the acetone peak areas, to the internal standard peak areas for the lower standard, higher standard, and the material under test, respectively: the content of C_2H_5OH is between 62.0% and 68.0%, and the content of acetone (C_3H_6O) is between 9.0% and 11.0%.

Assay—Transfer 50.0 mL of Benzethonium Chloride Tincture to a 150-mL beaker, and add, with continuous stirring, 10 mL of sodium tetraphenylboron solution (1 in 40). Cover, and allow to stand for 16 hours. Decant the supernatant liquid into a tared sintered-glass crucible, applying vacuum filtration. Suspend the precipitate in 20 mL of water, and transfer the precipitate to the crucible, washing well with water. Dry the precipitate and the crucible at 105° for 1 hour, cool, and weigh. The weight of the precipitate so obtained, multiplied by 0.6122, represents its equivalent of $C_{27}H_{42}ClNO_2$.

Benzocaine

$NH_2-C_6H_4-COOC_2H_5$

$C_9H_{11}NO_2$ 165.19
Benzoic acid, 4-amino-, ethyl ester.
Ethyl *p*-aminobenzoate [*94-09-7*].

» Benzocaine, dried over phosphorus pentoxide for 3 hours, contains not less than 98.0 percent and not more than 101.0 percent of $C_9H_{11}NO_2$.

Packaging and storage—Preserve in well-closed containers.

Reference standard—*USP Benzocaine Reference Standard*—Dry over phosphorus pentoxide for 3 hours before using.

Identification—

A: The infrared absorption spectrum of a potassium bromide dispersion of it, previously dried over phosphorus pentoxide for 3 hours, exhibits maxima only at the same wavelengths as that of a similar preparation of USP Benzocaine RS.

B: The ultraviolet absorption spectrum of a 1 in 200,000 solution in chloroform exhibits maxima and minima at the same wavelengths as that of a similar preparation of USP Benzocaine RS, concomitantly measured, and the respective absorptivities, calculated on the dried basis, at the wavelength of maximum absorbance at about 278 nm, do not differ by more than 3.0%.

C: Dissolve about 20 mg in 10 mL of water with the aid of a few drops of 3 *N* hydrochloric acid, and add 5 drops of a solution of sodium nitrite (1 in 10), followed by 2 mL of a solution of 100 mg of 2-naphthol in 5 mL of 1 *N* sodium hydroxide: an orange-red precipitate is formed.

Melting range, *Class I* ⟨741⟩: between 88° and 92°, but the range between beginning and end of melting does not exceed 2°.

Reaction—Dissolve 1.0 g in 10 mL of neutralized alcohol: a clear solution results. Dilute this solution with 10 mL of water, and add 2 drops of phenolphthalein TS and 1 drop of 0.10 *N* sodium hydroxide: a red color is produced.

Loss on drying ⟨731⟩—Dry it over phosphorus pentoxide for 3 hours: it loses not more than 1.0% of its weight.

Residue on ignition ⟨281⟩: not more than 0.1%.

Chloride—To a solution of 200 mg in 5 mL of alcohol, previously acidified with a few drops of diluted nitric acid, add a few drops of silver nitrate TS: no turbidity is produced immediately.

Heavy metals, *Method II* ⟨231⟩: 0.001%.

Readily carbonizable substances ⟨271⟩—Dissolve 500 mg in 5 mL of sulfuric acid TS: the solution has no more color than Matching Fluid A.

Ordinary impurities ⟨466⟩—

Test solution: dehydrated alcohol.

Standard solution: dehydrated alcohol.

Application volume: 10 µL.

Eluant: chloroform containing about 0.75% of dehydrated alcohol as a preservative, in a nonequilibrated chamber.

Visualization: 1.

Limit—The total of any ordinary impurities observed does not exceed 1%.

Assay—Dissolve about 300 mg of Benzocaine, previously dried and accurately weighed, in a mixture of 100 mL of water and 15 mL of hydrochloric acid. Cool the solution in an ice bath to about 10°, and titrate with 0.1 *M* sodium nitrite VS until a blue color is produced immediately when the titrated solution is applied on starch iodide paper. When the titration is complete, the end-point is reproducible after the mixture has been allowed to stand for 5 minutes. Perform a blank determination, and make any necessary correction. Each mL of 0.1 *M* sodium nitrite is equivalent to 16.52 mg of $C_9H_{11}NO_2$.

Benzocaine, and Phenylephrine Hydrochloride Otic Solution, Antipyrine,—*see* Antipyrine, Benzocaine, and Phenylephrine Hydrochloride Otic Solution

Benzocaine Topical Aerosol

» Benzocaine Topical Aerosol is a solution of Benzocaine in a pressurized container. It contains not less than 90.0 percent and not more than 110.0 percent of the labeled amount of $C_9H_{11}NO_2$.

Packaging and storage—Preserve in pressurized containers, and avoid exposure to excessive heat.

Identification—Spray a portion of Topical Aerosol into a beaker, and heat on a steam bath for a few minutes to expel residual propellant. A portion of this solution, equivalent to about 5 mg of benzocaine, responds to the *Identification test* under *Benzocaine Topical Solution.*

Other requirements—It meets the requirements for *Leak Testing* under *Aerosols* ⟨601⟩.

Assay—Spray a portion of Benzocaine Topical Aerosol into a beaker, and heat on a steam bath for a few minutes to expel residual propellant. Cool, and transfer an accurately weighed portion of the solution, equivalent to about 200 mg of benzocaine, to a 250-mL beaker. Add 50 mL of water and 5 mL of hydrochloric acid, and stir. Cool the solution in an ice bath to about 10°, and titrate slowly with 0.1 *M* sodium nitrite VS, determining the end-point potentiometrically, using a calomel-platinum electrode system. Perform a blank determination, and make any necessary correction. Each mL of 0.1 *M* sodium nitrite is equivalent to 16.52 mg of $C_9H_{11}NO_2$.

Benzocaine Cream

» Benzocaine Cream contains not less than 90.0 percent and not more than 110.0 percent of the labeled amount of $C_9H_{11}NO_2$ in a suitable cream base.

Packaging and storage—Preserve in tight containers, protected from light, and avoid prolonged exposure to temperatures exceeding 30°.

Identification—Place an amount of Cream, equivalent to about 50 mg of benzocaine, in a small beaker, add 20 mL of 0.5 *N* hydrochloric acid, warm gently to disperse the cream, cool, and filter. To 10 mL of the filtrate add 5 drops of a solution of sodium nitrite (1 in 10) and 2 drops of methyl red TS, and neutralize with 1 *N* sodium hydroxide. Add 2 mL of a solution of 100 mg of 2-naphthol in 5 mL of 1 *N* sodium hydroxide: an orange-red precipitate is formed.

Microbial limits—It meets the requirements of the tests for absence of *Staphylococcus aureus* and *Pseudomonas aeruginosa* under *Microbial Limit Tests* ⟨61⟩.

Minimum fill ⟨755⟩: meets the requirements.

Assay—Weigh accurately an amount of Benzocaine Cream, equivalent to about 200 mg of benzocaine, in a tared 250-mL beaker. Add 50 mL of water and 5 mL of hydrochloric acid, and stir by mechanical means, with gentle warming, until solution is effected. Cool the solution in an ice bath to about 10°, and titrate slowly with 0.1 *M* sodium nitrite VS, determining the end-point potentiometrically using a calomel-platinum electrode system. Each mL of 0.1 *M* sodium nitrite is equivalent to 16.52 mg of $C_9H_{11}NO_2$.

Benzocaine Ointment

» Benzocaine Ointment contains not less than 90.0 percent and not more than 110.0 percent of the labeled amount of $C_9H_{11}NO_2$ in a suitable ointment base.

Packaging and storage—Preserve in tight containers, protected from light, and avoid prolonged exposure to temperatures exceeding 30°.

Identification—

Ointments having water-soluble bases—Transfer an amount of Ointment, equivalent to about 5 mg of benzocaine, to a small beaker, add 20 mL of 0.5 *N* hydrochloric acid, warm gently to disperse the ointment, cool, and filter, if necessary. To 10 mL of the filtrate add 5 drops of a solution of sodium nitrite (1 in 10) and 2 drops of methyl red TS, and neutralize with 1 *N* sodium hydroxide. Add 2 mL of a solution of 100 mg of 2-naphthol in 5 mL of sodium hydroxide TS: an orange-red precipitate is formed.

Ointments having water-insoluble bases—Transfer an amount of Ointment, equivalent to about 2.5 mg of benzocaine, to a small separator, and dissolve in 20 mL of ether. Extract with 10 mL of 0.5 *N* hydrochloric acid, and filter the aqueous phase through paper into a small beaker. Add 5 drops of a solution of sodium nitrite (1 in 10) and 2 drops of methyl red TS, and neutralize with 1 *N* sodium hydroxide. Add 2 mL of a solution of 100 mg of 2-naphthol in 5 mL of 1 *N* sodium hydroxide: an orange-red precipitate is formed.

Microbial limits—It meets the requirements of the tests for absence of *Staphylococcus aureus* and *Pseudomonas aeruginosa* under *Microbial Limit Tests* ⟨61⟩.

Minimum fill ⟨755⟩: meets the requirements.

Assay—

Ointments having water-soluble bases—Weigh accurately an amount of Benzocaine Ointment, equivalent to about 200 mg of benzocaine, in a tared 250-mL beaker. Add 50 mL of water and 5 mL of hydrochloric acid, and stir by mechanical means until solution is effected. Cool the solution in an ice bath to about 10°, and titrate slowly with 0.1 *M* sodium nitrite VS, determining the end-point potentiometrically using a calomel-platinum electrode system. Perform a blank determination, and make any necessary correction. Each mL of 0.1 *M* sodium nitrite is equivalent to 16.52 mg of $C_9H_{11}NO_2$.

Ointments having water-insoluble bases—Weigh accurately an amount of Benzocaine Ointment, equivalent to about 200 mg of benzocaine, and transfer to a 125-mL separator. Suspend the ointment in 50 mL of ether by shaking, and extract with three successive 25-mL portions of 1 *N* hydrochloric acid, filtering each portion into a 250-mL beaker, and proceed as directed under *Ointments having water-soluble bases*, beginning with "Cool the solution."

Benzocaine Otic Solution

» Benzocaine Otic Solution contains not less than 90.0 percent and not more than 110.0 percent of the labeled amount of $C_9H_{11}NO_2$.

Packaging and storage—Preserve in tight, light-resistant containers.

Identification—Transfer a volume of Otic Solution, equivalent to about 2.5 mg of benzocaine, to a small beaker, add 10 mL of water, and dissolve with the aid of a few drops of 3 *N* hydrochloric acid. Add 5 drops of sodium nitrite solution (1 in 10), then add 2 mL of a solution of 100 mg of 2-naphthol in 5 mL of 1 *N* sodium hydroxide: an orange-red precipitate is formed.

Microbial limits ⟨61⟩—It meets the requirements of the tests for absence of *Salmonella* species and *Escherichia coli* and for absence of *Staphylococcus aureus* and *Pseudomonas aeruginosa;* the total aerobic microbial count is less than 100 per mL.

Assay—Transfer an accurately measured volume of Benzocaine Otic Solution, equivalent to about 400 mg of benzocaine, to a beaker, add 150 mL of cold 1.6 *N* hydrochloric acid, and stir until a fine dispersion is obtained. Cool the solution in an ice bath to about 10°, and titrate slowly with 0.1 *M* sodium nitrite VS determining the end-point potentiometrically, using a calomel-platinum electrode system. Perform a blank determination, and make any necessary correction. Each mL of 0.1 *M* sodium nitrite is equivalent to 16.52 mg of $C_9H_{11}NO_2$.

Benzocaine Otic Solution, Antipyrine and—*see* Antipyrine and Benzocaine Otic Solution

Benzocaine Topical Solution

» Benzocaine Topical Solution is a solution of Benzocaine in a suitable solvent. It contains not less than 90.0 percent and not more than 110.0 percent of the labeled amount of $C_9H_{11}NO_2$. It contains a suitable antimicrobial agent.

Packaging and storage—Preserve in tight containers, protected from light, and avoid prolonged exposure to temperatures exceeding 30°.

Identification—Transfer an amount of Topical Solution, equivalent to about 5 mg of benzocaine, to a small beaker, add 20 mL of 0.5 *N* hydrochloric acid, warm gently to disperse, cool, and filter. To 10 mL of the filtrate add 5 drops of sodium nitrite solution (1 in 10) and 2 drops of methyl red TS, and neutralize with 1 *N* sodium hydroxide. Add 2 mL of a solution of 100 mg of 2-naphthol in 5 mL of 1 *N* sodium hydroxide: an orange-red precipitate is formed.

Microbial limits—It meets the requirements of the tests for absence of *Staphylococcus aureus* and *Pseudomonas aeruginosa* under *Microbial Limit Tests* ⟨61⟩.

Assay—Transfer an accurately weighed portion of Benzocaine Topical Solution, equivalent to about 200 mg of benzocaine, to a 250-mL beaker. Add 50 mL of water and 5 mL of hydrochloric acid, and stir. Cool the solution in an ice bath to about 10°, and titrate slowly with 0.1 *M* sodium nitrite VS, determining the end-point potentiometrically, using a calomel-platinum electrode system. Perform a blank determination, and make any necessary correction. Each mL of 0.1 *M* sodium nitrite is equivalent to 16.52 mg of $C_9H_{11}NO_2$.

Benzoic Acid

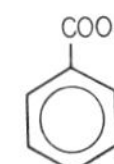

$C_7H_6O_2$ 122.12
Benzoic acid.
Benzoic acid [*65-85-0*].

» Benzoic Acid contains not less than 99.5 percent and not more than 100.5 percent of $C_7H_6O_2$, calculated on the anhydrous basis.

Packaging and storage—Preserve in well-closed containers.

Identification—It responds to the tests for *Benzoate* ⟨191⟩.

Congealing range ⟨651⟩: between 121° and 123°.

Water, *Method I* ⟨921⟩: not more than 0.7%, a 1 in 2 solution of methanol in pyridine being used as the solvent.

Residue on ignition ⟨281⟩: not more than 0.05%.

Arsenic, *Method II* ⟨211⟩: 3 ppm.

Heavy metals ⟨231⟩—Dissolve 2.0 g in 25 mL of acetone, and add 2 mL of water and 10 mL of hydrogen sulfide TS: any color produced is not darker than that of a control made with 25 mL of acetone, 2.0 mL of the standard lead solution, and 10 mL of hydrogen sulfide TS (0.001%).

Readily carbonizable substances ⟨271⟩—Dissolve 500 mg in 5 mL of sulfuric acid TS: the solution has no more color than Matching Fluid Q.

Readily oxidizable substances—Add 1.5 mL of sulfuric acid to 100 mL of water, heat to boiling, and add 0.1 *N* potassium permanganate, dropwise, until the pink color persists for 30 seconds. Dissolve 1.00 g of Benzoic Acid in the hot solution, and titrate with 0.1 *N* potassium permanganate VS to a pink color that persists for 15 seconds: not more than 0.50 mL of 0.10 *N* potassium permanganate is consumed.

Assay—Dissolve about 500 mg of Benzoic Acid, accurately weighed, in 25 mL of diluted alcohol that previously has been neutralized with 0.1 *N* sodium hydroxide, add phenolphthalein TS, and titrate with 0.1 *N* sodium hydroxide VS to a pink color. Each mL of 0.1 *N* sodium hydroxide is equivalent to 12.21 mg of $C_7H_6O_2$.

Benzoic and Salicylic Acids Ointment

» Benzoic and Salicylic Acids Ointment is Benzoic Acid and Salicylic Acid, present in a ratio of about 2 to 1, in a suitable ointment base. It contains not less than 90.0 percent and not more than 110.0 percent of the labeled amounts of benzoic acid ($C_7H_6O_2$) and salicylic acid ($C_7H_6O_3$).

Packaging and storage—Preserve in well-closed containers, and avoid exposure to temperatures exceeding 30°.

Labeling—Label Ointment to indicate the concentrations of Benzoic Acid and Salicylic Acid and to indicate whether the ointment base is water-soluble or water-insoluble.

Reference standards—*USP Benzoic Acid Reference Standard*—Dry over silica gel for 3 hours before using. *USP Salicylic Acid Reference Standard*—Dry over silica gel for 3 hours before using.

Identification—Dissolve an amount of Ointment, equivalent to about 60 mg of benzoic acid and 30 mg of salicylic acid, in 25 mL of a solvent consisting of equal volumes of chloroform and methanol. Prepare separate Standard solutions in the same solvent containing, respectively, 60 mg of USP Benzoic Acid RS in 25 mL and 30 mg of USP Salicylic Acid RS in 25 mL. Apply 5 µL of each solution at separate points 2.5 cm from the bottom edge of a 20- × 20-cm thin-layer chromatographic plate (see *Chromatography* ⟨621⟩) coated with a 0.25-mm layer of chromatographic silica gel mixture, and allow the spots to dry. Develop the chromatogram in a solvent system consisting of a mixture of chloroform, acetone, isopropyl alcohol, methanol, and ammonium hydroxide (30:30:15:15:10) until the solvent front has moved about three-fourths of the length of the plate. Remove the plate from the chromatographic chamber, mark the solvent front, and allow the solvent to evaporate. View the chromatogram under short-wavelength (254 nm) ultraviolet radiation: the two major fluorescent spots from the test solution correspond in color and in R_f value to those from the respective Standard solutions.

Minimum fill ⟨755⟩: meets the requirements.

Assay—

Ferric chloride–urea reagent—On the day of use, dissolve, without heating, 18 g of urea in a mixture of 2.5 mL of ferric chloride solution (6 in 10) and 12.5 mL of 0.05 *N* hydrochloric acid.

Column A—Insert a small pledget of glass wool above the stem constriction of a 20- × 2.5-cm chromatographic tube. Mix 1 g of chromatographic siliceous earth with 0.5 mL of dilute phosphoric acid (3 in 10) to form a uniform, fluffy mixture, transfer to the chromatographic tube, and pack evenly over the glass wool, exerting gentle pressure. Similarly, mix 4 g of chromatographic siliceous earth with 3 mL of *Ferric chloride–urea reagent*, and pack uniformly over the first layer. Cover the column with a pad of glass wool.

Column B—Insert a small pledget of glass wool above the stem constriction of a second 20- × 2.5-cm chromatographic tube. Mix 4 g of chromatographic siliceous earth with 2 mL of sodium bicarbonate solution (1 in 12), prepared just prior to use, to a uniform, fluffy mixture, and pack evenly over the glass wool. Cover the column with a pad of glass wool.

Assay preparation—Transfer an accurately weighed amount of Benzoic and Salicylic Acids Ointment, equivalent to about 100 mg of benzoic acid and 50 mg of salicylic acid, to a 250-mL volumetric flask, and dissolve in about 150 mL of chloroform by warming on a steam bath. Cool, dilute with chloroform to volume, and mix.

Procedure—Mount *Column A* directly over *Column B*, then pipet 10 mL of *Assay preparation* onto *Column A*, and allow it to pass into the column. Wash the columns with two 40-mL portions of chloroform, allowing the first portion to recede to the top of each column before adding the second portion. Discard the eluates, and separate the columns.

Salicylic acid content—Elute *Column A* with 95 mL of a 3 in 100 solution of glacial acetic acid in chloroform, collecting the eluate in a 100-mL volumetric flask. Dilute the contents of the flask with the same solvent to volume, and mix. Dissolve an accurately weighed quantity of USP Salicylic Acid RS in the same solvent, and dilute quantitatively and stepwise with this solvent to obtain a Standard solution having a known concentration of about 20 µg per mL. Concomitantly determine the absorbances of both solutions in 1-cm cells at the wavelength of maximum absorbance at about 311 nm, with a suitable spectrophotometer, using a 3 in 100 solution of glacial acetic acid in chloroform as the blank. Calculate the quantity, in mg, of salicylic acid ($C_7H_6O_3$) in the portion of Ointment taken by the formula:

$$2.5C(A_U / A_S),$$

in which C is the concentration, in µg per mL, of USP Salicylic Acid RS in the Standard solution, and A_U and A_S are the absorbances of the diluted eluate from *Column A* and the Standard solution, respectively.

Benzoic acid content—Elute *Column B* with 95 mL of a 3 in 100 solution of glacial acetic acid in chloroform, collecting the eluate in a 100-mL volumetric flask. Dilute the contents of the flask with the same solvent to volume, and mix. Dissolve an accurately weighed quantity of USP Benzoic Acid RS in the same solvent, and dilute quantitatively and stepwise with this solvent to obtain a Standard solution having a known concentration of about 40 µg per mL. Concomitantly determine the absorbances of both solutions in 1-cm cells at the wavelength of maximum absorbance at about 275 nm, with a suitable spectrophotometer, using a 3 in 100 solution of glacial acetic acid in chloroform as the blank. Calculate the quantity, in mg, of benzoic acid ($C_7H_6O_2$) in the portion of Ointment taken by the formula:

$$2.5C(A_U/A_S),$$

in which C is the concentration, in µg per mL, of USP Benzoic Acid RS in the Standard solution, and A_U and A_S are the absorbances of the diluted eluate from *Column B* and the Standard solution, respectively.

Benzoin

» Benzoin is the balsamic resin obtained from *Styrax benzoin* Dryander or *Styrax paralleloneurus* Perkins, known in commerce as Sumatra Benzoin, or from *Styrax tonkinensis* (Pièrre) Craib ex Hartwich, or other species of the Section *Anthostyrax* of the genus *Styrax*, known in commerce as Siam Benzoin (Fam. Styraceae).

Sumatra Benzoin yields not less than 75.0 percent of alcohol-soluble extractive, and Siam Benzoin yields not less than 90.0 percent of alcohol-soluble extractive.

Packaging and storage—Preserve in well-closed containers.

Labeling—Label it to indicate whether it is Sumatra Benzoin or Siam Benzoin.

Botanic characteristics—

Sumatra Benzoin—Blocks or lumps of varying size, made up of tears, compacted together, with a reddish brown, reddish gray, or grayish brown resinous mass. The tears are externally yellowish or rusty brown, milky white on fresh fracture; hard and brittle at ordinary temperatures but softened by heat and becoming gritty on chewing.

Siam Benzoin—Pebble-like tears of variable size and shape, compressed, yellowish brown to rusty brown externally, milky white on fracture, separate or very slightly agglutinated, hard and brittle at ordinary temperatures but softened by heat and becoming plastic on chewing.

Identification—

A: A solution in alcohol becomes milky upon the addition of water, and the mixture is acid to litmus paper.

B: Heat a few fragments in a test tube: Sumatra Benzoin evolves a sublimate consisting of plates and small, rod-like crystals of cinnamic acid and its esters that strongly polarize light. Siam Benzoin evolves a sublimate directly above the melted mass, consisting of numerous long, rod-shaped crystals of benzoic acid that do not strongly polarize light.

C: Heat about 500 mg in a test tube with 10 mL of potassium permanganate TS: only the Sumatra variety develops a faint odor of benzaldehyde.

Benzoic acid—Treat about 1 g of powdered Benzoin with 15 mL of warm carbon disulfide, filter through a small pledget of cotton, wash the cotton with an additional 5 mL of carbon disulfide, and allow the filtrate to evaporate spontaneously: the weight of the residue is not less than 6.0% (Sumatra Benzoin), or not less than 12.0% (Siam Benzoin) of the weight of Benzoin taken. This residue responds to the tests for *Benzoate* ⟨191⟩.

Acid-insoluble ash ⟨561⟩: not more than 1.0% in Sumatra Benzoin; not more than 0.5% in Siam Benzoin.

Foreign organic matter ⟨561⟩: not more than 1.0% in Siam Benzoin.

Assay—Place about 2 g of Benzoin, accurately weighed, in a tared extraction thimble, and insert the thimble in a continuous-extraction apparatus. Place about 100 mg of sodium hydroxide in the receiving flask of the apparatus, and extract the Benzoin with alcohol for 5 hours, or until completely extracted. Dry the extraction thimble containing the insoluble residue at 105° for 2 hours. On a separate portion of Benzoin, determine the water content by the *Azeotropic (Toluene Distillation) Method* (see *Water Determination* ⟨921⟩). Calculate the weight of water in the quantity of the Benzoin taken for assay, and subtract it from the original weight of the Benzoin taken. The difference between this result and the weight of the residue in the extraction thimble represents the alcohol-soluble extractive.

Compound Benzoin Tincture

» Prepare Compound Benzoin Tincture as follows.

Benzoin, in moderately coarse powder	100 g
Aloe, in moderately coarse powder	20 g
Storax	80 g
Tolu Balsam	40 g
To make	1000 mL

Prepare a *Tincture* by Process M (see *Pharmaceutical Dosage Forms* ⟨1151⟩), using alcohol as the menstruum.

Packaging and storage—Preserve in tight, light-resistant containers, and avoid exposure to direct sunlight and to excessive heat.

Labeling—Label it to indicate that it is flammable.

Specific gravity ⟨841⟩: between 0.870 and 0.885.

Alcohol content, *Method II* ⟨611⟩: between 74.0% and 80.0% of C_2H_5OH, the dilution to approximately 2% alcohol being made with methanol instead of with water.

Nonvolatile residue—Evaporate 3 mL of Tincture in a suitable tared dish on a steam bath, and dry the residue at 100° for 2 hours: the weight of the residue is between 525 mg and 675 mg.

Benzonatate

$CH_3(CH_2)_2CH_2NH-C_6H_4-COOCH_2CH_2(OCH_2CH_2)_nOCH_3$

$C_{30}H_{53}NO_{11}$ (av.) 603 (av.)

Benzoic acid, 4-(butylamino)-, 2,5,8,11,14,17,20,23,26-nonaoxaoctacos-28-yl ester.

2,5,8,11,14,17,20,23,26-Nonaoxaoctacosan-28-yl *p*-(butylamino)benzoate [*104-31-4*].

» Benzonatate contains not less than 95.0 percent and not more than 105.0 percent of $C_{30}H_{53}NO_{11}$.

Packaging and storage—Preserve in tight, light-resistant containers.

Reference standard—*USP Benzonatate Reference Standard*—After opening ampul, store in tightly closed, light-resistant container.

Identification—

A: The infrared absorption spectrum of a thin film of it exhibits maxima only at the same wavelengths as that of a similar preparation of USP Benzonatate RS.

B: The ultraviolet absorption spectrum of a solution (1 in 67,000) exhibits maxima and minima at the same wavelengths as that of a similar solution of USP Benzonatate RS, concomitantly measured.

Refractive index ⟨831⟩: between 1.509 and 1.511 at 20°.

Water, *Method I* ⟨921⟩: not more than 0.3%.

Residue on ignition ⟨281⟩: not more than 0.1%.

Arsenic, *Method I* ⟨211⟩: 1.5 ppm.

Chloride ⟨221⟩—Mix 20 mL of a solution (1 in 10) with 20 mL of water and 1 mL of nitric acid, shake for 1 hour, and allow to stand for 1 hour. Filter through a filter having a porosity of 0.2 µm, and to the filtrate add 1 mL of silver nitrate TS. Dilute with water to 50 mL, mix, and allow to stand protected from light for 10 minutes: the turbidity does not exceed that produced by 0.10 mL of 0.020 *N* hydrochloric acid (0.0035%).

Sulfate ⟨221⟩—Mix 5 mL of a solution (1 in 20) with 5 mL of water and 1 mL of 3 *N* hydrochloric acid, shake for 1 hour, and allow to stand for 1 hour. Filter through a filter having a porosity of 0.2 μm, and to the filtrate add 1 mL of barium chloride TS. Mix, and allow to stand for 10 minutes: the turbidity does not exceed that produced by 0.10 mL of 0.020 *N* sulfuric acid (0.04%).

Heavy metals, *Method II* ⟨231⟩: 0.001%.

Assay—Weigh accurately about 5 g of Benzonatate, and reflux with 25.0 mL of 0.5 *N* sodium hydroxide VS for 1 hour. Cool, add 25 mL of water and 10 drops of bromothymol blue TS, and titrate the excess alkali with 0.5 *N* hydrochloric acid VS. Perform a blank determination (see *Residual Titrations* under *Titrimetry* ⟨541⟩). Each mL of 0.5 *N* sodium hydroxide is equivalent to 301.5 mg of $C_{30}H_{53}NO_{11}$.

Benzonatate Capsules

» Benzonatate Capsules contain not less than 90.0 percent and not more than 110.0 percent of the labeled amount of benzonatate ($C_{30}H_{53}NO_{11}$ av.).

Packaging and storage—Preserve in tight, light-resistant containers.

Reference standard—*USP Benzonatate Reference Standard*—After opening ampul, store in tightly closed, light-resistant container.

Identification—

A: The contents of the Capsules meet the requirements of *Identification test A* under *Benzonatate.* If a difference is observed, or if excipients are present, use an amount of the contents of Capsules equivalent to about 100 mg of benzonatate, mixed with 25 mL of 0.01 *N* hydrochloric acid, and proceed as directed under *Identification—Organic Nitrogenous Bases* ⟨181⟩, beginning with "Transfer the liquid to a separator."

B: The contents of the Capsules respond to *Identification test B* under *Benzonatate.*

Uniformity of dosage units ⟨905⟩: meet the requirements.

Assay—

Standard preparation—Transfer about 50 mg of USP Benzonatate RS, accurately weighed, to a 100-mL volumetric flask, dilute with water to volume, and mix.

Assay preparation—Mix a number of Benzonatate Capsules, equivalent to about 500 mg of benzonatate, with 40 mL of chloroform in a suitable high-speed blender, and dilute with chloroform to 100.0 mL. Transfer 10.0 mL of this solution, equivalent to about 50 mg of benzonatate, to a 100-mL volumetric flask, and evaporate the chloroform on a steam bath with the aid of a current of air. Dissolve the residue in water, dilute with water to volume, and mix.

Procedure—Transfer 4.0 mL each of the *Standard preparation,* the *Assay preparation,* and water to provide the blank, to separate test tubes. To each tube add in succession 1.0 mL of 1 *M* hydroxylamine hydrochloride and 1.0 mL of 3.5 *N* sodium hydroxide, mixing after each addition. Allow to stand for 10 minutes, accurately timed, then add 1.0 mL of 3.5 *N* hydrochloric acid, mix, add 1.0 mL of ferric chloride solution (8 in 100), and mix. Allow to stand for 30 minutes, accurately timed. Gently swirl the tubes for 1 minute to remove any gas bubbles present, then concomitantly determine the absorbances of the solutions in 1-cm cells, at the wavelength of maximum absorbance at about 500 nm, with a suitable spectrophotometer, using the blank to set the instrument. Calculate the quantity, in mg, of benzonatate ($C_{30}H_{53}NO_{11}$ av.) in the number of Capsules taken by the formula:

$$C(A_U / A_S),$$

in which C is the concentration, in μg per mL, of USP Benzonatate RS in the *Standard preparation,* and A_U and A_S are the absorbances of the solutions from the *Assay preparation* and the *Standard preparation,* respectively.

Hydrous Benzoyl Peroxide

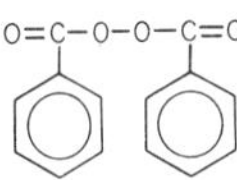

$C_{14}H_{10}O_4$ (anhydrous) 242.23
Peroxide, dibenzoyl.
Benzoyl peroxide [*94-36-0*].

» Hydrous Benzoyl Peroxide contains not less than 65.0 percent and not more than 82.0 percent of $C_{14}H_{10}O_4$. It contains about 26 percent of water for the purpose of reducing flammability and shock sensitivity.

Caution—Hydrous Benzoyl Peroxide may explode at temperatures higher than 60° or cause fires in the presence of reducing substances. Store it in the original container, treated to reduce static charges.

Packaging and storage—Store in the original container, at room temperature. [NOTE—Do not transfer Hydrous Benzoyl Peroxide to metal or glass containers fitted with friction tops. Do not return unused material to its original container, but destroy it by treatment with sodium hydroxide solution (1 in 10) until addition of a crystal of potassium iodide results in no release of free iodine.]

Identification—

A: Prepare a solution in methanol to contain 10 mg of benzoyl peroxide per mL. Apply 5 μL of this solution and 5 μL of a freshly prepared Standard solution of Hydrous Benzoyl Peroxide, previously subjected to the *Assay,* in methanol containing 10 mg per mL on a line parallel to and about 2.5 cm from the bottom of a thin-layer chromatographic plate (see *Chromatography* ⟨621⟩) coated with a 0.25-mm layer of chromatographic silica gel mixture. Place the plate in a developing chamber containing, and equilibrated with, a mixture of toluene, dichloromethane, and glacial acetic acid (50:2:1). Develop the chromatogram until the solvent front has moved about three-fourths of the length of the plate. Remove the plate, and allow the solvent to evaporate. Observe the plate under short-wavelength ultraviolet light: the R_f value of the principal spot obtained from the solution under test corresponds to that obtained from the Standard solution.

B: Dissolve an accurately weighed quantity in acetonitrile to obtain a solution containing 0.32 mg of benzoyl peroxide per mL. Chromatograph this test solution and a freshly prepared Standard solution of Hydrous Benzoyl Peroxide, previously subjected to the *Assay,* in acetonitrile containing 0.32 mg per mL as directed in the *Assay* under *Benzoyl Peroxide Gel:* the solution under test exhibits a major peak for benzoyl peroxide, the retention time of which corresponds to that exhibited by the Standard solution.

Chromatographic purity—Calculate the area percentage of each peak in the chromatogram of the test solution prepared as directed in *Identification test B:* the sum of the areas of all peaks other than the principal peak does not exceed 2.0% of the total area, and the area of any individual peak other than the principal peak does not exceed 1.5% of the total area.

Assay—Place about 500 mg of previously mixed Hydrous Benzoyl Peroxide in an accurately weighed conical flask fitted with a ground-glass stopper, and weigh again to obtain the weight of the test specimen. Add 30 mL of acetone, and swirl the flask gently to effect solution. Add 5 mL of potassium iodide solution (1 in 5), and mix. Wash the sides of the flask with a few mL of acetone, and allow the solution to stand for 1 minute. Titrate the liberated iodine with 0.1 *N* sodium thiosulfate VS. As the endpoint is approached add 1 drop of starch iodide paste TS, and continue the titration to the discharge of the blue color. Perform a blank determination, and make any necessary correction. Each mL of 0.1 *N* sodium thiosulfate is equivalent to 12.11 mg of $C_{14}H_{10}O_4$.

Benzoyl Peroxide Gel

» Benzoyl Peroxide Gel is benzoyl peroxide in a suitable gel base. It contains not less than 90.0 percent and not more than 125.0 percent of the labeled amount of $C_{14}H_{10}O_4$.

Packaging and storage—Preserve in tight containers.

Identification—Prepare a *Standard preparation* and an *Assay preparation* as directed in the *Assay*, except to omit the *Internal standard solution*, and chromatograph as directed in the *Assay:* the *Assay preparation* exhibits a major peak for benzoyl peroxide the retention time of which corresponds with that exhibited by the *Standard preparation.*

pH ⟨791⟩: between 3.5 and 6.0.

Assay—

Mobile phase—Prepare a solution of acetonitrile in water (about 5 in 10) such that the retention times for ethyl benzoate and benzoyl peroxide are about 7 and 14 minutes, respectively.

Internal standard solution—Dissolve ethyl benzoate in acetonitrile to obtain a solution having a concentration of about 3.6 mg per mL.

Standard preparation—Place a suitable quantity of hydrous benzoyl peroxide, recently subjected to the *Assay* under *Hydrous Benzoyl Peroxide*, in an accurately weighed conical flask fitted with a glass stopper, weigh again to obtain the weight of the specimen, and dissolve quantitatively in acetonitrile to obtain a solution containing a known concentration of about 0.8 mg of benzoyl peroxide per mL. Pipet 10 mL of this solution and 5 mL of *Internal standard solution* into a 25-mL volumetric flask, dilute with acetonitrile to volume, and mix. This *Standard preparation* contains about 0.32 mg of benzoyl peroxide per mL.

Assay preparation—Transfer an accurately weighed quantity of Benzoyl Peroxide Gel, equivalent to about 40 mg of benzoyl peroxide, to a 50-mL volumetric flask. Add 40 mL of acetonitrile and shake until the material is thoroughly dispersed. Sonicate the mixture for 5 minutes, dilute with acetonitrile to volume, mix, and filter. Pipet 10 mL of the filtrate and 5 mL of *Internal standard solution* into a 25-mL volumetric flask, dilute with acetonitrile to volume, and mix.

Chromatographic system (see *Chromatography* ⟨621⟩)—The high-pressure liquid chromatograph has a detector set at 254 nm and a 4-mm × 30-cm stainless steel column that contains packing L1, and is operated at room temperature. The flow rate is about 1 mL per minute. Chromatograph three replicate injections of the *Standard preparation*, and record the peak responses as directed under *Procedure:* the lowest and highest peak response ratios (R_S) agree within 2.0%, the resolution factor for the ethyl benzoate and benzoyl peroxide peaks is not less than 2.0, and the tailing factors for the ethyl benzoate and benzoyl peroxide peaks are not more than 2.0.

Procedure—Using a microsyringe or sampling valve, chromatograph 10 µL of the *Standard preparation*, and adjust the specimen size and other operating parameters, if necessary, until satisfactory chromatography and peak responses are obtained. Chromatograph equal volumes of the *Standard preparation* and the *Assay preparation*, and measure the peak responses. Calculate the quantity, in mg, of $C_{14}H_{10}O_4$ in the portion of Gel taken by the formula:

$$125C(R_U/R_S),$$

in which C is the concentration, in mg per mL, of benzoyl peroxide in the *Standard preparation*, and R_U and R_S are the ratios of benzoyl peroxide peak response to ethyl benzoate peak response obtained from the *Assay preparation* and the *Standard preparation*, respectively.

Benzoyl Peroxide Topical Gel, Erythromycin and—
see Erythromycin and Benzoyl Peroxide Topical Gel

Benzoyl Peroxide Lotion

» Benzoyl Peroxide Lotion is benzoyl peroxide in a suitable lotion base. It contains not less than 90.0 percent and not more than 110.0 percent of the labeled amount of $C_{14}H_{10}O_4$.

Packaging and storage—Preserve in tight containers.

Identification—

A: Dilute a quantity of Lotion with acetone to obtain a solution having a concentration of benzoyl peroxide equivalent to 10 mg per mL, and proceed with the solution so obtained as directed in the *Identification test A* under *Hydrous Benzoyl Peroxide*, beginning with "Apply 5 µL of this solution." The solution responds to the test.

B: It responds to the *Identification test* under *Benzoyl Peroxide Gel.*

pH ⟨791⟩: between 2.8 and 6.6.

Assay—

Mobile phase, Internal standard solution, Standard preparation, and *Chromatographic system*—Proceed as directed in the *Assay* under *Benzoyl Peroxide Gel.*

Assay preparation—Prepare as directed for *Assay preparation* in the *Assay* under *Benzoyl Peroxide Gel*, using Benzoyl Peroxide Lotion.

Procedure—Proceed as directed for *Procedure* in the *Assay* under *Benzoyl Peroxide Gel.* Calculate the quantity, in mg, of $C_{14}H_{10}O_4$ in the portion of Lotion taken by the formula:

$$125C(R_U/R_S),$$

in which C is the concentration, in mg per mL, of benzoyl peroxide in the *Standard preparation*, and R_U and R_S are the ratios of benzoyl peroxide peak response to ethyl benzoate peak response obtained from the *Assay preparation* and the *Standard preparation*, respectively.

Benzthiazide

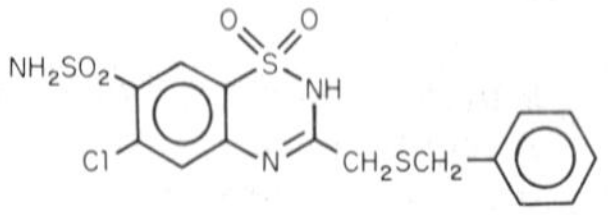

$C_{15}H_{14}ClN_3O_4S_3$ 431.93

2*H*-1,2,4-Benzothiadiazine-7-sulfonamide, 6-chloro-3-[[(phenylmethyl)thio]methyl]-, 1,1-dioxide.

3-[(Benzylthio)methyl]-6-chloro-2*H*-1,2,4-benzothiadiazine-7-sulfonamide 1,1-dioxide [*91-33-8*].

» Benzthiazide contains not less than 98.0 percent and not more than 101.5 percent of $C_{15}H_{14}ClN_3O_4S_3$, calculated on the dried basis.

Packaging and storage—Preserve in tight containers.

Reference standards—*USP Benzthiazide Reference Standard*—Dry in vacuum at 65° for 2 hours before using. *USP 4-Amino-6-chloro-1,3-benzenedisulfonamide Reference Standard*—Keep container tightly closed and protected from light. Dry over silica gel for 4 hours before using.

Identification—

A: The infrared absorption spectrum of a potassium bromide dispersion of it, previously dried, exhibits maxima only at the same wavelengths as that of a similar preparation of USP Benzthiazide RS.

B: The ultraviolet absorption spectrum of a 1 in 62,500 solution in 0.01 *N* alcoholic hydrochloric acid exhibits maxima and minima at the same wavelengths as that of a similar solution of USP Benzthiazide RS, concomitantly measured.

Loss on drying ⟨731⟩—Dry it in vacuum at 65° for 2 hours: it loses not more than 1.0% of its weight.

Residue on ignition ⟨281⟩: not more than 0.2%.

Selenium ⟨291⟩—The absorbance of the solution from the *Test Solution* prepared with 100 mg of Benzthiazide and 100 mg of magnesium oxide, is not greater than one-half that from the *Standard Solution* (0.003%).

Heavy metals, *Method II* ⟨231⟩: 0.0025%.

Diazotizable substances—

Standard preparation—Weigh accurately 10.0 mg of USP 4-Amino-6-chloro-1,3-benzenedisulfonamide RS, and dissolve in acetone to make 50.0 mL. Transfer 4.0 mL of the solution to a 100-mL volumetric flask, dilute with acetone to volume, and mix.

Test preparation—Dissolve 80.0 mg of Benzthiazide, accurately weighed, in sufficient acetone to make 100.0 mL, and mix.

Procedure—Transfer 1.0 mL each of the *Standard preparation* and of the *Test preparation* to separate tubes, and transfer 1.0 mL of acetone to a third tube to provide the blank. To each tube add 9.0 mL of dilute hydrochloric acid (1 in 12) and 0.10 mL of freshly prepared sodium nitrite solution (1 in 25), mix, and allow to stand for 1 minute. Add 0.80 mL of ammonium sulfamate solution (1 in 50), mix, and allow to stand for 3 minutes. Add 0.80 mL of a 1 in 50 solution of *N*-(1-naphthyl)ethylenediamine dihydrochloride in diluted alcohol to each tube, and mix. After 2 minutes, concomitantly determine the absorbances of the solutions in 1-cm cells at the wavelength of maximum absorbance at about 518 nm, with a suitable spectrophotometer, using the blank to set the instrument: the absorbance of the solution from the *Test preparation* does not exceed that of the solution from the *Standard preparation*, corresponding to not more than 1.0% of diazotizable substances.

Assay—Transfer about 100 mg of Benzthiazide, accurately weighed, to a 250-mL volumetric flask, add 200 mL of dilute alcoholic hydrochloric acid (1 in 1200), and stir for 1 hour to dissolve. Add dilute alcoholic hydrochloric acid (1 in 1200) to volume, and mix. Transfer 2.0 mL of this solution to a 50-mL volumetric flask, add dilute alcoholic hydrochloric acid (1 in 1200) to volume, and mix. Concomitantly determine the absorbances of this solution and of a Standard solution of USP Benzthiazide RS, in the same medium having a known concentration of about 16 µg per mL in 1-cm cells at the wavelength of maximum absorbance at about 283 nm, with a suitable spectrophotometer, using dilute alcoholic hydrochloric acid (1 in 1200) as the blank. Calculate the quantity, in mg, of $C_{15}H_{14}ClN_3O_4S_3$ in the Benzthiazide taken by the formula:

$$6.25C(A_U/A_S),$$

in which C is the concentration, in µg per mL, of USP Benzthiazide RS in the Standard solution, and A_U and A_S are the absorbances of the assay solution and the Standard solution, respectively.

Benzthiazide Tablets

» Benzthiazide Tablets contain not less than 90.0 percent and not more than 110.0 percent of the labeled amount of $C_{15}H_{14}ClN_3O_4S_3$.

Packaging and storage—Preserve in tight containers.

Reference standard—*USP Benzthiazide Reference Standard*—Dry in vacuum at 65° for 2 hours before using.

Identification—The solution of benzthiazide, obtained from the Tablets as directed in the *Assay*, exhibits an absorbance maximum at 295 ± 2 nm.

Disintegration ⟨701⟩: 15 minutes, the use of disks being omitted.

Uniformity of dosage units ⟨905⟩: meet the requirements.

Procedure for content uniformity—Transfer 1 finely powdered Tablet to a 100-mL volumetric flask, add 1 mL of dilute hydrochloric acid (1 in 120) and 50 mL of alcohol, and heat on a steam bath for 15 minutes. Cool, dilute with alcohol to volume, mix, and filter, discarding the first 20 mL of the filtrate. Dilute a portion of the subsequent filtrate with dilute alcoholic hydrochloric acid (1 in 1200) to provide a solution containing approximately 20 µg of benzthiazide per mL. Concomitantly determine the absorbances of this solution and a Standard solution of USP Benzthiazide RS in the same medium having a known concentration of about 20 µg per mL in 1-cm cells at the wavelength of maximum absorbance at about 283 nm, with a suitable spectrophotometer, using dilute alcoholic hydrochloric acid (1 in 1200) as the blank. Calculate the quantity, in mg, of $C_{15}H_{14}ClN_3O_4S_3$ in the Tablet by the formula:

$$(TC/D)(A_U/A_S),$$

in which T is the labeled quantity, in mg, of benzthiazide in the Tablet, C is the concentration, in µg per mL, of USP Benzthiazide RS in the Standard solution, D is the concentration, in µg per mL, of benzthiazide in the test solution, based upon the labeled quantity per Tablet and the extent of dilution, and A_U and A_S are the absorbances of the solution from the Tablet and the Standard solution, respectively.

Assay—Weigh and finely powder not less than 20 Benzthiazide Tablets. Transfer an accurately weighed portion of the powder, equivalent to about 50 mg of benzthiazide, to a 100-mL volumetric flask, add sodium hydroxide solution (1 in 125) to volume, and mix. Filter through fast filter paper, discarding the first 20 mL of the filtrate, and transfer 10.0 mL of the subsequent filtrate to a separator. To the separator add 3 mL of 3 *N* hydrochloric acid, shake vigorously for 2 minutes, and then extract successively with one 50-mL portion and two 25-mL portions of ether. Evaporate the combined ether extracts on a steam bath to a volume of about 5 mL. Add 10 mL of sodium hydroxide solution (1 in 125) while washing down the sides of the container, and continue the evaporation until all traces of ether are expelled. Cool, and dissolve the residual liquid in sodium hydroxide solution (1 in 125). Transfer the solution with the aid of the sodium hydroxide solution to a 100-mL volumetric flask, dilute with the sodium hydroxide solution to volume, and mix. Transfer 20.0 mL of this solution to a 50-mL volumetric flask, dilute with sodium hydroxide solution (1 in 125) to volume, and mix. Concomitantly determine the absorbances of this solution and a Standard solution of USP Benzthiazide RS in the same medium having a known concentration of about 20 µg per mL in 1-cm cells at the wavelength of maximum absorbance at about 295 nm, with a suitable spectrophotometer, using sodium hydroxide solution (1 in 125) as the blank. Calculate the quantity, in mg, of $C_{15}H_{14}ClN_3O_4S_3$ in the portion of Tablets taken by the formula:

$$2.5C(A_U/A_S),$$

in which C is the concentration, in µg per mL, of USP Benzthiazide RS in the Standard solution, and A_U and A_S are the absorbances of the solution from the Tablets and the Standard solution, respectively.

Benztropine Mesylate

CH3 N H H OCH H · CH3SO3H

$C_{21}H_{25}NO.CH_4O_3S$ 403.54

8-Azabicyclo[3.2.1]octane, 3-(diphenylmethoxy)-, *endo*-, methanesulfonate.

3α-(Diphenylmethoxy)-1αH,5αH-tropane methanesulfonate [*132-17-2*].

» Benztropine Mesylate contains not less than 98.0 percent and not more than 100.5 percent of $C_{21}H_{25}NO.CH_4O_3S$, calculated on the dried basis.

Packaging and storage—Preserve in tight containers.

Reference standard—*USP Benztropine Mesylate Reference Standard*—Dry at 105° for 2 hours before using.

Identification—The infrared absorption spectrum of a potassium bromide dispersion of it, previously dried, exhibits maxima only at the same wavelengths as that of a similar preparation of USP Benztropine Mesylate RS.

Melting range ⟨741⟩: between 141° and 145°.

Loss on drying ⟨731⟩—Dry it at 105° for 2 hours: it loses not more than 5.0% of its weight.

Residue on ignition ⟨281⟩: not more than 0.1%.

Assay—Transfer about 60 mg of Benztropine Mesylate, accurately weighed, to a 125-mL separator. Dissolve in 25 mL of water, add 5 mL of sodium carbonate TS, and extract with four 10-mL portions of chloroform. Wash the combined chloroform extracts with about 10 mL of water, and extract the wash solution with 5 mL of chloroform. Filter the combined chloroform extracts through a tightly packed pledget of cotton, and wash the cotton with about 5 mL of chloroform. Add methyl red TS, and titrate the chloroform solution with 0.01 *N* perchloric acid in dioxane VS. Perform a blank determination, and make any necessary correction. Each mL of 0.01 *N* perchloric acid is equivalent to 4.035 mg of $C_{21}H_{25}NO.CH_4O_3S$.

Benztropine Mesylate Injection

» Benztropine Mesylate Injection is a sterile solution of Benztropine Mesylate in Water for Injection. It contains not less than 90.0 percent and not more than 110.0 percent of the labeled amount of $C_{21}H_{25}NO.CH_4O_3S$.

Packaging and storage—Preserve in single-dose or in multiple-dose containers, preferably of Type I glass.

Reference standard—*USP Benztropine Mesylate Reference Standard*—Dry at 105° for 2 hours before using.

Identification—Dissolve about 10 mg of USP Benztropine Mesylate RS in 50 mL of water in a separator to obtain a solution containing about 0.2 mg per mL. Transfer a volume of Injection, equivalent to about 10 mg of benztropine mesylate, to a second separator, and dilute with water to 50 mL. Separately add 2 mL of 1 *N* sodium hydroxide to each separator, and mix. Extract each solution with three 10-mL portions of chloroform, collecting the chloroform extracts from both solutions in separate 50-mL beakers. Evaporate both chloroform extracts with the aid of gentle heat and a current of air to dryness, and separately dissolve each residue in 1 mL of chloroform. Apply separately 1 µL of the test solution and the Standard solution to a suitable thin-layer chromatographic plate (see *Chromatography* ⟨621⟩) coated with a 0.25-mm layer of chromatographic silica gel. Allow the applications to dry, and develop the chromatogram in a solvent system consisting of a mixture of chloroform, methanol, and a 1 in 4 solution of ammonium hydroxide (40:10:1) until the solvent front has moved about three-fourths of the length of the plate. Remove the plate from the developing chamber, mark the solvent front, and allow the solvent to evaporate. Locate the spots on the plate by lightly spraying with potassium iodoplatinate TS: the R_f value of the principal spot obtained from the test solution corresponds to that obtained from the Standard solution.

pH ⟨791⟩: between 5.0 and 8.0.

Other requirements—It meets the requirements under *Injections* ⟨1⟩.

Assay—

0.005 M Octylamine phosphate buffer—Transfer 0.83 mL of octylamine to a 1-liter volumetric flask, dilute with water to volume, adjust with phosphoric acid to a pH of 3.0, and filter the solution through a membrane filter.

Mobile phase—Mix 350 mL of *0.005 M Octylamine phosphate buffer* and 650 mL of acetonitrile, and degas the solution.

Standard preparation—Dissolve an accurately weighed quantity of USP Benztropine Mesylate RS in water to obtain a solution having a known concentration of about 1 mg per mL.

Assay preparation—Dilute with water an accurately measured volume of Benztropine Mesylate Injection, equivalent to about 10 mg of benztropine mesylate, to obtain a solution having a concentration of about 1 mg of benztropine mesylate per mL.

Chromatographic system (see *Chromatography* ⟨621⟩)—The liquid chromatograph is equipped with a 259-nm detector and a 4.6-mm × 25-cm column that contains packing L7. The flow rate is about 1.3 mL per minute. Adjust the flow rate to obtain a retention time of about 7 minutes for benztropine mesylate. Chromatograph five replicate injections of the *Standard preparation*, and record the peak responses as directed under *Procedure:* the relative standard deviation is not more than 2.0%.

Procedure—[NOTE—Use peak areas where peak responses are indicated.] Separately inject equal volumes (about 25 µL) of the *Standard preparation* and the *Assay preparation* into the chromatograph by means of a suitable microsyringe or sampling valve, record the chromatograms, and measure the responses for the major peaks. Calculate the quantity, in mg, of $C_{21}H_{25}NO.CH_4O_3S$ in each mL of the Injection taken by the formula:

$$10(C/V)(r_U/r_S),$$

in which C is the concentration, in mg per mL, of USP Benztropine Mesylate RS in the *Standard preparation*, V is the volume, in mL, of Injection taken, and r_U and r_S are the peak responses obtained for benztropine mesylate from the *Assay preparation* and the *Standard preparation*, respectively.

Benztropine Mesylate Tablets

» Benztropine Mesylate Tablets contain not less than 90.0 percent and not more than 110.0 percent of the labeled amount of $C_{21}H_{25}NO.CH_4O_3S$.

Packaging and storage—Preserve in well-closed containers.

Reference standard—*USP Benztropine Mesylate Reference Standard*—Dry at 105° for 2 hours before using.

Identification—Dissolve about 10 mg of USP Benztropine Mesylate RS in 50 mL of water in a separator to obtain a solution containing about 0.2 mg per mL. Transfer a portion of finely powdered Tablets, equivalent to about 10 mg of benztropine mesylate, to a suitable flask, add 50 mL of water, shake by mechanical means for 30 minutes, and filter into a second separator. Proceed as directed in the *Identification test* under *Benztropine Mesylate Injection*, beginning with "Separately add 2 mL of 1 *N* sodium hydroxide."

Dissolution ⟨711⟩—

Medium: 0.1 *N* hydrochloric acid; 900 mL.

Apparatus 2: 50 rpm.

Time: 30 minutes.

Bromophenol blue solution—Dissolve 50 mg of bromophenol blue in 100 mL of chloroform.

Standard preparation—Dissolve an accurately weighed quantity of USP Benztropine Mesylate RS in 0.1 *N* hydrochloric acid to obtain a solution having a known concentration of about 2 µg per mL.

Procedure—Pipet an aliquot of the filtered solution under test, containing about 66 µg of benztropine mesylate, into a 250-mL separator. Into a second separator pipet an equivalent volume of 0.1 *N* hydrochloric acid to provide a reagent blank, and into a third separator pipet 30 mL of *Standard preparation*. Treat each separator as follows: Add 1 mL of 1 *N* sulfuric acid and 10 mL of *Bromophenol blue solution*, and shake for 2 minutes. Determine the absorbances of the chloroform layers at the wavelength of maximum absorbance at about 415 nm, with a suitable spectrophotometer, against the reagent blank. Calculate the percentage of the Tablet dissolved by the formula:

$$(2700)(C/WV)(A_U/A_S),$$

in which C is the concentration, in µg per mL, of USP Benztropine Mesylate RS in the *Standard preparation*, W is the labeled quantity, in mg, of benztropine mesylate in the Tablet, V is the volume, in mL, of the aliquot of the solution under test used, and A_U and A_S are the absorbances of the solutions from the solution under test and the *Standard preparation*, respectively.

Tolerances—Not less than 80% (*Q*) of the labeled amount of $C_{21}H_{25}NO.CH_4O_3S$ is dissolved in 30 minutes.

Uniformity of dosage units ⟨905⟩: meet the requirements.

Assay—

0.005 M Octylamine phosphate buffer and *Mobile solvent*—Prepare as directed in the *Assay* under *Benztropine Mesylate Injection.*

Aqueous phosphoric acid–isopropyl alcohol solution—Mix 600 mL of water, 400 mL of isopropyl alcohol, and 1.0 mL of phosphoric acid.

Standard preparation—Dissolve an accurately weighed quantity of USP Benztropine Mesylate RS in water to obtain a solution having a known concentration of about 0.25 mg per mL.

Assay preparation—Weigh and finely powder not less than 20 Benztropine Mesylate Tablets. Transfer an accurately weighed portion of the powder, equivalent to about 10 mg of benztropine mesylate, to a stoppered, 50-mL centrifuge tube. Pipet 40 mL of *Aqueous phosphoric acid–isopropyl alcohol solution* into each centrifuge tube. Shake by mechanical means for not less than 60 minutes, then centrifuge for 5 minutes, and filter through a suitable polyfluoroethylene membrane filter.

Chromatographic system—Prepare as directed for *Chromatographic system* in the *Assay* under *Benztropine Mesylate Injection.*

Procedure—[NOTE—Use peak areas where peak responses are indicated.] Separately inject equal volumes (about 50 µL) of the *Standard preparation* and the *Assay preparation* into the chromatograph by means of a suitable microsyringe or sampling valve, record the chromatograms, and measure the responses for the major peaks. Calculate the quantity, in mg, of $C_{21}H_{25}NO \cdot CH_4O_3S$ in the portion of Tablets taken by the formula:

$$40C(r_U/r_S),$$

in which C is the concentration, in mg per mL, of USP Benztropine Mesylate RS in the *Standard preparation*, and r_U and r_S are the peak responses obtained for benztropine mesylate from the *Assay preparation* and the *Standard preparation*, respectively.

Benzyl Alcohol—*see* Benzyl Alcohol NF

Benzyl Benzoate

C_6H_5–COO–CH_2–C_6H_5

$C_{14}H_{12}O_2$ 212.25
Benzoic acid, phenylmethyl ester.
Benzyl benzoate [*120-51-4*].

» Benzyl Benzoate contains not less than 99.0 percent and not more than 100.5 percent of $C_{14}H_{12}O_2$.

Packaging and storage—Preserve in tight, well-filled, light-resistant containers, and avoid exposure to excessive heat.

Reference standard—*USP Benzyl Benzoate Reference Standard*—After opening the ampul, store in a tightly closed container, protected from light.

Identification—The infrared absorption spectrum of a thin film of it between sodium chloride plates exhibits maxima only at the same wavelengths as that of a similar preparation of USP Benzyl Benzoate RS.

Specific gravity ⟨841⟩: between 1.116 and 1.120.

Congealing temperature ⟨651⟩: not lower than 18.0°. Congelation may be brought about by addition of a fragment of previously congealed Benzyl Benzoate when the temperature has reached the expected congealing temperature.

Refractive index ⟨831⟩: between 1.568 and 1.570 at 20°.

Aldehyde—Transfer 10.0 g to a 125-mL conical flask containing 50 mL of alcohol and 5 mL of hydroxylamine hydrochloride solution (3.5 in 100), mix, and allow to stand for 10 minutes. Add 1 mL of bromophenol blue TS, and titrate with 0.1 *N* sodium hydroxide VS to a light green end-point. Perform a blank determination, and match the color of the end-point with that of the titrated test solution. The net volume of 0.1 *N* sodium hydroxide consumed does not exceed 0.50 mL (0.05% as benzaldehyde).

Acidity—Add 2 drops of phenolphthalein TS to 25 mL of alcohol, and add 0.020 *N* sodium hydroxide until a pink color is produced. Add 5.0 g of Benzyl Benzoate, mix, and titrate with 0.020 *N* sodium hydroxide: not more than 1.5 mL of 0.020 *N* sodium hydroxide is required to restore the pink color.

Assay—Transfer about 2 g of Benzyl Benzoate, accurately weighed, to a conical flask fitted with a reflux condenser, add 50.0 mL of 0.5 *N* alcoholic potassium hydroxide VS, and boil gently for 1 hour. Cool, add phenolphthalein TS, and titrate with 0.5 *N* hydrochloric acid VS. Perform a blank determination (see *Residual Titrations* ⟨541⟩). Each mL of 0.5 *N* alcoholic potassium hydroxide is equivalent to 106.1 mg of $C_{14}H_{12}O_2$.

Benzyl Benzoate Lotion

» Benzyl Benzoate Lotion contains not less than 26.0 percent and not more than 30.0 percent (w/w) of $C_{14}H_{12}O_2$.

Benzyl Benzoate	250 mL
Triethanolamine	5 g
Oleic Acid	20 g
Purified Water	750 mL
To make about	1000 mL

Mix the Triethanolamine with the Oleic Acid, add the Benzyl Benzoate, and mix. Transfer the mixture to a suitable container of about 2000-mL capacity, add 250 mL of Purified Water, and shake the mixture thoroughly. Finally add the remaining Purified Water, and again shake thoroughly.

Packaging and storage—Preserve in tight containers.

pH ⟨791⟩: between 8.5 and 9.2.

Assay—Place about 5 g of Benzyl Benzoate Lotion, accurately weighed, in a conical flask. Add 25 mL of alcohol and 2 drops of phenolphthalein TS. Cool the solution to about 15°, and titrate quickly with 0.1 *N* sodium hydroxide to a slight pink color. Add 50.0 mL of 0.5 *N* alcoholic potassium hydroxide VS, connect the flask to a reflux condenser, and boil gently for 1 hour. Cool, promptly add phenolphthalein TS, and titrate with 0.5 *N* hydrochloric acid VS. Perform a blank determination (see *Residual Titrations* under *Titrimetry* ⟨541⟩). Each mL of 0.5 *N* alcoholic potassium hydroxide is equivalent to 106.1 mg of $C_{14}H_{12}O_2$.

Benzylpenicilloyl Polylysine Concentrate

» Benzylpenicilloyl Polylysine Concentrate has a molar concentration of benzylpenicilloyl moiety ($C_{16}H_{19}N_2O_5S$) of not less than 0.0125 *M* and not more than 0.020 *M*. It contains one or more suitable buffers.

Packaging and storage—Preserve in tight containers.

Reference standard—*USP L-Lysine Hydrochloride Reference Standard*—Dry at 105° for 3 hours before using.

pH ⟨791⟩: between 6.5 and 8.5, the undiluted Concentrate being used.

Penicillenate and penamaldate—Transfer 1 mL of Concentrate to a 50-mL volumetric flask, dilute with *Saline phosphate buffer* to volume, and mix. Using a suitable spectrophotometer and using *Saline phosphate buffer* as a blank, determine the absorbances at the wavelengths of maximum absorption at about 322 nm and 282 nm. Calculate the molar concentration of penicillenate by the formula:

$$50A_{322}/26{,}600b,$$

in which A_{322} is the absorbance at 322 nm, 26,600 is the molar absorptivity of the penicillenate moiety at pH 7.6, and b is the length of the cell, in cm: not more than 0.00020 M is found. Calculate the molar concentration of penamaldate by the formula:

$$50A_{282}/22{,}325b,$$

in which A_{282} is the absorbance at 282 nm, 22,325 is the molar absorptivity of the penamaldate moiety at pH 7.6, and b is the length of the cell, in cm: not more than 0.00060 M is found.

Benzylpenicilloyl substitution—

Citrate buffer—Dissolve 19.69 g of sodium citrate dihydrate, 0.1 mL of pentachlorophenol, and 5 mL of 2,2′-thiodiethanol in 900 mL of 0.2 N hydrochloric acid, adjust with hydrochloric acid to a pH of 2.2, dilute with water to 1000 mL, and mix.

Ninhydrin reagent—Dissolve 18 g of ninhydrin and 0.7 g of hydrindantin in 675 mL of dimethyl sulfoxide, add 225 mL of 4 M lithium acetate solution previously adjusted with glacial acetic acid to a pH of 5.2, and mix.

Standard preparation—Dissolve an accurately weighed quantity of USP L-Lysine Hydrochloride RS in *Citrate buffer* to obtain a solution having a known concentration of about 91 µg per mL (5×10^{-4} M).

Test preparation—Transfer 1.0 mL of Benzylpenicilloyl Polylysine Concentrate to a 10-mL volumetric flask, dilute with water to volume, and mix. Transfer 1.0 mL of this solution to an ampul, add 1.5 mL of 6 N hydrochloric acid, and seal the ampul under nitrogen. Heat the ampul at 110° for 22 hours. Transfer the contents of the ampul to a round-bottom, 50-mL flask, and dry by vacuum rotary evaporation. Dissolve the residue three times, using 5-mL portions of water, evaporating to dryness after each dissolution. Dissolve the residue in 10 mL of *Citrate buffer*.

Chromatographic system (see *Chromatography* ⟨621⟩)—The liquid chromatograph is equipped with a 1.75-mm × 50-cm column that contains a packing of 8-µm 8 percent cross-linked sulfonated divinylbenzene polystyrene cation-exchange resin. The column effluent is mixed continuously with flowing *Ninhydrin reagent*, and the flowing mixture is heated at 130° for 1.5 minutes in a reaction coil. The absorbance of the reaction mixture is measured continuously by a 570-nm detector. Chromatograph the *Standard preparation*, and record the peak responses as directed under *Procedure:* the column efficiency determined from the analyte peak is not less than 1800 theoretical plates, and the relative standard deviation for replicate injections is not more than 4.0%.

Procedure—Separately inject equal volumes (about 20 µL) of the *Standard preparation* and the *Test preparation* into the chromatograph, record the chromatograms, and measure the responses for the major peaks. The retention time is about 57 minutes for L-lysine. Calculate the molar concentration of lysine in the Concentrate by the formula:

$$(0.1C/182.65)(r_U/r_S),$$

in which C is the concentration, in µg per mL, of USP L-Lysine Hydrochloride RS in the *Standard preparation*, 182.65 is the molecular weight of anhydrous lysine hydrochloride, and r_U and r_S are the peak responses obtained from the *Test preparation* and the *Standard preparation*, respectively. Calculate the percentage of benzylpenicilloyl substitution by the formula:

$$100(B/L),$$

in which B is the molar concentration of benzylpenicilloyl moiety in the Concentrate, as determined in the *Assay*, and L is the molar concentration of lysine in the Concentrate: not less than 50% and not more than 70% is found.

Assay—

Saline phosphate buffer—Dissolve 9 g of sodium chloride and 1.38 g of monobasic sodium phosphate in 900 mL of water, adjust with 5 N sodium hydroxide or phosphoric acid to a pH of 7.6, dilute with water to 1000 mL, and mix.

Mercuric chloride solution—Dissolve 35 mg of mercuric chloride in 500 mL of water, and mix.

Assay preparation—Transfer 1.0 mL of Benzylpenicilloyl Polylysine Concentrate to a 500-mL volumetric flask, dilute with *Saline phosphate buffer* to volume, and mix.

Procedure—Transfer 3.0 mL of *Assay preparation* to a spectrophotometric cell. Using a suitable spectrophotometer and using *Saline phosphate buffer* as the blank, determine the initial absorbance at the wavelength of maximum absorbance at about 282 nm. Add 0.02 mL of *Mercuric chloride solution* to the *Assay preparation* in the spectrophotometric cell, mix, and determine the absorbance at the same wavelength after 1 and 3 minutes. Repeat the addition of 0.02-mL portions of *Mercuric chloride solution* until a maximum absorbance reading is obtained. Calculate the molar concentration of benzylpenicilloyl moiety in the Concentrate by the formula:

$$500\{[3A_m/(3 + 0.02n)] - A_i\}/22{,}325b,$$

in which A_m is the highest absorbance observed, A_i is the initial absorbance, n is the number of 0.02-mL portions of *Mercuric chloride solution* added to the *Assay preparation* to obtain the maximum absorbance, 22,325 is the molar absorptivity of the penamaldate formed by the reaction of benzylpenicilloyl with mercuric chloride at pH 7.6, and b is the length of the cell, in cm: between 0.0125 M and 0.020 M is found.

Benzylpenicilloyl Polylysine Injection

» Benzylpenicilloyl Polylysine Injection has a molar concentration of benzylpenicilloyl moiety ($C_{16}H_{19}N_2O_5S$) of not less than 5.4×10^{-5} M and not more than 7.0×10^{-5} M. It contains one or more suitable buffers.

Packaging and storage—Preserve in single-dose or in multiple-dose containers, preferably of Type I glass, in a refrigerator.

Pyrogen—Combine the contents of a sufficient number of containers to obtain not less than 1.5 mL of Injection. Dilute 1.5 mL of Injection with pyrogen-free saline TS to 50 mL: it meets the requirements of the *Pyrogen Test* ⟨151⟩, the test dose being 1.0 mL per kg.

Sterility—It meets the requirements under *Sterility Tests* ⟨71⟩, when tested as directed in the section, *Test Procedures Using Membrane Filtration.*

pH ⟨791⟩: between 6.5 and 8.5.

Assay—

Saline phosphate buffer and *Mercuric chloride solution*—Prepare as directed in the *Assay* under *Benzylpenicilloyl Polylysine Concentrate.*

Assay preparation—Combine the contents of a sufficient number of containers to obtain not less than 3 mL of Benzylpenicilloyl Polylysine Injection. Transfer 3.0 mL of Injection to a 10-mL volumetric flask, dilute with *Saline phosphate buffer* to volume, and mix.

Procedure—Proceed as directed for *Procedure* in the *Assay* under *Benzylpenicilloyl Polylysine Concentrate.* Calculate the molar concentration of benzylpenicilloyl moiety in the Injection by the formula:

$$(10/3)\{[3A_m/(3 + 0.02n)] - A_i\}/22{,}325b.$$

Beta Carotene

H_3C CH_3 CH_3 CH_3 H_3C CH_3 CH_3 CH_3 CH_3 H_3C

$C_{40}H_{56}$ 536.88

β,β-Carotene.

all-trans-β-Carotene.

(*all-E*)-1,1′-(3,7,12,16-Tetramethyl-1,3,5,7,9,11,13,15,17-octadecanonaene-1,18-diyl)bis[2,6,6-trimethylcyclohexene] [*7235-40-7*].

» Beta Carotene contains not less than 96.0 percent and not more than 101.0 percent of $C_{40}H_{56}$.

Packaging and storage—Preserve in tight, light-resistant containers.

Identification—

A: Determine the absorbances of *Assay preparation B* at 455 nm and at 483 nm: the ratio of the absorbance at 455 nm to that at 483 nm is between 1.14 and 1.18.

B: Determine the absorbance of *Assay preparation B* at 455 nm and that of *Assay preparation A* at 340 nm: the ratio of the absorbance at 455 nm to that at 340 nm is not less than 1.5.

Melting range ⟨741⟩: between 176° and 182°, with decomposition.

Loss on drying ⟨731⟩—Dry it in vacuum over phosphorus pentoxide at 40° for 4 hours: it loses not more than 0.2% of its weight.

Residue on ignition ⟨281⟩: not more than 0.2%, 2 g of specimen being used.

Heavy metals, *Method II* ⟨231⟩: 0.001%.

Arsenic, *Method II* ⟨211⟩: 3 ppm.

Assay—[NOTE—Perform this procedure in subdued light, using low-actinic glassware.]

Assay preparation A—Transfer about 50 mg of Beta Carotene, accurately weighed, to a 100-mL volumetric flask, dissolve in 10 mL of acid-free chloroform, dilute with cyclohexane to volume, and mix. Pipet 5 mL of this solution into a 100-mL volumetric flask, dilute with cyclohexane to volume, and mix.

Assay preparation B—Pipet 5 mL of *Assay preparation A* into a 50-mL volumetric flask, dilute with cyclohexane to volume, and mix.

Procedure—Determine the absorbance of *Assay preparation B* at the wavelength of maximum absorbance at about 455 nm, with a suitable spectrophotometer, using cyclohexane as the blank. Calculate the quantity, in mg, of $C_{40}H_{56}$ in the portion of Beta Carotene taken by the formula:

$$20{,}000A_X/250,$$

in which A_X is the absorbance of *Assay preparation B* and 250 is the absorptivity of pure beta carotene.

Beta Carotene Capsules

» Beta Carotene Capsules contain not less than 90.0 percent and not more than 125.0 percent of the labeled amount of $C_{40}H_{56}$.

Packaging and storage—Preserve in tight, light-resistant containers.

Identification—Grind a portion of the Capsule contents, equivalent to about 10 mg of beta carotene, transfer to a centrifuge tube, add 5 mL of chloroform, shake for 1 minute, and centrifuge for 3 minutes. Filter the supernatant layer, collecting about 2 mL of the filtrate in a 25-mL conical flask. Add 5 mL of antimony trichloride TS to the filtrate: a transient purple or blue color forms.

Uniformity of dosage units ⟨905⟩: meet the requirements.

Assay—[NOTE—Perform this procedure in subdued light, using low-actinic glassware.]

Cyclohexane—Spectrophotometric grade, or material that has been purified by being passed through a column of activated silica gel and distilled.

Iodine solution—Transfer about 10 mg of iodine into a 100-mL volumetric flask. Dissolve in cyclohexane, dilute with cyclohexane to volume, and mix. Dilute 10.0 mL of this solution with cyclohexane to 100 mL, and mix. Prepare this solution fresh daily.

Assay preparation—Combine the contents of not less than 20 Beta Carotene Capsules, and grind, using a freezer mill, to a fine powder of uniform color. Transfer an accurately weighed quantity of the finely ground capsule contents, equivalent to about 75 mg of beta carotene, to a 1000-mL volumetric flask, add 500 mL of water, and heat at 60° for 15 minutes. Cool to ambient temperature, dilute with water to volume, and mix (*Assay solution I*). Transfer 5.0 mL of *Assay solution I* to a glass-stoppered, 50-mL centrifuge tube. Add 3 g of sodium sulfate decahydrate, 2 mL of 1 *N* hydrochloric acid, and 20.0 mL of chloroform. Shake by mechanical means for 10 minutes, centrifuge for 5 minutes, and remove the aqueous layer without disturbing the chloroform layer. Add 2 g of anhydrous sodium sulfate to the chloroform layer, shake vigorously, and allow to settle (*Assay solution II*). Transfer 5.0 mL of *Assay solution II* to a 50-mL volumetric flask, add 30 mL of cyclohexane, and mix. Add 0.05 mL of *Iodine solution*, dilute with cyclohexane to volume, mix, and allow to stand for 3 hours (*Assay solution III*). Transfer 20 mL of *Assay solution III* to a centrifuge tube, and centrifuge for 2 minutes.

Standard preparation—Accurately weigh about 17 mg of Beta Carotene, previously subjected to the *Assay* and previously dried in vacuum over phosphorus pentoxide at 40° for 4 hours, and transfer to a 1000-mL volumetric flask. Add 10 mL of water, heat at 60° for 15 minutes, and cool to room temperature. Add 3 g of sodium sulfate decahydrate and 2 mL of 1 *N* hydrochloric acid, and shake by mechanical means for 10 minutes. Add 200 mL of chloroform, and shake for 10 minutes. Add 750 mL of chloroform, shake, and dilute with chloroform to volume, disregarding the aqueous layer (*Standard solution I*). Shake vigorously, and allow the layers to separate completely. Transfer about 20 mL of the chloroform layer to a centrifuge tube, add 2 g of anhydrous sodium sulfate, shake vigorously, and allow to settle. Transfer 5.0 mL to a 50-mL volumetric flask, add 30 mL of cyclohexane, and mix. Add 0.05 mL of *Iodine solution*, dilute with cyclohexane to volume, mix, and allow to stand for 3 hours (*Standard solution II*). Transfer about 20 mL of *Standard solution II* to a centrifuge tube, and centrifuge for 2 minutes.

Procedure—Concomitantly measure the absorbances of *Assay solution III* and *Standard solution II* at 452 nm, with a suitable spectrophotometer, using cyclohexane as the blank. Calculate the quantity, in mg, of $C_{40}H_{56}$ in the portion of Capsule contents taken by the formula:

$$40C(A_U/A_S),$$

in which C is the concentration, in µg per mL, of Beta Carotene in the *Standard preparation*, and A_U and A_S are the absorbances of *Assay solution III* and *Standard solution II*, respectively.

Betaine Hydrochloride

$$[(CH_3)_3N^+{-}CH_2{-}COOH]\ Cl^-$$

$C_5H_{11}NO_2.HCl$ 153.61

Methanaminium, 1-carboxy-*N*,*N*,*N*-trimethyl-, chloride.

Betaine hydrochloride.

(Carboxymethyl)trimethylammonium chloride [*590-46-5*].

» Betaine Hydrochloride contains not less than 98.0 percent and not more than 100.5 percent of $C_5H_{11}NO_2.HCl$, calculated on the anhydrous basis.

Packaging and storage—Preserve in well-closed containers.

Reference standard—*USP Betaine Hydrochloride Reference Standard.*

Identification—

A: The infrared absorption spectrum of a potassium bromide dispersion of it exhibits maxima only at the same wavelengths as that of a similar preparation of USP Betaine Hydrochloride RS.

B: A solution (1 in 20) responds to the tests for *Chloride* ⟨191⟩.

pH ⟨791⟩: between 0.8 and 1.2, in a solution (1 in 4).

Water, *Method I* ⟨921⟩: not more than 0.5%.

Residue on ignition ⟨281⟩: not more than 0.1%.

Arsenic ⟨211⟩: 2 ppm.

Heavy metals ⟨231⟩: 0.001%.

Assay—Transfer about 400 mg of Betaine Hydrochloride, accurately weighed, to a conical flask, add 50 mL of glacial acetic acid, and heat gently with swirling until solution is complete. Add 25 mL of mercuric acetate TS, cool, add 2 drops of crystal violet TS, and titrate with 0.1 *N* perchloric acid VS to a green end-point. Perform a blank determination, and make any necessary correction. Each mL of 0.1 *N* perchloric acid is equivalent to 15.36 mg of $C_5H_{11}NO_2 \cdot HCl$.

Betamethasone

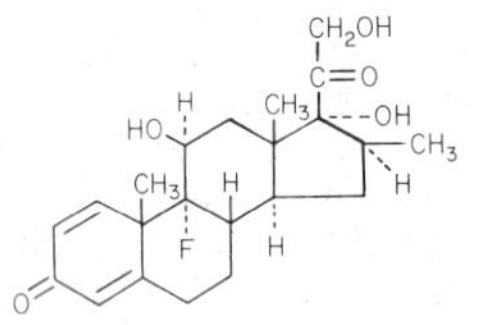

$C_{22}H_{29}FO_5$ 392.47

Pregna-1,4-diene-3,20-dione, 9-fluoro-11,17,21-trihydroxy-16-methyl-, (11β,16β)-.

9-Fluoro-11β,17,21-trihydroxy-16β-methylpregna-1,4-diene-3,20-dione [*378-44-9*].

» Betamethasone contains not less than 97.0 percent and not more than 103.0 percent of $C_{22}H_{29}FO_5$, calculated on the dried basis.

Packaging and storage—Preserve in well-closed containers.

Reference standard—*USP Betamethasone Reference Standard*—Dry at 105° for 3 hours before using.

Identification—

A: The infrared absorption spectrum of a mineral oil dispersion of it, previously dried, exhibits maxima only at the same wavelengths as that of a similar preparation of USP Betamethasone RS.

B: Prepare a solution in dehydrated alcohol containing 0.5 mg per mL. On a suitable thin-layer chromatographic plate (see *Chromatography* ⟨621⟩), coated with a 0.25-mm layer of chromatographic silica gel, apply 10 µL of this solution and 10 µL of a dehydrated alcohol solution of USP Betamethasone RS containing 0.5 mg per mL. Allow the spots to dry, and develop the chromatogram in a solvent system consisting of a mixture of chloroform and diethylamine (2:1) until the solvent front has moved about three-fourths of the length of the plate. Remove the plate from the developing chamber, mark the solvent front, and allow the solvent to evaporate. Locate the spots on the plate by lightly spraying with dilute sulfuric acid (1 in 2) and heating on a hot plate or under a lamp until spots appear: the R_f value of the principal spot obtained from the test solution corresponds to that obtained from the Standard solution.

Specific rotation ⟨781⟩: between +112° and +120°, calculated on the dried basis, determined in a solution in dioxane containing 100 mg in each 10 mL.

Loss on drying ⟨731⟩—Dry it at 105° for 3 hours: it loses not more than 1.0% of its weight.

Residue on ignition ⟨281⟩: not more than 0.2%, a platinum crucible being used.

Ordinary impurities ⟨466⟩—

Test solution: methanol.

Standard solution: methanol.

Application volume: 10 µL.

Eluant: a mixture of toluene, acetone, methyl ethyl ketone, and formic acid (55:20:20:5), in a nonequilibrated chamber.

Visualization: 5.

Assay—

Mobile solvent—Prepare a suitable aqueous acetonitrile solution (approximately 37 in 100), degassed for 5 to 10 minutes, such that the retention time of betamethasone is about 3.3 minutes and that of propylparaben is approximately 5.4 minutes. [NOTE—Do not leave the mobile solvent in the column overnight, but flush the system after use with water for 15 minutes, followed by methanol for 15 minutes.]

Internal standard solution—Prepare a solution of Propylparaben in alcohol having a known concentration of about 0.25 mg per mL.

Standard preparation—Dissolve an accurately weighed quantity of USP Betamethasone RS in alcohol to obtain a solution having a known concentration of about 0.2 mg per mL. Transfer 10.0 mL of this solution to a suitable vial, and add 10.0 mL of *Internal standard solution*, to obtain a *Standard preparation* having known concentrations of about 0.1 mg of betamethasone and about 0.125 mg of propylparaben per mL.

Assay preparation—Using about 80 mg of Betamethasone, accurately weighed, prepare as directed for *Standard preparation*.

Procedure—Introduce equal volumes (between 5 µL and 25 µL) of the *Assay preparation* and the *Standard preparation* into a high-pressure liquid chromatograph (see *Chromatography* ⟨621⟩) operated at room temperature, by means of a suitable microsyringe or sampling valve, adjusting the specimen size and other operating parameters such that the peak obtained with the internal standard in the *Standard preparation* is about 0.6 full scale. Typically, the apparatus is fitted with a 4-mm × 30-cm column that contains packing L1 and is equipped with an ultraviolet detector capable of monitoring absorption at 254 nm (or at the wavelength of maximum absorbance at about 240 nm if a detector with variable wavelengths is used) and a suitable recorder, and is capable of operating at a column pressure of up to 3500 psi. In a suitable chromatogram, the lowest and highest peak area ratios (R_S) of three successive injections of the *Standard preparation* agree within 2.0%. Determine the ratio of the peak heights, at equivalent retention times, obtained with the *Assay preparation* and the *Standard preparation*. Calculate the quantity, in mg, of $C_{22}H_{29}FO_5$ in the portion of Betamethasone taken by the formula:

$$800C(R_U / R_S),$$

in which *C* is the concentration, in mg per mL, of USP Betamethasone RS in the *Standard preparation*, and R_U and R_S are the peak height ratios of the betamethasone peak and the internal standard peak obtained from the *Assay preparation* and the *Standard preparation*, respectively.

Betamethasone Cream

» Betamethasone Cream contains not less than 90.0 percent and not more than 115.0 percent of the labeled amount of $C_{22}H_{29}FO_5$ in a suitable cream base.

Packaging and storage—Preserve in collapsible tubes or in tight containers.

Reference standard—*USP Betamethasone Reference Standard*—Dry at 105° for 3 hours before using.

Identification—Concentrate 10 mL of the *Assay preparation* on a steam bath to 1 mL, and apply 10 µL of the solution to a thin-layer chromatographic plate (see *Chromatography* ⟨621⟩), coated with a 0.25-mm layer of chromatographic silica gel. Apply to the same plate 10 µL of a solution containing 1 mg per mL of USP Betamethasone RS in dehydrated alcohol. Allow the spots to dry, and develop the chromatogram in a solvent system consisting of a mixture of chloroform and diethylamine (2:1) until the solvent front has moved about three-fourths of the length of the plate. Remove the plate from the developing chamber, mark the solvent front, and allow the solvent to evaporate. Spray the plate with a mixture of sulfuric acid, methanol, and nitric acid (10:10:1), and heat at 105° for 10 minutes: the R_f value of the principal spot obtained from the test solution corresponds to that obtained from the Standard solution.

Microbial limits—It meets the requirements of the tests for absence of *Staphylococcus aureus* and *Pseudomonas aeruginosa* under *Microbial Limit Tests* ⟨61⟩.

Minimum fill ⟨755⟩: meets the requirements.

Assay—

Mobile solvent, Internal standard solution, and *Standard preparation*—Prepare as directed in the *Assay* under *Betamethasone.*

Assay preparation—Transfer an accurately weighed quantity of Betamethasone Cream, equivalent to about 2 mg of betamethasone, into a capped 50-mL centrifuge tube. Add 10.0 mL of *Internal standard solution* and 10.0 mL of alcohol. Mix by rotation for about 20 minutes. Centrifuge at 2500 rpm for about 10 minutes. Transfer a portion of the supernatant solution to a suitable vial.

Procedure—Proceed as directed for *Procedure* in the *Assay* under *Betamethasone.* Calculate the quantity, in mg, of $C_{22}H_{29}FO_5$ in the portion of Cream taken by the formula:

$$20C(R_U/R_S),$$

in which C is the concentration, in mg per mL, of USP Betamethasone RS in the *Standard preparation*, and R_U and R_S are the peak height ratios of the betamethasone peak and the internal standard peak obtained from the *Assay preparation* and the *Standard preparation*, respectively.

Betamethasone Syrup

» Betamethasone Syrup contains not less than 90.0 percent and not more than 115.0 percent of the labeled amount of $C_{22}H_{29}FO_5$.

Packaging and storage—Preserve in well-closed containers.

Reference standard—*USP Betamethasone Reference Standard*—Dry at 105° for 3 hours before using.

Identification—Evaporate 25 mL of the *Assay preparation*, prepared as directed in the *Assay*, on a steam bath just to dryness, and dissolve the residue in 0.5 mL of alcohol. Proceed as directed for *Identification test B* under *Betamethasone*, beginning with "On a suitable thin-layer chromatographic plate."

Assay—

Developing solvent—Prepare a mixture of chloroform, methanol, and ammonium hydroxide (175:20:1).

Tetramethylammonium hydroxide reagent—Dilute 20 mL of tetramethylammonium hydroxide TS with alcohol to make 100 mL.

Standard preparation—Dissolve a suitable quantity of USP Betamethasone RS, accurately weighed, in a mixture of chloroform and methanol (1:1) to obtain a solution having a known concentration of about 0.6 mg per mL.

Test preparation—Use a pipet calibrated to contain a suitable volume, and transfer to a 50-mL centrifuge tube an accurately measured volume of Betamethasone Syrup, equivalent to about 1.2 mg of Betamethasone. Rinse the pipet with 15 mL of 0.1 *N* hydrochloric acid, then with 20 mL of ethyl acetate, and add the rinsings to the centrifuge tube. Rotate for about 10 minutes, or shake manually for about 1 minute. (Do not use a mechanical shaker.) Centrifuge to separate the phases. Transfer the upper phase (ethyl acetate) to a small, pear-shaped flask. Extract the aqueous phase twice more with 20-mL portions of ethyl acetate, and add the extracts to the pear-shaped flask. Evaporate the combined extracts on a steam bath under a gentle stream of nitrogen to dryness. Allow to cool to room temperature. Dissolve the residue in about 0.5 mL of chloroform and methanol (1:1) with a vortex mixer. Transfer the solution to a 2-mL volumetric flask with small portions of chloroform and methanol (1:1), dilute with chloroform and methanol (1:1) to volume, and mix.

Procedure—On a suitable thin-layer chromatographic plate (see *Chromatography* ⟨621⟩), coated with a 0.25-mm layer of chromatographic silica gel, apply 200-µL portions of the *Test preparation* and the *Standard preparation.* Allow the spots to dry, and develop the chromatogram, using the *Developing solvent*, until the solvent front has moved about 15 cm. Remove the plates from the developing chamber, and allow them to dry for about 15 minutes. Mark the betamethasone bands, using short-wavelength ultraviolet light, to include similar zones of silica gel for the *Test preparation*, the *Standard preparation*, and a zone containing no betamethasone for the blank. Scrape off these zones, and transfer them to separate, 50-mL centrifuge tubes. Add 15.0 mL of alcohol to each, and rotate for 20 minutes. Centrifuge to clarify. Transfer 10.0-mL portions of the supernatant solutions to separate, stoppered tubes. To each tube add 1.0 mL of blue tetrazolium TS, followed by 1.0 mL of *Tetramethylammonium hydroxide reagent*, and mix. Heat in a 35° water bath for about 1 hour. Remove from the water bath, add 1.0 mL of glacial acetic acid to each tube, and mix. Cool to room temperature. Concomitantly determine the absorbances of the solutions in 1-cm cells at the wavelength of maximum absorbance at about 525 nm, with a suitable spectrophotometer. Calculate the quantity, in mg, of $C_{22}H_{29}FO_5$ in each mL of the Syrup taken by the formula:

$$2(C/V)(A_U - A_B)/(A_S - A_B),$$

in which C is the concentration, in mg per mL, of USP Betamethasone RS in the *Standard preparation*, V is the volume, in mL, of Syrup taken, and A_U, A_S, and A_B are the absorbances of the solutions from the *Assay preparation*, the *Standard preparation*, and the blank, respectively.

Betamethasone Tablets

» Betamethasone Tablets contain not less than 90.0 percent and not more than 110.0 percent of the labeled amount of $C_{22}H_{29}FO_5$.

Packaging and storage—Preserve in well-closed containers.

Reference standard—*USP Betamethasone Reference Standard*—Dry at 105° for 3 hours before using.

Identification—Evaporate 50 mL of the *Assay preparation*, prepared as directed in the *Assay*, on a steam bath just to dryness, and dissolve the residue in 1 mL of chloroform. Proceed as directed for *Identification test B* under *Betamethasone*, beginning with "On a suitable thin-layer chromatographic plate."

Dissolution ⟨711⟩—

Medium: water; 900 mL. Add 1.0 mL of *Internal standard solution* to each vessel.

Apparatus 2: 50 rpm.

Time: 45 minutes.

Mobile phase—Prepare a filtered and degassed mixture of methanol and water (60:40), making adjustments if necessary (see *System Suitability* under *Chromatography* ⟨621⟩).

Internal standard solution—Prepare a solution in methanol of testosterone having a final concentration of about 0.5 mg per mL.

Standard solution—Prepare a solution of USP Betamethasone RS, in methanol, having an accurately known concentration of about 0.5 mg per mL. Pipet 1 mL of this solution and 1 mL of the *Internal standard solution* into a container, and dilute quantitatively with water to 900 mL.

Chromatographic system (see *Chromatography* ⟨621⟩)—The liquid chromatograph is equipped with a 254-nm detector and a 3.9-mm × 30-cm column that contains packing L1. The flow rate is about 2 mL per minute. Chromatograph replicate injections of the *Standard preparation*, and record the peak responses as directed under *Procedure:* the resolution, R, between betamethasone and testosterone is not less than 1.5, and the relative standard deviation for replicate injections is not more than 3.0%.

Procedure—Separately inject equal volumes (about 200 µL) of the *Standard solution* and filtered portions of the solution under test into the chromatograph, record the chromatograms, and measure the responses for the major peaks. The relative retention times are about 0.5 for betamethasone and 1.0 for testosterone. Calculate the quantity of $C_{22}H_{29}FO_5$ dissolved in comparison with the *Standard solution*, similarly chromatographed.

Tolerances—Not less than 75% (Q) of the labeled amount of $C_{22}H_{29}FO_5$ is dissolved in 45 minutes.

Uniformity of dosage units ⟨905⟩: meet the requirements.

Procedure for content uniformity—

Standard preparation—Prepare as directed under *Assay for Steroids* ⟨351⟩, using USP Betamethasone RS, to obtain a solution having a known concentration of about 12 µg per mL instead of 10 µg per mL.

Test preparation—Weigh and finely powder 1 Tablet. Transfer to a 125-mL separator, add 20 mL of water, and shake. Extract the betamethasone completely, using three 15-mL portions of chloroform, filtering each extract through chloroform-washed cotton into a 50-mL volumetric flask. Dilute with chloroform to volume, and mix. Transfer 20.0 mL of this solution to a glass-stoppered, 50-mL conical flask, evaporate the chloroform on a steam bath just to dryness, cool, and dissolve the residue in 20.0 mL of alcohol.

Procedure—Proceed as directed under *Assay for Steroids* ⟨351⟩, except to keep the flasks in a constant-temperature bath at 45 ± 1° for 90 minutes, then add 1.0 mL of glacial acetic acid, and cool. Calculate the quantity, in mg, of $C_{22}H_{29}FO_5$ in the Tablet by the formula:

$$(TC/D)(A_U/A_S),$$

in which T is the labeled quantity, in mg, of betamethasone in the Tablet, C is the concentration, in µg per mL, of USP Betamethasone RS in the *Standard preparation*, D is the concentration, in µg per mL, of betamethasone in the *Test preparation*, based upon the labeled quantity per Tablet and the extent of dilution, and A_U and A_S are the absorbances of the solutions from the *Test preparation* and the *Standard preparation*, respectively.

Assay—

Mobile phase—Prepare a filtered and degassed mixture of water and acetonitrile (2:1). Make adjustments if necessary (see *System Suitability* under *Chromatography* ⟨621⟩).

Internal standard solution—Transfer about 25 mg of beclomethasone to a 200-mL volumetric flask, add methanol to volume, and mix.

Standard preparation—Dissolve an accurately weighed quantity of USP Betamethasone RS in methanol, and dilute quantitatively and stepwise, if necessary, with methanol to obtain a solution having a known concentration of about 0.1 mg per mL. Mix equal volumes, accurately measured, of this solution and the *Internal standard solution* to obtain a *Standard preparation* having a final known concentration of about 0.05 mg of USP Betamethasone RS per mL.

Assay preparation—Weigh and finely powder not less than 20 Betamethasone Tablets. Transfer an accurately weighed portion of the powder, equivalent to about 0.5 mg of betamethasone, to a 125-mL separator. Add 25 mL of water, and shake by mechanical means for about 15 minutes. Add 5.0 mL of *Internal standard solution*. Extract with four 25-mL portions of chloroform. Filter the chloroform extracts through about 4 g of chloroform-washed anhydrous sodium sulfate, collecting the extracts in a 150-mL beaker. Evaporate the extracts on a steam bath with the aid of a stream of nitrogen to dryness, taking care to avoid overheating. Dissolve the residue in 2 mL of methanol, and transfer to a 10-mL volumetric flask. Rinse the beaker with small portions of methanol, transferring the rinses to the same flask. Dilute with methanol to volume, and mix.

Chromatographic system (see *Chromatography* ⟨621⟩)—The liquid chromatograph is equipped with a 254-nm detector and a 4-mm × 30-cm column that contains packing L1. The flow rate is about 1.2 mL per minute. Chromatograph the *Standard preparation*, and record the peak heights as directed under *Procedure:* the resolution, R, between the analyte and internal standard peaks is not less than 1.7, and the relative standard deviation for replicate injections is not more than 2.0%.

Procedure—Separately inject equal volumes (about 10 µL) of the *Standard preparation* and the *Assay preparation* into the chromatograph, record the chromatograms, and measure the heights of the major peaks. The relative retention times are about 1.4 for beclomethasone and 1.0 for betamethasone. Calculate the quantity, in mg, of $C_{23}H_{29}FO_5$ in the portion of Tablets taken by the formula:

$$10C(R_U/R_S),$$

in which C is the concentration, in mg per mL, of USP Betamethasone RS in the *Standard preparation*, and R_U and R_S are the peak height ratios obtained from the *Assay preparation* and the *Standard preparation*, respectively.

Betamethasone Acetate

$C_{24}H_{31}FO_6$ 434.50

Pregna-1,4-diene-3,20-dione, 9-fluoro-11,17-dihydroxy-16-methyl-21-(acetyloxy)-, (11β,16β)-.

9-Fluoro-11β,17,21-trihydroxy-16β-methylpregna-1,4-diene-3,20-dione 21-acetate [*987-24-6*].

» Betamethasone Acetate contains not less than 97.0 percent and not more than 103.0 percent of $C_{24}H_{31}FO_6$, calculated on the anhydrous basis.

Packaging and storage—Preserve in tight containers.

Reference standard—*USP Betamethasone Acetate Reference Standard*—Do not dry; determine the *Water* content by *Method I* before using.

Identification—

A: The infrared absorption spectrum of a mineral oil dispersion of it, previously dried at 105° for 3 hours, exhibits maxima only at the same wavelengths as that of a similar preparation of USP Betamethasone Acetate RS.

B: It responds to *Identification test B* under *Betamethasone*, USP Betamethasone Acetate RS being used instead of USP Betamethasone RS.

C: To about 50 mg in a test tube add 2 mL of alcoholic potassium hydroxide TS, and heat in a boiling water bath for 5 minutes. Cool, add 2 mL of dilute sulfuric acid (1 in 3.5), and boil gently for about 1 minute: the odor of ethyl acetate is perceptible.

Specific rotation ⟨781⟩: between +120° and +128°, calculated on the anhydrous basis, determined in a solution in dioxane containing 100 mg of Betamethasone Acetate in each 10 mL.

Water, *Method I* ⟨921⟩: not more than 4.0%.

Residue on ignition ⟨281⟩: not more than 0.2%, a platinum crucible being used.

Ordinary impurities ⟨466⟩—

Test solution: methanol.

Standard solution: methanol.

Application volume: 10 µL.

Eluant: a mixture of toluene and isopropyl alcohol (90:10), in a nonequilibrated chamber.

Visualization: 5.

Assay—

Mobile phase—Prepare a filtered and degassed mixture of water and acetonitrile (8:7). Make adjustments if necessary (see *System Suitability* under *Chromatography* ⟨621⟩).

Internal standard solution—Transfer about 35 mg of progesterone to a 50-mL volumetric flask, add methanol to volume, and mix.

Standard preparation—Dissolve an accurately weighed quantity of USP Betamethasone Acetate RS in methanol, and dilute quantitatively with methanol to obtain a solution containing about 0.5 mg per mL. Transfer 10.0 mL of the resulting solution to a 50-mL volumetric flask, add 10.0 mL of *Internal standard solution*, dilute with methanol to volume, and mix to obtain a solution having a known concentration of about 0.1 mg of USP Betamethasone Acetate RS per mL.

Assay preparation—Transfer about 50 mg of Betamethasone Acetate, accurately weighed, to a 100-mL volumetric flask, add methanol to volume, and mix. Transfer 10.0 mL of the resulting solution to a 50-mL volumetric flask, add 10.0 mL of *Internal standard solution*, dilute with methanol to volume, and mix.

Chromatographic system (see *Chromatography* ⟨621⟩)—The liquid chromatograph is equipped with a 254-nm detector and a 4-mm × 30-cm column that contains packing L1. The flow rate is about 1 mL per minute. Chromatograph the *Standard preparation*, and record the peak responses as directed under *Procedure:* the resolution, R, between the analyte and internal standard peaks is not less than 2, and the relative standard deviation for replicate injections is not more than 2.0%.

Procedure—Separately inject equal volumes (about 25 µL) of the *Standard preparation* and the *Assay preparation* into the chromatograph, record the chromatograms, and measure the responses for the major peaks. The relative retention times are about 3 for progesterone and 1.0 for betamethasone acetate. Calculate the quantity, in mg, of $C_{24}H_{31}FO_6$ in the portion of betamethasone acetate taken by the formula:

$$500C(R_U/R_S),$$

in which C is the concentration, in mg per mL, of USP Betamethasone Acetate RS in the *Standard preparation*, and R_U and R_S are the peak response ratios obtained from the *Assay preparation* and the *Standard preparation*, respectively.

Betamethasone Acetate Suspension, Sterile, Betamethasone Sodium Phosphate and—*see* Betamethasone Sodium Phosphate and Betamethasone Acetate Suspension, Sterile

Betamethasone Benzoate

$C_{29}H_{33}FO_6$ 496.57

Pregna-1,4-diene-3,20-dione, 17-(benzoyloxy)-9-fluoro-11,21-dihydroxy-16-methyl-, (11β,16β)-.

9-Fluoro-11β,17,21-trihydroxy-16β-methylpregna-1,4-diene-3,20-dione 17-benzoate [*22298-29-9*].

» Betamethasone Benzoate contains not less than 98.0 percent and not more than 102.0 percent of $C_{29}H_{33}FO_6$, calculated on the dried basis.

Packaging and storage—Preserve in tight containers.

Reference standard—*USP Betamethasone Benzoate Reference Standard*—Dry at 105° for 3 hours before using.

Identification—The infrared absorption spectrum of a mineral oil dispersion of it exhibits maxima only at the same wavelengths as that of a similar preparation of USP Betamethasone Benzoate RS.

Specific rotation ⟨781⟩: between +60° and +66°, calculated on the dried basis, determined in a solution in dioxane containing 80 mg in each 2 mL.

Loss on drying ⟨731⟩—Dry about 200 mg, accurately weighed, at 105° for 3 hours: it loses not more than 0.5% of its weight.

Related steroids—Dissolve 100.0 mg in 5.0 mL of methanol to obtain the *Test solution.* Dissolve a suitable quantity of USP Betamethasone Benzoate RS in methanol to obtain a *Standard solution* having a known concentration of about 5 mg per mL. Dilute a portion of this solution quantitatively and stepwise with methanol to obtain a *Diluted standard solution* having a concentration of about 100 µg per mL. Apply separate 10-µL portions of the three solutions on the starting line of a 20- × 20-cm thin-layer chromatographic plate coated with a 0.25-mm layer of chromatographic silica gel. Develop the chromatogram in a suitable chamber previously equilibrated with a solvent system consisting of a mixture of benzene, acetone, and methanol (75:25:4), until the solvent front has moved about 15 cm from the origin. Remove the plate from the chamber, air-dry, and view under short-wavelength ultraviolet light: the principal spot from the *Test solution* corresponds in R_f value to that of the *Standard solution,* and the *Test solution* shows not more than 3 additional spots, the intensity and size of which do not exceed those of the spot from the *Diluted standard solution.*

Assay—

Mobile phase—Prepare a suitable filtered solution of acetonitrile and water (60:40).

Internal standard solution—Prepare a solution of betamethasone dipropionate in methanol containing 0.6 mg per mL.

Standard preparation—Using an accurately weighed quantity of USP Betamethasone Benzoate RS, prepare a solution in methanol having a known concentration of about 0.6 mg per mL. Mix 5.0 mL of this solution and 10.0 mL of the *Internal standard solution* to obtain a *Standard preparation* having a known concentration of about 0.2 mg of betamethasone benzoate per mL.

Assay preparation—Transfer about 60 mg of Betamethasone Benzoate, accurately weighed, to a 100-mL volumetric flask. Dilute with methanol to volume, and mix. Mix 5.0 mL of this solution and 10.0 mL of the *Internal standard solution.*

Chromatographic system (see *Chromatography* ⟨621⟩)—The liquid chromatograph is equipped with a 254-nm detector and a 4-mm × 30-cm column that contains 5-µm packing L1. The flow rate is about 1 mL per minute. Chromatograph three replicate injections of the *Standard preparation,* and record the peak responses as directed under *Procedure:* the relative standard deviation is not more than 2.0%, and the resolution factor between betamethasone benzoate and the internal standard is not less than 3.

Procedure—Separately inject equal volumes (about 15 µL) of the *Standard preparation* and the *Assay preparation* into the chromatograph by means of a suitable microsyringe or sampling valve, record the chromatograms, and measure the responses for the major peaks. The retention times are about 7 and 5 minutes for betamethasone dipropionate and betamethasone benzoate, respectively. Calculate the quantity, in mg, of $C_{29}H_{33}FO_6$ in the portion of Betamethasone Benzoate taken by the formula:

$$300C(R_U/R_S),$$

in which C is the concentration, in mg per mL, of USP Betamethasone Benzoate RS in the *Standard preparation,* and R_U and R_S are the peak response ratios of the betamethasone benzoate peak and the internal standard peak obtained from the *Assay preparation* and the *Standard preparation,* respectively.

Betamethasone Benzoate Gel

» Betamethasone Benzoate Gel contains an amount of betamethasone benzoate ($C_{29}H_{33}FO_6$) equivalent to not less than 90.0 percent and not more than 110.0 percent of the labeled amount of $C_{29}H_{33}FO_6$.

Packaging and storage—Preserve in collapsible tubes or in tight containers.

Reference standard—*USP Betamethasone Benzoate Reference Standard*—Dry at 105° for 3 hours before using.

Identification—Place 2 mL each of the *Assay preparation* and the *Standard preparation,* obtained as directed in the *Assay,* in separate test tubes, evaporate in a current of nitrogen to dryness, and dissolve the residues in 0.1-mL portions of a mixture consisting of chloroform and methanol (7:3). Apply separate 10-µL portions of both solutions at points about 2 cm from the bottom edge of a 5- × 20-cm thin-layer chromatographic plate coated with a 0.25-mm layer of chromatographic silica gel mixture and previously activated by being heated at 105° for 30 minutes. Develop the plate in a suitable chamber, previously equilibrated with a solvent mixture consisting of benzene, acetone, and methanol (75:25:4) until the solvent front has moved to about 15 cm from the line of application. Remove the plate, air-dry, and examine under short-wavelength ultraviolet light: the principal spots from both solutions appear as ultraviolet light-absorbing zones at about R_f 0.6.

Microbial limits—It meets the requirements of the tests for absence of *Staphylococcus aureus* and *Pseudomonas aeruginosa* under *Microbial Limit Tests* ⟨61⟩.

Assay—

Standard preparation—Dissolve an accurately weighed quantity of USP Betamethasone Benzoate RS in methanol to obtain a solution having a known concentration of about 50 µg per mL.

Assay preparation—Transfer an accurately weighed quantity of Betamethasone Benzoate Gel, equivalent to about 0.5 mg of betamethasone benzoate, to a 125-mL separator containing 20 mL of water. Add 1 mL of saturated sodium acetate solution, and extract the mixture with three 25-mL portions of chloroform, shaking for about 2 minutes with each portion. Combine the chloroform extracts in a second separator, wash with 10 mL of water, and filter the chloroform through paper into a conical flask with the aid of several small portions of chloroform. Evaporate on a steam bath nearly to dryness, with the aid of a current of nitrogen, and remove the last traces of solvent without heating. Dissolve the residue in 10.0 mL of methanol.

Procedure—Transfer 1.0 mL each of the *Assay preparation*, the *Standard preparation*, and methanol to provide the reagent blank, to separate test tubes. Add 0.1 mL of a 1 in 1000 solution of cupric acetate in methanol to each tube, mix, and allow to stand for 10 minutes with occasional mixing. Evaporate the contents of each tube in a current of nitrogen at about 45° to dryness. Dissolve each residue in 0.2 mL of phenylhydrazine reagent (prepared by dissolving 100 mg of phenylhydrazine hydrochloride in 1.5 mL of water, adding 8.5 mL of glacial acetic acid, and mixing) with the aid of a vibrating mixing device. Allow to stand at room temperature for 5 minutes, then add 5.0 mL of alcohol, and mix. Concomitantly determine the absorbances of the solutions from the *Assay preparation* and the *Standard preparation* in 1-cm cells at the wavelength of maximum absorbance at about 364 nm, with a suitable spectrophotometer, against the reagent blank. Calculate the quantity, in mg, of $C_{29}H_{33}FO_6$ in the portion of Gel taken by the formula:

$$0.01C(A_U/A_S),$$

in which C is the concentration, in µg per mL, of USP Betamethasone Benzoate RS in the *Standard preparation*, and A_U and A_S are the absorbances of the solutions from the *Assay preparation* and the *Standard preparation*, respectively.

Betamethasone Dipropionate

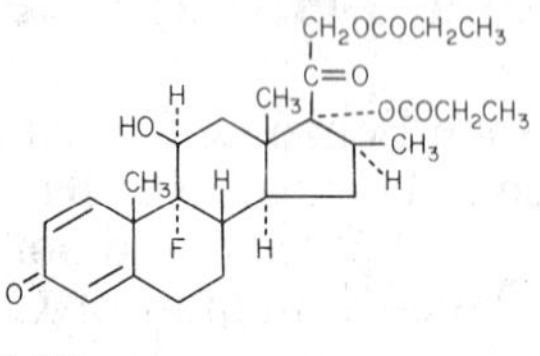

$C_{28}H_{37}FO_7$ 504.59

Pregna-1,4-diene-3,20-dione, 9-fluoro-11-hydroxy-16-methyl-17,21-bis(1-oxopropoxy)-, (11β,16β).

9-Fluoro-11β,17,21-trihydroxy-16β-methylpregna-1,4-diene-3,20-dione 17,21-dipropionate [*5593-20-4*].

» Betamethasone Dipropionate contains not less than 97.0 percent and not more than 103.0 percent of $C_{28}H_{37}FO_7$, calculated on the dried basis.

Packaging and storage—Preserve in well-closed containers.

Reference standards—*USP Betamethasone Dipropionate Reference Standard*—Dry at 105° for 3 hours before using. *USP Beclomethasone Dipropionate Reference Standard*—Dry at 105° for 3 hours before using.

Identification—

A: The infrared absorption spectrum of a mineral oil dispersion of it, previously dried at 105° for 3 hours, exhibits maxima only at the same wavelengths as that of a similar preparation of USP Betamethasone Dipropionate RS.

B: Prepare a solution of it in chloroform containing 1 mg per mL. On a suitable thin-layer chromatographic plate coated with a 0.25-mm layer of chromatographic silica gel mixture, apply 10 µL of this solution and 10 µL of a solution of USP Betamethasone Dipropionate RS in chloroform containing 1 mg per mL. Allow the spots to dry, and develop the chromatogram in a solvent system consisting of a mixture of chloroform and acetone (7:1) until the solvent has moved about three-fourths of the length of the plate. Remove the plate from the developing chamber, mark the solvent front, and allow the solvent to evaporate. Observe the dried spots under a short-wavelength ultraviolet light source: the R_f value of the principal spot obtained from the test solution corresponds to that obtained from the Standard solution.

Specific rotation ⟨781⟩: between +63° and +70°, calculated on the dried basis, determined in a solution in dioxane containing 100 mg in each 10 mL.

Loss on drying ⟨731⟩: Dry it at 105° for 3 hours: it loses not more than 1.0% of its weight.

Residue on ignition ⟨281⟩: not more than 0.2%, a platinum crucible being used.

Ordinary impurities ⟨466⟩—

Test solution: methanol.

Standard solution: methanol.

Application volume: 10 µL.

Eluant: a mixture of toluene and isopropyl alcohol (90:10), in a nonequilibrated chamber.

Visualization: 5.

Assay—

Mobile solvent—Prepare a suitable acetonitrile solution (approximately 1 in 2), degassed by ultrasonic vibration for 5 to 10 minutes, such that the retention time of betamethasone dipropionate is approximately 14 minutes and that of beclomethasone dipropionate is approximately 18 minutes. [NOTE—Do not leave the mobile solvent in the column overnight, but flush the system after use with water for 15 minutes, followed by methanol for 15 minutes.]

Internal standard solution—Prepare a solution of USP Beclomethasone Dipropionate RS in methanol having a known concentration of about 0.9 mg per mL.

Standard preparation—Prepare a solution of USP Betamethasone Dipropionate RS in methanol having a known concentration of about 0.6 mg per mL. Transfer 5.0 mL of this solution to a suitable vial, and add 5.0 mL of *Internal standard solution*, to obtain a *Standard preparation* having known concentrations of about 0.3 mg of betamethasone dipropionate and about 0.45 mg of beclomethasone dipropionate per mL.

Assay preparation—Weigh accurately about 60 mg of Betamethasone Dipropionate. Dilute quantitatively and stepwise with methanol to obtain a solution containing about 0.6 mg per mL. Transfer 5.0 mL of this solution to a suitable vial, and add 5.0 mL of *Internal standard solution*.

Procedure—Introduce equal volumes (between 5 µL and 25 µL) of the *Assay preparation* and the *Standard preparation* into a high-pressure liquid chromatograph (see *Chromatography* ⟨621⟩) operated at room temperature, by means of a suitable microsyringe or sampling valve, adjusting the specimen size and other operating parameters such that the peak obtained from the internal standard in the *Standard preparation* is about 0.6 full scale. Typically, the apparatus is fitted with a 4-mm × 30-cm column that contains packing L1, and is equipped with an ultraviolet detector capable of monitoring absorption at 254 nm or 240 nm and a suitable recorder, and is capable of operating at a column pressure of up to 3500 psi. In a suitable chromatogram, the lowest and highest peak area ratios (R_S) of three successive injections of the *Standard preparation* agree within 2.0%. Determine the ratio of the peak heights, at equivalent retention times, obtained with the *Assay preparation* and the *Standard preparation*, and calculate the quantity, in mg, of $C_{28}H_{37}FO_7$ in the portion of Betamethasone Dipropionate taken by the formula:

$$200C(R_U/R_S),$$

in which C is the concentration, in mg per mL, of USP Betamethasone Dipropionate RS in the *Standard preparation*, and R_U and R_S are the peak height ratios of the betamethasone dipropionate peak and the internal standard peak obtained from the *Assay preparation* and the *Standard preparation*, respectively.

Betamethasone Dipropionate Topical Aerosol

» Betamethasone Dipropionate Topical Aerosol is a solution, in suitable propellants in a pressurized container, of betamethasone dipropionate ($C_{28}H_{37}FO_7$) equivalent to not less than 90.0 percent and not more than 110.0 percent of the labeled amount of betamethasone ($C_{22}H_{29}FO_5$).

Packaging and storage—Preserve in tight, pressurized containers, and avoid exposure to excessive heat.

Reference standards—*USP Betamethasone Dipropionate Reference Standard*—Dry at 105° for 3 hours before using. *USP Beclomethasone Dipropionate Reference Standard*—Dry at 105° for 3 hours before using.

Identification—Proceed as directed in the *Identification test* under *Betamethasone Valerate Topical Aerosol*, except to transfer about 3 mL of the residue to a 50-mL centrifuge tube, and except to use 25 µL of a solution of USP Betamethasone Dipropionate RS in methanol containing about 3.2 mg per mL as the standard solution.

Other requirements—It meets the requirements for *Leak Testing* and *Pressure Testing* under *Aerosols* ⟨601⟩.

Assay—

Mobile solvent—Prepare as directed for *Mobile solvent* in the *Assay* under *Betamethasone Dipropionate*.

Internal standard solution—Prepare a solution of USP Beclomethasone Dipropionate RS, having a known concentration of about 0.90 mg per mL, in isopropanol containing acetic acid (1 in 1000).

Standard preparation—Prepare a solution of USP Betamethasone Dipropionate RS, having a known concentration of about 0.642 mg per mL, in isopropanol containing acetic acid (1 in 1000). Transfer 10.0 mL of this solution and 10.0 mL of *Internal standard solution* to a 100-mL volumetric flask, add isopropanol containing acetic acid (1 in 1000) to volume, and mix, to obtain a solution having known concentrations of about 0.09 mg of beclomethasone dipropionate and about 0.0642 mg of betamethasone dipropionate per mL.

Assay preparation—Proceed as directed for *Assay preparation* in the *Assay* under *Betamethasone Valerate Topical Aerosol*.

Procedure—Proceed as directed for *Procedure* in the *Assay* under *Betamethasone Dipropionate*. Calculate the quantity, in mg, of betamethasone ($C_{22}H_{29}FO_5$) equivalent to the quantity of betamethasone dipropionate ($C_{28}H_{37}FO_7$) in the container of the Topical Aerosol by the formula:

$$(392.47/504.59)(100C)(R_U/R_S),$$

in which 392.47 and 504.59 are the molecular weights of betamethasone and betamethasone dipropionate, respectively, *C* is the concentration, in mg per mL, of USP Betamethasone Dipropionate RS in the *Standard preparation*, and R_U and R_S are the peak height ratios of the betamethasone dipropionate and beclomethasone dipropionate peaks in the *Assay preparation* and the *Standard preparation*, respectively.

Betamethasone Dipropionate Cream

» Betamethasone Dipropionate Cream contains an amount of betamethasone dipropionate ($C_{28}H_{37}FO_7$) equivalent to not less than 90.0 percent and not more than 110.0 percent of the labeled amount of betamethasone ($C_{22}H_{29}FO_5$), in a suitable cream base.

Packaging and storage—Preserve in collapsible tubes or in tight containers.

Reference standards—*USP Betamethasone Dipropionate Reference Standard*—Dry at 105° for 3 hours before using. *USP Beclomethasone Dipropionate Reference Standard*—Dry at 105° for 3 hours before using.

Identification—Transfer about 1.5 g of Cream to a glass-stoppered, 50-mL centrifuge tube. Add 15 mL of methanol–hydrochloric acid solution prepared by mixing 1 volume of dilute hydrochloric acid (1 in 120) with 4 volumes of methanol. Shake to obtain a homogeneous mixture. Add 30 mL of solvent hexane, mix for 10 minutes, and centrifuge. Using a suitable syringe, transfer the lower aqueous phase to a second centrifuge tube, add about 20 mL of water, and mix. Extract this aqueous mixture with chloroform by shaking, centrifuging, and removing the lower, chloroform phase with a syringe. Evaporate the chloroform on a steam bath with the aid of a stream of nitrogen to dryness, cool, and dissolve the residue in chloroform to obtain a solution containing about 150 µg of betamethasone dipropionate per mL. On a suitable thin-layer chromatographic plate coated with a 0.25-mm layer of chromatographic silica gel mixture, apply 40 µL of this solution and 40 µL of a Standard solution of USP Betamethasone Dipropionate RS in chloroform containing 150 µg per mL. Allow the spots to dry, and develop the chromatograph in a solvent system consisting of a mixture of chloroform and acetone (7:1) until the solvent has moved about three-fourths of the length of the plate. Remove the plate from the developing chamber, mark the solvent front, and allow the solvent to evaporate. Observe the dried spots under a short-wavelength (254 nm) ultraviolet light source: the R_f value of the principal spot obtained from the test solution corresponds to that obtained from the Standard solution.

Minimum fill ⟨755⟩: meets the requirements.

Assay—

Mobile solvent—Prepare as directed in the *Assay* under *Betamethasone Dipropionate*.

Internal standard solution—Prepare a solution of USP Beclomethasone Dipropionate RS in methanol having a known concentration of about 0.45 mg per mL.

Standard preparation—Prepare a solution of USP Betamethasone Dipropionate RS in methanol having a known concentration of about 0.2 mg per mL. Transfer 10.0 mL of this solution to a suitable vial, and add 5.0 mL of *Internal standard solution*, to obtain a *Standard preparation* having known concentrations of about 0.133 mg of betamethasone dipropionate and about 0.15 mg of beclomethasone dipropionate per mL.

Assay preparation—Transfer an accurately weighed quantity of Betamethasone Dipropionate Cream, equivalent to about 2 mg of betamethasone dipropionate, into a capped 50-mL centrifuge tube. Add 5.0 mL of *Internal standard solution* and 10.0 mL of methanol. Heat in a water bath at 60°, shaking intermittently, until the assay specimen melts. Remove from the bath, and shake vigorously until the specimen has resolidified. Repeat the heating and shaking. Freeze in an ice-methanol bath for about 15 minutes, and centrifuge at 2500 rpm for about 5 minutes. Transfer a portion of the supernatant solution to a suitable vial.

Procedure—Proceed as directed for *Procedure* in the *Assay* under *Betamethasone Dipropionate*. Calculate the quantity, in mg, of $C_{22}H_{29}FO_5$ in the portion of Cream taken by the formula:

$$(392.47/504.59)(15C)(R_U/R_S),$$

in which 392.47 and 504.59 are the molecular weights of betamethasone and betamethasone dipropionate, respectively, *C* is the concentration, in mg per mL, of USP Betamethasone Dipropionate RS in the *Standard preparation*, and R_U and R_S are the peak height ratios of the betamethasone dipropionate peak and the internal standard peak obtained from the *Assay preparation* and the *Standard preparation*, respectively.

Betamethasone Dipropionate Cream, Clotrimazole and—*see* Clotrimazole and Betamethasone Dipropionate Cream

Betamethasone Dipropionate Lotion

» Betamethasone Dipropionate Lotion contains an amount of betamethasone dipropionate ($C_{28}H_{37}FO_7$) equivalent to not less than 90.0 percent and not more than 110.0 percent of the labeled amount of betamethasone ($C_{22}H_{29}FO_5$), in a suitable lotion base.

Packaging and storage—Preserve in tight containers.

Reference standards—*USP Betamethasone Dipropionate Reference Standard*—Dry at 105° for 3 hours before using. *USP Beclomethasone Dipropionate Reference Standard*—Dry at 105° for 3 hours.

Identification—Transfer a quantity of Lotion, equivalent to about 0.6 mg of betamethasone dipropionate, to a 50-mL vial. Add 10 mL of 0.1 *N* hydrochloric acid, then add 4 mL of chloroform. Disperse on a vortex mixer for about 1 minute, then shake vigorously for 10 minutes, and centrifuge at 2000 rpm for about 5 minutes. Transfer the chloroform layer to a suitable vial. Proceed as directed in the *Identification test* under *Betamethasone Dipropionate Cream*, beginning with "On a suitable thin-layer chromatographic plate."

Assay—

Mobile solvent—Prepare as directed in the *Assay* under *Betamethasone Dipropionate.*

Internal standard solution and *Standard preparation*—Prepare as directed in the *Assay* under *Betamethasone Dipropionate*, except to use chloroform as the solvent.

Assay preparation—Transfer an accurately weighed quantity of Betamethasone Dipropionate Lotion, equivalent to about 1.2 mg of betamethasone dipropionate, to a capped 50-mL centrifuge tube. Add 10.0 mL of 0.1 *N* hydrochloric acid, shake to disperse the specimen under assay, then add 2.0 mL of *Internal standard solution* and 2.0 mL of chloroform. To 10.0 mL of 0.1 *N* hydrochloric acid in a second capped 50-mL centrifuge tube add 4.0 mL of *Standard preparation*, and treat this solution and the *Assay preparation* as follows: Cap, and shake vigorously for about 2 minutes, or disperse on a vortex mixer for about 1 minute. Centrifuge at 2500 rpm for about 3 minutes. Transfer the chloroform phase to a suitable vial. Evaporate the chloroform under a stream of nitrogen at slightly elevated temperature to dryness. Cool the vials to room temperature, add 4.0 mL of methanol to each, and swirl to dissolve the residues.

Procedure—Using these as the *Standard preparation* and the *Assay preparation*, proceed as directed for *Procedure* in the *Assay* under *Betamethasone Dipropionate*. Calculate the quantity, in mg, of betamethasone ($C_{22}H_{29}FO_5$) in the portion of Lotion taken by the formula:

$$(392.47/504.59)(4C)(R_U/R_S),$$

in which 392.47 and 504.59 are the molecular weights of betamethasone and betamethasone dipropionate, respectively, C is the concentration, in mg per mL, of USP Betamethasone Dipropionate RS in the *Standard preparation*, and R_U and R_S are the peak height ratios of the betamethasone dipropionate peak and the internal standard peak obtained from the *Assay preparation* and the *Standard preparation*, respectively.

Betamethasone Dipropionate Ointment

» Betamethasone Dipropionate Ointment contains an amount of betamethasone dipropionate ($C_{28}H_{37}FO_7$) equivalent to not less than 90.0 percent and not more than 110.0 percent of the labeled amount of betamethasone ($C_{22}H_{29}FO_5$), in a suitable ointment base.

Packaging and storage—Preserve in collapsible tubes or in well-closed containers.

Reference standards—*USP Betamethasone Dipropionate Reference Standard*—Dry at 105° for 3 hours before using. *USP Beclomethasone Dipropionate Reference Standard*—Dry at 105° for 3 hours before using.

Identification—It responds to the *Identification test* under *Betamethasone Dipropionate Cream.*

Minimum fill ⟨755⟩: meets the requirements.

Assay—

Mobile solvent—Prepare as directed in the *Assay* under *Betamethasone Dipropionate.*

Internal standard solution and *Standard preparation*—Prepare as directed in the *Assay* under *Betamethasone Dipropionate Cream*, except to use alcohol as the solvent.

Assay preparation—Transfer an accurately weighed quantity of Betamethasone Dipropionate Ointment, equivalent to about 2 mg of betamethasone dipropionate, to a capped 50-mL centrifuge tube. Add 5.0 mL of *Internal standard solution* and 10.0 mL of alcohol. Heat in a water bath at 70°, shaking intermittently until the assay specimen melts. Remove from the bath, and shake vigorously until the ointment has solidified. Repeat the heating and shaking operation. Proceed as directed for *Assay preparation* in the *Assay* under *Betamethasone Dipropionate Cream*, beginning with "Freeze in an ice-methanol bath."

Procedure—Proceed as directed for *Procedure* in the *Assay* under *Betamethasone Dipropionate Cream.*

Betamethasone Sodium Phosphate

$C_{22}H_{28}FNa_2O_8P$ 516.41

Pregna-1,4-diene-3,20-dione, 9-fluoro-11,17-dihydroxy-16-methyl-21-(phosphonooxy)-, disodium salt, (11β,16β)-.

9-Fluoro-11β,17,21-trihydroxy-16β-methylpregna-1,4-diene-3,20-dione 21-(disodium phosphate) [*151-73-5*].

» Betamethasone Sodium Phosphate contains not less than 97.0 percent and not more than 103.0 percent of $C_{22}H_{28}FNa_2O_8P$, calculated on the anhydrous basis.

Packaging and storage—Preserve in tight containers.

Reference standards—*USP Betamethasone Sodium Phosphate Reference Standard*—Determine the *Water* content by *Method I* before using. *USP Butylparaben Reference Standard*—Dry over silica gel for 5 hours before using.

Identification—

A: The infrared absorption spectrum of a mineral oil dispersion of it, previously dried at 105° for 3 hours, exhibits maxima only at the same wavelengths as that of a similar preparation of USP Betamethasone Sodium Phosphate RS.

B: Prepare a solution in methanol containing 1 mg per mL. On a suitable thin-layer chromatographic plate (see *Chromatography* ⟨621⟩), coated with a 0.25-mm layer of chromatographic silica gel, apply 10 µL of this solution and 10 µL of a solution of USP Betamethasone Sodium Phosphate RS in methanol containing 1 mg per mL. Allow the spots to dry, and develop the chromatogram in a solvent system consisting of a mixture of butyl alcohol saturated with dilute hydrochloric acid (1 in 12) until the solvent front has moved about three-fourths of the length of the plate. Proceed as directed in the *Identification test* under *Betamethasone Cream*, beginning with "Remove the plate."

C: Ignite it at 800° (see *Residue on Ignition* ⟨281⟩): the residue responds to the tests for *Sodium* ⟨191⟩ and for *Phosphate* ⟨191⟩.

Specific rotation ⟨781⟩: between +99° and +105°, calculated on the anhydrous basis, determined in a solution in water containing 100 mg in each 10 mL.

Water, *Method I* ⟨921⟩: not more than 10.0%.

Phosphate ions—

Standard phosphate solution and *Phosphate reagent A*—Prepare as directed under *Phosphate in Reagents*, in the General Tests subsection in the section, *Reagents, Indicators, and Solutions.*

Phosphate reagent B—Dissolve 350 mg of *p*-methylaminophenol sulfate in 50 mL of water, add 20 g of sodium bisulfite, mix to dissolve, and dilute with water to 100 mL.

Procedure—Dissolve about 50 mg of Betamethasone Sodium Phosphate, accurately weighed, in a mixture of 10 mL of water and 5 mL of 2 *N* sulfuric acid contained in a 25-mL volumetric flask, by warming if necessary. Add 1 mL each of *Phosphate reagent A* and *Phosphate reagent B*, dilute with water to 25.0 mL, mix, and allow to stand at room temperature for 30 minutes. Similarly and concomitantly prepare a Standard solution, using 5.0 mL of *Standard phosphate solution* instead of the 50 mg of the substance under test. Concomitantly determine the absorbances of both solutions in 1-cm cells at the wavelength of maximum absorbance at about 730 nm, with a suitable spectrophotometer, using water as the blank. The absorbance of the test solution is not more than that of the Standard solution. The limit is 1.0% of phosphate (PO_4).

Free betamethasone—Dissolve 25.0 mg in water to make 25.0 mL. Transfer 5.0 mL of the solution to a separator, and extract with three 25-mL portions of chloroform. Filter each extract through a chloroform-saturated cotton pledget, combining the filtrates in a conical flask. Evaporate the chloroform on a steam bath with the aid of a current of air to dryness, and dissolve the residue in 25.0 mL of methanol. Determine the absorbance (A) of this solution in a 1-cm cell at the wavelength of maximum absorbance at about 239 nm, with a suitable spectrophotometer, using methanol as the blank. Calculate the quantity, in mg, of free betamethasone in the portion of Betamethasone Sodium Phosphate taken by the formula:

$$3.125A.$$

The limit is 250 µg (1.0%).

Assay—

Mobile solvent—Prepare a suitable mixture (approximately 3:2) of methanol and 0.07 *M* monobasic potassium phosphate (the latter having been previously filtered through a membrane filter), degassed by ultrasonic vibration for 5 minutes, such that the retention time of betamethasone sodium phosphate is approximately 1.2 minutes and that of butylparaben is approximately 2.5 minutes.

Internal standard solution—Transfer about 50 mg of USP Butylparaben RS, accurately weighed, to a 200-mL volumetric flask, and dilute with acetone to volume.

Standard preparation—Dissolve about 45 mg of USP Betamethasone Sodium Phosphate RS, accurately weighed, in water in a 100-mL volumetric flask, and add water to volume. Transfer 3.0 mL to a suitable vial, and add 2.0 mL of *Internal standard solution* and 3.0 mL of water, to obtain a solution having a known concentration of about 0.17 mg of betamethasone sodium phosphate per mL and 0.06 mg of butylparaben per mL.

Assay preparation—Weigh accurately about 45 mg of Betamethasone Sodium Phosphate, and prepare as directed under *Standard preparation.*

Procedure—Introduce equal volumes (between 5 µL and 25 µL) of the *Assay preparation* and the *Standard preparation* into a high-pressure liquid chromatograph (see *Chromatography* ⟨621⟩) operated at room temperature, by means of a suitable microsyringe or sampling valve, adjusting the specimen size and other operating parameters such that the peak obtained from the internal standard in the *Standard preparation* is about 0.6 full scale. Typically, the apparatus is fitted with a 4-mm × 30-cm column packed with chromatographic column packing L1 and equipped with an ultraviolet detector capable of monitoring absorption at 254 nm and a suitable recorder, and capable of operating at a column pressure of up to 3500 psi. In a suitable chromatogram, the coefficient of variation for five replicate injections of the *Standard preparation* is not more than 3.0%. Determine the ratio of the peak heights, at equivalent retention times, obtained from the *Assay preparation* and the *Standard preparation*, and calculate the quantity, in mg, of $C_{22}H_{28}FNa_2O_8P$ in the portion of Betamethasone Sodium Phosphate taken by the formula:

$$267C(R_U/R_S),$$

in which C is the concentration, in mg per mL, of USP Betamethasone Sodium Phosphate RS in the *Standard preparation*, and R_U and R_S are the peak height ratios of the betamethasone sodium phosphate peak to the internal standard peak obtained from the *Assay preparation* and the *Standard preparation*, respectively.

Betamethasone Sodium Phosphate Injection

» Betamethasone Sodium Phosphate Injection is a sterile solution of Betamethasone Sodium Phosphate in Water for Injection. It contains an amount of betamethasone sodium phosphate ($C_{22}H_{28}FNa_2O_8P$) equivalent to not less than 90.0 percent and not more than 110.0 percent of the labeled amount of betamethasone ($C_{22}H_{29}FO_5$).

Packaging and storage—Preserve in single-dose or in multiple-dose containers, preferably of Type I glass.

Reference standard—*USP Betamethasone Sodium Phosphate Reference Standard*—Determine the *Water* content by *Method I* before using.

Identification—Dilute the Injection with methanol, if necessary, to obtain a solution containing about 2 mg of betamethasone sodium phosphate per mL. Separately apply 10 µL of this test solution and 10 µL of a solution of USP Betamethasone Sodium Phosphate RS in methanol containing 2 mg per mL to a thin-layer chromatographic plate (see *Chromatography* ⟨621⟩) coated with chromatographic silica gel mixture. Develop the chromatogram in an equilibrated chamber containing *n*-butyl alcohol previously shaken with 1 *N* hydrochloric acid, until the solvent front has moved about three-fourths of the length of the plate. Remove the plate from the developing chamber, air-dry, then spray with a mixture of sulfuric acid, methanol, and nitric acid (10:10:1). Heat the plate at 105° for 10 minutes: the R_f value of the principal spot from the test solution corresponds to that obtained from the Standard solution.

Pyrogen—It meets the requirements of the *Pyrogen Test* ⟨151⟩.

pH ⟨791⟩: between 8.0 and 9.0.

Particulate matter ⟨788⟩: meets the requirements under *Small-volume Injections.*

Other requirements—It meets the requirements under *Injections* ⟨1⟩.

Assay—

Mobile phase—Prepare a filtered and degassed mixture of methanol and 0.05 *M* monobasic potassium phosphate (1:1). Make adjustments if necessary (see *System Suitability* under *Chromatography* ⟨621⟩).

Internal standard solution—Transfer about 100 mg of butylparaben to a 100-mL volumetric flask, add methanol to volume, and mix.

Standard preparation—Using an accurately weighed quantity of USP Betamethasone Sodium Phosphate RS, prepare a solution in water containing 4 mg per mL. Transfer 3.0 mL of this solution to a 25-mL volumetric flask, add 5.0 mL of *Internal standard solution*, dilute with water to volume, and mix to obtain a solution having a known concentration of about 0.5 mg of USP Betamethasone Sodium Phosphate RS per mL.

Assay preparation—Transfer an accurately measured volume of Betamethasone Sodium Phosphate Injection, equivalent to about 9 mg of betamethasone, to a 25-mL volumetric flask. Add 5.0 mL of the *Internal standard solution*, dilute with water to volume, and mix.

Chromatographic system (see *Chromatography* ⟨621⟩)—The liquid chromatograph is equipped with a 254-nm detector and a 3.9-mm × 30-cm column that contains packing L1. The flow rate is about 2 mL per minute. Chromatograph the *Standard preparation*, and record the peak responses as directed under *Procedure:* the resolution, R, between the analyte and internal standard peaks is not less than 5, and the relative standard deviation for replicate injections is not more than 2.0%.

Procedure—Separately inject equal volumes (about 20 µL) of the *Standard preparation* and the *Assay preparation* into the chromatograph, record the chromatograms, and measure the responses for the major peaks. The relative retention times are about 2.4 for butylparaben and 1.0 for betamethasone sodium phosphate. Calculate the quantity, in mg, of $C_{22}H_{29}FO_5$ in each mL of Betamethasone Sodium Phosphate Injection taken by the formula:

$$(392.47/516.41)(25C/V)(R_U/R_S),$$

in which 392.47 and 516.41 are the molecular weights of betamethasone and betamethasone sodium phosphate, respectively, C is the concentration, in mg per mL, of USP Betamethasone Sodium Phosphate RS in the *Standard preparation*, V is the volume, in mL, of Injection taken, and R_U and R_S are the peak response ratios obtained from the *Assay preparation* and the *Standard preparation*, respectively.

Sterile Betamethasone Sodium Phosphate and Betamethasone Acetate Suspension

» Sterile Betamethasone Sodium Phosphate and Betamethasone Acetate Suspension is a sterile preparation of Betamethasone Sodium Phosphate in solution and Betamethasone Acetate in suspension in Water for Injection. It contains an amount of betamethasone sodium phosphate ($C_{22}H_{28}FNa_2O_8P$) equivalent to not less than 90.0 percent and not more than 115.0 percent of the labeled amount of betamethasone ($C_{22}H_{29}FO_5$), and not less than 90.0 percent and not more than 115.0 percent of the labeled amount of betamethasone acetate ($C_{24}H_{31}FO_6$).

Packaging and storage—Preserve in multiple-dose containers, preferably of Type I glass.

Reference standards—*USP Betamethasone Sodium Phosphate Reference Standard*—Determine the *Water* content by *Method I* before using. *USP Betamethasone Acetate Reference Standard*—Do not dry; determine the *Water* content by *Method I* before using.

Identification—

A: Dilute 2 mL with 2 mL of methanol. On a suitable thin-layer chromatographic plate (see *Chromatography* ⟨621⟩), coated with a 0.25-mm layer of chromatographic silica gel, apply 10 µL of this solution and 10 µL of a methanol-water solution (1:1) of USP Betamethasone Sodium Phosphate RS containing 2 mg per mL. Proceed as directed for *Identification test B* under *Betamethasone Sodium Phosphate*, beginning with "Allow the spots to dry."

B: On a suitable thin-layer chromatographic plate (see *Chromatography* ⟨621⟩), coated with a 0.25-mm layer of chromatographic silica gel, apply 10 µL of the solution prepared for *Identification test A* and 10 µL of a methanol-water solution (1:1) of USP Betamethasone Acetate RS containing 1.5 mg per mL. Proceed as directed for *Identification test B* under *Betamethasone*, beginning with "Allow the spots to dry."

pH ⟨791⟩: between 6.8 and 7.2.

Other requirements—It meets the requirements under *Injections* ⟨1⟩.

Assay—

Mobile phase—Prepare a filtered and degassed mixture of methanol and 0.075 *M* monobasic potassium phosphate (7:5). Make adjustments if necessary (see *System Suitability* under *Chromatography* ⟨621⟩).

Internal standard solution—Transfer about 50 mg of methyltestosterone to a 50-mL volumetric flask, add methanol to volume, and mix.

Standard preparation—Transfer about 63 mg of USP Betamethasone Sodium Phosphate RS, accurately weighed, to a 25-mL volumetric flask, add *Mobile phase* to volume, and mix (*Solution 1*). Transfer about 45 mg of USP Betamethasone Acetate RS, accurately weighed, to a 25-mL volumetric flask, add methanol to volume, and mix (*Solution 2*). Pipet 5 mL each of *Solution 1* and *Solution 2* into a 100-mL volumetric flask. Add 10.0 mL of *Internal standard solution*, dilute with *Mobile phase* to volume, and mix to obtain a *Standard preparation* having known concentrations of about 126 µg of USP Betamethasone Sodium Phosphate RS per mL and 90 µg of USP Betamethasone Acetate RS per mL.

Assay preparation—Using a "To contain" pipet transfer an accurately measured volume of the well-mixed Sterile Suspension, equivalent to about 9 mg of betamethasone acetate, to a 100-mL volumetric flask. Rinse the pipet with about 10 mL of *Mobile phase*, collecting the rinse in the volumetric flask. Add 10.0 mL of *Internal standard solution*, dilute with *Mobile phase* to volume, and mix.

Chromatographic system (see *Chromatography* ⟨621⟩)—The liquid chromatograph is equipped with a 254-nm detector and a 3.9-mm × 30-cm column that contains packing L1. The flow rate is about 1.2 mL per minute. Chromatograph the *Standard preparation*, and record the peak responses as directed under *Procedure:* the resolution, R, between the betamethasone phosphate and betamethasone acetate peaks is not less than 5.0, and the resolution, R, between the betamethasone acetate and internal standard peaks is not less than 3.0, and the relative standard deviation for replicate injections is not more than 2.0%.

Procedure—Separately inject equal volumes (about 20 µL) of the *Standard preparation* and the *Assay preparation* into the chromatograph, record the chromatograms, and measure the responses for the major peaks. The relative retention times are about 0.5 for betamethasone phosphate, 1.7 for methyltestosterone, and 1.0 for betamethasone acetate. Calculate the quantity, in mg, of betamethasone acetate ($C_{24}H_{31}FO_6$) in each mL of the Suspension taken by the formula:

$$0.1C/V(R_U/R_S),$$

in which C is the concentration, in µg per mL, of USP Betamethasone Acetate RS in the *Standard preparation*, V is the volume, in mL, of Suspension taken, and R_U and R_S are the peak response ratios obtained for betamethasone acetate and methyltestosterone from the *Assay preparation* and the *Standard preparation*, respectively. Calculate the quantity, in mg, of betamethasone ($C_{22}H_{29}FO_5$) equivalent to the quantity of betamethasone sodium phosphate ($C_{22}H_{28}FNa_2O_8P$), in each mL of the Suspension taken by the formula:

$$(392.47/516.41)(0.1C/V)(R_U/R_S),$$

in which 392.47 and 516.41 are the molecular weights of betamethasone and betamethasone sodium phosphate, respectively, C is the concentration, in µg per mL, of USP Betamethasone Sodium Phosphate RS in the *Standard preparation*, V is the volume, in mL, of Suspension taken, and R_U and R_S are the peak response ratios obtained for betamethasone phosphate and methyltestosterone from the *Assay preparation* and the *Standard preparation*, respectively.

Betamethasone Valerate

$C_{27}H_{37}FO_6$ 476.58

Pregna-1,4-diene-3,20-dione, 9-fluoro-11,21-dihydroxy-16-methyl-17-[(1-oxopentyl)oxy]-, (11β,16β)-.

9-Fluoro-11β,17,21-trihydroxy-16β-methylpregna-1,4-diene-3,20-dione 17-valerate [*2152-44-5*].

» Betamethasone Valerate contains not less than 97.0 percent and not more than 103.0 percent of $C_{27}H_{37}FO_6$, calculated on the dried basis.

Packaging and storage—Preserve in tight containers.

Reference standards—*USP Betamethasone Valerate Reference Standard*—Dry at 105° for 3 hours before using. *USP Beclomethasone Dipropionate Reference Standard*—Dry at 105° for 3 hours before using.

Identification—

A: The infrared absorption spectrum of a mineral oil dispersion of it, previously dried, exhibits maxima only at the same wavelengths as that of a similar preparation of USP Betamethasone Valerate RS.

B: Prepare a solution in alcohol containing 1 mg per mL. On a suitable thin-layer chromatographic plate (see *Chromatography* ⟨621⟩), coated with a 0.25-mm layer of chromatographic silica gel mixture, apply 10 µL of this solution and 10 µL of a solution of USP Betamethasone Valerate RS in alcohol containing 1 mg per mL. Allow the spots to dry, and develop the chromatogram in a solvent system consisting of a mixture of toluene and ethyl acetate (1:1) until the solvent front has moved about three-fourths of the length of the plate. Remove the plate from the developing chamber, mark the solvent front, and allow the solvent to evaporate. Spray the plate with a mixture of sulfuric acid, methanol, and nitric acid (10:10:1), and heat at 105° for 15 minutes: the R_f value of the principal spot obtained from the test solution corresponds to that obtained from the Standard solution.

Specific rotation ⟨781⟩: between +75° and +82°, calculated on the dried basis, determined in a solution in dioxane containing 100 mg in each 10 mL.

Loss on drying ⟨731⟩—Dry it at 105° for 3 hours: it loses not more than 0.5% of its weight.

Residue on ignition ⟨281⟩: not more than 0.2%, a platinum crucible being used.

Assay—

Mobile phase—Prepare a filtered and degassed mixture of acetonitrile and water (3:2). Make adjustments if necessary (see *System Suitability* under *Chromatography* ⟨621⟩).

Internal standard solution—Transfer about 40 mg of beclomethasone dipropionate to a 100-mL volumetric flask, add a 1 in 1000 solution of glacial acetic acid in methanol to volume, and mix.

Standard preparation—Transfer about 30 mg of USP Betamethasone Valerate RS, accurately weighed, to a 50-mL volumetric flask, add a 1 in 1000 solution of glacial acetic acid in methanol to volume, and mix. Transfer 5.0 mL of this solution to a suitable stoppered vial, add 10.0 mL of *Internal standard solution*, and mix to obtain a solution having a known concentration of about 0.2 mg of USP Betamethasone Valerate RS per mL.

Assay preparation—Transfer about 60 mg of Betamethasone Valerate, accurately weighed, to a 100-mL volumetric flask, add a 1 in 1000 solution of glacial acetic acid in methanol to volume, and mix. Transfer 5.0 mL of this solution to a suitable stoppered vial, add 10.0 mL of *Internal standard solution*, and mix.

Chromatographic system (see *Chromatography* ⟨621⟩)—The liquid chromatograph is equipped with a 254-nm detector and a 4-mm × 30-cm column that contains packing L1. The flow rate is about 1.2 mL per minute. Chromatograph the *Standard preparation*, and record the peak responses as directed under *Procedure:* the resolution, *R*, between the betamethasone valerate and beclomethasone dipropionate peaks is not less than 4.5, and the relative standard deviation for replicate injections is not more than 2.0%.

Procedure—Separately inject equal volumes (about 10 µL) of the *Standard preparation* and the *Assay preparation* into the chromatograph, record the chromatograms, and measure the responses for the major peaks. The relative retention times are about 1.7 for beclomethasone dipropionate and 1.0 for betamethasone valerate. Calculate the quantity, in mg, of $C_{27}H_{37}FO_6$ in the portion of Betamethasone Valerate taken by the formula:

$$300C(R_U/R_S),$$

in which *C* is the concentration, in mg per mL, of USP Betamethasone Valerate RS in the *Standard preparation*, and R_U and R_S are the peak response ratios obtained from the *Assay preparation* and the *Standard preparation*, respectively.

Betamethasone Valerate Cream

» Betamethasone Valerate Cream contains an amount of betamethasone valerate ($C_{27}H_{37}FO_6$) equivalent to not less than 90.0 percent and not more than 110.0 percent of the labeled amount of betamethasone ($C_{22}H_{29}FO_5$), in a suitable cream base.

Packaging and storage—Preserve in collapsible tubes or in tight containers.

Reference standards—*USP Betamethasone Valerate Reference Standard*—Dry at 105° for 3 hours before using. *USP Beclomethasone Dipropionate Reference Standard*—Dry at 105° for 3 hours before using.

Identification—Transfer an amount of Cream, equivalent to about 2 mg of betamethasone, to a separator, add 20 mL of water and 2 mL of dilute hydrochloric acid (1 in 120), and mix. Extract with four 50-mL portions of chloroform, and combine the extracts. Filter through a cotton pledget, previously layered over with anhydrous sodium sulfate. Evaporate the filtrates on a steam bath under a stream of dry nitrogen to dryness. Dissolve the residue in alcohol to obtain a solution containing about 1 mg per mL. Proceed as directed in *Identification test B* under *Betamethasone Valerate*, beginning with "On a suitable thin-layer chromatographic plate."

Microbial limits—It meets the requirements of the tests for absence of *Staphylococcus aureus* and *Pseudomonas aeruginosa* under *Microbial Limit Tests* ⟨61⟩.

Minimum fill ⟨755⟩: meets the requirements.

Assay—

Mobile phase, Internal standard solution, Standard preparation, and *Chromatographic system*—Proceed as directed in the *Assay* under *Betamethasone Valerate.*

Assay preparation—Transfer an accurately weighed portion of Betamethasone Valerate Cream, equivalent to about 2.5 mg of betamethasone, to a 50-mL centrifuge tube. Add 10.0 mL of the *Internal standard solution* and 5.0 mL of a 1 in 1000 solution of glacial acetic acid in methanol. Insert the stopper into the tube, and place in a water bath held at 60° until the specimen melts. Remove from the bath, and shake vigorously until the specimen resolidifies. Repeat the heating and shaking two more times. Place the tube in an ice-methanol bath for 20 minutes, then centrifuge to separate the phases. Decant the clear supernatant solution into a suitable stoppered flask, and allow to warm to room temperature.

Procedure—Proceed as directed for *Procedure* in the *Assay* under *Betamethasone Valerate.* Calculate the quantity, in mg, of $C_{22}H_{29}FO_5$ in the portion of Cream taken by the formula:

$$(392.47/476.58)(15C)(R_U/R_S),$$

in which 392.47 and 476.58 are the molecular weights of betamethasone and betamethasone valerate, respectively, *C* is the concentration, in mg per mL, of USP Betamethasone Valerate RS in the *Standard preparation*, and R_U and R_S are the peak response ratios obtained from the *Assay preparation* and the *Standard preparation*, respectively.

Betamethasone Valerate Lotion

» Betamethasone Valerate Lotion contains an amount of betamethasone valerate ($C_{27}H_{37}FO_6$) equivalent to not less than 95.0 percent and not more than 115.0 percent of the labeled amount of betamethasone ($C_{22}H_{29}FO_5$).

Packaging and storage—Preserve in tight, light-resistant containers, and store at controlled room temperature.

Reference standards—*USP Betamethasone Valerate Reference Standard*—Dry at 105° for 3 hours before using. *USP Beclomethasone Dipropionate Reference Standard*—Dry at 105° for 3 hours before using.

Identification—Mix an amount of Lotion, equivalent to about 5 mg of betamethasone, with a mixture of methanol and chloroform (2:1) to make 10 mL. On a suitable thin-layer chromatographic plate (see *Chromatography* ⟨621⟩), coated with a 0.25-mm layer of chromatographic silica gel mixture, apply 20 µL of this solution and 20 µL of a Standard solution of USP Betamethasone Valerate RS in a mixture of methanol and chloroform (2:1) containing

0.6 mg per mL. Allow the spots to dry, and develop the chromatogram in a solvent system consisting of a mixture of chloroform and ethyl acetate (1:1), until the solvent front has moved about three-fourths of the length of the plate. Remove the plate from the developing chamber, mark the solvent front, and allow the solvent to evaporate. View the spots under ultraviolet light: the R_f value of the principal spot obtained from the test solution corresponds to that obtained from the Standard solution.

Microbial limits—It meets the requirements of the tests for absence of *Staphylococcus aureus* and *Pseudomonas aeruginosa* under *Microbial Limit Tests* ⟨61⟩.

pH ⟨791⟩: between 4.0 and 6.0.

Assay—

Mobile phase and *Chromatographic system*—Proceed as directed in the *Assay* under *Betamethasone Valerate.*

Internal standard solution—Transfer about 50 mg of beclomethasone dipropionate to a 25-mL volumetric flask, add chloroform to volume, and mix.

Standard preparation—Transfer about 40 mg of USP Betamethasone Valerate RS, accurately weighed, to a 25-mL volumetric flask, add chloroform to volume, and mix. Pipet 2 mL of this solution into a 50-mL centrifuge tube, add 10 mL of 0.1 *N* hydrochloric acid, then add 2.0 mL of *Internal standard solution.* Insert the stopper into the tube, shake vigorously for about 2 minutes, and centrifuge to separate the phases. Using a syringe, transfer the lower, chloroform phase to a small stoppered vial. Evaporate the chloroform on a steam bath, at low heat, with the aid of a stream of nitrogen. Add 4.0 mL of a 1 in 1000 solution of glacial acetic acid in methanol, and swirl to dissolve the residue.

Assay preparation—Transfer an accurately weighed portion of Betamethasone Valerate Lotion, equivalent to about 2.5 mg of betamethasone, to a stoppered, 50-mL centrifuge tube. Add 10.0 mL of 0.1 *N* hydrochloric acid, insert the stopper, and shake to disperse the specimen. Add 2.0 mL of chloroform and 2.0 mL of *Internal standard solution*, insert the stopper, and proceed as directed for *Standard preparation*, beginning with "shake vigorously for about 2 minutes."

Procedure—Proceed as directed for *Procedure* in the *Assay* under *Betamethasone Valerate.* Calculate the quantity, in mg, of betamethasone ($C_{22}H_{29}FO_5$) in the portion of Lotion taken by the formula:

$$(392.47/476.58)(4C)(R_U/R_S),$$

in which 392.47 and 476.58 are the molecular weights of betamethasone and betamethasone valerate, respectively, C is the concentration, in mg per mL, of USP Betamethasone Valerate RS in the Standard solution, and R_U and R_S are the peak response ratios obtained from the *Assay preparation* and the *Standard preparation*, respectively.

Betamethasone Valerate Ointment

» Betamethasone Valerate Ointment contains an amount of betamethasone valerate ($C_{27}H_{37}FO_6$) equivalent to not less than 90.0 percent and not more than 110.0 percent of the labeled amount of betamethasone ($C_{22}H_{29}FO_5$), in a suitable ointment base.

Packaging and storage—Preserve in collapsible tubes or in tight containers, and avoid exposure to excessive heat.

Reference standards—*USP Betamethasone Valerate Reference Standard*—Dry at 105° for 3 hours before using. *USP Beclomethasone Dipropionate Reference Standard*—Dry at 105° for 3 hours before using.

Identification—It responds to the *Identification test* under *Betamethasone Valerate Cream.*

Microbial limits—It meets the requirements of the tests for absence of *Staphylococcus aureus* and *Pseudomonas aeruginosa* under *Microbial Limit Tests* ⟨61⟩.

Minimum fill ⟨755⟩: meets the requirements.

Assay—

Mobile phase and *Chromatographic system*—Proceed as directed in the *Assay* under *Betamethasone Valerate.*

Internal standard solution—Transfer about 20 mg of beclomethasone dipropionate to a 50-mL volumetric flask, add a 1 in 1000 solution of glacial acetic acid in alcohol to volume, and mix.

Standard preparation—Transfer about 30 mg of USP Betamethasone Valerate RS, accurately weighed, to a 50-mL volumetric flask, add a 1 in 1000 solution of glacial acetic acid in alcohol to volume, and mix. Transfer 5.0 mL of this solution to a suitable stoppered vial, add 10.0 mL of *Internal standard solution*, and mix to obtain a solution having a known concentration of about 0.2 mg of USP Betamethasone Valerate RS per mL.

Assay preparation—Transfer an accurately weighed portion of Betamethasone Valerate Ointment, equivalent to about 2.5 mg of betamethasone, to a 50-mL centrifuge tube. Add 10.0 mL of the *Internal standard solution* and 5.0 mL of a 1 in 1000 solution of glacial acetic acid in alcohol. Insert the stopper into the tube, and place in a water bath held at 70° until the specimen melts. Remove from the bath, and shake vigorously until the specimen resolidifies. Repeat the heating and shaking two more times. Place the tube in an ice-methanol bath for 20 minutes, then centrifuge to separate the phases. Decant the clear supernatant solution into a suitable stoppered flask, and allow to warm to room temperature.

Procedure—Proceed as directed for *Procedure* in the *Assay* under *Betamethasone Valerate.* Calculate the quantity, in mg, of $C_{22}H_{29}FO_5$ in the portion of Ointment taken by the formula:

$$(392.47/476.58)(15C)(R_U/R_S),$$

in which 392.47 and 476.58 are the molecular weights of betamethasone and betamethasone valerate, respectively, C is the concentration, in mg per mL, of USP Betamethasone Valerate RS in the *Standard preparation*, and R_U and R_S are the peak response ratios obtained from the *Assay preparation* and the *Standard preparation*, respectively.

Bethanechol Chloride

$$\left[\begin{matrix}CH_3CHCH_2N^+(CH_3)_3\\ |\\ OCONH_2\end{matrix}\right] Cl^-$$

$C_7H_{17}ClN_2O_2$ 196.68

1-Propanaminium, 2-[(aminocarbonyl)oxy]-*N,N,N*-trimethyl-, chloride.

(2-Hydroxypropyl)trimethylammonium chloride carbamate [*590-63-6*].

» Bethanechol Chloride contains not less than 98.0 percent and not more than 101.5 percent of $C_7H_{17}ClN_2O_2$, calculated on the dried basis.

Packaging and storage—Preserve in tight containers.

Reference standard—*USP Bethanechol Chloride Reference Standard*—Dry at 105° for 2 hours before using.

Identification—

A: The infrared absorption spectrum of a mineral oil dispersion of it, previously dried, exhibits maxima only at the same wavelengths as that of a similar preparation of USP Bethanechol Chloride RS.

B: Dissolve about 50 mg in 2 mL of water, add 0.1 mL of cobaltous chloride solution (1 in 100), then add 0.1 mL of potassium ferrocyanide TS: an emerald-green color is produced, and almost entirely fades in 5 to 10 minutes (*distinction from choline chloride, which gives the same reaction but the color does not fade*).

C: To 1 mL of a solution (1 in 100) add 0.1 mL of iodine TS: a brown precipitate is formed, and it rapidly changes to a dark olive-green color.

D: A solution of it responds to the tests for *Chloride* ⟨191⟩.

pH ⟨791⟩: between 5.5 and 6.5, in a solution (1 in 100).

Loss on drying ⟨731⟩—Dry it at 105° for 2 hours: it loses not more than 1.0% of its weight.

Residue on ignition ⟨281⟩: not more than 0.1%.

Chloride content—Dissolve about 400 mg, previously dried and accurately weighed, in 30 mL of water. Add 40.0 mL of 0.1 *N* silver nitrate VS, then add 3 mL of nitric acid and 5 mL of nitrobenzene, shake well for a few minutes, add 2 mL of ferric ammonium sulfate TS, and titrate the excess silver nitrate with 0.1 *N* ammonium thiocyanate VS. Each mL of 0.1 *N* silver nitrate is equivalent to 3.545 mg of Cl: the content of Cl is between 17.7% and 18.3%.

Heavy metals, *Method I* ⟨231⟩—Dissolve 667 mg in 10 mL of water, add 2 mL of 1 *N* acetic acid, and dilute with water to 25 mL: the limit is 0.003%.

Assay—Dissolve about 500 mg of Bethanechol Chloride, accurately weighed, in a mixture of 10 mL of glacial acetic acid and 10 mL of mercuric acetate TS. Add 2 drops of crystal violet TS, and titrate with 0.1 *N* perchloric acid VS. Perform a blank determination, and make any necessary correction. Each mL of 0.1 *N* perchloric acid is equivalent to 19.67 mg of $C_7H_{17}ClN_2O_2$.

Bethanechol Chloride Injection

» Bethanechol Chloride Injection is a sterile solution of Bethanechol Chloride in Water for Injection. It contains not less than 95.0 percent and not more than 105.0 percent of the labeled amount of $C_7H_{17}ClN_2O_2$.

Packaging and storage—Preserve in single-dose containers, preferably of Type I glass.

Identification—It responds to *Identification tests B, C,* and *D* under *Bethanechol Chloride.*

pH ⟨791⟩: between 5.5 and 7.5.

Other requirements—It meets the requirements under *Injections* ⟨1⟩.

Assay—Transfer an accurately measured volume of Bethanechol Chloride Injection, equivalent to about 50 mg of bethanechol chloride, to a 50-mL beaker, and add 10 mL of dilute hydrochloric acid (1 in 1000). Add, with stirring, 20 mL of freshly prepared sodium tetraphenylboron solution (6 in 1000). Allow to stand for 10 minutes, filter through a tared crucible, dry the precipitate at 105° for 1 hour, cool, and weigh. The weight, in mg, of the precipitate, multiplied by 0.4094, represents the weight of $C_7H_{17}ClN_2O_2$ in the volume of the Injection taken.

Bethanechol Chloride Tablets

» Bethanechol Chloride Tablets contain not less than 90.0 percent and not more than 110.0 percent of the labeled amount of $C_7H_{17}ClN_2O_2$.

Packaging and storage—Preserve in tight containers.

Reference standard—*USP Bethanechol Chloride Reference Standard*—Dry at 105° for 2 hours before using.

Identification—Pulverize a quantity of Tablets, equivalent to about 100 mg of bethanechol chloride, add 15 mL of ether, and allow to digest for 15 minutes. Decant the ether, again extract the residue with 10 mL of ether, and discard the ether extracts. To the residue add 30 mL of alcohol, shake for 10 minutes, and allow to stand for 1 hour with frequent agitation. Filter with suction, and evaporate the filtrate on a steam bath to dryness: the bethanechol chloride so obtained, recrystallized from alcohol and dried at 105° for 2 hours, responds to *Identification test A* under *Bethanechol Chloride.*

Dissolution ⟨711⟩—

Medium: 0.1 *N* hydrochloric acid; 900 mL.

Apparatus 2: 50 rpm.

Time: 30 minutes.

Dipicrylamine reagent—Dissolve by swirling, without the aid of heat, 10 mg of dipicrylamine in 100 mL of methylene chloride. Prepare fresh.

Tetrabutylammonium iodide reagent—Dissolve 100 mg of tetrabutylammonium iodide in 20 mL of methylene chloride. Prepare fresh.

Standard preparation—Dissolve an accurately weighed quantity of USP Bethanechol Chloride RS in water to obtain a solution having a known concentration of about 33 µg per mL.

Procedure—Pipet an aliquot of the filtered test solution, estimated to contain about 66 µg of bethanechol chloride, into a 125-mL separator, and add an equivalent volume of 0.1 *N* sodium hydroxide. Into a similar separator pipet a volume of water equivalent to the total specimen volume to provide a reagent blank, and into a third 125-mL separator pipet 2 mL of *Standard preparation.* Treat each separator as follows: Add by pipet 0.5 mL of 0.1 *N* sodium hydroxide, and mix. Extract with four 5-mL portions of *Dipicrylamine reagent*, filter through a small pledget of cotton, and collect the methylene chloride extracts in 100-mL volumetric flasks. Dilute the contents of each flask with methylene chloride to about 75 mL. Pipet 1 mL of *Tetrabutylammonium iodide reagent* into each flask, dilute with methylene chloride to volume, and mix. Determine the absorbances of both solutions in 5-cm cells at the wavelength of maximum absorbance at about 420 nm, with a suitable spectrophotometer, against the reagent blank. Calculate the percentage dissolution of the Tablet by the formula:

$$180(C/WV)(A_U/A_S),$$

in which C is the concentration, in µg per mL, of USP Bethanechol Chloride RS in the *Standard preparation*, W is the labeled quantity, in mg, of bethanechol chloride in the Tablet, V is the volume, in mL, of the aliquot of test solution used, and A_U and A_S are the absorbances of the solutions from the test solution and the *Standard preparation*, respectively.

Tolerances—Not less than 80% (Q) of the labeled amount of $C_7H_{17}ClN_2O_2$ is dissolved in 30 minutes.

Uniformity of dosage units ⟨905⟩: meet the requirements.

Assay—

Sodium hypochlorite solution, 0.13%—Dilute sodium hypochlorite solution with water to obtain a 0.26% solution, and allow to stand for 30 minutes. Mix 75 mL of this solution with 75 mL of 1 *N* sodium hydroxide. Prepare this solution fresh daily.

Standard preparation—Dissolve an accurately weighed quantity of USP Bethanechol Chloride RS in 0.05 *N* hydrochloric acid, and dilute quantitatively with 0.05 *N* hydrochloric acid to obtain a solution having a known concentration of about 0.125 mg per mL.

Assay preparation—Weigh and finely powder not less than 20 Tablets. Transfer an accurately weighed portion of the powder, equivalent to about 25 mg of bethanechol chloride, to a 500-mL flask. Add 200.0 mL of 0.05 *N* hydrochloric acid, shake by mechanical means for 30 minutes, and centrifuge. Use the clear supernatant liquid as the *Assay preparation.*

Procedure—Pipet 2 mL each of the *Standard preparation* and the *Assay preparation* into separate 50-mL volumetric flasks. Transfer 3.0 mL of 0.05 *N* hydrochloric acid to another 50-mL volumetric flask to provide a reagent blank, and treat the contents of each flask as follows: [NOTE—The timing of the addition of reagents is critical. Use a stopwatch, and begin timing at the addition of the first reagent to the first flask.] At timed intervals, add 4 mL of *Sodium hypochlorite solution, 0.13%*, mix, allow to react for 15 minutes ± 30 seconds, and add 2.5 mL of phenol solution (7.5 in 1000). Rinse the inside walls of the flask with 1 to 2 mL of water, and mix. Allow to react for 5 minutes ± 30 seconds. Add 2.5 mL of 3.5 *N* hydrochloric acid, swirl to mix, and immediately add 1 mL of potassium iodide solution (3 in 1000). Mix, and allow to react for 5 minutes ± 30 seconds. Add 3 mL of previously centrifuged starch solution (1 in 100), dilute with water to volume, and mix. Immediately determine the absorbances of the *Standard preparation* and the *Assay preparation* at the wavelength of maximum absorbance at about 590 nm, with a suitable spectrophotometer, against the reagent blank. Calculate the quantity, in mg, of $C_7H_{17}ClN_2O_2$ in the portion of Tablets taken by the formula:

$$200C(A_U/A_S),$$

in which C is the concentration, in mg per mL, of USP Bethanechol Chloride RS in the *Standard preparation*, and A_U and A_S

are the absorbances of the solutions from the *Assay preparation* and the *Standard preparation*, respectively.

Biological Indicator for Dry-heat Sterilization, Paper Strip

» Biological Indicator for Dry-heat Sterilization, Paper Strip, is a preparation of viable spores made from a culture derived from a specified strain of *Bacillus subtilis* subspecies *niger*, on a suitable grade of paper[1] carrier, individually packaged in a suitable container readily penetrable by dry heat, and characterized for predictable resistance to dry-heat sterilization. The packaged Biological Indicator for Dry-heat Sterilization, Paper Strip, has a particular labeled spore count per carrier of not less than 10^4 and not more than 10^9 spores. When labeled for and subjected to dry-heat sterilization conditions at a particular temperature, it has a survival time and kill time appropriate to the labeled spore count and to the decimal reduction value (D value, in minutes) of the preparation, specified by:

Survival time (in minutes) = not less than (labeled D value) × ($\log_{10}$ labeled spore count per carrier − 2); and

Kill time (in minutes) = not more than (labeled D value) × ($\log_{10}$ labeled spore count per carrier + 4).

NOTE—See *Biological Indicators* ⟨1035⟩ for directives on selection of suitable indicators and on their applicability for different sterilization cycles, and for the characteristics of the basic or prototype article.

Packaging and storage—Preserve in the original package under the conditions recommended on the label, and protect from light, toxic substances, excessive heat, and moisture.

Expiration date—The expiration date is determined on the basis of stability studies and is not less than 18 months from the date of manufacture, the date of manufacture being the date on which the first determination of the total viable population was made.

Labeling—Label it to state that it is a Biological Indicator for Dry-heat Sterilization on a paper carrier, to indicate its D value, the method used to determine such D value, i.e., by spore count or fraction negative procedure after graded exposures to the sterilization conditions, survival time and kill time under the sterilization conditions stated on the label, its particular viable spore count, with a statement that such count has been determined after preliminary heat treatment, and its recommended storage conditions. State in the labeling the size of the paper carrier, the strain and ATCC number from which the spores were derived, and instructions for spore recovery and for its safe disposal. Indicate in the labeling that the stated D value is reproducible only under the exact conditions under which it was determined, that the user would not necessarily obtain the same result, and that the user would need to determine its suitability for the particular use.

Identification—The biological indicator organism complies substantially with the morphological, cultural, and biochemical characteristics of the strain of *Bacillus subtilis*, ATCC No. 9372, designated subspecies *niger*, detailed for that biological indicator organism under *Biological Indicator for Ethylene Oxide Sterilization, Paper Strip*.

Resistance performance tests—[NOTE—For all tests described in this section and in the section, *Purity*, handle each specimen with aseptic precautions, using sterilized equipment where applicable. For survival time, kill time, and D value determination, use apparatus of known thermodynamic characteristics, that has been validated for compliance with the requirements for safety[2] and performance,[3] that consists of a sterilizing chamber equipped with a means of heating the contained air, preferably electrically rather than gas fired, and so that there is adequate movement of the air, through forced ventilation (by mechanical devices such as blowers), with sensing and control devices for temperature and timing capable of indicating with an accuracy of not more than 0.5° and 1-second intervals, respectively. The geometrical pattern of the heat source(s) is such as to enable the biological indicators under test to be uniformly heated under the specified conditions. The temperature profile in the chamber is known, and cold spots, hot spots, and slow heat zones identified. The chamber has the capability of working within a temperature range of 40° to 300°, with an accuracy at any particular setting of not less than ±2.0°. The apparatus is equipped with a suitable additional access door or port so as to enable the entry and insertion (or removal) of specimens within 6 seconds and will enable the temperature to return to the set temperature within 0.5 minute where the specified temperature is 120° to 190° and within 1.0 minute where such temperature is 220° and above.]

Carry out the tests for *D value* and for *Survival time and kill time* at each of the temperatures for which the packaged Biological Indicator for Dry-heat Sterilization, Paper Strip, is labeled for use.

D value—Take a sufficient number of groups of specimens of Biological Indicator for Dry-heat Sterilization, Paper Strip, in their original individual containers, each group consisting of 10 specimens. The number of groups shall be such as to provide a range of observations from not less than one labeled D value below the labeled survival time through not less than one labeled D value above the labeled kill time. Place each group of 10 on a separate suitable specimen holder that permits each specimen to be exposed to the prescribed sterilizing condition at a specific location in the sterilizing chamber. Check the apparatus for operating parameters using specimen holders without specimens. Preheat the chamber for 30 minutes. Open the access door or port, place one of the holders with the group of 10 specimens in the sterilizing chamber, close the access door or port, and continue to operate the apparatus. Commence timing the heat exposure when the chamber temperature returns to 2° below the specified temperature. After the contents have been subjected to the sterilizing condition for a predetermined time, selected from a series of time increments, remove the holder with the heated specimens, and replace it with another containing 10 more specimens. Repeat the sterilizing procedure similarly, but for another predetermined time, and continue with successive groups until all have been appropriately heated. The heating time applied to the successive groups from the shortest selected should be in increments to form a series. The differences in sterilizing times over the series shall be constant and the difference between adjacent times shall not be greater than 75% of the labeled D value.

After completion of the sterilizing procedure, and within 2 hours, aseptically remove and add each strip to 30 mL of *Soybean Casein Digest Medium* in a suitable tube. Incubate each tube as directed for *General Procedure* under *Sterility Tests* ⟨71⟩,

[1] A satisfactory grade of paper is preferably one designed to hold spores, for example, a pure cotton-derived, medium-speed filter paper, such as Grade 740E or 591A, available from Schleicher and Schuell, Inc. For temperatures above 250° and for long periods of heating, Grade 740E may be preferable. It has been shown, by exposure at 250° and 280° for 1.5 hours, to lose about 2% and 5%, respectively, of moisture, with some browning. There was no ignition in 24 hours at those temperatures. Ignition temperature was not below 450°. For either grade browning without crumbling need not interfere with usage or testing.

[2] Safety includes design to prevent electric shock or gas explosion and burns, where operators can wear protective clothing and gloves against burns from touching hot surfaces.

[3] Descriptions of different types of dry-heat sterilizing equipment and detailed guidelines for determining, monitoring, and controlling the operating parameters have been published by the Health Industry Manufacturers Association (HIMA) in Report No. 78-4.7, *Operator Training for Dry Heat Sterilizing Equipment*, and by the Parenteral Drug Association, Inc. (PDA) in Technical Report No. 3, *Validation of Dry Heat Processes used for Sterilization and Depyrogenation.*

but at an incubation temperature of 30° to 35°. Observe each inoculated medium-containing tube at 24 hours and 48 hours, and every 1 or 2 days thereafter for a total of 7 days after inoculation.[4] Note the number of specimens showing no evidence of growth at any time.

Calculation—Designate the number of specimens taken for each group (i.e., 10) by n, and the difference between adjacent times (in minutes) by d. Designate for each group of the series the number of specimens showing no growth by:

$$f_1, f_2 \ldots\ldots f_k,$$

in which f_1 is the response of all 10 specimens showing growth (0/10 inactivated) in the group held for the shortest time for such result which is adjacent to an intermediate mortality, and f_k is the response of all 10 specimens of the group showing no growth (10/10 inactivated) in the group held for the longest time for such result which is adjacent to an intermediate mortality. Disregard all other observations for groups at the ends of the series, at f_1 and f_k, giving results that are not adjacent to an intermediate mortality. The test is valid if there is available a valid result (0/10) from a group held for a shorter time than that for the selected shortest time result (f_1) and there is available a valid result (10/10) from a group held for a longer time than that for the selected longest time result (f_k).

Subtract from 10 each f, from f_1 through f_k, and record the numbers. Multiply each such number by the corresponding f and sum the products, and designate that sum by $\Sigma f(10 - f)$. Determine the sum of the numbers showing no growth for the results f_1 through f_{k-1}, and designate that sum by Σf. Calculate the heating time (T) for achieving complete kill by the formula:

$$T = (\text{time for achieving the result } f_k) - d/2 - (d/10 \times \Sigma f).$$

Calculate the D value by the formula:

$$D = (T)/(\text{Log } N_o + 0.2507),$$

in which N_o is the average spore count per carrier determined by *Total viable spore count* (see below) at the time of making this test.

Calculate the variance of T (i.e., V_T) by the formula:

$$V_T = \frac{D^2}{n^2(n-1)} \times \Sigma f(10 - f),$$

and its standard deviation (SD) as:

$$\sqrt{V_T}.$$

Calculate the lower and upper confidence limits, respectively, for D as:

$$\frac{\text{lower confidence limit for } T}{\log N_o + 0.2507}$$

and

$$\frac{\text{upper confidence limit for } T}{\log N_o + 0.2507}$$

(The confidence limits for T are $T \pm$ 2SD; $p = 0.95$).

Replacement of missing values—If not more than 1 specimen from a group and not more than 2 specimens from all of the groups giving the results f_1 through f_k are lost, replace each missing value by adding 0 to the number showing no growth, if the number showing no growth in the remaining 9 specimens of that group is 4 or less, and adding 1 if the number showing no growth in the remaining 9 specimens of that group is 5 or more.

Interpretation—The test result, to meet the requirements of the test, is not outside ±20 percent of the D value stated on the label or labeling for the selected sterilizing temperature. The test is valid if the confidence limits of the estimate are within ±10 percent of the test result for the D value.

Survival time and kill time—Take 2 groups each consisting of 20 specimens of Biological Indicator for Dry-heat Sterilization, Paper Strip, in their original, individual containers. Place the specimens of a group in suitable specimen holders that permit each specimen to be exposed to the sterilizing conditions at a specific location in the sterilizing chamber. Check the chamber for operating parameters by preheating it to the selected temperature ±2°, using specimen holders without specimens. Preheat the unit to temperature, and equilibrate the heat chamber. Open the access door or port, and place the holder(s) in the chamber, close the access door or port, and continue to operate the apparatus. Commence timing the heat exposure when the chamber temperature returns to the lower limit of the selected temperature. Expose the specimens for the required survival time, enter the chamber, and remove the holder(s) containing the 20 specimens. Repeat the above procedure immediately, or preheat if a substantial interval has elapsed, so as to subject the second holder(s) containing 20 specimens similarly to the first, but for the required kill time. After completion of the sterilizing procedure, and at a noted time within 2 hours, aseptically remove and add each carrier to 30 mL of *Soybean Casein Digest Medium* in a suitable tube. Incubate each tube as directed for *General Procedure* under *Sterility Tests* ⟨71⟩, but at an incubation temperature of 30° to 35°. Observe each inoculated medium–containing tube at 24 hours and 48 hours, and every 1 or 2 days thereafter for a total of 7 days after inoculation.[4] Note the specimens showing no evidence of growth at any time. If all of the specimens subjected to dry-heat sterilization for the survival time show evidence of growth, while none of the specimens subjected to dry-heat sterilization for the kill time shows growth, the requirements of the tests are met. If either for the survival time test or for the kill time test, not more than 1 specimen out of both groups fails the survival requirement or the kill requirement (whichever is applicable) continue the corresponding test with 4 further groups each consisting of 20 specimens, according to the procedure described above. If all of the further specimens subjected to dry-heat sterilization *either* for the survival time meet the survival requirement *or* for the kill time meet the kill requirement, whichever is applicable, the requirements of the test are met.

Total viable spore count—Take 3 specimens of Biological Indicator for Dry-heat Sterilization, Paper Strip, out of their original individual containers, and proceed as directed for *Total viable spore count* under *Biological Indicator for Steam Sterilization, Paper Strip*, beginning with "Pulp the paper into component fibers," except to heat the vessel containing the suspension in a water bath at 65° to 70° for 15 minutes, and starting the timing when the temperature reaches 65°, and to incubate the inoculated media at 30° to 35°, and calculate the average number per specimen as directed therein. The average number is not less than 50% of the number of viable spores of the specified strain per carrier stated on the label, and not more than 300% of that number.

Purity—

Presence of contamination by other microorganisms—By examination of the spores on a suitable plate culture medium, there is no evidence of contamination with other microorganisms.

Stability—Where an additional test for stability, by spore count, is made during the shelf-life of the article, proceed as directed under *Total viable spore count.* The average number per specimen is not less than 50% of the highest such number obtained from any previous determinations made in the same dating period.

NOTE—

Disposal—Prior to destruction or discard, sterilize it by steam at 121° for not less than 30 minutes, or by not less than an equivalent method recommended by the manufacturer. This includes a strip used in test procedures for strips themselves.

Biological Indicator for Ethylene Oxide Sterilization, Paper Strip

» Biological Indicator for Ethylene Oxide Sterilization, Paper Strip is a preparation of viable spores made from a culture derived from a specified strain of *Bacillus subtilis* subspecies *niger* on a suitable grade

[4] Where growth is observed at any particular observation time, further incubation of the specimen(s) concerned is unnecessary.

of paper[1] carrier, individually packaged in a suitable container readily penetrable by ethylene oxide sterilizing gas mixture, and characterized for predictable resistance to sterilization with such gas. The packaged Biological Indicator for Ethylene Oxide Sterilization, Paper Strip, has a particular labeled spore count per carrier of not less than 10^4 and not more than 10^9 spores. Where labeled for and subjected to particular ethylene oxide sterilization conditions of a gaseous mixture, temperature, and relative humidity, it has a survival time and kill time appropriate to the labeled spore count and to the decimal reduction value (D value, in minutes) of the preparation, specified by:

Survival time (in minutes) = not less than (labeled D value) × ($\log_{10}$ labeled spore count per carrier − 2), and

Kill time (in minutes) = not more than (labeled D value) × ($\log_{10}$ labeled spore count per carrier + 4).

NOTE—See *Biological Indicators* ⟨1035⟩ for directives on selection of suitable indicators and on their applicability for different sterilization cycles, and for the characteristics of the basic or prototype article.

Packaging and storage—Preserve in the original package under the conditions recommended on the label, and protect it from light, toxic substances, excessive heat, and moisture.

Expiration date—The expiration date is determined on the basis of stability studies and is not less than 18 months from the date of manufacture, the date of manufacture being the date on which the first determination of the total viable population was made.

Labeling—Label it to state that it is a Biological Indicator for Ethylene Oxide Sterilization on a paper carrier, to indicate its D value, the method used to determine such D value, i.e., by spore count or fraction negative procedure after graded exposures to the sterilization conditions, survival time and kill time under the sterilization conditions stated on the label, its particular viable spore count, with a statement that such count has been determined after preliminary heat treatment, and its recommended storage conditions. State in the labeling the size of the paper carrier, the strain and ATCC number from which the spores were derived, and instructions for spore recovery and for its safe disposal. Indicate in the labeling that the stated D value is reproducible only under the exact conditions under which it was determined, that the user would not necessarily obtain the same result and that the user would need to determine its suitability for the particular use.

Identification—The biological indicator organism complies substantially with the morphological, cultural, and biochemical characteristics of the strain of *Bacillus subtilis*, ATCC No. 9372, designated subspecies *niger:* under microscopic examination it consists of Gram-positive rods of width 0.7 to 0.8 μm, and length 2 to 3 μm; the endospores are oval and central and the cells are not swollen; when incubated aerobically in appropriate media at 30° to 35°, growth occurs within 24 hours, and similar inoculated media incubated concomitantly at 55° to 60° show no evidence of growth in the same period; agar colonies have a dull appearance and may be cream or brown-colored; when incubated in nutrient broth it develops a pellicle, and shows little or no turbidity; when examined under conventional biochemical tests for microbial characterization, it develops a black pigment with tyrosine, it liquefies gelatin, utilizes citrate but not propionate or hippurate, reduces nitrate, and hydrolyzes both starch and glucose with no gas production; it shows a positive catalase reaction and gives a positive result with the Voges-Proskauer test.

[1] A satisfactory grade of paper is preferably one designed to hold spores, for example, a pure cotton-derived, medium-speed filter paper, such as Grade 740E or 591A, available from Schleicher and Schuell, Inc.

Resistance performance tests—[NOTE—For all tests described in this section and in the section, *Purity*, handle each specimen with aseptic precautions, using sterilized equipment where applicable. For survival time, kill time, and D value determination, use the apparatus that consists of a test chamber with a means of ensuring adequate mixing of the sterilant gas and a means for heating the sterilant gas to not higher than the preselected operating temperature so that no liquid enters the test chamber, equipped with temperature control and monitoring, pressure control, and humidification, and gas concentration monitoring devices, so that under the specified conditions of temperature, pressure, and humidification, the time to reach the specified gas concentration is not more than 1.0 minute, the time to exhaust the test chamber is not more than 1.0 minute, and the quantity of gas admitted to the test chamber is such as to enable the gaseous mixture to attain within ±5.0% of the gas concentration. The humidity-control device is capable of maintaining the relative humidity within 10 RH percent. The temperature-control device is capable of achieving ±2.0° in the chamber and the temperature-monitoring device has the capability of determining the uniformity of temperature to within ±0.5° at the exposure side. The apparatus has a suitable pressure-monitoring device (or devices) to indicate the vessel pressure with a precision of within ±0.84 kPa (±0.122 psia, ±6.35 mm of mercury). The specified times are applied and observed with the use of suitable operating devices and timing devices capable of indicating 1-second intervals.]

Carry out the tests for *D value* and for *Survival time and kill time* at each of the stated sterilization conditions for which the packaged Biological Indicator for Ethylene Oxide Sterilization, Paper Strip, is labeled for use.

STERILIZING PROCEDURE—

Place the loaded holder(s) in the test chamber, and

(1) Evacuate the test chamber to a pressure of not more than 100 ± 3 mm of mercury.

(2) Inject sufficient water vapor (e.g., saturated steam) to enable the chamber contents to attain within 10 RH percent of the required humidification condition and allow the chamber to equilibrate with moisture and to temperature for approximately 30 minutes.

(3) Inject a sufficient quantity of temperature-equilibrated ethylene oxide gas to attain the appropriate concentration ±30 mg of ethylene oxide per liter.

(4) Subject the specimens to the appropriate temperature, humidification and gas concentration conditions for the required time.

(5) Evacuate the test chamber to a pressure of 100 ± 3 mm of mercury, and release the vacuum with sterile filtered air. Repeat this until not less than 99% of the remaining gas has been removed, and remove the holder(s) with the exposed specimens.

If exposing further specimens to the sterilization conditions, proceed with steps (6) and (7).

(6) Flush the test chamber 5 times with filtered air after evacuation each time to a pressure of not more than 100 ± 3 mm of mercury.

(7) Repeat the entire sterilizing procedure above, steps (1) through (6), for other unexposed specimens, but maintain the specified conditions of step (4) for another required time.

CULTIVATION—

After completion of the sterilizing procedure, and at a noted time within 2 hours, aseptically remove and add each carrier to 30 mL of *Soybean Casein Digest Medium* in a suitable tube, and incubate as directed for *General Procedure* under *Sterility Tests* ⟨71⟩, but at an incubation temperature of 30° to 35°. Observe each inoculated medium–containing tube at 24 and 48 hours, and every 1 or 2 days thereafter, for a total of 7 days after inoculation.[2] Note the specimens showing no evidence of growth at any time, and examine material from any specimens that show evidence of growth for characteristic appearance of the spores.

D value—Take a sufficient number of groups of specimens of Biological Indicator for Ethylene Oxide Sterilization, Paper Strip, in their original individual containers, each group consisting of 10 specimens. The number of groups shall be such as to provide a range of observations from not less than one labeled D value

[2] Where growth is observed at any particular observation time, further incubation of the specimen(s) concerned is unnecessary.

below the labeled survival time through not less than one labeled D value above the labeled kill time. Place each group of 10 on a separate suitable specimen holder that permits each specimen to be exposed to the prescribed sterilizing conditions at a specific location in the test chamber. Check the apparatus for operating parameters using specimen holder without specimens. Place a group of specimens in the test chamber, and proceed with the steps as directed for *Sterilizing Procedure* including the evacuation and flushing described in step (6). Repeat steps (1) through (6) with successive groups of specimens, but subjecting each group to one of a series of exposure times. The differences in sterilizing times over the series shall be constant and the difference between adjacent times shall not be greater than 75% of the labeled D value. Treat the exposed specimens as directed under *Cultivation*.

Calculation—Designate the number of specimens taken for each group (i.e., 10) by n, and the difference between adjacent times (in minutes) by d. Designate for each group of the series the number of specimens showing no growth by:

$$f_1, f_2 \ldots\ldots f_k,$$

in which f_1 is the response of all 10 specimens showing growth (0/10 inactivated) in the group held for the shortest time for such result which is adjacent to an intermediate mortality, and f_k is the response of all 10 specimens of the group showing no growth (10/10 inactivated) in the group held for the longest time for such result which is adjacent to an intermediate mortality. Disregard all other observations for groups at the ends of the series, at f_1 and f_k, giving results that are not adjacent to an intermediate mortality. The test is valid if there is available a valid result (0/10) from a group held for a shorter time than that for the selected shortest time result (f_1) and there is available a valid result (10/10) from a group held for a longer time than that for the selected longest time result (f_k).

Subtract from 10 each f, from f_1 through f_k, and record the numbers. Multiply each such number by the corresponding f and sum the products, and designate that sum by $\Sigma f(10 - f)$. Determine the sum of the numbers showing no growth for the results f_1 through f_{k-1}, and designate the sum by Σf. Calculate the heating time (T) for achieving complete kill by the formula:

$$T = (\text{time for achieving the result } f_k) - d/2 - (d/10 \times \Sigma f).$$

Calculate the D value by the formula:

$$D = (T)/(\text{Log } N_o + 0.2507),$$

in which N_o is the average spore count per carrier determined by *Total viable spore count* (see below) at the time of making this test.

Calculate the variance of T (i.e., V_T) by the formula:

$$V_T = \frac{D^2}{n^2(n-1)} \times \Sigma f(10 - f),$$

and its standard deviation (SD) as:

$$\sqrt{V_T}.$$

Calculate the lower and upper confidence limits, respectively, for D as:

$$\frac{\text{lower confidence limit for } T}{\log N_o + 0.2507}$$

and

$$\frac{\text{upper confidence limit for } T}{\log N_o + 0.2507}$$

(The confidence limits for T are $T \pm 2\text{SD}$; $p = 0.95$).

Replacement of missing values—If not more than 1 specimen from a group and not more than 2 specimens from all of the groups giving the results f_1 through f_k are lost, replace each missing value by adding 0 to the number showing no growth, if the number showing no growth in the remaining 9 specimens of that group is 4 or less, and adding 1 if the number showing no growth in the remaining 9 specimens of that group is 5 or more.

Interpretation—The test result, to meet the requirements of the test, is not outside ±20 percent of the D value stated on the label or labeling for the selected sterilizing temperature. The test is valid if the confidence limits of the estimate are within ±10 percent of the test result for the D value.

Survival time and kill time—Take 2 groups each consisting of 20 specimens of Biological Indicator for Ethylene Oxide Sterilization, Paper Strip, in their original individual containers. Place the specimens of a group in suitable specimen holders that permit each specimen to be exposed to the sterilizing conditions at a specific location in the test chamber. Pre-heat the chamber to equilibrium at the selected temperature ±2°, and initiate and monitor the operating steps (1) through (6) as described under *Sterilizing Procedure* appropriate for the combination of gas concentration, temperature, and relative humidity, using a gassing time in step (4) appropriate to the survival time. Repeat the above procedure with 2 or more groups each consisting of 20 specimens, but using a gassing time in step (4) appropriate to the kill time. Treat the exposed specimens as directed under *Cultivation*. If all of the specimens subjected to the ethylene oxide sterilization conditions for the survival time show evidence of growth, while none of the specimens subjected to the ethylene oxide sterilization conditions for the kill time show evidence of growth, the requirements of the tests are met. If either for the survival time test or for the kill time test, not more than 1 specimen out of both groups fails the survival requirement or the kill requirement (whichever is applicable), continue the corresponding test with 4 further groups each consisting of 20 specimens, according to the procedure described above. If all of the further specimens subjected to ethylene oxide sterilization *either* for the survival time meet the survival requirement *or* for the kill time meet the kill requirement, whichever is applicable, the requirements of the test are met.

Total viable spore count—Take 3 specimens of Biological Indicator for Ethylene Oxide Sterilization, Paper Strip, out of their original, individual containers, and proceed as directed for *Total viable spore count* under *Biological Indicator for Steam Sterilization, Paper Strip*, beginning with "Pulp the paper into component fibers," except to heat the vessel containing the suspension in a water bath at 65° to 70° for 15 minutes, starting the timing when the temperature reaches 65°, and to incubate the inoculated media at 30° to 35°, and calculate the average number per specimen as directed therein. The average number is not less than 50% of the number of viable spores of the specified strain per carrier stated on the label, and not more than 300% of that number.

Purity—

Presence of contamination by other microorganisms—By examination of the spores on a suitable plate culture medium, there is no evidence of contamination with other microorganisms.

Stability—Where an additional test for stability, by spore count, is made during the shelf-life of the article, proceed as directed under *Total viable spore count*. The average number per specimen is not less than 50% of the highest such number obtained from any previous determinations made in the same dating period.

NOTE—

Disposal—Prior to destruction or discard, sterilize it by steam at 121° for not less than 30 minutes, or by not less than an equivalent method recommended by the manufacturer. This includes a strip used in test procedures for strips themselves.

Biological Indicator for Steam Sterilization, Paper Strip

» Biological Indicator for Steam Sterilization, Paper Strip, is a preparation of viable spores made from a culture derived from a specified strain of *Bacillus stearothermophilus*, on a suitable grade of paper[1]

[1] A satisfactory grade of paper is preferably one designed to hold spores, for example, a pure cotton-derived, medium-speed filter paper, such as Grade 740E or 591A, available from Schleicher and Schuell, Inc.

carrier, individually packaged in a suitable container readily penetrable by steam, and characterized for predictable resistance to steam sterilization. The packaged Biological Indicator for Steam Sterilization, Paper Strip, has a particular labeled spore count per carrier of not less than 10^4 and not more than 10^9 spores. When labeled for and subjected to steam sterilization conditions at a particular temperature, it has a survival time and kill time appropriate to the labeled spore count and to the decimal reduction value (D value, in minutes) of the preparation, specified by:

Survival time (in minutes) = not less than (labeled D value) × ($\log_{10}$ labeled spore count per carrier − 2); and

Kill time (in minutes) = not more than (labeled D value) × ($\log_{10}$ labeled spore count per carrier + 4).

NOTE—See *Biological Indicators* ⟨1035⟩ for directives on selection of suitable indicators and on their applicability for different sterilization cycles, and for the characteristics of the basic or prototype article.

Packaging and storage—Preserve in the original package under the conditions recommended on the label, and protect it from light, toxic substances, excessive heat, and moisture.

Expiration date—The expiration date is determined on the basis of stability studies and is not less than 18 months from the date of manufacture, the date of manufacture being the date on which the first determination of the total viable population was made.

Labeling—Label it to state that it is a Biological Indicator for Steam Sterilization on a paper carrier, to indicate its D value, the method used to determine such D value, i.e., by spore count or fraction negative procedure after graded exposures to the sterilization conditions, survival time and kill time under the sterilization conditions stated on the label, its particular viable spore count with a statement that such count has been determined after preliminary heat treatment, and its recommended storage conditions. State in the labeling the size of the paper carrier, the strain and ATCC number from which the spores were derived, and instructions for spore recovery and for its safe disposal. Indicate in the labeling that the stated D value is reproducible only under the exact conditions under which it was determined, that the user would not necessarily obtain the same result and that the user would need to determine its suitability for the particular use.

Identification—The biological indicator organism complies substantially with the morphological, cultural, and biochemical characteristics of the strain of *Bacillus stearothermophilus*, ATCC No. 7953 or 12980: under microscopic examination it consists of Gram-positive rods with oval endospores in subterminally swollen cells; when incubated in nutrient broth for 17 hours and used to inoculate appropriate solid media, growth occurs when the inoculated media are incubated aerobically for 24 hours at 55° to 60°, and similar inoculated media incubated concomitantly at 30° to 35° show no evidence of growth in the same period; when examined under conventional biochemical tests for microbial characterization, it shows a delayed weak positive catalase reaction, it does not utilize citrate, propionate or hippurate, but it reduces nitrate, it does not liquefy gelatin, it gives a negative result with the Voges-Proskauer test. Organisms derived from ATCC strain No. 7953 show negative egg yolk and starch hydrolysis reactions, while those derived from ATCC strain No. 12980 show positive reactions in both tests.

Resistance performance tests—[NOTE—For all tests described in this section and in the section, *Purity*, handle each specimen with aseptic precautions, using sterilized equipment where applicable. For survival time, kill time, and D value determination, use apparatus that consists of a chamber equipped with heating, temperature, and steam control and monitoring devices so that the contained air can be completely replaced by the steam admitted, the target temperature ±0.5° (with corresponding saturated steam pressure) is reached in not more than 10.0 seconds, and the chamber is capable of being exhausted to atmospheric pressure in not more than 5.0 seconds. The chamber can be entered and the contents removed within an additional 5 seconds. The apparatus is calibrated with appropriately distributed thermocouples or other heat-indicating devices to show that the temperature within the chamber has a uniformity of ±0.5°. The apparatus has a suitable pressure control and monitoring device (or devices) to indicate the target saturated steam pressure with a precision of within ±3.45 kPa (±0.5 psia, ±25.88 mm of mercury). The specified times are observed with the use of timing devices, capable of indicating 1-second intervals.] Carry out the tests for *D value* and for *Survival time and kill time* at each of the temperatures for which the packaged Biological Indicator for Steam Sterilization, Paper Strip, is labeled for use.

D value—Take a sufficient number of groups of specimens of Biological Indicator for Steam Sterilization, Paper Strip, in their original individual containers, each group consisting of 10 specimens. The number of groups shall be such as to provide a range of observations from not less than one labeled D value below the labeled survival time through not less than one labeled D value above the labeled kill time. Place each group of 10 on a separate suitable specimen holder that permits each specimen to be exposed to the prescribed sterilizing condition at a specific location in the sterilizing chamber. Check the apparatus for operating parameters using specimen holders without specimens. Exhaust the chamber, and within 15 seconds of opening the door, place one of the holders with the group of 10 specimens in the sterilizing chamber, and operate the apparatus to heat up the chamber contents as quickly as possible. After the contents have been subjected to the sterilizing condition for a predetermined time, selected from a series of time increments, exhaust the chamber as quickly as possible. Remove the holder with the heated specimens, and replace it with another containing 10 more specimens. Repeat the sterilizing procedure similarly, but for another predetermined time, and continue with successive groups until all have been appropriately heated. The heating time applied to the successive groups from the shortest selected should be in increments to form a series. The differences in sterilizing times over the series shall be constant and the difference between adjacent times shall not be greater than 75% of the labeled D value.

After completion of the sterilizing procedure, and within 2 hours, aseptically remove and add each strip to 30 mL of *Soybean Casein Digest Medium* in a suitable tube. Incubate each tube as directed for *General Procedure* under *Sterility Tests* ⟨71⟩, but at an incubation temperature of 55° to 60°. Observe each inoculated medium-containing tube at 24 hours and 48 hours, and every 1 or 2 days thereafter for a total of 7 days after inoculation.[2] Note the number of specimens showing no evidence of growth at any time.

Calculation—Designate the number of specimens taken for each group (i.e., 10) by n, and the difference between adjacent times (in minutes) by d. Designate for each group of the series the number of specimens showing no growth by:

$$f_1, f_2 \ldots\ldots f_k,$$

in which f_1 is the response of all 10 specimens showing growth (0/10 inactivated) in the group held for the shortest time for such result which is adjacent to an intermediate mortality, and f_k is the response of all 10 specimens of the group showing no growth (10/10 inactivated) in the group held for the longest time for such result which is adjacent to an intermediate mortality. Disregard all other observations for groups at the ends of the series, at f_1 and f_k, giving results that are not adjacent to an intermediate mortality. The test is valid if there is available a valid result (0/10) from a group held for a shorter time than that for the selected shortest time result (f_1) and there is available a valid result (10/10) from a group held for a longer time than that for the selected longest time result (f_k).

Subtract from 10 each f, from f_1 through f_k, and record the numbers. Multiply each such number by the corresponding f and sum the products, and designate that sum by $\Sigma f(10 - f)$. Determine the sum of the numbers showing no growth for the results

---

[2] Where growth is observed at any particular observation time, further incubation of the specimen(s) concerned is unnecessary.

f_1 through f_{k-1}, and designate that sum by Σf. Calculate the heating time (T) for achieving complete kill by the formula:

$$T = (\text{time for achieving the result } f_k) - d/2 - (d/10 \times \Sigma f).$$

Calculate the D value by the formula:

$$D = (T)/(\text{Log } N_o + 0.2507),$$

in which N_o is the average spore count per carrier determined by *Total viable spore count* (see below) at the time of making this test.

Calculate the variance of T (i.e., V_T) by the formula:

$$V_T = \frac{D^2}{n^2(n-1)} \times \Sigma f(10 - f),$$

and its standard deviation (SD) as:

$$\sqrt{V_T}.$$

Calculate the lower and upper confidence limits, respectively, for D as:

$$\frac{\text{lower confidence limit for } T}{\log N_o + 0.2507}$$

and

$$\frac{\text{upper confidence limit for } T}{\log N_o + 0.2507}$$

(The confidence limits for T are $T \pm 2\text{SD}$; $p = 0.95$).

Replacement of missing values—If not more than 1 specimen from a group and not more than 2 specimens from all of the groups giving the results f_1 through f_k are lost, replace each missing value by adding 0 to the number showing no growth, if the number showing no growth in the remaining 9 specimens of that group is 4 or less, and adding 1 if the number showing no growth in the remaining 9 specimens of that group is 5 or more.

Interpretation—The test result, to meet the requirements of the test, is not outside ±20 percent of the D value stated on the label or labeling for the selected sterilizing temperature. The test is valid if the confidence limits of the estimate are within ±10 percent of the test result for the D value.

Survival time and kill time—Take 2 groups each consisting of 20 specimens of Biological Indicator for Steam Sterilization, Paper Strip, in their original, individual containers. Place the specimens of a group in suitable specimen holders that permit each specimen to be exposed to the sterilizing conditions at a specific location in the sterilizing chamber. Check the chamber for operating parameters by preheating it to the selected temperature ±0.5°, using specimen holders without specimens. Exhaust the steam chamber, open the door as quickly as possible, and within 15 seconds of opening the door place the loaded holder(s) into the chamber, and operate the apparatus to heat the chamber contents as quickly as possible. Expose the specimens for the required survival time, counting the exposure from the time when the temperature record shows that the chamber has reached the required temperature. Exhaust the chamber as quickly as possible at the end of the exposure period. When the chamber can be safely entered, remove the holder(s) containing the specimens. Repeat the above procedure immediately, or preheat if a substantial interval has elapsed, so as to subject the holder(s) containing 20 specimens similarly to the first exposure. Repeat the above procedure with 2 more groups each consisting of 20 specimens, similarly, but expose the specimens for the required kill time. After completion of the sterilizing procedure, and within 2 hours, aseptically remove and add each carrier to 30 mL of *Soybean Casein Digest Medium* in a suitable tube. Incubate each tube as directed for *General Procedure* under *Sterility Tests* ⟨71⟩, but at an incubation temperature of 55° to 60°. Observe each inoculated medium-containing tube at 24 hours and 48 hours, and every 1 or 2 days thereafter for a total of 7 days after inoculation.[2] Note the specimens showing no evidence of growth at any time. If all of the specimens subjected to the steam sterilization for the survival time show evidence of growth, while none of the specimens subjected to the steam sterilization for the kill time shows growth, the requirements of the tests are met. If either for the survival time test or for the kill time test, not more than 1 specimen out of both groups fails the survival requirement or the kill requirement (whichever is applicable) continue the corresponding test with 4 further groups each consisting of 20 specimens, according to the procedure described above. If all of the further specimens subjected to steam sterilization *either* for the survival time meet the survival requirement *or* for the kill time meet the kill requirement, whichever is applicable, the requirements of the test are met.

Total viable spore count—Take 3 specimens of Biological Indicator for Steam Sterilization, Paper Strip, out of their original, individual containers. Pulp the paper into component fibers by placing them in a sterile 250-mL cup of a suitable blender containing 100 mL of chilled sterilized Purified Water and blending for 3 to 5 minutes to achieve a homogeneous suspension, remove and transfer a 10-mL aliquot of the suspension, and place it in a sterile, screw-capped, 16- × 125-mm vessel. Heat the vessel containing the suspension in a water bath at 95° to 100° for 15 minutes, starting the timing when the temperature reaches the temperature of 95°. Cool rapidly in an ice water bath (0° to 4°). Transfer two 1-mL aliquots to suitable vessels, and make appropriate serial dilutions in similar Purified Water, the dilutions being selected as calculated to yield 30 to 300 colonies on each of a pair of plates when treated as described below. Where the biological indicator has a low spore concentration, it may be necessary to modify the dilution series and to use more plates at each dilution. Prepare a separate series of plates for each aliquot. Place 1.0 mL of each selected dilution in each of two 15- × 100-mm petri dishes. Within 20 minutes, add to each plate 20 mL of *Soybean Casein Digest Agar Medium* (see *Microbial Limit Tests* ⟨61⟩) which has been melted and cooled to 45° to 50°. Swirl to attain a homogeneous suspension, and allow to solidify. Incubate the plates in an inverted position at 55° to 60°, and examine them after 24 and 48 hours, recording for each plate the number of colonies, and using the number of colonies after 48 hours for calculating the results. The test is valid if such number of spores is equal to or more than the number after 24 hours in each case. Calculate the average number per specimen from the results, using the appropriate dilution factor. The average number is not less than 50% of the number of viable spores of the specified strain per carrier stated on the label, and not more than 300% of that number.

Purity—

Presence of contamination by other microorganisms—By examination of the spores on a suitable plate culture medium, there is no evidence of contamination with other microorganisms.

Stability—Where an additional test for stability, by spore count, is made during the shelf-life of the article, proceed as directed under *Total viable spore count*. The average number per specimen is not less than 50% of the highest such number obtained from any previous determinations made in the same dating period.

NOTE—

Disposal—Prior to destruction or discard, sterilize it by steam at 121° for not less than 30 minutes, or by not less than an equivalent method recommended by the manufacturer. This includes a strip used in test procedures for strips themselves.

Biotin

$C_{10}H_{16}N_2O_3S$ 244.31

1*H*-Thieno[3,4-*d*]imidazole-4-pentanoic acid, hexahydro-2-oxo-, [3a*S*-(3aα,4β,6aα)]-.

(3a*S*,4*S*,6a*R*)-Hexahydro-2-oxo-1*H*-thieno[3,4-*d*]imidazole-4-valeric acid [*58-85-5*].

» Biotin contains not less than 97.5 percent and not more than 100.5 percent of $C_{10}H_{16}N_2O_3S$.

Packaging and storage—Store in tight containers.

Reference standard—*USP Biotin Reference Standard.*

Identification—The infrared absorption spectrum of a potassium bromide dispersion of it exhibits maxima only at the same wavelengths as that of a similar preparation of USP Biotin RS.

Specific rotation ⟨781⟩: between +89° and +93°, determined in a solution in 0.1 *N* sodium hydroxide containing 500 mg in each 25 mL.

Assay—Mix about 500 mg, accurately weighed, of Biotin with 100 mL of water, add phenolphthalein TS, and titrate the suspension slowly with 0.1 *N* sodium hydroxide VS, while heating and stirring continuously. Each mL of 0.1 *N* sodium hydroxide is equivalent to 24.43 mg of $C_{10}H_{16}N_2O_3S$.

Biperiden

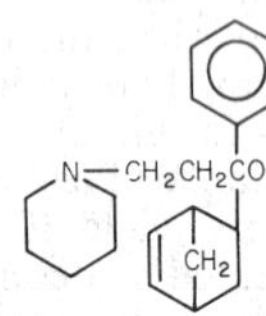

$C_{21}H_{29}NO$ 311.47
1-Piperidinepropanol, α-bicyclo[2.2.1]hept-5-en-2-yl-α-phenyl-.
α-5-Norbornen-2-yl-α-phenyl-1-piperidinepropanol
[*514-65-8*].

» Biperiden contains not less than 98.0 percent and not more than 101.0 percent of $C_{21}H_{29}NO$, calculated on the dried basis.

Packaging and storage—Preserve in well-closed, light-resistant containers.

Reference standard—*USP Biperiden Reference Standard*—Dry at 105° for 3 hours before using.

Identification—

A: The infrared absorption spectrum of a potassium bromide dispersion of it, previously dried, exhibits maxima only at the same wavelengths as that of a similar preparation of USP Biperiden RS.

B: Transfer about 180 mg of it, accurately weighed, to a 200-mL volumetric flask, add 1 mL of lactic acid, dilute with water to volume, and mix: the ultraviolet absorption spectrum of this solution exhibits maxima and minima at the same wavelengths as that of a similar solution of USP Biperiden RS, concomitantly measured, and the respective absorptivities, calculated on the dried basis, at the wavelength of maximum absorbance at about 257 nm do not differ by more than 3.0%.

C: Dissolve about 20 mg in 5 mL of phosphoric acid: a green color is produced.

D: Dissolve 200 mg in 80 mL of water with the aid of 0.5 mL of 3 *N* hydrochloric acid, warming, if necessary, to effect solution, and then cool. To 5 mL of the solution add 1 drop of hydrochloric acid and several drops of mercuric chloride TS: a white precipitate is formed. To a second 5-mL portion of the solution add bromine TS dropwise: a yellow precipitate forms which redissolves on shaking, and finally, upon the addition of more bromine TS, a permanent precipitate is formed.

Melting range, *Class I* ⟨741⟩: between 112° and 116°.

Loss on drying ⟨731⟩—Dry it at 105° for 3 hours: it loses not more than 1.0% of its weight.

Residue on ignition ⟨281⟩: not more than 0.1%.

Ordinary impurities ⟨466⟩—

Test solution: methanol.

Standard solution: methanol.

Eluant: a mixture of methanol and ammonium hydroxide (100:1.5).

Visualization: 3.

Assay—Dissolve about 500 mg of Biperiden, accurately weighed, in 20 mL of benzene, add 2 drops of crystal violet TS, and titrate with 0.1 *N* perchloric acid VS to a blue end-point. Perform a blank determination, and make any necessary correction. Each mL of 0.1 *N* perchloric acid is equivalent to 31.15 mg of $C_{21}H_{29}NO$.

Biperiden Hydrochloride

$C_{21}H_{29}NO.HCl$ 347.93
1-Piperidinepropanol, α-bicyclo[2.2.1]hept-5-en-2-yl-α-phenyl-, hydrochloride.
α-5-Norbornen-2-yl-α-phenyl-1-piperidinepropanol hydrochloride [*1235-82-1*].

» Biperiden Hydrochloride contains not less than 98.0 percent and not more than 101.0 percent of $C_{21}H_{29}NO.HCl$, calculated on the dried basis.

Packaging and storage—Preserve in well-closed, light-resistant containers.

Reference standard—*USP Biperiden Hydrochloride Reference Standard*—Dry at 105° for 3 hours before using.

Identification—

A: The infrared absorption spectrum of a potassium bromide dispersion of it, previously dried, exhibits maxima only at the same wavelengths as that of a similar preparation of USP Biperiden Hydrochloride RS.

B: The ultraviolet absorption spectrum of a 1 in 1000 solution of it in methanol exhibits maxima and minima at the same wavelengths as that of a similar preparation of USP Biperiden Hydrochloride RS, concomitantly measured, and the respective absorptivities, calculated on the dried basis, at the wavelength of maximum absorbance at about 257 nm do not differ by more than 3.0%.

C: Dissolve about 20 mg in 5 mL of phosphoric acid: a green color is produced.

D: To 5 mL of a solution (1 in 500) add bromine TS dropwise: a yellow precipitate, which dissolves on shaking, is formed. Addition of more bromine TS produces a precipitate which does not dissolve on shaking.

E: A 5-mL portion of a solution (1 in 500) responds to the tests for *Chloride* ⟨191⟩.

Loss on drying ⟨731⟩—Dry it at 105° for 3 hours: it loses not more than 0.5% of its weight.

Ordinary impurities ⟨466⟩—

Test solution: a mixture of alcohol and water (1:1).

Standard solution: a mixture of alcohol and water (1:1).

Eluant: a mixture of toluene, methanol, and glacial acetic acid (45:8:4).

Visualization: 5, followed by viewing under longwavelength ultraviolet light.

Assay—Weigh accurately about 500 mg of Biperiden Hydrochloride, and dissolve in 80 mL of glacial acetic acid, warming slightly, if necessary, to effect solution. Cool, add 1 drop of crystal violet TS and 10 mL of mercuric acetate TS, and titrate with 0.1 *N* perchloric acid VS to a blue end-point. Perform a blank determination, and make any necessary correction. Each mL of 0.1 *N* perchloric acid is equivalent to 34.79 mg of $C_{21}H_{29}NO.HCl$.

Biperiden Hydrochloride Tablets

» Biperiden Hydrochloride Tablets contain not less than 93.0 percent and not more than 107.0 percent of the labeled amount of $C_{21}H_{29}NO.HCl$.

Packaging and storage—Preserve in tight containers.

Reference standard—*USP Biperiden Hydrochloride Reference Standard*—Dry at 105° for 3 hours before using.

Identification—Transfer a quantity of finely powdered Tablets, equivalent to about 10 mg of biperiden hydrochloride, to a 10-mL volumetric flask. Pipet 5 mL of water into the flask, mix, and sonicate to disperse the powder. Pipet 5 mL of methanol into the flask, mix, and sonicate for 15 minutes. Filter the solution into a separator, add 2 mL of 1 *N* sodium hydroxide and 10 mL of chloroform, and shake for 3 minutes. Filter the chloroform layer into a stoppered flask, and use the chloroform filtrate as the test solution. In a similar manner, using about 10 mg of USP Biperiden Hydrochloride RS in place of the powdered Biperiden Hydrochloride Tablets, prepare the Standard solution. Condition a suitable thin-layer chromatographic plate (see *Chromatography* ⟨621⟩), coated with a 0.25-mm layer of chromatographic silica gel mixture, by heating the plate at 105° for 1 hour. Allow the plate to cool, and separately apply 20 µL each of the test solution and the Standard solution to the plate. Allow the applications to dry, and develop the chromatogram in a solvent system consisting of a mixture of methanol and ammonium hydroxide (100:1.5) until the solvent front has moved about three-fourths of the length of the plate. Remove the plate from the developing chamber, mark the solvent front, and allow the solvent to evaporate. Locate the spots on the plate by exposing the plate for 10 minutes to iodine vapors in a pre-equilibrated closed chamber, on the bottom of which there are iodine crystals: the R_f value of the principal spot obtained from the test solution corresponds to that obtained from the Standard solution.

Dissolution ⟨711⟩—

Medium: 0.1 *N* hydrochloric acid; 500 mL.

Apparatus 2: 50 rpm.

Time: 45 minutes.

Phosphate buffer–bromocresol purple solution—Prepare as directed in the *Assay*.

Standard preparation—Dissolve an accurately weighed quantity of USP Biperiden Hydrochloride RS in methanol, and dilute quantitatively with methanol to obtain a solution having a known concentration of about 0.8 mg per mL. Pipet 5 mL of this solution into a 500-mL volumetric flask, add 0.1 *N* hydrochloric acid to volume, and mix. Pipet 25 mL of this solution into a suitable beaker, and adjust with 0.1 *N* sodium hydroxide to a pH of 5.3. Transfer this solution to a 100-mL volumetric flask with the aid of water, dilute with water to volume, and mix to obtain a *Standard preparation* having a known concentration of about 2 µg per mL.

Test preparation—Filter 75 mL of the solution under test, pipet 50 mL of the clear filtrate into a suitable beaker, and adjust with 0.1 *N* sodium hydroxide to a pH of 5.3. Transfer this solution to a 100-mL volumetric flask with the aid of water, dilute with water to volume, and mix.

Procedure—Pipet 20 mL each of the *Standard preparation*, the *Test preparation*, and water to provide the blank, into individual separators, each containing 10.0 mL of *Phosphate buffer–bromocresol purple solution*. Extract the solution in each separator with 40.0 mL of chloroform for 10 minutes. After the layers have separated, filter each chloroform extract through filter paper into separate, glass-stoppered containers, discarding the first 10 mL of each filtrate. Determine the amount of $C_{21}H_{29}NO.HCl$ dissolved from absorbances at the wavelength of maximum absorbance at about 408 nm (10-cm cells) of the extract from the *Test preparation* in comparison with that of the extract from the *Standard preparation*, using the blank to set the instrument.

Tolerances—Not less than 75% (*Q*) of the labeled amount of $C_{21}H_{29}NO.HCl$ is dissolved in 45 minutes.

Uniformity of dosage units ⟨905⟩: meet the requirements.

Assay—

Phosphate buffer–bromocresol purple solution—Dissolve 38 g of monobasic sodium phosphate and 2 g of anhydrous dibasic sodium phosphate in water to make 1000 mL. Adjust the pH of the solution to 5.3 ± 0.1, if necessary (*Solution A*). Dissolve 400 mg of bromocresol purple in 30 mL of water, add 6.3 mL of 0.1 *N* sodium hydroxide, and dilute with water to 500 mL (*Solution B*). Mix equal volumes of *Solution A*, *Solution B*, and chloroform, shake in a separator, and discard the chloroform. If appreciable color is extracted, repeat with additional portions of chloroform until no color is extracted.

Standard preparation—Transfer about 80 mg of USP Biperiden Hydrochloride RS, accurately weighed, to a 100-mL volumetric flask, add methanol to volume, and mix. Transfer 5.0 mL of this solution to a second 100-mL volumetric flask, add 25 mL of water, dilute with methanol to volume, and mix. The concentration of USP Biperiden Hydrochloride RS in the *Standard preparation* is about 40 µg per mL.

Assay preparation—Weigh and finely powder not less than 20 Biperiden Hydrochloride Tablets. Transfer an accurately weighed portion of the powder, equivalent to about 2 mg of biperiden hydrochloride, to a 50-mL volumetric flask, add 12.5 mL of water, and heat on a steam bath for 15 minutes. Cool, dilute with methanol to volume, and mix.

Procedure—Transfer 5.0 mL each of the *Standard preparation*, the *Assay preparation*, and a methanol-water mixture (3:1) to provide the blank, to individual separators each, containing 10.0 mL of *Phosphate buffer–bromocresol purple solution*. Extract the solution in each separator with 20.0 mL of chloroform for 2 minutes. After the layers have separated, filter each chloroform extract through filter paper (Whatman No. 31 or equivalent) into separate glass-stoppered, 50-mL volumetric flasks. In the same manner, extract the solution in each separator with another 20.0-mL portion of chloroform, filter, and wash each filter with 8 mL of chloroform, collecting each combined filtrate and washing, respectively, in the 50-mL volumetric flask containing the first extract. Dilute with chloroform to volume, and mix. Concomitantly determine the absorbances of the solutions in 1-cm cells at the wavelength of maximum absorbance at about 408 nm, with a suitable spectrophotometer, using the blank to set the instrument. Calculate the quantity, in mg, of $C_{21}H_{29}NO.HCl$ in the portion of Tablets taken by the formula:

$$0.05C(A_U/A_S),$$

in which *C* is the concentration, in µg per mL, of USP Biperiden Hydrochloride RS in the *Standard preparation*, and A_U and A_S are the absorbances of the solutions from the *Assay preparation* and the *Standard preparation*, respectively.

Biperiden Lactate Injection

$C_{21}H_{29}NO.C_3H_6O_3$ 401.54

1-Piperidinepropanol, α-bicyclo[2.2.1]hept-5-en-2-yl-α-phenyl-, compd. with 2-hydroxypropanoic acid (1:1).

α-5-Norbornen-2-yl-α-phenyl-1-piperidinepropanol lactate (salt) [*7085-45-2*].

» Biperiden Lactate Injection is a sterile solution of biperiden lactate ($C_{21}H_{29}NO.C_3H_6O_3$) in Water for Injection, prepared from Biperiden with the aid of Lactic Acid. It contains not less than 95.0 percent and not more than 105.0 percent of the labeled amount of $C_{21}H_{29}NO.C_3H_6O_3$.

Packaging and storage—Preserve in single-dose containers, preferably of Type I glass, protected from light.

Reference standard—*USP Biperiden Reference Standard*—Dry at 105° for 3 hours before using.

Identification—Using a volume of Injection, equivalent to about 50 mg of biperiden lactate, and using a solution of 50 mg of USP Biperiden RS in 25 mL of 0.01 *N* hydrochloric acid, proceed as directed under *Identification—Organic Nitrogenous Bases* ⟨181⟩, beginning with "Transfer the liquid to a separator": the Injection meets the requirements of the test.

pH ⟨791⟩: between 4.8 and 5.8.

Other requirements—It meets the requirements under *Injections* ⟨1⟩.

Assay—

Phosphate buffer–bromocresol purple solution—Prepare as directed in the *Assay* under *Biperiden Hydrochloride Tablets*.

Standard preparation—Transfer about 80 mg of USP Biperiden RS, accurately weighed, to a 100-mL volumetric flask, add methanol to volume, and mix. Transfer 5.0 mL of this solution to a second 100-mL volumetric flask, add 25 mL of water, dilute with methanol to volume, and mix to obtain a *Standard preparation* having a known concentration of about 40 µg per mL.

Assay preparation—Transfer an accurately measured volume of Biperiden Lactate Injection, equivalent to about 5 mg of biperiden lactate, to a 100-mL volumetric flask, add 25 mL of water, dilute with methanol to volume, and mix.

Procedure—Proceed as directed in the *Assay* under *Biperiden Hydrochloride Tablets.* Calculate the quantity, in mg, of $C_{21}H_{29}NO.C_3H_6O_3$ in each mL of the Injection taken by the formula:

$$(401.54/311.47)(0.1C/V)(A_U/A_S),$$

in which 401.54 and 311.47 are the molecular weights of biperiden lactate and biperiden, respectively, C is the concentration, in µg per mL, of USP Biperiden RS in the *Standard preparation,* V is the volume, in mL, of Injection taken, and A_U and A_S are the absorbances of the solutions from the *Assay preparation* and the *Standard preparation* respectively.

Bisacodyl

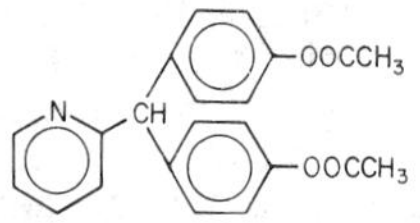

$C_{22}H_{19}NO_4$ 361.40

Phenol, 4,4′-(2-pyridinylmethylene)bis-, diacetate (ester).

4,4′-(2-Pyridylmethylene)diphenol diacetate (ester) [*603-50-9*].

» Bisacodyl contains not less than 98.0 percent and not more than 101.0 percent of $C_{22}H_{19}NO_4$, calculated on the dried basis.

Caution—Avoid inhalation and contact with the eyes, skin, and mucous membranes.

Packaging and storage—Preserve in well-closed containers.

Reference standard—*USP Bisacodyl Reference Standard*—Dry at 105° for 2 hours before using.

Identification—

A: The infrared absorption spectrum, determined in a 1.0-mm cell, of a 1 in 200 solution of it in chloroform, the specimen of Bisacodyl having been previously dried, exhibits maxima only at the same wavelengths as that of a similar preparation of USP Bisacodyl RS.

B: The ultraviolet absorption spectrum of a 1 in 50,000 solution in 0.05 *N* hydrochloric acid exhibits maxima and minima at the same wavelengths as that of a similar preparation of USP Bisacodyl RS, concomitantly measured, and the respective absorptivities, calculated on the dried basis, at the wavelength of maxima absorption at about 263 nm do not differ by more than 3.0%.

Melting range ⟨741⟩: between 131° and 135°.

Loss on drying ⟨731⟩—Dry it at 105° for 2 hours: it loses not more than 0.5% of its weight.

Residue on ignition ⟨281⟩: not more than 0.1%.

Heavy metals, *Method II* ⟨231⟩: 0.001%.

Assay—Dissolve about 250 mg of Bisacodyl, accurately weighed, in 70 mL of glacial acetic acid, add 3 drops of *p*-naphtholbenzein TS, and titrate with 0.1 *N* perchloric acid VS. Perform a blank determination, and make any necessary correction. Each mL of 0.1 *N* perchloric acid is equivalent to 36.14 mg of $C_{22}H_{19}NO_4$.

Bisacodyl Suppositories

» Bisacodyl Suppositories contain not less than 90.0 percent and not more than 110.0 percent of the labeled amount of $C_{23}H_{19}NO_4$.

Packaging and storage—Preserve in well-closed containers at a temperature not exceeding 30°.

Reference standard—*USP Bisacodyl Reference Standard*—Dry at 105° for 2 hours before using.

Identification—Transfer a quantity of Suppositories, equivalent to about 150 mg of bisacodyl, to a 500-mL conical flask, add 150 mL of solvent hexane, and heat on a steam bath until they are melted. Filter the solution, with the aid of vacuum, through a medium-porosity, sintered-glass funnel, and wash the residue with about 100 mL of warm solvent hexane until it is free from fat. Continue the vacuum until the residue appears dry. Dissolve the residue by rinsing the filter with about 50 mL of warm acetone, collecting the filtrate in a 150-mL beaker, and evaporate the filtrate on a steam bath to a volume of about 5 mL. To the residual liquid add about 75 mL of water, heat on a steam bath for 15 minutes, and cool. Scratch the sides of the beaker to induce crystallization, filter the crystals, and dry at 100° for about 15 minutes: the bisacodyl so obtained melts between 129° and 135°, and responds to *Identification test A* under *Bisacodyl.*

Assay—

Standard preparation—Transfer about 20 mg of USP Bisacodyl RS, accurately weighed, to a 100-mL volumetric flask, add 1 *N* hydrochloric acid to volume, and mix. Transfer 10.0 mL of the resulting solution to a 100-mL volumetric flask, dilute with water to volume, and mix.

Assay preparation—Place a number of Bisacodyl Suppositories, equivalent to about 50 mg of bisacodyl, in a 250-mL beaker, add 40 mL of anhydrous ethyl ether, and warm gently on a steam bath until they are melted. Transfer the mixture with the aid of 25 mL of ether to a 125-mL separator, and extract with a mixture of 15 mL of sodium hydroxide solution (1 in 250) and 5 mL of saturated sodium chloride solution. Collect the aqueous layer in a small separator, and extract it with 10 mL of ether. Combine the ether solutions, and extract with three 25-mL portions of 1 *N* hydrochloric acid. Combine the hydrochloric acid extracts in a 100-mL volumetric flask, add 1 *N* hydrochloric acid to volume, and mix. Transfer 20.0 mL of this solution to a 50-mL volumetric flask, add 1 *N* hydrochloric acid to volume, and mix. Transfer 10.0 mL of the resulting solution to a 100-mL volumetric flask, dilute with water to volume, and mix.

Procedure—Concomitantly determine the absorbances of the *Assay preparation* and the *Standard preparation* in 1-cm cells at the wavelength of maximum absorbance at about 263 nm, with a suitable spectrophotometer, using 0.1 *N* hydrochloric acid as the blank. Calculate the quantity, in mg, of $C_{22}H_{19}NO_4$ in the Suppositories taken by the formula:

$$2.5C(A_U/A_S),$$

in which C is the concentration, in µg per mL, of USP Bisacodyl RS in the *Standard preparation,* and A_U and A_S are the absorbances of the *Assay preparation* and the *Standard preparation,* respectively.

Bisacodyl Tablets

» Bisacodyl Tablets contain not less than 90.0 percent and not more than 110.0 percent of the labeled amount of $C_{22}H_{19}NO_4$. Bisacodyl Tablets are enteric-coated.

Packaging and storage—Preserve in well-closed containers at a temperature not exceeding 30°.

Labeling—Label Tablets to indicate that they are enteric-coated.

Reference standard—*USP Bisacodyl Reference Standard*—Dry at 105° for 2 hours before using.

Identification—Macerate a portion of powdered Tablets, equivalent to about 300 mg of bisacodyl, with 100 mL of acetone. Heat on a steam bath to boiling, filter, and evaporate to about 20 mL. Add 200 mL of water, and warm the mixture on the steam bath, passing a stream of nitrogen over the surface to evaporate the acetone. After 30 minutes, cool the mixture, and filter through a sintered-glass funnel. Discard the filtrate, and dissolve the crystals in 50 mL of acetone. Evaporate the solution to about 15 mL, add about 75 mL of water, heat on a steam bath for 15 minutes, and then cool. Scratch the sides of the beaker to induce crystallization, filter the crystals, and dry at 100° for about 15 minutes: the bisacodyl so obtained responds to *Identification test A* under *Bisacodyl.*

Disintegration ⟨701⟩—Proceed as directed for *Enteric-coated Tablets:* the tablets do not disintegrate after 1 hour of agitation in simulated gastric fluid TS, but then disintegrate within 45 minutes in simulated intestinal fluid TS.

Uniformity of dosage units ⟨905⟩: meet the requirements.

Assay—Weigh and finely powder not less than 20 Bisacodyl Tablets. Transfer an accurately weighed portion of the powder, equivalent to about 25 mg of bisacodyl, to a 50-mL conical flask, add 20 mL of chloroform, and shake by mechanical means for 15 minutes. Filter the mixture through a fine-porosity, sintered-glass funnel into a 250-mL separator, rinsing the flask and funnel with three 5-mL portions of chloroform and collecting the rinsings in the separator. To the separator add 75 mL of ether, and extract with three 75-mL portions of 1 *N* hydrochloric acid. Collect the extracts in a 250-mL volumetric flask, add 1 *N* hydrochloric acid to volume, and mix. Transfer 20.0 mL of this solution to a 100-mL volumetric flask, add 1 *N* hydrochloric acid to volume, and mix. Concomitantly determine the absorbances of this solution and a Standard solution prepared from USP Bisacodyl RS in the same medium having a known concentration of about 20 µg per mL, in 1-cm cells at the wavelength of maximum absorbance at about 263 nm, with a suitable spectrophotometer, using 1 *N* hydrochloric acid as the blank. Calculate the quantity, in mg, of $C_{22}H_{19}NO_4$ in the portion of Tablets taken by the formula:

$$1.25C(A_U/A_S),$$

in which *C* is the concentration, in µg per mL, of USP Bisacodyl RS in the Standard solution, and A_U and A_S are the absorbances of the solution from the Tablets and the Standard solution, respectively.

Milk of Bismuth

» Milk of Bismuth contains bismuth hydroxide and bismuth subcarbonate in suspension in water, and yields not less than 5.2 percent and not more than 5.8 percent (w/w) of bismuth trioxide (Bi_2O_3).

Bismuth Subnitrate	80 g
Nitric Acid	120 mL
Ammonium Carbonate	10 g
Strong Ammonia Solution,	
Purified Water, each, a sufficient quantity, to make	1000 mL

Mix the Bismuth Subnitrate with 60 mL of Purified Water and 60 mL of the Nitric Acid in a suitable container, and agitate, warming gently until solution is effected. Pour this solution, with constant stirring, into 5000 mL of Purified Water containing 60 mL of the Nitric Acid. Dilute 160 mL of Strong Ammonia Solution with 4300 mL of Purified Water in a glazed or glass vessel of at least 12,000-mL capacity. Dissolve the Ammonium Carbonate in this solution, and then pour the bismuth solution quickly into it with constant stirring. Add sufficient 6 *N* ammonium hydroxide, if necessary, to render the mixture distinctly alkaline, allow to stand until the precipitate has settled, then pour or siphon off the supernatant liquid, and wash the precipitate twice with Purified Water, by decantation. Transfer the magma to a strainer of close texture, so as to provide continuous washing with Purified Water, the outlet tube being elevated to prevent the surface of the magma from becoming dry. When the washings no longer yield a pink color with phenolphthalein TS, drain the moist preparation, transfer to a graduated vessel, add sufficient Purified Water to make 1000 mL, and mix.

NOTE—This method of preparation may be varied, provided the product meets the following requirements.

Packaging and storage—Preserve in tight containers, and protect from freezing.

Identification—

A: It responds to the tests for *Bismuth* ⟨191⟩ and for *Carbonate* ⟨191⟩.

B: Add 1 mL of 3 *N* hydrochloric acid to 1 mL of Milk of Bismuth: a clear solution is produced. Pour the clear solution into 10 volumes of water: a white precipitate is formed.

Microbial limits ⟨61⟩—The total bacterial count does not exceed 100 per mL and the test for *Escherichia coli* is negative.

Water-soluble substances—Boil 10 mL with 90 mL of water for 10 minutes, cool, add water to make the total volume 100 mL, mix, and filter. Evaporate 50 mL of the filtrate to dryness, and ignite it gently: the weight of the residue does not exceed 5 mg (0.1%).

Alkalies and alkaline earths—Dissolve 2.0 mL in 5 mL of hydrochloric acid, dilute with water to 100 mL, add hydrogen sulfide to precipitate the bismuth completely, and filter. To 50 mL of the clear filtrate add 5 drops of sulfuric acid, evaporate to dryness, and ignite: the weight of the residue does not exceed 3 mg (0.3%).

Arsenic, *Method I* ⟨211⟩—Evaporate 3.75 mL on a steam bath to dryness, add 2 mL of sulfuric acid, and heat until copious fumes of sulfur trioxide are evolved. The limit is 0.8 ppm.

Lead—To 5 mL add warm nitric acid, dropwise, until it is just dissolved, and pour the solution into 50 mL of water: a white precipitate may form. Filter, if necessary, evaporate the filtrate on a steam bath to 15 mL, again filter, and to 10 mL of the filtrate add an equal volume of 2 *N* sulfuric acid: no precipitate is formed.

Assay—Evaporate an accurately weighed quantity of Milk of Bismuth to dryness, and ignite the residue to constant weight. From the weight of the Bi_2O_3 so obtained determine the percentage in the assay specimen.

Bismuth Subgallate

COOH
HO O
O—BiOH

$C_7H_5BiO_6$ 394.09
Gallic acid bismuth basic salt [*99-26-3*].

» Bismuth Subgallate is a basic salt which, when dried at 105° for 3 hours, contains the equivalent of not less than 52.0 percent and not more than 57.0 percent of Bi_2O_3.

Packaging and storage—Preserve in tight, light-resistant containers.

Identification—

A: When heated to redness, it at first chars, leaving finally a yellow residue. This residue responds to the tests for *Bismuth* ⟨191⟩.

B: Agitate thoroughly about 100 mg with an excess of hydrogen sulfide TS, filter, and boil the filtrate to expel the dissolved gas. Cool, and add 1 drop of ferric chloride TS: a purplish blue mixture is produced.

Loss on drying ⟨731⟩—Dry it at 105° for 3 hours. It loses not more than 7.0% of its weight.

Nitrate—Mix about 100 mg with 5 mL of 2 *N* sulfuric acid and 5 mL of ferrous sulfate TS, filter the mixture, and carefully superimpose the filtrate, without mixing, on 5 mL of sulfuric acid, in a test tube: no reddish brown color appears at the zone of contact of the two liquids.

Alkalies and alkaline earths—Boil 1.0 g with 20 mL of a mixture of equal volumes of 6 *N* acetic acid and water, cool, and filter. Precipitate the bismuth from the filtrate by the addition of hydrogen sulfide, boil the mixture, and filter. Add 5 drops of sulfuric acid to the filtrate, evaporate to dryness, and ignite to constant weight: the weight of the residue does not exceed 5 mg (0.5%).

Arsenic—Triturate 200 mg with an equal weight of calcium hydroxide, and ignite. Dissolve the residue in 5 mL of 3 *N* hydrochloric acid: the solution without further treatment meets the requirements of the test for *Arsenic* ⟨211⟩.

Copper, Lead, and Silver—Ignite 3 g in a porcelain crucible, cool, and cautiously add, dropwise, just sufficient nitric acid to dissolve the residue upon warming. Evaporate the solution to dryness, again ignite, and cool. Cautiously dissolve the residue in just sufficient nitric acid with the aid of gentle heat, concentrate the solution to about 4 mL, pour it into 100 mL of water, filter, evaporate the filtrate on a steam bath to 20 mL, again filter, and divide this filtrate into portions of 5 mL each. Using these several portions as the test liquid, proceed as directed for *Copper*, *Lead*, and *Silver* under *Bismuth Subnitrate*. The specified results are obtained.

Free gallic acid—Shake 1.0 g with 20 mL of alcohol for 1 minute, filter and evaporate the filtrate to dryness on a steam bath, and dry the residue at 105° for 1 hour: the weight of the residue does not exceed 5 mg (0.5%).

Assay—Dry about 1 g of Bismuth Subgallate at 105° for 3 hours, then weigh accurately and ignite in a porcelain crucible. Allow it to cool and add nitric acid to the residue, dropwise, warming until complete solution has been effected. Evaporate the solution to dryness and carefully ignite the residue to constant weight. From the weight of the residue so obtained, determine the percentage of Bi_2O_3 in the portion of Bismuth Subgallate taken.

Bismuth Subnitrate

$Bi_5O(OH)_9(NO_3)_4$ 1461.99
Bismuth hydroxide nitrate oxide [$Bi_5O(OH)_9(NO_3)_4$].
Bismuth hydroxide nitrate oxide [$Bi_5O(OH)_9(NO_3)_4$] [*1304-85-4*].

» Bismuth Subnitrate is a basic salt that contains the equivalent of not less than 79.0 percent of bismuth trioxide (Bi_2O_3), calculated on the dried basis.

Packaging and storage—Preserve in well-closed containers.

Identification—It responds to the tests for *Bismuth* ⟨191⟩ and for *Nitrate* ⟨191⟩.

Loss on drying ⟨731⟩—Dry it at 105° for 2 hours: it loses not more than 3.0% of its weight.

Carbonate—Add 3 g to 3 mL of warm nitric acid: no effervescence occurs. Pour the solution into 100 mL of water: a white precipitate forms. Filter, evaporate the filtrate on a steam bath to 30 mL, again filter the liquid, divide the latter filtrate into portions of 5 mL each, and use these several portions in the tests for *Chloride*, *Sulfate*, *Copper*, *Lead*, and *Silver*.

Chloride ⟨221⟩—A 10-mL portion of the test liquid retained in the test for *Carbonate* shows no more chloride than corresponds to 0.50 mL of 0.020 *N* hydrochloric acid (0.035%).

Sulfate ⟨221⟩—To a 5-mL portion of the test liquid retained in the test for *Carbonate* add 5 drops of barium nitrate TS: no turbidity is produced immediately.

Alkalies and alkaline earths—Boil 1.0 g with 20 mL of a mixture of equal volumes of 6 *N* acetic acid and water, cool, and filter. Add 2 mL of 3 *N* hydrochloric acid, precipitate the bismuth by the addition of hydrogen sulfide, boil the mixture, and filter it. Add 5 drops of sulfuric acid to the filtrate, evaporate to dryness, and ignite to constant weight: the weight of the residue does not exceed 5 mg (0.5%).

Ammonium salts—Boil about 100 mg with 5 mL of 1 *N* sodium hydroxide: the vapor does not turn moistened red litmus paper blue.

Arsenic, *Method I* ⟨211⟩—Mix 375 mg with 5 mL of water, cautiously add 2 mL of sulfuric acid, and heat the mixture until fumes of sulfur trioxide are copiously evolved. Cool, cautiously add 10 mL of water, and again evaporate to strong fuming, repeating, if necessary, to remove any trace of nitric acid. The limit is 8 ppm.

Copper—To a 5-mL portion of the test liquid retained in the test for *Carbonate* add a slight excess of 6 *N* ammonium hydroxide: the liquid does not exhibit a bluish color.

Lead—Mix a 5-mL portion of the test liquid retained in the test for *Carbonate* with an equal volume of 2 *N* sulfuric acid: the liquid does not become cloudy.

Silver—To a 5-mL portion of the test liquid retained in the test for *Carbonate* add hydrochloric acid, dropwise: no precipitate is formed that is insoluble in a slight excess of hydrochloric acid, but that is soluble in 6 *N* ammonium hydroxide.

Assay—Transfer about 400 mg of Bismuth Subnitrate, accurately weighed, to a 250-mL beaker. Add 5 mL of water, then add 2 mL of nitric acid, and warm, if necessary, to effect solution. Dilute with water to 100 mL, add 0.3 mL of xylenol orange TS, and titrate with 0.05 *M* disodium ethylenediaminetetraacetate VS to a yellow end-point. Each mL of 0.05 *M* disodium ethylenediaminetetraacetate is equivalent to 11.65 mg of Bi_2O_3.

Sterile Bleomycin Sulfate

» Sterile Bleomycin Sulfate is the sulfate salt of bleomycin, a mixture of basic cytotoxic glycopeptides produced by the growth of *Streptomyces verticillus*, or produced by other means. It has a potency of not less than 1.5 Bleomycin Units and not more than 2.0 Bleomycin Units per mg. In addition, where packaged for dispensing, it contains not less than 90.0 percent and not more than 120.0 percent of the labeled amount of bleomycin.

Packaging and storage—Preserve in *Containers for Sterile Solids* as described under *Injections* ⟨1⟩.

Reference standard—*USP Bleomycin Sulfate Reference Standard*—Do not dry before using.

Constituted solution—At the time of use, the constituted solution prepared from Sterile Bleomycin Sulfate meets the requirements for *Constituted Solutions* under *Injections* ⟨1⟩.

Identification—The infrared absorption spectrum of a potassium bromide dispersion of it, previously dried, exhibits maxima only at the same wavelengths as that of a similar preparation of USP Bleomycin Sulfate RS.

Depressor substances—It meets the requirements of the *Depressor Substances Test* ⟨101⟩, the test dose being 1.0 mL per kg of a solution in sterile saline TS containing 0.5 Bleomycin Unit per mL.

Pyrogen—It meets the requirements of the *Pyrogen Test* ⟨151⟩, the test dose being 1.0 mL per kg of a solution in pyrogen-free saline TS containing 0.5 Bleomycin Unit per mL.

Sterility—It meets the requirements under *Sterility Tests* ⟨71⟩, when tested as directed in the section, *Test Procedures Using Membrane Filtration*, the entire contents of each container being used.

pH ⟨791⟩: between 4.5 and 6.0, in a solution containing 10 Bleomycin Units per mL.

Loss on drying ⟨731⟩—Dry the total contents of 2 containers in vacuum at a pressure not exceeding 5 mm of mercury at 60° for 3 hours: it loses not more than 6.0% of its weight.

Content of bleomycins—

Mobile phase—Dissolve 960 mg of sodium 1-pentanesulfonate in 1000 mL of deaerated 0.08 *N* acetic acid, adjust with ammonium hydroxide to a pH of 4.3, filter, and degas. [NOTE—1.86 g of disodium ethylenediaminetetraacetate may be included if needed to obtain satisfactory chromatography.] Use a linear gradient of 10% to 40% methanol mixed with this solution, with a gradient mixing time of 60 minutes, and allow chromatography to proceed with the final gradient mixture for a further 20 minutes or until demethylbleomycin A_2 has been eluted.

Test preparation—Dissolve Sterile Bleomycin Sulfate in deaerated water to obtain a solution having a concentration of about 2.5 Bleomycin Units per mL. Store this solution in a refrigerator until just prior to use.

Chromatographic system (see *Chromatography* ⟨621⟩)—The liquid chromatograph is equipped with a 254-nm detector and a 4.6-mm × 250-mm stainless steel column containing packing L1. The flow rate is about 1.2 mL per minute.

Procedure—Inject about 10 µL of the *Test preparation* into the chromatograph by means of a suitable microsyringe or sampling valve, record the chromatogram, and measure the peak responses for all peaks. The elution order is bleomycinic acid, bleomycin A_2 (major peak), bleomycin A_5, bleomycin B_2 (major peak), bleomycin B_4, and demethylbleomycin A_2. Calculate the percentage contents of bleomycin A_2, bleomycin B_2, and bleomycin B_4 by the formula:

$$100r_f/r_t,$$

in which r_f is the peak response corresponding to the particular bleomycin and r_t is the total of the responses of all peaks: the content of bleomycin A_2 is between 55% and 70%; the content of bleomycin B_2 is between 25% and 32%; the content of bleomycin B_4 is not more than 1%; and the combined percentage of bleomycin A_2 and bleomycin B_2 is not less than 85%.

Copper—

Standard preparation—Dissolve an accurately weighed quantity of cupric sulfate in 0.1 *N* hydrochloric acid, and dilute quantitatively with 0.1 *N* hydrochloric acid to obtain a solution containing the equivalent of 1.5 µg of copper per mL.

Test preparation—Dissolve about 10 mg of Sterile Bleomycin Sulfate, accurately weighed, in 10 mL of 0.1 *N* hydrochloric acid.

Procedure—Transfer the *Test preparation* and 10.0 mL of the *Standard preparation* to separate 60-mL separators, add 10.0 mL of a 1 in 10,000 solution of zinc dibenzyldithiocarbamate in carbon tetrachloride to each, and shake vigorously for 1 minute. Allow the phases to separate, then drain the lower phase from each separator through separate 1-g portions of anhydrous sodium sulfate. Concomitantly determine the absorbances of the solutions from the *Test preparation* and the *Standard preparation* at the wavelength of maximum absorbance at about 435 nm, with a suitable spectrophotometer, using carbon tetrachloride as the blank. Calculate the percentage of copper in the portions of Sterile Bleomycin Sulfate taken by the formula:

$$(1.5/W)(A_U/A_S),$$

in which W is the weight, in mg, of the portion of Sterile Bleomycin Sulfate taken, and A_U and A_S are the absorbances of the solutions from the *Test preparation* and the *Standard preparation*, respectively: not more than 0.1% is found.

Other requirements—Where it is packaged for dispensing, it meets the requirements for *Uniformity of Dosage Units* ⟨905⟩ and for *Labeling* under *Injections* ⟨1⟩.

Assay—

Assay preparation 1—Dissolve a suitable quantity of Sterile Bleomycin Sulfate, accurately weighed, in *Buffer No. 16*, and dilute quantitatively with *Buffer No. 16* to obtain a solution having a convenient concentration.

Assay preparation 2 (where it is packaged for dispensing)—Constitute Sterile Bleomycin Sulfate as directed in the labeling. Withdraw all of the withdrawable contents, using a suitable hypodermic needle and syringe, and dilute quantitatively with *Buffer No. 16* to obtain a solution having a convenient concentration.

Procedure—Proceed as directed under *Antibiotics—Microbial Assays* ⟨81⟩, using an accurately measured volume of *Assay preparation* diluted quantitatively and stepwise with *Buffer No. 16* to obtain a *Test Dilution* having a concentration assumed to be equal to the median dose level of the Standard.

Anti-A Blood Grouping Serum

» Anti-A Blood Grouping Serum conforms to the regulations of the federal Food and Drug Administration concerning biologics (660.20 to 660.29) (see *Biologics* ⟨1041⟩). It is a sterile, liquid or dried preparation containing the particular blood group antibodies derived from high-titered blood plasma or serum of human subjects, with or without stimulation by the injection of Blood Group Specific Substance A (or AB). It agglutinates human red cells containing A-antigens, i.e., blood groups A and AB (including subgroups A_1, A_2, A_1B, and A_2B but not necessarily weaker subgroups). It contains a suitable antimicrobial preservative. It meets the requirements to the test for potency, in parallel with, and not less than equivalent to, the U.S. Reference Blood Grouping Serum Anti-A, in agglutinating red blood cells from Group A_1 and Group A_2B donors. It meets the requirements of the tests for specificity with Group A_1, A_2B, B, and O cells and confirms the absence of contaminating antibodies reactive with M^g, Wr^a antigens as well as other antigens having an incidence of 1 percent or greater in the general population (see under *Blood Grouping Serums Anti-D, Anti-C, Anti-E, Anti-c, Anti-e*). It meets the requirements of the tests for avidity with Group A_1 and A_2B cells. All fresh or frozen red blood cell suspensions used for these tests are prepared under specified conditions and meet specified criteria. Anti-A Blood Grouping Serum may be artificially colored blue.

Packaging and storage—Preserve at a temperature between 2° and 8°.

Expiration date—The expiration date for liquid Serum is not later than 1 year, and for dried Serum not later than 5 years after date of issue from manufacturer's cold storage (5°, 1 year; or 0°, 2 years), provided that the expiration date for dried Serum is not later than 1 year after constitution.

Labeling—Label it to state that the source material was not reactive for hepatitis B surface antigen, but that no known test method offers assurance that products derived from human blood will not transmit hepatitis. Label it also to state that it is for in-vitro diagnostic use. [NOTE—The labeling is in black lettering imprinted on paper that is white or is colored completely or in part to match the specified blue color standard.]

Anti-B Blood Grouping Serum

» Anti-B Blood Grouping Serum conforms to the regulations of the federal Food and Drug Administration concerning biologics (660.20 to 660.29) (see *Biologics* ⟨1041⟩). It is a sterile, liquid or dried preparation containing the particular blood group antibodies derived from high-titered blood plasma or serum of human subjects, with or without stimulation by the injection of Blood Group Specific Substance B (or AB). It agglutinates human red cells containing B-antigens, i.e., blood groups B and AB (including subgroups A_1B and A_2B). It contains a suitable antimicrobial preservative. It meets the requirements of the test for potency, in parallel with, and not less than equivalent to, the U.S. Reference Blood Grouping Serum Anti-B, in agglutinating red blood cells from Group B donors. It meets the requirements of the tests for specificity with Group A_1, B, and O cells and confirms the absence of contaminating antibodies reactive with M^g, Wr^a antigens as well as other antigens having an incidence of 1 percent or greater in the general population (see under *Blood Grouping Serums Anti-D, Anti-C, Anti-E, Anti-c, Anti-e*). It meets the requirements of the test for avidity with Group B cells. All fresh or frozen red blood cell suspensions used for these tests are prepared under specified conditions and meet specified criteria. Anti-B Blood Grouping Serum may be artificially colored yellow.

Packaging and storage—Preserve at a temperature between 2° and 8°.

Expiration date—The expiration date for liquid Serum is not later than 1 year and for dried Serum not later than 5 years after date of issue from manufacturer's cold storage (5°, 1 year; or 0°, 2 years), provided that the expiration date for dried Serum is not later than 1 year after constitution.

Labeling—Label it to state that the source material was not reactive for hepatitis B surface antigen, but that no known test method offers assurance that products derived from human blood will not transmit hepatitis. Label it also to state that it is for in-vitro diagnostic use. [NOTE—The labeling is in black lettering imprinted on paper that is white or is colored completely or in part to match the specified yellow color standard.]

Blood Grouping Serums

» NOTE—This monograph deals with those Blood Grouping Serums for which there are no individual monographs and which are not routinely used or required for the testing of blood or blood products for transfusion.

Blood Grouping Serums conform to the regulations of the federal Food and Drug Administration concerning biologics (660.20 to 660.29) (see *Biologics* ⟨1041⟩). Each is a sterile, liquid or dried preparation containing one or more of the particular blood group antibodies derived from high-titered blood plasma or serum of human subjects, with or without stimulation by the injection of red cells or other substances, or of animals after stimulation by substances that cause such antibody production. It causes either directly, or indirectly by the antiglobulin test, the visible agglutination of human red cells containing the particular antigen(s) for which it is specific. It contains a suitable antimicrobial preservative. It meets the requirements of the test for potency, (1) in the case of tube test reagents, when tested by the specified method, of agglutinating red blood cells containing the specified antigens with the specified degree of reactivity, as defined, as follows: not less than a 1+ reaction (i.e., agglutinated cells dislodged into finely granular, but definite, small clumps) with a 1:8 dilution of Serum for Anti-K, Anti-k, Anti-Jk^a, Anti-Fy^a, Anti-C^w; not less than a 1+ reaction with a 1:4 dilution of Serum for Anti-S, Anti-s, Anti-P_1, Anti-M, Anti-I, Anti-e (saline), Anti-c (saline) and Anti-A_1; and not less than a 2+ reaction (i.e., agglutinated cells dislodged into many small clumps of equal size) with undiluted Serum for Anti-U, Anti-Kp^a, Anti-Kp^b, Anti-Js^a, Anti-Fy^b, Anti-N, Anti-Le^a, Anti-Le^b, Anti-Di^a, Anti-M^g, Anti-Jk^b, and Anti-Xg^a; and (2) in the case of reagents recommended for slide test methods, of agglutinating red blood cells, with the specified degree of reactivity, as defined, with both undiluted Serum and with a 1:2 dilution of Serum, when tested by the manufacturer's recommended method(s) using cells heterozygous for the corresponding antigen(s). It meets the requirements of the tests for specificity by the most sensitive method recommended by the manufacturer, in which not less than 4 positive and 4 negative phenotypes are included, and confirms the absence of contaminating antibodies reactive with M^g, Wr^a antigens as well as other antigens having an incidence of 1 percent or greater in the general population (see under *Blood Grouping Serums Anti-D, Anti-C, Anti-E, Anti-c, Anti-e*). It meets the requirements of the tests for avidity with the manufacturer's recommended method, red blood cells heterozygous for the corresponding antigen(s) being used. All fresh or frozen red blood cell suspensions used for these tests are prepared under specified conditions and meet specified criteria.

Packaging and storage—Preserve at a temperature between 2° and 8°.

Expiration date—The expiration date for liquid Serum is not later than 1 year and for dried Serum not later than 5 years after date of issue from manufacturer's cold storage (5°, 1 year; or 0°, 2 years), provided that the expiration date for dried Serum is not later than 1 year after constitution.

Labeling—Label each to state the source of the product if other than human and, if of human origin, to state that the source material was not reactive for hepatitis B surface antigen, but that no known test method offers assurance that products derived from human blood will not transmit hepatitis. Label each also to state that it is for in-vitro diagnostic use.

Blood Grouping Serums Anti-D, Anti-C, Anti-E, Anti-c, Anti-e

Anti-Rh Blood Grouping Serums.

» Blood Grouping Serums Anti-D, Anti-C, Anti-E, Anti-c, Anti-e (Anti-Rh Group) conform to the regulations of the federal Food and Drug Administration concerning biologics (660.20 to 660.29) (see *Biologics* ⟨1041⟩). They are sterile, liquid or dried preparations derived from the blood plasma or serum of human subjects who have developed specific Rh antibodies. They are free from agglutinins for the A or B antigens and from alloantibodies other than those for which claims are made in the labeling. They contain a suitable antimicrobial preservative. Liquid serums are not artificially colored. Two varieties of Anti-Rh Blood Grouping Serums are recognized, i.e., (1) saline agglutinating "complete" antiserums, which specifically agglutinate human red blood cells suspended in saline TS, and (2) "blocking or incomplete" antiserums, which contain protein or other macromolecular substances, usually require the cells to be suspended in serum or plasma, and generally are for slide or rapid tube tests. The most commonly used of these blood grouping serums are listed in Table 1, each reacting with the antigen(s) designated by the corresponding letter(s) with the alternative nomenclature indicated parenthetically.

Each Serum meets the requirements of the test for potency in the case of serums for saline tube test in parallel with, and not less than equivalent to, the U.S. Reference Blood Grouping Serum for Anti-D, Anti-C, or Anti-E, whichever is applicable, or, in the case of Anti-c and Anti-e for saline tube test which have no reference preparations, the test for minimum agglutination reactivity at a specified dilution; and in the case of serums for slide or rapid tube test in parallel with, and not less than equivalent to, the U.S. Reference Blood Grouping Serum for Anti-D, Anti-C, Anti-E, Anti-c, or Anti-e, whichever is applicable, in agglutinating as a minimum red blood cells from the donors indicated in Table 2 (which may be from Group A, B, AB, or O).

Each serum for slide or rapid tube test meets the requirements of the tests for avidity with the cells as indicated under tests for potency above.

Each serum meets the requirements of the tests for specificity by the most sensitive method recommended by the manufacturer, in which not less than 4 positive and 4 negative phenotypes are included (see Table 3), and confirms the absence of contaminating antibodies reactive with M^g, Wr^a antigens as well as other antigens having an incidence of 1 percent or greater in the general population, except where some of these confirmatory tests cannot be done by the manufacturer, in which event such omissions are noted. Antigens having such population incidence in the United States [other than low-incidence antigens, serum proteins (e.g., Gm, Km), leukocyte factors, drugs, chemicals or polyagglutinable cells, which are not necessarily excluded] are the following: A, B, H, Le^a, Le^b, Le^c, Le^d, I, K, k, Kp^a, Kp^b, Js^b, P_1, D, C, E, c, e, C^w, M, N, S, s, U, Lu^a, Lu^b, Jk^a, Jk^b, Fy^a, Fy^b, Xg^a, Do^a, Do^b, Yt^a, Yt^b, Lan, Co^a, Co^b, M^g, Wr^a, and Sd^a.

All fresh or frozen red blood cell suspensions used for these tests are prepared under specified conditions and meet specified criteria.

Packaging and storage—Preserve at a temperature between 2° and 8°.

Expiration date—The expiration date for liquid Serums is not later than 1 year and for dried Serums not later than 5 years after date of issue from manufacturer's cold storage (5°, 1 year; or 0°, 2 years), provided that the expiration date for dried Serums is not later than 1 year after constitution.

Labeling—Label each to state that the source material was not reactive for hepatitis B surface antigen, but that no known test method offers assurance that products derived from human blood will not transmit hepatitis. Label each to state that it is for in-vitro diagnostic use.

Table 1

Serum	Antigen(s) Reacting
Anti-D (Anti-Rh_o)	D (Rh_o)
Anti-C (Anti-rh′)	C (rh′)
Anti-E (Anti-rh″)	E (rh″)
Anti-CD (Anti-Rh_o′)	D (Rh_o) and C (rh′)
Anti-DE (Anti-Rh_o″)	D (Rh_o) and E (rh″)
Anti-CDE (Anti-Rh_o‴)	D (Rh_o), C (rh′), and E (rh″)
Anti-c (Anti-hr′)	c (hr′)
Anti-e (Anti-hr″)	e (hr″)

Table 2

Serum	Phenotype of Cells
Anti-D	cDe
Anti-C	Ccde
Anti-E	cdEe
Anti-CD	cDe, Ccde
Anti-DE	cDe, cdEe
Anti-CDE	cdEe, cDe, Ccde
Anti-c	CcDEe
Anti-e	cdEe

Leukocyte Typing Serum

» Leukocyte Typing Serum conforms to the regulations of the federal Food and Drug Administration concerning biologics (660.10 to 660.15) (see *Biologics* ⟨1041⟩). It is a dried or liquid preparation of serum derived from plasma or blood obtained from animals or from human donors containing an antibody or antibodies for identification of leukocyte antigens. It meets the requirements of the test for potency with reacting lymphocyte cell suspensions in producing not less than 80 percent increase in cell death over the negative control with not less than 70 percent of the cell samples bearing the corresponding antigen from the manufacturer's panel of cells of approved composition, when tested by all of the methods and at the dilution recommended in the labeling. It meets the requirements for specificity with the antibody or not more than three antibodies indicated on the label in producing not more than 3 percent of false positive reactions with cells lacking the corre-

Table 3

Serum	Cells
Anti-D	CcDe, cDe, Ccde, cdEe, A_1 cde, B cde, O cde, and where recommended for use by indirect antiglobulin technique, cde Bg(a+) cells from 3 different donors
Anti-C	cDe, Ccde, cdEe, C + rh_i neg. cells, A_1 cde, B cde, O cde
Anti-E	cDe, Ccde, cdEe, A_1 cde, B cde, O cde
Anti-CD	cDe, Ccde, cdEe, A_1 cde, B cde, O cde, and where recommended for detection of the G antigen, r^Gr
Anti-DE	cDe, Ccde, cdEe, A_1 cde, B cde, O cde
Anti-CDE	cDe, Ccde, cdEe, A_1 cde, B cde, O cde, and where recommended for detection of the G antigen, r^Gr
Anti-c	Ccde, A_1 CDe, B CDe, O CDe, and CDEe or CDE or CdE
Anti-e	cdEe, A_1 cDE, B cDE, O cDE, and CcDE or CDE or CdE

sponding antigen and not more than 14 percent of false negative reactions with cells possessing the corresponding antigen when tested by all of the methods recommended in the directions for use.

Packaging and storage—Preserve at the temperature recommended by the manufacturer.

Labeling—Label it to state the source of the product if other than human and, if of human origin, to state either that the source material was found reactive for hepatitis B surface antigen and that the product may transmit hepatitis, or that the source material was not reactive for hepatitis B surface antigen but that no known test method offers assurance that products derived from human blood will not transmit hepatitis. Label it also to include the name of the specific antibody or antibodies present and the requirements for potency in relation to the antigen(s) of the corresponding specificity with which it complies; the permissible limits of error for specificity reactions with which it complies; the name of the test method or methods recommended for the product; and a statement that it is for in-vitro diagnostic use. Provide with the package enclosure the following: adequate directions for performing the tests, including a description of all recommended test methods; descriptions of all supplementary reagents, including one of a suitable complement source; description of precautions in use, including a warning against exposure of the product to carbon dioxide; a caution statement to the effect that more than one antiserum is to be used for each specificity, that the antiserum is not to be diluted, and that cross-reacting antigens exist; and directions for constitution of the product, including instructions for the use, storage, and labeling of the constituted product.

Expiration date—The expiration date for dried Serum is not later than 2 years after date of issue from manufacturer's cold storage (5°, 1 year; or 0°, 2 years) and for liquid Serum not later than 1 year after such issue (5°, 1 year).

Blood Group Specific Substances A, B, and AB

» Blood Group Specific Substance A, or B, or AB conforms to the regulations of the federal Food and Drug Administration concerning biologics (680.20 to 680.26 Blood Group Substances) (see *Biologics* ⟨1041⟩). It is a sterile, pyrogen-free, nonanaphylactic isotonic solution of the polysaccharide-amino acid complexes that are capable of neutralizing the anti-A and the anti-B isoagglutinins of group O blood, and are used in the immunization of plasma donors for the production of in-vitro diagnostic reagents. It contains no added preservative. Blood Group Specific Substance A is prepared from hog stomach (gastric mucin), and Blood Group Specific Substances B and AB are prepared from horse stomach (gastric mucosa). It has a total nitrogen content of not more than 8 percent, calculated on the moisture- and ash-free basis. It meets the requirements of the test for potency (including identity) by inhibition of agglutination of appropriate Red Blood Cells as follows: for Blood Group Specific Substance A, an inhibition titer value not less than that of Reference Blood Group Substance A; for Blood Group Specific Substance B, an inhibition titer value not less than that of Reference Blood Group Substance B, but since it may contain some Blood Group Specific Substance A activity, such activity is less than that of Reference Blood Group Substance A; and for Blood Group Specific Substance AB, an inhibition titer value not less than that of Reference Blood Group Substance A and not less than that of Reference Blood Group Substance B. The tests are made in comparison with the specified Reference Blood Group Substance(s) and with both Reference Blood Grouping Serums Anti-A and Anti-B and produce neutralizing activity as shown in the accompanying table.

Packaging and storage—Preserve in single-dose containers, each containing a volume of not more than 1 mL consisting of a solution containing not more than 1.25 mg of Blood Group Specific Substance powder, at a temperature between 2° and 8°. Dispense

Recognized Leukocyte (HLA) Specificities

(April 1979)

Locus	HLA-A	Locus	HLA-B	Locus HLA-C	Locus HLA-D* (Mixed Leukocyte Culture)	Locus HLA-DR* (B lymphocytes)
A1	Aw19	B5	Bw4	Cw1	Dw1	DRw1
A2	Aw23	B7	Bw6	Cw2	Dw2	DRw2
A3	Aw24	B8	Bw16	Cw3	Dw3	DRw3
A9	Aw30	B12	Bw21	Cw4	Dw4	DRw4
A10	Aw31	B13	Bw22	Cw5	Dw5	DRw5
A11	Aw32	B14	Bw35	Cw6	Dw6	DRw6
A25	Aw33	B15	Bw38		Dw7	DRw7
A26	Aw34	B17	Bw39		Dw8	
A28	Aw36	B18	Bw41		Dw9	
A29	Aw43	B27	Bw42		Dw10	
		B37	Bw44		Dw11	
		B40	Bw45			
			Bw46			
			Bw47			
			Bw48			
			Bw49			
			Bw50			
			Bw51			
			Bw52			
			Bw53			
			Bw54			

*Not all antisera for these specificities may be generally available.

Substance	Tested with: Reference Grouping Serum	Tested with: Cell Suspension	Neutralizing Activity
Blood Group Specific Substance A			
Substance under test	Anti-A	A_1	+
Substance under test	Anti-B	B	−
Reference Substance A	Anti-A	A_1	+
Blood Group Specific Substance B			
Substance under test	Anti-B	B	+
Substance under test	Anti-A	A_1	−
Reference Substance B	Anti-B	B	+
Blood Group Specific Substance AB			
Substance under test	Anti-B	B	+
Substance under test	Anti-A	A_1	+
Reference Substance B	Anti-B	B	+
Reference Substance A	Anti-A	A_1	+

it in the unopened container in which it was placed by the manufacturer.

Expiration date—The expiration date is not more than 2 years after date of issue from manufacturer's cold storage (5°, 1 year, or 0°, 2 years).

Labeling—Label it to state that it was derived from porcine or equine stomachs, whichever is applicable, and that it contains a single dose consisting of the stated content of dry weight of powder dissolved in the stated volume of product. Label it also to state the route of administration, and to state that it is not to be administered intravenously nor to fertile women. Label Blood Group Specific Substance B with a warning that it may contain immunogenic A activity.

pH: between 6.0 and 6.8.

Other requirements—It meets the requirements of the tests for safety (guinea pigs and mice being used) and for anaphylaxis (by intraperitoneal injection into guinea pigs and subsequent challenge by intravenous injection of the same product).

Red Blood Cells

» Red Blood Cells conforms to the regulations of the federal Food and Drug Administration concerning biologics (640.10 to 640.18) (see *Biologics* ⟨1041⟩). It is the remaining red blood cells of whole human blood that has been collected from suitable whole blood donors, and from which plasma has been removed. It may be prepared at any time during the dating period of the whole blood from which it is derived, by centrifuging or undisturbed sedimentation for the separation of plasma and cells, not later than 21 days after the blood has been drawn, except that when acid citrate dextrose adenine solution has been used as the anticoagulant, such preparation may be made within 35 days therefrom. It contains a portion of the plasma sufficient to ensure optional cell preservation or contains a cryophylactic substance if it is for extended manufacturers' storage at −65° or colder.

Packaging and storage—Preserve in a hermetic container, which is of colorless, transparent, sterile, pyrogen-free Type I or Type II glass, or of a suitable plastic material (see *Transfusion and Infusion Assemblies* ⟨161⟩), in which it was placed by the processor. Store if unfrozen at a temperature between 1° and 6°, held constant within a 2° range except during shipment when the temperature may be between 1° and 10°, and store if for extended manufacturers' storage in frozen form at −65° or colder. The container of Red Blood Cells is accompanied by a securely attached smaller container holding an original pilot sample of blood taken from the donor at the same time as the whole human blood, or a pilot sample of Red Blood Cells removed at the time of its preparation.

Expiration date—The expiration date for unfrozen Red Blood Cells is not later than that of the whole human blood from which it is derived if plasma has not been removed, except that if the hermetic seal of the container is broken during preparation, the expiration date is not later than 24 hours after the seal is broken. The expiration date for frozen Red Blood Cells is not later than 3 years after the date of collection of the source blood when stored at −65° or colder and not later than 24 hours after removal from −65° storage provided it is then stored at the temperature for unfrozen Red Blood Cells.

Labeling—In addition to labeling requirements of Whole Blood applicable to this product, label it to indicate the approved variation to which it conforms, such as "Frozen," or "Deglycerolized." Label it also with the instruction to use a filter in the administration equipment.

Whole Blood

ACD Whole Blood.
CPD Whole Blood.
CPDA-1 Whole Blood.
Heparin Whole Blood.

» Whole Blood conforms to the regulations of the federal Food and Drug Administration concerning biologics (640.1 to 640.7) (see *Biologics* ⟨1041⟩). It is blood that has been collected from suitable whole blood human donors under rigid aseptic precautions, for transfusion to human recipients. It contains citrate ion (acid citrate dextrose or citrate phosphate dextrose or citrate phosphate dextrose with adenine) or Heparin Sodium as an anticoagulant. It may consist of blood from which the antihemophilic factor has been removed, in which case it is termed "Modified." It meets the requirements of tests made on a pilot sample in non-reacting in a serologic test for syphilis; for ABO blood group designation; and for classification in regard to Rh type, including those tests specified for variants and other related factors. Containers of Whole Blood shall not be entered for sterility testing prior to use of the blood for transfusion. [NOTE—Whole Blood may be issued prior to the results of testing, under the specified provisions.]

Packaging and storage—Preserve in the container into which it was originally drawn. Use pyrogen-free, sterile containers of colorless, transparent, Type I or Type II glass, or of a suitable plastic material (see *Transfusion and Infusion Assemblies* ⟨161⟩). The container is provided with a hermetic contamination-proof closure. Accessory equipment supplied with the blood is sterile and pyrogen-free (see *Transfusion and Infusion Assemblies* ⟨161⟩). Store at a temperature between 1° and 6° held constant within a 2° range, except during shipment, when the temperature may be between 1° and 10°. The container of Whole Blood is accompanied by at least one securely attached smaller container holding an original pilot sample of blood, for test purposes, taken at the same time from the same donor, with the same anticoagulant. Both containers bear the donor's identification symbol or number.

Expiration date—Its expiration date is not later than 21 days after the date of bleeding the donor, if it contains anticoagulant citrate dextrose solution or anticoagulant citrate phosphate dextrose solution, as the anticoagulant; or not later than 35 days if it contains anticoagulant citrate phosphate dextrose adenine solution as the anticoagulant; or not later than 48 hours after date of bleeding the donor, if it contains heparin ion as the anticoagulant.

Labeling—Label it to indicate the donor classification, quantity and kind of anticoagulant used and the corresponding volume of blood, the designation of ABO blood group and Rh factors, and in the case of Group O blood, whether or not isoagglutinin titers or other tests for exclusion of specified Group O bloods were performed and to indicate any group classification of the blood resulting therefrom. If an ABO blood group color scheme is used, the labeling color used shall be: Group A (yellow), Group B (pink), Group O (blue), and Group AB (white). Label it also with the type and result of a serologic test for syphilis, or to indicate that it was non-reactive in such test; and with the type and result of a test for hepatitis B surface antigen, or to indicate that it was non-reactive in such test. If it has been issued prior to determination of test results, label it also with a warning not to use it until the test results have been received and to specify that a crossmatch be performed. Where applicable, label it as "Modified," and indicate that antihemophilic factor has been removed and that it should not be used for patients requiring that factor.

Boluses, Ampicillin—*see* Ampicillin Boluses
Boluses, Dihydrostreptomycin Sulfate—*see* Dihydrostreptomycin Sulfate Boluses
Boluses, Tetracycline—*see* Tetracycline Boluses
Boric Acid—*see* Boric Acid NF

Botulism Antitoxin

» Botulism Antitoxin conforms to the regulations of the federal Food and Drug Administration concerning biologics (see *Biologics* ⟨1041⟩). It is a sterile, non-pyrogenic solution of the refined and concentrated antitoxic antibodies, chiefly globulins, obtained from the blood of healthy horses that have been immunized against the toxins produced by the type A and type B and/or type E strains of *Clostridium botulinum*. Its potency is determined with the U. S. Standard Botulism Antitoxin of the relevant type, tested by neutralizing activity in mice of the corresponding U. S. Control Botulism Test Toxin. It contains not more than 20.0 percent of solids, and contains a suitable antimicrobial agent.

Packaging and storage—Preserve in single-dose containers only, at a temperature between 2° and 8°.

Expiration date—The expiration date for Antitoxin containing a 20% excess of potency is not later than 5 years after date of issue from manufacturer's cold storage (5°, 1 year; or 0°, 2 years).

Labeling—Label it to state that it was prepared from horse blood.

Bromocriptine Mesylate

$(CH_3)_2CH$ OH H CONH O N H N O N H $CH_2CH(CH_3)_2$ · CH_3SO_3H N CH_3 H HN Br

$C_{32}H_{40}BrN_5O_5.CH_4SO_3$ 750.70

Ergotaman-3′,6′,18-trione, 2-bromo-12′-hydroxy-2′-(1-methylethyl)-5′-(2-methylpropyl)-, monomethanesulfonate (salt), (5′α)-.

2-Bromoergocryptine monomethanesulfonate (salt) [*22260-51-1*].

» Bromocriptine Mesylate contains not less than 98.0% and not more than 102.0% of $C_{32}H_{40}BrN_5O_5.CH_4SO_3$, calculated on the dried basis.

Packaging and storage—Preserve in tight, light-resistant containers, in a cold place.

Reference standard—*USP Bromocriptine Mesylate Reference Standard*—Determine the *Loss on drying* by thermogravimetric analysis, heating a 5- to 10-mg portion at 10° per minute from 25° to 160° under nitrogen flowing at about 45 mL per minute.

Identification—

A: The infrared absorption spectrum of a mineral oil dispersion of it exhibits maxima only at the same wavelengths as that of a similar preparation of undried USP Bromocriptine Mesylate RS.

B: Dissolve about 10 mg in 200 mL of 0.1 *M* methanolic methanesulfonic acid solution: the ultraviolet absorption spectrum of the solution so obtained exhibits maxima and minima at the same wavelengths as that of a solution of USP Bromocriptine Mesylate RS similarly prepared, having a concentration of 0.05 mg per mL, 0.1 *M* methanolic methanesulfonic acid being used as the blank, concomitantly measured.

Color of solution ⟨631⟩—

Matching solutions—Prepare three solutions, *A*, *B*, and *C*, containing, respectively, the following parts of cobaltous chloride CS, ferric chloride CS, cupric sulfate CS, and dilute hydrochloric acid (1 in 40):

A—3.0:3.0:2.4:31.6;
B—1.0:2.4:0.4:36.2;
C—0.6:2.4:0:37.0

Procedure—Prepare a test solution by dissolving 100 mg of Bromocriptine Mesylate in 10.0 mL of methanol, and compare this solution with 10-mL portions of the *Matching solutions* in suitable matched tubes: the solution is clear and not darker in color than *Matching solutions A*, *B*, and *C*.

Specific rotation ⟨781⟩: between +95° and +105°, calculated on the dried basis, determined in a mixture of methylene chloride and methanol (1:1) containing 100 mg in each 10 mL.

Loss on drying (see *Thermal Analysis* ⟨891⟩)—Determine the percentage of volatile substances by thermogravimetric analysis on an appropriately calibrated instrument, using about 10 mg of Bromocriptine Mesylate, accurately weighed. Heat the specimen under test at the rate of 10° per minute in an atmosphere of nitrogen at a flow rate of about 45 mL per minute. Record the thermogram from 25° to 160°: it loses not more than 6.0% of its weight.

Residue on ignition ⟨281⟩: not more than 0.1%.

Heavy metals, *Method II* ⟨231⟩: 0.002%.

Related substances—[NOTE—Conduct this test without exposure to daylight and with minimum exposure to artificial light. Perform the test rapidly, preparing and spotting the test solution last.] Prepare a solution in a mixture of chloroform and methanol (1:1) containing 10 mg of bromocriptine per mL. Similarly prepare three Standard solutions of USP Bromocriptine Mesylate RS in a mixture of chloroform and methanol (1:1) with final concentrations equivalent to 0.10 mg, 0.05 mg, and 0.025 mg of bromocriptine per mL (equivalent to 1.0%, 0.5%, and 0.25%, respectively). Separately apply, as 1-cm bands, 10 µL of each Standard solution and of the test solution on a suitable thin-layer chromatographic plate (see *Chromatography* ⟨621⟩), coated with a 0.25-mm layer of chromatographic silica gel mixture. Develop in a tank lined with filter paper, previously equilibrated for 20 minutes, using a solvent system consisting of a mixture of methylene chloride, dioxane, alcohol, and ammonium hydroxide (180:15:5:0.1) until the solvent front has moved a distance of 10 cm on the plate. Dry the plate well under vacuum at room temperature for 30 minutes. Spray first with Dragendorff's reagent (dissolve 850 mg of bismuth subnitrate in a mixture of 10 mL of glacial acetic acid and 40 mL of water; prepare a second solution consisting of 8 g of potassium iodide in 20 mL of water; just prior to use, transfer 5 mL of each of these two solutions to a 100-mL volumetric flask, add 20 mL of glacial acetic acid, dilute with water to volume, and mix), and then spray thoroughly with hydrogen peroxide solution, cover with a glass plate to prevent fading, and immediately visualize in daylight: any spot, other than the principal spot, obtained from the test solution is not greater in size and intensity than the spot obtained from the 1.0% Standard solution. Any other spots are not greater in size and intensity than the spot obtained from the 0.5% Standard solution. The sum of the related substances is not greater than 1.5%.

Methanesulfonic acid content—Transfer about 400 mg, accurately weighed, to a titration vessel, dissolve in 70 mL of methanol, and titrate under nitrogen with 0.1 *N* methanolic potassium hydroxide VS, determining the end-point potentiometrically. Perform a blank determination, and make any necessary correction. Each mL of 0.1 *N* methanolic potassium hydroxide is equivalent to 9.61 mg of CH_3SO_3H. Not less than 12.5% and not more than 13.4% of CH_3SO_3H, calculated on the dried basis, is found.

Assay—Transfer about 600 mg of Bromocriptine Mesylate, accurately weighed, to a titration vessel, dissolve in 80 mL of a mixture of acetic anhydride and glacial acetic acid (7:1), and titrate with 0.1 *N* perchloric acid VS, determining the end-point potentiometrically. Perform a blank determination, and make any necessary correction. Each mL of 0.1 *N* perchloric acid is equivalent to 75.07 mg of $C_{32}H_{40}BrN_5O_5 . CH_4SO_3$.

Bromocriptine Mesylate Tablets

» Bromocriptine Mesylate Tablets contain bromocriptine mesylate ($C_{32}H_{40}BrN_5O_5 . CH_4SO_3$) equivalent to not less than 90.0 percent and not more than 110.0 percent of the labeled amount of bromocriptine ($C_{32}H_{40}BrN_5O_5$).

Packaging and storage—Preserve in tight, light-resistant containers.

Reference standard—*USP Bromocriptine Mesylate Reference Standard*—Determine the *Loss on drying* by thermogravimetric analysis, heating a 5- to 10-mg portion, accurately weighed, at 10° per minute over a 200° range under nitrogen flowing at about 45 mL per minute.

Identification—Examine the chromatograms obtained in the test for *Related substances:* the principal spot obtained from the test solution is similar in R_f value, and color to that obtained from the Standard solution.

Related substances—[NOTE—Conduct this test without exposure to daylight and with minimum exposure to artificial light. Perform the test rapidly, preparing and applying the test solution last.] Transfer a quantity of finely powdered Tablets, equivalent to 20 mg of bromocriptine, to a conical flask. Add 10.0 mL of methanol, and mix for 20 minutes. Centrifuge the suspension at 4000 rpm for 10 minutes. The clear supernatant liquid is the test solution. Similarly prepare a Standard solution of USP Bromocriptine Mesylate RS in methanol containing 1 mg of bromocriptine per mL and three Standard dilutions having final concentrations equivalent to 0.10 mg, 0.30 mg, and 0.50 mg of bromocriptine per mL (equivalent to 1.0%, 3.0%, and 5.0%, respectively). Separately apply, as 1.5-cm bands, 10-µL portions of the Standard solution and of the three Standard dilutions, and 50 µL of the test solution, on a suitable thin-layer chromatographic plate (see *Chromatography* ⟨621⟩), coated with a 0.25-mm layer of chromatographic silica gel mixture. Dry the plate for 5 minutes in a stream of cold air. Develop in a tank lined with filter paper, previously equilibrated for 20 minutes, using a solvent system consisting of a mixture of methylene chloride, dioxane, alcohol, and ammonium hydroxide (180:15:5:0.1) until the solvent front has moved a distance of 10 cm on the plate. Dry the plate well under vacuum at room temperature for 15 minutes. Spray first with Dragendorff's reagent (prepared as directed in the test for *Related substances* under *Bromocriptine Mesylate*), and then spray thoroughly with hydrogen peroxide solution, cover with a glass plate to prevent fading, and immediately visualize in daylight. Do not evaluate the first 2 cm above the line of application (spots of the excipients). Any spot, other than the principal spot, obtained from the test solution is not greater in size and intensity than the spot obtained from the 3.0% Standard solution and any remaining spots are not greater in size and intensity than the spot obtained from the 1.0% Standard solution. The sum of the related substances is not greater than 5.0%.

Dissolution ⟨711⟩—

Medium: 0.1 *N* hydrochloric acid; 500 mL.

Apparatus 1: 120 rpm.

Time: 60 minutes.

Procedure—Determine the amount in solution in portions of the solution under test that previously have been filtered through a glass-fiber filter from fluorometric measurements at an excitation wavelength of 315 nm and an emission wavelength of 445 nm, using *Dissolution Medium* as the blank, in comparison with a Standard solution having a known concentration of USP Bromocriptine Mesylate RS in the same medium. A volume of alcohol not to exceed 5% of the total volume of the Standard solution may be used to bring the standard into solution prior to dilution with 0.1 *N* hydrochloric acid.

Tolerances—Not less than 80% (*Q*) of the labeled amount of bromocriptine ($C_{32}H_{40}BrN_5O_5$) is dissolved in 60 minutes.

Uniformity of dosage units ⟨905⟩: meet the requirements.

Procedure for content uniformity—Transfer 1 Tablet to a 50-mL volumetric flask, add about 25 mL of a solution of alcohol and water (1:1), and shake by mechanical means for 30 minutes. Dilute with a mixture of alcohol and water (1:1) to volume, mix, and filter through a 0.7-µm glass fiber filter, discarding the first few mL of the filtrate. Concomitantly determine the absorbances of this solution and of a Standard solution of USP Bromocriptine Mesylate RS in the same medium having a known concentration of about 57 µg per mL in 1-cm cells at the wavelength of maximum absorbance at about 306 nm, with a suitable spectrophotometer, using a mixture of alcohol and water (1:1) as the blank. Calculate the quantity, in mg, of bromocriptine ($C_{32}H_{40}BrN_5O_5$) in the Tablet by the formula:

$$(654.60/750.71)(TC/D)(A_U/A_S),$$

in which 654.60 and 750.71 are the molecular weights of bromocriptine and bromocriptine mesylate, respectively, T is the labeled quantity, in mg, of bromocriptine in the Tablet, C is the concentration, in µg per mL, of USP Bromocriptine Mesylate RS in the Standard solution, D is the concentration, in µg per mL, of bromocriptine in the solution from the Tablet, based upon the labeled quantity per Tablet and the extent of dilution, and A_U and A_S are the absorbances of the solution from the Tablet and the Standard solution, respectively.

Assay—

Mobile phase—Mix 650 mL of acetonitrile with 350 mL of 0.01 *M* ammonium carbonate.

Standard preparation—Transfer about 11 mg of USP Bromocriptine Mesylate RS, accurately weighed, to a 50-mL volumetric flask, dissolve in methanol, dilute with methanol to volume, and mix.

Assay preparation—Weigh and finely powder not less than 20 Bromocriptine Mesylate Tablets. Transfer an accurately weighed portion of the powder, equivalent to about 10 mg of bromocriptine, to an appropriate container, add 40 mL of methanol, and stir for 20 minutes, protected from light. Quantitatively filter through a fine glass filtering funnel into a 50-mL volumetric flask. Rinse the filter with methanol, adding the rinsing to the filtrate, dilute with methanol to volume, and mix.

Procedure—Using a sampling valve or suitable microsyringe, inject 50-μL aliquots of the *Standard preparation* and the *Assay preparation* into a high-performance liquid chromatograph equipped with a 250-mm × 4-mm stainless steel column containing packing L1, an ultraviolet detector capable of monitoring absorption at 300 nm, and suitable recorder and integrator. The column is maintained at ambient temperature, and the flow rate is adjusted to approximately 2 mL per minute. The coefficient of variation for three replicate injections of the *Standard preparation* is not more than 3.0%. Calculate the quantity, in mg, of bromocriptine ($C_{32}H_{40}BrN_5O_5$) in the portion of Tablets taken by the formula:

$$(654.60/750.71)50C(A_U/A_S),$$

in which 654.60 and 750.71 are the molecular weights of bromocriptine and bromocriptine mesylate, respectively, C is the concentration, in mg per mL, of USP Bromocriptine Mesylate RS in the *Standard preparation*, and A_U and A_S are the peak areas obtained from the *Assay preparation* and the *Standard preparation*, respectively.

Bromodiphenhydramine Hydrochloride

$C_{17}H_{20}BrNO.HCl$ 370.72

Ethanamine, 2-[(4-bromophenyl)phenylmethoxy]-*N,N*-dimethyl-, hydrochloride.

2-[(*p*-Bromo-α-phenylbenzyl)oxy]-*N,N*-dimethylethylamine hydrochloride [*1808-12-4*].

» Bromodiphenhydramine Hydrochloride contains not less than 98.0 percent and not more than 101.0 percent of $C_{17}H_{20}BrNO.HCl$, calculated on the dried basis.

Packaging and storage—Preserve in tight containers.

Reference standard—*USP Bromodiphenhydramine Hydrochloride Reference Standard*—Dry at 105° for 3 hours before using.

Identification—

A: The infrared absorption spectrum of a potassium bromide dispersion of it, previously dried, exhibits maxima only at the same wavelengths as that of a similar preparation of USP Bromodiphenhydramine Hydrochloride RS.

B: The ultraviolet absorption spectrum of a 1 in 67,000 solution in 0.1 *N* sulfuric acid exhibits maxima and minima at the same wavelengths as that of a similar solution of USP Bromodiphenhydramine Hydrochloride RS, concomitantly measured, and the respective absorptivities, calculated on the dried basis, at the wavelength of maximum absorbance at about 228 nm do not differ by more than 3.0%.

Melting range ⟨741⟩: between 148° and 152°.

Loss on drying ⟨731⟩—Dry it at 105° for 3 hours: it loses not more than 0.5% of its weight.

Assay—Dissolve about 700 mg of Bromodiphenhydramine Hydrochloride, accurately weighed, in 50 mL of glacial acetic acid, and add 10 mL of benzene and 15 mL of mercuric acetate TS. Add 2 drops of crystal violet TS, and titrate with 0.1 *N* perchloric acid VS to a green end-point. Perform a blank determination, and make any necessary correction. Each mL of 0.1 *N* perchloric acid is equivalent to 37.07 mg of $C_{17}H_{20}BrNO.HCl$.

Bromodiphenhydramine Hydrochloride Capsules

» Bromodiphenhydramine Hydrochloride Capsules contain not less than 93.0 percent and not more than 107.0 percent of the labeled amount of $C_{17}H_{20}BrNO.HCl$.

Packaging and storage—Preserve in tight containers.

Reference standard—*USP Bromodiphenhydramine Hydrochloride Reference Standard*—Dry at 105° for 3 hours before using.

Identification—Transfer the final solution obtained in the *Assay* to a separator, add about 1 mL of 0.1 *N* sulfuric acid, and shake with 25 mL of ether. (Methyl red enters the ether phase.) Drain the aqueous layer into another separator, add 5 mL of 1 *N* sodium hydroxide, and shake with 10 mL of chloroform. Drain the chloroform layer into a small flask containing 2 g of anhydrous sodium sulfate, and swirl. Pour the chloroform solution through a small cotton pledget, pre-rinsed with chloroform, into a beaker, and evaporate to about 5 mL. Apply a few drops of the solution directly to a potassium bromide plate, and completely remove the chloroform by warming for 2 to 3 minutes under an infrared lamp: the infrared absorption spectrum of the bromodiphenhydramine so obtained exhibits maxima only at the same wavelengths as that of a similar preparation of USP Bromodiphenhydramine Hydrochloride RS.

Dissolution ⟨711⟩—

Medium: water; 900 mL.

Apparatus 1: 100 rpm.

Time: 45 minutes.

Procedure—Determine the amount of $C_{17}H_{20}BrNO.HCl$ dissolved from ultraviolet absorbances at the wavelength of maximum absorbance at about 228 nm of filtered portions of the solution under test, suitably diluted with *Dissolution Medium*, if necessary, in comparison with a Standard solution having a known concentration of USP Bromodiphenhydramine Hydrochloride RS in the same medium.

Tolerances—Not less than 75% (*Q*) of the labeled amount of $C_{17}H_{20}BrNO.HCl$ is dissolved in 45 minutes.

Uniformity of dosage units ⟨905⟩: meet the requirements.

Procedure for content uniformity—Transfer, as completely as possible, the contents of 1 Capsule to a suitable small flask, add a quantity of alcohol, accurately measured, to yield a concentration of about 50 mg of bromodiphenhydramine hydrochloride per 100 mL, insert the stopper securely, and mix by swirling intermittently for a few minutes. Centrifuge a portion of the mixture at about 2000 rpm until the solution becomes clear. Transfer 3.0 mL of the supernatant liquid to a 100-mL volumetric flask containing 10 mL of 1 *N* sulfuric acid, dilute with alcohol to volume, and mix. Concomitantly determine the absorbances of this solution and of a solution of USP Bromodiphenhydramine Hydrochloride RS in a 1 in 10 mixture of 1 *N* sulfuric acid in alcohol having a known concentration of about 15 μg per mL, in 1-cm cells at the wavelength of maximum absorbance at about 228 nm, with a suitable spectrophotometer, using the 1 in 10 mixture of 1 *N* sulfuric acid in alcohol as the blank. Calculate the quantity, in mg, of $C_{17}H_{20}BrNO.HCl$ in the Capsule by the formula:

$$(TC/D)(A_U/A_S),$$

in which T is the labeled quantity, in mg, of bromodiphenhydramine hydrochloride in the Capsule, C is the concentration, in μg per mL, of USP Bromodiphenhydramine Hydrochloride RS in the Standard solution, D is the concentration, in μg per mL, of bromodiphenhydramine hydrochloride in the test solution, based upon the labeled quantity per Capsule and the extent of dilution, and A_U and A_S are the absorbances of the test solution and the Standard solution, respectively.

Assay—

Chromatographic column—Proceed as directed under *Column Partition Chromatography* ⟨621⟩, packing a chromatographic tube, equipped with a stopcock, with two segments of packing material. The lower segment is a mixture of 7.5 g of *Solid Support* and 5 mL of water, and the upper segment is a mixture prepared as directed under *Assay preparation*.

Assay preparation—Remove, as completely as possible, the contents of not less than 20 Bromodiphenhydramine Hydrochloride Capsules, and weigh. Mix, and transfer an accurately weighed portion of the powder, equivalent to about 250 mg of bromodiphenhydramine hydrochloride, to a 250-mL beaker. Add 4 mL of water, mix, add 1 mL of sodium hydroxide solution (2 in 25), and mix. Add 10 g of *Solid Support*, mix as directed under *Preparation of Chromatographic Column*, and transfer to the column, using a mixture of 3 g of *Solid Support* and 2 mL of water to rinse the beaker.

Procedure—Pass ether through the column at a rate of 2 to 4 mL per minute, collecting 75 mL of eluate in a 125-mL separator. Extract the eluate with 10.0 mL of 0.1 *N* sulfuric acid VS followed by two 5-mL portions of water, and combine the extracts in a conical flask. Add methyl red TS, and titrate the excess acid with 0.02 *N* sodium hydroxide VS. Perform a blank determination (see *Residual Titrations* under *Titrimetry* ⟨541⟩). Each mL of 0.1 *N* sulfuric acid is equivalent to 37.07 mg of $C_{17}H_{20}BrNO.HCl$.

Bromodiphenhydramine Hydrochloride Elixir

» Bromodiphenhydramine Hydrochloride Elixir contains not less than 93.0 percent and not more than 107.0 percent of the labeled amount of $C_{17}H_{20}BrNO.HCl$.

Packaging and storage—Preserve in tight, light-resistant containers.

Reference standard—*USP Bromodiphenhydramine Hydrochloride Reference Standard*—Dry at 105° for 3 hours before using.

Identification—It responds to the *Identification test* under *Bromodiphenhydramine Hydrochloride Capsules*.

Alcohol content, *Method I* ⟨611⟩: between 12.0% and 15.0% of C_2H_5OH.

Assay—Evaporate an accurately measured volume of Bromodiphenhydramine Hydrochloride Elixir, equivalent to about 250 mg of bromodiphenhydramine hydrochloride, to about half the original volume, using a suitable vacuum evaporator. Transfer the concentrated solution to a 250-mL separator, with the aid of sufficient warm water to bring the volume to the original volume. Add 20 g of sodium chloride, and shake until dissolved. Add 5 mL of 1 *N* sodium hydroxide, shake with 100 mL of ether, and drain the aqueous layer into a second separator containing 50 mL of ether. Shake, and discard the aqueous layer. Wash the ether solutions with two 20-mL portions of water, shaking each aqueous portion successively in the two separators, and then discard the aqueous solutions. Extract the ether solutions successively with 10.0 mL of 0.1 *N* sulfuric acid VS, followed by two 5-mL portions of water, and collect the aqueous extracts in a conical flask. Add methyl red TS to the solution in the flask, and titrate the excess acid with 0.02 *N* sodium hydroxide VS. Perform a blank determination (see *Residual Titrations* under *Titrimetry* ⟨541⟩). Each mL of 0.1 *N* sulfuric acid is equivalent to 37.07 mg of $C_{17}H_{20}BrNO.HCl$.

Brompheniramine Maleate

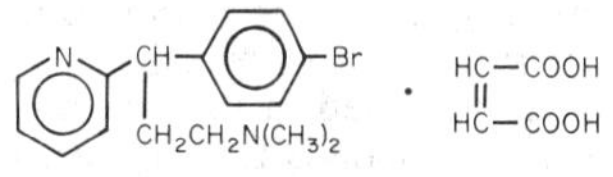

$C_{16}H_{19}BrN_2.C_4H_4O_4$ 435.32

2-Pyridinepropanamine, γ-(4-bromophenyl)-*N*,*N*-dimethyl-, (*Z*)-butenedioate (1:1).

2-[*p*-Bromo-α-[2-(dimethylamino)ethyl]benzyl]pyridine maleate (1:1) [*980-71-2*].

» Brompheniramine Maleate, dried at 105° for 3 hours, contains not less than 98.0 percent and not more than 100.5 percent of $C_{16}H_{19}BrN_2.C_4H_4O_4$.

Packaging and storage—Preserve in tight, light-resistant containers.

Reference standard—*USP Brompheniramine Maleate Reference Standard*—Dry at 105° for 3 hours before using.

Identification—

A: The infrared absorption spectrum of a potassium bromide dispersion of it, previously dried, exhibits maxima only at the same wavelengths as that of a similar preparation of USP Brompheniramine Maleate RS.

B: The ultraviolet absorption spectrum of a 1 in 30,000 solution in methanol exhibits maxima and minima at the same wavelengths as that of a similar solution of USP Brompheniramine Maleate RS, concomitantly measured.

Melting range ⟨741⟩: between 130° and 135°.

pH ⟨791⟩: between 4.0 and 5.0, in a solution (1 in 100).

Loss on drying ⟨731⟩—Dry it at 105° for 3 hours: it loses not more than 0.5% of its weight.

Residue on ignition ⟨281⟩: not more than 0.2%.

Related compounds—

Test preparation—Dissolve about 200 mg of Brompheniramine Maleate in 5 mL of methylene chloride, and mix.

Chromatographic system (see *Chromatography* ⟨621⟩)—The gas chromatograph is equipped with a flame-ionization detector and a 1.2-m × 4-mm glass column containing 3 percent phase G3 on support S1AB. The column temperature is maintained at about 190°, and the injection port and detector temperatures are both maintained at about 250°. The carrier gas is dry helium, flowing at a rate adjusted to obtain a retention time of 6 to 7 minutes for the main peak. Chromatograph the *Test preparation*, record the chromatogram, and determine the peak area as directed under *Procedure:* the tailing factor for the brompheniramine maleate peak is not more than 1.8.

Procedure—Inject a volume (about 1 µL) of the *Test preparation* into the chromatograph. Record the chromatogram for a total time of not less than twice the retention time of the brompheniramine peak, and measure the areas of the peaks. The total relative area of all extraneous peaks (except that of the solvent peak) does not exceed 2.0%.

Assay—Dissolve about 425 mg of Brompheniramine Maleate, previously dried and accurately weighed, in 50 mL of glacial acetic acid. Add 1 drop of crystal violet TS, and titrate with 0.1 *N* perchloric acid VS to a green end-point. Perform a blank determination, and make any necessary correction. Each mL of 0.1 *N* perchloric acid is equivalent to 21.77 mg of $C_{16}H_{19}BrN_2.C_4H_4O_4$.

Brompheniramine Maleate Elixir

» Brompheniramine Maleate Elixir contains not less than 95.0 percent and not more than 105.0 percent of the labeled amount of $C_{16}H_{19}BrN_2.C_4H_4O_4$.

Packaging and storage—Preserve in well-closed, light-resistant containers.

Reference standard—*USP Brompheniramine Maleate Reference Standard*—Dry at 105° for 3 hours before using.

Identification—Transfer a volume of Elixir, equivalent to about 50 mg of brompheniramine maleate, to a separator, render distinctly alkaline with 1 *N* sodium hydroxide, and extract with two 50-mL portions of chloroform, shaking gently to avoid emulsification. Wash the combined chloroform extracts with 10 mL of water, and discard the aqueous phase. Filter the combined chloroform extracts into a conical flask, and evaporate the solvent on a steam bath, with the aid of a current of air. To the residue add 25 mL of dilute hydrochloric acid (1 in 1200), and proceed as directed under *Identification—Organic Nitrogenous Bases* ⟨181⟩, beginning with "Transfer the liquid to a separator." The Elixir meets the requirements of the test.

pH ⟨791⟩: between 2.5 and 3.5.

Alcohol content, *Method I* ⟨611⟩: between 2.7% and 3.3% of C_2H_5OH.

Assay—Transfer an accurately measured volume of Brompheniramine Maleate Elixir, equivalent to about 20 mg of brompheniramine maleate, to a separator, render distinctly alkaline with 1 *N* sodium hydroxide, and extract with ten 10-mL portions of chloroform, shaking gently to avoid emulsification. Wash the combined chloroform extracts with 10 mL of water, wash the latter with 20 mL of chloroform, and discard the aqueous phase. Quantitatively filter the combined chloroform extracts and washings into a conical flask, and evaporate the solvent on a steam bath, with the aid of a current of air. To the residue add 25 mL of glacial acetic acid and 5 mL of acetic anhydride, agitate, and allow to stand for about 15 minutes. Add 1 drop of crystal violet TS, and titrate with 0.01 *N* perchloric acid VS to a blue-green end-point. Perform a blank determination, and make any necessary correction. Each mL of 0.01 *N* perchloric acid is equivalent to 2.177 mg of $C_{16}H_{19}BrN_2.C_4H_4O_4$.

Brompheniramine Maleate Injection

» Brompheniramine Maleate Injection is a sterile solution of Brompheniramine Maleate in Water for Injection. It contains not less than 90.0 percent and not more than 110.0 percent of the labeled amount of $C_{16}H_{19}BrN_2.C_4H_4O_4$.

Packaging and storage—Preserve in single-dose or in multiple-dose containers, preferably of Type I glass, protected from light.

Reference standard—*USP Brompheniramine Maleate Reference Standard*—Dry at 105° for 3 hours before using.

Identification—Dilute a volume of Injection, equivalent to about 50 mg of brompheniramine maleate, with dilute hydrochloric acid (1 in 1200) to 25 mL, and proceed as directed under *Identification—Organic Nitrogenous Bases* ⟨181⟩, beginning with "Transfer the liquid to a separator." The Injection meets the requirements of the test.

pH ⟨791⟩: between 6.3 and 7.3.

Other requirements—It meets the requirements under *Injections* ⟨1⟩.

Assay—Proceed with Brompheniramine Maleate Injection as directed under *Salts of Organic Nitrogenous Bases* ⟨501⟩, to prepare the solution employed for the determination of the absorbance, A_U, at 262 nm. For the determination of A_S, dissolve about 25 mg of USP Brompheniramine Maleate RS, accurately weighed, in 20 mL of dilute sulfuric acid (1 in 350), and treat this solution the same as the portion of Injection being assayed. Calculate the quantity, in mg, of $C_{16}H_{19}BrN_2.C_4H_4O_4$ in each mL of the Injection taken by the formula:

$$(W/V)(A_U/A_S),$$

in which *W* is the weight, in mg, of USP Brompheniramine Maleate RS in the *Standard Preparation*, and *V* is the volume, in mL, of Injection taken.

Brompheniramine Maleate Tablets

» Brompheniramine Maleate Tablets contain not less than 95.0 percent and not more than 105.0 percent of the labeled amount of $C_{16}H_{19}BrN_2.C_4H_4O_4$.

Packaging and storage—Preserve in tight containers.

Reference standard—*USP Brompheniramine Maleate Reference Standard*—Dry at 105° for 3 hours before using.

Identification—Tablets meet the requirements under *Identification—Organic Nitrogenous Bases* ⟨181⟩.

Dissolution ⟨711⟩—

Medium: water; 500 mL.

Apparatus 1: 100 rpm.

Time: 45 minutes.

Procedure—Determine the amount of $C_{16}H_{19}BrN_2.C_4H_4O_4$ dissolved from ultraviolet absorbances at the wavelength of maximum absorbance at about 264 nm of filtered portions of the solution under test, suitably diluted with 3 *N* hydrochloric acid, in comparison with a Standard solution having a known concentration of USP Brompheniramine Maleate RS in the same medium.

Tolerances—Not less than 75% (*Q*) of the labeled amount of $C_{16}H_{19}BrN_2.C_4H_4O_4$ is dissolved in 45 minutes.

Uniformity of dosage units ⟨905⟩: meet the requirements.

Assay—

Standard preparation—Dissolve an accurately weighed quantity of USP Brompheniramine Maleate RS in water, and dilute quantitatively with water to obtain a solution having a known concentration of about 160 µg per mL. Transfer 25.0 mL of this solution to a separator containing 25 mL of water, mix, and proceed as directed under *Assay preparation*, beginning with "adjust with sodium hydroxide solution (1 in 10) to a pH of 11." The concentration of USP Brompheniramine Maleate RS in the *Standard preparation* is about 20 µg per mL.

Assay preparation—Weigh and finely powder not less than 20 Brompheniramine Maleate Tablets. Weigh accurately a portion of the powder, equivalent to about 4 mg of brompheniramine maleate, mix with 50 mL of water for 10 minutes, adjust with sodium hydroxide solution (1 in 10) to a pH of 11, and cool to room temperature. Extract the mixture with two 75-mL portions of solvent hexane, and combine the extracts in a second separator. Extract the solvent hexane solution with three 50-mL portions of dilute hydrochloric acid (1 in 120), combining the acid extracts in a 200-mL volumetric flask. Add dilute hydrochloric acid (1 in 120) to volume, and mix.

Procedure—Concomitantly determine the absorbances of the *Assay preparation* and the *Standard preparation*, in 1-cm cells at the wavelength of maximum absorbance at about 264 nm, with a suitable spectrophotometer, using dilute hydrochloric acid (1 in 120) as the blank. Calculate the quantity, in mg, of $C_{16}H_{19}BrN_2.C_4H_4O_4$ in the portion of Tablets taken by the formula:

$$0.2C(A_U/A_S),$$

in which *C* is the concentration, in µg per mL, of USP Brompheniramine Maleate RS in the *Standard preparation*, and A_U and A_S are the absorbances of the *Assay preparation* and the *Standard preparation*, respectively.

Brompheniramine Maleate and Pseudoephedrine Sulfate Syrup

» Brompheniramine Maleate and Pseudoephedrine Sulfate Syrup contains not less than 90.0 percent and not more than 110.0 percent of the labeled amounts of Brompheniramine Maleate ($C_{10}H_{15}BrN_2.C_4H_4O_4$) and Pseudoephedrine Sulfate $(C_{10}H_{15}NO)_2.H_2SO_4$.

Reference standards—*USP Brompheniramine Maleate Reference Standard*—Dry at 105° for 3 hours before using. *USP Pseudoephedrine Sulfate Reference Standard*—Dry at 105° for 2 hours before using.

Identification—

A: The retention times of the major peaks in the chromatogram of the *Assay preparation* correspond to those of the *Standard preparation*, as obtained in the *Assay*.

B: A solution of it responds to the test for *Sulfate* ⟨191⟩.

Assay—

Mobile phase—Prepare a mixture of water, acetonitrile, methanol, and tetrahydrofuran (550:320:80:50). Transfer 1.0 mL of phosphoric acid, followed by 4.33 g of sodium lauryl sulfate to this mixture, and mix. Adjust with ammonium hydroxide to a pH of 3.50 ± 0.05, filter, and degas. Make adjustments if necessary (see *System Suitability* under *Chromatography* ⟨621⟩). [NOTE—The pH of the *Mobile phase* is critical and may cause 1 to 4 minutes of differences in the retention times of internal standard and brompheniramine maleate.]

Internal standard solution—Transfer about 50 mg of naphazoline hydrochloride to a 100-mL volumetric flask, add *Mobile phase* to volume, and mix.

Standard preparation—Dissolve an accurately weighed quantity of USP Brompheniramine Maleate RS in *Mobile phase*, and dilute quantitatively with *Mobile phase* to obtain a solution having a known concentration of about 6000J µg per mL, J being the ratio of the labeled amount, in mg, of brompheniramine maleate to the labeled amount, in mg, of pseudoephedrine sulfate per mL (*Solution P*). Transfer about 30 mg of USP Pseudoephedrine Sulfate RS, accurately weighed, to a 25-mL volumetric flask, add 5.0 mL each of *Solution P* and *Internal standard solution*, dilute with *Mobile phase* to volume, and mix to obtain a *Standard preparation* having known concentrations of about 1200J µg of USP Brompheniramine Maleate RS per mL and about 1.2 mg of USP Pseudoephedrine Sulfate RS per mL.

Assay preparation—Using a "To contain" pipet transfer an accurately measured volume of Brompheniramine Maleate and Pseudoephedrine Sulfate Syrup, equivalent to about 30 mg of Pseudoephedrine Sulfate, to a 25-mL volumetric flask. Rinse the pipet with about 5 mL of *Mobile phase*, collecting the rinse in the volumetric flask. Add 5.0 mL of *Internal standard solution*, dilute with *Mobile phase* to volume, and mix.

Chromatographic system (see *Chromatography* ⟨621⟩)—The liquid chromatograph is equipped with a 254-nm detector and a 4-mm × 30-cm column that contains packing L11. The flow rate is about 1.5 mL per minute. Chromatograph the *Standard preparation*, and record the peak responses as directed under *Procedure:* the resolution, R, between the pseudoephedrine sulfate and naphazoline hydrochloride peaks is not less than 3, and the resolution, R, between the brompheniramine maleate and naphazoline hydrochloride peaks is not less than 3, and the relative standard deviation for replicate injections is not more than 2.0%.

Procedure—Separately inject equal volumes (about 10 µL) of the *Standard preparation* and the *Assay preparation* into the chromatograph, record the chromatograms, and measure the responses for the major peaks. The relative retention times are about 1.0 for pseudoephedrine sulfate, 1.5 for naphazoline hydrochloride, and 2.5 for brompheniramine maleate. Calculate the quantity, in mg, of brompheniramine maleate ($C_{16}H_{19}BrN_2 \cdot C_4H_4O_4$) in each mL of the syrup taken by the formula:

$$25CV(R_U/R_S),$$

in which C is the concentration, in mg per mL, of USP Brompheniramine Maleate RS in the *Standard preparation*, V is the volume, in mL, of Syrup taken, and R_U and R_S are the peak response ratios obtained for brompheniramine maleate and naphazoline hydrochloride from the *Assay preparation* and the *Standard preparation*, respectively. Calculate the quantity, in mg, of pseudoephedrine sulfate $(C_{10}H_{15}NO)_2 \cdot H_2SO_4$ in each mL of the Syrup taken by the same formula, changing the terms to refer to pseudoephedrine sulfate.

Buffered Aspirin Tablets—*see* Aspirin Tablets, Buffered

Bumetanide

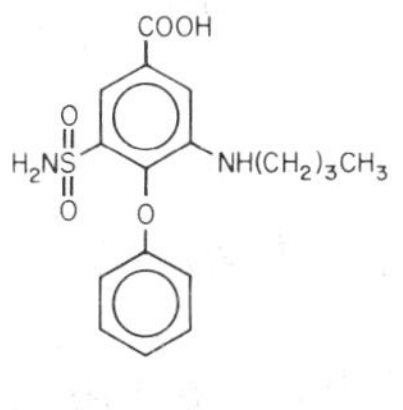

COOH
H_2NO_2S … $NH(CH_2)_3CH_3$

$C_{17}H_{20}N_2O_5S$ 364.42

Benzoic acid, 3-(aminosulfonyl)-5-(butylamino)-4-phenoxy-.
3-(Butylamino)-4-phenoxy-5-sulfamoylbenzoic acid
[*28395-03-1*].

» Bumetanide contains not less than 98.0 percent and not more than 102.0 percent of $C_{17}H_{20}N_2O_5S$, calculated on the dried basis.

Packaging and storage—Preserve in tight, light-resistant containers.

Reference standards—*USP Bumetanide Reference Standard*—Do not dry before using. *USP 3-Nitro-4-phenoxy-5-sulfamoylbenzoic Acid Reference Standard*—Do not dry before using. *USP 3-Amino-4-phenoxy-5-sulfamoylbenzoic Acid Reference Standard*—Do not dry before using. *USP Butyl 3-(butylamino)-4-phenoxy-5-sulfamoylbenzoate Reference Standard*—Do not dry before using.

Identification—

A: The infrared absorption spectrum of a mineral oil dispersion of it exhibits maxima only at the same wavelengths as that of a similar preparation of USP Bumetanide RS.

B: The ultraviolet absorption spectrum of a 1 in 20,000 solution in isopropyl alcohol exhibits maxima and minima at the same wavelengths as that of a similar solution of USP Bumetanide RS, concomitantly measured.

C: The principal spot obtained from the chromatogram of the *Test preparation* exhibits an R_f value corresponding to that of *Standard preparation A* in the test for *Chromatographic purity*.

Loss on drying ⟨731⟩—Dry it at 105° for 4 hours: it loses not more than 0.5% of its weight.

Residue on ignition ⟨281⟩: not more than 0.1%, a 1-g specimen being used.

Heavy metals, *Method II* ⟨231⟩: 0.002%.

Chromatographic purity—

Standard preparation A—Dissolve an accurately weighed quantity of USP Bumetanide RS in methanol to obtain a solution containing 25 mg per mL.

Standard preparation B—Dilute a volume of *Standard preparation A* quantitatively, and stepwise if necessary, with methanol to obtain a solution containing 0.05 mg per mL.

Standard preparation C—Dissolve an accurately weighed quantity of USP 3-Nitro-4-phenoxy-5-sulfamoylbenzoic Acid RS in methanol, and dilute quantitatively, and stepwise if necessary, with methanol to obtain a solution containing 0.05 mg per mL.

Standard preparation D—Dissolve an accurately weighed quantity of USP 3-Amino-4-phenoxy-5-sulfamoylbenzoic Acid RS in methanol, and dilute quantitatively, and stepwise if necessary, with methanol to obtain a solution containing 0.025 mg per mL.

Standard preparation E—Dissolve an accurately weighed quantity of USP Butyl 3-(butylamino)-4-phenoxy-5-sulfamoylbenzoate RS in methanol, and dilute quantitatively, and stepwise if necessary, with methanol to obtain a solution containing 0.025 mg per mL.

Test preparation—Dissolve an accurately weighed quantity of Bumetanide in methanol to obtain a solution containing 25 mg per mL.

Procedure—On a suitable thin-layer chromatographic plate (see *Chromatography* ⟨621⟩), coated with a 0.25-mm layer of chromatographic silica gel mixture, apply separately 20 µL of the *Test preparation* and 20 µL of each *Standard preparation*. Allow the spots to dry, place the plate in an unlined and unsaturated chromatographic chamber, and develop the chromatograms in a solvent system consisting of a mixture of chloroform, cyclohexane, glacial acetic acid, and methanol (80:10:10:2.5) until the solvent front has moved about three-fourths of the length of the plate. Remove the plate from the developing chamber, mark the solvent front, and allow the solvent to evaporate. Examine the plate under short-wavelength ultraviolet light: any secondary spots from the chromatogram of the *Test preparation* having R_f values corresponding to the R_f values of the principal spots obtained from the chromatograms of *Standard preparations C, D,* and *E* are not larger or more intense than the principal spots obtained from the chromatograms of *Standard preparations C, D,* and *E*, corresponding to not more than 0.2% of 3-nitro-4-phenoxy-5-sulfamoylbenzoic acid, not more than 0.1% of 3-amino-4-phenoxy-5-sulfamoylbenzoic acid, and not more than 0.1% of butyl 3-

(butylamino)-4-phenoxy-5-sulfamoylbenzoate, respectively. No other individual secondary spots in the chromatogram of the *Test preparation* are larger or more intense than the principal spot produced by *Standard preparation B*, corresponding to not more than 0.2%, and the total of any such spots observed is not more than 0.4%.

Assay—Dissolve about 1 g of Bumetanide, accurately weighed, in 150 mL of alcohol in a 250-mL conical flask. Add phenol red TS, and titrate with 0.1 *N* sodium hydroxide VS. Perform a blank determination (see *Titrimetry* ⟨541⟩), and make any necessary correction. Each mL of 0.1 *N* sodium hydroxide is equivalent to 36.44 mg of $C_{17}H_{20}N_2O_5S$.

Bumetanide Injection

» Bumetanide Injection is a sterile solution of Bumetanide in Water for Injection. It contains not less than 90.0 percent and not more than 110.0 percent of the labeled amount of $C_{17}H_{20}N_2O_5S$.

Packaging and storage—Preserve in single-dose or in multiple-dose containers, preferably of Type I glass, protected from light.

Reference standards—*USP Bumetanide Reference Standard*—Do not dry before using. *USP 3-Amino-4-phenoxy-5-sulfamoylbenzoic Acid Reference Standard*—Do not dry before using.

Identification—

A: The retention time of the major peak in the chromatogram of the *Assay preparation* corresponds to that of the *Standard preparation*, both relative to the internal standard, as obtained in the *Assay*.

B: The principal spot obtained from the chromatogram of the *Test preparation* exhibits an R_f value corresponding to that of the *Identification preparation* in the test for *Chromatographic purity*.

Pyrogen—It meets the requirements of the *Pyrogen Test* ⟨151⟩, the test dose being 1 mL per kg of Bumetanide Injection containing 0.25 mg of bumetanide in each mL.

pH ⟨791⟩: between 6.8 and 7.8.

Chromatographic purity—

Identification preparation—Dissolve USP Bumetanide RS in methanol to obtain a solution containing 10 mg per mL.

Standard preparations—Dilute a volume of the *Identification preparation* quantitatively, and stepwise if necessary, with methanol to obtain a solution containing 0.08 mg of USP Bumetanide RS per mL. Dilute quantitatively with methanol to obtain *Standard preparations* having the following compositions:

Standard preparation	Dilution	Concentration (μg RS per mL)	Percentage (%, for comparison with test specimen)
A	(undiluted)	80	0.8
B	(3 in 4)	60	0.6
C	(1 in 2)	40	0.4
D	(1 in 4)	20	0.2
E	(1 in 8)	10	0.1

Standard preparation F—Dissolve an accurately weighed quantity of USP 3-Amino-4-phenoxy-5-sulfamoylbenzoic Acid RS in methanol, and dilute quantitatively, and stepwise if necessary, with methanol to obtain a solution containing 0.02 mg per mL.

Test preparation—Pipet a volume of Bumetanide Injection, equivalent to 5 mg of bumetanide, into a 125-mL separator, and adjust with 0.1 *N* sodium hydroxide to a pH of 12. Extract with two 20-mL portions of ether, discard the ether extracts, and adjust the aqueous layer with 1 *N* acetic acid to a pH of 4. Extract with two 20-mL portions of ether, passing the extracts through anhydrous sodium sulfate. Wash the sodium sulfate with about 5 mL of ether. Evaporate the combined ether extracts with the aid of a stream of nitrogen to dryness, and dissolve the residue in 0.5 mL of methanol.

Procedure—On a suitable thin-layer chromatographic plate (see *Chromatography* ⟨621⟩), coated with a 0.25-mm layer of chromatographic silica gel mixture, apply separately 50 μL of the *Test preparation*, 50 μL of the *Identification preparation*, and 50 μL of each *Standard preparation*. Allow the spots to dry, place the plate in a chromatographic chamber, and develop the chromatograms in a solvent system consisting of a mixture of chloroform, cyclohexane, glacial acetic acid, and methanol (80:10:10:2.5) until the solvent front has moved about three-fourths of the length of plate. Remove the plate from the developing chamber, mark the solvent front, allow the solvent to evaporate, and examine the plate under short-wavelength ultraviolet light. Any secondary spot from the chromatogram of the *Test preparation* having an R_f value corresponding to the R_f value of the principal spot obtained from the chromatogram of *Standard preparation F* is not larger or more intense than the principal spot obtained from the chromatogram of *Standard preparation F* (0.2%), and the sum of the intensities of all other secondary spots obtained from the chromatogram of the *Test preparation*, compared to the intensities of the principal spots obtained from the chromatograms of *Standard preparations A* through *E*, corresponds to not more than 0.8%.

Other requirements—It meets the requirements under *Injections* ⟨1⟩.

Assay—

Mobile phase—Prepare a filtered and degassed mixture of methanol, water, tetrahydrofuran, and glacial acetic acid (50:45:5:2). Make adjustments if necessary (see *System Suitability* under *Chromatography* ⟨621⟩).

Internal standard solution—Transfer about 50 mg of 4-ethylbenzaldehyde to a 100-mL volumetric flask. Dissolve in methanol, dilute with methanol to volume, and mix. Transfer 10.0 mL of the resulting solution to a 100-mL volumetric flask, add 10.0 mL of tetrahydrofuran and 4.0 mL of glacial acetic acid, dilute with methanol to volume, and mix.

Standard preparation—Dissolve an accurately weighed quantity of USP Bumetanide RS in *Internal standard solution*, and dilute quantitatively with *Internal standard solution* to obtain a solution having a known concentration of about 250 μg per mL. Transfer 5.0 mL of the resulting solution to a 10-mL volumetric flask, dilute with water to volume, and mix to obtain a solution having a known concentration of about 125 μg of USP Bumetanide RS per mL.

Assay preparation—Transfer an accurately measured volume of Bumetanide Injection, equivalent to about 0.25 mg of bumetanide, to a flask. Add an equal volume of *Internal standard solution*, accurately measured, insert the stopper, and mix.

Chromatographic system (see *Chromatography* ⟨621⟩)—The liquid chromatograph is equipped with a 254-nm detector and a 3.9-mm × 30-cm column that contains packing L1. The flow rate is about 1 mL per minute. Chromatograph the *Standard preparation*, and record the peak responses as directed under *Procedure:* the tailing factor for the analyte peak is not more than 1.4, the resolution, *R*, between the analyte and internal standard peaks is not less than 1.5, and the relative standard deviation for replicate injections is not more than 2.0%.

Procedure—Separately inject equal volumes (about 20 μL) of the *Standard preparation* and the *Assay preparation* into the chromatograph, record the chromatograms, and measure the responses for the major peaks. The relative retention times are about 0.7 for 4-ethylbenzaldehyde and 1.0 for bumetanide. Calculate the quantity, in mg, of $C_{17}H_{20}N_2O_5S$ in each mL of the Injection taken by the formula:

$$(2C/V)(R_U/R_S),$$

in which *C* is the concentration, in mg per mL, of USP Bumetanide RS in the *Standard preparation*, *V* is the volume, in mL, of Injection taken, and R_U and R_S are the peak response ratios obtained from the *Assay preparation* and the *Standard preparation*, respectively.

Bumetanide Tablets

» Bumetanide Tablets contain not less than 90.0 percent and not more than 110.0 percent of the labeled amount of $C_{17}H_{20}N_2O_5S$.

Packaging and storage—Preserve in tight, light-resistant containers.

Reference standards—*USP Bumetanide Reference Standard*—Do not dry before using. *USP 3-Amino-4-phenoxy-5-sulfamoylbenzoic Acid Reference Standard*—Do not dry before using.

Identification—

A: The retention time of the major peak in the chromatogram of the *Assay preparation* corresponds to that of the *Standard preparation*, both relative to the internal standard, as obtained in the *Assay*.

B: The principal spot obtained from the chromatogram of the *Test preparation* exhibits an R_f value corresponding to that of the *Identification preparation* in the test for *Chromatographic purity*.

Dissolution ⟨711⟩—

Medium: water; 900 mL.

Apparatus 2: 50 rpm.

Time: 30 minutes.

pH 2.9 glycine buffer—Dissolve 7.505 g of glycine and 5.85 g of sodium chloride in water to make 1000 mL (stock solution). Dilute 80.0 mL of the stock solution and 20.0 mL of 0.1 *N* hydrochloric acid with water to 1000 mL. Adjust, if necessary, with 0.1 *N* hydrochloric acid or 0.1 *N* sodium hydroxide to a pH of 2.9.

Procedure—Determine the amount of $C_{17}H_{20}N_2O_5S$ dissolved, using a suitable fluorometer having an excitation wavelength of about 350 nm and a fluorescence emission of about 450 nm of filtered portions of the solution under test, suitably diluted with *pH 2.9 glycine buffer*, in comparison with a Standard solution having a known concentration of USP Bumetanide RS in the same medium.

Tolerances—Not less than 85% (*Q*) of the labeled amount of $C_{17}H_{20}N_2O_5S$ is dissolved in 30 minutes.

Chromatographic purity—

Identification preparation—Dissolve USP Bumetanide RS in methanol to obtain a solution containing 20 mg per mL.

Standard preparations—Dilute a volume of the *Identification preparation* quantitatively, and stepwise if necessary, with methanol to obtain a solution containing 0.16 mg of USP Bumetanide RS per mL. Dilute quantitatively with methanol to obtain *Standard preparations* having the following compositions:

Standard preparation	Dilution	Concentration (µg RS per mL)	Percentage (%, for comparison with test specimen)
A	(undiluted)	160	0.8
B	(3 in 4)	120	0.6
C	(1 in 2)	80	0.4
D	(1 in 4)	40	0.2
E	(1 in 8)	20	0.1

Standard preparation F—Dissolve an accurately weighed quantity of USP 3-Amino-4-phenoxy-5-sulfamoylbenzoic Acid RS in methanol, and dilute quantitatively, and stepwise if necessary, with methanol to obtain a solution containing 0.04 mg per mL.

Test preparation—Transfer an accurately weighed portion of finely powdered Bumetanide Tablets, equivalent to 10 mg of bumetanide, to a 50-mL centrifuge tube, add 20 mL of acetone (spectrophotometric or HPLC quality), and shake by mechanical means for 10 minutes. Centrifuge for 10 minutes, decant the supernatant liquid into a glass-stoppered, 25-mL conical flask, and evaporate with the aid of a stream of nitrogen to dryness. Dissolve the residue in 0.5 mL of methanol.

Procedure—On a suitable thin-layer chromatographic plate (see *Chromatography* ⟨621⟩), coated with a 0.25-mm layer of chromatographic silica gel mixture, apply separately 25 µL of the *Test preparation*, 25 µL of the *Identification preparation*, and 25 µL of each *Standard preparation*. Allow the spots to dry, place the plate in a chromatographic chamber, and develop the chromatograms in a solvent system consisting of a mixture of chloroform, cyclohexane, glacial acetic acid, and methanol (80:10:10:2.5) until the solvent front has moved about three-fourths of the length of the plate. Remove the plate from the developing chamber, mark the solvent front, allow the solvent to evaporate, and examine the plate under short-wavelength ultraviolet light. Any secondary spot from the chromatogram of the *Test preparation* having an R_f value corresponding to the R_f value of the principal spot obtained from the chromatogram of *Standard preparation F* is not larger or more intense than the principal spot obtained from the chromatogram of *Standard preparation F* (0.2%), and the sum of the intensities of all other secondary spots obtained from the chromatogram of the *Test preparation*, compared to the intensities of the principal spots obtained from the chromatograms of *Standard preparations A* through *E*, corresponds to not more than 0.8%.

Uniformity of dosage units ⟨905⟩: meet the requirements.

Assay—

Mobile phase, Internal standard solution, Standard preparation, and *Chromatographic system*—Prepare as directed in the *Assay* under *Bumetanide Injection.*

Assay preparation—Weigh and finely powder not less than 20 Bumetanide Tablets. Transfer an accurately weighed portion of the powder, equivalent to about 0.5 mg of bumetanide, to a 10-mL volumetric flask, add 2.0 mL of *Internal standard solution*, and sonicate for 5 minutes. Add 2.0 mL of water, and mix. Cool, and filter, discarding the first 1 mL of the filtrate.

Procedure—Separately inject equal volumes (about 20 µL) of the *Standard preparation* and the *Assay preparation* into the chromatograph, record the chromatograms, and measure the responses for the major peaks. The relative retention times are about 0.7 for 4-ethylbenzaldehyde and 1.0 for bumetanide. Calculate the quantity, in mg, of $C_{17}H_{20}N_2O_5S$ in the portion of Tablets taken by the formula:

$$4C(R_U / R_S),$$

in which *C* is the concentration, in mg per mL, of USP Bumetanide RS in the *Standard preparation*, and R_U and R_S are the peak response ratios obtained from the *Assay preparation* and the *Standard preparation*, respectively.

Bupivacaine in Dextrose Injection

» Bupivacaine in Dextrose Injection is a sterile solution of Bupivacaine Hydrochloride and Dextrose in Water for Injection. It contains not less than 93.0 percent and not more than 107.0 percent of the labeled amounts of bupivacaine hydrochloride ($C_{18}H_{28}N_2O.HCl$) and dextrose ($C_6H_{12}O_6$). It contains no preservative.

Packaging and storage—Preserve in single-dose containers, preferably of Type I glass.

Reference standards—*USP Bupivacaine Hydrochloride Reference Standard*—Do not dry; determine the water content at the time of use. *USP Dextrose Reference Standard*—Dry at 105° for 16 hours before using.

Identification—

A (see *Thin-layer Chromatographic Identification Test* ⟨201⟩):

Adsorbant: chromatographic silica gel mixture; 0.25 mm.

Developing solvent: mixture of butyl alcohol, water, dehydrated alcohol, and glacial acetic acid (6:2:1:1).

Test preparation: Bupivacaine in Dextrose Injection.

Standard preparations A, B, and *C*—Separately prepare (*A*) a solution of USP Bupivacaine Hydrochloride in water, (*B*) a solution of USP Dextrose RS in water, and (*C*) a solution of USP Bupivacaine Hydrochloride RS in (*B*) to obtain solutions having concentrations corresponding to the labeled concentrations of bupivacaine hydrochloride and dextrose in the Injection.

Naphthalenediol reagent—Dissolve 20 mg of 1,3-naphthalenediol in 10 mL of dehydrated alcohol containing 0.2 mL of sulfuric acid.

Iodoplatinate reagent—Mix equal volumes of platinic chloride solution (3 in 1000) and potassium iodide solution (6 in 100).

Procedure—Separately apply 10 µL each of the *Test preparation* and *Standard preparations A* and *C* on a portion of the chromatographic plate, and separately apply 1 µL each of the *Test preparation* and *Standard preparation B* on the remaining portion of the plate. Dry the applications in a current of warm air, develop the chromatograms, remove the plate from the developing chamber, and mark the solvent front. Dry the plate in warm circulating air, and examine the plate under short-wavelength ultraviolet light: the R_f value of the principal spot obtained from the *Test preparation* corresponds to the spots obtained from the adjacent chromatograms of *Standard preparations A* and *C*. Spray the plate with *Naphthalenediol reagent*, heat at 90° for 5 minutes, and examine the plate: the R_f value of the principal blue-purple spot obtained from the *Test preparation* corresponds to that obtained in the adjacent chromatogram of *Standard preparation B*. Cool the plate, spray it with *Iodoplatinate reagent*, and examine the plate: bupivacaine appears as a blue-purple spot on a salmon-colored background, and the dextrose spots fade slightly: the R_f value of the bupivacaine spot obtained from the *Test preparation* corresponds to those obtained from the adjacent chromatograms of *Standard preparations A* and *C*.

B: It responds to *Identification test B* under *Bupivacaine Hydrochloride Injection.*

Other requirements—It meets the requirements under *Injections* ⟨1⟩.

Assay for bupivacaine hydrochloride—

pH 6.8 phosphate buffer, Mobile phase, Internal standard solution, Standard preparation, Chromatographic system, and *Procedure*—Proceed as directed in the *Assay* under *Bupivacaine Hydrochloride Injection.*

Assay preparation—Transfer an accurately measured volume of Bupivacaine in Dextrose Injection, equivalent to about 50 mg of bupivacaine hydrochloride, to a 100-mL volumetric flask, add 10.0 mL of *Internal standard solution*, dilute with methanol to volume, and mix.

Assay for dextrose—Determine the angular rotation of Bupivacaine in Dextrose Injection in a suitable polarimeter tube (see *Optical Rotation* ⟨781⟩). The observed rotation, in degrees, multiplied by 9.452*A*, in which *A* is the ratio of 200 divided by the length, in mm, of the polarimeter tube employed, represents the weight, in mg, of dextrose ($C_6H_{12}O_6$) in each mL of the Injection.

Bupivacaine and Epinephrine Injection

» Bupivacaine and Epinephrine Injection is a sterile solution of Bupivacaine Hydrochloride and Epinephrine or Epinephrine Bitartrate in Water for Injection. It contains not less than 93.0 percent and not more than 107.0 percent of the labeled amount of bupivacaine hydrochloride ($C_{18}H_{28}N_2O.HCl$). The content of epinephrine ($C_9H_{13}NO_3$) does not exceed 0.001 percent (1 in 100,000). It contains the equivalent of not less than 90.0 percent and not more than 115.0 percent of the labeled amount of epinephrine ($C_9H_{13}NO_3$).

Packaging and storage—Preserve in single-dose or in multiple-dose containers, preferably of Type I glass, protected from light. Injection labeled to contain 0.5% or less of bupivacaine hydrochloride may be packaged in 50-mL multiple-dose containers.

Labeling—The label indicates that the Injection is not to be used if it is discolored or contains a precipitate.

Reference standards—*USP Bupivacaine Hydrochloride Reference Standard*—Do not dry; determine the water content at the time of use. *USP Epinephrine Bitartrate Reference Standard*—Dry in vacuum over silica gel for 3 hours before using.

Identification—

A: It responds to the *Identification tests* under *Bupivacaine Hydrochloride Injection.*

B: Pipet a volume of Injection, equivalent to about 50 µg of epinephrine, into a suitable container, add 0.1 mL of *Ferro-citrate solution* and 2.0 mL of *Buffer solution* (prepared as directed under *Epinephrine Assay* ⟨391⟩), mix, and allow the solution to stand for 10 minutes. Filter the solution: the filtrate is violet in color and may turn brownish.

pH ⟨791⟩: between 3.3 and 5.5.

Other requirements—It meets the requirements under *Injections* ⟨1⟩.

Assay for bupivacaine hydrochloride—

Mobile phase, Internal standard solution, Standard preparation, Chromatographic system, and *Procedure*—Proceed as directed in the *Assay* under *Bupivacaine Hydrochloride Injection.*

Assay preparation—Transfer an accurately measured volume of Bupivacaine and Epinephrine Injection equivalent to about 50 mg of bupivacaine hydrochloride, to a 100-mL volumetric flask, add 10.0 mL of *Internal standard solution*, dilute with methanol to volume, and mix.

Assay for epinephrine—

Mobile phase—Prepare a suitably filtered and degassed mixture of water, methanol, and 2 *M* monobasic sodium phosphate (900:50:50), containing in each 1000 mL, 40 mg of disodium ethylenediaminetetraacetate, 0.4 mL of phosphoric acid, and 0.4 g of sodium 1-octanesulfonate. Make adjustments, if necessary, to obtain a retention time of not less than 11 minutes for the epinephrine peak (see *System Suitability* under *Chromatography* ⟨621⟩).

Standard preparation—Dissolve an accurately weighed quantity of USP Epinephrine Bitartrate RS in *Mobile phase* to obtain a solution having a concentration of about 2 µg per mL.

Resolution solution—Dissolve suitable quantities of epinephrine bitartrate and dopamine hydrochloride in *Mobile phase* to obtain a solution containing about 2 µg of each per mL.

Assay preparation—Transfer an accurately measured volume of Bupivacaine and Epinephrine Injection, equivalent to about 25 µg of epinephrine, to a 25-mL volumetric flask, dilute with *Mobile phase* to volume, and mix.

Chromatographic system (see *Chromatography* ⟨621⟩)—The liquid chromatograph is equipped with an electrochemical detector held at a potential of +0.75 V and a 4.6-mm × 25-cm column that contains packing L1. The flow rate is about 1.2 mL per minute. Chromatograph the *Resolution solution*, and record the peak responses as directed under *Procedure:* the resolution, *R*, between the epinephrine and dopamine peaks is not less than 6.0 (the relative retention times are about 2 for dopamine and 1.0 for epinephrine). Chromatograph the *Standard preparation*, and record the peak responses as directed under *Procedure:* the relative standard deviation for replicate injections is not more than 2.0%.

Procedure—Separately inject equal volumes (about 20 µL) of the *Standard preparation* and the *Assay preparation* into the chromatograph, record the chromatograms, and measure the responses for the major peaks. Calculate the quantity, in µg, of epinephrine ($C_9H_{13}NO_3$) in each mL of the Injection taken by the formula:

$$(183.21/333.29)(25)(C/V)(r_U/r_S),$$

in which 183.21 and 333.29 are the molecular weights of epinephrine and epinephrine bitartrate, respectively, *C* is the concentration, in µg per mL, of USP Epinephrine Bitartrate RS in the *Standard preparation*, *V* is the volume, in mL, of Injection taken, and r_U and r_S are the peak responses obtained from the *Assay preparation* and the *Standard preparation*, respectively.

Bupivacaine Hydrochloride

$CH_3CH_2CH_2CH_2$ — N (piperidine) — CONH — (2,6-dimethylphenyl, CH_3, CH_3) · HCl · H_2O

$C_{18}H_{28}N_2O.HCl.H_2O$ 342.91

2-Piperidinecarboxamide, 1-butyl-*N*-(2,6-dimethylphenyl)-, monohydrochloride, monohydrate.

1-Butyl-2′,6′-pipecoloxylidide monohydrochloride, monohydrate [*14252-80-3*].

Anhydrous 324.89 [*18010-40-7*].

» Bupivacaine Hydrochloride contains not less than 98.5 percent and not more than 101.5 percent of $C_{18}H_{28}N_2O.HCl$, calculated on the anhydrous basis.

Packaging and storage—Preserve in well-closed containers.

Reference standard—*USP Bupivacaine Hydrochloride Reference Standard*—Do not dry; determine the water content at the time of use.

Identification—

A: Dissolve about 230 mg in 15 mL of water in a separator, add 1 mL of 6 *N* ammonium hydroxide, and extract with three 30-mL portions of chloroform. Evaporate the chloroform at room temperature with the aid of a stream of nitrogen, and dry the residue in vacuum. Add 2 mL of chloroform to the residue, and dissolve: the infrared absorption spectrum of this solution exhibits maxima only at the same wavelengths as that of a similar preparation of USP Bupivacaine Hydrochloride RS.

B: The ultraviolet absorption spectrum of a 1 in 2000 solution in 0.01 *N* hydrochloric acid exhibits maxima and minima at the same wavelengths as that of a similar preparation of USP Bupivacaine Hydrochloride RS, concomitantly measured, and the respective absorptivities, calculated on the anhydrous basis, at the wavelength of maximum absorbance at about 271 nm do not differ by more than 3.0%.

C: Dissolve about 50 mg in 10 mL of water in a small separator, render alkaline with 6 *N* ammonium hydroxide, and extract with 10 mL of ether: the aqueous layer responds to the tests for *Chloride* ⟨191⟩.

pH ⟨791⟩: between 4.5 and 6.0, in a solution (1 in 100).

Water, *Method I* ⟨921⟩: between 4.0% and 6.0%.

Residue on ignition ⟨281⟩: not more than 0.1%.

Heavy metals, *Method II* ⟨231⟩: not more than 0.001%.

Residual solvents—

Alcohol standard solution—Pipet 2 mL of dehydrated alcohol into a 100-mL volumetric flask, dilute with water to volume, and mix. Transfer 2.0 mL of this solution to a 50-mL volumetric flask, dilute with water to volume, and mix. The resulting solution contains 0.08% of alcohol.

Isopropyl alcohol standard solution—Pipet 2 mL of isopropyl alcohol into a 1000-mL volumetric flask, dilute with water to volume, and mix. Transfer 2.0 mL of this solution to a 100-mL volumetric flask, dilute with water to volume, and mix. The resulting solution contains 0.004% of isopropyl alcohol.

Test preparation—Transfer 1.0 g of Bupivacaine Hydrochloride, accurately weighed, to a 25-mL volumetric flask, dilute with water to volume, and mix.

Chromatographic system—Under typical conditions, the instrument is equipped with a flame-ionization detector and contains a 2-m × 6-mm column containing packing S3. The injection port is maintained at a temperature of about 200°, the column at about 175°, the detector at about 280°, and nitrogen is used as the carrier gas at a flow rate of about 40 mL per minute.

Procedure—Inject equal volumes (about 5 µL) of the *Test preparation*, the *Alcohol standard solution*, and the *Isopropyl alcohol standard solution* successively into the gas chromatograph. Measure the responses of the alcohol peak and the isopropyl alcohol peak in each chromatogram. Determine the percentage of alcohol by the formula:

$$2(r_U/r_S),$$

and determine the percentage of isopropyl alcohol by the formula:

$$0.1(r_U/r_S),$$

in which r_U and r_S are the responses of the respective analytes in the *Test preparation* and of the corresponding analytes in the *Alcohol standard solution* and the *Isopropyl alcohol standard solution*, respectively. The sum of the content of alcohol and the content of isopropyl alcohol does not exceed 2%.

Chromatographic purity—Dissolve a suitable quantity of Bupivacaine Hydrochloride in a mixture of chloroform and isopropylamine (99:1) to obtain a *Test solution* containing 20.0 mg per mL. Dissolve a suitable quantity of USP Bupivacaine Hydrochloride RS, accurately weighed, in the same solvent to obtain a *Standard solution* containing 20.0 mg per mL. Dilute a portion of this solution quantitatively with the same solvent to obtain a *Diluted standard solution* having a concentration of 100 µg per mL. Apply separate 10-µL portions of the three solutions on the starting line of a suitable thin-layer chromatographic plate (see *Chromatography* ⟨621⟩), coated with a 0.25-mm layer of chromatographic silica gel mixture. Develop the chromatogram in a suitable chamber with a solvent system consisting of a mixture of hexanes and isopropylamine (97:3) until the solvent has moved about three-fourths of the length of the plate. Remove the plate from the chamber, mark the solvent front, and dry it in warm air. Place the plate in a closed chamber with a dish containing 1 g of iodine in a shallow layer, and allow to remain for about 5 minutes. Remove the plate from the chamber, spray it with 7 *N* sulfuric acid, and examine the chromatogram: the R_f value of the principal spot from the *Test solution* corresponds to that of the *Standard solution*, and the estimated size and intensity of any other spot obtained from the *Test solution* does not exceed that of the principal spot obtained from the *Diluted standard solution* (0.5%), and the total of the estimated sizes and intensities of all of the other spots obtained from the *Test solution* does not exceed four times that of the principal spot obtained from the *Diluted standard solution* (2.0%).

Assay—Transfer about 600 mg of Bupivacaine Hydrochloride, accurately weighed, to a 250-mL conical flask, and dissolve in 20 mL of glacial acetic acid. Add 10 mL of mercuric acetate TS and 3 drops of crystal violet TS, and titrate with 0.1 *N* perchloric acid VS to a green end-point. Perform a blank determination, and make any necessary correction. Each mL of 0.1 *N* perchloric acid is equivalent to 32.49 mg of $C_{18}H_{28}N_2O.HCl$.

Bupivacaine Hydrochloride Injection

» Bupivacaine Hydrochloride Injection is a sterile solution of Bupivacaine Hydrochloride in Water for Injection. It contains not less than 93.0 percent and not more than 107.0 percent of the labeled amount of $C_{18}H_{28}N_2O.HCl$.

Packaging and storage—Preserve in single-dose or in multiple-dose containers, preferably of Type I glass. Injection labeled to contain 0.5% or less of bupivacaine hydrochloride may be packaged in 50-mL multiple-dose containers.

Reference standard—*USP Bupivacaine Hydrochloride Reference Standard*—Do not dry; determine the water content at the time of use.

Identification—

A: Dilute a volume of Injection, equivalent to about 50 mg of bupivacaine hydrochloride, with 0.01 *N* hydrochloric acid to 25 mL, and proceed as directed under *Identification—Organic Nitrogenous Bases* ⟨181⟩, beginning with "Transfer the liquid to a separator." The Injection meets the requirements of the test.

B: Place a volume of Injection, equivalent to about 50 mg of bupivacaine hydrochloride in a separator, render alkaline with 6 *N* ammonium hydroxide, and extract with 10 mL of ether: the aqueous layer responds to the tests for *Chloride* ⟨191⟩.

pH ⟨791⟩: between 4.0 and 6.5.

Other requirements—It meets the requirements under *Injections* ⟨1⟩.

Assay—

pH 6.8 phosphate buffer—Dissolve 1.94 g of monobasic potassium phosphate and 2.48 g of dibasic potassium phosphate in 1000 mL of water. Adjust, if necessary, with 1 *N* potassium hydroxide or 1 *M* phosphoric acid to a pH of 6.8.

Mobile phase—Prepare a fresh solution of acetonitrile and *pH 6.8 phosphate buffer* (65:35). Adjust, if necessary, with 1 *M* phosphoric acid to a pH of 7.7 ± 0.2. Filter the solution through a membrane filter of 1-µm or finer porosity, and degas.

Internal standard solution—Prepare a solution of dibutyl phthalate in methanol containing about 1.3 mg per mL.

Standard preparation—Dissolve about 50 mg of USP Bupivacaine Hydrochloride RS, accurately weighed, in 10.0 mL of water, using sonication if necessary, in a 100-mL volumetric flask.

Add 10 mL of *Internal standard solution*, dilute with methanol to volume, and mix.

Assay preparation—Transfer an accurately measured volume of Bupivacaine Hydrochloride Injection, equivalent to about 50 mg of bupivacaine hydrochloride, to a 100-mL volumetric flask, add 10.0 mL of *Internal standard solution*, dilute with methanol to volume, and mix.

Chromatographic system (see *Chromatography* ⟨621⟩)—The liquid chromatograph is equipped with a 263-nm detector and a 4-mm × 30-cm column that contains packing L1. The flow rate is about 2 mL per minute. Chromatograph three replicate injections of the *Standard preparation*, and record the peak responses as directed under *Procedure:* the relative standard deviation of the ratios of the bupivacaine hydrochloride peak to the dibutyl phthalate peak is not more than 1.0%, and the resolution factor between bupivacaine hydrochloride and dibutyl phthalate is not less than 2.0.

Procedure—Separately inject equal volumes (about 20 µL) of the *Standard preparation* and the *Assay preparation* into the chromatograph, record the chromatograms, and measure the responses for the major peaks. The relative retention times are about 1.2 for dibutyl phthalate and 1.0 for bupivacaine hydrochloride. Calculate the quantity, in mg, of $C_{18}H_{28}N_2O.HCl$ in the volume of Injection taken by the formula:

$$W(R_U/R_S),$$

in which W is the weight, in mg, of USP Bupivacaine Hydrochloride RS, calculated on the anhydrous basis, in the *Standard preparation*, and R_U and R_S are the ratios of the peak responses of bupivacaine hydrochloride to those of the internal standard obtained from the *Assay preparation* and the *Standard preparation*, respectively.

Busulfan

$CH_3SO_2O(CH_2)_4OSO_2CH_3$

$C_6H_{14}O_6S_2$ 246.29

1,4-Butanediol, dimethanesulfonate.

1,4-Butanediol dimethanesulfonate [*55-98-1*].

» Busulfan contains not less than 98.0 percent and not more than 100.5 percent of $C_6H_{14}O_6S_2$, calculated on the dried basis.

Packaging and storage—Preserve in tight containers.

Labeling—The label bears a warning that great care should be taken to prevent inhaling particles of Busulfan and exposing the skin to it.

Identification—

A: Fuse about 100 mg with about 100 mg of potassium nitrate and a pellet of potassium hydroxide weighing approximately 250 mg. Cool, dissolve the residue in water, acidify with 3 *N* hydrochloric acid, and add a few drops of barium chloride TS: a white precipitate is formed.

B: To 100 mg add 10 mL of water and 5 mL of 1 *N* sodium hydroxide. Heat until a clear solution is obtained: an odor characteristic of methanesulfonic acid is perceptible.

C: Cool the solution obtained in *Identification test B*, and divide it into two equal portions. To one portion add 1 drop of potassium permanganate TS: the purple color changes to violet, then to blue, and finally to emerald-green. Acidify the second portion of the solution with 2 *N* sulfuric acid, and add 1 drop of potassium permanganate TS: the color of the permanganate is not discharged.

Melting range ⟨741⟩: between 115° and 118°.

Loss on drying ⟨731⟩—Dry it in vacuum at 60° to constant weight: it loses not more than 2.0% of its weight.

Residue on ignition ⟨281⟩: not more than 0.1%.

Assay—Transfer about 80 mg of Busulfan, accurately weighed, to a 250-mL conical flask. Add about 30 mL of water, swirl, add phenolphthalein TS, and neutralize with 0.05 *N* sodium hydroxide. Connect the flask to a reflux air condenser, and boil the mixture gently for not less than 30 minutes, adding water occasionally to maintain the volume. Cool to room temperature, add phenolphthalein TS, and titrate with 0.05 *N* sodium hydroxide VS. Each mL of 0.05 *N* sodium hydroxide is equivalent to 6.158 mg of $C_6H_{14}O_6S_2$.

Busulfan Tablets

» Busulfan Tablets contain not less than 93.0 percent and not more than 107.0 percent of the labeled amount of $C_6H_{14}O_6S_2$.

Packaging and storage—Preserve in well-closed containers.

Identification—Pulverize a suitable number of Tablets, and extract the powder with several portions of acetone. Evaporate the combined acetone extracts, with the aid of a current of air, on a steam bath: the dry residue responds to the *Identification tests* under *Busulfan*, and melts at about 115°.

Disintegration ⟨701⟩: 30 minutes, the use of disks being omitted.

Uniformity of dosage units ⟨905⟩: meet the requirements.

Assay—Weigh and finely powder not less than 40 Busulfan Tablets. [*Caution: Guard against accidental inhalation of fine powder.*] Weigh accurately a portion of the powder, equivalent to about 80 mg of busulfan, and transfer to a 100-mL beaker. Extract with four 20-mL portions of acetone, each time stirring the mixture well, then allowing the insoluble matter to settle, and finally decanting the supernatant liquid through a sintered-glass filter into a 250-mL conical flask. Evaporate the combined acetone extracts to about 10 mL, add phenolphthalein TS, and neutralize with 0.05 *N* sodium hydroxide. Evaporate to dryness, add about 30 mL of water, and proceed as directed in the *Assay* under *Busulfan*, beginning with "Connect the flask." Each mL of 0.05 *N* sodium hydroxide is equivalent to 6.158 mg of $C_6H_{14}O_6S_2$.

Butabarbital

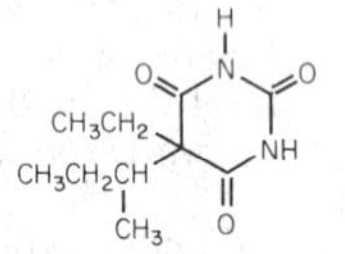

$C_{10}H_{16}N_2O_3$ 212.25

2,4,6(1*H*,3*H*,5*H*)-Pyrimidinetrione, 5-ethyl-5-(1-methylpropyl)-.

5-*sec*-Butyl-5-ethylbarbituric acid [*125-40-6*].

» Butabarbital contains not less than 98.5 percent and not more than 101.0 percent of $C_{10}H_{16}N_2O_3$, calculated on the dried basis.

Packaging and storage—Preserve in tight containers.

Reference standard—*USP Butabarbital Reference Standard*—Keep tightly closed. Dry at 105° for 4 hours before using.

Identification—The infrared absorption spectrum of a potassium bromide dispersion of it, previously dried, exhibits maxima only at the same wavelengths as that of a similar preparation of USP Butabarbital RS.

Melting range, *Class Ia* ⟨741⟩: between 164° and 167°.

Loss on drying ⟨731⟩—Dry it at 105° for 2 hours: it loses not more than 1.0% of its weight.

Residue on ignition ⟨281⟩: not more than 0.1%.

Chromatographic purity—

Standard solutions—Dissolve a quantity of USP Butabarbital RS in a mixture of chloroform and methanol (1:1) to obtain a solution having a concentration of 4.0 mg per mL (*Standard solution A*). Dilute 1.0 mL of *Standard solution A* with a mixture of chloroform and methanol (1:1) to 10.0 mL, and mix (*Standard solution B*).

Test solution—Dissolve a quantity of Butabarbital in a mixture of chloroform and methanol (1:1) to obtain a solution having a concentration of 40 mg per mL.

Procedure—Proceed as directed for *Procedure* in the test for *Chromatographic purity* under *Butabarbital Sodium.*

Assay—

Internal standard solution—Transfer about 400 mg of tetracosane to a 200-mL volumetric flask, add chloroform to volume, and mix.

Standard preparation—Transfer about 200 mg of USP Butabarbital RS, accurately weighed, to a 100-mL volumetric flask, add chloroform to volume, and mix. Transfer 10.0 mL of the resulting solution to a 50-mL volumetric flask, add 10.0 mL of *Internal standard solution*, and mix to obtain a solution having a known concentration of about 1.0 mg of USP Butabarbital RS per mL.

Assay preparation—Transfer about 200 mg of Butabarbital, accurately weighed, to a 100-mL volumetric flask, add chloroform to volume, and mix. Transfer 10.0 mL of the resulting solution to a 50-mL volumetric flask, add 10.0 mL of *Internal standard solution*, and mix.

Chromatographic system (see *Chromatography* ⟨621⟩)—The gas chromatograph is equipped with a flame-ionization detector and a 4-mm × 1.8-m column packed with 10 percent phase G37 on support S1AB. The column is maintained at about 260°, the injection port at about 260°, and the detector block at about 300°. Dry nitrogen is used as the carrier gas at a flow rate of about 50 mL per minute. Chromatograph the *Standard preparation*, and record the peak responses as directed under *Procedure:* the tailing factors for the analyte and internal standard peaks are not more than 1.3 and 1.2, respectively, the resolution, *R*, between the analyte and internal standard peaks is not less than 3.0, and the relative standard deviation for replicate injections is not more than 1.0%.

Procedure—Separately inject equal volumes (about 2 µL) of the *Standard preparation* and the *Assay preparation* into the chromatograph, record the chromatograms, and measure the responses for the major peaks. The relative retention times are about 0.6 for butabarbital and 1.0 for tetracosane. Calculate the quantity, in mg, of $C_{10}H_{16}N_2O_3$ in the portion of butabarbital taken by the formula:

$$200C(R_U/R_S),$$

in which *C* is the concentration, in mg per mL, of USP Butabarbital RS in the *Standard preparation*, and R_U and R_S are the peak response ratios obtained from the *Assay preparation* and the *Standard preparation*, respectively.

Butabarbital Sodium

$C_{10}H_{15}N_2NaO_3$ 234.23

2,4,6(1*H*,3*H*,5*H*)-Pyrimidinetrione, 5-ethyl-5-(1-methylpropyl)-, monosodium salt.

Sodium 5-*sec*-butyl-5-ethylbarbiturate [*143-81-7*].

» Butabarbital Sodium contains not less than 98.2 percent and not more than 100.5 percent of $C_{10}H_{15}N_2NaO_3$, calculated on the dried basis.

Packaging and storage—Preserve in tight containers.

Reference standard—*USP Butabarbital Reference Standard*—Keep container tightly closed. Dry at 105° for 4 hours before using.

Completeness of solution—Dissolve 1.0 g in 10 mL of carbon dioxide-free water: after 1 minute, the solution is clear and free from undissolved solid.

Identification—

A: Transfer about 150 mg to a suitable separator, dissolve in 10 mL of water, and add 15 mL of 3 *N* hydrochloric acid. Extract with three 20-mL portions of chloroform, filter the extracts through anhydrous sodium sulfate, and collect the extracts in a suitable beaker. Evaporate the combined chloroform extracts on a steam bath with the aid of a current of air to dryness, and dry the residue at 105° for 2 hours: the infrared absorption spectrum of a potassium bromide dispersion of the residue so obtained exhibits maxima only at the same wavelengths as that of a similar preparation of USP Butabarbital RS.

B: The ultraviolet absorption spectrum of a 1 in 100,000 solution in pH 9.6 alkaline borate buffer (see under *Buffer Solutions* in the section, *Reagents, Indicators, and Solutions*) exhibits maxima and minima at the same wavelengths as that of a similar solution of USP Butabarbital RS, concomitantly measured, and the respective molar absorptivities, calculated on the dried basis, at the wavelength of maximum absorbance at about 240 nm do not differ by more than 3.0%. [NOTE—The molecular weight of butabarbital ($C_{10}H_{16}N_2O_3$) is 212.25.]

C: Ignite about 100 mg: the residue responds to the tests for *Sodium* ⟨191⟩.

pH ⟨791⟩: between 10.0 and 11.2, in the solution prepared for the test for *Completeness of solution.*

Loss on drying ⟨731⟩—Dry it at 150° to constant weight: it loses not more than 5.0% of its weight.

Heavy metals, *Method II* ⟨231⟩: 0.003%.

Chromatographic purity—

Standard solutions—Dissolve a quantity of USP Butabarbital RS in a mixture of chloroform and methanol (1:1) to obtain a solution having a final concentration of 4.0 mg per mL (*Standard solution A*). Dilute 1.0 mL of *Standard solution A* with a mixture of chloroform and methanol (1:1) to 10.0 mL, and mix (*Standard solution B*).

Test solution—Dissolve a quantity of Butabarbital Sodium in a mixture of chloroform and methanol (1:1) to obtain a solution having a final concentration of 44 mg per mL.

Procedure—On a suitable thin-layer chromatographic plate (see *Chromatography* ⟨621⟩), coated with a 0.25-mm layer of chromatographic silica gel mixture, apply 10 µL of the *Test solution* and 10 µL each of *Standard solution A* and *Standard solution B.* Develop the chromatogram in a solvent system consisting of acetone, methylene chloride, methanol, and ammonium hydroxide (5:3:1:1) until the solvent front has moved about three-fourths of the length of the plate. Remove the plate from the developing chamber, and dry the plate in a current of air. Spray the plate with a 1 in 100 solution of mercurous nitrate in 0.15 *N* nitric acid and immediately estimate the intensities of any spots in the chromatogram of the *Test solution*, other than the principal spot, in comparison with *Standard solution B:* the R_f value of the principal spot obtained from the *Test solution* corresponds to that obtained from *Standard solution A*, and the sum of the intensities of any secondary spots observed in the chromatogram of the *Test solution* is not greater than the intensity of the principal spot produced by *Standard solution B*, corresponding to not more than a total of 1% of impurities.

Assay—

Standard preparation—Transfer about 25 mg of USP Butabarbital RS, accurately weighed, to a 200-mL volumetric flask, dissolve in pH 9.6 alkaline borate buffer (see under *Solutions* in the section, *Reagents, Indicators, and Solutions*), and dilute with the same solvent to volume.

Assay preparation—Transfer about 28 mg of Butabarbital Sodium, accurately weighed, to a 200-mL volumetric flask, dissolve in pH 9.6 alkaline borate buffer to volume, and dilute with the same solvent to volume.

Procedure—Transfer 10.0 mL each of the *Standard preparation* and the *Assay preparation* to separate 100-mL volumetric flasks, dilute each with pH 9.6 alkaline borate buffer to volume, and mix. Concomitantly determine the absorbances of the solutions at the wavelength of maximum absorbance at about 240 nm, with a suitable spectrophotometer, using pH 9.6 alkaline borate buffer as the blank. Calculate the quantity, in mg, of $C_{10}H_{15}N_2NaO_3$ in the portion of Butabarbital Sodium taken by the formula:

$$(234.23/212.25)(0.2C)(A_U/A_S),$$

in which 234.23 and 212.25 are the molecular weights of butabarbital sodium and butabarbital, respectively, *C* is the concentration, in µg per mL, of USP Butabarbital RS in the *Standard preparation*, and A_U and A_S are the absorbances of the solutions from the *Assay preparation* and the *Standard preparation*, respectively.

Butabarbital Sodium Capsules

» Butabarbital Sodium Capsules contain not less than 90.0 percent and not more than 110.0 percent of the labeled amount of $C_{10}H_{15}N_2NaO_3$.

Packaging and storage—Preserve in well-closed containers.

Reference standard—*USP Butabarbital Reference Standard*—Keep container tightly closed. Dry at 105° for 4 hours before using.

Identification—Mix a quantity of the Capsule contents, equivalent to about 150 mg of butabarbital sodium, with 1 mL of dimethyl sulfoxide and 1 mL of water, add hydrochloric acid dropwise until the preparation is just acid to litmus, and mix. Add 3 g of chromatographic siliceous earth, and mix. Proceed as directed for *Column Partition Chromatography* under *Chromatography* ⟨621⟩, packing the chromatographic tube as follows: The lower layer consists of 4 g of chromatographic siliceous earth mixed with 3 mL of sodium carbonate solution (1 in 10), and the upper layer is the test preparation. Wash the column with 75 mL of a water-saturated mixture of isooctane and ether (4:1), and discard the washing. Elute the butabarbital with 200 mL of water-saturated ether, collecting the eluate in a suitable vessel. Evaporate the eluate on a steam bath under a current of air to dryness, and dry the residue at 105° for 2 hours: the infrared absorption spectrum of a potassium bromide dispersion of the residue so obtained exhibits maxima only at the same wavelengths as that of a similar preparation of USP Butabarbital RS.

Dissolution ⟨711⟩—

Medium: water; 900 mL.

Apparatus 1: 100 rpm.

Time: 45 minutes.

Procedure—Determine the amount of $C_{10}H_{15}N_2NaO_3$ dissolved from ultraviolet absorbances at the wavelength of maximum absorbance at about 239 nm of filtered portions of the solution under test, mixed with sufficient ammonium hydroxide to provide a concentration of 0.5 *N* ammonium hydroxide, and suitably diluted with *Dissolution Medium*, if necessary, in comparison with a Standard solution having a known concentration of USP Butabarbital RS in the same medium.

Tolerances—Not less than 75% (*Q*) of the labeled amount of $C_{10}H_{15}N_2NaO_3$ is dissolved in 45 minutes.

Uniformity of dosage units ⟨905⟩: meet the requirements.

Procedure for content uniformity—

Acid-methanol mixture—Mix 1 volume of 1 *N* hydrochloric acid and 9 volumes of methanol.

Standard preparation—Transfer about 45 mg of USP Butabarbital RS, accurately weighed, to a 100-mL volumetric flask, dissolve in *Acid-methanol mixture*, dilute with the same solvent to volume, and mix.

Test preparation—Transfer the contents of 1 Capsule to a 25-mL volumetric flask, add *Acid-methanol mixture* to volume, and mix. Filter, discarding the first 5 mL of filtrate, and dilute the subsequent filtrate quantitatively and stepwise, if necessary, with *Acid-methanol mixture* to obtain a solution containing 500 to 600 µg of butabarbital sodium per mL.

Procedure—Transfer 2.0 mL each of the *Standard preparation* and the *Test preparation* to separate 100-mL volumetric flasks, and transfer 2.0 mL of *Acid-methanol mixture* to a third 100-mL volumetric flask to provide a blank. Dilute each flask with pH 9.6 alkaline borate buffer (see under *Solutions*, in the section, *Reagents, Indicators, and Solutions*), and mix. Concomitantly determine the absorbances of the solutions in 1-cm cells at the wavelength of maximum absorbance at about 240 nm, with a suitable spectrophotometer, using the blank to set the instrument. Calculate the quantity, in mg, of $C_{10}H_{15}N_2NaO_3$ in the Capsule by the formula:

$$(234.23/212.25)(TC/D)(A_U/A_S),$$

in which 234.23 and 212.25 are the molecular weights of butabarbital sodium and butabarbital, respectively, *T* is the labeled quantity, in mg, of butabarbital sodium in the Capsule, *C* is the concentration, in µg per mL, of USP Butabarbital RS in the *Standard preparation, D* is the concentration, in µg per mL, of butabarbital sodium in the *Test preparation*, based upon the labeled quantity per Capsule and the extent of dilution, and A_U and A_S are the absorbances of the solutions from the *Test preparation* and the *Standard preparation*, respectively.

Assay—

Internal standard: Secobarbital.

Internal standard solution—Dissolve an accurately weighed quantity of Secobarbital in chloroform, and dilute quantitatively with chloroform to obtain a solution having a known concentration of about 1.2 mg per mL.

Standard preparation—Dissolve accurately weighed quantities of USP Butabarbital RS and Secobarbital in chloroform, and dilute quantitatively with chloroform to obtain a solution that contains, in each mL, known amounts of about 0.8 mg of USP Butabarbital RS and about 1.2 mg of Secobarbital.

Assay preparation—Mix the contents of not less than 20 Butabarbital Sodium Capsules. Transfer an accurately weighed portion of the powder, equivalent to about 50 mg of butabarbital sodium, to a 50-mL volumetric flask, add 35 mL of dilute ammonium hydroxide (1 in 25), dilute with water to volume, and mix. Filter, if necessary, discarding the first 15 mL of the filtrate, and transfer 25.0 mL of the clear solution to a separator. Add 2 mL of hydrochloric acid, and extract with three 25-mL portions of chloroform. Filter the extracts through about 15 g of anhydrous sodium sulfate that is supported on a funnel by a small pledget of glass wool. Collect the combined filtrate in a 100-mL volumetric flask, wash the sodium sulfate with 15 mL of chloroform, collecting the washing with the filtrate, dilute with chloroform to volume, and mix. Combine 4.0 mL of this solution with 1.0 mL of *Internal standard solution* in a suitable container, and reduce the volume to about 1 mL by evaporation, with the aid of a stream of dry nitrogen, at room temperature.

Chromatographic system and *System suitability*—Proceed as directed for *Chromatographic System* and *System Suitability* under *Barbiturate Assay* ⟨361⟩, the resolution, *R*, between butabarbital and secobarbital being not less than 3.0. [NOTE—Relative retention times are, approximately, 0.6 for butabarbital and 1.0 for secobarbital.]

Procedure—Proceed as directed for *Procedure* under *Barbiturate Assay* ⟨361⟩. Calculate the quantity, in mg, of $C_{10}H_{15}N_2NaO_3$ in the portion of Capsule contents taken by the formula:

$$(234.23/212.25)(50)(R_U)(Q_S)(C_i)/(R_S),$$

in which 234.23 and 212.25 are the molecular weights of butabarbital sodium and butabarbital, respectively.

Butabarbital Sodium Elixir

» Butabarbital Sodium Elixir contains not less than 90.0 percent and not more than 110.0 percent of the labeled amount of $C_{10}H_{15}N_2NaO_3$.

Packaging and storage—Preserve in tight containers.

Reference standard—*USP Butabarbital Reference Standard*—Keep container tightly closed. Dry at 105° for 4 hours before using.

Identification—Place a volume of Elixir, equivalent to about 150 mg of butabarbital sodium, in a separator, render it distinctly alkaline by the addition of 1 *N* sodium hydroxide, and saturate it with sodium chloride. Extract the mixture with two 15-mL portions of ether, and discard the ether. Acidify the solution with hydrochloric acid, and render it just alkaline to litmus by adding small portions of sodium bicarbonate (carbonate-free). Extract the liberated acid barbiturate, using five 20-mL portions of chloroform. Wash the combined chloroform extracts with 10 mL of water acidified with 1 drop of hydrochloric acid, then extract the water with 10 mL of chloroform, adding the latter to the main chloroform solution. Filter the chloroform solution through a pledget of cotton or other suitable filter, previously washed with chloroform, into a tared beaker, and finally wash the separator and the filter with three 5-mL portions of chloroform. Evaporate the combined chloroform solution and washings on a steam bath with the aid of a current of air to dryness, and dry the residue at 105° for 2 hours: the infrared absorption spectrum of a potassium bromide dispersion of the residue so obtained exhibits maxima only at the same wavelengths as that of a similar preparation of USP Butabarbital RS.

Alcohol content, *Method II* ⟨611⟩: between 95.0% and 115.0% of the labeled amount of C_2H_5OH.

Assay—

Internal standard: Secobarbital.

Internal standard solution—Dissolve an accurately weighed quantity of Secobarbital in chloroform, and dilute quantitatively with chloroform to obtain a solution having a known concentration of about 0.7 mg per mL.

Standard preparation—Dissolve accurately weighed quantities of USP Butabarbital RS and Secobarbital in chloroform, and dilute quantitatively with chloroform to obtain a solution that contains, in each mL, known amounts of about 1 mg of USP Butabarbital RS and about 1.4 mg of Secobarbital.

Assay preparation—[NOTE—This preparation includes a bromination step for elimination of parabens and a carbonate-chloroform extraction for elimination of benzoic acid.] Transfer an accurately measured volume of Butabarbital Sodium Elixir, equivalent to about 30 mg of butabarbital sodium, to a separator, add 1 mL of bromine water (prepared by dissolving 2.0 mL of bromine and 10 g of potassium bromide in 60 mL of water), and swirl. Allow to stand for 5 minutes, add 1 mL of sodium bisulfite solution (1 in 10), and swirl. Add 300 mg of sodium bicarbonate in small portions, with mixing, and extract with four 10-mL portions of chloroform. Filter the extracts through about 15 g of anhydrous sodium sulfate that is supported on a funnel by a small pledget of glass wool. Collect the combined filtrates in a 50-mL volumetric flask, wash the sodium sulfate with 5 mL of chloroform, collecting the washing with the filtrate, dilute with chloroform to volume, and mix. Combine 2.0 mL of this solution with 2.0 mL of *Internal standard solution* in a suitable container, and reduce the volume to about 1 mL by evaporation, with the aid of a stream of dry nitrogen, at room temperature.

Chromatographic system and *System suitability*—Proceed as directed for *Chromatographic System* and *System Suitability* under *Barbiturate Assay* ⟨361⟩, the resolution, *R*, between butabarbital and secobarbital being not less than 3.0. [NOTE—Relative retention times are, approximately, 0.6 for butabarbital and 1.0 for secobarbital.]

Procedure—Proceed as directed for *Procedure* under *Barbiturate Assay* ⟨361⟩. Calculate the quantity, in mg, of $C_{10}H_{15}N_2NaO_3$ in each mL of the Elixir taken by the formula:

$$(234.23/212.25)(50)(R_U)(Q_S)(C_i)/V(R_S),$$

in which 234.23 and 212.25 are the molecular weights of butabarbital sodium and butabarbital, respectively, and *V* is the volume, in mL, of Elixir taken.

Butabarbital Sodium Tablets

» Butabarbital Sodium Tablets contain not less than 90.0 percent and not more than 110.0 percent of the labeled amount of $C_{10}H_{15}N_2NaO_3$.

Packaging and storage—Preserve in well-closed containers.

Reference standard—*USP Butabarbital Reference Standard*—Keep container tightly closed. Dry at 105° for 4 hours before using.

Identification—Using a quantity of ground Tablets equivalent to about 150 mg of butabarbital sodium, proceed as directed in the *Identification test* under *Butabarbital Sodium Capsules.*

Dissolution ⟨711⟩—

Medium: water; 900 mL.

Apparatus 1: 100 rpm.

Time: 45 minutes.

Procedure—Determine the amount of $C_{10}H_{15}N_2NaO_3$ dissolved from ultraviolet absorbances at the wavelength of maximum absorbance at about 239 nm of filtered portions of the solution under test, mixed with sufficient ammonium hydroxide to provide a concentration of 0.5 *N* ammonium hydroxide, and suitably diluted with *Dissolution Medium*, if necessary, in comparison with a Standard solution having a known concentration of USP Butabarbital RS in the same medium.

Tolerances—Not less than 75% (*Q*) of the labeled amount of $C_{10}H_{15}N_2NaO_3$ is dissolved in 45 minutes.

Uniformity of dosage units ⟨905⟩: meet the requirements.

Procedure for content uniformity—

Acid-methanol mixture and *Standard preparation*—Prepare as directed in the test for *Content uniformity* under *Butabarbital Sodium Capsules.*

Test preparation—Transfer 1 finely powdered Tablet to a 25-mL volumetric flask, and proceed as directed for *Test preparation* in the test for *Content uniformity* under *Butabarbital Sodium Capsules*, beginning with "add *Acid-methanol mixture* to volume."

Procedure—Procced as directed for *Procedure* in the test for *Content uniformity* under *Butabarbital Sodium Capsules.* Calculate the quantity, in mg, of $C_{10}H_{15}N_2NaO_3$ in the Tablet by the formula given therein, in which 234.23 and 212.25 are the molecular weights of butabarbital sodium and butabarbital, respectively, *T* is the labeled quantity, in mg, of butabarbital sodium in the Tablet, *C* is the concentration, in µg per mL, of USP Butabarbital RS in the *Standard preparation, D* is the concentration, in µg per mL, of butabarbital sodium in the *Test preparation*, based upon the labeled quantity per Tablet and the extent of dilution, and A_U and A_S are the absorbances of the solutions from the *Test preparation* and the *Standard preparation*, respectively.

Assay—

Internal standard: Secobarbital.

Internal standard solution—Dissolve an accurately weighed quantity of Secobarbital in chloroform, and dilute quantitatively with chloroform to obtain a solution having a known concentration of about 1.2 mg per mL.

Standard preparation—Dissolve accurately weighed quantities of USP Butabarbital RS and Secobarbital in chloroform, and dilute quantitatively with chloroform to obtain a solution that contains, in each mL, known amounts of about 0.8 mg of USP Butabarbital RS and about 1 mg of Secobarbital.

Assay preparation—Weigh and finely powder not less than 20 Butabarbital Sodium Tablets. Transfer an accurately weighed portion of the powder, equivalent to about 50 mg of butabarbital sodium, to a 50-mL volumetric flask, add 35 mL of dilute ammonium hydroxide (1 in 25), dilute with water to volume, and mix. Filter, if necessary, discarding the first 15 mL of the filtrate, and transfer 25.0 mL of the clear solution to a separator. Add 2 mL of hydrochloric acid, and extract with three 25-mL portions of chloroform. Filter the extracts through about 15 g of anhydrous sodium sulfate that is supported on a funnel by a small pledget of glass wool. Collect the combined filtrate in a 100-mL volumetric flask, wash the sodium sulfate with 15 mL of chloroform, collecting the washing with the filtrate, dilute with chloroform to volume, and mix. Combine 4.0 mL of this solution with 1.0 mL of *Internal standard solution* in a suitable container, and reduce the volume to about 1 mL by evaporation, with the aid of a stream of dry nitrogen, at room temperature.

Chromatographic system and *System suitability*—Proceed as directed for *Chromatographic System* and *System Suitability* under *Barbiturate Assay* ⟨361⟩, the resolution, *R*, between butabarbital and secobarbital being not less than 3.0. [NOTE—Relative retention times are, approximately, 0.6 for butabarbital and 1.0 for secobarbital.]

Procedure—Proceed as directed for *Procedure* under *Barbiturate Assay* ⟨361⟩. Calculate the quantity, in mg, of $C_{10}H_{15}N_2NaO_3$ in the portion of Tablets taken by the formula:

$$(234.23/212.25)(50)(R_U)(Q_S)(C_i)/(R_S),$$

in which 234.23 and 212.25 are the molecular weights of butabarbital sodium and butabarbital, respectively.

Butalbital

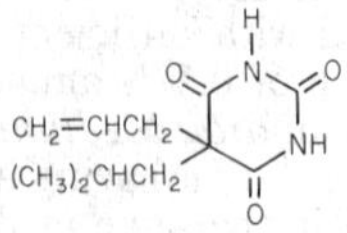

$C_{11}H_{16}N_2O_3$ 224.26

2,4,6(1*H*,3*H*,5*H*)-Pyrimidinetrione, 5-(2-methylpropyl)-5-(2-propenyl)-.

5-Allyl-5-isobutylbarbituric acid [*77-26-9*].

» Butalbital contains not less than 98.0 percent and not more than 102.0 percent of $C_{11}H_{16}N_2O_3$, calculated on the dried basis.

Packaging and storage—Preserve in well-closed containers.

Reference standard—*USP Butalbital Reference Standard*—Dry in vacuum at room temperature to constant weight before using.

Identification—

A: The infrared absorption spectrum of a potassium bromide dispersion of it, previously dried, exhibits maxima only at the same wavelengths as that of a similar preparation of USP Butalbital RS.

B: The ultraviolet absorption spectrum of a 1 in 67,000 solution in 0.1 *N* sodium hydroxide exhibits maxima and minima at the same wavelengths as that of a similar solution of USP Butalbital RS, concomitantly measured, and the respective absorptivities, calculated on the dried basis, at the wavelength of maximum absorbance at about 246 nm do not differ by more than 2.5%.

Melting range ⟨741⟩: between 138° and 141°.

Loss on drying ⟨731⟩—Dry it in vacuum at room temperature to constant weight: it loses not more than 0.2% of its weight.

Residue on ignition ⟨281⟩: not more than 0.1%.

Heavy metals, *Method II* ⟨231⟩: 0.002%.

Chromatographic purity—

Chloroform-methanol—Mix equal volumes of chloroform and methanol.

Standard preparations—Dissolve a quantity of USP Butalbital RS in *Chloroform-methanol* to obtain a solution having a concentration of 40 mg per mL (*Solution A*). Dilute 1.0 mL of *Solution A* with *Chloroform-methanol* to 100 mL, and mix (*Solution B*); mix 5.0 mL of *Solution B* with 5.0 mL of *Chloroform-methanol* (*Solution C*); and mix 5.0 mL of *Solution C* with 5.0 mL of *Chloroform-methanol* (*Solution D*).

Test preparation—Dissolve a quantity of Butalbital in *Chloroform-methanol* to obtain a solution having a concentration of 40 mg per mL.

Procedure—In a suitable chromatographic chamber, arranged for thin-layer chromatography and lined with filter paper, place a volume of a developing solvent consisting of a mixture of acetone, dichloromethane, methanol, and ammonium hydroxide (50:30:10:10) sufficient to develop the chromatogram. Cover the chamber, and allow it to equilibrate for 30 minutes. On a suitable thin-layer chromatographic plate (see *Chromatography* ⟨621⟩), coated with a 0.25-mm layer of chromatographic silica gel, apply 10 µL each of the *Test preparation* and *Solutions A*, *B*, *C*, and *D*. Develop the chromatogram until the solvent front has moved about three-fourths of the length of the plate. Remove the plate from the developing chamber, mark the solvent front, and dry the plate in a current of air. Spray the plate with a reagent prepared by dissolving 5 g of potassium hydroxide in a mixture of 25 mL of water and 75 mL of alcohol. Allow the plate to dry in warm air for 10 minutes, and examine the chromatograms under ultraviolet light: the chromatograms show principal spots at about the same R_f value, and the sum of the intensities of any secondary spots, if present in the chromatogram from the *Test preparation*, is not greater than 1% of that of the principal spot from *Solution A*. [NOTE—The relative intensities of the principal spots from the *Standard preparations* are: *Solution A*, 1; *Solution B*, 0.01; *Solution C*, 0.005; and *Solution D*, 0.0025.]

Assay—Dissolve about 180 mg of Butalbital, accurately weighed, in a mixture of 25 mL of alcohol and 25 mL of sodium carbonate solution (3 in 100), and titrate with 0.1 *N* silver nitrate VS, determining the end-point electrometrically, using a silver electrode, either with a suitable reference electrode containing a saturated aqueous solution of potassium nitrate, or a combination electrode in which the reference portion of the electrode contains a saturated aqueous solution of potassium nitrate. Each mL of 0.1 *N* silver nitrate is equivalent to 22.43 mg of $C_{11}H_{16}N_2O_3$.

Butalbital and Aspirin Tablets

» Butalbital and Aspirin Tablets contain not less than 90.0 percent and not more than 110.0 percent of the labeled amounts of butalbital ($C_{11}H_{16}N_2O_3$) and aspirin ($C_9H_8O_4$).

Packaging and storage—Preserve in tight containers.

Reference standards—*USP Butalbital Reference Standard*—Dry in vacuum at room temperature to constant weight before using. *USP Aspirin Reference Standard*—Keep container tightly closed. Dry over silica gel for 5 hours before using. *USP Salicylic Acid Reference Standard*—Dry over silica gel for 3 hours before using.

Identification—The retention times of the butalbital peak and the aspirin peak in the chromatogram of the *Assay preparation* correspond to those of the butalbital peak and the aspirin peak in the chromatograms of the *Butalbital and salicylic acid standard preparation* and the *Aspirin standard preparation*, as obtained in the *Assay for butalbital and aspirin and limit of free salicylic acid*.

Dissolution ⟨711⟩—

Medium: water; 900 mL.

Apparatus 1: 100 rpm.

Time: 60 minutes.

Determination of dissolved butalbital—

MOBILE PHASE—Prepare a suitable filtered and degassed mixture of water, acetonitrile, and phosphoric acid (3100:725:4). Adjust the ratio as necessary.

STANDARD PREPARATION—Dissolve accurately weighed quantities of USP Butalbital RS and of salicylic acid in *Mobile phase* to obtain a solution containing known concentrations of about 1 µg of butalbital per mL for each mg of the labeled amount per Tablet and about 30 µg of salicylic acid per mL.

CHROMATOGRAPHIC SYSTEM (see *Chromatography* ⟨621⟩)—The liquid chromatograph is equipped with a 214-nm detector and a 3.9-mm × 30-cm column that contains 10-µm packing L1. The flow rate is about 1.5 mL per minute. Chromatograph the *Standard preparation* and record the peak responses as directed under *Procedure:* the resolution, *R*, between the butalbital and salicylic acid peaks is not less than 3.0, and the relative standard deviation of butalbital responses for replicate injections is not more than 3.0%.

PROCEDURE—Filter a portion of the solution under test through a 0.5-µm filter. Separately inject equal volumes (about 10 to 25 µL) of the filtrate and the *Standard preparation* into the chromatograph, record the chromatograms, and measure the responses for the major peaks. The relative retention times are about 0.6 for aspirin, 0.85 for salicylic acid, and 1.0 for butalbital. [NOTE—After use, the column may be regenerated by passing through it at least 50 mL of a mixture of acetonitrile, methanol, and water (1:1:1), followed by 50 mL of a mixture of acetonitrile and water (1:1).] Calculate the amount, in mg, of butalbital ($C_{11}H_{16}N_2O_3$) dissolved by the formula:

$$0.9C(r_U / r_S),$$

in which *C* is the concentration, in µg per mL, of USP Butalbital RS in the *Standard preparation*, and r_U and r_S are the butalbital peak responses obtained from the solution under test and the *Standard preparation*, respectively.

Determination of dissolved aspirin—

pH 4.5 BUFFER—Dissolve 5.98 g of sodium acetate trihydrate in 500 mL of water, add 2.5 mL of glacial acetic acid, dilute with water to 1000 mL, and mix. Adjust this solution with glacial acetic acid to a pH of 4.5 ± 0.05, and mix.

PROCEDURE—Determine the amount of aspirin ($C_9H_8O_4$) dissolved from ultraviolet absorbances at the wavelength of the isosbestic point of aspirin and salicylic acid at 265 ± 2 nm of filtered portions of the solution under test, diluted with 4 volumes of *pH 4.5 buffer*, in comparison with a Standard solution having a known concentration of USP Aspirin RS in the same medium. [NOTE—Prepare the Standard solution at the time of use. An amount of alcohol not to exceed 1% of the total volume of the Standard solution may be used to bring the Reference Standard into solution prior to dilution first with water and then with 4 volumes of *pH 4.5 buffer*.]

Tolerances—Not less than 75% (*Q*) of the labeled amounts of butalbital ($C_{11}H_{16}N_2O_3$) and aspirin ($C_8H_9O_4$) are dissolved in 60 minutes.

Uniformity of dosage units ⟨905⟩: meet the requirements for *Content Uniformity* with respect to butalbital and for *Weight Variation* with respect to aspirin.

Assay for butalbital and aspirin and limit of free salicylic acid—

Mobile phase—Prepare a suitable filtered and degassed mixture of water, acetonitrile, and phosphoric acid (3100:725:4). Adjust the ratio as necessary.

Solvent mixture—Mix 40 mL of formic acid and 4000 mL of acetonitrile.

Butalbital standard stock solution—Dissolve an accurately weighed quantity of USP Butalbital RS in *Solvent mixture* to obtain a solution having a known concentration of about 3250*J* µg per mL, *J* being the ratio of the labeled amount, in mg, of butalbital to the labeled amount, in mg, of aspirin per tablet.

Salicylic acid standard stock solution—Dissolve an accurately weighed quantity of USP Salicylic Acid RS in *Solvent mixture* to obtain a solution having a known concentration of about 200 µg per mL.

Butalbital and salicylic acid standard preparation—Transfer 25.0 mL of *Butalbital standard stock solution* and 3.0 mL of *Salicylic acid standard stock solution* to a 250-mL volumetric flask, dilute with *Solvent mixture* to volume, and mix. This solution contains about 325*J* µg of butalbital and 2.4 µg of salicylic acid per mL.

Aspirin standard preparation—Dissolve an accurately weighed quantity of USP Aspirin RS in *Solvent mixture* to obtain a solution having a known concentration of about 325 µg per mL.

Resolution solution—Transfer 4.0 mL of *Butalbital standard stock solution* and 3.0 mL of *Salicylic acid standard stock solution* to a 50-mL volumetric flask, dilute with *Solvent mixture* to volume, and mix.

Assay preparation—Weigh and finely powder not less than 20 Butalbital and Aspirin Tablets. Transfer an accurately weighed portion of the powder, equivalent to about 80 mg of aspirin, to a 250-mL volumetric flask, dilute with *Solvent mixture* to volume, sonicate for 15 minutes, and mix. Filter a portion of this solution through a 0.5-µm porosity filter before use.

Chromatographic system (see *Chromatography* ⟨621⟩)—The liquid chromatograph is equipped with a 214-nm detector and a 3.9-mm × 30-cm column that contains 10-µm packing L1. The flow rate is about 1.5 mL per minute. Chromatograph the *Butalbital and salicylic acid standard preparation*, the *Aspirin standard preparation*, and the *Resolution solution* as directed under *Procedure:* the resolution, *R*, between the butalbital and salicylic acid peaks is not less than 3.0, and the relative standard deviation for replicate injections of the Standard preparations is not more than 3.0% for butalbital and aspirin, and not more than 6.0% for salicylic acid.

Procedure—Separately inject equal volumes (about 10 µL) of the *Standard preparations* and the *Assay preparation* into the chromatograph, record the chromatograms, and measure the responses for the major peaks, and for the minor peak corresponding to salicylic acid. The relative retention times are about 0.6 for aspirin, 0.85 for salicylic acid, and 1.0 for butalbital. [NOTE—After use, the column may be regenerated by passing through it at least 50 mL of a mixture of acetonitrile, methanol, and water (1:1:1), followed by a mixture of acetonitrile and water (1:1).] Calculate the quantity, in mg, of butalbital ($C_{11}H_{16}N_2O_3$) in the portion of Tablets taken by the formula:

$$0.25C(r_U/r_S),$$

in which *C* is the concentration, in µg per mL, of USP Butalbital RS in the *Butalbital and salicylic acid standard preparation*, and r_U and r_S are the butalbital peak responses obtained from the *Assay preparation* and the *Butalbital and salicylic acid standard preparation*, respectively. Calculate the quantity, in mg, of aspirin ($C_9H_8O_4$) in the portion of Tablets taken by the formula:

$$0.25C(r_U/r_S),$$

in which *C* is the concentration, in µg per mL, of USP Aspirin RS in the *Aspirin standard preparation*, and r_U and r_S are the aspirin peak responses obtained from the *Assay preparation* and the *Aspirin standard preparation*, respectively. Calculate the percentage of free salicylic acid in the Tablets taken by the formula:

$$25(C/a)(r_U/r_S),$$

in which *C* is the concentration, in µg per mL, of the USP Salicylic Acid RS in the *Butalbital and salicylic acid standard preparation*, *a* is the quantity, in mg, of aspirin in the portion of Tablets taken, based on the labeled amount, and r_U and r_S are the salicylic acid peak responses obtained from the *Assay preparation* and the *Butalbital and salicylic acid standard preparation*, respectively: not more than 3.0% is found.

Butamben

$H_2N-C_6H_4-COOCH_2CH_2CH_2CH_3$

$C_{11}H_{15}NO_2$ 193.24

Benzoic acid, 4-amino-, butyl ester.

Butyl *p*-aminobenzoate [*94-25-7*].

» Butamben, dried over phosphorus pentoxide for 3 hours, contains not less than 98.0 percent and not more than 101.0 percent of $C_{11}H_{15}NO_2$.

Packaging and storage—Preserve in well-closed containers.

Reference standard—*USP Butamben Reference Standard*—Dry over phosphorus pentoxide for 3 hours before using.

Completeness and color of solution—One g dissolves completely in 30 mL of alcohol and in 30 mL of ether, and the solutions are colorless.

Identification—The infrared absorption spectrum of a potassium bromide dispersion of it, previously dried, exhibits maxima only at the same wavelengths as that of a similar preparation of USP Butamben RS.

Melting range, *Class I* ⟨741⟩: between 57° and 59°.

Reaction—Dissolve 1 g in 10 mL of neutralized alcohol: a clear solution results. Dilute this solution with 10 mL of water, and add 2 drops of phenolphthalein TS and 1 drop of 0.1 *N* sodium hydroxide: a red color is produced.

Loss on drying ⟨731⟩—Dry it over phosphorus pentoxide for 3 hours: it loses not more than 1.0% of its weight.

Residue on ignition ⟨281⟩: not more than 0.2%.

Chloride—To a solution of 200 mg in 10 mL of alcohol add 1 mL of 2 *N* nitric acid and a few drops of silver nitrate TS: no opalescence is produced.

Heavy metals, *Method I* ⟨231⟩—Dissolve 2 g in 2 mL of 1 *N* acetic acid and sufficient alcohol to make 25 mL: the limit is 0.001%.

Assay—Dissolve about 400 mg of Butamben, previously dried and accurately weighed, in a mixture of 100 mL of water and 20 mL of hydrochloric acid. Cool the solution in an ice bath to about 10°, and titrate with 0.1 *M* sodium nitrite VS until a blue color is produced immediately when the titrated solution is spotted on starch iodide paper. When the titration is complete, the end-point is reproducible after the mixture has been allowed to stand for 5 minutes. Perform a blank determination, and make any necessary correction. Each mL of 0.1 *M* sodium nitrite is equivalent to 19.32 mg of $C_{11}H_{15}NO_2$.

Butoconazole Nitrate

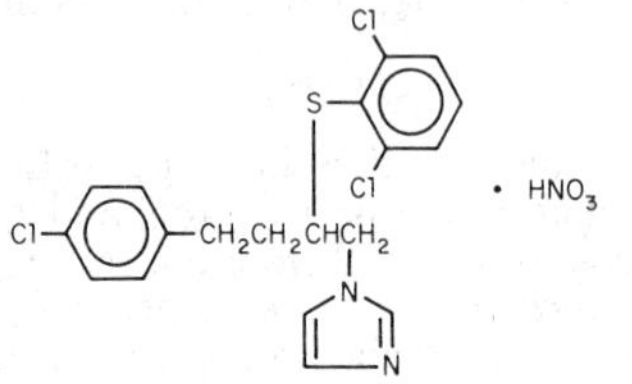

$C_{19}H_{17}Cl_3N_2S.HNO_3$ 474.79

1*H*-Imidazole, 1-[4-(4-chlorophenyl)-2-[(2,6-dichlorophenyl)-thio]butyl]-, mononitrate, (±)-.

(±)-1-[4-(*p*-Chlorophenyl)-2-[(2,6-dichlorophenyl)-thio]butyl]-imidazole mononitrate [*64872-77-1*].

» Butoconazole Nitrate contains not less than 98.0 percent and not more than 102.0 percent of $C_{19}H_{17}Cl_3N_2S.HNO_3$, calculated on the dried basis.

Packaging and storage—Preserve in well-closed, light-resistant containers.

Reference standard—*USP Butoconazole Nitrate Reference Standard*—Dry in vacuum at 60° for 3 hours before using.

Identification—The infrared absorption spectrum of a potassium bromide dispersion of it, previously dried, exhibits maxima only at the same wavelengths as that of a similar preparation of USP Butoconazole Nitrate RS.

Loss on drying ⟨731⟩—Dry it in vacuum at 60° for 3 hours: it loses not more than 1.0% of its weight.

Residue on ignition ⟨281⟩: not more than 0.1%.

Ordinary impurities ⟨466⟩—

Test solution: a mixture of methylene chloride and methanol (2:1).

Standard solutions: a mixture of methylene chloride and methanol (2:1).

Adsorbant: a 0.25-mm layer of chromatographic silica gel.

Eluant: a mixture of chloroform, tetrahydrofuran, cyclohexane, and ammonium hydroxide (18:18:13:1).

Visualization: 22.

Assay—

Phosphate buffer—Dissolve 2.18 g of monobasic potassium phosphate and 4.18 g of dibasic potassium phosphate in 900 mL of water, dilute with water to 1000 mL, and mix.

Mobile phase—Prepare a filtered and degassed mixture of methanol and *Phosphate buffer* (3:1), making adjustments if necessary (see *System Suitability* under *Chromatography* ⟨621⟩).

Standard preparation—Dissolve an accurately weighed quantity of USP Butoconazole Nitrate RS in *Mobile phase*, and dilute quantitatively with *Mobile phase* to obtain a solution having a known concentration of about 0.2 mg per mL.

Assay preparation—Transfer about 20 mg of Butoconazole Nitrate, accurately weighed, to a 100-mL volumetric flask, and dissolve in *Mobile phase*. Dilute with *Mobile phase* to volume, mix, and filter.

Chromatographic system (see *Chromatography* ⟨621⟩)—The liquid chromatograph is equipped with a 229-nm detector and a 4.6-mm × 25-cm column that contains packing L1 and is maintained at 40°. The flow rate is about 2 mL per minute. Chromatograph the *Standard preparation*, and record the peak responses as directed under *Procedure:* the column efficiency determined from the analyte peak is not less than 2800 theoretical plates, the tailing factor for the analyte peak is not more than 1.5, and the relative standard deviation for replicate injections is not more than 1.5%.

Procedure—Separately inject equal volumes (about 10 µL) of the *Standard preparation* and the *Assay preparation* into the chromatograph, record the chromatograms, and measure the responses for the major peaks. Calculate the quantity, in mg, of $C_{19}H_{17}Cl_3N_2S.HNO_3$ in the portion of Butoconazole Nitrate taken by the formula:

$$100C(r_U/r_S),$$

in which *C* is the concentration, in mg per mL, of USP Butoconazole Nitrate RS in the *Standard preparation*, and r_U and r_S are the peak responses obtained from the *Assay preparation* and the *Standard preparation*, respectively.

Butoconazole Nitrate Cream

» Butoconazole Nitrate Cream is Butoconazole Nitrate in a suitable cream base. It contains not less than 90.0 percent and not more than 110.0 percent of the labeled amount of $C_{19}H_{17}Cl_3N_2S.HNO_3$.

Packaging and storage—Preserve in collapsible tubes or in tight containers. Avoid excessive heat and avoid freezing.

Reference standard—*USP Butoconazole Nitrate Reference Standard*—Dry in vacuum at 60° for 3 hours before using.

Labeling—Cream that is intended for use as a vaginal preparation may be labeled Butoconazole Nitrate Vaginal Cream.

Identification—Prepare a mixture of the *Standard preparation* and the *Assay preparation* (1:1), prepared as directed in the *Assay*, and chromatograph as directed in the *Assay:* the chromatogram so obtained exhibits two main peaks, corresponding to butoconazole nitrate and the internal standard.

Minimum fill ⟨755⟩: meets the requirements.

Assay—

Acetate buffer—Dissolve 1.4 g of potassium acetate in 980 mL of water, adjust with about 2 mL of glacial acetic acid to a pH of 4.3 ± 0.1, dilute with water to 1000 mL, and mix. Adjust the buffer molarity (0.018–0.072 *M*) as necessary to obtain suitable chromatographic performance. Increased retention time may be achieved by a decrease in the buffer molarity.

Mobile phase—Prepare a filtered and degassed mixture of methanol and *Acetate buffer* (65:35). Make adjustments if necessary (see *System Suitability* under *Chromatography* ⟨621⟩).

Diluent—Prepare a mixture of methanol and *Acetate buffer* (6:4).

Internal standard solution—Dissolve 1-benzylimidazole in methanol to obtain a solution containing about 1.6 mg per mL.

Standard preparation—Dissolve an accurately weighed quantity of USP Butoconazole Nitrate RS in methanol, and dilute quantitatively, and stepwise if necessary, with methanol to obtain a stock solution having a known concentration of about 400 µg per mL. Transfer 2.0 mL of this stock solution and 3.0 mL of *Internal standard solution* to a 50-mL flask, add 35.0 mL of *Diluent*, and mix.

Assay preparation—Transfer to a 250-mL volumetric flask an accurately weighed quantity of Butoconazole Nitrate Cream, equivalent to about 100 mg of butoconazole nitrate. Add about 200 mL of methanol, and sonicate until the Cream is dissolved completely. Cool to room temperature, dilute with methanol to volume, and mix. Transfer 2.0 mL of this solution to a 50-mL flask, add 3.0 mL of *Internal standard solution* and 35.0 mL of *Diluent*, and mix. Allow the precipitated excipients which form to rise to the top of the solution, remove them by aspiration, and discard. Centrifuge or filter the remaining solution.

Chromatographic system (see *Chromatography* ⟨621⟩)—The liquid chromatograph is equipped with a 225-nm detector and a 4.6-mm × 25-cm column that contains packing L9 which has been converted to the potassium form by the use of 0.555 *M* potassium acetate solution. The flow rate is about 1 mL per minute. Chromatograph the *Standard preparation*, and record the peak responses as directed under *Procedure:* the column efficiency determined from the analyte peak is not less than 1100 theoretical plates, the tailing factor for the analyte peak is not more than 2.1, the resolution, *R*, between the analyte and internal standard peaks is not less than 4.0, and the relative standard deviation for replicate injections is not more than 1.5%.

Procedure—Separately inject equal volumes (about 20 µL) of the *Standard preparation* and the *Assay preparation* into the

chromatograph, record the chromatograms, and measure the responses for the major peaks. The relative retention times are about 0.6 for butoconazole nitrate and 1.0 for 1-benzylimidazole. Calculate the quantity, in mg, of $C_{19}H_{17}Cl_3N_2S.HNO_3$ in the portion of Cream taken by the formula:

$$0.25C(R_U/R_S),$$

in which C is the concentration, in µg per mL, of USP Butoconazole Nitrate RS in the stock solution used to prepare the *Standard preparation*, and R_U and R_S are the ratios of the butoconazole nitrate peak response to the internal standard peak response obtained from the *Assay preparation* and the *Standard preparation*, respectively.

Butane—*see* Butane NF

Butorphanol Tartrate

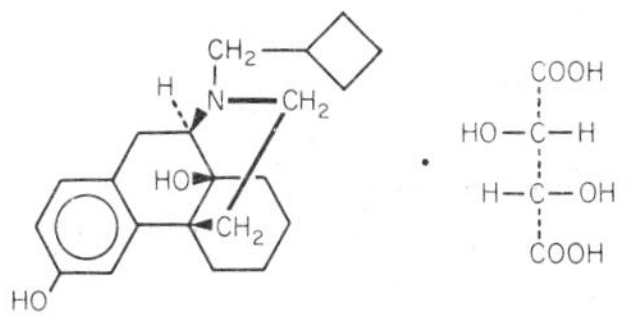

$C_{21}H_{29}NO_2.C_4H_6O_6$ 477.55

Morphinan-3,14-diol, 17-(cyclobutylmethyl)-, (−)-, [*S*-(*R**,*R**)]-2,3-dihydroxybutanedioate (1:1) (salt).

(−)-17-(Cyclobutylmethyl)morphinan-3,14-diol D-(−)-tartrate (1:1) (salt) [*58786-99-5*].

» Butorphanol Tartrate contains not less than 98.0 percent and not more than 102.0 percent of $C_{21}H_{29}NO_2.C_4H_6O_6$, calculated on the anhydrous basis.

Packaging and storage—Preserve in tight containers.

Reference standard—*USP Butorphanol Tartrate Reference Standard*—Do not dry; determine the *Water* content by *Method I* ⟨921⟩ before using.

Identification—

A: The infrared absorption spectrum of a potassium bromide dispersion of it exhibits maxima only at the same wavelengths as that of a similar preparation of USP Butorphanol Tartrate RS.

B: The R_f value of the principal spot obtained from the *Test preparation* in test *A* for *Chromatographic purity* corresponds to that obtained from the *Standard preparation*.

C: Add a few mg of Butorphanol Tartrate to a mixture of 15 mL of pyridine and 5 mL of acetic anhydride: an emerald-green color is produced (presence of tartrate).

Specific rotation ⟨781⟩: between −60° and −66°, calculated on the anhydrous basis, determined in a methanol solution containing 40 mg in each 10 mL.

Water, *Method I* ⟨921⟩: not more than 2.0%.

Residue on ignition ⟨281⟩: not more than 0.1%.

Heavy metals, *Method II* ⟨231⟩: 0.003%.

Chromatographic purity—

A (TLC):

Standard preparation—Prepare a solution in methanol containing 1 mg of USP Butorphanol Tartrate RS per mL.

Test preparation—Transfer 100 mg of Butorphanol Tartrate to a 10-mL volumetric flask. Dissolve in methanol, dilute with methanol to volume, and mix.

Iodoplatinate spray reagent—Prepare a 1 in 10 solution of chloroplatinic acid in water. To 0.5 mL of this solution add 33 mL of water and 1 g of potassium iodide to obtain the spray reagent. Prepare fresh daily.

Procedure—Apply 50 µL of the *Test preparation*, containing 500 µg of butorphanol tartrate, and 5 µL and 10 µL of the *Standard preparation*, containing 5 µg and 10 µg of USP Butorphanol Tartrate RS, respectively, about 2 cm apart on a line parallel to and about 2 cm from the bottom of a thin-layer chromatographic plate (see *Chromatography* ⟨621⟩), coated with a 0.25-mm layer of chromatographic silica gel mixture. Place the plate in a developing chamber containing, and equilibrated with, a mixture of chloroform, methanol, benzene, and ammonium hydroxide (85:25:20:5). Develop the chromatogram until the solvent front has moved about 10 cm above the line of application. Remove the plate, mark the solvent front, and allow the solvent to evaporate. Spray the plate with *Iodoplatinate spray reagent*. Estimate the percentage of the impurities present in the *Test preparation* by comparing the intensities of secondary spots, if present, with the intensities of the principal spots obtained from the chromatograms of the *Standard preparation*. The sum of the impurities observed is not greater than 2.0%.

B (GC): Dissolve a suitable quantity of Butorphanol Tartrate in methanol to obtain a solution containing about 10 mg per mL. Inject 1 µL of this solution into a suitable gas chromatograph equipped with a flame-ionization detector and a 1.8-m × 4-mm glass column containing 3 percent liquid phase G3 on support S1. The temperatures of the injection port, column, and detector are maintained at about 280°, 250°, and 290°, respectively. The carrier gas is nitrogen. Record a 30-minute chromatogram. Preferably using an electronic integrator, determine the areas of all peaks in the chromatogram excluding the area of the solvent. In a suitable chromatogram, the retention time for the alpha isomer of Butorphanol Tartrate is 1.2 relative to 1.0 for Butorphanol Tartrate, and the retention time of Butorphanol Tartrate is not less than 15 minutes. Calculate the percentage of synthesis precursors in the test specimen by the formula $100A_V/A_S$, in which A_V represents the sum of the areas of all minor peaks, and A_S is the sum of the areas of the major and minor peaks. The limit is 2.0%.

Assay—Dissolve about 500 mg of Butorphanol Tartrate, accurately weighed, in 75 mL of glacial acetic acid. Add crystal violet TS, and titrate with 0.1 *N* perchloric acid VS. Perform a blank determination, and make any necessary correction. Each mL of 0.1 *N* perchloric acid is equivalent to 47.76 mg of $C_{21}H_{29}NO_2.C_4H_6O_6$.

Butorphanol Tartrate Injection

» Butorphanol Tartrate Injection is a sterile solution of Butorphanol Tartrate in Water for Injection. It contains not less than 90.0 percent and not more than 110.0 percent of the labeled amount of $C_{21}H_{29}NO_2.C_4H_6O_6$. It may contain a suitable preservative and a buffer.

Packaging and storage—Preserve in single-dose or in multiple-dose containers, preferably of Type I glass, protected from light.

Reference standard—*USP Butorphanol Tartrate Reference Standard*—Do not dry; determine the *Water* content by *Method I* ⟨921⟩ before using.

Identification—Apply 10-µL portions of the Injection and a Standard solution of USP Butorphanol Tartrate RS having the same concentration about 2 cm apart on a line parallel to and about 2 cm from the bottom of a thin-layer chromatographic plate (see *Chromatography* ⟨621⟩) coated with a 0.25-mm layer of chromatographic silica gel mixture. Place the plate in a developing chamber containing a mixture of chloroform, ethyl acetate, and methanol (40:10:9), and develop the chromatogram until the solvent front has moved about 10 cm above the line of application. Remove the plate, mark the solvent front, and allow the solvent to evaporate. Examine the plate under short-wavelength ultraviolet light: the R_f value of the principal spot obtained from the solution under test corresponds to that obtained from the Standard solution. Benzethonium chloride, if present, is observed as a streaked zone near the point of application. Visualize the butorphanol spots by lightly spraying the plate with a 1 in 250 solution of bromocresol purple in dehydrated alcohol: butorphanol appears as a blue spot against a light yellow background.

pH ⟨791⟩: between 3.0 and 5.5.

Other requirements—It meets the requirements under *Injections* ⟨1⟩.

Assay—

Mobile solvent—Prepare a mixture of 0.05 *M* ammonium acetate and acetonitrile (3:1) adjusted by the addition of glacial acetic acid to a pH of 4.1. The mixture is appropriately filtered and degassed.

Internal standard solution—Dissolve about 50 mg of propylparaben in 5.0 mL of methanol contained in a 250-mL volumetric flask. Add water to volume, and mix.

Standard preparation—Transfer about 50 mg of USP Butorphanol Tartrate RS, accurately weighed, to a 25-mL volumetric flask containing 1.0 mL of 1 *N* sulfuric acid. Swirl the flask to dissolve the powder completely, add water to volume, and mix. Pipet 5 mL of the resulting solution into a 50-mL volumetric flask containing 10.0 mL of *Internal standard solution.* Add water to volume, mix, and filter through a microporous filter, discarding the first 5 mL of the filtrate and collecting the remainder in a suitable container.

Assay preparation—Transfer an accurately measured volume of Butorphanol Tartrate Injection, equivalent to about 10 mg of butorphanol tartrate, to a 50-mL volumetric flask. Add 10.0 mL of *Internal standard solution,* mix, add water to volume, and mix. Filter through a microporous filter, discarding the first 5 mL of the filtrate and collecting the remainder in a suitable container.

Chromatographic system (see *Chromatography* ⟨621⟩)—The liquid chromatograph is equipped with a 280-nm detector and a 4-mm × 30-cm column that contains packing L11. The flow rate is about 2 mL per minute. Chromatograph five replicate injections of the *Standard preparation,* and record the peak responses as directed under *Procedure:* the relative standard deviation is not more than 1.5%, and the capacity factor for butorphanol tartrate is not less than 2.0.

Procedure—Separately inject equal volumes (about 20 µL) of the *Standard preparation* and the *Assay preparation* into the chromatograph, adjusting the flow rate and other operating parameters, if necessary, until satisfactory chromatography and peak responses are obtained. Record the chromatograms, and measure the responses for the major peaks. The relative retention times are about 1.7 for propylparaben and 1.0 for butorphanol tartrate. Calculate the quantity, in mg, of $C_{21}H_{29}NO_2 . C_4H_6O_6$ in each mL of the Injection taken by the formula:

$$50(C/V)(R_U/R_S),$$

in which *C* is the concentration, in mg per mL, of USP Butorphanol Tartrate RS in the *Standard preparation, V* is the volume, in mL, of Injection taken, and R_U and R_S are the peak response ratios of the butorphanol tartrate peak and the internal standard peak obtained from the *Assay preparation* and the *Standard preparation,* respectively.

Butyl Alcohol—*see* Butyl Alcohol NF

Butylated Hydroxyanisole—*see* Butylated Hydroxyanisole NF

Butylated Hydroxytoluene—*see* Butylated Hydroxytoluene NF

Butylparaben—*see* Butylparaben NF

Caffeine

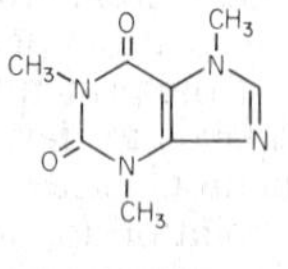

$C_8H_{10}N_4O_2$ (anhydrous) 194.19
1*H*-Purine-2,6-dione, 3,7-dihydro-1,3,7-trimethyl-.
1,3,7-Trimethylxanthine [*58-08-2*].
Monohydrate 212.21 [*5743-12-4*].

» Caffeine is anhydrous or contains one molecule of water of hydration. It contains not less than 98.5 percent and not more than 101.0 percent of $C_8H_{10}N_4O_2$, calculated on the anhydrous basis.

Packaging and storage—Preserve hydrous Caffeine in tight containers. Preserve anhydrous Caffeine in well-closed containers.

Labeling—Label it to indicate whether it is anhydrous or hydrous.

Reference standard—*USP Caffeine Reference Standard*—Dry at 80° for 4 hours before using.

Identification—

A: The infrared absorption spectrum of a mineral oil dispersion of it, previously dried, exhibits maxima only at the same wavelengths as that of a similar preparation of USP Caffeine RS.

B: Dissolve about 5 mg in 1 mL of hydrochloric acid in a porcelain dish, add 50 mg of potassium chlorate, and evaporate on a steam bath to dryness. Invert the dish over a vessel containing a few drops of 6 *N* ammonium hydroxide: the residue acquires a purple color, which disappears upon the addition of a solution of a fixed alkali.

Melting range ⟨741⟩: between 235° and 237.5°, determined after drying at 80° for 4 hours.

Water, *Method III* ⟨921⟩—Dry it at 80° for 4 hours: the anhydrous form loses not more than 0.5%, and the hydrous form not more than 8.5% of its weight.

Residue on ignition ⟨281⟩: not more than 0.1%.

Arsenic, *Method II* ⟨211⟩: 3 ppm.

Heavy metals ⟨231⟩—Mix 2.0 g with 5 mL of 0.1 *N* hydrochloric acid and 45 mL of water, warm gently until solution is complete, and cool to room temperature. Use 25 mL of this solution for the test: the limit is 0.001%.

Readily carbonizable substances ⟨271⟩—Dissolve 0.5 g in 5 mL of sulfuric acid TS: the solution has no more color than Matching Fluid D.

Other alkaloids—To 5 mL of a solution (1 in 50) add mercuric-potassium iodide TS: no precipitate is formed.

Assay—Dissolve about 400 mg, accurately weighed, of finely powdered Caffeine, with warming, in 40 mL of acetic anhydride. Cool, add 80 mL of benzene, and titrate with 0.1 *N* perchloric acid VS, determining the end-point potentiometrically. Each mL of 0.1 *N* perchloric acid is equivalent to 19.42 mg of $C_8H_{10}N_4O_2$.

Caffeine Capsules, Acetaminophen, Aspirin, and—*see* Acetaminophen, Aspirin, and Caffeine Capsules

Caffeine Capsules, Propoxyphene Hydrochloride, Aspirin, and—*see* Propoxyphene Hydrochloride, Aspirin, and Caffeine Capsules

Caffeine Suppositories, Ergotamine Tartrate and—*see* Ergotamine Tartrate and Caffeine Suppositories

Caffeine Tablets, Acetaminophen, Aspirin, and—*see* Acetaminophen, Aspirin, and Caffeine Tablets

Caffeine Tablets, Ergotamine Tartrate and—*see* Ergotamine Tartrate and Caffeine Tablets

Caffeine and Sodium Benzoate Injection

» Caffeine and Sodium Benzoate Injection is a sterile solution of Caffeine and Sodium Benzoate in Water for Injection. It contains an amount of anhydrous caffeine ($C_8H_{10}N_4O_2$) equivalent to not less than 45.0 percent and not more than 52.0 percent, and an amount of sodium benzoate ($C_7H_5NaO_2$) equivalent to not less than 47.5 percent and not more than 55.5

percent, of the labeled amounts of caffeine and sodium benzoate.

Packaging and storage—Preserve in single-dose containers, preferably of Type I glass.

Reference standard—*USP Caffeine Reference Standard*—Dry at 80° for 4 hours before using.

Identification—

A: The caffeine obtained in the *Assay for caffeine* herein responds to *Identification test A* under *Caffeine*.

B: Dip the end of a platinum wire into a portion of Injection, and then introduce it into a nonluminous flame: the flame is colored intensely yellow.

C: To about 0.5 mL of Injection add a few drops of ferric chloride TS: a salmon-colored precipitate is formed. To another portion of Injection add 3 *N* hydrochloric acid: a white precipitate is formed.

pH ⟨791⟩: between 6.5 and 8.5.

Other requirements—It meets the requirements under *Injections* ⟨1⟩.

Assay for caffeine—Measure accurately a volume of Caffeine and Sodium Benzoate Injection, equivalent to about 250 mg each of caffeine and sodium benzoate, transfer it completely with the aid of about 5 mL of water to a small separator, add 1 drop of phenolphthalein TS, and add 0.1 *N* sodium hydroxide, dropwise, until a permanent pink color is just produced. Shake the mixture with three or more 20-mL portions of chloroform to effect complete extraction of the caffeine, passing each chloroform extract through a small filter previously moistened with chloroform into a tared dish. (Retain the water layer for the *Assay for sodium benzoate*.) Wash the stem of the separator, the filter, and the funnel with 10 mL of hot chloroform, adding the washings to the dish. Evaporate the combined chloroform solutions on a steam bath, adding 2 mL of alcohol just before the last trace of chloroform is expelled. Complete the evaporation of the solvent, dry the residue, consisting of $C_8H_{10}N_4O_2$, at 80° for 4 hours, cool, and weigh.

Assay for sodium benzoate—To the water layer obtained in the *Assay for caffeine* add 75 mL of ether and 5 drops of methyl orange TS. Titrate with 0.1 *N* hydrochloric acid VS, mixing the liquids by vigorous shaking, until a permanent pink color is produced in the water layer. Each mL of 0.1 *N* hydrochloric acid is equivalent to 14.41 mg of $C_7H_5NaO_2$.

Calamine

Iron oxide (Fe_2O_3), mixture with zinc oxide.
Calamine (pharmaceutical preparation) [*8011-96-9*].

» Calamine is Zinc Oxide with a small proportion of ferric oxide, and contains, after ignition, not less than 98.0 percent and not more than 100.5 percent of zinc oxide (ZnO).

Packaging and storage—Preserve in well-closed containers.

Identification—

A: Treat 1 g with 10 mL of 3 *N* hydrochloric acid, and filter: the filtrate responds to the tests for *Zinc* ⟨191⟩.

B: Treat 1 g with 10 mL of 3 *N* hydrochloric acid, heat to boiling, and filter: the filtrate assumes a reddish color upon the addition of ammonium thiocyanate TS.

Microbial limits—It meets the requirements of the tests for absence of *Staphylococcus aureus* and *Pseudomonas aeruginosa* under *Microbial Limit Tests* ⟨61⟩.

Loss on ignition ⟨733⟩—Weigh accurately about 2 g, and ignite at 500° to constant weight: it loses not more than 2.0% of its weight.

Acid-insoluble substances—Dissolve 2.0 g in 50 mL of 3 *N* hydrochloric acid. If an insoluble residue remains, collect it on a tared filter, wash with water, dry at 105° for 1 hour, cool, and weigh: the weight of the residue does not exceed 40 mg (2.0%).

Alkaline substances—Digest 1.0 g with 20 mL of water on a steam bath for 15 minutes, filter, and add 2 drops of phenolphthalein TS: if a red color is produced, not more than 0.20 mL of 0.10 *N* sulfuric acid is required to discharge it.

Arsenic, *Method I* ⟨211⟩: 8 ppm.

Calcium—Digest 1 g in 25 mL of 3 *N* hydrochloric acid for 30 minutes, filter to remove the insoluble ferric oxide, and add 6 *N* ammonium hydroxide to the filtrate until the precipitate first formed is redissolved, then add 5 mL more of 6 *N* ammonium hydroxide. To 10 mL of this solution add 2 mL of ammonium oxalate TS: not more than a slight turbidity is produced.

Calcium or magnesium—To another 10-mL portion of the solution prepared for the test for *Calcium* add 2 mL of dibasic sodium phosphate TS: not more than a slight turbidity is produced.

Lead—To 1 g add 15 mL of water, stir well, then add 3 mL of glacial acetic acid, and warm on a steam bath until dissolved. Filter, and add 5 drops of potassium chromate TS: no turbidity is produced.

Assay—Digest about 1.5 g of freshly ignited Calamine, accurately weighed, with 50.0 mL of 1 *N* sulfuric acid VS, applying gentle heat until no further solution occurs. Filter the mixture, and wash the residue on the filter with hot water until the last washing is neutral to litmus paper. To the combined filtrate and washings add 2.5 g of ammonium chloride, cool, add methyl orange TS, and titrate with 1 *N* sodium hydroxide VS. Each mL of 1 *N* sulfuric acid is equivalent to 40.69 mg of ZnO.

Calamine Lotion

» Prepare Calamine Lotion as follows.

Calamine	80 g
Zinc Oxide	80 g
Glycerin	20 mL
Bentonite Magma	250 mL
Calcium Hydroxide Topical Solution, a sufficient quantity, to make	1000 mL

Dilute the Bentonite Magma with an equal volume of Calcium Hydroxide Topical Solution. Mix the powders intimately with the Glycerin and about 100 mL of the diluted magma, triturating until a smooth, uniform paste is formed. Gradually incorporate the remainder of the diluted magma. Finally add enough Calcium Hydroxide Solution to make 1000 mL, and shake well.

If a more viscous consistency in the Lotion is desired, the quantity of Bentonite Magma may be increased to not more than 400 mL.

NOTE—Shake Calamine Lotion well before dispensing.

Packaging and storage—Preserve in tight containers.

Microbial limits—It meets the requirements of the tests for absence of *Staphylococcus aureus* and *Pseudomonas aeruginosa* under *Microbial Limit Tests* ⟨61⟩.

Phenolated Calamine Lotion

» Prepare Phenolated Calamine Lotion as follows.

Liquefied Phenol	10 mL
Calamine Lotion	990 mL
To make	1000 mL

Mix the ingredients.

NOTE—Shake Phenolated Calamine Lotion well before dispensing.

Packaging and storage—Preserve in tight containers.

Calcifediol

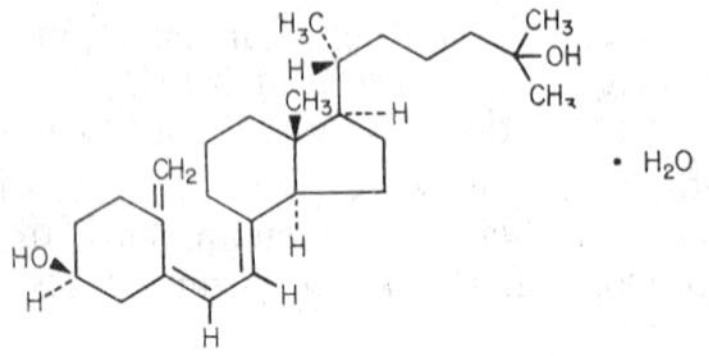

$C_{27}H_{44}O_2.H_2O$ 418.66
9,10-Secocholesta-5,7,10(19)-triene-3,25-diol monohydrate, (3β,5*Z*,7*E*)-.
25-Hydroxycholecalciferol monohydrate [*63283-36-3*].

» Calcifediol contains not less than 97.0 percent and not more than 103.0 percent of $C_{27}H_{44}O_2.H_2O$.

Packaging and storage—Preserve in tight, light-resistant containers at controlled room temperature.

Reference standard—*USP Calcifediol Reference Standard*—Do not dry before using.

Identification—The infrared absorption spectrum of a mineral oil dispersion of it exhibits maxima only at the same wavelengths as that of a similar preparation of USP Calcifediol RS.

Water, *Method Ia* ⟨921⟩: between 3.8% and 5.0%, determined on a 0.2-g specimen.

Assay—

Internal standard solution—Dissolve testosterone in ethyl acetate to obtain a solution having a concentration of about 0.10 mg per mL.

Mobile phase—Prepare a suitable degassed solution of about 6 volumes of heptane, 6 volumes of water-saturated heptane, 3 volumes of methylene chloride, and 5 volumes of ethyl acetate.

Standard preparation—Dissolve an accurately weighed quantity of USP Calcifediol RS in *Internal standard solution*, and dilute quantitatively and stepwise with *Internal standard solution* to obtain a solution having a known concentration of about 20 µg per mL.

Assay preparation—Transfer about 10 mg of Calcifediol, accurately weighed, to a 50-mL volumetric flask, dissolve in *Internal standard solution*, dilute with *Internal standard solution* to volume, and mix. Transfer 5.0 mL of this solution to a 50-mL volumetric flask, dilute with *Internal standard solution* to volume, and mix.

Chromatographic system (see *Chromatography* ⟨621⟩)—The liquid chromatograph is equipped with a 4-mm × 30-cm column that contains packing L3, a detector that monitors absorption at 254 nm, and a pump capable of providing constant flow up to a minimum of 2000 psi.

System suitability—The relative standard deviation for peak response ratios for four replicate injections of *Standard preparation* is not more than 3.0%, and the resolution factor is not less than 3.0.

Procedure—Introduce equal volumes of the *Standard preparation* and the *Assay preparation* into the chromatograph (see *Chromatography* ⟨621⟩), and measure the peak responses obtained. Calculate the quantity, in mg, of $C_{27}H_{44}O_2.H_2O$ in the portion of Calcifediol taken by the formula:

$$0.5C(R_U/R_S),$$

in which C is the concentration, in µg per mL, of USP Calcifediol RS in the *Standard preparation*, and R_U and R_S are the ratios of the peak responses of calcifediol to testosterone obtained from the *Assay preparation* and the *Standard preparation*, respectively.

Calcifediol Capsules

» Calcifediol Capsules contain not less than 90.0 percent and not more than 120.0 percent of the labeled amount of $C_{27}H_{44}O_2.H_2O$.

Packaging and storage—Preserve in tight, light-resistant containers.

Reference standard—*USP Calcifediol Reference Standard*—Do not dry before using.

Identification—Transfer the contents of a number of Capsules, equivalent to about 150 µg of calcifediol, to a suitable container, add 1 mL of methanol, and shake vigorously for 1 minute. Separate the layers by centrifugation, and transfer as much of the top, methanol layer as possible to a second container. Evaporate this extract to dryness, and dissolve the residue in about 1 mL of chloroform. Proceed as directed under *Thin-layer Chromatographic Identification Test* ⟨201⟩, applying 20 µL of this solution and 20 µL of a solution containing about the same concentration of USP Calcifediol RS in chloroform, and using a solvent system consisting of 60 parts of cyclohexane and 40 parts of ethyl acetate.

Disintegration ⟨701⟩—Proceed as directed for *Hard Gelatin Capsules*, using 600 mL of water as the disintegration fluid so that the top of the basket does not descend beneath the surface of the fluid. The time limit is 30 minutes. A capsule is considered to be disintegrated when oil is visible on the surface of the fluid in the tube.

Uniformity of dosage units ⟨905⟩: meet the requirements.

Assay—

Internal standard solution—Dissolve testosterone in ethyl acetate to obtain a solution having a concentration of about 35 µg per mL.

Mobile phase—Prepare as directed in the *Assay* under *Calcifediol*.

Standard preparation—Dissolve an accurately weighed quantity of USP Calcifediol RS in *Internal standard solution*, and dilute quantitatively and stepwise with *Internal standard solution* to obtain a solution having a known concentration of about 7 µg of USP Calcifediol RS per mL.

Assay preparation—Transfer a number of Calcifediol Capsules to a suitable container. Using a suitable implement, shear open a number of Capsules inside the container. Wash the implement with an accurately measured volume of *Internal standard solution* that will yield a solution having a concentration of about 7 µg of calcifediol per mL. Collect the rinsings in the container, and mix to obtain a homogeneous solution of the Capsule contents.

Chromatographic system and *System suitability*—Proceed as directed in the *Assay* under *Calcifediol*.

Procedure—Proceed as directed for *Procedure* in the *Assay* under *Calcifediol*. Calculate the quantity, in µg, of $C_{27}H_{44}O_2.H_2O$ in the portion of Capsule contents taken by the formula:

$$CV_U(R_U/R_S),$$

in which C is the concentration, in µg per mL, of USP Calcifediol RS in the *Standard preparation*, V_U is the volume, in mL, of *Internal standard solution* taken for the *Assay preparation*, and R_U and R_S are the peak response ratios of calcifediol to testosterone obtained from the *Assay preparation* and the *Standard preparation*, respectively.

Calcium Acetate

$(CH_3COO—)_2Ca$

$C_4H_6CaO_4$ 158.17

Acetic acid, calcium salt.
Calcium acetate [*543-90-8*].

» Calcium Acetate contains not less than 99.0 percent and not more than 100.5 percent of $C_4H_6CaO_4$, calculated on the anhydrous basis.

Packaging and storage—Preserve in tight containers.

Labeling—Where Calcium Acetate is intended for use in hemodialysis or peritoneal dialysis it is so labeled.

Identification—A solution (1 in 20) responds to the tests for *Calcium* ⟨191⟩ and for *Acetate* ⟨191⟩.

pH ⟨791⟩: 7.8 to 8.2, in a solution (1 in 20).

Water, *Method I* ⟨921⟩: not more than 7.0%, determined in a 0.7-g specimen, 20 mL of glacial acetic acid being added to the titration vessel in addition to the methanol.

Fluoride—Proceed as directed in the test for *Fluoride* under *Dibasic Calcium Phosphate:* the limit is 0.005%.

Arsenic, *Method I* ⟨211⟩: 3 ppm.

Heavy metals, *Method I* ⟨231⟩: 0.0025%.

Lead ⟨251⟩: 0.001%.

Chloride ⟨221⟩—A 0.15-g portion shows no more chloride than corresponds to 0.75 mL of 0.020 *N* hydrochloric acid (0.05%).

Sulfate ⟨221⟩—A 0.25-g portion shows no more sulfate than corresponds to 0.15 mL of 0.020 *N* sulfuric acid (0.06%).

Nitrate—Dissolve 1.0 g of it in 10 mL of water, add 5 mg of sodium chloride, 0.05 mL of indigo carmine TS, and, with stirring, 10 mL of nitrogen-free sulfuric acid: the blue color persists for not less than 10 minutes.

Readily oxidizable substances—Dissolve 2.0 g of it in 100 mL of boiling water, add a few glass beads, 6 mL of 10 *N* sulfuric acid, and 0.3 mL of 1 *N* potassium permanganate, mix, boil gently for 5 minutes, and allow the precipitate to settle: the pink color in the supernatant liquid is not completely discharged.

Aluminum (where it is labeled as intended for parenteral use or for use in hemodialysis or peritoneal dialysis)—Proceed as directed in the test for *Aluminum* under *Sodium Bicarbonate.*

Barium (where it is labeled as intended for use in hemodialysis or peritoneal dialysis)—[NOTE—The *Standard preparation* and the test solutions may be modified, if necessary, to obtain solutions, of suitable concentrations, adaptable to the linear or working range of the instrument.]

Standard preparation—Dissolve an accurately weighed quantity of anhydrous barium chloride in water to obtain a solution having a concentration of 1.516 mg per mL. This solution contains 1000 µg of barium per mL. Dilute an accurately measured volume of this solution quantitatively and stepwise, if necessary, with water to obtain a solution containing 12.5 µg of barium per mL.

Test preparation—Transfer 5.0 g of Calcium Acetate to a 100-mL volumetric flask, dissolve in water, dilute with water to volume, and mix.

Procedure—To three separate 25-mL volumetric flasks add 0 mL, 2.0 mL, and 4.0 mL of the *Standard preparation.* To each flask add 20.0 mL of the *Test preparation*, dilute with water to volume, and mix. These test solutions contain, respectively, 0, 1.0, and 2.0 µg per mL of barium from the *Standard preparation.* Concomitantly determine the absorbances of the test solutions at the barium emission line at 455.5 nm with a suitable atomic absorption spectrophotometer (see *Spectrophotometry and Light-scattering* ⟨851⟩) equipped with a nitrous oxide–acetylene flame, using water as the blank. Plot the absorbances of the test solutions versus their contents, in µg of barium per mL, as furnished by the *Standard preparation*, draw the straight line best fitting the three points, and extrapolate the line until it intercepts the concentration axis. From the intercept determine the amount, in µg, of barium in each mL of the test solution containing 0 mL of the *Standard preparation.* Calculate the percentage of Ba in the specimen by multiplying this value by 0.0025: the limit is 0.005%.

Magnesium (where it is labeled as intended for use in hemodialysis or peritoneal dialysis)—[NOTE—The *Standard preparation* and the test solutions may be modified, if necessary, to obtain solutions, of suitable concentrations, adaptable to the linear or working range of the instrument.]

Standard preparation—Dissolve an accurately weighed quantity of magnesium oxide in 1 *N* nitric acid to obtain a solution having a concentration of 1.516 mg per mL. This solution contains 1000 µg of magnesium per mL. Dilute an accurately measured volume of this solution quantitatively and stepwise, if necessary, with water to obtain a solution containing 5.0 µg of magnesium per mL.

Test preparation—Transfer 200 mg of Calcium Acetate to a 100-mL volumetric flask, dissolve in water, dilute with water to volume, and mix.

Procedure—To three separate 25-mL volumetric flasks add 0 mL, 2.0 mL, and 4.0 mL of the *Standard preparation.* To each flask add 20.0 mL of the *Test preparation*, dilute with water to volume, and mix. These test solutions contain, respectively, 0, 0.4, and 0.8 µg per mL of magnesium from the *Standard preparation.* Concomitantly determine the absorbances of the test solutions at the magnesium emission line at 285.2 nm with a suitable atomic absorption spectrophotometer (see *Spectrophotometry and Light-scattering* ⟨851⟩) equipped with an air-acetylene flame, using water as the blank. Plot the absorbances of the test solutions versus their contents of magnesium, in µg per mL, as furnished by the *Standard preparation*, draw the straight line best fitting the three points, and extrapolate the line until it intercepts the concentration axis. From the intercept determine the amount, in µg, of magnesium in each mL of the test solution containing 0 mL of the *Standard preparation.* Calculate the percentage of magnesium in the specimen by multiplying this value by 0.0625: the limit is 0.05%.

Potassium (where it is labeled as intended for use in hemodialysis or peritoneal dialysis)—[NOTE—The *Standard preparation* and the test solutions may be modified, if necessary, to obtain solutions, of suitable concentrations, adaptable to the linear or working range of the instrument.]

Standard preparation—Transfer 5.959 g of potassium chloride, previously dried at 105° for 2 hours and accurately weighed, to a 250-mL volumetric flask, diluted with water to volume, and mix. This solution contains 12.5 mg of potassium per mL. Dilute an accurately measured volume of this solution quantitatively and stepwise, if necessary, with water to obtain a solution containing 31.25 µg of potassium per mL.

Test preparation—Transfer 1.25 g of Calcium Acetate to a 100-mL volumetric flask, dissolve in water, dilute with water to volume, and mix.

Procedure—To three separate 25-mL volumetric flasks add 0 mL, 2.0 mL, and 4.0 mL of the *Standard preparation.* To each flask add 20.0 mL of the *Test preparation*, dilute with water to volume, and mix. These test solutions contain, respectively, 0, 2.5, and 5.0 µg per mL of potassium from the *Standard preparation.* Concomitantly determine the absorbances of the test solutions at the potassium emission line at 766.7 nm with a suitable atomic absorption spectrophotometer (see *Spectrophotometry and Light-scattering* ⟨851⟩) equipped with a nitrous oxide–acetylene flame, using water as the blank. Plot the absorbances of the test solutions versus their contents of potassium, in µg per mL, as furnished by the *Standard preparation*, draw the straight line best fitting the three points, and extrapolate the line until it intercepts the concentration axis. From the intercept determine the amount, in µg, of potassium in each mL of the test solution containing 0 mL of the *Standard preparation.* Calculate the percentage of potassium in the specimen by multiplying this value by 0.01: the limit is 0.05%.

Sodium (where it is labeled as intended for use in hemodialysis or peritoneal dialysis)—[NOTE—The *Standard preparation* and the test solutions may be modified, if necessary, to obtain solutions, of suitable concentrations, adaptable to the linear or working range of the instrument.]

Standard preparation—Transfer 6.355 g of sodium chloride, previously dried at 105° for 2 hours and accurately weighed, to a 250-mL volumetric flask, dilute with water to volume, and mix. This solution contains 10.0 mg of sodium per mL. Dilute an accurately measured volume of this solution quantitatively and stepwise, if necessary, with water to obtain a solution containing 250 µg of sodium per mL.

Test preparation—Transfer 1.0 g of Calcium Acetate to a 100-mL volumetric flask, dissolve in water, dilute with water to volume, and mix.

Procedure—To three separate 25-mL volumetric flasks add 0 mL, 2.0 mL, and 4.0 mL of the *Standard preparation.* To each flask add 20.0 mL of the *Test preparation,* dilute with water to volume, and mix. These test solutions contain, respectively, 0, 20.0, and 40.0 µg per mL of sodium from the *Standard preparation.* Concomitantly determine the absorbances of the test solutions at the sodium emission line at 589.0 nm with a suitable atomic absorption spectrophotometer (see *Spectrophotometry and Light-scattering* ⟨851⟩) equipped with a nitrous oxide–acetylene flame, using water as the blank. Plot the absorbances of the test solutions versus their contents of sodium, in µg per mL, as furnished by the *Standard preparation,* draw the straight line best fitting the three points, and extrapolate the line until it intercepts the concentration axis. From the intercept determine the amount, in µg, of sodium in each mL of the test solution containing 0 mL of the *Standard preparation.* Calculate the percentage of sodium in the specimen by multiplying this value by 0.0125: the limit is 0.5%.

Strontium (where it is labeled as intended for use in hemodialysis or peritoneal dialysis)—[NOTE—The *Standard preparation* and the test solutions may be modified, if necessary, to obtain solutions, of suitable concentrations, adaptable to the linear or working range of the instrument.]

Standard preparation—Dissolve an accurately weighed quantity of strontium acetate in water to obtain a solution having a concentration of 2.45 mg per mL. This solution contains 1000 µg of strontium per mL. Dilute an accurately measured volume of this solution quantitatively and stepwise, if necessary, with water to obtain a solution containing 50.0 µg of strontium per mL.

Test preparation—Transfer 2.0 g of Calcium Acetate to a 100-mL volumetric flask, dissolve in water, dilute with water to volume, and mix.

Procedure—To three separate 25-mL volumetric flasks add 0 mL, 2.0 mL, and 4.0 mL of the *Standard preparation.* To each flask add 20.0 mL of the *Test preparation,* dilute with water to volume, and mix. These test solutions contain, respectively, 0, 4.0, and 8.0 µg per mL of strontium from the *Standard preparation.* Concomitantly determine the absorbances of the test solutions at the strontium emission line at 460.7 nm with a suitable atomic absorption spectrophotometer (see *Spectrophotometry and Light-scattering* ⟨851⟩) equipped with a nitrous oxide–acetylene flame, using water as the blank. Plot the absorbances of the test solutions versus their contents of strontium, in µg per mL, as furnished by the *Standard preparation,* draw the straight line best fitting the three points, and extrapolate the line until it intercepts the concentration axis. From the intercept determine the amount, in µg, of strontium in each mL of the test solution containing 0 mL of the *Standard preparation.* Calculate the percentage of strontium in the specimen by multiplying this value by 0.00625: the limit is 0.05%.

Assay—Weigh accurately about 300 mg of Calcium Acetate, and dissolve in 150 mL of water containing 2 mL of 3 *N* hydrochloric acid. While stirring, preferably with a magnetic stirrer, add about 30 mL of 0.05 *M* disodium ethylenediaminetetraacetate VS from a 50-mL buret, then add 15 mL of 1 *N* sodium hydroxide and 300 mg of hydroxy naphthol blue indicator, and continue the titration to a blue end-point. Each mL of 0.05 *M* disodium ethylenediaminetetraacetate is equivalent to 7.909 mg of $C_4H_6CaO_4$.

Precipitated Calcium Carbonate

$CaCO_3$ 100.09

Carbonic acid, calcium salt (1:1).

Calcium carbonate (1:1) [*471-34-1*].

» Precipitated Calcium Carbonate, dried at 200° for 4 hours, contains calcium equivalent to not less than 98.0 percent and not more than 100.5 percent of $CaCO_3$.

Packaging and storage—Preserve in well-closed containers.

Identification—The addition of acetic acid to it produces effervescence, and the resulting solution, after boiling, responds to the tests for *Calcium* ⟨191⟩.

Loss on drying ⟨731⟩—Dry it at 200° for 4 hours: it loses not more than 2.0% of its weight.

Acid-insoluble substances—Mix 5.0 g with 10 mL of water, and add hydrochloric acid, dropwise, with agitation, until it ceases to cause effervescence, then add water to make the mixture measure 200 mL, and filter. Wash the insoluble residue with water until the last washing shows no chloride, and ignite: the weight of the residue does not exceed 10 mg (0.2%).

Fluoride—[NOTE—Prepare and store all solutions in plastic containers.]

Buffer solution, Standard solution, and *Electrode system*—Proceed as directed in the test for *Fluoride* under *Dibasic Calcium Phosphate.*

Standard response line—Proceed as directed in the test for *Fluoride* under *Dibasic Calcium Phosphate,* except to use 4.0 mL of hydrochloric acid, instead of 2.0 mL.

Procedure—Proceed as directed in the test for *Fluoride* under *Dibasic Calcium Phosphate,* except to use 4.0 mL of hydrochloric acid, instead of 2.0 mL. The limit is 0.005%.

Arsenic, *Method I* ⟨211⟩—Slowly dissolve 1.0 g in 15 mL of hydrochloric acid, and dilute with water to 55 mL: the resulting solution meets the requirements of the test, the addition of 20 mL of 7 *N* sulfuric acid specified under *Procedure* being omitted. The limit is 3 ppm.

Barium—A platinum wire, dipped in the filtrate obtained in the test for *Acid-insoluble substances* and held in a nonluminous flame, does not impart a green color.

Lead ⟨251⟩—Mix 1.0 g with 5 mL of water, slowly add 8 mL of 3 *N* hydrochloric acid, evaporate on a steam bath to dryness, and dissolve the residue in 5 mL of water: the limit is 0.001%.

Iron—Dissolve 80 mg in 5 mL of 2 *N* hydrochloric acid, transfer to a comparison tube with the aid of water, and dilute with water to 10 mL. Prepare a Standard solution by transferring 4.0 mL of *Standard Iron Solution,* prepared as directed under *Iron* ⟨241⟩, to a similar comparison tube and diluting with water to 10 mL. To each tube add 2 mL of citric acid solution (1 in 5) and 2 drops of thioglycolic acid, render just alkaline with ammonia TS, dilute with water to 20 mL, mix, and allow to stand for 5 minutes: any pink color produced by the specimen under test is not more intense than that produced by the Standard solution (0.05%).

Heavy metals ⟨231⟩—Mix 670 mg with 5 mL of water, slowly add 8 mL of 3 *N* hydrochloric acid, and evaporate on a steam bath to dryness. Dissolve the residue in 20 mL of water, filter, and add water to the filtrate to make 25 mL: the limit is 0.003%.

Magnesium and alkali salts—Mix 1.0 g with 35 mL of water, carefully add 3 mL of hydrochloric acid, heat the solution, and boil for 1 minute. Rapidly add 40 mL of oxalic acid TS, and stir vigorously until precipitation is well established. Add immediately to the warm mixture 2 drops of methyl red TS and then 6 *N* ammonium hydroxide, dropwise, until the mixture is just alkaline. Cool to room temperature, transfer to a 100-mL graduated cylinder, dilute with water to 100 mL, mix, and allow to stand for 4 hours or overnight. Filter, and to 50 mL of the clear filtrate in a platinum dish add 0.5 mL of sulfuric acid, and evaporate the mixture on a steam bath to a small volume. Carefully heat over a free flame to dryness, and continue heating to complete decomposition and volatilization of ammonium salts. Finally ignite the residue to constant weight. The weight of the residue does not exceed 5 mg (1.0%).

Assay—Weigh accurately about 200 mg of Precipitated Calcium Carbonate, previously dried at 200° for 4 hours, and transfer to a 250-mL beaker. Moisten thoroughly with a few mL of water, then add, dropwise, sufficient 3 *N* hydrochloric acid to effect complete solution. Add 100 mL of water, 15 mL of 1 *N* sodium hydroxide, and 300 mg of hydroxy naphthol blue trituration, and titrate with 0.05 *M* disodium ethylenediaminetetraacetate VS until the solution is a distinct blue in color. Each mL of 0.05 *M* disodium ethylenediaminetetraacetate is equivalent to 5.004 mg of $CaCO_3$.

Calcium Carbonate Oral Suspension

» Calcium Carbonate Oral Suspension contains not less than 90.0 percent and not more than 110.0 percent of the labeled amount of $CaCO_3$.

Packaging and storage—Preserve in tight containers, and avoid freezing.

Identification—The addition of acetic acid to it produces effervescence, and the resulting solution, after being boiled, responds to the tests for *Calcium* ⟨191⟩.

Microbial limits—Its total aerobic microbial count does not exceed 100 per mL, and it meets the requirements of the test for absence of *Escherichia coli* and *Pseudomonas aeruginosa* under *Microbial Limit Tests* ⟨61⟩.

pH ⟨791⟩: between 7.5 and 8.7.

Other requirements—It meets the requirements of the tests for *Fluoride*, *Arsenic*, *Lead*, and *Heavy metals* under *Precipitated Calcium Carbonate*, portions of Oral Suspension adequate to provide quantities of calcium carbonate equivalent to the stated amounts of test specimen being used.

Assay—Transfer an accurately measured quantity of Calcium Carbonate Oral Suspension, previously well shaken in its original container, equivalent to about 1 g of calcium carbonate, to a beaker with the aid of 25 mL of water, and add 20 mL of 1 *N* hydrochloric acid. Heat on a steam bath for 30 minutes, allow to cool, transfer with the aid of water to a 100-mL volumetric flask, dilute with water to volume, mix, and filter. Transfer 20.0 mL of the filtrate to a suitable container, dilute with water to 100 mL, add 15 mL of 1 *N* sodium hydroxide, 5 mL of triethanolamine, and 100 mg of hydroxy naphthol blue trituration, and titrate with 0.05 *M* disodium ethylenediaminetetraacetate VS until the solution is deep blue. Each mL of 0.05 *M* disodium ethylenediaminetetraacetate is equivalent to 5.004 mg of $CaCO_3$.

Calcium Carbonate Oral Suspension, Alumina, Magnesia, and—*see* Alumina, Magnesia, and Calcium Carbonate Oral Suspension

Calcium Carbonate Tablets

» Calcium Carbonate Tablets contain not less than 92.5 percent and not more than 107.5 percent of the labeled amount of $CaCO_3$.

Packaging and storage—Preserve in well-closed containers.

Identification—The addition of 6 *N* acetic acid to the Tablets produces effervescence, and the resulting solution, after being boiled to expel carbon dioxide and neutralized with 6 *N* ammonium hydroxide, responds to the tests for *Calcium* ⟨191⟩.

Disintegration ⟨701⟩: Where Tablets are labeled solely for antacid use, the disintegration time is 10 minutes.

Dissolution ⟨711⟩—For Tablets labeled for any indication other than, or in addition to, antacid use.

Medium: 0.1 *N* hydrochloric acid; 900 mL.
Apparatus 2: 75 rpm.
Time: 30 minutes.

DETERMINATION OF DISSOLVED CALCIUM CARBONATE—

5% Lanthanum chloride solution—Prepare a solution of lanthanum chloride in 0.1 *N* hydrochloric acid having a concentration of about 50 mg per mL.

Blank—Pipet 25 mL of *5% Lanthanum chloride solution* into a 250-mL volumetric flask, dilute with 0.1 *N* hydrochloric acid to volume, and mix.

Standard stock solution—Dissolve an accurately weighed quantity of calcium carbonate in 0.1 *N* hydrochloric acid, and dilute quantitatively, and stepwise if necessary, with 0.1 *N* hydrochloric acid to obtain a solution having a known concentration of about 100 µg of calcium per mL.

Standard solutions—Into four 100-mL volumetric flasks, each containing 10.0 mL of *5% Lanthanum chloride solution*, separately pipet 3-, 4-, 5-, and 6-mL portions of *Standard stock solution*. Dilute each with 0.1 *N* hydrochloric acid to volume, and mix to obtain *Standard solutions* having known concentrations of about 3, 4, 5, and 6 µg of calcium per mL, respectively.

Test solution—Filter a portion of the solution under test. Pipet a volume of the filtrate, estimated to contain 1 mg of calcium, into a 250-mL volumetric flask, add 25.0 mL of *5% Lanthanum chloride solution*, dilute with 0.1 *N* hydrochloric acid to volume, and mix.

Procedure—Concomitantly determine the absorbances of the *Standard solutions* and the *Test solution* at the calcium emission wavelength of 422.8 nm, with a suitable atomic absorption spectrophotometer (see *Spectrophotometry and Light-scattering* ⟨851⟩), equipped with a calcium hollow-cathode lamp and an air-acetylene flame, against the *Blank*. Construct a standard curve by plotting absorbances versus calcium concentrations of the *Standard solutions*, then from it obtain the concentration, *C*, in µg of calcium per mL, of the *Test solution*, and calculate the quantity, in mg, of $CaCO_3$ dissolved by the formula:

$$(100.09/40.08)(225C/v),$$

in which 100.09 is the molecular weight of calcium carbonate, 40.08 is the atomic weight of calcium, and v is the volume of the filtrate taken to prepare the *Test solution*.

Tolerances—Not less than 75% (*Q*) of the labeled amount of $CaCO_3$ is dissolved in 30 minutes.

Uniformity of dosage units ⟨905⟩: meet the requirements.

Assay—Weigh and finely powder not less than 20 Calcium Carbonate Tablets. Weigh accurately a portion of the powder, equivalent to about 200 mg of calcium carbonate, transfer to a suitable crucible, and ignite to constant weight. Cool the crucible, add 10 mL of water, and dissolve the residue by adding sufficient 3 *N* hydrochloric acid, dropwise, to achieve complete solution. Transfer the solution completely to a suitable container, dilute with water to 150 mL, add 15 mL of 1 *N* sodium hydroxide and 300 mg of hydroxy naphthol blue trituration, and titrate with 0.05 *M* disodium ethylenediaminetetraacetate VS until the solution is deep blue. Each mL of 0.05 *M* disodium ethylenediaminetetraacetate is equivalent to 5.004 mg of $CaCO_3$.

Calcium Carbonate and Magnesia Tablets

» Calcium Carbonate and Magnesia Tablets contain not less than 90.0 percent and not more than 110.0 percent of the labeled amounts of $CaCO_3$ and $Mg(OH)_2$.

Packaging and storage—Preserve in well-closed containers.

Labeling—Label the Tablets to indicate that they are to be chewed before being swallowed.

Identification—The addition of 3 *N* hydrochloric acid to the Tablets produces effervescence, and the resulting solution, after being boiled to expel carbon dioxide and neutralized with 6 *N* ammonium hydroxide, responds to the tests for *Calcium* ⟨191⟩ and for *Magnesium* ⟨191⟩.

Disintegration ⟨701⟩: 30 minutes, simulated gastric fluid TS being used as the test medium.

Uniformity of dosage units ⟨905⟩: meet the requirements for *Weight Variation* with respect to calcium carbonate and to magnesia.

Assay for calcium carbonate—Weigh and finely powder not less than 20 Calcium Carbonate and Magnesia Tablets. Weigh accurately a portion of the powder, equivalent to about 400 mg of calcium carbonate, transfer to a beaker with 25 mL of water, and add 10 mL of 1 *N* hydrochloric acid. Heat on a steam bath for 30 minutes, allow to cool, transfer to a 100-mL volumetric flask with the aid of water, dilute with water to volume, mix, and filter. Transfer 20.0 mL of the filtrate to a suitable container, dilute with water to 100 mL, add 15 mL of 1 *N* sodium hydroxide, 5 mL of triethanolamine, and 100 mg of hydroxy naphthol blue trituration, and titrate with 0.05 *M* disodium ethylenediaminetetraacetate VS until the solution is deep blue in color. Each mL of 0.05 *M* disodium ethylenediaminetetraacetate is equivalent to 5.004 mg of $CaCO_3$.

Assay for magnesium hydroxide—Transfer an accurately measured portion of the filtrate remaining from the *Assay for calcium carbonate*, equivalent to about 120 mg of calcium carbonate and magnesium hydroxide combined, to a suitable container, dilute with water to 100 mL, add 10 mL of ammonia–ammonium chloride buffer TS, 5 mL of triethanolamine, and 0.3 mL of eriochrome black TS, and titrate with 0.05 *M* disodium ethylenediaminetetraacetate VS to a blue end-point. The volume, in mL, of 0.05 *M* disodium ethylenediaminetetraacetate consumed, less the volume of 0.05 *M* disodium ethylenediaminetetraacetate corresponding to the content of calcium carbonate in the volume, in mL, of the filtrate taken, represents the volume, in mL, of 0.05 *M* disodium ethylenediaminetetraacetate equivalent to the quantity of magnesium hydroxide present. Each mL of 0.05 *M* disodium ethylenediaminetetraacetate is equivalent to 2.916 mg of $Mg(OH)_2$.

Calcium and Magnesium Carbonates Tablets

» Calcium and Magnesium Carbonates Tablets contain not less than 90.0 percent and not more than 110.0 percent of the labeled amount of calcium carbonate ($CaCO_3$) and not less than 85.0 percent and not more than 115.0 percent of the labeled amount of magnesium carbonate ($MgCO_3$).

Packaging and storage—Preserve in well-closed containers.

Identification—

A: The addition of 1 *N* hydrochloric acid to 1 Tablet produces effervescence, and the resulting solution, after having been filtered, responds to the tests for *Calcium* ⟨191⟩.

B: Heat 2 Tablets in 20 mL of 1 *N* sulfuric acid. Cool, add 20 mL of alcohol, mix, and allow to stand for 30 minutes. Filter this solution, and add 2 mL of 1 *N* hydrochloric acid to the filtrate: this solution responds to the tests for *Magnesium* ⟨191⟩.

Disintegration ⟨701⟩: 10 minutes, simulated gastric fluid TS being substituted for water in the test.

Uniformity of dosage units ⟨905⟩: meet the requirements for *Weight Variation* with respect to calcium carbonate and to magnesium carbonate.

Assay for calcium carbonate—Weigh and finely powder not less than 20 Calcium and Magnesium Carbonates Tablets. Weigh accurately a portion of the powder, equivalent to about 400 mg of calcium carbonate, transfer to a beaker with the aid of 25 mL of water, and add 10 mL of 1 *N* hydrochloric acid. Heat on a steam bath for 30 minutes, allow to cool, transfer with the aid of water to a 100-mL volumetric flask, dilute with water to volume, mix, and filter. Transfer 20.0 mL of the filtrate to a suitable container, dilute with water to 100 mL, add 15 mL of 1 *N* sodium hydroxide, 5 mL of triethanolamine, and 100 mg of hydroxy naphthol blue trituration, and titrate with 0.05 *M* disodium ethylenediaminetetraacetate VS until the solution is deep blue. Each mL of 0.05 *M* disodium ethylenediaminetetraacetate is equivalent to 5.004 mg of $CaCO_3$.

Assay for magnesium carbonate—Transfer an accurately measured portion of the filtrate remaining from the *Assay for calcium carbonate*, equivalent to about 120 mg of calcium carbonate and magnesium carbonate combined, to a suitable container, dilute with water to 100 mL, add 10 mL of ammonia–ammonium chloride buffer TS, 5 mL of triethanolamine, and 0.3 mL of eriochrome black TS, and titrate with 0.05 *M* disodium ethylenediaminetetraacetate VS to a blue end-point. From the volume of 0.05 *M* disodium ethylenediaminetetraacetate consumed, deduct the volume of 0.05 *M* disodium ethylenediaminetetraacetate corresponding to the content of calcium carbonate in the volume of the filtrate taken for the assay. The difference is the volume of 0.05 *M* disodium ethylenediaminetetraacetate equivalent to the amount of magnesium carbonate present. Each mL of 0.05 *M* disodium ethylenediaminetetraacetate is equivalent to 4.216 mg of $MgCO_3$.

Calcium Carbonate Tablets, Alumina, Magnesia, and—*see* Alumina, Magnesia, and Calcium Carbonate Tablets

Calcium Chloride

$CaCl_2 . 2H_2O$ 147.02
Calcium chloride, dihydrate.
Calcium chloride dihydrate [*10035-04-8*].
Anhydrous 110.99 [*10043-52-4*].

» Calcium Chloride contains an amount of $CaCl_2$ equivalent to not less than 99.0 percent and not more than 107.0 percent of $CaCl_2 . 2H_2O$.

Packaging and storage—Preserve in tight containers.

Identification—A solution (1 in 10) responds to the tests for *Calcium* ⟨191⟩ and for *Chloride* ⟨191⟩.

pH ⟨791⟩: between 4.5 and 9.2, in a solution (1 in 20).

Arsenic, *Method I* ⟨211⟩: 3 ppm.

Heavy metals ⟨231⟩—Dissolve 2.0 g in 25 mL of water: the limit is 0.001%.

Iron, aluminum, and phosphate—To a solution (1 in 20) add 2 drops of 3 *N* hydrochloric acid and 1 drop of phenolphthalein TS. Then add ammonium chloride–ammonium hydroxide TS, dropwise, until the solution is faintly pink, add 2 drops in excess, and heat the liquid to boiling: no turbidity or precipitate is produced.

Magnesium and alkali salts—Dissolve 1 g in about 50 mL of water, add 500 mg of ammonium chloride, and proceed as directed in the test for *Magnesium and alkali salts* under *Precipitated Calcium Carbonate*, beginning with "heat the solution, and boil for 1 minute": the weight of the residue does not exceed 5 mg (1.0%).

Assay—Weigh accurately about 1 g of Calcium Chloride, transfer to a 250-mL beaker, and dissolve in a mixture of 100 mL of water and 5 mL of 3 *N* hydrochloric acid. Transfer the solution to a 250-mL volumetric flask, dilute with water to volume, and mix. Pipet 50 mL of the solution into a suitable container, add 100 mL of water, 15 mL of 1 *N* sodium hydroxide, and 300 mg of hydroxy naphthol blue indicator, and titrate with 0.05 *M* disodium ethylenediaminetetraacetate VS until the solution is deep blue in color. Each mL of 0.05 *M* disodium ethylenediaminetetraacetate is equivalent to 7.351 mg of $CaCl_2 . 2H_2O$.

Calcium Chloride Injection

» Calcium Chloride Injection is a sterile solution of Calcium Chloride in Water for Injection. It contains not less than 95.0 percent and not more than 105.0 percent of the labeled amount of $CaCl_2 . 2H_2O$.

Packaging and storage—Preserve in single-dose containers, preferably of Type I glass.

Labeling—The label states the total osmolar concentration in mOsmol per liter. Where the contents are less than 100 mL, or where the label states that the Injection is not for direct injection but is to be diluted before use, the label alternatively may state the total osmolar concentration in mOsmol per mL.

Identification—It responds to the tests for *Calcium* and *Chloride* ⟨191⟩.

pH ⟨791⟩: between 5.5 and 7.5 in the undiluted Injection, except where the concentration is greater than 1 in 20, in which case this range applies to the Injection diluted with water to yield a concentration of 1 in 20.

Particulate matter ⟨788⟩: meets the requirements under *Small-volume Injections.*

Other requirements—It meets the requirements under *Injections* ⟨1⟩.

Assay—Transfer an accurately measured volume of Calcium Chloride Injection, equivalent to about 1 g of calcium chloride, to a 250-mL volumetric flask, add 5 mL of 3 *N* hydrochloric acid, dilute with water to volume, and mix. Pipet 50 mL of the resulting solution into a suitable container, add 100 mL of water, 15 mL of 1 *N* sodium hydroxide, and 300 mg of hydroxy naphthol blue, and titrate with 0.05 *M* disodium ethylenediaminetetraacetate VS until the solution is deep blue. Each mL of 0.05 *M* disodium ethylenediaminetetraacetate is equivalent to 7.351 mg of $CaCl_2 . 2H_2O$.

Calcium Citrate

$$\left[\begin{array}{c} CH_2COO- \\ | \\ HO-C-COO- \\ | \\ CH_2COO- \end{array} \right]_2 Ca_3 \cdot 4H_2O$$

$C_{12}H_{10}Ca_3O_{14}.4H_2O$ 570.50

1,2,3-Propanetricarboxylic acid, 2-hydroxy-, calcium salt (2:3), tetrahydrate.

Calcium citrate (3:2), tetrahydrate [*5785-44-4*].

» Calcium Citrate contains four molecules of water of hydration. When dried at 150° to constant weight, it contains not less than 97.5 percent and not more than 100.5 percent of $Ca_3(C_6H_5O_7)_2$.

Packaging and storage—Preserve in well-closed containers.

Identification—

A: Dissolve 0.5 g in a mixture of 10 mL of water and 2.5 mL of 2 *N* nitric acid. Add 1 mL of mercuric sulfate TS, heat to boiling, and add 1 mL of potassium permanganate TS: a white precipitate is formed.

B: Ignite 0.5 g completely at as low a temperature as possible, cool, and dissolve the residue in a mixture of 10 mL of water and 1 mL of glacial acetic acid. Filter, and add 10 mL of ammonium oxalate TS to the filtrate: a voluminous white precipitate is formed, and it is soluble in hydrochloric acid.

Loss on drying ⟨731⟩—Dry it at 150° for 4 hours: it loses between 10.0% and 13.3% of its weight.

Arsenic, *Method I* ⟨211⟩—Dissolve 1 g in 5 mL of 3 *N* hydrochloric acid, and dilute with water to 35 mL: the limit is 3 ppm.

Fluoride—Place 2.0 g together with 5 mL of perchloric acid, 15 mL of water, and a few glass beads in a 50-mL distilling flask connected to a condenser. The flask is equipped with a thermometer and a capillary tube, both of which extend into the liquid. Connect a small dropping funnel, filled with water, to the capillary tube. Support the flask on an insulating board having a hole that exposes about one-third of the flask to the flame. Distil until the temperature reaches 135°, receiving the distillate under the surface of a few mL of water, and then maintain at 135° to 140° by adding water from the funnel. Continue the distillation until 70 mL has been collected, dilute the distillate with water to 80 mL, and mix. Transfer 40 mL of the solution to a 50-mL color-comparison tube. In a matched tube place 40 mL of water as a control. Add to each tube 0.10 mL of sodium alizarinsulfonate solution (1 in 1000), and mix. Add 0.05 *N* sodium hydroxide dropwise, with stirring, to the tube containing the distillate until the color just matches that of the control, which is faintly pink. Then add to each tube 1.0 mL of 0.1 *N* hydrochloric acid, and mix. From a buret, graduated in 0.05-mL increments, add slowly to the tube containing the distillate sufficient thorium nitrate solution, prepared by dissolving 250 mg of thorium nitrate in 1000 mL of water, so that, after mixing, the color of the liquid just changes to a faint pink. Add the same volume of thorium nitrate solution, accurately measured, to the control, and mix. Then add sodium fluoride TS from a buret to the control to make the colors of the two tubes match after dilution to the same volume. Mix, and allow all air bubbles to escape before making the final color comparison. Check the end-point by adding 1 or 2 drops of sodium fluoride TS to the control: a distinct change in color takes place. Not more than 3.0 mL of the sodium fluoride TS is required (0.003%). Each mL of sodium fluoride TS is equivalent to 10 µg of fluoride (F).

Acid-insoluble substances—Dissolve 5 g by heating with a mixture of 10 mL of hydrochloric acid and 50 mL of water for 30 minutes: the residue so obtained, filtered, washed, and dried at 105° for 2 hours, weighs not more than 10 mg (0.2%).

Lead ⟨251⟩—Prepare a *Test Preparation* by dissolving 0.5 g in 20 mL of 3 *N* hydrochloric acid, evaporating on a steam bath to about 10 mL, diluting with water to about 20 mL, and cooling. Use 5 mL of *Diluted Standard Lead Solution* (5 µg of Pb) for the test: the limit is 0.001%.

Heavy metals, *Method I* ⟨231⟩—Dissolve 1 g in a mixture of 20 mL of water and 2 mL of hydrochloric acid, add 1.5 mL of ammonium hydroxide, and dilute with water to 25 mL: the limit is 0.002%.

Assay—Dissolve about 350 mg of Calcium Citrate, previously dried at 150° to constant weight and accurately weighed, in a mixture of 10 mL of water and 2 mL of 3 *N* hydrochloric acid, and dilute with water to about 100 mL. While stirring, add about 30 mL of 0.05 *M* disodium ethylenediaminetetraacetate VS from a 50-mL buret, then add 15 mL of 1 *N* sodium hydroxide and 300 mg of hydroxy naphthol blue indicator, and continue the titration to a blue end-point. Each mL of 0.05 *M* disodium ethylenediaminetetraacetate is equivalent to 8.307 mg of $Ca_3(C_6H_5O_7)_2$.

Calcium Glubionate Syrup

» Calcium Glubionate Syrup is a solution containing equimolar amounts of Calcium Gluconate and Calcium Lactobionate or with Calcium Lactobionate predominating. It contains not less than 95.0 percent and not more than 105.0 percent of the labeled amount of Calcium (Ca).

Packaging and storage—Preserve in tight containers, at a temperature not exceeding 30°, and avoid freezing.

Identification—

A: A dilution of the Syrup with water (1 in 10) responds to the tests for *Calcium* ⟨191⟩.

B: Dilute a portion of the Syrup with water to obtain a test solution containing about 0.4 mg of calcium per mL. Prepare a Standard solution in water containing 2 mg of calcium gluconate and 4 mg of calcium lactobionate per mL. On a suitable thin-layer chromatographic plate coated with 0.25-mm layer of chromatographic silica gel (see *Chromatography* ⟨621⟩), apply separately 5 µL of the test solution and 5 µL of the Standard solution. Dry the plate in a current of cool air. Place the plate in a suitable chromatographic chamber lined with filter paper and previously equilibrated with a solvent system consisting of a mixture of alcohol, water, ethyl acetate, and ammonium hydroxide (50:30:10:10). Develop the chromatogram until the solvent front has moved about three-fourths of the length of the plate. Remove the plate from the chamber, and dry at 100° for 20 minutes. Allow to cool, and spray with a spray reagent prepared as follows: Dissolve 2.5 g of ammonium molybdate in about 50 mL of 2 *N* sulfuric acid in a 100-mL volumetric flask, add 1.0 g of ceric sulfate, swirl to dissolve, dilute with 2 *N* sulfuric acid to volume, and mix. Heat the plate at 110° for about 10 minutes: the two principal spots obtained from the test solution correspond in color, size, and R_f value to those obtained from the Standard solution.

pH ⟨791⟩: between 3.4 and 4.5.

Assay—Transfer an accurately measured portion of Calcium Glubionate Syrup, equivalent to about 70 mg of Ca, to a suitable beaker. Add 2 mL of 3 *N* hydrochloric acid, and dilute with water to about 150 mL. While stirring with a magnetic stirrer, add from a 50-mL buret about 20 mL of 0.05 *M* disodium ethylenediaminetetraacetate VS, then add 15 mL of 1 *N* sodium hydroxide and 300 mg of hydroxy naphthol blue indicator, and continue the titration to a blue end-point. Each mL of 0.05 *M* disodium ethylenediaminetetraacetate is equivalent to 2.004 mg of Ca.

Calcium Gluceptate

$$\left[HOCH_2-\overset{H}{\underset{OH}{C}}-\overset{H}{\underset{OH}{C}}-\overset{OH}{\underset{H}{C}}-\overset{H}{\underset{OH}{C}}-CH(OH)-COO^- \right]_2 Ca$$

$C_{14}H_{26}CaO_{16}$ (anhydrous) 490.43
Glucoheptonic acid, calcium salt (2:1).
Calcium glucoheptonate (1:2) [*29039-00-7*].

» Calcium Gluceptate is anhydrous or contains varying amounts of water of hydration. It consists of the calcium salt of the alpha epimer of glucoheptonic acid or of a mixture of the alpha and beta epimers of glucoheptonic acid. It contains not less than 95.0 percent and not more than 102.0 percent of $C_{14}H_{26}CaO_{16}$, calculated on the dried basis.

Packaging and storage—Preserve in well-closed containers.

Labeling—Label it to indicate whether it is hydrous or anhydrous; if hydrous, label it to indicate also the degree of hydration.

Reference standard—*USP Calcium Gluceptate Reference Standard*—Dry in vacuum at 60° for 16 hours before using.

Identification—

A: The infrared absorption spectrum of a potassium bromide dispersion of it, previously dried, exhibits maxima only at the same wavelengths as that of a similar preparation of USP Calcium Gluceptate RS.

B: A solution (1 in 50) responds to the tests for *Calcium* ⟨191⟩.

pH ⟨791⟩: between 6.0 and 8.0, in a solution (1 in 10).

Loss on drying ⟨731⟩ (see *Thermal Analysis* ⟨891⟩)—[NOTE—The quantity taken for the determination may be adjusted, if necessary, for instrument sensitivity. Weight loss occurring at temperatures above about 160°, indicative of decomposition, is not to be interpreted as *Loss on drying*.] Determine the percentage of volatile substances by thermogravimetric analysis on an appropriately calibrated instrument, using 10 to 25 mg of Calcium Gluceptate accurately weighed. Heat the specimen under test at a rate of 5° per minute in an atmosphere of nitrogen, at a flow rate of 40 mL per minute. Record the thermogram to 150°: the anhydrous form loses not more than 1.0%, the $2H_2O$ form not more than 6.9%, and the $3\frac{1}{2}H_2O$ form not more than 11.4%, of its weight.

Chloride ⟨221⟩—A 1.0-g portion shows no more chloride than corresponds to 1.0 mL of 0.020 *N* hydrochloric acid (0.07%).

Sulfate ⟨221⟩—A 2.0-g portion shows no more sulfate than corresponds to 1.0 mL of 0.020 *N* sulfuric acid (0.05%).

Arsenic, *Method I* ⟨211⟩: 1 ppm.

Heavy metals ⟨231⟩—Dissolve 1 g in 25 mL of water: the limit is 0.002%.

Reducing sugars—Dissolve 0.50 g in 10 mL of hot water, add 2 mL of 3 *N* hydrochloric acid, boil for about 2 minutes, and cool. Add 5 mL of sodium carbonate TS, allow to stand for 5 minutes, dilute with water to 20 mL, and filter. Add 5 mL of the clear filtrate to about 2 mL of alkaline cupric tartrate TS, and boil for 1 minute: no red precipitate is formed immediately.

Assay—Weigh accurately about 800 mg of Calcium Gluceptate, and dissolve in 150 mL of water containing 2 mL of 3 *N* hydrochloric acid. While stirring, preferably with a magnetic stirrer, add from a 50-mL buret about 25 mL of 0.05 *M* disodium ethylenediaminetetraacetate VS, then add 15 mL of 1 *N* sodium hydroxide and 300 mg of hydroxy naphthol blue indicator, and continue the titration to a blue end-point. Each mL of 0.05 *M* disodium ethylenediaminetetraacetate is equivalent to 24.52 mg of $C_{14}H_{26}CaO_{16}$.

Calcium Gluceptate Injection

» Calcium Gluceptate Injection is a sterile solution of Calcium Gluceptate in Water for Injection. It contains not less than 95.0 percent and not more than 105.0 percent of the labeled amount of Ca.

Packaging and storage—Preserve in tight, single-dose containers, preferably of Type I or Type II glass.

Labeling—The label states the total osmolar concentration in mOsmol per liter. Where the contents are less than 100 mL, or where the label states that the Injection is not for direct injection but is to be diluted before use, the label alternatively may state the total osmolar concentration in mOsm per mL.

Reference standard—*USP Calcium Gluceptate Reference Standard*—Dry in vacuum at 60° for 16 hours before using.

Identification—

A: Transfer 5 mL of Injection to a separator, add 10 mL of chloroform, shake well, and allow the layers to separate. Draw off and discard the chloroform layer, and repeat the extraction with a second 10-mL portion of chloroform. Drain the water layer into a small beaker, evaporate to dryness, and dry in vacuum at 60° for 16 hours: the infrared absorption spectrum of a potassium bromide dispersion of the residue so obtained exhibits maxima only at the same wavelengths as that of a similar preparation of USP Calcium Gluceptate RS.

B: A dilution of the Injection with water (1 in 5) responds to the tests for *Calcium* ⟨191⟩.

Pyrogen—It meets the requirements of the *Pyrogen Test* ⟨151⟩, the test dose being the equivalent of 29 mg of calcium per kg.

pH ⟨791⟩: between 5.6 and 7.0.

Particulate matter ⟨788⟩: meets the requirements under *Small-volume Injections*.

Other requirements—It meets the requirements under *Injections* ⟨1⟩.

Assay—To an accurately measured volume of Calcium Gluceptate Injection, equivalent to about 45 mg of calcium, add 2 mL of 3 *N* hydrochloric acid and 148 mL of water. While stirring, preferably with a magnetic stirrer, add from a 50-mL buret about 15 mL of 0.05 *M* disodium ethylenediaminetetraacetate VS, then add 15 mL of 1 *N* sodium hydroxide and 300 mg of hydroxy naphthol blue indicator, and continue the titration to a blue end-point. Each mL of 0.05 *M* disodium ethylenediaminetetraacetate is equivalent to 2.004 mg of Ca.

Calcium Gluconate

$$\left[HOCH_2-\overset{H}{\underset{OH}{C}}-\overset{H}{\underset{OH}{C}}-\overset{OH}{\underset{H}{C}}-\overset{H}{\underset{OH}{C}}-COO^- \right]_2 Ca$$

$C_{12}H_{22}CaO_{14}$ (anhydrous) 430.38
D-Gluconic acid, calcium salt (2:1).
Calcium D-gluconate (1:2) [*18016-24-5*].
Monohydrate 448.39 [*299-28-5*].

» Calcium Gluconate is anhydrous or contains one molecule of water of hydration. The anhydrous form contains not less than 98.0 percent and not more than 102.0 percent of $C_{12}H_{22}CaO_{14}$, calculated on the dried basis, and the monohydrate form contains not less than 95.0 percent and not more than 98.9 percent of $C_{12}H_{22}CaO_{14}$, calculated on the dried basis, corresponding to not less than 99.0 percent and not more than 103.0 percent of $C_{12}H_{22}CaO_{14}.H_2O$.

Packaging and storage—Preserve in well-closed containers.

Labeling—Label it to indicate whether it is anhydrous or is the monohydrate. Where the quantity of calcium gluconate is indicated in the labeling of any preparation containing Calcium Gluconate, this shall be understood to be in terms of anhydrous calcium gluconate ($C_{12}H_{22}CaO_{14}$).

Reference standard—*USP Potassium Gluconate Reference Standard*—Dry in vacuum at 105° for 4 hours before using.

Identification—

A: A solution (1 in 50) responds to the test for *Calcium* ⟨191⟩.

B: Dissolve a quantity of it in water to obtain a test solution containing 10 mg per mL, heating in a water bath at 60° if necessary. Similarly, prepare a Standard solution of USP Potassium Gluconate RS in water containing 10 mg per mL. Apply separate 5-µL portions of the test solution and the Standard solution on a suitable thin-layer chromatographic plate (see *Chromatography* ⟨621⟩), coated with a 0.25-mm layer of chromatographic silica gel, and allow to dry. Develop the chromatogram in a solvent system consisting of a mixture of alcohol, water, ammonium hydroxide, and ethyl acetate (50:30:10:10) until the solvent front has moved about three-fourths of the length of the plate. Remove the plate from the chamber, and dry at 110° for 20 minutes. Allow to cool, spray with a spray reagent prepared as follows: Dissolve 2.5 g of ammonium molybdate in about 50 mL of 2 *N* sulfuric acid in a 100-mL volumetric flask, add 1.0 g of ceric sulfate, swirl to dissolve, dilute with 2 *N* sulfuric acid to volume, and mix. Heat the plate at 110° for about 10 minutes: the principal spot obtained from the test solution corresponds in color, size, and R_f value to that obtained from the Standard solution.

Loss on drying ⟨731⟩—Dry it at 105° for 16 hours: it loses not more than 3.0% of its weight.

Chloride ⟨221⟩—A 1.0-g portion shows no more chloride than corresponds to 1 mL of 0.020 *N* hydrochloric acid (0.07%).

Sulfate ⟨221⟩—A 2.0-g portion dissolved in boiling water shows no more sulfate than corresponds to 1 mL of 0.020 *N* sulfuric acid (0.05%).

Arsenic, *Method I* ⟨211⟩—Dissolve 1.0 g in a mixture of 10 mL of hydrochloric acid and 20 mL of water, and dilute with water to 55 mL: the resulting solution meets the requirements of the test, the addition of 20 mL of 7 *N* sulfuric acid specified under *Procedure* being omitted. The limit is 3 ppm.

Heavy metals ⟨231⟩—Mix 1 g with 4 mL of 1.2 *N* hydrochloric acid, add water to make 25 mL, warm gently until dissolved, and cool to room temperature: the limit is 0.002%.

Reducing substances—Transfer 1.0 g to a 250-mL conical flask, dissolve in 20 mL of hot water, cool, and add 25 mL of alkaline cupric citrate TS. Cover the flask, boil gently for 5 minutes, accurately timed, and cool rapidly to room temperature. Add 25 mL of 0.6 *N* acetic acid, 10.0 mL of 0.1 *N* iodine VS, and 10 mL of 3 *N* hydrochloric acid, and titrate with 0.1 *N* sodium thiosulfate VS, adding 3 mL of starch TS as the end-point is approached. Perform a blank determination, omitting the specimen, and note the difference in volumes required. Each mL of the difference in volume of 0.1 *N* sodium thiosulfate consumed is equivalent to 2.7 mg of reducing substances (as dextrose): the limit is 1.0%.

Assay—Weigh accurately about 800 mg of Calcium Gluconate, and dissolve in 150 mL of water containing 2 mL of 3 *N* hydrochloric acid. While stirring, preferably with a magnetic stirrer, add about 30 mL of 0.05 *M* disodium ethylenediaminetetraacetate VS from a 50-mL buret, then add 15 mL of 1 *N* sodium hydroxide and 300 mg of hydroxy naphthol blue indicator, and continue the titration to a blue end-point. Each mL of 0.05 *M* disodium ethylenediaminetetraacetate is equivalent to 21.52 mg of $C_{12}H_{22}CaO_{14}$.

Calcium Gluconate Injection

» Calcium Gluconate Injection is a sterile solution of Calcium Gluconate in Water for Injection. It contains not less than 95.0 percent and not more than 105.0 percent of the labeled amount of total Ca. The calcium is in the form of calcium gluconate, except that a small amount may be replaced with an equal amount of calcium in the form of Calcium Saccharate, or other suitable calcium salts, for the purpose of stabilization. It may require warming before use if crystallization has occurred. It may contain sodium hydroxide added for adjustment of the pH.

NOTE—If crystallization has occurred, warming may dissolve the precipitate. The Injection must be clear at the time of use.

Packaging and storage—Preserve in single-dose containers, preferably of Type I glass.

Labeling—Label the Injection to indicate its content, if any, of added calcium salts, calculated as percentage of calcium in the Injection. The label states the total osmolar concentration in mOsmol per liter. Where the contents are less than 100 mL, or where the label states that the Injection is not for direct injection but is to be diluted before use, the label alternatively may state the total osmolar concentration in mOsmol per mL.

Reference standard—*USP Potassium Gluconate Reference Standard*—Dry in vacuum at 105° for 4 hours before using.

Identification—

A: A volume of Injection diluted, if necessary, with water to obtain a test solution of calcium gluconate (1 in 100) responds to *Identification test B* under *Calcium Gluconate.*

B: A dilution of the Injection with water (1 in 5) responds to the tests for *Calcium* ⟨191⟩.

Pyrogen—It meets the requirements of the *Pyrogen Test* ⟨151⟩, the test dose being the equivalent of 300 mg of calcium gluconate per kg.

pH ⟨791⟩: between 6.0 and 8.2.

Particulate matter ⟨788⟩: meets the requirements under *Small-volume Injections.*

Other requirements—It meets the requirements under *Injections* ⟨1⟩.

Assay—To an accurately measured volume of Calcium Gluconate Injection, equivalent to about 500 mg of calcium gluconate, add 2 mL of 3 *N* hydrochloric acid, and dilute with water to 150 mL. While stirring, preferably with a magnetic stirrer, add about 20 mL of 0.05 *M* disodium ethylenediaminetetraacetate VS from a 50-mL buret, then add 15 mL of 1 *N* sodium hydroxide and 300 mg of hydroxy naphthol blue indicator, and continue the titration to a blue end-point. Each mL of 0.05 *M* disodium ethylenediaminetetraacetate is equivalent to 2.004 mg of Ca.

Calcium Gluconate Tablets

» Calcium Gluconate Tablets contain not less than 95.0 percent and not more than 105.0 percent of the labeled amount of $C_{12}H_{22}CaO_{14}$.

Packaging and storage—Preserve in well-closed containers.

Reference standard—*USP Potassium Gluconate Reference Standard*—Dry in vacuum at 105° for 4 hours before using.

Identification—A warm, filtered solution of Tablets, equivalent to calcium gluconate solution (1 in 10), diluted with water, where necessary, responds to the *Identification tests* under *Calcium Gluconate.*

Dissolution ⟨711⟩—

Medium: water; 900 mL.
Apparatus 2: 50 rpm.

Time: 45 minutes.

Procedure—Determine the amount of $C_{12}H_{22}CaO_{14}$ dissolved, employing atomic absorption spectrophotometry at a wavelength of about 422.8 nm in filtered portions of the solution under test, suitably diluted with water, in comparison with a Standard solution having a known concentration of calcium in the same medium.

Tolerances—Not less than 75% (*Q*) of the labeled amount of $C_{12}H_{22}CaO_{14}$ is dissolved in 45 minutes.

Uniformity of dosage units ⟨905⟩: meet the requirements.

Assay—Weigh and finely powder not less than 20 Calcium Gluconate Tablets. Weigh accurately a portion of the powder, equivalent to about 500 mg of calcium gluconate, transfer to a suitable crucible, and ignite, gently at first, until free from carbon. Cool the crucible, add 10 mL of water, and dissolve the residue by adding sufficient 3 *N* hydrochloric acid, dropwise, to achieve complete solution. Transfer the solution to a suitable container, dilute with water to 150 mL, and while stirring, preferably with a magnetic stirrer, add about 20 mL of 0.05 *M* disodium ethylenediaminetetraacetate VS from a 50-mL buret, then add 15 mL of 1 *N* sodium hydroxide and 300 mg of hydroxy naphthol blue indicator, and continue the titration to a blue end-point. Each mL of 0.05 *M* disodium ethylenediaminetetraacetate is equivalent to 21.52 mg of $C_{12}H_{22}CaO_{14}$.

Calcium Hydroxide

$Ca(OH)_2$ 74.09
Calcium hydroxide.
Calcium hydroxide [*1305-62-0*].

» Calcium Hydroxide contains not less than 95.0 percent and not more than 100.5 percent of $Ca(OH)_2$.

Packaging and storage—Preserve in tight containers.

Identification—

A: When mixed with from three to four times its weight of water, it forms a smooth magma. The clear, supernatant liquid from the magma is alkaline to litmus.

B: Mix 1 g with 20 mL of water, and add sufficient 6 *N* acetic acid to effect solution: the resulting solution responds to the tests for *Calcium* ⟨191⟩.

Acid-insoluble substances—Dissolve 2.0 g in 30 mL of hydrochloric acid, and heat to boiling. Filter the mixture, wash the residue with hot water, and ignite: the weight of the residue does not exceed 10 mg (0.5%).

Carbonate—Mix 2 g with 50 mL of water: the addition of an excess of 3 *N* hydrochloric acid to the mixture does not cause more than a slight effervescence.

Arsenic, *Method I* ⟨211⟩—Cautiously dissolve 1.0 g in 15 mL of hydrochloric acid, and dilute with water to 55 mL: the resulting solution meets the requirements of the test, the addition of 20 mL of 7 *N* sulfuric acid specified under *Procedure* being omitted. The limit is 3 ppm.

Heavy metals ⟨231⟩—Dissolve 1 g in 10 mL of 3 *N* hydrochloric acid, and evaporate on a steam bath to dryness. Dissolve the residue in 20 mL of water, and filter. Dilute the filtrate with water to 40 mL, and to 20 mL of the resulting solution add 1 mL of 0.1 *N* hydrochloric acid, then add water to make 25 mL: the limit is 0.004%.

Magnesium and alkali salts—Dissolve 0.50 g in a mixture of 30 mL of water and 10 mL of 3 *N* hydrochloric acid, and proceed as directed in the test for *Magnesium and alkali salts* under *Precipitated Calcium Carbonate*, beginning with "heat the solution, and boil for 1 minute." The weight of the residue does not exceed 12 mg (4.8%).

Assay—Weigh accurately about 1.5 g of Calcium Hydroxide, transfer to a beaker, and gradually add 30 mL of 3 *N* hydrochloric acid. When solution is complete, transfer the solution to a 500-mL volumetric flask, rinse the beaker thoroughly, adding the rinsings to the flask, dilute with water to volume, and mix. Pipet 50 mL of the solution into a suitable container, add 100 mL of water, 15 mL of 1 *N* sodium hydroxide, and 300 mg of hydroxy naphthol blue trituration, and titrate with 0.05 *M* disodium ethylenediaminetetraacetate VS to a blue end-point. Each mL of 0.05 *M* disodium ethylenediaminetetraacetate is equivalent to 3.705 mg of $Ca(OH)_2$.

Calcium Hydroxide Topical Solution

» Calcium Hydroxide Topical Solution is a solution containing, in each 100 mL, not less than 140 mg of $Ca(OH)_2$.

Calcium Hydroxide Topical Solution may be prepared as follows.

Calcium Hydroxide	3 g
Purified Water	1000 mL

Add the Calcium Hydroxide to 1000 mL of cool Purified Water, and agitate the mixture vigorously and repeatedly during 1 hour. Allow the excess calcium hydroxide to settle. Dispense only the clear, supernatant liquid.

NOTE—The solubility of calcium hydroxide varies with the temperature at which the solution is stored, being about 170 mg per 100 mL at 15°, and less at a higher temperature. The official concentration is based upon a temperature of 25°.

The undissolved portion of the mixture is not suitable for preparing additional quantities of Calcium Hydroxide Topical Solution.

Packaging and storage—Preserve in well-filled, tight containers, at a temperature not exceeding 25°.

Identification—

A: It absorbs carbon dioxide from the air, a film of calcium carbonate forming on the surface of the liquid.

B: When heated, it becomes turbid, owing to the separation of calcium hydroxide.

C: It responds to the tests for *Calcium* ⟨191⟩.

Alkalies and their carbonates—A portion of it, saturated with carbon dioxide and subsequently boiled, is neutral in reaction.

Assay—Pipet 100 mL of Calcium Hydroxide Topical Solution into a suitable container, add 50 mL of water, 15 mL of 1 *N* sodium hydroxide, and 300 mg of hydroxy naphthol blue trituration, and titrate with 0.05 *M* disodium ethylenediaminetetraacetate VS to a blue end-point. Each mL of 0.05 *M* disodium ethylenediaminetetraacetate is equivalent to 3.705 mg of $Ca(OH)_2$.

Calcium Lactate

$$\left[\begin{matrix}CH_3CHCOO- \\ \quad | \\ \quad OH\end{matrix}\right]_2 Ca \cdot xH_2O$$

$C_6H_{10}CaO_6 . xH_2O$ (anhydrous) 218.22
Propanoic acid, 2-hydroxy-, calcium salt (2:1), hydrate.
Calcium lactate (1:2) hydrate [*41372-22-9*].
Calcium lactate (1:2) pentahydrate 308.30 [*63690-56-2*].
Anhydrous [*814-80-2*].

» Calcium Lactate contains not less than 98.0 percent and not more than 101.0 percent of $C_6H_{10}CaO_6$, calculated on the dried basis.

Packaging and storage—Preserve in tight containers.

Labeling—The label indicates whether it is the dried form or is hydrous; if the latter, the label indicates the degree of hydration. Where the quantity of Calcium Lactate is indicated in the labeling of any preparation containing Calcium Lactate, this shall be understood to be in terms of calcium lactate pentahydrate ($C_6H_{10}CaO_6 . 5H_2O$).

Identification—

A: A solution (1 in 20) responds to the tests for *Calcium* ⟨191⟩.

B: To 10 mg add 1 mL of sulfuric acid, and heat for 2 minutes in a water bath maintained at a temperature of 85°. Cool the solution to room temperature, add about 10 mg of 4-phenylphenol crystals, swirl, and allow to stand for about 20 minutes: a violet color develops that deepens with the passage of time.

Acidity—Titrate 20 mL of a solution (1 in 20) with 0.10 *N* sodium hydroxide, using phenolphthalein TS as the indicator: not more than 0.50 mL is required for neutralization (0.45% as lactic acid).

Loss on drying ⟨731⟩—Distribute a 1-g to 2-g portion evenly in a suitable weighing dish to a depth of not more than 3 mm, and dry at 120° for 4 hours: the pentahydrate loses between 22.0% and 27.0% of its weight; the trihydrate loses between 15.0% and 20.0% of its weight; the monohydrate loses between 5.0% and 8.0% of its weight; and the dried form loses not more than 3.0% of its weight.

Heavy metals ⟨231⟩—Dissolve 1 g in 2.5 mL of 1 *N* acetic acid, and dilute with water to 25 mL: the limit is 0.002%.

Magnesium and alkali salts—Mix 1.0 g with 40 mL of water, carefully add 1 mL of hydrochloric acid, and heat the solution to boiling. Proceed as directed in the test for *Magnesium and alkali salts* under *Precipitated Calcium Carbonate*, beginning with "Rapidly add 40 mL of oxalic acid TS": the weight of the residue does not exceed 5.0 mg (1.0%).

Volatile fatty acid—Stir about 500 mg with 1 mL of sulfuric acid, and warm: the mixture does not emit an odor of volatile fatty acid.

Assay—Weigh accurately an amount of Calcium Lactate, equivalent to about 350 mg of $C_6H_{10}CaO_6$, transfer to a suitable container, and dissolve in a mixture of 150 mL of water and 2 mL of 3 *N* hydrochloric acid. While stirring, preferably with a magnetic stirrer, add about 30 mL of 0.05 *M* disodium ethylenediaminetetraacetate VS from a 50-mL buret, then add 15 mL of 1 *N* sodium hydroxide and 300 mg of hydroxy naphthol blue indicator, and continue the titration to a blue end-point. Each mL of 0.05 *M* disodium ethylenediaminetetraacetate is equivalent to 10.91 mg of $C_6H_{10}CaO_6$.

Calcium Lactate Tablets

» Calcium Lactate Tablets contain not less than 94.0 percent and not more than 106.0 percent of the labeled amount of $C_6H_{10}CaO_6 . 5H_2O$.

NOTE—An equivalent amount of Calcium Lactate with less water of hydration may be used in place of $C_6H_{10}CaO_6 . 5H_2O$ in preparing Calcium Lactate Tablets.

Packaging and storage—Preserve in tight containers.

Labeling—The quantity of calcium lactate stated in the labeling is in terms of calcium lactate pentahydrate.

Identification—

A: A filtered solution of Tablets, equivalent to calcium lactate solution (1 in 20), responds to the tests for *Calcium* ⟨191⟩.

B: A filtered solution of Tablets, equivalent to calcium lactate solution (1 in 20), responds to *Identification test B* under *Calcium Lactate*, 1 drop of the solution being used.

Dissolution ⟨711⟩—

Medium: water; 500 mL.

Apparatus 1: 100 rpm.

Time: 45 minutes.

Procedure—Determine the amount of $C_6H_{10}CaO_6 . 5H_2O$ dissolved, employing the procedure set forth in the *Assay*, making any necessary modifications.

Tolerances—Not less than 75% (*Q*) of the labeled amount of $C_6H_{10}CaO_6 . 5H_2O$ is dissolved in 45 minutes.

Uniformity of dosage units ⟨905⟩: meet the requirements.

Assay—Weigh and finely powder not less than 20 Calcium Lactate Tablets. Weigh accurately a portion of the powder, equivalent to about 350 mg of $C_6H_{10}CaO_6$, transfer to a suitable container, and add 150 mL of water and 2 mL of 3 *N* hydrochloric acid. Stir, using a magnetic stirrer, for 3 to 5 minutes. While stirring, add about 30 mL of 0.05 *M* disodium ethylenediaminetetraacetate VS from a 50-mL buret, then add 15 mL of 1 *N* sodium hydroxide and 300 mg of hydroxy naphthol blue indicator, and continue the titration to a blue end-point. Each mL of 0.05 *M* disodium ethylenediaminetetraacetate is equivalent to 15.42 mg of $C_6H_{10}CaO_6 . 5H_2O$.

Calcium Lactobionate

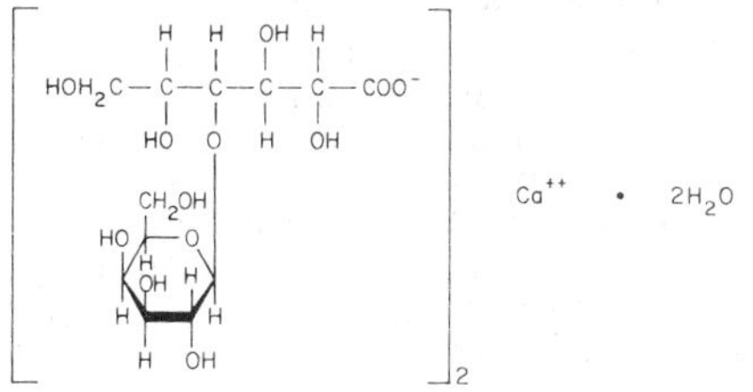

$C_{24}H_{42}CaO_{24} . 2H_2O$ 790.69

D-Gluconic acid, 4-*O*-β-D-galactopyranosyl-, calcium salt (2:1), dihydrate.

Lactobionic acid, calcium salt (2:1), dihydrate.

Calcium lactobionate (1:2), dihydrate [*110638-68-1*].

» Calcium Lactobionate contains not less than 96.0 percent and not more than 102.0 percent of $C_{24}H_{42}CaO_{24} . 2H_2O$.

Packaging and storage—Preserve in well-closed containers.

Reference standard—*USP Calcium Lactobionate Reference Standard*—Dry at 105° for 8 hours before using.

Identification—

A: A solution (1 in 50) responds to the tests for *Calcium* ⟨191⟩.

B: The infrared absorption spectrum of a potassium bromide dispersion of it, previously dried at 105° for 8 hours, exhibits maxima only at the same wavelengths as that of a similar preparation of USP Calcium Lactobionate RS.

C: Dissolve a quantity of it in water to obtain a test solution containing 10 mg per mL. Similarly prepare a Standard solution of USP Calcium Lactobionate RS in water containing 10 mg per mL. On a suitable thin-layer chromatographic plate (see *Chromatography* ⟨621⟩), coated with a 0.25-mm layer of chromatographic silica gel apply separately 5 µL of the test solution and 5 µL of the Standard solution. Dry the plate in a current of cool air. Place the plate in a suitable chromatographic chamber lined with filter paper and previously equilibrated with a solvent system consisting of a mixture of alcohol, water, ethyl acetate, and ammonium hydroxide (50:30:10:10). Develop the chromatogram until the solvent front has moved about three-fourths of the length of the plate. Remove the plate from the chamber, and dry at 100° for 20 minutes. Allow to cool, and spray with a spray reagent prepared as follows: Dissolve 2.5 g of ammonium molybdate in about 50 mL of 2 *N* sulfuric acid in a 100-mL volumetric flask, add 1.0 g of ceric sulfate, swirl to dissolve, dilute with 2 *N* sulfuric acid to volume, and mix. Heat the plate at 110° for about 10 minutes: the principal spot obtained from the test solution corresponds in color, size, and R_f value to that obtained from the Standard solution.

Specific rotation ⟨781⟩: between +22.0° and +26.5°, determined in a solution containing 1 g in each 10 mL.

pH ⟨791⟩: between 5.4 and 7.4, in a solution (1 in 20).

Halides—A 1.2-g portion tested as directed under *Chloride* ⟨221⟩ shows no more turbidity than corresponds to 0.7 mL of 0.020 *N* hydrochloric acid (0.04%).

Sulfate ⟨221⟩—A 2.0-g portion dissolved in boiling water shows no more sulfate than corresponds to 1 mL of 0.020 *N* sulfuric acid (0.05%).

Arsenic, *Method II* ⟨211⟩: 3 ppm.

Heavy metals ⟨231⟩—Mix 1 g with 4 mL of 1.2 *N* hydrochloric acid, add water to make 25 mL, warm gently until dissolved, and cool to room temperature: the limit is 0.002%.

Reducing substances—Transfer 1.0 g to a 250-mL conical flask, dissolve in 20 mL of water, and add 25 mL of alkaline cupric citrate TS. Cover the flask, boil gently for 5 minutes, accurately timed, and cool rapidly to room temperature. Add 25 mL of 0.6 *N* acetic acid, 10.0 mL of 0.1 *N* iodine VS, and 10 mL of 3 *N* hydrochloric acid, and titrate with 0.1 *N* sodium thiosulfate VS, adding 3 mL of starch TS as the end-point is approached. Perform a blank determination, omitting the specimen, and note the difference in volumes required. Each mL of the difference in volume of 0.1 *N* sodium thiosulfate consumed is equivalent to 2.7 mg of reducing substances (as dextrose): the limit is 1.0%.

Assay—Weigh accurately about 0.8 g of Calcium Lactobionate, and dissolve in a mixture of 150 mL of water and 2 mL of 3 *N* hydrochloric acid. While stirring with a magnetic stirrer, add from a 50-mL buret about 15 mL of 0.05 *M* disodium ethylenediaminetetraacetate VS, then add 15 mL of 1 *N* sodium hydroxide and 300 mg of hydroxy naphthol blue indicator, and continue the titration to a blue end-point. Each mL of 0.05 *M* disodium ethylenediaminetetraacetate is equivalent to 39.53 mg of $C_{24}H_{42}CaO_{24} . 2H_2O$.

Calcium Levulinate

$[CH_3COCH_2CH_2COO-]_2 Ca \cdot 2H_2O$

$C_{10}H_{14}CaO_6 . 2H_2O$ 306.33
Pentanoic acid, 4-oxo-, calcium salt (2:1), dihydrate.
Calcium levulinate (1:2) dihydrate [*5743-49-7*].
Anhydrous 270.30 [*591-64-0*].

» Calcium Levulinate contains not less than 97.5 percent and not more than 100.5 percent of $C_{10}H_{14}CaO_6$, calculated on the dried basis.

Packaging and storage—Preserve in well-closed containers.

Identification—

A: A solution (1 in 10) responds to the tests for *Calcium* ⟨191⟩.

B: To a solution of 0.5 g in 5 mL of water add 5 mL of 1 *N* sodium hydroxide, and filter. To the filtrate add 5 mL of iodine TS: a precipitate of iodoform is produced.

C: To a solution of 0.1 g in 2 mL of water add 5 mL of dinitrophenylhydrazine TS, and allow the mixture to stand in an ice bath for 1 hour. Collect the precipitate on a filter, wash well with cold water, and dry at 105° for 1 hour: the hydrazone so obtained melts between 198° and 206°.

Melting range, *Class I* ⟨741⟩: between 119° and 125°.

pH ⟨791⟩: between 7.0 and 8.5, in a solution (1 in 10).

Loss on drying ⟨731⟩—Dry it at a pressure not exceeding 5 mm of mercury at 60° for 5 hours: it loses between 10.5% and 12.0% of its weight.

Chloride ⟨221⟩—A 1.0-g portion shows no more chloride than corresponds to 1.0 mL of 0.020 *N* hydrochloric acid (0.07%).

Sulfate ⟨221⟩—A 2.0-g portion shows no more sulfate than corresponds to 1.0 mL of 0.020 *N* sulfuric acid (0.05%).

Arsenic, *Method I* ⟨211⟩: 3 ppm.

Heavy metals ⟨231⟩: 0.002%.

Reducing sugars—Dissolve 0.50 g in 10 mL of water, add 2 mL of 3 *N* hydrochloric acid, boil for about 2 minutes, and cool. Add 5 mL of sodium carbonate TS, allow to stand for 5 minutes, dilute with water to 20 mL, and filter. Add 5 mL of the clear filtrate to about 2 mL of alkaline cupric tartrate TS, and boil for 1 minute: no red precipitate is formed immediately.

Assay—Weigh accurately about 600 mg of Calcium Levulinate, and dissolve in 150 mL of water containing 2 mL of 3 *N* hydrochloric acid. While stirring with a magnetic stirrer, add from a 50-mL buret 30 mL of 0.05 *M* disodium ethylenediaminetetraacetate VS, then add 15 mL of 1 *N* sodium hydroxide and 300 mg of hydroxy naphthol blue indicator, and continue the titration to the blue end-point. Each mL of 0.05 *M* disodium ethylenediaminetetraacetate is equivalent to 13.51 mg of $C_{10}H_{14}CaO_6$.

Calcium Levulinate Injection

» Calcium Levulinate Injection is a sterile solution of Calcium Levulinate in Water for Injection. It contains not less than 95.0 percent and not more than 105.0 percent of the labeled amount of $C_{10}H_{14}CaO_6 . 2H_2O$.

Packaging and storage—Preserve in single-dose containers, preferably of Type I glass.

Labeling—The label states the total osmolar concentration in mOsmol per liter. Where the contents are less than 100 mL, or where the label states that the Injection is not for direct injection but is to be diluted before use, the label alternatively may state the total osmolar concentration in mOsmol per mL.

Identification—It responds to the *Identification tests* under *Calcium Levulinate.*

Pyrogen—It meets the requirements of the *Pyrogen Test* ⟨151⟩, the test dose being 200 mg of calcium levulinate per kg.

pH ⟨791⟩: between 6.0 and 8.0.

Particulate matter ⟨788⟩: meets the requirements under *Small-volume Injections.*

Other requirements—It meets the requirements under *Injections* ⟨1⟩.

Assay—Transfer an accurately measured volume of Calcium Levulinate Injection, equivalent to about 600 mg of calcium levulinate, to a 400-mL beaker, add 2 mL of hydrochloric acid, and proceed as directed in the *Assay* under *Calcium Levulinate*, beginning with "While stirring with a magnetic stirrer." Each mL of 0.05 *M* disodium ethylenediaminetetraacetate is equivalent to 15.32 mg of $C_{10}H_{14}CaO_6 . 2H_2O$.

Calcium Pantothenate

$[HOCH_2C(CH_3)_2-C(OH)(H)-CONH(CH_2)_2COO-]_2 Ca$

$C_{18}H_{32}CaN_2O_{10}$ 476.54
β-Alanine, *N*-(2,4-dihydroxy-3,3-dimethyl-1-oxobutyl)-, calcium salt (2:1), (*R*)-.
Calcium D-pantothenate (1:2) [*137-08-6*].

» Calcium Pantothenate is the calcium salt of the dextrorotatory isomer of pantothenic acid. It contains not less than 5.7 percent and not more than 6.0 percent of nitrogen (N), and not less than 8.2 percent and not more than 8.6 percent of calcium (Ca), both calculated on the dried basis.

Packaging and storage—Preserve in tight containers.

Reference standard—*USP Calcium Pantothenate Reference Standard*—Dry at 105° for 3 hours before using.

Identification—

A: The infrared absorption spectrum of a potassium bromide dispersion of it, previously dried, exhibits maxima only at the same wavelengths as that of a similar preparation of USP Calcium Pantothenate RS.

B: A solution (1 in 20) responds to the tests for *Calcium* ⟨191⟩.

Specific rotation ⟨781⟩: between +25.0° and +27.5°, calculated on the dried basis, determined in a solution containing 500 mg in each 10 mL.

Alkalinity—Dissolve 1.0 g in 15 mL of carbon dioxide–free water in a small flask. As soon as solution is complete, add 1.0 mL of 0.10 *N* hydrochloric acid, then add 0.05 mL of phenolphthalein TS, and mix: no pink color is produced within 5 seconds.

Loss on drying ⟨731⟩—Dry it at 105° for 3 hours: it loses not more than 5.0% of its weight.

Heavy metals ⟨231⟩—Dissolve 1.0 g in 25 mL of water: the limit is 0.002%.

Ordinary impurities ⟨466⟩—

Test solution: water.

Standard solution: water. Use 3-aminopropionic acid, in place of USP Calcium Pantothenate RS, as the standard.

Eluant: a mixture of alcohol and water (65:35).

Visualization: 4.

Limit: 1.0%.

Nitrogen content—Weigh accurately about 500 mg, transfer to a Kjeldahl flask, and proceed as directed under *Nitrogen Determination, Method I* ⟨461⟩.

Calcium content—Weigh accurately about 800 mg, and dissolve in 150 mL of water containing 2 mL of 3 *N* hydrochloric acid. Add 15 mL of 1 *N* sodium hydroxide and 300 mg of hydroxy naphthol blue indicator, and titrate with 0.05 *M* disodium ethylenediaminetetraacetate VS until the solution is a distinct blue in color. Each mL of 0.05 *M* disodium ethylenediaminetetraacetate is equivalent to 2.004 mg of Ca.

Calcium Pantothenate Tablets

» Calcium Pantothenate Tablets contain not less than 95.0 percent and not more than 115.0 percent of the labeled amount of the dextrorotatory isomer of $C_{18}H_{32}CaN_2O_{10}$.

Packaging and storage—Preserve in tight containers.

Labeling—Label Tablets to indicate the content of dextrorotatory calcium pantothenate.

Reference standard—*USP Calcium Pantothenate Reference Standard*—Dry at 105° for 3 hours before using.

Identification—

A: Digest a quantity of powdered Tablets, equivalent to about 150 mg of calcium pantothenate, with 15 mL of 1 *N* sodium hydroxide, and filter: the filtrate responds to *Identification test B* under *Calcium Pantothenate.*

B: To 5 mL of the filtrate obtained in *Identification test A* add 5 mL of 1 *N* hydrochloric acid and 2 drops of ferric chloride TS: a strong yellow color is produced.

Disintegration ⟨701⟩: 30 minutes.

Uniformity of dosage units ⟨905⟩: meet the requirements.

Calcium content—Weigh and finely powder not less than 20 Tablets. Weigh accurately a portion of the powder, equivalent to about 500 mg of calcium pantothenate, transfer to a suitable crucible, and ignite, gently at first, until free from carbon. Cool the crucible, add 10 mL of water, and dissolve the residue by adding sufficient 3 *N* hydrochloric acid, dropwise, to achieve complete solution. Transfer the solution to a suitable container, dilute with water to 150 mL, add 15 mL of 1 *N* sodium hydroxide and 300 mg of hydroxy naphthol blue indicator, and titrate with 0.05 *M* disodium ethylenediaminetetraacetate VS until the solution is deep blue in color. Each mL of 0.05 *M* disodium ethylenediaminetetraacetate is equivalent to 2.004 mg of Ca. The content of Ca found is not less than 7.9% and not more than 9.7% of the weight of $C_{18}H_{32}CaN_2O_{10}$ in the Tablets as determined by the *Assay*.

Assay—Weigh and finely powder not less than 20 Calcium Pantothenate Tablets. Weigh accurately a portion of the powder, equivalent to about 25 mg of calcium pantothenate, and transfer, with the aid of about 50 mL of water, to a 1000-mL volumetric flask. Add 2 mL of 1 *N* acetic acid and 100 mL of sodium acetate solution (1 in 60), then dilute with water to volume, and mix. Make further accurate dilutions, with water, to a concentration between 0.01 µg and 0.04 µg of calcium pantothenate per mL. Using this solution as the *Assay Preparation*, proceed as directed under *Calcium Pantothenate Assay* ⟨91⟩.

Racemic Calcium Pantothenate

$C_{18}H_{32}CaN_2O_{10}$ 476.54

β-Alanine, *N*-(2,4-dihydroxy-3,3-dimethyl-1-oxobutyl)-, calcium salt (2:1), (±)-.

Calcium DL-pantothenate (1:2) [*6381-63-1*].

» Racemic Calcium Pantothenate is a mixture of the calcium salts of the dextrorotatory and levorotatory isomers of pantothenic acid. It contains not less than 5.7 percent and not more than 6.0 percent of nitrogen (N), and not less than 8.2 percent and not more than 8.6 percent of calcium (Ca), both calculated on the dried basis.

NOTE—The physiological activity of Racemic Calcium Pantothenate is approximately one-half that of Calcium Pantothenate.

Packaging and storage—Preserve in tight containers.

Labeling—Label preparations containing it in terms of the equivalent amount of dextrorotatory calcium pantothenate.

Reference standard—*USP Calcium Pantothenate Reference Standard*—Dry at 105° for 3 hours before using.

Specific rotation ⟨781⟩: between −0.05° and +0.05°, calculated on the dried basis, determined in a solution containing 500 mg in each 10 mL.

Alkalinity—Dissolve 1.0 g in 15 mL of carbon dioxide–free water in a small flask. As soon as solution is complete, add 1.6 mL of 0.10 *N* hydrochloric acid, then add 0.05 mL of phenolphthalein TS, and mix: no pink color is produced within 5 seconds.

Other requirements—It responds to the *Identification tests* and meets the requirements of the tests for *Loss on drying, Heavy metals, Nitrogen content*, and *Calcium content* under *Calcium Pantothenate.*

Dibasic Calcium Phosphate

$CaHPO_4$ 136.06

Phosphoric acid, calcium salt (1:1).

Calcium phosphate (1:1) [*7757-93-9*].

Dihydrate 172.09 [*7789-77-7*].

» Dibasic Calcium Phosphate is anhydrous or contains two molecules of water of hydration. It contains not less than 30.0 percent and not more than 31.7 percent of calcium (Ca), calculated on the ignited basis.

Packaging and storage—Preserve in well-closed containers.

Labeling—Label it to indicate whether it is anhydrous or hydrous.

Identification—

A: Dissolve about 100 mg by warming with a mixture of 5 mL of 3 *N* hydrochloric acid and 5 mL of water, add 2.5 mL of 6 *N* ammonium hydroxide dropwise, with shaking, and then add 5 mL of ammonium oxalate TS: a white precipitate is formed.

B: To 10 mL of a warm solution (1 in 100) in a slight excess of nitric acid add 10 mL of ammonium molybdate TS: a yellow precipitate of ammonium phosphomolybdate is formed.

Loss on ignition ⟨733⟩—Ignite it at 800° to 825° to constant weight: anhydrous Dibasic Calcium Phosphate loses between 6.6% and 8.5% of its weight, and hydrous Dibasic Calcium Phosphate loses between 24.5% and 26.5% of its weight.

Acid-insoluble substances—Heat 5.0 g with a mixture of 40 mL of water and 10 mL of hydrochloric acid until no more dissolves, and dilute with water to 100 mL. If an insoluble residue remains, filter, wash with hot water until the last washing does not give a reaction for chloride, and dry the residue at 105° for 1 hour: the weight of the residue does not exceed 10 mg (0.2%).

Carbonate—Mix 1.0 g with 5 mL of water, and add 2 mL of hydrochloric acid: no effervescence occurs.

Chloride ⟨221⟩—To 0.30 g add 10 mL of water and 2 mL of nitric acid, and warm gently, if necessary, until no more dissolves. Dilute to 25 mL, filter, if necessary, and add 1 mL of silver nitrate TS: the turbidity does not exceed that produced by 1.0 mL of 0.020 *N* hydrochloric acid (0.25%).

Fluoride—[NOTE—Prepare and store all solutions in plastic containers.]

Buffer solution—Dissolve 73.5 g of sodium citrate in water to make 250 mL of solution.

Standard solution—Dissolve an accurately weighed quantity of USP Sodium Fluoride RS quantitatively in water to obtain a solution containing 1.1052 mg per mL. Transfer 20.0 mL of the resulting solution to a 100-mL volumetric flask containing 50 mL of *Buffer solution*, dilute with water to volume, and mix. Each mL of this solution contains 100 µg of fluoride ion.

Electrode system—Use a fluoride-specific, ion-indicating electrode and a silver–silver chloride reference electrode connected to a pH meter capable of measuring potentials with a minimum reproducibility of ±0.2 mV (see *pH* ⟨791⟩).

Standard response line—Transfer 50.0 mL of *Buffer solution* and 2.0 mL of hydrochloric acid to a beaker, and add water to make 100 mL. Add a plastic-coated stirring bar, insert the electrodes into the solution, stir for 15 minutes, and read the potential, in mV. Continue stirring, and at 5-minute intervals add 100 µL, 100 µL, 300 µL, and 500 µL of *Standard solution*, reading the potential 5 minutes after each addition. Plot the logarithms of the cumulative fluoride ion concentrations (0.1, 0.2, 0.5, and 1.0 µg per mL) versus potential, in mV.

Procedure—Transfer 2.0 g of the specimen under test to a beaker containing a plastic-coated stirring bar, add 20 mL of water and 2.0 mL of hydrochloric acid, and stir until dissolved. Add 50.0 mL of *Buffer solution* and sufficient water to make 100 mL of test solution. Rinse and dry the electrodes, insert them into the test solution, stir for 5 minutes, and read the potential, in mV. From the measured potential and the *Standard response line* determine the concentration, *C*, in µg per mL, of fluoride ion in the test solution. Calculate the percentage of fluoride in the specimen taken by multiplying *C* by 0.005: the limit is 0.005%.

Sulfate ⟨221⟩—Dissolve 1.0 g in the smallest possible amount of 3 *N* hydrochloric acid, dilute with water to 100 mL, and filter, if necessary. To 20 mL of the filtrate add 1 mL of barium chloride TS: the turbidity does not exceed that produced by 1.0 mL of 0.020 *N* sulfuric acid (0.5%).

Arsenic, *Method I* ⟨211⟩—Prepare the *Test Preparation* by dissolving 1.0 g in 25 mL of 3 *N* hydrochloric acid and diluting with water to 55 mL: the resulting solution meets the requirements of the test, the addition of 20 mL of 7 *N* sulfuric acid specified under *Procedure* being omitted. The limit is 3 ppm.

Barium—Heat 0.50 g with 10 mL of water, and add hydrochloric acid dropwise, stirring after each addition, until no more dissolves. Filter, and to the filtrate add 2 mL of potassium sulfate TS: no turbidity is produced within 10 minutes.

Heavy metals, *Method I* ⟨231⟩—Warm 1.3 g with 3 mL of 3 *N* hydrochloric acid until no more dissolves, dilute with water to 50 mL, and filter: the limit is 0.003%.

Assay—Weigh accurately a quantity of Dibasic Calcium Phosphate, equivalent to about 250 mg of dibasic calcium phosphate (dihydrate), dissolve, with the aid of gentle heat if necessary, in a mixture of 5 mL of hydrochloric acid and 3 mL of water contained in a 250-mL beaker equipped with a magnetic stirrer, and cautiously add 125 mL of water. With constant stirring, add, in the order named, 0.5 mL of triethanolamine, 300 mg of hydroxy naphthol blue indicator, and, from a 50-mL buret, about 23 mL of 0.05 *M* disodium ethylenediaminetetraacetate VS. Add sodium hydroxide solution (45 in 100) until the initial red color changes to clear blue, then continue to add it dropwise until the color changes to violet, then add an additional 0.5 mL. The pH is between 12.3 and 12.5. Continue the titration dropwise with the 0.05 *M* disodium ethylenediaminetetraacetate VS to the appearance of a clear blue end-point that persists for not less than 60 seconds. Each mL of 0.05 *M* disodium ethylenediaminetetraacetate is equivalent to 2.004 mg of Ca.

Dibasic Calcium Phosphate Tablets

» Dibasic Calcium Phosphate Tablets contain not less than 92.5 percent and not more than 107.5 percent of the labeled amount of $CaHPO_4 . 2H_2O$.

NOTE—An equivalent amount of Dibasic Calcium Phosphate with less water of hydration may be used in place of $CaHPO_4 . 2H_2O$ in preparing Dibasic Calcium Phosphate Tablets.

Packaging and storage—Preserve in well-closed containers.

Labeling—The quantity of dibasic calcium phosphate stated in the labeling is in terms of $CaHPO_4 . 2H_2O$.

Identification—

A: A filtered portion of the solution prepared for the *Assay* responds to the tests for *Calcium* ⟨191⟩.

B: A filtered portion of the solution prepared for the *Assay*, neutralized with ammonium hydroxide, responds to the tests for *Phosphate* ⟨191⟩.

Disintegration ⟨701⟩: 30 minutes.

Uniformity of dosage units ⟨905⟩: meet the requirements.

Assay—Weigh and finely powder not less than 20 Dibasic Calcium Phosphate Tablets. Weigh accurately a portion of the powder, equivalent to about 1 g of dibasic calcium phosphate (dihydrate), and transfer to a 100-mL volumetric flask containing 15 mL of hydrochloric acid and 10 mL of water. Heat on a steam bath, with occasional mixing to dissolve the dibasic calcium phosphate, but not longer than 30 minutes. Cool, add water to volume, and mix. If the preparation is not clear, filter, discarding the first 10 mL of the filtrate (reserve portions of the solution for the *Identification tests*). Transfer 25.0 mL of the solution to a 250-mL beaker equipped with a magnetic stirrer. Proceed as directed in the *Assay* under *Dibasic Calcium Phosphate*, beginning with "With constant stirring." Each mL of 0.05 *M* disodium ethylenediaminetetraacetate is equivalent to 8.604 mg of $CaHPO_4 . 2H_2O$.

Calcium Phosphate, Tribasic—*see* Calcium Phosphate, Tribasic NF

Calcium Polycarbophil

Calcium polycarbophil [*9003-97-8*].

» Calcium Polycarbophil is the calcium salt of polyacrylic acid cross-linked with divinyl glycol.

Packaging and storage—Preserve in tight containers.

Identification—When tested as directed in the test for *Absorbing power*, it absorbs about 35 times its original weight.

Loss on drying ⟨731⟩—Dry it in vacuum at 130° for 4 hours: it loses not more than 10.0% of its weight.

Absorbing power—Transfer about 250 mg, accurately weighed, to a tared 50-mL centrifuge tube fitted with a tight closure. Add 35 mL of 0.1 *N* hydrochloric acid to the tube, seal the tube, and shake by mechanical means for 30 minutes. Centrifuge at 2000 rpm for 20 minutes, and decant and discard the supernatant liquid. [NOTE—Exercise care to avoid any loss of particles.] Add 35 mL of 0.1 *N* hydrochloric acid, and shake for 30 minutes. Centrifuge, decanting and discarding the supernatant liquid. Repeat the foregoing steps, using water instead of acid. Add 35 mL of a sodium bicarbonate solution (15 in 1000), and shake, venting as necessary to release any carbon dioxide liberated. Shake for 1 hour, centrifuge, and decant the supernatant liquid. Add 35 mL of sodium bicarbonate solution (15 in 1000), and shake for 1 hour. Allow the tube and contents to stand overnight or until the contents have settled, and centrifuge. Withdraw the supernatant liquid, and weigh the tube and contents. Calculate the weight of sodium bicarbonate solution absorbed: not less than 35.0 g of the sodium bicarbonate solution is absorbed by 1.0 g of Calcium Polycarbophil, calculated on the dried basis.

Calcium content—Transfer about 2 g, accurately weighed, to a 250-mL beaker, add 100.0 mL of dilute hydrochloric acid (1 in 5), stir for 15 minutes, and filter, discarding the first 15 mL of the filtrate. Pipet 15 mL of the filtrate into a 250-mL beaker, and add, while stirring on a magnetic stirrer, 100 mL of water, 20.0 mL of 0.05 *M* disodium ethylenediaminetetraacetate VS, 30 mL of 1 *N* sodium hydroxide, and 300 mg of hydroxy naphthol blue trituration. Add additional sodium hydroxide dropwise until the color of the solution changes from red to blue and then to violet, and add an additional 0.5 mL. Continue titrating with 0.05 *M* disodium ethylenediaminetetraacetate VS to a persistent blue end-point. Each mL of 0.05 *M* disodium ethylenediaminetetraacetate is equivalent to 2.004 mg of Ca. The content of Ca found is not less than 18.0% and not more than 22.0%, calculated on the dried basis.

Calcium Saccharate

H H OH H
[—OOC—C—C—C—C—COO—] Ca · $4H_2O$
OH OH H OH

$C_6H_8CaO_8 \cdot 4H_2O$ 320.27
D-Glucaric acid, calcium salt (1:1) tetrahydrate.
Calcium D-glucarate (1:1), tetrahydrate [*5793-89-5*].

» Calcium Saccharate is the calcium salt of D-saccharic acid. It contains not less than 98.5 percent and not more than 102.0 percent of $C_6H_8CaO_8 \cdot 4H_2O$.

Packaging and storage—Preserve in well-closed containers.

Identification—

A: Dissolve about 0.2 g in 10 mL of water by the addition of 2 mL of hydrochloric acid: the solution so obtained responds to the tests for *Calcium* ⟨191⟩.

B: To 0.5 g add 10 mL of water, 1 mL of glacial acetic acid, and hydrochloric acid dropwise until it dissolves. Add 1 mL of freshly distilled phenylhydrazine, and heat the mixture on a steam bath for 30 minutes. Cool the solution, induce crystallization by scratching the inner surface of the test tube with a glass rod, and collect the crystals of the phenylhydrazide of saccharic acid: the crystals so obtained, after being thoroughly washed with water and dried at 105° for 1 hour, melt between 200° and 205°.

Specific rotation ⟨781⟩: between +18.5° and +22.5°, determined in a solution in 4.8 *N* hydrochloric acid containing 600 mg in each 10 mL that has been allowed to stand for 1 hour.

Arsenic, *Method I* ⟨211⟩—Dissolve 1.0 g in a mixture of 10 mL of hydrochloric acid and 20 mL of water, and dilute with water to 55 mL: the solution so obtained meets the requirements of the test, the addition of 20 mL of 7 *N* sulfuric acid specified under *Procedure* being omitted. The limit is 3 ppm.

Chloride ⟨221⟩—A 0.50-g portion dissolved in 10 mL of water by the addition of 2 mL of nitric acid shows no more chloride than corresponds to 0.50 mL of 0.020 *N* hydrochloric acid (0.07%).

Sulfate ⟨221⟩—A 0.50-g portion dissolved in 10 mL of water by the addition of 2 mL of hydrochloric acid shows no more sulfate than corresponds to 0.60 mL of 0.020 *N* sulfuric acid (0.12%).

Heavy metals, *Method II* ⟨231⟩: 0.002%.

Sucrose and reducing sugars—Dissolve 0.5 g in 10 mL of water by the addition of 2 mL of hydrochloric acid, and boil the solution for about 2 minutes. Cool, add 15 mL of sodium carbonate TS, allow to stand for 5 minutes, and filter. Add 5 mL of the clear filtrate to about 2 mL of alkaline cupric tartrate TS, and boil for 1 minute: no red precipitate is formed immediately.

Assay—Weigh accurately about 600 mg of Calcium Saccharate, and dissolve in 150 mL of water with the aid of a sufficient volume of hydrochloric acid. While stirring, preferably with a magnetic stirrer, add about 30 mL of 0.05 *M* disodium ethylenediaminetetraacetate VS from a 50-mL buret, then add 15 mL of 1 *N* sodium hydroxide and 300 mg of hydroxy naphthol blue indicator, and continue the titration to a blue end-point. Each mL of 0.05 *M* disodium ethylenediaminetetraacetate is equivalent to 16.01 mg of $C_6H_8CaO_8 \cdot 4H_2O$.

Calcium Silicate—*see* Calcium Silicate NF
Calcium Stearate—*see* Calcium Stearate NF
Calcium Sulfate—*see* Calcium Sulfate NF

Camphor

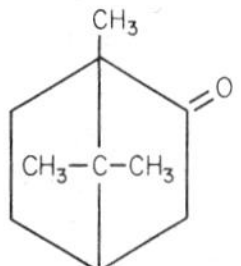

$C_{10}H_{16}O$ 152.24
Bicyclo[2.2.1]heptane-2-one, 1,7,7-trimethyl-.
Camphor.
2-Bornanone [*76-22-2*].

» Camphor is a ketone obtained from *Cinnamomum camphora* (Linné) Nees et Ebermaier (Fam. Lauraceae) (Natural Camphor) or produced synthetically (Synthetic Camphor).

Packaging and storage—Preserve in tight containers, and avoid exposure to excessive heat.

Labeling—Label it to indicate whether it is obtained from natural sources or is prepared synthetically.

Melting range ⟨741⟩: between 174° and 179°.

Specific rotation ⟨781⟩: between +41° and +43° for natural Camphor, determined in a solution in alcohol containing 1 g in each 10 mL. Synthetic Camphor is the optically inactive, racemic form.

Water—A 1 in 10 solution in solvent hexane is clear.

Nonvolatile residue—Heat 2.0 g in a tared dish on a steam bath until sublimation is complete. Then dry the residue at 120° for 3 hours, cool, and weigh: the weight of the residue does not exceed 1.0 mg (0.05%).

Halogens—Mix 100 mg of finely divided Camphor with 200 mg of sodium peroxide in a clean, dry, hard glass test tube of about 25-mm internal diameter and 20-cm length. Suspend the tube at an angle of about 45° by means of a clamp placed at the upper end, and gently heat the tube, starting near the upper end, but not heating the clamp, and gradually bringing the heat toward the lower part of the tube until incineration is complete. Dissolve the residue in 25 mL of warm water, acidify with nitric acid, and filter the solution into a comparison tube. Wash the test tube and the filter with two 10-mL portions of hot water, adding the washings to the filtered solution. To the filtrate add 0.50 mL of 0.10 *N* silver nitrate, dilute with water to 50 mL, and mix: the turbidity does not exceed that produced in a blank test with the same quantities of the same reagents and 0.050 mL of 0.020 *N* hydrochloric acid (0.035%).

Camphor Spirit

» Camphor Spirit is an alcohol solution containing, in each 100 mL, not less than 9.0 g and not more than 11.0 g of camphor ($C_{10}H_{16}O$).

Camphor................................	100 g
Alcohol, a sufficient quantity, to make..................................	1000 mL

Dissolve the camphor in about 800 mL of the alcohol, and add alcohol to make 1000 mL. Filter, if necessary.

Packaging and storage—Preserve in tight containers.

Alcohol content, *Method II* ⟨611⟩: between 80.0% and 87.0% of C_2H_5OH, the dilution to approximately 2% alcohol being made with methanol instead of with water.

Assay—Transfer 2.0 mL of Camphor Spirit to a suitable pressure bottle containing 50 mL of freshly prepared dinitrophenylhydrazine TS. Close the pressure bottle, immerse it in a water bath, and maintain at about 75° for 16 hours. Cool to room temperature, and transfer the contents to a beaker with the aid of 100 mL of 3 *N* sulfuric acid. Allow to stand at room temperature for not less than 12 hours, transfer the precipitate to a tared filter crucible, and wash with 100 mL of 3 *N* sulfuric acid followed by 75 mL of cold water in divided portions. Continue the suction until the excess water is removed, dry the crucible and precipitate at 80° for 2 hours, cool, and weigh. The weight of the precipitate so obtained, multiplied by 0.4581, represents the weight of $C_{10}H_{16}O$ in the specimen taken.

Camphorated Parachlorophenol—*see* Parachlorophenol, Camphorated

Candicidin

Candicidin.
Candicidin [*1403-17-4*].

» Candicidin is a substance produced by the growth of *Streptomyces griseus* Waksman et Henrici (Fam. Streptomycetaceae). It has a potency of not less than 1000 µg per mg, calculated on the dried basis.

Packaging and storage—Preserve in tight containers, in a refrigerator.

Reference standard—*USP Candicidin Reference Standard*—Dry in vacuum at a pressure not exceeding 5 mm of mercury at 40° for 3 hours before using.

Identification—The ultraviolet absorption spectrum of a solution containing 50 µg of candicidin activity per mL of a mixture of alcohol and water (53:47), exhibits maxima and minima at the same wavelengths as that of a similar solution of USP Candicidin RS, concomitantly measured.

pH ⟨791⟩: between 8.0 and 10.0, in an aqueous suspension containing 10 mg per mL.

Loss on drying ⟨731⟩—Dry about 100 mg in a capillary-stoppered bottle in vacuum at 60° for 3 hours: it loses not more than 4.0% of its weight.

Assay—Proceed with Candicidin as directed under *Antibiotics—Microbial Assays* ⟨81⟩.

Candicidin Ointment

» Candicidin Ointment contains not less than 90.0 percent and not more than 140.0 percent of the labeled amount of candicidin.

Packaging and storage—Preserve in well-closed containers, in a refrigerator.

Reference standard—*USP Candicidin Reference Standard*—Dry in vacuum at a pressure not exceeding 5 mm of mercury at 40° for 3 hours before using.

Minimum fill ⟨755⟩: meets the requirements.

Water, *Method I* ⟨921⟩: not more than 0.1%, 20 mL of a mixture of carbon tetrachloride, chloroform, and methanol (2:2:1) being used in place of methanol in the titration vessel.

Assay—Transfer an accurately weighed portion of Candicidin Ointment, equivalent to about 1800 µg of candicidin, to a separator containing 50 mL of a mixture of hexane and butylated hydroxyanisole (1000:1), shake to dissolve, extract with two 15-mL portions of a mixture of dimethylsulfoxide and butylated hydroxyanisole (1000:1), collect the dimethylsulfoxide extracts in a suitable volumetric flask, dilute with water to volume, and mix (*Stock Solution*). Proceed as directed under *Antibiotics—Microbial Assays* ⟨81⟩, diluting the *Stock Solution* quantitatively with water to obtain a *Test Dilution* having a concentration assumed to be equal to the median dose level of the Standard.

Candicidin Vaginal Tablets

» Candicidin Vaginal Tablets contain not less than 90.0 percent and not more than 150.0 percent of the labeled amount of candicidin.

Packaging and storage—Preserve in tight containers, in a refrigerator.

Reference standard—*USP Candicidin Reference Standard*—Dry in vacuum at a pressure not exceeding 5 mm of mercury at 40° for 3 hours before using.

Disintegration ⟨701⟩: 30 minutes.

Loss on drying ⟨731⟩—Grind not less than 4 Tablets to a fine powder, and dry about 100 mg of the powder in a capillary-stoppered bottle at 60° for 3 hours: it loses not more than 1.0% of its weight.

Assay—Weigh and finely powder not less than 5 Candicidin Vaginal Tablets. Suspend an accurately weighed portion of the powder, equivalent to about 6000 µg of candicidin, in 10 mL of dimethylsulfoxide, and mix. Centrifuge the mixture for 5 minutes, and decant the supernatant solution into a 250-mL volumetric flask. Wash the residue with three 5-mL portions of dimethylsulfoxide, collect the washings in the volumetric flask, dilute with water to volume, and mix (*Stock Solution*). Proceed as directed under *Antibiotics—Microbial Assays* ⟨81⟩, diluting the *Stock Solution* quantitatively with water to obtain a *Test Di-*

lution having a concentration assumed to be equal to the median dose level of the Standard.

Sterile Capreomycin Sulfate

· $2H_2SO_4$

	R	
Capreomycin IA	OH	$C_{25}H_{44}N_{14}O_8$
Capreomycin IB	H	$C_{25}H_{44}N_{14}O_7$

Capreomycin, sulfate.
Capreomycin sulfate [*1405-37-4*].

» Sterile Capreomycin Sulfate is the disulfate salt of capreomycin, a polypeptide mixture produced by the growth of *Streptomyces capreolus*, suitable for parenteral use. It has a potency equivalent to not less than 700 µg and not more than 1050 µg of capreomycin per mg. In addition, where packaged for dispensing, it contains the equivalent of not less than 90.0 percent and not more than 115.0 percent of the labeled amount of capreomycin.

Packaging and storage—Preserve in *Containers for Sterile Solids* as described under *Injections* ⟨1⟩. The constituted solution may be stored for 48 hours at room temperature, and up to 14 days in a refrigerator.

Reference standard—*USP Capreomycin Sulfate Reference Standard*—Dry in vacuum at a pressure not exceeding 5 mm of mercury at 100° for 4 hours before using.

Constituted solution—At the time of use, the constituted solution prepared from Sterile Capreomycin Sulfate meets the requirements for *Constituted Solutions* under *Injections* ⟨1⟩.

Depressor substances—It meets the requirements of the *Depressor Substances Test* ⟨101⟩, the test dose being 1.0 mL per kg of a solution in sterile saline TS containing 3.0 mg of capreomycin per mL.

Pyrogen—It meets the requirements of the *Pyrogen Test* ⟨151⟩, the test dose being 1.0 mL per kg of a solution in pyrogen-free saline TS containing 10 mg of capreomycin per mL.

pH ⟨791⟩: between 4.5 and 7.5, in a solution containing 30 mg per mL (or, where packaged for dispensing, in the solution constituted as directed in the labeling).

Loss on drying ⟨731⟩—Dry about 100 mg in vacuum at a pressure not exceeding 5 mm of mercury at 100° for 4 hours: it loses not more than 10.0% of its weight.

Residue on ignition ⟨281⟩: not more than 3.0%, the charred residue being moistened with 2 mL of nitric acid and 5 drops of sulfuric acid.

Heavy metals, *Method II* ⟨231⟩: 0.003%.

Capreomycin I content—

Citrate buffer—Dissolve 21.0 g of citric acid monohydrate in 1000 mL of water, and adjust with 12.5 *N* sodium hydroxide to a pH of 6.2.

Test preparation—Dissolve about 200 mg of Sterile Capreomycin Sulfate in water, dilute with water to 10 mL, and mix. Store this solution in a refrigerator.

Diluted test preparation—Pipet 1.0 mL of the *Test preparation* into a 100-mL volumetric flask, dilute with *Citrate buffer* to volume, and mix. Pipet 3 mL of this solution into a 50-mL volumetric flask, dilute with water to volume, and mix.

Procedure—Use a chromatography chamber suitable for descending chromatography (see *Chromatography* ⟨621⟩). Fill the chamber to a depth of 4 cm with a mixture of *n*-propyl alcohol and water (7:3), and allow to equilibrate for 2 days. On the starting line of a 20- × 50-cm strip of filter paper (Whatman No. 1 or equivalent) apply 100 µL of the *Test preparation* in a streak about 7 cm long, and allow to dry. Place this strip and an untreated blank strip in the chamber, and fill the solvent trough with a mixture of *n*-propyl alcohol, water, triethylamine, and glacial acetic acid (75:33:8:8). Develop the chromatograms until the solvent front reaches the bottom of the paper (about 16 hours). Remove the sheets from the chamber, and allow to air-dry for 1 hour. Examine the dried chromatograms under short-wavelength ultraviolet light, and locate the main zone (capreomycin I) at an R_f value of about 0.5. (A minor zone may be located at an R_f value of about 0.6.) Cut out the section of the paper containing the main zone, and cut it into small pieces (about 1.5 cm square). Cut a corresponding section from the blank sheet, and cut it into small pieces. On blank filter paper apply 100 µL of the *Test preparation* in a streak about 7 cm long, allow to dry, cut out the section containing the streak, and cut it into small squares. Place the pieces of paper from the chromatographed *Test preparation*, the chromatographed blank paper, and the unchromatographed *Test preparation* into three separate glass-stoppered, 50-mL conical flasks, each containing 10.0 mL of *Citrate buffer*. Shake by mechanical means for 1 hour, and filter. Pipet 3 mL of each filtrate into separate 50-mL volumetric flasks, dilute each with water to volume, and mix. Determine the absorbances of these solutions and the *Diluted test preparation* in 1-cm cells at the wavelength of maximum absorbance at about 268 nm, with a suitable spectrophotometer, using water as the blank. Calculate the recovery of capreomycin from the unchromatographed *Test preparation* by the formula:

$$100(A_U/A_D),$$

in which A_U and A_D are the absorbances of the solution from the unchromatographed *Test preparation* and the *Diluted test preparation*, respectively. If the recovery is between 98.0% and 102.0%, calculate the capreomycin I content, in percent, by the formula:

$$100(A_T - A_B)/A_D,$$

in which A_T and A_B are the absorbances of the solutions from the chromatographed *Test preparation* and the chromatographed blank paper, respectively. The capreomycin I content is not less than 90.0%.

Other requirements—It meets the requirements under *Injections* ⟨1⟩.

Assay—Proceed with Sterile Capreomycin Sulfate as directed under *Antibiotics—Microbial Assays* ⟨81⟩.

Capsules—*see complete list in index*

Captopril

$C_9H_{15}NO_3S$ 217.28
L-Proline, 1-[(2*S*)-3-mercapto-2-methyl-1-oxopropyl]-.
1-[(2*S*)-3-Mercapto-2-methylpropionyl]-L-proline [*62571-86-2*].

» Captopril contains not less than 97.5 percent and not more than 102.0 percent of $C_9H_{15}NO_3S$, calculated on the dried basis.

Packaging and storage—Preserve in tight containers.

Reference standards—*USP Captopril Reference Standard*—Do not dry before using. *USP 3-Mercapto-2-methylpropanoic Acid 1,2-Diphenylethylamine Salt Reference Standard*—Do not dry before using.

Identification—The infrared absorption spectrum of a potassium bromide dispersion of it exhibits maxima only at the same wavelengths as that of a similar preparation of USP Captopril RS.

Specific rotation ⟨781⟩: not less than −125° and not more than −134°, calculated on the dried basis, determined in a solution in absolute alcohol containing 10 mg per mL.

Loss on drying ⟨731⟩—Dry it in vacuum at 60° for 3 hours: it loses not more than 1.0% of its weight.

Residue on ignition ⟨281⟩: not more than 0.2%.

Heavy metals, *Method II* ⟨231⟩: 0.003%.

Assay—

0.1 N Potassium iodate titrant—Dissolve 3.567 g of potassium iodate, previously dried at 110° to constant weight, in water to make 1000.0 mL.

Procedure—Dissolve about 300 mg of Captopril, accurately weighed, in 100 mL of water in a suitable glass-stoppered flask, add 10 mL of 3.6 *N* sulfuric acid, 1 g of potassium iodide, and 2 mL of starch TS. Titrate with 0.1 *N Potassium iodate titrant* to a faint blue end-point that persists for not less than 30 seconds. Perform a blank determination (see *Titrimetry* ⟨541⟩), and make any necessary correction. Each mL of 0.1 *N Potassium iodate titrant* is equivalent to 21.73 mg of $C_9H_{15}NO_3S$.

Caramel—*see* Caramel NF

Carbamazepine

CONH₂

$C_{15}H_{12}N_2O$ 236.27
5*H*-Dibenz[*b,f*]azepine-5-carboxamide.
5*H*-Dibenz[*b,f*]azepine-5-carboxamide [*298-46-4*].

» Carbamazepine contains not less than 98.0 percent and not more than 102.0 percent of $C_{15}H_{12}N_2O$, calculated on the dried basis.

Packaging and storage—Preserve in tight containers.

Reference standard—*USP Carbamazepine Reference Standard*—Dry at 105° for 2 hours before using.

Identification—The infrared absorption spectrum of a mineral oil dispersion of it, previously dried, exhibits maxima only at the same wavelengths as that of a similar preparation of USP Carbamazepine RS.

X-ray diffraction ⟨941⟩—The X-ray diffraction pattern conforms to that of USP Carbamazepine RS, similarly determined.

Acidity—Add 2.0 g to 40.0 mL of water, mix for 15 minutes, and filter through paper. To a 10.0-mL aliquot of the solution so obtained add 1 drop of phenolphthalein TS, and titrate with 0.01 *N* sodium hydroxide VS from a 10-mL buret. Perform a blank determination, and make any necessary correction. Not more than 1.0 mL of 0.010 *N* sodium hydroxide is required for each 1.0 g of Carbamazepine.

Alkalinity—To a 10.0-mL aliquot of the solution prepared in the test for *Acidity* add 1 drop of methyl red TS, and titrate with 0.01 *N* hydrochloric acid VS from a 10-mL buret. Perform a blank determination, and make any necessary correction. Not more than 1.0 mL of 0.010 *N* hydrochloric acid is required for each 1.0 g of Carbamazepine.

Loss on drying ⟨731⟩—Dry it at 105° for 2 hours: it loses not more than 0.5% of its weight.

Residue on ignition ⟨281⟩: not more than 0.1%, a 2.0-g test specimen being used.

Chloride ⟨221⟩—Boil 1.0 g with 20.0 mL of water for 10 minutes, cool, again adjust the volume, and filter: a 10.0-mL portion of the filtrate shows no more chloride than corresponds to 0.10 mL of 0.020 *N* hydrochloric acid (0.014%).

Heavy metals, *Method II* ⟨231⟩: 0.001%.

Chromatographic purity: not more than 2.0% total impurities are found, the methods in both Part I and Part II being used.

Part I—

Mobile phase—Prepare as directed in the *Assay*.

Resolution solution—Dissolve suitable quantities of phenytoin and USP Carbamazepine RS in methanol to obtain a solution containing about 0.6 mg and 0.2 mg per mL, respectively. Dilute this solution quantitatively, and stepwise if necessary, with *Mobile phase* to obtain a solution containing about 60 µg of phenytoin per mL and about 20 µg of USP Carbamazepine RS per mL.

Test preparation—Dissolve an accurately weighed quantity of Carbamazepine quantitatively in methanol to obtain a solution containing 4.0 mg of specimen per mL.

Chromatographic system (see *Chromatography* ⟨621⟩)—The liquid chromatograph is equipped with a 230-nm detector and a 3.9-mm × 30-cm column that contains packing L1. [NOTE—Wash the column with 50 to 100 mL of methanol before and after use.] The flow rate is about 2 mL per minute. Chromatograph the *Resolution solution*, and record the peak responses as directed under *Procedure:* the resolution, *R*, between the phenytoin and carbamazepine peaks is not less than 2.8, and the relative standard deviation for replicate injections is not more than 2.0%. The relative retention times are about 0.7 for phenytoin and 1.0 for carbamazepine.

Procedure—Inject about 10 µL of the *Test preparation* into the chromatograph, record the chromatogram, and measure the peak responses. Calculate the percentage of each peak, other than the solvent peak and the carbamazepine peak, in the specimen of Carbamazepine taken by the same formula:

$$100r_i/r_t,$$

in which r_i is the response of each peak and r_t is the sum of the responses of all of the peaks, excluding that of the solvent peak.

Part II—

Mobile phase—Prepare a filtered and degassed mixture of water, methanol, and acetonitrile (10:7:3). Make adjustments if necessary (see *System Suitability* under *Chromatography* ⟨621⟩).

Resolution solution—Dissolve suitable quantities of iminostilbene and USP Carbamazepine RS in methanol to obtain a solution containing about 12.5 µg and 5.0 µg per mL, respectively.

Test preparation—Dissolve an accurately weighed quantity of Carbamazepine quantitatively in methanol to obtain a solution containing 4.0 mg of specimen per mL.

Chromatographic system (see *Chromatography* ⟨621⟩)—The liquid chromatograph is equipped with a 230-nm detector and a 3.9-mm × 30-cm column that contains packing L1. [NOTE—Wash the column with 50 to 100 mL of methanol before and after use.] The flow rate is about 2 mL per minute. Chromatograph the *Resolution solution*, and record the peak responses as directed under *Procedure* (except to inject 10 µL of the *Resolution solution*): the resolution, *R*, between the carbamazepine and iminostilbene peaks is not less than 10.0, and the relative standard deviation for replicate injections is not more than 2.0%. The relative retention times are about 0.3 for carbamazepine and 1.0 for iminostilbene.

Procedure—Inject about 40 µL of the *Test preparation* into the chromatograph, record the chromatogram, and measure the peak responses. Calculate the percentage of each peak, other than the solvent peak and the carbamazepine peak, in the specimen of Carbamazepine taken by the same formula:

$$100r_i/r_t,$$

in which r_i is the response of each peak and r_t is the sum of the responses of all of the peaks, excluding that of the solvent peak.

Assay—

Mobile phase—Prepare a filtered and degassed mixture of water, methanol, and methylene chloride (40:30:3). Make adjustments if necessary (see *System Suitability* under *Chromatography* ⟨621⟩).

Internal standard solution—Transfer about 60 mg of phenytoin to a 100-mL volumetric flask, dissolve in about 80 mL of methanol, add methanol to volume, and mix.

Standard preparation—Dissolve an accurately weighed quantity of USP Carbamazepine RS in methanol, and dilute quantitatively, and stepwise if necessary, with methanol to obtain a solution having a known concentration of about 0.2 mg per mL. Transfer 10.0 mL of this solution to a 100-mL volumetric flask, add 10.0 mL of *Internal standard solution*, dilute with *Mobile phase* to volume, and mix to obtain a solution having a known concentration of about 20 µg of USP Carbamazepine RS per mL.

Assay preparation—Transfer about 200 mg of Carbamazepine, accurately weighed, to a 100-mL volumetric flask, dissolve in methanol, dilute with methanol to volume, and mix. Transfer 10.0 mL of this solution to a 100-mL volumetric flask, dilute with methanol to volume, and mix. Transfer 10.0 mL of this solution to a 100-mL volumetric flask, add 10.0 mL of *Internal standard solution*, dilute with *Mobile phase* to volume, and mix.

Chromatographic system (see *Chromatography* ⟨621⟩)—The liquid chromatograph is equipped with a 230-nm detector and a 3.9-mm × 30-cm column that contains packing L1. [NOTE—Wash the column with 50 to 100 mL of methanol before and after use.] The flow rate is about 2 mL per minute. Chromatograph the *Standard preparation*, and record the peak responses as directed under *Procedure:* the resolution, R, between the analyte and internal standard peaks is not less than 2.8, and the relative standard deviation for replicate injections is not more than 2.0%.

Procedure—Separately inject equal volumes (about 10 µL) of the *Standard preparation* and the *Assay preparation* into the chromatograph, record the chromatograms, and measure the responses for the major peaks. The relative retention times are about 0.7 for phenytoin and 1.0 for carbamazepine. Calculate the quantity, in mg, of $C_{15}H_{12}N_2O$ in the Carbamazepine taken by the formula:

$$10C(R_U/R_S),$$

in which C is the concentration, in µg per mL, of USP Carbamazepine RS in the *Standard preparation*, and R_U and R_S are the ratios of the analyte peak response to the internal standard peak response obtained from the *Assay preparation* and the *Standard preparation*, respectively.

Carbamazepine Tablets

» Carbamazepine Tablets contain not less than 90.0 percent and not more than 110.0 percent of the labeled amount of $C_{15}H_{12}N_2O$.

Packaging and storage—Preserve in tight containers.

Reference standard—*USP Carbamazepine Reference Standard*—Dry at 105° for 2 hours before using.

Identification—Boil, in a 50-mL beaker, a quantity of powdered Tablets, equivalent to about 250 mg of carbamazepine, with 15 mL of acetone for 5 minutes. Filter while hot into a second beaker, using two 5-mL portions of hot acetone to effect transfer. Evaporate with the aid of nitrogen to about 5 mL, and cool in an ice bath until crystals are formed. Filter the crystals, wash with 3 mL of cold acetone, and dry in vacuum at 70° for 30 minutes: the crystals so obtained respond to the *Identification test* under *Carbamazepine.*

Dissolution ⟨711⟩—

Medium: water containing 1% sodium lauryl sulfate; 900 mL.

Apparatus 2: 75 rpm.

Time: 60 minutes.

Procedure—Determine the amount of $C_{15}H_{12}N_2O$ dissolved from ultraviolet absorbances at the wavelength of maximum absorbance at about 285 nm of filtered portions of the solution under test, suitably diluted with *Dissolution Medium*, if necessary, in comparison with a Standard solution having a known concentration of USP Carbamazepine RS in the same medium. [NOTE—A volume of methanol not exceeding 1% of the final total volume of the Standard solution may be used to dissolve the carbamazepine.]

Tolerances—Not less than 75% (Q) of the labeled amount of $C_{15}H_{12}N_2O$ is dissolved in 60 minutes.

Uniformity of dosage units ⟨905⟩: meet the requirements.

Assay—

Mobile phase, Internal standard solution, Standard preparation, and *Chromatographic system*—Prepare as directed in the *Assay* under *Carbamazepine.*

Assay preparation—Weigh and finely powder not less than 20 Carbamazepine Tablets. Transfer an accurately weighed portion of the powder, equivalent to about 200 mg of carbamazepine, to a 100-mL volumetric flask, add about 70 mL of methanol, shake by mechanical means for about 30 minutes, sonicate for about 2 minutes, dilute with methanol to volume, and mix. Allow the solution to stand for about 10 minutes, transfer 10.0 mL of the clear solution to a 100-mL volumetric flask, dilute with methanol to volume, and mix. Transfer 10.0 mL of this solution to a 100-mL volumetric flask, add 10.0 mL of *Internal standard solution*, dilute with *Mobile phase* to volume, and mix.

Procedure—Proceed as directed for *Procedure* in the *Assay* under *Carbamazepine.* Calculate the quantity, in mg, of $C_{15}H_{12}N_2O$ in the portion of Tablets taken by the formula:

$$10C(R_U/R_S),$$

in which the terms are as defined therein.

Carbamide Peroxide

$CO(NH_2)_2 \cdot H_2O_2$

$CH_6N_2O_3$ 94.07

Urea, compd. with hydrogen peroxide (1:1).

Urea compound with hydrogen peroxide (1:1) [*124-43-6*].

» Carbamide Peroxide contains not less than 98.0 percent and not more than 102.0 percent of $CH_6N_2O_3$.

Packaging and storage—Preserve in tight, light-resistant containers, and avoid exposure to excessive heat.

Identification—

A: Mix 1 mL of a solution (1 in 10) of it with 1 mL of nitric acid: a white, crystalline precipitate is formed.

B: A solution of it (1 in 10) responds to the tests for *Peroxide* ⟨191⟩.

Assay—Transfer about 100 mg of Carbamide Peroxide, accurately weighed, to a 500-mL iodine flask with the aid of 25 mL of water, add 5 mL of glacial acetic acid, and mix. Add 2 g of potassium iodide and 1 drop of ammonium molybdate TS, insert the stopper, and allow to stand in the dark for 10 minutes. Titrate the liberated iodine with 0.1 N sodium thiosulfate VS, adding 3 mL of starch TS as the end-point is approached. Each mL of 0.1 N sodium thiosulfate is equivalent to 4.704 mg of $CH_6N_2O_3$.

Carbamide Peroxide Topical Solution

$CO(NH_2)_2 \cdot H_2O_2$

$CH_6N_2O_3$ 94.07

Urea, compd. with hydrogen peroxide (1:1).

Urea compound with hydrogen peroxide (1:1) [*124-43-6*].

» Carbamide Peroxide Topical Solution is a solution in anhydrous glycerin of Carbamide Peroxide or of carbamide peroxide prepared from hydrogen peroxide and Urea. It contains not less than 78.0 percent and not more than 110.0 percent, by weight, of the labeled amount of $CH_6N_2O_3$.

Packaging and storage—Preserve in tight, light-resistant containers, and avoid exposure to excessive heat.

Identification—

A: Mix 1 mL with 1 mL of nitric acid: a white, crystalline precipitate is formed.

B: It responds to the tests for *Peroxide* ⟨191⟩.

Specific gravity ⟨841⟩: between 1.245 and 1.272.

pH ⟨791⟩: between 4.0 and 7.5.

Assay—Transfer an accurately weighed quantity of Carbamide Peroxide Topical Solution, equivalent to about 100 mg of carbamide peroxide, to a 500-mL iodine flask with the aid of 25 mL of water, add 5 mL of glacial acetic acid, and mix. Add 2 g of potassium iodide and 1 drop of ammonium molybdate TS, and allow to stand in the dark for 10 minutes. Titrate the liberated iodine with 0.1 *N* sodium thiosulfate VS, adding 3 mL of starch TS as the end-point is approached. Each mL of 0.1 *N* sodium thiosulfate is equivalent to 4.704 mg of $CH_6N_2O_3$.

Sterile Carbenicillin Disodium

$C_{17}H_{16}N_2Na_2O_6S$ (anhydrous) 422.36

4-Thia-1-azabicyclo[3.2.0]heptane-2-carboxylic acid, 6-[(carboxyphenylacetyl)amino]-3,3-dimethyl-7-oxo-, disodium salt, [2*S*-(2α,5α,6β)]-.

N-(2-Carboxy-3,3-dimethyl-7-oxo-4-thia-1-azabicyclo[3.2.0]-hept-6-yl)-2-phenylmalonamic acid disodium salt [*4800-94-6*].

» Sterile Carbenicillin Disodium has a potency equivalent to not less than 770 μg of carbenicillin ($C_{17}H_{18}N_2O_6S$) per mg, calculated on the anhydrous basis. In addition, where packaged for dispensing, it contains the equivalent of not less than 90.0 percent and not more than 120.0 percent of the labeled amount of carbenicillin.

Packaging and storage—Preserve in *Containers for Sterile Solids* as described under *Injections* ⟨1⟩.

Reference standard—*USP Carbenicillin Monosodium Monohydrate Reference Standard*—Do not dry before using.

Constituted solution—At the time of use, the constituted solution prepared from Sterile Carbenicillin Disodium meets the requirements for *Constituted Solutions* under *Injections* ⟨1⟩.

Pyrogen—It meets the requirements of the *Pyrogen Test* ⟨151⟩, the test dose being 1 mL per kg of a solution in Sterile Water for Injection containing 200 mg of carbenicillin per mL.

Sterility—It meets the requirements under *Sterility Tests* ⟨71⟩, when tested as directed in the section, *Test Procedures Using Membrane Filtration*, 6 g being aseptically dissolved in 200 mL of *Fluid A*.

pH ⟨791⟩: between 6.5 and 8.0, in a solution containing 10 mg of carbenicillin per mL (or, where packaged for dispensing, in the solution constituted as directed in the labeling).

Water, *Method I* ⟨921⟩: not more than 6.0%.

Particulate matter ⟨788⟩: meets the requirements under *Small-volume Injections*.

Other requirements—It meets the requirements for *Uniformity of Dosage Units* ⟨905⟩ and *Constituted Solutions* and *Labeling* under *Injections* ⟨1⟩.

Assay—

Assay preparation 1—Dissolve a suitable quantity of Sterile Carbenicillin Disodium, accurately weighed, in *Buffer No. 1*, and dilute quantitatively with *Buffer No. 1* to obtain a solution having a convenient concentration of carbenicillin.

Assay preparation 2 (where it is packaged for dispensing and where the package is represented as being a single-dose container)—Constitute Sterile Carbenicillin Disodium as directed in the labeling. Withdraw all of the withdrawable contents, and dilute quantitatively with *Buffer No. 1* to obtain a solution having a convenient concentration of carbenicillin.

Assay preparation 3 (where the label states the quantity of carbenicillin in a given volume of constituted solution)—Constitute Sterile Carbenicillin Disodium as directed in the labeling. Dilute an accurately measured volume of the constituted solution quantitatively with *Buffer No. 1* to obtain a solution having a convenient concentration of carbenicillin.

Procedure—Proceed as directed under *Antibiotics—Microbial Assays* ⟨81⟩, using an accurately measured volume of *Assay preparation* diluted quantitatively with *Buffer No. 1* to yield a *Test Dilution* having a concentration assumed to be equal to the median dose level of the Standard.

Carbenicillin Indanyl Sodium

$C_{26}H_{25}N_2NaO_6S$ 516.54

4-Thia-1-azabicyclo[3.2.0]heptane-2-carboxylic acid, 6-[[3-[(2,3-dihydro-1*H*-inden-5-yl)oxy]-1,3-dioxo-2-phenylpropyl]amino]-3,3-dimethyl-7-oxo-, monosodium salt, [2*S*-(2α,5α,6β)]-.

1-(5-Indanyl)(2*S*,5*R*,6*R*)-*N*-(2-carboxy-3,3-dimethyl-7-oxo-4-thia-1-azabicyclo[3.2.0]hept-6-yl)-2-phenylmalonamate monosodium salt [*26605-69-6*].

» Carbenicillin Indanyl Sodium has a potency equivalent to not less than 630 μg and not more than 769 μg of carbenicillin per mg, calculated on the anhydrous basis.

Packaging and storage—Preserve in tight containers. For periods up to 18 months, store at controlled room temperature.

Reference standard—*USP Carbenicillin Indanyl Sodium Reference Standard*—Do not dry before using.

Identification—The infrared absorption spectrum of a potassium bromide dispersion of it exhibits maxima only at the same wavelengths as that of a similar preparation of USP Carbenicillin Indanyl Sodium RS.

pH ⟨791⟩: between 5.0 and 8.0, in a solution containing 100 mg per mL.

Water, *Method I* ⟨921⟩: not more than 2.0%.

Assay—

Phosphate-citrate buffer—Dissolve 61.0 g of dibasic sodium phosphate (anhydrous) and 11.0 g of citric acid (anhydrous) in 950 mL of water, adjust with 6 *N* hydrochloric acid to a pH of 6.0, dilute with water to 1000 mL, and mix.

Standard preparation—Transfer about 125 mg of USP Carbenicillin Indanyl Sodium RS, accurately weighed, to a 25-mL volumetric flask, dissolve in water, dilute with water to volume, and mix. Transfer 5.0 mL of this solution to a glass-stoppered, 50-mL centrifuge tube, and add 15 mL of *Phosphate-citrate buffer* and 20.0 mL of methyl isobutyl ketone. Insert the stopper, and shake the tube for 10 seconds. Centrifuge for 10 minutes, and use the upper phase as the *Standard preparation*.

Assay preparation—Transfer about 125 mg of Carbenicillin Indanyl Sodium, accurately weighed, to a 25-mL volumetric flask, and proceed as directed under *Standard preparation*, beginning with "Dissolve in water."

Blank preparation—Pipet 5 mL of water into a glass-stoppered, 50-mL centrifuge tube, and proceed as directed under *Standard preparation*, beginning with "add 15 mL of *Phosphate-citrate buffer*."

Procedure—Determine the angular rotation of the *Assay preparation* in a suitable polarimeter equipped with a light source with a wavelength of 365 nm, using the *Blank preparation* to set the instrument to zero rotation (see *Optical Rotation* ⟨781⟩). Concomitantly determine the angular rotation of the *Standard preparation.* Calculate the potency, in µg of carbenicillin ($C_{17}H_{18}N_2O_6S$) per mg, of the Carbenicillin Indanyl Sodium taken by the formula:

$$P(W_S/W_U)(a_U/a_S),$$

in which P is the potency, in µg of carbenicillin per mg, of the USP Carbenicillin Indanyl Sodium RS, W_S and W_U are the quantities of USP Carbenicillin Indanyl Sodium RS and Carbenicillin Indanyl Sodium taken, respectively, and a_U and a_S are the angular rotations of the *Assay preparation* and the *Standard preparation*, respectively.

Carbenicillin Indanyl Sodium Tablets

» Carbenicillin Indanyl Sodium Tablets contain the equivalent of not less than 90.0 percent and not more than 120.0 percent of the labeled amount of carbenicillin.

Packaging and storage—Preserve in tight containers.

Reference standard—*USP Carbenicillin Indanyl Sodium Reference Standard*—Do not dry before using.

Identification—Triturate a quantity of finely powdered Tablets, equivalent to about 100 mg of carbenicillin, with 10 mL of a solvent mixture consisting of acetone, ethyl acetate, water, pyridine, and glacial acetic acid (200:100:75:25:1.5). Shake the mixture for 5 minutes, and dilute 1 volume of it with 9 volumes of the solvent mixture. On a suitable thin-layer chromatographic plate (see *Chromatography* ⟨621⟩), coated with a 0.25-mm layer of chromatographic silica gel mixture, apply 10 µL each of this solution and of a solution of USP Carbenicillin Indanyl Sodium RS in the same solvent mixture containing 1 mg of carbenicillin per mL. Allow the spots to dry, and develop the chromatogram in a solvent system consisting of a mixture of acetone, ethyl acetate, water, pyridine, and glacial acetic acid (400:300:75:25:2) until the solvent front has moved about three-fourths of the length of the plate. Remove the plate from the chamber, mark the solvent front, and heat the plate at 80° for 30 minutes. Allow the plate to cool, and expose it to iodine vapors in a closed chamber for about 30 seconds. Spray the plate with a reagent consisting of a 1 in 100 solution of ferric chloride in 0.1 *N* hydrochloric acid, potassium ferricyanide solution (1 in 100), and methanol (4:4:3): the principal spots from the test solution and the Standard solution are blue on a yellow-green background, and the R_f value of the principal spot obtained from the test solution corresponds to that obtained from the Standard solution (R_f about 0.5).

Dissolution ⟨711⟩—

Medium: water; 900 mL.

Apparatus 1: 100 rpm.

Time: 45 minutes.

Procedure—Determine the amount of carbenicillin ($C_{17}H_{18}N_2O_6S$) equivalent dissolved from the difference between ultraviolet absorbances at the wavelengths of maximum and minimum absorbance at about 267 nm and 254 nm, respectively, of filtered portions of the solution under test, suitably diluted with water, in comparison with a Standard solution having a known concentration of USP Carbenicillin Indanyl Sodium RS in the same medium.

Tolerances—Not less than 75% (*Q*) of the labeled amount of $C_{17}H_{18}N_2O_6S$ equivalent is dissolved in 45 minutes.

Uniformity of dosage units ⟨905⟩: meet the requirements.

Water, *Method I* ⟨921⟩: not more than 2.0%.

Assay—

Phosphate-citrate buffer, *Standard preparation*, and *Blank preparation*—Prepare as directed in the *Assay* under *Carbenicillin Indanyl Sodium.*

Assay preparation—Weigh and finely powder not less than 20 Carbenicillin Indanyl Sodium Tablets. Transfer an accurately weighed portion of the powder, equivalent to about 400 mg of carbenicillin, to a 100-mL volumetric flask. Add about 70 mL of water, shake for 5 minutes, dilute with water to volume, and mix. Transfer 5.0 mL of this solution to a glass-stoppered, 50-mL centrifuge tube, and proceed as directed for *Standard preparation* in the *Assay* under *Carbenicillin Indanyl Sodium,* beginning with "add 15 mL of *Phosphate-citrate buffer.*"

Procedure—Proceed as directed for *Procedure* in the *Assay* under *Carbenicillin Indanyl Sodium.* Calculate the quantity, in mg, of carbenicillin ($C_{17}H_{18}N_2O_6S$) in the portion of Tablets taken by the formula:

$$(PW_S/250)(a_U/a_S).$$

Carbidopa

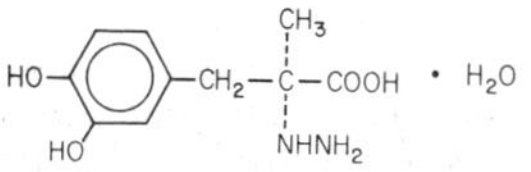

$$HO(HO)C_6H_3-CH_2-C(CH_3)(NHNH_2)-COOH \cdot H_2O$$

$C_{10}H_{14}N_2O_4 \cdot H_2O$ 244.25

Benzenepropanoic acid, α-hydrazino-3,4-dihydroxy-α-methyl-, monohydrate, (*S*)-.

(−)-L-α-Hydrazino-3,4-dihydroxy-α-methylhydrocinnamic acid monohydrate [*38821-49-7*].

Anhydrous 226.23 [*28860-95-9*].

» Carbidopa contains not less than 98.0 percent and not more than 101.0 percent of $C_{10}H_{14}N_2O_4 \cdot H_2O$.

Packaging and storage—Preserve in well-closed, light-resistant containers.

Reference standards—*USP Carbidopa Reference Standard*—Do not dry before using. Dry a portion separate from the analytical specimen at a pressure not exceeding 5 mm of mercury at 100° to constant weight, and apply a correction for *Loss on drying* for quantitative analyses. *USP Methyldopa Reference Standard*—Do not dry; determine the *Water* content by *Method I* before using. *USP 3-O-Methylcarbidopa Reference Standard*—Use as is.

Identification—

A: The infrared absorption spectrum of a mineral oil dispersion of it exhibits maxima only at the same wavelengths as that of a similar, undried, preparation of USP Carbidopa RS.

B: The ultraviolet absorption spectrum of a 1 in 25,000 solution of it in a 1 in 100 solution of hydrochloric acid in methanol exhibits maxima and minima at the same wavelengths as that of a similar solution of USP Carbidopa RS, concomitantly measured, and the respective absorptivities, calculated on the dried basis, at the wavelength of maximum absorbance at about 282 nm do not differ by more than 3.0%.

Specific rotation ⟨781⟩: between −21.0° and −23.5°, calculated as the monohydrate, determined in a solution containing 100 mg in each 10 mL, the solvent being aluminum chloride solution (2 in 3) that has been filtered and then adjusted with sodium hydroxide solution (1 in 100) to a pH of 1.5.

Loss on drying—Heat 1 g, accurately weighed, in a suitable vacuum drying apparatus at 100° and at a pressure of not more than 5 mm of mercury, to constant weight. Cool, and weigh: it loses not less than 6.9% and not more than 7.9% of its weight.

Residue on ignition ⟨281⟩: not more than 0.1%.

Heavy metals, *Method II* ⟨231⟩: 0.001%.

Methyldopa and 3-*O*-methylcarbidopa—

Mobile phase, *Resolution solution*, *Standard preparation*, *Assay preparation*, and *Chromatographic system*—Proceed as directed in the *Assay.*

Impurity standard preparation—Dissolve accurately weighed quantities of USP Methyldopa RS and USP 3-*O*-Methylcarbidopa RS in *Mobile phase* to obtain a solution having a known concentration of about 2.5 µg of each per mL.

Procedure—Separately inject equal volumes (about 20 µL) of the *Impurity standard preparation* and the *Assay preparation* into the chromatograph by means of a suitable sampling valve, and measure the peak responses. The retention times for carbidopa, methyldopa, and 3-*O*-methylcarbidopa are about 6, 5, and 11 minutes, respectively. Calculate the percentage of methyldopa in the portion of Carbidopa taken by the formula:

$$10(C/W)(r_U/r_S),$$

in which C is the concentration, in µg per mL, of USP Methyldopa RS in the *Impurity standard preparation, W* is the weight, in mg, of Carbidopa taken for the *Assay preparation*, and r_U and r_S are the peak responses for methyldopa obtained from the *Assay preparation* and the *Impurity standard preparation*, respectively. The limit is 0.5%. Calculate the percentage of 3-*O*-methylcarbidopa in the portion of Carbidopa taken by the formula:

$$10(C/W)(r_U/r_S),$$

in which C is the concentration, in µg per mL, of USP 3-*O*-Methylcarbidopa RS in the *Impurity standard preparation, W* is the weight, in mg, of carbidopa taken for the *Assay preparation*, and r_U and r_S are the peak responses for 3-*O*-methylcarbidopa obtained from the *Assay preparation* and the *Impurity standard preparation*, respectively. The limit is 0.5%.

Assay—

Mobile phase—Prepare a solution containing 5 volumes of alcohol and 95 volumes of 0.05 *M* monobasic sodium phosphate, previously adjusted with phosphoric acid to a pH of 2.7, and degas.

Standard preparation—Dissolve an accurately weighed quantity of USP Carbidopa RS in *Mobile phase* to obtain a solution having a known concentration of about 0.5 mg per mL, using gentle heat and ultrasonification, if necessary, to aid dissolution.

Resolution solution—Prepare a solution in *Mobile phase* containing, in each mL, 0.1 mg of USP Carbidopa RS and 0.1 mg of USP Methyldopa RS.

Assay preparation—Transfer about 50 mg of Carbidopa, accurately weighed, to a 100-mL volumetric flask, dilute with *Mobile phase* to volume, and mix.

Chromatographic system (see *Chromatography* ⟨621⟩)—The liquid chromatograph is equipped with a 280-nm detector and a 4-mm × 30-cm column that contains packing L1. The flow rate is about 1 mL per minute. Chromatograph three replicate injections of the *Standard preparation*, and record the peak responses as directed under *Procedure:* the relative standard deviation is not more than 1.5%. Chromatograph the *Resolution solution:* the resolution factor between carbidopa and methyldopa is not less than 0.9.

Procedure—Separately inject equal volumes (about 20 µL) of the *Standard preparation* and the *Assay preparation* into the chromatograph by means of a suitable sampling valve, record the chromatograms, and measure the responses for the major peaks. The retention time for carbidopa is about 6 minutes. Calculate the quantity, in mg, of $C_{10}H_{14}N_2O_4 . H_2O$ in the portion of Carbidopa taken by the formula:

$$100C(r_U/r_S),$$

in which C is the concentration, in mg per mL, of USP Carbidopa RS, as the monohydrate, in the *Standard preparation*, and r_U and r_S are the peak responses of the major peaks obtained from the *Assay preparation* and the *Standard preparation*, respectively.

Carbidopa and Levodopa Tablets

» Carbidopa and Levodopa Tablets contain not less than 90.0 percent and not more than 110.0 percent of the labeled amounts of carbidopa ($C_{10}H_{14}N_2O_4$) and of levodopa ($C_9H_{11}NO_4$).

Packaging and storage—Preserve in well-closed, light-resistant containers.

Reference standards—*USP Carbidopa Reference Standard*—Do not dry before using. Dry a portion separate from the analytical specimen at a pressure not exceeding 5 mm of mercury at 100° to constant weight, and apply a correction for *Loss on drying* for quantitative analyses. *USP Levodopa Reference Standard*—Dry at 105° for 4 hours before using.

Identification—Transfer a portion of powdered Tablets, equivalent to about 10 mg of carbidopa, to a 100-mL volumetric flask containing about 50 mL of 0.05 *N* hydrochloric acid. Agitate for 20 minutes, add methanol to volume, mix, and filter or centrifuge. Separately prepare 2 Standard solutions containing 2 mg per mL of USP Carbidopa RS and USP Levodopa RS, respectively, in a solvent prepared by mixing equal volumes of 0.05 *N* hydrochloric acid and methanol. Apply 20 µL of the test solution and 20 µL of each Standard solution at separate points on a thin-layer chromatographic plate (see *Chromatography* ⟨621⟩) coated with a 0.25-mm layer of chromatographic silica gel. Develop the chromatogram using a solvent system consisting of a mixture of acetone, chloroform, *n*-butanol, glacial acetic acid, and water (60:40:40:40:35) until the solvent front has moved about 15 cm. Air-dry, spray uniformly with about 0.5 mL of triketohydrindene reagent (prepared by dissolving 0.3 g of triketohydrindene in 100 mL of *n*-butanol acidified with 3 mL of glacial acetic acid), and heat at 105° for about 10 minutes: the solution under test exhibits two spots (reddish brown for levodopa and yellow-orange for carbidopa) having R_f values that correspond to those exhibited by the Standard solutions.

Dissolution ⟨711⟩—

Medium: 0.1 *N* hydrochloric acid; 750 mL.

Apparatus 1: 50 rpm.

Time: 30 minutes.

Procedure—Determine the amounts of carbidopa and levodopa in solution in filtered portions of the solution under test, in comparison with a Standard solution having known concentrations of USP Carbidopa RS and USP Levodopa RS in the same medium, as directed for *Procedure* in the *Assay*.

Tolerances—Not less than 80% (*Q*) of the labeled amounts of carbidopa ($C_{10}H_{14}N_2O_4$) and levodopa ($C_9H_{11}NO_4$) are dissolved in 30 minutes.

Uniformity of dosage units ⟨905⟩: meet the requirements.

Assay—

Decane sodium sulfonate solution—Dissolve 0.24 g of decane sodium sulfonate in 1 liter of water.

Mobile phase—Mix 11.04 g of monobasic sodium phosphate and 950 mL of water in a beaker. Add 1.3 mL of *Decane sodium sulfonate solution*, and adjust with phosphoric acid to a pH of 2.8. Transfer to a 1-liter volumetric flask, dilute with water to volume, and filter through a membrane filter.

Standard preparation—Transfer about 50 mg of USP Levodopa RS, accurately weighed, to a 100-mL volumetric flask. Add an accurately weighed quantity of USP Carbidopa RS, which is in a ratio with the USP Levodopa RS that corresponds with the ratio of carbidopa to levodopa in the Tablets. Add 10 mL of 0.1 *N* phosphoric acid. Warm gently to dissolve the standards. Dilute with water to volume, and mix.

Assay preparation—Weigh and finely powder not less than 20 Carbidopa and Levodopa Tablets. Transfer an accurately weighed portion of the powder, equivalent to about 50 mg of levodopa, to a 100-mL volumetric flask, add 10 mL of 0.1 *N* phosphoric acid, dilute with water to volume, and mix.

Chromatographic system (see *Chromatography* ⟨621⟩)—The liquid chromatograph is equipped with a 280-nm detector and a 3.9-mm × 30-cm column that contains packing L1. The flow rate, about 2 mL per minute, is adjusted until the retention times for levodopa and carbidopa are about 4 minutes and 11 minutes, respectively. Chromatograph five replicate injections of the *Standard preparation*, and record the peak responses as directed under *Procedure:* the relative standard deviation is not more than 2.0%, and the resolution factor between levodopa and carbidopa is not less than 6.

Procedure—Separately inject equal volumes (about 20 µL) of the *Standard preparation* and the *Assay preparation* into the chromatograph by means of a suitable microsyringe or sampling valve, record the chromatograms, and measure the responses for the major peaks. Calculate the quantity, in mg, of carbidopa ($C_{10}H_{14}N_2O_4$) in the portion of Tablets taken by the formula:

$$(100C)(r_U/r_S),$$

in which C is the concentration, in mg per mL, of USP Carbidopa RS in the *Standard preparation*, and r_U and r_S are the responses of the carbidopa peak obtained from the *Assay preparation* and the *Standard preparation*, respectively. Calculate the quantity, in mg, of levodopa ($C_9H_{11}NO_4$) by the same formula, reading the terms to refer to levodopa instead of carbidopa.

Carbinoxamine Maleate

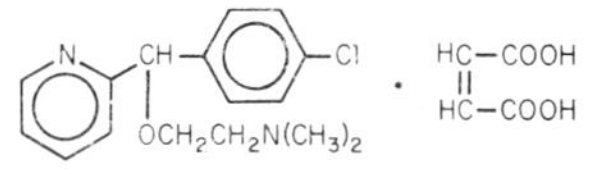

$C_{16}H_{19}ClN_2O.C_4H_4O_4$ 406.87

Ethanamine, 2-[(4-chlorophenyl)-2-pyridinylmethoxy]-*N*,*N*-dimethyl-, (*Z*)-2-butenedioate (1:1).

2-[*p*-Chloro-α-[2-(dimethylamino)ethoxy]benzyl]pyridine maleate (1:1) [*3505-38-2*].

» Carbinoxamine Maleate, dried at 105° for 2 hours, contains not less than 98.0 percent and not more than 102.0 percent of $C_{16}H_{19}ClN_2O.C_4H_4O_4$.

Packaging and storage—Preserve in tight, light-resistant containers.

Reference standard—*USP Carbinoxamine Maleate Reference Standard*—Dry at 105° for 2 hours before using.

Identification—

A: The infrared absorption spectrum of a mineral oil dispersion of it, previously dried, exhibits maxima only at the same wavelengths as that of a similar preparation of USP Carbinoxamine Maleate RS.

B: The ultraviolet absorption spectrum of a 1 in 20,000 solution in methanol exhibits maxima and minima at the same wavelengths as that of a similar solution of USP Carbinoxamine Maleate RS, concomitantly measured, and the respective absorptivities, calculated on the dried basis, at the wavelength of maximum absorbance at about 260 nm do not differ by more than 3.0%.

Melting range ⟨741⟩: between 116° and 121°, determined after drying.

pH ⟨791⟩: between 4.6 and 5.1, in a solution (1 in 100).

Loss on drying ⟨731⟩—Dry it at 105° for 2 hours: it loses not more than 0.5% of its weight.

Residue on ignition ⟨281⟩: not more than 0.1%.

Ordinary impurities ⟨466⟩—

Test solution: chloroform.

Standard solution: chloroform.

Eluant: a mixture of cyclohexane, chloroform, and diethylamine (75:15:10).

Visualization: 1.

Assay—Dissolve about 400 mg of Carbinoxamine Maleate, previously dried and accurately weighed, in 50 mL of glacial acetic acid, add 1 drop of crystal violet TS, and titrate with 0.1 *N* perchloric acid VS to a blue-green end-point. Perform a blank determination, and make any necessary correction. Each mL of 0.1 *N* perchloric acid is equivalent to 20.34 mg of $C_{16}H_{19}ClN_2O.C_4H_4O_4$.

Carbinoxamine Maleate Tablets

» Carbinoxamine Maleate Tablets contain not less than 93.0 percent and not more than 107.0 percent of the labeled amount of $C_{16}H_{19}ClN_2O.C_4H_4O_4$.

Packaging and storage—Preserve in tight, light-resistant containers.

Reference standard—*USP Carbinoxamine Maleate Reference Standard*—Dry at 105° for 2 hours before using.

Identification—It responds to the *Identification test* under *Carbinoxamine Maleate Elixir.*

Dissolution ⟨711⟩—

Medium: water; 900 mL.

Apparatus 2: 50 rpm.

Time: 45 minutes.

Procedure—Determine the amount of $C_{16}H_{19}ClN_2O.C_4H_4O_4$ dissolved from ultraviolet absorbances at the wavelength of maximum absorbance at about 260 nm of filtered portions of the solution under test, suitably diluted with *Dissolution Medium,* if necessary, in comparison with a Standard solution having a known concentration of USP Carbinoxamine Maleate RS in the same medium.

Tolerances—Not less than 75% (*Q*) of the labeled amount of $C_{16}H_{19}ClN_2O.C_4H_4O_4$ is dissolved in 45 minutes.

Uniformity of dosage units ⟨905⟩: meet the requirements.

Procedure for content uniformity—Place 1 Tablet in a 100-mL volumetric flask, add 10.0 mL of water, and shake by mechanical means for 15 minutes. Dilute with methanol to volume, and filter, discarding the first 20 mL of the filtrate. Dilute a portion of the subsequent filtrate quantitatively and stepwise, if necessary, with a mixture of methanol and water (9:1) to obtain a solution containing about 40 µg of carbinoxamine maleate per mL. Concomitantly determine the absorbances of this solution and of a Standard solution of USP Carbinoxamine Maleate RS, in the same medium having a known concentration of about 40 µg per mL, in 1-cm cells, at the wavelength of maximum absorbance at about 260 nm, with a suitable spectrophotometer, using a mixture of methanol and water (9:1) as the blank. Calculate the quantity, in mg, of $C_{16}H_{19}ClN_2O.C_4H_4O_4$ in the Tablet taken by the formula:

$$(TC/D)(A_U/A_S),$$

in which T is the labeled quantity, in mg, of carbinoxamine maleate in the Tablet, C is the concentration, in µg per mL, of USP Carbinoxamine Maleate RS in the Standard solution, D is the concentration, in µg per mL, of carbinoxamine maleate in the solution from the Tablet, based upon the labeled quantity per Tablet and the extent of dilution, and A_U and A_S are the absorbances of the solution from the Tablet and the Standard solution, respectively.

Assay—Weigh and finely powder not less than 30 Carbinoxamine Maleate Tablets. Transfer an accurately weighed portion of the powder, equivalent to about 100 mg of carbinoxamine maleate, to a separator, add 35 mL of water and 3 g of sodium bicarbonate, and mix. Extract with five 20-mL portions of chloroform, filtering the extracts through a pledget of cotton. Evaporate the combined chloroform extracts on a steam bath just to dryness, dissolve the residue in 50 mL of glacial acetic acid, add 1 drop of crystal violet TS, and titrate with 0.05 *N* perchloric acid VS to a blue-green end-point. Perform a blank determination, and make any necessary correction. Each mL of 0.05 *N* perchloric acid is equivalent to 10.17 mg of $C_{16}H_{19}ClN_2O.C_4H_4O_4$.

Carbol-Fuchsin Topical Solution

» Prepare Carbol-Fuchsin Topical Solution as follows.

Basic Fuchsin	3 g
Phenol	45 g
Resorcinol	100 g
Acetone	50 mL
Alcohol	100 mL
Purified Water, a sufficient quantity, to make	1000 mL

Dissolve the Basic Fuchsin in a mixture of the Acetone and Alcohol, and add to this solution the Phenol and Resorcinol previously dissolved in 725 mL of Purified Water. Then add sufficient Purified Water to make the product measure 1000 mL, and mix.

Packaging and storage—Preserve in tight, light-resistant containers.

Specific gravity ⟨841⟩: not less than 0.990 and not more than 1.050.
Alcohol content ⟨611⟩: between 7.0% and 10.0% of C_2H_5OH.

Carbomer 910—*see* Carbomer 910 NF
Carbomer 934—*see* Carbomer 934 NF
Carbomer 934P—*see* Carbomer 934P NF
Carbomer 940—*see* Carbomer 940 NF
Carbomer 941—*see* Carbomer 941 NF
Carbomer 1342—*see* Carbomer 1342 NF

Carbon Dioxide

CO_2 44.01
Carbon dioxide.
Carbon dioxide [*124-38-9*].

» Carbon Dioxide contains not less than 99.0 percent, by volume, of CO_2.

Packaging and storage—Preserve in cylinders.

NOTES—The following tests are designed to reflect the quality of Carbon Dioxide in both its vapor and liquid phases, which are present in previously unopened cylinders. Reduce the container pressure by means of a regulator. Withdraw the specimens for the tests with the least possible release of Carbon Dioxide consistent with proper purging of the sampling apparatus. Measure the gases with a gas volume meter downstream from the detector tubes in order to minimize contamination or change of the specimens. Perform tests in the sequence in which they are listed.

The various detector tubes called for in the respective tests are listed under *Reagents* in the section, *Reagents, Indicators, and Solutions*.

Identification—Pass 100 ± 5 mL, released from the vapor phase of the contents of the container, through a carbon dioxide detector tube at the rate specified for the tube: the indicator change extends throughout the entire indicating range of the tube.

Carbon monoxide—Pass 1050 ± 50 mL, released from the vapor phase of the contents of the container, through a carbon monoxide detector tube at the rate specified for the tube: the indicator change corresponds to not more than 0.001%.

Hydrogen sulfide—Pass 1050 ± 50 mL, released from the vapor phase, through a hydrogen sulfide detector tube at the rate specified for the tube: the indicator change corresponds to not more than 1 ppm.

Nitric oxide—Pass 550 ± 50 mL, released from the vapor phase, through a nitric oxide–nitrogen dioxide detector tube at the rate specified for the tube: the indicator change corresponds to not more than 2.5 ppm.

Nitrogen dioxide—Arrange the container so that when its valve is opened, a portion of the liquid phase of the contents is released through a piece of tubing of sufficient length to allow all of the liquid to vaporize during passage through it, and to prevent frost from reaching the inlet of the detector tube. Release into the tubing a flow of liquid sufficient to provide 550 mL of the vaporized specimen plus any excess necessary to assure adequate flushing of air from the system. Pass 550 ± 50 mL of this gas through a nitric oxide–nitrogen dioxide detector tube at the rate specified for the tube: the indicator change corresponds to not more than 2.5 ppm.

Ammonia—Proceed with Carbon Dioxide as directed in the test for *Nitrogen dioxide*, except to pass 1050 ± 50 mL of this gas through an ammonia detector tube at the rate specified for the tube: the indicator change corresponds to not more than 0.0025%.

Sulfur dioxide—Proceed with Carbon Dioxide as directed in the test for *Nitrogen dioxide*, except to pass 1050 ± 50 mL through a sulfur dioxide detector tube at the rate specified for the tube: the indicator change corresponds to not more than 5 ppm.

Water—Flush the regulator that has been flushed with 5 liters or more of the gas specimen. Pass 50 ± 5 liters, released from the vapor phase, through a water vapor detector tube connected to the regulator with a minimum length of metal or polyethylene tubing. Measure the gas passing through the detector tube with a gas flowmeter set at a flow rate of 2 liters per minute. The corrected indicator change corresponds to not more than 150 mg per cubic meter.

Assay—[NOTE—Sampling for this assay may be done from the vapor phase for convenience, but this results in more residual volume. If the specification of 1 mL is exceeded from the vapor phase, a liquid specimen may be taken.] Assemble a 100-mL gas buret provided with a leveling bulb and two-way stopcock to a gas absorption pipet of suitable capacity by connecting the pipet to one of the buret outlets. Fill the buret with slightly acidified water (turned pink with methyl orange), and fill the pipet with potassium hydroxide solution (1 in 2). By manipulation of the leveling bulb and leveling water, draw the potassium hydroxide solution to fill the pipet and capillary connection up to the stopcock, and then fill the buret with the leveling water and draw it through the other stopcock opening in such a manner that all gas bubbles are eliminated from the system. Draw into the buret 100.0 mL of specimen taken from the liquid phase as directed in the test for *Nitrogen dioxide*. By raising the leveling bottle, force the measured specimen into the pipet. The absorption may be facilitated by rocking the pipet or by flowing the specimen between pipet and buret. Draw any residual gas into the buret, and measure its volume: not more than 1.0 mL of gas remains.

Carbon Tetrachloride—*see* Carbon Tetrachloride NF

Carboprost Tromethamine

HO COOH H H H CH3 · $(HOCH_2)_3CNH_2$ H H_3C OH OH

$C_{21}H_{36}O_5 . C_4H_{11}NO_3$ 489.65
Prosta-5,13-dien-1-oic acid, 9,11,15-trihydroxy-15-methyl-, (5*Z*,9α,11α,13*E*,15*S*)-, compound with 2-amino-2-(hydroxymethyl)-1,3-propanediol (1:1).
(*Z*)-7-[(1*R*,2*R*,3*R*,5*S*)-3,5-Dihydroxy-2-[(*E*)-(3*S*)-3-hydroxy-3-methyl-1-octenyl]cyclopentyl]-5-heptenoic acid compound with 2-amino-2-(hydroxymethyl)-1,3-propanediol (1:1).
(15*S*)-15-Methylprostaglandin $F_{2\alpha}$ tromethamine [*58551-69-2*].

» Carboprost Tromethamine contains not less than 95.0 percent and not more than 105.0 percent of $C_{25}H_{47}NO_8$, calculated on the dried basis.

Caution—Great care should be taken to prevent inhaling particles of Carboprost Tromethamine and exposing the skin to it.

Packaging and storage—Preserve in well-closed containers, in a freezer.

Reference standard—*USP Carboprost Tromethamine Reference Standard*—Dry in vacuum at a pressure not exceeding 5 mm of mercury at 50° for 16 hours before using.

Identification—The infrared absorption spectrum of a mineral oil dispersion of it exhibits maxima only at the same wavelengths as that of a similar preparation of USP Carboprost Tromethamine RS.

Specific rotation ⟨781⟩: between +18° and +24°, calculated on the dried basis, determined in an alcohol solution containing 10 mg per mL.

Loss on drying ⟨731⟩: Dry it in vacuum at a pressure not exceeding 5 mm of mercury at 50° for 16 hours: it loses not more than 1.0% of its weight.

Residue on ignition ⟨281⟩: not more than 0.5%.

Limit of 15*R*-epimer and 5-*trans* isomer—Using the conditions described in the *Assay*, inject a volume of the *Assay preparation* that is about 2½ times the volume used in the *Assay*. Typical retention times are approximately 7, 8, 11, and 13 minutes for guaifenesin, the 2-naphthacyl ester of the 15*R*-epimer, the 2-naphthacyl ester of carboprost, and the 2-naphthacyl ester of the 5-*trans* isomer, respectively. Measure the peak areas for the 2-naphthacyl esters of the 15*R*-epimer (*Aa*), carboprost (*Ab*), and 5-*trans* isomer (*Ac*). Calculate the percentage of 15*R*-epimer (as the tromethamine salt), in the specimen taken, by the formula:

$$100Aa/(Aa + Ab + Ac).$$

The limit is not more than 2.0%. Calculate the percentage of 5-*trans* isomer (as the tromethamine salt), in the specimen taken, by the formula:

$$100Ac/(Aa + Ab + Ac).$$

The limit is not more than 3.0%.

Assay—

Mobile solvent—Add 0.5 mL of water to 7 mL of 1,3-butanediol, and mix with 992 mL of methylene chloride.

Internal standard preparation—Using the *Mobile solvent*, prepare a solution containing approximately 7 mg of guaifenesin per mL.

Citrate buffer—Dissolve 10.5 g of citric acid monohydrate in about 75 mL of water. Add 5 *N* sodium hydroxide slowly to adjust to a pH of 4.0, and dilute with water to 100 mL.

Standard preparation—Accurately weigh about 5 mg of USP Carboprost Tromethamine RS, and transfer to a stoppered, 50-mL centrifuge tube. Add 20.0 mL of methylene chloride and 2 mL of *Citrate buffer*. Shake the stoppered tube for about 10 minutes, and centrifuge. Remove and discard the top (aqueous) layer, and transfer a 4.0-mL aliquot of the lower (methylene chloride) layer to a suitable vial. Evaporate with the aid of a stream of nitrogen to dryness. Add 100 µL of a freshly prepared 1 in 50 solution of α-bromo-2′-acetonaphthone in acetonitrile. Swirl to wash down the sides of the vial. Add 50 µL of a freshly prepared 1 in 100 solution of diisopropylethylamine in acetonitrile, swirl again, and place the vial in a suitable heating device maintained at a temperature of 30° to 35° for not less than 15 minutes. Evaporate the acetonitrile from the vial with the aid of a stream of nitrogen, add 2.0 mL of *Internal standard preparation*, mix, and filter the resulting solution through a fine-porosity filter. Protect the filtered solution from light prior to injection to prevent degradation of the naphthacyl ester of carboprost.

Assay preparation—Using Carboprost Tromethamine, proceed as directed under *Standard preparation*.

Procedure—As a system suitability test, chromatograph different volumes of the *Standard preparation* using a suitable microsyringe or sampling valve to determine appropriate volume and other operating parameters. The retention times for guaifenesin and the 2-naphthacyl ester of carboprost are about 7 minutes and 11 minutes, respectively. In a suitable chromatogram, the resolution factor between these two peaks is not less than 4.0 and the relative standard deviation for four replicate injections of the *Standard preparation* show a relative standard deviation of not more than 2.0%. Use a suitable high-pressure liquid chromatograph of the general type (see *Chromatography* ⟨621⟩) capable of providing column pressure up to about 1500 psig operated at room temperature and equipped with an ultraviolet detector capable of monitoring absorption at 254 nm, a suitable recorder, and a 4-mm × 30-cm stainless steel column that contains 10-µm packing L3. Chromatograph equal volumes of the *Standard preparation* and the *Assay preparation*. Calculate the quantity, in mg, of $C_{25}H_{47}NO_8$, in the portion of Carboprost Tromethamine taken by the formula:

$$W_S(R_U/R_S),$$

in which W_S is the weight, in mg, of USP Carboprost Tromethamine RS used in the *Standard preparation*, and R_U and R_S are the ratios of the peak responses of the 2-naphthacyl ester of carboprost and the internal standard obtained from the *Assay preparation* and the *Standard preparation*, respectively.

Carboprost Tromethamine Injection

» Carboprost Tromethamine Injection is a sterile solution of Carboprost Tromethamine in aqueous solution, which may contain also benzyl alcohol, sodium chloride, and tromethamine. It contains not less than 90.0 percent and not more than 110.0 percent of the labeled amount of carboprost ($C_{21}H_{36}O_5$).

Packaging and storage—Preserve in single-dose or in multiple-dose containers, preferably of Type I glass, in a refrigerator.

Reference standard—*USP Carboprost Tromethamine Reference Standard*—Dry in vacuum at a pressure not exceeding 5 mm of mercury at 50° for 16 hours before using.

Identification—Extract a volume of Injection equivalent to about 2.5 mg of carboprost tromethamine with 1.5 to 2 times its volume of chloroform. Discard the chloroform layer, and acidify the aqueous layer with 3 to 5 drops of hydrochloric acid. Extract the acidified solution with an equivalent volume of chloroform. Filter the chloroform layer through a pledget of cotton, and concentrate it to a volume of less than 1 mL. Combine the resulting solution with 150 to 180 mg of potassium bromide. Dry the potassium bromide mixture in vacuum overnight, and prepare a potassium bromide pellet from the dried mixture: the infrared absorption spectrum of the resulting pellet exhibits maxima at the same wavelengths as that of a similar preparation of USP Carboprost Tromethamine RS.

Pyrogen—When diluted with Water for Injection to obtain a solution containing 10 µg of carboprost per mL and administered as a test dose of 1 mL per kg, it meets the requirements of the *Pyrogen Test* ⟨151⟩.

pH ⟨791⟩: between 7.0 and 8.0.

Other requirements—It meets the requirements under *Injections* ⟨1⟩.

Assay—

Mobile solvent and *Citrate buffer*—Prepare as directed for *Mobile solvent* and *Citrate buffer* in the *Assay* under *Carboprost Tromethamine*.

Internal standard preparation—Using the *Mobile solvent*, prepare a solution containing approximately 3 mg of guaifenesin per mL.

Standard preparation—Using an accurately weighed quantity of the Reference Standard, prepare an aqueous Standard solution containing approximately 0.332 mg of USP Carboprost Tromethamine RS and 9 mg of benzyl alcohol per mL. Pipet 2 mL of the resulting solution into a stoppered centrifuge tube. Add 20.0 mL of methylene chloride and 1.0 mL of *Citrate buffer*, shake the stoppered tube for about 10 minutes, and centrifuge. Remove and discard the top (aqueous) layer, transfer an 8.0 mL-aliquot of the lower (methylene chloride) layer to a suitable vial, and evaporate the solution with the aid of a stream of nitrogen. (The residue does not evaporate to dryness because of the presence of benzyl alcohol.) Add 100 µL of a freshly prepared 1 in 50 solution of α-bromo-2′acetonaphthone in acetonitrile, and swirl to wash down the sides of the vial. Add 50 µL of a freshly prepared 1 in 100 solution of diisopropylethylamine in acetonitrile. Swirl again, and place the vial in a suitable heating device maintained at a temperature of 30° to 35° for not less than 15 minutes. Evaporate the acetonitrile from the vial with the aid of a stream of nitrogen, add 1.0 mL of *Internal standard solution*, mix, and filter the resulting solution through a fine-porosity filter. Protect the filtered solution from light prior to injection to prevent degradation of the naphthacyl ester of carboprost.

Assay preparation—Pipet a volume of Carboprost Tromethamine Injection, equivalent to about 500 µg of carboprost, to a stoppered, 50-mL centrifuge tube. Proceed as directed for *Standard preparation*, beginning with "Add 20.0 mL of methylene chloride."

Procedure—Proceed as directed for *Procedure* in the *Assay* under *Carboprost Tromethamine.* Calculate the quantity, in µg, of carboprost in each mL of the Injection taken by the formula:

$$(368.51/489.65)(C/V)(R_U/R_S),$$

in which 368.51 and 489.65 are the molecular weights of carboprost and carboprost tromethamine, respectively, C is the concentration, in µg per mL, of USP Carboprost Tromethamine RS in the Standard solution used to prepare the naphthacyl ester, V is the volume, in mL, of Injection taken, and R_U and R_S are the ratios of the peak responses of the 2-naphthacyl ester of carboprost and the internal standard obtained from the *Assay preparation*, and the *Standard preparation*, respectively.

Carboxymethylcellulose Calcium—*see* Carboxymethylcellulose Calcium NF

Carboxymethylcellulose Sodium

Cellulose, carboxymethyl ether, sodium salt.
Cellulose carboxymethyl ether sodium salt [*9004-32-4*].

» Carboxymethylcellulose Sodium is the sodium salt of a polycarboxymethyl ether of cellulose. It contains not less than 6.5 percent and not more than 9.5 percent of sodium (Na), calculated on the dried basis.

Packaging and storage—Preserve in tight containers.

Labeling—Label it to indicate the viscosity in solutions of stated concentrations of either 1% (w/w) or 2% (w/w).

Identification—Add about 1 g of powdered Carboxymethylcellulose Sodium to 50 mL of water, while stirring to produce a uniform dispersion. Continue the stirring until a clear solution is produced, and use the solution for the following tests.

A: To 1 mL of the solution, diluted with an equal volume of water, in a small test tube, add 5 drops of 1-naphthol TS. Incline the test tube, and carefully introduce down the side of the tube 2 mL of sulfuric acid so that it forms a lower layer: a red-purple color develops at the interface.

B: To 5 mL of the solution add an equal volume of barium chloride TS: a fine, white precipitate is formed.

C: A portion of the solution responds to the tests for *Sodium* ⟨191⟩.

pH ⟨791⟩: between 6.5 and 8.5 in a solution (1 in 100).

Viscosity ⟨911⟩—Determine the viscosity in a water solution at the concentration stated on the label. Using undried Carboxymethylcellulose Sodium, weigh accurately the amount which, on the dried basis, will provide 200 g of solution of the stated concentration. Add the substance in small amounts to about 180 mL of stirred water contained in a tared, wide-mouth bottle, continue stirring rapidly until the powder is well wetted, add sufficient water to make the mixture weigh 200 g, and allow to stand, with occasional stirring, until solution is complete. Adjust the temperature to 25 ± 0.2°, and determine the viscosity, using a rotational type of viscosimeter, making certain that the system reaches equilibrium before taking the final reading. The viscosity of solutions of 2 percent concentration is not less than 80.0% and not more than 120.0% of that stated on the label; the viscosity of solutions of 1 percent concentration is not less than 75.0% and not more than 140.0% of that stated on the label.

Loss on drying ⟨731⟩—Dry it at 105° for 3 hours: it loses not more than 10.0% of its weight.

Heavy metals—Determine as directed in the test for *Heavy metals* under *Methylcellulose*, using a 500-mg specimen: the limit is 0.004%.

Assay—Transfer to a beaker about 500 mg of Carboxymethylcellulose Sodium, accurately weighed, add 80 mL of glacial acetic acid, heat the mixture on a boiling water bath for 2 hours, cool to room temperature, and titrate with 0.1 *N* perchloric acid VS, determining the end-point potentiometrically. Each mL of 0.1 *N* perchloric acid is equivalent to 2.299 mg of Na.

Carboxymethylcellulose Sodium Paste

» Carboxymethylcellulose Sodium Paste contains not less than 16.0 percent and not more than 17.0 percent of carboxymethylcellulose sodium.

Packaging and storage—Preserve in well-closed containers, and avoid prolonged exposure to temperatures exceeding 30°.

Identification—Digest a quantity of Paste, equivalent to about 1 g of carboxymethylcellulose sodium, with 50 mL of water until solution is virtually complete, and filter: the filtrate responds to the following tests.

A: To about 30 mL of the solution add 3 mL of hydrochloric acid: a white precipitate is formed.

B: To the remainder of the solution add an equal volume of barium chloride TS: a fine, white, precipitate is formed.

C: The filtrate from *Identification test A* responds to the tests for *Sodium* ⟨191⟩.

Microbial limits ⟨61⟩—The total bacterial count does not exceed 1000 per g, and the tests for *Salmonella* species and *Escherichia coli* are negative.

Loss on drying ⟨731⟩—Dry it at 105° for 3 hours: it loses not more than 2.0% of its weight.

Heavy metals ⟨231⟩—Determine as directed in the test for *Heavy metals* under *Methylcellulose*, using a 400-mg specimen: the limit is 0.005%.

Consistency—Determine as directed in the test for *Consistency* under *White Petrolatum:* the final average of the trials is not less than 30.0 mm and not more than 36.0 mm, indicating a consistency value between 300 and 360.

Assay—Transfer about 2 g of Carboxymethylcellulose Sodium Paste, accurately weighed, to a glass-stoppered, 250-mL conical flask. Add 75 mL of glacial acetic acid, attach a condenser, and reflux for 2 hours. Cool, transfer the mixture to a 250-mL beaker with the aid of small volumes of glacial acetic acid, and titrate with 0.1 *N* perchloric acid in dioxane VS, determining the end-point potentiometrically. Each mL of 0.1 *N* perchloric acid is equivalent to 29.67 mg of carboxymethylcellulose sodium.

Carboxymethylcellulose Sodium Tablets

» Carboxymethylcellulose Sodium Tablets contain an amount of sodium (Na) equivalent to not less than 6.5 percent and not more than 9.5 percent of the labeled amount of carboxymethylcellulose sodium.

Packaging and storage—Preserve in tight containers.

Identification—Digest a quantity of powdered Tablets, equivalent to about 1 g of carboxymethylcellulose sodium, with 50 mL of water until solution is virtually complete, and filter: the filtrate responds to the following tests.

A: To about 30 mL of the solution add 3 mL of hydrochloric acid: a white precipitate is formed.

B: To the remainder of the solution add an equal volume of barium chloride TS: a fine, white precipitate is formed.

C: The filtrate from *Identification test A* responds to the tests for *Sodium* ⟨191⟩.

Disintegration ⟨701⟩: 2 hours.

Uniformity of dosage units ⟨905⟩: meet the requirements.

Assay—Weigh and finely powder not less than 20 Carboxymethylcellulose Sodium Tablets. Weigh accurately a portion of the powder, equivalent to about 500 mg of carboxymethylcellulose sodium, add 80 mL of glacial acetic acid, heat the mixture on a steam bath for 2 hours, cool to room temperature, and titrate with 0.1 *N* perchloric acid VS, determining the end-point potentiometrically. Each mL of 0.1 *N* perchloric acid is equivalent to 2.299 mg of Na.

Carboxymethylcellulose Sodium and Microcrystalline Cellulose—*see* Cellulose, Microcrystalline and Carboxymethylcellulose Sodium NF

Carboxymethylcellulose Sodium 12—*see* Carboxymethylcellulose Sodium 12 NF

Carisoprodol

$$(CH_3)_2CHNHCOOCH_2C(CH_3)(CH_2CH_2CH_3)CH_2OOCNH_2$$

$C_{12}H_{24}N_2O_4$ 260.33

2-Methyl-2-propyl-1,3-propanediol carbamate isopropylcarbamate [*78-44-4*].

» Carisoprodol contains not less than 98.0 percent and not more than 102.0 percent of $C_{12}H_{24}N_2O_4$, calculated on the dried basis.

Packaging and storage—Preserve in tight containers.

Reference standards—*USP Carisoprodol Reference Standard*—Dry in vacuum at 60° for 3 hours before using. *USP Meprobamate Reference Standard*—Dry in vacuum at 60° for 3 hours before using.

Identification—

A: The infrared absorption spectrum of a potassium bromide dispersion of it exhibits maxima only at the same wavelengths as that of a similar preparation of USP Carisoprodol RS.

B: The R_f value of the principal spot due to carisoprodol in the chromatogram of the test solution obtained in the test for *Meprobamate* corresponds to that obtained in a similarly prepared chromatogram of a Standard solution of USP Carisoprodol RS in chloroform containing 100 mg per mL.

Melting range, *Class I* ⟨741⟩: between 91° and 94°.

Loss on drying ⟨731⟩—Dry it in vacuum at 60° for 3 hours: it loses not more than 0.5% of its weight.

Heavy metals, *Method II* ⟨231⟩: 0.001%.

Meprobamate—Dissolve 100 mg of Carisoprodol in 1.0 mL of chloroform. Dissolve USP Meprobamate RS in chloroform to obtain a Standard solution containing about 1 mg per mL. On a suitable thin-layer chromatographic plate (see *Chromatography* ⟨621⟩), coated with a 0.25-mm layer of chromatographic silica gel, apply a 10-µL portion of the test solution and a separate 5-µL portion of the Standard solution. Allow the spots to dry in a current of air, and develop the chromatogram in a solvent system consisting of a mixture of chloroform and acetone (4:1) until the solvent front has moved about three-fourths of the length of the plate. Remove the plate from the developing chamber, mark the solvent front, allow the solvent to evaporate, and spray the plate alternately with antimony trichloride TS and a 3 in 100 solution of furfural in chloroform until one or more black spots appear, heat the plate at 110° for 15 minutes, and examine the plate: any spot in the chromatogram of the test solution having an R_f value corresponding to that of meprobamate in the chromatogram of the Standard solution is not darker in color than the meprobamate spot in the chromatogram of the Standard solution (0.5%).

Assay—

Sodium methoxide titrant—Prepare and standardize as directed for 0.1 *N* sodium methoxide (in toluene) (see *Volumetric Solutions* under *Reagents, Indicators, and Solutions*), except to use 200 mL of methanol to dissolve the sodium metal, to add 750 mL of toluene, and to dilute with methanol to 1000 mL.

Procedure—Transfer about 400 mg of Carisoprodol, accurately weighed, to a suitable boiling flask, add 10 mL of pyridine, and mix. Add 1 drop of phenolphthalein TS, and titrate with *Sodium methoxide titrant* to a permanent pink end-point. Add 25.0 mL of *Sodium methoxide titrant*, connect the flask to a water-cooled condenser, and reflux on a hot plate for 30 minutes. Allow to cool, add 40 mL of alcohol and 7 drops of phenolphthalein TS, and titrate the excess alkali with 0.1 *N* hydrochloric acid VS until the pink color disappears. Perform a blank determination (see *Residual Titrations* ⟨541⟩). Each mL of *Sodium methoxide titrant* is equivalent to 26.03 mg of $C_{12}H_{24}N_2O_4$.

Carisoprodol Tablets

» Carisoprodol Tablets contain not less than 90.0 percent and not more than 110.0 percent of the labeled amount of $C_{12}H_{24}N_2O_4$.

Packaging and storage—Preserve in well-closed containers.

Reference standard—*USP Carisoprodol Reference Standard*—Dry in vacuum at 60° for 3 hours before using.

Identification—The retention time of the carisoprodol peak in the chromatogram of the *Assay preparation* corresponds to that of the carisoprodol peak in the chromatogram of the *Standard preparation,* obtained as directed in the *Assay.*

Uniformity of dosage units ⟨905⟩: meet the requirements.

Assay—

Mobile phase—Prepare a filtered and degassed mixture of water and acetonitrile (60:40). Make adjustments if necessary (see *System Suitability* under *Chromatography* ⟨621⟩).

Diluent—Prepare a mixture of methanol and 0.01 *N* sulfuric acid (60:40).

Standard preparation—Dissolve an accurately weighed quantity of USP Carisoprodol RS in *Diluent*, using sonication, if necessary, to obtain a solution having a known concentration of about 3.5 mg per mL.

Resolution solution—Prepare a solution in *Mobile phase* containing about 2.4 mg of 2-methyl-2-propyl-1,3-propanediol and 3.4 mg of carisoprodol per mL.

Assay preparation—Weigh and finely powder not less than 20 Carisoprodol Tablets. Transfer an accurately weighed portion of the powder, equivalent to about 350 mg of carisoprodol, to a 100-mL volumetric flask. Add about 50 mL of *Diluent*, place in an ultrasonic bath for 30 minutes, and shake by mechanical means for 60 minutes. Dilute with *Diluent* to volume, and mix. Filter a portion of this solution through a membrane filter of 0.5-µm or finer porosity, and use the filtrate as the *Assay preparation.*

Chromatographic system (see *Chromatography* ⟨621⟩)—The liquid chromatograph is equipped with a refractive index detector and a 3.9-mm × 30-cm column that contains packing L1. Maintain the detector and the column at 30 ± 1°. The flow rate is about 2 mL per minute. Chromatograph the *Resolution solution* and the *Standard preparation*, and record the peak responses as directed under *Procedure:* the resolution, *R*, between the 2-methyl-2-propyl-1,3-propanediol and carisoprodol peaks is not less than 2.0, and the relative standard deviation for three replicate injections of the *Standard preparation* is not more than 2.0%. The relative retention times are about 0.5 for 2-methyl-2-propyl-1,3-propanediol and 1.0 for carisoprodol.

Procedure—[NOTE—Use peak heights where peak responses are indicated.] Separately inject equal volumes (about 35 µL) of the *Standard preparation* and the *Assay preparation* into the chromatograph, record the chromatograms, and measure the responses for the major peaks. Calculate the quantity, in mg, of $C_{12}H_{24}N_2O_4$ in the portion of Tablets taken by the formula:

$$100C(r_U/r_S),$$

in which *C* is the concentration, in mg per mL, of USP Carisoprodol RS in the *Standard preparation*, and r_U and r_S are the peak responses obtained for carisoprodol from the *Assay preparation* and the *Standard preparation*, respectively.

Carisoprodol and Aspirin Tablets

» Carisoprodol and Aspirin Tablets contain not less than 90.0 percent and not more than 110.0 percent of the labeled amounts of carisoprodol ($C_{12}H_{24}N_2O_4$) and aspirin ($C_9H_8O_4$).

Packaging and storage—Preserve in well-closed containers.

Reference standards—*USP Carisoprodol Reference Standard*—Dry in vacuum at 60° for 3 hours before using. *USP Aspirin Reference Standard*—Dry over silica gel for 5 hours before using. *USP Salicylic Acid Reference Standard*—Dry over silica gel for 3 hours before using.

Identification—The retention times of the aspirin peak and the carisoprodol peak in the chromatogram of the *Assay preparation* correspond to those of the aspirin peak and the carisoprodol peak in the chromatogram of the *Aspirin and carisoprodol standard preparation*, obtained as directed under *Assay for aspirin and carisoprodol and limit of free salicylic acid.*

Dissolution ⟨711⟩—

Medium: water; 900 mL.

Apparatus 2: 75 rpm.

Time: 45 minutes.

Mobile phase—Mix 510 mL of methanol, previously filtered through a membrane filter of 0.5-µm or finer porosity, and 490 mL of glacial acetic acid solution (1 in 50), similarly filtered, and degas.

Standard preparation—Transfer about 90 mg of USP Aspirin RS and 90*J* mg of USP Carisoprodol RS, both accurately weighed, to a 250-mL volumetric flask, *J* being the ratio of the labeled amount, in mg, of carisoprodol to that of aspirin. Add 5 mL of acetonitrile, previously filtered through a membrane filter of 0.5-µm or finer porosity, swirl to dissolve, dilute with water to volume, and mix.

Resolution solution—Dissolve salicylic acid in *Standard preparation* to obtain a solution containing about 0.36 mg of salicylic acid per mL.

Chromatographic system (see *Chromatography* ⟨621⟩)—The liquid chromatograph is equipped with a refractive index detector and a 3.9-mm × 30-cm column that contains packing L1. Maintain the detector and the column at 30 ± 1°. The flow rate is about 2 mL per minute. Chromatograph the *Resolution solution* and the *Standard preparation*, and record the peak responses as directed under *Procedure:* the resolution, *R*, between the aspirin and salicylic acid peaks, and between the carisoprodol and salicylic acid peaks, is not less than 1.5, and the relative standard deviation for replicate injections of the *Standard preparation* is not more than 2.0%.

Procedure—Separately inject equal volumes (about 300 µL) of the *Standard preparation* and the solution under test, previously filtered, into the chromatograph, record the chromatograms, and measure the responses for the major peaks. The relative retention times are about 0.4 for aspirin and 1.0 for carisoprodol. Calculate the quantity, in mg, of aspirin ($C_9H_8O_4$) dissolved by the formula:

$$0.9C(r_U/r_S),$$

in which *C* is the concentration, in µg per mL, of USP Aspirin RS in the *Standard preparation*, and r_U and r_S are the peak responses obtained for aspirin from the solution under test and the *Standard preparation*, respectively. Calculate the quantity, in mg, of carisoprodol ($C_{12}H_{24}N_2O_4$) dissolved by the same formula, except to read "USP Carisoprodol RS" where "USP Aspirin RS" is specified, and "carisoprodol" where "aspirin" is specified.

Tolerances—Not less than 75% (*Q*) of the labeled amounts of $C_9H_8O_4$ and $C_{12}H_{24}N_2O_4$ are dissolved in 45 minutes.

Uniformity of dosage units ⟨905⟩: meet the requirements for *Content Uniformity* with respect to aspirin and to carisoprodol.

Assay for aspirin and carisoprodol and limit of free salicylic acid—

Mobile phase—Mix 5 mL of glacial acetic acid and 500 mL of water, and filter through a membrane filter of 0.5-µm or finer porosity. Add 360 mL of the filtrate to 640 mL of methanol, similarly filtered, mix, and degas. Make adjustments if necessary (see *System Suitability* under *Chromatography* ⟨621⟩).

Solvent mixture—Prepare a mixture of water, acetonitrile, and glacial acetic acid (59:40:1).

Aspirin and carisoprodol standard preparation—Transfer about 80 mg of USP Aspirin RS and 80*J* mg of USP Carisoprodol RS, both accurately weighed, to a 25-mL volumetric flask, *J* being the ratio of the labeled amount, in mg, of carisoprodol to that of aspirin. Add about 15 mL of *Solvent mixture*, swirl for 5 minutes, and sonicate for 25 to 30 seconds. Dilute with *Solvent mixture* to volume, and mix.

Salicylic acid standard preparation—Dissolve an accurately weighed quantity of USP Salicylic Acid RS in *Solvent mixture* to obtain a solution having a known concentration of about 16 µg per mL.

Resolution solution—Dissolve salicylic acid in *Aspirin and carisoprodol standard preparation* to obtain a solution containing about 0.5 mg of salicylic acid per mL.

Assay preparation—Weigh and finely powder not less than 20 Carisoprodol and Aspirin Tablets. Transfer an accurately weighed portion of the powder, equivalent to about 325 mg of aspirin, to a 100-mL volumetric flask. Add about 50 mL of *Solvent mixture*, swirl for 5 minutes, sonicate for 25 to 30 seconds, shake by mechanical means for 30 minutes, dilute with *Solvent mixture* to volume, and mix. Filter a portion of this solution through a membrane filter of 0.5-µm or finer porosity, and use the filtrate as the *Assay preparation.* [NOTE—Use within 8 hours.]

Chromatographic system (see *Chromatography* ⟨621⟩)—The liquid chromatograph is equipped with a refractive index detector, a 313-nm detector, and a 4.6-mm × 25-cm column that contains packing L7. Maintain the refractive index detector and the column at 30 ± 1°. The flow rate is about 1 mL per minute. Chromatograph the *Aspirin and carisoprodol standard preparation* and the *Resolution solution*, and record the peak responses as directed under *Procedure*, using the refractive index detector: the resolution, *R*, between the solvent and aspirin peaks in the chromatogram of the *Resolution solution* is not less than 1.2, the resolution, *R*, between the aspirin and salicylic acid peaks is not less than 1.5, and the relative standard deviation for replicate injections of the *Aspirin and carisoprodol standard preparation* is not more than 2.0%. Chromatograph the *Salicylic acid standard preparation*, and record the peak responses as directed under *Procedure:* the relative standard deviation for replicate injections is not more than 5.0%.

Procedure—Separately inject equal volumes (about 50 µL) of the *Aspirin and carisoprodol standard preparation*, the *Salicylic acid standard preparation*, and the *Assay preparation* into the chromatograph, record the chromatograms, and measure the responses for the major peaks, using the refractive index detector for the *Aspirin and carisoprodol standard preparation*, the 313-nm detector for the *Salicylic acid standard preparation*, and both detectors for the *Assay preparation.* The relative retention times are about 0.6 for aspirin, 0.7 for salicylic acid, and 1.0 for carisoprodol. Calculate the quantity, in mg, of aspirin ($C_9H_8O_4$) in the portion of Tablets taken by the formula:

$$100C(r_U/r_S),$$

in which *C* is the concentration, in mg per mL, of USP Aspirin RS in the *Aspirin and carisoprodol standard preparation*, and r_U and r_S are the peak responses, with the use of the refractive index detector, obtained for aspirin from the *Assay preparation* and the *Aspirin and carisoprodol standard preparation*, respectively. Calculate the quantity, in mg, of carisoprodol ($C_{12}H_{24}N_2O_4$) in the portion of Tablets taken by the same formula, except to read "USP Carisoprodol RS" where "USP Aspirin RS" is specified, and "carisoprodol" where "aspirin" is specified. Calculate the percentage of free salicylic acid in the Tablets taken by the formula:

$$10(C/a)(r_U/r_S),$$

in which *C* is the concentration, in µg per mL, of USP Salicylic Acid RS in the *Salicylic acid standard preparation*, *a* is the quantity, in mg, of aspirin in the portion of Tablets taken, based on the labeled amount, and r_U and r_S are the peak responses, with the use of the 313-nm detector, obtained for salicylic acid from the *Assay preparation* and the *Salicylic acid standard preparation*, respectively: not more than 3.0% is found.

Carisoprodol, Aspirin, and Codeine Phosphate Tablets

» Carisoprodol, Aspirin, and Codeine Phosphate Tablets contain not less than 90.0 percent and not more than 110.0 percent of the labeled amounts of carisoprodol ($C_{12}H_{24}N_2O_4$), aspirin ($C_9H_8O_4$), and codeine phosphate ($C_{18}H_{21}NO_3 . H_3PO_4 . \frac{1}{2}H_2O$).

Packaging and storage—Preserve in well-closed containers.

Reference standards—*USP Aspirin Reference Standard*—Dry over silica gel for 5 hours before using. *USP Carisoprodol Reference Standard*—Dry in vacuum at 60° for 3 hours before using. *USP Codeine Phosphate Reference Standard*—This is the hemihydrate form of codeine phosphate. Dry at 105° for 18 hours before using. *USP Codeine N-Oxide Reference Standard*—Store in a tightly closed container, protected from light. Do not dry before using. *USP Salicylic Acid Reference Standard*—Dry over silica gel for 3 hours before using.

Identification—The retention times of the aspirin, carisoprodol, and codeine phosphate peaks in the chromatograms of the *Assay preparations* correspond to those of the *Standard preparations* obtained as directed in the *Assay for aspirin and carisoprodol and limit of free salicylic acid* and the *Assay for codeine phosphate.*

Dissolution ⟨711⟩—

Medium: water; 900 mL.

Apparatus 2: 75 rpm.

Time: 45 minutes.

Determination of dissolved aspirin and carisoprodol—Proceed as directed in the test for *Dissolution* under *Aspirin and Carisoprodol Tablets.*

Determination of dissolved codeine phosphate—

MOBILE PHASE—Dissolve 2.2 g of docusate sodium and 0.8 g of ammonium nitrate in 550 mL of water, and filter through a membrane filter of 0.5-µm or finer porosity. Add 450 mL of similarly filtered acetonitrile to the filtrate, mix, and degas.

STANDARD PREPARATION—Dissolve an accurately weighed quantity of USP Codeine Phosphate RS in water to obtain a solution having a known concentration of about 18 µg per mL.

CHROMATOGRAPHIC SYSTEM (see *Chromatography* ⟨621⟩)—The liquid chromatograph is equipped with a 254-nm detector and a 3.9-mm × 30-cm column that contains packing L1. The flow rate is about 2 mL per minute. Chromatograph the *Standard preparation*, and record the peak responses as directed under *Procedure:* the relative standard deviation for replicate injections is not more than 2.0%.

PROCEDURE—Separately inject equal volumes (about 50 µL) of the *Standard preparation* and the solution under test, previously filtered, into the chromatograph, record the chromatograms, and measure the responses for the major peaks. Calculate the quantity, in mg, of codeine phosphate ($C_{18}H_{21}NO_3 . H_3PO_4 . \frac{1}{2}H_2O$) dissolved by the formula:

$$(406.37/397.36)(0.9C)(r_U/r_S),$$

in which 406.37 and 397.36 are the molecular weights of codeine phosphate hemihydrate and anhydrous codeine phosphate, respectively, C is the concentration, in µg per mL, of USP Codeine Phosphate RS in the *Standard preparation*, and r_U and r_S are the peak responses for the codeine phosphate peaks from the solution under test and the *Standard preparation*, respectively.

Tolerances—Not less than 75% (Q) of the labeled amounts of $C_9H_8O_4$, $C_{12}H_{24}N_2O_4$, and $C_{18}H_{21}NO_3 . H_3PO_4 . \frac{1}{2}H_2O$ are dissolved in 45 minutes.

Uniformity of dosage units ⟨905⟩: meet the requirements for *Content Uniformity* with respect to aspirin, to carisoprodol, and to codeine phosphate.

Assay for aspirin and carisoprodol and limit of free salicylic acid—

Mobile phase, Solvent mixture, Aspirin and carisoprodol standard preparation, Salicylic acid standard preparation, and *Resolution solution*—Prepare as directed in the *Assay for aspirin and carisoprodol and limit of free salicylic acid* under *Aspirin and Carisoprodol Tablets.*

Assay preparation—Weigh and finely powder not less than 20 Carisoprodol, Aspirin, and Codeine Phosphate Tablets. Transfer an accurately weighed portion of the powder, equivalent to about 325 mg of aspirin, to a 100-mL volumetric flask. Add about 50 mL of *Solvent mixture*, swirl for 5 minutes, sonicate for 25 to 30 seconds, shake by mechanical means for 30 minutes, dilute with *Solvent mixture* to volume, and mix. Filter a portion of this solution through a membrane filter of 0.5-µm or finer porosity, and use the filtrate as the *Assay preparation.*

Chromatographic system and *Procedure*—Proceed as directed in the *Assay for aspirin and carisoprodol and limit of free salicylic acid* under *Aspirin and Carisoprodol Tablets.*

Assay for codeine phosphate—

Mobile phase—Dissolve 2.2 g of docusate sodium in 600 mL of methanol. Dissolve 0.8 g of ammonium nitrate in 400 mL of water. Mix these two solutions, adjust with glacial acetic acid to a pH of 3.3 ± 0.05, filter through a membrane filter of 0.5-µm or finer porosity, and degas. Make adjustments if necessary (see *System Suitability* under *Chromatography* ⟨621⟩).

Solvent mixture—Mix equal volumes of methanol and 0.01 *N* sulfuric acid.

Standard preparation—Dissolve accurately weighed quantities of USP Codeine Phosphate RS and USP Aspirin RS in *Solvent mixture*, with the aid of swirling for 5 minutes and sonication for 25 to 30 seconds, to obtain a solution having known concentrations of about 0.16 mg of codeine phosphate and 0.16J mg of aspirin per mL, J being the ratio of the labeled amount, in mg, of aspirin to that of codeine phosphate.

Resolution solution—Transfer about 8 mg of USP Codeine Phosphate RS to a 50-mL volumetric flask containing about 6 mg of USP Codeine *N*-oxide RS, dilute with *Solvent mixture* to volume, and mix.

Assay preparation—Weigh and finely powder not less than 20 Carisoprodol, Aspirin, and Codeine Phosphate Tablets. Transfer an accurately weighed portion of the powder, equivalent to about 16 mg of codeine phosphate, to a 100-mL volumetric flask. Add about 50 mL of *Solvent mixture*, sonicate for 30 minutes, shake by mechanical means for about 30 minutes, dilute with *Solvent mixture* to volume, and mix.

Chromatographic system (see *Chromatography* ⟨621⟩)—The liquid chromatograph is equipped with a 254-nm detector and a 3.9-mm × 30-cm column that contains packing L1. The flow rate is about 1.5 mL per minute. Chromatograph the *Resolution solution* and the *Standard preparation*, and record the peak responses as directed under *Procedure:* the resolution, *R*, between the codeine phosphate and codeine *N*-oxide peaks is not less than 1.2, and the relative standard deviation for replicate injections of the *Standard preparation* is not more than 2.0%.

Procedure—Separately inject equal volumes (about 50 µL) of the *Standard preparation* and the *Assay preparation* into the chromatograph, record the chromatograms, and measure the responses for the major peaks. The relative retention times are about 0.9 for codeine *N*-oxide and 1.0 for codeine phosphate. Calculate the quantity, in mg, of codeine phosphate ($C_{18}H_{21}NO_3 . H_3PO_4 . \frac{1}{2}H_2O$) in the portion of Tablets taken by the formula:

$$(406.37)(397.36)(100C)(r_U/r_S),$$

in which 406.37 and 397.36 are the molecular weights of codeine phosphate hemihydrate and anhydrous codeine phosphate, respectively, C is the concentration, in mg per mL, of USP Codeine Phosphate RS in the *Standard preparation*, and r_U and r_S are the peak responses for the codeine phosphate peak obtained from the *Assay preparation* and the *Standard preparation*, respectively.

Carnauba Wax—*see* Wax, Carnauba NF

Carphenazine Maleate

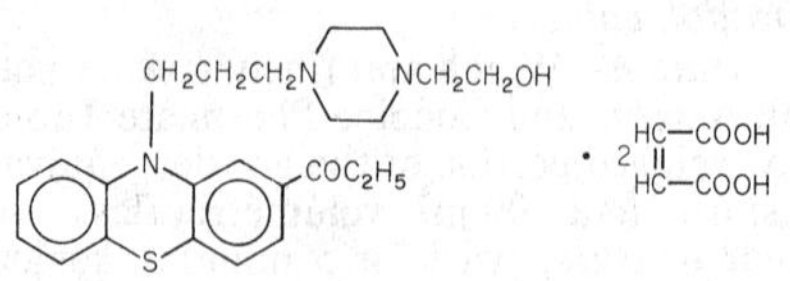

$C_{24}H_{31}N_3O_2S.2C_4H_4O_4$ 657.73

1-Propanone, 1-[10-[3-[4-(2-hydroxyethyl)-1-piperazinyl]propyl]-10*H*-phenothiazin-2-yl]-, (*Z*)-2-butenedioate (1:2).

1-[10-[3-[4-(2-Hydroxyethyl)-1-piperazinyl]propyl]phenothiazin-2-yl]-1-propanone maleate (1:2) [*2975-34-0*].

» Carphenazine Maleate contains not less than 98.0 percent and not more than 102.0 percent of $C_{24}H_{31}N_3O_2S.2C_4H_4O_4$, calculated on the anhydrous basis.

Packaging and storage—Preserve in tight, light-resistant containers.

Reference standard—*USP Carphenazine Maleate Reference Standard*—Dry in vacuum over phosphorus pentoxide for 18 hours before using.

NOTE—Throughout the following procedures, protect test or assay specimens, the Reference Standard, and solutions containing them, by conducting the procedures without delay, under subdued light, or using low-actinic glassware.

Identification—

A: The infrared absorption spectrum of a potassium bromide dispersion of it, previously dried in vacuum over phosphorus pentoxide for 18 hours, exhibits maxima only at the same wavelengths as that of a similar preparation of USP Carphenazine Maleate RS.

B: The ultraviolet absorption spectrum of a 1 in 50,000 solution in 0.1 *N* hydrochloric acid exhibits maxima and minima at the same wavelengths as that of a similar solution of USP Carphenazine Maleate RS, concomitantly measured, and the respective absorptivities, calculated on the anhydrous basis, at the wavelength of maximum absorbance at about 243 nm do not differ by more than 3.0%.

Melting range ⟨741⟩: between 176° and 185°, with decomposition, but the range between beginning and end of melting does not exceed 3°.

pH ⟨791⟩: between 2.5 and 3.5, in a suspension (1 in 100).

Water, *Method I* ⟨921⟩: not more than 1.0%.

Residue on ignition ⟨281⟩: not more than 0.2%.

Heavy metals, *Method II* ⟨231⟩: 0.0025%.

Ordinary impurities ⟨466⟩—

Test solution: a mixture of methanol and chloroform (1:1).

Standard solution: a mixture of methanol and chloroform (1:1).

Eluant: a mixture of chloroform, toluene, alcohol, and ammonium hydroxide (10:10:10:0.1).

Visualization: 1.

Assay—

Buffered palladous chloride solution—Transfer 500 mg of palladous chloride to a 250-mL beaker, add 5.0 mL of hydrochloric acid, and warm on a steam bath. Add 200 mL of hot water in small portions, and continue heating until solution is complete. Cool, transfer to a 500-mL volumetric flask, dilute with water to volume, and mix. Transfer 25.0 mL of this solution to a 500-mL volumetric flask, add 50.0 mL of 1.0 *N* sodium acetate and 50.0 mL of 1.0 *N* hydrochloric acid, dilute with water to volume, and mix.

Standard preparation—Transfer about 50 mg of USP Carphenazine Maleate RS, accurately weighed, to a 250-mL volumetric flask, add diluted alcohol to volume, and mix.

Assay preparation—Transfer about 50 mg of Carphenazine Maleate, accurately weighed, to a 250-mL volumetric flask, add diluted alcohol to volume, and mix.

Procedure—Transfer 3.0 mL each of the *Standard preparation*, the *Assay preparation*, and diluted alcohol to provide the blank to separate glass-stoppered tubes. To each tube add 4.0 mL of *Buffered palladous chloride solution*, and mix. Within 10 minutes, concomitantly determine the absorbances of the solutions in 1-cm cells at the wavelength of maximum absorbance at about 485 nm, with a suitable spectrophotometer, using the blank to set the instrument. Calculate the quantity, in mg, of $C_{24}H_{31}N_3O_2S.2C_4H_4O_4$ in the Carphenazine Maleate taken by the formula:

$$0.25C(A_U/A_S),$$

in which C is the concentration, in µg per mL, of USP Carphenazine Maleate RS in the *Standard preparation*, and A_U and A_S are the absorbances of the solutions from the *Assay preparation* and the *Standard preparation*, respectively.

Carphenazine Maleate Oral Solution

» Carphenazine Maleate Oral Solution contains not less than 95.0 percent and not more than 110.0 percent of the labeled amount of $C_{24}H_{31}N_3O_2S.2C_4H_4O_4$.

Packaging and storage—Preserve in tight, light-resistant containers.

Reference standard—*USP Carphenazine Maleate Reference Standard*—Dry in vacuum over phosphorus pentoxide for 18 hours before using.

NOTE—Throughout the following procedures, protect test or assay specimens, the Reference Standard, and solutions containing them, by conducting the procedures without delay, under subdued light, or using low-actinic glassware.

Identification—

A: Dilute a volume of Oral Solution, equivalent to about 50 mg of carphenazine maleate, with 0.01 *N* hydrochloric acid to 25 mL, and proceed as directed under *Identification—Organic Nitrogenous Bases* ⟨181⟩, beginning with "Transfer the liquid to a separator": the Oral Solution meets the requirements of the test.

B: Transfer a volume of Oral Solution, equivalent to about 100 mg of carphenazine maleate, to a separator. Add 2.5 *N* sodium hydroxide to render the solution alkaline to litmus, extract the carphenazine with four 25-mL portions of ether, and combine the ether extracts in a second separator. Extract the ether solution with three 10-mL portions of 0.1 *N* hydrochloric acid, collecting the acid extracts in a 100-mL volumetric flask. Dilute with 0.1 *N* hydrochloric acid to volume, mix, and transfer 2 mL of this solution to a second 100-mL volumetric flask. Dilute with 0.1 *N* hydrochloric acid to volume, and determine the absorbance of this solution in a 1-cm cell at 243 and 277 nm, with a suitable spectrophotometer, using 0.1 *N* hydrochloric acid as the blank: the ratio A_{243}/A_{277} is between 1.27 and 1.41.

pH ⟨791⟩: between 5.8 and 6.8.

Assay—

Buffered palladous chloride solution and *Standard preparation*—Prepare as directed in the *Assay* under *Carphenazine Maleate*.

Assay preparation—Transfer an accurately measured volume of Carphenazine Maleate Oral Solution, equivalent to about 50 mg of carphenazine maleate, to a 250-mL volumetric flask, add diluted alcohol to volume, and mix.

Procedure—Proceed as directed for *Procedure* in the *Assay* under *Carphenazine Maleate*. Calculate the quantity, in mg, of $C_{24}H_{31}N_3O_2S.2C_4H_4O_4$ in each mL of the Oral Solution taken by the formula:

$$(0.25C/V)(A_U/A_S),$$

in which V is the volume, in mL, of the Oral Solution taken, and C, A_U, and A_S are as defined therein.

Carrageenan—*see* Carrageenan NF

Cascara Sagrada

» Cascara Sagrada is the dried bark of *Rhamnus purshiana* De Candolle (Fam. Rhamnaceae). It yields not less than 7.0 percent of total hydroxyanthracene derivatives, calculated as cascaroside A, and calculated on the dried basis. Not less than 60 percent of the total hydroxyanthracene derivatives consists of cascarosides, calculated as cascaroside A.

Note—Collect Cascara Sagrada not less than one year prior to use.

Reference standard—*USP Danthron Reference Standard*—Dry over silica gel for 4 hours before using.

Botanic characteristics—

Cascara Sagrada—Usually in flattened or transversely curved pieces, occasionally in quills of variable length and from 1 to 5 mm in thickness. The outer surface is brown, purplish brown, or brownish red, longitudinally ridged, with or without grayish or whitish lichen patches, sometimes with numerous lenticels and occasionally with moss attached. The inner surface is longitudinally striate, light yellow, weak reddish brown, or moderate yellowish brown. The fracture is short with projections of phloem fiber bundles in the inner bark.

Histology—It shows a yellowish brown, purple, or reddish brown cork of up to 10 or more rows of small cells; stone cells in yellowish, tangentially elongated groups of 20 to 50 cells in the cortex, pericycle, and outer phloem regions; phloem rays 1 to 4 cells wide, 15 to 25 cells deep, frequently diagonal or curved, forming converging groups; phloem fibers in small bundles, more or less surrounded by crystal fibers and located between the phloem rays; parenchyma with brown walls and containing starch grains and calcium oxalate crystals.

Powdered Cascara Sagrada—Moderate yellowish brown to dusky yellowish orange. It shows numerous broken phloem fiber bundles with accompanying crystal fibers containing monoclinic prisms of calcium oxalate; stone cells more or less adherent, in small groups with thick, finely lamellated and porous walls; fragments of reddish brown to yellow cork; masses of parenchyma and phloem ray cells colored reddish brown to orange upon the addition of a solution of an alkali; starch grains spheroidal, up to 8 µm in diameter; calcium oxalate in monoclinic prisms or rosette aggregates from 6 to 20 µm in diameter, occasionally up to 45 µm in diameter.

Identification—

A: Add 100 mg of powdered Cascara Sagrada to 10 mL of hot water, shake the mixture occasionally until it is cold, filter, dilute the filtrate with water to 10 mL, and add 10 mL of 6 *N* ammonium hydroxide: an orange color is produced.

B: It becomes red to reddish brown in color when treated with 6 *N* ammonium hydroxide.

C: Macerate 100 mg of powdered Cascara Sagrada with 1 mL of alcohol, add 10 mL of water, boil the mixture, then cool, filter, and shake the filtrate with 10 mL of ether: a greenish yellow ether solution separates. Shake 3 mL of this ether solution with 3 mL of 6 *N* ammonium hydroxide, and dilute the separated ammonia solution with 20 mL of water: a distinct orange-pink color remains.

Water, *Method III—Procedure for Vegetable Drugs* ⟨921⟩—Dry it at 105° for 5 hours: it loses not more than 12.0% of its weight.

Foreign organic matter ⟨561⟩: not more than 4.0%.

Assay for cascarosides—[NOTE 1—Perform all extractions by shaking vigorously, and allow all phases to separate completely before transferring. Entrainment of aglycones into the aqueous phase, as indicated by a value of less than 2.7 for the ratio of the absorbance of the final solution at 515 nm to that at 440 nm, may lead to false results. NOTE 2—Throughout this assay, use 1 *N* sodium hydroxide that is prepared without added barium ions as directed for *Volumetric Solutions* in the section, *Reagents, Indicators, and Solutions.*]

Ferric chloride solution—Dissolve 100 g of ferric chloride in water to make 100 mL.

Standard preparation—Dissolve about 5 mg of USP Danthron RS, accurately weighed, in 50 mL of a 1 in 200 solution of magnesium acetate in methanol in a 500-mL volumetric flask, dilute with the same solution to volume, and mix.

Assay solution—Add about 1 g of Cascara Sagrada, accurately weighed, to about 70 mL of boiling water, boil for several minutes, with stirring, allow to cool, and transfer with the aid of water to a 100-mL volumetric flask. Dilute with water to volume, mix, and filter through suitable filter paper.

Assay preparation—Pipet 10 mL of *Assay solution* into a separator containing 5 mL of water and 2 drops of 1 *N* hydrochloric acid. Extract with 40 mL of carbon tetrachloride, and transfer the lower layer to a second separator. Add 10 mL of water to the second separator, and shake. Allow to separate, discard the lower layer, and transfer the water layer to the first separator. Extract the combined water layers with 40 mL of carbon tetrachloride, and transfer the lower layer to the second separator. Add 10 mL of water to the second separator, and shake. Allow to separate, discard the lower layer, and transfer the water layer to the first separator. Extract the combined aqueous phase with 30 mL of clear, freshly prepared, water-saturated ethyl acetate, and transfer the water layer to another separator. Repeat the extraction with two additional 30-mL portions of the freshly prepared, water-saturated ethyl acetate. Add 5 mL of water to the combined ethyl acetate extracts, shake, allow the phases to separate, discard the ethyl acetate extracts, and add 30 mL of the freshly prepared, water-saturated ethyl acetate to the water wash. Shake, allow the phases to separate, and discard the ethyl acetate phase. Transfer the combined aqueous phases, with the aid of water, to a 50-mL volumetric flask. Dilute with water to volume, and mix.

Procedure—Pipet 15 mL of *Assay preparation* into a suitable flask containing 2 mL of *Ferric chloride solution* and 12 mL of hydrochloric acid. Attach a condenser arranged for refluxing, and heat for 3 hours by keeping the flask immersed in boiling water or continuously exposed to steam heat. Cool, wash down the condenser, and transfer to a separator with the aid of 4 mL of 1 *N* sodium hydroxide and five 6-mL portions of water. Extract with 20 mL of carbon tetrachloride, and transfer the lower layer to another separator. Repeat the extraction with three additional 20-mL portions of carbon tetrachloride, wash the combined carbon tetrachloride extracts with two 10-mL portions of water, shaking each time for 2 minutes, and discard the water washings. Transfer the washed carbon tetrachloride extract to a 100-mL volumetric flask, dilute with carbon tetrachloride to volume, and mix. Evaporate a 20.0-mL portion carefully on a water bath to dryness, and dissolve the residue in 10.0 mL of a 1 in 200 solution of magnesium acetate in methanol. Determine the absorbance, against methanol as a reference, in 1-cm cells, with a suitable spectrophotometer, at the wavelength of maximum absorbance at about 515 nm. Concomitantly determine, in a similar manner, the absorbance of the *Standard preparation.* Calculate the quantity, in mg, of cascarosides in the portion of Cascara Sagrada taken by the formula:

$$1.667(2.88C)(A_U/A_S),$$

in which 2.88 is a composite of the ratio of the molecular weights and assay response factors of cascaroside A relative to danthron, C is the concentration, in µg per mL, of USP Danthron RS in the *Standard preparation,* and A_U and A_S are the absorbances of the solutions from the *Assay preparation* and the *Standard preparation,* respectively.

Assay for total hydroxyanthracene derivatives—[NOTE 1—Perform all extractions by shaking vigorously, and allow all phases to separate completely before transferring. Entrainment of aglycones into the aqueous phase, as indicated by a value of less than 2.6 for the ratio of the absorbance of the final solution at 515 nm to that at 440 nm, may lead to false results. NOTE 2—Throughout this assay, use 1 *N* sodium hydroxide that is prepared without added barium ions as directed for *Volumetric Solutions* in the section, *Reagents, Indicators, and Solutions.*]

Ferric chloride solution, Standard preparation, and *Assay solution*—Prepare as directed in the *Assay for cascarosides.*

Assay preparation—Pipet 10 mL of *Assay solution* into a separator containing 5 mL of water and 2 drops of 1 *N* hydrochloric acid. Extract with 40 mL of carbon tetrachloride, and transfer the lower layer to a second separator. Add 10 mL of water to the second separator, and shake. Allow to separate, discard the lower layer, and transfer the water layer to the first separator. Extract the combined water layers with 40 mL of carbon tetrachloride, and transfer the lower layer to the second separator. Add 10 mL of water to the second separator, and shake. Allow to separate, and discard the lower layer. Transfer the combined water layers, with the aid of water, to a 50-mL volumetric flask, dilute with water to volume, and mix.

Procedure—Proceed as directed for *Procedure* in the *Assay for cascarosides*, except to evaporate a 15.0-mL portion of the carbon tetrachloride solution instead of 20.0 mL. Calculate the quantity, in mg, of total hydroxyanthracene derivatives in the portion of Cascara Sagrada taken by the formula:

$$2.222(2.88C)(A_U/A_S),$$

in which 2.88 is a composite of the ratio of the molecular weights and assay response factors of cascaroside A relative to danthron, C is the concentration, in µg per mL, of USP Danthron RS in the *Standard preparation*, and A_U and A_S are the absorbances of the solutions from the *Assay preparation* and the *Standard preparation*, respectively.

Cascara Sagrada Extract

» Cascara Sagrada Extract contains, in each 100 g, not less than 10.0 g and not more than 12.0 g of hydroxyanthracene derivatives, of which not less than 50 percent consists of cascarosides, both calculated as cascaroside A.

Mix 900 g of Cascara Sagrada, in coarse powder, with 4000 mL of boiling water, and macerate the mixture for 3 hours. Then transfer it to a percolator, allow it to drain, exhaust it by percolation, using boiling water as the menstruum, and collect about 5000 mL of percolate. Evaporate the percolate to dryness, reduce the extract to a fine powder, and, after assaying, add sufficient starch, dried at 100°, or other inert, non-toxic diluents to make the product contain, in each 100 g, 11 g of hydroxyanthracene derivatives. Mix the powders, and pass the Extract through a number 60 sieve.

Packaging and storage—Preserve in tight, light-resistant containers, at a temperature not exceeding 30°.

Reference standard—*USP Danthron Reference Standard*—Dry over silica gel for 4 hours before using.

Assay for cascarosides—[NOTE 1—Perform all extractions by shaking vigorously, and allow all phases to separate completely before transferring. Entrainment of aglycones into the aqueous phase, as indicated by a value of less than 2.7 for the ratio of the absorbance of the final solution at 515 nm to that at 440 nm, may lead to false results. NOTE 2—Throughout this assay, use 1 *N* sodium hydroxide that is prepared without added barium ions as directed for *Volumetric Solutions* under *Reagents, Indicators, and Solutions.*]

Ferric chloride solution—Dissolve 100 g of ferric chloride in water to make 100 mL.

Standard preparation—Prepare as directed in the *Assay for cascarosides* under *Cascara Sagrada.*

Assay solution—Accurately weigh about 1 g of Cascara Sagrada Extract, and transfer to a 100-mL volumetric flask. Add about 60 mL of 70% alcohol, swirl and/or sonicate for 15 to 20 minutes, several times, allow to stand overnight, sonicate or swirl for 10 to 15 minutes, dilute with 70% alcohol to volume, mix, and filter through suitable filter paper.

Assay preparation—Prepare as directed for *Assay preparation* in the *Assay for cascarosides* under *Cascara Sagrada.*

Procedure—Pipet 25 mL of *Assay preparation* into a suitable flask containing 2 mL of *Ferric chloride solution* and 12 mL of hydrochloric acid. Proceed as directed for *Procedure* in the *Assay for cascarosides* under *Cascara Sagrada*, beginning with "Attach a condenser arranged for refluxing, and heat for 3 hours." Calculate the quantity, in mg, of cascarosides in the portion of Cascara Sagrada Extract taken by the formula:

$$2.88C(A_U/A_S),$$

in which 2.88 is a composite of the ratio of the molecular weights and assay response factors of cascaroside A and relative to danthron, C is the concentration, in µg per mL, of USP Danthron RS in the *Standard preparation*, and A_U and A_S are the absorbances of the solutions from the *Assay preparation* and the *Standard preparation*, respectively.

Assay for total hydroxyanthracene derivatives—[NOTE 1—Perform all extractions by shaking vigorously, and allow all phases to separate completely before transferring. Entrainment of aglycones into the aqueous phase, as indicated by a value of less than 2.6 for the ratio of the absorbance of the final solution at 515 nm to that at 440 nm, may lead to false results. NOTE 2—Throughout this assay, use 1 *N* sodium hydroxide that is prepared without added barium ions as directed for *Volumetric Solutions* under *Reagents, Indicators, and Solutions.*]

Ferric chloride solution, Standard preparation, and *Assay solution*—Prepare as directed in the *Assay for cascarosides.*

Assay preparation—Prepare as directed for *Assay preparation* in the *Assay for total hydroxyanthracene derivatives* under *Cascara Sagrada.*

Procedure—Pipet 10 mL of *Assay preparation* into a suitable flask containing 2 mL of *Ferric chloride solution* and 12 mL of hydrochloric acid. Proceed as directed for *Procedure* in the *Assay for cascarosides* under *Cascara Sagrada*, beginning with "Attach a condenser arranged for refluxing, and heat for 3 hours." Calculate the quantity, in mg, of total hydroxyanthracene derivatives in the portion of Cascara Sagrada Extract taken by the formula:

$$2.5(2.88C)(A_U/A_S),$$

in which 2.88 is a composite of the ratio of the molecular weights and assay response factors of cascaroside A relative to danthron, C is the concentration, in µg per mL, of USP Danthron RS in the *Standard preparation*, and A_U and A_S are the absorbances of the solutions from the *Assay preparation* and the *Standard preparation*, respectively.

Cascara Tablets

» Cascara Tablets are prepared from Cascara Sagrada Extract. They contain an amount of hydroxyanthracene derivatives, calculated as cascaroside A, not less than 9.35 percent and not more than 12.65 percent of the labeled amount of Cascara Sagrada Extract. Not less than 50 percent of the hydroxyanthracene derivatives are cascarosides, calculated as cascaroside A.

Packaging and storage—Preserve in tight containers; if the Tablets are coated, well-closed containers may be used.

Reference standard—*USP Danthron Reference Standard*—Dry over silica gel for 4 hours before using.

Disintegration ⟨701⟩: 60 minutes.

Uniformity of dosage units ⟨905⟩: meet the requirements.

Assay for cascarosides—[NOTE 1—Perform all extractions by shaking vigorously, and allow all phases to separate completely before transferring. Entrainment of aglycones into the aqueous phase, as indicated by a value of less than 2.7 for the ratio of the absorbance of the final solution at 515 nm to that at 440 nm, may lead to false results. NOTE 2—Throughout this assay, use 1 *N* sodium hydroxide that is prepared without added barium ions as directed for *Volumetric Solutions* under *Reagents, Indicators, and Solutions.*]

Ferric chloride solution—Dissolve 100 g of ferric chloride in water to make 100 mL.

Standard preparation—Prepare as directed in the *Assay for cascarosides* under *Cascara Sagrada.*

Assay solution—Weigh and finely powder not less than 20 Cascara Tablets. Transfer an accurately weighed portion of the powder, equivalent to about 1.0 g of Cascara Sagrada Extract, to a 100-mL volumetric flask. Add about 60 mL of 70% alcohol, swirl and/or sonicate for 15 to 20 minutes, several times, allow to stand overnight, sonicate or swirl for 10 to 15 minutes, dilute with 70% alcohol to volume, mix, and filter through suitable filter paper.

Assay preparation—Prepare as directed in the *Assay for cascarosides* under *Cascara Sagrada.*

Procedure—Pipet 20 mL of *Assay preparation* into a suitable flask containing 2 mL of *Ferric chloride solution* and 12 mL of hydrochloric acid. Proceed as directed for *Procedure* in the *Assay for cascarosides* under *Cascara Sagrada,* beginning with "Attach a condenser arranged for refluxing, and heat for 3 hours," except to evaporate a 15.0-mL portion of the carbon tetrachloride solution instead of 20.0 mL. Calculate the quantity, in mg, of cascarosides in the portion of Tablets taken by the formula:

$$1.667(2.88C)(A_U/A_S),$$

in which 2.88 is a composite of the ratio of the molecular weights and assay response factors of cascaroside A relative to danthron, C is the concentration, in µg per mL, of USP Danthron RS in the *Standard preparation,* and A_U and A_S are the absorbances of the solutions from the *Assay preparation* and the *Standard preparation,* respectively.

Assay for total hydroxyanthracene derivatives—[NOTE 1—Perform all extractions by shaking vigorously, and allow all phases to separate completely before transferring. Entrainment of aglycones into the aqueous phase, as indicated by a value of less than 2.6 for the ratio of the absorbance of the final solution at 515 nm to that at 440 nm, may lead to false results. NOTE 2—Throughout this assay, use 1 *N* sodium hydroxide that is prepared without added barium ions as directed for *Volumetric Solutions* in the section, *Reagents, Indicators, and Solutions.*]

Ferric chloride solution, Standard preparation, and *Assay solution*—Prepare as directed in the *Assay for cascarosides.*

Assay preparation—Prepare as directed for *Assay preparation* in the *Assay for total hydroxyanthracene derivatives* under *Cascara Sagrada.*

Procedure—Pipet 10 mL of *Assay preparation* into a suitable flask containing 2 mL of *Ferric chloride solution* and 12 mL of hydrochloric acid. Proceed as directed for *Procedure* in the *Assay for cascarosides* under *Cascara Sagrada,* beginning with "Attach a condenser arranged for refluxing, and heat for 3 hours," except to evaporate a 15.0-mL portion of the carbon tetrachloride solution instead of 20.0 mL. Calculate the quantity, in mg, of total hydroxyanthracene derivatives in the portion of Tablets taken by the formula:

$$3.333(2.88C)(A_U/A_S),$$

in which 2.88 is a composite of the ratio of the molecular weights and assay response factors of cascaroside A relative to danthron, C is the concentration, in µg per mL, of USP Danthron RS in the *Standard preparation,* and A_U and A_S are the absorbances of the solutions from the *Assay preparation* and the *Standard preparation,* respectively.

Cascara Sagrada Fluidextract

» Cascara Sagrada Fluidextract contains in each 100 mL not less than 3.5 g and not more than 4.0 g of total hydroxyanthracene derivatives, of which not less than 50 percent consists of cascarosides, both calculated as cascaroside A.

Packaging and storage—Preserve in tight, light-resistant containers, and avoid exposure to direct sunlight and to excessive heat.

Alcohol content, *Method I* ⟨611⟩: between 18.0% and 20.0% of C_2H_5OH.

Assay—Transfer 5.0 mL of the Fluidextract to a 100-mL volumetric flask, dilute with water to volume, and mix. Proceed as directed in the *Assay* under *Aromatic Cascara Fluidextract,* beginning with "Using this as the *Assay solution.*"

Aromatic Cascara Fluidextract

» Aromatic Cascara Fluidextract contains in each 100 mL not less than 0.50 g and not more than 0.75 g of total hydroxyanthracene derivatives, of which not less than 50 percent consists of cascarosides, both calculated as cascaroside A.

Packaging and storage—Preserve in tight, light-resistant containers, and avoid exposure to direct sunlight and to excessive heat.

Alcohol content ⟨611⟩: between 18% and 20% of C_2H_5OH, determined by the gas-liquid chromatographic method, acetone being used as the internal standard.

Assay—Transfer 10.0 mL of the Fluidextract to a 150-mL beaker. Place the beaker on a steam bath for about 45 minutes to evaporate oils and alcohol. Take up the residue in about 80 mL of boiling water, and transfer to a 100-mL volumetric flask with the aid of several small boiling water rinses. Allow the solution to cool, dilute with water to volume, and mix. Using this as the *Assay solution,* proceed as directed in the *Assay for cascarosides* and in the *Assay for total hydroxyanthracene derivatives* under *Cascara Sagrada.* Calculate the weight, in g, of cascarosides in 100 mL of Fluidextract taken by the formula:

$$(1/60)(2.88C)(A_U/A_S),$$

in which 2.88 is a composite of the ratio of the molecular weights and assay response factors of cascaroside A relative to danthron, C is the concentration, in µg per mL, of USP Danthron RS in the *Standard preparation,* and A_U and A_S are the absorbances of the solutions from the cascarosides *Assay preparation* and the *Standard preparation,* respectively. Calculate the quantity, in mg, of total hydroxyanthracene derivatives in 100 mL of Fluidextract taken by the formula:

$$(1/45)(2.88C)(A_U/A_S),$$

in which A_U is the absorbance of the solution from the total hydroxyanthracene derivatives *Assay preparation,* and the other symbols are as defined above.

Castor Oil

» Castor Oil is the fixed oil obtained from the seed of *Ricinus communis* Linné (Fam. Euphorbiaceae). It contains no added substances.

Packaging and storage—Preserve in tight containers, and avoid exposure to excessive heat.

Specific gravity ⟨841⟩: between 0.957 and 0.961.

Distinction from most other fixed oils—It is only partly soluble in solvent hexane (distinction from *most other fixed oils*), but it yields a clear liquid with an equal volume of alcohol (*foreign fixed oils*).

Heavy metals, *Method II* ⟨231⟩: 0.001%.

Free fatty acids ⟨401⟩—The free fatty acids in 10 g require for neutralization not more than 3.5 mL of 0.10 *N* sodium hydroxide.

Hydroxyl value—Transfer 2 g, accurately weighed, to a glass-stoppered, 250-mL conical flask, add 5.0 mL of a freshly prepared mixture of 1 volume of acetic anhydride and 3 volumes of pyridine, and swirl to mix. Connect the flask to a reflux condenser, and heat on a steam bath for 1 hour. Add 10 mL of water through the condenser, swirl to mix, heat on a steam bath for an additional 10 minutes, and allow to cool to room temperature. Add through the condenser 15 mL of normal butyl alcohol that previously has been neutralized to phenolphthalein, remove the condenser, and wash the tip of the condenser and the sides of the flask with an additional 10 mL of neutralized normal butyl alcohol. Add 1 mL of phenolphthalein TS, and titrate with 0.5 *N* alcoholic potassium hydroxide VS to a faint pink end-point. Perform a blank determination on a 5.0-mL portion of the acetic anhydride–pyridine mixture. To determine the amount of free acid in the Castor Oil, weigh accurately 10 g into a 250-mL conical flask, add 10 mL of pyridine that previously has been neutralized to phenolphthalein, swirl to mix, add 1 mL of phenolphthalein TS, and titrate with 0.5 *N* alcoholic potassium hydroxide VS to a faint pink end-point. Calculate the hydroxyl value by the formula:

$$56.1N[A + (BW/D) - C]/W,$$

in which N is the normality determined for the alcoholic potassium hydroxide solution, A is the volume, in mL, of 0.5 *N* alcoholic potassium hydroxide consumed by the blank, B is the volume, in mL, consumed in the free-acid titration, W is the weight, in g, of Oil taken, D is the weight, in g, of Oil used in the free-acid titration, and C is the volume, in mL, consumed in the sample titration. The hydroxyl value is between 160 and 168.

Iodine value ⟨401⟩: between 83 and 88.

Saponification value ⟨401⟩: between 176 and 182.

Castor Oil Capsules

» Castor Oil Capsules contain not less than 90.0 percent and not more than 110.0 percent of the labeled amount of castor oil, calculated from the tests for *Weight variation* and *Specific gravity*.

Packaging and storage—Preserve in tight containers, preferably at controlled room temperature.

Identification—The infrared absorption spectrum of a solution prepared by dissolving the contents of Capsules in chloroform to a concentration of about 40 mg of castor oil per mL, 0.1-mm sodium chloride cells being used and chloroform being used as the blank, exhibits maxima and minima only at the same wavelengths as that of a similar preparation of Castor Oil.

Uniformity of dosage units ⟨905⟩: meet the requirements.

Other requirements—The contents of Capsules meet the requirements of the tests for *Specific gravity*, *Hydroxyl value*, *Iodine value*, and *Saponification value* under *Castor Oil*.

Castor Oil Emulsion

» Castor Oil Emulsion contains not less than 90.0 percent and not more than 120.0 percent of the labeled amount of Castor Oil.

Packaging and storage—Preserve in tight containers.

Identification—Shake well, and place about 10 mL in a 125-mL separator. Add 10 mL of 1 *N* hydrochloric acid and about 20 mL of solvent hexane. Shake vigorously for 2 to 3 minutes, allow the layers to separate, discard the aqueous phase, and filter the upper layer through anhydrous sodium sulfate into a small beaker. Evaporate the solvent on a steam bath, and to the residue add 1 to 2 drops of sulfuric acid: a red color indicates the presence of castor oil.

Assay—

Internal standard solution—In a 100-mL volumetric flask dissolve about 1.2 g of di(2-ethylhexyl) phthalate in chloroform, dilute with chloroform to volume, and mix.

Chromatographic system (see *Chromatography* ⟨621⟩)—Under typical conditions, the instrument is equipped with a flame-ionization detector and contains a 1.8-m × 4-mm column packed with liquid phase G25 on support S1, and conditioned by flushing with helium for 2 to 5 minutes, then heating without further flushing at 250° for not less than 30 minutes, then cooling to room temperature, and finally heating while helium is flowing through it at 250° for not less than 60 minutes. During the chromatographic separation, the column is maintained at 245°, and the injection port and detector block are maintained at about 300°. The carrier gas is helium, and its flow rate is adjusted to obtain a peak due to castor oil about 5.5 minutes after introduction of the specimen and an internal standard peak about 8 minutes after introduction of the specimen.

Procedure—Transfer an accurately weighed amount of the well-shaken Castor Oil Emulsion, equivalent to about 100 mg of castor oil, to a long-neck, round-bottom, 100-mL boiling flask equipped with a suitable reflux condenser connected by a ground-glass joint. To a similar flask transfer about 100 mg of castor oil, accurately weighed, to provide the standard. Carry out the following steps on each: Add 30 mL of a mixture of 300 mL of methanol and 3.7 mL of sulfuric acid, reflux in a water bath maintained at 75° to 80° for 2.5 hours, cool, and rinse down the condenser with 10 mL of water. Transfer the contents of the flask to a 125-mL separator with the aid of 10 mL of water. Rinse the condenser and the flask with 25 mL of solvent hexane, and transfer to the separator. Shake the separator for 2 minutes, and draw off the aqueous layer into a second 125-mL separator. Add 20 mL of solvent hexane to the second separator, shake for 2 minutes, discard the aqueous layer, and transfer the solvent hexane layer to the first separator with the aid of 10 mL of solvent hexane. Wash the combined extracts with three 5-mL portions of water, discarding the washings, and transfer the washed extract to a 125-mL conical flask, through a funnel containing anhydrous sodium sulfate, with the aid of 25 mL of solvent hexane. Place the flask in a hot water bath, and evaporate with the aid of a current of air to dryness. To the residue add 10.0 mL of *Internal standard solution*, and mix until solution is complete. Inject about 5 µL into the gas chromatograph, and measure the heights of the peaks due to castor oil and internal standard. Calculate the quantity, in mg, of castor oil in the portion of Emulsion taken by the formula:

$$q_r(R_A/R_S),$$

in which q_r is the weight, in mg, of castor oil taken for the standard, and R_A and R_S are the ratios of the heights of the peaks due to castor oil and internal standard obtained from the Emulsion and the standard, respectively.

Castor Oil, Hydrogenated—*see* Castor Oil, Hydrogenated NF

Castor Oil, Polyoxyl 35—*see* Polyoxyl 35 Castor Oil NF

Castor Oil, Polyoxyl 40 Hydrogenated—*see* Polyoxyl 40 Hydrogenated Castor Oil NF

Aromatic Castor Oil

» Aromatic Castor Oil is Castor Oil containing suitable flavoring agents. It contains not less than 95.0 percent of castor oil.

Packaging and storage—Preserve in tight containers.

Alcohol content, *Method I* ⟨611⟩: not more than 4.0% of C_2H_5OH.

Assay—Proceed as directed in the *Assay* under *Castor Oil Emulsion*, using about 100 mg of Aromatic Castor Oil, accurately weighed. Calculate the quantity, in mg, of castor oil in the portion of Aromatic Castor Oil taken by the formula:

$$q_r(R_A/R_S),$$

in which q_r is the weight, in mg, of castor oil taken for the standard, and R_A and R_S are the ratios of the peak responses due to castor oil and internal standard obtained from the Aromatic Castor Oil and the standard, respectively.

Cefaclor

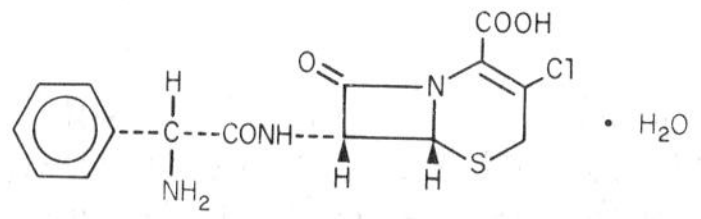

$C_{15}H_{14}ClN_3O_4S.H_2O$ 385.82

5-Thia-1-azabicyclo[4.2.0]oct-2-ene-2-carboxylic acid, 7-[(aminophenylacetyl)amino]-3-chloro-8-oxo-, monohydrate, [6*R*-[6α,7β(*R**)]]-.

(6*R*,7*R*)-7-[(*R*)-2-Amino-2-phenylacetamido]-3-chloro-8-oxo-5-thia-1-azabicyclo[4.2.0]oct-2-ene-2-carboxylic acid monohydrate.

3-Chloro-7-D-(2-phenylglycinamido)-3-cephem-4-carboxylic acid monohydrate [*70356-03-5*].

Anhydrous 367.81 [*53994-73-3*].

» Cefaclor has a potency of not less than 860 μg and not more than 1050 μg of $C_{15}H_{14}ClN_3O_4S$ per mg.

Packaging and storage—Preserve in tight containers.

Reference standard—*USP Cefaclor Reference Standard*—Do not dry before using.

Identification—

A: The infrared absorption spectrum of a mineral oil dispersion of it exhibits maxima only at the same wavelengths as that of a similar preparation of USP Cefaclor RS.

B: Place a suitable thin-layer chromatographic plate (see *Chromatography* ⟨621⟩) coated with a 0.25-mm layer of binder-free silica gel in a chamber containing a mixture of *n*-hexane and tetradecane (95:5) to a depth of about 1 cm, allow the solvent front to move the length of the plate, remove the plate from the chamber, and allow the solvent to evaporate. On this plate, apply 20 μL each of a solution of Cefaclor in water containing 2 mg per mL and a similarly prepared Standard solution of USP Cefaclor RS. Allow the spots to dry, and develop the chromatogram in a solvent system consisting of a mixture of 0.1 *M* citric acid, 0.1 *M* dibasic sodium phosphate, and a 1 in 15 solution of ninhydrin in acetone (60:40:1.5) until the solvent front has moved about three-fourths of the length of the plate. Remove the plate from the developing chamber, mark the solvent front, dry the plate for 10 minutes at 110°, and examine the chromatogram: the R_f value of the principal spot obtained from the test solution corresponds to that obtained from the Standard solution.

Crystallinity ⟨695⟩: meets the requirements.

pH ⟨791⟩: between 3.0 and 4.5, in an aqueous suspension containing 25 mg per mL.

Water, *Method I* ⟨921⟩: between 3.0% and 8.0%.

Assay—

Standard preparation—Dissolve a suitable quantity of USP Cefaclor RS, accurately weighed, in *Buffer No. 4* (see *Phosphate Buffers and Other Solutions* under *Antibiotics—Microbial Assays* ⟨81⟩) to obtain a solution having a known concentration of about 1 mg per mL.

Assay preparation—Using a suitable quantity of Cefaclor, accurately weighed, proceed as directed under *Standard preparation.*

Procedure—Proceed as directed for *Procedure* in the section, *Antibiotics—Hydroxylamine Assay*, under *Automated Methods of Analysis* ⟨16⟩. Calculate the potency, in μg, of $C_{15}H_{14}ClN_3O_4S$ in each mg of Cefaclor taken by the formula:

$$(CP/W)(A_U/A_S),$$

in which *W* is the weight, in mg, of Cefaclor taken in each mL of the *Assay preparation.*

Cefaclor Capsules

» Cefaclor Capsules contain the equivalent of not less than 90.0 percent and not more than 120.0 percent of the labeled amount of $C_{15}H_{14}ClN_3O_4S$.

Packaging and storage—Preserve in tight containers.

Reference standard—*USP Cefaclor Reference Standard*—Do not dry before using.

Identification—Mix the contents of 1 Capsule with water to obtain a concentration of about 2 mg of cefaclor per mL, and filter: the filtrate so obtained responds to *Identification test B* under *Cefaclor.*

Uniformity of dosage units ⟨905⟩: meet the requirements.

Water, *Method I* ⟨921⟩: not more than 8.0%.

Assay—

Standard preparation—Prepare as directed for *Standard preparation* in the *Assay* under *Cefaclor.*

Assay preparation—Place 1 Cefaclor Capsule in a high-speed glass blender jar containing an accurately measured volume of *Buffer No. 4* (see *Phosphate Buffers and Other Solutions* under *Antibiotics—Microbial Assays* ⟨81⟩) to obtain a solution containing about 1 mg per mL, blend for 4 ± 1 minutes, and filter.

Procedure—Proceed as directed for *Procedure* in the section, *Antibiotics—Hydroxylamine Assay*, under *Automated Methods of Analysis* ⟨16⟩. Calculate the quantity, in mg, of $C_{15}H_{14}ClN_3O_4S$ in the Capsule taken by the formula:

$$(T/D)(CP/1000)(A_U/A_S),$$

in which *T* is the labeled quantity, in mg, of cefaclor in the Capsule, and *D* is the concentration, in mg per mL, of cefaclor in the *Assay preparation*, based on the labeled quantity per Capsule and the extent of dilution.

Cefaclor for Oral Suspension

» Cefaclor for Oral Suspension is a dry mixture of Cefaclor and one or more suitable buffers, colors, diluents, and flavors. It contains the equivalent of not less than 90.0 percent and not more than 120.0 percent of the labeled amount of $C_{15}H_{14}ClN_3O_4S$.

Packaging and storage—Preserve in tight containers.

Reference standard—*USP Cefaclor Reference Standard*—Do not dry before using.

Identification—Constitute 1 container of Cefaclor for Oral Suspension as directed in the labeling. Mix a portion of the resulting suspension with water to obtain a concentration of about 2 mg of cefaclor per mL, and filter: the filtrate so obtained responds to *Identification test B* under *Cefaclor.*

Uniformity of dosage units ⟨905⟩—

FOR SOLID PACKAGED IN SINGLE-UNIT CONTAINERS: meets the requirements.

pH ⟨791⟩: between 2.5 and 5.0, in the suspension constituted as directed in the labeling.

Water, *Method I* ⟨921⟩: not more than 2.0%.

Assay—

Standard preparation—Prepare as directed for *Standard preparation* in the *Assay* under *Cefaclor.*

Assay preparation—Constitute 1 container of Cefaclor for Oral Suspension as directed in the labeling. Dilute 5.0 mL of the suspension, freshly mixed and free from air bubbles, quantitatively with *Buffer No. 4* (see *Phosphate Buffers and Other Solutions* under *Antibiotics—Microbial Assays* ⟨81⟩) to obtain a solution containing about 1 mg per mL.

Procedure—Proceed as directed for *Procedure* in the section, *Antibiotics—Hydroxylamine Assay*, under *Automated Methods of Analysis* ⟨16⟩. Calculate the quantity, in mg, of $C_{15}H_{14}ClN_3O_4S$ in each mL of the Oral Suspension taken by the formula:

$$(T/D)(CP)(A_U/A_S),$$

in which T is the labeled quantity, in mg, of cefaclor in each mL of the Oral Suspension, and D is the concentration, in mg per mL, of cefaclor in the *Assay preparation*, based on the labeled quantity per mL and the extent of dilution.

Cefadroxil

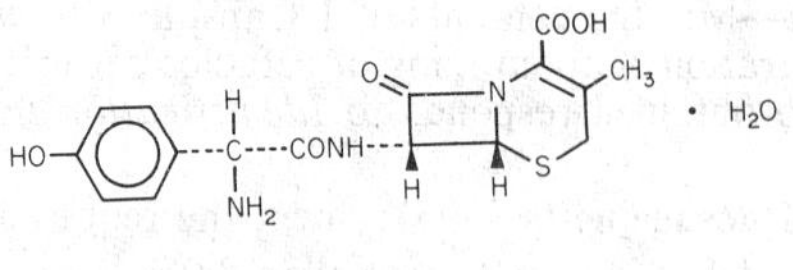

$C_{16}H_{17}N_3O_5S.H_2O$ 381.40

5-Thia-1-azabicyclo[4.2.0]oct-2-ene-2-carboxylic acid, 7-[[amino(4-hydroxyphenyl)acetyl]amino]-3-methyl-8-oxo-, monohydrate [6*R*-[6α,7β(*R**)]]-.

(6*R*,7*R*)-7-[(*R*)-2-Amino-2-(*p*-hydroxyphenyl)acetamido]-3-methyl-8-oxo-5-thia-1-azabicyclo[4.2.0]oct-2-ene-2-carboxylic acid monohydrate [*66592-87-8*].

Anhydrous 363.39 [*50370-12-2*].

» Cefadroxil has a potency equivalent to not less than 900 µg and not more than 1050 µg of $C_{16}H_{17}N_3O_5S$ per mg, calculated on the anhydrous basis.

Packaging and storage—Preserve in tight containers.

Reference standard—*USP Cefadroxil Reference Standard*—Do not dry before using.

Identification—

A: The ultraviolet absorption spectrum of a solution (1 in 50,000) exhibits maxima and minima at the same wavelengths as that of a similar solution of USP Cefadroxil RS, concomitantly measured, and the absorptivity, calculated on the anhydrous basis, at the wavelength of maximum absorbance at about 264 nm is between 95.0% and 104.0% of that of the USP Cefadroxil RS, the potency of the Reference Standard being taken into account.

B: Place a suitable thin-layer chromatographic plate (see *Chromatography* ⟨621⟩) coated with a 0.25-mm layer of binder-free silica gel in a chamber containing a mixture of *n*-hexane and tetradecane (95:5) to a depth of about 1 cm, allow the solvent front to move the length of the plate, remove the plate from the chamber, and allow the solvent to evaporate. On this plate apply 20 µL each of a solution of Cefadroxil in water containing 2 mg per mL and a similarly prepared Standard solution of USP Cefadroxil RS. Allow the spots to dry, and develop the chromatogram in a solvent system consisting of 0.1 *M* citric acid, 0.1 *M* dibasic sodium phosphate, and a 1 in 15 solution of ninhydrin in acetone (60:40:1.5) until the solvent front has moved about three-fourths of the length of the plate. Remove the plate from the developing chamber, mark the solvent front, and allow to air-dry. Spray the plate with a 1 in 500 solution of ninhydrin in dehydrated alcohol [NOTE—Protect this solution from light], dry for 10 minutes at 110°, and examine the chromatogram: the R_f value of the principal spot obtained from the test solution corresponds to that obtained from the Standard solution.

Crystallinity ⟨695⟩: meets the requirements.

pH ⟨791⟩: between 4.0 and 6.0, in a solution containing 50 mg per mL.

Water, *Method I* ⟨921⟩: between 4.2% and 6.0%.

Assay—

Standard preparation—Dissolve a suitable quantity of USP Cefadroxil RS, accurately weighed, in water to obtain a solution having a known concentration of about 1 mg per mL.

Assay preparation—Using a suitable quantity of Cefadroxil, accurately weighed, proceed as directed under *Standard preparation.*

Procedure—Proceed as directed for *Procedure* in the section, *Antibiotics—Hydroxylamine Assay*, under *Automated Methods of Analysis* ⟨16⟩. Calculate the potency, in µg, of $C_{16}H_{17}N_3O_5S$ in each mg of Cefadroxil taken by the formula:

$$(CP/W)(A_U/A_S),$$

in which W is the weight, in mg, of Cefadroxil taken in each mL of the *Assay preparation.*

Cefadroxil Capsules

» Cefadroxil Capsules contain the equivalent of not less than 90.0 percent and not more than 120.0 percent of the labeled amount of $C_{16}H_{17}N_3O_5S$.

Packaging and storage—Preserve in tight containers.

Reference standard—*USP Cefadroxil Reference Standard*—Do not dry before using.

Identification—Mix the contents of 1 Capsule with water to obtain a concentration of about 2 mg of cefadroxil per mL, and filter: the filtrate so obtained responds to *Identification test B* under *Cefadroxil.*

Uniformity of dosage units ⟨905⟩: meet the requirements.

Water, *Method I* ⟨921⟩: not more than 7.0%.

Assay—

Standard preparation—Prepare as directed for *Standard preparation* in the section, *Antibiotics—Hydroxylamine Assay*, under *Automated Methods of Analysis* ⟨16⟩, using USP Cefadroxil RS.

Assay preparation—Place not less than 5 Cefadroxil Capsules in a high-speed glass blender jar containing an accurately measured volume of water, so that the stock solution obtained after blending contains a convenient concentration of cefadroxil. Dilute an accurately measured volume of this stock solution quantitatively and stepwise with water to obtain an *Assay preparation* containing about 1 mg of cefadroxil per mL.

Procedure—Proceed as directed for *Procedure* in the section, *Antibiotics—Hydroxylamine Assay*, under *Automated Methods of Analysis* ⟨16⟩. Calculate the quantity, in mg, of $C_{16}H_{17}N_3O_5S$ in each Capsule taken by the formula:

$$(T/D)(C)(A_U/A_S),$$

in which T is the labeled quantity, in mg, of cefadroxil in each Capsule, and D is the concentration, in mg per mL, of cefadroxil in the *Assay preparation*, based on the labeled quantity per Capsule and the extent of dilution.

Cefadroxil for Oral Suspension

» Cefadroxil for Oral Suspension is a dry mixture of Cefadroxil and one or more suitable buffers, colors, diluents, and flavors. It contains the equivalent of not less than 90.0 percent and not more than 120.0 percent of the labeled amount of $C_{16}H_{17}N_3O_5S$.

Packaging and storage—Preserve in tight containers.

Reference standard—*USP Cefadroxil Reference Standard*—Do not dry before using.

Identification—Constitute 1 container of Cefadroxil for Oral Suspension as directed in the labeling. Mix a portion of the resulting suspension with water to obtain a concentration of about 2 mg of cefadroxil per mL, and filter: the filtrate so obtained responds to *Identification test B* under *Cefadroxil.*

Uniformity of dosage units ⟨905⟩—

FOR SOLID PACKAGED IN SINGLE-UNIT CONTAINERS: meets the requirements.

pH ⟨791⟩: between 4.5 and 6.0, in the suspension constituted as directed in the labeling.

Water, *Method I* ⟨921⟩: not more than 2.0%, 20 mL of a mixture of carbon tetrachloride, chloroform, and methanol (2:2:1) being used in place of methanol in the titration vessel.

Assay—

Standard preparation—Prepare as directed for *Standard preparation* in the *Assay* under *Cefadroxil.*

Assay preparation—Constitute 1 container of Cefadroxil for Oral Suspension as directed in the labeling. Dilute 5.0 mL of the Oral Suspension, freshly mixed and free from air bubbles, quantitatively with water to obtain a solution containing about 1 mg of cefadroxil per mL.

Procedure—Proceed as directed for *Procedure* in the section, *Antibiotics—Hydroxylamine Assay*, under *Automated Methods of Analysis* ⟨16⟩. Calculate the quantity, in mg, of $C_{16}H_{17}N_3O_5S$ in each mL of the Oral Suspension taken by the formula:

$$(T/D)(CP)(A_U/A_S),$$

in which T is the labeled quantity, in mg, of cefadroxil in each mL of Oral Suspension, and D is the concentration, in mg per mL, of cefadroxil in the *Assay preparation*, based on the labeled quantity per mL and the extent of dilution.

Cefadroxil Tablets

» Cefadroxil Tablets contain not less than 90.0 percent and not more than 120.0 percent of the labeled amount of $C_{16}H_{17}N_3O_5S$.

Packaging and storage—Preserve in tight containers.

Reference standard—*USP Cefadroxil Reference Standard*—Do not dry before using.

Identification—Mix a quantity of powdered Tablets, equivalent to about 250 mg of cefadroxil, with water to obtain a concentration of about 2 mg of cefadroxil per mL, and filter: the filtrate so obtained responds to *Identification test B* under *Cefadroxil.*

Disintegration ⟨701⟩: 15 minutes, simulated gastric fluid TS being used as the test medium.

Uniformity of dosage units ⟨905⟩: meet the requirements.

Water, *Method I* ⟨921⟩: not more than 8.0%.

Assay—

Standard preparation—Prepare as directed for *Standard Preparation* in the section, *Antibiotics—Hydroxylamine Assay*, under *Automated Methods of Analysis* ⟨16⟩, using USP Cefadroxil RS.

Assay preparation—Place not less than 5 Cefadroxil Tablets in a high-speed glass blender jar containing an accurately measured volume of water, and blend for 4 ± 1 minutes. Dilute an accurately measured volume of this stock solution quantitatively with water to obtain an *Assay preparation* containing about 1 mg of cefadroxil per mL.

Procedure—Proceed as directed for *Procedure* in the section, *Antibiotics—Hydroxylamine Assay*, under *Automated Methods of Analysis* ⟨16⟩. Calculate the quantity, in mg, of $C_{16}H_{17}N_3O_5S$ in each Tablet taken by the formula:

$$(T/D)(C)(A_U/A_S),$$

in which T is the labeled quantity, in mg, of cefadroxil in each Tablet, and D is the concentration, in mg, of cefadroxil in the *Assay preparation*, based on the labeled quantity per Tablet and the extent of dilution.

Cefamandole Nafate for Injection

» Cefamandole Nafate for Injection is a sterile mixture of Sterile Cefamandole Nafate or cefamandole nafate and one or more suitable buffers. It has a potency equivalent to not less than 810 µg and not more than 1000 µg of cefamandole ($C_{18}H_{18}N_6O_5S_2$) per mg, calculated on the anhydrous and sodium carbonate-free basis. It contains the equivalent of not less than 90.0 percent and not more than 115.0 percent of the labeled amount of cefamandole ($C_{18}H_{18}N_6O_5S_2$).

Packaging and storage—Preserve in *Containers for Sterile Solids* as described under *Injections* ⟨1⟩.

Reference standard—*USP Cefamandole Nafate Reference Standard*—Do not dry before using.

Constituted solution—At the time of use, the constituted solution prepared from Cefamandole Nafate for Injection meets the requirements for *Constituted Solutions* under *Injections* ⟨1⟩.

Identification—It responds to the *Identification test* under *Sterile Cefamandole Nafate.*

Pyrogen—When diluted with Sterile Water for Injection to a concentration of 50 mg of cefamandole ($C_{18}H_{18}N_6O_5S_2$) per mL, it meets the requirements of the *Pyrogen Test* ⟨151⟩, the test dose being 1.0 mL per kg.

Sterility—It meets the requirements under *Sterility Tests* ⟨71⟩, when tested as directed in the section, *Test Procedures Using Membrane Filtration.*

pH ⟨791⟩: between 6.0 and 8.0, determined after 30 minutes in a solution containing 100 mg per mL.

Water, *Method I* ⟨921⟩: not more than 3.0%.

Particulate matter ⟨788⟩: meets the requirements under *Small-volume Injections.*

Other requirements—It meets the requirements under *Injections* ⟨1⟩ and *Uniformity of Dosage Units* ⟨905⟩.

Assay—

Assay preparation 1 (where the article is represented as being in a single-dose container)—Constitute Cefamandole Nafate for Injection in a volume of water, accurately measured, corresponding to the volume of solvent specified in the labeling. Withdraw all of the withdrawable contents, using a suitable hypodermic needle and syringe, and dilute quantitatively with water to obtain a solution having a concentration of about 2 mg of cefamandole per mL. Transfer 5.0 mL of this solution to a 50-mL volumetric flask, add 30.0 mL of *pH 2.3 buffer*, dilute with water to volume, and mix.

Assay preparation 2 (where the label states the quantity of cefamandole in a given volume of constituted solution)—Constitute Cefamandole Nafate for Injection in a volume of water, accurately measured, corresponding to the volume of solvent specified in the labeling. Dilute an accurately measured volume of the constituted solution quantitatively with water to obtain a solution containing about 2 mg of cefamandole per mL. Transfer 5.0 mL of this solution to a 50-mL volumetric flask, add 30.0 mL of *pH 2.3 buffer*, dilute with water to volume, and mix.

Assay preparation 3—Using an accurately weighed quantity of Cefamandole Nafate for Injection, prepare as directed for *Standard preparation* under *Sterile Cefamandole Nafate.* Determine the sodium carbonate content of a separate, accurately weighed, 1-g portion of Cefamandole Nafate for Injection dissolved in 100 mL of water. Add methyl orange TS, and titrate with 0.2 *N* sulfuric acid VS. Each mL of 0.2 *N* sulfuric acid is equivalent to 10.60 mg of Na_2CO_3.

Procedure—Proceed as directed for *Procedure* in the *Assay* under *Sterile Cefamandole Nafate.* Calculate the quantity, in mg, of $C_{18}H_{18}N_6O_5S_2$ in the portion of constituted solution taken by the formula:

$$(CP)(L/1000D)(i_U/i_S),$$

in which C is the concentration, in mg per mL, of USP Cefamandole Nafate RS in the *Standard preparation*, L is the labeled quantity, in mg, in the portion of constituted solution taken, D is the concentration, in mg per mL, of cefamandole in *Assay preparation 1* or in *Assay preparation 2*, based on the volume of constituted solution taken and the extent of dilution, and the other terms are as defined therein. Calculate the potency, in µg of cefamandole ($C_{18}H_{18}N_6O_5S_2$) per mg, of the Cefamandole Nafate for Injection taken by the formula:

$$(CP/W)(i_U/i_S),$$

in which W is the weight, in mg, of the Cefamandole Nafate for Injection taken in each mL of *Assay preparation 3*, and the other terms are as defined therein.

Sterile Cefamandole Nafate

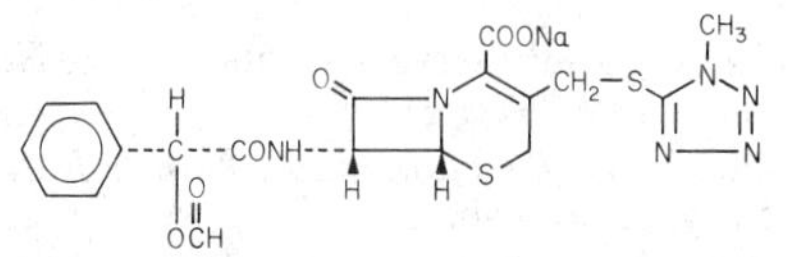

$C_{19}H_{17}N_6NaO_6S_2$ 512.50

5-Thia-1-azabicyclo[4.2.0]oct-2-ene-2-carboxylic acid, 7-[[(formyloxy)phenylacetyl]amino]-3-[[(1-methyl-1*H*-tetrazol-5-yl)thio]methyl]-8-oxo-, monosodium salt, [6*R*-[6α,7β(*R**)]]-.

Sodium (6*R*,7*R*)-7-(*R*)-mandelamido-3-[[(1-methyl-1*H*-tetrazol-5-yl)thio]methyl]-8-oxo-5-thia-1-azabicyclo[4.2.0]oct-2-ene-2-carboxylate formate (ester) [*42540-40-9*].

» Sterile Cefamandole Nafate is cefamandole nafate suitable for parenteral use. It has a potency equivalent to not less than 810 µg and not more than 1000 µg of cefamandole ($C_{18}H_{18}N_6O_5S_2$) per mg, calculated on the anhydrous basis.

Packaging and storage—Preserve in *Containers for Sterile Solids* as described under *Injections* ⟨1⟩.

Reference standard—*USP Cefamandole Nafate Reference Standard*—Do not dry before using.

Identification—Prepare a mixture of ethyl acetate, acetone, glacial acetic acid, and water (5:2:1:1) (*Developing solvent*). Prepare a solution of the specimen in *Developing solvent* containing 10 mg per mL. Prepare a Standard solution of USP Cefamandole Nafate RS in *Developing solvent* containing 10 mg per mL. Use these solutions promptly after preparation. Apply separately 10 µL of each solution on a suitable thin-layer chromatographic plate (see *Chromatography* ⟨621⟩), coated with a 0.25-mm layer of chromatographic silica gel mixture. Place the plate in a suitable chromatographic chamber, previously equilibrated with *Developing solvent* for not less than 30 minutes, and develop the chromatogram until the solvent front has moved about three-fourths of the length of the plate. Remove the plate from the developing chamber, mark the solvent front, and allow to air-dry. Locate the spots on the plate by examination under short-wavelength ultraviolet light: the R_f value of the principal spot obtained from the test solution corresponds to that obtained from the Standard solution.

Pyrogen—When diluted with Sterile Water for Injection to a concentration of 50 mg of cefamandole ($C_{18}H_{18}N_6O_5S_2$) per mL, it meets the requirements of the *Pyrogen Test* ⟨151⟩, the test dose being 1.0 mL per kg.

Sterility—It meets the requirements under *Sterility Tests* ⟨71⟩, when tested as directed in the section, *Test Procedures Using Membrane Filtration.*

pH ⟨791⟩: between 3.5 and 7.0, in a solution containing 100 mg per mL.

Water, *Method I* ⟨921⟩: not more than 2.0%.

Assay—

pH 2.3 buffer—Dissolve 3.6 g of anhydrous dibasic sodium phosphate, 39.4 g of citric acid monohydrate, and 70.8 g of potassium chloride in water to make 1000 mL.

Standard preparation—Transfer about 12 mg USP Cefamandole Nafate RS, accurately weighed, to a 50-mL volumetric flask containing 4 mL of water. Immediately before use, add 30.0 mL of *pH 2.3 buffer*, dilute with water to volume, and mix.

Assay preparation—Using Sterile Cefamandole Nafate, prepare as directed under *Standard preparation.*

Procedure—Transfer a portion of the *Assay preparation* to a suitable polarographic cell. Deaerate by bubbling scrubbed nitrogen through the solution for 5 minutes, and redirect the nitrogen flow to the surface outlet. Insert the dropping mercury electrode of a suitable polarograph (see *Polarography* ⟨801⟩) capable of measuring a current of 0.5 microampere or appropriate current to maintain on-scale response, using an average capillary, and a drop rate of 1 per second. Record the polarogram in the differential pulse mode from −0.3 volt to −1.05 volts, using a saturated calomel reference electrode and platinum wire counter electrode. Determine the peak height obtained, in microamperes, where the peak height is defined as the perpendicular distance from the extrapolated baseline to the highest point of the peak as compared to the full-scale current range. Similarly, determine the peak current of the *Standard preparation.* Calculate the quantity, in µg, of $C_{18}H_{18}N_6O_5S_2$ in each mg of the Sterile Cefamandole Nafate taken by the formula:

$$P(W_S/W_U)(i_U/i_S),$$

in which *P* is the potency, in µg of cefamandole per mg, of the USP Cefamandole Nafate RS, W_S and W_U are the quantities, in mg, of USP Cefamandole Nafate RS and Sterile Cefamandole Nafate taken to prepare the *Standard preparation* and the *Assay preparation*, respectively, and i_U and i_S are the peak currents, in microamperes, from the *Assay preparation* and the *Standard preparation*, respectively.

Cefamandole Sodium for Injection

$C_{18}H_{17}N_6NaO_5S_2$ 484.48

5-Thia-1-azabicyclo[4.2.0]oct-2-ene-2-carboxylic acid, 7-[(hydroxyphenylacetyl)amino]-3-[[(1-methyl-1*H*-tetrazol-5-yl)thio]methyl]-8-oxo-, [6*R*-[6α,7β(*R**)]]-, monosodium salt.

Monosodium (6*R*,7*R*)-7-(*R*)-mandelamido-3-[[(1-methyl-1*H*-tetrazol-5-yl)thio]methyl]-8-oxo-5-thia-1-azabicyclo[4.2.0]-oct-2-ene-2-carboxylate [*30034-03-8*].

» Cefamandole Sodium for Injection is a sterile mixture of Sterile Cefamandole Sodium and one or more suitable buffers. It contains not less than 90.0 percent and not more than 115.0 percent of the labeled amount of cefamandole ($C_{18}H_{18}N_6O_5S_2$).

Packaging and storage—Preserve in *Containers for Sterile Solids* as described under *Injections* ⟨1⟩.

Reference standards—*USP Cefamandole Lithium Reference Standard*—Do not dry before using. *USP Cefamandole Sodium Reference Standard*—Do not dry before using.

Constituted solution—At the time of use, the constituted solution prepared from Cefamandole Sodium for Injection meets the requirements for *Constituted Solutions* under *Injections* ⟨1⟩.

Identification—Prepare a mixture of ethyl acetate, acetone, glacial acetic acid, and water (5:2:1:1) (*Developing solvent*). Prepare a solution of the specimen in *Developing solvent* containing 10 mg per mL. Prepare a Standard solution of USP Cefamandole Sodium RS in *Developing solvent* containing 10 mg per mL. Use these solutions promptly after preparation. Apply separately 10 µL of each solution on a suitable thin-layer chromatographic plate (see *Chromatography* ⟨621⟩), coated with a 0.25-mm layer of chromatographic silica gel mixture. Place the plate in a suitable chromatographic chamber, previously equilibrated with *Developing solvent* for not less than 30 minutes, and develop the chromatogram until the solvent front has moved about three-fourths of the length of the plate. Remove the plate from the developing chamber, mark the solvent front, and allow to air-dry. Locate the spots on the plate by examination under short-wavelength ultraviolet light: the R_f value of the principal spot obtained from the test solution corresponds to that obtained from the Standard solution.

Pyrogen—It meets the requirements of the *Pyrogen Test* ⟨151⟩, the test dose being 1.0 mL per kg of a solution prepared by diluting Cefamandole Sodium for Injection with Sterile Water for Injection to a concentration of 50 mg of cefamandole per mL.

Sterility—It meets the requirements under *Sterility Tests* ⟨71⟩, when tested as directed in the section, *Test Procedures Using Membrane Filtration.*

pH ⟨791⟩: between 6.0 and 8.5, in a solution containing 100 mg of cefamandole per mL.

Water, *Method I* ⟨921⟩: not more than 3.0%.

Particulate matter ⟨788⟩: meets the requirements under *Small-volume Injections.*

Other requirements—It meets the requirements for *Uniformity of Dosage Units* ⟨905⟩ and *Labeling* under *Injections* ⟨1⟩.

Assay—

pH 2.3 buffer solution—Prepare as directed in the *Assay* under *Sterile Cefamandole Nafate.*

Standard preparation—Transfer about 12 mg of USP Cefamandole Lithium RS, accurately weighed, to a 50-mL volumetric flask containing 4 mL of water. Immediately before use, add 30.0 mL of *pH 2.3 buffer solution*, dilute with water to volume, and mix.

Assay preparation 1 (where it is represented as being in a single-dose container)—Constitute Cefamandole Sodium for Injection in a volume of water, accurately measured, corresponding to the volume of solvent specified in the labeling. Withdraw all of the withdrawable contents using a suitable hypodermic needle and syringe, and dilute quantitatively with water to obtain a solution having a concentration of 2 mg of cefamandole per mL. Transfer 5.0 mL of this solution to a 50-mL volumetric flask, add 30.0 mL of *pH 2.3 buffer solution*, dilute with water to volume, and mix.

Assay preparation 2 (where the label states the quantity of cefamandole in a given volume of constituted solution)—Constitute Cefamandole Sodium for Injection in a volume of water, accurately measured, corresponding to the volume of solvent specified in the labeling. Dilute an accurately measured volume of the constituted solution quantitatively with water to obtain a solution containing 2 mg of cefamandole per mL. Transfer 5.0 mL of this solution to a 50-mL volumetric flask, add 30.0 mL of *pH 2.3 buffer solution*, dilute with water to volume, and mix.

Procedure—Proceed as directed for *Procedure* in the *Assay* for *Sterile Cefamandole Nafate.* Calculate the quantity, in mg, of $C_{18}H_{18}N_6O_5S_2$ in the portion of constituted solution taken by the formula:

$$(CP)(L/1000D)(i_U/i_S),$$

in which C is the concentration, in mg per mL, of USP Cefamandole Lithium RS in the *Standard preparation*, P is the potency, in µg of cefamandole per mg, of USP Cefamandole Lithium RS, L is the labeled quantity, in mg, of cefamandole in the portion of constituted solution taken, D is the concentration, in mg per mL, of the *Assay preparation*, based on the volume of constituted solution taken and the extent of dilution, and the other terms are as defined therein.

Sterile Cefamandole Sodium

» Sterile Cefamandole Sodium has a potency equivalent to not less than 860 µg and not more than 1000 µg of cefamandole ($C_{18}H_{18}N_6O_5S_2$) per mg, calculated on the anhydrous basis.

Packaging and storage—Preserve in *Containers for Sterile Solids* as described under *Injections* ⟨1⟩.

Reference standards—*USP Cefamandole Lithium Reference Standard*—Do not dry before using. *USP Cefamandole Sodium Reference Standard*—Do not dry before using.

Identification—The infrared absorption spectrum of a mineral oil dispersion of it exhibits maxima only at the same wavelengths as that of a similar preparation of USP Cefamandole Sodium RS.

Pyrogen—It meets the requirements of the *Pyrogen Test* ⟨151⟩, the test dose being 1.0 mL per kg of a solution prepared by diluting Sterile Cefamandole Sodium with Sterile Water for Injection to a concentration of 50 mg of cefamandole per mL.

Sterility—It meets the requirements under *Sterility Tests* ⟨71⟩, when tested as directed in the section, *Test Procedures Using Membrane Filtration*, 5 g of specimen aseptically dissolved in 200 mL of *Fluid A* being used.

pH ⟨791⟩: between 3.5 and 7.0, in a solution (1 in 10).

Water, *Method I* ⟨921⟩: not more than 3.0%.

Particulate matter ⟨788⟩: meets the requirements under *Small-volume Injections.*

Assay—

pH 2.3 buffer solution—Prepare as directed in the *Assay* under *Sterile Cefamandole Nafate.*

Standard preparation—Transfer about 12 mg of USP Cefamandole Lithium RS, accurately weighed, to a 50-mL volumetric flask containing 4 mL of water. Immediately before use, add 30.0 mL of *pH 2.3 buffer solution*, dilute with water to volume, and mix.

Assay preparation—Using Sterile Cefamandole Sodium, prepare as directed under *Standard preparation.*

Procedure—Proceed as directed for *Procedure* in the *Assay* under *Sterile Cefamandole Nafate.* Calculate the quantity, in µg, of cefamandole ($C_{18}H_{18}N_6O_5S_2$) in each mg of the Sterile Cefamandole Sodium taken by the formula:

$$P(W_S/W_U)(i_U/i_S),$$

in which P is the potency, in µg of cefamandole per mg, of USP Cefamandole Lithium RS, W_S and W_U are the quantities, in mg, of USP Cefamandole Lithium RS and Sterile Cefamandole Sodium taken to prepare the *Standard preparation* and the *Assay preparation*, respectively, and the other terms are as defined therein.

Cefazolin

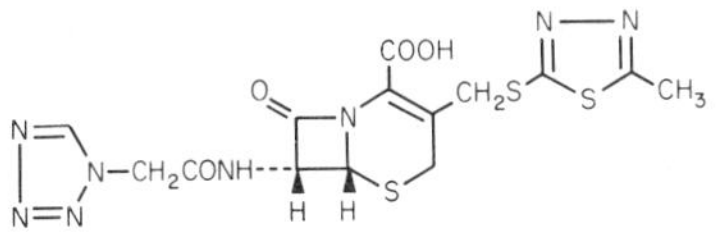

$C_{14}H_{14}N_8O_4S_3$ 454.50

5-Thia-1-azabicyclo[4.2.0]oct-2-ene-2-carboxylic acid, 3-[[(5-methyl-1,3,4-thiadiazol-2-yl)thio]methyl]-8-oxo-7-[[(1*H*-tetrazol-1-yl)acetyl]amino]-, (6*R*-*trans*).

(6*R*,7*R*)-3-[[(5-methyl-1,3,4-thiadiazol-2-yl)thio]methyl]-8-oxo-7-[2-(1*H*-tetrazol-1-yl)acetamido]-5-thia-1-azabicyclo[4.2.0]oct-2-ene-2-carboxylic acid [*25953-19-9*].

» Cefazolin contains not less than 950 µg and not more than 1030 µg of $C_{14}H_{14}N_8O_4S_3$ per mg, calculated on the anhydrous basis.

Packaging and storage—Preserve in tight containers.

Reference standard—*USP Cefazolin Reference Standard*—Do not dry before using.

Identification—The chromatogram of the *Assay preparation* obtained as directed in the *Assay* exhibits a major peak for cefazolin the retention time of which corresponds with that exhibited in the chromatogram of the *Standard preparation* obtained as directed in the *Assay.*

Water, *Method I* ⟨921⟩: not more than 2.0%.

Heavy metals, *Method II* ⟨231⟩: 0.002%.

Assay—

pH 3.6 buffer—Dissolve 0.900 g of anhydrous dibasic sodium phosphate and 1.298 g of citric acid monohydrate in water to make 1000 mL.

pH 7.0 buffer—Dissolve 5.68 g of anhydrous dibasic sodium phosphate and 3.63 g of monobasic potassium phosphate in water to make 1000 mL.

Mobile phase—Prepare a suitable mixture of *pH 3.6 buffer* and acetonitrile (9:1). Filter through a membrane filter (1-µm or finer porosity), and degas.

Internal standard solution—Transfer 750 mg of salicylic acid to a 100-mL volumetric flask, dissolve in 5 mL of methanol, dilute with *pH 7.0 buffer* to volume, and mix.

Standard preparation—Transfer about 50 mg of USP Cefazolin RS, accurately weighed, to a 50-mL volumetric flask, dissolve in *pH 7.0 buffer*, dilute with *pH 7.0 buffer* to volume, and mix. Transfer 5.0 mL of this solution to a 100-mL volumetric flask, add 5.0 mL of *Internal standard solution*, dilute with *pH 7.0 buffer* to volume, and mix.

Assay preparation—Using about 50 mg of Cefazolin, accurately weighed, proceed as directed under *Standard preparation.*

Chromatographic system (see *Chromatography* ⟨621⟩)—The liquid chromatograph is equipped with a 254-nm detector and a 4.0-mm × 30-cm column that contains 10-μm packing L1. The flow rate is about 2 mL per minute. Chromatograph the *Standard preparation*, and record the peak responses as directed under *Procedure:* the column efficiency determined from the analyte peak is not less than 1500 theoretical plates, the tailing factor for the analyte peak is not more than 1.5, the resolution, *R*, between the analyte and internal standard peaks is not less than 4.0, and the relative standard deviation for replicate injections is not more than 2.0%.

Procedure—Separately inject equal volumes (about 10 μL) of the *Standard preparation* and the *Assay preparation* into the chromatograph, record the chromatograms, and measure the responses for the major peaks. The relative retention times are about 0.7 for salicylic acid and 1.0 for cefazolin. Calculate the quantity, in μg, of $C_{14}H_{14}N_8O_4S_3$ in each mg of the Cefazolin taken by the formula:

$$(W_S/W_U)(P)(R_U/R_S),$$

in which W_S and W_U are the weights, in mg, of USP Cefazolin RS and Cefazolin taken to prepare the *Standard preparation* and the *Assay preparation*, respectively, *P* is the designated purity, in μg of cefazolin per mg, of USP Cefazolin RS, and R_U and R_S are the peak response ratios of the cefazolin peak to the internal standard peak obtained from the *Assay preparation* and the *Standard preparation*, respectively.

Cefazolin Sodium Injection

» Cefazolin Sodium Injection is a sterile solution of Cefazolin and Sodium Bicarbonate diluted with a suitable isoosmotic diluent. It contains not less than 90.0 percent and not more than 115.0 percent of the labeled amount of cefazolin ($C_{14}H_{14}N_8O_4S_3$).

Packaging and storage—Preserve in *Containers for Injections* as described under *Injections* ⟨1⟩. Maintain in the frozen state.

Labeling—It meets the requirements for *Labeling* under *Injections* ⟨1⟩. The label states that it is to be thawed just prior to use, describes conditions for proper storage of the resultant solution, and directs that the solution is not to be refrozen.

Reference standard—*USP Cefazolin Reference Standard*—Do not dry before using.

Identification—The chromatogram of the *Assay preparation* obtained as directed in the *Assay* exhibits a major peak for cefazolin the retention time of which corresponds to that exhibited in the chromatogram of the *Standard preparation* obtained as directed in the *Assay*.

Pyrogen—It meets the requirements of the *Pyrogen Test* ⟨151⟩, the test dose being a volume of undiluted Injection providing the equivalent of 50 mg of cefazolin per kg.

Sterility—It meets the requirements under *Sterility Tests* ⟨71⟩, when tested as directed in the section, *Test Procedures Using Membrane Filtration.*

pH ⟨791⟩: between 4.5 and 7.0.

Particulate matter ⟨788⟩: meets the requirements under *Small-volume Injections.*

Assay—

pH 3.6 buffer, pH 7.0 buffer, Mobile phase, Internal standard solution, Standard preparation, and *Chromatographic system*—Prepare as directed in the *Assay* under *Cefazolin.*

Assay preparation—Allow 1 container of Cefazolin Sodium Injection to thaw, and mix. Transfer an accurately measured volume of the Injection, equivalent to about 50 mg of cefazolin, to a 50-mL volumetric flask, dilute with *pH 7.0 buffer* to volume, and mix. Transfer 5.0 mL of this solution to a 100-mL volumetric flask, add 5.0 mL of *Internal standard solution*, dilute with *pH 7.0 buffer* to volume, and mix.

Procedure—Proceed as directed for *Procedure* in the *Assay* under *Cefazolin.* Calculate the quantity, in mg, of cefazolin ($C_{14}H_{14}N_8O_4S_3$) in each mL of the Injection taken by the formula:

$$(W_S/V)(P/1000)(R_U/R_S),$$

in which *V* is the volume, in mL, of Injection taken, and the other terms are as defined therein.

Sterile Cefazolin Sodium

$C_{14}H_{13}N_8NaO_4S_3$ 476.48

5-Thia-1-azabicyclo[4.2.0]oct-2-ene-2-carboxylic acid, 3-[[(5-methyl-1,3,4-thiadiazol-2-yl)thio]methyl]-8-oxo-7-[[(1*H*-tetrazol-1-yl)acetyl]amino]-, monosodium salt (6*R*-*trans*).

Monosodium (6*R*,7*R*)-3-[[(5-methyl-1,3,4-thiadiazol-2-yl)thio]-methyl]-8-oxo-7-[2-(1*H*-tetrazol-1-yl)acetamido]-5-thia-1-azabicyclo[4.2.0]oct-2-ene-2-carboxylate [*27164-46-1*].

» Sterile Cefazolin Sodium has a potency equivalent to not less than 850 μg and not more than 1050 μg of cefazolin ($C_{14}H_{14}N_8O_4S_3$) per mg, calculated on the anhydrous basis. In addition, where packaged for dispensing, it contains the equivalent of not less than 90.0 percent and not more than 115.0 percent of the labeled amount of cefazolin.

Packaging and storage—Preserve in *Containers for Sterile Solids* as described under *Injections* ⟨1⟩.

Reference standard—*USP Cefazolin Reference Standard*—Do not dry before using.

Constituted solution—At the time of use, the constituted solution prepared from Sterile Cefazolin Sodium meets the requirements for *Constituted Solutions* under *Injections* ⟨1⟩.

Identification—

A: The ultraviolet absorption spectrum of a 1 in 50,000 solution of it in 0.1 *M* sodium bicarbonate exhibits maxima and minima at the same wavelengths as that of a similar solution of USP Cefazolin RS, concomitantly measured.

B: The chromatogram of the *Assay preparation* obtained as directed in the *Assay* exhibits a major peak for cefazolin the retention time of which corresponds to that exhibited in the chromatogram of the *Standard preparation* obtained as directed in the *Assay.*

Specific rotation ⟨781⟩: between −24° and −10°, in a 0.1 *M* sodium bicarbonate solution containing 55 mg of specimen per mL, calculated on the anhydrous basis.

Pyrogen—It meets the requirements of the *Pyrogen Test* ⟨151⟩, the test dose being 1.0 mL per kg of a solution in Sterile Water for Injection containing 50 mg of cefazolin ($C_{14}H_{14}N_8O_4S_3$) per mL.

Sterility—It meets the requirements under *Sterility Tests* ⟨71⟩, when tested as directed in the section, *Test Procedures Using Membrane Filtration.*

pH ⟨791⟩: between 4.5 and 6.0, in a solution containing 100 mg of cefazolin per mL.

Water, *Method I* ⟨921⟩: not more than 6.0%.

Particulate matter ⟨788⟩: meets the requirements under *Small-volume Injections.*

Other requirements—It meets the requirements for *Uniformity of Dosage Units* ⟨905⟩ and *Labeling* under *Injections* ⟨1⟩.

Assay—

pH 3.6 buffer, pH 7.0 buffer, Mobile phase, Internal standard solution, Standard preparation, and *Chromatographic system*—Prepare as directed in the *Assay* under *Cefazolin.*

Assay preparation 1—Transfer about 50 mg of Sterile Cefazolin Sodium, accurately weighed, to a 50-mL volumetric flask, dissolve in *pH 7.0 buffer*, dilute with *pH 7.0 buffer* to volume, and mix. Transfer 4.0 mL of this solution to a 200-mL volumetric flask, add 5.0 mL of *Internal standard solution*, dilute with *pH 7.0 buffer* to volume, and mix.

Assay preparation 2 (where it is packaged for dispensing and is represented as being in a single-dose container)—Constitute Sterile Cefazolin Sodium in a volume of water, accurately measured, corresponding to the volume of solvent specified in the labeling. Withdraw all of the withdrawable contents, using a suitable hypodermic needle and syringe, and dilute quantitatively with *pH buffer* to obtain a stock solution containing about 1 mg of cefazolin per mL. Transfer 4.0 mL of this solution to a 200-mL volumetric flask, add 5.0 mL of *Internal standard solution*, dilute with *pH 7.0 buffer* to volume, and mix.

Assay preparation 3 (where the label states the quantity of cefazolin in a given volume of constituted solution)—Constitute Sterile Cefazolin Sodium in a volume of water, accurately measured, corresponding to the volume of solvent specified in the labeling. Dilute an accurately measured volume of the constituted solution quantitatively with *pH 7.0 buffer* to obtain a stock solution containing about 1 mg of cefazolin per mL. Transfer 4.0 mL of this solution to a 200-mL volumetric flask, add 5.0 mL of *Internal standard solution*, dilute with *pH 7.0 buffer* to volume, and mix.

Procedure—Proceed as directed for *Procedure* in the *Assay* under *Cefazolin*. Calculate the quantity, in µg, of cefazolin ($C_{14}H_{14}N_8O_4S_3$) in each mg of Sterile Cefazolin Sodium taken to prepare *Assay preparation 1* by the formula:

$$(W_S/W_U)(P)(R_U/R_S),$$

in which W_U is the weight, in mg, of Sterile Cefazolin Sodium taken, and the other terms are as defined therein. Calculate the quantity, in mg, of cefazolin ($C_{14}H_{14}N_8O_4S_3$) in the container, and in the volume of constituted solution taken, by the formula:

$$(L/D)(W_S/50{,}000)(P)(R_U/R_S),$$

in which L is the labeled quantity, in mg, of cefazolin in the container, or in the volume of constituted solution taken, D is the concentration, in mg per mL, of cefazolin in the stock solution used in preparing *Assay preparation 2* or *Assay preparation 3*, on the basis of the labeled quantity in the container, or in the volume of constituted solution taken, respectively, and the extent of dilution, and the other terms are as defined therein.

Sterile Cefonicid Sodium

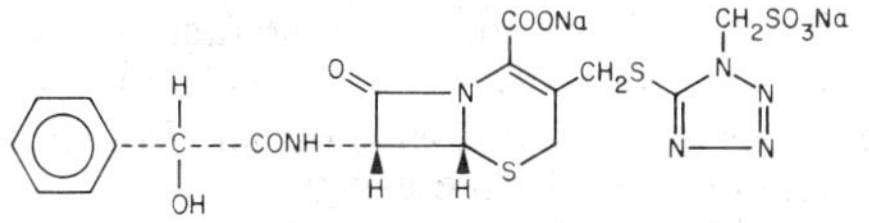

$C_{18}H_{16}N_6Na_2O_8S_3$ 586.52

5-Thia-1-azabicyclo[4.2.0]oct-2-ene-2-carboxylic acid, 7-[(hydroxyphenylacetyl)amino]-8-oxo-3-[[[1-(sulfomethyl)-1*H*-tetrazol-5-yl]thio]methyl]disodium salt, [6*R*-[6α,7β(*R**)]]-.

(6*R*,7*R*)-7-[(*R*)-Mandelamido]-8-oxo-3-[[[1-(sulfomethyl)-1-*H*-tetrazol-5-yl]thio]methyl]-5-thia-1-azabicyclo[4.2.0]oct-2-ene-2-carboxylic acid, disodium salt [*61270-78-8*].

» Sterile Cefonicid Sodium is cefonicid sodium suitable for parenteral use. It contains the equivalent of not less than 832 µg and not more than 970 µg of cefonicid ($C_{18}H_{18}N_6O_8S_3$) per mg, calculated on the anhydrous basis. In addition, where packaged for dispensing, it contains the equivalent of not less than 90.0 percent and not more than 120.0 percent of the labeled amount of cefonicid ($C_{18}H_{18}N_6O_8S_3$).

Packaging and storage—Preserve in *Containers for Sterile Solids* as described under *Injections* ⟨1⟩.

Reference standard—*USP Cefonicid Sodium Reference Standard*—Do not dry; determine the water content titrimetrically at the time of use.

Constituted solution—At the time of use, the constituted solution prepared from Sterile Cefonicid Sodium meets the requirements for *Constituted Solutions* under *Injections* ⟨1⟩.

Identification—The chromatogram of the *Assay preparation* obtained as directed in the *Assay* exhibits a major peak for cefonicid, the retention time of which corresponds to that exhibited in the chromatogram of the *Standard preparation* obtained as directed in the *Assay*.

Pyrogen—It meets the requirements of the *Pyrogen Test* ⟨151⟩, the test dose being 1 mL per kg of a solution prepared by diluting Sterile Cefonicid Sodium with Sterile Water for Injection to a concentration of 50 mg of cefonicid per mL.

Sterility—It meets the requirements under *Sterility Tests* ⟨71⟩, when tested as directed in the section, *Test Procedures Using Membrane Filtration*.

Specific rotation ⟨781⟩: between −37° and −47°, calculated on the anhydrous basis, determined in a solution in methanol containing 100 mg in each 10 mL.

pH ⟨791⟩: between 3.5 and 6.5, in a solution (1 in 20).

Water, *Method I* ⟨921⟩: not more than 5.0%.

Particulate matter ⟨788⟩: meets the requirements under *Small-volume Injections*.

Other requirements—It meets the requirements for *Uniformity of Dosage Units* ⟨905⟩ and for *Labeling* under *Injections* ⟨1⟩.

Assay—

Mobile phase—Prepare a mixture of water, methanol, and 0.2 *M* monobasic ammonium phosphate (33:5:2). Filter through a filter (0.5-µm or finer porosity), and degas. Adjust the composition, if necessary, to meet the performance requirements under *Chromatographic system*.

Standard preparation—Dissolve a suitable quantity of USP Cefonicid Sodium RS, accurately weighed, in *Mobile phase* to obtain a solution having a known concentration of about 20 µg of cefonicid ($C_{18}H_{18}N_6O_8S_3$) per mL.

Assay preparation 1—Transfer about 40 mg of Sterile Cefonicid Sodium, accurately weighed, to a 200-mL volumetric flask, dissolve in *Mobile phase*, dilute with *Mobile phase* to volume, and mix. Transfer 10.0 mL of the resulting solution to a 100-mL volumetric flask, dilute with *Mobile phase* to volume, and mix.

Assay preparation 2 (where it is represented as being in a single-dose container)—Constitute Sterile Cefonicid Sodium in a volume of water, accurately measured, corresponding to the volume of solvent specified in the labeling. Withdraw all of the withdrawable contents, using a suitable hypodermic needle and syringe, and dilute quantitatively with *Mobile phase* to obtain a solution containing about 20 µg of cefonicid per mL.

Assay preparation 3 (where the label states the quantity of cefonicid in a given volume of constituted solution)—Constitute Sterile Cefonicid Sodium in a volume of water, accurately measured, corresponding to the volume of solvent specified in the labeling. Dilute an accurately measured volume of the constituted solution quantitatively with *Mobile phase* to obtain a solution containing about 20 µg of cefonicid per mL.

Resolution solution—Dissolve a quantity of USP Cefonicid Sodium RS in *Mobile phase* to obtain a solution containing about 0.2 mg per mL. Heat on a steam bath for 30 minutes, and cool. This *Resolution solution* contains a mixture of cefonicid and desacetyl cefonicid.

Chromatographic system (see *Chromatography* ⟨621⟩)—The liquid chromatograph is equipped with a 254-nm detector and a 4-mm × 30-cm column that contains packing L1. The flow rate is about 2 mL per minute. Chromatograph the *Standard preparation* and the *Resolution solution*, and record the peak responses as directed under *Procedure:* the column efficiency determined from the analyte peak is not less than 1500 theoretical plates, the tailing factor for the analyte peak is not more than 1.3, the resolution, *R*, between the cefonicid and desacetyl cefonicid peaks is not less than 1.1, and the relative standard deviation for replicate injections of the *Standard preparation* is not more than 2%.

Procedure—Separately inject equal volumes (about 10 µL) of the *Standard preparation* and the *Assay preparation* into the chromatograph, record the chromatograms, and measure the responses for the major peaks. Calculate the quantity, in µg, of cefonicid ($C_{18}H_{18}N_6O_8S_3$) per mg of the Sterile Cefonicid Sodium taken by the formula:

$$2000(C/M)(r_U/r_S),$$

in which C is the concentration, in µg of cefonicid ($C_{18}H_{18}N_6O_8S_3$) per mL, of the *Standard preparation, M* is the quantity, in mg, of Sterile Cefonicid Sodium taken to prepare *Assay preparation 1*, and r_U and r_S are the peak responses obtained from *Assay preparation 1* and the *Standard preparation*, respectively. Calculate the quantity, in mg of cefonicid ($C_{18}H_{18}N_6O_8S_3$), withdrawn from the container, or in the portion of constituted solution taken by the formula:

$$(L/D)(C)(r_U/r_S),$$

in which L is the labeled quantity, in mg of cefonicid ($C_{18}H_{18}N_6O_8S_3$), in the container, or in the volume of constituted solution taken, and D is the concentration, in µg of cefonicid ($C_{18}H_{18}N_6O_8S_3$) per mL, of *Assay preparation 2* or *Assay preparation 3*, based on the labeled quantity in the container or in the portion of constituted solution taken, respectively, and the extent of dilution, and r_U and r_S are the peak responses obtained from the relevant *Assay preparation* and the *Standard preparation*, respectively.

Cefoperazone Sodium

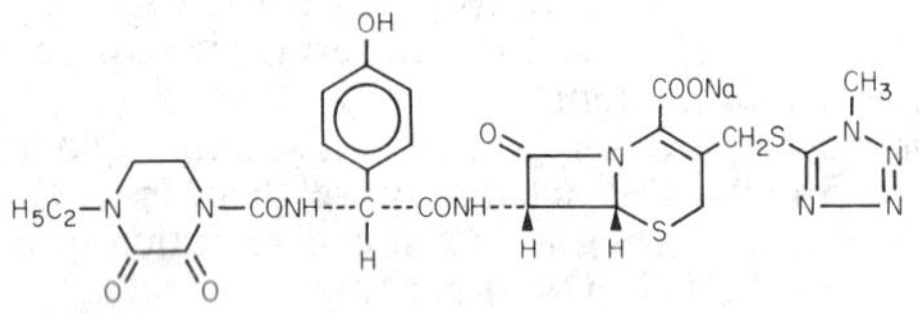

$C_{25}H_{26}N_9NaO_8S_2$ 667.65

5-Thia-1-azabicyclo[4.2.0]oct-2-ene-2-carboxylic acid, 7-[[[[(4-ethyl-2,3-dioxo-1-piperazinyl)carbonyl]amino](4-hydroxyphenyl)acetyl]amino]-3-[[(1-methyl-1*H*-tetrazol-5-yl)thio]methyl]-8-oxo-, monosodium salt, [6*R*-[6α,7β(*R**)]]-.

Sodium (6*R*,7*R*)-7-[(*R*)-2-(4-ethyl-2,3-dioxo-1-piperazinecarboxamido)-2-(*p*-hydroxyphenyl)acetamido-3-[[(1-methyl-*H*-tetrazol-5-yl)thio]methyl]-8-oxo-5-thia-1-azabicyclo[4.2.0]oct-2-ene-2-carboxylate [*62893-20-3*].

» Cefoperazone Sodium contains the equivalent of not less than 870 µg and not more than 1015 µg of cefoperazone ($C_{25}H_{27}N_9O_8S_2$) per mg, calculated on the anhydrous basis.

Packaging and storage—Preserve in tight containers.

Reference standard—*USP Cefoperazone Dihydrate Reference Standard*—Do not dry before using.

Identification—The chromatogram of the *Assay preparation* obtained as directed in the *Assay* exhibits a major peak for cefoperazone the retention time of which corresponds to that exhibited in the chromatogram of the *Standard preparation* obtained as directed in the *Assay*.

Crystallinity ⟨695⟩: meets the requirements. [NOTE—Cefoperazone Sodium in the freeze-dried form is exempt from this requirement.]

pH ⟨791⟩: between 4.5 and 6.5, in a solution (1 in 4).

Water, *Method I* ⟨921⟩: not more than 5.0%, except that where it is in the freeze-dried form, the limit is not more than 2.0%.

Assay—

Mobile phase—Place 14 mL of triethylamine and 5.7 mL of glacial acetic acid in a 100-mL volumetric flask, dilute with water to volume, and mix. Prepare a suitable mixture of this solution, 1 *N* acetic acid, acetonitrile, and water (1.2:2.8:120:876). Filter through a membrane filter (1-µm or finer porosity), and degas.

Standard preparation—Dissolve a suitable quantity of USP Cefoperazone Dihydrate RS, accurately weighed, in *Mobile phase* to obtain a solution having a known concentration of about 0.16 mg of cefoperazone ($C_{25}H_{27}N_9O_8S_2$) per mL.

Assay preparation—Using a suitable quantity of Cefoperazone Sodium, accurately weighed, proceed as directed under *Standard preparation.*

Chromatographic system (see *Chromatography* ⟨621⟩)—The liquid chromatograph is equipped with a 254-nm detector and a 4.0-mm × 30-cm column that contains packing L1. The flow rate is about 2 mL per minute. Chromatograph the *Standard preparation*, and record the peak responses as directed under *Procedure:* the relative standard deviation for replicate injections is not more than 2.0%, and the tailing factor is not more than 1.5.

Procedure—Separately inject equal volumes (about 10 µL) of the *Standard preparation* and the *Assay preparation* into the chromatograph, record the chromatograms, and measure the responses for the major peaks. Calculate the quantity, in µg of cefoperazone per mg, of the Cefoperazone Sodium taken by the formula:

$$1000(C/M)(r_U/r_S),$$

in which C is the concentration, in mg of cefoperazone ($C_{25}H_{27}N_9O_8S_2$) per mL, of the *Standard preparation, M* is the concentration, in mg per mL, of the *Assay preparation* based on the weight of Cefoperazone Sodium taken and the extent of dilution, and r_U and r_S are the peak responses from the *Assay preparation* and the *Standard preparation*, respectively.

Cefoperazone Sodium Injection

» Cefoperazone Sodium Injection is a sterile solution of Cefoperazone Sodium and a suitable osmolality adjusting substance in Water for Injection. It may contain a suitable buffer. It contains the equivalent of not less than 90.0 percent and not more than 120.0 percent of the labeled amount of cefoperazone ($C_{25}H_{27}N_9O_8S_2$).

Packaging and storage—Preserve in *Containers for Injections* as described under *Injections* ⟨1⟩. Maintain in the frozen state.

Labeling—It meets the requirements for *Labeling* under *Injections* ⟨1⟩. The label states that it is to be thawed just prior to use, describes conditions for proper storage of the resultant solution, and directs that the solution is not to be refrozen.

Reference standard—*USP Cefoperazone Dihydrate Reference Standard*—Do not dry before using.

Identification—The chromatogram of the *Assay preparation* obtained as directed in the *Assay* exhibits a major peak for cefoperazone the retention time of which corresponds to that exhibited in the chromatogram of the *Standard preparation* obtained as directed in the *Assay*.

Pyrogen—It meets the requirements of the *Pyrogen Test* ⟨151⟩, the test dose being a volume of undiluted Injection providing the equivalent of 10 mg of cefoperazone per kg.

Sterility—It meets the requirements under *Sterility Tests* ⟨71⟩, when tested as directed in the section, *Test Procedures Using Membrane Filtration.*

pH ⟨791⟩: between 4.5 and 6.5.

Particulate matter ⟨788⟩: meets the requirements under *Small-volume Injections.*

Assay—

Mobile phase, Standard preparation, and *Chromatographic system*—Prepare as directed in the *Assay* under *Cefoperazone Sodium.*

Assay preparation—Dilute an accurately measured volume of Injection quantitatively with *Mobile phase* to obtain a solution containing about 0.16 mg of cefoperazone per mL.

Procedure—Proceed as directed for *Procedure* in the *Assay* under *Cefoperazone Sodium.* Calculate the quantity, in mg, of cefoperazone ($C_{25}H_{27}N_9O_8S_2$), in the volume of Injection taken by the formula:

$$(L/D)(C)(r_U/r_S),$$

in which L is the labeled quantity, in mg, of cefoperazone ($C_{25}H_{27}N_9O_8S_2$), in the volume of Injection taken, D is the concentration, in mg of cefoperazone ($C_{25}H_{27}N_9O_8S_2$) per mL, of the *Assay preparation*, based on the labeled quantity in the portion of Injection taken and the extent of dilution, and the other terms are as defined therein.

Sterile Cefoperazone Sodium

» Sterile Cefoperazone Sodium is Cefoperazone Sodium suitable for parenteral use. It contains the equivalent of not less than 870 µg and not more than 1015 µg of cefoperazone ($C_{25}H_{27}N_9O_8S_2$) per mg, calculated on the anhydrous basis. In addition, where packaged for dispensing, it contains the equivalent of not less than 90.0 percent and not more than 120.0 percent of the labeled amount of cefoperazone ($C_{25}H_{27}N_9O_8S_2$).

Packaging and storage—Preserve in *Containers for Sterile Solids* as described under *Injections* ⟨1⟩.

Reference standard—*USP Cefoperazone Dihydrate Reference Standard*—Do not dry before using.

Constituted solution—At the time of use, the constituted solution prepared from Sterile Cefoperazone Sodium meets the requirements for *Constituted Solutions* under *Injections* ⟨1⟩.

Identification—The chromatogram of the *Assay preparation* obtained as directed in the *Assay* exhibits a major peak for cefoperazone the retention time of which corresponds to that exhibited in the chromatogram of the *Standard preparation* obtained as directed in the *Assay*.

Crystallinity ⟨695⟩: meets the requirements. [NOTE—Sterile Cefoperazone Sodium packaged for dispensing in the freeze-dried form is exempt from this requirement.]

Pyrogen—It meets the requirements of the *Pyrogen Test* ⟨151⟩, the test dose being 1.0 mL per kg of a solution prepared by diluting Sterile Cefoperazone Sodium with Sterile Water for Injection to a concentration of 10 mg of cefoperazone per mL.

Sterility—It meets the requirements under *Sterility Tests* ⟨71⟩, when tested as directed in the section, *Test Procedures Using Membrane Filtration.*

pH ⟨791⟩: between 4.5 and 6.5, in a solution (1 in 4).

Water, *Method I* ⟨921⟩: not more than 5.0%, except that where it is packaged for dispensing in the freeze-dried form, the limit is not more than 2.0%.

Particulate matter ⟨788⟩: meets the requirements under *Small-volume Injections.*

Other requirements—It meets the requirements for *Uniformity of Dosage Units* ⟨905⟩ and for *Labeling* under *Injections* ⟨1⟩.

Assay—

Mobile phase, Standard preparation, and *Chromatographic system*—Prepare as directed in the *Assay* under *Cefoperazone Sodium.*

Assay preparation 1—Dissolve an accurately weighed quantity of Sterile Cefoperazone Sodium quantitatively in *Mobile phase* to obtain a solution having a concentration of about 0.16 mg of cefoperazone ($C_{25}H_{27}N_9O_8S_2$) per mL.

Assay preparation 2 (where it is represented as being in a single-dose container)—Constitute Sterile Cefoperazone Sodium in a volume of water, accurately measured, corresponding to the volume of solvent specified in the labeling. Withdraw all of the withdrawable contents, using a suitable hypodermic needle and syringe, and dilute quantitatively with *Mobile phase* to obtain a solution containing about 0.16 mg of cefoperazone per mL.

Assay preparation 3 (where the label states the quantity of cefoperazone in a given volume of constituted solution)—Constitute Sterile Cefoperazone Sodium in a volume of water, accurately measured, corresponding to the volume of solvent specified in the labeling. Dilute an accurately measured volume of the constituted solution quantitatively with *Mobile phase* to obtain a solution containing about 0.16 mg of cefoperazone per mL.

Procedure—Separately inject equal volumes (about 10 µL) of the *Standard preparation* and the *Assay preparation* into the chromatograph, record the chromatograms, and measure the responses for the major peaks. Calculate the quantity, in µg of cefoperazone per mg, of the Sterile Cefoperazone Sodium taken by the formula:

$$1000(C/M)(r_U/r_S),$$

in which C is the concentration, in mg of cefoperazone ($C_{25}H_{27}N_9O_8S_2$) per mL, of the *Standard preparation, M* is the concentration, in mg per mL, of the *Assay preparation* based on the weight of Sterile Cefoperazone Sodium taken and the extent of dilution, and r_U and r_S are the peak responses from the *Assay preparation* and the *Standard preparation,* respectively. Calculate the quantity, in mg, of cefoperazone ($C_{25}H_{27}N_9O_8S_2$), withdrawn from the container, or in the portion of constituted solution taken by the formula:

$$(L/D)(C)(r_U/r_S),$$

in which L is the labeled quantity, in mg, of cefoperazone ($C_{25}H_{27}N_9O_8S_2$), in the container, or in the volume of constituted solution taken, and D is the concentration, in mg of cefoperazone ($C_{25}H_{27}N_9O_8S_2$) per mL, of *Assay preparation 2* or *Assay preparation 3,* based on the labeled quantity in the container or in the portion of constituted solution taken, respectively, and the extent of dilution.

Ceforanide for Injection

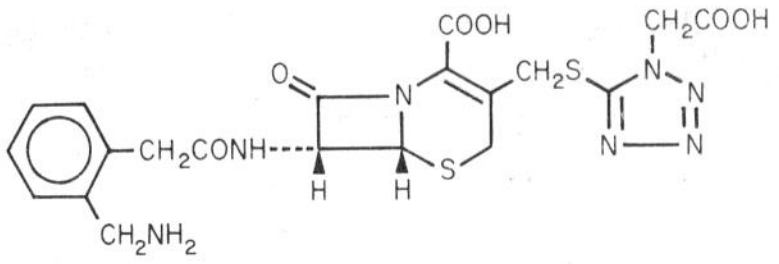

» Ceforanide for Injection is a sterile mixture of Sterile Ceforanide and L-Lysine. It contains not less than 900 µg and not more than 1050 µg of ceforanide ($C_{20}H_{21}N_7O_6S_2$) per mg on the L-Lysine-free basis, and not less than 90.0 percent and not more than 115.0 percent of the labeled amount of $C_{20}H_{21}N_7O_6S_2$.

Packaging and storage—Preserve in *Containers for Sterile Solids* as described under *Injections* ⟨1⟩.

Reference standard—*USP Ceforanide Reference Standard*—Do not dry before using.

Identification—

A: The chromatogram of the *Test preparation* obtained as directed in the test for *L-Lysine content* exhibits a major peak for L-lysine, the retention time of which corresponds to that exhibited in the chromatogram of the *Standard preparation* obtained as directed in the test for *L-Lysine content.*

B: The chromatogram of the *Assay preparation* obtained as directed in the *Assay* exhibits a major peak for ceforanide, the retention time of which corresponds to that exhibited in the chromatogram of the *Standard preparation* obtained as directed in the *Assay.*

Pyrogen—It meets the requirements of the *Pyrogen Test* ⟨151⟩, the test dose being 1 mL per kg of a solution prepared by diluting Ceforanide for Injection with Sterile Water for Injection to a concentration of 50 mg of ceforanide per mL.

Sterility—It meets the requirements under *Sterility Tests* ⟨71⟩, when tested as directed in the section, *Test Procedures Using Membrane Filtration,* except to constitute each container with 3 mL of Fluid A for each g of ceforanide contained therein, and to rinse the membrane with three 100-mL portions of *Fluid D* and one 100-mL portion of *Fluid A.*

pH ⟨791⟩: between 5.5 and 8.5, when constituted as directed in the labeling.

Water, *Method I* ⟨921⟩: not more than 3.0%.

Particulate matter ⟨788⟩: meets the requirements under *Small-volume Injections.*

L-Lysine content—

Mobile phase—Mix 62 volumes of methanol and 38 volumes of water, and adjust with glacial acetic acid to a pH of 3.0, making adjustments if necessary (see *System Suitability* under *Chromatography* ⟨621⟩).

Stock standard solution—Transfer about 36 mg of L-lysine, accurately weighed, to a 100-mL volumetric flask, dilute with water to volume, and mix.

Standard preparation—Transfer 2.0 mL of *Stock standard solution* to a glass-stoppered, 10-mL volumetric flask, add 2.0 mL of a 1.4% solution of tris(hydroxymethyl)aminomethane and 3.0 mL of a 1.5% solution of 1-fluoro-2,4-dinitrobenzene in dehydrated alcohol, insert the stopper tightly, and mix. Heat at 50° in a water bath for 30 minutes. Remove the flask from the water bath, allow to cool, dilute with methanol to volume, and mix.

Test preparation—Transfer about 150 mg of Ceforanide for Injection, accurately weighed, to a 100-mL volumetric flask, add water to volume, and mix. Transfer 2.0 mL of the resulting solution to a glass-stoppered, 10-mL volumetric flask, and proceed as directed under *Standard preparation,* beginning with "add 2.0 mL of a 1.4% solution of tris(hydroxymethyl)aminomethane."

Chromatographic system (see *Chromatography* ⟨621⟩)—The liquid chromatograph is equipped with a 254-nm detector and a 4.6-mm × 25-cm column that contains 5- to 10-μm packing L1. The flow rate is about 1.5 mL per minute. Chromatograph the *Standard preparation,* and record the peak responses as directed under *Procedure:* the column efficiency determined from the derivatized L-lysine peak is not less than 1500 theoretical plates, the tailing factor for the same peak is not more than 1.3, the resolution, *R,* between the derivatized L-lysine peak and the 1-fluoro-2,4-dinitrobenzene peak is not less than 4.5, the capacity factor, *k′*, is not less than 4 and not more than 6, and the relative standard deviation for replicate injections is not more than 1.5%.

Procedure—Separately inject equal volumes (about 20 μL) of the *Standard preparation* and the *Test preparation* into the chromatograph, record the chromatograms, and measure the responses for the major peaks. Calculate the percentage of L-lysine in the Ceforanide for Injection by the formula:

$$10(C/M)(r_U/r_S),$$

in which *C* is the concentration, in μg per mL, of L-lysine in the *Stock standard solution,* *M* is the quantity, in mg, of Ceforanide for Injection taken, and r_U and r_S are the peak responses obtained from the *Test preparation* and the *Standard preparation,* respectively. Use this percentage to calculate, on an L-lysine–free basis, the result from *Assay preparation 1* obtained as directed in the *Assay.*

Other requirements—It meets the requirements for *Uniformity of Dosage Units* ⟨905⟩ and for *Labeling* under *Injections* ⟨1⟩.

Assay—

Mobile phase, Standard preparation, and *Chromatographic system*—Prepare as directed in the *Assay* under *Ceforanide.*

Assay preparation 1—Dissolve a suitable quantity of Ceforanide for Injection, accurately weighed, in *Mobile phase,* and dilute quantitatively and stepwise with *Mobile phase* to obtain a solution having a concentration of about 1 mg of ceforanide per mL. Use this solution within 5 minutes.

Assay preparation 2 (where it is represented as being in a single-dose container)—Constitute Ceforanide for Injection in a volume of water, accurately measured, corresponding to the volume of solvent specified in the labeling. Withdraw all of the withdrawable contents, using a suitable hypodermic needle and syringe, and dilute quantitatively and stepwise with *Mobile phase* to obtain a solution containing about 1 mg of ceforanide per mL. Use this solution within 5 minutes.

Assay preparation 3 (where the label states the quantity of ceforanide in a given volume of constituted solution)—Constitute Ceforanide for Injection in a volume of water, accurately measured, corresponding to the volume of solvent specified in the labeling. Dilute an accurately measured volume of the constituted solution quantitatively and stepwise with *Mobile phase* to obtain a solution containing about 1 mg of ceforanide per mL. Use this solution within 5 minutes.

Procedure—Proceed as directed for *Procedure* in the *Assay* under *Ceforanide.* Calculate the quantity, in μg, of ceforanide ($C_{20}H_{21}N_7O_6S_2$) in each mg of the Ceforanide for Injection taken by the formula:

$$(CP/M)(r_U/r_S),$$

in which *M* is the concentration, in mg per mL, of *Assay preparation 1* based on the weight of Ceforanide for Injection taken and the extent of dilution, and the other terms are as defined therein. Calculate the quantity, in mg, of $C_{20}H_{21}N_7O_6S_2$ withdrawn from the container, or in the portion of constituted solution taken by the formula:

$$(L/D)(CP/1000)(r_U/r_S),$$

in which *L* is the labeled quantity, in mg, of ceforanide in the container, or in the volume of constituted solution taken, *D* is the concentration, in mg of ceforanide per mL, of *Assay preparation 2* or *Assay preparation 3,* based on the labeled quantity in the container or in the portion of constituted solution taken, respectively, and the extent of dilution, and the other terms are as defined therein.

Sterile Ceforanide

$C_{20}H_{21}N_7O_6S_2$ 519.55

5-Thia-1-azabicyclo[4.2.0]oct-2-ene-2-carboxylic acid, 7-[[[2-(aminomethyl)phenyl]acetyl]amino]-3-[[[1-(carboxymethyl)-1*H*-tetrazol-5-yl]thio]methyl]-8-oxo-, (6*R*-trans)-.

(6*R*,7*R*)-7-[2-(α-Amino-*o*-tolyl)acetamido]-3-[[[1-(carboxymethyl)-1*H*-tetrazol-5-yl]thio]methyl]-8-oxo-5-thia-1-azabicyclo[4.2.0]oct-2-ene-2-carboxylic acid.

7-[*o*-(Aminomethyl)phenylacetamido]-3-[[[1-(carboxymethyl)-1*H*-tetrazol-5-yl]thio]methyl]-3-cephem-4-carboxylic acid [*60925-61-3*].

» Sterile Ceforanide is ceforanide suitable for parenteral use. It contains not less than 900 μg and not more than 1050 μg of ceforanide ($C_{20}H_{21}N_7O_6S_2$) per mg.

Packaging and storage—Preserve in *Containers for Sterile Solids* as described under *Injections* ⟨1⟩.

Reference standard—*USP Ceforanide Reference Standard*—Do not dry before using.

Identification—

A: The infrared absorption spectrum of a mineral oil dispersion of it exhibits maxima only at the same wavelengths as that of a similar preparation of USP Ceforanide RS.

B: The chromatogram of the *Assay preparation* obtained as directed in the *Assay* exhibits a major peak for ceforanide, the retention time of which corresponds to that exhibited in the chromatogram of the *Standard preparation* obtained as directed in the *Assay.*

Pyrogen—Suspend 1.0 g of Sterile Ceforanide in 12.5 mL of Sterile Water for Injection, add 320 mg of L-lysine, previously tested and shown to meet the requirements of the *Pyrogen Test* ⟨151⟩, and mix. If the Sterile Ceforanide does not dissolve completely, add an additional amount of the L-lysine, not to exceed 20 mg, shake to dissolve, and dilute with Sterile Water for Injection to obtain 20 mL of test solution. This solution meets the requirements of the *Pyrogen Test* ⟨151⟩, the test dose being 1 mL per kg.

Sterility—It meets the requirements under *Sterility Tests* ⟨71⟩, when tested as directed in the section, *Test Procedures Using Membrane Filtration,* except to dissolve 6 g of Sterile Ceforanide in *Fluid A* to each 1000 mL of which has been added 10 g of sterile L-lysine, and to rinse the membrane with three 100-mL portions of *Fluid D* and one 100-mL portion of *Fluid A.*

pH ⟨791⟩: between 2.5 and 4.5, in a suspension containing 50 mg per mL.

Water, *Method I* ⟨921⟩: not more than 5.0%.

Assay—

Mobile phase—Mix 18 mL of tetrabutylammonium hydroxide solution (1 in 10) and 8.6 mL of 11 *N* potassium hydroxide, and add the mixture to 700 mL of water. Add 200 mL of methanol, adjust with phosphoric acid to a pH of 7.0, and add water to obtain 1000 mL of solution, making adjustments if necessary (see *System Suitability* under *Chromatography* ⟨621⟩). Filter, using a filter having a porosity of 1 µm or finer, and degas.

Standard preparation—Dissolve an accurately weighed quantity of USP Ceforanide RS in *Mobile phase* to obtain a solution having a known concentration of about 1 mg per mL. Use this solution within 5 minutes.

Assay preparation—Using a suitable quantity of Sterile Ceforanide, accurately weighed, proceed as directed under *Standard preparation.* Use this solution within 5 minutes.

Chromatographic system (see *Chromatography* ⟨621⟩)—The liquid chromatograph is equipped with a 254-nm detector and a 4-mm × 30-cm column that contains 5- to 10-µm packing L1. The flow rate is about 1 mL per minute. Chromatograph the *Standard preparation,* and record the peak responses as directed under *Procedure:* the column efficiency determined from the analyte peak is not less than 1900 theoretical plates, the tailing factor for the analyte peak is not more than 1.2, the capacity factor, k', is not less than 1.8 and not more than 5.0, and the relative standard deviation for replicate injections is not more than 1.5%.

Procedure—Separately inject equal volumes (about 10 µL) of the *Standard preparation* and the *Assay preparation* into the chromatograph, record the chromatograms, and measure the responses for the major peaks. Calculate the quantity, in µg, of $C_{20}H_{21}N_7O_6S_2$ in each mg of the Sterile Ceforanide taken by the formula:

$$(CP/M)(r_U/r_S),$$

in which C is the concentration, in mg per mL, of USP Ceforanide RS in the *Standard preparation, P* is the potency, in µg per mg, of the USP Ceforanide RS, M is the concentration, in mg per mL, of the *Assay preparation,* based on the amount of Sterile Ceforanide taken and the extent of dilution, and r_U and r_S are the peak responses obtained from the *Assay preparation* and the *Standard preparation,* respectively.

Cefotaxime Sodium

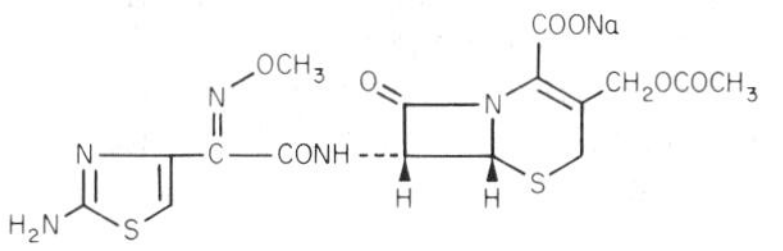

$C_{16}H_{16}N_5NaO_7S_2$ 477.44

5-Thia-1-azabicyclo[4.2.0]oct-2-ene-2-carboxylic acid, 3-[(acetyloxy)methyl]-7-[[(2-amino-4-thiazolyl)(methoxyimino)-acetyl]amino]-8-oxo, monosodium salt, [6*R*-[6α,7β(*Z*)]]-.

Sodium (6*R*,7*R*)-7-[2-(2-amino-4-thiazolyl)glyoxylamido]-3-(hydroxymethyl)-8-oxo-5-thia-1-azabicyclo[4.2.0]oct-2-ene-2-carboxylate 7^2-(*Z*)-(*O*-methyloxime), acetate (ester) [*64485-93-4*].

» Cefotaxime Sodium contains the equivalent of not less than 855 µg and not more than 1002 µg of cefotaxime ($C_{16}H_{17}N_5O_7S_2$) per mg, calculated on the anhydrous basis.

Packaging and storage—Preserve in tight containers.

Reference standard—*USP Cefotaxime Sodium Reference Standard*—Do not dry before using.

Identification—The chromatogram of the *Assay preparation* obtained as directed in the *Assay* exhibits a major peak for cefotaxime the retention time of which corresponds to that exhibited in the chromatogram of the *Standard preparation* obtained as directed in the *Assay.*

pH ⟨791⟩: between 4.5 and 6.5, in a solution (1 in 10).

Water, *Method I* ⟨921⟩: not more than 6.0%.

Assay—

Mobile phase—Dissolve 60 mg of monobasic potassium phosphate and 1.2 g of dibasic sodium phosphate in 1000 mL of water, add 120 mL of methanol, and mix. Filter through a membrane filter (0.5-µm or finer porosity), and degas. Make adjustments if necessary (see *System Suitability* under *Chromatography* ⟨621⟩).

Standard preparation—Dissolve a suitable quantity of USP Cefotaxime Sodium RS, accurately weighed, in water to obtain a stock solution having a known concentration of about 1000 µg of cefotaxime per mL. Transfer 10.0 mL of this stock solution to a 100-mL volumetric flask, dilute with water to volume, and mix.

Resolution preparation—Mix 1 mL of the stock solution used to prepare the *Standard preparation* and 9 mL of *Mobile phase,* add 10 mg of sodium carbonate, mix, and heat at 40° for 5 minutes. Cool in an ice bath, add 1 drop of glacial acetic acid and 1 mL of the stock solution used to prepare the *Standard preparation,* and mix.

Assay preparation—Dissolve a suitable quantity of Cefotaxime Sodium, accurately weighed, in water to obtain a solution having a concentration of about 0.1 mg of cefotaxime per mL.

Chromatographic system (see *Chromatography* ⟨621⟩)—The liquid chromatograph is equipped with a 254-nm detector and a 3.9-mm × 30-cm column that contains packing L1. The flow rate is about 1.5 mL per minute. Chromatograph 10 µL of the *Resolution preparation,* and record the peak responses as directed under *Procedure:* the relative retention times are about 0.4 for desacetylcefotaxime and 1.0 for cefotaxime, and the resolution, R, between the desacetylcefotaxime peak and the cefotaxime peak is not less than 5.0. Chromatograph the *Standard preparation,* and record the peak responses as directed under *Procedure:* the column efficiency is not less than 1200 theoretical plates when calculated by the formula:

$$5.545(t_r/W_{h/2})^2,$$

the tailing factor is not less than 0.8 and not more than 2, when calculated by the formula:

$$W_{0.1}/2f,$$

in which $W_{0.1}$ is width of peak at 10% height, and the relative standard deviation for replicate injections is not more than 2.0%.

Procedure—Separately inject equal volumes (about 10 µL) of the *Standard preparation* and the *Assay preparation* into the chromatograph, record the chromatograms, and measure the responses for the major peaks. Calculate the quantity, in µg, of cefotaxime ($C_{16}H_{17}N_5O_7S_2$) in each mg of the Cefotaxime Sodium taken by the formula:

$$(C/M)(r_U/r_S),$$

in which C is the concentration, in µg of cefotaxime per mL, of the *Standard preparation, M* is the weight, in mg, of Cefotaxime Sodium taken in each mL of the *Assay preparation,* based on the weight of Cefotaxime Sodium taken and the extent of dilution, and r_U and r_S are the peak responses from the *Assay preparation* and the *Standard preparation,* respectively.

Cefotaxime Sodium Injection

» Cefotaxime Sodium Injection is a sterile solution of Cefotaxime Sodium in Water for Injection. It contains one or more suitable buffers, and it may contain Dextrose or Sodium Chloride as a tonicity-adjusting agent. It contains the equivalent of not less than 90.0 percent and not more than 110.0 percent of the labeled amount of cefotaxime ($C_{16}H_{17}N_5O_7S_2$).

Packaging and storage—Preserve in single-dose containers, as described under *Injections* ⟨1⟩. Maintain in the frozen state.

Labeling—It meets the requirements for *Labeling* under *Injections* ⟨1⟩. The label states that it is to be thawed just prior to use, describes conditions for proper storage of the resultant solution, and directs that the solution is not to be refrozen.

Reference standard—*USP Cefotaxime Sodium Reference Standard*—Do not dry before using.

Identification—The chromatogram of the *Assay preparation* obtained as directed in the *Assay* exhibits a major peak for cefotaxime the retention time of which corresponds to that exhibited in the chromatogram of the *Standard preparation* obtained as directed in the *Assay*.

Pyrogen—It meets the requirements of the *Pyrogen Test* ⟨151⟩, the test dose being a volume of undiluted Injection providing the equivalent of 50 mg of cefotaxime per kg.

Sterility—It meets the requirements under *Sterility Tests* ⟨71⟩, when tested as directed in the section, *Test Procedures Using Membrane Filtration*.

pH ⟨791⟩: between 5.0 and 7.5.

Particulate matter ⟨788⟩: meets the requirements under *Small-volume Injections*.

Assay—

Mobile phase, Standard preparation, Resolution preparation, and *Chromatographic system*—Prepare as directed in the *Assay* under *Cefotaxime Sodium*.

Assay preparation—Allow 1 container of Cefotaxime Sodium Injection to thaw, and mix. Dilute an accurately measured volume of the Injection quantitatively and stepwise with water to obtain a solution containing the equivalent of about 0.1 mg of cefotaxime per mL.

Procedure—Proceed as directed for *Procedure* in the *Assay* under *Cefotaxime Sodium*. Calculate the quantity, in mg, of cefotaxime ($C_{16}H_{17}N_5O_7S_2$) in each mL of the Injection taken by the formula:

$$(L/D)(C/1000V)(r_U/r_S),$$

in which L is the labeled quantity, in mg, of cefotaxime per mL of Injection, D is the concentration, in mg per mL, of cefotaxime in the *Assay preparation*, based on the labeled quantity in the portion of Injection taken and the extent of dilution, C is the concentration, in µg of cefotaxime per mL, of the *Standard preparation*, V is the volume, in mL, of Injection taken, and r_U and r_S are the peak responses from the *Assay preparation* and the *Standard preparation*, respectively.

Sterile Cefotaxime Sodium

» Sterile Cefotaxime Sodium is Cefotaxime Sodium suitable for parenteral use. It contains the equivalent to not less than 855 µg and not more than 1002 µg of cefotaxime ($C_{16}H_{17}N_5O_7S_2$) per mg, calculated on the anhydrous basis. In addition, where packaged for dispensing, it contains the equivalent of not less than 90.0 percent and not more than 110.0 percent of the labeled amount of cefotaxime ($C_{16}H_{17}N_5O_7S_2$).

Packaging and storage—Preserve in *Containers for Sterile Solids* as described under *Injections* ⟨1⟩.

Reference standard—*USP Cefotaxime Sodium Reference Standard*—Do not dry before using.

Constituted solution—At the time of use, the constituted solution prepared from Sterile Cefotaxime Sodium meets the requirements for *Constituted Solutions* under *Injections* ⟨1⟩.

Pyrogen—It meets the requirements of the *Pyrogen Test* ⟨151⟩, the test dose being 1 mL per kg of a solution prepared by diluting Sterile Cefotaxime Sodium with Sterile Water for Injection to a concentration of 50 mg of cefotaxime per mL.

Sterility—It meets the requirements under *Sterility Tests* ⟨71⟩, when tested as directed in the section, *Test Procedures Using Membrane Filtration*.

Particulate matter ⟨788⟩: meets the requirements under *Small-volume Injections*.

Other requirements—It responds to the *Identification test* and meets the requirements for *pH* and *Water* under *Cefotaxime Sodium*. In addition, where packaged for dispensing, it meets the requirements for *Uniformity of Dosage Units* ⟨905⟩ and for *Labeling* under *Injections* ⟨1⟩.

Assay—

Mobile phase, Standard preparation, Resolution preparation, and *Chromatographic system*—Prepare as directed in the *Assay* under *Cefotaxime Sodium*.

Assay preparation 1—Dissolve a suitable quantity of Sterile Cefotaxime Sodium, accurately weighed, in water to obtain a solution having a concentration of about 0.1 mg of cefotaxime per mL.

Assay preparation 2 (where it is packaged for dispensing and is represented as being in a single-dose container)—Constitute 1 container of Sterile Cefotaxime Sodium in a volume of water, accurately measured, corresponding to the largest volume of solvent specified in the labeling. Withdraw all of the withdrawable contents, using a suitable hypodermic needle and syringe, and dilute quantitatively with water to obtain a solution having a concentration of about 0.1 mg of cefotaxime per mL.

Assay preparation 3 (where the label states the quantity of cefotaxime in a given volume of constituted solution)—Constitute 1 container of Sterile Cefotaxime Sodium in a volume of water, accurately measured, corresponding to the largest volume of solvent specified in the labeling. Dilute an accurately measured portion of the constituted solution quantitatively with water to obtain a solution having a concentration of about 0.1 mg of cefotaxime per mL.

Procedure—Proceed as directed for *Procedure* in the *Assay* under *Cefotaxime Sodium*. Calculate the quantity, in µg, of cefotaxime ($C_{16}H_{17}N_5O_7S_2$) in each mg of the Sterile Cefotaxime Sodium taken by the formula:

$$(C/M)(r_U/r_S),$$

in which M is the weight, in mg, of the Sterile Cefotaxime Sodium taken in each mL of *Assay preparation 1*, and the other terms are as defined therein. Calculate the quantity, in mg, of cefotaxime in the container, and in the portion of constituted solution taken, by the formula:

$$(LC/1000D)(r_U/r_S),$$

in which L is the labeled quantity, in mg, of cefotaxime in the container, or in the volume of constituted solution taken, and D is the concentration, in mg per mL, of cefotaxime in *Assay preparation 2* or in *Assay preparation 3*, on the basis of the labeled quantity in the container, or in the portion of constituted solution taken, respectively, and the extent of dilution, and the other terms are as defined therein.

Sterile Cefotetan Disodium

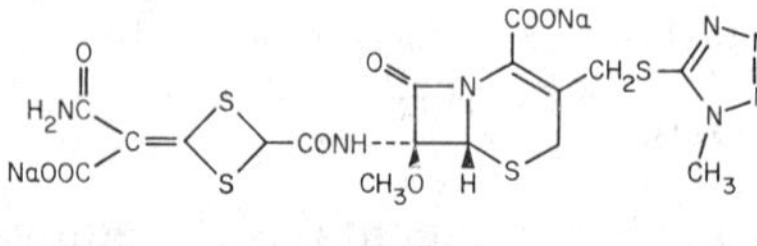

$C_{17}H_{15}N_7Na_2O_8S_4$ 619.57

5-Thia-1-azabicyclo[4.2.0]oct-2-ene-2-carboxylic acid, 7-[[[4-(2-amino-1-carboxy-2-oxoethylidene)-1,3-dithietan-2-yl]carbonyl]amino]-7-methoxy-3-[[(1-methyl-1*H*-tetrazol-5-yl)thio]methyl]-8-oxo-, disodium salt, [6*R*-(6α,7α)]-.

(6*R*,7*S*)-4-[[2-Carboxy-7-methoxy-3-[[(1-methyl-1*H*-tetrazol-5-yl)thio]methyl]-8-oxo-5-thia-1-azabicyclo[4.2.0]oct-2-en-7-yl]carbamoyl]-1,3-dithietane-$\Delta^{2,\alpha}$-malonamic acid, disodium salt.

(6*R*,7*S*)-7-[4-(Carbamoylcarboxymethylene)-1,3-dithietane-2-carboxamido]-7-methoxy-3-[[(1-methyl-1*H*-tetrazol-5-yl)thio]methyl]-8-oxo-5-thia-1-azabicyclo[4.2.0]oct-2-ene-2-carboxylic acid, disodium salt [*74356-00-6*].

» Sterile Cefotetan Disodium is cefotetan disodium suitable for parenteral use. It contains the equivalent of not less than 830 µg and not more than 970 µg of cefotetan ($C_{17}H_{17}N_7O_8S_4$) per mg, calculated on the anhydrous basis. In addition, where packaged for dispensing, it contains the equivalent of not less than 90.0 percent and not more than 120.0 percent of the labeled amount of cefotetan ($C_{17}H_{17}N_7O_8S_4$).

Packaging and storage—Preserve in *Containers for Sterile Solids* as described under *Injections* ⟨1⟩.

Reference standard—*USP Cefotetan Reference Standard*—Do not dry before using.

Constituted solution—At the time of use, the constituted solution prepared from Sterile Cefotetan Disodium meets the requirements for *Constituted Solutions* under *Injections* ⟨1⟩.

Identification—The retention time of the major peak in the chromatogram of the *Assay preparation* corresponds to that of the *Standard preparation*, obtained as directed in the *Assay*.

Bacterial endotoxins—When tested as directed under *Bacterial Endotoxins Test* ⟨85⟩, it contains not more than 0.1 USP Endotoxin Unit per mg of cefotetan.

Sterility—It meets the requirements under *Sterility Tests* ⟨71⟩, when tested as directed in the section, *Test Procedures Using Membrane Filtration*, 6 g being aseptically dissolved in 200 mL of *Fluid A*.

pH ⟨791⟩: between 4.0 and 6.5, in a solution (1 in 10).

Water, *Method I* ⟨921⟩: not more than 1.5%.

Particulate matter ⟨788⟩: meets the requirements under *Small-volume Injections*.

Other requirements—It meets the requirements for *Uniformity of Dosage Units* ⟨905⟩ and for *Labeling* under *Injections* ⟨1⟩.

Assay—[NOTE—Protect the *Standard preparation*, the *Resolution solution*, and the *Assay preparations* from light, and use within 2 hours.]

Mobile phase—Prepare a suitable filtered and degassed mixture of 0.1 *M* phosphoric acid, methanol, acetonitrile, and glacial acetic acid (1700:105:105:100). Make adjustments if necessary (see *System Suitability* under *Chromatography* ⟨621⟩).

Solvent—Prepare a mixture of water, methanol, and acetonitrile (90:5:5).

Standard preparation—Transfer about 40 mg of USP Cefotetan RS, accurately weighed, to a 200-mL volumetric flask, add 10 mL of methanol, swirl for several minutes, add 10 mL of acetonitrile, and swirl until dissolved. Dilute with water to volume, and mix.

Resolution solution—Place 10 mL of *Standard preparation* in a glass-stoppered flask containing a few mg of magnesium carbonate, and sonicate for 10 minutes. If the solution is not turbid, add a few more mg of magnesium carbonate, and repeat the sonication. Filter the turbid solution through a suitable filter of 0.5-µm or finer porosity. Collect the clear filtrate, and use as the *Resolution solution*.

Assay preparation 1—Using a suitable quantity of Sterile Cefotetan Disodium, accurately weighed, proceed as directed under *Standard preparation*.

Assay preparation 2 (where it is packaged for dispensing and where the package is represented as being in a single-dose container)—Constitute Sterile Cefotetan Disodium as directed in the labeling. Withdraw all of the withdrawable contents, and dilute quantitatively with *Solvent* to obtain a solution containing the equivalent of about 200 µg of cefotetan per mL.

Assay preparation 3 (where the label states the quantity of cefotetan in a given volume of constituted solution)—Constitute Sterile Cefotetan Disodium as directed in the labeling. Dilute an accurately measured volume of the constituted solution quantitatively with *Solvent* to obtain a solution containing the equivalent of about 200 µg of cefotetan per mL.

Chromatographic system (*see Chromatography* ⟨621⟩)—The liquid chromatograph is equipped with a 254-nm detector and a 4.6-mm × 25-cm column that contains packing L1. The flow rate is about 1.6 mL per minute. Chromatograph 10 µL of the *Resolution solution*, and record the peak responses as directed under *Procedure*. The relative retention times are about 0.75 for cefotetan and 1.0 for cefotetan tautomer, and the resolution, *R*, between the cefotetan peak and the cefotetan tautomer peak is not less than 2.0. Chromatograph the *Standard preparation*, and record the peak responses as directed under *Procedure:* the column efficiency is not less than 1500 theoretical plates when calculated by the formula:

$$5.545(t_r/W_{h/2})^2,$$

the tailing factor is not more than 1.3 when calculated by the formula:

$$W_{0.1}/2f,$$

in which $W_{0.1}$ is width of peak at 10% height, and the relative standard deviation for replicate injections is not more than 2.0%.

Procedure—[NOTE—Use peak areas where peak responses are indicated.] Separately inject equal volumes (about 20 µL) of the *Standard preparation* and the *Assay preparation* into the chromatograph, record the chromatograms, and measure the responses for the major peaks. Calculate the quantity, in µg, of cefotetan ($C_{17}H_{17}N_7O_8S_4$) per mg of the Sterile Cefotetan Disodium taken by the formula:

$$200(C/M)(r_U/r_S),$$

in which *C* is the concentration, in µg of cefotetan ($C_{17}H_{17}N_7O_8S_4$) per mL, of the *Standard preparation*, *M* is the weight, in mg, of Sterile Cefotetan Disodium taken to prepare *Assay preparation 1*, and r_U and r_S are the peak responses from the *Assay preparation* and the *Standard preparation*, respectively. Calculate the quantity, in mg, of cefotetan ($C_{17}H_{17}N_7O_8S_4$) withdrawn from the container, or in the portion of constituted solution taken by the formula:

$$(L/D)(C)(r_U/r_S),$$

in which *L* is the labeled quantity, in mg, of cefotetan in the container, or in the volume of constituted solution taken, and *D* is the concentration, in µg of cefotetan per mL, of *Assay preparation 2* or *Assay preparation 3*, based on the labeled quantity in the container or in the portion of constituted solution taken, respectively, and the extent of dilution.

Cefoxitin Sodium

COONa
CH₂OCONH₂
CH₂CONH
OCH₃ H

$C_{16}H_{16}N_3NaO_7S_2$ 449.43

5-Thia-1-azabicyclo[4.2.0]oct-2-ene-2-carboxylic acid, 3-[[(aminocarbonyl)oxy]methyl]-7-methoxy-8-oxo-7-[(2-thienylacetyl)amino]-, sodium salt (6*R-cis*)-.

Sodium (6*R*,7*S*)-3-(hydroxymethyl)-7-methoxy-8-oxo-7-[2-(2-thienyl)acetamido]-5-thia-1-azabicyclo-[4.2.0]oct-2-ene-2-carboxylate carbamate (ester) [*33564-30-6; 35607-66-0*].

» Cefoxitin Sodium contains the equivalent of not less than 927 µg and not more than 970 µg of cefoxitin ($C_{16}H_{17}N_3O_7S_2$) per mg, corresponding to not less than 97.5 percent and not more than 102.0 percent of cefoxitin sodium ($C_{16}H_{16}N_3NaO_7S_2$), calculated on the anhydrous and acetone- and methanol-free basis.

Packaging and storage—Preserve in tight containers.

Reference standard—*USP Cefoxitin Reference Standard*—Keep container tightly closed and store in a cold place, protected from light. Do not dry before using.

Identification—

A: The chromatogram of the *Assay preparation* obtained as directed in the *Assay* exhibits a major peak for cefoxitin the retention time of which corresponds to that exhibited in the chromatogram of the *Standard preparation* obtained as directed in the *Assay*.

B: The ultraviolet absorption spectrum of a 1 in 50,000 solution in phosphate buffer, prepared by dissolving 1.0 g of monobasic potassium phosphate and 1.8 g of anhydrous dibasic sodium phosphate in water to make 1000 mL, exhibits maxima and minima at the same wavelengths as that of a similar solution of USP Cefoxitin RS, concomitantly measured.

C: A solution (1 in 20) responds to the tests for *Sodium* ⟨191⟩.

Specific rotation ⟨781⟩: between +206° and +214°, calculated on the anhydrous and acetone- and methanol-free basis, determined in a solution in methanol containing 10 mg in each mL.

Crystallinity ⟨695⟩: meets the requirements.

pH ⟨791⟩: between 4.2 and 7.0, in a solution containing 100 mg per mL.

Water, *Method I* ⟨921⟩: not more than 1.0%, a mixture of ethylene glycol and pyridine (3:1) being used in place of methanol in the titration vessel.

Heavy metals, *Method II* ⟨231⟩: 0.002%.

Acetone and methanol—

Standard preparation—Transfer 5.0 mL of acetone to a 1000-mL volumetric flask, dilute with water to volume, and mix (*Solution A*). Transfer 5.0 mL of methanol to a 1000-mL volumetric flask, dilute with water to volume, and mix (*Solution B*). Transfer 50.0 mL of *Solution A* and 5.0 mL of *Solution B* to a 500-mL volumetric flask, dilute with water to volume, and mix to obtain a solution having concentrations of acetone and methanol of 0.050% and 0.005% (v/v), respectively.

Test preparation—Transfer 5.0 g of Cefoxitin Sodium to a 50-mL volumetric flask, dissolve in water, dilute with water to volume, and mix. Transfer 3.0 mL of the resulting solution to a 15-mL centrifuge tube, cool in an ice-water bath for 2 minutes, and add 3.0 mL of 0.24 *N* hydrochloric acid while swirling vigorously. Centrifuge to obtain a clear solution (*Test preparation*).

Chromatographic system (see *Chromatography* ⟨621⟩)—The gas chromatograph is equipped with a flame-ionization detector, and contains a 1.8-m × 6.3-mm glass column containing support S9, and a pre-column packed with 60- to 80-mesh silane-treated glass beads. The injection port is maintained at 100°, the columns are maintained at 110°, the detector is maintained at 200°, and nitrogen is used as the carrier gas at a flow rate of about 50 mL per minute. Chromatograph the *Standard preparation*, and record the peak responses as directed under *Procedure:* the column efficiency determined from the acetone and methanol peaks is not less than 160 and 200 theoretical plates, respectively, the tailing factors for the acetone and methanol peaks are not more than 1.3 and 2.3, respectively, and the relative standard deviation for replicate injections is not more than 5%.

Procedure—[NOTE—Use peak areas where peak responses are indicated.] Separately inject equal volumes (about 2 µL) of the *Standard preparation* and the *Test preparation* into the chromatograph, record the chromatograms, and measure the acetone and methanol peak responses. Calculate the percentages of acetone and methanol in the Cefoxitin Sodium taken, by the same formula:

$$15.8P(r_U/r_S),$$

in which P is the percentage (v/v) of acetone or methanol in the *Standard preparation*, and r_U and r_S are the acetone or methanol peak responses of the *Test preparation* and the *Standard preparation*, respectively: not more than 0.7% of acetone and 0.1% of methanol are found.

Assay—

Mobile phase—Prepare a suitable mixture of water, acetonitrile, and glacial acetic acid (840:160:10), filter through a membrane filter (1-µm or finer porosity), and degas. Make adjustments if necessary (see *System Suitability* under *Chromatography* ⟨621⟩).

Phosphate buffer—Dissolve 1.0 g of monobasic potassium phosphate and 1.8 g of dibasic sodium phosphate in 900 mL of water, adjust with phosphoric acid or 10 *N* sodium hydroxide to a pH of 7.1 ± 0.1, dilute with water to make 1000 mL, and mix. Filter through a membrane filter of 1-µm or finer porosity.

Standard preparation—Dissolve an accurately weighed quantity of USP Cefoxitin RS in *Phosphate buffer* to obtain a solution having a known concentration of about 0.3 mg per mL. [NOTE—Sonicate, if necessary to dissolve the specimen.] Use this solution within 5 hours.

Assay preparation—Transfer about 150 mg of Cefoxitin Sodium, accurately weighed, to a 500-mL volumetric flask, dissolve in *Phosphate buffer*, dilute with *Phosphate buffer* to volume, and mix. Use this solution within 5 hours.

Chromatographic system (see *Chromatography* ⟨621⟩)—The liquid chromatograph is equipped with a 254-nm detector and a 3.9-mm × 30-cm column that contains 5- to 10-µm packing L1. The flow rate is about 1 mL per minute. Chromatograph the *Standard preparation*, and record the peak responses as directed under *Procedure:* the column efficiency determined from the analyte peak is not less than 2800 theoretical plates, the tailing factor for the analyte peak is not more than 1.5, and the relative standard deviation for replicate injections is not more than 1.0%.

Procedure—Separately inject equal volumes (about 10 µL) of the *Standard preparation* and the *Assay preparation* into the chromatograph, record the chromatograms, and measure the responses for the major peaks. Calculate the quantity, in µg, of cefoxitin ($C_{16}H_{17}N_3O_7S_2$) per mg of the Cefoxitin Sodium taken by the formula:

$$500(CP/W)(r_U/r_S),$$

in which C is the concentration, in mg per mL, of USP Cefoxitin RS the *Standard preparation*, P is the potency, in µg per mg, of the USP Cefoxitin RS, W is the quantity, in mg, of Cefoxitin Sodium taken to prepare the *Assay preparation*, and r_U and r_S are the peak responses obtained from the *Assay preparation* and the *Standard preparation*, respectively.

Cefoxitin Sodium Injection

» Cefoxitin Sodium Injection is a sterile solution of Cefoxitin Sodium and one or more suitable buffer substances in Water for Injection. It contains Dextrose or Sodium Chloride as a tonicity-adjusting agent. It contains the equivalent of not less than 90.0 percent and not more than 120.0 percent of the labeled amount of cefoxitin ($C_{16}H_{17}N_3O_7S_2$).

Packaging and storage—Preserve in *Containers for Injections* as described under *Injections* ⟨1⟩. Maintain in the frozen state.

Labeling—It meets the requirements for *Labeling* under *Injections* ⟨1⟩. The label states that it is to be thawed just prior to use, describes conditions for proper storage of the resultant solution, and directs that the solution is not to be refrozen.

Reference standard—*USP Cefoxitin Reference Standard*—Keep container tightly closed and store in a cold place, protected from light. Do not dry before using.

Identification—The chromatogram of the *Assay preparation* obtained as directed in the *Assay* exhibits a major peak for cefoxitin the retention time of which corresponds to that exhibited in the chromatogram of the *Standard preparation* obtained as directed in the *Assay*.

Pyrogen—It meets the requirements of the *Pyrogen Test* ⟨151⟩, the test dose being a volume of undiluted Injection providing the equivalent of 50 mg of cefoxitin per kg.

Sterility—It meets the requirements under *Sterility Tests* ⟨71⟩, when tested as directed in the section, *Test Procedures Using Membrane Filtration.*

pH ⟨791⟩: between 4.5 and 8.0.

Particulate matter ⟨788⟩: meets the requirements under *Small-volume Injections.*

Assay—

Mobile phase, Phosphate buffer, Standard preparation, and *Chromatographic system*—Prepare as directed in the *Assay* under *Cefoxitin Sodium.*

Assay preparation—Allow 1 container of Cefoxitin Sodium Injection to thaw, and mix. Dilute an accurately measured volume of Injection quantitatively with *Phosphate buffer* to obtain a solution containing about 0.3 mg of cefoxitin per mL. Use this solution within 5 hours.

Procedure—Proceed as directed in the *Assay* under *Cefoxitin Sodium*. Calculate the quantity, in mg, of cefoxitin ($C_{16}H_{17}N_3O_7S_2$) in each mL of the Injection taken by the formula:

$$(CP/1000)(L/D)(r_U/r_S),$$

in which L is the labeled quantity, in mg, of cefoxitin ($C_{16}H_{17}N_3O_7S_2$) in each mL of Injection taken, D is the concentration, in mg per mL, of the *Assay preparation*, based on the volume of Injection taken and the extent of dilution, and the other terms are as defined therein.

Sterile Cefoxitin Sodium

» Sterile Cefoxitin Sodium contains the equivalent of not less than 927 µg and not more than 970 µg of cefoxitin ($C_{16}H_{17}N_3O_7S_2$) per mg, corresponding to not less than 97.5 percent and not more than 102.0 percent of cefoxitin sodium ($C_{16}H_{16}N_3NaO_7S_2$), calculated on the anhydrous and acetone- and methanol-free basis. In addition, where packaged for dispensing, it contains the equivalent of not less than 90.0 percent and not more than 120.0 percent of the labeled amount of cefoxitin.

Packaging and storage—Preserve in *Containers for Sterile Solids* as described under *Injections* ⟨1⟩.

Reference standard—*USP Cefoxitin Reference Standard*—Keep container tightly closed and store in a cold place, protected from light. Do not dry before using.

Constituted solution—At the time of use, the constituted solution prepared from Sterile Cefoxitin Sodium meets the requirements for *Constituted Solutions* under *Injections* ⟨1⟩.

Pyrogen—When diluted with Sterile Water for Injection to a concentration of 50 mg of cefoxitin ($C_{16}H_{17}N_3O_7S_2$) per mL, it meets the requirements of the *Pyrogen Test* ⟨151⟩, the test dose being 1.0 mL per kg.

Sterility—It meets the requirements under *Sterility Tests* ⟨71⟩, when tested as directed in the section, *Test Procedures Using Membrane Filtration*.

Particulate matter ⟨788⟩: meets the requirements under *Small-volume Injections*.

Other requirements—It responds to the *Identification tests* and meets the requirements for *Specific rotation*, *Crystallinity*, *pH*, *Water*, *Heavy metals*, and *Acetone and methanol* under *Cefoxitin Sodium*. In addition, where packaged for dispensing it meets the requirements for *Uniformity of Dosage Units* ⟨905⟩ and for *Labeling* under *Injections* ⟨1⟩.

Assay—

Mobile phase, *Phosphate buffer*, *Standard preparation*, and *Chromatographic system*—Prepare as directed in the *Assay* under *Cefoxitin Sodium*.

Assay preparation 1—Transfer about 150 mg of Sterile Cefoxitin Sodium, accurately weighed, to a 500-mL volumetric flask, dissolve in *Phosphate buffer*, dilute with *Phosphate buffer* to volume, and mix. Use this solution within 5 hours.

Assay preparation 2 (where it is packaged for dispensing and is represented as being in a single-dose container)—Constitute Sterile Cefoxitin Sodium in a volume of water, accurately measured, corresponding to the volume of solvent specified in the labeling. Withdraw all of the withdrawable contents, using a suitable hypodermic needle and syringe, and dilute quantitatively with water to obtain a solution having a concentration of about 0.3 mg of cefoxitin per mL. Use this solution within 5 hours.

Assay preparation 3 (where it is packaged for dispensing and the label states the quantity of cefoxitin in a given volume of constituted solution)—Constitute Sterile Cefoxitin Sodium in a volume of water, accurately measured, corresponding to the volume of solvent specified in the labeling. Dilute an accurately measured volume of the constituted solution quantitatively with water to obtain a solution containing about 0.3 mg of cefoxitin per mL. Use this solution within 5 hours.

Procedure—Proceed as directed in the *Assay* under *Cefoxitin Sodium*. Calculate the potency, in µg of cefoxitin ($C_{16}H_{17}N_3O_7S_2$) per mg, of Sterile Cefoxitin Sodium taken by the formula:

$$500(CP/W)(r_U/r_S),$$

in which W is the quantity, in mg, of Sterile Cefoxitin Sodium taken to prepare *Assay preparation 1*, and the other terms are as defined therein. Calculate the quantity, in mg, of cefoxitin ($C_{16}H_{17}N_3O_7S_2$) in the portion of constituted solution taken by the formula:

$$0.001(CP)(L/D)(r_U/r_S),$$

in which L is the labeled quantity, in mg, in the portion of constituted solution taken, D is the concentration, in mg per mL, of *Assay preparation 2* or *Assay preparation 3*, based on the volume of constituted solution taken and the extent of dilution, and the other terms are as defined therein.

Ceftazidime for Injection

» Ceftazidime for Injection is a sterile mixture of Sterile Ceftazidime and Sodium Carbonate. It contains not less than 90.0 percent and not more than 105.0 percent of ceftazidime ($C_{22}H_{22}N_6O_7S_2$) on the dried and sodium carbonate–free basis, and not less than 90.0 percent and not more than 120.0 percent of the labeled amount of ceftazidime ($C_{22}H_{22}N_6O_7S_2$).

Packaging and storage—Preserve in *Containers for Sterile Solids* as described under *Injections* ⟨1⟩, protected from light.

Reference standards—*USP Ceftazidime Pentahydrate Reference Standard*—Do not dry before using. *USP High Molecular Weight Ceftazidime Polymer Reference Standard*—Do not dry before using. *USP Dextran V_o Marker Reference Standard*—Do not dry before using. *USP Delta-2-Ceftazidime Isomer Reference Standard*—Do not dry before using.

Identification—

A: The chromatogram of the *Assay preparation* obtained as directed in the *Assay* exhibits a major peak for ceftazidime the retention time of which corresponds to that exhibited in the chromatogram of the *Standard preparation* obtained as directed in the *Assay*.

B: It dissolves in 1 *N* hydrochloric acid with effervescence, evolving a colorless gas, which when passed into calcium hydroxide TS produces a white precipitate immediately.

Pyrogen—It meets the requirements of the *Pyrogen Test* ⟨151⟩, the test dose being 1.0 mL per kg of a solution prepared by diluting Ceftazidime for Injection with Sterile Water for Injection to a concentration of 80 mg of ceftazidime per mL.

Sterility—It meets the requirements under *Sterility Tests* ⟨71⟩, when tested as directed in the section, *Test Procedures Using Membrane Filtration*.

pH ⟨791⟩: between 5.0 and 7.5, in a solution constituted in the sealed container, taking care to relieve the pressure inside the container during constitution, containing 100 mg of ceftazidime per mL.

Loss on drying ⟨731⟩—Dry about 300 mg, accurately weighed, in vacuum at a pressure not exceeding 5 mm of mercury at 25° for 4 hours: it loses not more than 13.5% of its weight. Heat the residue in vacuum at a pressure not exceeding 5 mm of mercury at 100° for 3 hours, and calculate the total percentage of weight loss. Use this percentage, m, to calculate, on the dried and sodium carbonate–free basis, the result from *Assay preparation 1* obtained as directed in the *Assay*.

Particulate matter ⟨788⟩: meets the requirements under *Small-volume Injections*.

Sodium carbonate—

Potassium chloride solution—Dissolve 19.07 g of potassium chloride in water to make 1000 mL of solution.

Standard preparation—Dissolve a suitable quantity of sodium chloride, previously dried at 105° for 2 hours and accurately weighed, in water to obtain a solution having a known concentration of about 14 µg per mL. Transfer 10 mL of this solution to a 100-mL volumetric flask, add 10.0 mL of *Potassium chloride solution*, dilute with water to volume, and mix.

Test preparation—Use the stock solution used to prepare *Assay preparation 1* in the *Assay*, diluting it if necessary quantitatively and stepwise with water to obtain a solution containing about 12.5 µg of sodium carbonate per mL. Transfer 10.0 mL of this solution to a 100-mL volumetric flask, add 10.0 mL of *Potassium chloride solution*, dilute with water to volume, and mix.

Blank solution—Transfer 10.0 mL of *Potassium chloride solution* to a 100-mL volumetric flask, dilute with water to volume, and mix.

Procedure—Concomitantly determine the absorbances of the *Standard preparation* and the *Test preparation* at the sodium emission line of 589.0 nm, with a suitable atomic absorption spectrophotometer (see *Spectrophotometry and Light-scattering* ⟨851⟩) equipped with a sodium hollow-cathode lamp and an air-acetylene flame, using the *Blank solution* as the blank. Calculate the percentage of sodium carbonate (Na_2CO_3) in the portion of Ceftazidime for Injection taken by the formula:

$$(105.99/116.89)(0.1C/M)(A_U/A_S),$$

in which 105.99 is the molecular weight of sodium carbonate, 116.89 is twice the molecular weight of sodium chloride, C is the concentration, in µg per mL, of sodium chloride in the *Standard preparation*, M is the quantity, in mg, of Ceftazidime for Injection in each mL of the *Test preparation*, based on the quantity taken to prepare the stock solution and the extent of dilution, and A_U and A_S are the absorbances of the *Test preparation* and the *Standard preparation*, respectively. Use this percentage to calculate, on the dried and sodium carbonate–free basis, the result from *Assay preparation 1* obtained as directed in the *Assay*.

High molecular weight ceftazidime polymer—

Mobile phase, System suitability preparation, Standard preparation, and *Chromatographic system*—Proceed as directed in the test for *High molecular weight ceftazidime polymer* under *Sterile Ceftazidime*.

Test preparation—Transfer an accurately weighed quantity of Ceftazidime for Injection, equivalent to about 100 mg of ceftazidime ($C_{22}H_{22}N_6O_7S_2$), to a 100-mL volumetric flask and add 80 mL of *Mobile phase*. Shake until dissolved, dilute with *Mobile phase* to volume, and mix. Use this solution immediately.

Procedure—Proceed as directed for *Procedure* in the test for *High molecular weight ceftazidime polymer* under *Sterile Ceftazidime*. Calculate the percentage of high molecular weight ceftazidime polymer in the Ceftazidime for Injection taken by the formula:

$$0.1(C/D)(r_U/r_S),$$

in which D is the concentration, in mg per mL, of ceftazidime in the *Test preparation*, based on the results of the *Assay* using *Assay preparation 1*, and the other terms are as defined in the *Procedure* in the test for *High molecular weight ceftazidime polymer* under *Sterile Ceftazidime*: not more than 0.4% is found.

Other requirements—It meets the requirements for *Uniformity of Dosage Units* ⟨905⟩ and for *Labeling* under *Injections* ⟨1⟩.

Assay—

pH 7 buffer, Mobile phase, Standard preparation, Resolution solution, and *Chromatographic system*—Proceed as directed in the *Assay* under *Sterile Ceftazidime*.

Assay preparation 1—Transfer an accurately weighed quantity of Ceftazidime for Injection, equivalent to about 250 mg of ceftazidime ($C_{22}H_{22}N_6O_7S_2$), to a 250-mL volumetric flask, dilute with water to volume, and mix to obtain a stock solution. [NOTE—Protect this solution from light.] Immediately prior to chromatography, transfer 5.0 mL of this solution to a 50-mL volumetric flask, dilute with water to volume, and mix.

Assay preparation 2 (where it is represented as being in a single-dose container)—Constitute a container of Ceftazidime for Injection in a volume of water, accurately measured, corresponding to the volume of solvent specified in the labeling. Withdraw all of the withdrawable contents, using a suitable hypodermic needle and syringe, and dilute quantitatively with water to obtain a solution containing about 1 mg of ceftazidime ($C_{22}H_{22}N_6O_7S_2$) per mL. [NOTE—Protect this solution from light.] Immediately prior to chromatography, transfer 5.0 mL of this solution to a 50-mL volumetric flask, dilute with water to volume, and mix.

Assay preparation 3 (where the label states the quantity of ceftazidime in a given volume of constituted solution)—Constitute a container of Ceftazidime for Injection in a volume of water, accurately measured, corresponding to the volume of solvent specified in the labeling. Dilute an accurately measured volume of the constituted solution quantitatively with water to obtain a solution containing about 1 mg of ceftazidime ($C_{22}H_{22}N_6O_7S_2$) per mL. [NOTE—Protect this solution from light.] Immediately prior to chromatography, transfer 5.0 mL of this solution to a 50-mL volumetric flask, dilute with water to volume, and mix.

Procedure—Proceed as directed for *Procedure* in the *Assay* under *Sterile Ceftazidime*. Calculate the percentage of ceftazidime ($C_{22}H_{22}N_6O_7S_2$) on the dried and sodium carbonate-free basis in the portion of Ceftazidime for Injection taken by the formula:

$$25{,}000[C/W(100 - m - s)](r_U/r_S),$$

in which C is the concentration, in µg per mL, of ceftazidime ($C_{22}H_{22}N_6O_7S_2$) in the *Standard preparation*, W is the quantity, in mg, of Ceftazidime for Injection taken to prepare *Assay preparation 1*, m is the total percentage of loss on drying, s is the percentage of sodium carbonate in the Ceftazidime for Injection taken, and r_U and r_S are the peak responses obtained from the *Assay preparation* and the *Standard preparation*, respectively. Calculate the quantity, in mg of ceftazidime ($C_{22}H_{22}N_6O_7S_2$), withdrawn from the container, or in the portion of constituted solution taken by the formula:

$$(L/D)(C)(r_U/r_S),$$

in which L is the labeled quantity, in mg of ceftazidime ($C_{22}H_{22}N_6O_7S_2$), in the container, or in the volume of constituted solution taken, and D is the concentration, in µg of ceftazidime ($C_{22}H_{22}N_6O_7S_2$) per mL, of *Assay preparation 2* or *Assay preparation 3*, based on the labeled quantity in the container or in the portion of constituted solution taken, respectively, and the extent of dilution.

Sterile Ceftazidime

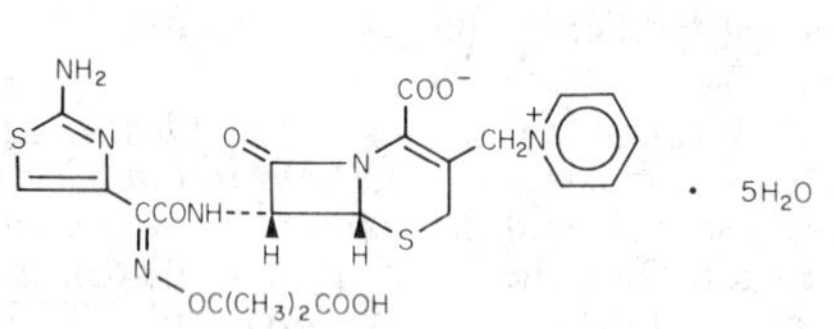

$C_{22}H_{22}N_6O_7S_2.5H_2O$ 636.65

Pyridinium, 1-[[7-[[(2-amino-4-thiazolyl)[(1-carboxy-1-methylethoxy)imino]acetyl]amino]-2-carboxy-8-oxo-5-thia-1-azabicyclo[4.2.0]oct-2-en-3-yl]methyl]-, hydroxide, inner salt, pentahydrate, [6*R*-[6α,7β(*Z*)]]-.

1-[[(6*R*,7*R*)-7-[2-(2-Amino-4-thiazolyl)glyoxylamido]-2-carboxy-8-oxo-5-thia-1-azabicyclo[4.2.0]oct-2-en-3-yl]methyl]-pyridinium hydroxide, inner salt, 7^2-(*Z*)-[*O*-(1-carboxy-1-methylethyl)oxime] pentahydrate [*78439-06-2*].

Anhydrous 546.57

» Sterile Ceftazidime is ceftazidime pentahydrate suitable for parenteral use. It contains not less than 95.0 percent and not more than 105.0 percent of ceftazidime ($C_{22}H_{22}N_6O_7S_2$), calculated on the dried basis.

Packaging and storage—Preserve in *Containers for Sterile Solids* as described under *Injections* ⟨1⟩, protected from light.

Reference standards—*USP Ceftazidime Pentahydrate Reference Standard*—Do not dry before using. *USP High Molecular Weight Ceftazidime Polymer Reference Standard*—Do not dry before using. *USP Dextran V_o Marker Reference Standard*—Do not dry before using. *USP Delta-2-Ceftazidime Isomer Reference Standard*—Do not dry before using.

Identification—The chromatogram of the *Assay preparation* obtained as directed in the *Assay* exhibits a major peak for ceftazidime the retention time of which corresponds to that exhibited in the chromatogram of the *Standard preparation* obtained as directed in the *Assay*.

Crystallinity ⟨695⟩: meets the requirements.

Pyrogen—It meets the requirements of the *Pyrogen Test* ⟨151⟩, the test dose being 1.0 mL per kg of a solution in pyrogen-free sodium carbonate solution (prepared by dissolving 9.9 g of pyrogen-free sodium carbonate in 1000 mL of Sterile Water for Injection) containing 80 mg of ceftazidime per mL.

Sterility—It meets the requirements under *Sterility Tests* ⟨71⟩, when tested as directed in the section, *Test Procedures Using Membrane Filtration*, except to use *Fluid A* to each 1000 mL of which has been added 10 g of sodium bicarbonate before sterilization.

pH ⟨791⟩: between 3.0 and 4.0, in a solution containing 5 mg per mL.

Loss on drying ⟨731⟩—Dry about 300 mg, accurately weighed, in vacuum at a pressure not exceeding 5 mm of mercury at 60° for 3 hours: it loses between 13.0% and 15.0% of its weight.

High molecular weight ceftazidime polymer—

Mobile phase—Use 0.1 *M* dibasic potassium phosphate adjusted with phosphoric acid to a pH of 7.0 ± 0.1.

System suitability preparation—Dissolve USP Dextran V_o Marker RS in *Mobile phase* to obtain a solution containing 100 µg per mL.

Standard preparation—Dissolve an accurately weighed quantity of USP High Molecular Weight Ceftazidime Polymer RS in *Mobile phase* to obtain a solution having a known concentration of about 4 µg of high molecular weight ceftazidime polymer per mL. Use this solution within 1 hour.

Test preparation—Transfer about 400 mg of Sterile Ceftazidime, accurately weighed, to a 100-mL volumetric flask, and add 80 mL of *Mobile phase*. Shake until dissolved, dilute with *Mobile phase* to volume, and mix. Use this solution within 5 minutes.

Chromatographic system (see *Chromatography* ⟨621⟩)—The liquid chromatograph is equipped with a 235-nm detector and a column that contains a 9-mm × 50-cm bed of packing L24, and is operated at a temperature between 20° and 25° maintained at ±1.0° of the selected temperature. The flow rate is about 1 mL per minute. Chromatograph 100 µL of the *System suitability preparation*, and record the peak responses as directed under *Procedure:* the column efficiency determined from the main peak is not less than 1500 theoretical plates when calculated by the formula:

$$5.545(t/W_{h/2})^2,$$

the tailing factor for the analyte peak is not less than 0.75 and not more than 1.5, and the relative standard deviation for replicate injections is not more than 4%.

Procedure—Separately inject equal volumes (about 100 µL) of the *Standard preparation* and the *Test preparation* into the chromatograph, record the chromatograms, and measure the responses for the high molecular weight ceftazidime polymer peaks. Calculate the percentage of high molecular weight ceftazidime polymer in the portion of Sterile Ceftazidime taken by the formula:

$$10(C/W)(r_U/r_S),$$

in which C is the concentration, in µg per mL, of high molecular weight ceftazidime polymer in the *Standard preparation*, W is the quantity, in mg, of Sterile Ceftazidime taken to prepare the *Test preparation*, and r_U and r_S are the peak responses obtained from the *Test preparation* and the *Standard preparation*, respectively: not more than 0.05% is found.

Assay—

pH 7 buffer—Dissolve 42.59 g of anhydrous dibasic sodium phosphate and 27.22 g of monobasic potassium phosphate in water to make 1000 mL of solution.

Mobile phase—Mix 40 mL of acetonitrile and 200 mL of *pH 7 buffer*, and dilute with water to obtain 2000 mL of solution. Filter, using a filter having a porosity of 1 µm or finer, and degas. Make adjustments if necessary (see *System Suitability* under *Chromatography* ⟨621⟩).

Standard preparation—Transfer about 29 mg of USP Ceftazidime Pentahydrate RS, accurately weighed, to a 25-mL volumetric flask containing 2.5 mL of *pH 7 buffer*, and shake until dissolved. Dilute with water to volume, and mix. [NOTE—Protect this solution from light.] Immediately prior to chromatography, transfer 5.0 mL of this stock solution to a 50-mL volumetric flask, dilute with water to volume, and mix. This solution contains about 100 µg of ceftazidime ($C_{22}H_{22}N_6O_7S_2$) per mL.

Assay preparation—Transfer about 115 mg of Sterile Ceftazidime, accurately weighed, to a 100-mL volumetric flask containing 10.0 mL of *pH 7 buffer*, and shake until dissolved. Dilute with water to volume, and mix. [NOTE—Protect this solution from light.] Immediately prior to chromatography, transfer 5.0 mL of this solution to a 50-mL volumetric flask, dilute with water to volume, and mix.

Resolution solution—Prepare a solution of USP Delta-2-Ceftazidime Isomer RS in *pH 7 buffer* containing about 0.1 mg per mL. Immediately prior to chromatography, mix 1 mL of this solution with 8 mL of water and 1 mL of the stock solution used to prepare the *Standard preparation*.

Chromatographic system (see *Chromatography* ⟨621⟩)—The liquid chromatograph is equipped with a 254-nm detector and a 4.6-mm × 15-cm column that contains 5-µm packing L1. The flow rate is about 2 mL per minute. Chromatograph the *Resolution solution*, and record the peak responses as directed under *Procedure:* the resolution, *R*, between the ceftazidime and the delta-2-ceftazidime isomer peaks is not less than 2.0. Chromatograph the *Standard preparation*, and record the responses as directed under *Procedure:* the tailing factor for the analyte peak is not less than 0.75 and not more than 1.5, and the relative standard deviation for replicate injections is not more than 1.0%.

Procedure—Separately inject equal volumes (about 20 µL) of the *Standard preparation* and the *Assay preparation* into the chromatograph, record the chromatograms, and measure the responses for the major peaks. Calculate the quantity, in mg, of ceftazidime ($C_{22}H_{22}N_6O_7S_2$) in the portion of Sterile Ceftazidime taken by the formula:

$$C(r_U/r_S),$$

in which C is the concentration, in µg per mL, of ceftazidime ($C_{22}H_{22}N_6O_7S_2$) in the *Standard preparation*, and r_U and r_S are the peak responses obtained from the *Assay preparation* and the *Standard preparation*, respectively.

Ceftizoxime Sodium

$C_{13}H_{12}N_5NaO_5S_2$ 405.38

5-Thia-1-azabicyclo[4.2.0]oct-2-ene-2-carboxylic acid, 7-[[(2,3-dihydro-2-imino-4-thiazoly)(methoxyimino)acetyl]amino]-8-oxomonosodium salt, [6*R*-[6α,7β(*Z*)]]-.

Sodium (6*R*,7*R*)-7-[2-(2-imino-4-thiazolin-4-yl)glyoxylamido]-8-oxo-5-thia-1-azabicyclo[4.2.0]oct-2-ene-2-carboxylate 7[2]-(*Z*)-(*O*-methyloxime) [*68401-82-1*].

» Ceftizoxime Sodium contains the equivalent of not less than 850 µg and not more than 995 µg of ceftizoxime ($C_{13}H_{13}N_5O_5S_2$) per mg, calculated on the anhydrous basis.

Packaging and storage—Preserve in tight containers.

Reference standard—*USP Ceftizoxime Reference Standard*—Do not dry before using.

Identification—The chromatogram of the *Assay preparation* obtained as directed in the *Assay* exhibits a major peak for ceftizoxime the retention time of which corresponds to that exhibited in the chromatogram of the *Standard preparation* obtained as directed in the *Assay*.

Crystallinity ⟨695⟩: meets the requirements.

pH ⟨791⟩: between 6.0 and 8.0, in a solution (1 in 10).

Water, *Method I* ⟨921⟩: not more than 8.5%.

Assay—

pH 3.6 buffer—Dissolve 1.42 g of citric acid monohydrate and 1.73 g of dibasic sodium phosphate in water to obtain 1000 mL of solution.

pH 7.0 buffer—Dissolve 3.63 g of monobasic potassium phosphate and 10.73 g of dibasic sodium phosphate in water to obtain 1000 mL of solution.

Mobile phase—Prepare a mixture of *pH 3.6 buffer* and acetonitrile (about 9:1). Filter through a filter (1-µm or finer porosity), and degas. Adjust the composition, if necessary, to meet the performance requirements under *Chromatographic system*.

Internal standard solution—Dissolve 1.2 g of salicylic acid in 10 mL of methanol, and dilute with *pH 7.0 buffer* to obtain 200 mL of solution.

Standard preparation—Dissolve a suitable quantity of USP Ceftizoxime RS, accurately weighed, in *pH 7.0 buffer* to obtain a solution having a known concentration of about 1 mg of ceftizoxime ($C_{13}H_{13}N_5O_5S_2$) per mL. Transfer 2.0 mL of this solution to a 100-mL volumetric flask, add 5.0 mL of *Internal standard solution*, dilute with *pH 7.0 buffer* to volume, and mix. This *Standard preparation* contains about 0.02 mg of ceftizoxime per mL.

Assay preparation—Using a suitable quantity of Ceftizoxime Sodium, accurately weighed, proceed as directed under *Standard preparation*.

Chromatographic system (see *Chromatography* ⟨621⟩)—The liquid chromatograph is equipped with a 254-nm detector and a 4.0-mm × 30-cm column that contains 5- to 10-µm packing L1. The flow rate is about 2 mL per minute. Chromatograph the *Standard preparation*, and record the peak responses as directed under *Procedure:* the column efficiency determined from the analyte peak is not less than 2000 theoretical plates, the tailing factor for the analyte peak is not more than 2, the resolution, *R*, between the analyte and internal standard peaks is not less than 4, and the relative standard deviation for replicate injections is not more than 2%.

Procedure—Separately inject equal volumes (about 10 µL) of the *Standard preparation* and the *Assay preparation* into the chromatograph, record the chromatograms, and measure the responses for the major peaks. The relative retention times are about 0.6 for ceftizoxime and 1.0 for salicylic acid. Calculate the quantity, in µg, of ceftizoxime per mg of the Ceftizoxime Sodium taken by the formula:

$$1000(C/M)(R_U/R_S),$$

in which C is the concentration, in mg of ceftizoxime ($C_{13}H_{13}N_5O_5S_2$) per mL, of the *Standard preparation*, M is the concentration, in mg per mL, of the *Assay preparation* based on the weight of Ceftizoxime Sodium taken and the extent of dilution, and R_U and R_S are the peak response ratios of the ceftizoxime peak to the internal standard peak obtained from the *Assay preparation* and the *Standard preparation*, respectively.

Ceftizoxime Sodium Injection

» Ceftizoxime Sodium Injection is a sterile solution of Ceftizoxime Sodium in a suitable isoosmotic diluent. It contains the equivalent of not less than 90.0 percent and not more than 115.0 percent of the labeled amount of ceftizoxime ($C_{13}H_{13}N_5O_5S_2$).

Packaging and storage—Preserve in *Containers for Injections* as described under *Injections* ⟨1⟩. Maintain in the frozen state.

Labeling—It meets the requirements for *Labeling* under *Injections* ⟨1⟩. The label states that it is to be thawed just prior to use, describes conditions for proper storage of the resultant solution, and directs that the solution is not to be refrozen.

Reference standard—*USP Ceftizoxime Reference Standard*—Do not dry before using.

Identification—The chromatogram of the *Assay preparation* obtained as directed in the *Assay* exhibits a major peak for ceftizoxime the retention time of which corresponds to that exhibited in the chromatogram of the *Standard preparation* obtained as directed in the *Assay*.

Pyrogen—It meets the requirements of the *Pyrogen Test* ⟨151⟩, the test dose being a volume of undiluted Injection providing the equivalent of 50 mg of ceftizoxime per kg.

Sterility—It meets the requirements under *Sterility Tests* ⟨71⟩, when tested as directed in the section, *Test Procedures Using Membrane Filtration*.

pH ⟨791⟩: between 5.5 and 8.0.

Particulate matter ⟨788⟩: meets the requirements under *Small-volume Injections*.

Assay—

pH 3.6 buffer, pH 7.0 buffer, Mobile phase, Internal standard solution, Standard preparation, and *Chromatographic system*—Prepare as directed in the *Assay* under *Ceftizoxime Sodium*.

Assay preparation—Allow 1 container of Ceftizoxime Sodium Injection to thaw, and mix. Transfer an accurately measured volume of the Injection, equivalent to about 40 mg of ceftizoxime, to a 100-mL volumetric flask, dilute with *pH 7.0 buffer* to volume, and mix. Transfer 10.0 mL of this solution to a 200-mL volumetric flask, add 5.0 mL of *Internal standard solution*, dilute with *pH 7.0 buffer* to volume, and mix.

Procedure—Proceed as directed for *Procedure* in the *Assay* under *Ceftizoxime Sodium*. Calculate the quantity, in mg, of ceftizoxime ($C_{13}H_{13}N_5O_5S_2$) in each mL of the Injection taken by the formula:

$$2000(C/V)(R_U/R_S),$$

in which V is the volume, in mL, of Injection taken, and the other terms are as defined therein.

Sterile Ceftizoxime Sodium

» Sterile Ceftizoxime Sodium is ceftizoxime sodium suitable for parenteral use. It contains the equivalent of not less than 850 µg and not more than 995 µg of ceftizoxime ($C_{13}H_{13}N_5O_5S_2$) per mg, calculated on the anhydrous basis. In addition, where packaged for dispensing, it contains the equivalent of not less than 90.0 percent and not more than 115.0 percent of the labeled amount of ceftizoxime ($C_{13}H_{13}N_5O_5S_2$).

Packaging and storage—Preserve in *Containers for Sterile Solids* as described under *Injections* ⟨1⟩.

Reference standard—*USP Ceftizoxime Reference Standard*—Do not dry before using.

Constituted solution—At the time of use, the constituted solution prepared from Sterile Ceftizoxime Sodium meets the requirements for *Constituted Solutions* under *Injections* ⟨1⟩.

Identification—The chromatogram of the *Assay preparation* obtained as directed in the *Assay* exhibits a major peak for ceftizoxime the retention time of which corresponds to that exhibited in the chromatogram of the *Standard preparation* obtained as directed in the *Assay*.

Crystallinity ⟨695⟩: meets the requirements.

Pyrogen—It meets the requirements of the *Pyrogen Test* ⟨151⟩, the test dose being 1 mL per kg of a solution prepared by diluting Sterile Ceftizoxime Sodium with Sterile Water for Injection to a concentration of 50 mg of ceftizoxime per mL.

Sterility—It meets the requirements under *Sterility Tests* ⟨71⟩, when tested as directed in the section, *Test Procedures Using Membrane Filtration*.

pH ⟨791⟩: between 6.0 and 8.0 in a solution (1 in 10).

Water, *Method I* ⟨921⟩: not more than 8.5%.

Particulate matter ⟨788⟩: meets the requirements under *Small-volume Injections*.

Other requirements—It meets the requirements for *Uniformity of Dosage Units* ⟨905⟩ and for *Labeling* under *Injections* ⟨1⟩.

Assay—

pH 3.6 buffer, *pH 7.0 buffer*, *Mobile phase*, *Internal standard solution*, and *Chromatographic system*—Prepare as directed in the *Assay* under *Ceftizoxime Sodium*.

Standard preparation—Dissolve a suitable quantity of USP Ceftizoxime RS, accurately weighed, in *pH 7.0 buffer* to obtain a solution having a known concentration of about 1 mg of ceftizoxime ($C_{13}H_{13}N_5O_5S_2$) per mL. Transfer 2.0 mL of this solution to a 100-mL volumetric flask, add 5.0 mL of *Internal standard solution*, dilute with *pH 7.0 buffer* to volume, and mix. This *Standard preparation* contains about 0.02 mg of ceftizoxime per mL.

Assay preparation 1—Using a suitable quantity of Sterile Ceftizoxime Sodium, accurately weighed, proceed as directed under *Standard preparation*.

Assay preparation 2 (where it is represented as being in a single-dose container)—Constitute Sterile Ceftizoxime Sodium in a volume of water, accurately measured, corresponding to the volume of solvent specified in the labeling. Withdraw all of the withdrawable contents, using a suitable hypodermic needle and syringe, and dilute quantitatively with *pH 7.0 buffer* to obtain a solution containing about 1 mg of ceftizoxime per mL. Transfer 2.0 mL of this solution to a 100-mL volumetric flask, add 5.0 mL of *Internal standard solution*, dilute with *pH 7.0 buffer* to volume, and mix.

Assay preparation 3 (where the label states the quantity of ceftizoxime in a given volume of constituted solution)—Constitute Sterile Ceftizoxime Sodium in a volume of water, accurately measured, corresponding to the volume of solvent specified in the labeling. Dilute an accurately measured volume of the constituted solution quantitatively with *pH 7.0 buffer* to obtain a solution containing about 1 mg of ceftizoxime per mL. Transfer 2.0 mL of this solution to a 100-mL volumetric flask, add 5.0 mL of *Internal standard solution*, dilute with *pH 7.0 buffer* to volume, and mix.

Procedure—Proceed with Sterile Ceftizoxime Sodium as directed for *Procedure* in the *Assay* under *Ceftizoxime Sodium*. Calculate the quantity, in µg, of ceftizoxime per mg of the Sterile Ceftizoxime Sodium taken by the formula:

$$1000(C/M)(R_U/R_S),$$

in which C is the concentration, in mg of ceftizoxime ($C_{13}H_{13}N_5O_5S_2$) per mL, of the *Standard preparation*, M is the concentration, in mg per mL, of the *Assay preparation* based on the weight of Sterile Ceftizoxime Sodium taken and the extent of dilution, and R_U and R_S are the peak response ratios of the ceftizoxime peak to the internal standard peak obtained from the *Assay preparation* and the *Standard preparation*, respectively. Calculate the quantity, in mg, of ceftizoxime ($C_{13}H_{13}N_5O_5S_2$) withdrawn from the container, or in the portion of constituted solution taken by the formula:

$$(L/D)(C)(R_U/R_S),$$

in which L is the labeled quantity, in mg of ceftizoxime ($C_{13}H_{13}N_5O_5S_2$), in the container, or in the volume of constituted solution taken, and D is the concentration, in mg of ceftizoxime ($C_{13}H_{13}N_5O_5S_2$) per mL, of *Assay preparation 2* or *Assay preparation 3*, based on the labeled quantity in the container or in the portion of constituted solution taken, respectively, and the extent of dilution.

Sterile Ceftriaxone Sodium

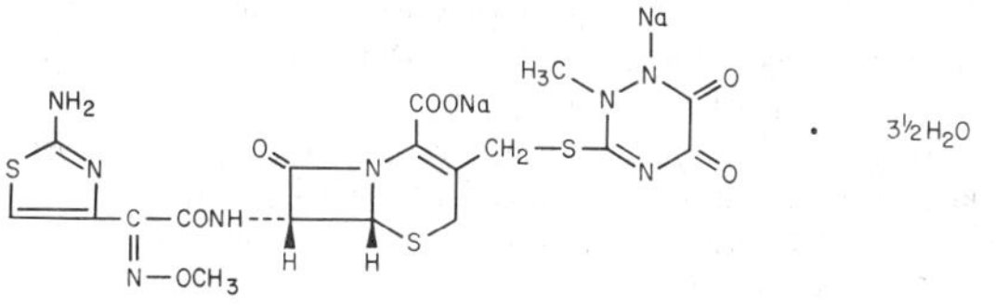

$C_{18}H_{16}N_8Na_2O_7S_3.3\frac{1}{2}H_2O$ 661.59

5-Thia-1-azabicyclo[4.2.0]oct-2-ene-2-carboxylic acid, 7-[[(2-amino-4-thiazolyl)(methoxyimino)acetyl]amino]-8-oxo-3-[[(1,2,5,6-tetrahydro-2-methyl-5,6-dioxo-1,2,4-triazin-3-yl)thio]methyl]-, disodium salt, trisesquihydrate [6*R*-[6α,7β(*Z*)]]-.

(6*R*,7*R*)-7-[2-(2-Amino-4-thiazolyl)glyoxylamido]-8-oxo-3-[[(1,2,5,6-tetrahydro-2-methyl-5,6-dioxo-*as*-triazin-3-yl)-thio]methyl]-5-thia-1-azabicyclo[4.2.0]oct-2-ene-2-carboxylic acid, 7^2-(*Z*)-(*O*-methyloxime), disodium salt, trisesquihydrate [*104376-79-6*].

Anhydrous 598.53

» Sterile Ceftriaxone Sodium is ceftriaxone sodium suitable for parenteral use. Where it is not packaged for dispensing, it contains the equivalent of not less than 795 µg of ceftriaxone ($C_{18}H_{18}N_8O_7S_3$) per mg, calculated on the anhydrous basis. Where it is packaged for dispensing, it contains the equivalent of not less than 776 µg of ceftriaxone ($C_{18}H_{18}N_8O_7S_3$) per mg, calculated on the anhydrous basis, and the equivalent of not less than 90.0 percent and not more than 115.0 percent of the labeled amount of ceftriaxone ($C_{18}H_{18}N_8O_7S_3$).

Packaging and storage—Preserve in *Containers for Sterile Solids* as described under *Injections* ⟨1⟩.

Reference standards—*USP Ceftriaxone Sodium Reference Standard*—Do not dry before using. *USP Ceftriaxone Sodium E-Isomer Reference Standard*—Keep container tightly closed and protected from light. Do not dry before using.

Constituted solution—At the time of use, the constituted solution prepared from Sterile Ceftriaxone Sodium meets the requirements for *Constituted Solutions* under *Injections* ⟨1⟩.

Identification—

A: The infrared absorption spectrum of a potassium bromide dispersion of it exhibits maxima only at the same wavelengths as that of a similar preparation of USP Ceftriaxone Sodium RS.

B: The chromatogram of the *Assay preparation* obtained as directed in the *Assay* exhibits a major peak for ceftriaxone, the retention time of which corresponds to that exhibited in the chromatogram of the *Standard preparation* obtained as directed in the *Assay*.

Crystallinity ⟨695⟩: meets the requirements.

Pyrogen—It meets the requirements of the *Pyrogen Test* ⟨151⟩, the test dose being 1 mL per kg of a solution prepared by diluting Sterile Ceftriaxone Sodium with Sterile Water for Injection to a concentration of 40 mg of ceftriaxone per mL.

Sterility—It meets the requirements under *Sterility Tests* ⟨71⟩, when tested as directed in the section, *Test Procedures Using Membrane Filtration*.

pH ⟨791⟩: between 6.0 and 8.0 in a solution (1 in 10).

Water, *Method I* ⟨921⟩: between 8.0% and 11.0%.

Particulate matter ⟨788⟩: meets the requirements under *Small-volume Injections*.

Other requirements—It meets the requirements for *Uniformity of Dosage Units* ⟨905⟩ and for *Labeling* under *Injections* ⟨1⟩.

Assay—

pH 7.0 buffer—Dissolve 13.6 g of dibasic potassium phosphate and 4.0 g of monobasic potassium phosphate in water to obtain 1000 mL of solution. Adjust this solution with phosphoric acid or 10 *N* potassium hydroxide to a pH of 7.0 ± 0.1.

pH 5.0 buffer—Dissolve 25.8 g of sodium citrate in 500 mL of water, adjust with citric acid solution (1 in 5) to a pH of 5.0 ± 0.1, and dilute with water to a volume of 1000 mL.

Mobile phase—Dissolve 3.2 g of tetraheptylammonium bromide in 400 mL of acetonitrile, add 44 mL of *pH 7.0 buffer* and 4 mL of *pH 5.0 buffer*, and add water to make 1000 mL. Filter through a membrane filter of 0.5-µm or finer porosity, and degas. Make adjustments if necessary (see *System Suitability* under *Chromatography* ⟨621⟩).

Standard preparation—Dissolve an accurately weighed quantity of USP Ceftriaxone Sodium RS in *Mobile phase*, to obtain a solution having a known concentration of about 0.2 mg per mL. Use this solution promptly after preparation.

Resolution solution—Dissolve a suitable quantity of USP Ceftriaxone Sodium E-Isomer RS in *Standard preparation*, and dilute with *Mobile phase* to obtain a solution containing about 160 µg of USP Ceftriaxone Sodium E-Isomer RS per mL and 160 µg of USP Ceftriaxone Sodium RS per mL. Use this solution promptly after preparation.

Assay preparation 1—Transfer about 40 mg of Sterile Ceftriaxone Sodium, accurately weighed, to a 200-mL volumetric flask, dissolve in *Mobile phase*, dilute with *Mobile phase* to volume, and mix. Use this solution promptly after preparation.

Assay preparation 2 (where it is represented as being in a single-dose container)—Constitute Sterile Ceftriaxone Sodium in a volume of water, accurately measured, corresponding to the volume of solvent specified in the labeling. Withdraw all of the withdrawable contents, using a suitable hypodermic needle and syringe, and dilute quantitatively with *Mobile phase* to obtain a solution containing about 180 µg of ceftriaxone per mL. Use this solution promptly after preparation.

Assay preparation 3 (where the label states the quantity of ceftriaxone in a given volume of constituted solution)—Constitute Sterile Ceftriaxone Sodium in a volume of water, accurately measured, corresponding to the volume of solvent specified in the labeling. Dilute an accurately measured volume of the constituted solution quantitatively with *Mobile phase* to obtain a solution containing about 180 µg of ceftriaxone per mL. Use this solution promptly after preparation.

Chromatographic system (see *Chromatography* ⟨621⟩)—The liquid chromatograph is equipped with a 270-nm detector and a 4.0-mm × 15-cm column that contains 5-µm packing L1. The flow rate is about 2 mL per minute. Chromatograph the *Resolution solution*, and record the peak responses as directed under *Procedure:* the resolution, *R*, between the ceftriaxone E-isomer and ceftriaxone peaks is not less than 3. Chromatograph the *Standard preparation*, and record the peak responses as directed under *Procedure:* the column efficiency determined from the analyte peak is not less than 1500 theoretical plates, the tailing factor for the analyte peak is not more than 2, and the relative standard deviation for replicate injections is not more than 2%.

Procedure—Separately inject equal volumes (about 20 µL) of the *Standard preparation* and the *Assay preparation* into the chromatograph, record the chromatograms, and measure the responses for the major peaks. Calculate the quantity, in µg, of ceftriaxone ($C_{18}H_{18}N_8O_7S_3$) per mg of the Sterile Ceftriaxone Sodium taken by the formula:

$$200(CP/W)(r_U/r_S),$$

in which *C* is the concentration, in mg per mL, of USP Ceftriaxone Sodium RS in the *Standard preparation*, *P* is the designated potency, in µg of ceftriaxone per mg, of USP Ceftriaxone Sodium RS, *W* is the quantity, in mg, of the Sterile Ceftriaxone Sodium taken to prepare *Assay preparation 1*, and r_U and r_S are the ceftriaxone peak responses obtained from the *Assay preparation* and the *Standard preparation*, respectively. Calculate the quantity, in mg, of ceftriaxone ($C_{18}H_{18}N_8O_7S_3$) withdrawn from the container, or in the portion of constituted solution taken by the formula:

$$(L/D)(CP)(r_U/r_S),$$

in which *L* is the labeled quantity, in mg, of ceftriaxone ($C_{18}H_{18}N_8O_7S_3$) in the container, or in the volume of constituted solution taken, and *D* is the concentration, in µg per mL, of ceftriaxone in *Assay preparation 2* or *Assay preparation 3*, based on the labeled quantity in the container or in the portion of constituted solution taken, respectively, and the extent of dilution, and the other terms are as defined above.

Sterile Cefuroxime Sodium

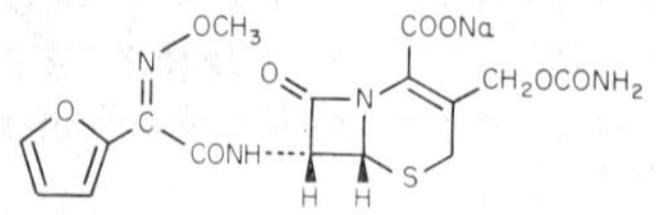

$C_{16}H_{15}N_4NaO_8S$ 446.36

5-Thia-1-azabicyclo[4.2.0]oct-2-ene-2-carboxylic acid, 3-[[(aminocarbonyl)oxy]methyl]-7-[[2-furanyl(methoxyimino)-acetyl]amino]-8-oxo-, monosodium salt [6*R*-[6α,7β(*Z*)]]-.

Sodium (6*R*,7*R*)-7-[2-(2-furyl)glyoxylamido]-3-(hydroxymethyl)-8-oxo-5-thia-1-azabicyclo[4.2.0]-oct-2-ene-2-carboxylate, 7[2]-(*Z*)-(*O*-methyloxime), carbamate (ester) [*56238-63-2*].

» Sterile Cefuroxime Sodium is cefuroxime sodium suitable for parenteral use. It contains the equivalent of not less than 855 µg and not more than 1000 µg of cefuroxime ($C_{16}H_{16}N_4O_8S$) per mg, calculated on the anhydrous basis. In addition, where it is packaged for dispensing, it contains the equivalent of not less than 90.0 percent and not more than 120.0 percent of the labeled amount of cefuroxime ($C_{16}H_{16}N_4O_8S$).

Packaging and storage—Preserve in *Containers for Sterile Solids* as described under *Injections* ⟨1⟩.

Reference standard—*USP Cefuroxime Sodium Reference Standard*—Do not dry before using.

Constituted solution—At the time of use, the constituted solution for intravenous administration prepared from Sterile Cefuroxime Sodium meets the requirements for *Constituted Solutions* under *Injections* ⟨1⟩.

Identification—The chromatogram of the *Assay preparation* obtained as directed in the *Assay* exhibits a major peak for cefuroxime the retention time of which corresponds to that exhibited in the chromatogram of the *Standard preparation* obtained as directed in the *Assay*.

Pyrogen—It meets the requirements of the *Pyrogen Test* ⟨151⟩, the test dose being 1 mL per kg of a solution prepared by diluting Sterile Cefuroxime Sodium with Sterile Water for Injection to a concentration of 50 mg of cefuroxime per mL.

Sterility—It meets the requirements under *Sterility Tests* ⟨71⟩, when tested as directed in the section, *Test Procedures Using Membrane Filtration*.

pH ⟨791⟩: between 6.0 and 8.5, in a solution (1 in 10).

Water, *Method I* ⟨921⟩: not more than 3.5%.

Particulate matter ⟨788⟩: meets the requirements under *Small-volume Injections*.

Other requirements—It meets the requirements for *Uniformity of Dosage Units* ⟨905⟩ and for *Labeling* under *Injections* ⟨1⟩.

Assay—

pH 3.4 acetate buffer—Transfer 50 mL of 0.1 *M* sodium acetate to a 1000-mL volumetric flask, dilute with 0.1 *N* acetic acid to volume, and mix.

Mobile phase—Prepare a suitable mixture of pH 3.4 acetate buffer and acetonitrile (about 10:1). Filter through a membrane filter (1-µm or finer porosity), and degas.

Internal standard solution—Prepare a solution of orcinol in water containing 1.5 mg per mL.

Standard preparation—Dissolve a suitable quantity of USP Cefuroxime Sodium RS, accurately weighed, in water to obtain a solution having a known concentration of about 1 mg of cefuroxime ($C_{16}H_{16}N_4O_8S$) per mL. Immediately transfer 5.0 mL of the resulting solution to a 100-mL volumetric flask, add 20.0 mL of *Internal standard solution*, dilute with water to volume, and mix. This *Standard preparation* contains 0.05 mg of cefuroxime per mL.

Assay preparation 1—Using a suitable quantity of Sterile Cefuroxime Sodium, accurately weighed, proceed as directed in the first sentence under *Standard preparation.* Immediately transfer 5.0 mL of the resulting solution to a 100-mL volumetric flask, add 20.0 mL of *Internal standard solution,* dilute with water to volume, and mix.

Assay preparation 2 (where it is represented as being in a single-dose container)—Constitute Sterile Cefuroxime Sodium in a volume of water, accurately measured, corresponding to the volume of solvent specified in the labeling. Withdraw all of the withdrawable contents, using a suitable hypodermic needle and syringe, and dilute quantitatively with water to obtain a solution containing about 1 mg of cefuroxime per mL. Immediately transfer 5.0 mL of the resulting solution to a 100-mL volumetric flask, add 20.0 mL of *Internal standard solution,* dilute with water to volume, and mix.

Assay preparation 3 (where the label states the quantity of cefuroxime in a given volume of constituted solution or suspension)—Constitute Sterile Cefuroxime Sodium in a volume of water, accurately measured, corresponding to the volume of solvent specified in the labeling. Dilute an accurately measured volume of the constituted solution or suspension quantitatively with water to obtain a solution containing about 1 mg of cefuroxime per mL. Immediately transfer 5.0 mL of the resulting solution to a 100-mL volumetric flask, add 20.0 mL of *Internal standard solution,* dilute with water to volume, and mix.

Chromatographic system (see *Chromatography* ⟨621⟩)—The liquid chromatograph is equipped with a 254-nm detector and a 4.6-mm × 15-cm column that contains 5-µm packing L15. The flow rate is about 2 mL per minute. Chromatograph the *Standard preparation,* and record the peak responses as directed under *Procedure:* the column efficiency determined from the analyte peak is not less than 1300 theoretical plates, the tailing factor for the analyte peak is not more than 2.0, the resolution, *R,* between the analyte and internal standard peaks is not less than 3.5, and the relative standard deviation for replicate injections is not more than 2.0%.

Procedure—Separately inject equal volumes (about 10 µL) of the *Standard preparation* and the *Assay preparation* into the chromatograph, record the chromatograms, and measure the responses for the major peaks. The relative retention times are about 0.5 for cefuroxime and 1.0 for orcinol. Calculate the quantity, in µg, of cefuroxime per mg of the Sterile Cefuroxime Sodium taken by the formula:

$$1000(C/M)(R_U/R_S),$$

in which C is the concentration, in mg of cefuroxime ($C_{16}H_{16}N_4O_8S$) per mL, of the *Standard preparation,* M is the concentration, in mg per mL, of the *Assay preparation* based on the weight of Sterile Cefuroxime Sodium taken and the extent of dilution, and R_U and R_S are the peak response ratios of the cefuroxime peak to the internal standard peak obtained from the *Assay preparation* and the *Standard preparation,* respectively. Calculate the quantity, in mg, of cefuroxime ($C_{16}H_{16}N_4O_8S$) withdrawn from the container, or in the portion of constituted solution or suspension taken by the formula:

$$(L/D)(C)(R_U/R_S),$$

in which L is the labeled quantity, in mg of cefuroxime ($C_{16}H_{16}N_4O_8S$), in the container, or in the volume of constituted solution or suspension taken, and D is the concentration, in mg of cefuroxime ($C_{16}H_{16}N_4O_8S$) per mL, of *Assay preparation 2* or *Assay preparation 3,* based on the labeled quantity in the container or in the portion of constituted solution or suspension taken, respectively, and the extent of dilution.

Cellulose, Microcrystalline—*see* Cellulose, Microcrystalline NF

Cellulose, Microcrystalline, and Carboxymethylcellulose Sodium—*see* Cellulose, Microcrystalline, and Carboxymethylcellulose Sodium NF

Oxidized Cellulose

» Oxidized Cellulose contains not less than 16.0 percent and not more than 24.0 percent of carboxyl groups (COOH), calculated on the dried basis. It is sterile.

Packaging and storage—Preserve in *Containers for Sterile Solids* as described under *Injections* ⟨1⟩, protected from direct sunlight. Store in a cold place.

Labeling—The package bears a statement to the effect that the sterility of Oxidized Cellulose cannot be guaranteed if the package bears evidence of damage, or if the package has been previously opened. Oxidized Cellulose meets the requirements for *Labeling* under *Injections* ⟨1⟩.

Identification—To about 200 mg add 10 mL of 0.25 *N* sodium hydroxide, and shake for 1 minute. Add 10 mL of water, and shake: the solution so obtained shows no more than a slight haze and is substantially free from fibers and from foreign particles. Allow to stand for 10 minutes: any swollen fibers initially present are no longer visible. Acidify with 3 *N* hydrochloric acid: a flocculent white precipitate is formed.

Sterility—It meets the requirements under *Sterility Tests* ⟨71⟩, the test specimen weighing approximately 250 mg and 0.5 mL of 0.1 *N* sodium hydroxide being added to the portions of media used.

Loss on drying ⟨731⟩—Dry it in vacuum over phosphorus pentoxide for 18 hours: it loses not more than 15.0% of its weight.

Residue on ignition ⟨281⟩: not more than 0.15%.

Nitrogen as nitrate or nitrite—Transfer about 1 g, previously dried in vacuum over phosphorus pentoxide for 18 hours and accurately weighed, to a 500-mL Kjeldahl flask. Arrange a 125-mL conical flask, containing 30 mL of boric acid solution (1 in 25) and 6 drops of mixed indicator (1 part of methyl red TS and 4 parts of bromocresol green TS), beneath the condenser of the distillation apparatus so that the tip of the condenser is well below the surface of the boric acid solution. To the Kjeldahl flask containing the sample add 1 g of Devarda's alloy, 100 mL of recently boiled water, a small lump of paraffin, and 100 mL of 1 *N* sodium hydroxide. Connect the Kjeldahl flask to the condenser by a suitable trap bulb. Heat the mixture in the flask until 45 to 50 mL of distillate has collected in the receiver. Rinse the condenser, and titrate the boric acid solution with 0.02 *N* sulfuric acid VS to a pale pink end-point. Perform a blank determination, and make any necessary correction. Each mL of 0.02 *N* sulfuric acid is equivalent to 0.2801 mg of nitrogen. The nitrogen content does not exceed 0.5%.

Formaldehyde—Weigh accurately about 500 mg, and transfer to a 500-mL iodine flask. Add 250 mL of water, and allow to stand for not less than 2 hours with intermittent shaking. Pipet 0.50 mL of the supernatant liquid into a glass-stoppered test tube, and add 10 mL of chromotropic acid TS. Stopper the tube loosely, and heat in a boiling water bath for 30 minutes. Cool, and determine the absorbance of the solution at 570 nm, with a suitable spectrophotometer, using a mixture of 0.5 mL of water and 10 mL of chromotropic acid TS as the blank: the absorbance does not exceed that produced when 0.50 mL of dilute formaldehyde solution (1 in 40,000) is treated in the same manner (0.5%).

Assay—Transfer about 500 mg of Oxidized Cellulose, previously dried over phosphorus pentoxide in vacuum for 18 hours and accurately weighed, to a 125-mL conical flask. Add 50.0 mL of calcium acetate solution (1 in 50), swirl until the sample is completely covered, allow the mixture to stand for 30 minutes, then add phenolphthalein TS, and titrate the solution with 0.1 *N* sodium hydroxide VS. Perform a blank determination by titrating 50.0 mL of the calcium acetate solution, and make any necessary correction. Each mL of 0.1 *N* sodium hydroxide is equivalent to 4.502 mg of carboxyl groups (COOH).

Oxidized Regenerated Cellulose

» Oxidized Regenerated Cellulose contains not less than 18.0 percent and not more than 24.0 percent of carboxyl groups (COOH), calculated on the dried basis. It is sterile.

Packaging and storage—Preserve in *Containers for Sterile Solids* as described under *Injections* ⟨1⟩, protected from direct sunlight. Store at controlled room temperature.

Labeling—The package bears a statement to the effect that the sterility of Oxidized Regenerated Cellulose cannot be guaranteed if the package bears evidence of damage, or if the package has been previously opened. Oxidized Regenerated Cellulose meets the requirements for *Labeling* under *Injections* ⟨1⟩.

Identification—To about 200 mg add 10 mL of 0.25 *N* sodium hydroxide, and shake for 1 minute. Add 10 mL of water, and shake: the solution so obtained shows no more than a slight haze and is substantially free from fibers and from foreign particles. Allow to stand for 10 minutes: any swollen fibers initially present are no longer visible. Acidify with 3 *N* hydrochloric acid: a flocculent white precipitate is formed.

Sterility—It meets the requirements under *Sterility Tests* ⟨71⟩, the test specimen weighing approximately 250 mg and 0.5 mL of 0.1 *N* sodium hydroxide being added to the portions of media used.

Loss on drying ⟨731⟩—Dry about 150 mg at 90° for 2 hours: it loses not more than 15% of its weight.

Residue on ignition ⟨281⟩: not more than 0.15%.

Nitrogen content—Transfer about 1 g, previously dried and accurately weighed, to a 500-mL Kjeldahl flask. Arrange a 125-mL conical flask, containing 30 mL of boric acid solution (1 in 25) and 6 drops of mixed indicator (1 part of methyl red TS and 4 parts of bromocresol green TS), beneath the condenser of the distillation apparatus so that the tip of the condenser is well below the surface of the boric acid solution. To the Kjeldahl flask containing the sample add 1 g of Devarda's alloy, 100 mL of recently boiled water, a small lump of paraffin, and 100 mL of 1 *N* sodium hydroxide. Connect the Kjeldahl flask to the condenser by a suitable trap bulb. Heat the mixture in the flask until 45 to 50 mL of distillate has collected in the receiver. Rinse the condenser, and titrate the boric acid solution with 0.02 *N* sulfuric acid VS to a pale pink end-point that persists for 30 seconds. Perform a blank determination, and make any necessary correction. Each mL of 0.02 *N* sulfuric acid is equivalent to 0.2801 mg of nitrogen. The nitrogen content does not exceed 0.5%.

Formaldehyde—Transfer 500 mg to a 500-mL iodine flask. Add 250 mL of water, and allow to stand for not less than 2 hours with intermittent shaking. Pipet 0.5 mL of the supernatant liquid into a glass-stoppered test tube, and add 10 mL of chromotropic acid TS. Stopper the tube loosely, and heat in a boiling water bath for 30 minutes. Cool, and determine the absorbance of the solution at 570 nm, with a suitable spectrophotometer, using a mixture of 0.5 mL of water and 10 mL of chromotropic acid TS as the blank: the absorbance does not exceed that produced when 0.5 mL of dilute formaldehyde solution (1 in 40,000) is treated in the same manner (0.5% CH_2O).

Assay—Transfer about 1 g of Oxidized Regenerated Cellulose, previously dried at 90° for 2 hours and accurately weighed, to a 250-mL conical flask. Pipet 10 mL of 0.5 *N* sodium hydroxide VS into the flask, swirl to dissolve, and add 100 mL of water. Immediately titrate with 0.1 *N* hydrochloric acid VS to a phenolphthalein end-point. Perform a blank determination, and note the difference in volumes required. Each mL of the difference in volumes of 0.1 *N* hydrochloric acid consumed is equivalent to 4.50 mg of carboxyl (COOH).

Cellulose, Powdered—*see* Cellulose, Powdered NF

Cellulose Acetate—*see* Cellulose Acetate NF

Cellulose Acetate Phthalate—*see* Cellulose Acetate Phthalate NF

Cellulose, Hydroxyethyl—*see* Hydroxyethyl Cellulose NF

Cellulose, Hydroxypropyl—*see* Hydroxypropyl Cellulose NF

Cellulose Sodium Phosphate

» Cellulose Sodium Phosphate is prepared by phosphorylation of alpha cellulose. It has an inorganic bound phosphate content of not less than 31.0 percent and not more than 36.0 percent, calculated on the dried basis.

Packaging and storage—Preserve in well-closed containers.

pH ⟨791⟩—Place 3 g of it in a 100-mL beaker, add 60 mL of water, and stir occasionally for 5 minutes. Filter through a sintered-glass crucible. The pH of the filtrate is between 6.0 and 9.0.

Loss on drying ⟨731⟩—Dry it at 105° for 3 hours: it loses not more than 10.0% of its weight.

Nitrogen, *Method I* ⟨461⟩: not more than 1.0%.

Heavy metals, *Method III* ⟨231⟩: 0.004%.

Calcium binding capacity—

Standard calcium solution—Transfer about 0.33 g of dried calcium carbonate, primary standard grade, accurately weighed, to a 250-mL beaker with the aid of a few mL of water; add more water to bring the volume to about 50 mL. Carefully and dropwise add 2 *N* hydrochloric acid until all of the solid dissolves; add 2 drops in excess. Heat the solution to boiling, and boil for 5 minutes. Cool the solution, and transfer to a 1000-mL volumetric flask. Dilute with water to volume, and mix. Calculate the molarity, M, of the solution by the formula:

$$g/100.09,$$

in which g is the weight, in g, of calcium carbonate taken.

Standard ethylenediaminetetraacetate titrant—Dissolve 10 g of disodium ethylenediaminetetraacetate in 100 mL of water. Slowly add alcohol until the first permanent precipitate is formed. Filter, and discard the solid. Add an equal volume of alcohol to the filtrate. Filter the resulting precipitate, discard the filtrate, and wash the residue on the filter, first with acetone, then with ether. Dry at 80° for 4 days at about 50% relative humidity. Transfer about 3.72 g of this purified disodium ethylenediaminetetraacetate, accurately weighed, to a 1000-mL volumetric flask, and dissolve with water. Dilute with water to volume, and mix. Calculate the molarity, M_S, of the solution by the formula:

$$w/372.24,$$

in which w is the weight, in g, of the purified disodium ethylenediaminetetraacetate taken.

Procedure—Transfer 0.15 ± 0.02 g of Cellulose Sodium Phosphate, accurately weighed, to a 250-mL beaker. Add 150.0 mL of *Standard calcium solution*, and stir the mixture for 5 minutes on a magnetic stirrer. Filter, discarding the first few mL of the filtrate. To 50.0 mL of the filtrate add about 50 mL of water, 15 mL of 1 *N* sodium hydroxide, and 300 mg of hydroxy naphthol blue trituration. Titrate with *Standard ethylenediaminetetraacetate titrant* to a permanent deep blue color and designate the number of mL consumed as V_S. Calculate the calcium binding capacity of the undried Cellulose Sodium Phosphate, in mmol per g, by the formula:

$$(150M_S - 3V_SM_S)/W,$$

in which W is the weight, in g, of Cellulose Sodium Phosphate taken. The calcium binding capacity, calculated on the dried basis, is not less than 1.8 mmol per g.

Sodium content—

Standard stock solution—Dissolve 508.5 mg of sodium chloride, previously dried at 105° for 2 hours, in 100 mL of water, transfer to a 1000-mL volumetric flask, dilute with water to volume, and mix. Each mL of this solution contains 200 μg of sodium.

Standard preparations—Transfer 5.0, 10.0, 15.0, and 20.0 mL of *Standard stock solution* to separate 100-mL volumetric flasks, dilute each with water to volume, and mix.

Test preparation—Dissolve about 250 mg of Cellulose Sodium Phosphate, accurately weighed, in 10 mL of a mixture of 20 mL of perchloric acid and 15 mL of nitric acid. Heat cautiously to the production of dense, white fumes, cool, transfer to a 1000-mL volumetric flask, dilute with water to volume, and mix.

Procedure—Concomitantly determine the emission of the *Test preparation* and of each *Standard preparation* at the sodium emission line of 589 nm, with a suitable flame photometer. Plot the emissions of the *Standard preparations* versus their concentration of sodium, and draw a straight line best fitting the four plotted points. From the graph so obtained determine the concentration of sodium in the *Test preparation.* Calculate the percentage of sodium in the undried Cellulose Sodium Phosphate. The content of sodium, calculated on the dried basis, is not less than 9.5 percent and not more than 13.0 percent.

Free phosphate—

Standard preparation—Prepare as directed under *Inorganic bound phosphate.*

Test preparation—Transfer about 2 g of Cellulose Sodium Phosphate, accurately weighed, to a 250-mL beaker. Add 100 mL of water, accurately measured, stir, allow to stand for 5 minutes, stir again, and filter through moderately retentive filter paper, collecting the filtrate in a dry flask.

Procedure—Transfer 2.0 mL of the *Standard preparation* and 5.0 mL of the *Test preparation* to separate 100-mL volumetric flasks. Proceed as directed in the *Procedure* under *Inorganic bound phosphate*, beginning with "Treat each of these." Calculate the percentage of free phosphate by the formula:

$$(4000/W)(A_U/A_S),$$

in which W is the weight, in mg, of undried Cellulose Sodium Phosphate taken: not more than 3.5 percent, calculated on the dried basis, is found.

Inorganic bound phosphate—

Standard preparation—Transfer 358.2 mg of monobasic potassium phosphate, primary standard grade, to a 250-mL volumetric flask, dissolve in water, dilute with water to volume, and mix. Each mL of this solution contains 1.0 mg of phosphate.

Test preparation—Transfer about 250 mg of Cellulose Sodium Phosphate, accurately weighed, to a 250-mL conical flask. Rinse to the bottom with a few mL of water. Add 10 mL of a mixture of 20 mL of perchloric acid and 15 mL of nitric acid. Heat cautiously to the production of dense, white fumes, cool the clear, almost colorless, solution, and transfer to a 100-mL volumetric flask with the aid of water. Dilute with water to volume, and mix.

Procedure—Transfer 2.0-mL portions of the *Standard preparation* and the *Test preparation* to separate 100-mL volumetric flasks. Treat each of these and a third flask, providing the blank, as follows: Add 10 mL of 5 *N* nitric acid, 10.0 mL of ammonium vanadate TS, and about 60 mL of water. Swirl, and add 10.0 mL of a freshly prepared solution of 2.5 g of ammonium molybdate in 50 mL of warm water. Dilute with water to volume, and mix. Concomitantly determine the absorbances, A_U and A_S, of the solutions from the *Standard preparation* and the *Test preparation*, respectively, at 400 nm with a suitable spectrophotometer, using the reagent blank to set the instrument. Calculate the percentage of total phosphate by the formula:

$$(10{,}000/W)(A_U/A_S),$$

in which W is the weight, in mg, of Cellulose Sodium Phosphate taken. Calculate the percentage of inorganic bound phosphate in the undried Cellulose Sodium Phosphate by subtracting from this result the percentage of *Free phosphate.*

Cephalexin

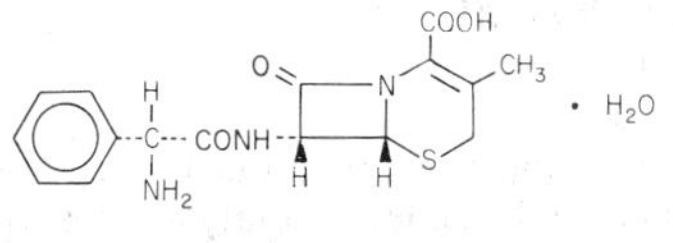

$C_{16}H_{17}N_3O_4S.H_2O$ 365.40

5-Thia-1-azabicyclo[4.2.0]oct-2-ene-2-carboxylic acid, 7-[(aminophenylacetyl)amino]-3-methyl-8-oxo-, monohydrate, [6*R*-[6α,7β(*R**)]]-.

(6*R*,7*R*)-7-[(*R*)-2-Amino-2-phenylacetamido]-3-methyl-8-oxo-5-thia-1-azabicyclo[4.2.0]oct-2-ene-2-carboxylic acid monohydrate [*23325-78-2*].

Anhydrous 347.39 [*15686-71-2*].

» Cephalexin has a potency of not less than 900 μg of $C_{16}H_{17}N_3O_4S$ per mg, calculated on the anhydrous basis.

Packaging and storage—Preserve in tight containers.

Reference standard—*USP Cephalexin Reference Standard*—Do not dry before using.

Identification—

A: The infrared absorption spectrum of a potassium bromide dispersion of it exhibits maxima only at the same wavelengths as that of a similar preparation of USP Cephalexin RS.

B: The ultraviolet absorption spectrum of a solution (1 in 50,000) exhibits maxima and minima at the same wavelengths as that of a similar solution of USP Cephalexin RS, concomitantly measured, and the absorptivity, calculated on the anhydrous basis, at the wavelength of maximum absorbance at about 262 nm is not less than 95.0% and not more than 104.0% of that of USP Cephalexin RS, the potency of the Reference Standard being taken into account.

Crystallinity ⟨695⟩: meets the requirements.

pH ⟨791⟩: between 3.0 and 5.5, in an aqueous suspension containing 50 mg per mL.

Water, *Method I* ⟨921⟩: between 4.0% and 8.0%.

Assay—Proceed with Cephalexin as directed under *Antibiotics—Microbial Assays* ⟨81⟩.

Cephalexin Capsules

» Cephalexin Capsules contain the equivalent of not less than 90.0 percent and not more than 120.0 percent of the labeled amount of cephalexin ($C_{16}H_{17}N_3O_4S$).

Packaging and storage—Preserve in tight containers.

Reference standard—*USP Cephalexin Reference Standard*—Do not dry before using.

Identification—Mix the contents of 1 Capsule with water to obtain a concentration of about 3 mg of cephalexin per mL, and filter (test solution). Place a suitable thin-layer chromatographic plate (see *Chromatography* ⟨621⟩) coated with a 0.25-mm layer of binder-free silica gel in a chamber containing a mixture of *n*-hexane and tetradecane (95:5) to a depth of about 1 cm, allow the solvent front to move the length of the plate, remove the plate from the chamber, and allow the solvent to evaporate. On this plate apply 10 μL each of the test solution and a Standard solution containing 3 mg of USP Cephalexin RS per mL. Allow the spots to dry, and develop the chromatogram in a solvent system consisting of a mixture of 0.1 *M* citric acid, 0.1 *M* dibasic sodium phosphate, and a 1 in 15 solution of ninhydrin in acetone (60:40:1.5) until the solvent front has moved about three-fourths of the length of the plate. Remove the plate from the developing chamber, mark the solvent front, dry the plate for 10 minutes at 110°, and examine the chromatogram: the R_f value of the principal spot obtained from the test solution corresponds to that obtained from the Standard solution.

Dissolution ⟨711⟩—

Medium: water; 900 mL.

Apparatus 1: 100 rpm.

Time: 45 minutes.

Procedure—Determine the amount of $C_{16}H_{17}N_3O_4S$ by a suitable validated spectrophotometric analysis of a filtered portion of the solution under test, suitably diluted with *Dissolution Medium*, if necessary, in comparison with a Standard solution having a known concentration of USP Cephalexin RS in the same medium.

Tolerances—Not less than 75% (*Q*) of the labeled amount of $C_{16}H_{17}N_3O_4S$ is dissolved in 45 minutes.

Uniformity of dosage units ⟨905⟩: meet the requirements.

Water, *Method I* ⟨921⟩: not more than 10.0%.

Assay—Transfer, as completely as possible, the contents of not less than 20 Cephalexin Capsules to a suitable tared container, determine the average content weight per capsule, and mix the combined contents. Transfer an accurately weighed quantity of the powder, equivalent to about 2000 mg of cephalexin, to a high-speed blender jar containing 200.0 mL of *Buffer No. 1* (see *Phosphate Buffers and Other Solutions* under *Antibiotics—Microbial Assays* ⟨81⟩), and blend for 3 to 5 minutes. Proceed as directed under *Antibiotics—Microbial Assays* ⟨81⟩, using an accurately measured volume of this stock solution diluted quantitatively and stepwise with *Buffer No. 1* to yield a *Test Dilution* having a concentration assumed to be equal to the median dose level of the Standard.

Cephalexin for Oral Suspension

» Cephalexin for Oral Suspension is a dry mixture of Cephalexin and one or more suitable buffers, colors, diluents, and flavors. It contains the equivalent of not less than 90.0 percent and not more than 120.0 percent of the labeled amount of $C_{16}H_{17}N_3O_4S$ per mL when constituted as directed in the labeling.

Packaging and storage—Preserve in tight containers.

Reference standard—*USP Cephalexin Reference Standard*—Do not dry before using.

Identification—Constitute 1 container of Cephalexin for Oral Suspension as directed in the labeling. Mix a portion of the resulting suspension with water to obtain a concentration of about 3 mg of cephalexin per mL, and filter (test solution). Proceed as directed in the *Identification test* under *Cephalexin Capsules*, beginning with "Place a suitable thin-layer chromatographic plate": the R_f value of the principal spot obtained from the test solution corresponds to that obtained from the Standard solution.

Uniformity of dosage units ⟨905⟩—

FOR SOLID PACKAGED IN SINGLE-UNIT CONTAINERS: meets the requirements.

pH ⟨791⟩: between 3.0 and 6.0, in the suspension constituted as directed in the labeling.

Water, *Method I* ⟨921⟩: not more than 2.0%.

Assay—Constitute Cephalexin for Oral Suspension as directed in the labeling. Proceed as directed under *Antibiotics—Microbial Assays* ⟨81⟩, using an accurately measured volume of the suspension so obtained, freshly mixed and free from air bubbles, diluted quantitatively and stepwise with *Buffer No. 1* to yield a *Test Dilution* having a concentration assumed to be equal to the median dose level of the Standard.

Cephalexin Tablets

» Cephalexin Tablets contain the equivalent of not less than 90.0 percent and not more than 120.0 percent of the labeled amount of $C_{16}H_{17}N_3O_4S$.

Packaging and storage—Preserve in tight containers.

Reference standard—*USP Cephalexin Reference Standard*—Do not dry before using.

Identification—Mix a quantity of finely powdered Tablets with water to obtain a concentration of about 3 mg of cephalexin per mL, and filter (test solution). Proceed as directed in the *Identification test* under *Cephalexin Capsules,* beginning with "Place a suitable thin-layer chromatographic plate": the R_f value of the principal spot obtained from the test solution corresponds to that obtained from the Standard solution.

Dissolution ⟨711⟩—

Medium: water; 900 mL.

Apparatus 1: 100 rpm.

Time: 45 minutes.

Procedure—Determine the amount of $C_{16}H_{17}N_3O_4S$ by a suitable validated spectrophotometric analysis of a filtered portion of the solution under test, suitably diluted with *Dissolution Medium*, if necessary, in comparison with a Standard solution having a known concentration of USP Cephalexin RS in the same medium.

Tolerances—Not less than 75% (*Q*) of the labeled amount of $C_{16}H_{17}N_3O_4S$ is dissolved in 45 minutes.

Uniformity of dosage units ⟨905⟩: meet the requirements.

Water, *Method I* ⟨921⟩: not more than 9.0%.

Assay—Weigh and finely powder not less than 20 Cephalexin Tablets. Transfer an accurately weighed quantity of the powder, equivalent to about 2000 mg of cephalexin, to a high-speed blender jar containing 200.0 mL of *Buffer No. 1* (see *Phosphate Buffers and Other Solutions* under *Antibiotics—Microbial Assays* ⟨81⟩), and blend for 3 to 5 minutes. Proceed as directed under *Antibiotics—Microbial Assays* ⟨81⟩, using an accurately measured volume of this stock solution diluted quantitatively and stepwise with *Buffer No. 1* to yield a *Test Dilution* having a concentration assumed to be equal to the median dose level of the Standard.

Cephalothin Sodium

$C_{16}H_{15}N_2NaO_6S_2$ 418.41

5-Thia-1-azabicyclo[4.2.0]oct-2-ene-2-carboxylic acid, 3-[(acetyloxy)methyl]-8-oxo-7-[(2-thienylacetyl)amino]-, monosodium salt, (6*R-trans*)-.

Monosodium (6*R*,7*R*)-3-(hydroxymethyl)-8-oxo-7-[2-(2-thienyl)acetamido]-5-thia-1-azabicyclo[4.2.0]oct-2-ene-2-carboxylate acetate (ester) [*58-71-9*].

» Cephalothin Sodium contains the equivalent of not less than 850 μg of cephalothin ($C_{16}H_{16}N_2O_6S_2$) per mg, calculated on the dried basis.

Packaging and storage—Preserve in tight containers.

Reference standard—*USP Cephalothin Sodium Reference Standard*—Dry in vacuum at a pressure not exceeding 5 mm of mercury at 60° for 3 hours before using.

Identification—The ultraviolet absorption spectrum of a solution (1 in 40,000) exhibits maxima and minima at the same wavelengths as that of a similar solution of USP Cephalothin Sodium RS, concomitantly measured.

Specific rotation ⟨781⟩: between +124° and +134°, determined in a solution containing a known amount of specimen, equivalent to about 50 mg of cephalothin, per mL, calculated on the dried basis.

Crystallinity ⟨695⟩: meets the requirements.

Pyrogen—When diluted with Sterile Water for Injection to a concentration of 50 mg of cephalothin ($C_{16}H_{16}N_2O_6S_2$) per mL, it meets the requirements of the *Pyrogen Test* ⟨151⟩, the test dose being 1.0 mL per kg.

pH ⟨791⟩: between 4.5 and 7.0, in a solution containing 250 mg per mL, or, where packaged for dispensing, in the solution constituted as directed in the labeling.

Loss on drying ⟨731⟩—Dry about 100 mg, accurately weighed, in a capillary-stoppered bottle in vacuum at a pressure not exceeding 5 mm of mercury at 60° for 3 hours: it loses not more than 1.5% of its weight.

Assay—Dissolve a suitable quantity of Cephalothin Sodium, accurately weighed, in *Buffer No. 1* (see *Phosphate Buffers and Other Solutions* under *Antibiotics—Microbial Assays* ⟨81⟩), and dilute quantitatively with *Buffer No. 1* to obtain a solution having a convenient concentration of cephalothin. Proceed as directed under *Antibiotics—Microbial Assays* ⟨81⟩, using an accurately measured volume of this solution diluted quantitatively with *Buffer No. 1* to yield a *Test Dilution* having a concentration assumed to be equal to the median dose level of the Standard.

Cephalothin Sodium Injection

» Cephalothin Sodium Injection is a sterile solution of Cephalothin Sodium, or Sterile Cephalothin Sodium, in Water for Injection. It may contain Dextrose or Sodium Chloride or one or more suitable buffers. It contains the equivalent of not less than 90.0 percent and not more than 115.0 percent of the labeled amount of cephalothin ($C_{16}H_{16}N_2O_6S_2$).

Packaging and storage—Preserve in *Containers for Injections* as described under *Injections* ⟨1⟩. Maintain in the frozen state.

Labeling—It meets the requirements for *Labeling* under *Injections* ⟨1⟩. The label states that it is to be thawed just prior to use, describes conditions for proper storage of the resultant solution, and directs that the solution is not to be refrozen.

Reference standard—*USP Cephalothin Sodium Reference Standard*—Dry in vacuum at a pressure not exceeding 5 mm of mercury at 60° for 3 hours before using.

pH ⟨791⟩: between 6.0 and 8.5.

Particulate matter ⟨788⟩: meets the requirements under *Small-volume Injections.*

Other requirements—It responds to the *Identification test* and meets the requirements of the tests for *Pyrogen* and *Sterility* under *Sterile Cephalothin Sodium.*

Assay—Proceed with Cephalothin Sodium Injection as directed under *Antibiotics—Microbial Assays* ⟨81⟩, using an accurately measured volume of Injection diluted quantitatively and stepwise with *Buffer No. 1* to yield a *Test Dilution* having a concentration assumed to be equal to the median dose level of the Standard.

Cephalothin Sodium for Injection

» Cephalothin Sodium for Injection is a sterile mixture of Sterile Cephalothin Sodium or cephalothin sodium and one or more suitable buffers. It has a potency equivalent to not less than 850 μg of cephalothin ($C_{16}H_{16}N_2O_6S_2$) per mg, calculated on the dried and sodium bicarbonate–free basis. In addition, where packaged for dispensing, it contains the equivalent of not less than 90.0 percent and not more than 115.0 percent of the labeled amount of cephalothin ($C_{16}H_{16}N_2O_6S_2$).

Packaging and storage—Preserve in *Containers for Sterile Solids* as described under *Injections* ⟨1⟩.

Reference standard—*USP Cephalothin Sodium Reference Standard*—Dry in vacuum at a pressure not exceeding 5 mm of mercury at 60° for 3 hours before using.

Constituted solution—At the time of use, the constituted solution prepared from Cephalothin Sodium for Injection meets the requirements for *Constituted Solutions* under *Injections* ⟨1⟩.

Specific rotation ⟨781⟩: between +124° and +134°, in a solution containing a known amount of specimen, equivalent to about 50 mg of cephalothin, per mL, calculated on the dried and sodium bicarbonate–free basis. Determine its sodium bicarbonate content in a separate, accurately weighed, 1-g portion dissolved in 50 mL of water. Add methyl orange TS, and titrate with 0.1 *N* sulfuric acid VS. Each mL of 0.1 *N* sulfuric acid is equivalent to 8.401 mg of $NaHCO_3$.

pH ⟨791⟩: between 6.0 and 8.5, in the solution constituted as directed in the labeling.

Particulate matter ⟨788⟩: meets the requirements under *Small-volume Injections.*

Other requirements—It responds to the *Identification test* and meets the requirements for *Pyrogen, Sterility,* and *Loss on drying* under *Sterile Cephalothin Sodium.* It meets also the requirements for *Uniformity of Dosage Units* ⟨905⟩ and *Labeling* under *Injections* ⟨1⟩.

Assay—

Standard preparation—Dissolve a suitable quantity of USP Cephalothin Sodium RS, accurately weighed, in water to obtain a solution having a known concentration of about 1 mg of cephalothin per mL.

Assay preparation 1—Using a suitable quantity of Cephalothin Sodium for Injection, accurately weighed, proceed as directed under *Standard preparation.*

Assay preparation 2 (where it is packaged for dispensing and is represented as being in a single-dose container)—Constitute Cephalothin Sodium for Injection in a volume of water, accurately measured, corresponding to the volume of solvent specified in the labeling. Withdraw all of the withdrawable contents, using a suitable hypodermic needle and syringe, and dilute quantitatively with water to obtain a solution having a concentration of about 1 mg of cephalothin per mL.

Assay preparation 3 (where the label states the quantity of cephalothin in a given volume of constituted solution)—Constitute 1 container of Cephalothin Sodium for Injection in a volume of water, accurately measured, corresponding to the volume of solvent specified in the labeling. Dilute an accurately measured portion of the constituted solution quantitatively with water to obtain a solution having a concentration of about 1 mg of cephalothin per mL.

Procedure—Proceed as directed for *Procedure* in the section, *Antibiotics—Hydroxylamine Assay*, under *Automated Methods of Analysis* ⟨16⟩. Calculate the potency, in μg of cephalothin ($C_{16}H_{16}N_2O_6S_2$) per mg, of the Cephalothin Sodium for Injection taken by the formula:

$$(CP/W)(A_U/A_S),$$

in which *W* is the weight, in mg, of the Cephalothin Sodium for Injection taken in each mL of *Assay preparation 1*, and the other terms are as defined therein. Calculate the quantity, in mg, of cephalothin in the container, and in the portion of constituted solution taken, by the formula:

$$(L/D)(CP/1000)(A_U/A_S),$$

in which *L* is the labeled quantity of cephalothin in the container, or in the volume of constituted solution taken, and *D* is the concentration, in mg per mL, of cephalothin in *Assay preparation 2* or in *Assay preparation 3,* on the basis of the labeled quantity in the container, or in the portion of constituted solution taken, respectively, and the extent of dilution.

Sterile Cephalothin Sodium

» Sterile Cephalothin Sodium is cephalothin sodium suitable for parenteral use. It has a potency equivalent to not less than 850 μg of cephalothin ($C_{16}H_{16}N_2O_6S_2$) per mg, calculated on the dried basis. In addition, where packaged for dispensing, it contains the equivalent of not less than 90.0 percent and not more than 115.0 percent of the labeled amount of cephalothin ($C_{16}H_{16}N_2O_6S_2$).

Packaging and storage—Preserve in *Containers for Sterile Solids* as described under *Injections* ⟨1⟩.

Reference standard—*USP Cephalothin Sodium Reference Standard*—Dry in vacuum at a pressure not exceeding 5 mm of mercury at 60° for 3 hours before using.

Constituted solution—At the time of use, the constituted solution prepared from Sterile Cephalothin Sodium meets the requirements for *Constituted Solutions* under *Injections* ⟨1⟩.

Sterility—It meets the requirements under *Sterility Tests* ⟨71⟩, when tested as directed in the section, *Test Procedures Using Membrane Filtration.*

pH ⟨791⟩: between 4.5 and 7.0, in a solution containing 250 mg per mL, or, where packaged for dispensing, in the solution constituted as directed in the labeling.

Particulate matter ⟨788⟩: meets the requirements under *Small-volume Injections.*

Other requirements—It responds to the *Identification test* and meets the requirements of the tests for *Specific rotation, Crystallinity, Pyrogen,* and *Loss on drying* under *Cephalothin Sodium.* In addition, where packaged for dispensing it meets the requirements for *Injections* ⟨1⟩ and for *Uniformity of Dosage Units* ⟨905⟩.

Assay—

Assay preparation 1—Dissolve a suitable quantity of Sterile Cephalothin Sodium, accurately weighed, in *Buffer No. 1* (see *Phosphate Buffers and Other Solutions* under *Antibiotics—Microbial Assays* ⟨81⟩), and dilute quantitatively with *Buffer No. 1* to obtain a solution having a convenient concentration of cephalothin.

Assay preparation 2 (where it is packaged for dispensing and is represented as being in a single-dose container)—Constitute Sterile Cephalothin Sodium in a volume of water, accurately measured, corresponding to the volume of solvent specified in the labeling. Withdraw all of the withdrawable contents using a suitable hypodermic needle and syringe, and dilute quantitatively with *Buffer No. 1* to obtain a solution having a convenient concentration of cephalothin.

Assay preparation 3 (where the label states the quantity of cephalothin in a given volume of constituted solution)—Constitute Sterile Cephalothin Sodium in a volume of water, accurately measured, corresponding to the volume of solvent specified in the labeling. Dilute an accurately measured volume of the constituted solution quantitatively with *Buffer No. 1* to obtain a solution having a convenient concentration of cephalothin.

Procedure—Proceed as directed under *Antibiotics—Microbial Assays* ⟨81⟩, using an accurately measured volume of *Assay preparation* diluted quantitatively with *Buffer No. 1* to obtain a *Test Dilution* having a concentration assumed to be equal to the median dose level of the Standard.

Sterile Cephapirin Sodium

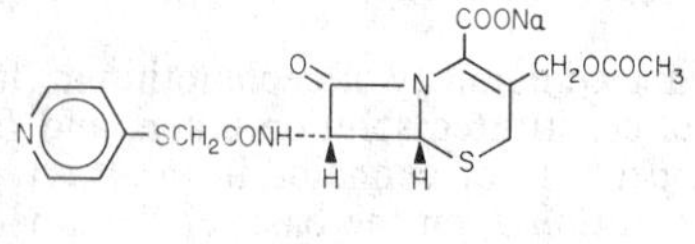

$C_{17}H_{16}N_3NaO_6S_2$ 445.44

5-Thia-1-azabicyclo[4.2.0]oct-2-ene-2-carboxylic acid, 3-[(acetyloxy)methyl]-8-oxo-7-[[(4-pyridinylthio)acetyl]amino]-, monosodium salt, (6*R-trans*)-.

Monosodium (6*R*,7*R*)-3-(hydroxymethyl)-8-oxo-7-[2-(4-pyridylthio)acetamido]-5-thia-1-azabicyclo[4.2.0]oct-2-ene-2-carboxylate acetate (ester) [*24356-60-3*].

» Sterile Cephapirin Sodium has a potency equivalent to not less than 855 μg and not more than 1000 μg of cephapirin ($C_{17}H_{17}N_3O_6S_2$) per mg. In addition, where packaged for dispensing, it contains the equivalent of not less than 90.0 percent and not more than 115.0 percent of the labeled amount of cephapirin.

Packaging and storage—Preserve in *Containers for Sterile Solids* as described under *Injections* ⟨1⟩.

Reference standard—*USP Cephapirin Sodium Reference Standard*—Do not dry before using.

Constituted solution—At the time of use, the constituted solution prepared from Sterile Cephapirin Sodium meets the requirements for *Constituted Solutions* under *Injections* ⟨1⟩.

Identification—

A: The infrared absorption spectrum of a potassium bromide dispersion of it exhibits maxima only at the same wavelengths as that of a similar preparation of USP Cephapirin Sodium RS.

B: It responds to the tests for *Sodium* ⟨191⟩.

Crystallinity ⟨695⟩: meets the requirements.

Pyrogen—When diluted with Sterile Water for Injection to a concentration of 100 mg of cephapirin ($C_{17}H_{17}N_3O_6S_2$) per mL, it meets the requirements of the *Pyrogen Test* ⟨151⟩, the test dose being 1.0 mL per kg.

Sterility—It meets the requirements under *Sterility Tests* ⟨71⟩, when tested as directed in the section, *Test Procedures Using Membrane Filtration.*

pH ⟨791⟩: between 6.5 and 8.5, in a solution containing 10 mg of cephapirin per mL.

Water, *Method I* ⟨921⟩: not more than 2.0%.

Particulate matter ⟨788⟩: meets the requirements under *Small-volume Injections.*

Other requirements—It meets the requirements for *Uniformity of Dosage Units* ⟨905⟩ and *Labeling* under *Injections* ⟨1⟩.

Assay—

Mobile phase—Prepare a filtered and degassed mixture of water, dimethylformamide, glacial acetic acid, and 11.7 *N* potassium hydroxide (1894:100:4:2). Make adjustments if necessary (see *System Suitability* under *Chromatography* ⟨621⟩). Increase the proportion of dimethylformamide to decrease the retention time of cephapirin.

Internal standard solution—Transfer about 500 mg of acetanilide to a 200-mL volumetric flask, add 5.0 mL of dimethylformamide, swirl to dissolve the acetanilide, dilute with *Mobile phase* to volume, and mix.

Standard preparation—Transfer about 21 mg of USP Cephapirin Sodium RS, accurately weighed, to a 100-mL volumetric flask, add 5.0 mL of *Internal standard solution,* dilute with *Mobile phase* to volume, and mix.

Assay preparation 1—Transfer about 42 mg of Sterile Cephapirin Sodium, accurately weighed, to a 200-mL volumetric flask, add 10.0 mL of *Internal standard solution,* dilute with *Mobile phase* to volume, and mix.

Assay preparation 2 (where it is packaged for dispensing and is represented as being in a single-dose container)—Constitute Sterile Cephapirin Sodium in a volume of water, accurately measured, corresponding to the volume of solvent specified in the labeling. Withdraw all of the withdrawable contents, using a suitable hypodermic needle and syringe, and dilute quantitatively and stepwise, if necessary, with *Mobile phase* to obtain a solution containing the equivalent of about 4 mg of cephapirin per mL. Transfer 10.0 mL of this solution to a 200-mL volumetric flask, add 10.0 mL of *Internal standard solution,* dilute with *Mobile phase* to volume, and mix. This solution contains the equivalent of about 0.2 mg of cephapirin per mL.

Assay preparation 3 (where the label states the quantity of cephapirin in a given volume of constituted solution)—Constitute Sterile Cephapirin Sodium in a volume of water, accurately measured, corresponding to the volume of solvent specified in the labeling. Dilute an accurately measured volume of the constituted solution quantitatively and stepwise, if necessary, with *Mobile phase* to obtain a solution containing the equivalent of about 4 mg of cephapirin per mL. Transfer 10.0 mL of this solution to a 200-mL volumetric flask, add 10.0 mL of *Internal standard solution*, dilute with *Mobile phase* to volume, and mix. This solution contains the equivalent of about 0.2 mg of cephapirin per mL. [NOTE—Use the *Standard preparation* and the *Assay preparation*, within 1 hour.]

Chromatographic system (see *Chromatography* ⟨621⟩)—The liquid chromatograph is equipped with a 254-nm detector and a 4-mm × 30-cm column that contains packing L1. The flow rate is about 2 mL per minute. Chromatograph the *Standard preparation*, and record the peak responses as directed under *Procedure:* the capacity factor, k', for the cephapirin peak is between 7.1 and 10.1, the column efficiency determined from the cephapirin peak is not less than 1900 theoretical plates, the tailing factor for the cephapirin peak is not more than 2.5, the resolution, R, between the cephapirin and internal standard peaks is not less than 2.1, and the relative standard deviation for replicate injections is not more than 2.0%.

Procedure—Separately inject equal volumes (about 20 µL) of the *Standard preparation* and the *Assay preparation* into the chromatograph, record the chromatograms, and measurQthe responses for the major peaks. The relative retention times are about 0.75 for cephapirin and 1.0 for acetanilide. Calculate the quantity, in µg, of cephapirin ($C_{17}H_{17}N_3O_6S_2$) per mg of the Sterile Cephapirin Sodium taken by the formula:

$$200(CP/W)(R_U/R_S),$$

in which C is the concentration, in mg per mL, of USP Cephapirin Sodium RS in the *Standard preparation*, P is the potency, in µg of cephapirin per mg, of USP Cephapirin Sodium RS, W is the quantity, in mg, of Sterile Cephapirin Sodium taken to prepare *Assay preparation 1*, and R_U and R_S are the peak response ratios of the cephapirin peak to the internal standard peak obtained from the *Assay preparation* and the *Standard preparation*, respectively. Calculate the quantity, in mg, of cephapirin ($C_{17}H_{17}N_3O_6S_2$) withdrawn from the container, or in the portion of constituted solution taken by the formula:

$$(L/D)(CP/1000)(R_U/R_S),$$

in which L is the labeled quantity, in mg, of cephapirin in the single-dose container, or in the volume of constituted solution taken, and D is the concentration, in mg, per mL, of cephapirin in *Assay preparation 2* or in *Assay preparation 3*, on the basis of the labeled quantity in the container, or in the portion of constituted solution taken, respectively, and the extent of dilution.

Cephradine

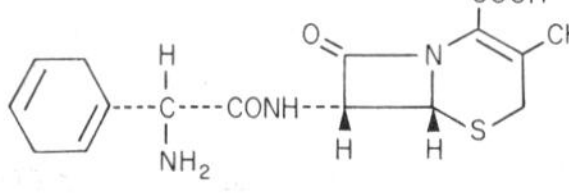

$C_{16}H_{19}N_3O_4S$ 349.40

5-Thia-1-azabicyclo[4.2.0]oct-2-ene-2-carboxylic acid, 7-[(amino-1,4-cyclohexadien-1-ylacetyl)amino]-3-methyl-8-oxo-, [6*R*-[6α,7β(*R**)]]-.

(6*R*,7*R*)-7-[(*R*)-2-Amino-2-(1,4-cyclohexadien-1-yl)acetamido]-3-methyl-8-oxo-5-thia-1-azabicyclo[4.2.0]oct-2-ene-2-carboxylic acid [*38821-53-3* (anhydrous)].

Monohydrate 367.42 [*31828-50-9* (non-stoichiometric hydrate)].

Dihydrate 385.43 [*58456-86-3*].

» Cephradine has a potency of not less than 900 µg and not more than 1050 µg of $C_{16}H_{19}N_3O_4S$ per mg, calculated on the anhydrous basis.

Packaging and storage—Preserve in tight containers.

Labeling—Where it is the dihydrate form, the label so indicates. Where the quantity of cephradine is indicated in the labeling of any preparation containing Cephradine, this shall be understood to be in terms of anhydrous cephradine ($C_{16}H_{19}N_3O_4S$).

Reference standards—*USP Cephradine Reference Standard*—Do not dry before using. *USP Cephalexin Reference Standard*—Do not dry before using.

Identification—The infrared absorption spectrum of a potassium bromide dispersion of it exhibits maxima only at the same wavelengths as that of a similar preparation of USP Cephradine RS (see *Spectrophotometry and Light-Scattering* ⟨851⟩). If a difference appears, dissolve equal portions of both the test specimen and the Reference Standard in methanol, evaporate the solution to dryness under identical conditions, and repeat the test on the residues.

Crystallinity ⟨695⟩: meets the requirements.

pH ⟨791⟩: between 3.5 and 6.0, in a solution containing 10 mg per mL.

Water, *Method I* ⟨921⟩: not more than 6.0%, except that if it is the dihydrate form, the limit is between 8.5% and 10.5%.

Cephalexin—

Mobile phase—Prepare a suitable mixture of water, methanol, 0.5 *M* sodium acetate, and 0.7 *N* acetic acid (782:200:15:3). Make adjustments if necessary (see *System Suitability* under *Chromatography* ⟨621⟩). Filter this solution through a membrane filter of 1-µm or finer porosity, and degas before use.

Standard preparation—Transfer about 20 mg of USP Cephalexin RS, accurately weighed, to a 50-mL volumetric flask, add 30 mL of *Mobile phase*, and sonicate. Dilute with *Mobile phase* to volume, and mix. Transfer 5.0 mL of this stock solution to a 100-mL volumetric flask, dilute with *Mobile phase* to volume, and mix.

Test preparation—Transfer about 40 mg of Cephradine, accurately weighed, to a 50-mL volumetric flask. Add 30 mL of *Mobile phase*, and sonicate. Dilute with *Mobile phase* to volume, and mix.

Resolution solution—Prepare a mixture of the *Test preparation* and the stock solution used to prepare the *Standard preparation* (10:1).

Chromatographic system (see *Chromatography* ⟨621⟩)—The liquid chromatograph is equipped with a 254-nm detector and a 4.6-mm × 25-cm column that contains 10-µm packing L1. The flow rate is about 1.2 mL per minute. Chromatograph the *Standard preparation* and the *Resolution solution*, and record the peak responses as directed under *Procedure:* the resolution, R, between the cephradine and cephalexin peaks is not less than 2.0, and the relative standard deviation for replicate injections of the *Standard preparation* is not more than 2.0%.

Procedure—Separately inject equal volumes (about 10 µL) of the *Standard preparation* and the *Test preparation* into the chromatograph, record the chromatograms, and measure the responses for the major peaks. Calculate the percentage of cephalexin ($C_{16}H_{17}N_3O_4S$) in the portion of Cephradine taken by the formula:

$$5(CP/W)(r_U/r_S),$$

in which C is the concentration, in mg per mL, of USP Cephalexin RS in the *Standard preparation*, P is the potency, in µg per mg, of the USP Cephalexin RS, W is the quantity, in mg, of Cephradine taken to prepare the *Test preparation*, and r_U and r_S are the cephalexin peak responses of the *Test preparation* and the *Standard preparation*, respectively: not more than 5.0%, calculated on the anhydrous basis, is found.

Assay—

Standard preparation—Dissolve a suitable quantity of USP Cephradine RS, accurately weighed, in water to obtain a solution having a known concentration of about 1 mg per mL.

Assay preparation—Using a suitable quantity of Cephradine, accurately weighed, proceed as directed under *Standard preparation*.

Procedure—Proceed as directed for *Procedure* in the section, *Antibiotics—Hydroxylamine Assay*, under *Automated Methods of Analysis* ⟨16⟩. Calculate the potency, in µg of $C_{16}H_{19}N_3O_4S$ in each mg of Cephradine taken by the formula:

$$(CP/M)(A_U/A_S),$$

in which M is the quantity, in mg, of Cephradine taken in each mL of the *Assay preparation*.

Cephradine Capsules

» Cephradine Capsules contain not less than 90.0 percent and not more than 120.0 percent of the labeled amount of $C_{16}H_{19}N_3O_4S$.

Packaging and storage—Preserve in tight containers.

Labeling—The quantity of cephradine stated in the labeling is in terms of anhydrous cephradine ($C_{16}H_{19}N_3O_4S$).

Reference standard—*USP Cephradine Reference Standard*—Do not dry before using.

Identification—Mix the contents of 1 Capsule with water to obtain a solution having a concentration of about 3 mg of cephradine per mL, and filter (test solution). Place a suitable thin-layer chromatographic plate (see *Chromatography* ⟨621⟩) coated with a 0.25-mm layer of binder-free silica gel in a chamber containing a mixture of *n*-hexane and tetradecane (95:5) to a depth of about 1 cm, allow the solvent front to move the length of the plate, remove the plate from the chamber, and allow the solvent to evaporate. On this plate apply 10 µL each of the test solution and a Standard solution containing 3 mg of USP Cephradine RS per mL. Allow the spots to dry, and develop the chromatogram in a solvent system consisting of a mixture of 0.1 *M* citric acid, 0.1 *M* dibasic sodium phosphate, and a 1 in 15 solution of ninhydrin in acetone (60:40:1.5) until the solvent front has moved about three-fourths of the length of the plate. Remove the plate from the developing chamber, mark the solvent front, dry the plate for 10 minutes at 110°, and examine the chromatogram: the R_f value of the principal spot obtained from the test solution corresponds to that obtained from the Standard solution.

Dissolution ⟨711⟩—

Medium: 0.12 *N* hydrochloric acid; 900 mL.

Apparatus 1: 100 rpm.

Time: 45 minutes.

Procedure—Determine the amount of $C_{16}H_{19}N_3O_4S$ dissolved from ultraviolet absorbances at the wavelength of maximum absorbance at about 255 nm of filtered portions of the solution under test, suitably diluted with *Dissolution Medium*, if necessary, in comparison with a Standard solution having a known concentration of USP Cephradine RS in the same medium.

Tolerances—Not less than 75% (*Q*) of the labeled amount of $C_{16}H_{19}N_3O_4S$ is dissolved in 45 minutes.

Uniformity of dosage units ⟨905⟩: meet the requirements.

Loss on drying ⟨731⟩—Dry about 100 mg, accurately weighed, of the mixed contents of 4 Cephradine Capsules in a capillary-stoppered bottle in vacuum at a pressure not exceeding 5 mm of mercury at 60° for 3 hours: the Capsule contents lose not more than 7.0% of their weight.

Assay—

Standard preparation—Dissolve a suitable quantity of USP Cephradine RS, accurately weighed, in water to obtain a solution having a known concentration of about 1 mg per mL.

Assay preparation—Place not less than 5 Cephradine Capsules in a high-speed glass blender jar containing an accurately measured volume of water, so that the stock solution obtained after blending for 3 to 5 minutes contains a convenient concentration of cephradine. Dilute an accurately measured volume of this stock solution quantitatively and stepwise with water to obtain an *Assay preparation* containing about 1 mg of cephradine per mL.

Procedure—Proceed as directed for *Procedure* in the section, *Antibiotics—Hydroxylamine Assay*, under *Automated Methods of Analysis* ⟨16⟩. Calculate the quantity, in mg, of $C_{16}H_{19}N_3O_4S$ in each Capsule taken by the formula:

$$(T/D)(C)(A_U/A_S),$$

in which T is the labeled quantity, in mg, of cephradine in each Capsule, and D is the concentration, in mg per mL, of cephradine in the *Assay preparation*, based on the labeled quantity per Capsule and the extent of dilution.

Cephradine for Injection

» Cephradine for Injection is a dry mixture of Cephradine and one or more suitable buffers and solubilizers. It contains not less than 90.0 percent and not more than 115.0 percent of the labeled amount of $C_{16}H_{19}N_3O_4S$.

Packaging and storage—Preserve in *Containers for Sterile Solids* as described under *Injections* ⟨1⟩.

Reference standard—*USP Cephradine Reference Standard*—Do not dry before using.

Identification—Dilute the contents of 1 container of Cephradine for Injection with water to obtain a test solution containing about 3 mg of cephradine per mL. Proceed as directed in the *Identification test* under *Cephradine Capsules*, beginning with "Place a suitable thin-layer chromatographic plate": the R_f value of the principal spot obtained from the test solution corresponds to that obtained from the Standard solution.

pH ⟨791⟩: between 8.0 and 9.6, in a solution containing 10 mg per mL.

Loss on drying ⟨731⟩—Dry about 100 mg, accurately weighed, in a capillary-stoppered bottle in vacuum at a pressure not exceeding 5 mm of mercury at 60° for 3 hours: it loses not more than 5.0% of its weight.

Particulate matter ⟨788⟩: meets the requirements under *Small-volume Injections*.

Other requirements—It meets the requirements for *Pyrogen*, *Sterility*, and *Assay* (*Assay preparation 2* and *Assay preparation 3* being used) under *Sterile Cephradine*. It meets also the requirements for *Uniformity of Dosage Units* ⟨905⟩ and *Labeling* under *Injections* ⟨1⟩.

Sterile Cephradine

» Sterile Cephradine has a potency of not less than 900 µg and not more than 1050 µg of $C_{16}H_{19}N_3O_4S$ per mg, calculated on the anhydrous basis. In addition, where packaged for dispensing, it contains not less than 90.0 percent and not more than 115.0 percent of the labeled amount of $C_{16}H_{19}N_3O_4S$.

Packaging and storage—Preserve in *Containers for Sterile Solids* as described under *Injections* ⟨1⟩.

Reference standards—*USP Cephradine Reference Standard*—Do not dry before using. *USP Cephalexin Reference Standard*—Do not dry before using.

Constituted solution—At the time of use, the constituted solution prepared from Sterile Cephradine meets the requirements for *Constituted Solutions* under *Injections* ⟨1⟩.

Pyrogen—When diluted with pyrogen-free sodium carbonate solution, prepared by dissolving 2.56 g of sodium carbonate, previously heated at 170° for not less than 4 hours, in 100 mL of Sterile Water for Injection, to a concentration of 80 mg of cephradine ($C_{16}H_{19}N_3O_4S$) per mL, it meets the requirements of the *Pyrogen Test* ⟨151⟩, the test dose being 1.0 mL per kg.

Sterility—It meets the requirements under *Sterility Tests* ⟨71⟩, when tested as directed in the section, *Test Procedures Using Membrane Filtration*.

Other requirements—It conforms to the definition, responds to the *Identification test*, and meets the requirements for *pH*, *Water*, *Crystallinity*, and *Cephalexin* under *Cephradine*. It meets also the requirements for *Uniformity of Dosage Units* ⟨905⟩ and *Labeling* under *Injections* ⟨1⟩.

Assay—

Standard preparation—Dissolve a suitable quantity of USP Cephradine RS, accurately weighed, in water to obtain a solution having a known concentration of about 1 mg per mL.

Assay preparation 1—Dissolve an accurately weighed quantity of Sterile Cephradine in water, and dilute quantitatively with water to obtain a solution containing about 1 mg per mL.

Assay preparation 2 (where it is packaged for dispensing and is represented as being in a single-dose container)—Constitute Sterile Cephradine in a volume of water, accurately measured, corresponding to the volume of solvent specified in the labeling. Withdraw all of the withdrawable contents, using a suitable hypodermic needle and syringe, and dilute quantitatively with water to obtain a solution containing about 1 mg per mL.

Assay preparation 3 (where the label states the quantity of cephradine in a given volume of constituted solution)—Constitute Sterile Cephradine in a volume of water, accurately measured, corresponding to the volume of solvent specified in the labeling. Dilute an accurately measured volume of the constituted solution quantitatively with water to obtain a solution containing about 1 mg per mL.

Procedure—Proceed as directed for *Procedure* in the section, *Antibiotics—Hydroxylamine Assay* under *Automated Methods of Analysis* ⟨16⟩. Calculate the quantity, in µg of cephradine ($C_{16}H_{17}N_3O_4S$) per mg, of the Sterile Cephradine taken by the formula:

$$(CP/M)(A_U/A_S),$$

in which M is the concentration, in mg per mL, of *Assay preparation 1* based on the weight of Sterile Cephradine taken and the extent of dilution. Calculate the quantity, in mg, of cephradine withdrawn from the container, or in the portion of constituted solution taken by the formula:

$$(CP)(L/1000D)(A_U/A_S),$$

in which L is the labeled quantity, in mg of cephradine in the container, or in the volume of constituted solution taken, and D is the concentration, in mg of cephradine per mL, of *Assay preparation 2* or *Assay preparation 3*, based on the labeled quantity in the container or in the portion of constituted solution taken, respectively, and the extent of dilution.

Cephradine for Oral Suspension

» Cephradine for Oral Suspension is a dry mixture of Cephradine and one or more suitable buffers, colors, diluents, and flavors. It contains not less than 90.0 percent and not more than 125.0 percent of the labeled amount of $C_{16}H_{19}N_3O_4S$.

Packaging and storage—Preserve in tight containers.

Reference standard—*USP Cephradine Reference Standard*—Do not dry before using.

Identification—Constitute 1 container of Cephradine for Oral Suspension as directed in the labeling. Mix a portion of the resulting suspension with water to obtain a concentration of about 3 mg of cephradine per mL, and filter (test solution). Proceed as directed in the *Identification test* under *Cephradine Capsules*, beginning with "Place a suitable thin-layer chromatographic plate": the R_f value of the principal spot obtained from the test solution corresponds to that obtained from the Standard solution.

Uniformity of dosage units ⟨905⟩—

FOR SOLID PACKAGED IN SINGLE-UNIT CONTAINERS: meets the requirements.

pH ⟨791⟩: between 3.5 and 6.0, in the suspension constituted as directed in the labeling.

Water, *Method I* ⟨921⟩: not more than 1.5%.

Assay—

Standard preparation—Dissolve a suitable quantity of USP Cephradine RS, accurately weighed, in water to obtain a solution having a known concentration of about 1 mg per mL.

Assay preparation—Constitute Cephradine for Oral Suspension as directed in the labeling. Dilute an accurately measured volume of the suspension so obtained, freshly mixed and free from air bubbles, quantitatively with water to obtain a solution containing about 1 mg of cephradine per mL.

Procedure—Proceed as directed for *Procedure* in the section, *Antibiotics—Hydroxylamine Assay* under *Automated Methods of Analysis* ⟨16⟩. Calculate the quantity, in mg, of $C_{16}H_{19}N_3O_4S$ in each mL of the constituted Cephradine for Oral Suspension taken by the formula:

$$(T/D)(CP)(A_U/A_S),$$

in which T is the labeled quantity, in mg, of cephradine in each mL of the constituted Cephradine for Oral Suspension, and D is the concentration, in mg per mL, of cephradine in the *Assay preparation*, based on the labeled quantity per mL and the extent of dilution.

Cephradine Tablets

» Cephradine Tablets contain not less than 90.0 percent and not more than 120.0 percent of the labeled amount of $C_{16}H_{19}N_3O_4S$.

Packaging and storage—Preserve in tight containers.

Reference standard—*USP Cephradine Reference Standard*—Do not dry before using.

Identification—Mix a quantity of finely powdered Tablets with water to obtain a concentration of about 3 mg of cephradine per mL, and filter (test solution). Proceed as directed in the *Identification test* under *Cephradine Capsules,* beginning with "Place a suitable thin-layer chromatographic plate": the R_f value of the principal spot obtained from the test solution corresponds to that obtained from the Standard solution.

Dissolution ⟨711⟩—

Medium: 0.12 N hydrochloric acid; 900 mL.

Apparatus 2: 75 rpm.

Time: 60 minutes.

Procedure—Determine the amount of $C_{16}H_{19}N_3O_4S$ dissolved from ultraviolet absorbances at the wavelength of maximum absorbance at about 255 nm of filtered portions of the solution under test, suitably diluted with *Dissolution Medium*, in comparison with a Standard solution having a known concentration of USP Cephradine RS in the same medium.

Tolerances—Not less than 85% (Q) of the labeled amount of $C_{16}H_{19}N_3O_4S$ is dissolved in 60 minutes.

Uniformity of dosage units ⟨905⟩: meet the requirements.

Water, *Method I* ⟨921⟩: not more than 6.0%, 20 mL of a mixture of carbon tetrachloride, chloroform, and methanol (2:2:1) being used in place of methanol in the titration vessel.

Assay—

Standard preparation—Prepare as directed for *Standard preparation* in the section, *Antibiotics—Hydroxylamine Assay*, under *Automated Methods of Analysis* ⟨16⟩, using USP Cephradine RS.

Assay preparation—Place not less than 5 Cephradine Tablets in a high-speed glass blender jar containing an accurately measured volume of water and blend for 4 ± 1 minutes. Dilute an accurately measured volume of this stock solution quantitatively and stepwise with water to obtain an *Assay preparation* containing about 1 mg of cephradine per mL.

Procedure—Proceed as directed for *Procedure* in the section, *Antibiotics—Hydroxylamine Assay*, under *Automated Methods of Analysis* ⟨16⟩. Calculate the quantity, in mg, of $C_{16}H_{19}N_3O_4S$ in each Tablet taken by the formula:

$$(T/D)(C)(A_U/A_S),$$

in which T is the labeled quantity, in mg, of cephradine in each

Tablet, and D is the concentration, in mg per mL, of cephradine in the *Assay preparation*, based on the labeled quantity per Tablet and the extent of dilution.

Cetostearyl Alcohol—*see* Cetostearyl Alcohol NF
Cetyl Alcohol—*see* Cetyl Alcohol NF
Cetyl Esters Wax—*see* Cetyl Esters Wax NF

Cetylpyridinium Chloride

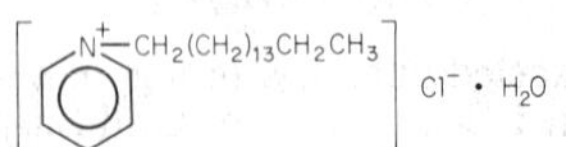

$C_{21}H_{38}ClN.H_2O$ 358.01
Pyridinium, 1-hexadecyl-, chloride, monohydrate.
1-Hexadecylpyridinium chloride monohydrate [*6004-24-6*].
Anhydrous 339.99 [*123-03-5*].

» Cetylpyridinium Chloride contains not less than 99.0 percent and not more than 102.0 percent of $C_{21}H_{38}ClN$, calculated on the anhydrous basis.

Packaging and storage—Preserve in well-closed containers.

Reference standard—*USP Cetylpyridinium Chloride Reference Standard*—Do not dry before using.

Identification—

A: The infrared absorption spectrum of a potassium bromide dispersion of it exhibits maxima only at the same wavelengths as that of a similar preparation of USP Cetylpyridinium Chloride RS.

B: The ultraviolet absorption spectrum of a 1 in 25,000 solution of it exhibits maxima and minima at the same wavelengths as that of a similar solution of USP Cetylpyridinium Chloride RS, concomitantly measured.

C: Dissolve 100 mg in 50 mL of water: a 10-mL portion of the solution responds to the tests for *Chloride* ⟨191⟩, except that a turbidity is produced, rather than a curdy white precipitate, when the silver nitrate TS is added.

Melting range, *Class I* ⟨741⟩: between 80° and 84°, the preliminary drying treatment being omitted.

Acidity—Dissolve 500 mg, accurately weighed, in 50 mL of water, add phenolphthalein TS, and titrate with 0.020 N sodium hydroxide: not more than 2.5 mL is required for neutralization.

Water, *Method I* ⟨921⟩: between 4.5% and 5.5%.

Residue on ignition ⟨281⟩: not more than 0.2%, calculated on the anhydrous basis.

Heavy metals, *Method II* ⟨231⟩: 0.002%.

Pyridine—Dissolve 1 g in 10 mL of sodium hydroxide solution (1 in 10) without heating: the odor of pyridine is not immediately perceptible.

Assay—Transfer about 200 mg of Cetylpyridinium Chloride, accurately weighed, to a glass-stoppered, 250-mL graduated cylinder containing 75 mL of water. Add 10 mL of chloroform, 0.4 mL of bromophenol blue solution (1 in 2000), and 5 mL of a freshly prepared solution of sodium bicarbonate (4.2 in 1000), and titrate with 0.02 M sodium tetraphenylboron VS until the blue color disappears from the chloroform layer. Add the last portions of the sodium tetraphenylboron solution dropwise, agitating vigorously after each addition. Each mL of 0.02 M sodium tetraphenylboron is equivalent to 6.800 mg of $C_{21}H_{38}ClN$.

Cetylpyridinium Chloride Lozenges

» Cetylpyridinium Chloride Lozenges contain not less than 90.0 percent and not more than 125.0 percent of the labeled amount of $C_{21}H_{38}ClN.H_2O$ in a suitable molded base.

Packaging and storage—Preserve in well-closed containers.

Reference standard—*USP Cetylpyridinium Chloride Reference Standard*—Do not dry before using.

Identification—

Chromatographic column—Pack a pledget of fine glass wool in the base of a 10-mm × 200-mm chromatographic tube. Add styrene-divinylbenzene cation-exchange resin (strong acid form), to form a uniform column 12 cm in height, and top the column with a pledget of fine glass wool.

Procedure—Weigh and finely powder not less than 20 Lozenges. Dissolve a portion of the powder, equivalent to about 500 µg of cetylpyridinium chloride, in about 50 mL of water, and immediately transfer this solution to the *Chromatographic column*. Discard the eluate, wash the column, successively, with 200 mL of water, 100 mL of alcohol, 100 mL of water, and 100 mL of 3 N hydrochloric acid, and discard the washings. Elute the column with 80 mL of a solvent consisting of a mixture of 7 volumes of alcohol and 3 volumes of 1.2 N hydrochloric acid. Collect the eluate in a 100-mL volumetric flask, dilute with the eluting solvent to volume, and mix. The ultraviolet absorption spectrum of this solution, measured between 225 and 300 nm, exhibits maxima and minima at the same wavelengths as that of a solution of USP Cetylpyridinium Chloride RS, in the same medium at a concentration of 5 µg per mL, concomitantly measured.

Assay—

0.004 M Sodium lauryl sulfate—[NOTE—Sulfuric acid is included in this solution to inhibit precipitate formation. If a precipitate forms under storage, discard the solution, and prepare and standardize a fresh *0.004 M Sodium lauryl sulfate* solution.] Dissolve 1.15 g of sodium lauryl sulfate in 500 mL of water, add 2 mL of sulfuric acid, dilute with water to 1000 mL, and mix. Determine the molarity of the solution as follows: Transfer to a glass-stoppered, 100-mL cylinder 10.0 mL of 0.004 M cetylpyridinium chloride (1.432 mg of USP Cetylpyridinium Chloride RS per mL in water), add 5 mL of 2 N sulfuric acid, 20 mL of chloroform, and 1 mL of methyl yellow TS, and titrate with the sodium lauryl sulfate solution, with frequent vigorous shaking, until the chloroform layer acquires the first permanent orange-pink color. Calculate the molarity, and restandardize before each use.

Procedure—Dissolve an accurately determined number (about 100) of Cetylpyridinium Chloride Lozenges in about 400 mL of water in a 500-mL volumetric flask, dilute with water to volume, and mix. Transfer an accurately measured aliquot of this solution, equivalent to about 10 mg of cetylpyridinium chloride, to a glass-stoppered, 100-mL cylinder, and add 5 mL of 2 N sulfuric acid, 20 mL of chloroform, and 1 mL of methyl yellow TS. Insert the stopper, shake until the chloroform layer develops a bright yellow color, and titrate with *0.004 M Sodium lauryl sulfate*, shaking thoroughly after each addition, until the chloroform layer develops the first permanent orange-pink color. Each mL of *0.004 M Sodium lauryl sulfate* is equivalent to 1.432 mg of $C_{21}H_{38}ClN.H_2O$.

Cetylpyridinium Chloride Topical Solution

» Cetylpyridinium Chloride Topical Solution contains not less than 95.0 percent and not more than 105.0 percent of the labeled amount of $C_{21}H_{38}ClN.H_2O$.

Packaging and storage—Preserve in tight containers.

Reference standard—*USP Cetylpyridinium Chloride Reference Standard*—Do not dry before using.

Identification—

A: Dilute a volume of Topical Solution to a concentration of about 40 µg of cetylpyridinium chloride per mL: the ultraviolet absorption spectrum of the resulting solution exhibits maxima and minima at the same wavelengths as that of a similar solution of USP Cetylpyridinium Chloride RS, concomitantly measured.

B: Evaporate a volume of Topical Solution, equivalent to about 500 mg of cetylpyridinium chloride, on a steam bath to one-half of its original volume: The resulting solution responds to *Identification tests D, E,* and *F* under *Cetylpyridinium Chloride.*

Assay—Transfer a volume of Cetylpyridinium Chloride Topical Solution, equivalent to about 150 mg of cetylpyridinium chloride, to a glass-stoppered, 500-mL graduated cylinder, and proceed as directed in the *Assay* under *Cetylpyridinium Chloride,* beginning with "Add 10 mL of chloroform." Each mL of 0.02 *M* sodium tetraphenylboron is equivalent to 7.160 mg of $C_{21}H_{38}ClN.H_2O$.

Activated Charcoal

» Activated Charcoal is the residue from the destructive distillation of various organic materials, treated to increase its adsorptive power.

Packaging and storage—Preserve in well-closed containers.

Microbial limits—It meets the requirements of the tests for absence of *Salmonella* species and *Escherichia coli* under *Microbial Limit Tests* ⟨61⟩.

Reaction—Boil 3.0 g with 60 mL of water for 5 minutes, allow to cool, restore the original volume by the addition of water, and filter: the filtrate is colorless and is neutral to litmus.

Loss on drying ⟨731⟩—Dry it at 120° for 4 hours: it loses not more than 15.0% of its weight.

Residue on ignition ⟨281⟩: not more than 4.0%, a 0.50-g test specimen being used.

Acid-soluble substances—Boil 1.0 g with a mixture of 20 mL of water and 5 mL of hydrochloric acid for 5 minutes, filter into a tared porcelain crucible, and wash the residue with 10 mL of hot water, adding the washing to the filtrate. To the combined filtrate and washing add 1 mL of sulfuric acid, evaporate to dryness, and ignite to constant weight: the residue weighs not more than 35 mg (3.5%).

Chloride ⟨221⟩—A 10-mL portion of the filtrate obtained in the test for *Reaction* shows no more chloride than is contained in 1.5 mL of 0.020 *N* hydrochloric acid (0.2%).

Sulfate ⟨221⟩—A 10-mL portion of the filtrate obtained in the test for *Reaction* shows no more sulfate than is contained in 1.0 mL of 0.020 *N* sulfuric acid (0.2%).

Sulfide—Place 0.50 g in a small conical flask, add 20 mL of water and 5 mL of hydrochloric acid, and boil gently: the escaping vapors do not darken paper moistened with lead acetate TS.

Cyanogen compounds—Place a mixture of 5 g of Activated Charcoal, 50 mL of water, and 2 g of tartaric acid in a distilling flask connected to a condenser provided with a tightly fitting adapter, the end of which dips below the surface of a mixture of 2 mL of 1 *N* sodium hydroxide and 10 mL of water, contained in a small flask surrounded by ice. Heat the mixture in the distilling flask to boiling, and distil about 25 mL. Dilute the distillate with water to 50 mL, and mix. To 25 mL of the diluted distillate add 12 drops of ferrous sulfate TS, heat the mixture almost to boiling, cool, and add 1 mL of hydrochloric acid: no blue color is produced.

Heavy metals ⟨231⟩—Boil 1.0 g with a mixture of 20 mL of 3 *N* hydrochloric acid and 5 mL of bromine TS for 5 minutes, filter, and wash the charcoal and the filter with 50 mL of boiling water. Evaporate the filtrate and washing to dryness, and to the residue add 1 mL of 1 *N* hydrochloric acid, 20 mL of water, and 5 mL of sulfurous acid. Boil the solution until all of the sulfur dioxide is expelled, filter if necessary, and dilute with water to 50 mL. To 20 mL of the solution add water to make 25 mL: the limit is 0.005%.

Uncarbonized constituents—Boil 0.25 g with 10 mL of 1 *N* sodium hydroxide for 5 seconds, and filter: the filtrate is colorless.

Adsorptive power—

Alkaloids—Shake 1 g of Activated Charcoal, previously dried at 120° for 4 hours, with a solution of 100 mg of strychnine sulfate in 50 mL of water for 5 minutes, and filter through a dry filter, rejecting the first 10 mL of filtrate. To a 10-mL portion of the subsequent filtrate add 1 drop of hydrochloric acid and 5 drops of mercuric-potassium iodide TS: no turbidity is produced.

Dyes—Pipet 50 mL of methylene blue solution (1 in 1000) into each of two glass-stoppered, 100-mL flasks. Add to one flask 250 mg, accurately weighed, of Activated Charcoal, insert the stopper in the flask, and shake for 5 minutes. Filter the contents of each flask through a dry filter, rejecting the first 20 mL of each filtrate. Pipet 25-mL portions of the remaining filtrates into two 250-mL volumetric flasks. Add to each flask 50 mL of sodium acetate solution (1 in 10), mix, and add from a buret 35.0 mL of 0.1 *N* iodine VS, swirling the mixture during the addition. Insert the stoppers in the flasks, and allow them to stand for 50 minutes, shaking them vigorously at 10-minute intervals. Dilute each mixture with water to volume, mix, allow to stand for 10 minutes, and filter through dry filters, rejecting the first 30 mL of each filtrate. Titrate the excess iodine in a 100-mL aliquot of each subsequent filtrate with 0.1 *N* sodium thiosulfate VS, adding 3 mL of starch TS as the end-point is approached. Calculate the number of mL of 0.1 *N* iodine consumed in each titration: the difference between the two volumes is not less than 0.7 mL.

Chewable Tablets, Isosorbide Dinitrate—*see* Isosorbide Dinitrate Chewable Tablets

Chloral Hydrate

$CCl_3CH(OH)_2$

$C_2H_3Cl_3O_2$ 165.40
1,1-Ethanediol, 2,2,2-trichloro-.
Chloral hydrate [*302-17-0*].

» Chloral Hydrate contains not less than 99.5 percent and not more than 102.5 percent of $C_2H_3Cl_3O_2$.

Packaging and storage—Preserve in tight containers.

Identification—Transfer to a 125-mL conical flask a portion of a solution in water equivalent to about 1 mg of chloral hydrate, and add water to bring the volume to about 10 mL. Add 10 mL of 1-ethylquinaldinium iodide solution (15 in 1000), which has been filtered through a 0.45-µm filter. Add 60 mL of isopropyl alcohol, 5 mL of an aqueous 0.1 *M* monoethanolamine solution, and 15 mL of water. Mix, and heat in a water bath at 60° for 15 minutes: a blue color develops.

Acidity—A 1 in 20 solution in alcohol does not at once redden moistened blue litmus paper.

Residue on ignition ⟨281⟩: not more than 0.1%.

Chloride—To a 1 in 10 solution in alcohol add a few drops of silver nitrate TS: any opalescence produced does not exceed that of a control containing 0.10 mL of 0.020 *N* hydrochloric acid (0.007%).

Readily carbonizable substances ⟨271⟩—Shake 500 mg, at intervals of 5 minutes during 1 hour, with 5 mL of sulfuric acid TS in a glass-stoppered cylinder that previously has been rinsed with sulfuric acid TS, and transfer the mixture to a comparison vessel: the mixture has no more color than Matching Fluid P.

Assay—Dissolve about 4 g of Chloral Hydrate, accurately weighed, in 10 mL of water, add 30.0 mL of 1 *N* sodium hydroxide VS, and allow the mixture to stand for 2 minutes. Add a few drops of phenolphthalein TS, and titrate the residual alkali at once with 1 *N* sulfuric acid VS. Each mL of 1 *N* sodium hydroxide corresponds to 165.4 mg of $C_2H_3Cl_3O_2$.

Chloral Hydrate Capsules

» Chloral Hydrate Capsules contain not less than 95.0 percent and not more than 110.0 percent of the labeled amount of $C_2H_3Cl_3O_2$.

Packaging and storage—Preserve in tight containers, preferably at controlled room temperature.

Identification—The contents of the Capsules respond to the *Identification test* under *Chloral Hydrate.*

Uniformity of dosage units ⟨905⟩: meet the requirements.

Assay—Place a counted number of Chloral Hydrate Capsules, equivalent to about 2.5 g of chloral hydrate, in a glass-stoppered, 250-mL flask, add 25 mL of water, insert the stopper in the flask, and heat on a steam bath with frequent swirling until the Capsules are dissolved. Cool to room temperature, add 25 mL of neutralized alcohol and 20.0 mL of 1 *N* sodium hydroxide VS, mix, and allow the mixture to stand for 4 minutes. Add phenolphthalein TS, and titrate the excess alkali with 1 *N* sulfuric acid VS. Each mL of 1 *N* sodium hydroxide is equivalent to 165.4 mg of $C_2H_3Cl_3O_2$.

Chloral Hydrate Syrup

» Chloral Hydrate Syrup contains not less than 95.0 percent and not more than 110.0 percent of the labeled amount of $C_2H_3Cl_3O_2$.

Packaging and storage—Preserve in tight, light-resistant containers.

Identification—It responds to the *Identification* test under *Chloral Hydrate.*

Assay—Transfer 25.0 mL of Chloral Hydrate Syrup to a 250-mL conical flask with the aid of several portions of water. Add 30.0 mL of 1 *N* sodium hydroxide VS, and mix. After the mixture has stood for 2 minutes, add 5 drops of phenolphthalein TS, and immediately titrate the excess sodium hydroxide with 1 *N* sulfuric acid VS. Designate the volume of 1 *N* sodium hydroxide VS consumed as *A*. Transfer 5.0 mL of Syrup to a second 250-mL conical flask with the aid of several portions of water. Add 10 drops of phenolphthalein TS, and titrate with 0.1 *N* sodium hydroxide VS. Designate the volume of 0.1 *N* sodium hydroxide VS consumed as *B*. Calculate the weight, in mg, of $C_2H_3Cl_3O_2$ in the amount of Syrup taken by the first titration by the formula:

$$165.4(A - 0.5B).$$

Chlorambucil

$(ClCH_2CH_2)_2N-C_6H_4-CH_2CH_2CH_2COOH$

$C_{14}H_{19}Cl_2NO_2$ 304.22
Benzenebutanoic acid, 4-[bis(2-chloroethyl)amino]-.
4-[*p*-[Bis(2-chloroethyl)amino]phenyl]butyric acid
[*305-03-3*].

» Chlorambucil contains not less than 98.0 percent and not more than 101.0 percent of $C_{14}H_{19}Cl_2NO_2$, calculated on the anhydrous basis.

Caution—Great care should be taken to prevent inhaling particles of Chlorambucil and exposing the skin to it.

Packaging and storage—Preserve in tight, light-resistant containers.

Reference standard—*USP Chlorambucil Reference Standard*—[*Caution—Avoid contact.*] Dry over silica gel for 24 hours before using.

Identification—

A: The infrared absorption spectrum of a 1 in 125 solution in carbon disulfide, in a 1-mm cell, exhibits maxima only at the same wavelengths as that of a similar solution of USP Chlorambucil RS.

B: Dissolve 50 mg in 5 mL of acetone, and dilute with water to 10 mL. Add 1 drop of 2 *N* sulfuric acid, then add 4 drops of silver nitrate TS: no opalescence is observed immediately (*absence of chloride ion*). Warm the solution on a steam bath: opalescence develops (*presence of ionizable chlorine*).

Melting range ⟨741⟩: between 65° and 69°.

Water, *Method I* ⟨921⟩: not more than 0.5%.

Assay—Dissolve about 200 mg of Chlorambucil, accurately weighed, in 10 mL of acetone, add 10 mL of water, and titrate with 0.1 *N* sodium hydroxide VS, using phenolphthalein TS as the indicator. Each mL of 0.1 *N* sodium hydroxide is equivalent to 30.42 mg of $C_{14}H_{19}Cl_2NO_2$.

Chlorambucil Tablets

» Chlorambucil Tablets contain not less than 85.0 percent and not more than 110.0 percent of the labeled amount of $C_{14}H_{19}Cl_2NO_2$.

Packaging and storage—Preserve coated Tablets in well-closed containers; preserve uncoated Tablets in well-closed, light-resistant containers.

Reference standard—*USP Chlorambucil Reference Standard—[Caution—Avoid contact.]* Dry over silica gel for 24 hours before using.

Identification—Shake a quantity of finely powdered Tablets, equivalent to about 16 mg of chlorambucil, with 20 mL of carbon disulfide. Filter, evaporate to dryness, and dissolve the residue in 2 mL of carbon disulfide: the resulting solution responds to *Identification test A* under *Chlorambucil.*

Disintegration ⟨701⟩: 15 minutes, the use of disks being omitted.

Uniformity of dosage units ⟨905⟩: meet the requirements.

Assay—

Internal standard solution—Transfer about 20 mg of USP Propylparaben RS to a 50-mL volumetric flask, dissolve in alcohol, dilute with alcohol to volume, and mix.

Standard preparation—Dissolve in alcohol a suitable quantity of USP Chlorambucil RS, accurately weighed, and prepare, by quantitative dilution, a solution in alcohol having a known concentration of about 1 mg per mL. Transfer 2.0 mL of the solution to a 100-mL volumetric flask containing about 50 mL of alcohol and, while gently swirling, add 5.0 mL of 0.1 *N* hydrochloric acid and 2.0 mL of *Internal standard solution.* Dilute with alcohol to volume, and mix.

Assay preparation—Weigh and finely powder not less than 20 Chlorambucil Tablets. Transfer an accurately weighed portion of the powder, equivalent to about 2 mg of chlorambucil, to a 100-mL volumetric flask containing about 50 mL of alcohol and, while gently swirling, add 5.0 mL of 0.1 *N* hydrochloric acid and 2.0 mL of *Internal standard solution.* Sonicate for 5 minutes, dilute with alcohol to volume, and mix. Filter through a medium-porosity, sintered-glass filtering funnel, maintaining reduced pressure for the minimum necessary time in order to avoid solvent loss of evaporation.

Mobile phase—Mix 500 mL of alcohol with 1.0 mL of glacial acetic acid in a 1-liter volumetric flask, dilute with water to volume, and mix. The alcohol concentration may be varied to meet system suitability requirements and to provide a suitable elution time for chlorambucil. Degas the solution at a pressure of approximately 250 mm of mercury for 2 minutes.

Chromatographic system—Typically, a high-pressure liquid chromatograph, operated at room temperature, is fitted with a

25- or 30-cm × 2-mm stainless steel column packed with spherical silica microbeads, 5 µm to 10 µm in diameter, to which is bonded a nominal 10% or 20% (w/w) octadecyl silane. The mobile phase is maintained at a flow rate capable of giving the required resolution (see *System suitability test*) and a suitable elution time. An ultraviolet detector that monitors absorption at the 254-nm wavelength is used.

System suitability test—Chromatograph 6 to 10 injections of the *Standard preparation*, and measure the peak responses as directed under *Procedure*. Calculate the response factor, P_S, and the resolution factor, *R* (see *Chromatography* ⟨621⟩). The response factor for 6 to 8 injections does not vary by more than 2.0% relative standard deviation and the resolution factor is not less than 2.0.

Procedure—Introduce equal volumes (10 to 12 µL) of the *Standard preparation* and the *Assay preparation* into the high-pressure liquid chromatograph by means of a suitable sampling valve or high-pressure microsyringe, and measure the peak responses at identical retention times obtained with each preparation. Calculate the quantity, in mg, of $C_{14}H_{19}Cl_2NO_2$ in the portion of Tablets taken by the formula:

$$0.1C(P_U/P_S),$$

in which *C* is the concentration, in µg per mL, of USP Chlorambucil RS in the *Standard preparation*, and P_U and P_S are the response factors of the *Assay preparation* and the *Standard preparation*, respectively.

Chloramphenicol

$C_{11}H_{12}Cl_2N_2O_5$ 323.13

Acetamide, 2,2-dichloro-*N*-[2-hydroxy-1-(hydroxymethyl)-2-(4-nitrophenyl)ethyl]-, [*R*-(*R**,*R**)]-.

D-*threo*-(−)-2,2-Dichloro-*N*-[β-hydroxy-α-(hydroxymethyl)-*p*-nitrophenethyl]acetamide [*56-75-7*].

» Chloramphenicol contains not less than 97.0 percent and not more than 103.0 percent of $C_{11}H_{12}Cl_2N_2O_5$.

Packaging and storage—Preserve in tight containers.

Reference standard—*USP Chloramphenicol Reference Standard*—Do not dry before using.

Identification—

A: The infrared absorption spectrum of a potassium bromide dispersion of it exhibits maxima only at the same wavelengths as that of a similar preparation of USP Chloramphenicol RS.

B: The retention time of the major peak in the chromatogram of the *Assay preparation* corresponds to that in the chromatogram of the *Standard preparation* as obtained in the *Assay*.

Melting range ⟨741⟩: between 149° and 153°.

Specific rotation ⟨781⟩: between +17.0° and +20.0°, determined in a solution in dehydrated alcohol containing 1.25 g in each 25 mL.

Crystallinity ⟨695⟩: meets the requirements.

pH ⟨791⟩: between 4.5 and 7.5, in an aqueous suspension containing 25 mg per mL.

Chromatographic purity—Dissolve an accurately weighed quantity of Chloramphenicol in methanol to obtain a test solution containing 10 mg per mL. Prepare a solution of USP Chloramphenicol RS in methanol containing 10 mg per mL (*Standard solution A*). Dilute portions of *Standard solution A* quantitatively with methanol to obtain *Standard solution B* containing 100 µg per mL and *Standard solution C* containing 50 µg per mL. Apply separate 20-µL portions of the test solution and *Standard solutions B* and *C* on a suitable thin-layer chromatographic plate (see *Chromatography* ⟨621⟩), coated with a 0.25-mm layer of chromatographic silica gel mixture. Develop the chromatogram in a solvent system consisting of a mixture of chloroform, methanol, and glacial acetic acid (79:14:7) until the solvent front has moved about three-fourths of the length of the plate. Remove the plate from the chamber, air-dry, and examine under short-wavelength ultraviolet light: any spot other than the principal spot obtained from the test solution does not exceed in size or intensity the principal spot obtained from *Standard solution B* (1%), and the sum of the impurities represented by all of the spots other than the principal spot, based on a comparison of the intensities of such spots with the intensities of the principal spots obtained from *Standard solutions B* and *C*, does not exceed 2%.

Assay—

Mobile phase—Prepare a suitable filtered mixture of water, methanol, and glacial acetic acid (55:45:0.1). Make adjustments if necessary (see *System Suitability* under *Chromatography* ⟨621⟩).

Standard preparation—Dissolve an accurately weighed quantity of USP Chloramphenicol RS in *Mobile phase*, and dilute quantitatively, and stepwise if necessary, with *Mobile phase* to obtain a solution having a known concentration of about 80 µg per mL. Filter a portion of this solution through a 0.5-µm or finer porosity filter, and use the clear filtrate as the *Standard preparation.*

Assay preparation—Transfer about 200 mg of Chloramphenicol, accurately weighed, to a 100-mL volumetric flask, add *Mobile phase* to volume, and mix. Transfer 4.0 mL of the resulting solution to a 100-mL volumetric flask, dilute with *Mobile phase* to volume, and mix. Filter a portion of this solution through a 0.5-µm or finer porosity filter, and use the clear filtrate as the *Assay preparation.*

Chromatographic system (see *Chromatography* ⟨621⟩)—The liquid chromatograph is equipped with a 280-nm detector and a 4.6-mm × 10-cm column that contains 5-µm packing L1. The flow rate is about 1 mL per minute. Chromatograph the *Standard preparation*, and record the peak responses as directed under *Procedure:* the column efficiency determined from the analyte peak is not less than 1800 theoretical plates, the tailing factor is not more than 2.0, and the relative standard deviation for replicate injections is not more than 1.0%.

Procedure—[NOTE—Use peak heights where peak responses are indicated.] Separately inject equal volumes (about 10 µL) of the *Standard preparation* and the *Assay preparation* into the chromatograph, record the chromatograms, and measure the responses for the major peaks. Calculate the quantity, in mg, of $C_{11}H_{12}Cl_2N_2O_5$ in the portion of Chloramphenicol taken by the formula:

$$2.5C(r_U/r_S),$$

in which *C* is the concentration, in µg per mL, of USP Chloramphenicol RS in the *Standard preparation*, and r_U and r_S are the peak responses obtained from the *Assay preparation* and the *Standard preparation*, respectively.

Chloramphenicol Capsules

» Chloramphenicol Capsules contain not less than 90.0 percent and not more than 120.0 percent of the labeled amount of $C_{11}H_{12}Cl_2N_2O_5$.

Packaging and storage—Preserve in tight containers.

Reference standard—*USP Chloramphenicol Reference Standard*—Do not dry before using.

Identification—The retention time of the major peak in the chromatogram of the *Assay preparation* corresponds to that in the chromatogram of the *Standard preparation* as obtained in the *Assay*.

Dissolution ⟨711⟩—

Medium: 0.1 *N* hydrochloric acid; 900 mL.

Apparatus 1: 100 rpm.

Time: 30 minutes.

Procedure—Determine the amount of $C_{11}H_{12}Cl_2N_2O_5$ dissolved from ultraviolet absorbances at the wavelength of maximum absorbance at about 278 nm of filtered portions of the solution under test, suitably diluted with *Dissolution Medium*, if necessary, in comparison with a Standard solution having a known concentration of USP Chloramphenicol RS in the same medium.

Tolerances—Not less than 85% (*Q*) of the labeled amount of $C_{11}H_{12}Cl_2N_2O_5$ is dissolved in 30 minutes.

Uniformity of dosage units ⟨905⟩: meet the requirements.

Assay—

Mobile phase and *Chromatographic system*—Proceed as directed in the *Assay* under *Chloramphenicol.*

Standard preparation—Transfer about 25 mg of USP Chloramphenicol RS, accurately weighed, to a 200-mL volumetric flask, add 10 mL of water, and heat on a steam bath until completely dissolved. Cool to room temperature, dilute with *Mobile phase* to volume, and mix. Filter a portion of this solution through a 0.5-μm or finer porosity filter, and use the clear filtrate as the *Standard preparation.*

Assay preparation—Transfer an accurately counted number of Chloramphenicol Capsules, equivalent to about 2500 mg of chloramphenicol, to a 1000-mL volumetric flask, add 100 mL of water, and heat on a steam bath until the Capsules have disintegrated. Add 300 mL of water, and heat on a steam bath for 20 minutes, with occasional mixing. Cool to room temperature, dilute with water to volume, and mix. Transfer 5.0 mL of the resulting solution to a 100-mL volumetric flask, dilute with *Mobile phase* to volume, and mix. Filter a portion of this solution through a 0.5-μm or finer porosity filter, and use the clear filtrate as the *Assay preparation.*

Procedure—Proceed as directed for *Procedure* in the *Assay* under *Chloramphenicol.* Calculate the quantity, in mg, of $C_{11}H_{12}Cl_2N_2O_5$ in each Capsule taken by the formula:

$$20(C/N)(r_U/r_S),$$

in which *N* is the number of Capsules taken, and the other terms are as defined therein.

Chloramphenicol Cream

» Chloramphenicol Cream contains not less than 90.0 percent and not more than 130.0 percent of the labeled amount of $C_{11}H_{12}Cl_2N_2O_5$.

Packaging and storage—Preserve in collapsible tubes or in tight containers.

Reference standard—*USP Chloramphenicol Reference Standard*—Do not dry before using.

Identification—The retention time of the major peak in the chromatogram of the *Assay preparation* corresponds to that in the chromatogram of the *Standard preparation* as obtained in the *Assay.*

Minimum fill ⟨755⟩: meets the requirements.

Assay—

Mobile phase and *Chromatographic system*—Proceed as directed in the *Assay* under *Chloramphenicol.*

Standard preparation—Transfer about 40 mg of USP Chloramphenicol RS, accurately weighed, to a 100-mL volumetric flask, dissolve in methanol, dilute with methanol to volume, and mix. Transfer 10.0 mL of the resulting solution to a 50-mL volumetric flask, dilute with *Mobile phase* to volume, and mix. Filter a portion of this solution through a 0.5-μm or finer porosity filter, and use the clear filtrate as the *Standard preparation.*

Assay preparation—Transfer an accurately weighed quantity of Chloramphenicol Cream, equivalent to about 40 mg of chloramphenicol, to a 100-mL volumetric flask, add about 80 mL of methanol, and sonicate for about 10 minutes. Cool to room temperature, dilute with methanol to volume, and mix. Transfer 10.0 mL of the resulting solution to a 50-mL volumetric flask, dilute with *Mobile phase* to volume, and mix. Filter a portion of this solution through a 0.5-μm or finer porosity filter, and use the clear filtrate as the *Assay preparation.*

Procedure—Proceed as directed for *Procedure* in the *Assay* under *Chloramphenicol.* Calculate the quantity, in mg, of $C_{11}H_{12}Cl_2N_2O_5$ in the portion of Cream taken by the formula:

$$0.5C(r_U/r_S),$$

in which the terms are as defined therein.

Chloramphenicol Injection

» Chloramphenicol Injection is a sterile solution of Chloramphenicol in one or more suitable solvents. It contains not less than 90.0 percent and not more than 115.0 percent of the labeled amount of $C_{11}H_{12}Cl_2N_2O_5$. It may contain suitable buffers.

Packaging and storage—Preserve in single-dose or in multiple-dose containers.

Labeling—Label it to indicate that it is for veterinary use only.

Reference standard—*USP Chloramphenicol Reference Standard*—Do not dry before using.

Identification—The retention time of the major peak in the chromatogram of the *Assay preparation* corresponds to that in the chromatogram of the *Standard preparation* as obtained in the *Assay.*

Pyrogen—When diluted, if necessary, with pyrogen-free saline TS to a concentration of 5.0 mg of chloramphenicol per mL, it meets the requirements of the *Pyrogen Test* ⟨151⟩, the test dose being 1 mL per kg.

Sterility—It meets the requirements under *Sterility Tests* ⟨71⟩, when tested as directed in the section, *Test Procedures Using Membrane Filtration,* 1 mL from each container being transferred directly to the membrane filter.

pH ⟨791⟩: between 5.0 and 8.0, in a solution diluted with water (1:1).

Other requirements—It meets the requirements under *Injections* ⟨1⟩.

Assay—

Mobile phase, Standard preparation, and *Chromatographic system*—Prepare as directed in the *Assay* under *Chloramphenicol.*

Assay preparation—Transfer an accurately measured volume of Chloramphenicol Injection, equivalent to about 200 mg of chloramphenicol, to a 100-mL volumetric volume flask, add *Mobile phase* to volume, and mix. Transfer 4.0 mL of the resulting solution to a 100-mL volumetric flask, dilute with *Mobile phase* to volume, and mix. Filter this solution through a 0.5-μm or finer porosity filter.

Procedure—Proceed as directed for *Procedure* in the *Assay* under *Chloramphenicol.* Calculate the quantity, in mg, of $C_{11}H_{12}Cl_2N_2O_5$ in each mL of the Injection taken by the formula:

$$2.5(C/V)(r_U/r_S),$$

in which *V* is the volume, in mL, of Injection taken, and the other terms are as defined therein.

Chloramphenicol Ophthalmic Ointment

» Chloramphenicol Ophthalmic Ointment contains not less than 90.0 percent and not more than 130.0 percent of the labeled amount of $C_{11}H_{12}Cl_2N_2O_5$.

Packaging and storage—Preserve in collapsible ophthalmic ointment tubes.

Reference standard—*USP Chloramphenicol Reference Standard*—Do not dry before using.

Identification—The retention time of the major peak in the chromatogram of the *Assay preparation* corresponds to that in the chromatogram of the *Standard preparation* as obtained in the *Assay.*

Sterility—It meets the requirements for *Ophthalmic Ointments* under *Sterility Tests* ⟨71⟩.

Minimum fill ⟨755⟩: meets the requirements.

Metal particles—It meets the requirements under *Metal Particles in Ophthalmic Ointments* ⟨751⟩.

Assay—

Mobile phase and *Chromatographic system*—Proceed as directed in the *Assay* under *Chloramphenicol.*

Standard preparation—Transfer about 25 mg of USP Chloramphenicol RS, accurately weighed, to a 100-mL volumetric flask, dissolve in methanol, dilute with methanol to volume, and mix. Transfer 10.0 mL of the resulting solution to a 25-mL volumetric flask, dilute with *Mobile phase* to volume, and mix. Filter a portion of this solution through a 0.5-μm or finer porosity filter, and use the clear filtrate as the *Standard preparation.*

Assay preparation—Transfer an accurately weighed quantity of Chloramphenicol Ophthalmic Ointment, equivalent to about 25 mg of chloramphenicol, to a suitable conical flask, add 20 mL of cyclohexane, mix, and sonicate for about 2 minutes. Add 60 mL of methanol, and mix. Filter this mixture, collecting the filtrate in a 100-mL volumetric flask. Wash the filter with methanol, collecting the washings in the volumetric flask. Dilute with methanol to volume, and mix. Transfer 50.0 mL of the resulting solution to a suitable round-bottom flask, and evaporate to dryness by rotating the flask under vacuum in a water bath at 35°. Dissolve the residue in 50.0 mL of methanol. Transfer 10.0 mL of the resulting solution to a 25-mL volumetric flask, dilute with *Mobile phase* to volume, and mix. Filter a portion of this solution through a 0.5-μm or finer porosity filter, and use the clear filtrate as the *Assay preparation.*

Procedure—Proceed as directed for *Procedure* in the *Assay* under *Chloramphenicol.* Calculate the quantity, in mg, of $C_{11}H_{12}Cl_2N_2O_5$ in the portion of Ophthalmic Ointment taken by the formula:

$$0.25C(r_U/r_S),$$

in which the terms are as defined therein.

Chloramphenicol Ophthalmic Solution

» Chloramphenicol Ophthalmic Solution is a sterile solution of Chloramphenicol. It contains not less than 90.0 percent and not more than 130.0 percent of the labeled amount of $C_{11}H_{12}Cl_2N_2O_5$.

Packaging and storage—Preserve in tight containers. The containers or individual cartons are sealed and tamper-proof so that sterility is assured at time of first use.

Reference standard—*USP Chloramphenicol Reference Standard*—Do not dry before using.

Identification—The retention time of the major peak in the chromatogram of the *Assay preparation* corresponds to that in the chromatogram of the *Standard preparation* as obtained in the *Assay.*

Sterility—It meets the requirements under *Sterility Tests* ⟨71⟩, when tested as directed in the section, *Test Procedures Using Membrane Filtration.*

pH ⟨791⟩: between 7.0 and 7.5, except that in the case of Ophthalmic Solution that is unbuffered or is labeled for veterinary use it is between 3.0 and 6.0.

Assay—

Mobile phase and *Chromatographic system*—Proceed as directed in the *Assay* under *Chloramphenicol.*

Standard preparation—Dissolve an accurately weighed quantity of USP Chloramphenicol RS in *Mobile phase,* and dilute quantitatively, and stepwise if necessary, with *Mobile phase* to obtain a solution having a known concentration of about 100 μg per mL. Filter a portion of this solution through a 0.5-μm or finer porosity filter, and use the clear filtrate as the *Standard preparation.*

Assay preparation—Transfer an accurately measured volume of Chloramphenicol Ophthalmic Solution, equivalent to about 50 mg of chloramphenicol, to a 100-mL volumetric flask, dilute with *Mobile phase* to volume, and mix. Transfer 5.0 mL of the resulting solution to a 25-mL volumetric flask, dilute with *Mobile phase* to volume, and mix. Filter a portion of this solution through a 0.5-μm or finer porosity filter, and use the clear filtrate as the *Assay preparation.*

Procedure—Proceed as directed for *Procedure* in the *Assay* under *Chloramphenicol.* Calculate the quantity, in mg, of $C_{11}H_{12}Cl_2N_2O_5$ in each mL of the Ophthalmic Solution taken by the formula:

$$0.5(C/V)(r_U/r_S),$$

in which V is the volume, in mL, of Ophthalmic Solution taken, and the other terms are as defined therein.

Chloramphenicol for Ophthalmic Solution

» Chloramphenicol for Ophthalmic Solution is a sterile, dry mixture of Chloramphenicol with or without one or more suitable buffers, diluents, and preservatives. It contains not less than 90.0 percent and not more than 130.0 percent of the labeled amount of $C_{11}H_{12}Cl_2N_2O_5$, when constituted as directed.

Packaging and storage—Preserve in tight containers.

Labeling—If packaged in combination with a container of solvent, label it with a warning that it is not for injection.

Reference standard—*USP Chloramphenicol Reference Standard*—Do not dry before using.

Identification—The retention time of the major peak in the chromatogram of the *Assay preparation* corresponds to that in the chromatogram of the *Standard preparation* as obtained in the *Assay.*

Sterility—It meets the requirements under *Sterility Tests* ⟨71⟩, when tested as directed in the section, *Test Procedures Using Membrane Filtration.*

pH ⟨791⟩: between 7.1 and 7.5, in an aqueous solution containing 5 mg of chloramphenicol per mL.

Assay—

Mobile phase and *Chromatographic system*—Proceed as directed in the *Assay* under *Chloramphenicol.*

Standard preparation—Dissolve an accurately weighed quantity of USP Chloramphenicol RS in *Mobile phase,* and dilute quantitatively, and stepwise if necessary, with *Mobile phase* to obtain a solution having a known concentration of about 100 μg per mL. Filter a portion of this solution through a 0.5-μm or finer porosity filter, and use the clear filtrate as the *Standard preparation.*

Assay preparation—Transfer the contents of 1 container of Chloramphenicol for Ophthalmic Solution to a suitable container with the aid of *Mobile phase,* and dilute quantitatively, and stepwise if necessary, with *Mobile phase* to obtain a solution having a concentration of about 100 μg of chloramphenicol per mL. Filter a portion of this solution through a 0.5-μm or finer porosity filter, and use the clear filtrate as the *Assay preparation.*

Procedure—Proceed as directed for *Procedure* in the *Assay* under *Chloramphenicol.* Calculate the quantity, in mg, of $C_{11}H_{12}Cl_2N_2O_5$ in the container of Chloramphenicol for Ophthalmic Solution taken by the formula:

$$(L/D)C(r_U/r_S),$$

in which L is the labeled quantity, in mg, of chloramphenicol in the container, D is the concentration, in μg per mL, of chloramphenicol in the *Assay preparation,* based on the labeled quantity and the extent of dilution, and the other terms are as defined therein.

Chloramphenicol Oral Solution

» Chloramphenicol Oral Solution is a solution of Chloramphenicol in a suitable solvent. It contains one or more suitable buffers and preservatives. It has a potency of not less than 90.0 percent and not more than 120.0 percent of the labeled amount of $C_{11}H_{12}Cl_2N_2O_5$.

Packaging and storage—Preserve in tight containers.

Labeling—Label it to indicate that it is for veterinary use only and that it is not to be used in animals raised for food production.

Reference standard—*USP Chloramphenicol Reference Standard*—Do not dry before using.

Identification—A volume of Oral Solution, equivalent to about 250 mg of chloramphenicol, responds to the *Identification test* under *Chloramphenicol Capsules.*

pH ⟨791⟩: between 5.0 and 8.5, when diluted with an equal volume of water.

Assay—Proceed as directed under *Antibiotics—Microbial Assays* ⟨81⟩, using an accurately measured volume of Oral Solution diluted quantitatively and stepwise with water to yield a *Test Dilution* having a concentration assumed to be equal to the median dose level of the Standard.

Chloramphenicol Otic Solution

» Chloramphenicol Otic Solution is a sterile solution of Chloramphenicol in a suitable solvent. It contains not less than 90.0 percent and not more than 130.0 percent of the labeled amount of $C_{11}H_{12}Cl_2N_2O_5$.

Packaging and storage—Preserve in tight containers.

Reference standard—*USP Chloramphenicol Reference Standard*—Do not dry before using.

Identification—The retention time of the major peak in the chromatogram of the *Assay preparation* corresponds to that in the chromatogram of the *Standard preparation* as obtained in the *Assay.*

Sterility—It meets the requirements under *Sterility Tests* ⟨71⟩, when tested as directed in the section, *Test Procedures Using Membrane Filtration.*

pH ⟨791⟩: between 4.0 and 8.0, when diluted with an equal volume of water.

Water, *Method I* ⟨921⟩: not more than 2.0%.

Assay—

Mobile phase and *Chromatographic system*—Proceed as directed in the *Assay* under *Chloramphenicol.*

Standard preparation—Dissolve an accurately weighed quantity of USP Chloramphenicol RS in *Mobile phase,* and dilute quantitatively, and stepwise if necessary, with *Mobile phase* to obtain a solution having a known concentration of about 100 µg per mL. Filter a portion of this solution through a 0.5-µm or finer porosity filter, and use the clear filtrate as the *Standard preparation.*

Assay preparation—Transfer an accurately measured volume of Chloramphenicol Otic Solution, equivalent to about 50 mg of chloramphenicol, to a 100-mL volumetric flask, dilute with *Mobile phase* to volume, and mix. Transfer 5.0 mL of the resulting solution to a 25-mL volumetric flask, dilute with *Mobile phase* to volume, and mix. Filter a portion of this solution through a 0.5-µm or finer porosity filter, and use the clear filtrate as the *Assay preparation.*

Procedure—Proceed as directed for *Procedure* in the *Assay* under *Chloramphenicol.* Calculate the quantity, in mg, of $C_{11}H_{12}Cl_2N_2O_5$ in each mL of the Otic Solution taken by the formula:

$$0.5(C/V)(r_U/r_S),$$

in which V is the volume, in mL, of Otic Solution taken, and the other terms are as defined therein.

Sterile Chloramphenicol

» Sterile Chloramphenicol is Chloramphenicol suitable for parenteral use. It contains not less than 97.0 percent and not more than 103.0 percent of $C_{11}H_{12}Cl_2N_2O_5$.

Packaging and storage—Preserve in *Containers for Sterile Solids* as described under *Injections* ⟨1⟩.

Reference standard—*USP Chloramphenicol Reference Standard*—Do not dry before using.

Pyrogen—It meets the requirements of the *Pyrogen Test* ⟨151⟩, the test dose being 1.0 mL per kg of a solution in pyrogen-free saline TS, containing 5.0 mg per mL, warmed, if necessary, to effect solution.

Sterility—It meets the requirements under *Sterility Tests* ⟨71⟩, when tested as directed in the section, *Test Procedures Using Membrane Filtration,* except to use 1 g of solid specimen, instead of 6 g.

Other requirements—It conforms to the definition, responds to the *Identification test,* and meets the requirements for *Melting range, Specific rotation, pH, Crystallinity,* and *Assay* under *Chloramphenicol.*

Chloramphenicol Tablets

» Chloramphenicol Tablets contain not less than 90.0 percent and not more than 120.0 percent of the labeled amount of $C_{11}H_{12}Cl_2N_2O_5$.

Packaging and storage—Preserve in tight containers.

Labeling—Label Tablets to indicate that they are for veterinary use only and are not to be used in animals raised for food production.

Reference standard—*USP Chloramphenicol Reference Standard*—Do not dry before using.

Identification—The retention time of the major peak in the chromatogram of the *Assay preparation* corresponds to that in the chromatogram of the *Standard preparation* as obtained in the *Assay.*

Disintegration ⟨701⟩: 60 minutes.

Uniformity of dosage units ⟨905⟩: meet the requirements.

Assay—

Mobile phase and *Chromatographic system*—Proceed as directed in the *Assay* under *Chloramphenicol.*

Standard preparation—Transfer about 25 mg of USP Chloramphenicol RS, accurately weighed, to a 200-mL volumetric flask, add 10 mL of water, and heat on a steam bath until completely dissolved. Cool to room temperature, dilute with *Mobile phase* to volume, and mix. Filter a portion of this solution through a 0.5-µm or finer porosity filter, and use the clear filtrate as the *Standard preparation.*

Assay preparation—Weigh and finely powder not less than 20 Chloramphenicol Tablets. Transfer an accurately weighed portion of the powder, equivalent to about 500 mg of chloramphenicol, to a 200-mL volumetric flask, add 80 mL of water, and heat on a steam bath for 20 minutes, with occasional mixing. Cool to room temperature, dilute with water to volume, and mix. Transfer 5.0 mL of the resulting solution to a 100-mL volumetric flask, dilute with *Mobile phase* to volume, and mix. Filter a portion of this solution through a 0.5-µm or finer porosity filter, and use the clear filtrate as the *Assay preparation.*

Procedure—Proceed as directed for *Procedure* in the *Assay* under *Chloramphenicol.* Calculate the quantity, in mg, of $C_{11}H_{12}Cl_2N_2O_5$ in the portion of Tablets taken by the formula:

$$4C(r_U/r_S),$$

in which the terms are as defined therein.

Chloramphenicol and Hydrocortisone Acetate for Ophthalmic Suspension

» Chloramphenicol and Hydrocortisone Acetate for Ophthalmic Suspension is a sterile, dry mixture of Chloramphenicol and Hydrocortisone Acetate with or without one or more suitable buffers, diluents, and preservatives. It contains not less than 90.0 percent and not more than 130.0 percent of the labeled amount of chloramphenicol ($C_{11}H_{12}Cl_2N_2O_5$), and not less than 90.0 percent and not more than 115.0 percent of the labeled amount of hydrocortisone acetate ($C_{23}H_{32}O_6$), when constituted as directed.

Labeling—If packaged in combination with a container of solvent, label it with a warning that it is not for injection.

Reference standards—*USP Chloramphenicol Reference Standard*—Do not dry before using. *USP Hydrocortisone Acetate Reference Standard*—Dry at 105° for 3 hours before using.

Identification—The retention times of the major peaks in the chromatogram of the *Assay preparation* correspond to that in the chromatogram of the *Standard preparation* as obtained in the *Assay.*

Sterility—It meets the requirements under *Sterility Tests* ⟨71⟩, when tested as directed in the section, *Test Procedures Using Membrane Filtration.*

pH ⟨791⟩: between 7.1 and 7.5, in an aqueous suspension containing 5 mg of chloramphenicol per mL.

Assay—

Mobile phase and *Chromatographic system*—Proceed as directed in the *Assay* under *Chloramphenicol.*

Standard preparation—Transfer about 37.5 mg of USP Chloramphenicol RS and 37.5*J* mg of USP Hydrocortisone Acetate RS, both accurately weighed, *J* being the ratio of the labeled amount, in mg, of hydrocortisone acetate to the labeled amount, in mg, of chloramphenicol in the Chloramphenicol and Hydrocortisone Acetate for Ophthalmic Solution, to a 100-mL volumetric flask, add 15 mL of water and 75 mL of methanol, sonicate for a few seconds, dilute with methanol to volume, and mix. Transfer 5.0 mL of the resulting solution to a 25-mL volumetric flask, dilute with *Mobile phase* to volume, and mix. Filter a portion of this solution through a 0.5-µm or finer porosity filter, and use the clear filtrate as the *Standard preparation.*

Assay preparation—Transfer the contents of an accurately counted number of containers of Chloramphenicol and Hydrocortisone Acetate for Ophthalmic Solution, equivalent to about 37.5 mg of chloramphenicol, to a 100-mL volumetric flask with the aid of 5 mL of water for each 12.5 mL of chloramphenicol contained therein. Wash each container with methanol, and add the washings to the volumetric flask. Dilute with methanol to volume, and mix. Transfer 5.0 mL of the resulting solution to a 25-mL volumetric flask, dilute with *Mobile phase* to volume, and mix. Filter a portion of this solution through a 0.5-µm or finer porosity filter, and use the clear filtrate as the *Assay preparation.*

Procedure—Proceed as directed for *Procedure* in the *Assay* under *Chloramphenicol.* Calculate the quantity, in mg, of chloramphenicol ($C_{11}H_{12}Cl_2N_2O_5$) in each container taken by the formula:

$$0.5(C/N)(r_U/r_S),$$

in which *N* is the number of containers taken, and the other terms are as defined therein. Calculate the quantity, in mg, of hydrocortisone acetate ($C_{23}H_{32}O_6$) in each container taken by the formula:

$$500(C/N)(r_U/r_S),$$

in which *C* is the concentration, in mg per mL, of USP Hydrocortisone Acetate RS in the *Standard preparation, N* is the number of containers taken, and the other terms are as defined therein.

Chloramphenicol and Polymyxin B Sulfate Ophthalmic Ointment

» Chloramphenicol and Polymyxin B Sulfate Ophthalmic Ointment contains not less than 90.0 percent and not more than 120.0 percent of the labeled amount of chloramphenicol ($C_{11}H_{12}Cl_2N_2O_5$) and not less than 90.0 percent and not more than 125.0 percent of the labeled amount of polymyxin B.

Packaging and storage—Preserve in collapsible ophthalmic ointment tubes.

Reference standards—*USP Chloramphenicol Reference Standard*—Do not dry before using. *USP Polymyxin B Sulfate Reference Standard*—Dry in vacuum at a pressure not exceeding 5 mm of mercury at 60° for 3 hours before using.

Identification—The retention time of the major peak in the chromatogram of the *Assay preparation* corresponds to that in the chromatogram of the *Standard preparation* as obtained in the *Assay for chloramphenicol.*

Sterility—It meets the requirements for *Ophthalmic Ointments* under *Sterility Tests* ⟨71⟩.

Metal particles—It meets the requirements of the test for *Metal Particles in Ophthalmic Ointments* ⟨751⟩.

Assay for chloramphenicol—

Mobile phase and *Chromatographic system*—Proceed as directed in the *Assay* under *Chloramphenicol.*

Standard preparation—Proceed as directed for *Standard preparation* in the *Assay* under *Chloramphenicol Ophthalmic Ointment.*

Assay preparation—Using Chloramphenicol and Polymyxin B Sulfate Ophthalmic Ointment, proceed as directed for *Assay preparation* in the *Assay* under *Chloramphenicol Ophthalmic Ointment.*

Procedure—Proceed as directed for *Procedure* in the *Assay* under *Chloramphenicol.* Calculate the quantity, in mg, of $C_{11}H_{12}Cl_2N_2O_5$ in the portion of Ophthalmic Ointment taken by the formula:

$$0.25C(r_U/r_S),$$

in which the terms are as defined therein.

Assay for polymyxin—Proceed as directed for polymyxin under *Antibiotics—Microbial Assays* ⟨81⟩, using an accurately weighed portion of Chloramphenicol and Polymyxin B Sulfate Ophthalmic Ointment, equivalent to about 5000 Polymyxin B Units, shaken in a separator containing about 50 mL of ether and extracted with four 20-mL portions of *Buffer No. 6.* Combine the aqueous extracts in a 100-mL volumetric flask, dilute with *Buffer No. 6* to volume, and mix. Dilute an accurately measured portion of this solution quantitatively with *Buffer No. 6* to obtain a *Test*

Dilution having a concentration assumed to be equal to the median dose level of the Standard.

Chloramphenicol, Polymyxin B Sulfate, and Hydrocortisone Acetate Ophthalmic Ointment

» Chloramphenicol, Polymyxin B Sulfate, and Hydrocortisone Acetate Ophthalmic Ointment contains not less than 90.0 percent and not more than 120.0 percent of the labeled amount of chloramphenicol ($C_{11}H_{12}Cl_2N_2O_5$), not less than 90.0 percent and not more than 125.0 percent of the labeled amount of polymyxin B, and not less than 90.0 percent and not more than 115.0 percent of the labeled amount of hydrocortisone acetate ($C_{23}H_{32}O_6$).

Packaging and storage—Preserve in collapsible ophthalmic ointment tubes.

Reference standards—*USP Chloramphenicol Reference Standard*—Do not dry before using. *USP Polymyxin B Sulfate Reference Standard*—Dry in vacuum at 60° and at a pressure not exceeding 5 mm of mercury for 3 hours before using. *USP Hydrocortisone Acetate Reference Standard*—Dry at 105° for 3 hours before using.

Identification—The retention times of the major peaks in the chromatogram of the *Assay preparation* correspond to those in the chromatogram of the *Standard preparation* as obtained in the *Assay for chloramphenicol and hydrocortisone acetate.*

Sterility—It meets the requirements for *Ophthalmic Ointments* under *Sterility Tests* ⟨71⟩.

Minimum fill ⟨755⟩: meets the requirements.

Metal particles—It meets the requirements under *Metal Particles in Ophthalmic Ointments* ⟨751⟩.

Assay for polymyxin—Proceed with Chloramphenicol, Polymyxin B Sulfate, and Hydrocortisone Acetate Ophthalmic Ointment as directed for polymyxin under *Antibiotics—Microbial Assays* ⟨81⟩, using an accurately weighed portion of Ophthalmic Ointment, equivalent to about 5000 Polymyxin B Units, shaken in a separator containing about 50 mL of ether and extracted with four 20-mL portions of *Buffer No. 6.* Combine the aqueous extracts in a 100-mL volumetric flask, dilute, if necessary, with *Buffer No. 6* to volume, and mix. Dilute an accurately measured portion of the resulting solution quantitatively with *Buffer No. 6* to obtain a *Test Dilution* having a concentration assumed to be equal to the median dose level of the Standard.

Assay for chloramphenicol and hydrocortisone acetate—

Mobile phase and *Chromatographic system*—Proceed as directed in the *Assay* under *Chloramphenicol.*

Standard preparation—Transfer about 25 mg of USP Chloramphenicol RS and 25*J* mg of USP Hydrocortisone Acetate RS, both accurately weighed, *J* being the ratio of the labeled amount, in mg, of hydrocortisone acetate to the labeled amount, in mg, of chloramphenicol per g of Ophthalmic Ointment, to a 100-mL volumetric flask, dissolve in methanol, dilute with methanol to volume, and mix. Transfer 10.0 mL of the resulting solution to a 25-mL volumetric flask, dilute with *Mobile phase* to volume, and mix. Filter a portion of this solution through a 0.5-µm or finer porosity filter, and use the clear filtrate as the *Standard preparation.*

Assay preparation—Using Chloramphenicol, Polymyxin B Sulfate, and Hydrocortisone Acetate Ophthalmic Ointment, proceed as directed for *Assay preparation* in the *Assay* under *Chloramphenicol Ophthalmic Ointment.*

Procedure—Proceed as directed for *Procedure* in the *Assay* under *Chloramphenicol.* Calculate the quantity, in mg, of chloramphenicol ($C_{11}H_{12}Cl_2N_2O_5$) in the portion of Ophthalmic Ointment taken by the formula:

$$0.25C(r_U / r_S),$$

in which the terms are as defined therein. Calculate the quantity, in mg, of hydrocortisone acetate ($C_{23}H_{32}O_6$) in the portion of Ophthalmic Ointment taken by the formula:

$$250C(r_U / r_S),$$

in which *C* is the concentration, in mg per mL, of USP Hydrocortisone Acetate RS in the *Standard preparation*, and the other terms are as defined therein.

Chloramphenicol and Prednisolone Ophthalmic Ointment

» Chloramphenicol and Prednisolone Ophthalmic Ointment contains not less than 90.0 percent and not more than 130.0 percent of the labeled amount of chloramphenicol ($C_{11}H_{12}Cl_2N_2O_5$), and not less than 90.0 percent and not more than 115.0 percent of the labeled amount of prednisolone ($C_{21}H_{28}O_5$).

Packaging and storage—Preserve in collapsible ophthalmic ointment tubes.

Reference standards—*USP Chloramphenicol Reference Standard*—Do not dry before using. *USP Prednisolone Reference Standard*—Dry at 105° for 3 hours before using.

Identification—

A: Transfer a quantity of Ophthalmic Ointment, equivalent to about 20 mg of chloramphenicol, to a screw-capped test tube, add 5 mL of 5 *N* sodium hydroxide and 2 mL of pyridine, and shake. Place the tube in a water bath at 50° for 20 minutes: a reddish brown color develops in the pyridine layer.

B: The retention time of the major peak in the chromatogram of the *Assay preparation* corresponds to that in the chromatogram of the *Standard preparation* as obtained in the *Assay for chloramphenicol.*

C: Transfer a quantity of Ophthalmic Ointment, equivalent to about 1.5 mg of prednisolone, to a screw-capped test tube, add 10 mL of methylene chloride, and shake to disperse. Heat at 60° for 15 minutes, and allow to cool while shaking for about 30 minutes. Allow to separate, draw off the upper ointment layer, and retain the lower methylene chloride layer. On a suitable thin-layer chromatographic plate (see *Chromatography* ⟨621⟩), coated with a 0.25-mm layer of chromatographic silica gel mixture, apply, in portions, 0.4 mL each of the methylene chloride test solution and a Standard solution of USP Prednisolone RS in chloroform containing 0.5 mg per mL. Allow each portion to dry before adding the next portion to the same spot. Develop the chromatogram in a chromatographic chamber lined with paper and equilibrated with a solvent system consisting of a mixture of chloroform and acetone (4:1) until the solvent front has moved about three-fourths of the length of the plate. Remove the plate from the developing chamber, mark the solvent front, and allow the solvent to evaporate. Locate the spots on the plate by examination under short-wavelength ultraviolet light: the R_f value of the principal spot obtained from the test solution corresponds to that obtained from the Standard solution.

Sterility—It meets the requirements for *Ophthalmic Ointments* under *Sterility Tests* ⟨71⟩.

Minimum fill ⟨755⟩: meets the requirements.

Metal particles—It meets the requirements under *Metal Particles in Ophthalmic Ointments* ⟨751⟩.

Assay for chloramphenicol—

Methanol-water solution and *Mobile phase*—Proceed as directed in the *Assay for prednisolone.*

Standard preparation—Dissolve an accurately weighed quantity of USP Chloramphenicol RS in *Methanol-water solution* to obtain a solution having a known concentration of about 0.3 mg per mL.

Assay preparation—Transfer an accurately weighed portion of Chloramphenicol and Prednisolone Ophthalmic Ointment, equivalent to about 3.0 mg of chloramphenicol, to a screw-capped test tube. Add 10 mL of *n*-heptane, and shake by mechanical means until the substance is dissolved. Add 10.0 mL of *Methanol-water solution*, and shake by mechanical means for 30 seconds. Allow the layers to separate, and carefully remove the upper phase. Centrifuge the lower phase for 15 minutes, and use the clear portion as the *Assay preparation*.

Chromatographic system (see *Chromatography* ⟨621⟩)—The liquid chromatograph is equipped with a 280-nm detector and a 3.9-mm × 30-cm column that contains packing L1. The flow rate is about 1 mL per minute. Chromatograph the *Standard preparation*, and record the peak responses as directed under *Procedure:* the peak for chloramphenicol obtained from the Ophthalmic Ointment, at a retention time corresponding to that of the peak obtained from the Reference Standard, exhibits baseline separation from the adjacent prednisolone peak, and the relative standard deviation of replicate injections is not more than 1.5%.

Procedure—Separately inject equal volumes (about 20 µL) of the *Standard preparation* and the *Assay preparation* into the chromatograph, record the chromatograms, and measure the responses for the major peaks. Calculate the quantity, in mg, of $C_{11}H_{12}Cl_2N_2O_5$ in the portion of Ophthalmic Ointment taken by the formula:

$$10C(r_U/r_S),$$

in which C is the concentration, in mg per mL, of USP Chloramphenicol RS in the *Standard preparation*, and r_U and r_S are the peak responses obtained from the *Assay preparation* and the *Standard preparation*, respectively.

Assay for prednisolone—

Methanol-water solution—Mix 4 volumes of methanol with 1 volume of water, and mix.

Mobile phase—Dissolve 0.68 g of sodium acetate trihydrate in 400 mL of water in a 1000-mL graduated cylinder, adjust with glacial acetic acid to a pH of 4.0, and dilute with water to 500 mL. Dilute with methanol to 1000 mL, and mix. Filter this solution through a membrane filter (1-µm or finer porosity), and degas. Make adjustments if necessary (see *System Suitability* under *Chromatography* ⟨621⟩).

Standard preparation—Dissolve an accurately weighed quantity of USP Prednisolone RS in *Methanol-water solution* to obtain a solution having a known concentration of about 0.2 mg per mL.

Assay preparation—Transfer an accurately weighed portion of Chloramphenicol and Prednisolone Ophthalmic Ointment, equivalent to about 2.0 mg of prednisolone, to a screw-capped test tube. Add 10 mL of *n*-heptane, and shake by mechanical means until the substance is dissolved. Add 10.0 mL of *Methanol-water solution*, and shake by mechanical means for 30 seconds. Allow the layers to separate, and carefully remove the upper phase. Centrifuge the lower phase for 15 minutes, and use the clear portion as the *Assay preparation*.

Chromatographic system (see *Chromatography* ⟨621⟩)—The liquid chromatograph is equipped with a 254-nm detector and a 3.9-mm × 30-cm column that contains packing L1. The flow rate is about 1.5 mL per minute. Chromatograph the *Standard preparation*, and record the peak responses as directed under *Procedure:* the peak for prednisolone obtained from the Ophthalmic Ointment, at a retention time corresponding to that of the peak obtained from the *Reference Standard*, exhibits baseline separation from the adjacent chloramphenicol peak, and the relative standard deviation of replicate injections is not more than 1.5%.

Procedure—Separately inject equal volumes (about 15 µL) of the *Standard preparation* and the *Assay preparation* into the chromatograph, record the chromatograms, and measure the responses for the major peaks. Calculate the quantity, in mg, of $C_{21}H_{28}O_5$ in the portion of Ophthalmic Ointment taken by the formula:

$$10C(r_U/r_S),$$

in which C is the concentration, in mg per mL, of USP Prednisolone RS in the *Standard preparation*, and r_U and r_S are the peak responses obtained from the *Assay preparation* and the *Standard preparation*, respectively.

Chloramphenicol Palmitate

$C_{27}H_{42}Cl_2N_2O_6$ 561.54

Hexadecanoic acid, 2-[(2,2-dichloroacetyl)amino]-3-hydroxy-3-(4-nitrophenyl)propyl ester, [*R*-(*R**,*R**)]-.

D-*threo*-(−)-2,2-Dichloro-*N*-[β-hydroxy-α-(hydroxymethyl)-*p*-nitrophenethyl]acetamide α-palmitate [*530-43-8*].

» Chloramphenicol Palmitate has a potency equivalent to not less than 555 µg and not more than 595 µg of chloramphenicol ($C_{11}H_{12}Cl_2N_2O_5$) per mg.

Packaging and storage—Preserve in tight containers.

Reference standard—*USP Chloramphenicol Palmitate Reference Standard*—Do not dry before using.

Identification—The retention time of the chloramphenicol palmitate peak in the chromatogram of the *Assay preparation* corresponds to that of the chloramphenicol palmitate peak in the chromatogram of the *Standard preparation*, as obtained in the *Assay*.

Melting range ⟨741⟩: between 87° and 95°.

Specific rotation ⟨781⟩: between +21° and +25°, determined in a solution in dehydrated alcohol containing 50 mg in each mL.

Crystallinity ⟨695⟩: meets the requirements.

Loss on drying ⟨731⟩—Dry it to constant weight over phosphorus pentoxide in vacuum at a pressure not exceeding 5 mm of mercury: it loses not more than 0.5% of its weight.

Acidity—Dissolve 1.0 g by heating at 35° with 5 mL of a 1:1 mixture of 80 percent alcohol and ether, previously neutralized using phenolphthalein TS. Titrate with 0.1 *N* sodium hydroxide VS, using phenolphthalein TS, until on gentle shaking a pink color persists for not less than 30 seconds: not more than 0.4 mL is consumed.

Free chloramphenicol—Dissolve 1.0 g in 80 mL of xylene with the aid of gentle warming. Cool, and extract with three 15-mL portions of water, combining the aqueous extracts and discarding the xylene. Dilute the combined aqueous extracts with water to 50 mL, extract with 10 mL of toluene, allow to separate, and discard the toluene. Centrifuge a portion of the aqueous solution, and determine the absorbance of the clear solution at the wavelength of maximum absorbance at about 278 nm, using a suitable spectrophotometer, and using as a reagent blank to set the instrument to zero the solution obtained by the same procedure without the specimen: the absorbance is not more than 0.268 (0.045%).

Assay—

Mobile phase—Prepare a suitable degassed mixture of methanol, water, and glacial acetic acid (172:27:1).

Standard preparation—Transfer about 65 mg of USP Chloramphenicol Palmitate RS to a 50-mL volumetric flask, add about 40 mL of methanol and 1 mL of glacial acetic acid, and sonicate for a few minutes. Dilute with methanol to volume, and mix. Transfer 10.0 mL of this solution to a 25-mL volumetric flask, dilute with *Mobile phase* to volume, and mix.

Assay preparation—Using about 65 mg of Chloramphenicol Palmitate, accurately weighed, prepare as directed under *Standard preparation*.

Chromatographic system (see *Chromatography* ⟨621⟩)—The liquid chromatograph is equipped with a 280-nm detector and a 3.9-mm × 30-cm column that contains 10-µm packing L1. The flow rate is about 2 mL per minute. Chromatograph the *Standard preparation*, and record the peak responses as directed under *Procedure:* the column efficiency determined from the analyte peak is not less than 2400 theoretical plates, and the relative standard deviation for replicate injections is not more than 0.5%.

Procedure—Separately inject equal volumes (about 10 µL) of the *Standard preparation* and the *Assay preparation* into the chromatograph, record the chromatograms, and measure the responses for the major peaks. Calculate the quantity, in µg, of chloramphenicol ($C_{11}H_{12}Cl_2N_2O_5$) equivalent in each mg of specimen taken by the formula:

$$(W_S/W_U)(P_S)(r_U/r_S),$$

in which W_S and W_U are the quantities, in mg, of USP Chloramphenicol Palmitate RS and Chloramphenicol Palmitate taken, respectively, P_S is the designated chloramphenicol equivalent, in µg per mg, of USP Chloramphenicol Palmitate RS, and r_U and r_S are the peak responses obtained from the *Assay preparation* and the *Standard preparation*, respectively.

Chloramphenicol Palmitate Oral Suspension

» Chloramphenicol Palmitate Oral Suspension contains the equivalent of not less than 90.0 percent and not more than 120.0 percent of the labeled amount of chloramphenicol ($C_{11}H_{12}Cl_2N_2O_5$). It contains one or more suitable buffers, colors, flavors, preservatives, and suspending agents.

Packaging and storage—Preserve in tight, light-resistant containers.

Reference standards—*USP Chloramphenicol Palmitate Reference Standard*—Do not dry before using. *USP Chloramphenicol Palmitate Polymorph A Reference Standard*—Do not dry before using. *USP Chloramphenicol Palmitate Nonpolymorph A Reference Standard*—Do not dry before using.

Identification—The retention time of the chloramphenicol palmitate peak in the chromatogram of the *Assay preparation* corresponds to that of the chloramphenicol palmitate peak in the chromatogram of the *Standard preparation*, as obtained in the *Assay*.

Uniformity of dosage units ⟨905⟩—

FOR SUSPENSION PACKAGED IN SINGLE-UNIT CONTAINERS: meets the requirements.

pH ⟨791⟩: between 4.5 and 7.0.

Polymorph A—

Standard preparation—Prepare a freshly mixed portion, free from air bubbles, of a dry mixture of 1 part by weight of USP Chloramphenicol Palmitate Polymorph A RS and 9 parts by weight of USP Chloramphenicol Palmitate Nonpolymorph A RS. Prepare a 1 in 3 mineral oil dispersion of this mixture, and place a portion of it between two sodium chloride plates, taking care not to allow air bubbles to form.

Test preparation—Place 20 mL of previously mixed Oral Suspension in a 50-mL centrifuge tube, add 20 mL of water, and mix. Centrifuge, and discard the supernatant solution. Add 20 mL of water to the residue in the centrifuge tube, mix, centrifuge, and discard the supernatant solution. Repeat this washing two times. Dry the residue in vacuum over silica gel for not less than 14 hours. Prepare a 1 in 3 mineral oil dispersion of the dried residue, and place a portion of it between two sodium chloride plates, taking care not to allow air bubbles to form.

Procedure—Concomitantly record the absorption spectra of the *Standard preparation* and the *Test preparation* from about 11 µm to about 13 µm, with a suitable infrared absorption spectrophotometer, using an empty cell to set the instrument to 100 percent transmittance. Adjust the cell thickness of the *Standard preparation* and of the *Test preparation* so that transmittances of 20 percent to 30 percent are obtained at 12.3 µm. On each spectrum, draw a straight baseline between the absorption minima at wavelengths of about 11.3 µm and 12.65 µm. Draw straight lines, perpendicular to the wavelength scale, at the wavelengths of maximum absorption at about 11.65 µm and 11.86 µm, intersecting both the baseline and the spectrum. Determine the absorbance ratio:

$$(A_{11.65a} - A_{11.65b})/(A_{11.86a} - A_{11.86b}),$$

in which the parenthetic expressions are the differences in absorbance values obtained at the wavelengths indicated by the subscripts for the spectrum (*a*) and at the point of intersection of the perpendicular line with the baseline (*b*). The absorbance ratio of the *Test preparation* is greater than that of the *Standard preparation.*

Assay—

Mobile phase, Standard preparation, and *Chromatographic system*—Proceed as directed in the *Assay* under *Chloramphenicol Palmitate.*

Assay preparation—Transfer an accurately measured volume of Chloramphenicol Palmitate Oral Suspension, well-shaken and free from air bubbles, equivalent to about 160 mg of chloramphenicol, to a 200-mL volumetric flask containing about 20 mL of methanol, add 4 mL of glacial acetic acid, dilute with methanol to volume, and mix. Filter about 20 mL of this solution through glass-fiber filter paper. Transfer 10.0 mL of the filtrate to a 25-mL volumetric flask, dilute with *Mobile phase* to volume, and mix.

Procedure—Proceed as directed for *Procedure* in the *Assay* under *Chloramphenicol Palmitate.* Calculate the quantity, in mg, of chloramphenicol ($C_{11}H_{12}Cl_2N_2O_5$) equivalent in each mL of Oral Suspension taken by the formula:

$$0.004(W_S/V)(P_S)(r_U/r_S),$$

in which *V* is the volume, in mL, of Oral Suspension taken, and the other terms are as defined therein.

Sterile Chloramphenicol Sodium Succinate

$C_{15}H_{15}Cl_2N_2NaO_8$ 445.19

Butanedioic acid, mono[2-[(2,2-dichloroacetyl)amino]-3-hydroxy-3-(4-nitrophenyl)propyl] ester, monosodium salt, [*R*-(*R**,*R**)]-.

D-*threo*-(−)-2,2-Dichloro-*N*-[β-hydroxy-α-(hydroxymethyl)-*p*-nitrophenethyl]acetamide α-(sodium succinate) [*982-57-0*].

» Sterile Chloramphenicol Sodium Succinate has a potency equivalent to not less than 650 µg and not more than 765 µg of chloramphenicol ($C_{11}H_{12}Cl_2N_2O_5$) per mg. In addition, where packaged for dispensing and constituted as directed in the labeling, it contains the equivalent of not less than 90.0 percent and not more than 115.0 percent of the labeled amount of $C_{11}H_{12}Cl_2N_2O_5$.

Packaging and storage—Preserve in *Containers for Sterile Solids* as described under *Injections* ⟨1⟩.

Reference standard—*USP Chloramphenicol Reference Standard*—Do not dry before using.

Identification—The *Assay preparation*, prepared as directed in the *Assay*, exhibits an absorption maximum at a wavelength of about 276 nm, when tested as directed for *Procedure* in the *Assay*.

Specific rotation ⟨781⟩: between +5.0° and +8.0°, calculated on the anhydrous basis, determined in a solution containing 50 mg per mL.

Pyrogen—When diluted, if necessary, with sterile saline TS to a concentration of 5.0 mg of chloramphenicol per mL, it meets the requirements of the *Pyrogen Test* ⟨151⟩, the test dose being 1 mL per kg.

Sterility—It meets the requirements under *Sterility Tests* ⟨71⟩, when tested as directed in the section, *Test Procedures Using Membrane Filtration*.

pH ⟨791⟩: between 6.4 and 7.0, in a solution containing the equivalent of 250 mg of chloramphenicol per mL.

Water, *Method I* ⟨921⟩: not more than 5.0%.

Particulate matter ⟨788⟩: meets the requirements under *Small-volume Injections*.

Free chloramphenicol—
Mobile phase—Prepare a suitably filtered and degassed mixture of 0.05 *M* monobasic ammonium phosphate, previously adjusted with 10% phosphoric acid and methanol (60:40) to a pH of 2.5 ± 0.1. Make adjustments if necessary (see *System Suitability* under *Chromatography* ⟨621⟩).
Standard preparation—Dissolve an accurately weighed quantity of USP Chloramphenicol RS in *Mobile phase* to obtain a solution containing a known concentration of about 6 µg per mL. Filter this solution through a 0.5-µm or finer porosity filter, and use the filtrate as the *Standard preparation.*
Test preparation—Dissolve the contents of 1 container in a volume of *Mobile phase* to obtain a solution containing the equivalent of about 100 mg of chloramphenicol per mL. Dilute this solution quantitatively and stepwise, if necessary, with *Mobile phase* to obtain a solution containing the equivalent of about 0.5 mg of chloramphenicol per mL. Filter a portion of this solution through a 0.5-µm or finer porosity filter, and use the filtrate as the *Test preparation.*
Chromatographic system (see *Chromatography* ⟨621⟩)—The liquid chromatograph is equipped with a 275-nm detector and a 4.6-mm × 10-cm column that contains 5-µm packing L1. The flow rate is about 1 mL per minute. Chromatograph the *Test preparation*, and record the peak responses as directed under *Procedure:* the column efficiency determined from the two major peaks, chloramphenicol-1-succinate and chloramphenicol-3-succinate, is not less than 1750 theoretical plates, the resolution, *R*, between the two peaks is not less than 2.0, and the tailing factor is not more than 1.2. Chromatograph the *Standard preparation*, and record the peak responses as directed under *Procedure:* the relative standard deviation for replicate injections is not less than 2.0%.
Procedure—[NOTE—Use peak areas where peak responses are indicated.] Separately inject equal volumes (about 10 µL) of the *Standard preparation* and the *Test preparation* into the chromatograph, record the chromatograms, and measure the responses for the free chloramphenicol peaks. Calculate the percentage of free chloramphenicol ($C_{11}H_{12}Cl_2N_2O_5$) in the specimen taken by the formula:

$$0.1(C/D)(r_U/r_S),$$

in which *C* is the concentration, in µg per mL, of USP Chloramphenicol RS in the *Standard preparation, D* is the concentration, in mg per mL, of chloramphenicol equivalent in the *Test preparation*, based on the labeled quantity in the container and the extent of dilution, and r_U and r_S are the chloramphenicol peak responses obtained from the *Test preparation* and the *Standard preparation*, respectively. Not more than 2.0% is found.

Assay—
Standard preparation—Dissolve an accurately weighed quantity of USP Chloramphenicol RS in water, and dilute quantitatively with water to obtain a solution having a known concentration of about 20 µg per mL.
Assay preparation 1—Dissolve an accurately weighed quantity of Sterile Chloramphenicol Sodium Succinate in water, and dilute quantitatively with water to obtain a solution having a concentration equivalent to about 20 µg of chloramphenicol per mL.
Assay preparation 2 (where it is packaged for dispensing)—Constitute 1 container of Sterile Chloramphenicol Sodium Succinate as directed in the labeling. Dilute an accurately measured volume of the constituted solution quantitatively with water to obtain a solution having a concentration of about 20 µg of chloramphenicol per mL.
Procedure—Concomitantly determine the absorbances of the *Standard preparation* in 1-cm cells at the wavelength of maximum absorbance at about 278 nm, with a suitable spectrophotometer, and the absorbance of the *Assay preparation* at the wavelength of maximum absorbance at about 276 nm, using water as the blank. Calculate the potency, in µg of chloramphenicol ($C_{11}H_{12}Cl_2N_2O_5$) per mg, of the Sterile Chloramphenicol Sodium Succinate taken by the formula:

$$(CP/W)(A_U/A_S),$$

in which *C* is the concentration, in µg per mL, of USP Chloramphenicol RS in the *Standard preparation, P* is the potency, in µg per mg, of USP Chloramphenicol RS, *W* is the weight, in µg, of Sterile Chloramphenicol Sodium Succinate taken in each mL of *Assay preparation 1*, and A_U and A_S are the absorbances of the *Assay preparation* and the *Standard preparation*, respectively. Calculate the quantity, in mg, of chloramphenicol ($C_{11}H_{12}Cl_2N_2O_5$), in each mL of the constituted Sterile Chloramphenicol Sodium Succinate taken by the formula:

$$(L/D)(CP/1000)(A_U/A_S),$$

in which *L* is the labeled quantity of chloramphenicol in each mL of constituted solution, and *D* is the concentration, in µg per mL, of chloramphenicol in *Assay preparation 2*, on the basis of the labeled quantity of chloramphenicol in each mL of constituted solution and the extent of dilution.

Chlordiazepoxide

NHCH3 N N O Cl

$C_{16}H_{14}ClN_3O$ 299.76
3*H*-1,4-Benzodiazepin-2-amine, 7-chloro-*N*-methyl-5-phenyl-, 4-oxide.
7-Chloro-2-(methylamino)-5-phenyl-3*H*-1,4-benzodiazepine 4-oxide [*58-25-3*].

» Chlordiazepoxide contains not less than 98.0 percent and not more than 102.0 percent of $C_{16}H_{14}ClN_3O$, calculated on the dried basis.

Packaging and storage—Preserve in tight, light-resistant containers.

Reference standards—*USP Chlordiazepoxide Reference Standard*—Dry at 105° for 3 hours before using. *USP 7-Chloro-1,3-dihydro-5-phenyl-2H-1,4-benzodiazepin-2-one 4-Oxide Reference Standard*—Keep container tightly closed and protected from light. Dry over silica gel for 4 hours before using. *USP 2-Amino-5-chlorobenzophenone Reference Standard*—Keep container tightly closed and protected from light. Dry over silica gel for 4 hours before using.

Identification—
A: The infrared absorption spectrum of a potassium bromide dispersion of it, previously dried, exhibits maxima only at the same wavelengths as that of a similar preparation of USP Chlordiazepoxide RS.
B: The retention time of the major peak in the chromatogram of the *Assay preparation* is the same as that of the *Standard preparation*, both relative to the internal standard, as obtained in the *Assay.*
C: To about 20 mg add 5 mL of hydrochloric acid and 10 mL of water, and heat to boiling to effect hydrolysis. To the cooled solution add 2 mL of sodium nitrite solution (1 in 1000), shake, add 1 mL of ammonium sulfamate solution (1 in 200), then shake for 2 minutes, and add 1 mL of *N*-1-naphthylethylenediamine dihydrochloride solution (1 in 1000): a reddish violet color is produced.

Loss on drying ⟨731⟩—Dry it at 105° for 3 hours: it loses not more than 0.3% of its weight.

Residue on ignition ⟨281⟩: not more than 0.1%.

Heavy metals, *Method II* ⟨231⟩: 0.002%.

Related compounds—Transfer 50.0 mg to a 10-mL conical flask, add 2.5 mL of acetone, and shake. Allow any undissolved particles to settle, and apply 50 µL of the supernatant solution to a suitable thin-layer chromatographic plate (see *Chromatography* ⟨621⟩), coated with a 0.25-mm layer of chromatographic silica gel. Apply to the same plate 10 µL of an acetone solution containing 100 µg per mL of USP 7-Chloro-1,3-dihydro-5-phenyl-2*H*-1,4-benzodiazepin-2-one 4-Oxide RS and 10 µL of an acetone solution containing 10 µg per mL of USP 2-Amino-5-chlorobenzophenone RS. Develop the chromatogram in a chromatographic chamber (not previously saturated with the developing solvent) in ethyl acetate until the solvent front has moved about three-fourths of the length of the plate. Remove the plate from the developing chamber, mark the solvent front, and allow the solvent to evaporate. Locate the spots on the plate by lightly spraying with 2 *N* sulfuric acid, drying at 105° for 15 minutes, and then spraying in succession with sodium nitrite solution (1 in 1000), ammonium sulfamate solution (1 in 200), and *N*-1-naphthylethylenediamine dihydrochloride solution (1 in 1000). Any spots from the test solution are not greater in size or intensity than the spots at the respective R_f values produced by the Standard solutions, corresponding to not more than 0.1% of 7-chloro-1,3-dihydro-5-phenyl-2*H*-1,4-benzodiazepin-2-one 4-oxide and to not more than 0.01% of 2-amino-5-chlorobenzophenone.

Assay—[NOTE—Use low-actinic glassware in this procedure.]

Mobile phase—Prepare a mixture consisting of water, tetrahydrofuran, and methanol (5:4:1).

Internal standard solution—Dissolve an amount of sulfanilamide in *Mobile phase* to obtain a solution having a concentration of about 0.6 mg per mL.

Standard preparation—Dissolve an accurately weighed quantity of USP Chlordiazepoxide RS in *Mobile phase* and dilute quantitatively with *Mobile phase* to obtain a solution having known concentrations of about 1 mg of USP Chlordiazepoxide RS per mL. Transfer 10.0 mL of this solution and 10.0 mL of *Internal standard solution* to a 100-mL volumetric flask, dilute with *Mobile phase* to volume, and mix.

Assay preparation—Transfer about 50 mg of Chlordiazepoxide, accurately weighed, to a 50-mL volumetric flask, dissolve in *Mobile phase*, dilute with *Mobile phase* to volume, and mix. Transfer 10.0 mL of the resulting solution and 10.0 mL of *Internal standard solution* to a 100-mL volumetric flask, dilute with *Mobile phase* to volume, and mix.

Chromatographic system—The chromatograph is equipped with a 254-nm detector and a 3.9-mm × 30-cm column that contains packing L1. The flow rate is about 1 mL per minute. Chromatograph five replicate injections of the *Standard preparation* and record the peak responses as directed under *Procedure:* the relative standard deviation is not more than 2.0%, and the resolution factor is not less than 1.5.

Procedure—Separately inject equal volumes (about 20 µL) of the *Standard preparation* and the *Assay preparation* into the chromatograph, and record the chromatograms. Measure the peak responses at approximate retention times of 3 minutes for sulfanilamide and 6 minutes for chlordiazepoxide. Calculate the quantity, in mg, of $C_{16}H_{14}ClN_3O$ in the portion of Chlordiazepoxide taken by the formula:

$$0.5C(R_U / R_S),$$

in which C is the concentration, in µg per mL, of USP Chlordiazepoxide RS in the *Standard preparation*, and R_U and R_S are the ratios of the peak responses of chlordiazepoxide and sulfanilamide from the *Assay preparation* and the *Standard preparation*, respectively.

Chlordiazepoxide Tablets

» Chlordiazepoxide Tablets contain not less than 90.0 percent and not more than 110.0 percent of the labeled amount of $C_{16}H_{14}ClN_3O$.

Packaging and storage—Preserve in tight, light-resistant containers.

Reference standards—*USP Chlordiazepoxide Reference Standard*—Dry at 105° for 3 hours before using. *USP 7-Chloro-1,3-dihydro-5-phenyl-2H-1,4-benzodiazepin-2-one 4-Oxide Reference Standard*—Keep container tightly closed and protected from light. Dry over silica gel for 4 hours before using. *USP 2-Amino-5-chlorobenzophenone Reference Standard*—Keep container tightly closed and protected from light. Dry over silica gel for 4 hours before using.

Identification—

A: The retention time of the major peak in the chromatogram of the *Assay preparation* is the same as that of the *Standard preparation*, both relative to the internal standard, as obtained in the *Assay*.

B: A portion of finely powdered Tablets, equivalent to about 20 mg of chlordiazepoxide, responds to *Identification test C* under *Chlordiazepoxide*.

Dissolution ⟨711⟩—

Medium: simulated gastric fluid TS, prepared without pepsin; 900 mL.

Apparatus 1: 100 rpm.

Time: 30 minutes.

Procedure—Determine the amount of $C_{16}H_{14}ClN_3O$ dissolved from ultraviolet absorbances at the wavelength of maximum absorbance at about 309 nm of filtered portions of the solution under test, suitably diluted with *Dissolution Medium*, if necessary, in comparison with a Standard solution having a known concentration of USP Chlordiazepoxide RS in the same medium.

Tolerances—Not less than 85% (Q) of the labeled amount of $C_{16}H_{14}ClN_3O$ is dissolved in 30 minutes.

Uniformity of dosage units ⟨905⟩: meet the requirements.

Related compounds—Transfer an accurately weighed portion of finely powdered Tablets, equivalent to about 25 mg of chlordiazepoxide, to a 10-mL conical flask, and proceed as directed in the test for *Related compounds* under *Chlordiazepoxide*, beginning with "add 2.5 mL of acetone," except to use 20 µL of an acetone solution containing 1 mg per mL of USP 7-Chloro-1,3-dihydro-5-phenyl-2*H*-1,4-benzodiazepin-2-one 4-Oxide RS instead of 10 µL of an acetone solution containing 100 µg per mL of the Reference Standard, and except to use 5 µL of an acetone solution containing 100 µg per mL of USP 2-Amino-5-chlorobenzophenone RS instead of 10 µL of an acetone solution containing 10 µg per mL of the Reference Standard. Any spots from the test solution are not greater in size or intensity than the spots at the respective R_f values produced by the Standard solutions, corresponding to not more than 4.0% of 7-chloro-1,3-dihydro-5-phenyl-2*H*-1,4-benzodiazepin-2-one 4-oxide and to not more than 0.1% of 2-amino-5-chlorobenzophenone.

Assay—

Mobile phase, *Internal standard solution*, *Standard preparation*, and *Chromatographic system*—Proceed as directed in the *Assay* under *Chlordiazepoxide*.

Assay preparation—Weigh and finely powder not less than 20 Chlordiazepoxide Tablets. Transfer an accurately weighed portion of the powder, equivalent to about 50 mg of chlordiazepoxide, to a 50-mL volumetric flask, add *Mobile phase* to volume, sonicate with continuous swirling for 5 minutes to disperse the mixture, shake by mechanical means for 10 minutes, and allow undissolved particles to settle. Transfer 10.0 mL of the resulting solution and 10.0 mL of *Internal standard solution* to a 100-mL volumetric flask, dilute with *Mobile phase* to volume, and mix. Filter a portion of this solution through a 5-µm porosity filter, and use the filtrate as the *Assay preparation*.

Procedure—Proceed as directed for *Procedure* in the *Assay* under *Chlordiazepoxide*. Calculate the quantity, in mg, of $C_{16}H_{14}ClN_3O$ in the portion of Tablets taken by the formula:

$$0.5C(R_U / R_S).$$

Chlordiazepoxide and Amitriptyline Hydrochloride Tablets

» Chlordiazepoxide and Amitriptyline Hydrochloride Tablets contain not less than 90.0 percent and not more than 110.0 percent of the labeled amount of chlordiazepoxide ($C_{16}H_{14}ClN_3O$) and an amount of amitriptyline hydrochloride equivalent to not less than 90.0 percent and not more than 110.0 percent of the labeled amount of amitriptyline ($C_{20}H_{23}N$).

Packaging and storage—Preserve in tight, light-resistant containers.

Reference standards—*USP Amitriptyline Hydrochloride Reference Standard*—Dry at a pressure not exceeding 5 mm of mercury at 60° to constant weight before using. *USP Chlordiazepoxide Reference Standard*—Dry at 105° for 3 hours before using. *USP 7-Chloro-1,3-dihydro-5-phenyl-2H-1,4-benzodiazepin-2-one 4-Oxide Reference Standard*—Keep container tightly closed and protected from light. Dry over silica gel for 4 hours before using. *USP 2-Amino-5-chlorobenzophenone Reference Standard*—Keep container tightly closed and protected from light. Dry over silica gel for 4 hours before using.

Identification—The retention times of the major peaks in the chromatogram of the *Assay preparation* are the same as those of the *Standard preparation*, all relative to the internal standard, as obtained in the *Assay*.

Dissolution ⟨711⟩—

Medium: simulated gastric fluid TS, prepared without pepsin; 900 mL.

Apparatus 1: 100 rpm.

Time: 30 minutes.

Procedure—Determine the absorbances of filtered portions of the solution under test, suitably diluted with *Dissolution Medium*, if necessary, and Standard solutions having known concentrations of USP Chlordiazepoxide RS and USP Amitriptyline Hydrochloride RS in the same medium, at wavelengths of 239 nm and 309 nm, using *Dissolution Medium* as the blank. Calculate the percentages of chlordiazepoxide ($C_{16}H_{14}ClN_3O$) and amitriptyline ($C_{20}H_{23}N$) dissolved by the formulas:

$$100[(CD_{309}/T)(A_U/A_S)_{309}]_Z$$

and

$$100(277.41/313.87)[(CD_{239}/T)(A_X/A_S)_{239}]_Y,$$

respectively, in which Z denotes chlordiazepoxide and Y denotes amitriptyline hydrochloride, C is the concentration, in mg per mL, of the Reference Standard in the respective Standard solution, D_{239} and D_{309} are dilution factors, T is the respective labeled amount, in mg, in the Tablet, 277.41 is the molecular weight of amitriptyline, 313.87 is the molecular weight of amitriptyline hydrochloride, and A_U and A_S are the absorbances of the solutions obtained from the solution under test and the Standard solution, respectively, at the wavelengths indicated by the subscripts, except the $(A_X)_{239}$ is obtained by calculation using the formula:

$$(A_U)_{239} - D_{239}[(A_U)_{309}][(a)_{239}/(a)_{309}]_Z,$$

in which a is the absorptivity of the Standard solution of USP Chlordiazepoxide RS at the wavelength indicated by the subscript.

Tolerances—Not less than 85% (Q) of the labeled amount of chlordiazepoxide ($C_{16}H_{14}ClN_3O$) and an amount of amitriptyline hydrochloride equivalent to not less than 85% (Q) of the labeled amount of amitriptyline ($C_{20}H_{23}N$) is dissolved in 30 minutes.

Uniformity of dosage units ⟨905⟩: meet the requirements for *Content Uniformity* with respect to chlordiazepoxide and to amitriptyline.

Related compounds—Transfer an accurately weighed portion of finely powdered Tablets, equivalent to about 25 mg of chlordiazepoxide, to a 10-mL conical flask, add 2.5 mL of acetone, and shake. Allow any undissolved particles to settle, and apply 50 µL of the supernatant solution to a thin-layer chromatographic plate (see *Chromatography* ⟨621⟩), coated with a 0.25-mm layer of chromatographic silica gel. Apply to the same plate 20 µL of an acetone solution containing 1 mg per mL of USP 7-Chloro-1,3-dihydro-5-phenyl-2*H*-1,4-benzodiazepin-2-one 4-Oxide RS and 10 µL of an acetone solution containing 50 µg per mL of USP 2-Amino-5-chlorobenzophenone RS. Develop the chromatogram in a chromatographic chamber (not previously saturated with the developing solvent) in ethyl acetate until the solvent front has moved about three-fourths of the length of the plate. Remove the plate from the developing chamber, mark the solvent front, and allow the solvent to evaporate. Locate the spots on the plate by lightly spraying with 2 *N* sulfuric acid, drying at 105° for 15 minutes, and then spraying in succession with sodium nitrite solution (1 in 1000), ammonium sulfamate solution (1 in 200), and *N*-1-naphthylethylenediamine dihydrochloride solution (1 in 1000): any spots from the test solution are not greater in size or intensity than the spots at the respective R_f values produced by the Standard solutions, corresponding to not more than 4.0% of 7-chloro-1,3-dihydro-5-phenyl-2*H*-1,4-benzodiazepin-2-one 4-oxide and not more than 0.1% of 2-amino-5-chlorobenzophenone.

Assay—[NOTE—Use low-actinic glassware in this procedure.]

Solvent mixture—Prepare a mixture consisting of water, tetrahydrofuran, and methanol (5:4:1).

pH 2.5 buffer—Mix 10.5 mL of 0.20 *N* sodium hydroxide with 100 mL of a solution consisting of 0.04 *M* acetic acid, 0.04 *M* phosphoric acid, and 0.04 *M* boric acid (prepared by dissolving 2.402 g of glacial acetic acid, 4.612 g of phosphoric acid, and 2.473 g of boric acid in sufficient water to obtain 1000 mL of solution).

Internal standard solution—Dissolve an amount of sulfanilamide in *Solvent mixture* to obtain a solution having a concentration of about 1 mg per mL.

Standard preparation—Dissolve accurately weighed quantities of USP Chlordiazepoxide RS and USP Amitriptyline Hydrochloride RS in *Solvent mixture*, and dilute quantitatively with *Solvent mixture* to obtain a solution having known concentrations of about 1 mg of USP Chlordiazepoxide RS per mL and about 2.8 mg of USP Amitriptyline Hydrochloride RS per mL. Transfer 10.0 mL of this solution and 10.0 mL of *Internal standard solution* to a 100-mL volumetric flask, dilute with *Solvent mixture* to volume, and mix.

Assay preparation—Weigh and finely powder not less than 20 Chlordiazepoxide and Amitriptyline Hydrochloride Tablets. Transfer an accurately weighed portion of the powder, equivalent to about 50 mg of chlordiazepoxide and about 125 mg of amitriptyline, to a 50-mL volumetric flask, add *Solvent mixture* to volume, sonicate to disperse the mixture, and allow undissolved particles to settle. Transfer 10.0 mL of the resulting solution and 10.0 mL of *Internal standard solution* to a 100-mL volumetric flask, dilute with *Solvent mixture* to volume, and mix. Filter a portion of this solution through a 5-µm porosity filter, and use the filtrate as the *Assay preparation*.

Chromatographic system (see *Chromatography* ⟨621⟩)—The chromatograph is equipped with a 254-nm detector and a 3.9-mm × 30-cm column that contains packing L1. The mobile phase is a solution of 0.01 *M* sodium lauryl sulfate in a mixture of *pH 2.5 buffer*, tetrahydrofuran, and methanol (5:4:1). The flow rate is about 1 mL per minute. Chromatograph five replicate injections of the *Standard preparation*, and record the peak responses as directed under *Procedure:* the relative standard deviation is not more than 2.0%, and the resolution factors are not less than 2.0.

Procedure—Separately inject equal volumes (about 20 µL) of the *Standard preparation* and the *Assay preparation* into the chromatograph, and record the chromatogram. Measure the peak responses at approximate retention times of 3 minutes for sulfanilamide, 5 minutes for chlordiazepoxide, and 7 minutes for amitriptyline hydrochloride. Calculate the quantity, in mg, of chlordiazepoxide ($C_{16}H_{14}ClN_3O$) in the portion of Tablets taken by the formula:

$$0.5[C(R_U/R_S)]_Z,$$

and calculate the quantity, in mg, of amitriptyline ($C_{20}H_{23}N$) in the portion of Tablets taken by the formula:

$$(277.41/313.87)[(0.5C)(R_U/R_S)]_Y,$$

in which Z denotes chlordiazepoxide and Y denotes amitriptyline hydrochloride, 277.41 is the molecular weight of amitriptyline, 313.87 is the molecular weight of amitriptyline hydrochloride, C is the concentration, in µg per mL, of the Reference Standard in the *Standard preparation*, and R_U and R_S are the ratios of the peak responses of the analyte and sulfanilamide from the *Assay preparation* and the *Standard preparation*, respectively.

Chlordiazepoxide Hydrochloride

$C_{16}H_{14}ClN_3O.HCl$ 336.22

3*H*-1,4-Benzodiazepin-2-amine, 7-chloro-*N*-methyl-5-phenyl-, 4-oxide, monohydrochloride.

7-Chloro-2-(methylamino)-5-phenyl-3*H*-1,4-benzodiazepine 4-oxide monohydrochloride [*438-41-5*].

» Chlordiazepoxide Hydrochloride contains not less than 98.0 percent and not more than 102.0 percent of $C_{16}H_{14}ClN_3O.HCl$, calculated on the dried basis.

Packaging and storage—Preserve in tight, light-resistant containers.

Reference standards—*USP Chlordiazepoxide Hydrochloride Reference Standard*—Dry in vacuum over phosphorus pentoxide at 60° for 4 hours before using. *USP 7-Chloro-1,3-dihydro-5-phenyl-2H-1,4-benzodiazepin-2-one 4-Oxide Reference Standard*—Keep container tightly closed and protected from light. Dry over silica gel for 4 hours before using. *USP 2-Amino-5-chlorobenzophenone Reference Standard*—Keep container tightly closed and protected from light. Dry over silica gel for 4 hours before using.

Identification—

A: The infrared absorption spectrum of a potassium bromide dispersion of it, previously dried, exhibits maxima only at the same wavelengths as that of a similar preparation of USP Chlordiazepoxide Hydrochloride RS.

B: The relative retention time of the major peak in the chromatogram of the *Assay preparation* corresponds to that of the *Standard preparation*, both relative to the internal standard, obtained as directed in the *Assay*.

C: To about 20 mg add 5 mL of hydrochloric acid and 10 mL of water, and heat to boiling to effect hydrolysis. To the cooled solution add 2 mL of sodium nitrite solution (1 in 1000), 1 mL of ammonium sulfamate solution (1 in 200), and 1 mL of *N*-1-naphthylethylenediamine dihydrochloride solution (1 in 1000): a reddish violet color is produced.

Melting range, *Class I* ⟨741⟩: between 212° and 218°, with decomposition.

Loss on drying ⟨731⟩—Dry it in vacuum over phosphorus pentoxide at 60° for 4 hours: it loses not more than 0.5% of its weight.

Residue on ignition ⟨281⟩: not more than 0.1%.

Heavy metals, *Method II* ⟨231⟩: 0.002%.

Related compounds—It meets the requirements of the test for *Related compounds* under *Chlordiazepoxide*.

Assay—[NOTE—Use low-actinic glassware in this procedure.]

Mobile phase—Prepare a mixture consisting of water, tetrahydrofuran, and methanol (5:4:1).

Internal standard solution—Dissolve an amount of sulfanilamide in *Mobile phase* to obtain a solution having a concentration of about 0.6 mg per mL.

Standard preparation—Dissolve an accurately weighed quantity of USP Chlordiazepoxide Hydrochloride RS in *Mobile phase*, and dilute quantitatively with *Mobile phase* to obtain a solution having known concentrations of about 1 mg of USP Chlordiazepoxide Hydrochloride RS per mL. Transfer 10.0 mL of the resulting solution and 10.0 mL of *Internal standard solution* to a 100-mL volumetric flask, dilute with *Mobile phase* to volume, and mix.

Assay preparation—Transfer about 50 mg of Chlordiazepoxide Hydrochloride, accurately weighed, to a 50-mL volumetric flask, dissolve in *Mobile phase*, dilute with *Mobile phase* to volume, and mix. Transfer 10.0 mL of the resulting solution and 10.0 mL of *Internal standard solution* to a 100-mL volumetric flask, dilute with *Mobile phase* to volume, and mix.

Chromatographic system (see *Chromatography* ⟨621⟩)—The liquid chromatograph is equipped with a 254-nm detector and a 3.9-mm × 30-cm column that contains packing L1. The flow rate is about 1 mL per minute. Chromatograph five replicate injections of the *Standard preparation*, and record the peak responses as directed under *Procedure:* the relative standard deviation is not more than 2.0%, and the resolution factor is not less than 4.0.

Procedure—Separately inject equal volumes (about 20 µL) of the *Standard preparation* and the *Assay preparation* into the chromatograph, record the chromatograms, and measure the responses for the major peaks. Relative retention times are about 0.5 for sulfanilamide and 1.0 for chlordiazepoxide hydrochloride. Calculate the quantity, in mg, of $C_{16}H_{14}ClN_3O.HCl$ in the portion of Chlordiazepoxide Hydrochloride taken by the formula:

$$0.5C(R_U/R_S),$$

in which C is the concentration, in µg per mL, of USP Chlordiazepoxide Hydrochloride RS in the *Standard preparation*, and R_U and R_S are the ratios of the peak responses of chlordiazepoxide hydrochloride and sulfanilamide obtained from the *Assay preparation* and the *Standard preparation*, respectively.

Chlordiazepoxide Hydrochloride Capsules

» Chlordiazepoxide Hydrochloride Capsules contain not less than 90.0 percent and not more than 110.0 percent of the labeled amount of $C_{16}H_{14}ClN_3O.HCl$.

Packaging and storage—Preserve in tight, light-resistant containers.

Reference standards—*USP Chlordiazepoxide Hydrochloride Reference Standard*—Dry in vacuum over phosphorus pentoxide at 60° for 4 hours before using. *USP 7-Chloro-1,3-dihydro-5-phenyl-2H-1,4-benzodiazepin-2-one 4-Oxide Reference Standard*—Keep container tightly closed and protected from light. Dry over silica gel for 4 hours before using. *USP 2-Amino-5-chlorobenzophenone Reference Standard*—Keep container tightly closed and protected from light. Dry over silica gel for 4 hours before using.

Identification—

A: The solution employed for measurement of absorbance in the *Assay* exhibits maxima at 245 ± 2 nm and 311 ± 2 nm, and the ratio A_{245}/A_{311} is between 2.90 and 3.45.

B: A portion of the contents of Capsules responds to *Identification test C* under *Chlordiazepoxide Hydrochloride*.

Dissolution ⟨711⟩—

Medium: water; 900 mL.

Apparatus 1: 100 rpm.

Time: 30 minutes.

Procedure—Measure the amount in solution in filtered portions of the *Dissolution Medium*, suitably diluted, in 1-cm cells at the wavelength of maximum absorbance at about 245 nm, with a suitable spectrophotometer, in comparison with a Standard solution of known concentration of USP Chlordiazepoxide Hydrochloride RS. Remove the contents of 12 Capsules as completely as possible with the aid of a current of air. Dissolve the empty capsule shells in 900 mL of *Dissolution Medium*. Filter a portion of the solution, and determine the absorbance at the same dilution and in the same manner as for the Capsules, making any necessary modifications.

Tolerances—Not less than 85% (*Q*) of the labeled amount of $C_{16}H_{14}ClN_3O.HCl$ is dissolved in 30 minutes.

Uniformity of dosage units ⟨905⟩: meet the requirements.

Procedure for content uniformity—[NOTE—Use low-actinic glassware in this procedure.] Transfer the contents of 1 Capsule to a 200-mL volumetric flask, dissolve in water, dilute with water to volume, and filter, discarding the first 20 mL of the filtrate. Dilute a portion of the filtrate quantitatively and stepwise with 0.1 *N* hydrochloric acid to obtain a solution having a concentration of about 6 µg of chlordiazepoxide hydrochloride per mL. Dissolve a suitable quantity of USP Chlordiazepoxide Hydrochloride RS, accurately weighed, in 0.1 *N* hydrochloric acid to obtain a Standard solution having a known concentration of about 6 µg per mL. Concomitantly determine the absorbances of the two solutions in 1-cm cells at the wavelength of maximum absorbance at about 245 nm, with a suitable spectrophotometer, using 0.1 *N* hydrochloric acid as the blank. Calculate the quantity, in mg, of $C_{16}H_{14}ClN_3O.HCl$ in the Capsule by the formula:

$$(T/D)C(A_U/A_S),$$

in which T is the labeled quantity, in mg, of chlordiazepoxide hydrochloride in the Capsule, D is the concentration, in µg per mL, of chlordiazepoxide hydrochloride in the test solution, based on the labeled quantity per Capsule and the extent of dilution, C is the concentration, in µg per mL, of USP Chlordiazepoxide Hydrochloride RS in the Standard solution, and A_U and A_S are the absorbances of the solution from the contents of the Capsule and the Standard solution, respectively.

Related compounds—Proceed with Capsules as directed in the test for *Related compounds* under *Chlordiazepoxide Hydrochloride*, but use an accurately weighed portion of Capsule contents, equivalent to about 25 mg of chlordiazepoxide hydrochloride, and use 15 µL of a 1 in 1000 solution of USP 7-Chloro-1,3-dihydro-5-phenyl-2*H*-1,4-benzodiazepin-2-one 4-Oxide RS in acetone, and 10 µL of a 1 in 20,000 solution of USP 2-Amino-5-chlorobenzophenone RS in acetone. Not more than 3.0% of 7-chloro-1,3-dihydro-5-phenyl-2*H*-1,4-benzodiazepin-2-one 4-oxide is found, and not more than 0.1% of 2-amino-5-chlorobenzophenone is found.

Assay—[NOTE—Use low-actinic glassware in this procedure.] Weigh the contents of not less than 20 Chlordiazepoxide Hydrochloride Capsules, and determine the average weight per Capsule. Mix the combined contents, and transfer an accurately weighed portion of the powder, equivalent to about 60 mg of chlordiazepoxide hydrochloride, to a 100-mL volumetric flask. Add methanol to volume, mix, and filter, discarding the first 15 mL of the filtrate. Pipet 5 mL of the clear filtrate into a 100-mL volumetric flask, and add a 1 in 360 solution of sulfuric acid in dehydrated alcohol to volume. Pipet 10 mL of this solution into a 50-mL volumetric flask, and dilute with the same acidified alcohol to volume. Dissolve an accurately weighed quantity of USP Chlordiazepoxide Hydrochloride RS in methanol to obtain a solution having a known concentration of about 600 µg per mL. Dilute this solution quantitatively and stepwise with the acidified alcohol to obtain a Standard solution having a known concentration of about 6 µg per mL. Concomitantly determine the absorbances of both solutions in 1-cm cells at the wavelength of maximum absorbance at about 245 nm, with a suitable spectrophotometer, using the acidified alcohol as the blank. Calculate the quantity, in mg, of $C_{16}H_{14}ClN_3O.HCl$ in the portion of Capsules taken by the formula:

$$10C(A_U/A_S),$$

in which C is the concentration, in µg per mL, of USP Chlordiazepoxide Hydrochloride RS in the Standard solution, and A_U and A_S are the absorbances of the solution from the Capsules and the Standard solution, respectively.

Sterile Chlordiazepoxide Hydrochloride

» Sterile Chlordiazepoxide Hydrochloride is Chlordiazepoxide Hydrochloride suitable for parenteral use.

Packaging and storage—Preserve in *Containers for Sterile Solids* as described under *Injections* ⟨1⟩, protected from light.

Reference standards—*USP Chlordiazepoxide Hydrochloride Reference Standard*—Dry in vacuum over phosphorus pentoxide at 60° for 4 hours before using. *USP 7-Chloro-1,3-dihydro-5-phenyl-2H-1,4-benzodiazepin-2-one 4-Oxide Reference Standard*—Keep container tightly closed and protected from light. Dry over silica gel for 4 hours before using. *USP 2-Amino-5-chlorobenzophenone Reference Standard*—Keep container tightly closed and protected from light. Dry over silica gel for 4 hours before using.

Completeness of solution ⟨641⟩—It dissolves in the solvent and in the concentration recommended in the labeling to yield a clear solution.

Constituted solution—At the time of use, the constituted solution prepared from Sterile Chlordiazepoxide Hydrochloride meets the requirements for *Constituted Solutions* under *Injections* ⟨1⟩.

Pyrogen—Both Sterile Chlordiazepoxide Hydrochloride and the accompanying solvent, when diluted with Sodium Chloride Injection to concentrations of 400 mg in 100 mL and 8 mL in 100 mL, respectively, meet the requirements of the *Pyrogen Test* ⟨151⟩, the test dose for each dilution being 1 mL per kg.

pH ⟨791⟩: between 2.5 and 3.5, in a solution (1 in 100).

Other requirements—It responds to the *Identification tests* and meets the requirements of the tests for *Loss on drying* and *Heavy metals* under *Chlordiazepoxide Hydrochloride*, and the test for *Related compounds* under *Chlordiazepoxide*. It meets also the requirements for *Sterility Tests* ⟨71⟩, *Uniformity of Dosage Units* ⟨905⟩, and *Labeling* under *Injections* ⟨1⟩.

Assay—Proceed with Sterile Chlordiazepoxide Hydrochloride as directed in the *Assay* under *Chlordiazepoxide Hydrochloride*.

Chlordiazepoxide Hydrochloride and Clidinium Bromide Capsules

» Chlordiazepoxide Hydrochloride and Clidinium Bromide Capsules contain not less than 90.0 percent and not more than 110.0 percent of the labeled amounts of chlordiazepoxide hydrochloride ($C_{16}H_{14}ClN_3O.HCl$) and clidinium bromide ($C_{22}H_{26}BrNO_3$).

Packaging and storage—Preserve in tight, light-resistant containers.

Reference standards—*USP Chlordiazepoxide Hydrochloride Reference Standard*—Dry in vacuum over phosphorus pentoxide at 60° for 4 hours before using. *USP 7-Chloro-1,3-dihydro-5-phenyl-2H-1,4-benzodiazepin-2-one 4-Oxide Reference Standard*—Keep container tightly closed and protected from light. Dry over silica gel for 4 hours before using. *USP 2-Amino-5-chlorobenzophenone Reference Standard*—Keep container tightly closed and protected from light. Dry over silica gel for 4 hours before using. *USP Clidinium Bromide Reference Standard*—Dry at 105° for 3 hours before using. *USP 3-Quinuclidinyl Benzilate Reference Standard*—Dry over silica gel for 4 hours before using. Keep container tightly closed and protected from light. [*Caution—Avoid contact; work in a well-ventilated hood.*] *USP 3-Hydroxy-1-methylquinuclidinium Bromide Reference Standard*—Dry over silica gel for 4 hours before using. Keep container tightly closed and protected from light.

Identification—The retention times of the major peaks in the chromatogram of the *Assay preparation* correspond to those of the *Standard preparation*, as obtained in the *Assay*.

Dissolution ⟨711⟩—

Medium: water; 900 mL.

Apparatus 1: 100 rpm.

Time: 30 minutes.

Determine the amounts of chlordiazepoxide hydrochloride and clidinium bromide dissolved using the following method.

Solution A—Dissolve 1.92 g of sodium 1-pentanesulfonate in 900 mL of water. Adjust the solution with dilute sulfuric acid (1 in 100) to a pH of 3.8 ± 0.1, dilute with water to 1.0 liter, and mix.

Mobile phase—Prepare a filtered and degassed mixture of *Solution A*, tetrahydrofuran, and methanol (75:18:6). Make adjustments if necessary (see *System Suitability* under *Chromatography* ⟨621⟩).

Chromatographic system—The liquid chromatograph is equipped with a 212-nm detector and a 4-mm × 25-cm column that contains packing L1. The flow rate is about 2 mL per minute. Chromatograph replicate injections of the Standard solution, and record the peak responses as directed under *Procedure:* the resolution between the two components is not less than 5.0, and the relative standard deviation for replicate injections is not more than 2.0%.

Procedure—Separately inject equal volumes (about 100 µL) of a Standard solution and a filtered portion of the solution under test into the chromatograph, record the chromatograms, and measure the responses for the major peaks. The relative retention times are about 0.6 for clidinium bromide and 1.0 for chlordiazepoxide hydrochloride. Calculate the amounts of chlordiazepoxide hydrochloride ($C_{16}H_{14}ClN_3O.HCl$) and clidinium bromide ($C_{22}H_{26}BrNO_3$) dissolved in comparison with a Standard solution having known concentrations of USP Chlordiazepoxide Hydrochloride RS and USP Clidinium Bromide RS, similarly prepared and chromatographed.

Tolerances—Not less than 75% (*Q*) each of the labeled amounts of $C_{16}H_{14}ClN_3O.HCl$ and $C_{22}H_{26}BrNO_3$ are dissolved in 30 minutes.

Uniformity of dosage units ⟨905⟩: meet the requirements.

Related compounds—

A: *7-Chloro-1,3-dihydro-5-phenyl-2H-1,4-benzodiazepin-2-one 4-oxide* and *2-amino-5-chlorobenzophenone*—Proceed with Capsules as directed in the test for *Related compounds* under *Chlordiazepoxide Hydrochloride Capsules.* Not more than 3.0% of 7-chloro-1,3-dihydro-5-phenyl-2*H*-1,4-benzodiazepin-2-one 4-oxide is found, and not more than 0.1% of 2-amino-5-chlorobenzophenone is found.

B: *3-Quinuclidinyl benzilate*—Proceed with Capsules as directed in *Related compounds test A* under *Clidinium Bromide Capsules.* Not more than 0.03% of 3-quinuclidinyl benzilate is found.

C: *3-Hydroxy-1-methylquinuclidinium bromide*—Proceed with Capsules as directed in *Related compounds test B* under *Clidinium Bromide Capsules*, except to apply 20 µL each, instead of 40 µL each, of the *Reference solution* and the *Test preparation* to the chromatographic plate. Not more than 1.0% of 3-hydroxy-1-methylquinuclidinium bromide is found.

Assay—[NOTE—Use low-actinic glassware in this procedure.]

Sodium 1-pentanesulfonate solution—Dissolve 1.92 g of sodium 1-pentanesulfonate in about 900 mL of water in a 1-liter volumetric flask. Adjust with 1 *N* sulfuric acid to a pH of 3.8 ± 0.1, dilute with water to volume, mix, and filter.

Mobile phase—Prepare a filtered and degassed mixture of *Sodium 1-pentanesulfonate solution*, tetrahydrofuran, and methanol (70:24:6). Make adjustments if necessary (see *System Suitability* under *Chromatography* ⟨621⟩).

Solvent mixture—Prepare a mixture of water and methanol (3:1).

Standard stock solution A—Dissolve an accurately weighed quantity of USP Chlordiazepoxide Hydrochloride RS in *Solvent mixture*, and dilute quantitatively with *Solvent mixture* to obtain a solution having a known concentration of about 0.5 mg per mL.

Standard stock solution B—Dissolve an accurately weighed quantity of USP Clidinium Bromide RS in *Solvent mixture*, and dilute quantitatively with *Solvent mixture* to obtain a solution having a known concentration of about 0.25 mg per mL.

Standard stock solution C—Transfer about 3 mg of USP 7-Chloro-1,3-dihydro-5-phenyl-2*H*-1,4-benzodiazepin-2-one 4-Oxide RS to a 10-mL volumetric flask. Add about 2 mL of methanol, shake to dissolve, dilute with methanol to volume, and mix. Transfer 1.0 mL of this solution to a 100-mL volumetric flask, dilute with *Solvent mixture* to volume, and mix to obtain a solution having a known concentration of about 3 µg per mL.

Standard preparation—Transfer 10.0 mL each of *Standard stock solution A* and *Standard stock solution B* to a 25-mL volumetric flask, dilute with water to volume, and mix. This solution contains about 0.2 mg of USP Chlordiazepoxide Hydrochloride RS and 0.1 mg of USP Clidinium Bromide RS per mL.

Resolution solution—Transfer 10.0 mL each of *Standard stock solution B* and *Standard stock solution C* to a 25-mL volumetric flask, dilute with water to volume, and mix.

Assay preparation—Weigh the contents of not less than 20 Chlordiazepoxide Hydrochloride and Clidinium Bromide Capsules, and calculate the average weight per Capsule. Mix the combined contents of the Capsules, and transfer an accurately weighed portion, equivalent to about 50 mg of chlordiazepoxide hydrochloride ($C_{16}H_{14}ClN_3O.HCl$), to a 100-mL volumetric flask. Add about 50 mL of methanol, sonicate for 3 to 4 minutes, and shake by mechanical means for 15 minutes. Dilute with water to volume, mix, and filter, discarding the first 20 mL of the filtrate. Transfer 10.0 mL of the filtrate to a 25-mL volumetric flask, dilute with water to volume, and mix.

Chromatographic system (see *Chromatography* ⟨621⟩)—The liquid chromatograph is equipped with a 254-nm detector and a 4-mm × 30-cm column that contains packing L1. The flow rate is about 2 mL per minute. Chromatograph the *Resolution preparation*, and record the peak responses as directed under *Procedure:* the resolution, *R*, between the clidinium bromide and 7-chloro-1,3-dihydro-5-phenyl-2*H*-1,4-benzodiazepin-2-one 4-oxide peaks is not less than 1.5. The relative retention times are about 0.45 for clidinium bromide, 0.65 for 7-chloro-1,3-dihydro-5-phenyl-2*H*-1,4-benzodiazepin-2-one 4-oxide, and 1.0 for chlordiazepoxide hydrochloride. Chromatograph the *Standard preparation*, and record the peak responses as directed under *Procedure:* the relative standard deviation for replicate injections is not more than 2.0%.

Procedure—Separately inject equal volumes (about 50 µL) of the *Assay preparation* and the *Standard preparation* into the chromatograph, record the chromatograms, and measure the responses for the major peaks. Calculate the quantity, in mg, of chlordiazepoxide hydrochloride ($C_{16}H_{14}ClN_3O.HCl$) in the portion of Capsules taken by the formula:

$$250C(r_U / r_S),$$

in which *C* is the concentration, in mg per mL, of USP Chlordiazepoxide Hydrochloride RS in the *Standard preparation*, and r_U and r_S are the chlordiazepoxide hydrochloride peak responses obtained from the *Assay preparation* and the *Standard preparation*, respectively. Calculate the quantity, in mg, of clidinium bromide ($C_{22}H_{26}BrNO_3$) in the portion of Capsules taken by the same formula, reading clidinium bromide instead of chlordiazepoxide hydrochloride.

Chlorobutanol—*see* Chlorobutanol NF

Chlorocresol—*see* Chlorocresol NF

Chloroform—*see* Chloroform NF

Chloroprocaine Hydrochloride

NH_2–(ring with Cl)–$COOCH_2CH_2N(C_2H_5)_2$ · HCl

$C_{13}H_{19}ClN_2O_2.HCl$ 307.22

Benzoic acid, 4-amino-2-chloro-, 2-(diethylamino)ethyl ester, monohydrochloride.

2-(Diethylamino)ethyl 4-amino-2-chlorobenzoate monohydrochloride [*3858-89-7*].

» Chloroprocaine Hydrochloride contains not less than 98.0 percent and not more than 102.0 percent of $C_{13}H_{19}ClN_2O_2.HCl$, calculated on the dried basis.

Packaging and storage—Preserve in well-closed containers.

Reference standard—*USP Chloroprocaine Hydrochloride Reference Standard*—Dry at 105° for 2 hours before using.

Identification—

A: The infrared absorption spectrum of a potassium bromide dispersion of it, previously dried, exhibits maxima only at the same wavelengths as that of a similar preparation of USP Chloroprocaine Hydrochloride RS.

B: The ultraviolet absorption spectrum of a 1 in 100,000 solution in pH 4.5 buffer solution [prepared by dissolving 13.61 g of monobasic potassium phosphate in 750 mL of water, adjusting to a pH of 4.5 ± 0.1 with potassium hydroxide solution (1 in 180) and diluting with water to 1000 mL] exhibits maxima and minima at the same wavelengths as that of a similar solution of USP Chloroprocaine Hydrochloride RS, concomitantly measured, and the respective absorptivities, calculated on the dried basis, at the wavelength of maximum absorbance at about 290 nm do not differ by more than 3.0%.

C: It responds to the tests for *Chloride* ⟨191⟩.

Melting range, *Class I* ⟨741⟩: between 173° and 176°.

Acidity—Dissolve 1.0 g in 25 mL of water, add 2 drops of methyl red TS, and titrate with 0.020 *N* sodium hydroxide: not more than 1.8 mL is required to produce a yellow color.

Loss on drying ⟨731⟩—Dry about 500 mg, accurately weighed, at 105° for 2 hours: it loses not more than 1.0% of its weight.

Residue on ignition ⟨281⟩: not more than 0.2%.

4-Amino-2-chlorobenzoic acid—

Standard preparation—[NOTE—Prepare this solution on the day of use.] Transfer 50.0 mg of recrystallized 4-amino-2-chlorobenzoic acid to a 100-mL volumetric flask, dissolve in 4 mL of alcohol, dilute with water to volume, and mix. Transfer 5.0 mL of this solution to a second 100-mL volumetric flask, dilute with water to volume, and mix.

Test preparation—Transfer 100.0 mg of Chloroprocaine Hydrochloride to a 25-mL volumetric flask, dissolve in water, dilute with water to volume, and mix.

Procedure—Transfer 1.0 mL each of the *Standard preparation* and of the *Test preparation* to individual separators. To each separator add 1 mL of dilute hydrochloric acid (1 in 120), 1 mL of water, and 2 mL of sodium nitrite solution (1 in 1000), mix, and allow to stand for 3 minutes. Add 2 mL of ammonium sulfamate solution (1 in 200), mix, and allow to stand for 2 minutes. Add 5 mL of *N*-(1-naphthyl)ethylenediamine dihydrochloride solution (1 in 1000), mix, and allow to stand for 10 minutes. Add 7 mL of chloroform, shake vigorously for 30 seconds, and add rapidly 0.8 mL of sodium hydroxide solution (35 in 100) with continuous agitation. Shake immediately for 2 minutes, and allow the layers to separate. Remove most of the chloroform, and repeat the extraction with 5-mL portions of chloroform until the chloroform phase remains colorless. Allow the aqueous layer to stand until clear: the color of the solution from the *Test preparation* is not darker than that of the solution from the *Standard preparation* when viewed downward over a white surface in matched color-comparison tubes (0.625%).

Assay—Transfer about 400 mg of Chloroprocaine Hydrochloride, accurately weighed, to a 200-mL beaker, add 80 mL of glacial acetic acid and 2 mL of acetic anhydride, and heat on a steam bath for 10 minutes. Cool, add 10 mL of mercuric acetate TS, and titrate with 0.1 *N* perchloric acid VS, determining the endpoint potentiometrically. Perform a blank determination, and make any necessary correction. Each mL of 0.1 *N* perchloric acid is equivalent to 30.72 mg of $C_{13}H_{19}ClN_2O_2 \cdot HCl$.

Chloroprocaine Hydrochloride Injection

» Chloroprocaine Hydrochloride Injection is a sterile solution of Chloroprocaine Hydrochloride in Water for Injection. It contains not less than 95.0 percent and not more than 105.0 percent of the labeled amount of $C_{13}H_{19}ClN_2O_2 \cdot HCl$.

Packaging and storage—Preserve in single-dose or in multiple-dose containers, preferably of Type I glass.

Reference standard—*USP Chloroprocaine Hydrochloride Reference Standard*—Dry at 105° for 2 hours before using.

Identification—Dissolve 60 mg of USP Chloroprocaine Hydrochloride RS in 10 mL of water in a 60-mL separator, and in a second 60-mL separator mix a volume of Chloroprocaine Hydrochloride Injection, equivalent to 60 mg of chloroprocaine hydrochloride, with sufficient water to obtain 10 mL of solution. Add 5 mL of dilute ammonium hydroxide (4 in 10) to each, mix, and immediately extract each with four 10-mL portions of chloroform, passing the extracts from the Reference Standard and the test specimen through cotton filters into separate 50-mL volumetric flasks. Dilute each with chloroform to volume, and mix. Add a mixture of chloroform and methanol (4:1) to a suitable chromatographic chamber arranged for thin-layer chromatography (see *Chromatography* ⟨621⟩), cover the chamber, and allow the system to equilibrate for 15 minutes. On a suitable thin-layer chromatographic plate, coated with a 0.25-mm layer of chromatographic silica gel mixture, separately apply 10-µL portions of the chloroform solutions obtained from the Reference Standard and the test specimen. Allow the spots to dry, and develop the chromatogram until the solvent front has moved about three-fourths of the length of the plate. Remove the plate from the developing chamber, mark the solvent front, and allow the solvent to evaporate. Locate the spots on the plate by examination under short-wavelength ultraviolet light: the R_f value of the principal spot obtained from the test solution corresponds to that obtained from the Standard solution (R_f about 0.40). After viewing, spray the chromatogram with iodoplatinate TS: violet-blue colored spots, characteristic of tertiary nitrogen compounds, are visible.

pH ⟨791⟩: between 2.7 and 4.0.

4-Amino-2-chlorobenzoic acid—

Standard preparation—Prepare as directed in the test for *4-Amino-2-chlorobenzoic acid* under *Chloroprocaine Hydrochloride.*

Test preparation—Transfer an accurately measured volume of Injection, equivalent to 100 mg of $C_{13}H_{19}ClN_2O_2 \cdot HCl$, as determined by the *Assay*, to a 25-mL volumetric flask, dilute with water to volume, and mix.

Procedure—Transfer 5.0 mL of the *Standard preparation* to a separator, transfer 1.0 mL of the *Test preparation* to another separator, and proceed as directed for *Procedure* in the test for *4-Amino-2-chlorobenzoic acid* under *Chloroprocaine Hydrochloride:* the color of the solution from the *Test preparation* is not darker than that from the *Standard preparation*, corresponding to not more than 3.0% of 4-amino-2-chlorobenzoic acid, based upon the content of $C_{13}H_{19}ClN_2O_2 \cdot HCl$ as determined by the *Assay.*

Other requirements—It meets the requirements under *Injections* ⟨1⟩.

Assay—

Standard preparation—Transfer about 50 mg of USP Chloroprocaine Hydrochloride RS, accurately weighed, to a 125-mL separator, add 20 mL of water, and mix.

Assay preparation—Transfer an accurately measured volume of Chloroprocaine Hydrochloride Injection, equivalent to about 50 mg of chloroprocaine hydrochloride, to a 125-mL separator, and dilute with water to about 20 mL.

Procedure—To the *Standard preparation* and to the *Assay preparation* add 5 mL of 6 *N* ammonium hydroxide, then treat each as follows: Extract with five 25-mL portions of chloroform, and filter the combined extracts through about 1 g of anhydrous sodium sulfate supported on a pledget of glass wool. Collect the filtrate in a 200-mL volumetric flask, dilute with chloroform to volume, and mix. Transfer 3.0 mL of this solution to a 100-mL volumetric flask, dilute with chloroform to volume, and mix. Concomitantly determine the absorbances of the solutions in 1-cm cells at the wavelength of maximum absorbance at about 278 nm, with a suitable spectrophotometer, using chloroform as the blank. Calculate the quantity, in mg, of $C_{13}H_{19}ClN_2O_2 \cdot HCl$ in each mL of the Injection taken by the formula:

$$(W/V)(A_U/A_S),$$

in which W is the weight, in mg, of USP Chloroprocaine Hydrochloride RS in the *Standard preparation*, V is the volume, in mL, of Injection taken, and A_U and A_S are the absorbances of the solutions from the *Assay preparation* and the *Standard preparation*, respectively.

Chloroquine

Cl, N, HN—CH(CH2)3N(C2H5)2, CH3

$C_{18}H_{26}ClN_3$ 319.88

1,4-Pentanediamine, N^4-(7-chloro-4-quinolinyl)-N^1,N^1-diethyl-.

7-Chloro-4-[[4-(diethylamino)-1-methylbutyl]amino]quinoline [*54-05-7*].

» Chloroquine contains not less than 98.0 percent and not more than 102.0 percent of $C_{18}H_{26}ClN_3$, calculated on the dried basis.

Packaging and storage—Preserve in well-closed containers.

Reference standard—*USP Chloroquine Phosphate Reference Standard*—Dry at 105° for 16 hours before using.

Identification—

A: Dissolve 35 mg in 4 mL of chloroform, and filter through a dry filter: the infrared absorption spectrum of the solution so obtained exhibits maxima only at the same wavelengths as that of a solution of USP Chloroquine Phosphate RS prepared as directed in *Identification test A* under *Chloroquine Phosphate.*

B: The ultraviolet absorption spectrum of a 1 in 100,000 solution in dilute hydrochloric acid (1 in 1000) exhibits maxima and minima at the same wavelengths as that of a similar solution of USP Chloroquine Phosphate RS, concomitantly measured. The ratio A_{343}/A_{329} is between 1.00 and 1.15.

Melting range ⟨741⟩: between 87° and 92°.

Loss on drying ⟨731⟩—Dry it at 105° for 2 hours: it loses not more than 2.0% of its weight.

Residue on ignition ⟨281⟩: not more than 0.2%.

Assay—Dissolve about 250 mg of Chloroquine, accurately weighed, in 50 mL of glacial acetic acid TS, add crystal violet TS, and titrate with 0.1 *N* perchloric acid VS. Perform a blank determination, and make any necessary correction. Each mL of 0.1 *N* perchloric acid is equivalent to 15.99 mg of $C_{18}H_{26}ClN_3$.

Chloroquine Hydrochloride Injection

$C_{18}H_{26}ClN_3 . 2HCl$ 392.80

1,4-Pentanediamine, N^4-(7-chloro-4-quinolinyl)-N^1,N^1-diethyl-, dihydrochloride.

7-(Chloro-4-[[4-diethylamino)-1-methylbutyl]amino]quinoline dihydrochloride [*3545-67-3*].

» Chloroquine Hydrochloride Injection is a sterile solution of Chloroquine in Water for Injection prepared with the aid of Hydrochloric Acid. It contains, in each mL, not less than 47.5 mg and not more than 52.5 mg of $C_{18}H_{26}ClN_3 . 2HCl$.

Packaging and storage—Preserve in single-dose containers, preferably of Type I glass.

Reference standard—*USP Chloroquine Phosphate Reference Standard*—Dry at 105° for 16 hours before using.

Identification—

A: The ultraviolet absorption spectrum of the solution employed for measurement of absorbance in the *Assay* exhibits maxima and minima at the same wavelengths as that of a similar solution of USP Chloroquine Phosphate RS, concomitantly measured. The ratio A_{343}/A_{329} is between 1.00 and 1.15.

B: To 20 mL of a dilution of Injection with water, containing about 20 mg of chloroquine hydrochloride, add 5 mL of trinitrophenol TS: a yellow precipitate is formed. Filter, wash the precipitate with water until the last washing is colorless, and dry over silica gel: the precipitate melts between 205° and 210°. [*Caution—Picrates may explode.*]

C: It responds to the tests for *Chloride* ⟨191⟩.

pH ⟨791⟩: between 5.5 and 6.5.

Other requirements—It meets the requirements under *Injections* ⟨1⟩.

Assay—Transfer an accurately measured volume of Chloroquine Hydrochloride Injection, equivalent to about 150 mg of chloroquine hydrochloride, to a 1000-mL volumetric flask, and dilute with water to volume. Transfer 5.0 mL of this solution to a 100-mL volumetric flask, add 10 mL of dilute hydrochloric acid (1 in 100), and dilute with water to volume. Dissolve an accurately weighed quantity of USP Chloroquine Phosphate RS in dilute hydrochloric acid (1 in 1000), and dilute quantitatively and stepwise with dilute hydrochloric acid (1 in 1000) to obtain a Standard solution having a known concentration of about 10 µg per mL. Concomitantly determine the absorbances of both solutions at the wavelength of maximum absorbance at about 343 nm, with a suitable spectrophotometer, using water as the blank. Calculate the quantity, in mg, of $C_{18}H_{26}ClN_3 . 2HCl$ in the portion of Injection taken by the formula:

$$15.23C(A_U/A_S),$$

in which C is the concentration, in µg per mL, of USP Chloroquine Phosphate RS in the Standard solution, and A_U and A_S are the absorbances of the solution from the Injection and the Standard solution, respectively.

Chloroquine Phosphate

$C_{18}H_{26}ClN_3 . 2H_3PO_4$ 515.87

1,4-Pentanediamine, N^4-(7-chloro-4-quinolinyl)-N^1,N^1-diethyl-, phosphate (1:2).

7-Chloro-4-[[4-(diethylamino)-1-methylbutyl]amino]quinoline phosphate (1:2) [*50-63-5*].

» Chloroquine Phosphate contains not less than 98.0 percent and not more than 102.0 percent of $C_{18}H_{26}ClN_3 . 2H_3PO_4$, calculated on the dried basis.

Packaging and storage—Preserve in well-closed containers.

Reference standard—*USP Chloroquine Phosphate Reference Standard*—Dry at 105° for 16 hours before using.

Identification—

A: It meets the requirements under *Identification—Organic Nitrogenous Bases* ⟨181⟩, chloroform being substituted for carbon disulfide in the test.

B: The ultraviolet absorption spectrum of a 1 in 100,000 solution in dilute hydrochloric acid (1 in 1000) exhibits maxima and minima at the same wavelengths as that of a similar solution of USP Chloroquine Phosphate RS, concomitantly measured, and the ratio A_{343}/A_{329} is between 1.00 and 1.15.

Loss on drying ⟨731⟩—Dry it at 105° for 16 hours: it loses not more than 2.0% of its weight.

Assay—Dissolve about 100 mg of Chloroquine Phosphate, accurately weighed, in about 5 mL of water, and dilute quantitatively and stepwise with dilute hydrochloric acid (1 in 1000) to obtain a solution containing about 10 µg per mL. Similarly prepare a Standard solution of USP Chloroquine Phosphate RS. Concomitantly determine the absorbances of both solutions in 1-cm cells at the wavelength of maximum absorbance at about 343 nm, with a suitable spectrophotometer, using dilute hydrochloric acid (1 in 1000) as the blank. Calculate the quantity, in mg, of $C_{18}H_{26}ClN_3 . 2H_3PO_4$ in the portion of Chloroquine Phosphate taken by the formula:

$$10C(A_U/A_S),$$

in which C is the concentration, in µg per mL, of USP Chloroquine Phosphate RS in the Standard solution, and A_U and A_S are the absorbances of the solution of Chloroquine Phosphate and the Standard solution, respectively.

Chloroquine Phosphate Tablets

» Chloroquine Phosphate Tablets contain not less than 93.0 percent and not more than 107.0 percent of the labeled amount of $C_{18}H_{26}ClN_3 . 2H_3PO_4$.

Packaging and storage—Preserve in well-closed containers.

Reference standard—*USP Chloroquine Phosphate Reference Standard*—Dry at 105° for 16 hours before using.

Identification—A filtered solution of the Tablets responds to *Identification tests A* and *B* under *Chloroquine Hydrochloride Injection.*

Dissolution ⟨711⟩—

Medium: water; 900 mL.

Apparatus 2: 100 rpm.

Time: 45 minutes.

Procedure—Determine the amount of $C_{18}H_{26}ClN_3 . 2H_3PO_4$ dissolved from ultraviolet absorbances at the wavelength of maximum absorbance at about 343 nm of filtered portions of the solution under test, suitably diluted with *Dissolution Medium*, if necessary, in comparison with a Standard solution having a known concentration of USP Chloroquine Phosphate RS in the same medium.

Tolerances—Not less than 75% (*Q*) of the labeled amount of $C_{18}H_{26}ClN_3 . 2H_3PO_4$ is dissolved in 45 minutes.

Uniformity of dosage units ⟨905⟩: meet the requirements.

Assay—Weigh and finely powder not less than 20 Chloroquine Phosphate Tablets. Transfer an accurately weighed portion of the powder, equivalent to about 800 mg of chloroquine phosphate, to a 200-mL volumetric flask, add about 100 mL of water, and shake by mechanical means for about 20 minutes. Add water to volume, mix, and filter, discarding the first 50 mL of the filtrate. Pipet 50 mL of the clear filtrate into a 250-mL separator, add 5 mL of 6 *N* ammonium hydroxide, agitate, and extract the liberated chloroquine with five 25-mL portions of chloroform. Wash the combined chloroform extracts with 10 mL of water, and extract the water washing with 10 mL of chloroform. Evaporate the combined chloroform extracts on a steam bath to about 10 mL, then add 50 mL of dilute hydrochloric acid (1 in 250), and continue heating on the steam bath until the odor of chloroform is no longer perceptible. Transfer the solution to a 200-mL volumetric flask, wash the evaporating vessel with portions of dilute hydrochloric acid (1 in 1000), adding the washings to the volumetric flask, add more of the same dilute hydrochloric acid to volume, and mix. Dilute this solution quantitatively and stepwise with dilute hydrochloric acid (1 in 1000) to obtain an estimated concentration of 10 µg per mL. Dissolve an accurately weighed quantity of USP Chloroquine Phosphate RS in dilute hydrochloric acid (1 in 1000), and dilute quantitatively and stepwise with the same dilute hydrochloric acid to obtain a Standard solution having a known concentration of about 10 µg per mL. Concomitantly determine the absorbances of both solutions in 1-cm cells at the wavelength of maximum absorbance at about 343 nm, with a suitable spectrophotometer, using dilute hydrochloric acid (1 in 1000) as the blank. Calculate the quantity, in mg, of $C_{18}H_{26}ClN_3 . 2H_3PO_4$ in the portion of Tablets taken by the formula:

$$80C(A_U / A_S),$$

in which *C* is the concentration, in µg per mL, of USP Chloroquine Phosphate RS in the Standard solution, and A_U and A_S are the absorbances of the solution from Chloroquine Phosphate Tablets and the Standard solution, respectively.

Chlorothiazide

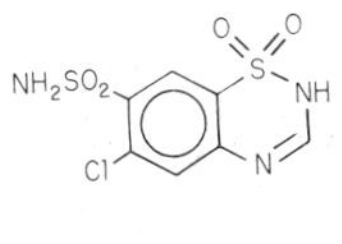

$C_7H_6ClN_3O_4S_2$ 295.72

2*H*-1,2,4-Benzothiadiazine-7-sulfonamide, 6-chloro-, 1,1-dioxide.

6-Chloro-2*H*-1,2,4-benzothiadiazine-7-sulfonamide 1,1-dioxide [*58-94-6*].

» Chlorothiazide contains not less than 98.0 percent and not more than 102.0 percent of $C_7H_6ClN_3O_4S_2$, calculated on the dried basis.

Packaging and storage—Preserve in well-closed containers.

Reference standards—*USP Chlorothiazide Reference Standard*—Dry at 105° for 1 hour before using. *USP 4-Amino-6-chloro-1,3-benzenedisulfonamide Reference Standard*—Keep container tightly closed and protected from light. Dry over silica gel for 4 hours before using.

Identification—

A: The infrared absorption spectrum of a mineral oil dispersion of it, previously dried at 105° for 1 hour, exhibits maxima only at the same wavelengths as that of a similar preparation of USP Chlorothiazide RS.

B: The ultraviolet absorption spectrum of a 1 in 100,000 solution of it in sodium hydroxide solution (1 in 250) exhibits maxima and minima at the same wavelengths as that of a similar solution of USP Chlorothiazide RS, concomitantly measured, and the respective absorptivities, calculated on the dried basis, at the wavelength of maximum absorbance at about 292 nm do not differ by more than 3.0%.

Loss on drying ⟨731⟩—Dry it at 105° for 1 hour: it loses not more than 1.0% of its weight.

Residue on ignition ⟨281⟩: not more than 0.1%.

Chloride—Dissolve 1.00 g in a mixture of 10 mL of water and 10 mL of sodium hydroxide solution (1 in 10). Cool in an ice bath, and add 20 mL of water and 5 mL of nitric acid. A flocculent, white precipitate is formed. Titrate potentiometrically with 0.050 *N* silver nitrate, using a silver–silver chloride electrode system: not more than 0.28 mL is required (0.05%).

Selenium ⟨291⟩: 0.003%.

Heavy metals, *Method II* ⟨231⟩: 0.001%.

4-Amino-6-chloro-1,3-benzenedisulfonamide—

Mobile phase, *Standard preparation*, and *Chromatographic system*—Proceed as directed in the *Assay*.

Test preparation—Proceed as directed for *Assay preparation* in the *Assay*.

Procedure—Separately inject equal volumes (about 20 µL) of the *Standard preparation* and the *Test preparation* into the chromatograph, record the chromatograms, and measure the responses for the major peaks. The relative retention times are about 0.9 for 4-amino-6-chloro-1,3-benzenedisulfonamide and 1.0 for chlorothiazide. Calculate the quantity, in mg, of 4-amino-6-chloro-1,3-benzenedisulfonamide in the portion of Chlorothiazide taken by the formula:

$$0.2C(r_U / r_S),$$

in which *C* is the concentration, in µg per mL, of USP 4-Amino-6-chloro-1,3-benzenedisulfonamide RS in the *Standard preparation*, and r_U and r_S are the peak responses of 4-amino-6-chloro-1,3-benzenedisulfonamide obtained from the *Test preparation* and the *Standard preparation*, respectively: not more than 1.0% is present.

Assay—

Mobile phase—Prepare a suitable degassed mixture of 0.1 *M* monobasic sodium phosphate and acetonitrile (9:1), adjust with phosphoric acid to a pH of 3.0 ± 0.1, and filter. Make adjustments if necessary (see *System Suitability* under *Chromatography* ⟨621⟩).

Standard preparation—[NOTE—A volume of acetonitrile not exceeding 10% of the total volume of solution may be used to dissolve the reference standards.] Dissolve accurately weighed quantities of USP Reference Standards in *Mobile phase* to obtain a solution having known concentrations of about 0.15 mg per mL of USP Chlorothiazide RS and about 1.5 µg per mL of USP 4-Amino-6-chloro-1,3-benzenedisulfonamide RS.

Assay preparation—Transfer about 30 mg of Chlorothiazide, accurately weighed, to a 200-mL volumetric flask, dissolve in a small volume of acetonitrile, not exceeding 10% of the total volume of the solution, dilute with *Mobile phase* to volume, and mix.

Chromatographic system (see *Chromatography* ⟨621⟩)—The liquid chromatograph is equipped with a 254-nm detector and a 4.6-mm × 25-cm column that contains packing L1. The flow rate is about 1.2 mL per minute. Chromatograph replicate injections of the *Standard preparation*, and record the peak responses as directed under *Procedure:* the relative standard deviation is not more than 1.5%, and the resolution, *R*, between 4-amino-6-chloro-1,3-benzenedisulfonamide and chlorothiazide is not less than 3.5.

Procedure—Separately inject equal volumes (about 20 µL) of the *Standard preparation* and the *Assay preparation* into the chromatograph, record the chromatograms, and measure the responses for the major chlorothiazide peaks. The relative retention times are about 0.9 for 4-amino-6-chloro-1,3-benzenedisulfonamide and 1.0 for chlorothiazide. Calculate the quantity, in mg, of $C_7H_6ClN_3O_4S_2$ in the portion of Chlorothiazide taken by the formula:

$$200C(r_U/r_S),$$

in which C is the concentration, in mg per mL, of USP Chlorothiazide RS in the *Standard preparation*, and r_U and r_S are the peak responses of chlorothiazide obtained from the *Assay preparation* and the *Standard preparation*, respectively.

Chlorothiazide Oral Suspension

» Chlorothiazide Oral Suspension contains not less than 90.0 percent and not more than 110.0 percent of the labeled amount of $C_7H_6ClN_3O_4S_2$.

Packaging and storage—Preserve in tight containers.

Reference standard—*USP Chlorothiazide Reference Standard*—Dry at 105° for 1 hour before using.

Identification —The ultraviolet absorption spectrum of the solution of chlorothiazide prepared from Oral Suspension as directed in the *Assay* exhibits maxima and minima at the same wavelengths as that of a solution of USP Chlorothiazide RS, prepared as directed in the *Assay*, concomitantly measured.

pH ⟨791⟩: between 3.2 and 4.0.

Assay—Transfer to a 250-mL volumetric flask an accurately measured volume of Chlorothiazide Oral Suspension, equivalent to about 250 mg of chlorothiazide, dilute with sodium hydroxide solution (1 in 250) to volume, and mix. Transfer 10.0 mL of this solution to a 100-mL volumetric flask, add dilute hydrochloric acid (1 in 100) to volume, and mix. Transfer 50.0 mL of the resulting solution to a 125-mL separator, and wash with two 25-mL portions of chloroform, discarding the washings. Transfer 10.0 mL of the washed solution to a 100-mL volumetric flask, dilute with sodium hydroxide solution (1 in 250) to volume, and mix. Dissolve an accurately weighed quantity of USP Chlorothiazide RS in sodium hydroxide solution (1 in 250) to obtain a Standard solution having a known concentration of about 10 µg per mL. Concomitantly determine the absorbances of both solutions in 1-cm cells at the wavelength of maximum absorbance at about 292 nm, with a suitable spectrophotometer, using sodium hydroxide solution (1 in 250) as the blank. Calculate the quantity, in mg, of $C_7H_6ClN_3O_4S_2$ in each mL of the Suspension taken by the formula:

$$25(C/V)(A_U/A_S),$$

in which C is the concentration, in µg per mL, of USP Chlorothiazide RS in the Standard solution, V is the volume, in mL, of Suspension taken, and A_U and A_S are the absorbances of the solution from the Suspension and the Standard solution, respectively.

Chlorothiazide Tablets

» Chlorothiazide Tablets contain not less than 90.0 percent and not more than 110.0 percent of the labeled amount of $C_7H_6ClN_3O_4S_2$.

Packaging and storage—Preserve in well-closed containers.

Reference standard—*USP Chlorothiazide Reference Standard*—Dry at 105° for 1 hour before using.

Identification—

A: The chromatogram of the *Assay preparation* obtained as directed in the *Assay* exhibits a peak for chlorothiazide, the retention time of which corresponds to that exhibited by the *Standard preparation*.

B: Powder 1 Tablet, and fuse it with a pellet of sodium hydroxide: the ammonia fumes produced turn moistened red litmus paper blue, and the residue responds to the test for *Sulfite* ⟨191⟩.

Dissolution ⟨711⟩—

Medium: 0.05 *M* pH 8.0 phosphate buffer (see *Buffer Solutions* in the section, *Reagents, Indicators, and Solutions*); 900 mL.

Apparatus 2: 75 rpm.

Time: 60 minutes.

Procedure—Determine the amount of $C_7H_6ClN_3O_4S_2$ dissolved from ultraviolet absorbances at the wavelength of maximum absorbance at about 294 nm of filtered portions of the solution under test, suitably diluted with *Dissolution Medium*, if necessary, in comparison with a Standard solution having a known concentration of USP Chlorothiazide RS in the same medium.

Tolerances—Not less than 75% (Q) of the labeled amount of $C_7H_6ClN_3O_4S_2$ is dissolved in 60 minutes.

Uniformity of dosage units ⟨905⟩: meet the requirements.

Assay—

Mobile phase—Prepare a filtered and degassed solution of 0.08 *M* monobasic sodium phosphate (adjusted with phosphoric acid to a pH of 2.9 ± 0.1) and methanol (95:5), making adjustments if necessary (see *System Suitability* under *Chromatography* ⟨621⟩).

Standard preparation—Transfer an accurately weighed quantity of about 25 mg of USP Chlorothiazide RS to a 50-mL volumetric flask, add 5 mL of 0.05 *M* monobasic sodium phosphate solution, followed by 10 mL of acetonitrile to the flask, and sonicate with occasional shaking for about 3 minutes. Dilute with water to volume, mix, and filter to obtain a *Standard preparation* having a known concentration of about 0.5 mg of USP Chlorothiazide RS per mL.

Assay preparation—[NOTE—Prepare fresh daily.] Weigh and finely powder not less than 20 Chlorothiazide Tablets. Weigh accurately a portion of the powder, equivalent to about 250 mg of chlorothiazide, and transfer to a 500-mL volumetric flask. Add 50 mL of 0.05 *M* monobasic sodium phosphate solution, and shake by mechanical means for about 15 minutes followed by sonication for about 2 minutes. Add 100 mL of acetonitrile, sonicate for about 3 minutes, dilute with water to volume, mix, and filter.

Chromatographic system (see *Chromatography* ⟨621⟩)—The liquid chromatograph is equipped with a 254-nm detector and a 3.9-mm × 30-cm column that contains packing L1 and is fitted with a guard column. The flow rate is about 2 mL per minute. Chromatograph the *Standard preparation*, and record the peak responses as directed under *Procedure:* the capacity factor (k′) is not less than 4.3, the tailing factor (T) for chlorothiazide is not more than 2.0, the theoretical plate count (N) for chlorothiazide is not less than 1300, and the relative standard deviation for replicate injections is not more than 2.0%.

Procedure—Separately inject equal volumes (about 15 µL) of the *Standard preparation* and the *Assay preparation* into the chromatograph, record the chromatograms, and measure the responses for the major peaks. Calculate the quantity, in mg, of $C_7H_6ClN_3O_4S_2$ in the portion of Chlorothiazide Tablets taken by the formula:

$$500C(r_U/r_S),$$

in which C is the concentration, in mg per mL, of USP Chlorothiazide RS in the *Standard preparation*, and r_U and r_S are the peak responses obtained from the *Assay preparation* and the *Standard preparation*, respectively.

Chlorothiazide Tablets, Methyldopa and—*see* Methyldopa and Chlorothiazide Tablets

Chlorothiazide Tablets, Reserpine and—*see* Reserpine and Chlorothiazide Tablets

Chlorothiazide Sodium for Injection

$C_7H_5ClN_3NaO_4S_2$ 317.70

2*H*-1,2,4-Benzothiadiazine-7-sulfonamide, 6-chloro-, 1,1-dioxide, monosodium salt.

6-Chloro-2*H*-1,2,4-benzothiadiazine-7-sulfonamide 1,1-dioxide monosodium salt [*7085-44-1*].

» Chlorothiazide Sodium for Injection is a sterile, freeze-dried mixture of Chlorothiazide Sodium (prepared by the neutralization of Chlorothiazide with the aid of Sodium Hydroxide) and Mannitol. It contains chlorothiazide sodium ($C_7H_5ClN_3NaO_4S_2$) equivalent to not less than 93.0 percent and not more than 107.0 percent of the labeled amount of chlorothiazide ($C_7H_6ClN_3O_4S_2$).

Packaging and storage—Preserve in *Containers for Sterile Solids* as described under *Injections* ⟨1⟩.

Reference standard—*USP Chlorothiazide Reference Standard*—Dry at 105° for 1 hour before using.

Constituted solution—At the time of use, the constituted solution prepared from Chlorothiazide Sodium for Injection meets the requirements for *Constituted Solutions* under *Injections* ⟨1⟩.

Identification—

A: The ultraviolet absorption spectrum of the solution prepared for measurement of absorbance in the *Assay* exhibits maxima and minima at the same wavelengths as that of a 1 in 100,000 solution of USP Chlorothiazide RS in sodium hydroxide solution (1 in 250), concomitantly measured.

B: Fuse an amount of Chlorothiazide Sodium for Injection, equivalent to about 100 mg of chlorothiazide, with a pellet of sodium hydroxide: the ammonia fumes produced turn moistened red litmus paper blue. The fusion mixture responds to the test for *Sulfite* ⟨191⟩.

pH ⟨791⟩: between 9.2 and 10.0, in a solution prepared as directed in the labeling.

Uniformity of dosage units ⟨905⟩: meet the requirements.

Other requirements—It meets the requirements under *Injections* ⟨1⟩.

Assay—Transfer an accurately weighed portion of Chlorothiazide Sodium for Injection, equivalent to about 500 mg of chlorothiazide, to a 1000-mL volumetric flask, add sodium hydroxide solution (1 in 250) to volume, and mix. Transfer 5.0 mL of this solution to a 250-mL volumetric flask, dilute with sodium hydroxide solution (1 in 250) to volume, and mix. Concomitantly determine the absorbances of this solution and a Standard solution of USP Chlorothiazide RS in the same medium having a known concentration of about 10 µg per mL in 1-cm cells at the wavelength of maximum absorbance at about 292 nm, with a suitable spectrophotometer, using sodium hydroxide solution (1 in 250) as the blank. Calculate the quantity, in mg, of $C_7H_6ClN_3O_4S_2$ in the portion of Chlorothiazide Sodium for Injection taken by the formula:

$$50C(A_U/A_S),$$

in which C is the concentration, in µg per mL, of USP Chlorothiazide RS in the Standard solution, and A_U and A_S are the absorbances of the assay solution and the Standard solution, respectively.

Chloroxylenol

C_8H_9ClO 156.61

Phenol, 4-chloro-3,5-dimethyl-.

4-Chloro-3,5-xylenol [*88-04-0*].

» Chloroxylenol contains not less than 98.5 percent of C_8H_9ClO.

Packaging and storage—Preserve in well-closed containers.

Reference standards—*USP Chloroxylenol Reference Standard*—Do not dry before using. *USP 2-Chloro-3,5-dimethylphenol Reference Standard*—Do not dry before using.

Identification—The infrared absorption spectrum of a potassium bromide dispersion of it exhibits maxima only at the same wavelengths as that of a similar preparation of USP Chloroxylenol RS.

Melting range ⟨741⟩: between 114° and 116°.

Residue on ignition ⟨281⟩: not more than 0.1%.

Iron ⟨241⟩—Transfer 0.10 g to a suitable crucible, add 5 drops of sulfuric acid, and ignite at a low heat until thoroughly ashed. Add to the carbonized mass 10 drops of sulfuric acid, and heat cautiously until white fumes are no longer evolved. Ignite, preferably in a muffle furnace, at 500° to 600°, until the carbon is completely burned off. Cool, add 4 mL of 6 *N* hydrochloric acid, cover, digest on a steam bath for 15 minutes, uncover, and slowly evaporate on a steam bath to dryness. Moisten the residue with 1 drop of hydrochloric acid, add 10 mL of hot water, and digest for 2 minutes. Dilute with water to about 25 mL. Filter, if necessary, rinse the crucible and the filter with 10 mL of water, combining the filtrate and rinsing in a 50-mL color-comparison tube, add 2 mL of hydrochloric acid, dilute with water to 45 mL, and mix. The limit is 0.01%.

Water, *Method I* ⟨921⟩: not more than 0.5%.

Assay and chromatographic purity—

Standard preparation—Dissolve accurately weighed quantities of 3,5-dimethylphenol and USP 2-Chloro-3,5-dimethylphenol RS quantitatively in chloroform to obtain a solution containing 0.08 mg of each per mL.

Assay preparation—Dissolve an accurately weighed quantity of Chloroxylenol quantitatively in chloroform to obtain a solution containing 40.0 mg of specimen per mL.

Chromatographic system (see *Chromatography* ⟨621⟩)—The gas chromatograph is equipped with a flame-ionization detector and a 1.8-m × 4-mm column packed with 3 percent phase G16 on packing S1A. The column is maintained at 180°, and the injection port and detector block are maintained at 200°. Dry nitrogen, flowing at a rate of about 30 mL per minute, is used as the carrier gas. Chromatograph the *Standard preparation*, and record the peak responses as directed under *Procedure:* the resolution, *R*, between the 3,5-dimethylphenol and 2-chloro-3,5-dimethylphenol peaks is not less than 4.5, and the relative standard deviation for replicate injections is not more than 10%.

Procedure—[NOTE—Use peak areas where peak responses are indicated.] Separately inject equal volumes (about 1 µL) of the *Standard preparation* and the *Assay preparation* into the gas chromatograph, record the chromatograms, and measure the responses for all of the peaks. Calculate the percentage of chloroxylenol (C_8H_9ClO) in the specimen of Chloroxylenol taken by the formula:

$$100r_m/r_t,$$

in which r_m is the response of the main peak and r_t is the sum of the responses of all of the peaks, excluding that of the solvent peak. Calculate the percentages of 3,5-dimethylphenol ($C_8H_{10}O$) and 2-chloro-3,5-dimethylphenol (C_8H_9ClO) in the specimen of Chloroxylenol taken by the same formula:

$$0.2r_U/r_S,$$

in which r_U and r_S are the peak responses of the corresponding analytes obtained from the *Assay preparation* and the *Standard preparation*, respectively: not more than 0.2% of either analyte is found. Calculate the percentage of each peak, other than the solvent peak, the main peak, the 3,5-dimethylphenol peak, and the 2-chloro-3,5-dimethylphenol peak, by the same formula:

$$100r_i/r_t,$$

in which r_i is the response of each peak: not more than 0.5% of any individual peak is found.

Chlorpheniramine Maleate

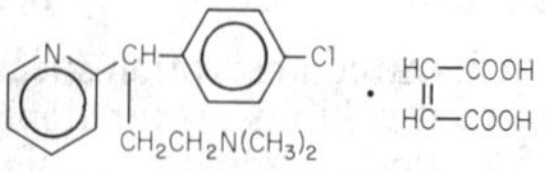

$C_{16}H_{19}ClN_2 . C_4H_4O_4$ 390.87

2-Pyridinepropanamine, γ-(4-chlorophenyl)-*N*,*N*-dimethyl-, (*Z*)-2-butenedioate (1:1).

2-[*p*-Chloro-α-[2-(dimethylamino)ethyl]benzyl]pyridine maleate (1:1) [*113-92-8*].

» Chlorpheniramine Maleate contains not less than 98.0 percent and not more than 100.5 percent of $C_{16}H_{19}ClN_2 . C_4H_4O_4$, calculated on the dried basis.

Packaging and storage—Preserve in tight, light-resistant containers.

Reference standard—*USP Chlorpheniramine Maleate Reference Standard*—Dry at 105° for 3 hours before using.

Identification—The infrared absorption spectrum of a potassium bromide dispersion of it exhibits maxima only at the same wavelengths as that of a similar preparation of USP Chlorpheniramine Maleate RS.

Melting range ⟨741⟩: between 130° and 135°.

Loss on drying ⟨731⟩—Dry it at 105° for 3 hours: it loses not more than 0.5% of its weight.

Residue on ignition ⟨281⟩: not more than 0.2%.

Related compounds—

Test preparation—Dissolve about 200 mg of Chlorpheniramine Maleate in 5 mL of methylene chloride, and mix.

Chromatographic system (see *Chromatography* ⟨621⟩)—The gas chromatograph is equipped with a flame-ionization detector and a 1.2-m × 4-mm glass column containing 3 percent phase G3 on support S1AB. The column temperature is maintained at about 190°, and the injection port and detector temperatures are both maintained at about 250°. The carrier gas is dry helium, flowing at a rate adjusted to obtain a retention time of 4 to 5 minutes for the main peak. Chromatograph the *Test preparation*, record the chromatogram, and determine the peak area as directed under *Procedure:* the tailing factor for the chlorpheniramine maleate peak is not more than 1.8.

Procedure—Inject a volume (about 1 μL) of the *Test preparation* into the chromatograph. Record the chromatogram for a total time of not less than twice the retention time of the chlorpheniramine peak, and measure the areas of the peaks. The total relative area of all extraneous peaks (except that of the solvent peak) does not exceed 2.0%.

Assay—Dissolve about 500 mg of Chlorpheniramine Maleate, accurately weighed, in 20 mL of glacial acetic acid. Add 2 drops of crystal violet TS, and titrate with 0.1 *N* perchloric acid VS. Perform a blank determination, and make any necessary correction. Each mL of 0.1 *N* perchloric acid is equivalent to 19.54 mg of $C_{16}H_{19}ClN_2 . C_4H_4O_4$.

Chlorpheniramine Maleate, and Dexamethasone Suspension, Sterile Penicillin G Procaine, Dihydrostreptomycin Sulfate,—*see* Sterile Penicillin G Procaine, Dihydrostreptomycin Sulfate, Chlorpheniramine Maleate, and Dexamethasone Suspension

Chlorpheniramine Maleate Injection

» Chlorpheniramine Maleate Injection is a sterile solution of Chlorpheniramine Maleate in Water for Injection. It contains not less than 90.0 percent and not more than 110.0 percent of the labeled amount of $C_{16}H_{19}ClN_2 . C_4H_4O_4$.

Packaging and storage—Preserve in single-dose or in multiple-dose containers, preferably of Type I glass, protected from light.

Reference standard—*USP Chlorpheniramine Maleate Reference Standard*—Dry at 105° for 3 hours before using.

Identification—

A: Dilute a volume of Injection, equivalent to about 50 mg of chlorpheniramine maleate, with dilute hydrochloric acid (1 in 1000) to 25 mL, and proceed as directed under *Identification—Organic Nitrogenous Bases* ⟨181⟩, beginning with "Transfer the liquid to a separator." The Injection meets the requirements of the test.

B: Evaporate a volume of Injection, equivalent to about 25 mg of chlorpheniramine maleate, on a steam bath to dryness, and dry the residue at 105° for 1 hour: it melts between 128° and 135°.

pH ⟨791⟩: between 4.0 and 5.2.

Other requirements—It meets the requirements under *Injections* ⟨1⟩.

Assay—Proceed with Chlorpheniramine Maleate Injection as directed under *Salts of Organic Nitrogenous Bases* ⟨501⟩, to prepare the solution employed for the determination of the absorbance, A_U, at 264 nm. For the determination of A_S, dissolve about 25 mg of USP Chlorpheniramine Maleate RS, accurately weighed, in 20 mL of dilute sulfuric acid (1 in 350), and treat this solution the same as the portion of Injection being assayed. Calculate the quantity, in mg, of $C_{16}H_{19}ClN_2 . C_4H_4O_4$ in each mL of the Injection taken by the formula:

$$(C/V)(A_U/A_S),$$

in which C is the weight, in mg, of USP Chlorpheniramine Maleate RS in the *Standard preparation*, and V is the volume, in mL, of Injection taken.

Chlorpheniramine Maleate Syrup

» Chlorpheniramine Maleate Syrup contains not less than 90.0 percent and not more than 110.0 percent of the labeled amount of $C_{16}H_{19}ClN_2 . C_4H_4O_4$.

Packaging and storage—Preserve in tight, light-resistant containers.

Reference standard—*USP Chlorpheniramine Maleate Reference Standard*—Dry at 105° for 3 hours before using.

Identification—

A: Evaporate the remaining extract from the *Assay* on a steam bath to a small volume, then transfer it to a smaller, more suitable vessel, and evaporate just to the point where hexane vapors are no longer perceptible. Transfer the oily residue, with the aid of four 3-mL portions of dimethylformamide, to a suitable glass-stoppered graduated cylinder, dilute with dimethylformamide to 15.0 mL, and mix: the optical rotation of the solution so obtained, in a 100-mm tube, after correcting for the blank, is not more than +0.01° (*distinction from dexchlorpheniramine maleate*).

B: The ultraviolet absorption spectrum of the solution employed for measurement of absorbance in the *Assay* exhibits maxima and minima at the same wavelengths as that of the solution obtained from USP Chlorpheniramine Maleate RS used in the *Assay*.

Alcohol content ⟨611⟩: between 6.0% and 8.0% of C_2H_5OH.

Assay—Transfer 10 mL of Chlorpheniramine Maleate Syrup, accurately measured, to a separator. Transfer about 40 mg of USP Chlorpheniramine Maleate RS, accurately weighed, to a 100-mL volumetric flask, dilute with water to volume, mix, and pipet 10 mL of this Standard solution into a separator similar to that containing the Syrup. Treat each solution as follows: Add 10 mL of sodium hydroxide solution (1 in 10), and extract with two 50-mL portions of solvent hexane. Combine the extracts in a second separator, wash with 10 mL of sodium hydroxide solution (1 in 250), and discard the washing. Extract the hexane solution with two 40-mL portions of dilute hydrochloric acid (1 in 100), collect the extracts in a 100-mL volumetric flask, add the same dilute acid to volume, and mix. Wash 50-mL portions of each solution, and of dilute hydrochloric acid (1 in 100), respectively, with three 30-mL portions of chloroform and then with 50 mL of solvent hexane, and discard the washings. Filter the acid phases through paper, discarding the first few mL of each filtrate, and determine the absorbances of the solutions obtained from the Syrup and the Standard solution in 1-cm cells at the wavelength of maximum absorbance at about 264 nm, with a suitable spectrophotometer, using the extracted acid as the blank. Calculate the quantity, in µg, of $C_{16}H_{19}ClN_2.C_4H_4O_4$ in each mL of the Syrup taken by the formula:

$$C(A_U/A_S),$$

in which C is the concentration, in µg per mL, of USP Chlorpheniramine Maleate RS in the Standard solution, and A_U and A_S are the absorbances of the solutions from Chlorpheniramine Maleate Syrup and the Standard solution, respectively.

Chlorpheniramine Maleate Tablets

» Chlorpheniramine Maleate Tablets contain not less than 93.0 percent and not more than 107.0 percent of the labeled amount of $C_{16}H_{19}ClN_2.C_4H_4O_4$.

Packaging and storage—Preserve in tight containers.

Reference standard—*USP Chlorpheniramine Maleate Reference Standard*—Dry at 105° for 3 hours before using.

Identification—Disperse a portion of powdered Tablets, equivalent to about 25 mg of chlorpheniramine maleate, in about 20 mL of dilute hydrochloric acid (1 in 100). Dissolve about 25 mg of USP Chlorpheniramine Maleate RS in 20 mL of dilute hydrochloric acid (1 in 100). Treat each solution as follows: Render alkaline, to a pH of about 11, with sodium hydroxide solution (1 in 10). Extract with two 50-mL portions of solvent hexane, collect the extracts in a beaker, and evaporate to dryness. Prepare a mineral oil dispersion of the residue so obtained and determine the infrared absorption spectrum of the preparation in the region between 2 µm and 12 µm: the spectrum of the test preparation exhibits maxima only at the same wavelengths as that of the Standard preparation.

Dissolution ⟨711⟩—

Medium: water; 500 mL.

Apparatus 2: 50 rpm.

Time: 45 minutes.

Procedure—Determine the amount of $C_{16}H_{19}ClN_2.C_4H_4O_4$ dissolved from ultraviolet absorbances at the wavelength of maximum absorbance at about 262 nm in filtered portions of the solution under test, suitably diluted with 3 *N* hydrochloric acid, in comparison with a Standard solution having a known concentration of USP Chlorpheniramine Maleate RS in the same medium.

Tolerances—Not less than 75% (*Q*) of the labeled amount of $C_{16}H_{19}ClN_2.C_4H_4O_4$ is dissolved in 45 minutes.

Uniformity of dosage units ⟨905⟩: meet the requirements.

Assay—Using a portion of powdered Chlorpheniramine Maleate Tablets equivalent to 4 mg of chlorpheniramine maleate, proceed as directed under *Salts of Organic Nitrogenous Bases* ⟨501⟩, but using dilute hydrochloric acid (1 in 100) instead of the dilute sulfuric acid (1 in 350), and dilute sulfuric acid (1 in 70), and using solvent hexane instead of the ether, and diluting 10 mL of the *Assay preparation* with dilute hydrochloric acid (1 in 100) to 25.0 mL to prepare the solution employed for the determination of the absorbance, A_U, at 264 nm. For the determination of A_S, prepare a solution containing about 40 mg of USP Chlorpheniramine Maleate RS, accurately weighed, in 200.0 mL of dilute hydrochloric acid (1 in 100), and treat 20.0 mL of this solution the same as the solution in dilute hydrochloric acid (1 in 100) of the portion of Tablets taken. Calculate the quantity, in mg, of $C_{16}H_{19}ClN_2.C_4H_4O_4$ in the portion of Tablets taken by the formula:

$$C(A_U/A_S),$$

in which C is the weight, in mg, of USP Chlorpheniramine Maleate RS in the 20.0-mL portion of the *Standard preparation.*

Chlorpheniramine Maleate Extended-release Capsules

» Chlorpheniramine Maleate Extended-release Capsules contain not less than 90.0 percent and not more than 110.0 percent of the labeled amount of $C_{16}H_{19}ClN_2.C_4H_4O_4$.

Packaging and storage—Preserve in tight containers.

Labeling—The labeling indicates steady-state blood level profiles and confidence intervals and states the in-vitro *Drug release* test conditions of times and tolerances.

Reference standard—*USP Chlorpheniramine Maleate Reference Standard*—Dry at 105° for 3 hours before using.

Identification—The retention time of the chlorpheniramine peak in the chromatogram of the *Assay preparation* corresponds to that of the *Standard preparation,* as obtained in the *Assay.*

Uniformity of dosage units ⟨905⟩—meet the requirements.

Assay—

Mobile phase—Dissolve 2.0 g of sodium perchlorate in 350 mL of water. Add 650 mL of methanol and 2.0 mL of triethylamine, and mix. Filter, and degas this solution prior to use. Make adjustments if necessary (see *Chromatography* ⟨621⟩).

Standard preparation—Dissolve an accurately weighed quantity of USP Chlorpheniramine Maleate RS in water to obtain a solution having a known concentration of about 0.12 mg per mL.

Assay preparation—Weigh and mix the contents of not less than 20 Chlorpheniramine Maleate Extended-release Capsules. Transfer an accurately weighed portion of the mixture, equivalent to about 120 mg of chlorpheniramine maleate, to a 200-mL volumetric flask. Add about 100 mL of water, bring to a boil on a hot plate, and continue boiling moderately for 5 minutes. Cool, dilute with water to volume, mix, and filter. Transfer 10.0 mL of the filtrate to a 50-mL volumetric flask, dilute with water to volume, and mix.

Chromatographic system (see *Chromatography* ⟨621⟩)—The liquid chromatograph is equipped with a 261-nm detector and a 3.9-mm × 15-cm column that contains 10-µm packing L1. The flow rate is about 1 mL per minute. Chromatograph the *Standard preparation,* and record the peak responses as directed under *Procedure:* the column efficiency is not less than 900 theoretical plates, the tailing factor is not greater than 2.0, and the relative standard deviation for replicate injections is not more than 1.5%.

Procedure—Separately inject equal volumes (about 20 µL) of the *Standard preparation* and the *Assay preparation* into the chromatograph, record the chromatograms, and measure the responses for the major peaks. Calculate the quantity, in mg, of $C_{16}H_{19}ClN_2.C_4H_4O_4$ in the portion of Capsules taken by the formula:

$$(1000C)(r_U/r_S),$$

in which C is the concentration, in mg per mL, of USP Chlorpheniramine Maleate RS in the *Standard preparation,* and r_U and r_S are the peak responses obtained from the *Assay preparation* and the *Standard preparation,* respectively.

Chlorpromazine

$C_{17}H_{19}ClN_2S$ 318.86
10*H*-Phenothiazine-10-propanamine, 2-chloro-*N,N*-dimethyl-.
2-Chloro-10-[3-(dimethylamino)propyl]phenothiazine [*50-53-3*].

» Chlorpromazine contains not less than 98.0 percent and not more than 101.0 percent of $C_{17}H_{19}ClN_2S$, calculated on the dried basis.

Packaging and storage—Preserve in tight, light-resistant containers.

Reference standard—*USP Chlorpromazine Hydrochloride Reference Standard*—Dry at 105° for 2 hours before using.

NOTE—Throughout the following procedures, protect test or assay specimens, the Reference Standard, and solutions containing them, by conducting the procedures without delay, under subdued light, or using low-actinic glassware.

Identification—

A: The infrared absorption spectrum of a 1 in 100 solution in carbon disulfide, in a 1.0-mm cell between 7 µm and 15 µm, exhibits maxima only at the same wavelengths as that of a solution prepared by dissolving 55 mg of USP Chlorpromazine Hydrochloride RS in 3 mL of 1 *N* sodium hydroxide and extracting the resulting solution with 5.0 mL of carbon disulfide.

B: The principal spot found in the test for *Other alkylated phenothiazines* corresponds in R_f with the spot from the *Standard solution*.

Loss on drying ⟨731⟩—Dry it in vacuum at room temperature for 3 hours: it loses not more than 1.0% of its weight.

Other alkylated phenothiazines—Dissolve 45.0 mg in 10 mL of methanol. Dissolve a suitable quantity of USP Chlorpromazine Hydrochloride RS in methanol to obtain a concentration of 5 mg per mL (*Standard solution*), and dilute it quantitatively and stepwise with methanol to obtain a concentration of 25 µg per mL (*Diluted standard solution*). Apply separately 10 µL of each of the three solutions on the starting line of a thin-layer chromatographic plate coated with chromatographic silica gel mixture. Develop the chromatogram, using as the solvent system a freshly prepared mixture of equal volumes of ether and ethyl acetate saturated with ammonium hydroxide, until the solvent front has moved about 10 cm from the origin. Remove the plate from the chamber, and air-dry for 20 minutes. View under short-wavelength ultraviolet light: the area and intensity of any spot, other than the principal spot, from the solution of Chlorpromazine are not greater than those of the spot from the *Diluted standard solution* (0.5%).

Assay—Place about 750 mg of Chlorpromazine, accurately weighed, in a 250-mL conical flask, and dissolve in 25 mL of glacial acetic acid, warming gently on a steam bath to effect solution. Cool, add crystal violet TS, and titrate with 0.1 *N* perchloric acid VS to a blue end-point. Perform a blank determination, and make any necessary correction. Each mL of 0.1 *N* perchloric acid is equivalent to 31.89 mg of $C_{17}H_{19}ClN_2S$.

Chlorpromazine Suppositories

» Chlorpromazine Suppositories contain not less than 90.0 percent and not more than 110.0 percent of the labeled amount of $C_{17}H_{19}ClN_2S$.

Packaging and storage—Preserve in well-closed, light-resistant containers, at controlled room temperature.

Reference standard—*USP Chlorpromazine Hydrochloride Reference Standard*—Dry at 105° for 2 hours before using.

NOTE—Throughout the following procedures, protect test or assay specimens, the Reference Standard, and solutions containing them, by conducting the procedures without delay, under subdued light, or using low-actinic glassware.

Identification—Suppositories respond to *Identification test B* under *Chlorpromazine*.

Other alkylated phenothiazines—Transfer a portion of Suppositories, equivalent to 45 mg of chlorpromazine, to a stoppered centrifuge tube, add 10 mL of methanol, shake vigorously to disperse the solid, warming gently if necessary, and centrifuge. Proceed as directed in the test for *Other alkylated phenothiazines* under *Chlorpromazine*, beginning with "Dissolve a suitable quantity of USP Chlorpromazine Hydrochloride RS." The area and intensity of any spot, other than the principal spot, from the solution from the Suppositories are not greater than those of the spot from the *Diluted standard solution* (0.5%).

Assay—Place not less than 10 Chlorpromazine Suppositories in a 250-mL beaker, reduce the mass to the consistency of a paste by crushing with a spatula, and mix. Weigh accurately a portion of the mass, equivalent to about 50 mg of chlorpromazine, place in a beaker, and dissolve in about 40 mL of ether. Transfer to a 250-mL separator with the aid of three 25-mL portions of ether, and extract with four 75-mL portions of 0.1 *N* hydrochloric acid, collecting the aqueous extracts in a 500-mL volumetric flask. Add 0.1 *N* hydrochloric acid to volume, and mix. Transfer 10.0 mL of this solution to a 200-mL volumetric flask, add 0.1 *N* hydrochloric acid to volume, and mix. Dissolve an accurately weighed quantity of USP Chlorpromazine Hydrochloride RS in 0.1 *N* hydrochloric acid, and dilute quantitatively and stepwise with the same solvent to obtain a Standard solution having a known concentration of about 5.5 µg of chlorpromazine hydrochloride per mL. Concomitantly determine the absorbances of both solutions in 1-cm cells at the wavelength of maximum absorbance at about 254 nm and at 277 nm, with a suitable spectrophotometer, using 0.1 *N* hydrochloric acid as the blank. Calculate the quantity, in mg, of chlorpromazine ($C_{17}H_{19}ClN_2S$) in the portion of Suppositories taken by the formula:

$$10(0.897C)(A_{254} - A_{277})_U/(A_{254} - A_{277})_S,$$

in which 0.897 is the ratio of the molecular weight of chlorpromazine to that of chlorpromazine hydrochloride, *C* is the concentration, in µg per mL, of USP Chlorpromazine Hydrochloride RS in the Standard solution, and the parenthetic expressions are the differences in the absorbances of the two solutions at the wavelengths indicated by the subscripts, for the solution from the Suppositories (*U*) and the Standard solution (*S*), respectively.

Chlorpromazine Hydrochloride

$C_{17}H_{19}ClN_2S.HCl$ 355.32
10*H*-Phenothiazine-10-propanamine, 2-chloro-*N,N*-dimethyl-, monohydrochloride.
2-Chloro-10-[3-(dimethylamino)propyl]phenothiazine monohydrochloride [*69-09-0*].

» Chlorpromazine Hydrochloride contains not less than 98.0 percent and not more than 101.5 percent of $C_{17}H_{19}ClN_2S.HCl$, calculated on the dried basis.

Packaging and storage—Preserve in tight, light-resistant containers.

Reference standard—*USP Chlorpromazine Hydrochloride Reference Standard*—Dry at 105° for 2 hours before using.

NOTE—Throughout the following procedures, protect test or assay specimens, the Reference Standard, and solutions containing them, by conducting the procedures without delay, under subdued light, or using low-actinic glassware.

Identification—

A: The infrared absorption spectrum of a potassium bromide dispersion of it exhibits maxima only at the same wavelengths as that of a similar preparation of USP Chlorpromazine Hydrochloride RS.

B: The principal spot found in the test for *Other alkylated phenothiazines* corresponds in R_f with the spot from the *Standard solution.*

C: A solution (1 in 10) responds to the tests for *Chloride* ⟨191⟩.

Melting range ⟨741⟩: between 195° and 198°.

Loss on drying ⟨731⟩—Dry it at 105° for 2 hours: it loses not more than 0.5% of its weight.

Residue on ignition ⟨281⟩: not more than 0.1%.

Other alkylated phenothiazines—Dissolve 50 mg, previously dried, in methanol to make 10 mL, and mix. Proceed as directed in the test for *Other alkylated phenothiazines* under *Chlorpromazine*, beginning with "Dissolve a suitable quantity of USP Chlorpromazine Hydrochloride RS." The area and intensity of any spot, other than the principal spot, from the solution of Chlorpromazine Hydrochloride are not greater than those of the spot from the *Diluted standard solution* (0.5%).

Assay—Transfer to a beaker about 700 mg of Chlorpromazine Hydrochloride, accurately weighed, and dissolve in 75 mL of glacial acetic acid. Add 10 mL of mercuric acetate TS, and titrate with 0.1 *N* perchloric acid VS, determining the end-point potentiometrically. Each mL of 0.1 *N* perchloric acid is equivalent to 35.53 mg of $C_{17}H_{19}ClN_2S.HCl$.

Chlorpromazine Hydrochloride Oral Concentrate

» Chlorpromazine Hydrochloride Oral Concentrate contains not less than 90.0 percent and not more than 110.0 percent of the labeled amount of $C_{17}H_{19}ClN_2S.HCl$.

Packaging and storage—Preserve in tight, light-resistant containers.

Labeling—Label it to indicate that it must be diluted prior to administration.

Reference standard—*USP Chlorpromazine Hydrochloride Reference Standard*—Dry at 105° for 2 hours before using.

NOTE—Throughout the following procedures, protect test or assay specimens, the Reference Standard, and solutions containing them, by conducting the procedures without delay, under subdued light, or using low-actinic glassware.

Identification—

A: It responds to *Identification test A* under *Chlorpromazine Hydrochloride Syrup*.

B: Dilute a portion of the Oral Concentrate with an equal volume of water: the resulting solution responds to the tests for *Chloride* ⟨191⟩.

Microbial limits—It meets the requirements of the tests for the absence of *E. coli* under *Microbial Limit Tests* ⟨61⟩.

pH ⟨791⟩: between 2.9 and 4.1.

Chlorpromazine sulfoxide—Proceed as directed in the test *Chlorpromazine sulfoxide* under *Chlorpromazine Hydrochloride Syrup*.

Assay—Transfer an accurately measured volume of Chlorpromazine Hydrochloride Oral Concentrate, previously diluted if necessary, equivalent to about 10 mg of chlorpromazine hydrochloride, to a 50-mL volumetric flask, add 0.1 *N* hydrochloric acid to volume, and mix. Proceed as directed in the *Assay* under *Chlorpromazine Hydrochloride Injection*, beginning with "Pipet 10 mL of the solution." Calculate the quantity, in mg, of $C_{17}H_{19}ClN_2S.HCl$ in each mL of the Oral Solution taken by the formula:

$$1.25C(A_{254} - A_{277})_U / V(A_{254} - A_{277})_S,$$

in which *C* is the concentration, in µg per mL, of USP Chlorpromazine Hydrochloride RS in the Standard solution, *V* is the volume, in mL, of Oral Concentrate taken, and the parenthetic expressions are the differences in the absorbances of the two solutions at the wavelengths indicated by the subscripts, for the solution from the Oral Concentrate (*U*) and the Standard solution (*S*), respectively.

Chlorpromazine Hydrochloride Injection

» Chlorpromazine Hydrochloride Injection is a sterile solution of Chlorpromazine Hydrochloride in Water for Injection. It contains, in each mL, not less than 23.75 mg and not more than 26.25 mg of $C_{17}H_{19}ClN_2S.HCl$.

Packaging and storage—Preserve in single-dose or in multiple-dose containers, preferably of Type I glass, protected from light.

Reference standard—*USP Chlorpromazine Hydrochloride Reference Standard*—Dry at 105° for 2 hours before using.

NOTE—Throughout the following procedures, protect test or assay specimens, the Reference Standard, and solutions containing them, by conducting thc procedures without delay, under subdued light, or using low-actinic glassware.

Identification—

A: Transfer a volume of Injection, equivalent to about 25 mg of chlorpromazine hydrochloride, to a 10-mL volumetric flask, dilute with methanol to volume, and mix (test solution). Dissolve a suitable quantity of USP Chlorpromazine Hydrochloride RS in dilute methanol (9 in 10) to obtain a Standard solution having a known concentration of 2.5 mg per mL. Apply separately 5-µL portions of each of the two solutions on the starting line of a thin-layer chromatographic plate (see *Chromatography* ⟨621⟩) coated with chromatographic silica gel mixture. Develop the chromatogram in a solvent system consisting of a freshly prepared mixture of equal volumes of ether and ethyl acetate saturated with ammonium hydroxide until the solvent front has moved about 10 cm from the origin. Remove the plate from the developing chamber, air-dry for 20 minutes, then view under short-wavelength ultraviolet light: the R_f value of the principal spot obtained from the test solution corresponds to that obtained from the Standard solution.

B: It responds to the tests for *Chloride* ⟨191⟩.

pH ⟨791⟩: between 3.4 and 5.4.

Other requirements—It meets the requirements under *Injections* ⟨1⟩.

Chlorpromazine sulfoxide—[NOTE—Conduct this test without exposure to daylight, and with the minimum necessary exposure to artificial light.]

Test preparation—Pipet 4 mL of the test solution prepared with methanol as directed in *Identification test A* into a 10-mL volumetric flask, dilute with methanol to volume, and mix.

Standard preparation—Dissolve a suitable quantity of USP Chlorpromazine Hydrochloride RS in methanol to obtain a solution having a concentration of 50 µg per mL.

Procedure—Apply separate 10-µL portions of the *Standard preparation* and the *Test preparation* on the starting line of a thin-layer chromatographic plate coated with a 0.25-mm layer of chromatographic silica gel mixture. Dry the applied solutions with the aid of a stream of nitrogen. Develop the chromatogram, using as the solvent system a freshly prepared mixture of equal volumes of ether and ethyl acetate saturated with ammonium hydroxide, until the solvent front has moved about 13 cm from the origin. Remove the plate from the chamber, and air-dry for 30 minutes. Examine under short-wavelength ultraviolet light: the area and intensity of the only other spot in the test specimen chromatogram, other than the principal spot, are not greater than those of the spot from the *Standard preparation* (5.0%).

Assay—Transfer an accurately measured volume of Chlorpromazine Hydrochloride Injection, equivalent to about 100 mg of chlorpromazine hydrochloride, to a 500-mL volumetric flask, add 0.1 *N* hydrochloric acid to volume, and mix. Pipet 10 mL of the solution into a 250-mL separator, add about 20 mL of water, render alkaline with ammonium hydroxide, and extract with four 25-mL portions of ether. Extract the combined ether extracts with four 25-mL portions of 0.1 *N* hydrochloric acid, collecting the aqueous extracts in a 250-mL volumetric flask. Aerate to remove residual ether, add 0.1 *N* hydrochloric acid to volume, and mix. Dissolve a suitable quantity, accurately weighed, of USP Chlorpromazine Hydrochloride RS in 0.1 *N* hydrochloric acid, and dilute quantitatively and stepwise with the same acid to obtain a Standard solution having a known concentration of about 8 µg per mL. Concomitantly determine the absorbances of both solutions in 1-cm cells at the wavelength of maximum absorbance at about 254 nm and at 277 nm, with a suitable spectrophotometer, using 0.1 *N* hydrochloric acid as the blank. Calculate the quantity, in mg, of $C_{17}H_{19}ClN_2S.HCl$ in each mL of the Injection taken by the formula:

$$12.5C(A_{254} - A_{277})_U / V(A_{254} - A_{277})_S,$$

in which *C* is the concentration, in µg per mL, of USP Chlorpromazine Hydrochloride RS in the Standard solution, *V* is the volume, in mL, of Injection taken, and the parenthetic expressions are the differences in the absorbances of the two solutions at the wavelengths indicated by the subscripts, for the solution from the Injection (*U*) and the Standard solution (*S*), respectively.

Chlorpromazine Hydrochloride Syrup

» Chlorpromazine Hydrochloride Syrup contains, in each 100 mL, not less than 190 mg and not more than 210 mg of $C_{17}H_{19}ClN_2S.HCl$.

Packaging and storage—Preserve in tight, light-resistant containers.

Reference standard—*USP Chlorpromazine Hydrochloride Reference Standard*—Dry at 105° for 2 hours before using.

NOTE—Throughout the following procedures, protect test or assay specimens, the Reference Standard, and solutions containing them, by conducting the procedures without delay, under subdued light, or using low-actinic glassware.

Identification—

A: Transfer a volume of it, equivalent to about 20 mg of chlorpromazine hydrochloride, to a 125-mL separator. Add 10 mL of water, 2 mL of sodium hydroxide solution (1 in 2), and mix. Extract with three 30-mL portions of ether. Filter the combined ether extracts through anhydrous sodium sulfate. With the aid of a stream of nitrogen evaporate the ether to about 5 mL. Quantitatively transfer the solution to a 40-mL centrifuge tube. Evaporate with a stream of nitrogen and mild heat to dryness. Dissolve the residue in 1.0 mL of methanol to obtain the Test solution. Separately apply 15 µL of this Test solution and 15 µL of a Standard solution, containing 0.2 mg of USP Chlorpromazine Hydrochloride RS per mL of methanol, on a thin-layer chromatographic plate (see *Chromatography* ⟨621⟩), coated with a 0.25-mm layer of chromatographic silica gel. Develop the chromatogram in a chamber containing a freshly prepared mixture of ethyl acetate that has been saturated with ammonium hydroxide, ether, and methanol (75:25:20) until the solvent front has moved about three-fourths of the length of the plate. Remove the plate from the chamber, air-dry, and spray with Iodoplatinate reagent prepared by dissolving 100 mg of platinic chloride in 10 mL of 0.1 *N* hydrochloric acid, adding 25 mL of potassium iodide solution (1 in 25), 0.5 mL of formic acid, and diluting with water to 100 mL. The R_f value of the principal spot from the test solution corresponds to that obtained from the Standard solution.

B: Dilute a portion of the Syrup with an equal volume of water: the resulting solution responds to the tests for *Chloride* ⟨191⟩.

Chlorpromazine sulfoxide—

Preparation of chlorpromazine sulfoxide solution—Transfer 5 mL of a solution in dilute hydrochloric acid (1 in 100) of USP Chlorpromazine Hydrochloride RS containing 10.6 mg per mL to a 50-mL volumetric flask. Add 2 mL of 30% hydrogen peroxide and heat at 60° for 10 minutes. Cool, dilute with 1 *M* sodium bisulfite to volume, and mix. Transfer 10.0 mL to a 60-mL separator, add 2 mL of sodium hydroxide solution (1 in 2), and mix. Extract with three 30-mL portions of ether. Filter the extracts through ether-wetted anhydrous sodium sulfate into a 250-mL conical flask. Cautiously evaporate the extracts to dryness. Dissolve the residue in 10.0 mL of methanol, and filter if necessary. Each mL of this solution contains 1 mg of chlorpromazine sulfoxide.

Procedure—Transfer an accurately measured volume of the Syrup, equivalent to about 20 mg of chlorpromazine hydrochloride, to a 125-mL separator. Add 10 mL of water and 2 mL of sodium hydroxide solution (1 in 2), and mix. Extract with three 30-mL portions of ether. Filter the combined ether extracts through anhydrous sodium sulfate. With the aid of a stream of nitrogen evaporate the ether to about 5 mL. Quantitatively transfer the solution to a 40-mL centrifuge tube. Evaporate with a stream of nitrogen and mild heat to dryness. Dissolve the residue in 1.0 mL of methanol to obtain the Test solution. Separately apply 15 µL of this Test solution and 15 µL of a Standard solution, containing 0.2 mg of USP Chlorpromazine Hydrochloride RS per mL of methanol, on a thin-layer chromatographic plate (see *Chromatography* ⟨621⟩), coated with a 0.25-mm layer of chromatographic silica gel. Develop the chromatogram in a chamber containing a freshly prepared mixture of ethyl acetate that has been saturated with ammonium hydroxide, ether, and methanol (75:25:20) until the solvent front has moved about three-fourths of the length of the plate. Remove the plate from the chamber, air-dry, and spray with Iodoplatinate reagent prepared by dissolving 100 mg of platinic chloride in 10 mL of 0.1 *N* hydrochloric acid, adding 25 mL of potassium iodide solution (1 in 25) and 0.5 mL of formic acid, and diluting with water to 100 mL. The chromatogram from the Test solution may exhibit a secondary spot whose R_f value corresponds to, and whose size and intensity are not greater than, those of the spot from the chlorpromazine sulfoxide solution (5.0%).

Assay—Transfer an accurately measured volume of Chlorpromazine Hydrochloride Syrup, equivalent to about 10 mg of chlorpromazine hydrochloride, to a 50-mL volumetric flask, add 0.1 *N* hydrochloric acid to volume, and mix. Proceed as directed in the *Assay* under *Chlorpromazine Hydrochloride Injection*, beginning with "Pipet 10 mL of the solution." Calculate the quantity, in mg, of $C_{17}H_{19}ClN_2S.HCl$ in each mL of the Syrup taken by the formula:

$$1.25C(A_{254} - A_{277})_U / V(A_{254} - A_{277})_S,$$

in which *C* is the concentration, in µg per mL, of USP Chlorpromazine Hydrochloride RS in the Standard solution, *V* is the volume, in mL, of Syrup taken, and the parenthetic expressions are the differences in the absorbances of the two solutions at the wavelengths indicated by the subscripts, for the solution from the Syrup (*U*) and the Standard solution (*S*), respectively.

Chlorpromazine Hydrochloride Tablets

» Chlorpromazine Hydrochloride Tablets contain not less than 95.0 percent and not more than 105.0 percent of the labeled amount of $C_{17}H_{19}ClN_2S.HCl$.

Packaging and storage—Preserve in well-closed, light-resistant containers.

Reference standard—*USP Chlorpromazine Hydrochloride Reference Standard*—Dry at 105° for 2 hours before using.

NOTE—Throughout the following procedures, protect test or assay specimens, the Reference Standard, and solutions containing them, by conducting the procedures without delay, under subdued light, or using low-actinic glassware.

Identification—

A: Tablets respond to *Identification test B* under *Chlorpromazine Hydrochloride*.

B: Digest a quantity of powdered Tablets, equivalent to about 25 mg of chlorpromazine hydrochloride, with 25 mL of water, and filter: the solution so obtained responds to *Identification test C* under *Chlorpromazine Hydrochloride*.

Dissolution ⟨711⟩—

Medium: 0.1 *N* hydrochloric acid; 900 mL.
Apparatus 1: 50 rpm.
Time: 30 minutes.

Procedure—Determine the amount of $C_{17}H_{19}ClN_2S.HCl$ dissolved from ultraviolet absorbances at the wavelength of maximum absorbance at about 254 nm of filtered portions of the solution under test, suitably diluted with *Dissolution Medium*, in comparison with a Standard solution having a known concentration of USP Chlorpromazine Hydrochloride RS in the same medium.

Tolerances—Not less than 80% (*Q*) of the labeled amount of $C_{17}H_{19}ClN_2S.HCl$ is dissolved in 30 minutes.

Uniformity of dosage units ⟨905⟩: meet the requirements.

Other alkylated phenothiazines—Transfer a portion of finely powdered Tablets, equivalent to 50 mg of chlorpromazine hydrochloride, to a stoppered centrifuge tube, add 10 mL of methanol, shake vigorously, and centrifuge (remove any sugar coating by prior washing with water). Proceed as directed in the test for *Other alkylated phenothiazines* under *Chlorpromazine*, beginning with "Dissolve a suitable quantity of USP Chlorpromazine Hydrochloride RS." The area and intensity of any spot, other than the principal spot, from the solution from the Tablets are not greater than those of the spot from the *Diluted standard solution* (0.5%).

Assay—Weigh and finely powder not less than 20 Chlorpromazine Hydrochloride Tablets. Transfer an accurately weighed portion of the powder, equivalent to about 100 mg of chlorpromazine hydrochloride, to a 500-mL volumetric flask. Add about 200 mL of water and 5 mL of hydrochloric acid, insert the stopper, and shake for about 10 minutes. Dilute with water to volume, and mix. Filter a portion of the solution, discarding the first 50 mL of the filtrate. Treat 10.0 mL of the filtrate as directed in the *Assay* under *Chlorpromazine Hydrochloride Injection*, beginning with "Pipet 10 mL of the solution." Calculate the quantity, in mg, of $C_{17}H_{19}ClN_2S.HCl$ in the portion of Tablets taken by the formula:

$$12.5C(A_{254} - A_{277})_U/(A_{254} - A_{277})_S,$$

in which *C* is the concentration, in µg per mL, of USP Chlorpromazine Hydrochloride RS in the Standard solution, and the parenthetic expressions are the differences in the absorbances of the two solutions at the wavelengths indicated by the subscripts, for the solution from the Tablets (*U*) and the Standard solution (*S*), respectively.

Chlorpropamide

Cl–C6H4–SO2–NH–C(=O)–NH–CH2CH2CH3

$C_{10}H_{13}ClN_2O_3S$ 276.74

Benzenesulfonamide, 4-chloro-*N*-[(propylamino)carbonyl]-.
1-[(*p*-Chlorophenyl)sulfonyl]-3-propylurea [*94-20-2*].

» Chlorpropamide contains not less than 97.0 percent and not more than 103.0 percent of $C_{10}H_{13}ClN_2O_3S$, calculated on the dried basis.

Packaging and storage—Preserve in well-closed containers.

Reference standard—*USP Chlorpropamide Reference Standard*—Dry in vacuum at 60° for 2 hours before using.

Identification—

A: The infrared absorption spectrum of a potassium bromide dispersion of it, previously dried, exhibits maxima only at the same wavelengths as that of a similar preparation of USP Chlorpropamide RS.

B: It responds to the *Thin-layer Chromatographic Identification Test* ⟨201⟩. Prepare the test solution by dissolving an accurately weighed quantity of Chlorpropamide in acetone to obtain a solution containing 1 mg per mL. Develop the chromatogram in a solvent system consisting of a mixture of methylene chloride, methanol, cyclohexane, and ammonium hydroxide (100:50:30:10).

Melting range ⟨741⟩: between 126° and 129°.

Loss on drying ⟨731⟩—Dry it in vacuum at 60° for 2 hours: it loses not more than 1.0% of its weight.

Selenium ⟨291⟩: 0.003%.

Heavy metals, *Method II* ⟨231⟩: 0.003%.

Residue on ignition ⟨281⟩: not more than 0.4%.

Assay—

Mobile phase—Prepare a filtered and degassed mixture of equal volumes of acetonitrile and dilute glacial acetic acid (1 in 100). [NOTE—Do not exceed 50% of acetonitrile.] Make adjustments if necessary (see *System Suitability* under *Chromatography* ⟨621⟩).

Standard preparation—Dissolve an accurately weighed quantity of USP Chlorpropamide RS in *Mobile phase*, and dilute quantitatively, and stepwise if necessary, with *Mobile phase* to obtain a solution having a known concentration of about 0.05 mg per mL.

Assay preparation—Transfer about 50 mg of Chlorpropamide, accurately weighed, to a 100-mL volumetric flask, add *Mobile phase* to volume, and mix. Transfer 10 mL of this solution to a second 100-mL volumetric flask, add *Mobile phase* to volume, and mix.

Chromatographic system (see *Chromatography* ⟨621⟩)—The liquid chromatograph is equipped with a 240-nm detector and a 4.6-mm × 25-cm column that contains packing L1. The flow rate is about 1.5 mL per minute. Chromatograph the *Standard preparation*, and record the peak responses as directed under *Procedure:* the tailing factor for the analyte peak is not more than 1.5, and the relative standard deviation for replicate injections is not more than 2.0%.

Procedure—Separately inject equal volumes (about 20 µL) of the *Standard preparation* and the *Assay preparation* into the chromatograph, record the chromatograms, and measure the responses for the major peaks. Calculate the quantity, in mg, of $C_{10}H_{13}ClN_2O_3S$ in the portion of Chlorpropamide taken by the formula:

$$1000C(r_U / r_S),$$

in which *C* is the concentration, in mg per mL, of USP Chlorpropamide RS in the *Standard preparation*, and r_U and r_S are the peak responses obtained from the *Assay preparation* and the *Standard preparation*, respectively.

Chlorpropamide Tablets

» Chlorpropamide Tablets contain not less than 90.0 percent and not more than 110.0 percent of the labeled amount of $C_{10}H_{13}ClN_2O_3S$.

Packaging and storage—Preserve in well-closed containers.

Reference standard—*USP Chlorpropamide Reference Standard*—Dry in vacuum at 60° for 2 hours before using.

Identification—Shake a quantity of finely powdered Tablets, equivalent to about 100 mg of chlorpropamide, with 20 mL of 1 *N* hydrochloric acid, and extract with 50 mL of chloroform. Filter the chloroform through chloroform-washed cotton into a suitable beaker, and evaporate the chloroform on a steam bath with the aid of a current of dry air to dryness. Dry the residue at 105° for 1 hour: the residue so obtained responds to the *Identification tests* under *Chlorpropamide*.

Dissolution ⟨711⟩—

Medium: water; 900 mL.

Apparatus 2: 50 rpm.

Time: 60 minutes.

Procedure—Determine the amount of $C_{10}H_{13}ClN_2O_3S$ dissolved from ultraviolet absorbances at the wavelength of maximum absorbance at about 230 nm of filtered portions of the solution under test, suitably diluted with 0.1 *N* hydrochloric acid in comparison with a Standard solution having a known concentration of USP Chlorpropamide RS in 0.1 *N* hydrochloric acid. [NOTE—A volume of alcohol not exceeding 10% of the final volume of the Standard solution may be used to dissolve the USP Chlorpropamide RS.]

Tolerances—Not less than 75% (Q) of the labeled amount of $C_{10}H_{13}ClN_2O_3S$ is dissolved in 60 minutes.

Uniformity of dosage units ⟨905⟩: meet the requirements.

Assay—

Mobile phase, Standard preparation, and *Chromatographic system*—Proceed as directed in the *Assay* under *Chlorpropamide.*

Assay preparation—Weigh and finely powder not less than 20 Chlorpropamide Tablets. Transfer an accurately weighed portion of the powder, equivalent to about 50 mg of chlorpropamide, to a 100-mL volumetric flask. Add *Mobile phase* to volume, mix, and filter, discarding the first 10 mL of the filtrate. Pipet 10 mL of the filtrate into a second 100-mL volumetric flask, add *Mobile phase* to volume, and mix.

Procedure—Proceed as directed for *Procedure* in the *Assay* under *Chlorpropamide.* Calculate the quantity, in mg, of $C_{10}H_{13}ClN_2O_3S$ in the portion of Tablets taken by the formula:

$$1000C(r_U/r_S),$$

in which C is the concentration, in mg per mL, of USP Chlorpropamide RS in the *Standard preparation,* and r_U and r_S are the peak responses obtained from the *Assay preparation* and the *Standard preparation,* respectively.

Chlorprothixene

H—C—CH₂CH₂N(CH₃)₂
Cl
S

$C_{18}H_{18}ClNS$ 315.87

1-Propanamine, 3-(2-chloro-9*H*-thioxanthen-9-ylidene)-*N*,*N*-dimethyl-, (*Z*)-.

(*Z*)-2-Chloro-*N*,*N*-dimethylthioxanthene-$\Delta^{9,\gamma}$-propylamine [*113-59-7*].

» Chlorprothixene contains not less than 99.0 percent and not more than 101.0 percent of $C_{18}H_{18}ClNS$, calculated on the dried basis.

Packaging and storage—Preserve in tight, light-resistant containers.

Reference standards—*USP Chlorprothixene Reference Standard*—Dry over silica gel to constant weight before using. *USP (E)-Chlorprothixene Reference Standard*—Keep container tightly closed and protected from light. Dry over silica gel to constant weight before using.

Identification—

A: The infrared absorption spectrum of a 1 in 10 solution of previously dried Chlorprothixene in carbon disulfide exhibits maxima only at the same wavelengths as that of a similar solution of USP Chlorprothixene RS.

B: A 1 in 20,000 solution in 0.1 *N* hydrochloric acid exhibits an absorbance maximum at 324 ± 2 nm, and its absorptivity at that maximum is within 2.0% of the absorptivity of a similar solution of USP Chlorprothixene RS, concomitantly measured. A 1 in 100,000 solution in 0.1 *N* hydrochloric acid exhibits an absorbance maximum at 267 ± 2 nm, and its absorptivity at that maximum is within 2.0% of the absorptivity of a similar solution of USP Chlorprothixene RS, concomitantly measured.

C: Dissolve about 10 mg in 2 mL of nitric acid: a light red solution is produced. Dilute the solution with 5 mL of water: the solution exhibits a green fluorescence under ultraviolet light.

Melting range, *Class I* ⟨741⟩: between 96.5° and 101.5°.

Loss on drying ⟨731⟩—Dry it over silica gel to constant weight: it loses not more than 0.1% of its weight.

Residue on ignition ⟨281⟩: not more than 0.1%.

Heavy metals, *Method II* ⟨231⟩: 0.002%.

Limit of (*E*)-chlorprothixene [(*E*)-2-chloro-*N*,*N*-dimethylthioxanthene-$\Delta^{9,\gamma}$-propylamine]—Prepare a solution in dehydrated alcohol containing 10 mg of chlorprothixene per mL. On a suitable thin-layer chromatographic plate (see *Chromatography* ⟨621⟩), coated with a 0.5-mm layer of chromatographic silica gel, apply 10 µL of this solution and 10 µL of a solution in dehydrated alcohol containing 300 µg of USP (*E*)-Chlorprothixene RS per mL. Allow the spots to dry, and develop the chromatogram in a low-actinic chamber, not pre-equilibrated with the developing solvent, with a developing solvent consisting of a mixture of dehydrated alcohol, benzene, and water (20:20:1), until the solvent front has moved about three-fourths of the length of the plate. Remove the plate from the developing chamber, mark the solvent front, and allow the solvent to evaporate. Locate the spots by spraying the plate cautiously with sulfuric acid in a spraying cabinet and immediately viewing under short-wavelength ultraviolet light: the spot from the test solution is not greater in size or intensity than the spot at the same R_f value produced by the Standard solution, corresponding to not more than 3.0% of (*E*)-chlorprothixene.

Assay—Dissolve about 800 mg of Chlorprothixene, accurately weighed, in 80 mL of chloroform, add methanolic methyl red TS, and titrate with 0.1 *N* perchloric acid in dioxane VS. Perform a blank determination, and make any necessary correction. Each mL of 0.1 *N* perchloric acid is equivalent to 31.59 mg of $C_{18}H_{18}ClNS$.

Chlorprothixene Injection

» Chlorprothixene Injection is a sterile solution of Chlorprothixene in Water for Injection, prepared with the aid of Hydrochloric Acid. It contains not less than 95.0 percent and not more than 105.0 percent of the labeled amount of $C_{18}H_{18}ClNS$.

Packaging and storage—Preserve in single-dose, low-actinic containers, protected from light.

Reference standard—*USP Chlorprothixene Reference Standard*—Dry over silica gel to constant weight before using.

Identification—

A: A volume of Injection, equivalent to about 12.5 mg of chlorprothixene, responds to *Identification test C* under *Chlorprothixene.*

B: The ultraviolet absorption spectrum of the solution employed for measurement of absorbance in the *Assay* exhibits maxima and minima at the same wavelengths as that of a 1 in 16,000 solution of USP Chlorprothixene RS in 0.1 *N* hydrochloric acid, concomitantly measured.

Pyrogen—It meets the requirements of the *Pyrogen Test* ⟨151⟩, the test dose being 1 mL per kg of a solution containing 1 mg per 10 mL.

pH ⟨791⟩: between 3.0 and 4.0.

Other requirements—It meets the requirements under *Injections* ⟨1⟩.

Assay—[NOTE—Use low-actinic glassware.] Transfer an accurately measured volume of Chlorprothixene Injection, equivalent to about 63 mg of chlorprothixene, to a 100-mL volumetric flask, dilute with 0.1 *N* hydrochloric acid to volume, and mix. Transfer 10.0 mL of this solution to a second 100-mL volumetric flask, dilute with the same acid to volume, and mix. Concomitantly determine the absorbances of this solution and of a solution of USP Chlorprothixene RS in the same medium having a known concentration of about 63 µg per mL in 1-cm cells at the wavelength of maximum absorbance at about 324 nm, with a suitable spectrophotometer, using 0.1 *N* hydrochloric acid as the blank. Calculate the quantity, in mg, of $C_{18}H_{18}ClNS$ in each mL of the Injection taken by the formula:

$$(C/V)(A_U/A_S),$$

in which C is the concentration, in µg per mL, of USP Chlorprothixene RS in the Standard solution, V is the volume, in mL, of Injection taken, and A_U and A_S are the absorbances of the solution from the Injection and the Standard solution, respectively.

Chlorprothixene Oral Suspension

» Chlorprothixene Oral Suspension contains not less than 90.0 percent and not more than 110.0 percent of the labeled amount of $C_{18}H_{18}ClNS$.

Packaging and storage—Preserve in tight, light-resistant containers.

Reference standard—*USP Chlorprothixene Reference Standard*—Dry over silica gel to constant weight before using.

Identification—On a suitable thin-layer chromatographic plate (see *Chromatography* ⟨621⟩), coated with a 0.25-mm layer of chromatographic silica gel, apply 10 µL of the heptane solution from the *Assay preparation* and 10 µL of a solution of USP Chlorprothixene RS in *Purified n-heptane* (see *Assay*) containing 1 mg per mL. Allow the spots to dry, and develop the chromatogram in a chamber, not pre-equilibrated with the developing solvent, in a solvent system consisting of a mixture of dehydrated alcohol, benzene, and water (20:20:1) until the solvent front has moved about three-fourths of the length of the plate. Remove the plate from the developing chamber, mark the solvent front, and allow the solvent to evaporate. Locate the spots on the plate by cautiously spraying with sulfuric acid in a spraying cabinet: the R_f value of the principal (orange) spot obtained from the test solution corresponds to that obtained from the Standard solution.

pH ⟨791⟩: between 3.5 and 4.5.

Assay—[NOTE—Use low-actinic glassware.]

Purified n-heptane—Pass chromatographic *n*-heptane through a column of purified siliceous earth.

Standard preparation—Dissolve a suitable quantity of USP Chlorprothixene RS, accurately weighed, in 0.1 *N* hydrochloric acid, and dilute quantitatively and stepwise, if necessary, with the same solvent to obtain a solution having a known concentration of about 50 µg per mL.

Assay preparation—Transfer an accurately measured volume of well-mixed Chlorprothixene Oral Suspension, equivalent to about 200 mg of chlorprothixene, to a glass-stoppered, 50-mL centrifuge tube with the aid of about 30 mL of water. Insert the stopper, shake, centrifuge, and discard the clear supernatant liquid. Repeat the washing with three 30-mL portions of water, and discard the washings. Add 5 mL of sodium hydroxide solution (1 in 10) and 15 mL of dimethylformamide, and shake for 15 minutes. Add 15 mL of *Purified n-heptane*, shake, centrifuge, and pass the clear supernatant heptane extract through a filter of about 2 g of anhydrous granular sodium sulfate supported on glass wool into a 200-mL volumetric flask. Repeat the extraction with five more 15-mL portions of *Purified n-heptane*, passing each extract through the same filter into the volumetric flask containing the main extract. Dilute with *Purified n-heptane* to volume, and retain a portion of this solution for the *Identification test*. Transfer 5.0 mL of this solution to a 100-mL volumetric flask, add about 20 mL of 0.1 *N* hydrochloric acid, and evaporate the heptane on a steam bath with the aid of a stream of nitrogen. Cool, dilute with 0.1 *N* hydrochloric acid to volume, and mix.

Procedure—Concomitantly determine the absorbances of the solutions in 1-cm cells at the wavelength of maximum absorbance at about 324 nm, with a suitable spectrophotometer, using 0.1 *N* hydrochloric acid as the blank. Calculate the quantity, in mg, of $C_{18}H_{18}ClNS$ in each mL of the Suspension taken by the formula:

$$(4C/V)(A_U/A_S),$$

in which C is the concentration, in µg per mL, of USP Chlorprothixene RS in the *Standard preparation*, V is the volume, in mL, of Suspension taken, and A_U and A_S are the absorbances of the *Assay preparation* and the *Standard preparation*, respectively.

Chlorprothixene Tablets

» Chlorprothixene Tablets contain not less than 93.0 percent and not more than 107.0 percent of the labeled amount of $C_{18}H_{18}ClNS$.

Packaging and storage—Preserve in well-closed, light-resistant containers.

Reference standard—*USP Chlorprothixene Reference Standard*—Dry over silica gel to constant weight before using.

Identification—

A: A portion of finely powdered Tablets, equivalent to about 10 mg of chlorprothixene, responds to *Identification test C* under *Chlorprothixene*.

B: The ultraviolet absorption spectrum of the solution employed for measurement of absorbance in the *Assay* exhibits maxima and minima at the same wavelengths as that of a 1 in 20,000 solution of USP Chlorprothixene RS in 0.5 *N* hydrochloric acid, concomitantly measured.

Dissolution ⟨711⟩—

Medium: 0.1 *N* hydrochloric acid; 900 mL.

Apparatus 1: 100 rpm.

Time: 30 minutes.

Procedure—Determine the amount of $C_{18}H_{18}ClNS$ dissolved from ultraviolet absorbances at the wavelength of maximum absorbance at about 324 nm of filtered portions of the solution under test, suitably diluted with *Dissolution Medium*, in comparison with a Standard solution having a known concentration of USP Chlorprothixene RS in the same medium.

Tolerances—Not less than 75% (*Q*) of the labeled amount of $C_{18}H_{18}ClNS$ is dissolved in 30 minutes.

Uniformity of dosage units ⟨905⟩: meet the requirements.

Procedure for content uniformity—[NOTE—Use low-actinic glassware.] Crush 1 Tablet in a suitable container, add 5 mL of dimethylformamide, and allow to stand for about 20 to 30 minutes. Dilute this solution quantitatively and stepwise, if necessary, with 0.1 *N* sulfuric acid, filtering prior to final dilution, to obtain a solution containing approximately 50 µg of chlorprothixene per mL. Concomitantly determine the absorbances of this solution and of a solution of USP Chlorprothixene RS in the same medium having a known concentration of about 50 µg per mL, in 1-cm cells at the wavelength of maximum absorbance at about 324 nm, with a suitable spectrophotometer, using 0.1 *N* sulfuric acid as the blank. Calculate the quantity, in mg, of $C_{18}H_{18}ClNS$ in the Tablet by the formula:

$$(TC/D)(A_U/A_S),$$

in which T is the labeled quantity, in mg, of chlorprothixene in the Tablet, C is the concentration, in µg per mL, of USP Chlorprothixene RS in the Standard solution, D is the concentration, in µg per mL, of chlorprothixene in the solution from the Tablet, based on the labeled quantity per Tablet and the extent of dilution, and A_U and A_S are the absorbances of the solution from the Tablet and the Standard solution, respectively.

Assay—[NOTE—Use low-actinic glassware.] Weigh and finely powder not less than 20 Chlorprothixene Tablets. Transfer an accurately weighed portion of the powder, equivalent to about 50 mg of chlorprothixene, to a glass-stoppered, 50-mL centrifuge tube, add 25 mL of dilute ammonium hydroxide (1 in 100), and shake. Extract with four 15-mL portions of ether, combining the ether extracts in a 250-mL separator. Extract the ether solution with four 30-mL portions of 0.5 *N* hydrochloric acid, collecting the acid extracts in a 200-mL volumetric flask. Dilute with 0.5 *N* hydrochloric acid to volume, mix, and filter, discarding the first 20 mL of the filtrate. Transfer 10.0 mL of the clear filtrate to a 50-mL volumetric flask, dilute with water to volume, and mix. Concomitantly determine the absorbances of this solution and of a solution of USP Chlorprothixene RS in the same medium having a known concentration of about 50 µg per mL in 1-cm cells at the wavelength of maximum absorbance at about 324 nm, with a suitable spectrophotometer, using water containing 10 mL of 0.5 *N* hydrochloric acid per 50 mL of solution as the blank. Calculate the quantity, in mg, of $C_{18}H_{18}ClNS$ in the portion of Tablets taken by the formula:

$$C(A_U/A_S),$$

in which C is the concentration, in µg per mL, of USP Chlorprothixene RS in the Standard solution, and A_U and A_S are the absorbances of the solution from the Tablets and the Standard solution, respectively.

Chlortetracycline Bisulfate

» Chlortetracycline Bisulfate has a potency equivalent to not less than 760 µg of chlortetracycline hydrochloride ($C_{22}H_{23}ClN_2O_8.HCl$) per mg, calculated on the dried and butyl alcohol–free basis.

Packaging and storage—Preserve in tight, light-resistant containers.

Labeling—Label it to indicate that it is intended for veterinary use only.

Reference standard—*USP Chlortetracycline Hydrochloride Reference Standard*—Do not dry before using.

Identification—The ultraviolet absorption spectrum of a 1 in 25,000 solution of it in 0.1 *N* hydrochloric acid exhibits maxima and minima at the same wavelengths as that of a similar solution of USP Chlortetracycline Hydrochloride RS, concomitantly measured, and the absorptivity, calculated on the dried and butyl alcohol–free basis, at the wavelength of maximum absorbance at about 368 nm is between 83.0% and 95.0% of that of the USP Chlortetracycline Hydrochloride RS, the potency of the Reference Standard being taken into account.

Crystallinity ⟨695⟩: meets the requirements.

Safety—Dissolve a quantity, accurately weighed, in sterile water to obtain a solution containing the equivalent of 2.0 mg of chlortetracycline hydrochloride per mL: it meets the requirements for antibiotics under *Safety Tests—General* ⟨157⟩, the test dose being 0.4 mL administered intravenously.

Loss on drying ⟨731⟩—Dry it in vacuum at a pressure not exceeding 5 mm of mercury at 60° for 3 hours: it loses not more than 2.0% of its weight.

Sulfate content—Transfer about 1 g, accurately weighed, to a 250-mL beaker, and dissolve in about 100 mL of water. Neutralize the solution with 7.5 *N* ammonium hydroxide to litmus paper, and warm. Filter, and wash the filter with warm water. Neutralize the filtrate with 6 *N* hydrochloric acid to litmus, and add an additional 4 mL of 6 *N* hydrochloric acid. Heat the solution to boiling, and add, with constant stirring, sufficient boiling barium chloride TS to precipitate all of the sulfate. Add an additional 2 mL of barium chloride TS, and digest on a steam bath for 1 hour. Filter the mixture through ashless filter paper, transferring the residue quantitatively to the filter, and wash the residue with hot water until no precipitate is obtained when 1 mL of silver nitrate TS is added to 5 mL of washing. Transfer the paper containing the residue to a tared crucible. Char the paper, without burning, and ignite the crucible and its contents to constant weight. Perform a blank determination concurrently with the test specimen determination, and subtract the weight of residue obtained from that obtained in the test specimen determination to obtain the weight of residue attributable to the sulfate content of the specimen: not less than 15.0% is found, calculated on the dried and butyl alcohol–free basis.

Butyl alcohol—

Ceric ammonium nitrate solution—Dissolve 20 g of ceric ammonium nitrate in 4 *N* nitric acid to obtain 100 mL of solution.

Standard preparations—Transfer about 3 g of butyl alcohol, accurately weighed, to a 1000-mL volumetric flask containing 800 mL of water, shake to dissolve, dilute with water to volume, and mix (*Standard preparation 1*). Transfer 10.0 mL of *Standard preparation 1* and 1 drop of dimethicone to a 50-mL distilling flask equipped with a condenser and an extension that reaches into a collecting tube maintained in an ice-water bath. Distil slowly, and collect about 8 mL of distillate. Warm the distillate to ambient temperature, and transfer with the aid of water to a 10-mL volumetric flask. Dilute with water to volume, and mix (*Standard preparation 2*).

Test preparation—Transfer an accurately weighed specimen, equivalent to about 30 mg of butyl alcohol, to a 50-mL distilling flask equipped with a condenser and an extension that reaches into a collecting tube maintained in an ice bath. Add 25 mL of water and 1 drop of dimethicone to the distilling flask. Distil slowly, and collect about 8 mL of the distillate. Warm the distillate to ambient temperature, and transfer with the aid of water to a 10-mL volumetric flask. Dilute with water to volume, and mix.

Procedure—To four separate test tubes add, respectively, 5.0 mL of *Standard preparation 1*, 5.0 mL of *Standard preparation 2*, 5.0 mL of *Test preparation*, and 5.0 mL of water to provide a blank. To each add 2.0 mL of *Ceric ammonium nitrate solution*, and mix. Concomitantly determine the absorbances of the solutions from the *Standard preparations* and the *Test preparation* at the wavelength of maximum absorbance at about 475 nm, with a suitable spectrophotometer, using the blank to set the instrument to zero. In a suitable determination, the absorbance of the solution from *Standard preparation 2* is not less than 98.0% of the absorbance of the solution from *Standard preparation 1*. Calculate the percentage of butyl alcohol in the specimen by the formula:

$$1000(W_S/W_U)(A_U/A_S),$$

in which W_S is the weight, in g, of butyl alcohol taken to prepare *Standard preparation 1*, W_U is the weight, in mg, of specimen taken, and A_U and A_S are the absorbances of the solutions from the *Test preparation* and *Standard preparation 2*, respectively: not more than 15.0% is found.

Assay—Proceed with Chlortetracycline Bisulfate as directed for chlortetracycline under *Antibiotics—Microbial Assays* ⟨81⟩.

Chlortetracycline and Sulfamethazine Bisulfates Soluble Powder

» Chlortetracycline and Sulfamethazine Bisulfates Soluble Powder is a dry mixture of Chlortetracycline Bisulfate and Sulfamethazine Bisulfate and one or more suitable buffers and diluents. It contains the equivalent of not less than 85.0 percent and not more than 125.0 percent of the labeled amounts of chlortetracycline hydrochloride ($C_{22}H_{24}N_2O_8.HCl$) and sulfamethazine ($C_{12}H_{14}N_4O_2S$).

Packaging and storage—Preserve in tight, light-resistant containers.

Labeling—Label it to indicate that it is intended for veterinary use only.

Reference standard—*USP Chlortetracycline Hydrochloride Reference Standard*—Do not dry before using.

Loss on drying ⟨731⟩—Dry about 100 mg, accurately weighed, in a capillary-stoppered bottle in vacuum at a pressure not exceeding 5 mm of mercury at 60° for 3 hours: it loses not more than 2.0% of its weight.

Assay for chlortetracycline hydrochloride—Proceed as directed for chlortetracycline under *Antibiotics—Microbial Assays* ⟨81⟩, using an accurately weighed quantity of Chlortetracycline and Sulfamethazine Bisulfates Soluble Powder, equivalent to about 100 mg of chlortetracycline hydrochloride, dissolved in an accurately measured volume of 0.01 *N* hydrochloric acid to obtain a stock solution having a convenient concentration. Dilute an accurately measured volume of this stock solution quantitatively and stepwise with water to obtain a *Test Dilution* having a concentration assumed to be equal to the median dose level of the Standard.

Assay for sulfamethazine—Proceed as directed under *Nitrite Titration* ⟨451⟩, using an accurately weighed quantity of Chlortetracycline and Sulfamethazine Bisulfates Soluble Powder, equivalent to about 500 mg of sulfamethazine. Each mL of 0.1 *M* sodium nitrite is equivalent to 27.83 mg of sulfamethazine ($C_{12}H_{14}N_4O_2S$).

Chlortetracycline Hydrochloride

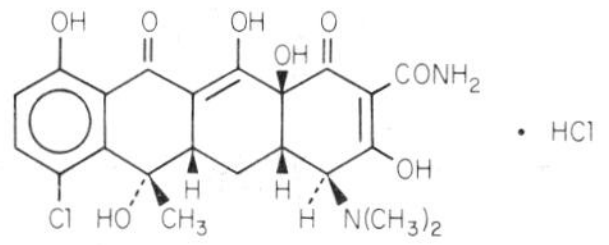

$C_{22}H_{23}ClN_2O_8 \cdot HCl$ 515.35

2-Naphthacenecarboxamide, 7-chloro-4-(dimethylamino)-1,4,4a,5,5a,6,11,12a-octahydro-3,6,10,12,12a-pentahydroxy-6-methyl-1,11-dioxo-, monohydrochloride [4*S*-(4α,4aα,5aα,6β,12aα)]-.

7-Chloro-4-(dimethylamino)-1,4,4a,5,5a,6,11,12a-octahydro-3,6,10,12,12a-pentahydroxy-6-methyl-1,11-dioxo-2-naphthacenecarboxamide monohydrochloride [*64-72-2*].

» Chlortetracycline Hydrochloride has a potency of not less than 900 μg of $C_{22}H_{23}ClN_2O_8 \cdot HCl$ per mg.

NOTE—Chlortetracycline Hydrochloride labeled solely for use in preparing oral veterinary dosage forms has a potency of not less than 820 μg of $C_{22}H_{23}ClN_2O_8 \cdot HCl$ per mg.

Packaging and storage—Preserve in tight, light-resistant containers.

Reference standards—*USP Chlortetracycline Hydrochloride Reference Standard*—Do not dry before using. *USP Oxytetracycline Reference Standard*—Do not dry before using. *USP Tetracycline Hydrochloride Reference Standard*—Do not dry before using.

Identification—

A: Proceed as directed for *Method II* under *Identification—Tetracyclines* ⟨193⟩, using a methanol solution containing 0.5 mg per mL as the *Test solution* and a methanol solution containing in each mL 0.5 mg of USP Chlortetracycline Hydrochloride RS, 0.5 mg of USP Oxytetracycline RS, and 0.5 mg of Tetracycline Hydrochloride RS as the *Resolution solution.*

B: A solution (1 in 20) responds to the tests for *Chloride* ⟨191⟩.

Specific rotation ⟨781⟩: between −235° and −250°, calculated on the dried basis, determined in a solution containing 5 mg per mL that has been allowed to stand in the dark for 30 minutes.

Crystallinity ⟨695⟩: meets the requirements.

pH ⟨791⟩: between 2.3 and 3.3, in a solution containing 10 mg per mL.

Loss on drying ⟨731⟩—Dry about 100 mg, accurately weighed, in a capillary-stoppered bottle in vacuum at a pressure not exceeding 5 mm of mercury at 60° for 3 hours: it loses not more than 2.0% of its weight.

Assay—Proceed with Chlortetracycline Hydrochloride as directed under *Antibiotics—Microbial Assays* ⟨81⟩.

Chlortetracycline Hydrochloride Capsules

» Chlortetracycline Hydrochloride Capsules contain not less than 90.0 percent and not more than 120.0 percent of the labeled amount of $C_{22}H_{23}ClN_2O_8 \cdot HCl$.

Packaging and storage—Preserve in tight, light-resistant containers.

Reference standards—*USP Chlortetracycline Hydrochloride Reference Standard*—Do not dry before using. *USP Oxytetracycline Reference Standard*—Do not dry before using. *USP Tetracycline Hydrochloride Reference Standard*—Do not dry before using.

Identification—Shake a suitable quantity of Capsule contents with methanol to obtain a solution containing about 0.5 mg of chlortetracycline hydrochloride per mL, and filter. Using the filtrate so obtained as the *Test solution*, and a methanol solution containing in each mL 0.5 mg of USP Chlortetracycline Hydrochloride RS, 0.5 mg of USP Oxytetracycline RS, and 0.5 mg of USP Tetracycline Hydrochloride RS as the *Resolution solution*, proceed as directed for *Method II* under *Identification—Tetracyclines* ⟨193⟩.

Dissolution ⟨711⟩—

Medium: water; 900 mL.

Apparatus 2: 75 rpm.

Time: 45 minutes.

Procedure—Determine the amount of $C_{22}H_{23}ClN_2O_8 \cdot HCl$ dissolved from ultraviolet absorbances at the wavelength of maximum absorbance at about 270 nm of filtered portions of the solution under test, suitably diluted with *Dissolution Medium*, if necessary, in comparison with a Standard solution having a known concentration of USP Chlortetracycline Hydrochloride RS in the same medium.

Tolerances—Not less than 75% (*Q*) of the labeled amount of $C_{22}H_{23}ClN_2O_8 \cdot HCl$ is dissolved in 45 minutes.

Uniformity of dosage units ⟨905⟩: meet the requirements.

Loss on drying ⟨731⟩—Dry about 100 mg, accurately weighed, in a capillary-stoppered bottle in vacuum at a pressure not exceeding 5 mm of mercury at 110° for 3 hours: it loses not more than 1.0% of its weight.

Assay—Proceed as directed under *Antibiotics—Microbial Assays* ⟨81⟩, using not less than 5 Chlortetracycline Hydrochloride Capsules blended for 4 ± 1 minutes in a high-speed glass blender jar containing an accurately measured volume of 0.01 *N* hydrochloric acid, so that the stock solution so obtained contains not less than 1 mg of chlortetracycline hydrochloride ($C_{22}H_{23}ClN_2O_8 \cdot HCl$) per mL. Dilute an accurately measured volume of the stock solution quantitatively and stepwise with water to obtain a *Test Dilution* having a concentration assumed to be equal to the median dose level of the Standard.

Chlortetracycline Hydrochloride Ointment

» Chlortetracycline Hydrochloride Ointment contains not less than 90.0 percent and not more than 125.0 percent of the labeled amount of $C_{22}H_{23}ClN_2O_8 \cdot HCl$ in a suitable ointment base.

Packaging and storage—Preserve in collapsible tubes or in well-closed, light-resistant containers.

Reference standard—*USP Chlortetracycline Hydrochloride Reference Standard*—Do not dry before using.

Water, *Method I* ⟨921⟩: not more than 0.5%, 20 mL of a mixture of carbon tetrachloride, chloroform, and methanol (2:2:1) being used in place of methanol in the titration vessel.

Minimum fill ⟨755⟩: meets the requirements.

Assay—Proceed as directed under *Antibiotics—Microbial Assays* ⟨81⟩, using an accurately weighed quantity of Chlortetracycline Hydrochloride Ointment, equivalent to about 30 mg of chlortetracycline hydrochloride, shaken in a separator with about 50 mL of ether, and extracted with four 20-mL portions of 0.01 *N* hydrochloric acid. Combine the aqueous extracts in a 100-mL volumetric flask, dilute with 0.01 *N* hydrochloric acid to volume, and mix. Dilute this stock solution quantitatively and stepwise with water to obtain a *Test Dilution* having a concentration assumed to be equal to the medium dose level of the Standard.

Chlortetracycline Hydrochloride Ophthalmic Ointment

» Chlortetracycline Hydrochloride Ophthalmic Ointment contains not less than 90.0 percent and not more than 125.0 percent of the labeled amount of $C_{22}H_{23}ClN_2O_8 . HCl$.

Packaging and storage—Preserve in collapsible ophthalmic ointment tubes.

Reference standard—*USP Chlortetracycline Hydrochloride Reference Standard*—Do not dry before using.

Sterility—It meets the requirements under *Sterility Tests* ⟨71⟩.

Minimum fill ⟨755⟩: meets the requirements.

Water, *Method I* ⟨921⟩: not more than 0.5%, 20 mL of a mixture of carbon tetrachloride, chloroform, and methanol (2:2:1) being used in place of methanol in the titration vessel.

Metal particles—It meets the requirements of the test for *Metal Particles in Ophthalmic Ointments* ⟨751⟩.

Assay—Proceed as directed under *Antibiotics—Microbial Assays* ⟨81⟩, using an accurately weighed quantity of Chlortetracycline Hydrochloride Ophthalmic Ointment, equivalent to about 10 mg of chlortetracycline hydrochloride, shaken in a separator with about 50 mL of ether, and extracted with four 20-mL portions of 0.01 *N* hydrochloric acid. Combine the aqueous extracts in a 100-mL volumetric flask, dilute with 0.01 *N* hydrochloric acid to volume, and mix. Dilute this stock solution quantitatively and stepwise with water to obtain a *Test Dilution* having a concentration assumed to be equal to the median dose level of the Standard.

Chlortetracycline Hydrochloride Soluble Powder

» Chlortetracycline Hydrochloride Soluble Powder contains not less than 90.0 percent and not more than 125.0 percent of the labeled amount of $C_{22}H_{23}ClN_2O_8 . HCl$.

Packaging and storage—Preserve in tight containers, protected from light.

Labeling—Label it to indicate that it is intended for oral veterinary use only.

Reference standard—*USP Chlortetracycline Hydrochloride Reference Standard*—Do not dry before using.

Loss on drying ⟨731⟩—Dry about 100 mg, accurately weighed, in a capillary-stoppered bottle in vacuum at a pressure not exceeding 5 mm of mercury at 60° for 3 hours: it loses not more than 2.0% of its weight.

Assay—

Assay preparation 1 (where it is labeled on a weight basis)—Dissolve about 3 g of Chlortetracycline Hydrochloride Soluble Powder in an accurately measured volume of 0.01 *N* hydrochloric acid sufficient to obtain a solution containing not less than 1000 µg of chlortetracycline hydrochloride ($C_{22}H_{23}ClN_2O_8 . HCl$) per mL.

Assay preparation 2 (where the label states the amount of chlortetracycline in the immediate container)—Transfer the contents of 1 container of Chlortetracycline Hydrochloride Soluble Powder to an accurately measured volume of 0.01 *N* hydrochloric acid sufficient to obtain a solution containing not less than 1000 µg of chlortetracycline hydrochloride ($C_{22}H_{23}ClN_2O_8 . HCl$) per mL.

Procedure—Proceed with Chlortetracycline Hydrochloride Soluble Powder as directed for chlortetracycline under *Antibiotics—Microbial Assays* ⟨81⟩, using an accurately measured volume of *Assay preparation* diluted quantitatively and stepwise with water to yield a *Test Dilution* having a concentration assumed to be equal to the median dose level of the Standard.

Sterile Chlortetracycline Hydrochloride

» Sterile Chlortetracycline Hydrochloride is Chlortetracycline Hydrochloride suitable for parenteral use. It has a potency of not less than 900 µg of $C_{22}H_{23}ClN_2O_8 . HCl$ per mg.

Packaging and storage—Preserve in *Containers for Sterile Solids* as described under *Injections* ⟨1⟩, protected from light.

Reference standards—*USP Chlortetracycline Hydrochloride Reference Standard*—Do not dry before using. *USP Oxytetracycline Reference Standard*—Do not dry before using. *USP Tetracycline Hydrochloride Reference Standard*—Do not dry before using.

Depressor substances—It meets the requirements of the *Depressor Substances Test* ⟨101⟩, the test dose being 0.6 mL per kg of a solution in sterile water containing 5.0 mg of chlortetracycline hydrochloride per mL.

Pyrogen—It meets the requirements of the *Pyrogen Test* ⟨151⟩, the test dose being 1 mL per kg of a solution in Sterile Water for Injection containing 5.0 mg of chlortetracycline hydrochloride per mL.

Sterility—It meets the requirements under *Sterility Tests* ⟨71⟩, when tested as directed in the section, *Test Procedures Using Membrane Filtration*, 6 g of specimen aseptically dissolved in 200 mL of *Fluid D* being used.

Other requirements—It conforms to the definition, responds to the *Identification tests*, and meets the requirements for *pH*, *Specific rotation*, *Loss on drying*, *Crystallinity*, and *Assay* under *Chlortetracycline Hydrochloride*.

Chlortetracycline Hydrochloride Tablets

» Chlortetracycline Hydrochloride Tablets contain not less than 90.0 percent and not more than 120.0 percent of the labeled amount of $C_{22}H_{23}ClN_2O_8 . HCl$.

Packaging and storage—Preserve in tight containers, protected from light.

Labeling—Label Tablets to indicate that they are intended for veterinary use only.

Reference standards—*USP Chlortetracycline Hydrochloride Reference Standard*—Do not dry before using. *USP Oxytetracycline Reference Standard*—Do not dry before using. *USP Tetracycline Hydrochloride Reference Standard*—Do not dry before using.

Identification—Shake a suitable quantity of finely ground Tablet powder with methanol to obtain a solution containing about 0.5 mg of chlortetracycline hydrochloride per mL, and filter. Using the filtrate so obtained as the *Test solution*, and a methanol solution containing in each mL 0.5 mg of USP Chlortetracycline Hydrochloride RS, 0.5 mg of Oxytetracycline RS, and 0.5 mg of USP Tetracycline Hydrochloride RS as the *Resolution solution*, proceed as directed for *Method II* under *Identification—Tetracyclines* ⟨193⟩.

Disintegration ⟨701⟩: 1 hour, simulated gastric fluid TS being used as the test medium in place of water.

Uniformity of dosage units ⟨905⟩: meet the requirements for *Weight Variation*.

Water, *Method I* ⟨921⟩: not more than 3.0%, or where the Tablets have a diameter of greater than 15 mm, not more than 6.0%, a quantity of finely ground Tablet powder, accurately weighed, being used.

Assay—Transfer not less than 5 Chlortetracycline Hydrochloride Tablets to a high-speed glass blender jar containing an accurately measured volume of 0.01 *N* hydrochloric acid, so that after blending for 3 to 5 minutes a solution of convenient concentration is obtained. Proceed as directed for chlortetracycline under *Antibiotics—Microbial Assays* ⟨81⟩, using an accurately measured volume of this solution diluted quantitatively and stepwise with water to obtain a *Test Dilution* having a concentration assumed to be equal to the median dose level of the Standard.

Chlorthalidone

$C_{14}H_{11}ClN_2O_4S$ 338.76

Benzenesulfonamide, 2-chloro-5-(2,3-dihydro-1-hydroxy-3-oxo-1*H*-isoindol-1-yl)-.

2-Chloro-5-(1-hydroxy-3-oxo-1-isoindolinyl)benzenesulfonamide [*77-36-1*].

» Chlorthalidone contains not less than 98.0 percent and not more than 102.0 percent of $C_{14}H_{11}ClN_2O_4S$, calculated on the dried basis.

Packaging and storage—Preserve in well-closed containers.

Reference standards—*USP Chlorthalidone Reference Standard*—Dry at 105° for 4 hours before using. *USP 4′-Chloro-3′-sulfamoyl-2-benzophenone Carboxylic Acid Reference Standard*—Dry at 105° for 4 hours before using.

Identification—

A: The infrared absorption spectrum of a mineral oil dispersion of it, previously dried, exhibits maxima only at the same wavelengths as that of a similar preparation of USP Chlorthalidone RS.

B: The ultraviolet absorption spectrum of a 1 in 10,000 solution in a 1 in 50 solution of 2 *N* hydrochloric acid in methanol exhibits maxima and minima at the same wavelengths as that of a similar solution of USP Chlorthalidone RS, concomitantly measured, and the respective absorptivities, calculated on the dried basis, at the wavelength of maximum absorbance at about 275 nm do not differ by more than 4.0%.

C: Dissolve about 50 mg in 3 mL of sulfuric acid: an intense yellow color develops.

Loss on drying ⟨731⟩—Dry about 2 g, accurately weighed, at 105° for 4 hours: it loses not more than 0.4% of its weight.

Residue on ignition ⟨281⟩: not more than 0.1%.

Chloride ⟨221⟩—Shake 1.0 g with 40 mL of water for 5 minutes, and filter through chloride-free filter paper previously rinsed with water: the filtrate shows no more chloride than corresponds to 0.50 mL of 0.020 *N* hydrochloric acid (0.035%).

Heavy metals, *Method II* ⟨231⟩: 0.001%.

Limit of 4′-chloro-3′-sulfamoyl-2-benzophenone carboxylic acid (CCA)—Proceed as directed in the *Assay*, except to calculate the quantity, in mg, of CCA in the portion of Chlorthalidone taken by the formula:

$$(0.5)C(R_U/R_S),$$

in which *C* is the concentration, in µg per mL, of USP 4′-Chloro-3′-sulfamoyl-2-benzophenone Carboxylic Acid RS in the *Standard preparation*, and R_U and R_S are the peak response ratios of CCA and the internal standard obtained from the *Assay preparation* and the *Standard preparation*, respectively: not more than 1.0% is present.

Assay—

Mobile phase—Prepare a suitable degassed mixture of 0.01 *M* dibasic ammonium phosphate and methanol (3:2), adjust dropwise with phosphoric acid to a pH of 5.5 ± 0.1, and filter.

Internal standard solution—Prepare a solution of 2,7-naphthalenediol in methanol having a concentration of about 1.0 mg per mL.

CCA solution—Prepare a solution of USP 4′-Chloro-3′-sulfamoyl-2-benzophenone Carboxylic Acid RS in methanol having a known concentration of about 5 µg per mL.

Standard preparation—Prepare a solution of USP Chlorthalidone RS in methanol having a known concentration of about 1 mg per mL. Pipet 5 mL of this solution into a 50-mL volumetric flask containing 5.0 mL of *Internal standard solution* and 10.0 mL of CCA solution. Dilute with water to volume, and mix.

Assay preparation—Transfer about 50 mg of Chlorthalidone, accurately weighed, to a 50-mL volumetric flask, dissolve in methanol, dilute with methanol to volume, and mix. Pipet 5 mL of this solution into a 50-mL volumetric flask containing 5.0 mL of *Internal standard solution* and 10.0 mL of methanol. Dilute with water to volume, and mix.

Chromatographic system (see *Chromatography* ⟨621⟩)—The liquid chromatograph is equipped with a 254-nm detector and a 4.6-mm × 25-cm column that contains packing L7. The flow rate is about 1.0 mL per minute. Chromatograph five replicate injections of the *Standard preparation*, and record the peak responses as directed under *Procedure:* the relative standard deviation is not more than 2.0%, and the resolution factors between Chlorthalidone and CCA, and between chlorthalidone and the internal standard, are not less than 1.5. The tailing factors for the Chlorthalidone and CCA peaks are not more than 2.0.

Procedure—Separately inject equal volumes (about 25 µL) of the *Standard preparation* and the *Assay preparation* into the chromatograph, record the chromatograms, and measure the responses for the major peaks. The relative retention times are about 0.5 for CCA, 0.8 for Chlorthalidone and 1.0 for the internal standard. Calculate the quantity, in mg, of $C_{14}H_{11}ClN_2O_4S$ in the portion of Chlorthalidone taken by the formula:

$$500C(R_U/R_S),$$

in which *C* is the concentration, in mg per mL, of USP Chlorthalidone RS in the *Standard preparation*, and R_U and R_S are the peak response ratios of chlorthalidone and the internal standard obtained from the *Assay preparation* and the *Standard preparation*, respectively.

Chlorthalidone Tablets

» Chlorthalidone Tablets contain not less than 92.0 percent and not more than 108.0 percent of the labeled amount of $C_{14}H_{11}ClN_2O_4S$.

Packaging and storage—Preserve in well-closed containers.

Reference standard—*USP Chlorthalidone Reference Standard*—Dry at 105° for 4 hours before using.

Identification—

A: Digest a quantity of powdered Tablets, equivalent to about 100 mg of chlorthalidone, with 10 mL of acetone on a steam bath for about 5 minutes. Filter the solution into a 50-mL beaker, add 20 mL of water, and boil on the steam bath for about 5 minutes, passing a gentle current of air above the solution to remove the acetone. Cool in an ice bath, filter, and dry the crystals at 105° for 4 hours: the crystals so obtained respond to *Identification test A* under *Chlorthalidone*.

B: The retention time of the major peak in the chromatogram of the *Assay preparation* corresponds to that of the *Standard preparation*, both relative to the internal standard, as obtained in the *Assay*.

Dissolution ⟨711⟩—

Medium: water; 900 mL.

Apparatus 2: 100 rpm.

Time: 60 minutes.

Standard preparation—Dissolve an accurately weighed quantity of USP Chlorthalidone RS in methanol to obtain a solution having a known concentration of about 5 mg per mL.

Procedure—Determine the amount of $C_{14}H_{11}ClN_2O_4S$ dissolved from ultraviolet absorbances at the wavelength of maximum absorbance at about 275 nm of filtered portions of the solution under test, suitably diluted with water, in comparison with a quantitative dilution in water of the *Standard preparation* having a known concentration of USP Chlorthalidone RS comparable to the concentration of the solution under test.

Tolerances—Not less than 50% (*Q*) of the labeled amount of $C_{14}H_{11}ClN_2O_4S$ is dissolved in 60 minutes.

Uniformity of dosage units ⟨905⟩: meet the requirements.

Assay—

Mobile phase and *Internal standard solution*—Prepare as directed in the *Assay* under *Chlorthalidone.*

Standard preparation—Prepare as directed in the *Assay* under *Chlorthalidone*, except to substitute 10.0 mL of methanol for the *CCA solution.*

Assay preparation—Weigh and finely powder not less than 20 Chlorthalidone Tablets. Transfer an accurately weighed portion of the powder, equivalent to about 100 mg of Chlorthalidone, to a 100-mL volumetric flask. Dissolve in about 50 mL of methanol, shake for 30 minutes, dilute with methanol to volume, and mix. Transfer about 30 mL of this solution to a 50-mL centrifuge tube, and centrifuge for 10 minutes. Pipet 5 mL of the clear supernatant solution into a 50-mL volumetric flask containing 5.0 mL of *Internal standard solution* and 10.0 mL of methanol. Dilute with water to volume, and mix.

Chromatographic system (see *Chromatography* ⟨621⟩)—The liquid chromatograph is equipped with a 254-nm detector and a 4.6-mm × 25-cm column that contains packing L7. The flow rate is about 1.0 mL per minute. Chromatograph five replicate injections of the *Standard preparation*, and record the peak responses as directed under *Procedure:* the relative standard deviation is not more than 2.0%, and the resolution factor between chlorthalidone and the internal standard is not less than 1.5. The tailing factors for the chlorthalidone and internal standard peaks are not more than 2.0.

Procedure—Separately inject equal volumes (about 25 µL) of the *Standard preparation* and the *Assay preparation* into the chromatograph, record the chromatograms, and measure the responses for the major peaks. The relative retention times are about 0.8 for chlorthalidone and 1.0 for the internal standard. Calculate the quantity, in mg, of $C_{14}H_{11}ClN_2O_4S$ in the portion of Tablets taken by the formula:

$$1000C(R_U/R_S),$$

in which *C* is the concentration, in mg per mL, of USP Chlorthalidone RS in the *Standard preparation*, and R_U and R_S are the peak response ratios of chlorthalidone and the internal standard obtained from the *Assay preparation* and the *Standard preparation*, respectively.

Chlorthalidone Tablets, Clonidine Hydrochloride and—*see* Clonidine Hydrochloride and Chlorthalidone Tablets

Chlorzoxazone

$C_7H_4ClNO_2$ 169.57

2(3*H*)-Benzoxazolone, 5-chloro-.

5-Chloro-2-benzoxazolinone [*95-25-0*].

» Chlorzoxazone contains not less than 98.0 percent and not more than 102.0 percent of $C_7H_4ClNO_2$, calculated on the dried basis.

Packaging and storage—Preserve in tight containers.

Reference standards—*USP Chlorzoxazone Reference Standard*—Dry at 105° for 2 hours before using. *USP 2-Amino-4-chlorophenol Reference Standard*—Dry at 105° for 2 hours before using.

Identification—

A: The infrared absorption spectrum of a potassium bromide dispersion of it, previously dried, exhibits maxima only at the same wavelengths as that of a similar preparation of USP Chlorzoxazone RS.

B: The ultraviolet absorption spectrum of a 1 in 50,000 solution in methanol exhibits maxima and minima at the same wavelengths as that of a similar preparation of USP Chlorzoxazone RS, concomitantly measured.

Melting range ⟨741⟩: between 189° and 194°.

Loss on drying ⟨731⟩—Dry it at 105° for 2 hours: it loses not more than 0.5% of its weight.

Heavy metals, *Method II* ⟨231⟩: 0.002%.

Residue on ignition ⟨281⟩: not more than 0.15%.

Chromatographic impurities—Prepare a *Test solution* in methanol containing 20 mg per mL. Dissolve a suitable quantity of USP 2-Amino-4-chlorophenol RS in methanol to obtain a solution containing 100 µg per mL (*Standard solution A*). Dissolve a suitable quantity of *p*-chlorophenol in methanol to obtain a solution containing 50 µg per mL (*Standard solution B*). Apply separate 10 µL portions of the three solutions on the starting line of a suitable thin-layer chromatographic plate (see *Chromatography* ⟨621⟩), coated with a 0.25-mm layer of chromatographic silica gel mixture. Allow the spots to dry, and develop the chromatogram in a solvent system consisting of hexane and dioxane (63:37) until the solvent front has moved about three-fourths of the length of the plate. Remove the plate from the developing chamber, mark the solvent front, and allow the solvent to evaporate. Locate the spots on the plate by examination under short-wavelength ultraviolet light: any spot obtained from the *Test solution*, other than one corresponding to chlorzoxazone, does not exceed, in size or intensity, the principal spot obtained from *Standard solution A*, corresponding to not more than 0.5% of any individual impurity. Expose the plate to iodine vapors in a closed chamber, and locate the spots: any spot obtained from the *Test solution*, other than one corresponding to chlorzoxazone, does not exceed, in size or intensity, the principal spot obtained from *Standard solution B*, corresponding to not more than 0.25% of any individual impurity.

Chlorine content—Dissolve about 300 mg, accurately weighed, in 10 mL of alcohol in a suitable flask. Add 3.5 g of Raney's nickel-aluminum catalyst, and connect to a suitable reflux condenser. Chill the flask in an ice bath, and add through the condenser 75 mL of 2.5 *N* sodium hydroxide. When the reaction has subsided, remove the ice bath. After 10 minutes, heat the flask gently, gradually increasing the heat until the mixture refluxes rapidly. After 90 minutes from the time of the addition of the alkali, discontinue heating, cool, and rinse the condenser with water, collecting the rinsings in the flask. Transfer the liquid to a 200-mL volumetric flask, wash the residue with water, and add the washing to the volumetric flask. Dilute with water to volume, and mix. Transfer 100.0 mL of this solution to a beaker, neutralize, then acidify (using congo red as the indicator) by adding nitric acid dropwise with mixing. Titrate with 0.1 *N* silver nitrate VS, determining the end-point potentiometrically, using silver and calomel electrodes. Each mL of 0.1 *N* silver nitrate is equivalent to 3.545 mg of Cl: the content of Cl, calculated on the dried basis, is between 20.6% and 21.2%.

Assay—Transfer about 50 mg of Chlorzoxazone, accurately weighed, to a 100-mL volumetric flask. Dissolve in methanol, dilute with methanol to volume, and mix. Transfer 4.0 mL of this solution to a 100-mL volumetric flask, dilute with methanol to volume, and mix. Concomitantly determine the absorbances of this solution and a Standard solution of USP Chlorzoxazone RS in methanol at a concentration of about 20 µg per mL in 1-cm cells at the wavelength of maximum absorbance at about 282 nm, with a suitable spectrophotometer, using methanol as the blank. Calculate the quantity, in mg, of $C_7H_4ClNO_2$ in the Chlorzoxazone taken by the formula:

$$2.5C(A_U/A_S),$$

in which *C* is the concentration, in µg per mL, of USP Chlorzoxazone RS in the Standard solution, and A_U and A_S are the absorbances of the solution of Chlorzoxazone and the Standard solution, respectively.

Chlorzoxazone Tablets

» Chlorzoxazone Tablets contain not less than 90.0 percent and not more than 110.0 percent of the labeled amount of $C_7H_4ClNO_2$.

Packaging and storage—Preserve in tight containers.

Reference standard—*USP Chlorzoxazone Reference Standard*—Dry at 105° for 2 hours before using.

Identification—The ultraviolet absorption spectrum of the solution employed for measurement of absorbance in the *Assay* exhibits maxima and minima at the same wavelengths as that of a similar solution of USP Chlorzoxazone RS, concomitantly measured.

Dissolution ⟨711⟩—

Medium: simulated gastric fluid TS; 900 mL.

Apparatus 2: 100 rpm.

Time: 60 minutes.

Procedure—Determine the amount of $C_7H_4ClNO_2$ dissolved, using filtered portions of the solution under test and the following procedure. Transfer an accurately measured portion of the filtrate containing about 0.5 mg of chlorzoxazone to a suitable flask containing 20.0 mL of chloroform. Shake by mechanical means for 15 minutes, centrifuge, and discard the aqueous layer. Determine the absorbance of the chloroform layer at the wavelength of maximum absorbance at about 282 nm in comparison with a Standard solution having a known concentration of USP Chlorzoxazone RS, similarly prepared and extracted.

Tolerances—Not less than 65% (*Q*) of the labeled amount of $C_7H_4ClNO_2$ is dissolved in 60 minutes.

Uniformity of dosage units ⟨905⟩: meet the requirements.

Assay—

Standard preparation—Dissolve a suitable quantity of USP Chlorzoxazone RS, accurately weighed, in methanol, and dilute quantitatively and stepwise with methanol to obtain a solution having a known concentration of about 20 µg per mL.

Assay preparation—Weigh and finely powder not less than 20 Chlorzoxazone Tablets. Transfer an accurately weighed portion of the powder, equivalent to about 100 mg of chlorzoxazone, to a 100-mL volumetric flask. Add about 80 mL of warm methanol, and shake by mechanical means for about 15 minutes. Dilute with methanol to volume, and mix. Filter, discarding the first 20 mL of the filtrate, and pipet 2 mL of the filtrate into a 100-mL volumetric flask. Dilute with methanol to volume, and mix.

Procedure—Concomitantly determine the absorbances of the *Assay preparation* and the *Standard preparation* in 1-cm cells at the wavelength of maximum absorbance at about 282 nm, with a suitable spectrophotometer, using methanol as the blank. Calculate the quantity, in mg, of $C_7H_4ClNO_2$ in the portion of Tablets taken by the formula:

$$5C(A_U/A_S),$$

in which C is the concentration, in µg per mL, of USP Chlorzoxazone RS in the *Standard preparation*, and A_U and A_S are the absorbances of the *Assay preparation* and the *Standard preparation*, respectively.

Chlorzoxazone and Acetaminophen Capsules

» Chlorzoxazone and Acetaminophen Capsules contain not less than 90.0 percent and not more than 110.0 percent of the labeled amounts of chlorzoxazone ($C_7H_4ClNO_2$) and acetaminophen ($C_8H_9NO_2$).

Packaging and storage—Preserve in tight containers.

Reference standards—*USP Chlorzoxazone Reference Standard*—Dry at 105° for 2 hours before using. *USP Acetaminophen Reference Standard*—Dry over silica gel for 18 hours before using.

Identification—Transfer a portion of Capsule contents, equivalent to about 50 mg of chlorzoxazone, to a 100-mL volumetric flask, add about 80 mL of warm methanol, and shake by mechanical means for 15 minutes. Dilute with methanol to volume, and mix. Filter, and discard the first 20 mL of the filtrate. Transfer 2.0 mL of the filtrate to a 100-mL volumetric flask, dilute with methanol to volume, and mix: the solution so obtained exhibits absorbance maxima at 247 ± 3 nm and at 282 ± 3 nm.

Uniformity of dosage units ⟨905⟩: meet the requirements for *Content Uniformity* with respect to chlorzoxazone and to acetaminophen.

Assay—

Mobile phase A, Mobile phase B, Internal standard solution, Standard preparation, Resolution solution, and *Chromatographic system*—Prepare as directed in the *Assay* under *Chlorzoxazone and Acetaminophen Tablets.*

Assay preparation—Remove, as completely as possible, the contents of not less than 20 Chlorzoxazone and Acetaminophen Capsules, and weigh accurately. Mix the combined contents, and transfer an accurately weighed portion of the powder, equivalent to about 50 mg of chlorzoxazone, to a 100-mL volumetric flask. Add about 70 mL of methanol, and shake by mechanical means for 1 hour. Dilute with methanol to volume, and mix. Filter a portion through filter paper and then through a polytef membrane filter of 1-µm or finer porosity. Transfer 7.0 mL of the filtrate to a suitable vial containing 7.0 mL of *Internal standard solution,* and mix.

Procedure—Proceed as directed for *Procedure* in the *Assay* under *Chlorzoxazone and Acetaminophen Tablets.* Calculate the quantities, in mg, of chlorzoxazone ($C_7H_4ClNO_2$) and acetaminophen ($C_8H_9NO_2$), respectively, in the portion of Capsules taken by the same formula:

$$200C(R_U/R_S),$$

in which C is the concentration, in mg per mL, of the appropriate USP Reference Standard in the *Standard preparation,* and R_U and R_S are the ratios of the peak responses of the corresponding analyte to those of the internal standard obtained from the *Assay preparation* and the *Standard preparation,* respectively.

Cholecalciferol

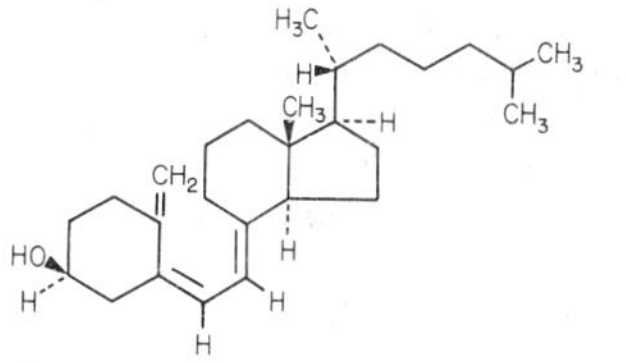

$C_{27}H_{44}O$ 384.64

9,10-Secocholesta-5,7,10(19)-trien-3-ol, (3β,5*Z*,7*E*)-.

Cholecalciferol [*67-97-0*].

» Cholecalciferol contains not less than 97.0 percent and not more than 103.0 percent of $C_{27}H_{44}O$.

Packaging and storage—Preserve in hermetically sealed containers under nitrogen, in a cool place and protected from light.

Reference standards—*USP Cholecalciferol Reference Standard*—Store in a cold place, protected from light. Allow it to attain room temperature before opening ampul. Use the material promptly, and discard the unused portion. *USP Vitamin D Assay System Suitability Reference Standard*—Store in a cool place, protected from light. Allow it to attain room temperature before opening ampul. Do not dry. Transfer unused contents of ampul to a tightly closed container, and store under nitrogen in the dark, in a cool place.

Identification—

A: The infrared absorption spectrum of a potassium bromide dispersion of it, in the range of 2 µm to 12 µm, exhibits maxima only at the same wavelengths as that of a similar preparation of USP Cholecalciferol RS.

B: The ultraviolet absorption spectrum of a 1 in 100,000 solution in alcohol exhibits maxima and minima at the same wavelengths as that of a similar solution of USP Cholecalciferol RS, concomitantly measured, and the respective absorptivities at the wavelength of maximum absorbance at about 265 nm do not differ by more than 3.0%.

C: To a solution of about 0.5 mg in 5 mL of chloroform add 0.3 mL of acetic anhydride and 0.1 mL of sulfuric acid, and shake vigorously: a bright red color is produced, and it rapidly changes through violet and blue to green.

D: Prepare without heating, and handle without delay, a 1 in 100 solution of squalane in chloroform containing 50 mg of cholecalciferol per mL, and prepare a Standard solution of USP Cholecalciferol RS in the same solvent and having the same concentration. Apply 10 µL of the test solution and 10 µL of the Standard solution on a line parallel to and about 2.5 cm from the bottom edge of a thin-layer chromatographic plate (see *Chromatography* ⟨621⟩) coated with a 0.25-mm layer of chromatographic silica gel mixture. Place the plate in a developing chamber containing and equilibrated with a mixture of equal volumes of cyclohexane and diethyl ether. Develop the chromatogram until the solvent front has moved about 15 cm above the line of application. Perform the development and subsequent operations in the dark. Remove the plate, allow the solvent to evaporate, and spray with a 1 in 50 solution of acetyl chloride in antimony trichloride TS: the chromatogram obtained from the test solution shows a yellowish orange area (cholecalciferol) having the same R_f value as the area of the Standard solution, and may show below the cholecalciferol area a violet area attributed to 7-dehydrocholesterol.

Specific rotation ⟨781⟩: between +105° and +112°, determined in a solution in alcohol containing 50 mg in each 10 mL. Prepare the solution without delay, using Cholecalciferol from a container opened not longer than 30 minutes, and determine the rotation within 30 minutes after the solution has been prepared.

Assay—

Dehydrated hexane—Prepare a chromatographic column by packing a chromatographic tube, 60-cm × 8-cm in diameter, with 500 g of 50-µm to 250-µm chromatographic siliceous earth, activated by drying at 150° for 4 hours (see *Column adsorption chromatography* under *Chromatography* ⟨621⟩). Pass 500 mL of hexanes through the column, and collect the eluate in a glass-stoppered flask.

Standard preparation—[NOTE—Use low-actinic glassware, and prepare solutions fresh daily.] Transfer about 30 mg of USP Cholecalciferol RS, accurately weighed, to a 50-mL volumetric flask, dissolve without heat in toluene, add toluene to volume, and mix. Pipet 10 mL of this stock solution into a 50-mL volumetric flask, dilute with *Mobile phase* to volume, and mix to obtain a solution having a known concentration of about 120 µg per mL.

Assay preparation—[NOTE—Use low-actinic glassware, and prepare solutions fresh daily.] Transfer about 30 mg of Cholecalciferol, accurately weighed, to a 50-mL volumetric flask, and proceed as directed for *Standard preparation*, beginning with "dissolve without heat in toluene," to obtain a solution having a concentration of about 120 µg per mL.

Mobile phase—Prepare a 3 in 1000 mixture of *n*-amyl alcohol in *Dehydrated hexane*. The ratio of components and the flow rate may be varied to meet system suitability requirements.

Chromatographic system (see *Chromatography* ⟨621⟩)—The liquid chromatograph is equipped with a 254-nm detector and a 4.6-mm × 25-cm column that contains packing L3.

System suitability preparation—Dissolve about 250 mg of USP Vitamin D Assay System Suitability RS in 10 mL of a mixture of equal volumes of toluene and *Mobile phase*. Heat this solution, under reflux, at 90° for 45 minutes, and cool. This solution contains cholecalciferol, pre-cholecalciferol, and *trans*-cholecalciferol.

System suitability test—Chromatograph five injections of the *System suitability preparation*, and measure the peak responses as directed under *Procedure:* the relative standard deviation for the peak response for cholecalciferol does not exceed 2.0%, and the resolution between *trans*-cholecalciferol and pre-cholecalciferol is not less than 1.0. [NOTE—Chromatograms obtained as directed for this test exhibit relative retention times of approximately 0.4 for pre-cholecalciferol, 0.5 for *trans*-cholecalciferol, and 1.0 for cholecalciferol.]

Procedure—Introduce equal volumes (5 to 10 µL) of the *Standard preparation* and the *Assay preparation* into the chromatograph (see *Chromatography* ⟨621⟩) by means of a suitable sampling valve. Measure the peak responses for the major peaks obtained, at corresponding retention times, from the *Assay preparation* and the *Standard preparation*. Calculate the quantity, in mg, of $C_{27}H_{44}O$ in the portion of Cholecalciferol taken by the formula:

$$0.25C(r_U/r_S),$$

in which C is the concentration, in µg per mL, of USP Cholecalciferol RS in the *Standard preparation*, and r_U and r_S are the peak responses for cholecalciferol obtained from the *Assay preparation* and the *Standard preparation*, respectively.

Cholera Vaccine

» Cholera Vaccine conforms to the regulations of the federal Food and Drug Administration concerning biologics (620.30 to 620.36) (see *Biologics* ⟨1041⟩). It is a sterile suspension, in Sodium Chloride Injection or other suitable diluent, of killed cholera vibrios (*Vibrio cholerae*) of a strain or strains selected for high antigenic efficiency, shown to yield a vaccine not less potent than vaccines prepared from Inaba strain 35A3 and Ogawa strain 41. It is prepared from equal portions of suspensions of cholera vibrios of the Inaba and Ogawa strains. It has a labeled potency of 8 units per serotype per mL. Its potency, determined by the specific mouse potency test based on the U. S. Standard Cholera Vaccines for the respective serotypes, is not less than 4.4 units per serotype per mL. It meets the requirements of the specific mouse toxicity test and of the test for nitrogen content. It contains a suitable antimicrobial agent.

Packaging and storage—Preserve at a temperature between 2° and 8°.

Expiration date—Its expiration date is not later than 18 months after date of issue from manufacturer's cold storage (5°, 1 year).

Labeling—Label it to state that it is to be well shaken before use and that it is not to be frozen.

Cholesterol—*see* Cholesterol NF

Cholestyramine Resin

Cholestyramine.
Cholestyramine [*11041-12-6*].

» Cholestyramine Resin is a strongly basic anion-exchange resin in the chloride form, consisting of styrene-divinylbenzene copolymer with quaternary ammonium functional groups. Each g exchanges not less than 1.8 g and not more than 2.2 g of sodium glycocholate, calculated on the dried basis.

Packaging and storage—Preserve in tight containers.

Reference standard—*USP Cholestyramine Resin Reference Standard*—Dry at a pressure not exceeding 50 mm of mercury at 70° for 16 hours before using.

Identification—The infrared absorption spectrum of a potassium bromide dispersion of it, previously dried, exhibits maxima only at the same wavelengths as that of a similar preparation of USP Cholestyramine Resin RS.

pH ⟨791⟩: between 4.0 and 6.0, in a slurry (1 in 100).

Loss on drying ⟨731⟩—Dry it at a pressure not exceeding 50 mm of mercury at 70° for 16 hours: it loses not more than 12.0% of its weight.

Residue on ignition ⟨281⟩: not more than 0.1%.

Heavy metals, *Method II* ⟨231⟩: 0.002%.

Dialyzable quaternary amines—

pH 9.2 buffer—Transfer 3.80 g of sodium borate to a 1000-mL volumetric flask, dissolve in water, dilute with water to volume, and mix.

Bromothymol blue solution—Transfer 150 mg of bromothymol blue and 405 mg of sodium carbonate decahydrate to a 100-mL volumetric flask, dilute with water to volume, and mix.

Standard preparation—Dilute 1 mL of 60% benzyltrimethylammonium chloride solution, accurately pipeted, quantitatively and stepwise with water to obtain a *Standard solution* having a concentration of 0.2 ± 0.01 mg per mL (prepare this solution fresh). Cut a 20- to 25-cm piece of cellulose dialysis tubing* having a molecular weight cut-off of 6,000 to 14,000 and a dry flat width of 5 to 9 cm, and place it in water to hydrate until pliable, appropriately sealing one end. Pipet 5 mL of the *Standard solution* into the tubing, add 5 mL of water, appropriately seal the open end, place the tube in a suitable vessel containing 100 mL of water so that is completely immersed in the water, and stir the fluid for 16 hours to effect dialysis.

Test preparation—Cut a 20- to 25-cm piece of cellulose dialysis tubing* having a molecular weight cut-off of 6,000 to 14,000 and a dry flat width of 5 to 9 cm, and place it in water to hydrate until pliable, appropriately sealing one end. Weigh 2 ± 0.01 g of Cholestyramine Resin, and carefully transfer the specimen into the tubing, taking care to ensure that none adheres to the upper walls of the tubing. Add 10 mL of water to the contents of the tube, appropriately seal the open end, and place the tube in a suitable vessel containing 100 mL of water so that it is completely immersed in the water. Stir the fluid for 16 hours to effect dialysis.

Procedure—Pipet the following into each of three separators: separator 1: 5 mL of *Standard preparation*, 5 mL of *pH 9.2 buffer*, 1 mL of *Bromothymol blue solution*, and 10 mL of chloroform; separator 2: 5 mL of *Test preparation*, 5 mL of *pH 9.2 buffer*, 1 mL of *Bromothymol blue solution*, and 10 mL of chloroform; separator 3: 5 mL of water, 5 mL of *pH 9.2 buffer*, 1 mL of *Bromothymol blue solution*, and 10 mL of chloroform. Shake each separator, vigorously, for 1 minute, allow the phases to separate until the chloroform phase is clear, and collect the chloroform extracts in separate 25-mL volumetric flasks. Repeat the extraction process with a second 10-mL portion of chloroform, and combine with the previous extracts. Dilute each solution with chloroform to volume, if necessary, and mix. Concomitantly determine the absorbances of the *Test preparation* and the *Standard preparation* at the wavelength of maximum absorbance at about 420 nm, with a suitable spectrophotometer, using the solution from separator 3 as the blank: the absorbance of the *Test preparation* does not exceed that of the *Standard preparation* (0.05% as benzyltrimethylammonium chloride).

Chloride content—To about 750 mg, accurately weighed, add 100 mL of water and 50 mg of potassium nitrate. Add, with stirring, 2 mL of nitric acid, and titrate with 0.1 *N* silver nitrate VS, determining the end-point potentiometrically, and using a silver-glass electrode system. Each mL of 0.1 *N* silver nitrate is equivalent to 3.545 mg of Cl. Not less than 13.0% and not more than 17.0% of Cl, calculated on the dried basis, is found.

* A suitable tubing is Visking No. C65, available from Union Carbide Corp., Films-Packaging Div., 6733 West 65th St., Chicago, IL 60638, or Spectrapor 1, available from various laboratory supply houses, or equivalent.

Exchange capacity—

Sodium glycocholate solution—Dissolve 1.50 g of sodium glycocholate in 40 mL of hot water, cool, and dilute with water to 50.0 mL.

Sulfuric acid solution—Cautiously add 100 mL of sulfuric acid to 25 mL of water, mix, and cool to room temperature.

Procedure—Weigh accurately a quantity of Cholestyramine Resin, equivalent to about 100 mg of anhydrous cholestyramine resin, into a glass-stoppered, 25-mL conical flask. Into a similar flask weigh accurately about 100 mg of USP Cholestyramine Resin RS. Into each flask pipet 10 mL of *Sodium glycocholate solution*, insert the stopper, and shake by mechanical means for 2 hours. Centrifuge a portion of each suspension, transfer 2.0-mL aliquots of the clear, supernatant liquids to separate 200-mL volumetric flasks, dilute with water to volume, and mix. Dilute also 2.0 mL of *Sodium glycocholate solution* with water to volume in a 200-mL volumetric flask, and mix. Pipet 1 mL each of the diluted test solution, the diluted Standard solution, and the diluted *Sodium glycocholate solution*, respectively, into separate, 10-mL volumetric flasks. To each flask add 4 mL of *Sulfuric acid solution*, insert the stopper, and mix. Loosen the stoppers, and heat the flasks in a water bath maintained at 60° for 15 minutes. Remove the flasks from the bath, cool to room temperature, and dilute with *Sulfuric acid solution* to volume. Concomitantly determine the absorbances of the three solutions at the wavelength of maximum absorbance at about 318 nm, with a suitable spectrophotometer, using water as the blank. Calculate the quantity, in g, of sodium glycocholate absorbed on each g of the resin by the formula:

$$M(A_R - A_U)(W_S)/(A_R - A_S)W_U,$$

in which M is the stated value, in g, of sodium glycocholate absorbed per g of the USP Cholestyramine Resin RS, A_R, A_U, and A_S are the absorbances of the treated sodium glycocholate solution, the treated test solution, and the treated Standard solution, respectively, W_U is the weight, in mg, of Cholestyramine Resin, on the dried basis, and W_S is the weight, in mg, of USP Cholestyramine Resin RS used.

Cholestyramine for Oral Suspension

» Cholestyramine for Oral Suspension is a mixture of Cholestyramine Resin with suitable excipients and coloring and flavoring agents. It contains not less than 85.0 percent and not more than 115.0 percent of the labeled amount of dried cholestyramine resin.

Packaging and storage—Preserve in tight containers.

Reference standard—*USP Cholestyramine Resin Reference Standard*—Dry at a pressure not exceeding 50 mm of mercury at 70° for 16 hours before using.

Identification—Transfer a quantity of Cholestyramine for Oral Suspension, equivalent to about 500 mg of dried cholestyramine resin, to a suitable flask, add 100 mL of 0.1 *N* hydrochloric acid, stir to suspend the solid, and heat on a steam bath for 10 minutes. Filter, wash the residue with three 50-mL portions of water, and dry at 70° and at a pressure not exceeding 50 mm of mercury for 16 hours: the infrared absorption spectrum of a potassium bromide dispersion of the residue so obtained exhibits maxima only at the same wavelengths as that of a similar preparation of USP Cholestyramine Resin RS.

Uniformity of dosage units ⟨905⟩: meets the requirements.

Water, *Method I* ⟨921⟩: not more than 12.5%.

Assay—

Sodium glycocholate solution and Sulfuric acid solution—Prepare as directed in the test for *Exchange capacity* under *Cholestyramine Resin*.

Procedure—Weigh accurately a portion of Cholestyramine for Oral Suspension, equivalent to about 100 mg of dried cholestyramine resin, into a glass-stoppered, 30-mL centrifuge tube. Add

25 mL of water, shake for 15 minutes, centrifuge, and decant the water. Add another 25-mL portion of water, shake, centrifuge and decant as before. Heat the centrifuge tube containing the washed specimen at 100° for 2 hours. Into a similar centrifuge tube weigh accurately about 100 mg of USP Cholestyramine Resin RS. Into each tube pipet 10 mL of *Sodium glycocholate solution*, insert the stopper, and shake by mechanical means for 2 hours. Centrifuge each suspension, transfer 2.0-mL aliquots of the clear, supernatant liquids to separate 200-mL volumetric flasks, dilute with water to volume, and mix. Proceed as directed for *Procedure* in the test for *Exchange capacity* under *Cholestyramine Resin*, beginning with "Dilute also 2.0 mL of *Sodium glycocholate solution*." Calculate the quantity, in mg, of cholestyramine resin in the portion of Cholestyramine for Oral Suspension taken by the formula:

$$(M/2)(A_R - A_U)W/(A_R - A_S),$$

in which M is the labeled value, in g, of sodium glycocholate absorbed per g of USP Cholestyramine Resin RS, A_R, A_U, and A_S are the absorbances of the treated sodium glycocholate solution, the treated test solution, and the treated Standard solution, respectively, and W is the weight, in mg, of USP Cholestyramine Resin RS used.

Chorionic Gonadotropin—*see* Gonadotropin, Chorionic

Chromic Chloride

$CrCl_3.6H_2O$ 266.48
Chromium chloride ($CrCl_3$) hexahydrate.
Chromium(3+) chloride hexahydrate [*10060-12-5*].
Anhydrous 158.36 [*10025-73-7*].

» Chromic Chloride contains not less than 98.0 percent and not more than 101.0 percent of $CrCl_3.6H_2O$.

Packaging and storage—Preserve in tight containers.

Identification—

A: To 5 mL of a solution (1 in 250) in a test tube add 1 mL of 5 *N* sodium hydroxide and 10 drops of 30 percent hydrogen peroxide, and heat gently for about 2 minutes: a yellow color develops.

B: To 5 mL of a solution (1 in 250) in a test tube add 5 drops of silver nitrate TS: a white, curdy precipitate is formed, and it is insoluble in nitric acid.

Insoluble matter—Transfer 10 g to a 250-mL beaker, add 100 mL of water, cover the beaker, and heat to boiling. Digest the hot solution on a steam bath for 30 minutes, and filter through a tared filtering crucible of fine porosity. Rinse the beaker with hot water, passing the rinsings through the filter, and wash the filter with hot water until the last washing is colorless. Dry the filter at 105°: the weight of the residue does not exceed 1 mg (0.01%).

Substances not precipitated by ammonium hydroxide—Dissolve 2.0 g in 80 mL of water, heat the solution to boiling, and add a slight excess of ammonium hydroxide. Continue heating to remove the excess ammonia, cool, dilute with water to 100.0 mL, and mix. Filter through a retentive filter, and transfer 50.0 mL of the clear filtrate to an evaporating dish that previously has been ignited and tared. Add 0.5 mL of sulfuric acid to the filtrate, evaporate on a steam bath to dryness, heat gently to remove the excess acid, and ignite gently: the weight of the residue does not exceed 2.0 mg (0.20% as sulfate).

Sulfate ⟨221⟩—Prepare a test solution by dissolving 2.0 g in 10 mL of water. Add 1 mL of 3 *N* hydrochloric acid, filter if necessary to obtain a clear solution, wash the filter with two 5-mL portions of water, and dilute with water to 40 mL. Prepare a control solution in a similar manner, but use 1.0 g of the substance under test, and after the filtration step add 0.10 mL of 0.020 *N* sulfuric acid. To each solution add 3 mL of barium chloride TS, mix, and allow to stand overnight. Decant most of the supernatant liquids, without disturbing the precipitates, but leaving twice the volume of liquid in the control solution as in the test solution. Dilute each solution with water to 25 mL, and sonicate for 1 minute: any turbidity in the test solution does not exceed that in the control solution (0.01%).

Iron ⟨241⟩—Dissolve 1.0 g in 100 mL of water, and mix. Transfer 10 mL of this solution to a 100-mL color comparison tube, dilute with water to 45 mL, add 2 mL of hydrochloric acid, and mix (*Test Preparation*). Proceed as directed for *Procedure*, except to add 15 mL of butyl alcohol to the *Test Preparation* and the *Standard Preparation* at the same time that the *Ammonium Thiocyanate Solution* is added. Shake for 30 seconds, and allow the layers to separate: the color in the upper butyl alcohol layer from the *Test Preparation* is not darker than that from the *Standard Preparation*. The limit is 0.01%.

Assay—Dissolve about 0.4 g of Chromic Chloride, accurately weighed, in 100 mL of water contained in a glass-stoppered, 500-mL conical flask, add 5 mL of 5 *N* sodium hydroxide, and mix. Pipet, slowly, 4 mL of 30 percent hydrogen peroxide into the flask, and boil the solution for 5 minutes. Cool the solution slightly, and add 5 mL of nickel sulfate solution (1 in 20). Boil the solution until no more oxygen is evolved, cool, and add 2 *N* sulfuric acid dropwise until the color of the solution changes from yellow to orange. Add to the flask a freshly prepared solution of 4 g of potassium iodide and 2 g of sodium bicarbonate in 100 mL of water, then add 6 mL of hydrochloric acid. Immediately insert the stopper in the flask, and allow to stand in the dark for 10 minutes. Rinse the stopper and the sides of the flask with a few mL of water, and titrate the liberated iodine with 0.1 *N* sodium thiosulfate VS to an orange color. Add 3 mL of starch TS, and continue the titration to a blue-green end-point. Each mL of 0.1 *N* sodium thiosulfate is equivalent to 8.882 mg of $CrCl_3.6H_2O$.

Chromic Chloride Injection

» Chromic Chloride Injection is a sterile solution of Chromic Chloride in Water for Injection. It contains not less than 95.0 percent and not more than 105.0 percent of the labeled amount of chromium (Cr).

Packaging and storage—Preserve in single-dose or in multiple-dose containers, preferably of Type I or Type II glass.

Labeling—Label the Injection to indicate that it is to be diluted to the appropriate strength with Sterile Water for Injection or other suitable fluid prior to administration.

Identification—The *Assay preparation*, prepared as directed in the *Assay*, exhibits an absorption maximum at about 360 nm when tested as directed for *Procedure* in the *Assay*.

Pyrogen—When diluted with Sodium Chloride Injection to contain 0.1 µg of chromium per mL, it meets the requirements of the *Pyrogen Test* ⟨151⟩.

pH ⟨791⟩: between 1.5 and 2.5.

Other requirements—It meets the requirements under *Injections* ⟨1⟩.

Assay—

Sodium chloride solution—Dissolve 54 g of sodium chloride in water, dilute with water to 2000 mL, and mix.

Chromium stock solution—Transfer 2.829 g of potassium dichromate, accurately weighed, to a 1000-mL volumetric flask, dissolve in water, dilute with water to volume, and mix. This solution contains 1000 µg of chromium per mL. Store in a polyethylene bottle.

Standard preparations—Pipet 10 mL of the *Chromium stock solution* into a 1000-mL volumetric flask, dilute with water to volume, and mix. Transfer 10.0 mL and 20.0 mL, respectively, of this stock solution to separate 100-mL volumetric flasks, and transfer 15.0 mL and 20.0 mL, respectively, of the stock solution to separate 50-mL volumetric flasks. Add 20 mL of *Sodium chloride solution* to each 100-mL volumetric flask, and 10 mL of *Sodium chloride solution* to each 50-mL volumetric flask, dilute the contents of each flask with water to volume, and mix.

These *Standard preparations* contain, respectively, 1.0, 2.0, 3.0, and 4.0 μg of chromium per mL.

Assay preparation—Transfer an accurately measured volume of Chromic Chloride Injection, equivalent to about 60 μg of chromium, to a 25-mL volumetric flask. From the labeled amount of sodium chloride, if any, in the Injection, calculate the amount, in mg, of sodium chloride in the volume of Injection taken, and add sufficient *Sodium chloride solution* to bring the total sodium chloride content of the flask to 135 mg. Dilute with water to volume, and mix.

Procedure—Concomitantly determine the absorbances of the *Standard preparations* and the *Assay preparation* at the chromium emission line of 357.9 nm, with a suitable atomic absorption spectrophotometer (see *Spectrophotometry and Light-scattering* ⟨851⟩) equipped with a chromium hollow-cathode lamp and an air-acetylene flame, using a 1:5 dilution of the *Sodium chloride solution* as the blank. Plot the absorbances of the *Standard preparations* versus concentration, in μg per mL, of chromium, and draw the straight line best fitting the four plotted points. From the graph so obtained, determine the concentration, in μg per mL, of chromium in the *Assay preparation*. Calculate the quantity, in μg, of chromium in each mL of the Injection taken by the formula:

$$25C/V,$$

in which C is the concentration, in μg per mL, of chromium in the *Assay preparation*, and V is the volume, in mL, of Injection taken.

Chromic Phosphate P 32 Suspension—*see* Phosphate P 32 Suspension, Chromic

Sodium Chromate Cr 51 Injection

Chromic acid ($H_2{}^{51}CrO_4$), disodium salt.
Disodium chromate ($Na_2{}^{51}CrO_4$) [*7775-11-3*].

» Sodium Chromate Cr 51 Injection is a sterile solution of radioactive chromium (^{51}Cr) processed in the form of sodium chromate in Water for Injection. For those uses where an isotonic solution is required, Sodium Chloride may be added in appropriate amounts as provided under *Injections* ⟨1⟩. Chromium 51 is produced by the neutron bombardment of enriched chromium 50.

Sodium Chromate Cr 51 Injection contains not less than 90.0 percent and not more than 110.0 percent of the labeled amount of ^{51}Cr as sodium chromate expressed in megabecquerels (millicuries) per mL at the time indicated in the labeling. The sodium chromate content is not less than 90.0 percent and not more than 110.0 percent of the labeled amount. The specific activity is not less than 370 megabecquerels (10 millicuries) per mg of sodium chromate at the end of the expiry period. Other chemical forms of radioactivity do not exceed 10.0 percent of the total radioactivity.

Packaging and storage—Preserve in single-dose or in multiple-dose containers.

Labeling—Label it to include the following, in addition to the information specified for *Labeling* under *Injections* ⟨1⟩: the time and date of calibration; the amount of sodium chromate expressed in μg per mL; the amount of ^{51}Cr as sodium chromate expressed as total megabecquerels (millicuries) and as megabecquerels (millicuries) per mL at the time of calibration; a statement to indicate whether the contents are intended for diagnostic or therapeutic use; the expiration date; and the statement, "Caution—Radioactive Material." The labeling indicates that in making dosage calculations, correction is to be made for radioactive decay and the quantity of chromium, and also indicates that the radioactive half-life of ^{51}Cr is 27.8 days.

Reference standard—*USP Endotoxin Reference Standard.*

Radionuclide identification (see *Radioactivity* ⟨821⟩)—Its gamma-ray spectrum is identical to that of a specimen of ^{51}Cr of known purity that exhibits a photopeak having an energy of 0.320 MeV.

Bacterial endotoxins—It meets the requirements of the *Bacterial Endotoxins Test* ⟨85⟩, the limit of endotoxin content being not more than $175/V$ USP Endotoxin Unit per mL of the Injection, when compared with the USP Endotoxin RS, in which V is the maximum recommended total dose, in mL, at the expiration date or time.

pH ⟨791⟩: between 7.5 and 8.5.

Radiochemical purity—Place a volume of Injection, appropriately diluted such that it provides a count rate of about 20,000 counts per minute, about 25 mm from one end of a 25- × 300-mm strip of chromatographic paper (see *Chromatography* ⟨621⟩), and immediately develop with a mixture of 5 parts of water, 2 parts of dilute alcohol (9.5 in 10), and 1 part of ammonium hydroxide. Air-dry the chromatogram, and determine the radioactivity distribution by scanning the chromatogram with a suitable collimated radiation detector. The radioactivity of the chromate band is not less than 90.0% of the total radioactivity. The R_f value for the chromate band falls within ±10% of the value found for a known sodium chromate specimen when determined under identical conditions.

Other requirements—It meets the requirements under *Injections* ⟨1⟩, except that it is not subject to the recommendation on *Volume in Container*.

Assay for sodium chromate—Prepare a Standard solution of sodium chromate adjusted with sodium bicarbonate solution (1 in 100) to a pH of 8.0 ± 0.5 and containing 1.4 μg of sodium chromate per mL. Determine the absorbances of the Standard solution and of Sodium Chromate Cr 51 Injection, respectively, in 5-cm cells at the wavelength of maximum absorbance at about 370 nm, with a suitable spectrophotometer, using water as the blank. If the absorbance of the Injection is not within 10% of that of the Standard solution, appropriately dilute either the Injection or the Standard solution. If the Injection is diluted, calculate the quantity, in μg, of Na_2CrO_4 per mL of the Injection taken by the formula:

$$1.4D_U(A_U/A_S),$$

in which D_U is the dilution factor for the Injection and A_U and A_S are the absorbances of the Injection and the Standard solution, respectively. If the Standard solution is diluted, use $1/D_S$ in which D_S is the dilution factor for the Standard, in place of D_U.

Assay for radioactivity—Using a suitable counting assembly (see *Selection of a Counting Assembly* under *Radioactivity* ⟨821⟩), determine the radioactivity, in MBq (μCi) per mL, of Sodium Chromate Cr 51 Injection by use of a calibrated system as directed under *Radioactivity* ⟨821⟩.

Chymotrypsin

Chymotrypsin.
Chymotrypsin [*9004-07-3*].

» Chymotrypsin is a proteolytic enzyme crystallized from an extract of the pancreas gland of the ox, *Bos taurus* Linné (Fam. Bovidae). It contains not less than 1000 USP Chymotrypsin Units in each mg, calculated on the dried basis, and not less than 90.0 percent and not more than 110.0 percent of the labeled potency, as determined by the *Assay*.

Packaging and storage—Preserve in tight containers, and avoid exposure to excessive heat.

Reference standards—*USP Chymotrypsin Reference Standard*—Keep container tightly closed, and store in a refrigerator. Allow contents to reach room temperature before opening, and do not dry before using. *USP Trypsin Crystallized Reference Standard*—Keep container tightly closed, and store in a refrigerator. Allow contents to reach room temperature before opening, and do not dry before using.

Microbial limits—It meets the requirements of the tests for absence of *Pseudomonas aeruginosa* and *Salmonella* species and *Staphylococcus aureus* under *Microbial Limit Tests* ⟨61⟩.

Loss on drying ⟨731⟩—Dry it in a vacuum oven at 60° for 4 hours: it loses not more than 5.0% of its weight.

Residue on ignition ⟨281⟩: not more than 2.5%.

Trypsin—

Chymotrypsin solution—Dissolve 100 mg in 10.0 mL of water.

pH 8.1 Tris(hydroxymethyl)aminomethane buffer, 0.08 M—Dissolve 294 mg of calcium chloride in 40 mL of 0.20 *M* tris(hydroxymethyl)aminomethane, adjust with 1 *N* hydrochloric acid to a pH of 8.1, and dilute with water to 100 mL.

Substrate solution—Transfer 98.5 mg of *p*-toluenesulfonyl-L-arginine methyl ester hydrochloride, suitable for use in assaying trypsin, to a 25-mL volumetric flask. Add 5 mL of *pH 8.1 Tris(hydroxymethyl)aminomethane buffer, 0.08 M*, and swirl until the substrate dissolves. Add 0.25 mL of methyl red–methylene blue TS, and dilute with water to volume.

Procedure—[NOTE—Determine the suitability of the substrate by performing the *Procedure* using the appropriate amount of USP Trypsin Crystallized RS in place of the test specimen.] By means of a micropipet, transfer 50 µL of *Chymotrypsin solution* to a depression on a white spot plate. Add 0.2 mL of *Substrate solution:* no purple color develops within 3 minutes (not more than 1% of trypsin).

Assay—

pH 7.0 phosphate buffer, fifteenth-molar—Dissolve 4.54 g of monobasic potassium phosphate in water to make 500 mL of solution. Dissolve 4.73 g of anhydrous dibasic sodium phosphate in water to make 500 mL of solution. Mix 38.9 mL of the monobasic potassium phosphate solution with 61.1 mL of dibasic sodium phosphate solution. If necessary, adjust to a pH of 7.0 by the dropwise addition of dibasic sodium phosphate solution.

Substrate solution—Dissolve 23.7 mg of *N*-acetyl-L-tyrosine ethyl ester, suitable for use in assaying Chymotrypsin, in about 50 mL of *pH 7.0 phosphate buffer, fifteenth-molar*, with warming. When the solution is cool, dilute with additional pH 7.0 buffer to 100 mL. [NOTE—*Substrate solution* may be stored in the frozen state and used after thawing, but it is important to freeze it immediately after preparation.]

Chymotrypsin solution—Dissolve a sufficient quantity of Chymotrypsin, accurately weighed, in 0.0012 *N* hydrochloric acid to yield a solution containing between 12 and 16 USP Chymotrypsin Units per mL. The dilution is correct if, during the conduct of the assay, there is a change in absorbance of between 0.008 and 0.012 in each 30-second interval.

Procedure—[NOTE—Determine the suitability of the substrate and check the adjustment of the spectrophotometer by performing the *Procedure* using USP Chymotrypsin RS in place of the assay specimen.] Conduct the assay in a suitable spectrophotometer equipped to maintain a temperature of 25 ± 0.1° in the cell compartment. Determine the temperature in the reaction cell before and after the measurement of absorbance in order to assure that the temperature does not change by more than 0.5°. Pipet 0.2 mL of 0.0012 *N* hydrochloric acid and 3.0 mL of *Substrate solution* into a 1-cm cell. Place this cell in the spectrophotometer, and adjust the instrument so that the absorbance will read 0.200 at 237 nm. Pipet 0.2 mL of *Chymotrypsin solution* into another 1-cm cell, add 3 mL of *Substrate solution*, and place the cell in the spectrophotometer. [NOTE—Carefully follow this order of addition, and begin timing the reaction from the addition of the *Substrate solution*.] Read the absorbance at 30-second intervals for not less than 5 minutes. Repeat the procedure on the same dilution at least once. Absolute absorbance values are less important than a constant rate of absorbance change. If the rate of change fails to remain constant for not less than 3 minutes, repeat the test and, if necessary, use a lower concentration. The duplicate determination at the same dilution matches the first determination in rate of absorbance change. Determine the average absorbance change per minute, using only the values within the 3-minute portion of the curve where the rate of absorbance change is constant. Plot a curve of absorbance against time. One USP Chymotrypsin Unit is the activity causing a change in absorbance of 0.0075 per minute under the conditions specified in this assay. Calculate the number of USP Chymotrypsin Units per mg taken by the formula:

$$(A_2 - A_1)/(0.0075TW),$$

in which A_2 is the absorbance straight-line initial reading, A_1 is the absorbance straight-line final reading, T is the elapsed time, in minutes, between the initial and final readings, and W is the weight, in mg, of Chymotrypsin in the volume of solution used in determining the absorbance.

Chymotrypsin for Ophthalmic Solution

» Chymotrypsin for Ophthalmic Solution is sterile Chymotrypsin. When constituted as directed in the labeling, it yields a solution containing not less than 80.0 percent and not more than 120.0 percent of the labeled potency.

Packaging and storage—Preserve in single-dose containers, preferably of Type I glass, and avoid exposure to excessive heat.

Completeness of solution ⟨641⟩—It dissolves in the solvent and in the concentration recommended in the labeling to yield a clear solution.

Identification—Prepare a *Substrate solution* as follows: Transfer 237.0 mg of *N*-acetyl-L-tyrosine ethyl ester, suitable for use in assaying chymotrypsin, to a 100-mL volumetric flask, add 2 mL of alcohol, and swirl until solution is effected. Add 20 mL of *pH 7.0 phosphate buffer, fifteenth-molar*, prepared as directed in the *Assay* under *Chymotrypsin*, add 1 mL of methyl red–methylene blue TS, and dilute with water to volume. If necessary, adjust to a pH of 7.0 by the dropwise addition of monobasic potassium phosphate solution, prepared by dissolving 4.54 g of monobasic potassium phosphate in sufficient water to yield 500 mL of solution. Dissolve the contents of 1 vial of Chymotrypsin for Ophthalmic Solution in 1 mL of saline TS, transfer 0.2 mL to a suitable dish, and add 0.2 mL of *Substrate solution:* a purple color is produced within 3 minutes (*distinction from trypsin, which produces no purple color within 3 minutes*).

pH ⟨791⟩: between 4.3 and 8.7, in the solution constituted as directed in the labeling.

Uniformity of dosage units ⟨905⟩: meets the requirements.

Procedure for content uniformity—Assay 10 individual units as directed in the *Assay*, and calculate the average of the 10 results. The average is not less than 80.0 percent and not more than 120.0 percent of the labeled amount. The contents of not more than 2 vials deviate by more than 10 percent from the average content. The contents of none of the vials deviate by more than 15 percent from the average.

Other requirements—It meets the requirements of the test for *Trypsin* under *Chymotrypsin*. It meets also the requirements for *Sterility Tests* ⟨71⟩.

Assay—Proceed with Chymotrypsin for Ophthalmic Solution as directed in the *Assay* under *Chymotrypsin*, but use the following as the *Chymotrypsin solution:* Dissolve the contents of 1 vial of Chymotrypsin for Ophthalmic Solution in 5.0 mL of 0.0012 *N* hydrochloric acid. Dilute an accurately measured volume (V, in mL) of this solution, equivalent to about 300 USP Chymotrypsin Units, with 0.0012 *N* hydrochloric acid to 25.0 mL. Calculate the number of USP Chymotrypsin Units per vial by the formula:

$$300(5/V)(A_2 - A_1)/[T(2.4)(0.0075)],$$

in which A_2 is the absorbance straight-line initial reading, A_1 is the absorbance straight-line final reading, T is the elapsed time in minutes between the initial and final readings, and 2.4 is the number of USP Chymotrypsin Units in the solution on which the absorbance was determined.

Ciclopirox Olamine

$C_{12}H_{17}NO_2.C_2H_7NO$ 268.36
2(1*H*)-Pyridinone, 6-cyclohexyl-1-hydroxy-4-methyl-, compound with 2-aminoethanol (1:1).
6-Cyclohexyl-1-hydroxy-4-methyl-2(1*H*)-pyridone compound with 2-aminoethanol (1:1) [*41621-49-2*].

» Ciclopirox Olamine contains not less than 75.7 percent and not more than 78.0 percent of ciclopirox ($C_{12}H_{17}NO_2$), calculated on the dried basis.

Packaging and storage—Preserve in well-closed containers.

Reference standard—*USP Ciclopirox Olamine Reference Standard.*

Identification—

A: The infrared absorption spectrum of a potassium bromide dispersion of it exhibits maxima only at the same wavelengths as that of a similar preparation of USP Ciclopirox Olamine RS.

B: Prepare a test solution by dissolving a suitable quantity of it in methanol to obtain a concentration of about 40 mg per mL. Similarly prepare a Standard solution, using USP Ciclopirox Olamine RS. Separately apply 10-µL portions of the test solution and the Standard solution on a suitable thin-layer chromatographic plate (see *Chromatography* ⟨621⟩) coated with a 0.25-mm layer of octadecylsilanized chromatographic silica gel mixture, and allow the spots to dry. Place the plate in a suitable chromatographic chamber saturated with a solvent system consisting of a mixture of acetonitrile, isopropyl alcohol, water, 1 *M* methanolic tetrabutylammonium hydroxide, and glacial acetic acid (50:40:10:0.8:0.6), and develop the chromatogram with the same solvent system. When the solvent front has moved about three-fourths of the length of the plate, remove the plate from the chamber, and allow to dry. Expose the plate to iodine vapors for 3 hours, and locate the spots on the plate: the R_f values of the principal spots obtained from the test solution correspond to those obtained from the Standard solution.

Loss on drying ⟨731⟩—Dry it in vacuum to constant weight: it loses not more than 1.5% of its weight.

Residue on ignition ⟨281⟩: not more than 0.1%.

Heavy metals, *Method II* ⟨231⟩: not more than 0.002%.

pH ⟨791⟩: between 8.0 and 9.0, in a solution (1 in 100).

Monoethanolamine content—Dissolve about 300 mg, accurately weighed, in 25 mL of glacial acetic acid. Titrate with 0.1 *N* perchloric acid VS, determining the end-point potentiometrically. Perform a blank determination and make any necessary correction. Each mL of 0.1 *N* perchloric acid is equivalent to 6.108 mg of C_2H_7NO. The content of monoethanolamine (C_2H_7NO), calculated on the anhydrous basis, is not less than 289 mg and not more than 298 mg per g of $C_{12}H_{17}NO_2$ found in the *Assay*.

Assay—Dissolve about 300 mg of Ciclopirox Olamine, accurately weighed, in 80 mL of dimethylformamide. Pass nitrogen gas over the solution to protect it from atmospheric carbon dioxide, and titrate with 0.1 *N* tetrabutylammonium hydroxide VS, determining the end-point potentiometrically, using a calomel-glass electrode system. Perform a blank determination, and make any necessary correction. Each mL of 0.1 *N* tetrabutylammonium hydroxide is equivalent to 20.73 mg of $C_{12}H_{17}NO_2$.

Ciclopirox Olamine Cream

» Ciclopirox Olamine Cream contains not less than 90.0 percent and not more than 110.0 percent of the labeled amount of $C_{12}H_{17}NO_2.C_2H_7NO$.

Packaging and storage—Preserve in collapsible tubes, at controlled room temperature.

Reference standard—*USP Ciclopirox Olamine Reference Standard.*

Identification—

A: Dilute 4 mL of the *Assay preparation* obtained as directed in the *Assay* with a mixture of methanol and 6.25 *N* sodium hydroxide (123:2) to make 100 mL: the ultraviolet absorption spectrum of the solution so obtained exhibits maxima and minima at the same wavelengths as that of a similar solution prepared from the *Standard preparation* obtained as directed in the *Assay*, concomitantly measured.

B: Transfer a quantity of Cream, equivalent to about 40 mg of ciclopirox olamine, to a 50-mL centrifuge tube containing 10 mL of 0.5 *N* sodium hydroxide and about 1 g of sodium chloride. Shake to disperse, and extract with three 10-mL portions of chloroform, centrifuging each time to separate the layers. Discard the chloroform extracts. Add 10 *N* hydrochloric acid to the aqueous layer until just acid to litmus, then add 1 mL of 0.5 *N* hydrochloric acid. Extract with two 15-mL portions of chloroform. Combine the two chloroform extracts, evaporate to dryness, and dissolve the residue in 1 mL of methanol (*Test solution*). Similarly prepare a Standard solution, using USP Ciclopirox Olamine RS. Proceed as directed in *Identification test B* under *Ciclopirox Olamine*, beginning with "Separately apply 10-µL portions of the test solution and the Standard solution." The specified result is obtained.

Minimum fill ⟨755⟩: meets the requirements.

pH ⟨791⟩: Add 15 mL of boiling water, previously adjusted with 0.1 *N* hydrochloric acid or 0.1 *N* sodium hydroxide to a pH of 6 to 7, to 3.5 g of Cream in a 50-mL centrifuge tube. Place a cap on the tube, and shake vigorously until an emulsion is formed. Loosen the cap, and heat the tube on a steam bath for 10 minutes. Allow to cool, centrifuge, and determine the pH of the aqueous phase: the pH is between 5.0 and 8.0.

Benzyl alcohol content (if present)—

Solvent mixture—Mix chloroform and methanol (4:1).

Internal standard solution—Prepare a solution of 1-nonyl alcohol in *Solvent mixture* containing about 1.75 mg per mL.

Standard preparation—Dilute an accurately weighed quantity of benzyl alcohol quantitatively and stepwise with *Solvent mixture* to obtain a solution having a known concentration of about 2 mg per mL. Transfer 5.0 mL of this solution and 5.0 mL of *Internal standard solution* to a 50-mL volumetric flask, dilute with *Solvent mixture* to volume, and mix.

Test preparation—Transfer 1.0 g of Cream to a 50-mL volumetric flask, add about 30 mL of *Solvent mixture*, and mix. Add 5.0 mL of *Internal standard solution*, dilute with *Solvent mixture* to volume, and mix to obtain a clear solution.

Chromatographic system (see *Chromatography* ⟨621⟩)—The gas chromatograph is equipped with a flame-ionization detector and contains a 2-m × 4-mm glass column packed with 3 percent phase G3 on 100- to 120-mesh support S1AB. The column is maintained at a temperature of about 100° and the injection port and detector are maintained at about 315°, and nitrogen is used as the carrier gas at a flow rate of about 45 mL per minute.

Procedure—Separately inject equal volumes (about 2 µL) of the *Standard preparation* and the *Assay preparation* into the chromatograph, record the chromatograms, and measure the responses for the major peaks. In a suitable chromatogram, the resolution, *R*, between the peaks is not less than 1.6, the tailing factor for the benzyl alcohol peak and the internal standard peak are not greater than 3.5, and the relative standard deviation of the peak response ratios, R_S, from replicate injections of the *Standard preparation* is not more than 3%. [Note—After 6 injections, raise the column temperature to about 300° for about 5 minutes, then cool to 100°.] Calculate the percentage of benzyl alcohol in the Cream by the formula:

$$5C(R_U/R_S),$$

in which *C* is the concentration, in mg per mL, of benzyl alcohol in the *Standard preparation*, and R_U and R_S are the peak response ratios of the benzyl alcohol peak to the internal standard peak obtained from the *Test preparation* and the *Standard preparation*, respectively: between 90.0% and 110.0% of the claimed amount is present.

Assay—

Ferrous sulfate solution—Transfer 600 mg of ferrous sulfate to a 25-mL volumetric flask. Add 0.6 mL of glacial acetic acid, dilute with water to volume, and mix.

Standard preparation—Dissolve an accurately weighed quantity of USP Ciclopirox Olamine RS in methanol to obtain a solution having a known concentration of about 0.2 mg per mL.

Assay preparation—Transfer an accurately weighed quantity of Ciclopirox Olamine Cream, equivalent to about 10 mg of ciclopirox olamine, to a 50-mL volumetric flask, add 25 mL of methanol, and shake by mechanical means for about 10 minutes. Dilute with methanol to volume, mix, centrifuge, and use the supernatant liquid.

Procedure—Transfer 4.0 mL of the *Standard preparation*, 4.0 mL of the *Assay preparation*, and 4.0 mL of methanol to provide a blank, to separate 25-mL volumetric flasks. Add 15 mL of methanol to each flask, and mix. Then to each flask add 1.0 mL of *Ferrous sulfate solution*, mix, dilute with methanol to volume, and mix. Store the flasks in the dark for 1 hour. Concomitantly determine the absorbances of the solutions from the *Assay preparation* and the *Standard preparation* against the blank in 1-cm cells at the wavelength of maximum absorbance at about 440 nm, with a suitable spectrophotometer. Calculate the quantity, in mg, of $C_{12}H_{17}NO_2 . C_2H_7NO$ in each g of the Cream taken by the formula:

$$50(C/W)(A_U/A_S),$$

in which C is the concentration, in mg per mL, of USP Ciclopirox Olamine RS in the *Standard preparation*, W is the weight, in g, of Cream taken, and A_U and A_S are the absorbances of the solutions from the *Assay preparation* and the *Standard preparation*, respectively.

Cimetidine

$C_{10}H_{16}N_6S$ 252.34

Guanidine, *N''*-cyano-*N*-methyl-*N'*-[2-[[(5-methyl-1*H*-imidazol-4-yl)methyl]thio]ethyl]-.

2-Cyano-1-methyl-3-[2-[[(5-methylimidazol-4-yl)methyl]-thio]ethyl]guanidine [*51481-61-9*].

» Cimetidine contains not less than 99.0 percent and not more than 101.5 percent of $C_{10}H_{16}N_6S$, calculated on the dried basis.

Packaging and storage—Preserve in tight, light-resistant containers, at controlled room temperature.

Reference standard—*USP Cimetidine Reference Standard*—Dry at 110° for 2 hours before using.

Identification—

A: The infrared absorption spectrum of a potassium bromide dispersion of it, previously dried, exhibits maxima only at the same wavelengths as that of a similar preparation of USP Cimetidine RS.

B: The ultraviolet absorption spectrum of a solution (1 in 80,000) in 0.1 *N* sulfuric acid exhibits maxima and minima at the same wavelengths as that of a similar solution of USP Cimetidine RS, concomitantly measured.

Selenium ⟨291⟩: 0.002%, a 300-mg test specimen being used.

Melting range ⟨741⟩: between 139° and 144°.

Loss on drying ⟨731⟩—Dry it at 110° for 2 hours: it loses not more than 1.0% of its weight.

Residue on ignition ⟨281⟩: not more than 0.2%.

Heavy metals, *Method II* ⟨231⟩: 0.002%.

Chromatographic purity—

Mobile phase—Mix 240 mL of methanol, 0.3 mL of phosphoric acid (85%), 940 mg of sodium 1-hexanesulfonate, and sufficient water to make 1 liter. Filter before use. Make adjustments if necessary (see *System Suitability* under *Chromatography* ⟨621⟩).

Standard preparation—Prepare a solution of USP Cimetidine RS in *Mobile phase* having a concentration of 0.80 µg per mL.

Test preparation—Transfer 100.0 mg of Cimetidine, accurately weighed, to a 250-mL volumetric flask, dissolve in about 50 mL of *Mobile phase*, and dilute with *Mobile phase* to volume. Mix, sonicate for 15 minutes, and again mix.

Chromatographic system (see *Chromatography* ⟨621⟩)—The liquid chromatograph is equipped with a 220-nm detector and 4.6-mm × 25-cm column that contains packing L1. The flow rate is about 2.0 mL per minute. Chromatograph the *Standard preparation*, and record the peak response as directed under *Procedure:* the capacity factor, k', is not less than 3.0, the number of theoretical plates, n, is not less than 2000, and the relative standard deviation of the response for replicate injections is not more than 2.0%.

Procedure—Separately inject equal volumes (about 50 µL) of the *Standard preparation* and the *Test preparation* into the chromatograph, record the chromatograms, and measure the peak responses. The sum of all the peak responses, excluding the cimetidine response, from the *Test preparation* is not more than 5 times the cimetidine response from the *Standard preparation*, and no single peak response is greater than that of the cimetidine response from the *Standard preparation*.

Assay—Dissolve about 240 mg of Cimetidine, accurately weighed, in 75 mL of glacial acetic acid, and titrate with 0.1 *N* perchloric acid VS, determining the end-point potentiometrically. Perform a blank determination, and make any necessary correction. Each mL of 0.1 *N* perchloric acid is equivalent to 25.23 mg of $C_{10}H_{16}N_6S$.

Cimetidine Tablets

» Cimetidine Tablets contain not less than 90.0 percent and not more than 110.0 percent of the labeled amount of $C_{10}H_{16}N_6S$.

Packaging and storage—Preserve in tight, light-resistant containers, at controlled room temperature.

Reference standard—*USP Cimetidine Reference Standard*—Dry at 110° for 2 hours before using.

Identification—The retention time of the major peak in the chromatogram of the *Assay preparation* corresponds to that of the *Standard preparation* obtained in the *Assay*.

Dissolution ⟨711⟩—

Medium: water; 900 mL.

Apparatus 1: 100 rpm.

Time: 15 minutes.

Procedure—Determine the amount of $C_{10}H_{16}N_6S$ dissolved from ultraviolet absorbances at about 218 nm of filtered portions of the solution under test, suitably diluted with 0.1 *N* sulfuric acid, in comparison with a Standard solution having a known concentration of USP Cimetidine RS in the same medium.

Tolerances—Not less than 75% (*Q*) of the labeled amount of $C_{10}H_{16}N_6S$ is dissolved in 15 minutes.

Uniformity of dosage units ⟨905⟩: meet the requirements.

Assay—

Buffer solution, Mobile phase, and *Chromatographic system*—Prepare as directed in the test for *Chromatographic purity* under *Cimetidine*, using the *Standard preparation* described below as the *System suitability preparation*.

Internal standard solution—Dissolve diphenhydramine hydrochloride in methanol to obtain a solution having a concentration of about 0.1 mg per mL.

Standard preparation—Dissolve a suitable quantity of USP Cimetidine RS, accurately weighed, in *Internal standard solution* to obtain a solution having a known concentration of about 0.05 mg per mL.

Assay preparation—Weigh and finely powder not less than 20 Cimetidine Tablets. Transfer an accurately weighed portion of the powder, equivalent to about 25 mg of cimetidine, to a 25-mL centrifuge tube, and add 10.0 mL of *Internal standard solution*.

Shake the tube on a horizontal shaker for 10 minutes, and centrifuge. Pipet 2 mL of the clear supernatant solution into a 100-mL volumetric flask, add *Internal standard solution* to volume, and mix.

Procedure—Separately inject equal volumes (about 10 μL) of the *Standard preparation* and the *Assay preparation*, and measure the responses for the major peaks. The relative retention times are about 0.63 for diphenhydramine and 1.0 for cimetidine. Calculate the quantity, in mg, of $C_{10}H_{16}N_6S$ in the portion of Tablets taken by the formula:

$$500C(R_U/R_S),$$

in which C is the concentration, in mg per mL, of USP Cimetidine RS in the *Standard preparation*, and R_U and R_S are the ratios of the response of the cimetidine peak to that of the diphenhydramine peak obtained from the *Assay preparation* and the *Standard preparation*, respectively.

Cinoxacin

$C_{12}H_{10}N_2O_5$ 262.22

[1,3]Dioxolo[4,5-*g*]cinnoline-3-carboxylic acid, 1-ethyl-1,4-dihydro-4-oxo-.

1-Ethyl-1,4-dihydro-4-oxo[1,3]dioxolo[4,5-*g*]cinnoline-3-carboxylic acid [*28657-80-9*].

» Cinoxacin contains not less than 97.0 percent and not more than 102.0 percent of $C_{12}H_{10}N_2O_5$, calculated on the dried basis.

Packaging and storage—Preserve in tight containers.

Reference standard—*USP Cinoxacin Reference Standard*—Dry in vacuum at 60° for 3 hours before using.

Identification—

A: The infrared absorption spectrum of a potassium bromide dispersion of it, previously dried, exhibits maxima only at the same wavelengths as that of a similar preparation of USP Cinoxacin RS.

B: The R_f value of the principal spot obtained from the *Test preparation* corresponds to that obtained from *Solution A* in the chromatogram prepared as directed in the test for *Related substances*.

Loss on drying ⟨731⟩—Dry it in vacuum at 60° for 3 hours: it loses not more than 1.0% of its weight.

Related substances—

Standard preparations—Prepare a solution of USP Cinoxacin RS in a mixed solvent prepared by mixing equal volumes of chloroform, dimethylformamide, dimethyl sulfoxide, and nitromethane containing 5 mg per mL (*Solution A*). Prepare a second solution by diluting 1.0 volume of *Solution A* with the same mixed solvent to obtain 100 volumes of solution (*Solution B*).

Test preparation—Prepare a solution of Cinoxacin in the same mixed solvent used for the *Standard preparations*, containing 5 mg per mL.

Procedure—In a suitable chromatographic chamber arranged for thin-layer chromatography and lined with paper, place a volume of a solvent system consisting of a mixture of acetonitrile, water, and ammonium hydroxide (105:30:7.5) sufficient to develop the chromatogram, cover, and allow to equilibrate for 30 minutes. On a suitable thin-layer chromatographic plate (see *Chromatography* ⟨621⟩), coated with a 0.25-mm layer of chromatographic silica gel mixture, apply 10-μL portions of *Solution A*, *Solution B*, and the *Test preparation*. Dry the plate, and apply three additional 10-μL portions of each solution at the corresponding initial locations. Dry the plate thoroughly after each application, and develop the chromatogram until the solvent front has moved to the top of the plate. Remove the plate from the developing chamber, and allow the solvent to evaporate. View the plate under short- and long-wavelength ultraviolet light: the R_f value of the principal spot obtained from the *Test preparation* corresponds to that obtained from *Solution A*, and no spot obtained from the *Test preparation*, other than the principal spot, is larger or more intense than the principal spot obtained from *Solution B* (1.0%).

Assay—

Sodium borate solution—Dissolve 38.1 g of sodium borate in water to make 1000 mL.

Internal standard solution—Prepare an aqueous solution containing 2 mg of sulfanilic acid per mL and 5.0 mL of *Sodium borate solution* in each 100 mL.

Mobile phase—Dilute 100.0 mL of *Sodium borate solution* and 0.426 g of sodium sulfate with water to 1000 mL, mix, and degas. [NOTE—The quantity of sodium sulfate may be varied to meet *System suitability* requirements, and to provide a suitable elution time.]

Standard preparation—Dissolve an accurately weighed quantity of USP Cinoxacin RS in *Sodium borate solution* to obtain a solution having a known concentration of about 1 mg per mL. Transfer 5.0 mL of this solution and 5.0 mL of the *Internal standard solution* to a 100-mL volumetric flask, dilute with water to volume, and mix. The *Standard preparation* contains about 50 μg of USP Cinoxacin RS per mL.

Assay preparation—Transfer about 50 mg of Cinoxacin, accurately weighed, to a 50-mL volumetric flask, dissolve in *Sodium borate solution*, dilute with the same solvent to volume, and mix. Transfer 5.0 mL of this solution and 5.0 mL of the *Internal standard solution* to a 100-mL volumetric flask, dilute with water to volume, and mix.

Chromatographic system (see *Chromatography* ⟨621⟩)—The chromatograph is equipped with a 254-nm detector and a 1.8-mm × 1-m column that contains packing L12. The flow rate is about 1 mL per minute. Chromatograph five replicate injections of the *Standard preparation*, and record the peak responses as directed under *Procedure:* the relative standard deviation is not more than 2.0%, the resolution factor between cinoxacin and sulfanilic acid is not less than 4.4, and the tailing factor for the cinoxacin peak is not more than 2.1.

Procedure—Separately inject equal volumes (about 1.0 μL) of the *Standard preparation* and the *Assay preparation* into the chromatograph, record the chromatograms, and measure the responses for the major peaks. The relative retention times are about 2.2 for sulfanilic acid and 1.0 for cinoxacin. Calculate the quantity, in mg, of $C_{12}H_{10}N_2O_5$ in the portion of Cinoxacin taken by the formula:

$$C(R_U/R_S),$$

in which C is the concentration, in μg per mL, of USP Cinoxacin RS in the *Standard preparation*, and R_U and R_S are the ratios of the peak response of cinoxacin to the peak response of sulfanilic acid obtained from the *Assay preparation* and the *Standard preparation*, respectively.

Cinoxacin Capsules

» Cinoxacin Capsules contain not less than 90.0 percent and not more than 110.0 percent of the labeled amount of $C_{12}H_{10}N_2O_5$.

Packaging and storage—Preserve in well-closed containers.

Reference standard—*USP Cinoxacin Reference Standard*—Dry in vacuum at 60° for 3 hours before using.

Identification—Using a portion of Capsule contents to prepare the *Test preparation*, proceed as directed in the test for *Related substances* under *Cinoxacin:* the R_f value of the principal spot obtained from the *Test preparation* corresponds to that obtained from *Solution A*.

Dissolution ⟨711⟩—

Medium: pH 6.5 phosphate buffer (see *Buffer Solutions* in the section, *Reagents, Indicators, and Solutions*); 500 mL for Capsules containing 250 mg or less of cinoxacin; 1000 mL for Capsules containing more than 250 mg of cinoxacin.

Apparatus 1: 100 rpm.
Time: 30 minutes.

Standard preparation—Dissolve an accurately weighed quantity of USP Cinoxacin RS in *Dissolution Medium* to obtain a solution having a known concentration of about 0.35 mg per mL.

Procedure—Determine the amount of $C_{12}H_{10}N_2O_5$ dissolved from ultraviolet absorbances at the wavelength of maximum absorbance at about 270 nm of filtered portions of the solution under test, suitably diluted with 0.1 *N* sodium hydroxide, in comparison with the *Standard preparation*, similarly diluted.

Tolerances—Not less than 60% (*Q*) of the labeled amount of $C_{12}H_{10}N_2O_5$ is dissolved in 30 minutes.

Uniformity of dosage units ⟨905⟩: meet the requirements.

Assay—Transfer the contents of not less than 20 Cinoxacin Capsules to a suitable tared container, and weigh. Transfer an accurately weighed portion of the mixed powder, equivalent to about 250 mg of cinoxacin, to a 100-mL volumetric flask. Dilute with 0.1 *M* sodium borate to volume, and mix. Filter the solution, discarding the first 20 mL of the filtrate, transfer 2.0 mL of the filtrate to a 500-mL volumetric flask, dilute with water to volume, and mix. Concomitantly determine the absorbances of this solution and of a Standard solution of USP Cinoxacin RS in the same medium having a known concentration of about 10 µg per mL, in 1-cm cells at the wavelength of maximum absorbance at about 352 nm, using 2 mL of 0.1 *M* sodium borate diluted with water to 500 mL as the blank. Calculate the quantity, in mg, of $C_{12}H_{10}N_2O_5$ in the portion of Capsules taken by the formula:

$$25C(A_U/A_S),$$

in which *C* is the concentration, in µg per mL, of USP Cinoxacin RS in the Standard solution, and A_U and A_S are the absorbances of the solution from the Capsules and the Standard solution, respectively.

Cinoxate

$CH_3O-C_6H_4-CH{=}CHC(O)OCH_2CH_2OCH_2CH_3$

$C_{14}H_{18}O_4$ 250.29
Propenoic acid, 3-(4-methoxyphenyl)-, 2-ethoxyethyl ester.
2-Ethoxyethyl *p*-methoxycinnamate [*104-28-9*].

» Cinoxate contains not less than 98.0 percent and not more than 101.0 percent of $C_{14}H_{18}O_4$.

Packaging and storage—Preserve in tight, light-resistant containers.

Reference standard—*USP Cinoxate Reference Standard*—Do not dry before using.

Identification—

A: The infrared absorption spectrum of a film of it exhibits maxima only at the same wavelengths as that of a similar preparation of USP Cinoxate RS.

B: The ultraviolet absorption spectrum of a 1 in 200,000 solution in isopropyl alcohol exhibits maxima and minima at the same wavelengths as that of a similar solution of USP Cinoxate RS, concomitantly measured.

Specific gravity ⟨841⟩: between 1.100 and 1.105.

Refractive index ⟨831⟩: between 1.564 and 1.569.

Acidity—Pipet 5 mL into a suitable container, and add 50 mL of neutralized alcohol. Add 4 drops of phenolphthalein TS, and titrate with 0.10 *N* sodium hydroxide: not more than 0.8 mL of 0.10 *N* sodium hydroxide is consumed.

Assay—Transfer about 1.5 g of Cinoxate, accurately weighed, to a 125-mL flask, and pipet 30 mL of 0.5 *N* alcoholic potassium hydroxide into the flask. Heat the mixture under reflux, using an air-cooled reflux condenser, on a boiling water bath for 1 hour. Cool, add 5 drops of phenolphthalein TS, and titrate the excess potassium hydroxide with 0.5 *N* hydrochloric acid VS. Perform a blank determination (see *Titrimetry* ⟨541⟩), and make any necessary correction. Each mL of 0.5 *N* alcoholic potassium hydroxide is equivalent to 125.1 mg of $C_{14}H_{18}O_4$.

Cinoxate Lotion

» Cinoxate Lotion is Cinoxate in a suitable hydroalcoholic vehicle. It contains not less than 90.0 percent and not more than 110.0 percent of the labeled amount of $C_{14}H_{18}O_4$.

Packaging and storage—Preserve in tight, light-resistant containers.

Reference standard—*USP Cinoxate Reference Standard*—Do not dry before using.

Identification—The ultraviolet absorption spectrum of the *Assay preparation*, prepared as directed in the *Assay*, exhibits maxima and minima at the same wavelengths as that of the solution obtained from the *Standard preparation*, concomitantly measured.

pH ⟨791⟩: between 5.4 and 6.4.

Alcohol content, *Method II* ⟨611⟩: between 47% and 57% of C_2H_5OH.

Assay—

Standard preparation—Dissolve an accurately weighed quantity of USP Cinoxate RS in alcohol, and dilute quantitatively with alcohol to obtain a solution having a known concentration of about 0.25 mg per mL.

Assay preparation—Transfer an accurately weighed quantity of Cinoxate Lotion, equivalent to about 25 mg of cinoxate, to a 100-mL volumetric flask. Dissolve in alcohol, and dilute with alcohol to volume.

Procedure—Concomitantly determine the absorbances of the *Assay preparation* and the *Standard preparation* at the wavelength of maximum absorbance at about 308 nm, with a suitable spectrophotometer, using alcohol as the blank. Calculate the quantity, in mg, of $C_{14}H_{18}O_4$ in the portion of Lotion taken by the formula:

$$100C(A_U/A_S),$$

in which *C* is the concentration, in mg per mL, of USP Cinoxate RS, and A_U and A_S are the absorbances of the *Assay preparation* and the *Standard preparation*, respectively.

Cisplatin

Cl^- NH_3 / Pt^{2+} / Cl^- NH_3

$Cl_2H_6N_2Pt$ 300.06
Platinum, diamminedichloro-, (*SP*-4-2)-.
cis-Diamminedichloroplatinum [*15663-27-1*].

» Cisplatin contains not less than 98.0 percent and not more than 102.0 percent of $Cl_2H_6N_2Pt$, calculated on the anhydrous basis.

Caution—Cisplatin is potentially cytotoxic. Great care should be taken to prevent inhaling particles and exposing the skin to it.

Packaging and storage—Preserve in tight containers. Protect from light.

Reference standards—*USP Cisplatin Reference Standard*—Do not dry before using. *USP Transplatin Reference Standard*—Do not dry before using. *USP Potassium Trichloroammineplatinate Reference Standard.*

Identification—

A: The retention time of the major peak in the chromatogram of the *Assay preparation* corresponds to that of the *Standard preparation*, as obtained in the *Assay*.

B: The infrared absorption spectrum of a potassium bromide dispersion of it exhibits maxima only at the same wavelengths as that of a similar preparation of USP Cisplatin RS. [NOTE—Hand grinding with mortar and pestle is recommended for consistent results.]

C: *Spray reagent*—Add 5.6 g of stannous chloride to 10 mL of hydrochloric acid, and stir for 5 minutes. [NOTE—It is not necessary that all of the solids dissolve.] Dissolve 0.2 g of potassium iodide in 90 mL of water. Mix the two solutions together. Disregard any precipitate that is formed. Store in the dark. The solution is usable for at least 1 week.

Procedure—Prepare a test solution containing 1 mg of Cisplatin per mL and a Standard solution containing 1 mg of USP Cisplatin RS per mL, both in dimethylformamide. Apply separately 5-µL quantities of each solution to a thin-layer chromatographic plate coated with a 0.25-mm layer of chromatographic silica gel mixture (see *Chromatography* ⟨621⟩). Place the plate in a suitable chromatographic chamber containing a filter paper lining and equilibrated for 30 minutes with a developer consisting of a mixture of acetone and 1 *N* nitric acid (180:20). Develop the plate for a distance of about 8 cm from the origin. Remove the plate, and allow it to air-dry. Complete the drying by heating in a forced-air oven at about 100° for 1 minute. Spray the plate with *Spray reagent*, heat it in an oven at about 100° for 5 minutes, cool, and spray with a 1 in 50 solution of potassium iodide in water, to bring out the full color of the spots: the principal spot from the test solution corresponds in appearance and R_f value to that produced by the Standard solution.

Crystallinity ⟨695⟩: meets the requirements.

Water, *Method I* ⟨921⟩: not more than 1.0%.

UV purity ratio—[NOTE—Cleanse all glassware with a mixture of hydrochloric acid and nitric acid (3:1), rinse thoroughly with water, and dry before use. Do not use dichromate for cleaning. Do not use acetone or pressurized air for drying. Protect the test solution from light, and use within 1 hour after its preparation.] Transfer 98.5 ± 0.5 mg of ground Cisplatin to a 100-mL volumetric flask, and add 0.1 *N* hydrochloric acid to volume. Using a clean magnetic stir bar, alternately stir at a high speed for 5 minutes and sonicate for 10 seconds until complete solution is effected, inverting the flask frequently to remove particles that may cling to the neck. Obtain the ultraviolet absorption spectrum, using thoroughly rinsed 2-cm cells, with 0.1 *N* hydrochloric acid in the reference cell: the ratio of the absorbance at the maximum near 301 nm to that at the minimum near 246 nm is not less than 4.5.

Platinum content—[NOTE—Thoroughly cleanse all glassware with nitric acid, and rinse with *Purified Water*, to prevent "mirroring" of the platinum precipitate.] Transfer about 0.5 g of Cisplatin, accurately weighed, to a 600-mL beaker. Add 300 mL of 0.1 *N* hydrochloric acid, and slowly dissolve by heating nearly to boiling on a hot plate covered with an insulating pad, and stirring frequently with a glass stirring rod. When solution is complete, remove the insulating pad, and boil for about 10 minutes. Remove the beaker from the hot plate, allow to cool for 1 minute without stirring, and filter through quantitative, fine-porosity, smooth, dense, ashless filter paper, collecting the filtrate in a 600-mL beaker, completing the transfer to the filter with hot water. Wash the filter with hot water. Place the beaker containing the combined filtrate and washings on a hot plate, and evaporate to a volume of about 300 mL. Place a glass stirring rod in the beaker, and heat the solution to boiling. Slowly add to the center of the beaker, by dropwise additions, 10.0 mL of hydrazine hydrate, 85 percent. [*Caution—Hydrazine is toxic.*] Add 2 drops of 10 *N* sodium hydroxide, boil for 10 minutes to coagulate the precipitate for ease of filtration, cool, and filter through quantitative, medium-porosity, smooth, ashless filter paper. Rinse the beaker with hot water, and pour the rinsings onto the filter. Wipe the beaker and the stirring rod with small pieces of the same kind of paper used for this filtration, and place these and the filter containing the precipitate in a No. 1 porcelain crucible, previously ignited to constant weight. Dry on a hot plate covered with an insulating pad, slowly increase the heat to char, and ignite for 1 hour at 800°. Cool in a desiccator, and again weigh: the weight of the platinum so obtained is between 64.42% and 65.22% of the weight of Cisplatin taken, on the anhydrous basis.

Trichloroammineplatinate—

Mobile phase—Transfer 0.8 g of ammonium sulfate to a 2-liter volumetric flask, dissolve in water, and dilute with water to volume. Degas, and filter through a membrane filter prior to use. The pH of this solution is 5.9 ± 0.1.

Standard preparation—[NOTE—Use low-actinic glassware.] Dissolve a suitable quantity of USP Potassium Trichloroammineplatinate RS, accurately weighed, in saline TS, and dilute quantitatively with saline TS to obtain a solution having a known concentration of about 6 µg per mL. Use within 4 hours.

Test preparation—[NOTE—Use low-actinic glassware.] Transfer about 50 mg of Cisplatin, accurately weighed, to a 100-mL volumetric flask, and dilute with saline TS to volume. Completely dissolve by stirring by mechanical means for 30 minutes. Use within 4 hours.

Chromatographic system (see *Chromatography* ⟨621⟩)—The liquid chromatograph is equipped with a 4.6-mm × 25-cm column that contains packing L14 and a 209-nm detector. The flow rate is about 2 mL per minute. Chromatograph the *Standard preparation*, and record the peak responses as directed under *Procedure*. If necessary, adjust the flow rate so that the retention time of the trichloroammineplatinate is between 5 and 10 minutes: the relative standard deviation for replicate injections is not more than 3.0%.

Procedure—Separately inject equal volumes (about 20 µL) of the *Standard preparation* and the *Test preparation* into the chromatograph, record the chromatograms, and measure the areas for the peaks due to trichloroammineplatinate. The relative retention times are about 1.0 for cisplatin (in the void volume) and 5.0 for trichloroammineplatinate. Calculate the percentage of trichloroammineplatinate by the formula:

$$10(318.48/357.58)(r_U/r_S)(C/W),$$

in which 318.48 and 357.58 are the formula weights of trichloroammineplatinate and potassium trichloroammineplatinate, respectively, r_U and r_S are the peak areas obtained from the *Test preparation* and the *Standard preparation*, respectively, C is the concentration, in µg per mL, of the *Standard preparation*, and W is the weight, in mg, of Cisplatin taken: not more than 1.0% is found.

Transplatin—

Buffer solution—Mix 85.5 mL of 8 *N* potassium hydroxide with 800 mL of water, and add phosphoric acid to adjust to a pH of 3.20. Allow the solution to cool to room temperature, and readjust to a pH of 3.20, if necessary. Transfer to a 1-liter volumetric flask, dilute with water to volume, and mix.

Mobile phase—Dilute 180 mL of *Buffer solution* in a 1-liter volumetric flask with water to volume, and mix. Filter, and degas, before use. If necessary, modify by adding water to increase the retention time or *Buffer solution* to decrease it.

Stock standard solution—Transfer about 10 mg of USP Transplatin RS, accurately weighed, to a 200-mL volumetric flask, dilute with saline TS to volume, and dissolve by stirring by mechanical means for 30 minutes.

Working standard solution—Pipet 5 mL of *Stock standard solution* into a 25-mL volumetric flask containing about 12 mg of USP Cisplatin RS. Dilute with saline TS to volume, and stir by mechanical means for 30 minutes to dissolve.

Standard preparation—Pipet 10 mL of *Working standard solution* into a 50-mL volumetric flask. Add 5.0 mL of a 1 in 200 solution of thiourea, prepared fresh daily, and 5.0 mL of 1 *N* hydrochloric acid. Dilute with saline TS to volume, and mix. Place about 10 mL of this solution in a suitable serum vial, seal with a polytef-lined closure, and heat in a heating block at 60 ± 0.5° for 60 minutes. Remove, and cool to room temperature.

Test solution—Transfer about 50 mg of Cisplatin, accurately weighed, to a 100-mL volumetric flask, dilute with saline TS to volume, and dissolve by stirring by mechanical means for 30 minutes.

Test preparation—Pipet 10 mL of *Test solution* into a 50-mL volumetric flask, and proceed as directed for *Standard preparation*, beginning with "add 5.0 mL of a 1 in 200 solution of thiourea."

Resolution solution—Place about 10 mg of USP Cisplatin RS in a 200-mL volumetric flask, dilute with saline TS to volume, and stir by mechanical means for 30 minutes to dissolve. Pipet 10 mL of this solution and 10 mL of *Stock standard solution* into a 50-mL volumetric flask, and proceed as directed for *Standard preparation*, beginning with "add 5.0 mL of a 1 in 200 solution of thiourea."

Chromatographic system (see *Chromatography* ⟨621⟩)—The liquid chromatograph is equipped with a 254-nm detector and a 4.6-mm × 25-cm column that contains packing L9. The column is maintained throughout at a temperature of 45°. The flow rate is about 2.0 mL per minute. Condition the column by pumping *Mobile phase* at a flow rate of 2.0 mL per minute for 30 minutes, then at 0.5 mL per minute for 30 minutes, and then again at 2.0 mL per minute for 30 minutes. Chromatograph the *Standard preparation*. The retention time of the derivatized transplatin is between 5.0 and 9.0 minutes; or, if it is not, modify the *Mobile phase* as necessary, and recondition the column. The column efficiency, n, is not less than 2500. Chromatograph the *Resolution solution*. The resolution, R, is not less than 1.7. Chromatograph the *Standard preparation* as directed for *Procedure*. The relative standard deviation of replicate injections is not more than 4.0%.

Procedure—Separately inject equal volumes (about 20 µL) of the *Test preparation* and the *Standard preparation* into the chromatograph, record the chromatograms, and measure the areas of the transplatin peaks. The relative retention times are about 1.0 for cisplatin and 1.3 for transplatin. Calculate the percentage of transplatin by the formula:

$$10(C/W)(r_U/r_S),$$

in which C is the concentration, in µg per mL, of USP Transplatin RS in the *Working standard solution*, W is the weight, in mg, of Cisplatin taken to prepare the *Test solution*, and r_U and r_S are the peak areas obtained from the *Test preparation* and the *Standard preparation*, respectively. Not more than 2.0% is found.

Assay—

Mobile phase—Prepare a suitable solution by mixing ethyl acetate, methanol, dimethylformamide, and degassed water (25:16:5:5), and degas.

Standard preparation—Dissolve an accurately weighed quantity of USP Cisplatin RS quantitatively in dimethylformamide to obtain a solution having a known concentration of about 1 mg per mL. Use within 1 hour.

Assay preparation—Dissolve about 100 mg of Cisplatin, accurately weighed, in dimethylformamide in a 100-mL volumetric flask, dilute with dimethylformamide to volume, and mix.

Chromatographic system (see *Chromatography* ⟨621⟩)—The liquid chromatograph is equipped with a 4.0-mm × 30-cm column that contains packing L8, and a 310-nm detector. The flow rate is about 2.0 mL per minute. Chromatograph the *Standard preparation*, and record the peak response as directed under *Procedure:* the relative standard deviation for replicate injections is not more than 2.0%.

Procedure—Separately inject equal volumes (about 40 µL) of the *Standard preparation* and the *Assay preparation* into the chromatograph, record the chromatograms, and measure the responses for the major peaks. Calculate the quantity, in mg, of $Cl_2H_6N_2Pt$ in the portion of Cisplatin taken by the formula:

$$100C(r_U/r_S),$$

in which C is the concentration, in mg per mL, of USP Cisplatin RS in the *Standard preparation*, and r_U and r_S are the peak responses obtained from the *Assay preparation* and the *Standard preparation*, respectively.

Cisplatin for Injection

» Cisplatin for Injection is a sterile, lyophilized mixture of Cisplatin, Mannitol, and Sodium Chloride. It contains not less than 90.0 percent and not more than 110.0 percent of the labeled amount of cisplatin ($Cl_2H_6N_2Pt$).

Caution—Cisplatin is potentially cytotoxic. Great care should be taken in handling the powder and preparing solutions.

Packaging and storage—Preserve in *Containers for Sterile Solids* as described under *Injections* ⟨1⟩. Protect from light.

Reference standards—*USP Cisplatin Reference Standard*—Do not dry before using. *USP Transplatin Reference Standard*—Do not dry before using. *USP Potassium Trichloroammineplatinate Reference Standard.*

Identification—

Spray reagent—Prepare as directed for *Spray reagent* in *Identification test C* under *Cisplatin.*

Standard preparation—Prepare a solution containing 1.0 mg of USP Cisplatin RS per mL, 9 mg of sodium chloride per mL, and 10 mg of D-mannitol per mL, in water.

Test preparation—Dissolve the contents of 1 container in water to provide a Cisplatin concentration of 1.0 mg per mL, based on label claim.

Procedure—Proceed as directed for *Procedure* in *Identification test C* under *Cisplatin*, beginning with "Apply separately 5-µL quantities." The principal spot from the *Test preparation* corresponds in appearance and R_f value to that from the *Standard preparation.*

Constituted solution—At the time of use, the constituted solution prepared from Cisplatin for Injection meets the requirements for *Constituted Solutions* under *Injections* ⟨1⟩.

Pyrogen—It meets the requirements of the *Pyrogen Test* ⟨151⟩, the test dose being 10 mL per kg of a solution in sterile, pyrogen-free saline TS containing 0.27 mg of Cisplatin per mL.

Sterility—It meets the requirements under *Sterility Tests* ⟨71⟩, when tested as directed in the section, *Test Procedures Using Membrane Filtration.*

pH ⟨791⟩: between 3.5 and 6.2, in the solution constituted as directed in the labeling, using Sterile Water for Injection.

Water, *Method I* ⟨921⟩—Use anhydrous formamide as the extraction solvent, and use the following procedure. Introduce about 50 mL of anhydrous formamide into the titration vessel, and titrate with the *Reagent* to the electrometric end-point. Use the formamide thus dried to rinse a suitable glass syringe equipped with a 22-gauge needle, about 8 cm long. Add the rinse back to the titration vessel, and again titrate the vessel contents, if necessary. Via the syringe, withdraw 5 mL of the formamide thus titrated, and, through the closure of the container, expel the contents into the container. Shake the container to obtain a solution. With the same syringe, withdraw all of the contents of the container, and transfer to the titration vessel. Titrate to the end-point, adjusting the feeding speed control to the lowest setting, to avoid over-titration. The amount of water found is not more than 2.0%.

Uniformity of dosage units ⟨905⟩: meets the requirements.

Trichloroammineplatinate—

Mobile phase, Standard preparation, and *Chromatographic system*—Proceed as directed in the test for *Trichloroammineplatinate* under *Cisplatin.*

Test preparation—Using low-actinic volumetric glassware, quantitatively dissolve with water the contents of 1 container to yield a 0.5-mg per mL solution of Cisplatin.

Procedure—Proceed as directed for *Procedure* in the test for *Trichloroammineplatinate* under *Cisplatin.* Calculate the percentage of trichloroammineplatinate by the formula:

$$0.1(318.48/357.58)(r_U/r_S)(CV/W),$$

in which 318.48 and 357.58 are the formula weights of trichloroammineplatinate and potassium trichloroammineplatinate, respectively, r_U and r_S are the peak areas obtained from the *Test preparation* and the *Standard preparation*, respectively, C is the concentration, in µg per mL, of the *Standard preparation*, V is the volume, in mL, of the constituted container contents, and W is the labeled amount, in mg, of Cisplatin per container. Not more than 1.0% is found.

Transplatin—

Buffer solution, Mobile phase, Stock standard solution, Working standard solution, Standard preparation, Resolution

solution, and *Chromatographic system*—Proceed as directed in the test for *Transplatin* under *Cisplatin.*

Test solution—Quantitatively dissolve the contents of 1 container with water to yield a 0.5-mg per mL solution of Cisplatin.

Test preparation—Prepare as directed for *Test preparation* in the test for *Transplatin* under *Cisplatin.*

Procedure—Proceed as directed for *Procedure* in the test for *Transplatin* under *Cisplatin.* Calculate the percentage of transplatin by the formula:

$$0.1(CV/W)(r_U/r_S),$$

in which C is the concentration, in µg per mL, of the *Standard preparation*, V is the volume, in mL, of the constituted container contents, W is the labeled amount, in mg, of Cisplatin per container, and r_U and r_S are the peak areas obtained from the *Test preparation* and the *Standard preparation*, respectively. Not more than 2.0% is found.

Other requirements—It meets the requirements for *Labeling* under *Injections* ⟨1⟩.

Assay—

Mobile phase, *Standard preparation*, and *Chromatographic system*—Proceed as directed in the *Assay* under *Cisplatin.*

Assay preparation—Quantitatively dissolve the Cisplatin in 1 container by sonicating for 5 minutes with dimethylformamide to yield a Cisplatin concentration of about 1.0 mg per mL. Filter 5 mL through a suitable membrane filter, and collect the filtrate after discarding the first mL passing through the filter.

Procedure—Proceed as directed for *Procedure* in the *Assay* under *Cisplatin.* Calculate the quantity, in mg, of $Cl_2H_6N_2Pt$ in the container taken by the formula:

$$CV(r_U/r_S),$$

in which C is the concentration, in mg per mL, of USP Cisplatin RS in the *Standard preparation*, V is the volume, in mL, of the constituted container contents, and r_U and r_S are the peak responses obtained from the *Assay preparation* and the *Standard preparation*, respectively.

Citrate Dextrose Solution, Anticoagulant—*see* Anticoagulant Citrate Dextrose Solution

Citrate Phosphate Dextrose Adenine Solution, Anticoagulant—*see* Anticoagulant Citrate Phosphate Dextrose Adenine Solution

Citrate Phosphate Dextrose Solution, Anticoagulant—*see* Anticoagulant Citrate Phosphate Dextrose Solution

Citric Acid

$CH_2(COOH)C(OH)(COOH)CH_2COOH$

$C_6H_8O_7$ 192.12

1,2,3-Propanetricarboxylic acid, 2-hydroxy-.

Citric acid [*77-92-9*].

Monohydrate 210.14 [*5949-29-1*].

» Citric Acid is anhydrous or contains one molecule of water of hydration. It contains not less than 99.5 percent and not more than 100.5 percent of $C_6H_8O_7$, calculated on the anhydrous basis.

Packaging and storage—Preserve in tight containers.

Labeling—Label it to indicate whether it is anhydrous or hydrous.

Identification—A solution responds to the tests for *Citrate* ⟨191⟩.

Water, *Method I* ⟨921⟩: not more than 0.5% (anhydrous form) and not more than 8.8% (hydrous form).

Residue on ignition ⟨281⟩: not more than 0.05%.

Oxalate—Neutralize 10 mL of a solution (1 in 10) with 6 *N* ammonium hydroxide, add 5 drops of 3 *N* hydrochloric acid, cool, and add 2 mL of calcium chloride TS: no turbidity is produced.

Sulfate—To 10 mL of a solution (1 in 100) add 1 mL of barium chloride TS to which has been added 1 drop of hydrochloric acid: no turbidity is produced.

Arsenic, *Method I* ⟨211⟩: 3 ppm.

Heavy metals ⟨231⟩: 0.001%.

Readily carbonizable substances—Transfer 1.0 g, powdered for the test, to a 22- × 175-mm test tube previously rinsed with 10 mL of sulfuric acid TS and allowed to drain for 10 minutes. Add 10 mL of sulfuric acid TS, agitate until solution is complete, and immerse in a water bath at 90 ± 1° for 60 ± 0.5 minutes, keeping the level of the acid below the level of the water during the entire period. Cool the tube in running water, and transfer the acid to a color-comparison tube: the color of the acid is not darker than that of a similar volume of Matching Fluid K (see *Color and Achromicity* ⟨631⟩) in a matching tube, the tubes being observed vertically against a white background.

Assay—Place about 3 g of Citric Acid in a tared flask, and weigh accurately. Dissolve in 40 mL of water, add phenolphthalein TS, and titrate with 1 *N* sodium hydroxide VS. Each mL of 1 *N* sodium hydroxide is equivalent to 64.04 mg of $C_6H_8O_7$.

Citric Acid, Magnesium Oxide, and Sodium Carbonate Irrigation

» Citric Acid, Magnesium Oxide, and Sodium Carbonate Irrigation is a sterile solution of Citric Acid, Magnesium Oxide, and Sodium Carbonate in Water for Injection. It contains not less than 95.0 percent and not more than 105.0 percent of the labeled amounts of citric acid ($C_6H_8O_7.H_2O$), magnesium oxide (MgO), and sodium carbonate (Na_2CO_3).

Packaging and storage—Preserve in single-dose containers, preferably of Type I or Type II glass.

Identification—

A: It responds to the tests for *Sodium* ⟨191⟩ and for *Magnesium* ⟨191⟩.

B: To 10 mL of the Irrigation add 1 mL of mercuric sulfate TS, heat to boiling, and add a few drops of potassium permanganate TS: a white precipitate is formed.

Pyrogen—Adjust a measured portion of Irrigation with sodium hydroxide to a pH of 7.0, and dilute with Sodium Chloride Injection to obtain a solution containing 4 mg of citric acid monohydrate per mL: it meets the requirements of the *Pyrogen Test* ⟨151⟩, the test dose being 10 mL per kg.

pH ⟨791⟩: between 3.8 and 4.2.

Other requirements—It meets the requirements under *Injections* ⟨1⟩, except that the container may be designed to empty rapidly, and may exceed 1000 mL in capacity.

Assay for citric acid—

Mobile phase—Add 2.0 mL of sulfuric acid to 800 mL of water, mix, and dilute with water to 1000 mL. Filter through a membrane filter, heat to 40°, and degas. Maintain the temperature of the *Mobile phase* at 40° throughout the analysis.

Standard preparation—Dissolve a suitable quantity of anhydrous citric acid, accurately weighed, in water to obtain a solution having a known concentration of about 1.2 mg per mL.

Resolution solution—Prepare a solution in water containing about 1 mg of citric acid and about 2 mg of boric acid per mL.

Assay preparation—Transfer an accurately measured volume of Citric Acid, Magnesium Oxide, and Sodium Carbonate Irrigation, equivalent to about 130 mg of citric acid monohydrate, to a 100-mL volumetric flask, dilute with water to volume, and mix.

Chromatographic system (see *Chromatography* ⟨621⟩)—The liquid chromatograph is equipped with a refractive index detector and a 7.8-mm × 30-cm column that contains packing L17. The column temperature is maintained at 40°. The flow rate is about 0.6 mL per minute. Chromatograph the *Standard preparation* and the *Resolution solution*, and record the peak responses as directed under *Procedure:* the relative standard deviation for replicate injections of the *Standard preparation* is not more than 1.5%, and the resolution, *R*, between the citric acid and boric acid peaks is not less than 4.0.

Procedure—Separately inject equal volumes (about 50 µL) of the *Standard preparation* and the *Assay preparation* into the chromatograph, record the chromatograms, and measure the responses for the major peaks. Calculate the quantity, in mg, of $C_6H_8O_7 . H_2O$ in each mL of the Irrigation taken by the formula:

$$(210.14/192.12)(100C/V)(r_U/r_S),$$

in which 210.14 and 192.12 are the molecular weights of citric acid monohydrate and anhydrous citric acid, respectively, C is the concentration, in mg per mL, of anhydrous citric acid in the *Standard preparation*, V is the volume, in mL, of Irrigation taken, and r_U and r_S are the peak responses for citric acid obtained from the *Assay preparation* and the *Standard preparation*, respectively.

Assay for magnesium oxide—Transfer an accurately measured volume of Citric Acid, Magnesium Oxide, and Sodium Carbonate Irrigation, equivalent to about 40 mg of magnesium oxide, to a beaker containing 130 mL of water heated to 75 ± 5°, and add 4 mL of ammonium chloride TS and then 5 mL of ammonium hydroxide. Mix, and add slowly, with stirring, 8 mL of 8-hydroxyquinoline TS. After allowing to stand for 30 minutes at 75°, filter through a sintered-glass crucible, previously dried and weighed, and wash the precipitate with 50 mL of a warm mixture of water and 6 *N* ammonium hydroxide (45:15), followed by 50 mL of cool water. Dry the crucible and contents at 105° for 3 hours, cool, and weigh. Determine the equivalent of MgO in the portion of Irrigation taken by multiplying the weight of $C_{18}H_{12}MgN_2O_2 . 2H_2O$ so obtained by 0.1156.

Assay for sodium carbonate—

Sodium chloride stock solution—Transfer 475 mg of sodium chloride, previously dried at 105° for 2 hours and accurately weighed, to a 100-mL volumetric flask. Dissolve in water, dilute with water to volume, and mix.

Internal standard solution—Transfer 636 mg of lithium chloride to a 1000-mL volumetric flask, dissolve in water, dilute with water to volume, and mix.

Standard preparation—Quantitatively prepare a mixture of *Internal standard solution* and *Sodium chloride stock solution* (99:1).

Assay preparation—Dilute an accurately measured volume of Citric Acid, Magnesium Oxide, and Sodium Carbonate Irrigation quantitatively with water to obtain a stock solution containing about 4.4 mg of sodium carbonate per mL. Quantitatively prepare a mixture of *Internal standard solution* and this stock solution (99:1).

Procedure—Concomitantly determine the emittances of the *Standard preparation* and the *Assay preparation* at 591 nm and 671 nm with a suitable flame photometer, adjusting the instrument to zero emittance with *Internal standard solution.* Calculate the quantity, in mg, of Na_2CO_3 in each mL of the Irrigation taken by the formula:

$$(105.99/116.88)(C)(L/D)(R_{U,591}/R_{U,671})(R_{S,671}/R_{S,591}),$$

in which 105.99 is the molecular weight of sodium carbonate, 116.88 is two times the molecular weight of sodium chloride, C is the concentration, in mg per mL, of sodium chloride in the *Sodium chloride stock solution*, L is the labeled quantity, in mg per mL, of sodium carbonate in the Irrigation, D is the concentration, in mg per mL, of sodium carbonate in the stock solution used to prepare the *Assay preparation*, on the basis of the labeled quantity in each mL, and the extent of dilution, $R_{U,591}$ and $R_{U,671}$ are the emittance readings obtained from the *Assay preparation* at the wavelengths indicated by the subscripts, and $R_{S,671}$ and $R_{S,591}$ are the emittance readings obtained from the *Standard preparation* at the wavelengths indicated by the subscripts.

Citric Acid Oral Solution, Potassium Citrate and—*see* Potassium Citrate and Citric Acid Oral Solution

Citric Acid Oral Solution, Sodium Citrate and—*see* Sodium Citrate and Citric Acid Oral Solution

Citric Acid Effervescent Tablets for Oral Solution, Potassium and Sodium Bicarbonates and—*see* Potassium and Sodium Bicarbonates and Citric Acid Effervescent Tablets for Oral Solution

Clavulanate Potassium

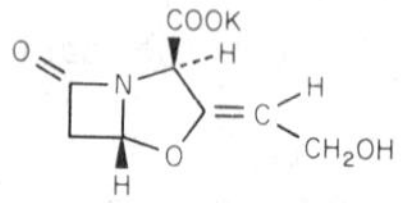

$C_8H_8KNO_5$ 237.25

4-Oxa-1-azabicyclo[3.2.0]heptane-2-carboxylic acid, 3-(2-hydroxyethylidene)-7-oxo-, monopotassium salt, [2*R*-(2α,3*Z*,5α)]-.

Potassium (*Z*)-(2*R*,5*R*)-3-(2-hydroxyethylidene)-7-oxo-4-oxa-1-azabicyclo[3.2.0]heptane-2-carboxylate [*61177-45-5*].

» Clavulanate Potassium contains the equivalent of not less than 75.5 percent and not more than 92.0 percent of clavulanic acid ($C_8H_9NO_5$), calculated on the anhydrous basis.

Packaging and storage—Preserve in tight containers.

Reference standards—*USP Clavulanate Lithium Reference Standard*—Do not dry before using. *USP Clavam-2-Carboxylate Potassium Reference Standard*—Keep container tightly closed and protected from light. Do not dry before using.

Identification—

A: The chromatogram of the *Assay preparation* obtained as directed in the *Assay* exhibits a major peak for clavulanic acid, the retention time of which corresponds to that exhibited in the chromatogram of the *Standard preparation* obtained as directed in the *Assay*.

B: A solution of it responds to the tests for *Potassium* ⟨191⟩.

pH ⟨791⟩: between 5.5 and 8.0, in a solution (1 in 100).

Water, *Method I* ⟨921⟩: not more than 1.5%.

Clavam-2-carboxylate potassium—

Mobile phase—Prepare 0.1 *M* monobasic sodium phosphate, adjust with phosphoric acid to a pH of 4.0 ± 0.1, and filter through a membrane filter of 0.5-µm or finer porosity. Make adjustments if necessary (see *System Suitability* under *Chromatography* ⟨621⟩).

Standard preparation—Dissolve a suitable quantity of USP Clavam-2-Carboxylate Potassium RS, accurately weighed, in water to obtain a solution having a known concentration of about 0.03 mg per mL.

Test preparation—Transfer about 100 mg of Clavulanate Potassium, accurately weighed, to a 10-mL volumetric flask, dissolve in water, dilute with water to volume, and mix.

Resolution solution—Dissolve a suitable quantity of Clavulanate Potassium in *Standard preparation* to obtain a solution containing about 1 mg of clavulanate potassium and 0.03 mg of clavam-2-carboxylate potassium per mL.

Chromatographic system (see *Chromatography* ⟨621⟩)—The liquid chromatograph is equipped with a 210-nm detector and a 4-mm × 30-cm column that contains 3- to 10-µm packing L1. The flow rate is about 0.5 mL per minute. Chromatograph the *Standard preparation*, and record the peak responses as directed under *Procedure:* the column efficiency determined from the analyte peak is not less than 4000 theoretical plates, the tailing factor for the analyte peak is not more than 1.5, and the relative standard deviation for replicate injections is not more than 2.0%. Chromatograph the *Resolution solution*, and record the peak responses as directed under *Procedure:* the resolution, *R*, between the clavam-2-carboxylic acid and clavulanic acid peaks is not less than 1.0.

Procedure—Separately inject equal volumes (about 20 µL) of the *Standard preparation* and the *Test preparation* into the chromatograph, record the chromatograms, and measure the responses for the major peaks. The relative retention times are about 0.7 for clavam-2-carboxylic acid and 1.0 for clavulanic acid. Calculate the percentage of clavam-2-carboxylate potassium in the specimen taken by the formula:

$$(1000CP/W)(r_U/r_S),$$

in which C is the concentration, in mg per mL, of USP Clavam-2-Carboxylate Potassium RS in the *Standard preparation*, P is the percentage of clavam-2-carboxylate potassium in the USP Clavam-2-Carboxylate Potassium RS, W is the quantity, in mg, of Clavulanate Potassium taken to prepare the *Test preparation*, and r_U and r_S are the clavam-2-carboxylic acid peak responses obtained from the *Test preparation* and the *Standard preparation*, respectively: not more than 0.01% is found.

Assay—

pH 4.4 sodium phosphate buffer—Dissolve 7.8 g of monobasic sodium phosphate in 900 mL of water, adjust with phosphoric acid or 10 *N* sodium hydroxide to a pH of 4.4 ± 0.1, dilute with water to make 1000 mL, and mix.

Mobile phase—Prepare a suitable mixture of *pH 4.4 sodium phosphate buffer* and methanol (95:5), and filter through a membrane filter of 0.5-µm or finer porosity. Make adjustments if necessary (see *System Suitability* under *Chromatography* ⟨621⟩).

Standard preparation—Dissolve an accurately weighed quantity of USP Clavulanate Lithium RS in water to obtain a solution having a known concentration of about 0.25 mg per mL.

Assay preparation—Transfer about 50 mg of Clavulanate Potassium, accurately weighed, to a 200-mL volumetric flask, dissolve in water, dilute with water to volume, and mix.

Resolution solution—Dissolve a suitable quantity of amoxicillin in *Standard preparation* to obtain a solution containing about 0.5 mg of amoxicillin and 0.25 mg of clavulanate lithium per mL.

Chromatographic system (see *Chromatography* ⟨621⟩)—The liquid chromatograph is equipped with a 220-nm detector and a 4-mm × 30-cm column that contains 3- to 10-µm packing L1. The flow rate is about 2 mL per minute. Chromatograph the *Standard preparation*, and record the peak responses as directed under *Procedure:* the column efficiency determined from the analyte peak is not less than 550 theoretical plates, the tailing factor for the analyte peak is not more than 1.5, and the relative standard deviation for replicate injections is not more than 2.0%. Chromatograph the *Resolution solution*, and record the peak responses as directed under *Procedure:* the relative retention times are about 0.5 for clavulanic acid and 1.0 for amoxicillin, and the resolution, R, between the amoxicillin and clavulanic acid peaks is not less than 3.5.

Procedure—Separately inject equal volumes (about 20 µL) of the *Standard preparation* and the *Assay preparation* into the chromatograph, record the chromatograms, and measure the responses for the major peaks. Calculate the quantity, in µg, of clavulanic acid ($C_8H_9NO_5$) in each mg of the Clavulanate Potassium taken by the formula:

$$(200CP)(r_U/r_S),$$

in which C is the concentration, in mg per mL, of USP Clavulanate Lithium RS in the *Standard preparation*, P is the designated potency, in µg of clavulanic acid per mg, of the USP Clavulanate Lithium RS, and r_U and r_S are the peak responses obtained from the *Assay preparation* and the *Standard preparation*, respectively.

Clavulanate Potassium, Sterile Ticarcillin Disodium and—*see* Sterile Ticarcillin Disodium and Clavulanate Potassium

Clavulanate Potassium for Oral Suspension, Amoxicillin and—*see* Amoxicillin and Clavulanate Potassium for Oral Suspension

Clavulanate Potassium Tablets, Amoxicillin and—*see* Amoxicillin and Clavulanate Potassium Tablets

Sterile Clavulanate Potassium

» Sterile Clavulanate Potassium is clavulanate potassium suitable for parenteral use. It contains the equivalent of not less than 75.5 percent and not more than 92.0 percent of clavulanic acid ($C_8H_9NO_5$), calculated on the anhydrous basis.

Packaging and storage—Preserve in *Containers for Sterile Solids* as described under *Injections* ⟨1⟩.

Reference standards—*USP Clavulanate Lithium Reference Standard*—Do not dry before using. *USP Clavam-2-Carboxylate Potassium Reference Standard*—Keep container tightly closed and protected from light. Do not dry before using.

Pyrogen—It meets the requirements of the *Pyrogen Test* ⟨151⟩, the test dose being 1.0 mL per kg of a solution in Sterile Water for Injection containing 10 mg of it per mL.

Sterility—It meets the requirements under *Sterility Tests* ⟨71⟩, when tested as directed in the section, *Test Procedures Using Membrane Filtration.*

Other requirements—It responds to the *Identification tests*, and meets the requirements for *pH, Water, Clavam-2-carboxylate potassium*, and *Assay* under *Clavulanate Potassium.*

Cleansing Emulsion, Hexachlorophene—*see* Hexachlorophene Cleansing Emulsion

Clemastine Fumarate

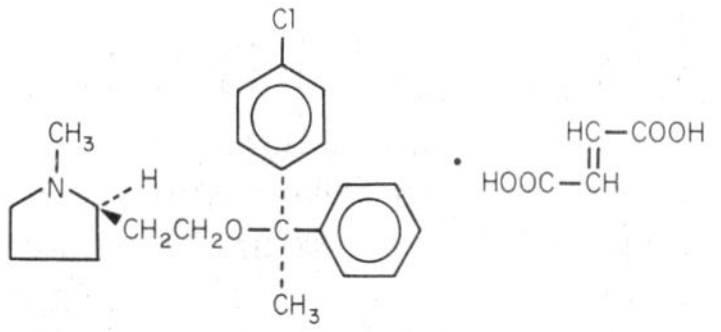

$C_{21}H_{26}ClNO.C_4H_4O_4$ 459.97

Pyrrolidine, 2-[2-[1-(4-chlorophenyl)-1-phenylethoxy]ethyl]-1-methyl-, [*R*-(*R**,*R**)]-, (*E*)-2-butenedioate (1:1).

(+)-(2*R*)-2-[2-[[(*R*)-*p*-Chloro-α-methyl-α-phenylbenzyl]-oxy]ethyl]-1-methylpyrrolidine fumarate (1:1) [*14976-57-9*].

» Clemastine Fumarate contains not less than 98.0 percent and not more than 102.0 percent of $C_{21}H_{26}ClNO.C_4H_4O_4$, calculated on the dried basis.

Packaging and storage—Preserve in tight, light-resistant containers, at a temperature not exceeding 25°.

Reference standard—*USP Clemastine Fumarate Reference Standard*—Dry at 105° to constant weight before using.

Clarity and color of solution—Dissolve 100 mg of Clemastine Fumarate in 10.0 mL of methanol, and mix to obtain the *Test solution.* Prepare a *Comparison solution* by mixing 2.5 mL of 0.00002 *M* sodium chloride, 2.5 mL of water, 5.0 mL of 2.5 *N* nitric acid, and 1.0 mL of 0.1 *N* silver nitrate, and use this solution within 5 minutes. Prepare a *Color matching fluid* by mixing 1 volume of *Matching Fluid C* (see *Color and Achromicity* ⟨631⟩) with 3 volumes of water. Transfer the *Test solution*, the *Comparison solution*, and 10 mL of *Color matching fluid* to separate test tubes having the same nominal diameter (about 12 mm). View the *Test solution* and the *Comparison solution* horizontally against a dull black background: the *Test solution* is clear or not more opalescent than the *Comparison solution.* View the *Test solution* and *Color matching fluid* horizontally against a dull white background: the *Test solution* is colorless or not more intensely colored than *Color matching fluid.*

Identification—

A: The infrared absorption spectrum of a mineral oil dispersion of it, previously dried, exhibits maxima only at the same wavelengths as that of a similar preparation of USP Clemastine Fumarate RS.

B: Prepare a *Test preparation* by dissolving 40 mg of Clemastine Fumarate in 2.0 mL of dilute alcohol (8 in 10) with slight warming. Similarly prepare a *Standard preparation* by dissolving 50 mg of fumaric acid in 10.0 mL of dilute alcohol (8 in 10). Separately apply 5-μL portions of the *Test preparation* and the *Standard preparation* on a suitable thin-layer chromatographic plate (see *Chromatography* ⟨621⟩), coated with a 0.25-mm layer of chromatographic silica gel mixture, and dry the spots with the aid of a current of air. Develop the chromatogram in a solvent system consisting of a mixture of diisopropyl ether, formic acid, and water (70:25:5) until the solvent front has moved about three-fourths of the length of the plate. Remove the plate from the developing chamber, mark the solvent front, dry at 100° for 30 minutes, cool, and spray the plate with 0.1 *M* potassium permanganate. Dry briefly with the aid of a current of warm air, and examine the chromatogram: the principal spot obtained from the *Test preparation* corresponds in R_f value, color, and intensity to that obtained from the *Standard preparation.*

Specific rotation ⟨781⟩: between +15.0° and +18.0°, calculated on the dried basis, determined at a temperature of 20 ± 0.5° in a solution in methanol containing 100 mg in each 10 mL.

pH ⟨791⟩: between 3.2 and 4.2, in a suspension (1 in 10).

Loss on drying ⟨731⟩—Dry it at 105° to constant weight: it loses not more than 0.5% of its weight.

Heavy metals, *Method II* ⟨231⟩: 0.002%.

Chromatographic purity—

Spray reagent—Dissolve 850 mg of bismuth subnitrate in a mixture of 10 mL of glacial acetic acid and 40 mL of water, and mix (*Solution A*). Dissolve 8 g of potassium iodide in 20 mL of water (*Solution B*). Mix 5.0 of *Solution A*, 5.0 mL of *Solution B*, and 20 mL of glacial acetic acid in a 100-mL volumetric flask, dilute with water to volume, and mix.

Standard preparation—Dissolve a suitable quantity of USP Clemastine Fumarate RS in a mixture of chloroform and methanol (1:1) to obtain a solution having a known concentration of 20 mg per mL. Dilute portions of this solution quantitatively with the mixture of chloroform and methanol (1:1) to prepare 5 *Comparison solutions* having known concentrations of 0.10, 0.08, 0.06, 0.04, and 0.02 mg per mL, respectively (0.5%, 0.4%, 0.3%, 0.2%, and 0.1% of the *Standard preparation,* respectively).

Test preparation—Dissolve 100 mg of Clemastine Fumarate in 5.0 mL of a mixture of chloroform and methanol (1:1), and mix.

Procedure—On a suitable thin-layer chromatographic plate (see *Chromatography* ⟨621⟩), coated with a 0.25-mm layer of chromatographic silica gel mixture, separately apply 5-μL portions of the *Standard preparation*, each of the 5 *Comparison solutions*, and the *Test preparation.* Allow the spots to dry, and develop the chromatogram in a solvent system consisting of a mixture of chloroform, methanol, and ammonium hydroxide (90:10:1) until the solvent front has moved about three-fourths of the length of the plate. Remove the plate from the developing chamber, mark the solvent front, and dry the plate at room temperature with the aid of a current of air. Locate the spots on the plate by spraying first with *Spray reagent*, then with 3 percent hydrogen peroxide: the principal spot obtained from the *Test preparation* corresponds in R_f value, color, and intensity to that obtained from the *Standard preparation;* the sum of the intensities of any secondary spots, if present in the chromatogram from the *Test preparation*, corresponds to not more than 1.0%, and the intensities of any secondary spots do not exceed 0.5% of that of the principal spot in the chromatogram from the *Standard preparation* on the basis of comparison with spots obtained from the *Comparison solutions.*

Assay—Transfer about 350 mg of Clemastine Fumarate, accurately weighed, to a small conical flask, and dissolve in 60 mL of glacial acetic acid. Titrate with 0.1 *N* perchloric acid VS, determining the end-point potentiometrically. Perform a blank determination, and make any necessary correction. Each mL of 0.1 *N* perchloric acid is equivalent to 46.00 mg of $C_{21}H_{26}ClNO \cdot C_4H_4O_4$.

Clemastine Fumarate Tablets

» Clemastine Fumarate Tablets contain not less than 90.0 percent and not more than 110.0 percent of the labeled amount of $C_{21}H_{26}ClNO \cdot C_4H_4O_4$.

Packaging and storage—Preserve in tight, light-resistant containers, at a temperature not exceeding 25°.

Reference standard—*USP Clemastine Fumarate Reference Standard*—Dry at 105° to constant weight before using.

Identification—

Spray reagent—Prepare as directed in the test for *Chromatographic purity* under *Clemastine Fumarate.*

Standard preparation—Prepare a solution in a mixture of chloroform and methanol (1:1) having a concentration of about 2.5 mg of USP Clemastine Fumarate RS per mL.

Test preparation—Place a portion of powdered Tablets, equivalent to about 2.5 mg of clemastine fumarate, in a glass-stoppered flask. Add 10 mL of a mixture of chloroform and methanol (1:1), and shake for 20 minutes. Filter, wash the residue with two 5-mL portions of the mixture of chloroform and methanol (1:1), and evaporate the combined filtrate and washings to dryness under vacuum. Dissolve the residue so obtained in 1 mL of the mixture of chloroform and methanol (1:1), and mix.

Procedure—Proceed as directed for *Procedure* in the test for *Chromatographic purity* under *Clemastine Fumarate*, applying 5-μL portions of the *Standard preparation* and the *Test preparation* on the thin-layer chromatographic plate: the R_f value of the principal spot obtained from the *Test preparation* corresponds to that obtained from the *Standard preparation.*

Dissolution ⟨711⟩—

Medium: pH 4.0 citrate buffer, prepared by dissolving 20.0 g of citric acid monohydrate in about 1000 mL of water, adding 22.0 mL of sodium hydroxide solution (3 in 10) and 8.8 mL of hydrochloric acid, and mixing with water to make 2000 mL of solution; 500 mL.

Apparatus 2: 50 rpm.

Time: 30 minutes.

Procedure—Centrifuge 60 mL of the solution under test for 20 minutes at 4000 rpm, and transfer 50.0 mL of the supernatant solution to a 125-mL separator. To a second 125-mL separator transfer 50.0 mL of a Standard preparation that is prepared by dissolving an accurately weighed quantity of USP Clemastine Fumarate RS in *Dissolution Medium* and diluting quantitatively and stepwise with *Dissolution Medium* to yield a solution having a known concentration comparable with that of the solution under test. To a third 125-mL separator transfer 50.0 mL of *Dissolution Medium* to provide a blank. Treat each of the solutions in the three separators as follows: Add 10 mL of methyl orange solution (2 in 10,000), mix, add 20.0 mL of chloroform, shake simultaneously by mechanical means for 10 minutes, remove the chloroform layer, and centrifuge the chloroform layer for 10 minutes at 4000 rpm. Determine the amount of $C_{21}H_{26}ClNO \cdot C_4H_4O_4$ dissolved from absorbances at the wavelength of maximum absorbance at about 420 nm of the chloroform solutions obtained from the solution under test and the Standard preparation, using the chloroform solution obtained from the blank to set the instrument.

Tolerances—Not less than 75% (*Q*) of the labeled amount of $C_{21}H_{26}ClNO \cdot C_4H_4O_4$ is dissolved in 30 minutes.

Uniformity of dosage units ⟨905⟩: meet the requirements.

Procedure for content uniformity—

Dye solution—Dissolve 100 mg of bromocresol purple in 1000 mL of 0.33 *N* acetic acid, and mix.

Acetous methanol—Dilute 100 mL of methanol with sufficient 0.33 *N* acetic acid to prepare 1000 mL of solution, and mix.

Standard preparation—Transfer about 27 mg of USP Clemastine Fumarate RS, accurately weighed, to a 100-mL volumetric flask, dissolve in 10 mL of methanol, dilute with 0.33 *N* acetic acid to volume, and mix. Transfer 10.0 mL of this solution to a 100-mL volumetric flask, dilute with *Acetous methanol* to volume, and mix.

Test preparation—Mix 1 finely powdered Tablet with an accurately measured volume of *Acetous methanol*, sufficient to obtain a solution having a concentration of about 27 µg of clemastine fumarate per mL. Shake for 30 minutes, and filter, discarding the first few mL of the filtrate.

Procedure—Transfer 15.0 mL each of the *Standard preparation*, the *Test preparation*, and *Acetous methanol* to provide the blank to individual 125-mL separators. Add 25 mL of *Dye solution* and 50.0 mL of chloroform to each, and shake by mechanical means for 15 minutes. Allow the layers to separate, and filter the chloroform layers. Concomitantly determine the absorbances of the filtered solutions obtained from the *Test preparation* and the *Standard preparation* at the wavelength of maximum absorbance at about 406 nm, using the blank to set the instrument. Calculate the quantity, in mg, of $C_{21}H_{26}ClNO.C_4H_4O_4$ in the Tablet by the formula:

$$(TC/D)(A_U/A_S),$$

in which *T* is the labeled quantity, in mg, of clemastine fumarate in the Tablet, *C* is the concentration, in µg per mL, of USP Clemastine Fumarate RS in the *Standard preparation*, *D* is the concentration, in µg per mL, of clemastine fumarate in the *Test preparation*, based on the labeled quantity per Tablet and the extent of dilution, and A_U and A_S are the absorbances of the solutions from the *Test preparation* and the *Standard preparation*, respectively.

Assay—

pH 7 phosphate buffer—To flask *A* transfer 9.47 g of anhydrous dibasic sodium phosphate to a 1000-mL volumetric flask, dilute with water to volume, and mix. To flask *B* transfer 9.08 g of monobasic potassium phosphate to a 1000-mL volumetric flask, dilute with water to volume, and mix. Mix 612 mL of *A* with 388 mL of *B*.

Dilute phosphate buffer—Prepare a mixture of 1 volume of *pH 7 phosphate buffer* and 3 volumes of water.

Mobile phase—Prepare a suitable and degassed solution of methanol and *Dilute phosphate buffer* (83:17).

Standard preparation—Dissolve an accurately weighed quantity of USP Clemastine Fumarate RS in a mixture of methanol and water (1:1) to obtain a solution having a known concentration of about 0.14 mg per mL.

Assay preparation—Weigh and finely powder not less than 20 Clemastine Fumarate Tablets. Transfer an accurately weighed quantity of the powder, equivalent to about 14 mg of clemastine fumarate, to a 200-mL conical flask. Pipet 100 mL of a mixture of methanol and water (1:1) into the flask, shake for 30 minutes, centrifuge, and filter the supernatant layer.

Chromatographic system (see *Chromatography* ⟨621⟩)—The liquid chromatograph is equipped with a 220-nm detector and a 4.6-mm × 25-cm column that contains packing L7. The flow rate is about 4 mL per minute. Chromatograph five replicate injections of the *Standard preparation*, and record the peak responses as directed under *Procedure:* the relative standard deviation is not more than 1.5%.

Procedure—Separately inject equal volumes (about 100 µL) of the *Standard preparation* and the *Assay preparation* into the chromatograph, record the chromatograms, and measure the responses for the major peaks. Calculate the quantity, in mg, of $C_{21}H_{26}ClNO.C_4H_4O_4$ in the portion of Tablets taken by the formula:

$$100C(r_U/r_S),$$

in which *C* is the concentration, in mg per mL, of USP Clemastine Fumarate RS in the *Standard preparation*, and r_U and r_S are the peak responses of clemastine fumarate obtained from the *Assay preparation* and the *Standard preparation*, respectively.

Clidinium Bromide

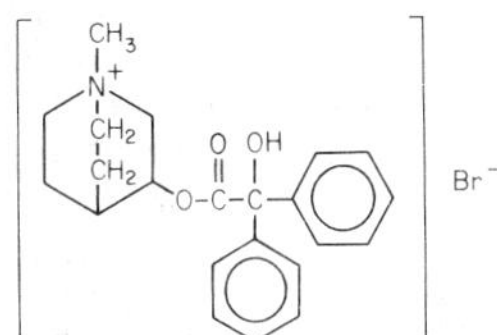

$C_{22}H_{26}BrNO_3$ 432.36

1-Azoniabicyclo[2.2.2]octane, 3-[(hydroxydiphenylacetyl)oxy]-1-methyl-, bromide.

3-Hydroxy-1-methylquinuclidinium bromide benzilate [*3485-62-9*].

» Clidinium Bromide contains not less than 99.0 percent and not more than 100.5 percent of $C_{22}H_{26}BrNO_3$, calculated on the dried basis.

Packaging and storage—Preserve in tight, light-resistant containers.

Reference standards—*USP Clidinium Bromide Reference Standard*—Dry at 105° for 3 hours before using. *USP 3-Quinuclidinyl Benzilate Reference Standard*—Dry over silica gel for 4 hours before using. Keep container tightly closed and protected from light. [*Caution—Avoid contact; work in a well-ventilated hood.*] *USP 3-Hydroxy-1-methylquinuclidinium Bromide Reference Standard*—Dry over silica gel for 4 hours before using. Keep container tightly closed and protected from light.

Identification—

A: The infrared absorption spectrum of a potassium bromide dispersion of it, previously dried, exhibits maxima only at the same wavelengths as that of a similar preparation of USP Clidinium Bromide RS.

B: Dissolve about 250 mg in 5 mL of water in a test tube, cool in an ice bath, add 5 mL of trinitrophenol TS, and scratch the inner surface of the tube with a glass rod to induce crystallization. Collect the precipitate on a filter, wash well with cold water, and dry at 105° for 1 hour: the picrate so obtained melts between 184° and 189°, when tested by the method for *Class I* ⟨741⟩. [*Caution—Picrates may explode.*]

C: To a solution of 100 mg in 2 mL of water add a few drops of 2 *N* nitric acid and 1 mL of silver nitrate TS: a yellowish white precipitate is formed.

Loss on drying ⟨731⟩—Dry it at 105° for 3 hours: it loses not more than 0.5% of its weight.

Residue on ignition ⟨281⟩: not more than 0.1%.

Heavy metals ⟨231⟩—Dissolve 1.0 g in 25 mL of water: the limit is 0.002%.

Related compounds—

Solvent mixture—Mix 70 volumes of acetone, 20 volumes of methanol, 5 volumes of water, and 5 volumes of hydrochloric acid.

Chromatographic plates—Use suitable thin-layer chromatographic plates (see *Chromatography* ⟨621⟩) coated with a 0.25-mm layer of chromatographic silica gel. Pre-develop the plates by placing in a chromatographic chamber saturated with *Solvent mixture*, and allow the *Solvent mixture* to move about 15 cm. Remove the plates from the chamber, dry at 105° for 15 minutes, and cool.

Spray reagent—Dissolve 850 mg of bismuth subnitrate in a mixture of 10 mL of glacial acetic acid and 40 mL of water. In a separate container, dissolve 20 g of potassium iodide in 50 mL of water. Mix the two solutions, and dilute with dilute sulfuric acid (1 in 10) to 500 mL. Add 7.5 ± 2.5 g of iodine, and mix until solution is complete.

Reference solution A—Dissolve 3.0 mg of USP 3-Quinuclidinyl Benzilate RS in 100 mL of 0.1 *N* methanolic hydrochloric acid, and mix.

Reference solution B—Dissolve 100 mg of USP Clidinium Bromide RS in 1.0 mL of 0.1 *N* methanolic hydrochloric acid, and add 20 µL of a solution of 25.0 mg of USP 3-Hydroxy-1-methylquinuclidinium Bromide RS in 1.0 mL of 0.1 *N* methanolic hydrochloric acid.

Standard preparation—Dissolve 100 mg of USP Clidinium Bromide RS in 1.0 mL of 0.1 *N* methanolic hydrochloric acid.

Test preparation—Dissolve 100 mg of Clidinium Bromide in 1.0 mL of 0.1 *N* methanolic hydrochloric acid.

Procedure A (*3-quinuclidinyl benzilate*)—Apply 20 µL each of the *Standard preparation, Reference solution A*, and the *Test preparation* to separate points 2 cm from the bottom of one of the *Chromatographic plates.* Place the plate in an unsaturated chromatographic chamber containing freshly prepared *Solvent mixture*, and allow the solvent front to move 10 cm. Remove the plate, dry at 105° for 10 minutes, cool, and spray with potassium iodoplatinate TS: the R_f value of the principal spot from the *Test preparation* corresponds to that from the *Standard preparation*, and the *Test preparation* shows no spot at an R_f value (about 0.8) corresponding to that of 3-quinuclidinyl benzilate.

Procedure B (*3-hydroxy-1-methylquinuclidinium bromide*)—Apply 20 µL each of *Reference solution B* and the *Test preparation* to separate points 2 cm from the bottom of a second *Chromatographic plate.* Place the plate in an unsaturated chromatographic chamber containing freshly prepared *Solvent mixture*, and allow the solvent front to move 15 cm. Remove the plate, dry at 105° for 10 minutes, cool, and spray with *Spray reagent:* any spot from the *Test preparation* at an R_f value (about 0.4) corresponding to that of the minor spot from *Reference solution B* is not greater in size or intensity than that minor spot (0.5%).

Assay—Dissolve about 1.2 g of Clidinium Bromide, accurately weighed, in 80 mL of glacial acetic acid, warming if necessary to effect solution. Cool, add 15 mL of mercuric acetate TS, and titrate with 0.1 *N* perchloric acid in dioxane VS, determining the end-point potentiometrically. Perform a blank determination, and make any necessary correction. Each mL of 0.1 *N* perchloric acid is equivalent to 43.24 mg of $C_{22}H_{26}BrNO_3$.

Clidinium Bromide Capsules

» Clidinium Bromide Capsules contain not less than 90.0 percent and not more than 110.0 percent of the labeled amount of $C_{22}H_{26}BrNO_3$.

Packaging and storage—Preserve in tight, light-resistant containers.

Reference standards—*USP Clidinium Bromide Reference Standard*—Dry at 105° for 3 hours before using. *USP 3-Quinuclidinyl Benzilate Reference Standard*—Dry over silica gel for 4 hours before using. Keep container tightly closed and protected from light. [*Caution—Avoid contact; work in a well-ventilated hood.*] *USP 3-Hydroxy-1-methylquinuclidinium Bromide Reference Standard*—Dry over silica gel for 4 hours before using. Keep container tightly closed and protected from light.

Identification—To the contents of 1 Capsule add 0.5 mL of sulfuric acid, and shake: a yellowish solution that decolorizes and finally turns a light violet color is produced.

Dissolution ⟨711⟩—

Medium: 0.1 *N* hydrochloric acid; 900 mL.

Apparatus 1: 100 rpm.

Time: 15 minutes.

Procedure—Determine the amount of $C_{22}H_{26}BrNO_3$ dissolved, employing the procedure set forth in the *Assay*, making any necessary modifications.

Tolerances—Not less than 80% (*Q*) of the labeled amount of $C_{22}H_{26}BrNO_3$ is dissolved in 15 minutes.

Uniformity of dosage units ⟨905⟩: meet the requirements.

Related compounds—

A: *3-Quinuclidinyl benzilate*—

REFERENCE SOLUTION—Dissolve 3.0 mg of USP 3-Quinuclidinyl Benzilate RS in 100 mL of methanol, and mix.

SPECIMEN MIXTURE—Empty a number of Capsules, equivalent to 15 mg of clidinium bromide, into a 100-mL beaker, add 3 mL of 1 *N* hydrochloric acid, and swirl to dissolve. Add 4 g of chromatographic siliceous earth, and mix with a spatula.

TEST PREPARATION—Place a glass wool plug in the bottom of a 2.5-cm × 35 ± 5-cm glass chromatographic tube, and add a mixture of 2 g of chromatographic siliceous earth triturated with 1 mL of 1 *N* hydrochloric acid. Pack lightly with a glass tamping rod, add the *Specimen mixture*, and dry-wash the beaker with an additional 0.5 to 1.0 g of chromatographic siliceous earth, adding the washing to the top of the column. Pack lightly with the tamping rod, and overlay the column with glass wool. Insert the lower exit tube of the column in a 125-mL separator, and elute the column with 100 mL of chloroform previously distilled over 1 *N* sulfuric acid and saturated with water. Extract the chloroform eluate with 20 mL of freshly prepared ascorbic acid solution (1 in 20), reserving the extract. Extract the eluate with a second, 15-mL portion of the ascorbic acid solution, combine the extracts in the separator, and discard the chloroform layer. Neutralize the acid extracts by the addition of sufficient sodium bicarbonate until the solution is slightly alkaline to pH paper. Extract the slightly alkaline solution with two 25-mL portions of chloroform, combine the chloroform extracts, and filter through dry, fluted filter paper into a 100-mL beaker. Evaporate the chloroform with the aid of a stream of nitrogen to dryness, and transfer the residue to a glass-stoppered, graduated, 1.0-mL micro-tube, using methanol to facilitate the transfer, and dilute with methanol to volume.

PROCEDURE—Apply 15 µL of the *Reference solution* and 100 µL of the *Test preparation* to separate points on a suitable thin-layer chromatographic plate (see *Chromatography* ⟨621⟩) coated with a 0.25-mm layer of chromatographic silica gel. Place the plate in a paper-lined, methanol-saturated chromatographic chamber, and develop the chromatogram with methanol until the solvent front has moved about 15 cm. Remove the plate, air-dry, spray with potassium iodoplatinate TS, and allow the spots to develop for 10 minutes: any spot from the *Test preparation* occurring at an R_f value of about 0.3 is not greater in size or intensity than the corresponding spot obtained from the *Reference solution* (0.03%).

B: *3-Hydroxy-1-methylquinuclidinium bromide*—

SOLVENT MIXTURE, CHROMATOGRAPHIC PLATE, and SPRAY REAGENT—Prepare as directed in the test for *Related compounds* under *Clidinium Bromide.*

REFERENCE SOLUTION—Dissolve 50 mg of USP Clidinium Bromide RS in 1.0 mL of 0.1 *N* methanolic hydrochloric acid, and add 40 µL of a solution of 12.5 mg of USP 3-Hydroxy-1-methylquinuclidinium Bromide RS in 1.0 mL of 0.1 *N* methanolic hydrochloric acid.

TEST PREPARATION—Empty a number of Capsules, equivalent to 25 mg of clidinium bromide, into a glass-stoppered centrifuge tube, and add 5 mL of a mixture of dehydrated alcohol, cyclohexane, and hydrochloric acid (125:125:2.1). Heat the tube gently, with shaking, to 50°, centrifuge, and decant the clear supernatant liquid to a second tube. Repeat the addition of the solvent mixture twice, heating, centrifuging, and decanting as before, and combine the three extracts in a single tube. Evaporate the combined extracts with the aid of gentle heat and a stream of nitrogen to dryness. Dissolve the residue in 0.5 mL of 0.1 *N* methanolic hydrochloric acid.

PROCEDURE—Apply 40 µL each of the *Reference solution* and the *Test preparation* to separate points 2 cm from the bottom of the *Chromatographic plate.* Place the plate in an unsaturated chromatographic chamber containing freshly prepared *Solvent mixture*, and allow the solvent front to move for about 10 cm. Remove the plate, dry at 105° for 10 minutes, cool, and spray with *Spray reagent:* any spot from the *Test preparation* occurring at an R_f value of about 0.4 is not greater in size or intensity than the corresponding spot obtained from the *Reference solution* (1%).

Assay—

Buffer solution—Dissolve 26.8 g of dibasic sodium phosphate in water, and dilute with water to 200 mL. Determine the pH of the solution, and adjust, if necessary, to a pH of 8.8 by the addition of small amounts of phosphoric acid or 1 *N* sodium hydroxide.

Thymol blue solution—Triturate 100 mg of thymol blue with 2.15 mL of 0.1 *N* sodium hydroxide. Dilute with water to 100 mL, and adjust to a pH of 8.8 by the addition of 0.1 *N* hydrochloric acid or 0.1 *N* sodium hydroxide, as necessary.

Standard preparation—Dissolve an accurately weighed quantity of USP Clidinium Bromide RS in water, and dilute quantitatively and stepwise with water to obtain a solution having a known concentration of about 100 µg per mL.

Assay preparation—Weigh the contents of not less than 20 Clidinium Bromide Capsules, and determine the average weight per capsule. Mix the combined contents to obtain a homogeneous specimen. Transfer an accurately weighed portion of the powder, equivalent to about 10 mg of clidinium bromide, to a 100-mL volumetric flask. Add 50 mL of water, shake for 5 minutes, dilute with water to volume, mix, and filter, discarding the first 20 mL of the filtrate and retaining the rest.

Procedure—Into separate 125-mL separators pipet 10 mL each of the *Standard preparation*, the *Assay preparation*, and water to serve as the blank. To each separator add 10 mL of *Buffer solution* and 5 mL of *Thymol blue solution*, and mix. Extract the contents of each separator with four 20-mL portions of chloroform, shaking each time for 1 minute and allowing the phases to separate for not less than 2 minutes. Combine the extracts from each separator into separate 100-mL volumetric flasks, dilute with chloroform to volume, and mix. Concomitantly determine the absorbances of the extracts from the *Standard preparation* and the *Assay preparation* in 1-cm cells at 410 nm, with a suitable spectrophotometer, using the blank to set the instrument. Calculate the quantity, in mg, of $C_{22}H_{26}BrNO_3$ in the portion of Capsule contents taken by the formula:

$$0.1C(A_U/A_S),$$

in which C is the concentration, in µg per mL, of USP Clidinium Bromide RS in the *Standard preparation*, and A_U and A_S are the absorbances of the extracts from the *Assay preparation* and the *Standard preparation*, respectively.

Clindamycin Hydrochloride

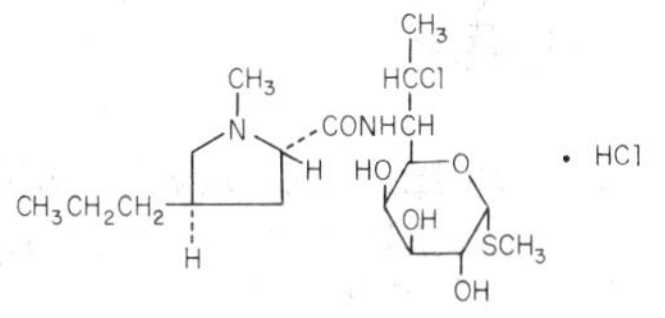

$C_{18}H_{33}ClN_2O_5S.HCl$ 461.44

L-*threo*-α-D-*galacto*-Octopyranoside, methyl 7-chloro-6,7,8-trideoxy-6-[[(1-methyl-4-propyl-2-pyrrolidinyl)-carbonyl]-amino]-1-thio-, (2*S*-*trans*)-, monohydrochloride.

Methyl 7-chloro-6,7,8-trideoxy-6-(1-methyl-*trans*-4-propyl-L-2-pyrrolidinecarboxamido)-1-thio-L-*threo*-α-D-*galacto*-octopyranoside monohydrochloride [*21462-39-5*].

Monohydrate 479.46 [*58207-19-5*].

» Clindamycin Hydrochloride is the hydrated hydrochloride salt of clindamycin, a substance produced by the chlorination of lincomycin. It has a potency equivalent to not less than 800 µg of clindamycin ($C_{18}H_{33}ClN_2O_5S$) per mg.

Packaging and storage—Preserve in tight containers.

Reference standard—*USP Clindamycin Hydrochloride Reference Standard*—Do not dry before using.

Identification—The infrared absorption spectrum of a mineral oil dispersion of it exhibits maxima only at the same wavelengths as that of a similar preparation of USP Clindamycin Hydrochloride RS.

Crystallinity ⟨695⟩: meets the requirements.

pH ⟨791⟩: between 3.0 and 5.5, in a solution containing 100 mg per mL.

Water, *Method I* ⟨921⟩: between 3.0% and 6.0%.

Assay—

Mobile phase—Add 2 g of *dl*-10-camphorsulfonic acid, 1 g of ammonium acetate, and 1 mL of glacial acetic acid to 200 mL of water in a 500-mL volumetric flask, and mix to dissolve. Dilute with methanol to volume, and mix. Adjust if necessary, with hydrochloric acid or sodium hydroxide solution (1 in 2) to a pH of 6.0 ± 0.1. Make adjustments, if necessary (see *System Suitability* under *Chromatography* ⟨621⟩).

Internal standard solution—Add 0.5 mL of phenylethyl alcohol to a 100-mL volumetric flask, dilute with *Mobile phase* to volume, and mix.

Standard preparation—Transfer about 90 mg of USP Clindamycin Hydrochloride RS, accurately weighed, to a suitable container. Add 5.0 mL of *Internal standard solution*, and swirl to dissolve.

Assay preparation—Transfer about 90 mg of Clindamycin Hydrochloride, accurately weighed, to a suitable container. Add 5.0 mL of *Internal standard solution*, and swirl to dissolve.

Chromatographic system (see *Chromatography* ⟨621⟩)—The liquid chromatograph is equipped with a refractive index detector and a 4-mm × 30-cm stainless steel column containing packing L1. The flow rate is about 1 mL per minute. Chromatograph the *Standard preparation*, and record the peak responses as directed under *Procedure:* the resolution, *R*, between the analyte and internal standard peaks is not less than 5.0, and the relative standard deviation for replicate injections is not more than 2.0%.

Procedure—Separately inject equal volumes (about 25 µL) of the *Standard preparation* and the *Assay preparation* into the chromatograph, record the chromatograms, and measure the responses for the major peaks. The relative retention times are about 0.6 for the internal standard and 1.0 for clindamycin. Calculate the potency, in µg of clindamycin ($C_{18}H_{33}ClN_2O_5S$) per mg, of the Clindamycin Hydrochloride taken by the formula:

$$(W_S/W_U)(P)(R_U/R_S),$$

in which W_S and W_U are the amounts, in mg, of USP Clindamycin Hydrochloride RS and Clindamycin Hydrochloride taken to prepare the *Standard preparation* and the *Assay preparation*, respectively, P is the potency, in µg of clindamycin per mg, of the USP Clindamycin Hydrochloride RS, and R_U and R_S are the ratios of the response of the clindamycin peak to that of the internal standard peak obtained from the *Assay preparation* and the *Standard preparation*, respectively.

Clindamycin Hydrochloride Capsules

» Clindamycin Hydrochloride Capsules contain the equivalent of not less than 90.0 percent and not more than 120.0 percent of the labeled amount of $C_{18}H_{33}ClN_2O_5S$.

Packaging and storage—Preserve in tight containers.

Reference standard—*USP Clindamycin Hydrochloride Reference Standard*—Do not dry before using.

Identification—The retention time of the major peak in the chromatogram of the *Assay preparation* obtained as directed in the *Assay* corresponds to that of the *Standard preparation*, relative to the internal standard.

Dissolution ⟨711⟩—

Medium: water; 900 mL.

Apparatus 1: 100 rpm.

Time: 30 minutes.

Procedure—Determine the amount of clindamycin hydrochloride dissolved using the following method.

Mobile phase—Dissolve 16 g of *dl*-10-camphorsulfonic acid, 8 g of ammonium acetate, and 8 mL of glacial acetic acid in 1600 mL of water, and mix. Add 2400 mL of methanol to this solution, mix, and adjust with hydrochloric acid or 5 *N* sodium hydroxide to a pH of 6.0 ± 0.05.

Standard solution—Prepare a solution of USP Clindamycin Hydrochloride RS in water having an accurately known concentration similar to that expected in the *Test solution.*

Test solution—Use a filtered portion of the solution under test, diluted with water if necessary.

Chromatographic system (see *Chromatography* ⟨621⟩)—The liquid chromatograph is equipped with a refractive index detector and a column that contains 3-µm packing L1. The flow rate is about 2 mL per minute. Chromatograph the *Standard solution*, and record the peak responses as directed under *Procedure:* the tailing factor is not more than 2.0, and the relative standard deviation for replicate injections is not more than 3.0%.

Procedure—Separately inject equal volumes (about 50 µL) of the *Standard solution* and the *Test solution* into the chromatograph, record the chromatograms, and measure the responses for the major peaks. Calculate the amount of $C_{18}H_{33}ClN_2O_5S$ dissolved.

Tolerances—Not less than 80% (*Q*) of the labeled amount of $C_{18}H_{33}ClN_2O_5S$ is dissolved in 30 minutes.

Uniformity of dosage units ⟨905⟩: meet the requirements.

Water, *Method I* ⟨921⟩: not more than 7.0%.

Assay—

Mobile phase, Internal standard solution, Standard preparation, and *Chromatographic system*—Proceed as directed in the *Assay* under *Clindamycin Hydrochloride.*

Assay preparation—Remove as completely as possible the contents of not less than 20 Clindamycin Hydrochloride Capsules, accurately counted, weigh, and mix. Transfer an accurately weighed portion of the powder, equivalent to about 75 mg of clindamycin, to a suitable container. Add 5.0 mL of *Internal standard solution,* and shake for about 30 minutes. Centrifuge or filter, if necessary, to obtain a clear solution.

Procedure—Proceed as directed in the *Assay* under *Clindamycin Hydrochloride.* Calculate the quantity, in mg, of $C_{18}H_{33}ClN_2O_5S$ in the portion of Capsules taken by the formula:

$$W_S(P/1000)(R_U/R_S),$$

in which the terms are as defined therein.

Clindamycin Palmitate Hydrochloride

$C_{34}H_{63}ClN_2O_6S.HCl$ 699.86

L-*threo*-α-D-*galacto*-Octopyranoside, methyl 7-chloro-6,7,8-trideoxy-6-[[(1-methyl-4-propyl-2-pyrrolidinyl)carbonyl]-amino]-1-thio-2-hexadecanoate, monohydrochloride, (2*S*-*trans*)-.

Methyl 7-chloro-6,7,8-trideoxy-6-(1-methyl-*trans*-4-propyl-L-2-pyrrolidinecarboxamido)-1-thio-L-*threo*-α-D-*galacto*-octopyranoside 2-palmitate monohydrochloride [*25507-04-4*].

» Clindamycin Palmitate Hydrochloride has a potency equivalent to not less than 540 µg of clindamycin ($C_{18}H_{33}ClN_2O_5S$) per mg.

Packaging and storage—Preserve in tight containers.

Reference standard—*USP Clindamycin Palmitate Hydrochloride Reference Standard*—Dry in vacuum at 60° for 16 hours before using.

Identification—The infrared absorption spectrum of a mineral oil dispersion of it exhibits maxima only at the same wavelengths as that of a similar preparation of USP Clindamycin Palmitate Hydrochloride RS.

pH ⟨791⟩: between 2.8 and 3.8, in a solution containing 10 mg per mL.

Water, *Method I* ⟨921⟩: not more than 3.0%.

Residue on ignition ⟨281⟩: not more than 0.5%.

Assay—

Internal standard solution—Dissolve cholesteryl benzoate in chloroform to obtain a solution containing about 5 mg per mL.

Standard preparation—Transfer about 150 mg of USP Clindamycin Palmitate Hydrochloride RS, accurately weighed, to a glass-stoppered, 15-mL conical centrifuge tube. Add 5 mL of water, 5.0 mL of *Internal standard solution,* and 1 mL of sodium carbonate solution (3 in 10), and mix. Insert the stopper, shake vigorously for not less than 10 minutes, and centrifuge. Remove the upper aqueous layer, and transfer 1.0 mL of the lower chloroform layer to a 15-mL centrifuge tube. Add 1.0 mL of pyridine and 1.0 mL of acetic anhydride. Agitate the tube to ensure complete mixing, cover the top of the centrifuge tube with a plastic cap through which a small hole has been punched, heat at 100° for 2.5 hours, and allow to cool. Mix, and centrifuge, if necessary. Use the clear solution.

Assay preparation—Transfer about 150 mg of Clindamycin Palmitate Hydrochloride, accurately weighed, to a glass-stoppered, 15-mL conical centrifuge tube, and proceed as directed for *Standard preparation,* beginning with "Add 5 mL of water."

Chromatographic system (see *Chromatography* ⟨621⟩)—The gas chromatograph is equipped with a flame-ionization detector and contains a 0.6-m × 3-mm glass column packed with 1 percent phase G36 on support S1AB. The column and detector are maintained at about 290° and 320°, respectively. Dry helium is used as the carrier gas at a flow rate of about 60 mL per minute.

Procedure—Separately inject equal volumes of about 1.0 µL of the *Standard preparation* and the *Assay preparation* into the chromatograph, record the chromatograms, and measure the responses for the major peaks. In a suitable chromatogram the resolution of the peaks is complete. The elution order is: cholesteryl benzoate, clindamycin palmitate. Calculate the potency, in µg of clindamycin ($C_{18}H_{33}ClN_2O_5S$) per mg, in the Clindamycin Palmitate Hydrochloride taken by the formula:

$$F(R_U/R_S)(W_S/W_U),$$

in which F is the potency, in µg of clindamycin per mg, of the USP Clindamycin Palmitate Hydrochloride RS, R_U and R_S are the ratios of the peak response of clindamycin palmitate to that of cholesteryl benzoate obtained from the *Assay preparation* and the *Standard preparation,* respectively, and W_S and W_U are the amounts, in mg, of USP Clindamycin Palmitate Hydrochloride RS and Clindamycin Palmitate Hydrochloride taken, respectively.

Clindamycin Palmitate Hydrochloride for Oral Solution

» Clindamycin Palmitate Hydrochloride for Oral Solution is a dry mixture of Clindamycin Palmitate Hydrochloride and one or more suitable buffers, colors, diluents, flavors, and preservatives. It contains the equivalent of not less than 90.0 percent and not more than 120.0 percent of the labeled amount of clindamycin ($C_{18}H_{33}ClN_2O_5S$), the labeled amount being 15 mg per mL when constituted as directed in the labeling.

Packaging and storage—Preserve in tight containers.

Reference standard—*USP Clindamycin Palmitate Hydrochloride Reference Standard*—Dry in vacuum at 60° for 16 hours before using.

Uniformity of dosage units ⟨905⟩—

FOR SOLID PACKAGED IN SINGLE-UNIT CONTAINERS: meets the requirements.

pH ⟨791⟩: between 2.5 and 5.0, in the solution constituted as directed in the labeling.

Water, *Method I* ⟨921⟩: not more than 3.0%.

Assay—

Internal standard solution and *Standard preparation*—Prepare as directed in the *Assay* under *Clindamycin Palmitate Hydrochloride.*

Assay preparation—Constitute the Clindamycin Palmitate Hydrochloride for Oral Solution as directed in the labeling, and transfer 5.0 mL of the constituted solution to a glass-stoppered, 15-mL conical centrifuge tube. Add 5.0 mL of *Internal standard solution* and 1 mL of sodium carbonate solution (3 in 10), and proceed as directed for *Standard preparation,* in the *Assay* under *Clindamycin Palmitate Hydrochloride,* beginning with "Insert the stopper, shake vigorously."

Chromatographic system—Proceed as directed in the *Assay* under *Clindamycin Palmitate Hydrochloride.*

Procedure—Proceed as directed in the *Assay* under *Clindamycin Palmitate Hydrochloride.* Calculate the quantity, in mg, of $C_{18}H_{33}ClN_2O_5S$ in each mL of the solution constituted from Clindamycin Palmitate Hydrochloride for Oral Solution taken by the formula:

$$(F/1000)(W_S/V)(R_U/R_S),$$

in which V is the volume, in mL, of constituted solution from Clindamycin Palmitate Hydrochloride for Oral Solution taken and the other terms are as defined therein.

Clindamycin Phosphate

$C_{18}H_{34}ClN_2O_8PS$ 504.96

L-*threo*-α-D-*galacto*-Octopyranoside, methyl 7-chloro-6,7,8-trideoxy-6-[[(1-methyl-4-propyl-2-pyrrolidinyl)carbonyl]-amino]-1-thio-, 2-(dihydrogen phosphate), (2*S*-*trans*)-.
Methyl 7-chloro-6,7,8-trideoxy-6-(1-methyl-*trans*-4-propyl-L-2-pyrrolidinecarboxamido)-1-thio-L-*threo*-α-D-*galacto*-octopyranoside 2-(dihydrogen phosphate) [*24729-96-2*].

» Clindamycin Phosphate has a potency equivalent to not less than 758 µg of clindamycin ($C_{18}H_{33}ClN_2O_5S$) per mg, calculated on the anhydrous basis.

Packaging and storage—Preserve in tight containers.

Reference standard—*USP Clindamycin Phosphate Reference Standard*—Do not dry before using.

Identification—The infrared absorption spectrum of a mineral oil dispersion of it, previously dried at 100° for 2 hours, exhibits maxima only at the same wavelengths as that of a similar preparation of USP Clindamycin Phosphate RS, previously dried at 100° for 2 hours.

Crystallinity ⟨695⟩: meets the requirements.

pH ⟨791⟩: between 3.5 and 4.5, in a solution containing 10 mg per mL.

Water, *Method I* ⟨921⟩: not more than 6.0%.

Other requirements—Clindamycin Phosphate intended for use in making Clindamycin Phosphate Injection complies with the requirements for *Depressor substances* and *Pyrogen* under *Sterile Clindamycin Phosphate.*

Assay—

Mobile phase—Dissolve 10.54 g of monobasic potassium phosphate in 775 mL of water, and adjust with phosphoric acid to a pH of 2.5. Add 225 mL of acetonitrile, mix, and filter. Make adjustments if necessary (see *System Suitability* under *Chromatography* ⟨621⟩). [NOTE—Ensure that the concentration of acetonitrile in the *Mobile phase* is not less than 22% and not more than 25%, in order to retain the correct elution order.]

Internal standard solution—Prepare a solution of 4′-hydroxyacetophenone in acetonitrile containing about 4 mg per mL. Dilute a volume of this solution with *Mobile phase* to obtain a solution having a concentration of about 0.04 mg per mL.

Standard preparation—Transfer about 24 mg of USP Clindamycin Phosphate RS, accurately weighed, to a 100-mL volumetric flask. Add 25.0 mL of *Internal standard solution*, dilute with *Mobile phase* to volume, and mix.

Assay preparation—Transfer about 24 mg of Clindamycin Phosphate, accurately weighed, to a 100-mL volumetric flask, add 25.0 mL of *Internal standard solution*, dilute with *Mobile phase* to volume, and mix.

Chromatographic system (see *Chromatography* ⟨621⟩)—The liquid chromatograph is equipped with a 210-nm detector and a 4.6-mm × 25-cm column that contains packing L7. The flow rate is about 1 mL per minute. Chromatograph the *Standard preparation*, and record the peak responses as directed under *Procedure:* the resolution, R, between the analyte and internal standard peaks is not less than 2.0, and the relative standard deviation for replicate injections is not more than 2.5%.

Procedure—Separately inject equal volumes (about 20 µL) of the *Standard preparation* and the *Assay preparation* into the chromatograph, record the chromatograms, and measure the responses for the major peaks. The relative retention times are about 1.0 for clindamycin phosphate and 1.2 for 4′-hydroxyacetophenone. Calculate the quantity, in µg, of $C_{18}H_{33}ClN_2O_5S$ in the portion of Clindamycin Phosphate taken by the formula:

$$100CP(R_U/R_S),$$

in which C is the concentration, in mg per mL, of USP Clindamycin Phosphate RS in the *Standard preparation*, P is the potency, in µg of $C_{18}H_{33}ClN_2O_5S$ per mg of the USP Clindamycin Phosphate RS, and R_U and R_S are the ratios of the response of the clindamycin phosphate peak to the response of the internal standard peak obtained from the *Assay preparation* and the *Standard preparation*, respectively.

Clindamycin Phosphate Injection

» Clindamycin Phosphate Injection is a sterile solution of Sterile Clindamycin Phosphate or Clindamycin Phosphate in Water for Injection with one or more suitable preservatives and sequestering agents. It contains the equivalent of not less than 90.0 percent and not more than 120.0 percent of the labeled amount of clindamycin ($C_{18}H_{33}ClN_2O_5S$).

Packaging and storage—Preserve in single-dose or in multiple-dose containers, preferably of Type I glass.

Reference standard—*USP Clindamycin Phosphate Reference Standard*—Do not dry before using.

Identification—The retention time of the major peak in the chromatogram of the *Assay preparation* corresponds to that of the *Standard preparation*, both relative to the internal standard, as obtained in the *Assay*.

Pyrogen—When diluted with sterile pyrogen-free saline TS to a concentration of 24 mg of clindamycin ($C_{18}H_{33}ClN_2O_5S$) per mL, it meets the requirements of the *Pyrogen Test* ⟨151⟩, the test dose being 1.0 mL per kg.

pH ⟨791⟩: between 5.5 and 7.0.

Particulate matter ⟨788⟩: meets the requirements under *Small-volume Injections.*

Other requirements—It meets the requirements under *Injections* ⟨1⟩.

Assay—

Mobile phase—Dissolve 10.54 g of monobasic potassium phosphate in 775 mL of water, and adjust with phosphoric acid to a pH of 2.5. Add 225 mL of acetonitrile, mix, and filter. Make adjustments if necessary (see *System Suitability* under *Chromatography* ⟨621⟩). [NOTE—Ensure that the concentration of acetonitrile in the *Mobile phase* is not less than 22% and not more than 25% in order to retain the correct elution order.]

Internal standard solution—Prepare a solution of methylparaben in acetonitrile containing about 6 mg per mL. Dilute a volume of this solution with *Mobile phase* to obtain a solution having a concentration of about 0.06 mg per mL.

Standard preparation—Transfer about 24 mg of USP Clindamycin Phosphate RS, accurately weighed, to a 100-mL volumetric flask, add 25.0 mL of *Internal standard solution*, dilute with *Mobile phase* to volume, and mix.

Assay preparation—Transfer an accurately measured volume of Clindamycin Phosphate Injection, equivalent to about 300 mg of clindamycin, to a 100-mL volumetric flask, dilute with *Mobile phase* to volume, and mix. Transfer 7.0 mL of the resulting solution to a 100-mL volumetric flask, add 25.0 mL of *Internal standard solution*, dilute with *Mobile phase* to volume, and mix.

Resolution solution—Prepare a solution of 4′-hydroxyacetophenone in acetonitrile containing about 4 mg per mL. Dilute a portion of the resulting solution with *Mobile phase* to obtain a solution having a concentration of about 0.04 mg per mL. Add about 25 mL of this solution to a 100-mL volumetric flask containing about 25 mg of USP Clindamycin Phosphate RS, dilute with *Mobile phase* to volume, and mix.

Chromatographic system (see *Chromatography* ⟨621⟩)—The liquid chromatograph is equipped with a 210-nm detector and a 4.6-mm × 25-cm column that contains packing L7. The flow rate is about 1 mL per minute. Chromatograph the *Resolution solution*, and record the peak responses as directed under *Procedure:* the resolution, R, between clindamycin phosphate and 4′-hydroxyacetophenone is not less than 2.0. The relative retention times are 1.0 for clindamycin phosphate and about 1.2 for 4′-hydroxyacetophenone. Chromatograph the *Standard prepa-*

ration, and record the peak responses as directed under *Procedure:* the relative standard deviation for replicate injections is not more than 2.5%.

Procedure—Separately inject equal volumes (about 20 µL) of the *Standard preparation* and the *Assay preparation* into the chromatograph, record the chromatograms, and measure the responses for the major peaks. Calculate the quantity, in mg, of $C_{18}H_{33}ClN_2O_5S$ in each mL of the Injection taken by the formula:

$$(10/7)(CP/V)(R_U/R_S),$$

in which C is the concentration, in mg per mL, of USP Clindamycin Phosphate RS in the *Standard preparation*, P is the potency, in µg of $C_{18}H_{33}ClN_2O_5S$ per mg of the USP Clindamycin Phosphate RS, V is the volume, in mL, of Injection taken, and R_U and R_S are the ratios of the response of the clindamycin phosphate peak to the response of the internal standard peak obtained from the *Assay preparation* and the *Standard preparation*, respectively.

Clindamycin Phosphate Topical Solution

» Clindamycin Phosphate Topical Solution contains the equivalent of not less than 90.0 percent and not more than 110.0 percent of the labeled amount of clindamycin ($C_{18}H_{33}ClN_2O_5S$).

Packaging and storage—Preserve in tight containers.

Reference standard—*USP Clindamycin Phosphate Reference Standard*—Do not dry before using.

Identification—The retention time of the major peak in the chromatogram of the *Assay preparation* corresponds to that of the *Standard preparation*, both relative to the internal standard, as obtained in the *Assay*.

pH ⟨791⟩: between 4.0 and 7.0.

Assay—

Mobile phase, Internal standard solution, Standard preparation, and *Chromatographic system*—Proceed as directed in the *Assay* under *Clindamycin Phosphate*.

Assay preparation—Transfer an accurately measured volume of Clindamycin Phosphate Topical Solution, equivalent to about 20 mg of clindamycin, to a 100-mL volumetric flask, add 25.0 mL of *Internal standard solution*, dilute with *Mobile phase* to volume, and mix.

Procedure—Proceed as directed for *Procedure* in the *Assay* under *Clindamycin Phosphate*. Calculate the quantity, in mg, of $C_{18}H_{33}ClN_2O_5S$ in each mL of the Topical Solution taken by the formula:

$$0.1(CP/V)(R_U/R_S),$$

in which V is the volume, in mL, of Topical Solution taken, and the other terms are as defined therein.

Sterile Clindamycin Phosphate

» Sterile Clindamycin Phosphate is Clindamycin Phosphate suitable for parenteral use. It has a potency equivalent to not less than 758 µg of clindamycin ($C_{18}H_{33}ClN_2O_5S$) per mg, calculated on the anhydrous basis.

Packaging and storage—Preserve in *Containers for Sterile Solids* as described under *Injections* ⟨1⟩.

Reference standard—*USP Clindamycin Phosphate Reference Standard*—Do not dry before using.

Depressor substances—It meets the requirements of the *Depressor Substances Test* ⟨101⟩, the test dose being 1.0 mL per kg of a solution prepared to contain 5.0 mg of clindamycin ($C_{18}H_{33}ClN_2O_5S$) per mL in sterile saline TS.

Pyrogen—It meets the requirements of the *Pyrogen Test* ⟨151⟩, the test dose being 1.0 mL per kg of a solution prepared to contain 24 mg of clindamycin ($C_{18}H_{33}ClN_2O_5S$) per mL in pyrogen-free saline TS.

Sterility—It meets the requirements under *Sterility Tests* ⟨71⟩, when tested as directed in the section, *Test Procedure Using Membrane Filtration*, 6 g of specimen aseptically dissolved in 200 mL of *Fluid A* being used.

Other requirements—It conforms to the definition, responds to the *Identification test*, and meets the requirements for *pH*, *Water*, *Crystallinity*, and *Assay* under *Clindamycin Phosphate*.

Clioquinol

OH I N Cl

C_9H_5ClINO 305.50

8-Quinolinol, 5-chloro-7-iodo-.

5-Chloro-7-iodo-8-quinolinol [*130-26-7*].

» Clioquinol, dried over phosphorus pentoxide for 5 hours, contains not less than 93.0 percent and not more than 100.5 percent of C_9H_5ClINO (the 5-chloro-7-iodo-8-quinolinol isomer).

Packaging and storage—Preserve in tight, light-resistant containers.

Reference standard—*USP Clioquinol Reference Standard*—Dry over phosphorus pentoxide for 5 hours before using.

Identification—

A: Prepare a Standard solution as directed for *Standard preparation* in the *Assay*, except to use 1.0 mL of pyridine instead of the *Internal standard solution*, and chromatograph as directed in the *Assay:* the chromatogram of the *Assay preparation* obtained in the *Assay* exhibits a peak for clioquinol, the retention time of which corresponds with that exhibited by the Standard solution.

B: The ultraviolet absorption spectrum of a 1 in 200,000 solution in 3 N hydrochloric acid exhibits maxima and minima at the same wavelengths as that of a similar solution of USP Clioquinol RS, concomitantly measured, and the respective absorptivities, calculated on the dried basis, at the wavelength of maximum absorbance at about 267 nm do not differ by more than 3.0%.

C: Heat 100 mg with 5 mL of sulfuric acid: copious violet vapors of iodine are evolved.

Loss on drying ⟨731⟩—Dry it over phosphorus pentoxide for 5 hours: it loses not more than 0.5% of its weight.

Residue on ignition ⟨281⟩: not more than 0.5%.

Free iodine and iodide—Shake 1.0 g with 20 mL of water for 30 seconds, allow to stand for 5 minutes, and filter. To 10 mL of the filtrate add 1 mL of 2 N sulfuric acid, then add 2 mL of chloroform, and shake: no violet color appears in the chloroform (*free iodine*). To the mixture add 5 mL of 2 N sulfuric acid and 1 mL of potassium dichromate TS, and shake for 15 seconds: the color in the chloroform layer is no deeper than that produced in a control test made in the following manner: Dilute 2.0 mL of potassium iodide solution (1 in 6000) with water to 10 mL, add 6 mL of 2 N sulfuric acid, 1 mL of potassium dichromate TS, and 2 mL of chloroform, and shake for 15 seconds (*0.05% of iodide*).

Assay—

Internal standard solution—Prepare a solution of pyrene in pyridine containing 2 mg per mL.

Standard preparation—Dissolve an accurately weighed quantity of USP Clioquinol RS in a mixture of pyridine and *n*-hexane (4:1) to obtain a Standard solution having a known concentration of about 3 mg per mL. Transfer 1.0 mL of the Standard solution to a screw-capped glass vial fitted with a septum, add 1.0 mL of bis(trimethylsilyl)acetamide and 1.0 mL of *Internal standard solution*, attach the cap, and mix. Heat in a water bath at 50° for 15 minutes, and then cool to ambient temperature.

Assay preparation—Transfer about 75 mg of Clioquinol, previously dried and accurately weighed, to a 25-mL volumetric flask, dissolve in a mixture of pyridine and *n*-hexane (4:1), dilute with the same solvent to volume, and mix. Transfer 1.0 mL of this solution to a screw-capped glass vial fitted with a septum, add 1.0 mL of bis(trimethylsilyl)acetamide and 1.0 mL of *Internal standard solution*, attach the cap, and mix. Heat in a water bath at 50° for 15 minutes, then cool to ambient temperature.

Chromatographic system (see *Chromatography* ⟨621⟩)—The gas chromatograph is equipped with a flame-ionization detector, and contains a 1.83-m × 2-mm glass column packed with 3 percent liquid phase G3 on 80- to 100-mesh support S1AB. The injection port and detector temperatures are maintained at 170° and 250°, respectively, and the initial column temperature is 200° for a conditioning period of not less than 16 hours (not connected to the detector) and is then reduced to 165°. Helium is used as the carrier gas at a flow rate of about 30 mL per minute, and hydrogen and air are introduced into the detector at rates of 25 mL and 500 mL per minute, respectively. Chromatograph the *Standard preparation*, and record the peak responses as directed under *Procedure:* the resolution, *R*, between the clioquinol and the internal standard peaks is not less than 3.

Procedure—Separately inject equal volumes (about 1 µL) of the *Standard preparation* and the *Assay preparation* into the chromatograph, record the chromatograms, and measure the responses for the major peaks. The relative retention times for clioquinol and pyrene are about 0.6 and 1.0, respectively. Calculate the quantity, in mg, of C_9H_5ClINO in the Clioquinol taken by the formula:

$$25C(R_U/R_S),$$

in which *C* is the concentration, in mg per mL, of USP Clioquinol RS in the Standard solution used to prepare the *Standard preparation*, and R_U and R_S are the ratios of the peak responses of the clioquinol peak to the internal standard peak obtained from the *Assay preparation* and the *Standard preparation*, respectively.

Clioquinol Cream

» Clioquinol Cream contains not less than 90.0 percent and not more than 110.0 percent of the labeled amount of C_9H_5ClINO in a suitable cream base.

Packaging and storage—Preserve in collapsible tubes or tight, light-resistant containers.

Reference standard—*USP Clioquinol Reference Standard*—Dry over phosphorus pentoxide for 5 hours before using.

Identification—

A: Prepare a Standard solution as directed for *Standard preparation* in the *Assay*, except to use 1.0 mL of pyridine instead of the *Internal standard solution*, and chromatograph as directed in the *Assay:* the chromatogram of the *Assay preparation* obtained in the *Assay* exhibits a peak for clioquinol, the retention time of which corresponds with that exhibited by the Standard solution.

B: Place a quantity of Cream, equivalent to about 25 mg of clioquinol, in a 100-mL volumetric flask, add about 75 mL of dilute hydrochloric acid (1 in 4), and heat on a steam bath to melt the cream, shaking vigorously to extract the clioquinol. Cool under running water, and add dilute hydrochloric acid (1 in 4) to volume. Filter through paper, and dilute 3 mL of the filtrate with dilute hydrochloric acid (1 in 4) to 100 mL: the ultraviolet absorption spectrum of this solution exhibits maxima and minima at the same wavelengths as that of a similar solution of USP Clioquinol RS, concomitantly measured.

Assay—

Internal standard solution, Standard preparation, and *Chromatographic system*—Proceed as directed in the *Assay* under *Clioquinol.*

Assay preparation—Transfer an accurately weighed portion of Clioquinol Cream, equivalent to about 150 mg of clioquinol, to a 60-mL separator. Place the separator on its side in a vacuum oven at a pressure of about 10 mm of mercury at 45° for 4 hours. Remove the separator from the oven, allow to cool, add 15 mL of a mixture of pyridine and *n*-hexane (4:1), insert a polytef stopper, and mix. Transfer the mixture to a 50-mL volumetric flask, and rinse the separator with two 15-mL portions of the same solvent, shaking each time for 30 seconds. Transfer both rinsings to the volumetric flask, dilute with the same solvent to volume, and mix. Transfer 1.0 mL to a screw-capped glass vial fitted with a septum, add 1.0 mL of bis(trimethylsilyl)acetamide and 1.0 mL of *Internal standard solution*, attach the cap, and mix.

Procedure—Proceed as directed for *Procedure* in the *Assay* under *Clioquinol.* Calculate the quantity, in mg, of C_9H_5ClINO in the portion of Cream taken by the formula:

$$50C(R_U/R_S),$$

in which *C* is the concentration, in mg per mL, of USP Clioquinol RS in the Standard solution used to prepare the *Standard preparation*, and R_U and R_S are the ratios of the peak responses of the clioquinol peak to the internal standard peak obtained from the *Assay preparation* and the *Standard preparation*, respectively.

Clioquinol Ointment

» Clioquinol Ointment contains not less than 90.0 percent and not more than 110.0 percent of the labeled amount of C_9H_5ClINO in a suitable ointment base.

Packaging and storage—Preserve in collapsible tubes or tight, light-resistant containers.

Reference standard—*USP Clioquinol Reference Standard*—Dry over phosphorus pentoxide for 5 hours before using.

Identification—It responds to the *Identification tests* under *Clioquinol Cream.*

Assay—

Internal standard solution, Standard preparation, and *Chromatographic system*—Proceed as directed in the *Assay* under *Clioquinol.*

Assay preparation—Transfer an accurately weighed portion of Clioquinol Ointment, equivalent to about 150 mg of clioquinol, to a 125-mL separator. Add 75 mL of *n*-hexane, and mix. Add 15 mL of dimethylformamide, and mix for 1 minute. Allow the layers to separate, and transfer the lower layer to a 50-mL volumetric flask. Repeat the extraction with separate 15-mL and 10-mL portions of dimethylformamide, and transfer the lower layers to the 50-mL volumetric flask. Dilute with dimethylformamide to volume, and mix. Transfer 1.0 mL of this solution to a screw-capped glass vial fitted with a septum, and evaporate at about 60° under a stream of nitrogen to dryness. Add 1.0 mL of a mixture of pyridine and *n*-hexane (4:1) to the residue, add 1.0 mL of bis(trimethylsilyl)acetamide and 1.0 mL of *Internal standard solution*, attach the cap, and mix. Heat in a water bath at 50° for 15 minutes, then cool to ambient temperature.

Procedure—Proceed as directed for *Procedure* under *Clioquinol.* Calculate the quantity, in mg, of C_9H_5ClINO in the portion of Ointment taken by the formula:

$$50C(R_U/R_S),$$

in which *C* is the concentration, in mg per mL, of USP Clioquinol RS in the Standard solution used to prepare the *Standard preparation*, and R_U and R_S are the ratios of the peak responses of the clioquinol peak to the internal standard peak obtained from the *Assay preparation* and the *Standard preparation*, respectively.

Compound Clioquinol Powder

» Compound Clioquinol Powder contains not less than 22.5 percent and not more than 27.5 percent of C_9H_5ClINO.

Clioquinol	250 g
Lactic Acid	25 g
Zinc Stearate	200 g
Lactose	525 g
To make	1000 g

Mix the Lactic Acid with the Lactose, then add the Clioquinol and the Zinc Stearate, and mix.

Packaging and storage—Preserve in well-closed, light-resistant containers.

Reference standard—*USP Clioquinol Reference Standard*—Dry over phosphorus pentoxide for 5 hours before using.

Identification—Place a quantity of Powder, equivalent to about 30 mg of clioquinol, in a glass-stoppered, 50-mL conical flask, add 20 mL of 1 *N* sulfuric acid, and shake for 5 minutes. Filter, transfer 5 mL of the filtrate to a glass-stoppered test tube, add 5 drops of potassium dichromate TS and 2 mL of chloroform, and shake well: a red-violet color develops in the chloroform layer.

Assay—Transfer an accurately weighed quantity of Compound Clioquinol Powder, equivalent to about 50 mg of clioquinol, to a 200-mL volumetric flask, add 100 mL of 3 *N* hydrochloric acid, and shake by mechanical means for 15 minutes. Dilute with 3 *N* hydrochloric acid to volume, and mix. Filter a portion of the solution, dilute 4.0 mL of the filtrate with 3 *N* hydrochloric acid to 200.0 mL, and mix. Concomitantly determine the absorbances of this solution and of a Standard solution of USP Clioquinol RS in the same medium having a known concentration of about 5 µg per mL in 1-cm cells at the wavelength of maximum absorbance at about 267 nm, with a suitable spectrophotometer, using 3 *N* hydrochloric acid as the blank. Calculate the quantity, in mg, of C_9H_5ClINO in the portion of Powder taken by the formula:

$$10C(A_U/A_S),$$

in which C is the concentration, in µg per mL, of USP Clioquinol RS in the Standard solution, and A_U and A_S are the absorbances of the solution from the Powder and the Standard solution, respectively.

Clocortolone Pivalate

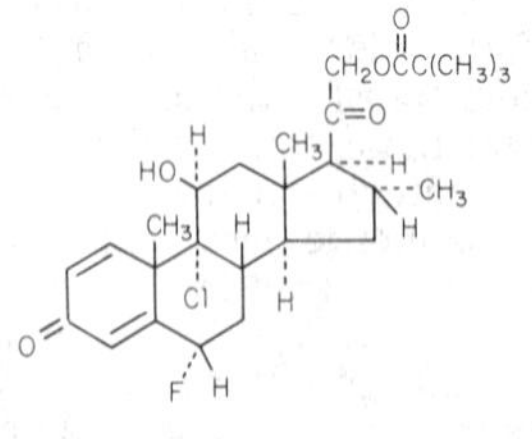

$C_{27}H_{36}ClFO_5$ 495.03

Pregna-1,4-diene-3,20-dione, 9-chloro-21-(2,2-dimethyl-1-oxopropoxy)-6-fluoro-11-hydroxy-16-methyl-, (6α,11β,16α)-.

9-Chloro-6α-fluoro-11β,21-dihydroxy-16α-methylpregna-1,4-diene-3,20-dione 21-pivalate [*34097-16-0*].

» Clocortolone Pivalate contains not less than 97.0 percent and not more than 103.0 percent of $C_{27}H_{36}ClFO_5$, calculated on the dried basis.

Packaging and storage—Preserve in tight, light-resistant containers.

Reference standard—*USP Clocortolone Pivalate Reference Standard*—Dry at 105° for 3 hours before using.

Color and clarity of solution—A 1 in 100 solution in chloroform is clear and practically colorless.

Identification—

A: The infrared absorption spectrum of a mineral oil dispersion of it, previously dried, exhibits maxima only at the same wavelength as that of a similar preparation of USP Clocortolone Pivalate RS.

B: The ultraviolet absorption spectrum of a 3 in 200,000 solution in methanol exhibits maxima and minima at the same wavelengths as that of a similar solution of USP Clocortolone Pivalate RS, concomitantly measured, and the respective absorptivities, calculated on the dried basis, at the wavelength of maximum absorbance at about 238 nm, do not differ by more than 3.0%.

Specific rotation ⟨781⟩: between +125° and +135°, calculated on the dried basis, determined in a solution in chloroform containing 400 mg in each 10 mL.

Loss on drying ⟨731⟩—Dry it at 105° for 3 hours: it loses not more than 1.0% of its weight.

Residue on ignition ⟨281⟩: not more than 0.2%, a 100-mg test specimen being used.

Chromatographic impurities—

Test preparation—Accurately weigh about 100 mg of Clocortolone Pivalate, and transfer to a 25-mL volumetric flask. Dissolve in a mixture of chloroform and methanol (1:1), and dilute with the same solvent to volume.

Standard preparation—Using an accurately weighed quantity of USP Clocortolone Pivalate RS, prepare a solution in a mixture of chloroform and methanol (1:1) having a known concentration of about 4 mg per mL.

Procedure—Score a 20- × 20-cm thin-layer chromatographic plate (see *Chromatography* ⟨621⟩) coated with a 0.25-mm layer of chromatographic silica gel mixture into three equal sections to be used for the *Test preparation*, the blank, and the *Standard preparation*, respectively. Activate the plate at 105° for 30 minutes before use. Apply 100 µL each of the *Test preparation* and the *Standard preparation* as streaks 2.5 cm from the bottom of the appropriate section of the plate, and dry the streaks with a gentle stream of air. Using a solvent system consisting of cyclohexane and ethyl acetate (2:1), develop the chromatogram in a suitable chromatographic chamber lined with absorbent paper and previously equilibrated, until the solvent front has moved 15 cm above the line of application. Air-dry the plate, and develop the chromatogram a second time using the same chromatographic system. Air-dry the plate, and locate the principal band occupied by the *Standard preparation* by viewing under ultraviolet light. Mark this band as well as corresponding bands in the blank and *Test preparation* sections of the plate. Quantitatively remove the silica gel from each band, and transfer to separate glass-stoppered, 50-mL centrifuge tubes. Add 25.0 mL of methanol to each tube, shake for not less than 20 minutes, and centrifuge. Concomitantly determine the absorbances of the supernatant solutions from the *Test preparation* and the *Standard preparation* against the blank at the wavelength of maximum absorbance at about 238 nm, with a suitable spectrophotometer. Calculate the percentage of chromatographic impurities in the *Test preparation* by the formula:

$$100 - [100(C_S/C_U)(A_U/A_S)],$$

in which C_S is the concentration, in mg per mL, of USP Clocortolone Pivalate RS in the *Standard preparation*, C_U is the concentration, in mg per mL, of the *Test preparation*, and A_U and A_S are the absorbances of the solutions from the *Test preparation* and the *Standard preparation*, respectively. The requirements of the test are met if the amount of impurities does not exceed 3.0%.

Assay—

Standard preparation—Dissolve an accurately weighed quantity of USP Clocortolone Pivalate RS in chloroform to obtain a solution having a known concentration of about 0.75 mg per mL. Dilute an accurately measured volume of this solution with methanol, and mix to obtain a *Standard preparation* having a known concentration of about 30 µg per mL.

Assay preparation—Accurately weigh about 75 mg of Clocortolone Pivalate, and transfer to a 100-mL volumetric flask. Dissolve in chloroform, dilute with chloroform to volume, and mix. Transfer 4.0 mL of this solution to a 100-mL volumetric flask, dilute with methanol to volume, and mix.

Procedure—Transfer 10.0-mL portions of the *Standard preparation* and the *Assay preparation* to separate glass-stoppered, 50-mL conical flasks, and evaporate on a steam bath to dryness. To each flask, and to a third flask to provide the blank, add 15.0 mL of a solution containing 250 mg of isoniazid and 0.3 mL of hydrochloric acid in 500 mL of methanol. Swirl the contents of the flasks to dissolve the residues. Insert the stoppers securely in the flasks, and place in a water bath at 60° for 2.5 hours. Cool to room temperature. Concomitantly determine the absorbances of the *Assay preparation* and the *Standard preparation* in 1-cm cells against the blank at the wavelength of maximum absorbance at about 405 nm, with a suitable spectrophotometer. Calculate the quantity, in mg, of $C_{27}H_{36}ClFO_5$ in the portion of Clocortolone Pivalate taken by the formula:

$$2.5C(A_U/A_S),$$

in which C is the concentration, in µg per mL, of USP Clocortolone Pivalate RS in the *Standard preparation*, and A_U and A_S are the absorbances of the *Assay preparation* and the *Standard preparation*, respectively.

Clocortolone Pivalate Cream

» Clocortolone Pivalate Cream contains not less than 90.0 percent and not more than 110.0 percent of $C_{27}H_{36}ClFO_5$ in a suitable cream base. It may contain suitable preservatives.

Packaging and storage—Preserve in collapsible tubes or in tight, light-resistant containers.

Reference standard—*USP Clocortolone Pivalate Reference Standard*—Dry at 105° for 3 hours before using.

Identification—Place a portion of Cream, equivalent to about 1 mg of clocortolone pivalate, in a suitable separator. Add 5 mL of water, and extract with 10 mL of chloroform. Evaporate the chloroform layer to dryness, and dissolve the residue in 2 mL of methanol. Apply 20 µL of this test solution and 20 µL of a Standard solution of USP Clocortolone Pivalate RS in chloroform containing about 0.5 mg per mL about 1.5 cm from the bottom of a suitable thin-layer chromatographic plate (see *Chromatography* ⟨621⟩) coated with a 0.25-mm layer of chromatographic silica gel mixture. Allow the spots to dry, and develop the chromatogram in a solvent system consisting of cyclohexane and ethyl acetate (2:1) until the solvent front has moved about three-fourths of the length of the plate. Remove the plate from the developing chamber, mark the solvent front, and allow the solvent to evaporate. Locate the spots on the plate by viewing under short-wavelength ultraviolet light: the R_f value of the principal spot obtained from the test solution corresponds to that obtained from the Standard solution.

Minimum fill ⟨755⟩: meets the requirements.

pH ⟨791⟩: between 5.0 and 7.0, in a 1 in 10 aqueous dispersion.

Particle size determination—Place a small portion of Cream on a microscope slide, apply a cover slide, press slightly, and examine under 40× objective magnification using a suitable microscope equipped with polarized light. Scan the complete slide preparation, and record the size of the largest crystal found in reference to a calibrated grid: no particle in the Cream is greater than 50 microns when measured in the longitudinal axis.

Assay—

Standard preparation—Dissolve an accurately weighed quantity of USP Clocortolone Pivalate RS in methanol to obtain a solution having a known concentration of about 0.06 mg per mL.

Assay preparation—Using a plastic syringe equipped with a suitable cannula, transfer an accurately weighed quantity of Clocortolone Pivalate Cream, equivalent to about 3 mg of clocortolone pivalate, to a 50-mL volumetric flask. Add about 25 mL of methanol, and warm the flask in a 60° water bath for about 10 minutes, with occasional swirling, to disperse the Cream. Cool to room temperature, dilute with methanol to volume, and mix. Allow any insoluble material to settle.

Procedure—Transfer 10.0 mL of the *Standard preparation* to a 25-mL volumetric flask. Transfer 10.0-mL portions of the *Assay preparation* into two separate 25-mL volumetric flasks labeled *Assay preparation* and *Assay blank*, respectively. Evaporate the contents of the three flasks with the aid of a stream of air or nitrogen to dryness. Transfer 20.0 mL of a solution containing 250 mg of isoniazid and 0.3 mL of hydrochloric acid in 500 mL of methanol to the flasks containing the *Standard preparation*, the *Assay preparation*, and a fourth 25-mL volumetric flask labeled *Reagent blank*. Pipet 20 mL of acidified methanol solution (0.3 mL of hydrochloric acid diluted with methanol to 500 mL) into the flask labeled *Assay blank*. Insert the stoppers securely in the flasks, and place in a water bath at 60° for 2.5 hours, occasionally swirling the contents of each flask. Cool the flasks to room temperature, dilute with the acidified methanol solution to volume, and mix. Centrifuge the *Assay preparation* at high speed for 10 minutes. Concomitantly determine the absorbances of the *Standard preparation*, the *Assay preparation*, the *Assay blank*, and the *Reagent blank* against acidified methanol solution as the solvent blank in 1-cm cells at the wavelength of maximum absorbance at about 390 nm, with a suitable spectrophotometer. Calculate the quantity, in mg, of $C_{27}H_{36}ClFO_5$ in the portion of Cream taken by formula:

$$50C(A_U/A_S),$$

in which C is the concentration, in mg per mL, of USP Clocortolone Pivalate RS in the *Standard preparation*, A_U is the absorbance of the *Assay preparation*, corrected for the *Assay blank* and the *Reagent blank*, and A_S is the absorbance of the *Standard preparation* corrected for the *Reagent blank*.

Clofibrate

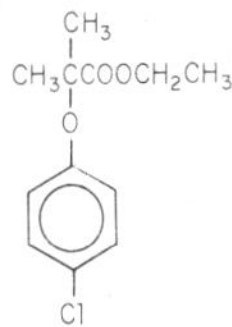

$C_{12}H_{15}ClO_3$ 242.70

Propanoic acid, 2-(4-chlorophenoxy)-2-methyl-, ethyl ester.

Ethyl 2-(*p*-chlorophenoxy)-2-methylpropionate [*637-07-0*].

» Clofibrate contains not less than 97.0 percent and not more than 103.0 percent of $C_{12}H_{15}ClO_3$, calculated on the anhydrous basis.

Packaging and storage—Preserve in tight, light-resistant containers.

Reference standard—*USP Clofibrate Reference Standard*—After opening ampul, store in a tight, light-resistant container. Do not dry before using.

Identification—

A: The infrared absorption spectrum, determined in a 0.1-mm cell, of a 1 in 10 solution of it in carbon disulfide exhibits maxima only at the same wavelengths as that of a similar preparation of USP Clofibrate RS.

B: The ultraviolet absorption spectrum of a 1 in 50,000 solution in methanol exhibits maxima and minima at the same wavelengths as that of a similar preparation of USP Clofibrate RS, concomitantly measured.

Refractive index ⟨831⟩: between 1.500 and 1.505, at 20°.

Acidity—Mix 10.0 g with 100 mL of neutralized alcohol, add 3 drops of phenolphthalein TS, and titrate with 0.10 *N* sodium hydroxide: not more than 0.90 mL is required for neutralization.

Water, *Method I* ⟨921⟩: not more than 0.2%.

Chromatographic impurities—Extract 10.0 g with 20 mL of 1 *N* sodium hydroxide in a separator. Wash the lower layer with 5 mL of water. Dry the organic layer with anhydrous sodium sulfate (*Liquid A*). Dissolve an accurately weighed portion of *Liquid A* in chloroform, mix, and dilute quantitatively with chloroform to obtain a solution containing 100 μg per mL (*Liquid B*). Inject a portion of *Liquid A* into a suitable gas chromatograph equipped with a flame-ionization detector and a recorder. Under typical conditions, the instrument contains a 4-mm × 1.5-m glass column packed with 30 percent phase G2 on packing S1AB. Maintain the column, injection port, and detector at 185°. Use nitrogen as the carrier gas. Record the chromatogram for a total time of not less than four times the retention time of the main peak, and measure the peak area response of each component. Similarly inject a portion of *Liquid B*, record the chromatogram, and measure the peak area response of the main peak: the area of any peak other than the main peak obtained from *Liquid A* is not greater than the area of the main peak obtained from *Liquid B* (0.01%), and the total of the areas of all extraneous peaks obtained from *Liquid A* does not exceed 12 times the area from *Liquid B* (0.12%).

Limit of *p*-chlorophenol—

Standard preparation—Dissolve 50.0 mg of *p*-chlorophenol in 500.0 mL of sodium hydroxide solution (1 in 250). Dilute 4.0 mL of this solution with sodium hydroxide solution (1 in 250) to 100.0 mL, and mix.

Test preparation—Transfer 2.0 g to a 250-mL separator, and dissolve in 50 mL of solvent hexane that previously has been washed with two 25-mL portions of sodium hydroxide solution (1 in 250) followed by successive quantities of water until the last washing is not alkaline to phenolphthalein TS. Extract this solution successively with two 20-mL portions and one 5-mL portion of sodium hydroxide solution (1 in 250). Allow the phases to separate as completely as possible, and collect the aqueous phases in a 50-mL volumetric flask, discarding the solvent hexane phase. Dilute the aqueous alkaline solution with sodium hydroxide solution (1 in 250) to volume, and mix.

Test preparation blank—Prepare as directed under *Test preparation*, omitting only the test specimen.

Procedure—Transfer 6.0 mL of *Standard preparation* to a 250-mL separator, and dilute with sodium hydroxide solution (1 in 250) to 20 mL. To a second 250-mL separator transfer 20.0 mL of sodium hydroxide solution (1 in 250) to provide the *Standard preparation blank*. To the third and fourth 250-mL separators transfer 20.0 mL each of *Test preparation* and *Test preparation blank*, respectively. To each separator add 20.0 mL of a buffer solution prepared by dissolving 24.7 g of boric acid in sufficient sodium hydroxide solution (1 in 250) to make 1000 mL. Add successively 0.1 mL of 4-aminoantipyrine solution (3 in 100) and 0.2 mL of potassium ferricyanide solution (1 in 10) to each separator, shaking after each addition. Allow the mixtures to stand for 3 minutes, add 10 mL of chloroform to each separator, and shake. Allow the layers to separate as completely as possible, and transfer the chloroform layer from each separator to separate 25-mL volumetric flasks. Wash the contents remaining in each separator with 2 mL of chloroform without shaking, and add the chloroform washings to the respective volumetric flasks. Add 5 mL of chloroform to each separator, and shake. Allow the layers to separate, and add the chloroform layer from each separator to the contents of the respective volumetric flasks. Finally, wash the contents in each of the separators with 3 mL of chloroform, adding the chloroform washings to the contents of the respective volumetric flasks. Dilute each volumetric flask with chloroform to volume, add about 0.5 g of anhydrous sodium sulfate, shake, and allow to stand for 5 minutes. Concomitantly determine the absorbances of the solutions in 5-cm cells at the wavelength of maximum absorbance at about 455 nm, with a suitable spectrophotometer, using chloroform as the blank. Record the absorbance of the solution from the *Standard preparation* as A_1, that from the *Standard preparation blank* as A_2, that from the *Test preparation* as A_3, and that from the *Test preparation blank* as A_4. The corrected absorbance of the solution from the *Test preparation* ($A_3 - A_4$) is not greater than the corrected absorbance of the solution from the *Standard preparation* ($A_1 - A_2$), corresponding to not more than 0.003% of *p*-chlorophenol.

Assay—

Ion-exchange resin—To a beaker containing 75 mL of 1 *N* sodium hydroxide add about 3 g of a strongly basic polystyrene anion-exchange resin, and allow the mixture to stand for about 15 minutes, with occasional stirring. Wash the resin with water until the last washing is neutral to litmus paper, then wash with three 50-mL portions of methanol.

Ion-exchange column—Place a plug of glass wool in the base of a 1- × 15-cm ion-exchange tube, and transfer to the tube a sufficient amount of *Ion-exchange resin*, slurried in methanol, to produce a column bed height of from 6 cm to 8 cm.

Standard preparation—Dissolve an accurately weighed quantity of USP Clofibrate RS in methanol, and dilute quantitatively and stepwise with methanol to obtain a solution having a known concentration of about 20 μg per mL.

Assay preparation—Transfer about 200 mg of Clofibrate, accurately weighed, to a 100-mL volumetric flask, add methanol to volume, and mix. Transfer 10.0 mL of this solution to the *Ion-exchange column*, and collect the eluate in a 100-mL volumetric flask. Rinse the column with 25 mL of methanol, collect the rinsing in the volumetric flask, dilute with methanol to volume, and mix. Transfer 5.0 mL of this solution to a 50-mL volumetric flask, dilute with methanol to volume, and mix.

Procedure—Concomitantly determine the absorbances of the *Standard preparation* and the *Assay preparation* in 1-cm cells at the wavelength of maximum absorbance at about 226 nm, with a suitable spectrophotometer, using methanol as the blank. Calculate the quantity, in mg, of $C_{12}H_{15}ClO_3$ in the portion of Clofibrate taken by the formula:

$$10C(A_U / A_S),$$

in which C is the concentration, in μg per mL, of USP Clofibrate RS in the *Standard preparation*, and A_U and A_S are the absorbances of the *Assay preparation* and the *Standard preparation*, respectively.

Clofibrate Capsules

» Clofibrate Capsules contain not less than 90.0 percent and not more than 110.0 percent of the labeled amount of $C_{12}H_{15}ClO_3$.

Packaging and storage—Preserve in well-closed, light-resistant containers.

Reference standard—*USP Clofibrate Reference Standard*—After opening ampul, store in a tight, light-resistant container. Do not dry before using.

Identification—Capsules respond to *Identification test A* under *Clofibrate*.

Uniformity of dosage units ⟨905⟩: meet the requirements.

Assay—Proceed with Clofibrate Capsules as directed in the *Assay* under *Clofibrate*, using the following as the *Assay preparation:* Weigh accurately not less than 20 Clofibrate Capsules in a tared weighing bottle. With a sharp blade, carefully open the Capsules, without loss of shell material, and transfer the contents to a 100-mL beaker. Remove any liquid from the emptied capsules by washing with several small portions of ether. Discard the washings, and allow the capsules to dry in a jet of dry air until the odor of ether no longer is perceptible. Weigh the empty capsules in the tared weighing bottle, and calculate the average net weight per capsule. Transfer an accurately weighed amount of capsule contents, equivalent to about 200 mg of clofibrate, to a 100-mL volumetric flask, add methanol to volume, and mix. Transfer 10.0 mL of this solution to the *Ion-exchange column*, and collect the eluate in a 100-mL volumetric flask. Rinse the column with 25 mL of methanol, collect the rinsings in the volumetric flask, dilute with methanol to volume, and mix. Transfer 5.0 mL of this solution to a 50-mL volumetric flask, dilute with methanol to volume, and mix. Calculate the quantity, in mg, of $C_{12}H_{15}ClO_3$ in the portion of Capsules taken by the formula:

$$10C(A_U / A_S),$$

in which C is the concentration, in μg per mL, of USP Clofibrate RS in the *Standard preparation*, and A_U and A_S are the absorbances of the *Assay preparation* and the *Standard preparation*, respectively.

Clomiphene Citrate

$C_{26}H_{28}ClNO.C_6H_8O_7$ 598.09

Ethanamine, 2-[4-(2-chloro-1,2-diphenylethenyl)phenoxy]-*N*,*N*-diethyl-, 2-hydroxy-1,2,3-propanetricarboxylate (1:1).

2-[*p*-(2-Chloro-1,2-diphenylvinyl)phenoxy]triethylamine citrate (1:1) [*50-41-9*].

» Clomiphene Citrate contains not less than 98.0 percent and not more than 101.0 percent of a mixture of the (*E*)- and (*Z*)- geometric isomers of $C_{26}H_{28}ClNO.C_6H_8O_7$, calculated on the anhydrous basis. It contains not less than 30.0 percent and not more than 50.0 percent of the *Z*-isomer, [(*Z*)-2-[4-(2-chloro-1,2-diphenylethenyl)phenoxy]-*N*,*N*-diethylethanamine 2-hydroxy-1,2,3-propanetricarboxylate (1:1).

Packaging and storage—Preserve in well-closed containers.

Reference standard—*USP Clomiphene Citrate Reference Standard*—Do not dry; determine the *Water* content by *Method I* before using.

Identification—

A: It meets the requirements under *Identification—Organic Nitrogenous Bases* ⟨181⟩.

B: The ultraviolet absorption spectrum of a 1 in 50,000 solution in 0.1 *N* hydrochloric acid exhibits maxima and minima at the same wavelengths as that of a similar solution of USP Clomiphene Citrate RS, concomitantly measured.

C: It responds to the tests for *Citrate* ⟨191⟩.

Water, *Method I* ⟨921⟩: not more than 1.0%.

Heavy metals, *Method II* ⟨231⟩: 0.002%.

***Z*-isomer**—

Mobile phase—Mix *n*-hexane, alcohol-free chloroform, and triethylamine (80:20:0.1). The amount of triethylamine may be varied to meet system suitability requirements.

Standard preparation—Weigh accurately about 25 mg of USP Clomiphene Citrate RS into a 125-mL separator, and dissolve in 25 mL of 0.1 *N* hydrochloric acid. Add 5 mL of 1 *N* sodium hydroxide, and extract with three 25-mL portions of alcohol-free chloroform. Dry the extract with anhydrous sodium sulfate, and add alcohol-free chloroform to make 100.0 mL. Transfer 20 mL of this solution to a 100-mL volumetric flask, add 0.1 mL of triethylamine, dilute with *n*-hexane to volume, and mix.

Test preparation—Using about 25 mg of Clomiphene Citrate, accurately weighed, proceed as directed for *Standard preparation*.

Chromatographic system (see *Chromatography* ⟨621⟩)—The liquid chromatograph is equipped with a 302-nm detector and a 4-mm × 30-cm column that contains 10-μm packing L3. The flow rate is about 2 mL per minute. Chromatograph the *Standard preparation*, and record the peak response as directed under *Procedure:* the resolution, *R*, between the *E*-isomer, eluting first, and the *Z*-isomer is not less than 1.5, and the relative standard deviation for replicate injections is not more than 2.0%.

Procedure—Separately inject equal volumes (about 50 μL) of the *Standard preparation* and the *Test preparation* into the chromatograph, record the chromatograms, and measure the responses for the *E*- and *Z*-isomers. Measure the areas, R_E and R_Z, of the (*E*)- and (*Z*)-isomer peaks, respectively, obtained for the *Standard preparation* and the *Test preparation*. Calculate the area percentages, $\%A_S$ and $\%A_U$, of the (*Z*)-isomer in the *Standard preparation* and the *Test preparation*, respectively, by the formula:

$$100R_Z/(R_E + R_Z).$$

Calculate the percentage of (*Z*)-isomer in the *Test preparation* by the formula:

$$\%W_S(\%A_U/\%A_S),$$

in which $\%W_S$ is the labeled weight percentage of (*Z*)-isomer in USP Clomiphene Citrate RS, and $\%A_S$ and $\%A_U$ are the area percentages of the (*Z*)-isomer peaks in chromatograms of the *Standard preparation* and the *Test preparation*, respectively. Between 30.0% and 50.0% of (*Z*)-isomer is found.

Limit of related compounds—Disperse about 300 mg, accurately weighed, in about 40 mL of water, add 10 mL of 1 *N* sodium hydroxide, and extract the precipitated base with 10 mL of chloroform. Transfer the chloroform extract to a stoppered centrifuge tube, and centrifuge. Inject between 1 μL and 2 μL of the clear chloroform extract into a suitable gas chromatograph equipped with a flame-ionization detector and a recorder. Under typical conditions, the instrument contains a 2-m × 4-mm glass column packed with 0.5 percent phase G10 and 1.3 percent phase G2 on packing S1A. Maintain the column and injection port at 230° and the detector at 250°. Using a suitable carrier gas, arrange the rate of flow so that the elution time of the clomiphene isomers is about 20 minutes. Record the chromatogram for a total time of not less than twice the retention time of the two separated clomiphene base isomers. Not more than 1.0% of any single extraneous volatile substance and not more than 2.0% of total extraneous volatile substances are found, determined by area normalization.

Assay—Dissolve about 1 g of Clomiphene Citrate, accurately weighed, in 70 mL of glacial acetic acid, add crystal violet TS, and titrate with 0.1 *N* perchloric acid VS to an emerald-green end-point. Perform a blank determination, and make any necessary corrections. Each mL of 0.1 *N* perchloric acid is equivalent to 59.81 mg of $C_{26}H_{28}ClNO.C_6H_8O_7$.

Clomiphene Citrate Tablets

» Clomiphene Citrate Tablets contain not less than 93.0 percent and not more than 107.0 percent of the labeled amount of $C_{26}H_{28}ClNO.C_6H_8O_7$.

Packaging and storage—Preserve in well-closed containers, protected from light.

Reference standard—*USP Clomiphene Citrate Reference Standard*—Do not dry; determine the *Water* content by *Method I* before using.

Identification—Place a portion of finely powdered Tablets, equivalent to about 30 mg of clomiphene citrate, in a centrifuge tube containing about 30 mL of a 1 in 2 solution of methanol in 0.1 *N* hydrochloric acid. Insert the stopper, and place the tube in a water bath at about 37° for 15 minutes. Shake occasionally. Centrifuge, and place the clear supernatant liquid in a separator. Extract with one 40-mL and two 25-mL portions of hexanes, and discard the extract. Render the aqueous solution alkaline with 1 *N* sodium hydroxide, and proceed as directed in *Identification test A* under *Clomiphene Citrate*, beginning with "extract the precipitated base," except to use 1.0 mL of carbon disulfide to dissolve the residue.

Dissolution ⟨711⟩—

Medium: water; 900 mL.

Apparatus 1: 100 rpm.

Time: 60 minutes.

Procedure—Determine the amount of $C_{26}H_{28}ClNO.C_6H_8O_7$ dissolved from ultraviolet absorbances at the wavelength of maximum absorbance at about 232 nm of filtered portions of the solution under test, suitably diluted with 0.1 *N* hydrochloric acid, in comparison with a Standard solution having a known concentration of USP Clomiphene Citrate RS in the same medium.

Tolerances—Not less than 75% (*Q*) of the labeled amount of $C_{26}H_{28}ClNO.C_6H_8O_7$ is dissolved in 60 minutes.

Uniformity of dosage units ⟨905⟩: meet the requirements.

Procedure for content uniformity—Transfer 1 finely powdered Tablet to a 100-mL volumetric flask, add 80 mL of 0.1 *N* isopropanolic hydrochloric acid [prepared by diluting hydrochloric acid with dilute isopropyl alcohol (3 in 10) to obtain the 0.1 *N* acid], and shake vigorously by mechanical means for 20 minutes. Add the 0.1 *N* isopropanolic hydrochloric acid to volume, and mix. Filter, discarding the first 20 mL of the filtrate. Concomitantly determine the absorbances of this solution, suitably diluted, if necessary, and a Standard solution of USP Clomiphene Citrate RS in the same medium having a known concentration of about 25 µg per mL, in 1-cm cells at the wavelength of maximum absorbance at about 292 nm, with a suitable spectrophotometer, using the 0.1 *N* isopropanolic hydrochloric acid as the blank. Calculate the quantity, in mg, of $C_{26}H_{28}ClNO.C_6H_8O_7$ in the Tablet by the formula:

$$(TC/D)(A_U/A_S),$$

in which *T* is the labeled quantity, in mg, of clomiphene citrate in the Tablet, *D* is the concentration, in µg per mL, of clomiphene citrate in the solution from the Tablet, based on the labeled quantity per Tablet and the extent of dilution, *C* is the concentration, in µg per mL, of USP Clomiphene Citrate RS in the Standard solution, and A_U and A_S are the absorbances of the solution from the Tablet and the Standard solution, respectively.

Assay—

Standard preparation—Dissolve about 50 mg of USP Clomiphene Citrate RS, accurately weighed, in 100 mL of dilute isopropyl alcohol (1 in 2), and dilute with a 1 in 10 solution of hydrochloric acid in isopropyl alcohol to obtain a solution having a known concentration of about 12.5 µg per mL.

Assay preparation—Weigh and finely powder not less than 20 Clomiphene Citrate Tablets. Transfer an accurately weighed portion of the powder, equivalent to about 100 mg of clomiphene citrate, to a 200-mL volumetric flask, add dilute methanol (1 in 2) to volume, and mix for not less than 30 minutes. Transfer about 25 mL of the solution to a glass-stoppered centrifuge tube, and centrifuge until clear.

Chromatographic column—Use a 25-cm × 15-mm column, made with a constriction and fitted with a stopcock. Place a pledget of glass wool at the point of constriction. To the column add, in aqueous slurry, sufficient polystyrene cation-exchange resin (strong acid) to a height of 2.5 cm. Wash the column, in the order named, with 150 mL of water and 25 mL of dilute methanol (1 in 2). Add the last wash solution in two equal portions, and deaerate the column after the first portion has been added. Proceed immediately with the *Procedure*.

Procedure—Pipet 10 mL of the supernatant liquid from the *Assay preparation* onto the prepared *Chromatographic column*, and adjust the flow rate to about 1 mL per minute. After the specimen has passed through the column, wash it, at a flow rate of about 2 mL per minute, with 50 mL of dilute methanol (1 in 2) and then with 50 mL of isopropyl alcohol. Elute the clomiphene from the column, at a flow rate of about 2 mL per minute, with 100 mL of a 1 in 10 solution of hydrochloric acid in isopropyl alcohol, collecting the eluate in a 100-mL volumetric flask, add the same solution to volume, and mix. Pipet 25 mL of the eluate into a 100-mL volumetric flask, and add the 1 in 10 solution of hydrochloric acid in isopropyl alcohol to volume, and mix. Concomitantly determine the absorbances of the *Assay preparation* and the *Standard preparation* in 1-cm cells at 232.5 nm, with a suitable spectrophotometer, using a 1 in 10 solution of hydrochloric acid in isopropyl alcohol as the blank. Calculate the quantity, in mg, of $C_{26}H_{28}ClNO.C_6H_8O_7$ in the portion of Tablets taken by the formula:

$$8C(A_U/A_S),$$

in which *C* is the concentration, in µg per mL, of USP Clomiphene Citrate RS in the *Standard preparation*, and A_U and A_S are the absorbances of the *Assay preparation* and the *Standard preparation*, respectively.

Clonazepam

$C_{15}H_{10}ClN_3O_3$ 315.72

2*H*-1,4-Benzodiazepin-2-one, 5-(2-chlorophenyl)-1,3-dihydro-7-nitro-.

5-(*o*-Chlorophenyl)-1,3-dihydro-7-nitro-2*H*-1,4-benzodiazepin-2-one [*1622-61-3*].

» Clonazepam contains not less than 99.0 percent and not more than 101.0 percent of $C_{15}H_{10}ClN_3O_3$, calculated on the dried basis.

Packaging and storage—Preserve in tight, light-resistant containers, at room temperature.

Reference standards—*USP Clonazepam Reference Standard*—Dry at 105° for 4 hours before using. *USP 3-Amino-4-(2-chlorophenyl)-6-nitrocarbostyril Reference Standard* and *USP 2-Amino-2′-chloro-5-nitrobenzophenone Reference Standard*—Do not dry before using.

Identification—The infrared absorption spectrum of a potassium bromide dispersion of it, previously dried, exhibits maxima only at the same wavelengths as that of a similar preparation of USP Clonazepam RS.

Loss on drying ⟨731⟩—Dry it at 105° for 4 hours: it loses not more than 0.5% of its weight.

Residue on ignition ⟨281⟩: not more than 0.1%.

Heavy metals, *Method II* ⟨231⟩: 0.002%.

Related compounds—Transfer 250 mg to a 10-mL volumetric flask, add acetone to volume, and mix. Apply 20 µL of this solution to a thin-layer chromatographic plate (see *Chromatography* ⟨621⟩) coated with a 0.25-mm layer of chromatographic silica gel mixture. Apply to the same plate 20 µL of an acetone solution containing 125 µg per mL of USP 3-Amino-4-(2-chlorophenyl)-6-nitrocarbostyril RS and 20 µL of an acetone solution containing 125 µg per mL of USP 2-Amino-2′-chloro-5-nitrobenzophenone RS. Develop the chromatogram in a solvent system consisting of ethyl acetate and carbon tetrachloride (1:1) until the solvent front has moved about three-fourths of the length of the plate. Remove the plate from the developing chamber, mark the solvent front, and air-dry it. Locate the spots on the plate by heavily spraying with 3 *N* sulfuric acid, dry at 105° for 15 to 30 minutes, and spray successively with sodium nitrite solution (1 in 1000), ammonium sulfamate solution (1 in 200), and *N*-1-naphthylethylenediamine dihydrochloride (1 in 1000), drying the plate with a stream of cool air after each spraying: any spots obtained from the solution under test are not greater in size or intensity than the spots at the respective R_f values produced by the Standard solutions, corresponding to not more than 0.5% of 3-amino-4-(2-chlorophenyl)-6-nitrocarbostyril, and not more than 0.5% of 2-amino-2′-chloro-5-nitrobenzophenone.

Assay—Dissolve about 700 mg of Clonazepam, accurately weighed, in 100 mL of acetic anhydride by stirring by mechanical means for about 20 minutes, add 5 drops of a 1 in 100 solution of Nile blue hydrochloride in glacial acetic acid (1 in 100), and titrate with 0.1 *N* perchloric acid VS to a yellow-green end-point. Perform a blank determination, and make any necessary correction. Each mL of 0.1 *N* perchloric acid is equivalent to 31.57 mg of $C_{15}H_{10}ClN_3O_3$.

Clonazepam Tablets

» Clonazepam Tablets contain not less than 90.0 percent and not more than 110.0 percent of the labeled amount of $C_{15}H_{10}ClN_3O_3$.

Packaging and storage—Preserve in tight, light-resistant containers, at room temperature.

Reference standards—*USP Clonazepam Reference Standard*—Dry at 105° for 4 hours before using. *USP 3-Amino-4-(2-chlorophenyl)-6-nitrocarbostyril Reference Standard*—Do not dry before using. *USP 2-Amino-2′-chloro-5-nitrobenzophenone Reference Standard*—Do not dry before using.

Identification—Place an amount of finely powdered Tablets, equivalent to about 10 mg of clonazepam, in a 125-mL separator. Add 25 mL of water, shake for 2 minutes, and extract with two 40-mL portions of chloroform. Pass the extracts through anhydrous sodium sulfate, combine them, and evaporate at room temperature with the aid of a stream of nitrogen to dryness. Wash the residue with three 10-mL portions of solvent hexane: the infrared absorption spectrum of a potassium bromide dispersion of the residue so obtained exhibits maxima only at the same wavelengths as that of a similar preparation of USP Clonazepam RS.

Dissolution ⟨711⟩—

Medium: degassed water; 900 mL.

Apparatus 2: 100 rpm.

Time: 60 minutes.

Determine the amount of Clonazepam dissolved, using the following method.

Mobile phase—Prepare a filtered and degassed mixture of water, methanol, and acetonitrile (40:30:30). Make adjustments if necessary (see *System Suitability* under *Chromatography* ⟨621⟩).

Standard solution—Prepare a solution of USP Clonazepam RS in methanol having a known concentration of about 0.05 mg per mL. Quantitatively dilute a portion of this solution with *Dissolution Medium* to obtain a *Standard solution* having a known concentration similar to the expected concentration in the solution under test.

Chromatographic system (see *Chromatography* ⟨621⟩)—The liquid chromatograph is equipped with a 254-nm detector and a 4-mm × 30-cm column that contains packing L1. The flow rate is about 1 mL per minute. Chromatograph replicate injections of the *Standard solution*, and record the peak responses as directed under *Procedure:* the relative standard deviation is not more than 2.0%, and the tailing factor is not more than 2.0.

Procedure—Separately inject equal volumes (about 100 µL) of the *Standard solution* and the solution under test into the chromatograph, record the chromatograms, and measure the responses for the major peaks. Calculate the quantity of $C_{15}H_{10}ClN_3O_3$ dissolved by comparison of the peak responses obtained from the *Standard solution* and the test solution.

Tolerances—Not less than 80% (*Q*) of the labeled amount of $C_{15}H_{10}ClN_3O_3$ is dissolved in 60 minutes.

Uniformity of dosage units ⟨905⟩: meet the requirements.

Procedure for content uniformity—Transfer 1 intact Tablet to a suitable container, add an amount of hot water (approximately 60°) equal to one-fifth of the nominal capacity of the container, and shake by mechanical means for 30 minutes. Add an amount of *N,N′*-dimethylacetamide equal to two-fifths of the nominal capacity of the container, and mix. Heat on a steam bath for 5 minutes, and sonicate for an additional 5 minutes. Cool to room temperature, and add an accurately measured volume of a 1 in 25 solution of *o*-dichlorobenzene in acetonitrile, sufficient to provide about 4 µg of *o*-dichlorobenzene for each 40 µg of clonazepam that is present. Dilute quantitatively and stepwise, if necessary, with acetonitrile to obtain a Tablet preparation containing about 40 µg of clonazepam per mL. Similarly prepare a Standard solution having a known concentration of about 40 µg of USP Clonazepam RS per mL, and having the same concentration of *o*-dichlorobenzene as the Tablet preparation, in the same medium. Proceed as directed for *Procedure* in the *Assay*.

Related compounds—Accurately weigh a portion of ground Tablets, equivalent to 10 mg of clonazepam, into a suitable flask. Add about 20 mL of acetone, insert the stopper, and shake for 1 minute. Filter the solution, and evaporate the filtrate on a steam bath to dryness. Dissolve the residue in 0.5 mL of acetone, and apply 25 µL to a thin-layer chromatographic plate (see *Chromatography* ⟨621⟩) coated with a 0.25-mm layer of chromatographic silica gel mixture. Apply to the same plate 25 µL of a 1 in 5000 solution of USP 3-Amino-4-(2-chlorophenyl)-6-nitrocarbostyril RS in acetone and 25 µL of a 1 in 5000 solution of USP 2-Amino-2′-chloro-5-nitrobenzophenone RS in acetone. Develop the plate in a mixture of ethyl acetate and carbon tetrachloride (1:1) until the solvent front has moved about 15 cm from the origin. Remove the plate from the chromatographic chamber, allow the solvent to evaporate, and locate the bands by viewing under short-wavelength ultraviolet light. Spray the plate heavily with dilute sulfuric acid (1 in 10) and heat in an oven at 105° for 15 minutes. Cool to room temperature, spray with sodium nitrite solution (1 in 1000), and dry under a stream of hot air. Spray with ammonium sulfamate solution (1 in 200), and dry under a stream of hot air. Spray with *N*-1-naphthylenediamine dihydrochloride solution (1 in 1000), and dry under a stream of hot air: any bands produced by the Tablet solution are not greater in size or intensity than the spots at the respective R_f values produced by the Standard solutions, corresponding to not more than 1.0% of 3-Amino-4-(2-chlorophenyl)-6-nitrocarbostyril and not more than 1.0% of 2-Amino-2′-chloro-5-nitrobenzophenone.

Assay—

Internal standard solution—Pipet 4 mL of *o*-dichlorobenzene into a 100-mL volumetric flask. Dilute with acetonitrile to volume, and mix.

Standard preparation—Dissolve an accurately weighed quantity of USP Clonazepam RS in acetonitrile, and dilute quantitatively and stepwise with acetonitrile to obtain a solution having a known concentration of about 0.5 mg per mL. Pipet 4 mL of this solution and 4 mL of *Internal standard solution* into a 50-mL volumetric flask, dilute with acetonitrile to volume, and mix. The final solution contains 40 µg of clonazepam per mL.

Assay preparation—Weigh and finely powder 20 Clonazepam Tablets. Transfer an accurately weighed portion of the powder, equivalent to about 2 mg of clonazepam, to a 50-mL volumetric flask. Pipet 20 mL of *N,N′*-dimethylacetamide into the flask, and sonicate for 5 minutes. Cool to room temperature, add, by pipet, 4 mL of *Internal standard solution*, dilute with acetonitrile to volume, mix, and filter if necessary.

Chromatographic system—The chromatograph is equipped with a 254-nm detector and a 4-mm × 30-cm column that contains packing L1. The mobile phase consists of a mixture of water, methanol, and acetonitrile (about 4:3:3). Chromatograph five replicate injections of the *Standard preparation*, and record the peak responses as directed under *Procedure:* the relative standard deviation is not more than 2.0%, and the resolution factor is not less than 10.0.

Procedure—Separately inject equal volumes (about 25 µL) of the *Standard preparation* and the *Assay preparation* into the chromatograph, and record the chromatograms. Measure the peak responses at approximate retention times of 6 minutes for clonazepam and 14.5 minutes for *o*-dichlorobenzene. Calculate the quantity, in mg, of $C_{15}H_{10}ClN_3O_3$ in the portion of Tablets taken by the formula:

$$0.05C(R_U/R_S),$$

in which C is the concentration, in µg per mL, of USP Clonazepam RS in the *Standard preparation*, and R_U and R_S are the ratios of peak responses of clonazepam and *o*-dichlorobenzene obtained from the *Assay preparation* and the *Standard preparation*, respectively.

Clonidine Hydrochloride

$C_9H_9Cl_2N_3 \cdot HCl$ 266.56

Benzenamine, 2,6-dichloro-*N*-2-imidazolidinylidene-, monohydrochloride.
2-[(2,6-Dichlorophenyl)imino]imidazolidine monohydrochloride [*4205-91-8*].

» Clonidine Hydrochloride contains not less than 98.5 percent and not more than 101.0 percent of $C_9H_9Cl_2N_3 . HCl$, calculated on the dried basis.

Packaging and storage—Preserve in tight containers.

Reference standard—*USP Clonidine Hydrochloride Reference Standard*—Dry at 105° to constant weight before using.

Identification—

A: The infrared absorption spectrum of a potassium bromide dispersion of it, previously dried, exhibits maxima only at the same wavelengths as that of a similar preparation of USP Clonidine Hydrochloride RS.

B: The ultraviolet absorption spectrum of a 1 in 3000 solution in 0.01 *N* hydrochloric acid exhibits maxima and minima at the same wavelengths as that of a similar solution of USP Clonidine Hydrochloride RS, concomitantly measured, and the respective absorptivities, calculated on the dried basis, at the wavelength of maximum absorbance at about 272 nm do not differ by more than 3.0%.

C: It responds to the tests for *Chloride* ⟨191⟩.

pH ⟨791⟩: between 3.5 and 5.5, in a solution (1 in 20).

Loss on drying ⟨731⟩—Dry it at 105° to constant weight: it loses not more than 0.5% of its weight.

Residue on ignition ⟨281⟩: not more than 0.1%.

Chromatographic purity—Dissolve 200 mg in methanol, and dilute with methanol to 2.0 mL to obtain the *Test solution.* Dissolve a suitable quantity of USP Clonidine Hydrochloride RS in methanol to obtain a *Standard solution* having a known concentration of 100 mg per mL. Dilute a portion of this solution quantitatively and stepwise with methanol to obtain a *Diluted standard solution* having a concentration of 100 μg per mL. On a suitable thin-layer chromatographic plate (see *Chromatography* ⟨621⟩), coated with a 0.25-mm layer of chromatographic silica gel mixture, apply separate 2-μL portions of the *Test solution*, the *Standard solution*, and the *Diluted standard solution.* Develop the chromatogram in a freshly prepared solvent system consisting of a mixture of toluene, dioxane, dehydrated alcohol, and ammonium hydroxide (10:8:2:1), until the solvent front has moved three-fourths of the length of the plate. Remove the plate from the chamber, allow the solvent to evaporate, and dry it at 100° for 1 hour. Dip the plate into a dipping chamber filled to three-fourths of its height with sodium hypochlorite solution, diluted to contain 0.5% available chlorine, dry in a fume hood with a current of air for 1 hour, and spray with starch–potassium iodide TS: the R_f value of the principal spot from the *Test solution* corresponds to that of the *Standard solution.* Any other spot obtained from the *Test solution* does not exceed, in size or intensity, the principal spot obtained from the *Diluted standard solution* (0.1%), and the total of any spots does not exceed 0.2%.

Assay—Dissolve about 200 mg of Clonidine Hydrochloride, accurately weighed, in about 80 mL of glacial acetic acid, add 15 mL of mercuric acetate TS, and titrate with 0.1 *N* perchloric acid VS, determining the end-point potentiometrically, using a glass electrode and a sleeve-type calomel electrode containing 0.1 *N* lithium perchlorate in glacial acetic acid (see *Titrimetry* ⟨541⟩). Perform a blank determination, and make any necessary correction. Each mL of 0.1 *N* perchloric acid is equivalent to 26.66 mg of $C_9H_9Cl_2N_3.HCl$.

Clonidine Hydrochloride Tablets

» Clonidine Hydrochloride Tablets contain not less than 90.0 percent and not more than 110.0 percent of the labeled amount of $C_9H_9Cl_2N_3 . HCl$.

Packaging and storage—Preserve in well-closed containers.

Reference standard—*USP Clonidine Hydrochloride Reference Standard*—Dry at 105° to constant weight before using.

Identification—

A: Transfer a quantity of finely powdered Tablets, equivalent to about 500 μg of clonidine hydrochloride, to a separator containing 30 mL of water and 5 mL of 1 *N* sodium hydroxide. Swirl gently to dissolve the test specimen, and extract with 20 mL of chloroform. Allow the layers to separate, centrifuge the chloroform layer, dry the chloroform with anhydrous sodium sulfate, and filter. Evaporate the filtrate to dryness, and dissolve the residue in 8 mL of 0.01 *N* hydrochloric acid: the ultraviolet absorption spectrum of the resulting solution exhibits maxima only at the same wavelengths as that of a similar solution of USP Clonidine Hydrochloride RS, concomitantly measured.

B: Transfer a quantity of finely powdered Tablets, equivalent to about 1 mg of clonidine hydrochloride, to a separator containing 30 mL of water and 5 mL of 1 *N* sodium hydroxide. Swirl gently to dissolve the test specimen, and extract with 20 mL of chloroform. Allow the layers to separate, centrifuge the chloroform layer, dry the chloroform with anhydrous sodium sulfate, and filter. Evaporate the filtrate to dryness, and dissolve the residue in 0.1 mL of methanol to obtain the test solution. Proceed as directed in the test for *Chromatographic purity* under *Clonidine Hydrochloride*, beginning with "Dissolve a suitable quantity of USP Clonidine Hydrochloride RS," except to omit the *Diluted standard solution:* the R_f value of the principal spot obtained from the test solution corresponds to that obtained from the Standard solution.

Dissolution ⟨711⟩—

Medium: water; 500 mL.

Apparatus 2: 50 rpm.

Time: 30 minutes.

Procedure—Determine the amount of $C_9H_9Cl_2N_3 . HCl$ dissolved, employing the procedure set forth in the *Assay*, making any necessary modifications.

Tolerances—Not less than 75% (*Q*) of the labeled amount of $C_9H_9Cl_2N_3 . HCl$ is dissolved in 30 minutes.

Uniformity of dosage units ⟨905⟩: meet the requirements.

Assay—

Mobile phase—Dissolve 1.1 g of sodium 1-octanesulfonate in 500 mL of water, add 500 mL of methanol, and 1 mL of phosphoric acid. Mix, and adjust with 1 *N* sodium hydroxide to a pH of 3.0. Mix, filter through a 0.45-μm or finer porosity filter, and degas. Make adjustments if necessary (see *System Suitability* under *Chromatography* ⟨621⟩).

Standard clonidine hydrochloride stock solution—Dissolve an accurately weighed quantity of USP Clonidine Hydrochloride RS in *Mobile phase*, and dilute quantitatively, and stepwise if necessary, with *Mobile phase* to obtain a solution having a known concentration of about 100 μg per mL.

2,6-Dichloroaniline stock solutions—Transfer about 12 mg of 2,6-dichloroaniline to a 100-mL volumetric flask, add *Mobile phase* to volume, and mix. Dilute an accurately measured volume of this solution quantitatively with *Mobile phase* to obtain *2,6-Dichloroaniline stock solution A* having a known concentration of about 12 μg per mL. Dilute an accurately measured volume of *2,6-Dichloroaniline stock solution A* quantitatively with *Mobile phase* to obtain *2,6-Dichloroaniline stock solution B* having a known concentration of about 1.2 μg per mL.

Standard preparation—Transfer 2.0 mL of *Standard clonidine hydrochloride stock solution* and 20.0 mL of *2,6-Dichloroaniline stock solution B* to a 200-mL volumetric flask, dilute with *Mobile phase* to volume, and mix. This solution contains about 1 μg of clonidine hydrochloride and about 0.12 μg of 2,6-dichloroaniline per mL.

Assay preparation—Weigh and finely powder not less than 20 Clonidine Hydrochloride Tablets. Transfer an accurately weighed portion of the powder, equivalent to about 0.1 mg of clonidine hydrochloride, to a 100-mL volumetric flask. Add about 60 mL of *Mobile phase*, shake by mechanical means for 15 to 30 minutes, dilute with *Mobile phase* to volume, and mix. Centrifuge a portion of this solution to obtain a clear solution (*Assay preparation*).

System suitability preparation—Transfer 2.0 mL of *Standard clonidine hydrochloride stock solution* and 20.0 mL of *2,6-Di-*

chloroaniline stock solution A to a 100-mL volumetric flask, dilute with *Mobile phase* to volume, and mix.

Chromatographic system (see *Chromatography* ⟨621⟩)—The liquid chromatograph is equipped with a 220-nm detector and a 4.6-mm × 15-cm column that contains packing L7 which has been deactivated for basic compounds. The flow rate is about 1.5 mL per minute. Chromatograph the *Standard preparation*, and record the peak responses as directed under *Procedure:* the relative standard deviation for replicate injections is not more than 2.0% for the clonidine hydrochloride peak and not more than 4.0% for the 2,6-dichloroaniline peak. Chromatograph the *System suitability preparation*, and record the peak responses as directed under *Procedure:* the column efficiency determined from the clonidine hydrochloride peak is not less than 3500 theoretical plates, the tailing factor for the clonidine hydrochloride peak is not more than 1.5, the column efficiency determined from the 2,6-dichloroaniline peak is not less than 5000 theoretical plates, the tailing factor for the 2,6-dichloroaniline peak is not more than 1.3, and the resolution, *R*, between the clonidine hydrochloride and 2,6-dichloroaniline peaks is not less than 10.0. The relative retention times are about 0.5 for clonidine hydrochloride and 1.0 for 2,6-dichloroaniline.

Procedure—[NOTE—Use peak areas where peak responses are indicated.] Separately inject equal volumes (about 50 µL) of the *Standard preparation* and the *Assay preparation* into the chromatograph, record the chromatograms, and measure the responses for the major peaks. Calculate the quantity, in mg, of $C_9H_9Cl_2N_3 . HCl$ in the portion of Tablets taken by the formula:

$$0.1C(r_U/r_S),$$

in which *C* is the concentration, in µg per mL, of USP Clonidine Hydrochloride RS in the *Standard preparation*, and r_U and r_S are the peak responses obtained from the *Assay preparation* and the *Standard preparation*, respectively.

Clonidine Hydrochloride and Chlorthalidone Tablets

» Clonidine Hydrochloride and Chlorthalidone Tablets contain not less than 90.0 percent and not more than 110.0 percent of the labeled amount of chlorthalidone ($C_{14}H_{11}ClN_2O_4S$) and not less than 90.0 percent and not more than 110.0 percent of the labeled amount of clonidine hydrochloride ($C_9H_9Cl_2N_3 . HCl$).

Packaging and storage—Preserve in well-closed containers.

Reference standards—*USP Chlorthalidone Reference Standard*—Dry at 105° for 4 hours before using. *USP Clonidine Hydrochloride Reference Standard*—Dry at 105° to constant weight before using.

Identification—

A: Transfer an amount of powdered Tablets, equivalent to about 3 mg of clonidine hydrochloride, to a beaker, add 30 mL of water, stir for 5 minutes, and filter, using a medium-porosity, sintered-glass funnel. Transfer the filtrate to a separator, add 5 mL of 0.1 *N* sodium hydroxide, and extract with 20 mL of chloroform, collecting the chloroform extract in a separator. Extract the chloroform phase with 15 mL of 0.01 *N* hydrochloric acid, collecting the acid extract in a beaker. Remove any residual chloroform from the acid extract by heating on a steam bath: the ultraviolet absorption spectrum of the solution so obtained exhibits maxima and minima at the same wavelengths as that of a similar solution of USP Clonidine Hydrochloride RS, concomitantly measured.

B: Transfer 10 powdered Tablets to a 50-mL beaker, add 10 mL of methanol, boil on a steam bath for 5 minutes, and filter. Add 20 mL of water to the filtrate, and boil on a steam bath for 5 minutes under a current of air. Cool, with stirring, in ice until crystals form, filter the crystals, and dry at 105° for 1 hour: the chlorthalidone so isolated melts between 215° and 222° (see *Melting Range or Temperature* ⟨741⟩).

C: The retention times of the chlorthalidone and clonidine hydrochloride peaks in the chromatogram of the *Assay preparation* correspond to those of the *Standard preparation*, as obtained in the *Assay*.

Dissolution ⟨711⟩—

Medium: water; 900 mL.

Apparatus 2: 100 rpm.

Time: 60 minutes.

Procedure—Pipet 20 mL of a centrifuged portion of the solution under test into a 25-mL volumetric flask, and dilute with 0.5% monobasic ammonium phosphate solution to volume. Use the resulting solution as the *Assay preparation*. Determine the amounts of chlorthalidone ($C_{14}H_{11}ClN_2O_4S$) and clonidine hydrochloride ($C_9H_9Cl_2N_3 . HCl$) dissolved, employing the procedure set forth in the *Assay*, making any necessary volumetric adjustments.

Tolerances—Not less than 50% (*Q*) of the labeled amount of $C_{14}H_{11}ClN_2O_4S$ and not less than 80% (*Q*) of the labeled amount of $C_9H_9Cl_2N_3 . HCl$ are dissolved in 60 minutes.

Uniformity of dosage units ⟨905⟩: meet the requirements for *Content Uniformity* with respect to both chlorthalidone and clonidine hydrochloride.

Assay—

Mobile phase—Dissolve 800 mg of monobasic ammonium phosphate in 800 mL of water, add 100 mL of methanol and 100 mL of acetonitrile, mix, filter, and degas. Make adjustments if necessary (see *System Suitability* under *Chromatography* ⟨621⟩).

Standard preparation—Dissolve an accurately weighed quantity of USP Clonidine Hydrochloride RS in 0.1% monobasic ammonium phosphate solution, and dilute quantitatively with the same solvent to obtain a solution having a known concentration of about 1500*J* µg per mL, *J* being the ratio of the labeled amount, in mg, of clonidine hydrochloride to the labeled amount, in mg, of chlorthalidone per Tablet (*Solution P*). Transfer about 15 mg of USP Chlorthalidone RS, accurately weighed, to a 100-mL volumetric flask, dissolve in 10 mL of methanol, add 25 mL of 0.1% monobasic ammonium phosphate solution and 10.0 mL of *Solution P*, dilute with 0.1% monobasic ammonium phosphate solution to volume, and mix to obtain a solution having known concentrations of about 150*J* µg of USP Clonidine Hydrochloride RS per mL and about 150 µg of USP Chlorthalidone RS per mL.

Assay preparation—Weigh and finely powder not less than 20 Clonidine Hydrochloride and Chlorthalidone Tablets. Transfer an accurately weighed portion of the powder, equivalent to about 15 mg of chlorthalidone, to a 100-mL volumetric flask, add 10 mL of methanol, and sonicate for 5 minutes. Add 40 mL of 0.1% monobasic ammonium phosphate solution, sonicate until the solution is free from agglomerates, allow to cool to ambient temperature, dilute with 0.1% monobasic ammonium phosphate solution to volume, mix, and centrifuge.

Chromatographic system (see *Chromatography* ⟨621⟩)—The liquid chromatograph is equipped with a 220-nm detector and a 4.6-mm × 10-cm column that contains packing L7. The flow rate is about 2 mL per minute. Chromatograph the *Standard preparation*, and record the peak responses as directed under *Procedure:* the resolution, *R*, between the clonidine hydrochloride and chlorthalidone peaks is not less than 3, and the relative standard deviation for replicate injections is not more than 2%.

Procedure—Separately inject equal volumes (about 20 µL) of the *Standard preparation* and the *Assay preparation* into the chromatograph, record the chromatograms, and measure the responses for the major peaks. The relative retention times are about 0.2 for clonidine hydrochloride and 1.0 for chlorthalidone. Calculate the quantity, in mg, of clonidine hydrochloride ($C_9H_9Cl_2N_3 . HCl$) in the portion of Tablets taken by the formula:

$$0.1C(r_U/r_S),$$

in which *C* is the concentration, in µg per mL, of USP Clonidine Hydrochloride RS in the *Standard preparation*, and r_U and r_S are the peak responses of clonidine hydrochloride obtained from the *Assay preparation* and the *Standard preparation*, respectively. Calculate the quantity, in mg, of chlorthalidone ($C_{14}H_{11}ClN_2O_4S$) in the portion of Tablets taken by the same formula, changing the terms to refer to chlorthalidone.

Clorazepate Dipotassium

$C_{16}H_{11}ClK_2N_2O_4$ 408.92

1*H*-1,4-Benzodiazepine-3-carboxylic acid, 7-chloro-2,3-dihydro-2-oxo-5-phenyl-, potassium salt compound with potassium hydroxide (1:1).

Potassium 7-chloro-2,3-dihydro-2-oxo-5-phenyl-1*H*-1,4-benzodiazepine-3-carboxylate compound with potassium hydroxide (1:1) [*57109-90-7*].

» Clorazepate Dipotassium contains not less than 98.5 percent and not more than 101.5 percent of $C_{16}H_{11}ClK_2N_2O_4$, calculated on the dried basis.

Packaging and storage—Preserve under nitrogen in tight, light-resistant containers.

Reference standards—*USP Clorazepate Dipotassium Reference Standard*—Dry in vacuum at 60° for 1 hour before using. *USP 2-Amino-5-chlorobenzophenone Reference Standard*—Keep container tightly closed and protected from light. Dry over silica gel for 4 hours before using.

Identification—

A: The infrared absorption spectrum of a mineral oil dispersion of it, previously dried, exhibits maxima only at the same wavelengths as that of a similar preparation of USP Clorazepate Dipotassium RS.

B: The ultraviolet absorption spectrum of a 1 in 150,000 solution in sodium hydroxide solution (1 in 2500) exhibits maxima and minima at the same wavelengths as that of a similar solution of USP Clorazepate Dipotassium RS, concomitantly measured.

Loss on drying ⟨731⟩—Dry it in vacuum at 60° for 1 hour: it loses not more than 0.5% of its weight.

Heavy metals, *Method II* ⟨231⟩: 0.002%.

Related compounds—

Diluting solution—Prepare a solution by dissolving 4 g of potassium carbonate in 100 mL of water.

Standard preparation A—Dissolve an accurately weighed quantity of USP Clorazepate Dipotassium RS in *Diluting solution* to obtain a solution having a concentration of 20 mg per mL.

Standard preparation B—Transfer 1.0 mL of *Standard preparation A* to a 200-mL volumetric flask, dilute with *Diluting solution* to volume, and mix.

Standard preparation C—Transfer an accurately weighed quantity of about 10 mg of USP 2-Amino-5-chlorobenzophenone RS to a 100-mL volumetric flask, dissolve in acetone, dilute with acetone to volume, and mix. Dilute 10.0 mL of this solution with acetone to 50.0 mL.

Test preparation—Dissolve an accurately weighed quantity of about 100 mg of Clorazepate Dipotassium in 5.0 mL of *Diluting solution*.

Procedure—On a suitable thin-layer chromatographic plate (see *Chromatography* ⟨621⟩), coated with a 0.25-mm layer of chromatographic silica gel mixture, apply separately 5 µL of the *Test preparation* and 5 µL of each *Standard preparation*. Allow the spots to dry, and develop the chromatograms in a solvent system consisting of a mixture of chloroform and acetone (85:15) until the solvent front has moved about three-fourths of the length of the plate. Remove the plate from the developing chamber, mark the solvent front, and allow the solvent to evaporate. Examine the plate under short-wavelength ultraviolet light, and compare the intensities of any secondary spots observed in the chromatogram of the *Test preparation* with those of the principal spots in the chromatograms of *Standard preparations A* and *B*. No secondary spot from the chromatogram of the *Test preparation* is larger or more intense than the principal spot obtained from *Standard preparation B* (0.5%). Successively spray the plate with a freshly prepared 1% solution of sodium nitrite in 1 *N* hydrochloric acid. Dry the plate in a current of air and spray with a 0.4% (w/v) solution of *N*-(1-Napthyl)ethylenediamine dihydrochloride in alcohol. Any violet-colored spot present in the chromatogram of the *Test preparation* is not greater in size or intensity than the principal spot obtained from *Standard preparation C* (0.1%). The sum of the intensities of all secondary spots obtained by both visualization techniques from the *Test preparation* corresponds to not more than 1%.

Assay—Transfer about 150 mg of Clorazepate Dipotassium, accurately weighed, to a 250-mL beaker, add 100 mL of glacial acetic acid, and stir until dissolved. Titrate with 0.1 *N* perchloric acid VS, determining the end-point potentiometrically, using a glass electrode and a calomel electrode containing a 1 in 100 solution of lithium perchlorate in glacial acetic acid. Perform a blank determination (see *Titrimetry* ⟨521⟩), and make any necessary correction. Each mL of 0.1 *N* perchloric acid is equivalent to 13.63 mg of $C_{16}H_{11}ClK_2N_2O_4$.

Clotrimazole

$C_{22}H_{17}ClN_2$ 344.84

1*H*-Imidazole, 1-[(2-chlorophenyl)diphenylmethyl]-.

1-(*o*-Chloro-α,α-diphenylbenzyl)imidazole [*23593-75-1*].

» Clotrimazole contains not less than 98.0 percent and not more than 102.0 percent of $C_{22}H_{17}ClN_2$, calculated on the dried basis.

Packaging and storage—Preserve in tight containers.

Reference standards—*USP Clotrimazole Reference Standard*—Dry at 105° for 2 hours before using. *USP (o-Chlorophenyl)diphenylmethanol Reference Standard*—Do not dry before using. *USP Imidazole Reference Standard*—Do not dry before using.

Identification—

A: The infrared absorption spectrum of a mineral oil dispersion of it, previously dried, exhibits maxima only at the same wavelengths as that of a similar preparation of USP Clotrimazole RS.

B: It responds to the *Thin-layer Chromatographic Identification Test* ⟨201⟩, a solution containing about 20 mg per mL of Clotrimazole in chloroform being used as the test solution, and a solvent system consisting of a mixture of xylene, *n*-propyl alcohol, and ammonium hydroxide (180:20:1) being used.

Loss on drying ⟨731⟩—Dry it at 105° for 2 hours: it loses not more than 0.5% of its weight.

Residue on ignition ⟨281⟩: not more than 0.1%.

Heavy metals, *Method II* ⟨231⟩: 0.001%.

Imidazole—Dissolve 500 mg, accurately weighed, in 5.0 mL of chloroform. On a suitable thin-layer chromatographic plate (see *Chromatography* ⟨621⟩), coated with a 0.25-mm layer of chromatographic silica gel mixture, apply 5 µL of this solution and 5 µL of a Standard solution of USP Imidazole RS in chloroform containing 500 µg per mL. Allow the spots to dry, and develop the chromatogram in a solvent system consisting of a mixture of methanol and chloroform (3:2) until the solvent front has moved about three-fourths of the length of the plate. Remove the plate from the developing chamber, mark the solvent front, allow it to air-dry for 5 minutes, place it in a closed container with a dish containing 100 g of iodine in a shallow layer, and allow to remain for 60 minutes. Remove the plate from the container, and observe the chromatogram: any brown spot obtained from the solution under test, at an R_f value corresponding to the principal spot from the Standard solution, is not greater in size or intensity than the principal spot obtained from the Standard solution (0.5% of imidazole).

(o-Chlorophenyl)diphenylmethanol—

Dibasic potassium phosphate solution, Mobile phase, Resolution solution, and *Chromatographic system*—Prepare as directed in the *Assay.*

Standard preparation—Transfer about 12.5 mg of USP (o-Chlorophenyl)diphenylmethanol RS, accurately weighed, to a 25-mL volumetric flask, add 10 mL of methanol to dissolve it, add 6.25 mL of *Dibasic potassium phosphate solution,* dilute with methanol to volume, and mix (*Stock solution*). Transfer 5.0 mL of this *Stock solution* to a 50-mL volumetric flask, dilute with *Mobile phase* to volume, and mix.

Test preparation—Use the *Stock assay solution* used to prepare the *Assay preparation* in the *Assay.*

Procedure—Separately inject equal volumes (about 20 µL) of the *Standard preparation* and the *Test preparation* into the chromatograph, record the chromatograms, and measure the responses for the major peaks. Calculate the percentage of (o-chlorophenyl)diphenylmethanol in the specimen taken by the formula:

$$1000(C/W)(r_U/r_S),$$

in which C is the concentration, in mg per mL, of USP (o-Chlorophenyl)diphenylmethanol RS in the *Standard preparation, W* is the weight, in mg of Clotrimazole taken, and r_U and r_S are the responses for (o-chlorophenyl)diphenylmethanol obtained from the *Test preparation* and the *Standard preparation,* respectively: not more than 0.5% is found.

Assay—

Dibasic potassium phosphate solution—Dissolve 4.35 g of dibasic potassium phosphate in water to make 1000 mL of solution.

Mobile phase—Mix methanol and *Dibasic potassium* phosphate solution (3:1), filter through a membrane filter of 0.2-µm or finer porosity, and degas. The ratio of volumes may be changed to obtain the required resolution.

Internal standard solution—Transfer 33 mg of testosterone propionate to a 200-mL volumetric flask, add 125 mL of methanol to dissolve it, add 50 mL of *Dibasic potassium phosphate solution,* dilute with methanol to volume, and mix.

Standard preparation—Transfer about 50 mg of USP Clotrimazole RS, accurately weighed, to a 50-mL volumetric flask. Add 25 mL of methanol to dissolve it, add 12.5 mL of *Dibasic potassium phosphate solution,* dilute with methanol to volume, and mix (*Stock standard solution*). Transfer 10.0 mL of this *Stock standard solution* to a 100-mL volumetric flask, add 4.0 mL of *Internal standard solution,* dilute with *Mobile phase* to volume, and mix.

Resolution solution—Transfer 3 mL of the *Stock standard solution* used to prepare the *Standard preparation* and 5 mL of the *Stock solution* used to prepare the *Standard preparation* in the *(o-Chlorophenyl)diphenylmethanol* test to a 25-mL volumetric flask, dilute with *Mobile phase* to volume, and mix.

Assay preparation—Transfer about 100 mg of Clotrimazole, accurately weighed, to a 10-mL volumetric flask, add 5 mL of methanol to dissolve it, add 2.5 mL of *Dibasic potassium phosphate solution,* dilute with methanol to volume, and mix (*Stock assay solution*). Transfer 1.0 mL of this *Stock assay solution* to a 100-mL volumetric flask, add 4.0 mL of *Internal standard solution,* dilute with *Mobile phase* to volume, and mix.

Chromatographic system (see *Chromatography* ⟨621⟩)—The liquid chromatograph is equipped with a 254-nm detector and a 4.6-mm × 25-cm column that contains 10-µm packing L1. The flow rate is about 1.5 mL per minute. Chromatograph the *Resolution solution* and the *Standard preparation,* and record the peak responses as directed under *Procedure:* the resolution, R, between the clotrimazole and (o-chlorophenyl)diphenylmethanol peaks is not less than 1.9, and the relative standard deviation for replicate injections of the *Standard preparation* is not more than 2.0%. The relative retention times are about 0.7 for (o-chlorophenyl)diphenylmethanol, 1.0 for clotrimazole, and 1.5 for testosterone propionate.

Procedure—Separately inject equal volumes (about 20 µL) of the *Standard preparation* and the *Assay preparation* into the chromatograph, record the chromatograms, and measure the responses for the major peaks. Calculate the quantity, in mg, of clotrimazole ($C_{22}H_{17}ClN_2$) in the portion of Clotrimazole taken to prepare the *Assay preparation* by the formula:

$$1000C(R_U/R_S),$$

in which C is the concentration, in mg per mL, of USP Clotrimazole RS in the *Standard preparation,* and R_U and R_S are the ratios of the peak responses of the clotrimazole peak and the testosterone propionate peak obtained from the *Assay preparation* and the *Standard preparation,* respectively.

Clotrimazole Cream

» Clotrimazole Cream contains not less than 90.0 percent and not more than 110.0 percent of the labeled amount of $C_{22}H_{17}ClN_2$.

Packaging and storage—Preserve in collapsible tubes or in tight containers, at a temperature between 2° and 30°.

Reference standards—*USP Clotrimazole Reference Standard*—Dry at 105° for 2 hours before using. *USP (o-Chlorophenyl)diphenylmethanol Reference Standard*—Do not dry before using.

Labeling—Cream that is intended for use as a vaginal preparation may be labeled Clotrimazole Vaginal Cream.

Identification—In a suitable chromatographic chamber, arranged for thin-layer chromatography (see *Chromatography* ⟨621⟩), containing 200 mL of ether, place a beaker containing 25 mL of ammonium hydroxide. Cover the chamber, and allow to equilibrate for 2 hours. Place a portion of Cream, equivalent to about 5 mg of clotrimazole, in a 50-mL centrifuge tube, add 5 mL of chloroform, mix, and centrifuge to obtain a clear test solution. On a suitable thin-layer chromatographic plate (see *Chromatography* ⟨621⟩), coated with a 0.25-mm layer of chromatographic silica gel mixture, apply 20 µL of the lower, chloroform phase and 20 µL of a solution of USP Clotrimazole RS in chloroform containing 1 mg per mL. Develop the chromatogram until the solvent front has moved about three-fourths of the length of the plate. Remove the plate from the developing chamber, mark the solvent front, and allow the solvent to evaporate. Locate the spots on the plate by examination under short-wavelength ultraviolet light: the R_f value of the principal spot from the test solution corresponds to that obtained from the Standard solution. Dissolve 3 g of bismuth subnitrate and 30 g of potassium iodide in 10 mL of dilute hydrochloric acid (1 in 4), dilute with water to 100 mL, mix, and prepare a spray reagent by diluting 10 mL of this solution and 5 mL of dilute hydrochloric acid (1 in 4) with water to 200 mL, and mixing. Spray the plate evenly with this spray reagent: the principal spots from the test solution and the Standard solution are orange.

Assay—

Dibasic potassium phosphate solution and *Mobile phase*—Prepare as directed in the *Assay* under *Clotrimazole.*

Internal standard solution—Prepare a solution of testosterone propionate in dehydrated alcohol having a concentration of about 0.07 mg per mL.

Standard stock solution—Transfer about 50 mg of USP Clotrimazole RS, accurately weighed, to a 50-mL volumetric flask, dilute with dehydrated alcohol to volume, and mix.

Standard preparation—Mix 10.0 mL of *Standard stock solution* and 10.0 mL of *Internal standard solution.*

Resolution solution—Prepare a solution of USP (o-Chlorophenyl)diphenylmethanol RS in dehydrated alcohol having a concentration of about 0.12 mg per mL. Mix 7 mL of this solution and 1 mL of *Standard stock solution.*

Assay preparation—Weigh accurately a portion of Clotrimazole Cream, equivalent to about 10 mg of clotrimazole, and transfer to a 50-mL screw-capped centrifuge tube. Add 10.0 mL of *Internal standard solution* and heat at 50° in a water bath for 5 minutes, with occasional shaking. Remove the tube from the bath and shake vigorously for 5 minutes. Cool in a methanol ice bath for 15 minutes, and promptly centrifuge. Transfer the supernatant liquid to a test tube. Add 10.0 mL of dehydrated alcohol to the residue in the centrifuge tube, and repeat the extraction starting with "heat at 50° in a water bath." Transfer the supernatant liquid to the test tube containing the supernatant liquid from the first extraction, and mix. Use this solution as the *Assay preparation.*

Chromatographic system (see *Chromatography* ⟨621⟩)—The liquid chromatograph is equipped with a 254-nm detector and a 2.1-mm × 6-cm guard column that contains 10-μm packing L7, and a 3.9-mm × 30-cm analytical column that contains 10-μm packing L1. The flow rate is about 1 mL per minute. Chromatograph the *Resolution solution* and the *Standard preparation,* and record the peak responses as directed under *Procedure:* the resolution, *R*, between the (*o*-chlorophenyl)diphenylmethanol and clotrimazole peaks is not less than 1.2, and the relative standard deviation for replicate injections of the *Standard preparation* is not more than 2.0%. The relative retention times are about 0.9 for (*o*-chlorophenyl)diphenylmethanol, 1.0 for clotrimazole, and 1.5 for testosterone propionate.

Procedure—Separately inject equal volumes (about 20 μL) of the *Standard preparation* and the *Assay preparation* into the chromatograph, record the chromatograms, and measure the responses for the major peaks. Calculate the quantity, in mg, of clotrimazole ($C_{22}H_{17}ClN_2$) in each g of the Cream taken by the formula:

$$20(C/W)(R_U/R_S),$$

in which *C* is the concentration, in mg per mL, of USP Clotrimazole RS in the *Standard preparation, W* is the weight, in g, of Cream taken, and R_U and R_S are the ratios of the peak responses of the clotrimazole peak and the testosterone propionate peak obtained from the *Assay preparation* and the *Standard preparation,* respectively.

Clotrimazole Lotion

» Clotrimazole Lotion contains not less than 90.0 percent and not more than 110.0 percent of the labeled amount of $C_{22}H_{17}ClN_2$.

Packaging and storage—Preserve in tight containers, at a temperature between 2° and 30°.

Reference standards—*USP Clotrimazole Reference Standard*—Dry at 105° for 2 hours before using. *USP (o-Chlorophenyl)diphenylmethanol Reference Standard*—Do not dry before using.

Identification—The chromatogram of the *Assay preparation* obtained as directed in the *Assay* exhibits a major peak for clotrimazole, the retention time of which corresponds to that exhibited in the chromatogram of the *Standard preparation* obtained as directed in the *Assay.*

pH ⟨791⟩: between 5.0 and 7.0.

Microbial limits—It meets the requirements of the tests for absence of *Staphylococcus aureus* and *Pseudomonas aeruginosa* under *Microbial Limit Tests* ⟨61⟩.

(*o*-Chlorophenyl)diphenylmethanol—Using the chromatograms of the *Assay preparation* and the *Standard preparation* obtained as directed in the *Assay,* calculate the percentage of (*o*-chlorophenyl)diphenylmethanol in the portion of Lotion taken by the formula:

$$2000(C/Q_C)(R_U/R_S),$$

in which *C* is the concentration, in mg per mL, of USP (*o*-Chlorophenyl)diphenylmethanol RS in the *Standard preparation,* Q_C is the quantity, in mg, of clotrimazole ($C_{22}H_{17}ClN_2$) in the portion of Lotion taken, and R_U and R_S are the ratios of the peak responses of the (*o*-chlorophenyl)diphenylmethanol peak to the testosterone propionate peak obtained from the *Assay preparation* and the *Standard preparation,* respectively: not more than 5% is found.

Assay—

Dibasic potassium phosphate solution—Dissolve 4.35 g of dibasic potassium phosphate in water to make 1000 mL.

Mobile phase—Mix methanol and *Dibasic Potassium phosphate solution* (3:1), filter through a membrane filter of 0.5-μm or finer porosity, and degas.

Internal standard solution—Prepare a solution of testosterone propionate in dehydrated alcohol having a concentration of about 0.07 mg per mL.

Clotrimazole standard stock solution—Transfer about 50 mg of USP Clotrimazole RS, accurately weighed, to a 25-mL volumetric flask, add dehydrated alcohol to volume, and mix.

(o-Chlorophenyl)diphenylmethanol standard stock solution—Prepare a solution of USP (*o*-Chlorophenyl)diphenylmethanol RS in dehydrated alcohol having a known concentration of about 0.1 mg per mL.

Standard preparation—Mix 5.0 mL of *Clotrimazole standard stock solution,* 5.0 mL of *(o-Chlorophenyl)diphenylmethanol standard stock solution,* and 10.0 mL of *Internal standard solution.*

Assay preparation—Weigh accurately a portion of freshly mixed Clotrimazole Lotion, equivalent to about 10 mg of clotrimazole, and transfer to a screw-capped, 50-mL centrifuge tube. Add 10.0 mL of *Internal standard solution,* place the cap on the tube, and heat at 50° in a water bath for 5 minutes, with occasional shaking. Remove the tube from the bath, and shake vigorously for 5 minutes. Cool in a methanol-ice bath for 15 minutes, and promptly centrifuge. Transfer the supernatant liquid to a test tube. Add 10.0 mL of dehydrated alcohol to the residue in the centrifuge tube, and repeat the extraction as directed above, beginning with "place the cap on the tube." Transfer the supernatant liquid to the test tube containing the supernatant liquid from the first extraction, mix, and use this solution as the *Assay preparation.*

Chromatographic system (see *Chromatography* ⟨621⟩)—The liquid chromatograph is equipped with a 254-nm detector and a 2.1-mm × 6-cm guard column that contains 10-μm packing L2, and a 3.9-mm × 30-cm analytical column that contains 10-μm packing L1. The flow rate is about 1 mL per minute. Chromatograph the *Standard preparation,* and record the peak responses as directed under *Procedure:* the resolution, *R*, between the (*o*-chlorophenyl)diphenylmethanol and clotrimazole peaks is not less than 1.2, the resolution, *R*, between the clotrimazole and testosterone propionate peaks is not less than 1.9, and the relative standard deviation for replicate injections of the *Standard preparation* is not more than 2.0%. The relative retention times are about 0.9 for (*o*-chlorophenyl)diphenylmethanol, 1.0 for clotrimazole, and 1.5 for testosterone propionate.

Procedure—Separately inject equal volumes (about 20 μL) of the *Standard preparation* and the *Assay preparation* into the chromatograph, record the chromatograms, and measure the responses for the major peaks. Calculate the quantity, in mg, of clotrimazole ($C_{22}H_{17}ClN_2$) in each g of the Lotion taken by the formula:

$$20(C/W)(R_U/R_S),$$

in which *C* is the concentration, in mg per mL, of USP Clotrimazole RS in the *Standard preparation, W* is the weight, in g, of Lotion taken, and R_U and R_S are the ratios of the peak responses of the clotrimazole peak and the testosterone propionate peak obtained from the *Assay preparation* and the *Standard preparation,* respectively.

Clotrimazole Topical Solution

» Clotrimazole Topical Solution is a solution of Clotrimazole in a suitable nonaqueous, hydrophilic solvent. It contains not less than 90.0 percent and not more than 115.0 percent of the labeled amount of $C_{22}H_{17}ClN_2$.

Packaging and storage—Preserve in tight containers, at a temperature between 2° and 30°.

Reference standards—*USP Clotrimazole Reference Standard*—Dry at 105° for 2 hours before using. *USP (o-Chlorophenyl)diphenylmethanol Reference Standard*—Do not dry before using.

Identification—Transfer a volume of Solution, equivalent to about 10 mg of Clotrimazole, to a screw-capped, 50-mL centrifuge tube, and add 5 mL of dilute ammonium hydroxide (1 in 100) and 10 mL of chloroform. Shake vigorously, centrifuge to obtain a clear chloroform phase, and proceed as directed in the *Identification test* under *Clotrimazole Cream*.

Assay—

Dibasic potassium phosphate solution, Mobile phase, and *Chromatographic system*—Prepare as directed in the *Assay* under *Clotrimazole Cream.*

Internal standard solution—Transfer about 66 mg of testosterone propionate to a 100-mL volumetric flask, dissolve in 75 mL of methanol, dilute with *Dibasic potassium phosphate solution* to volume, and mix.

Standard preparation—Transfer about 50 mg of USP Clotrimazole RS to a 50-mL volumetric flask, add 5.0 mL of *Internal standard solution*, dilute with *Mobile phase* to volume, and mix.

Resolution solution—Prepare a solution of USP (*o*-Chlorophenyl)diphenylmethanol RS in methanol having a concentration of about 0.2 mg per mL. Transfer 12 mL of this solution to a 25-mL volumetric flask, add 4 mL of *Dibasic potassium phosphate solution* and 3 mL of *Standard preparation*, dilute with *Mobile phase* to volume, and mix.

Assay preparation—Transfer an accurately measured volume of Clotrimazole Topical Solution, equivalent to about 50 mg of clotrimazole, to a 50-mL volumetric flask, add 5.0 mL of *Internal standard solution*, dilute with *Mobile phase* to volume, and mix.

Procedure—Separately inject equal volumes (about 20 μL) of the *Standard preparation* and the *Assay preparation* into the chromatograph, record the chromatograms, and measure the responses for the major peaks. Calculate the quantity, in mg, of $C_{22}H_{17}ClN_2$ in each mL of the Topical Solution taken by the formula:

$$50(C/V)(R_U/R_S),$$

in which C is the concentration, in mg per mL, of USP Clotrimazole RS in the *Standard preparation*, V is the volume, in mL, of Topical Solution taken, and R_U and R_S are the ratios of the peak responses of the clotrimazole peak to the testosterone propionate peak obtained from the *Assay preparation* and the *Standard preparation*, respectively.

Clotrimazole Vaginal Tablets

» Clotrimazole Vaginal Tablets contain not less than 90.0 percent and not more than 110.0 percent of the labeled amount of $C_{22}H_{17}ClN_2$.

Packaging and storage—Preserve in well-closed containers.

Reference standards—*USP Clotrimazole Reference Standard*—Dry at 105° for 2 hours before using. *USP (o-Chlorophenyl)diphenylmethanol Reference Standard*—Do not dry before using.

Identification—Place an amount of finely powdered Tablets, equivalent to about 50 mg of clotrimazole, in a screw-capped, 50-mL centrifuge tube. Add 10 mL of chloroform, and shake vigorously for about 2 minutes. Centrifuge to clarify [NOTE—The supernatant liquid may remain slightly turbid]. Proceed as directed in the *Identification test* under *Clotrimazole Cream*, except to use a Standard solution of USP Clotrimazole RS in chloroform containing 5 mg per mL.

Disintegration ⟨701⟩: 20 minutes.

Uniformity of dosage units ⟨905⟩: meet the requirements.

Assay—

Dibasic potassium phosphate solution, Mobile phase, and *Chromatographic system*—Prepare as directed in the *Assay* under *Clotrimazole Cream.*

Internal standard solution, Standard preparation, and *Resolution solution*—Prepare as directed in the *Assay* under *Clotrimazole Topical Solution.*

Assay preparation—Weigh and finely powder not less than 10 Clotrimazole Vaginal Tablets. Transfer an accurately weighed portion of the powder, equivalent to about 100 mg of clotrimazole, to a 50-mL, screw-capped centrifuge tube, add 10.0 mL of *Internal standard solution* and 15 mL of *Mobile phase*, rotate for 15 minutes, and centrifuge for 10 minutes. Using a suitable syringe, transfer the supernatant liquid to a 100-mL volumetric flask. Rinse the syringe with 25 mL of *Mobile phase*, adding the rinsings to the centrifuge tube. Rotate the centrifuge tube for 15 minutes, and centrifuge for 10 minutes. Using a suitable syringe, transfer the supernatant liquid to the 100-mL volumetric flask. Rinse the syringe with 25 mL of *Mobile phase*, and add the washings to the 100-mL volumetric flask. Dilute with *Mobile phase* to volume, and mix.

Procedure—Separately inject equal volumes (about 20 μL) of the *Standard preparation* and the *Assay preparation* into the chromatograph, record the chromatograms, and measure the responses for the major peaks. Calculate the quantity, in mg, of $C_{22}H_{17}ClN_2$ in the portion of Tablets taken by the formula:

$$100C(R_U/R_S),$$

in which C is the concentration, in mg per mL, of USP Clotrimazole RS in the *Standard preparation*, and R_U and R_S are the ratios of the peak responses of the clotrimazole peak to the testosterone propionate peak obtained from the *Assay preparation* and the *Standard preparation*, respectively.

Clotrimazole and Betamethasone Dipropionate Cream

» Clotrimazole and Betamethasone Dipropionate Cream contains not less than 90.0 percent and not more than 110.0 percent of the labeled amount of clotrimazole ($C_{22}H_{17}ClN_2$) and an amount of betamethasone dipropionate equivalent to not less than 90.0 percent and not more than 110.0 percent of the labeled amount of betamethasone ($C_{22}H_{29}FO_5$), in a suitable cream base.

Packaging and storage—Preserve in collapsible tubes or in tight containers.

Reference standards—*USP Clotrimazole Reference Standard*—Dry at 105° for 2 hours before using. *USP Betamethasone Dipropionate Reference Standard*—Dry at 105° for 3 hours before using. *USP (o-Chlorophenyl)diphenylmethanol Reference Standard*—Do not dry before using.

Identification—The chromatogram of the *Assay preparation* obtained as directed in the *Assay for clotrimazole and betamethasone and limit of (o-chlorophenyl)diphenylmethanol* exhibits major peaks for clotrimazole and betamethasone dipropionate, the retention times of which correspond to those exhibited in the chromatogram of the *Standard preparation* obtained as directed in the *Assay for clotrimazole and betamethasone and limit of (o-chlorophenyl)diphenylmethanol.*

Microbial limits—It meets the requirements of the tests for absence of *Staphylococcus aureus* and *Pseudomonas aeruginosa* under *Microbial Limit Tests* ⟨61⟩.

Minimum fill ⟨755⟩: meets the requirements.

Assay for clotrimazole and betamethasone and limit of (*o*-chlorophenyl)diphenylmethanol—

Dibasic ammonium phosphate solution—Dissolve 6.6 g of dibasic ammonium phosphate in water to make 1000 mL of solution.

Mobile phase—Mix 7 volumes of methanol and 3 volumes of *Dibasic ammonium phosphate solution*, and adjust with phosphoric acid to a pH of 7.0 ± 0.2. Filter through a membrane filter of 0.45-μm or finer porosity, and degas. Make adjustments if necessary (see *System Suitability* under *Chromatography* ⟨621⟩).

Internal standard solution—Prepare a solution of progesterone in alcohol having a concentration of about 0.15 mg per mL.

Clotrimazole stock standard solution—Prepare a solution of USP Clotrimazole RS in alcohol having a known concentration of about 5 mg per mL.

Betamethasone dipropionate stock standard solution—Prepare a solution of USP Betamethasone Dipropionate RS in alcohol having a known concentration of about 6.4*J* mg per mL, *J* being the ratio of the labeled amount, in mg, of betamethasone to the labeled amount, in mg, of clotrimazole in each g of Cream.

(o-Chlorophenyl)diphenylmethanol stock standard solution—Prepare a solution of USP (*o*-Chlorophenyl)diphenylmethanol RS in methanol having a known concentration of about 0.5 mg per mL.

Standard preparation—Transfer 1.0 mL of *(o-Chlorophenyl)diphenylmethanol stock standard solution* to a suitable container, and evaporate in a water bath at room temperature under a stream of nitrogen to dryness. To the residue add 2.0 mL each of the *Clotrimazole stock standard solution*, the *Betamethasone dipropionate stock standard solution*, and the *Internal standard solution*, and mix.

Assay preparation—Weigh accurately a portion of Clotrimazole and Betamethasone Dipropionate Cream, equivalent to about 10 mg of clotrimazole, and transfer to a screw-capped, 50-mL centrifuge tube. Add 2.0 mL of *Internal standard solution* and 4.0 mL of alcohol, place the cap on the tube, and heat at 60° in a water bath for 10 minutes, with occasional shaking. Remove the tube from the bath, cool in an ice bath for 20 minutes, and promptly centrifuge. Transfer a portion of the supernatant liquid to a test tube, and use this solution as the *Assay preparation*.

Chromatographic system (see *Chromatography* ⟨621⟩)—The liquid chromatograph is equipped with a 254-nm detector and a 4.6-mm × 25-cm column that contains 10-μm packing L1. The flow rate is about 1.7 mL per minute. Chromatograph the *Standard preparation*, and record the peak responses as directed under *Procedure:* the resolution, *R*, between the betamethasone dipropionate and (*o*-chlorophenyl)diphenylmethanol peaks is not less than 1.0, between the (*o*-chlorophenyl)diphenylmethanol and progesterone peaks is not less than 1.5, and between the progesterone and clotrimazole peaks is not less than 1.8. The relative standard deviation for replicate injections of the *Standard preparation* is not more than 2.0% for the clotrimazole and betamethasone dipropionate responses, and not more than 4.0% for the (*o*-chlorophenyl)diphenylmethanol responses. The relative retention times are about 1.0 for betamethasone dipropionate, 1.2 for (*o*-chlorophenyl)diphenylmethanol, 1.4 for progesterone, and 1.7 for clotrimazole.

Procedure—Separately inject equal volumes (about 20 μL) of the *Standard preparation* and the *Assay preparation* into the chromatograph, record the chromatograms, and measure the responses for the major peaks. Calculate the quantity, in mg, of clotrimazole ($C_{22}H_{17}ClN_2$) in each g of the Cream taken by the formula:

$$2(C/W)(R_U/R_S),$$

in which *C* is the concentration, in mg per mL, of USP Clotrimazole RS in the *Clotrimazole stock standard solution*, *W* is the weight, in g, of Cream taken, and R_U and R_S are the ratios of the peak responses of the clotrimazole peak and the progesterone peak obtained from the *Assay preparation* and the *Standard preparation*, respectively. Calculate the quantity, in mg, of betamethasone ($C_{22}H_{29}FO_5$) in each g of the Cream taken by the formula:

$$(392.47/504.59)(2)(C/W)(R_U/R_S),$$

in which 392.47 and 504.59 are the molecular weights of betamethasone and betamethasone dipropionate, respectively, *C* is the concentration, in mg per mL, of USP Betamethasone Dipropionate RS in the *Betamethasone dipropionate stock standard solution*, *W* is the weight, in g, of Cream taken, and R_U and R_S are the ratios of the peak responses of the betamethasone dipropionate peak and the progesterone peak obtained from the *Assay preparation* and the *Standard preparation*, respectively. Calculate the quantity, in mg, of (*o*-chlorophenyl)diphenylmethanol in each g of the Cream taken by the formula:

$$(C/W)(R_U/R_S),$$

in which *C* is the concentration, in mg per mL, of USP (*o*-Chlorophenyl)diphenylmethanol RS in the *(o-Chlorophenyl)diphenylmethanol stock standard solution*, *W* is the weight, in g, of Cream taken, and R_U and R_S are the ratios of the peak responses of the (*o*-chlorophenyl)diphenylmethanol peak and the progesterone peak obtained from the *Assay preparation* and the *Standard preparation*, respectively: the quantity of (*o*-chlorophenyl)diphenylmethanol found is not more than 5.0% of the labeled quantity of clotrimazole in the Cream.

Cloxacillin Benzathine

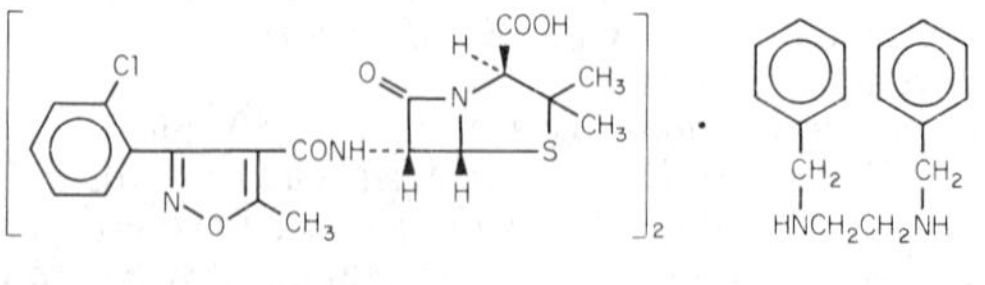

$(C_{19}H_{18}ClN_3O_5S)_2 \cdot C_{16}H_{20}N_2$ 1112.11

4-Thia-1-azabicyclo[3.2.0]heptane-2-carboxylic acid, 6-[[[3-(2-chlorophenyl)-5-methyl-4-isoxazolyl]carbonyl]amino]-3,3-dimethyl-7-oxo-, [2*S*-(2α,5α,6β)]-, cmpd. with *N,N′*-bis(phenylmethyl)-1,2-ethanediamine (2:1).

(2*S*,5*R*,6*R*)-6-[3-(*o*-Chlorophenyl)-5-methyl-4-isoxazolecarboxamido]-3,3-dimethyl-7-oxo-4-thia-1-azabicyclo[3.2.0]-heptane-2-carboxylic acid compound with *N,N′*-dibenzylethylenediamine (2:1) [*23736-58-5*].

» Cloxacillin Benzathine has a potency equivalent to not less than 704 μg and not more than 821 μg of cloxacillin ($C_{19}H_{18}ClN_3O_5S$) per mg, calculated on the anhydrous basis.

Packaging and storage—Preserve in tight containers.

Labeling—Label it to indicate that it is for veterinary use only.

Reference standards—*USP Cloxacillin Sodium Reference Standard*—Do not dry before using. *USP Cloxacillin Benzathine Reference Standard*—Do not dry before using.

Identification—

A: The infrared absorption spectrum of a potassium bromide dispersion of it exhibits maxima only at the same wavelengths as that of a similar preparation of USP Cloxacillin Benzathine RS.

B: To about 20 mg in a 50-mL conical flask add 5 mL of 5 *N* sodium hydroxide, and heat on a steam bath for 20 minutes. Cool, transfer 1 mL of this solution to a separator containing 10 mL of 1.2 *N* sulfuric acid, and extract with 50 mL of ether. Wash the ether extract with 30 mL of water, and extract the ether layer with 50 mL of 0.1 *N* sodium hydroxide: the ultraviolet absorption spectrum of the alkaline extract so obtained exhibits maxima and minima at the same wavelengths as that of a similar solution prepared from about 15 mg of USP Cloxacillin Sodium RS, concomitantly measured.

Crystallinity ⟨695⟩: meets the requirements.

pH ⟨791⟩: between 3.0 and 6.5, in a suspension containing 10 mg per mL.

Water, *Method I* ⟨921⟩: not more than 5.0%.

Assay—Proceed as directed for cloxacillin under *Antibiotics—Microbial Assays* ⟨81⟩, using an accurately weighed quantity of Cloxacillin Benzathine dissolved quantitatively in methanol to yield a stock solution having a convenient concentration. Promptly dilute an accurately measured volume of this stock solution quantitatively and stepwise with *Buffer No. 1* to obtain a *Test Dilution* having a concentration assumed to be equal to the median dose level of the Standard.

Cloxacillin Benzathine Intramammary Infusion

» Cloxacillin Benzathine Intramammary Infusion is a suspension of Cloxacillin Benzathine or Sterile Cloxacillin Benzathine in a suitable oil vehicle. It has a potency equivalent to not less than 90.0 percent and not more than 120.0 percent of the labeled amount of cloxacillin ($C_{19}H_{18}ClN_3O_5S$).

Packaging and storage—Preserve in disposable syringes that are well-closed containers, except that where the Infusion is labeled as sterile, the individual syringes or cartons are sealed and tamper-proof so that sterility is assured at time of use.

Labeling—Label it to indicate that it is for veterinary use only. Infusion that is sterile may be so labeled.

Reference standards—*USP Cloxacillin Sodium Reference Standard*—Do not dry before using. *USP Cloxacillin Benzathine Reference Standard*—Do not dry before using.

Identification—Transfer a quantity of Intramammary Infusion, equivalent to about 500 mg of cloxacillin, to a 50-mL centrifuge tube, add 25 mL of toluene, mix, and centrifuge. Decant and discard the toluene. Wash the residue with four 25-mL portions of toluene, sonicating for about 30 seconds after each addition of toluene. Dry the residue in vacuum over silica gel: the infrared absorption spectrum of a potassium bromide dispersion of the residue so obtained exhibits maxima only at the same wavelengths as that of a similar preparation of USP Cloxacillin Benzathine RS.

Sterility (where labeled as being sterile)—It meets the requirements under *Sterility Tests* ⟨71⟩, when tested as directed in the section, *Test Procedures for Direct Transfer to Test Media*, except to use Fluid Thioglycollate Medium containing polysorbate 80 solution (1 in 200) and an amount of sterile penicillinase sufficient to inactivate the cloxacillin in each tube, to use Soybean-Casein Digest Medium containing polysorbate 80 solution (1 in 200) and an amount of sterile penicillinase sufficient to inactivate the cloxacillin in each tube, and to shake the tubes once daily.

Water, *Method I* ⟨921⟩: not more than 1.0%, 20 mL of a mixture of carbon tetrachloride, chloroform, and methanol (2:2:1) being used in place of methanol in the titration vessel.

Assay—Proceed as directed for cloxacillin under *Antibiotics—Microbial Assays* ⟨81⟩, expelling the contents of 1 syringe of Cloxacillin Benzathine Intramammary Infusion into a high-speed glass blender jar containing sufficient methanol to yield a volume of 500.0 mL, and blend for 3 to 5 minutes. Promptly dilute an accurately measured volume of this solution quantitatively and stepwise with *Buffer No. 1* to obtain a *Test Dilution* having a concentration assumed to be equal to the median dose level of the Standard.

Sterile Cloxacillin Benzathine

» Sterile Cloxacillin Benzathine has a potency equivalent to not less than 704 µg and not more than 821 µg of cloxacillin ($C_{19}H_{18}ClN_3O_5S$) per mg, calculated on the anhydrous basis.

Packaging and storage—Preserve in tight containers.

Labeling—Label it to indicate that it is for veterinary use only.

Reference standards —*USP Cloxacillin Sodium Reference Standard*—Do not dry before using. *USP Cloxacillin Benzathine Reference Standard*—Do not dry before using.

Sterility—It meets the requirements under *Sterility Tests* ⟨71⟩, when tested as directed in the section, *Test Procedures for Direct Transfer to Test Media*, except to use Fluid Thioglycollate Medium containing polysorbate 80 solution (1 in 200) and an amount of sterile penicillinase sufficient to inactivate the cloxacillin in each tube, to use Soybean-Casein Digest Medium containing polysorbate 80 solution (1 in 200) and an amount of sterile penicillinase sufficient to inactivate the cloxacillin in each tube, and to shake the tubes once daily.

Other requirements—It responds to the *Identification test* and meets the requirements for *Crystallinity*, *pH*, *Water*, and *Assay* under *Cloxacillin Benzathine*.

Cloxacillin Sodium

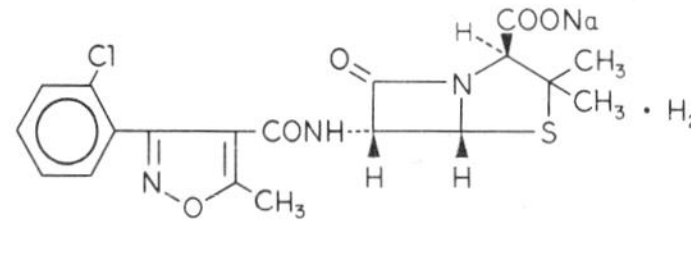

$C_{19}H_{17}ClN_3NaO_5S.H_2O$ 475.88

4-Thia-1-azabicyclo[3.2.0]heptane-2-carboxylic acid, 6-[[[3-(2-chlorophenyl)-5-methyl-4-isoxazolyl]carbonyl]-amino]-3,3-dimethyl-7-oxo-, monosodium salt, monohydrate, [2*S*-(2α,5α,6β)]-.

Monosodium (2*S*,5*R*,6*R*)-6-[3-(*o*-chlorophenyl)-5-methyl-4-isoxazolecarboxamido]-3,3-dimethyl-7-oxo-4-thia-1-azabicyclo[3.2.0]heptane-2-carboxylate monohydrate [*7081-44-9*].

Anhydrous 457.86 [*642-78-4*].

» Cloxacillin Sodium contains the equivalent of not less than 825 µg of cloxacillin ($C_{19}H_{18}ClN_3O_5S$) per mg.

Packaging and storage—Preserve in tight containers, at a temperature not exceeding 25°.

Reference standard—*USP Cloxacillin Sodium Reference Standard*—Do not dry before using.

Identification—

A: The infrared absorption spectrum of a potassium bromide dispersion of it exhibits maxima only at the same wavelengths as that of a similar preparation of USP Cloxacillin Sodium RS.

B: It responds to the tests for *Sodium* ⟨191⟩.

Crystallinity ⟨695⟩: meets the requirements.

pH ⟨791⟩: between 4.5 and 7.5, in a solution containing 10 mg per mL.

Water, *Method I* ⟨921⟩: between 3.0% and 5.0%.

Assay—

Diluent—Dissolve 5.44 g of monobasic potassium phosphate in water to make 2000 mL of solution, and adjust with 8 *N* potassium hydroxide to a pH of 5.0 ± 0.1.

Mobile phase—Prepare a suitable filtered mixture of *Diluent* and acetonitrile (75:25). Make adjustments if necessary (see *System Suitability* under *Chromatography* ⟨621⟩). Increasing the acetonitrile concentration decreases the retention time of cloxacillin.

Standard preparation—Dissolve an accurately weighed quantity of USP Cloxacillin Sodium RS in *Diluent* to obtain a solution having a known concentration of about 1.1 mg per mL.

Assay preparation—Transfer about 220 mg of Cloxacillin Sodium, accurately weighed, to a 200-mL volumetric flask, dilute with *Diluent* to volume, and mix. Stir with the aid of a magnetic stirrer for 5 minutes to assure dissolution of the specimen.

Chromatographic system (see *Chromatography* ⟨621⟩)—The liquid chromatograph is equipped with a 225-nm detector and a 4.6-mm × 25-cm column that contains packing L1. The flow rate is about 2 mL per minute. Chromatograph the *Standard preparation*, and record the peak responses as directed under *Procedure:* the capacity factor, k', for cloxacillin is between 2.2 and 5.7, the column efficiency determined from the analyte peak is not less than 1000 theoretical plates, the tailing factor for the analyte peak is not more than 1.5, and the relative standard deviation for replicate injections is not more than 2.0%.

Procedure—Separately inject equal volumes (about 10 µL) of the *Standard preparation* and the *Assay preparation* into the chromatograph, record the chromatograms, and measure the responses for the major peaks. Calculate the quantity, in µg, of cloxacillin ($C_{19}H_{18}ClN_3O_5S$) in each mg of the Cloxacillin Sodium taken by the formula:

$$200(CE/W)(r_U/r_S),$$

in which C is the concentration, in mg per mL, of USP Cloxacillin Sodium RS in the *Standard preparation*, E is the cloxacillin equivalent, in µg per mg, of the USP Cloxacillin RS, W is the weight, in mg, of the portion of Cloxacillin Sodium taken, and

r_U and r_S are the peak responses obtained from the *Assay preparation* and the *Standard preparation*, respectively.

Cloxacillin Sodium Capsules

» Cloxacillin Sodium Capsules contain the equivalent of not less than 90.0 percent and not more than 120.0 percent of the labeled amount of cloxacillin ($C_{19}H_{18}ClN_3O_5S$).

Packaging and storage—Preserve in tight containers.

Reference standard—*USP Cloxacillin Sodium Reference Standard*—Do not dry before using.

Dissolution ⟨711⟩—

Medium: water; 900 mL.

Apparatus 1: 100 rpm.

Time: 45 minutes.

Procedure—Determine the amount of cloxacillin ($C_{19}H_{18}ClN_3O_5S$) by a suitable validated spectrophotometric analysis of a filtered portion of the solution under test, suitably diluted with *Dissolution Medium*, if necessary, in comparison with a Standard solution having a known concentration of USP Cloxacillin Sodium RS in the same medium.

Tolerances—Not less than 75% (*Q*) of the labeled amount of cloxacillin ($C_{19}H_{18}ClN_3O_5S$) is dissolved in 45 minutes.

Uniformity of dosage units ⟨905⟩: meet the requirements.

Water, *Method I* ⟨921⟩: not more than 5.0%.

Assay—

Diluent, Mobile phase, Standard preparation, and *Chromatographic system*—Proceed as directed in the *Assay* under *Cloxacillin Sodium.*

Assay preparation—Weigh and finely powder the contents of not less than 10 Cloxacillin Sodium Capsules. Transfer an accurately weighed quantity of the powder, equivalent to about 200 mg of cloxacillin, to a 200-mL volumetric flask, dilute with *Diluent* to volume, mix, and stir for 10 minutes. Filter about 25 mL of this mixture, discarding the first 5 mL of the filtrate. Use the clear filtrate as the *Assay preparation.*

Procedure—Proceed as directed for *Procedure* in the *Assay* under *Cloxacillin Sodium.* Calculate the quantity, in mg, of cloxacillin ($C_{19}H_{18}ClN_3O_5S$) in the portion of Capsule contents taken by the formula:

$$0.2CE(r_U / r_S).$$

Cloxacillin Sodium Intramammary Infusion

» Cloxacillin Sodium Intramammary Infusion is a suspension of Sterile Cloxacillin Sodium in a suitable natural or chemically modified vegetable oil vehicle with a suitable dispersing agent. It has a potency equivalent to not less than 90.0 percent and not more than 120.0 percent of the labeled amount of cloxacillin ($C_{19}H_{18}ClN_3O_5S$).

Packaging and storage—Preserve in disposable syringes that are well-closed containers, except that where the Infusion is labeled as sterile, the individual syringes or cartons are sealed and tamper-proof so that sterility is assured at time of use.

Reference standard—*USP Cloxacillin Sodium Reference Standard*—Do not dry before using.

Labeling—Label it to indicate that it is for veterinary use only. Infusion that is sterile may be so labeled.

Identification—Transfer a quantity of Intramammary Infusion, equivalent to about 500 mg of cloxacillin, to a 50-mL centrifuge tube, add 15 mL of isooctane, mix, and centrifuge. Decant and discard the isooctane. Wash the residue with two 15-mL portions of isooctane and two 15-mL portions of ethyl ether, and discard the washings. Dry the residue in a current of air: the infrared absorption spectrum of a potassium bromide dispersion of the residue so obtained exhibits maxima only at the same wavelengths as that of a similar preparation of USP Cloxacillin Sodium RS.

Sterility (where labeled as being sterile)—It meets the requirements under *Sterility Tests* ⟨71⟩, when tested as directed in the section, *Test Procedures for Direct Transfer to Test Media,* except to use Fluid Thioglycollate Medium containing polysorbate 80 solution (1 in 200) and an amount of sterile penicillinase sufficient to inactivate the cloxacillin in each tube, to use Soybean-Casein Digest Medium containing polysorbate 80 solution (1 in 200) and an amount of sterile penicillinase sufficient to inactivate the cloxacillin in each tube, and to shake the tubes once daily.

Water, *Method I* ⟨921⟩: not more than 1.0%, 20 mL of a mixture of carbon tetrachloride, chloroform, and methanol (2:2:1) being used in place of methanol in the titration vessel.

Assay—Proceed as directed for cloxacillin under *Antibiotics—Microbial Assays* ⟨81⟩, expelling the contents of 1 syringe of Cloxacillin Intramammary Infusion into a high-speed glass blender jar containing 499.0 mL of *Buffer No. 1* and 1.0 mL of polysorbate 80, and blending for 3 to 5 minutes. Allow to stand for 10 minutes, and dilute an accurately measured volume of the aqueous phase quantitatively and stepwise with *Buffer No. 1* to obtain a *Test Dilution* having a concentration assumed to be equal to the median dose level of the Standard.

Cloxacillin Sodium for Oral Solution

» Cloxacillin Sodium for Oral Solution is a dry mixture of Cloxacillin Sodium and one or more suitable buffers, colors, flavors, and preservatives. It contains the equivalent of not less than 90.0 percent and not more than 120.0 percent of the labeled amount of cloxacillin ($C_{19}H_{18}ClN_3O_5S$).

Packaging and storage—Preserve in tight containers.

Reference standard—*USP Cloxacillin Sodium Reference Standard*—Do not dry before using.

Uniformity of dosage units ⟨905⟩—

FOR SOLID PACKAGED IN SINGLE-UNIT CONTAINERS: meets the requirements.

pH ⟨791⟩: between 5.0 and 7.5, in the solution constituted as directed in the labeling.

Water, *Method I* ⟨921⟩: not more than 1.0%.

Assay—

Diluent, Mobile phase, Standard preparation, and *Chromatographic system*—Proceed as directed in the *Assay* under *Cloxacillin Sodium.*

Assay preparation—Constitute Cloxacillin Sodium for Oral Solution as directed in the labeling. Dilute an accurately measured volume of the resulting solution, equivalent to about 250 mg of cloxacillin, to a 250-mL volumetric flask, dilute with *Diluent* to volume, mix, and stir for 15 minutes. Filter about 25 mL of this mixture, discarding the first 5 mL of the filtrate. Use the clear filtrate as the *Assay preparation.*

Procedure—Proceed as directed for *Procedure* in the *Assay* under *Cloxacillin Sodium.* Calculate the quantity, in mg, of cloxacillin ($C_{19}H_{18}ClN_3O_5S$) in the portion of Oral Solution taken by the formula:

$$0.25CE(r_U / r_S).$$

Sterile Cloxacillin Sodium

» Sterile Cloxacillin Sodium contains the equivalent of not less than 825 µg of cloxacillin ($C_{19}H_{18}ClN_3O_5S$) per mg.

Packaging and storage—Preserve in tight containers.

Reference standard—*USP Cloxacillin Sodium Reference Standard*—Do not dry before using.

Labeling—Label it to indicate that it is for veterinary use only.

Pyrogen—It meets the requirements of the *Pyrogen Test* ⟨151⟩, the test dose being 1 mL per kg of a solution in Sterile Water for Injection containing 20 mg of cloxacillin ($C_{19}H_{18}ClN_3O_5S$) per mL.

Sterility—It meets the requirements under *Sterility Tests* ⟨71⟩, when tested as directed in the section, *Test Procedures Using Membrane Filtration.*

Other requirements—It responds to the *Identification tests*, and meets the requirements for *Crystallinity, pH, Water*, and *Assay* under *Cloxacillin Sodium.*

Coal Tar

» Coal Tar is the tar obtained as a by-product during the destructive distillation of bituminous coal.

Packaging and storage—Preserve in tight containers.

Residue on ignition ⟨281⟩: not more than 2.0%, from 100 mg.

Coal Tar Ointment

» Prepare Coal Tar Ointment as follows.

Coal Tar	10 g
Polysorbate 80	5 g
Zinc Oxide Paste	985 g
To make	1000 g

Blend the Coal Tar with the Polysorbate 80, and incorporate the mixture with the Zinc Oxide Paste.

Packaging and storage—Preserve in tight containers.

Coal Tar Topical Solution

» Prepare Coal Tar Topical Solution as follows.

Coal Tar	200 g
Polysorbate 80	50 g
Alcohol, a sufficient quantity, to make	1000 mL

Mix the Coal Tar with 500 g of washed sand (see under *Reagents* in the section, *Reagents, Indicators, and Solutions*), and add the Polysorbate 80 and 700 mL of Alcohol. Macerate the mixture for 7 days in a closed vessel with frequent agitation. Filter, and rinse the vessel and the filter with sufficient Alcohol to make the product measure 1000 mL.

Packaging and storage—Preserve in tight containers.

Alcohol content ⟨611⟩: between 81.0% and 86.0% of C_2H_5OH.

Cyanocobalamin Co 57 Capsules

Vitamin B_{12}-^{57}Co.
Vitamin B_{12}-^{57}Co [*41559-38-0; 13115-03-2*].

» Cyanocobalamin Co 57 Capsules contain Cyanocobalamin in which a portion of the molecules contain radioactive cobalt (^{57}Co) in the molecular structure. Each Capsule contains not less than 90.0 percent and not more than 110.0 percent of the labeled amount of ^{57}Co as cyanocobalamin expressed in megabecquerels (microcuries) at the time indicated in the labeling. The cyanocobalamin content is not less than 90.0 percent and not more than 110.0 percent of the labeled amount. The specific activity is not less than 0.02 megabecquerel (0.5 microcurie) per μg of cyanocobalamin.

Packaging and storage—Preserve in well-closed, light-resistant containers.

Labeling—Label Capsules to include the following: the date of calibration; the amount of cyanocobalamin expressed in μg per Capsule; the amount of ^{57}Co as cyanocobalamin expressed in megabecquerels (microcuries) per Capsule at the time of calibration; the expiration date; and the statement, "Caution—Radioactive Material." The labeling indicates that in making dosage calculations, correction is to be made for radioactive decay, and also indicates that the radioactive half-life of ^{57}Co is 270.9 days.

Reference standard—*USP Cyanocobalamin Reference Standard*—Keep container tightly closed and protected from light, and store in a cool place. Dry over silica gel for 4 hours before using.

Radionuclide identification—A solution of 1 or more Capsules in water responds to the test for *Radionuclide identification* under *Cyanocobalamin Co 57 Solution.*

Uniformity of dosage units: meet the requirements.

Procedure for content uniformity—Determine the instrument response of each of 20 Capsules by measurement in a suitable counting assembly and under identical geometric conditions. Calculate the mean radioactivity value per Capsule. The requirements are met if not less than 19 of the Capsules are within the limits of 96.5% and 103.5% of the mean radioactivity value.

Radiochemical purity—A solution of 1 or more Capsules in water responds to the test for *Radiochemical purity* under *Cyanocobalamin Co 57 Oral Solution.*

Cyanocobalamin content—Determine the content, in μg per Capsule, of cyanocobalamin as directed under *Vitamin B_{12} Activity Assay* ⟨171⟩.

Assay for radioactivity—Using a suitable counting assembly (see *Selection of a Counting Assembly* under *Radioactivity* ⟨821⟩), determine the radioactivity, in MBq (μCi) per Capsule, of Cyanocobalamin Co 57 Capsules by use of a calibrated system as directed under *Radioactivity* ⟨821⟩.

Cyanocobalamin Co 57 Oral Solution

Vitamin B_{12}-^{57}Co.
Vitamin B_{12}-^{57}Co [*41559-38-0; 13115-03-2*].

» Cyanocobalamin Co 57 Oral Solution is a solution suitable for oral administration, containing Cyanocobalamin in which a portion of the molecules contain radioactive cobalt (^{57}Co) in the molecular structure. Cyanocobalamin Co 57 Oral Solution contains not less than 90.0 percent and not more than 110.0 percent of the labeled amount of ^{57}Co as cyanocobalamin expressed in megabecquerels (microcuries) per mL at the time indicated in the labeling. The cyanocobalamin content is not less than 90.0 percent and not more than 110.0 percent of the labeled amount. The specific activity is not less than 0.02 megabecquerel (0.5 microcurie) per μg of cyanocobalamin. Cyanocobalamin Co 57 Oral Solution contains a suitable antimicrobial agent.

Packaging and storage—Preserve in tight containers, and protect from light.

Labeling—Label it to include the following: the date of calibration; the amount of ^{57}Co as cyanocobalamin expressed as total megabecquerels (microcuries) and as megabecquerels (microcuries) per mL at the time of calibration; the amount of cyanocobalamin expressed in μg per mL; the name and quantity of the added preservative; the expiration date; and the statement, "Caution—Radioactive Material." The labeling indicates that in making dosage calculations, correction is to be made for radioactive decay, and also indicates that the radioactive half-life of ^{57}Co is 270.9 days, and directs that the Oral Solution be protected from light.

Reference standard—*USP Cyanocobalamin Reference Standard*—Keep container tightly closed and protected from light, and store in a cool place. Dry over silica gel for 4 hours before using.

Radionuclide identification (see *Radioactivity* ⟨821⟩)—Its gamma-ray spectrum is identical to that of a specimen of cobalt 57 of known purity that exhibits a major photopeak having an energy of 0.122 MeV.

pH ⟨791⟩: between 4.0 and 5.5.

Radiochemical purity—[NOTE—Solutions of cyanocobalamin are light-sensitive. Prepare the strips in diffuse light, and perform the development and drying steps in the absence of light.] Place 0.01 mL of a solution containing 1 mg of cyanocobalamin per mL about 45 mm from one end of a 25-mm × 300-mm strip of chromatographic paper (see *Chromatography* ⟨621⟩), and allow to dry. To the same area add a measured volume, such that it provides a count rate of not less than 20,000 counts per minute, of appropriately diluted Cyanocobalamin Co 57 Oral Solution, and allow to dry. Develop the chromatogram over a period of about 24 hours by descending chromatography, using a freshly prepared, homogeneous solution prepared by mixing 1 liter of secondary butyl alcohol, 1 mL of ammonium hydroxide, 20 mL of sodium cyanide solution (3.5 in 1000), and 300 mL of water (if phases separate, add 10-mL increments of secondary butyl alcohol, and shake until the mixture becomes homogeneous). Remove the paper strip from the apparatus when the cyanocobalamin spot has moved at least 75 mm from the point of application. Dry the chromatogram in air, and determine the radioactivity distribution by scanning with a suitable collimated radiation detector, or divide the strip horizontally into sections not exceeding 65 mm in width, and determine the radioactivity of the individual sections. The radioactivity of the cyanocobalamin band is not less than 95.0% of the total radioactivity.

Cyanocobalamin content—Determine the content, in μg per mL, of cyanocobalamin as directed under *Vitamin B_{12} Activity Assay* ⟨171⟩.

Assay for radioactivity—Using a suitable counting assembly (see *Selection of a Counting Assembly* under *Radioactivity* ⟨821⟩), determine the radioactivity, in MBq (μCi) per mL, of Cyanocobalamin Co 57 Oral Solution by use of a calibrated system as directed under *Radioactivity* ⟨821⟩.

Cyanocobalamin Co 60 Capsules

Vitamin B_{12}-^{60}Co.
Vitamin B_{12}-^{60}Co [*13422-53-2*].

» Cyanocobalamin Co 60 Capsules contain Cyanocobalamin in which a portion of the molecules contain radioactive cobalt (^{60}Co) in the molecular structure. Cyanocobalamin Co 60 Capsules contain not less than 90.0 percent and not more than 110.0 percent of the labeled amount of ^{60}Co as cyanocobalamin expressed in megabecquerels (microcuries) on the date indicated in the labeling. The cyanocobalamin content is not less than 90.0 percent and not more than 110.0 percent of the labeled amount. The specific activity is not less than 0.02 megabecquerel (0.5 microcurie) per μg of cyanocobalamin.

Packaging and storage—Preserve in well-closed, light-resistant containers.

Labeling—Label Capsules to include the following: the date of calibration; the amount of cyanocobalamin expressed in μg per Capsule; the amount of ^{60}Co as cyanocobalamin expressed in megabecquerels (microcuries) per Capsule on the date of calibration; the expiration date; and the statement, "Caution—Radioactive Material." The labeling indicates that in making dosage calculations, correction is to be made for radioactive decay, and also indicates that the radioactive half-life of ^{60}Co is 5.27 years.

Reference standard—*USP Cyanocobalamin Reference Standard*—Keep container tightly closed and protected from light, and store in a cool place. Dry over silica gel for 4 hours before using.

Radionuclide identification—A solution of one or more Capsules in water responds to the test for *Radionuclide identification* under *Cyanocobalamin Co 60 Solution.*

Uniformity of dosage units—The Capsules meet the requirements of the test for *Content uniformity* under *Cyanocobalamin Co 57 Capsules.*

Cyanocobalamin content—Determine the content, in μg per Capsule, of cyanocobalamin as directed under *Vitamin B_{12} Activity Assay* ⟨171⟩.

Assay for radioactivity—Cyanocobalamin Co 60 Capsules meet the requirements of the *Assay for radioactivity* under *Cyanocobalamin Co 57 Capsules.*

Cyanocobalamin Co 60 Oral Solution

Vitamin B_{12}-^{60}Co.
Vitamin B_{12}-^{60}Co [*13422-53-2*].

» Cyanocobalamin Co 60 Oral Solution is a solution suitable for oral administration, containing Cyanocobalamin in which a portion of the molecules contain radioactive cobalt (^{60}Co) in the molecular structure. Cyanocobalamin Co 60 Oral Solution contains not less than 90.0 percent and not more than 110.0 percent of the labeled amount of ^{60}Co as cyanocobalamin expressed in megabecquerels (microcuries) per mL on the date indicated in the labeling. The cyanocobalamin content is not less than 90.0 percent and not more than 110.0 percent of the labeled amount per mL. The amount of cobalt 60 as cyanocobalamin is not more than 0.04 megabecquerel (1 microcurie) per mL. The specific activity is not less than 0.02 megabecquerel (0.5 microcurie) per μg of cyanocobalamin. Cyanocobalamin Co 60 Oral Solution contains a suitable antimicrobial agent.

Packaging and storage—Preserve in single-dose or in multiple-dose containers, protected from light.

Labeling—Label it to include the following: the date of calibration; the amount of ^{60}Co as cyanocobalamin expressed as total megabecquerels (microcuries) and as megabecquerels (microcuries) per mL on the date of calibration; the amount of cyanocobalamin expressed in μg per mL; the name and quantity of the added preservative; the expiration date; and the statement, "Caution—Radioactive Material." The labeling indicates that in making dosage calculations, correction is to be made for radioactive decay, and also indicates that the radioactive half-life of ^{60}Co is 5.27 years.

Reference standard—*USP Cyanocobalamin Reference Standard*—Keep container tightly closed and protected from light, and store in a cool place. Dry over silica gel for 4 hours before using.

Radionuclide identification (see *Radioactivity* ⟨821⟩)—Its gamma-ray spectrum is identical to that of a specimen of cobalt 60 of known purity that exhibits photoelectric peaks having energies of 1.172 and 1.332 MeV.

pH ⟨791⟩: between 4.0 and 5.5.

Radiochemical purity—Cyanocobalamin Co 60 Oral Solution meets the requirements of the test for *Radiochemical purity* under *Cyanocobalamin Co 57 Oral Solution.*

Cyanocobalamin content—Determine the content, in µg per mL, of cyanocobalamin as directed under *Vitamin B_{12} Activity Assay* ⟨171⟩.

Assay for radioactivity—Cyanocobalamin Co 60 Oral Solution meets the requirements of the *Assay for radioactivity* under *Cyanocobalamin Co 57 Oral Solution.*

Cocaine

$C_{17}H_{21}NO_4$ 303.36

8-Azabicyclo[3.2.1]octane-2-carboxylic acid, 3-(benzoyloxy)-8-methyl-, methyl ester, [1*R*-(*exo,exo*)]-.

Methyl 3β-hydroxy-1αH,5αH-tropane-2β-carboxylate benzoate (ester) [*50-36-2*].

» Cocaine, dried over phosphorus pentoxide for 3 hours, contains not less than 99.0 percent and not more than 101.0 percent of $C_{17}H_{21}NO_4$.

Packaging and storage—Preserve in well-closed, light-resistant containers.

Reference standard—*USP Cocaine Hydrochloride Reference Standard*—Dry over silica gel for 3 hours before using.

Identification—

A: The ultraviolet absorption spectrum of a 1 in 75,000 solution in dilute hydrochloric acid (1 in 120) exhibits maxima and minima at the same wavelengths as that of a similar solution of USP Cocaine Hydrochloride RS, concomitantly measured, and the respective molar absorptivities, calculated on the dried basis, at the wavelength of maximum absorbance at about 233 nm do not differ by more than 3.0%. [NOTE—The molecular weight of cocaine hydrochloride ($C_{17}H_{21}NO_4 \cdot HCl$) is 339.82.]

B: It meets the requirements under *Identification—Organic Nitrogenous Bases* ⟨181⟩, USP Cocaine Hydrochloride RS being used, and sodium carbonate TS being used in place of sodium hydroxide TS.

C: Dissolve about 100 mg in a mixture of 0.4 mL of dilute hydrochloric acid (1 in 12) and water to make 5 mL, and add 5 drops of chromium trioxide solution (1 in 20): a yellow precipitate is formed, and it quickly redissolves when the mixture is shaken. Add 1 mL of hydrochloric acid: a permanent, orange-colored, crystalline precipitate is formed.

D: Dissolve about 10 mg in 1 mL of dilute hydrochloric acid (1 in 600), and evaporate on a steam bath just to dryness. Dissolve the residue in 2 drops of water, and add 1 mL of potassium permanganate solution (1 in 300): a violet, crystalline precipitate is formed, and it appears brownish violet when collected on a filter, and shows characteristic violet-red crystalline aggregates under the low power of a microscope, similar to those obtained from USP Cocaine Hydrochloride RS.

Melting range, *Class I* ⟨741⟩: between 96° and 98°.

Loss on drying ⟨731⟩—Dry it over phosphorus pentoxide for 3 hours: it loses not more than 1.0% of its weight.

Residue on ignition ⟨281⟩: not more than 0.1%.

Readily carbonizable substances ⟨271⟩—Dissolve about 500 mg in 5 mL of sulfuric acid TS: the solution has no more color than Matching Fluid A.

Cinnamyl-cocaine and other reducing substances—Dissolve about 300 mg of finely powdered Cocaine in 1 mL of dilute hydrochloric acid (1 in 12) with the aid of heat, if necessary, and dilute with water to 15 mL. Mix 5 mL of this solution with 0.3 mL of dilute sulfuric acid (1 in 35) and 0.1 mL of potassium permanganate solution (1 in 300): the violet color does not disappear entirely within 30 minutes.

Isoatropyl-cocaine—Dilute in a beaker 5 mL of the solution of Cocaine prepared in the test for *Cinnamyl-cocaine and other reducing substances* with 80 mL of water, add 0.2 mL of 6 *N* ammonium hydroxide, and stir the solution vigorously for 5 minutes, occasionally rubbing the inner wall of the beaker with a stirring rod: a crystalline precipitate of cocaine is formed, and the supernatant liquid is clear.

Assay—Dissolve about 600 mg of Cocaine, previously dried and accurately weighed, in 50 mL of glacial acetic acid, add 1 drop of crystal violet TS, and titrate with 0.1 *N* perchloric acid VS to a green end-point. Perform a blank determination, and make any necessary correction. Each mL of 0.1 *N* perchloric acid is equivalent to 30.34 mg of $C_{17}H_{21}NO_4$.

Cocaine Hydrochloride

$C_{17}H_{21}NO_4 \cdot HCl$ 339.82

8-Azabicyclo[3.2.1]octane-2-carboxylic acid, 3-(benzoyloxy)-8-methyl-, methyl ester, hydrochloride, [1*R*-(*exo,exo*)]-.

Methyl 3β-hydroxy-1αH,5αH-tropan-2β-carboxylate, benzoate (ester) hydrochloride [*53-21-4*].

» Cocaine Hydrochloride contains not less than 99.0 percent and not more than 101.0 percent of $C_{17}H_{21}NO_4 \cdot HCl$, calculated on the dried basis.

Packaging and storage—Preserve in well-closed, light-resistant containers.

Reference standard—*USP Cocaine Hydrochloride Reference Standard*—Dry over silica gel for 3 hours before using.

Identification—

A: It meets the requirements under *Identification—Organic Nitrogenous Bases* ⟨181⟩, sodium carbonate TS being used in place of 1 *N* sodium hydroxide.

B: To 5 mL of a solution (1 in 50) add 5 drops of chromium trioxide solution (1 in 20): a yellow precipitate is formed, and it quickly redissolves when the mixture is shaken gently. Add 1 mL of hydrochloric acid: a permanent, orange-colored crystalline precipitate is formed.

C: To a solution of about 10 mg in 2 drops of water add 1 mL of 0.1 *N* potassium permanganate: a violet, crystalline precipitate is formed, and it appears brownish violet when collected on a filter, and shows characteristic, violet-red crystalline aggregates under the low power of a microscope.

D: It responds to the tests for *Chloride* ⟨191⟩.

Specific rotation ⟨781⟩: between −71° and −73°, determined in a solution containing the equivalent of 200 mg in each 10 mL, the test specimen previously having been dried over silica gel for 3 hours.

Acidity—Dissolve 500 mg in 10 mL of water, add 1 drop of methyl red TS, and titrate with 0.020 *N* sodium hydroxide: not more than 0.50 mL is required to produce a yellow color.

Loss on drying ⟨731⟩—Dry it over silica gel for 3 hours: it loses not more than 1.0% of its weight.

Residue on ignition ⟨281⟩: not more than 0.1%.

Readily carbonizable substances ⟨271⟩—Dissolve 500 mg in 5 mL of sulfuric acid TS: the solution has no more color than Matching Fluid F.

Cinnamyl-cocaine and other reducing substances—To 5 mL of a solution (1 in 50) add 0.3 mL of 1 *N* sulfuric acid and 0.10 mL of 0.10 *N* potassium permanganate: the violet color does not disappear entirely within 30 minutes.

Isoatropyl-cocaine—Dilute 5 mL of a solution (1 in 50) in a beaker with 80 mL of water, add 0.2 mL of 6 *N* ammonium hydroxide, stir the solution vigorously during 5 minutes, occasionally rubbing the inner wall of the beaker with a stirring rod: a crystalline precipitate of cocaine is formed, and the supernatant liquid is clear.

Assay—Dissolve about 500 mg of Cocaine Hydrochloride, accurately weighed, in a mixture of 40 mL of glacial acetic acid and 10 mL of mercuric acetate TS. Add 2 drops of quinaldine red TS, and titrate with 0.1 *N* perchloric acid VS. Perform a blank determination, and make any necessary correction. Each mL of 0.1 *N* perchloric acid is equivalent to 33.98 mg of $C_{17}H_{21}NO_4.HCl$.

Cocaine Hydrochloride Tablets for Topical Solution

» Cocaine Hydrochloride Tablets for Topical Solution contain not less than 91.0 percent and not more than 109.0 percent of the labeled amount of $C_{17}H_{21}NO_4.HCl$.

Packaging and storage—Preserve in well-closed, light-resistant containers.

Reference standard—*USP Cocaine Hydrochloride Reference Standard*—Dry over silica gel for 3 hours before using.

Identification—

A: Add 5 drops of chromium trioxide solution (1 in 20) to 5 mL of a filtered solution of Tablets, equivalent to cocaine hydrochloride solution (1 in 50): a yellow precipitate is formed and it redissolves when the mixture is shaken. On the addition of 1 mL of hydrochloric acid, a permanent, yellowish orange, crystalline precipitate is formed.

B: Dissolve a portion of powdered Tablets, equivalent to about 10 mg of cocaine hydrochloride, in 1 mL of water, filter, and add 2 mL of 0.1 *N* potassium permanganate: a red-purple, crystalline precipitate, which appears brown when collected on a filter, is formed, and it shows characteristic, crystalline aggregates under the low power of a microscope.

C: Add silver nitrate TS, dropwise, to a filtered solution of Tablets, equivalent to cocaine hydrochloride solution (1 in 20): a white precipitate is formed, and it is insoluble in nitric acid.

Disintegration ⟨701⟩: 15 minutes.

Uniformity of dosage units ⟨905⟩: meet the requirements.

Procedure for content uniformity—Place 1 Tablet in a 100-mL volumetric flask, add 50 mL of water, and shake the flask until the tablet is dissolved. Dilute with water to volume, mix, and filter, discarding the first 20 mL of the filtrate. Dilute a portion of the subsequent filtrate, if necessary, with water to provide a solution containing approximately 80 µg of cocaine hydrochloride per mL. Concomitantly determine the absorbances of this test solution and a Standard solution of USP Cocaine Hydrochloride RS in the same medium having a known concentration of about 80 µg per mL, in 1-cm cells at the wavelength of maximum absorbance at about 275 nm, with a suitable spectrophotometer, using water as the blank. Calculate the quantity, in mg, of $C_{17}H_{21}NO_4.HCl$ in the Tablet by the formula:

$$(T/D)C(A_U/A_S),$$

in which *T* is the labeled quantity, in mg, of cocaine hydrochloride in the Tablet, *D* is the concentration, in µg per mL, of cocaine hydrochloride in the test solution, based upon the labeled quantity per Tablet and the extent of dilution, *C* is the concentration, in µg per mL, of USP Cocaine Hydrochloride RS in the Standard solution, and A_U and A_S are the absorbances of the solution from the Tablet and the Standard solution, respectively.

Assay—Weigh and finely powder not less than 20 Cocaine Hydrochloride Tablets for Topical Solution. Dissolve an accurately weighed portion of the powder, equivalent to about 60 mg of cocaine hydrochloride, in 10 mL of water, render the solution slightly alkaline with 6 *N* ammonium hydroxide, and completely extract the cocaine with small successive portions of ether. Evaporate the combined ether extracts on a steam bath to one-half their volume, transfer the remaining liquid to a separator, and wash with three 5-mL portions of water. Shake the water washings with a small portion of ether, and add the ether washing to the combined ether extracts. Add 10.0 mL of 0.05 *N* sulfuric acid VS to the ether solution, agitate the mixture thoroughly, and draw off the acidified water layer into a beaker. Wash the ether with two small portions of water, add the washings to the acid liquid, and titrate the excess acid with 0.02 *N* sodium hydroxide VS, using methyl red TS as the indicator. Each mL of 0.05 *N* sulfuric acid is equivalent to 16.99 mg of $C_{17}H_{21}NO_4.HCl$.

Coccidioidin

» Coccidioidin conforms to the regulations of the federal Food and Drug Administration concerning biologics (see *Biologics* ⟨1041⟩). It is a sterile solution containing the antigens obtained from the by-products of mycelial growth or from the spherules of the fungus *Coccidioides immitis.* It has a potency such that the 1:100 dilution is bioequivalent to the U.S. Reference Coccidioidin 1:100. It contains a suitable antimicrobial agent.

Packaging and storage—Preserve at a temperature between 2° and 8°.

Expiration date—The expiration date is not later than 3 years after date of issue from manufacturer's cold storage (5°, 1 year) for the mycelial product and not later than 18 months after date of issue from manufacturer's cold storage (5°, 18 months) for the spherule-derived product.

Labeling—Label it to state that any dilutions made of the product should be stored in a refrigerator and used within 24 hours. Label it also to state that a separate syringe and needle shall be used for each individual injection.

Cocoa Butter—*see* Cocoa Butter NF

Cod Liver Oil

» Cod Liver Oil is the partially destearinated fixed oil obtained from fresh livers of *Gadus morrhua* Linné and other species of Fam. Gadidae. Cod Liver Oil contains, in each g, not less than 255 µg (850 USP Units) of vitamin A and not less than 2.125 µg (85 USP Units) of vitamin D.

Cod Liver Oil may be flavored by the addition of not more than 1 percent of a suitable flavor or a mixture of flavors.

Packaging and storage—Preserve in tight containers. It may be bottled or otherwise packaged in containers from which air has been expelled by the production of a vacuum or by an inert gas.

Labeling—The vitamin A potency and vitamin D potency, when designated on the label, are expressed in USP Units per g of oil. The potencies may be expressed also in metric units, on the basis that 1 USP Vitamin A Unit = 0.3 µg and 40 USP Vitamin D Units = 1 µg.

Reference standard—*USP Cholecalciferol Reference Standard*—Store in a cold place, protected from light. Allow it to attain room temperature before opening ampul. Use the material promptly and discard the unused portion.

Identification for vitamin A—To 1 mL of a 1 in 40 solution in chloroform add 10 mL of antimony trichloride TS: a blue color results immediately.

Specific gravity ⟨841⟩: between 0.918 and 0.927.

Color—When viewed transversely in a tall, cylindrical, standard oil-specimen bottle of colorless glass of about 120-mL capacity, the color is not more intense than that of a mixture of 11 mL of cobaltous chloride CS, 76 mL of ferric chloride CS, and 33 mL of water, in a similar bottle of the same internal diameter.

Nondestearinated cod liver oil—Fill a tall, cylindrical, standard oil-specimen bottle of about 120-mL capacity with Cod Liver Oil at a temperature between 23° and 28°, insert the stopper, and immerse the bottle in a mixture of ice and water for 3 hours: the oil remains clear and does not deposit stearin.

Unsaponifiable matter ⟨401⟩: not more than 1.30%.

Acid value ⟨401⟩—Mix 15 mL of alcohol with 15 mL of ether, add 5 drops of phenolphthalein TS, and neutralize with 0.1 *N* sodium hydroxide. Dissolve 2.0 g of Cod Liver Oil in the mixture, and boil the oil solution gently under a reflux condenser for 10 minutes. Cool, and titrate the mixture with 0.1 *N* sodium hydroxide VS to the production of a pink color that persists after shaking for 30 seconds: not more than 1.0 mL of 0.10 sodium hydroxide is required.

Iodine value ⟨401⟩: between 145 and 180.

Saponification value ⟨401⟩—Its saponification value is between 180 and 192. If carbon dioxide has been used as a preservative, expose the Oil in a shallow dish in a vacuum desiccator for 24 hours before weighing the specimen for determination of the saponification value.

Assay for vitamin A—Using 500 mg to 1 g, accurately weighed, of Cod Liver Oil, proceed as directed under *Vitamin A Assay* ⟨571⟩.

Assay for vitamin D—Proceed with Cod Liver Oil as directed for *Biological Method* under *Vitamin D Assay* ⟨581⟩.

Codeine

$C_{18}H_{21}NO_3 . H_2O$ 317.38

Morphinan-6-ol, 7,8-didehydro-4,5-epoxy-3-methoxy-17-methyl-, monohydrate, (5α,6α)-.

7,8-Didehydro-4,5α-epoxy-3-methoxy-17-methylmorphinan-6α-ol monohydrate   *[6059-47-8]*.

Anhydrous   299.37   *[76-57-3]*.

» Codeine, dried at 80° for 4 hours, contains not less than 98.5 percent and not more than 100.5 percent of $C_{18}H_{21}NO_3$.

Packaging and storage—Preserve in tight, light-resistant containers.

Reference standard—*USP Codeine Sulfate Reference Standard*—Dry at 105° for 3 hours before using.

Identification—

A: The infrared absorption spectrum of a potassium bromide dispersion of it, previously dried, exhibits maxima and minima at the same wavelengths as that of the codeine obtained from 50 mg of USP Codeine Sulfate RS dissolved in 15 mL of water, then rendered alkaline with 6 *N* ammonium hydroxide and extracted with several 10-mL portions of chloroform, followed by evaporation of the combined chloroform extracts on a steam bath to dryness, and drying at 80° for 4 hours.

B: The ultraviolet absorption spectrum of a 1 in 10,000 solution in 0.1 *N* sulfuric acid exhibits maxima and minima at the same wavelengths as that of a similar solution of USP Codeine Sulfate RS, concomitantly measured, and the respective molar absorptivities, calculated on the dried basis, at the wavelength of maximum absorbance at about 284 nm do not differ by more than 3.0%. [NOTE—The molecular weight of anhydrous codeine sulfate, $(C_{18}H_{21}NO_3)_2 . H_2SO_4$, is 696.81.]

Melting range ⟨741⟩—When previously dried, it melts between 154° and 158°, but the range between beginning and end of melting does not exceed 2°.

Loss on drying ⟨731⟩—Dry it at 80° for 4 hours: it loses not more than 6.0% of its weight.

Residue on ignition ⟨281⟩: not more than 0.1%.

Readily carbonizable substances ⟨271⟩—Dissolve 10 mg in 5 mL of sulfuric acid TS: the solution has no more color than Matching Fluid S.

Chromatographic purity—Prepare a solution of it in dehydrated alcohol containing 40 mg per mL (*Solution A*). Dilute 2.0 mL of *Solution A* with dehydrated alcohol to 100 mL (*Solution B*). Dilute 1.0 mL of *Solution A* with dehydrated alcohol to 100 mL (*Solution C*). On a suitable thin-layer chromatographic plate (see *Chromatography* ⟨621⟩), coated with a 0.25-mm layer of chromatographic silica gel, apply separate 10-µL volumes of *Solution A*, *Solution B*, and *Solution C*. Allow the spots to dry, and develop the chromatogram in a solvent system consisting of a mixture of dehydrated alcohol, cyclohexane, and ammonium hydroxide (72:30:6) until the solvent front has moved three-fourths of the length of the plate. Remove the plate from the developing chamber, and allow the solvent to evaporate. Spray the plate with a reagent prepared by mixing 3 mL of chloroplatinic acid solution (1 in 10) with 97 mL of water, followed by the addition of 100 mL of potassium iodide solution (6 in 100), and examine the chromatogram: no spot obtained from *Solution A*, other than the principal spot and any spot observed at the origin, is more intense than the principal spot obtained from *Solution B* (2%), and not more than one such spot having an R_f greater than that of the principal spot is more intense than the principal spot obtained from *Solution C* (1%).

Morphine—Dissolve about 50 mg of potassium ferricyanide in 10 mL of water, and add 1 drop of ferric chloride TS and 1 mL of a neutral or slightly acid solution of Codeine (1 in 100) prepared with the aid of sulfuric acid: no blue color is produced immediately.

Assay—Dissolve about 400 mg of Codeine, previously dried and accurately weighed, by warming it in 30.0 mL of 0.1 *N* sulfuric acid VS. Cool, and add 10 mL of water. Add methyl red TS, and titrate the excess acid with 0.1 *N* sodium hydroxide VS. Perform a blank determination (see *Residual Titrations* under *Titrimetry* ⟨541⟩). Each mL of 0.1 *N* sulfuric acid is equivalent to 29.94 mg of $C_{18}H_{21}NO_3$.

Codeine Elixir, Terpin Hydrate and—*see* Terpin Hydrate and Codeine Elixir

Codeine Phosphate

$C_{18}H_{21}NO_3 . H_3PO_4 . \frac{1}{2}H_2O$ 406.37

Morphinan-6-ol, 7,8-didehydro-4,5-epoxy-3-methoxy-17-methyl-, (5α,6α)-, phosphate (1:1) (salt), hemihydrate.

7,8-Didehydro-4,5α-epoxy-3-methoxy-17-methylmorphinan-6α-ol phosphate (1:1) (salt) hemihydrate   *[41444-62-6]*.

Anhydrous   397.36   *[52-28-8]*.

» Codeine Phosphate contains not less than 99.0 percent and not more than 101.5 percent of $C_{18}H_{21}NO_3 . H_3PO_4$, calculated on the anhydrous basis.

Packaging and storage—Preserve in tight, light-resistant containers.

Reference standard—*USP Codeine Phosphate Reference Standard*—This is the hemihydrate form of codeine phosphate. Dry at 105° for 18 hours before using.

Identification—

A: The infrared absorption spectrum of a potassium bromide dispersion of it, previously dried at 105° for 18 hours, exhibits maxima only at the same wavelengths as that of a similar preparation of USP Codeine Phosphate RS.

B: Neutralize a solution (1 in 50) with 6 *N* ammonium hydroxide, and add silver nitrate TS: a yellow precipitate of silver phosphate is formed, and it is soluble in 2 *N* nitric acid and in 6 *N* ammonium hydroxide.

Acidity—Dissolve 100 mg in 20 mL of water, and titrate with 0.010 *N* sodium hydroxide to a pH of 5.4, using a pH meter: not more than 1.0 mL of 0.010 *N* sodium hydroxide is required.

Water, *Method I* ⟨921⟩: not more than 3.0%.

Chloride—To 10 mL of a solution (1 in 100), acidified with nitric acid, add a few drops of silver nitrate TS: no opalescence is produced immediately.

Sulfate—To 10 mL of a solution (1 in 100) add a few drops of barium chloride TS: no turbidity is produced immediately.

Morphine—Dissolve about 50 mg of potassium ferricyanide in 10 mL of water, and add 1 drop of ferric chloride TS and 1 mL of a solution of Codeine Phosphate (1 in 100): no blue color is produced immediately.

Chromatographic purity—Using Codeine Phosphate, proceed as directed in the test for *Chromatographic purity* under *Codeine*, except to use a mixture of 0.01 *N* hydrochloric acid and dehydrated alcohol (4:1), instead of dehydrated alcohol, to prepare *Solution A*, *Solution B*, and *Solution C.*

Assay—Dissolve about 1 g of Codeine Phosphate, accurately weighed, in 50 mL of glacial acetic acid, warming slightly if necessary to effect solution, and titrate with 0.1 *N* perchloric acid VS, determining the end-point potentiometrically. Perform a blank determination, and make any necessary correction. Each mL of 0.1 *N* perchloric acid is equivalent to 39.74 mg of $C_{18}H_{21}NO_3 . H_3PO_4$.

Codeine Phosphate, Alumina, and Magnesia Tablets, Aspirin,—*see* Aspirin, Codeine Phosphate, Alumina, and Magnesia Tablets

Codeine Phosphate Elixir, Acetaminophen and—*see* Acetaminophen and Codeine Phosphate Elixir

Codeine Phosphate Injection

» Codeine Phosphate Injection is a sterile solution of Codeine Phosphate in Water for Injection. It contains not less than 93.0 percent and not more than 107.0 percent of the labeled amount of $C_{18}H_{21}NO_3 . H_3PO_4 . \frac{1}{2}H_2O$.

NOTE—Do not use the Injection if it is more than slightly discolored or contains a precipitate.

Packaging and storage—Preserve in single-dose or in multiple-dose containers, preferably of Type I glass, protected from light.

Reference standard—*USP Codeine Phosphate Reference Standard*—This is the hemihydrate form of codeine phosphate. Dry at 105° for 18 hours before using.

Identification—

A: Dilute a volume of Injection, equivalent to about 90 mg of codeine phosphate, with water to about 10 mL, add 1 drop of hydrochloric acid, and extract with three 10-mL portions of chloroform, discarding the chloroform extracts. Add 6 *N* ammonium hydroxide until the solution is alkaline, and extract with several 10-mL portions of chloroform. Evaporate the combined chloroform extracts on a steam bath to dryness, and dry at 80° for 4 hours: the infrared absorption spectrum of a potassium bromide dispersion of the residue so obtained exhibits maxima at the same wavelengths as that of the codeine obtained by similarly treating 1 mL of a solution of USP Codeine Phosphate RS (1 in 100).

B: A volume of Injection, equivalent to about 60 mg of codeine phosphate, responds to *Identification test B* under *Codeine Phosphate.*

pH ⟨791⟩: between 3.0 and 6.0.

Morphine—Diluted with water to a concentration of 5 mg of codeine phosphate per mL, it meets the requirements of the test for *Morphine* under *Codeine Phosphate.*

Other requirements—It meets the requirements under *Injections* ⟨1⟩.

Assay—Transfer an accurately measured volume of Codeine Phosphate Injection, equivalent to about 75 mg of codeine phosphate, to a small separator, and add about 15 mL of water. Add 2 drops of phosphoric acid, and extract with four 10-mL portions of chloroform, collecting the chloroform extracts in a separator. Wash the combined chloroform extracts with 10 mL of water, and add the water wash to the first separator containing the sample. Discard the chloroform extracts. Proceed as directed in the *Assay* under *Codeine Phosphate Tablets*, beginning with "render the solution alkaline with 6 *N* ammonium hydroxide." Each mL of 0.02 *N* sulfuric acid is equivalent to 8.128 mg of $C_{18}H_{21}NO_3 . H_3PO_4 . 4H_2O$.

Codeine Phosphate Tablets

» Codeine Phosphate Tablets contain not less than 93.0 percent and not more than 107.0 percent of the labeled amount of $C_{18}H_{21}NO_3 . H_3PO_4 . \frac{1}{2}H_2O$.

Packaging and storage—Preserve in well-closed, light-resistant containers.

Reference standard—*USP Codeine Phosphate Reference Standard*—This is the hemihydrate form of codeine phosphate. Dry at 105° for 18 hours before using.

Identification—

A: Digest a quantity of finely powdered Tablets, equivalent to about 100 mg of codeine phosphate, with 15 mL of water and 5 mL of 2 *N* sulfuric acid for 1 hour. Filter, if necessary, and wash any undissolved residue with a few mL of water. Render the filtrate alkaline with 6 *N* ammonium hydroxide, extract with several small portions of chloroform, and proceed as directed in *Identification test A* under *Codeine Phosphate Injection*, beginning with "Evaporate the combined chloroform extracts." The specified results are observed.

B: To a quantity of finely powdered Tablets, equivalent to about 100 mg of codeine phosphate, add 10 mL of water and 2 drops of 2 *N* sulfuric acid. Digest, with frequent shaking, for 15 minutes, and filter. Neutralize 5 mL of the filtrate with 6 *N* ammonium hydroxide, and add silver nitrate TS: a yellow precipitate of silver phosphate is formed, and it is soluble in diluted nitric acid and in 6 *N* ammonium hydroxide.

Dissolution ⟨711⟩—

Medium: water; 900 mL.

Apparatus 2: 50 rpm.

Time: 45 minutes.

Procedure—Determine the amount of $C_{18}H_{21}NO_3 . H_3PO_4 . \frac{1}{2}H_2O$ dissolved from ultraviolet absorbances at the wavelength of maximum absorbance at about 284 nm of filtered portions of the solution under test, suitably diluted with *Dissolution Medium*, if necessary, in comparison with a Standard solution having a known concentration of USP Codeine Phosphate RS in the same medium.

Tolerances—Not less than 75% (*Q*) of the labeled amount of $C_{18}H_{21}NO_3 . H_3PO_4 . \frac{1}{2}H_2O$ is dissolved in 45 minutes.

Uniformity of dosage units ⟨905⟩: meet the requirements.

Procedure for content uniformity—Transfer 1 Tablet, previously crushed or finely powdered, to a 50-mL volumetric flask, add 25 mL of water, and shake to dissolve. Dilute with water to volume, and filter, if necessary, discarding the first 20 mL of the filtrate. Transfer an aliquot of the filtrate, equivalent to about 6 mg of codeine phosphate, to a 50-mL volumetric flask containing 2 mL of 3 *N* hydrochloric acid, and dilute with water to volume. Dissolve an accurately weighed quantity of USP Codeine Phosphate RS in 0.1 *N* hydrochloric acid, and dilute quantitatively and stepwise with the same solvent to obtain a Standard solution having a known concentration of about 120 µg per mL. Concomitantly determine the absorbances of both solutions in 1-cm cells at the wavelength of maximum absorbance at about 284 nm, with a suitable spectrophotometer, using water as the blank. Calculate the quantity, in mg, of $C_{18}H_{21}NO_3 . H_3PO_4 . \frac{1}{2}H_2O$ in the Tablet by the formula:

$$2.5(C/V)(A_U/A_S)(406.37/397.36),$$

in which C is the concentration, in µg per mL, of USP Codeine Phosphate RS in the Standard solution, V is the volume, in mL, of the aliquot taken of the solution of the Tablet, and A_U and A_S are the absorbances of the solution from the Tablet and the Standard solution, respectively, and 406.37 and 397.36 are the molecular weights of codeine phosphate hemihydrate and anhydrous codeine phosphate, respectively.

Morphine—A 1-mL portion of the filtrate from *Identification test B* meets the requirements of the test for *Morphine* under *Codeine Phosphate.*

Assay—Weigh and finely powder not less than 20 Codeine Phosphate Tablets. Weigh accurately a portion of the powder, equivalent to about 150 mg of codeine phosphate, and transfer to a 100-mL volumetric flask. Add 20 mL of 0.5 *N* sulfuric acid, and shake the mixture occasionally during 2 hours. Add water to volume, mix, and filter through a filtering crucible. Transfer to a separator an accurately measured portion of the filtrate, equivalent to not less than 75 mg of codeine phosphate, render the solution alkaline with 6 *N* ammonium hydroxide, and completely extract the alkaloid with successive 15-mL portions of chloroform. Evaporate the combined chloroform solution on a steam bath nearly to dryness. Dissolve the residue in about 2 mL of methanol, heating, if necessary, add methyl red TS, and titrate with 0.02 *N* sulfuric acid VS to a faint pink color. Add about 40 mL of freshly boiled, cooled water, and complete the titration with 0.02 *N* sulfuric acid VS. Each mL of 0.02 *N* sulfuric acid is equivalent to 8.128 mg of $C_{18}H_{21}NO_3 . H_3PO_4 . \frac{1}{2}H_2O$.

Codeine Phosphate Tablets, Aspirin and—*see* Aspirin and Codeine Phosphate Tablets

Codeine Phosphate Tablets, Carisoprodol, Aspirin, and—*see* Carisoprodol, Aspirin, and Codeine Phosphate Tablets

Codeine Sulfate

$(C_{18}H_{21}NO_3)_2 . H_2SO_4 . 3H_2O$ 750.86

Morphinan-6-ol, 7,8-didehydro-4,5-epoxy-3-methoxy-17-methyl-, (5α,6α)-, sulfate (2:1) (salt), trihydrate.

7,8-Didehydro-4,5α-epoxy-3-methoxy-17-methylmorphinan-6α-ol sulfate (2:1) (salt) trihydrate [*6854-40-6*].

Anhydrous 696.81 [*1420-53-7*].

» Codeine Sulfate, dried at 105° for 3 hours, contains not less than 98.5 percent and not more than 100.5 percent of $(C_{18}H_{21}NO_3)_2 . H_2SO_4$.

Packaging and storage—Preserve in tight, light-resistant containers.

Reference standard—*USP Codeine Sulfate Reference Standard*—Dry at 105° for 3 hours before using.

Identification—

A: The infrared absorption spectrum of a potassium bromide dispersion of it, previously dried, exhibits maxima only at the same wavelengths as that of a similar preparation of USP Codeine Sulfate RS.

B: The ultraviolet absorption spectrum of a solution (1 in 10,000) exhibits maxima and minima at the same wavelengths as that of a similar solution of USP Codeine Sulfate RS, concomitantly measured, and the respective absorptivities, calculated on the dried basis, at the wavelength of maximum absorbance at about 284 nm do not differ by more than 3.0%.

C: It responds to the tests for *Sulfate* ⟨191⟩.

Specific rotation ⟨781⟩: between −112.5° and −115.0°, calculated on the dried basis, determined in a solution containing the equivalent of 2 g of Codeine Sulfate in each 100 mL.

Acidity—Dissolve 500 mg in 15 mL of water, add 1 drop of methyl red TS, and titrate with 0.020 *N* sodium hydroxide: not more than 0.30 mL is required for neutralization.

Water, *Method III* ⟨921⟩—Dry about 500 mg, accurately weighed, at 105° for 3 hours: it loses between 6.0% and 7.5% of its weight.

Residue on ignition ⟨281⟩: not more than 0.1%.

Readily carbonizable substances ⟨271⟩—Dissolve 10 mg in 5 mL of sulfuric acid TS: the solution has no more color than Matching Fluid S.

Chromatographic purity—Using Codeine Sulfate, proceed as directed in the test for *Chromatographic purity* under *Codeine,* except to use a mixture of 0.01 *N* hydrochloric acid and dehydrated alcohol (4:1), instead of dehydrated alcohol, to prepare *Solution A, Solution B,* and *Solution C.*

Morphine—Dissolve about 50 mg of potassium ferricyanide in 10 mL of water, and add 1 drop of ferric chloride TS and 1 mL of a solution of Codeine Sulfate (1 in 100): no blue color is produced immediately.

Assay—Dissolve about 1.4 g of Codeine Sulfate, previously dried and accurately weighed, in 50 mL of glacial acetic acid, warming, if necessary, to effect solution. Titrate with 0.1 *N* perchloric acid VS, determining the end-point potentiometrically. Perform a blank determination, and make any necessary correction. Each mL of 0.1 *N* perchloric acid is equivalent to 69.68 mg of $(C_{18}H_{21}NO_3)_2 . H_2SO_4$.

Codeine Sulfate Tablets

» Codeine Sulfate Tablets contain not less than 93.0 percent and not more than 107.0 percent of the labeled amount of $(C_{18}H_{21}NO_3)_2 . H_2SO_4 . 3H_2O$.

Packaging and storage—Preserve in well-closed containers.

Reference standard—*USP Codeine Sulfate Reference Standard*—Dry at 105° for 3 hours before using.

Identification—

A: Digest a quantity of finely powdered Tablets, equivalent to about 50 mg of codeine sulfate, with 15 mL of water and 5 mL of 2 *N* sulfuric acid for 1 hour. Filter, if necessary, and wash any undissolved residue with a few mL of water. Render the filtrate alkaline with 6 *N* ammonium hydroxide, extract with several small portions of chloroform, and evaporate the chloroform solution on a steam bath to dryness: the resulting residue of codeine responds to *Identification test A* under *Codeine.*

B: A filtered solution of Tablets responds to the tests for *Sulfate* ⟨191⟩.

Dissolution ⟨711⟩—

Medium: water; 500 mL.

Apparatus 1: 100 rpm.

Time: 45 minutes.

Procedure—Determine the amount of $(C_{18}H_{21}NO_3)_2 . H_2SO_4 . 3H_2O$ dissolved from ultraviolet absorbances at the wavelength of maximum absorbance at about 284 nm of filtered portions of the solution under test, suitably diluted with water, if necessary, in comparison with a Standard solution having a known concentration of USP Codeine Sulfate RS in the same medium.

Tolerances—Not less than 75% (Q) of the labeled amount of $(C_{18}H_{21}NO_3)_2 . H_2SO_4 . 3H_2O$ is dissolved in 45 minutes.

Uniformity of dosage units ⟨905⟩: meet the requirements.

Procedure for content uniformity—Transfer 1 Tablet to a 50-mL volumetric flask. Add 20 mL of 0.5 *N* sulfuric acid and 10 mL of water, shake until the Tablet is disintegrated, and allow to stand for 16 hours. Dilute with water to volume, and filter, discarding the first few mL of the filtrate. Dilute a portion of the subsequent filtrate quantitatively and stepwise, if necessary, with 0.2 *N* sulfuric acid to obtain a solution containing approximately 120 µg of codeine sulfate (trihydrate) per mL. Concomitantly determine the absorbances of this solution and a Standard solution of USP Codeine Sulfate RS in the same medium having a known concentration of about 110 µg per mL, in 1-cm cells, at the wavelength of maximum absorbance at about 284 nm, with a suitable spectrophotometer, using 0.2 *N* sulfuric acid as the blank. Calculate the quantity, in mg, of $(C_{18}H_{21}NO_3)_2 . H_2SO_4 . 3H_2O$ in the Tablet by the formula:

$$(750.86/696.81)(TC/D)(A_U/A_S),$$

in which 750.86 and 696.81 are the molecular weights of codeine sulfate trihydrate and anhydrous codeine sulfate, respectively, *T* is the labeled quantity, in mg, of codeine sulfate in the Tablet, *C* is the concentration, in µg per mL, of USP Codeine Sulfate RS in the Standard solution, *D* is the concentration, in µg per mL, of codeine sulfate in the test solution, based upon the labeled quantity per Tablet and the extent of dilution, and A_U and A_S are the absorbances of the test solution and the Standard solution, respectively.

Assay—Weigh and finely powder not less than 20 Codeine Sulfate Tablets. Weigh accurately a portion of the powder, equivalent to about 150 mg of codeine sulfate, and transfer it completely to a 100-mL volumetric flask. Add sufficient water to make a thin suspension, then add 20 mL of 0.5 *N* sulfuric acid, shake the mixture occasionally for 2 hours, and allow to stand for 16 hours. Dilute with water to volume, mix, and filter through a filtering crucible. Transfer an accurately measured portion of the filtrate, equivalent to not less than 75 mg of codeine sulfate, to a separator, render the solution alkaline with 6 *N* ammonium hydroxide, and completely extract the alkaloid with successive 15-mL portions of chloroform. Evaporate the combined chloroform solution on a steam bath nearly to dryness, add 25.0 mL of 0.02 *N* sulfuric acid VS, and heat gently to dissolve the codeine and expel all of the chloroform. Cool, add methyl red TS, and titrate the excess acid with 0.02 *N* sodium hydroxide VS. Each mL of 0.02 *N* sulfuric acid is equivalent to 7.509 mg of $(C_{18}H_{21}NO_3)_2 . H_2SO_4 . 3H_2O$.

Colchicine

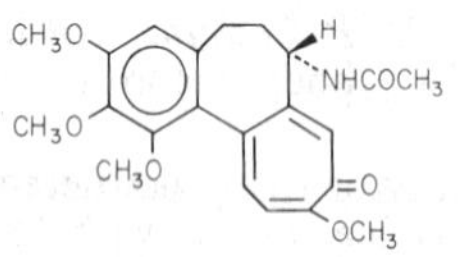

$C_{22}H_{25}NO_6$ 399.44

Acetamide, *N*-(5,6,7,9-tetrahydro-1,2,3,10-tetramethoxy-9-oxobenzo[*a*]heptalen-7-yl)-, (*S*)-.

Colchicine [*64-86-8*].

» Colchicine is an alkaloid obtained from various species of *Colchicum*. It contains not less than 94.0 percent and not more than 101.0 percent of $C_{22}H_{25}NO_6$, calculated on the anhydrous, solvent-free basis.

Caution—Colchicine is extremely poisonous.

Packaging and storage—Preserve in tight, light-resistant containers.

Reference standard—*USP Colchicine Reference Standard*—Dry at 105° for 3 hours before using.

Identification—The infrared absorption spectrum of a potassium bromide dispersion prepared from a portion of Colchicine, previously dried at 105° for 3 hours, exhibits maxima only at the same wavelengths as that of a similar preparation of USP Colchicine RS.

Specific rotation ⟨781⟩: between −240° and −250°, calculated on the anhydrous, solvent-free basis, determined in a solution containing 100 mg in each 10 mL of alcohol.

Water, *Method I* ⟨921⟩: not more than 3.0%.

Colchiceine—To 5 mL of a solution (1 in 100) add 2 drops of ferric chloride TS: no definite green color is produced.

Chloroform and ethyl acetate—

Internal standard—Dilute 1.0 mL of *n*-propyl alcohol with water to 100.0 mL.

Standard preparation—Pipet 1 mL of chloroform, 1 mL of ethyl acetate, and 1 mL of *n*-propyl alcohol into a 1000-mL volumetric flask, add water to volume, and mix. Each mL of *Standard preparation* contains 1.48 mg of chloroform and 0.90 mg of ethyl acetate.

Test preparation—Place about 250 mg of Colchicine, accurately weighed, in a 10-mL volumetric flask, dissolve in about 8 mL of water, and add 1.0 mL of *Internal standard*. Add water to volume, and mix.

Procedure—Determine appropriate sensitivity settings on a suitable gas chromatograph (see *Chromatography* ⟨621⟩) fitted with a 1.5-m × 4-mm column packed with 20 percent (w/v) phase G-14 on support S-1, maintaining the column temperature at 75°, using nitrogen as the carrier gas, and using a flame-ionization detector. Inject the *Standard preparation* and *Test preparation*, determine the corrected peak heights for chloroform and ethyl acetate relative to the peak height for *n*-propyl alcohol, and calculate the percentage, by weight, of chloroform and of ethyl acetate in the portion of Colchicine taken. The sum of the percentages of chloroform and ethyl acetate and of water, as determined in the test for *Water*, is not greater than 10.0%.

Chromatographic purity—The sum of the responses of any peaks other than that due to colchicine, eluting within 1.5 times the retention time for colchicine, is not more than 5.0% of the sum of all responses, obtained as directed in the *Assay*.

Assay—[NOTE—Perform all dilutions in low-actinic glassware.]

Mobile phase—Dilute 45 mL of 0.5 *M* monobasic potassium phosphate with water to 450 mL. Add about 530 mL of methanol, cool to room temperature, and add methanol to bring the volume to 1000 mL. Adjust with 0.5 *M* phosphoric acid to a pH of 5.5 ± 0.05, and filter through a 0.45-µm membrane filter.

Standard preparation—Dissolve an accurately weighed quantity of USP Colchicine RS in a mixture of methanol and water (1:1), and dilute quantitatively and stepwise with the same mixture to obtain a solution having a known concentration of about 6 µg per mL. This solution is stable for 4 months when stored tightly stoppered and in the dark.

Assay preparation—[NOTE—Prepare immediately before use.] Transfer about 60 mg of Colchicine, accurately weighed, to a 500-mL volumetric flask, dissolve in a mixture of methanol and water (1:1), dilute with the same mixture to volume, and mix. Pipet 5 mL of this solution into a 100-mL volumetric flask, dilute with the same mixture to volume, and mix.

Chromatographic system (see *Chromatography* ⟨621⟩)—The liquid chromatograph is equipped with a 254-nm detector and a 4.6-mm × 25-cm column that contains packing L7. The flow rate is about 1 mL per minute. Chromatograph the *Standard preparation*, and record the peak responses as directed under *Procedure:* the column efficiency is not less than 4500 theoretical plates, the retention time for colchicine is between 5.5 and 9.5 minutes, and the relative standard deviation for replicate injections is not more than 2%.

Procedure—Separately inject equal volumes (about 20 µL) of the *Standard preparation* and the *Assay preparation* into the chromatograph, record the chromatograms, and measure the responses for all peaks recorded during 1.5 times the retention time for colchicine. Calculate the quantity, in mg, of $C_{22}H_{25}NO_6$ in the Colchicine taken by the formula:

$$10C(r_U/r_S),$$

in which *C* is the concentration, in µg per mL, of the *Standard preparation*, and r_U and r_S are the colchicine peak responses obtained from the *Assay preparation* and the *Standard preparation*, respectively.

Colchicine Injection

» Colchicine Injection is a sterile solution of $C_{22}H_{25}NO_6$ in Water for Injection, prepared from

Colchicine with the aid of Sodium Hydroxide. It contains not less than 90.0 percent and not more than 110.0 percent of the labeled amount of $C_{22}H_{25}NO_6$.

Caution—Colchicine is extremely poisonous.

Packaging and storage—Preserve in single-dose containers, preferably of Type I glass, protected from light.

Reference standard—*USP Colchicine Reference Standard*—Dry at 105° for 3 hours before using.

Identification—

A: Transfer a volume of Injection, equivalent to about 2 mg of colchicine, to a separator. Add 5 mL of water, and extract with 15 mL of chloroform. Evaporate the chloroform extract, using mild heat, to dryness: the infrared absorption spectrum of a potassium bromide dispersion of the residue so obtained exhibits maxima only at the same wavelengths as that of a similar preparation of USP Colchicine RS.

B: The ultraviolet absorption spectrum of the Injection, diluted with water to a concentration of about 10 µg of colchicine per mL, exhibits maxima and minima at the same wavelengths as that of a similar solution of USP Colchicine RS, concomitantly measured.

pH ⟨791⟩: between 6.0 and 7.2, in a solution of Injection containing 1.0 mg of potassium chloride in each mL.

Other requirements—It meets the requirements under *Injections* ⟨1⟩.

Assay—[NOTE—Perform all dilutions in low-actinic glassware.]

Mobile phase, Standard preparation, and *Chromatographic system*—Prepare as directed in the *Assay* under *Colchicine.*

Assay preparation—[NOTE—Prepare immediately before use.] Transfer an accurately measured volume, *V* mL, of Colchicine Injection, equivalent to about 1 mg of colchicine, to a 50-mL volumetric flask, dilute with a mixture of methanol and water (1:1) to volume, and mix. Pipet 30 mL of this solution into a 100-mL volumetric flask, dilute with the same mixture to volume, and mix.

Procedure—Proceed as directed for *Procedure* in the *Assay* under *Colchicine*, and measure the responses for the colchicine peaks. Calculate the quantity, in mg, of $C_{22}H_{25}NO_6$ in each mL of the Injection taken by the formula:

$$(C/6V)(r_U/r_S),$$

in which *C* is the concentration, in µg per mL, of the *Standard preparation*, and r_U and r_S are the peak responses obtained from the *Assay preparation* and the *Standard preparation*, respectively.

Colchicine Tablets

» Colchicine Tablets contain not less than 90.0 percent and not more than 110.0 percent of the labeled amount of $C_{22}H_{25}NO_6$.

Packaging and storage—Preserve in well-closed, light-resistant containers.

Reference standard—*USP Colchicine Reference Standard*—Dry at 105° for 3 hours before using.

Identification—Weigh a portion of ground Tablets, equivalent to about 20 mg of colchicine, triturate with 20 mL of water, allow the solids to settle, and filter the supernatant liquid into a separator. Extract with 30 mL of chloroform. Evaporate the chloroform extract, using mild heat, to dryness: the infrared absorption spectrum of a potassium bromide dispersion of the residue so obtained exhibits maxima only at the same wavelengths as that of a similar preparation of USP Colchicine RS.

Dissolution ⟨711⟩—[NOTE—Conduct this procedure without delay, under subdued light, and using low-actinic glassware.]

Medium: water; 500 mL.

Apparatus 1: 100 rpm.

Time: 30 minutes.

Procedure—Extract a filtered portion of the solution under test with three 15-mL portions of chloroform, collecting the chloroform extracts in a suitable flask. Evaporate the combined extracts to a small volume, transfer to a 10-mL volumetric flask, add chloroform to volume, and mix. Determine the amount of $C_{22}H_{25}NO_6$ dissolved from absorbances at the wavelength of maximum absorbance at about 350 nm, in comparison with a Standard solution having a known concentration of USP Colchicine RS in chloroform, using chloroform as the blank.

Tolerances—Not less than 75% (*Q*) of the labeled amount of $C_{22}H_{25}NO_6$ is dissolved in 30 minutes.

Uniformity of dosage units ⟨905⟩: meet the requirements.

Procedure for content uniformity—[NOTE—Conduct this procedure without delay, under subdued light, using low-actinic glassware.] Powder 1 Tablet, and transfer to a separator. Add 5 mL of water, and mix. Extract with three 10-mL portions of chloroform, filtering the extracts through a layer of anhydrous sodium sulfate into a 50-mL volumetric flask. Dilute with chloroform to volume, and mix. Dissolve an accurately weighed quantity of USP Colchicine RS in chloroform, and dilute quantitatively and stepwise with chloroform to obtain a *Standard solution* having a known concentration of about 10 µg per mL. Concomitantly determine the absorbances of both solutions in 1-cm cells at the wavelength of maximum absorbance at about 350 nm, with a suitable spectrophotometer, using chloroform as the blank. Calculate the quantity, in mg, of $C_{22}H_{25}NO_6$ in the Tablet by the formula:

$$(TC/D)(A_U/A_S),$$

in which *T* is the labeled quantity, in mg, of colchicine in the Tablet, *C* is the concentration, in µg per mL, of USP Colchicine RS in the *Standard solution, D* is the concentration, in µg per mL, of colchicine in the solution from the Tablet, on the basis of the labeled quantity per Tablet and the extent of dilution, and A_U and A_S are the absorbances of the solution from the Tablet and the *Standard solution*, respectively.

Assay—[NOTE—Perform all dilutions in low-actinic glassware.]

Mobile phase, Standard preparation, and *Chromatographic system*—Prepare as directed in the *Assay* under *Colchicine.*

Assay preparation—[NOTE—Prepare immediately before use.] Weigh and finely powder not less than 20 Colchicine Tablets. Transfer an accurately weighed portion of the powder, equivalent to about 0.6 mg of colchicine, to a 100-mL volumetric flask, add about 50 mL of a mixture of methanol and water (1:1), and shake by mechanical means for 15 minutes, rinsing down the walls of the flask at about 8 minutes. Dilute with the same mixture to volume, and filter through a 0.45-µm membrane filter.

Procedure—Proceed as directed for *Procedure* in the *Assay* under *Colchicine*, and measure the responses for the colchicine peaks. Calculate the quantity, in mg, of $C_{22}H_{25}NO_6$ in the portion of Tablets taken by the formula:

$$0.1C(r_U/r_S),$$

in which *C* is the concentration, in µg per mL, of the *Standard preparation*, and r_U and r_S are the peak responses obtained from the *Assay preparation* and the *Standard preparation*, respectively.

Colchicine Tablets, Probenecid and—*see* Probenecid and Colchicine Tablets

Colestipol Hydrochloride

» Colestipol Hydrochloride is an insoluble, high molecular weight basic anion-exchange copolymer of diethylenetriamine and 1-chloro-2,3-epoxypropane with approximately one out of five amino nitrogens protonated. Each g binds not less than 1.1 mEq and not more than 1.6 mEq of sodium cholate, calculated as cholate binding capacity.

Packaging and storage—Preserve in tight containers.

Reference standard—*USP Colestipol Hydrochloride Reference Standard*—Do not dry before using.

Identification—Determine by pyrolysis gas chromatography, using the following procedure.

Standard preparation—Transfer an appropriate amount of USP Colestipol Hydrochloride RS into the probe. If necessary to keep the colestipol hydrochloride in the probe, mix 4 parts with 1 part of *n*-eicosane. Grind in a mortar with chloroform until the colestipol hydrochloride is uniformly coated with the *n*-eicosane. This preparation is stable indefinitely, but may require wetting with a small amount of chloroform before each use.

Test preparation—Proceed as directed under *Standard preparation*, using an appropriate amount of the test specimen.

Chromatographic system (see *Chromatography* ⟨621⟩)—The gas chromatograph is equipped with a flame-ionization detector and a 180-cm × 3-mm column packed with 80- to 100-mesh support S1A which has been coated with 0.25 percent potassium hydroxide and 5 percent phase G16. The carrier gas is helium at a flow rate of about 60 mL per minute. Maintain the detector and column temperatures at about 270° and about 85°, respectively. The pyrolysis unit is capable of reaching 1100° when equipped with a platinum probe, and pyrolysis time is not less than 10 seconds.

Procedure—Install the pyrolysis unit on the chromatograph, and position the probe so that it is just above but not touching the column packing. Set the pyrolysis temperature to about 1100°. Separately pyrolyze the *Standard preparation* and the *Test preparation*. After the completion of each pyrolysis cycle, remove and clean the probe. The pyrogram of the *Test preparation* corresponds to that of the *Standard preparation* obtained the same day.

pH ⟨791⟩—Prepare a 10% (w/w) suspension of it in deionized water in a clean vial. Insert the stopper, shake at approximately 10-minute intervals for 1 hour, and centrifuge. Transfer a portion of the clear supernatant liquid to a suitable container, and record the pH as soon as the reading has stabilized: the pH is between 6.0 and 7.5.

Loss on drying ⟨731⟩—Dry it in vacuum at a pressure of about 5 mm of mercury at 75° for 16 hours: it loses not more than 1.0% of its weight.

Residue on ignition ⟨281⟩: not more than 0.3%.

Heavy metals, *Method II* ⟨231⟩: not more than 0.002%.

Chloride content—

Test preparation—Using about 20 mg of Colestipol Hydrochloride, accurately weighed, proceed as directed under *Oxygen Flask Combustion* ⟨471⟩, 10 mL of 0.05 *N* sodium hydroxide being used as the absorbing liquid. Do not allow the paper specimen wrapper to come in contact with the liquid, and ignite the paper with an infrared igniter. After combustion is complete, shake the flask vigorously, and allow to stand, with frequent shaking, for about 40 minutes or until no cloudiness is present. Transfer the solution to a 50-mL beaker. Wash the flask with one 10-mL portion of water and one 20-mL portion of alcohol, adding each washing to the beaker, and add 0.2 mL of nitric acid.

Reagent blank preparation I—Using a paper specimen wrapper, complete the combustion, and allow the mixture to stand for about 40 minutes as directed under *Test preparation*. Transfer the solution so obtained to a 50-mL beaker containing 10.0 mL of 0.0075 *N* sodium chloride. Wash the combustion flask with two 10-mL portions of alcohol, adding the washings to the beaker, and add 0.2 mL of nitric acid.

Reagent blank preparation II—Mix 10.0 mL of 0.0075 *N* sodium chloride, 10 mL of water, 20 mL of alcohol, and 0.2 mL of nitric acid in a 50-mL beaker.

Procedure—Titrate the *Test preparation* and the *Reagent blank preparations I* and *II* with 0.05 *N* silver nitrate VS, determining the end-point potentiometrically, using a silver–silver chloride electrode and a glass reference electrode (see *Titrimetry* ⟨541⟩). Determine the volume of 0.05 *N* silver nitrate VS consumed by the test specimen by the formula:

$$V - (V_I - V_{II}),$$

in which V, V_I, and V_{II}, are the volumes, in mL, of titrant used for the *Test preparation*, *Reagent blank preparation I*, and *Reagent blank preparation II*, respectively. Each mL of 0.05 *N* silver nitrate is equivalent to 1.773 mg of Cl: the chloride content is between 6.5% and 9.0%, calculated on the dried basis.

Water absorption—Transfer about 5 g of Colestipol Hydrochloride, accurately weighed, to a dry, 100-mL plastic container, and add about 80 g of water, accurately weighed. Cover the container, and allow the suspension to equilibrate for 72 hours. With the aid of vacuum, filter the slurry transferred to a medium-porosity, fritted-glass funnel, and collect the filtrate in a tared plastic container. Disconnect the vacuum 2 minutes after the collection of the last portion of the filtrate. Weigh the container and the filtrate, and determine the weight, in g, of the filtrate. Determine the amount of water absorbed by subtracting the weight of the filtrate from the weight of water taken for the test, and divide the weight, in g, of the absorbed water by the weight, in g, of Colestipol Hydrochloride taken: each g absorbs between 3.3 g and 5.3 g of water.

Cholate binding capacity—

Cholate solution—Dissolve accurately weighed quantities of sodium cholate and sodium chloride in water, and dilute quantitatively with water to obtain a solution having known concentrations of 10.0 mg of sodium cholate per mL and 9.0 mg of sodium chloride per mL. Adjust the solution by the dropwise addition of hydrochloric acid to a pH of 6.4 ± 0.1.

Test preparation—Transfer 1.0 ± 0.01 g of Colestipol Hydrochloride to a glass-stoppered, 125-mL conical flask. Add 100.0 mL of *Cholate solution*, insert the stopper securely in the flask, shake vigorously for 90 minutes with the flask positioned horizontally on a platform shaker, remove the flask from the shaker, and allow the solids to settle for 5 minutes.

Procedure—Transfer 20.0 mL of supernatant liquid from the *Test preparation* to a 40-mL beaker, transfer 20.0 mL of *Cholate solution* to a second 40-mL beaker, and adjust both solutions by the dropwise addition of 1 *N* sodium hydroxide to a pH of 10.5 ± 0.5. Titrate both solutions with 0.1 *N* hydrochloric acid VS, determining the end-points potentiometrically, and measure the titrant volume corresponding to the difference between the midpoints of the two inflections in the titration curves obtained for each solution (see *Titrimetry* ⟨541⟩). Determine the volume of titrant equivalent to the bound cholate by subtracting the volume of 0.1 *N* hydrochloric acid VS used in titrating the *Test preparation* from that used in titrating the *Cholate solution*. Calculate the *Cholate binding capacity*, in mEq per g, by the formula:

$$5VN/W,$$

in which V is the volume, in mL, of titrant equivalent to the bound cholate, N is the normality of the 0.1 *N* hydrochloric acid VS, and W is the weight, in g, of Colestipol Hydrochloride taken for the *Test preparation*. The *Cholate binding capacity* is between 1.1 mEq per g and 1.6 mEq per g.

Water-soluble substances—Transfer 5.0 g of Colestipol Hydrochloride, accurately weighed, to a glass-stoppered, 125-mL conical flask, add 80.0 mL of water, insert the stopper in the flask, and mount the flask in a water-bath shaker maintained at 37 ± 1°. Operate the shaker for 72 hours, remove the flask from the shaker, and filter the contents through a fine-porosity, fritted-glass funnel, collecting the filtrate in a tared 125-mL conical flask. Rinse any residual test material in the flask with two 5-mL portions of water, pass the washings through the filter, and combine the filtrates from the washings with the filtrate from the test mixture. Evaporate the filtrate to dryness, filtered air or nitrogen being used, if necessary, to aid in the evaporation. Dry the residue in a vacuum oven maintained at 75° for 1 hour, allow to cool in a desiccator, and weigh: not more than 0.5% of water-soluble substances is found in the portion of Colestipol Hydrochloride taken.

Colestipol exchange capacity—

Resin base preparation—Combine not less than 2 g of Colestipol Hydrochloride and 100 mL of 1 *N* sodium hydroxide in a 125-mL conical flask, insert a stopper in the flask, secure the flask on a platform shaker, and shake the mixture for 3 to 4 hours. Filter the suspension through a coarse-porosity, fritted-glass funnel, and wash the resin with 500 mL of water. Transfer the resin to a 1000-mL beaker, add 200 mL of water, allow to stand for 10 minutes, filter the suspension, and check the pH of the filtrate. Repeat the washing procedure with 200-mL portions of water until the pH of the filtrate is below 8 (as much as 5000 mL of water may be required). Dry the colestipol base resin so obtained and the funnel at a pressure of about 5 mm of mercury at 60° for 16 hours. Break up any aggregates, and store the *Resin base preparation* in a desiccator.

Procedure—Transfer about 1.0 g of the *Resin base preparation* to a 125-mL conical flask, add 100.0 mL of 0.20 *N* hydrochloric acid, insert a stopper in the flask, and shake the mixture by mechanical means for 2.5 hours. Filter a portion of the suspension through a pledget of glass wool, and transfer 8.0 mL of the filtrate (test preparation) to a 25-mL beaker. Transfer 5.0 mL of the same 0.20 *N* hydrochloric acid that was used to equilibrate the resin to a second 25-mL beaker, and add 5.0 mL of water. Titrate both solutions with 0.2 *N* sodium hydroxide VS, determining the end-points potentiometrically (see *Titrimetry* ⟨541⟩), and calculate the *Colestipol exchange capacity*, in mEq per g, by the formula:

$$(100N/W)[(V_b/5) - (V_a/8)],$$

in which N is the normality of the sodium hydroxide VS, W is the weight, in g, of the *Resin base preparation* taken, V_b is the volume, in mL, of titrant used to neutralize the 5.0-mL aliquot of 0.20 *N* hydrochloric acid, and V_a is the volume, in mL, of titrant used to neutralize the residual acid in the test preparation. Each g exchanges not less than 9.0 mEq and not more than 11.0 mEq of sodium hydroxide, calculated as colestipol exchange capacity.

Colestipol Hydrochloride for Oral Suspension

» Colestipol Hydrochloride for Oral Suspension is a mixture of Colestipol Hydrochloride with a suitable flow-promoting agent. Each g binds not less than 1.1 mEq and not more than 1.6 mEq of sodium cholate, calculated as the cholate binding capacity.

Packaging and storage—Preserve in tight, single-dose or multiple-dose containers.

Reference standard—*USP Colestipol Hydrochloride Reference Standard*—Do not dry before using.

Minimum fill ⟨755⟩: meets the requirements for powders.

Water-soluble substances—Transfer 5.0 g of Colestipol Hydrochloride for Oral Suspension, accurately weighed, to a glass-stoppered, 125-mL conical flask, add 80.0 mL of water, insert the stopper in the flask, and mount the flask in a water-bath shaker maintained at 37 ± 1°. Operate the shaker for 72 hours, remove the flask from the shaker, and filter the contents through a fine-porosity, fritted-glass funnel, collecting the filtrate in a tared 100-mL fused quartz crucible. Rinse any residual test material in the flask with two 5-mL portions of water, pass the washings through the filter, and combine the filtrates from the washings with the filtrate from the test mixture. Evaporate the filtrate to dryness, filtered air or nitrogen being used, if necessary, to aid in the evaporation. Dry the residue in a vacuum oven maintained at 75° for 1 hour, allow to cool in a desiccator, and weigh. Calculate the initial percentage of water-soluble substances in the portion of Colestipol Hydrochloride for Oral Suspension taken. Again heat the residue in a muffle furnace maintained at 800 ± 25° for 4 hours, allow to cool in a desiccator, and weigh. Calculate the percentage of inert ingredients present. Calculate the actual percentage of water-soluble substances in the portion of Colestipol Hydrochloride for Oral Suspension taken by subtracting the percentage of inert ingredients from the initial percentage of water-soluble substances found. Not more than 0.5% of water-soluble substances is found.

Other requirements—Meets the requirements of the tests for *Cholate binding capacity*, *Identification*, *Water absorption*, and *pH* under *Colestipol Hydrochloride*.

Sterile Colistimethate Sodium

R—C(=O)—Dbu-Thr-Dbu-Dbu-Dbu-DLeu-Leu-Dbu—Dbu-Thr (cyclic, ring closed through N^{γ} of the fourth Dbu); side chains N^{γ}-R′ on Dbu residues

(Dbu is L-α,γ-diaminobutyric acid; *R* is $CH_3CH_2CH(CH_3)(CH_2)_4$ in the colistin A component and $CH_3CH(CH_3)(CH_2)_4$ in the colistin B component; *R′* is CH_2SO_3Na)

$C_{58}H_{105}N_{16}Na_5O_{28}S_5$ (colistin A component) 1749.81
$C_{57}H_{103}N_{16}Na_5O_{28}S_5$ (colistin B component) 1735.78
Colistimethate sodium.
Pentasodium colistinmethanesulfonate
[*8068-28-8; 21362-08-3*].

» Sterile Colistimethate Sodium is colistimethate sodium suitable for parenteral use. It has a potency equivalent to not less than 390 µg of colistin per mg. In addition, where packaged for dispensing, it contains the equivalent of not less than 90.0 percent and not more than 120.0 percent of the labeled amount of colistin.

Packaging and storage—Preserve in *Containers for Sterile Solids* as described under *Injections* ⟨1⟩.

Reference standard—*USP Colistimethate Sodium Reference Standard*—Dry in vacuum at a pressure not exceeding 5 mm of mercury at 60° for 3 hours before using.

Constituted solution—At the time of use, the constituted solution prepared from Sterile Colistimethate Sodium meets the requirements for *Constituted Solutions* under *Injections* ⟨1⟩.

Identification—The infrared absorption spectrum of a potassium bromide dispersion of it, previously dried, exhibits maxima only at the same wavelengths as that of a similar preparation of USP Colistimethate Sodium RS.

Pyrogen—It meets the requirements of the *Pyrogen Test* ⟨151⟩, the test dose being 1.0 mL per kg of a solution in Sterile Water for Injection containing the equivalent of 10 mg of colistin base per mL.

Sterility—It meets the requirements under *Sterility Tests* ⟨71⟩, when tested as directed in the section, *Test Procedures Using Membrane Filtration.*

pH ⟨791⟩: between 6.5 and 8.5, in a solution containing 10 mg per mL.

Loss on drying ⟨731⟩—Dry about 100 mg, accurately weighed, in a capillary-stoppered bottle in vacuum at a pressure not exceeding 5 mm of mercury at 60° for 3 hours: it loses not more than 7.0% of its weight.

Heavy metals, *Method II* ⟨231⟩: not more than 0.003%.

Free colistin—Dissolve 80 mg in 3 mL of water, and add 0.05 mL of silicotungstic acid solution (1 in 10): no immediate precipitate is formed.

Other requirements—Where packaged for dispensing, it meets the requirements for *Uniformity of Dosage Units* ⟨905⟩ and for *Constituted Solutions* and *Labeling* under *Injections* ⟨1⟩.

Assay—

Assay preparation 1—Dissolve a suitable quantity of Sterile Colistimethate Sodium, accurately weighed, in 2.0 mL of water, add a sufficient accurately measured volume of *Buffer No. 6* to obtain a solution having a convenient concentration.

Assay preparation 2 (where it is packaged for dispensing and is represented as being in a single-dose container)—Constitute Sterile Colistimethate Sodium in a volume of water, accurately measured, corresponding to the volume of diluent specified in the labeling. Withdraw all of the withdrawable contents, using a suitable hypodermic needle and syringe, and dilute quantitatively with *Buffer No. 6* to obtain a solution having a convenient concentration.

Assay preparation 3 (where the label states the quantity of colistin equivalent in a given volume of constituted solution)—Constitute Sterile Colistimethate Sodium in a volume of water, accurately measured, corresponding to the volume of diluent specified in the labeling. Dilute an accurately measured volume of the constituted solution quantitatively with *Buffer No. 6* to obtain a solution having a convenient concentration.

Procedure—Proceed as directed for Colistimethate Sodium under *Antibiotics—Microbial Assays* ⟨81⟩, using an accurately measured volume of *Assay preparation* diluted quantitatively with *Buffer No. 6* to yield a *Test Dilution* having a concentration assumed to be equal to the median dose level of the Standard.

Colistin Sulfate

O
‖
RC—Dbu-Thr-Dbu-Dbu-Dbu-DLeu-Leu-Dbu-Dbu-Thr (N^γ ring) · $2.5H_2SO_4$

(Dbu is L-α,γ-diaminobutyric acid; *R* is 5-methylheptyl in Colistin A and 5-methylhexyl in Colistin B)

Colistin, sulfate.
Colistins sulfate [*1264-72-8*].

» Colistin Sulfate is the sulfate salt of an antibacterial substance produced by the growth of *Bacillus polymyxa* var. *colistinus.* It has a potency equivalent to not less than 500 μg of colistin per mg.

Packaging and storage—Preserve in tight containers.

Reference standard—*USP Colistin Sulfate Reference Standard*—Dry in vacuum at a pressure not exceeding 5 mm of mercury at 60° for 3 hours before using.

Identification—Dissolve about 20 mg in 2 mL of buffer solution, prepared by adjusting 50 mL of 1 *M* monobasic potassium phosphate with 1 *N* sodium hydroxide to a pH of 7.0, diluting with water to 100 mL, and mixing. Add 0.2 mL of ninhydrin solution (1 in 200) and boil: a purple color is produced.

pH ⟨791⟩: between 4.0 and 7.0, in a solution containing 10 mg per mL.

Loss on drying ⟨731⟩—Dry about 100 mg, accurately weighed, in a capillary-stoppered bottle in vacuum at a pressure not exceeding 5 mm of mercury at 60° for 3 hours: it loses not more than 7.0% of its weight.

Assay—Proceed as directed for Colistin under *Antibiotics—Microbial Assays* ⟨81⟩.

Colistin Sulfate for Oral Suspension

» Colistin Sulfate for Oral Suspension is a dry mixture of Colistin Sulfate with or without one or more suitable buffers, colors, diluents, dispersants, and flavors. It contains the equivalent of not less than 90.0 percent and not more than 120.0 percent of the labeled amount of colistin.

Packaging and storage—Preserve in tight containers, protected from light.

Reference standard—*USP Colistin Sulfate Reference Standard*—Dry in vacuum at a pressure not exceeding 5 mm of mercury at 60° for 3 hours before using.

Uniformity of dosage units ⟨905⟩—

FOR SOLID PACKAGED IN SINGLE-UNIT CONTAINERS: meets the requirements.

pH ⟨791⟩: between 5.0 and 6.0, in the suspension constituted as directed in the labeling.

Loss on drying ⟨731⟩—Dry about 100 mg, accurately weighed, in a capillary-stoppered bottle in vacuum at a pressure not exceeding 5 mm of mercury at 60° for 3 hours: it loses not more than 3.0% of its weight.

Assay—Constitute Colistin Sulfate for Oral Suspension as directed in the labeling. Proceed as directed under *Antibiotics—Microbial Assays* ⟨81⟩ using an accurately measured volume of this suspension, freshly mixed and free from air bubbles, diluted quantitatively with *Buffer No. 6* to yield a *Test Dilution* having a concentration assumed to be equal to the median dose level of the Standard.

Colistin and Neomycin Sulfates and Hydrocortisone Acetate Otic Suspension

» Colistin and Neomycin Sulfates and Hydrocortisone Acetate Otic Suspension is a sterile suspension containing the equivalent of not less than 90.0 percent and not more than 135.0 percent of the labeled amount of colistin, not less than 90.0 percent and not more than 125.0 percent of the labeled amount of neomycin, and not less than 90.0 percent and not more than 110.0 percent of the labeled amount of hydrocortisone acetate ($C_{23}H_{32}O_6$). It contains one or more suitable buffers, detergents, dispersants, and preservatives.

NOTE—Where Colistin and Neomycin Sulfates and Hydrocortisone Acetate Otic Suspension is prescribed, without reference to the quantity of colistin, neomycin, or hydrocortisone acetate contained therein, a product containing 3.0 mg of colistin, 3.3 mg of neomycin, and 10 mg of hydrocortisone acetate per mL shall be dispensed.

Packaging and storage—Preserve in tight containers.

Reference standards—*USP Colistin Sulfate Reference Standard*—Dry in vacuum at a pressure not exceeding 5 mm of mercury at 60° for 3 hours before using. *USP Neomycin Sulfate Reference Standard*—Dry in vacuum at a pressure not exceeding 5 mm of mercury at 60° for 3 hours before using. *USP Hydrocortisone Acetate Reference Standard*—Dry at 105° for 3 hours before using.

Sterility—It meets the requirements under *Sterility Tests* ⟨71⟩, 0.25 mL from each container being transferred directly to 90 mL of each medium.

pH ⟨791⟩: between 4.8 and 5.2.

Assay for colistin—Proceed as directed under *Antibiotics—Microbial Assays* ⟨81⟩, using a freshly mixed, accurately measured volume of Colistin and Neomycin Sulfates and Hydrocortisone Acetate Otic Suspension diluted quantitatively and stepwise with *Buffer No. 6* to yield a *Test Dilution* having a concentration assumed to be equal to the median dose level of the Standard.

Assay for neomycin—Proceed as directed under *Antibiotics—Microbial Assays* ⟨81⟩, using a freshly mixed, accurately measured volume of Colistin and Neomycin Sulfates and Hydrocortisone Acetate Otic Suspension diluted quantitatively and stepwise with *Buffer No. 3* to yield a *Test Dilution* having a concentration assumed to be equal to the median dose level of the Standard.

Assay for hydrocortisone acetate—

Reagent blank—Dilute 200 mL of 22 *N* sulfuric acid with 100 mL of dehydrated alcohol.

Phenylhydrazine reagent—Dissolve 43.33 mg of phenylhydrazine hydrochloride in 100 mL of *Reagent blank.*

Standard preparation—Dissolve a suitable quantity of USP Hydrocortisone Acetate RS, accurately weighed, in chloroform, and dilute quantitatively and stepwise with chloroform to obtain a solution having a known concentration of about 10 μg per mL.

Assay preparation—Transfer 5.0 mL of freshly mixed Colistin and Neomycin Sulfates and Hydrocortisone Acetate Otic Suspension to a 125-mL separator. Extract with three 20-mL portions of chloroform, filtering each chloroform extract through a pledget of cotton previously saturated with chloroform, collect the filtrates in a 100-mL volumetric flask, dilute with chloroform to volume, and mix. Pipet 10 mL of this solution into a 100-mL volumetric flask, dilute with chloroform to volume, and mix. Pipet 20 mL of this solution into a 100-mL volumetric flask, dilute with chloroform to volume, and mix.

Procedure—Pipet 50 mL each of the *Standard preparation* and the *Assay preparation* into separate 125-mL separators, add 2 mL of 0.1 *N* sodium hydroxide to each separator, shake, and allow the layers to separate. Filter both chloroform layers through glass wool, and collect the filtrates in separate beakers. Pipet two 20-mL portions of each chloroform filtrate into separate 125-mL separators. Add 25.0 mL of *Phenylhydrazine reagent* to one separator each of the filtrates from the *Standard preparation* and the *Assay preparation*, respectively, and add 25.0 mL of *Reagent blank* to the remaining two separators. Shake all four separators well, allow the layers to separate, and discard the chloroform layers. Drain the aqueous layers into separate centrifuge tubes, and centrifuge for 2 minutes. Pipet 10 mL of each solution into separate glass-stoppered test tubes. Place the tubes in a water bath maintained at a temperature of 60° for 30 minutes, then cool the solution to room temperature. Concomitantly determine the absorbances of the solutions at the wavelength of maximum absorbance at about 410 nm, with a suitable spectrophotometer, using water to set the instrument. Calculate the quantity, in mg, of hydrocortisone acetate ($C_{23}H_{32}O_6$) in each mL of the Otic Suspension taken by the formula:

$$C(A_U - A_{UB}/A_S - A_{SB}),$$

in which C is the concentration, in µg per mL, of USP Hydrocortisone Acetate RS in the *Standard preparation*, A_U and A_S are the absorbances of the solutions from the *Assay preparation* and the *Standard preparation* treated with *Phenylhydrazine reagent*, respectively, and A_{UB} and A_{SB} are the absorbances of the solution from the *Assay preparation* and the *Standard preparation* treated with the *Reagent blank*, respectively.

Collodion

» Collodion contains not less than 5.0 percent, by weight, of pyroxylin.

Pyroxylin	40 g
Ether	750 mL
Alcohol	250 mL
To make about	1000 mL

Add the Alcohol and the Ether to the Pyroxylin contained in a suitable container, and insert the stopper into the container well. Shake the mixture occasionally until the Pyroxylin is dissolved.

Caution—Collodion is highly flammable.

Packaging and storage—Preserve in tight containers, at a temperature not exceeding 30°, remote from fire.

Labeling—The label bears a caution statement to the effect that Collodion is highly flammable.

Identification—

A: When exposed to air in a thin layer, it leaves a transparent, tenacious film. The film of pyroxylin so obtained burns rapidly, with a yellow flame.

B: When mixed with an equal volume of water, it yields a viscid, stringy mass of pyroxylin.

Specific gravity ⟨841⟩: between 0.765 and 0.775.

Acidity—Add 5 mL of it to 5 mL of water: the liquid separated from the pyroxylin is not acid to litmus.

Alcohol content ⟨611⟩—

Internal standard solution—Place 20 mL of acetone in a glass-stoppered, 100-mL graduated cylinder, dilute with 1,2-dichloroethane to 100 mL, and mix.

Standard solutions—Pipet 10-, 20-, and 30-mL portions of dehydrated alcohol into separate 100-mL volumetric flasks, dilute with 1,2-dichloroethane to volume, and mix.

Standard preparations—Pipet 10 mL of each *Standard solution* into separate glass-stoppered, 50-mL graduated cylinders. To each add 15 mL of 1,2-dichloroethane, 10 mL of hexane, and 10.0 mL of *Internal standard solution*, and mix.

Test preparation—Pipet 10 mL of Collodion into a glass-stoppered, 50-mL graduated cylinder containing 15 mL of 1,2-dichloroethane, 10 mL of hexane, and 10.0 mL of *Internal standard solution*, mix, and allow the precipitate to settle.

Chromatographic system—Under typical conditions, the instrument is equipped with a thermal conductivity detector, and it contains a 1.8-m × 3.5-mm glass column containing packing S3; the injection port, the detector, and the column are maintained at 200°, 250°, and 150°, respectively, and helium is used as the carrier gas, at a flow rate of about 50 mL per minute.

Procedure—Inject a suitable volume (about 4 µL) of each *Standard preparation* and of the *Test preparation* into a suitable gas chromatograph, and record the chromatograms (so as to obtain about 25% of maximum recorder response for the least concentrated solution). Calculate the relative response factor, F, for each *Standard solution* by the formula:

$$F = C_S/R_S,$$

in which C_S is the concentration of C_2H_5OH in the *Standard solution*, in percentage (V/V), and R_S is the area-ratio of alcohol to acetone observed for the respective *Standard preparation*. Calculate the percentage of alcohol, C_U, in the test specimen by the formula:

$$C_U = F_a R_U,$$

in which F_a is the average of the individual F values, and R_U is the area-ratio of alcohol to acetone observed for the *Test preparation*. The content of C_2H_5OH is between 22.0% and 26.0%.

Assay—Pour quickly about 10 mL of Collodion into a tared flask, insert the stopper, weigh the assay charge accurately, remove the stopper, warm on a steam bath, and add 10 mL of water dropwise, with constant stirring. Evaporate the mixture on a steam bath, and dry the residue at 105° for 4 hours. The weight of the pyroxylin so obtained corresponds to not less than 5.0%, by weight, of the assay charge.

Flexible Collodion

» Prepare Flexible Collodion as follows.

Camphor	20 g
Castor Oil	30 g
Collodion, a sufficient quantity, to make	1000 g

Weigh the ingredients, successively, into a dry, tared bottle, insert the stopper in the bottle, and shake the mixture until the camphor is dissolved.

Packaging and storage—Preserve in tight containers, at a temperature not exceeding 30°, remote from fire.

Labeling—The label bears a caution statement to the effect that Flexible Collodion is highly flammable.

Identification—

A: When exposed to air in a thin layer, it leaves a transparent tenacious film. The film exhibits a distinct odor of camphor. The film of pyroxylin so obtained burns rapidly, with a yellow flame.

B: When mixed with an equal volume of water, it yields a viscid, stringy mass of pyroxylin.

Specific gravity ⟨841⟩: between 0.770 and 0.790.

Alcohol content—Proceed as directed in the test for *Alcohol content* under *Collodion*. The content of C_2H_5OH is between 21.0% and 25.0%.

Collodion, Salicylic Acid—*see* Salicylic Acid Collodion

Colloidal Silicon Dioxide—*see* Silicon Dioxide, Colloidal NF

Complex, Factor IX—*see* Factor IX Complex

Complex, Tetracycline Phosphate—*see* Tetracycline Phosphate Complex

Compound Benzaldehyde Elixir—*see* Benzaldehyde Elixir, Compound NF

Compound Benzoin Tincture—*see* Benzoin Tincture, Compound

Compound Clioquinol Powder—*see* Clioquinol Powder, Compound

Compound Resorcinol Ointment—*see* Resorcinol Ointment, Compound

Compound Undecylenic Acid Ointment—*see* Undecylenic Acid Ointment, Compound

Compressed Air—*see* Air, Compressed

Compressible Sugar—*see* Sugar, Compressible NF

Concentrate, Glutaral—*see* Glutaral Concentrate

Concentrate, Hydrogen Peroxide—*see* Hydrogen Peroxide Concentrate

Concentrate, Platelet—*see* Platelet Concentrate

Confectioner's Sugar—*see* Sugar, Confectioner's NF

Conjugated Estrogens—*see* Estrogens, Conjugated

Copper Gluconate

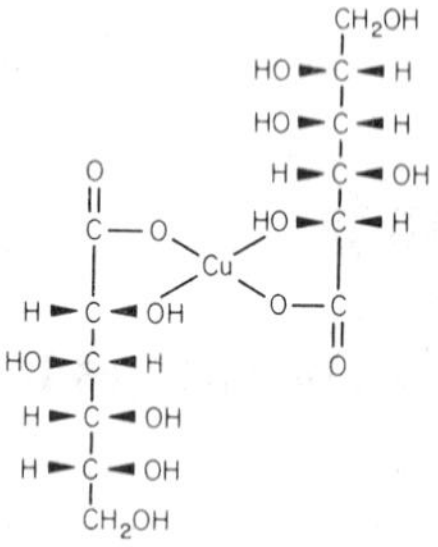

$C_{12}H_{22}CuO_{14}$ 453.84
Copper, bis(D-gluconato-O^1,O^2)-.
Copper D-gluconate (1:2) [*527-09-3*].

» Copper Gluconate contains not less than 98.0 percent and not more than 102.0 percent of $C_{12}H_{22}CuO_{14}$.

Packaging and storage—Preserve in well-closed containers.

Reference standard—*USP Potassium Gluconate Reference Standard*—Dry in vacuum at 105° for 4 hours before using.

Identification—

A: A solution (1 in 20) responds to the test for *Copper* ⟨191⟩.

B: It responds to *Identification test B* under *Calcium Gluconate*.

Chloride ⟨221⟩—A 1.0-g portion shows no more chloride than corresponds to 1 mL of 0.020 *N* hydrochloric acid (0.07%).

Sulfate ⟨221⟩—A 2.0-g portion dissolved in boiling water shows no more sulfate than corresponds to 1 mL of 0.020 *N* sulfuric acid (0.05%).

Arsenic, *Method I* ⟨211⟩—Dissolve 1.0 g in 35 mL of water: the limit is 3 ppm.

Lead—[NOTE—For the preparation of all aqueous solutions and for the rinsing of glassware before use, employ water that has been passed through a strong-acid, strong-base, mixed-bed ion-exchange resin before use. Select all reagents to have as low a content of lead as practicable, and store all reagent solutions in containers of borosilicate glass. Cleanse glassware before use by soaking in warm 8 *N* nitric acid for 30 minutes and by rinsing with deionized water.]

Ascorbic acid–sodium iodide solution—Dissolve 20 g of ascorbic acid and 38.5 g of sodium iodide in water in a 200-mL volumetric flask, dilute with water to volume, and mix.

Trioctylphosphine oxide solution—[*Caution—This solution causes irritation. Avoid contact with eyes, skin, and clothing. Take special precautions in disposing of unused portions of solutions to which this reagent is added.*] Dissolve 5.0 g of trioctylphosphine oxide in 4-methyl-2-pentanone in a 100-mL volumetric flask, dilute with the same solvent to volume, and mix.

Standard preparation and *Blank*—Transfer 5.0 mL of *Lead Nitrate Stock Solution*, prepared as directed in the test for *Heavy Metals* ⟨231⟩, to a 100-mL volumetric flask, dilute with water to volume, and mix. Transfer 2.0 mL of the resulting solution to a 50-mL volumetric flask. To this volumetric flask and to a second, empty 50-mL volumetric flask (*Blank*) add 10 mL of 9 *N* hydrochloric acid and about 10 mL of water. To each flask add 20 mL of *Ascorbic acid–sodium iodide solution* and 5.0 mL of *Trioctylphosphine oxide solution*, shake for 30 seconds, and allow to separate. Add water to bring the organic solvent layer into the neck of each flask, shake again, and allow to separate. The organic solvent layers are the *Blank* and the *Standard preparation*, and they contain 0.0 and 2.0 µg of lead per mL, respectively.

Test preparation—Add 1.0 g of Copper Gluconate, 10 mL of 9 *N* hydrochloric acid, about 10 mL of water, 20 mL of *Ascorbic acid–sodium iodide solution*, and 5.0 mL of *Trioctylphosphine oxide solution* to a 50-mL volumetric flask, shake for 30 seconds, and allow to separate. Add water to bring the organic solvent layer into the neck of the flask, shake again, and allow to separate. The organic solvent layer is the *Test preparation*.

Procedure—Concomitantly determine the absorbances of the *Blank*, the *Standard preparation*, and the *Test preparation* at the lead emission line at 283.3 nm, with a suitable atomic absorption spectrophotometer (see *Spectrophotometry and Light-scattering* ⟨851⟩) equipped with a lead hollow-cathode lamp and an air-acetylene flame, using 4-methyl-2-pentanone to set the instrument to zero. In a suitable analysis, the absorbance of the *Blank* is not greater than 20% of the difference between the absorbance of the *Standard preparation* and the absorbance of the *Blank*: the absorbance of the *Test preparation* does not exceed that of the *Standard preparation* (0.001%).

Reducing substances—Transfer 1.0 g to a 250-mL conical flask, dissolve in 10 mL of water, and add 25 mL of alkaline cupric citrate TS. Cover the flask, boil gently for 5 minutes, accurately timed, and cool rapidly to room temperature. Add 25 mL of 0.6 *N* acetic acid, 10.0 mL of 0.1 *N* iodine VS, and 10 mL of 3 *N* hydrochloric acid, and titrate with 0.1 *N* sodium thiosulfate VS, adding 3 mL of starch TS as the end-point is approached. Perform a blank determination, omitting the specimen, and note the difference in volumes required. Each mL of the difference in volume of 0.1 *N* sodium thiosulfate consumed is equivalent to 2.7 mg of reducing substances (as dextrose): the limit is 1.0%.

Assay—Dissolve about 1.5 g of Copper Gluconate, accurately weighed, in 100 mL of water. Add 2 mL of glacial acetic acid and 5 g of potassium iodide, mix, and titrate with 0.1 *N* sodium thiosulfate VS to a light yellow color. Add 2 g of ammonium thiocyanate, mix, add 3 mL of starch TS, and continue titrating to a milk-white end-point. Each mL of 0.1 *N* sodium thiosulfate is equivalent to 45.38 mg of $C_{12}H_{22}CuO_{14}$.

Corn Oil—*see* Corn Oil NF

Corticotropin Injection

Corticotropin.
Corticotropin [*9002-60-2*].

» Corticotropin Injection is a sterile solution, in a suitable diluent, of the material containing the polypeptide hormone having the property of increasing the rate of secretion of adrenal corticosteroids, which is obtained from the anterior lobe of the pituitary of mammals used for food by man. Its potency is not less than 80.0 percent and not more than 125.0 percent of the potency stated on the label in USP Corticotropin Units. It may contain a suitable antimicrobial agent.

Packaging and storage—Preserve in single-dose or in multiple-dose containers, preferably of Type I glass. Store in a cold place.

Labeling—If the labeling of Corticotropin Injection recommends intravenous administration, include specific information on dosage.

Reference standards—*USP Corticotropin Reference Standard*—Do not dry before using. Store at a temperature of 0° or below. *USP Posterior Pituitary Reference Standard*—Do not dry before using. Store at a temperature of 0° or below. *USP Ascorbic Acid Reference Standard*—Dry over silica gel for 18 hours before using. Keep container tightly closed and protected from light.

Vasopressin activity—Prepare a rat as directed in the *Assay* under *Vasopressin Injection.* Determine the sensitivity of the animal to vasopressin activity by injecting, at uniform intervals of time of 12 to 15 minutes, doses of a solution of USP Posterior Pituitary RS containing 0.1 USP Unit per mL. Dilute the Corticotropin Injection with saline TS so that each mL contains not less than the equivalent of 2.0 USP Corticotropin Units.

Inject the selected dose of the Standard solution, and repeat at regular intervals of not less than 12 minutes thereafter. At the mid-point of the time interval following each dose of Standard, inject a dose of the diluted Corticotropin Injection equal in volume to that of the Standard solution. Measure and record the responses observed: the response to each of the 2 doses of the Injection does not exceed that of the average of the doses of Standard solution that immediately precede and follow each respective dose of the Injection.

pH ⟨791⟩: between 3.0 and 7.0.

Particulate matter ⟨788⟩: meets the requirements under *Small-volume Injections.*

Other requirements—It meets the requirements under *Injections* ⟨1⟩.

Assay—

Standard preparation—Pipet 2.5 mL of gelatin TS into an opened container of USP Corticotropin RS, and mix, to obtain a solution having a concentration of 2.0 USP Corticotropin Units per mL. Using gelatin TS as a diluent, prepare three *Diluted standard solutions* such that the respective concentrations of corticotropin constitute a geometric series such as 1:2:4 or 1:3:9 and such that the quantity of corticotropin in each 0.5 mL lies within the range of 10 to 300 milli-units.

Assay preparation—In the same manner, using the same diluent, dilute the Corticotropin Injection to give three *Test solutions* corresponding to those of the standard.

The animals—Select healthy rats, of the same but either sex, that have been raised on a diet fully adequate with respect to vitamin and mineral content. Anesthetize the rats with ether, and remove the hypophysis from each by application of gentle suction through a fine-tipped tube. Between 16 and 48 hours after the operation, select those rats weighing between 80 and 180 g, but restrict the selection so that no rat is more than 30% heavier than the lightest, and the number of rats is an exact multiple of 6. Separate the selected rats into 6 groups, equal in size, of not less than 6 rats each, and assign at random one of the three *Diluted standard solutions* or of the three *Test solutions* to each group.

Procedure—Inject all rats of each group subcutaneously with the assigned test doses. Three hours after the injection, anesthetize the rats, and remove both adrenal glands from each rat, free them from adhering tissue, and promptly weigh each pair on a suitable balance to the nearest 0.2 mg. Place the weighed glands from each rat in suitable vessels each containing 8.0 mL of metaphosphoric acid solution (1 in 40), and comminute the glands as by grinding with a small quantity of washed sand. Cover each vessel, and proceed similarly until all glands have been extracted.

Ascorbic acid determination—Filter the metaphosphoric acid extracts, and pipet 4 mL of each filtrate into suitable vessels each containing 4.0 mL of indophenol-acetate TS. Mix by shaking, and read the absorbance at 520 nm, with a suitable spectrophotometer. From the observed absorbance and the standard curve prepared as directed in the next paragraph, calculate the amount of ascorbic acid in terms of mg of ascorbic acid in each 100 g of adrenal gland tissue.

Prepare a standard concentration-absorbance curve, using three standard solutions containing in each mL, respectively, 6.0, 8.0, and 10.0 µg of USP Ascorbic Acid RS in metaphosphoric acid solution (1 in 40). Pipet into each of 3 suitable vessels, preferably spectrophotometer cells, 4 mL of indophenol-acetate TS. Add 4.0 mL of one of the three standard ascorbic acid solutions to one of the cells, mix, and promptly read the absorbance in the same instrument and under the same conditions as for the adrenal gland extracts. Repeat the process for the other two standard ascorbic acid solutions, plot the concentration-absorbance values, and draw the straight line best fitting the 3 plotted points.

Calculation—Tabulate the observed concentration of ascorbic acid in the adrenal glands of each rat, designated by the symbol y, for each dosage group of f rats. If the data from one or more rats are missing, adjust to groups of equal size by suitable means (see *Replacement of Missing Values* ⟨111⟩). Total the values of y in each group, and designate each total as T, subscripts 1 to 3 for the three successive dosage levels and subscripts S and U for the Standard and the Injection, respectively. Test both the agreement in slope of the dosage-response lines for the Standard and for the Injection, and the lack of curvature, as directed for a 3-dose balanced assay (see *Tests of Assay Validity* ⟨111⟩). If the combined discrepancy as measured by F_3 exceeds its tabular value in Table 9 (see *Combination of Independent Assays* ⟨111⟩), regard these data as preliminary and repeat the assay.

Determine the logarithm of potency of the Injection by the formula:

$$M = (4iT_a/3T_b) + \log R,$$

in which $T_a = \Sigma(T_U - T_S)$, $T_b = \Sigma(T_3 - T_1)$, i is the interval between successive log doses of both the *Standard preparation* and the *Assay preparation*, and $R = v_S/v_U$ is the ratio of the high dose of the Standard in USP Units (v_S) to the high dose of the Injection in mL (v_U).

Compute the log confidence interval L (see *Confidence Intervals for Individual Assays* ⟨111⟩).

Replication—Repeat the entire determination at least once. Test the agreement among the two or more independent determinations, and compute the weight for each (see *Combination of Independent Assays* ⟨111⟩). Calculate the weighted mean log-potency $\overline{M}$ and its confidence interval, L_c (see *Confidence Intervals for Individual Assays* ⟨111⟩). The potency, P_*, is satisfactory if $P_* = \text{antilog } \overline{M}$ is not less than 80% and not more than 125% of the labeled potency and if the confidence interval does not exceed 0.40.

Corticotropin for Injection

Corticotropin.
Corticotropin [*9002-60-2*].

» Corticotropin for Injection is the sterile, dry material containing the polypeptide hormone having the property of increasing the rate of secretion of adrenal corticosteroids, which is obtained from the anterior lobe of the pituitary of mammals used for food by man. Its potency is not less than 80.0 percent and not more than 125.0 percent of the potency stated on the label in USP Corticotropin Units. It may contain a suitable antimicrobial agent and suitable diluents and buffers.

Packaging and storage—Preserve in *Containers for Sterile Solids* as described under *Injections* ⟨1⟩.

Labeling—If the labeling of Corticotropin for Injection recommends intravenous administration, include specific information on dosage.

Reference standards—*USP Corticotropin Reference Standard*—Do not dry before using. Store at a temperature of 0° or below. *USP Posterior Pituitary Reference Standard*—Do not dry before using. Store at a temperature of 0° or below. *USP Ascorbic Acid Reference Standard*—Dry over silica gel for 18 hours before using. Keep container tightly closed and protected from light.

Vasopressin activity—Prepare a solution of Corticotropin for Injection in saline TS so that each mL contains not less than the equivalent of 2.0 USP Corticotropin Units. The solution meets the requirements of the test for *Vasopressin activity* under *Corticotropin Injection.*

pH ⟨791⟩: between 2.5 and 6.0, in a solution constituted as directed in the labeling supplied by the manufacturer.

Particulate matter ⟨788⟩: meets the requirements under *Small-volume Injections.*

Other requirements—It meets the requirements for *Sterility Tests* ⟨71⟩, *Uniformity of Dosage Units* ⟨905⟩, and *Constituted Solutions* and *Labeling* under *Injections* ⟨1⟩.

Assay—Proceed with the constituted solution of Corticotropin for Injection as directed in the *Assay* under *Corticotropin Injection.*

Repository Corticotropin Injection

Corticotropin.
Corticotropin [*9002-60-2*].

» Repository Corticotropin Injection is corticotropin in a solution of partially hydrolyzed gelatin. Its potency is not less than 80.0 percent and not more than 125.0 percent of the potency stated on the label in USP Corticotropin Units. It may contain a suitable antimicrobial agent.

Packaging and storage—Preserve in single-dose or in multiple-dose containers, preferably of Type I glass.

Reference standards—*USP Corticotropin Reference Standard*—Do not dry before using. Store at a temperature of 0° or below. *USP Posterior Pituitary Reference Standard*—Do not dry before using. Store at a temperature of 0° or below. *USP Ascorbic Acid Reference Standard*—Dry over silica gel for 18 hours before using.

Vasopressin activity, pH, and Assay—Repository Corticotropin Injection meets the requirements for *Vasopressin activity*, *pH*, and *Assay* under *Corticotropin Injection.*

Other requirements—It meets the requirements under *Injections* ⟨1⟩.

Sterile Corticotropin Zinc Hydroxide Suspension

Corticotropin zinc hydroxide.
Corticotropin zinc hydroxide [*9050-75-3*].

» Sterile Corticotropin Zinc Hydroxide Suspension is a sterile suspension of corticotropin adsorbed on zinc hydroxide. Its potency is not less than 80.0 percent and not more than 125.0 percent of the potency stated on the label in USP Corticotropin Units. It contains not less than 1800 µg and not more than 2200 µg of zinc, and not less than 604 µg and not more than 776 µg of anhydrous dibasic sodium phosphate, for each 40 USP Corticotropin Units.

Packaging and storage—Preserve in single-dose or in multiple-dose containers, preferably of Type I glass. Store at controlled room temperature.

Labeling—Label it to indicate that it is not recommended for intravenous use and that the suspension is to be well shaken before use. The container label and the package label state the potency in USP Corticotropin Units in each mL.

Reference standards—*USP Corticotropin Reference Standard*—Do not dry before using. Store at a temperature of 0° or below. *USP Ascorbic Acid Reference Standard*—Dry over silica gel for 18 hours before using.

pH ⟨791⟩: between 7.5 and 8.5, determined potentiometrically.

Zinc—Pipet a volume of the well-shaken Suspension, equivalent to about 6 mg of zinc, into a 125-mL conical flask, and add 2 mL of a buffer mixture containing 5.4 g of ammonium chloride and 26 mL of stronger ammonia water in each 100 mL. Add 10 mL of water and 2 drops of eriochrome black TS, and titrate with 0.005 *M* disodium ethylenediaminetetraacetate VS to a clear blue end-point. Perform a blank titration, and make any necessary correction. Each mL of 0.005 *M* disodium ethylenediaminetetraacetate is equivalent to 0.327 mg of Zn.

Anhydrous dibasic sodium phosphate—

Ammonium molybdate reagent—Dissolve 6.4 g of ammonium molybdate in 40 mL of water, and add 50 mL of 10 *N* sulfuric acid. Mix, and dilute with water to 100 mL. This solution is stable for about 2 weeks.

Stannous chloride reagent—Dissolve 1 g of stannous chloride in 5 mL of hydrochloric acid. Just prior to use, dilute 1 mL of this solution with water to 100 mL.

Standard preparation—Weigh accurately about 275 mg of anhydrous dibasic sodium phosphate (Na_2HPO_4) into a 1000-mL volumetric flask. Dissolve in water, and dilute with water to volume. Dilute 10 mL of this solution with water to 100 mL.

Test preparation—Pipet 1 mL of the well-shaken Suspension into a 25-mL volumetric flask, add 0.1 mL of 10 *N* sulfuric acid, mix, dilute with water to volume, and again mix.

Procedure—Into separate 100-mL volumetric flasks pipet duplicate 10-mL portions of *Test preparation*, duplicate 10-mL portions of *Standard preparation*, and 10 mL of water to provide a blank. Treat each flask as follows: Dilute the contents with water to 60 mL, then add 10 mL of 10 *N* sulfuric acid, and mix. Add 10 mL of *Ammonium molybdate reagent*, mix, add 10 mL of water, and again mix. Add slowly, with mixing, 5 mL of *Stannous chloride reagent*, dilute with water to volume, and mix. Measure the absorbances of the solutions at 10 minutes, accurately timed, after the first addition of the *Stannous chloride reagent*, at a wavelength of 710 nm, with a suitable spectrophotometer, relative to the blank. Calculate the quantity, in µg, of anhydrous dibasic sodium phosphate in the portion of Suspension taken by the formula:

$$25C(A_U/A_S),$$

in which C is the concentration, in µg per mL, of Na_2HPO_4 in the *Standard preparation*, and A_U and A_S are the absorbances of the solutions from the *Test preparation* and the *Standard preparation*, respectively.

Other requirements—It meets the requirements under *Injections* ⟨1⟩.

Assay—Add sufficient 0.1 *N* hydrochloric acid to Sterile Corticotropin Zinc Hydroxide Suspension to effect complete solution, and using this in making the *Assay preparation*, proceed as directed in the *Assay* under *Corticotropin Injection.*

Cortisol and its dosage forms—*see* Hydrocortisone and its dosage forms

Cortisone Acetate

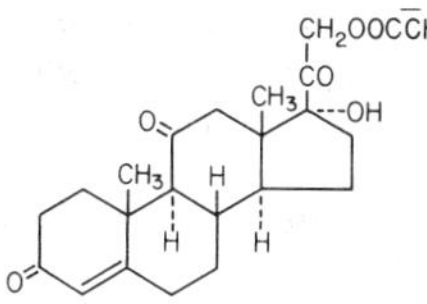

$C_{23}H_{30}O_6$ 402.49
Pregn-4-ene-3,11,20-trione, 21-(acetyloxy)-17-hydroxy-.
17,21-Dihydroxypregn-4-ene-3,11,20-trione 21-acetate
[50-04-4].

» Cortisone Acetate contains not less than 97.0 percent and not more than 102.0 percent of $C_{23}H_{30}O_6$, calculated on the dried basis.

Packaging and storage—Preserve in well-closed containers.

Reference standard—*USP Cortisone Acetate Reference Standard*—Dry at 105° for 30 minutes before using.

Identification—

A: Dissolve a portion of Cortisone Acetate in methanol in a beaker, evaporate the methanol on a steam bath with the aid of a current of air, then dry the residue at 105° for 30 minutes: the infrared absorption spectrum of a potassium bromide dispersion of the residue so obtained exhibits maxima only at the same wavelengths as that of a similar preparation of USP Cortisone Acetate RS.

B: The ultraviolet absorption spectrum of a 1 in 100,000 solution in methanol exhibits maxima and minima at the same wavelengths as that of a similar solution of USP Cortisone Acetate RS, concomitantly measured, and the respective absorptivities, calculated on the dried basis, at the wavelength of maximum absorbance at about 238 nm do not differ by more than 3.0%.

Specific rotation ⟨781⟩: between +208° and +217°, calculated on the dried basis, determined in a solution in dioxane containing 100 mg in each 10 mL.

Loss on drying ⟨731⟩—Dry it at 105° for 30 minutes: it loses not more than 1.0% of its weight.

Residue on ignition ⟨281⟩: negligible, from 100 mg.

Ordinary impurities ⟨466⟩—

Test solution: ethyl acetate.

Standard solution: ethyl acetate.

Eluant: a mixture of toluene, ethyl acetate, and glacial acetic acid (50:45:5), in a nonequilibrated chamber.

Visualization: 1.

Assay—

Mobile phase—Prepare a solution containing butyl chloride, water-saturated butyl chloride, tetrahydrofuran, methanol, and glacial acetic acid (95:95:14:7:6). Make adjustments if necessary (see *System Suitability* under *Chromatography* ⟨621⟩).

Internal standard solution—Prepare a solution of methylparaben in *Mobile phase* having a concentration of about 0.04 mg per mL.

Standard preparation—Transfer about 12 mg of USP Cortisone Acetate RS, accurately weighed, to a glass-stoppered, 50-mL conical flask. Add 25.0 mL of *Internal standard solution*, sonicate for 5 minutes, and combine approximately 1 mL of this solution and 3 mL of *Mobile phase* to obtain the *Standard preparation*.

Resolution solution—Dissolve a quantity of hydrocortisone acetate in the *Standard preparation* to obtain a solution containing 0.1 mg of hydrocortisone acetate per mL.

Assay preparation—Weigh accurately about 12 mg of Cortisone Acetate, and proceed as directed under *Standard preparation.*

Chromatographic system (see *Chromatography* ⟨621⟩)—The liquid chromatograph is equipped with a 254-nm detector and a 4.6-mm × 25-cm column that contains packing L3. The flow rate is about 1 mL per minute. Chromatograph the *Standard preparation* and the *Resolution solution*, and record the peak responses as directed under *Procedure:* the resolution, *R*, between cortisone acetate and hydrocortisone acetate is not less than 2.2 (if necessary add equal parts of butyl chloride and water-saturated butyl chloride to the *Mobile phase* to meet this requirement), and the relative standard deviation for replicate injections is not more than 2.0%.

Procedure—Separately inject equal volumes (about 15 µL) of the *Standard preparation* and the *Assay preparation* into the chromatograph, record the chromatograms, and measure the responses for the major peaks. The relative retention times are about 0.7 for methylparaben and 1.0 for cortisone acetate. Calculate the quantity, in mg, of $C_{23}H_{30}O_6$ in the portion of Cortisone Acetate taken by the formula:

$$W(R_U / R_S),$$

in which *W* is the weight, in mg, of USP Cortisone Acetate RS taken for the *Standard preparation*, and R_U and R_S are the peak response ratios of the cortisone acetate peak and the internal standard peak obtained from the *Assay preparation* and the *Standard preparation*, respectively.

Sterile Cortisone Acetate Suspension

» Sterile Cortisone Acetate Suspension is a sterile suspension of Cortisone Acetate in a suitable aqueous medium. It contains not less than 90.0 percent and not more than 110.0 percent of the labeled amount of $C_{23}H_{30}O_6$.

Packaging and storage—Preserve in single-dose or in multiple-dose containers, preferably of Type I glass.

Reference standard—*USP Cortisone Acetate Reference Standard*—Dry at 105° for 30 minutes before using.

Identification—Mix 25 mL of water with a volume of Suspension equivalent to about 25 mg of cortisone acetate. Centrifuge, or allow the insoluble material to settle, then decant and discard the supernatant liquid. Add 20 mL of methanol and, using agitation and warming as necessary, dissolve the residue. Evaporate the solvent on a steam bath with the aid of a current of air, then dry the residue at 105° for 30 minutes: the residue so obtained responds to *Identification test A* under *Cortisone Acetate.*

pH ⟨791⟩: between 5.0 and 7.0.

Other requirements—It meets the requirements under *Injections* ⟨1⟩.

Assay—

Mobile phase, Resolution solution, and *Chromatographic system*—Prepare as directed in the *Assay* under *Cortisone Acetate.*

Internal standard solution—Prepare a solution of prednisone in *Mobile phase* having a concentration of 0.5 mg per mL.

Standard preparation—Transfer about 12 mg of USP Cortisone Acetate RS, accurately weighed, to a stoppered, 50-mL conical flask. Add 20.0 mL of *Internal standard solution*, and sonicate for 5 minutes. Filter a portion through a polytef syringe filter, then combine 1 mL of the filtrate and 4 mL of *Mobile phase* to obtain the *Standard preparation.*

Assay preparation—Using a pipet calibrated "to contain," transfer 2.0 mL of freshly mixed Sterile Cortisone Acetate Suspension to a volumetric flask of a size to give a cortisone acetate concentration of 2 mg per mL when diluted to volume. Rinse the suspension remaining in the pipet into the flask with isopropyl alcohol, dilute with isopropyl alcohol to volume, and sonicate for 3 minutes. Deliver a 3.0-mL aliquot of this solution to a stoppered, 25-mL conical flask, and evaporate on a steam bath with the aid of a current of air to dryness. Add 10.0 mL of *Internal standard solution*, insert the stopper, and sonicate for 5 minutes. Filter a portion through a polytef syringe filter, then combine approximately 1 mL of the filtrate and 4 mL of *Mobile phase* to obtain the *Assay preparation.*

Procedure—Proceed as directed for *Procedure* in the *Assay* under *Cortisone Acetate.* The relative retention times are 0.6 for cortisone acetate and 1.0 for prednisone. Calculate the quantity, in mg, of $C_{23}H_{30}O_6$ in each mL of suspension taken by the formula:

$$W(V/12)(R_U/R_S),$$

in which W is the weight, in mg, of USP Cortisone Acetate RS taken for the *Standard preparation*, V is the capacity, in mL, of the volumetric flask used for the *Assay preparation*, and the other terms are as defined in the *Assay* under *Cortisone Acetate*.

Cortisone Acetate Tablets

» Cortisone Acetate Tablets contain not less than 90.0 percent and not more than 110.0 percent of the labeled amount of $C_{23}H_{30}O_6$.

Packaging and storage—Preserve in well-closed containers.

Reference standard—*USP Cortisone Acetate Reference Standard*—Dry at 105° for 30 minutes before using.

Identification—Powder a number of Tablets, equivalent to about 25 mg of cortisone acetate, add 25 mL of solvent hexane, and extract for 15 minutes with occasional agitation. Decant and discard the supernatant liquid, then extract the residue with 5 mL of chloroform, with frequent agitation, for 5 minutes. Filter, add 10 mL of methanol to the filtrate, mix, evaporate the solvent on a steam bath with the aid of a current of air, then dry the residue at 105° for 30 minutes: the residue so obtained responds to *Identification test A* under *Cortisone Acetate*.

Dissolution ⟨711⟩—

Medium: a mixture containing three volumes of isopropyl alcohol and 7 volumes of dilute hydrochloric acid (1 in 100); 900 mL.

Apparatus 1: 100 rpm.

Time: 30 minutes.

Standard solution—Dissolve a suitable quantity of USP Cortisone Acetate RS, accurately weighed, in the *Dissolution Medium* to obtain a solution having a known concentration of about 5.55 µg of cortisone acetate per mL.

Procedure—Determine the amount of $C_{23}H_{30}O_6$ in solution in filtered portions of the *Dissolution Medium*, suitably diluted, in 1-cm cells, at the wavelength of maximum absorbance at about 242 nm in comparison with the *Standard solution*.

Tolerances—Not less than 60% (*Q*) of the labeled amount of $C_{23}H_{30}O_6$ is dissolved in 30 minutes.

Uniformity of dosage units ⟨905⟩: meet the requirements.

Procedure for content uniformity—

Mobile phase, *Internal standard solution*, *Standard preparation*, *Resolution solution*, and *Chromatographic system*—Proceed as directed in the *Assay* under *Cortisone Acetate*.

Test preparation—Place 1 Tablet in a stoppered, 50-mL conical flask, deposit 0.25 mL of water on the tablet, insert the stopper in the flask, and allow to stand for 30 minutes. Add 2.5 mL of isopropyl alcohol, and place the unstoppered flask on a steam bath. Boil gently, if necessary, until the tablet disintegrates, then evaporate the solvent with the aid of a current of air. Remove from the steam bath, add 10.0 mL of *Internal standard solution* for each 5 mg of cortisone acetate declared, insert the stopper, and sonicate vigorously for 10 minutes. Proceed as directed under *Assay preparation*, beginning with "Filter a portion," to obtain the *Test preparation*.

Procedure—Proceed as directed for *Procedure* in the *Assay* under *Cortisone Acetate*. Calculate the quantity, in mg, of $C_{23}H_{30}O_6$ in each tablet tested by the formula:

$$W(V/25)(R_U/R_S),$$

in which W is the weight, in mg, of USP Cortisone Acetate RS taken for the *Standard preparation*, V is the volume, in mL, of *Internal standard solution* added to the *Test preparation*, and the other terms are as defined therein.

Assay—

Mobile phase, *Internal standard solution*, *Standard preparation*, *Resolution solution*, and *Chromatographic system*—Proceed as directed in the *Assay* under *Cortisone Acetate*.

Assay preparation—Accurately weigh, then finely powder, not less than 20 Cortisone Acetate Tablets. Transfer an accurately weighed portion of the powder, equivalent to about 12 mg of cortisone acetate, to a stoppered conical flask. Add 25.0 mL of *Internal standard solution*, insert the stopper in the flask, and sonicate vigorously for 5 minutes. Filter a portion through a polytef syringe filter, then combine approximately 1 mL of the filtrate and 3 mL of *Mobile phase* to obtain the *Assay preparation*.

Procedure—Proceed as directed for *Procedure* in the *Assay* under *Cortisone Acetate*. Calculate the quantity, in mg, of $C_{23}H_{30}O_6$ in the portion of Tablets taken by the formula given therein.

Purified Cotton

» Purified Cotton is the hair of the seed of cultivated varieties of *Gossypium hirsutum* Linné, or of other species of *Gossypium* (Fam. Malvaceae), freed from adhering impurities, deprived of fatty matter, bleached, and sterilized in its final container.

Packaging and storage—Package it in rolls of not more than 500 g of a continuous lap, with a light-weight paper running under the entire lap, the paper being of such width that it may be folded over the edges of the lap to a distance of at least 25 mm, the two together being tightly and evenly rolled, and enclosed and sealed in a well-closed container. It may be packaged also in other types of containers if these are so constructed that the sterility of the product is maintained.

Labeling—Its label bears a statement to the effect that the sterility cannot be guaranteed if the package bears evidence of damage or if the package has been opened previously.

Alkalinity or acidity—Thoroughly saturate about 10 g with 100 mL of recently boiled and cooled water, then with the aid of a glass rod press out two 25-mL portions of the water into white porcelain dishes. To one portion add 3 drops of phenolphthalein TS, and to the other portion add 1 drop of methyl orange TS: no pink color develops in either portion.

Residue on ignition ⟨281⟩—Place about 5 g, accurately weighed, in a porcelain or platinum dish, and moisten with 2 *N* sulfuric acid. Gently heat the cotton until it is charred, then ignite more strongly until the carbon is completely consumed: not more than 0.20% of residue remains.

Water-soluble substances—Place 10 g, accurately weighed, in a beaker containing 1000 mL of water, and boil gently for 30 minutes, adding water as required to maintain the volume. Pour the water through a funnel into another vessel, and press out the excess water from the cotton with a glass rod. Wash the cotton in the funnel with two 250-mL portions of boiling water, pressing the cotton after each washing. Filter the combined extract and washings, and wash the filter thoroughly with hot water. Evaporate the combined extract and washings to a small volume, transfer to a tared porcelain or platinum dish, evaporate to dryness, and dry the residue at 105° to constant weight: the residue weighs not more than 35 mg (0.35%).

Fatty matter—Pack 10 ± 0.01 g in a Soxhlet extractor provided with a tared receiver, and extract with ether for 5 hours at a rate such that the ether siphons over not less than four times per hour. The ether solution in the flask shows no trace of blue, green, or brownish color. Evaporate the extract to dryness, and dry at 105° for 1 hour: the weight of the residue does not exceed 70 mg (0.7%).

Dyes—Pack about 10 g in a narrow percolator, and extract slowly with alcohol until the percolate measures 50 mL: when observed downward through a column 20 cm in depth, the percolate may show a yellowish color, but neither a blue nor a green tint.

Other foreign matter—The pinches of it taken for the determination of *Fiber length* contain no oil stains or metallic particles.

Fiber length and Absorbency—Remove it from its wrappings, and condition it for not less than 4 hours in a standard atmosphere of 65 ± 2% relative humidity at 21 ± 1.1° (70 ± 2°F), before determining the *Fiber length* and *Absorbency*.

Fiber length—Determine the fiber length of Purified Cotton as directed under *Cotton—Fiber Length* ⟨691⟩: not less than 60% of the fibers, by weight, are 12.5 mm or greater in length, and not more than 10% of the fibers, by weight, are 6.25 mm or less in length.

Absorbency—Proceed as directed under *Cotton—Absorbency Test* ⟨691⟩: submersion is complete in 10 seconds at a temperature of 25°, and the cotton retains not less than 24 times its weight of water.

Sterility—It meets the requirements under *Sterility Tests* ⟨71⟩.

Cottonseed Oil—*see* Cottonseed Oil NF

Creams—*see complete list in index*

Cresol—*see* Cresol NF

Cromolyn Sodium

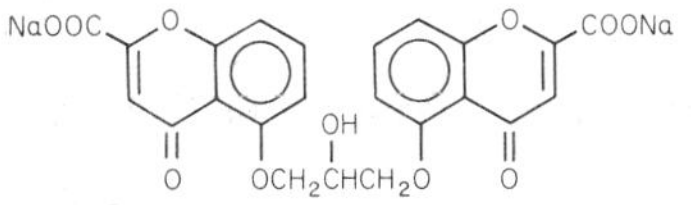

$C_{23}H_{14}Na_2O_{11}$ 512.34

4*H*-1-Benzopyran-2-carboxylic acid, 5,5′-[(2-hydroxy-1,3-propanediyl)bis(oxy)]bis[4-oxo-, disodium salt].

Disodium 5,5′-[(2-hydroxytrimethylene)dioxy]bis[4-oxo-4*H*-1-benzopyran-2-carboxylate] [*15826-37-6*].

» Cromolyn Sodium contains not less than 98.0 percent and not more than 101.0 percent of $C_{23}H_{14}Na_2O_{11}$, calculated on the dried basis.

Packaging and storage—Preserve in tight containers.

Reference standard—*USP Cromolyn Sodium Reference Standard*—Dry in vacuum at 105° to constant weight before using.

Identification—

A: The infrared absorption spectrum of a potassium bromide dispersion of it, previously dried in vacuum at 105° to constant weight, exhibits maxima only at the same wavelengths as that of a similar preparation of USP Cromolyn Sodium RS.

B: The ultraviolet absorption spectrum of a 1 in 40,000 solution in *pH 7.4 sodium phosphate buffer* prepared as directed in the *Assay* exhibits maxima at the same wavelengths as that of a similar solution of USP Cromolyn Sodium RS, concomitantly measured.

Acidity or alkalinity—Dissolve 1.0 g in 25 mL of carbon dioxide-free water, and add two drops of bromothymol blue TS. If the solution is yellow, not more than 0.25 mL of 0.1 *N* sodium hydroxide is required to produce a blue color. If the solution is blue, not more than 0.25 mL of 0.1 *N* hydrochloric acid is required to produce a yellow color.

Loss on drying ⟨731⟩—Dry it in vacuum at 105° to constant weight: it loses not more than 10.0% of its weight.

Related compounds—Dissolve 100 mg of Cromolyn Sodium in 10.0 mL of a mixture of water, stabilizer-free tetrahydrofuran, and acetone (6:4:1). Similarly prepare a solution of USP Cromolyn Sodium RS in the same solvent mixture having a concentration of 10 mg per mL (*Standard solution A*). Quantitatively dilute a volume of *Standard solution A* with the same solvent mixture to obtain a diluted standard solution having a concentration of 0.05 mg per mL (*Standard solution B*). Apply 10-µL portions of the test solution, *Standard solution A*, and *Standard solution B* to a suitable thin-layer chromatographic plate (see *Chromatography* ⟨621⟩), coated with a 0.25-mm layer of chromatographic silica gel mixture. Allow the spots to dry, and develop the chromatogram in a solvent system consisting of a mixture of chloroform, methanol, and glacial acetic acid (9:9:2) until the solvent front has moved about three-fourths of the length of the plate. Remove the plate from the developing chamber, mark the solvent front, and allow the solvent to evaporate. Locate the spots on the plate by viewing under short-wavelength ultraviolet light: the R_f value of the principal spot obtained from the test solution corresponds to that obtained from *Standard solution A*. Any spot in the chromatogram obtained from the test solution moving ahead of the principal spot is not more intense than the spot in the chromatogram obtained from *Standard solution B* (0.5%).

Oxalate—Dissolve 100 mg in 20 mL of water, add 5.0 mL of iron salicylate TS, and dilute with water to 50 mL. Determine the absorbance of the solution at 480 nm against a reagent blank. The absorbance is not less than that of a solution containing 350 µg of oxalic acid prepared in the same manner (0.35%).

Assay—

pH 7.4 sodium phosphate buffer—Dissolve 70 g of anhydrous dibasic sodium phosphate in 900 mL of water. Adjust to a pH of 7.4 by the addition of dilute phosphoric acid (1 in 10). Dilute with water to 1000 mL, and mix. Transfer 10 mL of this solution to a 100-mL volumetric flask, dilute with water to volume, and mix.

Standard preparation—Dissolve a suitable quantity of USP Cromolyn Sodium RS, previously dried in vacuum at 105° to constant weight and accurately weighed, in water to obtain a solution having a known concentration of about 250 µg per mL. Transfer 10 mL of this solution to a 100-mL volumetric flask, add 1 mL of *pH 7.4 sodium phosphate buffer*, dilute with water to volume, and mix. The final concentration is about 25 µg per mL.

Assay preparation—Transfer about 100 mg of Cromolyn Sodium, accurately weighed, to a 1000-mL volumetric flask, dissolve in about 100 mL of water, dilute with water to volume, and mix. Pipet 25 mL of this solution into a 100-mL volumetric flask, add 1 mL of *pH 7.4 sodium phosphate buffer*, dilute with water to volume, and mix.

Procedure—Concomitantly determine the absorbances of the *Standard preparation* and the *Assay preparation* in 1-cm cells at the wavelength of maximum absorbance at about 326 nm, with a suitable spectrophotometer, using a 1 in 100 solution of *pH 7.4 sodium phosphate buffer* as the blank. Calculate the quantity, in mg, of $C_{23}H_{14}Na_2O_{11}$ in the Cromolyn Sodium taken by the formula:

$$4C(A_U/A_S),$$

in which C is the concentration, in µg per mL, of USP Cromolyn Sodium RS in the *Standard preparation*, and A_U and A_S are the absorbances of the solutions from the *Assay preparation* and the *Standard preparation*, respectively.

Cromolyn Sodium Inhalation

» Cromolyn Sodium Inhalation is a sterile, aqueous solution of Cromolyn Sodium. It contains not less than 90.0 percent and not more than 110.0 percent of the labeled amount of $C_{23}H_{14}Na_2O_{11}$. It contains no added substances.

Packaging and storage—Preserve in single-dose, double-ended glass ampuls.

Reference standard—*USP Cromolyn Sodium Reference Standard*—Dry in vacuum at 105° to constant weight before using.

Identification—The ultraviolet absorption spectrum of the *Assay preparation* prepared as directed in the *Assay* exhibits maxima and minima at the same wavelengths as that of a similar solution of USP Cromolyn Sodium RS, concomitantly measured.

Related compounds—Apply 10-µL portions of Cromolyn Sodium Inhalation and Standard solutions of USP Cromolyn Sodium RS in a mixture of water, stabilizer-free tetrahydrofuran, and acetone (6:4:1) containing 10 mg per mL (*Standard solution A*) and 0.1 mg per mL (*Standard solution B*) to a suitable thin-layer chromatographic plate (see *Chromatography* ⟨621⟩) coated with a 0.25-mm layer of chromatographic silica gel mixture. Allow the spots to dry, and develop the chromatogram in a solvent system consisting of a mixture of chloroform, methanol, and glacial acetic acid (9:9:2) until the solvent front has moved about three-fourths of the length of the plate. Remove the plate from the developing chamber, mark the solvent front, and allow the solvent to evaporate. Locate the spots on the plate by viewing under short-wavelength ultraviolet light: the R_f value of the principal spot obtained from the Inhalation corresponds to that obtained from *Standard solution A*. Any spot in the chromatogram obtained from the Inhalation moving ahead of the principal spot is not more intense than the spot in the chromatogram obtained from *Standard solution B* (1.0%).

pH ⟨791⟩: between 4.0 and 7.0.

Sterility—It meets the requirements under *Sterility Tests* ⟨71⟩.

Uniformity of dosage units ⟨905⟩: meets the requirements.

Assay—

pH 7.4 sodium phosphate buffer and *Standard preparation*—Prepare as directed in the *Assay* under *Cromolyn Sodium*.

Assay preparation—Dilute with water an accurately measured volume of Cromolyn Sodium Inhalation, equivalent to about 25 mg of cromolyn sodium, to obtain a solution having a concentration of about 250 µg per mL. Pipet 10 mL of this solution into a 100-mL volumetric flask, add 1 mL of *pH 7.4 sodium phosphate buffer*, dilute with water to volume, and mix.

Procedure—Concomitantly determine the absorbances of the *Standard preparation* and the *Assay preparation* in 1-cm cells at the wavelength of maximum absorbance at about 326 nm, with a suitable spectrophotometer, using a 1 in 100 aqueous solution of *pH 7.4 sodium phosphate buffer* as the blank. Calculate the quantity, in mg, of $C_{23}H_{14}Na_2O_{11}$ in each mL of the Inhalation taken by the formula:

$$(C/V)(A_U/A_S),$$

in which C is the concentration, in µg per mL, of USP Cromolyn Sodium RS in the *Standard preparation*, V is the volume, in mL, of Inhalation taken, and A_U and A_S are the absorbances of the solutions from the *Assay preparation* and the *Standard preparation*, respectively.

Cromolyn Sodium for Inhalation

» Cromolyn Sodium for Inhalation is a mixture of equal parts of Lactose and Cromolyn Sodium contained in a hard gelatin capsule. It contains not less than 95.0 percent and not more than 125.0 percent of the labeled amount of $C_{23}H_{14}Na_2O_{11}$.

Packaging and storage—Preserve in tight, light-resistant containers. Avoid excessive heat.

Reference standard—*USP Cromolyn Sodium Reference Standard*—Dry in vacuum at 105° to constant weight before using.

Identification—It responds to *Identification test B* under *Cromolyn Sodium*, and to the *Identification test* under *Lactose*.

Uniformity of dosage units ⟨905⟩: meets the requirements.

Assay—

pH 7.4 sodium phosphate buffer—Prepare as directed in the *Assay* under *Cromolyn Sodium*.

Assay preparation—Remove and accurately weigh the contents of not less than 20 capsules of Cromolyn Sodium for Inhalation, and transfer the combined contents to a 250-mL volumetric flask. Dissolve in 100 mL of water, dilute with water to volume, and mix. Transfer an aliquot of this solution, equivalent to 8 mg of cromolyn sodium, to a 250-mL volumetric flask, add 1 mL of *pH 7.4 sodium phosphate buffer*, dilute with water to volume, and mix.

Standard preparation—Dissolve a suitable quantity of USP Cromolyn Sodium RS, previously dried in vacuum at 105° to constant weight and accurately weighed, in water to obtain a solution having a known concentration of about 350 µg per mL. Transfer 10 mL of this solution to a 100-mL volumetric flask, add 1 mL of *pH 7.4 sodium phosphate buffer*, dilute with water to volume, and mix. The final concentration is about 35 µg per mL.

Procedure—Concomitantly determine the absorbances of the *Standard preparation* and the *Assay preparation* in 1-cm cells at the wavelength of maximum absorbance at about 326 nm, with a suitable spectrophotometer, using a 1 in 250 solution of *pH 7.4 sodium phosphate buffer* as the blank. Calculate the quantity, in mg, of $C_{23}H_{14}Na_2O_{11}$ in the aliquot taken by the formula:

$$0.25C(A_U/A_S),$$

in which C is the concentration, in µg per mL, of USP Cromolyn Sodium RS in the *Standard preparation*, and A_U and A_S are the absorbances of the *Assay preparation* and the *Standard preparation*, respectively.

Cromolyn Sodium Nasal Solution

» Cromolyn Sodium Nasal Solution is an aqueous solution of Cromolyn Sodium. It contains not less than 90.0 percent and not more than 110.0 percent of the labeled amount of $C_{23}H_{14}Na_2O_{11}$. It may contain suitable stabilizers.

Packaging and storage—Preserve in tight, light-resistant containers.

Reference standard—*USP Cromolyn Sodium Reference Standard*—Dry in vacuum at 105° to constant weight before using.

Identification—It responds to *Identification test B* under *Cromolyn Sodium*.

pH ⟨791⟩: between 4.0 and 7.0.

Related compounds—It meets the requirements of the test for *Related compounds* under *Cromolyn Sodium Inhalation*, "Nasal Solution" being read in place of "Inhalation."

Assay—

pH 7.4 sodium phosphate buffer—Prepare as directed in the *Assay* under *Cromolyn Sodium*.

Assay preparation—Transfer 4 mL of Cromolyn Sodium Nasal Solution to a 100-mL volumetric flask, dilute with water to volume, and mix. Transfer an aliquot of this solution, equivalent to 8 mg of cromolyn sodium, to a 250-mL volumetric flask. Add 1 mL of *pH 7.4 sodium phosphate buffer*, dilute with water to volume, and mix.

Standard preparation—Prepare as directed in the *Assay* under *Cromolyn Sodium*.

Procedure—Proceed as directed for *Procedure* in the *Assay* under *Cromolyn Sodium for Inhalation*.

Cromolyn Sodium Ophthalmic Solution

» Cromolyn Sodium Ophthalmic Solution is a sterile, aqueous solution of Cromolyn Sodium. It contains not less than 90.0 percent and not more than 110.0 percent of the labeled amount of $C_{23}H_{14}Na_2O_{11}$. It may contain suitable antimicrobial and stabilizing agents.

Packaging and storage—Preserve in tight, light-resistant, single-dose or multiple-dose containers. Ophthalmic Solution that is packaged in multiple-dose containers contains a suitable antimicrobial agent.

Reference standard—*USP Cromolyn Sodium Reference Standard*—Dry in vacuum at 105° to constant weight before using.

Identification—It responds to *Identification test B* under *Cromolyn Sodium*.

Sterility—It meets the requirements under *Sterility Tests* ⟨71⟩.

pH ⟨791⟩: between 4.0 and 7.0.

Related compounds—It meets the requirements of the test for *Related compounds* under *Cromolyn Sodium Inhalation*, "Ophthalmic Solution" being read in place of "Inhalation."

Assay—

pH 7.4 sodium phosphate buffer—Prepare as directed in the *Assay* under *Cromolyn Sodium*.

Assay preparation—Transfer 4 mL of Cromolyn Sodium Ophthalmic Solution to a 100-mL volumetric flask, dilute with water to volume, and mix. Transfer an aliquot of this solution, equivalent to 8 mg of cromolyn sodium, to a 250-mL volumetric flask. Add 1 mL of *pH 7.4 sodium phosphate buffer*, dilute with water to volume, and mix.

Standard preparation—Prepare as directed in the *Assay* under *Cromolyn Sodium*.

Procedure—Proceed as directed for *Procedure* in the *Assay* under *Cromolyn Sodium for Inhalation*.

Croscarmellose Sodium—*see* Croscarmellose Sodium NF

Crospovidone—*see* Crospovidone NF

Crotamiton

$CH_3CH{=}CHCON(C_6H_4CH_3)CH_2CH_3$

$C_{13}H_{17}NO$ 203.28

2-Butenamide, *N*-ethyl-*N*-(2-methylphenyl)-.

N-Ethyl-*o*-crotonotoluidide [*483-63-6*].

» Crotamiton is a mixture of *cis* and *trans* isomers containing not less than 97.0 percent and not more than 103.0 percent of $C_{13}H_{17}NO$.

Packaging and storage—Preserve in tight, light-resistant containers.

Reference standard—*USP Crotamiton Reference Standard*—After opening the ampul, store in a tightly closed container, protected from light. Do not dry before using.

Identification—

A: The infrared absorption spectrum of a thin film of it exhibits maxima only at the same wavelengths as that of a similar preparation of USP Crotamiton RS.

B: The ultraviolet absorption spectrum of a 1 in 50,000 solution in cyclohexane exhibits maxima and minima at the same wavelengths as that of a similar preparation of USP Crotamiton RS, concomitantly measured.

C: To about 10 mL of a saturated solution in water add a few drops of potassium permanganate TS: a brown color is produced, and a brown precipitate is formed on standing.

Specific gravity ⟨841⟩: between 1.008 and 1.011 at 20°.

Refractive index ⟨831⟩: between 1.540 and 1.543 at 20°.

Residue on ignition ⟨281⟩: not more than 0.1%.

Bound halogen—Place 4 drops in a 3-mm (ID) test tube, and add calcium oxide to a height of 1 cm. Heat the tube in a flame, starting from the top, until the reaction is complete, then ignite for a short time. Transfer the contents to a beaker containing 10 mL of water, acidify with nitric acid, and filter. To the filtrate add 0.2 mL of silver nitrate solution (1 in 60): any opalescence obtained is not more than that obtained from a blank solution treated in the same manner.

Assay—Transfer about 50 mg of Crotamiton, accurately weighed, to a 100-mL volumetric flask, add cyclohexane to volume, and mix. Transfer 10.0 mL of this solution to a 250-mL volumetric flask, dilute with cyclohexane to volume, and mix. Determine the absorbance of this solution and of a solution of USP Crotamiton RS in the same medium having a known concentration of about 20 µg per mL in 1-cm cells at the wavelength of maximum absorbance at about 242 nm, with a suitable spectrophotometer, using cyclohexane as the blank. Calculate the quantity, in mg, of $C_{13}H_{17}NO$ in the Crotamiton taken by the formula:

$$2.5C(A_U/A_S),$$

in which *C* is the concentration, in µg per mL, of USP Crotamiton RS in the Standard solution, and A_U and A_S are the absorbances of the assay solution and the Standard solution, respectively.

Crotamiton Cream

» Crotamiton Cream contains not less than 93.0 percent and not more than 107.0 percent of the labeled amount of $C_{13}H_{17}NO$.

Packaging and storage—Preserve in collapsible tubes or in tight, light-resistant containers.

Reference standard—*USP Crotamiton Reference Standard*—After opening the ampul, store in a tightly closed container, protected from light. Do not dry before using.

Identification—The retention time of the major peak in the chromatogram of the *Assay preparation* corresponds to that of the *Standard preparation*, both relative to the internal standard, as obtained in the *Assay*.

Minimum fill ⟨755⟩: meets the requirements.

Assay—

Internal standard solution—Dissolve butyl benzoate in methanol to obtain a solution containing about 17.5 mg per mL.

Mobile phase—Prepare a suitable degassed and filtered mixture of acetonitrile and water (3:2).

Standard solution—Dissolve a suitable quantity of USP Crotamiton RS, accurately weighed, in methanol to obtain a solution having a known concentration of about 1 mg per mL.

Standard preparation—Pipet 10 mL of *Standard solution* and 5 mL of *Internal standard solution* into a 50-mL volumetric flask, dilute with methanol to volume, and mix.

Assay preparation—Transfer an accurately weighed portion of Crotamiton Cream, equivalent to about 50 mg of crotamiton, to a tared 50-mL volumetric flask. Add about 25 mL of methanol, and shake and sonicate to disperse the cream. Dilute with methanol to volume, and mix. Filter about 20 mL through moderately retentive filter paper. Pipet 10 mL of the clear filtrate and 5 mL of *Internal standard solution* into a 50-mL volumetric flask, dilute with methanol to volume, and mix.

Procedure—Inject equal volumes of the *Standard preparation* and the *Assay preparation* into a liquid chromatograph (see *Chromatography* ⟨621⟩) equipped with a 254-nm detector and a 4.6-mm × 25-cm stainless steel column that contains packing L1. In a suitable chromatogram, the resolution factor is not less than 3.0 between peaks for crotamiton and butyl benzoate, and the lowest and highest peak response ratios (R_S) of three replicate injections of *Standard preparation* agree within 2.0%. Calculate the quantity, in mg, of $C_{13}H_{17}NO$ in the portion of Cream taken by the formula:

$$250C(R_U/R_S),$$

in which *C* is the concentration, in mg per mL, of USP Crotamiton RS in the *Standard preparation*, and R_U and R_S are the peak response ratios of the crotamiton peak and the butyl benzoate peak obtained from the *Assay preparation* and the *Standard preparation*, respectively.

Cryoprecipitated Antihemophilic Factor—*see* Antihemophilic Factor, Cryoprecipitated

Crystallized Trypsin—*see* Trypsin, Crystallized

Cupric Chloride

$CuCl_2 . 2H_2O$ 170.48

Copper chloride ($CuCl_2$) dihydrate.

Copper(2+) chloride dihydrate [*10125-13-0*].

Anhydrous 134.45 [*7447-39-4*].

» Cupric Chloride contains not less than 99.0 percent and not more than 100.5 percent of $CuCl_2$, calculated on the dried basis.

Packaging and storage—Preserve in tight containers.

Identification—A solution (1 in 20) responds to the tests for *Chloride* ⟨191⟩ and for *Copper* ⟨191⟩.

Loss on drying ⟨731⟩—Dry it at 105° for 16 hours: it loses between 20.9% and 21.4% of its weight.

Insoluble matter—Transfer 10 g to a 250-mL beaker, add 100 mL of water, cover the beaker, and heat to boiling. Digest the hot solution on a steam bath for 1 hour, and filter through a tared, fine-porosity filtering crucible. Rinse the beaker with hot water, passing the rinsings through the filter, and finally wash the filter with additional hot water. Dry the filter at 105°: the residue weighs not more than 1.0 mg (0.01%). (Retain the combined filtrate and washings for the test for *Sulfate*.)

Sulfate—Heat to boiling the combined filtrate and washings retained from the test for *Insoluble matter*, add 5 mL of barium chloride TS, digest for 2 hours on a steam bath, and allow to stand overnight. Filter the solution through paper, wash the residue with hot water, and transfer the paper and residue to a tared crucible. Char the paper over a low flame, then ignite at 800 ± 25° to constant weight: the weight of the residue, corrected for the weight obtained in a blank test, does not exceed 1.2 mg (0.005%).

Substances not precipitated by ammonium sulfide—Dissolve 2.0 g in about 90 mL of water, add 1 mL of sulfuric acid, heat the solution to about 70°, and pass a stream of hydrogen sulfide through the solution to precipitate the copper completely. Dilute with water to 100 mL, mix, and allow the precipitate to settle. Filter, without washing, transfer 50.0 mL of the filtrate to a tared evaporating dish, and evaporate on a steam bath to dryness. Ignite the dish, gently at first, then at 800 ± 25° to constant weight: the residue weighs not more than 1.0 mg (0.1%). (Retain the residue for the test for *Iron*.)

Iron ⟨241⟩—To the residue retained from the test for *Substances not precipitated by ammonium sulfide* add 2 mL of hydrochloric acid, 2 mL of water, and 0.05 mL of nitric acid. Evaporate slowly on a steam bath to dryness, then dissolve the residue in a mixture of 1 mL of hydrochloric acid and 10 mL of water. Dilute with water to 100 mL, and mix (*Solution A*). To 10 mL of *Solution A* add 10 mL of water, and mix (*Solution B*): 10 mL of *Solution B* conforms to the test for *Iron* (0.015%). (Retain the remainder of *Solution A* for the test for *Other metals*.)

Other metals—To 20 mL of the *Solution A* retained from the test for *Iron* add a slight excess of ammonium hydroxide, boil the solution for 1 minute, filter, and wash the residue with water until the combined volume of filtrate and washing is 25 mL. Neutralize with 3 *N* hydrochloric acid, dilute with water to 30 mL, and add 0.15 mL of ammonium hydroxide and 1 mL of hydrogen sulfide TS: any brown color produced is not darker than that in a control containing 0.15 mL of ammonium hydroxide, 1 mL of hydrogen sulfide TS, and 0.02 mg of added nickel ion (Ni) (0.01% as Ni).

Assay—Transfer about 400 mg of Cupric Chloride, accurately weighed, to a beaker, and dissolve in 50 mL of water. Add 4 mL of acetic acid and 3 g of potassium iodide, mix, and titrate the liberated iodine with 0.1 *N* sodium thiosulfate VS, adding about 2 g of potassium thiocyanate and 3 mL of starch TS as the end-point is approached. Each mL of 0.1 *N* sodium thiosulfate is equivalent to 13.45 mg of $CuCl_2$.

Cupric Chloride Injection

» Cupric Chloride Injection is a sterile solution of Cupric Chloride in Water for Injection. It contains not less than 95.0 percent and not more than 105.0 percent of the labeled amount of copper (Cu).

Packaging and storage—Preserve in single-dose or in multiple-dose containers, preferably of Type I or Type II glass.

Labeling—Label the Injection to indicate that it is to be diluted to the appropriate strength with Sterile Water for Injection or other suitable fluid prior to administration.

Identification—The *Assay preparation*, prepared as directed in the *Assay*, exhibits an absorption maximum at about 325 nm when tested as directed for *Procedure* in the *Assay*.

Pyrogen—When diluted with Sodium Chloride Injection to contain 20 µg of copper per mL, it meets the requirements of the *Pyrogen Test* ⟨151⟩.

pH ⟨791⟩: between 1.5 and 2.5.

Particulate matter ⟨788⟩: meets the requirements under *Small-volume Injections*.

Other requirements—It meets the requirements under *Injections* ⟨1⟩.

Assay—

Copper stock solution—Transfer 1.000 g of copper to a 1000-mL volumetric flask, dissolve in 20 mL of nitric acid, dilute with 0.2 *N* nitric acid to volume, and mix. This solution contains 1000 µg of copper per mL. Store in a polyethylene bottle.

Standard preparations—Pipet 10 mL of *Copper stock solution* into a 500-mL volumetric flask, dilute with water to volume, and mix. Transfer 5.0, 10.0, 15.0, and 20.0 mL, respectively, of this solution to separate 100-mL volumetric flasks, dilute the contents of each flask with water to volume, and mix. These *Standard preparations* contain, respectively, 1.0, 2.0, 3.0, and 4.0 µg of copper per mL.

Assay preparation—Transfer an accurately measured volume of Cupric Chloride Injection, equivalent to about 2 mg of copper, to a 100-mL volumetric flask, dilute with water to volume, and mix. Pipet 15 mL of this solution into a 100-mL volumetric flask, dilute with water to volume, and mix.

Procedure—Concomitantly determine the absorbances of the *Standard preparations* and the *Assay preparation* at the copper emission line of 324.8 nm, with a suitable atomic absorption spectrophotometer (see *Spectrophotometry and Light-scattering* ⟨851⟩) equipped with a copper hollow-cathode lamp and an air-acetylene flame using water as the blank. Plot the absorbances of the *Standard preparations* versus concentration, in µg per mL, of copper, and draw the straight line best fitting the four plotted points. From the graph so obtained, determine the concentration, in µg per mL, of copper in the *Assay preparation*. Calculate the quantity, in mg, of copper in each mL of the Injection taken by the formula:

$$2C/3V,$$

in which C is the concentration, in µg per mL, of copper in the *Assay preparation*, and V is the volume, in mL, of Injection taken.

Cupric Sulfate

$CuSO_4 \cdot 5H_2O$ 249.68
Sulfuric acid, copper(2+) salt (1:1), pentahydrate.
Copper(2+) sulfate (1:1) pentahydrate [*7758-99-8*].
Anhydrous 159.60 [*7758-98-7*].

» Cupric Sulfate, dried at 250° to constant weight, contains not less than 98.5 percent and not more than 100.5 percent of $CuSO_4$.

Packaging and storage—Preserve in tight containers.

Identification—A solution (1 in 10) responds to the tests for *Copper* ⟨191⟩ and for *Sulfate* ⟨191⟩.

Loss on drying ⟨731⟩—Dry it at 250° to constant weight: it loses between 33.0% and 36.5% of its weight.

Alkalies and alkaline earths—Dissolve 2.0 g in 100 mL of water, add 1 mL of 3 *N* hydrochloric acid, and pass hydrogen sulfide through the solution until all of the copper is precipitated. Filter the mixture, evaporate 50 mL of the filtrate to dryness, and ignite: the weight of the residue does not exceed 3 mg (0.3%).

Assay—Place about 650 mg of Cupric Sulfate in an accurately weighed container fitted with a ground-glass stopper, dry, allow to cool in a desiccator, and weigh again to obtain the weight of the specimen. Dissolve in 50 mL of water, add 4 mL of 6 *N* acetic acid and 3 g of potassium iodide, and titrate the liberated iodine with 0.1 *N* sodium thiosulfate VS, adding about 2 g of potassium thiocyanate and 3 mL of starch TS as the end-point is approached. Perform a blank determination, and make any necessary correction. Each mL of 0.1 *N* sodium thiosulfate is equivalent to 15.96 mg of $CuSO_4$.

Cupric Sulfate Injection

» Cupric Sulfate Injection is a sterile solution of Cupric Sulfate in Water for Injection. It contains not less than 95.0 percent and not more than 105.0 percent of the labeled amount of copper (Cu).

Packaging and storage—Preserve in single-dose or in multiple-dose containers, preferably of Type I or Type II glass.

Labeling—Label the Injection to indicate that it is to be diluted to the appropriate strength with Sterile Water for Injection or other suitable fluid prior to administration.

Identification—The *Assay preparation*, prepared as directed in the *Assay*, exhibits an absorption maximum at about 325 nm when tested as directed for *Procedure* in the *Assay*.

Pyrogen—When diluted with Sodium Chloride Injection to contain 20 µg of copper per mL, it meets the requirements of the *Pyrogen Test* ⟨151⟩.

pH ⟨791⟩: between 2.0 and 3.5.

Particulate matter ⟨788⟩: meets the requirements under *Small-volume Injections*.

Other requirements—It meets the requirements under *Injections* ⟨1⟩.

Assay—

Copper stock solution and *Standard preparations*—Prepare as directed in the *Assay* under *Cupric Chloride Injection*.

Assay preparation—Transfer an accurately measured volume of Cupric Sulfate Injection, equivalent to about 2 mg of copper, to a 100-mL volumetric flask, dilute with water to volume, and mix. Pipet 15 mL of this solution into a 100-mL volumetric flask, dilute with water to volume, and mix.

Procedure—Proceed as directed for *Procedure* in the *Assay* under *Cupric Chloride Injection*.

Cyanocobalamin

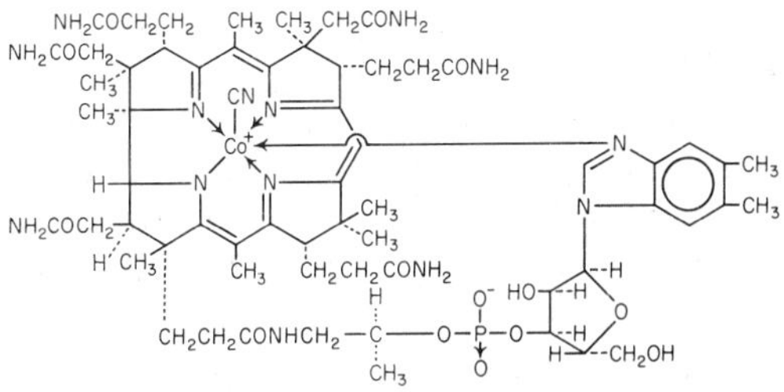

$C_{63}H_{88}CoN_{14}O_{14}P$ 1355.38

Vitamin B_{12}.

Vitamin B_{12} [*68-19-9*].

» Cyanocobalamin contains not less than 96.0 percent and not more than 100.5 percent of $C_{63}H_{88}CoN_{14}O_{14}P$, calculated on the dried basis.

Packaging and storage—Preserve in tight, light-resistant containers.

Reference standard—*USP Cyanocobalamin Reference Standard*—Dry over silica gel for 4 hours before using.

Identification—

A: The absorption spectrum of the solution employed for measurement of absorbance in the *Assay* exhibits maxima within ±1 nm at 278 nm and 361 nm and within ±2 nm at 550 nm. The ratio A_{361}/A_{278} is between 1.70 and 1.90, and the ratio A_{361}/A_{550} is between 3.15 and 3.40.

B: Fuse about 1 mg with about 50 mg of potassium pyrosulfate in a porcelain crucible. Cool, break up the mass with a glass rod, add 3 mL of water, and dissolve by boiling. Add 1 drop of phenolphthalein TS, and add sodium hydroxide solution (1 in 10), dropwise, until just pink. Add 500 mg of sodium acetate, 0.5 mL of 1 *N* acetic acid, and 0.5 mL of nitroso R salt solution (1 in 500): a red or orange-red color appears at once. Add 0.5 mL of hydrochloric acid, and boil for 1 minute: the red color persists.

C: Dissolve about 5 mg in 5 mL of water in a 50-mL distilling flask connected with a short, water-cooled, condenser. To the flask add 2.5 mL of hypophosphorous acid, close, boil gently but short of distillation for 10 minutes, then distil 1 mL into a test tube containing 1 mL of sodium hydroxide solution (1 in 50). To the test tube add 4 drops of cold saturated ferrous ammonium sulfate solution, shake gently, then add about 30 mg of sodium fluoride, and bring the contents to a boil. Immediately add, dropwise, 5 *N* sulfuric acid until a clear solution results, then add 3 to 5 drops more of the acid: a blue or blue-green color develops within a few minutes.

Loss on drying ⟨731⟩—Heat about 25 mg, accurately weighed, in a suitable vacuum drying apparatus at 105° and at a pressure of not more than 5 mm of mercury for 2 hours, cool, and weigh: it loses not more than 12.0% of its weight.

Pseudo cyanocobalamin—Dissolve 1.0 mg in 20 mL of water contained in a small separator, add 5 mL of a mixture of equal volumes of carbon tetrachloride and cresol, and shake well for about 1 minute. Allow to separate, draw off the lower layer into a second small separator, add 5 mL of 5 *N* sulfuric acid, shake well, and allow to separate completely (the complete separation of the layer may be facilitated by centrifuging): the separated upper layer is colorless or has no more color than a mixture of 0.15 mL of 0.10 *N* potassium permanganate and 250 mL of water.

Assay—Transfer about 30 mg of Cyanocobalamin, accurately weighed, with the aid of water to a 1-liter volumetric flask, dilute with water to volume, and mix. Dissolve an accurately weighed quantity of USP Cyanocobalamin RS in water, and dilute quantitatively and stepwise with water to obtain a Standard solution having a known concentration of about 30 µg per mL. Concomitantly determine the absorbances of both solutions in 1-cm cells at the wavelength of maximum absorbance at about 361 nm, with a suitable spectrophotometer, using water as the blank. Calculate the quantity, in mg, of $C_{63}H_{88}CoN_{14}O_{14}P$ in the Cyanocobalamin taken by the formula:

$$C(A_U/A_S),$$

in which *C* is the concentration, in µg per mL, of USP Cyanocobalamin RS in the Standard solution, and A_U and A_S are the absorbances of the solution of Cyanocobalamin and the Standard solution, respectively.

Cyanocobalamin Injection

» Cyanocobalamin Injection is a sterile solution of Cyanocobalamin in Water for Injection, or in Water for Injection rendered isotonic by the addition of Sodium Chloride. It contains not less than 95.0 percent and not more than 115.0 percent of the labeled amount of anhydrous cyanocobalamin ($C_{63}H_{88}CoN_{14}O_{14}P$).

Packaging and storage—Preserve in light-resistant, single-dose or multiple-dose containers, preferably of Type I glass.

Reference standard—*USP Cyanocobalamin Reference Standard*—Dry over silica gel for 4 hours before using.

Identification—The absorption spectrum, in the range of 300 nm to 550 nm, of the solution employed for measurement of absorbance in the *Assay* exhibits maxima at the same wavelengths as that of a similar solution of USP Cyanocobalamin RS, concomitantly measured, and the ratio A_{361}/A_{550} is between 3.15 and 3.40.

pH ⟨791⟩: between 4.5 and 7.0.

Other requirements—It meets the requirements under *Injections* ⟨1⟩.

Assay—Dilute, if necessary, an accurately measured volume of Cyanocobalamin Injection, equivalent to not less than 300 µg of cyanocobalamin, quantitatively and stepwise with water to a concentration of about 30 µg per mL. Dissolve an accurately weighed quantity of USP Cyanocobalamin RS in water, and dilute quantitatively and stepwise with water to obtain a Standard solution having a known concentration of about 30 µg per mL. Concomitantly determine the absorbances of both solutions in 1-cm cells at the wavelength of maximum absorbance at about 361 nm, with a suitable spectrophotometer, using water as the blank. Calculate the quantity, in µg, of $C_{63}H_{88}CoN_{14}O_{14}P$ in each mL of the Injection taken by the formula:

$$10(C/V)(A_U/A_S),$$

in which C is the concentration, in µg per mL, of USP Cyanocobalamin RS in the Standard solution, V is the volume, in mL, of Injection taken, and A_U and A_S are the absorbances of the solution from the Injection and the Standard solution, respectively.

Cyanocobalamin Co 57 Capsules—*see under "cobalt"*

Cyanocobalamin Co 57 Oral Solution—*see under "cobalt"*

Cyanocobalamin Co 60 Capsules—*see under "cobalt"*

Cyanocobalamin Co 60 Oral Solution—*see under "cobalt"*

Cyclacillin

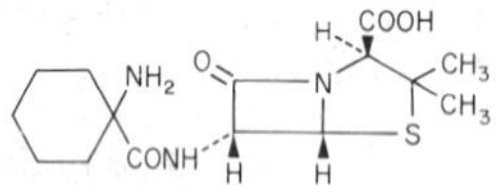

$C_{15}H_{23}N_3O_4S$ 341.42

4-Thia-1-azabicyclo[3.2.0]heptane-2-carboxylic acid, 6-[[(1-aminocyclohexyl)carbonyl]amino]-3,3-dimethyl-7-oxo-, [2*S*-(2α,5α,6β)]-.

6-(1-Aminocyclohexanecarboxamido)-3,3-dimethyl-7-oxo-4-thia-1-azabicyclo[3.2.0]heptane-2-carboxylic acid [*3485-14-1*].

» Cyclacillin contains not less than 90.0 percent of $C_{15}H_{23}N_3O_4S$, calculated on the anhydrous basis. It has a potency of not less than 900 µg and not more than 1050 µg of $C_{15}H_{23}N_3O_4S$ per mg.

Packaging and storage—Preserve in tight containers.

Reference standard—*USP Cyclacillin Reference Standard*—Do not dry before using.

Identification—The infrared absorption spectrum of a potassium bromide dispersion of it exhibits maxima only at the same wavelengths as that of a similar preparation of USP Cyclacillin RS.

Crystallinity ⟨695⟩: meets the requirements.

pH ⟨791⟩: between 4.0 and 6.5, in a solution containing 10 mg per mL.

Water, *Method I* ⟨921⟩: not more than 1.0%, 20 mL of a mixture of carbon tetrachloride, chloroform, and methanol (2:2:1) being used in place of methanol in the titration vessel.

Concordance—Calculate the *Acid titration–base titration concordance* by the formula:

$$A - B,$$

the *Acid titration–potency concordance* by the formula:

$$A - (P_A/10),$$

and the *Base titration–potency concordance* by the formula:

$$B - (P_A/10),$$

in which A and B are the *Content of cyclacillin*, in percentage, as determined by the *Titration of acidic function* and the *Titration of basic function*, respectively, under the test for *Content of cyclacillin*, and P_A is the potency, in µg per mg, as determined in the *Assay:* none of the concordance values exceed 6.0%.

Content of cyclacillin—

Titration of acidic function—Dissolve about 100 mg of Cyclacillin, accurately weighed, in 20 mL of dimethylsulfoxide and 30 mL of methanol. Stir by mechanical means, and titrate with 0.02 *N* lithium methoxide in methanol VS, determining the endpoint potentiometrically using a glass electrode and a calomel electrode containing potassium chloride–saturated methanol (see *Titrimetry* ⟨541⟩). Perform a blank determination and make any necessary correction. Each mL of 0.02 *N* lithium methoxide is equivalent to 6.828 mg of $C_{15}H_{23}N_3O_4S$.

Titration of basic function—Dissolve about 100 mg of Cyclacillin, accurately weighed, in 50 mL of glacial acetic acid. Titrate with 0.02 *N* perchloric acid VS, determining the end-point potentiometrically using a glass electrode and a calomel electrode containing potassium chloride–saturated methanol (see *Titrimetry* ⟨541⟩). Perform a blank determination and make any necessary correction. Each mL of 0.02 *N* perchloric acid is equivalent to 6.828 mg of $C_{15}H_{23}N_3O_4S$.

Assay—

Standard preparation—Prepare as directed for *Standard preparation* under *Iodometric Assay—Antibiotics* ⟨425⟩, using USP Cyclacillin RS.

Assay preparation—Proceed as directed for *Assay preparation* under *Iodometric Assay—Antibiotics* ⟨425⟩.

Procedure—Proceed as directed for *Procedure* under *Iodometric Assay—Antibiotics* ⟨425⟩. Calculate the potency, in µg of $C_{15}H_{23}N_3O_4S$ per mg, of the Cyclacillin taken by the formula:

$$(F)(B - I)/(2D),$$

in which D is the concentration, in mg per mL, of the *Assay preparation*, based on the weight of Cyclacillin taken and the extent of dilution.

Cyclacillin for Oral Suspension

» Cyclacillin for Oral Suspension is a dry mixture of Cyclacillin with one or more suitable buffers, colors, flavors, preservatives, sweeteners, and suspending agents. It contains not less than 90.0 percent and not more than 120.0 percent of the labeled amount of $C_{15}H_{23}N_3O_4S$.

Packaging and storage—Preserve in tight containers.

Reference standard—*USP Cyclacillin Reference Standard*—Do not dry before using.

Identification—Prepare a solution of it in 0.1 *N* sodium hydroxide containing 1 mg of cyclacillin per mL, and allow to stand for 45 minutes before use. Prepare a solution of USP Cyclacillin RS in 0.1 *N* sodium hydroxide containing 1 mg per mL, and allow to stand for 15 minutes before use. Apply separately 5 µL of each solution on a thin-layer chromatographic plate coated with a 0.25-mm layer of chromatographic silica gel mixture (see *Chromatography* ⟨621⟩). Place the plate in a suitable chromatographic chamber, and develop the chromatogram with ammonium formate solution (1 in 100). When the solvent front has moved about three-fourths of the length of the plate, remove the plate from the chamber, and dry at 80° for 30 minutes. Locate the spots on the plate by spraying lightly with a mixture of starch iodide paste TS, water, glacial acetic acid, and 0.1 *N* iodine (25:25:3:1): the spots are white against a blue background, and the R_f value of the principal spot obtained from the test solution corresponds to that obtained from the Standard solution.

Uniformity of dosage units ⟨905⟩—

FOR SOLID PACKAGED IN SINGLE-UNIT CONTAINERS: meets the requirements.

pH ⟨791⟩: between 4.5 and 6.5, in the suspension constituted as directed in the labeling.

Water, *Method I* ⟨921⟩: not more than 1.5%, 20 mL of a mixture of carbon tetrachloride, chloroform, and methanol (2:2:1) being used in place of methanol in the titration vessel.

Assay—

Standard preparation—Prepare as directed for *Standard preparation* under *Iodometric Assay—Antibiotics* ⟨425⟩, using USP Cyclacillin RS.

Assay preparation—Constitute Cyclacillin for Oral Suspension as directed in the labeling. Dilute an accurately measured volume of the suspension, freshly mixed and free from air bubbles, quantitatively and stepwise with *Buffer No. 1* (see *Media and Diluents* under *Antibiotics—Microbial Assays* ⟨81⟩) to obtain a solution having a concentration of 1 mg of cyclacillin per mL.

Procedure—Proceed as directed for *Procedure* under *Iodometric Assay—Antibiotics* ⟨425⟩. Calculate the quantity, in mg, of $C_{15}H_{23}N_3O_4S$ in each mL of the Suspension taken by the formula:

$$(T/D)(F/2000)(B - I),$$

in which *T* is the labeled quantity, in mg per mL, of cyclacillin in the constituted Cyclacillin for Oral Suspension, and *D* is the concentration, in mg per mL, of cyclacillin in the *Assay preparation* on the basis of the labeled quantity in the constituted Cyclacillin for Oral Suspension and the extent of dilution.

Cyclacillin Tablets

» Cyclacillin Tablets contain not less than 90.0 percent and not more than 120.0 percent of the labeled amount of $C_{15}H_{23}N_3O_4S$.

Packaging and storage—Preserve in tight containers.

Reference standard—*USP Cyclacillin Reference Standard*—Do not dry before using.

Identification—To a portion of powdered Tablets add 0.1 *N* sodium hydroxide to make a solution containing 1 mg per mL. Allow the solution to stand for 15 minutes before use. The solution responds to the *Identification test* under *Cyclacillin for Oral Suspension.*

Dissolution ⟨711⟩—

Medium: Water; 500 mL.

Apparatus 2: 50 rpm.

Time: 45 minutes.

Procedure—Filter the solution under test, dilute with *Dissolution Medium,* if necessary, and proceed as directed for *Procedure* in the section, *Antibiotics—Hydroxylamine Assay,* under *Automated Methods of Analysis* ⟨16⟩. Determine the amount of $C_{15}H_{23}N_3O_4S$ dissolved by comparison with a Standard solution having a known concentration of USP Cyclacillin RS in the same medium.

Tolerances—Not less than 75% (*Q*) of the labeled amount of $C_{15}H_{23}N_3O_4S$ is dissolved in 45 minutes.

Water, *Method I* ⟨921⟩: not more than 5.0%, 20 mL of a mixture of carbon tetrachloride, chloroform, and methanol (2:2:1) being used in place of methanol in the titration vessel.

Assay—

Standard preparation—Prepare as directed for *Standard preparation* under *Iodometric Assay—Antibiotics* ⟨425⟩, using USP Cyclacillin RS.

Assay preparation—Place not less than 5 Cyclacillin Tablets in a high-speed glass blender jar containing an accurately measured volume of water, and blend for 4 ± 1 minutes. Dilute an accurately measured volume of this stock solution quantitatively and stepwise with water to obtain an *Assay preparation* containing about 1 mg of cyclacillin per mL.

Procedure—Proceed as directed for *Procedure* under *Iodometric Assay—Antibiotics* ⟨425⟩. Calculate the quantity, in mg, of $C_{15}H_{23}N_3O_4S$ in each Tablet taken by the formula:

$$(T/D)(F/2000)(B - I),$$

in which *T* is the labeled quantity, in mg, of cyclacillin in each Tablet, and *D* is the concentration, in mg per mL, of cyclacillin in the *Assay preparation* on the basis of the labeled quantity in each Tablet and the extent of dilution.

Cyclizine

$C_{18}H_{22}N_2$ 266.39

Piperazine, 1-(diphenylmethyl)-4-methyl-.

1-(Diphenylmethyl)-4-methylpiperazine [*82-92-8*].

» Cyclizine contains not less than 98.0 percent and not more than 100.5 percent of $C_{18}H_{22}N_2$, calculated on the anhydrous basis.

Packaging and storage—Preserve in tight, light-resistant containers.

Reference standard—*USP Cyclizine Reference Standard*—Do not dry; determine the water content titrimetrically at the time of use.

Clarity and color of solution—A 1 in 100 solution in 3 *N* hydrochloric acid is clear and colorless.

Identification—

A: The infrared absorption spectrum of a potassium bromide dispersion of it, previously dried in vacuum at 60° for 5 hours, exhibits maxima only at the same wavelengths as that of a similar preparation of USP Cyclizine RS.

B: The ultraviolet absorption spectrum of a 1 in 5000 solution in dilute sulfuric acid (1 in 70) exhibits maxima and minima at the same wavelengths as that of a similar solution of USP Cyclizine RS, concomitantly measured, and the respective absorptivities, calculated on the anhydrous basis, at the wavelength of maximum absorbance at about 262 nm do not differ by more than 3.0%.

Melting range, *Class I* ⟨741⟩: between 106° and 109°.

pH ⟨791⟩: between 7.6 and 8.6, in a saturated solution.

Water, *Method I* ⟨921⟩: not more than 1.0%.

Residue on ignition ⟨281⟩: not more than 0.1%.

Chloride ⟨221⟩—Shake 1.0 g of it with 25 mL of water, filter, and add 1 mL of 2 *N* nitric acid: the acidified filtrate shows no more chloride than corresponds to 0.20 mL of 0.020 *N* hydrochloric acid (0.014%).

Ordinary impurities ⟨466⟩—

Test solution: methanol.

Standard solution: methanol.

Eluant: a mixture of chloroform, methanol, and ammonium hydroxide (80:20:1).

Visualization: 2.

Assay—Dissolve about 300 mg of Cyclizine, accurately weighed, in 75 mL of glacial acetic acid in a 250-mL beaker, warming, if necessary, to effect solution. Cool, add 1 drop of crystal violet TS, and titrate with 0.1 *N* perchloric acid VS to a blue-green end-point. Perform a blank determination, and make any necessary correction. Each mL of 0.1 *N* perchloric acid is equivalent to 13.32 mg of $C_{18}H_{22}N_2$.

Cyclizine Hydrochloride

$C_{18}H_{22}N_2.HCl$ 302.85
Piperazine, 1-(diphenylmethyl)-4-methyl-, monohydrochloride.
1-(Diphenylmethyl)-4-methylpiperazine monohydrochloride [*303-25-3*].

» Cyclizine Hydrochloride contains not less than 98.0 percent and not more than 100.5 percent of $C_{18}H_{22}N_2.HCl$, calculated on the dried basis.

Packaging and storage—Preserve in tight, light-resistant containers.

Reference standard—*USP Cyclizine Hydrochloride Reference Standard*—Dry at 120° for 3 hours before using.

Identification—

A: The infrared absorption spectrum of a potassium bromide dispersion of it, previously dried, exhibits maxima only at the same wavelengths as that of a similar preparation of USP Cyclizine Hydrochloride RS.

B: Dissolve 500 mg in 10 mL of a mixture of 3 volumes of alcohol and 2 volumes of water, warming if necessary. Cool the solution in an ice bath, add 1 mL of 1 *N* sodium hydroxide and 20 mL of water, stir well, and filter: the precipitate of the base so obtained, washed with water and dried in vacuum at 60° for 2 hours, melts between 106° and 109°.

C: It responds to the tests for *Chloride* ⟨191⟩.

pH ⟨791⟩: between 4.5 and 5.5, determined potentiometrically in a 1 in 50 solution, the solvent being a mixture of 2 volumes of alcohol and 3 volumes of water.

Loss on drying ⟨731⟩—Dry it at 120° for 3 hours: it loses not more than 1.0% of its weight.

Residue on ignition ⟨281⟩: not more than 0.2%.

Ordinary impurities ⟨466⟩—

Test solution: methanol.

Standard solution: methanol.

Eluant: a mixture of chloroform, methanol, and ammonium hydroxide (80:20:1).

Visualization: 2.

Assay—Transfer to a beaker about 400 mg of Cyclizine Hydrochloride, accurately weighed, and dissolve in 80 mL of glacial acetic acid. Add 10 mL of mercuric acetate TS, and titrate with 0.1 *N* perchloric acid VS, determining the end-point potentiometrically. Perform a blank determination, and make any necessary correction. Each mL of 0.1 *N* perchloric acid is equivalent to 15.14 mg of $C_{18}H_{22}N_2.HCl$.

Cyclizine Hydrochloride Tablets

» Cyclizine Hydrochloride Tablets contain not less than 93.0 percent and not more than 107.0 percent of the labeled amount of $C_{18}H_{22}N_2.HCl$.

Packaging and storage—Preserve in tight, light-resistant containers.

Reference standard—*USP Cyclizine Hydrochloride Reference Standard*—Dry at 120° for 3 hours before using.

Identification—

A: Shake a quantity of finely powdered Tablets, equivalent to about 500 mg of cyclizine hydrochloride, with 25 mL of water for 5 minutes, and filter the mixture. Cool the filtrate in an ice bath, add a slight excess of 1 *N* sodium hydroxide, and stir well: the precipitate so obtained responds to *Identification test B* under *Cyclizine Hydrochloride*.

B: Tablets meet the requirements under *Identification—Organic Nitrogenous Bases* ⟨181⟩.

Dissolution ⟨711⟩—

Medium: water; 900 mL.

Apparatus 2: 50 rpm.

Time: 45 minutes.

Procedure—Determine the amount of $C_{18}H_{22}N_2.HCl$ dissolved, employing the procedure set forth in the *Assay*, making any necessary modifications.

Tolerances—Not less than 75% (*Q*) of the labeled amount of $C_{18}H_{22}N_2.HCl$ is dissolved in 45 minutes.

Uniformity of dosage units ⟨905⟩: meet the requirements.

Assay—Proceed with Cyclizine Hydrochloride Tablets as directed under *Salts of Organic Nitrogenous Bases* ⟨501⟩, diluting the *Standard Preparation* and the *Assay Preparation*, respectively, with an equal volume of dilute sulfuric acid (1 in 100), and determining the absorbance at the wavelength of maximum absorbance at about 264 nm. Calculate the quantity, in mg, of $C_{18}H_{22}N_2.HCl$ in the portion of Tablets taken by the formula:

$$50C(A_U/A_S),$$

in which *C* is the concentration, in mg per mL, of USP Cyclizine Hydrochloride RS in the *Standard Preparation*.

Cyclizine Lactate Injection

$C_{18}H_{22}N_2.C_3H_6O_3$ 356.46
Piperazine, 1-(diphenylmethyl)-4-methyl-, mono(2-hydroxypropanoate).
1-(Diphenylmethyl)-4-methylpiperazine monolactate [*5897-19-8*].

» Cyclizine Lactate Injection is a sterile solution of cyclizine lactate ($C_{18}H_{22}N_2.C_3H_6O_3$) in Water for Injection, prepared from Cyclizine with the aid of Lactic Acid. It contains not less than 95.0 percent and not more than 105.0 percent of the labeled amount of $C_{18}H_{22}N_2.C_3H_6O_3$.

Packaging and storage—Preserve in single-dose containers, preferably of Type I glass, protected from light.

Reference standard—*USP Cyclizine Reference Standard*—Do not dry; determine the water content titrimetrically at the time of use.

Identification—

A: Dilute a volume of Injection, equivalent to about 500 mg of cyclizine lactate, with alcohol to obtain a solution containing about 20 mg per mL. Cool the solution in an ice bath, add 1 mL of 1 *N* sodium hydroxide and 20 mL of water, stir, and filter. Wash the residue on the filter with 2 mL of ice-cold water, and dry in vacuum at 75° for 1 hour: the cyclizine so obtained melts between 106° and 109°, the procedure for *Class I* being used (see *Melting Range or Temperature* ⟨741⟩).

B: Dilute a volume of Injection, equivalent to about 50 mg of cyclizine lactate, with dilute hydrochloric acid (1 in 1200) to 25 mL, and proceed as directed under *Identification—Organic Nitrogenous Bases* ⟨181⟩, beginning with "Transfer the liquid to a separator": the Injection meets the requirements of the test.

C: To a volume of Injection, equivalent to 5 mg of cyclizine lactate, add 2 mL of sulfuric acid, and heat for 2 minutes in a water bath maintained at a temperature of 85°. Cool the solution to room temperature, and add 5 to 10 drops of the solution to about 10 mg of 4-phenylphenol crystals: a violet color is produced in 2 to 3 minutes.

pH ⟨791⟩: between 3.2 and 4.7.

Other requirements—It meets the requirements under *Injections* ⟨1⟩.

Assay—

Standard preparation—Transfer about 50 mg of USP Cyclizine RS, accurately weighed, to a glass-stoppered, 25-mL conical flask, add 10.0 mL of cyclohexane, insert the stopper, and mix.

Assay preparation—Transfer an accurately measured volume of Cyclizine Lactate Injection, equivalent to about 100 mg of cyclizine lactate, to a separator, add 10 mL of water, and mix. Add 1 mL of sodium hydroxide solution (1 in 2), mix, add 15.0 mL of cyclohexane, and shake the mixture for 1 minute. Allow the layers to separate, discard the aqueous layer, transfer the cyclohexane layer to a glass-stoppered, 25-mL conical flask containing 2 g of sodium chloride, and mix.

Procedure—Concomitantly determine the absorbances of the solutions in 0.4-mm cells at the wavelength of maximum absorbance at about 14.2 µm, with a suitable spectrophotometer, using cyclohexane as the blank. Calculate the quantity, in mg, of $C_{18}H_{22}N_2 . C_3H_6O_3$ in each mL of the Injection taken by the formula:

$$(356.46/266.39)(15C/V)(A_U/A_S),$$

in which 356.46 and 266.39 are the molecular weights of cyclizine lactate and cyclizine, respectively, C is the concentration, in mg per mL, of USP Cyclizine RS in the *Standard preparation*, V is the volume, in mL, of Injection taken, and A_U and A_S are the absorbances of the *Assay preparation* and the *Standard preparation*, respectively.

Cyclobenzaprine Hydrochloride

· HCl

$CHCH_2CH_2N(CH_3)_2$

$C_{20}H_{21}N . HCl$ 311.85

1-Propanamine, 3-(5*H*-dibenzo[*a,d*]cyclohepten-5-ylidene)-*N,N*-dimethyl-, hydrochloride.

N,N-Dimethyl-5*H*-dibenzo[*a,d*]cycloheptene-$\Delta^{5,\gamma}$-propylamine hydrochloride [*6202-23-9*].

» Cyclobenzaprine Hydrochloride contains not less than 99.0 percent and not more than 101.0 percent of $C_{20}H_{21}N . HCl$, calculated on the dried basis.

Packaging and storage—Preserve in well-closed containers.

Reference standard—*USP Cyclobenzaprine Hydrochloride Reference Standard*—Dry at 105° to constant weight before using.

Identification—

A: The infrared absorption spectrum of a mineral oil dispersion of it, previously dried, exhibits maxima only at the same wavelengths as that of a similar preparation of USP Cyclobenzaprine Hydrochloride RS.

B: The ultraviolet absorption spectrum of a 1 in 65,000 solution in methanol exhibits maxima and minima at the same wavelengths as that of a similar solution of USP Cyclobenzaprine Hydrochloride RS, concomitantly measured, and the respective absorptivities, calculated on the dried basis, at the wavelength of maximum absorbance at about 290 nm do not differ by more than 3.0%.

C: A solution (1 in 50) responds to the tests for *Chloride* ⟨191⟩.

Melting range ⟨741⟩: between 215° and 219°, but the range between beginning and end of melting does not exceed 2°.

Loss on drying ⟨731⟩—Dry it at 105° to constant weight: it loses not more than 1.0% of its weight.

Residue on ignition ⟨281⟩: not more than 0.1%.

Heavy metals, *Method II* ⟨231⟩: 0.001%.

Chromatographic purity—Dissolve 100 mg in methanol, and dilute with methanol to 5.0 mL to obtain the *Test solution*. Dissolve a suitable quantity of USP Cyclobenzaprine Hydrochloride RS in methanol to obtain a *Standard solution* having a known concentration of about 20 mg per mL. Dilute a portion of this solution quantitatively and stepwise with methanol to obtain a *Diluted standard solution* having a concentration of about 100 µg per mL. Apply separate 5-µL portions of the three solutions on the starting line of a suitable thin-layer chromatographic plate (see *Chromatography* ⟨621⟩), coated with a 0.25-mm layer of chromatographic silica gel mixture and previously washed with methanol. Develop the chromatogram in a suitable chamber with a freshly prepared solvent system consisting of a mixture of acetone, toluene, and ammonium hydroxide (75:25:1) until the solvent front has moved about three-fourths of the length of the plate. Remove the plate from the chamber, air-dry, and view under short-wavelength ultraviolet light: the R_f value of the principal spot from the *Test solution* corresponds to that of the *Standard solution*, and any other spot obtained from the *Test solution* does not exceed, in size or intensity, the principal spot obtained from the *Diluted standard solution* (0.5%).

Assay—Dissolve about 400 mg of Cyclobenzaprine Hydrochloride, accurately weighed, in about 80 mL of glacial acetic acid, add 15 mL of mercuric acetate TS, and titrate with 0.1 *N* perchloric acid VS, determining the end-point potentiometrically, using a platinum ring electrode and a sleeve-type calomel electrode containing 0.1 *N* lithium perchlorate in glacial acetic acid (see *Titrimetry* ⟨541⟩). Perform a blank determination, and make any necessary correction. Each mL of 0.1 *N* perchloric acid is equivalent to 31.19 mg of $C_{20}H_{21}N . HCl$.

Cyclobenzaprine Hydrochloride Tablets

» Cyclobenzaprine Hydrochloride Tablets contain not less than 90.0 percent and not more than 110.0 percent of the labeled amount of $C_{20}H_{21}N . HCl$.

Packaging and storage—Preserve in well-closed containers.

Reference standard—*USP Cyclobenzaprine Hydrochloride Reference Standard*—Dry at 105° to constant weight before using.

Identification—Transfer a quantity of finely powdered Tablets, equivalent to about 50 mg of cyclobenzaprine hydrochloride, to a small flask, add 10 mL of methylene chloride, swirl to dissolve the test specimen, and filter. Evaporate the clear filtrate to about 5 mL, transfer to a suitable centrifuge tube, and add 1 to 2 mL of ether. Evaporate with the aid of a stream of air to about 1 mL, and agitate until crystallization occurs. Wash the crystals with several portions of ether, and air-dry: the infrared absorption spectrum of a mineral oil dispersion of the dried crystals so obtained exhibits maxima and minima at the same wavelengths as that of a similar preparation of USP Cyclobenzaprine Hydrochloride RS.

Dissolution ⟨711⟩—

Medium: 0.1 *N* hydrochloric acid; 900 mL.

Apparatus 1: 50 rpm.

Time: 30 minutes.

Procedure—Determine the amount of $C_{20}H_{21}N . HCl$ dissolved from ultraviolet absorbances at the wavelength of maximum absorbance at about 290 nm of filtered portions of the solution under test, suitably diluted with *Dissolution Medium*, if necessary, in comparison with a Standard solution having a known concentration of USP Cyclobenzaprine Hydrochloride RS in the same medium.

Tolerances—Not less than 75% (Q) of the labeled amount of $C_{20}H_{21}N . HCl$ is dissolved in 30 minutes.

Uniformity of dosage units ⟨905⟩: meet the requirements.

Assay—

Standard preparation—Dissolve a suitable quantity of USP Cyclobenzaprine Hydrochloride RS, accurately weighed, in 0.1 *N* sulfuric acid, and dilute quantitatively and stepwise with 0.1 *N* sulfuric acid to obtain a solution having a known concentration of about 20 µg per mL.

Assay preparation—Weigh and finely powder not less than 20 Cyclobenzaprine Hydrochloride Tablets. Transfer an accurately weighed portion of the powder, equivalent to about 20 mg of cyclobenzaprine hydrochloride, to a 100-mL volumetric flask containing about 80 mL of 0.1 *N* hydrochloric acid. Insert the stopper, shake the flask for 30 minutes, dilute with 0.1 *N* hydrochloric acid to volume, mix, and filter.

Into one of three separators (*A*) add 2 g of sodium chloride and 20.0 mL of the filtered *Assay preparation*. Shake to dissolve the sodium chloride. To each of the other separators (*B* and *C*) add 50 mL of 0.1 *N* sulfuric acid. To separator *A* add 45 mL of methylene chloride, and shake vigorously for 1 minute. Allow the phases to separate, and filter the methylene chloride layer through cotton into separator *B*. Shake separator *B* vigorously for 1 minute, allow the phases to separate, and transfer the methylene chloride layer into separator *C*. Shake separator *C* vigorously for 1 minute, allow the phases to separate, and draw off and discard the methylene chloride. Repeat the separations in the same manner, using a second 45-mL portion of methylene chloride. Combine the acid solutions in separators *B* and *C* into a 200-mL volumetric flask, rinsing the separators with 50 mL of 0.1 *N* sulfuric acid and adding it to the flask, dilute with 0.1 *N* sulfuric acid to volume, and mix.

Procedure—Concomitantly determine the absorbances of the *Standard preparation* and the *Assay preparation* in 1-cm cells at the wavelength of maximum absorbance at about 290 nm, with a suitable spectrophotometer, using 0.1 *N* sulfuric acid as the blank. Calculate the quantity, in mg, of $C_{20}H_{21}N.HCl$ in the portion of Tablets taken by the formula:

$$C(A_U/A_S),$$

in which *C* is the concentration, in µg per mL, of USP Cyclobenzaprine Hydrochloride RS in the *Standard preparation*, and A_U and A_S are the absorbances of the solutions from the *Assay preparation* and the *Standard preparation*, respectively.

Cyclomethicone—*see* Cyclomethicone NF

Cyclopentolate Hydrochloride

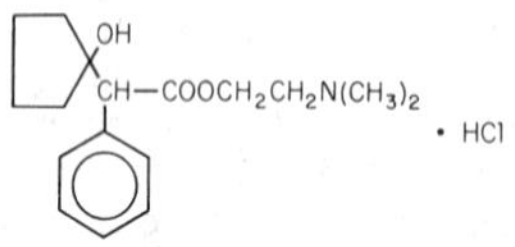

$C_{17}H_{25}NO_3.HCl$ 327.85

Benzeneacetic acid, α-(1-hydroxycyclopentyl)-, 2-(dimethylamino)ethyl ester, hydrochloride.

2-(Dimethylamino)ethyl 1-hydroxy-α-phenylcyclopentaneacetate hydrochloride [*5870-29-1*].

» Cyclopentolate Hydrochloride contains not less than 98.0 percent and not more than 100.5 percent of $C_{17}H_{25}NO_3.HCl$, calculated on the dried basis.

Packaging and storage—Preserve in tight containers, and store in a cold place.

Reference standard—*USP Cyclopentolate Hydrochloride Reference Standard*—Dry at 105° for 4 hours before using.

Identification—

A: The infrared absorption spectrum of a potassium bromide dispersion of it exhibits maxima only at the same wavelengths as that of a similar preparation of USP Cyclopentolate Hydrochloride RS.

B: Dissolve about 50 mg of Cyclopentolate Hydrochloride, accurately weighed, in 1 mL of methanol. Dissolve an accurately weighed quantity of USP Cyclopentolate Hydrochloride RS in methanol to obtain a Standard solution having a known concentration of about 50 mg per mL. Apply 20 µL each of the test solution and the Standard solution on a thin-layer chromatographic plate coated with chromatographic silica gel mixture, and subject to ascending chromatography in an unsaturated tank, using cyclohexane-diethylamine (95:5) as the developing solvent. After the solvent front has moved not less than 10 cm, air-dry the plate, and view under short-wavelength ultraviolet light: the R_f value of the principal spot obtained from the test solution corresponds to that obtained from the Standard solution.

C: A solution (1 in 500) responds to the tests for *Chloride* ⟨191⟩.

pH ⟨791⟩: between 4.5 and 5.5, in a solution (1 in 100).

Loss on drying ⟨731⟩—Dry it at 105° for 4 hours: it loses not more than 0.5% of its weight.

Residue on ignition ⟨281⟩: not more than 0.05%.

Assay—Dissolve about 1 g of Cyclopentolate Hydrochloride, accurately weighed, in 80 mL of glacial acetic acid, and add 25 mL of mercuric acetate TS. Titrate with 0.1 *N* perchloric acid VS, using crystal violet TS as the indicator. Perform a blank determination, and make any necessary correction. Each mL of 0.1 *N* perchloric acid is equivalent to 32.79 mg of $C_{17}H_{25}NO_3.HCl$.

Cyclopentolate Hydrochloride Ophthalmic Solution

» Cyclopentolate Hydrochloride Ophthalmic Solution is a sterile, aqueous solution of Cyclopentolate Hydrochloride. It may contain suitable buffers and other additives. It contains not less than 90.0 percent and not more than 105.0 percent of the labeled amount of $C_{17}H_{25}NO_3.HCl$.

Packaging and storage—Preserve in tight containers, and store at controlled room temperature.

Reference standard—*USP Cyclopentolate Hydrochloride Reference Standard*—Dry at 105° for 4 hours before using.

Identification—Place in a 125-mL separator a volume of Ophthalmic Solution, equivalent to about 50 mg of cyclopentolate hydrochloride, and place in a second separator about 50 mg of USP Cyclopentolate Hydrochloride RS dissolved in 5 mL of water. Treat each solution as follows: Add 1 g of potassium carbonate, and extract with two 10-mL portions of ether. Filter the ether extracts through ether-washed filter paper, collect the filtrate in a small beaker, and evaporate to dryness: the residue so obtained responds to *Identification test A* under *Cyclopentolate Hydrochloride*.

Sterility—It meets the requirements under *Sterility Tests* ⟨71⟩.

pH ⟨791⟩: between 3.0 and 5.5.

Assay—

Standard preparation—Dissolve an accurately weighed quantity of USP Cyclopentolate Hydrochloride RS in 0.1 *N* sulfuric acid, and dilute quantitatively and stepwise with the same acid to obtain a solution having a known concentration of about 600 µg per mL.

Assay preparation—Dilute a volume of Cyclopentolate Hydrochloride Ophthalmic Solution quantitatively and stepwise with 0.1 *N* sulfuric acid to obtain a solution having a concentration of about 600 µg per mL.

Procedure—Pipet 10 mL each of the *Standard preparation*, the *Assay preparation*, and 0.1 *N* sulfuric acid to provide a blank, into individual, 60-mL separators. Wash each solution with four 10-mL portions of chloroform (saturated with water), and discard the washings.

Pipet 4 mL each of the washed *Standard preparation*, *Assay preparation*, and blank into separate test tubes. Add 1.0 mL of hydroxylamine hydrochloride solution (7 in 100) and 1.0 mL of sodium hydroxide solution (7 in 50) to each tube, mix, and allow the tubes to stand at room temperature for 10 minutes. Add 1.0 mL of 4 *N* hydrochloric acid and 1.0 mL of a filtered solution of a 2 in 25 solution of ferric chloride in 0.1 *N* hydrochloric acid, mix, and allow the tubes to stand at room temperature for 10 minutes. Concomitantly determine the absorbances of the three solutions in 1-cm cells at the wavelength of maximum absorbance at about 500 nm, with a suitable spectrophotometer, using water in the reference cell. Subtract the absorbance of the blank from the absorbances due to the solutions from the *Assay preparation* and the *Standard preparation*. Calculate the quantity, in µg, of $C_{17}H_{25}NO_3.HCl$ in each mL of the *Assay preparation* taken by the formula:

$$C(A_U/A_S),$$

in which *C* is the concentration, in µg per mL, of USP Cyclopentolate Hydrochloride RS in the *Standard preparation*, and A_U and A_S are the corrected absorbances of the solutions from the *Assay preparation* and the *Standard preparation*, respectively.

Cyclophosphamide

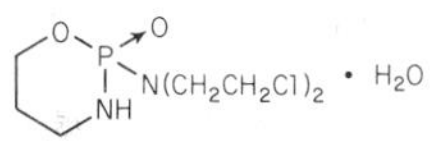

$C_7H_{15}Cl_2N_2O_2P.H_2O$ 279.10

2*H*-1,3,2-Oxazaphosphorin-2-amine, *N,N*-bis(2-chloroethyl)-tetrahydro-, 2-oxide, monohydrate.

2-[Bis(2-chloroethyl)amino]tetrahydro-2*H*-1,3,2-oxazaphosphorine 2-oxide monohydrate [*6055-19-2*].

Anhydrous 261.09 [*50-18-0*].

» Cyclophosphamide contains not less than 97.0 percent and not more than 103.0 percent of $C_7H_{15}Cl_2N_2O_2P$, calculated on the anhydrous basis.

Caution—Great care should be taken in handling Cyclophosphamide, as it is a potent cytotoxic agent.

Packaging and storage—Preserve in tight containers, at a temperature between 2° and 30°.

Reference standard—*USP Cyclophosphamide Reference Standard*—Do not dry; determine the water content titrimetrically when used for quantitative analyses.

Identification—

A: The infrared absorption spectrum of a potassium bromide dispersion of it exhibits maxima only at the same wavelengths as that of a similar preparation of USP Cyclophosphamide RS.

B: The retention time of the major peak in the chromatogram of the *Assay preparation* corresponds to that of the *Standard preparation*, both relative to the internal standard, as obtained in the *Assay*.

pH ⟨791⟩: between 3.9 and 7.1, in a solution (1 in 100), determined 30 minutes after its preparation.

Water, *Method I* ⟨921⟩: between 5.7% and 6.8%.

Heavy metals ⟨231⟩—Dissolve 1.0 g in 25 mL of water, and filter if necessary: the limit is 0.002%.

Assay—

Mobile phase—Prepare a suitable, degassed solution of water and acetonitrile (70:30).

Internal standard solution—Dissolve about 185 mg of ethylparaben in 250 mL of alcohol in a 1000-mL volumetric flask, dilute with water to volume, and mix.

Standard preparation—Transfer an accurately weighed quantity of USP Cyclophosphamide RS, equivalent to about 25 mg of anhydrous cyclophosphamide, to a 50-mL volumetric flask, add about 25 mL of water, and shake to dissolve the Reference Standard. Add 5.0 mL of *Internal standard solution*, dilute with water to volume, and mix to obtain a *Standard preparation* having a known concentration of about 0.5 mg of anhydrous cyclophosphamide per mL.

Assay preparation—Transfer an accurately weighed quantity of Cyclophosphamide, equivalent to about 200 mg of anhydrous cyclophosphamide, to a 200-mL volumetric flask, add about 50 mL of water, shake for about 5 minutes, dilute with water to volume, and mix. Pipet 25 mL of this solution and 5 mL of *Internal standard solution* into a 50-mL volumetric flask, dilute with water to volume, and mix.

Chromatographic system (see *Chromatography* ⟨621⟩)—The liquid chromatograph is equipped with a 195-nm detector and a 3.9-mm × 30-cm column that contains packing L1. The flow rate is about 1.5 mL per minute. Chromatograph six replicate injections of the *Standard preparation*, and record the peak responses as directed under *Procedure:* the relative standard deviation is not more than 2% and the resolution factor between cyclophosphamide and ethylparaben is not less than 2.

Procedure—Separately inject equal volumes (about 25 µL) of the *Standard preparation* and the *Assay preparation* into the chromatograph, record the chromatograms, and measure the responses for the major peaks. The relative retention times are about 0.7 for cyclophosphamide and 1.0 for ethylparaben. Calculate the quantity, in mg, of $C_7H_{15}Cl_2N_2O_2P$ in the Cyclophosphamide taken by the formula:

$$400C(R_U/R_S),$$

in which C is the concentration, in mg per mL, of anhydrous cyclophosphamide in the *Standard preparation*, as determined from the concentration of USP Cyclophosphamide RS corrected for moisture content by a titrimetric water determination, and R_U and R_S are the ratios of the peak responses of cyclophosphamide to those of ethylparaben in the *Assay preparation* and the *Standard preparation*, respectively.

Cyclophosphamide for Injection

» Cyclophosphamide for Injection is a sterile mixture of Cyclophosphamide with a suitable diluent. It contains not less than 90.0 percent and not more than 110.0 percent of the labeled amount of anhydrous cyclophosphamide ($C_7H_{15}Cl_2N_2O_2P$).

Packaging and storage—Preserve in *Containers for Sterile Solids* as described under *Injections* ⟨1⟩. Storage at a temperature not exceeding 25° is recommended. It will withstand brief exposure to temperatures up to 30°, but is to be protected from temperatures above 30°.

Reference standard—*USP Cyclophosphamide Reference Standard*—Do not dry; determine the water content titrimetrically when used for quantitative analyses.

Constituted solution—At the time of use, the constituted solution prepared from Cyclophosphamide for Injection meets the requirements for *Constituted Solutions* under *Injections* ⟨1⟩.

Identification—

A: It responds to the *Thin-layer Chromatographic Identification Test* ⟨201⟩, a solution of it in chloroform, equivalent to 20 mg of cyclophosphamide per mL, filtered if necessary, being used as the test solution. Apply 5 µL of the test solution and Standard solution, use a solvent system consisting of a mixture of chloroform, methanol, and ammonium hydroxide (75:20:5), and visualize the spots by placing the plate in an iodine chamber.

B: The retention time of the major peak in the chromatogram of the *Assay preparation* corresponds to that of the *Standard preparation*, both relative to the internal standard, as obtained in the *Assay*.

Pyrogen—It meets the requirements of the *Pyrogen Test* ⟨151⟩, the test dose being 1 mL per kg, of a solution in water for injection containing the equivalent of 10 mg of anhydrous cyclophosphamide per mL.

pH ⟨791⟩: between 3.0 and 7.5, in a solution prepared as directed in the test for *Completeness of solution*, determined 30 minutes after its preparation.

Other requirements—It meets the requirements for *Sterility Tests* ⟨71⟩, *Uniformity of Dosage Units* ⟨905⟩, and *Labeling* under *Injections* ⟨1⟩.

Assay—

Mobile phase, Internal standard solution, and *Standard preparation*—Prepare as directed in the *Assay* under *Cyclophosphamide*.

Assay preparation—Weigh accurately a portion of Cyclophosphamide for Injection, equivalent to about 200 mg of anhydrous cyclophosphamide, and proceed as directed for *Standard preparation* in the *Assay* under *Cyclophosphamide*.

Chromatographic system—Proceed as directed for *Chromatographic system* in the *Assay* under *Cyclophosphamide*.

Procedure—Proceed as directed for *Procedure* in the *Assay* under *Cyclophosphamide*. Calculate the quantity, in mg, of $C_7H_{15}Cl_2N_2O_2P$ in the portion of Cyclophosphamide for Injection taken by the formula:

$$400C(R_U/R_S),$$

in which the terms are as defined therein.

Cyclophosphamide Tablets

» Cyclophosphamide Tablets contain not less than 90.0 percent and not more than 110.0 percent of the labeled amount of anhydrous cyclophosphamide ($C_7H_{15}Cl_2N_2O_2P$).

Packaging and storage—Preserve in tight containers. Storage at a temperature not exceeding 25° is recommended. Tablets will withstand brief exposure to temperatures up to 30°, but are to be protected from temperatures above 30°.

Reference standard—*USP Cyclophosphamide Reference Standard*—Do not dry; determine the water content titrimetrically when used for quantitative analyses.

Identification—

A: Extract a portion of finely powdered Tablets, equivalent to about 50 mg of cyclophosphamide, with 25 mL of chloroform, filter about 2 mL of the chloroform solution, mix the filtrate with 500 mg of potassium bromide, evaporate the chloroform, carefully removing the last trace of solvent in a small vacuum flask, and use the residue to prepare a potassium bromide dispersion: the infrared absorption spectrum of the potassium bromide dispersion so obtained exhibits maxima, between 6.5 μm and 14 μm, only at the same wavelengths as that of a similar preparation of USP Cyclophosphamide RS.

B: The retention time of the major peak in the chromatogram of the *Assay preparation* corresponds to that of the *Standard preparation*, as obtained in the *Assay*.

Disintegration ⟨701⟩: 30 minutes, determined as directed under *Plain Coated Tablets*.

Uniformity of dosage units ⟨905⟩: meet the requirements.

Procedure for content uniformity—

Perchloric acid solution—Dissolve 23.5 mL of perchloric acid in water, and dilute with water to 1 liter.

4-(p-Nitrobenzyl)pyridine solution—Dissolve 1.5 g of 4-(*p*-nitrobenzyl)pyridine in 200 mL of ethylene glycol.

Sodium hydroxide solution—Dissolve 20 g of sodium hydroxide in 1000 mL of diluted alcohol.

Procedure—Place 1 Tablet in a volumetric flask of suitable size so that the final concentration is about 500 μg per mL. Fill the flask about two-thirds full of water, shake until the tablet is completely disintegrated, dilute with water to volume, and filter, discarding the first 10 mL of the filtrate. Place in separate 27-mm × 170-mm test tubes 2.0 mL of the filtrate, 2.0 mL of water to provide a blank, and 2.0 mL of the Standard solution, prepared by dissolving an accurately weighed quantity of USP Cyclophosphamide RS in water and diluting quantitatively and stepwise with water to obtain a solution having a known concentration of about 500 μg per mL. Treat each tube as follows: Add 0.7 mL of *Perchloric acid solution*, mix, and heat at 95° for 10 minutes. Cool, add 1.0 mL of sodium acetate TS, mix, add 1.6 mL of *4-(p-Nitrobenzyl)pyridine solution*, mix, and heat at 95° for 10 minutes. Cool, add 8.0 mL of *Sodium hydroxide solution*, and mix. Within 4 minutes, determine the absorbances of the solutions in 1-cm cells at the wavelength of maximum absorbance at 560 nm, with a suitable spectrophotometer, against the blank. Calculate the quantity, in mg, of $C_7H_{15}Cl_2N_2O_2P$ in the Tablet by the formula:

$$(T/500)C(A_U/A_S) \times 0.935,$$

in which T is the labeled quantity, in mg, of anhydrous cyclophosphamide in the Tablet, C is the concentration, in μg per mL, of USP Cyclophosphamide RS in the Standard solution, A_U and A_S are the absorbances of the solution from the Tablet and the Standard solution, respectively, and 0.935 is the ratio of the molecular weight of anhydrous cyclophosphamide to that of the monohydrate form.

Assay—

Mobile phase, Internal standard solution, and *Standard preparation*—Prepare as directed in the *Assay* under *Cyclophosphamide*.

Assay preparation—Transfer not less than 10 Cyclophosphamide Tablets to a volumetric flask of suitable size so that the final concentration is about 1 mg of anhydrous cyclophosphamide per mL. Fill about half full with water, shake for 30 minutes, dilute with water to volume and mix. Filter through fast, fluted filter paper, discarding the first 40 to 50 mL of the filtrate. Pipet 25 mL of the filtrate and 5 mL of *Internal standard solution* into a 50-mL volumetric flask, dilute with water to volume, and mix.

Chromatographic system—Proceed as directed for *Chromatographic system* in the *Assay* under *Cyclophosphamide*.

Procedure—Proceed as directed for *Procedure* in the *Assay* under *Cyclophosphamide*. Calculate the quantity, in mg, of $C_7H_{15}Cl_2N_2O_2P$ per Tablet taken by the formula:

$$(2CV/N)(R_U/R_S),$$

in which C is the concentration, in mg per mL, of anhydrous cyclophosphamide in the *Standard preparation*, as determined from the concentration of USP Cyclophosphamide RS corrected for moisture by a titrimetric water determination, V is the volume, in mL, of the volumetric flask to which the N Tablets were transferred, N is the number of Tablets taken, and R_U and R_S are the ratios of the peak responses of cyclophosphamide to those of the internal standard in the *Assay preparation* and the *Standard preparation*, respectively.

Cyclopropane

H2
C
H2C—CH2

C_3H_6 42.08
Cyclopropane.
Cyclopropane [*75-19-4*].

» Cyclopropane contains not less than 99.0 percent, by volume, of C_3H_6.

Caution—Cyclopropane is highly flammable. Do not use where it may be ignited.

Packaging and storage—Preserve in cylinders. [NOTE—Maintain cylinders of Cyclopropane at 25 ± 2° for not less than 6 hours prior to withdrawing specimens for the tests and assay, and correct the results to 25° and 760 mm of mercury.]

Labeling—The label bears a warning that cyclopropane is highly flammable and is not to be used where it may be ignited.

Acidity or alkalinity—Add 0.3 mL of methyl red TS and 0.3 mL of bromothymol blue TS to 400 mL of boiling water, and boil the solution for 5 minutes. Pour 100 mL of the boiling solution into each of three color-comparison tubes marked *A*, *B*, and *C*, respectively. To tube *B* add 0.20 mL of 0.012 *N* hydrochloric acid, and to tube *C* add 0.40 mL of 0.012 *N* hydrochloric acid. Insert the stopper in each of the tubes, and cool them to room temperature. Pass 2000 mL of Cyclopropane through the solution in tube *B* at a rate requiring about 30 minutes for the passage of the gas: the color of the solution in tube *B* is no deeper orange-red than that in tube *C* and no deeper yellow-green than that in tube *A*.

NOTE—The various detector tubes called for in the respective tests are listed under *Reagents* in the section, *Reagents, Indicators, and Solutions*.

Carbon dioxide—Place the container so that when its valve is opened, the gaseous phase can be sampled. Connect one end of a carbon dioxide detector tube to the container valve, and the other end to a gas flow meter. Pass 1000 mL of the Cyclopropane through the tube at a suitable rate: the indicator change corresponds to not more than 0.03%.

Halogens—Provide a 500-mL flask with a tightly fitting two-hole stopper. Through one opening pass a delivery tube bent at right angles and extending just beyond the lower surface of the stopper. Through the other opening insert a capillary tube bent at right angles and having a bore of 1 ± 0.2 mm, in the same manner. Place in a 50-mL cylinder, having an internal diameter of 2 ± 0.25 cm, 40 mL of a solution containing 850 mg of sodium carbonate in 1000 mL of water. Provide the cylinder with a two-hole stopper, and through one opening pass a right-angle delivery tube, having a bore of 3 ± 0.5 mm, to within 2 mm of the bottom of the cylinder. The end of the delivery tube that extends out of the cylinder is provided with an enlargement 8 ± 0.5 cm long having an internal diameter of 2 ± 0.25 cm. Through the other opening in the stopper pass another right-angle delivery tube, having it extend just below the surface of the stopper. Collect 500 mL of Cyclopropane in the flask. By means of hydrostatic pressure, applied through the delivery tube, force the gas through the capillary tube, the water used being previously saturated with Cyclopropane. Ignite the gas, place the enlarged end of the delivery tube, connected with the cylinder, around the flame, extending the flame one-third of the way into the enlargement. Apply suction to the shorter delivery tube connected with the cylinder, thus drawing the spent gases through the sodium carbonate solution, the period of ignition of the 500 mL of Cyclopropane requiring approximately 30 minutes. Make any necessary correction for the amount of halogen in the volume of air used for the ignition of the gas. Transfer the sodium carbonate solution to a 500-mL volumetric flask, and rinse the cylinder thoroughly, collecting the rinsings in the flask. Dilute the solution with water to volume, and mix. To a 50-mL aliquot add sufficient nitric acid to make it acid to litmus paper, and then add 1 mL of acid in excess. Prepare a blank containing 0.50 mL of 0.0012 *N* hydrochloric acid and 4 mL of the sodium carbonate solution in 46 mL of water, acidify to litmus with nitric acid, then add 1 mL of acid in excess and 1 mL of silver nitrate TS to each solution: after 5 minutes any opalescence in the solution representing the Cyclopropane does not exceed that in the blank (0.02% as chloride).

Propylene, allene, and other unsaturated hydrocarbons—Place the container so that when its valve is opened, the gaseous phase can be sampled. Connect one end of an olefin detector tube to the container valve, and the other end to a gas flow meter. Pass the Cyclopropane through the detector tube at a suitable rate: the color of the indicating layer of the tube contents matches the color standard after the passage of not less than 400 mL of Cyclopropane (0.9% as propylene).

Assay—Place the container so that when its valve is opened, the gaseous phase can be sampled. Withdraw 100 mL of Cyclopropane, accurately measured in a gas buret previously filled with mercury and equipped with a leveling bulb at the lower end. Connect one arm of the buret stopcock to a pipet that previously has been filled with sulfuric acid. By appropriate manipulation of the stopcock and the leveling bulb, transfer the gas between the pipet and the buret, bringing about sufficient contact of the gas with the acid to reduce the volume of unabsorbed gas to a minimum as measured in the buret. Not more than 1.0 mL of gas remains.

Cycloserine

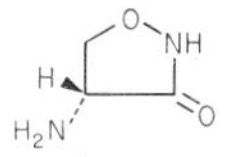

$C_3H_6N_2O_2$ 102.09

3-Isoxazolidinone, 4-amino-, (*R*)-.

(+)-4-Amino-3-isoxazolidinone [*68-41-7*].

» Cycloserine has a potency of not less than 900 µg of $C_3H_6N_2O_2$ per mg.

Packaging and storage—Preserve in tight containers.

Reference standard—*USP Cycloserine Reference Standard*—Dry in vacuum at a pressure not exceeding 5 mm of mercury at 60° for 3 hours before using.

Identification—Dissolve about 1 mg in 10 mL of 0.1 *N* sodium hydroxide. To 1 mL of the resulting solution add 3 mL of 1 *N* acetic acid and 1 mL of a mixture, prepared 1 hour before use, of equal parts of sodium nitroprusside solution (1 in 25) and 4 *N* sodium hydroxide: a blue color gradually develops.

Crystallinity ⟨695⟩: meets the requirements.

pH ⟨791⟩: between 5.5 and 6.5, in a solution (1 in 10).

Loss on drying ⟨731⟩—Dry about 100 mg in a capillary-stoppered bottle in vacuum at 60 for 3 hours: it loses not more than 1.0% of its weight.

Residue on ignition ⟨281⟩: not more than 0.5%, the charred residue being moistened with 2 mL of nitric acid and 5 drops of sulfuric acid.

Assay—Proceed with Cycloserine as directed under *Antibiotics—Microbial Assays* ⟨81⟩.

Cycloserine Capsules

» Cycloserine Capsules contain not less than 90.0 percent and not more than 120.0 percent of the labeled amount of $C_3H_6N_2O_2$.

Packaging and storage—Preserve in tight containers.

Reference standard—*USP Cycloserine Reference Standard*—Dry in vacuum at a pressure not exceeding 5 mm of mercury at 60° for 3 hours before using.

Identification—Shake a quantity of the contents of Capsules, equivalent to about 10 mg of cycloserine, with 100 mL of 0.1 *N* sodium hydroxide, and filter: 1 mL of the filtrate so obtained responds to the *Identification test* under *Cycloserine*.

Dissolution ⟨711⟩—

Medium: water; 900 mL.

Apparatus 1: 100 rpm.

Time: 45 minutes.

Procedure—Determine the amount of $C_3H_6N_2O_2$ dissolved from ultraviolet absorbances at the wavelength of maximum absorbance at about 219 nm of filtered portions of the solution under test, suitably diluted with *Dissolution Medium*, if necessary, in comparison with a Standard solution having a known concentration of USP Cycloserine RS in the same medium.

Tolerances—Not less than 75% (*Q*) of the labeled amount of $C_3H_6N_2O_2$ is dissolved in 45 minutes.

Uniformity of dosage units ⟨905⟩: meet the requirements.

Loss on drying ⟨731⟩—Dry about 100 mg of the contents of Capsules in a capillary-stoppered bottle in vacuum at 60° for 3 hours: it loses not more than 1.0% of its weight.

Assay—Place not less than 5 Cycloserine Capsules in a high-speed glass blender jar containing an accurately measured volume of water to obtain a suitable stock test solution, and blend for 4 ± 1 minutes. Proceed as directed under *Antibiotics—Microbial Assays* ⟨81⟩, using an accurately measured volume of this stock test solution diluted quantitatively with water to obtain a *Test Dilution* having a concentration assumed to be equal to the median dose level of the Standard.

Cyclosporine

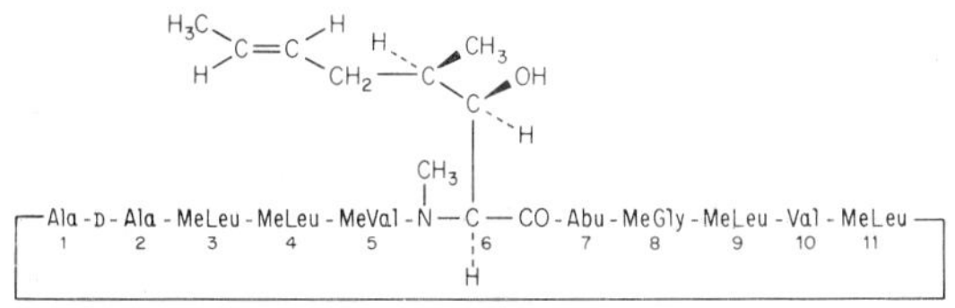

$C_{62}H_{111}N_{11}O_{12}$ 1202.63

Cyclo[[(*E*)-(2*S*,3*R*,4*R*)-3-hydroxy-4-methyl-2-(methylamino)-6-octenoyl]-L-2-aminobutyryl-*N*-methylglycyl-*N*-methyl-L-leucyl-L-valyl-*N*-methyl-L-leucyl-L-alanyl-D-alanyl-*N*-methyl-L-leucyl-*N*-methyl-L-leucyl-*N*-methyl-L-valyl].

[*R*-[*R**,*R**-(*E*)]]-Cyclic(L-alanyl-D-alanyl-*N*-methyl-L-leucyl-*N*-methyl-L-leucyl-*N*-methyl-L-valyl-3-hydroxy-*N*,4-dimethyl-L-2-amino-6-octenoyl-L-α-aminobutyryl-*N*-methylglycyl-*N*-methyl-L-leucyl-L-valyl-*N*-methyl-L-leucyl) [*59865-13-3*].

» Cyclosporine contains not less than 975 μg and not more than 1020 μg of $C_{62}H_{111}N_{11}O_{12}$ per mg, calculated on the dried basis.

Packaging and storage—Preserve in tight, light-resistant containers.

Reference standard—*USP Cyclosporine Reference Standard.*

Identification—The chromatogram of the *Assay preparation* obtained as directed in the *Assay* exhibits a major peak for cyclosporine, the retention time of which corresponds to that exhibited in the chromatogram of the *Standard preparation* obtained as directed in the *Assay.*

Loss on drying ⟨731⟩—Dry about 100 mg, accurately weighed, in a capillary-stoppered bottle in vacuum at a pressure not exceeding 5 mm of mercury at 60° for 3 hours: it loses not more than 2.0% of its weight.

Heavy metals, *Method II* ⟨231⟩: not less than 0.002%.

Assay—

Mobile phase—Prepare a suitable filtered and degassed mixture of acetonitrile, water, methanol, and phosphoric acid (900:525:75:0.075), making adjustments if necessary (see *System Suitability* under *Chromatography* ⟨621⟩).

Standard preparation—Dissolve an accurately weighed quantity of USP Cyclosporine RS in dehydrated alcohol to obtain a solution having a known concentration of about 0.5 mg per mL. Shake for 15 minutes to ensure dissolution. Use this solution promptly after preparation.

Assay preparation—Using an accurately weighed quantity of Cyclosporine, prepare as directed under *Standard preparation.* Use this solution promptly after preparation.

Chromatographic system (see *Chromatography* ⟨621⟩)—The liquid chromatograph is equipped with a 210-nm detector, a 0.25-mm × 1-m stainless steel tube connected to a 4.6-mm × 5-cm guard column that contains 40- to 60-μm packing L7, and a 4-mm × 25-cm analytical column that contains 3- to 10-μm packing L7. The tube and the columns are maintained at 70°. The flow rate is about 2 mL per minute. Chromatograph the *Standard preparation,* and record the peak responses as directed under *Procedure:* the capacity factor, k', is not less than 3 and not more than 10, the column efficiency determined from the analyte peak is not less than 1500 theoretical plates, the tailing factor for the analyte peak is not more than 1.5, and the relative standard deviation for replicate injections is not more than 1.5%.

Procedure—Separately inject equal volumes (about 20 μL) of the *Standard preparation* and the *Assay preparation* into the chromatograph, record the chromatograms, and measure the responses for the major peaks. Calculate the quantity, in μg, of $C_{62}H_{111}N_{11}O_{12}$ in each mg of Cyclosporine taken by the formula:

$$(CP/U)(r_U/r_S),$$

in which C is the concentration, in mg per mL, of USP Cyclosporine RS in the *Standard preparation,* P is the purity, in μg per mg, of USP Cyclosporine RS, U is the concentration, in mg per mL, of specimen in the *Assay preparation,* and r_U and r_S are the peak responses obtained from the *Assay preparation* and the *Standard preparation,* respectively.

Cyclosporine Concentrate for Injection

» Cyclosporine Concentrate for Injection is a sterile solution of Cyclosporine in a mixture of alcohol and a suitable vegetable oil. It contains not less than 90.0 percent and not more than 110.0 percent of the labeled amount of cyclosporine ($C_{62}H_{111}N_{11}O_{12}$).

Packaging and storage—Preserve in single-dose or in multiple-dose containers.

Labeling—Label it to indicate that it is to be diluted with a suitable parenteral vehicle prior to intravenous infusion.

Reference standards—*USP Cyclosporine Reference Standard. USP Endotoxin Reference Standard.*

Identification—

A: Prepare a solution of it in methanol containing about 0.5 mg of cyclosporine per mL (test solution). Prepare a Standard solution containing 0.5 mg per mL of USP Cyclosporine RS in methanol. On a suitable thin-layer chromatographic plate (see *Chromatography* ⟨621⟩), coated with a 0.25-mm layer of chromatographic silica gel mixture, separately apply 10-μL portions of the test solution and the Standard solution. Allow the spots to dry in a current of air, place the plate in a suitable chromatographic chamber, and develop the chromatogram, using ethyl ether as the developing solvent, until the solvent front has moved about three-fourths of the length of the plate. Remove the plate from the chamber, mark the solvent front, and allow it to dry. Place the plate in a second chromatographic chamber, and develop the chromatogram in a solvent system consisting of a mixture of ethyl acetate, methyl ethyl ketone, water, and formic acid (60:40:2:1) until the solvent front has moved about three-fourths of the length of the plate. Remove the plate from the chamber, and allow it to dry. Spray the plate with a freshly prepared mixture of 5 mL of *Solution A* (340 mg of bismuth subnitrate dissolved in 20 mL of 20% acetic acid), 5 mL of *Solution B* (8 g of potassium iodide dissolved in 20 mL of water), 20 mL of glacial acetic acid, and water to make 100 mL. Immediately again spray the plate with hydrogen peroxide TS. Cyclosporine appears as a brown spot having an R_f value of about 0.45 on the chromatograms: the R_f value of the principal spot obtained from the test solution corresponds to that obtained from the Standard solution. [NOTE—Disregard any spots at the origin.]

B: The chromatogram of the *Assay preparation* obtained as directed in the *Assay* exhibits a major peak for cyclosporine, the retention time of which corresponds to that exhibited in the chromatogram of the *Standard preparation* obtained as directed in the *Assay.*

Bacterial endotoxins—Proceed as directed under *Bacterial Endotoxin Test* ⟨85⟩, USP Endotoxin RS being used, preparing the test specimen as follows: Make a 1:10 dilution of the Concentrate with Water for Injection. Add 0.1 mL of the resulting suspension and 0.1 mL of appropriately constituted LAL reagent to a suitable pyrogen-free test tube, and vortex for about 5 seconds: the article under test contains not more than 0.84 USP Endotoxin Unit per mg of cyclosporine.

Sterility—It meets the requirements under *Sterility Tests* ⟨71⟩.

Alcohol content—

Internal standard solution—Mix 3 mL of *n*-propyl alcohol and 50 mL of butyl alcohol.

Standard stock solution—Transfer about 1.6 g of dehydrated alcohol, accurately weighed, to a 25-mL volumetric flask, dilute with butyl alcohol to volume, and mix.

Standard preparation—Transfer 5.0 mL of *Standard stock solution* and 6.0 mL of *Internal standard solution* to a 25-mL volumetric flask, dilute with butyl alcohol to volume, and mix.

Test preparation—Transfer an accurately weighed portion of Cyclosporine Concentrate for Injection, equivalent to about 320 mg of C_2H_5OH, to a 25-mL volumetric flask, add 6.0 mL of *Internal standard solution,* dilute with butyl alcohol to volume, and mix.

Chromatographic system—The gas chromatograph is equipped with a flame-ionization detector and contains a 2-mm × 2-m glass column packed with support S3. The injection port is maintained at a temperature of about 280°, the detector is maintained at about 290°, and the column is maintained at 145° for 8 minutes and is programmed thereafter to rise to 270° at a rate of 32° per minute. Nitrogen is used as the carrier gas, flowing at a rate of about 35 mL per minute. [NOTE—Make adjustments if necessary to obtain satisfactory chromatography.]

System suitability—Chromatograph the *Standard preparation,* and record the peak responses as directed under *Procedure:* the relative standard deviation for replicate injections is not greater than 2.0%.

Procedure—[NOTE—Use peak areas where peak responses are indicated.] Inject separate suitable portions (about 1 μL) of the *Standard preparation* and the *Test preparation* into the chromatograph, record the chromatograms, and measure the responses for the major peaks. The elution order is: alcohol, *n*-

propyl alcohol, and butyl alcohol. Calculate the quantity, in mg, of C_2H_5OH in the portion of Cyclosporine Concentrate for Injection taken by the formula:

$$25C(R_U/R_S),$$

in which C is the concentration, in mg per mL, of C_2H_5OH in the *Standard preparation*, and R_U and R_S are the peak response ratios of the alcohol peak to the *n*-propyl alcohol internal standard peak obtained from the *Test preparation* and the *Standard preparation*, respectively: it contains between 80.0% and 120.0% of the labeled amount of C_2H_5OH.

Assay—

Mobile phase—Prepare a suitable filtered and degassed mixture of acetonitrile, water, methanol, and phosphoric acid (550: 400:50:0.5), making adjustments if necessary (see *System Suitability* under *Chromatography* ⟨621⟩).

Standard preparation—Dissolve an accurately weighed quantity of USP Cyclosporine RS in methanol to obtain a solution having a known concentration of about 0.5 mg per mL. Use this solution promptly after preparation.

Assay preparation 1 (where it is represented as being in a single-dose container)—Using a suitable hypodermic needle and syringe, withdraw all of the withdrawable contents from 1 container of Cyclosporine Concentrate for Injection, and dilute quantitatively with methanol to obtain a solution containing about 0.5 mg of cyclosporine per mL. Use this solution promptly after preparation.

Assay preparation 2 (where this label states the quantity of cyclosporine in a given volume)—Dilute an accurately measured volume of Cyclosporine Concentrate for Injection quantitatively with methanol to obtain a solution containing about 0.5 mg of cyclosporine per mL. Use this solution promptly after preparation.

Chromatographic system (see *Chromatography* ⟨621⟩)—The liquid chromatograph is equipped with a 210-nm detector and a 4.6-mm × 25-cm analytical column that contains packing L16. The column is maintained at 70°. The flow rate is about 1 mL per minute. Chromatograph the *Standard preparation*, and record the peak responses as directed under *Procedure:* the capacity factor, k', is not less than 3 and not more than 10, the column efficiency determined from the analyte peak is not less than 700 theoretical plates, the tailing factor for the analyte peak is not more than 1.5, and the relative standard deviation for replicate injections is not more than 1.5%.

Procedure—Separately inject equal volumes (about 20 µL) of the *Standard preparation* and the *Assay preparation* into the chromatograph, record the chromatograms, and measure the responses for the major peaks. Calculate the quantity, in mg, of cyclosporine ($C_{26}H_{111}N_{11}O_{12}$) withdrawn from the container or in each mL of Cyclosporine Concentrate for Injection taken by the same formula:

$$(L/D)(CP/1000)(r_U/r_S),$$

in which L is the labeled quantity, in mg, of cyclosporine in the container or in each mL of Cyclosporine Concentrate for Injection, D is the concentration, in mg of cyclosporine per mL, of *Assay preparation 1* or *Assay preparation 2* based on the labeled quantity in the container or in the volume of Cyclosporine Concentrate for Injection taken and the extent of dilution, respectively, C is the concentration, in mg per mL, of USP Cyclosporine RS in the *Standard preparation*, P is the purity, in µg per mg, of USP Cyclosporine RS, and r_U and r_S are the peak responses obtained from *Assay preparation 1* or *Assay preparation 2* and the *Standard preparation*, respectively.

Cyclosporine Oral Solution

» Cyclosporine Oral Solution is a solution of Cyclosporine in a mixture of alcohol and a suitable vegetable oil. It contains not less than 90.0 percent and not more than 110.0 percent of the labeled amount of cyclosporine ($C_{62}H_{111}N_{11}O_{12}$).

Packaging and storage—Preserve in tight containers.

Reference standard—*USP Cyclosporine Reference Standard.*

Identification—

A: Using a solution of it in a mixture of methanol and chloroform (4:1) containing about 1 mg of cyclosporine per mL (test solution) and a Standard solution containing 1 mg of USP Cyclosporine RS in the same solvent mixture, proceed as directed in *Identification test A* under *Cyclosporine Concentrate for Injection*, beginning with "On a suitable thin-layer chromatographic plate": the Oral Solution meets the requirements of the test.

B: The chromatogram of the *Assay preparation* obtained as directed in the *Assay* exhibits a major peak for cyclosporine, the retention time of which corresponds to that exhibited in the chromatogram of the *Standard preparation* obtained as directed in the *Assay*.

Alcohol content—

Internal standard solution, Chromatographic system, and *System suitability*—Proceed as directed in the test for *Alcohol content* under *Cyclosporine Concentrate for Injection.*

Standard stock solution—Transfer about 2.5 g of dehydrated alcohol, accurately weighed, to a 50-mL volumetric flask, dilute with butyl alcohol to volume, and mix.

Standard preparation—Transfer 5.0 mL of *Standard stock solution* and 6.0 mL of *Internal standard solution* to a 25-mL volumetric flask, dilute with butyl alcohol to volume, and mix.

Test preparation—Transfer an accurately weighed portion of Cyclosporine Oral Solution, equivalent to about 250 mg of C_2H_5OH, to a 25-mL volumetric flask, add 6.0 mL of *Internal standard solution*, dilute with butyl alcohol to volume, and mix.

Procedure—Proceed as directed for *Procedure* in the test for *Alcohol content* under *Cyclosporine Concentrate for Injection.* Calculate the quantity, in mg, of C_2H_5OH in the portion of Cyclosporine Oral Solution taken by the formula:

$$25C(R_U/R_S).$$

It contains between 80.0% and 120.0% of the labeled amount of C_2H_5OH.

Assay—

Mobile phase—Prepare as directed in the *Assay* under *Cyclosporine Concentrate for Injection.*

Solvent mixture—Prepare a mixture of methanol and chloroform (4:1).

Standard preparation—Dissolve an accurately weighed quantity of USP Cyclosporine RS in *Solvent mixture* to obtain a solution having a known concentration of about 1 mg per mL. Use this solution promptly after preparation.

Assay preparation—Dilute an accurately measured volume of Cyclosporine Oral Solution quantitatively with *Solvent mixture* to obtain a solution containing about 1 mg of cyclosporine per mL. Use this solution promptly after preparation.

Chromatographic system—Proceed as directed for *Chromatographic system* in the *Assay* under *Cyclosporine Concentrate for Injection*, except to maintain the column at 50°, instead of at 70°.

Procedure—Separately inject equal volumes (about 10 µL) of the *Standard preparation* and the *Assay preparation* into the chromatograph, record the chromatograms, and measure the responses for the major peaks. Calculate the quantity, in mg, of $C_{62}H_{111}N_{11}O_{12}$ in each mL of the Oral Solution taken by the formula:

$$(L/D)(CP/1000)(r_U/r_S),$$

in which L is the labeled quantity, in mg, of cyclosporine in each mL of Oral Solution taken, D is the concentration, in mg of cyclosporine per mL, of the *Assay preparation*, based on the labeled quantity of cyclosporine in the volume of Oral Solution taken and the extent of dilution, C is the concentration, in mg per mL, of USP Cyclosporine RS in the *Standard preparation*, P is the purity, in µg per mg, of USP Cyclosporine RS, and r_U and r_S are the peak responses obtained from the *Assay preparation* and the *Standard preparation*, respectively.

Cyproheptadine Hydrochloride

· HCl · $1.5H_2O$

$C_{21}H_{21}N.HCl.1½H_2O$ 350.89

Piperidine, 4-(5*H*-dibenzo[*a,d*]cyclohepten-5-ylidene)-1-methyl-, hydrochloride, sesquihydrate.

4-(5*H*-Dibenzo[*a,d*]cyclohepten-5-ylidene)-1-methylpiperidine hydrochloride sesquihydrate [*41354-29-4*].

Anhydrous 323.86 [*969-33-5*].

» Cyproheptadine Hydrochloride, previously dried, contains not less than 98.5 percent and not more than 100.5 percent of $C_{21}H_{21}N.HCl$.

Packaging and storage—Preserve in well-closed containers.

Reference standard—*USP Cyproheptadine Hydrochloride Reference Standard*—Dry at a pressure between 1 mm and 5 mm of mercury at 100° to constant weight before using.

Identification—

A: The infrared absorption spectrum of a mineral oil dispersion of it, previously dried, exhibits maxima only at the same wavelengths as that of a similar preparation of USP Cyproheptadine Hydrochloride RS.

B: The ultraviolet absorption spectrum of a 1 in 62,500 solution in alcohol exhibits maxima and minima at the same wavelengths as that of a similar solution of USP Cyproheptadine Hydrochloride RS, concomitantly measured, and the respective absorptivities, calculated on the dried basis, at the wavelength of maximum absorbance at about 286 nm do not differ by more than 3.0%.

C: Dissolve 100 mg in 10 mL of methanol. Place 1 drop of the solution on a filter paper, dry, and view under short-wavelength ultraviolet light: a bright blue fluorescence is observed.

Acidity—Dissolve 1.0 g in 25 mL of water, add methyl red TS, and titrate with 0.10 *N* sodium hydroxide: not more than 0.15 mL is required (0.05% as HCl).

Loss on drying ⟨731⟩—Dry it at a pressure between 1 mm and 5 mm of mercury at 100° to constant weight: it loses between 7.0% and 9.0% of its weight.

Residue on ignition ⟨281⟩: not more than 0.1%.

Heavy metals, *Method II* ⟨231⟩: 0.003%.

Assay—Dissolve about 650 mg of Cyproheptadine Hydrochloride, previously dried and accurately weighed, in 50 mL of glacial acetic acid, heating to effect solution. Cool, add 10 mL of mercuric acetate TS, 0.5 mL of acetic anhydride, and 1 drop of crystal violet TS, and titrate with 0.1 *N* perchloric acid VS to a green end-point. Perform a blank determination, and make any necessary correction. Each mL of 0.1 *N* perchloric acid is equivalent to 32.39 mg of $C_{21}H_{21}N.HCl$.

Cyproheptadine Hydrochloride Syrup

» Cyproheptadine Hydrochloride Syrup contains not less than 90.0 percent and not more than 110.0 percent of the labeled amount of $C_{21}H_{21}N.HCl$.

Packaging and storage—Preserve in tight containers.

Reference standard—*USP Cyproheptadine Hydrochloride Reference Standard*—Dry at a pressure between 1 mm and 5 mm of mercury at 100° to constant weight before using.

Identification—Place about 50 mL of Syrup in a separator, add 25 mL of sodium bicarbonate solution (2 in 100), and extract with three 15-mL portions of isooctane. Wash the combined isooctane extracts with 15 mL of sodium bicarbonate solution (2 in 100), and discard the washing. Evaporate the isooctane solution on a steam bath to dryness, and dissolve the residue in 1 mL of carbon disulfide, filtering if necessary. Determine the infrared absorption spectrum as directed under *Identification—Organic Nitrogenous Bases* ⟨181⟩, obtaining the spectrum of USP Cyproheptadine Hydrochloride RS as directed: the Syrup meets the requirements of the test.

pH ⟨791⟩: between 3.5 and 4.5.

Assay—

Standard preparation—Transfer about 10 mg of USP Cyproheptadine Hydrochloride RS, accurately weighed, to a 500-mL volumetric flask, dissolve in 1 mL of methanol, dilute with 0.1 *N* hydrochloric acid to volume, and mix to obtain a *Standard preparation* having a known concentration of about 20 µg of anhydrous cyproheptadine hydrochloride per mL.

Assay preparation—Transfer an accurately measured volume of Cyproheptadine Hydrochloride Syrup, equivalent to about 2 mg of cyproheptadine hydrochloride, to a 125-mL separator, add 20 mL of sodium bicarbonate solution (1 in 100), and extract with two 25-mL portions of isooctane. Wash the combined isooctane extracts with 15 mL of sodium bicarbonate solution (1 in 100), and discard the washing. Extract the isooctane solution with one 50-mL portion and one 25-mL portion of 0.1 *N* hydrochloric acid, collecting the aqueous acid extract in a 100-mL volumetric flask. Dilute with 0.1 *N* hydrochloric acid to volume, and mix. Filter a portion of the solution through dry filter paper, discarding the first 20 mL of the filtrate.

Procedure—Concomitantly determine the absorbances of the *Standard preparation* and the *Assay preparation* in 1-cm cells at the wavelength of maximum absorbance at about 286 nm, with a suitable spectrophotometer, using 0.1 *N* hydrochloric acid as the blank. Calculate the quantity, in mg, of $C_{21}H_{21}N.HCl$ in each mL of the Syrup taken by the formula:

$$(0.1C/V)(A_U/A_S),$$

in which C is the concentration, in µg per mL, of USP Cyproheptadine Hydrochloride RS in the *Standard preparation*, V is the volume, in mL, of Syrup taken, and A_U and A_S are the absorbances of the *Assay preparation* and the *Standard preparation*, respectively.

Cyproheptadine Hydrochloride Tablets

» Cyproheptadine Hydrochloride Tablets contain not less than 90.0 percent and not more than 110.0 percent of the labeled amount of $C_{21}H_{21}N.HCl$.

Packaging and storage—Preserve in well-closed containers.

Reference standard—*USP Cyproheptadine Hydrochloride Reference Standard*—Dry at a pressure between 1 mm and 5 mm of mercury at 100° to constant weight before using.

Identification—Tablets meet the requirements under *Identification—Organic Nitrogenous Bases* ⟨181⟩.

Dissolution ⟨711⟩—

Medium: 0.1 *N* hydrochloric acid; 900 mL.

Apparatus 2: 50 rpm.

Time: 30 minutes.

Procedure—Determine the amount of $C_{21}H_{21}N.HCl$ dissolved from ultraviolet absorbances at the wavelength of maximum absorbance at about 285 nm of filtered portions of the solution under test, suitably diluted with *Dissolution Medium*, if necessary, in comparison with a Standard solution having a known concentration of USP Cyproheptadine Hydrochloride RS in the same medium.

Tolerances—Not less than 80% (Q) of the labeled amount of $C_{21}H_{21}N.HCl$ is dissolved in 30 minutes.

Uniformity of dosage units ⟨905⟩: meet the requirements.

Assay—

Mixed solvent—Mix 10 mL of methanol, 2 drops of hydrochloric acid, and sufficient chloroform (saturated with water) to make 100 mL.

Standard preparation—Dissolve a suitable quantity of USP Cyproheptadine Hydrochloride RS, accurately weighed, in the

Mixed solvent, and prepare, by quantitative and stepwise dilution, a solution in *Mixed solvent* containing about 20 µg per mL.

Chromatographic column—Proceed as directed under *Column Partition Chromatography* ⟨621⟩, packing a chromatographic tube with two segments of packing material. The lower segment is a mixture of 2 g of *Solid Support* and 1 mL of sulfamic acid solution (1 in 20), and the upper segment is a mixture prepared as directed under *Assay preparation.*

Assay preparation—Weigh and finely powder not less than 20 Cyproheptadine Hydrochloride Tablets. Transfer an accurately weighed portion of the powder, equivalent to about 4 mg of cyproheptadine hydrochloride, to a 100-mL beaker, add 1 mL of spectrophotometric grade dimethyl sulfoxide, and mix to wet the powder evenly. Add 1 mL of sulfamic acid solution (1 in 20), mix, add 3 g of *Solid Support*, mix as directed under *Chromatographic column*, and transfer to the column.

Procedure—Wash the column with 100 mL of ether (saturated with water), and discard the washing. Place under the column, as a receiver, a 200-mL volumetric flask containing 20 mL of methanol and 4 drops of hydrochloric acid. Pass through the column 150 mL of chloroform (saturated with water). Mix the eluate, allow to cool to room temperature, dilute with chloroform (saturated with water) to volume, and mix. Concomitantly determine the absorbances of this solution and of the *Standard preparation* in 1-cm cells at the wavelength of maximum absorbance at about 286 nm, with a suitable spectrophotometer, using the *Mixed solvent* as the blank. Calculate the quantity, in mg, of $C_{21}H_{21}N.HCl$ in the portion of Tablets taken by the formula:

$$0.2C(A_U/A_S),$$

in which C is the concentration, in µg per mL, of USP Cyproheptadine Hydrochloride RS in the *Standard preparation*, and A_U and A_S are the absorbances of the solution from the *Assay preparation* and the *Standard preparation*, respectively.

Cysteine Hydrochloride

$$HSCH_2-\underset{NH_2}{\overset{H}{C}}-COOH \cdot HCl \cdot H_2O$$

$C_3H_7NO_2S.HCl.H_2O$ 175.63

L-Cysteine hydrochloride monohydrate.

L-Cysteine hydrochloride monohydrate [*7048-04-6*].

Anhydrous 157.61 [*52-89-1*].

» Cysteine Hydrochloride contains not less than 98.5 percent and not more than 101.5 percent of $C_3H_7NO_2S.HCl$, as L-cysteine hydrochloride, calculated on the dried basis.

Packaging and storage—Preserve in well-closed containers.

Reference standard—*USP L-Cysteine Hydrochloride Reference Standard*—Dry at a pressure of 5 mm of mercury for 24 hours before using.

Identification—The infrared absorption spectrum of a potassium bromide dispersion of it, previously dried, exhibits maxima only at the same wavelengths as that of a similar preparation of USP L-Cysteine Hydrochloride RS.

Specific rotation ⟨781⟩: between +5.7° and +6.8°, calculated on the dried basis, determined in a solution in 6 *N* hydrochloric acid containing 800 mg in each 10 mL.

Loss on drying ⟨731⟩—Dry it at a pressure of 5 mm of mercury for 24 hours: it loses between 8.0% and 12.0% of its weight.

Residue on ignition ⟨281⟩: not more than 0.4%.

Sulfate ⟨221⟩—A solution containing 0.33 g shows no more sulfate than corresponds to 0.10 mL of 0.020 *N* sulfuric acid (0.03%).

Arsenic ⟨211⟩: 1.5 ppm.

Iron ⟨241⟩: 0.003%.

Heavy metals, *Method I* ⟨231⟩: 0.0015%.

Chloride content ⟨221⟩—Dissolve about 350 mg, accurately weighed, in 10 mL of dilute nitric acid (1:1). Add 30.0 mL of 0.1 *N* silver nitrate VS. Add 40 mL of saturated potassium permanganate solution. Heat for 30 minutes on a steam bath, and cool. Add 30 percent hydrogen peroxide dropwise until the solution becomes colorless. Add 8 mL of ferric ammonium sulfate TS and 1 mL of nitrobenzene. Titrate with 0.1 *N* ammonium thiocyanate VS. Each mL of 0.1 *N* silver nitrate is equivalent to 3.545 mg of chloride: between 19.8% and 20.8% is found.

Assay—Weigh accurately about 250 mg of Cysteine Hydrochloride into an iodine flask. Add 20 mL of water and 4 g of potassium iodide, and mix to dissolve. Cool the solution in an ice bath, and add 5 mL of 3 *N* hydrochloric acid and 25.0 mL of 0.1 *N* iodine VS. Insert the stopper, and allow to stand in the dark for 20 minutes. Titrate the excess iodine with 0.1 *N* sodium thiosulfate VS, adding 3 mL of starch TS as the end-point is approached. Perform a blank determination, and make any necessary correction. Each mL of 0.1 *N* iodine is equivalent to 15.76 mg of $C_3H_7NO_2S.HCl$.

Cysteine Hydrochloride Injection

» Cysteine Hydrochloride Injection is a sterile solution of Cysteine Hydrochloride in Water for Injection. It contains not less than 85.0 percent and not more than 115.0 percent of the labeled amount of $C_3H_7NO_2S.HCl.H_2O$.

Packaging and storage—Preserve in single-dose or in multiple-dose containers, preferably of Type I glass.

Reference standard—*USP L-Cysteine Hydrochloride Reference Standard*—Dry at a pressure of 5 mm of mercury for 24 hours before using.

Identification—

A: To 2 mL of Injection add 3 mL of water, and mix. Add 10 mL of cupric sulfate TS: a bluish gray precipitate is formed.

B: To 2 mL of Injection add 3 mL of water, and mix. Add 2 mL of 3 *N* sodium hydroxide and 2 drops of sodium nitroferricyanide solution (1 in 20): a red-purple color is produced, and it rapidly changes to yellow.

C: It responds to the tests for *Chloride* ⟨191⟩.

Pyrogen—It meets the requirements of the *Pyrogen Test* ⟨151⟩.

pH ⟨791⟩: between 1.0 and 2.5.

Heavy metals, *Method II* ⟨231⟩: 2 ppm.

Other requirements—It meets the requirements under *Injections* ⟨1⟩.

Assay—

Standard solution—Dissolve an accurately weighed quantity of USP L-Cysteine Hydrochloride RS in nitrogen-saturated water to obtain a solution having a known concentration of about 1 mg per mL.

Standard preparation—Transfer 20.0 mL of *Standard solution* to a 200-mL volumetric flask, dilute with nitrogen-saturated 1.0 *N* sodium hydroxide to volume, and mix.

Assay preparation—Dilute an accurately measured volume of Cysteine Hydrochloride Injection, equivalent to about 250 mg of cysteine hydrochloride, quantitatively and stepwise with nitrogen-saturated 1.0 *N* sodium hydroxide, to obtain a solution having a concentration of about 0.1 mg per mL.

Procedure—Transfer a suitable amount of *Standard preparation* to a polarographic cell. With mercury dropping from the electrode, lower the dropping mercury electrode of a polarograph so that the end is submerged in the liquid. Bubble oxygen-free, water-saturated nitrogen through the liquid for 15 minutes. Record the polarogram from −0.2 volt to −1.10 volts, using a saturated calomel electrode as the reference electrode. In a similar manner, record the polarograms obtained using portions of the *Assay preparation* and of the nitrogen-saturated 1.0 *N* sodium hydroxide. Determine the height of the diffusion current wave at −0.4 volt. Calculate the quantity, in mg, of $C_3H_7NO_2S.HCl.H_2O$ in each mL of the Injection taken by the formula:

$$2500(C/V)[(i_d)_U/(i_d)_S],$$

in which *C* is the concentration, in mg per mL, of USP L-Cysteine Hydrochloride RS in the *Standard preparation*, *V* is the volume, in mL, of Injection taken, and $(i_d)_U$ and $(i_d)_S$ are the observed diffusion currents, corrected for the diffusion current of the 0.1 *N* sodium hydroxide, of the *Assay preparation* and the *Standard preparation*, respectively.

Cytarabine

$C_9H_{13}N_3O_5$ 243.22
2(1*H*)-Pyrimidinone, 4-amino-1-β-D-arabinofuranosyl-.
1-β-D-Arabinofuranosylcytosine [*147-94-4*].

» Cytarabine contains not less than 95.0 percent and not more than 105.0 percent of $C_9H_{13}N_3O_5$, calculated on the dried basis.

Packaging and storage—Preserve in well-closed, light-resistant containers.

Reference standards—*USP Cytarabine Reference Standard*—[*Caution—Avoid contact.*] Dry at a pressure not exceeding 5 mm of mercury at 60° for 3 hours before using. *USP Uracil Arabinoside Reference Standard*—Do not dry before using.

Identification—

A: The infrared absorption spectrum of a mineral oil dispersion of it, previously dried at a pressure of not more than 5 mm of mercury at 60° for 3 hours, exhibits maxima only at the same wavelengths as that of a similar preparation of USP Cytarabine RS.

B: The retention time of the major peak in the chromatogram of the *Assay preparation* corresponds to that in the chromatogram of the *Standard preparation*, both relative to the internal standard, as obtained in the *Assay*.

Specific rotation ⟨781⟩: between +154° and +160°, calculated on the dried basis, determined in a solution in water containing 100 mg in each 10 mL.

Loss on drying ⟨731⟩—Dry it in vacuum at a pressure not exceeding 5 mm of mercury at 60° for 3 hours: it loses not more than 1.0% of its weight.

Residue on ignition ⟨281⟩: not more than 0.5%.

Assay—

Mobile phase—Dissolve 0.69 g of monobasic sodium phosphate and 1.34 g of dibasic sodium phosphate in about 950 mL of water. Add 50 mL of methanol, mix, filter, and degas. Make adjustments if necessary (see *System Suitability* under *Chromatography* ⟨621⟩).

Internal standard solution—Dissolve a suitable quantity of *p*-toluic acid in methanol to obtain a solution having a concentration of about 1.4 mg per mL.

Standard preparation—Dissolve an accurately weighed quantity of USP Cytarabine RS in *Internal standard solution* to obtain a *Stock standard solution* having a known concentration of about 0.6 mg per mL. Dilute an aliquot of this solution with *Mobile phase* to a final volume of about 6 times the volume of the aliquot of the *Stock standard solution* taken.

Resolution solution—Dissolve suitable quantities of USP Uracil Arabinoside RS and USP Cytarabine RS in water to obtain a solution containing about 0.02 mg and 0.6 mg per mL, respectively. Dilute an aliquot of this solution with an equal volume of *Internal standard solution*, and mix. Dilute this solution with a volume of *Mobile phase* equivalent to about 4 times the volume of *Internal standard solution* taken, and mix.

Assay preparation—Dissolve an accurately weighed quantity of Cytarabine in an accurately measured volume of *Internal standard solution* to obtain a *Stock assay solution* having a concentration of about 0.6 mg per mL. Dilute an aliquot of this solution with *Mobile phase* to a final volume of about 6 times the volume of the aliquot of the *Stock assay solution* taken.

Chromatographic system (see *Chromatography* ⟨621⟩)—The liquid chromatograph is equipped with a 254-nm detector and a 4- to 4.6-mm × 25- to 30-cm column that contains packing L1. The flow rate is about 1 mL per minute. Chromatograph the *Resolution solution*, and record the peak responses as directed under *Procedure:* the resolution, *R*, between the cytarabine and the uracil arabinoside peaks is not less than 2.5. The relative retention times are about 0.5 for cytarabine, 0.65 for uracil arabinoside, and 1.0 for *p*-toluic acid. Chromatograph the *Standard preparation*, and record the peak responses as directed under *Procedure:* the relative standard deviation for replicate injections is not more than 2.0%. [NOTE—After chromatography has been completed, flush the column with a solution obtained by mixing water and methanol (7:3). Failure to flush the column can result in hydrolysis of the stationary phase or microbiological growth in the column.]

Procedure—Separately inject equal volumes (about 10 µL) of the *Standard preparation* and the *Assay preparation* into the chromatograph, record the chromatograms, and measure the responses for the major peaks. Calculate the quantity, in mg, of $C_9H_{13}N_3O_5$ in the portion of Cytarabine taken by the formula:

$$CV(R_U/R_S),$$

in which *C* is the concentration, in mg per mL, of USP Cytarabine RS, in the *Stock standard solution*, *V* is the volume of *Internal standard solution* taken, in mL, to dissolve the test specimen in the *Stock assay solution*, and R_U and R_S are the peak response ratios obtained from the *Assay preparation* and the *Standard preparation*, respectively.

Sterile Cytarabine

» Sterile Cytarabine is Cytarabine suitable for parenteral use. It contains not less than 90.0 percent and not more than 110.0 percent of the labeled amount of $C_9H_{13}N_3O_5$.

Packaging and storage—Preserve in *Containers for Sterile Solids* as described under *Injections* ⟨1⟩.

Reference standards—*USP Cytarabine Reference Standard*—[*Caution—Avoid contact.*] Dry at a pressure not exceeding 5 mm of mercury at 60° for 3 hours before using. *USP Uracil Arabinoside Reference Standard*—Do not dry before using.

Constituted solution—At the time of use, the constituted solution prepared from Sterile Cytarabine meets the requirements for *Constituted Solutions* under *Injections* ⟨1⟩.

Identification—The retention time of the major peak in the chromatogram of the *Assay preparation* corresponds to that in the chromatogram of the *Standard preparation*, both relative to the internal standard, as obtained in the *Assay*.

pH ⟨791⟩: between 4.0 and 8.0, in a solution containing the equivalent of 10 mg of cytarabine per mL.

Water, *Method I* ⟨921⟩: not more than 3.0%.

Other requirements—It meets the requirements for *Sterility Tests* ⟨71⟩, *Uniformity of Dosage Units* ⟨905⟩, and *Labeling* under *Injections* ⟨1⟩.

Assay—

Mobile phase, Internal standard solution, Standard preparation, Resolution solution, and *Chromatographic system*—Prepare as directed in the *Assay* under *Cytarabine.*

Assay preparation—Constitute each of 10 vials of *Sterile Cytarabine* in a volume of water corresponding to the volume specified in the labeling. Transfer quantitatively, with rinsing, the contents of all 10 vials to a suitable volumetric flask, dilute with water to volume, and mix to obtain a solution having a concentration of about 10 mg per mL. Transfer 5.0 mL of this solution to a 50-mL volumetric flask, dilute with water to volume, and mix. Transfer 3.0 mL of the resulting solution to a suitable container, add 5.0 mL of *Internal standard solution*, dilute with about 20 mL of *Mobile phase*, and mix.

Procedure—Proceed as directed for *Procedure* in the *Assay* under *Cytarabine*. Calculate the quantity, in mg, of $C_9H_{13}N_3O_5$ in the portion of the constituted solution taken by the formula:

$$(500/3)C(R_U/R_S),$$

in which the terms are as defined therein.

Dacarbazine

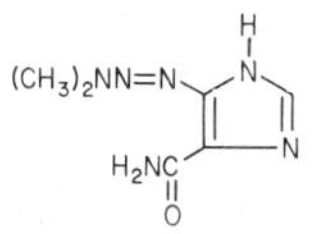

$C_6H_{10}N_6O$ 182.18

1*H*-Imidazole-4-carboxamide, 5-(3,3-dimethyl-1-triazenyl)-.
5-(3,3-Dimethyl-1-triazeno)imidazole-4-carboxamide [*4342-03-04*].

» Dacarbazine contains not less than 97.0 percent and not more than 102.0 percent of $C_6H_{10}N_6O$.

Caution—Great care should be taken in handling Dacarbazine, as it is a potent cytotoxic agent.

Packaging and storage—Preserve in tight, light-resistant containers, in a refrigerator.

Reference standards—*USP Dacarbazine Reference Standard*—Keep container tightly closed and protected from light, and store in a refrigerator. Dry in vacuum over phosphorus pentoxide at 60° for 2 hours before using. *USP 5-Aminoimidazole-4-carboxamide Hydrochloride Reference Standard*—Do not dry before using. Keep container tightly closed and protected from light, and store in a refrigerator. *USP 2-Azahypoxanthine Reference Standard*—This is the monohydrate form of 2-azahypoxanthine. Do not dry before using. Keep container tightly closed and protected from light, and store in a refrigerator.

Identification—The infrared absorption spectrum of a potassium bromide dispersion of it exhibits maxima only at the same wavelengths as that of a similar preparation of USP Dacarbazine RS.

Residue on ignition ⟨281⟩: not more than 0.1%.

Related compounds—Dissolve an accurately weighed quantity of Dacarbazine in 1.0 *N* hydrochloric acid to obtain a solution having a concentration of 40 mg per mL, and apply 5 µL of the solution to a suitable thin-layer chromatographic plate (see *Chromatography* ⟨621⟩), coated with a 0.25-mm layer of chromatographic silica gel mixture. Apply, separately, 5 µL of an aqueous solution containing 0.40 mg of USP 5-Aminoimidazole-4-carboxamide Hydrochloride RS per mL, and 5 µL of an aqueous solution containing 0.40 mg of USP 2-Azahypoxanthine RS per mL. Develop the chromatogram in pH 5 acetate buffer (prepared by mixing 74 mL of 0.20 *M* acetic acid and 176 mL of 0.20 *M* sodium acetate in a 500-mL volumetric flask, diluting with water to volume, and mixing) until the solvent front has moved about three-fourths of the length of the plate. Remove the plate from the developing chamber, mark the solvent front, and allow the solvent to evaporate. Locate the spots on the plate by viewing under short-wavelength ultraviolet light: any spots obtained from the test solution are not greater in size or intensity than the spots, occurring at the respective R_f values, produced by the Standard solutions, corresponding to not more than 1.0% of 5-aminoimidazole-4-carboxamide hydrochloride and not more than 1.0% of 2-azahypoxanthine.

Assay—[NOTE—Throughout this procedure, avoid exposing Dacarbazine and its solutions to light.]

Standard preparations—Transfer about 30 mg of USP Dacarbazine RS, accurately weighed, to a 50-mL volumetric flask, add 0.1 *N* hydrochloric acid to volume, and mix (*Standard stock solution*). Dilute a portion of *Standard stock solution* quantitatively and stepwise with 0.1 *N* hydrochloric acid to obtain an *Acidic standard preparation* having a known concentration of about 6 µg per mL. Dilute a portion of *Standard stock solution* quantitatively and stepwise with pH 7.0 phosphate buffer (see *Buffer Solutions* in the section, *Reagents, Indicators, and Solutions*) to obtain a *Neutral standard preparation* having a known concentration of about 6 µg per mL.

Assay preparations—Prepare as directed under *Standard preparations*, except to use about 30 mg of Dacarbazine, accurately weighed.

Procedure—Concomitantly determine the absorbances of the *Acidic standard preparation* and the *Acidic assay preparation* in 1-cm cells at the wavelength of maximum absorbance at about 323 nm, with a suitable spectrophotometer, using 0.1 *N* hydrochloric acid as the blank. Concomitantly determine the absorbances of the *Neutral standard preparation* and the *Neutral assay preparation* in 1-cm cells at the wavelength of maximum absorbance at about 329 nm, using pH 7.0 phosphate buffer (see *Buffer Solutions* in the section, *Reagents, Indicators, and Solutions*) as the blank. Calculate the quantity, in mg, of $C_6H_{10}N_6O$ in the portion of Dacarbazine taken by the formula:

$$5C[(A_{323} + A_{329})_U/(A_{323} + A_{329})_S],$$

in which *C* is the concentration, in µg per mL, of USP Dacarbazine RS in the *Standard preparations*, and the parenthetic expressions are the sums of the absorbances of the *Assay preparations* (*U*) and the *Standard preparations* (*S*), respectively, measured at the wavelengths indicated by the subscripts.

Dacarbazine for Injection

» Dacarbazine for Injection is a sterile, freeze-dried mixture of Dacarbazine and suitable buffers or diluents. It contains not less than 95.0 percent and not more than 105.0 percent of the labeled amount of $C_6H_{10}N_6O$.

Caution—Great care should be taken to prevent inhaling particles of Dacarbazine for Injection and exposing the skin to it.

Packaging and storage—Preserve in single-dose or multiple-dose *Containers for Sterile Solids* as described under *Injections* ⟨1⟩, preferably of Type I glass, protected from light.

Reference standards—*USP Dacarbazine Reference Standard*—Keep container tightly closed and protected from light, and store in a refrigerator. Dry in vacuum over phosphorus pentoxide at 60° for 2 hours before using. *USP 2-Azahypoxanthine Reference Standard*—This is the monohydrate form of 2-azahypoxanthine. Do not dry before using. Keep container tightly closed and protected from light, and store in a refrigerator.

Completeness of solution—When dissolved as directed in the labeling, it yields a clear, pale yellow to yellow solution.

Constituted solution—At the time of use, the constituted solution prepared from Dacarbazine for Injection meets the requirements for *Constituted Solutions* under *Injections* ⟨1⟩.

Identification—

A: Dissolve a suitable quantity of Dacarbazine for Injection in water to obtain a solution having a concentration of 10 mg of dacarbazine per mL. Apply, separately, 1 µL of the freshly prepared solution and 1 µL of an aqueous solution containing 10 mg each of USP Dacarbazine RS and citric acid per mL to a suitable thin-layer chromatographic plate (see *Chromatography* ⟨621⟩), coated with a 0.25-mm layer of chromatographic silica gel. Develop the chromatogram in a solvent system consisting of a mixture of isopropyl alcohol and 1 *N* ammonium hydroxide (3:1) until the solvent front has moved about three-fourths of the length of the plate. Remove the plate from the developing chamber, mark the solvent front, and allow the solvent to evaporate. Spray the plate evenly with a freshly prepared solution containing 1% of ferric chloride and 1% of potassium ferricyanide (prepared by mixing 5 mL of a 10% aqueous solution of ferric chloride with 5 mL of a 10% aqueous solution of potassium ferricyanide and diluting with water to 50 mL). Dacarbazine appears as an intense blue spot on a light yellow background: the R_f value of the spot obtained from the test solution corresponds to that obtained from the Standard solution.

B: To 1 mL of a solution (1 in 100) in a test tube add a few crystals of periodic acid and 4 drops of methanol. Shake, and after 1 minute add 5 mL of a 0.2% acetylacetone reagent solution (prepared by mixing 15.0 g of ammonium acetate, 0.30 mL of glacial acetic acid, and 0.20 mL of acetylacetone in a 100-mL volumetric flask, adding water to volume, and mixing). Shake, and place in a water bath maintained at a temperature of 60°: an intense yellow color develops in a few minutes (*presence of mannitol*).

C: To 2 drops of an aqueous solution (1 in 100) in a 15-mL test tube add 10 mL of a solution prepared by mixing 10 mL of acetic anhydride with 30 mL of pyridine: an intense yellow color is produced immediately and after a few minutes becomes red-violet (*presence of citric acid*).

Pyrogen—It meets the requirements of the *Pyrogen Test* ⟨151⟩, the test dose being 1 mL per kg of a solution in Sodium Chloride Injection containing 5.0 mg per mL.

pH ⟨791⟩: between 3.0 and 4.0, in a solution containing an amount of Dacarbazine for Injection equivalent to about 1 g of dacarbazine in 100 mL of water.

Water, *Method I* ⟨921⟩: not more than 1.5%.

2-Azahypoxanthine—[NOTE—The *Mobile phase* employed in this procedure is corrosive. The system should be rinsed well with methanol following completion of analysis.]

Mobile phase—Transfer 2.2 g of docusate sodium to a 1000-mL volumetric flask, dissolve in a mixture of 100 mL of water and 15 mL of glacial acetic acid, and dilute with water to volume. Filter the solution through a 0.5-μm porosity filter. Prepare this solution fresh daily.

Standard preparation—Prepare a solution of USP 2-Azahypoxanthine RS to contain 0.04 mg per mL.

Test preparation—Constitute the contents of 1 vial of Dacarbazine for Injection. Using the contents of the constituted vial, dilute quantitatively with water to obtain a solution containing 4 mg of dacarbazine per mL.

Chromatographic system (see *Chromatography* ⟨621⟩)—The liquid chromatograph is equipped with a 254-nm detector and a 3.9-mm × 30-cm column that contains packing L1. The flow rate is about 1.2 mL per minute. Chromatograph five replicate injections of the *Standard preparation*, and record the peak responses as directed under *Procedure:* the relative standard deviation is not more than 2.0%.

Procedure—Separately inject equal volumes (about 20 μL) of the *Standard preparation* and the *Test preparation* into the chromatograph by means of a suitable sampling valve or high-pressure microsyringe. Measure the peak responses at corresponding retention times obtained from the *Standard preparation* and the *Test preparation*, and calculate the quantity, in mg, of 2-azahypoxanthine monohydrate in the dacarbazine taken by the formula:

$$(CV)(r_U / r_S),$$

in which C is the concentration, in mg per mL, of USP 2-Azahypoxanthine RS in the *Standard preparation*, V is the final volume, in mL, of the *Test preparation*, and r_U and r_S are the peak responses obtained from the *Test preparation* and the *Standard preparation*, respectively: not more than 1.0% is found.

Other requirements—It meets the requirements for *Sterility Tests* ⟨71⟩, *Uniformity of Dosage Units* ⟨905⟩, and *Labeling* under *Injections* ⟨1⟩.

Assay—Transfer an accurately weighed quantity of Dacarbazine for Injection, equivalent to about 100 mg of dacarbazine, to a 250-mL volumetric flask, dissolve in 0.1 *N* hydrochloric acid, dilute with 0.1 *N* hydrochloric acid to volume, and mix. Transfer 2.0 mL of this solution to a 250-mL volumetric flask, dilute with 0.1 *N* hydrochloric acid to volume, and mix. Dissolve an accurately weighed quantity of USP Dacarbazine RS in 0.1 *N* hydrochloric acid, and dilute quantitatively and stepwise with the same solvent to obtain a Standard solution having a known concentration of about 3.2 μg per mL. Concomitantly determine the absorbances of both solutions in 1-cm cells at the wavelength of maximum absorbance at about 323 nm, with a suitable spectrophotometer, using 0.1 *N* hydrochloric acid as the blank. Calculate the quantity, in mg, of $C_6H_{10}N_6O$ in the portion of Dacarbazine for Injection taken by the formula:

$$31.25C(A_U / A_S),$$

in which C is the concentration, in μg per mL, of USP Dacarbazine RS in the Standard solution, and A_U and A_S are the absorbances of the solution of Dacarbazine for Injection and the Standard solution, respectively.

Dactinomycin

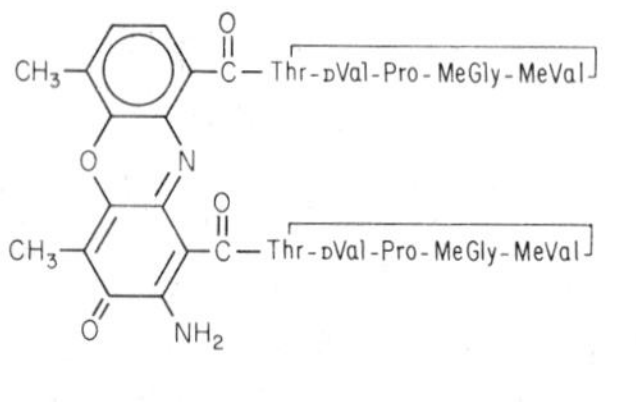

$C_{62}H_{86}N_{12}O_{16}$ 1255.43

Actinomycin D.

Actinomycin D [*50-76-0*].

» Dactinomycin has a potency of not less than 900 μg of $C_{62}H_{86}N_{12}O_{16}$ per mg, calculated on the dried basis.

Caution—Great care should be taken to prevent inhaling particles of Dactinomycin and exposing the skin to it.

Packaging and storage—Preserve in tight containers, protected from light and excessive heat.

Reference standard—*USP Dactinomycin Reference Standard*—Dry in vacuum at a pressure not exceeding 5 mm of mercury at 60° for 3 hours before using.

Identification—

A: The ultraviolet absorption spectrum of a 1 in 40,000 solution in methanol exhibits maxima and minima at the same wavelengths as that of a similar solution of USP Dactinomycin RS, concomitantly measured, and the absorptivity, calculated on the dried basis, at the wavelength of maximum absorbance at about 445 nm is not less than 95.0% and not more than 103.0% of that of USP Dactinomycin RS, the potency of the Reference Standard being taken into account. The ratio A_{240}/A_{445} is between 1.30 and 1.50.

B: The chromatogram obtained from the *Assay preparation* in the *Assay* exhibits a major peak for dactinomycin the retention time of which corresponds to that exhibited by the *Standard preparation*, and the chromatogram compares qualitatively to that obtained from the *Standard preparation.*

Crystallinity ⟨695⟩: meets the requirements.

Pyrogen—It meets the requirements of the *Pyrogen Test* ⟨151⟩, the test dose being 1.0 mL per kg of a solution prepared to contain 0.2 mg of dactinomycin per mL of Sterile Water for Injection.

Loss on drying ⟨731⟩—Dry it in vacuum at a pressure not exceeding 5 mm of mercury at 60° for 3 hours: it loses not more than 15.0% of its weight.

Assay—[NOTE—In this procedure, use freshly prepared *Standard preparation* and *Assay preparation*, protected from light.]

Mobile phase—Prepare a suitable mixture of acetonitrile, 0.04 *M* sodium acetate, and 0.07 *M* acetic acid (approximately 46:25:25), filter through a membrane filter (1-μm or finer porosity), and degas. [NOTE—The acetonitrile concentration may be varied to provide appropriate *Chromatographic system* performance and to provide a suitable elution time.]

Standard preparation—Dissolve an accurately weighed quantity of USP Dactinomycin RS in *Mobile phase*, and dilute quantitatively with *Mobile phase* to obtain a solution having a known concentration of about 1200 μg of dactinomycin per mL.

Assay preparation—Dissolve about 30 mg of Dactinomycin, accurately weighed, in *Mobile phase* to make 25.0 mL, and mix.

Chromatographic system (see *Chromatography* ⟨621⟩)—The liquid chromatograph is equipped with a 254-nm detector and a 3.9-mm × 30-cm column that contains packing L1. The flow rate is about 1.0 mL per minute. Chromatograph three replicate injections of the *Standard preparation*, and record the peak responses as directed under *Procedure:* the relative standard deviation is not more than 1.0%.

Procedure—Separately inject equal volumes (about 20 µL) of the *Standard preparation* and the *Assay preparation* into the chromatograph, record the chromatograms, and measure the responses for the major peaks. The retention time is about 25 minutes for dactinomycin. Calculate the potency, in µg, of $C_{62}H_{86}N_{12}O_{16}$ per mg by the formula:

$$25(C/W)(r_U/r_S),$$

in which C is the concentration, in µg, of dactinomycin in each mL of the *Standard preparation*, W is the weight, in mg, of Dactinomycin taken, and r_U and r_S are the peak responses of the *Assay preparation* and the *Standard preparation*, respectively.

Dactinomycin for Injection

» Dactinomycin for Injection is a sterile mixture of Dactinomycin and Mannitol. It contains not less than 90.0 percent and not more than 120.0 percent of the labeled amount of $C_{62}H_{86}N_{12}O_{16}$, the labeled amount being 0.5 mg in each container.

Caution—Great care should be taken to prevent inhaling particles of Dactinomycin and exposing the skin to it.

Packaging and storage—Preserve in light-resistant *Containers for Sterile Solids* as described under *Injections* ⟨1⟩.

Labeling—Label it to include the statement, "Protect from light."

Reference standard—*USP Dactinomycin Reference Standard*—Dry in vacuum at a pressure not exceeding 5 mm of mercury at 60° for 3 hours before using.

Constituted solution—At the time of use, the constituted solution prepared from Dactinomycin for Injection meets the requirements for *Constituted Solutions* under *Injections* ⟨1⟩.

Identification—

A: The ultraviolet absorption spectrum of a methanol solution containing about 25 µg of dactinomycin per mL exhibits maxima and minima at the same wavelengths as that of a similar solution of USP Dactinomycin RS, concomitantly measured, and the ratio A_{240}/A_{445} is between 1.30 and 1.50.

B: The chromatogram obtained from the *Assay preparation* in the *Assay* exhibits a major peak for dactinomycin the retention time of which corresponds to that exhibited by the *Standard preparation*, and the chromatogram compares qualitatively to that obtained from the *Standard preparation*.

Pyrogen—It meets the requirements of the *Pyrogen Test* ⟨151⟩, the test dose being 1.0 mL per kg of a solution prepared to contain 0.2 mg of dactinomycin per mL of Sterile Water for Injection.

Sterility—It meets the requirements under *Sterility Tests* ⟨71⟩, when tested as directed in the section, *Test Procedures Using Membrane Filtration*, each container being constituted aseptically by injecting Sterile Water for Injection through the stopper, and the entire contents of all the containers being collected aseptically with the aid of 200 mL of *Fluid A* before filtering.

pH ⟨791⟩: between 5.5 and 7.5, in the solution constituted as directed in the labeling.

Loss on drying ⟨731⟩—Dry it in vacuum at a pressure not exceeding 5 mm of mercury at 60° for 3 hours: it loses not more than 4.0% of its weight.

Other requirements—It meets the requirements under *Injections* ⟨1⟩.

Assay—[Note—In this procedure, use freshly prepared *Standard preparation* and *Assay preparation*, protected from light.]

Mobile phase—Mix 6 volumes of acetonitrile and 4 volumes of water, filter through a membrane filter (1-µm or finer porosity), and degas. [Note—The acetonitrile concentration may be varied to provide appropriate *Chromatographic system* performance and to provide a suitable elution time.]

Standard preparation—Dissolve an accurately weighed quantity of USP Dactinomycin RS in *Mobile phase*, and dilute quantitatively with *Mobile phase* to obtain a solution having a known concentration of about 250 µg of dactinomycin per mL.

Assay preparation—Add an accurately measured volume of *Mobile phase* to 1 container of Dactinomycin for Injection to obtain a solution containing about 250 µg of dactinomycin per mL, and filter, if necessary, to obtain a clear solution.

Chromatographic system (see *Chromatography* ⟨621⟩)—The liquid chromatograph is equipped with a 254-nm detector and a 3.9-mm × 30-cm column that contains packing L1. The flow rate is about 2.5 mL per minute. Chromatograph the *Standard preparation*, and record the peak response as directed under *Procedure:* the column efficiency is not less than 1200 theoretical plates, the tailing factor is not more than 2, and the relative standard deviation for replicate injections is not more than 3.0%.

Procedure—Separately inject equal volumes (about 10 µL) of the *Standard preparation* and the *Assay preparation* into the chromatograph, record the chromatograms, and measure the responses for the major peaks. The retention time is about 6 minutes for dactinomycin. Calculate the quantity, in mg, of $C_{62}H_{86}N_{12}O_{16}$ in the container of Dactinomycin for Injection taken by the formula:

$$(CV/1000)(r_U/r_S),$$

in which C is the concentration, in µg, of dactinomycin in each mL of the *Standard preparation*, V is the volume, in mL, of the *Assay preparation*, and r_U and r_S are the peak responses of the *Assay preparation* and the *Standard preparation*, respectively.

Danazol

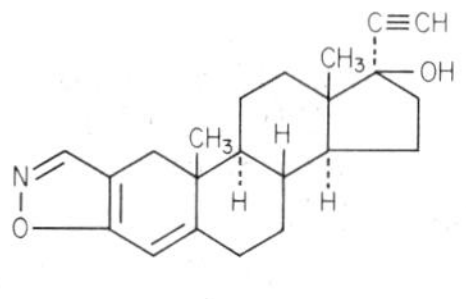

$C_{22}H_{27}NO_2$ 337.46

Pregna-2,4-dien-20-yno[2,3-*d*]isoxazol-17-ol, (17α)-.
17α-Pregna-2,4-dien-20-yno[2,3-*d*]isoxazol-17-ol
[*17230-88-5*].

» Danazol contains not less than 97.0 percent and not more than 102.0 percent of $C_{22}H_{27}NO_2$, calculated on the dried basis.

Packaging and storage—Preserve in tight, light-resistant containers.

Reference standard—*USP Danazol Reference Standard*—Dry at a pressure not exceeding 5 mm of mercury at 60° to constant weight before using.

Identification—

A: The infrared absorption spectrum of a potassium bromide dispersion of it, previously dried, exhibits maxima only at the same wavelengths as that of a similar preparation of USP Danazol RS.

B: The ultraviolet absorption spectrum of the solution prepared as directed in the *Assay* exhibits maxima and minima at the same wavelengths as that of a similar solution of USP Danazol RS, concomitantly measured.

Specific rotation ⟨781⟩: between +21° and +27°, calculated on the dried basis, determined in a solution in chloroform containing 100 mg in each 10 mL.

Loss on drying ⟨731⟩—Dry it at a pressure not exceeding 5 mm of mercury at 60° to constant weight: it loses not more than 2.0% of its weight.

Chromatographic impurities—

Solvent—Prepare a mixture of chloroform and methanol (9:1).

Standard preparations—Dissolve an accurately weighed quantity of USP Danazol RS in *Solvent* to obtain a solution having a known concentration of 1 mg per mL. Dilute quantitatively with *Solvent* to obtain Standard preparations having the following compositions:

Standard Preparation	Dilution	Concentration (μg RS per mL)	Percentage (%, for comparison with test specimen)
A	(1 in 2)	500	1.0
B	(1 in 4)	250	0.5
C	(1 in 10)	100	0.2
D	(1 in 20)	50	0.1

Test preparation—Dissolve an accurately weighed quantity of Danazol in *Solvent* to obtain a solution containing 50 mg per mL.

Procedure—On a suitable thin-layer chromatographic plate (see *Chromatography* ⟨621⟩), coated with a 0.25-mm layer of chromatographic silica gel mixture, apply separately 5 μL of the *Test preparation* and 5 μL of each *Standard preparation.* Position the plate in a chromatographic chamber and develop the chromatograms in a solvent system consisting of a mixture of cyclohexane and ethyl acetate (7:3) until the solvent front has moved about three-fourths of the length of the plate. Remove the plate from the developing chamber, mark the solvent front, and allow the solvent to evaporate in warm, circulating air. Examine the plate under short-wavelength ultraviolet light. Expose the plate to iodine vapors for 5 minutes. Compare the intensities of any secondary spots observed in the chromatogram of the *Test preparation* with those of the principal spots in the chromatograms of the *Standard preparations:* the sum of the intensities of secondary spots obtained from the *Test preparation* corresponds to not more than 1.0% of related compounds, with no single impurity corresponding to more than 0.5%.

Assay—Dissolve about 100 mg of Danazol, accurately weighed and previously dried, in about 50 mL of alcohol in a 100-mL volumetric flask, swirl until dissolved, dilute with alcohol to volume, and mix. Transfer 2.0 mL of this solution to a 100-mL volumetric flask, dilute with alcohol to volume, and mix. Similarly, dissolve an accurately weighed quantity of USP Danazol RS in alcohol to obtain a Standard solution having a known concentration of about 20 μg per mL. Concomitantly determine the absorbances of both solutions in 1-cm cells at the wavelength of maximum absorbance at about 285 nm, using alcohol as the blank. Calculate the quantity, in mg, of $C_{22}H_{27}NO_2$ in the portion of Danazol taken by the formula:

$$5C(A_U / A_S),$$

in which C is the concentration, in μg per mL, of USP Danazol RS in the Standard solution, and A_U and A_S are the absorbances of the solution of Danazol and the Standard solution, respectively.

Danazol Capsules

» Danazol Capsules contain not less than 90.0 percent and not more than 110.0 percent of the labeled amount of $C_{22}H_{27}NO_2$.

Packaging and storage—Preserve in well-closed containers.

Reference standard—*USP Danazol Reference Standard*—Dry at a pressure not exceeding 5 mm of mercury at 60° to constant weight before using.

Identification—Shake the contents of a sufficient number of Capsules, equivalent to about 50 mg of Danazol, with 50 mL of chloroform, and filter. Evaporate the filtrate on a steam bath with the aid of a stream of nitrogen to dryness. The infrared absorption spectrum of a potassium bromide dispersion of the residue, previously dried, exhibits maxima at the same wavelengths as that of a similar preparation of USP Danazol RS.

Dissolution ⟨711⟩—

Medium: isopropyl alcohol in 0.1 *N* hydrochloric acid (4 in 10); 900 mL.

Apparatus 2: 80 rpm.

Time: 30 minutes.

Procedure—Determine the amount of $C_{22}H_{27}NO_2$ dissolved as follows. Remove an aliquot from the solution under test at a point midway between the stirring shaft and the wall of the vessel and approximately midway in depth. Measure the amount in solution in filtered portions of the *Dissolution Medium*, suitably diluted with the *Dissolution Medium*, at the wavelength of maximum absorbance at about 286 nm, with a suitable spectrophotometer, in comparison with a solution of known concentration of USP Danazol RS prepared as follows. Transfer 10 mg of USP Danazol RS, accurately weighed, to a 10-mL volumetric flask, and dissolve in isopropyl alcohol. Transfer 2.0 mL to a 100-mL volumetric flask, dilute with *Dissolution Medium* to volume, and mix.

Tolerances—Not less than 65% (Q) of the labeled amount of $C_{22}H_{27}NO_2$ is dissolved in 30 minutes.

Uniformity of dosage units ⟨905⟩: meet the requirements.

Assay—Remove, as completely as possible, the contents of not less than 20 Danazol Capsules, and weigh accurately. Mix the combined contents thoroughly, and transfer an accurately weighed quantity of the powder, equivalent to about 100 mg of Danazol, to a glass-stoppered, 125-mL conical flask. Add 100.0 mL of chloroform, and shake for 3 minutes. Filter, discarding the first 10 to 15 mL of filtrate. Dilute 10.0 mL of the subsequent filtrate with chloroform to 100.0 mL, and mix. Transfer 10.0 mL of this solution to a 50-mL volumetric flask, dilute with chloroform to volume, and mix. Dissolve an accurately weighed quantity of USP Danazol RS in chloroform to obtain a solution having a known concentration of about 20 μg per mL. Concomitantly determine the absorbances of both solutions in 1-cm cells at the wavelength of maximum absorbance at about 287 nm, with a suitable spectrophotometer, against chloroform as a blank. Calculate the quantity, in mg, of $C_{22}H_{27}NO_2$ in the portion of Capsules taken by the formula:

$$5C(A_U / A_S),$$

in which C is the concentration, in μg per mL, of USP Danazol RS in the Standard solution, and A_U and A_S are the absorbances of the solution from the Capsules and the Standard solution, respectively.

Dapsone

$H_2N-C_6H_4-SO_2-C_6H_4-NH_2$

$C_{12}H_{12}N_2O_2S$ 248.30

Benzenamine, 4,4′-sulfonylbis-.

4,4′-Sulfonyldianiline [*80-08-0*].

» Dapsone contains not less than 99.0 percent and not more than 101.0 percent of $C_{12}H_{12}N_2O_2S$, calculated on the dried basis.

Packaging and storage—Preserve in well-closed, light-resistant containers.

Reference standard—*USP Dapsone Reference Standard*—Dry at 105° for 3 hours before using.

Identification—

A: The infrared absorption spectrum of a mineral oil dispersion of it, previously dried, exhibits maxima only at the same wavelengths as that of a similar preparation of USP Dapsone RS.

B: The ultraviolet absorption spectrum of a 1 in 200,000 solution in methanol exhibits maxima and minima at the same wavelengths as that of a similar solution of USP Dapsone RS, concomitantly measured.

Melting range ⟨741⟩: between 175° and 181°.

Loss on drying ⟨731⟩—Dry it at 105° for 3 hours: it loses not more than 1.5% of its weight.

Residue on ignition ⟨281⟩: not more than 0.1%.

Selenium ⟨291⟩: 0.003%, a 100-mg test specimen, mixed with 100 mg of magnesium oxide, being used.

Assay—

Mobile phase—Transfer 100 mL of isopropyl alcohol, 100 mL of acetonitrile, and 100 mL of ethyl acetate to a 1000-mL volumetric flask. Add pentane to volume without mixing, then mix, and allow the mixture to cool to room temperature.

Standard preparation—Dissolve an accurately weighed quantity of USP Dapsone RS in methanol to obtain a solution having a known concentration of about 250 µg per mL. Pipet 5 mL of this solution into a 50-mL volumetric flask, dilute with *Mobile phase* to volume, and mix to obtain a *Standard preparation* having a known concentration of about 25 µg per mL.

Assay preparation—Transfer about 50 mg of Dapsone, accurately weighed, to a 200-mL volumetric flask. Dissolve in methanol, dilute with methanol to volume, and mix. Pipet 5 mL of this solution into a 50-mL volumetric flask, dilute with *Mobile phase* to volume, and mix.

Chromatographic system—The chromatograph (see *Chromatography* ⟨621⟩) is equipped with a 254-nm detector and a 4-mm × 30-cm column that contains 10-µm diameter packing L3. Chromatograph a sufficient number of injections of the *Standard preparation* as directed under *Procedure* to ensure that the relative standard deviation is not more than 2%.

Procedure—Separately introduce equal volumes (about 10 µL) of the *Standard preparation* and the *Assay preparation* into the chromatograph by means of a suitable microsyringe or sampling valve, adjusting the specimen size and other operating parameters to obtain satisfactory chromatograms. Measure the responses for the major peaks obtained at corresponding retention times with the *Assay preparation* and the *Standard preparation*. Calculate the quantity, in mg, of $C_{12}H_{12}N_2O_2S$ in the portion of Dapsone taken by the formula:

$$2C(P_U/P_S),$$

in which C is the concentration, in µg per mL, of USP Dapsone RS in the *Standard preparation*, and P_U and P_S are the peak responses obtained from the *Assay preparation* and the *Standard preparation*, respectively.

Dapsone Tablets

» Dapsone Tablets contain not less than 92.5 percent and not more than 107.5 percent of the labeled amount of $C_{12}H_{12}N_2O_2S$.

Packaging and storage—Preserve in well-closed, light-resistant containers.

Reference standard—*USP Dapsone Reference Standard*—Dry at 105° for 3 hours before using.

Identification—

A: Transfer a quantity of finely powdered Tablets, equivalent to about 100 mg of dapsone, to a suitable container, add 5 mL of acetone, shake for 5 minutes, filter, and evaporate the filtrate to dryness. Dry this residue at 105° for 1 hour: the residue so obtained responds to *Identification test A* under *Dapsone*.

B: Triturate a quantity of finely powdered Tablets, equivalent to about 100 mg of dapsone, with 50 mL of methanol, and filter. Dilute a portion of the filtrate with methanol to make approximately a 1 in 200,000 solution: this solution responds to *Identification test B* under *Dapsone*.

Dissolution ⟨711⟩—

Medium: dilute hydrochloric acid (2 in 100); 1000 mL.

Apparatus 1: 100 rpm.

Time: 60 minutes.

Procedure—Withdraw and filter a portion of the solution under test. Transfer an accurately measured portion of the filtrate, estimated to contain about 0.2 mg of dapsone, to a 25-mL volumetric flask, add 5 mL of 1 *N* sodium hydroxide, dilute with water to volume, and mix. Determine the amount of $C_{12}H_{12}N_2O_2S$ dissolved from ultraviolet absorbances at the wavelength of maximum absorbance at about 290 nm of the solutions so obtained from the solution under test in comparison with a Standard solution having a known concentration of USP Dapsone RS in the same medium.

Tolerances—Not less than 75% (*Q*) of the labeled amount of $C_{12}H_{12}N_2O_2S$ is dissolved in 60 minutes.

Uniformity of dosage units ⟨905⟩: meet the requirements.

Procedure for content uniformity—Transfer 1 Tablet to a 100-mL volumetric flask, add 2.0 mL of water, and allow to stand for 30 minutes, swirling occasionally. Add about 70 mL of methanol, and place the flask in an ultrasonic bath until the specimen is completely dispersed. Add methanol to volume, mix, and centrifuge a portion of the mixture. Quantitatively dilute an accurately measured volume of the clear supernatant liquid with methanol to obtain a solution having a concentration of about 8 µg of dapsone per mL. Dissolve an accurately weighed quantity of USP Dapsone RS in methanol to obtain a Standard solution having a known concentration of about 8 µg per mL. Concomitantly determine the absorbances of the test solution and the Standard solution in 1-cm cells at the wavelength of maximum absorbance at about 296 nm, with a suitable spectrophotometer, using methanol as the blank. Calculate the quantity, in mg, of $C_{12}H_{12}N_2O_2S$ in the Tablet by the formula:

$$(TC/D)(A_U/A_S),$$

in which T is the labeled quantity, in mg, of dapsone in the Tablet, C is the concentration, in µg per mL, of USP Dapsone RS in the Standard solution, D is the concentration, in µg per mL, of dapsone in the solution from the Tablet, based upon the labeled quantity per Tablet and the extent of dilution, and A_U and A_S are the absorbances of the solution from the Tablet and the Standard solution, respectively.

Assay—

Mobile phase, *Standard preparation*, and *Chromatographic system*—Prepare as directed in the *Assay* under *Dapsone*.

Assay preparation—Weigh and finely powder not less than 20 Dapsone Tablets. Transfer an accurately weighed portion of the powder, equivalent to about 50 mg of dapsone, to a 200-mL volumetric flask. Add 150 mL of methanol, and place the flask in an ultrasonic bath at a temperature of 35° for 15 minutes, with occasional shaking. Allow to cool to room temperature, add methanol to volume, and mix. Centrifuge a portion of the mixture until clear. Transfer 5.0 mL of the clear supernatant liquid to a 50-mL volumetric flask, dilute with *Mobile phase* to volume, and mix.

Procedure—Proceed as directed for *Procedure* in the *Assay* under *Dapsone*. Calculate the quantity, in mg, of $C_{12}H_{12}N_2O_2S$ in the portion of Tablets taken by the formula:

$$2C(P_U/P_S).$$

Daunorubicin Hydrochloride

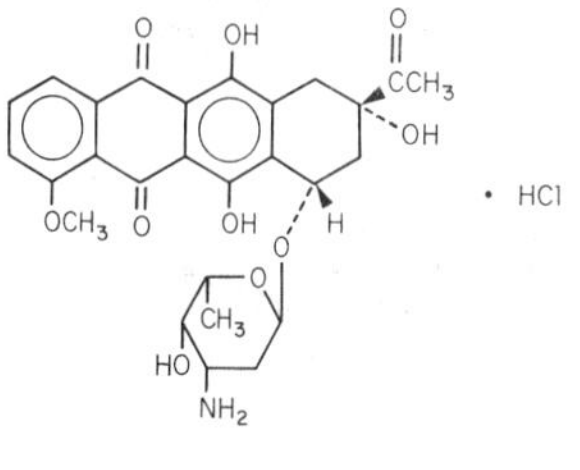

$C_{27}H_{29}NO_{10}\cdot HCl$ 563.99

5,12-Naphthacenedione, 8-acetyl-10-[(3-amino-2,3,6-trideoxy-α-L-*lyxo*-hexopyranosyl)]oxy]-7,8,9,10-tetrahydro-6,8,11-trihydroxy-1-methoxy-, (8*S*-*cis*)-, hydrochloride.

(1*S*,3*S*)-3-Acetyl-1,2,3,4,6,11-hexahydro-3,5,12-trihydroxy-10-methoxy-6,11-dioxo-1-naphthacenyl 3-amino-2,3,6-trideoxy-α-L-*lyxo*-hexopyranoside hydrochloride [*23541-50-6*].

» Daunorubicin Hydrochloride has a potency equivalent to not less than 842 µg and not more than 1030 µg of $C_{27}H_{29}NO_{10}$ per mg.

Caution—Great care should be taken to prevent inhaling particles of daunorubicin hydrochloride and exposing the skin to it.

Packaging and storage—Preserve in tight containers, protected from light and excessive heat.

Reference standard—*USP Daunorubicin Hydrochloride Reference Standard*—Do not dry before using.

Identification—The retention time of the main peak obtained with the *Assay preparation* corresponds to that obtained with the *Standard preparation* as directed in the *Assay*.

Crystallinity ⟨695⟩: meets the requirements.

pH ⟨791⟩: between 4.5 and 6.5, in a solution containing 5 mg per mL.

Water, *Method I* ⟨921⟩: not more than 3.0%, 20 mL of a mixture of carbon tetrachloride, chloroform, and methanol (2:2:1) being used in place of methanol in the titration vessel.

Assay—

Mobile solvent—Mix 62 volumes of water and 38 volumes of acetonitrile, and adjust with phosphoric acid to a pH of 2.2 ± 0.2. The acetonitrile concentration may be varied to meet system suitability requirements and to provide a suitable elution time for daunorubicin. Filter the solution through a membrane filter (1-µm or finer porosity), and degas.

Internal standard solution—Prepare a solution of 2-naphthalenesulfonic acid in a mixture of water and acetonitrile, in the same proportions as in the *Mobile solvent*, containing about 2 mg per mL.

Standard preparation—Dissolve a suitable quantity of USP Daunorubicin Hydrochloride RS, accurately weighed, in the *Internal standard solution* to obtain a solution having a known concentration of about 1 mg of daunorubicin per mL.

Assay preparation—Dissolve about 25 mg of Daunorubicin Hydrochloride, accurately weighed, in the *Internal standard solution* to make 25.0 mL, and mix.

Chromatographic system (see *Chromatography* ⟨621⟩)—The chromatograph is equipped with a 254-nm detector and a 4.6-mm × 30-cm column that contains packing L1. The flow rate is about 1.5 mL per minute. Chromatograph replicate injections of the *Standard preparation*, and record the peak responses as directed under *Procedure:* the resolution factor between daunorubicin and 2-naphthalenesulfonic acid is not less than 2.0.

Procedure—Separately inject equal volumes (about 5 µL) of the *Standard preparation* and the *Assay preparation* into the chromatograph, record the chromatograms, and measure the responses for the major peaks. Calculate the potency, in µg of $C_{27}H_{29}NO_{10}$ per mg, by the formula:

$$(25C/W)(R_U/R_S),$$

in which C is the concentration, in µg, of daunorubicin in each mL of the *Standard preparation*, W is the weight, in mg, of Daunorubicin Hydrochloride taken, and R_U and R_S are the ratios of peak responses of daunorubicin peak to 2-naphthalenesulfonic acid peak obtained from the *Assay preparation* and the *Standard preparation*, respectively.

Daunorubicin Hydrochloride for Injection

» Daunorubicin Hydrochloride for Injection is a sterile mixture of Daunorubicin Hydrochloride and Mannitol. It contains the equivalent of not less than 90.0 percent and not more than 115.0 percent of the labeled amount of $C_{27}H_{29}NO_{10}$.

Packaging and storage—Preserve in light-resistant *Containers for Sterile Solids* as described under *Injections* ⟨1⟩.

Reference standard—*USP Daunorubicin Hydrochloride Reference Standard*—Do not dry before using.

Constituted solution—At the time of use, the constituted solution prepared from Daunorubicin Hydrochloride for Injection meets the requirements for *Constituted Solutions* under *Injections* ⟨1⟩.

Identification—The retention time of the main peak obtained with the *Assay preparation* corresponds to that obtained with the *Standard preparation* as directed in the *Assay*.

Depressor substances—It meets the requirements of the *Depressor Substances Test* ⟨101⟩, the test dose being 1.0 mL per kg of a solution prepared to contain 1.5 mg of daunorubicin per mL in sterile saline TS.

Pyrogen—It meets the requirements of the *Pyrogen Test* ⟨151⟩, the test dose being 1.0 mL per kg of a solution prepared to contain 2.25 mg of daunorubicin per mL in pyrogen-free saline TS.

pH ⟨791⟩: between 4.5 and 6.5, in the solution constituted as directed in the labeling.

Water, *Method I* ⟨921⟩: not more than 3.0%, the *Test preparation* being prepared as directed for a hygroscopic specimen.

Other requirements—It meets the requirements under *Injections* ⟨1⟩.

Assay—

Mobile solvent, Internal standard solution, Standard preparation, and *Chromatographic system*—Prepare as directed in the *Assay* under *Daunorubicin Hydrochloride.*

Assay preparation—Transfer the contents of 1 vial of Daunorubicin Hydrochloride for Injection with the aid of *Internal standard solution* to an appropriate volumetric flask, and dilute with *Internal standard solution* to volume to obtain a solution containing about 1 mg of daunorubicin per mL.

Procedure—Proceed as directed for *Procedure* in the *Assay* under *Daunorubicin Hydrochloride.* Calculate the quantity, in mg, of $C_{27}H_{29}NO_{10}$ in the vial of Daunorubicin Hydrochloride for Injection taken by the formula:

$$(CV/1000)(R_U/R_S),$$

in which V is the volume, in mL, of the *Assay preparation*, and the other terms are as defined therein.

Deferoxamine Mesylate

$$H_2N(CH_2)_5N(OH)C(=O)(CH_2)_2C(=O)NH(CH_2)_5N(OH)C(=O)(CH_2)_2C(=O)NH(CH_2)_5N(OH)C(=O)CH_3 \cdot CH_3SO_3H$$

$C_{25}H_{48}N_6O_8.CH_4O_3S$ 656.79

Butanediamide, *N′*-[5-[[4-[[5-(acetylhydroxyamino)pentyl]-amino]-1,4-dioxobutyl]hydroxyamino]pentyl]-*N*-(5-aminopentyl)-*N*-hydroxy-, monomethanesulfonate.

N-[5-[3-[(5-Aminopentyl)hydroxycarbamoyl]propionamido]-pentyl]-3-[[5-(*N*-hydroxyacetamido)pentyl]carbamoyl]-propionohydroxamic acid monomethanesulfonate (salt) [*138-14-7*].

» Deferoxamine Mesylate contains not less than 98.0 percent and not more than 102.0 percent of $C_{25}H_{48}N_6O_8.CH_4O_3S$, calculated on the anhydrous basis.

Packaging and storage—Preserve in tight containers.

Reference standard—*USP Deferoxamine Mesylate Reference Standard*—Do not dry; determine the water content titrimetrically at the time of use.

Identification—Dissolve 5 mg in 5 mL of water, add 2 mL of tribasic sodium phosphate solution (1 in 200), mix, then add 10 drops of β-naphthoquinone-4-sodium sulfonate solution (1 in 40): a blackish brown color is produced.

pH ⟨791⟩: between 4.0 and 6.0, in a solution (1 in 100).

Water, *Method I* ⟨921⟩: not more than 2.0%.

Residue on ignition ⟨281⟩: not more than 0.1%, 2.0 g being used for the test.

Chloride ⟨221⟩—A 1.2-g portion shows no more chloride than corresponds to 0.20 mL of 0.020 *N* hydrochloric acid (0.012%).

Sulfate ⟨221⟩—A 0.5-g portion shows no more sulfate than corresponds to 0.20 mL of 0.020 *N* sulfuric acid (0.04%).

Heavy metals, *Method II* ⟨231⟩: 0.001%.

Assay—

Ferric chloride solution—Dissolve 6.7 g of ferric chloride in dilute hydrochloric acid (1 in 100) in a 100-mL volumetric flask. Add dilute hydrochloric acid (1 in 100) to volume, mix, and filter.

Standard preparation—Dissolve a suitable quantity of USP Deferoxamine Mesylate RS, accurately weighed, in water to obtain a solution having a known concentration of about 1000 µg per mL.

Assay preparation—Dissolve about 50 mg of Deferoxamine Mesylate, accurately weighed, in water to make 50.0 mL, and mix.

Procedure—Pipet 2 mL each of the *Standard preparation*, the *Assay preparation*, and water to provide a blank, into separate 25-mL volumetric flasks. To each flask add 3 mL of *Ferric chloride solution*, dilute with water to volume, and mix. Concomitantly determine the absorbances of the solutions from the *Standard preparation* and the *Assay preparation* against the blank, in 1-cm cells, at the wavelength of maximum absorbance at about 485 nm, with a suitable spectrophotometer. Calculate the quantity, in mg, of $C_{25}H_{48}N_6O_8 \cdot CH_4O_3S$ in the Deferoxamine Mesylate taken by the formula:

$$0.05C(A_U / A_S),$$

in which C is the concentration, in µg per mL, of USP Deferoxamine Mesylate RS in the *Standard preparation*, and A_U and A_S are the absorbances of the solutions from the *Assay preparation* and *Standard preparation*, respectively.

Sterile Deferoxamine Mesylate

» Sterile Deferoxamine Mesylate is Deferoxamine Mesylate suitable for parenteral use. It contains not less than 90.0 percent and not more than 110.0 percent of the labeled amount of $C_{25}H_{48}N_6O_8 \cdot CH_4O_3S$.

Packaging and storage—Preserve in single-dose or in multiple-dose containers, preferably of Type I glass.

Reference standard—*USP Deferoxamine Mesylate Reference Standard*—Do not dry; determine the water content titrimetrically at the time of use.

Constituted solution—At the time of use, the constituted solution prepared from Sterile Deferoxamine Mesylate meets the requirements for *Constituted Solutions* under *Injections* ⟨1⟩.

Identification—It responds to the *Identification test* under *Deferoxamine Mesylate.*

pH ⟨791⟩: between 4.0 and 6.0, in a solution (1 in 100).

Pyrogen—It meets the requirements of the *Pyrogen Test* ⟨151⟩, the test dose being 0.3 mL per kg of a solution containing 100 mg of deferoxamine mesylate per mL.

Water, *Method I* ⟨921⟩: not more than 1.5%.

Other requirements—It meets the requirements under *Injections* ⟨1⟩ and *Uniformity of Dosage Units* ⟨905⟩.

Assay—Proceed with Sterile Deferoxamine Mesylate as directed in the *Assay* under *Deferoxamine Mesylate.* Calculate the quantity, in mg, of $C_{25}H_{48}N_6O_8 \cdot CH_4O_3S$ in the Sterile Deferoxamine Mesylate taken by the formula:

$$0.05C(A_U / A_S).$$

Dehydrated Alcohol—*see* Alcohol, Dehydrated

Dehydroacetic Acid—*see* Dehydroacetic Acid NF

Dehydrocholic Acid

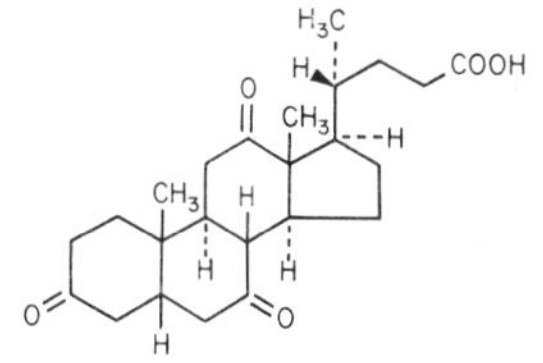

$C_{24}H_{34}O_5$ 402.53

Cholan-24-oic acid, 3,7,12-trioxo-, (5β)-.

3,7,12-Trioxo-5β-cholan-24-oic acid [*81-23-2*].

» Dehydrocholic Acid contains not less than 98.5 percent and not more than 101.0 percent of $C_{24}H_{34}O_5$, calculated on the dried basis. Dehydrocholic Acid for parenteral use melts between 237° and 242°.

Packaging and storage—Preserve in well-closed containers.

Reference standard—*USP Dehydrocholic Acid Reference Standard*—Dry at 105° for 2 hours before using.

Identification—The infrared absorption spectrum of a potassium bromide dispersion of it, previously dried, exhibits maxima only at the same wavelengths as that of a similar preparation of USP Dehydrocholic Acid RS.

Melting range ⟨741⟩: between 231° and 242°, but the range between beginning and end of melting does not exceed 3°.

Specific rotation ⟨781⟩: between +29.0° and +32.5°, calculated on the dried basis, determined in a solution in dioxane containing 200 mg in each 10 mL.

Microbial limit—It meets the requirements of the test for absence of *Salmonella* species under *Microbial Limit Tests* ⟨61⟩.

Loss on drying ⟨731⟩—Dry it at 105° for 2 hours: it loses not more than 1.0% of its weight.

Residue on ignition ⟨281⟩: not more than 0.3%.

Odor on boiling—Boil 2 g with 100 mL of water for 2 minutes: the mixture is odorless.

Barium—Add to the mixture obtained in the test for *Odor on boiling* 2 mL of hydrochloric acid, and again boil for 2 minutes. Cool, filter, and wash the filter with water until the filtrate measures 100 mL. To 10 mL of the filtrate add 1 mL of 2 *N* sulfuric acid: no turbidity is produced.

Heavy metals, *Method II* ⟨231⟩: 0.002%.

Assay—Transfer about 500 mg of Dehydrocholic Acid, accurately weighed, to a 300-mL conical flask, add 60 mL of neutralized alcohol, and warm on a steam bath until solution is effected. Cool, add phenolphthalein TS and 20 mL of water, and titrate with 0.1 *N* sodium hydroxide VS, adding 100 mL of water shortly before the end-point is reached. Each mL of 0.1 *N* sodium hydroxide is equivalent to 40.25 mg of $C_{24}H_{34}O_5$.

Dehydrocholic Acid Tablets

» Dehydrocholic Acid Tablets contain not less than 94.0 percent and not more than 106.0 percent of the labeled amount of $C_{24}H_{34}O_5$.

Packaging and storage—Preserve in well-closed containers.

Identification—Mix a quantity of finely powdered Tablets, equivalent to about 500 mg of dehydrocholic acid, with 15 mL of water, and add slowly, with stirring, 2 mL of sodium carbonate TS. Filter, and add to the filtrate 3 *N* hydrochloric acid (about 2 mL), dropwise, until no more precipitate is formed. Filter the precipitate, wash with small portions of cold water until free from chloride, and dry at 105° for 2 hours: the dehydrocholic acid so obtained responds to the *Identification test* and meets the requirements of the test for *Melting range* under *Dehydrocholic Acid.*

Microbial limit—Tablets meet the requirements of the test for absence of *Salmonella* species under *Microbial Limit Tests* ⟨61⟩.

Disintegration ⟨701⟩: 30 minutes.

Uniformity of dosage units ⟨905⟩: meet the requirements.

Assay—Weigh and finely powder not less than 20 Dehydrocholic Acid Tablets. Transfer an accurately weighed portion of the powder, equivalent to about 500 mg of dehydrocholic acid, to a 300-mL conical flask, add 60 mL of neutralized alcohol, and warm on a steam bath for 10 minutes. Cool, add phenolphthalein TS and 20 mL of water, and titrate with 0.1 *N* sodium hydroxide VS, adding 100 mL of water shortly before the end-point is reached. Each mL of 0.1 *N* sodium hydroxide is equivalent to 40.25 mg of $C_{24}H_{34}O_5$.

Delayed-release Capsules, Aspirin—*see* Aspirin Delayed-release Capsules

Delayed-release Capsules, Doxycycline Hyclate—*see* Doxycycline Hyclate Delayed-release Capsules

Delayed-release Tablets, Aspirin—*see* Aspirin Delayed-release Tablets

Delayed-release Tablets, Erythromycin—*see* Erythromycin Delayed-release Tablets

Demecarium Bromide

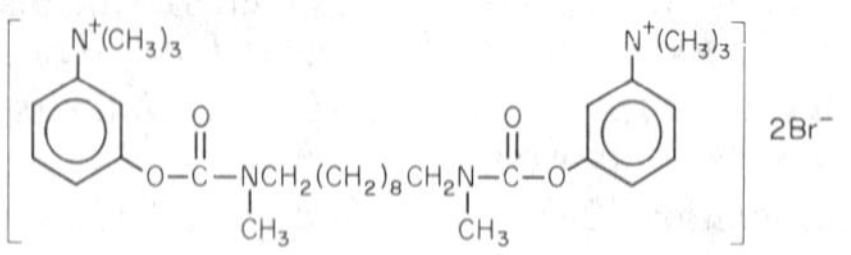

$C_{32}H_{52}Br_2N_4O_4$ 716.60

Benzenaminium, 3,3′-[1,10-decanediylbis[(methylimino)-carbonyloxy]]bis[*N,N,N*-trimethyl-, dibromide.

(*m*-Hydroxyphenyl)trimethylammonium bromide decamethylenebis[methylcarbamate] (2:1) [*56-94-0*].

» Demecarium Bromide contains not less than 95.0 percent and not more than 100.5 percent of $C_{32}H_{52}Br_2N_4O_4$, calculated on the anhydrous basis.

Packaging and storage—Preserve in tight, light-resistant containers.

Reference standard—*USP Demecarium Bromide Reference Standard.*

Identification—

A: The infrared absorption spectrum of a potasim bromide dispersion of it exhibits maxima only at the same wavelengths as that of a similar preparation of USP Demecarium Bromide RS.

B: Dissolve about 100 mg in 50 mL of 1 *N* sodium hydroxide, and reflux for 15 minutes. Cool, and add 3 mL of the refluxed solution to 25 mL of saturated sodium bicarbonate solution. Add, with mixing, 4 mL of *N,N*-dimethyl-*p*-phenylenediamine dihydrochloride solution (1.5 in 10,000) and 2 mL of sodium hypochlorite solution (1.5 in 20,000): a violet-blue color is produced.

C: Dissolve about 50 mg in 20 mL of water, add 10 mL of a 1 in 50 solution of ammonium reineckate in methanol, and allow to stand for 30 minutes with occasional swirling: a pink reineckate of demecarium forms, and it melts between 131° and 136°, with decomposition.

D: A solution of it responds to the tests for *Bromide* ⟨191⟩.

pH ⟨791⟩: between 5.0 and 7.0, in a solution (1 in 100).

Water, *Method I* ⟨921⟩: not more than 2.0%.

Residue on ignition ⟨281⟩: not more than 0.1%.

Heavy metals, *Method I* ⟨231⟩: 0.002%.

***m*-Trimethylammoniophenol bromide**—

Control preparation—Dissolve 100 mg of *m*-dimethylaminophenol in 10 mL of alcohol in a 1000-mL volumetric flask, dilute with water to volume, and mix. Pipet 1 mL of this solution into a 500-mL volumetric flask, dilute with water to volume, and mix.

Test preparation—Transfer 100 mg of Demecarium Bromide to a 100-mL volumetric flask, add water to volume, and mix.

Procedure—Pipet 25 mL of the *Test preparation* into a glass-stoppered, 50-mL centrifuge tube, and pipet 25 mL of the *Control preparation* into a second, similar tube. To each tube add 3 mL of pH 7.0 phosphate buffer (see under *Solutions* in the section, *Reagents, Indicators, and Solutions*), 1 mL of *N,N*-dimethyl-*p*-phenylenediamine dihydrochloride solution (1.5 in 10,000), 5 mL of isobutyl alcohol, and 1 mL of sodium hypochlorite solution (1.5 in 20,000). Insert the stoppers in the tubes, shake the mixtures for 5 minutes, and centrifuge: any blue color produced in the upper layer obtained from the *Test preparation* is not more intense than that obtained from the *Control preparation.*

Assay—Dissolve about 0.8 g of Demecarium Bromide, accurately weighed, in a mixture of 75 mL of glacial acetic acid and 15 mL of mercuric acetate TS, warming slightly, if necessary, to effect solution. Add 2 drops of crystal violet TS, and titrate with 0.1 *N* perchloric acid VS. Perform a blank determination, and make any necessary correction. Each mL of 0.1 *N* perchloric acid is equivalent to 35.83 mg of $C_{32}H_{52}Br_2N_4O_4$.

Demecarium Bromide Ophthalmic Solution

» Demecarium Bromide Ophthalmic Solution is a sterile, aqueous solution of Demecarium Bromide. It contains not less than 92.0 percent and not more than 108.0 percent of the labeled amount of $C_{32}H_{52}Br_2N_4O_4$. It contains a suitable antimicrobial agent.

Packaging and storage—Preserve in tight, light-resistant containers.

Reference standard—*USP Demecarium Bromide Reference Standard.*

Identification—

A: Mix about 10 mL with 12.5 mL of 1 *N* sodium hydroxide, and proceed as directed in *Identification test B* under *Demecarium Bromide*, beginning with "reflux for 15 minutes": a violet-blue color is produced.

B: To 10 mL add 5 mL of a 1 in 50 solution of ammonium reineckate in methanol, and allow to stand for 30 minutes with occasional swirling: a pink reineckate of demecarium forms, and it melts between 131° and 136°, with decomposition.

C: It responds to the tests for *Bromide* ⟨191⟩.

Sterility—It meets the requirements under *Sterility Tests* ⟨71⟩.

Assay—Into each of two 50-mL volumetric flasks marked *1* and *2* pipet a volume of Demecarium Bromide Ophthalmic Solution, equivalent to about 2.5 mg of demecarium bromide. Into each of two additional 50-mL volumetric flasks, marked *3* and *4*, pipet 5.0 mL of a freshly prepared Standard solution made by dissolving about 50 mg of USP Demecarium Bromide RS, accurately weighed, in 100.0 mL of water. To flasks *1* and *3*, add 1 *N* sodium hydroxide to volume, and mix. Transfer 10.0 mL of each of these solutions to separate glass-stoppered tubes, insert the stoppers loosely, heat the tubes in a water bath for 15 minutes, then cool to room temperature. The concentration of USP Demecarium Bromide RS in the Standard solution is about 50 µg per mL. To flasks *2* and *4*, add pH 10.0 alkaline borate buffer (see under *Solutions* in the section, *Reagents, Indicators, and Solutions*) to volume, and mix. Concomitantly determine the absorbances of the two sodium hydroxide solutions at the wavelength of maximum absorbance at about 292 nm, with a suitable spectrophotometer, using the corresponding borate buffer solutions as the respective solvent blanks. Calculate the quantity, in mg, of $C_{32}H_{52}Br_2N_4O_4$ in each mL of the Ophthalmic Solution taken by the formula:

$$(50C/V)(A_U/A_S),$$

in which C is the concentration, in mg per mL, of USP Demecarium Bromide RS in the Standard solution, V is the volume, in mL, of Ophthalmic Solution taken, and A_U and A_S are the absorbances of the sodium hydroxide solution from the Demecarium Bromide Ophthalmic Solution and the Standard solution, respectively.

Demeclocycline

$C_{21}H_{21}ClN_2O_8$ 464.86

2-Naphthacenecarboxamide, 7-chloro-4-(dimethylamino)-1,4,4a,5,5a,6,11,12a-octahydro-3,6,10,12,12a-pentahydroxy-1,11-dioxo-, [4*S*-(4α,4aα,5aα,6β,12aα)]-.

7-Chloro-4-(dimethylamino)-1,4,4a,5,5a,6,11,12a-octahydro-3,6,10,12,12a-pentahydroxy-1,11-dioxo-2-naphthacenecarboxamide [*127-33-3*].

Sesquihydrate 491.88 [*13215-10-6*].

» Demeclocycline has a potency equivalent to not less than 970 μg of demeclocycline hydrochloride ($C_{21}H_{21}ClN_2O_8.HCl$) per mg, calculated on the anhydrous basis.

Packaging and storage—Preserve in tight, light-resistant containers.

Reference standard—*USP Demeclocycline Hydrochloride Reference Standard*—Dry in vacuum at a pressure not exceeding 5 mm of mercury at 60° for 3 hours before using.

Identification—

A: Transfer about 40 mg, accurately weighed, to a 250-mL volumetric flask, add 2 mL of 0.1 *N* hydrochloric acid to dissolve, dilute with water to volume, and mix. Transfer 10.0 mL of this solution to a 100-mL volumetric flask, add about 75 mL of water and 5.0 mL of 5 *N* sodium hydroxide, dilute with water to volume, and mix: the ultraviolet absorption spectrum of this solution, measured at 6 minutes, accurately timed, after the addition of the sodium hydroxide, exhibits maxima and minima at the same wavelengths as that of a similar solution of USP Demeclocycline Hydrochloride RS, concomitantly measured, and the absorptivity, calculated on the anhydrous basis, at the wavelength of maximum absorbance at about 380 nm is between 103.5% and 111.3% of that of the USP Demeclocycline Hydrochloride RS, the potency of the Reference Standard being taken into account.

B: Transfer 40 mg, accurately weighed, to a 200-mL volumetric flask, add 100 mL of 0.1 *N* hydrochloric acid, shake to dissolve, dilute with 0.1 *N* hydrochloric acid to volume, and mix. Transfer 5.0 mL of this solution into each of two 50-mL volumetric flasks (*Solutions 1* and *2*). Prepare similar solutions of USP Demeclocycline Hydrochloride RS (*Solutions 3* and *4*). To *Solutions 1* and *3*, add 10 mL of 6 *N* hydrochloric acid, and to *Solutions 2* and *4*, add 10 mL of 3 *N* hydrochloric acid. Heat the four flasks in a water bath for 20 minutes, cool, dilute the contents with water to volume, and mix. Determine the absorbances of *Solutions 1* and *3* at the wavelength of maximum absorbance at about 368 nm, with a suitable spectrophotometer, using *Solutions 2* and *4*, respectively, as the blanks. Determine the absorbances of *Solutions 2* and *4* at the wavelength of maximum absorbance at about 430 nm, using *Solutions 1* and *3*, respectively, as the blanks. Calculate the ratio:

$$(W_SP/1000)(A_{368} + A_{430})_U/W_U(A_{368} + A_{430})_S,$$

in which W_S is the weight, in mg, of USP Demeclocycline Hydrochloride RS taken, calculated on the dried basis, P is the potency, in μg per mg, of the USP Demeclocycline Hydrochloride RS, W_U is the weight, in mg, of specimen taken, calculated on the anhydrous basis, and the final two parenthetic expressions are the absorbances of the four solutions at the wavelengths indicated by the subscripts for the specimen (U) and the Standard (S): the ratio is between 0.97 and 1.17.

Crystallinity ⟨695⟩: meets the requirements.

pH ⟨791⟩: between 4.0 and 5.5, in a solution containing 10 mg per mL.

Water, *Method I* ⟨921⟩: between 4.3% and 6.7%.

Assay—Proceed with Demeclocycline as directed under *Antibiotics—Microbial Assays* ⟨81⟩.

Demeclocycline Oral Suspension

» Demeclocycline Oral Suspension contains the equivalent of not less than 90.0 percent and not more than 125.0 percent of the labeled amount of demeclocycline hydrochloride ($C_{21}H_{21}ClN_2O_8.HCl$). It may contain one or more suitable buffers, preservatives, stabilizers, and suspending agents.

Packaging and storage—Preserve in tight containers, protected from light.

Reference standard—*USP Demeclocycline Hydrochloride Reference Standard*—Dry in vacuum at a pressure not exceeding 5 mm of mercury at 60° for 3 hours before using.

pH ⟨791⟩: between 4.0 and 5.8.

Assay—Transfer an accurately measured quantity of Demeclocycline Oral Suspension, freshly mixed and free from air bubbles, equivalent to about 150 mg of demeclocycline hydrochloride ($C_{21}H_{21}N_2O_8.HCl$), to a 1000-mL volumetric flask, dilute with 0.1 *N* hydrochloric acid to volume, and mix. Proceed as directed under *Antibiotics—Microbial Assays* ⟨81⟩, using an accurately measured volume of this stock solution diluted quantitatively and stepwise with water to obtain a *Test Dilution* having a concentration assumed to be equal to the median dose level of the Standard.

Demeclocycline Hydrochloride

$C_{21}H_{21}ClN_2O_8.HCl$ 501.32

2-Naphthacenecarboxamide, 7-chloro-4-(dimethylamino)-1,4,-4a,5,5a,6,11,12a-octahydro-3,6,10,12,12a-pentahydroxy-1,11-dioxo-, monohydrochloride, [4*S*-(4α,4aα,5aα,6β,12aα)]-.

7-Chloro-4-(dimethylamino)-1,4,4a,5,5a,6,11,12a-octahydro-3,6,10,12,12a-pentahydroxy-1,11-dioxo-2-naphthacenecarboxamide monohydrochloride [*64-73-3*].

» Demeclocycline Hydrochloride has a potency of not less than 900 μg of $C_{21}H_{21}ClN_2O_8.HCl$ per mg, calculated on the dried basis.

Packaging and storage—Preserve in tight, light-resistant containers.

Reference standard—*USP Demeclocycline Hydrochloride Reference Standard*—Dry in vacuum at a pressure not exceeding 5 mm of mercury at 60° for 3 hours before using.

Identification—

A: It responds to *Identification test A* under *Demeclocycline*, except that its absorptivity, calculated on the dried basis, is between 95.8% and 104.2% of that of the USP Demeclocycline Hydrochloride RS, the potency of the Reference Standard being taken into account.

B: It responds to *Identification test B* under *Demeclocycline*.

Crystallinity ⟨695⟩: meets the requirements.

pH ⟨791⟩: between 2.0 and 3.0, in a solution containing 10 mg per mL.

Loss on drying ⟨731⟩—Dry about 100 mg, accurately weighed, in a capillary-stoppered bottle in vacuum at a pressure not exceeding 5 mm of mercury at 60° for 3 hours: it loses not more than 2.0% of its weight.

Assay—Proceed with Demeclocycline Hydrochloride as directed under *Antibiotics—Microbial Assays* ⟨81⟩.

Demeclocycline Hydrochloride Capsules

» Demeclocycline Hydrochloride Capsules contain not less than 90.0 percent and not more than 125.0 percent of the labeled amount of demeclocycline hydrochloride ($C_{21}H_{21}ClN_2O_8.HCl$).

Packaging and storage—Preserve in tight, light-resistant containers.

Reference standard—*USP Demeclocycline Hydrochloride Reference Standard*—Dry in vacuum at a pressure not exceeding 5 mm of mercury at 60° for 3 hours before using.

Dissolution ⟨711⟩—

Medium: water; 900 mL.

Apparatus 2: 75 rpm.

Time: 45 minutes.

Procedure—Determine the amount of $C_{21}H_{21}ClN_2O_8$ dissolved from ultraviolet absorbances at the wavelength of maximum absorbance at about 270 nm of filtered portions of the solution under test, suitably diluted with *Dissolution Medium*, if necessary, in comparison with a Standard solution having a known concentration of USP Demeclocycline Hydrochloride RS in the same medium.

Tolerances—Not less than 75% (*Q*) of the labeled amount of demeclocycline ($C_{21}H_{21}ClN_2O_8$) is dissolved in 45 minutes.

Uniformity of dosage units ⟨905⟩: meet the requirements.

Loss on drying ⟨731⟩—Dry about 100 mg of Capsule contents, accurately weighed, in a capillary-stoppered bottle in vacuum at 60° for 3 hours: the material loses not more than 2.0% of its weight, except that if the Capsules contain starch the material loses not more than 8.0% of its weight.

Assay—Place not less than 5 Demeclocycline Hydrochloride Capsules in a high-speed glass blender jar containing an accurately measured volume of 0.1 *N* hydrochloric acid, so that the stock solution so obtained contains not less than 150 µg of demeclocycline hydrochloride ($C_{21}H_{21}ClN_2O_8.HCl$) per mL, and blend for 4 ± 1 minutes. Proceed as directed under *Antibiotics—Microbial Assays* ⟨81⟩, using an accurately measured volume of this stock solution diluted quantitatively and stepwise with water to obtain a *Test Dilution* having a concentration assumed to be equal to the median dose level of the Standard.

Demeclocycline Hydrochloride Tablets

» Demeclocycline Hydrochloride Tablets contain not less than 90.0 percent and not more than 125.0 percent of the labeled amount of demeclocycline hydrochloride ($C_{21}H_{21}ClN_2O_8.HCl$).

Packaging and storage—Preserve in tight, light-resistant containers.

Reference standard—*USP Demeclocycline Hydrochloride Reference Standard*—Dry in vacuum at a pressure not exceeding 5 mm of mercury at 60° for 3 hours before using.

Dissolution ⟨711⟩—

Medium: water; 900 mL.

Apparatus 2: 75 rpm.

Time: 45 minutes.

Procedure—Determine the amount of $C_{21}H_{21}ClN_2O_8$ dissolved from ultraviolet absorbances at the wavelength of maximum absorbance at about 270 nm of filtered portions of the solution under test, suitably diluted with *Dissolution Medium*, if necessary, in comparison with a Standard solution having a known concentration of USP Demeclocycline Hydrochloride RS in the same medium.

Tolerances—Not less than 75% (*Q*) of the labeled amount of demeclocycline ($C_{21}H_{21}ClN_2O_8$) is dissolved in 45 minutes.

Uniformity of dosage units ⟨905⟩: meet the requirements.

Loss on drying ⟨731⟩—Dry about 100 mg of finely ground Tablet powder, accurately weighed, in a capillary-stoppered bottle in vacuum at 60° for 3 hours: it loses not more than 2.0% of its weight.

Assay—Place not less than 5 Demeclocycline Hydrochloride Tablets in a high-speed glass blender jar containing an accurately measured volume of 0.1 *N* hydrochloric acid, so that the stock solution so obtained contains not less than 150 µg of demeclocycline hydrochloride ($C_{21}H_{21}ClN_2O_8.HCl$) per mL, and blend for 4 ± 1 minutes. Proceed as directed under *Antibiotics—Microbial Assays* ⟨81⟩, using an accurately measured volume of this stock solution diluted quantitatively and stepwise with water to obtain a *Test Dilution* having a concentration assumed to be equal to the median dose level of the Standard.

Demeclocycline Hydrochloride and Nystatin Capsules

» Demeclocycline Hydrochloride and Nystatin Capsules contain not less than 90.0 percent and not more than 125.0 percent of the labeled amount of demeclocycline hydrochloride ($C_{21}H_{21}ClN_2O_8.HCl$) and not less than 90.0 percent and not more than 135.0 percent of the labeled amount of USP Nystatin Units.

Packaging and storage—Preserve in tight, light-resistant containers.

Reference standards—*USP Demeclocycline Hydrochloride Reference Standard*—Dry in vacuum at a pressure not exceeding 5 mm of mercury at 60° for 3 hours before using. *USP Nystatin Reference Standard*—Dry in vacuum at a pressure not exceeding 5 mm of mercury at 40° for 2 hours before using.

Identification—Shake a suitable quantity of Capsule contents with methanol to obtain a *Test Solution* containing about 1 mg of demeclocycline hydrochloride per mL, and proceed as directed under *Identification—Tetracyclines* ⟨193⟩.

Dissolution ⟨711⟩—

Medium: water; 900 mL.

Apparatus 2: 75 rpm.

Time: 45 minutes.

Procedure—Determine the amount of $C_{21}H_{21}ClN_2O_8.HCl$ dissolved from ultraviolet absorbances at the wavelength of maximum absorbance at about 270 nm of filtered portions of the solution under test, suitably diluted with *Dissolution Medium*, if necessary, in comparison with a Standard solution having a known concentration of USP Demeclocycline Hydrochloride RS in the same medium.

Tolerances—Not less than 75% (*Q*) of the labeled amount of demeclocycline hydrochloride ($C_{21}H_{21}ClN_2O_8.HCl$) is dissolved in 45 minutes.

Loss on drying ⟨731⟩—Dry about 100 mg of Capsule contents, accurately weighed, in a capillary-stoppered bottle in vacuum at a pressure not exceeding 5 mm of mercury at 60° for 3 hours: it loses not more than 5.0% of its weight.

Assay for demeclocycline hydrochloride—Place not less than 5 Demeclocycline Hydrochloride and Nystatin Capsules in a high-speed glass blender jar containing an accurately measured volume of 0.1 *N* hydrochloric acid, and blend for 4 ± 1 minutes, so that the stock solution so obtained contains not less than 150 µg of demeclocycline hydrochloride ($C_{21}H_{21}ClN_2O_8.HCl$) per mL. Proceed as directed under *Antibiotics—Microbial Assays* ⟨81⟩, using an accurately measured volume of this stock solution diluted quantitatively and stepwise with water to obtain a *Test Dilution* having a concentration of demeclocycline hydrochloride assumed to be equal to the median dose level of the Standard.

Assay for nystatin—Proceed as directed for Nystatin under *Antibiotics—Microbial Assays* ⟨81⟩, blending not less than 5 Demeclocycline Hydrochloride and Nystatin Capsules for 4 ± 1 minutes in a high-speed blender with a sufficient accurately measured volume of dimethylformamide to obtain a solution of convenient concentration. Dilute an accurately measured portion of this solution quantitatively with dimethylformamide to obtain a stock solution containing about 400 USP Nystatin Units per mL. Dilute this stock solution quantitatively with *Buffer No. 6* to obtain a *Test Dilution* having a concentration of nystatin assumed to be equal to the median dose level of the Standard.

Demeclocycline Hydrochloride and Nystatin Tablets

» Demeclocycline Hydrochloride and Nystatin Tablets contain not less than 90.0 percent and not more than 125.0 percent of the labeled amount of demeclocycline hydrochloride ($C_{21}H_{21}ClN_2O_8.HCl$) and not less than 90.0 percent and not more than 135.0 percent of the labeled amount of USP Nystatin Units.

Packaging and storage—Preserve in tight, light-resistant containers.

Reference standards—*USP Demeclocycline Hydrochloride Reference Standard*—Dry in vacuum at a pressure not exceeding 5 mm of mercury at 60° for 3 hours before using. *USP Nystatin Reference Standard*—Dry in vacuum at a pressure not exceeding 5 mm of mercury at 40° for 2 hours before using.

Identification—Shake a suitable quantity of powdered Tablets with methanol to obtain a *Test Solution* containing about 1 mg of demeclocycline hydrochloride per mL, and proceed as directed under *Identification—Tetracyclines* ⟨193⟩.

Dissolution ⟨711⟩—

Medium: water; 900 mL.

Apparatus 2: 75 rpm.

Time: 45 minutes.

Procedure—Determine the amount of $C_{21}H_{21}ClN_2O_8.HCl$ dissolved from ultraviolet absorbances at the wavelength of maximum absorbance at about 270 nm of filtered portions of the solution under test, suitably diluted with *Dissolution Medium*, if necessary, in comparison with a Standard solution having a known concentration of USP Demeclocycline Hydrochloride RS in the same medium.

Tolerances—Not less than 75% (*Q*) of the labeled amount of demeclocycline ($C_{21}H_{21}ClN_2O_8.HCl$) is dissolved in 45 minutes.

Loss on drying ⟨731⟩—Dry about 100 mg of powdered Tablets, accurately weighed, in a capillary-stoppered bottle in vacuum at a pressure not exceeding 5 mm of mercury at 60° for 3 hours: it loses not more than 4.0% of its weight.

Assay for demeclocycline hydrochloride—Place not less than 5 Demeclocycline Hydrochloride and Nystatin Tablets in a high-speed glass blender jar containing an accurately measured volume of 0.1 *N* hydrochloric acid and blend for 4 ± 1 minutes, so that the stock solution so obtained contains not less than 150 µg of demeclocycline hydrochloride ($C_{21}H_{21}ClN_2O_8.HCl$) per mL. Proceed as directed under *Antibiotics—Microbial Assays* ⟨81⟩, using an accurately measured volume of this stock solution diluted quantitatively and stepwise with water to obtain a *Test Dilution* having a concentration of demeclocycline hydrochloride assumed to be equal to the median dose level of the Standard.

Assay for nystatin—Proceed as directed for Nystatin under *Antibiotics—Microbial Assays* ⟨81⟩, blending not less than 5 Demeclocycline Hydrochloride and Nystatin Tablets for 4 ± 1 minutes in a high-speed blender with a sufficient accurately measured volume of dimethylformamide to obtain a solution of convenient concentration. Dilute an accurately measured portion of this solution quantitatively with dimethylformamide to obtain a stock solution containing about 400 USP Nystatin Units per mL. Dilute this stock solution quantitatively with *Buffer No. 6* to obtain a *Test Dilution* having a concentration of nystatin assumed to be equal to the median dose level of the Standard.

Denatonium Benzoate—*see* Denatonium Benzoate NF

Dental Paste, Triamcinolone Acetonide—*see* Triamcinolone Acetonide Dental Paste

Desipramine Hydrochloride

$CH_2CH_2CH_2NHCH_3$ · HCl

$C_{18}H_{22}N_2.HCl$ 302.85

5*H*-Dibenz[*b,f*]azepine-5-propanamine, 10,11-dihydro-*N*-methyl-, monohydrochloride.

10,11-Dihydro-5-[3-(methylamino)propyl]-5*H*-dibenz[*b,f*]-azepine monohydrochloride [*58-28-6*].

» Desipramine Hydrochloride, dried in vacuum at 105° for 2 hours, contains not less than 98.0 percent and not more than 100.5 percent of $C_{18}H_{22}N_2.HCl$.

Packaging and storage—Preserve in tight containers.

Reference standards—*USP Desipramine Hydrochloride Reference Standard*—Dry in vacuum at 105° for 2 hours before using. *USP Iminodibenzyl Reference Standard*—Keep container tightly closed and protected from light. Dry over silica gel for 4 hours before using.

Identification—

A: The infrared absorption spectrum of a mineral oil dispersion of it, previously dried, exhibits maxima only at the same wavelengths as that of a similar preparation of USP Desipramine Hydrochloride RS.

B: The ultraviolet absorption spectrum of a 3 in 100,000 solution of it in 0.1 *N* hydrochloric acid exhibits maxima and minima at the same wavelengths as that of a similar solution of USP Desipramine Hydrochloride RS, concomitantly measured, and the respective absorptivities, calculated on the dried basis, at the wavelength of maximum absorbance at about 251 nm do not differ by more than 2.0%.

C: Add about 5 mg to 2 mL of nitric acid on a spot plate: an intense blue color is produced.

D: A 1 in 20 solution in alcohol responds to the tests for *Chloride* ⟨191⟩.

Loss on drying ⟨731⟩—Dry it in vacuum at 105° for 2 hours: it loses not more than 0.5% of its weight.

Residue on ignition ⟨281⟩: not more than 0.1%.

Heavy metals, *Method II* ⟨231⟩: 0.001%.

Iminodibenzyl—

Standard preparation—Dissolve an accurately weighed quantity of USP Iminodibenzyl RS in alcohol, and dilute quantitatively and stepwise with alcohol to obtain a solution having a concentration of 50 µg per mL. Transfer 1.0 mL of this solution to a low-actinic, 25-mL volumetric flask, add 10 mL of a mixture of equal volumes of hydrochloric acid and alcohol, and mix.

Test preparation—Transfer 50.0 mg of Desipramine Hydrochloride to a low-actinic, 25-mL volumetric flask, add 10 mL of a mixture of equal volumes of hydrochloric acid and alcohol, and mix.

Procedure—To the two flasks containing the *Standard preparation* and the *Test preparation*, and to a third flask containing 10 mL of a mixture of equal volumes of hydrochloric acid and alcohol to provide the blank, add slowly 5 mL of a 0.4% (v/v) solution of furfural in alcohol, mix, then add 5 mL of hydrochloric acid, and allow the flasks to stand in a constant-temperature bath at 25° for 3 hours. Dilute each flask to volume with a mixture of equal volumes of hydrochloric acid and alcohol, and mix. Concomitantly determine the absorbances of the solutions in 1-cm cells at the wavelength of maximum absorbance at about 565 nm, with a suitable spectrophotometer, using the blank to set the instrument: the absorbance of the solution from the *Test preparation* is not greater than that from the *Standard preparation* (0.1%).

Assay—Dissolve about 0.6 g of Desipramine Hydrochloride, previously dried and accurately weighed, in 100 mL of glacial acetic acid, add 10 mL of mercuric acetate TS, and titrate with 0.1 *N* perchloric acid VS, determining the end-point potentiometrically, using a calomel-glass electrode system. Perform a blank determination, and make any necessary correction. Each mL of 0.1 *N* perchloric acid is equivalent to 30.29 mg of $C_{18}H_{22}N_2.HCl$.

Desipramine Hydrochloride Capsules

» Desipramine Hydrochloride Capsules contain not less than 92.0 percent and not more than 108.0 percent of the labeled amount of $C_{18}H_{22}N_2.HCl$.

Packaging and storage—Preserve in tight containers.

Reference standard—*USP Desipramine Hydrochloride Reference Standard*—Dry in vacuum at 105° for 2 hours before using.

Identification—Mix the contents of Capsules, equivalent to about 125 mg of desipramine hydrochloride, with 10 mL of chloroform, stir for 5 minutes, and filter. Evaporate the chloroform on a steam bath to a volume of about 4 mL, add ether until the solution becomes turbid, then heat slightly on the steam bath to remove the turbidity, and cool: the crystals of desipramine hydrochloride so obtained, when filtered and dried at 100° for 30 minutes, respond to *Identification test A* under *Desipramine Hydrochloride*.

Dissolution ⟨711⟩—

Medium: water; 900 mL.

Apparatus 1: 100 rpm.

Time: 45 minutes.

Procedure—Determine the amount of $C_{18}H_{22}N_2.HCl$ dissolved from ultraviolet absorbances at the wavelength of maximum absorbance at about 251 nm of filtered portions of the solution under test, suitably diluted with *Dissolution Medium*, if necessary, in comparison with a Standard solution having a known concentration of USP Desipramine Hydrochloride RS in the same medium.

Tolerances—Not less than 75% (*Q*) of the labeled amount of $C_{18}H_{22}N_2.HCl$ is dissolved in 45 minutes.

Uniformity of dosage units ⟨905⟩: meet the requirements.

Procedure for content uniformity—Transfer the contents of 1 Capsule and the capsule shell to a 100-mL volumetric flask with the aid of 70 mL of 30 percent alcohol, and shake for 30 minutes, using a mechanical shaker that provides vigorous agitation. Dilute with 30 percent alcohol to volume, mix, and filter, discarding the first 20 mL of the filtrate. Dilute a portion of the subsequent filtrate quantitatively and stepwise with 30 percent alcohol to provide a solution containing approximately 25 µg of desipramine hydrochloride per mL. Concomitantly determine the absorbances of this solution and a Standard solution of USP Desipramine Hydrochloride RS in the same medium having a known concentration of about 25 µg per mL in 1-cm cells at the wavelength of maximum absorbance at about 251 nm, with a suitable spectrophotometer, using 30 percent alcohol as the blank. Calculate the quantity, in mg, of $C_{18}H_{22}N_2.HCl$ in the Capsule by the formula:

$$(TC/D)(A_U/A_S),$$

in which T is the labeled quantity, in mg, of desipramine hydrochloride in the Capsule, C is the concentration, in µg per mL, of USP Desipramine Hydrochloride RS in the Standard solution, D is the concentration, in µg per mL, of desipramine hydrochloride in the test solution, based upon the labeled quantity per Capsule and the extent of dilution, and A_U and A_S are the absorbances of the solution from the Capsule and the Standard solution, respectively.

Assay—Transfer, as completely as possible, the contents of not less than 20 Desipramine Hydrochloride Capsules to a tared weighing bottle, and weigh. Transfer an accurately weighed portion of the powder, equivalent to about 250 mg of desipramine hydrochloride, to a 250-mL volumetric flask, add about 100 mL of 0.01 *N* hydrochloric acid, and shake by mechanical means for 1 hour. Dilute with 0.01 *N* hydrochloric acid to volume, mix, and filter, discarding the first 20 mL of the filtrate. Transfer 5.0 mL of the subsequent filtrate to a separator, add 50 mL of 2.5 *N* sodium hydroxide, and extract with four 20-mL portions of ether, shaking each portion for 2 minutes and collecting the extracts in a second separator. Extract the combined ether extracts with four 30-mL portions of 0.01 *N* hydrochloric acid, combine the acid extracts in a 200-mL volumetric flask, dilute with 0.01 *N* hydrochloric acid to volume, and mix. Concomitantly determine the absorbances of this solution and of a Standard solution of USP Desipramine Hydrochloride RS in the same medium having a known concentration of about 25 µg per mL in 1-cm cells at the wavelength of maximum absorbance at about 251 nm, with a suitable spectrophotometer, using 0.01 *N* hydrochloric acid as the blank. Calculate the quantity, in mg, of $C_{18}H_{22}N_2.HCl$ in the portion of Capsules taken by the formula:

$$10C(A_U/A_S),$$

in which C is the concentration, in µg per mL, of USP Desipramine Hydrochloride RS in the Standard solution, and A_U and A_S are the absorbances of the solution from the Capsules and the Standard solution, respectively.

Desipramine Hydrochloride Tablets

» Desipramine Hydrochloride Tablets contain not less than 95.0 percent and not more than 105.0 percent of the labeled amount of $C_{18}H_{22}N_2.HCl$.

Packaging and storage—Preserve in tight containers.

Reference standard—*USP Desipramine Hydrochloride Reference Standard*—Dry in vacuum at 105° for 2 hours before using.

Identification—Finely powder a number of Tablets, equivalent to about 350 mg of desipramine hydrochloride, and triturate the powder with 15 mL of chloroform. Filter the chloroform extract through paper into a wide-mouth test tube, and evaporate the filtrate to about 3 mL. Carefully add ether until the liquid becomes turbid, heat cautiously to produce a clear solution, then cool, and allow to stand. Collect the crystals, wash with ether, and dry in vacuum at 80° for 30 minutes: the desipramine hydrochloride so obtained responds to *Identification test A* under *Desipramine Hydrochloride*.

Dissolution ⟨711⟩—

Medium: 0.1 *N* hydrochloric acid; 900 mL.

Apparatus 2: 50 rpm.

Time: 60 minutes.

Procedure—Determine the amount of $C_{18}H_{22}N_2.HCl$ dissolved from ultraviolet absorbances at the wavelength of maximum absorbance at about 251 nm of filtered portions of the solution under test, suitably diluted with *Dissolution Medium*, if necessary, in comparison with a Standard solution having a known concentration of USP Desipramine Hydrochloride RS in the same medium.

Tolerances—Not less than 75% (*Q*) of the labeled amount of $C_{18}H_{22}N_2.HCl$ is dissolved in 60 minutes.

Uniformity of dosage units ⟨905⟩: meet the requirements.

Procedure for content uniformity—Transfer 1 finely powdered Tablet to a 100-mL volumetric flask, add 50 mL of 0.1 *N* hydrochloric acid, and shake by mechanical means for 15 minutes. Dilute with 0.1 *N* hydrochloric acid to volume, mix, and filter, discarding the first 20 mL of the filtrate. Dilute a portion of the subsequent filtrate quantitatively and stepwise with 0.1 *N* hydrochloric acid to provide a solution containing approximately 25 µg of desipramine hydrochloride per mL. Concomitantly determine the absorbances of this solution and a Standard solution of USP Desipramine Hydrochloride RS in the same medium having a known concentration of about 25 µg per mL in 1-cm cells at the wavelength of maximum absorbance at about 251 nm, with a suitable spectrophotometer, using 0.1 *N* hydrochloric acid as the blank. Calculate the quantity, in mg, of $C_{18}H_{22}N_2.HCl$ in the tablet by the formula:

$$(TC/D)(A_U/A_S),$$

in which T is the labeled quantity, in mg, of desipramine hydrochloride in the Tablet, C is the concentration, in µg per mL, of USP Desipramine Hydrochloride RS in the Standard solution, D is the concentration, in µg per mL, of desipramine hydrochloride in the test solution, based upon the labeled quantity per Tablet and the extent of dilution, and A_U and A_S are the absorbances of the solution from the Tablet and the Standard solution, respectively.

Assay—

Standard preparation—Dissolve a suitable quantity of USP Desipramine Hydrochloride RS, accurately weighed, in 0.1 *N* hydrochloric acid, and dilute quantitatively and stepwise with the same solvent to obtain a solution having a known concentration of about 25 µg per mL.

Assay preparation—Finely powder a number of Desipramine Hydrochloride Tablets, equivalent to about 0.5 g of desipramine hydrochloride, and transfer the powder to a 500-mL volumetric flask with the aid of 100 mL of 0.1 *N* hydrochloric acid. Add 150 mL of the 0.1 *N* acid to the mixture, insert the stopper, and shake by mechanical means for 30 minutes. Dilute with 0.1 *N* hydrochloric acid to volume, mix, and filter, discarding the first 10 mL of the filtrate. Transfer 5.0 mL of the subsequent filtrate to a 200-mL volumetric flask, dilute with 0.1 *N* hydrochloric acid to volume, and mix.

Procedure—Transfer 15.0 mL each of the *Standard preparation* and the *Assay preparation*, respectively, to separate glass-stoppered, 50-mL centrifuge tubes, and to each tube add 2 mL of 2.5 *N* sodium hydroxide and 15.0 mL of cyclohexane suitable for use in ultraviolet spectrophotometry. Insert the stoppers, shake the tubes by mechanical means for 30 minutes, and centrifuge at about 1500 rpm until the clear phases separate (5 to 10 minutes). Concomitantly determine the absorbances of the cyclohexane extracts in 1-cm cells at the wavelength of maximum absorbance at about 255 nm, with a suitable spectrophotometer, using cyclohexane suitable for use in ultraviolet spectrophotometry as the blank. Calculate the quantity, in mg, of $C_{18}H_{22}N_2.HCl$ in the portion of Tablets taken by the formula:

$$20C(A_U/A_S),$$

in which C is the concentration, in µg per mL, of USP Desipramine Hydrochloride RS in the *Standard preparation*, and A_U and A_S are the absorbances of the solutions from the *Assay preparation* and the *Standard preparation*, respectively.

Desoximetasone

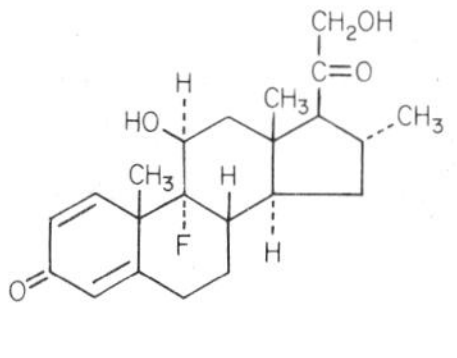

$C_{22}H_{29}FO_4$ 376.47

Pregna-1,4-diene-3,20-dione, 9-fluoro-11,21-dihydroxy-16-methyl-, (11β,16α)-.

9-Fluoro-11β,21-dihydroxy-16α-methylpregna-1,4-diene-3,20-dione [*382-67-2*].

» Desoximetasone contains not less than 97.0 percent and not more than 103.0 percent of $C_{22}H_{29}FO_4$, calculated on the dried basis.

Packaging and storage—Preserve in well-closed containers.

Reference standard—*USP Desoximetasone Reference Standard*—Dry at 105° to constant weight before using.

Identification—

A: The infrared absorption spectrum of a potassium bromide dispersion of it, previously dried at 105° for 3 hours, exhibits maxima only at the same wavelengths as that of a similar preparation of USP Desoximetasone RS.

B: Prepare a solution of it in a mixture of chloroform and alcohol (3:1) containing 10 mg per mL. Prepare a solution of USP Desoximetasone RS in the same mixture, containing 10 mg per mL. Apply separately 20 µL of each solution on a thin-layer chromatographic plate coated with a 0.25-mm layer of chromatographic silica gel mixture. Allow the spots to dry, and develop the chromatogram in a saturated chamber containing a mixture of chloroform and ethyl acetate (1:1). Allow the solvent front to move 10 cm beyond the application point. After drying, examine the plate under ultraviolet light at 254 nm. Spray the dried plate with a 1 in 5 solution of *p*-toluenesulfonic acid in alcohol. The major spot from the test solution corresponds in R_f value (about 0.25) and appearance to that obtained from the Standard solution.

Melting range ⟨741⟩: between 206° and 218°, but the range between beginning and end of melting does not exceed 4°.

Specific rotation ⟨781⟩: between +107° and +112°, calculated on the dried basis, determined in a solution in chloroform containing 50 mg in each 10 mL.

Loss on drying ⟨731⟩—Dry it at 105° to constant weight: it loses not more than 1.0% of its weight.

Residue on ignition ⟨281⟩: not more than 0.2%.

Heavy metals, *Method II* ⟨231⟩: 0.002%.

Assay—

Mobile phase—Prepare a suitable filtered and degassed solution of methanol, water, and glacial acetic acid (65:35:1), such that the retention time of desoximetasone is about 6 minutes.

Standard preparation—On the day of use, weigh accurately about 20 mg of USP Desoximetasone RS, and dissolve in methanol to obtain 50.0 mL. Dilute 10.0 mL of this solution with a mixture of methanol and acetonitrile (1:1) to 100.0 mL.

Assay preparation—Weigh accurately 40 mg of Desoximetasone, dissolve in 100.0 mL of methanol, and proceed as directed for *Standard preparation*, beginning with "Dilute 10.0 mL of this solution."

Procedure—Using an injection loop, inject 10-µL portions of the *Assay preparation* and the *Standard preparation* into a high-pressure liquid chromatograph equipped with an ultraviolet detector capable of monitoring absorption at 254 nm. The instrument is equipped with a 4.6-mm × 15-cm stainless steel column that contains packing L7 and is operated at a flow rate of about 1 mL per minute. In a suitable chromatogram, five replicate injections of the *Standard preparation* show a relative standard deviation of not more than 2.0%, and the tailing factor is not more than 1.5. Calculate the quantity, in mg, of $C_{22}H_{29}FO_4$ in the portion of Desoximetasone taken by the formula:

$$1000C(H_U/H_S),$$

in which C is the concentration, in mg per mL, of USP Desoximetasone RS in the *Standard preparation*, and H_U and H_S are the peak heights of desoximetasone obtained from the *Assay preparation* and the *Standard preparation*, respectively.

Desoximetasone Cream

» Desoximetasone Cream is Desoximetasone in an emollient cream base. It contains not less than 90.0 percent and not more than 110.0 percent of the labeled amount of $C_{22}H_{29}FO_4$.

Packaging and storage—Preserve in collapsible tubes, at controlled room temperature.

Reference standard—*USP Desoximetasone Reference Standard*—Dry at 105° to constant weight before using.

Identification—Evaporate 25 mL of the *Assay preparation*, prepared as directed in the *Assay*, on a steam bath just to dryness, and dissolve the residue in 2 mL of acetonitrile. This is the test solution. Prepare a Standard solution of USP Desoximetasone RS in acetonitrile containing 500 µg per mL. Using 10 µL instead of 20 µL of each solution, proceed as directed in *Identification* *test B* under *Desoximetasone*, beginning with "Apply separately 20 µL of each." The specified result is observed.

Minimum fill ⟨755⟩: meets the requirements.

pH ⟨791⟩: between 4.0 and 8.0, in a solution prepared in the following manner. Add 15 mL of boiling water to 3.5 g of the Cream in a 50-mL centrifuge tube, cap the tube, shake vigorously until the cream is uniformly dispersed, then place the tube in a steam bath until the water and oil layers separate completely. Cool, separate the layers, and determine the pH of the aqueous phase.

Assay—

Mobile phase—Prepare a filtered and degassed mixture of methanol, water, and glacial acetic acid (65:35:1). Make adjustments if necessary (see *System Suitability* under *Chromatography* ⟨621⟩).

Internal standard solution—Prepare a solution of ethylparaben in methanol having a concentration of about 0.04 mg per mL.

Standard preparation—Dissolve an accurately weighed quantity of USP Desoximetasone RS in methanol to obtain a solution having a known concentration of about 0.4 mg per mL. Pipet 5 mL of this solution into a 50-mL centrifuge tube. Add 10.0 mL of *Internal standard solution*, dilute with methanol quantitatively to 40.0 mL, and mix to obtain the *Standard preparation* having a known concentration of about 0.05 mg per mL.

Assay preparation—Transfer an accurately weighed amount of Desoximetasone Cream, equivalent to about 2 mg of desoximetasone, to a 50-mL centrifuge tube, and add a few 3-mm glass beads. Add 10.0 mL of *Internal standard solution* and about 30 mL of methanol, and mix. Tightly cap the centrifuge tube, and immerse it for 10 minutes in a bath maintained at a temperature of 65°. Remove the tube from the bath, and immediately vortex at high speed for 30 seconds. Return the tube to the hot water bath for 5 minutes, remove it from the bath, and immediately vortex for 30 seconds. Repeat the procedure one more time, then cool the tube in an ice-bath held at 10° until no further flocculent precipitation occurs. Centrifuge, and use the supernatant solution.

Chromatographic system (see *Chromatography* ⟨621⟩)—The liquid chromatograph is equipped with a 254-nm detector and a 4.6-mm × 15-cm column that contains packing L7. The flow rate is about 1 mL per minute. Chromatograph the *Standard preparation*, and record the peak responses as directed under *Procedure:* the tailing factor for the analyte peak is not more than 2.0, the resolution, *R*, between the analyte and internal standard peaks is not less than 2.0, and the relative standard deviation for replicate injections is not more than 2.0%.

Procedure—Separately inject equal volumes (about 10 µL) of the *Standard preparation* and the *Assay preparation* into the chromatograph, record the chromatograms, and measure the responses for the major peaks. The relative retention times are about 1 and 2 for ethylparaben and desoximetasone, respectively. Calculate the quantity, in mg, of $C_{22}H_{29}FO_4$ in the portion of Desoximetasone Cream taken by the formula:

$$40C(R_U/R_S),$$

in which *C* is the concentration, in mg per mL, of USP Desoximetasone RS in the *Standard preparation*, and R_U and R_S are the peak response ratios obtained from the *Assay preparation* and the *Standard preparation*, respectively.

Desoximetasone Gel

» Desoximetasone Gel contains not less than 90.0 percent and not more than 110.0 percent of the labeled amount of $C_{22}H_{29}FO_4$.

Packaging and storage—Preserve in collapsible tubes, at controlled room temperature.

Reference standard—*USP Desoximetasone Reference Standard*—Dry at 105° to constant weight before using.

Identification—Transfer an amount of Gel, equivalent to 100 µg of desoximetasone, to a 15-mL centrifuge tube. Add 3 mL of acetonitrile, sonicate for approximately 1 minute, centrifuge, and transfer the clear supernatant solution to another 15-mL centrifuge tube. Evaporate the solution under nitrogen at a temperature between 35° and 45° to dryness. Dissolve the residue in 100 µL of methanol, using a sonicator. Streak separately the entire test solution and 100 µL of a Standard solution of USP Desoximetasone RS in methanol containing 1 mg per mL on a thin-layer chromatographic plate (see *Chromatography* ⟨621⟩) coated with a 0.25-mm layer of chromatographic silica gel mixture and an area of preadsorbent material on which specimens are applied. Allow the streaks to dry, and develop the chromatogram in a saturated chamber containing a mixture of acetone and chloroform (1:1). Allow the solvent front to move not less than 10 cm beyond the origin. After drying, examine the plate under ultraviolet light at 254 nm: the R_f value of the principal spot obtained in the chromatogram of the test solution corresponds to that of the Standard solution.

Minimum fill ⟨755⟩: meets the requirements.

Alcohol content—Transfer about 2.5 g of Gel, accurately weighed, to a 50-mL volumetric flask. Dissolve in methanol, dilute with methanol to volume, and mix. Determine the alcohol content of the specimen thus prepared by the *Gas-liquid Chromatographic Method* (see *Alcohol Determination* ⟨611⟩), using isopropyl alcohol as the internal standard and using methanol in place of water as the solvent: between 18.0% and 24.0% (w/w) of C_2H_5OH is found.

Assay—

Mobile phase—Prepare a suitable filtered and degassed solution of methanol, water, and glacial acetic acid (65:35:1). Adjust the ratio, if necessary, so that the retention time of desoximetasone is about 8 minutes.

Standard preparation—Using an accurately weighed quantity of USP Desoximetasone RS, prepare a solution in methanol containing 0.5 mg per mL. Dilute an accurately measured volume of this solution with methanolic calcium chloride dihydrate solution (1.5 in 100) to obtain a *Standard preparation* having a known concentration of about 0.025 mg per mL.

Assay preparation—Transfer an accurately weighed quantity of Desoximetasone Gel, equivalent to about 1.25 mg of desoximetasone, to a 50-mL volumetric flask, add approximately 40 mL of methanolic calcium chloride dihydrate solution (1.5 in 100), and sonicate to disperse the gel. Dilute with the same solution to volume, mix, and centrifuge. Use the clear supernatant solution.

Chromatographic system (see *Chromatography* ⟨621⟩)—The liquid chromatograph is equipped with a 254-nm detector and a 3.9-mm × 30-cm column that contains packing L1. The flow rate is about 1.5 mL per minute. Chromatograph five replicate injections of the *Standard preparation*, and record the peak responses as directed under *Procedure:* the relative standard deviation is not more than 2.0%.

Procedure—Separately inject equal volumes (about 10 µL) of the *Standard preparation* and the *Assay preparation* into the chromatograph by means of a sampling valve, record the chromatograms, and measure the responses for the major peaks. Calculate the quantity, in mg, of $C_{22}H_{29}FO_4$ in the portion of Gel taken by the formula:

$$50C(r_U/r_S),$$

in which *C* is the concentration, in mg per mL, of USP Desoximetasone RS in the *Standard preparation*, and r_U and r_S are the peak responses of desoximetasone obtained from the *Assay preparation* and the *Standard preparation*, respectively.

Desoximetasone Ointment

» Desoximetasone Ointment contains not less than 90.0 percent and not more than 110.0 percent of the labeled amount of desoximetasone ($C_{22}H_{29}FO_4$).

Packaging and storage—Preserve in collapsible tubes, at controlled room temperature.

Reference standard—*USP Desoximetasone Reference Standard*—Dry at 105° to constant weight before using.

Identification—Transfer an accurately weighed quantity of Ointment, equivalent to about 5 mg of desoximetasone, to a 50-mL centrifuge tube. Add 20 mL of hexane, heat gently to 60°, and shake until the Ointment is completely dispersed. Add 8 mL of acetonitrile, insert the stopper in the tube, and shake vigorously for 5 minutes. Cool to room temperature, and centrifuge until the lower layer is clear. Transfer the lower layer to a 10-mL volumetric flask, dilute with acetonitrile to volume, and mix. Prepare a solution of USP Desoximetasone RS in acetonitrile containing 0.5 mg per mL. Separately apply 5 µL of each solution on a thin-layer chromatographic plate (see *Chromatography* ⟨621⟩) coated with a 0.25-mm layer of chromatographic silica gel mixture. Allow the spots to dry, and develop the plate in a saturated chamber containing a mixture of ethyl acetate and chloroform (4:1) until the solvent front has moved about three-fourths of the length of the plate. Remove the plate, and allow to air-dry. Examine under short-wavelength ultraviolet light. Spray the dried plate with a 1 in 5 solution of *p*-toluenesulfonic acid in alcohol. Heat the plate at 100° for 5 minutes, and examine under long-wavelength ultraviolet light: the R_f value and appearance (brownish yellow fluorescent spot) of the principal spot from the test solution, correspond to those of the principal spot from the Standard solution.

Minimum fill ⟨755⟩: meets the requirements.

Assay—

Mobile phase—Prepare a filtered and degassed mixture of methanol, water, and glacial acetic acid (65:35:1). Make adjustments if necessary (see *System Suitability* under *Chromatography* ⟨621⟩).

Standard preparation—Dissolve an accurately weighed quantity of USP Desoximetasone RS in methanol to obtain a solution having a known concentration of about 0.4 mg per mL. Quantitatively dilute 1 volume of this solution with 9 volumes of a 1:1 mixture of methanol and spectrophotometric acetonitrile that is saturated with *n*-heptane, and mix.

Assay preparation—Transfer an accurately weighed amount of Desoximetasone Ointment, equivalent to about 2 mg of desoximetasone, to a 50-mL centrifuge tube. Add 20 mL of *n*-heptane that has been previously saturated with spectrophotometric acetonitrile, and heat gently with occasional shaking until the Ointment is completely dispersed. Allow to cool slightly, and extract with a 10-mL portion of spectrophotometric acetonitrile. Shake vigorously, centrifuge, remove the bottom layer of acetonitrile with a syringe and needle, and transfer to a 50-mL volumetric flask. Using the same needle and syringe, extract the desoximetasone with successive 10-mL and 8-mL portions of acetonitrile, combining all acetonitrile layers in the 50-mL flask. Dilute with methanol nearly to volume, mix, and allow the solution to reach room temperature. Dilute with methanol to volume, and mix.

Chromatographic system (see *Chromatography* ⟨621⟩)—The liquid chromatograph is equipped with a 254-nm detector and a 4.6-mm × 15-cm column that contains packing L7. The flow rate is about 1 mL per minute. Chromatograph the *Standard preparation*, and record the peak responses as directed under *Procedure:* the tailing factor for the analyte peak is not more than 2.0, the resolution, *R*, between the analyte and solvent peaks is not less than 5.0, and the relative standard deviation for replicate injections is not more than 2.0%.

Procedure—Separately inject equal volumes (about 10 µL) of the *Standard preparation* and the *Assay preparation* into the chromatograph, record the chromatograms, and measure the responses for the major peaks. Calculate the quantity, in mg, of $C_{22}H_{29}FO_4$ in the portion of Desoximetasone Ointment taken by the formula:

$$50C(r_U/r_S),$$

in which *C* is the concentration, in mg per mL, of USP Desoximetasone RS in the *Standard preparation*, and r_U and r_S are the peak responses obtained from the *Assay preparation* and the *Standard preparation*, respectively.

Desoxycorticosterone Acetate

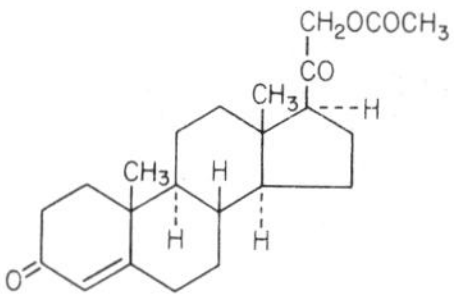

$C_{23}H_{32}O_4$ 372.50

Pregn-4-ene-3,20-dione, 21-(acetyloxy)-.

11-Deoxycorticosterone acetate [*56-47-3*].

» Desoxycorticosterone Acetate contains not less than 97.0 percent and not more than 103.0 percent of $C_{23}H_{32}O_4$, calculated on the dried basis.

Packaging and storage—Preserve in well-closed, light-resistant containers.

Reference standard—*USP Desoxycorticosterone Reference Standard*—Dry in vacuum over silica gel for 4 hours before using.

Identification—

A: The infrared absorption spectrum of a potassium bromide dispersion of it, previously dried in vacuum over silica gel for 4 hours, exhibits maxima only at the same wavelengths as that of a similar preparation of USP Desoxycorticosterone Acetate RS.

B: The ultraviolet absorption spectrum of a 1 in 100,000 solution in alcohol exhibits maxima and minima at the same wavelengths as that of a similar solution of USP Desoxycorticosterone Acetate RS, concomitantly measured, and the respective absorptivities, calculated on the dried basis, at the wavelength of maximum absorbance at about 240 nm, do not differ by more than 3.0%.

Melting range ⟨741⟩: between 155° and 161°.

Specific rotation ⟨781⟩: between +171° and +179°, determined in a solution in dioxane containing 100 mg in each 10 mL.

Loss on drying ⟨731⟩—Dry it in vacuum over silica gel for 4 hours: it loses not more than 0.5% of its weight.

Assay—

Standard preparation—Prepare as directed for *Standard Preparation* under *Assay for Steroids* ⟨351⟩, using USP Desoxycorticosterone Acetate RS.

Assay preparation—Weigh accurately about 100 mg of Desoxycorticosterone Acetate, dissolve in sufficient alcohol to make 200.0 mL, and mix. Pipet 5 mL of this solution into a 250-mL volumetric flask, add alcohol to volume, and mix. Pipet 20 mL of the resulting solution into a glass-stoppered, 50-mL conical flask.

Procedure—Proceed as directed for *Procedure* under *Assay for Steroids* ⟨351⟩. Calculate the quantity, in mg, of $C_{23}H_{32}O_4$ in the Desoxycorticosterone Acetate taken by the formula:

$$10C(A_U/A_S).$$

Desoxycorticosterone Acetate Injection

» Desoxycorticosterone Acetate Injection is a sterile solution of Desoxycorticosterone Acetate in vegetable oil. It contains not less than 90.0 percent and not more than 115.0 percent of the labeled amount of $C_{23}H_{32}O_4$.

Packaging and storage—Preserve in single-dose or in multiple-dose containers, preferably of Type I or Type III glass, protected from light.

Reference standard—*USP Desoxycorticosterone Reference Standard*—Dry in vacuum over silica gel for 4 hours before using.

Identification—Evaporate 25 mL of the *Assay preparation* from the *Assay* on a steam bath just to dryness, and dissolve the residue in 1 mL of chloroform. Using this as the test solution, proceed as directed under *Thin-layer Chromatographic Identification Test* ⟨201⟩.

Other requirements—It meets the requirements under *Injections* ⟨1⟩.

Assay—

Alcohol-isooctane and Isooctane-alcohol—Shake equal volumes of 90 percent alcohol and isooctane in a separator for 10 to 15 minutes, and allow to separate. Withdraw the layers into separate containers, designating the lower layer as "alcohol-isooctane" and the upper layer as "isooctane-alcohol."

Standard preparation—Prepare as directed for *Standard Preparation* under *Assay for Steroids* ⟨351⟩, using USP Desoxycorticosterone Acetate RS.

Assay preparation—Transfer an accurately measured volume of Desoxycorticosterone Acetate Injection, equivalent to about 5 mg of desoxycorticosterone acetate, to a separator containing 50 mL of *Isooctane-alcohol.* Extract with six 20-mL portions of *Alcohol-isooctane*, receiving the extracts in a 250-mL volumetric flask, dilute with *Alcohol-isooctane* to volume, and mix. Pipet 10 mL of this solution into a glass-stoppered, 50-mL conical flask, evaporate on a steam bath with the aid of a gentle current of air just to dryness, and dissolve the residue in 20.0 mL of alcohol.

Procedure—Proceed as directed for *Procedure* under *Assay for Steroids* ⟨351⟩. Calculate the quantity, in mg, of $C_{23}H_{32}O_4$ in each mL of the Injection taken by the formula:

$$0.5(C/V)(A_U/A_S),$$

in which V is the volume, in mL, of Injection taken.

Desoxycorticosterone Acetate Pellets

» Desoxycorticosterone Acetate Pellets are sterile pellets composed of Desoxycorticosterone Acetate in compressed form, without the presence of any binder, diluent, or excipient. They contain not less than 97.0 percent and not more than 103.0 percent of $C_{23}H_{32}O_4$.

Packaging and storage—Preserve in tight containers suitable for maintaining sterile contents, holding one pellet each.

Reference standard—*USP Desoxycorticosterone Acetate Reference Standard*—Dry in vacuum over silica gel for 4 hours before using.

Identification—

A: The infrared absorption spectrum of a potassium bromide dispersion of finely powdered Pellets exhibits maxima only at the same wavelengths as that of a similar preparation of USP Desoxycorticosterone Acetate RS.

B: The ultraviolet absorption of a 1 in 100,000 solution in alcohol exhibits maxima and minima at the same wavelengths as that of a similar solution of USP Desoxycorticosterone Acetate RS, concomitantly measured, and the respective absorptivities, calculated on the dried basis, at the wavelength of maximum absorbance at about 240 nm do not differ by more than 3.0%.

Solubility in alcohol—A solution of 25 mg of powdered Pellets in 1 mL of alcohol is clear and practically free from insoluble residue.

Melting range ⟨741⟩: between 155° and 161°.

Specific rotation ⟨781⟩: between +171° and +179°, determined in a solution in dioxane containing 100 mg in each 10 mL.

Sterility—Pellets meet the requirements under *Sterility Tests* ⟨71⟩.

Weight variation—Weigh 5 Pellets singly, and calculate the average weight. The average weight is not less than 95% and not more than 105% of the labeled weight of $C_{23}H_{32}O_4$, and each Pellet weighs not less than 90% and not more than 110% of the labeled weight of $C_{23}H_{32}O_4$.

Assay—

Standard preparation—Prepare as directed under *Assay for Steroids* ⟨351⟩, using USP Desoxycorticosterone Acetate RS.

Assay preparation—Weigh and finely powder not less than 10 Desoxycorticosterone Acetate Pellets. Weigh accurately about 100 mg of the powder, dissolve it in sufficient alcohol to make 200.0 mL, and mix. Transfer 5.0 mL of this solution to a 250-mL volumetric flask, dilute with alcohol to volume, and mix. Transfer 20.0 mL of this solution to a glass-stoppered, 50-mL conical flask.

Procedure—Proceed as directed for *Procedure* under *Assay for Steroids* ⟨351⟩. Calculate the quantity, in mg, of $C_{23}H_{32}O_4$ in the portion of Pellets taken by the formula:

$$10C(A_U/A_S).$$

Desoxycorticosterone Pivalate

$C_{26}H_{38}O_4$ 414.58

Pregn-4-ene-3,20-dione, 21-(2,2-dimethyl-1-oxopropoxy)-.

11-Deoxycorticosterone pivalate [*808-48-0*].

» Desoxycorticosterone Pivalate contains not less than 97.0 percent and not more than 103.0 percent of $C_{26}H_{38}O_4$, calculated on the dried basis.

Packaging and storage—Preserve in well-closed, light-resistant containers.

Reference standard—*USP Desoxycorticosterone Pivalate Reference Standard*—Dry at 105° for 2 hours before using.

Identification—

A: The infrared absorption spectrum of a potassium bromide dispersion of it, previously dried, exhibits maxima only at the same wavelengths as that of a similar preparation of USP Desoxycorticosterone Pivalate RS.

B: The ultraviolet absorption spectrum of a 1 in 50,000 solution in methanol exhibits maxima and minima at the same wavelengths as that of a similar solution of USP Desoxycorticosterone Pivalate RS, concomitantly measured, and the respective absorptivities, calculated on the dried basis, at the wavelength of maximum absorbance at about 241 nm do not differ by more than 3.0%.

Melting range ⟨741⟩: between 200° and 206°.

Specific rotation ⟨781⟩: between +155° and +163°, calculated on the dried basis, determined in a solution in dioxane containing 100 mg in each 10 mL.

Loss on drying ⟨731⟩—Dry it at 105° for 2 hours: it loses not more than 0.5% of its weight.

Assay—

Mobile phase—Prepare a filtered and degassed mixture of methanol and water (4:1). Make adjustments if necessary (see *System Suitability* under *Chromatography* ⟨621⟩).

Internal standard solution—Transfer about 100 mg of desoxycorticosterone acetate to a 50-mL volumetric flask, add methanol to volume, and mix.

Standard preparation—Transfer about 25 mg of USP Desoxycorticosterone Pivalate RS, accurately weighed, to a 50-mL volumetric flask, add 40 mL of methanol, and mix. Add 5.0 mL of *Internal standard solution*, dilute with methanol to volume, and mix to obtain a solution having a known concentration of about 0.5 mg of USP Desoxycorticosterone Pivalate RS per mL.

Assay preparation—Transfer about 50 mg of desoxycorticosterone pivalate, accurately weighed, to a 100-mL volumetric flask, add 80 mL of methanol, and mix. Add 10.0 mL of *Internal standard solution*, dilute with methanol to volume, and mix.

Chromatographic system (see *Chromatography* ⟨621⟩)—The liquid chromatograph is equipped with a 254-nm detector and a 4.6-mm × 25-cm column that contains packing L7. The flow rate is about 1.5 mL per minute. Chromatograph the *Standard preparation*, and record the peak responses as directed under *Procedure:* the resolution, R, between the analyte and internal standard peaks is not less than 2.0, and the relative standard deviation for replicate injections is not more than 1.5%.

Procedure—Separately inject equal volumes (about 25 µL) of the *Standard preparation* and the *Assay preparation* into the chromatograph, record the chromatograms, and measure the responses for the major peaks. The relative retention times are about 0.5 for desoxycorticosterone acetate and 1.0 for desoxycorticosterone pivalate. Calculate the quantity, in mg, of $C_{23}H_{32}O_4$ in the portion of Desoxycorticosterone Pivalate taken by the formula:

$$100C(R_U/R_S),$$

in which C is the concentration, in mg per mL, of USP Desoxycorticosterone Pivalate RS in the *Standard preparation*, and R_U and R_S are the peak response ratios obtained from the *Assay preparation* and the *Standard preparation*, respectively.

Sterile Desoxycorticosterone Pivalate Suspension

» Sterile Desoxycorticosterone Pivalate Suspension is a sterile suspension of Desoxycorticosterone Pivalate in an aqueous medium. It contains not less than 90.0 percent and not more than 110.0 percent of the labeled amount of $C_{26}H_{38}O_4$.

Packaging and storage—Preserve in single-dose or in multiple-dose containers, preferably of Type I glass, protected from light.

Reference standard—*USP Desoxycorticosterone Pivalate Reference Standard*—Dry at 105° for 2 hours before using.

Identification—Centrifuge a portion of Suspension, decant the supernatant liquid, wash the residue by stirring with several successive portions of water, centrifuging and decanting each time, and finally dry the residue at 105°: the desoxycorticosterone pivalate so obtained melts between 198° and 204°, and when about 5 mg of the residue is dissolved in 2 mL of sulfuric acid, the solution is yellowish, with a greenish fluorescence. Dilute the solution with 2 mL of water: the color changes to a dark red-blue, and on further dilution with 2 mL of water it is discharged.

pH ⟨791⟩: between 5.0 and 7.0.

Other requirements—It meets the requirements under *Injections* ⟨1⟩.

Assay—

Mobile phase, Internal standard solution, Standard preparation, and *Chromatographic system*—Proceed as directed in the *Assay* under *Desoxycorticosterone Pivalate.*

Assay preparation—Using a "to contain" pipet, transfer an accurately measured volume of the Sterile Suspension, equivalent to about 125 mg of desoxycorticosterone pivalate, to a 250-mL volumetric flask. Add about 200 mL of methanol, and sonicate to dissolve. Add 25.0 mL of the *Internal standard solution*, dilute with methanol to volume, and mix. Centrifuge a 20-mL portion at high speed for about 5 minutes. Filter the supernatant solution through a 5-µm disk, discarding the first 5 mL of the filtrate.

Procedure—Proceed as directed for *Procedure* in the *Assay* under *Desoxycorticosterone Pivalate.* Calculate the quantity, in mg, of $C_{23}H_{32}O_4$ in each mL of the Sterile Suspension taken by the formula:

$$250(C/V)(R_U/R_S),$$

in which V is the volume, in mL, of Sterile Suspension taken and the other terms are as previously defined.

Dexamethasone

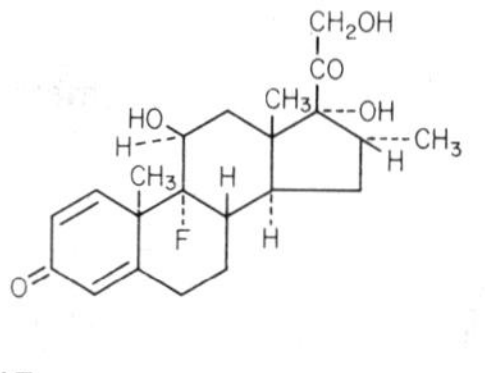

$C_{22}H_{29}FO_5$ 392.47

Pregna-1,4-diene-3,20-dione, 9-fluoro-11,17,21-trihydroxy-16-methyl-, (11β,16α)-.

9-Fluoro-11β,17,21-trihydroxy-16α-methylpregna-1,4-diene-3,20-dione [*50-02-2*].

» Dexamethasone contains not less than 97.0 percent and not more than 102.0 percent of $C_{22}H_{29}FO_5$, calculated on the dried basis.

Packaging and storage—Preserve in well-closed containers.

Reference standard—*USP Dexamethasone Reference Standard*—Dry at 105° for 3 hours before using.

Identification—

A: The infrared absorption spectrum of a potassium bromide dispersion of it, previously dried, exhibits maxima only at the same wavelengths as that of a similar preparation of USP Dexamethasone RS. If a difference appears, separately dissolve portions of both the test specimen and the Reference Standard in acetonitrile, evaporate each solution to dryness, and repeat the test on the residues.

B: The ultraviolet absorption spectrum of a 1 in 100,000 solution in methanol exhibits maxima and minima at the same wavelengths as that of a similar solution of USP Dexamethasone RS, concomitantly measured, and the respective absorptivities, calculated on the dried basis, at the wavelength of maximum absorbance at about 239 nm do not differ by more than 3.0%.

Specific rotation ⟨781⟩: between +72° and +80°, calculated on the dried basis, determined in a solution in dioxane containing 100 mg in each 10 mL.

Loss on drying ⟨731⟩—Dry it at 105° for 3 hours: it loses not more than 1.0% of its weight.

Residue on ignition ⟨281⟩: not more than 0.2% from 250 mg.

Assay—

Mobile phase—Prepare a suitable degassed solution of water and acetonitrile (about 7:3) such that at an approximate flow rate of 2 mL per minute, the retention time of Dexamethasone is about 7 minutes.

Standard preparation—Prepare a solution of USP Dexamethasone RS in methanol having a known concentration of about 7.5 mg per mL. Dilute an accurately measured volume of this solution with the *Mobile phase* to obtain a *Standard preparation* having a known concentration of about 0.3 mg per mL.

Assay preparation—Using 30 mg of Dexamethasone, proceed as directed for *Standard preparation.*

Procedure—Introduce equal volumes (between 15 µL and 30 µL) of the *Assay preparation* and the *Standard preparation* into a high-pressure liquid chromatograph (see *Chromatography* ⟨621⟩) operated at room temperature by means of a suitable microsyringe or sampling valve, adjusting the operating parameters such that the peak obtained with the *Standard preparation* is 60% full-scale. Typically, the apparatus is fitted with a 4-mm × 25-cm column containing packing L7, is equipped with an ultraviolet detector capable of monitoring absorption at 254 nm and a suitable recorder and is operated at about 1000 psi. Five replicate injections of the *Standard preparation* show a relative standard deviation of not more than 3.0%. Determine the peak responses, at equivalent retention times, obtained with the *Assay preparation* and the *Standard preparation*, and calculate the quantity, in mg, of $C_{22}H_{29}FO_5$ in the portion of Dexamethasone taken by the formula:

$$100C(r_U/r_S),$$

in which C is the concentration, in mg per mL, of USP Dexamethasone RS in the *Standard preparation*, and r_U and r_S are the peak responses obtained from the *Assay preparation* and the *Standard preparation*, respectively.

Dexamethasone Topical Aerosol

» Dexamethasone Topical Aerosol is Dexamethasone in a suitable lotion base mixed with suitable propellants in a pressurized container. Dexamethasone Topical Aerosol delivers not less than 90.0 percent and not more than 120.0 percent of the labeled amount of $C_{22}H_{29}FO_5$.

Packaging and storage—Preserve in pressurized containers, and avoid exposure to excessive heat.

Reference standard—*USP Dexamethasone Reference Standard*—Dry at 105° for 3 hours before using.

Identification—Evaporate 5 mL of the *Assay preparation*, obtained as directed in the *Assay*, on a steam bath just to dryness, and dissolve the residue in 1 mL of chloroform. On a suitable thin-layer chromatographic plate (see *Chromatography* ⟨621⟩), coated with a 0.25-mm layer of chromatographic silica gel, apply 100 µL of this solution and 10 µL of a solution of USP Dexamethasone RS in chloroform containing 500 µg per mL. Allow the spots to dry, and develop the chromatogram in a solvent system consisting of a mixture of chloroform and diethylamine (2:1) until the solvent front has moved about three-fourths of the length of the plate. Remove the plate from the developing chamber, mark the solvent front, and allow the solvent to evaporate. Locate the spots on the plate by lightly spraying with dilute sulfuric acid (1 in 2) and heating on a hot plate or under a heat lamp until spots appear: the R_f value of the principal spot obtained from the test solution corresponds to that obtained from the Standard solution.

Microbial limits—It meets the requirements of the tests for absence of *Staphylococcus aureus* and *Pseudomonas aeruginosa* under *Microbial Limit Tests* ⟨61⟩.

Other requirements—It meets the requirements for *Leak Testing* and *Pressure Testing* under *Aerosols* ⟨601⟩.

Assay—

Standard preparation—Prepare as directed under *Assay for Steroids* ⟨351⟩, using USP Dexamethasone RS.

Assay preparation—Shake the Dexamethasone Topical Aerosol container gently, and invert, immersing the valve in about 75 mL of alcohol contained in a 400-mL beaker. Actuate the valve by pushing against the bottom of the beaker. Remove the container at 15-second intervals, shake gently, and allow the container to warm to room temperature. Continue spraying until the contents of the container are exhausted. Transfer the alcohol solution to a 100-mL volumetric flask, dilute with alcohol to volume, and mix. Dilute an accurately measured volume of this solution, equivalent to about 1 mg of dexamethasone, with alcohol to 100.0 mL, and mix. Transfer 20.0 mL of this solution to a glass-stoppered, 50-mL flask.

Procedure—Proceed as directed for *Procedure* under *Assay for Steroids* ⟨351⟩, except to allow to stand in the dark for 45 minutes. Calculate the quantity, in mg, of $C_{22}H_{29}FO_5$ in each container by the formula:

$$(10C/V)(A_U/A_S),$$

in which V is the volume, in mL, of assay solution taken for the second dilution.

Dexamethasone Elixir

» Dexamethasone Elixir contains not less than 90.0 percent and not more than 110.0 percent of the labeled amount of $C_{22}H_{29}FO_5$.

Packaging and storage—Preserve in tight containers.

Reference standard—*USP Dexamethasone Reference Standard*—Dry at 105° for 3 hours before using.

Identification—Evaporate 9 mL of the *Assay preparation*, prepared as directed in the *Assay*, on a steam bath just to dryness, and dissolve the residue in 2 mL of a mixture of methylene chloride and methanol (1:1). Apply separately 5 µL of this solution and 5 µL of a solution of Dexamethasone RS in the mixture of methylene chloride and methanol (1:1), containing 0.5 mg per mL, on a thin-layer chromatographic plate coated with a 0.25-mm layer of chromatographic silica gel mixture (see *Chromatography* ⟨621⟩). Allow the spots to dry, and develop the chromatogram in a solvent system consisting of a mixture of chloroform, acetone, and glacial acetic acid (80:40:1) until the solvent front has moved about three-fourths of the length of the plate. Remove the plate from the developing chamber, mark the solvent front, and allow the solvent to evaporate. Locate the spots by viewing under short-wavelength ultraviolet light: the R_f value of the principal spot obtained from the solution under test corresponds to that obtained from the Standard solution.

Alcohol content ⟨611⟩ *Method II:* between 3.8% and 5.7% of C_2H_5OH, *n*-propyl alcohol being used as the internal standard.

Assay—

Mobile phase—Prepare a filtered and degassed mixture of water and acetonitrile (2:1). Make adjustments if necessary (see *System Suitability* under *Chromatography* ⟨621⟩).

Standard preparation—Dissolve an accurately weighed quantity of USP Dexamethasone RS in dilute methanol (1 in 2), and dilute quantitatively, and stepwise if necessary, with dilute methanol (1 in 2) to obtain a solution having a known concentration of about 0.1 mg per mL.

Assay preparation—Transfer an accurately measured volume of Dexamethasone Elixir, freshly mixed and free from air bubbles, equivalent to about 1 mg of dexamethasone, to a 10-mL volumetric flask, dilute with water to volume, mix, and filter through a suitable membrane filter.

Chromatographic system (see *Chromatography* ⟨621⟩)—The liquid chromatograph is equipped with a 254-nm detector and a 4.6-mm × 30-cm column that contains packing L1. The flow rate is about 1.5 mL per minute. Chromatograph the *Standard preparation*, and record the peak responses as directed under *Procedure:* the relative standard deviation for replicate injections is not more than 3.0%.

Procedure—Separately inject equal volumes (between 5 µL and 25 µL) of the *Standard preparation* and the *Assay preparation* into the chromatograph, record the chromatograms, and measure the responses for the major peaks. Calculate the quantity, in mg, of $C_{22}H_{29}FO_5$ in each mL of the Elixir taken by the formula:

$$10(C/V)(r_U/r_S),$$

in which C is the concentration, in mg per mL, of USP Dexamethasone RS in the *Standard preparation*, V is the volume, in mL, of Elixir taken, and r_U and r_S are the peak responses obtained from the *Assay preparation* and the *Standard preparation*, respectively.

Dexamethasone Gel

» Dexamethasone Gel contains not less than 90.0 percent and not more than 110.0 percent of the labeled amount of $C_{22}H_{29}FO_5$.

Packaging and storage—Preserve in collapsible tubes. Keep tightly closed. Avoid exposure to temperatures exceeding 30°.

Reference standard—*USP Dexamethasone Reference Standard*—Dry at 105° for 3 hours before using.

Identification—Evaporate 25 mL of the *Assay preparation*, prepared as directed in the *Assay*, on a steam bath just to dryness, and dissolve the residue in 0.5 mL of chloroform. The chloroform extract responds to the *Thin-layer Chromatographic Identification Test* ⟨201⟩, 10 µL of the chloroform extract and 10 µL of a Standard solution containing about 500 µg per mL of USP Dexamethasone RS being applied, and a mixture of chloroform and diethylamine (2:1) being used for development. Locate the spots on the plate by lightly spraying with dilute sulfuric acid (1 in 2) and heating.

Assay—

Mobile solvent—Dilute 100 mL of methylene chloride with isooctane to one liter.

Chromatographic columns—Tamp a pledget of glass wool at the constriction of a glass chromatographic tube measuring about 30-cm × 1.5-cm, equipped with a polytetrafluorethylene stopcock. Fill the tube about half-full with *Mobile solvent*. Mix 8 g of chromatographic siliceous earth with 8 mL of methanol and water (1:1). Transfer successive portions of the mixture to the column, emptying and adding *Mobile solvent* after each addition to pack the column. Finally drain the column to a layer of *Mobile solvent* about 1 cm above the absorbant. Pack a second tube to provide a blank and proceed as directed for *Assay preparation* but omit the specimen.

Standard preparation—Prepare a solution of USP Dexamethasone RS in alcohol to obtain a solution having a known concentration of about 10 µg per mL.

Assay preparation—Accurately weigh an amount of Dexamethasone Gel, equivalent to about 0.5 mg of Dexamethasone in a 100-mL beaker, add 1 g of chromatographic siliceous earth and mix. Transfer the mixture to the column, wash the beaker with small portions of *Mobile solvent*, adding them to the column. Adjust the flow rate to about 2 mL per minute, discarding the eluate. Elute with four additional 25-mL portions of *Mobile solvent*, and discard. Rinse the sample beaker with two 25-mL portions of methylene chloride and transfer to the column, collecting the eluate in a suitable beaker. Elute the column with six additional 25-mL portions of methylene chloride, combining the eluates and evaporating with a gentle stream of air to dryness, dissolve the residue in alcohol, and transfer to a 50-mL volumetric flask. Wash the beaker with successive 5-mL portions of alcohol, collecting the washings in the flask, dilute with alcohol to volume, and mix. Centrifuge or filter and then pipet 10 mL of this solution, 10 mL of the *Standard preparation*, 10 mL of the column blank solution, and 10 mL of alcohol to provide a reagent blank to separate flasks. Proceed as directed under *Assay for Steroids* ⟨351⟩, except to allow to stand in the dark for 45 minutes. Calculate the quantity, in mg, of $C_{22}H_{29}FO_5$ in the portion of Gel taken by the formula:

$$(0.05C)(A_U - A_{CB}/A_S - A_{RB}),$$

in which C is the concentration, in µg per mL, of USP Dexamethasone RS in the *Standard preparation*, and A_U, A_{CB}, A_S, and A_{RB} are the absorbances of the solutions from the *Assay preparation*, column blank preparation, *Standard preparation*, and reagent blank preparation, respectively.

Dexamethasone Ophthalmic Ointment, Neomycin and Polymyxin B Sulfates and—*see* Neomycin and Polymyxin B Sulfates and Dexamethasone Ophthalmic Ointment

Dexamethasone Ophthalmic Suspension

» Dexamethasone Ophthalmic Suspension is a sterile, aqueous suspension of dexamethasone containing a suitable antimicrobial preservative. It may contain suitable buffers, stabilizers, and suspending and viscosity agents. It contains not less than 90.0 percent and not more than 110.0 percent of the labeled amount of $C_{22}H_{29}FO_5$.

Packaging and storage—Preserve in tight containers.

Reference standard—*USP Dexamethasone Reference Standard*—Keep container tightly closed. Dry at 105° for 3 hours before using.

Identification—Transfer a volume of Suspension, equivalent to about 2.5 mg of dexamethasone, to a test tube, add 5 mL of chloroform, and shake. Centrifuge, and apply 10 µL of the chloroform layer and 10 µL of a Standard solution of USP Dexamethasone RS in chloroform containing 500 µg per mL on a thin-layer chromatographic plate (see *Chromatography* ⟨621⟩) coated with a 0.25-mm layer of chromatographic silica gel mixture. Develop the chromatogram in *Solvent A* as directed under *Single-steroid Assay* ⟨511⟩. Mark the solvent front, and locate the spots on the plate by spraying with a 1 in 5 solution of *p*-toluenesulfonic acid in a mixture of 9 volumes of alcohol and 1 volume of propylene glycol, and heating until spots appear. The R_f value of the principal spot obtained from the solution under test corresponds to that obtained from the Standard solution.

Sterility—It meets the requirements under *Sterility Tests* ⟨71⟩.

pH ⟨791⟩: between 5.0 and 6.0.

Assay—

Mobile phase—Prepare a filtered and degassed mixture of water and acetonitrile (60:40). Make adjustments if necessary (see *System Suitability* under *Chromatography* ⟨621⟩).

Standard preparation—Dissolve an accurately weighed quantity of USP Dexamethasone RS in *Mobile phase*, and dilute quantitatively, and stepwise if necessary, with *Mobile phase* to obtain a solution having a known concentration of about 0.12 mg per mL.

Assay preparation—Transfer an accurately measured volume of Dexamethasone Ophthalmic Suspension, freshly mixed and free from air bubbles, equivalent to about 3 mg of dexamethasone, to a 25-mL volumetric flask, dilute with *Mobile phase* to volume, and mix.

Chromatographic system (see *Chromatography* ⟨621⟩)—The liquid chromatograph is equipped with a 254-nm detector and a 4.6-mm × 25-cm column that contains packing L1. The flow rate is about 2 mL per minute. Chromatograph the *Standard preparation*, and record the peak response as directed under *Procedure:* the column efficiency determined from the analyte peak is not less than 1750 theoretical plates, the tailing factor for the analyte peak is not more than 3.0, and the relative standard deviation for replicate injections is not more than 3.0%.

Procedure—Separately inject equal volumes (about 10 µL) of the *Standard preparation* and the *Assay preparation* into the chromatograph, record the chromatograms, and measure the responses for the major peaks. Calculate the quantity, in mg, of $C_{22}H_{29}FO_5$ in each mL of the Ophthalmic Suspension taken by the formula:

$$25(C/V)(r_U/r_S),$$

in which C is the concentration, in mg per mL, of USP Dexamethasone RS in the *Standard preparation*, V is the volume, in mL, of Ophthalmic Suspension taken, and r_U and r_S are the peak responses obtained from the *Assay preparation* and the *Standard preparation*, respectively.

Dexamethasone Ophthalmic Suspension, Neomycin and Polymyxin B Sulfates and—*see* Neomycin and Polymyxin B Sulfates and Dexamethasone Ophthalmic Suspension

Dexamethasone Suspension, Sterile Penicillin G Procaine, Dihydrostreptomycin Sulfate, Chlorpheniramine Maleate, and—*see* Sterile Penicillin G Procaine, Dihydrostreptomycin Sulfate, Chlorpheniramine Maleate, and Dexamethasone Suspension

Dexamethasone Tablets

» Dexamethasone Tablets contain not less than 90.0 percent and not more than 110.0 percent of the labeled amount of $C_{22}H_{29}FO_5$.

Packaging and storage—Preserve in well-closed containers.

Reference standard—*USP Dexamethasone Reference Standard*—Dry at 105° for 3 hours before using.

Identification—Evaporate 10 mL of the methanol extract of Tablets obtained as directed under *Assay preparation* in the *Assay* on a steam bath just to dryness, and dissolve the residue in 1 mL of chloroform. Apply 10 µL of this solution and 20 µL of a solution of Dexamethasone RS in chloroform containing 500 µg per mL on a thin-layer chromatographic plate (see *Chromatography* ⟨621⟩) coated with a 0.25-mm layer of chromatographic silica gel mixture. Develop the chromatogram in *Solvent A* as directed under *Single-steroid Assay* ⟨511⟩. Mark the solvent front, and locate the spots on the plate by visualizing under short-wavelength ultraviolet light: the R_f value of the principal spot obtained from the solution under test corresponds to that obtained from the Standard solution.

Dissolution ⟨711⟩—

Medium: dilute hydrochloric acid (1 in 100); 500 mL.

Apparatus 1: 100 rpm.

Time: 45 minutes.

Standard solution—Prepare as directed for *Standard Preparation* under *Assay for Steroids* ⟨351⟩, using USP Dexamethasone RS.

Procedure—Extract a filtered aliquot of *Dissolution Medium,* equivalent to about 200 µg of dexamethasone, with three 15-mL portions of chloroform. Evaporate the combined chloroform extracts on a steam bath just to dryness, cool, and dissolve the residue in 20 mL of alcohol. Proceed as directed for *Procedure* under *Assay for Steroids* ⟨351⟩, except to allow to stand in the dark for 45 minutes. Calculate the portion, in mg, of $C_{22}H_{29}FO_5$ dissolved by the formula:

$$10(C/V)(A_U/A_S),$$

in which V is the volume, in mL, of the aliquot extracted with chloroform.

Tolerances—Not less than 70% (Q) of the labeled amount of $C_{22}H_{29}FO_5$ is dissolved in 45 minutes.

Uniformity of dosage units ⟨905⟩: meet the requirements.

Procedure for content uniformity—

Standard preparation—Prepare as directed for *Standard Preparation* under *Assay for Steroids* ⟨351⟩, using USP Dexamethasone RS.

Test preparation—Place 1 Tablet in a separator with 15 mL of water, and swirl to disintegrate the Tablet completely. Extract with four 10-mL portions of chloroform, filtering each portion through chloroform-washed cotton into a 50-mL volumetric flask, add chloroform to volume, and mix. Pipet a volume of this solution, equivalent to about 200 µg of dexamethasone into a glass-stoppered, 50-mL conical flask, evaporate the chloroform on a steam bath just to dryness, cool, and dissolve the residue in 20.0 mL of alcohol. Use this where *Assay Preparation* is specified in the *Procedure.*

Procedure—Proceed as directed for *Procedure* under *Assay for Steroids* ⟨351⟩, except to allow to stand in the dark for 45 minutes. Calculate the quantity, in mg, of total steroids, as $C_{22}H_{29}FO_5$, in the Tablet by the formula:

$$(C/V)(A_U/A_S),$$

in which V is the volume, in mL, of the aliquot taken to prepare the *Test preparation.*

Assay—

Mobile solvent—Prepare a suitable aqueous solution of acetonitrile, approximately 1 in 3, such that the retention time of dexamethasone is between 3 minutes and 6 minutes.

Standard preparation—Dissolve an accurately weighed quantity of USP Dexamethasone RS in dilute methanol (1 in 2) to obtain a solution having a known concentration of about 0.1 mg per mL.

Assay preparation—Weigh and finely powder not less than 10 Dexamethasone Tablets. Weigh accurately a portion of the powder, equivalent to about 5 mg of dexamethasone, transfer to a 50-mL volumetric flask, and add 30 mL of dilute methanol (1 in 2). Sonicate the flask for about 2 minutes, shake by mechanical means for 30 minutes, and dilute with the same solvent to volume. Filter a portion of the mixture through a suitable filter to obtain a clear filtrate.

Procedure—Introduce equal volumes (between 5 µL and 25 µL) of the *Assay preparation* and the *Standard preparation* into a high-pressure liquid chromatograph (see *Chromatography* ⟨621⟩) operated at room temperature, by means of a loop injector, adjusting the specimen size and other operating parameters such that the peak obtained with the *Standard preparation* is about 0.6 full scale. Typically, the apparatus is fitted with a 4.6-mm × 30-cm column packed with packing L1 and is equipped with an ultraviolet detector capable of monitoring absorption at 254 nm and a suitable recorder. In a suitable chromatogram, the coefficient of variation for five replicate injections of a single specimen is not more than 3.0%. Measure the responses of the peaks, at identical retention times, obtained with the *Assay preparation* and the *Standard preparation.* Calculate the quantity, in mg, of $C_{22}H_{29}FO_5$ in the portion of Tablets taken by the formula:

$$50C(r_U/r_S),$$

in which C is the concentration, in mg per mL, of USP Dexamethasone RS in the *Standard preparation,* and r_U and r_S are the peak responses obtained from the *Assay preparation* and the *Standard preparation,* respectively.

Dexamethasone Acetate

$C_{24}H_{31}FO_6 . H_2O$ 452.52

Pregna-1,4-diene-3,20-dione, 21-(acetyloxy)-9-fluoro-11,17-dihydroxy-16-methyl-,(11β,16α)-monohydrate.

9-Fluoro-11β,17,21-trihydroxy-16α-methylpregna-1,4-diene-3,20-dione 21-acetate monohydrate [*55812-90-3*].

Anhydrous 434.50 [*1177-87-3*].

» Dexamethasone Acetate contains one molecule of water of hydration or is anhydrous. It contains not less than 97.0 percent and not more than 102.0 percent of $C_{24}H_{31}FO_6$, calculated on the dried basis.

Packaging and storage—Preserve in well-closed containers.

Labeling—Label it to indicate whether it is hydrous or anhydrous.

Reference standard—*USP Dexamethasone Acetate Reference Standard*—Dry in vacuum at 105° for 3 hours.

Identification—

A: The infrared absorption spectrum of a mineral oil dispersion of it exhibits maxima only at the same wavelengths as that of a similar, undried preparation of USP Dexamethasone Acetate RS.

B: The ultraviolet absorption spectrum of a 1 in 70,000 solution in methanol exhibits maxima and minima at the same wavelengths as that of a similar solution of USP Dexamethasone Acetate RS, concomitantly measured, and the respective absorptivities, calculated on the dried basis, at the wavelength of maximum absorbance at about 239 nm do not differ by more than 3.0%.

Specific rotation ⟨781⟩: between +82° and +88°, calculated on the dried basis, determined in a solution in dioxane containing 100 mg in each 10 mL.

Loss on drying ⟨731⟩—Dry it in vacuum at 105° for 3 hours: the hydrous form loses between 3.5% and 4.5%, and the anhydrous form not more than 0.4%, of its weight.

Residue on ignition ⟨281⟩: not more than 0.1%.

Heavy metals, *Method II* ⟨231⟩: not more than 0.002%.

Assay—

Mobile phase—Prepare a suitable degassed solution of water and acetonitrile (about 3:2), making adjustments if necessary (see *System Suitability* under *Chromatography* ⟨621⟩).

Standard preparation—Dissolve an accurately weighed quantity of USP Dexamethasone Acetate RS in acetonitrile, and dilute quantitatively with acetonitrile to obtain a solution having a known concentration of about 0.3 mg per mL.

Assay preparation—Transfer about 30 mg of Dexamethasone Acetate, accurately weighed, to a 100-mL volumetric flask, add acetonitrile to volume, and mix.

Chromatographic system (see *Chromatography* ⟨621⟩)—The liquid chromatograph is equipped with a 254-nm detector and a 4-mm × 25-cm column that contains packing L7. The flow rate is about 2 mL per minute. Chromatograph the *Standard preparation,* and record the peak response as directed under *Procedure:* the relative standard deviation for replicate injections is not more than 3.0%.

Procedure—Separately inject equal volumes (between 15 µL and 30 µL) of the *Standard preparation* and the *Assay preparation* into the chromatograph, record the chromatograms, and measure the responses for the major peaks. Calculate the quantity, in mg, of $C_{24}H_{31}FO_6$ in the portion of Dexamethasone Acetate taken by the formula:

$$100C(r_U/r_S),$$

in which C is the concentration, in mg per mL, of USP Dexamethasone Acetate RS in the *Standard preparation*, and r_U and r_S are the peak responses obtained from the *Assay preparation* and the *Standard preparation*, respectively.

Sterile Dexamethasone Acetate Suspension

» Sterile Dexamethasone Acetate Suspension is a sterile suspension of Dexamethasone Acetate in Water for Injection. It contains an amount of dexamethasone acetate monohydrate ($C_{24}H_{31}FO_6 \cdot H_2O$) equivalent to not less than 90.0 percent and not more than 110.0 percent of the labeled amount of dexamethasone ($C_{22}H_{29}FO_5$).

Packaging and storage—Preserve in single-dose or in multiple-dose containers, preferably of Type I glass.

Reference standard—*USP Dexamethasone Acetate Reference Standard*—Dry in vacuum at 105° for 3 hours before using.

Identification—Transfer the contents of a well-shaken container of Suspension to a fine-porosity sintered-glass vacuum filter, filter, and wash with several 10-mL portions of water. Remove the powder from the filter and allow to air-dry. [NOTE—Do not use heat to dry the specimen. Total or partial dehydration may occur.] The infrared absorption spectrum of a mineral oil dispersion of the residue so obtained exhibits maxima and minima at the same wavelengths as that of a similar undried preparation of USP Dexamethasone Acetate RS.

pH ⟨791⟩: between 5.0 and 7.5.

Other requirements—It meets the requirements under *Injections* ⟨1⟩.

Assay—

Standard preparation—Dissolve an accurately weighed quantity of USP Dexamethasone Acetate RS in alcohol, and dilute quantitatively and stepwise with alcohol to obtain a solution having a known concentration of about 10 µg per mL.

Assay preparation—Transfer 5.0 mL of well-shaken Sterile Dexamethasone Acetate Suspension, equivalent to about 40 mg of dexamethasone, to a 125-mL separator containing 20 mL of water and 25 mL of methylene chloride. Shake carefully for about 1 minute, allow the layers to separate, and filter the methylene chloride into a 100-mL volumetric flask. Repeat the extraction with two additional 25-mL portions of methylene chloride adding each filtrate to the flask. Dilute with methylene chloride to volume, and mix. Transfer 5.0 mL of this solution to a 200-mL volumetric flask, evaporate with a stream of air to dryness, dissolve in alcohol, dilute with alcohol to volume, and mix.

Procedure—Pipet 10 mL each of the *Standard preparation*, the *Assay preparation*, and alcohol into separate glass-stoppered flasks. Proceed as directed for *Procedure* under *Assay for Steroids* ⟨351⟩, except to allow to stand in the dark for 45 minutes. Calculate the quantity, in mg, of $C_{22}H_{29}FO_5$ in each mL of the Suspension taken by the formula:

$$(0.903)(4C/V)(A_U/A_S),$$

in which C is the concentration, in µg per mL, of USP Dexamethasone Acetate RS in the *Standard preparation*, V is the volume, in mL, of Suspension taken, and A_U and A_S are the absorbances of the solutions from the *Assay preparation* and the *Standard preparation*, respectively, and 0.903 is the ratio of the molecular weight of dexamethasone to that of dexamethasone acetate.

Dexamethasone Sodium Phosphate

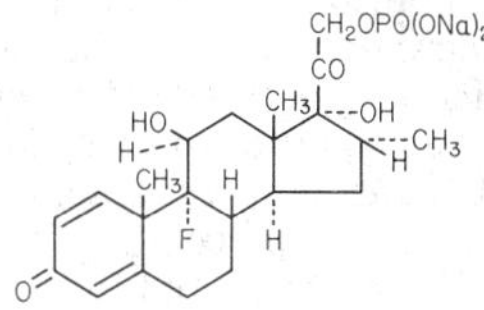

$C_{22}H_{28}FNa_2O_8P$ 516.41

Pregna-1,4-diene-3,20-dione, 9-fluoro-11,17-dihydroxy-16-methyl-21-(phosphonooxy)-, disodium salt, (11β,16α)-.

9-Fluoro-11β,17,21-trihydroxy-16α-methylpregna-1,4-diene-3,20-dione 21-(dihydrogen phosphate) disodium salt [*2392-39-4*].

» Dexamethasone Sodium Phosphate contains not less than 97.0 percent and not more than 102.0 percent of $C_{22}H_{28}FNa_2O_8P$, calculated on the water-free and alcohol-free basis.

Packaging and storage—Preserve in tight containers.

Reference standards—*USP Dexamethasone Reference Standard*—Dry at 105° for 3 hours before using. *USP Dexamethasone Phosphate Reference Standard*—Dry at a pressure of 5 mm of mercury at 40° to constant weight before using.

Identification—

A: *pH 9 buffer with magnesium*—Mix 3.1 g of boric acid and 500 mL of water in a 1-liter volumetric flask, add 21 mL of 1 *N* sodium hydroxide and 10 mL of 0.1 *M* magnesium chloride, dilute with water to volume, and mix.

Alkaline phosphatase solution—Transfer 95 ± 5 mg of alkaline phosphatase enzyme to a 50-mL volumetric flask, dissolve by adding *pH 9 buffer with magnesium* to volume, and mix. Prepare this solution fresh daily.

Standard preparation—Weigh 15 mg of USP Dexamethasone RS into a 5-mL volumetric flask. Dissolve in ethyl acetate, and dilute with ethyl acetate to volume. [NOTE—Sonication may be required to ensure dissolution.]

Test preparation—Weigh 20 mg of Dexamethasone Sodium Phosphate into a 15-mL centrifuge tube. Add 5.0 mL of *Alkaline phosphatase solution*, shake vigorously, and allow to stand for 30 minutes. Add 5.0 mL of ethyl acetate, shake vigorously, centrifuge, and use the upper, ethyl acetate layer.

Procedure—Apply 10-µL portions of the *Test preparation* and the *Standard preparation* to a thin-layer chromatographic plate (see *Chromatography* ⟨621⟩) coated with a 0.25-mm layer of chromatographic silica gel mixture. Develop the chromatogram in a mobile phase consisting of a mixture of chloroform, methanol, and water (180:15:1) to a distance of three-fourths of the length of the plate. Air-dry the plate and observe under short-wavelength ultraviolet light: the R_f value of the principal spot obtained from the *Test preparation* corresponds to that obtained from the *Standard preparation*.

B: The residue from the ignition of it responds to the tests for *Phosphate* ⟨191⟩ and for *Sodium* ⟨191⟩.

Specific rotation ⟨781⟩: between +74° and +82°, calculated on the water-free and alcohol-free basis, determined in a solution containing 10 mg per mL.

pH ⟨791⟩: between 7.5 and 10.5, in a solution (1 in 100).

Water, *Method I* ⟨921⟩—Determine the water content. The sum of the percentages of water content, and alcohol content, determined as directed in the test for *Alcohol*, does not exceed 16.0%.

Alcohol—Proceed as directed in the *Gas-liquid Chromatographic Method* under *Alcohol Determination* ⟨611⟩, except to use column packing S8 and to use the following modifications.

Internal standard solution—Pipet 1 mL of isopropyl alcohol into a 100-mL volumetric flask, add water to volume, and mix.

Standard solution—Prepare a 1 in 50 solution of alcohol in water. Determine the specific gravity at 25° (see *Specific Gravity* ⟨841⟩), and obtain the percentage of C_2H_5OH by reference to the *Alcoholometric Table* in the section, *Reference Tables*.

Standard preparation—Into a 10-mL volumetric flask pipet 4 mL of *Standard solution* and 5 mL of *Internal standard solution*.

Add water to volume, and mix. Inject 2 µL of this solution into the gas chromatograph.

Test preparation—Transfer about 500 mg of Dexamethasone Sodium Phosphate, accurately weighed, into a 10-mL volumetric flask. Pipet 5 mL of *Internal standard solution* into the flask, and mix to dissolve. Add water to volume, and mix. Inject 2 µL of this solution into the gas chromatograph.

Calculation—Calculate the percentage of alcohol in the Dexamethasone Sodium Phosphate by the formula:

$$4(S/W)(Z/Y),$$

in which S is the percentage of alcohol in the *Standard solution*, W is the weight, in g, of Dexamethasone Sodium Phosphate used in the *Test preparation*, and Y and Z are the ratios of the alcohol peak heights to the internal standard peak heights for the *Standard preparation* and the *Test preparation*, respectively. The content of C_2H_5OH is not more than 8.0%.

Phosphate ions—

Standard phosphate solution—Dissolve 143.3 mg of dried monobasic potassium phosphate, KH_2PO_4, in water to make 1000.0 mL. This solution contains the equivalent of 0.10 mg of phosphate (PO_4) in each mL.

Phosphate reagent A—Dissolve 5 g of ammonium molybdate in 1 *N* sulfuric acid to make 100 mL.

Phosphate reagent B—Dissolve 350 mg of *p*-methylaminophenol sulfate in 50 mL of water, add 20 g of sodium bisulfite, mix to dissolve, and dilute with water to 100 mL.

Procedure—Dissolve about 50 mg of Dexamethasone Sodium Phosphate, accurately weighed, in a mixture of 10 mL of water and 5 mL of 2 *N* sulfuric acid contained in a 25-mL volumetric flask, by warming if necessary. Add 1 mL each of *Phosphate reagent A* and *Phosphate reagent B*, dilute with water to 25 mL, mix, and allow to stand at room temperature for 30 minutes. Similarly and concomitantly, prepare a standard solution, using 5.0 mL of *Standard phosphate solution* instead of the 50 mg of the substance under test. Concomitantly determine the absorbances of both solutions in 1-cm cells at 730 nm, with a suitable spectrophotometer, using water as the blank. The absorbance of the test solution is not more than that of the standard solution. The limit is 1.0% of phosphate (PO_4).

Free dexamethasone—

Mobile phase, Standard preparation, Assay preparation, System suitability preparation, and *Chromatographic system*—Proceed as directed in the *Assay*.

Procedure—Separately inject equal volumes (about 20 µL) of the *Standard preparation* and the *Assay preparation* into the chromatograph, record the chromatograms, and measure the responses for the dexamethasone peaks. Calculate the quantity, in µg, of dexamethasone in the portion of Dexamethasone Sodium Phosphate taken by the formula:

$$1000C(r_U/r_S),$$

in which C is the concentration, in µg per mL, of USP Dexamethasone RS in the *Standard preparation*, and r_U and r_S are the peak responses for the dexamethasone peak obtained from the *Assay preparation* and the *Standard preparation*, respectively: not more than 1.0% is found.

Assay—

Mobile phase—Prepare a solution containing 7.5 mL of triethylamine in 1 liter of water. Adjust by the addition of phosphoric acid to a pH of 5.4. Prepare a filtered and degassed mixture of 74 parts of the resulting solution with 26 parts of methanol. Make adjustments if necessary (see *System Suitability* under *Chromatography* ⟨621⟩).

Standard preparation—Dissolve an accurately weighed quantity of USP Dexamethasone Phosphate RS in *Mobile phase* to obtain a solution containing about 0.5 mg per mL. Prepare a second solution by dissolving an accurately weighed quantity of USP Dexamethasone RS in a mixture of methanol and water (1:1) to obtain a solution containing about 50 µg per mL. Transfer 10.0 mL of the first solution and 1.0 mL of the second solution to a 100-mL volumetric flask. Dilute with *Mobile phase* to volume, and mix to obtain the *Standard preparation* having known concentrations of 50 µg of USP Dexamethasone Phosphate RS per mL and 0.5 µg of USP Dexamethasone RS per mL.

Assay preparation—Transfer about 50 mg of Dexamethasone Sodium Phosphate, accurately weighed, to a 100-mL volumetric flask, dissolve in *Mobile phase*, dilute with *Mobile phase* to volume, and mix. Further dilute 5.0 mL of this solution with *Mobile phase* to 50.0 mL.

System suitability preparation—Prepare a solution in *Mobile phase* containing in each mL 0.05 mg of USP Dexamethasone Phosphate RS and 0.02 mg of USP Dexamethasone RS.

Chromatographic system (see *Chromatography* ⟨621⟩)—The liquid chromatograph is equipped with a 254-nm detector and a 4.5-mm × 25-cm column that contains 5-µm packing L11. The flow rate is about 1.2 mL per minute. Chromatograph the *Standard preparation* and the *System suitability preparation*, record the peak responses as directed under *Procedure*, and determine the chromatographic characteristics from chromatograms obtained from the *System suitability preparation:* the column efficiency determined from the analyte peak is not less than 900 theoretical plates, the tailing factor for the analyte peak is not more than 1.6, the resolution, R, between the dexamethasone phosphate and dexamethasone peaks is not less than 1.8, and the relative standard deviation for replicate injections is not more than 1.0%.

Procedure—Separately inject equal volumes (about 20 µL) of the *Standard preparation* and the *Assay preparation* into the chromatograph, record the chromatograms, and measure the responses for the dexamethasone phosphate peaks. The relative retention times are about 0.7 for dexamethasone and 1.0 for dexamethasone phosphate. Calculate the quantity, in mg, of $C_{22}H_{28}FNa_2O_8P$ in the portion of Dexamethasone Sodium Phosphate taken by the formula:

$$(516.41/472.45)C(r_U/r_S),$$

in which 516.41 and 472.45 are the molecular weights of dexamethasone sodium phosphate and dexamethasone phosphate, respectively, C is the concentration, in µg per mL, of USP Dexamethasone Phosphate RS in the *Standard preparation*, and r_U and r_S are the peak responses for the analyte obtained from the *Assay preparation* and the *Standard preparation*, respectively.

Dexamethasone Sodium Phosphate Inhalation Aerosol

» Dexamethasone Sodium Phosphate Inhalation Aerosol is a suspension, in suitable propellants and alcohol, in a pressurized container, of dexamethasone sodium phosphate ($C_{22}H_{28}FNa_2O_8P$) equivalent to not less than 90.0 percent and not more than 110.0 percent of the labeled amount of dexamethasone phosphate ($C_{22}H_{30}FO_8P$).

Packaging and storage—Preserve in tight, pressurized containers, and avoid exposure to excessive heat.

Reference standard—*USP Dexamethasone Reference Standard*—Dry at 105° for 3 hours before using.

Identification—Prepare a pH 9.0 buffer solution by dissolving 3.1 g of boric acid, 203 mg of magnesium chloride, and 860 mg of sodium hydroxide in water to make 1000 mL. Dissolve 50 mg of alkaline phosphatase enzyme in 50 mL of the pH 9.0 buffer solution, and transfer 5 mL of the resulting solution to a glass-stoppered, 50-mL tube containing 5 mL of the *Assay preparation* prepared as directed in the *Assay*. Incubate at 37° for 45 minutes, add 25 mL of methylene chloride, and shake for 2 minutes. The methylene chloride extract so obtained responds to the *Identification test* under *Dexamethasone Sodium Phosphate Injection*, beginning with "Evaporate 15 mL of the methylene chloride extract."

Alcohol content, *Method II* ⟨611⟩: between 1.7% and 2.3% of C_2H_5OH.

Other requirements—It meets the requirements for *Leak Testing* and *Pressure Testing* under *Aerosols* ⟨601⟩.

Assay—

Standard preparation—Transfer about 25 mg of USP Dexamethasone RS, accurately weighed, to a 25-mL volumetric flask, dilute with alcohol to volume, and mix. Transfer 10.0 mL of the solution to a 100-mL volumetric flask, dilute with water to volume, and mix to obtain a solution containing about 100 µg per mL.

Assay preparation—Transfer 15 mL of methylene chloride into a glass-stoppered, 50-mL centrifuge tube, and place in a dry ice–acetone bath for several minutes. Accurately weigh the Dexamethasone Sodium Phosphate Inhalation Aerosol container (W_1), shake the container several times, and attach a sampling needle to the aerosol container. Remove the centrifuge tube from the bath, and fire the container 10 times into the methylene chloride. Pipet 10 mL of 0.1 *N* sulfuric acid through the sampling needle into the centrifuge tube. Again weigh the Aerosol container without the sampling needle (W_2). Insert the stopper in the centrifuge tube, and shake cautiously, releasing the pressure occasionally. Allow the phases to separate, and equilibrate to room temperature. The aqueous phase is the *Assay preparation.*

Procedure—Pipet 2 mL each of the *Assay preparation* and the *Standard preparation* into separate flasks. Add 10.0 mL of 0.1 *N* sulfuric acid to each, and swirl to mix. Concomitantly determine the absorbances of both solutions in 1-cm cells at the wavelength of maximum absorbance at about 239 nm, with a suitable spectrophotometer, using 0.1 *N* sulfuric acid as the blank. Calculate the quantity, in mg, of dexamethasone phosphate ($C_{22}H_{30}FO_8P$) in each g of Aerosol taken by the formula:

$$(472.45/392.47)(A_U/A_S)[C/(W_1 - W_2)],$$

in which 472.45 and 392.47 are the molecular weights of dexamethasone phosphate and dexamethasone, respectively, A_U and A_S are the absorbances of the solutions from the *Assay preparation* and the *Standard preparation*, respectively, *C* is the concentration, in mg per mL, of USP Dexamethasone RS taken, and W_1 and W_2 are the weights, in g, as previously defined.

Dexamethasone Sodium Phosphate Cream

» Dexamethasone Sodium Phosphate Cream contains an amount of dexamethasone sodium phosphate ($C_{22}H_{28}FNa_2O_8P$) equivalent to not less than 90.0 percent and not more than 115.0 percent of the labeled amount of dexamethasone phosphate ($C_{22}H_{30}FO_8P$).

Packaging and storage—Preserve in collapsible tubes or in tight containers.

Reference standards—*USP Dexamethasone Reference Standard*—Dry at 105° for 3 hours before using. *USP Dexamethasone Phosphate Reference Standard*—Dry at a pressure of 5 mm of mercury at 40° to constant weight before using.

Identification—Prepare a pH 9.0 buffer solution by dissolving 3.1 g of boric acid, 203 mg of magnesium chloride, and 860 mg of sodium hydroxide in water to make 1000 mL. Dissolve 50 mg of alkaline phosphatase enzyme in 50 mL of the pH 9.0 buffer solution, and transfer 5 mL of the resulting solution to a glass-stoppered, 50-mL tube containing 5 mL of the *Assay preparation*, prepared as directed in the *Assay*. Incubate at 37° for 45 minutes, then add 25 mL of methylene chloride, and shake for 2 minutes. The methylene chloride extract so obtained responds to the *Identification test* under *Dexamethasone Sodium Phosphate Injection*, beginning with "Evaporate 15 mL of the methylene chloride extract."

Microbial limits—It meets the requirements of the tests for absence of *Staphylococcus aureus* and *Pseudomonas aeruginosa* under *Microbial Limit Tests* ⟨61⟩.

Assay—

Alcohol–aqueous phosphate buffer—Dissolve 0.29 g of dibasic sodium phosphate in 450 mL of water, add 550 mL of alcohol, and mix.

0.05 M Phosphate buffer—In a 1-liter volumetric flask, dissolve 6.9 g of monobasic sodium phosphate in 500 mL of water, dilute with water to volume, and mix.

Mobile phase—Prepare a suitable degassed solution of methanol and *0.05 M Phosphate buffer* (52:48) which, at ambient temperature and at a flow rate of 1.5 mL per minute, gives a retention time of about 8.5 minutes for dexamethasone phosphate.

Standard preparation—Using an accurately weighed quantity of USP Dexamethasone Phosphate RS, prepare a solution in *Alcohol–aqueous phosphate buffer* having a known concentration of about 30 µg per mL. Prepare this solution fresh.

Assay preparation—Transfer an accurately weighed quantity of Dexamethasone Sodium Phosphate Cream, equivalent to about 3 mg of dexamethasone phosphate, to a 150-mL beaker. Add 65 mL of *Alcohol–aqueous phosphate buffer*, and heat just to boiling. Pour the contents of the beaker into a 125-mL separator containing 45 mL of isooctane. After shaking for 1 minute, decant the lower layer into a 100-mL volumetric flask with the aid of a glass funnel. Rinse the 150-mL beaker with two 15-mL portions of *Alcohol–aqueous phosphate buffer*, extracting the remaining isooctane in the separator with each portion and decanting the lower layer from each extraction into the 100-mL volumetric flask. Dilute with *Alcohol–aqueous phosphate buffer* to volume, and mix. Filter through a membrane filter before injecting.

Chromatographic system (see *Chromatography* ⟨621⟩)—The liquid chromatograph is equipped with a 254-nm detector and a 4-mm × 30-cm column that contains packing L1. Chromatograph five replicate injections of the *Standard preparation*, and record the peak responses as directed under *Procedure:* the relative standard deviation is not more than 1.5%.

Procedure—By means of a suitable sampling valve, separately inject equal volumes (about 20 µL) of the *Standard preparation* and the *Assay preparation* into the chromatograph, record the chromatograms, and measure the responses for the major peaks. Calculate the quantity, in mg, of $C_{22}H_{30}FO_8P$ in the portion of Cream taken by the formula:

$$0.1C(r_U/r_S),$$

in which *C* is the concentration, in µg per mL, of USP Dexamethasone Phosphate RS in the *Standard preparation*, and r_U and r_S are the peak responses at equivalent retention times obtained from the *Assay preparation* and the *Standard preparation*, respectively.

Dexamethasone Sodium Phosphate Cream, Neomycin Sulfate and—*see* Neomycin Sulfate and Dexamethasone Sodium Phosphate Cream

Dexamethasone Sodium Phosphate Injection

» Dexamethasone Sodium Phosphate Injection is a sterile solution of Dexamethasone Sodium Phosphate in Water for Injection. It contains not less than 90.0 percent and not more than 115.0 percent of the labeled amount of dexamethasone phosphate ($C_{22}H_{30}FO_8P$), present as the disodium salt.

Packaging and storage—Preserve in single-dose or in multiple-dose containers, preferably of Type I glass, protected from light.

Reference standards—*USP Dexamethasone Reference Standard*—Dry at 105° for 3 hours before using. *USP Dexamethasone Phosphate Reference Standard*—Dry at a pressure of 5 mm of mercury at 40° to constant weight before using.

Identification—Pipet a volume of Injection, equivalent to 10 mg of dexamethasone phosphate, into a 100-mL volumetric flask, add water to volume, and mix. Pipet 5 mL of this solution into a 125-mL separator, and wash with two 10-mL portions of water-washed methylene chloride, discarding the washings. Transfer the solution into a glass-stoppered, 50-mL tube, and add 5 mL of alkaline phosphatase solution, prepared by dissolving 50 mg of alkaline phosphatase enzyme in 50 mL of *pH 9 buffer with magnesium* (prepared as directed in *Identification test A* under *Dexamethasone Sodium Phosphate*). Allow to stand at 37° for 45 minutes, and extract with 25 mL of methylene chloride. Evaporate 15 mL of the methylene chloride extract on a steam bath to dryness, and dissolve the residue in 1 mL of methylene chloride. On a 20-cm × 20-cm, thin-layer chromatographic plate (see *Chromatography* ⟨621⟩) coated with a 0.25-mm layer of chromatographic silica gel, apply 5 µL of this solution and 5 µL of a solution of USP Dexamethasone RS in methylene chloride containing 300 µg per mL. Allow the spots to dry, and develop the chromatogram in a tank completely lined with a strip of filter paper, using a solvent system consisting of 50 parts of chloroform, 50 parts of acetone, and 1 part of water, until the solvent front has moved about three-fourths of the length of the plate. Remove the plate from the developing tank, mark the solvent front, and allow the spots to dry. Spray the plate with dilute sulfuric acid (1 in 2), and heat at 105° until brown or black spots appear: the R_f value of the principal spot obtained from the test specimen corresponds to that obtained from the Reference Standard.

pH ⟨791⟩: between 7.0 and 8.5.

Other requirements—It meets the requirements under *Injections* ⟨1⟩.

Assay—

Mobile phase—Prepare a suitable degassed solution of 0.01 *M* monobasic potassium phosphate in a mixture of methanol and water (1:1) which, at ambient temperature and at a flow rate of about 1.6 mL per minute, gives a retention time of about 5 minutes for dexamethasone phosphate.

Standard preparation—[NOTE—Prepare this solution at the time of use.] Dissolve an accurately weighed quantity of USP Dexamethasone Phosphate RS in *Mobile phase* to obtain a solution having a known concentration of about 80 µg per mL.

Assay preparation—Transfer an accurately measured volume of Dexamethasone Sodium Phosphate Injection, equivalent to about 8 mg of dexamethasone phosphate, to a 100-mL volumetric flask. Dilute with *Mobile phase* to volume, and mix.

Chromatographic system (see *Chromatography* ⟨621⟩)—The liquid chromatograph is equipped with a 254-nm detector and a 4-mm × 30-cm column that contains packing L1. Chromatograph five replicate injections of the *Standard preparation*, and record the peak responses as directed under *Procedure:* the relative standard deviation is not more than 1.5%.

Procedure—By means of a suitable sampling valve, separately inject equal volumes (about 20 µL) of the *Standard preparation* and the *Assay preparation* into the chromatograph, record the chromatograms, and measure the responses for the major peaks. Calculate the quantity, in mg, of $C_{22}H_{30}FO_8P$ in each mL of the Injection taken by the formula:

$$0.1(C/V)(r_U/r_S),$$

in which C is the concentration, in µg per mL, of USP Dexamethasone Phosphate RS in the *Standard preparation*, V is the volume, in mL, of Injection taken, and r_U and r_S are the peak responses at equivalent retention times obtained from the *Assay preparation* and the *Standard preparation*, respectively.

Dexamethasone Sodium Phosphate Ophthalmic Ointment

» Dexamethasone Sodium Phosphate Ophthalmic Ointment is a sterile ointment containing an amount of dexamethasone sodium phosphate ($C_{22}H_{28}FNa_2O_8P$) equivalent to not less than 90.0 percent and not more than 115.0 percent of the labeled amount of dexamethasone phosphate ($C_{22}H_{30}FO_8P$).

Packaging and storage—Preserve in collapsible ophthalmic ointment tubes.

Reference standards—*USP Dexamethasone Reference Standard*—Dry at 105° for 3 hours before using. *USP Dexamethasone Phosphate Reference Standard*—Dry at a pressure of 5 mm of mercury at 40° to constant weight before using.

Identification—The *Assay preparation*, prepared as directed in the *Assay*, responds to the *Identification test* under *Dexamethasone Sodium Phosphate Cream.*

Minimum fill ⟨755⟩: meets the requirements.

Sterility—It meets the requirements under *Sterility Tests* ⟨71⟩.

Metal particles—It meets the requirements of the test for *Metal Particles in Ophthalmic Ointments* ⟨751⟩.

Assay—

Alcohol–aqueous phosphate buffer, 0.05 M Phosphate buffer, Mobile phase, Standard preparation, and *Chromatographic system*—Prepare as directed in the *Assay* under *Dexamethasone Sodium Phosphate Cream.*

Assay preparation—Using an accurately weighed portion of Dexamethasone Sodium Phosphate Ophthalmic Ointment, prepare as directed in the *Assay* under *Dexamethasone Sodium Phosphate Cream.*

Procedure—Proceed as directed for *Procedure* in the *Assay* under *Dexamethasone Sodium Phosphate Cream.* Calculate the quantity, in mg, of $C_{22}H_{30}FO_8P$ in the portion of Ophthalmic Ointment taken by the formula:

$$0.1C(r_U/r_S).$$

Dexamethasone Sodium Phosphate Ophthalmic Ointment, Neomycin Sulfate and—*see* Neomycin Sulfate and Dexamethasone Sodium Phosphate Ophthalmic Ointment

Dexamethasone Sodium Phosphate Ophthalmic Solution

» Dexamethasone Sodium Phosphate Ophthalmic Solution is a sterile, aqueous solution of Dexamethasone Sodium Phosphate. It contains an amount of dexamethasone sodium phosphate ($C_{22}H_{28}FNa_2O_8P$) equivalent to not less than 90.0 percent and not more than 115.0 percent of the labeled amount of dexamethasone phosphate ($C_{22}H_{30}FO_8P$).

Packaging and storage—Preserve in tight, light-resistant containers.

Reference standards—*USP Dexamethasone Reference Standard*—Dry at 105° for 3 hours before using. *USP Dexamethasone Phosphate Reference Standard*—Dry at a pressure of 5 mm of mercury at 40° to constant weight before using.

Identification—The *Assay preparation*, prepared as directed in the *Assay*, responds to the *Identification test* under *Dexamethasone Sodium Phosphate Cream.*

pH ⟨791⟩: between 6.6 and 7.8.

Sterility—It meets the requirements under *Sterility Tests* ⟨71⟩.

Assay—

Mobile phase, Standard preparation, and *Chromatographic system*—Prepare as directed in the *Assay* under *Dexamethasone Sodium Phosphate Injection.*

Assay preparation—Transfer an accurately measured volume of Dexamethasone Sodium Phosphate Ophthalmic Solution, equivalent to about 8 mg of dexamethasone phosphate, to a 100-mL volumetric flask, dilute with *Mobile phase* to volume, and mix.

Procedure—Proceed as directed for *Procedure* in the *Assay* under *Dexamethasone Sodium Phosphate Injection.* Calculate the quantity, in mg, of $C_{22}H_{30}FO_8P$ in each mL of the Ophthalmic Solution taken by the formula:

$$0.1(C/V)(r_U/r_S),$$

in which V is the volume, in mL, of Ophthalmic Solution taken.

Dexamethasone Sodium Phosphate Ophthalmic Solution, Neomycin Sulfate and—*see* Neomycin Sulfate and Dexamethasone Sodium Phosphate Ophthalmic Solution

Dexbrompheniramine Maleate

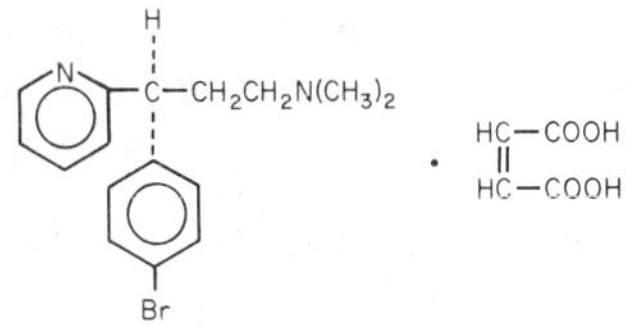

$C_{16}H_{19}BrN_2 . C_4H_4O_4$ 435.32

2-Pyridinepropanamine, γ-(4-bromophenyl)-*N*,*N*-dimethyl-, (*S*)-, (*Z*)-2-butenedioate (1:1).

(+)-2-[*p*-Bromo-α-[2-(dimethylamino)ethyl]benzyl]pyridine maleate (1:1) [*2391-03-9*].

» Dexbrompheniramine Maleate contains not less than 98.0 percent and not more than 100.5 percent of $C_{16}H_{19}BrN_2 . C_4H_4O_4$, calculated on the dried basis.

Packaging and storage—Preserve in tight, light-resistant containers.

Reference standard—*USP Dexbrompheniramine Maleate Reference Standard*—Dry at 65° for 4 hours before using.

Identification—

A: The infrared absorption spectrum of a potassium bromide dispersion of it, previously dried, exhibits maxima only at the same wavelengths as that of a similar preparation of USP Dexbrompheniramine Maleate RS.

B: The ultraviolet absorption spectrum of a 1 in 30,000 solution in methanol exhibits maxima and minima at the same wavelengths as that of a similar solution of USP Dexbrompheniramine Maleate RS, concomitantly measured, and the respective absorptivities, calculated on the dried basis, at the wavelength of maximum absorbance at about 261 nm do not differ by more than 3.0%.

Specific rotation ⟨781⟩: between +35.0° and +38.5°, calculated on the dried basis, determined in a solution in dimethylformamide containing 500 mg in each 10 mL.

Loss on drying ⟨731⟩—Dry it at 65° for 4 hours: it loses not more than 0.5% of its weight.

Residue on ignition ⟨281⟩: not more than 0.2%.

Related compounds—

Test preparation—Dissolve about 200 mg of Dexbrompheniramine Maleate in 5 mL of methylene chloride, and mix.

Chromatographic system (see *Chromatography* ⟨621⟩)—The gas chromatograph is equipped with a flame-ionization detector and a 1.2-m × 4-mm glass column containing 3 percent phase G3 on support S1AB. The column temperature is maintained at about 190°, and the injection port and detector temperatures are both maintained at about 250°. The carrier gas is dry helium, flowing at a rate adjusted to obtain a retention time of 6 to 7 minutes for the main peak. Chromatograph the *Test preparation*, record the chromatogram, and determine the peak area as directed under *Procedure*. The tailing factor for the dexbrompheniramine maleate peak is not more than 1.8.

Procedure—Inject a volume (about 1 µL) of the *Test preparation* into the chromatograph. Record the chromatogram for a total time of not less than twice the retention time of the dexbrompheniramine peak, and measure the areas of the peaks. The total relative area of all extraneous peaks (except that of the solvent peak) does not exceed 2.0%.

Assay—Dissolve about 400 mg of Dexbrompheniramine Maleate, accurately weighed, in 50 mL of glacial acetic acid, add 1 drop of crystal violet TS, and titrate with 0.1 *N* perchloric acid VS to a green end-point. Perform a blank determination, and make any necessary correction. Each mL of 0.1 *N* perchloric acid is equivalent to 21.77 mg of $C_{16}H_{19}BrN_2 . C_4H_4O_4$.

Dexchlorpheniramine Maleate

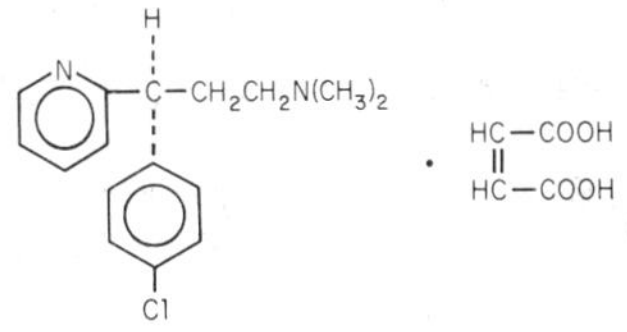

$C_{16}H_{19}ClN_2 . C_4H_4O_4$ 390.87

2-Pyridinepropanamine, γ-(4-chlorophenyl)-*N*,*N*-dimethyl-, (*S*)-, (*Z*)-2-butenedioate (1:1).

(+)-2-[*p*-Chloro-α-[2-(dimethylamino)ethyl]benzyl]pyridine maleate (1:1) [*2438-32-6*].

» Dexchlorpheniramine Maleate, dried at 65° for 4 hours, contains not less than 98.0 percent and not more than 100.5 percent of $C_{16}H_{19}ClN_2 . C_4H_4O_4$.

Packaging and storage—Preserve in tight, light-resistant containers.

Reference standard—*USP Dexchlorpheniramine Maleate Reference Standard*—Dry at 65° for 4 hours before using.

Identification—

A: The infrared absorption spectrum of a potassium bromide dispersion of it, previously dried, exhibits maxima only at the same wavelengths as that of a similar preparation of USP Dexchlorpheniramine Maleate RS.

B: The ultraviolet absorption spectrum of a solution (1 in 25,000) exhibits maxima and minima at the same wavelengths as that of a similar solution of USP Dexchlorpheniramine Maleate RS, concomitantly measured.

Melting range, *Class I* ⟨741⟩: between 110° and 115°.

Specific rotation ⟨781⟩: between +39.5° and +43.0°, calculated on the dried basis, determined in a solution in dimethylformamide containing 500 mg in each 10 mL.

pH ⟨791⟩: between 4.0 and 5.0, in a solution (1 in 100).

Loss on drying ⟨731⟩—Dry it at 65° for 4 hours: it loses not more than 0.5% of its weight.

Residue on ignition ⟨281⟩: not more than 0.2%.

Related compounds—

Test preparation—Dissolve about 200 mg of Dexchlorpheniramine Maleate in 5 mL of methylene chloride, and mix.

Chromatographic system (see *Chromatography* ⟨621⟩)—The gas chromatograph is equipped with a flame-ionization detector and a 1.2-m × 4-mm glass column containing 3 percent phase G3 on support S1AB. The column temperature is maintained at about 190°, and the injection port and detector temperatures are both maintained at about 250°. The carrier gas is dry helium, flowing at a rate adjusted to obtain a retention time of 4 to 5 minutes for the main peak. Chromatograph the *Test preparation*, record the chromatogram, and determine the peak area as directed under *Procedure*. The tailing factor for the dexchlorpheniramine maleate peak is not more than 1.8.

Procedure—Inject a volume (about 1 µL) of the *Test preparation* into the chromatograph. Record the chromatogram for a total time of not less than twice the retention time of the dexchlorpheniramine peak, and measure the areas of the peaks. The total relative area of all extraneous peaks (except that of the solvent peak) does not exceed 2.0%.

Assay—Dissolve about 400 mg of Dexchlorpheniramine Maleate, previously dried and accurately weighed, in 50 mL of glacial acetic acid, add 1 drop of crystal violet TS, and titrate with 0.1

N perchloric acid VS to a green end-point. Perform a blank determination, and make any necessary correction. Each mL of 0.1 *N* perchloric acid is equivalent to 19.54 mg of $C_{16}H_{19}ClN_2.C_4H_4O_4$.

Dexchlorpheniramine Maleate Syrup

» Dexchlorpheniramine Maleate Syrup contains not less than 90.0 percent and not more than 110.0 percent of the labeled amount of $C_{16}H_{19}ClN_2.C_4H_4O_4$.

Packaging and storage—Preserve in tight, light-resistant containers.

Reference standard—*USP Dexchlorpheniramine Maleate Reference Standard*—Dry at 65° for 4 hours before using.

Identification—

A: Evaporate the remaining extract from the *Assay* on a steam bath to a small volume, then transfer it to a smaller, more suitable vessel, and evaporate just to the point where hexane vapors are no longer perceptible. Transfer the oily residue, with the aid of four 3-mL portions of dimethylformamide, to a suitable glass-stoppered graduated cylinder, dilute with dimethylformamide to 15.0 mL, and mix: the optical rotation of the solution so obtained, in a 100-mm tube, after correcting for the blank, is between +0.06° and +0.11° (*distinction from chlorpheniramine maleate*).

B: The ultraviolet absorption spectrum of the solution employed for measurement of absorbance in the *Assay* exhibits maxima and minima at the same wavelengths as that of the solution obtained from USP Dexchlorpheniramine Maleate RS used in the *Assay*.

Alcohol content ⟨611⟩: between 5.0% and 7.0% of C_2H_5OH.

Assay—

Standard preparation—Transfer about 40 mg of USP Dexchlorpheniramine Maleate RS, accurately weighed, to a 100-mL volumetric flask, add water to volume, and mix. Transfer 10.0 mL of this solution to a separator, adjust with 1 *N* sodium hydroxide to a pH of 11, and cool. Extract with two 50-mL portions of solvent hexane, shaking each portion for 2 minutes before separating the phases, and combining the hexane extracts in a second separator. Extract the hexane solution with two 40-mL portions of dilute hydrochloric acid (1 in 120), combine the acid extracts in a 100-mL volumetric flask, add dilute hydrochloric acid (1 in 120) to volume, and mix. Filter the solution into a glass-stoppered conical flask, discarding the first few mL of the filtrate. The concentration of USP Dexchlorpheniramine Maleate RS in the *Standard preparation* is about 40 µg per mL.

Assay preparation—Transfer an accurately measured volume of Dexchlorpheniramine Maleate Syrup, equivalent to about 40 mg of dexchlorpheniramine maleate, to a 250-mL separator, using a pipet calibrated "to contain" the required volume. Rinse the pipet with small portions of water, add the rinsings to the separator, adjust with 1 *N* sodium hydroxide to a pH of 11, and cool. Extract with five 70-mL portions of solvent hexane, combine the hexane extracts in a 500-mL separator, and wash the hexane solution with two 10-mL portions of sodium hydroxide solution (1 in 250). Extract the combined alkaline washings with two 20-mL portions of solvent hexane, and add these extracts to the bulk of the alkali-washed hexane solution. Filter the hexane solution through a pledget of cotton that previously has been saturated with solvent hexane into a 500-mL volumetric flask, rinse the separator with portions of solvent hexane, pass the rinsings through the filter to add to volume, and mix. Transfer 50.0 mL of this solution to a separator (retain the remaining extract for *Identification test A*), and proceed as directed for *Standard preparation*, beginning with "Extract the hexane solution."

Procedure—Concomitantly determine the absorbances of the *Standard preparation* and the *Assay preparation* in 1-cm cells at the wavelength of maximum absorbance at about 264 nm, using dilute hydrochloric acid (1 in 120) as the blank. Calculate the quantity, in mg, of $C_{16}H_{19}ClN_2.C_4H_4O_4$ in each mL of the Syrup taken by the formula:

$$(C/V)(A_U/A_S),$$

in which C is the concentration, in µg per mL, of USP Dexchlorpheniramine Maleate RS in the *Standard preparation*, V is the volume, in mL, of Syrup taken, and A_U and A_S are the absorbances of the *Assay preparation* and the *Standard preparation*, respectively.

Dexchlorpheniramine Maleate Tablets

» Dexchlorpheniramine Maleate Tablets contain not less than 90.0 percent and not more than 110.0 percent of the labeled amount of $C_{16}H_{19}ClN_2.C_4H_4O_4$.

Packaging and storage—Preserve in tight containers.

Reference standard—*USP Dexchlorpheniramine Maleate Reference Standard*—Dry at 65° for 4 hours before using.

Identification—

A: Tablets meet the requirements under *Identification—Organic Nitrogenous Bases* ⟨181⟩.

B: Shake a quantity of finely powdered Tablets, equivalent to about 150 mg of dexchlorpheniramine maleate, with 100 mL of 1 *N* acetic acid for 10 minutes, filter through a sintered-glass funnel into a suitable vessel, adjust the filtrate with sodium hydroxide solution (1 in 10) to a pH of 11, and extract the solution with six 100-mL portions of solvent hexane. Concentrate the combined extracts on a steam bath to a small volume, transfer to a smaller, more suitable vessel, and evaporate just to the point where hexane vapors are no longer perceptible. Transfer the oily residue, with the aid of four 3-mL portions of dimethylformamide, to a suitable glass-stoppered graduated cylinder, dilute with dimethylformamide to 15.0 mL, and mix: the optical rotation of the solution so obtained, in a 100-mm tube, after correcting for the blank, is between +0.24° and +0.35° (*distinction from chlorpheniramine maleate*).

Dissolution ⟨711⟩—

Medium: water; 500 mL.

Apparatus 2: 50 rpm.

Time: 45 minutes.

Procedure—Determine the amount of $C_{16}H_{19}ClN_2.C_4H_4O_4$ dissolved, using the following procedure.

Internal standard solution—Prepare a solution of Dexbrompheniramine Maleate in water having a final concentration of about 90 µg per mL.

Standard preparation—Dissolve an accurately weighed quantity of USP Dexchlorpheniramine Maleate RS in water, and dilute quantitatively and stepwise with water to obtain a stock solution having a known concentration of about 12.5 µg per mL. Pipet 5 mL of this stock solution into a 50-mL centrifuge tube, add 10.0 mL of water and 1.0 mL of *Internal standard solution*, and mix. Adjust with sodium hydroxide solution (1 in 2) to a pH of 11 ± 0.1, and add 3.0 mL of chromatographic solvent hexane. Insert the stopper in the tube, shake by mechanical means for 3 minutes, centrifuge, and use the clear supernatant hexane layer.

Test preparation—Pipet 15 mL of a portion of the solution under test into a 50-mL centrifuge tube, add 1.0 mL of *Internal standard solution*, and mix. Proceed as directed for *Standard preparation*, beginning with "Adjust with sodium hydroxide solution (1 in 2)."

Chromatographic system (see *Chromatography* ⟨621⟩)—The gas chromatograph is equipped with a flame-ionization detector and a 2-mm × 1.8-m column that contains a packing consisting of 1.2 percent phase G16 and 0.5 percent potassium hydroxide on support S1AB. The carrier gas is helium maintained at a flow rate of about 60 mL per minute. The column, injector, and detector temperatures are maintained at 205°, 250°, and 250°, respectively. Chromatograph replicate injections of the *Standard preparation*, and record the peak response as directed under *Procedure:* the relative standard deviation is not more than 2.0%, and the resolution, *R*, between dexchlorpheniramine and dexbrompheniramine is not less than 1.9.

Procedure—Separately inject equal volumes (about 2 µL) of the *Standard preparation* and the *Test preparation* into the chromatograph, record the chromatograms, and measure the responses for the major peaks. The relative retention times are about 0.7 for dexchlorpheniramine and 1.0 for dexbrompheniramine. Calculate the amount of $C_{16}H_{19}ClN_2.C_4H_4O_4$ dissolved by comparison of the peak response ratios.

Tolerances—Not less than 75% (*Q*) of the labeled amount of $C_{16}H_{19}ClN_2.C_4H_4O_4$ is dissolved in 45 minutes.

Uniformity of dosage units ⟨905⟩: meet the requirements.

Assay—

Standard preparation—Prepare as directed in the *Assay* under *Dexchlorpheniramine Maleate Syrup*. The concentration of USP Dexchlorpheniramine Maleate RS in the *Standard preparation* is about 40 µg per mL.

Assay preparation—Weigh and finely powder not less than 20 Dexchlorpheniramine Maleate Tablets. Transfer an accurately weighed portion of the powder, equivalent to about 8 mg of dexchlorpheniramine maleate, to a 250-mL separator, mix with 50 mL of water for 10 minutes, adjust with sodium hydroxide solution (1 in 10) to a pH of 11, and cool to room temperature. Extract the mixture with two 75-mL portions of solvent hexane, and combine the extracts in a second separator. Extract the solvent hexane solution with three 50-mL portions of dilute hydrochloric acid (1 in 120), combining the acid extracts in a 200-mL volumetric flask. Add dilute hydrochloric acid (1 in 120) to volume, and mix.

Procedure—Concomitantly determine the absorbances of the *Assay preparation* and the *Standard preparation* in 1-cm cells at the wavelength of maximum absorbance at about 264 nm, with a suitable spectrophotometer, using dilute hydrochloric acid (1 in 120) as the blank. Calculate the quantity, in mg, of $C_{16}H_{19}ClN_2.C_4H_4O_4$ in the portion of Tablets taken by the formula:

$$0.2C(A_U/A_S),$$

in which *C* is the concentration, in µg per mL, of USP Dexchlorpheniramine Maleate RS in the *Standard preparation*, and A_U and A_S are the absorbances of the *Assay preparation* and the *Standard preparation*, respectively.

Dexpanthenol

$$HOCH_2-\underset{CH_3}{\overset{CH_3}{C}}-\underset{H}{\overset{OH}{C}}-CONHCH_2CH_2CH_2OH$$

$C_9H_{19}NO_4$ 205.25

Butanamide, 2,4-dihydroxy-*N*-(3-hydroxypropyl)-3,3-dimethyl-, (*R*)-.

D-(+)-2,4-Dihydroxy-*N*-(3-hydroxypropyl)-3,3-dimethylbutyramide [*81-13-0*].

» Dexpanthenol contains not less than 98.0 percent and not more than 102.0 percent of $C_9H_{19}NO_4$, calculated on the anhydrous basis.

Packaging and storage—Preserve in tight containers.

Reference standard—*USP Dexpanthenol Reference Standard.*

Identification—

A: The infrared absorption spectrum of a thin film of it exhibits maxima only at the same wavelengths as that of a similar preparation of USP Dexpanthenol RS.

B: To 1 mL of a 1 in 10 solution add 5 mL of 1 *N* sodium hydroxide and 1 drop of cupric sulfate TS, and shake vigorously: a deep blue color develops.

C: To 1 mL of a solution (1 in 100) add 1 mL of 1 *N* hydrochloric acid, and heat on a steam bath for about 30 minutes. Cool, add 100 mg of hydroxylamine hydrochloride, mix, and add 5 mL of 1 *N* sodium hydroxide. Allow to stand for 5 minutes, then adjust with 1 *N* hydrochloric acid to a pH of between 2.5 and 3.0, and add 1 drop of ferric chloride TS: a purplish red color develops.

Specific rotation ⟨781⟩: between +29.0° and +31.5°, calculated on the anhydrous basis, determined in a solution containing 500 mg in each 10 mL.

Refractive index ⟨831⟩: between 1.495 and 1.502, at 20°.

Water, *Method I* ⟨921⟩: not more than 1.0%.

Residue on ignition ⟨281⟩: not more than 0.1%.

Aminopropanol—Transfer about 5 g, accurately weighed, to a 50-mL flask, and dissolve in 10 mL of water. Add bromothymol blue TS, and titrate with 0.1 *N* sulfuric acid VS to a yellow endpoint. Each mL of 0.1 *N* sulfuric acid is equivalent to 7.5 mg of aminopropanol. Not more than 1.0% is found.

Assay—

Potassium biphthalate solution—Dissolve 20.42 g of potassium biphthalate in glacial acetic acid contained in a 1000-mL volumetric flask. If necessary, warm the mixture on a steam bath to achieve complete solution, observing precautions against absorption of moisture. Cool to room temperature, dilute with glacial acetic acid to volume, and mix.

Procedure—Transfer about 400 mg of Dexpanthenol, accurately weighed, to a 300-mL flask fitted to a reflux condenser by means of a standard-taper glass joint, add 50.0 mL of 0.1 *N* perchloric acid VS, and reflux for 5 hours. Cool, observing precautions to prevent atmospheric moisture from entering the condenser, and rinse the condenser with glacial acetic acid, collecting the rinsings in the flask. Add 5 drops of crystal violet TS, and titrate with *Potassium biphthalate solution* to a blue-green endpoint. Perform a blank determination, and note the differences in volumes required. Each mL of the difference in volumes of 0.1 *N* perchloric acid consumed is equivalent to 20.53 mg of $C_9H_{19}NO_4$.

Dexpanthenol Preparation

» Dexpanthenol Preparation contains not less than 94.5 percent and not more than 98.5 percent of dexpanthenol ($C_9H_{19}NO_4$), and not less than 2.7 percent and not more than 4.2 percent of pantolactone, both calculated on the anhydrous basis.

Packaging and storage—Preserve in tight containers.

Reference standards—*USP Dexpanthenol Reference Standard. USP Pantolactone Reference Standard.*

Identification—

A: The infrared absorption spectrum of a thin film of it exhibits maxima only at the same wavelengths as that of a similar preparation of USP Dexpanthenol RS, except that there is an additional maximum at about 5.6 µm due to pantolactone.

B: It responds to *Identification test B* under *Dexpanthenol.*

Specific rotation ⟨781⟩: between +27.5° and +30.0°, calculated on the anhydrous basis, determined in a solution containing 500 mg in each 10 mL.

Pantolactone—

Internal standard solution—Prepare a solution of 2,6-dimethylphenol in toluene to contain about 100 mg per mL.

Standard solution—Transfer about 100 mg of USP Pantolactone RS, accurately weighed, to a 10-mL volumetric flask, add methylene chloride to volume, and mix.

Standard preparation—Pipet 0.4 mL of *Standard solution* into a suitable small vial. Evaporate the solvent by means of a steady stream of dry air, and add 50 µL, accurately measured, of *Internal standard solution.* Add 1 mL of a mixture of pyridine, hexamethyldisilazane, and trimethylchlorosilane (9:3:1), immediately insert the stopper, and shake vigorously for 30 seconds.

Test preparation—Transfer about 100 mg of Dexpanthenol Preparation, accurately weighed, to a suitable small vial, and proceed as directed under *Standard preparation*, beginning with "add 50 µL, accurately measured, of *Internal standard solution.*"

Chromatographic system (see *Chromatography* ⟨621⟩)—Under typical conditions, the gas chromatograph is equipped with a flame-ionization detector and contains a suitable column, 1.8 m × 2.0 mm, packed with 5 percent liquid phase G2 on support S1A. The column and injection port are maintained isothermally at 170° and 180°, respectively. Using a suitable carrier gas, adjust the flow rate so that the derivatized pantolactone elutes in about 4 minutes.

System suitability—Chromatograph five injections of the *Standard preparation*, and record the peak responses as directed under *Procedure:* the relative standard deviation of the peak response ratios (R_S) of the five injections is not more than 2.0%. The retention time of the derivatized pantolactone is about 0.75 relative to that of the derivatized internal standard. In a suitable chromatogram, the resolution factor between the two peaks is not less than 2.0.

Procedure—Inject about 0.5 µL of the *Standard preparation* into the gas chromatograph, record the chromatogram to obtain not less than 40% of maximum recorder response, and measure the peak responses of the derivatized pantolactone and the derivatized internal standard. Similarly, inject about 0.5 µL of the *Test preparation*, record the chromatogram, and measure the peak responses of the corresponding components. Calculate the quantity, in mg, of pantolactone in the portion of Dexpanthenol Preparation taken by the formula:

$$0.4C_S(R_U/R_S),$$

in which C_S is the concentration, in mg per mL, of USP Pantolactone RS in the *Standard solution*, and R_U and R_S are the ratios of the peak response due to the pantolactone to that due to the internal standard obtained from the *Test preparation* and the *Standard preparation*, respectively.

Other requirements—It meets the requirements for *Refractive index*, *Water*, *Residue on ignition*, *Aminopropanol*, and *Assay* under *Dexpanthenol*.

Dextrates—*see* Dextrates NF
Dextrin—*see* Dextrin NF

Dextroamphetamine Sulfate

$(C_9H_{13}N)_2 . H_2SO_4$ 368.49

Benzeneethanamine, α-methyl-, (*S*)-, sulfate (2:1).
(+)-α-Methylphenethylamine sulfate (2:1) [*51-63-8*].

» Dextroamphetamine Sulfate, the dextrorotatory isomer of amphetamine sulfate, contains not less than 98.0 percent and not more than 101.0 percent of $(C_9H_{13}N)_2 . H_2SO_4$, calculated on the dried basis.

Packaging and storage—Preserve in well-closed containers.

Identification—

A: Dissolve about 100 mg in 5 mL of water, add 5 mL of 1 *N* sodium hydroxide, cool to 10° to 15°, add 1 mL of a mixture of 1 volume of benzoyl chloride and 2 volumes of anhydrous ethyl ether, insert the stopper, and shake for 3 minutes. Filter the precipitate, wash it with about 10 mL of cold water, and recrystallize it from diluted alcohol: the crystals of the benzoyl derivative of dextroamphetamine so obtained, after being dried at 105° for 1 hour, melt between 155° and 160°.

B: A solution (1 in 10) responds to the tests for *Sulfate* ⟨191⟩.

Specific rotation ⟨781⟩: between +20° and +23.5°, calculated on the dried basis, determined in a solution containing 400 mg in each 10 mL.

pH ⟨791⟩: between 5.0 and 6.0, in a solution (1 in 20).

Loss on drying ⟨731⟩—Dry it at 105° for 2 hours: it loses not more than 1.0% of its weight.

Residue on ignition ⟨281⟩: not more than 0.1%.

Ordinary impurities ⟨466⟩—

Test solution: methanol.

Standard solution: methanol.

Eluant: a mixture of methanol and ammonium hydroxide (50:1).

Visualization: 1.

Assay—Dissolve about 500 mg of Dextroamphetamine Sulfate, accurately weighed, in 50 mL of glacial acetic acid, and titrate with 0.1 *N* perchloric acid VS, determining the end-point potentiometrically. Perform a blank determination, and make any necessary correction. Each mL of 0.1 *N* perchloric acid is equivalent to 36.85 mg of $(C_9H_{13}N)_2 . H_2SO_4$.

Dextroamphetamine Sulfate Capsules

» Dextroamphetamine Sulfate Capsules contain not less than 90.0 percent and not more than 110.0 percent of the labeled amount of $(C_9H_{13}N)_2 . H_2SO_4$.

Packaging and storage—Preserve in tight containers.

Reference standard—*USP Dextroamphetamine Sulfate Reference Standard*—Dry at 105° for 2 hours before using.

Identification—

A: Mix an amount of the Capsule contents, equivalent to about 50 mg of dextroamphetamine sulfate, with about 10 mL of water for 30 minutes, and filter into a small flask. Cool the filtrate to about 15°, and proceed as directed in the *Identification test* under *Dextroamphetamine Sulfate Elixir*, beginning with "add 3 mL of 1 *N* sodium hydroxide."

B: The retention time of the major peak in the chromatogram of the *Assay preparation* is the same as that of the *Standard preparation* obtained in the *Assay*.

Dissolution ⟨711⟩—

Medium: water; 500 mL.

Apparatus 1: 100 rpm.

Time: 45 minutes.

Procedure—Determine the amount of $(C_9H_{13}N)_2 . H_2SO_4$ dissolved, employing the procedure set forth in the *Assay*, making any necessary modifications.

Tolerances—Not less than 75% (*Q*) of the labeled amount of $(C_9H_{13}N)_2 . H_2SO_4$ is dissolved in 45 minutes.

Uniformity of dosage units ⟨905⟩: meet the requirements.

Assay—

Mobile phase—Dissolve 1.1 g of sodium 1-heptanesulfonate in 525 mL of water. Add 25 mL of dilute glacial acetic acid (14 in 100) and 450 mL of methanol. Adjust dropwise, if necessary, with glacial acetic acid to a pH of 3.3 ± 0.1. Filter through a 0.5-µm membrane filter. The volume of methanol may be adjusted so that the retention time for dextroamphetamine is about 5 minutes.

Standard preparation—Dissolve an accurately weighed quantity of USP Dextroamphetamine Sulfate RS in 0.12 *N* phosphoric acid to obtain a solution having a known concentration of about 0.3 mg per mL.

Assay preparation—Remove, as completely as possible, the contents of not less than 20 Dextroamphetamine Sulfate Capsules, and weigh. Transfer an accurately weighed portion of the mixed powder, equivalent to about 15 mg of dextroamphetamine sulfate, to a 50-mL volumetric flask. Add 40 mL of 0.12 *N* phosphoric acid, and sonicate for 15 minutes. Dilute with 0.12 *N* phosphoric acid to volume, and mix. Filter through a 0.5-µm membrane filter, discarding the first 20 mL of the filtrate.

Chromatographic system (see *Chromatography* ⟨621⟩)—The liquid chromatograph is equipped with a 254-nm detector and a 3.9-mm × 30-cm column that contains packing L1. The flow rate is about 2 mL per minute. Chromatograph three replicate injections of the *Standard preparation*, and record the peak responses as directed under *Procedure:* the relative standard deviation is not more than 2.0%.

Procedure—Separately inject equal volumes (about 50 µL) of the *Standard preparation* and the *Assay preparation* into the chromatograph by means of a suitable automatic injector or sampling valve, record the chromatograms, and measure the responses for the major peaks. Calculate the quantity, in mg, of $(C_9H_{13}N)_2 . H_2SO_4$ in the portion of Capsules taken by the formula:

$$50C(r_U/r_S),$$

in which C is the concentration, in mg per mL, of USP Dextroamphetamine Sulfate RS in the *Standard preparation*, and r_U and r_S are the peak responses obtained from the *Assay preparation* and the *Standard preparation*, respectively.

Dextroamphetamine Sulfate Elixir

» Dextroamphetamine Sulfate Elixir contains, in each 100 mL, not less than 90.0 mg and not more than 110.0 mg of $(C_9H_{13}N)_2 \cdot H_2SO_4$.

Packaging and storage—Preserve in tight, light-resistant containers.

Reference standard—*USP Dextroamphetamine Sulfate Reference Standard*—Dry at 105° for 2 hours before using.

Identification—Transfer 25 mL of Elixir to a 250-mL separator, add 25 mL of water and 5 mL of 2.5 *N* sodium hydroxide, mix, and extract with 60 mL of ether. Wash the ether extract with two 5-mL portions of 0.25 *N* sodium hydroxide, and discard the washings. Filter the ether extract through a pledget of cotton, previously saturated with ether, into a 100-mL beaker, and evaporate on a steam bath in a current of air to about 1 mL. Dissolve the residue in 3 mL of alcohol, and transfer to a glass-stoppered, 125-mL conical flask containing 25 mL of water. Rinse the beaker with 3 mL of alcohol, and transfer to the flask. Cool to about 15°, add 3 mL of 1 *N* sodium hydroxide, then add 1 mL of a mixture of 1 volume of benzoyl chloride and 2 volumes of anhydrous ethyl ether, and shake for 2 minutes. Filter the precipitate, wash with about 15 mL of cold water, and recrystallize twice from diluted alcohol: the benzoyl derivative of dextroamphetamine so obtained, after being dried at 105° for 1 hour, melts between 154° and 160°.

Alcohol content ⟨611⟩: between 9.0% and 11.0% of C_2H_5OH.

Isomeric purity—Transfer 150 mL of Elixir to a 500-mL separator, add 15 mL of 2.5 *N* sodium hydroxide, and extract with one 60-mL and two 40-mL portions of ether. Wash the combined ether extracts with two 10-mL portions of 0.25 *N* sodium hydroxide. Wash the aqueous alkaline extracts with 20 mL of ether, adding the ether washing to the combined ether extracts. Filter the ether extracts through a pledget of cotton, previously saturated with ether, into a 250-mL beaker, rinse the cotton with a small amount of ether, and evaporate on a steam bath in a current of air to about 2 mL. Dissolve the residue in 20 mL of chloroform, and transfer to a separator containing 35 mL of 0.1 *N* sulfuric acid. Complete the transfer with two additional 20-mL portions of chloroform. Shake the separator vigorously for 1 minute, allow the layers to separate, and discard the chloroform. Add to the liquid in the separator 2.5 g of sodium bicarbonate, preventing it from coming in contact with the mouth of the separator, and swirl until most of the bicarbonate has dissolved. By means of a 1-mL syringe, rapidly inject 1.0 mL of acetic anhydride directly into the contents of the separator. Immediately insert the stopper in the separator, and shake vigorously until the evolution of carbon dioxide has ceased, releasing the pressure as necessary through the stopcock. Allow to stand for 5 minutes, and extract the solution with 50 mL of chloroform, shaking vigorously for 1 minute. Filter the chloroform extract through a pledget of filter cotton into a 100-mL beaker, rinse the cotton with a small amount of chloroform, and evaporate on a steam bath in a current of air or nitrogen to dryness. Heat and triturate the residue until the odor of chloroform is no longer perceptible. Allow the residue to cool, inducing it to crystallize. Reduce the crystals to a fine powder, heat at 80° for 30 minutes, and cool: the specific rotation of the acetylamphetamine so obtained, determined in a solution in chloroform containing 20 mg per mL, a 200-mm semimicro polarimeter tube being used, is between −37.5° and −44.0°.

Assay—

Chromatographic column—Proceed as directed for *Column Partition Chromatography* under *Chromatography* ⟨621⟩, packing a chromatographic tube with a mixture of 2 g of *Solid Support* and 1 mL of 0.06 *N* hydrochloric acid.

Standard preparation—Dissolve an accurately weighed quantity of USP Dextroamphetamine Sulfate RS in 1.8 *N* sulfuric acid (previously saturated with chloroform), and dilute quantitatively and stepwise with the same solvent to obtain a solution having a known concentration of about 0.5 mg per mL.

Assay preparation—Pipet 5 mL of Dextroamphetamine Sulfate Elixir into a 100-mL beaker, add 1 drop of 3 *N* hydrochloric acid, and swirl to mix. Add 6 g of purified siliceous earth, and mix with a glass rod until a fluffy mixture is obtained.

Procedure—Transfer the *Assay preparation* to the *Chromatographic column*, and complete the preparation of the column. Wash the prepared column with 100 mL of chloroform that previously has been saturated with water, and discard the washing. Arrange to collect the eluate in a separator containing 10.0 mL of 1.8 *N* sulfuric acid that previously has been saturated with chloroform. Pass through the column 60 mL of a freshly prepared ammoniacal chloroform solution, made by shaking 50 volumes of chloroform with 1 volume of ammonium hydroxide for 1 to 2 minutes and discarding the aqueous phase. Complete the elution with 60 mL of chloroform (previously saturated with water). Shake the separator vigorously for 1 minute, allow the layers to separate, and discard the chloroform. Concomitantly determine the absorbances of the *Standard preparation* and the *Assay preparation* in 1-cm cells at 280 nm and at the maximum at about 257 nm, with a suitable spectrophotometer, using 1.8 *N* sulfuric acid (previously saturated with chloroform) as the blank. Calculate the quantity, in mg, of $(C_9H_{13}N)_2 \cdot H_2SO_4$ in the portion of Elixir taken by the formula:

$$10C(A_{257} - A_{280})_U/(A_{257} - A_{280})_S,$$

in which C is the concentration, in mg per mL, of USP Dextroamphetamine Sulfate RS in the *Standard preparation*, and the parenthetic expressions are the differences in the absorbances of the two solutions at the wavelengths indicated by the subscripts, for the *Assay preparation* (*U*) and the *Standard preparation* (*S*), respectively.

Dextroamphetamine Sulfate Tablets

» Dextroamphetamine Sulfate Tablets contain not less than 93.0 percent and not more than 107.0 percent of the labeled amount of $(C_9H_{13}N)_2 \cdot H_2SO_4$.

Packaging and storage—Preserve in well-closed containers.

Reference standard—*USP Dextroamphetamine Sulfate Reference Standard*—Dry at 105° for 2 hours before using.

Identification—

A: Macerate a quantity of powdered Tablets, representing about 50 mg of dextroamphetamine sulfate, with about 10 mL of water for 30 minutes, and filter into a small flask. Cool the filtrate to about 15°, and proceed as directed in the *Identification test* under *Dextroamphetamine Sulfate Elixir*, beginning with "add 3 mL of 1 *N* sodium hydroxide."

B: The retention time of the major peak in the chromatogram of the *Assay preparation* corresponds to that in the chromatogram of the *Standard preparation*, as obtained in the *Assay*.

Dissolution ⟨711⟩—

Medium: water; 500 mL.

Apparatus 1: 100 rpm.

Time: 45 minutes.

Mobile phase—Dissolve 1.1 g of sodium 1-heptanesulfonate in 575 mL of water. Add 25 mL of dilute glacial acetic acid (14 in 100) and 400 mL of methanol. Adjust by the dropwise addition of glacial acetic acid to a pH of 3.3 ± 0.1, if necessary, filter, and degas the solution. Make adjustments if necessary (see *System Suitability* under *Chromatography* ⟨621⟩).

Chromatographic system (see *Chromatography* ⟨621⟩)—The liquid chromatograph is equipped with a 254-nm detector and a 3.9-mm × 30-cm column that contains packing L1. The flow rate is about 1 mL per minute. Chromatograph replicate injections of the Standard solution, and record the peak responses as directed under *Procedure:* the relative standard deviation is not more than 2.0%.

Procedure—Inject a volume (about 500 μL) of a filtered portion of the solution under test into the chromatograph, record the chromatogram, and measure the response for the major peak. Calculate the quantity of $(C_9H_{13}N)_2 . H_2SO_4$ dissolved in comparison with a Standard solution having a known concentration of USP Dextroamphetamine Sulfate RS in the same medium and similarly chromatographed.

Tolerances—Not less than 75% (*Q*) of the labeled amount of $(C_9H_{13}N)_2 . H_2SO_4$ is dissolved in 45 minutes.

Uniformity of dosage units ⟨905⟩: meet the requirements.

Isomeric purity—Pack a pledget of fine glass wool in the base of a 200- × 25-mm chromatographic tube, with the aid of a tamping rod. Add 5 g of chromatographic siliceous earth, and tamp firmly to compress the material to a uniform mass.

Finely powder a number of Tablets, equivalent to about 130 mg of dextroamphetamine sulfate, mix the powder in a mortar with 5 g of chromatographic siliceous earth, add 1 mL of methanol and 0.5 mL of ammonium hydroxide, and triturate to a uniform mixture. Transfer the mixture without delay to the chromatographic tube, and tamp as before. Wipe the mortar and pestle with a small amount of glass wool, and insert it into the tube on top of the column. Arrange a 125-mL separator containing 35 mL of 0.1 *N* sulfuric acid to receive the effluent. Pass 60 mL of chloroform through the column. Proceed as directed in the test for *Isomeric purity* under *Dextroamphetamine Sulfate Elixir*, beginning with "Shake the separator vigorously."

Assay—

Mobile phase—Dissolve 1.1 g of sodium 1-heptanesulfonate in 525 mL of water. Add 25 mL of dilute glacial acetic acid (14 in 100) and 450 mL of methanol. Adjust dropwise, if necessary, with glacial acetic acid to a pH of 3.3 ± 0.1. Filter through a 0.5-μm membrane filter. Make adjustments if necessary (see *System Suitability* under *Chromatography* ⟨621⟩).

Standard preparation—Dissolve an accurately weighed quantity of USP Dextroamphetamine Sulfate RS in 0.12 *N* phosphoric acid to obtain a solution having a known concentration of about 0.3 mg per mL.

Assay preparation—Weigh and finely powder not less than 20 Dextroamphetamine Sulfate Tablets. Transfer an accurately weighed portion of the mixed powder, equivalent to about 15 mg of dextroamphetamine sulfate, to a 50-mL volumetric flask. Add 40 mL of 0.12 *N* phosphoric acid, and sonicate for 15 minutes. Dilute with 0.12 *N* phosphoric acid to volume, and mix. Filter through a 0.5-μm membrane filter, discarding the first 20 mL of the filtrate.

Chromatographic system (see *Chromatography* ⟨621⟩)—The liquid chromatograph is equipped with a 254-nm detector and a 3.9-mm × 30-cm column that contains packing L1. The flow rate is about 2 mL per minute. Chromatograph replicate injections of the *Standard preparation*, and record the peak responses as directed under *Procedure:* the tailing factor is not more than 3, and the relative standard deviation is not more than 2.0%.

Procedure—Separately inject equal volumes (about 50 μL) of the *Standard preparation* and the *Assay preparation* into the chromatograph, record the chromatograms, and measure the responses for the major peaks. Calculate the quantity, in mg, of $(C_9H_{13}N)_2 . H_2SO_4$ in the portion of Tablets taken by the formula:

$$50C(r_U / r_S),$$

in which *C* is the concentration, in mg per mL, of USP Dextroamphetamine Sulfate RS in the *Standard preparation*, and r_U and r_S are the peak responses obtained from the *Assay preparation* and the *Standard preparation*, respectively.

Dextromethorphan

$C_{18}H_{25}NO$ 271.4

Morphinan, 3-methoxy-17-methyl-, (9α,13α,14α)-.

3-Methoxy-17-methyl-9α,13α,14α-morphinan [*125-71-3*].

» Dextromethorphan contains not less than 98.0 percent and not more than 101.0 percent of $C_{18}H_{25}NO$, calculated on the anhydrous basis.

Packaging and storage—Preserve in tight containers.

Reference standard—*USP Dextromethorphan Reference Standard*—Do not dry; determine the water content titrimetrically at the time of use.

Identification—

A: The infrared absorption spectrum of a potassium bromide dispersion of it exhibits maxima only at the same wavelengths as that of a similar preparation of USP Dextromethorphan RS.

B: The ultraviolet absorption spectrum of a 1 in 10,000 solution in dilute hydrochloric acid (1 in 120) exhibits maxima and minima at the same wavelengths as that of a similar solution of USP Dextromethorphan RS, concomitantly measured, and the respective absorptivities, calculated on the anhydrous basis, at the wavelength of maximum absorbance at about 278 nm do not differ by more than 3.0%.

Melting range, *Class I* ⟨741⟩: between 109.5° and 112.5°.

Specific rotation ⟨781⟩—The specific rotation of Dextromethorphan at 589 nm, calculated on the anhydrous basis, determined in a solution in chloroform containing 1 g of Dextromethorphan in each 10 mL, and the specific rotation of USP Dextromethorphan RS, similarly measured, do not differ by more than 1.0%.

Water, *Method Ia* ⟨921⟩: not more than 0.5%.

Residue on ignition ⟨281⟩: not more than 0.1%.

Heavy metals, *Method II* ⟨231⟩: 0.002%.

Dimethylaniline—Transfer about 500 mg to a 25-mL volumetric flask, add 19 mL of water and 1 mL of 3 *N* hydrochloric acid, and dissolve by warming on a steam bath. Cool, add 2 mL of 1 *N* acetic acid and 1 mL of sodium nitrite solution (1 in 100), dilute with water to volume, and mix: this solution shows no more color than corresponds to a solution, similarly prepared, containing 5 μg of *N,N*-dimethylaniline in 25 mL (0.001%).

Phenolic compounds—Dissolve about 10 mg in 2 mL of 3 *N* hydrochloric acid, and add 2 drops of ferric chloride TS. Mix, add 2 drops of potassium ferricyanide TS, and observe: no blue-green color develops after 2 minutes.

Assay—Dissolve about 700 mg of Dextromethorphan, accurately weighed, in 60 mL of glacial acetic acid, warming slightly, if necessary, to effect solution. Add 2 drops of crystal violet TS, and titrate with 0.1 *N* perchloric acid VS to a blue-green endpoint. Perform a blank determination, and make any necessary correction. Each mL of 0.1 *N* perchloric acid is equivalent to 27.14 mg of $C_{18}H_{25}NO$.

Dextromethorphan Hydrobromide

$C_{18}H_{25}NO . HBr . H_2O$ 370.33

Morphinan, 3-methoxy-17-methyl-, (9α,13α,14α)-, hydrobromide, monohydrate.

3-Methoxy-17-methyl-9α,13α,14α-morphinan hydrobromide monohydrate [*6700-34-1*].

Anhydrous 352.31 [*125-69-9*].

» Dextromethorphan Hydrobromide contains not less than 98.0 percent and not more than 102.0 percent of $C_{18}H_{25}NO . HBr$, calculated on the anhydrous basis.

Packaging and storage—Preserve in tight containers.

Reference standard—*USP Dextromethorphan Hydrobromide Reference Standard*—Do not dry; determine the *Water* content by *Method I* before using.

Identification—

A: The infrared absorption spectrum of a potassium bromide dispersion of it, previously dried in vacuum over phosphorus pentoxide for 4 hours, exhibits maxima only at the same wavelengths as that of a similar preparation of USP Dextromethorphan Hydrobromide RS.

B: The ultraviolet absorption spectrum of a 1 in 10,000 solution in 0.1 *N* hydrochloric acid exhibits maxima and minima at the same wavelengths as that of a similar solution of USP Dextromethorphan Hydrobromide RS, concomitantly measured, and the respective absorptivities, calculated on the anhydrous basis, at the wavelength of maximum absorbance at about 278 nm do not differ by more than 3.0%.

C: To 5 mL of a solution (1 in 200) add 5 drops of 2 *N* nitric acid and 2 mL of silver nitrate TS: a yellowish white precipitate is formed.

Specific rotation ⟨781⟩—The specific rotation of it, at 325 nm, determined photoelectrically in a solution containing 180 mg of Dextromethorphan Hydrobromide in each 10 mL (warming, if necessary, to obtain complete solution), calculated on the anhydrous basis, and the specific rotation of USP Dextromethorphan Hydrobromide RS, similarly measured, do not differ by more than 1.0%.

pH ⟨791⟩: between 5.2 and 6.5 in a solution (1 in 100).

Water, *Method I* ⟨921⟩: between 3.5% and 5.5%.

Residue on ignition ⟨281⟩: not more than 0.1%.

***N,N*-Dimethylaniline**—Transfer about 500 mg to a 25-mL volumetric flask, add 20 mL of water, and dissolve by warming on a steam bath. Cool, add 2 mL of 1 *N* acetic acid and 1 mL of sodium nitrite solution (1 in 100), dilute with water to volume, and mix: this solution shows no more color than corresponds to a solution, similarly prepared, containing 5 µg of *N,N*-dimethylaniline in 25 mL (0.001%).

Phenolic compounds—To about 5 mg add 1 drop of 3 *N* hydrochloric acid, 1 mL of water, and 2 drops of ferric chloride TS. Mix, add 2 drops of potassium ferricyanide TS, and observe after 2 minutes: no blue-green color develops.

Assay—

Mobile phase—Prepare a filtered and degassed solution containing 0.007 *M* docusate sodium and 0.007 *M* ammonium nitrate in a mixture of acetonitrile and water (70:30), and adjust the solution with glacial acetic acid to a pH of 3.4. [NOTE—Dissolve the docusate sodium in the acetonitrile and water mixture before adding the ammonium nitrate.]

Standard preparation—Dissolve an accurately weighed quantity of USP Dextromethorphan Hydrobromide RS in water, and dilute quantitatively with water to obtain a solution having a known concentration of about 0.10 mg per mL.

Assay preparation—Transfer about 10 mg of Dextromethorphan Hydrobromide, accurately weighed, to a 100-mL volumetric flask, add water to volume, and mix.

Chromatographic system (see *Chromatography* ⟨621⟩)—The liquid chromatograph is equipped with a 280-nm detector and a 4.6-mm × 25-cm column that contains 5-µm packing L1. The flow rate is about 1 mL per minute. Chromatograph replicate injections of the *Standard preparation*, and record the peak responses as directed under *Procedure:* the relative standard deviation is not more than 2.0% and the tailing factor for the major peak is not more than 2.5.

Procedure—Separately inject equal volumes (about 100 µL) of the *Standard preparation* and the *Assay preparation* into the chromatograph, record the chromatograms, and measure the responses for the major peaks. Calculate the quantity, in mg, of $C_{18}H_{25}NO.HBr$ in the portion of Dextromethorphan Hydrobromide taken by the formula:

$$100C(r_U/r_S),$$

in which *C* is the concentration, in mg per mL, of Dextromethorphan Hydrobromide RS, on the anhydrous basis, in the *Standard preparation*, and r_U and r_S are the peak responses obtained from the *Assay preparation* and the *Standard preparation*, respectively.

Dextromethorphan Hydrobromide Elixir, Terpin Hydrate and—*see* Terpin Hydrate and Dextromethorphan Hydrobromide Elixir

Dextromethorphan Hydrobromide Syrup

» Dextromethorphan Hydrobromide Syrup contains not less than 95.0 percent and not more than 105.0 percent of the labeled amount of $C_{18}H_{25}NO.HBr.H_2O$.

Packaging and storage—Preserve in tight, light-resistant containers.

Reference standard—*USP Dextromethorphan Hydrobromide Reference Standard*—Do not dry; determine the *Water* content by *Method I* before using.

Identification—

A: Transfer about 50 mL of Syrup to a 250-mL separator, add 20 mL of water, 5 mL of 2.5 *N* sodium hydroxide, and 40 mL of solvent hexane, and shake thoroughly. Remove the solvent hexane layer, and filter through anhydrous sodium sulfate into a 150-mL beaker. Repeat the solvent hexane extraction, using two 40-mL portions and collecting the extracts in the beaker after filtering. Evaporate the combined extracts at 50° under nitrogen to dryness, and dissolve and dilute the residue in 10 mL of chloroform: the solution is dextrorotatory (see *Optical Rotation* ⟨781⟩). Retain the chloroform solution for *Identification test B.*

B: Evaporate the chloroform solution from *Identification test A* on a steam bath to dryness, dissolve the residue in 2 mL of 2 *N* sulfuric acid, and add 1 mL of a freshly prepared solution of mercuric nitrate (prepared by dissolving 700 mg of mercuric nitrate in 4 mL of water, adding 100 mg of sodium nitrate, mixing, and filtering): no red color is produced immediately, but after heating, a yellow to red color develops in about 15 minutes.

Assay—

Mobile phase and *Standard preparation*—Prepare as directed in the *Assay* under *Dextromethorphan Hydrobromide.*

Assay preparation—Pipet, using a to-contain pipet, a volume of Dextromethorphan Hydrobromide Syrup, equivalent to about 10 mg of dextromethorphan hydrobromide, into a 100-mL volumetric flask, dilute with water to volume, and mix.

Chromatographic system and *Procedure* (see *Chromatography* ⟨621⟩)—Proceed as directed in the *Assay* under *Dextromethorphan Hydrobromide.* Calculate the quantity, in mg, of $C_{18}H_{25}NO.HBr.H_2O$ in the volume of Syrup taken by the formula:

$$(370.33/352.31)(100C)(r_U/r_S),$$

in which 370.33 and 352.31 are the molecular weights of dextromethorphan hydrobromide and anhydrous dextromethorphan hydrobromide, respectively, *C* is the concentration, in mg per mL, of USP Dextromethorphan Hydrobromide RS, on the anhydrous basis, in the *Standard preparation*, and r_U and r_S are the peak responses obtained from the *Assay preparation* and the *Standard preparation*, respectively.

Dextrose

CH₂OH, O, OH, OH, HO, OH · H₂O

D-Glucopyranose monohydrate

$C_6H_{12}O_6.H_2O$ 198.17
D-Glucose, monohydrate.
D-Glucose monohydrate [*5996-10-1*].
Anhydrous 180.16 [*50-99-7*].

» Dextrose is a sugar usually obtained by the hydrolysis of Starch. It contains one molecule of water of hydration or is anhydrous.

Packaging and storage—Preserve in well-closed containers.

Labeling—Label it to indicate whether it is hydrous or anhydrous.

Identification—Add a few drops of a solution (1 in 20) to 5 mL of hot alkaline cupric tartrate TS: a copious red precipitate of cuprous oxide is formed.

Color of solution—Dissolve 25 g in water to make 50.0 mL of solution: the solution has no more color than a solution prepared by mixing 1.0 mL of cobaltous chloride CS, 3.0 mL of ferric chloride CS, and 2.0 mL of cupric sulfate CS with water to make 10 mL, and diluting 3 mL of this solution with water to 50 mL. Make the comparison by viewing the solutions downward in matched color-comparison tubes against a white surface.

Specific rotation ⟨781⟩: between +52.6° and +53.2°, calculated on the anhydrous basis, determined in a solution containing 10 g of Dextrose and 0.2 mL of 6 *N* ammonium hydroxide in each 100 mL.

Acidity—Dissolve 5.0 g in 50 mL of carbon dioxide–free water. Add phenolphthalein TS, and titrate with 0.020 *N* sodium hydroxide to the production of a distinct pink color: not more than 0.30 mL is required for neutralization.

Water, *Method III* ⟨921⟩—Dry it at 105° for 16 hours: the hydrous form loses between 7.5% and 9.5% of its weight, and the anhydrous form loses not more than 0.5% of its weight.

Residue on ignition ⟨281⟩: not more than 0.1%.

Chloride ⟨221⟩—A 2.0-g portion shows no more chloride than corresponds to 0.50 mL of 0.020 *N* hydrochloric acid (0.018%).

Sulfate ⟨221⟩—A 2.0-g portion shows no more sulfate than corresponds to 0.50 mL of 0.020 *N* sulfuric acid (0.025%).

Arsenic, *Method I* ⟨211⟩: 1 ppm.

Heavy metals ⟨231⟩—Dissolve 4.0 g in water to make 25 mL of solution: the limit is 5 ppm.

Dextrin—Reflux 1 g of finely powdered Dextrose with 20 mL of alcohol: it dissolves completely.

Soluble starch, sulfites—To a solution of 1 g in 10 mL of water add 1 drop of iodine TS: the liquid is colored yellow.

Dextrose Excipient—*see* Dextrose Excipient NF

Dextrose Injection

» Dextrose Injection is a sterile solution of Dextrose in Water for Injection. It contains not less than 95.0 percent and not more than 105.0 percent of the labeled amount of $C_6H_{12}O_6 \cdot H_2O$. Dextrose Injection contains no antimicrobial agents.

Packaging and storage—Preserve in single-dose containers, preferably of Type I or Type II glass.

Labeling—The label states the total osmolar concentration in mOsmol per liter. Where the contents are less than 100 mL, or where the label states that the Injection is not for direct injection but is to be diluted before use, the label alternatively may state the total osmolar concentration in mOsmol per mL.

Identification—It responds to the *Identification test* under *Dextrose.*

Pyrogen—It meets the requirements of the *Pyrogen Test* ⟨151⟩. [NOTE—Dilute, with Water for Injection, Injections containing more than 10% of dextrose to give a concentration of 10% of dextrose.]

pH ⟨791⟩: between 3.5 and 6.5, determined on a portion to which 0.30 mL of saturated potassium chloride solution has been added for each 100 mL and which previously has been diluted with water, if necessary, to a concentration of not more than 5% of dextrose.

Particulate matter ⟨788⟩: meets the requirements under *Small-volume Injections.*

Heavy metals ⟨231⟩—Transfer a volume of Injection, equivalent to 4.0 g of dextrose, to a suitable vessel, and adjust the volume to 25 mL by evaporation or addition of water, as necessary: the limit is $0.0005C$%, in which C is the labeled amount, in g, of $C_6H_{12}O_6 \cdot H_2O$ per mL of Injection.

5-Hydroxymethylfurfural and related substances—Dilute an accurately measured volume of Injection, equivalent to 1.0 g of $C_6H_{12}O_6 \cdot H_2O$, with water to 250.0 mL. Determine the absorbance of this solution in a 1-cm cell at 284 nm, with a suitable spectrophotometer, using water as the blank: the absorbance is not more than 0.25.

Other requirements—It meets the requirements under *Injections* ⟨1⟩.

Assay—Transfer an accurately measured volume of Dextrose Injection, containing 2 to 5 g of dextrose, to a 100-mL volumetric flask. Add 0.2 mL of 6 *N* ammonium hydroxide, dilute with water to volume, and mix. Determine the angular rotation in a suitable polarimeter tube at 25° (see *Optical Rotation* ⟨781⟩). The observed rotation, in degrees, multiplied by $1.0425A$, in which A is the ratio 200 divided by the length, in mm, of the polarimeter tube employed, represents the weight, in g, of $C_6H_{12}O_6 \cdot H_2O$ in the volume of Injection taken.

Dextrose Injection, Alcohol and—*see* Alcohol and Dextrose Injection

Dextrose Injection, Dopamine Hydrochloride and—*see* Dopamine Hydrochloride and Dextrose Injection

Dextrose Injection, Lidocaine Hydrochloride and—*see* Lidocaine Hydrochloride and Dextrose Injection

Dextrose Injection, Potassium Chloride in—*see* Potassium Chloride in Dextrose Injection

Dextrose and Sodium Chloride Injection

» Dextrose and Sodium Chloride Injection is a sterile solution of Dextrose and Sodium Chloride in Water for Injection. It contains not less than 95.0 percent and not more than 105.0 percent of the labeled amount of $C_6H_{12}O_6 \cdot H_2O$ and of NaCl. It contains no antimicrobial agents.

Packaging and storage—Preserve in single-dose containers, preferably of Type I or Type II glass.

Labeling—The label states the total osmolar concentration in mOsmol per liter. Where the contents are less than 100 mL, or where the label states that the Injection is not for direct injection but is to be diluted before use, the label alternatively may state the total osmolar concentration in mOsmol per mL.

Identification—It responds to the *Identification test* under *Dextrose*, and to the tests for *Sodium* ⟨191⟩ and for *Chloride* ⟨191⟩.

Pyrogen—It meets the requirements of the test for *Pyrogen* under *Dextrose Injection.*

pH ⟨791⟩: between 3.5 and 6.5, determined on a portion diluted with water, if necessary, to a concentration of not more than 5% of dextrose.

5-Hydroxymethylfurfural and related substances—Dilute an accurately measured volume of Injection, equivalent to 1.0 g of $C_6H_{12}O_6 \cdot H_2O$ with water to 500.0 mL. Determine the absorbance of this solution in a 1-cm cell at 284 nm, with a suitable spectrophotometer, using water as the blank: the absorbance is not more than 0.25.

Other requirements—It meets the requirements under *Injections* ⟨1⟩.

Assay for dextrose—Transfer an accurately measured volume of Dextrose and Sodium Chloride Injection, containing from 2 to 5 g of dextrose, to a 100-mL volumetric flask. Add 0.2 mL of 6

N ammonium hydroxide, dilute with water to volume, and mix. Determine the angular rotation in a suitable polarimeter tube at 25° (see *Optical Rotation* ⟨781⟩). The observed rotation, in degrees, multiplied by 1.0425A, in which A is the ratio 200 divided by the length, in mm, of the polarimeter tube employed, represents the weight, in g, of $C_6H_{12}O_6 \cdot H_2O$ in the volume of Injection taken.

Assay for sodium chloride—Transfer an accurately measured volume of Dextrose and Sodium Chloride Injection, equivalent to about 90 mg of sodium chloride, into a porcelain casserole, and add 140 mL of water and 1 mL of dichlorofluorescein TS. Mix, and titrate with 0.1 N silver nitrate VS until the silver chloride flocculates and the mixture acquires a faint pink color. Each mL of 0.1 N silver nitrate is equivalent to 5.844 mg of NaCl.

Dextrose Tablets, Sodium Chloride and—*see* Sodium Chloride and Dextrose Tablets

Diacetylated Monoglycerides—*see* Diacetylated Monoglycerides NF

Diatrizoate Meglumine

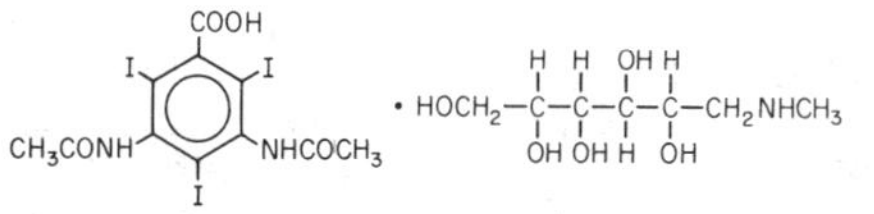

$C_{11}H_9I_3N_2O_4 \cdot C_7H_{17}NO_5$ 809.13

Benzoic acid, 3,5-bis(acetylamino)-2,4,6-triiodo-, compd. with 1-deoxy-1-(methylamino)-D-glucitol (1:1).

1-Deoxy-1-(methylamino)-D-glucitol 3,5-diacetamido-2,4,6-triiodobenzoate (salt) [*131-49-7*].

» Diatrizoate Meglumine contains not less than 98.0 percent and not more than 102.0 percent of $C_{11}H_9I_3N_2O_4 \cdot C_7H_{17}NO_5$, calculated on the dried basis.

Packaging and storage—Preserve in well-closed containers.

Reference standards—*USP 5-Acetamido-3-amino-2,4,6-triiodobenzoic Acid Reference Standard*—Dry at 105° for 4 hours before using. Keep container tightly closed and protected from light. *USP Diatrizoic Acid Reference Standard*—This material is the hydrous form of Diatrizoic Acid. Dry at 105° for 4 hours before using.

Identification—

A: It responds to the *Thin-layer Chromatographic Identification Test* ⟨201⟩, the test solution and the Standard solution of USP Diatrizoic Acid RS being prepared at a concentration of 1 mg per mL in an 0.8 in 1000 solution of sodium hydroxide in methanol, the solvent mixture being a mixture of chloroform, methanol, and ammonium hydroxide (20:10:2), and short-wavelength ultraviolet light being used to locate the spots.

B: Heat about 500 mg in a suitable crucible: violet vapors are evolved.

Specific rotation ⟨781⟩: between −5.65° and −6.37°, calculated on the dried basis, determined in a solution containing 1 g in each 10 mL.

Loss on drying ⟨731⟩—Dry it at 105° for 4 hours: it loses not more than 1.0% of its weight.

Residue on ignition ⟨281⟩: not more than 0.1%.

Free aromatic amine—Transfer 1.0 g to a 50-mL volumetric flask, and add 5 mL of water and 10 mL of 0.1 N sodium hydroxide. To a second 50-mL volumetric flask transfer 4 mL of water, 10 mL of 0.1 N sodium hydroxide, and 1.0 mL of a Standard solution prepared by dissolving a suitable quantity of USP 5-Acetamido-3-amino-2,4,6-triiodobenzoic Acid RS in 0.1 N sodium hydroxide. Use 0.2 mL of 0.1 N sodium hydroxide for each 5.0 mg of Standard, and dilute with water to obtain a known concentration of 500 µg per mL. To a third 50-mL volumetric flask add 5 mL of water and 10 mL of 0.1 N sodium hydroxide to provide a blank.

Treat each flask as follows: add 25 mL of methyl sulfoxide, insert the stopper, and mix by swirling gently. Chill in an ice bath in the dark for 5 minutes. [NOTE—In conducting the following steps, keep the flasks in the ice bath and in the dark as much as possible until all of the reagents have been added.] Add slowly 2 mL of hydrochloric acid, mix, and allow to stand for 5 minutes. Add 2 mL of sodium nitrite solution (1 in 50), mix, and allow to stand for 5 minutes. Add 1 mL of sulfamic acid solution (2 in 25), shake, and allow to stand for 5 minutes. [*Caution—Considerable pressure is produced.*] Add 2 mL of a 1 in 1000 solution of N-(1-naphthyl)ethylenediamine dihydrochloride in dilute propylene glycol (7 in 10), and mix.

Remove the flasks from the ice bath and from storage in the dark, and allow to stand in a water bath at 22° to 25° for 10 minutes. Shake gently and occasionally during this period, releasing the pressure by loosening the stopper. Dilute with water to volume, and mix.

Within 5 minutes from the time of diluting the solutions in all three flasks to 50 mL, concomitantly determine the absorbances of the solution from the substance under test and the Standard solution in 1-cm cells at the wavelength of maximum absorbance at about 465 nm, with a suitable spectrophotometer, versus the prepared blank. The absorbance of the solution from the Diatrizoate Meglumine is not greater than that of the Standard solution (0.05%).

Iodine and iodide—

Test preparation—Transfer 2.0 g to a 50-mL centrifuge tube provided with a stopper, dilute with water to 24 mL, and shake to dissolve.

Procedure—Add 5 mL of toluene and 5 mL of 2 N sulfuric acid, shake well, and centrifuge: the toluene layer shows no red color. Add 1 mL of sodium nitrite solution (1 in 50), shake, and centrifuge: any red color in the toluene layer is not darker than that obtained when a mixture of 2.0 mL of potassium iodide solution (1 in 4000) and 22 mL of water is substituted for the solution under test (0.02% of iodide).

Heavy metals ⟨231⟩—

Standard preparation—Transfer 2.0 mL of *Standard Lead Solution* (20 µg of Pb) to a 50-mL color-comparison tube, add 5 mL of 1 N sodium hydroxide, dilute with water to 40 mL, and mix.

Test preparation—Dissolve 1.0 g of Diatrizoate Meglumine in 20 mL of water and 5 mL of 1 N sodium hydroxide, transfer the solution to a 50-mL color-comparison tube, dilute with water to 40 mL, and mix.

Procedure—To each of the tubes containing the *Standard preparation* and the *Test preparation* add 10 mL of sodium sulfide TS, mix, allow to stand for 5 minutes, and view downward over a white surface: the color of the solution from the *Test preparation* is not darker than that of the solution from the *Standard preparation* (0.002%).

Assay—Transfer about 400 mg of Diatrizoate Meglumine, accurately weighed, to a glass-stoppered, 125-mL conical flask, add 30 mL of 1.25 N sodium hydroxide and 500 mg of powdered zinc, connect the flask to a reflux condenser, and reflux the mixture for 1 hour. Cool the flask to room temperature, rinse the condenser with 20 mL of water, disconnect the flask from the condenser, and filter the mixture. Rinse the flask and the filter thoroughly, adding the rinsings to the filtrate. Add 5 mL of glacial acetic acid and 1 mL of tetrabromophenolphthalein ethyl ester TS, and titrate with 0.05 N silver nitrate VS until the yellow precipitate just turns green. Each mL of 0.05 N silver nitrate is equivalent to 13.49 mg of $C_{11}H_9I_3N_2O_4 \cdot C_7H_{17}NO_5$.

Diatrizoate Meglumine Injection

» Diatrizoate Meglumine Injection is a sterile solution of Diatrizoate Meglumine in Water for Injection, or a sterile solution of Diatrizoic Acid in Water for Injection prepared with the aid of Meglumine. It contains not less than 95.0 percent and not more than

105.0 percent of the labeled amount of diatrizoate meglumine ($C_{11}H_9I_3N_2O_4.C_7H_{17}NO_5$). It may contain small amounts of suitable buffers and of Edetate Calcium Disodium or Edetate Disodium as a stabilizer. Diatrizoate Meglumine Injection intended for intravascular use contains no antimicrobial agents.

Packaging and storage—Preserve Injection intended for intravascular injection either in single-dose containers, preferably of Type I or Type III glass, protected from light or, where intended for administration with a pressure injector through a suitable transfer connection, in similar glass 500-mL or 1000-mL bottles, protected from light. Injection packaged for other than intravascular use may be packaged in 100-mL multiple-dose containers, preferably of Type I or Type III glass, protected from light.

Labeling—Label containers of Injection intended for intravascular injection, where packaged in single-dose containers, to direct the user to discard any unused portion remaining in the container or, where packaged in bulk bottles to state, "Bulk Container—only for sterile filling of pressure injectors," to state that it contains no antimicrobial preservatives, and to direct the user to discard any unused portion remaining in the container after 6 hours. Indicate also in the labeling of bulk bottles that a pressure injector is to be charged with a dose just prior to administration of the Injection. Label containers of Injection intended for other than intravascular injection to show that the contents are not intended for intravascular injection.

Reference standards—*USP 5-Acetamido-3-amino-2,4,6-triiodobenzoic Acid Reference Standard*—Dry at 105° for 4 hours before using. Keep container tightly closed and protected from light. *USP Diatrizoic Acid Reference Standard*—This material is the hydrous form of Diatrizoic Acid. Dry at 105° for 4 hours before using.

Identification—

A: Dilute a volume of Injection, if necessary, with an 0.8 in 1000 solution of sodium hydroxide in methanol to obtain a test solution having a concentration of 1 mg per mL. The test solution responds to the *Thin-layer Chromatographic Identification Test* ⟨201⟩, the Standard solution being prepared at a concentration of 1 mg of USP Diatrizoic Acid RS per mL in an 0.8 in 1000 solution of sodium hydroxide in methanol, the solvent mixture being a mixture of chloroform, methanol, and ammonium hydroxide (20:10:2), and short-wavelength ultraviolet light being used to locate the spots.

B: Evaporate a volume of Injection, equivalent to about 500 mg of diatrizoate meglumine, to dryness, and heat the residue so obtained in a suitable crucible: violet vapors are evolved.

Pyrogen—It meets the requirements of the *Pyrogen Test* ⟨151⟩, the test dose being the equivalent of 2.5 g of diatrizoate meglumine per kg.

pH ⟨791⟩: between 6.0 and 7.7.

Free aromatic amine—Transfer a volume of Injection, accurately measured and equivalent to 1 g of diatrizoate meglumine, to a glass-stoppered, 50-mL volumetric flask. Dilute with water to 5 mL, and add 10 mL of 0.1 *N* sodium hydroxide. Into a second 50-mL volumetric flask pipet a known volume of a Standard solution prepared by dissolving a suitable quantity of USP 5-Acetamido-3-amino-2,4,6-triiodobenzoic Acid RS in 0.1 *N* sodium hydroxide. Use 0.2 mL of 0.1 *N* sodium hydroxide for each 5.0 mg of standard, and dilute with water to obtain a known concentration of 500 µg per mL. The volume of Standard solution used contains a quantity of free aromatic amine corresponding to 0.05% of the weight of diatrizoate meglumine in the volume of Injection taken. Dilute with water to 5 mL, and add 10 mL of 0.1 *N* sodium hydroxide. Proceed as directed in the test for *Free aromatic amine* under *Diatrizoate Meglumine*, beginning with "To a third 50-mL volumetric flask add 5 mL of water."

Iodine and iodide—Dilute a volume of Injection, equivalent to 2.0 g of diatrizoate meglumine, with 24 mL of water in a 50-mL centrifuge tube provided with a stopper. Add 5 mL of toluene and 5 mL of 2 *N* sulfuric acid, shake, and centrifuge: the toluene layer shows no red color. Add 1 mL of sodium nitrite solution (1 in 50), shake, and centrifuge: any red color in the toluene layer is not darker than that obtained when a volume of potassium iodide solution (1 in 4000), containing a quantity of iodide corresponding to 0.02% of the weight of diatrizoate meglumine in the volume of Injection taken, is diluted with water to 24 mL and substituted for the solution under test (0.02% of iodide).

Heavy metals ⟨231⟩—In a 50-mL color-comparison tube, mix a volume of Injection, equivalent to 1.0 g of diatrizoate meglumine, with 5 mL of 1 *N* sodium hydroxide, dilute with water to 40 mL, and mix. Using this as the *Test preparation*, proceed as directed in the test for *Heavy metals* under *Diatrizoate Meglumine:* the limit is 0.002%.

Meglumine content—Determine the angular rotation (see *Optical Rotation* ⟨781⟩) of the Injection, using a 10-cm cell and a suitable polarimeter. Calculate the content, in mg per mL, of meglumine in the Injection by the formula:

$$1000a/24.9,$$

in which a is the observed angular rotation, in degrees, corrected for the blank, and the factor, 24.9, is the average specific rotation, in degrees, of meglumine. The meglumine content is between 22.9% and 25.3% of the labeled amount of diatrizoate meglumine.

Other requirements—It meets the requirements under *Injections* ⟨1⟩.

Assay—Pipet a volume of Diatrizoate Meglumine Injection, or a suitable dilution of it, equivalent to about 600 mg of diatrizoate meglumine, into a glass-stoppered, 125-mL conical flask, add 30 mL of 1.25 *N* sodium hydroxide and 500 mg of powdered zinc, connect the flask to a reflux condenser, and reflux the mixture for 1 hour. Cool the flask to room temperature, rinse the condenser with 20 mL of water, disconnect the flask from the condenser, and filter the mixture. Rinse the filter and the flask thoroughly, adding the rinsings to the filtrate. Add 5 mL of glacial acetic acid and 1 mL of tetrabromophenolphthalein ethyl ester TS, and titrate with 0.05 *N* silver nitrate VS until the yellow precipitate just turns green. Each mL of 0.05 *N* silver nitrate is equivalent to 13.49 mg of $C_{11}H_9I_3N_2O_4.C_7H_{17}NO_5$.

Diatrizoate Meglumine and Diatrizoate Sodium Injection

» Diatrizoate Meglumine and Diatrizoate Sodium Injection is a sterile solution of Diatrizoate Meglumine and Diatrizoate Sodium in Water for Injection, or a sterile solution of Diatrizoic Acid in Water for Injection prepared with the aid of Sodium Hydroxide and Meglumine. It contains not less than 95.0 percent and not more than 105.0 percent of the labeled amounts of diatrizoate meglumine ($C_{11}H_9I_3N_2O_4.$-$C_7H_{17}NO_5$) and of iodine (I). It may contain small amounts of suitable buffers and of Edetate Calcium Disodium or Edetate Disodium as a stabilizer. Diatrizoate Meglumine and Diatrizoate Sodium Injection intended for intravascular use contains no antimicrobial agents.

Packaging and storage—Preserve either in single-dose containers, preferably of Type I or Type III glass, protected from light or, where intended for administration with a pressure injector through a suitable transfer connection, in similar glass 500-mL or 1000-mL bottles, protected from light.

Labeling—Label containers of Injection intended for intravascular injection, where packaged in single-dose containers, to direct the user to discard any unused portion remaining in the container or, where packaged in bulk bottles to state, "Bulk Container—only for sterile filling of pressure injectors," to state that it contains no antimicrobial preservatives, and to direct the user to discard any unused portion remaining in the container after 6 hours. Indicate also in the labeling of bulk bottles that a pressure injector is to be charged with a dose just prior to administration of the Injection. Label containers of Injection intended for other than intravascular injection to show that the contents are not intended for intravascular injection.

Reference standards—*USP 5-Acetamido-3-amino-2,4,6-triiodobenzoic Acid Reference Standard*—Dry at 105° for 4 hours before using. Keep container tightly closed and protected from light. *USP Diatrizoic Acid Reference Standard*—This material is the hydrous form of Diatrizoic Acid. Dry at 105° for 4 hours before using.

Identification—

A: Dilute a volume of Injection, if necessary, with an 0.8 in 1000 solution of sodium hydroxide in methanol to obtain a test solution having a concentration of 1 mg per mL. The test solution responds to the *Thin-layer Chromatographic Identification Test* ⟨201⟩, the Standard solution being prepared at a concentration of 1 mg of USP Diatrizoic Acid RS per mL in an 0.8 in 1000 solution of sodium hydroxide in methanol, the solvent mixture being a mixture of chloroform, methanol, and ammonium hydroxide (20:10:2), and short-wavelength ultraviolet light being used to locate the spots.

B: Evaporate a volume of Injection, equivalent to about 500 mg of diatrizoate meglumine and diatrizoate sodium, to dryness, and heat the residue so obtained in a suitable crucible: violet vapors are evolved.

Pyrogen—It meets the requirements of the *Pyrogen Test* ⟨151⟩, the test dose being the equivalent of 2.5 g of the total of diatrizoate meglumine and diatrizoate sodium per kg.

pH ⟨791⟩: between 6.0 and 7.7.

Free aromatic amine—Transfer an accurately measured volume of Injection, equivalent to about 1 g of diatrizoate meglumine and diatrizoate sodium, to a 50-mL volumetric flask. Dilute with water to 5 mL, and add 10 mL of 0.1 *N* sodium hydroxide. To a second 50-mL volumetric flask transfer 4 mL of water, 10 mL of 0.1 *N* sodium hydroxide, and 1.0 mL of a Standard solution prepared by dissolving a suitable quantity of USP 5-Acetamido-3-amino-2,4,6-triiodobenzoic Acid RS in 0.1 *N* sodium hydroxide. Use 0.2 mL of 0.1 *N* sodium hydroxide for each 5.0 mg of standard, and dilute with water to obtain a known concentration of 500 µg per mL. Proceed as directed in the test for *Free aromatic amine* under *Diatrizoate Meglumine*, beginning with "To a third 50-mL volumetric flask add 5 mL of water."

Iodine and iodide—Transfer an accurately measured volume of Injection, equivalent to about 2.0 g of the total of diatrizoate meglumine and diatrizoate sodium, to a 50-mL centrifuge tube provided with a stopper. Dilute with water to 24 mL. Add 5 mL of toluene and 5 mL of 2 *N* sulfuric acid, shake, and centrifuge: the toluene layer shows no red color. Add 1 mL of sodium nitrite solution (1 in 50), shake well, and centrifuge: any red color in the toluene layer is not darker than that obtained when a volume of potassium iodide solution (1 in 4000) containing a quantity of iodide corresponding to 0.02% of the weight of diatrizoate meglumine and diatrizoate sodium in the volume of Injection taken, is diluted with water to 24 mL and substituted for the solution under test (0.02% of iodide).

Heavy metals ⟨231⟩—In a 50-mL color-comparison tube, mix a volume of Injection, equivalent to 1.0 g of the total of diatrizoate meglumine and diatrizoate sodium, with 5 mL of 1 *N* sodium hydroxide, dilute with water to 40 mL, and mix. Using this as the *Test preparation*, proceed as directed in the test for *Heavy metals* under *Diatrizoate Meglumine:* the limit is 0.002%.

Other requirements—It meets the requirements under *Injections* ⟨1⟩.

Assay for diatrizoate meglumine—Pipet 5 mL of Diatrizoate Meglumine and Diatrizoate Sodium Injection into a 10-mL volumetric flask, add water to volume, and mix. Determine the angular rotation (see *Optical Rotation* ⟨781⟩) of the diluted Injection, using a 100-mm tube. Calculate the content, in mg per mL, of $C_{11}H_9I_3N_2O_4 \cdot C_7H_{17}NO_5$ in the Injection by the formula:

$$2000a/6.01,$$

in which a is the observed angular rotation, in degrees, corrected for the blank, and the factor 6.01 is the specific rotation, in degrees, of diatrizoate meglumine.

Assay for iodine—Transfer an accurately measured volume of Diatrizoate Meglumine and Diatrizoate Sodium Injection, equivalent to about 4 g of the total of diatrizoate meglumine and diatrizoate sodium, to a 50-mL volumetric flask, dilute with water to volume, and mix. Pipet 5 mL of this solution into a glass-stoppered, 125-mL conical flask, add 30 mL of 1.25 *N* sodium hydroxide and 500 mg of powdered zinc, connect the flask to a reflux condenser, and reflux the mixture for 1 hour. Cool the flask to room temperature, rinse the condenser with 20 mL of water, disconnect the flask from the condenser, and filter the mixture. Rinse the flask and filter thoroughly, adding the rinsings to the filtrate. Add 5 mL of glacial acetic acid and 1 mL of tetrabromophenolphthalein ethyl ester TS, and titrate with 0.05 *N* silver nitrate VS until the yellow precipitate just turns green. Each mL of 0.05 *N* silver nitrate is equivalent to 6.345 mg of iodine.

Diatrizoate Meglumine and Diatrizoate Sodium Solution

» Diatrizoate Meglumine and Diatrizoate Sodium Solution is a solution of Diatrizoic Acid in Purified Water prepared with the aid of Meglumine and Sodium Hydroxide. It contains not less than 95.0 percent and not more than 105.0 percent of the labeled amounts of diatrizoate meglumine ($C_{11}H_9I_3N_2O_4 \cdot C_7H_{17}NO_5$) and of iodine (I). It may contain small amounts of suitable buffers, Edetate Disodium, and flavoring agents.

Packaging and storage—Preserve in tight, light-resistant containers.

Labeling—Label the container to indicate that the contents are not intended for parenteral use.

Reference standard—*USP Diatrizoic Acid Reference Standard*—This material is the hydrous form of Diatrizoic Acid. Dry at 105° for 4 hours before using.

Identification—

A: Dilute a volume of Solution, if necessary, with an 0.8 in 1000 solution of sodium hydroxide in methanol, to obtain a test solution having a concentration of 1 mg per mL. The test solution responds to the *Thin-layer Chromatographic Identification Test* ⟨201⟩, the Standard solution being prepared at a concentration of 1 mg of USP Diatrizoic Acid RS per mL in an 0.8 in 1000 solution of sodium hydroxide in methanol, the solvent mixture being a mixture of chloroform, methanol, and ammonium hydroxide (20:10:2), and short-wavelength ultraviolet light being used to locate the spots.

B: Evaporate a volume of Solution, equivalent to about 500 mg of diatrizoate meglumine and diatrizoate sodium, to dryness, and heat the residue so obtained in a crucible: violet vapors are evolved.

pH ⟨791⟩: between 6.0 and 7.6.

Iodine and iodide—Using as the *Test preparation* a volume of Solution equivalent to about 2 g of diatrizoate meglumine and diatrizoate sodium and diluting it with water to 24 mL in a 50-mL centrifuge tube provided with a stopper, proceed as directed for *Procedure* in the test for *Iodine and iodide* under *Diatrizoate Meglumine.*

Assay for diatrizoate meglumine—Pipet into a 250-mL volumetric flask a volume of Diatrizoate Meglumine and Diatrizoate Sodium Solution, equivalent to about 1.5 g of diatrizoate meglumine and diatrizoate sodium, add water to volume, and mix. Pipet 10 mL of this solution into a glass-stoppered, 250-mL flask. Add 4 mL of 2 *N* sulfuric acid and 20 mL of sodium periodate solution (1 in 200). Insert the stopper, and set aside in the dark for 1 hour, then add 50 mL of water, mix, and add 10 mL of potassium iodide TS. Insert the stopper quickly, mix by swirling for 20 seconds, and immediately titrate with 0.1 *N* sodium thiosulfate VS, using 3 mL of starch TS. Perform a blank determination, and make any necessary correction. Each mL of 0.1 *N* sodium thiosulfate is equivalent to 10.11 mg of $C_{11}H_9I_3N_2O_4 \cdot C_7H_{17}NO_5$.

Assay for iodine—Transfer an accurately measured volume of Diatrizoate Meglumine and Diatrizoate Sodium Solution, equivalent to about 4 g of the total of diatrizoate meglumine and diatrizoate sodium, to a 50-mL volumetric flask, dilute with water to volume, and mix. Pipet 5 mL of this solution into a glass-stoppered, 125-mL conical flask, add 30 mL of 1.25 *N* sodium hydroxide and 500 mg of powdered zinc, connect the flask to a reflux condenser, and reflux the mixture for 1 hour. Cool the flask to room temperature, rinse the condenser with 20 mL of water, disconnect the flask from the condenser, and filter the mixture. Rinse the flask and filter thoroughly, adding the rinsings to the filtrate. Add 5 mL of glacial acetic acid and 1 mL of tetrabromophenolphthalein ethyl ester TS, and titrate with 0.05 *N* silver nitrate VS until the yellow precipitate just turns green. Each mL of 0.05 *N* silver nitrate is equivalent to 6.345 mg of iodine.

Diatrizoate Sodium

$C_{11}H_8I_3N_2NaO_4$ 635.90

Benzoic acid, 3,5-bis(acetylamino)-2,4,6-triiodo-, monosodium salt.

Monosodium 3,5-diacetamido-2,4,6-triiodobenzoate [*737-31-5*].

» Diatrizoate Sodium contains not less than 98.0 percent and not more than 102.0 percent of $C_{11}H_8I_3N_2NaO_4$, calculated on the anhydrous basis.

Packaging and storage—Preserve in well-closed containers.

Reference standards—*USP 5-Acetamido-3-amino-2,4,6-triiodobenzoic Acid Reference Standard*—Dry at 105° for 4 hours before using. Keep container tightly closed and protected from light. *USP Diatrizoic Acid Reference Standard*—This material is the hydrous form of Diatrizoic Acid. Dry at 105° for 4 hours before using.

Identification—

A: It responds to the *Thin-layer Chromatographic Identification Test* ⟨201⟩, the test solution and the Standard solution being prepared at a concentration of 1 mg of USP Diatrizoic Acid RS per mL in an 0.8 in 1000 solution of sodium hydroxide in methanol, the solvent mixture being a mixture of chloroform, methanol, and ammonium hydroxide (20:10:2), and short-wavelength ultraviolet light being used to locate the spots.

B: Heat about 500 mg in a suitable crucible: violet vapors are evolved.

C: It responds to the flame test for *Sodium* ⟨191⟩.

Water, *Method I* ⟨921⟩: not more than 10.0%.

Free aromatic amine—Transfer 1.0 g to a 50-mL volumetric flask, and add 5 mL of water and 10 mL of 0.1 *N* sodium hydroxide. Proceed as directed in the test for *Free aromatic amine* under *Diatrizoate Meglumine*, beginning with "To a second 50-mL volumetric flask transfer 4 mL of water."

Iodine and iodide—

Test preparation—Transfer 2.0 g to a 50-mL centrifuge tube provided with a stopper, dilute with water to 24 mL, and shake to dissolve.

Procedure—Proceed as directed for *Procedure* in the test for *Iodine and iodide* under *Diatrizoate Meglumine*.

Heavy metals ⟨231⟩—Dissolve 1.0 g of Diatrizoate Sodium in 20 mL of water and 5 mL of 1 *N* sodium hydroxide, transfer the solution to a 50-mL color-comparison tube, dilute with water to 40 mL, and mix. Using this as the *Test preparation*, proceed as directed for *Heavy metals* under *Diatrizoate Meglumine:* the limit is 0.002%.

Assay—Transfer about 300 mg of Diatrizoate Sodium, accurately weighed, to a glass-stoppered, 125-mL conical flask, add 30 mL of 1.25 *N* sodium hydroxide and 500 mg of powdered zinc, connect the flask to a reflux condenser, and reflux the mixture for 1 hour. Cool the flask to room temperature, rinse the condenser with 20 mL of water, disconnect the flask from the condenser, and filter the mixture. Rinse the flask and filter thoroughly, adding the rinsings to the filtrate. Add 5 mL of glacial acetic acid and 1 mL of tetrabromophenolphthalein ethyl ester TS, and titrate with 0.05 *N* silver nitrate VS until the yellow precipitate just turns green. Each mL of 0.05 *N* silver nitrate is equivalent to 10.60 mg of $C_{11}H_8I_3N_2NaO_4$.

Diatrizoate Sodium Injection

» Diatrizoate Sodium Injection is a sterile solution of Diatrizoate Sodium in Water for Injection, or a sterile solution of Diatrizoic Acid in Water for Injection prepared with the aid of Sodium Hydroxide. It contains not less than 95.0 percent and not more than 105.0 percent of the labeled amount of diatrizoate sodium ($C_{11}H_8I_3N_2NaO_4$). It may contain small amounts of suitable buffers and of Edetate Calcium Disodium or Edetate Disodium as a stabilizer. Diatrizoate Sodium Injection intended for intravascular use contains no antimicrobial agents.

Packaging and storage—Preserve Injection intended for intravascular injection in single-dose containers, preferably of Type I or Type III glass, protected from light. Injection intended for other than intravascular use may be packaged in 100-mL multiple-dose containers, preferably of Type I or Type III glass, protected from light.

Labeling—Label containers of Injection intended for intravascular injection to direct the user to discard any unused portion remaining in the container. Label containers of Injection intended for other than intravascular injection to show that the contents are not intended for intravascular injection.

Reference standards—*USP Acetamido-3-amino-2,4,6-triiodobenzoic Acid Reference Standard*—Dry at 105° for 4 hours before using. Keep container tightly closed and protected from light. *USP Diatrizoic Acid Reference Standard*—This material is the hydrous form of Diatrizoic Acid. Dry at 105° for 4 hours before using.

Identification—

A: Dilute a volume of Injection, if necessary, with an 0.8 in 1000 solution of sodium hydroxide in methanol to obtain a test solution having a concentration of 1 mg per mL. The test solution responds to the *Thin-layer Chromatographic Identification Test* ⟨201⟩, the Standard solution being prepared at a concentration of 1 mg of USP Diatrizoic Acid RS per mL in an 0.8 in 1000 solution of sodium hydroxide in methanol, the solvent mixture being a mixture of chloroform, methanol, and ammonium hydroxide (20:10:2), and short-wavelength ultraviolet light being used to locate the spots.

B: Evaporate a volume of Injection, equivalent to about 500 mg of diatrizoate sodium, to dryness, and heat the residue so obtained in a crucible: violet vapors are evolved.

Pyrogen—It meets the requirements of the *Pyrogen Test* ⟨151⟩, the test dose being the equivalent of 2.5 g of diatrizoate sodium per kg.

pH ⟨791⟩: between 6.0 and 7.7.

Free aromatic amine—Transfer a volume of Injection, equivalent to 1.0 g of diatrizoate sodium, to a 50-mL volumetric flask, dilute with water to 5 mL, and add 10 mL of 0.1 *N* sodium hydroxide. Proceed as directed in the test for *Free aromatic amine* under *Diatrizoate Meglumine*, beginning with "To a second 50-mL volumetric flask transfer 4 mL of water."

Iodine and iodide—Using a volume of Injection equivalent to 2.0 g of diatrizoate sodium, and diluting it with water to 24 mL in a 50-mL centrifuge tube provided with a stopper, proceed as directed for *Procedure* in the test for *Iodine and iodide* under *Diatrizoate Meglumine*.

Heavy metals ⟨231⟩—In a 50-mL color-comparison tube mix a volume of Injection, equivalent to 1.0 g of diatrizoate sodium, with 5 mL of 1 *N* sodium hydroxide, dilute with water to 40 mL, and mix. Using this as the *Test preparation*, proceed as directed in the test for *Heavy metals* under *Diatrizoate Meglumine:* the limit is 0.002%.

Other requirements—It meets the requirements under *Injections* ⟨1⟩.

Assay—Pipet into a glass-stoppered 125-mL conical flask a volume of Diatrizoate Sodium Injection, equivalent to about 500 mg of diatrizoate sodium. Add 30 mL of 1.25 *N* sodium hydroxide and 500 mg of powdered zinc, and reflux the mixture for 1 hour. Cool to room temperature, wash the condenser with 20 mL of water, and filter the mixture. Wash the flask and the filter with small portions of water, adding the washings to the filtrate. Add to the filtrate 5 mL of glacial acetic acid and 1 mL of tetrabromophenolphthalein ethyl ester TS, and titrate with 0.05 *N* silver nitrate VS until the color of the yellow precipitate just changes to green. Each mL of 0.05 *N* silver nitrate is equivalent to 10.60 mg of $C_{11}H_8I_3N_2NaO_4$.

Diatrizoate Sodium Injection, Diatrizoate Meglumine and—*see* Diatrizoate Meglumine and Diatrizoate Sodium Injection

Diatrizoate Sodium Solution

» Diatrizoate Sodium Solution is a solution of Diatrizoate Sodium in Purified Water, or a solution of Diatrizoic Acid in Purified Water prepared with the aid of Sodium Hydroxide. It contains not less than 95.0 percent and not more than 105.0 percent of the labeled amount of diatrizoate sodium ($C_{11}H_8I_3N_2NaO_4$). It may contain a suitable preservative.

Packaging and storage—Preserve in tight, light-resistant containers.

Labeling—Label the container to indicate that the contents are not intended for parenteral use.

Reference standard—*USP Diatrizoic Acid Reference Standard*—This material is the hydrous form of Diatrizoic Acid. Dry at 105° for 4 hours before using.

Identification—

A: Dilute a volume of Solution, if necessary, with an 0.8 in 1000 solution of sodium hydroxide in methanol to obtain a test solution having a concentration of 1 mg per mL. The test solution responds to the *Thin-layer Chromatographic Identification Test* ⟨201⟩, the Standard solution being prepared at a concentration of 1 mg of USP Diatrizoic Acid RS per mL in an 0.8 in 1000 solution of sodium hydroxide in methanol, the solvent mixture being a mixture of chloroform, methanol, and ammonium hydroxide (20:10:2), and short-wavelength ultraviolet light being used to locate the spots.

B: Evaporate a volume of Solution, equivalent to about 500 mg of diatrizoate sodium, to dryness, and heat the residue so obtained in a suitable crucible: violet vapors are evolved.

pH ⟨791⟩: between 4.5 and 7.5.

Iodine and iodide—Using as the *Test preparation* a volume of Solution equivalent to 2.0 g of diatrizoate sodium and diluting it with water to 24 mL in a 50-mL centrifuge tube provided with a stopper, proceed as directed for *Procedure* in the test for *Iodine and iodide* under *Diatrizoate Meglumine.*

Assay—Pipet a volume of Diatrizoate Sodium Solution, equivalent to about 400 mg of diatrizoate sodium, into a 125-mL conical flask. Add 30 mL of 1.25 *N* sodium hydroxide and 500 mg of powdered zinc, connect the flask to a reflux condenser, and reflux the mixture for 1 hour. Cool the flask to room temperature, rinse the condenser with 20 mL of water, disconnect the flask from the condenser, and filter the mixture. Rinse the flask and filter thoroughly, adding the rinsings to the filtrate. Add 5 mL of glacial acetic acid and 1 mL of tetrabromophenolphthalein ethyl ester TS, and titrate with 0.05 *N* silver nitrate VS until the yellow precipitate just turns green. Each mL of 0.05 *N* silver nitrate is equivalent to 10.60 mg of $C_{11}H_8I_3N_2NaO_4$.

Diatrizoate Sodium Solution, Diatrizoate Meglumine and—*see* Diatrizoate Meglumine and Diatrizoate Sodium Solution

Diatrizoic Acid

COOH; I, I; CH_3CONH, $NHCOCH_3$; I

$C_{11}H_9I_3N_2O_4$ (anhydrous) 613.92
Benzoic acid, 3,5-bis(acetylamino)-2,4,6-triiodo-.
3,5-Diacetamido-2,4,6-triiodobenzoic acid [*117-96-4*].
Dihydrate 649.95 [*50978-11-5*].

» Diatrizoic Acid is anhydrous or contains two molecules of water of hydration. It contains not less than 98.0 percent and not more than 102.0 percent of $C_{11}H_9I_3N_2O_4$, calculated on the anhydrous basis.

Packaging and storage—Preserve in well-closed containers.

Labeling—Label it to indicate whether it is anhydrous or hydrous.

Reference standards—*USP 5-Acetamido-3-amino-2,4,6-triiodobenzoic Acid Reference Standard*—Dry at 105° for 4 hours before using. Keep container tightly closed and protected from light. *USP Diatrizoic Acid Reference Standard*—This material is the hydrous form of Diatrizoic Acid. Dry at 105° for 4 hours before using.

Identification—

A: It responds to the *Thin-layer Chromatographic Identification Test* ⟨201⟩, the test solution and the Standard solution being prepared at a concentration of 1 mg per mL in an 0.8 in 1000 solution of sodium hydroxide in methanol, the solvent mixture being a mixture of chloroform, methanol, and ammonium hydroxide (20:10:2), and short-wavelength ultraviolet light being used to locate the spots.

B: Heat about 500 mg in a suitable crucible: violet vapors are evolved.

Water, *Method I* ⟨921⟩: not more than 1.0% (anhydrous form), and between 4.5% and 7.0% (hydrous form).

Residue on ignition ⟨281⟩: not more than 0.1%.

Free aromatic amine—Transfer 1.0 g to a 50-mL volumetric flask, and add 12.5 mL of water and 2.5 mL of 1 *N* sodium hydroxide. Proceed as directed in the test for *Free aromatic amine* under *Diatrizoate Meglumine*, beginning with "To a second 50-mL volumetric flask transfer 4 mL of water." The absorbance of the solution from the Diatrizoic Acid is not greater than that of the Standard solution (0.05%).

Iodine and iodide—

Test preparation—Suspend 10.0 g in 10 mL of water, and add in small portions, with stirring, 1.5 mL of sodium hydroxide solution (2 in 5). When solution is complete, adjust to a pH between 7.0 and 7.5 with a dilute solution (1 in 125) of sodium hydroxide or hydrochloric acid, and dilute with water to 20 mL.

Procedure—Dilute 4.0 mL of *Test preparation* with 20 mL of water in a 50-mL centrifuge tube provided with a stopper, and proceed as directed for *Procedure* under *Diatrizoate Meglumine.*

Heavy metals ⟨231⟩—To a 50-mL color-comparison tube transfer 2.0 mL of solution prepared as directed for *Test preparation* in the test for *Iodine and iodide*, add 5 mL of 1 *N* sodium hydroxide, dilute with water to 40 mL, and mix. Using this as the *Test preparation*, proceed as directed in the test for *Heavy metals* under *Diatrizoate Meglumine:* the limit is 0.002%.

Assay—Transfer about 300 mg of Diatrizoic Acid, accurately weighed, to a glass-stoppered, 125-mL conical flask, and proceed as directed in the *Assay* under *Diatrizoate Meglumine*, beginning with "add 30 mL of 1.25 *N* sodium hydroxide." Each mL of 0.05 *N* silver nitrate is equivalent to 10.23 mg of $C_{11}H_9I_3N_2O_4$.

Diazepam

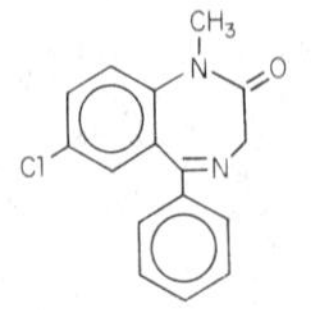

$C_{16}H_{13}ClN_2O$ 284.75
2*H*-1,4-Benzodiazepin-2-one, 7-chloro-1,3-dihydro-1-methyl-5-phenyl-.
7-Chloro-1,3-dihydro-1-methyl-5-phenyl-2*H*-1,4-benzodiazepin-2-one [*439-14-5*].

» Diazepam contains not less than 98.5 percent and not more than 101.0 percent of $C_{16}H_{13}ClN_2O$, calculated on the dried basis.

Packaging and storage—Preserve in tight, light-resistant containers.

Reference standards—*USP Diazepam Reference Standard*—Dry in vacuum over phosphorus pentoxide at 60° for 4 hours before using. *USP 3-Amino-6-chloro-1-methyl-4-phenylcarbostyril Reference Standard*—Do not dry before using. *USP 7-Chloro-1,3-dihydro-5-phenyl-2H-1,4-benzodiazepin-2-one Reference Standard*—Do not dry before using. *USP 2-Methylamino-5-chlorobenzophenone Reference Standard*—Do not dry before using.

Identification—

A: The infrared absorption spectrum of a potassium bromide dispersion of it, previously dried, exhibits maxima only at the same wavelengths as that of a similar preparation of USP Diazepam RS.

B: A 1 in 25,000 solution of it in a 1 in 360 solution of sulfuric acid in dehydrated alcohol exhibits an absorbance maximum at 368 ± 2 nm, and its absorptivity at that maximum is within 3.0% of the absorptivity of a similar solution of USP Diazepam RS, concomitantly measured. A 15 in 50 dilution of the 1 in 25,000 solution in the same alcoholic sulfuric acid solution exhibits a maximum at 285 ± 2 nm, and its absorptivity at that maximum is within 3.0% of the absorptivity of a similar solution of USP Diazepam RS, concomitantly measured.

C: On a suitable thin-layer chromatographic plate (see *Chromatography* ⟨621⟩) coated with a 0.25-mm layer of chromatographic silica gel mixture, apply 10 µL each of a solution of it in acetone containing 50 mg per mL and of a similar solution of USP Diazepam RS. Allow the spots to dry, and develop the chromatogram in an unsaturated chamber with a solvent system consisting of ethyl acetate and *n*-heptane (1:1), until the solvent front has moved about three-fourths of the length of the plate. Remove the plate from the developing chamber, mark the solvent front, and allow the solvent to evaporate. Locate the spots on the plate by viewing under short-wavelength ultraviolet light: the R_f value of the principal spot obtained from the test solution corresponds to that obtained from the Standard solution.

Melting range, *Class I* ⟨741⟩: between 131° and 135°.

Loss on drying ⟨731⟩—Dry it in vacuum over phosphorus pentoxide at 60° for 4 hours: it loses not more than 0.5% of its weight.

Residue on ignition ⟨281⟩: not more than 0.1%.

Heavy metals, *Method II* ⟨231⟩: 0.002%.

Related compounds—Place 1.0 g of Diazepam in a 10-mL volumetric flask. Dissolve in acetone, dilute with acetone to volume, and apply 10 µL of this Test solution to a thin-layer chromatographic plate (see *Chromatography* ⟨621⟩) coated with a 0.25-mm layer of chromatographic silica gel mixture. Apply to the same plate 10 µL of a 1 in 100,000 solution of USP 2-Methylamino-5-chlorobenzophenone RS in acetone, 10 µL of a 1 in 10,000 solution of USP 3-Amino-6-chloro-1-methyl-4-phenylcarbostyril RS in acetone, and 10 µL of a 3 in 10,000 solution of USP 7-Chloro-1,3-dihydro-5-phenyl-2*H*-1,4-benzodiazepin-2-one RS in acetone. Develop the plate in a solvent system consisting of ethyl acetate and *n*-heptane (1:1) until the solvent front has moved about 15 cm from the origin. Remove the plate from the chromatographic chamber, and allow the solvent to evaporate. Locate the bands by viewing under short-wavelength ultraviolet light and by spraying with a freshly prepared 1:1 solution of ferric chloride (1 in 10) and potassium ferricyanide (1 in 20): any spots produced by the solution other than the principal spot are not greater in size or intensity than the spots at the Test respective R_f values produced by the Standard solutions corresponding to not more than 0.01% of 2-methylamino-5-chlorobenzophenone, not more than 0.1% of 3-amino-6-chloro-1-methyl-4-phenylcarbostyril, and not more than 0.3% of 7-chloro-1,3-dihydro-5-phenyl-2*H*-1,4-benzodiazepin-2-one.

Assay—Dissolve about 800 mg of Diazepam, accurately weighed, in 75 mL of acetic anhydride in a 250-mL conical flask. Titrate with 0.1 *N* perchloric acid VS determining the end-point potentiometrically, using a glass electrode and a calomel electrode containing saturated lithium chloride in glacial acetic acid (see *Titrimetry* ⟨541⟩). Perform a blank determination, and make any necessary correction. Each mL of 0.1 *N* perchloric acid is equivalent to 28.48 mg of $C_{16}H_{13}ClN_2O$.

Diazepam Capsules

» Diazepam Capsules contain not less than 90.0 percent and not more than 110.0 percent of the labeled amount of $C_{16}H_{13}ClN_2O$.

Packaging and storage—Preserve in tight, light-resistant containers.

Reference standard—*USP Diazepam Reference Standard*—Dry in vacuum over phosphorus pentoxide at 60° for 4 hours before using.

Identification—

A: The retention time of the major peak in the chromatogram of the *Assay preparation* corresponds to that of the *Standard preparation*, both relative to the internal standard, as obtained in the *Assay*.

B: Transfer an accurately weighed amount of Capsule contents, equivalent to about 10 mg of diazepam, to a 50-mL centrifuge tube, and add 2 mL of acetone. Place the centrifuge tube in an ultrasonic bath for 5 minutes, and centrifuge. Using 100 µL of the supernatant liquid as the test solution, and 100 µL of a solution of USP Diazepam RS in acetone containing 5 mg per mL as the Standard solution, proceed as directed for *Identification test C* under *Diazepam:* the specified result is observed.

Dissolution ⟨711⟩—

Medium: 0.1 *N* hydrochloric acid; 900 mL.

Apparatus 1: 100 rpm.

Time: 45 minutes.

Procedure—Determine the amount of $C_{16}H_{13}ClN_2O$ dissolved from ultraviolet absorbances at the wavelength of maximum absorbance at about 284 nm of filtered portions of the solution under test, suitably diluted with *Dissolution Medium*, if necessary, in comparison with a Standard solution having a known concentration of USP Diazepam RS in the same medium.

Tolerances—Not less than 85% (*Q*) of the labeled amount of $C_{16}H_{13}ClN_2O$ is dissolved in 45 minutes.

Uniformity of dosage units ⟨905⟩: meet the requirements.

Procedure for content uniformity—Place the contents of 1 Capsule into a suitable volumetric flask. Add a suitable quantity of *Internal standard solution* (prepared as directed in the *Assay*), half-fill the flask with methanol, shake by mechanical means for 30 minutes, and dilute quantitatively and stepwise, if necessary, with methanol to obtain a final solution containing about 0.2 mg of diazepam per mL and 0.05 mg of ethylparaben per mL. Proceed as directed in the *Assay*.

Assay—

Mobile phase—Prepare a filtered and degassed mixture of methanol and water (65:35). Make adjustments if necessary (see *System Suitability* under *Chromatography* ⟨621⟩).

Internal standard solution—Transfer about 50 mg of ethylparaben to a 200-mL volumetric flask. Dissolve in methanol, dilute with methanol to volume, and mix.

Standard preparation—Dissolve an accurately weighed quantity of USP Diazepam RS in methanol, and dilute quantitatively, and stepwise if necessary, with methanol to obtain a solution having a known concentration of about 1 mg per mL. Pipet 5 mL of this solution and 5 mL of *Internal standard solution* into a 25-mL volumetric flask, dilute with methanol to volume, and mix to obtain a *Standard preparation* having a known concentration of about 0.2 mg of USP Diazepam RS per mL.

Assay preparation—Weigh and mix the contents of not less than 20 Diazepam Capsules. Transfer an accurately weighed portion of the capsule contents, equivalent to about 10 mg of diazepam, to a 50-mL volumetric flask. Add 10.0 mL of *Internal standard solution*, and half-fill the flask with methanol. Shake by mechanical means for 30 minutes, dilute with methanol to volume, and mix.

Chromatographic system (see *Chromatography* ⟨621⟩)—The liquid chromatograph is equipped with a 254-nm detector and a 3.9-mm × 30-cm column that contains packing L1. The flow rate is about 1.2 mL per minute. Chromatograph replicate injections of the *Standard preparation*, and record the peak responses as directed under *Procedure:* the relative standard deviation is not more than 2.0%, and the resolution, *R*, between ethylparaben and diazepam is not less than 4.5.

Procedure—Separately inject equal volumes (about 5 µL) of the *Standard preparation* and the *Assay preparation* into the chromatograph, record the chromatograms, and measure the responses for the major peaks. The relative retention times are about 1 for ethylparaben and 2 for diazepam. Calculate the quantity, in mg, of $C_{16}H_{13}ClN_2O$ in the portion of Capsules taken by the formula:

$$50C(R_U/R_S),$$

in which *C* is the concentration, in mg per mL, of USP Diazepam RS in the *Standard preparation*, and R_U and R_S are the peak response ratios obtained from the *Assay preparation* and the *Standard preparation*, respectively.

Diazepam Extended-release Capsules

» Diazepam Extended-release Capsules contain not less than 90.0 percent and not more than 110.0 percent of the labeled amount of $C_{16}H_{13}ClN_2O$.

Packaging and storage—Preserve in tight, light-resistant containers.

Reference standard—*USP Diazepam Reference Standard*—Dry in vacuum over phosphorus pentoxide at 60° for 4 hours before using.

Identification—

A: The retention time of the major peak in the chromatogram of the *Assay preparation* corresponds to that of the *Standard preparation*, both relative to the internal standard, as obtained in the *Assay*.

B: Accurately weigh a quantity of Capsule contents, equivalent to 10 mg of diazepam, place in a 50-mL centrifuge tube, and add 2 mL of acetone. Place the centrifuge tube in an ultrasonic bath for 5 minutes, remove from the ultrasonic bath, and centrifuge. Using 100 µL of the supernatant liquid as the test solution, 50 µL of a solution of USP Diazepam RS in acetone containing 10 mg per mL as the Standard solution, and a solvent system consisting of equal volumes of ethyl acetate and *n*-heptane, proceed as directed in *Identification test C* under *Diazepam*.

Drug release ⟨724⟩—

Medium: simulated gastric fluid TS, prepared without enzymes; 900 mL.

Apparatus 1: 100 rpm.

Time: 0.042*D* hours; 0.167*D* hours; 0.333*D* hours; 0.500*D* hours.

Mobile phase—Prepare a suitable degassed and filtered mixture of methanol and water (65:35). Make adjustments if necessary (see *System Suitability* under *Chromatography* ⟨621⟩).

Buffer solution—Dissolve 77.1 g of ammonium acetate in water to make 1000 mL of solution, and adjust with ammonium hydroxide to a pH of 8.7.

Standard preparation—Dissolve an accurately weighed quantity of USP Diazepam RS in *Dissolution Medium*, dilute quantitatively with *Dissolution Medium* to obtain a solution having a known concentration of about 0.15 mg per mL, and mix. Transfer 2.0-, 5.0-, 8.0-, and 10.0-mL aliquots of this solution to separate 100-mL volumetric flasks, add *Dissolution Medium* to volume, and mix. Pipet 1.0 mL of each solution and 1.0 mL of *Buffer solution* into individual small vials, mix, and allow to stand at room temperature for about 10 minutes.

Test preparation—Wrap each Capsule in a coil made from a 10-cm piece of 18-gauge copper wire weighing approximately 750 mg, so that the wire encircles the Capsule 4 times. The Capsule enclosed in the coil remains at the bottom of the basket (it should not float). Filter a portion of the solution under test, obtained at each time interval, through a suitable 0.6-µm porosity filter. Pipet 1.0 mL of each solution and 1.0 mL of *Buffer solution* into individual small vials, mix, and allow to stand at room temperature for about 10 minutes.

Chromatographic system (see *Chromatography* ⟨621⟩)—The liquid chromatograph is equipped with a 254-nm detector and an 8-mm × 10-cm column that contains packing L1. The flow rate is about 5.0 mL per minute. Chromatograph the appropriate *Standard preparation*, and record the peak responses as directed under *Procedure:* the relative standard deviation for replicate injections is not more than 2.0%, and the tailing factor is not greater than 1.7.

Procedure—Separately inject equal volumes (about 100 µL) of the *Standard preparation* and the *Test preparation* into the chromatograph, record the chromatograms, and measure the responses for the major peaks. Determine the amount of $C_{16}H_{13}ClN_2O$ dissolved from peak responses of diazepam obtained from the *Test preparation* and the *Standard preparation*.

Tolerances—The percentage of the labeled amount of $C_{16}H_{13}ClN_2O$ dissolved is within the range stated at each of the following times.

Time (hours)	*Amount dissolved*
0.042*D*	between 15% and 27%
0.167*D*	between 49% and 66%
0.333*D*	between 76% and 96%
0.500*D*	between 85% and 115%

Uniformity of dosage units ⟨905⟩: meet the requirements.

Assay—

Mobile phase—Prepare a suitable degassed and filtered mixture of methanol and water (65:35). Make adjustments if necessary (see *System Suitability* under *Chromatography* ⟨621⟩).

Internal standard solution—Transfer about 300 mg of ethylparaben to a 200-mL volumetric flask, dissolve in methanol, dilute with methanol to volume, and mix.

Standard preparation—Dissolve an accurately weighed quantity of USP Diazepam RS in methanol, dilute quantitatively with methanol to obtain a solution having a known concentration of about 1 mg per mL, and mix. Transfer 15.0 mL of this solution and 5.0 mL of *Internal standard solution* to a 100-mL volumetric flask, dilute with methanol to volume, and mix.

Assay preparation—Weigh and mix the contents of not less than 20 Diazepam Extended-release Capsules. Transfer an accurately weighed portion of the mixture, equivalent to about 15 mg of diazepam, to a 100-mL volumetric flask. Add 5.0 mL of *Internal standard solution* and about 45 mL of methanol. Shake by mechanical means for 30 minutes, dilute with methanol to volume, and mix. Centrifuge about 30 mL of this solution for 5 minutes, and filter.

Chromatographic system (see *Chromatography* ⟨621⟩)—The liquid chromatograph is equipped with a 254-nm detector and an 8-mm × 10-cm column that contains packing L1. The flow rate is about 5.0 mL per minute. Chromatograph the *Standard preparation*, and record the peak responses as directed under *Procedure:* the relative standard deviation for replicate injections is not more than 2.0%, and the resolution, R, between ethylparaben and diazepam is not less than 4.5.

Procedure—Separately inject equal volumes (about 10 µL) of the *Standard preparation* and the *Assay preparation* into the chromatograph, record the chromatograms, and measure the responses for the major peaks. The relative retention times are about 0.5 for ethylparaben and 1.0 for diazepam. Calculate the quantity, in mg, of $C_{16}H_{13}ClN_2O$ in the portion of Capsules taken by the formula:

$$100C(R_U/R_S),$$

in which C is the concentration, in mg per mL, of USP Diazepam RS in the *Standard preparation*, and R_U and R_S are the peak response ratios obtained from the *Assay preparation* and the *Standard preparation*, respectively.

Diazepam Injection

» Diazepam Injection is a sterile solution of Diazepam in a suitable medium. It contains not less than 90.0 percent and not more than 110.0 percent of the labeled amount of $C_{16}H_{13}ClN_2O$.

Packaging and storage—Preserve in single-dose or in multiple-dose containers, preferably of Type I glass, protected from light.

Reference standard—*USP Diazepam Reference Standard*—Dry in vacuum over phosphorus pentoxide at 60° for 4 hours before using.

Identification—

A: The retention time of the major peak in the chromatogram of the *Assay preparation* corresponds to that of the *Standard preparation*, both relative to the internal standard, as obtained in the *Assay*.

B: Transfer to a suitable separator a volume of Diazepam Injection, equivalent to about 10 mg of diazepam. Add 20 mL of pH 7 buffer solution (prepared by mixing 104.5 mL of 0.2 *M* dibasic sodium phosphate and 85.5 mL of 0.2 *M* monobasic potassium phosphate in a 1000-mL volumetric flask, diluting with water to volume, and adjusting, if necessary, to a pH of 7.0), and extract with four 20-mL portions of chloroform, passing each extract through about 5 g of anhydrous sodium sulfate and combining the extracts in a 100-mL volumetric flask. Add chloroform to volume, and mix. Evaporate 50 mL of this chloroform solution with the aid of a stream of nitrogen, add 1 mL of chloroform, and mix. Using 30 µL of this solution as the test solution, 30 µL of a solution of USP Diazepam RS in acetone containing 5 mg per mL as the Standard solution, and a solvent system consisting of equal volumes of ethyl acetate and *n*-heptane, proceed as directed in *Identification test C* under *Diazepam*. The specified result is observed.

Pyrogen—The Injection, undiluted, meets the requirements of the *Pyrogen Test* ⟨151⟩, the test dose being 0.25 mg per kg.

pH ⟨791⟩: between 6.2 and 6.9.

Other requirements—It meets the requirements under *Injections* ⟨1⟩.

Assay—

Internal standard preparation—Pipet 1.0 mL of tolualdehyde into a 50-mL volumetric flask and dilute with methanol to volume. Dilute 1.0 mL to 50.0 mL with methanol.

Standard preparation—Dissolve an accurately weighed quantity of USP Diazepam RS in methanol and dilute quantitatively with methanol to obtain a solution having a known concentration of about 1 mg per mL. Pipet 5 mL of this solution and 2 mL of *Internal standard preparation* into a 10-mL volumetric flask, dilute with methanol to volume, and mix to obtain the *Standard preparation*.

Assay preparation—Dilute an accurately measured volume of Diazepam Injection, equivalent to about 50 mg of diazepam, and a suitable quantity of *Internal standard preparation* quantitatively and stepwise with methanol to obtain a final solution containing approximately 0.5 mg per mL of diazepam and an amount of tolualdehyde identical to that in the *Standard preparation*.

Chromatographic system—Under typical conditions the high-pressure liquid chromatograph is equipped with an ultraviolet detector that monitors absorption at 254 nm and contains a 4.6-mm × 25-cm column containing packing L1. The mobile phase consists of methanol and water (70:30) and is maintained at a flow rate of about 1.4 mL per minute, giving a resolution factor of not less than 5.0. Five replicate injections of the *Standard preparation* show a relative standard deviation of not more than 2.0%.

Procedure—Inject equal volumes (about 10 µL) of the *Standard preparation* and the *Assay preparation* into the high-pressure liquid chromatograph and record the chromatograms. Measure the peak responses at approximate retention times of 5 minutes (tolualdehyde) and 10 minutes (diazepam). Calculate the quantity, in mg, of $C_{16}H_{13}ClN_2O$ in each mL of the Injection, taken by the formula:

$$100(C/V)(R_U/R_S),$$

in which C is the concentration of USP Diazepam RS, in mg per mL, in the *Standard preparation*, V is the volume of Diazepam Injection, in mL, taken for the *Assay preparation*, and R_U and R_S are the ratios of peak responses of diazepam and tolualdehyde obtained from the *Assay preparation* and the *Standard preparation*, respectively.

Diazepam Tablets

» Diazepam Tablets contain not less than 90.0 percent and not more than 110.0 percent of the labeled amount of $C_{16}H_{13}ClN_2O$.

Packaging and storage—Preserve in tight, light-resistant containers.

Reference standard—*USP Diazepam Reference Standard*—Dry in vacuum over phosphorus pentoxide at 60° for 4 hours before using.

Identification—

A: The retention time of the major peak in the chromatogram of the *Assay preparation* corresponds to that of the *Standard preparation*, both relative to the internal standard, as obtained in the *Assay*.

B: Accurately weigh an amount of Tablet mass, equivalent to 10 mg of diazepam, place in a 50-mL centrifuge tube, and add 2 mL of acetone. Place the centrifuge tube in an ultrasonic bath for 5 minutes, and centrifuge. Using 100 µL of the supernatant liquid as the test solution, 100 µL of a solution of USP Diazepam RS in acetone containing 5 mg per mL as the Standard solution, and a solvent system consisting of equal volumes of ethyl acetate and *n*-heptane, proceed as directed in *Identification test C* under *Diazepam*. The specified result is observed.

Dissolution ⟨711⟩—

Medium: 0.1 *N* hydrochloric acid; 900 mL.

Apparatus 1: 100 rpm.

Time: 30 minutes.

Procedure—Measure the amount in solution in filtered portions of the *Dissolution Medium* at the wavelength of maximum absorbance at about 242 nm with a suitable spectrophotometer against 0.1 *N* hydrochloric acid. Similarly measure the absorbance of a Standard solution of known concentration of USP Diazepam RS.

Tolerances—Not less than 85% (Q) of the labeled amount of $C_{16}H_{13}ClN_2O$ is dissolved in 30 minutes.

Uniformity of dosage units ⟨905⟩: meet the requirements.

Assay—

Mobile phase, Internal standard solution, Standard preparation, and *Chromatographic system*—Prepare as directed in the *Assay* under *Diazepam Capsules*.

Assay preparation—Weigh and finely powder not less than 20 Diazepam Tablets. Transfer an accurately weighed portion of the powder, equivalent to about 10 mg of diazepam, to a 50-mL volumetric flask. Pipet 10 mL of *Internal standard solution* into the flask, and add about 25 mL of methanol. Shake by mechanical means for 30 minutes, dilute with methanol to volume, and mix. Filter the solution, discarding the first 10 mL of the filtrate.

Procedure—Proceed as directed for *Procedure* in the *Assay* under *Diazepam Capsules*. Calculate the quantity, in mg, of $C_{16}H_{13}ClN_2O$ in the portion of Tablets taken by the formula:

$$50C(R_U/R_S),$$

in which C is the concentration, in mg per mL, of USP Diazepam RS in the *Standard preparation*, and R_U and R_S are the peak response ratios obtained from the *Assay preparation* and the *Standard preparation*, respectively.

Diazoxide

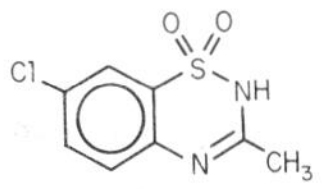

$C_8H_7ClN_2O_2S$ 230.67

2*H*-1,2,4-Benzothiadiazine, 7-chloro-3-methyl-, 1,1-dioxide.

7-Chloro-3-methyl-2*H*-1,2,4-benzothiadiazine 1,1-dioxide [*364-98-7*].

» Diazoxide contains not less than 97.0 percent and not more than 102.0 percent of $C_8H_7ClN_2O_2S$, calculated on the dried basis.

Packaging and storage—Preserve in well-closed containers.

Reference standard—*USP Diazoxide Reference Standard*—Dry at 105° for 4 hours before using.

Identification—

A: The infrared absorption spectrum of a mineral oil dispersion of it exhibits maxima only at the same wavelengths as that of a similar preparation of USP Diazoxide RS.

B: The retention time of the major peak in the chromatogram of the *Assay preparation* corresponds to that of the *Standard preparation*, both relative to the internal standard, as obtained in the *Assay*.

Loss on drying ⟨731⟩—Dry it at 105° for 4 hours: it loses not more than 0.5% of its weight.

Residue on ignition ⟨281⟩: not more than 0.1%.

Assay—

Mobile phase—Prepare a filtered and degassed mixture of 0.01 *M* sodium 1-pentanesulfonate, methanol, and glacial acetic acid (80:20:1). Make adjustments if necessary (see *System Suitability* under *Chromatography* ⟨621⟩).

Internal standard solution—Transfer about 50 mg of hydrochlorothiazide to a 25-mL volumetric flask, add methanol to volume, and mix.

Standard preparation—Dissolve an accurately weighed quantity of USP Diazoxide RS in methanol, and dilute quantitatively, and stepwise if necessary, with methanol to obtain a solution having a known concentration of about 1 mg per mL. Transfer 5.0 mL of this solution to a 100-mL volumetric flask, add 2.0 mL of *Internal standard solution*, dilute with a mixture of water and methanol (4:1) to volume, and mix to obtain a solution having a known concentration of about 50 µg of USP Diazoxide RS per mL.

Assay preparation—Transfer about 50 mg of Diazoxide, accurately weighed, to a 50-mL volumetric flask, add methanol to volume, and mix. Transfer 5.0 mL of this solution to a 100-mL volumetric flask, add 2.0 mL of *Internal standard solution*, dilute with a mixture of water and methanol (4:1) to volume, and mix.

Chromatographic system (see *Chromatography* ⟨621⟩)—The liquid chromatograph is equipped with a 254-nm detector and a 3.9-mm × 30-cm column that contains packing L11. The flow rate is about 2 mL per minute. Chromatograph the *Standard preparation*, and record the peak responses as directed under *Procedure:* the resolution, *R*, between the analyte and internal standard peaks is not less than 5, and the relative standard deviation for replicate injections is not more than 1.5%.

Procedure—Separately inject equal volumes (about 10 µL) of the *Standard preparation* and the *Assay preparation* into the chromatograph, record the chromatograms, and measure the responses for the major peaks. The relative retention times are about 0.4 for hydrochlorothiazide and 1.0 for diazoxide. Calculate the quantity, in mg, of $C_8H_7ClN_2O_2S$ in the portion of Diazoxide taken by the formula:

$$C(R_U/R_S),$$

in which C is the concentration, in µg per mL, of USP Diazoxide RS in the *Standard preparation*, and R_U and R_S are the peak response ratios obtained from the *Assay preparation* and the *Standard preparation*, respectively.

Diazoxide Capsules

» Diazoxide Capsules contain not less than 90.0 percent and not more than 110.0 percent of the labeled amount of $C_8H_7ClN_2O_2S$.

Packaging and storage—Preserve in well-closed containers.

Reference standard—*USP Diazoxide Reference Standard*—Dry at 105° for 4 hours before using.

Identification—

A: Place a portion of the contents of Capsules, equivalent to about 50 mg of diazoxide, in a suitable centrifuge tube, add 25 mL of 0.1 *N* sodium hydroxide, shake for 15 minutes, and centrifuge: the supernatant liquid so obtained responds to the *Thin-layer Chromatographic Identification Test* ⟨201⟩, a solvent system consisting of a mixture of ethyl acetate, methanol, and ammonium hydroxide (17:4:3) being used.

B: The retention time of the major peak in the chromatogram of the *Assay preparation* corresponds to that of the *Standard preparation*, both relative to the internal standard, as obtained in the *Assay*.

Dissolution ⟨711⟩—

Medium: pH 7.6 phosphate buffer (see *Buffer Solutions* in the section, *Reagents, Indicators, and Solutions*); 900 mL.

Apparatus 1: 100 rpm.

Time: 45 minutes.

Procedure—Determine the amount of $C_8H_7ClN_2O_2S$ dissolved from ultraviolet absorbances at the wavelength of maximum absorbance at about 268 nm of filtered portions of the solution under test in comparison with a solution of USP Diazoxide RS similarly prepared.

Tolerances—Not less than 75% (*Q*) of the labeled amount of $C_8H_7ClN_2O_2S$ is dissolved in 45 minutes.

Uniformity of dosage units ⟨905⟩: meet the requirements.

Assay—

Mobile phase, Internal standard solution, Standard preparation, and *Chromatographic system*—Prepare as directed in the *Assay* under *Diazoxide*.

Assay preparation—Remove, as completely as possible, the contents of not less than 20 Diazoxide Capsules, and weigh accurately. Mix the combined contents, and transfer an accurately weighed quantity of the powder, equivalent to about 100 mg of diazoxide, to a 100-mL volumetric flask, add 15 mL of water, and shake by mechanical means for 5 minutes. Add 60 mL of methanol, shake by mechanical means for 15 minutes, dilute with methanol to volume, and mix. Transfer 5.0 mL of this solution to a 100-mL volumetric flask, add 2.0 mL of *Internal standard solution*, dilute with a mixture of water and methanol (4:1) to volume, and mix.

Procedure—Proceed as directed for *Procedure* in the *Assay* under *Diazoxide*. Calculate the quantity, in mg, of $C_8H_7ClN_2O_2S$ in the portion of Capsules taken by the formula:

$$2C(R_U/R_S),$$

in which C is the concentration, in µg per mL, of USP Diazoxide RS in the *Standard preparation*, and R_U and R_S are the peak response ratios obtained from the *Assay preparation* and the *Standard preparation*, respectively.

Diazoxide Injection

» Diazoxide Injection is a sterile solution of Diazoxide in Water for Injection, prepared with the aid of Sodium Hydroxide. It contains not less than 90.0 percent and not more than 110.0 percent of the labeled amount of $C_8H_7ClN_2O_2S$.

Packaging and storage—Preserve in single-dose containers, preferably of Type I glass, protected from light.

Reference standard—*USP Diazoxide Reference Standard*—Dry at 105° for 4 hours before using.

Identification—

A: A dilution of the Injection in methanol containing 1 mg of diazoxide per mL responds to the *Thin-layer Chromatographic Identification Test* ⟨201⟩, a solution of USP Diazoxide RS in methanol containing 1 mg per mL being used as the Standard solution. Develop the chromatogram with a solvent system consisting of a mixture of ethyl acetate, methanol, and ammonium hydroxide (17:4:3).

B: The retention time of the major peak in the chromatogram of the *Assay preparation* corresponds to that of the *Standard preparation*, both relative to the internal standard, as obtained in the *Assay*.

pH ⟨791⟩: between 11.2 and 11.9.

Pyrogen—The Injection, undiluted, meets the requirements of the *Pyrogen Test* ⟨151⟩, the test dose being 6 mg per kg.

Other requirements—It meets the requirements under *Injections* ⟨1⟩.

Assay—

Mobile phase, Internal standard solution, Standard preparation, and *Chromatographic system*—Prepare as directed in the *Assay* under *Diazoxide*.

Assay preparation—Transfer an accurately measured volume of Diazoxide Injection, equivalent to about 45 mg of diazoxide, to a 50-mL volumetric flask, dilute with methanol to volume, and mix. Transfer 5.0 mL of this solution to a 100-mL volumetric flask, add 2.0 mL of *Internal standard solution*, dilute with a mixture of water and methanol (4:1) to volume, and mix.

Procedure—Proceed as directed for *Procedure* in the *Assay* under *Diazoxide*. Calculate the quantity, in mg, of $C_8H_7ClN_2O_2S$ in each mL of the Injection taken by the formula:

$$C/V(R_U/R_S),$$

in which C is the concentration, in µg per mL, of USP Diazoxide RS in the *Standard preparation*, V is the volume, in mL, of Injection taken, and R_U and R_S are the peak response ratios obtained from the *Assay preparation* and the *Standard preparation*, respectively.

Diazoxide Oral Suspension

» Diazoxide Oral Suspension contains not less than 90.0 percent and not more than 110.0 percent of the labeled amount of $C_8H_7ClN_2O_2S$.

Packaging and storage—Preserve in tight, light-resistant containers.

Reference standard—*USP Diazoxide Reference Standard*—Dry at 105° for 4 hours before using.

Identification—

A: Place a portion of Oral Suspension, equivalent to about 50 mg of diazoxide, in a 50-mL volumetric flask, add 30 mL of 0.1 *N* sodium hydroxide, shake for 30 minutes, dilute with 0.1 *N* sodium hydroxide to volume, and mix: the solution so obtained responds to the *Thin-layer Chromatographic Identification Test* ⟨201⟩, a solvent system consisting of a mixture of ethyl acetate, methanol, and ammonium hydroxide (17:4:3) being used.

B: The retention time of the major peak in the chromatogram of the *Assay preparation* corresponds to that of the *Standard preparation*, both relative to the internal standard, as obtained in the *Assay*.

Assay—

Mobile phase, Internal standard solution, Standard preparation, and *Chromatographic system*—Prepare as directed in the *Assay* under *Diazoxide*.

Assay preparation—Transfer an accurately measured volume of freshly mixed Diazoxide Oral Suspension, equivalent to about 100 mg of diazoxide, to a 50-mL centrifuge tube, add 2 mL of water and 35 mL of methanol, shake for 15 minutes, and centrifuge for 5 minutes. Transfer the supernatant liquid to a 200-mL volumetric flask. Repeat the extraction process two times, beginning with the addition of 2 mL of water, combine the extracts in the 200-mL volumetric flask, dilute with methanol to volume, and mix. Transfer 10.0 mL of this solution to a 100-mL volumetric flask, add 2.0 mL of *Internal standard solution*, dilute with a mixture of water and methanol (4:1) to volume, and mix.

Procedure—Proceed as directed for *Procedure* in the *Assay* under *Diazoxide*. Calculate the quantity, in mg, of $C_8H_7ClN_2O_2S$ in each mL of the Oral Suspension taken by the formula:

$$2(C/V)(R_U/R_S),$$

in which C is the concentration, in µg per mL, of USP Diazoxide RS in the *Standard preparation*, V is the volume, in mL, of Oral Suspension taken, and R_U and R_S are the peak response ratios obtained from the *Assay preparation* and the *Standard preparation*, respectively.

Dibasic Calcium Phosphate—*see* Calcium Phosphate, Dibasic

Dibasic Sodium Phosphate—*see* Sodium Phosphate, Dibasic

Dibucaine

$C_{20}H_{29}N_3O_2$ 343.47

4-Quinolinecarboxamide, 2-butoxy-*N*-[2-(diethylamino)ethyl]-.
2-Butoxy-*N*-[2-(diethylamino)ethyl]cinchoninamide [*85-79-0*].

» Dibucaine contains not less than 97.0 percent and not more than 100.5 percent of $C_{20}H_{29}N_3O_2$, calculated on the dried basis.

Packaging and storage—Preserve in tight, light-resistant containers.

Reference standard—*USP Dibucaine Hydrochloride Reference Standard*—Dry at 80° for 5 hours before using.

Identification—

A: The infrared absorption spectrum of a mineral oil dispersion of it, previously dried, exhibits maxima only at the same wavelengths as that of a similar dispersion of the residue prepared by dissolving 30 mg of USP Dibucaine Hydrochloride RS in 5 mL of 0.5 *N* sodium hydroxide, extracting the resulting solution with 5 mL of ether, evaporating the ether, and drying the residue over phosphorus pentoxide.

B: The ultraviolet absorption spectrum of a 1 in 100,000 solution in 1 *N* hydrochloric acid exhibits maxima and minima at the same wavelengths as that of a similar solution of USP Dibucaine Hydrochloride RS, concomitantly measured, and the respective molar absorptivities, calculated on the dried basis, at the wavelength of maximum absorbance at about 247 nm do not differ by more than 3.0%. [NOTE—The molecular weight of dibucaine hydrochloride, $C_{20}H_{29}N_3O_2 \cdot HCl$, is 379.93.]

Melting range ⟨741⟩: between 62.5° and 66.0°, determined after drying.

Loss on drying ⟨731⟩—Dry it over phosphorus pentoxide for 16 hours: it loses not more than 1.0% of its weight.

Residue on ignition ⟨281⟩: not more than 0.2%.

Chromatographic purity—Proceed as directed for *Chromatographic purity* under *Dibucaine Hydrochloride*, except to use a *Test solution* containing 36.2 mg of Dibucaine per mL: the principal spot obtained from the *Test solution* corresponds in R_f value, color, and intensity to that obtained from the *Standard solution;* the sum of the intensities of any secondary spots, if present in the chromatogram from the *Test solution*, corresponds to not more than 2.0% of that of the principal spot on the chromatogram from the *Standard solution* on the basis of comparison with the spots obtained from the *Comparison solutions*.

Assay—

Mobile phase—Dissolve 1.20 g of sodium lauryl sulfate, 0.20 g of sodium acetate, and 2.0 g of triethylamine in 300 mL of water. Adjust with glacial acetic acid to a pH of 5.6, add 700 mL of methanol, mix, and filter through a suitable filter of 0.5-µm or finer porosity. Make adjustments if necessary (see *System Suitability* under *Chromatography* ⟨621⟩).

Solvent mixture—Prepare a mixture of methanol and water (70:30).

Standard preparation—Dissolve an accurately weighed quantity of USP Dibucaine Hydrochloride RS in *Solvent mixture* to obtain a solution having a known concentration of about 1 mg per mL. Filter through a suitable filter of 0.5-µm or finer porosity.

Assay preparation—Transfer about 90 mg of Dibucaine, accurately weighed, to a 100-mL volumetric flask, add *Solvent mixture* to volume, and mix. Filter through a suitable filter of 0.5-µm or finer porosity.

Chromatographic system (see *Chromatography* ⟨621⟩)—The liquid chromatograph is equipped with a 254-nm detector and a 3.9-mm × 30-cm column that contains packing L1. The flow rate is about 1.5 mL per minute. Chromatograph the *Standard preparation*, and record the peak responses as directed under *Procedure:* the column efficiency, determined from the analyte peak, is not less than 1500 theoretical plates, the tailing factor for the analyte peak is not more than 3.0, and the relative standard deviation of replicate injections is not more than 2%.

Procedure—[NOTE—Use peak areas where peak responses are indicated.] Separately inject equal volumes (about 10 µL) of the *Standard preparation* and the *Assay preparation* into the chromatograph, record the chromatograms, and measure the responses for the major peaks. Calculate the quantity, in mg, of $C_{20}H_{29}N_3O_2$ in the portion of Dibucaine taken by the formula:

$$(343.47/379.93)(100C)(r_U/r_S),$$

in which 343.47 and 379.93 are the molecular weights of dibucaine and dibucaine hydrochloride, respectively, C is the concentration, in mg per mL, of USP Dibucaine Hydrochloride RS in the *Standard preparation*, and r_U and r_S are the responses of the dibucaine peaks obtained from the *Assay preparation* and the *Standard preparation*, respectively.

Dibucaine Cream

» Dibucaine Cream contains not less than 90.0 percent and not more than 110.0 percent of the labeled amount of $C_{20}H_{29}N_3O_2$ in a suitable cream base.

Packaging and storage—Preserve in collapsible tubes or in tight, light-resistant containers.

Reference standard—*USP Dibucaine Hydrochloride Reference Standard*—Dry at 80° for 5 hours before using.

Identification—The ultraviolet absorption spectrum of the solution employed for measurement of absorbance in the *Assay* exhibits maxima and minima at the same wavelengths as that of the solution of USP Dibucaine Hydrochloride RS prepared as directed in the *Assay*.

Microbial limits—It meets the requirements of the tests for absence of *Staphylococcus aureus* and *Pseudomonas aeruginosa* under *Microbial Limit Tests* ⟨61⟩.

Minimum fill ⟨755⟩: meets the requirements.

Assay—

Mobile phase and *Chromatographic system*—Proceed as directed in the *Assay* under *Dibucaine*.

Standard preparation—Transfer about 25 mg of USP Dibucaine Hydrochloride RS, accurately weighed, to a 100-mL volumetric flask, add 20.0 mL of 0.1 *N* hydrochloric acid, and swirl to dissolve. Dilute with methanol to volume, and mix. Filter through a suitable filter of 0.5-µm or finer porosity.

Assay preparation—Weigh accurately a portion of Dibucaine Cream, equivalent to about 22 mg of dibucaine, transfer to a separator containing 25 mL of ether, and mix to dissolve. Extract successively with two 9-mL portions of 0.1 *N* hydrochloric acid, combining the extracts in a 100-mL volumetric flask. Extract the ether phase in the separator with 2 mL of water, collecting the aqueous extract in the 100-mL volumetric flask. Dilute with methanol to volume, and mix. Filter through a suitable filter of 0.5-µm or finer porosity.

Procedure—[NOTE—Use peak areas where peak responses are indicated.] Separately inject equal volumes (about 20 µL) of the *Standard preparation* and the *Assay preparation* into the chromatograph, record the chromatograms, and measure the responses for the major peaks. Calculate the quantity, in mg, of $C_{20}H_{29}N_3O_2$ in the portion of Dibucaine Cream taken by the formula:

$$(343.47/379.93)(100C)(r_U/r_S),$$

in which 343.47 and 379.93 are the molecular weights of dibucaine and dibucaine hydrochloride, respectively, C is the concentration, in mg per mL, of USP Dibucaine Hydrochloride RS in the *Standard preparation*, and r_U and r_S are the responses of the dibucaine peaks obtained from the *Assay preparation* and the *Standard preparation*, respectively.

Dibucaine Ointment

» Dibucaine Ointment contains not less than 90.0 percent and not more than 110.0 percent of the labeled amount of $C_{20}H_{29}N_3O_2$ in a suitable ointment base.

Packaging and storage—Preserve in collapsible tubes or in tight, light-resistant containers.

Reference standard—*USP Dibucaine Hydrochloride Reference Standard*—Dry at 80° for 5 hours before using.

Identification—The ultraviolet absorption spectrum of the solution employed for measurement of absorbance in the *Assay* exhibits maxima and minima at the same wavelengths as that of the solution of USP Dibucaine Hydrochloride RS prepared as directed in the *Assay*.

Microbial limits—It meets the requirements of the tests for absence of *Staphylococcus aureus* and *Pseudomonas aeruginosa* under *Microbial Limit Tests* ⟨61⟩.

Minimum fill ⟨755⟩: meets the requirements.

Assay—

Mobile phase and *Chromatographic system*—Proceed as directed in the *Assay* under *Dibucaine*.

Standard preparation—Transfer about 55 mg of USP Dibucaine Hydrochloride RS, accurately weighed, to a 50-mL volumetric flask, add 24.0 mL of 0.1 *N* hydrochloric acid, and swirl to dissolve. Dilute with a mixture of methanol and 1.0 *N* hydrochloric acid (13:12) to volume, and mix. Transfer 10.0 mL of this solution to a 50-mL volumetric flask, dilute with methanol to volume, and mix. Filter through a suitable filter of 0.5-µm or finer porosity.

Assay preparation—Weigh accurately a portion of Dibucaine Ointment, equivalent to about 50 mg of dibucaine, transfer to a separator containing 50 mL of ether, and mix to dissolve. Extract successively with 50-mL, 40-mL, and 30-mL portions of 0.1 *N* hydrochloric acid, combining the extracts in a 250-mL volumetric flask. Dilute with methanol to volume, and mix. Filter through a suitable filter of 0.5-µm or finer porosity.

Procedure—[NOTE—Use peak areas where peak responses are indicated.] Separately inject equal volumes (about 10 µL) of the *Standard preparation* and the *Assay preparation* into the chromatograph, record the chromatograms, and measure the responses for the major peaks. Calculate the quantity, in mg, of $C_{20}H_{29}N_3O_2$ in the portion of Dibucaine Ointment taken by the formula:

$$(343.47/379.93)(250C)(r_U/r_S),$$

in which 343.47 and 379.93 are the molecular weights of dibucaine and dibucaine hydrochloride, respectively, C is the concentration, in mg per mL, of USP Dibucaine Hydrochloride RS in the *Standard preparation*, and r_U and r_S are the responses of the dibucaine peaks obtained from the *Assay preparation* and the *Standard preparation*, respectively.

Dibucaine Hydrochloride

$C_{20}H_{29}N_3O_2 \cdot HCl$ 379.93

4-Quinolinecarboxamide, 2-butoxy-*N*-[2-(diethylamino)ethyl]-, monohydrochloride.

2-Butoxy-*N*-[2-(diethylamino)ethyl]cinchoninamide monohydrochloride [*61-12-1*].

» Dibucaine Hydrochloride contains not less than 97.0 percent and not more than 100.5 percent of $C_{20}H_{29}N_3O_2 \cdot HCl$, calculated on the dried basis.

Packaging and storage—Preserve in tight, light-resistant containers.

Reference standard—*USP Dibucaine Hydrochloride Reference Standard*—Dry at 80° for 5 hours before using.

Identification—

A: The infrared absorption spectrum of a mineral oil dispersion of it, previously dried, exhibits maxima only at the same wavelengths as that of a similar preparation of USP Dibucaine Hydrochloride RS.

B: The ultraviolet absorption spectrum of a 1 in 100,000 solution in 1 *N* hydrochloric acid exhibits maxima and minima at the same wavelengths as that of a similar solution of USP Dibucaine Hydrochloride RS, concomitantly measured, and the respective absorptivities, calculated on the dried basis, at the wavelength of maximum absorbance at about 247 nm do not differ by more than 3.0%.

C: A solution of it responds to the tests for *Chloride* ⟨191⟩ when tested as specified for alkaloidal hydrochlorides.

Loss on drying ⟨731⟩—Dry it at 80° for 5 hours: it loses not more than 2.0% of its weight.

Residue on ignition ⟨281⟩: not more than 0.1%.

Chromatographic purity—Dissolve a suitable quantity of Dibucaine Hydrochloride, accurately weighed, in chloroform to obtain a *Test solution* having a concentration of 40.0 mg per mL. Dissolve a suitable quantity of USP Dibucaine Hydrochloride RS, accurately weighed, in chloroform to obtain a *Standard solution* having a known concentration of about 40 mg per mL. Dilute portions of this solution quantitatively and stepwise with chloroform to obtain three *Comparison solutions* having concentrations of 40, 120, and 200 µg per mL (0.1%, 0.3%, and 0.5%) of the *Standard solution*, respectively. Apply separate 5-µL portions of the five solutions on the starting line of a suitable thin-layer chromatographic plate (see *Chromatography* ⟨621⟩), coated with a 0.25-mm layer of chromatographic silica gel mixture. Develop the chromatogram in a solvent system consisting of a mixture of toluene, acetone, methanol, and ammonium hydroxide (50:30:5:1) until the solvent front has moved about three-fourths of the length of the plate. Remove the plate from the chamber, mark the solvent front, and air-dry. Spray the plate heavily with a 1 in 200 solution of potassium dichromate in dilute sulfuric acid (1 in 5). Place the plate in an oven at 140° for 10 minutes, and view under short-wavelength ultraviolet light: the principal spot obtained from the *Test solution* corresponds in R_f value, color, and intensity to that obtained from the *Standard solution;* the sum of the intensities of any secondary spots, if present in the chromatogram from the *Test solution*, corresponds to not more than 1.0%, and the intensity of any secondary spot does not exceed 0.5% of that of the principal spot on the chromatogram from the *Standard solution* on the basis of comparison with the spots obtained from the *Comparison solutions.*

Assay—

Mobile phase, Solvent mixture, Standard preparation, and *Chromatographic system*—Proceed as directed in the *Assay* under *Dibucaine.*

Assay preparation—Transfer about 100 mg of Dibucaine Hydrochloride, accurately weighed, to a 100-mL volumetric flask, add *Solvent mixture* to volume, and mix. Filter through a suitable filter of 0.5-µm or finer porosity.

Procedure—Proceed as directed for *Procedure* in the *Assay* under *Dibucaine.* Calculate the quantity, in mg, of $C_{20}H_{29}N_3O_2 \cdot HCl$ in the portion of Dibucaine Hydrochloride taken by the formula:

$$100C(r_U/r_S),$$

in which C is the concentration, in mg per mL, of USP Dibucaine Hydrochloride RS in the *Standard preparation*, and r_U and r_S are the responses of the dibucaine peaks obtained from the *Assay preparation* and the *Standard preparation*, respectively.

Dibucaine Hydrochloride Injection

» Dibucaine Hydrochloride Injection is a sterile solution of Dibucaine Hydrochloride in Water for Injection. It contains not less than 95.0 percent and not more than 105.0 percent of the labeled amount of $C_{20}H_{29}N_3O_2 \cdot HCl$.

Packaging and storage—Preserve in single-dose or in multiple-dose containers, preferably of Type I glass, and protect from light.

Reference standard—*USP Dibucaine Hydrochloride Reference Standard*—Dry at 80° for 5 hours before using.

Identification—

A: The ultraviolet absorption spectrum of the *Assay preparation*, prepared as directed in the *Assay*, exhibits maxima and minima at the same wavelengths as that of the *Standard preparation*, prepared as directed in the *Assay*, concomitantly measured.

B: Place a volume of Injection, equivalent to about 30 mg of dibucaine hydrochloride, in a suitable evaporating dish, and concentrate on a steam bath to a volume of about 10 mL. Transfer the solution to a separator, render distinctly alkaline with 1 *N* sodium hydroxide, and extract with four 20-mL portions of ether. Wash the combined ether extracts with 5 mL of water, discarding the washing. Evaporate the ether extracts with the aid of a current of air to dryness, and dry the residue over phosphorus pentoxide for 3 hours: the dibucaine so obtained melts between 62° and 65°.

pH ⟨791⟩: between 4.5 and 7.0.

Particulate matter ⟨788⟩: meets the requirements under *Small-volume Injections.*

Other requirements—It meets the requirements under *Injections* ⟨1⟩.

Assay—

Mobile phase and *Chromatographic system*—Proceed as directed in the *Assay* under *Dibucaine.*

Standard preparation—Transfer about 50 mg of USP Dibucaine Hydrochloride RS, accurately weighed, to a 100-mL volumetric flask, add an accurately measured volume of water, equivalent to the volume of Injection taken to prepare the *Assay preparation*, dilute with methanol to volume, and mix. Where the *Assay preparation* is prepared in a 50-mL volumetric flask, transfer about 25 mg of USP Dibucaine Hydrochloride RS, accurately weighed, to a 100-mL volumetric flask, add an accurately measured volume of water, equivalent to twice the volume of Injection taken to prepare the *Assay preparation*, dilute with methanol to volume, and mix. Filter through a suitable filter of 0.5-µm or finer porosity.

Assay preparation—Transfer an accurately measured volume of Dibucaine Hydrochloride Injection, equivalent to about 50 mg of dibucaine, to a 100-mL volumetric flask. Dilute with methanol to volume, and mix. Where the Injection is labeled to contain 1 mg or less of dibucaine hydrochloride per mL, transfer an accurately measured volume of Injection, equivalent to about 13 mg of dibucaine hydrochloride, to a 50-mL volumetric flask, dilute with methanol to volume, and mix. Filter through a suitable filter of 0.5-μm or finer porosity.

Procedure—[NOTE—Use peak areas where peak responses are indicated.] Separately inject equal volumes (about 10 μL, or 20 μL where the concentration of dibucaine hydrochloride is about 0.25 mg per mL) of the *Standard preparation* and the *Assay preparation* into the chromatograph, record the chromatograms, and measure the responses for the major peaks. Calculate the quantity, in mg, of $C_{20}H_{29}N_3O_2.HCl$ in each mL of the Dibucaine Hydrochloride Injection taken by the formula:

$$C(v/V)(r_U/r_S),$$

in which C is the concentration, in mg per mL, of USP Dibucaine Hydrochloride RS in the *Standard preparation*, v is the volume, in mL, of the *Assay preparation*, V is the volume, in mL, of Injection taken, and r_U and r_S are the responses of the dibucaine peaks obtained from the *Assay preparation* and the *Standard preparation*, respectively.

Dichlorodifluoromethane—*see* Dichlorodifluoromethane NF

Dichlorotetrafluoroethane—*see* Dichlorotetrafluoroethane NF

Dicloxacillin Sodium

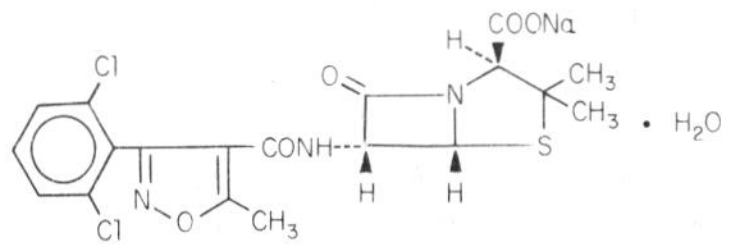

$C_{19}H_{16}Cl_2N_3NaO_5S.H_2O$ 510.32

4-Thia-1-azabicyclo[3.2.0]heptane-2-carboxylic acid, 6-[[[3-(2,6-dichlorophenyl)-5-methyl-4-isoxazolyl]carbonyl]-amino]-3,3-dimethyl-7-oxo-, monosodium salt, monohydrate, [2*S*-(2α,5α,6β)]-.

Monosodium (2*S*,5*R*,6*R*)-6-[3-(2,6-dichlorophenyl)-5-methyl-4-isoxazolecarboxamido]-3,3-dimethyl-7-oxo-4-thia-1-azabicyclo[3.2.0]heptane-2-carboxylate monohydrate [*13412-64-1*].

Anhydrous 492.31 [*343-55-5*].

» Dicloxacillin Sodium has a potency equivalent to not less than 850 μg of dicloxacillin ($C_{19}H_{17}Cl_2N_3O_5S$) per mg.

Packaging and storage—Preserve in tight containers.

Reference standard—*USP Dicloxacillin Sodium Reference Standard*—Do not dry before using.

Identification—

A: The infrared absorption spectrum of a potassium bromide dispersion of it exhibits maxima only at the same wavelengths as that of a similar preparation of USP Dicloxacillin Sodium RS.

B: Ignite about 100 mg: a 1 in 20 solution of the residue in acetic acid responds to the tests for *Sodium* ⟨191⟩.

Crystallinity ⟨695⟩: meets the requirements.

pH ⟨791⟩: between 4.5 and 7.5, in a solution containing 10 mg per mL.

Water, *Method I* ⟨921⟩: between 3.0% and 5.0%.

Organic chlorine content—Weigh accurately about 25 mg of Dicloxacillin Sodium, and proceed as directed under *Oxygen Flask Combustion* ⟨471⟩, using 10 mL of 0.1 *N* sodium hydroxide as the absorbing liquid. Transfer the resulting solution to a suitable titration vessel, heat on a steam bath for about 30 minutes, cool to room temperature, add 5 mL of 8 *N* nitric acid, and titrate with 0.01 *N* silver nitrate VS, determining the end-point potentiometrically, using silver and silver–silver chloride electrodes. Each mL of 0.01 *N* silver nitrate is equivalent to 0.3545 mg of Cl. Calculate the total chlorine content, in percentage, and subtract from it the *Free chloride:* between 13.0% and 14.2% is found.

Free chloride content—Transfer about 125 mg, accurately weighed, to a titration flask, dissolve in 10 mL of 0.1 *N* sodium hydroxide, add 20 mL of water, and heat on a steam bath for about 30 minutes. Cool to room temperature, add 5 mL of 8 *N* nitric acid, and titrate with 0.01 *N* silver nitrate VS, determining the end-point potentiometrically, using silver and silver–silver chloride electrodes. Each mL of 0.01 *N* silver nitrate is equivalent to 0.3545 mg of Cl: not more than 0.5% is found.

Assay—Proceed with Dicloxacillin Sodium as directed under *Antibiotics—Microbial Assays* ⟨81⟩.

Dicloxacillin Sodium Capsules

» Dicloxacillin Sodium Capsules contain not less than 90.0 percent and not more than 120.0 percent of the labeled amount of dicloxacillin ($C_{19}H_{17}Cl_2N_3O_5S$).

Packaging and storage—Preserve in tight containers.

Reference standard—*USP Dicloxacillin Sodium Reference Standard*—Do not dry before using.

Dissolution ⟨711⟩—

Medium: water; 900 mL.

Apparatus 1: 100 rpm.

Time: 45 minutes.

Procedure—Determine the amount of dicloxacillin ($C_{19}H_{17}Cl_2N_3O_5S$) by a suitable validated spectrophotometric analysis of a filtered portion of the solution under test, suitably diluted with *Dissolution Medium*, if necessary, in comparison with a Standard solution having a known concentration of USP Dicloxacillin Sodium RS in the same medium.

Tolerances—Not less than 75% (*Q*) of the labeled amount of $C_{19}H_{17}Cl_2N_3O_5S$ is dissolved in 45 minutes.

Uniformity of dosage units ⟨905⟩: meet the requirements.

Water, *Method I* ⟨921⟩: not more than 5.0%.

Assay—Proceed as directed under *Antibiotics—Microbial Assays* ⟨81⟩, using not less than 5 Dicloxacillin Sodium Capsules blended for 4 ± 1 minutes in a high-speed glass blender jar containing an accurately measured volume of *Buffer No. 1.* Dilute an accurately measured volume of this stock solution quantitatively with *Buffer No. 1* to obtain a *Test Dilution* having a concentration assumed to be equal to the median dose level of the Standard.

Sterile Dicloxacillin Sodium

» Sterile Dicloxacillin Sodium is Dicloxacillin Sodium suitable for parenteral use. It has a potency equivalent to not less than 850 μg of dicloxacillin ($C_{19}H_{17}Cl_2N_3O_5S$) per mg. In addition, where packaged for dispensing, it contains not less than 90.0 percent and not more than 120.0 percent of the labeled amount of dicloxacillin ($C_{19}H_{17}Cl_2N_3O_5S$).

Packaging and storage—Preserve in *Containers for Sterile Solids* as described under *Injections* ⟨1⟩.

Reference standard—*USP Dicloxacillin Sodium Reference Standard*—Do not dry before using.

Constituted solution—At the time of use, the constituted solution prepared from Sterile Dicloxacillin Sodium meets the requirements for *Constituted Solutions* under *Injections* ⟨1⟩.

Pyrogen—It meets the requirements of the *Pyrogen Test* ⟨151⟩, the test dose being 1 mL of a solution in pyrogen-free saline TS containing 20 mg of dicloxacillin per mL.

Sterility—It meets the requirements under *Sterility Tests* ⟨71⟩, when tested as directed in the section, *Test Procedures Using Membrane Filtration*, using 6 g of specimen aseptically dissolved in 200 mL of *Fluid A*.

pH ⟨791⟩: between 4.5 and 7.5, in a solution containing 10 mg per mL or, where packaged for dispensing, in the solution constituted as directed in the labeling.

Particulate matter ⟨788⟩: meets the requirements under *Small-volume Injections.*

Other requirements—It conforms to the definition, responds to the *Identification tests*, and meets the requirements for *Water*, *Crystallinity*, *Free chloride*, and *Organic chlorine content* under *Dicloxacillin Sodium.* It meets also the requirements for *Uniformity of Dosage Units* ⟨905⟩ and for *Labeling* under *Injections* ⟨1⟩.

Assay—Proceed as directed under *Antibiotics—Microbial Assays* ⟨81⟩. Where it is packaged for dispensing, constitute 1 container of Sterile Dicloxacillin Sodium as directed in the labeling. Remove an accurately measured portion, and dilute quantitatively with *Buffer No. 1* to obtain a *Test Dilution* having a concentration assumed to be equal to the median dose level of the Standard.

Dicloxacillin Sodium for Oral Suspension

» Dicloxacillin Sodium for Oral Suspension is a dry mixture of Dicloxacillin Sodium and one or more suitable buffers, colors, flavors, and preservatives. It contains not less than 90.0 percent and not more than 120.0 percent of the labeled amount of dicloxacillin ($C_{19}H_{17}Cl_2N_3O_5S$).

Packaging and storage—Preserve in tight containers.

Reference standard—*USP Dicloxacillin Sodium Reference Standard*—Do not dry before using.

Uniformity of dosage units ⟨905⟩—

FOR SOLID PACKAGED IN SINGLE-UNIT CONTAINERS: meets the requirements.

pH ⟨791⟩: between 4.5 and 7.5, in the suspension constituted as directed in the labeling.

Water, *Method I* ⟨921⟩: not more than 2.0%.

Assay—Proceed as directed under *Antibiotics—Microbial Assays* ⟨81⟩, using Dicloxacillin Sodium for Oral Suspension constituted as directed in the labeling. Transfer an accurately measured quantity of the constituted suspension, freshly mixed and free from air bubbles, equivalent to about 125 mg of dicloxacillin, to a 100-mL volumetric flask. Add 20 mL of dimethylformamide, and shake by mechanical means for 30 minutes. Dilute with *Buffer No. 1* to volume, and mix. Dilute a portion of this stock solution with *Buffer No. 1* to obtain a *Test Dilution* having a concentration assumed to be equal to the median dose level of the Standard.

Dicumarol

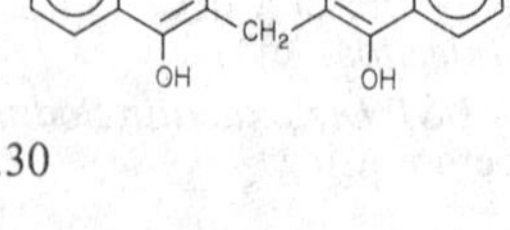

$C_{19}H_{12}O_6$ 336.30

2*H*-1-Benzopyran-2-one], 3,3′-methylenebis[4-hydroxy-.
3,3′-Methylenebis[4-hydroxycoumarin] [*66-76-2*].

» Dicumarol contains not less than 98.5 percent and not more than 101.0 percent of $C_{19}H_{12}O_6$, calculated on the dried basis.

Packaging and storage—Preserve in well-closed containers.

Reference standard—*USP Dicumarol Reference Standard*—Dry at 105° for 3 hours before using.

Identification—

A: The infrared absorption spectrum of a potassium bromide dispersion of it exhibits maxima only at the same wavelengths as that of a similar preparation of USP Dicumarol RS.

B: The ultraviolet absorption spectrum of a 1 in 100,000 solution in chloroform exhibits maxima and minima at the same wavelengths as that of a similar preparation of USP Dicumarol RS, concomitantly measured.

Acidity—Shake 500 mg with 10 mL of water for 1 minute, and filter: the filtrate requires for neutralization not more than 0.50 mL of 0.020 *N* sodium hydroxide, methyl red TS being used as the indicator.

Loss on drying ⟨731⟩—Dry it at 105° for 3 hours: it loses not more than 0.5% of its weight.

Residue on ignition ⟨281⟩: not more than 0.25%.

Ordinary impurities ⟨466⟩—

Test solution: 5 mg per mL in chloroform.

Standard solutions: 0.025 mg per mL, 0.05 mg per mL, and 0.1 mg per mL in chloroform.

Eluant: a mixture of chloroform, methanol, and glacial acetic acid (95:5:5).

Procedure—Proceed as directed for *Procedure* under *Ordinary Impurities* ⟨466⟩, except to employ the following instructions for drying the plate before and after developing the chromatograms. After applying the *Test solution* and *Standard solutions*, air-dry the plate at 105° for 10 minutes, and cool. Develop the chromatograms and remove the plate from the chamber. Air-dry the developed plate, dry it at 105° for 10 minutes, and cool.

Visualization: 16; then dry the plate and spray with starch TS.

Assay—Weigh accurately about 400 mg of Dicumarol, and dissolve in 50 mL of *n*-butylamine. Add 3 drops of a saturated solution of azo violet in methanol, and titrate with 0.1 *N* sodium methoxide VS to a blue end-point, observing precautions to prevent absorption of atmospheric carbon dioxide. Perform a blank determination, and make any necessary correction. Each mL of 0.1 *N* sodium methoxide is equivalent to 16.81 mg of $C_{19}H_{12}O_6$.

Dicumarol Tablets

» Dicumarol Tablets contain not less than 90.0 percent and not more than 110.0 percent of the labeled amount of $C_{19}H_{12}O_6$.

Packaging and storage—Preserve in well-closed containers.

Labeling—Label Tablets to state that Dicumarol Tablets may not be interchangeable with Dicumarol Capsules without retitration of the patient.

Reference standard—*USP Dicumarol Reference Standard*—Dry at 105° for 3 hours before using.

Identification—The retention time of the major peak in the chromatogram of the *Assay preparation* corresponds to that of the *Standard preparation*, as obtained in the *Assay*.

Dissolution ⟨711⟩—

Medium: 0.1 *M* tris buffer, prepared by dissolving 12.1 g of tris(hydroxymethyl)aminomethane in 900 mL of water in a 1000-mL volumetric flask, adjusting with hydrochloric acid to a pH of 9.0 ± 0.1, and diluting with water to volume; 1000 mL.

Apparatus 1: 100 rpm.

Time: 30 minutes.

Standard preparation—[NOTE—Prepare this solution fresh daily.] Dissolve an accurately weighed quantity of USP Dicumarol RS in 0.1 *N* sodium hydroxide, and dilute quantitatively with 0.1 *N* sodium hydroxide to obtain a solution having a known concentration of about 200 µg per mL. Transfer 3.0 mL of this solution to a 200-mL volumetric flask, add 0.1 *N* sodium hydroxide to volume, and mix.

Procedure—Dilute a filtered portion of the solution under test quantitatively with 0.1 *N* sodium hydroxide to obtain a solution having a concentration of about 3 µg of $C_{19}H_{12}O_6$ per mL. Determine the amount of $C_{19}H_{12}O_6$ dissolved from ultraviolet absorbances at the wavelength of maximum absorbance at about 314 nm of this solution in comparison with the *Standard preparation*, using 0.1 *N* sodium hydroxide as the blank.

Tolerances—Not less than 60% (*Q*) of the labeled amount of $C_{19}H_{12}O_6$ is dissolved in 30 minutes.

Uniformity of dosage units ⟨905⟩: meet the requirements.

Assay—

Mobile phase—Prepare a filtered and degassed mixture of water, tetrahydrofuran, methanol, and glacial acetic acid (65:35:10:0.1). Make adjustments if necessary (see *System Suitability* under *Chromatography* ⟨621⟩).

Standard preparation Dissolve an accurately weighed quantity of USP Dicumarol RS on 0.01 *N* sodium hydroxide with the aid of sonication to obtain a solution having a known concentration of about 0.25 mg per mL. Transfer 5.0 mL of this solution to a 25-mL volumetric flask, dilute with *Mobile phase* to volume, and mix to obtain a solution having a known concentration of about 0.05 mg of USP Dicumarol RS per mL.

Assay preparation—Weigh and finely powder not less than 20 Dicumarol Tablets. Transfer an accurately weighed portion of the powder, equivalent to about 25 mg of dicumarol, to a 100-mL volumetric flask, add about 50 mL of 0.01 *N* sodium hydroxide, sonicate to dissolve, dilute with 0.01 *N* sodium hydroxide to volume, mix, and filter, discarding the first 15 mL of the filtrate. Pipet 5 mL of the filtrate into a 25-mL volumetric flask, dilute with *Mobile phase* to volume, and mix.

Chromatographic system (see *Chromatography* ⟨621⟩)—The liquid chromatograph is equipped with a 311-nm detector and a 3.9-mm × 30-cm column that contains packing L1. The flow rate is about 1.5 mL per minute. Chromatograph the *Standard preparation*, and record the peak responses as directed under *Procedure:* the tailing factor for the analyte peak is not more than 2.0, and the relative standard deviation for replicate injections is not more than 2.0%.

Procedure—Separately inject equal volumes (about 20 µL) of the *Standard preparation* and the *Assay preparation* into the chromatograph, record the chromatograms, and measure the responses for the major peaks. Calculate the quantity, in mg, of $C_{19}H_{12}O_6$ in the portion of Tablets taken by the formula:

$$500C(r_U / r_S),$$

in which *C* is the concentration, in mg per mL, of USP Dicumarol RS in the *Standard preparation*, and r_U and r_S are the dicumarol peak responses obtained from the *Assay preparation* and the *Standard preparation*, respectively.

Dicyclomine Hydrochloride

· HCl
$COOCH_2CH_2N(C_2H_5)_2$

$C_{19}H_{35}NO_2 \cdot HCl$ 345.95

[Bicyclohexyl]-1-carboxylic acid, 2-(diethylamino)ethyl ester, hydrochloride.

2-(Diethylamino)ethyl [bicyclohexyl]-1-carboxylate hydrochloride [*67-92-5*].

» Dicyclomine Hydrochloride contains not less than 99.0 percent and not more than 102.0 percent of $C_{19}H_{35}NO_2 \cdot HCl$, calculated on the dried basis.

Packaging and storage—Preserve in well-closed containers.

Reference standard—*USP Dicyclomine Hydrochloride Reference Standard*—Dry at 105° for 4 hours before using.

Identification—

A: The infrared absorption spectrum of a potassium bromide dispersion of it, previously dried, exhibits maxima only at the same wavelengths as that of a similar preparation of USP Dicyclomine Hydrochloride RS.

B: Mix about 5 mL of a 1 in 500 solution of it with about 2 mL of 2 *N* nitric acid, and add about 2 mL of silver nitrate TS: a white precipitate is formed which is insoluble in nitric acid but soluble in a slight excess of 6 *N* ammonium hydroxide.

Melting range, *Class I* ⟨741⟩: between 169° and 174°.

pH ⟨791⟩: between 5.0 and 5.5, in a solution (1 in 100).

Loss on drying ⟨731⟩—Dry it at 105° for 4 hours: it loses not more than 1.0% of its weight.

Readily carbonizable substances ⟨271⟩—Dissolve 500 mg in 5 mL of sulfuric acid TS: the solution has no more color than Matching Fluid D.

Assay—Dissolve about 600 mg of Dicyclomine Hydrochloride, accurately weighed, in 70 mL of glacial acetic acid, add 10 mL of mercuric acetate TS and 1 drop of crystal violet TS, and titrate with 0.1 *N* perchloric acid VS to a blue end-point. Perform a blank determination, and make any necessary correction. Each mL of 0.1 *N* perchloric acid is equivalent to 34.60 mg of $C_{19}H_{35}NO_2 \cdot HCl$.

Dicyclomine Hydrochloride Capsules

» Dicyclomine Hydrochloride Capsules contain not less than 93.0 percent and not more than 107.0 percent of the labeled amount of $C_{19}H_{35}NO_2 \cdot HCl$.

Packaging and storage—Preserve in well-closed containers.

Reference standard—*USP Dicyclomine Hydrochloride Reference Standard*—Dry at 105° for 4 hours before using.

Identification—

A: Transfer a portion of the contents of the Capsules, equivalent to about 100 mg of dicyclomine hydrochloride, to a separator containing 10 mL of water and 1 mL of hydrochloric acid. Extract the aqueous acid solution with two 30-mL portions of chloroform, transfer the chloroform extracts to a second separator containing 20 mL of water and 1 mL of sodium hydroxide solution (1 in 10), and shake. Filter the chloroform layer through anhydrous sodium sulfate into a suitable container, and add 3 mL of a freshly prepared 1 in 20 solution of acetyl chloride in anhydrous methanol, prepared by cautiously adding acetyl chloride dropwise to anhydrous methanol with stirring. Evaporate under reduced pressure at room temperature until the residue has been thoroughly dried: the infrared absorption spectrum of a potassium bromide dispersion of the dicyclomine hydrochloride so obtained exhibits maxima and minima at the same wavelengths as that of a similar preparation of USP Dicyclomine Hydrochloride RS.

B: The retention time of the major peak in the chromatogram of the *Assay preparation* corresponds to that of the *Standard preparation*, as obtained in the *Assay*.

Dissolution ⟨711⟩—

Medium: 0.01 *N* hydrochloric acid; 500 mL.

Apparatus 2: 50 rpm.

Time: 45 minutes.

Picric acid solution—Transfer about 100 mg of picric acid and 14.5 g of anhydrous sodium acetate to a 500-mL volumetric flask and dissolve in 400 mL of water. Add 20.0 mL of glacial acetic acid, dilute with water to volume, and mix.

Standard solution—Prepare a solution in 0.01 *N* hydrochloric acid having a known concentration of about 18 µg per mL of USP Dicyclomine Hydrochloride RS.

Procedure—Determine the amount of $C_{19}H_{35}NO_2 \cdot HCl$ dissolved using the following procedure. Pipet 20 mL each of the solution under test, the *Standard solution*, and 0.01 *N* hydrochloric acid to provide a blank, into separate suitable separators. Add 5.0 mL of *Picric acid solution* and 25.0 mL of chloroform to each separator, and shake for one minute. Collect the chloroform layer in a suitable vessel containing 2.0 g of anhydrous sodium sulfate, shake and allow the solutions to stand for ten minutes. Concomitantly determine the absorbances of the solutions at the wavelength of maximum absorbance at about 405 nm, using the blank to set the spectrophotometer.

Tolerances—Not less than 75% (*Q*) of the labeled amount of $C_{19}H_{35}NO_2.HCl$ is dissolved in 45 minutes.

Uniformity of dosage units ⟨905⟩: meet the requirements.

Assay—

Mobile phase—Transfer 385 mg of ammonium acetate to a 1000-mL volumetric flask, add 100 mL of water, mix, and sonicate to dissolve. Dilute with methanol to volume, mix, filter through a membrane filter of 0.47-µm or finer porosity, and degas. Make adjustments if necessary (see *System Suitability* under *Chromatography* ⟨621⟩).

Standard preparation—Dissolve an accurately weighed quantity of USP Dicyclomine Hydrochloride RS in *Mobile phase*, and dilute quantitatively, and stepwise if necessary, with *Mobile phase* to obtain a solution having a known concentration of about 0.4 mg per mL.

Assay preparation—Remove, as completely as possible, the contents of not less than 20 Dicyclomine Hydrochloride Capsules, and weigh accurately. Mix the combined contents, and transfer an accurately weighed portion of the powder, equivalent to about 20 mg of dicyclomine hydrochloride, to a conical flask. Add 50.0 mL of *Mobile phase*, sonicate with occasional swirling for 5 minutes, and filter, discarding the first 10 mL of the filtrate.

Chromatographic system (see *Chromatography* ⟨621⟩)—The liquid chromatograph is equipped with a 215-nm detector and a 3.9-mm × 30-cm column that contains packing L1. The flow rate is about 1.5 mL per minute. Chromatograph the *Standard preparation*, and record the peak responses as directed under *Procedure:* the tailing factor for the analyte peak is not more than 2.5, and the relative standard deviation for replicate injections is not more than 1.5%.

Procedure—Separately inject equal volumes (about 50 µL) of the *Standard preparation* and the *Assay preparation* into the chromatograph, record the chromatograms, and measure the responses for the major peaks. Calculate the quantity, in mg, of $C_{19}H_{35}NO_2.HCl$ in the portion of Capsules taken by the formula:

$$50C(r_U/r_S),$$

in which *C* is the concentration, in mg per mL, of USP Dicyclomine Hydrochloride RS in the *Standard preparation*, and r_U and r_S are the peak responses obtained from the *Assay preparation* and the *Standard preparation*, respectively.

Dicyclomine Hydrochloride Injection

» Dicyclomine Hydrochloride Injection is a sterile, isotonic solution of Dicyclomine Hydrochloride in Water for Injection. It contains not less than 93.0 percent and not more than 107.0 percent of the labeled amount of $C_{19}H_{35}NO_2.HCl$.

Packaging and storage—Preserve in single-dose or in multiple-dose containers, preferably of Type I glass.

Reference standard—*USP Dicyclomine Hydrochloride Reference Standard*—Dry at 105° for 4 hours before using.

Identification—

A: Transfer a portion of Injection, equivalent to about 100 mg of dicyclomine hydrochloride, to a separator containing 10 mL of water and 1 mL of hydrochloric acid. Shake with 25 mL of ether, and discard the ether layer. Proceed as directed in *Identification test A* under *Dicyclomine Hydrochloride Capsules*, beginning with "Extract the aqueous acid solution."

B: The retention time of the major peak in the chromatogram of the *Assay preparation* corresponds to that of the *Standard preparation*, as obtained in the *Assay*.

Other requirements—It meets the requirements under *Injections* ⟨1⟩.

Assay—

Mobile phase, Standard preparation, and *Chromatographic system*—Prepare as directed in the *Assay* under *Dicyclomine Hydrochloride Capsules*.

Assay preparation—Transfer an accurately measured volume of Dicyclomine Hydrochloride Injection, equivalent to about 20 mg of dicyclomine hydrochloride, to a 50-mL volumetric flask, dilute with *Mobile phase* to volume, and mix.

Procedure—Proceed as directed for *Procedure* in the *Assay* under *Dicyclomine Hydrochloride Capsules*. Calculate the quantity, in mg, of $C_{19}H_{35}NO_2.HCl$ in each mL of the Injection taken by the formula:

$$50(C/V)(r_U/r_S),$$

in which *C* is the concentration, in mg per mL, of USP Dicyclomine Hydrochloride RS in the *Standard preparation*, *V* is the volume, in mL, of Injection taken, and r_U and r_S are the peak responses obtained from the *Assay preparation* and the *Standard preparation*, respectively.

Dicyclomine Hydrochloride Syrup

» Dicyclomine Hydrochloride Syrup contains not less than 95.0 percent and not more than 105.0 percent of the labeled amount of $C_{19}H_{35}NO_2.HCl$.

Packaging and storage—Preserve in tight containers.

Reference standard—*USP Dicyclomine Hydrochloride Reference Standard*—Dry at 105° for 4 hours before using.

Identification—

A: Transfer a portion of the Syrup, equivalent to about 100 mg of dicyclomine hydrochloride, to a separator containing 10 mL of water and 1 mL of hydrochloric acid. Extract with two 30-mL portions of ether, and discard the ether. Proceed as directed in *Identification test A* under *Dicyclomine Hydrochloride Capsules*, beginning with "Extract the aqueous acid solution."

B: The chromatogram of the *Assay preparation* exhibits a major peak at a retention time corresponding to that of the peak representing dicyclomine hydrochloride in the chromatogram of the *Standard preparation*, obtained as directed in the *Assay*.

Assay—

Mobile phase, Standard preparation, and *Chromatographic system*—Prepare as directed in the *Assay* under *Dicyclomine Hydrochloride Capsules*.

Assay preparation—Transfer an accurately measured volume of Dicyclomine Hydrochloride Syrup, equivalent to about 20 mg of dicyclomine hydrochloride, to a separator, add 30 mL of water, and extract with five 20-mL portions of chloroform, collecting the chloroform extracts in a 200-mL beaker. Evaporate the chloroform extracts on a steam bath under a current of air to dryness. Remove the beaker from the steam bath, and transfer the residue with the aid of small amounts of *Mobile phase* to a 50-mL volumetric flask. Dilute with *Mobile phase* to volume, and mix.

Procedure—Proceed as directed for *Procedure* in the *Assay* under *Dicyclomine Hydrochloride Capsules*. Calculate the quantity, in mg, of $C_{19}H_{35}NO_2.HCl$ in each mL of the Syrup taken by the formula:

$$50(C/V)(r_U/r_S),$$

in which *C* is the concentration, in mg per mL, of USP Dicyclomine Hydrochloride RS in the *Standard preparation*, *V* is the volume, in mL, of Syrup taken, and r_U and r_S are the peak responses obtained from the *Assay preparation* and the *Standard preparation*, respectively.

Dicyclomine Hydrochloride Tablets

» Dicyclomine Hydrochloride Tablets contain not less than 93.0 percent and not more than 107.0 percent of the labeled amount of $C_{19}H_{35}NO_2.HCl$.

Packaging and storage—Preserve in well-closed containers.

Reference standard—*USP Dicyclomine Hydrochloride Reference Standard*—Dry at 105° for 4 hours before using.

Identification—

A: Transfer a portion of finely powdered Tablets, equivalent to about 100 mg of dicyclomine hydrochloride, to a separator containing 10 mL of water and 1 mL of hydrochloric acid. Proceed as directed in *Identification test A* under *Dicyclomine Hydrochloride Capsules*, beginning with "Extract the aqueous acid solution."

B: The retention time of the major peak in the chromatogram of the *Assay preparation* corresponds to that of the *Standard preparation*, as obtained in the *Assay*.

Dissolution ⟨711⟩—

Medium: 0.01 *N* hydrochloric acid; 500 mL.

Apparatus 2: 50 rpm.

Time: 45 minutes.

Picric acid solution and *Standard solution*—Prepare as directed for *Dissolution* under *Dicyclomine Hydrochloride Capsules*.

Procedure—Proceed as directed for *Dissolution* under *Dicyclomine Hydrochloride Capsules* except to use 10.0 mL of the solution under test and to add 10.0 mL of 0.01 *N* hydrochloric acid to the separator containing the solution under test.

Tolerances—Not less than 75% (*Q*) of the labeled amount of $C_{19}H_{35}NO_2 . HCl$ is dissolved in 45 minutes.

Uniformity of dosage units ⟨905⟩: meet the requirements.

Assay—

Mobile phase, *Standard preparation*, and *Chromatographic system*—Prepare as driected in the *Assay* under *Dicyclomine Hydrochloride Capsules*.

Assay preparation—Weigh and finely powder not less than 20 Dicyclomine Hydrochloride Tablets. Transfer an accurately weighed portion of the powder, equivalent to about 20 mg of dicyclomine hydrochloride, to a conical flask. Add 50.0 mL of *Mobile phase*, sonicate with occasional swirling for 5 minutes, and filter, discarding the first 10 mL of the filtrate.

Procedure—Proceed as directed for *Procedure* in the *Assay* under *Dicyclomine Hydrochloride Capsules*. Calculate the quantity, in mg, of $C_{19}H_{35}NO_2 . HCl$ in the portion of Tablets taken by the formula:

$$50C(r_U / r_S),$$

in which *C* is the concentration, in mg per mL, of USP Dicyclomine Hydrochloride RS in the *Standard preparation*, and r_U and r_S are the peak responses obtained from the *Assay preparation* and the *Standard preparation*, respectively.

Dienestrol

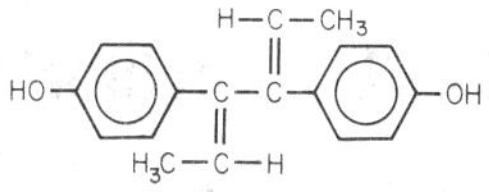

$C_{18}H_{18}O_2$ 266.34

Phenol, 4,4′-(1,2-diethylidene-1,2-ethanediyl)bis-, (*E*,*E*)-.

4,4′-(Diethylideneethylene)diphenol [*84-17-3; 13029-44-2*].

» Dienestrol contains not less than 98.0 percent and not more than 100.5 percent of $C_{18}H_{18}O_2$, calculated on the dried basis.

Packaging and storage—Preserve in well-closed containers.

Reference standard—*USP Dienestrol Reference Standard*—Dry at 105° for 2 hours before using.

Identification—

A: The infrared absorption spectrum of a potassium bromide dispersion of it, previously dried, exhibits maxima only at the same wavelengths as that of a similar preparation of USP Dienestrol RS.

B: The ultraviolet absorption spectrum of a 1 in 200,000 solution in alcohol exhibits maxima and minima at the same wavelengths as that of a similar solution of USP Dienestrol RS, concomitantly measured, and the respective absorptivities, calculated on the dried basis, at the wavelength of maximum absorbance at about 228 nm do not differ by more than 3.0%.

C: To a solution of about 10 mg in 0.5 mL of alcohol add 1 mL of hydrochloric acid and about 50 mg of vanillin: a blue color is produced immediately, and it persists on dilution with water but disappears on the addition of alkali (*distinction from diethylstilbestrol, which produces no color*).

Melting range ⟨741⟩: between 227° and 234°, but the range between beginning and end of melting does not exceed 3°.

Loss on drying ⟨731⟩—Dry it at 105° for 2 hours: it loses not more than 0.5% of its weight.

Residue on ignition ⟨281⟩: not more than 0.2%.

Assay—

Mobile phase, *Internal standard solution*, and *Standard preparation*—Prepare as directed in the *Assay* under *Dienestrol Cream*.

Assay preparation—Transfer about 25 mg of Dienestrol, accurately weighed, to a 100-mL volumetric flask. Add methanol to volume, and mix. Pipet 2 mL of this solution, 5 mL of *Internal standard solution*, and 5 mL of water into a 50-mL volumetric flask. Dilute with methanol to volume, and mix.

Procedure—Proceed as directed for *Procedure* in the *Assay* under *Dienestrol Cream*. Calculate the quantity, in mg, of $C_{18}H_{18}O_2$ in the portion of Dienestrol taken by the formula:

$$2.5C(R_U / R_S).$$

Dienestrol Cream

» Dienestrol Cream is Dienestrol in a suitable water-miscible base. It contains not less than 90.0 percent and not more than 110.0 percent of the labeled amount of $C_{18}H_{18}O_2$.

Packaging and storage—Preserve in collapsible tubes or in tight containers.

Reference standard—*USP Dienestrol Reference Standard*—Dry at 105° for 2 hours before using.

Identification—The chromatogram of the *Assay preparation* employed in the *Assay* exhibits two peaks, for dienestrol and the internal standard, whose retention times are identical to those exhibited by the *Standard preparation*.

Minimum fill ⟨755⟩: meets the requirements.

Assay—

Mobile phase—Prepare a suitable degassed solution of methanol (about 3 in 5) such that the retention time of dienestrol is about 8 to 10 minutes and that of methyltestosterone is about 11 to 14 minutes. Make adjustments, if necessary (see *System Suitability* under *Chromatography* ⟨621⟩).

Internal standard solution—Dissolve methyltestosterone in methanol to obtain a solution having a concentration of about 125 µg per mL.

Standard preparation—Accurately weigh a suitable quantity of USP Dienestrol RS, dissolve in methanol, and dilute quantitatively and stepwise with methanol to obtain a concentration of about 50 µg per mL. Pipet 10 mL of this solution, 5 mL of *Internal standard solution*, and 5 mL of water to a 50-mL volumetric flask, dilute with methanol to volume, and mix. The concentration of dienestrol in the *Standard preparation* is about 10 µg per mL.

Assay preparation—Pipet 20 mL of methanol and 5 mL of *Internal standard solution* into a 50-mL screw-capped tube. Using a 5-mL plastic syringe, transfer an accurately weighed quantity of Dienestrol Cream, equivalent to about 0.5 mg of dienestrol, to the tube. Cap the tube, and disperse the mixture with a suitable vibrating mixer at high speed for 3 minutes, then in an ultrasonic bath for 5 minutes. Add 20 mL of methanol, continue shaking by mechanical means for another 10 minutes, then chill in an ice bath for 5 minutes. Filter the mixture through paper, discarding the first 5 mL of the filtrate.

Procedure—Introduce equal volumes (about 50 µL) of the *Assay preparation* and the *Standard preparation* into a high-pressure liquid chromatograph operated at room temperature, by means of a suitable microsyringe or sampling valve. Typically, the apparatus is fitted with a 25-cm × 4.6-mm column containing packing L1 and equipped with an ultraviolet detector capable of monitoring absorption at 254 nm and a suitable recorder. The *Mobile phase* is maintained at a flow rate of about 2 mL per minute. In a suitable chromatographic system, six replicate injections of the *Standard preparation* show a relative standard deviation of not more than 2.0% and a resolution factor of not less than 2.0 between the peaks for dienestrol and the internal standard. Calculate the quantity, in mg, of $C_{18}H_{18}O_2$ in the portion of Cream taken by the formula:

$$0.05C(R_U/R_S),$$

in which *C* is the concentration, in µg per mL, of USP Dienestrol RS in the *Standard preparation*, and R_U and R_S are the peak area ratios of dienestrol to methyltestosterone obtained from the *Assay preparation* and the *Standard preparation*, respectively.

Diethanolamine—*see* Diethanolamine NF

Diethylcarbamazine Citrate

CH_3—N(piperazine)N—$CON(C_2H_5)_2$ · HO—C(CH_2COOH)(CH_2COOH)—COOH

$C_{10}H_{21}N_3O.C_6H_8O_7$ 391.42

1-Piperazinecarboxamide, *N,N*-diethyl-4-methyl-, 2-hydroxy-1,2,3-propanetricarboxylate.

N,N-Diethyl-4-methyl-1-piperazinecarboxamide citrate (1:1) [*1642-54-2*].

» Diethylcarbamazine Citrate contains not less than 98.0 percent and not more than 100.5 percent of $C_{10}H_{21}N_3O.C_6H_8O_7$, calculated on the anhydrous basis.

Packaging and storage—Preserve in tight containers.

Reference standard—*USP Diethylcarbamazine Citrate Reference Standard*—Do not dry before using.

Identification—

A: It meets the requirements under *Identification—Organic Nitrogenous Bases* ⟨181⟩.

B: It responds to the test for *Citrate* ⟨191⟩.

Water, *Method I* ⟨921⟩: not more than 0.5%.

Residue on ignition ⟨281⟩: not more than 0.1%.

Heavy metals ⟨231⟩—Dissolve 1.0 g in 20 mL of water. Add 1 mL of 0.1 *N* hydrochloric acid, dilute with water to 25 mL, and mix: the limit is 0.002%.

Ordinary impurities ⟨466⟩—

Test solution: methanol.

Standard solution: methanol.

Eluant: a mixture of methanol and ammonium hydroxide (100:1.5).

Visualization: 16.

Assay—Dissolve about 750 mg of Diethylcarbamazine Citrate, accurately weighed, in 50 mL of glacial acetic acid, warming slightly to effect solution. Cool the solution to room temperature, add 10 drops of *p*-naphtholbenzein TS, and titrate with 0.1 *N* perchloric acid VS. Perform a blank determination, and make any necessary correction. Each mL of 0.1 *N* perchloric acid is equivalent to 39.14 mg of $C_{10}H_{21}N_3O.C_6H_8O_7$.

Diethylcarbamazine Citrate Tablets

» Diethylcarbamazine Citrate Tablets contain not less than 95.0 percent and not more than 105.0 percent of the labeled amount of $C_{10}H_{21}N_3O.C_6H_8O_7$.

NOTE—Diethylcarbamazine Citrate Tablets labeled solely for veterinary use are exempt from the requirements of the test for *Dissolution.*

Packaging and storage—Preserve in tight containers.

Reference standard—*USP Diethylcarbamazine Citrate Reference Standard*—Do not dry before using.

Identification—Tablets meet the requirements under *Identification—Organic Nitrogenous Bases* ⟨181⟩.

Disintegration ⟨701⟩—

FOR TABLETS LABELED SOLELY FOR VETERINARY USE: 30 minutes.

Dissolution ⟨711⟩—

Medium: water; 900 mL.

Apparatus 2: 50 rpm.

Time: 45 minutes.

Procedure—Determine the amount of $C_{10}H_{21}N_3O.C_6H_8O_7$ dissolved by treating filtered portions of the solution under test, suitably diluted with *Dissolution Medium,* if necessary, as directed in the *Assay*, beginning with "Add 5 mL of 5 *N* sodium hydroxide," and using 0.01 *N* perchloric acid.

Tolerances—Not less than 75% (*Q*) of the labeled amount of $C_{10}H_{21}N_3O.C_6H_8O_7$ is dissolved in 45 minutes.

Uniformity of dosage units ⟨905⟩: meet the requirements.

Assay—Weigh and finely powder not less than 20 Diethylcarbamazine Citrate Tablets. Weigh accurately a portion of the powder, equivalent to about 750 mg of diethylcarbamazine citrate, transfer it to a centrifuge tube, and extract with six 10-mL portions of water. Centrifuge after each extraction, decanting the supernatant liquid through filter paper into a 250-mL separator. Wash the residue on the filter paper with two small portions of water. Add 5 mL of 5 *N* sodium hydroxide to the combined aqueous solutions, and extract with two 25-mL portions of chloroform. Add another 5-mL portion of the sodium hydroxide solution to the aqueous solution, and again extract with two 25-mL portions of chloroform. Filter the combined chloroform extracts through paper, wash the paper with two small portions of chloroform, and titrate with 0.1 *N* perchloric acid VS, determining the end-point potentiometrically. Each mL of 0.1 *N* perchloric acid is equivalent to 39.14 mg of $C_{10}H_{21}N_3O.C_6H_8O_7$.

Diethyl Phthalate—*see* Diethyl Phthalate NF

Diethylpropion Hydrochloride

$CH_3CH(N(C_2H_5)_2)C(=O)$—C_6H_5 · HCl

$C_{13}H_{19}NO.HCl$ 241.76

1-Propanone, 2-(diethylamino)-1-phenyl-, hydrochloride.

2-(Diethylamino)propiophenone hydrochloride [*134-80-5*].

» Diethylpropion Hydrochloride contains not less than 97.0 percent and not more than 103.0 percent of $C_{13}H_{19}NO.HCl$, calculated on the anhydrous basis. It may contain tartaric acid as a stabilizer.

Packaging and storage—Preserve in well-closed, light-resistant containers.

Labeling—The label indicates whether it contains tartaric acid as a stabilizer.

Reference standards—*USP Diethylpropion Hydrochloride Reference Standard*—Dry over silica gel for 4 hours before using.

USP 2-Ethylaminopropiophenone Hydrochloride Reference Standard—Do not dry before using. Keep container tightly closed, protected from light, and store in a cool place.

Identification—

A: The infrared absorption spectrum of a potassium bromide dispersion of it, previously dried over silica gel for 4 hours, exhibits maxima only at the same wavelengths as that of a similar preparation of USP Diethylpropion Hydrochloride RS.

B: The retention time of the major peak in the chromatogram of the *Assay preparation* corresponds to that of the *Standard preparation*, as obtained in the *Assay*.

C: A solution (1 in 100) responds to the tests for *Chloride* ⟨191⟩.

Water, *Method I* ⟨921⟩: not more than 0.5%.

Secondary amines—Dissolve 100 mg in 2 mL of methylene chloride in a centrifuge tube. Transfer to a second tube 2 mL of a Standard solution of diethylamine hydrochloride (dried at 105° for 2 hours before being used) in methylene chloride having a known concentration of 250 µg per mL. Treat each solution as follows: Extract with 2 mL of a buffer solution containing 5.7 g of sodium carbonate and 3.0 g of sodium bicarbonate per 100 mL of water. Centrifuge, if necessary, to clarify the upper phase, and immediately transfer 0.5 mL of it to a spot plate. Immediately add 2 drops of acetaldehyde TS, and then, in rapid succession, add 1 drop of sodium nitroferricyanide solution (1 in 100) to each spot. Immediately and simultaneously stir both spots to mix the reagents: any blue color produced within 3 minutes by the test solution is not more intense than that of the Standard solution (not more than 0.5% of secondary amines as diethylamine hydrochloride).

Free bromine—One drop of a solution (1 in 10) produces no discoloration when placed upon starch iodide paper.

Hydrobromic acid and bromide—To 10 mL of a solution (1 in 10) add 1 mL of sodium hydroxide solution (1 in 10), extract with about 25 mL of chloroform, and discard the chloroform extract. Add 1 mL of 6 *N* hydrochloric acid, 0.5 mL of chloroform, and 0.5 mL of freshly prepared chloramine T solution (1 in 10), and shake vigorously: no yellow or brown-red color is produced in the chloroform layer.

Chromatographic purity—

Phosphate buffer—Dissolve 136.1 g of monobasic potassium phosphate in 900 mL of water, add 3.2 mL of phosphoric acid, dilute with water to 1000 mL, and mix.

Diluent—Prepare a mixture of water, *Phosphate buffer*, and acetonitrile (8:1:1).

Mobile phase—Mix 100 mL of acetonitrile, 100 mL of *Phosphate buffer*, 7.0 mL of diethylamine, and sufficient water to make 1 liter. Filter, and degas before use. Make adjustments if necessary (see *System Suitability* under *Chromatography* ⟨621⟩).

Test preparation—Transfer 100 mg of Diethylpropion Hydrochloride, accurately weighed, to a 50-mL volumetric flask, dissolve in about 40 mL of *Diluent*, add *Diluent* to volume, and mix.

Standard preparation—Dissolve an accurately weighed quantity of USP Diethylpropion Hydrochloride RS in *Diluent*, and dilute quantitatively, and stepwise if necessary, with *Diluent* to obtain a solution having a known concentration of about 0.01 mg per mL.

System suitability solution—Prepare a solution in *Diluent* containing about 4 µg of USP 2-Ethylaminopropiophenone Hydrochloride RS and 2 mg of USP Diethylpropion Hydrochloride RS per mL.

Chromatographic system (see *Chromatography* ⟨621⟩)—The liquid chromatograph is equipped with a 254-nm detector and a 4.6-mm × 15-cm column that contains packing L11. The flow rate is about 1 mL per minute. Chromatograph the *System suitability solution*, and record the peak responses as directed under *Procedure:* the relative retention times are about 0.6 for 2-ethylaminopropiophenone hydrochloride and 1.0 for diethylpropion hydrochloride, and the resolution, *R*, between the 2-ethylaminopropiophenone hydrochloride and diethylpropion hydrochloride peaks is not less than 4.0. Chromatograph the *Standard preparation*, and record the peak responses as directed under *Procedure:* the relative standard deviation for replicate injections is not more than 1.0%.

Procedure—Separately inject equal volumes (about 20 µL) of the *Standard preparation* and the *Test preparation* into the chromatograph, record the chromatograms, and measure the peak responses. The sum of all of the peak responses, excluding the diethylpropion hydrochloride response, from the *Test preparation* is not greater than the diethylpropion hydrochloride response from the *Standard preparation* (0.5%).

Assay—

Phosphate buffer—Dissolve 136.1 g of monobasic potassium phosphate in 900 mL of water, add 4.3 mL of phosphoric acid, dilute with water to 1000 mL, and mix.

Mobile phase—Prepare a suitable mixture of water, acetonitrile, *Phosphate buffer*, and 1.0 *M* sodium nitrate (730:200:50:20), filter through a membrane filter (0.7-µm or finer porosity), and degas. Make adjustments if necessary (see *System Suitability* under *Chromatography* ⟨621⟩).

Standard preparation—Dissolve an accurately weighed quantity of USP Diethylpropion Hydrochloride RS in *Mobile phase*, and dilute quantitatively, and stepwise if necessary, with *Mobile phase* to obtain a solution having a known concentration of about 40 µg per mL.

Assay preparation—Transfer about 100 mg of Diethylpropion Hydrochloride, accurately weighed, to a 250-mL volumetric flask, dissolve in *Mobile phase*, dilute with *Mobile phase* to volume, and mix. Transfer 10.0 mL of this solution to a 100-mL volumetric flask, dilute with *Mobile phase* to volume, and mix.

System suitability preparation—Prepare a solution in *Mobile phase* containing about 200 µg of benzoic acid and 40 µg of USP Diethylpropion Hydrochloride RS per mL.

Chromatographic system (see *Chromatography* ⟨621⟩)—The liquid chromatograph is equipped with a 254-nm detector and a 4-mm × 30-cm column that contains packing L1. The flow rate is about 1.5 mL per minute. Chromatograph the *System suitability preparation*, and record the peak responses as directed under *Procedure:* the relative retention times are about 0.5 for diethylpropion hydrochloride and 1.0 for benzoic acid, and the resolution, *R*, between the diethylpropion hydrochloride and benzoic acid peaks is not less than 2.0. Chromatograph the *Standard preparation*, and record the peak responses as directed under *Procedure:* the relative standard deviation for replicate injections is not more than 1.0%.

Procedure—Separately inject equal volumes (about 50 µL) of the *Standard preparation* and the *Assay preparation* into the chromatograph, record the chromatograms, and measure the responses for the major peaks. Calculate the quantity, in mg, of $C_{13}H_{19}NO.HCl$ in the Diethylpropion Hydrochloride taken by the formula:

$$2.5C(r_U/r_S),$$

in which C is the concentration, in µg per mL, of USP Diethylpropion Hydrochloride RS in the *Standard preparation*, and r_U and r_S are the peak responses obtained from the *Assay preparation* and the *Standard preparation*, respectively.

Diethylpropion Hydrochloride Tablets

» Diethylpropion Hydrochloride Tablets contain not less than 90.0 percent and not more than 110.0 percent of the labeled amount of $C_{13}H_{19}NO.HCl$.

Packaging and storage—Preserve in well-closed containers.

Reference standard—*USP Diethylpropion Hydrochloride Reference Standard*—Dry over silica gel for 4 hours before using.

Identification—

A: The Tablets meet the requirements under *Identification—Organic Nitrogenous Bases* ⟨181⟩.

B: The retention time of the major peak in the chromatogram of the *Assay preparation* corresponds to that of the *Standard preparation*, as obtained in the *Assay*.

Dissolution ⟨711⟩—

Medium: water; 900 mL.

Apparatus 2: 50 rpm.

Time: 45 minutes.

Procedure—Determine the amount of $C_{13}H_{19}NO.HCl$ dissolved from ultraviolet absorbances at the wavelength of maximum absorbance at about 253 nm of filtered portions of the solution under test, suitably diluted with 0.1 *N* hydrochloric acid, in comparison with a Standard solution having a known concentration of USP Diethylpropion Hydrochloride RS in the same medium.

Tolerances—Not less than 75% (*Q*) of the labeled amount of $C_{13}H_{19}NO.HCl$ is dissolved in 45 minutes.

Uniformity of dosage units ⟨905⟩: meet the requirements.

Assay—

Phosphate buffer, *Mobile phase*, and *Chromatographic system*—Prepare as directed in the *Assay* under *Diethylpropion Hydrochloride*.

Standard preparation—Dissolve an accurately weighed quantity of USP Diethylpropion Hydrochloride RS in 0.1 *N* hydrochloric acid, and dilute quantitatively, and stepwise if necessary, with 0.1 *N* hydrochloric acid to obtain a stock solution having a known concentration of about 160 µg per mL. Transfer 5.0 mL of this stock solution to a 100-mL volumetric flask, dilute with *Mobile phase* to volume, and mix to obtain a solution having a known concentration of about 8 µg per mL.

Assay preparation—Weigh and finely powder not less than 20 Diethylpropion Hydrochloride Tablets. Transfer an accurately weighed portion of the powder, equivalent to about 40 mg of diethylpropion hydrochloride, to a 250-mL volumetric flask. Add 200 mL of 0.1 *N* hydrochloric acid, and stir with the aid of a stir bar for 45 minutes. Remove the stir bar, dilute with 0.1 *N* hydrochloric acid to volume, mix, and filter, discarding the first 25 mL of the filtrate. Transfer 5.0 mL of the filtrate to a 100-mL volumetric flask, dilute with *Mobile phase* to volume, and mix. If necessary, filter the solution through a 0.7-µm porosity membrane filter.

System suitability preparation—Dissolve benzoic acid in 0.1 *N* hydrochloric acid to obtain a solution having a concentration of about 1 mg per mL. Transfer 10.0 mL of this solution to a 100-mL volumetric flask, add 5.0 mL of the stock solution prepared as directed for the *Standard preparation*, dilute with *Mobile phase* to volume, and mix.

Procedure—Proceed as directed for *Procedure* in the *Assay* under *Diethylpropion Hydrochloride*. Calculate the quantity, in mg, of $C_{13}H_{19}NO.HCl$ in the portion of Tablets taken by the formula:

$$5C(r_U / r_S),$$

in which *C* is the concentration, in µg per mL, of USP Diethylpropion Hydrochloride RS in the *Standard preparation*, and r_U and r_S are the peak responses obtained from the *Assay preparation* and the *Standard preparation*, respectively.

Diethylstilbestrol

$C_{18}H_{20}O_2$ 268.35

Phenol 4,4′-(1,2-diethyl-1,2-ethenediyl)bis-, (*E*)-.

α,α′-Diethyl-(*E*)-4,4′-stilbenediol [*56-53-1*].

» Diethylstilbestrol contains not less than 97.0 percent and not more than 100.5 percent of $C_{18}H_{20}O_2$, calculated on the dried basis.

Packaging and storage—Preserve in tight, light-resistant containers.

Reference standard—*USP Diethylstilbestrol Reference Standard*—Dry at 105° for 2 hours before using.

Identification—

A: Dilute the alcohol solutions used to prepare the *Standard preparation* and the *Assay preparation*, in the *Assay*, with alcohol to contain 10 µg of USP Diethylstilbestrol RS and diethylstilbestrol, respectively, in each mL. Determine the absorbances of each solution in the range from 230 nm to 350 nm, using alcohol as the blank. The spectrum of Diethylstilbestrol exhibits a maximum and an additional inflection at the same wavelengths as that of the solution of USP Diethylstilbestrol RS, concomitantly measured, and the absorptivity of Diethylstilbestrol at the wavelength of maximum absorbance does not differ from that of the Reference Standard by more than 3.0%.

B: The absorption spectrum, in the range of 250 nm to 450 nm, of the yellow solution obtained in the *Assay* after irradiation of the *Assay preparation* exhibits inflections only at the same wavelengths as that of the solution obtained after irradiation of the *Standard preparation*.

Melting range ⟨741⟩: between 169° and 175°, but the range between beginning and end of melting does not exceed 4°.

Acidity or alkalinity—A solution of 100 mg in 5 mL of neutralized 70 percent alcohol is neutral to litmus.

Loss on drying ⟨731⟩—Dry it at 105° for 2 hours: it loses not more than 0.5% of its weight.

Residue on ignition ⟨281⟩: not more than 0.05%.

Assay—

Standard preparation—Dissolve in alcohol a suitable quantity of USP Diethylstilbestrol RS, accurately weighed, and prepare, by stepwise dilution with alcohol, a solution containing about 20 µg per mL. Mix 25.0 mL of this solution with an equal volume of dibasic potassium phosphate solution (1 in 55).

Assay preparation—Proceed with a suitable quantity, accurately weighed, of Diethylstilbestrol as directed under *Standard preparation*.

Procedure—[*Caution—Protect the eyes from direct rays of ultraviolet light throughout this procedure.*] Transfer 4 mL of the *Standard preparation* to a stoppered, 1-cm quartz cell, place about 5 cm from a low-pressure, short-wave mercury lamp, rated at from 2 to 20 watts, and irradiate for about 5 minutes. Place the cell in the sample compartment of a suitable spectrophotometer, and measure the absorbance at the wavelength of maximum absorbance at about 418 nm, using water as the blank. Continue irradiation for successive 1- to 3-minute intervals, measuring at 418 nm until the maximum absorbance (about 0.7) has been obtained. If necessary, adjust the geometry of the irradiation apparatus so as to obtain maximum, reproducible absorbance at 418 nm. Similarly, irradiate a 4-mL portion of the *Assay preparation*, recording the absorbance at 418 nm, at successive short intervals until maximum absorbance is obtained. Concomitantly determine the absorbances of the *Assay preparation* and the *Standard preparation* in 1-cm cells at 418 nm, using water as the blank, and subtract these values from those for the respective irradiated solutions, to obtain the corrected maximum absorbances. Calculate the quantity, in µg, of $C_{18}H_{20}O_2$ in each mL of the *Assay preparation* taken by the formula:

$$C(A_U / A_S),$$

in which *C* is the concentration, in µg per mL, of USP Diethylstilbestrol RS in the *Standard preparation*, and A_U and A_S are the corrected maximum absorbances of the irradiated *Assay preparation* and *Standard preparation*, respectively.

Diethylstilbestrol Injection

» Diethylstilbestrol Injection is a sterile solution of Diethylstilbestrol in a suitable vegetable oil. It contains not less than 90.0 percent and not more than 110.0 percent of the labeled amount of $C_{18}H_{20}O_2$.

Packaging and storage—Preserve in light-resistant, single-dose or multiple-dose containers, preferably of Type I glass.

Reference standard—*USP Diethylstilbestrol Reference Standard*—Dry at 105° for 2 hours before using.

Identification—It responds to *Identification test B* under *Diethylstilbestrol*.

Other requirements—It meets the requirements under *Injections* ⟨1⟩.

Assay—Transfer an accurately measured volume of Diethylstilbestrol Injection, equivalent to not less than 1 mg and not more than 50 mg of diethylstilbestrol, to a 125-mL separator containing 75 mL of isooctane. Extract the solution with one 20-mL and two 10-mL portions of 1 *N* sodium hydroxide. Wash the combined alkaline extracts with two 10-mL portions of chloroform. Transfer the alkaline solution to a 150-mL beaker, and with the careful addition of 2.5 *M* phosphoric acid adjust the pH of the solution to 9.5. Transfer the adjusted solution, with the aid of a small volume of water, to a 125-mL separator, and extract with four 20-mL portions of chloroform. Filter the extracts into a 100-mL volumetric flask through a chloroform-wetted pledget of cotton, washing the filter with several small portions of chloroform to adjust the solution to volume. If necessary, dilute the solution quantitatively and stepwise with chloroform so that it contains about 10 µg of diethylstilbestrol per mL. Transfer 20 mL of the solution to a 50-mL conical flask, and evaporate with the aid of gentle heating in a current of air to about 5 mL. Complete the evaporation of the solvent in the air current without further application of heat. Dissolve the residue in 10.0 mL of alcohol, add 10.0 mL of dibasic potassium phosphate solution (1 in 55), and mix. Using this clear solution as the *Assay preparation*, proceed as directed in the *Assay* under *Diethylstilbestrol.*

Diethylstilbestrol Tablets

» Diethylstilbestrol Tablets contain not less than 90.0 percent and not more than 110.0 percent of the labeled amount of $C_{18}H_{20}O_2$.

Packaging and storage—Preserve in well-closed containers.

Reference standard—*USP Diethylstilbestrol Reference Standard*—Dry at 105° for 2 hours before using.

Identification—The Tablets respond to *Identification test B* under *Diethylstilbestrol.*

Disintegration ⟨701⟩: 30 minutes.

Uniformity of dosage units ⟨905⟩: meet the requirements.

Procedure for content uniformity—[NOTE—Use the same lot of glacial acetic acid throughout the procedure.] Place 1 Tablet, previously crushed or finely powdered, in a volumetric flask of such size that a solution of the Tablet, or a quantitative and stepwise dilution thereof, contains about 10 µg of diethylstilbestrol per mL. Add to the flask a volume of water, accurately measured and equivalent to 20% of the total volume of the flask. Shake to disintegrate the tablet, and digest on a steam bath for 15 minutes. Cool, add a volume of glacial acetic acid equivalent to 50% of the total volume of the flask, and shake, by mechanical means, for 15 minutes. Dilute with glacial acetic acid to volume, and filter, discarding the first few mL of the filtrate. Dilute a portion of the subsequent filtrate quantitatively and stepwise, if necessary, with 14 *N* acetic acid to obtain a solution having a concentration of about 10 µg of diethylstilbestrol per mL (this is the *Assay preparation*). Dissolve a suitable quantity of USP Diethylstilbestrol RS, accurately weighed, in 14 *N* acetic acid, and dilute quantitatively and stepwise with the same solvent to obtain a *Standard preparation* having a known concentration of about 10 µg per mL. Proceed with the irradiation as directed for *Procedure* in the *Assay* under *Diethylstilbestrol.* Concomitantly determine the absorbance of each solution in a 1-cm cell at the wavelength of maximum absorbance at about 418 nm, with a suitable spectrophotometer, using water as the blank. Calculate the quantity, in mg, of $C_{18}H_{20}O_2$ in the Tablet by the formula:

$$(T/10)C(A_U/A_S),$$

in which T is the labeled quantity, in mg, of diethylstilbestrol in the Tablet, C is the concentration, in µg per mL, of USP Diethylstilbestrol RS in the *Standard preparation*, and A_U and A_S are the corrected maximum absorbances of the irradiated *Assay preparation* and *Standard preparation*, respectively.

Assay—Weigh and finely powder not less than 20 Diethylstilbestrol Tablets. Weigh accurately not less than 200 mg of the powder, equivalent to not less than 0.5 mg of diethylstilbestrol, and transfer to a small beaker. Add 10 mL of alcohol, cover the beaker with a watch glass, heat carefully to gentle simmering, and allow to simmer for 10 minutes, with frequent stirring. Cool, transfer the mixture to a separator, using 30 mL of chloroform and 20 mL of water, add 2 mL of 2 *N* sulfuric acid, and shake vigorously. Remove the clearly defined chloroform layer to a second separator, wash with 20 mL of water, and filter through a pledget of cotton moistened with chloroform into a 100-mL volumetric flask. Repeat the extraction in the two separators with three 20-mL portions of chloroform, filtering each chloroform extract into the volumetric flask, add chloroform to volume, and mix. If necessary, dilute this solution quantitatively and stepwise with chloroform to obtain a solution containing not more than 40 µg of diethylstilbestrol per mL. Transfer a portion of this solution, equivalent to about 200 µg of diethylstilbestrol, to a 50-mL conical flask, and evaporate with the aid of gentle heating in a current of air to about 5 mL. Complete the evaporation of the solvent in the air current without further application of heat. Dissolve the residue in 10.0 mL of alcohol, add 10.0 mL of dibasic potassium phosphate solution (1 in 55), and mix. Using this clear solution as the *Assay preparation*, proceed as directed in the *Assay* under *Diethylstilbestrol.*

Diethylstilbestrol Diphosphate

$C_{18}H_{22}O_8P_2$ 428.31

Phenol, 4,4′-(1,2-diethyl-1,2-ethenediyl)bis-, bis(dihydrogen phosphate), (*E*)-.

α,α′-Diethyl-(*E*)-4,4′-stilbenediol bis(dihydrogen phosphate) [*522-40-7*].

» Diethylstilbestrol Diphosphate contains not less than 95.0 percent and not more than 101.0 percent of $C_{18}H_{22}O_8P_2$, calculated on the dried basis.

Packaging and storage—Preserve in tight containers, at a temperature not exceeding 21°.

Reference standard—*USP Diethylstilbestrol Diphosphate Reference Standard*—Dry at 105° for 4 hours before using.

Identification—

A: The infrared absorption spectrum of a potassium bromide dispersion of it exhibits maxima only at the same wavelengths as that of a similar preparation of USP Diethylstilbestrol Diphosphate RS.

B: Dissolve about 15 mg in 1 mL of sulfuric acid: an orange color is produced, and it disappears on dilution with about 10 volumes of water.

Loss on drying ⟨731⟩—Dry it at 105° for 4 hours: it loses not more than 1.0% of its weight.

Chloride—Weigh accurately about 2.0 g into a conical flask, dissolve in 25 mL of alcohol, add 4 mL of nitric acid and 10 mL of water, then add 20 mL of silver nitrate TS, with stirring, and heat the mixture on a steam bath until the precipitate has coagulated. Transfer the precipitate to a tared filtering crucible, wash with three 10-mL portions of alcohol, dry at 105° for 30 minutes, cool in a desiccator, protected from light, and weigh: the weight of the precipitate is not more than 120 mg, corresponding to not more than 1.5% of chloride.

Free diethylstilbestrol—

Vanadyl sulfate solution—Dissolve about 100 mg of vanadyl sulfate in 20 mL of sulfuric acid in a 200-mL volumetric flask, cautiously add glacial acetic acid to volume, and mix.

Standard diethylstilbestrol solution—Dissolve about 10 mg of USP Diethylstilbestrol RS, accurately weighed, in diluted alcohol, and dilute with diluted alcohol to 100.0 mL.

Procedure—Transfer about 500 mg of Diethylstilbestrol Diphosphate, accurately weighed, to a beaker containing about 20 mL of water. Add 1 *N* sodium hydroxide to adjust to a pH of 8.0, determined potentiometrically. Transfer the solution to a separator, and extract with four 20-mL portions of peroxide-free ether. Evaporate the combined ether extracts on a steam bath to about 50 mL, transfer to a glass-stoppered, 50-mL conical flask, and evaporate to dryness. Transfer 5 mL of *Standard diethylstilbestrol solution* to a 50-mL conical flask, and evaporate on a steam bath to dryness. Pipet 10 mL of *Vanadyl sulfate solution* into each of the flasks containing the residues from the test solution and the *Standard diethylstilbestrol solution*, and into a third dry flask, to provide the blank. Heat the loosely stoppered flasks on a steam bath for 1 hour, allow the contents to cool, and determine the absorbances of the solutions from the test preparation and the Standard preparation, relative to that of the blank, in 1-cm cells at the wavelength of maximum absorbance at about 520 nm. Calculate the quantity, in mg, of free diethylstilbestrol in the portion of Diethylstilbestrol Diphosphate taken by the formula:

$$0.005C(A_U/A_S),$$

in which C is the concentration, in µg per mL, of USP Diethylstilbestrol RS in the *Standard diethylstilbestrol solution*, and A_U and A_S are the absorbances of the solution from Diethylstilbestrol Diphosphate and the Standard preparation, respectively. The limit of free diethylstilbestrol is 0.15%.

Diethylstilbestrol monophosphate—Transfer about 50 mg of Diethylstilbestrol Diphosphate, accurately weighed, to a beaker containing about 20 mL of water. Add 1 *N* sodium hydroxide or 1 *N* phosphoric acid to adjust to a pH of about 2.2, determined potentiometrically. Proceed as directed for *Procedure* in the test for *Free diethylstilbestrol*, beginning with "Transfer the solution to a separator." Calculate the total quantity, T, in mg, of free diethylstilbestrol and diethylstilbestrol monophosphate, as diethylstilbestrol, in the portion of Diethylstilbestrol Diphosphate taken by the formula:

$$0.005C(A_U/A_S),$$

in which the terms are as defined under *Procedure*. Calculate the quantity, in mg, of diethylstilbestrol monophosphate in the portion of Diethylstilbestrol Diphosphate taken by the formula:

$$1.298(T - F),$$

in which T is as defined above, and F is the content, in mg, of free diethylstilbestrol in the portion of Diethylstilbestrol Diphosphate taken, calculated from the result of the test for *Free diethylstilbestrol*. The limit of diethylstilbestrol monophosphate is 1.5%.

Pyridine—

Cyanogen bromide reagent—Dissolve 10 g of cyanogen bromide in water to make 100.0 mL.

Buffer solution—Transfer to a 1000-mL volumetric flask 87 g of dibasic potassium phosphate, 107 g of ammonium chloride, and 6.7 mL of ammonium hydroxide, dilute with water to volume, and mix.

Procedure—Transfer about 100 mg of Diethylstilbestrol Diphosphate, accurately weighed, to a 50-mL volumetric flask, add 25 mL of sodium hydroxide solution (1 in 500), shake until dissolved, add water to volume, and mix, to obtain the *Test preparation*. Similarly prepare a *Standard preparation* containing 10 µg of pyridine per mL of sodium hydroxide solution (1 in 1000). Pipet 2 mL of the *Test preparation* and of the *Standard preparation* into separate glass-stoppered, 10-mL centrifuge tubes, add 3.0 mL of *Buffer solution* and 5.0 mL of *Cyanogen bromide reagent* to each tube, and mix. At 2 minutes, accurately timed, after the addition of *Cyanogen bromide reagent* measure the absorbance of each mixture relative to that of water at about 420 nm, with a suitable spectrophotometer. The absorbance of the *Test preparation* is not greater than that of the *Standard preparation* (0.5%).

Assay—Transfer about 100 mg of Diethylstilbestrol Diphosphate, accurately weighed, to a 250-mL volumetric flask, add 25 mL of alcohol and 5 mL of 1 *N* sodium hydroxide, dilute with water to volume, and mix. Pipet 5 mL of this solution into a 100-mL volumetric flask, dilute with water to volume, and mix. Similarly prepare a Standard solution of USP Diethylstilbestrol Diphosphate RS having a known concentration of about 20 µg per mL. Concomitantly determine the absorbances of both solutions in 1-cm cells at the wavelength of maximum absorbance at about 241 nm, with a suitable spectrophotometer. Calculate the quantity, in mg, of $C_{18}H_{22}O_8P_2$ in the portion of Diethylstilbestrol Diphosphate taken by the formula:

$$5C(A_U/A_S),$$

in which C is the concentration, in µg per mL, of USP Diethylstilbestrol Diphosphate RS in the Standard solution, and A_U and A_S are the absorbances of the solution of Diethylstilbestrol Diphosphate and the Standard solution, respectively.

Diethylstilbestrol Diphosphate Injection

» Diethylstilbestrol Diphosphate Injection is a sterile, buffered solution of Diethylstilbestrol Diphosphate. It contains not less than 45.0 mg and not more than 55.0 mg of $C_{18}H_{22}O_8P_2$ in each mL.

Packaging and storage—Preserve in single-dose or in multiple-dose containers.

Reference standard—*USP Diethylstilbestrol Diphosphate Reference Standard*—Dry at 105° for 4 hours before using.

Identification—Add 1 mL of sulfuric acid to about 0.25 mL of Injection: an orange color is produced, and it disappears on dilution with 10 volumes of water.

pH ⟨791⟩: between 9.0 and 10.5.

Free diethylstilbestrol—Transfer 2.0 mL of the Injection to a beaker containing about 20 mL of water. Add 1 *N* phosphoric acid to adjust to a pH of 8.0, determined potentiometrically. Proceed as directed for *Procedure* in the test for *Free diethylstilbestrol* under *Diethylstilbestrol Diphosphate*, beginning with "Transfer the solution to a separator." The limit of free diethylstilbestrol is 0.2 mg per mL of Injection.

Diethylstilbestrol monophosphate—Transfer 3.0 mL of Injection to a 10-mL volumetric flask, dilute with water to volume, and mix. Transfer 1 mL of this solution to a beaker containing about 20 mL of water, and proceed as directed in the test for *Diethylstilbestrol monophosphate* under *Diethylstilbestrol Diphosphate*, beginning with "Add 1 *N* sodium hydroxide or 1 *N* phosphoric acid." The limit of diethylstilbestrol monophosphate is 2.0 mg per mL of Injection.

Other requirements—It meets the requirements under *Injections* ⟨1⟩.

Assay—Transfer an accurately measured volume of Diethylstilbestrol Diphosphate Injection, equivalent to about 150 mg of diethylstilbestrol diphosphate, to a 200-mL volumetric flask, add water to volume, and mix. Transfer 5.0 mL of this solution to a 250-mL volumetric flask, and dilute with water to volume, and mix. Concomitantly determine the absorbances of this solution and a Standard solution of USP Diethylstilbestrol Diphosphate RS having a known concentration of about 15 µg per mL, and containing 1 mL of 0.02 *N* sodium hydroxide per 100 mL of solution, in 1-cm cells at the wavelength of maximum absorbance at about 241 nm, with a suitable spectrophotometer, using water containing 0.5 mL of 0.02 *N* sodium hydroxide per 100 mL, as the blank. Calculate the quantity, in mg, of $C_{18}H_{22}O_8P_2$ in each mL of the Injection taken by the formula:

$$3.33C(A_U/A_S),$$

in which C is the concentration, in µg per mL, of USP Diethylstilbestrol Diphosphate RS in the Standard solution, and A_U and A_S are the absorbances of the solution from the Injection and the Standard solution, respectively.

Diethyltoluamide

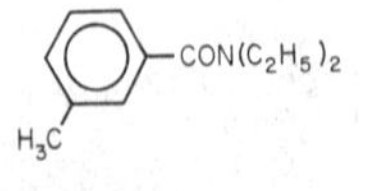

$C_{12}H_{17}NO$ 191.27

Benzamide, *N,N*-diethyl-3-methyl-.

N,N-Diethyl-*m*-toluamide [*134-62-3*].

» Diethyltoluamide contains not less than 95.0 percent and not more than 103.0 percent of the *meta*-isomer of $C_{12}H_{17}NO$, calculated on the anhydrous basis.

Packaging and storage—Preserve in tight containers.

Reference standard—*USP Diethyltoluamide Reference Standard*—Do not dry; determine the water content at the time of use.

Identification—The infrared absorption spectrum of the solution employed for measurement of absorbance in the *Assay*, in the spectral region between 8 and 15 µm, exhibits maxima only at the same wavelengths as that of a solution of USP Diethyltoluamide RS, prepared as directed in the *Assay*.

Specific gravity ⟨841⟩: between 0.996 and 1.002.

Refractive index ⟨831⟩: between 1.520 and 1.524.

Acidity—Dissolve 10.0 g in 50 mL of neutralized alcohol, add phenolphthalein TS, and titrate with 0.010 *N* sodium hydroxide: not more than 4.0 mL of 0.010 *N* sodium hydroxide is required.

Water, *Method I* ⟨921⟩: not more than 0.5%.

Assay—Transfer about 200 mg of Diethyltoluamide, accurately weighed, to a 10-mL volumetric flask, dilute with carbon disulfide to volume, and mix to obtain the *Assay preparation.* Concomitantly determine the absorbances of the *Assay preparation* and of a *Standard preparation* of USP Diethyltoluamide RS in carbon disulfide having a known concentration of about 20 mg per mL in 1-mm cells at the wavelength of maximum absorbance at about 14.1 µm and at the wavelength of minimum absorbance at about 14.4 µm, with a suitable infrared spectrophotometer, using carbon disulfide as the blank. Calculate the quantity, in mg, of the *meta*-isomer of $C_{12}H_{17}NO$ in the Diethyltoluamide taken by the formula:

$$10C(A_{U14.1} - A_{U14.4})/(A_{S14.1} - A_{S14.4}),$$

in which *C* is the concentration, in mg per mL, of USP Diethyltoluamide RS in the *Standard preparation,* and A_U and A_S are the absorbances of the *Assay preparation* and the *Standard preparation*, respectively, at the wavelengths indicated by the subscripts.

Diethyltoluamide Topical Solution

» Diethyltoluamide Topical Solution is a solution of Diethyltoluamide in Alcohol or Isopropyl Alcohol. It contains not less than 92.0 percent and not more than 108.0 percent of the labeled amount of the meta isomer of $C_{12}H_{17}NO$.

If it contains Alcohol, not less than 95.0 percent and not more than 105.0 percent of the labeled amount of C_2H_5OH is present.

Packaging and storage—Preserve in tight containers.

Reference standard—*USP Diethyltoluamide Reference Standard*—Do not dry; determine the water content at the time of use.

Identification—It responds to the *Identification test* under *Diethyltoluamide.*

Alcohol content (if present) ⟨611⟩: between 29.0% and 89.0% of C_2H_5OH.

Assay—Transfer an accurately weighed quantity of Diethyltoluamide Topical Solution, equivalent to about 200 mg of diethyltoluamide, to a beaker. Place the beaker in a vacuum oven containing silica gel and calcium chloride, and adjusted to a pressure of about 380 mm of mercury, and heat at 35° for 6 hours. Transfer the residue with the aid of carbon disulfide to a 10-mL volumetric flask, and carbon disulfide to volume, and mix. Transfer to a 10-mL volumetric flask about 200 mg of USP Diethyltoluamide RS, accurately weighed, add carbon disulfide to volume, and mix, to obtain the Standard solution. Concomitantly determine the absorbances of both solutions in 1-mm cells at the wavelength of maximum absorbance at about 14.1 µm and at the wavelength of minimum absorbance at about 14.4 µm, with a suitable infrared spectrophotometer, using carbon disulfide as the blank. Calculate the quantity, in mg, of the *meta*-isomer of $C_{12}H_{17}NO$ in the portion of Topical Solution taken by the formula:

$$10C(A_{U14.1} - A_{U14.4})/(A_{S14.1} - A_{S14.4}),$$

in which *C* is the concentration, in mg per mL, of USP Diethyltoluamide RS in the Standard solution, and A_U and A_S are the absorbances of the solution from Diethyltoluamide Solution and the Standard solution, respectively, at the wavelengths indicated by the subscripts.

Diflorasone Diacetate

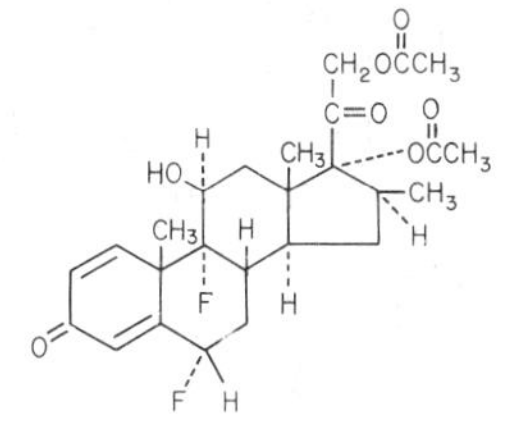

$C_{26}H_{32}F_2O_7$ 494.53

Pregna-1,4-diene-3,20-dione,17,21-bis(acetyloxy)-6,9-difluoro-11-hydroxy-16-methyl-, (6α,11β,16β)-.

6α,9-Difluoro-11β,17,21-trihydroxy-16β-methylpregna-1,4-diene-3,20-dione 17,21-diacetate [*33564-31-7*].

» Diflorasone Diacetate contains not less than 97.0 percent and not more than 103.0 percent of $C_{26}H_{32}F_2O_7$, calculated on the dried basis.

Packaging and storage—Preserve in tight containers.

Reference standard—*USP Diflorasone Diacetate Reference Standard*—Dry in vacuum at 60° and at a pressure not exceeding 5 mm of mercury for 16 hours before using.

Identification—The infrared absorption spectrum of a mineral oil dispersion of it exhibits maxima only at the same wavelengths as that of a similar preparation of USP Diflorasone Diacetate RS.

Specific rotation ⟨781⟩: between +58° and +68°, determined in a solution containing 200 mg in each 10 mL of chloroform.

Loss on drying ⟨731⟩—Dry it in vacuum at 60° and at a pressure not exceeding 5 mm of mercury for 16 hours: it loses not more than 0.5% of its weight.

Residue on ignition ⟨281⟩: not more than 0.5%.

Assay—

Mobile phase—Prepare a solution containing a mixture of water-saturated *n*-butyl chloride, water-saturated methylene chloride, glacial acetic acid, and tetrahydrofuran (350:125:15:10). Make adjustments if necessary (see *System Suitability* under *Chromatography* ⟨621⟩).

Internal standard solution—Using water-saturated chloroform, prepare a solution of isoflupredone acetate containing about 0.04 mg per mL.

Standard preparation—Dissolve an accurately weighed quantity of USP Diflorasone Diacetate RS in *Internal standard solution* to obtain a solution having a known concentration of about 33 µg per mL.

Assay preparation—Transfer about 15 mg of Diflorasone Diacetate, accurately weighed, to a 500-mL volumetric flask. Add *Internal standard solution* to volume, and mix.

Chromatographic system (see *Chromatography* ⟨621⟩)—The liquid chromatograph is equipped with a 254-nm detector and a 4.6-mm × 10-cm column that contains 3-µm packing L3. The flow rate is about 2.5 mL per minute. Chromatograph the *Standard preparation*, and record the peak areas as directed under *Procedure:* the resolution, *R*, between the analyte and internal standard peaks is not less than 12, and the relative standard deviation for not less than four replicate injections is not more than 2.0%.

Procedure—Separately inject equal volumes (about 10 µL) of the *Standard preparation* and the *Assay preparation* into the chromatograph, record the chromatograms, and measure the areas for the major peaks. The relative retention times are about 1.0 for diflorasone diacetate and 2.4 for the internal standard. Calculate the quantity, in mg, of $C_{26}H_{32}F_2O_7$ in the portion of diflorasone diacetate taken by the formula:

$$0.5C(R_U/R_S),$$

in which C is the concentration, in µg per mL, of USP Diflorasone Diacetate RS in the *Standard preparation*, and R_U and R_S are the ratios of the peak areas for diflorasone diacetate and the internal standard areas obtained from the *Assay preparation* and the *Standard preparation*, respectively.

Diflorasone Diacetate Cream

» Diflorasone Diacetate Cream contains not less than 90.0 percent and not more than 110.0 percent of the labeled amount of $C_{26}H_{32}F_2O_7$.

Packaging and storage—Preserve in collapsible tubes, preferably at controlled room temperature.

Reference standard—*USP Diflorasone Diacetate Reference Standard*—Dry in vacuum at 60° and at a pressure not exceeding 5 mm of mercury for 16 hours before using.

Identification—The chromatogram of the *Assay preparation* exhibits a major peak for diflorasone diacetate at a retention time that corresponds to that exhibited in the chromatogram of the *Standard preparation* obtained in the *Assay*.

Microbial limits—It meets the requirements of the tests for absence of *Staphylococcus aureus* and *Pseudomonas aeruginosa* under *Microbial Limit Tests* ⟨61⟩.

Minimum fill ⟨755⟩—meets the requirements.

Assay—

Mobile phase, Internal standard solution, Standard preparation, and *Chromatographic system*—Proceed as directed in the *Assay* under *Diflorasone Diacetate*.

Assay preparation—Transfer an accurately weighed amount of Diflorasone Diacetate Cream, equivalent to about 1 mg of diflorasone diacetate, to a suitable container. Add 30.0 mL of *Internal standard solution*, and shake for about 30 minutes. Centrifuge the solution, and remove and discard the top (excipient) layer. Use the lower, clear chloroform layer.

Procedure—Proceed as directed for *Procedure* in the *Assay* under *Diflorasone Diacetate*. Calculate the quantity, in mg, of $C_{26}H_{32}F_2O_7$ in the portion of Cream taken by the formula:

$$0.03C(R_U/R_S),$$

in which the terms are as defined therein.

Diflorasone Diacetate Ointment

» Diflorasone Diacetate Ointment contains not less than 90.0 percent and not more than 110.0 percent of the labeled amount of $C_{26}H_{32}F_2O_7$.

Packaging and storage—Preserve in collapsible tubes, preferably at controlled room temperature.

Reference standard—*USP Diflorasone Diacetate Reference Standard*—Dry in vacuum at 60° and at a pressure not exceeding 5 mm of mercury for 16 hours before using.

Identification—It responds to the *Identification test* under *Diflorasone Diacetate Cream*.

Microbial limits—Proceed with Diflorasone Diacetate Ointment as directed for *Microbial limits* under *Diflorasone Diacetate Cream*.

Minimum fill ⟨755⟩: meets the requirements.

Assay—

Mobile phase, Internal standard solution, Standard preparation, and *Chromatographic system*—Proceed as directed in the *Assay* under *Diflorasone Diacetate*.

Assay preparation—Using Diflorasone Diacetate Ointment instead of the Cream, proceed as directed for *Assay preparation* in the *Assay* under *Diflorasone Diacetate Cream*.

Procedure—Proceed as directed for *Procedure* in the *Assay* under *Diflorasone Diacetate*. Calculate the quantity, in mg, of $C_{26}H_{32}F_2O_7$ in the portion of Ointment taken by the formula:

$$0.03C(R_U/R_S),$$

in which the terms are as defined therein.

Diflunisal

COOH, F, OH, F

$C_{13}H_8F_2O_3$ 250.20

[1,1′-Biphenyl]-3-carboxylic acid, 2′,4′-difluoro-4-hydroxy-.
2′,4′-Difluoro-4-hydroxy-3-biphenylcarboxylic acid [*22494-42-4*].

» Diflunisal contains not less than 98.0 percent and not more than 101.5 percent of $C_{13}H_8F_2O_3$, calculated on the dried basis.

Packaging and storage—Preserve in well-closed containers.

Reference standard—*USP Diflunisal Reference Standard*—Dry in vacuum at 60° and at a pressure not exceeding 5 mm of mercury for 4 hours before using.

Identification—

A: The infrared absorption spectrum of a mineral oil dispersion of it exhibits maxima only at the same wavelengths as that of a similar preparation of USP Diflunisal RS.

B: The ultraviolet absorption spectrum of a 1 in 25,000 solution in 0.1 *N* methanolic hydrochloric acid exhibits maxima and minima at the same wavelengths as that of a similar solution of USP Diflunisal RS, concomitantly measured, and the respective absorptivities, calculated on the dried basis, at the wavelength of maximum absorbance at about 315 nm do not differ by more than 3.0%.

Loss on drying ⟨731⟩—Dry it in vacuum at a pressure not exceeding 5 mm of mercury at 60° for 4 hours: it loses not more than 0.3% of its weight.

Residue on ignition ⟨281⟩: not more than 0.1%.

Heavy metals, *Method II* ⟨231⟩: 0.001%.

Chromatographic purity—Prepare a solution of it in methanol containing about 10 mg per mL. Prepare solutions of USP Diflunisal RS in methanol having concentrations of 10 mg, 0.05 mg, and 0.02 mg per mL, respectively (*Standard solutions A, B,* and *C*). Apply 5-µL portions of all four solutions on a suitable thin-layer chromatographic plate (see *Chromatography* ⟨621⟩), coated with a 0.25-mm layer of chromatographic silica gel mixture and previously washed with methanol. Allow the spots to dry, and develop the chromatogram in a freshly prepared solvent system consisting of a mixture of *n*-hexane, dioxane, and glacial acetic acid (85:10:5) in a paper-lined, equilibrated tank, until the solvent front has moved about three-fourths of the length of the plate. Remove the plate from the developing chamber, mark the solvent front, allow to air-dry, and examine the plate under short-wavelength ultraviolet light: the chromatograms show principal spots at about the same R_f value. Estimate the concentration of any spot observed in the chromatogram of the test solution, other than the principal spot, by comparison with the spots in the chromatograms of *Standard solutions B* and *C*: the intensity of any individual spot is not greater than that of the principal spot obtained from *Standard solution C* (0.2%), and the sum of all additional spots is not greater than that of the principal spot obtained from *Standard solution B* (0.5%).

Assay—

Mobile phase—Prepare a suitable mixture of water, methanol, acetonitrile, and glacial acetic acid (55:23:10:2) such that the retention time of diflunisal is about 18 minutes.

Standard preparation—Dissolve an accurately weighed quantity of USP Diflunisal RS in a mixture of acetonitrile and water (4:1) to obtain a solution having a known concentration of about 1 mg per mL. Dilute an accurately measured volume of this solution with a mixture of acetonitrile and water (1:1) to obtain a solution having a known concentration of about 0.2 mg per mL.

Assay preparation—Transfer about 50 mg of Diflunisal, accurately weighed, to a 50-mL volumetric flask. Dilute with a mixture of acetonitrile and water (4:1) to volume, and mix. Transfer 5.0 mL of this solution to a 25-mL volumetric flask. Dilute with a mixture of acetonitrile and water (1:1) to volume, and mix.

Chromatographic system (see *Chromatography* ⟨621⟩)—The liquid chromatograph is equipped with a 254-nm detector and a 3.9-mm × 30-cm column that contains packing L1 and is maintained at a temperature of 40°. The flow rate is about 1.5 mL per minute. Chromatograph the *Standard preparation*, and record the peak responses as directed under *Procedure:* the column efficiency determined from the analyte peak is not less than 2500 theoretical plates, the tailing factor is not more than 2.0, the capacity factor is not less than 7.2, and the relative standard deviation for replicate injections is not more than 1%.

Procedure—Separately inject equal volumes (about 10 µL) of the *Standard preparation* and the *Assay preparation* into the chromatograph, record the chromatograms, and measure the responses for the major peaks. Calculate the quantity, in mg, of $C_{13}H_8F_2O_3$ in the portion of Diflunisal taken by the formula:

$$250C(r_U / r_S),$$

in which C is the concentration, in mg per mL, of USP Diflunisal RS in the *Standard preparation*, and r_U and r_S are the peak responses of the major peaks obtained from the *Assay preparation* and the *Standard preparation*, respectively.

Diflunisal Tablets

» Diflunisal Tablets contain not less than 90.0 percent and not more than 110.0 percent of the labeled amount of $C_{13}H_8F_2O_3$.

Packaging and storage—Preserve in well-closed containers.

Reference standard—*USP Diflunisal Reference Standard*—Dry in vacuum at a pressure not exceeding 5 mm of mercury at 60° for 4 hours before using.

Identification—

A: The retention time of the major peak in the chromatogram of the *Assay preparation* corresponds to that of the *Standard preparation*, obtained as directed in the *Assay*.

B: Transfer a quantity of finely ground Tablets, equivalent to about 100 mg of diflunisal, to a 10-mL volumetric flask, add 2 mL of water, and sonicate for 5 minutes. Dilute with methanol to volume, sonicate for an additional 5 minutes, mix, and filter. Separately apply 10 µL each of the filtrate and a Standard solution of USP Diflunisal RS in methanol solution (4 in 5) containing 10 mg per mL to a thin-layer chromatographic plate (see *Chromatography* ⟨621⟩) coated with a 0.25-mm layer of chromatographic silica gel mixture. Develop the chromatogram in a solvent system consisting of *n*-hexane, glacial acetic acid, and chloroform (17:3:2) until the solvent front has moved about three-fourths of the length of the plate. Remove the plate from the chamber, air-dry, and examine under long-wavelength ultraviolet light: the R_f value of the principal spot in the chromatogram of the test solution corresponds to that obtained from the Standard solution.

Dissolution ⟨711⟩—

pH 7.20, 0.1 M Tris buffer—Dissolve 121 g of tris(hydroxymethyl)aminomethane (THAM) in 9 liters of water. Adjust the solution with a 7 in 100 solution of anhydrous citric acid in water to a pH of 7.45, at 25°. Dilute with water to 10.0 liters, equilibrate to 37°, and adjust to a pH of 7.20, if necessary.

Medium: pH 7.20, 0.1 M Tris buffer; 900 mL.

Apparatus 2: 50 rpm.

Time: 30 minutes.

Procedure—Determine the amount of $C_{13}H_8F_2O_3$ dissolved from ultraviolet absorbances at the wavelength of maximum absorbance at about 306 nm of filtered portions of the solution under test, suitably diluted with *pH 7.20, 0.1 M Tris buffer*, in comparison with a Standard solution having a known concentration of USP Diflunisal RS in the same medium.

Tolerances—Not less than 80% (*Q*) of the labeled amount of $C_{13}H_8F_2O_3$ is dissolved in 30 minutes.

Uniformity of dosage units ⟨905⟩: meet the requirements.

Procedure for content uniformity—Transfer 1 finely powdered Tablet to a 200-mL volumetric flask, add 50 mL of water, shake by mechanical means for 30 minutes, and sonicate for 2 minutes. Add 100 mL of alcohol to the flask, shake by mechanical means for 15 minutes, and sonicate for 2 minutes. Dilute with alcohol to volume, mix, and centrifuge a portion of the solution. Quantitatively dilute an accurately measured volume of the resultant clear supernatant liquid with alcohol, if necessary, to obtain a test solution containing about 1.25 mg per mL. Transfer about 125 mg of USP Diflunisal RS, accurately weighed, to a 100-mL volumetric flask, add 75 mL of alcohol to dissolve, dilute with water to volume, and mix to obtain the Standard solution. Transfer 3.0 mL each of the Standard solution and the test solution to separate 50-mL volumetric flasks. To each flask add 5.0 mL of a solution containing 1 g of ferric nitrate in 100 mL of 0.08 *N* nitric acid, dilute with water to volume, and mix. Concomitantly determine the absorbances of the solutions at the wavelength of maximum absorbance at about 550 nm, with a suitable spectrophotometer, using water as the blank. Calculate the quantity, in mg, of $C_{13}H_8F_2O_3$, in the Tablet by the formula:

$$(TC / D)(A_U / A_S),$$

in which T is the labeled quantity, in mg, of diflunisal in the Tablet, C is the concentration, in µg per mL, of USP Diflunisal RS in the Standard solution, D is the concentration, in µg per mL, of diflunisal in the test solution, based upon the labeled quantity per Tablet and the extent of dilution, and A_U and A_S are the absorbances of the solutions from the test solution and the Standard solution, respectively.

Assay—

Mobile phase—Prepare a suitable degassed mixture of water, methanol, acetonitrile, and glacial acetic acid (45:40:17:6) such that the retention time of diflunisal is about 8 minutes.

Standard preparation—Dissolve a suitable quantity of USP Diflunisal RS in acetonitrile to obtain a solution having a known concentration of about 1.0 mg per mL.

Assay preparation—Weigh and finely powder not less than 20 Diflunisal Tablets. Transfer an accurately weighed portion of the powder, equivalent to about 100 mg of diflunisal, to a 100-mL volumetric flask containing about 5 mL of water. Sonicate for 5 minutes, add 60.0 mL of acetonitrile, sonicate for an additional 5 minutes, dilute with water to volume, mix, and filter.

Chromatographic system (see *Chromatography* ⟨621⟩)—The liquid chromatograph is equipped with a 254-nm detector and a 3.9-mm × 30-cm column that contains packing L1. The flow rate is about 2.0 mL per minute. Chromatograph the *Standard preparation*, and record the peak responses as directed under *Procedure:* the tailing factor for the analyte peak is not more than 2.0, and the relative standard deviation for replicate injections is not more than 2.0%.

Procedure—Separately inject equal volumes (about 20 µL) of the *Standard preparation* and the *Assay preparation* into the chromatograph, record the chromatograms, and measure the responses for the major peaks. Calculate the quantity, in mg, of $C_{13}H_8F_2O_3$ in the portion of Tablets taken by the formula:

$$100C(r_U / r_S),$$

in which C is the concentration, in mg per mL, of USP Diflunisal RS in the *Standard preparation*, and r_U and r_S are the peak responses obtained from the *Assay preparation* and the *Standard preparation*, respectively.

Digitalis

» Digitalis is the dried leaf of *Digitalis purpurea* Linné (Fam. Scrophulariaceae). The potency of Digitalis is such that, when assayed as directed, 100 mg is equivalent to not less than 1 USP Digitalis Unit.*

NOTE—When Digitalis is prescribed, Powdered Digitalis is to be dispensed.

Packaging, storage, and labeling—Preserve in containers that protect it from absorbing moisture. Digitalis labeled to indicate that it is to be used only in the manufacture of glycosides is exempt from the moisture and storage requirements.

Reference standard—*USP Digitalis Reference Standard*—Do not dry before using.

Botanic characteristics—

Unground Digitalis—This occurs as more or less crumpled or broken leaves. The leaf blades are ovate, oblong-ovate to ovate-lanceolate, mostly 10 cm to 35 cm in length and 4 cm to 11 cm in width and contracted into a winged petiole. The apex is obtuse; the margin irregularly crenate or serrate; the lower surface densely pubescent, the upper surface wrinkled and finely hairy. The venation is conspicuously reticulate, the mid-rib and principal veins broad and flat, and the lower veins are continued into the wings of the petiole. The color of the upper surface is dark green, of the lower surface grayish from the dense pubescence, the larger veins often purplish. The odor is slight when dry, peculiar and characteristic when moistened. The taste is very bitter.

Histology—Digitalis shows an upper epidermis whose cells possess slightly wavy anticlinal walls, numerous hairs, and no stomata; a lower epidermis with wavy anticlinal walls, numerous oval stomata, and many hairs, and frequently not attached over irregular areas to the cell layer within, especially near the veins; a broad chlorenchyma of a single layer of short palisade cells and several layers of spongy parenchyma; and numerous vascular bundles in the larger veins and petioles, separated by vascular rays one cell in width. On the apex of each marginal tooth one or two water stomata occur.

Ground Digitalis—This is dark green in color. Present are chiefly numerous irregular fragments of epidermis and chlorenchyma; nonglandular hairs that are frequently curved or crooked, up to 500 µm in length, uniseriate, two- to eight-celled, some of the cells collapsed so that the planes of adjoining cells may be at right angles, the terminal cell pointed or rounded; few, small glandular hairs, usually with a one- or two-celled stalk, and a one- or two-celled head; fragments of veins and petioles with annular, reticulate, spiral and simple pitted vessels and tracheids. Calcium oxalate is absent.

Acid-insoluble ash ⟨561⟩: not more than 5.0%.

Foreign organic matter ⟨561⟩—The proportion of stems, browned leaves, flowers, and other foreign organic matter does not exceed 2.0%.

Water, *Method III, Procedure for Vegetable Drugs* ⟨921⟩: not more than 6.0%.

Assay—

Standard preparation—Weigh the contents of 1 container of USP Digitalis RS to the nearest mg, either in the original container or in a weighing bottle, and transfer to a dry, hard-glass, glass-stoppered container or centrifuge tube of at least 50-mL capacity. Complete the weighing within 5 minutes after opening the ampul. Add a menstruum consisting of 4 volumes of alcohol and 1 volume of water so that the total volume of menstruum added corresponds to 10 mL for each g of powder. Insert the stopper, the upper third of which is greased lightly with petrolatum. Shake the mixture for 24 ± 2 hours at 25 ± 5° by mechanical means which continuously brings the solid material into fresh contact with the liquid phase. Immediately thereafter transfer, if necessary, to a centrifuge tube, centrifuge, and decant the supernatant tincture into a dry, hard-glass bottle having a tight closure. Preserve under refrigeration, and use within 30 days.

Assay preparation—Transfer about 5 g of Digitalis, reduced to a fine powder and accurately weighed, to a hard-glass, glass-stoppered container or centrifuge tube of at least 50-mL capacity. Proceed as directed under *Standard preparation*, beginning with "Add a menstruum." Preserve under refrigeration, and use within 30 days.

Pigeons—Employ adult pigeons free from gross evidence of disease or emaciation, and of such weight that the heaviest weighs less than twice the weight of the lightest. Divide the pigeons into groups as nearly alike as practicable with respect to breed and weight so that the average weight of the group assigned by random choice to the *Standard preparation* shall not differ by more than 30% from the average weight of the group assigned to the preparation to be assayed. Withhold food but not water during the period 16 to 28 hours prior to use. Preparatory to injection, lightly anesthetize the pigeon with ether, and immobilize it; expose an alar vein, and cannulate with a suitable cannula. Maintain the anesthesia during cannulation and throughout the subsequent injection period at such a level that pain is absent, the pupillary and corneal reflexes are present, and the voluntary musculature is not relaxed beyond permitting the pigeon to make some voluntary movement occasionally.

Preparation of test dilutions—On the day of the assay, dilute portions of the *Standard preparation* and of the preparation to be assayed (*Assay preparation*) with isotonic sodium chloride solution in such a way that the estimated fatal dose of each dilution will be 15 mL per kg of body weight.

Injection of test dilutions—Arrange to inject the appropriate test dilution by suitable means such as a small-bore buret calibrated to 0.05 mL. Start the injection after ensuring the absence of air bubbles from the injection apparatus, by infusing, within a few seconds, a volume of the test dilution equivalent to 1 mL per kg of body weight. Repeat this dose at 5-minute intervals thereafter until the pigeon dies of cardiac arrest.

Use a total of not less than 6 pigeons for the *Standard preparation* and not less than 6 pigeons for the preparation to be assayed. If the average number of doses for any given dilution required to produce death is less than 13 or greater than 19, or if the larger exceeds the smaller in the same assay by more than 4 doses, regard these data as preliminary. Use them as a guide, and repeat with a fresh, higher or lower dilution. Complete the assay within the period of 30 days for preservation of the *Standard preparation* and *Assay preparation.*

Calculation of potency—Tabulate and average the number of doses of the *Standard preparation*, designating the average $\bar{z}_S$, and likewise obtain the corresponding average, $\bar{z}_U$, for the *Assay preparation.* Compute the potency in USP Digitalis Units per mL (i.e., per 100 mg) of the *Assay preparation* as:

$$\text{Potency} = \bar{z}_S R/\bar{z}_U,$$

where R equals v_S/v_U, in which v_S is the number of USP Digitalis Units per mL of *Standard preparation* dilution, and v_U is the volume, in mL, of *Assay preparation* per mL of dilution. Compute the log confidence interval, L (see Equation (31) under *Confidence Intervals for Individual Assays* in *Design and Analysis of Biological Assays* ⟨111⟩). If L exceeds 0.30, repeat the assay or inject more pigeons with one or both preparations until the confidence interval is 0.30 or less.

The potency of Digitalis, calculated from that of the *Assay preparation*, is satisfactory if the result is not less than 0.85 USP Digitalis Unit per 100 mg.

Powdered Digitalis

» Powdered Digitalis is Digitalis dried at a temperature not exceeding 60°, reduced to a fine or a very fine powder, and adjusted, if necessary to conform to the official potency by admixture with sufficient Lac-

* One USP Digitalis Unit represents the potency of 100 mg of USP Digitalis RS.

tose, Starch, or exhausted marc of digitalis, or with Powdered Digitalis having either a lower or a higher potency.

The potency of Powdered Digitalis is such that, when assayed as directed, 100 mg is equivalent to 1 USP Digitalis Unit.*

NOTE—When Digitalis is prescribed, Powdered Digitalis is to be dispensed.

Packaging and storage—Preserve in tight, light-resistant containers. A package of a suitable desiccant may be enclosed in the container.

Reference standards—*USP Digitalis Reference Standard*—Do not dry before using. *USP Digitoxin Reference Standard*—Dry in vacuum at 105° for 1 hour before using. *USP Gitoxin Reference Standard*—[*Caution—Avoid contact.*] Do not dry before using. Keep container tightly closed and protected from light.

Identification—

A: It conforms to the description for *Ground Digitalis* in the section, *Botanic characteristics*, under *Digitalis*.

B: Transfer 100 mg to a 15-mL centrifuge tube containing 2.0 mL of diluted alcohol and 1.0 mL of lead acetate TS, mix, shake, and boil for 2 minutes. Centrifuge, decant the supernatant liquid into a second 15-mL centrifuge tube, add 2.0 mL of chloroform, and mix. Centrifuge, then remove the lower layer, and filter it through a chloroform-washed small column of anhydrous sodium sulfate (100 to 300 mg) into a 5-mL centrifuge tube. Evaporate the chloroform solution under a stream of nitrogen to dryness, and dissolve the residue in 100 µL of a mixture of methanol and chloroform (1:1). Prepare a Standard solution in the same manner, using 100 mg of USP Digitalis RS (*Standard solution A*). Prepare a second Standard solution by dissolving USP Digitoxin RS and USP Gitoxin RS in a mixture of methanol and chloroform (1:1) such that the final concentration of each is approximately 0.2 mg per mL (*Standard solution B*). On a suitable thin-layer chromatographic plate (see *Chromatography* ⟨621⟩), coated with a 0.25-mm layer of chromatographic silica gel mixture, apply 10 µL of the test solution, 10 µL of *Standard solution A*, and 10 µL of *Standard solution B*, each as a narrow band, about 15 mm long, and allow the bands to dry. Develop the chromatogram in a saturated chamber, using a solvent system consisting of a mixture of ethyl acetate, methanol, and water (30:4:3) until the solvent front has moved about 15 cm from the origin. Mix 10 mL of chloramine T solution (3 in 100) with 40 mL of a 1 in 4 solution of trichloroacetic acid in alcohol (store the mixture in a cool place, and use it within 1 week), and spray the air-dried chromatographic plate with this mixture. Heat the plate at 110° for 15 to 20 minutes, and examine it under long-wavelength ultraviolet light. Locate the 2 prominent bands obtained from *Standard solution A* corresponding in R_f value to the 2 bands obtained from *Standard solution B*. The chromatogram obtained from the solution under test shows bands corresponding to them, and also shows bands corresponding to the 3 other bands most prominent in the chromatogram from *Standard solution A* but of lower R_f value. Relative R_f values for the 5 bands are: 1.0 (digitoxin); 0.8 to 0.9 (gitoxin); 0.6 to 0.7; 0.4 to 0.5; and 0.3 to 0.4.

Microbial limit—It meets the requirements of the test for absence of *Salmonella* species under *Microbial Limit Tests* ⟨61⟩.

Acid-insoluble ash ⟨561⟩: not more than 5.0%.

Water, *Method III, Procedure for Vegetable Drugs* ⟨921⟩: not more than 5.0%.

Assay—Proceed with Powdered Digitalis as directed in the *Assay* under *Digitalis*.

The potency of Powdered Digitalis, calculated from that of the *Assay preparation*, is satisfactory if the result is not less than 0.85 USP Digitalis Unit and not more than 1.20 USP Digitalis Units per 100 mg.

* One USP Digitalis Unit represents the potency of 100 mg of USP Digitalis RS.

Digitalis Capsules

» Digitalis Capsules contain an amount of Powdered Digitalis equivalent to not less than 85.0 percent and not more than 120.0 percent of the labeled potency.

Packaging and storage—Preserve in tight containers.

Reference standard—*USP Digitalis Reference Standard*—Do not dry before using.

Microbial limit—It meets the requirements of the test for absence of *Salmonella* species under *Microbial Limit Tests* ⟨61⟩.

Uniformity of dosage units ⟨905⟩: meet the requirements.

Assay—

Standard preparation—Prepare as directed in the *Assay* under *Digitalis*.

Assay preparation—Empty the contents of not less than 20 Digitalis Capsules into a hard-glass, glass-stoppered container of not less than 50-mL capacity. Add a menstruum consisting of 4 volumes of alcohol and 1 volume of water so that the total volume of menstruum corresponds to 1 mL for each expected USP Digitalis Unit. Insert the stopper, the upper third of which is greased lightly with petrolatum. Shake the mixture at 25 ± 5° for 24 ± 2 hours by mechanical means, which continuously brings the solid material into fresh contact with the liquid phase. Immediately thereafter transfer to a centrifuge tube, centrifuge, and decant the supernatant tincture into a dry, hard-glass bottle having a tight closure. Preserve under refrigeration, and use within 30 days.

Pigeons, Preparation of test dilutions, Injection of test dilutions, and *Calculation of potency*—Proceed as directed in the *Assay* under *Digitalis*.

Digitalis Tablets

» Digitalis Tablets contain an amount of Powdered Digitalis equivalent to not less than 85.0 percent and not more than 120.0 percent of the labeled potency.

Packaging and storage—Preserve in tight containers.

Reference standard—*USP Digitalis Reference Standard*—Do not dry before using.

Disintegration ⟨701⟩: 30 minutes.

Microbial limit—It meets the requirements of the test for absence of *Salmonella* species under *Microbial Limit Tests* ⟨61⟩.

Uniformity of dosage units ⟨905⟩: meet the requirements.

Assay—

Standard preparation—Prepare as directed in the *Assay* under *Digitalis*.

Assay preparation—Weigh and finely powder not less than 25 Digitalis Tablets. Weigh accurately a portion of the powder, equivalent to not less than 20 Tablets. Transfer to a dry, hard-glass, glass-stoppered container of not less than 50-mL capacity. Proceed as directed for *Assay preparation—Capsules of dry powdered digitalis* in the *Assay* under *Digitalis Capsules*, beginning with "Add a menstruum."

Pigeons, Preparation of test dilutions, Injection of test dilutions, and *Calculation of potency*—Proceed as directed in the *Assay* under *Digitalis*.

Digitoxin

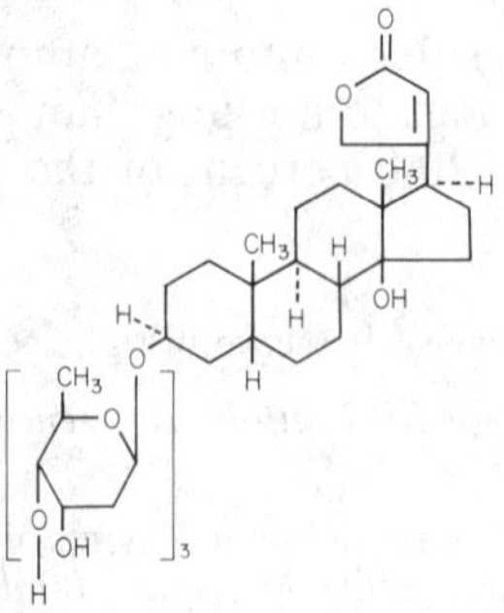

$C_{41}H_{64}O_{13}$ 764.95

Card-20(22)-enolide, 3-[(*O*-2,6-dideoxy-β-D-*ribo*-hexopyranosyl-(1→4)-*O*-2,6-dideoxy-β-D-*ribo*-hexopyranosyl-(1→4)-2,6-dideoxy-β-D-*ribo*-hexopyranosyl)oxy]-14-hydroxy, (3β,5β)-.

Digitoxin [*71-63-6*].

» Digitoxin is a cardiotonic glycoside obtained from *Digitalis purpurea* Linné, *Digitalis lanata* Ehrhart (Fam. Scrophulariaceae), and other suitable species of *Digitalis.* Digitoxin contains not less than 92.0 percent and not more than 103.0 percent of $C_{41}H_{64}O_{13}$, calculated on the dried basis.

Caution—Handle Digitoxin with exceptional care since it is highly potent.

Packaging and storage—Preserve in tight containers.

Reference standard—*USP Digitoxin Reference Standard*—Dry in vacuum at 105° for 1 hour before using.

Identification—

A: The infrared absorption spectrum of a potassium bromide dispersion of it exhibits maxima only at the same wavelengths as that of a similar preparation of USP Digitoxin RS.

B: Prepare a test solution in methanol containing 1 mg per mL. On a suitable thin-layer chromatographic plate (see *Chromatography* ⟨621⟩), coated with a 0.25-mm layer of chromatographic silica gel mixture, apply 1 µL of the test solution and 1 µL of a Standard solution of USP Digitoxin RS in methanol containing 1 mg per mL. Allow the applications to dry, and develop the chromatogram in a solvent system consisting of a mixture of methylene chloride and methanol (93:7) until the solvent front has moved about three-fourths of the length of the plate. Remove the plate from the developing chamber, mark the solvent front, and dry the plate at 100° to remove the solvent. Spray the plate with a 6 in 10 solution of sulfuric acid in methanol, heat at 105° for 10 minutes, and examine the chromatogram under long-wavelength ultraviolet light: the R_f value of the principal spot obtained from the test solution corresponds to that obtained from the Standard solution.

C: The retention time of the major peak in the chromatogram of the *Assay preparation* corresponds to that of the major peak in the chromatogram of the *Standard preparation* as obtained in the *Assay.*

Loss on drying ⟨731⟩—Dry it in vacuum at 105° for 1 hour: it loses not more than 1.5% of its weight.

Residue on ignition ⟨281⟩: negligible, from 100 mg.

Assay—

Mobile phase—Prepare a filtered and degassed mixture of water and acetonitrile (55:45). Make adjustments if necessary (see *System Suitability* under *Chromatography* ⟨621⟩).

Standard preparation—Dissolve an accurately weighed quantity of USP Digitoxin RS in *Mobile phase,* and dilute quantitatively, and stepwise if necessary, with *Mobile phase* to obtain a solution having a known concentration of about 40 µg per mL.

Assay preparation—Transfer about 50 mg of Digitoxin, accurately weighed, to a 200-mL volumetric flask. Dissolve in *Mobile phase,* dilute with *Mobile phase* to volume, and mix. Pipet 4 mL of this solution into a 25-mL volumetric flask, dilute with *Mobile phase* to volume, and mix.

System suitability preparation—Prepare a solution in *Mobile phase* containing about 40 µg each of digitoxin and digoxin per mL.

Chromatographic system (see *Chromatography* ⟨621⟩)—The liquid chromatograph is equipped with a 218-nm detector and a 3.9-mm × 30-cm column that contains packing L1. The flow rate is about 1 mL per minute. Chromatograph the *Standard preparation* and the *System suitability preparation,* and record the peak responses as directed under *Procedure:* the tailing factor for the analyte peak is not more than 2.0, the resolution, *R,* between the digoxin and digitoxin peaks is not less than 2.0, and the relative standard deviation for replicate injections of the *Standard preparation* is not more than 2.0%. The relative retention times are about 0.35 for digoxin and 1.0 for digitoxin.

Procedure—Separately inject equal volumes (about 50 µL) of the *Standard preparation* and the *Assay preparation* into the chromatograph, record the chromatograms, and measure the responses for the major peaks. Calculate the quantity, in mg, of $C_{41}H_{64}O_{13}$ in the portion of Digitoxin taken by the formula:

$$1.25C(r_U / r_S),$$

in which *C* is the concentration, in µg per mL, of USP Digitoxin RS in the *Standard preparation,* and r_U and r_S are the peak responses obtained from the *Assay preparation* and the *Standard preparation,* respectively.

Digitoxin Injection

» Digitoxin Injection is a sterile solution of Digitoxin in 5 to 50 percent (v/v) of alcohol, and may contain Glycerin or other suitable solubilizing agents. It contains not less than 90.0 percent and not more than 110.0 percent of the labeled amount of $C_{41}H_{64}O_{13}$.

Packaging and storage—Preserve in single-dose or in multiple-dose containers, preferably of Type I glass, protected from light.

Reference standard—*USP Digitoxin Reference Standard*—Dry in vacuum at 105° for 1 hour before using.

Identification—

A: To a portion of Injection, equivalent to about 1 mg of digitoxin, add 10 mL of water, and extract with 10 mL of chloroform. Evaporate the chloroform extract on a steam bath with the aid of a current of air to dryness. Dissolve the residue in 2 mL of a solution prepared by mixing 0.3 mL of ferric chloride TS and 50 mL of glacial acetic acid, and underlay with 2 mL of sulfuric acid: at the zone of contact of the two liquids a brown color, which gradually changes to light green, then to blue, is produced, and finally the entire acetic acid layer acquires a blue color.

B: To a portion of Injection, equivalent to about 0.2 mg of digitoxin, add 10 mL of water, and extract with 10 mL of chloroform. Evaporate the chloroform extract on a steam bath with the aid of a current of air to dryness. Add 2 mL of a freshly prepared 1 in 100 solution of *m*-dinitrobenzene in alcohol, and allow to stand for 10 minutes with frequent shaking. Add 2 mL of a 1 in 200 solution of tetramethylammonium hydroxide in alcohol, and mix: a red-violet color develops slowly and then fades.

C: The retention time of the major peak in the chromatogram of the *Assay preparation* corresponds to that of the major peak in the chromatogram of the *Standard preparation* as obtained in the *Assay.*

Alcohol content ⟨611⟩: between 90.0% and 110.0% of the labeled percentage of C_2H_5OH.

Other requirements—It meets the requirements under *Injections* ⟨1⟩.

Assay—

Mobile phase, Standard preparation, System suitability preparation, and *Chromatographic system*—Prepare as directed in the *Assay* under *Digitoxin.*

Assay preparation—Transfer an accurately measured volume of Digitoxin Injection, equivalent to about 1 mg of digitoxin, to a 25-mL volumetric flask. Dilute with *Mobile phase* to volume, and mix.

Procedure—Proceed as directed for *Procedure* in the *Assay* under *Digitoxin.* Calculate the quantity, in µg, of $C_{41}H_{64}O_{13}$ in each mL of the Injection taken by the formula:

$$25(C/V)(r_U/r_S),$$

in which C is the concentration, in µg per mL, of USP Digitoxin RS in the *Standard preparation*, V is the volume, in mL, of Injection taken, and r_U and r_S are the peak responses obtained from the *Assay preparation* and the *Standard preparation*, respectively.

Digitoxin Tablets

» Digitoxin Tablets contain not less than 90.0 percent and not more than 110.0 percent of the labeled amount of $C_{41}H_{64}O_{13}$.

NOTE—Avoid the use of strongly adsorbing substances, such as bentonite, in the manufacture of Digitoxin Tablets.

Packaging and storage—Preserve in well-closed containers.

Reference standard—*USP Digitoxin Reference Standard*—Dry in vacuum at 105° for 1 hour before using.

Identification—

A: Transfer a quantity of finely powdered Tablets, equivalent to not less than 1 mg of digitoxin, to a suitable flask, add 20 mL of chloroform, and sonicate. Filter, and evaporate the filtrate on a steam bath with the aid of a current of air to dryness. Dissolve the residue in 2 mL of a solution prepared by mixing 0.3 mL of ferric chloride TS and 50 mL of glacial acetic acid, and underlay with 2 mL of sulfuric acid: at the zone of contact of the two liquids a brown color, which gradually changes to light green, then to blue, is produced, and finally the entire acetic acid layer acquires a blue color.

B: The retention time of the major peak in the chromatogram of the *Assay preparation* corresponds to that of the major peak in the chromatogram of the *Standard preparation* as obtained in the *Assay*.

Dissolution ⟨711⟩—[NOTE—Throughout this procedure, use scrupulously clean glassware, which previously has been rinsed successively with hydrochloric acid, water, and alcohol, and carefully dried. Take precautions to prevent contamination from fluorescent particles and from metal and rubber surfaces.]

Medium: dilute hydrochloric acid (3 in 500); 500 mL. [NOTE—Use the same batch of *Dissolution Medium* throughout the test.]

Apparatus 1: 120 ± 5 rpm.

Times: 30 minutes; 60 minutes.

Standard solution—Weigh accurately about 30 mg of USP Digitoxin RS, dissolve in a minimum amount of alcohol in a 500-mL volumetric flask, add dilute alcohol (4 in 5) to volume, and mix.

Standard preparations—Just prior to use, dilute 5.0 mL of the *Standard solution* with *Dissolution Medium* to 500.0 mL, and mix. Transfer aliquots (2.0 to 10.0 mL) of this solution to individual separators to prepare standards equivalent to 20, 40, 60, 80, and 100% of the labeled amount of digitoxin in 500 mL. Add *Dissolution Medium* to make 10 mL, and proceed as directed under *Procedure*, beginning with "Extract with three 15-mL portions of chloroform."

Procedure—Proceed as directed for *Procedure* under *Dissolution* ⟨711⟩. After 30 minutes, accurately timed, withdraw a suitable aliquot of the solution under test from a point midway between the stirring shaft and the wall of the vessel, and approximately midway in depth. Filter the solution promptly after withdrawal, using a suitable membrane filter of not greater than 0.8-µm porosity, discarding the first 10 mL of the filtrate. Without replacing the *Dissolution Medium* withdrawn, continue to rotate the basket, and after an additional 30 minutes, accurately timed, similarly withdraw and filter another aliquot. Treat each of these solutions as follows: Assuming dissolution of 100% of the labeled amount of digitoxin, transfer aliquots, equivalent to 6 µg of digitoxin, to suitable separators. Extract with three 15-mL portions of chloroform, and combine the chloroform extracts in glass-stoppered flasks. Evaporate the combined extracts on a steam bath, with the aid of a current of air, to dryness. In a similar manner, prepare a blank using a suitable volume of *Dissolution Medium.*

Measurement of fluorescence—Begin with the *Standard preparations*, and keep all flasks in the same sequence throughout, so that the elapsed time from addition of reagents to reading of fluorescence is the same for each set. Treat 1 flask at a time as follows: Add 10 mL of a solution freshly prepared by dissolving 35 mg of ascorbic acid in 25 mL of methanol and cautiously adding the solution to 100 mL of hydrochloric acid. Mix, and add 1 mL of a solution freshly prepared by diluting 1 mL of 30 percent hydrogen peroxide with water to 500 mL and diluting 1 volume of the resulting solution with 20 volumes of water. Mix, and insert the stopper in the flask. After 45 minutes, measure the fluorescence at about 575 nm, the excitation wavelength being about 395 nm. Correct each reading for the blank, and plot a standard curve of fluorescence versus percentage dissolution. By calculation from the standard curve, determine the percentage dissolution of digitoxin in each Tablet within 30 minutes and the total percentage dissolution of digitoxin within 60 minutes, taking into account the volume of the solution under test removed after the first 30 minutes of the test.

Tolerances—Not less than 60% of the labeled amount of $C_{41}H_{64}O_{13}$ is dissolved within 30 minutes for each Tablet tested, and not less than 85% of the labeled amount is dissolved within 60 minutes for the average of the Tablets tested.

Uniformity of dosage units ⟨905⟩: meet the requirements.

Procedure for content uniformity—Place 1 Tablet in a suitable glass-stoppered conical flask. Add an accurately measured volume of *Mobile phase* (prepared as directed in the *Assay* under *Digitoxin*) sufficient to obtain a solution containing about 10 µg of digitoxin per mL, and shake by mechanical means until the Tablet has completely disintegrated (not less than 30 minutes). Centrifuge, and use the clear supernatant solution as the test solution. Dissolve an accurately weighed quantity of USP Digitoxin RS in *Mobile phase*, and dilute quantitatively, and stepwise if necessary, with *Mobile phase* to obtain a Standard solution having a known concentration of about 10 µg per mL. Proceed as directed in the *Assay*. Calculate the quantity, in mg, of $C_{41}H_{64}O_{13}$ in the Tablet by the formula:

$$(LC/D)(r_U/r_S),$$

in which L is the labeled quantity, in mg, of digitoxin in the Tablet, C is the concentration, in µg per mL, of USP Digitoxin RS in the Standard solution, D is the concentration, in µg per mL, of digitoxin in the test solution based on the labeled quantity in the Tablet and the extent of dilution, and r_U and r_S are the digitoxin peak responses obtained from the test solution and the Standard solution, respectively.

Assay—

Mobile phase, Standard preparation, System suitability preparation, and *Chromatographic system*—Prepare as directed in the *Assay* under *Digitoxin.*

Assay preparation—Weigh and finely powder not less than 20 Digitoxin Tablets. Transfer an accurately weighed quantity of powder, equivalent to about 1 mg of digitoxin, to a 25-mL volumetric flask. Add 15 mL of *Mobile phase*, and sonicate. Dilute with *Mobile phase* to volume, and mix. Filter a portion of this solution, discarding the first few mL of the filtrate. The filtrate is the *Assay preparation.*

Procedure—Proceed as directed for *Procedure* in the *Assay* under *Digitoxin.* Calculate the quantity, in µg, of $C_{41}H_{64}O_{13}$ in the portion of Tablets taken by the formula:

$$25C(r_U/r_S),$$

in which C is the concentration, in µg per mL, of USP Digitoxin RS in the *Standard preparation*, and r_U and r_S are the peak responses obtained from the *Assay preparation* and the *Standard preparation*, respectively.

Digoxin

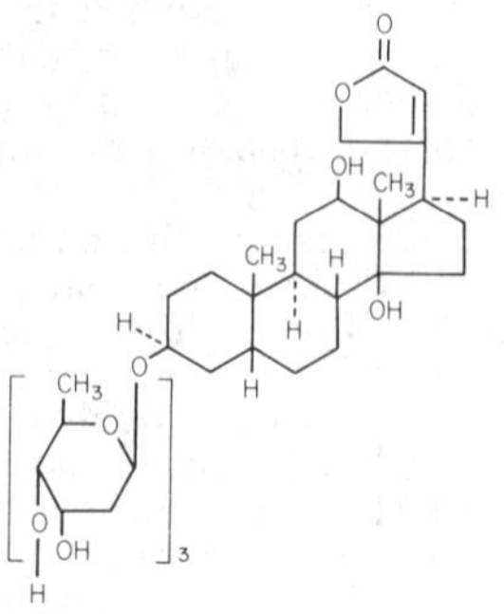

$C_{41}H_{64}O_{14}$ 780.95

Card-20(22)-enolide, 3-[(*O*-2,6-dideoxy-β-D-*ribo*-hexopyranosyl-(1→4)-*O*-2,6-dideoxy-β-D-*ribo*-hexopyranosyl-(1→4)-2,6-dideoxy-β-D-*ribo*-hexopyranosyl)oxy]-12,14-dihydroxy-, (3β,5β,12β)-.

Digoxin.

3β-[(*O*-2,6-Dideoxy-β-D-*ribo*-hexopyranosyl-(1→4)-*O*-2,6-dideoxy-β-D-*ribo*-hexopyranosyl-(1→4)-2,6-dideoxy-β-D-*ribo*-hexopyranosyl)oxy]-12β,14-dihydroxy-5β-card-20(22)-enolide [*20830-75-5*].

» Digoxin is a cardiotonic glycoside obtained from the leaves of *Digitalis lanata* Ehrhart (Fam. Scrophulariaceae). It contains not less than 95.0 percent and not more than 101.0 percent of $C_{41}H_{64}O_{14}$, calculated on the dried basis.

Caution—Handle Digoxin with exceptional care, since it is extremely poisonous.

Packaging and storage—Preserve in tight containers.

Reference standards—*USP Digoxin Reference Standard*—[*Caution—Poisonous—cardiotonic.*] Dry in vacuum at 105° for 1 hour before using. *USP Gitoxin Reference Standard*—[*Caution—Avoid contact.*] Do not dry before using. Keep container tightly closed and protected from light.

Identification—

A: The infrared absorption spectrum of a potassium bromide dispersion of it exhibits maxima only at the same wavelengths as that of a similar preparation of USP Digoxin RS.

B: The retention time of the major peak in the chromatogram of the *Assay preparation* corresponds to that of the *Standard preparation* as obtained in the *Assay*.

C: Examine in visible light the thin-layer chromatograph prepared as directed in the test for *Related glycosides:* the R_f value of the principal blue spot obtained from the *Test solution* corresponds to that obtained from the *Standard solution.*

Loss on drying ⟨731⟩—Dry it in vacuum at 105° for 1 hour: it loses not more than 1.0% of its weight.

Residue on ignition ⟨281⟩: not more than 0.5%, a 100-mg specimen being used.

Related glycosides—

Chloramine T-trichloroacetic acid reagent—Mix 10 mL of a freshly prepared solution of chloramine T (3 in 100) and 40 mL of a 1 in 4 solution of trichloroacetic acid in dehydrated alcohol.

Spotting solvent—Prepare a mixture of chloroform and methanol (2:1).

Standard solution—Dissolve an accurately weighed quantity of USP Digoxin RS in *Spotting solvent* to obtain a solution containing 10 mg per mL.

Gitoxin standard solution—Dissolve an accurately weighed quantity of USP Gitoxin RS in *Spotting solvent* to obtain a solution containing 0.30 mg per mL.

Test solution—Transfer 250.0 mg of Digoxin to a 25-mL volumetric flask, dissolve in *Spotting solvent*, dilute with *Spotting solvent* to volume, and mix.

Procedure—Apply 10 µL of the *Test solution*, 10 µL of the *Standard solution*, and 10 µL of the *Gitoxin standard solution* on a line parallel to and about 2.5 cm from the bottom edge of a reversed-phase thin-layer chromatographic plate, coated with a 0.25-mm layer of chromatographic silica gel mixture to which is permanently bonded octadecylsilane (C18). Allow the spots to dry, and place the plates in a developing chamber containing a mixture of methanol and water (7:3). Develop the chromatogram until the solvent front has moved about 15 cm above the line of application. Remove the plate, and allow the solvent to evaporate. Spray the plate with *Chloramine T-trichloroacetic acid reagent*, freshly mixed, and heat in an oven at 110° for 10 minutes. Examine the plate under long-wavelength ultraviolet light: no spot from the *Test solution* except that due to digoxin is more intense than the spot from the *Gitoxin standard solution* (not more than 3% of any related glycoside as gitoxin).

Assay—

Mobile phase—Prepare a suitable degassed and filtered mixture of water and acetonitrile (37:13), making adjustments if necessary (see *System Suitability* under *Chromatography* ⟨621⟩).

Standard preparation—Dissolve an accurately weighed quantity of USP Digoxin RS in diluted alcohol, and dilute quantitatively and stepwise with diluted alcohol to obtain a solution having a known concentration of about 250 µg per mL. Use a sonic bath to aid dissolution.

Assay preparation—Weigh accurately about 50 mg of Digoxin, and transfer to a 200-mL volumetric flask. Dissolve in about 150 mL of diluted alcohol by sonication, dilute with diluted alcohol to volume, and mix.

System suitability preparation—Prepare a solution in diluted alcohol of USP Digoxin RS and digoxigenin bisdigitoxoside having concentrations of about 40 µg of each per mL.

Chromatographic system (see *Chromatography* ⟨621⟩)—The liquid chromatograph is equipped with a 218-nm detector and a 4.2-mm × 25-cm column that contains packing L1. The flow rate is about 3.0 mL per minute. Chromatograph the *System suitability preparation*, and record the peak responses as directed under *Procedure:* the relative standard deviation for replicate injection is not more than 2.0%, the tailing factor for the digoxin peak is not more than 2.0, and the resolution, *R*, between digoxin and digoxigenin bisdigitoxoside is not less than 2.0.

Procedure—Separately inject equal volumes (about 10 µL) of the *Standard preparation* and the *Assay preparation* into the chromatograph, record the chromatograms, and measure the responses for the major peaks. Calculate the quantity, in mg, of $C_{41}H_{64}O_{14}$ in the portion of Digoxin taken by the formula:

$$0.2C(r_U / r_S),$$

in which *C* is the concentration, in µg per mL, of USP Digoxin RS in the *Standard preparation*, and r_U and r_S are the responses for the digoxin peaks obtained from the *Assay preparation* and the *Standard preparation*, respectively.

Digoxin Elixir

» Digoxin Elixir contains, in each 100 mL, not less than 4.50 mg and not more than 5.25 mg of $C_{41}H_{64}O_{14}$.

Packaging and storage—Preserve in tight containers, and avoid exposure to excessive heat.

Reference standard—*USP Digoxin Reference Standard*—[*Caution—Poisonous—cardiotonic.*] Dry in vacuum at 105° for 1 hour before using.

Identification—

A: The retention time of the major peak in the chromatogram of Digoxin Elixir corresponds to that of the *Standard preparation* as obtained in the *Assay*.

B: *Chloramine T-trichloroacetic acid reagent*—Mix 10 mL of a freshly prepared solution of chloramine T (3 in 100) and 40 mL of a 1 in 4 solution of trichloroacetic acid in dehydrated alcohol.

Spotting solvent—Prepare a mixture of chloroform and methanol (2:1).

Standard solution—Dissolve an accurately weighed quantity of USP Digoxin RS in *Spotting solvent* to obtain a solution containing 0.25 mg per mL.

Test solution—Pipet a volume of Elixir, equivalent to 0.5 mg of digoxin, into a separator, and add 5 mL of water and 20 mL of carbon tetrachloride. Shake, allow to separate, and discard the carbon tetrachloride layer. Extract the aqueous layer with three 10-mL portions of chloroform, combining the extracts in a conical flask. Evaporate the combined chloroform extracts on a steam bath with the aid of a current of air to dryness. Add 2 mL of *Spotting solvent* to the residue, and shake for 2 minutes.

Procedure—Proceed as directed for *Procedure* in the test for *Related glycosides* under *Digoxin*, except to omit the use of the *Gitoxin standard solution*. Examine the plate under long-wavelength ultraviolet light: the R_f value of the principal spot in the chromatogram of the *Test solution* corresponds to that of the *Standard solution*.

Alcohol content ⟨611⟩: between 9.0% and 11.5% of C_2H_5OH.

Assay—

Mobile phase—Prepare a filtered and degassed mixture of water, acetonitrile, and isopropyl alcohol (70:27.5:2.5). Make adjustments if necessary (see *System Suitability* under *Chromatography* ⟨621⟩).

Standard preparation—Dissolve an accurately weighed quantity of USP Digoxin RS in diluted alcohol, and dilute quantitatively and stepwise with diluted alcohol to obtain a solution having a known concentration of about 5 µg per mL.

Assay preparation—Transfer an accurately measured volume of Digoxin Elixir, equivalent to about 500 µg of digoxin, to a 100-mL volumetric flask, dilute with diluted alcohol to volume, and mix.

System suitability preparation—Prepare as directed in the *Assay* under *Digoxin*.

Chromatographic system (see *Chromatography* ⟨621⟩)—The liquid chromatograph is equipped with a 218-nm detector and a 4.6-mm × 15-cm column that contains packing L1. The flow rate is about 0.5 mL per minute. Chromatograph the *System suitability preparation*, and record the peak responses as directed under *Procedure*: the tailing factor for the analyte peak is not more than 2.0, the resolution, *R*, between the digoxin and digoxigenin bisdigitoxoside peaks is not less than 2.0, and the relative standard deviation for replicate injections is not more than 2.0%.

Procedure—Separately inject equal volumes (about 10 µL) of the *Standard preparation* and the *Assay preparation* into the chromatograph, record the chromatograms, and measure the responses for the major peaks. Calculate the quantity, in µg, of $C_{41}H_{64}O_{14}$ in each mL of the Elixir taken by the formula:

$$(100C/V)(r_U/r_S),$$

in which *C* is the concentration, in µg per mL, of USP Digoxin RS in the *Standard preparation*, *V* is the volume, in mL, of Elixir taken, and r_U and r_S are the digoxin peak responses obtained from the *Assay preparation* and the *Standard preparation*, respectively.

Digoxin Injection

» Digoxin Injection is a sterile solution of Digoxin in Water for Injection and Alcohol or other suitable solvents. It contains not less than 90.0 percent and not more than 105.0 percent of the labeled amount of $C_{41}H_{64}O_{14}$.

Packaging and storage—Preserve in single-dose containers, preferably of Type I glass. Avoid exposure to excessive heat.

Reference standard—*USP Digoxin Reference Standard*—[*Caution—Poisonous—cardiotonic.*] Dry in vacuum at 105° for 1 hour before using.

Identification—

A: Injection responds to *Identification test A* under *Digoxin Elixir*.

B: *Chloramine T-trichloroacetic acid reagent*, *Spotting solvent*, and *Standard solution*—Proceed as directed for *Identification test B* under *Digoxin Elixir*.

Test solution—Pipet a volume of Injection, equivalent to 0.5 mg of digoxin, into a separator, and add 5 mL of water. Extract with three 10-mL portions of chloroform, combining the extracts in a conical flask. Evaporate the combined chloroform extracts on a steam bath with the aid of a current of air to dryness. (If traces of water or propylene glycol remain, dry the flask in vacuum at 100° for 30 minutes.) Dissolve the residue in 2 mL of *Spotting solvent*.

Procedure—Proceed as directed for *Procedure* in the test for *Related glycosides* under *Digoxin*, except to omit the use of the *Gitoxin standard solution*. Examine the plate under long-wavelength ultraviolet light: the R_f value of the principal spot in the chromatogram of the *Test solution* corresponds to that of the *Standard solution*.

Alcohol content ⟨611⟩: between 9.0% and 11.0% of C_2H_5OH.

Other requirements—It meets the requirements under *Injections* ⟨1⟩.

Assay—

Mobile phase—Proceed as directed in the *Assay* under *Digoxin*.

Standard preparation—Dissolve an accurately weighed quantity of USP Digoxin RS in diluted alcohol, and dilute quantitatively with diluted alcohol to obtain a solution having a known concentration of about 250 µg per mL. Use a sonic bath to aid dissolution. If necessary, dilute quantitatively to match, approximately, the concentration of the Injection.

Chromatographic system and *System suitability preparation*—Proceed as directed in the *Assay* under *Digoxin*.

Procedure—Separately inject equal volumes (about 10 µL) of the *Standard preparation* and Digoxin Injection into the chromatograph, record the chromatograms, and measure the responses for the major peaks. Calculate the quantity, in µg, of $C_{41}H_{64}O_{14}$ in each mL of the Injection taken by the formula:

$$C(r_U/r_S),$$

in which *C* is the concentration, in µg per mL, of USP Digoxin RS in the *Standard preparation*, and r_U and r_S are the responses for the digoxin peaks obtained from the Injection and the *Standard preparation*, respectively.

Digoxin Tablets

» Digoxin Tablets contain not less than 90.0 percent and not more than 105.0 percent of the labeled amount of $C_{41}H_{64}O_{14}$.

Packaging and storage—Preserve in tight containers.

Reference standard—*USP Digoxin Reference Standard*—[*Caution—Poisonous—cardiotonic.*] Dry in vacuum at 105° for 1 hour before using.

Identification—

A: The retention time of the major peak in the chromatogram of the *Assay preparation* corresponds to that of the *Standard preparation* as obtained in the *Assay*.

B: *Chloramine T-trichloroacetic acid reagent*, *Spotting solvent*, and *Standard solution*—Proceed as directed for *Identification test B* under *Digoxin Elixir*.

Test solution—Transfer an accurately weighed portion of finely powdered Tablets, equivalent to 0.5 mg of digoxin, to a 10-mL centrifuge tube. Add 2 mL of *Spotting solvent*, shake for 10 minutes, and centrifuge. Decant the supernatant liquid, and use it as the *Test solution*.

Procedure—Proceed as directed for *Procedure* in the test for *Related glycosides* under *Digoxin*, except to omit the use of the *Gitoxin standard solution*. Examine the plate under long-wavelength ultraviolet light: the R_f value of the principal spot in the chromatogram of the *Test solution* corresponds to that of the *Standard solution*.

Dissolution ⟨711⟩—[NOTE—Throughout this procedure, use scrupulously clean glassware, which previously has been rinsed successively with hydrochloric acid, water, and alcohol, and carefully dried. Take precautions to prevent contamination from fluorescent particles and from metal and rubber surfaces.]

Medium: 0.1 *N* hydrochloric acid; 500 mL. [NOTE—Use the same batch of *Dissolution Medium* throughout the test.]

Ascorbic acid–methanol—Prepare a solution containing 2 mg of ascorbic acid per mL of methanol.

Hydrogen peroxide–methanol—On the day of use, dilute 2.0 mL of recently assayed 30 percent hydrogen peroxide with methanol to 100 mL. Store in a refrigerator. Just prior to use, dilute 2.0 mL of this solution with methanol to 100 mL.

Standard solutions—Weigh accurately about 25 mg of USP Digoxin RS, dissolve in a minimum amount of alcohol in a 500-mL volumetric flask, add dilute alcohol (4 in 5) to volume, and mix. Dilute 10.0 mL of this solution with dilute alcohol (4 in 5) to 100.0 mL, and mix. Just prior to use, dilute suitable aliquots of the resulting solution with *Dissolution Medium* to 50.0 mL to prepare *Standard solutions* equivalent to 20%, 40%, 60%, 80%, and 100%, respectively, of the labeled amount of digoxin in 500 mL.

Apparatus 1: 120 rpm.

Time: 60 minutes.

Procedure—Filter a portion of the solution under test promptly after withdrawal, using a suitable membrane filter of not greater than 0.8-μm porosity, discarding the first 10 mL of the filtrate. This is the *Test solution*.

Measurement of fluorescence—Transfer to individual glass-stoppered flasks duplicate 1.0-mL portions of the *Test solution*, 1.0-mL portions of each of the *Standard solutions*, and 1.0 mL of the *Dissolution Medium* to provide a blank. Begin with the *Standard solutions*, and keep all flasks in the same sequence throughout, so that the elapsed time from addition of reagents to reading of fluorescence is the same for each flask in the set. Treating one flask at a time, add the following three reagents, in the order named, in as rapid a sequence as possible, swirling after each addition: 1.0 mL of *Ascorbic acid–methanol*, 5.0 mL of hydrochloric acid, and 1.0 mL of *Hydrogen peroxide–methanol*. Insert the stoppers in the flasks, and after 2 hours, measure the fluorescence at about 485 nm, the excitation wavelength being about 372 nm. To check the stability of the fluorometer, repeat the measurement of fluorescence on one or more treated *Standard solutions*. Correct each reading for the blank, and plot a standard curve of fluorescence versus percentage dissolution. Determine the percentage dissolution of digoxin in the *Test solution* by reading from the standard graph.

Tolerances—Not less than 65% of the labeled amount of $C_{41}H_{64}O_{14}$ is dissolved in 60 minutes for not fewer than eleven-twelfths of the Tablets tested, and no individual Tablet tested is less than 55% dissolved in 60 minutes.

Uniformity of dosage units ⟨905⟩: meet the requirements.

Assay—

Mobile phase, *Chromatographic system*, and *System suitability preparation*—Proceed as directed in the *Assay* under *Digoxin*.

Standard preparation—Dissolve an accurately weighed quantity of USP Digoxin RS in diluted alcohol, and dilute quantitatively and stepwise with diluted alcohol to obtain a solution having a known concentration of about 40 μg per mL. Use a sonic bath to aid dissolution.

Assay preparation—Weigh and finely powder not less than 20 Digoxin Tablets. Transfer an accurately weighed portion of the powder, equivalent to about 1 mg of digoxin, to a glass-stoppered, 50-mL conical flask. Add 25.0 mL of diluted alcohol with swirling, sonicate for about 30 minutes, and cool. Filter a portion of this solution through a 0.8-μm porosity membrane filter, discarding the first 10 mL of the filtrate.

Procedure—Separately inject equal volumes (about 50 μL) of the *Standard preparation* and the *Assay preparation* into the chromatograph, record the chromatograms, and measure the responses for the major peaks. Calculate the quantity, in mg, of $C_{41}H_{64}O_{14}$ in the portion of Tablets taken by the formula:

$$25C(r_U / r_S),$$

in which C is the concentration, in mg per mL, of USP Digoxin RS in the *Standard preparation*, and r_U and r_S are the responses for the digoxin peaks obtained from the *Assay preparation* and the *Standard preparation*, respectively.

Dihydroergotamine Mesylate

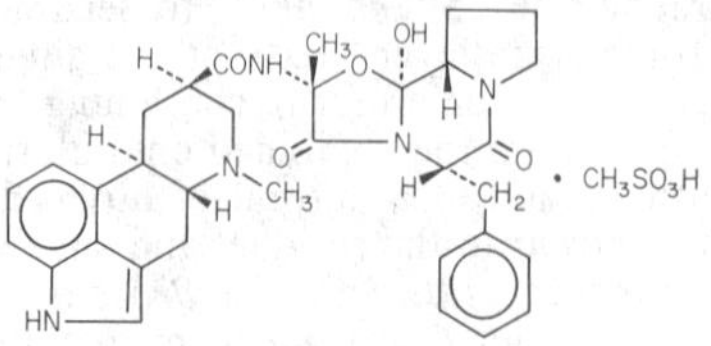

$C_{33}H_{37}N_5O_5 \cdot CH_4O_3S$ 679.79

Ergotaman-3′,6′,18-trione, 9,10-dihydro-12′-hydroxy-2′-methyl-5′-(phenylmethyl)-, (5′α)-, monomethanesulfonate (salt).

Dihydroergotamine monomethanesulfonate [*6190-39-2*].

» Dihydroergotamine Mesylate contains not less than 97.0 percent and not more than 103.0 percent of $C_{33}H_{37}N_5O_5 \cdot CH_4O_3S$, calculated on the dried basis.

Packaging and storage—Preserve in well-closed, light-resistant containers.

Reference standard—*USP Dihydroergotamine Mesylate Reference Standard*—Dry in vacuum at 100° to constant weight before using.

Identification—

A: The infrared absorption spectrum of a potassium bromide dispersion of it, previously dried, exhibits maxima at the same wavelengths as that of a similar preparation of USP Dihydroergotamine Mesylate RS.

B: The ultraviolet absorption spectrum of a 1 in 20,000 solution in 70 percent alcohol exhibits maxima and minima at the same wavelengths as that of a similar solution of USP Dihydroergotamine Mesylate RS, concomitantly measured, and the respective absorptivities, calculated on the dried basis, at the wavelength of maximum absorbance at about 280 nm do not differ by more than 3.0%.

C: The principal spot from the *Test preparation* found in the test for *Related alkaloids* corresponds in R_f value to that obtained from the *Standard preparation*.

Specific rotation ⟨781⟩: between −16.7° and −22.7°, calculated on the dried basis, determined in a solvent mixture consisting of chloroform, alcohol, and ammonium hydroxide (10:10:1) containing 250 mg in each 10 mL.

pH ⟨791⟩: between 4.4 and 5.4, in a solution (1 in 1000).

Loss on drying ⟨731⟩—Dry it in vacuum at 100° to constant weight: it loses not more than 4.0% of its weight.

Related alkaloids—

Solvent mixture—Mix 10 volumes of chloroform, 10 volumes of methanol, and 1 volume of ammonium hydroxide.

Test preparation—Prepare a solution of Dihydroergotamine Mesylate in *Solvent mixture* to contain 20 mg per mL.

Standard preparation and *Standard dilutions*—Prepare a solution of USP Dihydroergotamine Mesylate RS in *Solvent mixture* to contain 20 mg per mL (*Standard preparation*). Prepare a series of dilutions of *Standard preparation* in *Solvent mixture* to contain 0.40 mg, 0.20 mg, and 0.10 mg per mL (*Standard dilutions*).

Procedure—In a suitable chromatographic chamber arranged for thin-layer chromatography place a volume of a solvent system consisting of a mixture of chloroform and alcohol (9:1) sufficient to develop the chromatogram, cover, and allow to equilibrate for 30 minutes. On a suitable thin-layer chromatographic plate coated with a 0.25-mm layer of chromatographic silica gel apply 5-μL portions of *Test preparation*, *Standard preparation*, and each of the three *Standard dilutions*. Allow the spots to dry, and develop the chromatogram until the solvent front has moved about three-fourths of the length of the plate. Remove the plate from the developing chamber, mark the solvent front, and allow the solvent to evaporate. Locate the spots on the plate by lightly spraying with a solution prepared by dissolving 800 mg of *p*-dimethylaminobenzaldehyde in a cooled mixture of 80 g of alcohol and 20 g of sulfuric acid. The R_f value of the principal spot obtained from the *Test preparation* corresponds to that obtained from the *Standard preparation*. Estimate the concentration of any other spots observed in the lane for the *Test prepa-*

ration by comparison with the *Standard dilutions*. The spots from the 0.40-, 0.20-, and 0.10-mg-per-mL dilutions are equivalent to 2.0%, 1.0%, and 0.50% of impurities, respectively. The sum of the impurities is not greater than 2.0%.

Assay—

Standard preparation—Transfer about 10 mg of USP Dihydroergotamine Mesylate RS, accurately weighed, to a 200-mL volumetric flask, add 2 mL of methanol, dilute with tartaric acid solution (1 in 100) to volume, and mix.

Assay preparation—Using about 10 mg of Dihydroergotamine Mesylate, accurately weighed, prepare as directed for *Standard preparation*.

Procedure—Transfer 3.0 mL each of the *Standard preparation*, the *Assay preparation*, and tartaric acid solution (1 in 100) to provide the blank, to separate conical flasks. Add 6.0 mL of *p*-dimethylaminobenzaldehyde TS to each, shake, and allow to stand for 20 minutes. Concomitantly determine the absorbances of the solutions in 1-cm cells at the wavelength of maximum absorbance at about 585 nm, with a suitable spectrophotometer, against the blank. Calculate the quantity, in mg, of $C_{33}H_{37}N_5O_5.CH_4O_3S$ in the portion of Dihydroergotamine Mesylate taken by the formula:

$$0.2C(A_U/A_S),$$

in which C is the concentration, in µg per mL, of USP Dihydroergotamine Mesylate RS in the *Standard preparation*, and A_U and A_S are the absorbances of the solutions from the *Assay preparation* and the *Standard preparation*, respectively.

Dihydroergotamine Mesylate Injection

» Dihydroergotamine Mesylate Injection is a sterile solution of Dihydroergotamine Mesylate in Water for Injection. It contains not less than 90.0 percent and not more than 110.0 percent of the labeled amount of $C_{33}H_{37}N_5O_5.CH_4O_3S$.

Packaging and storage—Preserve in single-dose containers, preferably of Type I glass, protected from light.

Reference standard—*USP Dihydroergotamine Mesylate Reference Standard*—Dry in vacuum at 100° to constant weight before using.

Identification—Dilute 2 mL of Injection with water to 25 mL: the ultraviolet absorption spectrum of the solution so obtained exhibits maxima and minima at the same wavelengths as that of a similar solution of USP Dihydroergotamine Mesylate RS, concomitantly measured.

pH ⟨791⟩: between 3.2 and 4.0.

Other requirements—It meets the requirements under *Injections* ⟨1⟩.

Assay—

Reagent preparation—Dissolve 250 mg of *p*-dimethylaminobenzaldehyde in a cooled mixture of 130 mL of sulfuric acid and 70 mL of water, and add 0.40 mL of ferric chloride solution (1 in 20).

Standard preparation—Dissolve in tartaric acid solution (1 in 100) a suitable quantity of USP Dihydroergotamine Mesylate RS, accurately weighed, and dilute quantitatively and stepwise with the same solvent to obtain a solution having a known concentration of about 50 µg per mL.

Assay preparation—Transfer an accurately measured volume of Dihydroergotamine Mesylate Injection, equivalent to about 5 mg of dihydroergotamine mesylate, to a 100-mL volumetric flask, dilute with tartaric acid solution (1 in 100) to volume, and mix.

Procedure—Transfer 5.0 mL each of the *Standard preparation*, the *Assay preparation*, and tartaric acid solution (1 in 100) to provide the blank, to separate 50-mL conical flasks. Add 10.0 mL of the *Reagent preparation* to each, shake, and allow to stand for 30 minutes. Concomitantly determine the absorbances of the solutions in 1-cm cells at the wavelength of maximum absorbance at about 585 nm, with a suitable spectrophotometer, using the blank to set the instrument. Calculate the quantity, in mg, of $C_{33}H_{37}N_5O_5.CH_4O_3S$ in each mL of the Injection taken by the formula:

$$(0.1C/V)(A_U/A_S),$$

in which C is the concentration, in µg per mL, of USP Dihydroergotamine Mesylate RS in the *Standard preparation*, V is the volume, in mL, of Injection taken, and A_U and A_S are the absorbances of the solutions from the *Assay preparation* and the *Standard preparation*, respectively.

Dihydroergotamine Mesylate, Heparin Sodium, and Lidocaine Hydrochloride Injection

» Dihydroergotamine Mesylate, Heparin Sodium, and Lidocaine Hydrochloride Injection is a sterile solution of Dihydroergotamine Mesylate, Heparin Sodium, and Lidocaine Hydrochloride in Water for Injection. It contains not less than 90.0 percent and not more than 110.0 percent of the labeled amounts of dihydroergotamine mesylate ($C_{33}H_{37}N_5O_5.CH_4O_4S$) and lidocaine hydrochloride ($C_{14}H_{22}N_2O.HCl$) and exhibits a potency of not less than 90.0 percent and not more than 110.0 percent of the potency stated on the label in terms of USP Heparin Units.

NOTE—Heparin Units are consistently established on the basis of the *Assay* set forth herein, independently of International Units, and the respective units are not equivalent (see *General Notices*).

Packaging and storage—Preserve in single-dose or in multiple-dose containers, preferably of Type I glass.

Reference standards—*USP Dihydroergotamine Mesylate Reference Standard*—Dry in vacuum at 100° to constant weight before using. *USP Heparin Sodium Reference Standard*—Store in a cool place, and do not freeze. *USP Lidocaine Reference Standard*—Dry in vacuum over silica gel for 24 hours before using.

Identification—Dilute the contents of 1 vial with 10 mL of water, transfer to a 125-mL separator, and add 5 mL of ammonium hydroxide. Extract with six 15-mL portions of chloroform, collecting the chloroform extracts in a conical flask. Evaporate the chloroform solution on a rotary evaporator with the aid of a warm water bath to dryness. Dissolve the residue in 1.0 mL of dilute alcohol (1 in 2). Apply 10 µL of this solution and 10 µL each of a solution containing 0.7 mg per mL of USP Dihydroergotamine Mesylate RS in dilute alcohol (1 in 2) and 9.2 mg per mL of USP Lidocaine RS in dilute alcohol (1 in 2), on a suitable thin-layer chromatographic plate (see *Chromatography* ⟨621⟩), coated with a 0.25-mm layer of chromatographic silica gel mixture. Allow the spots to dry, and develop the plate in an equilibrated chamber containing a mixture of chloroform, methanol, and ammonium hydroxide (85:18:3) until the solvent front has moved about three-fourths of the length of the plate. Remove the plate from the chamber, and dry in a current of cold air. Examine the plate under short-wavelength ultraviolet light: the R_f value of fluorescent spot from the test solution corresponds to that obtained from the USP Lidocaine Hydrochloride RS solution. Spray the plate with a solution containing 800 mg of *p*-dimethylaminobenzaldehyde in 50 mL of ethanolic sulfuric acid (1 in 2): the test solution exhibits a blue spot that corresponds in color and R_f value to that obtained from the USP Dihydroergotamine Mesylate RS solution.

Pyrogen—When 0.7 mL of the Injection is diluted with *Water for Injection* to 10.0 mL, it meets the requirements of the *Pyrogen Test* ⟨151⟩.

pH ⟨791⟩: between 5.0 and 6.5, determined on a portion diluted with an equal volume of potassium nitrate solution (1 in 100).

Other requirements—It meets the requirements under *Injections* ⟨1⟩.

Assay for heparin sodium—Proceed as directed in the *Assay* under *Heparin Sodium*, substituting the Injection for the solution of heparin sodium prepared as directed under *Assay preparation*. Under *Calculation*, v_U in the equation for R is the volume, in mL, of the Injection per mL of the *Assay preparation*. The potency of the Injection in USP Heparin Units per mL is $P_* =$ antilog M.

Assay for dihydroergotamine mesylate and lidocaine hydrochloride—

Mobile phase—Prepare a filtered and degassed mixture of water and acetonitrile (65:35). To 1 liter of this mixture add 0.4 mL of triethylamine, and adjust the solution by the dropwise addition of about 1 mL of dilute phosphoric acid (1 in 2) to a pH of 8.2. Make adjustments if necessary (see *System Suitability* under *Chromatography* ⟨621⟩).

Standard preparation—Using accurately weighed quantities of USP Dihydroergotamine Mesylate RS and USP Lidocaine RS prepare a solution in dilute alcohol (1 in 2) having known concentrations of about 0.2 mg and 3 mg, respectively, per mL.

Assay preparation—Using a "To contain" pipet, transfer an accurately measured volume of Injection, equivalent to about 2 mg of dihydroergotamine mesylate, to a 10-mL volumetric flask. Rinse the pipet with dilute alcohol (1 in 2), collecting the rinse in the volumetric flask. Add dilute alcohol (1 in 2) to volume, and mix.

Chromatographic system (see *Chromatography* ⟨621⟩)—The liquid chromatograph is equipped with a 268-nm detector and 3.9-mm × 15-cm column that contains packing L1. The flow rate is about 3 mL per minute. Chromatograph the *Standard preparation*, and record the peak responses as directed under *Procedure:* the column efficiency determined from the dihydroergotamine peak is not less than 800 theoretical plates, the resolution, R, between the dihydroergotamine and lidocaine peaks is not less than 2.0, and the relative standard deviation for replicate injections is not more than 3.0%.

Procedure—Separately inject equal volumes (about 50 µL) of the *Standard preparation* and the *Assay preparation* into the chromatograph, record the chromatograms, and measure the responses for the major peaks. The relative retention times are 1.0 for lidocaine and 1.5 for dihydroergotamine. Calculate the quantity, in mg, of dihydroergotamine mesylate ($C_{33}H_{37}N_5O_5 \cdot CH_4O_4S$) in each mL of the Injection taken by the formula:

$$10(C/V)(r_U/r_S),$$

in which C is the concentration, in mg per mL of USP Dihydroergotamine Mesylate RS in the *Standard preparation*, V is the volume, in mL, of Injection taken, and r_U and r_S are the peak responses obtained from the *Assay preparation* and the *Standard preparation*, respectively. Calculate the quantity, in mg, of lidocaine hydrochloride ($C_{14}H_{22}N_2O.HCl$) in each mL of the Injection taken by the formula:

$$(270.80/234.34)(10)(C/V)(r_U/r_S),$$

in which 270.80 and 234.34 are the molecular weights of lidocaine hydrochloride and lidocaine, respectively, C is the concentration, in mg per mL, of USP Lidocaine RS in the *Standard preparation*, and V, r_U and r_S are as defined above.

Dihydrostreptomycin Sulfate

R	R′
CH_3	CH_2OH

$\cdot 3H_2SO_4$

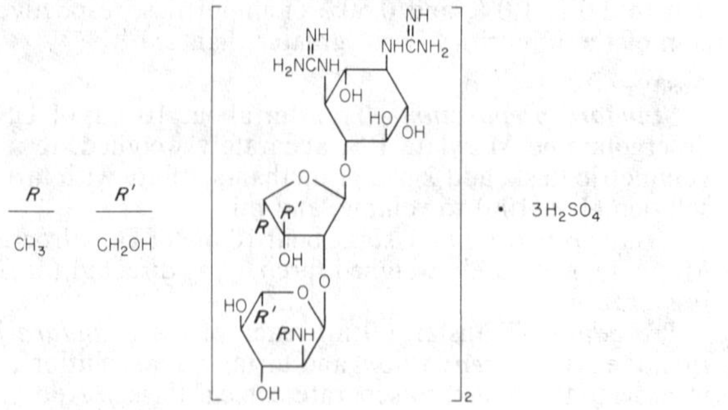

$(C_{21}H_{41}N_7O_{12})_2.3H_2SO_4$ 1461.41
Dihydrostreptomycin sulfate (2:3) (salt) [*5490-27-7*].

» Dihydrostreptomycin Sulfate has a potency equivalent to not less than 650 µg of dihydrostreptomycin ($C_{21}H_{41}N_7O_{12}$) per mg, except that if it is labeled as being crystalline, it has a potency equivalent to not less than 725 µg of dihydrostreptomycin per mg, or if it is labeled as being solely for oral use, it has a potency equivalent to not less than 450 µg of dihydrostreptomycin per mg.

Packaging and storage—Preserve in tight containers.

Labeling—Label it to indicate that it is intended for veterinary use only. If it is crystalline, it may be so labeled. If it is intended solely for oral use, it is so labeled.

Reference standards—*USP Dihydrostreptomycin Sulfate Reference Standard*—Dry in vacuum at a pressure not exceeding 5 mm of mercury at 100° for 4 hours before using. *USP Streptomycin Sulfate Reference Standard*—Dry in vacuum at a pressure not exceeding 5 mm of mercury at 60° for 3 hours before using.

Identification—

A: To a solution of 4 mg in 2 mL of water, add 0.5 mL of 1 *N* hydrochloric acid, and heat in a water bath for 20 minutes. Remove the tube from the bath, and add 1.0 mL of a 1 in 200 solution of 1-naphthol in 1 *N* sodium hydroxide. Heat again for 10 minutes, cool briefly in an ice bath, and add water to make 25 mL: a red color develops, intensifying during about 10 minutes.

B: A solution (1 in 50) responds to the tests for *Sulfate* ⟨191⟩.

Crystallinity ⟨695⟩ (where labeled as being crystalline): meets the requirements.

pH ⟨791⟩: between 4.5 and 7.0, in a solution containing 200 mg of dihydrostreptomycin per mL, except that if it is labeled as being solely for oral use, the pH is between 3.0 and 7.0.

Loss on drying ⟨731⟩—Dry about 100 mg in a capillary-stoppered bottle in vacuum at 60° for 3 hours: it loses not more than 5.0% of its weight, except that if it is labeled as being solely for oral use, it loses not more than 14.0% of its weight.

Streptomycin—

Ferric chloride stock solution—Dissolve 5 g of ferric chloride in 50 mL of 0.1 *N* hydrochloric acid.

Ferric chloride solution—Dilute 2.5 mL of *Ferric chloride stock solution* with sufficient 0.01 *N* hydrochloric acid to make 100 mL. Use this solution within 1 day.

Standard preparation—Dissolve an accurately weighed quantity of USP Streptomycin Sulfate RS in water to obtain a stock solution containing 1.0 mg of streptomycin ($C_{21}H_{39}N_7O_{12}$) per mL. Transfer 1.0, 2.0, 3.0, 4.0, and 5.0 mL, respectively, of this stock solution to each of five 25-mL volumetric flasks. Transfer 9.0, 8.0, 7.0, 6.0, and 5.0 mL of water to the flasks, respectively.

Test preparation—Transfer about 800 mg of Dihydrostreptomycin Sulfate, accurately weighed, to a 25-mL volumetric flask, dissolve in water, dilute with water to volume, and mix. Transfer 10.0 mL of this solution to a second 25-mL volumetric flask.

Procedure—To each of the flasks containing the *Standard preparations* and the *Test preparation*, and to a seventh 25-mL volumetric flask containing 10.0 mL of water to provide a blank, add 2.0 mL of 1 *N* sodium hydroxide, and heat in a water bath for 10 minutes. Cool the flasks in ice water for 3 minutes, and to each add 2.0 mL of 1.2 *N* hydrochloric acid and 5.0 mL of *Ferric chloride solution.* Dilute with water to volume, and mix. Concomitantly determine the absorbances of the solutions from the *Standard preparations* and the *Test preparation* at the wavelength of maximum absorbance at about 550 nm, with a suitable spectrophotometer, using the blank to set the instrument at zero. Plot the absorbance values of the solutions from the *Standard preparations* versus concentration, in µg per mL, of streptomycin, and draw the straight line best fitting the five plotted points. From the graph so obtained, determine the concentration, *C*, in µg per mL, of streptomycin in the solution from the *Test preparation.* Calculate the percentage of streptomycin in the portion of Dihydrostreptomycin Sulfate taken by the formula:

$$6250C/WP,$$

in which *W* is the weight, in mg, of Dihydrostreptomycin Sulfate taken, and *P* is the potency, in µg of dihydrostreptomycin per mg, of the Dihydrostreptomycin Sulfate taken as determined in the *Assay:* not more than 3.0% is found, except that if it is labeled as being crystalline, not more than 1.0% is found, or if it is labeled as being solely for oral use, not more than 5.0% is found.

Assay—Proceed with Dihydrostreptomycin Sulfate as directed for the turbidimetric assay of dihydrostreptomycin under *Antibiotics—Microbial Assays* ⟨81⟩.

Dihydrostreptomycin Sulfate Boluses

» Dihydrostreptomycin Sulfate Boluses contain the equivalent of not less than 85.0 percent and not more than 120.0 percent of the labeled amount of dihydrostreptomycin ($C_{21}H_{41}N_7O_{12}$).

Packaging and storage—Preserve in tight containers.

Labeling—Label Boluses to indicate that they are intended for veterinary use only.

Reference standard—*USP Dihydrostreptomycin Sulfate Reference Standard*—Dry in vacuum at a pressure not exceeding 5 mm of mercury at 100° for 4 hours before using.

Loss on drying ⟨731⟩—Dry about 100 mg, accurately weighed, of finely ground Boluses in a capillary-stoppered bottle in vacuum at a pressure not exceeding 5 mm of mercury at 60° for 3 hours: it loses not more than 10.0% of its weight.

Assay—Proceed as directed for the cylinder-plate assay of dihydrostreptomycin under *Antibiotics—Microbial Assays* ⟨81⟩, using 3 Dihydrostreptomycin Sulfate Boluses added to a mixture of 499.0 mL of *Buffer No. 3* and 1.0 mL of polysorbate 80 and blended for 5 minutes in a high-speed glass blender jar. Allow to stand for not less than 1 hour, and repeat the blending. While stirring this mixture, withdraw an accurately measured volume of it, and dilute quantitatively and stepwise with *Buffer No. 3* to obtain a *Test Dilution* having a concentration assumed to be equal to the median dose level of the Standard.

Dihydrostreptomycin Sulfate Injection

» Dihydrostreptomycin Sulfate Injection contains the equivalent of not less than 90.0 percent and not more than 120.0 percent of the labeled amount of dihydrostreptomycin ($C_{21}H_{41}N_7O_{12}$). It contains one or more suitable preservatives.

Packaging and storage—Preserve in single-dose or in multiple-dose containers.

Labeling—Label it to indicate that it is intended for veterinary use only.

Reference standard—*USP Dihydrostreptomycin Sulfate Reference Standard*—Dry in vacuum at a pressure not exceeding 5 mm of mercury at 100° for 4 hours before using.

Identification—It responds to the *Identification tests* under *Dihydrostreptomycin Sulfate.*

Depressor substances—It meets the requirements of the *Depressor Substances Test* ⟨101⟩, the test dose being 1.0 mL per kg of a solution in sterile saline TS, containing 3.0 mg of dihydrostreptomycin per mL.

Pyrogen—It meets the requirements of the *Pyrogen Test* ⟨151⟩, the test dose being 1.0 mL per kg of a solution in Sterile Water for Injection containing 10 mg of dihydrostreptomycin per mL.

Sterility—It meets the requirements under *Sterility Tests* ⟨71⟩, when tested as directed in the section, *Test Procedures Using Membrane Filtration.*

pH ⟨791⟩: between 5.0 and 8.0.

Assay—

Assay preparation 1 (where it is represented as being in a single-dose container)—Withdraw all of the withdrawable contents of Dihydrostreptomycin Sulfate Injection, using a suitable hypodermic needle and syringe, and dilute quantitatively with water to obtain a solution containing a convenient quantity of dihydrostreptomycin in each mL.

Assay preparation 2 (where the label states the quantity of dihydrostreptomycin in a given volume of solution)—Dilute an accurately measured volume of Dihydrostreptomycin Sulfate Injection quantitatively with water to obtain a solution containing a convenient quantity of dihydrostreptomycin in each mL.

Procedure—Proceed as directed for the turbidimetric assay of dihydrostreptomycin under *Antibiotics—Microbial Assays* ⟨81⟩, using an accurately measured volume of *Assay preparation* diluted quantitatively with water to yield a *Test Dilution* having a concentration assumed to be equal to the median dose level of the Standard.

Sterile Dihydrostreptomycin Sulfate

» Sterile Dihydrostreptomycin Sulfate is Dihydrostreptomycin Sulfate suitable for parenteral use. It has a potency equivalent to not less than 650 µg of dihydrostreptomycin ($C_{21}H_{41}N_7O_{12}$) per mg, except that if it is labeled as being crystalline, it has a potency equivalent to not less than 725 µg of dihydrostreptomycin per mg.

Packaging and storage—Preserve in *Containers for Sterile Solids* as described under *Injections* ⟨1⟩.

Labeling—Label it to indicate that it is intended for veterinary use only. If it is crystalline, it may be so labeled.

Reference standards—*USP Dihydrostreptomycin Sulfate Reference Standard*—Dry in vacuum at a pressure not exceeding 5 mm of mercury at 100° for 4 hours before using. *USP Streptomycin Sulfate Reference Standard*—Dry in vacuum at a pressure not exceeding 5 mm of mercury at 60° for 3 hours before using.

Depressor substances—It meets the requirements of the *Depressor Substances Test* ⟨101⟩, the test dose being 1.0 mL per kg of a solution in sterile saline TS, containing 3.0 mg of dihydrostreptomycin per mL.

Pyrogen—It meets the requirements of the *Pyrogen Test* ⟨151⟩, the test dose being 1.0 mL per kg of a solution in Sterile Water for Injection containing 10 mg of dihydrostreptomycin per mL.

Sterility—It meets the requirements under *Sterility Tests* ⟨71⟩, when tested as directed in the section, *Test Procedures Using Membrane Filtration.*

Other requirements—It responds to the *Identification tests*, and meets the requirements for *Crystallinity*, *pH*, *Loss on drying*, *Streptomycin*, and *Assay* under *Dihydrostreptomycin Sulfate.*

Dihydrostreptomycin Sulfate Intramammary Infusion, Penicillin G Procaine and—*see* Penicillin G Procaine and Dihydrostreptomycin Sulfate Intramammary Infusion

Dihydrostreptomycin Sulfate Suspension, Sterile Penicillin G Procaine and—*see* Sterile Penicillin G Procaine and Dihydrostreptomycin Sulfate Suspension

Dihydrostreptomycin Sulfate, Chlorpheniramine Maleate, and Dexamethasone Suspension, Sterile Penicillin G Procaine,—*see* Sterile Penicillin G Procaine, Dihydrostreptomycin Sulfate, Chlorpheniramine Maleate, and Dexamethasone Suspension

Dihydrostreptomycin Sulfate, and Prednisolone Suspension, Sterile Penicillin G Procaine,—*see* Sterile Penicillin G Procaine, Dihydrostreptomycin Sulfate, and Prednisolone Suspension

Dihydrotachysterol

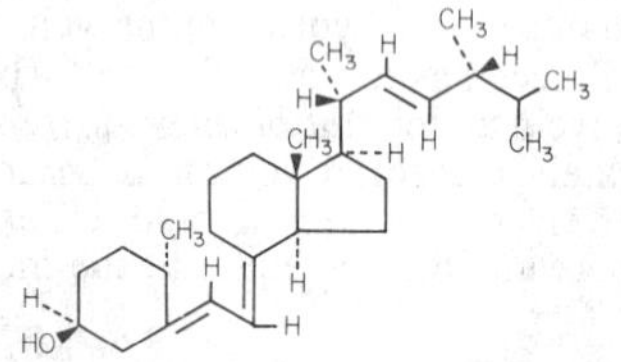

$C_{28}H_{46}O$ 398.67
9,10-Secoergosta-5,7,22-trien-3-ol, (3β,5*E*,7*E*,10α,22*E*)-.
Dihydrotachysterol.
9,10-Secoergosta-5,7,22-trien-3β-ol [*67-96-9*].

» Dihydrotachysterol contains not less than 97.0 percent and not more than 103.0 percent of $C_{28}H_{46}O$.

Packaging and storage—Preserve in light-resistant, hermetic glass containers from which air has been displaced by an inert gas.

Reference standard—*USP Dihydrotachysterol Reference Standard*—Store in a cold place, protected from light. Allow to reach room temperature before opening ampuls, and use the material promptly.

Identification—

A: The infrared absorption spectrum of a potassium bromide dispersion of it exhibits maxima only at the same wavelengths as that of a similar preparation of USP Dihydrotachysterol RS.

B: The ultraviolet absorption spectrum of a 1 in 125,000 solution in alcohol exhibits maxima and minima at the same wavelengths as that of a similar solution of USP Dihydrotachysterol RS.

Specific rotation ⟨781⟩: between +100° and +103°, determined in a solution in alcohol containing 200 mg in each 10 mL.

Residue on ignition ⟨281⟩: not more than 0.1%.

Assay—

Mobile phase—Prepare a degassed and filtered solution of isooctane and isopropyl alcohol (100:1). Make adjustments if necessary (see *System Suitability* under *Chromatography* ⟨621⟩).

Standard preparation—Dissolve an accurately weighed quantity of USP Dihydrotachysterol RS in *Mobile phase*, and dilute quantitatively and stepwise, if necessary, with *Mobile phase* to obtain a solution having a known concentration of about 10 µg per mL.

Assay preparation—Transfer about 50 mg of Dihydrotachysterol, accurately weighed, to a 500-mL volumetric flask, dissolve in *Mobile phase*, dilute with *Mobile phase* to volume, and mix. Pipet 10 mL of the resulting solution into a 100-mL volumetric flask, dilute with *Mobile phase* to volume, and mix.

System suitability preparation—Prepare a solution of ergocalciferol in *Mobile phase* having a concentration of about 0.7 mg per mL. Reflux under nitrogen for 20 minutes, and cool to ambient temperature (*Solution A*). Pipet 3 mL of *Solution A* and 2 mL of a solution in *Mobile phase* containing 0.1 mg of dihydrotachysterol per mL into a 25-mL volumetric flask, dilute with *Mobile phase* to volume, and mix.

Chromatographic system (see *Chromatography* ⟨621⟩)—The liquid chromatograph is equipped with a 254-nm detector and a 4.6-mm × 25-cm column that contains packing L3. The flow rate is about 1 mL per minute. Chromatograph replicate injections of the *Standard preparation*, and record the peak responses as directed under *Procedure:* the relative standard deviation is not more than 2.5%. Chromatograph the *System suitability preparation*, and record the peak responses as directed under *Procedure:* the resolution, *R*, between the pre-ergocalciferol and dihydrotachysterol peaks is not less than 1.5.

Procedure—Separately inject equal volumes (about 20 µL) of the *Standard preparation* and the *Assay preparation* into the chromatograph, record the chromatograms, and measure the responses for the major peaks. The relative retention times are about 0.6 for pre-ergocalciferol, 0.7 for dihydrotachysterol, and 1.0 for ergocalciferol. Calculate the quantity, in mg, of $C_{28}H_{46}O$ in the portion of Dihydrotachysterol taken by the formula:

$$5C(r_U / r_S),$$

in which *C* is the concentration, in µg per mL, of USP Dihydrotachysterol RS in the *Standard preparation*, and r_U and r_S are the peak responses for dihydrotachysterol obtained from the *Assay preparation* and the *Standard preparation*, respectively.

Dihydrotachysterol Capsules

» Dihydrotachysterol Capsules contain a solution of Dihydrotachysterol in a suitable vegetable oil. Capsules contain not less than 90.0 percent and not more than 110.0 percent of the labeled amount of $C_{28}H_{46}O$.

Packaging and storage—Preserve in well-closed, light-resistant containers.

Reference standard—*USP Dihydrotachysterol Reference Standard*—Store in a cold place, protected from light. Allow to reach room temperature before opening ampuls, and use the material promptly.

Identification—

A: Cut open 1 Capsule, and remove the contents: the Capsule contents respond to *Identification test A* under *Dihydrotachysterol Oral Solution.*

B: The retention time of the major peak in the chromatogram of the *Assay preparation* corresponds to that of the *Standard preparation* obtained as directed in the *Assay.*

Uniformity of dosage units ⟨905⟩: meet the requirements.

Assay—

Mobile phase, Standard preparation, System suitability preparation, and *Chromatographic system*—Prepare as directed in the *Assay* under *Dihydrotachysterol.*

Assay preparation—Combine the contents of not less than 20 Dihydrotachysterol Capsules. Transfer an accurately weighed quantity of Capsule contents, equivalent to about 500 µg of dihydrotachysterol, to a 50-mL volumetric flask, dissolve in 25 mL of *Mobile phase*, and mix. Dilute with *Mobile phase* to volume, and mix.

Procedure—Proceed as directed for *Procedure* in the *Assay* under *Dihydrotachysterol.* Calculate the quantity, in µg, of $C_{28}H_{46}O$ in the portion of Capsules taken by the formula:

$$50C(r_U / r_S),$$

in which *C* is the concentration, in µg per mL, of USP Dihydrotachysterol RS in the *Standard preparation*, and r_U and r_S are the peak responses obtained from the *Assay preparation* and the *Standard preparation*, respectively.

Dihydrotachysterol Oral Solution

» Dihydrotachysterol Oral Solution contains not less than 90.0 percent and not more than 110.0 percent of the labeled amount of $C_{28}H_{46}O$.

Packaging and storage—Preserve in tight, light-resistant glass containers.

Reference standard—*USP Dihydrotachysterol Reference Standard*—Store in a cold place, protected from light. Allow to reach room temperature before opening ampuls, and use the material promptly.

Identification—

A: Place 1 drop of Oral Solution on a thin-layer chromatographic plate coated with a 0.25-mm layer of chromatographic silica gel. Spray the plate with 3.5% phosphomolybdic acid solution (prepared by dissolving 3.5 g of phosphomolybdic acid in 100 mL of isopropyl alcohol), and immediately heat the plate over a hot plate: a dark blue spot appears on a yellow background.

B: The retention time of the major peak in the chromatogram of the *Assay preparation* corresponds to that of the *Standard preparation* obtained as directed in the *Assay*.

Assay—

Mobile phase, Standard preparation, System suitability preparation, and *Chromatographic system*—Prepare as directed in the *Assay* under *Dihydrotachysterol*.

Assay preparation [FOR ORAL SOLUTION IN AN OIL MEDIUM]—Transfer an accurately measured volume of Dihydrotachysterol Oral Solution, equivalent to about 500 µg of dihydrotachysterol, by means of a "to contain" pipet to a 50-mL volumetric flask. Rinse the pipet with *Mobile phase*, add the rinsing to the flask, dilute with *Mobile phase* to volume, and mix.

Assay preparation [FOR ORAL SOLUTION IN AN AQUEOUS MEDIUM]—Transfer an accurately measured volume of Dihydrotachysterol Oral Solution, equivalent to about 600 µg of dihydrotachysterol, to a separator containing about 30 mL of water. Add about 1 g of sodium chloride, mix, and extract with three 15-mL portions of chloroform, filtering each portion through absorbent cotton into a suitable glass-stoppered conical flask. Wash the cotton with about 5 mL of chloroform, collecting the washing in the glass-stoppered conical flask. Evaporate the chloroform extracts and washing with the aid of a current of air to dryness. Dissolve the residue in 50.0 mL of *Mobile phase*, and mix.

Procedure—Proceed as directed for *Procedure* in the *Assay* under *Dihydrotachysterol*. Calculate the quantity, in µg, of $C_{28}H_{46}O$ in the volume of Oral Solution taken by the formula:

$$50C(r_U / r_S),$$

in which C is the concentration, in µg per mL, of USP Dihydrotachysterol RS in the *Standard preparation*, and r_U and r_S are the peak responses obtained from the *Assay preparation* and the *Standard preparation*, respectively.

Dihydrotachysterol Tablets

» Dihydrotachysterol Tablets contain not less than 90.0 percent and not more than 110.0 percent of the labeled amount of $C_{28}H_{46}O$.

Packaging and storage—Preserve in well-closed, light-resistant containers.

Reference standard—*USP Dihydrotachysterol Reference Standard*—Store in a cold place, protected from light. Allow to reach room temperature before opening ampuls, and use the material promptly.

Identification—

A: Transfer a portion of powdered Tablets, equivalent to about 2 mg of dihydrotachysterol, to a glass-stoppered flask, add about 25 mL of methylene chloride, shake for 15 minutes, and filter. Evaporate the filtrate to dryness, and dissolve the residue in 0.4 mL of methylene chloride. Apply 10 µL of this solution and 10 µL of a Standard solution of USP Dihydrotachysterol RS in methylene chloride containing 5 mg per mL at separate points on a thin-layer chromatographic plate coated with a 0.25-mm layer of chromatographic silica gel. Develop the chromatogram using a solvent system consisting of ether and cyclohexane (1:1) until the solvent front has moved about three-fourths of the length of the plate. Air-dry, spray lightly with a 1 in 5 solution of *p*-toluenesulfonic acid in a mixture of alcohol and propylene glycol (9:1), and heat at 80° until reddish brown spots appear (about 10 minutes): the R_f value of the principal spot obtained from the solution under test corresponds to that obtained from the Standard solution.

B: The retention time of the major peak in the chromatogram of the *Assay preparation* corresponds to that of the *Standard preparation* obtained as directed in the *Assay*.

Disintegration ⟨701⟩: 10 minutes.

Uniformity of dosage units ⟨905⟩: meet the requirements.

Assay—

Mobile phase, Standard preparation, System suitability preparation, and *Chromatographic system*—Prepare as directed in the *Assay* under *Dihydrotachysterol*.

Assay preparation—Weigh and finely powder not less than 20 Dihydrotachysterol Tablets. Transfer an accurately weighed portion of the powder, equivalent to about 5 mg of dihydrotachysterol, to a 100-mL volumetric flask. Add about 80 mL of *Mobile phase*, and shake by mechanical means for 10 minutes. Dilute with *Mobile phase* to volume, mix, and filter, discarding the first 10 mL of the filtrate. Pipet 10 mL of the filtrate into a 50-mL volumetric flask, dilute with *Mobile phase* to volume, and mix.

Procedure—Proceed as directed for *Procedure* in the *Assay* under *Dihydrotachysterol*. Calculate the quantity, in µg, of $C_{28}H_{46}O$ in the portion of Tablets taken by the formula:

$$500C(r_U / r_S),$$

in which C is the concentration, in µg per mL, of USP Dihydrotachysterol RS in the *Standard preparation*, and r_U and r_S are the peak responses obtained from the *Assay preparation* and the *Standard preparation*, respectively.

Dihydroxyaluminum Aminoacetate

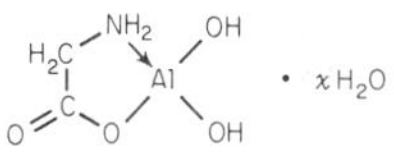

$C_2H_6AlNO_4 . xH_2O$

Aluminum, (glycinato-*N,O*)dihydroxy-, hydrate.

(Glycinato)dihydroxyaluminum hydrate [*41354-48-7*].

Anhydrous 135.06 [*13682-92-3*].

» Dihydroxyaluminum Aminoacetate yields not less than 94.0 percent and not more than 102.0 percent of dihydroxyaluminum aminoacetate ($C_2H_6AlNO_4$), calculated on the dried basis. It may contain small amounts of aluminum oxide and of Aminoacetic Acid.

Packaging and storage—Preserve in well-closed containers.

Identification—Suspend 1 g in 25 mL of water, add hydrochloric acid, dropwise, until a clear solution is formed, and divide it into two equal parts for the following tests.

A: One portion of the solution responds to the tests for *Aluminum* ⟨191⟩.

B: To the other portion of the solution add 1 drop of liquefied phenol and 5 mL of sodium hypochlorite TS: a blue color is produced.

pH ⟨791⟩: between 6.5 and 7.5, in a suspension of 1 g of it, finely powdered, in 25 mL of water.

Loss on drying ⟨731⟩—Dry it at 130° to constant weight: it loses not more than 14.5% of its weight.

Mercury, *Method IIa* ⟨261⟩—Transfer 2.0 g to a 100-mL beaker, and add 35 mL of 1 *N* sulfuric acid: the limit is 1 ppm.

Isopropyl alcohol—Transfer about 5 g to a flask provided with a reflux condenser, and add 100 mL of potassium permanganate solution (1 in 300) and 10 mL of sulfuric acid. Reflux the mixture for 30 minutes, distil, and collect 10 mL of the distillate. To 1 mL of the distillate add 5 drops of sodium nitroferricyanide TS and 2 mL of 1 *N* sodium hydroxide, then add a slight excess of 6 *N* acetic acid: no red color is produced.

Nitrogen—Determine the nitrogen content as directed under *Nitrogen Determination, Method II* ⟨461⟩, using about 100 mg, previously dried and accurately weighed. Each mL of 0.1 *N* sulfuric acid is equivalent to 1.401 mg of nitrogen. Not less than 9.90% and not more than 10.60% of nitrogen is found.

Assay—

Disodium ethylenediaminetetraacetate titrant—Prepare and standardize as directed in the *Assay* under *Alum.*

Procedure—Transfer about 2.5 g of Dihydroxyaluminum Aminoacetate, accurately weighed, to a 150-mL beaker, add 15 mL of hydrochloric acid, and warm, if necessary, to dissolve the specimen completely. Transfer the solution with the aid of water to a 500-mL volumetric flask, dilute with water to volume, and mix. Transfer 20.0 mL of this solution to a 250-mL beaker, and add, with continuous stirring, 25.0 mL of *Disodium ethylenediaminetetraacetate titrant* and then 20 mL of acetic acid–ammonium acetate buffer TS. Heat the solution near the boiling point for 5 minutes, cool, and add 50 mL of alcohol and 2 mL of dithizone TS. Titrate with 0.05 *M* zinc sulfate VS until the color changes from green-violet to rose pink. Perform a blank determination, substituting 20 mL of water for the assay solution, and make any necessary correction. Each mL of 0.05 *M Disodium ethylenediaminetetraacetate titrant* is equivalent to 6.753 mg of $C_2H_6AlNO_4$.

Dihydroxyaluminum Aminoacetate Capsules

» Dihydroxyaluminum Aminoacetate Capsules contain not less than 90.0 percent and not more than 110.0 percent of the labeled amount of $C_2H_6AlNO_4$.

Packaging and storage—Preserve in well-closed containers.

Identification—Weigh accurately the contents of not less than 20 Capsules, and mix. Suspend a portion of the powder, equivalent to about 1 g of dihydroxyaluminum aminoacetate, in 25 mL of water, add hydrochloric acid, dropwise, until all of the dihydroxyaluminum aminoacetate is dissolved, and filter: the solution so obtained responds to the *Identification tests* under *Dihydroxyaluminum Aminoacetate.* Retain the remaining portion of the powder for the following tests as indicated.

Disintegration ⟨701⟩: 10 minutes, simulated gastric fluid TS being substituted for water in the test.

Uniformity of dosage units ⟨905⟩: meet the requirements.

Acid-neutralizing capacity ⟨301⟩: not less than 5 mEq of acid is consumed by the minimum single dose recommended in the labeling.

pH ⟨791⟩: between 6.5 and 7.5, in a suspension of the powder retained from the *Identification test*, equivalent to about 1 g of dihydroxyaluminum aminoacetate in 25 mL of water.

Assay—

Disodium ethylenediaminetetraacetate titrant—Prepare and standardize as directed in the *Assay* under *Ammonium Alum.*

Procedure—Transfer an accurately weighed portion of the powder retained from the *Identification test*, equivalent to about 2.5 g of dihydroxyaluminum aminoacetate, to a flask, add 15 mL of hydrochloric acid and 10 mL of water, and boil gently for about 5 minutes. Cool, transfer the solution with the aid of water to a 500-mL volumetric flask, dilute with water to volume, mix, and filter to obtain a clear solution. Proceed as directed in the *Assay* under *Dihydroxyaluminum Aminoacetate*, beginning with "Transfer 20.0 mL of this solution."

Dihydroxyaluminum Aminoacetate Magma

» Dihydroxyaluminum Aminoacetate Magma is a suspension that contains not less than 90.0 percent and not more than 110.0 percent of the labeled amount of $C_2H_6AlNO_4$.

Packaging and storage—Preserve in tight containers, and protect from freezing.

Identification—Dilute a volume of Magma, equivalent to about 1 g of dihydroxyaluminum aminoacetate, with water to 25 mL, add hydrochloric acid, dropwise, until a solution results, and then filter: the filtered solution responds to the *Identification tests* under *Dihydroxyaluminum Aminoacetate.*

Microbial limits ⟨61⟩—The total bacterial count does not exceed 100 per mL and the test for *Escherichia coli* is negative.

pH ⟨791⟩: between 6.5 and 7.5, in a dilution in water, equivalent to about 1 g of dihydroxyaluminum aminoacetate in 25 mL.

Assay—

Disodium ethylenediaminetetraacetate titrant—Prepare and standardize as directed in the *Assay* under *Alum.*

Procedure—[NOTE—Shake the container by mechanical means for 1 hour before removing the specimen.] Weigh accurately in a tared beaker a quantity of Dihydroxyaluminum Aminoacetate Magma, equivalent to about 2.5 g of dihydroxyaluminum aminoacetate, add 15 mL of hydrochloric acid, and boil gently for about 5 minutes. Cool, transfer the solution with the aid of water to a 500-mL volumetric flask, dilute with water to volume, mix, and filter, if necessary, to obtain a clear solution. Proceed as directed in the *Assay* under *Dihydroxyaluminum Aminoacetate*, beginning with "Transfer 20.0 mL of this solution."

Dihydroxyaluminum Aminoacetate Tablets

» Dihydroxyaluminum Aminoacetate Tablets contain not less than 90.0 percent and not more than 110.0 percent of the labeled amount of $C_2H_6AlNO_4$.

Packaging and storage—Preserve in well-closed containers.

Identification—Weigh and finely powder not less than 20 Tablets. Suspend a portion of the powder, equivalent to about 1 g of dihydroxyaluminum aminoacetate, in 25 mL of water, add hydrochloric acid, dropwise, until all of the dihydroxyaluminum aminoacetate is dissolved, and filter. The solution responds to the *Identification tests* under *Dihydroxyaluminum Aminoacetate.* Retain the remaining portion of the powder for the following tests as indicated.

Disintegration ⟨701⟩: 10 minutes, simulated gastric fluid TS being substituted for water in the test.

Uniformity of dosage units ⟨905⟩: meet the requirements.

pH ⟨791⟩: between 6.5 and 7.5, in a suspension of the powder, retained in the *Identification test*, equivalent to about 1 g of dihydroxyaluminum aminoacetate in 25 mL of water.

Assay—

Disodium ethylenediaminetetraacetate titrant—Prepare and standardize as directed in the *Assay* under *Alum.*

Procedure—Transfer an accurately weighed portion of the powder, retained in the *Identification test*, equivalent to about 2.5 g of dihydroxyaluminum aminoacetate, to a flask, add 15 mL of hydrochloric acid and 10 mL of water, and boil gently for about 5 minutes. Cool, transfer the solution, with the aid of water, to a 500-mL volumetric flask, dilute with water to volume, mix, and filter to obtain a clear solution. Proceed as directed in the *Assay* under *Dihydroxyaluminum Aminoacetate*, beginning with "Transfer 20.0 mL of this solution."

Dihydroxyaluminum Sodium Carbonate

$NaAl(OH)_2CO_3$ 144.00
Aluminum, [carbonato(1-)-*O*]dihydroxy-, monosodium salt.
Sodium (*T*-4)-[carbonato(2-)-*O,O'*]dihydroxyaluminate(1-).
Sodium (carbonato)dihydroxyaluminate(1-) [*539-68-4; 16482-55-6*].

» Dihydroxyaluminum Sodium Carbonate, dried to constant weight at 130°, contains not less than 98.3 percent and not more than 107.9 percent of dihydroxyaluminum sodium carbonate (CH_2AlNaO_5).

Packaging and storage—Preserve in tight containers.

Identification—A 1-g portion, treated with 20 mL of 3 *N* hydrochloric acid, dissolves with effervescence, and the resulting solution responds to the tests for *Aluminum* ⟨191⟩ and for *Sodium* ⟨191⟩.

pH ⟨791⟩: between 9.9 and 10.2 in a suspension (1 in 25).

Loss on drying ⟨731⟩—Dry it at 130° to constant weight: it loses not more than 14.5% of its weight.

Isopropyl alcohol—

Standard isopropyl alcohol solution and *Factor determination*—Dilute about 500 mg of isopropyl alcohol, accurately weighed, with water to 500.0 mL, calculate the weight, in mg, of isopropyl alcohol per mL of the solution, and record this value as *W*. Transfer 5.0 mL of this solution to a 500-mL iodine flask, add 50 mL of dilute sulfuric acid (1 in 2) and 5.0 mL of 0.07 *N* potassium dichromate, insert the stopper loosely in the flask, and heat in a boiling water bath for 15 minutes. Remove the flask from the water bath, cool slightly, add 225 mL of water, 5 mL of phosphoric acid, 10.0 mL of 0.05 *N* ferrous ammonium sulfate VS, and 10 drops of diphenylamine TS, and titrate with 0.07 *N* potassium dichromate VS to a permanent purple color. Calculate the factor, *F*, by the formula:

$$W/[(V_1N_1) - (10N_2)],$$

in which V_1 is the total volume, in mL, N_1 is the normality of the potassium dichromate solution, and N_2 is the normality of the ferrous ammonium sulfate solution.

Procedure—Transfer a 500-mg portion to a 500-mL iodine flask, slowly add 10 mL of dilute sulfuric acid (1 in 2), swirl to dissolve, and then add 5.0 mL of 0.07 *N* potassium dichromate VS. Insert the stopper loosely in the flask, heat in a boiling water bath for 15 minutes, remove the flask from the water bath, and cool slightly. Add 225 mL of water, 5 mL of phosphoric acid, 10.0 mL of 0.05 *N* ferrous ammonium sulfate VS, and 10 drops of diphenylamine TS, and titrate with 0.07 *N* potassium dichromate VS to a permanent purple color. Calculate the percentage of isopropyl alcohol in the specimen taken by the formula:

$$F[(V_1N_1) - (10N_2)].$$

The limit is 1.0%.

Sodium content—

Potassium chloride solution—Prepare a solution of potassium chloride in water containing 38 mg per mL.

Sodium chloride stock solution—Dissolve a suitable quantity of sodium chloride, previously dried at 105° for 2 hours and accurately weighed, in water, and dilute quantitatively and stepwise with water to obtain a solution containing 25.42 µg per mL (10.0 µg of sodium per mL).

Standard preparations—On the day of use, transfer 4.0 mL of 1 *N* hydrochloric acid and 10.0 mL of *Potassium chloride solution* to each of two 100-mL volumetric flasks. To the respective flasks add 5.0 mL and 10.0 mL of *Sodium chloride stock solution*. Dilute with water to volume, and mix. These solutions contain about 0.5 µg and 1.0 µg of sodium per mL, respectively.

Test preparation—Transfer about 250 mg of Dihydroxyaluminum Sodium Carbonate, previously dried and accurately weighed, to a 200-mL volumetric flask, add 40 mL of 1 *N* hydrochloric acid, and boil for 1 minute. Cool, dilute with water to volume, and mix. Transfer 10.0 mL of this solution to a 100-mL volumetric flask, dilute with water to volume, and mix. Transfer 5.0 mL of this solution to a 100-mL volumetric flask containing 4.0 mL of 1 *N* hydrochloric acid and 10.0 mL of *Potassium chloride solution*, dilute with water to volume, and mix.

Procedure—Concomitantly determine the absorbances of the *Standard preparations* and the *Test preparation* at the sodium emission line at 589.0 nm with a suitable atomic absorption spectrophotometer (see *Spectrophotometry and Light-scattering* ⟨851⟩) equipped with a sodium hollow-cathode lamp and an air-acetylene flame, using as a blank a solution prepared by pipeting 4 mL of 1 *N* hydrochloric acid and 10.0 mL of *Potassium chloride solution* into a 100-mL volumetric flask, diluting with water to volume, and mixing. Plot the absorbances of the *Standard preparations* versus concentrations, in µg per mL of sodium, and draw a straight line between the plotted points. From the graph so obtained, determine the concentration, *C*, in µg per mL of sodium, in the *Test preparation*. Calculate the percentage of sodium in the portion of Dihydroxyaluminum Sodium Carbonate taken by the formula:

$$4000C/W,$$

in which *W* is the quantity, in mg, of Dihydroxyaluminum Sodium Carbonate taken: between 15.2% and 16.8% is found.

Mercury, *Method IIa* ⟨261⟩—Transfer 2.0 g to a 100-mL beaker, and add 35 mL of 1 *N* sulfuric acid: the limit is 1 ppm.

Assay—

Disodium ethylenediaminetetraacetate titrant—Dissolve 18.6 g of disodium ethylenediaminetetraacetate in water to make 500 mL, and standardize as directed in the *Assay* under *Alum*.

Procedure—Transfer about 200 mg of Dihydroxyaluminum Sodium Carbonate, previously dried and accurately weighed, to a 250-mL beaker, add 10 mL of 2 *N* nitric acid, cover the beaker, and boil for 1 minute. Add 25.0 mL of 0.1 *M* disodium ethylenediaminetetraacetate VS, again boil for 1 minute, cool, and then add 10 mL of acetic acid–ammonium acetate buffer TS, 50 mL of acetone, and 2 mL of dithizone TS. Using a pH meter, adjust with the addition of ammonium hydroxide to a pH of 4.5, and titrate with 0.05 *M* zinc sulfate VS, maintaining the pH at 4.5 by the addition of ammonium hydroxide as necessary, to an orange-pink color. Perform a blank determination, and make any necessary correction. Each mL of 0.1 *M Disodium ethylenediaminetetraacetate titrant* is equivalent to 14.40 mg of CH_2AlNaO_5.

Dihydroxyaluminum Sodium Carbonate Tablets

» Dihydroxyaluminum Sodium Carbonate Tablets contain not less than 90.0 percent and not more than 110.0 percent of the labeled amount of CH_2AlNaO_5.

Packaging and storage—Preserve in well-closed containers.

Labeling—Label the Tablets to indicate that they are to be chewed before swallowing.

Identification—A 1 in 10 suspension of powdered Tablets in 3 *N* hydrochloric acid responds to the tests for *Aluminum* ⟨191⟩ and for *Sodium* ⟨191⟩.

Uniformity of dosage units ⟨905⟩: meet the requirements.

Assay—

Disodium ethylenediaminetetraacetate titrant—Dissolve 18.6 g of disodium ethylenediaminetetraacetate in water to make 500 mL, and standardize as directed in the *Assay* under *Alum*.

Procedure—Weigh and finely powder not less than 20 Dihydroxyaluminum Sodium Carbonate Tablets. Transfer an accurately weighed portion of the powder, equivalent to about 200 mg of dihydroxyaluminum sodium carbonate, to a 250-mL beaker, and proceed as directed in the *Assay for aluminum oxide* under *Dihydroxyaluminum Sodium Carbonate*, beginning with "add 10 mL of 2 *N* nitric acid." Each mL of 0.1 *M Disodium ethylenediaminetetraacetate titrant* is equivalent to 14.40 mg of CH_2AlNaO_5.

Diltiazem Hydrochloride

$C_{22}H_{26}N_2O_4S.HCl$ 450.98

1,5-Benzothiazepin-4(5*H*)one, 3-(acetyloxy)-5-[2-(dimethylamino)ethyl]-2,3-dihydro-2-(4-methoxyphenyl)-, monohydrochloride, (+)-*cis*-.

(+)-5-[2-(Dimethylamino)ethyl]-*cis*-2,3-dihydro-3-hydroxy-2-(*p*-methoxyphenyl)-1,5-benzothiazepin-4(5*H*)-one acetate (ester) monohydrochloride [*33286-22-5*].

» Diltiazem Hydrochloride contains not less than 98.5 percent and not more than 101.5 percent of $C_{22}H_{26}N_2O_4S.HCl$, calculated on the dried basis.

Packaging and storage—Preserve in tight, light-resistant containers.

Reference standards—*USP Diltiazem Hydrochloride Reference Standard*—Dry at 105° for 3 hours before using. *USP Desacetyl Diltiazem Hydrochloride Reference Standard*—Dry at 105° for 3 hours before using.

Identification—

A: The infrared absorption spectrum of a potassium bromide dispersion of it, previously dried, exhibits maxima only at the same wavelengths as that of a similar preparation of USP Diltiazem Hydrochloride RS.

B: The retention time of the major peak in the chromatogram of the *Assay preparation* corresponds to that of the *Standard preparation*, obtained as directed in the *Assay*.

C: It responds to the tests for *Chloride* ⟨191⟩.

Specific rotation ⟨781⟩: between +110° and +116°, calculated on the dried basis, determined in a 1 in 100 solution in water.

Loss on drying ⟨731⟩—Dry it at 105° for 3 hours: it loses not more than 0.5% of its weight.

Heavy metals ⟨231⟩: not more than 20 ppm.

Related compounds—

Buffer and *Mobile phase*—Prepare as directed in the *Assay*.

Standard solution—Use the *System suitability preparation* prepared as directed under *Assay*.

Test solution—Prepare as directed for the *Assay preparation* in the *Assay*.

Chromatographic system—Prepare as directed under *Assay*. The relative standard deviation of the peak response for replicate injections of the *Standard solution* is not more than 10.0%.

Procedure—Separately inject equal volumes (about 10 µL) of the *Standard solution* and the *Test solution* into the chromatograph, record the chromatograms, and measure the responses for all of the peaks. The relative retention times are about 0.65 for desacetyl diltiazem and 1.0 for diltiazem. Calculate the percentage of desacetyl diltiazem hydrochloride in the specimen of Diltiazem Hydrochloride by the formula:

$$10(C/W)(r_U/r_S),$$

in which *C* is the concentration, in µg per mL, of USP Desacetyl Diltiazem Hydrochloride RS in the *Standard solution*, *W* is the weight, in mg, of Diltiazem Hydrochloride taken, and r_U and r_S are the desacetyl diltiazem peak responses obtained from the *Test solution* and the *Standard solution*, respectively: not more than 0.5% of desacetyl diltiazem hydrochloride is found. Calculate the percentage of each impurity peak, other than the main peak and the desacetyl diltiazem peak, by the formula:

$$10(C/W)(r_I/r_S),$$

in which r_I is the response of each impurity peak and all other quantities are as defined above: not more than 1.0% total impurities including desacetyl diltiazem hydrochloride with no individual impurity greater than 0.5%, is found.

Assay—

Buffer—Dissolve 1.16 g of *d*-10-camphorsulfonic acid in 1000 mL of 0.1 *M* sodium acetate, adjust this solution by the addition of 0.1 *N* sodium hydroxide to a pH of 6.2, and mix.

Mobile phase—Prepare a mixture of *Buffer*, acetonitrile, and methanol (50:25:25), filter, and degas. Make adjustments if necessary (see *System Suitability* under *Chromatography* ⟨621⟩).

Standard preparation—Prepare a solution in methanol having an accurately known concentration of about 1.2 mg of USP Diltiazem Hydrochloride per mL.

Assay preparation—Transfer about 120 mg of Diltiazem Hydrochloride, accurately weighed, to a 100-mL volumetric flask, dissolve in methanol, dilute with methanol to volume, and mix.

System suitability preparation—Prepare a solution in methanol containing 0.012 mg each of USP Diltiazem Hydrochloride RS and USP Desacetyl Diltiazem Hydrochloride RS per mL.

Chromatographic system—The liquid chromatograph is equipped with a 240-nm detector and a 3.9-mm × 30-cm column that contains packing L1. The flow rate is about 1.6 mL per minute. Chromatograph the *System suitability preparation*, and record the peak responses as directed under *Procedure:* the relative retention times are about 0.65 for desacetyl diltiazem and 1.0 for diltiazem, the resolution, *R*, between desacetyl diltiazem and diltiazem is not less than 3, and the number of theoretical plates, *n*, for the diltiazem peak is not less than 1200. Chromatograph the *Standard preparation*, and record the peak responses as directed under *Procedure:* the relative standard deviation for replicate injections is not more than 2.0%.

Procedure—Separately inject equal volumes (about 10 µL) of the *Standard preparation* and the *Assay preparation* into the chromatograph, record the chromatograms, and measure the responses for the major peaks. Calculate the quantity, in mg, of $C_{22}H_{26}N_2O_4S.HCl$ in the Diltiazem Hydrochloride taken by the formula:

$$100C(r_U/r_S),$$

in which *C* is the concentration, in mg per mL, of USP Diltiazem Hydrochloride RS in the *Standard preparation*, and r_U and r_S are the peak responses obtained from the *Assay preparation* and the *Standard preparation*, respectively.

Diltiazem Tablets

» Diltiazem Tablets contain not less than 90.0 percent and not more than 110.0 percent of the labeled amount of diltiazem hydrochloride ($C_{22}H_{26}N_2O_4S.HCl$).

Packaging and storage—Preserve in tight, light-resistant containers.

Labeling—Label Tablets to state both the content of the active moiety and the content of the salt used in formulating the article.

Reference standards—*USP Diltiazem Hydrochloride Reference Standard*—Dry at 105° for 3 hours before using. *USP Desacetyl Diltiazem Reference Standard*—Dry at 105° for 3 hours before using.

Identification—

A: The ultraviolet absorption spectrum of the solution employed for measurement of absorbance in the *Dissolution* exhibits maxima and minima at the same wavelengths as that of a similar solution of USP Diltiazem Hydrochloride RS, concomitantly measured.

B: The retention time of the major peak in the chromatogram of the *Assay preparation* corresponds to that in the chromatogram of the *Standard preparation* as obtained in the *Assay*.

Dissolution ⟨711⟩—

Medium: water; 900 mL.

Apparatus 2: 50 rpm.

Time: 45 minutes.

Procedure—Determine the amount of $C_{22}H_{26}N_2O_4S.HCl$ dissolved from ultraviolet absorbances at the wavelength of maximum absorbance at about 240 nm using filtered portions of the solution under test, suitably diluted with water, if necessary, in comparison with a Standard solution having a known concentration of USP Diltiazem Hydrochloride RS in the same medium.

Tolerances—Not less than 75% (*Q*) of the labeled amount of $C_{22}H_{26}N_2O_4S.HCl$ is dissolved in 45 minutes.

Uniformity of dosage units ⟨905⟩: meet the requirements.

Assay—

Buffer, Mobile phase, Standard preparation, System suitability preparation, Chromatographic system, and *Procedure*—Proceed as directed in the *Assay* under *Diltiazem Hydrochloride.*

Assay preparation—Weigh and finely powder not less than 20 Diltiazem Tablets. Transfer an accurately weighed portion of the powder, equivalent to about 600 mg of diltiazem hydrochloride to a 500-mL volumetric flask. Add 200 mL of methanol, sonicate for one hour, cool, dilute with methanol to volume, and mix. Centrifuge a 25-mL aliquot at 3500 rpm for 15 minutes and use the clear supernatant liquid for injection into the liquid chromatograph.

Procedure—Calculate the quantity, in mg, of $C_{22}H_{26}N_2O_4S.HCl$ in the portion of Tablets taken by the formula:

$$500C(r_U / r_S),$$

in which *C* is the concentration, in mg per mL, of USP Diltiazem Hydrochloride RS in the *Standard preparation*, and r_U and r_S are the diltiazem hydrochloride peak areas in the *Assay preparation* and the *Standard preparation*, respectively.

Diisopropanolamine—*see* Diisopropanolamine NF

Diluted Alcohol—*see* Alcohol, Diluted NF

Diluted Hydrochloric Acid—*see* Hydrochloric Acid, Diluted NF

Diluted Isosorbide Dinitrate—*see* Isosorbide Dinitrate, Diluted

Diluted Nitroglycerin—*see* Nitroglycerin, Diluted

Diluted Pentaerythritol Tetranitrate—*see* Pentaerythritol Tetranitrate, Diluted

Diluted Phosphoric Acid—*see* Phosphoric Acid, Diluted NF

Dimenhydrinate

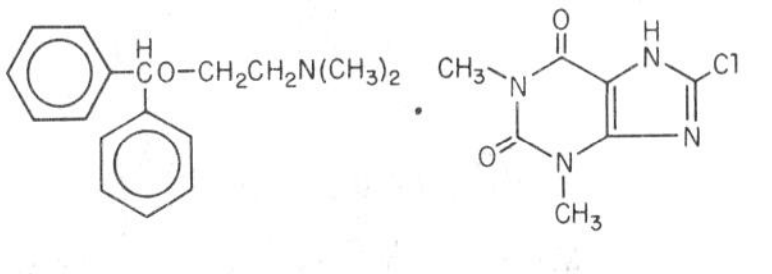

$C_{17}H_{21}NO.C_7H_7ClN_4O_2$ 469.97

1*H*-Purine-2,6-dione, 8-chloro-3,7-dihydro-1,3-dimethyl-, compd. with 2-(diphenylmethoxy)-*N,N*-dimethylethanamine (1:1).

8-Chlorotheophylline, compound with 2-(diphenylmethoxy)-*N,N*-dimethylethylamine (1:1) [*523-87-5*].

» Dimenhydrinate contains not less than 53.0 percent and not more than 55.5 percent of diphenhydramine ($C_{17}H_{21}NO$), and not less than 44.0 percent and not more than 47.0 percent of 8-chlorotheophylline ($C_7H_7ClN_4O_2$), calculated on the dried basis.

Packaging and storage—Preserve in well-closed containers.

Reference standard—*USP Dimenhydrinate Reference Standard*—Dry in vacuum over phosphorus pentoxide for 24 hours before using.

Identification—

A: It meets the requirements under *Identification—Organic Nitrogenous Bases* ⟨181⟩.

B: Dissolve about 250 mg in 15 mL of diluted alcohol, add 15 mL of water and 2 mL of 2 *N* sulfuric acid, and cool for 30 minutes. Scratch the inside of the container to facilitate crystallization. Filter the mixture, wash the crystals with a few mL of ice-cold water, and dry the crystals. The 8-chlorotheophylline melts between 300° and 305° with decomposition.

C: To about 10 mg of the 8-chlorotheophylline obtained in *Identification test B*, contained in a porcelain dish, add 1 mL of hydrochloric acid and 100 mg of potassium chlorate, evaporate on a steam bath to dryness, and invert the dish over a vessel containing a few drops of ammonia TS: the residue acquires a purple color, which is destroyed by solutions of fixed alkalies.

D: Mix about 50 mg of the 8-chlorotheophylline obtained in *Identification test B* with about 500 mg of sodium peroxide in a nickel crucible, and heat until the mass is well sintered. Dissolve the melt in 20 mL of water, acidify with 2 *N* nitric acid, filter if necessary, and add 1 mL of silver nitrate TS: a curdy, white precipitate is formed, and it is soluble in 6 *N* ammonium hydroxide and reappears upon acidification with nitric acid.

Melting range ⟨741⟩: between 102° and 107°.

Loss on drying ⟨731⟩—Dry it in vacuum over phosphorus pentoxide for 24 hours: it loses not more than 0.5% of its weight.

Residue on ignition ⟨281⟩: not more than 0.3%.

Chloride—When the ammoniacal filtrate from the precipitation of silver chlorotheophylline in the *Assay for 8-chlorotheophylline* is acidified preparatory to titration, the solution shows not more than a faint opalescence.

Bromide and iodide—Mix in a test tube 100 mg of Dimenhydrinate, 50 mg of sodium nitrite, and 10 mL of chloroform. Add 10 mL of 3 *N* hydrochloric acid, insert the stopper in the tube, and shake: the chloroform remains colorless.

Assay for diphenhydramine—Dissolve about 150 mg of Dimenhydrinate, accurately weighed, in 75 mL of glacial acetic acid, and titrate with 0.05 *N* perchloric acid VS, determining the endpoint potentiometrically. Perform a blank determination, and make any necessary correction. Each mL of 0.05 *N* perchloric acid is equivalent to 12.77 mg of diphenhydramine ($C_{17}H_{21}NO$).

Assay for 8-chlorotheophylline—Place about 800 mg of Dimenhydrinate, accurately weighed, in a 200-mL volumetric flask, add 50 mL of water, 3 mL of 6 *N* ammonium hydroxide, and 6 mL of ammonium nitrate solution (1 in 10), and warm the mixture on a steam bath for 5 minutes. Add 25.0 mL of 0.1 *N* silver nitrate VS, mix, and warm on a steam bath for 15 minutes with frequent shaking. Cool, dilute with water to volume, mix, and allow the precipitate to settle. Filter through a dry filter paper, discarding the first 20 mL of the filtrate. Pipet 100 mL of the filtrate into a 250-mL flask, acidify with nitric acid, and add an excess of 3 mL of the acid. Add 2 mL of ferric ammonium sulfate TS, and titrate the excess silver nitrate with 0.1 *N* ammonium thiocyanate VS. Each mL of 0.1 *N* silver nitrate is equivalent to 21.46 mg of $C_7H_7ClN_4O_2$.

Dimenhydrinate Injection

» Dimenhydrinate Injection is a solution of Dimenhydrinate in a mixture of Propylene Glycol and water. It contains not less than 95.0 percent and not more than 105.0 percent of the labeled amount of $C_{17}H_{21}NO.C_7H_7ClN_4O_2$.

Packaging and storage—Preserve in single-dose or in multiple-dose containers, preferably of Type I or Type III glass.

Reference standards—*USP Dimenhydrinate Reference Standard*—Dry in vacuum over phosphorus pentoxide for 24 hours before using. *USP Diphenhydramine Hydrochloride Reference Standard*—Dry at 105° for 3 hours before using.

Identification—Pipet 5 mL of Injection into a 250-mL separator containing 35 mL of water. Add 3 mL of 6 *N* ammonium hydroxide and 10 g of sodium chloride, shake the mixture until the sodium chloride dissolves, and extract with three successive 50-mL portions of ether. Wash the combined ether extracts with successive 25-mL portions of water until any color produced in the final washing by the addition of 2 drops of phenolphthalein TS is discharged by 1 drop of dilute hydrochloric acid (1 in 100). Extract with 10 mL of water containing 0.5 mL of 3 *N* hydrochloric acid, and aerate to remove residual ether: the aqueous extract so obtained meets the requirements under *Identification—Organic Nitrogenous Bases* ⟨181⟩, USP Diphenhydramine Hydrochloride RS being used as the reference material.

pH ⟨791⟩: between 6.4 and 7.2.

8-Chlorotheophylline content—

Mobile phase, Standard preparation, and *Chromatographic system*—Prepare as directed in the test for *8-Chlorotheophylline content* under *Dimenhydrinate Tablets.*

Test preparation—Transfer an accurately measured volume of Dimenhydrinate Injection, equivalent to about 50 mg of dimenhydrinate, to a 100-mL volumetric flask, dilute with methanol to volume, and mix to obtain a stock solution. Retain a portion of this stock solution for use in the *Assay.* Pipet 5 mL of this stock solution into a 50-mL volumetric flask, dilute with methanol to volume, mix, and filter through a membrane filter (0.5-μm or finer porosity).

Procedure—Proceed as directed for *Procedure* in the test for *8-Chlorotheophylline content* under *Dimenhydrinate Tablets.* Calculate the quantity, in mg, of $C_{17}H_{21}NO$ in each mL of the Injection taken by the formula:

$$(214.61/469.97)(1000C/V)(r_U/r_S),$$

in which V is the volume, in mL, of Injection taken, and the other terms are as defined therein: an amount of 8-chlorotheophylline that is between 43.4% and 47.9% of the amount of dimenhydrinate obtained in the *Assay* is found.

Other requirements—It meets the requirements under *Injections* ⟨1⟩.

Assay—

Mobile phase, Internal standard solution, and *Chromatographic system*—Prepare as directed in the *Assay* under *Dimenhydrinate Tablets.*

Standard preparation—Use a portion of the Standard stock solution that was prepared for *Standard preparation* in the test for *8-Chlorotheophylline content.* Mix 5.0 mL of this Standard stock solution and 5.0 mL of *Internal standard solution,* and filter through a membrane filter (0.5-μm or finer porosity).

Assay preparation—Use a portion of the stock solution that was prepared for *Test preparation* in the test for *8-Chlorotheophylline content.* Mix 5.0 mL of this stock solution and 5.0 mL of *Internal standard solution,* and filter through a membrane filter (0.5-μm or finer porosity).

Procedure—Proceed as directed for *Procedure* in the *Assay* under *Dimenhydrinate Tablets.* Calculate the quantity, in mg, of $C_{17}H_{21}NO.C_7H_7ClN_4O_2$ in each mL of the Injection taken by the formula:

$$(200C/V)(R_U/R_S),$$

in which V is the volume, in mL, of Injection taken, and the other terms are as defined therein.

Dimenhydrinate Syrup

» Dimenhydrinate Syrup contains not less than 94.0 percent and not more than 106.0 percent of the labeled amount of $C_{17}H_{21}NO.C_7H_7ClN_4O_2$.

Packaging and storage—Preserve in tight containers.

Reference standards—*USP Dimenhydrinate Reference Standard*—Dry in vacuum over phosphorus pentoxide for 24 hours before using. *USP Diphenhydramine Hydrochloride Reference Standard*—Dry at 105° for 3 hours before using.

Identification—Transfer to a separator 30 mL of Syrup, and add 50 mL of water, 3 mL of 6 *N* ammonium hydroxide, and 10 g of sodium chloride. Shake the mixture until the sodium chloride dissolves, and extract it first with 50 mL, then with three 40-mL portions, of ether. Wash the combined ether extracts with successive 25-mL portions of water until any color produced in the final washing by the addition of 2 drops of phenolphthalein TS is discharged by 1 drop of 0.1 *N* hydrochloric acid. Shake the ether extract with 25 mL of 0.01 *N* hydrochloric acid, separate the aqueous layer, and aerate it to remove residual ether: the aqueous solution so obtained meets the requirements under *Identification—Organic Nitrogenous Bases* ⟨181⟩, USP Diphenhydramine Hydrochloride RS being used as the reference material.

Alcohol content ⟨611⟩: between 4.0% and 6.0% of C_2H_5OH.

8-Chlorotheophylline content—

Mobile phase and *Chromatographic system*—Prepare as directed in the test for *8-Chlorotheophylline content* under *Dimenhydrinate Tablets.*

Standard preparation—Dissolve an accurately weighed quantity of USP Dimenhydrinate RS in a mixture of methanol and water (50:50), and dilute quantitatively and stepwise with a mixture of methanol and water (50:50) to obtain a solution having a known concentration of about 50 μg per mL.

Test preparation—Transfer an accurately measured volume of Syrup, equivalent to about 50 mg of dimenhydrinate, to a 100-mL volumetric flask, dilute with a mixture of methanol and water (50:50) to volume, and mix to obtain a stock solution. Retain a portion of this stock solution for use in the *Assay.* Pipet 5 mL of this stock solution to a 50-mL volumetric flask, dilute with a mixture of methanol and water (50:50) to volume, mix, and filter.

Procedure—Proceed as directed for *Procedure* in the test for *8-Chlorotheophylline content* under *Dimenhydrinate Tablets.* Calculate the quantity, in mg, of $C_{17}H_{21}NO$ in each mL of the Syrup taken by the formula:

$$(214.61/469.97)(1000C/V)(r_U/r_S),$$

in which V is the volume, in mL, of Syrup taken, and the other terms are as defined therein: an amount of 8-chlorotheophylline that is between 43.4% and 47.9% of the amount of dimenhydrinate obtained in the *Assay* is found.

Assay—

Mobile phase, Internal standard solution, and *Chromatographic system*—Prepare as directed in the *Assay* under *Dimenhydrinate Tablets.*

Standard preparation—Dissolve an accurately weighed quantity of USP Dimenhydrinate RS in a mixture of methanol and water (50:50) to obtain a Standard stock solution having a known concentration of about 0.5 mg per mL. Mix 5.0 mL of this Standard stock solution and 5.0 mL of *Internal standard solution,* and filter.

Assay preparation—Use a portion of the stock solution that was prepared for *Test preparation* in the test for *8-Chlorotheophylline content.* Mix 5.0 mL of this stock solution and 5.0 mL of *Internal standard solution,* and filter.

Procedure—Proceed as directed for *Procedure* in the *Assay* under *Dimenhydrinate Tablets.* Calculate the quantity, in mg, of $C_{17}H_{21}NO.C_7H_7ClN_4O_2$ in each mL of the Syrup taken by the formula:

$$(200C/V)(R_U/R_S),$$

in which V is the volume, in mL, of Syrup taken, and the other terms are as defined therein.

Dimenhydrinate Tablets

» Dimenhydrinate Tablets contain not less than 95.0 percent and not more than 105.0 percent of the labeled amount of $C_{17}H_{21}NO.C_7H_7ClN_4O_2$.

Packaging and storage—Preserve in well-closed containers.

Reference standard—*USP Dimenhydrinate Reference Standard*—Dry in vacuum over phosphorus pentoxide for 24 hours before using.

Identification—Triturate a number of Tablets, equivalent to about 500 mg of dimenhydrinate, with 25 mL of warm alcohol, and filter the mixture. Dilute the filtrate with 40 mL of water, and again filter: the filtrate responds to the *Identification tests* under *Dimenhydrinate*.

Dissolution ⟨711⟩—

Medium: water; 900 mL.

Apparatus 2: 50 rpm.

Time: 45 minutes.

Procedure—Determine the amount of $C_{17}H_{21}NO.C_7H_7ClN_4O_2$ dissolved from ultraviolet absorbances at the wavelength of maximum absorbance at about 276 nm of filtered portions of the solution under test, suitably diluted with water, if necessary, in comparison with a Standard solution having a known concentration of USP Dimenhydrinate RS in the same medium.

Tolerances—Not less than 75% (*Q*) of the labeled amount of $C_{17}H_{21}NO.C_7H_7ClN_4O_2$ is dissolved in 45 minutes.

Uniformity of dosage units ⟨905⟩: meet the requirements.

Procedure for content uniformity—Transfer 1 Tablet, previously crushed or finely powdered, to a 100-mL volumetric flask with the aid of methanol, add methanol to volume, mix, and filter, discarding the first portion of the filtrate. Dilute the filtrate quantitatively and stepwise with methanol to obtain a solution having a concentration of about 25 µg per mL. Prepare a solution of USP Dimenhydrinate RS to obtain a Standard solution in methanol having a known concentration of about 25 µg per mL. Concomitantly determine the absorbances of both solutions in 1-cm cells at the wavelength of maximum absorbance at about 276 nm, with a suitable spectrophotometer, using methanol as the blank. Calculate the quantity, in mg, of $C_{17}H_{21}NO.C_7H_7ClN_4O_2$ in the Tablet by the formula:

$$(T/25)C(A_U/A_S),$$

in which *T* is the labeled quantity, in mg, of dimenhydrinate in the Tablet, *C* is the concentration, in µg per mL, of Dimenhydrinate in the Standard solution, and A_U and A_S are the absorbances of the solution from the Tablet and the Standard solution, respectively.

8-Chlorotheophylline content—

Mobile phase—Dissolve 0.81 g of *dl*-10-camphorsulfonic acid and 0.70 g of sodium acetate trihydrate in 700 mL of water. Add 300 mL of methanol, mix, and filter through a membrane filter (0.5-µm or finer porosity).

Standard preparation—Dissolve an accurately weighed quantity of USP Dimenhydrinate RS in methanol to obtain a Standard stock solution having a known concentration of about 0.5 mg per mL. Retain a portion of this Standard stock solution for use in the *Assay*. Pipet 5 mL into a 50-mL volumetric flask, dilute with methanol to volume, mix, and filter through a membrane filter (0.5-µm or finer porosity).

Test preparation—Weigh and finely powder not less than 20 Dimenhydrinate Tablets. Transfer an accurately weighed portion of the powder, equivalent to about 50 mg of dimenhydrinate, to a 100-mL volumetric flask, add about 95 mL of methanol, shake well, and sonicate for 1 minute. Dilute with methanol to volume, and mix to obtain a stock solution. Retain a portion of this stock solution for use in the *Assay*. Pipet 5 mL of this stock solution into a 50-mL volumetric flask, dilute with methanol to volume, mix, and filter through a membrane filter (0.5-µm or finer porosity).

Chromatographic system (see *Chromatography* ⟨621⟩)—The liquid chromatograph is equipped with a 280-nm detector, a 2-mm × 12.5-cm guard column that contains packing L2, and a 4.6-mm × 25-cm analytical column that contains packing L1. The flow rate is about 2 mL per minute. Chromatograph three replicate injections of the *Standard preparation*, and record the peak responses as directed under *Procedure:* the relative standard deviation is not more than 1.0%.

Procedure—Separately inject equal volumes (about 25 µL) of the *Standard preparation* and the *Test preparation* into the chromatograph by means of a suitable microsyringe or sampling valve, record the chromatograms, and measure the responses for the major peaks. The retention time is about 5 minutes for 8-chlorotheophylline. At the completion of the analysis, rinse the pump and the columns with a mixture of water and methanol (70:30). Calculate the quantity, in mg, of 8-chlorotheophylline ($C_7H_7ClN_4O_2$) in the portion of Tablets taken by the formula:

$$(214.61/469.97)(1000C)(r_U/r_S),$$

in which 214.61 and 469.97 are the molecular weights of 8-chlorotheophylline and dimenhydrinate, respectively, *C* is the concentration, in mg per mL, of USP Dimenhydrinate RS in the *Standard preparation*, and r_U and r_S are the peak responses of the *Test preparation* and the *Standard preparation*, respectively: an amount of 8-chlorotheophylline that is between 43.4% and 47.9% of the amount of dimenhydrinate obtained in the *Assay* is found.

Assay—

Mobile phase—Dissolve 0.83 g of dibasic ammonium phosphate in 165 mL of water, slowly add 935 mL of methanol with constant stirring, and filter through a membrane filter (0.5-µm or finer porosity).

Internal standard solution—Dissolve bis(2-ethylhexyl) phthalate in methanol to obtain a solution containing 30 µL per 250 mL.

Standard preparation—Use a portion of the Standard stock solution that was prepared for *Standard preparation* in the test for *8-Chlorotheophylline content*. Mix 5.0 mL of this Standard stock solution and 5.0 mL of *Internal standard solution*, and filter through a membrane filter (0.5-µm or finer porosity).

Assay preparation—Use a portion of the stock solution that was prepared for *Test preparation* in the test for *8-Chlorotheophylline content*. Mix 5.0 mL of this stock solution and 5.0 mL of *Internal standard solution*, and filter through a membrane filter (0.5-µm or finer porosity).

Chromatographic system (see *Chromatography* ⟨621⟩)—The liquid chromatograph is equipped with a 254-nm detector and a 3.9-mm × 30-cm column that contains packing L1. The flow rate is about 2.2 mL per minute. Chromatograph three replicate injections of the *Standard preparation*, and record the peak responses as directed under *Procedure:* the relative standard deviation is not more than 1.0%, and the resolution, *R*, between the diphenhydramine peak and the bis(2-ethylhexyl) phthalate peak is not less than 2.5. [NOTE—After use, rinse the column briefly with methanol.]

Procedure—Separately inject equal volumes (about 25 µL) of the *Standard preparation* and the *Assay preparation* into the chromatograph by means of a suitable microsyringe or sampling valve, record the chromatograms, and measure the responses for the major peaks. The retention times are about 4 and 6 minutes for the diphenhydramine component of dimenhydrinate and for bis(2-ethylhexyl) phthalate, respectively. Calculate the quantity, in mg, of $C_{17}H_{21}NO.C_7H_7ClN_4O_2$ in the portion of Tablets taken by the formula:

$$200C(R_U/R_S),$$

in which *C* is the concentration, in mg per mL, of USP Dimenhydrinate RS in the *Standard preparation*, and R_U and R_S are the ratios of the peak responses of diphenhydramine to those of bis(2-ethylhexyl) phthalate obtained from the *Assay preparation* and *Standard preparation*, respectively.

Dimercaprol

CH_2CHCH_2OH (with SH on C_1 and C_2)

SH SH

$C_3H_8OS_2$ 124.22

1-Propanol, 2,3-dimercapto.

2,3-Dimercapto-1-propanol [*59-52-9*].

» Dimercaprol contains not less than 97.0 percent and not more than 100.5 percent of $C_3H_8OS_2$, and not more than 1.5 percent of 1,2,3-trimercaptopropane ($C_3H_8S_3$).

Packaging and storage—Preserve in tight containers, in a cold place.

Specific gravity ⟨841⟩: between 1.242 and 1.244.

Distilling range, *Method I* ⟨721⟩: between 66° and 68°, under a pressure of 0.2 mm of mercury.

Refractive index ⟨831⟩: between 1.567 and 1.573.

1,2,3-Trimercaptopropane and related impurities—

Adsorbant—Use a suitable chromatographic grade of 100-mesh silicic acid.

Standard buffer solution—Prepare 100 mL of pH 6.0 Phosphate Buffer (see *pH* ⟨791⟩), and dissolve in it 100 mg of sodium bisulfite.

Acid-washed solvent hexane—To 100 mL of solvent hexane contained in a separator add 10 mL of sulfuric acid, shake well for not less than 12 hours, and allow the layers to separate. Transfer the acid-washed solvent to a distilling flask, and distil slowly, retaining only that portion distilling between 35° and 50°. Use only freshly distilled material.

Diisopropyl ether—Place 100 mL of diisopropyl ether in a distilling flask, and distil, retaining only that portion distilling between 68° and 69°. Use only freshly distilled material. [*Caution—Do not evaporate to the point of near-dryness, since diisopropyl ether tends to form explosive peroxides.*]

Solvent hexane–diisopropyl ether mixture (mobile solvent)—Mix 50 mL of *Diisopropyl ether* with 50 mL of *Acid-washed solvent hexane.*

Chromatographic tube—Insert a small plug of glass wool at the juncture of the tube and the stem of a 600- × 13-mm chromatographic tube.

Chromatographic column—Mix 20 g of *Adsorbant* with 20 mL of *Standard buffer solution.* Make into a slurry by mixing with 100 mL of chloroform. Transfer successive portions of the slurry into the *Chromatographic tube*, packing firmly and evenly with a close-fitting, ground-glass tamper after each addition. Keep a layer of liquid above the packed column to prevent the formation of air spaces. Wash the column free from chloroform with *Solvent hexane–diisopropyl ether mixture*, and allow the solvent to fall to the level of the *Adsorbant.*

Procedure—Place about 250 mg of Dimercaprol, accurately weighed and demonstrated to be free from hydrogen sulfide as directed in the *Assay*, in a 5-mL volumetric flask, add *Solvent hexane–diisopropyl ether mixture* to volume, and mix. Transfer 2.0 mL of the resulting solution to the prepared *Chromatographic column.* When the liquid has passed into the column, wash the wall of the tube with a 2-mL portion of *Solvent hexane–diisopropyl ether mixture*, and allow the washing to fall to the level of the *Adsorbant.* Fill the *Chromatographic tube* with solvent, and collect two successive fractions: (*A*) a 20-mL fraction containing all of the 1,2,3-trimercaptopropane, and (*B*) a 3-mL fraction that serves as a check on the separation. To each fraction add an equal volume of alcohol, and titrate with 0.1 *N* iodine VS until a permanent yellow color is produced. Perform a blank titration on 20 mL of the solvent mixture that has been passed through the column prior to introduction of the test specimen, and make any necessary correction. Fraction (*B*) does not decolorize 1 drop of 0.1 *N* iodine VS. Each mL of 0.1 *N* iodine added is equivalent to 4.676 mg of $C_3H_8S_3$. Not more than 1.5% of 1,2,3-trimercaptopropane ($C_3H_8S_3$) is found.

Assay—Test the Dimercaprol for the presence of hydrogen sulfide by examining the vapor above the assay specimen with moistened lead acetate test paper. If the paper darkens, bubble dry, oxygen-free nitrogen or carbon dioxide through the assay specimen until a fresh strip of test paper gives a negative test. Transfer about 2 mL of hydrogen sulfide–free Dimercaprol to a tared, glass-stoppered, 100-mL volumetric flask, weigh accurately, add methanol to volume, and mix. Pipet 10 mL of the solution into a 50-mL conical flask, and titrate with 0.1 *N* iodine VS until a permanent yellow color is produced. Perform a blank titration, and make any necessary correction. Calculate the percentage of $C_3H_8OS_2$ taken by the formula:

$$0.6211V/W - 1.328T,$$

in which *V* is the volume, in mL, of 0.1 *N* iodine used, *W* is the weight, in g, of specimen in the aliquot taken, and *T* is the percentage of $C_3H_8S_3$ found in the determination of *1,2,3-Trimercaptopropane and related impurities.*

Dimercaprol Injection

» Dimercaprol Injection is a sterile solution of Dimercaprol in a mixture of Benzyl Benzoate and vegetable oil. It contains, in each 100 g, not less than 9.0 g and not more than 11.0 g of $C_3H_8OS_2$.

Packaging and storage—Preserve in single-dose or in multiple-dose containers, preferably of Type I or Type III glass.

1,2,3-Trimercaptopropane and related impurities—

Adsorbant, Standard buffer solution, Acid-washed solvent hexane, Diisopropyl ether, Solvent hexane–diisopropyl ether mixture, Chromatographic tube, and *Chromatographic column*—Prepare as directed in the test for *1,2,3-Trimercaptopropane and related impurities* under *Dimercaprol.*

Procedure—Place about 1 g of Dimercaprol Injection, accurately weighed and demonstrated to be free from hydrogen sulfide as directed in the *Assay* under *Dimercaprol*, in a 5-mL beaker, add 2 mL of *Solvent hexane–diisopropyl ether mixture*, and mix. Transfer the resulting solution to the prepared *Chromatographic column.* When the liquid has passed into the column, wash the beaker with two 2-mL portions of *Solvent hexane–diisopropyl ether mixture*, and allow the washings to fall to the level of the *Adsorbant.* Proceed as directed for *Procedure* in the test for *1,2,3-Trimercaptopropane and related impurities* under *Dimercaprol.* The limit of 1,2,3-trimercaptopropane ($C_3H_8S_3$) is not more than 4.5%, by weight, of the content of dimercaprol.

Other requirements—It meets the requirements under *Injections* ⟨1⟩, except that at times it may be turbid or contain small amounts of flocculent material.

Assay—Transfer about 2 mL of Dimercaprol Injection to a tared conical flask, and weigh accurately. Add 100 mL of a mixture of 1 volume of chloroform and 3 volumes of methanol, agitate to dissolve the Injection, and titrate with 0.1 *N* iodine VS to the production of a permanent yellow color. Perform a blank determination, and make any necessary correction. Calculate the percentage of $C_3H_8OS_2$ taken by the formula:

$$0.6211(V/W - v/w),$$

in which *V* is the volume, in mL, of 0.1 *N* iodine used, *W* is the weight, in g, of Injection taken, and *v* and *w* are the volume, in mL, of 0.1 *N* iodine and the weight, in g, of Injection, respectively, used in the test for *1,2,3-Trimercaptopropane and related impurities.*

Dimethicone—*see* Dimethicone NF

Dimethyl Sulfoxide

$(CH_3)_2S{=}O$

C_2H_6OS 78.13
Methane, sulfinylbis-.
Methyl sulfoxide [*67-68-5*].

» Dimethyl Sulfoxide contains not less than 99.9 percent of C_2H_6OS.

Packaging and storage—Preserve in tight, light-resistant containers.

Reference standard—*USP Dimethyl Sulfoxide Reference Standard*—After opening, store in a tightly closed, light-resistant container.

Identification—

A: The infrared absorption spectrum of a thin film of it between potassium bromide plates exhibits maxima at the same wavelengths as that of a similar preparation of USP Dimethyl Sulfoxide RS.

B: Add 1.5 mL cautiously and dropwise to 2.5 mL of hydriodic acid in a test tube cooled in ice. Filter the mixture rapidly, and collect the precipitate. Dry the precipitate in vacuum: a deep-violet, crystalline solid is obtained, and it is soluble in chloroform, yielding a red solution.

Specific gravity ⟨841⟩: between 1.095 and 1.097.

Congealing temperature ⟨651⟩: 18.3°, indicating not less than 99.9% of C_2H_6OS.

Refractive index ⟨831⟩: between 1.4755 and 1.4775.

Water, *Method I* ⟨921⟩: not more than 0.1%. [NOTE—Weigh and transfer the test specimen in an environment of low humidity to minimize absorption of atmospheric water.]

Acidity—Dissolve 50.0 g in 100 mL of water, and add phenolphthalein TS. If the solution remains colorless, titrate with 0.01 *N* sodium hydroxide until a pink color appears: not more than 5.0 mL of 0.01 *N* sodium hydroxide is consumed.

Nonvolatile residue—Evaporate 50 g in a rotary evaporator at a pressure of about 30 mm of mercury at 95°. Wash the residue from the evaporator flask into a tared dish with several 25-mL portions of glass-distilled methanol, and evaporate on a hot plate in an exhaust hood: the weight of the residue does not exceed 5.0 mg.

Ultraviolet absorbance—Maintain Dimethyl Sulfoxide in a water bath at a temperature of less than 20° [NOTE—Do not freeze.], and purge with dry nitrogen for 30 minutes. Record the ultraviolet absorption spectrum between 270 nm and 350 nm in a 1-cm cell, using water as the blank: the spectrum is smooth with no absorption maxima; the absorbance at 275 nm is not more than 0.20, and the absorbance ratios, A_{285}/A_{275} and A_{295}/A_{275}, at the wavelengths indicated by the subscripts, are not more than 0.65 and 0.45, respectively.

Substances darkened by potassium hydroxide—Add 0.5 mL of water and 1.0 g of solid potassium hydroxide to 25 mL of Dimethyl Sulfoxide in a glass-stoppered, 50-mL flask. [NOTE—Use only solid potassium hydroxide that is white with no discoloration.] Insert the stopper, and heat in a steam cone for 20 minutes. Cool to room temperature: the absorbance of the solution at 350 nm, measured in a 2-cm cell, water being used as the blank, does not exceed 0.046.

Dimethyl sulfone—

Resolution solution—Prepare a solution containing about 0.15 mg of dimethyl sulfone per mL and 0.1 mg of dibenzyl per mL in Dimethyl Sulfoxide.

Chromatographic system—The gas chromatograph is equipped with a flame-ionization detector and a suitable recorder, and contains a 1.5-m × 3-mm column packed with 10 percent liquid phase G25 on packing S1A (see *Chromatography* ⟨621⟩). The column is temperature-programmed at a rate of about 10° per minute from 100° to 170°, the injection port is maintained at a temperature of about 210°, and the detector block is maintained at a temperature of about 220°. Helium is used as the carrier gas, flowing at the rate of about 30 mL per minute. Chromatograph the *Resolution solution*, and record the peak responses as directed under *Procedure:* the column efficiency as determined from the dimethyl sulfoxide peak is not less than 1000 theoretical plates, and the resolution, *R*, between the dimethyl sulfone and dibenzyl peaks is not less than 5.0.

Procedure—By means of a suitable sampling valve or high-pressure microsyringe, inject about 1 µL of Dimethyl Sulfoxide, record the chromatograms, and measure the responses of the peaks: the response of any peak, other than that of dimethyl sulfoxide, is not greater than 0.03% of the total of the responses of all of the peaks, and the total of the responses of all secondary peaks is not greater than 0.1% of the total of the responses of all of the peaks.

Dimethyl Sulfoxide Irrigation

» Dimethyl Sulfoxide Irrigation is a sterile solution of Dimethyl Sulfoxide in Water for Injection. It contains not less than 95.0 percent and not more than 105.0 percent of the labeled amount of C_2H_6OS.

Packaging and storage—Preserve in single-dose containers, and store at controlled room temperature, protected from strong light.

Labeling—Label it to indicate prominently that it is not intended for injection.

Reference standard—*USP Dimethyl Sulfoxide Reference Standard*—After opening ampul, store in a tightly closed, light-resistant container.

Identification—A portion of the *Assay preparation*, chromatographed as directed in the *Assay*, exhibits a major peak for dimethyl sulfoxide, the retention time of which is identical with that exhibited by the *Standard preparation*.

Pyrogen—It meets the requirements of the *Pyrogen Test* ⟨151⟩, the test dose being 4 mL per kg of a solution prepared by diluting the Irrigation with pyrogen-free saline TS to a concentration of 5 mg of dimethyl sulfoxide per mL.

Sterility—It meets the requirements under *Sterility Tests* ⟨71⟩.

pH ⟨791⟩: between 5.0 and 7.0, when diluted with water to obtain a solution containing 50 mg of dimethyl sulfoxide per mL.

Assay—

Standard preparation—Dissolve an accurately weighed quantity of USP Dimethyl Sulfoxide RS in acetone, and dilute quantitatively with acetone to obtain a solution having a known concentration of about 80 mg per mL.

Assay preparation—Transfer an accurately measured volume of Dimethyl Sulfoxide Irrigation, equivalent to about 2 g of dimethyl sulfoxide, to a 25-mL volumetric flask, dilute with acetone to volume, and mix.

Chromatographic system (see *Chromatography* ⟨621⟩)—Proceed as directed in the test for *Dimethyl sulfone* under *Dimethyl Sulfoxide*. The relative standard deviation for replicate injections of the *Standard preparation* is not more than 2.0%.

Procedure—Introduce equal volumes (about 1.0 µL) of the *Standard preparation* and the *Test preparation* into the gas chromatograph. Measure the peak responses, at corresponding retention times, so obtained, and calculate the quantity, in mg, of C_2H_6OS in each mL of the Irrigation taken by the formula:

$$25(C/V)(r_U/r_S),$$

in which C is the concentration, in mg per mL, of USP Dimethyl Sulfoxide RS in the *Standard preparation*, V is the volume, in mL, of Irrigation taken, and r_U and r_S are the peak responses from the *Assay preparation* and the *Standard preparation*, respectively.

Dinoprost Tromethamine

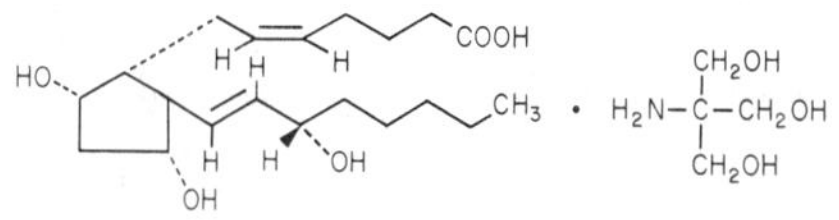

$C_{20}H_{34}O_5 . C_4H_{11}NO_3$ 475.62

Prosta-5,13-dien-1-oic acid, 9,11,15-trihydroxy-, (5*Z*,9α,11α,13*E*,15*S*)-, compd. with 2-amino-2-(hydroxymethyl)-1,3-propanediol (1:1).

(*E*,*Z*)-(1*R*,2*R*,3*R*,5*S*)-7-[3,5-Dihydroxy-2-[(3*S*)-(3-hydroxy-1-octenyl)]cyclopentyl]-5-heptenoic acid compound with 2-amino-2-(hydroxymethyl)-1,3-propanediol (1:1).

Prostaglandin $F_{2\alpha}$ tromethamine [*38562-01-5*].

» Dinoprost Tromethamine contains not less than 95.0 percent and not more than 105.0 percent of $C_{20}H_{34}O_5 . C_4H_{11}NO_3$, calculated on the dried basis.

Caution—Great care should be taken to prevent inhaling particles of Dinoprost Tromethamine and exposing the skin to it.

Packaging and storage—Preserve in tight containers.

Reference standard—*USP Dinoprost Tromethamine Reference Standard*—Do not dry before using.

Identification—The infrared absorption spectrum of a mineral oil dispersion of it exhibits maxima only at the same wavelengths as that of a similar preparation of USP Dinoprost Tromethamine RS.

Specific rotation ⟨781⟩: between +19° and +26°, calculated on the dried basis, determined in an alcohol solution containing 20 mg per mL.

Loss on drying ⟨731⟩—Dry it in vacuum at room temperature and at a pressure not exceeding 5 mm of mercury for 16 hours: it loses not more than 1.0% of its weight.

Residue on ignition ⟨281⟩: not more than 0.5%.

Limit of 5,6-*trans* Isomer and 15-*R* epimer—

Mobile phase, Internal standard solution, Reagent preparations, Standard preparation, Assay preparation, and *Chromatographic system*—Proceed as directed in the *Assay*.

Procedure—Proceed as directed for *Procedure* in the *Assay*. Calculate the percentage of the 5,6-*trans* isomer by the formula:

$$100[R_1/(R_1 + R_D + R_2)],$$

in which R_1, R_2, and R_D are the areas of the 5,6-*trans* isomer, 15-*R* epimer, and dinoprost tromethamine peaks, respectively. The limit is not more than 3.0%. Calculate the percentage of the 15-*R* epimer by the formula:

$$100[R_2/(R_2 + R_D + R_1)].$$

The limit is 1.5%.

Assay—

Mobile phase—Prepare a solution consisting of methylene chloride, 1,3-butanediol, and water (496:3.5:0.25).

Internal standard solution—Prepare a solution in *Mobile phase* containing about 0.75 mg of guaifenesin per mL.

Reagent preparations—

A—Prepare a solution containing about 10 mg of α-bromo-2′-acetonaphthone per mL of acetonitrile. Use a freshly prepared solution.

B—Prepare a solution containing 5 µL of diisopropylethylamine per mL of acetonitrile. Use a freshly prepared solution.

C—Prepare a citrate buffer solution by dissolving 10.5 g of citric acid monohydrate in about 75 mL of water and adding 5 *N* sodium hydroxide until a pH of 4.0 is obtained. Dilute with water to 100 mL, and mix.

Standard preparation—Accurately weigh about 0.5 mg of USP Dinoprost Tromethamine RS into a suitable container. Add about 2 mL of water, about 1 mL of *Reagent C*, and 20.0 mL of methylene chloride. Shake and centrifuge. Transfer 6.0 mL of the lower layer into a suitable container, and evaporate with the aid of nitrogen to dryness. Wash the inside of the container with 200 µL of *Reagent A*. Swirl to dissolve, add 100 µL of *Reagent B*, and mix. Allow the solution to stand for about 1 hour at room temperature, evaporate to dryness, add 4.0 mL of *Internal standard solution*, and mix to obtain a *Standard preparation* having a known concentration of about 0.0375 mg of USP Dinoprost Tromethamine RS per mL.

Assay preparation—Transfer about 0.5 mg of Dinoprost Tromethamine, accurately weighed, to a suitable container, and proceed as directed for *Standard preparation*.

Chromatographic system (see *Chromatography* ⟨621⟩)—The liquid chromatograph is equipped with a 254-nm detector and a 4.6-mm × 30-cm column that contains packing L3. The flow rate is about 1.5 mL per minute. Chromatograph the *Standard preparation*, and record the peak responses as directed under *Procedure:* the resolution, *R*, between dinoprost tromethamine and internal standard peaks is not less than 10, and the relative standard deviation for replicate injections is not more than 2.0%.

Procedure—Inject equal volumes (about 20 µL) of the *Assay preparation* and *Standard preparation* into the liquid chromatograph, record the chromatograms, and measure the peak responses at equivalent retention times. The relative retention times are about 0.4, 0.5, 1.0, and 1.2 for the internal standard, the 15-*R* epimer, dinoprost tromethamine, and the 5,6-*trans* isomer, respectively. Calculate the quantity, in mg, of $C_{20}H_{34}O_5 . C_4H_{11}NO_3$ in the portion of dinoprost tromethamine taken by the formula:

$$13.33C(R_U/R_S),$$

in which *C* is the concentration, in mg per mL, of USP Dinoprost Tromethamine RS, in the *Standard preparation*, and R_U and R_S are the ratios of the responses for the dinoprost tromethamine and internal standard peaks obtained from the *Assay preparation* and the *Standard preparation*, respectively.

Dinoprost Tromethamine Injection

» Dinoprost Tromethamine Injection is a sterile solution of Dinoprost Tromethamine in Water for Injection. It may contain a suitable preservative, such as benzyl alcohol. It contains not less than 90.0 percent and not more than 110.0 percent of the labeled amount of dinoprost ($C_{20}H_{34}O_5$).

Packaging and storage—Preserve in single-dose or in multiple-dose containers, preferably of Type I glass.

Reference standards—*USP Dinoprost Tromethamine Reference Standard*—Do not dry before using. *USP Endotoxin Reference Standard.*

Identification—The retention time of the derivatized dinoprost peak in the chromatogram of the *Assay preparation* corresponds to that of the derivatized dinoprost peak in the *Standard preparation*, both relative to the internal standard, obtained as directed in the *Assay*.

Bacterial endotoxin—It meets the requirements of the *Bacterial Endotoxins Test* ⟨85⟩, the limit of endotoxin content being not more than 5 USP Endotoxin Units per 0.6 mg of dinoprost.

pH ⟨791⟩: between 7.0 and 9.0.

Other requirements—It meets the requirements under *Injections* ⟨1⟩, and it meets the requirements under *Sterility Tests* ⟨71⟩, when tested as directed in the section, *Test Procedures Using Membrane Filtration.*

Assay—

Mobile phase, Internal standard solution, Reagent preparations, and *Chromatographic system*—Proceed as directed in the *Assay* under *Dinoprost Tromethamine.*

Diluting solution—Use *Sterile Water for Injection* containing 0.945% of benzyl alcohol.

Standard preparation—Accurately weigh about 6.7 mg of USP Dinoprost Tromethamine RS, and transfer to a suitable container. Add 10.0 mL of *Diluting solution*, and mix to dissolve. Transfer 1.0 mL of this solution to a suitable container, and add 1.0 mL of *Reagent C*, and 20.0 mL of methylene chloride. Shake and centrifuge. Transfer 5.0 mL of the lower layer into a suitable container, and evaporate with the aid of nitrogen to dryness. Wash the inside of the container with 200 µL of *Reagent A*. Swirl to dissolve, add 100 µL of *Reagent B*, and mix. Allow the solution to stand for about 1 hour at room temperature, evaporate to dryness, add 4.0 mL of *Internal standard solution*, and mix to obtain a *Standard preparation* having a known concentration of about 0.0419 mg of USP Dinoprost Tromethamine RS per mL.

Assay preparation—Transfer an accurately measured volume of Dinoprost Tromethamine Injection, equivalent to about 5 mg of dinoprost, to a suitable container, add 9.0 mL of *Diluting solution*, and mix. Proceed as directed for *Standard preparation*, beginning with "Transfer 1.0 mL of this solution."

Procedure—Proceed as directed for *Procedure* in the *Assay* under *Dinoprost Tromethamine.* Calculate the quantity, in mg, of $C_{20}H_{34}O_5$ in each mL of the Injection taken by the formula:

$$(354.49/475.62)(160C/V)(R_U/R_S),$$

in which 354.49 and 475.62 are the molecular weights of dinoprost and dinoprost tromethamine, respectively, *C* is the concentration, in mg per mL, of USP Dinoprost Tromethamine RS in the *Standard preparation*, *V* is the volume, in mL, of Injection taken, and R_U and R_S are the ratios of the responses for the dinoprost tromethamine and internal standard peaks obtained from the *Assay preparation* and the *Standard preparation*, respectively.

Dioxybenzone

$C_{14}H_{12}O_4$ 244.25
Methanone, (2-hydroxy-4-methoxyphenyl)(2-hydroxyphenyl)-.
2,2′-Dihydroxy-4-methoxybenzophenone [*131-53-3*].

» Dioxybenzone contains not less than 97.0 percent and not more than 103.0 percent of $C_{14}H_{12}O_4$, calculated on the dried basis.

Packaging and storage—Preserve in tight, light-resistant containers.

Reference standard—*USP Dioxybenzone Reference Standard*—Dry in vacuum at 40° for 2 hours before using.

Identification—

A: The infrared absorption spectrum of a potassium bromide dispersion of it, previously dried, exhibits maxima only at the same wavelengths as that of a similar preparation of USP Dioxybenzone RS.

B: The ultraviolet absorption spectrum of a 1 in 100,000 solution of it in methanol exhibits maxima and minima at the same wavelengths as that of a similar preparation of USP Dioxybenzone RS, concomitantly measured, and the respective absorptivities, calculated on the dried basis, at the wavelength of maximum absorbance at about 286 nm do not differ by more than 3.0%.

Congealing temperature ⟨651⟩: not lower than 68.0°.

Loss on drying ⟨731⟩—Dry it in vacuum at 40° for 2 hours: it loses not more than 2.0% of its weight.

Assay—Dissolve about 100 mg of Dioxybenzone, accurately weighed, in toluene in a 100-mL volumetric flask, dilute with toluene to volume, and mix. Pipet 1 mL of this solution into a second 100-mL volumetric flask, add methanol to volume, and mix. Similarly, prepare a Standard solution of USP Dioxybenzone RS, accurately weighed, having a known concentration of about 10 µg per mL. Concomitantly determine the absorbances of both solutions in 1-cm cells at the wavelength of maximum absorbance at about 325 nm, with a suitable spectrophotometer, using a 1 in 100 solution of toluene in methanol as the blank. Calculate the quantity, in mg, of $C_{14}H_{12}O_4$ in the Dioxybenzone taken by the formula:

$$10C(A_U/A_S),$$

in which *C* is the concentration, in µg per mL, of USP Dioxybenzone RS in the Standard solution, and A_U and A_S are the absorbances of the solution of Dioxybenzone and the Standard solution, respectively.

Dioxybenzone and Oxybenzone Cream

» Dioxybenzone and Oxybenzone Cream is a mixture of approximately equal parts of Dioxybenzone and Oxybenzone in a suitable cream base. It contains, in each 100 g, not less than 2.7 g and not more than 3.3 g each of dioxybenzone ($C_{14}H_{12}O_4$) and oxybenzone ($C_{14}H_{12}O_3$).

Packaging and storage—Preserve in tight containers.

Reference standards—*USP Dioxybenzone Reference Standard*—Dry in vacuum at 40° for 2 hours before using. *USP Oxybenzone Reference Standard*—Dry in vacuum at 40° for 2 hours before using.

Identification—The ultraviolet absorption spectra of the solutions prepared from the Cream for measurement of absorbance in the *Assay* exhibit maxima and minima at the same wavelengths as those of the respective *Standard preparations* employed in the *Assay*.

Minimum fill ⟨755⟩: meets the requirements.

Assay—

Standard preparations—Transfer about 30 mg of USP Dioxybenzone RS, accurately weighed, to a 50-mL volumetric flask, dissolve in methanol, dilute with the same solvent to volume, and mix. Pipet 2 mL of this solution into a 100-mL volumetric flask, add methanol to volume, and mix. Similarly, prepare a *Standard preparation* of USP Oxybenzone RS, accurately weighed, having a known concentration of about 12 µg per mL.

Assay preparation—Dissolve an accurately weighed portion of Dioxybenzone and Oxybenzone Cream, equivalent to about 25 mg each of dioxybenzone and oxybenzone, in methanol in a 100-mL volumetric flask, dilute with methanol to volume, and mix. Pipet 1 mL of this solution into a 15-mL conical test tube, evaporate on a water bath just to dryness, using a gentle stream of air, and dissolve the residue in about 200 µL of methanol.

Procedure—Prepare sheets of chromatographic paper (Whatman No. 1 or equivalent), each measuring about 23 × 28.5 cm, as follows: Immerse the sheets in a 1 in 20 solution of light mineral oil in solvent hexane, withdraw them immediately, and allow to dry in air. On one sheet mark a starting line about 2.5 cm from the long edge, and apply the entire *Assay preparation* as a uniform streak along the starting line, using a stream of air or an air blower, if necessary, to maintain the width of the streak between 5 mm and 10 mm. Rinse the conical test tube, which contained the *Assay preparation*, with about 100 µL of methanol, and apply the rinse to the starting line. Similarly, repeat the rinsing and streaking with two additional portions of methanol, and then allow the paper to dry in air for 5 minutes.

Staple together the short edges of the paper to form a cylinder, and place it in a 12- × 25-cm cylindrical chromatographic chamber containing about 40 mL of a mobile solvent consisting of a mixture of equal volumes of acetone and water. Seal the chamber, and allow the chromatogram to develop for 2 hours.

Remove the paper from the chamber, air-dry, then remove the staples, and view the chromatogram under short-wavelength (254 nm) ultraviolet radiation. Mark the two bands representing the separated dioxybenzone and oxybenzone, respectively. [NOTE—Determine the relative position of each benzone on the chromatogram by applying suitable aliquots of each *Standard preparation* to another prepared chromatographic sheet, and developing the chromatogram in a manner similar to that described for the *Assay preparation.*] Cut the marked bands from the sheet, and then, keeping the band segments separate, cut each into several pieces to facilitate extraction. Place the pieces from each band in separate glass-stoppered, 50-mL conical flasks, add 20.0 mL of methanol to each flask, and shake gently for 30 minutes.

To provide the chromatographic blank, treat one of the prepared chromatographic sheets in the same manner as described above, but omit the application of the *Assay preparation.* Cut from the chromatographed paper the areas corresponding to the bands produced by the benzones from the *Assay preparation*, and in the same manner extract the blank bands for 30 minutes with 20.0 mL of methanol.

Concomitantly determine the absorbance of each of the 4 solutions thus prepared, and of each of the *Standard preparations*, in a 1-cm cell at the wavelength of maximum absorbance at about 325 nm, with a suitable spectrophotometer, using methanol as the blank. Calculate the quantity, in mg, of dioxybenzone ($C_{14}H_{12}O_4$) in the portion of Cream taken by the formula:

$$2C(A_U - A_B)/A_S,$$

in which *C* is the concentration, in µg per mL, of USP Dioxybenzone RS in the dioxybenzone *Standard preparation*, and A_U, A_B, and A_S are the absorbances of the dioxybenzone solution from the *Assay preparation*, the dioxybenzone chromatographic blank solution, and the dioxybenzone *Standard preparation*, respectively. In a similar manner, calculate the quantity, in mg, of oxybenzone ($C_{14}H_{12}O_3$) in the portion of Cream taken, using as *C*, A_U, A_B, and A_S the respective values pertaining to the oxybenzone determination.

Diperodon

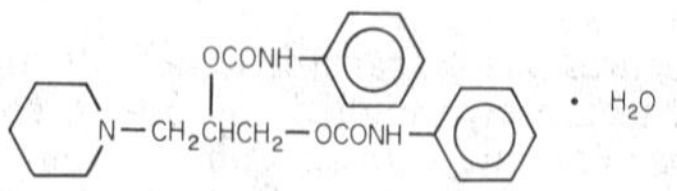

$C_{22}H_{27}N_3O_4.H_2O$ 415.49
1,2-Propanediol, 3-(1-piperidinyl)-, bis(phenylcarbamate) (ester), monohydrate.
3-Piperidino-1,2-propanediol dicarbanilate (ester) monohydrate [*51552-99-9*].
Anhydrous 397.47 [*101-08-6*].

» Diperodon contains not less than 98.0 percent and not more than 102.0 percent of $C_{22}H_{27}N_3O_4$, calculated on the anhydrous basis.

Packaging and storage—Preserve in well-closed containers.

Reference standard—*USP Diperodon Reference Standard*—Do not dry; determine the water content at the time of use.

Identification—

A: The infrared absorption spectrum of a mineral oil dispersion of it exhibits maxima only at the same wavelengths as that of a similar preparation of USP Diperodon RS.

B: Prepare a solution of it in methanol to contain approximately 100 mg in 10 mL. Prepare a similar solution of USP Diperodon RS. Apply 10 µL portions of each solution to a thin-layer chromatographic plate coated with a 0.25-mm layer of chromatographic silica gel, dry, and develop the chromatogram (see *Chromatography* ⟨621⟩), using a mixture of butanol, acetic acid, and water (4:2:1). Observe the dried chromatographic plate under short-wavelength ultraviolet light: the R_f value of the principal spot obtained from the solution of Diperodon corresponds to that obtained from the Standard solution.

Water, *Method I* ⟨921⟩: between 3.5% and 5.0%.

Residue on ignition ⟨281⟩: not more than 0.1%.

Chloride ⟨221⟩—A solution of 100 mg in 10 mL of alcohol, upon the addition, successively, of 5 mL of water, 1 mL of 2 *N* nitric acid, and 1 mL of silver nitrate TS, shows no more turbidity than that of a control prepared with 0.15 mL of 0.020 *N* hydrochloric acid (0.1%).

Heavy metals, *Method II* ⟨231⟩: 0.002%.

Assay—Dissolve about 350 mg of Diperodon, accurately weighed, in 40 mL of glacial acetic acid, warming, if necessary, to effect solution. Cool, add crystal violet TS, and titrate with 0.1 *N* perchloric acid VS to an emerald green end-point. Perform a blank determination, and make any necessary correction. Each mL of 0.1 *N* perchloric acid is equivalent to 39.75 mg of $C_{22}H_{27}N_3O_4$.

Diperodon Ointment

» Diperodon Ointment contains an amount of diperodon ($C_{22}H_{27}N_3O_4.H_2O$) equivalent to not less than 90.0 percent and not more than 110.0 percent of the labeled amount of $C_{22}H_{27}N_3O_4$ in a suitable ointment base.

Packaging and storage—Preserve in collapsible tubes or in tight containers.

Reference standard—*USP Diperodon Reference Standard*—Do not dry; determine the water content at the time of use.

Identification—

A: The ultraviolet absorption spectrum of the *Assay preparation*, prepared as directed in the *Assay*, exhibits maxima and minima at the same wavelengths as that of the *Standard preparation*, prepared as directed in the *Assay*.

B: Evaporate a 40-mL portion of the diperodon eluate obtained in the *Assay* on a water bath with the aid of a current of air just to dryness. Add 10 mL of methanol to dissolve the residue, and repeat the evaporation. Dissolve the residue in 1 mL of methanol: the resulting solution responds to *Identification test B* under *Diperodon*.

Minimum fill ⟨755⟩: meets the requirements.

Assay—

Chromatographic column—Pack a pledget of glass wool in the base of a 20- × 300-mm chromatographic tube fitted with a stopcock. Half fill the tube with carbon tetrachloride, and add 29 g of alumina at a rate that will allow the particles to be moistened with the solvent before impacting on the developing adsorbent bed. Drain the solvent, pass an additional 50 mL of carbon tetrachloride through the column at a moderate rate, and close the stopcock with the solvent level with the top of the alumina column.

Standard preparation—Transfer an accurately weighed quantity of USP Diperodon RS, equivalent to about 25 mg of anhydrous diperodon, to a 100-mL volumetric flask. Dissolve in alcohol, dilute with alcohol to volume, and mix. Transfer 3.0 mL of this solution to a second 100-mL volumetric flask, add 0.1 mL of hydrochloric acid, mix, and evaporate on a water bath with the aid of a current of air to dryness. Dissolve the residue in a freshly prepared solvent consisting of a mixture of equal volumes of isopropyl alcohol and *n*-hexane, dilute with the same solvent to volume, and mix. The concentration of anhydrous diperodon, equivalent to the USP Diperodon RS in the *Standard preparation*, is about 7.5 µg per mL.

Assay preparation—Transfer an accurately weighed portion of Diperodon Ointment, equivalent to about 50 mg of diperodon, to a 150-mL beaker. Add 10 mL of alcohol and 0.1 mL of hydrochloric acid, and evaporate on a water bath with the aid of a current of air to dryness. Dissolve the residue, while still warm, in a solvent prepared by mixing 56 volumes of chloroform, 37 volumes of *n*-hexane, and 7 volumes of alcohol. Transfer to a 100-mL volumetric flask with the aid of the same solvent, cool, dilute with the same solvent to volume, and mix. [NOTE—This solution may become cloudy on standing; if clouding occurs, warm slightly, mix, and cool.] Transfer 10.0 mL of this solution to the top of the *Chromatographic column*, and allow the solvent to drain through the column at a rate of about 1 mL per minute. Do not allow the solution to descend below the top of the alumina column at any time during this and succeeding elutions. Wash the inner wall of the tube with 3 mL of carbon tetrachloride, elute, repeat the washing and elution, and wash the column with 145 mL of carbon tetrachloride, eluting at a rate of 1 to 2 mL per minute. Discard all of the eluted solvent and washings. Elute the diperodon from the column into a 200-mL volumetric flask, at a flow rate of 1 to 2 mL per minute, with a 1 in 10 solution of methanol in chloroform until the flask is filled to volume, and mix. Transfer 30.0 mL of this solution to a suitable vessel, add 0.1 mL of hydrochloric acid, mix, and evaporate on a water bath with the aid of a current of air just to dryness. Immediately add 3 mL of a solvent consisting of a mixture of equal volumes of isopropyl alcohol and *n*-hexane, and again evaporate just to dryness. Dissolve the residue in 50 mL of the isopropyl alcohol–*n*-hexane solvent, and warm slightly to effect solution (about 1 minute) with the beaker covered with a watch glass. Transfer to a 100-mL volumetric flask, cool, dilute with the isopropyl alcohol–*n*-hexane solvent to volume, and mix.

Procedure—Concomitantly determine the absorbances of the solutions in 1-cm cells at the wavelength of maximum absorbance at about 235 nm and at 300 nm, with a suitable spectrophotometer, using a mixture of equal volumes of isopropyl alcohol and *n*-hexane as the blank. Calculate the quantity, in mg, of $C_{22}H_{27}N_3O_4$ in the portion of Ointment taken by the formula:

$$(6.667C)[(A_{U235} - A_{U300})/(A_{S235} - A_{S300})],$$

in which *C* is the concentration, in µg per mL, of diperodon, calculated on the anhydrous basis, in the *Standard preparation*, and the parenthetic expressions are the differences in the absorbances of the two solutions at the wavelengths indicated by the subscripts, for the solution from the *Assay preparation* (*U*) and the *Standard preparation* (*S*), respectively.

Diphemanil Methylsulfate

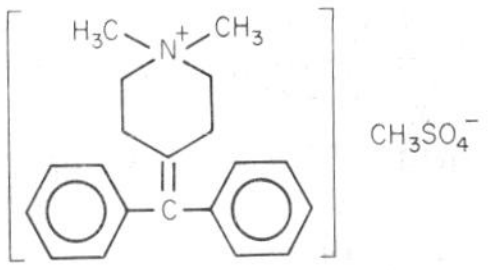

$C_{21}H_{27}NO_4S$ 389.51
Piperidinium, 4-(diphenylmethylene)-1,1-dimethyl-, methyl sulfate.
4-(Diphenylmethylene)-1,1-dimethylpiperidinium methyl sulfate [*62-97-5*].

» Diphemanil Methylsulfate contains not less than 97.0 percent and not more than 103.0 percent of $C_{21}H_{27}NO_4S$, calculated on the dried basis.

Packaging and storage—Preserve in tight containers.

Reference standard—*USP Diphemanil Methylsulfate Reference Standard*—Dry at 105° for 4 hours before using.

Identification—

A: The infrared absorption spectrum of a mineral oil dispersion of it, previously dried, exhibits maxima only at the same wavelengths as that of a similar preparation of USP Diphemanil Methylsulfate RS.

B: It responds to the *Thin-layer Chromatographic Identification Test* ⟨201⟩, the test solution and the Standard solution being prepared at a concentration of 1 mg per mL in methanol, and the solvent system being a mixture of methoxyethanol and hydrochloric acid (100:1).

Melting range ⟨741⟩: between 189° and 196°.

Loss on drying ⟨731⟩—Dry it at 105° for 4 hours: it loses not more than 0.5% of its weight.

Residue on ignition ⟨281⟩: not more than 0.1%.

Heavy metals, *Method II* ⟨231⟩: 0.002%.

Ordinary impurities ⟨466⟩—

Test solution: methanol.

Standard solution: methanol.

Eluant: a mixture of *n*-butanol, glacial acetic acid, and water (7:2:1).

Visualization: 3.

Assay—

Buffer solution—Dissolve 9.23 g of anhydrous dibasic sodium phosphate in 800 mL of water, adjust with saturated citric acid solution to a pH of 4, dilute with water to 1000 mL, and mix.

Standard preparation—Dissolve an accurately weighed quantity of USP Diphemanil Methylsulfate RS in water to obtain a solution having a known concentration of about 100 µg per mL.

Assay preparation—Transfer about 50 mg of Diphemanil Methylsulfate, accurately weighed, to a 500-mL volumetric flask, dissolve in water, dilute with water to volume, and mix.

Procedure—Transfer 10.0 mL each of the *Standard preparation*, the *Assay preparation*, and water to provide the blank, to individual 250-mL separators, and add to each 40 mL of *Buffer solution*, 10 mL of a solution of bromophenol blue in chloroform (1 in 1000), and 60 mL of chloroform. Shake the separators vigorously for 2 minutes, allow to stand for 15 minutes, and then withdraw the chloroform layers through chloroform-washed cotton into 100-mL volumetric flasks. Repeat the extraction with 20 mL of chloroform, adding the filtered chloroform extracts to the respective volumetric flasks, dilute with chloroform to volume, and mix. Without delay, concomitantly determine the absorbances of the solutions, in 1-cm cells, at the wavelength of maximum absorbance at about 412 nm, with a suitable spectrophotometer, using the blank to set the instrument. Calculate the quantity, in mg, of $C_{21}H_{27}NO_4S$ in the Diphemanil Methylsulfate taken by the formula:

$$0.5C(A_U/A_S),$$

in which C is the concentration, in µg per mL, of USP Diphemanil Methylsulfate RS in the *Standard preparation*, and A_U and A_S are the absorbances of the solutions from the *Assay preparation* and the *Standard preparation*, respectively.

Diphemanil Methylsulfate Tablets

» Diphemanil Methylsulfate Tablets contain not less than 92.5 percent and not more than 107.5 percent of the labeled amount of $C_{21}H_{27}NO_4S$.

Packaging and storage—Preserve in tight containers.

Reference standard—*USP Diphemanil Methylsulfate Reference Standard*—Dry at 105° for 4 hours before using.

Identification—Shake a quantity of finely powdered Tablets, equivalent to about 25 mg of diphemanil methylsulfate, with 25 mL of methanol for 15 minutes, and filter. Proceed as directed in the *Identification test* under *Diphemanil Methylsulfate*, beginning with "On a line parallel to."

Dissolution ⟨711⟩—

Medium: water; 900 mL.

Apparatus 1: 100 rpm.

Time: 30 minutes.

Procedure—Determine the amount of $C_{21}H_{27}NO_4S$ dissolved from ultraviolet absorbances at the wavelength of maximum absorbance at about 230 nm of filtered portions of the solution under test, suitably diluted with *Dissolution Medium*, if necessary, in comparison with a Standard solution having a known concentration of USP Diphemanil Methylsulfate RS in the same medium.

Tolerances—Not less than 80% (*Q*) of the labeled amount of $C_{21}H_{27}NO_4S$ is dissolved in 30 minutes.

Uniformity of dosage units ⟨905⟩: meet the requirements.

Assay—

Buffer solution and *Standard preparation*—Prepare as directed in the *Assay* under *Diphemanil Methylsulfate*.

Assay preparation—Weigh and finely powder not less than 10 Diphemanil Methylsulfate Tablets. Transfer an accurately weighed portion of the powder, equivalent to about 50 mg of diphemanil methylsulfate, to a 500-mL volumetric flask, add 150 mL of water, shake by mechanical means for 15 minutes, add water to volume, mix, and filter through a dry filter, discarding the first 25 mL of the filtrate.

Procedure—Proceed as directed for *Procedure* in the *Assay* under *Diphemanil Methylsulfate*. Calculate the quantity, in mg, of $C_{21}H_{27}NO_4S$ in the portion of Tablets taken by the formula:

$$0.5C(A_U/A_S),$$

in which C is the concentration, in µg per mL, of USP Diphemanil Methylsulfate RS in the *Standard preparation*.

Diphenhydramine Citrate

$C_{17}H_{21}NO \cdot C_6H_8O_7$ 447.49
Ethanamine, 2-(diphenylmethoxy)-*N*,*N*-dimethyl-, 2-hydroxy-1,2,3-propanetricarboxylate (1:1).
2-(Diphenylmethoxy)-*N*,*N*-dimethylethylamine citrate (1:1) [*88637-37-0*].

» Diphenhydramine Citrate contains not less than 98.0 percent and not more than 100.5 percent of $C_{17}H_{21}NO \cdot C_6H_8O_7$, calculated on the dried basis.

Packaging and storage—Preserve in tight, light-resistant containers.

Reference standard—*USP Diphenhydramine Citrate Reference Standard*.

Identification—

A: The infrared absorption spectrum of a potassium bromide dispersion of it, previously dried, exhibits maxima only at the same wavelengths as that of a similar preparation of USP Diphenhydramine Citrate RS.

B: The ultraviolet absorption spectrum of a solution (1 in 1400) exhibits maxima and minima at the same wavelengths as that of a similar solution of USP Diphenhydramine Citrate RS, concomitantly measured.

C: It responds to the test for *Citrate* ⟨191⟩.

Melting range ⟨741⟩: between 146° and 150°, but the range between beginning and end of melting does not exceed 2°.

Loss on drying ⟨731⟩—Dry it at 105° for 3 hours: it loses not more than 0.5% of its weight.

Residue on ignition ⟨281⟩—To about 8 g, accurately weighed, add 5 mL of sulfuric acid, and char. After the substance is thoroughly charred, add 4 mL of nitric acid and a few drops of sulfuric acid, heat gently until fumes are no longer evolved, and ignite at 800 ± 25° until the carbon is consumed. Place in a muffle furnace at 550 ± 50° for about 1 hour. Continue the ignition until constant weight is attained: not more than 0.1% remains.

Assay—Dissolve about 1.6 g of Diphenhydramine Citrate, accurately weighed, in a mixture of 100 mL of glacial acetic acid and 20 mL of xylene. Add 20 mL of mercuric acetate TS, and titrate with 0.1 *N* perchloric acid VS, determining the end-point potentiometrically. Perform a blank determination, and make any necessary correction. Each mL of 0.1 *N* perchloric acid is equivalent to 44.75 mg of $C_{17}H_{21}NO.C_6H_8O_7$.

Diphenhydramine Citrate Tablets, Acetaminophen and—*see* Acetaminophen and Diphenhydramine Citrate Tablets

Diphenhydramine Hydrochloride

H
C$_6$H$_5$—CO—$CH_2CH_2N(CH_3)_2$ · HCl
C$_6$H$_5$

$C_{17}H_{21}NO.HCl$ 291.82

Ethanamine, 2-(diphenylmethoxy)-*N*,*N*-dimethyl-, hydrochloride.

2-(Diphenylmethoxy)-*N*,*N*-dimethylethylamine hydrochloride [*147-24-0*].

» Diphenhydramine Hydrochloride contains not less than 98.0 percent and not more than 102.0 percent of $C_{17}H_{21}NO.HCl$, calculated on the dried basis.

Packaging and storage—Preserve in tight, light-resistant containers.

Reference standard—*USP Diphenhydramine Hydrochloride Reference Standard*—Dry at 105° for 3 hours before using.

Identification—

A: It meets the requirements under *Identification—Organic Nitrogenous Bases* ⟨181⟩.

B: The retention time of the major peak in the chromatogram of the *Assay preparation* corresponds to that of the *Standard preparation*, as obtained in the *Assay*.

C: It responds to the tests for *Chloride* ⟨191⟩.

Melting range ⟨741⟩: between 167° and 172°.

Loss on drying ⟨731⟩—Dry it at 105° for 3 hours: it loses not more than 0.5% of its weight.

Residue on ignition ⟨281⟩: not more than 0.1%.

Assay—

Mobile phase—Prepare a solution of acetonitrile, water, and triethylamine (50:50:0.5), adjust with glacial acetic acid to a pH of 6.5, filter, and degas. Make adjustments if necessary (see *System Suitability* under *Chromatography* ⟨621⟩).

Standard preparation—Dissolve an accurately weighed quantity of USP Diphenhydramine Hydrochloride RS in water to obtain a solution having a known concentration of about 0.5 mg per mL.

Assay preparation—Transfer about 25 mg of Diphenhydramine Hydrochloride, accurately weighed, to a 50-mL volumetric flask, dissolve in water, dilute with water to volume, mix, and filter.

System suitability solution—Dissolve about 5 mg of benzophenone in 5 mL of acetonitrile, dilute with water to 100 mL, and mix. Transfer 1.0 mL of this solution and 5 mg of diphenhydramine hydrochloride to a 10-mL volumetric flask, dilute with water to volume, and mix.

Chromatographic system (see *Chromatography* ⟨621⟩)—The liquid chromatograph is equipped with a 254-nm detector and a 4.6-mm × 25-cm column that contains packing L10. The flow rate is about 1 mL per minute. Chromatograph the *System suitability solution*, and record the peak responses as directed under *Procedure:* the resolution, *R*, between the benzophenone and diphenhydramine peaks is not less than 2.0. The relative retention times are about 0.8 for benzophenone and 1.0 for diphenhydramine hydrochloride. Chromatograph replicate injections of the *Standard preparation*, and record the peak responses as directed under *Procedure:* the relative standard deviation is not more than 2.0%, the tailing factor for the diphenhydramine hydrochloride peak is not more than 1.3, and the number of theoretical plates is not less than 3000.

Procedure—Separately inject equal volumes (about 10 µL) of the *Standard preparation* and the *Assay preparation* into the chromatograph, record the chromatograms, and measure the responses for the major peaks. Calculate the quantity, in mg, of $C_{17}H_{21}NO.HCl$ in the portion of Diphenhydramine Hydrochloride taken by the formula:

$$50C(r_U/r_S),$$

in which C is the concentration, in mg per mL, of USP Diphenhydramine Hydrochloride RS in the *Standard preparation*, and r_U and r_S are the peak responses obtained from the *Assay preparation* and the *Standard preparation*, respectively.

Diphenhydramine Hydrochloride Capsules

» Diphenhydramine Hydrochloride Capsules contain not less than 90.0 percent and not more than 110.0 percent of the labeled amount of $C_{17}H_{21}NO.HCl$.

Packaging and storage—Preserve in tight containers.

Reference standard—*USP Diphenhydramine Hydrochloride Reference Standard*—Dry at 105° for 3 hours before using.

Identification—

A: The contents of the Capsules meet the requirements under *Identification—Organic Nitrogenous Bases* ⟨181⟩.

B: The retention time of the major peak in the chromatogram of the *Assay preparation* corresponds to that of the *Standard preparation*, as obtained in the *Assay*.

Dissolution ⟨711⟩—

Medium: water; 500 mL.

Apparatus 1: 100 rpm.

Time: 45 minutes.

Mobile phase and *Chromatographic system*—Prepare as directed in the *Assay*.

Procedure—Inject a measured volume (about 50 µL) of a filtered portion of the solution under test into the chromatograph, record the chromatogram, and measure the response for the major peak. Determine the quantity of $C_{17}H_{21}NO.HCl$ dissolved in comparison with a Standard solution having a known concentration of USP Diphenhydramine Hydrochloride RS in the same medium and similarly chromatographed.

Tolerances—Not less than 75% (Q) of the labeled amount of $C_{17}H_{21}NO.HCl$ is dissolved in 45 minutes.

Uniformity of dosage units ⟨905⟩: meet the requirements.

Assay—

Mobile phase, Standard preparation, System suitability solution, and *Chromatographic system*—Prepare as directed in the *Assay* under *Diphenhydramine Hydrochloride.*

Assay preparation—Weigh and combine the contents of not less than 20 Diphenhydramine Hydrochloride Capsules. Transfer an accurately weighed portion of the combined Capsule contents, equivalent to about 50 mg of diphenhydramine hydrochloride, to a 100-mL volumetric flask. Dissolve in water, dilute with water to volume, mix, and filter.

Procedure—Proceed as directed for *Procedure* in the *Assay* under *Diphenhydramine Hydrochloride.* Calculate the quantity, in mg, of $C_{17}H_{21}NO.HCl$ in the portion of Capsule contents taken by the formula:

$$100C(r_U/r_S),$$

in which C is the concentration, in mg per mL, of USP Diphenhydramine Hydrochloride RS in the *Standard preparation,* and r_U and r_S are the peak responses obtained from the *Assay preparation* and the *Standard preparation,* respectively.

Diphenhydramine Hydrochloride Elixir

» Diphenhydramine Hydrochloride Elixir contains not less than 90.0 percent and not more than 110.0 percent of the labeled amount of $C_{17}H_{21}NO.HCl$.

Packaging and storage—Preserve in tight, light-resistant containers.

Reference standard—*USP Diphenhydramine Hydrochloride Reference Standard*—Dry at 105° for 3 hours before using.

Identification—

A: Place a portion of Elixir, equivalent to 50 mg of diphenhydramine hydrochloride, in a separator, add 0.5 mL of 2 N sulfuric acid, and extract with three 15-mL portions of ether, discarding the extracts. Add 5 mL of water. In a second separator dissolve 50 mg of USP Diphenhydramine Hydrochloride RS in 25 mL of water. Treat each solution as follows: Add 2 mL of 1 N sodium hydroxide, and extract with 75 mL of *n*-heptane. Wash the *n*-heptane extract with 10 mL of water, evaporate the extract to dryness, and dissolve the residue in 4 mL of carbon disulfide. Filter through a dry filter to clarify the solution, if necessary, and proceed as directed under *Identification—Organic Nitrogenous Bases* ⟨181⟩, beginning with "Determine the absorption spectra of the filtered solutions": the Elixir meets the requirements of the test.

B: The retention time of the major peak in the chromatogram of the *Assay preparation* corresponds to that of the *Standard preparation,* as obtained in the *Assay.*

Alcohol content ⟨611⟩: between 12.0% and 15.0% of C_2H_5OH.

Assay—

Mobile phase, Standard preparation, System suitability solution, and *Chromatographic system*—Prepare as directed in the *Assay* under *Diphenhydramine Hydrochloride.*

Assay preparation—Transfer an accurately measured volume of Diphenhydramine Hydrochloride Elixir, equivalent to about 50 mg of diphenhydramine hydrochloride, to a 100-mL volumetric flask, dilute with water to volume, and mix.

Procedure—Proceed as directed for *Procedure* in the *Assay* under *Diphenhydramine Hydrochloride.* Calculate the quantity, in mg, of $C_{17}H_{21}NO.HCl$ in each mL of the Elixir taken by the formula:

$$100(C/V)(r_U/r_S),$$

in which C is the concentration, in mg per mL, of USP Diphenhydramine Hydrochloride RS in the *Standard preparation,* V is the volume, in mL, of Elixir taken, and r_U and r_S are the peak responses obtained from the *Assay preparation* and the *Standard preparation,* respectively.

Diphenhydramine Hydrochloride Injection

» Diphenhydramine Hydrochloride Injection is a sterile solution of Diphenhydramine Hydrochloride in Water for Injection. It contains not less than 90.0 percent and not more than 110.0 percent of the labeled amount of $C_{17}H_{21}NO.HCl$.

Packaging and storage—Preserve in single-dose or in multiple-dose containers, preferably of Type I glass, protected from light.

Reference standard—*USP Diphenhydramine Hydrochloride Reference Standard*—Dry at 105° for 3 hours before using.

Identification—

A: Dilute a volume of Injection, equivalent to about 50 mg of diphenhydramine hydrochloride, with 0.03 N sulfuric acid to 25 mL, and proceed as directed under *Identification—Organic Nitrogenous Bases* ⟨181⟩, beginning with "Transfer the liquid to a separator": the Injection meets the requirements of the test.

B: The retention time of the major peak in the chromatogram of the *Assay preparation* corresponds to that of the *Standard preparation,* as obtained in the *Assay.*

pH ⟨791⟩: between 4.0 and 6.5.

Other requirements—It meets the requirements under *Injections* ⟨1⟩.

Assay—

Mobile phase, Standard preparation, System suitability solution, and *Chromatographic system*—Prepare as directed in the *Assay* under *Diphenhydramine Hydrochloride.*

Assay preparation—Transfer an accurately measured volume of Diphenhydramine Hydrochloride Injection, equivalent to about 50 mg of diphenhydramine hydrochloride, to a 100-mL volumetric flask, dilute with water to volume, and mix.

Procedure—Proceed as directed for *Procedure* in the *Assay* under *Diphenhydramine Hydrochloride.* Calculate the quantity, in mg, of $C_{17}H_{21}NO.HCl$ in each mL of the Injection taken by the formula:

$$100(C/V)(r_U/r_S),$$

in which C is the concentration, in mg per mL, of USP Diphenhydramine Hydrochloride RS in the *Standard preparation,* V is the volume, in mL, of Injection taken, and r_U and r_S are the peak responses obtained from the *Assay preparation* and the *Standard preparation,* respectively.

Diphenoxylate Hydrochloride

$C_{30}H_{32}N_2O_2.HCl$ 489.06

4-Piperidinecarboxylic acid, 1-(3-cyano-3,3-diphenylpropyl)-4-phenyl-, ethyl ester, monohydrochloride.

Ethyl 1-(3-cyano-3,3-diphenylpropyl)-4-phenylisonipecotate monohydrochloride [*3810-80-8*].

» Diphenoxylate Hydrochloride contains not less than 98.0 percent and not more than 102.0 percent of $C_{30}H_{32}N_2O_2.HCl$, calculated on the dried basis.

Packaging and storage—Preserve in well-closed containers.

Reference standard—*USP Diphenoxylate Hydrochloride Reference Standard*—Dry at 105° for 2 hours before using.

Identification—

A: The infrared absorption spectrum of a 1 in 10 solution in chloroform, determined in a 0.2-mm cell, exhibits maxima only at the same wavelengths as that of a similar solution of USP Diphenoxylate Hydrochloride RS.

B: The ultraviolet absorption spectrum of a 1 in 2000 solution of it in a 1 in 1000 solution of hydrochloric acid in methanol exhibits maxima and minima at the same wavelengths as that of a similar solution of USP Diphenoxylate Hydrochloride RS, concomitantly measured.

C: A saturated solution responds to the tests for *Chloride* ⟨191⟩.

Melting range ⟨741⟩: between 220° and 226°.

Loss on drying ⟨731⟩—Dry it at 105° for 2 hours: it loses not more than 0.5% of its weight.

Ordinary impurities ⟨466⟩—

Test solution: chloroform.

Standard solution: chloroform.

Eluant: a mixture of chloroform, cyclohexane, dehydrated alcohol, and formic acid (50:40:10:1).

Visualization: 17; then examine the plate immediately under short-wavelength ultraviolet light.

Limits—The sum of the intensities of all secondary spots obtained from the *Test solution* corresponds to not more than 1.0%.

Assay—Dissolve about 300 mg of Diphenoxylate Hydrochloride, accurately weighed, in 75 mL of glacial acetic acid, add 4 mL of mercuric acetate TS, and titrate with 0.1 *N* perchloric acid VS, determining the end-point potentiometrically. Each mL of 0.1 *N* perchloric acid is equivalent to 48.91 mg of $C_{30}H_{32}N_2O_2 . HCl$.

Diphenoxylate Hydrochloride and Atropine Sulfate Oral Solution

» Diphenoxylate Hydrochloride and Atropine Sulfate Oral Solution contains not less than 93.0 percent and not more than 107.0 percent of the labeled amount of diphenoxylate hydrochloride ($C_{30}H_{32}N_2O_2 . HCl$), and not less than 80.0 percent and not more than 120.0 percent of the labeled amount of atropine sulfate [$(C_{17}H_{23}NO_3)_2 . H_2SO_4 . H_2O$].

Packaging and storage—Preserve in tight, light-resistant containers.

Reference standards—*USP Atropine Sulfate Reference Standard*—[*Caution*—*Avoid contact.*] Dry at 120° for 4 hours before using. *USP Diphenoxylate Hydrochloride Reference Standard*—Dry at 105° for 2 hours before using.

Identification—Transfer a volume of Oral Solution, equivalent to about 100 mg of diphenoxylate hydrochloride, to a separator, add 1 mL of 3 *N* hydrochloric acid and sufficient water to make about 100 mL, and proceed as directed in the *Identification test* under *Diphenoxylate Hydrochloride and Atropine Sulfate Tablets*, beginning with "extract with five 25-mL portions."

pH ⟨791⟩: between 3.0 and 4.3, determined in a dilution of the Oral Solution with an equal volume of water.

Alcohol content ⟨611⟩: between 13.5% and 16.5% of C_2H_5OH.

Assay for diphenoxylate hydrochloride—Transfer an accurately measured volume of Diphenoxylate Hydrochloride and Atropine Sulfate Oral Solution, equivalent to about 100 mg of diphenoxylate hydrochloride, to a separator, add 4 mL of 3 *N* hydrochloric acid, and extract with six 30-mL portions of chloroform. Wash the combined chloroform extracts with 25 mL of water, and discard the washing. Transfer the chloroform to a beaker, and evaporate nearly to dryness. Add 100 mL of glacial acetic acid and 4 mL of mercuric acetate TS to the beaker, and titrate with 0.05 *N* perchloric acid in dioxane VS, determining the end-point potentiometrically. Each mL of 0.05 *N* perchloric acid is equivalent to 24.45 mg of $C_{30}H_{32}N_2O_2 . HCl$.

Assay for atropine sulfate—

pH 2.8 buffer and *Internal standard solution*—Prepare as directed in the *Assay for atropine sulfate* under *Diphenoxylate Hydrochloride and Atropine Sulfate Tablets.*

Standard preparation—Transfer about 25 mg of USP Atropine Sulfate RS, accurately weighed, to a 200-mL volumetric flask. Dissolve in water, dilute with water to volume, and mix. Pipet 2 mL of the resulting Standard solution into a 125-mL separator containing about 50 mL of water. Add 2.0 mL of *Internal standard solution*, 10.0 mL of *pH 2.8 buffer*, and 25 mL of water-saturated methylene chloride. Insert the stopper, shake for 2 minutes, and allow the layers to separate. Discard the lower, organic layer. Repeat the extraction four times, using 25-mL portions of water-saturated methylene chloride each time, allowing the layers to separate and discarding the organic layer each time. To the remaining aqueous layer add 3 mL of 0.1 *N* sodium hydroxide, and shake briefly. Using a pH meter, adjust the solution to a pH of 9.0 ± 0.3. Immediately add 10 mL of water-saturated methylene chloride, insert the stopper, and shake. Transfer the lower, organic layer to a 50-mL container. Repeat the extraction twice more with 10-mL portions of water-saturated methylene chloride, combining the organic extracts. Under a stream of nitrogen, evaporate the organic extracts to dryness. Dissolve the residue in 0.1 mL of methylene chloride.

Assay preparation—Transfer an accurately measured volume of Diphenoxylate Hydrochloride and Atropine Sulfate Oral Solution, equivalent to about 250 µg of atropine sulfate, to a 125-mL separator. Proceed as directed for *Standard preparation*, beginning with "Add 2.0 mL of *Internal standard solution.*"

Procedure—Proceed as directed for *Procedure* in the *Assay for atropine sulfate* under *Diphenoxylate Hydrochloride and Atropine Sulfate Tablets.* Calculate the quantity, in µg, of $(C_{17}H_{23}NO_3)_2 . H_2SO_4 . H_2O$ in each mL of the Oral Solution taken by the formula:

$$2(C/V)(R_U/R_S),$$

in which *C* is the concentration, in µg per mL, of USP Atropine Sulfate RS (corrected to the monohydrate) in the Standard solution, *V* is the volume, in mL, of Oral Solution taken, and R_U and R_S are the ratios of atropine to the internal standard obtained from the *Assay preparation* and the *Standard preparation*, respectively.

Diphenoxylate Hydrochloride and Atropine Sulfate Tablets

» Diphenoxylate Hydrochloride and Atropine Sulfate Tablets contain not less than 95.0 percent and not more than 105.0 percent of the labeled amount of diphenoxylate hydrochloride ($C_{30}H_{32}N_2O_2 . HCl$), and not less than 80.0 percent and not more than 120.0 percent of the labeled amount of atropine sulfate [$(C_{17}H_{23}NO_3)_2 . H_2SO_4 . H_2O$].

Packaging and storage—Preserve in well-closed, light-resistant containers.

Reference standards—*USP Diphenoxylate Hydrochloride Reference Standard*—Dry at 105° for 2 hours before using. *USP Atropine Sulfate Reference Standard*—Dry at 120° for 4 hours before using.

Identification—Transfer an amount of finely powdered Tablets, equivalent to about 100 mg of diphenoxylate hydrochloride, to a separator, add 100 mL of water and 1 mL of 3 *N* hydrochloric acid, and extract with five 25-mL portions of a mixture of 9 volumes of chloroform and 1 volume of isopropyl alcohol. After each extraction transfer the bottom chloroform layer to a second separator, filtering each portion through a sintered-glass filter. Wash the combined chloroform solutions with two 25-mL portions of water, discard the washings, and evaporate the chloroform on a steam bath to dryness. To the residue add 10 mL of ether previously saturated with hydrochloric acid, carefully evaporate to dryness, then add a second portion of the acid-saturated ether, and again evaporate to dryness. Slurry the residue with *n*-hexane, allow the solids to settle, and carefully decant the supernatant liquid. Dry the solids at 105° for 1 hour: the diphenoxylate hydrochloride so obtained responds to *Identification tests A* and *B* under *Diphenoxylate Hydrochloride.*

Dissolution ⟨711⟩—

Medium: 0.2 *M* acetic acid; 500 mL.

Apparatus 1: 150 rpm.

Time: 45 minutes.

Determine the amount of $C_{30}H_{32}N_2O_2.HCl$ dissolved, employing the following procedure.

Mobile phase—Prepare a suitable degassed mixture of acetonitrile and 0.05 *M* monobasic potassium phosphate (65:35).

Standard solution—Dissolve an accurately weighed quantity of USP Diphenoxylate Hydrochloride RS in methanol to obtain a solution having a known concentration of about 250 µg per mL. Pipet 10 mL of this solution into a 500-mL volumetric flask, dilute with *Dissolution Medium* to volume, mix, and filter.

Chromatographic system (see *Chromatography* ⟨621⟩)—The liquid chromatograph is equipped with a 210-nm detector and a 3.9-mm × 30-cm column that contains packing L11. The flow rate is about 1.0 mL per minute. Chromatograph replicate injections of the *Standard solution*, and record the peak responses as directed under *Procedure:* the relative standard deviation is not more than 2.0%, and the tailing factor is not more than 1.5.

Procedure—Separately inject equal volumes (about 50 µL) of the *Standard solution* and of filtered portions of the solution under test into the chromatograph, record the chromatograms, measure the response for the major peak, and determine the amount of $C_{30}H_{32}N_2O_2.HCl$ dissolved.

Tolerances—Not less than 75% (*Q*) of the labeled amount of $C_{30}H_{32}N_2O_2.HCl$ is dissolved in 45 minutes.

Uniformity of dosage units ⟨905⟩: meet the requirements.

Procedure for content uniformity—Place 1 Tablet in a glass-stoppered, 50-mL centrifuge tube, add an accurately measured volume of methanol to produce a preparation having a concentration of about 250 µg of diphenoxylate hydrochloride per mL, and crush the Tablet with a glass rod. Mix by shaking vigorously, and centrifuge at 2000 rpm for 10 minutes. Dissolve a quantity of USP Diphenoxylate Hydrochloride RS in the same medium to obtain a Standard solution having a known concentration of about 250 µg per mL. Concomitantly determine the absorbances of the supernatant liquid and of the Standard solution in 1-cm cells at the wavelength of maximum absorbance at about 257 nm, with a suitable spectrophotometer, using methanol as the blank. Calculate the quantity, in mg, of $C_{30}H_{32}N_2O_2.HCl$ in the Tablet taken by the formula:

$$(T/D)C(A_U/A_S),$$

in which *T* is the labeled quantity, in mg, of diphenoxylate hydrochloride in the Tablet, *D* is the concentration, in µg per mL, of diphenoxylate hydrochloride in the test solution, based on the labeled quantity per Tablet and the extent of dilution, *C* is the concentration, in µg per mL, of the Standard solution, and A_U and A_S are the absorbances of the test solution and the Standard solution, respectively.

Assay for diphenoxylate hydrochloride—Weigh and finely powder not less than 60 Diphenoxylate Hydrochloride and Atropine Sulfate Tablets. Transfer an accurately weighed portion of the powder, equivalent to about 150 mg of diphenoxylate hydrochloride, to a 250-mL separator, add 100 mL of water and 1 mL of 3 *N* hydrochloric acid, and extract with five 25-mL portions of a mixture of 9 volumes of chloroform and 1 volume of isopropyl alcohol, shaking each portion for not less than 2 minutes. After each extraction, transfer the bottom, chloroform layer to a second separator, filtering each portion through a sintered-glass filter. Rinse the filter with a few mL of chloroform, add the rinsing to the chloroform solution, and extract the combined chloroform solutions with two 25-mL portions of water. Discard the washings, and evaporate the chloroform in a beaker on a steam bath to a volume of about 5 mL. Add 5 mL of mercuric acetate TS and 100 mL of glacial acetic acid, and titrate with 0.05 *N* perchloric acid VS, determining the end-point potentiometrically. Each mL of 0.05 *N* perchloric acid is equivalent to 24.45 mg of $C_{30}H_{32}N_2O_2.HCl$.

Assay for atropine sulfate—

pH 2.8 buffer—Dissolve 1.9 g of aminoacetic acid and 1.5 g of sodium chloride in 250 mL of water. Adjust by the gradual addition of about 85 mL of 0.1 *N* hydrochloric acid to a pH of 2.8.

Internal standard solution—Transfer about 20 mg of homatropine hydrobromide to a 200-mL volumetric flask, dissolve in water, dilute with water to volume, and mix.

Standard preparation—Transfer about 25 mg of USP Atropine Sulfate RS, accurately weighed, to a 200-mL volumetric flask. Dissolve in water, dilute with water to volume, and mix. Pipet 2 mL of the resulting Standard solution into a 50-mL centrifuge tube. Add 2.0 mL of *Internal standard solution*, 10.0 mL of *pH 2.8 buffer*, and 10 mL of water-saturated chloroform. Insert the stopper in the tube, shake for 2 minutes, and centrifuge at 1000 rpm for 5 minutes. Allow the layers to separate, and discard the lower, organic layer. Repeat the extraction with a second 10-mL portion of water-saturated chloroform, centrifuge, and discard the organic layer. To the remaining aqueous layer add 3 mL of 0.1 *N* sodium hydroxide, and shake briefly. Using a pH meter, adjust the solution to a pH of 9.0 ± 0.3. Immediately add 10 mL of water-saturated chloroform, insert the stopper, shake, and centrifuge at 1000 rpm for 5 minutes. Transfer the lower, organic layer to a second 50-mL container. Repeat the extraction twice more with 10-mL portions of water-saturated chloroform, combining the organic extracts. Under a stream of nitrogen, evaporate the organic extracts to dryness. Dissolve the residue in 0.1 mL of chloroform.

Assay preparation—Weigh and finely powder not less than 100 Diphenoxylate Hydrochloride and Atropine Sulfate Tablets. Transfer a portion of the powder, equivalent to about 250 µg of atropine sulfate, to a 50-mL centrifuge tube. Proceed as directed for *Standard preparation*, beginning with "Add 2.0 mL of *Internal standard solution.*"

Procedure—Inject equal volumes (about 2 µL) of the *Assay preparation* and the *Standard preparation* into a suitable gas chromatograph equipped with a flame-ionization detector and a 1.2-m × 4-mm glass column containing 3 percent phase G3 on packing S1. Maintain the injection port and detector temperature at 250° and the column at 230°. Use nitrogen as the carrier gas at a flow rate of 40 mL per minute. In a suitable chromatogram, five replicate injections of the *Standard preparation* show a relative standard deviation of not more than 2.5% and a resolution factor of not less than 2.0 between the peaks for atropine and the internal standard. Calculate the quantity, in µg, of $(C_{17}H_{23}NO_3)_2.H_2SO_4.H_2O$ in the portion of Tablets taken by the formula:

$$2C(R_U/R_S),$$

in which *C* is the concentration, in µg per mL, of USP Atropine Sulfate RS (corrected to the monohydrate) in the Standard solution, and R_U and R_S are the ratios of atropine to the internal standard obtained from the *Assay preparation* and the *Standard preparation*, respectively.

Diphtheria Antitoxin

» Diphtheria Antitoxin conforms to the regulations of the federal Food and Drug Administration concerning biologics (see *Biologics* ⟨1041⟩). It is a sterile, non-pyrogenic solution of the refined and concentrated proteins, chiefly globulins, containing antitoxic antibodies obtained from the blood serum or plasma of healthy horses that have been immunized against diphtheria toxin or toxoid. It has a potency of not less than 500 antitoxin units per mL based on the U.S. Standard Diphtheria Antitoxin, and a diphtheria test toxin, tested in guinea pigs. It contains not more than 20.0 percent of solids.

Packaging and storage—Preserve at a temperature between 2° and 8°.

Expiration date—The expiration date for Antitoxin containing a 20% excess of potency is not later than 5 years after date of issue from manufacturer's cold storage.

Labeling—Label it to state that it was prepared from horse serum or plasma.

Diphtheria Toxin for Schick Test

» Diphtheria Toxin for Schick Test conforms to the regulations of the federal Food and Drug Administration concerning biologics (650.1 to 650.7) (see *Biologics* ⟨1041⟩). It is a sterile solution of the diluted, standardized toxic products of growth of the diphtheria bacillus (*Corynebacterium diphtheriae*) of which the parent toxin contains not less than 400 MLD (minimum lethal doses) per mL or 400,000 MRD (minimum skin reaction doses) per mL in guinea pigs. Its potency is determined in terms of the U.S. Standard Diphtheria Toxin for Schick Test, tested in guinea pigs.

Packaging and storage—Preserve at a temperature between 2° and 8°.

Expiration date—The expiration date is not later than 1 year after date of issue from manufacturer's cold storage (5°, 1 year).

Diphtheria Toxoid

» Diphtheria Toxoid conforms to the regulations of the federal Food and Drug Administration concerning biologics (see *Biologics* ⟨1041⟩). It is a sterile solution of the formaldehyde-treated products of growth of the diphtheria bacillus (*Corynebacterium diphtheriae*). It meets the requirements of the specific guinea pig potency and detoxification tests. It contains not more than 0.02 percent of residual free formaldehyde. It contains a preservative other than a phenoloid compound.

Packaging and storage—Preserve at a temperature between 2° and 8°.

Expiration date—The expiration date is not later than 2 years after date of issue from manufacturer's cold storage (5°, 1 year).

Labeling—Label it to state that it is not to be frozen.

Diphtheria Toxoid Adsorbed

» Diphtheria Toxoid Adsorbed conforms to the regulations of the federal Food and Drug Administration concerning biologics (see *Biologics* ⟨1041⟩). It is a sterile preparation of plain diphtheria toxoid that meets all of the requirements for that product with the exception of those for antigenicity, and that has been precipitated or adsorbed by alum, aluminum hydroxide, or aluminum phosphate adjuvants. It meets the requirements of the specific guinea pig antigenicity test in the production of not less than 2 units of antitoxin per mL based on the U.S. Standard Diphtheria Antitoxin and a diphtheria test toxin. It meets the requirements of the specific guinea pig detoxification test.

Packaging and storage—Preserve at a temperature between 2° and 8°.

Expiration date—The expiration date is not later than 2 years after date of issue from manufacturer's cold storage (5°, 1 year).

Labeling—Label it to state that it is to be well shaken before use and that it is not to be frozen.

Aluminum content—It contains not more than 0.85 mg per single injection, determined by analysis, or not more than 1.14 mg calculated on the basis of the amount of aluminum compound added.

Diphtheria Toxoids Adsorbed for Adult Use, Tetanus and—*see* Tetanus and Diphtheria Toxoids Adsorbed for Adult Use

Diphtheria and Tetanus Toxoids

» Diphtheria and Tetanus Toxoids conforms to the regulations of the federal Food and Drug Administration concerning biologics (see *Biologics* ⟨1041⟩). It is a sterile solution prepared by mixing suitable quantities of fluid diphtheria toxoid and fluid tetanus toxoid. The antigenicity or potency and the proportions of the toxoids are such as to provide an immunizing dose of each toxoid in the total dosage prescribed in the labeling, and each component meets the requirements for those products. It contains not more than 0.02 percent of residual free formaldehyde.

Packaging and storage—Preserve at a temperature between 2° and 8°.

Expiration date—The expiration date is not later than 2 years after date of issue from manufacturer's cold storage (5°, 1 year).

Labeling—Label it to state that it is not to be frozen.

Diphtheria and Tetanus Toxoids Adsorbed

» Diphtheria and Tetanus Toxoids Adsorbed conforms to the regulations of the federal Food and Drug Administration concerning biologics (see *Biologics* ⟨1041⟩). It is a sterile suspension prepared by mixing suitable quantities of plain or adsorbed diphtheria toxoid and plain or adsorbed tetanus toxoid, and an aluminum adsorbing agent if plain toxoids are used. The antigenicity or potency and the proportions of the toxoids are such as to provide an immunizing dose of each toxoid in the total dosage prescribed in the labeling, and each component meets the requirements for those products. It contains not more than 0.02 percent of residual free formaldehyde.

Packaging and storage—Preserve at a temperature between 2° and 8°.

Expiration date—The expiration date is not later than 2 years after date of issue from manufacturer's cold storage (5°, 1 year).

Labeling—Label it to state that it is to be well shaken before use and that it is not to be frozen.

Diphtheria and Tetanus Toxoids and Pertussis Vaccine

» Diphtheria and Tetanus Toxoids and Pertussis Vaccine conforms to the regulations of the federal Food and Drug Administration concerning biologics (see *Biologics* ⟨1041⟩). It is a sterile suspension prepared by mixing suitable quantities of pertussis vaccine component of killed pertussis bacilli (*Bordetella pertussis*), or a fraction of this organism, fluid diphtheria toxoid, and fluid tetanus toxoid. The antigenicity or potency and the proportions of the components are such as to provide an immunizing dose of each product in the total dosage prescribed in the labeling, and each component meets the requirements for those products.

Packaging and storage—Preserve at a temperature between 2° and 8°.

Expiration date—The expiration date is not later than 18 months after date of issue from manufacturer's cold storage (5°, 1 year).

Labeling—Label it to state that it is to be well shaken before use and that it is not to be frozen.

Diphtheria and Tetanus Toxoids and Pertussis Vaccine Adsorbed

» Diphtheria and Tetanus Toxoids and Pertussis Vaccine Adsorbed conforms to the regulations of the federal Food and Drug Administration concerning biologics (see *Biologics* ⟨1041⟩). It is a sterile suspension prepared by mixing suitable quantities of plain or adsorbed diphtheria toxoid, plain or adsorbed tetanus toxoid, plain or adsorbed pertussis vaccine, and an aluminum adsorbing agent if plain antigen components are used. The antigenicity or potency and the proportions of the components are such as to provide an immunizing dose of each product in the total dosage prescribed in the labeling, and each component meets the requirements for those products.

Packaging and storage—Preserve at a temperature between 2° and 8°.

Expiration date—The expiration date is not later than 18 months after date of issue from manufacturer's cold storage (5°, 1 year).

Labeling—Label it to state that it is to be well shaken before use and that it is not to be frozen.

Dipivefrin Hydrochloride

$(CH_3)_3CCOO$—[benzene ring]—$CHCH_2NHCH_3$ (OH) · HCl; $(CH_3)_3CCOO$

$C_{19}H_{29}NO_5.HCl$ 387.90

Propanoic acid, 2,2-dimethyl-, 4-[1-hydroxy-2-(methylamino)ethyl]-1,2-phenylene ester, hydrochloride, (±)-.

(±)-3,4-Dihydroxy-α-[(methylamino)methyl]benzyl alcohol 3,4-dipivalate hydrochloride [*64019-93-8*].

» Dipivefrin Hydrochloride contains not less than 98.5 percent and not more than 101.5 percent of $C_{19}H_{29}NO_5.HCl$, calculated on the dried basis.

Packaging and storage—Preserve in tight containers.

Reference standard—*USP Dipivefrin Hydrochloride Reference Standard*—Dry in a suitable vacuum drying tube over phosphorus pentoxide at 60° for 6 hours before using.

Identification—

A: The infrared absorption spectrum of a potassium bromide dispersion of it, previously dried, exhibits maxima only at the same wavelengths as that of a similar preparation of USP Dipivefrin Hydrochloride RS.

B: It responds to the *Thin-layer Chromatographic Identification Test* ⟨201⟩, the test solution and the Standard solution being prepared at a concentration of 2 mg per mL in water, chromatographic silica gel being used as the absorbant, the solvent mixture being chloroform, methanol, and formic acid (30:10:1) and 0.1% potassium ferricyanide in water, alcohol, and ethylenediamine (50:45:5), and heat being used to locate the spots.

C: A solution (1 in 100) responds to the tests for *Chloride* ⟨191⟩.

Melting range ⟨741⟩: between 155° and 165°, but the range between beginning and end of melting does not exceed 2°.

Loss on drying ⟨731⟩—Dry it in a suitable vacuum drying tube over phosphorus pentoxide at 60° for 6 hours: it loses not more than 1.0% of its weight.

Residue on ignition ⟨281⟩: not more than 0.3%.

Heavy metals, *Method I* ⟨231⟩: not more than 0.0015%.

Iron ⟨241⟩: not more than 5 ppm.

Assay—

Mobile phase—Prepare a mixture of acetonitrile, 0.014 *M* dodecyl sodium sulfate, and glacial acetic acid (24:15:1).

Standard preparation—Dissolve a suitable quantity of USP Dipivefrin Hydrochloride RS, accurately weighed, in 0.0015 *N* hydrochloric acid to obtain a solution having a known concentration of about 5 mg per mL.

Assay preparation—Prepare as directed under *Standard preparation*, using 500 mg of Dipivefrin Hydrochloride, accurately weighed, in place of the Reference Standard.

Chromatographic system (see *Chromatography* ⟨621⟩)—The liquid chromatograph is equipped with a 254-nm detector and a 4-mm × 30-cm column that contains packing L1. The flow rate is about 2 mL per minute. Chromatograph the *Standard preparation*, and record the peak responses as directed under *Procedure:* the column efficiency is not less than 500 theoretical plates, the tailing factor for the major peak is not more than 1.2, and the relative standard deviation for replicate injections is not more than 2.0%.

Procedure—Separately inject equal volumes (about 20 µL) of the *Standard preparation* and the *Assay preparation* into the chromatograph by means of a suitable microsyringe or sampling valve, record the chromatograms, and measure the responses for the major peaks. Calculate the quantity, in mg, of $C_{19}H_{29}NO_5.HCl$ in the portion of Dipivefrin Hydrochloride taken by the formula:

$$100C(r_U / r_S),$$

in which C is the concentration, in mg per mL, of USP Dipivefrin Hydrochloride RS in the *Standard preparation*, and r_U and r_S are the peak responses obtained from the *Assay preparation* and the *Standard preparation*, respectively.

Dipivefrin Hydrochloride Ophthalmic Solution

» Dipivefrin Hydrochloride Ophthalmic Solution is a sterile, aqueous solution of Dipivefrin Hydrochloride. It contains not less than 90.0 percent and not more than 115.0 percent of the labeled amount of $C_{19}H_{29}NO_5.HCl$. It contains a suitable antimicrobial agent and may contain stabilizers, suitable buffers, and chelating agents.

Packaging and storage—Preserve in tight, light-resistant containers.

Reference standard—*USP Dipivefrin Hydrochloride Reference Standard*—Dry in a suitable vacuum drying tube over phosphorus pentoxide at 60° for 6 hours before using.

Identification—Proceed as directed in *Identification test B* under *Dipivefrin Hydrochloride*, using 40-µL portions of the *Assay preparation* and the *Standard preparation* obtained as directed in the *Assay*, except to locate the spots under ultraviolet light.

Sterility—It meets the requirements under *Sterility Tests* ⟨71⟩.

pH ⟨791⟩: between 2.5 and 3.5.

Assay—

Mobile phase and *Chromatographic system*—Prepare as directed in the *Assay* under *Dipivefrin Hydrochloride.*

Standard preparation—Dissolve a suitable quantity of USP Dipivefrin Hydrochloride RS, accurately weighed, in 0.0015 *N* hydrochloric acid to obtain a solution having a known concentration of about 1 mg per mL.

Assay preparation—Transfer an accurately measured volume of Dipivefrin Hydrochloride Ophthalmic Solution, equivalent to about 25 mg of dipivefrin hydrochloride, to a 25-mL volumetric flask, dilute with 0.0015 *N* hydrochloric acid to volume, if necessary, and mix.

Procedure—Proceed as directed in the *Assay* under *Dipivefrin Hydrochloride.* Calculate the quantity, in mg, of $C_{19}H_{29}NO_5 \cdot HCl$ in each mL of the Ophthalmic Solution taken by the formula:

$$(25C/V)(r_U/r_S),$$

in which *C* is the concentration, in mg per mL, of USP Dipivefrin Hydrochloride RS in the *Standard preparation*, *V* is the volume, in mL, of Ophthalmic Solution taken, and r_U and r_S are the peak responses obtained from the *Assay preparation* and the *Standard preparation*, respectively.

Dipyridamole

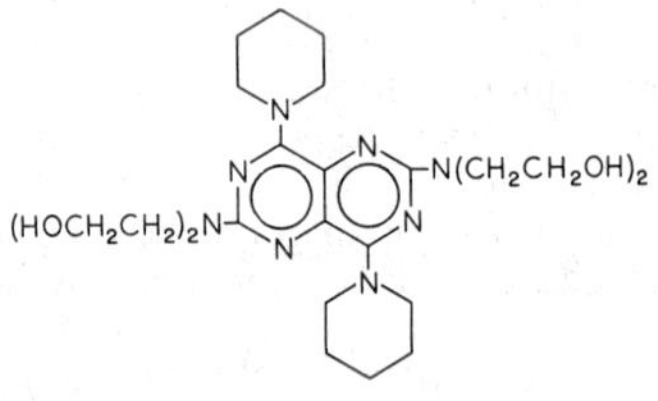

$C_{24}H_{40}N_8O_4$ 504.63

Ethanol, 2,2′,2″,2‴-[(4,8-di-1-piperidinylpyrimido[5,4-*d*]-pyrimidine-2,6-diyl)dinitrilo]tetrakis-.

2,2′,2″,2‴-[(4,8-Dipiperidinopyrimido[5,4-*d*]pyrimidine-2,6-diyl)dinitrilo]tetraethanol [*58-32-2*].

» Dipyridamole contains not less than 98.0 percent and not more than 102.0 percent of $C_{24}H_{40}N_8O_4$, calculated on the dried basis.

Packaging and storage—Preserve in tight, light-resistant containers.

Reference standard—*USP Dipyridamole Reference Standard*—Dry at 105° for 3 hours before using.

Identification—The infrared absorption spectrum of a potassium bromide dispersion of it, previously dried, exhibits maxima only at the same wavelengths as that of a similar preparation of USP Dipyridamole RS.

Melting range ⟨741⟩: between 162° and 168°, but the range between beginning and end of melting does not exceed 2°.

Loss on drying ⟨731⟩—Dry it at 105° for 3 hours: it loses not more than 0.2% of its weight.

Chloride—Dissolve 500 mg in 5 mL of alcohol and 2 mL of 2 *N* nitric acid, and add 1 mL of silver nitrate TS: no turbidity or precipitate is produced.

Residue on ignition ⟨281⟩: not more than 0.1%.

Heavy metals, *Method II* ⟨231⟩: 0.001%.

Chromatographic purity—

Mobile phase and *Chromatographic system*—Prepare as directed in the *Assay* under *Dipyridamole Tablets.*

Test preparation A—Prepare a solution of Dipyridamole in methanol having a known concentration of 1 mg per mL.

Test preparation B—Dilute 1.0 mL of *Test preparation A* with methanol to 100 mL, and mix.

Procedure—Inject 10 µL of *Test preparation B* into the chromatograph by means of a sampling valve, adjusting the operating parameters so that the response of the main peak (retention time about 6.5 minutes) obtained is about 5% full scale. Inject 10 µL of *Test preparation A*, and run the chromatograph for 10 minutes: the sum of responses of all secondary peaks obtained from *Test preparation A* is not greater than the response of the main peak obtained from *Test preparation B* (1.0%).

Assay—Transfer about 450 mg of Dipyridamole, accurately weighed, to a 250-mL beaker, and dissolve in 50 mL of glacial acetic acid. Stir for 30 minutes. Add 75 mL of acetone, and stir for an additional 15 minutes. Titrate with 0.1 *N* perchloric acid VS, determining the end-point potentiometrically, using a glass electrode and a silver–silver chloride reference electrode system. Perform a blank titration, and make any necessary correction. Each mL of 0.1 *N* perchloric acid is equivalent to 50.46 mg of $C_{24}H_{40}N_8O_4$.

Dipyridamole Tablets

» Dipyridamole Tablets contain not less than 90.0 percent and not more than 110.0 percent of the labeled amount of $C_{24}H_{40}N_8O_4$.

Packaging and storage—Preserve in tight, light-resistant containers.

Reference standard—*USP Dipyridamole Reference Standard*—Dry at 105° for 3 hours before using.

Identification—Triturate a quantity of finely powdered Tablets, equivalent to about 100 mg of dipyridamole, with 10 mL of 0.1 *N* hydrochloric acid, and filter, collecting the filtrate in a beaker. Add 0.1 *N* sodium hydroxide until the solution is basic and a precipitate forms. Heat the mixture on a steam bath for 1 minute, cool, and filter. Dry the residue at 105° for 1 hour: the residue so obtained responds to the *Identification test* under *Dipyridamole.*

Dissolution ⟨711⟩—

Medium: 0.1 *N* hydrochloric acid; 900 mL.

Apparatus 2: 50 rpm.

Time: 30 minutes.

Procedure—Determine the amount of $C_{24}H_{40}N_8O_4$ dissolved from ultraviolet absorbances at the wavelength of maximum absorbance at about 282 nm of filtered portions of the solution under test, suitably diluted with *Dissolution Medium*, if necessary, in comparison with a Standard solution having a known concentration of USP Dipyridamole RS in the same medium.

Tolerances—Not less than 70% (*Q*) of the labeled amount of $C_{24}H_{40}N_8O_4$ is dissolved in 30 minutes.

Uniformity of dosage units ⟨905⟩: meet the requirements.

Procedure for content uniformity—Transfer 1 Tablet to a 100-mL volumetric flask, add 50 mL of 1 *N* hydrochloric acid, heat in a steam bath for 5 minutes, and shake by mechanical means for 30 minutes. Cool to room temperature, dilute with 1 *N* hydrochloric acid to volume, and mix. Filter, discarding the first 25 mL of the filtrate. Dilute an accurately measured portion of the subsequent filtrate with water to provide a solution containing about 10 µg of dipyridamole per mL. Concomitantly determine the absorbances of this solution, and of a solution of USP Dipyridamole RS in the same medium having a known concentration of about 10 µg per mL, in 1-cm cells at the wavelength of maximum absorbance at about 282 nm, with a suitable spectrophotometer, using 0.02 *N* hydrochloric acid as the blank. Calculate the quantity, in mg, of $C_{24}H_{40}N_8O_4$ in the Tablet taken by the formula:

$$(TC/D)(A_U/A_S),$$

in which T is the labeled quantity, in mg, of dipyridamole in the Tablet, C is the concentration, in µg per mL, of USP Dipyridamole RS in the *Standard solution*, D is the concentration, in µg per mL, of dipyridamole in the solution from the Tablet based upon the labeled quantity per Tablet and the extent of dilution, and A_U and A_S are the absorbances of the solution from the Tablet and the *Standard solution*, respectively.

Assay—

Mobile phase—Dissolve 250 mg of dibasic sodium phosphate in 250 mL of water, and adjust with dilute phosphoric acid (1 in 3) to a pH of 4.6. Add 750 mL of methanol, mix, filter through a 0.5-µm membrane filter, and degas. Make adjustments if necessary (see *System Suitability* under *Chromatography* ⟨621⟩).

Standard preparation—Using an accurately weighed quantity of USP Dipyridamole RS, prepare a solution in methanol having a known concentration of about 15 µg per mL.

Assay preparation—Transfer not less than 20 Tablets to a 1000-mL volumetric flask, add 100 mL of water, and sonicate for 15 minutes. Add about 750 mL of methanol, and shake by mechanical means for 30 minutes. Dilute with methanol to volume, mix, and centrifuge. Dilute an accurately measured volume (V_S mL) of the clear supernatant solution quantitatively with *Mobile phase* to obtain a solution (V_A mL) containing about 15 µg of dipyridamole per mL.

Chromatographic system (see *Chromatography* ⟨621⟩)—The liquid chromatograph is equipped with a 288-nm detector and a 3.9-mm × 30-cm column that contains packing L1. The flow rate is about 1.5 mL per minute. Chromatograph the *Standard preparation*, and record the peak responses as directed under *Procedure:* the column efficiency determined from the analyte peak is not less than 1000 theoretical plates, the tailing factor for the analyte peak is not more than 2.0, and the relative standard deviation for replicate injections is not more than 2.0%.

Procedure—Separately inject equal volumes (about 50 µL) of the *Standard preparation* and the *Assay preparation* into the chromatograph, record the chromatograms, and measure the responses for the major peaks. Calculate the quantity, in mg, of $C_{24}H_{40}N_8O_4$ in the Tablets taken by the formula:

$$C(V_A/V_S)(r_U/r_S),$$

in which C is the concentration, in µg per mL, of USP Dipyridamole RS in the *Standard preparation*, V_A is the volume, in mL, of the *Assay preparation*, V_S is the volume, in mL, of supernatant solution taken for the *Assay preparation*, and r_U and r_S are the peak responses obtained from the *Assay preparation* and the *Standard preparation*, respectively.

Disopyramide Phosphate

$[(CH_3)_2CH]_2NCH_2CH_2CCONH_2 \cdot H_3PO_4$

$C_{21}H_{29}N_3O.H_3PO_4$ 437.47

2-Pyridineacetamide, α-[2-[bis(1-methylethyl)amino]ethyl]-α-phenyl-, phosphate (1:1).

α-[2-(Diisopropylamino)ethyl]-α-phenyl-2-pyridineacetamide phosphate (1:1) *[22059-60-5]*.

» Disopyramide Phosphate contains not less than 98.0 percent and not more than 102.0 percent of $C_{21}H_{29}N_3O.H_3PO_4$, calculated on the dried basis.

Packaging and storage—Preserve in tight, light-resistant containers.

Reference standard—*USP Disopyramide Phosphate Reference Standard*—Dry at 105° for 4 hours before using.

Identification—

A: The infrared absorption spectrum of a mineral oil dispersion of it exhibits maxima only at the same wavelengths as that of a similar preparation of USP Disopyramide Phosphate RS.

B: A solution (1 in 200) responds to the tests for *Phosphate* ⟨191⟩.

pH ⟨791⟩: between 4.0 and 5.0 in a solution (1 in 20).

Loss on drying ⟨731⟩—Dry it at 105° for 4 hours: it loses not more than 0.5% of its weight.

Heavy metals, *Method II* ⟨231⟩: 0.002%.

Chromatographic purity—

Standard solutions—Prepare solutions *A* and *B* of USP Disopyramide Phosphate RS in methanol having concentrations of about 50 and 100 µg per mL, respectively.

Test solution—Prepare a solution of Disopyramide Phosphate in methanol having a concentration of about 10 mg per mL.

Procedure—On a suitable thin-layer chromatographic plate (see *Chromatography* ⟨621⟩), coated with a 0.25-mm layer of chromatographic silica gel, separately apply 10-µL portions of the *Standard solutions A* and *B*, and the *Test solution.* Allow the spots to dry, and develop the chromatogram in a solvent system consisting of a mixture of toluene, dehydrated alcohol, and ammonium hydroxide (170:28:2) until the solvent front has moved about three-fourths of the length of the plate. Remove the plate from the chamber, allow to air-dry, and spray with Potassium Bismuth Iodide TS: the R_f value of the principal spot obtained from the *Test solution* corresponds to that obtained from *Standard solution B.* Estimate the levels of any additional spots observed in the chromatogram of the *Test preparation* by comparison with the principal spots in the chromatograms of *Standard solutions A* and *B:* the sum of the intensities of any additional spots observed is not greater than that obtained from *Standard solution B* (equivalent to 1%).

Assay—Dissolve about 160 mg of Disopyramide Phosphate, accurately weighed, in 50 mL of glacial acetic acid TS, and titrate with 0.1 *N* perchloric acid VS, determining the end-point potentiometrically. Perform a blank determination, and make any necessary correction. Each mL of 0.1 *N* perchloric acid is equivalent to 21.87 mg of $C_{21}H_{29}N_3O.H_3PO_4$.

Disopyramide Phosphate Capsules

» Disopyramide Phosphate Capsules contain an amount of Disopyramide Phosphate equivalent to not less than 90.0 percent and not more than 110.0 percent of the labeled amount of disopyramide ($C_{21}H_{29}N_3O$).

Packaging and storage—Preserve in well-closed containers.

Reference standard—*USP Disopyramide Phosphate Reference Standard*—Dry at 105° for 4 hours before using.

Identification—Transfer a portion of Capsule contents, equivalent to about 125 mg of disopyramide phosphate, to a 25-mL volumetric flask, add 20 mL of methanol, and shake by mechanical means for 20 minutes. Dilute with methanol to volume, mix, and filter through paper (Whatman No. 2 or equivalent), discarding the first 10 mL of the filtrate. On a suitable thin-layer chromatographic plate (see *Chromatography* ⟨621⟩), coated with a 0.25-mm layer of chromatographic silica gel mixture, apply 10 µL each of the subsequent filtrate and of a solution of USP Disopyramide Phosphate RS in methanol containing 6.2 mg per mL. Allow the spots to dry, and develop the chromatogram in a solvent system consisting of a mixture of toluene, alcohol, and ammonium hydroxide (170:28:2) until the solvent front has moved about three-fourths of the length of the plate. Remove the plate from the developing chamber, mark the solvent front, and allow the solvent to evaporate. Locate the spots on the plate by viewing under short-wavelength ultraviolet light: the R_f value of the principal spot obtained from the test solution corresponds to that obtained from the Standard solution.

Dissolution ⟨711⟩—

Medium: water; 1000 mL.

Apparatus 2: 50 rpm.

Time: 20 minutes.

Procedure—Filter 15 mL of the solution under test, and transfer 10.0 mL of the filtrate to a 25-mL volumetric flask. Dilute with 2 *N* sulfuric acid to volume, and mix. Determine the amount of disopyramide ($C_{21}H_{29}N_3O$) dissolved from ultraviolet absorbances at the wavelength of maximum absorbance at about 268 nm of this solution, using water as the blank, in comparison with a Standard solution having a known concentration of USP Disopyramide Phosphate RS in the same medium.

Tolerances—Not less than 80% (*Q*) of the labeled amount of $C_{21}H_{29}N_3O$ is dissolved in 20 minutes.

Uniformity of dosage units ⟨905⟩: meet the requirements.

Assay—

Methanolic sulfuric acid—Cautiously add 5.4 mL of sulfuric acid to about 1800 mL of methanol with stirring, dilute with methanol to 2000 mL, and mix.

Procedure—Weigh the contents of not less than 20 Disopyramide Phosphate Capsules, and calculate the average weight per Capsule. Mix the combined contents of the Capsules, and transfer an accurately weighed portion, equivalent to about 125 mg of disopyramide phosphate, to a glass-stoppered, 125-mL flask. Add 50 mL of *Methanolic sulfuric acid*, and stir for 30 minutes. Filter through a medium-porosity, sintered-glass filter, and rinse thoroughly with *Methanolic sulfuric acid.* Transfer the combined filtrate and rinsings to a 100-mL volumetric flask, dilute with *Methanolic sulfuric acid* to volume, and mix. Dilute an accurately measured portion of this solution quantitatively and stepwise with the same solvent to obtain a solution having a concentration of about 40 µg per mL. Dissolve an accurately weighed portion of USP Disopyramide Phosphate RS in *Methanolic sulfuric acid*, and dilute quantitatively and stepwise with the same solvent to obtain a solution having a known concentration of about 40 µg per mL. Concomitantly determine the absorbances of both solutions in 1-cm cells at the wavelength of maximum absorbance at about 268 nm, with a suitable spectrophotometer, using *Methanolic sulfuric acid* as the blank. Calculate the quantity, in mg, of $C_{21}H_{29}N_3O$ in the portion of Capsules taken by the formula:

$$3.125(339.48/437.47)C(A_U/A_S),$$

in which 339.48 and 437.47 are the molecular weights of disopyramide and disopyramide phosphate, respectively, *C* is the concentration, in µg per mL, of USP Disopyramide Phosphate RS in the Standard solution, and A_U and A_S are the absorbances of the solution from the Capsules and the Standard solution, respectively.

Disopyramide Phosphate Extended-release Capsules

» Disopyramide Phosphate Extended-release Capsules contain an amount of Disopyramide Phosphate equivalent to not less than 90.0 percent and not more than 110.0 percent of the labeled amount of disopyramide ($C_{21}H_{29}N_3O$).

Packaging and storage—Preserve in well-closed containers.

Reference standard—*USP Disopyramide Phosphate Reference Standard*—Dry at 105° for 4 hours before using.

Identification—Transfer a portion of Capsule contents, equivalent to about 195 mg of disopyramide phosphate, to a 25-mL volumetric flask, add 20 mL of methanol, and shake by mechanical means for 20 minutes. Dilute with methanol to volume, mix, and filter, discarding the first 10 mL of the filtrate. On a suitable thin-layer chromatographic plate (see *Chromatography* ⟨621⟩), coated with 0.25-mm layer of chromatographic silica gel mixture, apply 20 µL each of the subsequent filtrate and of a solution of USP Disopyramide Phosphate RS in methanol containing 7.7 mg per mL. Allow the spots to dry, and develop the chromatogram in a solvent system consisting of toluene, absolute alcohol, and ammonium hydroxide (170:28:2) until the solvent front has moved about three-fourths of the length of the plate. Remove the plate from the developing chamber, mark the solvent front, and allow the solvent to evaporate. Locate the spots on the plate by viewing under short-wavelength ultraviolet light: the R_f value of the principal spot obtained from the test solution corresponds to that obtained from the Standard solution.

Drug release ⟨724⟩—

pH 2.5, 0.1 M phosphate buffer—Dissolve 272 g of monobasic potassium phosphate in 20 liters of water, and adjust with hydrochloric acid to a pH of 2.50 ± 0.04. [NOTE—Do not adjust back to pH 2.50 with base if too much acid is added. It is imperative that the ionic strength of the buffer be controlled.]

Medium: pH 2.5, 0.1 M phosphate buffer; 1000 mL.

Apparatus 1: 100 rpm.

Time: 0.083*D* hours; 0.167*D* hours; 0.417*D* hours; 1.000*D* hours.

Procedure—Filter 10 mL of the solution under test at the required test points. Determine the amount of disopyramide ($C_{21}H_{29}N_3O$) dissolved from ultraviolet absorbances at the wavelength of maximum absorbance at about 261 nm of this solution, suitably diluted with *Dissolution Medium*, if necessary, using *Dissolution Medium* as the blank, in comparison with a Standard solution having a known concentration of USP Disopyramide Phosphate RS dissolved in *Dissolution Medium.*

Tolerances—The percentage of the labeled amount of disopyramide ($C_{21}H_{29}N_3O$) dissolved is within the range stated at each of the following times.

Time (hours)	Amount dissolved
0.083*D*	between 5% and 25%
0.167*D*	between 17% and 43%
0.417*D*	between 50% and 80%
1.000*D*	not less than 85%

Uniformity of dosage units ⟨905⟩: meet the requirements.

Assay—

Standard preparation—Dissolve an accurately weighed quantity of USP Disopyramide Phosphate RS in 0.1 *N* sulfuric acid, and dilute quantitatively and stepwise with the same solvent to obtain a solution having a known concentration of about 40 µg per mL.

Assay preparation—Grind the contents of not less than 20 Disopyramide Phosphate Extended-release Capsules to a powder fine enough to pass through a 40-mesh screen. Transfer an accurately weighed portion of the powder, equivalent to about 650 mg of disopyramide phosphate, to a 500-mL volumetric flask. Add about 400 mL of 0.1 *N* sulfuric acid, and shake for 30 minutes. Dilute with 0.1 *N* sulfuric acid to volume, mix, and filter. Dilute an accurately measured portion of the filtrate quantitatively and stepwise with 0.1 *N* sulfuric acid to obtain a solution having a concentration of about 40 µg per mL.

Procedure—Concomitantly determine the absorbances of the *Assay preparation* and the *Standard preparation* at the wavelength of maximum absorbance at about 261 nm, with a suitable spectrophotometer, using 0.1 *N* sulfuric acid as the blank. Calculate the quantity, in mg, of $C_{21}H_{29}N_3O$ in the portion of Capsules taken by the formula:

$$16.25(339.48/437.27)C(A_U/A_S),$$

in which 339.48 and 437.47 are the molecular weights of disopyramide and disopyramide phosphate, respectively, *C* is the concentration, in µg per mL, of USP Disopyramide Phosphate RS in the Standard solution, and A_U and A_S are the absorbances of the *Assay preparation* and the *Standard preparation*, respectively.

Disulfiram

$$(C_2H_5)_2N\overset{S}{\overset{\|}{C}}-S-S-\overset{S}{\overset{\|}{C}}N(C_2H_5)_2$$

$C_{10}H_{20}N_2S_4$ 296.52

Thioperoxydicarbonic diamide $[(H_2N)C(S)]_2S_2$, tetraethyl-.

Bis(diethylthiocarbamoyl) disulfide [*97-77-8*].

» Disulfiram contains not less than 98.0 percent and not more than 102.0 percent of $C_{10}H_{20}N_2S_4$.

Packaging and storage—Preserve in tight, light-resistant containers.

Reference standard—*USP Disulfiram Reference Standard*—Do not dry before using.

Identification—

A: The infrared absorption spectrum of a potassium bromide dispersion of it exhibits maxima only at the same wavelengths as that of a similar preparation of USP Disulfiram RS.

B: The retention time of the major peak in the chromatogram of the *Assay preparation* corresponds to that of the *Standard preparation*, as obtained in the *Assay*.

Melting range, *Class I* ⟨741⟩: between 69° and 72°.

Residue on ignition ⟨281⟩: not more than 0.1%.

Selenium ⟨291⟩: 0.003%.

Assay—

Buffer solution A—Dissolve 68 g of monobasic potassium phosphate in 1000 mL of water.

Buffer solution B—Dilute 100 mL of *Buffer solution A* with water to 1000 mL, and adjust with 45% potassium hydroxide solution to a pH of 7.0.

Mobile phase—Prepare a filtered and degassed mixture of methanol and *Buffer solution B* (70:30). Make adjustments if necessary (see *System Suitability* under *Chromatography* ⟨621⟩).

Standard preparation—Dissolve an accurately weighed quantity of USP Disulfiram RS in alcohol with sonication and swirling if necessary to obtain a solution having a known concentration of about 1 mg per mL. [NOTE—Discard this solution after 5 days.] Quantitatively dilute this solution with *Mobile phase* to obtain the *Standard preparation* having a known concentration of about 0.02 mg of USP Disulfiram RS per mL. [NOTE—Prepare the *Standard preparation* fresh daily.]

Assay preparation—Transfer about 50 mg of Disulfiram, accurately weighed, to a 50-mL volumetric flask, add about 40 mL of alcohol, sonicate for 5 minutes to completely dissolve, cool, dilute with alcohol to volume, and mix. [NOTE—Discard this solution after 5 days.] Quantitatively dilute this solution with *Mobile phase* to obtain the *Assay preparation* having a concentration of about 0.02 mg per mL. [NOTE—Prepare the *Assay preparation* fresh daily.]

Chromatographic system (see *Chromatography* ⟨621⟩)—The liquid chromatograph is equipped with a 250-nm detector and a 4-mm × 15-cm column that contains 5-µm packing L1. The flow rate is about 1 mL per minute. Chromatograph the *Standard preparation*, and record the peak responses as directed under *Procedure:* the column efficiency determined from the analyte peak is not less than 1800 plates, the tailing factor, *T*, for the analyte peak is not more than 2, and the relative standard deviation for replicate injections is not more than 2.0%.

Procedure—Separately inject equal volumes (about 20 µL) of the *Standard preparation* and the *Assay preparation* into the chromatograph, record the chromatograms, and measure the responses for the major peaks. Calculate the quantity, in mg, of $C_{10}H_{20}N_2S_4$ in the portion of Disulfiram taken by the formula:

$$2.5C(r_U / r_S),$$

in which *C* is the concentration, in µg per mL, of USP Disulfiram RS in the *Standard preparation*, and r_U and r_S are the peak responses obtained from the *Assay preparation* and the *Standard preparation*, respectively.

Disulfiram Tablets

» Disulfiram Tablets contain not less than 90.0 percent and not more than 110.0 percent of the labeled amount of $C_{10}H_{20}N_2S_4$.

Packaging and storage—Preserve in tight, light-resistant containers.

Reference standard—*USP Disulfiram Reference Standard*—Do not dry before using.

Identification—

A: The infrared absorption spectrum of a potassium bromide dispersion of a portion of finely powdered Tablets exhibits maxima only at the same wavelengths as that of a similar preparation of USP Disulfiram RS.

B: The retention time of the major peak in the chromatogram of the *Assay preparation* corresponds to that of the *Standard preparation*, as obtained in the *Assay*.

Disintegration ⟨701⟩: 15 minutes, the use of disks being omitted.

Uniformity of dosage units ⟨905⟩: meet the requirements.

Assay—

Buffer solution A, Buffer solution B, Mobile phase, Standard preparation, and *Chromatographic system*—Proceed as directed in the *Assay* under *Disulfiram.*

Assay preparation—Weigh and finely powder not less than 20 Disulfiram Tablets. Transfer an accurately weighed portion of the powder, equivalent to about 100 mg of disulfiram, to a 100-mL volumetric flask, add about 70 mL of alcohol and swirl, sonicate for 5 minutes, and shake by mechanical means for 30 minutes or until dissolved. Dilute with alcohol to volume, mix, and filter. [NOTE—Discard this solution after 5 days.] Quantitatively dilute this solution with *Mobile phase* to obtain the *Assay preparation* having a concentration of about 0.02 mg per mL. [NOTE—Prepare the *Assay preparation* fresh daily.]

Procedure—Proceed as directed for *Procedure* in the *Assay* under *Disulfiram.* Calculate the quantity, in mg, of $C_{10}H_{20}N_2S_4$ in the portion of Tablets taken by the formula:

$$5C(r_U / r_S),$$

in which *C* is the concentration, in µg per mL, of USP Disulfiram RS in the *Standard preparation*, and r_U and r_S are the peak responses obtained from the *Assay preparation* and the *Standard preparation*, respectively.

Dobutamine Hydrochloride

HO, HO—C6H3—$(CH_2)_2NHCH(CH_3)(CH_2)_2$—C6H4—OH · HCl

$C_{18}H_{23}NO_3 . HCl$ 337.85

1,2-Benzenediol, 4-[2-[[3-(4-hydroxyphenyl)-1-methylpropyl]amino]ethyl]-, hydrochloride, (±)-.

(±)-4-[2-[[3-(*p*-Hydroxyphenyl)-1-methylpropyl]amino]ethyl]-pyrocatechol hydrochloride [*49745-95-1*].

» Dobutamine Hydrochloride contains not less than 97.0 percent and not more than 103.0 percent of $C_{18}H_{23}NO_3 . HCl$, calculated on the anhydrous basis.

Caution—Great care should be taken to prevent inhaling particles of Dobutamine Hydrochloride and exposing the skin to it. Protect the eyes.

Packaging and storage—Preserve in tight containers, and store at controlled room temperature.

Reference standard—*USP Dobutamine Hydrochloride Reference Standard*—Determine the water content at the time of use for quantitative analyses.

Identification—

A: The infrared absorption spectrum of a potassium bromide dispersion of it, previously dried over a suitable desiccant, exhibits maxima only at the same wavelengths as that of a similar preparation of USP Dobutamine Hydrochloride RS.

B: A solution (1 in 100) responds to the tests for *Chloride* ⟨191⟩.

Water, *Method I* ⟨921⟩: not more than 1.0%.

Residue on ignition ⟨281⟩: not more than 0.2%.

Heavy metals, *Method II* ⟨231⟩: 0.003%.

Assay—

Internal standard solution—Dissolve an amount of *n*-triacontane in chloroform, and mix with an equal volume of pyridine to obtain a solution having a concentration of about 0.8 mg of *n*-triacontane per mL.

Standard preparation—Dissolve an accurately weighed quantity of USP Dobutamine Hydrochloride RS in pyridine, and dilute quantitatively with pyridine to obtain a solution having a known concentration of about 1 mg per mL.

Assay preparation—Transfer about 50 mg of Dobutamine Hydrochloride, accurately weighed, to a 50-mL volumetric flask, dissolve in pyridine, dilute with pyridine to volume, and mix.

Chromatographic system (see *Chromatography* ⟨621⟩)—The chromatograph is equipped with a flame-ionization detector and a 0.9-m × 3-mm glass column packed with 3 percent liquid phase (25 percent phenyl, 25 percent cyanopropylmethyl silicone) on 100- to 120-mesh, acid-washed, silanized support S1. The column is maintained at a temperature of about 230°, the injection port and detector block are maintained at a temperature of about 270°, and dry helium is used as the carrier gas. Chromatograph five injections of the solution obtained from the *Standard preparation*, and record the peak responses as directed under *Procedure:* the retention time for silylated dobutamine is about 1.9 relative to *n*-triacontane, and the resolution factor between the peaks is not less than 3.0. The relative standard deviation is not more than 1.5%.

Procedure—Transfer 2.0 mL of the *Standard preparation* and 2.0 mL of the *Assay preparation* to separate screw-capped tubes containing 2.0-mL portions of *Internal standard solution*, and mix. Add 0.40 mL of 1-trimethylsilylimidazole, quickly cap the tubes, mix, heat at 80° for 15 minutes, and allow to cool to room temperature. Inject a portion of the solution obtained from the *Standard preparation* into the chromatograph, record the chromatogram so as to obtain not less than 50% of maximum recorder response, and measure the responses obtained for silylated dobutamine and *n*-triacontane. Similarly inject a portion of the solution obtained from the *Assay preparation*, record the chromatogram, and measure the responses. Calculate the quantity, in mg, of $C_{18}H_{23}NO_3.HCl$ in the Dobutamine Hydrochloride taken by the formula:

$$50C(R_U/R_S),$$

in which C is the concentration, in mg per mL, of USP Dobutamine Hydrochloride RS in the *Standard preparation*, and R_U and R_S are the ratios of the responses of silylated dobutamine to *n*-triacontane from the *Assay preparation* and the *Standard preparation*, respectively.

Dobutamine Hydrochloride for Injection

» Dobutamine Hydrochloride for Injection is a sterile mixture of Dobutamine Hydrochloride with suitable diluents. It contains an amount of dobutamine hydrochloride equivalent to not less than 90.0 percent and not more than 110.0 percent of the labeled amount of dobutamine ($C_{18}H_{23}NO_3$).

Caution—Great care should be taken to prevent inhaling particles of Dobutamine Hydrochloride for Injection and exposing the skin to it. Protect the eyes.

Packaging and storage—Preserve in *Containers for Sterile Solids* as described under *Injections* ⟨1⟩, at controlled room temperature.

Reference standard—*USP Dobutamine Hydrochloride Reference Standard*—Determine the water content at the time of use for quantitative analyses.

Constituted solution—At the time of use, the constituted solution prepared from Dobutamine Hydrochloride for Injection meets the requirements for *Constituted Solutions* under *Injections* ⟨1⟩.

Identification—Prepare a solution in methanol, clarified by centrifugation, to contain 10 mg of dobutamine hydrochloride per mL. Apply 10 µL of this solution and 10 µL of a freshly prepared Standard solution of USP Dobutamine Hydrochloride RS in methanol containing 10 mg per mL on a thin-layer chromatographic plate (see *Chromatography* ⟨621⟩) coated with a 0.25-mm layer of chromatographic silica gel mixture. Allow the spots to dry, and develop the chromatogram in a solvent system consisting of ethyl acetate, *n*-propyl alcohol, water, and glacial acetic acid (100:40:15:5) until the solvent front has moved about three-fourths of the length of the plate. Remove the plate from the developing chamber, mark the solvent front, and allow the solvent to evaporate at room temperature. Observe the plate under short-wavelength ultraviolet light: the R_f value of the principal spot obtained from the solution under test corresponds to that obtained from the Standard solution.

Pyrogen—It meets the requirements of the *Pyrogen Test* ⟨151⟩, the test dose being an amount of Dobutamine Hydrochloride for Injection equivalent to 5.0 mg of dobutamine per kg.

Uniformity of dosage units ⟨905⟩: meets the requirements.

pH ⟨791⟩: between 2.5 and 5.5, the contents of 1 vial being dissolved in 10 mL of water.

Particulate matter ⟨788⟩: meets the requirements under *Small-volume Injections.*

Other requirements—It meets the requirements under *Injections* ⟨1⟩.

Assay—

Internal standard solution—Dissolve an amount of *n*-triacontane in chloroform, and mix with an equal volume of pyridine to obtain a solution having a concentration of about 0.5 mg of *n*-triacontane per mL.

pH 9.0 buffer—Dissolve about 5 g of sodium bicarbonate in 100 mL of water, and add increments of a saturated solution of sodium carbonate, with mixing, to obtain a solution having a pH of 9.0 ± 0.05, determined electrometrically.

Standard preparation—Transfer an accurately weighed quantity of USP Dobutamine Hydrochloride RS, equivalent to about 50 mg of dobutamine, to a separator, add 5 mL of 0.01 *N* hydrochloric acid, mix, and proceed as directed for *Assay preparation*, beginning with "Add 30 mL of *pH 9.0 buffer*."

Assay preparation—Dissolve the contents of not less than 10 containers of Dobutamine Hydrochloride for Injection in 0.01 *N* hydrochloric acid, and dilute quantitatively with 0.01 *N* hydrochloric acid to obtain a solution having a concentration of about 10 mg of dobutamine per mL. Transfer 5.0 mL of this solution to a separator. Add 30 mL of *pH 9.0 buffer*, and extract with two 40-mL portions of ethyl acetate, filtering the ethyl acetate extracts through anhydrous sodium sulfate into a 100-mL volumetric flask. Add ethyl acetate to volume, and mix.

Chromatographic system—Proceed as directed for *Chromatographic system* in the *Assay* under *Dobutamine Hydrochloride*.

Procedure—Transfer 2.0 mL of freshly prepared *Standard preparation* and 2.0 mL of freshly prepared *Assay preparation* to separate screw-capped tubes, and evaporate each with the aid of a gentle stream of dry nitrogen to dryness. Add 2.0 mL of *Internal standard solution* to each tube, and mix to dissolve the residues. Add 0.30 mL of *n*-trimethylsilylimidazole to each, and proceed as directed for *Procedure* in the *Assay* under *Dobutamine Hydrochloride*, beginning with "quickly cap the tubes." Calculate the quantity, in mg, of dobutamine ($C_{18}H_{23}NO_3$) in each container of Dobutamine Hydrochloride for Injection taken by the formula:

$$(301.38/337.85)(2VC)(R_U/R_S),$$

in which 301.38 is the molecular weight of dobutamine, 337.85 is the molecular weight of dobutamine hydrochloride, V is the volume, in mL, of the solution prepared to contain 10 mg of dobutamine per mL, C is the concentration, in mg per mL, of USP Dobutamine Hydrochloride RS in the *Standard preparation*, and R_U and R_S are the ratios of the responses of silylated dobutamine to *n*-triacontane from the *Assay preparation* and the *Standard preparation*, respectively.

Docusate Calcium

$$\left[\begin{array}{l} \quad\quad C_2H_5 \\ COOCH_2CH(CH_2)_3CH_3 \\ CH_2 \\ CH-SO_3- \\ COOCH_2CH(CH_2)_3CH_3 \\ \quad\quad C_2H_5 \end{array}\right]_2 Ca$$

$C_{40}H_{74}CaO_{14}S_2$ 883.22

Butanedioic acid, sulfo-, 1,4-bis(2-ethylhexyl) ester, calcium salt.

1,4-Bis(2-ethylhexyl) sulfosuccinate, calcium salt [*128-49-4*].

» Docusate Calcium contains not less than 91.0 percent and not more than 100.5 percent of $C_{40}H_{74}CaO_{14}S_2$, calculated on the anhydrous basis.

Packaging and storage—Preserve in well-closed containers.

Reference standards—*USP Docusate Calcium Reference Standard*—Do not dry; determine the water content at the time of use. *USP Bis(2-ethylhexyl) Maleate Reference Standard*—Do not dry before using.

Clarity of solution—Dissolve 25 g in 94 mL of alcohol: the solution does not develop a haze within 24 hours when maintained at a temperature of 25 ± 1°.

Identification—

A: Place a small piece of it on a salt plate, add 1 drop of acetone, and promptly cover with another salt plate. Rub the plates together to dissolve the specimen, slide the plates apart, and allow the acetone to evaporate: the infrared absorption spectrum of the film so obtained exhibits maxima only at the same wavelengths as that of a similar preparation of USP Docusate Calcium RS.

B: Dissolve 25 mg in 2 mL of acetone. Add 2 mL of water, mix, and add 2 drops of sulfuric acid: a white precipitate is formed.

C: Prepare a solution of it in isopropyl alcohol containing 10 mg per mL, and mix. On a suitable thin-layer chromatographic plate (see *Chromatography* ⟨621⟩), coated with a 0.25-mm layer of chromatographic silica gel, apply, with the aid of a stream of nitrogen, 10 µL of this solution and 10 µL of an isopropyl alcohol solution of USP Docusate Calcium RS containing 10 mg per mL. Allow the spots to dry, and develop the chromatogram in a solvent system consisting of a mixture of ethyl acetate, ammonium hydroxide, and alcohol (5:2:2) until the solvent front has moved three-fourths of the length of the plate. Remove the plate from the developing chamber, mark the solvent front, and allow the solvent to evaporate. Expose the plate to iodine vapors in a closed chamber for about 30 minutes, and locate the spots: the R_f value of the spot obtained from the test solution corresponds to that obtained from the Standard solution.

Water, *Method I* ⟨921⟩: not more than 2.0%.

Residue on ignition ⟨281⟩: between 14.5% and 16.5%, calculated on the anhydrous basis.

Arsenic, *Method II* ⟨211⟩: 2 ppm.

Heavy metals, *Method I* ⟨231⟩—Transfer 2.0 g to a platinum crucible, and ignite until free from carbon. Cool, moisten the residue with 1 mL of hydrochloric acid, and evaporate on a steam bath to dryness. Add 2 mL of 1 *N* acetic acid, digest on a steam bath for 5 minutes, filter into one of a pair of matched 50-mL, color-comparison tubes, and wash the residue with sufficient water to make 25 mL: the limit is 0.001%.

Bis(2-ethylhexyl) maleate—

Electrolyte solution—In a 250-mL borosilicate glass beaker dissolve 21.2 g of lithium perchlorate in 175 mL of water, and adjust by the dropwise addition of glacial acetic acid to a pH of 3.0. Transfer to a 200-mL volumetric flask, dilute with water to volume, and mix.

Sample stock solution—Transfer the equivalent of 12.5 g of anhydrous Docusate Calcium, accurately weighed, to a 150-mL beaker, add about 85 mL of isopropyl alcohol, stir to dissolve, and transfer to a 250-mL volumetric flask. Rinse the beaker with another 85-mL portion of isopropyl alcohol, stirring, if necessary, to dissolve any residual Docusate Calcium, combine the rinsing with the solution in the volumetric flask, dilute with isopropyl alcohol to volume, and mix.

Standard preparation—Transfer 25 mg of USP Bis(2-ethylhexyl) Maleate RS, accurately weighed, to a 25-mL volumetric flask, dissolve in isopropyl alcohol, dilute with isopropyl alcohol to volume, and mix. In a second 100-mL volumetric flask combine 10.0 mL of this solution, 50.0 mL of *Sample stock solution*, and 20.0 mL of *Electrolyte solution*, mix, and allow to stand for 2 minutes. Dilute with isopropyl alcohol to volume, and mix.

Test preparation—In a 100-mL volumetric flask combine 50.0 mL of *Sample stock solution* with 20.0 mL of *Electrolyte solution*, mix, and allow to stand for 2 minutes. Dilute with isopropyl alcohol to volume, and mix.

Procedure—Transfer a portion of the *Test preparation* to a polarographic cell, and deaerate by bubbling through the solution, for 15 minutes, nitrogen that has previously been passed through isopropyl alcohol. Continue to flush the surface of the solution with the nitrogen, insert the dropping mercury electrode of a suitable polarograph (see *Polarography* ⟨801⟩), and record the polarogram from −0.9 to −1.5 volts, using a saturated calomel electrode as the reference electrode. Determine the height of the polarogram at the half-wave potential, at about −1.2 volts, measuring from the baseline. From this height subtract the height, at the same potential, of the polarogram obtained from a blank solution prepared by diluting 20.0 mL of *Electrolyte solution* with isopropyl alcohol to 100 mL in a volumetric flask and mixing. Designate this corrected height as H_T. Concomitantly and similarly determine the height of the polarogram of the *Standard preparation*, and subtract the blank value from it. Designate the result as H_S. The value, H_T, is not greater than one-half H_S, corresponding to not more than 0.4% of bis(2-ethylhexyl) maleate.

Assay—

Tetra-n-butylammonium iodide solution—Transfer 1.250 g of tetra-*n*-butylammonium iodide to a 500-mL volumetric flask, dilute with water to volume, and mix.

Salt solution—Transfer 100 g of anhydrous sodium sulfate and 10 g of sodium carbonate to a 400-mL beaker, and add sufficient water to dissolve the two salts. Transfer the solution to a 1000-mL volumetric flask, dilute with water to volume, and mix.

Procedure—Dissolve about 50 mg of Docusate Calcium, accurately weighed, in 50 mL of chloroform in a glass-stoppered, 250-mL conical flask. Add 50 mL of *Salt solution* and 500 µL of bromophenol blue TS, and mix. Titrate with *Tetra-n-butylammonium iodide solution* until about 1 mL from the end-point, and shake the stoppered flask vigorously for about 2 minutes. Continue the titration in 2-drop increments, shaking vigorously for about 10 seconds after each addition, and then allow the flask to stand about 10 seconds. Continue the titration until the chloroform layer just assumes a blue color. Each mL of *Tetra-n-butylammonium iodide solution* is equivalent to 2.989 mg of $C_{40}H_{74}CaO_{14}S_2$.

Docusate Calcium Capsules

» Docusate Calcium Capsules contain not less than 85.0 percent and not more than 115.0 percent of the labeled amount of $C_{40}H_{74}CaO_{14}S_2$.

Packaging and storage—Preserve in tight containers, and store at controlled room temperature in a dry place.

Reference standard—*USP Docusate Calcium Reference Standard*—Do not dry; determine the water content at the time of use.

Identification—

A: Ash the contents of 1 Capsule, dissolve the residue in 10 mL of dilute hydrochloric acid (1 in 12), and filter: the filtrate responds to the tests for *Calcium* ⟨191⟩.

B: Empty the contents of 1 Capsule into a conical flask, add isopropyl alcohol to obtain a concentration of 10 mg of docusate calcium per mL, and mix. This solution responds to *Identification test C* under *Docusate Calcium*.

Uniformity of dosage units ⟨905⟩—

SOLID-FILLED CAPSULES: meet the requirements for *Content Uniformity*.

SOLUTION-FILLED CAPSULES: meet the requirements for *Weight Variation*.

Assay—

Basic fuchsin solution—Dissolve 100 mg of basic fuchsin, previously recrystallized from alcohol, in 3 mL of methanol in a 100-mL volumetric flask, dilute with water to volume, and mix. Filter prior to use.

Standard preparation—Weigh accurately about 50 mg of USP Docusate Calcium RS into a 10-mL volumetric flask, dissolve in isopropyl alcohol, dilute with isopropyl alcohol to volume, and mix. Transfer 5.0 mL of this solution to a 500-mL volumetric flask containing 90 mL of a mixture of water and isopropyl alcohol (1:1) with swirling, dilute with water to volume, and mix. The concentration of USP Docusate Calcium RS in the *Standard preparation* is about 50 µg per mL.

Assay preparation—To 10 Docusate Calcium Capsules in a 400-mL beaker add 300 mL of hot water. Heat on a steam bath, with occasional stirring, until the Capsules are completely disintegrated, cool, and transfer with the aid of 100 mL of warm water to a 1000-mL volumetric flask. Rinse the beaker with 100 mL of isopropyl alcohol, add the rinsing to the volumetric flask, and shake to dissolve any previously undissolved particles. Dilute with a mixture of water and isopropyl alcohol (1:1) to volume, and mix. Transfer 10.0 mL of this solution to a 100-mL volumetric flask, dilute with water to volume, and mix.

Procedure—Transfer 4.0 mL of the *Standard preparation* to a 125-mL separator. To a second 125-mL separator transfer an accurately measured volume of the *Assay preparation*, equivalent to 200 to 240 µg of docusate calcium, and add a solution of isopropyl alcohol (1 in 100), if necessary, to bring the volume of the solution in the separator to 4.0 mL. To each separator transfer 20.0 mL of pH 1.2 hydrochloric acid buffer (see under *Solutions* in the section, *Reagents, Indicators, and Solutions*) and 2.0 mL of *Basic fuchsin solution*, add 20 mL of chloroform to each, and shake for 1 minute. Allow the phases to separate, and transfer the chloroform extracts, through separate pledgets of absorbent cotton, into separate 100-mL volumetric flasks. Extract each of the solutions in the separators in the same manner with additional 20-mL portions of chloroform until no more color is visible in the extract, and pass each extract through the pledget of cotton used for the preceding extract into the flask containing the preceding extract. Dilute the contents of each flask to volume by passing chloroform through the cotton pledgets that had been used to filter the extracts, and mix. Concomitantly determine the absorbances of the solutions in 1-cm cells at the wavelength of maximum absorbance at about 545 nm, with a suitable spectrophotometer, using chloroform as the blank. Calculate the quantity, in mg, of $C_{40}H_{74}CaO_{14}S_2$ in each Capsule taken by the formula:

$$(4C/V)(A_U/A_S),$$

in which C is the concentration, in µg per mL, of anhydrous docusate calcium in the *Standard preparation*, V is the volume, in mL, of *Assay preparation* taken, and A_U and A_S are the absorbances of the solutions from the *Assay preparation* and the *Standard preparation*, respectively.

Docusate Potassium

C_2H_5
$COOCH_2CH(CH_2)_3CH_3$
CH_2
$CH{-}SO_3K$
$COOCH_2CH(CH_2)_3CH_3$
C_2H_5

$C_{20}H_{37}KO_7S$ 460.67

Butanedioic acid, sulfo-, 1,4-bis(2-ethylhexyl) ester, potassium salt.

Potassium 1,4-bis(2-ethylhexyl) sulfosuccinate [*7491-09-0*].

» Docusate Potassium contains not less than 95.0 percent and not more than 100.5 percent of $C_{20}H_{37}KO_7S$, calculated on the dried basis.

Packaging and storage—Preserve in well-closed containers.

Reference standards—*USP Docusate Potassium Reference Standard*—Do not dry; determine the water content at the time of use. *USP Bis(2-ethylhexyl) Maleate Reference Standard*—Do not dry before using.

Identification—

A: Place a small piece of it on a salt plate, add 1 drop of acetone, and promptly cover with another salt plate. Rub the plates together to dissolve the specimen, slide the plates apart, and allow the acetone to evaporate: the infrared absorption spectrum of the film so obtained exhibits maxima only at the same wavelengths as that of a similar preparation of USP Docusate Potassium RS.

B: It responds to the flame test for *Potassium* ⟨191⟩.

Loss on drying ⟨731⟩—Dry it in a glass container at 105° for 4 hours: it loses not more than 3.0% of its weight.

Residue on ignition ⟨281⟩: between 18.0% and 20.0%, calculated on the dried basis.

Arsenic, *Method II* ⟨211⟩: 2 ppm.

Heavy metals ⟨231⟩—It meets the requirements of the test for *Heavy metals* under *Docusate Sodium*.

Bis(2-ethylhexyl) maleate—Place 12.5 g of Docusate Potassium, previously dried and accurately weighed, in a 100-mL volumetric flask. Pipet 20 mL of pH 10 alkaline borate buffer (see under *Solutions* in the section, *Reagents, Indicators, and Solutions*) into the flask, and add about 70 mL of alcohol. Swirl the flask and contents, with gentle warming if necessary, until the solid dissolves. Adjust the reaction dropwise, if necessary, with 1 *N* sodium hydroxide to a pH of 10. Add alcohol to volume, and proceed as directed in the test for *Bis(2-ethylhexyl) maleate* under *Docusate Sodium*, beginning with "Transfer a portion of this solution to a polarographic cell." The diffusion current for the test solution of Docusate Potassium is not greater than one-half the diffusion current of the solution containing the added bis(2-ethylhexyl) maleate [not more than 0.4% of bis(2-ethylhexyl) maleate].

Assay—

Tetra-n-butylammonium iodide solution, and *Salt solution*—Prepare as directed in the *Assay* under *Docusate Sodium*.

Procedure—Dissolve about 100 mg of Docusate Potassium, accurately weighed, in 50 mL of chloroform in a glass-stoppered, 250-mL conical flask. Proceed as directed in the *Assay* under *Docusate Sodium*, beginning with "Add 50 mL of *Salt solution*." Each mL of *Tetra-n-butylammonium iodide solution* is equivalent to 3.118 mg of $C_{20}H_{37}KO_7S$.

Docusate Potassium Capsules

» Docusate Potassium Capsules contain not less than 90.0 percent and not more than 110.0 percent of the labeled amount of $C_{20}H_{37}KO_7S$.

Packaging and storage—Preserve in tight containers, and store at controlled room temperature.

Reference standard—*USP Docusate Potassium Reference Standard*—Do not dry; determine the water content at the time of use.

Identification—The chromatogram of the *Assay preparation* determined as directed in the *Assay* exhibits a major peak for docusate potassium, the retention time of which corresponds to that exhibited by the *Standard preparation*, similarly determined.

Uniformity of dosage units ⟨905⟩—

SOLID-FILLED CAPSULES: meet the requirements for *Content Uniformity*.

SOLUTION-FILLED CAPSULES: meet the requirements for *Weight Variation*.

Assay—

Methanol-water solution—Transfer 180 mL of water to a 1000-mL volumetric flask, dilute with methanol to volume, and mix.

Tetrabutylammonium hydroxide–methanol solution—Prepare a solution of tetrabutylammonium hydroxide in methanol containing 25 g per 100 mL.

Mobile phase—Mix 180 mL of water, 6.0 mL of glacial acetic acid, and 8.0 mL of *Tetrabutylammonium hydroxide–methanol solution* in a 1000-mL volumetric flask, dilute with methanol to volume, and mix. The water concentration may be varied to meet system suitability requirements and to provide a suitable elution time (about 5 minutes) for docusate potassium.

Standard preparation—Dissolve in *Methanol-water solution* a suitable quantity, accurately weighed, of USP Docusate Potassium RS, and dilute quantitatively with the same solvent to obtain a solution having a known concentration of about 4 mg per mL, calculated on the anhydrous basis.

Assay preparation for solid-filled capsules—Open and empty into a suitable container the contents of a counted number of Docusate Potassium Capsules, equivalent to about 1000 mg of docusate potassium. Place the capsule shells in the container, and add 250.0 mL of *Methanol-water solution.* Shake by mechanical means for 20 minutes, and clarify a portion of the mixture by centrifuging. Filter a portion of the clear supernatant liquid through a membrane filter (0.5-µm or finer porosity).

Assay preparation for solution-filled capsules—Cut open a counted number of Docusate Potassium Capsules, equivalent to about 2.5 g of docusate potassium, and place the shells and contents in a suitable container. Add about 25 mL of methanol, agitate for not less than 2 minutes, and decant the liquid into a 200-mL volumetric flask. Repeat the addition of methanol, agitation, and decanting not less than four times. Dilute with methanol to volume, and mix. Pipet 8 mL of this solution into a 25-mL volumetric flask, add 4.5 mL of water, dilute with methanol to volume, and mix.

Chromatographic system (see *Chromatography* ⟨621⟩)—The chromatograph is equipped with a refractive index detector and a 4.6-mm × 25-cm column that contains packing L1. The flow rate is about 1.8 mL per minute. Chromatograph five replicate injections of the *Standard preparation*, and record the peak responses as directed under *Procedure:* the relative standard deviation is not more than 3.0%.

Procedure—Separately inject equal volumes (about 100 µL) of the *Standard preparation* and the *Assay preparation* into the chromatograph by means of a suitable microsyringe or sampling valve, record the chromatograms, and measure the responses for the major peaks. Calculate the quantity, in mg, of $C_{20}H_{37}KO_7S$ in each solid-filled Capsule taken by the formula:

$$(250C/N)(H_U/H_S),$$

and calculate the quantity, in mg, of $C_{20}H_{37}KO_7S$ in each solution-filled Capsule taken by the formula:

$$(625C/N)(H_U/H_S),$$

in which C is the concentration, in mg per mL, of anhydrous USP Docusate Potassium RS in the *Standard preparation*, N is the number of capsules taken, and H_U and H_S are the peak responses from the *Assay preparation* and the *Standard preparation*, respectively.

Docusate Sodium

$C_{20}H_{37}NaO_7S$ 444.56

Butanedioic acid, sulfo-, 1,4-bis(2-ethylhexyl) ester, sodium salt.

Sodium 1,4-bis(2-ethylhexyl) sulfosuccinate [*577-11-7*].

» Docusate Sodium contains not less than 99.0 percent and not more than 100.5 percent of $C_{20}H_{37}NaO_7S$, calculated on the anhydrous basis.

Packaging and storage—Preserve in well-closed containers.

Reference standards—*USP Docusate Sodium Reference Standard*—Do not dry; determine the water content at time of use. *USP Bis(2-ethylhexyl) Maleate Reference Standard*—Do not dry before using.

Clarity of solution—Dissolve 25 g in 100 mL of alcohol: the solution does not develop a haze within 24 hours.

Identification—Place a small piece of it on a salt plate, add 1 drop of acetone, and promptly cover with another salt plate. Rub the plates together to dissolve the specimen, slide the plates apart, and allow the acetone to evaporate: the infrared absorption spectrum of the film so obtained exhibits maxima only at the same wavelengths as that of a similar preparation of USP Docusate Sodium RS.

Water, *Method I* ⟨921⟩: not more than 2.0%.

Residue on ignition ⟨281⟩: between 15.5% and 16.5%, calculated on the anhydrous basis.

Arsenic, *Method II* ⟨211⟩: 3 ppm.

Heavy metals ⟨231⟩—Transfer 2.0 g to a platinum crucible, and ignite until free from carbon. Cool, moisten the residue with 1 mL of hydrochloric acid, and evaporate on a steam bath to dryness. Add 2 mL of 6 *N* acetic acid, digest on a steam bath for 5 minutes, filter into one of a pair of matched 50-mL color-comparison tubes, and wash the residue with sufficient water to make 25 mL: the limit is 0.001%.

Bis(2-ethylhexyl) maleate—Place 12.5 g of Docusate Sodium, previously dried at 105° for 2 hours and accurately weighed, in a 100-mL volumetric flask. Pipet 20 mL of pH 10 alkaline borate buffer (see under *Solutions* in the section, *Reagents, Indicators, and Solutions*) into the flask, and add about 70 mL of alcohol. Swirl the flask and contents, with gentle warming, until the solid dissolves. Cool, and add alcohol to volume. Transfer a portion of this test solution to a polarographic cell that is immersed in a water bath regulated at 25 ± 0.5°, and deaerate by bubbling through the solution, for 10 minutes, purified nitrogen that previously has been passed through a solution prepared by mixing 4 volumes of alcohol and 1 volume of water. Insert the dropping mercury electrode of a suitable polarograph, and record the polarogram from −0.5 volt to −1.6 volts, using a saturated calomel electrode as the reference electrode (see *Polarography* ⟨801⟩). Determine the height of the diffusion current at the current plateau at about −1.45 volts. To a separate 50-mL portion of the solution add 25 mg of USP Bis(2-ethylhexyl) Maleate RS, accurately weighed, and when solution is complete, deaerate a portion of the resulting solution, and record the polarogram, as directed in the foregoing. The diffusion current for the test solution of Docusate Sodium is not greater than one-half the diffusion current of the solution containing the added USP Bis(2-ethylhexyl) Maleate RS [not more than 0.4% of bis(2-ethylhexyl) maleate].

Assay—

Tetra-n-butylammonium iodide solution—Transfer 1.250 g of tetra-*n*-butylammonium iodide to a 500-mL volumetric flask, dilute with water to volume, and mix.

Salt solution—Transfer 100 g of anhydrous sodium sulfate and 10 g of sodium carbonate to a 400-mL beaker, and add water to dissolve the two salts. Transfer the solution to a 1000-mL volumetric flask, dilute with water to volume, and mix.

Procedure—Dissolve about 50 mg of Docusate Sodium, accurately weighed, in 50 mL of chloroform in a glass-stoppered, 250-mL conical flask. Add 50 mL of *Salt solution* and 0.5 mL of bromophenol blue TS, and mix. Titrate with *Tetra-n-butylammonium iodide solution* until about 1 mL from the end-point, and shake the stoppered flask vigorously for about 2 minutes. Continue the titration in 2-drop increments, shaking vigorously for about 10 seconds after each addition, and then allow the flask to stand about 10 seconds. Continue the titration until the chloroform layer just assumes a blue color. Each mL of *Tetra-n-butylammonium iodide solution* is equivalent to 3.009 mg of $C_{20}H_{37}NaO_7S$.

Docusate Sodium Capsules

» Docusate Sodium Capsules contain not less than 90.0 percent and not more than 110.0 percent of the labeled amount of $C_{20}H_{37}NaO_7S$.

Packaging and storage—Preserve in tight containers, and store at controlled room temperature.

Reference standard—*USP Docusate Sodium Reference Standard*—Do not dry; determine the water content at the time of use.

Identification—The retention time of the docusate sodium peak in the chromatogram of the *Assay preparation* corresponds to that of the docusate sodium peak in the chromatogram of the *Standard preparation*, as obtained in the *Assay*.

Uniformity of dosage units ⟨905⟩—

SOLID-FILLED CAPSULES: meet the requirements for *Content Uniformity*.

SOLUTION-FILLED CAPSULES: meet the requirements for *Weight Variation*.

Assay—

Mobile phase—Mix 65 volumes of acetonitrile and 35 volumes of dilute phosphoric acid (1 in 2500), and degas.

Internal standard solution—Dissolve *p*-toluenesulfonic acid in a mixture of acetonitrile and water (70:30) to obtain a solution containing about 0.01 mg per mL.

Standard preparation—Transfer about 160 mg of USP Docusate Sodium RS, accurately weighed, to a 100-mL volumetric flask. Dissolve in a mixture of methanol and water (1:1), dilute with the same solvent to volume, and mix. Transfer 5.0 mL of this solution to a 125-mL separator, add 10 mL of 0.1 *N* hydrochloric acid, and swirl. Immediately extract with five 25-mL portions of chloroform, swirling gently for the first extraction and shaking each subsequent extraction vigorously for 30 seconds. Combine the extracts in a vacuum evaporation flask, and evaporate at 30° under vacuum to dryness. As soon as the residue is dry, add 10.0 mL of *Internal standard solution*, and swirl to dissolve the residue.

Assay preparation—To a number of Docusate Sodium Capsules, equivalent to about 1 g of docusate sodium, in a 400-mL beaker, add 100 mL of hot water. Heat on a steam bath, with occasional stirring, until the Capsules are completely disintegrated, cool, and transfer with the aid of methanol to a 500-mL volumetric flask. Dilute with a mixture of methanol and water (1:1) to volume, and mix. Transfer 4.0 mL of this solution to a 125-mL separator, and proceed as directed for *Standard preparation*, beginning with "Add 10 mL of 0.1 *N* hydrochloric acid."

Chromatographic system (see *Chromatography* ⟨621⟩)—The liquid chromatograph is equipped with a 214-nm detector and a 4.6-mm × 25-cm column that contains packing L14. The flow rate is about 1.5 mL per minute. Chromatograph five replicate injections of the *Standard preparation*, and record the peak responses as directed under *Procedure:* the relative standard deviation is not more than 2.0%, and the resolution, *R*, between docusate sodium and *p*-toluenesulfonic acid is not less than 9.0.

Procedure—Separately inject equal volumes (about 25 µL) of the *Standard preparation* and the *Assay preparation* into the chromatograph by means of a suitable microsyringe or sampling valve, record the chromatograms, and measure the responses for the major peaks. The relative retention times are about 0.4 for docusate sodium and 1.0 for *p*-toluenesulfonic acid. Calculate the quantity, in mg, of $C_{20}H_{37}NaO_7S$ in each Capsule taken by the formula:

$$1250(C/N)(R_U/R_S),$$

in which *C* is the concentration, in mg per mL, of anhydrous docusate sodium in the *Standard preparation*, as determined from the concentration of USP Docusate Sodium RS corrected for moisture by a titrimetric water determination, *N* is the number of Capsules taken, and R_U and R_S are the ratios of the peak responses of docusate sodium to those of *p*-toluenesulfonic acid obtained from the *Assay preparation* and the *Standard preparation*, respectively.

Docusate Sodium Solution

» Docusate Sodium Solution contains not less than 90.0 percent and not more than 110.0 percent of the labeled amount of $C_{20}H_{37}NaO_7S$.

Packaging and storage—Preserve in tight containers.

Reference standard—*USP Docusate Sodium Reference Standard*—Do not dry; determine the water content at time of use.

Identification—The retention time of the docusate sodium peak in the chromatogram of the *Assay preparation* corresponds to that of the docusate sodium peak in the chromatogram of the *Standard preparation*, as obtained in the *Assay*.

pH ⟨791⟩: between 4.5 and 6.9.

Assay—

Mobile phase, Internal standard solution, Standard preparation, and *Chromatographic system*—Proceed as directed in the *Assay* under *Docusate Sodium Capsules*.

Assay preparation—Dilute an accurately measured volume of Docusate Sodium Solution quantitatively with methanol to obtain a solution containing about 1.6 mg of docusate sodium per mL. Proceed as directed for *Standard preparation* in the *Assay* under *Docusate Sodium Capsules*, beginning with "Transfer 5.0 mL of this solution to a 125-mL separator."

Procedure—Proceed as directed for *Procedure* in the *Assay* under *Docusate Sodium Capsules*. Calculate the quantity, in mg, of $C_{20}H_{37}NaO_7S$ in each mL of the Solution taken by the formula:

$$(CL/D)(R_U/R_S),$$

in which *L* is the labeled quantity, in mg, of docusate sodium in each mL of Solution taken, *D* is the concentration, in mg of docusate sodium per mL, of the *Assay preparation* based on the labeled quantity in each mL and the extent of dilution, and the other terms are as defined therein.

Docusate Sodium Syrup

» Docusate Sodium Syrup contains not less than 90.0 percent and not more than 110.0 percent of the labeled amount of $C_{20}H_{37}NaO_7S$.

Packaging and storage—Preserve in tight, light-resistant containers.

Reference standard—*USP Docusate Sodium Reference Standard*—Do not dry; determine the water content at time of use.

Identification—Dilute a volume of Syrup, equivalent to about 10 mg of docusate sodium, with isopropyl alcohol to obtain a preparation containing about 2 mg per mL, and mix. On a suitable thin-layer chromatographic plate (see *Chromatography* ⟨621⟩), coated with a 0.25-mm layer of chromatographic silica gel mixture, apply, with the aid of a stream of nitrogen, 50 µL of the upper layer of this preparation and 50 µL of an isopropyl alcohol solution of USP Docusate Sodium RS containing about 2 mg per mL. Allow the spots to dry, and develop the chromatogram in a two-phase solvent system consisting of ethyl acetate, water, alcohol, and ammonium hydroxide (25:20:10:1) until the solvent front has moved about three-fourths of the length of the plate. Remove the plate from the developing chamber, mark the solvent front, and allow the solvent to evaporate. Expose the plate to iodine vapors in a closed chamber for about 30 minutes, and locate the spots: the test preparation produces a spot at the same R_f value and of approximately the same size as that obtained from the Standard solution.

pH ⟨791⟩: between 5.5 and 6.5.

Assay—

Standard preparation—Transfer about 100 mg of USP Docusate Sodium RS, accurately weighed, to a 100-mL volumetric flask, dissolve in 2.5 mL of alcohol, dilute with water to volume, and mix. Transfer 10.0 mL of this solution to a 1000-mL volumetric flask, dilute with water to volume, and mix. The concentration of USP Docusate Sodium RS in the *Standard preparation* is about 10 µg per mL.

Assay preparation—Transfer an accurately measured volume of Docusate Sodium Syrup, equivalent to about 100 mg of docusate sodium, to a 1000-mL volumetric flask, allowing the pipet to drain for 15 minutes. Dilute with water to volume, and mix.

Transfer 10.0 mL of the solution to a 100-mL volumetric flask, dilute with water to volume, and mix.

Procedure—Transfer 20.0 mL each of the *Standard preparation* and of the *Assay preparation* to individual separators, and place 20 mL of water in a third separator to provide the blank. To each separator add 5 drops of hydrochloric acid, mix by swirling, add 1.0 mL of methylene blue solution (1 in 1000), and mix by swirling. To each separator add 20.0 mL of chloroform, and shake vigorously for 5 minutes. Wash each chloroform solution, in clean separators, with 20 mL of water, shaking vigorously for 60 seconds. Discard the washings, and filter each chloroform solution through a layer of 3 g of anhydrous granular sodium sulfate, supported on glass wool, into a 100-mL volumetric flask, washing each separator with two 10-mL portions of chloroform and filtering the washings into each flask. Dilute each flask with chloroform to volume, and mix. Concomitantly determine the absorbances of the solutions in 1-cm cells at the wavelength of maximum absorbance at about 650 nm, with a suitable spectrophotometer, using the blank to set the instrument. Calculate the quantity, in mg, of $C_{20}H_{37}NaO_7S$ in each mL of Syrup taken by the formula:

$$(10C/V)(A_U/A_S),$$

in which C is the concentration, in µg per mL, calculated on the anhydrous basis, of USP Docusate Sodium RS in the *Standard preparation*, V is the volume, in mL, of Syrup taken, and A_U and A_S are the absorbances of the solutions from the *Assay preparation* and the *Standard preparation*, respectively.

Docusate Sodium Tablets

» Docusate Sodium Tablets contain not less than 90.0 percent and not more than 110.0 percent of the labeled amount of $C_{20}H_{37}NaO_7S$.

Packaging and storage—Preserve in well-closed containers.

Reference standard—*USP Docusate Sodium Reference Standard*—Do not dry; determine the water content at time of use.

Identification—Finely divide a suitable number of Tablets, extract with solvent hexane, filter, and evaporate the solvent hexane extract on a steam bath: the infrared absorption spectrum of a potassium bromide dispersion of a portion of the dry residue exhibits maxima and minima at the same wavelengths as that of a similar preparation of USP Docusate Sodium RS, concomitantly measured.

Disintegration ⟨701⟩: 1 hour, simulated gastric fluid TS being substituted for water in the test for *Uncoated Tablets*.

Uniformity of dosage units ⟨905⟩: meet the requirements.

Assay—

Basic fuchsin solution—Prepare as directed in the *Assay* under *Docusate Calcium Capsules*.

Standard preparation—Transfer an accurately weighed portion of USP Docusate Sodium RS, equivalent to about 50 mg of anhydrous docusate sodium, to a 10-mL volumetric flask. Add 5 mL of isopropyl alcohol, dissolve in isopropyl alcohol, dilute with the same solvent to volume, and mix. Pipet 5 mL of this solution into 300 mL of water contained in a 500-mL volumetric flask, dilute with water to volume, and mix.

Assay preparation—Weigh and finely powder not less than 20 Docusate Sodium Tablets. Weigh accurately a portion of the powder, equivalent to about 50 mg of docusate sodium, into a 10-mL volumetric flask. Add 8 mL of isopropyl alcohol, sonicate for 5 minutes, add the same solvent to volume, and mix. Centrifuge a portion of the mixture, transfer 5.0 mL of the clear solution to a 500-mL volumetric flask, dilute with water to volume, and mix.

Procedure—Pipet 4.0 mL each of the *Standard preparation* and the *Assay preparation* into separate 125-mL separators, and proceed as directed in the *Assay* under *Docusate Calcium Capsules*, beginning with "To each separator." Calculate the quantity, in mg, of $C_{20}H_{37}NaO_7S$ in the portion of Tablets taken by the formula:

$$C(A_U/A_S),$$

in which C is the concentration, in µg per mL, calculated on the anhydrous basis, of USP Docusate Sodium RS in the *Standard preparation*, and A_U and A_S are the absorbances of the solutions from the *Assay preparation* and the *Standard preparation*, respectively.

Dopamine Hydrochloride

HO
HO–⟨ring⟩–$CH_2CH_2NH_2$ · HCl

$C_8H_{11}NO_2.HCl$ 189.64

1,2-Benzenediol, 4-(2-aminoethyl)-, hydrochloride.

4-(2-Aminoethyl)pyrocatechol hydrochloride [*62-31-7*].

» Dopamine Hydrochloride contains not less than 98.0 percent and not more than 102.0 percent of $C_8H_{11}NO_2.HCl$, calculated on the dried basis.

Packaging and storage—Preserve in tight containers.

Reference standard—*USP Dopamine Hydrochloride Reference Standard*—Dry at 105° for 2 hours before using.

Clarity and color of solution—A solution of 400 mg in 10 mL of sodium bisulfite solution (1 in 1000) is clear and colorless or practically colorless.

Identification—

A: The infrared absorption spectrum of a potassium bromide dispersion of it exhibits maxima only at the same wavelengths as that of a similar preparation of USP Dopamine Hydrochloride RS.

B: The ultraviolet absorption spectrum of a 1 in 25,000 solution in sodium bisulfite solution (1 in 1000) exhibits maxima and minima at the same wavelengths as that of a similar solution of USP Dopamine Hydrochloride RS, concomitantly measured.

C: It responds to the tests for *Chloride* ⟨191⟩.

pH ⟨791⟩: between 3.0 and 5.5, in a solution (1 in 25).

Loss on drying ⟨731⟩—Dry it at 105° for 2 hours: it loses not more than 0.5% of its weight.

Residue on ignition ⟨281⟩: not more than 0.1%.

Heavy metals, *Method I* ⟨231⟩—Dissolve 1 g in 25 mL of water: the limit is 0.002%.

Sulfate ⟨221⟩—Dissolve 500 mg in 40 mL of water: any turbidity produced is not more than that produced in a solution containing 0.10 mL of 0.020 *N* sulfuric acid.

Readily carbonizable substances ⟨271⟩—Dissolve 100 mg in 5 mL of sulfuric acid TS: the solution has no more color than *Matching Fluid A*.

Chromatographic purity—

Standard preparation and *Standard dilutions*—Prepare a solution of USP Dopamine Hydrochloride RS in methanol to contain 30 mg per mL (*Standard preparation*). Prepare a series of dilutions of the *Standard preparation* in methanol to contain 0.6 mg, 0.3 mg, and 0.15 mg per mL (*Standard dilutions*), corresponding to 2.0%, 1.0%, and 0.5% of impurities, respectively.

Test preparation—Transfer 150 mg of Dopamine Hydrochloride to a 5-mL volumetric flask, dilute with methanol to volume, and mix.

Procedure—In a suitable chromatographic chamber arranged for thin-layer chromatography (see *Chromatography* ⟨621⟩), place a solvent system consisting of 13 volumes of chloroform, 9 volumes of methanol, and 4 volumes of dilute glacial acetic acid (3 in 10). Line the chamber with filter paper, and allow to equilibrate. On a 20-cm × 20-cm chromatographic plate coated with a 0.25-mm layer of chromatographic silica gel mixture, apply 10-µL portions of the *Standard preparation*, the *Standard dilutions*, and the *Test preparation*. Develop the chromatogram until the solvent front has moved about 15 cm. Remove the plate from the developing chamber, allow to dry at room temperature for several minutes, and spray evenly with a freshly prepared mixture containing equal volumes of ferric chloride solution (1 in 10) and potassium ferricyanide solution (1 in 20). [Dopamine and its related impurities appear as blue spots under visible light.] The *Test preparation* exhibits its principal spot at an R_f value corresponding to that of the *Standard preparation* and not more than three secondary spots. Estimate the concentration of any secondary spots exhibited by the *Test preparation* by comparison with the *Standard dilutions:* the sum of the impurities is not greater than 1.0%.

Assay—Dissolve about 300 mg of Dopamine Hydrochloride, accurately weighed, in 70 mL of glacial acetic acid, add 10 mL of mercuric acetate TS, and mix. Titrate with 0.1 *N* perchloric acid VS, determining the end-point potentiometrically. Perform a blank determination, and make any necessary correction. Each mL of 0.1 *N* perchloric acid is equivalent to 18.96 mg of $C_8H_{11}NO_2 \cdot HCl$.

Dopamine Hydrochloride Injection

» Dopamine Hydrochloride Injection is a sterile solution of Dopamine Hydrochloride in Water for Injection. It contains not less than 95.0 percent and not more than 105.0 percent of the labeled amount of $C_8H_{11}NO_2 \cdot HCl$. It may contain a suitable antioxidant.

NOTE—Do not use the Injection if it is darker than slightly yellow or discolored in any other way.

Packaging and storage—Preserve in single-dose containers of Type I glass.

Labeling—Label it to indicate that the Injection is to be diluted with a suitable parenteral vehicle prior to intravenous infusion.

Reference standard—*USP Dopamine Hydrochloride Reference Standard*—Dry at 105° for 2 hours before using.

Identification—Transfer a volume of Injection, equivalent to about 80 mg of dopamine hydrochloride, to a 50-mL volumetric flask, add 10 mL of methanol, dilute with water to volume, and mix. Apply 5-µL portions of this test solution and a Standard solution containing 1.6 mg of USP Dopamine Hydrochloride RS in each mL of dilute methanol (1 in 5) to a suitable thin-layer chromatographic plate (see *Chromatography* ⟨621⟩) coated with a 0.25-mm layer of chromatographic silica gel mixture. Allow the spots to dry, and develop the chromatogram in a solvent system consisting of a mixture of *n*-butyl alcohol, glacial acetic acid, and water (4:1:1) in a chamber until the solvent front has moved about three-fourths of the length of the plate. Remove the plate from the developing chamber, and allow to air-dry. Locate the spots under short-wavelength ultraviolet light: the principal spot obtained from the test solution corresponds in R_f value to that obtained from the Standard solution.

Pyrogen—It meets the requirements of the *Pyrogen Test* ⟨151⟩, the test dose being 2.8 mg of dopamine hydrochloride per kg.

pH ⟨791⟩: between 2.5 and 5.0.

Particulate matter ⟨788⟩: meets the requirements under *Small-volume Injections.*

Other requirements—It meets the requirements under *Injections* ⟨1⟩.

Assay—

Mobile phase—Mix 260 mL of acetonitrile and 1740 mL of a 1 in 1000 solution of sodium 1-octanesulfonate in dilute glacial acetic acid (1 in 100). Filter through a membrane filter (0.45 µm or finer porosity), and degas.

Standard preparation—Dissolve a suitable quantity of USP Dopamine Hydrochloride RS, accurately weighed, in *Mobile phase* to obtain a solution having a concentration of about 1.6 mg per mL. Pipet 10 mL of this solution into a 100-mL volumetric flask, dilute with *Mobile phase* to volume, and mix to obtain *Standard preparation* having a known concentration of about 0.16 mg of dopamine hydrochloride per mL.

Resolution solution—Prepare a solution of benzoic acid in methanol containing about 20 mg per mL. Dilute 1 volume of this solution with 3 volumes of the *Mobile phase* to obtain a solution having a final concentration of about 5 mg per mL. Transfer 10.0 mL of this solution and 10.0 mL of a Standard solution containing 1.6 mg of USP Dopamine Hydrochloride RS per mL to a 100-mL volumetric flask, dilute with *Mobile phase* to volume, and mix.

Assay preparation—Transfer an accurately measured volume of Dopamine Hydrochloride Injection, equivalent to about 16 mg of dopamine hydrochloride, to a 100-mL volumetric flask, dilute with *Mobile phase* to volume, and mix.

Chromatographic system (see *Chromatography* ⟨621⟩)—The liquid chromatograph is equipped with a 280-nm detector and a 30-cm × 4-mm stainless steel column packed with packing L1. The flow rate is about 1.5 mL per minute. Chromatograph the *Standard preparation*, and record the peak responses as directed under *Procedure:* the relative standard deviation for replicate injections is not more than 3.0%. In a similar manner chromatograph the *Resolution solution:* the resolution factor between the benzoic acid and dopamine hydrochloride peaks is not less than 4.0.

Procedure—Separately inject equal volumes (about 40 µL) of the *Standard preparation* and the *Assay preparation* into the chromatograph by means of a suitable microsyringe or sampling valve, record the chromatograms, and measure the responses for the major peaks. Calculate the quantity, in mg, of $C_8H_{11}NO_2 \cdot HCl$ in each mL of the Injection taken by the formula:

$$(100C/V)(r_U/r_S),$$

in which *C* is the concentration, in mg per mL, of USP Dopamine Hydrochloride RS in the *Standard preparation*, *V* is the volume, in mL, of Injection taken, and r_U and r_S are the responses of dopamine hydrochloride obtained from the *Assay preparation* and the *Standard preparation*, respectively.

Dopamine Hydrochloride and Dextrose Injection

» Dopamine Hydrochloride and Dextrose Injection is a sterile solution of Dopamine Hydrochloride and Dextrose in Water for Injection. It contains not less than 95.0 percent and not more than 105.0 percent of the labeled amount of dopamine hydrochloride ($C_8H_{11}NO_2 \cdot HCl$) and of dextrose ($C_6H_{12}O_6 \cdot H_2O$).

NOTE—Do not use the Injection if it is darker than slightly yellow or discolored in any other way.

Packaging and storage—Preserve in single-dose containers, preferably of Type I or Type II glass.

Labeling—The label states the total osmolar concentration in mOsmol per liter. Where the contents are less than 100 mL, or where the label states that the Injection is not for direct injection but is to be diluted before use, the label alternatively may state the total osmolar concentration in mOsm per mL.

Reference standards—*USP Dopamine Hydrochloride Reference Standard*—Dry at 105° for 2 hours before using. *USP Dextrose Reference Standard*—Dry at 105° for 16 hours before using.

Identification—

A: It responds to the *Identification test* under *Dextrose.*

B: The retention time of the major peak in the chromatogram of the *Assay preparation* corresponds to that of the *Standard preparation* obtained as directed in the *Assay for dopamine hydrochloride.*

Pyrogen—It meets the requirements of the *Pyrogen Test* ⟨151⟩.

pH ⟨791⟩: between 2.5 and 4.5.

Particulate matter ⟨788⟩: meets the requirements under *Small-volume Injections.*

Other requirements—It meets the requirements under *Injections* ⟨1⟩.

Assay for dopamine hydrochloride—Proceed with Dopamine Hydrochloride and Dextrose Injection as directed in the *Assay* under *Dopamine Hydrochloride Injection.*

Assay for dextrose—Transfer an accurately measured volume of Dopamine Hydrochloride and Dextrose Injection, containing 2 to 5 g of dextrose, to a 100-mL volumetric flask. Add 0.2 mL of 6 *N* ammonium hydroxide, dilute with water to volume, and mix. Determine the angular rotation in a suitable polarimeter tube at 25° (see *Optical Rotation* ⟨781⟩). The observed rotation, in degrees, multiplied by 1.0425*A*, in which *A* is the ratio 200 divided by the length, in mm, of the polarimeter tube employed, represents the weight, in g, of $C_6H_{12}O_6 . H_2O$ in the volume of Injection taken.

Doxapram Hydrochloride

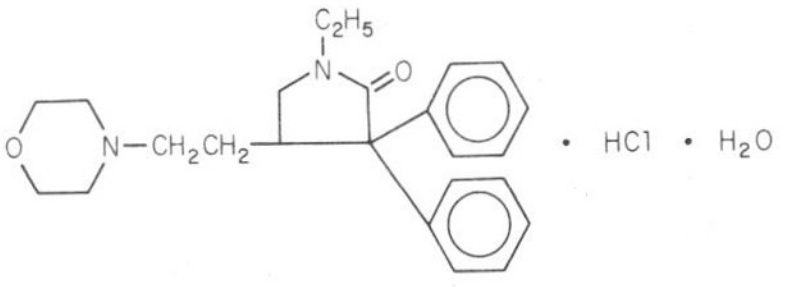

$C_{24}H_{30}N_2O_2 . HCl . H_2O$ 432.99

2-Pyrrolidinone, 1-ethyl-4-[2-(4-morpholinyl)ethyl]-3,3-diphenyl-, monohydrochloride, monohydrate.

1-Ethyl-4-(2-morpholinoethyl)-3,3-diphenyl-2-pyrrolidinone monohydrochloride monohydrate [*7081-53-0*].

Anhydrous 414.97 [*113-07-5*].

» Doxapram Hydrochloride, dried at 105° for 2 hours, contains not less than 98.0 percent and not more than 100.5 percent of $C_{24}H_{30}N_2O_2 . HCl$.

Packaging and storage—Preserve in tight containers.

Reference standard—*USP Doxapram Hydrochloride Reference Standard*—Dry at 105° for 2 hours before using.

Identification—

A: The infrared absorption spectrum of a potassium bromide dispersion of it, previously dried, exhibits maxima only at the same wavelengths as that of a similar preparation of USP Doxapram Hydrochloride RS.

B: The ultraviolet absorption spectrum of a solution (1 in 2500) exhibits maxima and minima at the same wavelengths as that of a similar solution of USP Doxapram Hydrochloride RS, concomitantly measured, and the respective absorptivities, calculated on the dried basis, at the wavelength of maximum absorbance at about 258 nm do not differ by more than 3.0%.

pH ⟨791⟩: between 3.5 and 5.0, in a solution (1 in 100).

Loss on drying ⟨731⟩—Dry it at 105° for 2 hours: it loses between 3.0% and 4.5% of its weight.

Residue on ignition ⟨281⟩: not more than 0.3%.

Arsenic, *Method II* ⟨211⟩: 5 ppm.

Heavy metals, *Method II* ⟨231⟩: 0.002%.

Chromatographic purity—

Dragendorff reagent—Dissolve 17 g of bismuth subnitrate and 200 g of tartaric acid in 800 mL of water (*Solution A*). Dissolve 160 g of potassium iodide in 400 mL of water (*Solution B*). Mix *Solution A* and *Solution B*. To 25 mL of this stock solution add 50 g of tartaric acid and 250 mL of water, and mix.

Test preparation—Dissolve 57 mg of Doxapram Hydrochloride in 0.5 mL of 0.1 *N* sodium hydroxide, add 1.0 mL of chloroform, and shake.

Standard preparation A—Dissolve 57 mg of USP Doxapram Hydrochloride RS in 0.5 mL of 0.1 *N* sodium hydroxide, add 1.0 mL of chloroform, and shake.

Standard preparation B—Dissolve 11.4 mg of USP Doxapram Hydrochloride RS in 0.5 mL of 0.1 *N* sodium hydroxide, add 100 mL of chloroform, and shake.

Procedure—On a suitable thin-layer chromatographic plate (see *Chromatography* ⟨621⟩), coated with a 0.25-mm layer of chromatographic silica gel mixture, apply 10-µL portions of the chloroform solutions obtained from the *Test preparation* and the *Standard preparations*. Allow the spots to dry, and develop the chromatogram in a chromatographic chamber lined with paper and equilibrated with a solvent system consisting of a mixture of isopropyl alcohol and 1 *N* ammonium hydroxide (4:1) until the solvent front has moved about three-fourths of the length of the plate. Remove the plate from the developing chamber, mark the solvent front, and allow the solvent to evaporate. Spray the plate with *Dragendorff reagent* in order to visualize the spots: the R_f value of the principal spot obtained from the *Test preparation* corresponds to that obtained from *Standard preparation A*, and no spot, other than the principal spot, in the chromatogram of the *Test preparation* is larger or more intense than the principal spot obtained from *Standard preparation B* (0.2%).

Assay—Dissolve about 800 mg of Doxapram Hydrochloride, previously dried and accurately weighed, in 50 mL of glacial acetic acid, add 1 drop of crystal violet TS and 10 mL of mercuric acetate TS, and titrate with 0.1 *N* perchloric acid VS to a blue-green end-point. Perform a blank determination, and make any necessary correction. Each mL of 0.1 *N* perchloric acid is equivalent to 41.50 mg of $C_{24}H_{30}N_2O_2 . HCl$.

Doxapram Hydrochloride Injection

» Doxapram Hydrochloride Injection is a sterile solution of Doxapram Hydrochloride in Water for Injection. It contains not less than 90.0 percent and not more than 110.0 percent of the labeled amount of $C_{24}H_{30}N_2O_2 . HCl . H_2O$.

Packaging and storage—Preserve in single-dose or in multiple-dose containers, preferably of Type I glass.

Reference standard—*USP Doxapram Hydrochloride Reference Standard*—Dry at 105° for 2 hours before using.

Identification—

A: The chromatogram obtained from the *Assay preparation* in the *Assay* exhibits a major peak for doxapram the retention time of which corresponds to that of the doxapram peak in the chromatogram of the *Standard preparation*.

B: Evaporate to dryness about 5 mL of the chloroform solution remaining from the preparation of the *Assay preparation* in the *Assay*. Dissolve the residue in 0.01 *N* sulfuric acid, dilute with the same solvent to 100 mL, and mix: the ultraviolet absorption spectrum of the solution so obtained exhibits maxima and minima at the same wavelengths as a similarly prepared solution from the chloroform solution remaining from the preparation of the *Standard preparation* in the *Assay*.

pH ⟨791⟩: between 3.5 and 5.0.

Other requirements—It meets the requirements under *Injections* ⟨1⟩.

Assay—

Internal standard solution—Prepare a solution of cholesterol in chloroform having a concentration of about 1.25 mg per mL.

Standard preparation—Transfer about 100 mg of USP Doxapram Hydrochloride RS, accurately weighed, to a separator containing 5 mL of water. Add 1 mL of a saturated solution of sodium chloride to the separator, insert the stopper, and mix. Add 5 mL of 2.5 *N* sodium hydroxide, and extract with three 25-mL portions of chloroform. Pass each extract through a pledget of glass wool, combine the filtrates in a 100-mL volumetric flask, dilute with chloroform to volume, and mix. Mix 6.0 mL of this solution and 2.0 mL of *Internal standard solution*.

Assay preparation—Transfer an acccurately measured volume of Doxapram Hydrochloride Injection, equivalent to about 100 mg of doxapram hydrochloride hydrate, to a separator and proceed as directed for *Standard preparation*, beginning with "Add 1 mL of a saturated solution of sodium chloride."

Procedure—Inject equal volumes (about 2 µL) of the *Standard preparation* and the *Assay preparation* successively into a gas chromatograph equipped with a flame-ionization detector and a 0.9-m × 2-mm glass column containing 3 percent liquid phase G19 on 80- to 100-mesh support S1A. Maintain the injection port and detector temperature at 260° and the column at 240°. Use dry helium as the carrier gas at a flow rate of about 30 mL per minute. [NOTE—Cure and condition the column as necessary (see *Chromatography* ⟨621⟩).] The relative retention times are about 0.4 for cholesterol and 1.0 for doxapram. In a suitable chromatogram, the relative standard deviation of the peak response ratios of replicate injections of the *Standard preparation* is not more than 2.0%, and the resolution, R, between cholesterol and doxapram is not less than 4.0. Calculate the quantity, in mg, of $C_{24}H_{30}N_2O_2 \cdot HCl \cdot H_2O$ in each mL of the Injection taken by the formula:

$$(432.99/414.97)(W/V)(R_U/R_S),$$

in which 432.99 and 414.97 are the molecular weights of the hydrated and anhydrous forms of doxapram hydrochloride, respectively, W is the weight, in mg, of USP Doxapram Hydrochloride RS taken, V is the volume, in mL, of Injection taken, and R_U and R_S are the ratios of the peak responses of doxapram and cholesterol obtained from the *Assay preparation* and the *Standard preparation*, respectively.

Doxepin Hydrochloride

$CHCH_2CH_2N(CH_3)_2$ · HCl

$C_{19}H_{21}NO \cdot HCl$ 315.84

1-Propanamine, 3-dibenz[*b,e*]oxepin-11(6*H*)ylidene-*N,N*-dimethyl-, hydrochloride.

N,N-Dimethyldibenz[*b,e*]oxepin-$\Delta^{11(6H),\gamma}$-propylamine hydrochloride [*1229-29-4; 4698-39-9* ((*E*)-isomer); *25127-31-5* ((*Z*)-isomer)].

» Doxepin Hydrochloride, an (*E*) and (*Z*) geometric isomer mixture, contains the equivalent of not less than 98.0 percent and not more than 102.0 percent of doxepin ($C_{19}H_{21}NO \cdot HCl$), calculated on the dried basis. It contains not less than 13.6 percent and not more than 18.1 percent of the (*Z*)-isomer, and not less than 81.4 percent and not more than 88.2 percent of the (*E*)-isomer.

Packaging and storage—Preserve in well-closed containers.

Reference standard—*USP Doxepin Hydrochloride Reference Standard*—Dry in vacuum at 60° for 3 hours before using.

Identification—

A: The infrared absorption spectrum of a potassium bromide dispersion of it exhibits maxima only at the same wavelengths as that of a similar preparation of USP Doxepin Hydrochloride RS.

B: The ultraviolet absorption spectrum of a 1 in 20,000 solution in alcohol exhibits maxima and minima at the same wavelengths as that of a similar preparation of USP Doxepin Hydrochloride RS, concomitantly measured, and the respective absorptivities, at the wavelength of maximum absorbance at about 296 nm, do not differ by more than 3.0%.

Melting range, *Class I* ⟨741⟩: between 185° and 191°.

Loss on drying—Dry it in vacuum at 60° for 3 hours: it loses not more than 0.5% of its weight.

Residue on ignition ⟨281⟩: not more than 0.2%.

Heavy metals, *Method II* ⟨231⟩: 0.002%.

Chloride content—Dissolve about 100 mg, accurately weighed, in a mixture of 100 mL of water and 100 mL of alcohol. Titrate with 0.05 *N* silver nitrate VS, determining the end-point potentiometrically using a silver–silver sulfide sensing electrode and a double-junction reference electrode containing potassium nitrate filling solution in the outer jacket and a standard filling solution in the inner jacket. Each mL of 0.05 *N* silver nitrate is equivalent to 1.773 mg of chloride. Not less than 10.9% and not more than 11.6% of chloride is found.

Assay—

Internal standard solution—Transfer about 250 mg of chlorpheniramine maleate to a 250-mL separator containing 50 mL of water. Add 10 mL of sodium hydroxide solution (1 in 10), and swirl. Extract with four 10-mL portions of chloroform, filtering each chloroform extract through cotton into a 50-mL volumetric flask. Dilute with chloroform to volume, and mix.

Standard preparation—Transfer about 55 mg of USP Doxepin Hydrochloride RS, accurately weighed, to a 50-mL volumetric flask. Dissolve in water, and dilute with water to volume. Pipet 25 mL of this solution into a 125-mL separator, add 5 mL of 1 *N* sodium hydroxide, and shake. Extract with three 25-mL portions of chloroform, filtering each chloroform extract through cotton into a 100-mL volumetric flask containing 5.0 mL of *Internal standard solution.* Dilute with chloroform to volume, and mix.

Assay preparation—Using 55 mg of Doxepin Hydrochloride, accurately weighed, prepare as directed for *Standard preparation.*

Procedure—Inject separately 2-µL portions of the *Standard preparation* and the *Assay preparation* into a gas chromatograph equipped with a flame-ionization detector and that contains a 3.1-m × 2-mm column packed with 3 percent liquid phase G7 on support S1A. The column is maintained at a temperature of about 200°, the injection port at 250°, and the detector at 250°. The carrier gas is helium, flowing at a rate of 30 mL per minute. The retention times of chlorpheniramine, the (*Z*)-isomer, and the (*E*)-isomer are about 2.3, about 4.8, and about 5.4 minutes, respectively. Three replicate injections of the *Standard preparation* show a resolution factor, R (see *Chromatography* ⟨621⟩), for the (*E*)- and (*Z*)-isomers of not less than 1.6. Determine the peak areas of the internal standard and the (*Z*)- and (*E*)-isomers in the *Standard preparation* and the *Assay preparation.* Designate the ratios of the peak areas of the (*Z*)-isomer to the internal standard obtained from the *Assay preparation* and the *Standard preparation* as $R_{U(Z)}$ and $R_{S(Z)}$, respectively. Similarly designate the ratios of the peak areas of the (*E*)-isomer to the internal standard obtained from the *Assay preparation* and the *Standard preparation* as $R_{U(E)}$ and $R_{S(E)}$, respectively. Calculate the quantity, in mg, of $C_{19}H_{21}NO \cdot HCl$ in the portion of Doxepin Hydrochloride taken by the formula:

$$200C[(R_{U(Z)} + R_{U(E)})/(R_{S(Z)} + R_{S(E)})],$$

in which C is the concentration, in mg per mL, of USP Doxepin Hydrochloride RS in the *Standard preparation.* Calculate the percentage of the (*Z*)-isomer in the *Assay preparation* by the formula:

$$(R_{U(Z)}/R_{S(Z)})(W_S/W_T)(P_Z),$$

in which W_S is the weight, in mg, of USP Doxepin Hydrochloride RS, W_T is the weight, in mg, of Doxepin Hydrochloride taken, and P_Z is the labeled percentage of (*Z*)-isomer in the USP Doxepin Hydrochloride RS. Similarly calculate the percentage of the (*E*)-isomer in the *Assay preparation* by the formula:

$$(R_{U(E)}/R_{S(E)})(W_S/W_T)(P_E),$$

in which P_E is the labeled percentage of (*E*)-isomer in the USP Doxepin Hydrochloride RS.

Doxepin Hydrochloride Capsules

» Doxepin Hydrochloride Capsules contain not less than 90.0 percent and not more than 110.0 percent of the labeled amount of doxepin ($C_{19}H_{21}NO$).

Packaging and storage—Preserve in well-closed containers.

Reference standard—*USP Doxepin Hydrochloride Reference Standard*—Dry in vacuum at 60° for 3 hours before using.

Identification—Transfer the contents of 2 Capsules to a suitable flask. Add about 30 mL of chloroform, and shake by mechanical means for 30 minutes. Dilute with chloroform to 50 mL, and mix. Quickly filter a portion of the solution through Whatman

No. 4 filter paper, rejecting the first 15 mL of the filtrate. Pipet an aliquot of the filtered solution containing about 10 mg of doxepin into a separator. Add, if necessary, additional chloroform to bring the volume to 25 mL. Extract with 50 mL of 1 *N* sodium hydroxide, and drain the chloroform into another separator. Discard the sodium hydroxide washing. Wash the chloroform solution with 20 mL of 1 *N* sodium hydroxide. Filter the chloroform layer through cotton into a 100-mL volumetric flask, extract the aqueous layer with 25 mL of chloroform, and add the chloroform layer to the volumetric flask. Dilute with chloroform to volume, and mix. Inject a portion of the chloroform solution, equivalent to 0.5 µg of doxepin, diluting the chloroform solution, if necessary, into a gas chromatograph equipped with a flame-ionization detector and a 3.1-m × 2-mm column packed with 3 percent cyanomethylphenyl silicone liquid phase on 80- to 110-mesh acid-washed, silanized diatomaceous earth. The carrier gas is helium flowing at a rate of 30 mL per minute. The column temperature is 220°, and the temperature of the injection port and the detector is 250°. The retention times obtained for the (*Z*)- and (*E*)-isomer peaks in the chromatogram of the solution from the Capsules are the same as those obtained for a 20-mg specimen of USP Doxepin Hydrochloride RS treated and chromatographed in a similar manner.

Dissolution ⟨711⟩—

Medium: water; 900 mL.

Apparatus 1: 50 rpm.

Time: 30 minutes.

Procedure—Determine the amount of $C_{19}H_{21}NO$ dissolved from ultraviolet absorbances at the wavelength of maximum absorbance at about 292 nm of filtered portions of the solution under test, suitably diluted with *Dissolution Medium*, if necessary, in comparison with a Standard solution having a known concentration of USP Doxepin Hydrochloride RS in the same medium.

Tolerances—Not less than 80% (*Q*) of the labeled amount of $C_{19}H_{21}NO$ is dissolved in 30 minutes.

Uniformity of dosage units ⟨905⟩: meet the requirements.

Water, *Method I* ⟨921⟩: not more than 9.0%, determined on the contents of 1 Capsule.

Assay—

Standard preparation—Transfer 11 mg, accurately weighed, of USP Doxepin Hydrochloride RS to a 10-mL volumetric flask. Dissolve in water, and dilute with water to volume. Transfer this solution into a 250-mL separator, add 2 mL of 1 *N* sodium hydroxide, and mix. Extract with three 30-mL portions of chloroform, combining the extracts in another 250-mL separator. Extract the combined chloroform extracts with three 50-mL portions of dilute sulfuric acid (3 in 1000). Filter the acid extracts through cotton, previously wetted with the same dilute sulfuric acid, into a 200-mL volumetric flask. Dilute with the same dilute sulfuric acid to volume, and mix.

Assay preparation—Transfer the contents of 20 Capsules to a 500-mL volumetric flask, add about 300 mL of chloroform, and shake by mechanical means for 30 minutes. Dilute with chloroform to volume, and mix. Quickly filter a portion of the solution through Whatman No. 4 filter paper, rejecting the first 25 mL of the filtrate. Pipet an aliquot of the filtered solution containing about 10 mg of doxepin into a separator. Add, if necessary, additional chloroform to bring the volume to 25 mL. Extract with 50 mL of 1 *N* sodium hydroxide, and drain the chloroform into another separator. Wash the sodium hydroxide layer with 50 mL of chloroform, adding the washing to the separator containing the chloroform layer. Discard the sodium hydroxide solution. Extract the combined chloroform washings with three 50-mL portions of dilute sulfuric acid (3 in 1000). Filter the extracts through cotton, previously wetted with the same dilute sulfuric acid, into a 200-mL volumetric flask, dilute with the same dilute sulfuric acid to volume, and mix.

Procedure—Concomitantly determine the absorbances of the *Standard preparation* and the *Assay preparation* in 1-cm cells at the wavelength of maximum absorbance at about 292 nm, with a suitable spectrophotometer, using dilute sulfuric acid (3 in 1000) as the blank. Calculate the quantity, in mg, of $C_{19}H_{21}NO$ in the aliquot taken by the formula:

$$0.885(200C)(A_U/A_S),$$

in which *C* is the concentration, in mg per mL, of USP Doxepin Hydrochloride RS in the *Standard preparation*, 0.885 is the ratio of the molecular weight of doxepin to that of doxepin hydrochloride, and A_U and A_S are the absorbances of the *Assay preparation* and the *Standard preparation*, respectively.

Doxepin Hydrochloride Oral Solution

» Doxepin Hydrochloride Oral Solution contains not less than 90.0 percent and not more than 110.0 percent of the labeled amount of doxepin ($C_{19}H_{21}NO$).

Packaging and storage—Preserve in tight, light-resistant containers.

Labeling—Label it to indicate that it is to be diluted with water or other suitable fluid to approximately 120 mL, just prior to administration.

Reference standard—*USP Doxepin Hydrochloride Reference Standard*—Dry in vacuum at 60° for 3 hours before using.

Identification—

Mobile solvent—Add 0.2 mL of diethylamine to a solution containing 250 mL of chloroform and 750 mL of acetonitrile in a vacuum flask. Prior to use, degas the contents of the flask by stirring vigorously with a magnetic stirrer, while applying vacuum, for 10 minutes.

Procedure—Transfer 5.0 mL of the Oral Solution to a 60-mL separator, add 1 mL of sodium hydroxide solution (1 in 25), 1 g of sodium chloride, and 5.0 mL of ethyl acetate, and shake the mixture vigorously for 1 minute. Allow the phases to separate, transfer 1.0 mL of the clear upper phase to a 25-mL volumetric flask, dilute with *Mobile solvent* to volume, and mix. Transfer about 22 mg of USP Doxepin Hydrochloride RS to a 50-mL volumetric flask, dilute with the same *Mobile solvent* to volume, and mix. Inject 4-µL portions of both solutions into a high-pressure liquid chromatograph (see *Chromatography* ⟨621⟩), fitted with a 50-cm × 2-mm column packed with silica microspheres and equipped with an ultraviolet detector capable of monitoring absorption at 254 nm and a suitable recorder. Adjust the operating parameters to obtain a flow rate of about 24 mL per hour. The chromatogram of the test solution exhibits two peaks having retention times that are identical with those obtained with the Standard solution.

pH ⟨791⟩: between 4.0 and 7.0, the test specimen being allowed to remain in contact with the electrodes for 15 minutes prior to the measurement.

Assay—Transfer an accurately measured volume of Doxepin Hydrochloride Oral Solution, equivalent to 100 mg of doxepin, to a 100-mL volumetric flask, dilute with dilute hydrochloric acid (1 in 120) to volume, and mix. Dilute 4.0 mL of this solution with the same solvent to 50.0 mL. Transfer 15.0 mL of the resulting solution to a 125-mL separator, and extract with two 20-mL portions of ether. Dilute 10.0 mL of the extracted aqueous phase with dilute hydrochloric acid (1 in 120) to 25.0 mL. Prepare a Standard solution from a suitable quantity of USP Doxepin Hydrochloride RS, by quantitative and stepwise dilution with dilute hydrochloric acid (1 in 120) to obtain a solution having a known concentration of about 1.1 mg per mL. Take a 4-mL aliquot of the Standard solution through the above-described procedure, beginning with "Dilute 4.0 mL of this solution." Concomitantly determine the absorbances of both solutions in 1-cm cells at the wavelength of maximum absorbance at about 292 nm, with a suitable spectrophotometer, using dilute hydrochloric acid (1 in 120) as the blank. Calculate the quantity, in mg, of doxepin ($C_{19}H_{21}NO$) in each mL of Doxepin Hydrochloride Oral Solution taken by the formula:

$$0.885(0.1C/V)(A_U/A_S),$$

in which 0.885 is the ratio of the molecular weight of doxepin to that of doxepin hydrochloride, *C* is the concentration, in µg per mL, of USP Doxepin Hydrochloride RS in the Standard solution, *V* is the volume, in mL, of Oral Solution taken, and A_U and A_S are the absorbances of the solution from the Oral Solution and the Standard solution, respectively.

Doxorubicin Hydrochloride

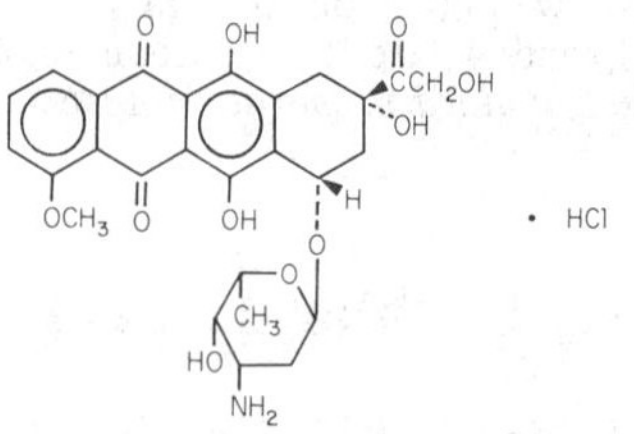

$C_{27}H_{29}NO_{11}.HCl$ 579.99

5,12-Naphthacenedione, 10-[(3-amino-2,3,6-trideoxy-α-L-*lyxo*-hexopyranosyl)oxy]-7,8,9,10-tetrahydro-6,8,11-trihydroxy-8-(hydroxylacetyl)-1-methoxy-, hydrochloride (8*S*-*cis*)-.

(8*S*,10*S*)-10-[(3-Amino-2,3,6-trideoxy-α-L-*lyxo*-hexopyranosyl)-oxy]-8-glycoloyl-7,8,9,10-tetrahydro-6,8,11-trihydroxy-1-methoxy-5,12-naphthacenedione hydrochloride [*25316-40-9*].

» Doxorubicin Hydrochloride contains not less than 97.0 percent and not more than 102.0 percent of $C_{27}H_{29}NO_{11}.HCl$, calculated on the anhydrous, solvent-free basis.

Caution—Great care should be taken to prevent inhaling particles of doxorubicin hydrochloride and exposing the skin to it.

Packaging and storage—Preserve in tight containers.

Reference standard—*USP Doxorubicin Hydrochloride Reference Standard*—Do not dry before using.

Identification—When chromatographed as directed in the *Assay*, the *Assay preparation* exhibits a major peak for doxorubicin, the retention time of which corresponds to that exhibited by the *Standard preparation*, and the chromatogram compares qualitatively to that obtained from the *Standard preparation.*

Crystallinity ⟨695⟩: meets the requirements.

Depressor substances—It meets the requirements of the *Depressor Substances Test* ⟨101⟩, the test dose being 1.0 mL per kg of a solution in sterile saline TS containing 1.5 mg of doxorubicin hydrochloride per mL.

pH ⟨791⟩: between 4.0 and 5.5, in a solution containing 5 mg per mL.

Water, *Method I* ⟨921⟩: not more than 4.0%.

Chromatographic purity—Using the chromatogram of the *Assay preparation* obtained as directed in the *Assay*, calculate the percentage of impurities by the formula:

$$100S/(S + r),$$

in which *S* is the sum of the responses of the minor component peaks, and *r* is the response of the major doxorubicin peak: the total of any impurities detected is not more than 3.0%.

Solvent residue (as acetone and alcohol)—

Standard preparation—Transfer to a 100-mL volumetric flask about 200 mg of acetone, 300 mg of dehydrated alcohol, and 1000 mg of dioxane, each accurately weighed, and mix. Dilute with water to volume, and mix. Transfer 5.0 mL of the resulting solution to a 50-mL volumetric flask, dilute with water to volume, and mix. This solution contains about 0.2 mg of acetone, 0.3 mg of C_2H_5OH, and 1 mg of dioxane per mL.

Solvent—Transfer about 100 mg of dioxane, accurately weighed, to a 100-mL volumetric flask, dilute with water to volume, and mix.

Test preparation—Dissolve about 200 mg of Doxorubicin Hydrochloride in 3.0 mL (3.0 g) of *Solvent.*

Chromatographic system (see *Chromatography* ⟨621⟩)—The gas chromatograph is equipped with a flame-ionization detector and a 4-mm × 2-m column packed with 8 percent liquid phase G16 on 100- to 120-mesh support S1AB (potassium hydroxide-washed). The column is maintained at about 60°, and helium is used as the carrier gas. Adjust the column temperature and carrier gas flow rate so that dioxane elutes in about 6 minutes. Chromatograph the *Standard preparation*, and record the peak responses as directed under *Procedure:* the resolution, *R*, between adjacent peaks is not less than 2.0, the relative standard deviations of the ratios of the peak responses of the acetone and dioxane peaks and of the alcohol and dioxane peaks for replicate injections is not more than 4.0%, and the tailing factor for the alcohol peak is not more than 1.5.

Procedure—[NOTE—Use peak areas where peak responses are indicated.] Separately inject equal volumes (about 1 µL) of the *Standard preparation* and the *Test preparation* into the chromatograph, record the chromatograms, and measure the responses for the major peaks. The relative retention times are about 0.2 for acetone, 0.5 for alcohol, and 1.0 for dioxane. Calculate the percentage, by weight, of acetone (CH_3COCH_3) and alcohol (C_2H_5OH), respectively, in the portion of Doxorubicin Hydrochloride taken by the same formula:

$$100(C_a/C_d)(D_U/W_U)(R_U/R_S),$$

in which C_a is the concentration, in mg per mL, of acetone or alcohol in the *Standard preparation*, C_d is the concentration, in mg per mL, of dioxane in the *Standard preparation*, D_U is the total quantity, in mg, of dioxane in the *Test preparation*, W_U is the quantity, in mg, of Doxorubicin Hydrochloride taken to prepare the *Test preparation*, and R_U and R_S are the response ratios of the analyte peak (acetone or alcohol) to the dioxane peak obtained from the *Test preparation* and the *Standard preparation*, respectively: the total of acetone and alcohol is not greater than 2.5%. Use the results obtained to calculate the result obtained as directed in the *Assay* on the solvent-free basis.

Assay—

Mobile phase—Mix equal volumes of acetonitrile and 0.01 *M* sodium lauryl sulfate in 0.02 *M* phosphoric acid. The acetonitrile concentration may be varied to meet system suitability requirements and to provide a suitable elution time for doxorubicin. Filter the solution through a membrane filter (1-µm or finer porosity), and degas.

Resolution solution—Prepare a solution of *n*-hexyl *p*-hydroxybenzoate in *Mobile phase* containing about 0.2 mg per mL.

Standard preparation—Dissolve an accurately weighed quantity of USP Doxorubicin Hydrochloride RS in *Mobile phase* to obtain a solution having a known concentration of about 0.2 mg per mL.

Assay preparation—Transfer about 20 mg of Doxorubicin Hydrochloride, accurately weighed, to a 100-mL volumetric flask, add *Mobile phase* to volume, and mix.

Chromatographic system (see *Chromatography* ⟨621⟩)—The chromatograph is equipped with a 254-nm detector and a 4.6-mm × 25-cm column that contains packing L1. The flow rate is about 1.5 mL per minute. Chromatograph the *Standard preparation*, and record the peak responses as directed under *Procedure:* the relative standard deviation of replicate injections is not more than 2.0%. Chromatograph the *Resolution solution*, and record the peak responses as directed under *Procedure:* the resolution, *R*, between doxorubicin and *n*-hexyl *p*-hydroxybenzoate is not less than 8.0.

Procedure—Separately inject equal volumes (about 20 µL) of the *Standard preparation* and the *Assay preparation* into the chromatograph, record the chromatograms, and measure the responses for the major peaks: the retention time for doxorubicin is about 0.3 relative to *n*-hexyl *p*-hydroxybenzoate. Calculate the quantity, in mg, of $C_{27}H_{29}NO_{11}.HCl$ in the Doxorubicin Hydrochloride taken by the formula:

$$100C(r_U/r_S),$$

in which *C* is the concentration, in mg, of doxorubicin hydrochloride in each mL of the *Standard preparation*, and r_U and r_S are the responses of the doxorubicin peak obtained from the *Assay preparation* and the *Standard preparation*, respectively.

Doxorubicin Hydrochloride for Injection

» Doxorubicin Hydrochloride for Injection is a sterile mixture of Doxorubicin Hydrochloride and Lactose. It contains not less than 90.0 percent and not more than 115.0 percent of the labeled amount of $C_{27}H_{29}NO_{11}.HCl$.

Caution—Great care should be taken to prevent inhaling particles of Doxorubicin Hydrochloride and exposing the skin to it.

Packaging and storage—Preserve in *Containers for Sterile Solids* as described under *Injections* ⟨1⟩.

Reference standard—*USP Doxorubicin Hydrochloride Reference Standard*—Do not dry before using.

Constituted solution—At the time of use, the constituted solution prepared from Doxorubicin Hydrochloride for Injection meets the requirements for *Constituted Solutions* under *Injections* ⟨1⟩.

Pyrogen—It meets the requirements of the *Pyrogen Test* ⟨151⟩, the test dose being 1.0 mL per kg of a solution in pyrogen-free saline TS containing 2.25 mg of doxorubicin hydrochloride per mL.

Sterility—It meets the requirements under *Sterility Tests* ⟨71⟩, when tested as directed in the section, *Test Procedures Using Membrane Filtration*, each container being constituted aseptically by injecting Sterile Water for Injection through the stopper, and the entire contents of all the containers being collected aseptically with the aid of 200 mL of *Fluid A* before filtering.

pH ⟨791⟩: between 4.5 and 6.5, in the solution constituted as directed in the labeling, except that water is used as the diluent.

Water, *Method I* ⟨921⟩: not more than 4.0%, the *Test Preparation* being prepared as directed for a hygroscopic specimen.

Other requirements—It responds to the *Identification test* under *Doxorubicin Hydrochloride*, and meets the requirements for *Uniformity of Dosage Units* ⟨905⟩ and *Labeling* under *Injections* ⟨1⟩.

Assay—

Mobile phase, Resolution solution, Standard preparation, and *Chromatographic system*—Proceed as directed in the *Assay* under *Doxorubicin Hydrochloride.*

Assay preparation—Dilute the contents of 1 container quantitatively with *Mobile phase* to obtain a solution containing about 0.2 mg of doxorubicin hydrochloride per mL.

Procedure—Proceed as directed for *Procedure* in the *Assay* under *Doxorubicin Hydrochloride.* Calculate the quantity, in mg, of $C_{27}H_{29}NO_{11}.HCl$ in the container of Doxorubicin Hydrochloride for Injection taken by the formula:

$$C(L/D)(r_U/r_S),$$

in which C is the concentration, in mg per mL, of doxorubicin hydrochloride in the *Standard preparation*, L is the labeled quantity of doxorubicin hydrochloride in the container, D is the concentration, in mg per mL, of doxorubicin hydrochloride in the *Assay preparation* on the basis of the labeled quantity in the container and the extent of dilution, and r_U and r_S are the responses of the doxorubicin peak obtained from the *Assay preparation* and the *Standard preparation*, respectively.

Doxycycline

$C_{22}H_{24}N_2O_8.H_2O$ 462.46

2-Naphthacenecarboxamide, 4-(dimethylamino)-1,4,4a,5,5a,6,-11,12a-octahydro-3,5,10,12,12a-pentahydroxy-6-methyl-1,11-dioxo-, [4*S*-(4α,4aα,5α,5aα,6α,12aα)]-, monohydrate.

4-(Dimethylamino)-1,4,4a,5,5a,6,11,12a-octahydro-3,5,10,-12,12a-pentahydroxy-6-methyl-1,11-dioxo-2-naphthacene-carboxamide monohydrate [*17086-28-1*].

Anhydrous 444.44 [*564-25-0*].

» Doxycycline has a potency equivalent to not less than 880 µg and not more than 980 µg of $C_{22}H_{24}N_2O_8$ per mg.

Packaging and storage—Preserve in tight, light-resistant containers.

Reference standard—*USP Doxycycline Hyclate Reference Standard*—Do not dry before using.

Identification—Dissolve a suitable quantity in methanol to obtain a *Test Solution* containing 1 mg of doxycycline per mL, and proceed as directed under *Identification—Tetracyclines* ⟨193⟩.

Crystallinity ⟨695⟩: meets the requirements.

pH ⟨791⟩: between 5.0 and 6.5, in an aqueous suspension containing 10 mg per mL.

Water, *Method I* ⟨921⟩: between 3.6% and 4.6%.

Assay—

Mobile phase and *Chromatographic system*—Prepare as directed in the *Assay* under *Doxycycline Hyclate.*

NOTE—Throughout the following sections, protect the *Standard preparation* and the *Assay preparation* from light.

Standard preparation—Transfer about 30 mg of USP Doxycycline Hyclate RS, accurately weighed, to a 50-mL volumetric flask, add 25 mL of methanol, dilute with water to volume, and mix. Filter through a filter of 0.5-µm or finer porosity.

Assay preparation—Transfer about 55 mg of Doxycycline, accurately weighed, to a 100-mL volumetric flask, add 50 mL of methanol and 1.2 mL of 0.1 *N* hydrochloric acid, swirl to dissolve, dilute with water to volume, and mix. Filter through a filter of 0.5-µm or finer porosity.

Procedure—Separately inject equal volumes (about 20 µL) of the *Standard preparation* and the *Assay preparation* into the chromatograph, record the chromatograms, and measure the responses for the major peaks. Calculate the quantity, in µg, of doxycycline ($C_{22}H_{24}N_2O_8$) in each mg of the Doxycycline taken by the formula:

$$2(PW_S/W_U)(r_U/r_S),$$

in which P is the potency, in µg of doxycycline per mg, of the USP Doxycycline Hyclate RS, W_S is the weight, in mg, of USP Doxycycline Hyclate RS taken to prepare the *Standard preparation*, W_U is the weight, in mg, of Doxycycline taken to prepare the *Assay preparation*, and r_U and r_S are the peak responses obtained from the *Assay preparation* and the *Standard preparation*, respectively.

Doxycycline for Oral Suspension

» Doxycycline for Oral Suspension contains the equivalent of not less than 90.0 percent and not more than 125.0 percent of the labeled amount of $C_{22}H_{24}N_2O_8$ when constituted as directed. It contains one or more suitable buffers, colors, diluents, flavors, and preservatives.

Packaging and storage—Preserve in tight, light-resistant containers.

Reference standard—*USP Doxycycline Hyclate Reference Standard*—Do not dry before using.

Identification—To a quantity of Doxycycline for Oral Suspension (powder), equivalent to about 50 mg of doxycycline, add 50 mL of methanol, shake, and allow to settle. Using the clear supernatant liquid as the *Test Solution*, proceed as directed under *Identification—Tetracyclines* ⟨193⟩.

Uniformity of dosage units ⟨905⟩—

FOR SOLID PACKAGED IN SINGLE-UNIT CONTAINERS: meets the requirements.

pH ⟨791⟩: between 5.0 and 6.5, in the suspension constituted as directed in the labeling.

Water, *Method I* ⟨921⟩: not more than 3.0%.

Assay—Constitute Doxycycline for Oral Suspension as directed in the labeling. Dilute an accurately measured quantity of the suspension, freshly mixed and free from air bubbles, quantitatively with 0.1 *N* hydrochloric acid to obtain a stock solution containing not less than 150 μg of doxycycline per mL. Proceed as directed for doxycycline under *Antibiotics—Microbial Assays* ⟨81⟩, using an accurately measured volume of stock solution diluted quantitatively and stepwise with water to yield a *Test Dilution* having a concentration assumed to be equal to the median dose level of the Standard.

Doxycycline Calcium Oral Suspension

» Doxycycline Calcium Oral Suspension is prepared from Doxycycline Hyclate, and contains one or more suitable buffers, colors, diluents, flavors, and preservatives. It contains the equivalent of not less than 90.0 percent and not more than 125.0 percent of the labeled amount of doxycycline ($C_{22}H_{24}N_2O_8$).

Packaging and storage—Preserve in tight, light-resistant containers.

Reference standard—*USP Doxycycline Hyclate Reference Standard*—Do not dry before using.

Identification—To an accurately measured volume of Oral Suspension, equivalent to about 50 mg of doxycycline, add 50 mL of methanol, shake, and allow to settle. Using the clear supernatant liquid as the *Test Solution*, proceed as directed under *Identification—Tetracyclines* ⟨193⟩.

Uniformity of dosage units ⟨905⟩—

FOR SUSPENSION PACKAGED IN SINGLE-UNIT CONTAINERS: meets the requirements.

pH ⟨791⟩: between 6.5 and 8.0.

Assay—Transfer an accurately measured quantity of Doxycycline Calcium Oral Suspension, freshly mixed and free from air bubbles, equivalent to about 150 mg of doxycycline to a 1000-mL volumetric flask, dilute with 0.1 *N* hydrochloric acid to volume, and mix. Proceed as directed for doxycycline under *Antibiotics—Microbial Assays* ⟨81⟩, using an accurately measured volume of this solution diluted quantitatively and stepwise with water to yield a *Test Dilution* having a concentration assumed to be equal to the median dose level of the Standard.

Doxycycline Hyclate

$(C_{22}H_{24}N_2O_8 . HCl)_2 . C_2H_6O . H_2O$ 1025.89

2-Naphthacenecarboxamide, 4-(dimethylamino)-1,4,4a,5,5a,-6,11,12a-octahydro-3,5,10,12,12a-pentahydroxy-6-methyl-1,11-dioxo-, monohydrochloride, compd. with ethanol (2:1), monohydrate, [4*S*-(4α,4aα,5α,5aα,6α,12aα)]-.

4-(Dimethylamino)-1,4,4a,5,5a,6,11,12a-octahydro-3,5,10,-12,12a-pentahydroxy-6-methyl-1,11-dioxo-2-naphthacene-carboxamide monohydrochloride, compound with ethyl alcohol (2:1), monohydrate [*24390-14-5*].

» Doxycycline Hyclate has a potency equivalent to not less than 800 μg and not more than 920 μg of doxycycline ($C_{22}H_{24}N_2O_8$) per mg.

Packaging and storage—Preserve in tight containers, protected from light.

Reference standard—*USP Doxycycline Hyclate Reference Standard*—Do not dry before using.

Identification—The infrared absorption spectrum of a potassium bromide dispersion of it exhibits maxima only at the same wavelengths as that of a similar preparation of USP Doxycycline Hyclate RS.

Crystallinity ⟨695⟩: meets the requirements.

pH ⟨791⟩: between 2.0 and 3.0, in a solution containing 10 mg of doxycycline per mL.

Water, *Method I* ⟨921⟩: between 1.4% and 2.75%.

Assay—

Mobile phase—Mix 450 mL of 0.1 *M* monobasic sodium phosphate and 550 mL of methanol. Add 3 mL of *N,N*-dimethyl-*n*-octylamine, adjust with phosphoric acid to a pH of 6.25 ± 0.05, filter through a membrane filter of 0.5-μm or finer porosity, and degas. Make adjustments if necessary (see *System Suitability* under *Chromatography* ⟨621⟩).

[NOTE—Throughout the following sections, protect the *Standard preparation* and the *Assay preparation* from light.]

Standard preparation—Dissolve an accurately weighed quantity of USP Doxycycline Hyclate RS in water to obtain a solution having a known concentration of about 1000 μg of doxycycline ($C_{22}H_{24}N_2O_8$) per mL. Filter through a membrane filter of 0.5-μm or finer porosity.

Assay preparation—Transfer about 120 mg of Doxycycline Hyclate, accurately weighed, to a 100-mL volumetric flask, dissolve in water, dilute with water to volume, and mix. Filter through a membrane filter of 0.5-μm or finer porosity.

Chromatographic system (see *Chromatography* ⟨621⟩)—The liquid chromatograph is equipped with a 280-nm detector, a 4.6-mm × 3-cm guard column that contains packing L7, and a 3.9-mm × 30-cm analytical column that contains 10-μm packing L1 of low acidity. The flow rate is about 1.5 mL per minute. Chromatograph the *Standard preparation*, and record the peak responses as directed under *Procedure:* the column efficiency determined from the analyte peak is not less than 1000 theoretical plates, the tailing factor for the analyte peak is not more than 2.0, and the relative standard deviation for replicate injections is not more than 2.0%.

Procedure—Separately inject equal volumes (about 10 μL) of the *Standard preparation* and the *Assay preparation* into the chromatograph, record the chromatograms, and measure the responses for the major peaks. Calculate the potency, in μg of doxycycline ($C_{22}H_{24}N_2O_8$) per mg, of the Doxycycline Hyclate taken by the formula:

$$100(C/W)(r_U/r_S),$$

in which C is the concentration, in μg per mL, of doxycycline ($C_{22}H_{24}N_2O_8$) in the *Standard preparation*, W is the quantity, in mg, of Doxycycline Hyclate taken to prepare the *Assay preparation*, and r_U and r_S are the peak responses obtained from the *Assay preparation* and the *Standard preparation*, respectively.

Doxycycline Hyclate Capsules

» Doxycycline Hyclate Capsules contain the equivalent of not less than 90.0 percent and not more than 120.0 percent of the labeled amount of doxycycline ($C_{22}H_{24}N_2O_8$).

Packaging and storage—Preserve in tight, light-resistant containers.

Reference standard—*USP Doxycycline Hyclate Reference Standard*—Do not dry before using.

Identification—Shake a suitable quantity of Capsule contents with methanol to obtain a solution containing the equivalent of 1 mg of doxycycline per mL, and filter. Using the filtrate as the *Test Solution*, proceed as directed under *Identification—Tetracyclines* ⟨193⟩.

Dissolution ⟨711⟩—

Medium: water; 900 mL.

Apparatus 2: 75 rpm, the distance between the blade and the inside bottom of the flask being maintained at 4.5 ± 0.5 cm during the test.

Time: 30 minutes.

Procedure—Determine the amount of $C_{22}H_{24}N_2O_8$ dissolved from ultraviolet absorbances at the wavelength of maximum absorbance at about 276 nm of filtered portions of the solution under test, suitably diluted with *Dissolution Medium*, if necessary, in comparison with a Standard solution having a known concentration of USP Doxycycline Hyclate RS in the same medium.

Tolerances—Not less than 80% (*Q*) of the labeled amount of $C_{22}H_{24}N_2O_8$ is dissolved in 30 minutes.

Uniformity of dosage units ⟨905⟩: meet the requirements.

Water, *Method I* ⟨921⟩: not more than 5.0%.

Assay—

Mobile phase, Standard preparation, and *Chromatographic system*—Proceed as directed in the *Assay* under *Doxycycline Hyclate.*

Assay preparation—Remove, as completely as possible, the contents of not less than 20 Doxycycline Hyclate Capsules, and weigh accurately. Mix the combined contents, and transfer an accurately weighed portion of the powder, equivalent to about 50 mg of doxycycline, to a 50-mL volumetric flask, add about 35 mL of water, shake for 15 minutes, dilute with water to volume, and mix. Filter through filter paper, discarding the first 10 mL of the filtrate, then filter through a membrane filter of 0.5-µm or finer porosity.

Procedure—Proceed as directed for *Procedure* in the *Assay* under *Doxycycline Hyclate.* Calculate the quantity, in mg, of doxycycline ($C_{22}H_{24}N_2O_8$) in the portion of Capsules taken by the formula:

$$0.05C(r_U / r_S),$$

in which *C* is the concentration, in µg per mL, of doxycycline ($C_{22}H_{24}N_2O_8$) in the *Standard preparation,* and r_U and r_S are the peak responses obtained from the *Assay preparation* and the *Standard preparation,* respectively.

Doxycycline Hyclate Delayed-release Capsules

» Doxycycline Hyclate Delayed-release Capsules contain the equivalent of not less than 90.0 percent and not more than 120.0 percent of the labeled amount of doxycycline ($C_{22}H_{24}N_2O_8$).

Packaging and storage—Preserve in tight, light-resistant containers.

Labeling—The label indicates that the contents of the Delayed-release Capsules are enteric-coated.

Reference standard—*USP Doxycycline Hyclate Reference Standard*—Do not dry before using.

Identification—Shake a suitable quantity of finely powdered Capsule contents with methanol to obtain a solution containing the equivalent of 1 mg of doxycycline per mL, and filter. Using the filtrate as the *Test Solution,* proceed as directed under *Identification—Tetracyclines* ⟨193⟩.

Drug release, *Method B* ⟨724⟩—

Acid stage—[NOTE—Conduct the test by transferring the contents of each Capsule to the individual basket units of the apparatus.]

Medium: 0.06 *N* hydrochloric acid; 900 mL.

Apparatus 1: 50 rpm.

Time: 20 minutes.

Diluting solvent: 0.1 *N* hydrochloric acid.

Procedure—Determine the amount of $C_{22}H_{24}N_2O_8$ dissolved from ultraviolet absorbances at the wavelength of maximum absorbance at about 345 nm of filtered portions of the solution under test, suitably diluted with *Diluting solvent,* in comparison with a Standard solution having a known concentration of about 0.01 mg of USP Doxycycline Hyclate RS per mL in *Diluting solvent.*

Tolerances—Level 1 (6 Capsules tested)—No individual value exceeds 50% dissolved. *Level 2* (6 Capsules tested)—Not more than 2 individual values of 12 tested are greater than 50% dissolved.

Buffer stage—[NOTE—Conduct this stage of testing on separate specimens, selecting Capsules that were not previously subjected to *Acid-stage* testing and transferring the contents of each Capsule to the individual basket units of the apparatus.]

Medium: pH 5.5 Neutralized Phthalate Buffer (see *Buffer solutions* in the section, *Reagents, Indicators, and Solutions*); 1000 mL.

Apparatus 1: 50 rpm.

Time: 30 minutes.

Diluting solvent: 0.1 *N* hydrochloric acid.

Procedure—Determine the amount of $C_{22}H_{24}N_2O_8$ dissolved from ultraviolet absorbances at the wavelength of maximum absorbance at about 345 nm of filtered portions of the solution under test, suitably diluted with *Diluting solvent,* in comparison with a Standard solution having a known concentration of about 0.01 mg of USP Doxycycline Hyclate RS per mL in *Diluting solvent.*

Tolerances—Not less than 85% (*Q*) of the labeled amount of $C_{22}H_{24}N_2O_8$ is dissolved in 30 minutes.

Uniformity of dosage units ⟨905⟩: meet the requirements.

Water, *Method I* ⟨921⟩: not more than 5.0%.

Assay—

Mobile phase, Standard preparation, and *Chromatographic system*—Proceed as directed in the *Assay* under *Doxycycline Hyclate.*

Assay preparation—Remove, as completely as possible, the contents of not less than 20 Doxycycline Hyclate Delayed-release Capsules, and weigh accurately. Mix the combined contents, and transfer an accurately weighed portion of the powder, equivalent to about 50 mg of doxycycline, to a 50-mL volumetric flask, add about 35 mL of water, shake for 15 minutes, dilute with water to volume, and mix. Filter through filter paper, discarding the first 10 mL of the filtrate, then filter through a membrane filter of 0.5-µm or finer porosity.

Procedure—Proceed as directed for *Procedure* in the *Assay* under *Doxycycline Hyclate.* Calculate the quantity, in mg, of doxycycline ($C_{22}H_{24}N_2O_8$) in the portion of Capsules taken by the formula:

$$0.05C(r_U / r_S),$$

in which *C* is the concentration, in µg per mL, of doxycycline ($C_{22}H_{24}N_2O_8$) in the *Standard preparation,* and r_U and r_S are the peak responses obtained from the *Assay preparation* and the *Standard preparation,* respectively.

Doxycycline Hyclate for Injection

» Doxycycline Hyclate for Injection is a sterile, dry mixture of Doxycycline Hyclate and a suitable buffer or a sterile-filtered and lyophilized mixture of Doxycycline Hyclate and a suitable buffer. It contains the equivalent of not less than 90.0 percent and not more than 120.0 percent of the labeled amount of doxycycline ($C_{22}H_{24}N_2O_8$).

Packaging and storage—Preserve in *Containers for Sterile Solids* as described under *Injections* ⟨1⟩, protected from light.

Reference standard—*USP Doxycycline Hyclate Reference Standard*—Do not dry before using.

Constituted solution—At the time of use, the constituted solution prepared from Doxycycline Hyclate for Injection meets the requirements for *Constituted Solutions* under *Injections* ⟨1⟩.

Identification—Dissolve a suitable quantity in methanol to obtain a solution containing the equivalent of 1 mg of doxycycline per mL, and filter. Using the filtrate as the *Test Solution,* proceed as directed under *Identification—Tetracyclines* ⟨193⟩.

Depressor substances—It meets the requirements of the *Depressor Substances Test* ⟨101⟩, the test dose being 1.0 mL per kg of a solution in sterile saline TS containing 5.0 mg of doxycycline per mL.

Pyrogen—It meets the requirements of the *Pyrogen Test* ⟨151⟩, the test dose being 1.0 mL per kg of a solution prepared to contain 7.5 mg of doxycycline per mL in Sodium Chloride Injection.

Sterility—It meets the requirements under *Sterility Tests* ⟨71⟩, when tested as directed in the section, *Test Procedures Using Membrane Filtration, Fluid D* being used instead of *Fluid A.*

pH ⟨791⟩: between 1.8 and 3.3, in the solution constituted as directed in the labeling.

Loss on drying ⟨731⟩—Dry about 100 mg, accurately weighed, in a capillary-stoppered bottle in vacuum at a pressure not exceeding 5 mm of mercury at 60° for 3 hours: it loses not more than 2.0% of its weight.

Particulate matter ⟨788⟩: meets the requirements under *Small-volume Injections.*

Assay—

Mobile phase, Standard preparation, and *Chromatographic system*—Proceed as directed in the *Assay* under *Doxycycline Hyclate.*

Assay preparation 1 (where it is represented as being in a single-dose container)—Constitute Doxycycline Hyclate for Injection as directed in the labeling. Withdraw all of the withdrawable contents, using a suitable hypodermic needle and syringe, and dilute quantitatively with water to obtain a solution containing the equivalent of about 1000 µg of doxycycline per mL.

Assay preparation 2 (where the label states the quantity of doxycycline equivalent in a given volume of constituted solution)—Constitute Doxycycline Hyclate for Injection as directed in the labeling. Dilute an accurately measured volume of the constituted solution quantitatively with water to obtain a solution containing the equivalent of about 1000 µg of doxycycline per mL.

Procedure—Proceed as directed for *Procedure* in the *Assay* under *Doxycycline Hyclate.* Calculate the quantity, in mg, of doxycycline ($C_{22}H_{24}N_2O_8$) withdrawn from the container, or in the portion of constituted solution taken, by the formula:

$$(L/D)(C)(r_U/r_S),$$

in which *L* is the labeled quantity, in mg, of doxycycline in the container, or in the volume of constituted solution taken, *D* is the concentration, in µg of doxycycline per mL, of *Assay preparation 1* or *Assay preparation 2,* based on the labeled quantity in the container or in the portion of constituted solution taken, respectively, and the extent of dilution, and the other terms are as defined therein.

Sterile Doxycycline Hyclate

» Sterile Doxycycline Hyclate is Doxycycline Hyclate suitable for parenteral use. It has a potency equivalent to not less than 800 µg and not more than 920 µg of doxycycline ($C_{22}H_{24}N_2O_8$) per mg.

Packaging and storage—Preserve in *Containers for Sterile Solids* as described under *Injections* ⟨1⟩, protected from light.

Reference standard—*USP Doxycycline Hyclate Reference Standard*—Do not dry before using.

Depressor substances—It meets the requirements of the *Depressor Substances Test* ⟨101⟩, the test dose being 1.0 mL per kg of a solution in sterile saline TS containing 5.0 mg of doxycycline per mL.

Pyrogen—It meets the requirements of the *Pyrogen Test* ⟨151⟩, the test dose being 1.0 mL per kg of a solution prepared to contain 7.5 mg of doxycycline per mL in Sodium Chloride Injection.

Sterility—It meets the requirements under *Sterility Tests* ⟨71⟩, when tested as directed in the section, *Test Procedures Using Membrane Filtration, Fluid D* being used instead of *Fluid A.*

Other requirements—It responds to the *Identification test* and meets the requirements for *pH, Water, Crystallinity,* and *Assay* under *Doxycycline Hyclate.*

Doxycycline Hyclate Tablets

» Doxycycline Hyclate Tablets contain the equivalent of not less than 90.0 percent and not more than 120.0 percent of the labeled amount of doxycycline ($C_{22}H_{24}N_2O_8$).

Packaging and storage—Preserve in tight, light-resistant containers.

Reference standard—*USP Doxycycline Hyclate Reference Standard*—Do not dry before using.

Identification—Shake a suitable quantity of finely ground Tablets with methanol to obtain a solution containing the equivalent of 1 mg of doxycycline per mL and filter. Using the filtrate as the *Test Solution,* proceed as directed under *Identification—Tetracyclines* ⟨193⟩.

Dissolution ⟨711⟩—

Medium: water; 900 mL.

Apparatus 2: 75 rpm, the distance between the blade and the inside bottom of the flask being maintained at 4.5 ± 0.5 cm during the test.

Time: 90 minutes.

Procedure—Determine the amount of $C_{22}H_{24}N_2O_8$ dissolved from ultraviolet absorbances at the wavelength of maximum absorbance at about 276 nm of filtered portions of the solution under test, suitably diluted with *Dissolution Medium,* if necessary, in comparison with a Standard solution having a known concentration of USP Doxycycline Hyclate RS in the same medium.

Tolerances—Not less than 85% (*Q*) of the labeled amount of $C_{22}H_{24}N_2O_8$ is dissolved in 90 minutes.

Uniformity of dosage units ⟨905⟩: meet the requirements.

Water, *Method I* ⟨921⟩: not more than 5.0%, 20 mL of a mixture of carbon tetrachloride, chloroform, and methanol (2:2:1) being used in place of methanol in the titration vessel.

Assay—

Mobile phase, Standard preparation, and *Chromatographic system*—Proceed as directed in the *Assay* under *Doxycycline Hyclate.*

Assay preparation—Weigh and finely powder not less than 20 Doxycycline Hyclate Tablets. Transfer an accurately weighed portion of the powder, equivalent to about 50 mg of doxycycline, to a 50-mL volumetric flask, add about 35 mL of water, shake for 15 minutes, dilute with water to volume, and mix. Filter through filter paper, discarding the first 10 mL of the filtrate, then filter through a membrane filter of 0.5-µm or finer porosity.

Procedure—Proceed as directed for *Procedure* in the *Assay* under *Doxycycline Hyclate.* Calculate the quantity, in mg, of doxycycline ($C_{22}H_{24}N_2O_8$) in the portion of Tablets taken by the formula:

$$0.05C(r_U/r_S),$$

in which *C* is the concentration, in µg per mL, of doxycycline ($C_{22}H_{24}N_2O_8$) in the *Standard preparation,* and r_U and r_S are the peak responses obtained from the *Assay preparation* and the *Standard preparation,* respectively.

Doxylamine Succinate

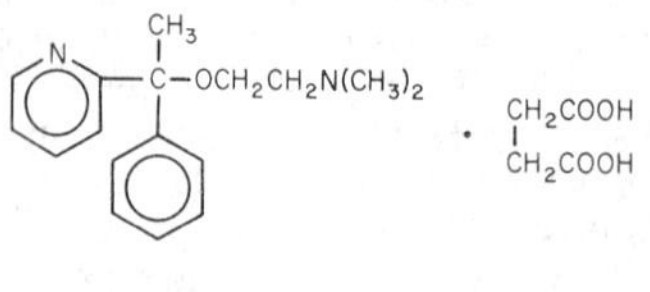

$C_{17}H_{22}N_2O.C_4H_6O_4$ 388.46

Ethanamine, *N,N*-dimethyl-2-[1-phenyl-1-(2-pyridinyl)ethoxy]-, butanedioate (1:1).

2-[α-[2-(Dimethylamino)ethoxy]-α-methylbenzyl]pyridine succinate (1:1) [*562-10-7*].

» Doxylamine Succinate contains not less than 98.0 percent and not more than 101.0 percent of $C_{17}H_{22}N_2O.C_4H_6O_4$, calculated on the dried basis.

Packaging and storage—Preserve in well-closed, light-resistant containers.

Reference standard—*USP Doxylamine Succinate Reference Standard*—Dry in vacuum over phosphorus pentoxide for 5 hours before using.

Identification—

A: The ultraviolet absorption spectrum of a 1 in 50,000 solution in 0.1 *N* hydrochloric acid exhibits maxima and minima at the same wavelengths as that of a similar solution of USP Doxylamine Succinate RS, concomitantly measured, and the respective absorptivities, calculated on the dried basis, at the wavelength of maximum absorbance at about 262 nm do not differ by more than 3.0%.

B: It meets the requirements under *Identification—Organic Nitrogenous Bases* ⟨181⟩.

C: Dissolve about 500 mg in 5 mL of water, and add a slight excess of 6 *N* ammonium hydroxide. Extract the liberated doxylamine with several portions of ether, discard the ether extracts, and evaporate the aqueous solution on a steam bath to dryness. Add 2 mL of 3 *N* hydrochloric acid, and again evaporate on the steam bath to dryness. Cool, add about 10 mL of ether, allow to stand for a few minutes, and decant the clear supernatant liquid. Evaporate the ether solution to dryness, and dry the residue at 105° for 30 minutes: the succinic acid so obtained melts between 184° and 188°, the procedure for *Class I* being used (see *Melting Range or Temperature* ⟨741⟩).

Melting range, *Class I* ⟨741⟩: between 103° and 108°, but the range between beginning and end of melting does not exceed 3°.

Loss on drying ⟨731⟩—Dry it in vacuum over phosphorus pentoxide for 5 hours: it loses not more than 0.5% of its weight.

Residue on ignition ⟨281⟩: not more than 0.1%.

Volatile related compounds—Dissolve 650 mg in 20 mL of 0.1 *N* hydrochloric acid in a separator. Render the solution alkaline with 2.5 *N* sodium hydroxide, and immediately extract with four 25-mL portions of ether, filtering each extract through an ether-saturated pledget of cotton. Evaporate the combined ether extracts on a water bath with the aid of a current of air to dryness at a temperature not exceeding 50°, and dissolve the residue in 5 mL of alcohol. Inject about 1 µL of this solution into a suitable gas chromatograph (see *Chromatography* ⟨621⟩), equipped with a flame-ionization detector. Under typical conditions, the instrument contains a 2-m × 4-mm glass column containing 5 percent packing G16 and 5 percent packing G12 on 60- to 80-mesh S1A. The column is maintained at about 212°, the injection port and detector block are maintained at about 250°, and dry helium is used as the carrier gas. The total relative area of all extraneous peaks (except that of the solvent peak) does not exceed 2.0%, and the relative area of any individual extraneous peak does not exceed 1.0%.

Assay—Dissolve about 500 mg of Doxylamine Succinate, accurately weighed, in 80 mL of glacial acetic acid. Add crystal violet TS, and titrate with 0.1 *N* perchloric acid VS to an emerald-green end-point. Perform a blank determination, and make any necessary correction. Each mL of 0.1 *N* perchloric acid is equivalent to 19.42 mg of $C_{17}H_{22}N_2O.C_4H_6O_4$.

Doxylamine Succinate Syrup

» Doxylamine Succinate Syrup contains not less than 92.0 percent and not more than 108.0 percent of the labeled amount of $C_{17}H_{22}N_2O.C_4H_6O_4$.

Packaging and storage—Preserve in tight, light-resistant containers.

Reference standard—*USP Doxylamine Succinate Reference Standard*—Dry in vacuum over phosphorus pentoxide for 5 hours before using.

Identification—Use a volume of Syrup equivalent to about 50 mg of doxylamine succinate, and proceed as directed under *Identification—Organic Nitrogenous Bases* ⟨181⟩, beginning with "Transfer the liquid to a separator." The Syrup meets the requirements of the test.

Assay—Proceed with Doxylamine Succinate Syrup as directed under *Salts of Organic Nitrogenous Bases* ⟨501⟩, determining the absorbance at the wavelength of maximum absorbance at about 262 nm. Calculate the quantity, in mg, of $C_{17}H_{22}N_2O.C_4H_6O_4$ in each mL of the Syrup taken by the formula:

$$(0.05C/V)(A_U/A_S),$$

in which *C* is the concentration, in µg per mL, of USP Doxylamine Succinate RS in the *Standard Preparation*, and *V* is the volume, in mL, of Syrup taken.

Doxylamine Succinate Tablets

» Doxylamine Succinate Tablets contain not less than 92.0 percent and not more than 108.0 percent of the labeled amount of $C_{17}H_{22}N_2O.C_4H_6O_4$.

Packaging and storage—Preserve in well-closed, light-resistant containers.

Reference standard—*USP Doxylamine Succinate Reference Standard*—Dry in vacuum over phosphorus pentoxide for 5 hours before using.

Identification—Tablets meet the requirements under *Identification—Organic Nitrogenous Bases* ⟨181⟩.

Dissolution ⟨711⟩—

Medium: 0.1 *N* hydrochloric acid; 900 mL.

Apparatus 2: 50 rpm.

Time: 30 minutes.

Procedure—Determine the amount of $C_{17}H_{22}N_2O.C_4H_6O_4$ dissolved from ultraviolet absorbances at the wavelength of maximum absorbance at about 262 nm of filtered portions of the solution under test, suitably diluted with 0.1 *N* hydrochloric acid, in comparison with a Standard solution having a known concentration of USP Doxylamine Succinate RS in the same medium.

Tolerances—Not less than 80% (*Q*) of the labeled amount of $C_{17}H_{22}N_2O.C_4H_6O_4$ is dissolved in 30 minutes.

Uniformity of dosage units ⟨905⟩: meet the requirements.

Procedure for content uniformity—Transfer 1 finely powdered Tablet to a 100-mL volumetric flask containing 65 mL of 0.1 *N* hydrochloric acid. Shake frequently during a 10-minute period, dilute with 0.1 *N* hydrochloric acid to volume, and mix. Allow the insoluble material to settle, and filter, discarding the first 20 mL of the filtrate. Dilute a portion of the subsequent filtrate quantitatively and stepwise, if necessary, with 0.1 *N* hydrochloric acid to provide a solution containing approximately 25 µg of doxylamine succinate per mL. Concomitantly determine the absorbances of this solution and of a Standard solution of USP Doxylamine Succinate RS in the same medium having a known concentration of about 25 µg per mL in 1-cm cells at the wavelength of maximum absorbance at about 262 nm, with a suitable spectrophotometer, using 0.1 *N* hydrochloric acid as the blank. Calculate the quantity, in mg, of $C_{17}H_{22}N_2O.C_4H_6O_4$ in the Tablet by the formula:

$$(TC/D)(A_U/A_S),$$

in which *T* is the labeled quantity, in mg, of doxylamine succinate in the Tablet, *C* is the concentration, in µg per mL, of USP Doxylamine Succinate RS in the Standard solution, *D* is the concentration, in µg per mL, of doxylamine succinate in the solution from the Tablet, based on the labeled quantity per Tablet and the extent of dilution, and A_U and A_S are the absorbances of the solution from the Tablet and the Standard solution, respectively.

Assay—Proceed with Doxylamine Succinate Tablets as directed under *Salts of Organic Nitrogenous Bases* ⟨501⟩, determining the absorbance at the wavelength of maximum absorbance at about 262 nm. Calculate the quantity, in mg, of $C_{17}H_{22}N_2O \cdot C_4H_6O_4$ in the portion of Tablets taken by the formula:

$$0.05C(A_U/A_S),$$

in which C is the concentration, in µg per mL, of USP Doxylamine Succinate RS in the *Standard preparation.*

Dried Aluminum Hydroxide Gel—*see* Aluminum Hydroxide Gel, Dried

Dried Aluminum Hydroxide Gel Capsules—*see* Aluminum Hydroxide Gel Capsules, Dried

Dried Basic Aluminum Carbonate Gel Capsules—*see* Aluminum Carbonate Gel Capsules, Dried Basic

Dried Basic Aluminum Carbonate Gel Tablets—*see* Aluminum Carbonate Gel Tablets, Dried Basic

Dried Ferrous Sulfate—*see* Ferrous Sulfate, Dried

Dried Sodium Phosphate—*see* Sodium Phosphate, Dried

Droperidol

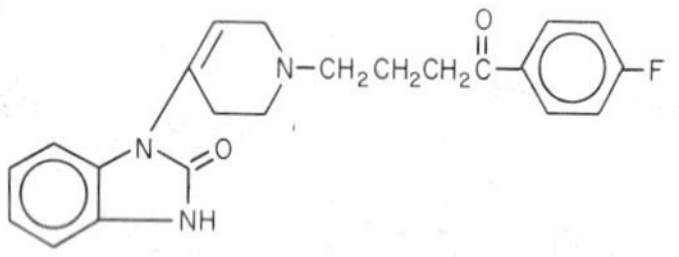

$C_{22}H_{22}FN_3O_2$ 379.43

2*H*-Benzimidazol-2-one, 1-[1-[4-(4-fluorophenyl)-4-oxobutyl]-1,2,3,6-tetrahydro-4-pyridinyl]-1,3-dihydro-.

1-[1-[3-(*p*-Fluorobenzoyl)propyl]-1,2,3,6-tetrahydro-4-pyridyl]-2-benzimidazolinone [*548-73-2*].

» Droperidol, dried in vacuum at 70° for 4 hours, contains not less than 98.0 percent and not more than 102.0 percent of $C_{22}H_{22}FN_3O_2$.

Packaging and storage—Preserve in tight, light-resistant containers, under nitrogen, in a cool place.

Reference standards—*USP Droperidol Reference Standard*—Dry in vacuum at 70° for 4 hours before using. *USP 4,4′-Bis[1,2,3,6-tetrahydro-4-(2-oxo-1-benzimidazolinyl)-1-pyridyl]butyrophenone Reference Standard*—Keep container tightly closed and protected from light. Dry in vacuum at 70° for 4 hours before using.

Identification—

A: The infrared absorption spectrum of a potassium bromide dispersion of it, previously dried, exhibits maxima only at the same wavelengths as that of a similar preparation of USP Droperidol RS.

B: In a 125-mL separator, combine 15.0 mL of a 1 in 6700 solution of Droperidol in chloroform, 10 mL of pH 6.0 phosphate buffer (see *Buffer Solutions* in the section, *Reagents, Indicators, and Solutions*), 15 mL of water, and 10 mL of freshly prepared sodium bisulfite solution (1 in 100). Shake for 15 minutes, and discard the aqueous layer. Transfer 5.0 mL of the chloroform solution to a glass-stoppered conical flask containing 50.0 mL of citric acid solution (1.9 in 100), insert the stopper, and shake for 30 minutes: the ultraviolet absorption spectrum of the citric acid layer exhibits maxima and minima at the same wavelengths as that of a similar solution of USP Droperidol RS, and the respective absorptivities, calculated on the dried basis, at the wavelength of maximum absorbance at about 246 nm do not differ by more than 3.0%.

Melting range, *Class I* ⟨741⟩: between 147° and 150°.

Loss on drying ⟨731⟩—Dry it in vacuum at 70° for 4 hours: it loses not more than 5.0% of its weight.

Residue on ignition ⟨281⟩: not more than 0.2%.

Heavy metals, *Method II* ⟨231⟩: 0.002%.

4,4′-Bis[1,2,3,6-tetrahydro-4-(2-oxo-1-benzimidazolinyl)-1-pyridyl]butyrophenone—Dissolve 30.0 mg in 70 mL of isopropyl alcohol in a 100-mL volumetric flask. Add 10.0 mL of 0.1 *N* hydrochloric acid, dilute with isopropyl alcohol to volume, and mix. Concomitantly determine the absorbances of this solution and a Standard solution of USP 4,4′-Bis[1,2,3,6-tetrahydro-4-(2-oxo-1-benzimidazolinyl)-1-pyridyl]butyrophenone RS in the same medium at a concentration of 4.5 µg per mL, in 1-cm cells at the wavelength of maximum absorbance at about 330 nm, with a suitable spectrophotometer, using a 1 in 10 solution of 0.1 *N* hydrochloric acid in isopropyl alcohol as the blank: the absorbance of the test solution does not exceed that of the Standard solution, corresponding to not more than 1.5%.

Assay—Dissolve about 240 mg of Droperidol, previously dried and accurately weighed, in 50 mL of glacial acetic acid, add *p*-naphtholbenzein TS, and titrate with 0.1 *N* perchloric acid VS. Perform a blank determination, and make any necessary correction. Each mL of 0.1 *N* perchloric acid is equivalent to 37.94 mg of $C_{22}H_{22}FN_3O_2$.

Droperidol Injection

» Droperidol Injection is a sterile solution of Droperidol in Water for Injection, prepared with the aid of Lactic Acid. It contains not less than 90.0 percent and not more than 110.0 percent of the labeled amount of $C_{22}H_{22}FN_3O_2$, as the lactate.

Packaging and storage—Preserve in single-dose or in multiple-dose containers, preferably of Type I glass, protected from light.

Reference standard—*USP Droperidol Reference Standard*—Dry in vacuum at 70° for 4 hours before using.

Identification—The ultraviolet absorption spectrum of the *Assay preparation*, prepared as directed in the *Assay*, exhibits maxima and minima at the same wavelengths as that of the *Standard preparation*, prepared as directed in the *Assay*, concomitantly measured.

Bacterial endotoxins—When tested as directed under *Bacterial Endotoxins test* ⟨85⟩, it contains not more than 35.7 USP Endotoxin Units per mg of droperidol.

pH ⟨791⟩: between 3.0 and 3.8.

Other requirements—It meets the requirements under *Injections* ⟨1⟩.

Assay—

Standard preparation—Dissolve a suitable quantity of USP Droperidol RS, accurately weighed, in chloroform, and dilute quantitatively with chloroform to obtain a solution having a known concentration of about 1.25 mg per mL. Transfer 10.0 mL of this solution to a 100-mL volumetric flask, dilute with chloroform to volume, and mix. Transfer 15.0 mL of the resulting solution to a suitable glass-stoppered flask containing 10 mL of pH 6.0 phosphate buffer (see *Buffer Solutions* in the section, *Reagents, Indicators, and Solutions*), 15 mL of water, and 10 mL of freshly prepared sodium bisulfite solution (1 in 100), and proceed as directed under *Assay preparation*, beginning with "Shake for 15 minutes." The concentration of USP Droperidol RS in the *Standard preparation* is about 12.5 µg per mL.

Assay preparation—Transfer to a 100-mL volumetric flask an accurately measured volume of Droperidol Injection, equivalent to about 12.5 mg of droperidol, dilute with water to volume, and mix. Transfer 15.0 mL of this solution to a suitable glass-stoppered flask containing 10 mL of pH 6.0 phosphate buffer, 15.0 mL of chloroform, and 10 mL of freshly prepared sodium bisulfite solution (1 in 100). Shake for 15 minutes, then centrifuge, and discard the aqueous layer. Transfer 5.0 mL of the chloroform layer to a suitable glass-stoppered flask containing 50.0 mL of citric acid solution (1.9 in 100), insert the stopper, and shake for 30 minutes. Centrifuge, and discard the chloroform layer.

Procedure—Concomitantly determine the absorbances of the *Assay preparation* and the *Standard preparation* in 1-cm cells at the wavelength of maximum absorbance at about 246 nm, with a suitable spectrophotometer, using as the blank citric acid solution (1.9 in 100) previously saturated with chloroform by shaking with one-tenth its volume of chloroform. Calculate the quantity, in mg, of $C_{22}H_{22}FN_3O_2$ in each mL of the Injection taken by the formula:

$$(C/V)(A_U/A_S),$$

in which C is the concentration, in µg per mL, of USP Droperidol RS in the *Standard preparation*, V is the volume, in mL, of Injection taken, and A_U and A_S are the absorbances obtained from the *Assay preparation* and the *Standard preparation*, respectively.

Dry-heat Sterilization, Paper Strip, Biological Indicator for—*see* Biological Indicator for Dry-heat Sterilization, Paper Strip

Absorbable Dusting Powder

» Absorbable Dusting Powder is an absorbable powder prepared by processing cornstarch and intended for use as a lubricant for surgical gloves. It contains not more than 2.0 percent of magnesium oxide.

Packaging and storage—Preserve in well-closed containers. It may be preserved in sealed paper packets.

Identification—A 1 in 10 suspension is colored purplish blue to deep blue by iodine TS.

Stability to autoclaving—Transfer about 2 g to a suitable paper packet, and seal or close the packet with a double fold. Wrap the paper packet in muslin, transfer to an autoclave, heat to 121° for 30 minutes, and cool: the powder is not caked, and any lumps are easily crushed between the fingers.

Sedimentation—Boil 100 mL of a 1 in 10 suspension in water for 20 minutes. Cool, transfer to a 100-mL graduated cylinder, dilute with water to volume, and allow to stand undisturbed for 24 hours: the volume occupied by the settled Absorbable Dusting Powder is between 45 mL and 75 mL.

pH ⟨791⟩: between 10.0 and 10.8, in a 1 in 10 suspension.

Loss on drying ⟨731⟩—Dry about 2 g, accurately weighed, at 105° to constant weight: it loses not more than 12% of its weight.

Residue on ignition ⟨281⟩—Heat about 1 g, accurately weighed, in a covered platinum crucible until most of the carbon is burned away, but do not ignite the sample. Remove the cover, and ignite to constant weight: not more than 3.0% of residue remains.

Heavy metals, *Method II* ⟨231⟩: 0.001%.

Assay for magnesium oxide—Weigh accurately about 2.5 g, and transfer to a beaker. Add 25 mL of water and 2 mL of 3 *N* hydrochloric acid, and stir the mixture for 5 minutes. Add 5 mL of hydroxylamine hydrochloride solution (1 in 10), 15 mL of ammonia–ammonium chloride TS, 5 mL of potassium cyanide solution (1 in 10), and 5 drops of eriochrome black TS, mix, and titrate with 0.05 *M* disodium ethylenediaminetetraacetate VS until the solution becomes distinctly blue in color. Each mL of 0.05 *M* disodium ethylenediaminetetraacetate is equivalent to 2.015 mg of MgO.

Dyclonine Hydrochloride

O
CH₂CH₂C—⟨benzene⟩—OCH₂CH₂CH₂CH₃ · HCl
N (piperidine)

$C_{18}H_{27}NO_2 \cdot HCl$ 325.88

1-Propanone, 1-(4-butoxyphenyl)-3-(1-piperidinyl)-.
4′-Butoxy-3-piperidinopropiophenone hydrochloride [*536-43-6*].

» Dyclonine Hydrochloride contains not less than 98.0 percent and not more than 102.0 percent of $C_{18}H_{27}NO_2 \cdot HCl$, calculated on the dried basis.

Packaging and storage—Preserve in tight, light-resistant containers.

Reference standard—*USP Dyclonine Hydrochloride Reference Standard*—Dry at 105° for 1 hour before using.

Identification—

A: The infrared absorption spectrum of a mineral oil dispersion of it, previously dried, exhibits maxima only at the same wavelengths as that of a similar preparation of USP Dyclonine Hydrochloride RS.

B: The ultraviolet absorption spectrum of a 1 in 40,000 solution in methanol exhibits maxima and minima at the same wavelengths as that of a similar solution of USP Dyclonine Hydrochloride RS, concomitantly measured.

C: Add 2 mL of silver nitrate TS to 10 mL of Dyclonine Hydrochloride solution (1 in 100): a white precipitate is formed. Add 2 mL of nitric acid, centrifuge, and discard the supernatant liquid. Wash the precipitate twice by adding 10 mL of 2 *N* nitric acid, centrifuging, and discarding the supernatant liquid: the precipitate so obtained is soluble in 6 *N* ammonium hydroxide.

Melting range ⟨741⟩: between 173° and 178°.

pH ⟨791⟩: between 4.0 and 7.0, in a solution (1 in 100).

Loss on drying ⟨731⟩—Dry it at 105° for 1 hour: it loses not more than 1.0% of its weight.

Residue on ignition ⟨281⟩: not more than 0.2%.

Assay—

Mobile phase, *Standard preparation*, and *Chromatographic system*—Proceed as directed in the *Assay* under *Dyclonine Hydrochloride Gel*.

Assay preparation—Transfer about 50 mg of Dyclonine Hydrochloride, accurately weighed, to a 500-mL volumetric flask, dissolve in 0.001 *N* phosphoric acid, dilute with 0.001 *N* phosphoric acid to volume, and mix.

Procedure—Separately inject equal volumes (about 20 µL) of the *Standard preparation* and the *Assay preparation* into the chromatograph, record the chromatograms, and measure the responses for the major peaks. Calculate the quantity, in mg, of dyclonine hydrochloride ($C_{18}H_{27}NO_2 \cdot HCl$) in the portion of Dyclonine Hydrochloride taken by the formula:

$$500C(r_U/r_S),$$

in which C is the concentration, in mg per mL, of USP Dyclonine Hydrochloride RS in the *Standard preparation*, and r_U and r_S are the peak responses obtained from the *Assay preparation* and the *Standard preparation*, respectively.

Dyclonine Hydrochloride Gel

» Dyclonine Hydrochloride Gel contains not less than 90.0 percent and not more than 110.0 percent of the labeled amount of $C_{18}H_{27}NO_2 \cdot HCl$. It may contain suitable stabilizers and antimicrobial agents.

Packaging and storage—Preserve in collapsible, opaque plastic tubes or in tight, light-resistant glass containers. [NOTE—Do not use aluminum or tin tubes.]

Reference standard—*USP Dyclonine Hydrochloride Reference Standard*—Dry at 105° for 1 hour before using.

Identification—Shake a portion of Gel, equivalent to about 400 mg of dyclonine hydrochloride, with 25 mL of chloroform, and allow the layers to separate. Remove the chloroform layer, evaporate on a steam bath to dryness, and dry the residue at 105° for 1 hour: the dyclonine hydrochloride so obtained responds to the *Identification tests* under *Dyclonine Hydrochloride*.

pH ⟨791⟩: between 2.0 and 4.0.

Assay—

Mobile phase—Dissolve 0.20 g of monobasic potassium phosphate and 0.45 mL of *n*-heptylamine in about 350 mL of water. Adjust with phosphoric acid to a pH of 3.0, dilute with water to 400 mL, add 600 mL of acetonitrile, and mix.

Standard preparation—Dissolve an accurately weighed quantity of USP Dyclonine Hydrochloride RS in 0.001 *N* phosphoric acid to obtain a solution having a known concentration of about 0.1 mg per mL.

Assay preparation—Transfer an accurately measured portion of Dyclonine Hydrochloride Gel, equivalent to about 5.0 mg of dyclonine hydrochloride, to a 50-mL volumetric flask. Add 10 mL of 0.001 *N* phosphoric acid, and sonicate to dissolve the gel. Dilute with 0.001 *N* phosphoric acid to volume, and mix.

Chromatographic system (see *Chromatography* ⟨621⟩)—The liquid chromatograph is equipped with a 254-nm detector and a 4-mm × 25-cm column that contains 5-μm diameter packing L13. The flow rate is about 1.2 mL per minute. Adjust the flow rate, if necessary, so that the retention time of dyclonine hydrochloride is not less than 5 minutes. Chromatograph five replicate injections of the *Standard preparation*, and record the peak responses as directed under *Procedure:* the relative standard deviation is not more than 3.0%. The tailing factor is not more than 2.0.

Procedure—Separately inject equal volumes (about 20 μL) of the *Standard preparation* and the *Assay preparation* into the chromatograph, record the chromatograms, and measure the responses for the major peaks. Calculate the quantity, in mg, of dyclonine hydrochloride ($C_{18}H_{27}NO_2.HCl$) in the portion of Gel taken by the formula:

$$50C(r_U/r_S),$$

in which C is the concentration, in mg per mL, of USP Dyclonine Hydrochloride RS in the *Standard preparation*, and r_U and r_S are the peak responses obtained from the *Assay preparation* and the *Standard preparation*, respectively.

Dyclonine Hydrochloride Topical Solution

» Dyclonine Hydrochloride Topical Solution is a sterile, aqueous solution of Dyclonine Hydrochloride. It contains not less than 92.0 percent and not more than 108.0 percent of the labeled amount of $C_{18}H_{27}NO_2.HCl$. It may contain suitable stabilizers and antimicrobial agents.

Packaging and storage—Preserve in tight, light-resistant containers.

Reference standard—*USP Dyclonine Hydrochloride Reference Standard*—Dry at 105° for 1 hour before using.

Identification—To a volume of Topical Solution, equivalent to about 300 mg of dyclonine hydrochloride, add 15 mL of chloroform, shake, and allow the layers to separate. Remove a portion of the chloroform solution, evaporate on a steam bath to dryness, and dry the residue at 105° for 1 hour: the dyclonine hydrochloride so obtained responds to the *Identification tests* under *Dyclonine Hydrochloride*.

Sterility—It meets the requirements under *Sterility Tests* ⟨71⟩.

pH ⟨791⟩: between 3.0 and 5.0.

Assay—

Mobile phase, Standard preparation, and *Chromatographic system*—Proceed as directed in the *Assay* under *Dyclonine Hydrochloride Gel*.

Assay preparation—Transfer an accurately measured volume of Dyclonine Hydrochloride Topical Solution, equivalent to about 50 mg of dyclonine hydrochloride, to a 500-mL volumetric flask. Dilute with water to volume, and mix.

Procedure—Separately inject equal volumes (about 20 μL) of the *Standard preparation* and the *Assay preparation* into the chromatograph, record the chromatograms, and measure the responses for the major peaks. Calculate the quantity, in mg, of $C_{18}H_{27}NO_2.HCl$ in each mL of the Dyclonine Hydrochloride Topical Solution taken by the formula:

$$500(C/V)(r_U/r_S),$$

in which C is the concentration, in mg per mL, of USP Dyclonine Hydrochloride RS in the *Standard preparation*, V is the volume, in mL, of the Topical Solution taken, and r_U and r_S are the peak responses obtained from the *Assay preparation* and the *Standard preparation*, respectively.

Dydrogesterone

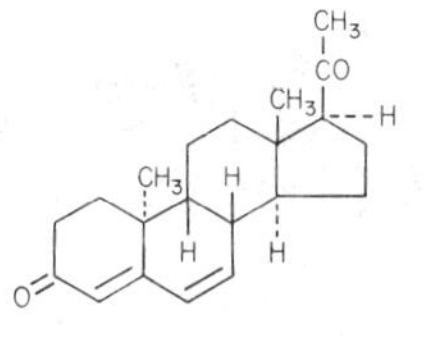

$C_{21}H_{28}O_2$ 312.45

Pregna-4,6-diene-3,20-dione, (9β,10α)-.

9β,10α-Pregna-4,6-diene-3,20-dione [*152-62-5*].

» Dydrogesterone contains not less than 98.0 percent and not more than 102.0 percent of $C_{21}H_{28}O_2$, calculated on the dried basis.

Packaging and storage—Preserve in well-closed containers.

Reference standard—*USP Dydrogesterone Reference Standard*—Dry in vacuum at 50° for 1 hour before using.

Identification—

A: The infrared absorption spectrum of a potassium bromide dispersion of it, previously dried, exhibits maxima only at the same wavelengths as that of a similar preparation of USP Dydrogesterone RS.

B: The ultraviolet absorption spectrum of a 1 in 167,000 solution in methanol exhibits maxima and minima at the same wavelengths as that of a similar solution of USP Dydrogesterone RS, concomitantly measured, and the respective absorptivities, calculated on the dried basis, at the wavelength of maximum absorbance at about 285 nm do not differ by more than 2.5%.

C: Dissolve about 50 mg in 5 mL of dehydrated alcohol, add about 50 mg each of hydroxylamine hydrochloride and anhydrous sodium acetate, and reflux gently for 2 hours. Concentrate the liquid to about 3 mL, then add 10 mL of water, mix, and centrifuge. Wash the precipitate with two 3-mL portions of warm methanol, and dry at 105° for 1 hour: the dydrogesterone dioxime so obtained melts between 275° and 282°, with decomposition (see *Melting Range or Temperature* ⟨741⟩).

Melting range ⟨741⟩: between 167° and 171°.

Specific rotation ⟨781⟩: between −470° and −500°, calculated on the dried basis, determined in a solution in chloroform containing 50 mg in each 10 mL.

Loss on drying ⟨731⟩—Dry it in vacuum at 50° for 1 hour: it loses not more than 0.5% of its weight.

Residue on ignition ⟨281⟩: not more than 0.1%.

Heavy metals, *Method II* ⟨231⟩: 0.002%.

Assay—

Standard preparation—Prepare as directed under *Single-steroid Assay* ⟨511⟩, using USP Dydrogesterone RS, except to obtain a solution having a final concentration of 8 mg per mL.

Assay preparation—Weigh accurately about 80 mg of Dydrogesterone, previously dried, dissolve in a sufficient quantity of a mixture of equal volumes of alcohol and chloroform to make 10.0 mL, and mix.

Procedure—Proceed as directed for *Procedure* under *Single-steroid Assay* ⟨511⟩, applying 20 μL each of the *Standard preparation* and the *Assay preparation* and using a solvent system consisting of a mixture of benzene and acetone (4:1), through the fourth sentence of the second paragraph under *Procedure*. Then centrifuge the tubes for 5 minutes, and determine the absorbances of the supernatant solutions in 1-cm cells at the wavelength of maximum absorbance at about 285 nm, with a suitable spectrophotometer, against the blank. Calculate the quantity, in mg, of $C_{21}H_{28}O_2$ in the portion of Dydrogesterone taken by the formula:

$$10C(A_U/A_S),$$

in which C is the concentration, in mg per mL, of USP Dydrogesterone RS in the *Standard preparation*, and A_U and A_S are the absorbances of the solutions from the *Assay preparation* and the *Standard preparation*, respectively.

Dydrogesterone Tablets

» Dydrogesterone Tablets contain not less than 90.0 percent and not more than 110.0 percent of the labeled amount of $C_{21}H_{28}O_2$.

Packaging and storage—Preserve in well-closed containers.

Reference standard—*USP Dydrogesterone Reference Standard*—Dry in vacuum at 50° for 1 hour before using.

Identification—Triturate a quantity of finely powdered Tablets, equivalent to about 5 mg of dydrogesterone, with 10 mL of chloroform, and filter. On a suitable thin-layer chromatographic plate (see *Chromatography* ⟨621⟩), coated with a 0.25-mm layer of chromatographic silica gel mixture, apply 10 µL each of this solution and of a solution of USP Dydrogesterone RS in chloroform containing 500 µg per mL. Allow the spots to dry, and develop the chromatogram in a solvent system consisting of a mixture of benzene and ether (3:1) until the solvent front has moved about three-fourths of the length of the plate. Remove the plate from the developing chamber, mark the solvent front, and allow the solvent to evaporate. Locate the spots on the plate by viewing under short-wavelength ultraviolet light: the R_f value of the principal spot obtained from the test solution corresponds to that obtained from the Standard solution.

Dissolution ⟨711⟩—

Medium: water:isopropyl alcohol (89:11); 900 mL.

Apparatus 2: 100 rpm.

Time: 45 minutes.

Procedure—Determine the amount of $C_{21}H_{28}O_2$ dissolved from ultraviolet absorbances at the wavelength of maximum absorbance at about 285 nm of filtered portions of the solution under test, suitably diluted with *Dissolution Medium*, if necessary, in comparison with a Standard solution having a known concentration of USP Dydrogesterone RS in the same medium.

Tolerances—Not less than 75% (Q) of the labeled amount of $C_{21}H_{28}O_2$ is dissolved in 45 minutes.

Uniformity of dosage units ⟨905⟩: meet the requirements.

Procedure for content uniformity—Transfer 1 finely powdered Tablet to a 100-mL volumetric flask, add 50 mL of methanol, and heat in a water bath at 60° for 10 minutes, with frequent agitation. Remove from the bath, shake by mechanical means for 20 minutes, and cool. Dilute with methanol to volume, mix, and centrifuge. Dilute a portion of the clear, supernatant liquid with methanol to provide a solution containing approximately 5 µg of dydrogesterone per mL. Concomitantly determine the absorbances of this solution and a solution of USP Dydrogesterone RS in the same medium having a known concentration of about 5 µg per mL, in 1-cm cells at the wavelength of maximum absorbance at about 285 nm, with a suitable spectrophotometer, using methanol as the blank. Calculate the quantity, in mg, of $C_{21}H_{28}O_2$ in the Tablet by the formula:

$$(TC/D)(A_U/A_S),$$

in which T is the labeled quantity, in mg, of dydrogesterone in the Tablet, C is the concentration, in µg per mL, of USP Dydrogesterone RS in the Standard solution, D is the concentration, in µg per mL, of dydrogesterone in the solution from the Tablet, based upon the labeled quantity per Tablet and the extent of dilution, and A_U and A_S are the absorbances of the solution from the Tablet and the Standard solution, respectively. Dydrogesterone Tablets meet the requirements for *Tablets*.

Assay—

Standard preparation—Dissolve a suitable quantity of USP Dydrogesterone RS, accurately weighed, in chloroform, and dilute quantitatively and stepwise with chloroform to obtain a solution having a concentration of about 20 µg per mL. Transfer 5.0 mL of this solution to a glass-stoppered, 50-mL conical flask.

Assay preparation—Weigh and finely powder not less than 10 Dydrogesterone Tablets. Transfer an accurately weighed portion of the powder, equivalent to about 20 mg of dydrogesterone, to a separator with the aid of about 15 mL of water, and extract with four 25-mL portions of chloroform, filtering each extract through chloroform-moistened cotton into a 250-mL volumetric flask. Dilute the contents of the flask with chloroform to volume, and mix. Transfer 25.0 mL of this solution to a 100-mL volumetric flask, dilute with chloroform to volume, and mix. Transfer 5.0 mL of this last solution to a glass-stoppered, 50-mL conical flask.

Procedure—To each of the flasks containing the *Standard preparation* and the *Assay preparation* and to a similar flask containing 5.0 mL of chloroform to provide the blank add 10.0 mL of a solution containing 375 mg of isoniazid and 0.47 mL of hydrochloric acid in 500 mL of methanol. Mix, allow to stand for 45 minutes, and concomitantly determine the absorbances of the solutions in 1-cm cells at the wavelength of maximum absorbance at about 420 nm, with a suitable spectrophotometer, against the blank. Calculate the quantity, in mg, of $C_{21}H_{28}O_2$ in the portion of Tablets taken by the formula:

$$C(A_U/A_S),$$

in which C is the concentration, in µg per mL, of USP Dydrogesterone RS in the *Standard preparation*, and A_U and A_S are the absorbances of the solutions from the *Assay preparation* and the *Standard preparation*, respectively.

Dyphylline

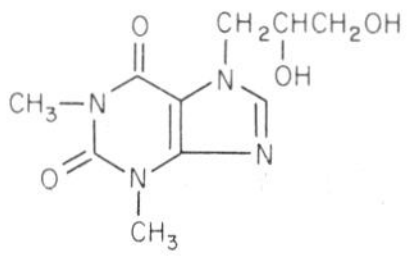

$C_{10}H_{14}N_4O_4$ 254.25

1*H*-Purine-2,6-dione, 7-(2,3-dihydroxypropyl)-3,7-dihydro-1,3-dimethyl-.

7-(2,3-Dihydroxypropyl)theophylline [*479-18-5*].

» Dyphylline contains not less than 98.0 percent and not more than 102.0 percent of $C_{10}H_{14}N_4O_4$, calculated on the dried basis.

Packaging and storage—Preserve in tight containers.

Reference standard—*USP Dyphylline Reference Standard*—Dry at 105° for 3 hours before using.

Identification—

A: The infrared absorption spectrum of a potassium bromide dispersion of it, previously dried, exhibits maxima only at the same wavelengths as that of a similar preparation of USP Dyphylline RS.

B: The ultraviolet absorption spectrum of a solution (1 in 100,000) exhibits maxima and minima at the same wavelengths as that of a similar solution of USP Dyphylline RS, concomitantly measured.

Melting range ⟨741⟩: between 160° and 164°.

pH ⟨791⟩: between 5.0 and 7.5, in a solution (1 in 100).

Loss on drying ⟨731⟩—Dry about 1 g, accurately weighed, at 105° for 3 hours: it loses not more than 0.5% of its weight.

Residue on ignition ⟨281⟩: not more than 0.15%.

Chloride ⟨221⟩: A 1.0-g portion shows no more chloride than corresponds to 0.50 mL of 0.020 *N* hydrochloric acid (0.035%).

Sulfate ⟨221⟩: A 1.0-g portion shows no more sulfate than corresponds to 0.10 mL of 0.020 *N* sulfuric acid (0.010%).

Heavy metals, *Method I* ⟨231⟩: 0.002%.

Limit of theophylline—Dissolve 2.0 g in 10 mL of hot water. Cool, add 10 drops of bromothymol blue TS and 5.0 mL of silver nitrate TS, and mix. Not more than 0.50 mL of 0.10 *N* sodium hydroxide is required to change the color of the solution to blue (0.5%).

Related substances—Dissolve about 200 mg in a mixture of 30 volumes of methanol and 20 volumes of water in a 10-mL volumetric flask, dilute with the same mixture to volume, and mix (*Test solution*). Dilute 1 mL of the *Test solution* with methanol to 100 mL (*Diluted test solution*). Separately apply 10 µL of each solution to a suitable thin-layer chromatographic plate (see *Chromatography* ⟨621⟩) coated with a 0.25-mm layer of binder-free chromatographic silica gel mixture. Develop the plate in a suitable chamber containing a mixture of chloroform, dehydrated alcohol, and ammonium hydroxide (90:10:1) until the solvent front has moved about 15 cm. Remove the plate, air-dry at room temperature, and examine under short-wavelength ultraviolet light: no spot in the chromatogram of the *Test solution* other than the principal spot is larger or more intense than the principal spot from the *Diluted test solution* (1.0%).

Assay—Dissolve about 200 mg of Dyphylline, accurately weighed, in 2 mL of formic acid, and carefully add, with stirring, 50 mL of acetic anhydride. Add 0.2 mL of Sudan IV TS, and titrate with 0.1 *N* perchloric acid VS to a deep violet end-point. Perform a blank determination, and make any necessary correction. Each mL of 0.1 *N* perchloric acid is equivalent to 25.42 mg of $C_{10}H_{14}N_4O_4$.

Dyphylline Elixir

» Dyphylline Elixir contains not less than 90.0 percent and not more than 110.0 percent of the labeled amount of $C_{10}H_{14}N_4O_4$.

Packaging and storage—Preserve in tight containers.

Reference standard—*USP Dyphylline Reference Standard*—Dry at 105° for 3 hours before using.

Identification—The retention time of the major peak in the chromatogram of the *Assay preparation* corresponds to that of the *Standard preparation*, obtained as directed in the *Assay*.

Alcohol content—

Standard preparation—Pipet 5 mL of dehydrated alcohol and 5 mL of acetone into a 200-mL volumetric flask containing 50 mL of water, add water to volume, and mix. Pipet 10 mL of this solution into a 200-mL volumetric flask, add water to volume, and mix.

Test preparation—Transfer an accurately measured volume of Elixir, equivalent to about 5 mL of alcohol, to a 200-mL volumetric flask containing about 30 mL of water. Pipet 5 mL of acetone into the flask, add water to volume, and mix. Pipet 10 mL of this solution into a 200-mL volumetric flask, add water to volume, and mix.

Chromatographic system—The gas chromatograph is equipped with a flame-ionization detector and contains a 75-cm × 4-mm column packed with 20 percent phase G20 on support S1AB, conditioned as directed (see *Chromatography* ⟨621⟩). The column is maintained at a temperature of about 85°, and the injection port and detector block are maintained at about 175° and 225°, respectively. Nitrogen is used as the carrier gas at a flow rate of about 18 mL per minute. Chromatograph the *Standard preparation*, and record the peak responses as directed under *Procedure:* the relative standard deviation for replicate injections does not exceed 2.0% in the ratio of the peak of alcohol to the peak of acetone, the resolution is not less than 2.0, and the tailing factor of the alcohol peak is not greater than 1.5.

Procedure—Separately inject equal volumes (about 4 µL) of the *Test preparation* and the *Standard preparation*, in duplicate, into the gas chromatograph, record the chromatograms, and measure the responses for the major peaks. Calculate the percentage of alcohol in the specimen by the formula:

$$(500/V)(R_U/R_S),$$

in which V is the volume, in mL, of Elixir taken, and R_U and R_S are the ratios of the response of the alcohol peak to that of the acetone peak obtained from the *Assay preparation* and the *Standard preparation*, respectively: the alcohol content, obtained as the average of the calculated results, is between 90.0% and 110.0% of the labeled amount of C_2H_5OH.

Assay—

Mobile phase and *Chromatographic system*—Prepare as directed in the *Assay* under *Dyphylline Tablets*.

Standard preparation—Dissolve an accurately weighed quantity of USP Dyphylline RS in *Mobile phase* to obtain a solution having a known concentration of about 500 µg per mL.

Assay preparation—Transfer an accurately measured volume of Dyphylline Elixir, equivalent to about 100 mg of dyphylline, to a 200-mL volumetric flask, add *Mobile phase* to volume, and mix.

Procedure—Separately inject equal volumes (about 10 µL), of the *Standard preparation* and the *Assay preparation* into the chromatograph, record the chromatograms, and measure the responses for the major peaks. Calculate the quantity, in mg, of $C_{10}H_{14}N_4O_4$ in the volume of Elixir taken by the formula:

$$0.2C(r_U/r_S),$$

in which C is the concentration, in µg per mL, of USP Dyphylline RS in the *Standard preparation*, and r_U and r_S are the dyphylline peak responses from the *Assay preparation* and the *Standard preparation*, respectively.

Dyphylline Injection

» Dyphylline Injection contains not less than 90.0 percent and not more than 110.0 percent of the labeled amount of $C_{10}H_{14}N_4O_4$.

Packaging and storage—Preserve in single-dose or in multiple-dose containers, preferably of Type I glass, protected from light. To avoid precipitation, store at a temperature of not below 15°, but avoid excessive heat.

Reference standard—*USP Dyphylline Reference Standard*—Dry at 105° for 3 hours before using.

Labeling—Label it to indicate that the Injection is not to be used if crystals have separated.

Identification—

A: Dilute about 5 mL with 20 mL of water, and add 1.0 mL of 2 *N* sodium hydroxide and 2 drops of potassium permanganate TS: a green color is produced.

B: The retention time of the major peak in the chromatogram of the *Assay preparation* corresponds to that of the *Standard preparation*, as obtained in the *Assay*.

C: Transfer a portion of Injection, equivalent to about 100 mg of dyphylline, to a 200-mL volumetric flask, dilute with water to volume, and mix. Transfer 2 mL to a 100-mL volumetric flask, dilute with water to volume, and mix: the ultraviolet absorption spectrum of the solution exhibits maxima and minima at the same wavelengths as that of a solution of USP Dyphylline RS containing 10 µg per mL, concomitantly measured.

Pyrogen—It meets the requirements of the *Pyrogen Test* ⟨151⟩.

pH ⟨791⟩: between 5.0 and 8.0.

Other requirements—It meets the requirements under *Injections* ⟨1⟩.

Assay—

Mobile phase, Standard preparation, and *Chromatographic system*—Prepare as directed in the *Assay* under *Dyphylline Tablets*.

Assay preparation—Transfer an accurately measured volume of Dyphylline Injection, equivalent to about 100 mg of dyphylline, to a 500-mL volumetric flask, add water to volume, and mix.

Procedure—Proceed as directed in the *Assay* under *Dyphylline Tablets*. Calculate the quantity, in mg, of $C_{10}H_{14}N_4O_4$ in each mL of the Injection taken by the formula:

$$(0.5C/V)(r_U/r_S),$$

in which V is the volume, in mL, of Injection taken.

Dyphylline Tablets

» Dyphylline Tablets contain not less than 90.0 percent and not more than 110.0 percent of the labeled amount of $C_{10}H_{14}N_4O_4$.

Packaging and storage—Preserve in tight containers.

Reference standard—*USP Dyphylline Reference Standard*—Dry at 105° for 3 hours before using.

Identification—

A: To a solution of 5 Tablets in 20 mL of water add 1 mL of 2 *N* sodium hydroxide and 2 drops of potassium permanganate TS: a green color is produced.

B: The retention time of the major peak in the chromatogram of the *Assay preparation* corresponds to that of the *Standard preparation*, as obtained in the *Assay*.

C: Weigh a portion of finely powdered Tablets, equivalent to about 100 mg of dyphylline, and transfer to a 200-mL volumetric flask. Add about 100 mL of water, shake by mechanical means for about 15 minutes, dilute with water to volume, and mix. Filter, and transfer 2 mL of the filtrate to a 100-mL volumetric flask, dilute with water to volume, and mix: the ultraviolet absorption spectrum of the solution exhibits maxima and minima at the same wavelengths as that of a solution of USP Dyphylline RS containing 10 µg per mL, concomitantly measured.

Dissolution ⟨711⟩—

Medium: water; 900 mL.

Apparatus 1: 100 rpm.

Time: 45 minutes.

Procedure—Determine the amount of $C_{10}H_{14}N_4O_4$ dissolved from ultraviolet absorbances at the wavelength of maximum absorbance at about 273 nm of filtered portions of the solution under test, suitably diluted with water, in comparison with a Standard solution having a known concentration of USP Dyphylline RS in the same medium.

Tolerances—Not less than 75% (*Q*) of the labeled amount of $C_{10}H_{14}N_4O_4$ is dissolved in 45 minutes.

Uniformity of dosage units ⟨905⟩: meet the requirements.

Assay—

Mobile phase—Dissolve 1.4 g of monobasic potassium phosphate in 1350 mL of water in a 2-liter volumetric flask, add methanol to volume, and mix. Filter through a 0.5-µm porosity membrane filter. [NOTE—The composition may be varied to meet system suitability requirements.]

Standard preparation—Dissolve an accurately weighed quantity of USP Dyphylline RS in water to obtain a solution having a known concentration of about 200 µg per mL.

Assay preparation—Weigh and finely powder not less than 20 Tablets. Transfer an accurately weighed portion of the powder, equivalent to about 100 mg of dyphylline, to a 500-mL volumetric flask, add 200 mL of water, insert the stopper, and shake by mechanical means for 10 minutes. Add water to volume, and mix. Just prior to injection, filter about 5 mL through a 1.2-µm porosity membrane filter.

Chromatographic system (see *Chromatography* ⟨621⟩)—The liquid chromatograph is equipped with a 254-nm detector and a 4-mm × 30-cm column that contains packing L1. The flow rate is about 1.0 mL per minute. Chromatograph five replicate injections of the *Standard preparation*, and record the peak responses as directed under *Procedure:* the relative standard deviation is not more than 2.0%. The tailing factor is not more than 2.0.

Procedure—Inject separately equal volumes, accurately measured (about 10 µL), of the *Standard preparation* and the *Assay preparation* into the chromatograph, record the chromatograms, and measure the peak responses. Calculate the quantity, in mg, of $C_{10}H_{14}N_4O_4$ in the portion of Tablets taken by the formula:

$$0.5C(r_U/r_S),$$

in which C is the concentration, in µg per mL, of USP Dyphylline RS in the *Standard preparation*, and r_U and r_S are the dyphylline peak responses obtained from the *Assay preparation* and the *Standard preparation*, respectively.

Dyphylline and Guaifenesin Elixir

» Dyphylline and Guaifenesin Elixir contains not less than 90.0 percent and not more than 110.0 percent of the labeled amounts of dyphylline ($C_{10}H_{14}N_4O_4$) and guaifenesin ($C_{10}H_{14}O_4$).

Packaging and storage—Preserve in tight containers.

Reference standards—*USP Dyphylline Reference Standard*—Dry at 105° for 3 hours before using. *USP Guaifenesin Reference Standard*—Dry in vacuum at a pressure not below 10 mm of mercury at 60° to constant weight before using.

Identification—

A: To an amount of the Elixir equivalent to about 100 mg of dyphylline, add water to make 20 mL, mix, and add 2.0 mL of 2 *N* sodium hydroxide and 2 drops of potassium permanganate TS: a green color is produced.

B: Transfer a volume of Elixir, equivalent to about 100 mg of guaifenesin, to a 60-mL separator, add 10 mL of chloroform, shake for 30 seconds, and allow the layers to separate. Decant the lower (chloroform) layer through chloroform-washed cotton into a small beaker. Evaporate 1 mL of the extract, on a watch glass, on a steam bath to dryness. To the residue add 1 drop of formaldehyde TS and a few drops of sulfuric acid: a deep cherry-red to purple color is produced.

C: The retention times of the major peaks in the chromatogram of the *Assay preparation* correspond to those of the *Standard preparation*, as obtained in the *Assay*.

pH ⟨791⟩: between 5.0 and 7.0.

Alcohol content—

Standard preparation, *Test preparation*, *Chromatographic system*, and *Procedure*—Proceed as directed in *Alcohol content* under *Dyphylline Elixir:* the alcohol content is between 90.0% and 110.0% of the labeled amount of C_2H_5OH.

Assay—

Mobile phase—Prepare a suitable filtered and degassed mixture of 0.01 *M* monobasic potassium phosphate and methanol (79:21). Make adjustments if necessary (see *System Suitability* under *Chromatography* ⟨621⟩).

Standard preparation—Dissolve accurately weighed quantities of USP Dyphylline RS and USP Guaifenesin RS in *Mobile phase*, and dilute quantitatively, and stepwise if necessary, with *Mobile phase* to obtain a solution having known concentrations of about 0.1 mg of guaifenesin and about 0.1*J* mg of dyphylline per mL, *J* being the ratio of the labeled amount of dyphylline to that of guaifenesin.

Resolution solution—Prepare a solution in *Mobile phase* containing in each mL about 0.1 mg each of dyphylline and guaifenesin and about 0.01 mg of guaiacol.

Assay preparation—Transfer an accurately measured volume of Dyphylline and Guaifenesin Elixir, equivalent to about 100 mg of guaifenesin, to a 100-mL volumetric flask, dilute with *Mobile phase* to volume, and mix. Transfer 5.0 mL of this solution to a 50-mL volumetric flask, dilute with *Mobile phase* to volume, and mix.

Chromatographic system (see *Chromatography* ⟨621⟩)—The liquid chromatograph is equipped with a 230-nm detector, a guard column that contains packing L1, and a 4.6-mm × 15-cm column that contains packing L1. The flow rate is about 2 mL per minute. Chromatograph the *Resolution solution*, and record the peak responses as directed under *Procedure:* the resolution, *R*, between the guaiacol and guaifenesin peaks is not less than 1.8, and the resolution, *R*, between the guaiacol and dyphylline peaks is not less than 9.0. The relative retention times are about 0.25 for dyphylline, 0.7 for guaiacol, and 1.0 for guaifenesin. Chromatograph the *Standard preparation*, and record the peak responses as directed under *Procedure:* the relative standard deviation for replicate injections is not more than 2.0% for both dyphylline and guaifenesin.

Procedure—Separately inject equal volumes (about 20 µL) of the *Standard preparation* and the *Assay preparation* into the chromatograph, record the chromatograms, and measure the responses for the major peaks. Calculate the quantities, in mg, of dyphylline ($C_{10}H_{14}N_4O_4$) and guaifenesin ($C_{10}H_{14}O_4$) in each mL of the Elixir taken by the formula:

$$1000C/V(r_U/r_S),$$

in which C is the concentration, in mg per mL, of the appropriate USP Reference Standard in the *Standard preparation*, V is the volume, in mL, of Elixir taken, and r_U and r_S are the peak responses of the corresponding analyte in the *Assay preparation* and the *Standard preparation*, respectively.

Dyphylline and Guaifenesin Tablets

» Dyphylline and Guaifenesin Tablets contain not less than 90.0 percent and not more than 110.0 percent of the labeled amounts of dyphylline ($C_{10}H_{14}N_4O_4$) and guaifenesin ($C_{10}H_{14}O_4$).

Packaging and storage—Preserve in tight containers.

Reference standards—*USP Dyphylline Reference Standard*—Dry at 105° for 3 hours before using. *USP Guaifenesin Reference Standard*—Dry in vacuum at a pressure not below 10 mm of mercury at 60° to constant weight before using.

Identification—

A: Triturate a quantity of finely powdered Tablets, equivalent to about 100 mg of dyphylline, with 20 mL of water and transfer to a test tube. Add 2.0 mL of 2 *N* sodium hydroxide and 2 drops of potassium permanganate TS: a green color is produced.

B: Triturate a quantity of finely powdered Tablets, equivalent to about 100 mg of guaifenesin, with 10 mL of chloroform, filter, and evaporate 1 mL of the filtrate, on a watch glass, on a steam bath to dryness. To the residue add 1 drop of formaldehyde TS and a few drops of sulfuric acid: a deep cherry-red to purple color is produced.

C: The retention times of the major peaks in the chromatogram of the *Assay preparation* correspond to those of the *Standard preparation*, as obtained in the *Assay*.

Uniformity of dosage units ⟨905⟩: meet the requirements.

Assay—

Mobile phase, Standard preparation, and Resolution solution—Prepare as directed in the *Assay* under *Dyphylline and Guaifenesin Elixir.*

Chromatographic system—Prepare as directed in the *Assay* under *Dyphylline and Guaifenesin Elixir*, except omit the use of the guard column.

Assay preparation—Weigh and finely powder not less than 20 Dyphylline and Guaifenesin Tablets. Transfer an accurately weighed portion of the powder, equivalent to about 100 mg of guaifenesin, to a 100-mL volumetric flask. Add about 80 mL of *Mobile phase*, sonicate for 30 minutes, and shake by mechanical means for an additional 30 minutes. Dilute with *Mobile phase* to volume, mix, and filter through a membrane filter of 0.45-µm or finer porosity, discarding the first 10 mL of the filtrate. Transfer 5.0 mL of the clear filtrate to a 50-mL volumetric flask, dilute with *Mobile phase* to volume, and mix.

Procedure—Proceed as directed for *Procedure* in the *Assay* under *Dyphylline and Guaifenesin Elixir.* Calculate the quantities, in mg, of dyphylline ($C_{10}H_{14}N_4O_4$) and guaifenesin ($C_{10}H_{14}O_4$) in the portion of Tablets taken by the formula:

$$1000C(r_U/r_S),$$

in which the terms are as defined therein.

Echothiophate Iodide

$$\left[(CH_3)_3\overset{+}{N}CH_2CH_2{-}S{-}\overset{O}{\overset{\uparrow}{P}}(OC_2H_5)_2\right] I^-$$

$C_9H_{23}INO_3PS$ 383.22

Ethanaminium, 2-[(diethoxyphosphinyl)thio]-*N,N,N*-trimethyl-, iodide.

(2-Mercaptoethyl)trimethylammonium iodide *S*-ester with *O,O*-diethyl phosphorothioate [*513-10-0*].

» Echothiophate Iodide contains not less than 95.0 percent and not more than 100.5 percent of $C_9H_{23}INO_3PS$, calculated on the dried basis.

Packaging and storage—Preserve in tight, light-resistant containers, preferably at a temperature below 0°.

Identification—

A: Dissolve about 100 mg in 2 mL of water in a test tube, and add 1 mL of nitric acid: a brown precipitate of iodine is formed. Transfer 1 drop of this mixture to another test tube, add 1 mL of carbon tetrachloride, and shake: the carbon tetrachloride takes on a pink color.

B: Heat the remainder of the brown reaction mixture from *Identification test A* over a flame until a colorless solution remains (about 3 minutes). Cool, and dilute with water to about 10 mL: a 2-mL portion of this solution responds to the tests for *Phosphate* ⟨191⟩.

C: A 2-mL portion of the solution obtained in *Identification test B* responds to the tests for *Sulfate* ⟨191⟩.

Loss on drying ⟨731⟩—Dry it in vacuum over phosphorus pentoxide at 50° for 3 hours: it loses not more than 1.0% of its weight.

Assay—[NOTE—In the preparation of all reagents, and throughout this procedure, wherever water is specified, use only water that has been distilled, boiled for 10 minutes, and cooled while protected from the atmosphere.]

pH 12 phosphate buffer—Transfer 5.44 g of anhydrous dibasic sodium phosphate to a 100-mL volumetric flask, and add a volume of 1 *N* sodium hydroxide VS that contains 36.5 mEq of sodium hydroxide. Add about 40 mL of water, shake to dissolve the sodium phosphate, dilute with water to volume, and mix.

0.004 N Iodine—Dilute 0.1 *N* iodine with water to 0.004 *N*, and standardize the solution on the day of use as follows: Weigh accurately about 150 mg of arsenic trioxide, and dissolve in 20 mL of 1 *N* sodium hydroxide, by warming if necessary, in a 500-mL volumetric flask. Dilute with 40 mL of water, add 2 drops of methyl orange TS, then add 3 *N* hydrochloric acid until the yellow color is changed to pink. Add 2 g of sodium bicarbonate, add water to volume, and mix. Transfer 5.0 mL of this solution to a titration vessel, and add 50 mL of pH 7 phosphate buffer (see under *Solutions* in the section, *Reagents, Indicators, and Solutions*). Titrate with 0.004 *N* iodine, determining the endpoint potentiometrically, using platinum and silver–silver chloride electrodes. Calculate the normality. Each 197.8 µg of arsenic trioxide is equivalent to 1 mL of 0.004 *N* iodine.

Procedure—Dissolve about 125 mg of Echothiophate Iodide, accurately weighed, in about 50 mL of water in a 100-mL volumetric flask, add water to volume, and mix. Transfer 10.0 mL of this solution to a titration vessel containing 30 mL of water. Add 10.0 mL of *pH 12 phosphate buffer*, mix, cover, and allow to stand for 20 minutes at 25 ± 3°. Add 2 mL of glacial acetic acid, rapidly and with mixing. Titrate with *0.004 N iodine*, determining the end-point potentiometrically, using platinum and silver–silver chloride electrodes. Correct for the amount of free thiol sulfur by repeating the procedure but adding the glacial acetic acid first, then the *pH 12 phosphate buffer*, and titrating immediately. Subtract the volume of *0.004 N iodine* used in the second titration from that used in the first. Each mL of *0.004 N iodine* is equivalent to 1.533 mg of $C_9H_{23}INO_3PS$.

Echothiophate Iodide for Ophthalmic Solution

» Echothiophate Iodide for Ophthalmic Solution is sterile Echothiophate Iodide. It may contain Mannitol or other suitable diluent. It contains not less than 95.0 percent and not more than 115.0 percent of the labeled amount of $C_9H_{23}INO_3PS$.

Packaging and storage—Preserve in tight containers, preferably of Type I glass, at controlled room temperature.

Completeness of solution ⟨641⟩—The contents of 1 container dissolve in 10 mL of carbon dioxide–free water to yield a clear solution.

Identification—

A: Dissolve a quantity, equivalent to about 12 mg of echothiophate iodide, in 10 mL of water. To 1 mL of this solution add 0.2 mL of 2 *N* hydrochloric acid and 0.2 mL of 30 percent hydrogen peroxide: the color turns brown. Add a few drops of carbon tetrachloride, and shake: the carbon tetrachloride acquires a pink color.

B: To 5 mL of the solution prepared for *Identification test A* add 0.5 mL of sodium hydroxide solution (1 in 2), heat at 50° for 2 minutes, cool to room temperature, then add 1 mL of sodium nitroferricyanide TS: a deep-red color is produced.

Sterility—It meets the requirements under *Sterility Tests* ⟨71⟩.

Water—[NOTE—Dry all glassware used in the following procedure at 105° for 4 hours, and store in a desiccator or dry box. Perform as many operations as possible in a dry box.]

Dry alcohol—Wash about 150 g of 8- to 12-mesh type 3A molecular sieve with several portions of dehydrated alcohol to remove the fine particles. Place the washed molecular sieve in a shallow borosilicate glass tray, heat in an oven at 350° for 2 hours, and cool in a dry box. Transfer the dry molecular sieve to a 1-liter conical flask, add about 700 mL of dehydrated alcohol, insert the stopper, mix, and allow to stand for not less than 48 hours before using.

Internal standard solution—[NOTE—Prepare this solution fresh daily.] Place 0.17 mL of methanol in a 100-mL volumetric flask, add *Dry alcohol* to volume, and mix.

Standard preparations—[NOTE—Prepare these solutions fresh daily.] Into three 25-mL volumetric flasks, each containing about 15 mL of *Internal standard solution*, transfer 5 µL, 40 µL, and 75 µL of water, respectively. Dilute with *Internal standard solution* to volume, and mix.

Test preparation—Carefully remove the protective retainer and cap from five vials of Echothiophate Iodide for Ophthalmic Solution without removing the elastomeric septum closure. Discard the separated parts, and weigh accurately each closed vial and contents. Inject through the septum of each vial 400 µL of *Internal standard solution*, accurately measured, using a suitable gas-tight syringe, and allow to stand for 1 hour, swirling occasionally to dissolve the residue. After 1 hour, using a gas-tight syringe, remove 300 µL of solution from each vial, transfer to a dry small-volume sample-collecting vial equipped with a sampling valve system,* and mix the combined solutions.

Chromatographic system (see *Chromatography* ⟨621⟩)—The chromatograph is equipped with a thermal conductivity detector and a 1.8-m × 2-mm silylated glass column packed with 80- to 100-mesh surface-silanized packing S3. The column is maintained at a temperature of about 115°, the injection port and detector block are maintained at temperatures of about 200° and 225°, respectively, and dry helium is used as the carrier gas at a flow rate of about 45 mL per minute. Chromatograph a sufficient number of injections of a *Standard preparation*, and record the peak responses as directed under *Procedure*. The resolution factor between the water and methanol peaks is not less than 2.0; and the relative standard deviation is not more than 5.0%.

* Suitable sample-collecting vials and sampling valve systems are available as catalog Nos. 13098 and 13099 3-mL and 5-mL vials, and catalog No. 10135 valve, from Pierce Chemical Co., Box 117, Rockford, Ill. 61105.

Procedure—Inject a portion (3 µL to 4 µL) of each *Standard preparation* into the chromatograph, record the chromatogram, and measure the responses for the first (water) and second (methanol) major peaks obtained for each. Plot the ratios of the peak responses of water to methanol versus the concentration, in mg per mL, of water in each *Standard preparation*. [NOTE—If the plot is not linear, discard it, and repeat the chromatography on additional portions of the *Standard preparations*.] Similarly inject a portion of the *Test preparation*, record the chromatogram, and measure the responses for the two major peaks. By comparison with the linear standard plot, determine the concentration, in mg per mL, of water in the *Test preparation* as that corresponding to the ratio of the peak responses of water to methanol from the *Test preparation*. Remove the elastomeric septum closure from each test vial, discard the contents, and rinse each vial and closure with several portions of methanol. Dry the vials and the closures in a stream of dry nitrogen, weigh accurately, and subtract this weight from that of the closed vials and contents obtained as directed under *Test preparation*. Calculate the water content, in percentage, by the formula:

$$100CV/W,$$

in which C is the concentration, in mg per mL, of water in the *Test preparation*, V is the volume, in mL, of *Internal standard solution* added to each test specimen vial as directed under *Test preparation*, and W is the average weight, in mg, of the test specimens in the vials: not more than 2.0% is found.

Assay—[NOTE—In the preparation of all reagents, and throughout this procedure, wherever water is specified, use only water that has been distilled, boiled for 10 minutes, and cooled while protected from the atmosphere.] Dissolve the contents of a counted number of vials of Echothiophate Iodide for Ophthalmic Solution, equivalent to not less than 30 mg of echothiophate iodide, by adding 5.0 mL of water to each vial. Combine the solutions, and mix. Dilute a portion of the mixture, equivalent to about 12 mg of echothiophate iodide, with water to 40 mL, and proceed as directed for *Procedure* in the *Assay* under *Echothiophate Iodide*, beginning with "Add 10.0 mL of *pH 12 phosphate buffer*."

Edetate Calcium Disodium

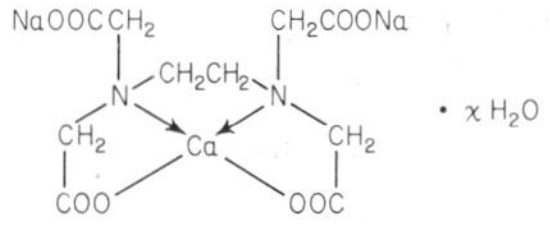

$C_{10}H_{12}CaN_2Na_2O_8 . xH_2O$ (anhydrous) 374.27

Calciate(2-), [[*N*,*N*′-1,2-ethanediylbis[*N*-(carboxymethyl)-glycinato]](4-)-*N*,*N*′,*O*,*O*′,O^N,$O^{N'}$]-, disodium, hydrate, (*OC*-6-21)-.

Disodium[(ethylenedinitrilo)tetraacetato]calciate(2-) hydrate [*23411-34-9*].

Anhydrous [*62-33-9*].

» Edetate Calcium Disodium is a mixture of the dihydrate and trihydrate of calcium disodium ethylenediaminetetraacetate (predominantly the dihydrate). It contains not less than 97.0 percent and not more than 102.0 percent of $C_{10}H_{12}CaN_2Na_2O_8$, calculated on the anhydrous basis.

Packaging and storage—Preserve in tight containers.

Reference standard—*USP Edetate Calcium Disodium Reference Standard*—Do not dry before using.

Identification—

A: The infrared absorption spectrum of a mineral oil dispersion of it exhibits maxima only at the same wavelengths as that of a similar preparation of USP Edetate Calcium Disodium RS.

B: A solution (1 in 20) responds to the oxalate test for *Calcium* ⟨191⟩, and to the flame test for *Sodium* ⟨191⟩.

C: To 5 mL of water add 2 drops of ammonium thiocyanate TS and 2 drops of ferric chloride TS. To the deep red solution add about 50 mg of Edetate Calcium Disodium, and mix: the deep red color disappears.

pH ⟨791⟩: between 6.5 and 8.0, in a solution (1 in 5).

Water, *Method I* ⟨921⟩: not more than 13.0%.

Heavy metals, *Method II* ⟨231⟩: 0.002%.

Magnesium-chelating substances—Weigh accurately 1.0 g, transfer to a small beaker, and dissolve in 5 mL of water. Add 5 mL of ammonia–ammonium chloride buffer TS. Then to the buffered solution add 5 drops of eriochrome black TS, and titrate with 0.10 *M* magnesium acetate to the appearance of a deep wine-red color: not more than 2.0 mL is required.

Assay—Weigh accurately about 1.2 g of Edetate Calcium Disodium, transfer to a 250-mL beaker, and dissolve in 75 mL of water. Add 25 mL of 1 *N* acetic acid and 1 mL of diphenylcarbazone TS, and titrate slowly with 0.1 *M* mercuric nitrate VS to the appearance of the first purplish color. Perform a blank determination, and make any necessary correction. Each mL of 0.1 *M* mercuric nitrate is equivalent to 37.43 mg of $C_{10}H_{12}CaN_2Na_2O_8$.

Edetate Calcium Disodium Injection

» Edetate Calcium Disodium Injection is a sterile solution of Edetate Calcium Disodium in Water for Injection. It contains, in each mL, not less than 180 mg and not more than 220 mg of $C_{10}H_{12}CaN_2Na_2O_8$.

Packaging and storage—Preserve in single-dose containers, preferably of Type I glass.

Reference standard—*USP Edetate Calcium Disodium Reference Standard*—Do not dry before using.

Identification—

A: Transfer a volume of Injection, equivalent to about 1 g of edetate calcium disodium, to an evaporating dish, and evaporate on a steam bath to dryness: the residue responds to *Identification test A* under *Edetate Calcium Disodium.*

B: It responds to *Identification test B* under *Edetate Calcium Disodium.*

Pyrogen—It meets the requirements of the *Pyrogen Test* ⟨151⟩, the test dose being 400 mg of edetate calcium disodium per kg.

pH ⟨791⟩: between 6.5 and 8.0.

Particulate matter ⟨788⟩: meets the requirements under *Small-volume Injections.*

Other requirements—It meets the requirements under *Injections* ⟨1⟩.

Assay—Dilute an accurately measured volume of Edetate Calcium Disodium Injection, equivalent to about 1 g of edetate calcium disodium with water to about 75 mL. Add 25 mL of 1 *N* acetic acid and 1 mL of diphenylcarbazone TS, mix, and titrate slowly with 0.1 *M* mercuric nitrate VS to the appearance of the first purplish color. Perform a blank determination, and make any necessary correction. Each mL of 0.1 *M* mercuric nitrate is equivalent to 37.43 mg of $C_{10}H_{12}CaN_2Na_2O_8$.

Edetate Disodium

$C_{10}H_{14}N_2Na_2O_8 . 2H_2O$ 372.24
Glycine, *N,N′*-1,2-ethanediylbis[*N*-(carboxymethyl)-, disodium salt, dihydrate.
Disodium (ethylenedinitrilo)tetraacetate dihydrate [*6381-92-6*].
Anhydrous 336.21 [*139-33-3*].

» Edetate Disodium contains not less than 99.0 percent and not more than 101.0 percent of $C_{10}H_{14}N_2Na_2O_8$, calculated on the dried basis.

Packaging and storage—Preserve in well-closed containers.

Reference standard—*USP Edetate Disodium Reference Standard.*

Identification—

A: The infrared absorption spectrum of a potassium bromide dispersion of it exhibits maxima only at the same wavelengths as that of a similar preparation of USP Edetate Disodium RS.

B: To 5 mL of water in a test tube add 2 drops of ammonium thiocyanate TS and 2 drops of ferric chloride TS, and mix. To the deep red solution add about 50 mg of Edetate Disodium, and mix: the red color is discharged, leaving a yellowish solution.

C: It responds to the flame test for *Sodium* ⟨191⟩.

pH ⟨791⟩: between 4.0 and 6.0, in a solution (1 in 20).

Loss on drying (see *Thermal Analysis* ⟨891⟩)—[NOTE—The quantity taken for the determination may be adjusted, if necessary, for instrument sensitivity. Weight loss occurring at temperatures above about 240°, indicative of decomposition, is not to be interpreted as *Loss on drying.*] Determine the percentage of volatile substances by thermogravimetric analysis on an appropriately calibrated instrument, using 10 to 25 mg of edetate disodium, accurately weighed. Heat the specimen under test at the rate of 5° per minute in an atmosphere of nitrogen, at a flow rate of 40 mL per minute. Record the thermogram over a 200° range: it loses not less than 8.7% and not more than 11.4% of its weight.

Calcium—To a solution (1 in 20) add 2 drops of methyl red TS, and neutralize with 6 *N* ammonium hydroxide. Add 3 *N* hydrochloric acid dropwise until the solution is just acid, and then add 1 mL of ammonium oxalate TS: no precipitate is formed.

Heavy metals, *Method II* ⟨231⟩: 0.005%.

Nitrilotriacetic acid—

Mobile phase—Add 10 mL of a 1 in 4 solution of tetrabutylammonium hydroxide in methanol to 200 mL of water, and adjust with 1 *M* phosphoric acid to a pH of 7.5 ± 0.1. Transfer the solution to a 1000-mL volumetric flask, add 90 mL of methanol, dilute with water to volume, mix, filter through a membrane filter (0.5-µm or finer porosity), and degas.

Cupric nitrate solution—Prepare a solution containing about 10 mg per mL.

Stock standard solution—Transfer about 100 mg of nitrilotriacetic acid, accurately weighed, to a 10-mL volumetric flask, add 0.5 mL of ammonium hydroxide, and mix. Dilute with water to volume, and mix.

Standard preparation—Transfer 1.0 g of Edetate Disodium to a 100-mL volumetric flask, add 100 µL of *Stock standard solution*, dilute with *Cupric nitrate solution* to volume, and mix. Sonicate, if necessary, to achieve complete solution.

Test preparation—Transfer 1.0 g of Edetate Disodium to a 100-mL volumetric flask, dilute with *Cupric nitrate solution* to volume, and mix. Sonicate, if necessary, to achieve complete solution.

Chromatographic system (see *Chromatography* ⟨621⟩)—The chromatograph is equipped with a 254-nm detector and a 4.6-mm × 15-cm column that contains packing L7. The flow rate is about 2 mL per minute. Chromatograph three replicate injections of the *Standard preparation*, and record the peak responses as directed under *Procedure:* the relative standard deviation is not more than 2.0%, and the resolution factor between nitrilotriacetic acid and edetate disodium is not less than 4.0.

Procedure—Separately inject equal volumes (about 50 µL) of the *Standard preparation* and the *Test preparation* into the chromatograph, record the chromatograms, and measure the responses for the major peaks. The retention times are about 3.5 minutes for nitrilotriacetic acid and 9 minutes for edetate disodium. The response of the nitrilotriacetic acid peak of the *Test preparation* does not exceed the difference between the nitrilotriacetic acid peak responses obtained from the *Standard preparation* and the *Test preparation* (0.1%).

Assay—

Assay preparation—Dissolve about 5 g of Edetate Disodium, accurately weighed, in about 100 mL of water contained in a 250-mL volumetric flask, add water to volume, and mix.

Procedure—Place about 200 mg of chelometric standard calcium carbonate, previously dried at 300° for 3 hours, cooled in a desiccator for 2 hours, and accurately weighed, in a 400-mL beaker, add 10 mL of water, and swirl to form a slurry. Cover the beaker with a watch glass, and without removing the latter, add 2 mL of 3 *N* hydrochloric acid from a pipet. Swirl the contents of the beaker, and dissolve the calcium carbonate. Wash down the sides of the beaker, the outer surface of the pipet, and the watch glass with water, and dilute with water to about 100 mL. While stirring the solution, preferably with a magnetic stirrer, add about 30 mL of *Assay preparation* from a 50-mL buret. Add 15 mL of 1 *N* sodium hydroxide and 0.30 g of hydroxy naphthol blue trituration, and continue the titration with the *Assay preparation* to a blue end-point. Calculate the weight, in mg, of $C_{10}H_{14}N_2Na_2O_8$ in the Edetate Disodium taken by the formula:

$$839.8(W/V),$$

in which *W* is the weight, in mg, of calcium carbonate and *V* is the volume, in mL, of the *Assay preparation* consumed in the titration.

Edetate Disodium Injection

» Edetate Disodium Injection is a sterile solution of Edetate Disodium in Water for Injection, which, as a result of pH adjustment, contains varying amounts of the disodium and trisodium salts. It contains the equivalent of not less than 90.0 percent and not more than 110.0 percent of the labeled amount of $C_{10}H_{14}N_2Na_2O_8$.

Packaging and storage—Preserve in single-dose containers, preferably of Type I glass.

Reference standard—*USP Edetate Disodium Reference Standard.*

Identification—Transfer a volume of Injection, equivalent to about 1 g of edetate disodium, to an evaporating dish, adjust with 3 *N* hydrochloric acid to a pH of 5.0, and evaporate on a steam bath to dryness: the residue responds to *Identification tests A* and *C* under *Edetate Disodium.*

Pyrogen—To a measured volume of Injection, representing 3.0 g of edetate disodium, add 37.5 mL of Calcium Gluconate Injection and sufficient Water for Injection to give a final concentration of 50 mg of edetate disodium per mL: the solution meets the requirements of the *Pyrogen Test* ⟨151⟩, the test dose being 5 mL of the solution per kg.

pH ⟨791⟩: between 6.5 and 7.5.

Other requirements—It meets the requirements under *Injections* ⟨1⟩.

Assay—

Assay preparation—Dilute an accurately measured volume of Edetate Disodium Injection, equivalent to about 2 g of edetate disodium, with water to volume in a 100-mL volumetric flask, and mix.

Procedure—Proceed as directed for *Procedure* in the *Assay* under *Edetate Disodium*, but use the formula:

$$335.9(W/V),$$

in calculating the weight of $C_{10}H_{14}N_2Na_2O_8$.

Edetic Acid—*see* Edetic Acid NF

Edrophonium Chloride

[Structure: HO–C6H4–N⁺(CH3)2(C2H5) Cl⁻]

$C_{10}H_{16}ClNO$ 201.70

Benzenaminium, *N*-ethyl-3-hydroxy-*N*,*N*-dimethyl-, chloride.

Ethyl(*m*-hydroxyphenyl)dimethylammonium chloride [*116-38-1*].

» Edrophonium Chloride contains not less than 98.0 percent and not more than 100.5 percent of $C_{10}H_{16}ClNO$, calculated on the dried basis.

Packaging and storage—Preserve in well-closed containers.

Reference standard—*USP Edrophonium Chloride Reference Standard*—Dry in vacuum over phosphorus pentoxide for 3 hours before using.

Identification—

A: The infrared absorption spectrum of a potassium bromide dispersion of it, previously dried, exhibits maxima only at the same wavelengths as that of a similar preparation of USP Edrophonium Chloride RS.

B: The ultraviolet absorption spectrum of a 1 in 20,000 solution in 0.1 *N* hydrochloric acid exhibits maxima and minima at the same wavelengths as that of a similar solution of USP Edrophonium Chloride RS, concomitantly measured, and the respective absorptivities, calculated on the dried basis, at the wavelength of maximum absorbance at about 273 nm, do not differ by more than 2.0%.

C: To 10 mL of a solution (1 in 10) add 1 drop of ferric chloride TS: a violet-blue color is produced.

D: It responds to the tests for *Chloride* ⟨191⟩.

Melting range ⟨741⟩: between 165° and 170°, with decomposition.

pH ⟨791⟩: between 4.0 and 5.0, in a solution (1 in 10).

Loss on drying ⟨731⟩—Dry it in a suitable vacuum drying tube, using phosphorus pentoxide as the desiccant, for 3 hours: it loses not more than 0.5% of its weight.

Residue on ignition ⟨281⟩: not more than 0.1%.

Heavy metals ⟨231⟩—Dissolve 1.0 g in 25 mL of water: the limit is 0.002%.

Dimethylaminophenol—Dissolve about 500 mg, accurately weighed, in 5 mL of water in a separator. Add 5 mL of pH 8.0 phosphate buffer (see *Buffer Solutions* in the section, *Reagents, Indicators, and Solutions*), and shake. Extract with five 20-mL portions of chloroform, pooling the extracts in a 100-mL volumetric flask, and add chloroform to volume. The absorbance of this solution, determined in a 1-cm cell at 252 nm, with a suitable spectrophotometer, is not more than that of a 1 in 200,000 solution of dimethylaminophenol in chloroform, similarly measured (0.1%).

Assay—Transfer about 175 mg of Edrophonium Chloride, accurately weighed, to a suitable flask, and dissolve in about 20 mL of glacial acetic acid. Add 5 mL of mercuric acetate TS, then add 1 drop of crystal violet TS, and titrate with 0.1 *N* perchloric acid VS. Perform a blank determination, and make any necessary correction. Each mL of 0.1 *N* perchloric acid is equivalent to 20.17 mg of $C_{10}H_{16}ClNO$.

Edrophonium Chloride Injection

» Edrophonium Chloride Injection is a sterile solution of Edrophonium Chloride in Water for Injection. It contains not less than 95.0 percent and not more than 105.0 percent of the labeled amount of $C_{10}H_{16}ClNO$.

Packaging and storage—Preserve in single-dose or in multiple-dose containers, preferably of Type I glass.

Labeling—Label Injection in multiple-dose containers to indicate an expiration date of not later than 3 years after date of manufacture, and label Injection in single-dose containers to indicate an expiration date of not later than 4 years after the date of manufacture.

Reference standard—*USP Edrophonium Chloride Reference Standard*—Dry in vacuum over phosphorus pentoxide for 3 hours before using.

Identification—

A: The ultraviolet absorption spectrum of the solution employed for measurement of absorbance in the *Assay* exhibits maxima and minima at the same wavelengths as that of the solution of USP Edrophonium Chloride RS used in the *Assay*.

B: Place in a small separator a volume of Injection, equivalent to about 30 mg of edrophonium chloride, add 15 mL of pH 9.6 alkaline borate buffer (see *Buffer Solutions* in the section, *Reagents, Indicators, and Solutions*) and 5 mL of a 1 in 1000 solution of thymol blue in pH 9.6 alkaline borate buffer, and mix. Add 10 mL of chloroform, shake thoroughly, and allow to settle: a yellow color is produced in the chloroform layer.

C: It responds to the tests for *Chloride* ⟨191⟩.

Pyrogen—It meets the requirements of the *Pyrogen Test* ⟨151⟩, the test dose being 1 mL per kg, of a solution prepared to contain 1 mg of edrophonium chloride per mL in pyrogen-free saline TS.

pH ⟨791⟩: between 5.0 and 5.8.

Other requirements—It meets the requirements under *Injections* ⟨1⟩.

Assay—Pipet a volume of Edrophonium Chloride Injection, equivalent to about 50 mg of edrophonium chloride, into a glass-stoppered, 50-mL centrifuge tube. Add 5 mL of pH 8.0 phosphate buffer (see *Buffer Solutions* in the section, *Reagents, Indicators, and Solutions*) and 5 g of sodium chloride. Wash the solution with four 20-mL portions of a mixture of equal volumes of solvent hexane and ether. Transfer the aqueous phase to a 100-mL volumetric flask, add 0.1 *N* hydrochloric acid to volume, and mix. Transfer a 5-mL aliquot of this solution to a 50-mL volumetric flask, add 0.1 *N* hydrochloric acid to volume, and mix. Dissolve an accurately weighed quantity of USP Edrophonium Chloride RS in 0.1 *N* hydrochloric acid, and dilute quantitatively and stepwise with the acid to obtain a Standard solution having a known concentration of about 50 µg per mL. Concomitantly determine the absorbances of both solutions in 1-cm cells at the wavelength of maximum absorbance at about 273 nm, with a suitable spectrophotometer, using 0.1 *N* hydrochloric acid as the blank. Calculate the quantity, in mg, of $C_{10}H_{16}ClNO$ in each mL of the Injection taken by the formula:

$$(C/V)(A_U/A_S),$$

in which C is the concentration, in µg per mL, of USP Edrophonium Chloride RS in the Standard solution, V is the volume, in mL, of Injection taken, and A_U and A_S are the absorbances of the solution from the Injection and the Standard solution, respectively.

Elixirs—*see complete list in index*

Emetine Hydrochloride

$C_{29}H_{40}N_2O_4 \cdot 2HCl$ 553.57

Emetan, 6′,7′,10,11-tetramethoxy-, dihydrochloride.

Emetine dihydrochloride [*316-42-7*].

» Emetine Hydrochloride is the hydrochloride of an alkaloid obtained from Ipecac, or prepared by methylation of cephaeline, or prepared synthetically. It contains not less than 98.0 percent and not more than 101.5 percent of $C_{29}H_{40}N_2O_4 \cdot 2HCl$, calculated on the anhydrous basis.

Packaging and storage—Preserve in tight, light-resistant containers.

Reference standards—*USP Emetine Hydrochloride Reference Standard*—Dry small portions only, well-exposed, at 105° to constant weight before using. *USP Cephaeline Hydrobromide Reference Standard*—Dry a portion at 105° to constant weight before using.

Identification—

A: The infrared absorption spectrum of a potassium bromide dispersion of it, previously dried at 105° for 2 hours, exhibits maxima only at the same wavelengths as that of a similar preparation of USP Emetine Hydrochloride RS.

B: The ultraviolet absorption spectrum of a 1 in 20,000 solution in 0.5 *N* sulfuric acid exhibits maxima and minima only at the same wavelengths as that of a similar solution of USP Emetine Hydrochloride RS, concomitantly measured.

C: A solution (1 in 20) responds to the tests for *Chloride* ⟨191⟩.

Water, *Method I* ⟨921⟩: between 15.0% and 19.0%.

Residue on ignition ⟨281⟩: not more than 0.2%.

Acidity—Dissolve 100 mg in 10 mL of water, add 1 drop of methyl red TS, and titrate with 0.020 *N* sodium hydroxide: not more than 0.5 mL is required to produce a yellow color.

Cephaeline—[NOTE—Conduct this test in subdued light until after the chromatogram has been completely developed.]

Standard preparation—Dissolve 23 mg of USP Cephaeline Hydrobromide RS in 100.0 mL of methanol.

Test preparation—Dissolve 100 mg of Emetine Hydrochloride in 10.0 mL of methanol.

Spray reagent—Dissolve 300 mg of *p*-nitroaniline in 25 mL of 2 *N* hydrochloric acid, and cool to about 4°. Slowly add 5 mL of sodium nitrite solution (1 in 25), maintaining the temperature at about 4°. Freshly prepare the solution for each test.

Procedure—Apply 10-µL portions of the *Standard preparation* and the *Test preparation*, respectively, on a suitable thin-layer chromatographic plate coated with a 0.25-mm layer of chromatographic silica gel. Place the plate in a chromatographic tank containing a mixture of 9 volumes of chloroform and 1 volume of diethylamine, and develop the chromatogram until the solvent front has moved about 12 cm. Remove the plate from the tank, and allow the plate to dry in air for 20 minutes. Spray the dried plate with 2.5 *N* sodium hydroxide solution, and dry at 50° for 5 minutes. Then spray the plate with *Spray reagent:* any cephaeline spot from the *Test preparation* is not larger or more intense than that produced by the *Standard preparation* (2%).

Assay—Dissolve about 150 mg of Emetine Hydrochloride, accurately weighed, in 5 mL of glacial acetic acid, warming, if necessary, to effect solution. Allow the solution to cool, add 10 mL of dioxane, 5 mL of mercuric acetate TS, and 3 drops of crystal violet TS, and titrate with 0.1 *N* perchloric acid VS. Perform a blank determination, and make any necessary correction. Each mL of 0.1 *N* perchloric acid is equivalent to 27.68 mg of $C_{29}H_{40}N_2O_4 \cdot 2HCl$.

Emetine Hydrochloride Injection

» Emetine Hydrochloride Injection is a sterile solution of Emetine Hydrochloride in Water for Injection. It contains an amount of anhydrous emetine hydrochloride ($C_{29}H_{40}N_2O_4 \cdot 2HCl$) equivalent to not less than 84.0 percent and not more than 94.0 percent of the labeled amount of emetine hydrochloride.

Packaging and storage—Preserve in single-dose, light-resistant containers, preferably of Type I glass.

Reference standards—*USP Emetine Hydrochloride Reference Standard*—Dry small portions only, well exposed, at 105° to constant weight before using. *USP Cephaeline Hydrobromide Reference Standard*—Dry a portion at 105° to constant weight before using.

Identification—

A: Evaporate 1 mL of Injection on a steam bath to dryness: the residue responds to *Identification test A* under *Emetine Hydrochloride.*

B: The Injection responds to *Identification test C* under *Emetine Hydrochloride.*

pH ⟨791⟩: between 3.0 and 5.0.

Cephaeline—[NOTE—Conduct this test in subdued light until after the chromatogram has been completely developed.]

Standard preparation and *Spray reagent*—Prepare as directed in the test for *Cephaeline* under *Emetine Hydrochloride.*

Test preparation—Evaporate 1.0 mL of Emetine Hydrochloride Injection on a steam bath with the aid of a current of nitrogen just to dryness. Dissolve the residue in a sufficient volume of methanol to obtain a *Test preparation* having a concentration of 10 mg of emetine hydrochloride per mL.

Procedure—Proceed as directed for *Procedure* in the test for *Cephaeline* under *Emetine Hydrochloride.*

Other requirements—It meets the requirements under *Injections* ⟨1⟩.

Assay—Transfer an accurately measured volume of Emetine Hydrochloride Injection, equivalent to about 120 mg of emetine hydrochloride, to a suitable extraction apparatus containing about 20 mL of water. Add 6 *N* ammonium hydroxide until the solution is strongly alkaline, and extract with ether until 0.5 mL of the water layer, slightly acidified with hydrochloric acid, remains unaffected by the addition of a few drops of mercuric-potassium iodide TS. Evaporate the combined ether extracts on a steam bath, allowing the last few mL of the ether to evaporate spontaneously. Add to the residue 2 mL of neutralized alcohol and 30.0 mL of 0.02 *N* sulfuric acid VS, and warm gently until the alkaloid is dissolved. Cool, add methyl red TS, and titrate the excess acid with 0.02 *N* sodium hydroxide VS. Each mL of 0.02 *N* sulfuric acid is equivalent to 5.536 mg of $C_{29}H_{40}N_2O_4 . 2HCl$.

Emulsifying Wax—*see* Wax, Emulsifying NF

Emulsion, Mineral Oil—*see* Mineral Oil Emulsion

Enalapril Maleate

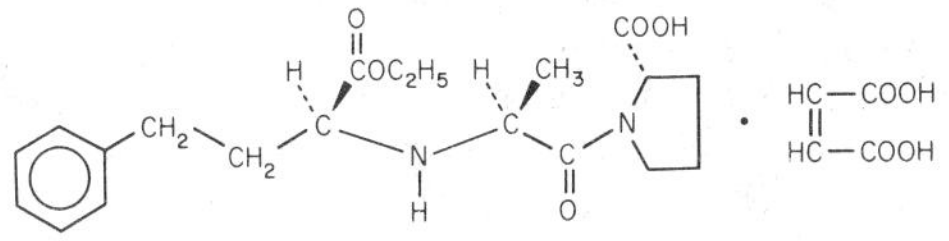

$C_{20}H_{28}N_2O_5 . C_4H_4O_4$ 492.52

L-Proline, 1-[*N*-[1-(ethoxycarbonyl)-3-phenylpropyl]-L-alanyl]-, (*S*)-, (*Z*)-2-butenedioate (1:1).

1-[*N*-[(*S*)-1-Carboxy-3-phenylpropyl]-L-alanyl]-L-proline 1′-ethyl ester, maleate (1:1) [*76095-16-4*].

» Enalapril Maleate contains not less than 98.0 percent and not more than 102.0 percent of $C_{20}H_{28}N_2O_5 . C_4H_4O_4$, calculated on the dried basis.

Packaging and storage—Preserve in well-closed containers.

Reference standard—*USP Enalapril Maleate Reference Standard*—Dry in vacuum at a pressure not exceeding 5 mm of mercury at 60° for 2 hours before using.

Identification—The infrared absorption spectrum of a mineral oil dispersion of it, previously dried, exhibits maxima only at the same wavelengths as that of a similar preparation of USP Enalapril Maleate RS.

Specific rotation ⟨781⟩: between −41.0° and −43.5°, calculated on the dried basis, determined in a solution in methanol containing 100 mg in each 10 mL.

Loss on drying ⟨731⟩—Dry it in vacuum at a pressure not exceeding 5 mm of mercury at 60° for 2 hours: it loses not more than 1.0% of its weight.

Residue on ignition ⟨281⟩: not more than 0.5%.

Heavy metals, *Method II* ⟨231⟩: 0.001%.

Assay—

pH 6.8 phosphate buffer—Dissolve 2.8 g of monobasic sodium phosphate in about 900 mL of water. Add a volume of sodium hydroxide solution (1 in 4) sufficient to adjust to a pH of 6.8, dilute with water to 1 liter, and mix.

Mobile phase—Prepare a filtered and degassed mixture of *pH 6.8 phosphate buffer* and acetonitrile (4:1). Make adjustments if necessary (see *System Suitability* under *Chromatography* ⟨621⟩).

Standard preparation—[NOTE—Prepare this solution fresh daily.] Dissolve an accurately weighed quantity of USP Enalapril Maleate RS in *Mobile phase*, and dilute quantitatively with *Mobile phase* to obtain a solution having a known concentration of about 0.3 mg per mL.

Assay preparation—Transfer about 30 mg of Enalapril Maleate, accurately weighed, to a 100-mL volumetric flask, dissolve in *Mobile phase*, dilute with *Mobile phase* to volume, and mix.

Chromatographic system (see *Chromatography* ⟨621⟩)—The liquid chromatograph is equipped with a 210-nm detector and a 4.1-mm × 25-cm column containing packing L21 and maintained at a temperature of 70°. The flow rate is about 1 mL per minute. Chromatograph the *Standard preparation*, and record the peak responses as directed under *Procedure:* the column efficiency determined from the analyte peak is not less than 300 theoretical plates, the tailing factor for the analyte peak is not more than 1.3, and the relative standard deviation for replicate injections is not more than 1.0%.

Procedure—Separately inject equal volumes (about 10 µL) of the *Standard preparation* and the *Assay preparation* into the chromatograph, record the chromatograms, and measure the responses for the major peaks. Calculate the quantity, in mg, of $C_{20}H_{28}N_2O_5 . C_4H_4O_4$ in the portion of Enalapril Maleate taken by the formula:

$$100C(r_U / r_S),$$

in which *C* is the concentration, in mg per mL, of USP Enalapril Maleate RS in the *Standard preparation*, and r_U and r_S are the enalapril peak responses obtained from the *Assay preparation* and the *Standard preparation*, respectively.

Enalapril Maleate Tablets

» Enalapril Maleate Tablets contain not less than 90.0 percent and not more than 110.0 percent of the labeled amount of $C_{20}H_{28}N_2O_5 . C_4H_4O_4$.

Packaging and storage—Preserve in well-closed containers.

Reference standard—*USP Enalapril Maleate Reference Standard*—Dry in vacuum at a pressure not exceeding 5 mm of mercury at 60° for 2 hours before using.

Identification—The retention time of the major peak in the chromatogram of the *Assay preparation* corresponds to that of the *Standard preparation*, as obtained in the *Assay.*

Dissolution ⟨711⟩—

Medium: water; 900 mL.

Apparatus 2: 50 rpm.

Time: 30 minutes.

Procedure—Determine the amount of $C_{20}H_{28}N_2O_5 . C_4H_4O_4$ dissolved, employing the procedure set forth for *Content Uniformity*, making any necessary modifications for appropriate sample and standard concentrations.

Tolerances—Not less than 80% (*Q*) of the labeled amount of $C_{20}H_{28}N_2O_5 . C_4H_4O_4$ is dissolved in 30 minutes.

Uniformity of dosage units ⟨905⟩: meet the requirements.

Procedure for content uniformity—

Buffer solution A—Dissolve 136 g of monobasic potassium phosphate in 800 mL of water, adjust with phosphoric acid to a pH of 4.0, dilute with water to 1000 mL, and mix.

Buffer solution B—Transfer 20.0 mL of *Buffer solution A* to a 1000-mL volumetric flask, dilute with water to volume, and mix.

Mobile phase—Prepare a filtered and degassed mixture of water, acetonitrile, and *Buffer solution A* (34:15:1). Make adjustments if necessary (see *System Suitability* under *Chromatography* ⟨621⟩).

Standard preparation—Dissolve an accurately weighed quantity of USP Enalapril Maleate RS in *Buffer solution B* to obtain a solution having a known concentration of about 100 µg per mL.

Test preparation—Transfer one finely powdered Tablet to a 50-mL volumetric flask, add about 30 mL of *Buffer solution B*, and sonicate for 15 minutes. Dilute with *Buffer solution B* to volume, sonicate for 15 minutes, mix, and filter, discarding the first portion of the filtrate. Dilute a portion of the filtrate with *Buffer solution B* to obtain a solution containing about 100 µg per mL.

Chromatographic system—Prepare as directed in the *Assay*.

Procedure—Separately inject equal volumes (about 50 µL) of the *Test preparation* and the *Standard preparation* into the chromatograph, record the chromatograms, and measure the responses for the major peaks. Calculate the quantity, in mg, of $C_{20}H_{28}N_2O_5 \cdot C_4H_4O_4$ in the Tablet taken by the formula:

$$(TC/D)(r_U/r_S),$$

in which T is the labeled quantity, in mg, of enalapril maleate in the Tablet, C is the concentration, in µg per mL, of USP Enalapril Maleate RS in the *Standard preparation*, D is the concentration, in µg per mL, of enalapril maleate in the *Test preparation*, based upon the labeled quantity per Tablet and the extent of dilution, and r_U and r_S are the enalapril peak responses obtained from the *Test preparation* and the *Standard preparation*, respectively.

Related compounds—

Mobile phase, *Buffer solution*, and *Chromatographic system*—Prepare as directed in the *Assay*.

Standard preparation—Dilute the *Standard preparation* prepared as directed in the *Assay* with *Buffer solution* to obtain a solution having a known concentration of about 20 µg per mL.

Test preparation—Prepare as directed for *Assay preparation* in the *Assay*.

Procedure—Proceed as directed in the *Assay*. Measure the responses of the enalapril peak obtained from the *Standard preparation* and of all peaks other than that of enalapril obtained from the *Test preparation*. Calculate the amount, in mg, of related compounds in the portion of Tablets taken by the formula:

$$0.25C(r_U/r_S),$$

in which C is the concentration, in µg per mL, of USP Enalapril Maleate RS in the *Standard preparation*, r_U is the sum of all peak responses other than that of enalapril in the *Test preparation*, and r_S is the enalapril peak response obtained from the *Standard preparation*: not more than 5.0%, calculated on the basis of the quantity, in mg, of enalapril maleate in the portion of Tablets taken, as determined in the *Assay*, is found.

Assay—

Buffer solution—Dissolve 136 mg of monobasic potassium phosphate in 800 mL of water, adjust with phosphoric acid to a pH of 2.0, dilute with water to 1000 mL, and mix.

Mobile phase—Prepare a filtered and degassed mixture of *Buffer solution* and acetonitrile (17:8). Make adjustments if necessary (see *System Suitability* under *Chromatography* ⟨621⟩).

Standard preparation—Dissolve an accurately weighed quantity of USP Enalapril Maleate RS in *Buffer solution*, and dilute quantitatively with *Buffer solution* to obtain a solution having a known concentration of about 200 µg per mL.

Assay preparation—Weigh and finely powder not less than 20 Enalapril Maleate Tablets. Transfer an accurately weighed portion of the powder, equivalent to about 50 mg of enalapril maleate, to a 250-mL volumetric flask. Add about 150 mL of *Buffer solution*, and sonicate for 15 minutes. Dilute with *Buffer solution* to volume, sonicate for 15 minutes, mix, and filter, discarding the first portion of the filtrate.

Chromatographic system (see *Chromatography* ⟨621⟩)—The liquid chromatograph is equipped with a 215-nm detector and a 4.6-mm × 20-cm column containing packing L7 and maintained at a temperature of 80°. The flow rate is about 2 mL per minute. Chromatograph the *Standard preparation*, and record the peak responses as directed under *Procedure:* the column efficiency determined from the analyte peak is not less than 1000 theoretical plates, the tailing factor for the analyte peak is not more than 2.0, and the relative standard deviation for replicate injections is not more than 2.0%.

Procedure—Separately inject equal volumes (about 50 µL) of the *Standard preparation* and the *Assay preparation* into the chromatograph, record the chromatograms, and measure the responses for the major peaks. Calculate the quantity, in mg, of $C_{20}H_{28}N_2O_5 \cdot C_4H_4O_4$ in the portion of Tablets taken by the formula:

$$0.25C(r_U/r_S),$$

in which C is the concentration, in µg per mL, of USP Enalapril Maleate RS in the *Standard preparation*, and r_U and r_S are the enalapril peak responses obtained from the *Assay preparation* and the *Standard preparation*, respectively.

Enemas—*see complete list in index*

Enflurane

```
    F     F  F
    |     |  |
  H-C--O--C--C-H
    |     |  |
    F     F  Cl
```

$C_3H_2ClF_5O$ 184.49

Ethane, 2-chloro-1-(difluoromethoxy)-1,1,2-trifluoro-.
2-Chloro-1,1,2-trifluoroethyl difluoromethyl ether [*13838-16-9*].

» Enflurane contains not less than 99.9 percent and not more than 100.0 percent of $C_3H_2ClF_5O$, calculated on the anhydrous basis.

Packaging and storage—Preserve in tight, light-resistant containers, and avoid exposure to excessive heat.

Reference standard—*USP Enflurane Reference Standard*—After opening ampul, store in tightly closed, light-resistant container.

Identification—The infrared absorption spectrum of a thin film of it exhibits maxima only at the same wavelengths as that of a similar preparation of USP Enflurane RS.

Specific gravity ⟨841⟩: not less than 1.516 and not more than 1.519.

Distilling range, *Method II* ⟨721⟩: between 55.5° and 57.5°, a correction factor of 0.041° per mm being applied as necessary.

Refractive index ⟨831⟩: not less than 1.3020 and not more than 1.3038 at 20°.

Acidity or alkalinity—Shake 20 mL with 20 mL of carbon dioxide–free water for 3 minutes, and allow the layers to separate: the aqueous layer requires not more than 0.10 mL of 0.010 *N* sodium hydroxide or not more than 0.60 mL of 0.010 *N* hydrochloric acid for neutralization, bromocresol purple TS being used as the indicator.

Water, *Method I* ⟨921⟩: not more than 0.14%.

Nonvolatile residue—Allow 10.0 mL to evaporate at room temperature in a tared evaporating dish, and dry the residue at 50° for 2 hours: the weight of the residue does not exceed 2 mg.

Chloride ⟨221⟩—Shake 25 mL with 25 mL of water for 5 minutes, and allow the liquids to separate completely. Draw off the water layer, and add to it 1 drop of nitric acid and 5 drops of silver nitrate TS: any turbidity produced is no greater than that produced in a solution containing 0.35 mL of 0.020 *N* hydrochloric acid.

Fluoride ions—[NOTE—Use plasticware throughout this test.]

pH 5.25 buffer—Dissolve 110 g of sodium chloride and 1 g of sodium citrate in 700 mL of water in a 2000-mL volumetric flask. Cautiously add 150 g of sodium hydroxide, and dissolve with shaking. Cool to room temperature, and, while stirring, cautiously add 450 mL of glacial acetic acid to the cooled solution. Cool, add 600 mL of isopropyl alcohol, dilute with water to volume, and mix: the pH of this solution is between 5.0 and 5.5.

Standard stock solution—Transfer 221 mg of sodium fluoride, previously dried at 150° for 4 hours, to a 100-mL volumetric flask, add about 20 mL of water, and mix to dissolve. Add 1.0 mL of sodium hydroxide solution (1 in 2500), dilute with water to volume, and mix. Each mL of this solution contains 1 mg of fluoride ions. Store in a tightly closed, plastic container.

Standard preparations—Dilute portions of the *Standard stock solution* quantitatively and stepwise with *pH 5.25 buffer* to obtain 100-mL solutions having concentrations of 1, 3, 5, and 10 µg per mL.

Test preparation—Shake 25 mL with 25 mL of water for 5 minutes, and allow the liquids to separate completely. Transfer 5.0 mL of the water layer to a 10-mL volumetric flask, dilute with *pH 5.25 buffer* to volume, and mix.

Procedure—Concomitantly measure the potential (see *Titrimetry* ⟨541⟩), in mV, of the *Standard preparations* and the *Test preparation*, with a pH meter capable of a minimum reproducibility of ±0.2 mV, equipped with a glass-sleeved calomel-fluoride specific-ion electrode system. [NOTE—When taking measurements, immerse the electrodes in the solution which has been transferred to a 150-mL beaker containing a polytetrafluoroethylene-coated stirring bar. Allow to stir on a magnetic stirrer having an insulated top until equilibrium is attained (1 to 2 minutes), and record the potential. Rinse and dry the electrodes between measurements, taking care to avoid damaging the crystal of the specific-ion electrode.] Plot the logarithm of the fluoride-ion concentrations, in µg per mL, of the *Standard preparations* versus potential, in millivolts. From the measured potential of the *Test preparation* and the standard curve, determine the concentration, in µg per mL, of fluoride ions in the *Test preparation:* not more than 10 µg per mL is found.

Assay—Inject a volume of Enflurane of suitable size, but not more than 30 µL, into a suitable gas chromatograph (see *Gas Chromatography* under *Chromatography* ⟨621⟩) equipped with a thermal conductivity detector. Under typical conditions, the instrument contains a 3-m × 4-mm stainless steel column packed with 20 percent liquid phase G4 on 60- to 80-mesh S1A, the column is temperature-programmed at about 6° per minute from 60° to 125°, and the injection port is maintained at about 200°. Dry helium is used as the carrier gas at a flow rate of about 60 mL per minute. Calculate the percentage purity by dividing 100 times the area under the Enflurane peak by the sum of all of the areas in the chromatogram.

Ephedrine

$C_{10}H_{15}NO$ 165.23
Benzenemethanol, α-[1-(methylamino)ethyl]-, [*R*-(*R**,*S**)]-.
(−)-Ephedrine [*299-42-3*].
Hemihydrate 174.24 [*50906-05-3*].

» Ephedrine is anhydrous or contains not more than one-half molecule of water of hydration. It contains not less than 98.5 percent and not more than 100.5 percent of $C_{10}H_{15}NO$, calculated on the anhydrous basis.

Packaging and storage—Preserve in tight, light-resistant containers, in a cold place.

Labeling—Label it to indicate whether it is hydrous or anhydrous. Where the quantity of Ephedrine is indicated in the labeling of any preparation containing Ephedrine, this shall be understood to be in terms of anhydrous Ephedrine.

Reference standard—*USP Ephedrine Sulfate Reference Standard*—Dry at 105° for 3 hours before using.

Identification—Weigh accurately about 100 mg, and add by buret the exact volume of 0.1 *N* sulfuric acid, determined in the *Assay*, to neutralize it. Dilute with water in a volumetric flask to 25 mL. Mix 2 mL with 10 mL of alcohol, and evaporate on a steam bath with the aid of a current of air to dryness: the residue so obtained responds to *Identification test A* under *Ephedrine Sulfate*.

Specific rotation ⟨781⟩—Dissolve about 600 mg in 10 mL of ether in a tared beaker, add 0.6 mL of hydrochloric acid, evaporate to dryness, and dry the residue at 105° for 3 hours: the specific rotation of the ephedrine hydrochloride so obtained, determined in a solution containing 500 mg in each 10 mL, is between −33.0° and −35.5°.

Water, *Method Ib* ⟨921⟩: between 4.5% and 5.5%, for hydrated Ephedrine; not more than 0.5% for anhydrous Ephedrine.

Residue on ignition ⟨281⟩: not more than 0.1%.

Chloride ⟨221⟩—A solution of 500 mg shows no more chloride than corresponds to 0.20 mL of 0.020 *N* hydrochloric acid (0.030%).

Sulfate—Dissolve 100 mg in 40 mL of water, and add 1 mL of 3 *N* hydrochloric acid and 1 mL of barium chloride TS: no turbidity develops within 10 minutes.

Ordinary impurities ⟨466⟩—
Test solution: methanol.
Standard solution: methanol.
Eluant: a mixture of *n*-butanol, glacial acetic acid, and water (5:3:2); the thin-layer chromatographic plate coating material is cellulose with a fluorescent indicator.
Visualization: 16.

Assay—Dissolve about 500 mg of Ephedrine, accurately weighed, in 10 mL of neutralized alcohol, and add 5 drops of methyl red TS and 40.0 mL of 0.1 *N* hydrochloric acid VS. Titrate the excess acid with 0.1 *N* sodium hydroxide VS. Perform a blank determination (see *Residual Titrations* ⟨541⟩). Each mL of 0.1 *N* hydrochloric acid is equivalent to 16.52 mg of $C_{10}H_{15}NO$.

Ephedrine Hydrochloride

$C_{10}H_{15}NO.HCl$ 201.70
Benzenemethanol, α-[1-(methylamino)ethyl]-, hydrochloride, [*R*-(*R**,*S**)]-.
(−)-Ephedrine hydrochloride [*50-98-6*].

» Ephedrine Hydrochloride contains not less than 98.0 percent and not more than 100.5 percent of $C_{10}H_{15}NO.HCl$, calculated on the dried basis.

Packaging and storage—Preserve in well-closed, light-resistant containers.

Reference standard—*USP Ephedrine Sulfate Reference Standard*—Dry at 105° for 3 hours before using.

Identification—

A: Dissolve 100 mg in 5 mL of water, add 1 mL of potassium carbonate solution (1 in 5), and extract with 2 mL of chloroform: the infrared absorption spectrum of the chloroform extract so obtained exhibits maxima only at the same wavelengths as that of a similar preparation of USP Ephedrine Sulfate RS.

B: A solution of it responds to the tests for *Chloride* ⟨191⟩.

Melting range, *Class I* ⟨741⟩: between 217° and 220°.

Specific rotation ⟨781⟩: between −33.0° and −35.5°, calculated on the dried basis, determined in a solution containing 500 mg in each 10 mL.

Acidity or alkalinity—Dissolve 1.0 g in 20 mL of water, and add 1 drop of methyl red TS. If the solution is yellow, it is changed to red by not more than 0.10 mL of 0.020 *N* sulfuric acid. If the solution is pink, it is changed to yellow by not more than 0.20 mL of 0.020 *N* sodium hydroxide.

Loss on drying ⟨731⟩—Dry it at 105° for 3 hours: it loses not more than 0.5% of its weight.

Residue on ignition ⟨281⟩: not more than 0.1%.

Sulfate—Dissolve 50 mg in 40 mL of water, and add 1 mL of 3 *N* hydrochloric acid and 1 mL of barium chloride TS: no turbidity develops within 10 minutes.

Assay—Dissolve about 500 mg of Ephedrine Hydrochloride, accurately weighed, in 25 mL of glacial acetic acid. Add 10 mL of mercuric acetate TS and 2 drops of crystal violet TS, and titrate with 0.1 *N* perchloric acid VS to an emerald-green endpoint. Perform a blank determination, and make any necessary correction. Each mL of 0.1 *N* perchloric acid is equivalent to 20.17 mg of $C_{10}H_{15}NO.HCl$.

Ephedrine Hydrochloride, and Phenobarbital Tablets, Theophylline,—*see* Theophylline, Ephedrine Hydrochloride, and Phenobarbital Tablets

Ephedrine Sulfate

$(C_{10}H_{15}NO)_2.H_2SO_4$ 428.54

Benzenemethanol, α-[1-(methylamino)ethyl]-, [*R*-(*R**,*S**)]-, sulfate (2:1) (salt).

(−)-Ephedrine sulfate (2:1) (salt) [*134-72-5*].

» Ephedrine Sulfate contains not less than 98.0 percent and not more than 101.0 percent of $(C_{10}H_{15}NO)_2.H_2SO_4$, calculated on the dried basis.

Packaging and storage—Preserve in well-closed, light-resistant containers.

Reference standard—*USP Ephedrine Sulfate Reference Standard*—Dry at 105° for 3 hours before using.

Identification—

A: The infrared absorption spectrum of a potassium bromide dispersion of it, previously triturated with a few drops of methanol and dried at 105° for 3 hours, exhibits maxima only at the same wavelengths as that of a similar preparation of USP Ephedrine Sulfate RS.

B: A solution of it responds to the tests for *Sulfate* ⟨191⟩.

Specific rotation ⟨781⟩: between −30.5° and −32.5°, calculated on the dried basis, determined in a solution containing 500 mg in each 10 mL.

Acidity or alkalinity—Dissolve 1.0 g in 20 mL of water, and add 1 drop of methyl red TS. If the solution is yellow, it is changed to red by not more than 0.10 mL of 0.020 *N* sulfuric acid. If the solution is pink, it is changed to yellow by not more than 0.20 mL of 0.020 *N* sodium hydroxide.

Loss on drying ⟨731⟩—Dry about 500 mg, accurately weighed, at 105° for 3 hours: it loses not more than 0.5% of its weight.

Residue on ignition ⟨281⟩: not more than 0.1%.

Chloride ⟨221⟩—A 200-mg portion shows no more chloride than corresponds to 0.40 mL of 0.020 *N* hydrochloric acid (0.14%).

Ordinary impurities ⟨466⟩—

Test solution: alcohol.

Standard solution: alcohol.

Eluant: a mixture of alcohol, glacial acetic acid, and water (10:3:1).

Visualization: 1.

Assay—Transfer about 300 mg of Ephedrine Sulfate, accurately weighed, to a separator, and dissolve in about 10 mL of water. Saturate the solution with sodium chloride (about 3 g), add 5 mL of 1 *N* sodium hydroxide, and extract with four 25-mL portions of chloroform. Wash the combined chloroform extracts by shaking with 10 mL of a saturated solution of sodium chloride, and filter through chloroform-saturated purified cotton into a beaker. Extract the wash solution with 10 mL of chloroform, and add to the main chloroform extract. Add methyl red TS, and titrate with 0.1 *N* perchloric acid in dioxane VS. Perform a blank determination, and make any necessary correction. Each mL of 0.1 *N* perchloric acid is equivalent to 21.43 mg of $(C_{10}H_{15}NO)_2.H_2SO_4$.

Ephedrine Sulfate Capsules

» Ephedrine Sulfate Capsules contain not less than 92.0 percent and not more than 108.0 percent of the labeled amount of $(C_{10}H_{15}NO)_2.H_2SO_4$.

Packaging and storage—Preserve in tight, light-resistant containers.

Reference standard—*USP Ephedrine Sulfate Reference Standard*—Dry at 105° for 3 hours before using.

Identification—Macerate the contents of a sufficient number of Capsules, equivalent to about 200 mg of ephedrine sulfate, with 15 mL of warm alcohol for 20 minutes, filter, and evaporate the filtrate on a steam bath to dryness: the residue so obtained responds to the *Identification tests* under *Ephedrine Sulfate*.

Dissolution ⟨711⟩—

Medium: water; 500 mL.

Apparatus 1: 100 rpm.

Time: 30 minutes.

Procedure—Dilute filtered portions of the solutions under test with water to a concentration of about 25 µg per mL. Transfer 5.0-mL portions to suitable tubes. Add 1 mL of a saturated sodium carbonate solution and 2 mL of sodium metaperiodate solution (2 in 100) to each, mix, and allow to stand for 10 minutes. Add 20.0 mL of hexanes, shake for 30 seconds, and allow the phases to separate. Measure the absorbances of the hexanes extract in 1-cm cells at the wavelength of maximum absorbance, at about 242 nm, with a suitable spectrophotometer, using hexanes as the blank. Determine the amount of $(C_{10}H_{15}NO)_2.H_2SO_4$ dissolved by comparison with a similarly treated Standard solution having a known concentration of USP Ephedrine Sulfate RS in water. Remove the contents of 1 Capsule as completely as possible, with the aid of a current of air, dissolve the empty capsule shell in the *Dissolution Medium*, determine the absorbance at the same dilution and in the same manner as for the Capsules, and make any necessary corrections.

Tolerances—Not less than 80% (*Q*) of the labeled amount of $(C_{10}H_{15}NO)_2.H_2SO_4$ is dissolved in 30 minutes.

Uniformity of dosage units ⟨905⟩: meet the requirements.

Assay—

Standard preparation—Weigh accurately about 25 mg of USP Ephedrine Sulfate RS, transfer to a 50-mL volumetric flask with the aid of 10 mL of water, add methanol to volume, and mix. Dilute 5.0 mL of this solution with water to 100.0 mL.

Assay preparation—Weigh accurately the contents of not less than 20 Ephedrine Sulfate Capsules, and mix. Transfer an accurately weighed portion of the mixture, equivalent to about 25 mg of ephedrine sulfate, to a glass-stoppered conical flask, and add by pipet 50 mL of a 1 in 5 mixture of water in methanol. Shake by mechanical means for 10 minutes, and filter. Dilute 5.0 mL of the filtrate with water to 100.0 mL.

Procedure—Transfer 5-mL portions of the *Assay preparation* and the *Standard preparation* to separate glass-stoppered, 50-mL centrifuge tubes. Add 1 mL of saturated sodium carbonate solution and 2 mL of sodium metaperiodate solution (1 in 50) to each tube, mix, and allow to stand for 10 minutes. Pipet 20 mL of *n*-hexane into each tube, shake for 30 seconds, and allow the phases to separate. Concomitantly determine the absorbances of the *n*-hexane extracts in 1-cm cells at the wavelength of maximum absorbance at about 242 nm, with a suitable spectrophotometer, using *n*-hexane as the blank. Calculate the quantity, in mg, of $(C_{10}H_{15}NO)_2.H_2SO_4$ in the portion of Capsule contents taken by the formula:

$$C(A_U/A_S),$$

in which *C* is the concentration, in µg per mL, of USP Ephedrine Sulfate RS in the *Standard preparation*, and A_U and A_S are the absorbances of the hexane extracts of the *Assay preparation* and the *Standard preparation*, respectively.

Ephedrine Sulfate Injection

» Ephedrine Sulfate Injection is a sterile solution of Ephedrine Sulfate in Water for Injection. It contains not less than 95.0 percent and not more than 105.0 percent of the labeled amount of $(C_{10}H_{15}NO)_2 \cdot H_2SO_4$.

Packaging and storage—Preserve in single-dose or in multiple-dose, light-resistant containers, preferably of Type I glass.

Reference standard—*USP Ephedrine Sulfate Reference Standard*—Dry at 105° for 3 hours before using.

Identification—

A: Mix 1 mL of Injection with 5 mL of alcohol, and evaporate on a steam bath with the aid of a current of air to dryness: the residue so obtained responds to *Identification test A* under *Ephedrine Sulfate*.

B: It responds to the tests for *Sulfate* ⟨191⟩.

C: The angular rotation of the Injection is levorotatory.

pH ⟨791⟩: between 4.5 and 7.0.

Other requirements—It meets the requirements under *Injections* ⟨1⟩.

Assay—Transfer an accurately measured volume of Ephedrine Sulfate Injection, equivalent to about 250 mg of ephedrine sulfate, to a separator, add water, if necessary, to make about 10 mL, and proceed as directed in the *Assay* under *Ephedrine Sulfate*, beginning with "Saturate the solution."

Ephedrine Sulfate Nasal Solution

» Ephedrine Sulfate Nasal Solution contains not less than 93.0 percent and not more than 107.0 percent of the labeled amount of $(C_{10}H_{15}NO)_2 \cdot H_2SO_4$.

Packaging and storage—Preserve in tight, light-resistant containers.

Reference standard—*USP Ephedrine Sulfate Reference Standard*—Dry at 105° for 3 hours before using.

Identification—It responds to the *Identification tests* under *Ephedrine Sulfate Injection.*

Microbial limits—It meets the requirements of the tests for absence of *Staphylococcus aureus* and *Pseudomonas aeruginosa* under *Microbial Limit Tests* ⟨61⟩.

Assay—

Standard preparation—Weigh accurately about 26 mg of USP Ephedrine Sulfate RS, transfer to a 50-mL volumetric flask with the aid of 10 mL of water, add methanol to volume, and mix. Pipet 5 mL of the resulting solution into a 100-mL volumetric flask, dilute with water to volume, and mix.

Assay preparation—Transfer an accurately measured volume of Ephedrine Sulfate Nasal Solution, equivalent to about 26 mg of ephedrine sulfate, to a 50-mL volumetric flask, dilute with a 1 in 5 mixture of water in methanol to volume, and mix. Pipet 5 mL of the resulting solution into a 100-mL volumetric flask, dilute with water to volume, and mix.

Procedure—Transfer 5-mL portions of the *Assay preparation* and the *Standard preparation* to separate glass-stoppered, 50-mL centrifuge tubes. Add 1 mL of saturated sodium carbonate solution and 2 mL of sodium metaperiodate solution (1 in 50) to each tube, mix, and allow to stand for 10 minutes. Pipet 20 mL of *n*-hexane into each tube, shake for 30 seconds, and allow the phases to separate. Concomitantly determine the absorbances of the *n*-hexane extracts in 1-cm cells at the wavelength of maximum absorbance at about 242 nm, with a suitable spectrophotometer, using *n*-hexane as the blank. Calculate the quantity, in mg, of $(C_{10}H_{15}NO)_2 \cdot H_2SO_4$ in each mL of the Nasal Solution taken by the formula:

$$(C/V)(A_U/A_S),$$

in which V is the volume, in mL, of Nasal Solution taken, C is the concentration, in µg per mL, of USP Ephedrine Sulfate RS in the *Standard preparation,* and A_U and A_S are the absorbances of the hexane extracts of the *Assay preparation* and the *Standard preparation*, respectively.

Ephedrine Sulfate Syrup

» Ephedrine Sulfate Syrup contains, in each 100 mL, not less than 360 mg and not more than 440 mg of $(C_{10}H_{15}NO)_2 \cdot H_2SO_4$.

Packaging and storage—Preserve in tight, light-resistant containers, and avoid exposure to excessive heat.

Reference standard—*USP Ephedrine Sulfate Reference Standard*—Dry at 105° for 3 hours before using.

Identification—The 0.1 *N* sulfuric acid extract of the chloroform solution obtained as directed under *Assay preparation* responds to *Identification test C* under *Ephedrine Sulfate Injection.*

Alcohol content ⟨611⟩: between 2.0% and 4.0% of C_2H_5OH.

Assay—

Standard preparation—Dissolve an accurately weighed quantity of USP Ephedrine Sulfate RS in 0.1 *N* sulfuric acid to obtain a solution having a known concentration of about 20 µg per mL.

Assay preparation—Transfer 5 mL of Ephedrine Sulfate Syrup to a separator, add 1 mL of 1 *N* sulfuric acid, and extract with 10 mL of chloroform. Discard the extract, and add 5 mL of potassium carbonate solution (1 in 5). After gas evolution has ceased, extract the solution with three 10-mL portions of chloroform, and combine the extracts in a second separator. Extract the chloroform solution with 50.0 mL of 0.1 *N* sulfuric acid. Filter the acid layer through paper, and dilute 5.0 mL of it with 0.1 *N* sulfuric acid to 100.0 mL.

Procedure—Proceed as directed for *Procedure* in the *Assay* under *Ephedrine Sulfate Capsules.* Calculate the quantity, in mg, of $(C_{10}H_{15}NO)_2 \cdot H_2SO_4$ in the portion of Syrup taken by the formula:

$$C(A_U/A_S),$$

in which C is the concentration, in µg per mL, of USP Ephedrine Sulfate RS in the *Standard preparation,* and A_U and A_S are the absorbances of the solutions from the *Assay preparation* and the *Standard preparation*, respectively.

Ephedrine Sulfate Tablets

» Ephedrine Sulfate Tablets contain not less than 93.0 percent and not more than 107.0 percent of the labeled amount of $(C_{10}H_{15}NO)_2 \cdot H_2SO_4$.

Packaging and storage—Preserve in well-closed containers.

Reference standard—*USP Ephedrine Sulfate Reference Standard*—Dry at 105° for 3 hours before using.

Identification—Triturate a quantity of finely powdered Tablets, equivalent to about 200 mg of ephedrine sulfate, with 15 mL of warm alcohol, filter, and evaporate the filtrate to dryness on a steam bath: the residue so obtained responds to the *Identification tests* under *Ephedrine Sulfate.*

Dissolution ⟨711⟩—

Medium: water; 500 mL.

Apparatus 2: 50 rpm.

Time: 45 minutes.

Procedure—Dilute filtered portions of the solutions under test with water to a concentration of about 25 µg per mL. Transfer 5.0-mL portions to suitable tubes. Add 1 mL of a saturated sodium carbonate solution and 2 mL of sodium metaperiodate solution (2 in 100) to each, mix, and allow to stand for 10 minutes. Add 20.0 mL of hexanes, shake for 30 seconds, and allow the phases to separate. Measure the absorbances of the hexanes extract in 1-cm cells at the wavelength of maximum absorbance, at about 242 nm, with a suitable spectrophotometer, using hexanes as the blank. Determine the amount of $(C_{10}H_{15}NO)_2 \cdot H_2SO_4$ dissolved by comparison with a similarly treated Standard solution having a known concentration of USP Ephedrine Sulfate RS in water.

Tolerances—Not less than 75% (*Q*) of the labeled amount of $(C_{10}H_{15}NO)_2 \cdot H_2SO_4$ is dissolved in 45 minutes.

Uniformity of dosage units ⟨905⟩: meet the requirements.

Assay—

Standard preparation and *Procedure*—Proceed with Ephedrine Sulfate Tablets as directed for *Standard preparation* and *Procedure* in the *Assay* under *Ephedrine Sulfate Capsules.*

Assay preparation—Weigh and finely powder not less than 20 Ephedrine Sulfate Tablets. Weigh accurately a portion of the powder, equivalent to about 25 mg of ephedrine sulfate, transfer to a glass-stoppered conical flask, and proceed as directed for *Assay preparation* in the *Assay* under *Ephedrine Sulfate Capsules,* beginning with "add by pipet."

Ephedrine Sulfate and Phenobarbital Capsules

» Ephedrine Sulfate and Phenobarbital Capsules contain not less than 91.0 percent and not more than 109.0 percent of the labeled amounts of ephedrine sulfate $(C_{10}H_{15}NO_2 \cdot H_2SO_4)$ and of phenobarbital $(C_{12}H_{12}N_2O_3)$.

Packaging and storage—Preserve in well-closed containers.

Reference standards—*USP Phenobarbital Reference Standard*—Dry at 105° for 2 hours before using. *USP Ephedrine Sulfate Reference Standard*—Dry at 105° for 3 hours before using.

Identification—

A: Mix the contents of a number of Capsules, equivalent to about 500 mg of phenobarbital, with 25 mL of water and 4 mL of 1 *N* sodium hydroxide. Filter into a separator, acidify with hydrochloric acid, and add 1 mL in excess. Extract with three 15-mL portions of chloroform, pass the chloroform solution through a chloroform-moistened filter, collect it in a beaker, and evaporate to dryness: the infrared absorption spectrum of a potassium bromide dispersion of the residue so obtained exhibits maxima only at the same wavelengths as that of a similar preparation of USP Phenobarbital RS.

B: Mix the contents of a number of Capsules, equivalent to about 200 mg of ephedrine sulfate, with 25 mL of water. Filter into a separator, add 2 mL of 6 *N* ammonium hydroxide, and extract with two 15-mL portions of chloroform. Filter the chloroform solutions through a chloroform-washed pledget of cotton into a beaker, and allow the chloroform to evaporate spontaneously: white crystals of ephedrine hydrochloride are formed. Dissolve the crystals in 4 mL of water: the angular rotation of the solution so obtained is levorotatory.

Dissolution ⟨711⟩—

Medium: water; 900 mL.

Apparatus 1: 100 rpm.

Time: 45 minutes.

Procedure—Determine the amounts of $(C_{10}H_{15}NO)_2 \cdot H_2SO_4$ and $C_{12}H_{12}N_2O_5$ dissolved, employing the procedures set forth in the *Assay for ephedrine sulfate* and in the *Assay for phenobarbital,* respectively, making any necessary modifications.

Tolerances—Not less than 75% (*Q*) of the labeled amount of $(C_{10}H_{15}NO)_2 \cdot H_2SO_4$ and not less than 75% (*Q*) of the labeled amount of $C_{12}H_{12}N_2O_5$ are dissolved in 45 minutes.

Uniformity of dosage units ⟨905⟩: meet the requirements for *Content Uniformity* with respect to ephedrine sulfate and to phenobarbital.

Assay for phenobarbital—Weigh accurately the contents of not less than 20 Ephedrine Sulfate and Phenobarbital Capsules. Transfer a portion of the powder, equivalent to about 30 mg of phenobarbital, to a separator, add 15 mL of water and 5 mL of 3 *N* hydrochloric acid, and extract with four 25-mL portions of chloroform, filtering each portion through chloroform-moistened cotton into a 250-mL volumetric flask. Dilute with chloroform to volume, and mix. Transfer an accurately measured volume of this solution, equivalent to about 1 mg of phenobarbital, to a beaker, and evaporate on a steam bath just to dryness. Transfer the residue to a 100-mL volumetric flask with the aid of 5 mL of alcohol and then with pH 9.6 alkaline borate buffer (see *Buffer Solutions* in the section, *Reagents, Indicators, and Solutions*), then dilute with the buffer solution to volume, and mix. Concomitantly determine the absorbances of this solution and of a Standard solution of USP Phenobarbital RS in the same medium having a known concentration of about 10 µg per mL in 1-cm cells at the wavelength of maximum absorbance at about 240 nm, with a suitable spectrophotometer, using pH 9.6 alkaline borate buffer as the blank. Calculate the quantity, in mg, of $C_{12}H_{12}N_2O_3$ in the portion of Capsule contents taken by the formula:

$$3C(A_U / A_S),$$

in which *C* is the concentration, in µg per mL, of USP Phenobarbital RS in the Standard solution, and A_U and A_S are the absorbances of the solution from the Capsules and the Standard solution, respectively.

Assay for ephedrine sulfate—Proceed with Ephedrine Sulfate and Phenobarbital Capsules as directed in the *Assay* under *Ephedrine Sulfate Capsules.*

Epinephrine

$C_9H_{13}NO_3$ 183.21

1,2-Benzenediol, 4-[1-hydroxy-2-(methylamino)ethyl]-, (*R*)-.

(−)-3,4-Dihydroxy-α-[(methylamino)methyl]benzyl alcohol [*51-43-4*].

» Epinephrine contains not less than 97.0 percent and not more than 100.5 percent of $C_9H_{13}NO_3$, calculated on the dried basis.

Packaging and storage—Preserve in tight, light-resistant containers.

Reference standards—*USP Epinephrine Bitartrate Reference Standard*—Dry in vacuum over silica gel for 3 hours before using. *USP Norepinephrine Bitartrate Reference Standard*—Do not dry before using.

Identification—To 5 mL of pH 4.0 acid phthalate buffer (see *Buffer Solutions* in the section, *Reagents, Indicators, and Solutions*) add 0.5 mL of a slightly acid solution of Epinephrine (1 in 1000) and 1.0 mL of 0.1 *N* iodine. Mix, and allow to stand for 5 minutes. Add 2 mL of sodium thiosulfate solution (1 in 40): a deep red color is produced.

Specific rotation ⟨781⟩: between −50.0° and −53.5°, calculated on the dried basis, determined in a 1 in 50 solution in dilute hydrochloric acid (1 in 20).

Loss on drying ⟨731⟩—Dry it in vacuum over silica gel for 18 hours: it loses not more than 2.0% of its weight.

Residue on ignition ⟨281⟩: negligible, from 100 mg.

Adrenalone—Its absorptivity (see *Spectrophotometry and Light-scattering* ⟨851⟩) at 310 nm, determined in a solution in dilute hydrochloric acid (1 in 200) containing 2 mg per mL, is not more than 0.2.

Limit of norepinephrine—

Epinephrine standard solution—Dilute with methanol an accurately measured volume of a solution of USP Epinephrine Bitartrate RS in formic acid containing about 364 mg per mL to obtain a solution having a concentration of about 20 mg per mL.

Norepinephrine standard solution—Dilute with methanol an accurately measured volume of a solution of USP Norepinephrine

Bitartrate RS in formic acid containing 16 mg per mL to obtain a solution having a known concentration of 1.6 mg per mL.

Test solution—Dissolve 200 mg of Epinephrine in 1.0 mL of formic acid, and dilute with methanol to 10.0 mL, and mix.

Procedure—On a suitable thin-layer chromatographic plate (see *Chromatography* ⟨621⟩), coated with a 0.25-mm layer of chromatographic silica gel mixture, apply 5-µL portions of *Epinephrine standard solution, Norepinephrine standard solution*, and *Test solution*. Allow the spots to dry, and develop the chromatogram in an unsaturated tank using a solvent system consisting of a mixture of *n*-butanol, water, and formic acid (7:2:1) until the solvent front has moved about three-fourths of the length of the plate. Remove the plate from the developing chamber, mark the solvent front, and allow the solvent to evaporate in warm circulating air. Spray with Folin-Ciocalteu Phenol TS, followed by sodium carbonate solution (1 in 10): the R_f value of the principal spot obtained from the *Test solution* corresponds to that obtained from the *Epinephrine standard solution*. Any spot obtained from the *Test solution* is not larger nor more intense than the spot with the same R_f value obtained from the *Norepinephrine standard solution*, corresponding to not more than 4.0% of norepinephrine.

Assay—Dissolve about 300 mg of Epinephrine, accurately weighed, in 50 mL of glacial acetic acid TS, warming slightly if necessary to effect solution. Add crystal violet TS, and titrate with 0.1 *N* perchloric acid VS. Perform a blank determination, and make any necessary correction. Each mL of 0.1 *N* perchloric acid is equivalent to 18.32 mg of $C_9H_{13}NO_3$.

Epinephrine Inhalation—*see* Epinephrine Inhalation Solution

Epinephrine Inhalation Aerosol

» Epinephrine Inhalation Aerosol is a solution of Epinephrine in propellants and Alcohol prepared with the aid of mineral acid, and with ascorbic acid as a preservative, in a pressurized container. It contains not less than 90.0 percent and not more than 115.0 percent of the labeled amount of $C_9H_{13}NO_3$, and it delivers not less than 75.0 percent and not more than 125.0 percent of the labeled amount of $C_9H_{13}NO_3$ per inhalation through an oral inhalation actuator.

Packaging and storage—Preserve in small, nonreactive, light-resistant aerosol containers equipped with metered-dose valves and provided with oral inhalation actuators.

Reference standard—*USP Epinephrine Bitartrate Reference Standard*—Dry in vacuum over silica gel for 3 hours before using.

Identification—Place 10 mL of water in a small beaker, and deliver 2 sprays from the Inhalation Aerosol under the surface of the water, actuating the valve by pressing the tip against the bottom of the beaker. To 5 mL of the solution add 1 drop of dilute sulfuric acid (1 in 200), add 0.5 mL of 0.1 *N* iodine, allow to stand for 5 minutes, and add 1 mL of 0.1 *N* sodium thiosulfate: a red-brown color is produced.

Unit spray content—

Unit spray sampling apparatus—Prepare as directed under *Aerosols* ⟨601⟩.

Ferro-citrate solution and *Buffer solution*—Prepare as directed under *Epinephrine Assay* ⟨391⟩.

Standard preparation—Dissolve an accurately weighed quantity of USP Epinephrine Bitartrate RS in a freshly prepared sodium bisulfite solution (1 in 500) to obtain a solution having a known concentration of about 100 µg per mL.

Test preparation—Transfer 20.0 mL of freshly prepared sodium bisulfite solution (1 in 500) to the collection chamber of the sampling apparatus, and clamp the intake tube in a horizontal position. Prime the valve of the Inhalation Aerosol by alternately shaking and firing it 10 times through its oral inhalation actuator. Couple the aerosol container and actuator to the intake tube of the apparatus, turn on the vacuum source, and set the flow regulator for an airflow rate of 12 ± 1 liters per minute. Turn off the vacuum, and remove the aerosol container, leaving the actuator in place. Shake the container, seat it in the actuator, turn on the vacuum, immediately deliver a single spray into the system, and turn off the vacuum. Deliver a total of 4 sprays in this manner. Uncouple the aerosol container and actuator from the assembly, clamp the intake tube in a vertical position, and scavenge the epinephrine remaining in the intake system as follows: Disconnect the collection chamber, transfer the test solution to a small beaker, then wash the inside of the intake tube with the test solution, collecting the washing in the collection chamber placed a few cm below its normal position. Recycle the test solution twice in this manner, then add 10 mL of chloroform to the collection chamber, insert the stopper, shake vigorously for 1 minute, centrifuge for 5 minutes, and use the clear supernatant liquid as the *Test preparation*.

Procedure—Transfer 5.0 mL each of the *Standard preparation* and the *Test preparation* to separate test tubes. To each tube add 100 µL of *Ferro-citrate solution* and 1.0 mL of *Buffer solution*, and mix. Concomitantly determine the absorbances of the solutions in 1-cm cells at the wavelength of maximum absorbance at about 530 nm, with a suitable spectrophotometer, using water as the blank. Calculate the quantity, in mg, of $C_9H_{13}NO_3$ delivered per spray by the formula:

$$(183.21/333.29)(0.02C/4)(A_U/A_S),$$

in which *C* is the concentration, in µg per mL, of USP Epinephrine Bitartrate RS in the *Standard preparation*, 183.21 and 333.29 are the molecular weights of epinephrine and epinephrine bitartrate, respectively, and A_U and A_S are the absorbances of the solutions from the *Test preparation* and the *Standard preparation*, respectively.

Other requirements—It meets the requirements for *Leak Testing* under *Aerosols* ⟨601⟩.

Assay—Weigh the Epinephrine Inhalation Aerosol, chill to a temperature below −30°, remove the valve by suitable means, and allow the Aerosol to warm slowly to room temperature to expel the more volatile propellant fractions. Transfer the residues in the aerosol container and valve to a 125-mL separator with the aid of six 5-mL portions of dilute sulfuric acid (1 in 1000), and extract the solution with three 25-mL portions of carbon tetrachloride. Proceed as directed in the *Assay* under *Epinephrine Nasal Solution*, beginning with "Rinse the stopper and mouth of the separator," but use 10.0 mL instead of 5.0 mL of chloroform in the determination of the specific rotation. Dry the empty aerosol container and valve, weigh them, and determine the net weight of the Aerosol. Calculate the quantity, in mg, of $C_9H_{13}NO_3$ in the Inhalation Aerosol taken by the formula:

$$(183.21/309.32)(W)(0.5 + 0.5R/93),$$

in which 183.21 and 309.32 are the molecular weights of epinephrine and triacetylepinephrine, respectively, and *W* is the weight, in mg, and *R* is the specific rotation (in degrees, without regard to the sign), of the isolated triacetylepinephrine.

Epinephrine Injection

» Epinephrine Injection is a sterile solution of Epinephrine in Water for Injection prepared with the aid of Hydrochloric Acid. It contains not less than 90.0 percent and not more than 115.0 percent of the labeled amount of $C_9H_{13}NO_3$.

Note—Do not use the Injection if it is brown or contains a precipitate.

Packaging and storage—Preserve in single-dose or in multiple-dose, light-resistant containers, preferably of Type I glass.

Reference standard—*USP Epinephrine Bitartrate Reference Standard*—Dry in vacuum over silica gel for 3 hours before using.

Identification—It responds to the *Identification test* under *Epinephrine Nasal Solution.*

pH ⟨791⟩: between 2.2 and 5.0.

Total acidity—Transfer 5.0 mL of Injection to a flask, add 10 mL of water, and titrate with 0.01 *N* sodium hydroxide VS to a pH of 7.40. Perform a blank determination, and make any necessary correction. Not more than 25.0 mL of 0.01 *N* sodium hydroxide is required.

Other requirements—It meets the requirements under *Injections* ⟨1⟩.

Assay—

Mobile phase—To 1 liter of 0.05 *M* monobasic sodium phosphate add about 519 mg of sodium 1-octanesulfonate and about 45 mg of disodium ethylenediaminetetraacetate, and mix. Adjust by the dropwise addition of phosphoric acid, if necessary, to a pH of 3.8. Mix 85 volumes of this solution with 15 volumes of methanol. Make adjustments if necessary (see *System Suitability* under *Chromatography* ⟨621⟩).

Standard preparation—Dissolve an accurately weighed quantity of USP Epinephrine Bitartrate RS in *Mobile phase*, and dilute quantitatively and stepwise, if necessary, with *Mobile phase* to obtain a solution having a known concentration of about 0.1 mg of epinephrine per mL.

Assay preparation—Transfer an accurately measured volume of Epinephrine Injection, equivalent to about 1 mg of epinephrine, to a 10-mL volumetric flask, dilute with *Mobile phase* to volume, and mix.

System suitability preparation—Dissolve 10 mg of dopamine hydrochloride in 100 mL of the *Standard preparation*, and mix.

Chromatographic system (see *Chromatography* ⟨621⟩)—The liquid chromatograph is equipped with a 280-nm detector and a 4.6-mm × 15-cm column that contains packing L7. The flow rate is about 2 mL per minute. Chromatograph the *Standard preparation* and the *System suitability preparation*, and record the peak responses as directed under *Procedure:* the resolution, *R*, between the epinephrine and dopamine hydrochloride peaks is not less than 3.5 and the relative standard deviation for replicate injections is not more than 2.0%.

Procedure—Separately inject equal volumes (about 20 µL) of the *Standard preparation* and the *Assay preparation* into the chromatograph, record the chromatograms, and measure the responses for the major peaks. The relative retention times are about 1.0 for epinephrine and 2.0 for dopamine hydrochloride. Calculate the quantity, in mg, of $C_9H_{13}NO_3$ in each mL of the Epinephrine Injection taken by the formula:

$$(183.21/333.29)10(C/V)(r_U/r_S),$$

in which 183.21 and 333.29 are the molecular weights of epinephrine and epinephrine bitartrate, respectively, *C* is the concentration, in mg per mL, of USP Epinephrine Bitartrate RS in the *Standard preparation*, *V* is the volume, in mL, of Epinephrine Injection taken, and r_U and r_S are the peak responses obtained from the *Assay preparation* and the *Standard preparation*, respectively.

Epinephrine Injection, Bupivacaine and—*see* Bupivacaine and Epinephrine Injection

Epinephrine Injection, Lidocaine and—*see* Lidocaine and Epinephrine Injection

Epinephrine Injection, Lidocaine Hydrochloride and—*see* Lidocaine Hydrochloride and Epinephrine Injection

Epinephrine Injection, Meprylcaine Hydrochloride and—*see* Meprylcaine Hydrochloride and Epinephrine Injection

Epinephrine Injection, Prilocaine and—*see* Prilocaine and Epinephrine Injection

Epinephrine Injection, Procaine Hydrochloride and—*see* Procaine Hydrochloride and Epinephrine Injection

Epinephrine Inhalation Solution

» Epinephrine Inhalation Solution is a solution of Epinephrine in Purified Water prepared with the aid of Hydrochloric Acid. It contains, in each 100 mL, not less than 0.9 g and not more than 1.15 g of $C_9H_{13}NO_3$.

NOTE—Do not use the Inhalation Solution if it is brown or contains a precipitate.

Packaging and storage—Preserve in small, well-filled, tight, light-resistant containers.

Identification—It responds to the *Identification test* under *Epinephrine Nasal Solution.*

Assay—Pipet 10 mL of Epinephrine Inhalation Solution into a 125-mL separator, and extract the solution with two 10-mL portions of carbon tetrachloride. Then proceed as directed in the *Assay* under *Epinephrine Nasal Solution*, beginning with "Rinse the stopper and mouth of the separator," but use for the acetylation 1.05 g of sodium bicarbonate and 0.50 mL of acetic anhydride, and extract the acetylated product with six 15-mL portions of chloroform instead of the 25-mL portions specified therein, and use 15.0 mL of chloroform instead of 5.0 mL in the determination of the specific rotation.

Epinephrine Nasal Solution

» Epinephrine Nasal Solution is a solution of Epinephrine in Purified Water prepared with the aid of Hydrochloric Acid. It contains, in each 100 mL, not less than 90 mg and not more than 115 mg of $C_9H_{13}NO_3$.

NOTE—Do not use the Solution if it is brown or contains a precipitate.

Packaging and storage—Preserve in small, well-filled, tight, light-resistant containers.

Identification—To 5 mL of pH 4.0 acid phthalate buffer (see *Buffer Solutions* in the section, *Reagents, Indicators, and Solutions*) add 0.5 mL of Nasal Solution and 1.0 mL of 0.1 *N* iodine. Mix, and allow to stand for 5 minutes. Add 2 mL of sodium thiosulfate solution (1 in 40): a deep red color is produced.

Assay—Pipet 30 mL of Epinephrine Nasal Solution into a 125-mL separator, add 25 mL of carbon tetrachloride, shake vigorously for 1 minute, allow the liquids to separate, and discard the carbon tetrachloride. Wash twice more with carbon tetrachloride, separating and discarding the lower layer as completely as possible each time. Rinse the stopper and mouth of the separator with a few drops of water. Add 0.2 mL of starch TS, then while swirling the separator add iodine and potassium iodide TS dropwise until the blue color formed persists, and immediately add just sufficient 0.1 *N* sodium thiosulfate to discharge the blue color. [NOTE—Proceed with the assay from this point without delay.]

Add to the liquid in the separator 2.10 g of sodium bicarbonate, preventing it from coming in contact with the mouth of the separator, and swirl until most of the bicarbonate has dissolved. By means of a 1-mL syringe that is not fitted with a needle, rapidly inject 1.0 mL of acetic anhydride directly into the contents of the separator. Immediately insert the stopper in the separator, and shake vigorously until the evolution of carbon dioxide has ceased (7 to 10 minutes), releasing the pressure as necessary through the stopcock. Allow to stand for 5 minutes, and extract the solution with six 25-mL portions of chloroform, filtering each extract through a small pledget of cotton, previously washed with chloroform, into a beaker.

Evaporate the combined chloroform extracts on a steam bath in a current of air to about 3 mL, transfer the residue by means of small portions of chloroform to a tared 50-mL beaker, and heat again to evaporate the solvent completely. Heat further at

105° for 30 minutes, cool in a desiccator, and weigh the residue of triacetylepinephrine. Add 5.0 mL of chloroform, cover the beaker well, gently swirl the contents until the residue has completely dissolved, and determine the specific rotation, R, using a 200-mm semi-micro polarimeter tube.

Calculate the quantity, in mg, of $C_9H_{13}NO_3$ in the volume of Nasal Solution taken by the formula:

$$(183.21/309.32)(W)(0.5 + 0.5R/93),$$

in which 183.21 and 309.32 are the molecular weights of epinephrine and triacetylepinephrine, respectively, and W is the weight, in mg, and R is the specific rotation (in degrees, without regard to the sign), of the isolated triacetylepinephrine.

Epinephrine Ophthalmic Solution

» Epinephrine Ophthalmic Solution is a sterile, aqueous solution of Epinephrine prepared with the aid of Hydrochloric Acid. It contains not less than 90.0 percent and not more than 115.0 percent of the labeled amount of $C_9H_{13}NO_3$. It contains a suitable antibacterial agent and may contain an anti-oxidant, suitable buffers, and chelating and tonicity-adjusting agents.

NOTE—Do not use the Solution if it is brown or contains a precipitate.

Packaging and storage—Preserve in tight, light-resistant containers.

Reference standard—*USP Epinephrine Bitartrate Reference Standard*—Dry in vacuum over silica gel for 3 hours before using.

Identification—

A: The ultraviolet absorption spectrum of the *Assay preparation* prepared as directed in the *Assay* exhibits maxima and minima at the same wavelengths as that of a similar solution of USP Epinephrine Bitartrate RS.

B: A solution (1 in 2) is levorotatory.

Sterility—It meets the requirements under *Sterility Tests* ⟨71⟩.

pH ⟨791⟩: between 2.2 and 4.5.

Assay—

pH 5.8 buffer—Mix 1 volume of 1 M dibasic potassium phosphate with 9 volumes of 1 M monobasic potassium phosphate. Adjust to a pH of 5.80 ± 0.05 by the addition of small volumes of either solution.

Standard preparation—Dissolve a suitable quantity of USP Epinephrine Bitartrate RS, accurately weighed, in 0.1 N hydrochloric acid to obtain a solution having a known concentration of about 40 µg of epinephrine per mL.

Assay preparation—Transfer an accurately measured volume of Epinephrine Ophthalmic Solution, equivalent to about 20 mg of epinephrine, to a 250-mL beaker containing 2.0 mL of *pH 5.8 buffer*. Add 9 g of chromatographic siliceous earth, and mix. Carefully transfer the mixture to a 45- × 2.2-cm chromatographic tube containing a pledget of glass wool at the bottom, and tap the column gently to effect packing. Dry-wash the beaker with about 1 g of chromatographic siliceous earth, add to the column, and plug the top with a pledget of glass wool. Wash the column with 100 mL of water-washed ether, and discard the eluant. Add 10.0 mL of 0.1 N hydrochloric acid to a 125-mL separator, and place the separator under the column. To about 100 mL of water-washed ether add 1 mL of bis(2-ethylhexyl) phosphoric acid, and elute the column with this solution, collecting the eluate in the separator. Extract the epinephrine into the aqueous acid layer, and carefully transfer the aqueous layer to a 500-mL volumetric flask. Shake the ether layer with two 50-mL portions of 0.1 N hydrochloric acid, add the acidic aqueous extracts to the volumetric flask, dilute with 0.1 N hydrochloric acid to volume, and mix.

Procedure—Concomitantly determine the absorbances of the *Assay preparation* and the *Standard preparation* at the wavelength of maximum absorbance at about 280 nm, with a suitable spectrophotometer, using 0.1 N hydrochloric acid as the blank. Calculate the quantity, in mg, of $C_9H_{13}NO_3$ in each mL of the Ophthalmic Solution taken by the formula:

$$0.5(C/V)(A_U/A_S),$$

in which C is the concentration, in µg per mL, of epinephrine in the *Standard preparation*, V is the volume, in mL, of Ophthalmic Solution taken, and A_U and A_S are the absorbances of the *Assay preparation* and the *Standard preparation*, respectively.

Sterile Epinephrine Oil Suspension

» Sterile Epinephrine Oil Suspension is a sterile suspension of Epinephrine in a suitable vegetable oil. It contains, in each mL, not less than 1.8 mg and not more than 2.4 mg of $C_9H_{13}NO_3$.

Packaging and storage—Preserve in single-dose, light-resistant containers, preferably of Type I or Type III glass.

Other requirements—It meets the requirements under *Injections* ⟨1⟩.

Assay—Transfer to a small separator an accurately measured volume of Sterile Epinephrine Oil Suspension, equivalent to about 30 mg of epinephrine, add 25 mL of solvent hexane, and swirl the separator until the oil has dissolved. Extract the epinephrine from this solution with 10 mL of dilute sulfuric acid (1 in 700) and two 10-mL portions of water, receiving the extracts in a 125-mL separator. Wash the combined extracts with two 10-mL portions of carbon tetrachloride, discarding the washings, and proceed as directed in the *Assay* under *Epinephrine Nasal Solution*, beginning with "Add to the liquid in the separator 2.10 g of sodium bicarbonate."

Epinephrine Bitartrate

$C_9H_{13}NO_3 . C_4H_6O_6$ 333.29

1,2-Benzenediol, 4-[1-hydroxy-2-(methylamino)ethyl]-, (R)-, [R-(R^*,R^*)-2,3-dihydroxybutanedioate (1:1) (salt).

(−)-3,4-Dihydroxy-α-[(methylamino)methyl]benzyl alcohol (+)-tartrate (1:1) salt [*51-42-3*].

» Epinephrine Bitartrate contains not less than 97.0 percent and not more than 102.0 percent of $C_9H_{13}NO_3 . C_4H_6O_6$, calculated on the dried basis.

Packaging and storage—Preserve in tight, light-resistant containers.

Reference standards—*USP Epinephrine Bitartrate Reference Standard*—Dry in vacuum over silica gel for 3 hours before using. *USP Norepinephrine Bitartrate Reference Standard*—Do not dry before using.

Identification—Dissolve about 500 mg in 20 mL of water containing about 100 mg of sodium bisulfite. Add 6 N ammonium hydroxide until the solution has a distinct odor of ammonia, and allow to stand in a refrigerator for 1 hour. Filter the precipitate, wash it with three 2-mL portions of cold water, then with 5 mL of cold alcohol, and finally with 5 mL of cold ether, and dry in vacuum over silica gel for 3 hours. The epinephrine so obtained responds to the *Identification test* under *Epinephrine*, and its specific rotation (see *Optical Rotation* ⟨781⟩), determined by dissolving 200 mg, accurately weighed, in sufficient dilute hydrochloric acid (1 in 20) to make 10.0 mL, is between −50° and −53.5°.

Melting range ⟨741⟩: between 147° and 152°, with decomposition.

Loss on drying ⟨731⟩—Dry it in vacuum over silica gel for 3 hours: it loses not more than 0.5% of its weight.

Residue on ignition ⟨281⟩: negligible, from 100 mg.

Adrenalone—Its absorptivity (see *Spectrophotometry and Light-scattering* ⟨851⟩) at 310 nm, determined in a solution in dilute hydrochloric acid (1 in 200) containing 4 mg per mL, is not more than 0.2.

Limit of norepinephrine bitartrate—

Epinephrine standard solution—Dilute with methanol an accurately measured volume of an aqueous solution of USP Epinephrine Bitartrate RS containing about 200 mg per mL to obtain a solution having a known concentration of about 20 mg per mL.

Norepinephrine standard solution—Dilute with methanol an accurately measured volume of an aqueous solution of USP Norepinephrine Bitartrate RS containing 8.0 mg per mL to obtain a solution having a known concentration of 0.80 mg per mL.

Test solution—Dissolve 200 mg of Epinephrine Bitartrate in 1.0 mL of water, dilute with methanol to 10.0 mL, and mix.

Procedure—On a suitable thin-layer chromatographic plate (see *Chromatography* ⟨621⟩), coated with a 0.25-mm layer of chromatographic silica gel mixture, apply 5-µL portions of *Epinephrine standard solution*, *Norepinephrine standard solution*, and *Test solution*. Allow the spots to dry, and develop the chromatogram in an unsaturated tank using a solvent system consisting of *n*-butanol, water, and formic acid (7:2:1) until the solvent front has moved about three-fourths of the length of the plate. Remove the plate from the developing chamber, mark the solvent front, and allow the solvent to evaporate in warm circulating air. Spray with Folin-Ciocalteu Phenol TS, followed by sodium carbonate solution (1 to 10): the R_f value of the principal spot obtained from the *Test solution* corresponds to that obtained from the *Epinephrine standard solution*. Any spot obtained from the *Test solution* is not larger nor more intense than the spot with the same R_f value obtained from *Norepinephrine standard solution*, corresponding to not more than 4.0% of norepinephrine bitartrate.

Assay—Dissolve about 500 mg of Epinephrine Bitartrate, accurately weighed, in 20 mL of glacial acetic acid, warming slightly if necessary to effect solution. Add crystal violet TS, and titrate with 0.1 *N* perchloric acid VS. Perform a blank determination, and make any necessary correction. Each mL of 0.1 *N* perchloric acid is equivalent to 33.33 mg of $C_9H_{13}NO_3 . C_4H_6O_6$.

Epinephrine Bitartrate Inhalation Aerosol

» Epinephrine Bitartrate Inhalation Aerosol is a suspension of microfine Epinephrine Bitartrate in propellants in a pressurized container. It contains not less than 90.0 percent and not more than 110.0 percent of the labeled amount of $C_9H_{13}NO_3 . C_4H_6O_6$, and it delivers not less than 75.0 percent and not more than 125.0 percent of the labeled amount of $C_9H_{13}NO_3 . C_4H_6O_6$ per inhalation through an oral inhalation actuator.

Packaging and storage—Preserve in small, nonreactive, light-resistant aerosol containers equipped with metered-dose valves and provided with oral inhalation actuators.

Reference standard—*USP Epinephrine Bitartrate Reference Standard*—Dry in vacuum over silica gel for 3 hours before using.

Identification—

A: Place 10 mL of water in a small beaker, and deliver 3 sprays from the Aerosol under the surface of the water, actuating the valve by pressing the tip against the bottom of the beaker. Filter, and to 5 mL of the filtrate add 1 drop of dilute hydrochloric acid (1 in 120). Add 0.5 mL of 0.1 *N* iodine, allow to stand for 5 minutes, and add 1 mL of 0.1 *N* sodium thiosulfate: a red-brown color is produced.

B: Actuate the valve of the Aerosol by pressing the tip against a station of a white porcelain spot plate. Cover the spot with 2 or 3 drops of a mixture of 3 volumes of pyridine and 1 volume of acetic anhydride: an emerald-green color is produced.

Unit spray content—

Unit spray sampling apparatus—Prepare as directed under *Aerosols* ⟨601⟩.

Ferro-citrate solution and *Buffer solution*—Prepare as directed under *Epinephrine Assay* ⟨391⟩.

Standard preparation—Dissolve an accurately weighed quantity of USP Epinephrine Bitartrate RS in a freshly prepared sodium bisulfite solution (1 in 500) to obtain a solution having a known concentration of about 100 µg per mL.

Test preparation—Transfer 20.0 mL of a freshly prepared sodium bisulfite solution (1 in 500) to the collection chamber of the *Unit spray sampling apparatus*, and clamp the intake tube in a horizontal position. Prime the valve of the Aerosol by alternately shaking and firing it 10 times through its oral inhalation actuator. Couple the aerosol container and actuator to the intake tube of the apparatus, turn on the vacuum source, and set the flow regulator for an airflow rate of 12 ± 1 liters per minute. Turn off the vacuum, and remove the aerosol container, leaving the actuator in place. Shake the container, seat it in the actuator, turn on the vacuum, immediately deliver a single spray into the system, and turn off the vacuum. Deliver a total of 6 sprays in this manner. Uncouple the aerosol container and actuator from the assembly, clamp the intake tube in a vertical position, and scavenge the epinephrine bitartrate remaining in the intake system as follows: Disconnect the collection chamber, transfer the test solution to a small beaker, and wash the inside of the intake tube with the test solution, collecting the washing in the collection chamber placed a few cm below its normal position. Recycle the test solution twice in this manner. Transfer the solution to a centrifuge tube, and add 10.0 mL of chloroform. Insert the stopper, shake vigorously for 1 minute, centrifuge for 5 minutes, and use the clear supernatant liquid as the *Test preparation*.

Procedure—Using the *Test preparation* in place of the *Assay preparation*, proceed as directed for *Procedure* in the *Assay*. Calculate the quantity, in mg, of $C_9H_{13}NO_3 . C_4H_6O_6$ delivered per spray by the formula:

$$(0.02C/6)(A_U/A_S),$$

in which C is the concentration, in µg per mL, of USP Epinephrine Bitartrate RS in the *Standard preparation*, and A_U and A_S are the absorbances of the solutions from the *Test preparation* and the *Standard preparation*, respectively.

Particle size—Proceed with Epinephrine Bitartrate Inhalation Aerosol as directed in the test for *Particle size* under *Isoproterenol Sulfate Inhalation Aerosol*. It meets the limits of the test.

Other requirements—It meets the requirements for *Leak Testing* under *Aerosols* ⟨601⟩.

Assay—

Ferro-citrate solution and *Buffer solution*—Prepare as directed under *Epinephrine Assay* ⟨391⟩.

Standard preparation—Prepare as directed under *Unit spray content*.

Assay preparation—[NOTE—A suitable specimen beaker is one having a small indentation formed on its inside bottom surface having dimensions adequate to accept the aerosol valve stem during actuation, thereby preventing particle entrapment and side-of-stem leakage during the delivery of the specimen.] Place 20 mL of chloroform in a suitable 100-mL beaker. Prime the valve of Epinephrine Bitartrate Inhalation Aerosol by alternately shaking and firing it 10 times through its oral inhalation actuator. Accurately weigh the Aerosol, shake it, and immediately deliver a single spray under the surface of the chloroform, actuating the valve by pressing the tip into the indentation in the bottom of the beaker. Raise the Aerosol above the surface of the chloroform, and shake it gently preparatory to delivering another spray similarly under the surface of the chloroform. Deliver a total of 3 sprays in this manner. Rinse the valve stem and ferrule with about 2 mL of chloroform, collecting the rinsing with the specimen in the beaker. Allow the Aerosol to dry, weigh it, and determine the total weight of the 3 sprays. Transfer the solution to a centrifuge tube with the aid of two 3-mL portions of chloroform, and add 10.0 mL of freshly prepared sodium bisulfite solution (1 in 500). Insert the stopper, shake vigorously for 1 minute, centrifuge for 5 minutes, and use the clear supernatant liquid as the *Assay preparation.*

Procedure—Transfer 5.0 mL each of the *Standard preparation* and the *Assay preparation* to separate test tubes. To each tube add 100 µL of *Ferro-citrate solution* and 1.0 mL of *Buffer solution*, and mix. Concomitantly determine the absorbances of the solutions in 1-cm cells at the wavelength of maximum absorbance at about 530 nm, with a suitable spectrophotometer, using water as the blank. Calculate the quantity, in mg, of $C_9H_{13}NO_3 . C_4H_6O_6$ in each mL of the Aerosol taken by the formula:

$$(0.01Cd/W)(A_U/A_S),$$

in which *C* is the concentration, in µg per mL, of USP Epinephrine Bitartrate RS in the *Standard preparation, d* is the density, in g per mL, of the Aerosol, determined as directed for *d* in the *Procedure* in the *Assay* under *Isoproterenol Sulfate Inhalation Aerosol, W* is the weight, in g, of the specimen taken, and A_U and A_S are the absorbances of the solutions from the *Assay preparation* and the *Standard preparation*, respectively.

Epinephrine Bitartrate Injection, Lidocaine Hydrochloride and—*see* Lidocaine Hydrochloride and Epinephrine Bitartrate Injection

Epinephrine Bitartrate Ophthalmic Solution

» Epinephrine Bitartrate Ophthalmic Solution is a sterile, buffered, aqueous solution of Epinephrine Bitartrate. It contains an amount of epinephrine bitartrate equivalent to not less than 90.0 percent and not more than 115.0 percent of the labeled amount of epinephrine ($C_9H_{13}NO_3$). It contains a suitable antibacterial agent and may contain suitable preservatives.

NOTE—*Do not use the Solution if it is brown or contains a precipitate.*

Packaging and storage—Preserve in small, well-filled, tight, light-resistant containers.

pH ⟨791⟩: between 3.0 and 3.8.

Other requirements—It responds to the *Identification test* under *Epinephrine Nasal Solution*, and meets the requirements under *Sterility Tests* ⟨71⟩.

Assay—Pipet a volume of Epinephrine Bitartrate Ophthalmic Solution, equivalent to 55 to 60 mg of epinephrine bitartrate, into a 125-mL separator, add water to make 10 mL, and extract the solution with two 10-mL portions of carbon tetrachloride. Then proceed as directed in the *Assay* under *Epinephrine Nasal Solution*, beginning with "Rinse the stopper and mouth of the separator," but use for the acetylation 1.05 g of sodium bicarbonate and 0.50 mL of acetic anhydride, and extract the acetylated product with six 15-mL portions of chloroform instead of the 25-mL portions specified therein. Calculate the quantity, in mg, of $C_9H_{13}NO_3$ in the portion of Ophthalmic Solution taken by the formula:

$$(183.21/309.32)(W)(0.5 + 0.5R/93),$$

in which 183.21 and 309.32 are the molecular weights of epinephrine and triacetylepinephrine, respectively, and *W* is the weight, in mg, and *R* is the specific rotation (in degrees, without regard to the sign), of the isolated triacetylepinephrine.

Epinephrine Bitartrate for Ophthalmic Solution

» Epinephrine Bitartrate for Ophthalmic Solution is a sterile, dry mixture of Epinephrine Bitartrate and suitable antioxidants, prepared by freeze-drying. It contains an amount of epinephrine bitartrate equivalent to not less than 90.0 percent and not more than 110.0 percent of the labeled amount of epinephrine ($C_9H_{13}NO_3$).

Packaging and storage—Preserve in *Containers for Sterile Solids* as described under *Injections* ⟨1⟩.

Completeness of solution ⟨641⟩—A 100-mg portion dissolves in 5 mL of water to yield a clear solution.

Constituted solution—At the time of use, the constituted solution prepared from Epinephrine Bitartrate for Ophthalmic Solution meets the requirements for *Constituted Solutions* under *Injections* ⟨1⟩.

Other requirements—A solution of it responds to the *Identification test* under *Epinephrine Nasal Solution*, and meets the requirements of the *Assay* under *Epinephrine Bitartrate Ophthalmic Solution.* It meets also the requirements under *Sterility Tests* ⟨71⟩ and *Uniformity of Dosage Units* ⟨905⟩.

Epinephryl Borate Ophthalmic Solution

H
CH3NHCH2—C—
OH
O B—OH O

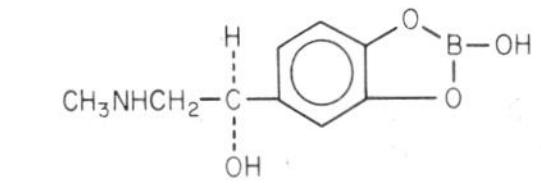

$C_9H_{12}BNO_4$ 209.01

1,3,2-Benzodioxaborole-5-methanol, 2-hydroxy-α-[(methylamino)methyl]-, (*R*)-.

(−)-3,4-Dihydroxy-α-[(methylamino)methyl]benzyl alcohol, cyclic 3,4-ester with boric acid [*5579-16-8*].

» Epinephryl Borate Ophthalmic Solution is a sterile solution in water of Epinephrine as a borate complex. It contains an amount of epinephryl borate ($C_9H_{12}BNO_4$) equivalent to not less than 90.0 percent and not more than 115.0 percent of the labeled amount of epinephrine ($C_9H_{13}NO_3$). It contains a suitable antibacterial agent and one or more suitable preservatives and buffering agents.

NOTE—*Do not use the Ophthalmic Solution if it is brown or contains a precipitate.*

Packaging and storage—Preserve in small, well-filled, tight, light-resistant containers.

Identification—

A: To 5 mL of pH 4.0 acid phthalate buffer (see *Buffer Solutions* in the section, *Reagents, Indicators, and Solutions*) add 0.5 mL of Ophthalmic Solution and 1 mL of 0.1 *N* iodine. Mix, allow to stand for 5 minutes, and add 2 mL of 0.1 *N* sodium thiosulfate: a deep red color is produced.

B: To 5 mL in a porcelain evaporating dish add 5 drops of sulfuric acid and 5 mL of methanol: the ignited mixture burns with a green-bordered flame.

Sterility—It meets the requirements under *Sterility Tests* ⟨71⟩.

pH ⟨791⟩: between 5.5 and 7.6.

Assay—Transfer an accurately measured volume of Epinephryl Borate Ophthalmic Solution, equivalent to about 100 mg of epinephrine, to a 250-mL separator. Dilute with water to 30 mL, and adjust with dilute hydrochloric acid (1 in 12) to a pH of 4.0 ± 0.2. Add 25 mL of carbon tetrachloride, shake vigorously for 1 minute, allow the phases to separate, and discard the carbon tetrachloride washing. In the same manner, wash with two additional 25-mL portions of carbon tetrachloride, and discard the washings. Rinse the stopper and the mouth of the separator with 2 to 3 mL of water such that the rinsings enter the separator and combine with the solution under assay. Add 0.2 mL of starch TS, and, while swirling the separator, add iodine and potassium iodide TS dropwise until the blue color persists. Immediately add a volume of 0.1 *N* sodium thiosulfate just sufficient to discharge the blue color. [NOTE—Proceed with the assay from this point without delay.] Add 2.10 g of sodium bicarbonate through a dry powder funnel to prevent the powder from coming in contact with the mouth of the separator, and swirl to dissolve most of the solid. By means of a syringe fitted with a suitable pipet, rapidly inject 1.0 mL of acetic anhydride directly into the contents of the separator. Swirl the unstoppered separator gently for 3 minutes to allow carbon dioxide to escape. Insert the stopper, and shake gently until the evolution of carbon dioxide has ceased (7 to 10 minutes), releasing the pressure through the stopcock as necessary. Allow to stand for 5 minutes. Extract with six 25-mL portions of chloroform, shaking for 1 minute each time, filtering each extract through a small pledget of chloroform-saturated cotton and collecting the extracts in a 400-mL beaker. Add several glass beads, and evaporate on a steam bath to about 3 mL. With the aid of 15 to 20 mL of chloroform, transfer the residue to a tared 50-mL beaker, and evaporate on the steam bath to dryness. Dry the residue at 105° for 30 minutes, cool in a desiccator, and weigh the triacetylepinephrine so obtained. Transfer 10.0 mL of chloroform to the beaker, and gently swirl to dissolve the residue, dislodging the semisolid residue from the glass surface, if necessary, with a small metal spatula. Determine the angular rotation of the solution in a 100-mm polarimeter tube. Calculate the quantity, in mg, of epinephrine ($C_9H_{13}NO_3$) in the volume of Ophthalmic Solution taken by the formula:

$$(183.21/309.32)(W)(0.5 + 0.5R/93),$$

in which 183.21 and 309.32 are the molecular weights of epinephrine and triacetylepinephrine, respectively, *W* is the weight, in mg, of the isolated triacetylepinephrine, and *R* is the specific rotation, in degrees, of the triacetylepinephrine solution.

Epitetracycline Hydrochloride

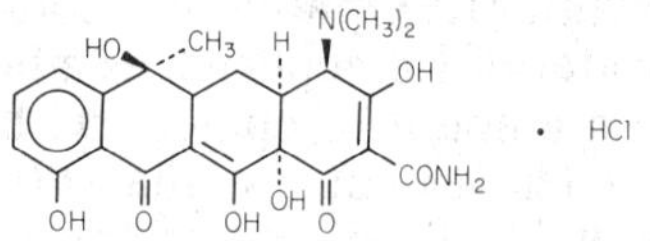

$C_{22}H_{24}N_2O_8 . HCl$ 480.90

2-Naphthacenecarboxamide, 4-(dimethylamino)-1,4,4a,5,5a,6,11,12a-octahydro-3,6,10,12,12a-pentahydroxy-6-methyl-1,11-dioxo-, monohydrochloride, 4-epimer [4*S*-(4α,4aα,5aα,6β,12aα)]-.

4-(Dimethylamino)-1,4,4a,5,5a,6,11,12a-octahydro-3,6,10,12,12a-pentahydroxy-6-methyl-1,11-dioxo-2-naphthacenecarboxamide monohydrochloride, 4-epimer.

» Epitetracycline Hydrochloride contains not less than 70.0 percent of $C_{22}H_{24}N_2O_8 . HCl$.

Packaging and storage—Preserve in tight, light-resistant containers.

Reference standard—*USP Tetracycline Hydrochloride Reference Standard*—Do not dry before using.

pH ⟨791⟩: between 2.3 and 4.0, in a solution containing 10 mg per mL.

Loss on drying ⟨731⟩—Dry about 100 mg in a capillary-stoppered bottle in vacuum at a pressure not exceeding 5 mm of mercury at 60° for 3 hours: it loses not more than 6.0% of its weight.

4-Epianhydrotetracycline ⟨226⟩—Dissolve about 250 mg, accurately weighed, in 10 mL of 0.1 *N* hydrochloric acid, and adjust with 6 *N* ammonium hydroxide to a pH of 7.8. Transfer this solution with the aid of *EDTA buffer* to a 50-mL volumetric flask, dilute with *EDTA buffer* to volume, and mix. Use this solution, without delay, as the *Test solution:* not more than 2.0% is found.

Assay—

Edetate disodium solution, Stationary phase, Alkaline methanol solution, Column support, Chromatographic column, and *Standard preparation*—Prepare as directed in the section, *Epitetracycline hydrochloride content and Assay for tetracycline hydrochloride,* under *Tetracycline Hydrochloride for Topical Solution.*

Assay preparation—Transfer about 22 mg of Epitetracycline Hydrochloride, accurately weighed, to a 25-mL volumetric flask, add 1 mL of methanol, and swirl to dissolve. Dilute with *Stationary phase* to volume, and mix. Transfer 2.0 mL of this solution to a 10-mL volumetric flask, dilute with *Stationary phase* to volume, and mix. Pipet 2.0 mL of this solution into the *Chromatographic column,* and allow it to penetrate the *Column support.* Add 20 mL of benzene to the solvent reservoir, and collect the eluate at the rate of about 1 mL per minute. When the benzene level reaches the top of the *Column support,* add 60 mL of chloroform to the solvent reservoir, and continue collecting eluate. When the chloroform level reaches the top of the *Column support,* discard the collected eluate, add 50 mL of a mixture of butyl alcohol and methanol (1:1), and collect 8 mL of eluate in a 10-mL graduated cylinder. Replace the 10-mL graduated cylinder with a low-actinic 50-mL volumetric flask, and continue collecting eluate until the column runs dry. The eluate in the 50-mL volumetric flask is the *Assay preparation.*

Procedure—Add 2.0 mL of *Alkaline methanol solution* to the *Standard preparation* and to the *Assay preparation,* dilute each with chloroform to volume, and mix. Concomitantly, within 10 minutes of preparation, determine the absorbances of these solutions at the wavelength of maximum absorbance at about 366 nm, with a suitable spectrophotometer, using chloroform as the blank. Calculate the quantity, in mg, of $C_{22}H_{24}N_2O_8 . HCl$ in the Epitetracycline Hydrochloride taken by the formula:

$$0.001(WP)(A_U/A_S),$$

in which *W* is the weight, in mg, of USP Tetracycline Hydrochloride RS taken, *P* is the potency, in µg per mg, of the USP Tetracycline Hydrochloride RS, and A_U and A_S are the absorbances of the solutions from the *Assay preparation* and the *Standard preparation,* respectively.

Equilin

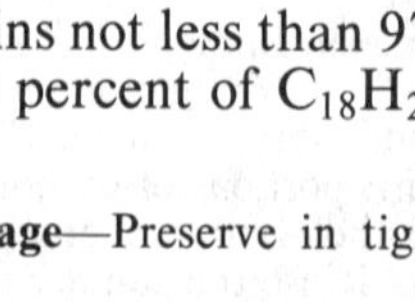

$C_{18}H_{20}O_2$ 268.35

Estra-1,3,5(10),7-tetraen-17-one, 3-hydroxy-.

3-Hydroxyestra-1,3,5(10),7-tetraen-17-one [*474-86-2*].

» Equilin contains not less than 97.0 percent and not more than 103.0 percent of $C_{18}H_{20}O_2$, calculated on the dried basis.

Packaging and storage—Preserve in tight, light-resistant containers.

Reference standard—*USP Equilin Reference Standard*—Do not dry before using. Store tightly closed, in a cool place.

Clarity of solution—Add 100 mg to 100 mL of 1 *N* sodium hydroxide in a 125-mL conical flask, heat on a steam bath until solution is complete, then cool, and transfer to a 100-mL color-comparison tube: the solution is clear.

Identification—

A: The infrared absorption spectrum of a potassium bromide dispersion of it exhibits maxima only at the same wavelengths as that of a similar preparation of undried USP Equilin RS.

B: The ultraviolet absorption spectrum of a 1 in 20,000 solution in alcohol exhibits maxima and minima at the same wavelengths as that of a similar solution of USP Equilin RS, concomitantly measured.

Specific rotation ⟨781⟩: between +300° and +316°, calculated on the dried basis, determined in a solution in dioxane containing 200 mg in each 10 mL.

Loss on drying ⟨731⟩—Dry it in vacuum at 105° for 1 hour: it loses not more than 0.5% of its weight.

Residue on ignition ⟨281⟩: not more than 0.5%.

Assay—

Mobile phase—Prepare a suitable and degassed solution containing 35 volumes of acetonitrile and 65 volumes of water.

Internal standard solution—Dissolve phenol in acetonitrile to obtain a solution having a concentration of about 35 µg per mL.

Standard preparation—Dissolve a suitable quantity of USP Equilin RS, accurately weighed, in *Internal standard solution* to obtain a solution having a known concentration of about 0.2 mg per mL.

Assay preparation—Transfer about 10 mg of Equilin, accurately weighed, to a 50-mL volumetric flask, dissolve in *Internal standard solution*, dilute with *Internal standard solution* to volume, and mix.

Chromatographic system (see *Chromatography* ⟨621⟩)—The liquid chromatograph is equipped with a 280-nm detector and a 3.9-mm × 30-cm column that contains packing L1. The flow rate is about 2 mL per minute. Chromatograph five replicate injections of the *Standard preparation*, and record the peak responses as directed under *Procedure:* the relative standard deviation is not more than 2.0%, and the resolution factor between equilin and phenol is not less than 5. Adjust the operating parameters such that the peak obtained from the *Standard preparation* is about 0.7 full-scale.

Procedure—Separately inject equal volumes (about 20 µL) of the *Standard preparation* and the *Assay preparation* into the chromatograph by means of a suitable microsyringe or sampling valve, record the chromatograms, and measure the responses for the major peaks: the retention times for equilin and phenol are about 14 and 3 minutes, respectively. Calculate the quantity, in mg, of $C_{18}H_{20}O_2$ in the portion of Equilin taken by the formula:

$$50C(R_U/R_S),$$

in which C is the concentration, in mg per mL, of USP Equilin RS in the *Standard preparation*, and R_U and R_S are the ratios of the peak responses of the equilin peak to the internal standard peak obtained from the *Assay preparation* and the *Standard preparation*, respectively.

Ergocalciferol

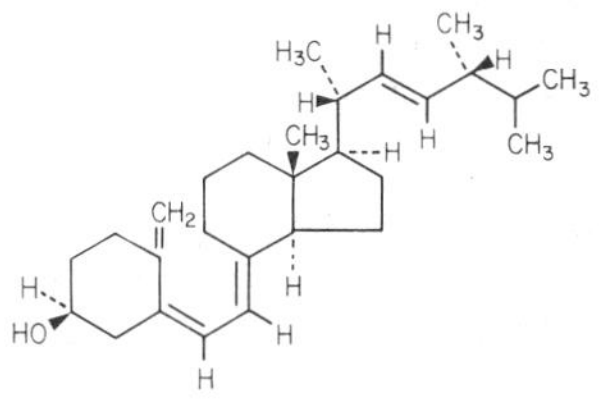

$C_{28}H_{44}O$ 396.65

9,10-Secoergosta-5,7,10(19),22-tetraen-3-ol, (3β,5*Z*,7*E*,22*E*)-.

Ergocalciferol [*50-14-6*].

» Ergocalciferol contains not less than 97.0 percent and not more than 103.0 percent of $C_{28}H_{44}O$.

Packaging and storage—Preserve in hermetically sealed containers under nitrogen, in a cool place and protected from light.

Reference standards—*USP Ergocalciferol Reference Standard*—Store in a cold place, protected from light. *USP Ergosterol Reference Standard*—Do not dry before using. Keep container tightly closed and protected from light, and store in a refrigerator. Allow to attain room temperature before opening container. *USP Vitamin D Assay System Suitability Reference Standard*—Store in a cool place, protected from light. Allow it to attain room temperature before opening ampul. Do not dry. Transfer unused contents of ampul to a tightly closed container, and store under nitrogen in the dark, in a cool place.

Identification—

A: The infrared absorption spectrum of a potassium bromide dispersion of it, in the range of 2 to 12 µm, exhibits maxima only at the same wavelengths as that of a similar preparation of USP Ergocalciferol RS.

B: The ultraviolet absorption spectrum of a 1 in 100,000 solution in alcohol exhibits maxima and minima at the same wavelengths as that of a similar solution of USP Ergocalciferol RS, concomitantly measured, and the respective absorptivities at the wavelength of maximum absorbance at about 265 nm do not differ by more than 3.0%.

C: To a solution of about 0.5 mg in 5 mL of chloroform add 0.3 mL of acetic anhydride and 0.1 mL of sulfuric acid, and shake vigorously: a bright red color is produced and rapidly changes through violet and blue to green.

D: Prepare without heating, and handle without delay, a 1 in 100 solution of squalane in chloroform containing 50 mg of Ergocalciferol per mL, and prepare a Standard solution of USP Ergocalciferol RS in the same solvent and of the same concentration. Prepare a 1 in 100 solution of squalane in chloroform containing 100 µg of USP Ergosterol RS per mL. Apply 10 µL of the test solution, 10 µL of the Standard solution, and 10 µL of the ergosterol solution on a line parallel to and about 2.5 cm from the bottom edge of a thin-layer chromatographic plate (see *Chromatography* ⟨621⟩) coated with a 0.25-mm layer of chromatographic silica gel mixture. Place the plate in a developing chamber containing, and equilibrated with, a mixture of equal volumes of cyclohexane and ether. Develop the chromatogram until the solvent front has moved about 15 cm above the line of application. Perform the development and subsequent operations in the dark. Remove the plate, allow the solvent to evaporate, and spray with a 1 in 50 solution of acetyl chloride in antimony trichloride TS. The chromatogram obtained with the test solution shows a yellowish orange area (ergocalciferol) having the same R_f value as the area of the Standard solution of ergocalciferol and may show a violet area below the ergocalciferol area. The color of the violet area is not more intense than that of the violet area in the chromatogram obtained from the solution of ergosterol.

Melting range, *Class Ib* ⟨741⟩: between 115° and 119°.

Specific rotation ⟨781⟩: between +103° and +106°, determined in a solution in alcohol containing 150 mg in each 10 mL. Prepare the solution without delay, using Ergocalciferol from a container opened not longer than 30 minutes, and determine the rotation within 30 minutes after the solution has been prepared.

Reducing substances—To 10 mL of a 1 in 100 solution in dehydrated alcohol add 0.5 mL of a 1 in 200 solution of blue tetrazolium in dehydrated alcohol. Then add 0.5 mL of a solution prepared by diluting 1 volume of tetramethylammonium hydroxide TS with dehydrated alcohol to make 10 volumes. Allow the mixture to stand for 5 minutes, accurately timed, then add 1 mL of glacial acetic acid. Prepare a blank by treating 10 mL of dehydrated alcohol in the same manner. Determine the absorbance of the solution at 525 nm, with a suitable spectrophotometer, against the blank: the absorbance is not greater than that obtained from a solution containing 0.2 µg per mL of hydroquinone in dehydrated alcohol, similarly treated.

Assay—

Dehydrated hexane, Mobile phase, Chromatographic system, System suitability preparation, and *System suitability test*—Proceed as directed in the *Assay* under *Cholecalciferol.*

Standard preparation—[NOTE—Use low-actinic glassware, and prepare solutions fresh daily.] Transfer about 30 mg of USP Ergocalciferol RS, accurately weighed, to a 50-mL volumetric flask, dissolve without heat in toluene, add toluene to volume, and mix. Pipet 10 mL of this stock solution into a 50-mL volumetric flask, dilute with *Mobile phase* to volume, and mix to obtain a solution having a known concentration of about 120 µg per mL.

Assay preparation—[NOTE—Use low-actinic glassware, and prepare solutions fresh daily.] Transfer about 30 mg of Ergocalciferol, accurately weighed, to a 50-mL volumetric flask, and proceed as directed for *Standard preparation*, beginning with "dissolve without heat in toluene," to obtain a solution having a concentration of about 120 μg per mL.

Procedure—Introduce equal volumes (5 to 10 μL) of the *Standard preparation* and the *Assay preparation* into the chromatograph (see *Chromatography* ⟨621⟩) by means of a suitable sampling valve. Measure the peak responses for the major peaks obtained, at corresponding retention times, with the *Assay preparation* and the *Standard preparation*. Calculate the quantity, in mg, of $C_{28}H_{44}O$ in the portion of Ergocalciferol taken by the formula:

$$0.25C(r_U/r_S),$$

in which C is the concentration, in μg per mL, of USP Ergocalciferol RS in the *Standard preparation*, and r_U and r_S are the peak responses for ergocalciferol obtained from the *Assay preparation* and the *Standard preparation*, respectively.

Ergocalciferol Capsules

» Ergocalciferol Capsules usually consist of an edible vegetable oil solution of Ergocalciferol, encapsulated with Gelatin. Ergocalciferol Capsules contain not less than 100.0 percent and not more than 120.0 percent of the labeled amount of $C_{28}H_{44}O$.

Packaging and storage—Preserve in tight, light-resistant containers.

Labeling—Label the Capsules to indicate the content of ergocalciferol in mg. The activity may be expressed also in terms of USP Units, on the basis that 40 USP Vitamin D Units = 1 μg.

Reference standards—*USP Ergocalciferol Reference Standard*—Store in a cold place, protected from light. Allow it to attain room temperature before opening ampul. Use the material promptly, and discard the unused portion. *USP Vitamin D Assay System Suitability Reference Standard*—Store in a cool place, protected from light. Allow it to attain room temperature before opening ampul. Do not dry. Transfer unused contents of ampul to a tightly closed container, and store under nitrogen, in the dark, in a cool place.

Uniformity of dosage units ⟨905⟩: meet the requirements.

Assay—[NOTE—Throughout this *Assay*, protect solutions containing, and derived from, the test specimen and the Reference Standard from the atmosphere and light, preferably by the use of a blanket of inert gas and low-actinic glassware.]

Ether—Use ethyl ether. Use within 24 hours after opening container.

Dehydrated hexane—Prepare a chromatographic column by packing a chromatographic tube, 60 cm × 8 cm in diameter, with 500 g of 50- to 250-μm chromatographic siliceous earth, activated by drying at 150° for 4 hours (see *Column adsorption chromatography* under *Chromatography* ⟨621⟩). Pass 500 mL of hexanes through the column, and collect the eluate in a glass-stoppered flask.

Butylated hydroxytoluene solution—Dissolve a quantity of butylated hydroxytoluene in chromatographic hexane to obtain a solution containing 10 mg per mL.

Aqueous potassium hydroxide solution—Dissolve 500 g of potassium hydroxide in 500 mL of freshly boiled water, mix, and cool. Prepare this solution fresh daily.

Alcoholic potassium hydroxide solution—Dissolve 3 g of potassium hydroxide in 50 mL of freshly boiled water, add 10 mL of alcohol, dilute with freshly boiled water to 100 mL, and mix. Prepare this solution fresh daily.

Sodium ascorbate solution—Dissolve 3.5 g of ascorbic acid in 20 mL of 1 *N* sodium hydroxide. Prepare this solution fresh daily.

Sodium sulfide solution—Dissolve 12 g of sodium sulfide in 20 mL of water, dilute with glycerin to 100 mL, and mix.

Mobile phase—Prepare a 3 in 1000 mixture of *n*-amyl alcohol in *Dehydrated hexane*. The ratio of components and the flow rate may be varied to meet system suitability requirements.

Standard preparation—Transfer about 25 mg of USP Ergocalciferol RS, accurately weighed, to a 50-mL volumetric flask, dissolve without heat in toluene, add toluene to volume, and mix. Prepare stock solution fresh daily.

Assay preparation—Reflux not less than 10 Capsules with a mixture of 10 mL of *Sodium ascorbate solution* and 2 drops of *Sodium sulfide solution* on a steam bath for 10 minutes, crush any remaining solids with a blunt glass rod, and continue heating for 5 minutes. Cool, add 25 mL of alcohol and 3 mL of *Aqueous potassium hydroxide solution*, and mix.

Reflux the mixture on a steam bath for 30 minutes. Cool rapidly under running water, and transfer the saponified mixture to a conical separator, rinsing the saponification flask with two 15-mL portions of water, 10 mL of alcohol, and two 50-mL portions of ether. Shake the combined saponified mixture and rinsings vigorously for 30 seconds, and allow to stand until both layers are clear. Transfer the aqueous phase to a second conical separator, add a mixture of 10 mL of alcohol and 50 mL of solvent hexane, and shake vigorously. Allow to separate, transfer the aqueous phase to a third conical separator, and transfer the hexane phase to the first separator, rinsing the second separator with two 10-mL portions of solvent hexane, adding the rinsings to the first separator. Shake the aqueous phase in the third separator with 50 mL of solvent hexane, and add the hexane phase to the first separator. Wash the combined ether-hexane extracts by shaking vigorously with three 50-mL portions of *Alcoholic potassium hydroxide solution*, and wash with 50-mL portions of water vigorously until the last washing is neutral to phenolphthalein. Drain any remaining drops of water from the combined ether-hexane extracts, add 2 sheets of 9-cm filter paper, in strips, to the separator, and shake. Transfer the washed ether-hexane extracts to a round-bottom flask, rinsing the separator and paper with solvent hexane. Combine the hexane rinsings with the ether-hexane extracts, add 100 μL of *Butylated hydroxytoluene solution*, and mix. Evaporate in vacuum to dryness by swirling in a water bath maintained at a temperature not higher than 40°. Cool under running water, and introduce nitrogen sufficient to restore atmospheric pressure. Without delay, dissolve the residue in a measured volume of a 1 in 5 mixture of toluene and *Mobile phase*, until the concentration of vitamin D is about 25 μg per mL, to obtain the *Assay preparation*.

Chromatographic system—Use a chromatograph, operated at room temperature, fitted with an ultraviolet detector that monitors absorption at 254 nm, and a 25-cm × 4.6-mm stainless steel column packed with column packing L3.

System suitability test—Transfer about 100 mg of USP Vitamin D Assay System Suitability RS to a 10-mL volumetric flask, add a 1 in 5 mixture of toluene and *Mobile phase* to volume, and mix. Heat a portion of this solution under reflux, at 90° for 45 minutes, and cool. Chromatograph five injections of this solution, and measure the peak responses as directed under *Procedure:* the relative standard deviation for the cholecalciferol peak response does not exceed 2.0%, and the resolution between *trans*-cholecalciferol and pre-cholecalciferol is not less than 1.0. [NOTE—Chromatograms obtained as directed for this test exhibit relative retention times of approximately 0.4 for pre-cholecalciferol, 0.5 for *trans*-cholecalciferol, and 1.0 for cholecalciferol.]

Vitamin D response factor—Transfer 4.0 mL of the *Standard preparation* to a 100-mL volumetric flask, dilute with *Mobile phase* to volume, and mix to obtain the *Working standard preparation*. Store this *Working standard preparation* at a temperature not above 0°, retaining the unused portion for the *Procedure*. Inject 20 μL of the *Working standard preparation* into the column, and measure the peak response for vitamin D. Calculate the response factor, F_D, by the formula:

$$C_S/r_S,$$

in which C_S is the concentration, in μg per mL, of ergocalciferol in the *Working standard preparation*, and r_S is the peak response of ergocalciferol.

Pre-vitamin D response factor—Pipet 4 mL of the *Standard preparation* into a round-bottom flask fitted with a reflux condenser, and add 2 or 3 crystals of butylated hydroxytoluene. Displace the air with nitrogen, and heat in a water bath maintained at a temperature of 90° in subdued light under a nitrogen atmosphere for 45 minutes, to obtain a solution containing vitamin D and pre-vitamin D. Cool, transfer with the aid of several portions of *Mobile phase* to a 100-mL volumetric flask, dilute with

Mobile phase to volume, and mix to obtain the *Working mixture.* Inject 20 µL of this *Working mixture* into the analytical column, and measure the peak responses for vitamin D and pre-vitamin D. Calculate the concentration, C'_S, in µg per mL, of vitamin D in the (heated) *Working mixture* by the formula:

$$F_D r'_S,$$

in which r'_S is the peak response for vitamin D. Calculate the concentration, C'_{pre}, in µg per mL, of pre-vitamin D, in the *Working mixture* by the formula:

$$C'_{pre} = C_S - C'_S.$$

Calculate the response factor, F_{pre}, for pre-vitamin D by the formula C'_{pre}/r'_{pre}, in which r'_{pre} is the peak response of pre-vitamin D. [NOTE—The value of F_{pre} determined in duplicate, on different days, can be used during the entire procedure.]

Procedure—Inject 20 µL of the *Assay preparation* into the column, and measure the peak responses for vitamin D and pre-vitamin D. Calculate the concentration, in µg per mL, of $C_{28}H_{44}O$ in the *Assay preparation* by the formula:

$$r''_D F_D + r''_{pre} F_{pre},$$

in which r''_D and r''_{pre} are the peak responses of vitamin D and pre-vitamin D, respectively.

Ergocalciferol Oral Solution

» Ergocalciferol Oral Solution is a solution of Ergocalciferol in an edible vegetable oil, in Polysorbate 80, or in Propylene Glycol. It contains not less than 100.0 percent and not more than 120.0 percent of the labeled amount of $C_{28}H_{44}O$.

Packaging and storage—Preserve in tight, light-resistant containers.

Labeling—Label the Oral Solution to indicate the concentration of ergocalciferol in mg. The activity may be expressed also in terms of USP Units, on the basis that 40 USP Vitamin D Units = 1 µg.

Reference standard—*USP Ergocalciferol Reference Standard*—Store in a cold place, protected from light. Allow it to attain room temperature before opening ampul. Use the material promptly, and discard the unused portion.

Assay—

Mobile phase—Use chloroform containing alcohol as preservative.

Standard solution A—Dissolve an accurately weighed quantity of USP Ergocalciferol RS in chloroform, and dilute quantitatively with chloroform to obtain a solution having a known concentration of about 500 µg per mL. [NOTE—Prepare this *Standard solution A* fresh daily.]

Standard solution B—Transfer 5.0 mL of *Standard solution A* to a 50-mL volumetric flask, dilute with chloroform to volume, and mix to obtain a solution having a known concentration of 50 µg per mL. [NOTE—Reserve a portion of this solution for the test for *Pre-ergocalciferol response factor*.]

Standard preparation—By quantitative dilution, with chloroform, of 5 mL of *Standard solution B*, prepare a solution having a known concentration of about 5 µg per mL. [NOTE—Store this *Standard preparation* at a temperature not above 0°.]

Assay preparation—Transfer an accurately measured volume of Ergocalciferol Oral Solution, equivalent to about 250 µg of ergocalciferol, to a 50-mL volumetric flask, dilute with chloroform to volume, and mix.

Chromatographic system (see *Chromatography* ⟨621⟩)—The liquid chromatograph is equipped with a 254-nm detector and a 4.6-mm × 25-cm column that contains 5-µm packing L3. The flow rate is about 1 mL per minute. The relative retention times of pre-ergocalciferol and ergocalciferol are about 0.8 and 1.0, respectively.

Calibration—Inject a suitable volume (10 µL to 20 µL) of the *Standard preparation* into the chromatograph, record the chromatogram, and measure the response for the major peak. Calculate the *Ergocalciferol response factor*, F_D, by the formula:

$$C_S/R_S,$$

in which C_S is the concentration, in µg per mL, of USP Ergocalciferol RS in the *Standard preparation*, and R_S is the peak response of ergocalciferol.

Pre-ergocalciferol response factor—Transfer 5.0 mL of *Standard solution B* into a round-bottomed flask fitted with a reflux condenser. Displace the air with nitrogen, and reflux for 1 hour in a water bath under a nitrogen atmosphere to obtain a solution containing ergocalciferol and pre-ergocalciferol. Cool, transfer, with the aid of several portions of chloroform, to a 50-mL volumetric flask, dilute with chloroform to volume, and mix to obtain the *Working mixture.* Inject a suitable volume (about 10 µL to 20 µL) of the *Working mixture* into the chromatograph, record the chromatogram, and measure the peak responses for ergocalciferol and pre-ergocalciferol. Calculate the concentration, C^1_S, in µg per mL, of ergocalciferol in the (heated) *Working mixture* by the formula:

$$F_D R_E,$$

in which R_E is the peak response for ergocalciferol. Calculate the concentration, C^1_{Pre}, in µg per mL, of pre-ergocalciferol in the (heated) *Working mixture* by the formula:

$$C_S - C^1_S.$$

Calculate the response factor, F_{Pre}, for pre-ergocalciferol by the formula:

$$C^1_{Pre}/R_P,$$

in which R_P is the peak response of pre-ergocalciferol.

System suitability—The resolution, *R*, between the pre-ergocalciferol peak and the ergocalciferol peak determined during the test for *Pre-ergocalciferol response factor* is not less than 1.0, and the relative standard deviation for the peak response of replicate injections of the *Standard preparation* is not more than 2.0%.

Procedure—Separately inject equal volumes (10 µL to 20 µL) of the *Standard preparation* and the *Assay preparation* into the chromatograph, record the chromatograms, and measure the responses for the ergocalciferol peaks. Calculate the quantity, in µg, of $C_{28}H_{44}O$ in the *Assay preparation* by the formula:

$$(F_D S_E) + (F_{Pre} S_{Pre}),$$

in which S_E and S_{Pre} are the peak responses of ergocalciferol and pre-ergocalciferol, respectively.

Ergocalciferol Tablets

» Ergocalciferol Tablets contain not less than 100.0 percent and not more than 120.0 percent of the labeled amount of $C_{28}H_{44}O$.

Packaging and storage—Preserve in tight, light-resistant containers.

Labeling—Label the Tablets to indicate the content of ergocalciferol in mg. The activity may be expressed also in terms of USP Units, on the basis that 40 USP Vitamin D Units = 1 µg.

Reference standard—*USP Ergocalciferol Reference Standard*—Store in a cold place, protected from light. Allow it to attain room temperature before opening ampul. Use the material promptly, and discard the unused portion.

Identification—

A: Evaporate under a stream of nitrogen to dryness 1 mL of the filtrate obtained in *Identification test B*, and dissolve the residue in 50 mL of alcohol: the ultraviolet absorption spectrum of the solution exhibits maxima and minima at the same wavelengths as that of a similar preparation of USP Ergocalciferol RS, concomitantly measured.

B: Triturate a quantity of powdered Tablets, equivalent to about 5 mg of ergocalciferol, with 10 mL of chloroform, and filter into a small flask: the filtrate so obtained responds to *Identification test C* under *Ergocalciferol.*

Disintegration ⟨701⟩: 30 minutes.

Uniformity of dosage units ⟨905⟩: meet the requirements.

Assay—Proceed with Ergocalciferol Tablets as directed for *Chemical Method* under *Vitamin D Assay* ⟨581⟩.

Ergoloid Mesylates

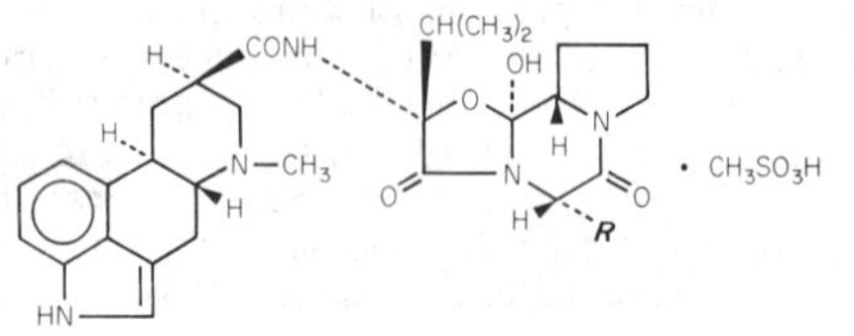

	R
Dihydroergocornine	$—CH(CH_3)_2$
Dihydroergocristine	$—CH_2C_6H_5$
Dihydro-α-ergocryptine	$—CH_2CH(CH_3)_2$
Dihydro-β-ergocryptine	$—CH(CH_3)CH_2CH_3$

$C_{31}H_{41}N_5O_5.CH_4O_3S$ (dihydroergocornine mesylate) 659.80

$C_{35}H_{41}N_5O_5.CH_4O_3S$ (dihydroergocristine mesylate) 707.84

$C_{32}H_{43}N_5O_5.CH_4O_3S$ (dihydro-α-ergocryptine mesylate) 673.82

$C_{32}H_{43}N_5O_5.CH_4O_3S$ (dihydro-β-ergocryptine mesylate) 673.82

Ergotaman-3′,6′,18-trione, 9,10-dihydro-12′-hydroxy-2′,5′-bis(1-methylethyl)-, (5′α,10α)-, monomethanesulfonate (salt) mixture with 9,10α-dihydro-12′-hydroxy-2′-(1-methylethyl)-5′α-(phenylmethyl)ergotaman-3′,6′,18-trione monomethanesulfonate (salt), 9,10α-dihydro-12′-hydroxy-2′-(1-methylethyl)-5′α-(2-methylpropyl)ergotaman-3′,6′,18-trione monomethanesulfonate (salt), and 9,10α-dihydro-12′-hydroxy-2′-(1-methylethyl)-5′α-(1-methylpropyl)ergotaman-3′,6′,18-trione monomethanesulfonate (salt).

Dihydroergotoxine monomethanesulfonate (salt).

An equiproportional mixture of dihydroergocornine mesylate, dihydroergocristine mesylate, and ratio of dihydro-α-ergocryptine mesylate to dihydro-β-ergocryptine mesylate is 1.5–2.5:1 [*8067-24-1*].

» Ergoloid Mesylates is a mixture of the methanesulfonate salts of the three hydrogenated alkaloids, dihydroergocristine ($C_{35}H_{41}N_5O_5.CH_4O_3S$), dihydroergocornine ($C_{31}H_{41}N_5O_5.CH_4O_3S$), and dihydroergocryptine ($C_{32}H_{43}N_5O_5.CH_4O_3S$), in an approximate weight ratio of 1:1:1. Ergoloid Mesylates contains not less than 97.0 percent and not more than 103.0 percent of the alkaloid methanesulfonate mixture, and not less than 30.3 percent and not more than 36.3 percent of the methanesulfonate salt of each of the individual alkaloids, calculated on the anhydrous basis. Dihydroergocryptine mesylate exists as a mixture of *alpha*- and *beta*- isomers. The ratio of *alpha*- to *beta*- isomers is not less than 1.5:1.0 and not more than 2.5:1.0.

Packaging and storage—Preserve in tight, light-resistant containers.

Reference standard—*USP Ergoloid Mesylates Reference Standard*—Do not dry; determine the *Water* content by *Method I* before using.

Identification—

A: The infrared absorption spectrum of a potassium bromide dispersion of it exhibits maxima only at the same wavelengths as that of a similar, undried preparation of USP Ergoloid Mesylates RS.

B: In a suitable chromatographic chamber, arranged for thin-layer chromatography, place a volume of a solvent system consisting of a mixture of acetone, *n*-butyl alcohol, ammonium hydroxide, and water (65:20:10:5) sufficient to develop the chromatogram. Prepare a test solution of Ergoloid Mesylates in a mixture of chloroform and methanol (9:1) containing 40 mg per mL. On a suitable thin-layer chromatographic plate (see *Chromatography* ⟨621⟩), coated with a 0.25-mm layer of chromatographic silica gel, apply 10 µL of this solution and 10 µL of a reference solution of methanesulfonic acid containing 0.4 mL in 100 mL of a mixture of chloroform and methanol (9:1). Develop the chromatogram until the solvent front has moved 10 cm. Remove the plate from the developing chamber, mark the solvent front, and dry in a current of cold air. Spray the plate with a 1 in 1000 solution of bromocresol purple in alcohol that previously has been adjusted to the purple color with 6 *N* ammonium hydroxide, then place in a stream of warm air until the spots appear: the R_f value of the methanesulfonic acid spot obtained from the test solution corresponds to that obtained from the reference solution.

Specific rotation ⟨781⟩: between +11.0° and +15.0°, calculated on the anhydrous basis, determined in a solution containing 100 mg of it in 10 mL of dilute alcohol (1 in 2).

pH ⟨791⟩: between 4.2 and 5.2 in a solution (1 in 200).

Water, *Method I* ⟨921⟩: not more than 5.0%.

Ergotamine—Prepare three solutions in a mixture of chloroform and methanol (9:1) containing 5 mg of Ergoloid Mesylates per mL, 5 mg of USP Ergoloid Mesylates RS per mL, and 5 mg of Ergotamine Tartrate per mL. Apply 5-µL volumes of the solutions at points about 2 cm from the bottom edge of a thin-layer chromatographic plate (see *Chromatography* ⟨621⟩) coated with a 0.25-mm layer of chromatographic silica gel mixture, and allow the spots to dry. Add the solvent system consisting of a mixture of chloroform and methanol (9:1) and a small beaker of ammonium hydroxide to a suitable chamber, seal, and allow to equilibrate for 30 minutes. Develop the chromatogram in the equilibrated chamber until the solvent front has moved about 15 cm from the points of application. Remove the plate, air-dry, and locate the spots, first by viewing under long-wavelength ultraviolet light, and then by spraying with a reagent prepared by dissolving 800 mg of *p*-(dimethylamino)benzaldehyde in a mixture of 80 mL of alcohol and 11 mL of sulfuric acid: the chromatogram from Ergoloid Mesylates shows primary spots that correspond in size and color to the spots obtained from the USP Ergoloid Mesylates RS solution, and shows no spot corresponding to the principal spot in the chromatogram of Ergotamine Tartrate.

Non-hydrogenated alkaloids—Prepare a solution in alcohol containing 0.4 mg of Ergoloid Mesylates per mL, and prepare a 1 in 10 dilution of the first solution. Determine the absorbances in 1-cm cells of the first solution at 317.5 nm and the dilution at 280 nm, using alcohol as the blank: the absorbance of the first solution is not more than 0.15 times that of the dilution.

Assay—

Mobile phase—Prepare a degassed solution containing a mixture of water, acetonitrile, and triethylamine (80:20:2.5). Adjust the ratio as necessary.

Standard preparation—Transfer about 10 mg of USP Ergoloid Mesylates RS, accurately weighed, to a 10-mL volumetric flask. Dissolve in a mixture of acetonitrile and water (1:1), dilute with the same solvent to volume, and mix. Prepare this solution fresh.

Assay preparation—Using about 10 mg of Ergoloid Mesylates, accurately weighed, proceed as directed for *Standard preparation.*

Chromatographic system (see *Chromatography* ⟨621⟩)—The liquid chromatograph is equipped with a 280-nm detector and a 4-mm × 30-cm column that contains packing L1. Chromatograph the *Standard preparation,* and record the peak responses as directed under *Procedure:* the column efficiency determined for the dihydro-β-ergocryptine mesylate peak is not less than 950 theoretical plates, the tailing factor for dihydro-β-ergocryptine mesylate is not more than 2.5, the resolution, *R,* between dihydro-α-ergocryptine mesylate and dihydroergocristine mesylate is not less than 1.35, the resolution, *R,* between dihydroergocristine and dihydro-β-ergocryptine is not less than 1.0, and the relative standard deviation of the sum of the four peaks for replicate injections is not more than 1.5%.

Procedure—Separately inject equal volumes (about 20 µL) of the *Standard preparation* and the *Assay preparation* into the chromatograph by means of a suitable microsyringe or sampling valve, record the chromatograms, and measure the responses for the major peaks. The order of elution is dihydroergocornine, dihydro-α-ergocryptine, dihydroergocristine, and dihydro-β-ergocryptine. Calculate the quantity, in mg, of dihydroergocristine mesylate ($C_{35}H_{41}N_5O_5 . CH_4O_3S$), dihydroergocryptine mesylate ($C_{32}H_{43}N_5O_5 . CH_4O_3S$), and dihydroergocornine mesylate ($C_{31}H_{41}N_5O_5 . CH_4O_3S$) in the Ergoloid Mesylates taken by the formula:

$$10C(r_U / r_S),$$

in which C is the concentration, in mg per mL, of the corresponding dihydrogenated alkaloid mesylate in USP Ergoloid Mesylates RS in the *Standard preparation*, and r_U and r_S are the peak responses of the corresponding analyte obtained from the *Assay preparation* and the *Standard preparation*, respectively.

Ergoloid Mesylates Oral Solution

» Ergoloid Mesylates Oral Solution contains not less than 90.0 percent and not more than 110.0 percent of the labeled amount of Ergoloid Mesylates, consisting of not less than 30.3 percent and not more than 36.3 percent of the methanesulfonate salt of each of the individual alkaloids (dihydroergocristine, dihydroergocornine, and dihydroergocryptine). The ratio of *alpha*- to *beta*-dihydroergocryptine mesylate is not less than 1.5:1.0 and not more than 2.5:1.0.

Packaging and storage—Preserve in tight, light-resistant containers at a temperature not exceeding 30°.

Reference standard—*USP Ergoloid Mesylates Reference Standard*—Do not dry; determine the *Water* content by *Method I* before using.

Identification—The retention times of the major peaks in the chromatogram of the *Assay preparation* are the same as those of the *Standard preparation* as obtained in the *Assay*.

Alcohol content, *Method II* ⟨611⟩: between 90.0% and 110.0% of the labeled amount of C_2H_5OH.

Assay—

Mobile phase and *Chromatographic system*—Prepare as directed in the *Assay* under *Ergoloid Mesylates*.

Standard preparation—Dissolve an accurately weighed quantity of USP Ergoloid Mesylates RS in acetonitrile solution (1 in 5) to obtain a solution having a known concentration of about 0.25 mg per mL.

Assay preparation—Transfer an accurately measured volume of Oral Solution, equivalent to about 25 mg of ergoloid mesylates, to a 100-mL volumetric flask. Rinse the pipet with small portions of acetonitrile solution (1 in 5), collecting the rinsings in the volumetric flask. Dilute with acetonitrile solution (1 in 5) to volume, and mix.

Procedure—Proceed as directed for *Procedure* in the *Assay* under *Ergoloid Mesylates*, except to inject about 60-µL volumes of the *Assay preparation* and the *Standard preparation*. Calculate the quantity, in mg, of dihydroergocristine mesylate ($C_{35}H_{41}N_5O_5 . CH_4O_3S$), dihydroergocryptine mesylate ($C_{32}H_{43}N_5O_5 . CH_4O_3S$), and dihydroergocornine mesylate ($C_{31}H_{41}N_5O_5 . CH_4O_3S$) in each mL of the Oral Solution taken by the formula:

$$100(C/V)(r_U / r_S),$$

in which V is the volume, in mL, of Oral Solution taken, and the other terms are as defined therein.

Ergoloid Mesylates Tablets

» Ergoloid Mesylates Tablets contain not less than 90.0 percent and not more than 110.0 percent of the labeled amount of ergoloid mesylates, consisting of not less than 30.3 percent and not more then 36.3 percent of the methanesulfonate salt of each of the individual alkaloids (dihydroergocristine, dihydroergocornine, and dihydroergocryptine). The ratio of *alpha*- to *beta*-dihydroergocryptine mesylate is not less than 1.5:1.0 and not more than 2.5:1.0.

Packaging and storage—Preserve in tight, light-resistant containers.

Labeling—Label Tablets to indicate whether they are intended for sublingual administration or for swallowing.

Reference standard—*USP Ergoloid Mesylates Reference Standard*—Do not dry; determine the *Water* content by *Method I* before using.

Identification—Mix a small amount of powdered Tablets, equivalent to about 5 mg of ergoloid mesylates, with 5 mL of water and 5 mL of a mixture of equal volumes of glacial acetic acid and sulfuric acid, and add 1 drop of freshly prepared ferric chloride solution (1 in 20): a violet-blue color develops within 5 minutes.

Disintegration ⟨701⟩ (for Tablets intended for sublingual use): 15 minutes.

Dissolution ⟨711⟩ (for Tablets intended to be swallowed)—

Medium: water; 500 mL.

Apparatus 2: 50 rpm, the distance between the paddle blade and the inside bottom of the vessel being maintained at 4.5 ± 0.2 cm during the test.

Time: 30 minutes.

Procedure—Determine the amount of Ergoloid Mesylates in solution in filtered portions of the *Dissolution Medium*, suitably diluted, in a fluorometer at an excitation wavelength of 280 nm and an emission wavelength of 360 nm, in comparison with a Standard solution of known concentration of USP Ergoloid Mesylates RS.

Tolerances—Not less than 75% (*Q*) of the labeled amount of Ergoloid Mesylates is dissolved in 30 minutes.

Uniformity of dosage units ⟨905⟩: meet the requirements.

Procedure for content uniformity—[NOTE—Protect the solutions from light throughout this procedure.]—Shake 1 Tablet for 30 minutes with sufficient tartaric acid solution (1 in 100) to obtain a final test solution containing about 60 µg of ergoloid mesylates per mL. Filter the mixture. To three separate flasks transfer 3-mL portions of the filtrate, a solution of USP Ergoloid Mesylates RS in tartaric acid solution (1 in 100) having a known concentration of about 60 µg per mL, and tartaric acid solution (1 in 100) to provide a blank. Add 6 mL of *p*-(dimethylamino)benzaldehyde TS to each flask, and allow to stand in the dark for 30 minutes. Concomitantly determine the absorbances of both solutions in 1-cm cells at the wavelength of maximum absorbance at about 580 nm, with a suitable spectrophotometer, against the blank. Calculate the quantity, in mg, of ergoloid mesylates in the Tablet by the formula:

$$(TC/D)(A_U / A_S),$$

in which T is the labeled quantity, in mg, of ergoloid mesylates in the Tablet, D is the concentration, in µg per mL, of ergoloid mesylates in the solution from the Tablet based on the labeled quantity per Tablet and the extent of dilution, C is the concentration, in µg per mL, of USP Ergoloid Mesylates RS in the Standard solution, and A_U and A_S are the absorbances of the test solution and Standard solution, respectively.

Assay—

Mobile phase, *Standard preparation*, and *Chromatographic system*—Prepare as directed in the *Assay* under *Ergoloid Mesylates*.

Assay preparation—Transfer a portion of finely powdered Ergoloid Mesylates Tablets, equivalent to about 50 mg of ergoloid mesylates, to a suitable container. Add 30 mL of a solvent mixture of acetonitrile and water (1:1). Stir the mixture for 15 minutes. Filter, and wash the residue with three successive 5-mL portions of the solvent mixture, collecting all of the filtrates in a 50-mL volumetric flask. Dilute with the solvent mixture to volume, and mix.

Procedure—Proceed as directed for *Procedure* in the *Assay* under *Ergoloid Mesylates*, except to inject about 50-µL volumes of the *Assay preparation* and the *Standard preparation*, and to use 50 instead of 10 in the calculation formula.

Ergonovine Maleate

$C_{19}H_{23}N_3O_2 \cdot C_4H_4O_4$ 441.48

Ergoline-8-carboxamide, 9,10-didehydro-*N*-(2-hydroxy-1-methylethyl)-6-methyl-, [8β(*S*)]-, (*Z*)-2-butenedioate (1:1) (salt).

9,10-Didehydro-*N*-[(*S*)-2-hydroxy-1-methylethyl]-6-methylergoline-8β-carboxamide maleate (1:1) (salt) [*129-51-1*].

» Ergonovine Maleate contains not less than 97.0 percent and not more than 103.0 percent of $C_{19}H_{23}N_3O_2 \cdot C_4H_4O_4$, calculated on the dried basis.

Packaging and storage—Preserve in tight, light-resistant containers, in a cold place.

Reference standard—*USP Ergonovine Maleate Reference Standard*—Dry in vacuum at 80° for 3 hours before using.

Identification—

A: The infrared absorption spectrum of a potassium bromide dispersion of it exhibits maxima only at the same wavelengths as that of a similar preparation of USP Ergonovine Maleate RS.

B: The ultraviolet absorption spectrum of a 1 in 50,000 solution in alcohol exhibits maxima and minima at the same wavelengths as that of a similar solution of USP Ergonovine Maleate RS, concomitantly measured, and the respective absorptivities, calculated on the dried basis, at the wavelength of maximum absorbance at about 311 nm do not differ by more than 3.0%.

C: The R_f value of the principal blue spot obtained from the *Test preparation* corresponds to that obtained from the *Standard preparation* in the chromatogram prepared as directed in the test for *Related alkaloids*.

Specific rotation ⟨781⟩: between +51° and +56°, calculated on the dried basis, determined in a solution containing 50 mg in each 10 mL, using 1-decimeter cells.

Loss on drying ⟨731⟩—Dry it in vacuum at 80° for 3 hours: it loses not more than 2.0% of its weight.

Related alkaloids—[NOTE—Conduct this test promptly, without exposure to daylight and with minimum exposure to artificial light.]

Solvent mixture—Mix 9 volumes of alcohol with 1 volume of ammonium hydroxide.

Test preparation—Immediately prior to use, prepare a solution of Ergonovine Maleate in *Solvent mixture* to contain 10 mg per mL.

Standard preparation and *Standard dilutions*—Prepare a solution of USP Ergonovine Maleate RS in *Solvent mixture* to contain 10 mg per mL (*Standard preparation*). Prepare a series of dilutions of the *Standard preparation* in *Solvent mixture* to contain 0.20 mg, 0.10 mg, and 0.05 mg per mL (*Standard dilutions*). Use immediately after preparation.

Procedure—In a suitable chromatographic chamber arranged for thin-layer chromatography place a volume of a solvent system consisting of chloroform, methanol, and water (75:25:3) sufficient to develop the chromatogram, cover, and allow to equilibrate for 30 minutes. On a suitable thin-layer chromatographic plate coated with a 0.25-mm layer of chromatographic silica gel, apply 5-µL portions of the *Standard preparation*, each of the three *Standard dilutions*, and finally the *Test preparation*. Without delay, develop the chromatogram until the solvent front has moved about three-fourths of the length of the plate. Remove the plate from the developing chamber, mark the solvent front, and allow the solvent to evaporate. Locate the spots on the plate by spraying thoroughly and evenly with a solution prepared by dissolving 1 g of *p*-dimethylaminobenzaldehyde in a cooled mixture of 50 mL of alcohol and 50 mL of hydrochloric acid. Immediately dry in a stream of nitrogen for about 2 minutes: the R_f value of the principal spot obtained from the *Test preparation* corresponds to that obtained from the *Standard preparation*. Estimate the concentration of any other spots observed in the lane for the *Test preparation* by comparison with the *Standard dilutions*. The spots from the 0.20-, 0.10-, and 0.05-mg-per-mL dilutions are equivalent to 2.0%, 1.0%, and 0.50% of impurities, respectively. The sum of the impurities is not greater than 2.0%.

Assay—

Standard preparation—Using a suitable quantity of USP Ergonovine Maleate RS, accurately weighed, prepare a solution in water having a known concentration of about 40 µg per mL.

Assay preparation—Transfer about 40 mg of Ergonovine Maleate, accurately weighed, to a 100-mL volumetric flask, dilute with water to volume, and mix. Dilute 10.0 mL of this solution with water to 100.0 mL.

Procedure—Transfer 5.0 mL each of the *Standard preparation*, the *Assay preparation*, and water to provide a blank, to separate conical flasks. Add 10.0 mL of *p*-dimethylaminobenzaldehyde TS with constant swirling to each, and allow to stand for 20 minutes. Concomitantly determine the absorbances of the solutions in 1-cm cells at the wavelength of maximum absorbance at about 555 nm, with a suitable spectrophotometer, against the blank. Calculate the quantity, in mg, of $C_{19}H_{23}N_3O_2 \cdot C_4H_4O_4$ taken by the formula:

$$C(A_U/A_S),$$

in which C is the concentration, in µg per mL, of USP Ergonovine Maleate RS in the *Standard preparation*, and A_U and A_S are the absorbances of the solutions from the *Assay preparation* and the *Standard preparation*, respectively.

Ergonovine Maleate Injection

» Ergonovine Maleate Injection is a sterile solution of Ergonovine Maleate in Water for Injection. It contains not less than 90.0 percent and not more than 110.0 percent of the labeled amount of $C_{19}H_{23}N_3O_2 \cdot C_4H_4O_4$.

Packaging and storage—Preserve in single-dose, light-resistant containers, preferably of Type I glass, and store in a cold place.

Reference standard—*USP Ergonovine Maleate Reference Standard*—Dry in vacuum at 80° for 3 hours before using.

Identification—The R_f value of the principal blue spot obtained from the *Test preparation* corresponds to that obtained from *Standard preparation A* in the chromatogram prepared as directed in the test for *Related alkaloids*.

pH ⟨791⟩: between 2.7 and 3.5.

Related alkaloids—[NOTE—Conduct this test promptly, without exposure to daylight and with minimum exposure to artificial light.]

Solvent mixture, Standard preparation, and *Standard dilutions*—Prepare as directed in the test for *Related alkaloids* under *Ergonovine Maleate*.

Test preparation—Immediately prior to use, transfer a volume of Injection, equivalent to about 5 mg of ergonovine maleate, to a separator, and extract with three 5-mL portions of chloroform. Discard the chloroform extracts. Render alkaline to litmus with 6 *N* ammonium hydroxide, and extract with three 5-mL portions of chloroform. Evaporate the combined extracts with the aid of a stream of nitrogen, but without heat, to dryness. Dissolve the residue so obtained in 0.5 mL of *Solvent mixture*.

Procedure—Proceed as directed for *Procedure* in the test for *Related alkaloids* under *Ergonovine Maleate*.

Other requirements—It meets the requirements under *Injections* ⟨1⟩.

Assay—

0.05 M Phosphate buffer—Dissolve 6.8 g of monobasic potassium phosphate in 600 mL of water and adjust with phosphoric acid to a pH of 2.1. Dilute with water to 1000 mL, and mix.

Mobile phase—Prepare a suitable and degassed solution of *0.05 M Phosphate buffer* and acetonitrile (80:20) such that the retention time is approximately 3 minutes with a flow rate of 1 mL per minute.

Standard preparation—Dissolve an accurately weighed quantity of USP Ergonovine Maleate RS in *Mobile phase*, adding sufficient water to equal 10% of the final volume, to obtain a solution having a known concentration of about 0.02 mg per mL.

Assay preparation—Quantitatively dilute an accurately measured volume of the Injection, equivalent to about 2 mg of ergonovine maleate, with *Mobile phase* and water, if necessary, to obtain a solution having a concentration of about 0.02 mg per mL in which the Injection volume plus any added water constitutes 10% of the final volume.

Chromatographic system (see *Chromatography* ⟨621⟩)—The liquid chromatograph is equipped with a 312-nm detector and a 3-mm × 30-cm column that contains packing L1. Chromatograph five replicate injections of the *Standard preparation*, and record the peak responses as directed under *Procedure:* the relative standard deviation is not more than 3.0%.

Procedure—By means of a suitable sampling valve, introduce equal volumes (about 100 µL) of the *Assay preparation* and the *Standard preparation* into the chromatograph. Measure the peak responses of Ergonovine Maleate, at corresponding retention times, obtained from the *Assay preparation* and the *Standard preparation*. Calculate the quantity, in mg, of $C_{19}H_{23}N_3O_2 . C_4H_4O_4$ in each mL of the Injection taken by the formula:

$$(CD/V)(r_U/r_S),$$

in which *C* is the concentration, in mg per mL, of USP Ergonovine Maleate RS in the *Standard preparation*, *V* is the volume, in mL, of Injection taken, *D* is the dilution factor, and r_U and r_S are the peak responses obtained from the *Assay preparation* and the *Standard preparation*, respectively.

Ergonovine Maleate Tablets

» Ergonovine Maleate Tablets contain not less than 90.0 percent and not more than 110.0 percent of the labeled amount of $C_{19}H_{23}N_3O_2 . C_4H_4O_4$.

Packaging and storage—Preserve in well-closed containers.

Reference standard—*USP Ergonovine Maleate Reference Standard*—Dry in vacuum at 80° for 3 hours before using.

Identification—The R_f value of the principal blue spot obtained from the *Test preparation* corresponds to that obtained from the *Standard preparation* in the chromatogram prepared as directed in the test for *Related alkaloids*.

Dissolution ⟨711⟩—

Medium: water; 900 mL.

Apparatus 1: 100 rpm.

Time: 45 minutes.

Procedure—Determine the amount of $C_{19}H_{23}N_3O_2 . C_4H_4O_4$ dissolved from fluorometric measurements, using 322 nm as the excitation wavelength and 428 nm as the emission wavelength, of filtered portions of the solution under test, suitably diluted with *Dissolution Medium*, if necessary, in comparison with a Standard solution having a known concentration of USP Ergonovine Maleate RS in the same medium.

Tolerances—Not less than 75% (*Q*) of the labeled amount of $C_{19}H_{23}N_3O_2 . C_4H_4O_4$ is dissolved in 45 minutes.

Uniformity of dosage units ⟨905⟩: meet the requirements.

Related alkaloids—[NOTE—Conduct this test promptly, without exposure to daylight and with minimum exposure to artificial light.]

Solvent mixture—Mix 75 volumes of chloroform, 25 volumes of methanol, and 1 volume of ammonium hydroxide.

Detecting reagent—Cautiously dissolve 800 mg of *p*-dimethylaminobenzaldehyde in a mixture of alcohol and sulfuric acid (101:11).

Standard stock solution—Transfer 25 mg of USP Ergonovine Maleate RS, accurately weighed, to a separator, shake with 10 mL of water, render the mixture alkaline to litmus with 6 *N* ammonium hydroxide, and extract with three 10-mL portions of chloroform. Evaporate the combined extracts with the aid of a stream of nitrogen, but without heat, to dryness. Dissolve and dilute the residue to 10.0 mL with the *Solvent mixture*.

Standard preparations A, B, C, and *D*—Dilute accurately measured volumes of *Standard stock solution* quantitatively with *Solvent mixture* (designated below as parts by volume of *Standard stock solution* in total parts by volume of the finished Standard preparation) to obtain Standard preparations, designated below by letter, having the following compositions:

Standard Solution	Dilution	Concentration (µg RS per mL)	Percentage (%, for comparison with test specimen)
A	(1 in 20)	125	5.0
B	(1 in 33)	75	3.0
C	(1 in 100)	25	1.0
D	(1 in 200)	12.5	0.5

Test preparation—Immediately prior to use, transfer a quantity of finely powdered Tablets, equivalent to about 5 mg of ergonovine maleate, to a separator, shake with 10 mL of water, render the mixture alkaline to litmus with 6 *N* ammonium hydroxide, and extract with three 10-mL portions of chloroform. Evaporate the combined extracts with the aid of a stream of nitrogen, but without heat, to dryness. Dissolve the residue obtained in 2.0 mL of *Solvent mixture*.

Procedure—On a suitable thin-layer chromatographic plate (see *Chromatography* ⟨621⟩), coated with a 0.25-mm layer of chromatographic silica gel mixture, apply separately 20 µL of the *Test preparation* and 20 µL of each *Standard preparation*. Dry the plate with the aid of a current of cool air. Position the plate in a chromatographic chamber, and develop the chromatograms in *Solvent mixture* until the solvent front has moved about three-fourths of the length of the plate. Remove the plate from the developing chamber, mark the solvent front, and allow the solvent to evaporate in a current of cool air. Examine the plate under long-wavelength ultraviolet light. Mark the principal and any secondary fluorescent spots. Spray the plate with *Detecting reagent*, and mark the principal and secondary blue spots. Compare the intensities of any secondary spots observed in the chromatogram of the *Test preparation* with those of the principal spots in the chromatograms of the *Standard preparations:* the sum of the intensities of secondary spots obtained from the *Test preparation* corresponds to not more than 5.0% of related compounds.

Assay—

0.05 M Phosphate buffer, *Mobile phase*, and *Chromatographic system*—Proceed as directed in the *Assay* under *Ergonovine Maleate Injection*.

Standard preparation—Dissolve an accurately weighed quantity of USP Ergonovine Maleate RS in *Mobile phase* to obtain a solution having a known concentration of about 0.02 mg per mL.

Assay preparation—Weigh and finely powder not less than 20 Ergonovine Maleate Tablets. Transfer an accurately weighed portion of the powder, equivalent to about 1 mg of ergonovine maleate, to a 50-mL volumetric flask, add 25 mL of *Mobile phase*, place in a sonic bath for 5 minutes, cool to room temperature, dilute with *Mobile phase* to volume, mix, and centrifuge. Use the supernatant liquid as directed in the *Procedure*.

Procedure—Proceed as directed for *Procedure* in the *Assay* under *Ergonovine Maleate Injection*. Calculate the quantity, in mg, of $C_{19}H_{23}N_3O_2 . C_4H_4O_4$ in the portion of Tablets taken by the formula:

$(50C)(r_U/r_S)$,

in which the terms are as defined therein.

Ergotamine Tartrate

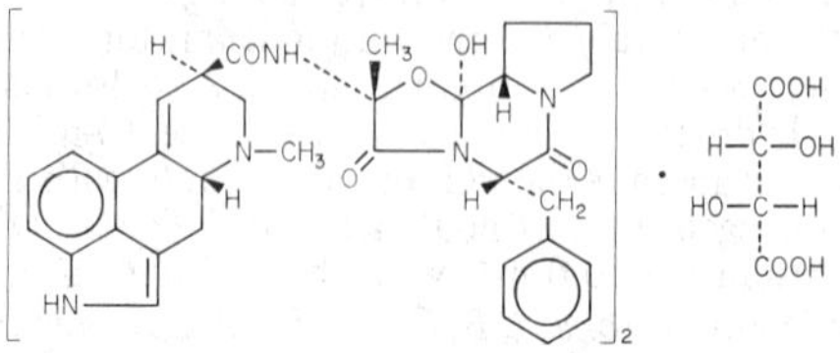

$(C_{33}H_{35}N_5O_5)_2 \cdot C_4H_6O_6$ 1313.43

Ergotaman-3′,6′,18-trione, 12′-hydroxy-2′-methyl-5′-(phenylmethyl)-, (5′α)-,[*R*-(*R**,*R**)]-2,3-dihydroxybutanedioate (2:1) (salt).

Ergotamine tartrate (2:1) (salt) [*379-79-3*].

» Ergotamine Tartrate contains not less than 97.0 percent and not more than 100.5 percent of $(C_{33}H_{35}N_5O_5)_2 \cdot C_4H_6O_6$, calculated on the dried basis.

Packaging and storage—Preserve in well-closed, light-resistant containers in a cold place.

Reference standard—*USP Ergotamine Tartrate Reference Standard*—Dry in vacuum at 60° for 4 hours before using.

Identification—The chromatogram of the test solution prepared as directed in the test for *Related alkaloids* exhibits its principal fluorescent spot and principal blue spot at the same R_f value as the principal spot of *Standard solution A.*

Specific rotation of ergotamine base ⟨781⟩—[NOTE—For this test, use chloroform from which any alcohol present has been removed by prior washing with water.] Dissolve about 350 mg in 25 mL of tartaric acid solution (1 in 100) contained in a separator, then add 500 mg of sodium bicarbonate, and mix gently but thoroughly. Add 10 mL of chloroform, shake vigorously, and after the layers have separated draw off the chloroform phase through a small filter, previously moistened with chloroform, into a 50-mL volumetric flask. Rapidly continue the extraction with three 10-mL portions of chloroform, passing the extracts through the same filter. Place the flask in a bath at 20° for 10 minutes. Adjust the volume of extract to 50.0 mL at 20° by the addition of chloroform. Mix the solution, and determine the angular rotation at 20°. Determine the concentration of ergotamine in the chloroform solution by evaporating a 25.0-mL aliquot of the solution on a rotary evaporator to dryness, maintaining the temperature of the bath below 45°. Dissolve the residue in 25 mL of glacial acetic acid, add 1 drop of crystal violet TS, and titrate with 0.05 *N* perchloric acid VS to an emerald-green end-point. Perform a blank determination, and make any necessary correction. Each mL of 0.05 *N* perchloric acid is equivalent to 29.08 mg of $C_{33}H_{35}N_5O_5$. From the angular rotation of the solution and the concentration of ergotamine base, calculate the specific rotation of the base: the specific rotation is between −155° and −165°.

Loss on drying ⟨731⟩—Dry about 100 mg, accurately weighed, in vacuum at 60° for 4 hours: it loses not more than 5.0% of its weight.

Related alkaloids—[NOTE—Conduct this test without exposure to daylight and with the minimum necessary exposure to artificial light.]

Solvent mixture—Mix 9 volumes of chloroform with 1 volume of methanol.

Standard preparation and *Standard dilutions*—Prepare a solution of USP Ergotamine Tartrate RS in *Solvent mixture* to contain 10.0 mg per mL (*Standard preparation*). Prepare a series of dilutions of the *Standard preparation* in *Solvent mixture* to contain 0.2 mg, 0.1 mg, 0.05 mg, and 0.025 mg per mL (*Standard dilutions*) corresponding to 2.0%, 1.0%, 0.5%, and 0.25% of the *Standard preparation*, respectively.

Test preparation—Dissolve 50.0 mg of Ergotamine Tartrate in 5.0 mL of *Solvent mixture.*

Procedure—In a suitable chromatographic chamber arranged for thin-layer chromatography (see *Chromatography* ⟨621⟩) place a volume of a solvent system consisting of ether, dimethylformamide, chloroform, and dehydrated alcohol (70:15:10:5). Line the chamber with filter paper, and allow it to equilibrate for 15 minutes. On a suitable chromatographic plate coated with a 0.25-mm layer of chromatographic silica gel, apply 5-µL portions of the *Test preparation*, the *Standard preparation*, and each of the *Standard dilutions*. Place each spot over an opened bottle of ammonium hydroxide for 20 seconds, then allow the plate to dry in a current of cold air for 20 seconds. Develop the chromatogram until the solvent front has moved about 17 cm. Remove the plate from the developing chamber, allow the solvent to evaporate in a current of cold air for approximately 2 minutes, and spray with a freshly prepared solution of 200 mg of *p*-(dimethylamino)benzaldehyde in a mixture of 5.5 mL of hydrochloric acid and 4.5 mL of water. Dry the plate at 60° for about 5 minutes and compare the chromatograms: the R_f value of the principal spot obtained from the *Test preparation* corresponds to that obtained from the *Standard preparation;* and the sum of the intensities of any secondary spots in the chromatogram from the *Test preparation* is not greater than the intensity of the principal spot from the 2.0% *Standard dilution*, and the intensity of not more than one of the secondary spots is greater than that of the principal spot from the 1.0% *Standard dilution.*

Assay—Transfer about 200 mg of Ergotamine Tartrate, accurately weighed, to a small conical flask, and dissolve in 15 mL of a mixture of 6 volumes of acetic anhydride and 100 volumes of glacial acetic acid. Add 1 drop of crystal violet TS, and titrate with 0.05 *N* perchloric acid VS from a 10-mL buret. Perform a blank determination, and make any necessary correction. Each mL of 0.05 *N* perchloric acid is equivalent to 32.84 mg of $(C_{33}H_{35}N_5O_5)_2 \cdot C_4H_6O_6$.

Ergotamine Tartrate Inhalation Aerosol

» Ergotamine Tartrate Inhalation Aerosol is a suspension of microfine Ergotamine Tartrate in propellants in a pressurized container. It contains not less than 90.0 percent and not more than 110.0 percent of the labeled amount of $(C_{33}H_{35}N_5O_5)_2 \cdot C_4H_6O_6$, and it delivers not less than 75.0 percent and not more than 125.0 percent of the labeled amount of $(C_{33}H_{35}N_5O_5)_2 \cdot C_4H_6O_6$ per inhalation through an oral inhalation actuator.

Packaging and storage—Preserve in small, non-reactive, light-resistant aerosol containers equipped with metered-dose valves and provided with oral inhalation actuators.

Reference standard—*USP Ergotamine Tartrate Reference Standard*—Dry in vacuum at 60° for 4 hours before using.

Identification—Place 10 mL of water in a small beaker, and deliver 2 sprays from the Inhalation Aerosol under the surface of the water, actuating the valve by pressing the tip against the bottom of the beaker. To one part of the resulting solution in a test tube add 2 parts of *p*-(dimethylamino)benzaldehyde TS, and mix: a blue color is produced.

Unit spray content—

Unit spray sampling apparatus—Prepare as directed under *Aerosols* ⟨601⟩.

Standard preparation—Dissolve an accurately weighed quantity of USP Ergotamine Tartrate RS in tartaric acid solution (1 in 500) to obtain a solution having a known concentration of about 80 µg per mL.

Test preparation—Transfer 20.0 mL of tartaric acid solution (1 in 500) to the collection chamber of the *Unit spray sampling apparatus* (see *Aerosols* ⟨601⟩), and clamp the intake tube in a horizontal position. Prime the valve of the Inhalation Aerosol by alternately shaking and firing it 10 times through its oral inhalation actuator. Couple the aerosol container and actuator to the intake tube of the apparatus, turn on the vacuum source, and set the flow regulator for an airflow rate of 12 ± 1 liters per minute. Turn off the vacuum, remove the aerosol container, leaving the actuator in place. Shake the container, seat it in the actuator, turn on the vacuum, immediately deliver a single spray into the system, and turn off the vacuum. Deliver a total of 4 sprays in this manner. Uncouple the aerosol container and actuator from the assembly, clamp the intake tube in a vertical position, and scavenge the ergotamine tartrate remaining in the intake system as follows: Disconnect the collection chamber, transfer the test solution to a small beaker, warm to 50°, and wash the inside of the intake tube with the warm test solution, collecting the washing in the collection chamber placed a few cm below its normal position. Recycle the test solution twice in this manner. Transfer the solution to a centrifuge tube, and add 10.0 mL of trichlorotrifluoroethane. Insert the stopper, shake vigorously for 1 minute, centrifuge for 5 minutes, and use the clear supernatant liquid as the *Test preparation.*

Procedure—Using the *Test preparation* in the place of the *Assay preparation*, proceed as directed for *Procedure* in the *Assay*. Calculate the quantity, in mg, of $(C_{33}H_{35}N_5O_5)_2 . C_4H_6O_6$ delivered per spray by the formula:

$$(0.02C/4)(A_U/A_S),$$

in which C is the concentration, in µg per mL, of USP Ergotamine Tartrate RS in the *Standard preparation*, and A_U and A_S are the absorbances of the solutions from the *Test preparation* and the *Standard preparation*, respectively.

Particle size—Prime the valve of the Inhalation Aerosol by alternately shaking and firing it several times through its oral inhalation actuator, and then actuate one measured spray onto a clean, dry microscope slide held 5 cm from the end of the actuator, perpendicular to the direction of the spray. Carefully rinse the slide with about 2 mL of carbon tetrachloride, and allow to dry. Examine the slide under a microscope, equipped with a calibrated ocular micrometer, using 450× magnification. Focus on the particles of 25 fields of view near the center of the test specimen pattern, and note the size of the great majority of individual particles: they are less than 5 µm in diameter. Record the number and size of all individual crystalline particles (not agglomerates) more than 10 µm in length measured along the longest axis: not more than 10 such particles are observed.

Other requirements—It meets the requirements for *Leak Testing* under *Aerosols* ⟨601⟩.

Assay—

Standard preparation—Prepare as directed under *Unit spray content.*

Assay preparation—[NOTE—A suitable specimen beaker is one having a small indentation formed on its inside bottom surface having dimensions adequate to accept the aerosol valve stem during actuation, thereby preventing particle entrapment and side-of-stem leakage during the delivery of the specimen.] Place 10 mL of trichlorotrifluoroethane in a suitable 100-mL beaker. Prime the valve of Ergotamine Tartrate Inhalation Aerosol by alternately shaking and firing it 10 times through its oral inhalation actuator. Accurately weigh the Aerosol, shake it, and immediately deliver a single spray under the surface of the trichlorotrifluoroethane, actuating the valve by pressing the tip into the indentation in the bottom of the beaker. Raise the Aerosol above the surface of the trichlorotrifluoroethane, and shake it gently preparatory to delivering another spray similarly under the surface of the trichlorotrifluoroethane. Deliver a total of 2 sprays in this manner. Rinse the valve stem and ferrule with about 2 mL of trichlorotrifluoroethane, collecting the rinsing with the specimen in the beaker. Allow the Aerosol to dry, weigh it, and determine the total weight of the 2 sprays. Transfer the solution to a centrifuge tube with the aid of two 3-mL portions of trichlorotrifluoroethane, and add 10.0 mL of tartaric acid solution (1 in 500). Insert the stopper, shake vigorously for 15 minutes, centrifuge for 5 minutes, and use the clear supernatant liquid as the *Assay preparation.*

Procedure—Transfer 5.0 mL each of the *Standard preparation*, the *Assay preparation* and a blank consisting of tartaric acid solution (1 in 500) to separate test tubes. To each tube add 10.0 mL of *p*-dimethylaminobenzaldehyde TS, mix, and allow to stand for 30 minutes. Concomitantly determine the absorbances of the solutions against the blank in 1-cm cells at the wavelength of maximum absorbance at about 546 nm, with a suitable spectrophotometer. Calculate the quantity, in mg, of $(C_{33}H_{35}N_5O_5)_2 . C_4H_6O_6$ in each mL of the Aerosol taken by the formula:

$$(0.01Cd/W)(A_U/A_S),$$

in which C is the concentration, in µg per mL, of USP Ergotamine Tartrate RS in the *Standard preparation*, d is the density, in g per mL, of Aerosol determined as directed for d in the *Procedure* in the *Assay* under *Isoproterenol Sulfate Inhalation Aerosol*, W is the weight, in g, of the specimen taken, and A_U and A_S are the absorbances of the solutions from the *Assay preparation* and the *Standard preparation*, respectively.

Ergotamine Tartrate Injection

» Ergotamine Tartrate Injection is a sterile solution of Ergotamine Tartrate and the tartrates of its epimer, ergotaminine, and of other related alkaloids, in Water for Injection to which Tartaric Acid or suitable stabilizers have been added. The total alkaloid content, in each mL, is not less than 450 µg and not more than 550 µg. The content of ergotamine tartrate $[(C_{33}H_{35}N_5O_5)_2 . C_4H_6O_6]$ is not less than 52.0 percent and not more than 74.0 percent of the content of total alkaloid; the content of ergotaminine tartrate is not more than 45.0 percent of the content of total alkaloid.

Packaging and storage—Preserve in single-dose, light-resistant containers, preferably of Type I glass.

Reference standard—*USP Ergotamine Tartrate Reference Standard*—Dry in vacuum at 60° for 4 hours before using.

pH ⟨791⟩: between 3.5 and 4.0.

Other requirements—It meets the requirements under *Injections* ⟨1⟩.

Assay—

Chloroform—Use chloroform that recently has been saturated with water.

Standard preparation—Dissolve about 10 mg of USP Ergotamine Tartrate RS, accurately weighed, with warming if necessary, in 50 mL of diluted alcohol and sufficient water to give a concentration of 50.0 µg per mL.

Ergotamine and Ergotaminine preparations—Pipet a volume of Ergotamine Tartrate Injection, equivalent to about 5 mg of ergotamine tartrate, into a beaker. Add 5 mL of *Chloroform* and a portion of sodium bicarbonate approximately equivalent in weight to one-tenth that of the portion of Injection taken. Mix, and add sufficient chromatographic siliceous earth to make a fluffy mixture (about 1 g for each mL of the Injection taken plus 3 g in addition). Pack the mixture in a chromatographic tube about 2.5 cm in diameter and about 30 cm in length. Rinse the sides of the beaker with 2 mL of *Chloroform.* Add sufficient chromatographic siliceous earth to make a fluffy mixture, and transfer it to the column.

Prepare a second column using a mixture of 9 g of the siliceous earth with 7 mL of citric acid solution (1 in 4). Place a mixture of 2 g of the siliceous earth and 2 mL of water on top of the second column. Insert a pledget of glass wool, and mount the tube containing the specimen so that the eluate from it will drain into the tube containing the citric acid solution. Add a total of 90 mL of *Chloroform* to the upper tube, and receive the eluate from the lower tube in a 200-mL volumetric flask. Rinse the tip of the upper tube with *Chloroform.* Pass sufficient *Chloroform* through the lower tube to dilute the eluate to volume. This eluate is the *Ergotaminine preparation.*

Extrude the adsorbant from the second column by means of slight air pressure into a 600-mL beaker containing 10 g of sodium bicarbonate, and mix. Cautiously add 50 mL of water, with continuous stirring. Wash the mixture with water into a 250-mL separator, and extract the ergotamine with four 15-mL portions of *Chloroform.* Pass the extracts through a glass wool filter, combining them in a 100-mL volumetric flask, wash the filter, and dilute with *Chloroform* to volume. This is the *Ergotamine preparation.*

Pipet 10 mL of the *Ergotamine preparation* and 20 mL of the *Ergotaminine preparation* into separate, small conical flasks, and evaporate with the aid of a current of air to dryness.

Total alkaloid preparation—Pipet a volume of Ergotamine Tartrate Injection, equivalent to about 2.5 mg of ergotamine tartrate, into a 50-mL volumetric flask, add 25 mL of alcohol, then add tartaric acid solution (1 in 100) to volume.

Procedure—Pipet 5 mL each of the *Standard preparation* and the *Total alkaloid preparation* into separate, small conical flasks. To the dried residues of *Ergotamine preparation* and *Ergotaminine preparation* add 5.0 mL of a freshly prepared solution of equal volumes of alcohol and tartaric acid solution (1 in 100). In turn, place each flask in an ice bath, and swirl continuously while adding, dropwise, 10.0 mL of *p*-dimethylaminobenzaldehyde TS. Allow to stand in subdued light at room temperature for not less than 90 minutes and not more than 2 hours. Concomitantly determine the absorbances of the four solutions at the wavelength of maximum absorbance at about 545 nm, with a suitable spectrophotometer, against a reagent blank.

Calculation—Calculate the quantity of total alkaloids in terms of mg of $(C_{33}H_{35}N_5O_5)_2 \cdot C_4H_6O_6$ in the volume of Injection taken by the formula:

$$0.05C(A_U/A_S),$$

in which C is the concentration, in µg per mL, of USP Ergotamine Tartrate RS in the *Standard preparation,* and A_U and A_S are the absorbances of the solutions from the *Total alkaloid preparation* and the *Standard preparation,* respectively.

Calculate the percentage of ergotamine tartrate by the formula:

$$(A'/A_U)50,$$

in which A' represents the absorbance of the solution from the *Ergotamine preparation* and A_U is as defined in the preceding paragraph. Calculate the percentage of ergotaminine tartrate by the formula:

$$(A''/A_U)50,$$

in which A'' represents the absorbance of the solution from the *Ergotaminine preparation.*

Ergotamine Tartrate Tablets

» Ergotamine Tartrate Tablets contain not less than 90.0 percent and not more than 110.0 percent of the labeled amount of $(C_{33}H_{35}N_5O_5)_2 \cdot C_4H_6O_6$.

Packaging and storage—Preserve in well-closed, light-resistant containers.

Labeling—Label Tablets to indicate whether they are intended for sublingual administration or for swallowing.

Reference standard—*USP Ergotamine Tartrate Reference Standard*—Dry in vacuum at 60° for 4 hours before using.

Identification—Triturate a quantity of finely powdered Tablets, equivalent to about 5 mg of ergotamine tartrate, with 10 mL of solvent hexane for a few minutes, allow to settle, and discard the solvent hexane extract. Add to the residue 10 mL of chloroform saturated with ammonia (prepared by shaking chloroform with ammonium hydroxide, then drawing off the chloroform layer), triturate for a few minutes, filter, and evaporate the filtrate on a steam bath to dryness. Dissolve the residue in a mixture of 4 mL of glacial acetic acid and 4 mL of ethyl acetate. To 1 mL of this solution add slowly, with continuous agitation and cooling, 1 mL of sulfuric acid: a blue color with a red tinge develops. Add 0.1 mL of ferric chloride TS, previously diluted with an equal volume of water: the red tinge becomes less apparent and the blue color more pronounced.

Disintegration ⟨701⟩ (for Tablets intended for sublingual use): 5 minutes.

Dissolution ⟨711⟩ (for Tablets intended to be swallowed)—

Medium: tartaric acid solution (1 in 100); 1000 mL.

Apparatus 2: 75 rpm.

Time: 30 minutes.

Procedure—Determine the amount of $(C_{33}H_{35}N_2O_5)_2 \cdot C_4H_6O_6$ dissolved from fluorescence intensities, using the maximum excitation wavelength at about 327 nm and the maximum emission wavelength at about 427 nm, of filtered portions of the solution under test, suitably diluted with *Dissolution Medium,* if necessary, in comparison with a Standard solution having a known concentration of USP Ergotamine Tartrate RS in the same medium.

Tolerances—Not less than 75% (Q) of the labeled amount of $(C_{33}H_{35}N_2O_5)_2 \cdot C_4H_6O_6$ is dissolved in 30 minutes.

Uniformity of dosage units ⟨905⟩: meet the requirements.

Assay—

Mobile phase—Prepare a filtered and degassed mixture of acetonitrile and 0.01 *M* monobasic potassium phosphate (55:45). Make adjustments if necessary (see *System Suitability* under *Chromatography* ⟨621⟩).

Internal standard solution—Transfer about 40 mg of ergonovine maleate to a 250-mL volumetric flask, add a mixture of acetonitrile and water (55:45) to volume, and mix.

Standard preparation—Transfer about 10 mg of USP Ergotamine Tartrate RS, accurately weighed, to a 50-mL volumetric flask, add a mixture of acetonitrile and water (55:45) to volume, and mix. Transfer 5.0 mL of this solution to a 50-mL volumetric flask, add 5.0 mL of *Internal standard solution,* dilute with the mixture of acetonitrile and water (55:45) to volume, and mix to obtain a solution having a known concentration of about 0.02 mg of USP Ergotamine Tartrate RS per mL.

Assay preparation—Transfer a number of whole Tablets, equivalent to about 10 mg of ergotamine tartrate, to a 500-mL volumetric flask. Add 50.0 mL of *Internal standard solution,* 300 mL of a mixture of acetonitrile and water (55:45), and sonicate for about 10 minutes. Dilute with the mixture of acetonitrile and water (55:45) to volume, and mix. Filter through a 0.45-µm membrane disk, discarding the first 25 mL of the filtrate.

Chromatographic system (see *Chromatography* ⟨621⟩)—The liquid chromatograph is equipped with a 254-nm detector and a 3.9-mm × 30-cm column that contains packing L1. The flow rate is about 1 mL per minute. Chromatograph the *Standard preparation,* and record the peak responses as directed under *Procedure:* the column efficiency determined from the analyte peak is not less than 3000 theoretical plates, the tailing factor for the analyte peak is not more than 2.0, the resolution, R, between the analyte and internal standard peaks is not less than 3.0, and the relative standard deviation for replicate injections is not more than 2.0%.

Procedure—Separately inject equal volumes (about 20 µL) of the *Standard preparation* and the *Assay preparation* into the chromatograph, record the chromatograms, and measure the responses for the major peaks. The relative retention times are about 0.7 for ergonovine maleate and 1.0 for ergotamine tartrate. Calculate the quantity, in mg, of $(C_{33}H_{35}N_5O_5)O_2 \cdot C_4H_6O_6$ in the portion of Ergotamine Tartrate Tablets taken by the formula:

$$500C(R_U/R_S),$$

in which C is the concentration, in mg per mL, of USP Ergotamine Tartrate RS in the *Standard preparation,* and R_U and R_S are the peak response ratios obtained from the *Assay preparation* and the *Standard preparation,* respectively.

Ergotamine Tartrate and Caffeine Suppositories

» Ergotamine Tartrate and Caffeine Suppositories contain not less than 90.0 percent and not more than 110.0 percent of the labeled amounts of ergotamine tartrate $[(C_{33}H_{35}N_5O_5)_2.C_4H_6O_6]$ and of caffeine $(C_8H_{10}N_4O_2)$.

Packaging and storage—Preserve in tight containers, at a temperature not above 25°. Do not expose unwrapped Suppositories to sunlight.

Reference standards—*USP Ergotamine Tartrate Reference Standard*—Dry in vacuum at 60° for 4 hours before using. *USP Caffeine Reference Standard*—Dry at 80° for 4 hours before using. *USP Ergotaminine Reference Standard*—Do not dry; use as is.

Identification—Melt 1 Suppository in 10 mL of hot tartaric acid solution (1 in 100), and mix. Chill the mixture until the layer of oil has hardened, then filter, divide the filtrate into two parts, and use this filtrate for the following tests.

A: To one part of the filtrate add 10 mL of *p*-dimethylaminobenzaldehyde TS: a blue color develops (*presence of ergotamine*).

B: Transfer the remaining part of the filtrate to a small evaporating dish, evaporate on a steam bath to dryness, add 1 mL of hydrochloric acid and 100 mg of potassium chlorate, and evaporate. Invert the dish over a vessel containing ammonium hydroxide: the residue acquires a purple color, which disappears upon the addition of a solution of fixed alkali (*presence of caffeine*).

Assay—[NOTE—Protect all solutions from light.]

Mobile phase A—Prepare a degassed mixture of water, acetonitrile, and triethylamine (850:150:0.5), and adjust by the dropwise addition of fluorometric grade sulfuric acid to a pH of 3.1 ± 0.1.

Mobile phase B—Prepare a degassed mixture of water, acetonitrile, and triethylamine (1380:620:1), and adjust by the dropwise addition of fluorometric grade sulfuric acid to a pH of 3.1 ± 0.1. Make any necessary adjustments in *pH* to meet relative retention times, and make other adjustments if necessary (see *System Suitability* under *Chromatography* ⟨621⟩).

Solvent mixture—Mix equal volumes of methanol and tartaric acid solution (1 in 100).

Ergotamine tartrate standard solution—Using an accurately weighed amount of USP Ergotamine Tartrate RS, prepare a solution in *Solvent mixture* having a known concentration of about 40 µg per mL.

Mixed standard preparation—Into a 10-mL volumetric flask pipet 5 mL of *Ergotamine tartrate standard solution*, and add about 10 mg of USP Caffeine RS, accurately weighed. Dilute with *Solvent mixture* to volume, and mix to obtain a solution having known concentrations of about 20 µg of USP Ergotamine Tartrate RS per mL and about 1 mg of USP Caffeine RS per mL.

System suitability preparation—Pipet 5 mL of the *Mixed standard preparation* and 1 mL of a solution containing about 20 µg of USP Ergotaminine RS per mL into a 10-mL volumetric flask. Dilute with *Solvent mixture* to volume, and mix.

Assay preparation—Weigh not less than 20 Ergotamine Tartrate and Caffeine Suppositories, and grind to a fine mesh. Transfer a portion of the ground mass, equivalent to about 2 mg of ergotamine tartrate, to a suitable glass-stoppered flask. Add 100.0 mL of *Solvent mixture*, insert the stopper in the flask, and place it in a water bath maintained at 40°. Shake vigorously for 5 minutes, or longer if necessary, until the specimen is completely melted. Sonicate for 30 minutes, and transfer to a freezer for 45 minutes. Filter through 0.7-µm glass fiber filter, discarding the first 5 to 10 mL of the filtrate.

Chromatographic system (see *Chromatography* ⟨621⟩)—The liquid chromatograph is equipped with a 4.6-mm × 25-cm column that contains packing L7. The chromatograph is equipped with a 244-nm detector in series with a fluorometric detector operating at an excitation wavelength of 229 nm and an emission wavelength of 435 nm. Equilibrate the system with *Mobile phase A*. The flow rate is about 2 mL per minute. At 3 minutes after injection of a specimen, or after caffeine has eluted, whichever occurs last, switch to *Mobile phase B*, and at 23 minutes after the initial injection, return to *Mobile phase A*. Allow not less than 2 minutes to elapse between injections. Chromatograph the *Mixed standard preparation*, and the *System suitability preparation*, and record the peak responses as directed under *Procedure*. The resolution, *R*, between the ergotamine and ergotaminine peaks is not less than 3.0, and the relative standard deviation for replicate injections of the *Mixed standard preparation* is not more than 2.0%.

Procedure—Separately inject equal volumes (about 10 µL) of the *Mixed standard preparation* and the *Assay preparation* into the chromatograph, record the chromatograms, and measure the responses for the major peaks. The relative retention times are about 4 for ergotamine, 4.5 for ergotaminine, and 1.0 for caffeine. Calculate the quantity, in mg, of caffeine $(C_8H_{10}N_4O_2)$ in the portion of Suppositories taken by the formula:

$$100C(r_U / r_S),$$

in which *C* is the concentration, in mg per mL, of USP Caffeine RS in the *Mixed standard preparation*, and r_U and r_S are the peak responses obtained from the *Assay preparation* and the *Mixed standard preparation*, respectively. Calculate the quantity, in mg, of ergotamine tartrate $[(C_{33}H_{35}N_5O_5)_2.C_4H_6O_6]$ in the portion of Suppositories taken by the formula:

$$0.1C(I_U / I_S),$$

in which *C* is the concentration, in µg per mL, of USP Ergotamine Tartrate RS in the *Mixed standard preparation*, and I_U and I_S are the fluorometric responses obtained from the *Assay preparation* and the *Mixed standard preparation*, respectively.

Ergotamine Tartrate and Caffeine Tablets

» Ergotamine Tartrate and Caffeine Tablets contain not less than 90.0 percent and not more than 110.0 percent of the labeled amounts of ergotamine tartrate $[(C_{33}H_{35}N_5O_5)_2.C_4H_6O_6]$ and of caffeine $(C_8H_{10}N_4O_2)$.

Packaging and storage—Preserve in well-closed, light-resistant containers.

Reference standards—*USP Ergotamine Tartrate Reference Standard*—Dry in vacuum at 60° for 4 hours before using. *USP Ergotaminine Reference Standard*—Do not dry; use as is. *USP Caffeine Reference Standard*—Keep container tightly closed. Dry at 80° for 4 hours before using.

Identification—Crush 1 Tablet, shake with 10 mL of chloroform and 3 drops of ammonium hydroxide, and filter. Divide the filtrate into two parts in small evaporating dishes, evaporate on a steam bath to dryness, and use the residues for the following tests.

A: Mix one of the residues with 5 mL of tartaric acid solution (1 in 100), and add 10 mL of *p*-dimethylaminobenzaldehyde TS: a blue color develops (*presence of ergotamine*).

B: To the other residue add 1 mL of hydrochloric acid and 100 mg of potassium chlorate, and evaporate on a steam bath to dryness. Invert the dish over a vessel containing ammonium hydroxide: the residue acquires a purple color, which disappears upon the addition of a solution of fixed alkali (*presence of caffeine*).

Dissolution ⟨711⟩—

Medium: tartaric acid solution (1 in 100); 900 mL.
Apparatus 2: 75 rpm.
Time: 30 minutes.

Standard preparation—Using suitable quantities of USP Ergotamine Tartrate RS and USP Caffeine RS, accurately weighed, prepare a solution, using the *Dissolution Medium*, having known concentrations of about 1 µg of ergotamine tartrate and 100 µg of caffeine in each mL.

Procedure—Measure the amount of caffeine in solution in filtered portions of the *Dissolution Medium*, suitably diluted, at the wavelength of maximum absorbance at about 273 nm, with a suitable spectrophotometer, in comparison with the *Standard preparation*. Similarly, measure the amount of ergotamine tartrate in solution in filtered portions of the *Dissolution Medium*, suitably diluted, in a suitable fluorometer, using 327 nm as the excitation wavelength and 427 nm as the emission wavelength, in comparison with the *Standard preparation.*

Tolerances—Not less than 70% (*Q*) of the labeled amount of ergotamine tartrate ($(C_{33}H_{35}N_5O_5)_2 \cdot C_4H_6O_6$) and not less than 75% (*Q*) of the labeled amount of caffeine ($C_8H_{10}N_4O_2$) is dissolved in 30 minutes.

Uniformity of dosage units ⟨905⟩: meet the requirements.

Assay—[NOTE—Protect all solutions from light.]

Mobile phase A—Prepare a degassed mixture of water, acetonitrile, and triethylamine (850:150:0.5), and adjust by the dropwise addition of fluorometric grade sulfuric acid to a pH of 2.7 ± 0.1.

Mobile phase B—Prepare a degassed mixture of water, acetonitrile, and triethylamine (1380:620:1), and adjust by the dropwise addition of fluorometric grade sulfuric acid to a pH of 2.7 ± 0.1. Make necessary adjustments in *pH* with sodium hydroxide solution (1 in 20) or with fluorometric grade sulfuric acid to meet relative retention times, and make other adjustments if necessary (see *System Suitability* under *Chromatography* ⟨621⟩).

Solvent mixture—Transfer 10 g of tartaric acid to a 1-liter volumetric flask, add 500 mL of water, and mix with shaking. Add 330 mL of alcohol, and mix. Dilute with water to volume, and mix. Use a freshly prepared mixture.

Ergotamine tartrate standard solution—Using an accurately weighed amount of USP Ergotamine Tartrate RS, prepare a solution in *Solvent mixture* having a known concentration of about 40 µg per mL.

Caffeine standard solution—Using an accurately weighed amount of USP Caffeine RS, prepare a solution in *Solvent mixture* having a known concentration of about 4 mg per mL.

Mixed standard preparation—Pipet 10 mL of *Ergotamine tartrate standard solution* and 10 mL of *Caffeine standard solution* into a 100-mL volumetric flask. Dilute with *Solvent mixture* to volume, and mix to obtain a solution having known concentrations of 4 µg of USP Ergotamine Tartrate RS per mL and 0.4 mg of USP Caffeine RS per mL.

Assay preparation—Weigh and finely powder not less than 20 Ergotamine Tartrate and Caffeine Tablets. Transfer an accurately weighed portion of the powder, equivalent to about 10 mg of ergotamine tartrate, to a 250-mL volumetric flask. Add 150 mL of *Solvent mixture* and 20 drops of benzalkonium chloride solution (1 in 2). Shake by mechanical means for 45 minutes. [NOTE—Two to 3 mL of methanol may be added, if necessary, to break up bubbles that form during shaking.] Dilute with *Solvent mixture* to volume, and mix. Filter through a 0.5-µm membrane, discarding the first 20 mL of the filtrate. Pipet 5 mL of the subsequent filtrate into a 50-mL volumetric flask, dilute with *Solvent mixture* to volume, and mix.

System suitability preparation—Pipet 20 mL of *Caffeine standard solution*, 20 mL of *Ergotamine tartrate standard solution*, and 4 mL of a solution containing 20 µg of USP Ergotaminine RS per mL of *Solvent mixture*, into a 200-mL volumetric flask. Dilute with *Solvent mixture* to volume, and mix.

Chromatographic system (see *Chromatography* ⟨621⟩)—The liquid chromatograph is equipped with a 4.6-mm × 25-cm column that contains packing L7. The chromatograph is equipped with a 254-nm detector in series with a fluorometric detector operating at an excitation wavelength of 325 nm and an emission wavelength of 435 nm. Equilibrate the system with *Mobile phase A*. The flow rate is about 2 mL per minute. At 3 minutes after injection of a specimen, or after caffeine has eluted, whichever occurs last, switch to *Mobile phase B*, and at 18 minutes after initial injection, return to *Mobile phase A*. Wait not less than 2 minutes between injections. Chromatograph the *Mixed standard preparation* and the *System suitability preparation*, and record the peak responses as directed under *Procedure:* the tailing factor for the ergotamine peak is not more than 2.0, the resolution, *R*, between the ergotamine and ergotaminine peaks is not less than 3.0, and the relative standard deviation for replicate injections of the *Mixed standard preparation* is not more than 2.0%.

Procedure—Separately inject equal volumes (about 20 µL) of the *Mixed standard preparation* and the *Assay preparation* into the chromatograph, record the chromatograms, and measure the responses for the major peaks. The relative retention times are about 3.5 for ergotamine, 4 for ergotaminine, and 1.0 for caffeine. Calculate the quantity, in mg, of caffeine ($C_8H_{10}N_4O_2$) in the portion of Tablets taken by the formula:

$$2500C(r_U / r_S),$$

in which *C* is the concentration, in mg per mL, of USP Caffeine RS in the *Mixed standard preparation*, and r_U and r_S are the 254-nm peak responses obtained from the *Assay preparation* and the *Mixed standard preparation*, respectively. Calculate the quantity, in mg, of ergotamine tartrate [$(C_{33}H_{35}N_5O_5)_2 \cdot C_4H_6O_6$] in the portion of Tablets taken by the formula:

$$2.5C(I_U / I_S),$$

in which *C* is the concentration, in µg per mL, of USP Ergotamine Tartrate RS in the *Mixed standard preparation*, and I_U and I_S are the fluorometric responses obtained from the *Assay preparation* and the *Mixed standard preparation*, respectively.

Diluted Erythrityl Tetranitrate

```
        CH2ONO2
           |
    H---C---ONO2
           |
    H---C---ONO2
           |
        CH2ONO2
```

$C_4H_6N_4O_{12}$ 302.11

1,2,3,4-Butanetetrol, tetranitrate, (*R**,*S**)-.

Erythritol tetranitrate [*7297-25-8*].

» Diluted Erythrityl Tetranitrate is a dry mixture of erythrityl tetranitrate ($C_4H_6N_4O_{12}$) with lactose or other suitable inert excipients to permit safe handling and compliance with federal Interstate Commerce Commission regulations pertaining to interstate shipment. It contains not less than 90.0 percent and not more than 110.0 percent of the labeled amount of $C_4H_6N_4O_{12}$.

Caution—Undiluted erythrityl tetranitrate is a powerful explosive, and proper precautions must be taken in handling. It can be exploded by percussion or by excessive heat. Only extremely small amounts should be isolated.

Packaging and storage—Preserve in tight containers, and avoid exposure to excessive heat.

Identification—Transfer an amount of Diluted Erythrityl Tetranitrate, equivalent to about 10 mg of erythrityl tetranitrate, to a small separator, add 5 mL of water and 1 mL of 2 *N* sulfuric acid, and extract with 25 mL of chloroform. Transfer the chloroform layer to a small flask containing 1 g of sodium chloride, shake, and allow to settle. Transfer 2 mL of the solution to the surface of 500 mg of potassium bromide contained in a small beaker, and remove the chloroform by evaporation, using gentle heat: the infrared absorption spectrum of a pressed disk [*Caution—Wear safety goggles, and work behind a safety shield*] prepared from the residue exhibits maxima at the following wavelengths, in µm: 3.39 (*sh*), 3.45 (*m*), 3.52 (*w*), 6.00 (*s*), 6.12 (*s*), 6.88 (*m*), 7.26 (*m*), 7.45 (*sh*), 7.68 (*s*), 7.81 (*s*), 7.92 (*s*), 8.13 (*sh*), 9.43 (*m*), 9.68 (*s*), 10.16 (*s*), 10.88 (*s*), 11.40 (*s*), 11.99 (*s*), 13.22 (*m*), and 14.25 (*m*).

Assay—

Standard preparation—Transfer about 130 mg of potassium nitrate, previously dried at 105° for 4 hours and accurately weighed, to a 10-mL volumetric flask, add water to volume, and mix. Transfer 1.0 mL of this solution to a 50-mL volumetric flask, dilute with glacial acetic acid to volume, and mix to obtain a *Standard preparation* having a known concentration of potassium nitrate of about 260 µg per mL.

Assay preparation—Transfer an accurately weighed portion of Diluted Erythrityl Tetranitrate, equivalent to about 10 mg of erythrityl tetranitrate, to a small separator, add 20 mL of chloroform and 10 mL of water, and shake. Drain the chloroform layer into a 50-mL volumetric flask. Extract the aqueous layer with two 10-mL portions of chloroform, adding the extracts to the volumetric flask. Dilute the contents of the flask with chloroform to volume, and mix.

Procedure—Transfer 1.0 mL each of the *Standard preparation* and of the *Assay preparation* to separate small beakers. Evaporate the chloroform from the *Assay preparation* with the aid of a current of air and very gentle heat, and then dissolve the residue in 1.0 mL of glacial acetic acid. To each beaker add 2 mL of phenoldisulfonic acid TS, mix, and allow to stand for 15 minutes. Add 8 mL of water to each beaker, mix, and carefully add 8 mL of ammonium hydroxide. Cool, transfer the solutions to separate 25-mL volumetric flasks, dilute each with water to volume, and mix. Concomitantly determine the absorbances of the solutions in 1-cm cells at the wavelength of maximum absorbance at about 405 nm, with a suitable spectrophotometer, using water as the blank. Calculate the quantity, in mg, of $C_4H_6N_4O_{12}$ in the Diluted Erythrityl Tetranitrate taken by the formula:

$$(302.11/101.10)(0.0125C)(A_U/A_S),$$

in which 302.11 and 101.10 are the molecular weights of erythrityl tetranitrate and potassium nitrate, respectively, C is the concentration, in µg per mL, of potassium nitrate in the *Standard preparation*, and A_U and A_S are the absorbances of the solutions from the *Assay preparation* and the *Standard preparation*, respectively.

Erythrityl Tetranitrate Tablets

» Erythrityl Tetranitrate Tablets are prepared from Diluted Erythrityl Tetranitrate. They contain not less than 90.0 percent and not more than 110.0 percent of the labeled amount of $C_4H_6N_4O_{12}$.

Caution—Undiluted erythrityl tetranitrate is a powerful explosive, and proper precautions must be taken in handling. It can be exploded by percussion or by excessive heat. Only extremely small amounts should be isolated.

Packaging and storage—Preserve in tight containers, and avoid exposure to excessive heat.

Identification—A portion of finely powdered Tablets, equivalent to about 10 mg of erythrityl tetranitrate, responds to the *Identification test* under *Diluted Erythrityl Tetranitrate.*

Disintegration ⟨701⟩: 10 minutes, determined without the use of disks.

Uniformity of dosage units ⟨905⟩: meet the requirements.

Assay—

Standard preparation—Prepare as directed in the *Assay* under *Diluted Erythrityl Tetranitrate.*

Assay preparation—Weigh and finely powder not less than 20 Erythrityl Tetranitrate Tablets. Transfer an accurately weighed portion of the powder, equivalent to about 10 mg of erythrityl tetranitrate, to a small separator, and proceed as directed for *Assay preparation* in the *Assay* under *Diluted Erythrityl Tetranitrate*, beginning with "Add 20 mL of chloroform."

Procedure—Proceed as directed for *Procedure* in the *Assay* under *Diluted Erythrityl Tetranitrate.* Calculate the quantity, in mg, of $C_4H_6N_4O_{12}$ in the portion of Tablets taken by the formula:

$$(302.11/101.10)(0.0125C)(A_U/A_S),$$

in which 302.11 and 101.10 are the molecular weights of erythrityl tetranitrate and potassium nitrate, respectively, C is the concentration, in µg per mL, of potassium nitrate in the *Standard preparation*, and A_U and A_S are the absorbances of the solutions from the *Assay preparation* and the *Standard preparation*, respectively.

Erythromycin

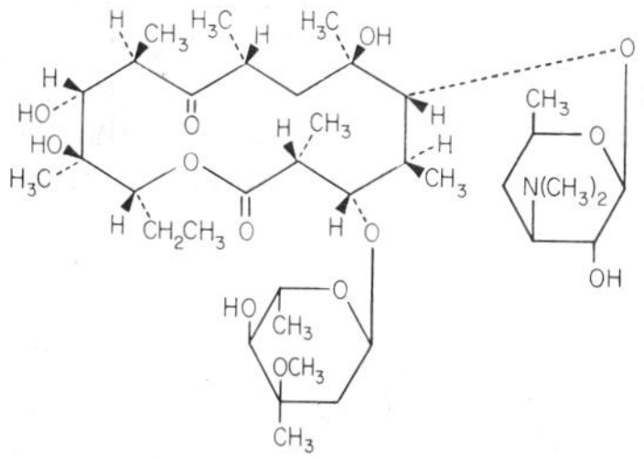

$C_{37}H_{67}NO_{13}$ 733.94

Erythromycin.

Erythromycin.

(3*R**,4*S**,5*S**,6*R**,7*R**,9*R**,11*R**,12*R**,13*S**,14*R**)-4-[(2,6-Dideoxy-3-*C*-methyl-3-*O*-methyl-α-L-*ribo*-hexopyranosyl)-oxy]-14-ethyl-7,12,13-trihydroxy-3,5,7,9,11,13-hexamethyl-6-[[3,4,6-trideoxy-3-(dimethylamino)-β-D-*xylo*-hexopyranosyl]oxy]oxacyclotetradecane-2,10-dione [114-07-8].

» Erythromycin contains not less than 850 µg of $C_{37}H_{67}NO_{13}$ per mg, calculated on the anhydrous basis.

Packaging and storage—Preserve in tight containers.

Reference standard—*USP Erythromycin Reference Standard*—Dry in vacuum at a pressure not exceeding 5 mm of mercury at 60° for 3 hours before using.

Identification—The infrared absorption spectrum of a 1 in 100 solution of it, previously dried at a pressure not exceeding 5 mm of mercury at 60° for 3 hours, in chloroform, determined in a 1.0-mm cell, exhibits maxima only at the same wavelengths as that of a similar preparation of USP Erythromycin RS.

Specific rotation ⟨781⟩: between −71° and −78°, calculated on the anhydrous basis, determined after standing for 30 minutes in a solution in dehydrated alcohol containing 20 mg per mL.

Crystallinity ⟨695⟩: meets the requirements.

pH ⟨791⟩: between 8.0 and 10.5, in a methanol-water solution prepared by diluting 1 volume of a methanol solution containing 40 mg per mL with 19 volumes of water.

Water, *Method I* ⟨921⟩: not more than 10.0%, 20 mL of a mixture of carbon tetrachloride, chloroform, and methanol (2:2:1) being used in place of methanol in the titration vessel.

Residue on ignition ⟨281⟩: not more than 2.0%, the charred residue being moistened with 2 mL of nitric acid and 5 drops of sulfuric acid.

Assay—Proceed with Erythromycin as directed under *Antibiotics—Microbial Assays* ⟨81⟩.

Erythromycin Delayed-release Capsules

» Erythromycin Delayed-release Capsules contain not less than 90.0 percent and not more than 115.0 percent of the labeled amount of $C_{37}H_{67}NO_{13}$.

Packaging and storage—Preserve in tight containers.

Reference standard—*USP Erythromycin Reference Standard*—Dry in vacuum at a pressure not exceeding 5 mm of mercury at 60° for 3 hours before using.

Identification—Prepare a test solution by mixing a quantity of finely ground Capsule contents with methanol to obtain a concentration of about 2.5 mg of erythromycin per mL. Prepare a Standard solution of USP Erythromycin RS in methanol containing 2.5 mg per mL. Apply separately 10 µL of each solution on a thin-layer chromatographic plate (see *Chromatography* ⟨621⟩), coated with a 0.25-mm layer of chromatographic silica gel. Place the plate in an unlined chromatographic chamber, and develop the chromatogram in a solvent system consisting of a mixture of methanol and chloroform (85:15) until the solvent front has moved about 7 cm. Remove the plate from the chamber, mark the solvent front, and allow the solvent to evaporate. Spray the plate with a mixture of alcohol, *p*-methoxybenzaldehyde, and sulfuric acid (90:5:5). Heat the plate at 100° for 10 minutes, and examine the chromatogram, in which erythromycin appears as a black-to-purple spot: the R_f value of the principal spot obtained from the test solution corresponds to that obtained from the Standard solution.

Drug release, *Method B* ⟨724⟩—

Apparatus 1: 50 rpm.

Times: 60 minutes for *Acid stage;* 45 minutes for *Buffer stage.*

Procedure—Transfer the contents of 1 Capsule to the apparatus. Proceed as directed for *Acid stage*, 900 mL of 0.06 *N* hydrochloric acid being placed in the vessel instead of 1000 mL of 0.1 *N* hydrochloric acid, and the apparatus being operated for 60 minutes instead of 2 hours. Do not perform an analysis at the end of the *Acid stage.* Continue as directed for *Buffer stage*, 900 mL of the pH 6.8 phosphate buffer being used instead of 1000 mL. Determine the amount of $C_{37}H_{67}NO_{13}$ dissolved after 105 minutes by assaying a filtered portion of the solution under test as directed under *Antibiotics—Microbial Assays* ⟨81⟩.

Tolerances—Not less than 80% (*Q*) of the labeled amount of $C_{37}H_{67}NO_{13}$ is dissolved in 105 minutes.

Water, *Method I* ⟨921⟩: not more than 7.5%.

Assay—Proceed as directed under *Antibiotics—Microbial Assays* ⟨81⟩, using not less than 5 Erythromycin Delayed-release Capsules blended for about 3 minutes in a high-speed glass blender jar containing 200 mL of methanol. Add 300 mL of *Buffer No. 3*, and blend again for about 3 minutes. Dilute an accurately measured volume of this stock solution quantitatively with *Buffer No. 3* to obtain a *Test Dilution* having a concentration assumed to be equal to the median dose level of the Standard.

Erythromycin Ointment

» Erythromycin Ointment is Erythromycin in a suitable ointment base. It contains not less than 90.0 percent and not more than 125.0 percent of the labeled amount of $C_{37}H_{67}NO_{13}$.

Packaging and storage—Preserve in collapsible tubes or in other tight containers, preferably at controlled room temperature.

Reference standard—*USP Erythromycin Reference Standard*—Dry in vacuum at a pressure not exceeding 5 mm of mercury at 60° for 3 hours before using.

Identification—Transfer a quantity of Ointment, equivalent to about 5 mg of erythromycin, to a separator containing 50 mL of solvent hexane. Shake until dissolved. Extract with three separate 20-mL portions of methanol. Combine the methanol extracts in a beaker, and evaporate to dryness. Dissolve the residue in 2 mL of methanol (test solution). Proceed as directed in the *Identification test* under *Erythromycin Delayed-release Capsules*, beginning with "Prepare a Standard solution of USP Erythromycin RS."

Minimum fill ⟨755⟩: meets the requirements.

Water, *Method I* ⟨921⟩: not more than 1.0%, 20 mL of a mixture of carbon tetrachloride, chloroform, and methanol (2:2:1) being used in place of methanol in the titration vessel.

Assay—Transfer an accurately weighed quantity of Erythromycin Ointment, equivalent to about 5 mg of erythromycin, to a separator containing 50 mL of solvent hexane. Shake until dissolved. Wash with four separate 20-mL portions of a mixture of methanol and water (4:1). Combine the washings in a 100-mL volumetric flask, dilute with the methanol-water solution to volume, and mix. Proceed as directed under *Antibiotics—Microbial Assays* ⟨81⟩, using an accurately measured volume of this stock test solution diluted quantitatively with *Buffer No. 3* to yield a *Test Dilution* having a concentration assumed to be equal to the median dose level of the Standard.

Erythromycin Ophthalmic Ointment

» Erythromycin Ophthalmic Ointment is a sterile preparation of Erythromycin in a suitable ointment base. It contains not less than 90.0 percent and not more than 120.0 percent of the labeled amount of $C_{37}H_{67}NO_{13}$.

Packaging and storage—Preserve in collapsible ophthalmic ointment tubes.

Reference standard—*USP Erythromycin Reference Standard*—Dry in vacuum at a pressure not exceeding 5 mm of mercury at 60° for 3 hours before using.

Identification—Transfer a quantity of Ophthalmic Ointment, equivalent to about 5 mg of erythromycin, to a separator containing 50 mL of solvent hexane. Shake until dissolved. Extract with three separate 20-mL portions of methanol. Combine the methanol extracts in a beaker, and evaporate to dryness. Dissolve the residue in 2 mL of methanol (test solution). Proceed as directed in the *Identification test* under *Erythromycin Delayed-release Capsules*, beginning with "Prepare a Standard solution of USP Erythromycin RS."

Sterility—It meets the requirements under *Sterility Tests* ⟨71⟩.

Minimum fill ⟨755⟩: meets the requirements.

Metal particles—It meets the requirements of the test for *Metal Particles in Ophthalmic Ointments* ⟨751⟩.

Other requirements—It meets the requirements for *Water* and *Assay* under *Erythromycin Ointment.*

Erythromycin Pledgets

» Erythromycin Pledgets are suitable absorbent pads impregnated with Erythromycin Topical Solution. Pledgets contain not less than 90.0 percent of the labeled volume of Erythromycin Topical Solution.

Packaging and storage—Preserve in tight containers.

Labeling—Label Pledgets to indicate that each Pledget is to be used once and then discarded. Label Pledgets also to indicate the volume, in mL, of Erythromycin Topical Solution contained in each Pledget, and the concentration, in mg of erythromycin per mL, of the Erythromycin Topical Solution.

Other requirements—The Erythromycin Topical Solution expressed from Erythromycin Pledgets meets the requirements for *Identification, Water*, and *Alcohol content* under *Erythromycin Topical Solution.*

Topical solution assay—Proceed as directed for erythromycin under *Antibiotics—Microbial Assays* ⟨81⟩, using the solution expressed from not less than 10 Erythromycin Pledgets. Dilute an accurately measured volume of this solution quantitatively with *Buffer No. 3* to obtain a *Test Dilution* having a concentration assumed to be equal to the median dose level of the Standard.

Assay and minimum volume—Proceed as directed for erythromycin under *Antibiotics—Microbial Assays* ⟨81⟩, assaying 10 Erythromycin Pledgets individually. Transfer 1 Pledget to a suitable container, add about 50 mL of *Buffer No. 3*, and shake for 2 minutes. Transfer the solution so obtained to a 100-mL volumetric flask. Wash the Pledget with two 20-mL portions of *Buffer No. 3*, add the washings to the 100-mL volumetric flask, dilute with *Buffer No. 3* to volume, and mix. Dilute an accurately measured volume of this solution quantitatively with *Buffer No. 3* to obtain a *Test Dilution* having a concentration assumed to be equal to the median dose level of the Standard. Calculate the volume, in mL, of Erythromycin Topical Solution in each Pledget taken by dividing the number of mg of erythromycin in each individual Pledget by the number of mg of erythromycin in each mL of Erythromycin Topical Solution as obtained in the *Topical Solution assay.* The volume of Erythromycin Topical Solution in each Pledget is not less than 90.0% of the labeled amount. If this requirement is not met, determine the volume of Erythromycin Topical Solution in each of 20 additional Erythromycin Pledgets. The volume of Erythromycin Topical Solution in not more than 1 of the 30 Pledgets is less than 90.0% of the labeled amount.

Erythromycin Topical Solution

» Erythromycin Topical Solution is a solution of Erythromycin in a suitable vehicle. It contains not less than 90.0 percent and not more than 125.0 percent of the labeled amount of $C_{37}H_{67}NO_{13}$.

Packaging and storage—Preserve in tight containers.

Reference standard—*USP Erythromycin Reference Standard*—Dry in vacuum at a pressure not exceeding 5 mm of mercury at 60° for 3 hours before using.

Identification—Prepare a test solution by mixing a portion of the Topical Solution with methanol to obtain a concentration of about 2.5 mg of erythromycin per mL. Proceed as directed in the *Identification test* under *Erythromycin Delayed-release Capsules*, beginning with "Prepare a Standard solution of USP Erythromycin RS."

Water, *Method I* ⟨921⟩: not more than 8.0% if it contains 20 mg per mL, or not more than 5.0% if it contains 15 mg per mL, or not more than 2.0% if it contains acetone, 20 mL of a mixture of pyridine and methanol (1:1) being used in place of methanol in the titration vessel.

Alcohol content, *Method II* ⟨611⟩: between 92.5% and 107.5% of the labeled amount of C_2H_5OH.

Assay—Proceed as directed under *Antibiotics—Microbial Assays* ⟨81⟩, using an accurately measured volume of Erythromycin Topical Solution diluted quantitatively with *Buffer No. 3* to yield a *Test Dilution* having a concentration assumed to be equal to the median dose level of the Standard.

Erythromycin Tablets

» Erythromycin Tablets contain not less than 90.0 percent and not more than 120.0 percent of the labeled amount of $C_{37}H_{67}NO_{13}$.

NOTE—Tablets that are enteric-coated meet the requirements for *Erythromycin Delayed-release Tablets.*

Packaging and storage—Preserve in tight containers.

Reference standard—*USP Erythromycin Reference Standard*—Dry in vacuum at a pressure not exceeding 5 mm of mercury at 60° for 3 hours before using.

Identification—Prepare a test solution by mixing a quantity of finely powdered Tablets with methanol to obtain a concentration of about 2.5 mg of erythromycin per mL. Proceed as directed in the *Identification test* under *Erythromycin Delayed-release Capsules*, beginning with "Prepare a Standard solution of USP Erythromycin RS."

Dissolution ⟨711⟩—

Medium: 0.05 *M* pH 6.8 phosphate buffer (see *Buffer solutions* in the section, *Reagents, Indicators, and Solutions*); 900 mL.

Apparatus 2: 50 rpm.

Time: 60 minutes.

Test solution—If necessary, dilute a filtered portion of the solution under test with *Dissolution Medium* to obtain a solution having a concentration of about 0.28 mg of erythromycin per mL, and mix.

Standard solution—Dissolve an accurately weighed quantity of USP Erythromycin RS in methanol (not more than 1 mL of methanol for each 14 mg of the Reference Standard), and dilute with water, quantitatively and with mixing, to obtain a stock solution containing about 0.56 mg per mL. Immediately prior to use, dilute the stock solution quantitatively with water to obtain a *Standard solution* having a known concentration of about 0.28 mg per mL.

Procedure—Transfer 5.0-mL portions of the *Test solution* and the *Standard solution* to separate 25-mL volumetric flasks, and treat each as follows: Add 2.0 mL of water, and allow to stand for 5 minutes with intermittent swirling. Add 15.0 mL of 0.25 *N* sodium hydroxide, dilute with *Dissolution Medium* to volume, and mix. Heat to 60° for 5 minutes, and allow to cool. Concomitantly determine the absorbances of these solutions at the wavelength of maximum absorbance at about 236 nm, with a suitable spectrophotometer, using blank solutions similarly prepared, except that 2.0 mL of 0.5 *N* sulfuric acid is substituted for the 2.0 mL of water. Calculate the amount of $C_{37}H_{67}NO_{13}$ dissolved.

Tolerances—Not less than 70% (*Q*) of the labeled amount of $C_{37}H_{67}NO_{13}$ is dissolved in 60 minutes.

Uniformity of dosage units ⟨905⟩: meet the requirements.

Loss on drying ⟨731⟩—Dry about 100 mg of powdered Tablets in a capillary-stoppered bottle in vacuum at 60° for 3 hours: it loses not more than 5.0% of its weight.

Assay—Place not less than 4 Erythromycin Tablets in a high-speed glass blender jar with 200 mL of methanol, and blend for 3 minutes. Add 300 mL of *Buffer No. 3*, and blend for 3 minutes. Proceed as directed under *Antibiotics—Microbial Assays* ⟨81⟩, using an accurately measured volume of this stock test solution diluted quantitatively with *Buffer No. 3* to yield a *Test Dilution* having a concentration assumed to be equal to the median dose level of the Standard.

Erythromycin Delayed-release Tablets

» Erythromycin Delayed-release Tablets contain not less than 90.0 percent and not more than 120.0 percent of the labeled amount of $C_{37}H_{67}NO_{13}$.

Packaging and storage—Preserve in tight containers.

Labeling—The label indicates that Erythromycin Delayed-release Tablets are enteric-coated.

Reference standard—*USP Erythromycin Reference Standard*—Dry in vacuum at a pressure not exceeding 5 mm of mercury at 60° for 3 hours before using.

Identification—Prepare a test solution by mixing a quantity of finely powdered Tablets with methanol to obtain a concentration of about 2.5 mg of erythromycin per mL. Proceed as directed in the *Identification test* under *Erythromycin Delayed-release Capsules*, beginning with "Prepare a Standard solution of USP Erythromycin RS."

Drug release, *Method B* ⟨724⟩—

Apparatus 1: 100 rpm.

Time: 60 minutes, Stage 1.
60 minutes, Stage 2.

Acid stage—Using 900 mL of simulated gastric fluid TS (prepared without pepsin) in place of 0.1 *N* hydrochloric acid, conduct this stage of the test for 1 hour, and do not perform an analysis of the medium.

Buffer stage—Using 900 mL of 0.05 *M* pH 6.8 phosphate buffer (see *Buffer solutions* in the section, *Reagents, Indicators, and Solutions*), conduct this stage of the test for 60 minutes.

Test solution—If necessary, dilute a filtered portion of the solution under test with *Dissolution Medium* to obtain a solution having a concentration of about 0.28 mg of erythromycin per mL, and mix.

Procedure—Transfer a 2.0-mL portion of the *Test solution* to a suitable separator. Add 6 mL of pH 1.2 buffer (see *Solutions* in the section, *Reagents, Indicators, and Solutions*), and 8 mL of a solution of bromocresol purple, prepared by dissolving 1 g of bromocresol purple in 1 liter of pH 4.5 phosphate buffer, and mix. Extract with 40.0 mL of chloroform. Determine the amount of $C_{37}H_{67}NO_{13}$ dissolved from ultraviolet absorbances at the wavelength of maximum absorbance at about 410 nm using the chloroform extracts. Similarly prepare a Standard solution, having a known concentration of USP Erythromycin RS, and treat similarly.

Uniformity of dosage units ⟨905⟩: meet the requirements.

Water, *Method I* ⟨921⟩: not more than 6.0%, 20 mL of a mixture of carbon tetrachloride, chloroform, and methanol (2:2:1) being used in place of methanol in the titration vessel.

Assay—Place not less than 4 Erythromycin Delayed-release Tablets in a high-speed glass blender jar with 200 mL of methanol, and blend for 3 minutes. Add 300 mL of *Buffer No. 3*, and blend for 3 minutes. Proceed as directed under *Antibiotics—Microbial Assays* ⟨81⟩, using an accurately measured volume of this stock test solution diluted quantitatively with *Buffer No. 3* to yield a *Test Dilution* having a concentration assumed to be equal to the median dose level of the Standard.

Erythromycin and Benzoyl Peroxide Topical Gel

» Erythromycin and Benzoyl Peroxide Topical Gel is a mixture of Erythromycin in a suitable gel vehicle containing benzoyl peroxide and one or more suitable dispersants, stabilizers, and wetting agents. It contains not less than 90.0 percent and not more than 125.0 percent of the labeled amounts of erythromycin ($C_{37}H_{67}NO_{13}$) and benzoyl peroxide ($C_{14}H_{10}O_4$).

Packaging and storage—Before mixing, preserve the Erythromycin and the vehicle containing benzoyl peroxide in separate, tight containers. After mixing, preserve the mixture in tight containers.

Reference standard—*USP Erythromycin Reference Standard*—Dry in vacuum at a pressure not exceeding 5 mm of mercury at 60° for 3 hours before using.

Identification—Prepare a *Standard preparation* and an *Assay preparation* as directed in the *Assay*, except to omit the *Internal standard solution*, and chromatograph as directed in the *Assay*: the *Assay preparation* exhibits a major peak for benzoyl peroxide the retention time of which corresponds to that exhibited by the *Standard preparation*.

Minimum fill ⟨755⟩: meets the requirements.

Assay for erythromycin—Proceed as directed for erythromycin under *Antibiotics—Microbial Assays* ⟨81⟩, using an accurately weighed portion of Erythromycin and Benzoyl Peroxide Topical Gel blended for 3 to 5 minutes in a high-speed glass blender jar containing 0.5 mL of polysorbate 80 and an accurately measured volume of *Buffer No. 3* sufficient to obtain a stock solution having a convenient concentration of erythromycin. Dilute an accurately measured volume of this stock solution quantitatively with *Buffer No. 3* to obtain a *Test Dilution* having a concentration of erythromycin assumed to be equal to the median dose level of the Standard.

Assay for benzoyl peroxide—

Mobile phase, Internal standard solution, Standard preparation, and *Chromatographic system*—Proceed as directed in the *Assay* under *Benzoyl Peroxide Gel.*

Assay preparation—Prepare as directed for *Assay preparation* in the *Assay* under *Benzoyl Peroxide Gel*, using Erythromycin and Benzoyl Peroxide Topical Gel.

Procedure—Proceed as directed for *Procedure* in the *Assay* under *Benzoyl Peroxide Gel.* Calculate the quantity, in mg, of benzoyl peroxide ($C_{14}H_{10}O_4$) in the portion of Topical Gel taken by the formula:

$$125C(R_U/R_S),$$

in which C is the concentration, in mg per mL, of benzoyl peroxide in the *Standard preparation*, and R_U and R_S are the ratios of benzoyl peroxide peak response to ethyl benzoate peak response obtained from the *Assay preparation* and the *Standard preparation*, respectively.

Erythromycin Estolate

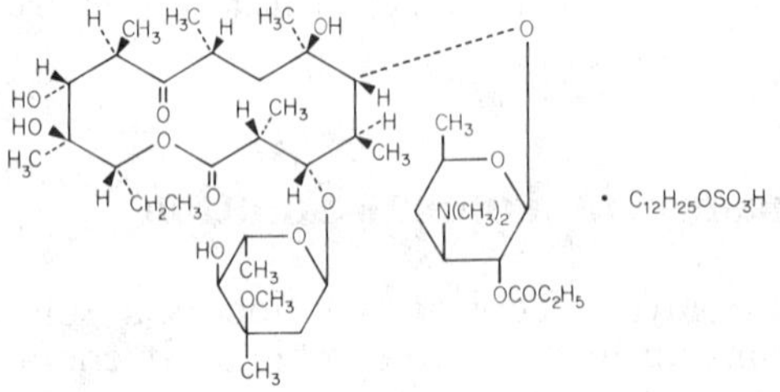

$C_{40}H_{71}NO_{14}.C_{12}H_{26}O_4S$ 1056.39

Erythromycin, 2′-propanoate, dodecyl sulfate (salt).

Erythromycin 2′-propionate dodecyl sulfate (salt) [*3521-62-8*].

» Erythromycin Estolate has a potency equivalent to not less than 600 µg of erythromycin ($C_{37}H_{67}NO_{13}$) per mg, calculated on the anhydrous basis.

Packaging and storage—Preserve in tight containers.

Reference standards—*USP Erythromycin Reference Standard*—Dry in vacuum at a pressure not exceeding 5 mm of mercury at 60° for 3 hours before using. *USP Erythromycin Estolate Reference Standard*—Do not dry before using.

Identification—The infrared absorption spectrum of a potassium bromide dispersion of it exhibits maxima only at the same wavelengths as that of a similar preparation of USP Erythromycin Estolate RS.

Crystallinity ⟨695⟩: meets the requirements.

pH ⟨791⟩: between 4.5 and 7.0, in an aqueous suspension containing 10 mg per mL.

Water, *Method I* ⟨921⟩: not more than 4.0%.

Free erythromycin—Prepare a test solution of it in methanol containing 10.0 mg per mL. Prepare a Standard solution of USP Erythromycin RS in methanol containing 0.3 mg per mL. [NOTE—Prepare these solutions immediately before use.] Prepare a suitable thin-layer chromatographic plate (see *Chromatography* ⟨621⟩), coated with a 0.25-mm layer of chromatographic silica gel mixture. Before using, place the plate in an unlined developing chamber containing about 100 mL of methanol, and allow the solvent front to travel to the top of the plate, marking the direction of travel. Remove the plate, and allow to dry. Apply separate 1-µL volumes of the test solution and the Standard solution on the plate, allow the spots to dry, and develop the chromatograms in a freshly prepared solvent system consisting of a mixture of methanol and chloroform (85:15) until the solvent front has moved about one-half of the length of the plate. Remove the plate from the developing chamber, mark the solvent front, and allow the plate to dry. Place the plate under a hood, and spray uniformly with a solution consisting of 150 mg of xanthydrol dissolved in a mixture of hydrochloric acid and glacial acetic acid (92.5:7.5). Heat the sprayed plate in an oven at 100° for 5 minutes. [*Caution—Avoid exposure to acid fumes when removing the plate from the oven.*] Examine the plate for reddish violet spots: free erythromycin has an R_f value of about 0.3, and erythromycin estolate has an R_f value of about 0.7. Any spot corresponding to free erythromycin obtained from the test solution does not exceed in size or intensity that of the principal spot obtained from the Standard solution (3.0%).

Assay—Proceed with Erythromycin Estolate as directed for erythromycin under *Antibiotics—Microbial Assays* ⟨81⟩, using an accurately weighed quantity of Erythromycin Estolate dissolved in methanol to obtain a solution containing the equivalent of 1.0 mg of erythromycin per mL. Immediately dilute quantitatively with 9 volumes of *Buffer No. 3*, and allow to stand at room temperature for 18 hours. Dilute a portion of this solution quantitatively with *Buffer No. 3* to obtain a *Test Dilution* having a concentration assumed to be equal to the median dose level of the Standard.

Erythromycin Estolate Capsules

» Erythromycin Estolate Capsules contain the equivalent of not less than 90.0 percent and not more than 115.0 percent of the labeled amount of erythromycin ($C_{37}H_{67}NO_{13}$).

Packaging and storage—Preserve in tight containers.

Reference standard—*USP Erythromycin Reference Standard*—Dry in vacuum at a pressure not exceeding 5 mm of mercury at 60° for 3 hours before using.

Identification—Prepare a test solution by mixing a quantity of Capsule contents with methanol to obtain a concentration equivalent to about 20 mg of erythromycin per mL. Prepare a Standard solution of USP Erythromycin Estolate RS in methanol containing the equivalent of 20 mg of erythromycin per mL. Apply separately 3 µL of each solution on a thin-layer chromatographic plate (see *Chromatography* ⟨621⟩), coated with a 0.25-mm layer of chromatographic silica gel. Proceed as directed in the *Identification test* under *Erythromycin Delayed-release Capsules*, beginning with "Place the plate in an unlined chromatographic chamber."

Uniformity of dosage units ⟨905⟩: meet the requirements.

Water, *Method I* ⟨921⟩: not more than 5.0%.

Assay—Place not less than 4 Erythromycin Estolate Capsules in a high-speed glass blender jar with 200.0 mL of methanol, and blend for 3 minutes. Add 300.0 mL of *Buffer No. 3,* and blend for 3 minutes. Allow this solution to stand at room temperature for 18 hours. Proceed as directed for erythromycin under *Antibiotics—Microbial Assays* ⟨81⟩, using an accurately measured volume of this stock test solution diluted quantitatively with *Buffer No. 3* to yield a *Test Dilution* having a concentration assumed to be equal to the median dose level of the Standard.

Erythromycin Estolate Oral Suspension

» Erythromycin Estolate Oral Suspension contains the equivalent of not less than 90.0 percent and not more than 115.0 percent of the labeled amount of erythromycin ($C_{37}H_{67}NO_{13}$). It contains one or more suitable buffers, colors, diluents, dispersants, and flavors.

Packaging and storage—Preserve in tight containers, in a cold place.

Reference standard—*USP Erythromycin Reference Standard*—Dry in vacuum at a pressure not exceeding 5 mm of mercury at 60° for 3 hours before using.

Identification—Transfer a quantity of Oral Suspension, equivalent to about 20 mg of erythromycin, to a separator. Add 15 mL of 0.02 *N* sodium hydroxide, and swirl to mix. Add 2 g of sodium chloride and 25 mL of chloroform, and shake for 3 minutes. Drain the chloroform phase through a small amount of chloroform-washed anhydrous sodium sulfate, and collect the chloroform extract in a beaker, rinsing the sodium sulfate with an additional 10 mL of chloroform. Evaporate the chloroform to dryness. Dissolve the residue in 1 mL of methanol (test solution). Prepare a Standard solution by transferring a quantity of USP Erythromycin Estolate RS, equivalent to 20 mg of erythromycin, to a separator and carrying out the extraction procedure described for preparation of the test solution. Apply separately 3 µL of each solution on a thin-layer chromatographic plate (see *Chromatography* ⟨621⟩), coated with a 0.25-mm layer of chromatographic silica gel. Proceed as directed in the *Identification test* under *Erythromycin Delayed-release Capsules*, beginning with "Place the plate in an unlined chromatographic chamber."

Uniformity of dosage units ⟨905⟩—

FOR SUSPENSION PACKAGED IN SINGLE-UNIT CONTAINERS: meets the requirements.

pH ⟨791⟩: between 3.5 and 6.5.

Assay—Dilute an accurately measured volume of Erythromycin Estolate Oral Suspension, freshly mixed and free from air bubbles, quantitatively with methanol to obtain a solution containing the equivalent of 2.5 mg of erythromycin per mL. Dilute with 1.5 volumes of *Buffer No. 3*, and allow to stand at room temperature for 18 hours. Proceed as directed for erythromycin under *Antibiotics—Microbial Assays* ⟨81⟩, using an accurately measured volume of this stock test solution diluted quantitatively with *Buffer No. 3* to yield a *Test Dilution* having a concentration assumed to be equal to the median dose level of the Standard.

Erythromycin Estolate for Oral Suspension

» Erythromycin Estolate for Oral Suspension is a dry mixture of Erythromycin Estolate with one or more suitable buffers, colors, diluents, dispersants, and flavors. It contains the equivalent of not less than 90.0 percent and not more than 115.0 percent of the labeled amount of erythromycin ($C_{37}H_{67}NO_{13}$).

Packaging and storage—Preserve in tight containers.

Reference standard—*USP Erythromycin Reference Standard*—Dry in vacuum at a pressure not exceeding 5 mm of mercury at 60° for 3 hours before using.

Identification—Prepare a test solution by mixing a quantity of Erythromycin Estolate for Oral Suspension with methanol to obtain a concentration equivalent to about 20 mg of erythromycin per mL. Prepare a Standard solution of USP Erythromycin Estolate RS in methanol containing the equivalent of 20 mg of erythromycin per mL. Apply separately 3 µL of each solution on a thin-layer chromatographic plate (see *Chromatography* ⟨621⟩), coated with a 0.25-mm layer of chromatographic silica gel. Proceed as directed in the *Identification test* under *Erythromycin Delayed-release Capsules*, beginning with "Place the plate in an unlined chromatographic chamber."

Uniformity of dosage units ⟨905⟩—

FOR SOLID PACKAGED IN SINGLE-UNIT CONTAINERS: meets the requirements.

pH ⟨791⟩: between 5.0 and 7.0 (if pediatric drops, between 5.0 and 5.5), in the suspension constituted as directed in the labeling.

Water, *Method I* ⟨921⟩: not more than 2.0%.

Assay—Constitute Erythromycin Estolate for Oral Suspension as directed in the labeling, and proceed as directed in the *Assay* under *Erythromycin Estolate Oral Suspension.*

Erythromycin Estolate Tablets

» Erythromycin Estolate Tablets contain the equivalent of not less than 90.0 percent and not more than 120.0 percent (115.0 percent, if chewable) of the labeled amount of erythromycin ($C_{37}H_{67}NO_{13}$).

Packaging and storage—Preserve in tight containers.

Labeling—Label Tablets to indicate whether they are to be chewed before swallowing.

Reference standard—*USP Erythromycin Reference Standard*—Dry in vacuum at a pressure not exceeding 5 mm of mercury at 60° for 3 hours before using.

Identification—Prepare a test solution by mixing a quantity of finely powdered Tablets with methanol to obtain a concentration equivalent to about 20 mg of erythromycin per mL. Prepare a Standard solution of USP Erythromycin Estolate RS in methanol containing the equivalent of 20 mg of erythromycin per mL. Apply separately 3 µL of each solution on a thin-layer chromatographic plate (see *Chromatography* ⟨621⟩), coated with a 0.25-mm layer of chromatographic silica gel. Proceed as directed in the *Identification test* under *Erythromycin Delayed-release Cap-*

sules, beginning with "Place the plate in an unlined chromatographic chamber." [NOTE—Use the following procedure for chewable Tablets: Transfer a quantity of finely powdered Tablets, equivalent to about 20 mg of erythromycin, to a separator, and proceed as directed in the *Identification test* under *Erythromycin Estolate Oral Suspension*, beginning with "Add 15 mL of 0.02 *N* sodium hydroxide."]

Disintegration ⟨701⟩: 30 minutes [NOTE—Chewable tablets are exempt from this requirement].

Uniformity of dosage units ⟨905⟩: meet the requirements.

Water, *Method I* ⟨931⟩: not more than 5.0%; or if chewable tablets, not more than 4.0%.

Assay—Proceed with Erythromycin Estolate Tablets as directed in the *Assay* under *Erythromycin Estolate Capsules*.

Erythromycin Ethylsuccinate

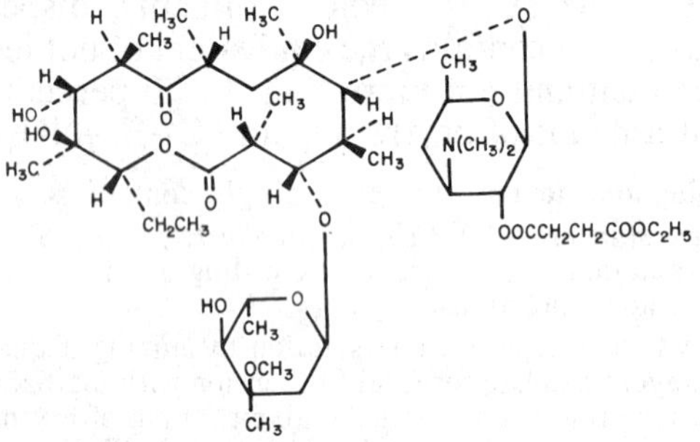

$C_{43}H_{75}NO_{16}$ 862.06

Erythromycin 2′-(ethyl butanedioate).

Erythromycin 2′-(ethyl succinate) [*41342-53-4; 1264-62-6*].

» Erythromycin Ethylsuccinate has a potency equivalent to not less than 765 µg of erythromycin ($C_{37}H_{67}NO_{13}$) per mg, calculated on the anhydrous basis.

Packaging and storage—Preserve in tight containers.

Reference standards—*USP Erythromycin Reference Standard*—Dry in vacuum at a pressure not exceeding 5 mm of mercury at 60° for 3 hours before using. *USP Erythromycin Ethylsuccinate Reference Standard*—Do not dry before using.

Identification—The infrared spectrum of a 1 in 100 solution in chloroform, determined in a 1.0-mm cell, exhibits maxima only at the same wavelengths as that of a similar preparation of USP Erythromycin Ethylsuccinate RS.

Crystallinity ⟨695⟩: meets the requirements.

pH ⟨791⟩: between 6.0 and 8.5, in a 1% aqueous suspension.

Water, *Method I* ⟨921⟩: not more than 3.0%.

Residue on ignition ⟨281⟩: not more than 1.0% after ignition at 550 ± 50°, the charred residue being moistened with 2 mL of nitric acid and 5 drops of sulfuric acid.

Assay—Proceed as directed for erythromycin under *Antibiotics—Microbial Assays* ⟨81⟩, using an accurately weighed quantity of Erythromycin Ethylsuccinate dissolved in methanol to yield a solution containing the equivalent of about 1 mg of erythromycin per mL. Dilute a portion of this solution quantitatively with *Buffer No. 3* to obtain a *Test Dilution* having a concentration assumed to be equal to the median dose level of the Standard.

Erythromycin Ethylsuccinate Injection

» Erythromycin Ethylsuccinate Injection is a sterile solution of Erythromycin Ethylsuccinate in Polyethylene Glycol 400, and contains 2 percent of butylaminobenzoate and a suitable preservative. It contains the equivalent of not less than 90.0 percent and not more than 115.0 percent of the labeled amount of erythromycin ($C_{37}H_{67}NO_{13}$).

Packaging and storage—Preserve in single-dose or in multiple-dose containers, preferably of Type I glass.

Reference standard—*USP Erythromycin Reference Standard*—Dry in vacuum at a pressure not exceeding 5 mm of mercury at 60° for 3 hours before using.

Sterility—It meets the requirements under *Sterility Tests* ⟨71⟩, when tested as directed in the section, *Test Procedures Using Membrane Filtration*, using a membrane filter resistant to the solvent effect of polyethylene glycol 400.

Water, *Method I* ⟨921⟩: not more than 1.5%.

Other requirements—It meets the requirements under *Injections* ⟨1⟩.

Assay—Proceed as directed for erythromycin under *Antibiotics—Microbial Assays* ⟨81⟩, using an accurately measured volume of Erythromycin Ethylsuccinate Injection diluted quantitatively with methanol to yield a solution containing the equivalent of about 1 mg of erythromycin per mL. Dilute a portion of this stock solution quantitatively with *Buffer No. 3* to obtain a *Test Dilution* having a concentration assumed to be equal to the median dose level of the Standard.

Sterile Erythromycin Ethylsuccinate

» Sterile Erythromycin Ethylsuccinate is Erythromycin Ethylsuccinate suitable for parenteral use. It has a potency equivalent to not less than 765 µg of erythromycin ($C_{37}H_{67}NO_{13}$) per mg, calculated on the anhydrous basis.

Packaging and storage—Preserve in *Containers for Sterile Solids* as described under *Injections* ⟨1⟩.

Reference standards—*USP Erythromycin Reference Standard*—Dry in vacuum at a pressure not exceeding 5 mm of mercury at 60° for 3 hours before using. *USP Erythromycin Ethylsuccinate Reference Standard*—Do not dry before using.

Sterility—It meets the requirements under *Sterility Tests* ⟨71⟩, when tested as directed in the section, *Test Procedures for Direct Transfer to Test Media*.

Heavy metals ⟨231⟩—Dissolve 1 g in 25 mL of water, and proceed as directed for *Method I*: the limit is 0.002%.

Other requirements—It conforms to the definition, responds to the *Identification test*, and meets the requirements for *pH*, *Water*, *Residue on ignition*, *Crystallinity*, and *Assay* under *Erythromycin Ethylsuccinate*.

Erythromycin Ethylsuccinate Oral Suspension

» Erythromycin Ethylsuccinate Oral Suspension is a suspension of Erythromycin Ethylsuccinate containing one or more suitable buffers, colors, dispersants, flavors, and preservatives. It contains the equivalent of not less than 90.0 percent and not more than 120.0 percent of the labeled amount of erythromycin ($C_{37}H_{67}NO_{13}$).

Packaging and storage—Preserve in tight containers, and store in a cold place.

Reference standard—*USP Erythromycin Reference Standard*—Dry in vacuum at a pressure not exceeding 5 mm of mercury at 60° for 3 hours before using.

Identification—To a quantity of the Oral Suspension add a volume of methanol sufficient to yield a solution having a concentration equivalent to about 2.5 mg of erythromycin per mL. Shake this mixture by mechanical means for about 30 minutes. Centrifuge a portion of this mixture, and use the clear supernatant liquid as the test solution. Prepare a Standard solution of USP Erythromycin Ethylsuccinate RS in methanol containing about 3 mg per mL. Apply separately 10 μL each of the test solution and the Standard solution on a suitable thin-layer chromatographic plate (see *Chromatography* ⟨621⟩), coated with a 0.25-mm layer of chromatographic silica gel mixture, and allow to dry. Place the plate in an unlined chromatographic chamber, and develop the chromatograms in a solvent system consisting of a mixture of methanol and chloroform (85:15) until the solvent front has moved about 9 cm. Remove the plate from the chamber, mark the solvent front, and allow the solvent to evaporate. Spray the plate with a mixture of dehydrated alcohol, *p*-methoxybenzaldehyde, and sulfuric acid (90:5:5). Heat the plate at 100° for 10 minutes, and examine the chromatograms, in which the erythromycin and succinic acid moieties appear as black-to-purple spots: the R_f values of the principal spots obtained from the test solution correspond to those obtained from the Standard solution.

Uniformity of dosage units ⟨905⟩—
FOR SUSPENSION PACKAGED IN SINGLE-UNIT CONTAINERS: meets the requirements.

pH ⟨791⟩: between 6.5 and 8.5.

Assay—Proceed as directed for erythromycin under *Antibiotics—Microbial Assays* ⟨81⟩, using an accurately measured volume of Erythromycin Ethylsuccinate Oral Suspension, freshly mixed and free from air bubbles, blended for 4 ± 1 minutes in a high-speed glass blender jar with sufficient methanol to give a stock solution containing the equivalent of about 1 mg of erythromycin per mL. Dilute this stock solution quantitatively with *Buffer No. 3* to obtain a *Test Dilution* having a concentration assumed to be equal to the median dose level of the Standard.

Erythromycin Ethylsuccinate for Oral Suspension

» Erythromycin Ethylsuccinate for Oral Suspension is a dry mixture of Erythromycin Ethylsuccinate with one or more suitable buffers, colors, diluents, dispersants, and flavors. It contains the equivalent of not less than 90.0 percent and not more than 120.0 percent of the labeled amount of erythromycin ($C_{37}H_{67}NO_{13}$).

Packaging and storage—Preserve in tight containers.

Reference standard—*USP Erythromycin Reference Standard*—Dry in vacuum at a pressure not exceeding 5 mm of mercury at 60° for 3 hours before using.

Identification—To a quantity of Erythromycin Ethylsuccinate for Oral Suspension add a volume of methanol sufficient to yield a solution containing the equivalent of about 2.5 mg of erythromycin per mL, and stir for 30 minutes. Centrifuge a portion of this mixture, and use the clear supernatant liquid as the test solution. Proceed as directed in the *Identification test* under *Erythromycin Ethylsuccinate Oral Suspension*, beginning with "Prepare a Standard solution."

Uniformity of dosage units ⟨905⟩—
FOR SOLID PACKAGED IN SINGLE-UNIT CONTAINERS: meets the requirements.

pH ⟨791⟩: between 7.0 and 9.0, in the suspension constituted as directed in the labeling.

Loss on drying ⟨731⟩—Dry about 100 mg in a capillary-stoppered bottle in vacuum at 60° for 3 hours: it loses not more than 1.0% of its weight.

Assay—Constitute Erythromycin Ethylsuccinate for Oral Suspension as directed in the labeling, and proceed as directed in the *Assay* under *Erythromycin Ethylsuccinate Oral Suspension*.

Erythromycin Ethylsuccinate Tablets

» Erythromycin Ethylsuccinate Tablets contain the equivalent of not less than 90.0 percent and not more than 120.0 percent of the labeled amount of erythromycin ($C_{37}H_{67}NO_{13}$).

Packaging and storage—Preserve in tight containers.

Labeling—Label chewable Tablets to indicate that they are to be chewed before swallowing.

Reference standard—*USP Erythromycin Reference Standard*—Dry in vacuum at a pressure not exceeding 5 mm of mercury at 60° for 3 hours before using.

Identification—To a quantity of powdered Tablets add a volume of methanol sufficient to yield a solution containing the equivalent of about 2.5 mg of erythromycin per mL. Shake this mixture by mechanical means for about 30 minutes. Centrifuge a portion of this mixture, and use the clear supernatant liquid as the test solution. Proceed as directed in the *Identification test* under *Erythromycin Ethylsuccinate Oral Suspension*, beginning with "Prepare a Standard solution."

Dissolution ⟨711⟩—
Medium: 0.1 *N* hydrochloric acid; 900 mL.
Apparatus 2: 50 rpm.
Time: 45 minutes.
Procedure—Determine the amount of $C_{37}H_{67}NO_{13}$ equivalent dissolved, using a suitable validated spectrophotometric procedure, of filtered portions of the solution under test, suitably diluted with 0.1 *N* hydrochloric acid, if necessary, in comparison with a Standard solution having a known concentration of USP Erythromycin RS in the same medium.
Tolerances—Not less than 75% (*Q*) of the labeled amount of $C_{37}H_{67}NO_{13}$ equivalent is dissolved in 45 minutes.

Uniformity of dosage units ⟨905⟩: meet the requirements.

Loss on drying ⟨731⟩—[NOTE—Chewable Tablets are exempt from this requirement.] Dry about 100 mg in vacuum at a pressure not exceeding 5 mm of mercury at 60° for 3 hours: it loses not more than 4.0% of its weight.

Water, *Method I* ⟨921⟩ (Chewable Tablets only): not more than 5.0%.

Assay—Proceed with Erythromycin Ethylsuccinate Tablets as directed in the *Assay* under *Erythromycin Tablets*.

Erythromycin Ethylsuccinate and Sulfisoxazole Acetyl for Oral Suspension

» Erythromycin Ethylsuccinate and Sulfisoxazole Acetyl for Oral Suspension is a dry mixture of Erythromycin Ethylsuccinate and Sulfisoxazole Acetyl with one or more suitable buffers, colors, flavors, surfactants, and suspending agents. It contains the equivalent of not less than 90.0 percent and not more than 120.0 percent of the labeled amount of erythromycin ($C_{37}H_{67}NO_{13}$) and the equivalent of not less than 90.0 percent and not more than 115.0 percent of the labeled amount of sulfisoxazole ($C_{11}H_{13}N_3O_3S$).

NOTE—Where Erythromycin Ethylsuccinate and Sulfisoxazole Acetyl for Oral Suspension is prescribed, without reference to the quantity of erythromycin or sulfisoxazole contained therein, a product containing 40 mg of erythromycin and 120 mg of sulfisoxazole per mL when constituted as directed in the labeling shall be dispensed.

Packaging and storage—Preserve in tight containers.

Reference standards—*USP Erythromycin Reference Standard*—Dry in vacuum at a pressure not exceeding 5 mm of mercury at

60° for 3 hours before using. *USP Sulfisoxazole Acetyl Reference Standard*—Dry at 105° for 3 hours before using.

Identification—To a quantity of the Erythromycin Ethylsuccinate and Sulfisoxazole Acetyl for Oral Suspension add a volume of methanol sufficient to yield a solution having a concentration equivalent to about 2.5 mg of erythromycin per mL. Shake this mixture by mechanical means for about 30 minutes. Centrifuge a portion of this mixture, and use the clear supernatant liquid as the test solution. Prepare a solution of USP Erythromycin Ethylsuccinate RS in methanol containing about 3 mg per mL (*Standard solution A*). Prepare a solution of USP Sulfisoxazole Acetyl RS in methanol containing about 8.7 mg per mL (*Standard solution B*). Apply separately 10 µL each of the test solution and the two Standard solutions on a suitable thin-layer chromatographic plate (see *Chromatography* ⟨621⟩), coated with a 0.25-mm layer of chromatographic silica gel mixture, and allow to dry. Place the plate in an unlined chromatographic chamber, and develop the chromatograms in a solvent system consisting of a mixture of methanol and chloroform (85:15) until the solvent front has moved about 9 cm. Remove the plate from the chamber, mark the solvent front, and allow the solvent to evaporate. Spray the plate with a mixture of dehydrated alcohol, *p*-methoxybenzaldehyde, and sulfuric acid (90:5:5). Heat the plate at 100° for 10 minutes, and examine the chromatograms, in which the erythromycin and succinic acid moieties appear as black-to-purple spots and the sulfisoxazole acetyl appears as a yellow spot: the R_f values of the principal black-to-purple spots obtained from the test solution correspond to those obtained from *Standard solution A*, and the R_f value of the principal yellow spot obtained from the test solution corresponds to that obtained from *Standard solution B*.

Uniformity of dosage units ⟨905⟩—

FOR SOLID PACKAGED IN SINGLE-UNIT CONTAINERS: meets the requirements for *Content Uniformity* with respect to erythromycin and sulfisoxazole.

pH ⟨791⟩: between 5.0 and 7.0, in the suspension constituted as directed in the labeling.

Loss on drying ⟨731⟩: Dry about 100 mg in a capillary-stoppered bottle in vacuum at 60° for 3 hours: it loses not more than 1.0% of its weight.

Assay for erythromycin—Constitute Erythromycin Ethylsuccinate and Sulfisoxazole Acetyl for Oral Suspension as directed in the labeling, and allow to stand for 1 hour. Gently shake the suspension, transfer 5.0 mL to a high-speed blender jar containing 195.0 mL of methanol, and blend for 4 ± 1 minutes. Proceed as directed under *Antibiotics—Microbial Assays* ⟨81⟩, using an accurately measured volume of this stock test solution diluted quantitatively and stepwise with *Buffer No. 3* to yield a *Test Dilution* having a concentration assumed to be equal to the median dose level of the Standard (1.0 µg of erythromycin per mL).

Assay for sulfisoxazole—

Mobile solvent—Mix 40 volumes of acetonitrile and 60 volumes of water. The acetonitrile concentration may be varied to meet system suitability requirements and to provide a suitable elution time for sulfisoxazole acetyl. Filter the solution through a membrane filter (1-µm or finer porosity).

Internal standard solution—Prepare a solution of benzanilide in acetonitrile having a concentration of about 0.33 mg per mL. Filter the solution through a membrane filter (1-µm or finer porosity).

Standard preparation—Prepare a solution of USP Sulfisoxazole Acetyl RS in *Internal standard solution* having a known concentration of about 1 mg per mL.

Assay preparation—Constitute Erythromycin Ethylsuccinate and Sulfisoxazole Acetyl for Oral Suspension as directed in the labeling, and allow to stand for 1 hour. Gently shake the suspension, transfer to a 125-mL separator an accurately measured volume of it, equivalent to about 600 mg of sulfisoxazole, and extract with three 75-mL portions of chloroform. Collect the chloroform extracts in a 250-mL volumetric flask, dilute with chloroform to volume, and mix. Filter a portion of this solution through a membrane filter (1-µm or finer porosity). Pipet 4 mL of the filtrate into a glass-stoppered, 25-mL conical flask, and evaporate with the aid of a current of dry air to dryness. Add 10.0 mL of *Internal standard solution*, and mix.

Chromatographic system (see *Chromatography* ⟨621⟩)—The liquid chromatograph is equipped with a 254-nm detector and a 4-mm × 30-cm column that contains packing L1. The flow rate is about 1.2 mL per minute. Chromatograph replicate injections of the *Standard preparation*, and record the peak responses as directed under *Procedure:* the resolution factor between sulfisoxazole acetyl and benzanilide is not less than 3.0.

Procedure—Separately inject equal volumes (about 5 µL) of the *Standard preparation* and the *Assay preparation* into the chromatograph, record the chromatograms, and measure the responses for the major peaks. Calculate the quantity, in mg, of sulfisoxazole ($C_{11}H_{13}N_3O_3S$) in each mL of the constituted suspension taken by the formula:

$$(267.30/309.34)(625C/V)(R_U/R_S),$$

in which 267.30 and 309.34 are the molecular weights of sulfisoxazole and sulfisoxazole acetyl, respectively, C is the concentration, in mg, of USP Sulfisoxazole Acetyl RS in each mL of the *Standard preparation*, V is the volume, in mL of constituted suspension taken, and R_U and R_S are the ratios of peak responses of sulfisoxazole acetyl peak to benzanilide peak obtained from the *Assay preparation* and the *Standard preparation*, respectively.

Sterile Erythromycin Gluceptate

$C_{37}H_{67}NO_{13}.C_7H_{14}O_8$ 960.12
Erythromycin monoglucoheptonate (salt).
Erythromycin glucoheptonate (1:1) (salt)
[*304-63-2; 23067-13-2*].

» Sterile Erythromycin Gluceptate is erythromycin gluceptate suitable for parenteral use. It has a potency equivalent to not less than 600 µg of erythromycin ($C_{37}H_{67}NO_{13}$) per mg, calculated on the anhydrous basis. In addition, where packaged for dispensing, it contains the equivalent of not less than 90.0 percent and not more than 115.0 percent of the labeled amount of erythromycin ($C_{37}H_{67}NO_{13}$).

Packaging and storage—Preserve in *Containers for Sterile Solids* as described under *Injections* ⟨1⟩.

Reference standards—*USP Erythromycin Reference Standard*—Dry in vacuum at a pressure not exceeding 5 mm of mercury at 60° for 3 hours before using. *USP Erythromycin Gluceptate Reference Standard*—Do not dry before using.

Identification—The infrared absorption spectrum of a mineral oil dispersion of it exhibits maxima only at the same wavelengths as that of a similar preparation of USP Erythromycin Gluceptate RS.

Pyrogen—It meets the requirements of the *Pyrogen Test* ⟨151⟩, the test dose being 1.0 mL per kg of a solution in Sterile Water for Injection containing the equivalent of 30.0 mg of erythromycin per mL.

Sterility—It meets the requirements under *Sterility Tests* ⟨71⟩, when tested as directed in the section, *Test Procedures Using Membrane Filtration.*

pH ⟨791⟩: between 6.0 and 8.0, in a solution containing 25 mg per mL.

Water, *Method I* ⟨921⟩: not more than 5.0%.

Particulate matter ⟨788⟩: meets the requirements under *Small-volume Injections.*

Other requirements—Where packaged for dispensing it meets the requirements for *Uniformity of Dosage Units* ⟨905⟩ and *Constituted Solutions* and *Labeling* under *Injections* ⟨1⟩.

Assay—Proceed as directed for erythromycin under *Antibiotics—Microbial Assays* ⟨81⟩, using an accurately weighed quantity of Sterile Erythromycin Gluceptate dissolved in methanol to yield a solution containing the equivalent of about 10 mg of erythromycin per mL. Dilute this solution quantitatively with 9 volumes of *Buffer No. 3* to obtain a stock solution containing the equivalent of about 1 mg of erythromycin per mL. Where it is packaged for dispensing, constitute Sterile Erythromycin Gluceptate as directed in the labeling. Withdraw all of the withdrawable contents where the package is represented as being a single-dose container; or where the labeling specifies the amount of potency in a given volume of the resultant preparation, withdraw an accurately measured volume. Dilute quantitatively with *Buffer No. 3* to obtain a stock solution having a convenient concentration. Dilute a portion of the stock solution quantitatively with *Buffer No. 3* to obtain a *Test Dilution* having a concentration assumed to be equal to the median dose level of the Standard.

Erythromycin Lactobionate for Injection

$C_{37}H_{67}NO_{13}.C_{12}H_{22}O_{12}$ 1092.23
Erythromycin mono(4-*O*-β-D-galactopyranosyl-D-gluconate) (salt).
Erythromycin lactobionate (1:1) (salt) [*3847-29-8*].

» Erythromycin Lactobionate for Injection is a sterile, dry mixture of erythromycin lactobionate and a suitable preservative. It contains the equivalent of not less than 90.0 percent and not more than 120.0 percent of the labeled amount of erythromycin ($C_{37}H_{67}NO_{13}$).

Packaging and storage—Preserve in *Containers for Sterile Solids* as described under *Injections* ⟨1⟩.

Reference standards—*USP Erythromycin Reference Standard*—Dry in vacuum at a pressure not exceeding 5 mm of mercury at 60° for 3 hours before using. *USP Erythromycin Lactobionate Reference Standard*—Dry in vacuum at a pressure not exceeding 5 mm of mercury at 60° for 3 hours before using.

Constituted solution—At the time of use, the constituted solution prepared from Erythromycin Lactobionate for Injection meets the requirements for *Constituted Solutions* under *Injections* ⟨1⟩.

Identification—The infrared absorption spectrum of a mineral oil dispersion of it, previously dried in vacuum at a pressure not exceeding 5 mm of mercury at 60° for 3 hours, exhibits maxima only at the same wavelengths as that of a similar preparation of USP Erythromycin Lactobionate RS.

Pyrogen—It meets the requirements of the *Pyrogen Test* ⟨151⟩, the test dose being 1.0 mL per kg of a solution in Sterile Water for Injection containing the equivalent of 30.0 mg of erythromycin per mL.

pH ⟨791⟩: between 6.5 and 7.5, in a solution containing the equivalent of 50 mg of erythromycin per mL.

Water, *Method I* ⟨921⟩: not more than 5.0%.

Particulate matter ⟨788⟩: meets the requirements under *Small-volume Injections* when the constituted solution is diluted with filtered water to a concentration of not more than 5 mg of erythromycin base per mL before the test is performed.

Heavy metals, *Method II* ⟨231⟩: 0.005%.

Other requirements—It meets the requirements under *Injections* ⟨1⟩.

Assay—Proceed as directed for erythromycin under *Antibiotics—Microbial Assays* ⟨81⟩, using Erythromycin Lactobionate for Injection constituted as directed in the labeling. Withdraw all of the withdrawable contents where the package is represented as being a single-dose container; or, where the labeling specifies the amount of erythromycin equivalent in a given volume of the resultant preparation, withdraw an accurately measured volume. Dilute quantitatively with water to obtain a stock solution containing the equivalent of about 10 mg of erythromycin per mL. Dilute this stock solution quantitatively with *Buffer No. 3* to obtain a *Test Dilution* having a concentration assumed to be equal to the median dose level of the Standard.

Sterile Erythromycin Lactobionate

» Sterile Erythromycin Lactobionate has a potency equivalent to not less than 525 µg of erythromycin ($C_{37}H_{67}NO_{13}$) per mg, calculated on the anhydrous basis. In addition, where packaged for dispensing, it contains the equivalent of not less than 90.0 percent and not more than 120.0 percent of the labeled amount of erythromycin ($C_{37}H_{67}NO_{13}$).

Packaging and storage—Preserve in *Containers for Sterile Solids* as described under *Injections* ⟨1⟩.

Reference standards—*USP Erythromycin Reference Standard*—Dry in vacuum at a pressure not exceeding 5 mm of mercury at 60° for 3 hours before using. *USP Erythromycin Lactobionate Reference Standard*—Dry in vacuum at a pressure not exceeding 5 mm of mercury at 60° for 3 hours before using.

Identification—The infrared absorption spectrum of a mineral oil dispersion of it, previously dried in vacuum at a pressure not exceeding 5 mm of mercury at 60° for 3 hours, exhibits maxima only at the same wavelengths as that of a similar preparation of USP Erythromycin Lactobionate RS.

Pyrogen—It meets the requirements of the *Pyrogen Test* ⟨151⟩, the test dose being 1.0 mL per kg of a solution in Sterile Water for Injection containing the equivalent of 30.0 mg of erythromycin per mL.

Sterility—It meets the requirements under *Sterility Tests* ⟨71⟩, when tested as directed in the section, *Test Procedures Using Membrane Filtration*.

pH ⟨791⟩: between 6.5 and 7.5, in a solution containing the equivalent of 50 mg of erythromycin per mL.

Water, *Method I* ⟨921⟩: not more than 5.0%.

Particulate matter ⟨788⟩: meets the requirements under *Small-volume Injections* when it is diluted with filtered water to a concentration of not more than 5 mg of erythromycin per mL before the test is performed.

Residue on ignition ⟨281⟩: not more than 2.0%, the charred residue being moistened with 2 mL of nitric acid and 5 drops of sulfuric acid.

Heavy metals, *Method II* ⟨231⟩: 0.005%.

Other requirements—Where packaged for dispensing, it meets the requirements for *Uniformity of Dosage Units* ⟨905⟩ and for *Constituted Solutions* and *Labeling* under *Injections* ⟨1⟩.

Assay—

Standard preparation—Prepare as directed for erythromycin under *Antibiotics—Microbial Assays* ⟨81⟩.

Assay preparation 1—Dissolve an accurately weighed quantity of Sterile Erythromycin Lactobionate quantitatively in methanol to obtain a stock solution containing the equivalent of about 10 mg of erythromycin per mL. Dilute this stock solution quantitatively with *Buffer No. 3* (see *Media and Diluents* under *Antibiotics—Microbial Assays* ⟨81⟩) to obtain a *Test Dilution* having a concentration assumed to be equal to the median dose level of the Standard.

Assay preparation 2 (where it is packaged for dispensing and is represented as being in a single-dose container)—Constitute Sterile Erythromycin Lactobionate in a volume of water, accurately measured, corresponding to the volume of solvent specified in the labeling. Withdraw all of the withdrawable contents, using a suitable hypodermic needle and syringe, and dilute quantitatively with *Buffer No. 3* to obtain a *Test Dilution* having a concentration assumed to be equal to the median dose level of the Standard.

Assay preparation 3 (where the label states the quantity of erythromycin in a given volume of constituted solution)—Constitute 1 container of Sterile Erythromycin Lactobionate in a volume of water, accurately measured, corresponding to the volume of solvent specified in the labeling. Dilute an accurately measured volume of the constituted solution quantitatively with *Buffer No. 3* to obtain a *Test Dilution* having a concentration assumed to be equal to the median dose level of the Standard.

Procedure—Proceed as directed for erythromycin under *Antibiotics—Microbial Assays* ⟨81⟩.

Erythromycin Stearate

$C_{37}H_{67}NO_{13}.C_{18}H_{36}O_2$ 1018.42
Erythromycin octadecanoate (salt).
Erythromycin stearate (salt) [*643-22-1*].

» Erythromycin Stearate is the stearic acid salt of Erythromycin, with an excess of Stearic Acid. It has a potency equivalent to not less than 550 µg of erythromycin ($C_{37}H_{67}NO_{13}$) per mg, calculated on the anhydrous basis.

Packaging and storage—Preserve in tight containers.

Reference standards—*USP Erythromycin Reference Standard*—Dry in vacuum at a pressure not exceeding 5 mm of mercury at 60° for 3 hours before using. *USP Erythromycin Stearate Reference Standard*—Do not dry before using.

Identification—The infrared absorption spectrum of a mineral oil dispersion of it exhibits maxima only at the same wavelengths as that of a similar preparation of USP Erythromycin Stearate RS.

Crystallinity ⟨695⟩: meets the requirements.

pH ⟨791⟩: between 6.0 and 11.0, in a 1% aqueous suspension.

Water, *Method I* ⟨921⟩: not more than 4.0%.

Residue on ignition ⟨281⟩: not more than 1.0%, the charred residue being moistened with 2 mL of nitric acid and 5 drops of sulfuric acid.

Assay—Proceed with Erythromycin Stearate as directed in the *Assay* under *Erythromycin Ethylsuccinate.*

Erythromycin Stearate for Oral Suspension

» Erythromycin Stearate for Oral Suspension is a dry mixture of Erythromycin Stearate with one or more suitable buffers, colors, diluents, dispersants, and flavors. It contains the equivalent of not less than 90.0 percent and not more than 120.0 percent of the labeled amount of erythromycin ($C_{37}H_{67}NO_{13}$).

Packaging and storage—Preserve in tight containers.

Reference standard—*USP Erythromycin Reference Standard*—Dry in vacuum at a pressure not exceeding 5 mm of mercury at 60° for 3 hours before using.

Identification—To a quantity of Erythromycin Stearate for Oral Suspension add a volume of methanol sufficient to yield a solution having a concentration equivalent to about 5 mg of erythromycin per mL. Shake this mixture by mechanical means for about 30 minutes. Centrifuge a portion of this mixture, and use the clear supernatant liquid as the test solution. Prepare a Standard solution of USP Erythromycin Stearate RS in methanol containing about 8 mg per mL. Apply separately 20 µL each of the test solution and the Standard solution on a suitable thin-layer chromatographic plate (see *Chromatography* ⟨621⟩), coated with a 0.25-mm layer of chromatographic silica gel mixture, and allow to dry. Place the plate in an unlined chromatographic chamber, and develop the chromatograms in a solvent system consisting of a mixture of methanol and chloroform (85:15) until the solvent front has moved about 9 cm. Remove the plate from the chamber, mark the solvent front, and allow the solvent to evaporate. Spray the plate with a methanolic solution of 2′,7′-dichlorofluorescein (1 in 500), and examine the plate under long-wavelength ultraviolet light: the R_f values of the principal fluorescent spots obtained from the test solution correspond to those obtained from the Standard solution. Then spray the plate with a mixture of dehydrated alcohol, *p*-methoxybenzaldehyde, and sulfuric acid (90:5:5). Heat the plate at 100° for 10 minutes, and examine the chromatograms, in which the erythromycin appears as a black-to-purple spot: the R_f value of the principal spot obtained from the test solution corresponds to that obtained from the Standard solution.

Uniformity of dosage units ⟨905⟩—

FOR SOLID PACKAGED IN SINGLE-UNIT CONTAINERS: meets the requirements.

pH ⟨791⟩: between 6.0 and 9.0, in the suspension constituted as directed in the labeling.

Water, *Method I* ⟨921⟩: not more than 2.0%.

Assay—Constitute Erythromycin Stearate for Oral Suspension as directed in the labeling, and proceed as directed in the *Assay* under *Erythromycin Ethylsuccinate Oral Suspension.*

Erythromycin Stearate Tablets

» Erythromycin Stearate Tablets contain the equivalent of not less than 90.0 percent and not more than 120.0 percent of the labeled amount of erythromycin ($C_{37}H_{67}NO_{13}$).

Packaging and storage—Preserve in tight containers.

Reference standard—*USP Erythromycin Reference Standard*—Dry in vacuum at a pressure not exceeding 5 mm of mercury at 60° for 3 hours before using.

Identification—To a quantity of powdered Tablets add a volume of methanol sufficient to yield a solution containing the equivalent of about 5 mg of erythromycin per mL. Shake this mixture by mechanical means for about 30 minutes. Centrifuge a portion of this mixture, and use the clear supernatant liquid as the test solution. Proceed as directed in the *Identification test* under *Erythromycin Stearate for Oral Suspension*, beginning with "Prepare a Standard solution."

Dissolution ⟨711⟩—

Medium: 0.05 *M* pH 6.8 phosphate buffer (see under *Solutions* in the section, *Reagents, Indicators, and Solutions*); 900 mL.

Apparatus 2: 100 rpm.

Time: 120 minutes.

Stock standard solution—Dissolve an accurately weighed quantity of USP Erythromycin RS in methanol to obtain a solution containing about 14 mg per mL. Dilute quantitatively with water, and mix to obtain a solution having a known concentration of about 0.56 mg of USP Erythromycin RS per mL.

Working standard solution—On the day of use, dilute 25.0 mL of *Stock standard solution* with water to 50.0 mL, and mix.

Test solution—After 120 minutes, withdraw a portion of the solution under test, filter, and dilute with *Dissolution Medium*, if necessary, to obtain a solution having an estimated concentration of about 0.28 mg of erythromycin per mL.

Procedure—Transfer 5.0-mL portions of the *Working standard solution* to two 25-mL volumetric flasks, one of which serves as a working standard blank. Similarly, transfer 5.0-mL portions of each *Test solution* to two 25-mL volumetric flasks, one of which serves as a blank for that *Test solution*. To each of the flasks designated as a blank add 2.0 mL of 0.5 *N* sulfuric acid and to the remaining flasks add 2.0 mL of water. Allow to stand for 5 minutes with intermittent swirling. To all flasks add 15.0 mL of 0.25 *N* sodium hydroxide, dilute with *Dissolution Medium* to volume, and mix. Heat the flasks in a water bath at 60 ± 0.5° for 5 minutes, and allow to cool. Using a suitable spectrophotometer, determine the absorbance of each solution, corrected for its blank solution, at the wavelength of maximum absorbance at about 236 nm. Determine the amount of $C_{37}H_{67}NO_{13}$ dissolved from the corrected absorbance of the solution obtained from the test specimen in comparison with that of the solution obtained from the Reference Standard.

Tolerances—Not less than 75% (*Q*) of the labeled amount of $C_{37}H_{67}NO_{13}$ is dissolved in 120 minutes.

Uniformity of dosage units ⟨905⟩: meet the requirements.

Loss on drying ⟨731⟩—Dry about 100 mg of powdered Tablets in a capillary-stoppered bottle in vacuum at 60° for 3 hours: it loses not more than 5.0% of its weight.

Assay—Proceed with Erythromycin Stearate Tablets as directed in the *Assay* under *Erythromycin Tablets*.

Erythrosine Sodium

$C_{20}H_6I_4Na_2O_5 . H_2O$ 897.88

Spiro[isobenzofuran-1(3*H*),9′-[9*H*]xanthen]-3-one, 3′,6′-dihydroxy-2′,4′,5′,7′-tetraiodo-, disodium salt, monohydrate.

2′,4′,5′,7′-Tetraiodofluorescein disodium salt monohydrate [*49746-10-3*].

Anhydrous 879.86 [*568-63-8; 16423-68-0*].

» Erythrosine Sodium is a dye consisting principally of the monohydrate of 2′,4′,5′,7′-tetraiodofluorescein disodium salt, with smaller amounts of lower iodinated fluoresceins. It contains not less than 87.0 percent of dye, calculated as $C_{20}H_6I_4Na_2O_5 . H_2O$. It conforms to the regulations of the federal Food and Drug Administration concerning certified dyes (21 CFR 74.1303).

Packaging and storage—Preserve in tight containers.

Other requirements—It complies with the volatile matter, chlorides and sulfates, water-insoluble matter, unhalogenated intermediates, sodium iodide, triiodoresorcinol, 2-(2′,4′-dihydroxy-3′,5′-diiodobenzoyl)benzoic acid, monoiodofluoresceins, lead, arsenic, and other requirements of the federal FDA concerning Erythrosine Sodium.

Erythrosine Sodium Topical Solution

» Erythrosine Sodium Topical Solution is a solution of Erythrosine Sodium in Purified Water. It contains not less than 90.0 percent and not more than 110.0 percent of the labeled amount of erythrosine sodium, calculated as $C_{20}H_6I_4Na_2O_5 . H_2O$. It contains one or more suitable flavoring and preservative agents.

Packaging and storage—Preserve in tight, light-resistant containers.

Reference standard—*USP Erythrosine Sodium Reference Standard*—Dry at 135° for 16 hours before using.

Identification—

A: It responds to *Identification test A* under *Erythrosine Sodium Soluble Tablets.*

B: Proceed with Erythrosine Sodium Topical Solution as directed in *Identification test B* under *Erythrosine Sodium Soluble Tablets*, beginning with "Filter through filter paper." The specified result is observed.

pH ⟨791⟩: between 6.8 and 8.0.

Assay—Dilute an accurately measured volume of Erythrosine Sodium Topical Solution, equivalent to about 100 mg of erythrosine sodium, quantitatively and stepwise with methanol to obtain a solution having a concentration of about 4 µg per mL. Prepare a Standard solution of USP Erythrosine Sodium RS as directed in the *Assay* under *Erythrosine Sodium Soluble Tablets.* Concomitantly determine the absorbances of both solutions in 1-cm cells at the wavelength of maximum absorbance at about 530 nm, with a suitable spectrophotometer, using methanol as the blank. Calculate the quantity, in mg, of $C_{20}H_6I_4Na_2O_5 . H_2O$ in each mL of the Topical Solution taken by the formula:

$$25(C/V)(A_U/A_S),$$

in which C is the concentration, in µg per mL, of USP Erythrosine Sodium RS in the Standard solution, V is the volume, in mL, of Topical Solution taken, and A_U and A_S are the absorbances of the solution prepared from the Erythrosine Sodium Topical Solution and the Standard solution, respectively.

Erythrosine Sodium Soluble Tablets

» Erythrosine Sodium Soluble Tablets contain not less than 90.0 percent and not more than 110.0 percent of the labeled amount of erythrosine sodium, calculated as $C_{20}H_6I_4Na_2O_5 . H_2O$.

Packaging and storage—Preserve in tight, moisture-resistant, light-resistant containers.

Reference standard—*USP Erythrosine Sodium Reference Standard*—Dry at 135° for 16 hours before using.

Identification—

A: The visible absorption spectrum of the solution employed for measurement of absorbance in the *Assay* exhibits maxima and minima at the same wavelengths as that of the Standard solution employed in the *Assay*, concomitantly measured.

B: Dissolve the powder mass remaining after removal of the aliquot specified in the *Assay* in 100 mL of methanol. Filter through filter paper (Whatman No. 5 or equivalent), and discard the filtrate. Dry the filter paper at 105° for 20 minutes. Place 1 drop of 2 *N* sulfuric acid on a section of the filter paper: a brownish yellow spot is observed. Place 1 drop of sodium hydroxide solution (1 in 10) on another section of the filter paper: the resulting spot remains red or becomes a more intense red color.

Uniformity of dosage units ⟨905⟩: meet the requirements.

Assay—Weigh and finely powder not less than 20 Erythrosine Sodium Soluble Tablets. Weigh accurately a portion of the powder, equivalent to about 0.8 mg of erythrosine sodium, place in a 200-mL volumetric flask, dilute with methanol to volume, and mix. Dissolve a suitable quantity of USP Erythrosine Sodium RS, accurately weighed, in methanol, and dilute quantitatively and stepwise with methanol to obtain a Standard solution having a known concentration of about 4 µg per mL. Concomitantly determine the absorbances of both solutions in 1-cm cells at the wavelength of maximum absorbance at about 530 nm, with a suitable spectrophotometer, using methanol as the blank. Calculate the quantity, in mg, of $C_{20}H_6I_4Na_2O_5 . H_2O$ in the portion of Soluble Tablets taken by the formula:

$$0.2C(A_U/A_S),$$

in which C is the concentration, in µg per mL, of USP Erythrosine Sodium RS in the Standard solution, and A_U and A_S are the absorbances of the solution from the Tablets and the Standard solution, respectively.

Esterified Estrogens—*see* Estrogens, Esterified

Estradiol

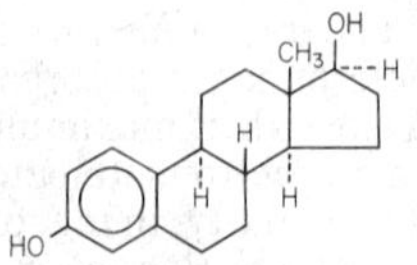

$C_{18}H_{24}O_2$ 272.39

Estra-1,3,5(10)-triene-3,17-diol, (17β)-.

Estra-1,3,5(10)-triene-3,17β-diol [*50-28-2*].

» Estradiol contains not less than 97.0 percent and not more than 103.0 percent of $C_{18}H_{24}O_2$, calculated on the anhydrous basis.

Packaging and storage—Preserve in tight, light-resistant containers.

Reference standard—*USP Estradiol Reference Standard*—Do not dry; determine the *Water* content by *Method I* before using.

Identification—

A: The infrared absorption spectrum of a mineral oil dispersion of it, exhibits maxima only at the same wavelengths as that of a similar preparation of USP Estradiol RS.

B: The ultraviolet absorption spectrum of a 1 in 20,000 solution in alcohol exhibits maxima and minima at the same wavelengths as that of a similar solution of USP Estradiol RS, concomitantly measured, and the respective absorptivities, calculated on the anhydrous basis, at the wavelength of maximum absorbance at about 280 nm do not differ by more than 3.0%.

Melting range, *Class I* ⟨741⟩: between 173° and 179°.

Specific rotation ⟨781⟩: between +76° and +83°, calculated on the anhydrous basis, determined in a solution in dioxane containing 100 mg in each 10 mL.

Water, *Method I* ⟨921⟩: not more than 3.5%.

Assay—

Mobile phase—Prepare a filtered and degassed mixture of acetonitrile and water (55:45). Make adjustments if necessary (see *System Suitability* under *Chromatography* ⟨621⟩).

Internal standard solution—Transfer about 300 mg of ethylparaben to a 500-mL volumetric flask, add methanol to volume, and mix.

Standard preparation—Dissolve accurately weighed quantities of USP Estradiol RS and USP Estrone RS in methanol to obtain a solution containing 0.40 mg and 0.24 mg, respectively, in each mL. Pipet 10 mL of this solution and 5 mL of the *Internal standard solution* into a 200-mL volumetric flask. Add 100 mL of methanol, dilute with water to volume, and mix to obtain a solution having a known concentration of about 20 µg of USP Estradiol RS per mL.

Assay preparation—Transfer about 100 mg of Estradiol, accurately weighed, to a 250-mL volumetric flask, add methanol to volume, and mix. Transfer 10.0 mL of this solution to a 200-mL volumetric flask, add 5.0 mL of *Internal standard solution* and 100 mL of methanol, dilute with water to volume, and mix.

Chromatographic system (see *Chromatography* ⟨621⟩)—The liquid chromatograph is equipped with a 205-nm detector and a 3.9-mm × 30-cm column that contains packing L1. The flow rate is about 1 mL per minute. Chromatograph the *Standard preparation*, and record the peak responses as directed under *Procedure:* the resolution, *R*, between the analyte and estrone peaks is not less than 2.0, and the relative standard deviation for replicate injections is not more than 2.0%.

Procedure—Separately inject equal volumes (about 25 µL) of the *Standard preparation* and the *Assay preparation* into the chromatograph, record the chromatograms, and measure the responses for the major peaks. The relative retention times are about 0.7 for the internal standard, about 1.3 for estrone, and 1.0 for estradiol. Calculate the quantity, in mg, of $C_{18}H_{24}O_2$ in the portion of Estradiol taken by the formula:

$$5C(R_U/R_S),$$

in which *C* is the concentration, in µg per mL, of USP Estradiol RS in the *Standard preparation*, and R_U and R_S are the peak response ratios obtained from the *Assay preparation* and the *Standard preparation*, respectively.

Estradiol Vaginal Cream

» Estradiol Vaginal Cream contains not less than 90.0 percent and not more than 110.0 percent of the labeled amount of $C_{18}H_{24}O_2$ in a suitable cream base.

Packaging and storage—Preserve in collapsible tubes or in tight containers.

Reference standards—*USP Estradiol Reference Standard*—Do not dry; determine the *Water* content by *Method I* before using. *USP Estrone Reference Standard*—Dry at 105° for 3 hours before using.

Identification—Transfer a portion of Vaginal Cream, equivalent to about 1 mg of estradiol, to a 150-mL beaker. Add 25 mL of acetonitrile, and gently heat to boiling. Boil for 45 seconds, and cool to room temperature. Add 25 mL of water, and swirl. Filter with the aid of suction. Transfer the filtrate to a 125-mL separator, add 50 mL of chloroform, and shake. Allow the layers to separate, drain the chloroform layer into a flask, and evaporate in a rotary evaporator to dryness. Dissolve the residue in 2 mL of chloroform to obtain the test solution. On a suitable thin-layer chromatographic plate (see *Chromatography* ⟨621⟩), coated with a 0.25-mm layer of chromatographic silica gel mixture, apply separately 50 µL of the test solution and 50 µL of a Standard solution of USP Estradiol RS in chloroform containing about 0.5 mg per mL, and dry the applications with the aid of a stream of nitrogen. Position the plate in a chromatographic chamber, and develop the chromatograms in a solvent system consisting of a mixture of toluene and acetone (4:1) until the solvent front has moved about three-fourths of the length of the plate. Remove the plate from the developing chamber, mark the solvent front, and allow the solvent to evaporate. Spray the plate with a fine mist of a mixture of sulfuric acid and methanol (1:1), then heat the plate for 3 to 5 minutes at 90°. Observe the plate under visible light: the R_f value and color of the principal spot obtained from the test solution correspond to those obtained from the Standard solution.

Microbial limits—It meets the requirements of the tests for absence of *Staphylococcus aureus* and *Pseudomonas aeruginosa* under *Microbial Limit Tests* ⟨61⟩.

Minimum fill ⟨755⟩: meets the requirements.

pH ⟨791⟩: between 3.5 and 6.5.

Assay—

Mobile phase—Prepare a filtered and degassed mixture of acetonitrile and water (1:1). Make adjustments if necessary (see *System Suitability* under *Chromatography* ⟨621⟩).

Internal standard solution—Dissolve a suitable quantity of dydrogesterone in acetonitrile to obtain a solution containing about 60 µg per mL. Use a freshly prepared solution.

Standard preparation—Transfer about 10 mg of USP Estradiol RS and about 7.5 mg of USP Estrone RS, both accurately weighed, to a 1000-mL volumetric flask. Add 50.0 mL of *Internal standard solution* and 450 mL of acetonitrile, and mix. Dilute with water to volume, and mix to obtain a solution having a known concentration of about 10 µg of USP Estradiol RS per mL.

Assay preparation—Transfer an accurately weighed portion of Estradiol Vaginal Cream, equivalent to about 0.5 mg of estradiol, to a 150-mL beaker. Add 2.5 mL of *Internal standard solution*, 22.5 mL of acetonitrile, and a few boiling chips. Cover with a watch glass, and heat gently until the Cream melts, swirling occasionally. Heat to boiling for about 45 seconds. Allow to cool to room temperature, add 25.0 mL of water, and mix. Filter first through paper and then through a micro disk filter.

Chromatographic system (see *Chromatography* ⟨621⟩)—The liquid chromatograph is equipped with a 280-nm detector and a 3.9-mm × 30-cm column that contains packing L1. The flow rate is about 1 mL per minute. Chromatograph the *Standard preparation*, and record the peak responses as directed under *Procedure:* the resolution, *R*, between the analyte and estrone peaks is not less than 1.9, and the relative standard deviation for replicate injections is not more than 3.0%.

Procedure—Separately inject equal volumes (about 50 µL) of the *Standard preparation* and the *Assay preparation* into the chromatograph, record the chromatograms, and measure the responses for the major peaks. The relative retention times are about 2.0 for the internal standard, 1.0 for estradiol, and 1.25 for estrone. Calculate the quantity, in mg, of $C_{18}H_{24}O_2$ in the portion of Cream taken by the formula:

$$0.05C(R_U/R_S),$$

in which C is the concentration, in µg per mL, of USP Estradiol RS in the *Standard preparation*, and R_U and R_S are the peak response ratios of estradiol and the internal standard obtained from the *Assay preparation* and the *Standard preparation*, respectively.

Estradiol Pellets

» Estradiol Pellets are sterile pellets composed of Estradiol in compressed form, without the presence of any binder, diluent, or excipient. They contain not less than 97.0 percent and not more than 103.0 percent of $C_{18}H_{24}O_2$.

Packaging and storage—Preserve in tight containers, suitable for maintaining sterile contents, that hold 1 Pellet each.

Reference standard—*USP Estradiol Reference Standard*—Do not dry; determine the *Water* content by *Method I* before using.

Solubility in chloroform—A solution of 25 mg of Pellets in 10 mL of chloroform is clear and practically free from insoluble residue.

Weight variation—Weigh 5 Pellets singly, and calculate the average weight. The average weight is between 95% and 105% of the labeled weight of $C_{18}H_{24}O_2$, and each Pellet weighs between 90% and 110% of the labeled weight of $C_{18}H_{24}O_2$.

Other requirements—Pellets meet the requirements under *Estradiol* and under *Sterility Tests* ⟨71⟩.

Assay—

Standard preparation—Prepare as directed in the *Assay* under *Estradiol Sterile Suspension.*

Assay preparation—Weigh and finely powder not less than 10 Estradiol Pellets. Transfer a portion of the powder, equivalent to about 100 mg of estradiol, to a suitable container, dissolve in a sufficient quantity of a mixture of equal volumes of alcohol and chloroform to make 5.0 mL, and mix.

Procedure—Proceed as directed for *Procedure* in the *Assay* under *Estradiol Sterile Suspension.* Calculate the quantity, in mg, of $C_{18}H_{24}O_2$ in the portion of Pellets taken by the formula:

$$5C(A_U/A_S),$$

in which all terms are as defined therein.

Sterile Estradiol Suspension

» Sterile Estradiol Suspension is a sterile suspension of Estradiol in Water for Injection. It contains not less than 90.0 percent and not more than 110.0 percent of the labeled amount of $C_{18}H_{24}O_2$.

Packaging and storage—Preserve in single-dose or in multiple-dose containers, preferably of Type I glass.

Reference standard—*USP Estradiol Reference Standard*—Do not dry; determine the *Water* content by *Method I* before using.

Identification—Transfer a volume of well-mixed Suspension, equivalent to about 10 mg of estradiol, to a flask, render it acid to bromophenol blue TS with dilute hydrochloric acid (1 in 12), mix thoroughly, and place in an ice bath for 15 minutes. Filter the acidified suspension with suction through a sintered-glass funnel. Wash the crystals of estradiol so isolated with five successive 5-mL portions of water, and dry the funnel and contents at 105° to constant weight. The estradiol so obtained responds to *Identification test A* and meets the requirements of the test for *Melting range* under *Estradiol.*

Uniformity of dosage units ⟨905⟩: meets the requirements.

Other requirements—It meets the requirements under *Injections* ⟨1⟩.

Assay—

Standard preparation—Dissolve a suitable quantity of USP Estradiol RS, accurately weighed, in methanol, and dilute quantitatively and stepwise, if necessary, with methanol to obtain a solution having a known concentration of about 40 µg per mL.

Assay preparation—Transfer an accurately measured volume of well-mixed Sterile Estradiol Suspension, equivalent to about 1 mg of estradiol, to a 100-mL beaker, and add water, if necessary, to obtain a volume of about 5 mL. Add 6 g of purified siliceous earth, mix, and pack the mixture tightly into a 20- × 200-mm chromatographic tube containing in its base a pledget of fine glass wool. Dry-rinse the beaker with about 1 g of purified siliceous earth, add the rinsing to the packed column, and wipe out the beaker with a pledget of glass wool used to top the column. Elute the column with 50 mL of ether that previously has been saturated with water, and collect the eluate in a glass-stoppered, 125-mL conical flask. Evaporate with the aid of gentle heat and a current of air to dryness, add 25.0 mL of methanol to the residue, and mix.

Procedure—Transfer 1.0 mL each of the *Standard preparation* and the *Assay preparation* to separate glass-stoppered, 16- × 150-mm test tubes, and evaporate with the aid of gentle heat and a current of air to dryness. Using a suitable syringe, add 1.0 mL of iron-phenol TS to each tube and to a third, similar tube to provide the blank. Suspend the tubes in a vigorously boiling water bath, mixing them simultaneously after heating for 5 minutes. Remove the tubes after heating in the water bath for a total of 35 minutes, and immediately cool in an ice-water bath. Remove from the ice bath, add 10.0 mL of dilute sulfuric acid (1 in 3) to each tube, mix to obtain homogeneous solutions, and allow to reach room temperature. Concomitantly determine the absorbances of the solutions in 1-cm cells at the wavelength of maximum absorbance at about 520 nm, with a suitable spectrophotometer, against the blank. Calculate the quantity, in mg, of $C_{18}H_{24}O_2$ in each mL of the Suspension taken by the formula:

$$(0.025C/V)(A_U/A_S),$$

in which C is the concentration, in µg per mL, of USP Estradiol RS in the *Standard preparation*, V is the volume, in mL, of Suspension taken, and A_U and A_S are the absorbances of the solutions from the *Assay preparation* and the *Standard preparation*, respectively.

Estradiol Tablets

» Estradiol Tablets contain not less than 90.0 percent and not more than 115.0 percent of the labeled amount of $C_{18}H_{24}O_2$.

Packaging and storage—Preserve in tight, light-resistant containers.

Reference standard—*USP Estradiol Reference Standard*—Do not dry; determine the *Water* content by *Method I* before using.

Identification—Place a quantity of finely powdered Tablets, equivalent to about 4 mg of estradiol, in a screw-capped, 20-mL vial. Add 10 mL of chloroform, and sonicate for 2 minutes. Filter through medium-porosity filter paper. On a suitable thin-layer chromatographic plate (see *Chromatography* ⟨621⟩), coated with a 0.25-mm layer of chromatographic silica gel mixture, apply 20 µL each of this solution and a Standard solution of USP Estradiol RS in chloroform containing 0.4 mg per mL. Allow the spots to dry, and develop the chromatogram in a lined chamber with a solvent system consisting of toluene and acetone (4:1) until the solvent front has moved 10 cm beyond the starting line. Remove the plate from the developing chamber, mark the solvent front, and allow to air-dry. Spray the plate with a mixture of methanol and sulfuric acid (1:1), and heat at 100° for about 5 minutes: the principal spots obtained from the test solution and the Standard solution have the same color and R_f value.

Dissolution ⟨711⟩—

Medium: 0.3% sodium lauryl sulfate in water; 500 mL.

Apparatus 2: 100 rpm.

Time: 60 minutes.

Mobile phase—Prepare a suitable degassed and filtered solution of water and acetonitrile (55:45).

Standard solution—Prepare a solution of USP Estradiol RS in methanol having an accurately known concentration of about 0.02 mg per mL. Dilute aliquots of this solution with *Dissolution Medium* to obtain a final solution having a concentration approximately equal to the expected concentration of drug in the dissolution medium, assuming 100% dissolution.

Test solution—Use a filtered portion of the solution under test from the dissolution vessel.

Chromatographic system (see *Chromatography* ⟨621⟩)—The liquid chromatograph is equipped with a 205-nm detector and a 4.6-mm × 7.5-cm column that contains packing L1. The flow rate is about 1.5 mL per minute. Chromatograph replicate injections of the *Standard preparation*, and record the peak areas as directed under *Procedure:* the relative standard deviation is not more than 2.0%, and the tailing factor is not more than 2.0.

Procedure—Separately inject equal volumes (about 100 µL) of the *Standard solution* and the *Test solution* into the chromatograph, record the chromatograms, and measure the areas for the major peaks. Calculate the quantity of $C_{18}H_{24}O_2$ dissolved by comparison of the peak areas obtained from the *Test solution* and the *Standard solution.*

Tolerances—Not less than 75% (*Q*) of the labeled amount of $C_{18}H_{24}O_2$ is dissolved in 60 minutes.

Uniformity of dosage units ⟨905⟩: meet the requirements.

Assay—

Mobile phase, Internal standard solution, Standard preparation, and *Chromatographic system*—Proceed as directed in the *Assay* under *Estradiol.*

Assay preparation—Weigh and finely powder not less than 10 Estradiol Tablets. Transfer a portion of the powder, equivalent to about 8 mg of estradiol, to a 100-mL volumetric flask. Add 4 mL of water, and swirl. Add 10.0 mL of *Internal standard solution* and about 60 mL of methanol. Shake by mechanical means for 15 minutes, dilute with methanol to volume, mix, and allow the solids to settle. Filter a portion, discarding the first 10 mL of the filtrate. Mix 5.0 mL of the subsequent filtrate with 5.0 mL of methanol and 10.0 mL of water.

Procedure—Proceed as directed for *Procedure* in the *Assay* under *Estradiol.* Calculate the quantity, in mg, of $C_{18}H_{24}O_2$ in the portion of Tablets taken by the formula:

$$0.4C(R_U/R_S),$$

in which the terms are as defined therein.

Estradiol Cypionate

$C_{26}H_{36}O_3$ 396.57

Estra-1,3,5(10)-triene-3,17-diol, (17β)-, 17-cyclopentanepropanoate.

Estradiol 17-cyclopentanepropionate [*313-06-4*].

» Estradiol Cypionate contains not less than 97.0 percent and not more than 103.0 percent of $C_{26}H_{36}O_3$, calculated on the dried basis.

Packaging and storage—Preserve in tight, light-resistant containers.

Reference standard—*USP Estradiol Cypionate Reference Standard*—Dry at 105° for 4 hours before using.

Identification—

A: The infrared absorption spectrum of a potassium bromide dispersion of it, previously dried, exhibits maxima only at the same wavelengths as that of a similar preparation of USP Estradiol Cypionate RS.

B: The ultraviolet absorption spectrum of a 1 in 10,000 solution in alcohol exhibits maxima and minima at the same wavelengths as that of a similar solution of USP Estradiol Cypionate RS, concomitantly measured, and the respective absorptivities, calculated on the dried basis, at the wavelength of maximum absorbance at about 280 nm do not differ by more than 3.0%.

Melting range ⟨741⟩: between 149° and 153°.

Specific rotation ⟨781⟩: between +39° and +44°, calculated on the dried basis, determined in a solution in dioxane containing 200 mg in each 10 mL.

Loss on drying ⟨731⟩—Dry it at 105° for 4 hours: it loses not more than 1.0% of its weight.

Residue on ignition ⟨281⟩: not more than 0.2%.

Assay—

Mobile solvent—Dissolve 0.8 g of ammonium nitrate in 300 mL of water, add 700 mL of acetonitrile, and mix.

Internal standard solution—Prepare a solution of testosterone benzoate in tetrahydrofuran containing 2.0 mg per mL.

Standard preparation—Accurately weigh about 10 mg of USP Estradiol Cypionate RS, and transfer to a 10-mL volumetric flask. Add *Internal standard solution* to volume, and shake vigorously to dissolve.

Assay preparation—Using 10 mg of Estradiol Cypionate, accurately weighed, proceed as directed under *Standard preparation.*

Procedure—Separately inject 10-µL aliquots of the *Assay preparation* and the *Standard preparation* into a suitable high-pressure liquid chromatograph fitted with a 280-nm detector, a 4-mm × 30-cm column containing packing L1 and operated at room temperature. The *Mobile phase* is maintained at a pressure and flow rate capable of giving the required resolution and a suitable elution time. In a suitable system, the resolution factor *R* (see *Chromatography* ⟨621⟩) is not less than 3.0 between the peaks for estradiol cypionate and the internal standard. Five replicate injections of the *Standard preparation* show a relative standard deviation that is not more than 1.5%. Calculate the quantity, in mg, of $C_{26}H_{36}O_3$ in the portion of Estradiol Cypionate taken by the formula:

$$10C(R_U/R_S),$$

in which *C* is the concentration, in mg per mL, of USP Estradiol Cypionate RS in the *Standard preparation*, and R_U and R_S are the ratios of the peak responses of the estradiol cypionate and internal standard peaks obtained from the *Assay preparation* and the *Standard preparation*, respectively.

Estradiol Cypionate Injection

» Estradiol Cypionate Injection is a sterile solution of Estradiol Cypionate in a suitable oil. It contains not less than 90.0 percent and not more than 110.0 percent of the labeled amount of $C_{26}H_{36}O_3$.

Packaging and storage—Preserve in single-dose or in multiple-dose, light-resistant containers, preferably of Type I glass.

Reference standard—*USP Estradiol Cypionate Reference Standard*—Dry at 105° for 4 hours before using.

Identification—Transfer a volume of Injection, equivalent to 5 mg of estradiol cypionate, to a glass-stoppered, 50-mL test tube, and add 30 mL of alcohol. Shake the mixture vigorously for 5 minutes, centrifuge until the two layers have separated, and transfer the alcohol layer, with the aid of a hypodermic syringe, to a 50-mL beaker. Evaporate on a steam bath to dryness, add 5 mL of potassium hydroxide solution (1 in 10), and heat on the steam bath for 15 minutes. Mix 50 mg of sulfanilic acid with 2 mL of 3 *N* hydrochloric acid, warm the mixture, then cool it in ice water, and slowly add, with agitation, 0.3 mL of sodium nitrite solution (1 in 10). Add this solution to the saponified estradiol cypionate: a red color is produced.

Other requirements—It meets the requirements under *Injections* ⟨1⟩.

Assay—

Mobile solvent—Prepare as directed in the *Assay* under *Estradiol Cypionate.*

Internal standard solution—Prepare a solution of testosterone benzoate in tetrahydrofuran containing 0.2 mg per mL.

Standard preparation—Accurately weigh about 10 mg of USP Estradiol Cypionate RS, and transfer to a 100-mL volumetric flask. Add *Internal standard solution* to volume, and shake vigorously to dissolve.

Assay preparation—Using a "to contain" pipet, transfer an accurately measured volume, in mL, of Estradiol Cypionate Injection, equivalent to about 10 mg of estradiol cypionate, to a 100-mL volumetric flask. Rinse the pipet with small portions of *Internal standard solution*, collecting the washings in the volumetric flask. Dilute with *Internal standard solution* to volume, and mix.

Procedure—Proceed as directed for *Procedure* in the *Assay* under *Estradiol Cypionate.* Calculate the quantity, in mg, of $C_{26}H_{36}O_3$ in each mL of the Injection taken by the formula:

$$(100C/V)(R_U/R_S),$$

in which C is the concentration, in mg per mL, of USP Estradiol Cypionate RS in the *Standard preparation*, V is the volume, in mL, of Injection taken, and R_U and R_S are the ratios of the peak responses of the estradiol cypionate and internal standard peaks obtained from the *Assay preparation* and the *Standard preparation*, respectively.

Estradiol Valerate

$C_{23}H_{32}O_3$ 356.50

Estra-1,3,5(10)-triene-3,17-diol(17β)-, 17-pentanoate.

Estradiol 17-valerate [*979-32-8*].

» Estradiol Valerate contains not less than 97.0 percent and not more than 101.0 percent of $C_{23}H_{32}O_3$.

Packaging and storage—Preserve in tight, light-resistant containers.

Reference standard—*USP Estradiol Valerate Reference Standard*—Do not dry; use as is.

Identification—The infrared absorption spectrum of a potassium bromide dispersion of it, prepared by adding chloroform to a mixture of Estradiol Valerate and potassium bromide, then grinding, and drying at 105°, exhibits absorption maxima only at the same wavelengths as that of a similar preparation of USP Estradiol Valerate RS.

Melting range, *Class Ia* ⟨741⟩: between 143° and 150°.

Specific rotation ⟨781⟩: between +41° and +47°, determined in a solution of dioxane containing 250 mg in each 10 mL.

Water, *Method I* ⟨921⟩: not more than 0.1%.

Limit of estradiol—Apply 5 µL of a solution of Estradiol Valerate in acetone, containing 5 mg per mL, and 5 µL of a solution of estradiol in acetone, containing 50 µg per mL, about 2.5 cm from the lower edge of a thin-layer chromatographic plate (see *Chromatography* ⟨621⟩) coated with a 0.25-mm layer of chromatographic silica gel. Develop the chromatogram in a solvent system consisting of a mixture of cyclohexane and ethyl acetate (7:3) in an unlined chamber until the solvent front has moved about 15 cm above the point of application. Remove the plate, dry at 90° for 30 minutes, and spray the plate lightly with a 3 in 10 solution of methanol in sulfuric acid, prepared by cautiously adding sulfuric acid to 30 mL of methanol in a 100-mL volumetric flask, in an ice bath, to volume. Heat the plate at 90° for 30 minutes: any spot in the chromatogram of Estradiol Valerate close to the origin and corresponding to the estradiol spot is not larger nor more intense than that produced by the standard. (The limit is 1.0% of estradiol.)

Free acid—Neutralize 25 mL of alcohol, in a conical flask, with 0.01 *N* sodium hydroxide VS to a faint blue color, using bromothymol blue TS. Accurately weigh 500 mg of Estradiol Valerate, and dissolve it in the neutralized alcohol. Titrate rapidly with 0.01 *N* sodium hydroxide VS to a faint blue color. Each mL of 0.01 *N* sodium hydroxide is equivalent to 1.021 mg of valeric acid. The free acid content, expressed as valeric acid, does not exceed 0.5%.

Ordinary impurities ⟨466⟩—

Test solution: acetone.

Standard solution: acetone.

Eluant: a mixture of cyclohexane and ether (4:1).

Visualization: 5 followed by 1.

Assay—

Mobile phase—Dissolve 0.8 g of ammonium nitrate in 300 mL of water, add 700 mL of acetonitrile, and mix. Filter, and degas. Make adjustments if necessary (see *System Suitability* under *Chromatography* ⟨621⟩).

Internal standard solution—Prepare a solution of testosterone benzoate in tetrahydrofuran having a concentration of about 2.0 mg per mL.

Standard preparation—Dissolve an accurately weighed quantity of USP Estradiol Valerate RS in *Internal standard solution*, and dilute quantitatively with *Internal standard solution* to obtain a solution having a known concentration of about 1 mg of USP Estradiol Valerate RS per mL.

Assay preparation—Transfer about 25 mg of Estradiol Valerate, accurately weighed, to a 25-mL volumetric flask, add *Internal standard solution* to volume, and mix.

Chromatographic system (see *Chromatography* ⟨621⟩)—The liquid chromatograph is equipped with a 280-nm detector and a 4-mm × 30-cm column that contains packing L1. The flow rate is about 2 mL per minute. Chromatograph the *Standard preparation*, and record the peak responses as directed under *Procedure:* the column efficiency determined from the analyte peak is not less than 1100 theoretical plates, the resolution, R, between the analyte and internal standard peaks is not less than 3.0, and the relative standard deviation for replicate injections is not more than 1.5%.

Procedure—Separately inject equal volumes (about 10 µL) of the *Standard preparation* and the *Assay preparation* into the chromatograph, record the chromatograms, and measure the responses for the major peaks. The relative retention times are about 1.2 for testosterone benzoate and 1.0 for estradiol valerate. Calculate the quantity, in mg, of $C_{23}H_{32}O_3$ in the portion of Estradiol Valerate taken by the formula:

$$25C(R_U/R_S),$$

in which C is the concentration, in mg per mL, of USP Estradiol Valerate RS in the *Standard preparation*, and R_U and R_S are the peak response ratios obtained from the *Assay preparation* and the *Standard preparation*, respectively.

Estradiol Valerate Injection

» Estradiol Valerate Injection is a sterile solution of Estradiol Valerate in a suitable vegetable oil. It contains not less than 90.0 percent and not more than 115.0 percent of the labeled amount of $C_{23}H_{32}O_3$.

Packaging and storage—Preserve in single-dose or in multiple-dose, light-resistant containers, preferably of Type I or Type III glass.

Reference standard—*USP Estradiol Valerate Reference Standard*—Do not dry; use as is.

Identification—

Phenol reagent (*Folin-Ciocalteu reagent*)—Dissolve 100 g of sodium tungstate ($Na_2WO_4 . 2H_2O$) and 25 g of sodium molybdate ($Na_2MoO_4 . 2H_2O$) in 700 mL of water, in a 1500-mL flask connected by a standard taper joint to a reflux condenser. Add

50 mL of phosphoric acid and 100 mL of hydrochloric acid, and reflux gently for 10 hours. Cool, and add 150 g of lithium sulfate, 50 mL of water, and 4 to 6 drops of bromine. Boil the mixture without the condenser for 15 minutes to remove the excess bromine, cool, transfer to a 1-liter volumetric flask, dilute with water to volume, and filter: the filtrate is golden yellow in color, and has no greenish tint. Store the filtrate in a tight container in a refrigerator. Dilute 1 volume of the filtrate with 2 volumes of water prior to use as the *Phenol reagent*.

Procedure—Transfer 0.5 mL of Injection to a separator containing 10 mL of solvent hexane and 10 mL of 80 percent methanol. Shake the contents for 2 minutes, and allow the phases to separate. Add 1 mL of *Phenol reagent* and 3 mL of sodium carbonate solution (1 in 5) to 1 mL of the bottom layer, and mix: a blue color develops.

Limit of estradiol—Prepare a solution of estradiol in acetone containing 30.0% of the labeled concentration of the Estradiol Valerate Injection, dilute 1.0 mL with the oil labeled as vehicle for the Injection to 10.0 mL, and mix. Apply 5 µL of Estradiol Valerate Injection as a spot 2.5 cm from the bottom edge of and in the center of one section of a thin-layer chromatographic plate (see *Chromatography* ⟨621⟩) coated with a 0.25-mm layer of chromatographic silica gel, and apply 5 µL of the estradiol solution at the corresponding point in the other section of the plate. Allow the applications to be absorbed by the layer without air-drying, and proceed as directed in the test for *Limit of estradiol* under *Estradiol Valerate*, beginning with "Develop the chromatogram in a solvent system." (The limit of estradiol is 3.0%.)

Other requirements—It meets the requirements under *Injections* ⟨1⟩.

Assay—

Mobile phase, and *Chromatographic system*—Prepare as directed in the *Assay* under *Estradiol Valerate*.

Internal standard solution—Prepare a solution of testosterone benzoate in tetrahydrofuran having a concentration of about 8.0 mg per mL.

Standard preparation—Transfer about 20 mg of USP Estradiol Valerate RS, accurately weighed, to a 25-mL volumetric flask. Add 5.0 mL of the *Internal standard solution*, dilute with tetrahydrofuran to volume, and mix to obtain a solution having a known concentration of about 0.8 mg of USP Estradiol Valerate RS per mL.

Assay preparation—Using a "to contain" pipet, transfer an accurately measured volume of Estradiol Valerate Injection, equivalent to about 20 mg of estradiol valerate, to a 25-mL volumetric flask. Rinse the pipet with small portions of tetrahydrofuran, collecting the washings in the volumetric flask. Add 5.0 mL of *Internal standard solution*, dilute with tetrahydrofuran to volume, and mix.

Procedure—Proceed as directed for *Procedure* in the *Assay* under *Estradiol Valerate*. Calculate the quantity, in mg, of $C_{23}H_{32}O_3$ in each mL of the Injection taken by the formula:

$$25(C/V)(R_U/R_S),$$

in which C is the concentration, in mg per mL, of USP Estradiol Valerate RS in the *Standard preparation*, V is the volume, in mL, of Injection taken, and R_U and R_S are the peak response ratios obtained from the *Assay preparation* and the *Standard preparation*, respectively.

Estriol

$C_{18}H_{24}O_3$ 288.39

Estra-1,3,5(10)-triene-3,16,17-triol, (16α,17β)-.

Estriol [*50-27-1*].

» Estriol contains not less than 97.0 percent and not more than 102.0 percent of $C_{18}H_{24}O_3$, calculated on the dried basis.

Packaging and storage—Preserve in tight containers.

Reference standard—*USP Estriol Reference Standard*—Dry at 105° for 3 hours before using.

Completeness of solution—Dissolve 500 mg in 10 mL of pyridine: the solution is clear and free from undissolved solid.

Identification—

A: The infrared absorption spectrum of a potassium bromide dispersion of it, previously dried, exhibits maxima only at the same wavelengths as that of a similar preparation of USP Estriol RS.

B: The ultraviolet absorption spectrum of a 1 in 10,000 solution in alcohol exhibits maxima and minima at the same wavelengths as that of a similar solution of USP Estriol RS, concomitantly measured.

Specific rotation ⟨781⟩: between +54° and +62°, calculated on the dried basis, determined in a solution in dioxane containing 40 mg in each 10 mL.

Loss on drying ⟨731⟩—Dry it at 105° for 3 hours: it loses not more than 0.5% of its weight.

Residue on ignition ⟨281⟩: not more than 0.1%.

Chromatographic impurities—

Test preparation—Prepare a solution of Estriol in dioxane-water (9:1) to obtain a solution containing 20.0 mg per mL.

Standard solution and Standard dilutions—Prepare a solution of USP Estriol RS in a mixture of dioxane and water (9:1) to obtain a solution containing 20 mg per mL (*Standard solution*). Prepare a series of dilutions of the *Standard solution* in a mixture of dioxane and water (9:1) to obtain solutions containing 0.40, 0.20, 0.10, and 0.05 mg per mL (*Standard dilutions*).

Chromatographic chamber—Line a suitable chamber (see *Chromatography* ⟨621⟩) with absorbent paper, and pour into the chamber 200 mL of developing solvent, prepared by mixing, just prior to use, 90 mL of chloroform, 5 mL of methanol, 5 mL of acetone, and 5 mL of acetic acid. Equilibrate the chamber for 15 minutes before using.

Procedure—Apply 5-µL volumes of the *Test preparation*, *Standard solution*, and each of the four *Standard dilutions* at equidistant points along a line 2.5 cm from one edge of a 20- × 20-cm thin-layer chromatographic plate (see *Chromatography* ⟨621⟩) coated with a 0.25-mm layer of chromatographic silica gel mixture. Place the plate in the *Chromatographic chamber*, seal the chamber, and allow the chromatogram to develop until the solvent front has moved 15 cm above the line of application. Remove the plate, and allow the solvent to evaporate. Spray the plate with a mixture of methanol and sulfuric acid (7:3), then heat the plate at 100° for 15 minutes. The lane of the *Test preparation* exhibits its principal spot at the same R_f value as the principal spot of the *Standard solution*. If spots other than the principal spot are observed in the lane of the *Test preparation*, estimate the concentration of each by comparison with the *Standard dilutions*. The spots from the 0.40-, 0.20-, 0.10-, and 0.05-mg-per-mL dilutions are equivalent to 2.0%, 1.0%, 0.5%, and 0.25% of impurities, respectively. The requirements of the test are met if the sum of impurities in the *Test preparation* is not greater than 2.0%.

Assay—Dissolve about 50 mg of Estriol, accurately weighed, in alcohol to make 100.0 mL, and mix. Dilute 10.0 mL of this solution with alcohol to 100.0 mL. Similarly, dissolve a suitable quantity of USP Estriol RS, accurately weighed, in alcohol to obtain a Standard solution having a known concentration of about 50 µg per mL. Concomitantly determine the absorbances of both solutions in 1-cm cells at the wavelength of maximum absorbance at about 281 nm. Calculate the quantity, in mg, of $C_{18}H_{24}O_3$ in the portion of Estriol taken by the formula:

$$C(A_U/A_S),$$

in which C is the concentration, in µg per mL, of USP Estriol RS in the Standard solution, and A_U and A_S are the absorbances of the solution of Estriol and the Standard solution, respectively.

Conjugated Estrogens

» Conjugated Estrogens is a mixture of sodium estrone sulfate and sodium equilin sulfate, derived wholly or in part from equine urine or synthetically from Estrone and Equilin. It may contain other conjugated estrogenic substances of the type excreted by pregnant mares. It is a dispersion of the estrogenic substances on a suitable powdered diluent.

Conjugated Estrogens contains not less than 50.0 percent and not more than 63.0 percent of sodium estrone sulfate and not less than 22.5 percent and not more than 32.5 percent of sodium equilin sulfate, and the total of sodium estrone sulfate and sodium equilin sulfate is not less than 80 percent of the labeled content of Conjugated Estrogens.

Packaging and storage—Preserve in tight containers.

Labeling—Label it to state the content of Conjugated Estrogens on a weight to weight basis.

Reference standards—*USP Estrone Reference Standard*—Dry at 105° for 3 hours before using. *USP Equilin Reference Standard*—Do not dry before using. Store tightly closed, in a cool place. *USP 17α-Dihydroequilin Reference Standard*—Do not dry before using. Store in a cold place, protected from light. Store the contents of the opened ampul in a tightly closed container, under nitrogen, protected from light, in a cold place. *USP Testosterone Reference Standard*—Dry in vacuum over phosphorus pentoxide for 4 hours before using.

Identification—The following results are obtained with respect to the *Assay preparation* treated as directed for *Procedure* in the *Assay*.

A: The chromatogram exhibits distinctive peaks for estrone and equilin at relative retention times corresponding to those exhibited in the chromatogram of the *Standard preparation.*

B: The chromatogram of Conjugated Estrogens derived wholly or in part from equine urine exhibits a peak for 17α-dihydroequilin at the same relative retention time as that exhibited in the chromatogram of the *Standard preparation.* Additional peaks or shoulders, if present, correspond to α-estradiol (0.24), 17β-dihydroequilin (0.35), equilenin (3.0), β-estradiol (0.28), and 9-dehydroestrone (1.8), the numbers in parentheses referring to the approximate retention times relative to that of testosterone.

17α-Dihydroequilin—Where a peak corresponding to 17α-dihydroequilin is detected in *Identification test B*, proceed as directed under *Assay*, but calculate r_U and r_S from the peak areas at the same relative retention times for this compound in the chromatograms from the *Assay preparation* and the *Standard preparation*, respectively: not more than 20.0% is found.

Free steroids—Proceed as directed in the *Assay*, but do not add the sulfatase enzyme preparation. Prepare a reagent blank in the same manner. Calculate the combined areas of all peaks other than testosterone and the ratio, r_{FS}, of the combined areas to the area of testosterone peak, correcting for any reagent blank peaks. The ratio r_{FS}/r_S relative to the estrone from the *Standard preparation* is not greater than 0.052 (3.0% of *free steroids*).

Assay—

Internal standard solution—Dissolve a quantity of USP Testosterone RS in alcohol to obtain a solution containing about 100 µg per mL.

Standard preparation—Prepare by quantitative and stepwise dilution an alcohol solution having a known concentration, in each mL, of about 180 µg of USP Estrone RS, about 90 µg of USP Equilin RS, and about 45 µg of USP 17α-Dihydroequilin RS. Pipet 1.0 mL of this solution and 1.0 mL of *Internal standard solution* into a suitable centrifuge tube fitted with a tight screw-cap or stopper. Evaporate the mixture with the aid of a stream of nitrogen to dryness, maintaining the temperature below 50°. To the dry residue add 15 µL of dried pyridine and 65 µL of trimethylchlorosilanebis(trimethylsilyl)trifluoroacetamide solution (1 in 100). Immediately cover the tube tightly, mix, and allow to stand for 15 minutes before injecting into the gas chromatograph.

Acetate buffer, pH 5.2—Mix 79 mL of sodium acetate TS with 21 mL of 1 *N* acetic acid, dilute with water to 500 mL, and mix. Use within 1 month.

Assay preparation—Transfer a quantity of Conjugated Estrogens, accurately weighed and equivalent to about 2 mg of total conjugated estrogens, to a 50-mL centrifuge tube, fitted with a polytef-lined screw cap, containing 15 mL of *Acetate buffer, pH 5.2*, and 1 g of barium chloride. Add 1 *N* acetic acid or *Acetate buffer, pH 5.2*, as necessary, to adjust the mixture to a pH of 5.2. Cap the tube tightly, and shake by mechanical means for 30 minutes. Place in a sonic bath for 30 seconds, then shake for an additional 30 minutes. Add a suitable sulfatase enzyme preparation equivalent to 2500 Units, and shake for 20 minutes in a water bath maintained at 50°. Add 15.0 mL of ethylene dichloride to the warm mixture, again cap the tube, and shake by mechanical means for 15 minutes. Centrifuge for 10 minutes or until the lower layer is clear. Transfer as much of the organic phase as possible, and dry by filtering rapidly through a filter consisting of a pledget of dry glass wool and about 5 g of anhydrous sodium sulfate in a small funnel. Protect from loss by evaporation. Transfer 3.0 mL of the solution to a suitable centrifuge tube fitted with a tight screw-cap or stopper. Add 1.0 mL of *Internal standard solution.* Proceed as directed under *Standard preparation*, beginning with "Evaporate the mixture."

Procedure—Inject about 3 µL of the *Standard preparation* into a suitable gas chromatograph equipped with a flame-ionization detector, preferably having a glass column measuring 1.8 m × 3.5 mm packed with 2 percent (w/w) liquid phase G4 on support S1AB. Where 1.8-m × 2-mm columns are used, dilute the *Internal standard solution*, the *Standard preparation*, and the *Assay preparation* 1 in 5 with the indicated solvents. Maintain the column at about 200°, and use moisture and oxygen-free carrier gas. Adjust the operating conditions to keep the elution time through equilin to between 32 and 40 minutes. The observed relative retention times of USP 17α-Dihydroequilin RS, USP Estrone RS, and USP Equilin RS estrogens versus testosterone are approximately 0.30, 1.45, and 1.61, respectively. In a suitable chromatogram, the resolution factor, *R* (see *Chromatography* ⟨621⟩), is not less than 1.2 between the peaks for estrone and equilin and the S_R (%) for R_S from the estrone peak area ratios of the *Standard preparation* is not greater than 3.0 on the basis of not less than four replicate injections. Inject about 3 µL of the *Assay preparation* under the same conditions. Calculate the peak area ratios R_U and R_S for both estrone and equilin relative to testosterone from both the *Assay preparation* and the *Standard preparation.* Calculate the amount, in mg, of each sodium estrogen sulfate (estrone, equilin) in the quantity of Conjugated Estrogens taken by the formula:

$$(0.005)(1.381)(C_i)(R_U/R_S),$$

in which C_i is the concentration, in µg per mL, of USP Estrogen RS (Estrone, Equilin) in the alcohol solution, and 1.381 is the factor converting free estrogen to the conjugate sodium salt.

Conjugated Estrogens Tablets

» Conjugated Estrogens Tablets contain not less than 73.0 percent and not more than 95.0 percent of the labeled amount of conjugated estrogens as the total of sodium estrone sulfate and sodium equilin sulfate. The ratio of sodium equilin sulfate to sodium estrone sulfate in the Tablets is not less than 0.35 and not more than 0.65.

Packaging and storage—Preserve in well-closed containers.

Reference standards—*USP Estrone Reference Standard*—Dry at 105° for 3 hours before using. *USP Equilin Reference Standard*—Do not dry before using. Store tightly closed, in a cool place. *USP 17α-Dihydroequilin Reference Standard*—Do not dry before using. Store in a cold place, protected from light. Store the contents of the opened ampul in a tightly closed container, under nitrogen, protected from light, in a cold place. *USP Testosterone Reference Standard*—Dry in vacuum over phosphorus pentoxide for 4 hours before using.

Identification—Tablets respond to the *Identification tests* under *Conjugated Estrogens.*

Dissolution ⟨711⟩—

Medium: simulated gastric fluid TS, without pepsin; 900 mL.

Apparatus—Proceed as directed under *Disintegration* ⟨701⟩ for *Uncoated Tablets,* except to use only one Tablet in the basket assembly, and to use a basket rack assembly that contains no metal parts other than the 10-mesh stainless steel wire cloth.

Time: 1 hour.

Mobile phase—Prepare a filtered and degassed mixture of 0.025 *M* monobasic potassium phosphate and acetonitrile (75:25). Make adjustments if necessary (see *System Suitability* under *Chromatography* ⟨621⟩).

Standard solution—Accurately weigh and finely grind not less than 20 Conjugated Estrogens Tablets, and mix the powder. Transfer an accurately weighed portion of the powder, equivalent to one average tablet weight, to a 900-mL volumetric flask, add 500 mL of *Dissolution Medium,* shake for 15 minutes, dilute with *Dissolution Medium* to volume, and mix. The quantity of conjugated estrogens in the portion taken is calculated with the *Assay* value.

System suitability solution—Prepare a solution containing about 0.6 mg of USP Estrone RS per mL in methanol. Dilute this solution 1000-fold with *Mobile phase.*

Chromatographic system (see *Chromatography* ⟨621⟩)—The liquid chromatograph is equipped with a 10-nm detector and a 4-mm × 30-cm column that contains packing L1. The flow rate is about 1 mL per minute. Chromatograph the *System suitability solution.* The chromatographic run time is not less than than 1.1 times the retention time of the estrone peak. Chromatograph replicate injections of the *Standard solution,* and measure the peak responses as directed under *Procedure:* the relative standard deviation for the estrone sulfate peak is not more than 1.5%, and the resolution between equilin sulfate and estrone sulfate is not less than 1.0.

Procedure—Separately inject equal volumes (about 200 µL) of the *Standard solution* and a filtered portion of the solution under test into the chromatograph, record the chromatograms, and measure the peak responses of the estrone sulfate peaks. The relative retention times are about 0.9 for equilin sulfate and 1.0 for estrone sulfate, the estrone sulfate peak being the last major peak in the chromatogram. Calculate the quantity of conjugated estrogens dissolved by the formula:

$$900C(r_U / r_S),$$

in which C is the concentration, in mg per mL, of conjugated estrogens in the *Standard solution,* and r_U and r_S are the responses of the estrone sulfate peaks obtained from the solution under test and the *Standard solution,* respectively.

Tolerance—Not less than 75% (Q) of the labeled amount of conjugated estrogens is dissolved in 1 hour.

Uniformity of dosage units—Assay 10 individual Tablets as directed in the *Assay,* and calculate the average content of conjugated estrogens, as the average of the total contents of sodium estrone sulfate and sodium equilin sulfate, in the 10 Tablets. The requirements are met if the content of each of the Tablets is not less than 85.0 percent and not more than 115.0 percent of the average content of conjugated estrogens. If the content of not more than 2 Tablets falls outside the range of 85.0 percent to 115.0 percent of the average content but not outside the range of 75.0 percent to 125.0 percent, assay an additional 20 Tablets. The requirements are met if the content of not more than 2 of the 30 Tablets falls outside the limits of 85.0 percent and 115.0 percent of that average, and no unit is outside the range of 75.0 percent to 125.0 percent of the average content.

Assay—If the Conjugated Estrogens Tablets are sugar-coated, carefully remove with water the color and sugar coatings, leaving the shellac coating intact, and dry under nitrogen. Weigh not less than 20 of the Tablets, and finely powder. Using a suitable portion of the powder, proceed as directed in the *Assay* under *Conjugated Estrogens.*

Esterified Estrogens

» Esterified Estrogens is a mixture of the sodium salts of the sulfate esters of the estrogenic substances, principally estrone. It is a dispersion of the estrogenic substances on a suitable powdered diluent. The content of total esterified estrogens is not less than 90.0 percent and not more than 110.0 percent of the labeled amount.

Esterified Estrogens contains not less than 75.0 percent and not more than 85.0 percent of sodium estrone sulfate, and not less than 6.0 percent and not more than 15.0 percent of sodium equilin sulfate, in such proportion that the total of these two components is not less than 90.0 percent, all percentages being calculated on the basis of the total esterified estrogens content.

Packaging and storage—Preserve in tight containers.

Labeling—Label it to state the content of Esterified Estrogens on a weight to weight basis.

Reference standards—*USP Estrone Reference Standard*—Dry at 105° for 3 hours before using. *USP Equilin Reference Standard*—Do not dry before using. Store tightly closed, in a cool place. *USP Testosterone Reference Standard*—Dry in vacuum over phosphorus pentoxide for 4 hours before using.

Identification—It responds to *Identification test A* under *Conjugated Estrogens.*

Free steroids—Proceed with Esterified Estrogens as directed in the test for *Free steroids* under *Conjugated Estrogens.* The limit is 3.0% of free steroids.

Assay—Proceed with Esterified Estrogens as directed in the *Assay* under *Conjugated Estrogens.*

Esterified Estrogens Tablets

» Esterified Estrogens Tablets contain not less than 90.0 percent and not more than 115.0 percent of the labeled amount of esterified estrogens as the total of sodium estrone sulfate and sodium equilin sulfate. The ratio of sodium equilin sulfate to sodium estrone sulfate is not less than 0.071 and not more than 0.20.

Packaging and storage—Preserve in well-closed containers.

Reference standards—*USP Estrone Reference Standard*—Dry at 105° for 3 hours before using. *USP Equilin Reference Standard*—Do not dry before using. Store tightly closed, in a cool place. *USP Testosterone Reference Standard*—Dry in vacuum over phosphorus pentoxide for 4 hours before using.

Identification—Tablets respond to the *Identification test* under *Esterified Estrogens.*

Disintegration ⟨701⟩: 1 hour.

Uniformity of dosage units—Assay 10 individual Tablets as directed in the *Assay,* and calculate the average content of esterified estrogens, as the average of the total contents of sodium estrone sulfate and sodium equilin sulfate, in the 10 Tablets. The requirements are met if the content of each of the Tablets is not less than 85.0 percent and not more than 115.0 percent of the average content of esterified estrogens. If the content of not more than 2 Tablets falls outside the range of 85.0 percent to 115.0 percent of the average content but not outside the range of 75.0 percent to 125.0 percent, assay an additional 20 Tablets. The requirements are met if the content of not more than 2 of the 30 Tablets falls outside the limits of 85.0 percent and 115.0 percent of the average, and no unit is outside the range of 75.0 percent to 125.0 percent of the average content.

Assay—Weigh and finely powder not less than 20 Esterified Estrogens Tablets. Using a suitable portion of the powder, proceed as directed in the *Assay* under *Conjugated Estrogens*.

Estrone

$C_{18}H_{22}O_2$ 270.37
Estra-1,3,5(10)-trien-17-one, 3-hydroxy-.
3-Hydroxyestra-1,3,5(10)-trien-17-one [*53-16-7*].

» Estrone contains not less than 97.0 percent and not more than 103.0 percent of $C_{18}H_{22}O_2$, calculated on the dried basis.

Packaging and storage—Preserve in tight, light-resistant containers.

Reference standard—*USP Estrone Reference Standard*—Dry at 105° for 3 hours before using.

Clarity of solution—Add 100 mg to 100 mL of 1 *N* sodium hydroxide in a 125-mL conical flask, heat on a steam bath until solution is complete, then cool, and transfer to a 100-mL color-comparison tube: the solution is clear.

Identification—

A: The infrared absorption spectrum of a potassium bromide dispersion of it, previously dried, exhibits maxima only at the same wavelengths as that of a similar preparation of USP Estrone RS.

B: Prepare a 1 in 20,000 solution in alcohol by heating on a steam bath and cooling to room temperature: the ultraviolet absorption spectrum of this solution exhibits maxima and minima at the same wavelengths as that of a similar solution of USP Estrone RS, concomitantly measured.

Specific rotation ⟨781⟩: between +158° and +165°, determined in a solution in dioxane containing 100 mg of Estrone, previously dried, in each 10 mL.

Loss on drying ⟨731⟩—Dry it at 105° for 3 hours: it loses not more than 0.5% of its weight.

Residue on ignition ⟨281⟩: not more than 0.5%.

Equilenin and equilin—Dissolve 10 mg in sufficient alcohol to make 50 mL. Transfer 5 mL of the solution to a small beaker. Add 5 mL of a buffer solution prepared by dissolving 2 mL of glacial acetic acid and 13.3 g of anhydrous sodium acetate in water to make 100 mL, warm to about 50°, and add 1 mL of a freshly prepared 1 in 200 solution of 2,6-dibromoquinone-chlorimide in alcohol. Mix, and allow to stand for 30 minutes. Transfer the solution to a small separator, add 10 mL of chloroform and 20 mL of 1 *N* sodium hydroxide, and shake vigorously for 2 minutes. Separate the chloroform layer, and filter rapidly through a dry filter paper into a dry test tube, discarding the first 2 mL of the filtrate. Viewed transversely against a white background, the chloroform filtrate shows no more red color than that produced by similarly treating 5 mL of an alcohol solution containing 20 µg of equilenin.

Ordinary impurities ⟨466⟩—

Test solution: acetone.

Standard solution: acetone.

Eluant: a mixture of toluene and ethyl acetate (70:30), in a nonequilibrated chamber.

Visualization: 5.

Assay—

Mobile phase—Prepare a filtered and degassed mixture of acetonitrile and 0.05 *M* monobasic potassium phosphate (1:1). Make adjustments if necessary (see *System Suitability* under *Chromatography* ⟨621⟩).

Standard preparation—Transfer about 20 mg of USP Estrone RS, accurately weighed, to a 100-mL volumetric flask, add methanol to volume, and mix. If necessary, sonicate to aid solution. Transfer 5 mL of this solution to a 25-mL volumetric flask, dilute with *Mobile phase* to volume, and mix to obtain a *Standard preparation* having a known concentration of about 40 µg of USP Estrone RS per mL.

Assay preparation—Transfer about 20 mg of Estrone, accurately weighed, to a 100-mL volumetric flask, add methanol to volume, and mix. If necessary, sonicate to aid solution. Transfer 5.0 mL of this solution to a 25-mL volumetric flask, dilute with *Mobile phase* to volume, and mix.

Chromatographic system (see *Chromatography* ⟨621⟩)—The liquid chromatograph is equipped with a 280-nm detector and a 4-mm × 15-cm column that contains 5-µm packing L1. The flow rate is about 1 mL per minute. Chromatograph the *Standard preparation*, and record the peak responses as directed under *Procedure:* the column efficiency determined from the analyte peak is not less than 1500 theoretical plates, the tailing factor for the analyte peak is not more than 2.0, and the relative standard deviation for replicate injections is not more than 2.0%.

Procedure—Separately inject equal volumes (about 50 µL) of the *Standard preparation* and the *Assay preparation* into the chromatograph, record the chromatograms, and measure the responses for the major peaks. Calculate the quantity, in mg, of $C_{18}H_{22}O_2$ in the portion of Estrone taken by the formula:

$$0.5C(r_U/r_S),$$

in which C is the concentration, in µg per mL, of USP Estrone RS in the *Standard preparation*, and r_U and r_S are the peak responses obtained from the *Assay preparation* and the *Standard preparation*, respectively.

Estrone Injection

» Estrone Injection is a sterile solution of Estrone in a suitable oil. It contains not less than 90.0 percent and not more than 115.0 percent of the labeled amount of $C_{18}H_{22}O_2$.

Packaging and storage—Preserve in single-dose or in multiple-dose containers, preferably of Type I glass.

Identification—Dissolve the residue obtained in the *Assay* in sufficient alcohol to obtain a solution containing 500 µg of estrone in each mL. Transfer to an acetylation flask, and evaporate to dryness. Add 10 mg of hydroxylamine hydrochloride, 0.20 mL of glacial acetic acid, and 5 mL of alcohol, and reflux for 5 hours. Dilute with 5 mL of water, filter, and recrystallize the precipitate from hot alcohol: the estrone oxime so obtained melts between 236° and 242°, the procedure for *Class I* being used (see *Melting Range or Temperature* ⟨741⟩).

Other requirements—It meets the requirements under *Injections* ⟨1⟩.

Assay—[NOTE—Use only water as a lubricant for the separators used in this assay, and complete the assay without interruption other than at the stage of obtaining the dry residue from the benzene extract.] Transfer a volume of Estrone Injection, equivalent to about 10 mg of estrone, to a suitable separator containing 25 mL, or not less than twice the volume of the Injection taken, of solvent hexane. Add 10 mL of sodium hydroxide solution (1 in 10), shake vigorously for 2 minutes, and allow the layers to separate completely. Transfer the aqueous layer to a second 125-mL separator, and repeat the extraction of the solvent hexane with two additional, successive 10-mL portions of the sodium hydroxide solution, adding each extract to the second separator. Complete the alkaline extraction as quickly as possible, since long standing in strongly alkaline solution may cause decomposition of the estrone. Wash the combined alkaline extracts with 25 mL of solvent hexane. Using dilute sulfuric acid (1 in 2), acidify the combined alkaline extracts until acid to litmus. Cool thoroughly, add 25 mL of benzene, shake carefully for 1 minute, and allow the layers to separate. Transfer the acid layer to another 125-mL separator, and extract with a second 25-mL portion of benzene. Discard the acid layer. Extract the benzene layers with two 5-mL portions of sodium carbonate TS and two 5-mL portions of water. Discard the aqueous layers. Transfer the benzene solutions to a beaker with the aid of benzene, and evaporate on a steam bath with the aid of a current of air to dryness.

Dissolve the residue from the benzene extract in a small quantity of chloroform, warming, if necessary, and completely transfer the solution, with the aid of a few mL of chloroform, to a 20- × 150-mm test tube. Carefully evaporate the chloroform on a steam bath with the aid of a current of air. Add 100 mg of trimethylacethydrazide ammonium chloride and 500 μL of glacial acetic acid to the test tube, insert the stopper loosely, and heat in a boiling water bath for 5 minutes. Cool the reaction mixture in an ice bath, dissolve in a small volume of cold water, and completely transfer, with the aid of a small volume of water, to a 125-mL separator containing 25 mL of cold water. Neutralize the solution to litmus with 1 *N* sodium hydroxide (approximately 6 mL), and wash at once with three 15-mL portions of chloroform. Combine the chloroform washings in another separator, and wash them with 5 mL of water. Discard the chloroform, and add the wash water to the first separator. Add 2 mL of dilute sulfuric acid (1 in 2), and allow to remain at room temperature for 2 hours. Add 15 mL of chloroform, shake vigorously for 1 minute, and allow the layers to separate. Transfer the chloroform layer to another separator, and repeat the extraction of the water layer with three additional, successive 15-mL portions of chloroform. Wash the combined chloroform extracts with 5 mL of water, filter through chloroform-washed cotton into a beaker, evaporate to a small volume, and transfer completely, with the aid of several small portions of chloroform, to a tared 25-mL beaker. Evaporate on a steam bath with the aid of a current of air to dryness, and dry the residue of estrone in a vacuum desiccator to constant weight: the weight of the residue, corrected for the residue of a reagent blank similarly prepared, indicates the amount of $C_{18}H_{22}O_2$ in the volume of Injection taken.

Sterile Estrone Suspension

» Sterile Estrone Suspension is a sterile suspension of Estrone in Water for Injection. It contains not less than 90.0 percent and not more than 115.0 percent of the labeled amount of $C_{18}H_{22}O_2$.

Packaging and storage—Preserve in single-dose or in multiple-dose containers, preferably of Type I glass.

Reference standards—*USP Estrone Reference Standard*—Dry at 105° for 3 hours before using. *USP Progesterone Reference Standard*—Dry in vacuum over silica gel for 4 hours before using.

Identification—Transfer a volume of Suspension, equivalent to about 5 mg of estrone, to a glass-stoppered centrifuge tube, and add 2.5 mL of a mixture of ether and benzene (1:1). Shake for 2 minutes, and allow insoluble matter to settle, centrifuging, if necessary, to obtain a clear supernatant solution. On a suitable thin-layer chromatographic plate (see *Chromatography* ⟨621⟩), coated with a 0.25-mm layer of chromatographic silica gel, apply 5 μL each of this supernatant solution and a 1 in 500 solution of USP Estrone RS in a mixture of ether and benzene (1:1). Allow the spots to dry, and develop the chromatogram in a solvent system consisting of a mixture of benzene and acetone (4:1) until the solvent front has moved about three-fourths of the length of the plate. Remove the plate from the developing chamber, mark the solvent front, and allow the solvent to evaporate. Spray the plate with a mixture of dehydrated alcohol and sulfuric acid (3:1), and heat in an oven at 105° for 10 minutes: the R_f value and appearance (pale orange to amber by direct observation in daylight, and fluorescing pale yellow-green under long-wavelength ultraviolet light) of the principal spot obtained from the test solution correspond to those obtained from the Standard solution.

Uniformity of dosage units ⟨905⟩: meets the requirements.

Other requirements—It meets the requirements under *Injections* ⟨1⟩.

Assay—

Mobile phase, *Standard preparation*, and *Chromatographic system*—Prepare as directed in the *Assay* under *Estrone*.

Assay preparation—Transfer an accurately measured volume of the well-mixed Suspension, equivalent to about 10 mg of estrone to a 50-mL volumetric flask. Add 30 mL of methanol and swirl for 5 minutes. Dilute with methanol to volume, and mix. Transfer 5.0 mL of this solution to a 25-mL volumetric flask, dilute with *Mobile phase* to volume, and mix.

Procedure—Proceed as directed for *Procedure* in the *Assay* under *Estrone*. Calculate the quantity, in mg, of $C_{18}H_{22}O_2$ in each mL of Sterile Estrone Suspension taken by the formula:

$$0.5(C/V)(r_U/r_S),$$

in which V is the volume, in mL, of the Sterile Suspension taken and the other terms are as defined therein.

Estropipate

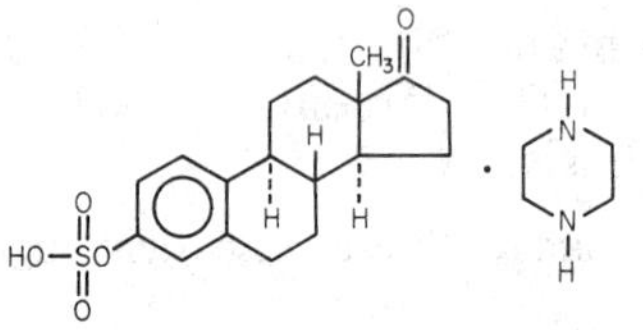

$C_{18}H_{22}O_5S.C_4H_{10}N_2$ 436.56

Estra-1,3,5(10)-trien-17-one, 3-(sulfooxy)-, compd. with piperazine (1:1).

Estrone hydrogen sulfate compound with piperazine (1:1) [*7280-37-7*].

» Estropipate contains not less than 97.0 percent and not more than 103.0 percent of $C_{18}H_{22}O_5S.C_4H_{10}N_2$, calculated on the dried basis.

Packaging and storage—Preserve in tight containers.

Reference standards—*USP Estropipate Reference Standard*—Dry at 105° for 1 hour before using. *USP Estrone Reference Standard*—Dry at 105° for 3 hours before using.

Identification—The infrared absorption spectrum of a potassium bromide dispersion of it, previously dried, exhibits maxima only at the same wavelengths as that of a similar preparation of USP Estropipate RS.

Loss on drying ⟨731⟩—Dry it at 105° for 1 hour: it loses not more than 1.0% of its weight.

Residue on ignition ⟨281⟩: not more than 0.5%.

Free estrone—

Stock impurity standard preparation—Weigh accurately 25.0 mg of USP Estrone RS into a 100-mL volumetric flask, dilute with spectrophotometric-grade methanol to volume, and sonicate to achieve complete solution.

Impurity standard preparation—Weigh accurately 25.0 mg of USP Estropipate RS into a 25-mL volumetric flask, add 2.0 mL of *Stock impurity standard preparation*, dilute with spectrophotometric-grade methanol to volume, and sonicate to achieve complete solution.

Standard preparation—Weigh accurately 25.0 mg of USP Estropipate RS into a 25-mL volumetric flask, dilute with spectrophotometric-grade methanol to volume, and sonicate to achieve complete solution.

Test preparation—Using a portion of Estropipate, accurately weighed, prepare as directed under *Standard preparation*.

Mobile phase—Mix 650 mL of 0.025 *M* potassium dihydrogen phosphate with 350 mL of spectrophotometric-grade acetonitrile. Filter the solution through a membrane filter having a porosity of 1 μm or less, and degas at a pressure of less than 100 mm of mercury until no further bubbles appear. The concentration of acetonitrile may be varied to meet system suitability requirements and to provide a suitable elution time for all components.

Chromatographic system—Typically, a high-pressure liquid chromatograph, operated at room temperature, is fitted with a 30-cm × 3.9-mm stainless steel column that contains packing L1. The mobile phase is maintained at a pressure and flow rate (approximately 1.5 mL per minute) capable of giving the required resolution (see *System suitability test*) and a suitable elution time. An ultraviolet detector that monitors absorption at a wavelength of 213 nm is used with a recorder adjusted such that approximately 0.04 absorbance unit gives a full-scale reading.

System suitability test—Chromatograph two injections of the *Impurity standard preparation*, and determine that after the injection front the small peak (estrone) after the major peak does not differ in peak response between the duplicate injections by more than 4%. Also determine that the small peak after the major component has a retention time relative to the major component of approximately 5.5. (For a particular column, resolution may be increased by decreasing the amount of acetonitrile in the *Mobile phase*.)

Procedure—Inject separately 5.0-µL portions of the *Standard preparation*, *Impurity standard preparation*, and *Test preparation* into the high-pressure liquid chromatograph by means of a suitable sampling valve or high-pressure microsyringe. Measure the peak responses for the estrone peak relative to the estropipate peak obtained with the *Standard preparation*, the *Impurity standard preparation*, and the *Test preparation*. Calculate the percentage of free estrone by the formula:

$$2.5(C/W)(H_U/H_S),$$

in which H_U and H_S are the measured peak heights of the impurity (estrone) in the *Test preparation* and the *Impurity standard preparation* corrected for the peak height of estrone in the *Standard preparation*, respectively, W is the weight, in mg, of estropipate in the *Test preparation*, and C is the concentration, in µg per mL, of USP Estrone RS in the *Impurity standard preparation*. Not more than 2.0% is found.

Assay—

Standard preparation—Prepare as directed under *Free estrone*.

Assay preparation—Prepare as directed for *Test preparation* under *Free estrone*.

Chromatographic system—Use the same system as in test for *Free estrone*. Adjust the recorder so that approximately 0.4 absorbance unit gives a full-scale reading.

System suitability test—Chromatograph two injections of the *Standard preparation*, and determine that only one major peak is observed after the injection front. The peak responses between the duplicate injections for the major peak do not differ by more than 3%.

Procedure—Inject 5.0 µL of the *Assay preparation* and the *Standard preparation* into the high-pressure liquid chromatograph by means of a suitable sampling valve or high-pressure microsyringe. Measure the peak heights for the respective estropipate peak (it is actually an estrone sulfate peak) obtained with the *Assay preparation* and the *Standard preparation*. Calculate the quantity, in mg, of $C_{18}H_{22}O_5S \cdot C_4H_{10}N_2$ in the portion of Estropipate taken by the formula:

$$25C(H_U/H_S),$$

in which H_U and H_S are the peak heights obtained with the *Assay preparation* and the *Standard preparation*, respectively, and C is the concentration, in mg per mL, of USP Estropipate RS in the *Standard preparation*.

Estropipate Vaginal Cream

» Estropipate Vaginal Cream contains not less than 90.0 percent and not more than 120.0 percent of the labeled amount of $C_{18}H_{22}O_5S \cdot C_4H_{10}N_2$, in a suitable cream base.

Packaging and storage—Preserve in collapsible tubes.

Reference standard—*USP Estropipate Reference Standard*—Dry at 105° for 1 hour before using.

Identification—

Arsenomolybdate spray reagent—Dissolve 25 g of ammonium molybdate in 450 mL of water, add 21 mL of sulfuric acid, and mix. Add 3 g of dibasic sodium arsenate heptahydrate dissolved in 25 mL of water, mix, and incubate at 37 ± 2° for 24 to 48 hours. Store, protected from light.

Procedure—Transfer a portion of Vaginal Cream, equivalent to about 8 mg of estropipate, to a container. Add 20 mL of methanol, stir to obtain a homogeneous mixture, and filter through filter paper. On a suitable thin-layer chromatographic plate (see *Chromatography* ⟨621⟩) coated with a 0.25-mm layer of chromatographic silica gel mixture, apply 30 µL of the filtrate and 30 µL of a Standard solution of USP Estropipate RS in methanol containing about 400 µg per mL. Allow the spots to dry, and develop the chromatogram in a solvent system consisting of a mixture of chloroform and methanol (10:8) until the solvent front has moved about 14 cm from the origin. Remove the plate from the developing chamber, mark the solvent front, and allow the solvent to evaporate. Locate the spots by lightly spraying with *Arsenomolybdate spray reagent* and heating at 105° for about 20 minutes: the R_f values of the principal spots obtained from the test solution correspond to those obtained from the Standard solution.

Minimum fill ⟨755⟩: meets the requirements.

Assay—

Internal standard concentrate, *Standard solution concentrate*, and *Internal standard solution*—Prepare as directed in the *Assay* under *Estropipate Tablets*.

Mobile phase—Dissolve 13.6 g of monobasic potassium phosphate in 1000 mL of water, and mix. Mix 645 mL of the resulting solution with 177 mL of methanol, add 177 mL of acetonitrile with stirring, and mix. Filter, and degas. Make adjustments if necessary (see *System Suitability* under *Chromatography* ⟨621⟩).

Piperazine solution—Dissolve about 4.4 g of piperazine in 1000 mL of water.

Internal standard identification solution—Transfer 5.0 mL of *Internal standard solution* to a 25-mL volumetric flask, dilute with *Piperazine solution* to volume, and mix.

Standard preparation—Transfer 5.0 mL of *Standard solution concentrate* to a 100-mL volumetric flask, dilute with *Piperazine solution* to volume, and mix. Transfer 5.0 mL of this solution to a 25-mL volumetric flask, add 5.0 mL of *Internal standard solution*, dilute with *Piperazine solution* to volume, and mix. Filter a portion of this solution through a 0.4-µm porosity filter. The concentration of estropipate in the *Standard preparation* is about 0.01 mg per mL.

Assay preparation—Transfer an accurately weighed portion of Estropipate Vaginal Cream, equivalent to about 1.5 mg of estropipate, to a 125-mL separator. Add 50 mL of chloroform and 20 mL of *Piperazine solution*, and extract for 3 minutes. Collect the chloroform layer in a second separator, and repeat the extraction of the aqueous phase with an additional 20-mL portion of chloroform. Combine the chloroform extracts and wash them with 20 mL of *Piperazine solution*, shaking for 3 minutes. Discard the chloroform layer, and transfer the two aqueous phases to a 100-mL volumetric flask, rinsing the separators with additional *Piperazine solution*. Dilute the aqueous solution with *Piperazine solution* to volume, and mix. Transfer 15.0 mL of this solution to a 25-mL volumetric flask, add 5.0 mL of *Internal standard solution*, dilute with *Piperazine solution* to volume, and mix. Filter a portion of this solution through a 0.4-µm porosity filter.

Chromatographic system (see *Chromatography* ⟨621⟩)—The liquid chromatograph is equipped with a 213-nm detector and a 4.6-mm × 25-cm column that contains packing L1. The flow rate is about 1 mL per minute to achieve baseline resolution of the estropipate peak from other detectable responses in the *Assay preparation*. Chromatograph the *Standard preparation*, and record the peak responses as directed under *Procedure:* the resolution, *R*, between the analyte and internal standard peaks is not less than 2.0, and the relative standard deviation for replicate injections is not more than 2.0%. [NOTE—Peaks may elute up to 1½ hours after injection of the *Assay preparation*. Injections made subsequent to the *Assay preparation* should be timed accordingly.]

Procedure—Proceed as directed in the *Assay* under *Estropipate Tablets.* Calculate the quantity, in mg, of $C_{18}H_{22}O_5S.$-$C_4H_{10}N_2$ in the portion of Vaginal Cream taken by the formula:

$$166.67C(R_U/R_S).$$

Estropipate Tablets

» Estropipate Tablets contain not less than 90.0 percent and not more than 110.0 percent of the labeled amount of $C_{18}H_{22}O_5S.C_4H_{10}N_2$.

Packaging and storage—Preserve in well-closed containers.

Reference standards—*USP Estropipate Reference Standard*—Dry at 105° for 1 hour before using. *USP Estrone Reference Standard*—Dry at 105° for 3 hours before using.

Identification—Place in a centrifuge tube an amount of powdered Tablets, equivalent to about 5 mg of estropipate, add 10 mL of alcohol, and shake by mechanical means for 15 minutes. Centrifuge, evaporate the clear supernatant liquid with the aid of a current of air to dryness, and dissolve the residue in 1 mL of alcohol. On a suitable thin-layer chromatographic plate (see *Chromatography* ⟨621⟩), coated with a 0.25-mm layer of chromatographic silica gel mixture, apply 50 µL of this solution and 50 µL of an alcohol solution of USP Estropipate RS containing 5 mg per mL. Allow the spots to dry, and develop the chromatogram in a solvent system consisting of a mixture of chloroform and methanol (1:1) until the solvent front has moved about three-fourths of the length of the plate. Remove the plate from the developing chamber, mark the solvent front, and allow the solvent to evaporate. Protect the lower half of the plate by suitable means, invert the plate, and spray the unprotected half with a reagent freshly prepared as follows: Slowly add, with cooling, 5 mL of sulfuric acid to 5 mL of acetic anhydride; with continued cooling, add 50 mL of dehydrated alcohol in small increments, and mix. Allow the plate to dry, return it to its normal position, protect the half of the plate previously subjected to spray treatment, and spray the unprotected half with a 1 in 50 solution of vanillin in isopropyl alcohol. Allow to dry, and heat at 100° for 20 minutes: the R_f value of the yellowish brown spot obtained from the test solution in the upper half of the plate corresponds to that obtained from the Standard solution (estrone moiety), and a transient yellow spot is visible at the point of origin of the test solution and of the Standard solution (piperazine moiety).

Dissolution ⟨711⟩—

Medium: water; 900 mL.

Apparatus 2: 75 rpm.

Time: 60 minutes.

Mobile phase, Internal standard concentrate, Standard solution concentrate, and *Chromatographic system*—Prepare as directed in the *Assay.*

Internal standard solution—Pipet 2 mL of *Internal standard concentrate* into a 250-mL volumetric flask, dilute with water to volume, and mix.

Standard preparation—Pipet 2 mL of *Standard solution concentrate* into a 100-mL volumetric flask, dilute with water to volume, and mix. Pipet 4 mL of the resulting solution into a second 100-mL volumetric flask, add 8.0 mL of *Internal standard solution,* dilute with water to volume, and mix to obtain a *Standard preparation* having a known concentration of about 0.8 µg of USP Estropipate RS per mL.

Test preparation—Pipet 2 mL of *Internal standard solution* into a 25-mL volumetric flask, and add an accurately measured volume of a filtered portion of the solution under test that is equivalent to about 20 µg of estropipate. Dilute with water to volume if the contents are less than the nominal volume of the flask, and mix.

Procedure—Inject a volume (about 300 µL) of the *Test preparation* into the chromatograph, record the chromatogram, and measure the responses for the major peaks. The relative retention times are about 0.78 for 4′-nitroacetophenone and 1.0 for estropipate. Calculate the quantity of $C_{18}H_{22}O_5S.C_4H_{10}N_2$ dissolved in comparison with the *Standard preparation,* similarly chromatographed.

Tolerances—Not less than 75% (*Q*) of the labeled amount of $C_{18}H_{22}O_5S.C_4H_{10}N_2$ is dissolved in 60 minutes.

Uniformity of dosage units ⟨905⟩: meet the requirements.

Procedure for content uniformity—

Mobile phase, Internal standard solution, Standard preparation, and *Chromatographic system*—Prepare as directed in the *Assay.*

Test preparation—Transfer 1 Tablet to a 50-mL volumetric flask, add 20 mL of water, insert the stopper in the flask, and shake by mechanical means for about 30 minutes or until the Tablet disintegrates. Add 20 mL of methanol, insert the stopper in the flask, and shake by mechanical means for about 60 minutes. Dilute with a mixture of water and methanol (1:1) to volume, and mix. Transfer an accurately measured volume of this stock solution, equivalent to about 0.25 mg of estropipate, to a 25-mL volumetric flask, add 5.0 mL of *Internal standard solution,* dilute with the mixture of water and methanol (1:1) to volume, and mix. Filter this solution through a solvent-resistant membrane filter of not more than 1-µm porosity, discarding the first portion of the filtrate.

Procedure—Proceed as directed for *Procedure* in the *Assay.*

Assay—

Mobile phase—Dissolve 13.6 g of monobasic potassium phosphate in 1000 mL of water, and mix. Mix 600 mL of the resulting solution with 200 mL of methanol, add 200 mL of acetonitrile with stirring, and mix. Filter, and degas. Make adjustments if necessary (see *System Suitability* under *Chromatography* ⟨621⟩).

Internal standard concentrate—Transfer about 100 mg of 4′-nitroacetophenone to a 100-mL volumetric flask, and dissolve in methanol (sonicate if necessary). Dilute with methanol to volume, and mix.

Internal standard solution—Pipet 5 mL of *Internal standard concentrate* into a 200-mL volumetric flask, dilute with a mixture of water and methanol (1:1) to volume, and mix.

Standard solution concentrate—Dissolve an accurately weighed quantity of USP Estropipate RS in a mixture of water and methanol (1:1), sonicating if necessary, and dilute quantitatively with the mixture of water and methanol (1:1) to obtain a solution having a known concentration of about 1 mg per mL.

Standard preparation—Pipet 5 mL of *Standard solution concentrate* into a 100-mL volumetric flask, dilute with a mixture of water and methanol (1:1) to volume, and mix. Pipet 5 mL of this solution and 5 mL of *Internal standard solution* into a 25-mL volumetric flask, dilute with the mixture of water and methanol (1:1) to volume, and mix. Filter this solution through a solvent-resistant membrane filter of not more than 1-µm porosity, discarding the first portion of the filtrate, to obtain a *Standard preparation* having a known concentration of about 0.01 mg per mL.

Assay preparation—Transfer 20 Estropipate Tablets to a 1000-mL volumetric flask, add 200 mL of water, insert the stopper in the flask, and shake by mechanical means for about 30 minutes or until the Tablets disintegrate completely. Add 200 mL of methanol, insert the stopper in the flask, and shake by mechanical means for about 60 minutes. Dilute with a mixture of water and methanol (1:1) to volume, and mix. Transfer an accurately measured volume of this stock solution, equivalent to about 0.25 mg of estropipate, to a 25-mL volumetric flask, add 5.0 mL of *Internal standard solution,* dilute with a mixture of water and methanol (1:1) to volume, and mix. Filter this solution through a solvent-resistant membrane filter of not more than 1-µm porosity, discarding the first portion of the filtrate.

Chromatographic system (see *Chromatography* ⟨621⟩)—The liquid chromatograph is equipped with a 213-nm detector and a 4.6-mm × 25-cm column that contains packing L1. The flow rate is about 2 mL per minute. Chromatograph the *Standard preparation,* and record the peak responses as directed under *Procedure:* the resolution, *R,* between the analyte and internal standard peaks is not less than 2.0, and the relative standard deviation for replicate injections is not more than 2.0%.

Procedure—Separately inject equal volumes (about 50 µL) of the *Standard preparation* and the *Assay preparation* into the chromatograph, record the chromatograms, and measure the responses for the major peaks. The relative retention times are about 0.85 for 4′-nitroacetophenone and 1.0 for estropipate. Calculate the quantity, in mg, of $C_{18}H_{22}O_5S.C_4H_{10}N_2$ in each Tablet taken by the formula:

$$(1250C/V)(R_U/R_S),$$

in which C is the concentration, in mg per mL, of USP Estropipate RS in the *Standard preparation,* V is the volume, in mL, of the stock solution taken to prepare the *Assay preparation,* and R_U and R_S are the peak response ratios obtained from the *Assay preparation* and the *Standard preparation,* respectively.

Ethacrynate Sodium for Injection

$C_{13}H_{11}Cl_2NaO_4$ 315.12
Acetic acid, [2,3-dichloro-4-(2-methylene-1-oxobutyl)phenoxy]-, sodium salt.
Sodium [2,3-dichloro-4-(2-methylenebutyryl)phenoxy]acetate [*6500-81-8*].

» Ethacrynate Sodium for Injection is a sterile, freeze-dried powder prepared by the neutralization of Ethacrynic Acid with the aid of Sodium Hydroxide. It contains an amount of ethacrynate sodium equivalent to not less than 90.0 percent and not more than 110.0 percent of the labeled amount of $C_{13}H_{12}Cl_2O_4$.

Packaging and storage—Preserve in *Containers for Sterile Solids* as described under *Injections* ⟨1⟩.

Labeling—Label it to indicate that it was prepared by freeze-drying, having been filled into its container in the form of a true solution.

Reference standard—*USP Ethacrynic Acid Reference Standard*—Dry at a pressure not exceeding 5 mm of mercury at 60° for 2 hours before using.

Constituted solution—At the time of use, the constituted solution prepared from Ethacrynate Sodium for Injection meets the requirements for *Constituted Solutions* under *Injections* ⟨1⟩.

Identification—The ultraviolet absorption spectrum of a 1 in 20,000 solution in acidified methanol (prepared by adding 9 mL of hydrochloric acid to 1000 mL of methanol) exhibits maxima and minima at the same wavelengths as that of a similar solution of USP Ethacrynic Acid RS, concomitantly measured.

pH ⟨791⟩: between 6.3 and 7.7, in a solution containing the equivalent of about 50 mg of ethacrynic acid in 50 mL.

Other requirements—It meets the requirements for *Sterility Tests* ⟨71⟩, *Uniformity of Dosage Units* ⟨905⟩, and *Labeling* under *Injections* ⟨1⟩.

Assay—

Triethylamine solution, Solvent mixture, Mobile phase, Standard preparation, and *Chromatographic system*—Prepare as directed in the *Assay* under *Ethacrynic Acid Tablets.*

Assay preparation—Select a number of containers of Ethacrynate Sodium for Injection, the combined contents of which, on the basis of the labeled amount, are equivalent to about 500 mg of ethacrynic acid. Add about 5 mL of *Solvent mixture* to each container, mix to dissolve the contents, and combine the resulting solutions in a 200-mL volumetric flask. Rinse each container with two additional 5 mL portions of *Solvent mixture,* add the rinsings to the solution in the volumetric flask, dilute with *Solvent mixture* to volume, and mix. Transfer 20.0 mL of this solution to a 100-mL volumetric flask, dilute with *Solvent mixture* to volume, and mix.

Procedure—Proceed as directed for *Procedure* in the *Assay* under *Ethacrynic Acid Tablets.* Calculate the quantity, in mg, of $C_{13}H_{12}Cl_2O_4$ in each container of Ethacrynate Sodium for Injection by the formula:

$$(1000C/N)(r_U/r_S),$$

in which N is the number of containers selected for the *Assay preparation.*

Ethacrynic Acid

$C_{13}H_{12}Cl_2O_4$ 303.14
Acetic acid, [2,3-dichloro-4-(2-methylene-1-oxobutyl)phenoxy]-.
[2,3-Dichloro-4-(2-methylenebutyryl)phenoxy]acetic acid [*58-54-8*].

» Ethacrynic Acid contains not less than 97.0 percent and not more than 102.0 percent of $C_{13}H_{12}Cl_2O_4$, calculated on the dried basis.

Caution—Use care in handling Ethacrynic Acid, since it irritates the skin, eyes, and mucous membranes.

Packaging and storage—Preserve in well-closed containers.

Reference standard—*USP Ethacrynic Acid Reference Standard*—Dry at a pressure not exceeding 5 mm of mercury at 60° for 2 hours before using.

Identification—

A: The infrared absorption spectrum of a mineral oil dispersion of it exhibits maxima only at the same wavelengths as that of a similar preparation of USP Ethacrynic Acid RS.

B: The ultraviolet absorption spectrum of a 1 in 20,000 solution in methanol exhibits maxima and minima at the same wavelengths as that of a similar solution of USP Ethacrynic Acid RS, concomitantly measured, and the respective absorptivities, calculated on the dried basis, at the wavelength of maximum absorbance at about 271 nm do not differ by more than 3.0%.

C: Add 2 mL of 1 *N* sodium hydroxide to about 25 mg of it, and heat for several minutes in a boiling water bath. Cool the solution, acidify with 0.25 mL of 18 *N* sulfuric acid, add 0.5 mL of chromotropic acid sodium salt solution (1 in 10), then add, cautiously, 2 mL of sulfuric acid TS: a deep violet color is produced.

Loss on drying ⟨731⟩—Dry it at a pressure not exceeding 5 mm of mercury at 60° for 2 hours: it loses not more than 0.25% of its weight.

Residue on ignition ⟨281⟩: not more than 0.1%.

Toluene extractives—Weigh accurately about 1 g into a glass-stoppered, 100-mL cylinder. Add 50 mL of sodium sulfite solution (2 in 25), and agitate until the solid dissolves. Allow to stand for 20 minutes, add 5 mL of hydrochloric acid, and mix. Divide the solution between two centrifuge tubes, each of which contains 15 mL of toluene. Close each tube tightly, using a polyethylene stopper, and shake vigorously during 2 minutes, occasionally relieving the pressure from the sulfur dioxide by loosening the stoppers. Centrifuge the tubes, withdraw most of the upper layer by means of a syringe, avoiding withdrawal of any of the lower, aqueous phase, and transfer the toluene extracts to a tared evaporating dish. Repeat the extraction twice with additional 15-mL portions of toluene, and evaporate the combined extracts on a steam bath to dryness. Dry the residue at a pressure not exceeding 5 mm of mercury at 60° for 2 hours. Cool, and weigh: not more than 2.0% of extractives is found.

Equivalent weight—Dissolve about 400 mg, accurately weighed, in 100 mL of methanol, add 5 mL of water, and titrate with 0.1 *N* sodium hydroxide VS, determining the end-point potentiometrically, using a calomel-glass electrode system. Perform a blank determination, and make any necessary correction. Calculate the equivalent weight on the dried basis: it is between 294 and 309.

Heavy metals, *Method II* ⟨231⟩: 0.001%.

Assay—Dissolve about 100 mg of Ethacrynic Acid, accurately weighed, in 20 mL of glacial acetic acid in an iodine flask. Pipet 20 mL of 0.1 *N* bromine VS into the flask, add 3 mL of hydrochloric acid, immediately stopper, and seal the flask with a few mL of water in the stopper well. Swirl the flask, and allow to stand in the dark for 1 hour. Add 50 mL of water and 15 mL of potassium iodide TS, and immediately titrate with 0.1 *N* sodium thiosulfate VS, adding 2.0 mL of starch TS as the endpoint is approached. Perform a blank determination (see *Residual Titrations* under *Titrimetry* ⟨541⟩). Each mL of 0.1 *N* bromine is equivalent to 15.16 mg of $C_{13}H_{12}Cl_2O_4$.

Ethacrynic Acid Tablets

» Ethacrynic Acid Tablets contain not less than 90.0 percent and not more than 110.0 percent of the labeled amount of $C_{13}H_{12}Cl_2O_4$.

Packaging and storage—Preserve in well-closed containers.

Reference standard—*USP Ethacrynic Acid Reference Standard*—Dry at a pressure not exceeding 5 mm of mercury at 60° for 2 hours before using.

Identification—

A: Weigh a portion of finely powdered Tablets, equivalent to about 50 mg of ethacrynic acid, and transfer to a separator containing 25 mL of 0.1 *N* hydrochloric acid. Extract with two 40-mL portions of methylene chloride, filter the extracts into a 100-mL volumetric flask, and dilute with methylene chloride to volume. Transfer 10.0 mL of this solution to a 100-mL volumetric flask, evaporate in a gentle current of air to dryness, and promptly dissolve the residue in a portion of a 9 in 1000 mixture of hydrochloric acid in anhydrous methanol, then dilute with the acidic methanol to volume: the ultraviolet absorption spectrum of the solution exhibits maxima and minima at the same wavelengths as that of a solution of USP Ethacrynic Acid RS in the acidic methanol containing 50 µg per mL, concomitantly measured.

B: A portion of powdered Tablets, equivalent to 25 mg of ethacrynic acid, responds to *Identification test C* under *Ethacrynic Acid.*

Dissolution ⟨711⟩—

Medium: 0.1 *M* phosphate buffer, prepared by mixing 13.6 g of monobasic potassium phosphate and 96.2 mL of 0.1 *N* sodium hydroxide with water to obtain 1000 mL of a solution having a pH of 8.0 ± 0.05; 900 mL.

Apparatus 2: 50 rpm.

Time: 45 minutes.

Procedure—Determine the amount of $C_{13}H_{12}Cl_2O_4$ dissolved from ultraviolet absorbances at the wavelength of maximum absorbance at about 277 nm of filtered portions of the solution under test, suitably diluted with *Dissolution Medium*, in comparison with a Standard solution having a known concentration of USP Ethacrynic Acid RS in the same medium.

Tolerances—Not less than 75% (*Q*) of the labeled amount of $C_{13}H_{12}Cl_2O_4$ is dissolved in 45 minutes.

Uniformity of dosage units ⟨905⟩: meet the requirements.

Procedure for content uniformity—Add 1 Tablet to a 100-mL volumetric flask containing 10 mL of water, and allow to stand for 15 minutes, shaking occasionally until the tablet is disintegrated. Add a 9 in 1000 mixture of hydrochloric acid in methanol to volume, and mix. Filter a portion of the mixture, and pipet a volume of the filtrate, equivalent to 5 mg of ethacrynic acid, into a 100-mL volumetric flask. Dilute with the acidic methanol to volume, and mix. Dissolve an accurately weighed quantity of USP Ethacrynic Acid RS in the acidic methanol, and dilute quantitatively and stepwise with the same solvent to obtain a Standard solution having a known concentration of about 50 µg per mL. Concomitantly determine the absorbances of both solutions in 1-cm cells at the wavelength of maximum absorbance at about 269 nm, with a suitable spectrophotometer, using the acidic methanol as the blank. Calculate the quantity, in mg, of $C_{13}H_{12}Cl_2O_4$ in the Tablet by the formula:

$$(T/D)C(A_U/A_S),$$

in which *T* is the labeled quantity, in mg, of ethacrynic acid in the Tablet, *D* is the concentration, in µg per mL, of ethacrynic acid in the solution from the Tablet, on the basis of the labeled quantity per Tablet and the extent of dilution, *C* is the concentration, in µg per mL, of USP Ethacrynic Acid RS in the Standard solution, and A_U and A_S are the absorbances of the solution from the Tablet and the Standard solution, respectively.

Assay—

Triethylamine solution—Mix 10 mL of triethylamine and about 900 mL of water in a 1-liter volumetric flask. Adjust with phosphoric acid to a pH of 6.8 ± 0.1, dilute with water to volume, mix, and filter.

Solvent mixture—Prepare a mixture of water and acetonitrile (3:2).

Mobile phase—Prepare a filtered and degassed mixture of *Triethylamine solution* and acetonitrile (3:2). Make adjustments if necessary (see *System Suitability* under *Chromatography* ⟨621⟩).

Standard preparation—Dissolve an accurately weighed quantity of USP Ethacrynic Acid RS in *Solvent mixture*, and dilute quantitatively with *Solvent mixture*, to obtain a solution having a known concentration of about 0.5 mg per mL.

Assay preparation—Weigh and finely powder not less than 20 Ethacrynic Acid Tablets. Transfer an accurately weighed portion of the powder, equivalent to about 50 mg of ethacrynic acid, to a 100-mL volumetric flask, add about 80 mL of *Solvent mixture*, and shake or sonicate to dissolve the ethacrynic acid. Dilute with *Solvent mixture* to volume, and mix.

Chromatographic system (see *Chromatography* ⟨621⟩)—The liquid chromatograph is equipped with a 254-nm detector and a 3.9-mm × 30-cm column that contains packing L1. The flow rate is about 1 mL per minute. Chromatograph the *Standard preparation*, and record the peak responses as directed under *Procedure:* the column efficiency determined from the analyte peak is not less than 1200 theoretical plates, the tailing factor for the analyte peak is not more than 2, the capacity factor, *k′*, is not less than 0.8, and the relative standard deviation for replicate injections is not more than 1.0.%.

Procedure—Separately inject equal volumes (about 10 µL) of the *Standard preparation* and the *Assay preparation* into the chromatograph, record the chromatograms, and measure the responses for the major peaks. Calculate the quantity, in mg, of $C_{13}H_{12}Cl_2O_4$ in the portion of Tablets taken by the formula:

$$100C(r_U/r_S),$$

in which *C* is the concentration, in mg per mL, of USP Ethacrynic Acid RS in the *Standard preparation*, and r_U and r_S are the peak responses obtained from the *Assay preparation* and the *Standard preparation*, respectively.

Ethambutol Hydrochloride

$CH_3CH_2-C(H)(CH_2OH)-NHCH_2CH_2NH-C(H)(CH_2OH)-CH_2CH_3 \cdot 2HCl$

$C_{10}H_{24}N_2O_2.2HCl$ 277.23

1-Butanol, 2,2′-(1,2-ethanediyldiimino)bis-, dihydrochloride, [*S*-(*R**,*R**)]-.

(+)-2,2′-(Ethylenediimino)-di-1-butanol dihydrochloride [*1070-11-7*].

» Ethambutol Hydrochloride contains not less than 98.0 percent and not more than 100.5 percent of $C_{10}H_{24}N_2O_2.2HCl$, calculated on the dried basis.

Packaging and storage—Preserve in well-closed containers.

Reference standards—*USP Aminobutanol Reference Standard*—This material is hygroscopic. After opening ampul, store in tightly closed container. Do not dry; determine the water content titrimetrically at the time of use. *USP Ethambutol Hydro-*

chloride Reference Standard—Dry at 105° for 2 hours before using.

Identification—

A: The infrared absorption spectrum of a potassium bromide dispersion of it, previously dried, exhibits maxima only at the same wavelengths as that of a similar preparation of USP Ethambutol Hydrochloride RS.

B: A solution (1 in 10) responds to the tests for *Chloride* ⟨191⟩.

Specific rotation ⟨781⟩: between +6.0° and +6.7°, calculated on the dried basis, determined in a solution containing 1.0 g in each 10 mL.

Loss on drying ⟨731⟩—Dry it at 105° for 2 hours: it loses not more than 0.5% of its weight.

Heavy metals, *Method II* ⟨231⟩: 0.002%.

Aminobutanol—

0.2 M Borate buffer—Dissolve 1.24 g of boric acid in 90 mL of water with stirring, and adjust with sodium hydroxide solution (1 in 5) to a pH of 9.0. Transfer to a 100-mL volumetric flask, dilute with water to volume, and mix.

Aminobutanol standard solution—Transfer about 50 mg of USP Aminobutanol RS, accurately weighed, to a 100-mL volumetric flask, dilute with water to volume, and mix. Transfer 1.0 mL of the resulting solution to a 100-mL volumetric flask, dilute with water to volume, and mix.

Fluorescamine solution—Dissolve 5 mg of fluorescamine in 50 mL of acetone in a glass-stoppered, graduated cylinder.

Test preparation—Transfer about 50 mg of Ethambutol Hydrochloride, accurately weighed, to a 100-mL volumetric flask, dilute with water to volume, and mix.

Procedure—Pipet a 10-mL portion of *Test preparation* into a glass-stoppered, 100-mL conical flask, and add 10 mL of water and 20 mL of *0.2 M Borate buffer*. To another 100-mL flask add 10.0 mL of *Test preparation*, 10.0 mL of *Aminobutanol standard solution*, and 20 mL of *0.2 M Borate buffer*. Place the flasks on a magnetic stirrer, and while the contents are being stirred rapidly, add 10 mL of *Fluorescamine solution* rapidly. Insert the stoppers in the flasks, invert, and shake briefly. After 1 minute, accurately timed, determine the relative fluorescence intensities of both solutions in 1-cm cells, with a suitable fluorometer, at about 485 nm, with the excitation wavelength at about 385 nm. The fluorescence intensity of the solution obtained from the *Test preparation* is not greater than the difference between the intensities of the two solutions (not more than 1.0%).

Ordinary impurities ⟨466⟩—

Test solution: methanol.

Standard solution: methanol.

Eluant: a mixture of methanol and ammonium hydroxide (18:1).

Visualization: 16.

Assay—Dissolve about 200 mg of Ethambutol Hydrochloride, accurately weighed, in a mixture of 100 mL of glacial acetic acid and 5 mL of mercuric acetate TS. Add crystal violet TS, and titrate with 0.1 *N* perchloric acid VS (the color change at the end-point is from blue to blue-green). Perform a blank determination, and make any necessary correction. Each mL of 0.1 *N* perchloric acid is equivalent to 13.86 mg of $C_{10}H_{24}N_2O_2 \cdot 2HCl$.

Ethambutol Hydrochloride Tablets

» Ethambutol Hydrochloride Tablets contain not less than 95.0 percent and not more than 105.0 percent of the labeled amount of $C_{10}H_{24}N_2O_2 \cdot 2HCl$.

Packaging and storage—Preserve in well-closed containers.

Reference standards—*USP Aminobutanol Reference Standard*—This material is hygroscopic. After opening ampul, store in tightly closed container. Do not dry; determine the water content titrimetrically at the time of use. *USP Ethambutol Hydrochloride Reference Standard*—Dry at 105° for 2 hours before using.

Identification—Triturate a quantity of finely ground Tablets, equivalent to about 100 mg of ethambutol, with 3 mL of methanol in a glass mortar. Add 5 mL of methanol to obtain a suspension, then filter through a funnel lined with a suitable filter paper (Whatman No. 42 or the equivalent) previously moistened with methanol, and collect the filtrate in a beaker containing 100 mL of acetone. Stir the mixture, and allow crystallization to proceed for 15 minutes. Decant the liquid, and gently dry the crystals with the aid of a current of air until the odor of methanol no longer is detectable: a portion of the crystals so obtained responds to the *Identification tests* under *Ethambutol Hydrochloride*.

Dissolution ⟨711⟩—

Medium: water; 900 mL.

Apparatus 1: 100 rpm.

Time: 45 minutes.

Phosphate buffer—Dissolve 38.0 g of monobasic sodium phosphate and 2.0 g of anhydrous dibasic sodium phosphate in water to obtain 1000 mL of solution.

Bromocresol green solution—Dissolve 200 mg of bromocresol green in 30 mL of water and 6.5 mL of 0.1 *N* sodium hydroxide. Dilute with *Phosphate buffer* to 500 mL, mix, and add 0.1 *N* hydrochloric acid to adjust to a pH of 4.6 ± 0.1.

Standard preparation—Dissolve an accurately weighed quantity of USP Ethambutol Hydrochloride RS in water, and dilute quantitatively with water to obtain a solution having a known concentration of about 0.1 mg per mL.

Procedure—Into 3 separate, glass-stoppered, 50-mL centrifuge tubes pipet (a) 1 mL of a filtered portion of the solution under test, (b) 1 mL of *Standard preparation*, and (c) 1 mL of water to provide a blank. Add 5.0 mL of *Bromocresol green solution* to each tube, mix, add 10.0 mL of chloroform to each, insert the stoppers, and shake the mixtures vigorously. Allow the mixtures to separate, discard the upper aqueous layers, and filter the 3 chloroform layers through separate pledgets of cotton. Determine the amount of $C_{10}H_{24}N_2O_2 \cdot 2HCl$ dissolved from absorbances, at the wavelength of maximum absorbance at about 415 nm, obtained from the test solution in comparison with those obtained from the Standard solution, using the blank to set the instrument.

Tolerances—Not less than 75% (*Q*) of the labeled amount of $C_{10}H_{24}N_2O_2 \cdot 2HCl$ is dissolved in 45 minutes.

Uniformity of dosage units ⟨905⟩: meet the requirements.

Aminobutanol—

0.2 M Borate buffer, Aminobutanol standard solution, and *Fluorescamine solution*—Prepare as directed in the test for *Aminobutanol* under *Ethambutol Hydrochloride*.

Test preparation—Place a number of Tablets, equivalent to 400 mg of ethambutol hydrochloride, in a beaker, cover with acetone, and allow to stand for 15 minutes. Decant the acetone, dry the tablets, and remove the coating. Grind the tablet cores in a mortar to a fine powder, moisten with methanol, and triturate to a fine paste. Transfer the mixture with the aid of methanol to a 100-mL volumetric flask, dilute with methanol to volume, and mix. Filter the mixture through a dry, folded filter paper. Pipet 25 mL of the filtrate into a 200-mL volumetric flask, and dilute with water to volume. Allow to stand for 15 minutes, and filter through a dry, folded filter paper, discarding the first cloudy portions of the filtrate. The clear filtrate is the *Test preparation*.

Procedure—Proceed as directed for *Procedure* in the test for *Aminobutanol* under *Ethambutol Hydrochloride*.

Assay—Weigh and finely powder not less than 20 Ethambutol Hydrochloride Tablets. Transfer an accurately weighed portion of the powder, equivalent to about 200 mg of ethambutol hydrochloride, to a 125-mL separator with the aid of 10 mL of sodium hydroxide solution (1 in 12), and swirl the contents to form a fine suspension. Extract with five 25-mL portions of chloroform, and filter the chloroform extracts through anhydrous sodium sulfate into a 400-mL beaker. Evaporate the combined chloroform extracts on a water bath almost to dryness, and remove the remainder of the chloroform with the aid of a current of air. Add 100 mL of glacial acetic acid and 5 mL of mercuric acetate TS, stir until completely dissolved, and proceed as directed in the *Assay* under *Ethambutol Hydrochloride*, beginning with "Add crystal violet TS."

Ethchlorvynol

```
        OH
        |
HC≡C—C—CH=CHCl
        |
        CH2CH3
```

C_7H_9ClO 144.60
1-Penten-4-yn-3-ol, 1-chloro-3-ethyl-.
1-Chloro-3-ethyl-1-penten-4-yn-3-ol [*113-18-8*].

» Ethchlorvynol contains not less than 98.0 percent and not more than 100.0 percent of *E*-ethchlorvynol (C_7H_9ClO), calculated on the anhydrous basis.

Packaging and storage—Preserve in tight, light-resistant glass or polyethylene containers, using polyethylene-lined closures.

Reference standard—*USP Ethchlorvynol Reference Standard*—Do not dry before using.

Identification—

A: The infrared absorption spectrum, obtained by spreading a capillary film of it between sodium chloride plates, exhibits maxima only at the same wavelengths as that of a similar preparation of USP Ethchlorvynol RS.

B: Dissolve about 1 g in 20 mL of methanol. To 1 mL of the solution, add about 4 drops of 6 *N* ammonium hydroxide, mix, then add silver nitrate TS, a few drops at a time: a yellowish white precipitate is formed, and it at first redissolves, but becomes insoluble when an excess of silver nitrate TS has been added.

C: To 10 mL of the solution prepared in *Identification test B* add 5 mL of freshly prepared *m*-phenylenediamine hydrochloride–oxalic acid solution (prepared by dissolving 1 g of *m*-phenylenediamine hydrochloride and 1 g of oxalic acid in 35 mL of water and filtering, if necessary): a reddish orange color is produced in about 3 minutes.

Refractive index ⟨831⟩: between 1.476 and 1.480.

Acidity—Dissolve 5.0 mL of Ethchlorvynol in 50 mL of a mixture of equal volumes of water and methanol that has been neutralized to the phenolphthalein end-point with 0.1 *N* sodium hydroxide. Add 1 mL of phenolphthalein TS, and titrate with 0.10 *N* sodium hydroxide to a pink end-point: not more than 1.7 mL of 0.10 *N* sodium hydroxide is required for neutralization.

Water, *Method I* ⟨921⟩: not more than 0.2%.

Assay and chromatographic purity—

Resolution solution—Add 2.5 µL of toluene to 0.5 mL of USP Ethchlorvynol RS, and mix.

Chromatographic system (see *Chromatography* ⟨621⟩)—The gas chromatograph is equipped with a thermal conductivity detector and a 1.8-m × 4-mm glass column (pretreated with 10% dimethyldichlorosilane in toluene) packed with 10 percent phase G16 on 60- to 80-mesh support S1AB. The column is maintained at about 160°, and the injector and the detector are maintained at about 200°. The carrier gas is dry helium, flowing at a rate of about 30 mL per minute. Chromatograph the *Resolution solution*, and record the peak responses as directed under *Procedure:* the resolution, *R*, between the *Z*- and *E*-ethchlorvynol peaks is not less than 1.0, and the relative standard deviation for replicate injections is not more than 2.0%. The relative retention times for toluene, β-chlorovinylethyl ketone (if present), *Z*-ethchlorvynol, and *E*-ethchlorvynol are about 0.1, 0.2, 0.8, and 1.0, respectively.

Procedure—[Note—Use peak areas where peak responses are indicated.] Inject about 3 µL of Ethchlorvynol into the chromatograph, record the chromatogram, and measure the responses for all of the peaks. Calculate the percentage of *E*-ethchlorvynol (C_7H_9ClO) in the specimen of Ethchlorvynol taken by the formula:

$$100r_e/r_t,$$

in which r_e is the response of the *E*-ethchlorvynol peak obtained in the chromatogram of the Ethchlorvynol and r_t is the sum of the responses of all of the peaks observed in the chromatogram. Calculate the percentages of toluene and β-chlorovinylethyl ketone in the specimen of Ethchlorvynol taken by the same formula:

$$0.5r_U/r_t,$$

in which r_U is the peak response of the relevant analyte obtained in the chromatogram of Ethchlorvynol: not more than 0.2% of toluene or 0.3% of β-chlorovinylethyl ketone is found. Calculate the percentage of each peak, other than the *E*-ethchlorvynol peak, the toluene peak, and the β-chlorovinylethyl ketone peak, observed in the chromatogram of the Ethchlorvynol by the same formula:

$$100r_i/r_t,$$

in which r_i is the response of each peak: the total of all observed impurities, including toluene and β-chlorovinylethyl ketone, is not more than 2.0%.

Ethchlorvynol Capsules

» Ethchlorvynol Capsules contain not less than 90.0 percent and not more than 110.0 percent of the labeled amount of *E*-ethchlorvynol (C_7H_9ClO).

Packaging and storage—Preserve in tight, light-resistant containers.

Reference standard—*USP Ethchlorvynol Reference Standard*—Do not dry before using.

Identification—Dissolve an amount of the contents of Capsules, equivalent to about 1 g of ethchlorvynol, in 20 mL of methanol. The solution responds to the *Identification tests* under *Ethchlorvynol*.

Uniformity of dosage units ⟨905⟩: meet the requirements, chloroform being used as the solvent in the procedure for *Weight Variation*.

Assay—

Methanol solution—Add 600 mL of methanol to 1400 mL of water in a suitable container, and mix. Allow the solution to equilibrate to room temperature.

Internal standard solution—Dissolve a quantity of chlorobutanol in *Methanol solution* to obtain a solution having a known concentration of about 6 mg per mL.

Standard preparation—Dissolve an accurately weighed quantity of USP Ethchlorvynol RS in *Methanol solution* to obtain a solution having a known concentration of about 4 mg per mL. Transfer 5.0 mL of this solution to a suitable flask, add 5.0 mL of *Internal standard solution*, and mix to obtain a *Standard preparation* having a known concentration of about 2 mg of USP Ethchlorvynol RS per mL.

Assay preparation—Transfer not less than 20 Ethchlorvynol Capsules, accurately counted, to a 500-mL volumetric flask. Add 250 mL of *Methanol solution*, heat at 70° to 80° for not more than 3 hours, and stir until the Capsules burst. Cool the solution to room temperature, add *Methanol solution* to volume, and mix. Quantitatively dilute a portion of this solution if necessary with *Methanol solution* to obtain a solution containing about 4 mg of *E*-ethchlorvynol per mL, and mix. Transfer 5.0 mL of this solution to a suitable flask, add 5.0 mL of *Internal standard solution*, and mix to obtain an *Assay preparation* containing about 2 mg of *E*-ethchlorvynol per mL.

Chromatographic system (see *Chromatography* ⟨621⟩)—The gas chromatograph is equipped with a flame-ionization detector and a 1.8-m × 4-mm glass column (pretreated with 10% dimethyldichlorosilane in toluene) packed with 10 percent phase G16 on 60- to 80-mesh support S1AB. The column is maintained at about 160°, and the injector and the detector are maintained at about 200°. The carrier gas is dry helium, flowing at a rate of about 30 mL per minute. Chromatograph the *Standard preparation*, and record the peak responses as directed under *Procedure:* the resolution, *R*, between the ethchlorvynol and chlorobutanol peaks is not less than 4.0, and the relative standard deviation for replicate injections is not more than 1.5%.

Procedure—Separately inject equal volumes (about 2 µL) of the *Standard preparation* and the *Assay preparation* into the gas chromatograph, record the chromatographs, and measure the responses for the major peaks. The relative retention times are about 0.55 for chlorobutanol and 1.0 for *E*-ethchlorvynol. Calculate the quantity, in mg, of *E*-ethchlorvynol (C_7H_9ClO) in each Capsule taken by the formula:

$$(CL/D)(R_U/R_S),$$

in which *C* is the concentration, in mg per mL, of USP Ethchlorvynol RS in the *Standard preparation*, *L* is the labeled quantity, in mg, of ethchlorvynol in each Capsule, *D* is the concentration, in mg per mL, of ethchlorvynol in each mL of the *Assay preparation* based on the number of Capsules taken, the labeled quantity, in mg, of ethchlorvynol in each Capsule, and the extent of dilution, and R_U and R_S are the peak response ratios obtained from the *Assay preparation* and the *Standard preparation*, respectively.

Ether

$C_2H_5.O.C_2H_5$

$C_4H_{10}O$ 74.12
Ethane, 1,1′-oxybis-.
Ethyl ether [*60-29-7*].

» Ether contains not less than 96.0 percent and not more than 98.0 percent of $C_4H_{10}O$, the remainder consisting of alcohol and water.

Caution—Ether is highly volatile and flammable. Its vapor, when mixed with air and ignited, may explode.

NOTE—Ether to be used for anesthesia must be preserved in tight containers of not more than 3-kg capacity, and is not to be used for anesthesia if it has been removed from the original container longer than 24 hours. Ether to be used for anesthesia may, however, be shipped in larger containers for repackaging in containers as directed above, provided the ether at the time of repackaging meets the requirements of the tests of this Pharmacopeia.

Packaging and storage—Preserve in partly filled, tight, light-resistant containers, at a temperature not exceeding 30°, remote from fire.

Specific gravity ⟨841⟩: between 0.713 and 0.716 (indicating 96.0% to 98.0% of $C_4H_{10}O$).

Acidity—To 10 mL of water in a glass-stoppered flask, add 0.10 mL of bromothymol blue TS and 0.010 *N* sodium hydroxide until a blue color persists after vigorous shaking. Add 25 mL of Ether, and shake briskly to mix the two layers. If no blue color remains, titrate with 0.010 *N* sodium hydroxide until the blue color is restored and persists for several minutes: not more than 0.80 mL of 0.010 *N* sodium hydroxide is required (0.003% as CH_3COOH). [NOTE—Exercise great care to avoid contamination from carbon dioxide when adding the Ether and titrating.]

Nonvolatile residue—Allow 50 mL to evaporate spontaneously from a tared dish, and dry at 105° for 1 hour: the weight of the residue does not exceed 1 mg (0.003%).

Foreign odor—Place 10 mL in a clean, dry, evaporating dish, and allow it to evaporate spontaneously to a volume of about 1 mL: no foreign odor is perceptible. Transfer this residue to a piece of clean, odorless, absorbent paper: no foreign odor is perceptible when the last traces of Ether evaporate from the paper.

Aldehyde—Place 20 mL in a glass-stoppered cylinder, and add 7 mL of a mixture of 1 mL of alkaline mercuric-potassium iodide TS and 17 mL of a saturated solution of sodium chloride. Insert the stopper in the cylinder, shake vigorously for 10 seconds, then set aside for 1 minute: the water layer shows no turbidity.

Peroxide—

Titanium tetrachloride solution—Cool separately, in small beakers surrounded by crushed ice, 10 mL of 6 *N* hydrochloric acid and 10 mL of water–white titanium tetrachloride. Add the titanium tetrachloride dropwise to the chilled acid. Allow the mixture to stand at ice-bath temperature until all of the yellow solid dissolves, dilute the solution with 6 *N* hydrochloric acid to 1000 mL, and mix.

Peroxide standard solution—Pipet 25 mL of hydrogen peroxide solution into a 1000-mL volumetric flask, dilute with water to volume, and mix. Pipet 15 mL of this solution into a 1000-mL volumetric flask, dilute with water to volume, and mix. Each mL contains 0.011 mg of H_2O_2.

Procedure—To 50 mL of Ether, in a separator, add 5.0 mL of *Titanium tetrachloride solution.* Shake vigorously, allow the layers to separate, and drain the lower layer into a glass-stoppered, 25-mL graduated cylinder. Dilute with water to 10.0 mL, and mix. Any yellow color in the solution does not exceed that of a solution prepared by adding 5.0 mL of *Titanium tetrachloride solution* and 1.0 mL of *Peroxide standard solution* to a glass-stoppered, 25-mL graduated cylinder, diluting with water to 10.0 mL, and mixing, the colors being determined in 1-cm cells, with a suitable spectrophotometer, at a wavelength of 410 nm. The limit is 0.3 ppm.

Low-boiling hydrocarbons—

Standard preparation—Transfer about 50 mL of anhydrous ethyl ether, previously tested for absence of hydrocarbons, as directed under *Procedure*, to a 100-mL volumetric flask. Add 0.20 mL of pentane, dilute with the same anhydrous ethyl ether to volume, and mix.

Apparatus—Under typical conditions, a gas chromatograph (see *Chromatography* ⟨621⟩) is equipped with a flame-ionization detector, and contains a 3.7-m × 2-mm (ID) stainless steel column packed with 30 percent phase G22 on 30- to 60-mesh support S1C. The injector port, column, and detector are maintained at temperatures of about 230°, 80°, and 250°, respectively, nitrogen being used as the carrier gas at a flow rate of about 30 mL per minute.

Procedure—Inject 1 µL of *Standard preparation* and Ether, respectively, into the apparatus, and measure the total area under the hydrocarbon peaks (under typical conditions the retention times are about 2.3 minutes for isopentane, 2.7 minutes for pentane, and 4.3 minutes for 2-methylpentane) in the Ether. The total area of the hydrocarbon peaks does not exceed that of the peak for pentane in the *Standard preparation* (0.2%).

Ether, Vinyl—*see* Vinyl Ether

Ethinamate

OOCNH2
C≡CH

$C_9H_{13}NO_2$ 167.21
Cyclohexanol, 1-ethynyl-, carbamate.
1-Ethynylcyclohexanol carbamate [*126-52-3*].

» Ethinamate, dried in vacuum at 50° for 4 hours, contains not less than 98.0 percent and not more than 100.5 percent of $C_9H_{13}NO_2$, calculated on the dried basis.

Packaging and storage—Preserve in tight containers.

Reference standard—*USP Ethinamate Reference Standard*—Do not dry before using.

Identification—

A: The infrared absorption spectrum of a potassium bromide dispersion of it, previously dried, exhibits maxima only at the same wavelengths as that of a similar preparation of USP Ethinamate RS.

B: Dissolve about 100 mg in 10 mL of alcohol, and divide the solution into three equal portions. To one portion add bromine TS, dropwise: it is instantly decolorized (*presence of unsaturation*). To the second portion add 10 mL of silver nitrate TS: a white precipitate forms. Add 1 mL of ammonium hydroxide to the reaction mixture: a yellow color, which slowly deepens, is produced. Add 10 mL of alcohol to this alkaline mixture: the precipitate dissolves, leaving a yellow to red solution. To the third portion add 10 mL of water: no precipitate forms.

Melting range ⟨741⟩: between 94° and 98°.

Loss on drying ⟨731⟩—Dry it in vacuum at 50° for 4 hours: it loses not more than 1.0% of its weight.

Residue on ignition ⟨281⟩: not more than 0.2%.

Assay—Transfer about 200 mg of Ethinamate, previously dried and accurately weighed, to a glass-stoppered, 250-mL conical flask, and dissolve in 25 mL of alcohol. Add 100 mL of silver nitrate solution (1 in 20), and shake or stir vigorously for 1 to 2 minutes. Using 8 to 10 drops of methyl red–methylene blue TS, titrate with 0.05 *N* sodium hydroxide VS to a gray end-point. Perform a blank determination, and make any necessary correction. Each mL of 0.05 *N* sodium hydroxide is equivalent to 8.360 mg of $C_9H_{13}NO_2$.

Ethinamate Capsules

» Ethinamate Capsules contain not less than 90.0 percent and not more than 110.0 percent of the labeled amount of $C_9H_{13}NO_2$.

Packaging and storage—Preserve in tight containers.

Reference standard—*USP Ethinamate Reference Standard*—Do not dry before using.

Identification—

A: Place an amount of the Capsule contents, equivalent to about 500 mg of ethinamate, in a beaker, add 50 mL of alcohol, mix, and filter. Proceed with a 10-mL portion of the filtrate as directed in the *Identification test* under *Ethinamate*, beginning with "divide the solution into three equal portions." The specified result is obtained.

B: Place an amount of the Capsule contents, equivalent to about 500 mg of ethinamate, in a beaker, add 10 mL of chloroform, stir well, and filter through a filter paper that previously has been rinsed with chloroform. Collect the filtrate in a suitable beaker, evaporate on a water bath at a temperature not higher than 70°, with the aid of a current of air to dryness, and dry the residue in vacuum at 50° for 1 hour: the infrared absorption spectrum of a potassium bromide dispersion of the residue so obtained exhibits maxima only at the same wavelengths as that of a similar preparation of USP Ethinamate RS.

Dissolution ⟨711⟩—

Medium: water; 900 mL.

Apparatus 2: 50 rpm.

Time: 45 minutes.

Procedure—Filter a portion of the solution under test. Transfer an accurately measured volume of the filtrate, equivalent to about 2 mg of ethinamate, to a 100-mL volumetric flask, and add water to obtain 20 mL of solution. Add 2 mL of 3.6 *N* sulfuric acid, swirl to mix, and place the flask in a bath maintained at a temperature of 87 ± 3°. Allow it to remain in the heating bath for 20 minutes, swirling the mixture frequently, initially, and occasionally thereafter. Add 25.0 mL of freshly prepared sodium phenolate solution (prepared by mixing 200 g of sodium hydroxide and 250 g of phenol with 500 mL of water), swirl to mix, and return the flask to the heating bath. Note the time, add 30.0 mL of sodium hypochlorite solution, and swirl the mixture frequently, initially, and occasionally thereafter. Remove the flask from the heating bath 20 minutes following the addition of the sodium hypochlorite solution, cool to room temperature, dilute with water to volume, and mix. Determine the amount of $C_9H_{13}NO_2$ dissolved from absorbances, at the wavelength of maximum absorbance at about 590 nm, measured 45 minutes, accurately timed, after adding the sodium hypochlorite solution, obtained from the test solution in comparison with those obtained from a Standard solution similarly prepared with a known quantity of USP Ethinamate RS, using water as the blank.

Tolerances—Not less than 75% (*Q*) of the labeled amount of $C_9H_{13}NO_2$ is dissolved in 45 minutes.

Uniformity of dosage units ⟨905⟩: meet the requirements.

Assay—

Standard preparation—Transfer about 100 mg of USP Ethinamate RS, accurately weighed, to a separator, dissolve in 50 mL of chloroform, and add 25 mL of water. Proceed as directed for *Assay preparation*, beginning with "Shake, and drain the lower, chloroform layer."

Assay preparation—Remove as completely as possible the contents of not less than 20 Ethinamate Capsules, and weigh. Transfer an accurately weighed portion of the contents, equivalent to about 500 mg of ethinamate, to a 250-mL volumetric flask, add 100 mL of chloroform, and shake on a mechanical shaker for 15 minutes. Dilute with chloroform to volume, and mix. Filter, and discard the first 20 mL of the filtrate. Transfer 50.0 mL of the subsequent filtrate to a separator containing 25 mL of water. Shake, and drain the lower, chloroform layer into a suitable beaker. (If an emulsion occurs, break it by adding 1 g of calcium chloride.) Extract with two additional 25-mL portions of chloroform, draining each chloroform layer into the same beaker. Evaporate the combined chloroform extracts to about 3 mL, transfer to a 10-mL volumetric flask, and dilute with chloroform to volume.

Procedure—Concomitantly determine the absorbances of the *Standard preparation* and the *Assay preparation* in 0.1-mm cells at the wavelength of maximum absorbance at about 5.80 µm, with a suitable spectrophotometer, using chloroform as the blank. Calculate the quantity, in mg, of $C_9H_{13}NO_2$ in the portion of Capsule contents taken by the formula:

$$5W(A_U/A_S),$$

in which W is the weight, in mg, of USP Ethinamate RS taken for the *Standard preparation*, and A_U and A_S are the absorbances of the *Assay preparation* and the *Standard preparation*, respectively.

Ethinyl Estradiol

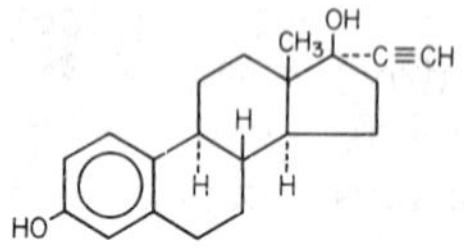

$C_{20}H_{24}O_2$ 296.41

19-Norpregna-1,3,5(10)-trien-20-yne-3,17-diol, (17α)-.

19-Nor-17α-pregna-1,3,5(10)-trien-20-yne-3,17-diol [*57-63-6*].

» Ethinyl Estradiol contains not less than 97.0 percent and not more than 102.0 percent of $C_{20}H_{24}O_2$, calculated on the dried basis.

Packaging and storage—Preserve in tight, non-metallic, light-resistant containers.

Reference standard—*USP Ethinyl Estradiol Reference Standard*—Dry in vacuum over silica gel for 4 hours before using.

Completeness of solution—Dissolve 100 mg in 5 mL of alcohol: the solution is clear and free from undissolved solid.

Identification—

A: The infrared absorption spectrum of a potassium bromide dispersion of it, previously dried, exhibits maxima only at the same wavelengths as that of a similar preparation of USP Ethinyl Estradiol RS.

B: The ultraviolet absorption spectrum of a 1 in 20,000 solution in alcohol exhibits maxima and minima at the same wavelengths as that of a similar solution of USP Ethinyl Estradiol RS, concomitantly measured, and the respective absorptivities, calculated on the dried basis, at the wavelength of maximum absorbance at about 281 nm do not differ by more than 3.0%.

Melting range ⟨741⟩: between 180° and 186°. It may exist also in a polymorphic modification, melting between 142° and 146°.

Specific rotation ⟨781⟩: between −28.0° and −29.5°, calculated on the dried basis, determined in a solution of colorless pyridine, from a freshly opened container, containing 40 mg in each 10 mL.

Loss on drying ⟨731⟩—Dry about 100 mg, accurately weighed, in vacuum over silica gel for 4 hours: it loses not more than 0.5% of its weight.

Assay—

Mobile phase—Prepare a filtered and degassed mixture of water and acetonitrile (1 : 1). Make adjustments if necessary (see *System Suitability* under *Chromatography* ⟨621⟩).

Internal standard solution—Prepare a solution of ethylparaben in a mixture of water and acetonitrile (1 : 1) containing about 0.5 mg per mL.

Standard preparation—Transfer about 10 mg of USP Ethinyl Estradiol RS, accurately weighed, to a 50-mL volumetric flask, and add 10 mL of *Mobile phase.* Add 5.0 mL of *Internal standard solution*, dilute with *Mobile phase* to volume, and mix to obtain a solution having a known concentration of about 0.2 mg of USP Ethinyl Estradiol RS per mL.

Assay preparation—Transfer about 25 mg of Ethinyl Estradiol, accurately weighed, to a 25-mL volumetric flask, add *Mobile phase* to volume, and mix. Transfer 10.0 mL of this solution to a 50-mL volumetric flask, add 5.0 mL to *Internal standard solution*, dilute with *Mobile phase* to volume, and mix.

Chromatographic system (see *Chromatography* ⟨621⟩)—The liquid chromatograph is equipped with a 280-nm detector and a 4.6-mm × 15-cm column that contains packing L1. The flow rate is about 1 mL per minute. Chromatograph the *Standard preparation*, and record the peak responses as directed under *Procedure:* the resolution, *R*, between the analyte and internal standard peaks is not less than 4.5, and the relative standard deviation for replicate injections is not more than 2.0%.

Procedure—Separately inject equal volumes (about 25 µL) of the *Standard preparation* and the *Assay preparation* into the chromatograph, record the chromatograms, and measure the responses for the major peaks. The relative retention times are about 0.6 for ethylparaben and 1.0 for ethinyl estradiol. Calculate the quantity, in mg, of $C_{20}H_{24}O_2$ in the portion of Ethinyl Estradiol taken by the formula:

$$125C(R_U / R_S),$$

in which C is the concentration, in mg per mL, of USP Ethinyl Estradiol RS in the *Standard preparation*, and R_U and R_S are the peak response ratios obtained from the *Assay preparation* and the *Standard preparation*, respectively.

Ethinyl Estradiol Tablets

» Ethinyl Estradiol Tablets contain not less than 90.0 percent and not more than 115.0 percent of the labeled amount of $C_{20}H_{24}O_2$.

Packaging and storage—Preserve in well-closed containers.

Reference standard—*USP Ethinyl Estradiol Reference Standard*—Dry in vacuum over silica gel for 4 hours before using.

Identification—Triturate a quantity of finely powdered Tablets, equivalent to about 1 mg of ethinyl estradiol, with four 20-mL portions of chloroform, and decant the chloroform through filter paper into a beaker. Evaporate the filtrate on a steam bath with the aid of a stream of nitrogen, and dissolve the residue in 0.5 mL of chloroform. Activate a thin-layer chromatographic plate (see *Chromatography* ⟨621⟩), coated with a 0.25-mm layer of chromatographic silica gel mixture, by heating it at 105° for 30 minutes. Apply 10 µL of the solution under test and 10 µL of a solution of USP Ethinyl Estradiol RS in chloroform containing 2 mg per mL at points about 3 cm from one end of the plate, and place the plate in a developing chamber containing a mixture of 29 volumes of chloroform and 1 volume of alcohol to a depth of 2 cm, the developing chamber previously having been equilibrated with the solvent mixture as described under *Thin-layer Chromatography* ⟨621⟩. Remove the plate when the solvent moves to about 15 cm above the initial spots, dry it at room temperature, and spray it with *Sulfuric acid–methanol*, prepared as directed in the *Assay:* the spots from the solution under test and the Standard solution appear yellow, changing to orange-yellow in room light, and they are intensely fluorescent when viewed in long-wavelength ultraviolet light. Heat the plate at 105° for 10 minutes: the spots from the solution under test and the Standard solution appear dark pink in room light and show an intense orange-red fluorescence when viewed under long-wavelength ultraviolet light. Both spots appear at the same R_f, about 0.3, and no other spots are visible.

Disintegration ⟨701⟩: 30 minutes.

Uniformity of dosage units ⟨905⟩: meet the requirements.

Assay—[NOTE—In this procedure, use scrupulously dry glassware and separators fitted with solvent-resistant stopcocks, and use isooctane that gives no color when shaken with an equal volume of sulfuric acid.]

Sulfuric acid–methanol—Cautiously add sulfuric acid to 30 mL of chilled anhydrous methanol in a 100-mL volumetric flask, in small increments and with mixing. Adjust to room temperature, dilute with sulfuric acid to volume, and mix.

Standard preparation—Dissolve about 30 mg of USP Ethinyl Estradiol RS in methanol, and dilute with methanol to 50 mL. Pipet 5 mL of this solution into a 100-mL volumetric flask, add isooctane to volume, and mix. Pipet 5 mL of the resulting solution into a 50-mL volumetric flask, add isooctane to volume, and mix.

Assay preparation—Weigh and finely powder not less than 20 Ethinyl Estradiol Tablets. Weigh accurately a portion of the powder, equivalent to about 150 µg of ethinyl estradiol, transfer to a 125-mL conical flask, and add 5 mL of water. Heat the mixture on a steam bath for 5 minutes, with occasional stirring. Add about 70 mL of methanol to the warm mixture, insert the stopper in the flask, and shake by mechanical means for 10 minutes. Filter the mixture into a 100-mL volumetric flask, rinsing the flask and the filter with several small portions of methanol, dilute the filtrate with methanol to volume, and mix. Transfer 10.0 mL of the solution to a 30-mL beaker, and evaporate on a steam bath with the aid of a stream of nitrogen to near dryness. Transfer the residue to a 125-mL separator with three 1-mL portions of 1 *N* sodium hydroxide. Rinse the beaker with 2 mL of dilute sulfuric acid (1 in 4), and add the rinsing to the separator. Rinse the beaker with three 25-mL portions of a 1 in 50 mixture of chloroform in isooctane, successively extracting the aqueous solution with each portion for 2 minutes. Combine the extracts in a second 125-mL separator through a cotton-pledget filter, and wash the filter with 5 mL of isooctane.

Procedure—Pipet 5 mL of the *Standard preparation* into a 125-mL separator, add 75 mL of a 1 in 50 mixture of chloroform in isooctane, and mix. Pipet 5 mL of *Sulfuric acid–methanol* into the separator containing the *Standard preparation* and 5 mL into that containing the *Assay preparation*, allowing the pipet to drain for not less than 2 minutes. Treat each separator as follows: Shake for 2 minutes, allow to separate completely, and transfer 4.0 mL of the fluorescent pink phase to a glass-stoppered, 15-mL centrifuge tube containing 0.50 mL of methanol. Insert the stopper, mix vigorously, and centrifuge to dispel air bubbles. Concomitantly determine the absorbances of both solutions in 1-cm cells at the wavelength of maximum absorbance at about 538 nm, with a suitable spectrophotometer, using water as the blank. Calculate the quantity, in µg, of $C_{20}H_{24}O_2$ in the portion of Tablets taken by the formula:

$$50C(A_U / A_S),$$

in which C is the concentration, in µg per mL, of USP Ethinyl Estradiol RS in the *Standard preparation*, and A_U and A_S are the absorbances of the solutions from the *Assay preparation* and the *Standard preparation*, respectively.

Ethinyl Estradiol Tablets, Ethynodiol Diacetate and—*see* Ethynodiol Diacetate and Ethinyl Estradiol Tablets

Ethinyl Estradiol Tablets, Levonorgestrel and—*see* Levonorgestrel and Ethinyl Estradiol Tablets
Ethinyl Estradiol Tablets, Norethindrone and—*see* Norethindrone and Ethinyl Estradiol Tablets
Ethinyl Estradiol Tablets, Norethindrone Acetate and—*see* Norethindrone Acetate and Ethinyl Estradiol Tablets
Ethinyl Estradiol Tablets, Norgestrel and—*see* Norgestrel and Ethinyl Estradiol Tablets

Ethiodized Oil Injection

» Ethiodized Oil Injection is an iodine addition product of the ethyl ester of the fatty acids of poppyseed oil, containing not less than 35.2 percent and not more than 38.9 percent of organically combined iodine. It is sterile.

Packaging and storage—Preserve in well-filled, light-resistant, single-dose or multiple-dose containers.

Identification—Place 1 drop in a test tube, and heat directly in a flame: the violet color of iodine vapors is observed.

Specific gravity ⟨841⟩: between 1.280 and 1.293, at 15°.

Viscosity ⟨911⟩: between 50 centipoises and 100 centipoises, at 15°.

Sterility—It meets the requirements under *Sterility Tests* ⟨71⟩.

Acidity—Dissolve 1.0 mL in 10 mL of chloroform in a glass-stoppered cylinder, add phenolphthalein TS and 0.30 mL of sodium hydroxide solution (1 in 250), insert the stopper, and shake vigorously: a red color is produced.

Free iodine—Dissolve 1.0 mL in 5 mL of chloroform, add 20 mL of potassium iodide solution (1 in 20), agitate vigorously, and add 2 drops of starch TS: no blue color is produced.

Assay—[*Caution—Observe rigorously the precautions set forth for Procedure under Oxygen Flask Combustion* ⟨471⟩.] Weigh accurately about 30 mg (1 drop) of Ethiodized Oil Injection in a tared cellulose acetate capsule, and proceed as directed for *Procedure* under *Oxygen Flask Combustion* ⟨471⟩, beginning with "Place the specimen," and using a thick-walled, 500-mL combustion flask. Use 10 mL of sodium hydroxide solution (1 in 100) and 1 mL of freshly prepared sodium bisulfite solution (1 in 100) as the absorbing liquid. Pipet 1 mL of bromine–sodium acetate TS into the cup of the flask, loosen the stopper, and allow the solution to be sucked into the flask. Wash down the cup and the ground joint with water, insert the stopper in the flask, shake it vigorously, then add 5 drops of formic acid, and again shake the flask. Remove the stopper, and rinse the stopper and the specimen holder with water, collecting the rinsings in the flask. Bubble nitrogen through the solution to displace all of the oxygen from the solution and the flask. Add 0.50 g of potassium iodide and 3 mL of 2 *N* sulfuric acid, allow the mixture to stand for 2 minutes, add 3 mL of starch TS, and titrate the liberated iodine with 0.05 *N* sodium thiosulfate VS. Each mL of 0.05 *N* sodium thiosulfate is equivalent to 1.058 mg of iodine (I).

Ethionamide

N C2H5 S=C−NH2

$C_8H_{10}N_2S$ 166.24
4-Pyridinecarbothioamide, 2-ethyl-.
2-Ethylthioisonicotinamide [*536-33-4*].

» Ethionamide contains not less than 98.0 percent and not more than 102.0 percent of $C_8H_{10}N_2S$, calculated on the anhydrous basis.

Packaging and storage—Preserve in tight containers.

Reference standard—*USP Ethionamide Reference Standard*—Do not dry before using.

Identification—

A: The infrared absorption spectrum of a potassium bromide dispersion of it, previously dried in vacuum over phosphorus pentoxide for 18 hours, exhibits maxima only at the same wavelengths as that of a similar preparation of USP Ethionamide RS.

B: The ultraviolet absorption spectrum of the solution of Ethionamide employed for measurement of absorbance in the *Assay* exhibits maxima and minima only at the same wavelengths as that of a similar solution of USP Ethionamide RS, concomitantly measured.

Melting range: ⟨741⟩: between 158° and 164°.

pH ⟨791⟩: between 6.0 and 7.0, in a 1 in 100 slurry in water.

Water, *Method I* ⟨921⟩: not more than 2.0%.

Residue on ignition ⟨281⟩: not more than 0.2%.

Selenium ⟨291⟩: 0.003%, a 200-mg test specimen being used.

Assay—Transfer about 100 mg of Ethionamide, accurately weighed, to a 250-mL volumetric flask, dissolve in about 100 mL of methanol, dilute with methanol to volume, and mix. Transfer a 5-mL aliquot to a 200-mL volumetric flask, dilute with methanol to volume, and mix. Dissolve a suitable quantity of USP Ethionamide RS, accurately weighed, in methanol, and dilute quantitatively and stepwise with methanol to obtain a Standard solution having a known concentration of about 10 µg per mL. Concomitantly determine the absorbances of both solutions in 1-cm cells at the wavelength of maximum absorbance at about 290 nm, with a suitable spectrophotometer, using methanol as the blank. Calculate the quantity, in mg, of $C_8H_{10}N_2S$ in the portion of Ethionamide taken by the formula:

$$10C(A_U / A_S),$$

in which C is the concentration, in µg per mL, of USP Ethionamide RS in the Standard solution, calculated on the anhydrous basis, and A_U and A_S are the absorbances of the solution of Ethionamide and the Standard solution, respectively.

Ethionamide Tablets

» Ethionamide Tablets contain not less than 95.0 percent and not more than 110.0 percent of the labeled amount of $C_8H_{10}N_2S$.

Packaging and storage—Preserve in tight containers.

Reference standard—*USP Ethionamide Reference Standard*—Do not dry before using.

Identification—

A: The solution of Tablets employed for measurement of absorbance in the *Assay* exhibits an absorbance maximum at 290 ± 2 nm.

B: Digest a quantity of powdered Tablets, equivalent to about 1 g of ethionamide, with 50 mL of methanol, and filter through a medium-porosity, sintered-glass funnel. Evaporate the filtrate on a steam bath to dryness: the residue so obtained melts between 155° and 164°.

Disintegration ⟨701⟩: 15 minutes, the use of disks being omitted.

Uniformity of dosage units ⟨905⟩: meet the requirements.

Assay—Weigh and finely powder not less than 20 Ethionamide Tablets. Transfer an accurately weighed portion of the powder, equivalent to about 100 mg of ethionamide, to a medium-porosity, sintered-glass funnel that is fitted into a 250-mL suction flask. Extract the specimen by stirring it with successive 10-mL portions of methanol, using a total of about 100 mL of solvent, drawing off each portion of liquid with gentle suction before adding the next portion. Transfer the combined methanol extracts to a 250-mL volumetric flask, dilute with methanol to volume, and mix. Proceed as directed in the *Assay* under *Ethionamide*, beginning with "Transfer a 5-mL aliquot." Calculate the quantity, in mg, of $C_8H_{10}N_2S$ in the portion of Tablets taken by the formula:

$$10C(A_U/A_S),$$

in which C is the concentration, in µg per mL, of USP Ethionamide RS in the Standard solution, calculated on the anhydrous basis, and A_U and A_S are the absorbances of the solution of Ethionamide Tablets and the Standard solution, respectively.

Ethopropazine Hydrochloride

N(C2H5)2
CH2CHCH3
N
S
· HCl

$C_{19}H_{24}N_2S.HCl$ 348.93

10*H*-Phenothiazine-10-ethanamine, *N*,*N*-diethyl-α-methyl-, monohydrochloride.

10-[2-(Diethylamino)propyl]phenothiazine monohydrochloride [*1094-08-2*].

» Ethopropazine Hydrochloride contains not less than 98.0 percent and not more than 101.5 percent of $C_{19}H_{24}N_2S.HCl$, calculated on the dried basis.

Packaging and storage—Preserve in tight, light-resistant containers.

Reference standard—*USP Ethopropazine Hydrochloride Reference Standard*—Dry at 105° for 2 hours before using.

NOTE—Throughout the following procedures, protect test or assay specimens, the Reference Standard, and solutions containing them, by conducting the procedures without delay, under subdued light, or using low-actinic glassware.

Identification—

A: The infrared absorption spectrum of a 1 in 10 solution of it in chloroform, measured in a 0.1-mm cell, exhibits maxima only at the same wavelengths as that of a similar preparation of USP Ethopropazine Hydrochloride RS.

B: The ultraviolet absorption spectrum of a 6 in 100,000 solution in alcohol exhibits maxima and minima at the same wavelengths as that of a similar solution of USP Ethopropazine Hydrochloride RS, concomitantly measured.

C: It responds to the tests for *Chloride* ⟨191⟩.

Loss on drying ⟨731⟩—Dry 1 g, accurately weighed, at 105° for 2 hours: it loses not more than 0.5% of its weight.

Heavy metals, *Method II* ⟨231⟩: 0.002%.

Ordinary impurities ⟨466⟩—

Test solution: methanol.

Standard solution: methanol.

Eluant: a mixture of chloroform, alcohol, and ammonium hydroxide (80:20:1).

Visualization: 16.

Assay—Weigh accurately about 500 mg into a 150-mL beaker, add 15 mL of glacial acetic acid, and dissolve by warming on a steam bath. Add 75 mL of acetone, 10 mL of mercuric acetate TS, and 1 mL of methyl orange TS, and titrate with 0.1 *N* perchloric acid VS to a pink end-point. Perform a blank determination, and make any necessary correction. Each mL of 0.1 *N* perchloric acid is equivalent to 34.89 mg of $C_{19}H_{24}N_2S.HCl$.

Ethopropazine Hydrochloride Tablets

» Ethopropazine Hydrochloride Tablets contain not less than 90.0 percent and not more than 110.0 percent of the labeled amount of $C_{19}H_{24}N_2S.HCl$.

Packaging and storage—Preserve in well-closed containers, protected from light.

Reference standard—*USP Ethopropazine Hydrochloride Reference Standard*—Dry at 105° for 2 hours before using.

NOTE—Throughout the following procedures, protect test or assay specimens, the Reference Standard, and solutions containing them, by conducting the procedures without delay, under subdued light, or using low-actinic glassware.

Identification—The principal spot found in the test for *Other alkylated phenothiazines* corresponds in R_f to the spot from the *Standard solution.*

Dissolution ⟨711⟩—

Medium: 0.1 *N* hydrochloric acid; 900 mL.

Apparatus 1: 100 rpm.

Time: 45 minutes.

Procedure—Determine the amount of $C_{19}H_{24}N_2S.HCl$ dissolved, employing the procedure set forth in the *Assay*, making any necessary modifications.

Tolerances—Not less than 75% (*Q*) of the labeled amount of $C_{19}H_{24}N_2S.HCl$ is dissolved in 45 minutes.

Uniformity of dosage units ⟨905⟩: meet the requirements.

Other alkylated phenothiazines—[NOTE—In this procedure, use low-actinic glassware; apply and develop the thin-layer plate in an area protected from direct sunlight and fluorescent light.] Transfer a quantity of powdered Tablets, equivalent to about 50 mg of ethopropazine hydrochloride, to a 25-mL volumetric flask, add about 15 mL of alcohol, and shake by mechanical means for 15 minutes. Dilute with alcohol to volume, and mix to obtain the *Test solution.* Dissolve a suitable quantity of USP Ethopropazine Hydrochloride RS in alcohol to obtain a *Standard solution* having a known concentration of about 2 mg per mL, and dilute quantitatively and stepwise with alcohol to obtain a *Diluted standard solution* having a known concentration of about 200 µg per mL. At five equidistant points about 2 cm from one edge of a 20-cm × 20-cm thin-layer plate coated with a 0.25-mm layer of chromatographic silica gel mixture, apply 50 µL of *Test solution* and 50 µL of *Standard solution,* in 10-µL increments, and apply 5-, 10-, and 25-µL portions of *Diluted standard solution,* overspotting these applications with alcohol to a total volume of 50 µL. Develop the plate for about 1 hour in a chamber lined on three sides with filter paper and previously equilibrated with a mixture of 90 volumes of chloroform, 10 volumes of alcohol, and 1 volume of ammonium hydroxide. Remove the plate, and air-dry. View under short-wavelength ultraviolet light: the area and intensity of any spot, other than the principal spot, from the *Test solution* are not greater than those of the middle spot from the *Diluted standard solution* (2.0%).

Assay—[NOTE—Use low-actinic glassware in this procedure.]

Standard preparation—Dissolve a suitable quantity of USP Ethopropazine Hydrochloride RS, accurately weighed, in 0.1 *N* hydrochloric acid, and dilute quantitatively and stepwise with the same solvent to obtain a solution having a known concentration of about 8 µg per mL.

Assay preparation—Weigh and finely powder not less than 20 Ethopropazine Hydrochloride Tablets. Transfer an accurately weighed portion of the powder, equivalent to about 100 mg of ethopropazine hydrochloride, to a 500-mL volumetric flask. Add about 200 mL of water and 5 mL of hydrochloric acid, and shake for about 10 minutes. Dilute with water to volume, and mix. Filter a portion of the solution, discarding the first 50 mL of the filtrate. Pipet 10 mL of the subsequent filtrate into a 250-mL separator, add about 20 mL of water, render alkaline with ammonium hydroxide, and extract with four 25-mL portions of ether. Extract the combined ether extracts with four 25-mL portions of 0.1 *N* hydrochloric acid, collecting the aqueous extracts in a 250-mL volumetric flask. Aerate to remove residual ether, add 0.1 *N* hydrochloric acid to volume, and mix.

Procedure—Concomitantly determine the absorbances of the *Standard preparation* and the *Assay preparation* in 1-cm cells at the wavelengths of maximum absorbance at about 254 nm and at 277 nm, with a suitable spectrophotometer, using 0.1 *N* hydrochloric acid as the blank. Calculate the quantity, in mg, of $C_{19}H_{24}N_2S.HCl$ in the portion of Tablets taken by the formula:

$$12.5C(A_{254} - A_{277})_U/(A_{254} - A_{277})_S,$$

in which *C* is the concentration, in µg per mL, of USP Ethopropazine Hydrochloride RS in the *Standard preparation*, and the parenthetic expressions are the differences in the absorbances of the two solutions at the wavelengths indicated by the subscripts, for the *Assay preparation* (*U*) and the *Standard preparation* (*S*), respectively.

Ethosuximide

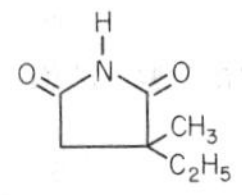

$C_7H_{11}NO_2$ 141.17
2,5-Pyrrolidinedione, 3-ethyl-3-methyl-.
2-Ethyl-2-methylsuccinimide [*77-67-8*].

» Ethosuximide contains not less than 98.0 percent and not more than 101.0 percent of $C_7H_{11}NO_2$, calculated on the anhydrous basis.

Packaging and storage—Preserve in tight containers.

Reference standard—*USP Ethosuximide Reference Standard*—Do not dry.

Identification—The infrared absorption spectrum of a 1 in 15 solution in chloroform, determined in a 0.1-mm cell, exhibits maxima only at the same wavelengths as that of a similar solution of USP Ethosuximide RS.

Melting range ⟨741⟩: between 47° and 52°.

Water, *Method I* ⟨921⟩: not more than 0.5%.

Residue on ignition ⟨281⟩: not more than 0.5%.

Cyanide—Dissolve 1 g in 10 mL of alcohol, and add 3 drops of ferrous sulfate TS, 1 mL of 1 *N* sodium hydroxide, and a few drops of ferric chloride TS. Warm gently, and acidify with 2 *N* sulfuric acid: no blue precipitate or blue color is formed within 15 minutes.

2-Ethyl-2-methylsuccinic anhydride and other impurities—Prepare a solution in chloroform containing 250 mg of ethosuximide per mL, and inject 1 µL of the solution into a suitable gas chromatographic apparatus equipped with a flame-ionization detector. The operating conditions of the apparatus may vary, depending upon the particular instrument used, but a suitable chromatogram is obtained with the following conditions.

Preferably, use an apparatus corresponding to the general type having a column that is about 1.8 m × 6.4 mm, and that is packed with 5 percent (w/w) G5 on 60- to 80-mesh S1A. The carrier gas is helium, flowing at the rate of 90 mL per minute as measured at the column inlet. The port temperature is 260°, the column is run isothermally at 140°, and the detector temperature is 280°. The hydrogen-flame detector is adjusted for a flow of hydrogen at 90 mL per minute, and of air at 450 mL per minute. Adjust the instrument to provide adequate sensitivity for the detection of the anhydride. Typically, a setting 32-fold more sensitive than that used for the detection of ethosuximide is satisfactory. Measure the peak area due to ethosuximide and the other peak areas due to the anhydride and other impurities, if present, and correct for differences in sensitivity settings. Calculate the percentage of 2-ethyl-2-methylsuccinic anhydride and other impurities present by the formula:

$$100A/B,$$

in which *A* is the sum of their corrected peak areas and *B* is the sum of the corrected peak areas due to ethosuximide, anhydride, and other impurities. Not more than 0.2% is found.

Assay—Dissolve about 200 mg of Ethosuximide, accurately weighed, in 50 mL of dimethylformamide, add 2 drops of a 1 in 1000 solution of azo violet in dimethylformamide, and titrate with 0.1 *N* sodium methoxide VS to a deep blue end-point, taking precautions to prevent absorption of atmospheric carbon dioxide. Perform a blank determination, and make any necessary correction. Each mL of 0.1 *N* sodium methoxide is equivalent to 14.12 mg of $C_7H_{11}NO_2$.

Ethosuximide Capsules

» Ethosuximide Capsules contain not less than 93.0 percent and not more than 107.0 percent of the labeled amount of $C_7H_{11}NO_2$, present in the form of a solution of Ethosuximide in Polyethylene Glycol 400 or other suitable solvent.

Packaging and storage—Preserve in tight containers.

Reference standard—*USP Ethosuximide Reference Standard*—Do not dry.

Identification—Place a portion of the Capsule contents, equivalent to about 300 mg of ethosuximide, in a separator containing 50 mL of ether. Shake with three 10-mL portions of water, discarding the aqueous extracts. Add about 5 g of anhydrous sodium sulfate, swirl for 3 minutes, and filter through a small pledget of cotton that previously has been washed with ether, into a small flask. Evaporate the ether solution at room temperature in a current of air to dryness, and dissolve the residue in 5 mL of chloroform: the infrared absorption spectrum of the solution, determined in a 0.1-mm cell, exhibits maxima only at the same wavelengths as that of a 1 in 15 solution of USP Ethosuximide RS in chloroform.

Uniformity of dosage units ⟨905⟩: meet the requirements.

Assay—Remove, pool, mix, and accurately weigh the contents of not less than 20 Ethosuximide Capsules. Transfer a weighed portion of the contents, equivalent to about 200 mg of ethosuximide, to a separator containing 50 mL of ether. Shake with 10 mL of water, and collect the ether in a flask, filtering through cotton that previously has been wetted with ether. Shake the aqueous solution successively with three additional 50-mL portions of ether, and combine the ether portions in the flask, filtering as before. Rinse the cotton, and evaporate the ether at room temperature in a current of air to about 10 mL. Add 3 to 4 g of anhydrous sodium sulfate, swirl for 2 to 3 minutes, and transfer the ether to a 125-mL conical flask, pouring through a small pledget of cotton. Rinse the sodium sulfate, and complete the transfer with the aid of anhydrous ethyl ether. Evaporate the ether at room temperature in a current of air, removing from the air stream as soon as the residue appears dry. Without delay, dissolve the residue in 50 mL of dimethylformamide, add 2 drops of a 1 in 1000 solution of azo violet in dimethylformamide, and titrate with 0.1 *N* sodium methoxide VS to a deep blue end-point, taking precautions to prevent absorption of atmospheric carbon dioxide. Perform a blank determination, and make any necessary correction. Each mL of 0.1 *N* sodium methoxide is equivalent to 14.12 mg of $C_7H_{11}NO_2$.

Ethotoin

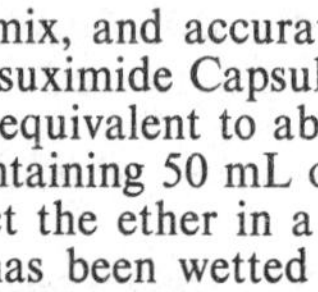

$C_{11}H_{12}N_2O_2$ 204.23
3-Ethyl-5-phenyl-imidazolidin-2,4-dione [*86-35-1*].

» Ethotoin contains not less than 97.5 percent and not more than 102.0 percent of $C_{11}H_{12}N_2O_2$, calculated on the dried basis.

Packaging and storage—Preserve in tight containers.

Reference standards—*USP Ethotoin Reference Standard*—Do not dry before using. *USP 5-Phenylhydantoin Reference Standard*—Do not dry before using.

Identification—

A: The ultraviolet absorption spectrum of a 1 in 1000 solution in alcohol exhibits maxima and minima at the same wavelengths as that of a similar preparation of USP Ethotoin RS, concomitantly measured.

B: The retention time of the major peak in the chromatogram of the *Assay preparation* corresponds to that of the *Standard preparation*, as obtained in the *Assay*.

Loss on drying ⟨731⟩—Dry it in vacuum at 60° for 4 hours: it loses not more than 1.0% of its weight.

Residue on ignition ⟨281⟩: not more than 0.1%.

Chloride—Transfer 1.0 g of Ethotoin to a suitable separator and dissolve in 50 mL of ether. Extract with three 15-mL portions of water, collect the combined extracts in a beaker, heat on a steam bath to expel any traces of ether, and allow to cool to room temperature. Transfer the solution to a 50-mL color comparison tube, add 2 *N* nitric acid until the solution is acidic, add 1 mL of 2 *N* nitric acid in excess, mix, add 1 mL of silver nitrate TS, dilute with water to 50 mL, and allow to stand for 5 minutes, protected from direct sunlight. The turbidity produced does not exceed that of a solution prepared by mixing 2 mL of freshly prepared 0.002 *N* hydrochloric acid, 1 mL of 2 *N* nitric acid, 1 mL of silver nitrate TS, and 46 mL of water (0.014%).

Heavy metals, *Method II* ⟨231⟩: 0.002%.

5-Phenylhydantoin and related compounds—

Mobile phase and *5-Phenylhydantoin stock solution*—Prepare as directed in the *Assay*.

Standard preparation—Prepare as directed for the *System suitability solution* in the *Assay*.

Test preparation—Prepare as directed for the *Assay preparation* in the *Assay*.

Chromatographic system—Use the system as directed under *Assay*. Determine its suitability for this test by chromatographing the *Standard preparation* as directed under *Procedure*. The resolution, *R*, between the 5-phenylhydantoin and ethotoin peaks is not less than 6.0, and the relative standard deviation of the 5-phenylhydantoin peak responses in replicate injections is not more than 3.0%. The relative retention times are about 0.4 for 5-phenylhydantoin and 1.0 for ethotoin.

Procedure—Separately inject equal volumes (about 20 µL) of the *Standard preparation* and the *Test preparation* into the chromatograph, record the chromatograms, and measure the responses for the corresponding peaks. Calculate the quantity, in mg, of 5-phenylhydantoin in the Ethotoin taken by the formula:

$$200C(r_U / r_S),$$

in which C is the concentration, in mg per mL, of USP 5-Phenylhydantoin RS in the *Standard preparation*, and r_U and r_S are the 5-phenylhydantoin peak responses obtained from the *Test preparation* and the *Standard preparation*, respectively: not more than 1.5% of 5-phenylhydantoin is found.

Similarly, calculate the quantity, in mg, of total unknown impurities in the Ethotoin taken by the formula:

$$200C(r_U / r_S),$$

in which C is the concentration, in mg per mL, of USP 5-phenylhydantoin in the *Standard preparation*, r_U is the total of all unknown impurity peak responses obtained from the *Test preparation*, and r_S is the 5-phenylhydantoin peak response obtained from the *Standard preparation:* not more than 1.0% of unknown impurities is found.

Assay—

Mobile phase—Dissolve 0.65 g of monobasic potassium phosphate in 600 mL of water, adjust with phosphoric acid solution (1 in 10) to a pH of 3.5 ± 0.1, and dilute with water to 650 mL. Add 350 mL of methanol, mix, filter through a membrane filter of 0.5-µm or finer porosity, and degas. Make adjustments if necessary (see *System Suitability* under *Chromatography* ⟨621⟩).

5-Phenylhydantoin stock solution—Dissolve, with the aid of sonication if necessary, an accurately weighed quantity of USP 5-Phenylhydantoin RS in *Mobile phase* to obtain a solution having a known concentration of about 0.37 mg per mL.

Standard preparation—Dissolve, with the aid of sonication if necessary, an accurately weighed quantity of USP Ethotoin RS in *Mobile phase* to obtain a solution having a known concentration of about 0.25 mg of ethotoin per mL.

System suitability solution—Transfer 25 mg of USP Ethotoin RS, accurately weighed, to a 100-mL volumetric flask, add about 1.0 mL of *5-Phenylhydantoin stock solution*, add *Mobile phase* to volume, and sonicate to dissolve.

Assay preparation—Transfer about 50 mg of Ethotoin, accurately weighed, to a 200-mL volumetric flask, dissolve in *Mobile phase*, dilute with *Mobile phase* to volume, and sonicate to dissolve.

Chromatographic system (see *Chromatography* ⟨621⟩)—The liquid chromatograph is equipped with a 210-nm detector and a 4.6-mm × 30-cm column that contains packing L1. The flow rate is about 1.5 mL per minute. Chromatograph the *System suitability solution*, and record the peak responses as directed under *Procedure:* the resolution, *R*, between the 5-phenylhydantoin and ethotoin peaks is not less than 6.0. The relative retention times are about 0.4 for 5-phenylhydantoin and 1.0 for ethotoin. Chromatograph the *Standard preparation*, and record the peak responses as directed under *Procedure:* the relative standard deviation for replicate injections is not more than 3.0%.

Procedure—Separately inject equal volumes (about 20 µL) of the *Standard preparation* and the *Assay preparation* into the chromatograph, record the chromatograms, and measure the responses for the major peaks. Calculate the quantity, in mg, of $C_{11}H_{12}N_2O_2$ in the portion of Ethotoin taken by the formula:

$$200C(r_U / r_S),$$

in which C is the concentration, in mg per mL, of USP Ethotoin RS in the *Standard preparation*, and r_U and r_S are the peak responses obtained from the *Assay preparation* and the *Standard preparation*, respectively.

Ethotoin Tablets

» Ethotoin Tablets contain not less than 90.0 percent and not more than 110.0 percent of the labeled amount of $C_{11}H_{12}N_2O_2$.

Packaging and storage—Preserve in tight containers.

Reference standard—*USP Ethotoin Reference Standard*—Do not dry before using.

Identification—The retention time of the major peak in the chromatogram of the *Assay preparation* corresponds to that of the *Standard preparation*, both relative to the internal standard, as obtained in the *Assay*.

Dissolution ⟨711⟩—

Medium: 0.1 *N* hydrochloric acid; 900 mL.

Apparatus 2: 100 rpm.

Time: 60 minutes.

Standard solution—Transfer about 100 mg of USP Ethotoin RS, accurately weighed, to a 25-mL volumetric flask. Dissolve in methanol, dilute with methanol to volume, and mix. Transfer 4.0 mL of this solution to a 50-mL volumetric flask, add *Dissolution Medium* to volume, and mix.

Procedure—Determine the amount of $C_{11}H_{12}N_2O_2$ dissolved from ultraviolet absorbances at the wavelength of maximum absorbance at about 257 nm on filtered portions of the solution under test, suitably diluted with *Dissolution Medium*, if necessary, in comparison with the *Standard solution*.

Tolerances—Not less than 80% (*Q*) of the labeled amount of $C_{11}H_{12}N_2O_2$ is dissolved in 60 minutes.

Uniformity of dosage units ⟨905⟩: meet the requirements.

Assay—

Mobile phase—Prepare a filtered and degassed mixture of water and acetonitrile (3:1). Make adjustments if necessary (see *System Suitability* under *Chromatography* ⟨621⟩).

Internal standard solution—Prepare a solution of ethylparaben in *Mobile phase* having a concentration of 0.02 mg per mL.

Standard preparation—Dissolve an accurately weighed quantity of USP Ethotoin RS in *Mobile phase*, and dilute quantitatively, and stepwise if necessary, with *Mobile phase* to obtain a solution having a known concentration of about 1.0 mg per mL. To 5.0 mL of this solution add 5.0 mL of *Internal standard solution*, and mix.

Assay preparation—Weigh and finely powder not less than 20 Ethotoin Tablets. Transfer an accurately weighed portion of the powder, equivalent to about 100 mg of ethotoin, to a 100-mL volumetric flask. Add 75 mL of *Mobile phase*, shake vigorously for 60 minutes, dilute with *Mobile phase* to volume, mix, and filter. To 5.0 mL of the filtrate add 5.0 mL of the *Internal standard solution*, and mix.

Chromatographic system (see *Chromatography* ⟨621⟩)—The liquid chromatograph is equipped with a 254-nm detector and a 3.9-mm × 30-cm column that contains packing L1. The flow rate is about 1.5 mL per minute. Chromatograph the *Standard preparation*, and record the peak responses as directed under *Procedure:* the resolution, *R*, between the analyte and internal standard peaks is not less than 2.0, and the relative standard deviation for replicate injections is not more than 2.0%.

Procedure—Separately inject equal volumes (about 50 µL) of the *Standard preparation* and the *Assay preparation* into the chromatograph, record the chromatograms, and measure the responses for the major peaks. The relative retention times are about 0.5 for ethotoin and 1.0 for ethylparaben. Calculate the quantity, in mg, of $C_{11}H_{12}N_2O_2$ in the portion of Tablets taken by the formula:

$$200C(R_U/R_S),$$

in which *C* is the concentration, in mg per mL, of USP Ethotoin RS in the *Standard preparation*, and R_U and R_S are the peak response ratios obtained from the *Assay preparation* and the *Standard preparation*, respectively.

Ethyl Acetate—*see* Ethyl Acetate NF

Ethyl Chloride

CH_3CH_2Cl

C_2H_5Cl 64.51
Ethane, chloro-.
Chloroethane [*75-00-3*].

» Ethyl Chloride contains not less than 99.5 percent and not more than 100.5 percent of C_2H_5Cl.

Caution—Ethyl Chloride is highly flammable. Do not use where it may be ignited.

Packaging and storage—Preserve in tight containers, preferably hermetically sealed, and remote from fire.

Reaction—Shake 10 mL with 10 mL of water, both having been previously cooled to 0°, and allow the supernatant layer of ethyl chloride to volatilize spontaneously: the remaining liquid is neutral to litmus. Retain the liquid for the test for *Alcohol*.

Alcohol—To the liquid obtained in the test for *Reaction* add a few drops of potassium dichromate TS and 2 mL of 2 *N* sulfuric acid, and boil the mixture: no odor of acetaldehyde is perceptible, and no greenish or purplish color is produced.

Nonvolatile residue and odor—Allow 5 mL to evaporate spontaneously from a tared, shallow dish: no foreign odor is perceptible while the last portions evaporate, and the weight of the residue is negligible.

Chloride—Add a few drops of silver nitrate TS to 10 mL of alcohol, cool to 0°, and add to the clear liquid about 500 µL of Ethyl Chloride cooled to the same temperature: no turbidity is produced immediately.

Assay—Introduce about 1.5 mL of cold Ethyl Chloride into a tared glass-stoppered pressure bottle containing 25.0 mL of 1 *N* alcoholic potassium hydroxide VS, rapidly replace the stopper, and weigh accurately. Tie down the stopper, insert the bottle in a wire basket, and immerse in a water bath at room temperature. [*Caution—Before raising the bath temperature, take adequate precautions to cover the bottle or erect a suitable safety shield to prevent injury in case the bottle should burst.*] Heat the water bath to boiling, maintain at this temperature for 30 minutes, and then cool gradually to room temperature before handling the bottle. Remove the stopper, add phenolphthalein TS, and titrate the excess alkali with 1 *N* hydrochloric acid VS. Perform a blank determination (see *Residual Titrations* ⟨541⟩). Each mL of 1 *N* alcoholic potassium hydroxide is equivalent to 64.51 mg of C_2H_5Cl.

Ethyl Oleate—*see* Ethyl Oleate NF

Ethyl Vanillin—*see* Ethyl Vanillin NF

Ethylcellulose—*see* Ethylcellulose NF

Ethylcellulose Aqueous Dispersion—*see* Ethylcellulose Aqueous Dispersion NF

Ethylene Oxide Sterilization, Paper Strip, Biological Indicator for—*see* Biological Indicator for Ethylene Oxide Sterilization, Paper Strip

Ethylenediamine

$H_2NCH_2CH_2NH_2$

$C_2H_8N_2$ 60.10
1,2-Ethanediamine.
Ethylenediamine [*107-15-3*].

» Ethylenediamine contains not less than 98.0 percent and not more than 100.5 percent, by weight, of $C_2H_8N_2$.

Caution—Use care in handling Ethylenediamine because of its caustic nature and the irritating properties of its vapor.

NOTE—Ethylenediamine is strongly alkaline and may readily absorb carbon dioxide from the air to form a nonvolatile carbonate. Protect Ethylenediamine against undue exposure to the atmosphere.

Packaging and storage—Preserve in well-filled, tight, glass containers.

Identification—To 2 mL of cupric sulfate solution (1 in 100) add 3 drops of a solution of Ethylenediamine (1 in 6), and shake: a purplish blue color is produced.

Heavy metals, *Method I* ⟨231⟩—Evaporate 5.0 mL on a steam bath to dryness, add to the residue 1 mL of hydrochloric acid and 0.5 mL of nitric acid, and evaporate again to dryness. Dissolve the residue in 20 mL of warm water, cool, dilute with water to 100 mL, mix, and use 20 mL of this solution for the test: the limit is 0.002%.

Assay—Weigh accurately about 1 mL of Ethylenediamine in a tared, glass-stoppered flask containing about 25 mL of water. Dilute with water to about 75 mL, add a mixed indicator of bromocresol green TS and methyl red TS (5 in 6), mix, and titrate with 1 *N* hydrochloric acid VS. Perform a blank determination, and make any necessary correction. Each mL of 1 *N* hydrochloric acid is equivalent to 30.05 mg of $C_2H_8N_2$.

Ethylnorepinephrine Hydrochloride

HO—C6H3(OH)—CH(OH)CH(NH2)CH2CH3 · HCl

$C_{10}H_{15}NO_3.HCl$ 233.69
1,2-Benzenediol, 4-(2-amino-1-hydroxybutyl)-, hydrochloride.
α-(1-Aminopropyl)-3,4-dihydroxybenzyl alcohol hydrochloride [*3198-07-0*].

» Ethylnorepinephrine Hydrochloride contains not less than 98.0 percent and not more than 101.0 percent of $C_{10}H_{15}NO_3.HCl$, calculated on the dried basis.

Packaging and storage—Preserve in well-closed, light-resistant containers.

Reference standard—*USP Ethylnorepinephrine Hydrochloride Reference Standard*—Dry in vacuum over phosphorus pentoxide for 24 hours before using.

Identification—

A: The infrared absorption spectrum of a potassium bromide dispersion of it exhibits maxima only at the same wavelengths as that of a similar preparation of USP Ethylnorepinephrine Hydrochloride RS.

B: The ultraviolet absorption spectrum of a solution (1 in 20,000) exhibits maxima and minima at the same wavelengths as that of a similar solution of USP Ethylnorepinephrine Hydrochloride RS, concomitantly measured, and the respective absorptivities, calculated on the dried basis, at the wavelengths of maximum absorbance at about 279 nm do not differ by more than 3.0%.

C: Dissolve about 50 mg in 100 mL of water. Mix 2 mL of this solution with 3 mL of mercuric acetate solution (1 in 25): a light pink color is produced (*distinction from epinephrine which gives, at the same concentration, a red color, and from nordefrin, which gives a yellow color, changing to red on heating*).

D: It responds to the tests for *Chloride* ⟨191⟩.

Loss on drying ⟨731⟩—Dry it in vacuum over phosphorus pentoxide for 24 hours: it loses not more than 1.0% of its weight.

Residue on ignition ⟨281⟩: not more than 0.1%.

Sulfate ⟨221⟩—A 50-mg portion shows no more sulfate than corresponds to 0.10 mL of 0.020 *N* sulfuric acid (0.2%).

Ordinary impurities ⟨466⟩—

Test solution: methanol.

Standard solution: methanol.

Eluant: a mixture of alcohol, glacial acetic acid, and water (5:3:2).

Visualization: 1.

Assay—Transfer about 500 mg of Ethylnorepinephrine Hydrochloride, accurately weighed, to a conical flask, add 60 mL of glacial acetic acid, 10 mL of mercuric acetate TS, and 6 drops of crystal violet TS, and titrate with 0.1 *N* perchloric acid VS. Perform a blank determination, and make any necessary correction. Each mL of 0.1 *N* perchloric acid is equivalent to 23.37 mg of $C_{10}H_{15}NO_3.HCl$.

Ethylnorepinephrine Hydrochloride Injection

» Ethylnorepinephrine Hydrochloride Injection is a sterile solution of Ethylnorepinephrine Hydrochloride in Water for Injection. It contains not less than 90.0 percent and not more than 115.0 percent of the labeled amount of $C_{10}H_{15}NO_3.HCl$.

Note—Do not use the Injection if it is brown or contains a precipitate.

Packaging and storage—Preserve in single-dose or in multiple-dose, light-resistant containers, preferably of Type I glass.

Reference standard—*USP Ethylnorepinephrine Hydrochloride Reference Standard*—Dry in vacuum over phosphorus pentoxide for 24 hours before using.

Identification—

A: Dilute a volume of Injection with water to obtain a solution having a concentration of about 50 µg of ethylnorepinephrine hydrochloride per mL: the ultraviolet absorption spectrum of this solution exhibits maxima and minima at the same wavelengths as that of a similar solution of USP Ethylnorepinephrine Hydrochloride RS, concomitantly measured.

B: *Reagent solution*—Dissolve 1.5 g of ferrous sulfate in 200 mL of water containing 1 mL of 1 *N* hydrochloric acid and 1 g of sodium bisulfite. To 5 mL of this solution add 250 mg of sodium citrate and 250 mg of sodium bisulfite. Dissolve, and mix.

Buffer solution—Dissolve 50.4 g of sodium bicarbonate in 480 mL of water containing 10 mL of ammonium hydroxide and 22.5 g of aminoacetic acid. Dilute with water to 500 mL, and mix.

Procedure—To 1 mL of Ethylnorepinephrine Hydrochloride Injection add 5 mL of water, 0.1 mL of *Reagent solution*, and 2 mL of *Buffer solution*, and mix: a violet color is produced.

pH ⟨791⟩: between 2.5 and 5.0.

Other requirements—It meets the requirements under *Injections* ⟨1⟩.

Assay—

Sodium acetate–ferricyanide reagent—Dissolve 50 mg of potassium ferricyanide in 100 mL of sodium acetate solution, prepared by mixing 33.3 g of sodium acetate with water to make 200 mL of solution.

Alkaline ascorbate solution—Just prior to use, mix 135 mL of sodium hydroxide solution (1 in 10) with 15 mL of ascorbic acid solution (1 in 50).

Standard preparation—Dissolve an accurately weighed quantity of USP Ethylnorepinephrine Hydrochloride RS in 0.1 *N* hydrochloric acid, and dilute quantitatively with 0.1 *N* hydrochloric acid to obtain a solution having a known concentration of about 100 µg per mL.

Assay preparation—Transfer an accurately measured volume of Ethylnorepinephrine Hydrochloride Injection, equivalent to about 10 mg of ethylnorepinephrine hydrochloride, to a 100-mL volumetric flask, dilute with 0.1 *N* hydrochloric acid to volume, and mix. Transfer 2.0 mL of this solution to a small beaker containing 5 mL of 0.1 *N* hydrochloric acid. Gently reduce the volume of the solution on a steam bath to about 3 mL. Cool to room temperature. Transfer, with the aid of small portions of 0.1 *N* hydrochloric acid, to a 25-mL volumetric flask, keeping the resulting volume at 5.0 mL. Proceed as directed under *Procedure*, beginning with "At accurately timed intervals."

Procedure—Transfer 2.0 mL each of the *Standard preparation*, and 0.1 *N* hydrochloric acid to provide a blank, to separate 25-mL volumetric flasks, and add 3.0 mL of 0.1 *N* hydrochloric acid to each flask. At accurately timed intervals, proceed as follows with each flask: Add 5.0 mL of *Sodium acetate–ferricyanide reagent*, and mix. After 8 minutes, accurately timed, quench the reaction by adding 10.0 mL of *Alkaline ascorbate solution*, dilute with water to volume, and mix. After 30 minutes, accurately timed, determine the absorbance of each solution in 1-cm cells relative to the blank, at the wavelength of maximum absorbance at about 402 nm, with a suitable spectrophotometer. Calculate the quantity, in mg, of $C_{10}H_{15}NO_3.HCl$ in each mL of the Injection taken by the formula:

$$0.1(C/V)(A_U/A_S),$$

in which C is the concentration, in µg per mL, of USP Ethylnorepinephrine Hydrochloride RS in the *Standard preparation*, V is the volume, in mL, of Injection taken, and A_U and A_S are the absorbances of the solutions from the *Assay preparation* and the *Standard preparation*, respectively.

Ethylparaben—*see* Ethylparaben NF

Ethynodiol Diacetate

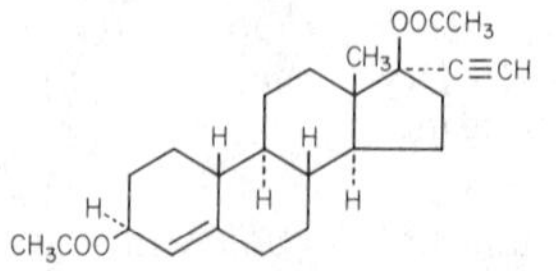

$C_{24}H_{32}O_4$ 384.51
19-Norpregn-4-en-20-yne-3,17-diol, diacetate, (3β,17α)-.
19-Nor-17α-pregn-4-en-20-yne-3β,17-diol diacetate [*297-76-7*].

» Ethynodiol Diacetate contains not less than 97.0 percent and not more than 102.0 percent of $C_{24}H_{32}O_4$.

Packaging and storage—Preserve in well-closed, light-resistant containers.

Reference standard—*USP Ethynodiol Diacetate Reference Standard*—Do not dry; use as is.

Identification—The infrared absorption spectrum of a potassium bromide dispersion of it exhibits maxima only at the same wavelengths as that of a similar preparation of USP Ethynodiol Diacetate RS.

Specific rotation ⟨781⟩: between −70° and −76°, determined in a solution in chloroform containing 100 mg in each 10 mL.

Limit of conjugated diene—The absorbance of a 1 in 2000 solution in methanol, determined in a 1-cm cell at the wavelength of maximum absorbance at about 236 nm, methanol being used as the blank, is not more than 0.500.

Assay—

Mobile solvent—Prepare a mixture of cyclohexane and ethyl acetate in the ratio of about 95 to 5, such that the retention time of ethynodiol diacetate is within 8 to 12 minutes.

Standard preparation—Dissolve about 40 mg of USP Ethynodiol Diacetate RS, accurately weighed, in *Mobile solvent*, and dilute with *Mobile solvent* to 5.0 mL.

Assay preparation—Dissolve about 40 mg of Ethynodiol Diacetate, accurately weighed, in *Mobile solvent*, and dilute with *Mobile solvent* to 5.0 mL.

Procedure—Introduce equal volumes (about 25 µL) of the *Assay preparation* and the *Standard preparation*, in triplicate, into a high-performance liquid chromatograph (see *Chromatography* ⟨621⟩) by means of a suitable microsyringe or sampling valve, adjusting the specimen size and other operating parameters such that the peak obtained with the *Standard preparation* is about 0.6 full-scale. Typically, the apparatus is fitted with a 2-mm × 25-cm column containing L3 having a mean surface area of about 300 m^2 per g, and is equipped with a high-sensitivity refractive index detector and a suitable recorder. Measure the height of the peaks, at identical retention times, obtained with the *Assay preparation* and the *Standard preparation*, and calculate the quantity, in mg, of $C_{24}H_{32}O_4$ in the portion of Ethynodiol Diacetate taken by the formula:

$$5C(H_U/H_S),$$

in which *C* is the concentration, in mg per mL, of USP Ethynodiol Diacetate RS in the *Standard preparation*, and H_U and H_S are the peak heights obtained from the *Assay preparation* and the *Standard preparation*, respectively.

Ethynodiol Diacetate and Ethinyl Estradiol Tablets

» Ethynodiol Diacetate and Ethinyl Estradiol Tablets contain not less than 93.0 percent and not more than 107.0 percent of the labeled amount of ethynodiol diacetate ($C_{24}H_{32}O_4$), and not less than 90.0 percent and not more than 110.0 percent of the labeled amount of ethinyl estradiol ($C_{20}H_{24}O_2$).

Packaging and storage—Preserve in well-closed containers.

Reference standards—*USP Ethynodiol Diacetate Reference Standard*—Do not dry; use as is. *USP Ethinyl Estradiol Reference Standard*—Dry in vacuum over silica gel for 4 hours before using.

Identification—Place a quantity of finely powdered Tablets, equivalent to about 10 mg of ethynodiol diacetate, in a stoppered 15-mL centrifuge tube. Add 10 mL of acetonitrile, insert the stopper in the tube, and mix by shaking and inversion for about 2 minutes. Centrifuge at about 1200 rpm for 10 minutes, and decant the supernatant liquid through filter paper into a suitable container. Evaporate a 5-mL aliquot of the filtrate on a steam bath with the aid of a stream of nitrogen, and dissolve the residue in 1 mL of chloroform. Apply 20 µL each of the solution under test, a Standard solution of USP Ethynodiol Diacetate RS in chloroform containing 5 mg per mL, and a Standard solution of USP Ethinyl Estradiol RS in chloroform containing 0.25 mg per mL at points about 3 cm from one end of a thin-layer chromatographic plate (see *Chromatography* ⟨621⟩) coated with a 0.25-mm layer of chromatographic silica gel mixture. Place the plate in a developing chamber containing a mixture of 3 volumes of ethyl acetate and 7 volumes of cyclohexane to a depth of 2 cm, the developing chamber previously having been equilibrated with the solvent mixture as described under *Thin-layer Chromatography* (see *Chromatography* ⟨621⟩). Remove the plate when the solvent moves to about 15 cm above the initial spots, allow it to dry in air at room temperature, spray it with a 1 in 10 solution of phosphomolybdic acid in alcohol, and heat it at 80° for 10 minutes: the spots from the solution under test and the Standard solutions appear dark on a light-green background. The ethynodiol diacetate and ethinyl estradiol spots from the solution under test have the same relative positions on the plate as the spots from the Standard solutions. The R_f values of ethynodiol diacetate and of ethinyl estradiol in this system are about 0.8 and about 0.4, respectively.

Disintegration ⟨701⟩: 15 minutes, the use of disks being omitted.

Uniformity of dosage units ⟨905⟩: meet the requirements for *Content Uniformity* with respect to ethynodiol diacetate and to ethinyl estradiol, 5-cm cells being used instead of 1-cm cells in the *Procedure* under *Assay for ethinyl estradiol*.

Assay for ethynodiol diacetate—

Standard preparation—Dissolve about 50 mg of USP Ethynodiol Diacetate RS, accurately weighed, in methanol in a volumetric flask to make 100.0 mL, and mix. Pipet a 10-mL aliquot of this solution into a 100-mL volumetric flask, add methanol to volume, and mix.

Assay preparation—Weigh and finely powder not less than 60 Ethynodiol Diacetate and Ethinyl Estradiol Tablets. Transfer an accurately weighed portion of the powder, equivalent to about 20 mg of ethynodiol diacetate, to a 200-mL volumetric flask, add 4 mL of water, and swirl gently until the substance becomes uniformly moistened. Add 4 mL of methanol, and warm gently on a steam bath. Add 100 mL of methanol, heat to boiling, with intermittent swirling, on a steam bath, cool in a water bath, add methanol to volume, and mix. Filter through dry filter paper, discarding the first 15 mL of the filtrate, and collect not less than 50 mL of the filtrate.

Procedure—Pipet 20 mL of the *Standard preparation* into a 100-mL volumetric flask. Pipet 10 mL of the *Assay preparation* into a 100-mL volumetric flask, add 10 mL of methanol, and mix. To each flask add 1.0 mL of dilute hydrochloric acid (3 in 5), boil gently on a steam bath for 5 minutes, cool, add methanol to volume, and mix. Determine the absorbance of the solution from the *Standard preparation* in a 1-cm cell at the wavelength of maximum absorbance at about 236 nm, with a suitable spectrophotometer, relative to a blank prepared by diluting 10.0 mL of *Standard preparation* with methanol to 50.0 mL. Determine the absorbance of the solution from the *Assay preparation* in a 1-cm cell at the wavelength of maximum absorbance at about 236 nm, with a suitable spectrophotometer, relative to a blank prepared by diluting 10.0 mL of *Assay preparation* with methanol to 100.0 mL. Calculate the quantity, in µg, of $C_{24}H_{32}O_4$ in the portion of Tablets taken by the formula:

$$(C/2.5)(A_U/A_S),$$

in which *C* is the concentration, in µg per mL, of USP Ethynodiol

Diacetate RS in the *Standard preparation*, and A_U and A_S are the absorbances of the solutions from the *Assay preparation* and the *Standard preparation*, respectively.

Assay for ethinyl estradiol—[NOTE—In this procedure, use scrupulously dry glassware and separators fitted with solvent-resistant stopcocks, use isooctane that gives no color when shaken with an equal volume of sulfuric acid, and use the same batch of sulfuric acid preparation for a given set of assays.]

Sulfuric acid preparation—Cautiously add 400 mL of sulfuric acid to 100 mL of chilled water, with mixing, and adjust to room temperature.

Benzene-isooctane—Mix 200 mL of benzene with 300 mL of isooctane in a suitable stoppered container.

Standard preparation—Dissolve about 50 mg of USP Ethinyl Estradiol RS, accurately weighed, in 100 mL of benzene in a 250-mL volumetric flask, warming slightly if necessary. Adjust to room temperature, add isooctane to volume, and mix. Pipet a 5-mL aliquot of this solution into a 100-mL volumetric flask, add *Benzene-isooctane* to volume, and mix.

Assay preparation—Weigh and finely powder not less than 60 Ethynodiol Diacetate and Ethinyl Estradiol Tablets. Transfer an accurately weighed portion of the powder, equivalent to about 200 µg of ethinyl estradiol, to a stoppered, 50-mL tapered centrifuge tube. Add 2 mL of water, mix on a vortex mixer to disperse the solids, and allow to stand for 5 minutes. Add 0.5 mL of 5 *N* hydrochloric acid, and again disperse the solids on a vortex mixer, dislodging any solids from the tip of the centrifuge tube. Add 20 mL of *Benzene-isooctane* to the centrifuge tube, insert the stopper, shake vigorously for 2 minutes, and centrifuge at 1200 rpm for 5 minutes. Carefully withdraw the aqueous layer into a 10-mL syringe fitted with a long needle, taking care not to withdraw any of the organic layer. Transfer the organic layer to a 125-mL separator, rinsing the centrifuge tube with 2 mL of *Benzene-isooctane*, and add the rinsing to the separator. Transfer the aqueous layer back to the centrifuge tube, rinsing the syringe with 1 mL of water, and add the rinsing to the centrifuge tube. Repeat the extraction, using 15 mL of *Benzene-isooctane*, and combine all *Benzene-isooctane* extracts and rinsings in the 125-mL separator.

Procedure—Pipet 20 mL of *Standard preparation* into a 125-mL separator, and add 20 mL of *Benzene-isooctane*. Pipet 5 mL of 1 *N* sodium hydroxide into the separator containing the *Assay preparation* and 5 mL into that containing the *Standard preparation*. Shake vigorously for 2 minutes, allow the phases to separate completely, and carefully drain the aqueous phases into separate 10-mL volumetric flasks. Repeat the extraction with 4.0 mL of 1 *N* sodium hydroxide, combine these extracts with the first extracts in the volumetric flasks, add water to volume, and mix. Pipet 25-mL aliquots of *Sulfuric acid preparation* into two separate 125-mL conical flasks, and place the flasks in ice baths. Slowly pipet 5-mL aliquots of the sodium hydroxide solutions of the *Assay preparation* and *Standard preparation*, respectively, into the corresponding chilled *Sulfuric acid preparation*, with stirring, regulating the rate of addition to maintain the temperature of the solution below 30°. Adjust to room temperature, and determine the absorbances of both solutions in 1-cm cells at the wavelength of maximum absorbance at about 536 nm, with a suitable spectrophotometer, relative to a water blank. Calculate the quantity, in µg, of $C_{20}H_{24}O_2$ in the portion of Tablets taken by the formula:

$$20C(A_U/A_S),$$

in which *C* is the concentration, in µg per mL, of USP Ethinyl Estradiol RS in the *Standard preparation*, and A_U and A_S are the absorbances of the solutions from the *Assay preparation* and the *Standard preparation*, respectively.

Ethynodiol Diacetate and Mestranol Tablets

» Ethynodiol Diacetate and Mestranol Tablets contain not less than 90.0 percent and not more than 110.0 percent of the labeled amounts of ethynodiol diacetate ($C_{24}H_{32}O_4$) and mestranol ($C_{21}H_{26}O_2$).

Packaging and storage—Preserve in well-closed containers.

Reference standards—*USP Ethynodiol Diacetate Reference Standard*—Do not dry; use as is. *USP Mestranol Reference Standard*—Dry at 105° for 3 hours before using.

Identification—Place a quantity of finely powdered Tablets, equivalent to about 10 mg of ethynodiol diacetate, in a stoppered, 15-mL centrifuge tube. Add 10 mL of acetonitrile, insert the stopper in the tube, and mix by shaking and inversion for about 2 minutes. Centrifuge at about 1200 rpm for 10 minutes, and decant the supernatant liquid through filter paper into a suitable container. Evaporate a 5-mL aliquot of the filtrate on a steam bath with the aid of a stream of nitrogen, and dissolve the residue in 1 mL of chloroform. Apply 5 µL of the solution under test, 5 µL of a Standard solution of USP Ethynodiol Diacetate RS in chloroform containing 5 mg per mL, and 5 µL of a Standard solution of USP Mestranol RS in chloroform containing 0.5 mg per mL at points about 3 cm from one end of a thin-layer chromatographic plate (see *Chromatography* ⟨621⟩), coated with a 0.25-mm layer of chromatographic silica gel mixture. Place the plate in a developing chamber containing a mixture of 30 volumes of ethyl acetate and 70 volumes of cyclohexane to a depth of 2 cm, the developing chamber previously having been equilibrated with the solvent mixture as described under *Thin-layer Chromatography* ⟨621⟩. Remove the plate when the solvent moves to about 15 cm above the initial spots, allow it to air-dry at room temperature, spray it with a 1 in 2 solution of sulfuric acid in water, and heat it at 80° for 10 minutes: the ethynodiol diacetate spots appear yellowish tan and the mestranol spots appear pink when viewed under white light. The ethynodiol diacetate and mestranol spots from the solution under test have the same relative positions on the plate as the spots from the respective Standard solutions: the R_f value of ethynodiol diacetate in this system is about 0.8, and the R_f value of mestranol in this system is about 0.6.

Disintegration ⟨701⟩: 15 minutes, the use of disks being omitted.

Uniformity of dosage units ⟨905⟩: meet the requirements for *Content Uniformity* with respect to ethynodiol diacetate and to mestranol.

Procedure for content uniformity—Proceed as directed in the *Assay*, except to reduce by half the concentrations in the *Standard preparation* and the *Assay preparation* and to inject twice the volume into the chromatograph.

Assay—

Mobile phase—Prepare a filtered and degassed mixture of methanol, water, and tetrahydrofuran (63:30:7), making adjustments if necessary (see *System Suitability* under *Chromatography* ⟨621⟩).

Standard preparation—Dissolve accurately weighed quantities of USP Ethynodiol Diacetate RS and USP Mestranol RS in a mixture of methanol and water (4:1) to obtain a solution having a known concentration of about 0.2 mg of ethynodiol diacetate per mL and 0.02 mg of mestranol per mL.

Assay preparation—Transfer 10 Tablets to a glass-stoppered flask, and pipet a sufficient volume of a mixture of methanol and water (4:1) into the flask to obtain a solution having final concentration of approximately 0.2 mg of ethynodiol diacetate per mL and 0.02 mg of mestranol per mL. Agitate vigorously until the tablets are completely disintegrated. Shake vigorously by mechanical means for 30 minutes. Allow the solids to settle for 10 minutes, and filter the solution through a 0.5-µm filter.

Chromatographic system (see *Chromatography* ⟨621⟩)—The liquid chromatograph is equipped with a 204-nm detector and a 4.6-mm × 25-cm column that contains packing L13. The flow rate is approximately 1.2 mL per minute. Chromatograph the *Standard preparation*, and record the peak responses as directed under *Procedure*: the relative retention times are about 0.6 for mestranol and 1.0 for ethynodiol diacetate. The mestranol is baseline separated from the ethynodiol diacetate peak, and the resolution, *R*, is not less than 4.0 between these two peaks. The relative standard deviation for replicate injections is not more than 2.0% for each analyte.

Procedure—Separately inject equal volumes (about 25 µL) of the *Standard preparation* and the *Assay preparation* into the

chromatograph, record the chromatograms, and measure the responses for the major peaks. Calculate the quantity, in mg, of ethynodiol diacetate ($C_{24}H_{32}O_4$) and mestranol ($C_{21}H_{26}O_2$) per Tablet taken by the formula:

$$0.1CV(r_U/r_S),$$

in which C is the concentration, in mg per mL, of the appropriate USP Reference Standard in the *Standard preparation*, V is the volume, in mL, of the methanol-water solvent added to obtain the *Assay preparation*, and r_U and r_S are the peak responses of the corresponding analyte obtained from the *Assay preparation* and the *Standard preparation*, respectively.

Etidronate Injection, Technetium Tc 99m—*see under "technetium"*

Etidronate Disodium

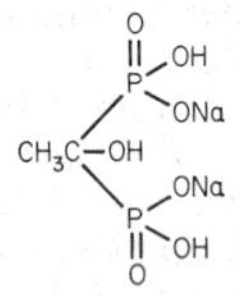

$C_2H_6Na_2O_7P_2$ 249.99

Phosphonic acid, (1-hydroxyethylidene)bis-, disodium salt.
Disodium dihydrogen (1-hydroxyethylidene)diphosphonate [*7414-83-7*].

» Etidronate Disodium contains not less than 97.0 percent and not more than 101.0 percent of $C_2H_6Na_2O_7P_2$, calculated on the dried basis.

Packaging and storage—Preserve in tight containers.

Reference standards—*USP Etidronate Disodium Reference Standard*—Do not dry before using. Keep container tightly closed, and store in a cool, dry place. *USP Etidronic Acid Monohydrate Reference Standard*—Do not dry before using. Keep container tightly closed, and store in a cool, dry place.

Identification—

A: The infrared absorption spectrum of a mineral oil dispersion of it exhibits maxima only at the same wavelengths as that of a similar preparation of USP Etidronate Disodium RS.

B: A solution (1 in 100) responds to the flame test for *Sodium* ⟨191⟩.

pH ⟨791⟩: between 4.2 and 5.2, in a solution (1 in 100).

Loss on drying ⟨731⟩—Dry it at 210° to constant weight: it loses not more than 5.0% of its weight.

Heavy metals, *Method II* ⟨231⟩—Use 0.5 g of Etidronate Disodium for the *Test Preparation* and 2.5 mL of *Standard Lead Solution* for the *Standard Preparation*. Dilute the *Test Preparation* and the *Standard Preparation* with water to 50 mL prior to adding the hydrogen sulfide TS. The limit is 0.005%.

Phosphite—

Buffer solution—Dissolve 6.9 g of monobasic sodium phosphate in 500 mL of water, add 400 mL of 0.1 *N* sodium hydroxide, and mix.

Test preparation—Transfer about 3.5 g of Etidronate Disodium, accurately weighed, to a glass-stoppered, 250-mL conical flask, and dissolve in 70 mL of water.

Procedure—Add 20 mL of *Buffer solution* to the *Test preparation*, and adjust with sodium hydroxide solution (25 in 100), using a pH meter, to a pH of 7.3 ± 0.2. Add 20 mL of iodine TS in divided portions, with mixing, and insert the stopper securely in the flask. [NOTE—If more than half of the iodine TS is absorbed (decolorized) by the portion of Etidronate Disodium taken, discard the mixture, and use a *Test preparation* containing a smaller quantity of Etidronate Disodium.] Allow to stand for 3 hours protected from light, and adjust with 6 *N* acetic acid to a pH of 4.5 ± 0.2. Titrate with 0.1 *N* sodium thiosulfate VS. When the iodine color becomes pale, add 2 mL of starch TS, and continue the titration until the blue color is discharged. Perform a blank determination (see *Titrimetry* ⟨541⟩), and make any necessary correction. Calculate the percentage of phosphite in the portion of Etidronate Disodium taken by the formula:

$$0.520V/W,$$

in which V is the volume, in mL, of titrant consumed, and W is the weight, in g, of Etidronate Disodium taken: not more than 1.0% is found.

Assay—

Titrant—Transfer 6.9 g of thorium nitrate, accurately weighed, to a 1000-mL beaker, and dissolve in 25 mL of 1 *N* nitric acid. Dissolve 4.7 g of (1,2-cyclohexylenedinitrilo)tetraacetic acid in 41 mL of 1 *N* sodium hydroxide, and transfer to the 1000-mL beaker. Add about 600 mL of water, and adjust with methenamine to a pH of between 5.0 and 5.5. Transfer to a 1000-mL volumetric flask, dilute with water to volume, and mix. Allow to stand for 1 week before use. This solution is about 0.0125 *M* in thorium-cyclohexylenedinitrilotetraacetic acid complex. Each mL is equivalent to approximately 1.5 mg of $C_2H_6Na_2O_7P_2$.

Standard preparation—Transfer 180 mg of USP Etidronic Acid Monohydrate RS to a 100-mL volumetric flask, dissolve in 50 mL of water, add 16 mL of 0.1 *N* sodium hydroxide, dilute with water to volume, and mix.

Assay preparation—Transfer about 400 mg of Etidronate Disodium, accurately weighed, to a 200-mL volumetric flask, dissolve in 100 mL of water, dilute with water to volume, and mix.

Procedure—Pipet 15 mL each of the *Standard preparation* and the *Assay preparation* into separate 100-mL beakers. To each beaker add 20 mL of water, 1 mL of xylenol orange TS, and 2 mL of methenamine solution (20 in 100), and mix. Adjust with 0.1 *N* nitric acid to a pH of 6.50 ± 0.05. Concomitantly titrate the resulting solutions to a reddish violet end-point. Calculate the quantity, in mg, of $C_2H_6Na_2O_7P_2$ in the portion of Etidronate Disodium taken by the formula:

$$2(249.99/224.05)W_S(V_U/V_S),$$

in which 249.99 and 224.05 are the molecular weights of etidronate disodium and etidronic acid monohydrate, respectively, W_S is the weight, in mg, of USP Etidronic Acid Monohydrate RS taken, and V_U and V_S are the volumes, in mL, of *Titrant* consumed by the *Assay preparation* and the *Standard preparation*, respectively.

Etidronate Disodium Tablets

» Etidronate Disodium Tablets contain not less than 90.0 percent and not more than 110.0 percent of the labeled amount of $C_2H_6Na_2O_7P_2$.

Packaging and storage—Preserve in tight containers.

Reference standards—*USP Etidronate Disodium Reference Standard*—Do not dry before using. Keep container tightly closed, and store in a cool, dry place. *USP Etidronic Acid Monohydrate Reference Standard*—Do not dry before using. Keep container tightly closed, and store in a cool, dry place.

Identification—The infrared absorption spectrum of a mineral oil dispersion prepared from finely powdered Tablets exhibits maxima only at the same wavelengths as that of a similar preparation of USP Etidronate Disodium RS.

Dissolution ⟨711⟩—

Medium: water; 900 mL.

Apparatus 1: 100 rpm.

Time: 30 minutes.

Procedure—Determine the amount of $C_2H_6Na_2O_7P_2$ dissolved, employing the procedure set forth in the *Assay* making any necessary volumetric adjustments.

Tolerances—Not less than 70% (*Q*) of the labeled amount of $C_2H_6Na_2O_7P_2$ is dissolved in 30 minutes.

Uniformity of dosage units ⟨905⟩: meet the requirements.

Assay—

Titrant and *Standard preparation*—Prepare as directed in the *Assay* under *Etidronate Disodium*.

Assay preparation—Weigh and finely powder not less than 20 Etidronate Disodium Tablets. Transfer an accurately weighed portion of the powder, equivalent to about 200 mg of etidronate disodium, to a 100-mL volumetric flask, and dilute with water to volume. Stir on a magnetic stirrer for about 10 minutes, and filter.

Procedure—Proceed as directed for *Procedure* in the *Assay* under *Etidronate Disodium*. Calculate the quantity, in mg, of $C_2H_6Na_2O_7P_2$ in the portion of Tablets taken by the formula:

$$200(V_U/V_S),$$

in which V_U and V_S are the volumes, in mL, of *Titrant* consumed by the *Assay preparation* and the *Standard preparation*, respectively.

Eucatropine Hydrochloride

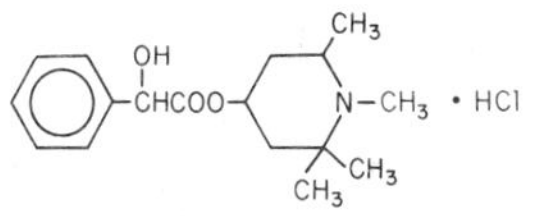

$C_{17}H_{25}NO_3.HCl$ 327.85

Benzeneacetic acid, α-hydroxy-, 1,2,2,6-tetramethyl-4-piperidinyl ester hydrochloride.

1,2,2,6-Tetramethyl-4-piperidyl mandelate hydrochloride [*536-93-6*].

» Eucatropine Hydrochloride contains not less than 99.0 percent and not more than 100.5 percent of $C_{17}H_{25}NO_3.HCl$, calculated on the dried basis.

Packaging and storage—Preserve in tight, light-resistant containers.

Identification—

A: Dissolve about 50 mg in 5 mL of water, render the solution alkaline with 6 *N* ammonium hydroxide, and extract with two 10-mL portions of ether. Evaporate the ether on a steam bath, and recrystallize the residue from solvent hexane: the eucatropine base so obtained melts between 111° and 114°.

B: A solution of it responds to the tests for *Chloride* ⟨191⟩.

Melting range ⟨741⟩: between 183° and 186°.

Loss on drying ⟨731⟩—Dry it over silica gel for 4 hours: it loses not more than 0.5% of its weight.

Residue on ignition ⟨281⟩: not more than 0.1%.

Assay—Weigh accurately about 500 mg of Eucatropine Hydrochloride, and dissolve it in 10 mL of water. Saturate the solution with sodium chloride. Render it alkaline with 6 *N* ammonium hydroxide, and extract the base completely with successive 15-mL portions of ether. Wash the combined ether extracts with 10 mL of water, and extract the water washing with 10 mL of ether. To the combined ether solutions add 25.0 mL of 0.1 *N* sulfuric acid VS, and stir well. Heat gently until the ether is expelled, cool, add methyl red TS, and titrate the excess acid with 0.1 *N* sodium hydroxide VS. Each mL of 0.1 *N* sulfuric acid is equivalent to 32.79 mg of $C_{17}H_{25}NO_3.HCl$.

Eucatropine Hydrochloride Ophthalmic Solution

» Eucatropine Hydrochloride Ophthalmic Solution is a sterile, isotonic, aqueous solution of Eucatropine Hydrochloride. It contains not less than 95.0 percent and not more than 105.0 percent of the labeled amount of $C_{17}H_{25}NO_3.HCl$. It may contain suitable antimicrobial agents.

Packaging and storage—Preserve in tight containers.

Reference standard—*USP Eucatropine Hydrochloride Reference Standard*—Dry over silica gel for 4 hours before using.

Identification—

A: It meets the requirements under *Identification—Organic Nitrogenous Bases* ⟨181⟩.

B: It responds to the tests for *Chloride* ⟨191⟩.

Sterility—It meets the requirements under *Sterility Tests* ⟨71⟩.

pH ⟨791⟩: between 4.0 and 5.0.

Assay—

Standard preparation—Weigh accurately about 50 mg of USP Eucatropine Hydrochloride RS, dissolve in water, and dilute with water in a volumetric flask to 100 mL. Dilute 10.0 mL of this solution with water to 50.0 mL to obtain a solution having a known concentration of about 100 µg per mL. Prepare this solution fresh.

Assay preparation—Transfer a portion of Eucatropine Hydrochloride Ophthalmic Solution, equivalent to 50 mg of eucatropine hydrochloride, to a 100-mL volumetric flask, and dilute with water to volume. Dilute 10.0 mL of this solution with water to 50.0 mL.

Procedure—Transfer duplicate 2-mL portions of the *Standard preparation* and of the *Assay preparation* to separate glass-stoppered, 40-mL centrifuge tubes. To one set of two tubes add 3 mL of water and 1 mL of sodium hydroxide solution (1 in 100). Heat these tubes in a boiling water bath for 10 minutes, and allow to cool to room temperature. To the remaining set of tubes, which provide the blanks for the *Standard preparation* and *Assay preparation*, respectively, add 4 mL of water. To each tube add 2 mL of approximately 0.2 *M* ceric sulfate in diluted sulfuric acid (prepared by dissolving 12.6 g of ceric ammonium sulfate in 50 mL of water and 3 mL of sulfuric acid, and diluting with water to 100 mL) and 20.0 mL of isooctane. Shake by mechanical means for 15 minutes, allow the layers to separate, and remove the isooctane from each tube. Concomitantly determine the absorbances of the isooctane solutions from the hydrolyzed aliquots in 1-cm cells at the wavelength of maximum absorbance at about 242 nm, with a suitable spectrophotometer, against the respective blanks. Calculate the quantity, in mg, of $C_{17}H_{25}NO_3.HCl$ in the portion of Ophthalmic Solution taken by the formula:

$$0.5C(A_U/A_S),$$

in which *C* is the concentration, in µg per mL, of USP Eucatropine Hydrochloride RS in the *Standard preparation*, and A_U and A_S are the absorbances of the solutions from the *Assay preparation* and the *Standard preparation*, respectively.

Eugenol

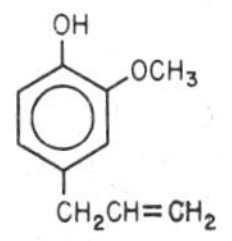

$C_{10}H_{12}O_2$ 164.20

Phenol, 2-methoxy-4-(2-propenyl)-.

4-Allyl-2-methoxyphenol [*97-53-0*].

» Eugenol is obtained from Clove Oil and from other sources.

Packaging and storage—Preserve in tight, light-resistant containers.

Solubility in 70 percent alcohol—One volume dissolves in 2 volumes of 70 percent alcohol.

Specific gravity ⟨841⟩: between 1.064 and 1.070.

Distilling range, *Method II* ⟨721⟩—Not less than 95% distils between 250° and 255°.

Refractive index ⟨831⟩: between 1.540 and 1.542 at 20°.

Heavy metals, *Method II* ⟨231⟩: 0.004%.

Hydrocarbons—Dissolve 1 mL in 20 mL of 0.5 *N* sodium hydroxide in a stoppered, 50-mL tube, add 18 mL of water, and mix: a clear mixture results immediately, but it may become turbid when exposed to air.

Phenol—Shake 1 mL with 20 mL of water, filter, and add 1 drop of ferric chloride TS to 5 mL of the clear filtrate: the mixture exhibits a transient grayish green color but not a blue or violet color.

Evans Blue

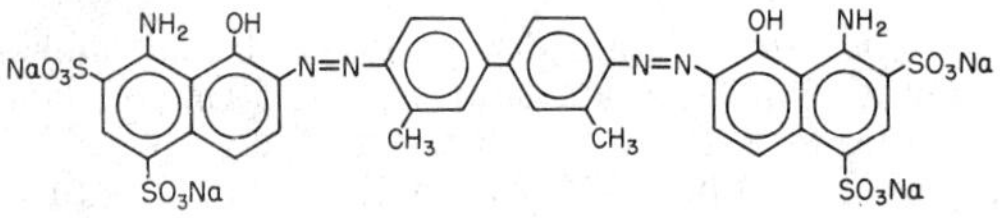

$C_{34}H_{24}N_6Na_4O_{14}S_4$ 960.79

1,3-Naphthalenedisulfonic acid, 6,6′-[(3,3′-dimethyl[1,1′-biphenyl]-4,4′-diyl)bis(azo)]bis[4-amino-5-hydroxy]-, tetrasodium salt.

C.I. direct blue 53 tetrasodium salt [*314-13-6*].

» Evans Blue contains not less than 95.0 percent and not more than 105.0 percent of $C_{34}H_{24}N_6Na_4O_{14}S_4$, calculated on the dried basis.

Packaging and storage—Preserve in tight containers.

Reference standard—*USP Evans Blue Reference Standard.*

Identification—

A: To a 50- × 8-mm test tube clamped in a vertical position add a piece of clean sodium metal about 8 mm in diameter, then add 50 mg of Evans Blue, and heat the lower part of the tube until the sodium melts and sodium vapors rise. Add another 50 mg of Evans Blue, and heat the bottom of the tube to a dull red heat. Allow to cool, and add 1 mL of alcohol to dissolve any unchanged sodium. Again heat the tube, and while it is still hot drop it into a small beaker containing 10 mL of water (*caution*). Break up the tube with a stirring rod, heat the solution to boiling, and filter: the filtrate is colorless. To 1 mL of the filtrate add 2 drops of sodium nitroferricyanide TS: a red-violet color appears. To 2 mL of the filtrate add a crystal of ferrous sulfate and 0.1 mL of 1 *N* sodium hydroxide, heat the mixture to boiling, filter, acidify the filtrate with 3 *N* hydrochloric acid, and add 1 drop of ferric chloride TS: a blue color appears.

B: The solution employed for measurement of absorbance in the *Assay* exhibits an absorbance maximum at about 610 nm, and its spectrum between 250 nm and 750 nm is similar to that of the Standard solution.

Loss on drying ⟨731⟩—Dry it at 105° for 2 hours: it loses not more than 15.0% of its weight.

Insoluble substances—Dissolve about 1 g, accurately weighed, in 100 mL of water. Filter through a tared, fine-porosity, sintered-glass crucible, wash with water until the dye color disappears from the crucible, and dry at 105° to constant weight: the weight of the residue is not more than 0.05% of the weight of test specimen taken.

Acetate—A 2-mL portion of the solution of it prepared in the test for *Chloride* gives no odor of acetic acid or of ethyl acetate when heated to boiling with 2 mL of sulfuric acid and 2 mL of alcohol.

Chloride—Dissolve 25 mg in 5 mL of water. To 2 mL of this solution add 2 mL of nitric acid, and heat to boiling. Add 9 mL of water to 1 mL of the boiled solution, then add 0.5 mL of silver nitrate TS: no turbidity develops.

Heavy metals, *Method II* ⟨231⟩: 0.007%.

Assay—Transfer to a 1000-mL volumetric flask an amount of Evans Blue, accurately weighed, equivalent to about 350 mg of Evans blue on the dried basis. Add about 700 mL of water, shake the flask until the dye dissolves, dilute with water to volume, and mix. Pipet 10 mL of the solution into a second 1000-mL volumetric flask, dilute with water to volume, and mix. Dissolve an accurately weighed quantity of USP Evans Blue RS in water, and dilute quantitatively and stepwise with water to obtain a Standard solution having a known concentration of about 3.5 µg per mL. Concomitantly determine the absorbances of both solutions in 1-cm cells at the wavelength of maximum absorbance at about 610 nm, with a suitable spectrophotometer, using water as the blank. Calculate the quantity, in mg, of $C_{34}H_{24}N_6Na_4O_{14}S_4$ in the Evans Blue taken by the formula:

$$100C(A_U/A_S),$$

in which C is the concentration, in µg per mL, of USP Evans Blue RS in the Standard solution, and A_U and A_S are the absorbances of the solution of Evans Blue and the Standard solution, respectively.

Evans Blue Injection

» Evans Blue Injection is a sterile solution of Evans Blue in Water for Injection. It contains, in each mL, not less than 4.30 mg and not more than 4.75 mg of $C_{34}H_{24}N_6Na_4O_{14}S_4$.

Packaging and storage—Preserve in single-dose containers, preferably of Type I glass.

Reference standard—*USP Evans Blue Reference Standard.*

Identification—Evaporate on a steam bath to dryness a volume of Injection equivalent to about 100 mg of Evans blue: the residue responds to the *Identification tests* under *Evans Blue.*

Pyrogen—It meets the requirements of the *Pyrogen Test* ⟨151⟩, the test dose being 1 mL per kg.

pH ⟨791⟩: between 5.5 and 7.5.

Other requirements—It meets the requirements under *Injections* ⟨1⟩.

Assay—Pipet into a 250-mL volumetric flask 10 mL of Evans Blue Injection, dilute with water to volume, and mix. Pipet 10 mL of this solution into a 500-mL volumetric flask, dilute with water to volume, and mix. Dissolve an accurately weighed quantity of USP Evans Blue RS in water, and dilute quantitatively and stepwise with water to obtain a Standard solution having a known concentration of about 3.6 µg per mL. Concomitantly determine the absorbances of both solutions in 1-cm cells at the wavelength of maximum absorbance at about 610 nm, with a suitable spectrophotometer, using water as the blank. Calculate the quantity, in mg, of $C_{34}H_{24}N_6Na_4O_{14}S_4$ in each mL of the Injection taken by the formula:

$$12.5(C/V)(A_U/A_S),$$

in which C is the concentration, in µg per mL, of USP Evans Blue RS in the Standard solution, V is the volume, in mL, of Injection taken, and A_U and A_S are the absorbances of the solution from the Injection and the Standard solution, respectively.

Extended Insulin Zinc Suspension—*see* Insulin Zinc Suspension, Extended

Extended Phenytoin Sodium Capsules—*see* Phenytoin Sodium Capsules, Extended

Extended-release Capsules—*see complete list in index*

Extended-release Tablets—*see complete list in index*

Factor IX Complex

» Factor IX Complex conforms to the regulations of the federal Food and Drug Administration concerning biologics (see *Biologics* ⟨1041⟩). It is a sterile, freeze-dried powder consisting of partially purified Factor IX fraction, as well as concentrated Factors II, VII, and X fractions, of venous plasma obtained from healthy human donors. It contains no preservative. It meets the requirements of the test for potency in having not less than 80 percent and not more than 120 percent of the potency stated on the label in Factor IX Units by comparison with the U.S. Factor IX Standard or with a working reference that has been calibrated with it.

Packaging and storage—Preserve in hermetic containers in a refrigerator.

Expiration date—The expiration date is not later than 2 years from the date of manufacture.

Labeling—Label it with a warning that it is to be used within 4 hours after constitution, and to state that it is for intravenous administration and that a filter is to be used in the administration equipment.

Famotidine

$(H_2N)_2C{=}N$–(2-thiazolyl, 4-position)–$CH_2SCH_2CH_2C(=NSO_2NH_2)NH_2$

$C_8H_{15}N_7O_2S_3$ 337.43

Propanimidamide, N'-(aminosulfonyl)-3-[[[2-[(diaminomethylene)amino]-4-thiazolyl]-methyl]thio]-.

[1-Amino-3-[[[2-[(diaminomethylene)amino]-4-thiazolyl]-methyl]thio]propylidene] [*76824-35-6*].

» Famotidine contains not less than 98.5 percent and not more than 101.0 percent of $C_8H_{15}N_7O_2S_3$, calculated on the dried basis.

Packaging and storage—Preserve in well-closed containers, protected from light.

Reference standard—*USP Famotidine Reference Standard*—Dry at a pressure between 1 mm and 5 mm of mercury at 80° for 5 hours before using.

Identification—

A: The ultraviolet absorption spectrum of a solution (1 in 40,000) in *Phosphate buffer* exhibits maximum and minimum at the same wavelengths as that of a similar solution of USP Famotidine RS concomitantly measured, and the respective absorptivities, calculated on the dried basis at the wavelength of maximum absorbance at about 265 nm do not differ by more than 3.0%. [NOTE—Prepare the *Phosphate buffer* as follows. Adjust 250 mL of 0.02 *M* phosphoric acid with sodium hydroxide solution (1 in 10) to a pH of 2.5, dilute with water to 500 mL, and mix.]

B: The infrared absorption spectrum of a potassium bromide dispersion of it, previously dried, exhibits maxima only at the same wavelengths as that of a similar preparation of USP Famotidine RS.

Loss on drying ⟨731⟩—Dry it at a pressure between 1 mm and 5 mm of mercury at 80° for 5 hours: it loses not more than 0.5% of its weight.

Residue on ignition ⟨281⟩: not more than 0.1%.

Heavy metals, *Method II* ⟨231⟩: not more than 0.001%.

Chromatographic purity—

Test preparation—Prepare a solution in glacial acetic acid containing 20 mg of Famotidine per mL.

Standard preparations—Dissolve USP Famotidine RS in glacial acetic acid to obtain a solution having a known concentration of 0.2 mg per mL. Dilute portions of this solution quantitatively with glacial acetic acid to obtain *Standard preparations A, B, C,* and *D,* containing 100 µg, 60 µg, 40 µg, and 20 µg of the *Reference standard* per mL, respectively.

Procedure—On a suitable high performance thin-layer chromatographic plate (see *Chromatography* ⟨621⟩), coated with a 0.25-mm layer of chromatographic silica gel mixture, apply separately 5 µL of the *Test preparation* and 5 µL of each *Standard preparation*, and dry under a stream of nitrogen. Position the plate in a chromatographic chamber, and develop the chromatograms in a solvent system consisting of a mixture of ethyl acetate, methanol, toluene, and ammonium hydroxide (40:25:20:2) until the solvent front has moved about three-fourths of the length of the plate. Remove the plate from the developing chamber, mark the solvent front, and air-dry the plate. Examine the plate under short-wavelength ultraviolet light, and compare the intensities of any secondary spots observed in the chromatogram of the *Test preparation* with those of the principal spots in the chromatograms of the *Standard preparations:* no secondary spot from the chromatogram of the *Test preparation* is larger in size or more intense than the principal spot obtained from *Standard preparation A* (0.5%) and the sum of the intensities of the secondary spots obtained from the *Test preparation* corresponds to not more than 2.0%.

Assay—Dissolve about 250 mg of Famotidine, accurately weighed, in 80 mL of glacial acetic acid, and titrate with 0.1 *N* perchloric acid VS (see *Titrimetry* ⟨541⟩), using a suitable anhydrous electrode system. Any aqueous electrolyte solution contained in the electrodes employed should be removed, the electrode rendered anhydrous and filled with 0.1 *N* lithium perchlorate in acetic anhydride. Perform a blank determination and make any necessary correction. Each mL of 0.1 *N* perchloric acid is equivalent to 16.87 mg of $C_8H_{15}N_7O_2S_3$.

Famotidine Tablets

» Famotidine Tablets contain not less than 90.0 percent and not more than 110.0 percent of the labeled amount of $C_8H_{15}N_7O_2S_3$.

Packaging and storage—Preserve in well-closed, light-resistant containers.

Reference standard—*USP Famotidine Reference Standard*—Dry at a pressure between 1 mm and 5 mm of mercury at 80° for 5 hours before using.

Identification—The retention time of the major peak in the chromatogram of the *Assay preparation* corresponds to that of the *Standard preparation*, as obtained with *Assay*.

Dissolution ⟨711⟩—

Medium: pH 4.5, 0.1 *M* phosphate buffer; prepared by dissolving 13.6 g of monobasic potassium phosphate in one liter of water; 900 mL.

Apparatus 2: 50 rpm.

Time: 30 minutes.

Procedure—Determine the amount of $C_8H_{15}N_7O_2S_3$ dissolved from ultraviolet absorbances at the wavelength of maximum absorbance at about 265 nm using filtered portions of the solution under test, suitably diluted with *Dissolution Medium*, if necessary, in comparison with a *Standard solution* having a known concentration of USP Famotidine RS in the same medium.

Tolerances—Not less than 75% (*Q*) of the labeled amount of $C_8H_{15}N_7O_2S_3$ is dissolved in 30 minutes.

Uniformity of dosage units ⟨905⟩: meet the requirements.

Assay—

Buffer solution—Dissolve 1.36 g of monobasic potassium phosphate in 800 mL of water, and adjust to a pH of 7.0, determined potentiometrically, by the addition of 1 *N* sodium hydroxide with mixing. Dilute with water to 1000 mL, and mix.

Mobile phase—Prepare a mixture of water, methanol, and *Buffer solution* (31:6:3), adjust to a pH of 5.0, determined potentiometrically, by the addition of 0.1 *N* sodium hydroxide, mix, filter, and degas. Make adjustments if necessary (see *System Suitability* under *Chromatography* ⟨621⟩).

Standard solution—[NOTE—Prepare fresh daily.] Dissolve a suitable quantity of USP Famotidine RS in *Buffer solution* to obtain a solution having a known concentration of about 80 µg of famotidine per mL.

Assay preparation—Weigh and finely powder not less than 20 Famotidine Tablets. Transfer an accurately weighed portion of the powder, equivalent to about 200 mg of famotidine, to a 500-mL volumetric flask. Add 50 mL of *Buffer solution* and 300 mL of water, sonicate for 5 minutes, and mechanically shake for 1 hour. Dilute with water to volume, mix, and filter. Quantitatively dilute a portion of the clear filtrate with *Buffer solution* to obtain a solution containing about 80 µg of famotidine per mL.

Chromatographic system (see *Chromatography* ⟨621⟩)—The liquid chromatograph is equipped with a 254-nm detector and a 4.6-mm × 25-cm column that contains packing L3. The flow rate is about 1.0 mL per minute. Chromatograph the *Standard preparation*, and record the peak responses as directed under *Procedure:* the capacity factor is not less than 4.0, the column efficiency, determined from the analyte peak, is not less than 3200 theoretical plates, the tailing factor for the famotidine peak is not more than 3.0, and the relative standard deviation for replicate injections is not more than 2.0%. [NOTE—When *System Suitability* cannot be achieved, the following wash sequence for the column may be used: wash with 0.05 *M* phosphoric acid at a flow rate of 1.3 mL per minute for 15 minutes, wash with water at a flow rate of 1.3 mL per minute for 15 minutes, wash with acetonitrile at a flow rate of 1.3 mL per minute for 15 minutes, wash with methanol at a flow rate of 1.3 mL per minute for 15 minutes, and finally precondition the column with *Mobile phase.*]

Procedure—Separately inject equal volumes (about 25 µL) of the *Standard preparation* and the *Assay preparation* into the chromatograph, record the chromatograms, and measure the responses for the major peaks. Calculate the quantity, in mg, of $C_8H_{15}N_7O_2S_3$ in the portion of Tablets by the formula:

$$2.5C(r_U/r_S),$$

in which *C* is the concentration, in µg per mL, of USP Famotidine RS in the *Standard preparation*, and r_U and r_S are the peak responses obtained from the *Assay preparation* and the *Standard preparation*, respectively.

Fat, Hard—*see* Hard Fat NF

Fenoprofen Calcium

$C_{30}H_{26}CaO_6.2H_2O$ 558.64

Benzeneacetic acid, α-methyl-3-phenoxy-, calcium salt dihydrate, (±)-.

Calcium (±)-*m*-phenoxyhydratropate dihydrate [*53746-45-5*].

Anhydrous 522.61 [*34597-40-5*].

» Fenoprofen Calcium contains not less than 97.0 percent and not more than 103.0 percent of $(C_{15}H_{13}O_3)_2Ca$, calculated on the anhydrous basis.

Packaging and storage—Preserve in tight containers.

Reference standard—*USP Fenoprofen Calcium Reference Standard*—Do not dry before using; determine the water content titrimetrically at the time of use for quantitative analyses.

Identification—

A: The infrared absorption spectrum of a potassium bromide dispersion of it exhibits maxima only at the same wavelengths as that of a similar preparation of USP Fenoprofen Calcium RS.

B: The R_f value of the principal spot from the *Test preparation*, obtained as directed in the test for *Chromatographic purity*, corresponds to that from the *Standard preparation* (*Solution A*), obtained as directed in the test for *Chromatographic purity*.

C: Heat a 1 in 50 mixture of it with acetic acid, filter, and add 2 mL of ammonium oxalate TS to the filtrate: a white precipitate, which is soluble in 3 *N* hydrochloric acid, is formed.

Water, *Method I* ⟨921⟩: between 5.0% and 8.0%.

Heavy metals, *Method II* ⟨231⟩: 0.001%.

Chromatographic purity—

Standard preparations—To 20.0 mg of USP Fenoprofen Calcium RS in a suitable vial add 1.0 mL of a mixed solvent prepared by mixing equal volumes of chloroform and methanol (*Solution A*). Prepare a second solution by diluting 1.0 volume of *Solution A* with the same mixed solvent to obtain 100 volumes of solution (*Solution B*).

Test preparation—To 20.0 mg of Fenoprofen Calcium in a suitable vial add 1.0 mL of the same mixed solvent used for the *Standard preparations*.

Procedure—On a suitable thin-layer chromatographic plate (see *Chromatography* ⟨621⟩), coated with a 0.25-mm layer of chromatographic silica gel mixture, apply 50-µL portions of *Solution A*, *Solution B*, and the *Test preparation*. Develop the chromatogram in a solvent system consisting of toluene and glacial acetic acid (10:1) until the solvent front has moved about three-fourths of the length of the plate. Remove the plate from the developing chamber, mark the solvent front, allow the solvent to evaporate, and examine the plate under short-wavelength ultraviolet light: the chromatograms from the *Standard preparations* and the *Test preparation* show principal spots at about the same R_f value. Spray the plate with bromocresol green spray reagent (prepared by dissolving 40 mg of bromocresol green in 100 mL of alcohol and adding 0.1 *N* sodium hydroxide until a blue coloration just appears), and examine the chromatogram: excluding the blue spot for calcium remaining at the point of application, no secondary spot, if present in the chromatogram from the *Test preparation*, is more intense than the principal spot obtained from *Solution B*.

Calcium content—Transfer about 750 mg of Fenoprofen Calcium, accurately weighed, to a 50-mL volumetric flask, dissolve in alcohol with the aid of heat, if necessary, cool, dilute with alcohol to volume, and mix to obtain the *Test preparation*. In a 150-mL beaker, mix 70 mL of water, 2 mL of sodium hydroxide solution (1 in 10), and about 0.3 g of hydroxy naphthol blue trituration. Add about 1 mL of *Test preparation*, and titrate to the blue end-point with 0.05 *M* disodium ethylenediaminetetraacetate. Transfer 10.0 mL of *Test preparation* to this solution, and titrate to the blue end-point with 0.05 *M* disodium ethylenediaminetetraacetate VS. Each mL of 0.05 *M* disodium ethylenediaminetetraacetate is equivalent to 2.004 mg of Ca: not less than 7.3% and not more than 8.0% of Ca, calculated on the anhydrous basis, is found.

Assay—

Internal standard solution—Prepare a solution of docosane in chloroform having a concentration of about 2.5 mg per mL.

Standard preparation—Transfer about 75 mg of USP Fenoprofen Calcium RS, accurately weighed, to a 125-mL separator containing 15 mL of water and 3 mL of 6 *N* hydrochloric acid. Add 15 mL of chloroform, mix for 3 minutes, and allow the phases to separate. Drain the chloroform extract through a layer of about 2 g of anhydrous sodium sulfate, supported on glass wool and previously washed with chloroform, into a 50-mL volumetric flask. Extract the aqueous phase in the separator with two additional 15-mL portions of chloroform, draining each chloroform extract through the same sodium sulfate into the volumetric flask. Rinse the sodium sulfate filter with about 2 mL of chloroform, collect the rinsing with the combined chloroform extracts, dilute with chloroform to volume, and mix.

Assay preparation—Proceed as directed for *Standard preparation*, using Fenoprofen Calcium instead of the Reference Standard.

Chromatographic system (see *Chromatography* ⟨621⟩)—Under typical conditions, the gas chromatograph is equipped with a flame-ionization detector, and contains a 0.6-m × 3-mm glass column packed with 1 percent liquid phase G9 on 80- to 100-mesh support S1AB. The column is maintained isothermally at about 140°, the injection port is maintained at about 200°, and the detector block is maintained at about 240°. Dry helium is used as the carrier gas at a flow rate of about 60 mL per minute. [NOTE—Cure and condition the column as necessary (see *Chromatography* ⟨621⟩).]

System suitability—Chromatograph five injections of the solution from the *Standard preparation*, and record the peak areas, as directed under *Procedure:* the retention time for fenoprofen is about 0.3 relative to docosane. In a suitable chromatogram, the resolution factor between the 2 peaks is not less than 1.0, and the relative standard deviation for five replicate injections of the solution from the *Standard preparation* is not more than 2.0.

Procedure—Transfer 5.0 mL of the *Standard preparation* and 5.0 mL of the *Assay preparation* to separate screw-capped tubes. To each add 3.0 mL of *Internal standard solution*, quickly cap the tubes, and mix. To each add 0.20 mL of bis(trimethylsilyl)trifluoroacetamide, quickly cap the tubes, mix, and heat at 60° for 30 minutes. Inject a portion of the solution obtained from the *Standard preparation* into a suitable gas chromatograph, record the chromatogram so as to obtain not less than 50% of maximum recorder response, and measure the peak area of each component. Similarly inject a portion of the solution obtained from the *Assay preparation*, record the chromatogram, and measure the peak area of each component. Calculate the quantity, in mg, of $(C_{15}H_{13}O_3)_2Ca$ in the Fenoprofen Calcium taken by the formula:

$$50C(R_U/R_S),$$

in which C is the concentration, in mg per mL, of anhydrous fenoprofen calcium in the *Standard preparation*, as determined from the concentration of USP Fenoprofen Calcium RS corrected for moisture content by a titrimetric water determination, and R_U and R_S are the ratios of the areas of the fenoprofen peak to the docosane peak obtained from the *Assay preparation* and the *Standard preparation*, respectively.

Fenoprofen Calcium Capsules

» Fenoprofen Calcium Capsules contain not less than 90.0 percent and not more than 110.0 percent of the labeled amount of fenoprofen ($C_{15}H_{14}O_3$).

Packaging and storage—Preserve in well-closed containers.

Reference standards—*USP Fenoprofen Calcium Reference Standard*—Do not dry before using; determine the water content titrimetrically at the time of use for quantitative analyses. *USP Fenoprofen Sodium Reference Standard*—Do not dry before using; determine the water content titrimetrically at the time of use for quantitative analyses.

Identification—

A: Transfer to suitable flasks the portions of the two solutions retained from the *Assay*, and evaporate both on a water bath with the aid of a current of air to dryness: the infrared absorption spectrum of a film of the liquid residue obtained from the *Assay preparation*, between sodium chloride plates, exhibits maxima only at the same wavelengths as that of the liquid residue obtained from the *Standard preparation*.

B: Place an amount of Capsule contents, equivalent to about 300 mg of fenoprofen, in a suitable container, and dissolve in 10 mL of acetone. Filter the solution through paper, and collect the filtrate in a crucible. Carefully evaporate to dryness, and ignite the crucible and its contents. Dissolve the residue in 10 mL of 1 *N* hydrochloric acid, transfer the solution to a beaker, add 2 drops of methyl red TS, neutralize with 6 *N* ammonium hydroxide, and add 3 *N* hydrochloric acid dropwise until the solution is acid to the indicator. Upon the addition of ammonium oxalate TS, a white precipitate is formed. The residue so obtained is insoluble in acetic acid but dissolves in hydrochloric acid.

Dissolution ⟨711⟩—

Medium: pH 7.0 phosphate buffer (see *Buffer Solutions* in the section, *Reagents, Indicators, and Solutions*); 1000 mL.

Apparatus 1: 10-mesh basket; 100 rpm.

Time: 60 minutes.

Procedure—Filter 20 mL of the solution under test, and transfer 5.0 mL of the filtrate to a 25-mL volumetric flask. Dilute with *Dissolution Medium* to volume, and mix. Determine the absorbances of this solution and a Standard solution prepared from USP Fenoprofen Sodium RS, in the same medium having a known concentration of about 60 µg per mL, at the wavelength of maximum absorbance at about 270 nm, using *Dissolution Medium* as the blank. Calculate the amount of $C_{15}H_{14}O_3$ dissolved, in mg, by the formula:

$$(242.27/264.26)(5C)(A_U/A_S),$$

in which 242.27 is the molecular weight of fenoprofen, 264.26 is the molecular weight of anhydrous fenoprofen sodium, C is the concentration of anhydrous fenoprofen sodium in the Standard solution, as determined from the concentration of USP Fenoprofen Sodium RS corrected for moisture content by a titrimetric water determination, and A_U and A_S are the absorbances of the solutions obtained from the substance under test and the Reference Standard, respectively.

Tolerances—Not less than 75% (Q) of the labeled amount of $C_{15}H_{14}O_3$ is dissolved in 60 minutes.

Uniformity of dosage units ⟨905⟩: meet the requirements.

Assay—

Internal standard solution—Prepare a solution of diphenamid in chloroform having a concentration of about 5 mg per mL.

Standard preparation—Transfer about 100 mg of USP Fenoprofen Calcium RS, accurately weighed, to a 125-mL separator containing 5 mL of acetone and 2 mL of 6 *N* hydrochloric acid. Swirl to dissolve, add 15 mL of water, and extract with three 15-mL portions of chloroform, draining each chloroform extract through a layer of about 2 g of anhydrous sodium sulfate, supported on glass wool and previously washed with chloroform, into a 50-mL volumetric flask. Rinse the sodium sulfate filter with about 2 mL of chloroform, collect the rinsing with the combined chloroform extracts, dilute with chloroform to volume, and mix. Transfer 3.0 mL of this solution to a 10-mL conical flask, reserving the remaining portion of the solution for *Identification test A*, and evaporate with the aid of gentle heat and a current of air to dryness. Cool, add 2.0 mL of *Internal standard solution*, swirl to dissolve the residue, add 0.20 mL of bis(trimethylsilyl)trifluoroacetamide, insert the stopper in the flask, and mix.

Assay preparation—Weigh the contents of not less than 20 Fenoprofen Calcium Capsules, and calculate the average weight per Capsule. Mix the combined contents of the Capsules, and transfer an accurately weighed portion, equivalent to about 85 mg of fenoprofen, to a 125-mL separator containing 5 mL of acetone and 2 mL of 6 *N* hydrochloric acid. Proceed as directed for *Standard preparation*, beginning with "Swirl to dissolve."

Chromatographic system (see *Chromatography* ⟨621⟩)—Under typical conditions, the gas chromatograph is equipped with a flame-ionization detector, and contains a 1.2-m × 3-mm glass column packed with 1 percent liquid phase G3 on 80- to 100-mesh support S1A. The column is maintained isothermally at about 180°, the injection port and detector block are maintained at about 240°, and dry helium is used as the carrier gas at a flow rate of about 60 mL per minute. [NOTE—Cure and condition the column as necessary (see *Chromatography* ⟨621⟩).]

System suitability—Chromatograph five injections of the *Standard preparation*, and record the peak areas, as directed under *Procedure:* the retention time for fenoprofen is about 0.5 relative to diphenamid. In a suitable chromatogram, the resolution factor between the two peaks is not less than 1.0, and the relative standard deviation for five replicate injections of the *Standard preparation* is not more than 2.0.

Procedure—Inject a portion of the *Standard preparation* into a suitable gas chromatograph, record the chromatogram so as to obtain not less than 50% of maximum recorder response, and measure the peak area of each component. Similarly inject a portion of the *Assay preparation*, record the chromatogram, and measure the peak area of each component. Calculate the quantity, in mg, of $C_{15}H_{14}O_3$ in the portion of Capsules taken by the formula:

$$(484.55/522.61)W(R_U/R_S),$$

in which 484.55 is two times the molecular weight of fenoprofen, 522.61 is the molecular weight of anhydrous fenoprofen calcium, W is the weight, in mg, of anhydrous fenoprofen calcium in the USP Fenoprofen Calcium RS taken, as corrected for moisture content by a titrimetric water determination, and R_U and R_S are the ratios of the areas of the fenoprofen peak to the diphenamid peak obtained from the *Assay preparation* and the *Standard preparation*, respectively.

Fenoprofen Calcium Tablets

» Fenoprofen Calcium Tablets contain not less than 90.0 percent and not more than 110.0 percent of the labeled amount of fenoprofen ($C_{15}H_{14}O_3$).

Packaging and storage—Preserve in well-closed containers.

Reference standards—*USP Fenoprofen Calcium Reference Standard*—Do not dry before using; determine the water content titrimetrically at the time of use for quantitative analyses. *USP Fenoprofen Sodium Reference Standard*—Do not dry before using; determine the water content titrimetrically at the time of use for quantitative analyses.

Identification—An amount of finely powdered Tablets, equivalent to about 600 mg of fenoprofen, responds to the *Identification tests* under *Fenoprofen Calcium Capsules.*

Dissolution ⟨711⟩—

Medium: pH 7.0 phosphate buffer (see *Buffer Solutions* in the section, *Reagents, Indicators, and Solutions*); 1000 mL.

Apparatus 1: 10-mesh basket; 100 rpm.

Time: 60 minutes.

Procedure—Filter 20 mL of the solution under test, and transfer 10.0 mL of the filtrate to a 100-mL volumetric flask. Dilute with *Dissolution Medium* to volume, and mix. Determine the absorbances of this solution and a Standard solution prepared from USP Fenoprofen Sodium RS, in the same medium having a known concentration of about 60 µg per mL at the wavelength of maximum absorbance at about 270 nm, using *Dissolution Medium* as the blank. Calculate the amount of $C_{15}H_{14}O_3$ dissolved, in mg, by the formula:

$$(242.27/264.26)(10C)(A_U/A_S),$$

in which 242.27 is the molecular weight of fenoprofen, 264.26 is the molecular weight of anhydrous fenoprofen sodium, C is the concentration of anhydrous fenoprofen sodium in the Standard solution, as determined from the concentration of USP Fenoprofen Sodium RS corrected for moisture content by a titrimetric water determination, and A_U and A_S are the absorbances of the solutions obtained from the substance under test and the Reference Standard, respectively.

Tolerances—Not less than 75% (Q) of the labeled amount of $C_{15}H_{14}O_3$ is dissolved in 60 minutes.

Uniformity of dosage units ⟨905⟩: meet the requirements.

Assay—

Internal standard solution, Standard preparation, Chromatographic system, and *System suitability*—Prepare as directed in the *Assay* under *Fenoprofen Calcium Capsules.*

Assay preparation—Weigh and finely powder not less than 20 Fenoprofen Calcium Tablets. Transfer an accurately weighed portion of the powder, equivalent to about 85 mg of fenoprofen, to a 125-mL separator containing 5 mL of acetone and 2 mL of 6 *N* hydrochloric acid. Proceed as directed for *Standard preparation* in the *Assay* under *Fenoprofen Calcium Capsules*, beginning with "Swirl to dissolve."

Procedure—Inject a portion of the *Standard preparation* into a suitable gas chromatograph, record the chromatogram so as to obtain not less than 50% of maximum recorder response, and measure the peak area of each component. Similarly inject a portion of the *Assay preparation*, record the chromatogram, and measure the peak area of each component. Calculate the quantity, in mg, of $C_{15}H_{14}O_3$ in the portion of Tablets taken by the formula:

$$(484.55/522.61)W(R_U/R_S),$$

in which 484.55 is two times the molecular weight of fenoprofen, 522.61 is the molecular weight of anhydrous fenoprofen calcium, W is the weight, in mg, of anhydrous fenoprofen calcium in the USP Fenoprofen Calcium RS taken, as corrected for moisture content by a titrimetric water determination, and R_U and R_S are the ratios of the areas of the fenoprofen peak to the diphenamid peak obtained from the *Assay preparation* and the *Standard preparation*, respectively.

Fentanyl Citrate

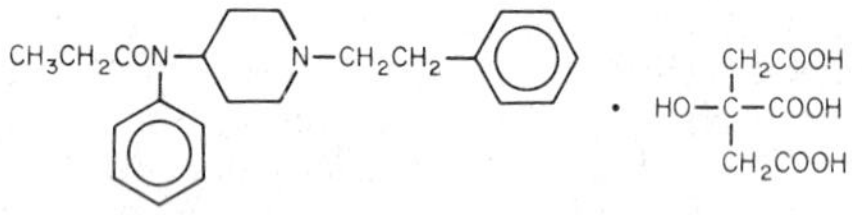

$C_{22}H_{28}N_2O \cdot C_6H_8O_7$ 528.60

Propanamide, *N*-phenyl-*N*-[1-(2-phenylethyl)-4-piperidinyl]-, 2-hydroxy-1,2,3-propanetricarboxylate (1:1).

N-(1-Phenethyl-4-piperidyl)propionanilide citrate (1:1) [*990-73-8*].

» Fentanyl Citrate contains not less than 98.0 percent and not more than 102.0 percent of $C_{22}H_{28}N_2O \cdot C_6H_8O_7$, calculated on the dried basis.

Caution—Great care should be taken to prevent inhaling particles of Fentanyl Citrate and exposing the skin to it.

Packaging and storage—Preserve in well-closed, light-resistant containers.

Reference standard—*USP Fentanyl Citrate Reference Standard*—Dry in vacuum at 60° for 2 hours before using.

Identification—

A: The infrared absorption spectrum of a potassium bromide dispersion of it exhibits maxima only at the same wavelengths as that of a similar preparation of USP Fentanyl Citrate RS.

B: The ultraviolet absorption spectrum of a 1 in 2000 solution of it in dilute hydrochloric acid (1 in 10) in methanol exhibits maxima and minima at the same wavelengths as that of a similar solution of USP Fentanyl Citrate RS, concomitantly measured.

Melting range ⟨741⟩: between 147° and 152°, determined after drying in vacuum at 60° for 2 hours.

Loss on drying ⟨731⟩—Dry it in vacuum at 60° for 2 hours: it loses not more than 0.5% of its weight.

Residue on ignition ⟨281⟩: not more than 0.5%.

Heavy metals, *Method II* ⟨231⟩: 0.002%.

Ordinary impurities ⟨466⟩—

Test solution: a mixture of chloroform and methanol (4:1).

Standard solution: a mixture of chloroform and methanol (4:1).

Eluant: a mixture of chloroform, methanol, and formic acid (85:10:5).

Visualization: iodine vapor.

Assay—Dissolve about 500 mg of Fentanyl Citrate, accurately weighed, in 30 mL of glacial acetic acid. Add 3 drops of *p*-naphtholbenzein TS, and titrate with 0.05 *N* perchloric acid VS. Perform a blank determination, and make any necessary correction. Each mL of 0.05 *N* perchloric acid is equivalent to 26.43 mg of $C_{22}H_{28}N_2O \cdot C_6H_8O_7$.

Fentanyl Citrate Injection

» Fentanyl Citrate Injection is a sterile solution of Fentanyl Citrate in Water for Injection. It contains the equivalent of not less than 90.0 percent and not more than 110.0 percent of the labeled amount of fentanyl ($C_{22}H_{28}N_2O$), present as the citrate.

Packaging and storage—Preserve in single-dose containers, preferably of Type I glass, protected from light.

Reference standards—*USP Fentanyl Citrate Reference Standard*—Dry in vacuum at 60° for 2 hours before using. *USP Endotoxin Reference Standard.*

Identification—The ultraviolet absorption spectrum of the Injection, measured in a 5-cm cell, exhibits maxima at the same wavelengths as that of a similar solution of USP Fentanyl Citrate RS containing 50 µg of fentanyl per mL, concomitantly measured.

Bacterial endotoxins—When tested as directed under *Bacterial Endotoxins Test* ⟨85⟩, it contains not more than 33.3 USP Endotoxin Units per mg.

pH ⟨791⟩: between 4.0 and 7.5.

Other requirements—It meets the requirements under *Injections* ⟨1⟩.

Assay—

Mobile phase—Prepare a filtered and degassed mixture containing 4 volumes of ammonium acetate solution (1 in 100) and 6 volumes of a mixture of methanol, acetonitrile, and glacial acetic acid (400:200:0.6). Adjust this solution to a pH of 6.6 ± 0.1 by the dropwise addition of glacial acetic acid, and make adjustments if necessary (see *System Suitability* under *Chromatography* ⟨621⟩), to obtain a retention time of about 5 minutes for the fentanyl peak.

Standard preparation—Dissolve an accurately weighed quantity of USP Fentanyl Citrate RS in water, and dilute quantitatively with water to obtain a solution having a known concentration of about 80 µg per mL.

Assay preparation—If necessary, dilute the Injection with water so that each mL contains the equivalent of about 50 µg of fentanyl.

Chromatographic system (see *Chromatography* ⟨621⟩)—The liquid chromatograph is equipped with a 230-nm detector and a 4.6-mm × 25-cm column that contains packing L1. The flow rate is about 2 mL per minute. Chromatograph the *Standard preparation*, and record the peak response as directed under *Procedure:* the tailing factor for the fentanyl peak is not more than 2.0, and the relative standard deviation for replicate injections is not more than 2.0%.

Procedure—Separately inject equal volumes (about 25 µL) of the *Standard preparation* and the *Assay preparation* into the chromatograph, record the chromatograms, and measure the responses for the major peaks. Calculate the quantity, in µg, of fentanyl ($C_{22}H_{28}N_3O$) in each mL of the Injection taken by the formula:

$$(336.48/528.60)CD(r_U/r_S),$$

in which 336.48 and 528.60 are the molecular weights of fentanyl and fentanyl citrate, respectively, C is the concentration, in µg per mL, of USP Fentanyl Citrate RS in the *Standard preparation*, D is the dilution factor used to obtain the *Assay preparation*, and r_U and r_S are the peak responses for the fentanyl peak obtained from the *Assay preparation* and the *Standard preparation*, respectively.

Ferric Oxide—*see* Ferric Oxide NF

Ferrous Citrate Fe 59 Injection—*see under "iron"*

Ferrous Fumarate

HC—C=O
O=C—CH O
O—Fe

$C_4H_2FeO_4$ 169.90

2-Butenedioic acid, (*E*)-, iron(2+) salt.

Iron(2+) fumarate [*141-01-5*].

» Ferrous Fumarate contains not less than 97.0 percent and not more than 101.0 percent of $C_4H_2FeO_4$, calculated on the dried basis.

Packaging and storage—Preserve in well-closed containers.

Reference standard—*USP Fumaric Acid Reference Standard.*

Identification—

A: To 1.5 g add 25 mL of dilute hydrochloric acid (1 in 2). Dilute with water to 50 mL, heat to effect complete solution, then cool, filter on a fine-porosity, sintered-glass crucible, wash the precipitate with dilute hydrochloric acid (3 in 100), saving the filtrate for *Identification test B*, and dry the precipitate at 105°: the infrared absorption of a potassium bromide dispersion of the dried precipitate so obtained exhibits maxima only at the same wavelengths as that of a similar preparation of USP Fumaric Acid RS.

B: A portion of the filtrate obtained in the preceding test responds to the tests for *Iron* ⟨191⟩.

Loss on drying ⟨731⟩—Dry it at 105° for 16 hours: it loses not more than 1.5% of its weight.

Sulfate—Transfer 1.0 g to a 250-mL beaker, add 100 mL of water, and heat on a steam bath, adding hydrochloric acid dropwise, until complete solution is effected (about 2 mL of the acid will be required). Filter the solution if necessary, and dilute the filtrate with water to 100 mL. Heat the filtrate to boiling, add 10 mL of barium chloride TS, warm on a steam bath for 2 hours, cover, and allow to stand for 16 hours. (If crystals of ferrous fumarate form, warm the solution on the steam bath to dissolve them.) Filter the solution through paper, wash the residue with hot water, and transfer the paper containing the residue to a tared crucible. Char the paper, without burning, and ignite the crucible and its contents at 600° to constant weight: each mg of residue is equivalent to 0.412 mg of SO_4. Not more than 0.2% is found.

Arsenic, *Method I* ⟨211⟩—Transfer 2.0 g to a beaker, and add 10 mL of water and 10 mL of sulfuric acid. Warm to precipitate the fumaric acid completely, cool, add 30 mL of water, and filter into a 100-mL volumetric flask. Wash the precipitate with water, adding the washings to the flask, add water to volume, and mix. Transfer 50.0 mL of this solution into the arsine generator flask, and dilute with water to 55 mL: the resulting solution meets the requirements of the test, the addition of 20 mL of 7 *N* sulfuric acid specified under *Procedure* being omitted. The limit is 3 ppm.

Ferric iron—Transfer 2.0 g, accurately weighed, to a glass-stoppered, 250-mL conical flask, add 25 mL of water and 4 mL of hydrochloric acid, and heat on a hot plate until solution is complete. Insert the stopper in the flask, and cool to room temperature. Add 3 g of potassium iodide, insert the stopper in the flask, swirl to mix, and allow to stand in the dark for 5 minutes. Remove the stopper, add 75 mL of water, and titrate with 0.1 *N* sodium thiosulfate VS, adding 3 mL of starch TS as the endpoint is approached. Not more than 7.16 mL of 0.1 *N* sodium thiosulfate is consumed (2.0%).

Lead—[NOTE—For the preparation of all aqueous solutions and for the rinsing of glassware before use, employ water that has been passed through a strong-acid, strong-base, mixed-bed ion-exchange resin before use. Select all reagents to have as low a content of lead as practicable, and store all reagent solutions in containers of borosilicate glass. Clean glassware before use by soaking in warm 8 *N* nitric acid for 30 minutes and by rinsing with deionized water.]

Ascorbic acid–sodium iodide solution—Dissolve 20 g of ascorbic acid and 38.5 g of sodium iodide in water in a 200-mL volumetric flask, dilute with water to volume, and mix.

Trioctylphosphine oxide solution—[*Caution—This solution causes irritation. Avoid contact with eyes, skin, and clothing. Take special precautions in disposing of unused portions of solutions to which this reagent is added.*] Dissolve 5.0 g of trioctylphosphine oxide in 4-methyl-2-pentanone in a 100-mL volumetric flask, dilute with the same solvent to volume, and mix.

Standard preparation and *Blank*—Transfer 5.0 mL of *Lead Nitrate Stock Solution*, prepared as directed in the test for *Heavy Metals* ⟨231⟩, to a 100-mL volumetric flask, dilute with water to volume, and mix. Transfer 2.0 mL, of the resulting solution to a 50-mL beaker. To this beaker and to a second, empty beaker (*Blank*) add 6 mL of nitric acid and 10 mL of perchloric acid, and evaporate in a hood to dryness. [*Caution—Use perchloric acid in a well-ventilated fume hood with proper precautions.*] Cool, dissolve the residues in 10 mL of 9 *N* hydrochloric acid, and transfer with the aid of about 10 mL of water to separate 50-mL volumetric flasks. To each flask add 20 mL of *Ascorbic acid–sodium iodide solution* and 5.0 mL of *Trioctylphosphine oxide solution*, shake for 30 seconds, and allow to separate. Add water to bring the organic solvent layer into the neck of each flask, shake again, and allow to separate. The organic solvent layers are the *Blank* and the *Standard preparation*, and they contain 0.0 and 2.0 µg of lead per mL, respectively.

Test preparation—Add 1.0 g of Ferrous Fumarate to a 50-mL beaker, and add 6 mL of nitric acid and 10 mL of perchloric acid. [*Caution—Use perchloric acid in a well-ventilated fume hood with proper precautions.*] Cover with a ribbed watch glass, and heat in a hood until completely dry. Cool, dissolve the residue in 10 mL of 9 *N* hydrochloric acid, and transfer with the aid of about 10 mL of water to a 50-mL volumetric flask. Add 20 mL of *Ascorbic acid–sodium iodide solution* and 5.0 mL of *Trioctylphosphine oxide solution*, shake for 30 seconds, and allow to separate. Add water to bring the organic solvent layer into the neck of the flask, shake again, and allow to separate. The organic solvent layer is the *Test preparation.*

Procedure—Concomitantly determine the absorbances of the *Blank*, the *Standard preparation*, and the *Test preparation* at the lead emission line at 283.3 nm with a suitable atomic absorption spectrophotometer (see *Spectrophotometry and Light-scattering* ⟨851⟩) equipped with a lead hollow-cathode lamp and an air-acetylene flame, using 4-methyl-2-pentanone to set the instrument to zero. In a suitable analysis, the absorbance of the *Blank* is not greater than 20% of the difference between the absorbance of the *Standard preparation* and the absorbance of the *Blank:* the absorbance of the *Test preparation* does not exceed that of the *Standard preparation* (0.001%).

Mercury—[NOTES—(1) Carry out this procedure in subdued light, since mercuric dithizonate is light-sensitive. (2) For preparation of solutions, see *Mercury* ⟨261⟩.] Dissolve about 1 g, accurately weighed, in 30 mL of dilute nitric acid (1 in 10), with the aid of heat, on a steam bath. Cool quickly by immersion in an ice bath, and filter through a filter that previously has been washed with dilute nitric acid (1 in 10) and water. To the filtrate add 20 mL of sodium citrate solution (1 in 4) and 1 mL of *Hydroxylamine Hydrochloride Solution.*

Prepare a control solution consisting of 3.0 mL of *Standard Mercury Solution*, 30 mL of dilute nitric acid (1 in 10), 5 mL of sodium citrate solution (1 in 4), and 1 mL of *Hydroxylamine Hydrochloride Solution.*

Treat the solution under test and the control solution in parallel as follows: Using ammonium hydroxide, adjust to a pH of 1.8, determined potentiometrically, and transfer to a separator. Extract with two 5-mL portions of *Dithizone Extraction Solution* and 5 mL of chloroform, pooling the chloroform extracts in a second separator. Add 10 mL of dilute hydrochloric acid (1 in 2), shake, allow the layers to separate, and discard the chloroform layer. Wash the acid extract with 3 mL of chloroform, and discard the washing. Add 0.1 mL of disodium ethylenediaminetetraacetate solution (1 in 50) and 2 mL of 6 *N* acetic acid, mix, and add slowly 5 mL of ammonium hydroxide. Close the separator, cool it under cold running water, and dry its outer surface. Remove the stopper, and pour the contents into a beaker. Adjust to a pH of 1.8 in the same manner as before, and return the solution to its separator. Add 5.0 mL of *Diluted Dithizone Extraction Solution*, shake vigorously, and allow the layers to separate. At this point, compare the colors developed in the chloroform layers of the two solutions that have been treated in parallel: the color developed by the solution under test is not more intense than that developed by the control solution (3 ppm).

Assay—Transfer 500 mg of Ferrous Fumarate, accurately weighed, to a 500-mL conical flask, and add 25 mL of dilute hydrochloric acid (2 in 5). Heat to boiling, and add a solution of 5.6 g of stannous chloride in 50 mL of dilute hydrochloric acid (3 in 10) dropwise until the yellow color disappears, then add 2 drops in excess. Cool the solution in an ice bath to room temperature, add 10 mL of mercuric chloride solution (1 in 20), and allow to stand for 5 minutes. Add 200 mL of water, 25 mL of dilute sulfuric acid (1 in 2), and 4 mL of phosphoric acid, then add 2 drops of orthophenanthroline TS, and titrate with 0.1 *N* ceric sulfate VS. Perform a blank determination, and make any necessary correction. Each mL of 0.1 *N* ceric sulfate is equivalent to 16.99 mg of $C_4H_2FeO_4$.

Ferrous Fumarate Tablets

» Ferrous Fumarate Tablets contain not less than 95.0 percent and not more than 110.0 percent of the labeled amount of $C_4H_2FeO_4$.

Packaging and storage—Preserve in tight containers.

Labeling—Label Tablets in terms of ferrous fumarate ($C_4H_2FeO_4$) and in terms of elemental iron.

Identification—To a portion of powdered Tablets, equivalent to 1 g of ferrous fumarate, add 25 mL of dilute hydrochloric acid (1 in 2), mix, and add 25 mL of water. Boil the solution for a few minutes, cool, and filter: the filtrate responds to the tests for *Iron* ⟨191⟩.

Disintegration ⟨701⟩: 30 minutes.

Uniformity of dosage units ⟨905⟩: meet the requirements.

Assay—Weigh and finely powder not less than 20 Ferrous Fumarate Tablets. Transfer an accurately weighed portion of the powder, equivalent to about 500 mg of ferrous fumarate, to a 250-mL beaker. Add 25 mL of water, 25 mL of nitric acid, and 7.5 mL of perchloric acid. Cover with a ribbed watch glass, and heat to the production of strong fumes. Cool, rinse the watch glass and the sides of the beaker with water, and evaporate in a hood to near-dryness. Wash down the watch glass and the sides of the beaker with 2 mL of hydrochloric acid and then with a small volume of water. Warm slightly, if necessary, to dissolve the residue. Transfer to a glass-stoppered, 250-mL conical flask. Repeat the washing with 2 mL of hydrochloric acid, and complete the transfer to the flask, using not more than 20 to 25 mL of water for the transfer. Add 4 g of potassium iodide to the flask, insert the stopper, and allow to stand in the dark for 5 minutes. Add 75 mL of water, and titrate with 0.1 *N* sodium thiosulfate VS, adding 3 mL of starch TS as the end-point is approached. Each mL of 0.1 *N* sodium thiosulfate is equivalent to 16.99 mg of $C_4H_2FeO_4$.

Ferrous Gluconate

$$\left[HOCH_2-\overset{H}{\underset{OH}{C}}-\overset{H}{\underset{OH}{C}}-\overset{OH}{\underset{H}{C}}-\overset{H}{\underset{OH}{C}}-COO-\right]_2 Fe \cdot 2H_2O$$

$C_{12}H_{22}FeO_{14} \cdot 2H_2O$ 482.17

D-Gluconic acid, iron(2+) salt (2:1), dihydrate.

Iron(2+) gluconate (1:2) dihydrate [*12389-15-0*].

Anhydrous 446.14 [*299-29-6*].

» Ferrous Gluconate contains not less than 97.0 percent and not more than 102.0 percent of $C_{12}H_{22}FeO_{14}$, calculated on the dried basis.

Packaging and storage—Preserve in tight containers.

Reference standard—*USP Potassium Gluconate Reference Standard*—Dry in vacuum at 105° for 4 hours before using.

Identification—

A: It responds to *Identification test B* under *Calcium Gluconate.*

B: A solution (1 in 200) yields a dark blue precipitate with potassium ferricyanide TS.

Loss on drying ⟨731⟩—Dry it at 105° for 16 hours: it loses between 6.5% and 10.0% of its weight.

Chloride ⟨221⟩—A 1.0-g portion shows no more chloride than corresponds to 1.0 mL of 0.020 *N* hydrochloric acid (0.07%).

Sulfate ⟨221⟩—A 1.0-g portion shows no more sulfate than corresponds to 1.0 mL of 0.020 *N* sulfuric acid (0.1%).

Oxalic acid—Dissolve 1.0 g in 10 mL of water, add 2 mL of hydrochloric acid, and transfer to a separator. Extract successively with 50 mL and 20 mL of ether. Combine the ether extracts, add 10 mL of water, and evaporate the ether on a steam bath. Add 1 drop of 6 *N* acetic acid and 1 mL of calcium acetate solution (1 in 20): no turbidity is produced within 5 minutes.

Arsenic, *Method I* ⟨211⟩—Transfer 1.0 g of Ferrous Gluconate to a 100-mL round-bottom flask fitted with a 24/40 standard-taper joint. Add 40 mL of 9 *N* sulfuric acid and 2 mL of potassium bromide solution (3 in 10). Immediately connect to a suitable distillation apparatus having a reservoir with a water jacket, cooled with circulating ice water, and heat to dissolve the test specimen. Distil, collect 25 mL of distillate, and transfer the distillate to the arsine generator flask. Wash the condenser and reservoir several times with small portions of water, add the washings to the distillate in the generator flask, add bromine TS until the solution is slightly yellow, and dilute with water to 35 mL. Proceed as directed under *Procedure:* the limit is 3 ppm.

Ferric iron—Dissolve about 5 g, accurately weighed, in a mixture of 100 mL of water and 10 mL of hydrochloric acid, and add 3 g of potassium iodide. Shake, and allow to stand in the dark for 5 minutes. Titrate any liberated iodine with 0.1 *N* sodium thiosulfate VS, adding 3 mL of starch TS as the end-point is approached. Perform a blank determination, and make any necessary correction. Each mL of 0.1 *N* sodium thiosulfate is equivalent to 5.585 mg of ferric iron. Ferrous Gluconate contains not more than 2.0% of ferric iron.

Lead—[NOTE—For the preparation of all aqueous solutions and for the rinsing of glassware before use, employ water that has been passed through a strong-acid, strong-base, mixed-bed ion-exchange resin before use. Select all reagents to have as low a content of lead as practicable, and store all reagent solutions in containers of borosilicate glass. Clean glassware before use by soaking in warm 8 *N* nitric acid for 30 minutes and by rinsing with deionized water.]

Ascorbic acid–sodium iodide solution—Dissolve 20 g of ascorbic acid and 38.5 g of sodium iodide in water in a 200-mL volumetric flask, dilute with water to volume, and mix.

Trioctylphosphine oxide solution—[*Caution—This solution causes irritation. Avoid contact with eyes, skin, and clothing. Take special precautions in disposing of unused portions of solutions to which this reagent is added.*] Dissolve 5.0 g of trioctylphosphine oxide in 4-methyl-2-pentanone in a 100-mL volumetric flask, dilute with the same solvent to volume, and mix.

Standard preparation and *Blank*—Transfer 5.0 mL of *Lead Nitrate Stock Solution*, prepared as directed in the test for *Heavy Metals* ⟨231⟩, to a 100-mL volumetric flask, dilute with water to volume, and mix. Transfer 2.0 mL of the resulting solution to a 50-mL volumetric flask. To this volumetric flask and to a second, empty 50-mL volumetric flask (*Blank*) add 10 mL of 9 *N* hydrochloric acid and about 10 mL of water. To each flask add 20 mL of *Ascorbic acid–sodium iodide solution* and 5.0 mL of *Trioctylphosphine oxide solution*, shake for 30 seconds, and allow to separate. Add water to bring the organic solvent layer into the neck of each flask, shake again, and allow to separate. The organic solvent layers are the *Blank* and the *Standard preparation*, and they contain 0.0 and 2.0 µg of lead per mL, respectively.

Test preparation—Add 1.0 g of Ferrous Gluconate, 10 mL of 9 *N* hydrochloric acid, about 10 mL of water, 20 mL of *Ascorbic acid–sodium iodide solution*, and 5.0 mL of *Trioctylphosphine oxide solution* to a 50-mL volumetric flask, shake for 30 seconds, and allow to separate. Add water to bring the organic solvent layer into the neck of the flask, shake again, and allow to separate. The organic solvent layer is the *Test preparation.*

Procedure—Concomitantly determine the absorbances of the *Blank*, the *Standard preparation*, and the *Test preparation* at the lead emission line at 283.3 nm, with a suitable atomic absorption spectrophotometer (see *Spectrophotometry and Light-scattering* ⟨851⟩) equipped with a lead hollow-cathode lamp and an air-acetylene flame, using 4-methyl-2-pentanone to set the instrument to zero. In a suitable analysis, the absorbance of the *Blank* is not greater than 20% of the difference between the absorbance of the *Standard preparation* and the absorbance of the *Blank:* the absorbance of the *Test preparation* does not exceed that of the *Standard preparation* (0.001%).

Mercury ⟨261⟩: 3 ppm.

Reducing sugars—Dissolve 500 mg in 10 mL of water, warm, and render alkaline with 1 mL of 6 *N* ammonium hydroxide. Pass hydrogen sulfide gas into the solution to precipitate the iron, and permit the solution to stand for 30 minutes to coagulate the precipitate. Filter, and wash the precipitate with two successive 5-mL portions of water. Acidify the combined filtrate and washings with hydrochloric acid, and add 2 mL of 3 *N* hydrochloric acid in excess. Boil the solution until the vapors no longer darken lead acetate paper, and continue to boil, if necessary, until it has been concentrated to about 10 mL. Cool, add 5 mL of sodium carbonate TS and 20 mL of water, filter, and adjust the volume of the filtrate to 100 mL. To 5 mL of the filtrate add 2 mL of alkaline cupric tartrate TS, and boil for 1 minute: no red precipitate is formed within 1 minute.

Assay—Dissolve about 1.5 g of Ferrous Gluconate, accurately weighed, in a mixture of 75 mL of water and 15 mL of 2 *N* sulfuric acid in a 300-mL conical flask. Add 250 mg of zinc dust, close the flask with a stopper containing a Bunsen valve, and allow to stand at room temperature for 20 minutes or until the solution becomes colorless. Filter the solution through a filtering crucible containing an asbestos mat coated with a thin layer of zinc dust, and wash the crucible and contents with 10 mL of 2 *N* sulfuric acid, followed by 10 mL of water. [NOTE—Prepare and use the filtering crucible in a well-ventilated hood.] Add orthophenanthroline TS, and titrate the filtrate in the suction flask immediately with 0.1 *N* ceric sulfate VS. Perform a blank determination, and make any necessary correction. Each mL of 0.1 *N* ceric sulfate is equivalent to 44.61 mg of $C_{12}H_{22}FeO_{14}$.

Ferrous Gluconate Capsules

» Ferrous Gluconate Capsules contain not less than 93.0 percent and not more than 107.0 percent of the labeled amount of $C_{12}H_{22}FeO_{14}.2H_2O$.

Packaging and storage—Preserve in tight containers.

Labeling—Label Capsules in terms of the content of ferrous gluconate ($C_{12}H_{22}FeO_{14}.2H_2O$) and in terms of the content of elemental iron.

Reference standard—*USP Potassium Gluconate Reference Standard*—Dry in vacuum at 105° for 4 hours before using.

Identification—Dissolve a quantity of the contents of Capsules, equivalent to about 1 g of ferrous gluconate, in 100 mL of water, and filter: the solution, diluted with water, where necessary, responds to the *Identification tests* under *Ferrous Gluconate.*

Dissolution ⟨711⟩—

Medium: 0.1 *N* hydrochloric acid; 900 mL.

Apparatus 1: 100 rpm.

Time: 45 minutes.

Procedure—Determine the amount of $C_{12}H_{22}FeO_{14}.2H_2O$ dissolved, employing atomic absorption spectrophotometry at a wavelength of about 248.3 nm in filtered portions of the solution under test, suitably diluted with *Dissolution Medium*, in comparison with a Standard solution having a known concentration of iron in the same medium.

Tolerances—Not less than 75% (*Q*) of the labeled amount of $C_{12}H_{22}FeO_{14}.2H_2O$ is dissolved in 45 minutes.

Uniformity of dosage units ⟨905⟩: meet the requirements.

Assay—

2,2′-Bipyridine solution—Dissolve 400 mg of 2,2′-bipyridine in 100 mL of water, using heat, if necessary, to effect solution. Cool, and filter.

pH 4.6 acetate buffer—Dissolve 3.0 g of sodium acetate in 50 mL of water, add 2.0 mL of glacial acetic acid, dilute with water to 200 mL, and mix.

Standard stock solution—Transfer an amount of ferrous ammonium sulfate, equivalent to 702.2 mg of ferrous ammonium sulfate hexahydrate $[Fe(NH_4)_2(SO_4)_2 \cdot 6H_2O]$ to a 100-mL volumetric flask, dissolve in water, dilute with water to volume, and mix. Transfer 10.0 mL of the solution to a second 100-mL volumetric flask, add 5 mL of 2 *N* sulfuric acid, dilute with water to volume, and mix. This solution contains 100 µg of iron (Fe) per mL.

Assay stock solution for hard gelatin capsules—Transfer, as completely as possible, the contents of not less than 20 Ferrous Gluconate Capsules to a suitable tared container. Mix and finely powder the combined contents, and transfer an accurately weighed portion of the powder, equivalent to about 430 mg of ferrous gluconate, to a 500-mL volumetric flask. Add about 300 mL of water, mix, heat on a steam bath, if necessary, to dissolve, cool, dilute with water to volume, and mix.

Assay stock solution for soft gelatin capsules—Place a number of Ferrous Gluconate Capsules, equivalent to about 430 mg of ferrous gluconate, in a 500-mL volumetric flask. Add about 300 mL of water, heat on a steam bath to dissolve the Capsules, cool, dilute with water to volume, and mix.

Standard preparation—Transfer 3.0 mL of the *Standard stock solution* to a 100-mL volumetric flask, and add, in the order named, 70 mL of *pH 4.6 acetate buffer*, 10.0 mL of sodium thiosulfate solution (1 in 10), and 5.0 mL of *2,2′-Bipyridine solution*, with mixing following each addition. Heat for 60 minutes on a steam bath, cool, dilute with *pH 4.6 acetate buffer* to volume, and mix. The concentration of Fe in the *Standard preparation* is about 3 µg per mL.

Standard preparation blank—Proceed as directed for *Standard preparation*, but omit the addition of 5.0 mL of *2,2′-Bipyridine solution.*

Assay preparation—Transfer 3.0 mL of the *Assay stock solution* to a 100-mL volumetric flask, and proceed as directed for *Standard preparation*, beginning with "add, in the order named, 70 mL of *pH 4.6 acetate buffer.*"

Assay preparation blank—Proceed as directed for *Assay preparation*, but omit the addition of 5.0 mL of *2,2′-Bipyridine solution.*

Procedure—Filter each solution, and concomitantly determine the absorbances of the *Standard preparation* and the *Assay preparation* in 1-cm cells at the wavelength of maximum absorbance at about 522 nm, with a suitable spectrophotometer, using the *Standard preparation blank* and the *Assay preparation blank*, respectively, to set the instrument. Calculate the quantity, in mg, of $C_{12}H_{22}FeO_{14} \cdot 2H_2O$ in the portion of Capsules taken by the formula:

$$(482.17/55.85)(16.67C)(A_U/A_S),$$

in which 482.17 is the molecular weight of ferrous gluconate (dihydrate), and 55.85 is the atomic weight of iron, *C* is the concentration, in µg per mL, of the *Standard preparation*, and A_U and A_S are the absorbances of the *Assay preparation* and the *Standard preparation*, respectively.

Ferrous Gluconate Elixir

» Ferrous Gluconate Elixir contains not less than 94.0 percent and not more than 106.0 percent of the labeled amount of $C_{12}H_{22}FeO_{14} \cdot 2H_2O$.

Packaging and storage—Preserve in tight, light-resistant containers.

Labeling—Label Elixir in terms of the content of ferrous gluconate ($C_{12}H_{22}FeO_{14} \cdot 2H_2O$) and in terms of the content of elemental iron.

Reference standard—*USP Potassium Gluconate Reference Standard*—Dry in vacuum at 105° for 4 hours before using.

Identification—A volume of Elixir diluted, where necessary, with water responds to the *Identification tests* under *Ferrous Gluconate.*

pH ⟨791⟩: between 3.4 and 3.8.

Alcohol content ⟨611⟩: between 6.3% and 7.7% of C_2H_5OH.

Assay—Transfer an accurately measured volume of Ferrous Gluconate Elixir, equivalent to about 1.2 g of ferrous gluconate, to a flask containing a cooled mixture of 80 mL of recently boiled water and 80 mL of 2 *N* sulfuric acid. Add orthophenanthroline TS, and titrate immediately with 0.1 *N* ceric sulfate VS. Perform a blank determination, and make any necessary correction. Each mL of 0.1 *N* ceric sulfate is equivalent to 48.22 mg of $C_{12}H_{22}FeO_{14} \cdot 2H_2O$.

Ferrous Gluconate Tablets

» Ferrous Gluconate Tablets contain not less than 93.0 percent and not more than 107.0 percent of the labeled amount of $C_{12}H_{22}FeO_{14} \cdot 2H_2O$.

Packaging and storage—Preserve in tight containers.

Labeling—Label Tablets in terms of the content of ferrous gluconate ($C_{12}H_{22}FeO_{14} \cdot 2H_2O$) and in terms of the content of elemental iron.

Reference standard—*USP Potassium Gluconate Reference Standard*—Dry in vacuum at 105° for 4 hours before using.

Identification—Dissolve a quantity of powdered Tablets, equivalent to about 1 g of ferrous gluconate, in 100 mL of water, and filter: the solution so obtained, diluted with water, where necessary, responds to the *Identification tests* under *Ferrous Gluconate.*

Dissolution ⟨711⟩—

Medium: simulated gastric fluid TS; 900 mL.

Apparatus 2: 150 rpm.

Time: 80 minutes.

Procedure—Determine the amount of $C_{12}H_{22}FeO_{14} \cdot 2H_2O$ dissolved, employing atomic absorption spectrophotometry at a wavelength of about 248.3 nm in filtered portions of the solution under test, suitably diluted with *Dissolution Medium*, in comparison with a Standard solution having a known concentration of iron in the same medium.

Tolerances—Not less than 80% (*Q*) of the labeled amount of $C_{12}H_{22}FeO_{14} \cdot 2H_2O$ is dissolved in 80 minutes.

Uniformity of dosage units ⟨905⟩: meet the requirements.

Assay—Weigh and finely powder not less than 20 Ferrous Gluconate Tablets. Weigh accurately a portion of the powder, equivalent to about 1.5 g of ferrous gluconate, and dissolve in a mixture of 75 mL of water and 15 mL of 2 *N* sulfuric acid contained in a 300-mL conical flask. Proceed as directed in the *Assay* under *Ferrous Gluconate*, beginning with "Add 250 mg of zinc dust." Each mL of 0.1 *N* ceric sulfate is equivalent to 48.22 mg of $C_{12}H_{22}FeO_{14} \cdot 2H_2O$.

Ferrous Sulfate

$FeSO_4 \cdot 7H_2O$ 278.01
Sulfuric acid, iron(2+) salt (1:1), heptahydrate.
Iron(2+) sulfate (1:1) heptahydrate [*7782-63-0*].
Anhydrous 151.90 [*7720-78-7*].

» Ferrous Sulfate contains an amount of $FeSO_4$ equivalent to not less than 99.5 percent and not more than 104.5 percent of $FeSO_4 \cdot 7H_2O$.

Packaging and storage—Preserve in tight containers.

Labeling—Label it to indicate that it is not to be used if it is coated with brownish yellow basic ferric sulfate.

Identification—It responds to the tests for *Ferrous Salts* ⟨191⟩ and for *Sulfate* ⟨191⟩.

Arsenic, *Method I* ⟨211⟩—Transfer 1.0 g to a round-bottom, 100-mL flask fitted with a glass joint, add 40 mL of 9 *N* sulfuric acid and 2 mL of potassium bromide solution (3 in 10), and immediately connect the flask to a condenser having a matching glass joint and a reservoir with a water jacket that is cooled by ice-water. Heat the flask gently over a low flame until the solid dissolves, then distil until 25 mL of distillate collects in the reservoir. Transfer this distillate to the arsine generator flask, and wash the condenser and the reservoir with several small portions of water, adding the washings to the generator flask. Swirl to mix, add bromine TS until the color of the solution is slightly yellow, and dilute with water to 35 mL. The limit is 3 ppm.

Lead—Using Ferrous Sulfate, proceed as directed in the test for *Lead* under *Ferrous Gluconate:* the limit is 0.001%.

Mercury—It meets the requirements of the test for *Mercury* under *Ferrous Fumarate.*

Assay—Dissolve about 1 g of Ferrous Sulfate, accurately weighed, in a mixture of 25 mL of 2 *N* sulfuric acid and 25 mL of freshly boiled and cooled water. Add orthophenanthroline TS, and immediately titrate with 0.1 *N* ceric sulfate VS. Perform a blank determination, and make any necessary correction. Each mL of 0.1 *N* ceric sulfate is equivalent to 15.19 mg of $FeSO_4$ or to 27.80 mg of $FeSO_4 . 7H_2O$.

Ferrous Sulfate Oral Solution

» Ferrous Sulfate Oral Solution contains not less than 94.0 percent and not more than 106.0 percent of the labeled amount of $FeSO_4 . 7H_2O$.

Packaging and storage—Preserve in tight, light-resistant containers.

Labeling—Label Oral Solution in terms of the content of ferrous sulfate ($FeSO_4 . 7H_2O$) and in terms of the content of elemental iron.

Identification—It responds to the tests for *Ferrous Salts* ⟨191⟩ and for *Sulfate* ⟨191⟩.

pH ⟨791⟩: between 1.8 and 5.3.

Assay—Transfer to a 200-mL flask an accurately measured volume of Ferrous Sulfate Oral Solution, equivalent to about 625 mg of $FeSO_4 . 7H_2O$. Add 25 mL of 2 *N* sulfuric acid and 75 mL of freshly boiled and cooled water, then add orthophenanthroline TS, and immediately titrate with 0.1 *N* ceric sulfate VS. Perform a blank determination, and make any necessary correction. Each mL of 0.1 *N* ceric sulfate is equivalent to 27.80 mg of $FeSO_4 . 7H_2O$.

Ferrous Sulfate Syrup

» Ferrous Sulfate Syrup contains, in each 100 mL, not less than 3.75 g and not more than 4.25 g of $FeSO_4 . 7H_2O$, equivalent to not less than 0.75 g and not more than 0.85 g of elemental iron.

Ferrous Sulfate	40	g
Citric Acid, hydrous	2.1	g
Peppermint Spirit	2	mL
Sucrose	825	g
Purified Water, a sufficient quantity, to make	1000	mL

Dissolve the Ferrous Sulfate, the Citric Acid, the Peppermint Spirit, and 200 g of the Sucrose in 450 mL of Purified Water, and filter the solution until clear. Dissolve the remainder of the Sucrose in the clear filtrate, and add Purified Water to make 1000 mL. Mix, and filter, if necessary, through a pledget of cotton.

Packaging and storage—Preserve in tight containers.

Labeling—Label Syrup in terms of the content of ferrous sulfate ($FeSO_4 . 7H_2O$) and in terms of the content of elemental iron.

Identification—Dissolve 1 mL of Syrup with water to about 100 mL: the solution responds to the tests for *Ferrous Salts* ⟨191⟩ and for *Sulfate* ⟨191⟩.

Assay—Transfer 25 mL of Ferrous Sulfate Syrup, accurately measured, to a 250-mL conical flask. Add 15 mL of 2 *N* sulfuric acid and 100 mL of water, and shake well. Add 3 drops of orthophenanthroline TS, and titrate with 0.1 *N* ceric sulfate VS. Perform a blank determination, and make any necessary correction. Each mL of 0.1 *N* ceric sulfate is equivalent to 27.80 mg of $FeSO_4 . 7H_2O$.

Ferrous Sulfate Tablets

» Ferrous Sulfate Tablets contain not less than 95.0 percent and not more than 110.0 percent of the labeled amount of $FeSO_4 . 7H_2O$.

NOTE—An equivalent amount of Dried Ferrous Sulfate may be used in place of $FeSO_4 . 7H_2O$ in preparing Ferrous Sulfate Tablets.

Packaging and storage—Preserve in tight containers.

Labeling—Label Tablets in terms of ferrous sulfate ($FeSO_4 . 7H_2O$) and in terms of elemental iron.

Identification—Dissolve a quantity of powdered Tablets, equivalent to about 250 mg of ferrous sulfate, in sufficient water, acidified with hydrochloric acid, to make 25 mL of solution: the solution responds to the tests for *Ferrous Salts* ⟨191⟩ and for *Sulfate* ⟨191⟩.

Disintegration ⟨701⟩: 30 minutes.

Uniformity of dosage units ⟨905⟩: meet the requirements.

Assay—Weigh and finely powder not less than 20 Ferrous Sulfate Tablets. Dissolve a portion of the powder, equivalent to about 500 mg of ferrous sulfate and accurately weighed, in a beaker containing a mixture of 20 mL of 2 *N* sulfuric acid and 80 mL of freshly boiled and cooled water. Filter the solution rapidly as soon as all soluble ingredients in the tablets are dissolved, and wash the container and the filter with small portions of a mixture of 20 mL of 2 *N* sulfuric acid and 80 mL of freshly boiled and cooled water. Add orthophenanthroline TS, and immediately titrate the combined filtrate and washings with 0.1 *N* ceric sulfate VS. Each mL of 0.1 *N* ceric sulfate is equivalent to 27.80 mg of $FeSO_4 . 7H_2O$.

Dried Ferrous Sulfate

$FeSO_4 . xH_2O$
Sulfuric acid, iron(2+) salt (1:1), hydrate.
Iron(2+) sulfate (1:1) hydrate [*13463-43-9*].
Anhydrous 151.90 [*7720-78-7*].

» Dried Ferrous Sulfate contains not less than 86.0 percent and not more than 89.0 percent of anhydrous ferrous sulfate ($FeSO_4$).

Packaging and storage—Preserve in well-closed containers.

Identification—It responds to the tests for *Ferrous Salts* ⟨191⟩ and for *Sulfate* ⟨191⟩.

Insoluble substances—Dissolve 2.0 g in 20 mL of freshly boiled dilute sulfuric acid (1 in 100), heat to boiling, and digest in a covered beaker on a steam bath for 1 hour. Filter through a tared filtering crucible, wash thoroughly, and dry at 105°: the weight of the insoluble residue does not exceed 1 mg (0.05%).

Arsenic—It meets the requirements of the test for *Arsenic* under *Ferrous Sulfate.*

Lead—Using Dried Ferrous Sulfate, proceed as directed in the test for *Lead* under *Ferrous Gluconate:* the limit is 0.001%.

Mercury—It meets the requirements of the test for *Mercury* under *Ferrous Fumarate.*

Assay—Dissolve about 800 mg of Dried Ferrous Sulfate, accurately weighed, in a mixture of 25 mL of 2 *N* sulfuric acid and 25 mL of freshly boiled and cooled water. Add orthophenanthroline TS, and immediately titrate with 0.1 *N* ceric sulfate VS. Perform a blank determination, and make any necessary correction. Each mL of 0.1 *N* ceric sulfate is equivalent to 15.19 mg of $FeSO_4$.

Film, Absorbable Gelatin—*see* Gelatin Film, Absorbable

Flexible Collodion—*see* Collodion, Flexible

Flucytosine

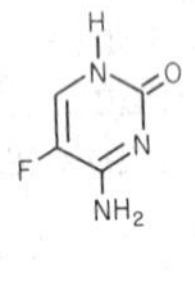

$C_4H_4FN_3O$ 129.09

Cytosine, 5-fluoro-.

5-Fluorocytosine [*2022-85-7*].

» Flucytosine contains not less than 98.5 percent and not more than 101.0 percent of $C_4H_4FN_3O$, calculated on the dried basis.

Packaging and storage—Preserve in tight, light-resistant containers.

Reference standards—*USP Flucytosine Reference Standard*—Dry at 105° for 4 hours before using. *USP Fluorouracil Reference Standard*—[*Caution—Avoid contact.*] Dry in a suitable vacuum drying tube, using phosphorus pentoxide as the desiccant, at 80° for 4 hours before using.

Identification—

A: The ultraviolet absorption spectrum of a 1 in 120,000 solution in dilute hydrochloric acid (1 in 100) exhibits a maximum at the same wavelength as that of a similar solution of USP Flucytosine RS, concomitantly measured, and the respective absorptivities, calculated on the dried basis, at the wavelength of maximum absorbance at about 285 nm do not differ by more than 2.0%.

B: The R_f value of the principal spot in the specimen chromatogram in the test for *Fluorouracil* corresponds to that obtained with the solution of USP Flucytosine RS.

Loss on drying ⟨731⟩—Dry it at 105° for 4 hours: it loses not more than 1.5% of its weight.

Residue on ignition ⟨281⟩: not more than 0.1%.

Heavy metals, *Method II* ⟨231⟩: 0.002%.

Fluoride ions—[NOTE—All glassware and/or plasticware used in this test should be scrupulously clean and even free from trace amounts of fluoride. The use of plasticware to contain the solutions while the potential is measured is recommended.]

Buffer solution—To 110 g of sodium chloride in a 2-liter volumetric flask add 1 g of sodium citrate and 700 mL of water, and dissolve with shaking. Carefully add 150 g of sodium hydroxide, and dissolve with shaking. Cool to room temperature, and while stirring, cautiously add 450 mL of glacial acetic acid. Cool to room temperature, add 600 mL of isopropyl alcohol, dilute with water to volume, and mix. The pH of this solution is between 5.0 and 5.5.

Standard stock solution—Weigh accurately 2.211 g of sodium fluoride, previously dried at 150° for 4 hours, into a 1-liter volumetric flask, and dissolve in about 200 mL of water. Add 1.0 mL of sodium hydroxide solution (1 in 250), dilute with water to volume, and mix. Each mL of this solution contains 1 mg of fluoride ion. Store the solution in a closed plastic container.

Standard preparations—Dilute a portion of *Standard stock solution* quantitatively and stepwise with *Buffer solution* to obtain a *Standard preparation* having a fluoride concentration of 1 µg per mL. Prepare the final dilution in a 100-mL volumetric flask. In the same manner, prepare additional *Standard preparations* having fluoride concentrations of 3, 5, and 10 µg per mL, respectively.

Test preparation—Place 1 g of Flucytosine, accurately weighed, in a 100-mL volumetric flask, dissolve in *Buffer solution*, and dilute with *Buffer solution* to volume.

Procedure—Concomitantly measure the potential (see *Titrimetry* ⟨541⟩), in mV, of the *Standard preparations* and the *Test preparation*, with a suitable pH meter equipped with a fluoride-specific ion electrode and a glass-sleeved calomel reference electrode that has been modified in the following manner: Mix 70 mL of freshly prepared saturated potassium chloride solution with 30 mL of isopropyl alcohol, fill the electrode with the clear supernatant liquid, and allow the electrode to remain in the mixture for not less than 2 hours prior to use, or preferably overnight.

When taking the measurements, transfer the solution to a 150-mL beaker, and immerse the electrodes. Insert a polytetrafluoroethylene-coated stirring bar into the beaker, place the beaker on a magnetic stirrer having an insulated top, and allow to stir until equilibrium is attained (about 1 to 2 minutes). Rinse and dry the electrodes between measurements, taking care not to scratch the crystal in the specific ion electrode.

Measure the potential of each *Standard preparation*, and plot the fluoride concentration, in mg per 100 mL, versus the potential, in mV, on semilogarithmic paper. Measure the potential of the *Test preparation*, and determine from the standard curve the fluoride concentration, in mg per 100 mL. Calculate the percentage of fluoride in the portion of Flucytosine taken by the formula:

$$C/10,$$

in which *C* is the fluoride concentration, in mg per 100 mL, from the standard curve: not more than 0.05% of fluoride is found.

Fluorouracil—Dissolve 250 mg in 10 mL of a mixture of glacial acetic acid and water (8:2). Apply 20 µL of this solution to a thin-layer chromatographic plate (see *Chromatography* ⟨621⟩) coated with a 0.5-mm layer of chromatographic silica gel mixture. To the same plate apply 20 µL, in 10-µL increments, of a 1 in 40,000 solution of USP Fluorouracil RS in a mixture of glacial acetic acid and water (4:1). Develop the chromatogram in a mixture of chloroform and glacial acetic acid (13:7) until the solvent front has moved not less than 14 cm from the origin. Remove the plate from the developing chamber, and allow the solvent to evaporate. Locate the spots on the plate by observing under short-wavelength ultraviolet radiation: any spot from the solution under test is not greater in size and intensity than the spot at the respective R_f produced by the Standard solution, corresponding to not more than 0.1% of fluorouracil.

Assay—Place about 400 mg of Flucytosine, accurately weighed, in a 250-mL beaker, add 150 mL of a mixture of 2 volumes of glacial acetic acid and 1 volume of acetic anhydride, and dissolve, warming gently if necessary. Titrate potentiometrically with 0.1 *N* perchloric acid VS, using a calomel-glass electrode system. Perform a blank determination, and make any necessary correction. Each mL of 0.1 *N* perchloric acid is equivalent to 12.91 mg of $C_4H_4FN_3O$.

Flucytosine Capsules

» Flucytosine Capsules contain not less than 90.0 percent and not more than 110.0 percent of the labeled amount of $C_4H_4FN_3O$.

Packaging and storage—Preserve in tight, light-resistant containers.

Reference standard—*USP Flucytosine Reference Standard*—Dry at 105° for 4 hours before using.

Identification—

A: The ultraviolet absorption spectrum of the solution from the Capsule contents obtained in the *Assay* exhibits maximum absorption at the same wavelength as that of the Standard solution, and the two spectra are similar between 260 nm and 350 nm.

B: Shake a portion of the contents of Capsules, equivalent to about 500 mg of flucytosine, with 10 mL of water. Filter, and to 2 mL of the filtrate add 1 mL of sodium pentacyanoaminoferrate reagent [prepared by dissolving 100 mg of sodium (tri)pentacyanoaminoferrate in 20 mL of sodium carbonate solution (1 in 100)] and 1 mL of 3 percent hydrogen peroxide: on standing, a darker green is produced than that produced by a blank.

Dissolution ⟨711⟩—

Medium: water; 900 mL.

Apparatus 2: 100 rpm.

Time: 45 minutes.

Procedure—Determine the amount of $C_4H_4FN_3O$ dissolved from ultraviolet absorbances at the wavelength of maximum absorbance at about 276 nm of filtered portions of the solution under test, suitably diluted with water, if necessary, in comparison with a Standard solution having a known concentration of USP Flucytosine RS in the same medium.

Tolerances—Not less than 75% (Q) of the labeled amount of $C_4H_4FN_3O$ is dissolved in 45 minutes.

Uniformity of dosage units ⟨905⟩: meet the requirements.

Assay—Weigh the contents of not less than 20 Flucytosine Capsules, and determine the average weight per Capsule. Mix the combined contents, and transfer an accurately weighed portion of the powder, equivalent to about 250 mg of flucytosine, to a 250-mL volumetric flask. Add about 50 mL of 0.1 *N* hydrochloric acid, shake by mechanical means for 30 minutes, then add 0.1 *N* hydrochloric acid to volume, mix, and filter, discarding the first 20 mL of filtrate. Dilute 10 mL of the clear filtrate with 0.1 *N* hydrochloric acid to 250 mL. Dilute 10.0 mL of this solution with 0.1 *N* hydrochloric acid to 50 mL. Dissolve an accurately weighed quantity of USP Flucytosine RS in 0.1 *N* hydrochloric acid, and dilute quantitatively and stepwise with the same solvent to obtain a Standard solution having a known concentration of about 8 µg per mL. Concomitantly determine the absorbances of both solutions in 1-cm cells at the wavelength of maximum absorbance at about 285 nm, with a suitable spectrophotometer, using 0.1 *N* hydrochloric acid as the blank. Calculate the quantity, in mg, of $C_4H_4FN_3O$ in the portion of Capsule contents taken by the formula:

$$31.25C(A_U/A_S),$$

in which C is the concentration, in µg per mL, of USP Flucytosine RS in the Standard solution, and A_U and A_S are the absorbances of the solution from the Capsule contents and the Standard solution, respectively.

Fludrocortisone Acetate

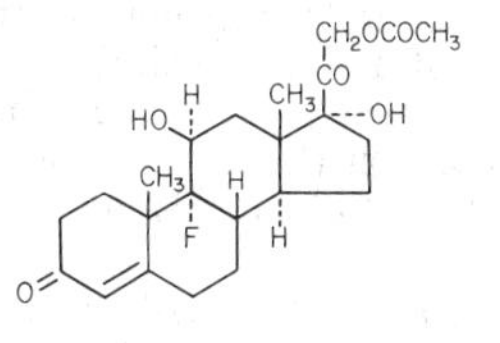

$C_{23}H_{31}FO_6$ 422.49

Pregn-4-ene-3,20-dione, 21-(acetyloxy)-9-fluoro-11,17-dihydroxy-, (11β)-.

9-Fluoro-11β,17,21-trihydroxypregn-4-ene-3,20-dione 21-acetate [*514-36-3*].

» Fludrocortisone Acetate contains not less than 97.0 percent and not more than 103.0 percent of $C_{23}H_{31}FO_6$, calculated on the dried basis.

Packaging and storage—Preserve in well-closed containers, protected from light.

Reference standard—*USP Fludrocortisone Acetate Reference Standard*—Dry in vacuum at 100° for 2 hours over magnesium perchlorate before using.

Identification—The infrared absorption spectrum of a mineral oil dispersion of it exhibits maxima only at the same wavelengths as that of a similar preparation of USP Fludrocortisone Acetate RS.

Specific rotation ⟨781⟩: between +126° and +138°, calculated on the dried basis, determined in a solution in acetone containing 50 mg in each 10 mL.

Loss on drying ⟨731⟩—Dry it in vacuum at 100° for 2 hours over magnesium perchlorate: it loses not more than 3.0% of its weight.

Residue on ignition ⟨281⟩: not more than 0.1%.

Chromatographic impurities—Dissolve about 100 mg of it in a mixture of 5 mL of chloroform and 1 mL of acetone in a 10-mL volumetric flask, and dilute with chloroform to volume. Dilute 1 mL of this solution with chloroform to 100 mL. Apply 10 µL, in 5-µL increments, of the test solution and of its dilution on a line parallel to and about 2.5 cm from the bottom of a thin-layer chromatographic plate (see *Chromatography* ⟨621⟩) coated with a 0.25-mm layer of chromatographic silica gel mixture. Develop the plate in a suitable chamber containing a mixture of chloroform, methanol, and water (85:14:1) until the solvent front has moved about 15 cm. Remove the plate, air-dry, and examine under short-wavelength ultraviolet light: no spot in the chromatogram of the more concentrated test solution, other than the principal spot, is larger or more intense than the spot from the diluted test solution (1.0%).

Assay—

Standard preparation—Dissolve about 25 mg of USP Fludrocortisone Acetate RS, accurately weighed, in chloroform to make 250 mL, and mix. Pipet 10 mL of this solution into a 50-mL volumetric flask, add chloroform to volume, and mix.

Assay preparation—Prepare as directed under *Standard preparation*, using Fludrocortisone Acetate instead of the Reference Standard.

Procedure—Pipet 10 mL of the *Assay preparation* and the *Standard preparation*, respectively, into separate 25-mL volumetric flasks, and pipet 10 mL of chloroform into a third flask to provide a blank. Treat each flask as follows: Add 1.0 mL of a solution prepared by dissolving 50 mg of blue tetrazolium in 10 mL of methanol, and mix. Add 1.0 mL of a mixture of 1 volume of tetramethylammonium hydroxide TS and 4 volumes of methanol, mix, and allow to stand for 10 minutes. Dilute with a 1 in 100 solution of hydrochloric acid in methanol to volume. Concomitantly determine the absorbances of the solutions from the *Assay preparation* and the *Standard preparation* in 1-cm cells at about 525 nm, with a suitable spectrophotometer, against the reagent blank. Calculate the quantity, in mg, of $C_{23}H_{31}FO_6$ in the portion of Fludrocortisone Acetate taken by the formula:

$$1.25C(A_U/A_S),$$

in which C is the concentration, in µg per mL, of USP Fludrocortisone Acetate RS in the *Standard preparation*, and A_U and A_S are the absorbances of the solutions from the *Assay preparation* and the *Standard preparation*, respectively.

Fludrocortisone Acetate Tablets

» Fludrocortisone Acetate Tablets contain not less than 90.0 percent and not more than 110.0 percent of the labeled amount of $C_{23}H_{31}FO_6$.

Packaging and storage—Preserve in well-closed containers.

Reference standards—*USP Fludrocortisone Acetate Reference Standard*—Dry in vacuum at 100° for 2 hours over magnesium perchlorate before using. *USP Norethindrone Reference Standard*—Dry in vacuum at 105° for 3 hours before using.

Identification—Transfer a portion of powdered Tablets, equivalent to about 1 mg of fludrocortisone acetate, to a glass-stoppered, 15-mL centrifuge tube, add 10 mL of acetone, and shake by mechanical means for 3 minutes. Centrifuge the mixture, and apply 20 µL, in 5-µL increments, of the clear solution and 20 µL of a solution of USP Fludrocortisone Acetate RS in acetone, containing about 100 µg per mL, at points along a line about 2.5 cm from the bottom of a thin-layer chromatographic plate (see *Chromatography* ⟨621⟩) coated with a 0.25-mm layer of chromatographic silica gel mixture. Develop the plate in a suitable chamber containing a mixture of chloroform, methanol, and water (85:14:1) until the solvent front has moved about 15 cm. Remove the plate, air-dry, and examine under short-wavelength ultraviolet light: the R_f value of the principal spot in the chromatogram of the test solution corresponds to that obtained with the Standard solution.

Disintegration ⟨701⟩: 30 minutes.

Uniformity of dosage units ⟨905⟩: meet the requirements.

Procedure for content uniformity—

Mobile solvent—Prepare as directed in the *Assay*.

Internal standard solution—Prepare a solution of USP Norethindrone RS in acetonitrile having a concentration of about 10 µg per mL.

Standard preparation—Dissolve a suitable quantity of USP Fludrocortisone Acetate RS, accurately weighed, in *Internal standard solution* to obtain a solution having a known concentration of about 0.20 mg per mL. Add 5.0 mL of this solution to 10.0 mL of water contained in a low-actinic 50-mL volumetric flask. Dilute with *Internal standard solution* to volume to obtain a solution having a known concentration of about 20 µg of fludrocortisone acetate per mL.

Test preparation—Add 1.0 mL of water to 1 Tablet in a 10-mL centrifuge tube, and mix on a vortex-type mixer for 1 minute or until disintegration is complete. Add 4.0 mL of *Internal standard solution*, mix on a vortex-type mixer for 1 minute, then shake by mechanical means for not less than 40 minutes. Centrifuge at 3600 rpm for 20 minutes, or until a clear supernatant solution is obtained. Use the clear supernatant solution.

Procedure—Proceed as directed for *Procedure* in the *Assay*. Calculate the quantity, in mg, of $C_{23}H_{31}FO_6$ in the Tablet by the formula:

$$(T/D)C(R_U/R_S),$$

in which C, R_U, and R_S are as defined in the *Assay*, T is the labeled quantity, in mg, of fludrocortisone acetate in the Tablet, and D is the concentration, in µg per mL, of fludrocortisone acetate in the *Test preparation*, based on the labeled quantity per Tablet and the extent of dilution.

Assay—

Mobile solvent—Prepare a suitable, degassed acetonitrile solution, 40% to 45% (v/v), such that the resolution factor, R, is not less than 2.5 between the peaks from fludrocortisone acetate and the internal standard.

Internal standard solution—Prepare a solution of USP Norethindrone RS in acetonitrile having a concentration of about 75 µg per mL.

Standard preparation—Dissolve a suitable quantity of USP Fludrocortisone Acetate RS, accurately weighed, in *Internal standard solution* to obtain a solution having a known concentration of about 0.50 mg per mL. Pipet 5 mL of this solution into a 25-mL volumetric flask containing 5.0 mL of water. Dilute with *Internal standard solution* to volume, to obtain a solution having a known concentration of about 0.1 mg of fludrocortisone acetate per mL.

Assay preparation—Weigh and finely powder not less than 35 Fludrocortisone Acetate Tablets. Transfer an accurately weighed portion of the powder, equivalent to about 2.5 mg of fludrocortisone acetate, to a low-actinic, glass-stoppered, 50-mL centrifuge tube. Add 5.0 mL of water, and mix for 1 minute. Add 20.0 mL of *Internal standard solution*, mix by mechanical means for 40 minutes, then centrifuge for 15 minutes or until a clear supernatant solution is obtained. Use the clear supernatant solution.

Procedure—Introduce equal volumes (about 20 µL) of the *Assay preparation* and the *Standard preparation* into a high-pressure liquid chromatograph (see *Chromatography* ⟨621⟩) operated at room temperature, by means of a suitable microsyringe or sampling valve, adjusting the specimen size and other operating parameters such that the peak obtained with the *Standard preparation* is about 0.7 full scale. Typically, the apparatus is fitted with a 3.9-mm (ID) × 30-cm stainless steel column packed with packing L1, and equipped with an ultraviolet detector capable of monitoring absorption at 254 nm and a suitable recorder. In a suitable chromatogram, the coefficient of variation for five replicate injections of the *Standard preparation* is not more than 3.0% and the resolution factor, R, is not less than 2.5 between the two peaks. Measure the height of the peaks, at identical retention times, obtained with the *Assay preparation* and the *Standard preparation*, and calculate the quantity, in mg, of $C_{23}H_{31}FO_6$ in the portion of Tablets taken by the formula:

$$25C(R_U/R_S),$$

in which C is the concentration, in mg per mL, of USP Fludrocortisone Acetate RS in the *Standard preparation*, and R_U and R_S are the ratios of the peak heights of the fludrocortisone acetate peak to the internal standard peak from the *Assay preparation* and the *Standard preparation*, respectively.

Fluidextract, Cascara Sagrada—*see* Cascara Sagrada Fluidextract

Fluidextract, Senna—*see* Senna Fluidextract

Flumethasone Pivalate

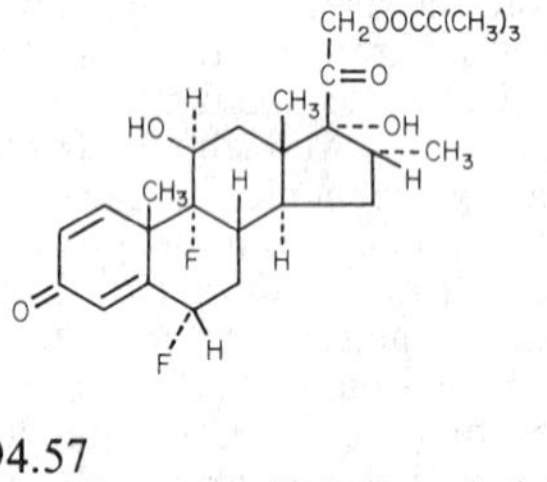

$C_{27}H_{36}F_2O_6$ 494.57

Pregna-1,4-diene-3,20-dione, 21-(2,2-dimethyl-1-oxopropoxy)-6,9-difluoro-11,17-dihydroxy-16-methyl-, (6α,11β,16α)-.

6α,9-Difluoro-11β,17,21-trihydroxy-16α-methylpregna-1,4-diene-3,20-dione 21-pivalate [*2002-29-1*].

» Flumethasone Pivalate contains not less than 97.0 percent and not more than 103.0 percent of $C_{27}H_{36}F_2O_6$, calculated on the dried basis.

Packaging and storage—Preserve in tight, light-resistant containers.

Reference standard—*USP Flumethasone Pivalate Reference Standard*—Dry at 105° for 4 hours before using.

Identification—

A: The infrared absorption spectrum of a mineral oil dispersion of it, previously dried, exhibits maxima only at the same wavelengths as that of a similar preparation of USP Flumethasone Pivalate RS.

B: The ultraviolet absorption spectrum of a 1 in 50,000 solution in methanol exhibits maxima and minima at the same wavelengths as that of a similar solution of USP Flumethasone Pivalate RS, concomitantly measured, and the respective absorptivities, calculated on the dried basis at the wavelength of maximum absorbance at about 237 nm, do not differ by more than 3.0%.

Specific rotation ⟨781⟩: between +71° and +82°, calculated on the dried basis, determined in a solution in dioxane containing 100 mg in each 10 mL.

Loss on drying ⟨731⟩—Dry it at 105° for 4 hours: it loses not more than 1.0% of its weight.

Chromatographic impurities—Prepare a solution in dioxane containing 20 mg per mL. On a suitable thin-layer chromatographic plate (see *Chromatography* ⟨621⟩), coated with a 0.25-mm layer of chromatographic silica gel, apply 5 µL of this solution and 5 µL each of three dioxane solutions containing in each mL, respectively, 200 (1%), 400 (2%), and 600 (3%) µg of USP Flumethasone Pivalate RS. Allow the spots to dry, and develop the chromatogram in a solvent system consisting of toluene and ethyl acetate (7:3) until the solvent front has moved about three-fourths of the length of the plate. Remove the plate from the developing chamber, mark the solvent front, and allow the solvent to evaporate. Locate the spots on the plate by lightly spraying with dilute sulfuric acid (1 in 2), heating at 100° for 30 minutes, and inspecting under long-wavelength ultraviolet light: the total content of any impurities detected, when compared to the Standard solutions, does not exceed 3.0%.

Assay—

Standard preparation—Dissolve a suitable quantity of USP Flumethasone Pivalate RS, accurately weighed, in alcohol, and dilute quantitatively and stepwise with alcohol to obtain a solution having a known concentration of about 20 µg per mL. Transfer 10.0 mL of this solution to a glass-stoppered, 20-mL conical flask.

Assay preparation—Transfer about 20 mg of Flumethasone Pivalate, accurately weighed, to a 100-mL volumetric flask, dissolve in alcohol, dilute with alcohol to volume, and mix. Transfer 10.0 mL of this solution to a second 100-mL volumetric flask, dilute with alcohol to volume, and mix. Transfer 10.0 mL of this solution to a glass-stoppered, 20-mL conical flask.

Procedure—To each of the flasks containing the *Standard preparation* and the *Assay preparation*, and to a similar flask containing 10.0 mL of alcohol to provide the blank, add 1.0 mL of tetramethylammonium hydroxide TS. Mix, allow to stand for 20 minutes, accurately timed, add 1.0 mL of blue tetrazolium TS to each flask, and mix. Allow to stand for 40 minutes, add 1.0 mL of glacial acetic acid to each flask, mix, and concomitantly determine the absorbances of the solutions in 1-cm cells at the wavelength of maximum absorbance at about 520 nm, with a suitable spectrophotometer, against the blank. Calculate the quantity, in mg, of $C_{27}H_{36}F_2O_6$ in the portion of Flumethasone Pivalate taken by the formula:

$$C(A_U / A_S),$$

in which C is the concentration, in µg per mL, of USP Flumethasone Pivalate RS in the *Standard preparation*, and A_U and A_S are the absorbances of the solutions from the *Assay preparation* and the *Standard preparation*, respectively.

Flumethasone Pivalate Cream

» Flumethasone Pivalate Cream contains not less than 90.0 percent and not more than 110.0 percent of the labeled amount of $C_{27}H_{36}F_2O_6$ in a suitable cream base.

Packaging and storage—Preserve in collapsible tubes.

Reference standard—*USP Flumethasone Pivalate Reference Standard*—Dry at 105° for 4 hours before using.

Identification—Place a quantity of Cream, equivalent to about 400 µg of flumethasone pivalate, in a 50-mL centrifuge tube, and treat as directed in the *Assay*, collecting the extracts in a 50-mL centrifuge tube. Evaporate the acetonitrile on a water bath (about 75°) with the aid of a stream of nitrogen. Dissolve the residue in 2.0 mL of acetonitrile with the aid of heat, allow to cool, and centrifuge. On a suitable thin-layer chromatographic plate (see *Chromatography* ⟨621⟩), coated with a 0.25-mm layer of chromatographic silica gel mixture, apply 50 µL of this solution and 50 µL of a solution of chloroform and methanol (1:1) of USP Flumethasone Pivalate RS containing 200 µg per mL. Allow the spots to dry, and develop the chromatogram in toluene until the solvent front has moved about three-fourths of the length of the plate. Remove the plate from the developing chamber, mark the solvent front, and allow the solvent to evaporate. Redevelop the plate in the same manner in a solvent system consisting of toluene and ethyl acetate (7:3), remove the plate from the developing chamber, and allow the solvent to evaporate. Locate the spots on the plate by examination under short-wavelength ultraviolet light: the R_f value of the principal spot obtained from the test solution corresponds to that obtained from the Standard solution.

Microbial limits—It meets the requirements of the tests for absence of *Staphylococcus aureus* and *Pseudomonas aeruginosa* under *Microbial Limit Tests* ⟨61⟩.

Minimum fill ⟨755⟩: meets the requirements.

Assay—

Solvent acetonitrile—Saturate acetonitrile with isooctane.

Solvent isooctane—Saturate isooctane with acetonitrile.

4-Aminoantipyrine solution—Dissolve about 200 mg of 4-aminoantipyrine in a 1 in 100 solution of hydrochloric acid in methanol to make 50 mL of solution, and mix. Prepare this solution on the day of use.

Standard preparation—Dissolve a suitable quantity of USP Flumethasone Pivalate RS, accurately weighed, in *Solvent acetonitrile*, and dilute quantitatively and stepwise with *Solvent acetonitrile* to obtain a solution having a known concentration of about 40 µg per mL.

Assay preparation—Transfer an accurately weighed portion of Flumethasone Pivalate Cream, equivalent to about 400 µg of flumethasone pivalate, to a tared glass-stoppered, 50-mL centrifuge tube. Place the tube in a vacuum desiccator, and dry the specimen in vacuum over silica gel until about 70% loss in weight is obtained.

Procedure—[NOTE—Perform the extractions in glass-stoppered, 50-mL centrifuge tubes with separation being effected by centrifugation, the portion to be retained being withdrawn into a hypodermic syringe fitted with a blunt-end, 14-gauge, 15-cm needle.] Add 10.0 mL of *Solvent acetonitrile* to the *Assay preparation*, and transfer 10.0 mL of the *Standard preparation* to a separate, glass-stoppered, 50-mL centrifuge tube. Add 25 mL of *Solvent isooctane* to each tube, shake by mechanical means until the cream is dispersed, and then shake both mixtures for 5 minutes. Separate and withdraw each acetonitrile layer, and filter through cotton pledgets, previously saturated with *Solvent acetonitrile*, into separate 25-mL volumetric flasks. Repeat the extraction, using a 10-mL portion of *Solvent acetonitrile* for each tube. Separate, withdraw, and filter each acetonitrile layer through the same respective filter, and combine the extracts with the main extracts. Dilute each with *Solvent acetonitrile* to volume, and mix. Transfer 10.0 mL of each solution to separate, glass-stoppered, 20-mL tubes, and to a third tube transfer 10.0 mL of *Solvent acetonitrile* to provide the blank. Evaporate the solvent on a water bath (about 75°) with the aid of a stream of nitrogen. Add 5.0 mL of *4-Aminoantipyrine solution* to each tube, insert the stopper, shake by mechanical means until the residue dissolves, and allow to stand for 1 hour. Concomitantly determine the absorbances of the solutions in 1-cm cells at the wavelength of maximum absorbance at about 390 nm, with a suitable spectrophotometer, against the blank. Calculate the quantity, in mg, of $C_{27}H_{36}F_2O_6$ in the portion of Cream taken by the formula:

$$0.01C(A_U / A_S),$$

in which C is the concentration, in µg per mL, of USP Flumethasone Pivalate RS in the *Standard preparation*, and A_U and A_S are the absorbances of the solutions from the *Assay preparation* and the *Standard preparation*, respectively.

Flunisolide

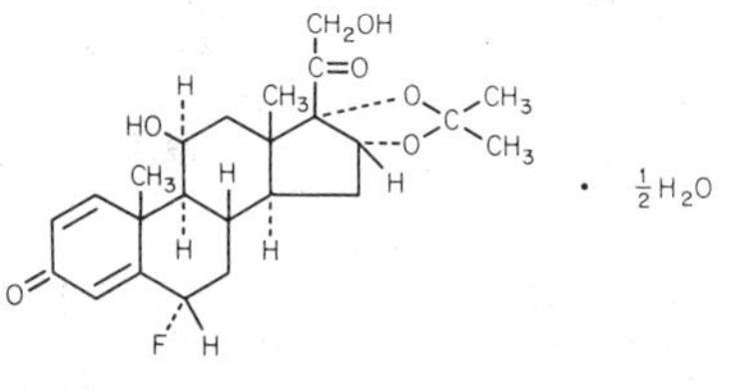

$C_{24}H_{31}FO_6 \cdot \frac{1}{2}H_2O$ 443.51

Pregna-1,4-diene-3,20-dione, 6-fluoro-11,21-dihydroxy-16,17-[(1-methylethylidene)bis(oxy)]-, hemihydrate, (6α,11β,16α)-.

6α-Fluoro-11β,16α,17,21-tetrahydroxypregna-1,4-diene-3,20-dione cyclic 16,17-acetal with acetone, hemihydrate [*77326-96-6*].

» Flunisolide contains not less than 97.0 percent and not more than 102.0 percent of $C_{24}H_{31}FO_6 \cdot \frac{1}{2}H_2O$, calculated on the dried basis.

Packaging and storage—Preserve in well-closed containers.

Reference standard—*USP Flunisolide Reference Standard*—Dry in vacuum at 60° for 3 hours before using. This dried standard is the hemihydrate of flunisolide.

Identification—

A: The infrared absorption spectrum of a potassium bromide dispersion of it, previously dried, exhibits maxima only at the same wavelengths as that of a similar preparation of USP Flunisolide RS.

B: The ultraviolet absorption spectrum of a 1 in 100,000 solution in methanol exhibits a maximum at the same wavelength as that of a similar solution of USP Flunisolide RS, concomitantly measured.

Specific rotation ⟨781⟩: between +103° and +111°, calculated on the dried basis, determined in a solution in chloroform containing 100 mg in each 10 mL.

Loss on drying ⟨731⟩—Dry it in vacuum at 60° for 3 hours: it loses not more than 1.0% of its weight.

Water, *Method I* ⟨921⟩: between 1.8% and 2.5% (determined on a dried specimen).

Residue on ignition ⟨281⟩: not more than 0.1% from 250 mg.

Chromatographic impurities—

Standard solutions—Prepare a solution of USP Flunisolide RS in acetone to contain 10 mg per mL (*Standard solution A*). Dilute 1 mL of *Standard solution A* with acetone to 100 mL (*Standard solution B*).

Test preparation—Prepare a solution of Flunisolide in acetone to contain 10 mg per mL.

Procedure—On a suitable thin-layer chromatographic plate (see *Chromatography* ⟨621⟩), coated with a 0.25-mm layer of chromatographic silica gel mixture, apply 10-µL volumes of *Standard solution A, Standard solution B,* and the *Test preparation.* Place the plate in a suitable chromatographic chamber previously equilibrated with a mixture of toluene and alcohol (90:10), seal the chamber, and develop the chromatogram until the solvent front has moved three-fourths of the length of the plate. Remove the plate, allow the solvent to evaporate, and examine the plate under short-wavelength ultraviolet light: the R_f value of the principal spot obtained from the *Test preparation* corresponds to that obtained from *Standard solution A.* No secondary spot exhibits an intensity greater than that of the principal spot from *Standard solution B.*

Assay—

Mobile phase—Prepare a suitable degassed solution of water and acetonitrile (3:2) such that at an approximate flow rate of 1.6 mL per minute, the retention time of Flunisolide is about 6 minutes.

Standard preparation—Dissolve an accurately weighed quantity of USP Flunisolide RS in *Mobile phase* to obtain a solution having a known concentration of about 0.2 mg per mL.

Assay preparation—Using 20 mg of Flunisolide, accurately weighed, proceed as directed for *Standard preparation.*

Chromatographic system (see *Chromatography* ⟨621⟩)—The liquid chromatograph is equipped with a 254-nm detector and a 4-mm × 25-cm column that contains 5- to 10-µm packing L7. The flow rate is about 1.6 mL per minute. Chromatograph the *Standard preparation,* and record the peak response as directed under *Procedure:* the column efficiency is not less than 2700 theoretical plates, the tailing factor for the flunisolide peak is not more than 1.7, and the relative standard deviation for replicate injections is not more than 1.0%.

Procedure—Separately inject equal volumes (between 15 µL and 30 µL) of the *Standard preparation* and the *Assay preparation* into the chromatograph, record the chromatograms, and measure the responses for the major peaks. Calculate the quantity, in mg, of $C_{24}H_{31}FO_6 \cdot \frac{1}{2}H_2O$ in the portion of Flunisolide taken by the formula:

$$100C(r_U/r_S),$$

in which C is the concentration, in mg per mL, of USP Flunisolide RS in the *Standard preparation,* and r_U and r_S are the peak responses obtained from the *Assay preparation* and the *Standard preparation,* respectively.

Flunisolide Nasal Solution

» Flunisolide Nasal Solution is an aqueous, buffered solution of Flunisolide. It is supplied in a form suitable for nasal administration. It contains not less than 90.0 percent and not more than 110.0 percent of the labeled amount of $C_{24}H_{31}FO_6$.

Packaging and storage—Preserve in tight containers, protected from light, at controlled room temperature.

Reference standard—*USP Flunisolide Reference Standard*—Dry in vacuum at 60° for 3 hours before using. This dried standard is the hemihydrate of flunisolide.

Identification—Proceed as directed in the *Assay,* except to inject 50 µL to 200 µL of a mixture of the *Assay preparation* and the *Standard preparation* (1:1) onto the column, adjusting the response to obtain a response that is between 50% and 90% full-scale: a single peak is observed in the chromatogram for the mixed solution.

pH ⟨791⟩: between 4.5 and 6.0, a silver–silver chloride (internal element) electrode being used in conjunction with a fiber junction calomel electrode.

Quantity delivered per spray—Prime the spray pump by delivering 10 sprays into a fume hood. Accurately weigh the entire assembly, record the weight, and deliver 8 more sprays into the hood. Again weigh the assembly, and record the weight. Calculate the quantity, in µg, of $C_{24}H_{31}FO_6$ delivered per spray by the formula:

$$[(W_1 - W_2)/8][A/1.04],$$

in which W_1 and W_2 are the first and second weights, respectively, in g, A is the quantity, in µg per mL, of $C_{24}H_{31}FO_6$ found in the *Assay,* and 1.04 is the density of Flunisolide Nasal Solution, in g per mL. The quantity delivered is between 17 µg and 33 µg per spray.

Assay—

Mobile phase—Prepare a suitable degassed solution of water, acetonitrile, and glacial acetic acid (69:30:1 to 64:35:1). Adjust the ratio as necessary to obtain suitable chromatographic performance.

Internal standard solution—Dissolve norethindrone in acetonitrile to obtain a solution containing about 300 µg per mL.

Standard preparation—Dissolve an accurately weighed quantity of USP Flunisolide RS in a mixture of acetonitrile and *Mobile phase* (1:1) to obtain a solution having a known concentration of about 250 µg per mL. Transfer 1.0 mL of this solution, and 1.0 mL of *Internal standard solution,* by means of "to contain" pipets, to a 50-mL volumetric flask. Rinse the pipets with *Mobile phase,* adding the rinsings to the flask, dilute with *Mobile phase* to volume, and mix. The final concentration of USP Flunisolide RS is about 35 µg per mL.

Assay preparation—Transfer an accurately measured volume of Flunisolide Nasal Solution, equivalent to about 250 µg of flunisolide, to a 50-mL volumetric flask, and add 1.0 mL of *Internal standard solution* by means of "to contain" pipets. Rinse the pipets with *Mobile phase,* adding the rinsing to the flask, dilute with *Mobile phase* to volume, and mix.

Chromatographic system (see *Chromatography* ⟨621⟩)—The liquid chromatograph is equipped with a 254-nm detector and a 4-mm × 25-cm column that contains 5- to 10-µm packing L1. The flow rate is about 2 mL per minute. Chromatograph the *Standard preparation,* and record the peak responses as directed under *Procedure:* the resolution, R, between the analyte and internal standard peaks is not less than 5.0, and the relative standard deviation for replicate injections is not more than 1.5%.

Procedure—Separately inject equal volumes (about 50 µL) of the *Standard preparation* and the *Assay preparation* into the chromatograph, record the chromatograms, and measure the responses for the major peaks. The relative retention times are about 0.6 for flunisolide and 1.0 for norethindrone. Calculate the quantity, in mg, of $C_{24}H_{31}FO_6$ in each mL of the Nasal Solution taken by the formula:

$$(434.5/443.5)(50C/V)(R_U/R_S),$$

in which 434.5 and 443.5 are the molecular weights of $C_{24}H_{31}FO_6$ and $C_{24}H_{31}FO_6 \cdot \frac{1}{2}H_2O$, respectively, *C* is the concentration, in mg per mL, of USP Flunisolide RS in the *Standard preparation*, *V* is the volume, in mL, of Nasal Solution taken, and R_U and R_S are the peak response ratios of the flunisolide peak and the norethindrone peak obtained from the *Assay preparation* and the *Standard preparation*, respectively.

Fluocinolone Acetonide

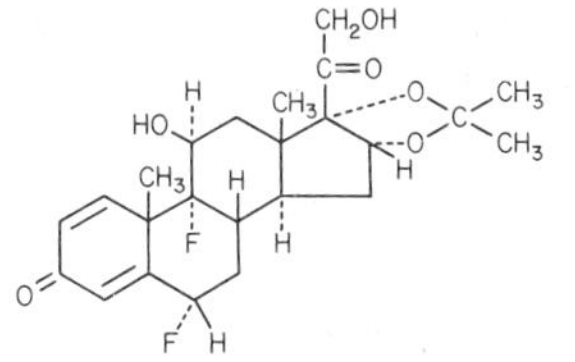

$C_{24}H_{30}F_2O_6$ (anhydrous) 452.50
Pregna-1,4-diene-3,20-dione, 6,9-difluoro-11,21-dihydroxy-16,17-[(1-methylethylidene)bis(oxy)]-, (6α,11β,16α)-.
6α,9-Difluoro-11β,16α,17,21-tetrahydroxypregna-1,4-diene-3,20-dione, cyclic 16,17-acetal with acetone [*67-73-2*].
Dihydrate 488.53

» Fluocinolone Acetonide is anhydrous or contains two molecules of water of hydration. It contains not less than 97.0 percent and not more than 102.0 percent of $C_{24}H_{30}F_2O_6$, calculated on the dried basis.

Packaging and storage—Preserve in well-closed containers.

Labeling—Label it to indicate whether it is anhydrous or hydrous.

Reference standard—*USP Fluocinolone Acetonide Reference Standard*—Dry in vacuum at 105° for 3 hours before using.

Identification—

A: The infrared absorption spectrum of a potassium bromide dispersion of it, previously dried exhibits maxima only at the same wavelengths as that of a similar preparation of USP Fluocinolone Acetonide RS. If a difference appears, dissolve portions of both the test specimen and the Reference Standard in ethyl acetate, evaporate the solutions to dryness, and repeat the test on the residues.

B: It responds to the *Thin-layer Chromatographic Identification Test* ⟨201⟩, the test solution and the Standard solution being prepared at a concentration of 5 mg per mL in acetone, chromatographic silica gel being used as the adsorbant, the solvent mixture being nitromethane, dichloromethane, and methanol (50:50:1), and ultraviolet light being used to locate the spots.

Specific rotation ⟨781⟩: between +98° and +108°, calculated on the dried basis, determined in a solution in methanol containing 100 mg in each 10 mL.

Loss on drying ⟨731⟩—Dry it in vacuum at 105° for 3 hours: anhydrous Fluocinolone Acetonide loses not more than 1.0% of its weight, and hydrous Fluocinolone Acetonide loses not more than 8.5% of its weight.

Assay—

Mobile phase—Prepare a suitable, degassed solution of water, acetonitrile, and tetrahydrofuran (77:13:10).

Standard preparation—Dissolve about 20 mg of USP Fluocinolone Acetonide RS, accurately weighed, in 23 mL of a mixture of acetonitrile and tetrahydrofuran (13:10), dilute with water to 100.0 mL, and mix.

Assay preparation—Transfer about 20 mg of Fluocinolone Acetonide, accurately weighed, to a 100-mL volumetric flask, dissolve in 23 mL of a mixture of acetonitrile and tetrahydrofuran (13:10), dilute with water to volume, and mix.

Chromatographic system (see *Chromatography* ⟨621⟩)—The liquid chromatograph is equipped with a 254-nm detector and a 4.5-mm × 10-cm column that contains packing L1. Adjust the flow rate so that the retention time for fluocinolone acetonide is between 9 and 13 minutes. Chromatograph the *Standard preparation*, and record the peak response as directed under *Procedure:* the column efficiency is not less than 3000 theoretical plates, and the relative standard deviation for replicate injections is not more than 3.0%.

Procedure—Separately inject equal volumes (about 20 µL) of the *Standard preparation* and the *Assay preparation* into the chromatograph, record the chromatograms, and measure the responses for the major peaks. Calculate the quantity, in mg, of $C_{24}H_{30}F_2O_6$ in the portion of Fluocinolone Acetonide taken by the formula:

$$100C(r_U/r_S),$$

in which *C* is the concentration, in mg per mL, of USP Fluocinolone Acetonide RS in the *Standard preparation*, and r_U and r_S are the peak responses obtained from the *Assay preparation* and the *Standard preparation*, respectively.

Fluocinolone Acetonide Cream

» Fluocinolone Acetonide Cream contains not less than 90.0 percent and not more than 110.0 percent of the labeled amount of $C_{24}H_{30}F_2O_6$.

Packaging and storage—Preserve in collapsible tubes or in tight containers.

Reference standards—*USP Fluocinolone Acetonide Reference Standard*—Dry at 105° for 3 hours before using. *USP Norethindrone Reference Standard*—Dry in vacuum at 105° for 3 hours before using.

Identification—Transfer a quantity of the Cream, equivalent to about 0.5 mg of fluocinolone acetonide, to a centrifuge tube, disperse it in 5 mL of water, add 10 mL of chloroform, shake, and centrifuge. Remove and discard the aqueous layer, add 10 mL of water to the tube, shake, and centrifuge. Dry about 2 mL of the chloroform extract over about 200 mg of anhydrous sodium sulfate: the dried extract responds to the *Thin-layer Chromatographic Identification Test* ⟨201⟩, 50 µL of the dried chloroform extract and 50 µL of a Standard solution containing about 50 µg per mL of USP Fluocinolone Acetonide RS being applied, and a mixture of chloroform and diethylamine (2:1) being used for development.

Microbial limits—It meets the requirements of the tests for absence of *Staphylococcus aureus* and *Pseudomonas aeruginosa* under *Microbial Limit Tests* ⟨61⟩.

Minimum fill ⟨755⟩: meets the requirements.

Assay—

Internal standard solution—Dissolve USP Norethindrone RS in acetonitrile to obtain a solution containing about 200 µg per mL.

Standard preparation—Dissolve an accurately weighed quantity of USP Fluocinolone Acetonide RS in acetonitrile to obtain a solution having a known concentration of about 300 µg per mL. Transfer 5.0 mL of this solution, 6.0 mL of *Internal standard solution*, and 15.0 mL of water to a 50-mL volumetric flask. Dilute with acetonitrile to volume, and mix. The *Standard preparation* contains 30 µg of USP Fluocinolone Acetonide RS per mL.

Mobile solvent—Prepare a mixture of acetonitrile and water (3:5). Adjust the ratio as necessary to obtain suitable chromatographic performance.

Assay preparation—Dissolve an accurately weighed portion of Fluocinolone Acetonide Cream, equivalent to about 0.75 mg of fluocinolone acetonide, in about 10 mL of acetonitrile by heating on a steam bath. Transfer the mixture to a 25-mL volumetric flask with the aid of three 2-mL portions of acetonitrile. Add 3.0 mL of *Internal standard solution* and 5.0 mL of water, cool, and mix. Dilute with acetonitrile to volume, mix, and cool in an ice bath. Centrifuge or filter the mixture to obtain a clear solution.

Apparatus—Use a high-pressure liquid chromatograph (see *Chromatography* ⟨621⟩) of the general type equipped with a detector for monitoring ultraviolet absorbance at about 254 nm, and capable of providing a flow rate of about 2 mL per minute for the *Mobile solvent*. Use a column that contains packing L1.

Procedure—Chromatograph equal volumes of the *Assay preparation* and the *Standard preparation*. Three replicate injections of the *Standard preparation* show a resolution factor of not less than 2.0 between the peaks for norethindrone and fluocinolone acetonide and a relative standard deviation of not more than 1.5%. Calculate the quantity, in mg, of $C_{24}H_{30}F_2O_6$ in the portion of Cream taken by the formula:

$$0.025C(R_U/R_S),$$

in which C is the concentration, in µg per mL, of USP Fluocinolone Acetonide RS in the *Standard preparation*, and R_U and R_S are the ratios of the peak areas of fluocinolone acetonide and norethindrone obtained from the *Assay preparation* and the *Standard preparation*, respectively.

Fluocinolone Acetonide Cream, Neomycin Sulfate and—*see* Neomycin Sulfate and Fluocinolone Acetonide Cream

Fluocinolone Acetonide Ointment

» Fluocinolone Acetonide Ointment contains not less than 90.0 percent and not more than 110.0 percent of the labeled amount of $C_{24}H_{30}F_2O_6$.

Packaging and storage—Preserve in collapsible tubes or in tight containers.

Reference standards—*USP Fluocinolone Acetonide Reference Standard*—Dry at 105° for 3 hours before using. *USP Norethindrone Reference Standard*—Dry in vacuum at 105° for 3 hours before using.

Identification—Evaporate 10.0 mL of the *Assay preparation* obtained in the *Assay* to dryness, and dissolve the residue in 1 mL of chloroform: it responds to the *Thin-layer Chromatographic Identification Test* ⟨201⟩, 50 µL of the test solution and 50 µL of the Standard solution, containing about 50 µg per mL of USP Fluocinolone Acetonide RS, being applied and a mixture of chloroform and diethylamine (2:1) being used for development.

Microbial limits—It meets the requirements of the tests for absence of *Staphylococcus aureus* and *Pseudomonas aeruginosa* under *Microbial Limit Tests* ⟨61⟩.

Assay—

Internal standard solution—Dissolve a suitable quantity of USP Norethindrone RS in methanol to obtain a solution containing about 850 µg per mL.

Diluted internal standard solution—Transfer 5.0 mL of *Internal standard solution* to a 250-mL flask. Dilute with methanol to volume, and mix.

Standard preparation—Dissolve an accurately weighed quantity of USP Fluocinolone Acetonide RS in acetonitrile to obtain a solution having a known concentration of about 200 µg per mL. Transfer 10.0 mL of this solution and 2.0 mL of *Internal standard solution* to a 100-mL volumetric flask. Dilute with methanol to volume, and mix. The concentration of USP Fluocinolone Acetonide RS in the *Standard preparation* is 20 µg per mL.

Mobile solvent—Prepare a mixture of acetonitrile and water (1:1). Adjust the ratio as necessary to obtain suitable chromatographic performance.

Assay preparation—Transfer an accurately weighed portion of Fluocinolone Acetonide Ointment, equivalent to about 0.7 mg of fluocinolone acetonide, to a 50-mL, round-bottom centrifuge tube. Add 35.0 mL of *Diluted internal standard solution*, emulsify using an ultrasonic probe, and centrifuge to bring the insoluble matter to the bottom. The clear supernatant liquid is the *Assay preparation*.

Apparatus—Use a suitable high-pressure liquid chromatograph (see *Chromatography* ⟨621⟩) of the general type equipped with a detector for monitoring ultraviolet absorbance at about 254 nm, and capable of providing a flow rate of about 2 mL per minute for the *Mobile solvent*. Use a 50-cm × 4-mm column that contains packing L1 so as to provide a resolution factor, *R* (see *Chromatography* ⟨621⟩), of at least 2.0 between peaks for norethindrone and fluocinolone acetonide. Three replicate injections of the *Standard preparation* show a relative standard deviation of not more than 1.5%.

Procedure—Chromatograph equal volumes of the *Assay preparation* and the *Standard preparation*, adjusting the system as necessary to obtain peaks of between about 50% and 90% of full-scale. Calculate the quantity, in mg, of $C_{24}H_{30}F_2O_6$ in the portion of Ointment taken by the formula:

$$0.035C(R_U/R_S),$$

in which C is the concentration, in µg per mL, of USP Fluocinolone Acetonide RS in the *Standard preparation*, and R_U and R_S are the ratios of the peak areas of fluocinolone acetonide and the internal standard obtained from the *Assay preparation* and the *Standard preparation*, respectively.

Fluocinolone Acetonide Topical Solution

» Fluocinolone Acetonide Topical Solution contains not less than 90.0 percent and not more than 110.0 percent of the labeled amount of $C_{24}H_{30}F_2O_6$.

Packaging and storage—Preserve in tight containers.

Reference standard—*USP Fluocinolone Acetonide Reference Standard*—Dry at 105° for 3 hours before using.

Identification—Transfer a quantity of Topical Solution, equivalent to about 0.5 mg of fluocinolone acetonide, to a separator, add 5 mL of water, and extract with 10 mL of chloroform. Withdraw the chloroform layer into a second separator, wash with 10 mL of water, and dry about 2 mL of the chloroform extract over about 200 mg of anhydrous sodium sulfate: the dried extract responds to the *Thin-layer Chromatographic Identification Test* ⟨201⟩, 50 µL of the dried chloroform extract and 50 µL of the Standard solution being applied, and a mixture of chloroform and diethylamine (2:1) being used for development.

Microbial limits—It meets the requirements of the tests for absence of *Staphylococcus aureus* and *Pseudomonas aeruginosa* under *Microbial Limit Tests* ⟨61⟩.

Assay—

Internal standard solution—Dissolve norethindrone in acetonitrile to obtain a solution containing about 200 µg per mL.

Standard preparation—Dissolve an accurately weighed quantity of USP Fluocinolone Acetonide RS in acetonitrile to obtain a solution having a known concentration of about 200 µg per mL. Transfer 5.0 mL of this solution, 4.0 mL of *Internal standard solution*, 10 mL of propylene glycol, and about 25 mL of acetonitrile to a 50-mL volumetric flask. Mix, cool to room temperature, dilute with acetonitrile to volume, and mix. The final concentration of USP Fluocinolone Acetonide RS is 20 µg per mL.

Mobile solvent—Prepare a mixture of acetonitrile and water (2:3). Adjust the ratio as necessary to obtain suitable chromatographic performance.

Assay preparation—Transfer an accurately measured volume of Fluocinolone Acetonide Topical Solution, equivalent to about 0.5 mg of fluocinolone acetonide, to a 25-mL volumetric flask. Add 2.0 mL of *Internal standard solution* and 10 mL of acetonitrile. Mix, cool to room temperature, dilute with acetonitrile to volume, and mix.

Apparatus—Use a suitable high-pressure liquid chromatograph (see *Chromatography* ⟨621⟩) of the general type equipped with a detector for monitoring ultraviolet absorbance at about 254 nm, and capable of providing a flow rate of about 2 mL per minute for the *Mobile solvent*. Use a column containing packing

L1 so as to provide a resolution factor, R, of at least 2.0 between peaks for norethindrone and fluocinolone acetonide.

Procedure—Chromatograph equal volumes of the *Assay preparation* and the *Standard preparation*, adjusting the system as necessary to obtain peaks of between about 50% and 90% full-scale. Calculate the quantity, in mg, of $C_{24}H_{30}F_2O_6$ in each mL of the Fluocinolone Acetonide Topical Solution taken by the formula:

$$0.025(C/V)(R_U/R_S),$$

in which C is the concentration, in µg per mL, of USP Fluocinolone Acetonide RS in the *Standard preparation*, V is the volume, in mL, of Solution taken, and R_U and R_S are the ratios of the areas of the fluocinolone acetonide peak to the internal standard peak obtained from the *Assay preparation* and the *Standard preparation*, respectively.

Fluocinonide

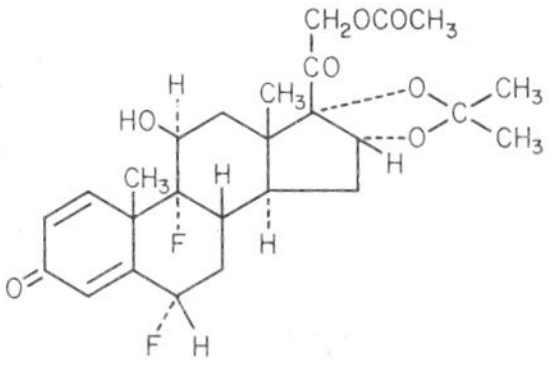

$C_{26}H_{32}F_2O_7$ 494.53

Pregna-1,4-diene-3,20-dione, 21-(acetyloxy)-6,9-difluoro-11-hydroxy-16,17-[(1-methylethylidene)bis(oxy)]-, (6α,11β,16α)-.

6α,9-Difluoro-11β,16α,17,21-tetrahydroxypregna-1,4-diene-3,20-dione, cyclic 16,17-acetal with acetone, 21-acetate [*356-12-7*].

» Fluocinonide contains not less than 97.0 percent and not more than 103.0 percent of $C_{26}H_{32}F_2O_7$, calculated on the dried basis.

Packaging and storage—Preserve in well-closed containers.

Reference standard—*USP Fluocinonide Reference Standard*—Dry at 105° for 3 hours before using.

Identification—

A: The infrared absorption spectrum of a potassium bromide dispersion of it exhibits maxima only at the same wavelengths as that of a similar preparation of USP Fluocinonide RS.

B: The ultraviolet absorption spectrum of a 1 in 100,000 solution in methanol exhibits maxima and minima at the same wavelengths as that of a similar solution of USP Fluocinonide RS, concomitantly measured, and the respective absorptivities, calculated on the dried basis, at the wavelength of maximum absorbance at about 238 nm do not differ by more than 3.0%.

C: Proceed as directed in the *Assay* to the point at which the chromatograph is viewed under ultraviolet light: the R_f value of the principal spot from the *Assay preparation* corresponds to that obtained from the *Standard preparation.*

Specific rotation ⟨781⟩: between +81° and +89°, calculated on the dried basis, determined in a solution in chloroform containing 100 mg in each 10 mL.

Loss on drying ⟨731⟩—Dry it at 105° for 3 hours: it loses not more than 1.0% of its weight.

Residue on ignition ⟨281⟩: negligible, from 100 mg.

Assay—

Standard preparation—Prepare as directed for *Standard Preparation* under *Single-steroid Assay* ⟨511⟩, using USP Fluocinonide RS.

Assay preparation—Weigh accurately about 100 mg of Fluocinonide, dissolve in a mixture of equal volumes of chloroform and alcohol to make 50.0 mL, and mix.

Procedure—Proceed as directed for *Procedure* under *Single-steroid Assay* ⟨511⟩, using chloroform and methanol solution (97:3) to develop the chromatogram. Calculate the quantity, in mg, of $C_{26}H_{32}F_2O_7$ in the Fluocinonide taken by the formula:

$$0.05C(A_U/A_S).$$

Fluocinonide Cream

» Fluocinonide Cream contains not less than 90.0 percent and not more than 110.0 percent of the labeled amount of $C_{26}H_{32}F_2O_7$.

Packaging and storage—Preserve in collapsible tubes or in tight containers.

Reference standard—*USP Fluocinonide Reference Standard*—Dry at 105° for 3 hours before using.

Identification—Weigh about 5 g of Fluocinonide Cream into a glass-stoppered, 50-mL centrifuge tube containing 5 mL of water and 10 mL of methanol. Add 20 mL of cyclohexane, shake vigorously, centrifuge, and discard the upper phase. Add 20 mL of water and 5 mL of chloroform, shake vigorously, centrifuge until the lower phase is clear, and discard the upper phase. The clear chloroform extract is the *Test preparation.* Proceed as directed in *Identification test C* under *Fluocinonide*, applying a volume of the *Test preparation* containing about 20 µg of fluocinonide and 10 µL of the *Standard preparation:* the R_f value of the principal spot from the *Test preparation* corresponds to that obtained from the *Standard preparation.*

Microbial limits—It meets the requirements of the tests for absence of *Staphylococcus aureus* and *Pseudomonas aeruginosa* under *Microbial Limit Tests* ⟨61⟩.

Minimum fill ⟨755⟩: meets the requirements.

Assay—

Internal standard solution—Dissolve norethindrone in acetonitrile to obtain a solution containing about 200 µg per mL.

Mobile solvent—Use a mixture of acetonitrile and water (1:1).

Standard preparation—Dissolve an accurately weighed quantity of USP Fluocinonide RS in acetonitrile to obtain a solution containing about 200 µg per mL. Transfer 10.0 mL of this solution, 4.0 mL of *Internal standard solution*, and 10.0 mL of water to a 100-mL volumetric flask. Dilute with acetonitrile to volume, and mix. The final concentration of USP Fluocinonide RS is 20 µg per mL.

Assay preparation—Transfer an accurately weighed quantity of Fluocinonide Cream, containing about 2 mg of fluocinonide, to a 100-mL volumetric flask. Add about 60 mL of acetonitrile, and dissolve the cream by heating on a steam bath. Add 4.0 mL of *Internal standard solution* and 10.0 mL of water, and allow to cool. Dilute with acetonitrile to volume, and mix. Filter the mixture through a fine-sintered glass funnel, using vacuum, and use the filtrate.

Apparatus—Use a suitable high-pressure liquid chromatograph (see *Chromatography* ⟨621⟩) of the general type equipped with a detector for monitoring ultraviolet absorbance at about 254 nm, and capable of providing a flow rate of about 2 mL per minute for the *Mobile solvent.* Use a column that contains packing L1 so as to provide a resolution factor, R (see *Chromatography* ⟨621⟩), of not less than 2.0, between peaks for norethindrone and fluocinonide.

Procedure—Chromatograph equal volumes of the *Assay preparation* and the *Standard preparation.* Calculate the quantity, in mg, of $C_{26}H_{32}F_2O_7$ in the portion of Cream taken by the formula:

$$0.1C(R_U/R_S),$$

in which C is the concentration, in µg per mL, of USP Fluocinonide RS in the *Standard preparation*, and R_U and R_S are the ratios of the areas of the fluocinonide peak to the internal standard peak obtained from the *Assay preparation* and the *Standard preparation*, respectively.

Fluocinonide Gel

» Fluocinonide Gel contains not less than 90.0 percent and not more than 110.0 percent of the labeled amount of $C_{26}H_{32}F_2O_7$.

Packaging and storage—Preserve in collapsible tubes or in tight containers.

Reference standards—*USP Fluocinonide Reference Standard*—Dry at 105° for 3 hours before using. *USP Norethindrone Reference Standard*—Dry in vacuum at 105° for 3 hours before using.

Identification—Weigh about 5 g of Gel into a glass-stoppered, 50-mL centrifuge tube containing 20 mL of sodium chloride solution (1 in 10). Add 5 mL of chloroform and 15 mL of methanol, and shake vigorously. Centrifuge to clarify the chloroform layer, and remove the solid material present at the interphase. Discard the upper phase. Dry a portion of the chloroform layer over anhydrous sodium sulfate. Using the dried extract as the *Test preparation*, proceed as directed in *Identification test C* under *Fluocinonide*, applying a volume of the *Test preparation* containing about 20 µg of fluocinonide and 10 µL of the *Standard preparation:* the R_f value of the principal spot from the *Test preparation* corresponds to that obtained from the *Standard preparation.*

Minimum fill ⟨755⟩: meets the requirements.

Assay—

Internal standard solution—Dissolve a portion of USP Norethindrone RS in acetonitrile to obtain a solution containing about 200 µg per mL.

Standard preparation—Dissolve an accurately weighed quantity of USP Fluocinonide RS in acetonitrile to obtain a solution having a known concentration of about 200 µg per mL. Transfer 10.0 mL of this solution and 4.0 mL of *Internal standard solution* to a 100-mL volumetric flask, dilute with acetonitrile to volume, and mix. The final concentration is 20 µg per mL.

Mobile solvent—Prepare a mixture of acetonitrile and water (1:1). Adjust the ratio as necessary to obtain suitable chromatographic performance.

Assay preparation—Transfer an accurately weighed quantity of Fluocinonide Gel, containing about 2 mg of fluocinonide, to a 100-mL volumetric flask. Add about 60 mL of acetonitrile, and dissolve the gel by heating on a steam bath. Cool to room temperature, add 4.0 mL of *Internal standard solution*, dilute with acetonitrile to volume, and mix. Centrifuge a portion at about 2500 rpm for about 5 minutes. Filter a portion of the centrifugate through an acetonitrile-insoluble membrane filter. The filtrate is the *Assay preparation.*

Apparatus—Use a suitable high-pressure liquid chromatograph (see *Chromatography* ⟨621⟩) of the general type equipped with a detector for monitoring ultraviolet absorbance at about 254 nm, and capable of providing a flow rate of about 2 mL per minute for the *Mobile solvent.* Use a 50-cm × 4-mm column that contains packing L1 so as to provide a resolution factor, *R* (see *Chromatography* ⟨621⟩), of at least 2.0 between peaks for norethindrone and fluocinonide. Three replicate injections of the *Standard preparation* show a relative standard deviation of not more than 1.5%.

Procedure—Chromatograph equal volumes of the *Assay preparation* and the *Standard preparation*, adjusting the system as necessary to obtain peaks of between about 50% and 90% of full-scale. Calculate the quantity, in mg, of $C_{26}H_{32}F_2O_7$ in the portion of Gel taken by the formula:

$$0.1C(R_U/R_S),$$

in which *C* is the concentration, in µg per mL, of USP Fluocinonide RS in the *Standard preparation*, and R_U and R_S are the ratios of the peak areas of fluocinonide and the internal standard obtained from the *Assay preparation* and the *Standard preparation*, respectively.

Fluocinonide Ointment

» Fluocinonide Ointment contains not less than 90.0 percent and not more than 110.0 percent of the labeled amount of $C_{26}H_{32}F_2O_7$.

Packaging and storage—Preserve in collapsible tubes or in tight containers.

Reference standard—*USP Fluocinonide Reference Standard*—Dry at 105° for 3 hours before using.

Identification—Weigh about 5 g of Ointment into a glass-stoppered, 50-mL centrifuge tube containing 20 mL of cyclohexane. Gently disperse to form a suspension. Add 5 mL of water and 10 mL of methanol. Shake vigorously, allow the phases to separate, and discard the upper phase. Add 20 mL of water and 5 mL of chloroform, shake vigorously, centrifuge, and transfer a portion of the chloroform layer to a small test tube containing about 200 mg of anhydrous sodium sulfate. Mix, and allow to stand until the extract is clear. Using the clear chloroform extract as the *Test preparation*, proceed as directed in *Identification test C* under *Fluocinonide*, applying a volume of the *Test preparation* containing about 20 µg of fluocinonide and 10 µL of the *Standard preparation:* the R_f value of the principal spot from the *Test preparation* corresponds to that obtained from the *Standard preparation.*

Minimum fill ⟨755⟩: meets the requirements.

Assay—

Internal standard solution—Dissolve norethindrone in methanol to obtain a solution containing about 850 µg per mL.

Diluted internal standard solution—Transfer 5.0 mL of *Internal standard solution* to a 250-mL flask, dilute with methanol to volume, and mix.

Mobile solvent—Use a mixture of acetonitrile and water (1:1). Adjust the ratio as necessary to obtain suitable chromatographic performance.

Standard preparation—Dissolve an accurately weighed quantity of USP Fluocinonide RS in acetonitrile to obtain a solution having a known concentration of about 400 µg per mL. Transfer 10.0 mL of this solution and 2.0 mL of *Internal standard solution* to a 100-mL volumetric flask. Dilute with methanol to volume, and mix. The final concentration of USP Fluocinonide RS is about 40 µg per mL.

Assay preparation—Transfer an accurately weighed quantity of Fluocinonide Ointment, containing about 1.35 mg of fluocinonide, to a round-bottom, 50-mL centrifuge tube. Add 35.0 mL of *Diluted internal standard solution.* Emulsify, using an ultrasonic probe, and centrifuge to bring the insoluble matter to the bottom. The clear supernatant solution is the *Assay preparation.*

Apparatus—Use a suitable high-pressure liquid chromatograph (see *Chromatography* ⟨621⟩) of the general type equipped with a detector for monitoring ultraviolet absorbance at about 254 nm, and capable of providing a flow rate of about 1 mL per minute for the *Mobile solvent.* Use a column that contains packing L1 so as to provide a resolution factor, *R* (see *Chromatography* ⟨621⟩), of not less than 2.0 between peaks for norethindrone and fluocinonide.

Procedure—Chromatograph equal volumes of the *Assay preparation* and the *Standard preparation*, adjusting the system as necessary to obtain peaks of between about 50% and 90% of full-scale. Calculate the quantity, in mg, of $C_{26}H_{32}F_2O_7$ in the portion of Fluocinonide Ointment taken by the formula:

$$0.035C(R_U/R_S),$$

in which *C* is the concentration, in µg per mL, of USP Fluocinonide RS in the *Standard preparation*, and R_U and R_S are the ratios of the areas of the fluocinonide peak to the internal standard peak obtained from the *Assay preparation* and the *Standard preparation*, respectively.

Fluocinonide Topical Solution

» Fluocinonide Topical Solution contains not less than 90.0 percent and not more than 110.0 percent of the labeled amount of $C_{26}H_{32}F_2O_7$.

Packaging and storage—Preserve in tight containers.

Reference standard—*USP Fluocinonide Reference Standard*—Dry at 105° for 3 hours before using.

Identification—Transfer 5.0 mL of Topical Solution to a glass-stoppered, 50-mL centrifuge tube containing 5 mL of water and 10 mL of methanol, add 20 mL of cyclohexane, shake vigorously, centrifuge, and discard the upper phase. Add 20 mL of water and 5 mL of chloroform, shake vigorously, centrifuge until the lower phase is clear, and discard the upper phase. The clear chloroform extract is the test solution. Proceed as directed under *Thin-layer Chromatographic Identification Test* ⟨201⟩, applying a volume of the test solution, containing about 20 μg of fluocinonide, and 10 μL of a Standard solution of USP Fluocinonide RS in a mixture of chloroform and alcohol (1:1) having a concentration of about 2 mg per mL. Use a mixture of chloroform and methanol (97:3) as the developing solvent: the R_f value of the principal spot from the test solution corresponds to that obtained from the Standard solution.

Minimum fill ⟨755⟩: meets the requirements.

Alcohol content—

Standard solution—Dilute 20.0 mL of USP Alcohol with methanol to volume in a 200-mL volumetric flask.

Internal standard solution—Dilute 20.0 mL of isopropyl alcohol with methanol to volume in a 100-mL volumetric flask.

Test preparation—Using a "to contain" pipet, transfer 2 mL of Fluocinonide Topical Solution to a 100-mL volumetric flask, rinsing the pipet 3 times with methanol and collecting the rinsings in the volumetric flask. Add 5.0 mL of *Internal standard solution*, dilute with methanol to volume, and mix.

Standard preparation—Pipet 6 mL of the *Standard solution* and 5 mL of the *Internal standard solution* into a 100-mL volumetric flask, dilute with methanol to volume, and mix.

Chromatographic system (see *Chromatography* ⟨621⟩)—The gas chromatograph is equipped with a flame-ionization detector and a 1.8-m × 2-mm glass column that is packed with 80- to 100-mesh packing S3. The carrier gas is helium, flowing at a rate of about 40 mL per minute. The injection port and detector temperatures are maintained at about 175° and 225°, respectively. The initial column temperature, 130°, is held for 8 minutes, then is raised at the rate of 30° per minute to 225°, and is held at 225° for 6 minutes. Chromatograph the *Standard preparation*, record the chromatogram, and determine the peak response ratio as directed under *Procedure:* the resolution, *R*, of the alcohol and isopropyl alcohol peaks is not less than 1.5, the tailing factor of the alcohol peak is not more than 1.25, and the relative standard deviation for peak response ratios from replicate injections is not more than 1.5%.

Procedure—Separately inject equal volumes (2 μL to 3 μL) of the *Test preparation* and the *Standard preparation* into the chromatograph, record the chromatograms, and measure the responses for the major peaks. Calculate the percentage (v/v) of C_2H_5OH in the Topical Solution by the formula:

$$(0.3)(95.45)(R_U/R_S),$$

in which 95.45 is the percentage (v/v) of C_2H_5OH in USP Alcohol, and R_U and R_S are the peak response ratios obtained from the *Test preparation* and the *Standard preparation*, respectively: between 28.4% and 39.0% of C_2H_5OH is present.

Assay—

Mobile phase—Use a mixture of acetonitrile and water (55:45). Make adjustments if necessary (see *System Suitability* under *Chromatography* ⟨621⟩).

Internal standard solution—Dissolve propylparaben in *Mobile phase* to obtain a solution containing about 32 μg per mL.

Standard preparation—Dissolve an accurately weighed quantity of USP Fluocinonide RS in acetonitrile to obtain a solution containing about 500 μg per mL. Transfer 2.0 mL of this solution and 5.0 mL of *Internal standard solution* to a 25-mL volumetric flask. Dilute with *Mobile phase* to volume, and mix to obtain a solution having a known concentration of about 40 μg of USP Fluocinonide RS per mL.

Assay preparation—Using a "to contain" pipet, transfer a volume of Fluocinonide Topical Solution, equivalent to about 1 mg of fluocinonide, to a 25-mL volumetric flask, rinsing the pipet with about 5 mL of *Mobile phase*, and adding the rinsings to the volumetric flask. Add 5.0 mL of *Internal standard solution*, dilute with *Mobile phase* to volume, and mix.

Chromatographic system (see *Chromatography* ⟨621⟩)—The liquid chromatograph is equipped with a 254-nm detector and a 3.9-mm × 30-cm column that contains packing L1. The flow rate is about 1 mL per minute. Chromatograph the *Standard preparation* and the *System suitability preparation*, and record the peak responses as directed under *Procedure:* the tailing factor for the analyte peak is not more than 1.5, the resolution, *R*, between the fluocinonide and the internal standard peaks is not less than 2.5, and the relative standard deviation for replicate injections of the *Standard preparation* is not more than 1.5%.

Procedure—Separately inject equal volumes (about 20 μL) of the *Standard preparation* and the *Assay preparation* into the chromatograph, record the chromatograms, and measure the responses for the major peaks. The relative retention times are 1.0 for fluocinonide and about 0.6 for propylparaben. Calculate the quantity, in mg, of $C_{26}H_{32}F_2O_7$ in each mL of the Topical Solution taken by the formula:

$$0.025(C/V)(R_U/R_S),$$

in which *C* is the concentration, in μg per mL, of USP Fluocinonide RS in the *Standard preparation*, *V* is the volume, in mL of Topical Solution taken, and R_U and R_S are the peak response ratios obtained from the *Assay preparation* and the *Standard preparation*, respectively.

Fluorescein

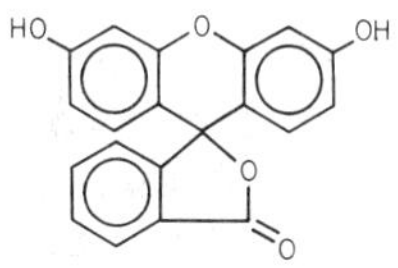

$C_{20}H_{12}O_5$ 332.31

Spiro[isobenzofuran-1(3*H*),9′-[9*H*]xanthen]-3-one, 3′,6′-dihydroxy-.

Fluorescein [*2321-07-5*].

» Fluorescein contains not less than 97.0 percent and not more than 102.0 percent of $C_{20}H_{12}O_5$, calculated on the anhydrous basis.

Packaging and storage—Preserve in tight containers.

Reference standards—*USP Diacetylfluorescein Reference Standard*—Keep container tightly closed. Dry at 105° for 4 hours before using. *USP Fluorescein Reference Standard*—Dry over silica gel for 16 hours before using.

Identification—The infrared absorption spectrum of a potassium bromide dispersion of it, previously dried over silica gel for 16 hours, exhibits maxima only at the same wavelengths as that of a similar preparation of USP Fluorescein RS.

Water, *Method I* ⟨921⟩: not more than 1.0%.

Zinc—Suspend 100 mg in 10 mL of a saturated solution of sodium chloride, add 2 mL of 3 *N* hydrochloric acid, mix, filter, and add 1 mL of potassium ferrocyanide TS to the filtrate: no turbidity is produced.

Acriflavine—Suspend 10 mg in 5 mL of water, swirl the mixture, and filter. To the filtrate add a few drops of sodium salicylate solution (1 in 10): no precipitate is formed.

Assay—

Standard preparation—Dissolve about 110 mg of USP Diacetylfluorescein RS, accurately weighed, in 10 mL of alcohol contained in a 100-mL volumetric flask. Add 2 mL of 2.5 *N* sodium hydroxide, and heat on a steam bath at about the boiling temperature for 20 minutes, with frequent swirling. Cool, dilute with water to volume, and mix. Dilute quantitatively and stepwise with water to obtain a solution having a known concentration of about 1.1 µg of diacetylfluorescein per mL. Transfer 3.0 mL of this solution to a 100-mL volumetric flask containing 20 mL of pH 9.0 alkaline borate buffer (see *Buffer Solutions* in the section, *Reagents, Indicators, and Solutions*), dilute with water to volume, and mix.

Assay preparation—Dissolve about 90 mg of Fluorescein, accurately weighed, in 10 mL of alcohol contained in a 100-mL volumetric flask. Add 2 mL of 2.5 *N* sodium hydroxide, and heat on a steam bath at about the boiling temperature for 20 minutes, with frequent swirling. Cool, dilute with water to volume, and mix. Dilute quantitatively and stepwise with water to obtain a solution having a concentration of 0.9 µg per mL. Transfer 3.0 mL of this solution to a 100-mL volumetric flask containing 20 mL of pH 9.0 alkaline borate buffer (see *Buffer Solutions* in the section, *Reagents, Indicators, and Solutions*), dilute with water to volume, and mix.

Procedure—Concomitantly determine the fluorescence intensities, *I*, of the *Standard preparation* and the *Assay preparation* in a fluorometer at an excitation wavelength of 485 nm and an emission wavelength of 515 nm. Calculate the quantity, in mg, of $C_{20}H_{12}O_5$ in the Fluorescein taken by the formula:

$$(332.31/416.39)(3333C)(I_U/I_S),$$

in which 332.31 and 416.39 are the molecular weights of fluorescein and diacetylfluorescein, respectively, *C* is the concentration, in µg per mL, of USP Diacetylfluorescein RS in the *Standard preparation*, and I_U and I_S are the fluorescence values observed for the *Assay preparation* and the *Standard preparation*, respectively.

Fluorescein Injection

» Fluorescein Injection is a sterile solution, in Water for Injection, of Fluorescein prepared with the aid of Sodium Hydroxide. It contains the equivalent of not less than 90.0 percent and not more than 110.0 percent of the labeled amount of fluorescein sodium ($C_{20}H_{10}Na_2O_5$). It may contain Sodium Bicarbonate.

Packaging and storage—Preserve in single-dose containers, preferably of Type I glass.

Reference standard—*USP Diacetylfluorescein Reference Standard*—Keep container tightly closed. Dry at 105° for 4 hours before using.

Identification—It responds to *Identification tests A* and *C* under *Fluorescein Sodium.*

Pyrogen—It meets the requirements of the *Pyrogen Test* ⟨151⟩, the test dose being the equivalent of 250 mg of fluorescein sodium per kg.

pH ⟨791⟩: between 8.0 and 9.8.

Other requirements—It meets the requirements under *Injections* ⟨1⟩.

Assay—

Standard preparation—Prepare as directed in the *Assay* under *Fluorescein Sodium.*

Assay preparation—Transfer an accurately measured volume of Fluorescein Injection, equivalent to about 100 mg of fluorescein sodium, and dilute quantitatively and stepwise with water to obtain a solution having a concentration of 1 µg per mL. Transfer 3.0 mL of this solution to a 100-mL volumetric flask containing 20 mL of pH 9.0 alkaline borate buffer (see *Buffer Solutions* in the section, *Reagents, Indicators, and Solutions*), dilute with water to volume, and mix.

Procedure—Proceed as directed for *Procedure* in the *Assay* under *Fluorescein Sodium.* Calculate the quantity, in mg, of $C_{20}H_{10}Na_2O_5$ equivalent in the Fluorescein Injection taken by the formula:

$$3333C(I_U/I_S),$$

in which *C* is the concentration, in µg per mL, of fluorescein sodium in the *Standard preparation*, and I_U and I_S are the fluorescence values observed for the *Assay preparation* and the *Standard preparation*, respectively.

Fluorescein Sodium

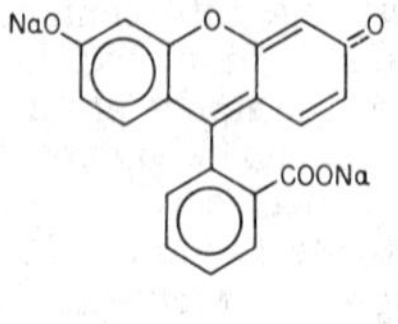

$C_{20}H_{10}Na_2O_5$ 376.28

Spiro[isobenzofuran-1(3*H*),9′-[9*H*]xanthene]-3-one, 3′6′-dihydroxy, disodium salt.

Fluorescein disodium salt [*518-47-8*].

» Fluorescein Sodium contains not less than 90.0 percent and not more than 102.0 percent of $C_{20}H_{10}Na_2O_5$, calculated on the anhydrous basis.

Packaging and storage—Preserve in tight containers.

Reference standard—*USP Diacetylfluorescein Reference Standard*—Keep container tightly closed. Dry at 105° for 4 hours before using.

Identification—

A: A solution of it is strongly fluorescent, even in extreme dilution. The fluorescence disappears when the solution is made acid, and reappears when the solution is again made alkaline.

B: The residue remaining after the incineration of it responds to the tests for *Sodium* ⟨191⟩.

C: Place 1 drop of a solution (1 in 2000) upon a piece of filter paper: a yellow spot is produced and, when exposed while moist to the vapor of bromine for 1 minute and then to ammonia vapor, it becomes deep pink in color.

Water, *Method I* ⟨921⟩: not more than 17.0%.

Zinc—Dissolve 100 mg in 10 mL of a saturated solution of sodium chloride, add 2 mL of 3 *N* hydrochloric acid, shake well, filter, and add 1 mL of potassium ferrocyanide TS to the filtrate: no turbidity is produced.

Acriflavine—Dissolve 10 mg in 5 mL of water, and add a few drops of sodium salicylate solution (1 in 10): no precipitate is formed.

Assay—

Standard preparation—Dissolve an accurately weighed quantity of USP Diacetylfluorescein RS in 10 mL of alcohol contained in a 100-mL volumetric flask. [NOTE—110.7 mg of anhydrous USP Diacetylfluorescein RS is equivalent to 100.0 mg of fluorescein sodium.] Add 2 mL of 2.5 *N* sodium hydroxide, and heat on a steam bath at about the boiling temperature for 20 minutes, with frequent swirling. Cool, dilute with water to volume, and mix. Transfer a suitable aliquot to a volumetric flask, and dilute with water to volume to obtain a solution having a concentration of 1 µg of fluorescein sodium per mL. Transfer a 3-mL aliquot to a 100-mL volumetric flask containing 20 mL of pH 9.0 alkaline borate buffer (see *Buffer Solutions* in the section, *Reagents, Indicators, and Solutions*), dilute with water to volume, and mix. The concentration of fluorescein sodium in the *Standard preparation* is 0.03 µg per mL.

Assay preparation—Dissolve about 100 mg of Fluorescein Sodium, accurately weighed, in water, and dilute quantitatively and stepwise with water to obtain a solution having a concentration of 1 µg per mL. Transfer 3.0 mL of this solution to a 100-mL volumetric flask containing 20 mL of pH 9.0 alkaline borate buffer (see *Buffer Solutions* in the section, *Reagents, Indicators, and Solutions*), dilute with water to volume, and mix.

Procedure—Concomitantly determine the fluorescence intensities, *I*, of the *Standard preparation* and the *Assay preparation* in a fluorometer at an excitation wavelength of 485 nm and an emission wavelength at 515 nm. Calculate the quantity, in mg, of $C_{20}H_{10}Na_2O_5$ in the Fluorescein Sodium taken by the formula:

$$3333C(I_U/I_S),$$

in which C is the concentration, in µg per mL, of fluorescein sodium in the *Standard preparation*, and I_U and I_S are the fluorescence values observed for the *Assay preparation* and the *Standard preparation*, respectively.

Fluorescein Sodium Ophthalmic Strips

» Fluorescein Sodium Ophthalmic Strips contain not less than 100.0 percent and not more than 160.0 percent of the labeled amount of $C_{20}H_{10}Na_2O_5$.

Packaging and storage—Package not more than 2 Strips in a single-unit container in such manner as to maintain sterility until the package is opened. Package individual packages in a second protective container.

Labeling—The label of the second protective container bears a statement that the contents may not be sterile if the individual package has been damaged or previously opened. The label states the amount of fluorescein sodium in each Strip.

Reference standard—*USP Diacetylfluorescein Reference Standard*—Keep container tightly closed. Dry at 105° for 4 hours before using.

Identification—Cut the colored tip from 1 Strip, place it in a small test tube containing 1 mL of water, and agitate for 1 minute: the resulting solution of fluorescein sodium responds to *Identification tests A* and *C* under *Fluorescein Sodium.*

Sterility—It meets the requirements under *Sterility Tests* ⟨71⟩.

Content uniformity—The content of $C_{20}H_{10}Na_2O_5$ in each of not less than 10 Strips, determined as directed in the *Assay*, is not less than 85.0% and not more than 175.0% of the labeled amount.

Assay—

Standard preparation—Prepare as directed in the *Assay* under *Fluorescein Sodium.*

Assay preparation—Remove 1 Fluorescein Sodium Ophthalmic Strip from its package, taking care not to allow any portion of the tip to adhere to the packaging material, transfer to a 100-mL volumetric flask, add 50 mL of water, shake the flask vigorously, and dilute with water to volume. Shake occasionally, and, after 1 hour, mix the contents of the flask. Transfer an aliquot (V) of this solution, equivalent to about 100 µg of fluorescein sodium, to a 100-mL volumetric flask, dilute with water to volume, and mix. Transfer 3 mL of the resulting solution to a 100-mL volumetric flask containing 20 mL of pH 9.0 alkaline borate buffer (see *Buffer Solutions* in the section, *Reagents, Indicators, and Solutions*), dilute with water to volume, and mix.

Procedure—Proceed as directed for *Procedure* in the *Assay* under *Fluorescein Sodium.* Calculate the quantity, in mg, of $C_{20}H_{10}Na_2O_5$ in the Strip taken by the formula:

$$(333)(C/V)(I_U/I_S),$$

in which C is the concentration, in µg per mL, of fluorescein sodium in the *Standard preparation*, V is the volume of the aliquot of solution taken for the *Assay preparation*, and I_U and I_S are the fluorescence intensities observed for the *Assay preparation* and the *Standard preparation*, respectively. Calculate the average content from the individual assays of not less than 10 Strips.

Fludeoxyglucose F 18 Injection

» Fludeoxyglucose F 18 Injection is a sterile, aqueous solution, suitable for intravenous administration, of 2-[^{18}F]fluoro-2-deoxy-D-glucose in which a portion of the molecules are labeled with radioactive ^{18}F. It contains not less than 90.0 percent and not more than 110.0 percent of the labeled amount of ^{18}F expressed in MBq (mCi) per mL at the time indicated in the labeling. It may contain suitable preservatives and/or stabilizing agents.

Specific activity: not less than 37 × 10^3 MBq (1 Ci) per mmol.

Packaging and storage—Preserve in single-dose or in multiple-dose containers that are adequately shielded.

Labeling—Label it to include the following, in addition to the information specified for *Labeling* under *Injection* ⟨1⟩: the time and date of calibration; the amount of ^{18}F as fluorodeoxyglucose expressed as total MBq (mCi) per mL, at time of calibration; the expiration date; the name and quantity of any added preservative or stabilizer; and the statement "Caution, Radioactive Material." The labeling indicates, that in making dosage calculations, correction is to be made for radioactive decay, and also indicates that the radioactive half-life of ^{18}F is 110 minutes. The label indicates "Do not use if cloudy or if it contains particulate matter."

Reference standard—*USP Endotoxin Reference Standard.*

Radionuclide identification (see *Radioactivity* ⟨821⟩)—Its gamma-ray spectrum is identical to that of a specimen of ^{18}F in that it exhibits a positron annihilation peak at 0.511 MeV and possibly a sum peak of 1.02 MeV dependent upon geometry and detector efficiency.

Bacterial endotoxin—It meets the requirements of the *Bacterial Endotoxin Test* ⟨85⟩; the limit of endotoxin content being not more than $175/V$ USP Endotoxin Unit per mL of the Injection, in which V is the maximum recommended total dose, in mL, at the expiration time.

pH ⟨791⟩: within a 0.4 pH unit range between 4.5 and 8.5.

Radiochemical purity—

[NOTE—This article may be synthesized by different methods and processes and may, therefore, contain different impurities. It may be necessary to employ additional validated limit tests relevant to the synthetic procedure used to assure radiochemical purity of the final product.]

Procedure A—Apply a volume of the Injection appropriately diluted, such that it provides a count rate of about 2 × 10^4 counts per minute to a 35- × 95-mm activated silica gel thin-layer chromatographic plate (see *Chromatography* ⟨621⟩). Develop the chromatogram in a solvent system consisting of a mixture of acetonitrile and water (95:5) until the solvent has moved about three-fourths of the length of the plate. Allow the chromatogram to dry. Determine the radioactivity distribution by scanning the chromatogram with a suitable collimated radiation detector. The radioactivity of the main peak at an R_f value of about 0.4 is not less than 90% of the total radioactivity.

Procedure B—Transfer a suitable volume of the Injection to a small reaction vial. Evaporate to dryness under reduced pressure. Add to the residue one ampul (1 mL) of silylating reagent[1] consisting of 3 parts of 1,1,1,3,3,3-hexamethyldisilazane, 1 part of trimethylchorosilane, and 9 parts of pyridine. Stopper the vial and shake vigorously, then allow to stand at room temperature for 6 minutes. Inject a suitable volume of the reaction mixture into a gas chromatograph (see *Chromatography* ⟨621⟩) equipped with a thermal conductivity detector, helium being used as the carrier gas flowing at about 15 mL per minute. The 1.8-m × 3-mm column is held isothermally at 150° and is packed with 4 percent G2[2] and 6 percent G6[3] on 80- to 100-mesh support S1A[4]. Record the chromatogram for 45 minutes and integrate all the peak areas. The main peaks observed at about 20 and 28 minutes are the silylated derivatives of the α and β anomers of 2-fluorodeoxyglucose and the secondary peaks observed at 23 and 36 minutes are the silylated derivatives of the α and β anomers of the fluorodeoxymannose. The sum of these four peaks is not less than 95.0% of the total observed peak areas.

[1] Available from Alltech-Supplied Science, 2051 Waukegan Ave., Deerfield, IL 60015 as Sil-Prep kit.
[2] SE-30, available from Ansper Co., Inc., Ann Arbor, MI.
[3] OV 210, available from Ansper Co., Inc., Ann Arbor, MI.
[4] Chromosorb W-HP, available from Ansper Co., Inc., Ann Arbor, MI.

Radionuclidic purity—Using a suitable gamma-ray spectrometer (see *Selection of a Counting Assembly* under *Radioactivity* ⟨821⟩), determine the absence of radiation, other than at 0.511 MeV, over a period of 2 hours.

Chemical purity—This article may be synthesized by different methods and processes and, therefore, contains different impurities. It is necessary to demonstrate the absence of any unlabeled starting ingredients and reagent chemicals employed in the synthetic process that may be present at levels of 2.0% or more, by the use of one or more validated limit tests using known separation techniques of thin-layer chromatography, gas-liquid chromatography and/or high-pressure liquid chromatography.

Other requirements—It meets the requirements under *Injections* ⟨1⟩, except that the Injection may be distributed or dispensed prior to completion of the test for *Sterility*, the latter test being started on the day following final manufacture, and except that it is not subject to the recommendation of *Volume in Container*.

Assay for radioactivity—Using a suitable counting assembly (see *Selection of a Counting Assembly* under *Radioactivity* ⟨821⟩), determine the radioactivity in MBq (mCi) per mL, of Fludeoxyglucose F 18 Injection by use of a calibrated system as directed under *Radioactivity* ⟨821⟩.

Fluorometholone

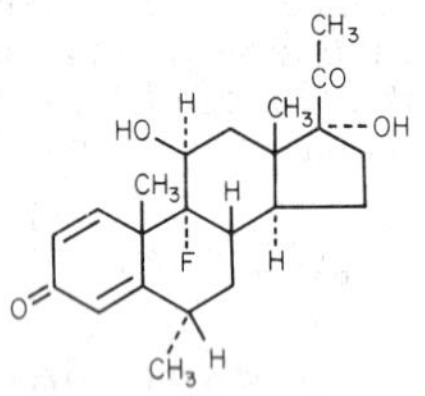

$C_{22}H_{29}FO_4$ 376.47

Pregna-1,4-diene-3,20-dione, 9-fluoro-11,17-dihydroxy-6-methyl-, (6α,11β)-.

9-Fluoro-11β,17-dihydroxy-6α-methylpregna-1,4-diene-3,20-dione [*426-13-1*].

» Fluorometholone contains not less than 97.0 percent and not more than 103.0 percent of $C_{22}H_{29}FO_4$, calculated on the dried basis.

Packaging and storage—Preserve in tight, light-resistant containers.

Reference standards—*USP Fluorometholone Reference Standard*—Dry in vacuum at 60° for 3 hours before using. *USP Fluoxymesterone Reference Standard*—Dry at 105° for 3 hours before using.

Identification—

A: The infrared absorption spectrum of a potassium bromide dispersion of it, previously dried, exhibits maxima only at the same wavelengths as that of a similar preparation of USP Fluorometholone RS.

B: The ultraviolet absorption spectrum of a 1 in 100,000 solution in methanol exhibits maxima and minima at the same wavelengths as that of a similar solution of USP Fluorometholone RS, concomitantly measured, and the respective absorptivities, calculated on the dried basis, at the wavelength of maximum absorbance at about 239 nm do not differ by more than 3.0%.

C: Prepare a solution in methanol containing 500 µg per mL. On a suitable thin-layer chromatographic plate (see *Chromatography* ⟨621⟩), coated with a 250-µm layer of chromatographic silica gel mixture, apply 100 µL of this solution and 100 µL of a methanol solution of USP Fluorometholone RS containing 500 µg per mL. Allow the spots to dry, and develop the chromatogram in a solvent system consisting of a mixture of methylene chloride and acetone (4:1) until the solvent front has moved about three-fourths of the length of the plate. Remove the plate from the developing chamber, mark the solvent front, and allow the solvent to evaporate. Locate the spots on the plate by examination under short-wavelength ultraviolet light: the R_f value of the principal spot obtained from the specimen solution corresponds to that obtained from the Standard solution.

Specific rotation ⟨781⟩: between +52° and +60°, calculated on the dried basis, determined in a solution in pyridine containing 100 mg of Fluorometholone in each 10 mL.

Loss on drying ⟨731⟩—Dry it in vacuum at 60° for 3 hours: it loses not more than 1.0% of its weight.

Residue on ignition ⟨281⟩: not more than 0.2%.

Assay—

Mobile phase—Prepare a suitable filtered solution of methanol and water (60:40) such that the retention time of fluorometholone is about 3 minutes.

Standard preparation—Dissolve an accurately weighed quantity of USP Fluorometholone RS in *Mobile phase* to obtain a solution having a known concentration of about 100 µg per mL.

Assay preparation—Transfer about 20 mg of Fluorometholone, accurately weighed, to a 200-mL volumetric flask, add *Mobile phase* to volume, and mix.

Chromatographic system (see *Chromatography* ⟨621⟩)—The liquid chromatograph is equipped with a 254-nm detector and a 4.6-mm × 25-cm column that contains packing L1. The flow rate is about 2 mL per minute. Chromatograph six replicate injections of the *Standard preparation*, and record the peak responses as directed under *Procedure:* the relative standard deviation is not more than 2.0%.

Procedure—Separately inject equal volumes (about 10 µL) of the *Standard preparation* and the *Assay preparation* into the chromatograph, using a suitable microsyringe or sampling valve, record the chromatograms, and measure the responses for the major peaks. Calculate the quantity, in mg, of $C_{22}H_{29}FO_4$ in the portion of Fluorometholone taken by the formula:

$$0.2C(r_U/r_S),$$

in which C is the concentration, in µg per mL, of USP Fluorometholone RS in the *Standard preparation*, and r_U and r_S are the peak responses obtained from the *Assay preparation* and the *Standard preparation*, respectively.

Fluorometholone Cream

» Fluorometholone Cream contains not less than 90.0 percent and not more than 110.0 percent of the labeled amount of $C_{22}H_{29}FO_4$.

Packaging and storage—Preserve in collapsible tubes.

Reference standards—*USP Fluorometholone Reference Standard*—Dry in vacuum at 60° for 3 hours before using. *USP Fluoxymesterone Reference Standard*—Dry at 105° for 3 hours before using.

Identification—The retention ratios of the main peak to the internal standard peak obtained with the *Standard preparation* and the *Assay preparation* as directed in the *Assay* do not differ by more than 2.0%.

Microbial limits—It meets the requirements of the tests for absence of *Staphylococcus aureus* and *Pseudomonas aeruginosa* under *Microbial Limit Tests* ⟨61⟩.

Minimum fill ⟨755⟩: meets the requirements.

Assay—

Internal standard solution—Dissolve USP Fluoxymesterone RS in acetonitrile to obtain a solution containing about 100 µg per mL.

Mobile solvent—Prepare a solution containing butyl chloride, water-saturated butyl chloride, tetrahydrofuran, methanol, and glacial acetic acid (95:95:14:7:6).

Standard preparation—Dissolve a suitable quantity of USP Fluorometholone RS, accurately weighed, in *Internal standard solution* to obtain a solution having a known concentration of about 50 µg per mL.

Assay preparation—Transfer an accurately weighed quantity of Fluorometholone Cream, equivalent to about 1 mg of fluorometholone, into a suitable container, add 20.0 mL of *Internal standard solution*, and mix.

Procedure—Treat 20.0 mL each of the *Standard preparation* and the *Assay preparation* in the following manner. To each add 10.0 mL of hexane, and shake for about 15 minutes, then allow the layers to separate, and centrifuge, if necessary. Using a suitable microsyringe or sampling valve, inject equal volumes of lower (acetonitrile) layers obtained from the *Standard preparation* and the *Assay preparation* into a suitable high-pressure liquid chromatograph (see *Chromatography* ⟨621⟩) of the general type equipped with a detector for monitoring ultraviolet light absorption at about 254 nm, equipped with a suitable recorder, capable of providing column pressure up to about 1000 psi. The instrument contains a 4-mm × 30-cm stainless steel column that contains packing L3. In a suitable chromatogram, the resolution factor, *R* (see *Chromatography* ⟨621⟩), is not less than 2.4 between peaks for fluorometholone and the internal standard, and the lowest and highest peak area ratios (R_S) of three replicate injections of the *Standard preparation* agree within 2.0%. Calculate the quantity, in mg, of $C_{22}H_{29}FO_4$ in the portion of Cream taken by the formula:

$$20C(R_U/R_S),$$

in which *C* is the concentration, in mg per mL, of USP Fluorometholone RS in the *Standard preparation*, and R_U and R_S are the peak area ratios of the fluorometholone peak and the internal standard peak obtained from the *Assay preparation* and the *Standard preparation*, respectively.

Fluorometholone Ointment, Neomycin Sulfate and—
see Neomycin Sulfate and Fluorometholone Ointment

Fluorometholone Ophthalmic Suspension

» Fluorometholone Ophthalmic Suspension is a sterile suspension of Fluorometholone in a suitable aqueous medium. It contains not less than 90.0 percent and not more than 110.0 percent of the labeled amount of $C_{22}H_{29}FO_4$. It may contain suitable stabilizers, buffers, and antimicrobial agents.

Packaging and storage—Preserve in tight containers.

Reference standard—*USP Fluorometholone Reference Standard*—Dry in vacuum at 60° for 3 hours before using.

Identification—Mix 1 mL of the well-shaken Suspension with 2 mL of a mixture of methanol and water (3:2) until a solution is obtained. Apply 20-µL portions of this solution and of a Standard solution of USP Fluorometholone RS in the same solvent containing 500 µg per mL on a suitable thin-layer chromatographic plate (see *Chromatography* ⟨621⟩) coated with a 0.25-mm layer of chromatographic silica gel mixture and previously activated by heating at 80° for 5 minutes. Allow the spots to dry, and develop the chromatogram in a solvent system consisting of methylene chloride and acetone (4:1) until the solvent front has moved not less than 15 cm. Remove the plate from the developing chamber, mark the solvent front, and allow to air-dry. Examine the plate under short-wavelength ultraviolet light: the R_f value and intensity of the principal spot obtained from the test solution correspond to those obtained from the Standard solution.

Sterility—It meets the requirements under *Sterility Tests* ⟨71⟩.

pH ⟨791⟩: between 6.0 and 7.5.

Assay—

Mobile phase—Prepare a suitable filtered solution of methanol and water (60:40) such that the retention time of fluorometholone is about 3 minutes.

Standard preparation—Using a suitable quantity of USP Fluorometholone RS, accurately weighed, prepare a solution in methanol containing 0.5 mg per mL. Pipet 10 mL of this solution and 5 mL of water into a 50-mL volumetric flask. Dilute with methanol to volume, and mix to obtain a *Standard preparation* having a known concentration of about 100 µg per mL.

Assay preparation—Pipet a volume of well-shaken Fluorometholone Suspension, equivalent to about 5 mg of fluorometholone, into a 50-mL volumetric flask, dilute with methanol to volume, and mix. Filter through a 5-µm membrane filter, and use the clear filtrate.

Chromatographic system (see *Chromatography* ⟨621⟩)—The liquid chromatograph is equipped with a 254-nm detector and a 4.6-mm × 25-cm column that contains packing L1. The flow rate is about 2 mL per minute. Chromatograph six replicate injections of the *Standard preparation*, and record the peak responses as directed under *Procedure:* the relative standard deviation is not more than 2.0%.

Procedure—Separately inject equal volumes (about 10 µL) of the *Standard preparation* and the *Assay preparation* into the chromatograph, using a suitable microsyringe or sampling valve, record the chromatograms, and measure the responses for the major peaks. Calculate the quantity, in mg, of $C_{22}H_{29}FO_4$ in the portion of Ophthalmic Suspension taken by the formula:

$$0.05C(r_U/r_S),$$

in which *C* is the concentration, in µg per mL, of USP Fluorometholone RS in the *Standard preparation*, and r_U and r_S are the peak responses obtained from the *Assay preparation* and the *Standard preparation*, respectively.

Fluorouracil

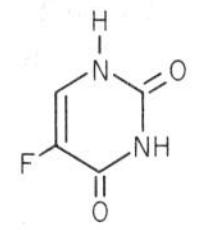

$C_4H_3FN_2O_2$ 130.08
2,4(1*H*,3*H*)-Pyrimidinedione, 5-fluoro-.
5-Fluorouracil [*51-21-8*].

» Fluorouracil contains not less than 98.5 percent and not more than 101.0 percent of $C_4H_3FN_2O_2$, calculated on the dried basis.

Caution—Great care should be taken to prevent inhaling particles of Fluorouracil and exposing the skin to it.

Packaging and storage—Preserve in tight, light-resistant containers.

Reference standard—*USP Fluorouracil Reference Standard*—Dry in vacuum over phosphorus pentoxide at 80° for 4 hours before using.

Identification—

A: The infrared absorption spectrum of a mineral oil dispersion of it exhibits maxima only at the same wavelengths as that of a similar preparation of USP Fluorouracil RS.

B: The ultraviolet absorption spectrum of a 1 in 100,000 solution in a pH 4.7 acetate buffer (prepared from 8.4 g of sodium acetate and 3.35 mL of glacial acetic acid mixed with water to make 1000 mL) exhibits maxima and minima at the same wavelengths as that of a similar solution of USP Fluorouracil RS, concomitantly measured, and the respective absorptivities, calculated on the dried basis at the wavelength of maximum absorbance at about 266 nm, do not differ by more than 3.0%.

C: To 5 mL of a solution (1 in 100) add 1 mL of bromine water TS: the bromine color is discharged.

Loss on drying ⟨731⟩—Dry it in vacuum over phosphorus pentoxide at 80° for 4 hours: it loses not more than 0.5% of its weight.

Residue on ignition ⟨281⟩: not more than 0.1%.

Heavy metals, *Method II* ⟨231⟩: 0.002%.

Fluorine content—[NOTE—All laboratory utensils used in this procedure should be scrupulously clean and free from even trace amounts of fluoride. The use of plasticware, wherever possible, in the preparation and storage of solutions and for measurement of potentials is recommended.]

Isopropyl alcohol solution—Dilute 295 mL of isopropyl alcohol with water to 500 mL.

Buffer solution—To 55 g of sodium chloride in a 1-liter volumetric flask add 500 mg of sodium citrate, 255 g of sodium acetate, and 300 mL of water. Shake to dissolve, and add 115 mL of glacial acetic acid. Cool to room temperature, add 300 mL of isopropyl alcohol, dilute with water to volume, and mix. The pH of the resulting solution is between 5.0 and 5.5.

Reagent blank—Pipet 15 mL of 1,2-dimethoxyethane into a flat-bottom, glass-joint, 500-mL flask, and proceed as directed under *Test preparation*, beginning with "add the contents of a 15-mL vial of sodium biphenyl solution."

Modified calomel reference electrode—Mix 70 mL of a freshly prepared saturated potassium chloride solution with 30 mL of isopropyl alcohol, fill the electrode with the clear supernatant liquid, and allow the electrode to soak in the remainder of the solution for a minimum of 2 hours before using. Store the electrode immersed in the potassium chloride–isopropyl alcohol solution when not in use.

Standard stock solution—Weigh accurately 2.211 g of sodium fluoride, previously dried at 150° for 4 hours, into a 1-liter volumetric flask, and dissolve in about 200 mL of water. Add 1 mL of sodium hydroxide solution (1 in 25), dilute with water to volume, and mix. Store this solution in plastic containers. One mL is equivalent to 1 mg of fluoride.

Standard curve—Dilute 10.0 mL of *Standard stock solution* with water to 100 mL. Into each of four 100-mL volumetric flasks pipet 0.8, 1.0, 1.2, and 1.6 mL, respectively, of the resulting solution. To each flask add 15 mL of *Reagent blank*, dilute with *Buffer solution* to volume, and mix. Use these dilutions, containing, respectively, 0.8, 1.0, 1.2, and 1.6 µg per mL, to construct the standard curve as follows. Determine the potentials of each solution as directed under *Procedure*. Plot the results of fluorine concentration, as the abscissa, in mg per 100 mL versus the potential, as the ordinate, on semilogarithmic graph paper, for each of the standards. Draw the best straight line through the plotted points.

Test preparation—Place 200 mg of Fluorouracil, accurately weighed, in a 250-mL volumetric flask, add about 150 mL of 1,2-dimethoxyethane, shake by mechanical means to dissolve, dilute with the same solvent to volume, and mix. Pipet 15 mL of this solution into a flat-bottom, glass-joint, 500-mL flask, add the contents of a 15-mL vial of sodium biphenyl solution through a long-stem funnel to prevent splattering, swirl the flask gently, and cover with a watch crystal. Allow to stand at room temperature for 20 minutes, then cautiously add 50.0 mL of isopropyl alcohol while swirling the flask. Add 10.0 mL of 30 percent hydrogen peroxide and 4.0 mL of 1 *N* sodium hydroxide, and connect the flask to a water-cooled reflux condenser that previously has been cleaned with water and isopropyl alcohol and dried. Place the flask on a hot plate, set at about 245°, and reflux for 1 hour. Cool to room temperature, rinse the condenser with 15 mL of *Isopropyl alcohol solution*, transfer the contents of the flask to a 250-mL volumetric flask using *Isopropyl alcohol solution* as a rinse, dilute with the same solvent to volume, and mix. Pipet 15 mL of this solution into a 100-mL volumetric flask, and dilute with *Buffer solution* to volume.

Procedure—Measure the potential, in mV, of the *Test preparation*, with a suitable pH meter having a minimum reproducibility of ±0.2 mV, and equipped with a fluoride-specific ion electrode and a glass-sleeved *Modified calomel reference electrode*. When taking a measurement, immerse the electrodes into the solution, which has been transferred to a 150-mL plastic beaker, insert a suitable plastic-coated stirring bar, place the beaker on a magnetic stirrer, taking adequate precautions to prevent heat transfer, and stir for 2 minutes before reading. Dry the electrodes between measurements, taking care not to scratch the crystal surface of the specific ion electrode. Determine the quantity of fluorine, in mg per 100 mL of *Test preparation*, from the *Standard curve*. Multiply the quantity by the factor 138.9 to express the result as percentage. Not less than 13.9% and not more than 15.0% of fluorine, calculated on the dried basis, is found.

Assay—

0.1 N Tetrabutylammonium hydroxide in methanol—Dilute with methanol a commercially available solution of tetrabutylammonium hydroxide in methanol, and standardize as directed under *Tetrabutylammonium Hydroxide, Tenth-Normal (0.1 N)*.

Procedure—Transfer about 400 mg of Fluorouracil, accurately weighed, to a 250-mL conical flask, add 80 mL of dimethylformamide, and warm gently to dissolve. Cool, add 5 drops of a 1 in 100 solution of thymol blue in dimethylformamide, and titrate with *0.1 N Tetrabutylammonium hydroxide in methanol* to a blue end-point, taking precautions to prevent absorption of atmospheric carbon dioxide (see *Titrimetry* ⟨541⟩). Perform a blank determination, and make any necessary correction. Each mL of 0.1 *N* tetrabutylammonium hydroxide is equivalent to 13.01 mg of $C_4H_3FN_2O_2$.

Fluorouracil Cream

» Fluorouracil Cream contains not less than 90.0 percent and not more than 110.0 percent of the labeled amount of $C_4H_3FN_2O_2$. It may contain Sodium Hydroxide to adjust the pH.

Packaging and storage—Preserve in tight containers, at controlled room temperature.

Reference standard—*USP Fluorouracil Reference Standard*—Dry in vacuum over phosphorus pentoxide at 80° for 4 hours before using.

Identification—Prepare a test solution by placing a quantity of Cream, equivalent to about 5 mg of fluorouracil, in a glass-stoppered conical flask, add 50 mL of alcohol, and shake until dissolved. Dissolve 5 mg of USP Fluorouracil RS in 50 mL of alcohol to obtain a Standard solution. In 20-µL increments, apply 100 µL each of the Standard solution and the test solution on a line about 3 cm from the bottom edge of a thin-layer chromatographic plate coated with a 0.25-mm layer of chromatographic silica gel mixture, which previously has been dried and activated at 105° for 5 minutes. Develop the chromatogram in a solvent system consisting of a mixture of ethyl acetate, methanol, and ammonium hydroxide (75:25:1), allowing the solvent front to move about 15 cm beyond the initial spotting line. Remove the plate, air-dry for 15 minutes, and examine under short-wavelength ultraviolet light: the R_f value of the principal spot obtained from the test solution corresponds to that obtained from the Standard solution.

Microbial limits—It meets the requirements of the tests for absence of *Staphylococcus aureus* and *Pseudomonas aeruginosa* under *Microbial Limit Tests* ⟨61⟩.

Minimum fill ⟨755⟩: meets the requirements.

Assay—

pH 4.7 buffer—To 8.4 g of sodium acetate in a 1-liter volumetric flask add 3.35 mL of glacial acetic acid, dilute with water to volume, and mix.

Procedure—Transfer an accurately weighed portion of Fluorouracil Cream, equivalent to about 45 mg of fluorouracil, to a 200-mL volumetric flask. Add 100 mL of *pH 4.7 buffer*, shake by mechanical means for 5 minutes, dilute with *pH 4.7 buffer* to volume, and mix. Filter a portion of this solution, discarding the first 20 mL of filtrate. Pipet 20 mL of the clear filtrate into a 100-mL volumetric flask, dilute with *pH 4.7 buffer* to volume, and mix. Pipet 20 mL of this solution into a 125-mL separator, and extract with five 20-mL portions of ethyl ether, combining the extracts in a 250-mL separator. Extract the remaining aqueous layer with four 20-mL portions of chloroform, and discard the chloroform extracts. Transfer the aqueous portion to a 100-mL volumetric flask. Extract the ethyl ether in the 250-mL separator with two 20-mL portions of *pH 4.7 buffer*, and combine the extracts with the aqueous solution in the 100-mL flask. Evaporate any residual solvent on a steam bath under a stream of nitrogen, cool to room temperature, and dilute with *pH 4.7 buffer* to volume. Dissolve an accurately weighed quantity of USP Fluorouracil RS in *pH 4.7 buffer*, and dilute quantitatively and stepwise with the same solvent to obtain a Standard solution having a known concentration of about 9 µg per mL. Concomitantly determine the absorbances of both solutions in 1-cm cells at the wavelength of maximum absorbance at about 266 nm, with a suitable spectrophotometer, using *pH 4.7 buffer* as the blank.

Calculate the quantity, in mg, of $C_4H_3FN_2O_2$ in the portion of Cream taken by the formula:

$$5C(A_U/A_S),$$

in which C is the concentration, in µg per mL, of USP Fluorouracil RS in the Standard solution, and A_U and A_S are the absorbances of the solution from the Cream and the Standard solution, respectively.

Fluorouracil Injection

» Fluorouracil Injection is a sterile solution of Fluorouracil in Water for Injection, prepared with the aid of Sodium Hydroxide. It contains, in each mL, not less than 45 mg and not more than 55 mg of $C_4H_3FN_2O_2$.

Note—If a precipitate is formed as a result of exposure to low temperatures, redissolve it by heating to 60° with vigorous shaking, and allow to cool to body temperature prior to use.

Packaging and storage—Preserve in single-dose containers, preferably of Type I glass, at controlled room temperature. Avoid freezing and exposure to light.

Labeling—Label it to indicate the expiration date, which is not more than 24 months after date of manufacture.

Reference standard—*USP Fluorouracil Reference Standard*—Dry in vacuum over phosphorus pentoxide at 80° for 4 hours before using.

Identification—

A: The ultraviolet absorption spectrum of the solution employed for measurement of absorbance in the *Assay* exhibits maxima and minima at the same wavelengths as that of the solution of USP Fluorouracil RS used in the *Assay*.

B: Carefully acidify a portion of Injection, equivalent to about 100 mg of fluorouracil, with glacial acetic acid. Stir and slightly chill the solution to precipitate the fluorouracil, collect the precipitate, wash with 1 mL of water, and then dry in vacuum over phosphorus pentoxide at 80° for 4 hours: the residue so obtained responds to *Identification test A* under *Fluorouracil*.

C: It responds to *Identification test C* under *Fluorouracil*.

Pyrogen—It meets the requirements of the *Pyrogen Test* ⟨151⟩, the test dose being 1 mL per kg of a solution prepared by dilution of the Injection with Sodium Chloride Injection to a concentration of 10 mg per mL.

pH ⟨791⟩: between 8.6 and 9.4.

Other requirements—It meets the requirements under *Injections* ⟨1⟩.

Assay—

Buffer—Prepare as directed in the *Assay* under *Fluorouracil Cream*.

Procedure—Pipet into a 500-mL volumetric flask a volume of Fluorouracil Injection, equivalent to 100 mg of fluorouracil, add *pH 4.7 buffer* to volume, and mix. Pipet 5 mL of this solution into a 100-mL volumetric flask, add *pH 4.7 buffer* to volume, and mix. Dissolve an accurately weighed quantity of USP Fluorouracil RS in *pH 4.7 buffer*, and dilute quantitatively and stepwise with *pH 4.7 buffer* to obtain a Standard solution having a known concentration of about 10 µg per mL. Concomitantly determine the absorbances of both solutions at the wavelength of maximum absorbance at about 266 nm, with a suitable spectrophotometer, using *pH 4.7 buffer* as the blank. Calculate the quantity, in mg, of $C_4H_3FN_2O_2$ in each mL of the Injection taken by the formula:

$$10(C/V)(A_U/A_S),$$

in which C is the concentration, in µg per mL, of USP Fluorouracil RS in the Standard solution, V is the volume, in mL, of Injection taken, and A_U and A_S are the absorbances of the solution from the Injection and the Standard solution, respectively.

Fluorouracil Topical Solution

» Fluorouracil Topical Solution contains not less than 90.0 percent and not more than 110.0 percent of the labeled amount of $C_4H_3FN_2O_2$.

Packaging and storage—Preserve in tight containers, at controlled room temperature.

Reference standard—*USP Fluorouracil Reference Standard*—Dry in vacuum over phosphorus pentoxide at 80° for 4 hours before using.

Identification—It responds to the *Identification test* under *Fluorouracil Cream*.

Microbial limits—It meets the requirements of the tests for absence of *Staphylococcus aureus* and *Pseudomonas aeruginosa* under *Microbial Limit Tests* ⟨61⟩.

Assay—

pH 4.7 buffer—Prepare as directed in the *Assay* under *Fluorouracil Cream*.

Procedure—Transfer an accurately weighed portion of Fluorouracil Topical Solution, equivalent to about 22 mg of fluorouracil, to a 100-mL volumetric flask, add 50 mL of *pH 4.7 buffer*, shake by mechanical means for 5 minutes, dilute with *pH 4.7 buffer* to volume, and mix. Proceed as directed for *Procedure* in the *Assay* under *Fluorouracil Cream*, beginning with "Filter a portion of this solution." Calculate the quantity, in mg, of $C_4H_3FN_2O_2$ in the portion of Topical Solution taken by the formula:

$$2.5C(A_U/A_S),$$

in which C is the concentration, in µg per mL, of USP Fluorouracil RS in the Standard solution, and A_U and A_S are the absorbances of the solution from the Topical Solution and the Standard solution, respectively.

Fluoxymesterone

$C_{20}H_{29}FO_3$ 336.45

Androst-4-en-3-one, 9-fluoro-11,17-dihydroxy-17-methyl-, (11β,17β)-.

9-Fluoro-11β,17β-dihydroxy-17-methylandrost-4-en-3-one [*76-43-7*].

» Fluoxymesterone contains not less than 97.0 percent and not more than 102.0 percent of $C_{20}H_{29}FO_3$, calculated on the dried basis.

Packaging and storage—Preserve in well-closed containers, protected from light.

Reference standard—*USP Fluoxymesterone Reference Standard*—Dry at 105° for 3 hours before using.

Identification—

A: The infrared absorption spectrum of a potassium bromide dispersion of it, previously dried, exhibits maxima only at the same wavelengths as that of a similar preparation of USP Fluoxymesterone RS. If a difference appears, dissolve portions of both the test specimen and the Reference Standard in dehydrated alcohol, evaporate the solutions to dryness, and repeat the test on the residues.

B: The ultraviolet absorption spectrum of a 1 in 100,000 solution in alcohol exhibits maxima and minima at the same wavelengths as that of a similar solution of USP Fluoxymesterone RS, concomitantly measured, and the respective absorptivities, calculated on the dried basis, at the wavelength of maximum absorbance at about 242 nm do not differ by more than 2.5%.

Specific rotation ⟨781⟩: between +104° and +112°, calculated on the dried basis, determined in a solution in alcohol containing 100 mg in each 10 mL.

Loss on drying ⟨731⟩—Dry it at 105° for 3 hours: it loses not more than 1.0% of its weight.

Ordinary impurities ⟨466⟩—

Test solution: methanol.

Standard solution: methanol.

Application volume: 10 µL.

Eluant: a mixture of toluene, ethyl acetate, and alcohol (6:2:2), in a nonequilibrated chamber.

Visualization: 1.

Assay—

Internal standard solution—Dissolve methylprednisolone in a mixture of chloroform and methanol (95:5) to obtain a solution containing about 200 µg per mL.

Mobile solvent—Prepare a solution containing butyl chloride, water-saturated butyl chloride, tetrahydrofuran, methanol, and glacial acetic acid (475:475:70:35:30).

Standard preparation—Dissolve an accurately weighed quantity of USP Fluoxymesterone RS in *Internal standard solution* to obtain a solution having a known concentration of about 0.25 mg per mL.

Assay preparation—Dissolve about 25 mg of Fluoxymesterone, accurately weighed, in 100.0 mL of *Internal standard solution* to obtain a solution having a concentration of about 0.25 mg per mL.

Procedure—Inject equal volumes of the *Assay preparation* and the *Standard preparation* into a suitable high-pressure liquid chromatograph (see *Chromatography* ⟨621⟩) of the general type equipped with a detector for monitoring ultraviolet light at 254 nm, equipped with a suitable recorder, and capable of providing column pressure up to about 2000 psi. The instrument contains a 4-mm × 30-cm stainless steel column that contains packing L3. In a suitable chromatogram, the resolution factor is not less than 3.0 between the peaks for fluoxymesterone and the internal standard, and the relative standard deviation of the peak response ratios of four replicate injections of the *Standard preparation* is not more than 2.0%. Calculate the quantity, in mg, of $C_{20}H_{29}FO_3$ in the portion of Fluoxymesterone taken by the formula:

$$100C(R_U/R_S),$$

in which C is the concentration, in mg per mL, of USP Fluoxymesterone RS in the *Standard preparation*, and R_U and R_S are the ratios of the peak responses for the fluoxymesterone peak and the internal standard peak obtained from the *Assay preparation* and the *Standard preparation*, respectively.

Fluoxymesterone Tablets

» Fluoxymesterone Tablets contain not less than 90.0 percent and not more than 110.0 percent of the labeled amount of $C_{20}H_{29}FO_3$.

Packaging and storage—Preserve in well-closed containers, protected from light.

Reference standard—*USP Fluoxymesterone Reference Standard*—Dry at 105° for 3 hours before using.

Identification—Triturate a quantity of powdered Tablets, equivalent to about 20 mg of fluoxymesterone, with 20 mL of hot chloroform, and decant the supernatant liquid through a filter. Repeat the extraction with two 20-mL portions of hot chloroform. Evaporate the combined chloroform solutions on a water bath to dryness, digest the residue with 5 mL of acetone, decant the supernatant liquid, add to it 20 mL of water, and filter off the precipitate. Dissolve the precipitate in 5 mL of acetone, add 20 mL of water, and filter: the precipitate, after being dried at 105° for 3 hours, responds to *Identification test A* under *Fluoxymesterone*.

Dissolution ⟨711⟩—

Medium: 0.1 *N* hydrochloric acid; 900 mL.

Apparatus 2: 75 rpm.

Time: 60 minutes.

Mobile phase—Prepare a degassed and filtered solution of water and acetonitrile (58:42). Make adjustments if necessary (see *Chromatography* ⟨621⟩).

Internal standard solution—Dissolve a quantity of USP Norethindrone RS in alcohol to obtain a solution having a final concentration of about 46 µg per mL.

Standard solution—Transfer about 28 mg of USP Fluoxymesterone RS, accurately weighed, to a 25-mL volumetric flask, dissolve in alcohol, dilute with alcohol to volume, and mix. Pipet 5 mL of the resulting solution into a 250-mL volumetric flask, dilute with 0.1 *N* hydrochloric acid to volume, and mix. Pipet 5 mL of this solution and 2 mL of *Internal standard solution* into a 25-mL volumetric flask, dilute with 0.1 *N* hydrochloric acid to volume, and mix.

Test solution—Pipet a filtered 20-mL aliquot of the solution under test and 2 mL of *Internal standard solution* into a 25-mL volumetric flask, dilute with 0.1 *N* hydrochloric acid to volume, and mix.

Chromatographic system (see *Chromatography* ⟨621⟩)—The liquid chromatograph is equipped with a 254-nm detector and a 4-mm × 30-cm column that contains packing L1. The flow rate is about 3 mL per minute. Chromatograph replicate injections of the *Standard solution*, and measure the peak responses as directed under *Procedure:* the relative standard deviation is not more than 2.0%, the resolution between fluoxymesterone and norethindrone is not less than 2, and the relative retention times are 0.5 for fluoxymesterone and 1.0 for norethindrone.

Procedure—Inject a volume (about 20 µL) of the *Test solution* into the chromatograph, record the chromatogram, and measure the responses for the major peaks. Calculate the amount of $C_{20}H_{29}FO_3$ dissolved by comparison with the *Standard solution*, similarly chromatographed.

Tolerances—Not less than 70% (Q) of the labeled amount of $C_{20}H_{29}FO_3$ is dissolved in 60 minutes.

Uniformity of dosage units ⟨905⟩: meet the requirements.

Procedure for content uniformity—

Internal standard solution, *Mobile solvent*, and *Standard preparation*—Prepare as directed in the *Assay* under *Fluoxymesterone.*

Test preparation—Transfer 1 Tablet to a suitable container, add 2 mL of water, and sonicate for 30 minutes or until the tablet completely disintegrates. Add an accurately measured volume of *Internal standard solution* (5.0 mL for each mg of fluoxymesterone in the Tablet), and shake the mixture for 15 minutes. Filter a portion of the chloroform layer, and use the clear filtrate.

Procedure—Proceed as directed for *Procedure* in the *Assay* under *Fluoxymesterone*, using the *Test preparation* in place of the *Assay preparation*. Calculate the quantity, in mg, of $C_{20}H_{29}FO_3$ in the Tablet by the formula:

$$(TC/D)(R_U/R_S),$$

in which T is the labeled quantity, in mg, of fluoxymesterone in the Tablet, D is the concentration, in mg per mL, of fluoxymesterone in the *Test preparation*, based on the labeled quantity per Tablet and the extent of dilution, and the other terms are as defined therein.

Assay—

Internal standard solution, *Mobile solvent*, and *Standard preparation*—Prepare as directed in the *Assay* under *Fluoxymesterone.*

Assay preparation—Accurately weigh 20 Tablets, and grind to a fine powder in a mortar and pestle. Accurately weigh a portion of the powder, equivalent to 5 mg of fluoxymesterone, and transfer to a suitable container. Add 20.0 mL of *Internal standard solution*, sonicate for 10 minutes, and shake for 15 minutes. Filter a portion of the liquid, and analyze the clear filtrate as directed for *Procedure*.

Procedure—Proceed as directed for *Procedure* in the *Assay* under *Fluoxymesterone*. Calculate the quantity, in mg, of $C_{20}H_{29}FO_3$ in the portion of Tablets taken by the formula:

$$20C(R_U/R_S),$$

in which the terms are as defined therein.

Fluphenazine Enanthate

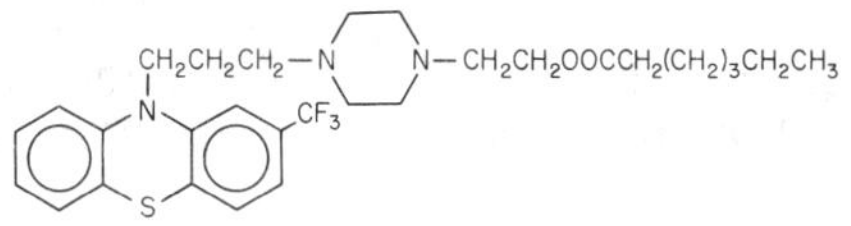

$C_{29}H_{38}F_3N_3O_2S$ 549.69
Heptanoic acid, 2-[4-[3-[2-(trifluoromethyl)-10*H*-phenothiazin-10-yl]propyl]-1-piperazinyl]ethyl ester.
2-[4-[3-[2-(Trifluoromethyl)phenothiazin-10-yl]propyl]-1-piperazinyl]ethyl heptanoate [*2746-81-8*].

» Fluphenazine Enanthate contains not less than 97.0 percent and not more than 103.0 percent of $C_{29}H_{38}F_3N_3O_2S$, calculated on the dried basis.

Packaging and storage—Preserve in tight, light-resistant containers.

Reference standard—*USP Fluphenazine Enanthate Dihydrochloride Reference Standard*—Do not dry; determine the *Water* content by *Method I* (see *Water Determination* ⟨921⟩) before using. Keep container tightly closed and protected from light.

NOTE—Throughout the following procedures, protect test or assay specimens, the Reference Standard, and solutions containing them, by conducting the procedures without delay, under subdued light, or using low-actinic glassware.

Identification—

A: Place about 50 mg of Fluphenazine Enanthate and about 50 mg of USP Fluphenazine Enanthate Dihydrochloride RS in separate, glass-stoppered, small centrifuge tubes, and treat each tube as follows: Add 1.5 mL of sodium hydroxide solution (1 in 250), and mix. Add 2 mL of carbon disulfide, shake vigorously for 2 minutes, and centrifuge. Dry the lower, clear layer by filtering it through 2 g of anhydrous sodium sulfate: the infrared absorption spectrum of the test preparation, determined in a 0.1-mm cell, exhibits maxima only at the same wavelengths as that of the Standard preparation, similarly measured.

B: The ultraviolet absorption spectrum of a 1 in 100,000 solution in methanolic hydrochloric acid (prepared by adding 8.5 mL of hydrochloric acid to an equal volume of water in a 1000-mL volumetric flask, diluting with methanol to volume, and mixing) exhibits maxima and minima at the same wavelengths as that of a similar solution of USP Fluphenazine Enanthate Dihydrochloride RS, concomitantly measured, and the respective molar absorptivities, calculated on the dried basis, at the wavelength of maximum absorbance at about 258 nm, do not differ by more than 2.5%. [NOTE—The molecular weight of fluphenazine enanthate dihydrochloride, $C_{29}H_{38}F_3N_3O_2S.2HCl$, is 622.62.]

Loss on drying ⟨731⟩—Dry it in vacuum at 60° for 3 hours: it loses not more than 1.0% of its weight.

Residue on ignition ⟨281⟩: not more than 0.2%.

Heavy metals, *Method II* ⟨231⟩: 0.003%.

Ordinary impurities ⟨466⟩—

Test solution: alcohol.

Standard solution: alcohol.

Eluant: a mixture of alcohol, glacial acetic acid, and water (3:1:1).

Visualization: 1.

Assay—Dissolve about 500 mg of Fluphenazine Enanthate, accurately weighed, in 50 mL of glacial acetic acid, add 1 drop of crystal violet TS, and titrate with 0.1 *N* perchloric acid VS to a blue-green end-point. Perform a blank determination, and make any necessary correction. Each mL of 0.1 *N* perchloric acid is equivalent to 27.49 mg of $C_{29}H_{38}F_3N_3O_2S$.

Fluphenazine Enanthate Injection

» Fluphenazine Enanthate Injection is a sterile solution of Fluphenazine Enanthate in a suitable vegetable oil. It contains not less than 90.0 percent and not more than 110.0 percent of the labeled amount of $C_{29}H_{38}F_3N_3O_2S$.

Packaging and storage—Preserve in single-dose or in multiple-dose containers, preferably of Type I or Type III glass, protected from light.

NOTE—Throughout the following procedures, protect test or assay specimens, the Reference Standard, and solutions containing them, by conducting the procedures without delay, under subdued light, or using low-actinic glassware.

Identification—To a volume of Injection, equivalent to about 50 mg of fluphenazine enanthate, add 2 mL of methanol and 3 mL of palladium chloride solution (1 in 1000): a rust-red color is produced. Add an excess of the palladium chloride solution: the color is intensified to a brownish red.

Other requirements—It meets the requirements under *Injections* ⟨1⟩.

Assay—Dissolve an accurately measured volume of Fluphenazine Enanthate Injection, equivalent to about 150 mg of fluphenazine enanthate, in 75 mL of glacial acetic acid, add 1 drop of crystal violet TS, and titrate with 0.1 *N* perchloric acid VS to a blue-green end-point. Perform a blank determination, and make any necessary correction. Each mL of 0.1 *N* perchloric acid is equivalent to 27.49 mg of $C_{29}H_{38}F_3N_3O_2S$.

Fluphenazine Hydrochloride

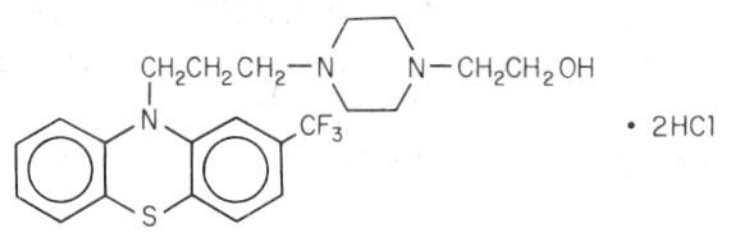

$C_{22}H_{26}F_3N_3OS.2HCl$ 510.44
1-Piperazineethanol, 4-[3-[2-(trifluoromethyl)-10*H*-phenothiazin-10-yl]propyl]-, dihydrochloride.
4-[3-[2-(Trifluoromethyl)phenothiazin-10-yl]propyl]-1-piperazineethanol dihydrochloride [*146-56-5*].

» Fluphenazine Hydrochloride contains not less than 97.0 percent and not more than 103.0 percent of $C_{22}H_{26}F_3N_3OS.2HCl$, calculated on the dried basis.

Packaging and storage—Preserve in tight, light-resistant containers.

Reference standard—*USP Fluphenazine Hydrochloride Reference Standard*—Dry at 65° for 3 hours before using.

NOTE—Throughout the following procedures, protect test or assay specimens, the Reference Standard, and solutions containing them, by conducting the procedures without delay, under subdued light, or using low-actinic glassware.

Identification—

A: The infrared absorption spectrum of a potassium bromide dispersion of it, previously dried, exhibits maxima only at the same wavelengths as that of a similar preparation of USP Fluphenazine Hydrochloride RS.

B: The ultraviolet absorption spectrum of a 1 in 100,000 solution in methanol exhibits maxima and minima at the same wavelengths as that of a similar preparation of USP Fluphenazine Hydrochloride RS, concomitantly measured, and the respective absorptivities, calculated on the dried basis, at the wavelength of maximum absorbance at about 259 nm do not differ by more than 2.5%.

C: A solution of Fluphenazine Hydrochloride responds to the tests for *Chloride* ⟨191⟩.

Loss on drying ⟨731⟩—Dry it at 65° for 3 hours: it loses not more than 1% of its weight.

Residue on ignition ⟨281⟩: not more than 0.5%.

Heavy metals, *Method II* ⟨231⟩: 0.003%.

Assay—Dissolve about 500 mg of Fluphenazine Hydrochloride, accurately weighed, in a mixture of 50 mL of glacial acetic acid and 10 mL of mercuric acetate TS, warming slightly if necessary to effect solution. Add 1 drop of crystal violet TS, and titrate with 0.1 *N* perchloric acid VS to a blue-green end-point. Perform a blank determination, and make any necessary correction. Each mL of 0.1 *N* perchloric acid is equivalent to 25.52 mg of $C_{22}H_{26}F_3N_3OS.2HCl$.

Fluphenazine Hydrochloride Elixir

» Fluphenazine Hydrochloride Elixir contains not less than 90.0 percent and not more than 110.0 percent of the labeled amount of $C_{22}H_{26}F_3N_3OS.2HCl$.

Packaging and storage—Preserve in tight containers, protected from light.

Reference standard—*USP Fluphenazine Hydrochloride Reference Standard*—Dry at 65° for 3 hours before using.

NOTE—Throughout the following procedures, protect test or assay specimens, the Reference Standard, and solutions containing them, by conducting the procedures without delay, under subdued light, or using low-actinic glassware.

Identification—Transfer a volume of Elixir, equivalent to about 10 mg of fluphenazine hydrochloride, to a separator, and to a second separator transfer 10 mg of USP Fluphenazine Hydrochloride RS. To each separator add 20 mL of 6 *N* sodium hydroxide, and extract each mixture with 20 mL of isooctane. Evaporate the isooctane solutions to dryness, and proceed as directed in the *Identification test* under *Fluphenazine Hydrochloride Tablets*, beginning with "dissolve the residues in 0.5-mL portions."

pH ⟨791⟩: between 5.3 and 5.8.

Alcohol content ⟨611⟩: between 13.5% and 15.0% of C_2H_5OH.

Assay—

Palladium chloride solution—Dissolve 500 mg of palladium chloride in a mixture of 250 mL of water and 5 mL of hydrochloric acid, by heating on a steam bath. Cool, dilute with water to 500 mL, and store in an amber bottle. Use within 30 days. On the day of use, transfer 6 mL to a 50-mL volumetric flask, and add dilute hydrochloric acid (1 in 50) to volume.

Standard preparation—Transfer about 50 mg of USP Fluphenazine Hydrochloride RS, accurately weighed, to a 100-mL volumetric flask, add 14 mL of alcohol, shake until dissolved, and dilute with water to volume. Pipet 10 mL of this solution into a 50-mL volumetric flask, add dilute hydrochloric acid (1 in 50) to volume, and mix.

Assay preparation—Transfer an accurately measured volume of Fluphenazine Hydrochloride Elixir, equivalent to about 5 mg of fluphenazine hydrochloride, to a 50-mL volumetric flask, add dilute hydrochloric acid (1 in 50) to volume, and mix.

Procedure—Transfer 10.0 mL each of the *Standard preparation*, *Assay preparation*, and 10.0 mL of dilute hydrochloric acid (1 in 50) to provide a blank, into separate, clean, dry 25-mL volumetric flasks. Treating each flask individually, dilute with *Palladium chloride solution* to volume, mix, and immediately measure the absorbance relative to that of water in 1-cm cells at about 485 nm, with a suitable spectrophotometer. Correct the recorded absorbances by subtracting the blank absorbance. Add 50 mg of anhydrous sodium sulfite to each of the flasks, shake to dissolve, and allow to stand for 3 minutes. Measure the absorbance of each solution as previously directed. Calculate the quantity, in mg, of $C_{22}H_{26}F_3N_3OS.2HCl$ in each mL of the Elixir taken by the formula:

$$0.05(C/V)(A_U/A_S),$$

in which C is the concentration, in µg per mL, of USP Fluphenazine Hydrochloride RS in the *Standard preparation*, V is the volume, in mL, of Elixir taken, and A_U and A_S are the differences in the corrected absorbances before and after the addition of anhydrous sodium sulfite for the *Assay preparation* and the *Standard preparation*, respectively.

Fluphenazine Hydrochloride Injection

» Fluphenazine Hydrochloride Injection is a sterile solution of Fluphenazine Hydrochloride in Water for Injection. It contains not less than 95.0 percent and not more than 110.0 percent of the labeled amount of $C_{22}H_{26}F_3N_3OS.2HCl$.

Packaging and storage—Preserve in single-dose or in multiple-dose containers, preferably of Type I glass, protected from light.

Reference standard—*USP Fluphenazine Hydrochloride Reference Standard*—Dry at 65° for 3 hours before using.

NOTE—Throughout the following procedures, protect test or assay specimens, the Reference Standard, and solutions containing them, by conducting the procedures without delay, under subdued light, or using low-actinic glassware.

Identification—Transfer a volume of Injection, equivalent to about 10 mg of fluphenazine hydrochloride, to a separator, and to a second separator transfer 10 mg of USP Fluphenazine Hydrochloride RS. To each separator add 20 mL of 6 *N* sodium hydroxide, and extract each mixture with 20 mL of isooctane. Evaporate the isooctane solutions to dryness, and proceed as directed in the *Identification test* under *Fluphenazine Hydrochloride Tablets*, beginning with "dissolve the residues in 0.5-mL portions."

pH ⟨791⟩: between 4.8 and 5.2.

Other requirements—It meets the requirements under *Injections* ⟨1⟩.

Assay—Proceed with Fluphenazine Hydrochloride Injection as directed in the *Assay* under *Fluphenazine Hydrochloride Elixir*.

Fluphenazine Hydrochloride Oral Solution

» Fluphenazine Hydrochloride Oral Solution is an aqueous solution of Fluphenazine Hydrochloride. It contains not less than 95.0 percent and not more than 110.0 percent of the labeled amount of $C_{22}H_{26}F_3N_3OS.2HCl$.

Packaging and storage—Preserve in tight containers, protected from light.

Labeling—Label it to indicate that it is to be diluted to appropriate strength with water or other suitable fluid prior to administration.

Reference standard—*USP Fluphenazine Hydrochloride Reference Standard*—Dry at 65° for 3 hours before using.

NOTE—Throughout the following procedures, protect test or assay specimens, the Reference Standard, and solutions containing them, by conducting the procedures without delay, under subdued light, or using low-actinic glassware.

Identification—Transfer a volume of Oral Solution, equivalent to about 10 mg of fluphenazine hydrochloride, to a separator, and to a second separator transfer 10 mg of USP Fluphenazine Hydrochloride RS. To each separator add 20 mL of sodium hydroxide solution (1 in 4), and extract each mixture with 20 mL of isooctane. Evaporate the isooctane solutions to dryness, and proceed as directed in the *Identification test* under *Fluphenazine Hydrochloride Tablets*, beginning with "dissolve the residues in 0.5-mL portions."

pH ⟨791⟩: between 4.0 and 5.0.

Alcohol content ⟨611⟩: not less than 90.0% and not more than 110.0% of the labeled amount, the labeled amount being not more than 15.0% of C_2H_5OH.

Assay—Proceed with Fluphenazine Hydrochloride Oral Solution as directed in the *Assay* under *Fluphenazine Hydrochloride Elixir*.

Fluphenazine Hydrochloride Tablets

» Fluphenazine Hydrochloride Tablets contain not less than 90.0 percent and not more than 110.0 percent of the labeled amount of $C_{22}H_{26}F_3N_3OS.2HCl$.

Packaging and storage—Preserve in tight, light-resistant containers.

Reference standard—*USP Fluphenazine Hydrochloride Reference Standard*—Dry at 65° for 3 hours before using.

NOTE—Throughout the following procedures, protect test or assay specimens, the Reference Standard, and solutions containing them, by conducting the procedures without delay, under subdued light, or using low-actinic glassware.

Identification—Transfer a portion of finely powdered Tablets, equivalent to about 10 mg of fluphenazine hydrochloride, to a separator, and to a second separator transfer 10 mg of USP Fluphenazine Hydrochloride RS. Add 5 mL of water and 20 mL of dilute hydrochloric acid (1 in 120) to each separator, shake for 10 minutes, and to each mixture add 20 mL of chloroform-saturated sodium carbonate solution (1 in 10). Extract each mixture with five 20-mL portions of chloroform, shaking gently to avoid emulsion formation, and pass the extracts through separate chloroform-washed cotton filters into separate 150-mL beakers. Evaporate the extracts on a steam bath to dryness, and dissolve the residues in 0.5-mL portions of a mixture of 4 volumes of methanol and 1 volume of water. On a suitable thin-layer chromatographic plate (see *Chromatography* ⟨621⟩), coated with a 0.25-mm layer of chromatographic silica gel mixture, apply 10 µL of each solution. Allow the spots to dry, and develop the chromatogram in a solvent system consisting of a mixture of benzene, methanol, and ammonium hydroxide (40:10:1) until the solvent front has moved about three-fourths of the length of the plate. Remove the plate from the developing chamber, mark the solvent front, and allow the solvent to evaporate. Locate the spots on the plate by lightly spraying with a 2 in 5 solution of sulfuric acid in methanol: the R_f value and color of the principal spot obtained from the test solution correspond to those obtained from the Standard solution.

Dissolution ⟨711⟩—

Medium: 0.1 *N* hydrochloric acid; 900 mL.

Apparatus 1: 100 rpm.

Time: 45 minutes.

Procedure—Determine the amount of $C_{22}H_{26}F_3N_3OS.2HCl$ dissolved, employing the procedure set forth in the *Assay*, making any necessary modifications.

Tolerances—Not less than 75% (*Q*) of the labeled amount of $C_{22}H_{26}F_3N_3OS.2HCl$ is dissolved in 45 minutes.

Uniformity of dosage units ⟨905⟩: meet the requirements.

Assay—

Palladium chloride solution—Dissolve 500 mg of palladium chloride in a mixture of 250 mL of water and 5 mL of hydrochloric acid, heating on a steam bath until solution is effected, cool, and dilute with water to 500 mL. Store in an amber bottle and use within 30 days. On the day of use, mix 5 mL of this solution with 10 mL of dilute hydrochloric acid (1 in 10), and dilute with water to 100 mL.

Standard preparation—Transfer about 25 mg of USP Fluphenazine Hydrochloride RS, accurately weighed, to a 250-mL volumetric flask, dilute with water to volume, and mix. Transfer 10.0 mL of this solution to a separator, add 1 mL of sodium hydroxide solution (1 in 2), and extract with four 20-mL portions of *n*-hexane. Filter the extracts through a pledget of glass wool into a 100-mL volumetric flask. Dilute with *n*-hexane to volume, and mix.

Assay preparation—Weigh and finely powder not less than 20 Fluphenazine Hydrochloride Tablets. Transfer an accurately weighed portion of the powder, equivalent to about 1 mg of fluphenazine hydrochloride, to a separator, add 10 mL of water, and proceed as directed under *Standard preparation*, beginning with "add 1 mL of sodium hydroxide solution (1 in 2)."

Procedure—Extract 50.0 mL each of the *Standard preparation* and the *Assay preparation* with 10.0 mL of *Palladium chloride solution*, in separate 250-mL separators. Shake vigorously for 3 minutes. Concomitantly determine the absorbances of the solutions in 1-cm cells at the wavelength of maximum absorbance at about 480 nm, with a suitable spectrophotometer, using *Palladium chloride solution* as the blank. Calculate the quantity, in mg, of $C_{22}H_{26}F_3N_3OS.2HCl$ in the portion of Tablets taken by the formula:

$$0.1C(A_U/A_S),$$

in which *C* is the concentration, in µg per mL, of USP Fluphenazine Hydrochloride RS in the *Standard preparation*, and A_U and A_S are the absorbances of the solutions from the *Assay preparation* and the *Standard preparation*, respectively.

Flurandrenolide

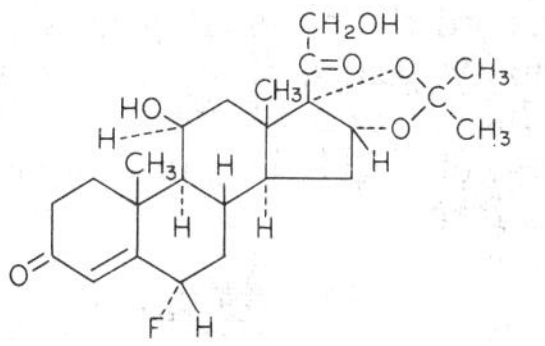

$C_{24}H_{33}FO_6$ 436.52

Pregn-4-ene-3,20-dione, 6-fluoro-11,21-dihydroxy-16,17-[(1-methylethylidene)bis(oxy)]-, (6α,11β,16α)-.

6α-Fluoro-11β,16α,17,21-tetrahydroxypregn-4-ene-3,20-dione, cyclic 16,17-acetal with acetone [*1524-88-5*].

» Flurandrenolide contains not less than 97.0 percent and not more than 102.0 percent of $C_{24}H_{33}FO_6$, calculated on the dried basis.

Packaging and storage—Preserve in tight containers in a cold place, protected from light.

Reference standards—*USP Flurandrenolide Reference Standard*—Dry in vacuum at 105° for 4 hours before using. *USP Prednisone Reference Standard*—Dry at 105° for 3 hours before using.

Identification—

A: The infrared absorption spectrum of a potassium bromide dispersion of it, previously dried, exhibits maxima only at the same wavelengths as that of a similar preparation of USP Flurandrenolide RS.

B: The ultraviolet absorption spectrum of a 1 in 50,000 solution in methanol exhibits maxima and minima at the same wavelengths as that of a similar preparation of USP Flurandrenolide RS, concomitantly measured, and the respective absorptivities, calculated on the dried basis, at the wavelength of maximum absorbance at about 237 nm do not differ by more than 3.0%.

Specific rotation ⟨781⟩: between +145° and +153°, calculated on the dried basis, determined in a solution in chloroform containing 100 mg in each 10 mL.

Loss on drying ⟨731⟩—Dry it in vacuum at 105° for 4 hours: it loses not more than 1.0% of its weight.

Ordinary impurities ⟨466⟩—

Test solution: methanol.

Standard solution: methanol.

Application volume: 10 µL.

Eluant: a mixture of toluene and isopropyl alcohol (90:10), in a nonequilibrated chamber.

Visualization: 1.

Assay—

Mobile phase—Prepare a filtered and degassed mixture of methanol and water (60:40). Make adjustments if necessary (see *System Suitability* under *Chromatography* ⟨621⟩).

Internal standard solution—Dissolve Prednisone in *Mobile phase*, with the aid of sonication, to obtain a solution containing about 1 mg per mL.

Standard preparation—Transfer about 5 mg of USP Flurandrenolide RS, accurately weighed, to a 10-mL volumetric flask, add 2.0 mL of *Internal standard solution*, dilute with *Mobile phase* to volume, sonicate to aid solution, and mix to obtain a solution having a known concentration of about 0.5 mg of USP Flurandrenolide RS per mL.

Assay preparation—Transfer about 5 mg of Flurandrenolide, accurately weighed, to a 10-mL volumetric flask, add 2.0 mL of *Internal standard solution*, dilute with *Mobile phase* to volume, sonicate to aid solution, and mix.

Chromatographic system (see *Chromatography* ⟨621⟩)—The liquid chromatograph is equipped with a 240-nm detector and a 4-mm × 25-cm column that contains packing L1. The flow rate is about 1 mL per minute. Chromatograph the *Standard preparation*, and record the peak responses as directed under *Procedure:* the order of elution is prednisone followed by flurandrenolide, the resolution, *R*, between the analyte and internal standard peaks is not less than 2.0, and the relative standard deviation for replicate injections is not more than 3.0%.

Procedure—Separately inject equal volumes (about 20 µL) of the *Standard preparation* and the *Assay preparation* into the chromatograph, record the chromatograms, and measure the responses for the major peaks. The relative retention times are about 0.5 for prednisone and 1.0 for flurandrenolide. Calculate the quantity, in mg, of $C_{24}H_{33}FO_6$ in the portion of Flurandrenolide taken by the formula:

$$10C(R_U/R_S),$$

in which *C* is the concentration, in mg per mL, of USP Flurandrenolide RS in the *Standard preparation*, and R_U and R_S are the peak response ratios obtained from the *Assay preparation* and the *Standard preparation*, respectively.

Flurandrenolide Cream

» Flurandrenolide Cream contains not less than 90.0 percent and not more than 110.0 percent of the labeled amount of $C_{24}H_{33}FO_6$.

Packaging and storage—Preserve in tight containers, protected from light.

Reference standard—*USP Flurandrenolide Reference Standard*—Dry in vacuum at 105° for 4 hours before using.

Identification—Extract a quantity of weighed Cream, equivalent to about 500 µg of flurandrenolide, as directed for the *Assay preparation*. Omit the addition of the internal standard, and evaporate the chloroform extracts on a steam bath under a stream of nitrogen to about 3 mL. Transfer the chloroform solution to a 10-mL flask, and evaporate with the aid of a stream of nitrogen to dryness. Dissolve the residue in 2 mL of chloroform to prepare the test solution. Flurandrenolide Cream meets the requirements of the *Thin-layer Chromatographic Identification Test* ⟨201⟩, 4.0 µL each of the test solution and Standard solution being applied and the solvent system consisting of a mixture of ethyl acetate and ether (70:30).

Microbial limits—It meets the requirements of the tests for absence of *Staphylococcus aureus* and *Pseudomonas aeruginosa* under *Microbial Limit Tests* ⟨61⟩.

Minimum fill ⟨755⟩: meets the requirements.

Assay—

Methanolic sodium chloride—Transfer 100 mL of sodium chloride solution (1 in 10) to a 500-mL volumetric flask. Dilute with methanol to volume, and mix.

Mobile phase—Prepare a filtered and degassed mixture of methanol and water (70:30). Make adjustments if necessary (see *System Suitability* under *Chromatography* ⟨621⟩).

Internal standard solution—Transfer about 10 mg of testosterone to a 100-mL volumetric flask, add methanol to volume, and mix.

Standard preparation—Transfer about 16 mg of USP Flurandrenolide RS, accurately weighed, to a 100-mL volumetric flask, add methanol to volume, and mix. Transfer 3.0 mL of this solution to a 10-mL volumetric flask, add 4.0 mL of *Internal standard solution*, dilute with water to volume, and mix to obtain a solution having a known concentration of about 48 µg of USP Flurandrenolide RS per mL.

Assay preparation—Transfer an accurately weighed quantity of Flurandrenolide Cream, equivalent to about 500 µg of flurandrenolide, to a 125-mL separator. Add 50 mL of hexane and 25 mL of *Methanolic sodium chloride*, and shake until the Cream is thoroughly dispersed. Allow the phases to separate, and drain the lower aqueous phase into a second 125-mL separator containing 15 mL of hexane. Shake vigorously, allow the phases to separate, and drain the lower aqueous phase into a 250-mL separator containing 75 mL of water. Serially extract the hexane phases remaining in the two 125-mL separators with two additional 25-mL portions of *Methanolic sodium chloride*, adding each aqueous phase to the 250-mL separator. Discard the hexane phases. Extract the combined aqueous phases with four 25-mL portions of chloroform. Filter each chloroform extract through 10 g of anhydrous sodium sulfate into a 125-mL conical beaker. Rinse the sodium sulfate with water-washed chloroform, and add the wash to the beaker. Add 4.0 mL of *Internal standard solution* to the beaker containing the chloroform extract. Evaporate the solution on a steam bath under a stream of nitrogen nearly to dryness. Remove the beaker from the steam bath, and evaporate the remaining solution with the aid of nitrogen to dryness. Add 10 mL of *Mobile phase* to the beaker, and place it in an ultrasonic bath to dissolve the residue. Filter the solution through a suitable 0.5-µm filter with a prefilter above the membrane filter to prevent clogging.

Chromatographic system (see *Chromatography* ⟨621⟩)—The liquid chromatograph is equipped with a 240-nm detector and a 4.6-mm × 25-cm column that contains packing L1. The flow rate is about 1 mL per minute. Chromatograph the *Standard preparation*, and record the peak responses as directed under *Procedure:* the resolution, *R*, between the analyte and internal standard peaks is not less than 2.0, and the relative standard deviation for replicate injections is not more than 3.0%.

Procedure—Separately inject equal volumes (about 20 µL) of the *Standard preparation* and the *Assay preparation* into the chromatograph, record the chromatograms, and measure the responses for the major peaks. The relative retention times are about 2 for testosterone and 1.0 for flurandrenolide. Calculate the quantity, in mg, of $C_{24}H_{33}FO_6$ in the portion of Cream taken by the formula:

$$10C(R_U/R_S),$$

in which *C* is the concentration, in mg per mL, of USP Flurandrenolide RS in the *Standard preparation*, and R_U and R_S are the peak response ratios obtained from the *Assay preparation* and the *Standard preparation*, respectively.

Flurandrenolide Cream, Neomycin Sulfate and—*see* Neomycin Sulfate and Flurandrenolide Cream

Flurandrenolide Lotion

» Flurandrenolide Lotion contains not less than 90.0 percent and not more than 110.0 percent of the labeled amount of $C_{24}H_{33}FO_6$.

Packaging and storage—Preserve in tight containers, protected from heat, light, and freezing.

Reference standard—*USP Flurandrenolide Reference Standard*—Dry in vacuum at 105° for 4 hours before using.

Identification—It responds to the *Identification test* under *Flurandrenolide Cream.*

Microbial limits—It meets the requirements of the tests for absence of *Staphylococcus aureus* and *Pseudomonas aeruginosa* under *Microbial Limit Tests* ⟨61⟩.

pH ⟨791⟩: between 3.5 and 6.0, determined in a 1 in 10 dilution of the Lotion in water containing 0.30 mL of saturated potassium chloride solution per 100 mL.

Assay—

Methanolic sodium chloride, Mobile phase, Internal standard solution, Standard preparation, and *Chromatographic system*—Prepare as directed in the *Assay* under *Flurandrenolide Cream.*

Assay preparation—Transfer an accurately weighed portion of Flurandrenolide Lotion, calculated from the density to contain about 500 µg of flurandrenolide, to a separator. (Determine the density by taring a 100-mL volumetric flask containing 50.0 mL of water, adding approximately 25 g of well-shaken Lotion, and again weighing, then carefully adjusting the contents of the volumetric flask with water from a buret to volume, and finally calculating the density by the formula:

$$A/B,$$

in which A is the weight, in g, of the Lotion taken, and B is 50.0 mL minus the volume, in mL, of water necessary to adjust the contents of the volumetric flask to volume.) Proceed as directed for *Assay preparation* in the *Assay* under *Flurandrenolide Cream,* beginning with "Add 50 mL of hexane and 25 mL of *Methanolic sodium chloride.*"

Procedure—Proceed as directed in the *Assay* under *Flurandrenolide Cream.* Calculate the quantity, in mg, of $C_{24}H_{33}FO_6$ in each mL of the Lotion taken by the formula:

$$10C(D/W)(R_U/R_S),$$

in which C is the concentration, in mg per mL, of USP Flurandrenolide RS in the *Standard preparation, D* is the density of the Lotion, W is the weight, in g, of Lotion taken, and R_U and R_S are the peak response ratios obtained from the *Assay preparation* and the *Standard preparation,* respectively.

Flurandrenolide Lotion, Neomycin Sulfate and—*see* Neomycin Sulfate and Flurandrenolide Lotion

Flurandrenolide Ointment

» Flurandrenolide Ointment contains not less than 90.0 percent and not more than 110.0 percent of the labeled amount of $C_{24}H_{33}FO_6$.

Packaging and storage—Preserve in tight containers, protected from light.

Reference standard—*USP Flurandrenolide Reference Standard*—Dry in vacuum at 105° for 4 hours before using.

Identification—It responds to the *Identification test* under *Flurandrenolide Cream.*

Microbial limits—It meets the requirements of the tests for absence of *Staphylococcus aureus* and *Pseudomonas aeruginosa* under *Microbial Limit Tests* ⟨61⟩.

Minimum fill ⟨755⟩: meets the requirements.

Assay—Proceed with Flurandrenolide Ointment as directed in the *Assay* under *Flurandrenolide Cream.* Calculate the quantity, in mg, of $C_{24}H_{33}FO_6$ in the portion of Ointment taken by the formula:

$$10C(R_U/R_S),$$

in which C is the concentration, in mg per mL, of USP Flurandrenolide RS in the *Standard preparation,* and R_U and R_S are the peak response ratios obtained from the *Assay preparation* and the *Standard preparation,* respectively.

Flurandrenolide Ointment, Neomycin Sulfate and—*see* Neomycin Sulfate and Flurandrenolide Ointment

Flurandrenolide Tape

» Flurandrenolide Tape is a non-porous, pliable, adhesive-type tape having Flurandrenolide impregnated in the adhesive material, the adhesive material on one side being transported on a removable, protective slit-paper liner. Flurandrenolide Tape contains not less than 80.0 percent and not more than 125.0 percent of the labeled amount of $C_{24}H_{33}FO_6$.

Packaging and storage—Preserve at controlled room temperature.

Reference standard—*USP Flurandrenolide Reference Standard*—Dry in vacuum at 105° for 4 hours before using.

Identification—Extract a portion of Flurandrenolide Tape, equivalent to about 200 µg of flurandrenolide, as directed for the *Assay preparation* in the *Assay.* Omit the addition of the internal standard, and evaporate the chloroform extracts on a steam bath under a stream of nitrogen to about 3 mL. Transfer the chloroform solution to a 10-mL flask, and evaporate with the aid of a stream of nitrogen to dryness. Dissolve the residue in 1.0 mL of a mixture of equal volumes of chloroform and methanol, warming gently to effect solution: it meets the requirements of the *Thin-layer Chromatographic Identification Test* ⟨201⟩, 20 µL each of the test solution and Standard solution being applied, and the solvent mixture consisting of equal volumes of benzene and ethyl acetate.

Microbial limits—It meets the requirements of the tests for absence of *Staphylococcus aureus* and *Pseudomonas aeruginosa* under *Microbial Limit Tests* ⟨61⟩.

Assay—

Methanolic sodium chloride, Mobile phase, and *Chromatographic system*—Prepare as directed in the *Assay* under *Flurandrenolide Cream.*

Flurandrenolide standard solution—Dissolve about 7 mg of USP Flurandrenolide RS, accurately weighed, in 50 mL of methanol in a 100-mL volumetric flask. Dilute with methanol to volume, and mix.

Internal standard solution—Dissolve about 4 mg of Testosterone in 50 mL of methanol in a 100-mL volumetric flask. Dilute with methanol to volume, and mix.

Standard preparation—Pipet 3.0 mL of *Flurandrenolide standard solution* and 4.0 mL of *Internal standard solution* into a 10-mL volumetric flask. Dilute with water to volume, and mix.

Assay preparation—Accurately measure and cut a portion of Flurandrenolide Tape, equivalent to about 200 µg of flurandrenolide. Remove and discard the paper liner from the portion of Tape. Touch the flattened end of a glass rod to the adhesive side of the Tape, and carefully transfer the tape to the bottom of a 600-mL beaker containing 15 mL of anhydrous methanol, taking care that the adhesive side of the tape does not adhere to the wall of the beaker. Remove the glass rod from the tape, and wash it with 5 mL of anhydrous methanol, adding the wash to the beaker. Place the beaker containing the Tape and the methanol in an ultrasonic bath for 3 minutes, rotating the beaker in such manner that the methanol is in contact with all portions of the Tape. Transfer the methanol to a 250-mL separator. Extract the Tape, using sonication, with two additional 20-mL portions of anhydrous methanol, adding each portion to the separator. To the combined methanol extract add 15 mL of sodium chloride solution (1 in 10) and 50 mL of hexane, and shake vigorously. Allow the phases to separate, and drain the lower aqueous phase into a second separator containing 15 mL of hexane. Shake vigorously, allow the phases to separate, and drain the lower phase into a third 250-mL separator containing 100 mL of water. Serially extract the hexane phases remaining in the two separators with one 25-mL portion of *Methanolic sodium chloride,* adding the extract to the third separator. Discard the hexane phases. Extract the combined aqueous phases with four 25-mL portions of chloroform. Filter each chloroform extract through 10 g of anhydrous sodium sulfate into a 125-mL conical beaker. Rinse the sodium sulfate with water-washed chloroform, and add the wash to the beaker. Add 4.0 mL of *Internal standard solution* to the beaker. Evaporate the solution on a steam bath under a stream of nitrogen to near dryness. Remove the beaker from the steam bath and evaporate the remaining solution with the aid of nitrogen to dryness. Add 10 mL of *Mobile solvent* to the beaker, and place it in an ultrasonic bath to dissolve the residue. Filter the solution through a suitable 0.5-μm filter with a prefilter above the membrane filter to prevent clogging.

Procedure—Separately inject equal volumes (about 50 μL) of the *Standard preparation* and the *Assay preparation* into the chromatograph, record the chromatograms, and measure the responses for the major peaks. The relative retention times are about 2 for testosterone and 1.0 for flurandrenolide. Calculate the quantity, in μg, of $C_{24}H_{33}FO_6$ in the portion of Tape taken by the formula:

$$10C(R_U/R_S),$$

in which C is the concentration, in μg per mL, of USP Flurandrenolide RS in the *Standard preparation*, and R_U and R_S are the peak response ratios obtained from the *Assay preparation* and the *Standard preparation*, respectively.

Flurazepam Hydrochloride

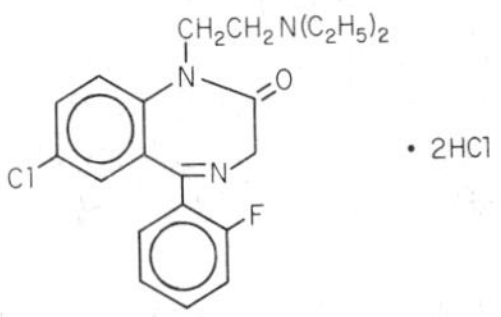

$C_{21}H_{23}ClFN_3O.2HCl$ 460.81

2*H*-1,4-Benzodiazepin-2-one, 7-chloro-1-[2-(diethylamino)ethyl]-5-(2-fluorophenyl)-1,3-dihydro-, dihydrochloride.

7-Chloro-1-[2-(diethylamino)ethyl]-5-(*o*-fluorophenyl)-1,3-dihydro-2*H*-1,4-benzodiazepin-2-one dihydrochloride [*1172-18-5*].

» Flurazepam Hydrochloride contains not less than 99.0 percent and not more than 101.0 percent of $C_{21}H_{23}ClFN_3O.2HCl$, calculated on the dried basis.

Packaging and storage—Preserve in tight, light-resistant containers.

Reference standards—*USP Flurazepam Hydrochloride Reference Standard*—Protect from light. Dry over silica gel for 4 hours before using. *USP Flurazepam Related Compound C Reference Standard*—Keep container tightly closed and protected from light. Dry at 105° for 4 hours before using. *USP Flurazepam Related Compound F Reference Standard*—Keep container tightly closed and protected from light. Dry in vacuum over silica gel at 60° for 4 hours before using.

Identification—

A: The infrared absorption spectrum of a potassium bromide dispersion of it, previously dried, exhibits maxima only at the same wavelengths as that of a similar preparation of USP Flurazepam Hydrochloride RS.

B: The ultraviolet absorption spectrum of a 1 in 100,000 solution of it in a 1 in 36 solution of sulfuric acid in methanol exhibits maxima and minima at the same wavelengths as that of Flurazepam Hydrochloride RS, concomitantly measured, and the respective absorptivities, calculated on the dried basis, at the wavelength of maximum absorbance at about 239 nm do not differ by more than 3.0%.

C: Prepare a solution of it in methanol containing 3 mg per mL. On a suitable thin-layer chromatographic plate (see *Chromatography* ⟨621⟩), coated with a 0.25-mm layer of chromatographic silica gel mixture, apply 10 μL of this solution and 10 μL of a methanol solution of USP Flurazepam Hydrochloride RS containing 3 mg per mL. Allow the spots to dry, and develop the chromatogram in a solvent system consisting of a mixture of ethyl acetate and ammonium hydroxide (200:1) until the solvent front has moved about three-fourths of the length of the plate. Remove the plate from the developing chamber, mark the solvent front, and allow the solvent to evaporate. Locate the spots on the plate by viewing under short-wavelength ultraviolet light: the R_f value of the principal spot in the chromatogram of the test solution corresponds to that obtained from the solution of the Reference Standard.

D: To 2 mL of a solution (1 in 20), add 1 mL of 2 *N* nitric acid: the solution responds to the tests for *Chloride* ⟨191⟩, 5 drops of silver nitrate TS being used.

Loss on drying ⟨731⟩—Protect from light. Dry it over silica gel in vacuum at 60° for 4 hours: it loses not more than 1.5% of its weight.

Residue on ignition ⟨281⟩: not more than 0.1%.

Heavy metals, *Method II* ⟨231⟩: 0.002%.

Limit of fluoride ion—[NOTE—Use plasticware throughout the procedure.]

pH 5.25 buffer—Dissolve 110 g of sodium chloride and 1 g of sodium citrate in 700 mL of water in a 2000-mL volumetric flask. Cautiously add 150 g of sodium hydroxide, and dissolve with shaking. Cool to room temperature, and, while stirring, cautiously add 450 mL of glacial acetic acid to the cooled solution. Cool, add 600 mL of isopropyl alcohol, dilute with water to volume, and mix: the pH of this solution is between 5.0 and 5.5.

Standard stock solution—Transfer 221 mg of sodium fluoride to a 100-mL volumetric flask, add about 20 mL of water, and mix to dissolve. Add 1.0 mL of sodium hydroxide solution (1 in 2500), dilute with water to volume, and mix. Each mL of this solution contains 1 mg of fluoride ions. Store in a tightly closed, plastic container.

Standard preparations—Dilute portions of the *Standard stock solution* quantitatively and stepwise with *pH 5.25 buffer* to obtain 100-mL solutions having concentrations of 1, 3, 5, and 10 μg per mL.

Test preparation—Transfer 1.0 g of Flurazepam Hydrochloride, accurately weighed, to a 100-mL volumetric flask, dissolve in *pH 5.25 buffer*, dilute with *pH 5.25 buffer* to volume, and mix.

Procedure—Concomitantly measure the potential (see *Titrimetry* ⟨541⟩), in mV, of the *Standard preparations* and of the *Test preparation*, with a pH meter capable of a minimum reproducibility of ±0.2 mV, equipped with a glass-sleeved calomel-fluoride specific-ion electrode system. [NOTE—When taking measurements, immerse the electrodes in the solution, which has been transferred to a 150-mL beaker containing a polytetrafluoroethylene-coated stirring bar. Allow to stir on a magnetic stirrer having an insulated top until equilibrium is attained (1 to 2 minutes), and record the potential. Rinse and dry the electrodes between measurements, being careful to avoid damaging the crystal of the specific-ion electrode.] Plot the logarithm of the fluoride-ion concentrations, in μg per mL, of the *Standard preparations* versus the potential in mV. From the measured potential of the *Test preparation* and the standard curve determine the concentration, in μg per mL, of fluoride ion in the *Test preparation:* not more than 0.05% is found.

Related compounds—Transfer 1.0 g of Flurazepam Hydrochloride to a 10-mL volumetric flask, add methanol to volume, and mix. Apply 10 μL of this solution to a suitable thin-layer chromatographic plate (see *Chromatography* ⟨621⟩), coated with a 0.25-mm layer of chromatographic silica gel mixture. Apply to the same plate 10 μL of a methanol solution containing 100 μg per mL of USP Flurazepam Related Compound C RS and 10 μL of a methanol solution containing 100 μg per mL of USP Flurazepam Related Compound F RS. Allow the spots to dry, and develop the chromatogram in a paper-lined chamber that has been pre-equilibrated with developing solvent. (Discard the solvent, and add fresh developing solvent just prior to inserting the thin-layer plate.) Use a solvent system consisting of ether and diethylamine (150:4) until the solvent front has moved about 16 cm. Remove the plate from the developing chamber, mark the solvent front, and allow the solvent to evaporate for not more than 5 minutes. Replace the solvent system in the developing chamber with a freshly prepared portion of solvent having the same composition, return the plate to the chamber, and redevelop the chromatogram in the same manner and in the same direction as before. Remove the plate from the developing chamber, and allow the solvent to evaporate. Locate the spots by viewing the plate under short-wavelength ultraviolet light. Any spots from the test solution are not greater in size or intensity than the spots, occurring at the respective R_f values, produced by the Standard solutions, corresponding to not more than 0.1% of 5-chloro-2-(2-diethylaminoethylamino)-2′-fluorobenzophenone hydrochloride (flurazepam related compound C), and not more than 0.1% of 7-chloro-5-(2-fluorophenyl)-1,3-dihydro-2*H*-1,4-benzodiazepin-2-one (flurazepam related compound F).

Assay—Transfer about 600 mg of Flurazepam Hydrochloride, accurately weighed, to a 250-mL beaker, dissolve in 80 mL of glacial acetic acid, and add 20 mL of mercuric acetate TS. Titrate with 0.1 *N* perchloric acid VS, determining the end-point potentiometrically, using a calomel-glass electrode system. Perform a blank determination, and make any necessary correction. Each mL of 0.1 *N* perchloric acid is equivalent to 23.04 mg of $C_{21}H_{23}ClFN_3O.2HCl$.

Flurazepam Hydrochloride Capsules

» Flurazepam Hydrochloride Capsules contain not less than 90.0 percent and not more than 110.0 percent of the labeled amount of $C_{21}H_{23}ClFN_3O.2HCl$.

Packaging and storage—Preserve in tight, light-resistant containers.

Reference standard—*USP Flurazepam Hydrochloride Reference Standard*—Protect from light. Dry over silica gel for 4 hours before using.

Identification—

A: Dissolve a portion of Capsules, equivalent to about 30 mg of flurazepam hydrochloride, in 10 mL of methanol, filter, and proceed as directed in *Identification test C* under *Flurazepam Hydrochloride*.

B: The solution prepared for measurement of ultraviolet absorbance in the *Assay* exhibits maxima at 239 ± 2 nm, 284 ± 2 nm, and 363 ± 2 nm. The ratio A_{239}/A_{284} is between 1.95 and 2.50.

C: Capsules meet the requirements under *Identification—Organic Nitrogenous Bases* ⟨181⟩.

Dissolution ⟨711⟩—

Medium: 0.1 *N* hydrochloric acid; 900 mL.

Apparatus 1: 100 rpm.

Time: 20 minutes.

Procedure—Determine the amount of $C_{21}H_{23}ClFN_3O.2HCl$ dissolved from ultraviolet absorbances at the wavelength of the isosbestic point at about 271 nm of filtered portions of the solution under test, suitably diluted with *Dissolution Medium*, if necessary, in comparison with a Standard solution having a known concentration of USP Flurazepam Hydrochloride RS.

Tolerances—Not less than 75% (*Q*) of the labeled amount of $C_{21}H_{23}ClFN_3O.2HCl$ is dissolved in 20 minutes.

Uniformity of dosage units ⟨905⟩: meet the requirements.

Assay—Transfer, as completely as possible, the contents of not less than 20 Flurazepam Hydrochloride Capsules to a tared weighing bottle, and weigh. Mix the combined contents, and transfer an accurately weighed quantity of the powder, equivalent to about 100 mg of flurazepam hydrochloride, to a 100-mL volumetric flask. Add 50 mL of a 1 in 36 solution of sulfuric acid in methanol, shake by mechanical means for 5 minutes, dilute with a 1 in 36 solution of sulfuric acid in methanol to volume, and mix. Filter, discarding the first 20 mL of the filtrate, transfer 10.0 mL of the subsequent filtrate to a second 100-mL volumetric flask, dilute with a 1 in 36 solution of sulfuric acid in methanol to volume, and mix. Transfer 10.0 mL of this solution to a third 100-mL volumetric flask, dilute with a 1 in 36 solution of sulfuric acid in methanol to volume, and mix. Concomitantly determine the absorbances of this solution and a Standard solution of USP Flurazepam Hydrochloride RS in the same medium having a known concentration of about 10 µg per mL, in 1-cm cells, at the wavelength of maximum absorbance at about 239 nm, with a suitable spectrophotometer, using a 1 in 36 solution of sulfuric acid in methanol as the blank. Calculate the quantity, in mg, of $C_{21}H_{23}ClFN_3O.2HCl$ in the portion of Capsules taken by the formula:

$$10C(A_U/A_S),$$

in which *C* is the concentration, in µg per mL, of USP Flurazepam Hydrochloride RS in the Standard solution, and A_U and A_S are the absorbances of the solution from the Capsules and the Standard solution, respectively.

Flurbiprofen Sodium

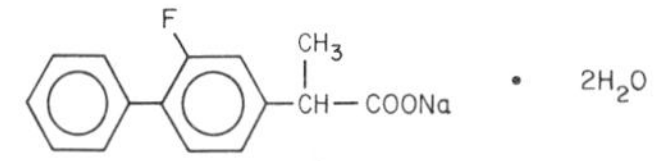

$C_{15}H_{12}FNaO_2.2H_2O$ 302.28

[1,1′-Biphenyl]-4-acetic acid, 2-fluoro-α-methyl, sodium salt dihydrate, (±)-.

Sodium (±)-2-[2-fluoro-4-biphenylyl)propionate dihydrate.

Anhydrous 266.25

» Flurbiprofen Sodium contains not less than 98.5 percent and not more than 101.5 percent of $C_{15}H_{12}FNaO_2.2H_2O$.

Packaging and storage—Preserve in well-closed containers.

Reference standards—*USP Flurbiprofen Reference Standard*—Dry in vacuum at a pressure not exceeding 5 mm of mercury over phosphorus pentoxide in a suitable drying tube at 55° for 2 hours before using. *USP Flurbiprofen Sodium Reference Standard*—Dry to constant weight in vacuum at a pressure not exceeding 1 mm of mercury over phosphorus pentoxide in a suitable drying tube at 60° before using. *USP 2-(4-Biphenylyl)Propionic Acid Reference Standard*—Do not dry before using.

Identification—

A: The infrared absorption spectrum of a mineral oil dispersion of it, previously dried, exhibits maxima only at the same wavelengths as that of a similar preparation of USP Flurbiprofen Sodium RS.

B: The ultraviolet absorption spectrum of a 1 in 100,000 solution of it in a pH 6.0 buffer consisting of 2.42 g of monobasic sodium phosphate and 0.66 g of dibasic sodium phosphate dissolved in water to make 1000 mL exhibits maxima and minima at the same wavelengths as that of a similar preparation of USP Flurbiprofen Sodium RS, concomitantly measured, and the respective absorptivities, calculated on the dried basis, at the wavelength of maximum absorbance at about 246 nm do not differ by more than 3.0%.

C: The residue obtained by igniting it responds to the tests for *Sodium* ⟨191⟩.

Specific rotation ⟨781⟩: between −0.45° and +0.45°, calculated on the dried basis, determined in a solution in methanol containing 500 mg in each 10 mL.

Loss on drying ⟨731⟩—Dry about 0.3 g of it in vacuum at a pressure not exceeding 1 mm of mercury over phosphorus pentoxide in a suitable drying tube at 60° for 18 hours: it loses not less than 11.3% and not more than 12.5% of its weight.

Heavy metals, *Method I* ⟨231⟩: 0.001%.

Assay and limit of 2-(4-biphenylyl)propionic acid—

Solvent—Mix 500 mL of methanol and 250 mL of water.

Mobile phase—Prepare a suitable mixture of acetonitrile, water, and glacial acetic acid (500:490:10), filter through a suitable filter having a porosity of 0.5 µm or less, and degas. Make adjustments if necessary (see *System Suitability* under *Chromatography* ⟨621⟩).

Standard flurbiprofen preparation—Dissolve an accurately weighed quantity of USP Flurbiprofen RS in methanol to obtain a stock solution having a known concentration of about 1 mg per mL. Transfer 5.0 mL of this solution to a 100-mL volumetric flask, dilute with *Solvent* to volume, and mix. This solution contains about 0.05 mg of flurbiprofen per mL.

Standard 2-(4-biphenylyl)propionic acid preparation—Dissolve an accurately weighed quantity of USP 2-(4-biphenylyl)Propionic Acid RS in methanol to obtain a stock solution having a known concentration of about 0.15 mg per mL. Transfer

1.0 mL of this solution to a 200-mL volumetric flask, dilute with *Solvent* to volume, and mix. This solution contains about 0.00075 mg of 2-(4-biphenylyl)propionic acid per mL.

Resolution solution—Transfer 5 mL of the stock solution used to prepare the *Standard flurbiprofen preparation* and 2 mL of the stock solution used to prepare the *Standard 2-(4-biphenylyl)propionic acid preparation* to a 100-mL volumetric flask, dilute with *Solvent* to volume, and mix.

Assay preparation—Transfer about 100 mg of Flurbiprofen Sodium, accurately weighed, to a 100-mL volumetric flask, dissolve in methanol, dilute with methanol to volume, and mix. Transfer 5.0 mL of this solution to a second 100-mL volumetric flask, dilute with *Solvent* to volume, and mix.

Chromatographic system (see *Chromatography* ⟨621⟩)—The liquid chromatograph is equipped with a 280-nm detector and a 4-mm × 30-cm column that contains packing L7. The flow rate is about 2 mL per minute. Chromatograph the *Resolution solution*, and record the peak responses as directed under *Procedure:* the resolution between the 2-(4-biphenylyl)propionic acid peak and the flurbiprofen peak is not less than 1.0. Chromatograph the *Standard flurbiprofen preparation*, and record the peak responses as directed under *Procedure:* the tailing factor for the analyte peak is not more than 2.5, and the relative standard deviation for replicate injections is not more than 1.0%.

Procedure—[NOTE—Use peak areas where peak responses are indicated.] Separately inject equal volumes (about 20 µL) of the *Standard flurbiprofen preparation*, the *Standard 2-(4-biphenylyl)propionic acid preparation*, and the *Assay preparation* into the chromatograph, record the chromatograms, and measure the responses for the major peaks. Calculate the percentage of $C_{15}H_{12}FNaO_2 . 2H_2O$ in the portion of Flurbiprofen Sodium taken by the formula:

$$(302.28/244.26)(200,000C/W)(r_U/r_S),$$

in which 302.28 and 244.26 are the molecular weights of flurbiprofen sodium dihydrate and anhydrous flurbiprofen, respectively, *C* is the concentration, in mg per mL, of USP Flurbiprofen RS in the *Standard flurbiprofen preparation*, *W* is the weight, in mg, of the portion of Flurbiprofen Sodium taken to prepare the *Assay preparation*, and r_U and r_S are the flurbiprofen peak responses obtained from the *Assay preparation* and the *Standard flurbiprofen preparation*, respectively. Calculate the percentage of 2-(4-biphenylyl)propionic acid in the portion of Flurbiprofen Sodium taken by the formula:

$$200,000(C/W)(r_U/r_S),$$

in which *C* is the concentration, in mg per mL, of USP 2-(4-Biphenylyl)Propionic Acid RS in the *Standard 2-(4-biphenylyl)propionic acid preparation*, *W* is the weight, in mg, of the portion of Flurbiprofen Sodium taken to prepare the *Assay preparation*, and r_U and r_S are the 2-(4-biphenylyl)propionic acid peak responses obtained from the *Assay preparation* and the *Standard 2-(4-biphenylyl)propionic acid preparation*, respectively: not more than 1.5% is found.

Flurbiprofen Sodium Ophthalmic Solution

» Flurbiprofen Sodium Ophthalmic Solution contains not less than 90.0 percent and not more than 110.0 percent of the labeled amount of $C_{15}H_{12}FNaO_2 . 2H_2O$.

Packaging and storage—Preserve in tight containers.

Reference standard—*USP Flurbiprofen Reference Standard*—Dry to constant weight in vacuum at a pressure not exceeding 5 mm of mercury over phosphorus pentoxide in a suitable drying tube at 55° before using.

Identification—The retention time of the major peak in the chromatogram of the *Assay preparation* corresponds to that of the major peak in the chromatogram of the *Standard preparation* as obtained in the *Assay*.

pH ⟨791⟩: between 6.0 and 7.0.

Antimicrobial preservatives—Effectiveness ⟨51⟩: meets the requirements.

Sterility—It meets the requirements under *Sterility Tests* ⟨71⟩, when tested as directed in the section, *Test Procedures Using Membrane Filtration.*

Assay—

Solvent, Mobile phase, Standard flurbiprofen preparation, Standard 2-(4-biphenylyl)propionic acid preparation, and *Resolution solution*—Proceed as directed for *Assay and limit of 2-(4-biphenylyl)propionic acid* under *Flurbiprofen Sodium.*

Assay preparation—Use the undiluted Flurbiprofen Sodium Ophthalmic Solution.

Chromatographic system—Proceed as directed for *Assay and limit of 2-(4-biphenylyl)propionic acid* under *Flurbiprofen Sodium*, using a 4-mm × 5-cm guard column that contains 5-µm packing L1.

Procedure—[NOTE—Use peak areas where peak responses are indicated.] Separately inject equal volumes (about 15 µL) of the *Standard flurbiprofen preparation* and the *Assay preparation* into the chromatograph, record the chromatograms, and measure the responses for the major peaks. Calculate the quantity of $C_{15}H_{12}FNaO_2 . 2H_2O$ in each mL of the Flurbiprofen Sodium Ophthalmic Solution taken by the formula:

$$(302.28/244.26)(C)(r_U/r_S),$$

in which 302.28 and 244.26 are molecular weights of flurbiprofen sodium dihydrate and anhydrous flurbiprofen, respectively, *C* is the concentration, in mg per mL, of USP Flurbiprofen Sodium RS in the *Standard flurbiprofen preparation*, and r_U and r_S are the peak responses obtained from the *Assay preparation* and the *Standard flurbiprofen preparation*, respectively.

Folic Acid

H
H_2N N N
N N CH_2NH CONH
O
$HOOCCH_2CH_2$---C---COOH
H

$C_{19}H_{19}N_7O_6$ 441.40

L-Glutamic acid, *N*-[4-[[(2-amino-1,4-dihydro-4-oxo-6-pteridinyl)methyl]amino]benzoyl]-.

Folic acid.

N-[*p*-[[(2-Amino-4-hydroxy-6-pteridinyl)methyl]amino]-benzoyl]-L-glutamic acid [*59-30-3*].

» Folic Acid contains not less than 95.0 percent and not more than 102.0 percent of $C_{19}H_{19}N_7O_6$, calculated on the anhydrous basis.

Packaging and storage—Preserve in well-closed, light-resistant containers.

Reference standard—*USP Folic Acid Reference Standard*—Do not dry; determine the water content at time of use.

Identification—The ultraviolet absorption spectrum of a 1 in 100,000 solution in sodium hydroxide solution (1 in 250) exhibits maxima and minima at the same wavelengths as that of a similar solution of USP Folic Acid RS, concomitantly measured. The ratio A_{256}/A_{365} for Folic Acid is between 2.80 and 3.00.

Water—Proceed as directed for *Method I* ⟨921⟩, except to stir the methanol solvent prior to and during the addition of the test specimen, and during the titration: not more than 8.5% is found.

Residue on ignition ⟨281⟩: not more than 0.3%.

Assay—

Standard preparation—Weigh accurately about 30 mg of USP Folic Acid RS, corrected for water content, and dissolve in an aqueous solvent containing 2 mL of ammonium hydroxide and 1 g of sodium perchlorate per 100 mL. Using the same solvent, adjust the volume quantitatively according to the injection size such that 5 to 20 µg of folic acid is chromatographed.

Assay preparation—Using Folic Acid in place of the Reference Standard, prepare as directed under *Standard preparation.* Adjust to the same volume as the *Standard preparation.*

Mobile phase—Add 35.1 g of sodium perchlorate, 1.40 g of monobasic potassium phosphate, 7.0 mL of 1 *N* potassium hydroxide, and 40 mL of methanol to a 1-liter volumetric flask, dilute with water to volume, and mix. Adjust with 1 *N* potassium hydroxide or phosphoric acid to a pH of 7.2. The methanol concentration may be varied to meet system suitability requirements and to provide a suitable elution time for folic acid.

Chromatographic system—Typically, a high-pressure liquid chromatograph, operated at room temperature, is fitted with a 25- to 30-cm × 4-mm stainless steel column packed with chromatographic column packing L1. The *Mobile phase* is maintained at a pressure and flow rate capable of giving the required resolution (see *System suitability test*) and a suitable elution time. An ultraviolet detector that monitors absorption at the 254-nm wavelength is used.

System suitability preparation—Prepare a solution containing about 1 mg per mL each of USP Folic Acid RS and USP Calcium Formyltetrahydrofolate Authentic Substance in an aqueous solvent containing 2 mL of ammonium hydroxide and 1 g of sodium perchlorate per 100 mL. Filter, before use, through a membrane filter (1-µm or finer porosity).

System suitability test—Chromatograph five injections of equal volume, up to 25 µL, of the *Standard preparation*, and measure the peak response as directed under *Procedure:* the relative standard deviation, calculated by the formula:

$$100 \times (\text{standard deviation}/\text{mean peak response}),$$

for the peak response does not exceed 2%. Inject a volume, up to 25 µL, of the *System suitability preparation:* the resolution factor between calcium formyltetrahydrofolate and folic acid, calculated by Equation 6 under *Chromatography* ⟨621⟩, is not less than 3.6. (For a particular column, resolution may be increased by decreasing the amount of methanol in the mobile phase.)

Procedure—Introduce equal volumes, up to 25 µL, of the *Assay preparation* and the *Standard preparation* into the high-pressure liquid chromatograph by means of a suitable sampling valve or high-pressure microsyringe. Measure the responses for the major peaks obtained, at corresponding retention times, from the *Assay preparation* and the *Standard preparation.* Calculate the quantity, in mg, of $C_{19}H_{19}N_7O_6$ in the portion of Folic Acid taken by the formula:

$$VC(P_U/P_S),$$

in which *V* is the volume, in mL, of the *Assay preparation*, *C* is the concentration, in mg per mL, of USP Folic Acid RS in the *Standard preparation*, and P_U and P_S are the peak responses of the solutions from the *Assay preparation* and the *Standard preparation*, respectively.

Folic Acid Injection

» Folic Acid Injection is a sterile solution of Folic Acid in Water for Injection prepared with the aid of Sodium Hydroxide or Sodium Carbonate. It contains not less than 95.0 percent and not more than 110.0 percent of the labeled amount of $C_{19}H_{19}N_7O_6$.

Packaging and storage—Preserve in single-dose or in multiple-dose containers, preferably of Type I glass, protected from light.

Reference standard—*USP Folic Acid Reference Standard*—Do not dry; determine the water content at time of use.

Identification—To a volume of the Injection equivalent to about 100 mg of folic acid add water to make about 25 mL. Adjust with hydrochloric acid to a pH of 3.0, cool to 5°, then filter, and wash the precipitate of folic acid with cold water until the last washing shows an absence of chloride. Then wash with acetone, and dry at 80° for 1 hour: the ultraviolet absorption spectrum of a 1 in 100,000 solution of the folic acid so obtained in sodium hydroxide solution (1 in 250) exhibits maxima and minima at the same wavelengths as that of a similar solution of USP Folic Acid RS, concomitantly measured. The ratio A_{256}/A_{365} is between 2.80 and 3.00.

pH ⟨791⟩: between 8.0 and 11.0.

Other requirements—It meets the requirements under *Injections* ⟨1⟩.

Assay—Dilute an accurately measured volume of Folic Acid Injection quantitatively and stepwise with an aqueous solvent containing 2 mL of ammonium hydroxide and 1 g of sodium perchlorate per 100 mL, to obtain a solution of such concentration that 5 µg to 20 µg of folic acid is chromatographed. Using this as the *Assay preparation*, proceed as directed for *Assay* under *Folic Acid*, and perform the necessary calculation to obtain the quantity, in mg, of $C_{19}H_{19}N_7O_6$ in each mL of the Injection.

Folic Acid Tablets

» Folic Acid Tablets contain not less than 90.0 percent and not more than 115.0 percent of the labeled amount of $C_{19}H_{19}N_7O_6$.

Packaging and storage—Preserve in well-closed containers.

Reference standard—*USP Folic Acid Reference Standard*—Do not dry; determine the water content at time of use.

Identification—Digest a quantity of powdered Tablets, equivalent to about 100 mg of folic acid, with 100 mL of sodium hydroxide solution (1 in 250), and filter. Proceed as directed in the *Identification test* under *Folic Acid Injection*, beginning with "Adjust with hydrochloric acid to a pH of 3.0."

Disintegration ⟨701⟩: 30 minutes.

Uniformity of dosage units ⟨905⟩: meet the requirements.

Assay—Weigh and finely powder not less than 20 Folic Acid Tablets. Transfer a portion of the powder, accurately weighed and equivalent to about 10 mg of folic acid, to a 50-mL volumetric flask with the aid of an aqueous solvent containing 2 mL of ammonium hydroxide and 1 g of sodium perchlorate per 100 mL. Shake gently until the folic acid has dissolved, dilute with the same solvent to volume, mix, and filter through a dry filter, rejecting the first portion of the filtrate. Dilute a portion of the clear filtrate quantitatively and stepwise with the same solvent to obtain a solution of such concentration that 5 µg to 20 µg of folic acid is chromatographed. Using this as the *Assay preparation*, proceed as directed for *Assay* under *Folic Acid*, and perform the necessary calculation to obtain the quantity, in mg, of $C_{19}H_{19}N_7O_6$ in the portion of powdered Tablets taken.

Formaldehyde Solution

CH_2O 30.03
Formaldehyde.
Formaldehyde [*50-00-0*].

» Formaldehyde Solution in bulk containers contains not less than 37.0 percent, by weight, of formaldehyde (CH_2O), with methanol added to prevent polymerization. Formaldehyde Solution in small containers (4 liters or less) contains not less than 36.5 percent, by weight, of formaldehyde (CH_2O), with methanol present to prevent polymerization.

Packaging and storage—Preserve in tight containers, preferably at a temperature not below 15°.

Labeling—The label of bulk containers of Formaldehyde Solution directs the drug repackager to demonstrate compliance with the USP *Assay* limit for formaldehyde of not less than 37.0%, by weight, immediately prior to repackaging.

Identification—

A: Dilute 2 mL with 10 mL of water in a test tube, and add 1 mL of silver-ammonia-nitrate TS: metallic silver is produced either in the form of a finely divided, gray precipitate, or as a bright, metallic mirror on the sides of the test tube.

B: Add 2 drops to 5 mL of sulfuric acid in which about 20 mg of salicylic acid has been dissolved, and warm the liquid very gently: a permanent, deep-red color appears.

Acidity—Measure 20.0 mL into a flask containing 20 mL of water, add 2 drops of bromothymol blue TS, and titrate with 0.1 *N* sodium hydroxide VS: not more than 10.0 mL of 0.1 *N* sodium hydroxide is consumed.

Assay—Transfer about 3 mL of Formaldehyde Solution to a tared flask containing 10 mL of water, insert the stopper in the flask tightly, and determine the exact weight of the Solution taken. Add 50.0 mL of 1 *N* sodium hydroxide VS, and add slowly, through a small funnel, 50 mL of hydrogen peroxide TS that has been previously neutralized to bromothymol blue TS with 1 *N* sodium hydroxide. Heat the mixture cautiously on a steam bath for 15 minutes, shaking it occasionally with a rotary motion. Allow the mixture to cool, rinse the funnel and the inner wall of the flask with water, and after allowing it to stand for 30 minutes, add 2 to 5 drops of bromothymol blue TS, and titrate the excess alkali with 1 *N* sulfuric acid VS. Perform a blank determination (see *Residual Titrations* under *Titrimetry* ⟨541⟩). Also make a correction based upon the acidity found in the test for *Acidity*. Each mL of 1 *N* sodium hydroxide is equivalent to 30.03 mg of CH_2O.

Fructose

β-D-Fructopyranose ⇌ β-D-Fructofuranose

$C_6H_{12}O_6$ 180.16

D-Fructose.

D-Fructose [*57-48-7*].

» Fructose, dried in vacuum at 70° for 4 hours, contains not less than 98.0 percent and not more than 102.0 percent of $C_6H_{12}O_6$.

Packaging and storage—Preserve in well-closed containers.

Reference standard—*USP Fructose Reference Standard*—Dry in vacuum at 70° for 4 hours before using.

Identification—The infrared absorption spectrum of a potassium bromide dispersion of it, previously dried, exhibits maxima only at the same wavelengths as that of a similar preparation of USP Fructose RS.

Color of solution—Dissolve 25 g in water to make 50 mL: the solution has no more color than a solution prepared by mixing 1.0 mL of cobaltous chloride CS, 3.0 mL of ferric chloride CS, and 2.0 mL of cupric sulfate CS with water to make 10 mL, and diluting 3 mL of this solution with water to make 50 mL. Make the comparison by viewing the solutions downward in matched color-comparison tubes against a white surface.

Acidity—Dissolve 5.0 g in 50 mL of carbon dioxide–free water, add phenolphthalein TS, and titrate with 0.02 *N* sodium hydroxide VS to a distinct pink color: not more than 0.50 mL of 0.02 *N* sodium hydroxide is required for neutralization.

Loss on drying ⟨731⟩—Dry it in vacuum at 70° for 4 hours: it loses not more than 0.5% of its weight.

Residue on ignition ⟨281⟩: not more than 0.5%.

Chloride ⟨221⟩—A 2.0-g portion shows no more chloride than corresponds to 0.50 mL of 0.020 *N* hydrochloric acid (0.018%).

Sulfate ⟨221⟩—A 2.0-g portion shows no more sulfate than corresponds to 0.50 mL of 0.020 *N* sulfuric acid (0.025%).

Arsenic, *Method II* ⟨211⟩—Prepare a *Test Preparation* by heating the acidified mixture nearly to dryness and cooling before 30% hydrogen peroxide is added. The limit is 1 ppm.

Calcium and magnesium (as calcium)—Dissolve about 20 g, accurately weighed, in 200 mL of water, and add 2 drops of hydrochloric acid, 5 mL of ammonia–ammonium chloride buffer TS, and 8 drops of eriochrome black TS. Mix, and titrate with 0.005 *M* disodium ethylenediaminetetraacetate VS to a blue endpoint. Each mL of 0.005 *M* disodium ethylenediaminetetraacetate is equivalent to 200.4 μg of Ca. Not more than 5.0 mL of 0.005 *M* disodium ethylenediaminetetraacetate is consumed (0.005% of Ca).

Heavy metals, *Method I* ⟨231⟩—Dissolve 4 g in 23 mL of water, and add 2 mL of 1 *N* acetic acid: the limit is 5 ppm.

Hydroxymethylfurfural—Transfer to a test tube 10 mL of a solution (1 in 10), add 5 mL of ether, and shake vigorously. Transfer 2 mL of the ether layer to a test tube, and add 1 mL of a 1 in 100 solution of resorcinol in hydrochloric acid: a slight pink color may develop, but no cherry-red color appears immediately.

Assay—Transfer about 10 g of Fructose, previously dried and accurately weighed, to a 100-mL volumetric flask, and dissolve in 50 mL of water. Add 0.2 mL of 6 *N* ammonium hydroxide, dilute with water to volume, and mix. After 30 minutes, determine the angular rotation in a 100-mm tube at 25° (see *Optical Rotation* ⟨781⟩). The observed rotation, in degrees, multiplied by −1.124, represents the weight, in g, of $C_6H_{12}O_6$ in the Fructose taken.

Fructose Injection

» Fructose Injection is a sterile solution of Fructose in Water for Injection. It contains not less than 95.0 percent and not more than 105.0 percent of the labeled amount of $C_6H_{12}O_6$.

Fructose Injection contains no antimicrobial agents.

Packaging and storage—Preserve in single-dose containers, preferably of Type I or Type II glass.

Labeling—The label states the total osmolar concentration in mOsmol per liter. Where the contents are less than 100 mL, or where the label states that the Injection is not for direct injection but is to be diluted before use, the label alternatively may state the total osmolar concentration in mOsmol per mL.

Reference standard—*USP Fructose Reference Standard*—Dry in vacuum at 70° for 4 hours before using.

Identification—

Silver nitrate solution—Dissolve 0.6 g of silver nitrate in 2.0 mL of water in a 100-mL volumetric flask, dilute with acetone to volume, and mix.

Sodium hydroxide solution—Dissolve 2.0 g of sodium hydroxide in 5.0 mL of water in a 100-mL volumetric flask, dilute with alcohol to volume, and mix.

Alcoholic monobasic sodium phosphate—Dissolve 4.0 g of monobasic sodium phosphate in 150 mL of water, and add, with mixing, 500 mL of alcohol. Use the entire mixture, even if there are two phases. Prepare the solution fresh daily.

Standard preparation—Transfer 25 mg, accurately weighed, of USP Fructose RS to a 10-mL volumetric flask, add water to volume, and mix.

Test preparation—Dilute an accurately measured volume of Injection with water to obtain a solution having a known concentration of about 2.5 mg of fructose per mL.

Procedure—Immerse a suitable thin-layer chromatographic plate coated with a 0.25-mm layer of chromatographic silica gel, coated side down, in *Alcoholic monobasic sodium biphosphate* for 2.0 minutes. Remove the plate from the solution, place it, coated side up, on a clean, absorbent towel, dry it in a current of warm air, and activate it at 105° for 30 minutes. Cool to room temperature, and apply 5 μL each of the *Standard preparation* and the *Test preparation*. Allow the spots to dry, and develop the chromatogram in a solvent system consisting of a mixture of acetone, *n*-butyl alcohol, and water (50:40:10). Remove the plate, allow the solvent to evaporate, and spray the plate with *Silver nitrate solution.* Allow the plate to dry for 30 seconds, then spray with *Sodium hydroxide solution:* the R_f value of the spot appearing within 3 minutes from the *Test preparation* corresponds to that obtained from the *Standard preparation.*

Pyrogen—Fructose Injection, diluted, if necessary, with Water for Injection to contain 10% of fructose, meets the requirements of the *Pyrogen Test* ⟨151⟩.

pH ⟨791⟩: between 3.0 and 6.0, determined on a portion to which 0.30 mL of a saturated potassium chloride solution has been added for each 100 mL and which previously has been diluted with water, if necessary, to a concentration of not more than 5% of fructose.

Heavy metals, *Method I* ⟨231⟩—Place a volume of Injection, equivalent to 4 g of fructose, in a porcelain dish, and evaporate to a volume of about 10 mL. Cool, and dilute with water to 25 mL: the limit is 5 ppm.

Hydroxymethylfurfural—It meets the requirements of the test for *Hydroxymethylfurfural* under *Fructose.*

Other requirements—It meets the requirements under *Injections* ⟨1⟩.

Assay—Transfer an accurately measured volume of Fructose Injection, equivalent to 5 g of fructose, to a 100-mL volumetric flask. Add 0.2 mL of 6 *N* ammonium hydroxide, dilute with water to volume, and mix. After 30 minutes determine the angular rotation (see *Optical Rotation* ⟨781⟩), and record the observed rotation, *a*, as an absolute number. Calculate the quantity, in mg, of $C_6H_{12}O_6$ in each mL of Injection taken by the formula:

$$1124a/lV,$$

in which *l* is the length, in dm, of the polarimeter tube, and *V* is the volume, in mL, of Injection taken.

Fructose and Sodium Chloride Injection

» Fructose and Sodium Chloride Injection is a sterile solution of Fructose and Sodium Chloride in Water for Injection. It contains not less than 95.0 percent and not more than 105.0 percent of the labeled amounts of $C_6H_{12}O_6$ (fructose) and of NaCl (sodium chloride).

Fructose and Sodium Chloride Injection contains no antimicrobial agents.

Packaging and storage—Preserve in single-dose containers, preferably of Type I or Type II glass.

Labeling—The label states the total osmolar concentration in mOsmol per liter. Where the contents are less than 100 mL, or where the label states that the Injection is not for direct injection but is to be diluted before use, the label alternatively may state the total osmolar concentration in mOsmol per mL.

Reference standard—*USP Fructose Reference Standard*—Dry at 70° for 4 hours before using.

Identification—It responds to the *Identification tests* under *Fructose Injection,* and to the tests for *Sodium* ⟨191⟩ and for *Chloride* ⟨191⟩.

Pyrogen—Fructose and Sodium Chloride Injection, diluted, if necessary, with Water for Injection to contain not more than 0.9% of sodium chloride and not more than 10% of fructose, meets the requirements of the *Pyrogen Test* ⟨151⟩.

pH ⟨791⟩: between 3.0 and 6.0.

Heavy metals, *Method I* ⟨231⟩—Proceed as directed in the test for *Heavy metals* under *Fructose Injection:* the limit is 5 ppm.

Hydroxymethylfurfural—It meets the requirements of the test for *Hydroxymethylfurfural* under *Fructose.*

Other requirements—It meets the requirements under *Injections* ⟨1⟩.

Assay for fructose—Transfer an accurately measured volume of Fructose and Sodium Chloride Injection, containing about 5 g of fructose, to a 100-mL volumetric flask, add 0.2 mL of 6 *N* ammonium hydroxide, dilute with water to volume, and mix. After 30 minutes determine the angular rotation (see *Optical Rotation* ⟨781⟩), and record the observed rotation, *a*, as an absolute number. Calculate the quantity, in mg, of $C_6H_{12}O_6$ in each mL of Injection taken by the formula:

$$1124a/lV,$$

in which *l* is the length, in dm, of the polarimeter tube, and *V* is the volume, in mL, of Injection taken.

Assay for sodium chloride—Transfer an accurately measured volume of Fructose and Sodium Chloride Injection, equivalent to about 90 mg of sodium chloride, to a porcelain casserole, and add 140 mL of water and 1 mL of dichlorofluorescein TS. Mix, and titrate with 0.1 *N* silver nitrate VS until the silver chloride flocculates and the mixture turns a faint pink color. Each mL of 0.1 *N* silver nitrate is equivalent to 5.844 mg of NaCl.

Basic Fuchsin

NH₂ R · HCl H₂N— —C= =NH

	R
Rosaniline	CH_3
Pararosaniline	H

Benzenamine, 4-[(4-aminophenyl)(4-imino-2,5-cyclohexadien-1-ylidene)methyl]-2-methyl-, monohydrochloride.
C.I. Basic Violet 14 monohydrochloride [*632-99-5*].

» Basic Fuchsin is a mixture of rosaniline and pararosaniline hydrochlorides. It contains the equivalent of not less than 88.0 percent of rosaniline hydrochloride ($C_{20}H_{19}N_3.HCl$), calculated on the dried basis.

Packaging and storage—Preserve in well-closed containers.

Identification—

A: To 5 mL of a solution (1 in 1000) add a few drops of hydrochloric acid: a yellow color is produced (*distinction from acid fuchsin*).

B: To 5 mL of a solution (1 in 500) add a few drops of tannic acid TS: a red precipitate is formed.

C: To 10 mL of a solution (1 in 500) add 10 mL of ammonia TS and 500 mg of zinc dust, and agitate the mixture: the solution becomes decolorized. Place a few drops of the decolorized solution on filter paper, and nearby on the same paper place a few drops of 3 *N* hydrochloric acid: a red color develops at the zone of contact.

Loss on drying ⟨731⟩—Dry it at 105° to constant weight: it loses not more than 5.0% of its weight.

Residue on ignition ⟨281⟩—Ignite 1 g with 0.5 mL of sulfuric acid: the weight of the residue is not more than 0.3%.

Alcohol-insoluble substances—Boil 1 g, accurately weighed, with 50 mL of alcohol under a reflux condenser for 15 minutes, filter through a tared filtering crucible, wash the residue on the filter with hot alcohol until the washings cease to be colored violet, and dry the crucible at 105° for 1 hour: the amount of insoluble residue is not more than 1.0%.

Arsenic, *Method II* ⟨211⟩: 8 ppm.

Lead ⟨251⟩: Place 1 g in a small Kjeldahl flask, add 5 mL of sulfuric acid, and insert a small funnel into the flask. Gently rotate the flask until the sulfuric acid has completely wetted the Basic Fuchsin, then heat with a small flame until carbonization is complete. Allow to cool, and add, in small quantities, 5 mL of nitric acid. Again heat gently until fumes of sulfur trioxide are evolved. Allow to cool, add another 5 mL of nitric acid, and heat to the evolution of sulfur trioxide. Allow to cool, add about 25 mL of water, and boil for a few minutes. Cool, neutralize with stronger ammonia water, using litmus paper as the indicator, and add 5 mL of nitric acid. Transfer the solution to a 100-mL volumetric flask, dilute to volume, and mix. A 20-mL portion of this solution contains not more than 30 ppm of lead.

Assay—Dissolve about 100 mg of Basic Fuchsin, accurately weighed, in 175 mL of water in a 500-mL closed system titration vessel fitted with a gas inlet tube, a gas outlet tube, an upright reflux condenser, and a buret. Add about 25 mL of sodium tartrate solution (30 in 100) and a polytef-coated magnetic stirring bar, and heat to boiling. Flush this titration vessel for 15 minutes with nitrogen that has been passed through two successive gas washing bottles each containing 500 mL of a mixture of water, titanium trichloride solution (20 in 100), and hydrochloric acid (400:40:40) to which about 10 mg of safranin O has been added. Continue the heating and nitrogen flow, and while stirring titrate with 0.05 *N* titanium trichloride VS to a yellow end-point. Each mL of 0.05 *N* titanium trichloride is equivalent to 3.379 mg of $C_{20}H_{19}N_3 . HCl$.

Fumaric Acid—*see* Fumaric Acid NF

Furazolidone

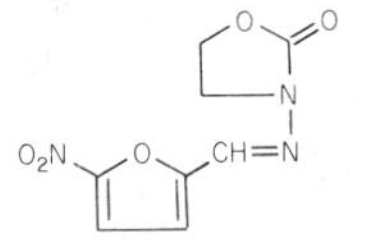

$C_8H_7N_3O_5$ 225.16
2-Oxazolidinone, 3-[[(5-nitro-2-furanyl)methylene]amino]-.
3-[(5-Nitrofurfurylidene)amino]-2-oxazolidinone
[*67-45-8*].

» Furazolidone contains not less than 97.0 percent and not more than 103.0 percent of $C_8H_7N_3O_5$, calculated on the dried basis.

Packaging and storage—Preserve in tight, light-resistant containers, and avoid exposure to direct sunlight.

Reference standard—*USP Furazolidone Reference Standard*—Dry at 100° for 1 hour before using.

Identification—

A: The infrared absorption spectrum of a potassium bromide dispersion of it, previously dried, exhibits maxima only at the same wavelengths as that of a similar preparation of USP Furazolidone RS.

B: The ultraviolet absorption spectrum of a 1 in 100,000 solution, prepared as directed in the *Assay*, exhibits maxima and minima at the same wavelengths as that of a similar solution of USP Furazolidone RS, concomitantly measured.

C: Add about 50 mg to 10 mL of a freshly prepared mixture of dimethylformamide and alcoholic potassium hydroxide TS (9:1): the solution becomes purple, immediately changes to deep blue, and, upon standing for 10 minutes, again turns purple.

Loss on drying ⟨731⟩—Dry it at 100° for 1 hour: it loses not more than 1.0% of its weight.

Residue on ignition ⟨281⟩: not more than 0.05%.

Assay—Transfer about 100 mg of Furazolidone, accurately weighed, to a 250-mL volumetric flask, dilute with dimethylformamide to volume, and mix. Transfer 5.0 mL of this solution to a 250-mL volumetric flask, dilute with water to volume, and mix (assay solution). Similarly, dissolve a suitable quantity of USP Furazolidone RS, accurately weighed, in dimethylformamide to obtain a Standard stock solution having a known concentration of about 400 µg per mL. Transfer 5.0 mL of this stock solution to a 250-mL volumetric flask, dilute with water to volume, and mix (Standard solution). Concomitantly determine the absorbances of the assay solution and the Standard solution at the wavelength of maximum absorbance at about 367 nm, with a suitable spectrophotometer, using dimethylformamide solution (1 in 50) as the blank. Calculate the quantity, in mg, of $C_8H_7N_3O_5$ in the Furazolidone taken by the formula:

$$12.5C(A_U/A_S),$$

in which *C* is the concentration, in µg per mL, of USP Furazolidone RS in the Standard solution, and A_U and A_S are the absorbances of the assay solution and the Standard solution, respectively.

Furazolidone Oral Suspension

» Furazolidone Oral Suspension is a suspension of Furazolidone in a suitable aqueous vehicle. It contains not less than 90.0 percent and not more than 110.0 percent of the labeled amount of $C_8H_7N_3O_5$.

Packaging and storage—Preserve in tight, light-resistant containers, and avoid exposure to excessive heat.

Reference standard—*USP Furazolidone Reference Standard*—Dry at 100° for 1 hour before using.

Identification—Add a quantity of Oral Suspension, equivalent to about 50 mg of furazolidone, to 10 mL of a freshly prepared mixture of dimethylformamide and alcoholic potassium hydroxide TS (9:1): the solution turns purple, immediately changes to deep blue, and, upon standing for 10 minutes, again turns purple.

pH ⟨791⟩: between 6.0 and 8.5.

Assay—Transfer an accurately measured volume of Furazolidone Oral Suspension, equivalent to about 160 mg of furazolidone, to a suitable flask. Add 5 mL of water, and mix. Transfer the mixture with the aid of dimethylformamide to a 1000-mL volumetric flask. Add about 500 mL of dimethylformamide, shake by mechanical means for 10 minutes, dilute with dimethylformamide to volume, and mix. Transfer 5.0 mL of this solution to a 100-mL volumetric flask, dilute with water to volume, and mix (assay solution). Similarly, dissolve a suitable quantity of USP Furazolidone RS, accurately weighed, in dimethylformamide to obtain a Standard stock solution having a known concentration of about 160 µg per mL. Transfer 5.0 mL of this stock solution to a 100-mL volumetric flask, dilute with water to volume, and mix (Standard solution). Concomitantly determine the absorbances of the assay solution and the Standard solution at the wavelength of maximum absorbance at about 367 nm, with a suitable spectrophotometer, using dimethylformamide solution (1 in 20) as the blank. Calculate the quantity, in mg, of $C_8H_7N_3O_5$ in each mL of the Oral Suspension taken by the formula:

$$20(C/V)(A_U/A_S),$$

in which *C* is the concentration, in µg per mL, of USP Furazolidone RS in the Standard solution, *V* is the volume, in mL, of Oral Suspension taken, and A_U and A_S are the absorbances of the assay solution and the Standard solution, respectively.

Furazolidone Tablets

» Furazolidone Tablets contain not less than 90.0 percent and not more than 110.0 percent of the labeled amount of $C_8H_7N_3O_5$.

Packaging and storage—Preserve in tight, light-resistant containers, and avoid exposure to excessive heat.

Reference standard—*USP Furazolidone Reference Standard*—Dry at 100° for 1 hour before using.

Identification—Add a quantity of powdered Tablets, equivalent to about 50 mg of furazolidone, to 10 mL of a freshly prepared mixture of dimethylformamide and alcoholic potassium hydroxide TS (9:1): the solution turns purple, immediately changes to deep blue, and, upon standing for 10 minutes, again turns purple.

Uniformity of dosage units ⟨905⟩: meet the requirements.

Assay—Weigh and finely powder not less than 20 Furazolidone Tablets. Transfer an accurately weighed portion of the powder, equivalent to about 100 mg of furazolidone, to a 250-mL volumetric flask. Add about 150 mL of dimethylformamide, warm to about 50°, and sonicate to aid in dissolving the furazolidone. Cool, dilute with dimethylformamide to volume, mix, and centrifuge a portion of the mixture. Transfer 5.0 mL of the clear solution so obtained to a 250-mL volumetric flask, dilute with water to volume, and mix (assay solution). Similarly, dissolve a suitable quantity of USP Furazolidone RS, accurately weighed, in dimethylformamide to obtain a Standard stock solution having a known concentration of about 400 µg per mL. Transfer 5.0 mL of this stock solution to a 250-mL volumetric flask, dilute with water to volume, and mix (Standard solution). Concomitantly determine the absorbances of the assay solution and the Standard solution at the wavelength of maximum absorbance at about 367 nm, with a suitable spectrophotometer, using dimethylformamide solution (1 in 50) as the blank. Calculate the quantity, in mg, of $C_8H_7N_3O_5$ in the portion of Tablets taken by the formula:

$$12.5(C/V)(A_U/A_S),$$

in which C is the concentration, in µg per mL, of USP Furazolidone RS in the Standard solution, and A_U and A_S are the absorbances of the assay solution and the Standard solution, respectively.

Furosemide

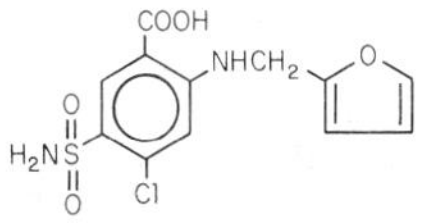

$C_{12}H_{11}ClN_2O_5S$ 330.74

Benzoic acid, 5-(aminosulfonyl)-4-chloro-2-[(2-furanylmethyl)amino]-.

4-Chloro-*N*-furfuryl-5-sulfamoylanthranilic acid [*54-31-9*].

» Furosemide contains not less than 98.0 percent and not more than 101.0 percent of $C_{12}H_{11}ClN_2O_5S$, calculated on the dried basis.

Packaging and storage—Preserve in well-closed, light-resistant containers.

Reference standards—*USP Furosemide Reference Standard*—Dry at 105° for 3 hours before using. *USP 4-Chloro-5-sulfamoylanthranilic Acid Reference Standard*—Keep container tightly closed and protected from light. Do not dry before using. *USP 2-Chloro-4-N-furfurylamino-5-sulfamoylbenzoic Acid Reference Standard*—Keep container tightly closed and protected from light. Do not dry before using.

Identification—

A: The infrared absorption spectrum of a potassium bromide dispersion of it exhibits maxima only at the same wavelengths as that of a similar preparation of USP Furosemide RS.

B: The ultraviolet absorption spectrum of a 1 in 125,000 solution in 0.02 *N* sodium hydroxide exhibits maxima and minima at the same wavelengths as that of a similar solution of USP Furosemide RS, concomitantly measured, and the respective absorptivities, calculated on the dried basis, at the wavelength of maximum absorbance at about 271 nm do not differ by more than 3.0%.

C: Dissolve about 5 mg in 10 mL of methanol. Transfer 1 mL of this solution to a flask, add 10 mL of 2.5 *N* hydrochloric acid, and reflux on a steam bath for 15 minutes. Cool, and add 15 mL of 1 *N* sodium hydroxide and 5 mL of sodium nitrite solution (1 in 1000). Allow the mixture to stand for 3 minutes, add 5 mL of ammonium sulfamate solution (1 in 200), mix, and add 5 mL of freshly prepared *N*-1-naphthylethylenediamine dihydrochloride solution (1 in 1000): a red to red-violet color is produced.

Loss on drying ⟨731⟩—Dry it at 105° for 3 hours: it loses not more than 1.0% of its weight.

Residue on ignition ⟨281⟩: not more than 0.1%.

Heavy metals, *Method II* ⟨231⟩: 0.002%.

Related compounds—[NOTE—Protect Furosemide solutions from exposure to light.]

Mobile phase—Prepare a filtered and degassed mixture of water, tetrahydrofuran, and glacial acetic acid (60:40:1). Make adjustments if necessary (see *System Suitability* under *Chromatography* ⟨621⟩).

Diluting solution—Dissolve 1.92 g of sodium 1-pentanesulfonate in 22 mL of glacial acetic acid contained in a 1000-mL volumetric flask. Dilute with a mixture of acetonitrile and water (50:50) to volume, and mix.

Standard preparation—Prepare a solution in *Diluting solution* containing 5.0 µg each of USP 4-Chloro-5-sulfamoylanthranilic Acid RS and USP 2-Chloro-4-*N*-furfurylamino-5-sulfamoylbenzoic Acid RS per mL.

Resolution solution—Dissolve suitable quantities of USP Furosemide RS and USP 2-Chloro-4-*N*-furfurylamino-5-sulfamoylbenzoic Acid RS in *Diluting solution* to obtain a solution containing about 1 mg and 5 µg per mL, respectively.

Test preparation—Transfer an accurately weighed quantity of Furosemide to a suitable volumetric flask, dissolve in *Diluting solution*, add *Diluting solution* to volume to obtain a solution having a known concentration of about 1.0 mg per mL, and mix.

Chromatographic system (see *Chromatography* ⟨621⟩)—The liquid chromatograph is equipped with a detector capable of recording at both 254 nm and 272 nm and a 4.6-mm × 25-cm column that contains packing L1. [NOTE—The 2,4-dichloro-5-sulfamoylbenzoic acid impurity does not respond at 272 nm and the 2,4-bis(furfurylamino)-5-sulfamoylbenzoic acid impurity has a very intense absorbance at 254 nm.] The flow rate is about 1.0 mL per minute. Chromatograph replicate injections of the *Resolution solution*, and record the peak responses as directed under *Procedure* [NOTE—The retention time for furosemide is about 10 minutes]: the relative standard deviation of the furosemide peak area is not more than 1.0%, and the resolution, *R*, between furosemide and 2-chloro-4-*N*-furfurylamino-5-sulfamoylbenzoic acid is not less than 1.6.

Procedure—[NOTE—Use peak areas where peak responses are indicated.] Separately inject equal volumes (about 20 µL) of the *Standard preparation* and the *Test preparation* into the chromatograph, and record the chromatograms. [NOTE—The chromatographic run time is not less than 2.5 times the retention time of the furosemide peak.] The sum of the responses at 254 nm of those peaks eluting before furosemide in the chromatogram obtained from the *Test preparation* is not more than the response at 254 nm of the 4-chloro-5-sulfamoylanthranilic acid peak in the chromatogram obtained from the *Standard preparation* (0.5%). The sum of the responses at 272 nm of those peaks eluting after furosemide in the chromatogram obtained from the *Test preparation* is not more than the response at 272 nm of the 2-chloro-4-*N*-furfurylamino-5-sulfamoylbenzoic acid peak in the chromatogram obtained from the *Standard preparation* (0.5%).

Assay—Dissolve about 600 mg of Furosemide, accurately weighed, in 50 mL of dimethylformamide to which has been added 3 drops of bromothymol blue TS, and which previously has been neutralized with 0.1 *N* sodium hydroxide. Titrate with 0.1 *N* sodium hydroxide VS to a blue end-point. Each mL of 0.1 *N* sodium hydroxide is equivalent to 33.07 mg of $C_{12}H_{11}ClN_2O_5S$.

Furosemide Injection

» Furosemide Injection is a sterile solution of Furosemide in Water for Injection prepared with the aid of Sodium Hydroxide. It contains not less than 90.0 percent and not more than 110.0 percent of the labeled amount of $C_{12}H_{11}ClN_2O_5S$.

Packaging and storage—Store in single-dose or in multiple-dose, light-resistant containers, of Type I glass.

Reference standards—*USP Furosemide Reference Standard*—Dry at 105° for 3 hours before using. *USP 4-Chloro-5-sulfamoylanthranilic Acid Reference Standard*—Keep container tightly closed and protected from light. Do not dry before using. *USP 2-Chloro-4-N-furfurylamino-5-sulfamoylbenzoic Acid Reference Standard*—Keep container tightly closed and protected from light. Do not dry before using.

Identification—Transfer to a 100-mL volumetric flask a volume of Furosemide Injection, equivalent to about 40 mg of furosemide, dilute with water to volume, and mix. Dilute 2.0 mL of this solution with 0.02 *N* sodium hydroxide in a second 100-mL volumetric flask to volume, and mix. Dissolve about 10 mg of USP Furosemide RS in 6.0 mL of 0.1 *N* sodium hydroxide in a 25-mL volumetric flask, and dilute with water to volume. Dilute 2.0 mL of the resulting solution quantitatively with 0.02 *N* sodium hydroxide to obtain a Standard solution having a concentration of about 8 µg per mL. Concomitantly determine the ultraviolet absorption spectra of both solutions: the ultraviolet absorption spectra so obtained exhibit maxima and minima at the same wavelengths.

Pyrogen—It meets the requirements of the *Pyrogen Test* ⟨151⟩, the test dose being 2.0 mg per kg.

pH ⟨791⟩: between 8.0 and 9.3.

Particulate matter ⟨788⟩: meets the requirements under *Small-volume Injections.*

4-Chloro-5-sulfamoylanthranilic acid—[NOTE—Protect furosemide solutions from exposure to light.]

Mobile phase, Diluting solution, Resolution solution, and *Chromatographic system*—Prepare as directed in the test for *Related compounds* under *Furosemide.*

Standard preparation—Prepare a solution in *Diluting solution* containing 10.0 µg of USP 4-Chloro-5-sulfamoylanthranilic Acid RS per mL.

Test preparation—Transfer an accurately measured volume of Injection, equivalent to about 10 mg of furosemide, to a 10-mL volumetric flask, add *Diluting solution* to volume, and mix.

Procedure—[NOTE—Use peak areas where peak responses are indicated.] Separately inject equal volumes (about 20 µL) of the *Standard preparation* and the *Test preparation* into the chromatograph, record the chromatograms, and measure the peak responses. The response at 254 nm obtained for any peak observed in the chromatogram of the *Test preparation* at a retention time corresponding to that of the Reference Standard in the *Standard preparation* is not greater than the response at 254 nm obtained for the peak in the chromatogram of the *Standard preparation,* corresponding to not more than 1.0% of 4-chloro-5-sulfamoylanthranilic acid.

Other requirements—It meets the requirements under *Injections* ⟨1⟩.

Assay—[NOTE—Protect furosemide solutions from exposure to light.]

Mobile phase, Diluting solution, Resolution solution, and *Chromatographic system*—Prepare as directed in the test for *Related compounds* under *Furosemide.*

Standard preparation—Dissolve an accurately weighed quantity of USP Furosemide RS in *Diluting solution* to obtain a solution having a known concentration of about 1.0 mg per mL.

Assay preparation—Transfer an accurately measured volume of Injection, equivalent to about 10 mg of furosemide, to a 10-mL volumetric flask, add *Diluting solution* to volume, and mix.

Procedure—Proceed as directed for *Procedure* in the limit test for *4-Chloro-5-sulfamoylanthranilic acid.* Calculate the quantity, in mg, of $C_{12}H_{11}ClN_2O_5S$ in each mL of the Injection taken by the formula:

$$10(C/V)(r_U/r_S),$$

in which C is the concentration, in mg per mL, of USP Furosemide RS in the *Standard preparation,* V is the volume, in mL, of Injection taken, and r_U and r_S are the peak responses obtained from the *Assay preparation* and the *Standard preparation,* respectively.

Furosemide Tablets

» Furosemide Tablets contain not less than 90.0 percent and not more than 110.0 percent of the labeled amount of $C_{12}H_{11}ClN_2O_5S$.

Packaging and storage—Preserve in well-closed, light-resistant containers.

Reference standards—*USP Furosemide Reference Standard*—Dry at 105° for 3 hours before using. *USP 4-Chloro-5-sulfamoylanthranilic Acid Reference Standard*—Keep container tightly closed and protected from light. Do not dry before using. *USP 2-Chloro-4-N-furfurylamino-5-sulfamoylbenzoic Acid Reference Standard*—Keep container tightly closed and protected from light. Do not dry before using.

Identification—Transfer a portion of finely powdered Tablets, equivalent to about 40 mg of furosemide, to a 100-mL volumetric flask. Add 25 mL of 0.1 *N* sodium hydroxide, and allow to stand for 30 minutes with occasional shaking. Dilute with water to volume, and mix. Filter the solution, discarding the first 10 mL of the filtrate, and transfer 2.0 mL to a second 100-mL volumetric flask. Add 0.02 *N* sodium hydroxide to volume, and mix. Proceed as directed in the *Identification test* under *Furosemide Injection,* beginning with "Dissolve about 10 mg of USP Furosemide RS."

Dissolution ⟨711⟩—

Medium: pH 5.8 phosphate buffer (see *Buffer Solutions* in the section, *Reagents, Indicators, and Solutions*); 900 mL.

Apparatus 2: 50 rpm.

Time: 30 minutes.

Procedure—Determine the amount of $C_{12}H_{11}ClN_2O_5S$ dissolved from ultraviolet absorbances at the isosbestic point at 274 nm of filtered portions of the solution under test, suitably diluted with pH 5.8 phosphate buffer, in comparison with a Standard solution having a known concentration of USP Furosemide RS in the same medium.

Tolerances—Not less than 65% (*Q*) of the labeled amount of $C_{12}H_{11}ClN_2O_5S$ is dissolved in 30 minutes.

Uniformity of dosage units ⟨905⟩: meet the requirements.

4-Chloro-5-sulfamoylanthranilic acid—[NOTE—Protect furosemide solutions from exposure to light.]

Mobile phase, Diluting solution, Resolution solution, and *Chromatographic system*—Prepare as directed in the test for *Related compounds* under *Furosemide.*

Standard preparation—Prepare a solution in *Diluting solution* containing 8.0 µg of USP 4-Chloro-5-sulfamoylanthranilic Acid RS per mL.

Test preparation—Transfer an accurately weighed portion of the finely powdered Tablets, equivalent to about 10 mg of furosemide, to a 10-mL volumetric flask, add *Diluting solution* to volume, and mix.

Procedure—[NOTE—Use peak areas where peak responses are indicated.] Separately inject equal volumes (about 20 µL) of the *Standard preparation* and the *Test preparation* into the chromatograph, record the chromatograms, and measure the peak responses. The response at 254 nm obtained for any peak observed in the chromatogram of the *Test preparation* at a retention time corresponding to that of the Reference Standard in the *Standard preparation* is not greater than the response at 254 nm obtained for the peak in the chromatogram of the *Standard preparation,* corresponding to not more than 0.8% of 4-chloro-5-sulfamoylanthranilic acid.

Assay—[NOTE—Protect furosemide solutions from exposure to light.]

Mobile phase, Diluting solution, Resolution solution, and *Chromatographic system*—Prepare as directed in the test for *Related compounds* under *Furosemide.*

Standard preparation—Dissolve an accurately weighed quantity of USP Furosemide RS in *Diluting solution* to obtain a solution having a known concentration of about 1.0 mg per mL.

Assay preparation—Weigh and finely powder not less than 20 Furosemide Tablets. Transfer an accurately weighed portion of the powder, equivalent to about 50 mg of furosemide, to a 50-mL volumetric flask, add 30 mL of *Diluting solution,* and sonicate for 10 minutes. Add *Diluting solution* to volume, mix, and filter, discarding the first 10 mL of the filtrate.

Procedure—Proceed as directed for *Procedure* in the limit test for *4-Chloro-5-sulfamoylanthranilic acid* under *Furosemide Injection.* Calculate the quantity, in mg, of $C_{12}H_{11}ClN_2O_5S$ in the portion of Tablets taken by the formula:

$$50C(r_U/r_S),$$

in which C is the concentration, in mg per mL of USP Furosemide RS in the *Standard preparation,* and r_U and r_S are the peak

responses obtained from the *Assay preparation* and the *Standard preparation*, respectively.

Gallamine Triethiodide

$C_{30}H_{60}I_3N_3O_3$ 891.54

Ethanaminium, 2,2′,2″-[1,2,3-benzenetriyltris(oxy)]tris-[*N,N,N*-triethyl]-, triiodide.

[*v*-Phenenyltris(oxyethylene)]tris[triethylammonium] triiodide [*65-29-2*].

» Gallamine Triethiodide contains not less than 98.0 percent and not more than 101.0 percent of $C_{30}H_{60}I_3N_3O_3$, calculated on the dried basis.

Packaging and storage—Preserve in tight containers, protected from light.

Reference standard—*USP Gallamine Triethiodide Reference Standard*—Dry at 100° for 4 hours before using.

Clarity and color of solution—A solution (1 in 50) is clear and colorless.

Identification—

A: The infrared absorption spectrum of a potassium bromide dispersion of it, previously dried, exhibits maxima only at the same wavelengths as that of a similar preparation of USP Gallamine Triethiodide RS.

B: A solution (1 in 100) responds to the tests for *Iodide* ⟨191⟩.

pH ⟨791⟩: between 5.3 and 7.0, in a solution (1 in 50).

Loss on drying ⟨731⟩—Dry it at 100° for 4 hours: it loses not more than 1.5% of its weight.

Residue on ignition ⟨281⟩: not more than 0.1%.

Heavy metals ⟨231⟩—Dissolve 1.0 g in 25 mL of water: the limit is 0.002%.

Assay—Dissolve about 200 mg of Gallamine Triethiodide, accurately weighed, in 10 mL of dimethylformamide, add 5 mL of mercuric acetate TS, 150 mL of dioxane, and 6 drops of a 1 in 100 solution of bromophenol blue indicator in dimethylformamide, and titrate with 0.1 *N* perchloric acid VS. Perform a blank determination, and make any necessary correction. Each mL of 0.1 *N* perchloric acid is equivalent to 29.72 mg of $C_{30}H_{60}I_3N_3O_3$.

Gallamine Triethiodide Injection

» Gallamine Triethiodide Injection is a sterile solution of Gallamine Triethiodide in Water for Injection. It contains not less than 95.0 percent and not more than 105.0 percent of the labeled amount of $C_{30}H_{60}I_3N_3O_3$.

Packaging and storage—Preserve in single-dose or in multiple-dose containers, preferably of Type I glass, protected from light.

Reference standard—*USP Gallamine Triethiodide Reference Standard*—Dry at 100° for 4 hours before using.

Identification—Evaporate a volume of Injection, equivalent to not less than 200 mg of gallamine triethiodide, to dryness. Take up the residue in warm alcohol, and filter through fine filter paper. Remove a portion of the filtrate, equivalent to about 100 mg of gallamine triethiodide, and evaporate to dryness: the residue responds to the *Identification tests* under *Gallamine Triethiodide*.

pH ⟨791⟩: between 6.5 and 7.5.

Other requirements—It meets the requirements under *Injections* ⟨1⟩.

Assay—Dilute an accurately measured volume of Gallamine Triethiodide Injection, equivalent to about 100 mg of gallamine triethiodide, in a 1000-mL volumetric flask with water to volume, and mix. Dissolve a suitable quantity of USP Gallamine Triethiodide RS, accurately weighed, in water to obtain a Standard solution having a known concentration of about 100 µg per mL. Pipet 5 mL of each solution into respective 125-mL separators. To each separator add 10 mL of pH 5.3 phosphate buffer (prepared by dissolving 38.0 g of monobasic sodium phosphate and 2.0 g of anhydrous dibasic sodium phosphate in 1000 mL of water) and 5 mL of a solution prepared by dissolving 200 mg of bromocresol green in 250 mL of water containing 6.4 mL of 0.1 *N* sodium hydroxide and washing the solution with five 10-mL portions of chloroform. Extract the contents of each separator with four 20-mL portions of chloroform, filtering each extract through a small pledget of cotton into a 100-mL volumetric flask. Dilute the combined extracts in each volumetric flask with chloroform to volume, and mix. Concomitantly determine the absorbances of both solutions in 1-cm cells at the wavelength of maximum absorbance at about 416 nm, with a suitable spectrophotometer, using chloroform as the blank. Calculate the quantity, in mg, of $C_{30}H_{60}I_3N_3O_3$ in each mL of the Injection taken by the formula:

$$(C/V)(A_U/A_S),$$

in which *C* is the concentration, in µg per mL, of USP Gallamine Triethiodide RS in the Standard solution, *V* is the volume, in mL, of Injection taken, and A_U and A_S are the absorbances of the solution from Gallamine Triethiodide Injection and the Standard solution, respectively.

Gallium Citrate Ga 67 Injection

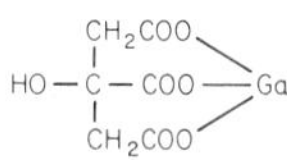

$C_6H_5{}^{67}GaO_7$

1,2,3-Propanetricarboxylic acid, 2-hydroxy-, gallium-^{67}*Ga* (1:1) salt.

Gallium-^{67}*Ga* citrate (1:1) [*41183-64-6; 52260-70-5*].

» Gallium Citrate Ga 67 Injection is a sterile aqueous solution of radioactive, essentially carrier-free, gallium citrate Ga 67 suitable for intravenous administration. It contains not less than 90.0 percent and not more than 110.0 percent of the labeled amount of ^{67}Ga as citrate expressed in megabecquerels (microcuries or millicuries) per mL at the time indicated in the labeling. Other chemical forms of radioactivity do not exceed 15.0 percent of the total radioactivity. It may contain a preservative or stabilizer.

Packaging and storage—Preserve in single-dose or in multiple-dose containers.

Labeling—Label it to include the following, in addition to the information specified for *Labeling* under *Injections* ⟨1⟩: the time and date of calibration; the amount of ^{67}Ga as labeled gallium citrate expressed as total megabecquerels (microcuries or millicuries) and concentration as megabecquerels (microcuries or millicuries) per mL at the time of calibration; the expiration date and time; and the statement, "Caution—Radioactive Material." The labeling indicates that in making dosage calibrations, correction is to be made for radioactive decay, and also indicates that the radioactive half-life of ^{67}Ga is 78.26 hours.

Reference standard—*USP Endotoxin Reference Standard.*

Bacterial endotoxins—It meets the requirements of the *Bacterial Endotoxins Test* ⟨85⟩, the limit of endotoxin content being not more than 175/*V* USP Endotoxin Unit per mL of the Injection, when compared with the USP Endotoxin RS, in which *V* is the maximum recommended total dose, in mL, at the expiration date or time.

pH ⟨791⟩: between 4.5 and 8.0.

Radiochemical purity—Place a measured volume of Injection, to provide a count rate of about 20,000 counts per minute, about 25 mm from one end of a 25-mm × 300-mm strip of chromatographic paper (see *Chromatography* ⟨621⟩); do not dry. If sample dilution is necessary, use sodium citrate solution (2 in 1000) adjusted with 3 *N* hydrochloric acid to a pH of 6.0. Develop the chromatogram over a period of 2½ hours (solvent movement of at least 10 cm) at 2° to 8° by ascending chromatography, using a solvent system consisting of a mixture of water, alcohol, and pyridine (4:2:1), adjusted with 3 *N* hydrochloric acid to a pH of 6.0 and equilibrated for not less than 18 hours at 2° to 8°, and air-dry. Determine the radioactivity distribution by scanning the chromatograph with a suitable collimated radiation detector: the radioactivity in the gallium citrate is not less than 85.0% of the total radioactivity when measured at an R_f value of approximately 0.7.

Radionuclide identification (see *Radioactivity* ⟨821⟩)—Its gamma-ray spectrum is identical to that of a specimen of ^{67}Ga of known purity that exhibits major photopeaks having energies of 93.3, 184.6, and 300.2 KeV.

Radionuclidic purity—Using a suitable counting assembly (see *Selection of a Counting Assembly* under *Radioactivity* ⟨821⟩), determine the radionuclidic purity of the Injection: not less than 99% of the total radioactivity is present as Ga 67 at the time of calibration.

Other requirements—It meets the requirements under *Injections* ⟨1⟩, except that the Injection may be distributed or dispensed prior to the completion of the test for *Sterility*, the latter test being started on the day of manufacture, and except that it is not subject to the recommendation on *Volume in Container*.

Assay for radioactivity—Using a suitable counting assembly (see *Selection of a Counting Assembly* under *Radioactivity* ⟨821⟩), determine the radioactivity in MBq (μCi or mCi) per mL of Gallium Ga 67 Injection by use of a calibrated system as directed under *Radioactivity* ⟨821⟩.

Absorbent Gauze

» Absorbent Gauze is cotton, or a mixture of cotton and not more than 53.0 percent, by weight, of rayon, and is in the form of a plain woven cloth conforming to the standards set forth herein. Absorbent Gauze that has been rendered sterile is packaged to protect it from contamination.

Note—Condition all Absorbent Gauze for not less than 4 hours in a standard atmosphere of 65 ± 2% relative humidity at 21 ± 1.1° before determining the weight, thread count, and absorbency. Remove the Absorbent Gauze from its wrappings before placing it in the conditioning atmosphere, and if it is in the form of bolts or rolls, cut the quantity necessary for the various tests from the piece, excluding the first two and the last two meters when the total quantity of Gauze available so permits.

Packaging and storage—Preserve in well-closed containers. Absorbent Gauze that has been rendered sterile is so packaged that the sterility of the contents of the package is maintained until the package is opened for use.

Labeling—Its type or thread count, length, and width, and the number of pieces contained, are stated on the container, and the designation "non-sterilized" or "not sterilized" appears prominently thereon unless the Gauze has been rendered sterile, in which case it may be labeled to indicate that it is sterile. The package label of sterile Gauze indicates that the contents may not be sterile if the package bears evidence of damage or has been previously opened.

The name of the manufacturer, packer, or distributor is stated on the package.

General characteristics—Absorbent Gauze is white cloth of various thread counts and weights. It may be supplied in various lengths and widths, and in the form of rolls or folds.

The accompanying table designates for each commercial type the thread count and weight in g per square meter.

Type	Threads per 2.54 cm		Average Count, Threads per 6.45 sq. cm	Weight,[1] g per sq. meter
	Warp	Filling		
I	41 to 47	33 to 39	76 to 84[2]	43.8 to 55.8
II	30 to 34	26 to 30	57 to 63	32.9 to 41.9
III	26 to 30	22 to 26	49 to 55	28.4 to 36.2
IV	22 to 26	18 to 22	41 to 47	24.5 to 31.1
V	20 to 24	16 to 20	37 to 43	22.5 to 28.8
VI	18 to 22	14 to 18	33 to 39	19.8 to 25.2
VII	18 to 22	8 to 14	27 to 35	18.1 to 23.1
VIII	12 to 16	8 to 12	21 to 27	12.1 to 15.5

[1] For Absorbent Gauze that contains purified rayon, increase these values by 2.5 percent.

[2] For Type I rolled gauze, the range is 75 to 85 threads per 6.45 sq. cm.

Thread count—If the dimensions of the piece permit, count the warp and filling threads of Absorbent Gauze in three separate 76.2-mm squares, not counting threads nearer any edge than one-tenth of the dimension of the fabric and not including the same threads in any two counts. For pieces not greater than 76.2 mm in either dimension, count all the threads in three different places in that dimension of the piece.

Average the three counts for the warp and filling, respectively: the average lies within the ranges tabulated under *General characteristics*.

For Absorbent Gauze packaged in rolls, count the number of warp and filling threads in areas of 1.27 cm square at 5 points evenly spread along the center line of the bandage, no point being within 30.5 cm of either end of the bandage.

Length—Unfold or unroll it, smooth it without stretching it, and measure its length along the center line: the length is not less than 98.0% of that stated on the label.

Width—Measure the width at each of the points selected for the *Thread count:* the average of the three measurements is within 1.6 mm of the width stated on the label.

Weight—Weigh a piece of gauze of stated size: the weight, expressed in terms of g per square meter, meets the requirements for weight under *General characteristics*.

Absorbency—Fold about 0.1 m² into a 10-cm section. For Absorbent Gauze packaged in rolls, use the entire roll. Hold the folded or rolled Gauze horizontally almost in contact with the surface of water at approximately 25°, and allow it to drop lightly upon the water: complete submersion takes place in not more than 30 seconds.

Sterility—Absorbent Gauze that has been rendered sterile meets the requirements under *Sterility Tests* ⟨71⟩.

Dried and ignited residue, Acid or alkali, and Dextrin or starch, in water extract—Place 20 ± 0.1 g in 500 mL of water, and boil the mixture for 15 minutes, adding boiling water as necessary to maintain the original volume. Pour the water through a funnel into a 1000-mL volumetric flask, transfer the Absorbent Gauze to the funnel, press out the excess water with a glass rod, and wash it with two 250-mL portions of boiling water, pressing the gauze after each washing. Cool the combined washings, dilute to volume, and mix. Then apply the following tests.

Dried residue—Evaporate 400 mL of the extract, filtering if necessary, in a suitable dish on a steam bath, and dry the residue at 105° to constant weight: the weight of the residue so obtained does not exceed an amount, in mg, calculated by the formula:

$$80 - 0.6C,$$

in which C is the corrected percentage of cotton (50 mg, maximum, or 0.6%).

Ignited residue—Ignite the dried residue in a muffle furnace at a dull-red heat to constant weight: the weight of the ignited residue does not exceed an amount, in mg, calculated by the formula:

$$20 - 0.14C,$$

in which C is the corrected percentage of cotton (13 mg maximum, or 0.16%).

Acid or alkali—To separate 200-mL portions of the extract add 3 drops of phenolphthalein TS and 1 drop of methyl orange TS, respectively: no pink color develops in either portion.

Dextrin or starch—To a 200-mL portion of the extract add 1 drop of iodine TS: no red, violet, or blue color develops.

Residue on ignition—Place about 5 g, accurately weighed, in a suitable dish, and moisten with 2 *N* sulfuric acid. Gently heat the mixture until it is charred, then ignite more strongly until the carbon is completely consumed: the weight of the residue corresponds to not more than the percentage of the weight of the Gauze, calculated by the formula:

$$0.002C + 0.015(100 - C),$$

in which C is the corrected percentage of cotton (0.89% maximum).

Fatty matter—Pack 10 ± 0.01 g in a continuous-extraction thimble with a tared flask, and extract with ether for 5 hours, adjusting the rate so that the ether siphons not less than four times per hour. The ether extract in the flask shows no trace of blue, green, or brownish color. Evaporate the extract to dryness, and dry at 105° to constant weight: the weight of the residue does not exceed an amount, in mg, calculated by the formula:

$$0.4C + 30,$$

in which C is the corrected percentage of cotton (70 mg maximum, or 0.7%).

Alcohol-soluble dyes—Pack 10 g in a narrow percolator, and extract slowly with alcohol until the percolate measures 50 mL: when observed downward in a column 20 cm in depth, the percolate may show a yellowish color, but neither a blue nor a green tint.

Cotton and rayon content—

Sulfuric acid solution (59.5 percent by weight)—Add sulfuric acid slowly to water until the specific gravity, determined at 20°, is between 1.4902 and 1.4956.

Procedure—Place about 500 mg of Absorbent Gauze, previously bleached and dried at 110° to constant weight and accurately weighed, in a glass-stoppered, 125-mL flask, add 50.0 mL of *Sulfuric acid solution*, and shake by mechanical means for 30 minutes. Filter the mixture through a tared sintered-glass crucible, using three 10-mL portions of *Sulfuric acid solution* to rinse the flask and applying suction each time to drain the acid. Wash the residue in the crucible with 50 mL of 2 *N* sulfuric acid, then wash it with water until the filtrate is neutral to litmus. Add 40 mL of 6 *N* ammonium hydroxide to the crucible, allow the residue to soak for 10 minutes, then apply suction to remove the liquid. Similarly wash the residue with three 50-mL portions of water, allowing the residue to soak for 15 minutes each time. Dry the residue at 105° to 110° to constant weight. Calculate C, the corrected percentage of cotton, by the formula:

$$[100(1.046J/G) - 1.6],$$

in which J is the weight, in mg, of the residue, G is the weight, in mg, of the portion of Absorbent Gauze taken, and 1.046 and 1.6 are empirical correction factors. Calculate R, the corrected percentage of rayon, by the formula:

$$100 - C.$$

Gauze Bandage—*see* Bandage, Gauze

Petrolatum Gauze

» Petrolatum Gauze is Absorbent Gauze saturated with White Petrolatum. The weight of the petrolatum in the gauze is not less than 70.0 percent and not more than 80.0 percent of the weight of petrolatum gauze. Petrolatum Gauze is sterile. It may be prepared by adding, under aseptic conditions, molten, sterile, White Petrolatum to dry, sterile, Absorbent Gauze, previously cut to size, in the ratio of 60 g of petrolatum to each 20 g of gauze.

Packaging and storage—Each Petrolatum Gauze unit is so packaged individually that the sterility of the unit is maintained until the package is opened for use.

Labeling—The package label bears a statement to the effect that the sterility of the Petrolatum Gauze cannot be guaranteed if the package bears evidence of damage or has been opened previously. The package label states the width, length, and type or thread count of the Gauze.

Sterility—It meets the requirements under *Sterility Tests* ⟨71⟩.

Other tests—The petrolatum recovered by draining in the *Assay* has the characteristics of and meets the requirements of the tests under *White Petrolatum.* The conditioned gauze obtained in the *Assay* meets the requirements of the tests for *Thread count*, *Length*, *Width*, and *Weight* under *Absorbent Gauze.*

Assay—Weigh not less than 20 units of Petrolatum Gauze, place them in a heated glass funnel, maintaining the temperature at approximately 75°, and allow the petrolatum to melt and drain from the funnel. Draining may be facilitated by pressing the gauze with a glass rod or porcelain spatula.

Wash the gauze on the funnel with successive portions of warm methyl chloroform until it is free from petrolatum, allow the residual methyl chloroform to evaporate spontaneously, condition the gauze in a standard atmosphere of 65 ± 2% relative humidity at 21 ± 1.1° for not less than 4 hours, and weigh. The difference between the weight of the gauze and that of the Petrolatum Gauze taken represents the weight of petrolatum.

Gelatin—*see* Gelatin NF

Absorbable Gelatin Film

» Absorbable Gelatin Film is Gelatin in the form of a sterile, absorbable, water-insoluble film.

Packaging and storage—Preserve in a hermetically sealed or other suitable container in such manner that the sterility of the product is maintained until the container is opened for use.

Labeling—The package bears a statement to the effect that the sterility of Absorbable Gelatin Film cannot be guaranteed if the package bears evidence of damage, or if the package has been previously opened.

Sterility—It meets the requirements under *Sterility Tests* ⟨71⟩.

Residue on ignition ⟨281⟩: not more than 2.0%.

Proteolytic digest—Place 150 mg (±5 mg) in a glass-stoppered, 150-mL flask containing 100 mL of a 1 in 100 solution of pepsin in 0.1 *N* hydrochloric acid, previously warmed to 37°. Maintain at 37 ± 1°, and agitate gently every 30 minutes until digestion is complete: the average time of three proteolytic digest determinations is between 4 and 8 hours.

Absorbable Gelatin Sponge

» Absorbable Gelatin Sponge is Gelatin in the form of a sterile, absorbable, water-insoluble sponge.

Packaging and storage—Preserve in a hermetically sealed or other suitable container in such manner that the sterility of the product is maintained until the container is opened for use.

Labeling—The package bears a statement to the effect that the sterility of Absorbable Gelatin Sponge cannot be guaranteed if the package bears evidence of damage, or if the package has been previously opened.

Sterility—It meets the requirements under *Sterility Tests* ⟨71⟩.

Residue on ignition ⟨281⟩: not more than 2.0%.

Digestibility—Place a 50-mg piece in a beaker of water. Knead gently between the fingers until thoroughly wet, and until all the air has been removed, taking care not to break the tissue. Lift from the water, and remove the excess water with absorbent paper. Place the wetted sample in a 150-mL flask that contains 100 mL of a 1 in 100 solution of pepsin in 0.1 *N* hydrochloric acid previously warmed to 37°. Maintain at a temperature of 37°, and agitate gently and continuously until digestion is complete: the average digestion time of three determinations is not more than 75 minutes.

Water absorption—Cut a portion of about 10 mg from 1 Absorbable Gelatin Sponge, weigh accurately, and place in a beaker of water. Knead gently between the fingers until thoroughly wet, and until all air has been removed, taking care not to break the tissue. Lift the portion of sponge from the water, and blot twice by pressing firmly between two pieces of absorbent paper. Drop the expressed sponge into a tared weighing bottle containing about 20 mL of water, and allow to stand for 2 minutes. Lift the sponge from the water with a suitable hooked instrument, allow to drain over the weighing bottle for 5 seconds, and discard the sponge. Again weigh the weighing bottle and water: the loss in weight represents the weight of water absorbed by the sponge. Absorbable Gelatin Sponge absorbs not less than 35 times its weight of water.

Gels—*see complete list in index*

Gemfibrozil

CH_3

$O(CH_2)_3C(CH_3)_2COOH$

CH_3

$C_{15}H_{22}O_3$ 250.34

Pentanoic acid, 5-(2,5-dimethylphenoxy)-2,2-dimethyl-.

2,2-Dimethyl-5-(2,5-xylyloxy)valeric acid [*25812-30-0*].

» Gemfibrozil contains not less than 98.0 percent and not more than 102.0 percent of $C_{15}H_{22}O_3$, calculated on the dried basis.

Packaging and storage—Preserve in tight containers.

Reference standard—*USP Gemfibrozil Reference Standard*—Dry over silica gel for 4 hours before using.

Identification—The infrared absorption spectrum of a potassium bromide dispersion of it exhibits maxima only at the same wavelengths as that of a similar preparation of USP Gemfibrozil RS.

Melting range ⟨741⟩: between 58° and 61°.

Water, *Method I* ⟨921⟩: not more than 0.25%.

Heavy metals, *Method II* ⟨231⟩: 0.002%.

Chromatographic purity—

Mobile phase—Add 10 mL of glacial acetic acid to 750 mL of methanol in a 1000-mL volumetric flask, dilute with water to volume, mix, and filter through a membrane filter.

Test solution—Transfer about 100 mg of Gemfibrozil, accurately weighed, to a 10-mL volumetric flask, dissolve in *Mobile phase*, dilute with *Mobile phase* to volume, and mix.

Diluted test solution—Transfer 2.0 mL of *Test solution* to a 100-mL volumetric flask, dilute with *Mobile phase* to volume, and mix.

Chromatographic system (see *Chromatography* ⟨621⟩)—The liquid chromatograph is equipped with a 276-nm detector and a 4.6-mm × 25-cm column that contains packing L1. The flow rate is about 1 mL per minute.

Procedure—Inject equal volumes (about 100 µL) of the *Test solution* and the *Diluted test solution* into the chromatograph by means of a suitable microsyringe or sampling valve, record the chromatograms, and measure the responses for the peaks. The total of all of the peak responses, other than the gemfibrozil peak, recorded for the *Test solution* over a time span of 1.3 × the retention time of gemfibrozil, is not greater than the response of the main peak obtained from the *Diluted test solution* (2.0%), and the response of any peak at a retention time of about 0.25 × that of gemfibrozil is not greater than one-tenth that of the main peak obtained from the *Diluted test solution* (0.2%).

Assay—

Mobile phase—Add 10 mL of glacial acetic acid to 800 mL of methanol in a 1000-mL volumetric flask, dilute with water to volume, mix, and filter through a membrane filter.

Standard preparation—Dissolve a suitable quantity of USP Gemfibrozil RS, accurately weighed, in methanol to obtain a solution having a known concentration of about 1 mg per mL. Transfer 5.0 mL of this solution to a 25.0-mL volumetric flask, dilute with *Mobile phase* to volume, and mix.

Assay preparation—Transfer about 100 mg of Gemfibrozil, accurately weighed, to a 100-mL volumetric flask, dissolve in methanol, dilute with methanol to volume, and mix. Transfer 5.0 mL of this solution to a 25.0-mL volumetric flask, dilute with *Mobile phase* to volume, and mix.

System suitability preparation—Prepare a solution in *Mobile phase* containing, in each mL, about 0.2 mg of gemfibrozil and about 0.05 mg of 2,5-xylenol.

Chromatographic system (see *Chromatography* ⟨621⟩)—The liquid chromatograph is equipped with a 276-nm detector and a 3.9-mm × 30-cm column that contains packing L1. The flow rate is about 0.8 mL per minute. Chromatograph the *Standard preparation*, and record the peak responses as directed under *Procedure:* the relative standard deviation for replicate injections is not more than 2.0%. Chromatograph about 10 µL of the *System suitability preparation:* the resolution, *R*, between gemfibrozil and 2,5-xylenol is not less than 8.0.

Procedure—Separately inject equal volumes (about 10 µL) of the *Standard preparation* and the *Assay preparation* into the chromatograph by means of a suitable microsyringe or sampling valve, record the chromatograms, and measure the responses for the major peaks. Calculate the quantity, in mg, of $C_{15}H_{22}O_3$ in the portion of Gemfibrozil taken by the formula:

$$500C(r_U/r_S),$$

in which *C* is the concentration, in µg per mL, of USP Gemfibrozil RS in the *Standard preparation*, and r_U and r_S are the peak responses obtained from the *Assay preparation* and the *Standard preparation*, respectively.

Gemfibrozil Capsules

» Gemfibrozil Capsules contain not less than 90.0 percent and not more than 110.0 percent of the labeled amount of $C_{15}H_{22}O_3$.

Packaging and storage—Preserve in tight containers.

Reference standard—*USP Gemfibrozil Reference Standard*—Dry over silica gel for 4 hours before using.

Identification—Shake a portion of Capsule contents, equivalent to about 100 mg of gemfibrozil, with 10 mL of 0.1 *N* sodium hydroxide. Filter the mixture into a 50-mL centrifuge tube, and acidify the filtrate with 3 *N* sulfuric acid to obtain a copious precipitate. Centrifuge, and discard the clear solution. Wash the precipitate with small portions of water, and allow it to air-dry: the infrared absorption spectrum of a potassium bromide dispersion of the precipitate, previously dried over silica gel for 4 hours, exhibits maxima only at the same wavelengths as that of a similar preparation of USP Gemfibrozil RS.

Dissolution ⟨711⟩—

Medium: 0.2 *M* pH 7.5 phosphate buffer prepared as directed under *Solutions* in the section, *Reagents, Indicators, and Solutions*, except that 0.8 *M* monobasic potassium phosphate solution and 0.8 *M* sodium hydroxide solution are used; 900 mL.

Apparatus 2: 50 rpm.

Time: 45 minutes.

Procedure—Determine the amount of $C_{15}H_{22}O_3$ dissolved from ultraviolet absorbances at the wavelength of maximum absorbance at about 276 nm of filtered portions of the solution under test, suitably diluted with 1 *N* sodium hydroxide, in comparison with a Standard solution having a known concentration of USP Gemfibrozil RS in the same medium.

Tolerances—Not less than 80% (*Q*) of the labeled amount of $C_{15}H_{22}O_3$ is dissolved in 45 minutes.

Uniformity of dosage units ⟨905⟩: meet the requirements.

Assay—

Mobile phase, Standard preparation, and *System suitability preparation*—Proceed as directed in the *Assay* under *Gemfibrozil.*

Assay preparation—Remove, as completely as possible, the contents of not less than 20 Gemfibrozil Capsules, weigh, and mix. Transfer an accurately weighed portion of the powder, equivalent to about 100 mg of gemfibrozil, to a 100-mL volumetric flask, add about 80 mL of methanol, and shake to dissolve. Dilute with methanol to volume, mix, and filter. Transfer 5.0 mL of this clear solution to a 25-mL volumetric flask, dilute with *Mobile phase* to volume, and mix.

Procedure—Proceed as directed for *Procedure* in the *Assay* under *Gemfibrozil.* Calculate the quantity, in mg, of $C_{15}H_{22}O_3$ in the portion of Capsules taken by the formula:

$$500C(r_U/r_S),$$

in which the terms are as defined therein.

Gentamicin Sulfate

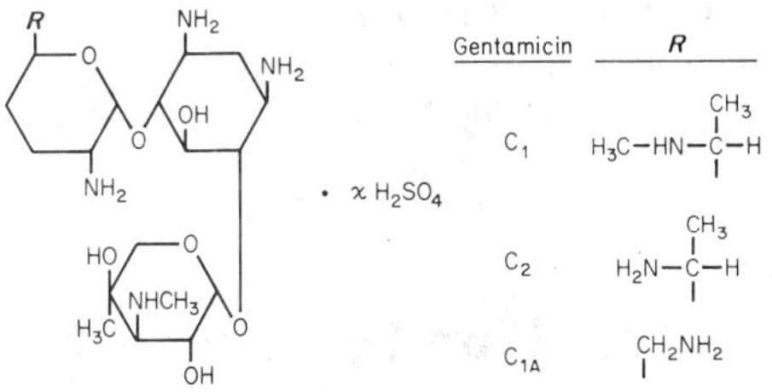

Gentamicin	R
C_1	$H_3C{-}HN{-}C(CH_3){-}H$
C_2	$H_2N{-}C(CH_3){-}H$
C_{1A}	CH_2NH_2

$\cdot\ x\ H_2SO_4$

Gentamicin sulfate (salt).
Gentamycin sulfate [*1405-41-0*].

» Gentamicin Sulfate is the sulfate salt, or a mixture of such salts, of the antibiotic substances produced by the growth of *Micromonospora purpurea.* It has a potency equivalent to not less than 590 µg of gentamicin per mg, calculated on the dried basis.

Packaging and storage—Preserve in tight containers.

Reference standard—*USP Gentamicin Sulfate Reference Standard*—Dry in vacuum at a pressure not exceeding 5 mm of mercury at 110° for 3 hours before using.

Identification—

A: The infrared absorption spectrum of a potassium bromide dispersion of it exhibits maxima only at the same wavelengths as that of a similar preparation of USP Gentamicin Sulfate RS.

B: It responds to the tests for *Sulfate* ⟨191⟩.

Specific rotation ⟨781⟩: between +107° and +121°, calculated on the dried basis, in a solution containing 10 mg per mL.

pH ⟨791⟩: between 3.5 and 5.5, in a solution (1 in 25).

Loss on drying ⟨731⟩—Dry it in vacuum at a pressure not exceeding 5 mm of mercury at 110° for 3 hours: it loses not more than 18.0% of its weight.

Residue on ignition ⟨281⟩: not more than 1.0%.

Content of gentamicins—

o-Phthalaldehyde solution—Dissolve 1.0 g of *o*-phthalaldehyde in 5 mL of methanol, and add 95 mL of 0.4 *M* boric acid, previously adjusted with 8 *N* potassium hydroxide to a pH of 10.4, and 2 mL of thioglycolic acid. Adjust the resulting solution with 8 *N* potassium hydroxide to a pH of 10.4.

Mobile phase—Mix 700 mL of methanol, 250 mL of water, and 50 mL of glacial acetic acid. Dissolve 5 g of sodium 1-heptanesulfonate in this solution. Make adjustments if necessary (see *System Suitability* under *Chromatography* ⟨621⟩).

Standard preparation—Prepare a solution of USP Gentamicin Sulfate RS in water containing about 0.65 mg per mL. Transfer 10 mL of this solution to a suitable test tube, add 5 mL of isopropyl alcohol and 4 mL of *o-Phthalaldehyde solution*, mix, and add isopropyl alcohol to obtain 25 mL of solution. Heat at 60° in a water bath for 15 minutes, and cool.

Test preparation—Using Gentamicin Sulfate, proceed as directed for *Standard preparation.*

Chromatographic system (see *Chromatography* ⟨621⟩)—The liquid chromatograph is equipped with a 330-nm detector and a 5-mm × 10-cm column that contains 5-µm packing L1. The flow rate is about 1.5 mL per minute. Chromatograph the *Standard preparation*, and record the peak responses as directed under *Procedure:* the capacity factor determined from the gentamicin C_1 peak is between 2 and 7, the column efficiency determined from the gentamicin C_2 peak is not less than 1200 theoretical plates, the resolution, *R*, between any two peaks is not less than 1.25, and the relative standard deviation for replicate injections is not more than 2.0%.

Procedure—[NOTE—Use peak heights where peak responses are indicated.] Separately inject equal volumes (about 20 µL) of the *Standard preparation* and the *Assay preparation* into the chromatograph, record the chromatograms, and measure the responses for the major peaks. The elution order is gentamicin C_1, gentamicin C_{1a}, gentamicin C_{2a}, and gentamicin C_2. Calculate the percentage contents of gentamicin C_1, gentamicin C_{1a}, gentamicin C_{2a}, and gentamicin C_2 by the formula:

$$100r_f/r_s,$$

in which r_f is the peak response corresponding to the particular gentamicin, and r_s is the sum of the responses of all four peaks: the content of gentamicin C_1 is between 25% and 50%, the content of gentamicin C_{1a} is between 10% and 35%, and the sum of the contents of gentamicin C_{2a} and gentamicin C_2 is between 25% and 55%.

Assay—Proceed with Gentamicin Sulfate as directed under *Antibiotics—Microbial Assays* ⟨81⟩.

Gentamicin Sulfate Cream

» Gentamicin Sulfate Cream contains the equivalent of not less than 90.0 percent and not more than 135.0 percent of the labeled amount of gentamicin.

Packaging and storage—Preserve in collapsible tubes or in other tight containers, and avoid exposure to excessive heat.

Reference standard—*USP Gentamicin Sulfate Reference Standard*—Dry in vacuum at a pressure not exceeding 5 mm of mercury at 110° for 3 hours before using.

Identification—Shake a quantity of Cream, equivalent to about 5 mg of gentamicin, with a mixture of 200 mL of chloroform and 5 mL of water. Allow to separate, and filter the aqueous phase: the filtrate so obtained meets the requirements of the *Identification test* under *Gentamicin Sulfate Injection.*

Minimum fill ⟨755⟩: meets the requirements.

Assay—Proceed with Gentamicin Sulfate Cream as directed in the *Assay* under *Gentamicin Sulfate Ointment.*

Gentamicin Sulfate Injection

» Gentamicin Sulfate Injection is a sterile solution of Gentamicin Sulfate in Water for Injection. It contains the equivalent of not less than 90.0 percent and not more than 125.0 percent of the labeled amount of gentamicin. It may contain suitable buffers, preservatives, and sequestering agents, unless it is intended for intrathecal use, in which case it contains only suitable tonicity agents.

Packaging and storage—Preserve in single-dose or in multiple-dose containers, preferably of Type I glass.

Reference standard—*USP Gentamicin Sulfate Reference Standard*—Dry in vacuum at a pressure not exceeding 5 mm of mercury at 110° for 3 hours before using.

Identification—On a suitable thin-layer chromatographic plate coated with a 0.25-mm layer of chromatographic silica gel having an average pore size of 6 nm (see *Chromatography* ⟨621⟩), apply separately a volume of Injection equivalent to 20 µg of gentamicin and the same volume of a similar preparation of USP Gentamicin Sulfate RS. [NOTE—Dilute the Injection with water, if necessary, to obtain a test solution containing 1000 µg of gentamicin per mL. Where the Injection contains less than 1000 µg per mL, apply a volume of it, equivalent to 20 µg of gentamicin, to the chromatographic plate, in separate portions of not more than 20 µL each, each application being allowed to dry before the next is applied.] Place the plate in a suitable chromatographic chamber, and develop the chromatogram with a solvent system consisting of the upper phase of a mixture of chloroform, dilute ammonium hydroxide (1 in 3.5), and methanol (2:1:1) until the solvent front has moved about three-fourths of the length of the plate. Remove the plate from the chamber, and air-dry. Spray the plate with a 1 in 200 solution of ninhydrin in a mixture of acetone and pyridine (1:1), and heat the plate at 105° for 2 minutes: the intensities and R_f values of the three principal spots obtained from the test solution correspond to those obtained from the Standard solution.

Pyrogen—It meets the requirements of the *Pyrogen Test* ⟨151⟩, the test dose being a volume of undiluted Injection providing the equivalent of 10 mg of gentamicin per kg, except that where the Injection is less concentrated than 1 mg per mL, the test dose is 10 mL of the Injection per kg.

pH ⟨791⟩: between 3.0 and 5.5.

Particulate matter ⟨788⟩: meets the requirements under *Small-volume Injections.*

Other requirements—It meets the requirements under *Injections* ⟨1⟩.

Assay—Proceed as directed under *Antibiotics—Microbial Assays* ⟨81⟩, using an accurately measured volume of Gentamicin Sulfate Injection diluted quantitatively and stepwise with *Buffer No. 3* to obtain a *Test Dilution* having a concentration assumed to be equal to the median dose level of the Standard (0.1 µg of gentamicin per mL).

Gentamicin Sulfate Ointment

» Gentamicin Sulfate Ointment contains the equivalent of not less than 90.0 percent and not more than 135.0 percent of the labeled amount of gentamicin.

Packaging and storage—Preserve in collapsible tubes or in other tight containers, and avoid exposure to excessive heat.

Reference standard—*USP Gentamicin Sulfate Reference Standard*—Dry in vacuum at a pressure not exceeding 5 mm of mercury at 110° for 3 hours before using.

Identification—Shake a quantity of Ointment, equivalent to about 5 mg of gentamicin, with a mixture of 200 mL of chloroform and 5 mL of water. Allow to separate, and filter the aqueous layer: the filtrate so obtained meets the requirements of the *Identification test* under *Gentamicin Sulfate Injection.*

Minimum fill ⟨755⟩: meets the requirements.

Water, *Method I* ⟨921⟩: not more than 1.0%, 20 mL of a mixture of carbon tetrachloride, chloroform, and methanol (2:2:1) being used in place of methanol in the titration vessel.

Assay—Proceed with Gentamicin Sulfate Ointment as directed under *Antibiotics—Microbial Assays* ⟨81⟩, using an accurately weighed quantity of Ointment, equivalent to about 1 mg of gentamicin, shaken with about 50 mL of ether in a separator, and extracted with four 20-mL portions of *Buffer No. 3.* Combine the aqueous extracts, and dilute quantitatively and stepwise with *Buffer No. 3* to obtain a *Test Dilution* having a concentration assumed to be equal to the median dose level of the Standard.

Gentamicin Sulfate Ophthalmic Ointment

» Gentamicin Sulfate Ophthalmic Ointment contains the equivalent of not less than 90.0 percent and not more than 135.0 percent of the labeled amount of gentamicin.

Packaging and storage—Preserve in collapsible ophthalmic ointment tubes, and avoid exposure to excessive heat.

Reference standard—*USP Gentamicin Sulfate Reference Standard*—Dry in vacuum at a pressure not exceeding 5 mm of mercury at 110° for 3 hours before using.

Identification—Shake a quantity of Ophthalmic Ointment, equivalent to about 5 mg of gentamicin, with a mixture of 200 mL of chloroform and 5 mL of water. Allow to separate, and filter the aqueous layer: the filtrate so obtained meets the requirements of the *Identification test* under *Gentamicin Sulfate Injection.*

Sterility—It meets the requirements under *Sterility Tests* ⟨71⟩.

Minimum fill ⟨755⟩: meets the requirements.

Metal particles—It meets the requirements of the test for *Metal Particles in Ophthalmic Ointments* ⟨751⟩.

Other requirements—It meets the requirements of the test for *Water* and of the *Assay* under *Gentamicin Sulfate Ointment.*

Gentamicin Sulfate Ophthalmic Solution

» Gentamicin Sulfate Ophthalmic Solution is a sterile, buffered solution of Gentamicin Sulfate with preservatives. It contains the equivalent of not less than 90.0 percent and not more than 135.0 percent of the labeled amount of gentamicin.

Packaging and storage—Preserve in tight containers, and avoid exposure to excessive heat.

Reference standard—*USP Gentamicin Sulfate Reference Standard*—Dry in vacuum at a pressure not exceeding 5 mm of mercury at 110° for 3 hours before using.

pH ⟨791⟩: between 6.5 and 7.5.

Other requirements—It meets the requirements of the *Identification test* under *Gentamicin Sulfate Injection,* and meets the requirements under *Sterility Tests* ⟨71⟩, when tested as directed in the section, *Test Procedures Using Membrane Filtration.*

Assay—Proceed with Gentamicin Sulfate Ophthalmic Solution as directed in the *Assay* under *Gentamicin Sulfate Injection.*

Sterile Gentamicin Sulfate

» Sterile Gentamicin Sulfate is Gentamicin Sulfate suitable for parenteral use. It has a potency equivalent to not less than 590 µg of gentamicin per mg, calculated on the dried basis.

Packaging and storage—Preserve in *Containers for Sterile Solids* as described under *Injections* ⟨1⟩.

Reference standard—*USP Gentamicin Sulfate Reference Standard*—Dry in vacuum at a pressure not exceeding 5 mm of mercury at 110° for 3 hours before using.

Pyrogen—It meets the requirements of the *Pyrogen Test* ⟨151⟩, the test dose being 1.0 mL per kg of a solution in pyrogen-free saline TS containing 10.0 mg of gentamicin per mL.

Sterility—It meets the requirements under *Sterility Tests* ⟨71⟩, when tested as directed in the section, *Test Procedures Using Membrane Filtration*, 6 g of specimen aseptically dissolved in 200 mL of *Fluid A* being used.

Other requirements—It conforms to the definition, responds to the *Identification test*, and meets the requirements for *Specific rotation*, *pH*, *Loss on drying*, *Content of gentamicins*, and *Assay* under *Gentamicin Sulfate*.

Gentian Violet

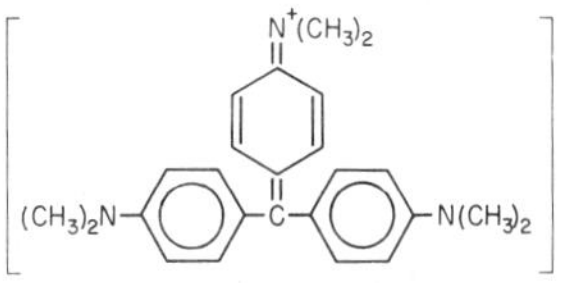

$C_{25}H_{30}ClN_3$ 407.99

Methanaminium, *N*-[4-[bis[4-(dimethylamino)phenyl]-methylene]-2,5-cyclohexadien-1-ylidene]-*N*-methyl-, chloride.

C. I. Basic violet 3.

[4-[Bis[*p*-(dimethylamino)phenyl]methylene]-2,5-cyclohexadien-1-ylidene]dimethylammonium chloride [*548-62-9*].

» Gentian Violet contains not less than 96.0 percent and not more than 100.5 percent of $C_{25}H_{30}ClN_3$, calculated on the anhydrous basis.

Packaging and storage—Preserve in well-closed containers.

Identification—

A: Sprinkle about 1 mg on 1 mL of sulfuric acid: it dissolves in the acid with an orange or brown-red color. When this solution is diluted cautiously with water, the color changes to brown, then to green, and finally to blue.

B: Dissolve about 20 mg in 10 mL of water, and add 5 drops of hydrochloric acid. To 5 mL of this solution add tannic acid TS dropwise: a deep blue precipitate is formed.

C: To the remainder of the solution prepared for *Identification test B* add about 500 mg of zinc dust, and warm the mixture: rapid decolorization occurs. Place a drop of the decolorized solution adjacent to a drop of 6 *N* ammonium hydroxide on a filter paper: a blue color is produced at the zone of contact.

Water, *Method I* ⟨921⟩: not more than 7.5%.

Residue on ignition ⟨281⟩: not more than 1.5%.

Alcohol-insoluble substances—Boil 1.0 g, accurately weighed, with 50 mL of alcohol under a reflux condenser for 15 minutes, filter through a tared filtering crucible, wash the residue on the filter with hot alcohol until the last washing is not colored violet, and dry the crucible at 105° for 1 hour: not more than 1.0% of insoluble residue remains.

Arsenic, *Method I* ⟨211⟩—Mix 300 mg with about 2.5 g each of powdered potassium nitrate and anhydrous sodium carbonate, and heat the mixture in a crucible until the organic matter is completely oxidized. Dissolve the cooled residue in 15 mL of 2 *N* sulfuric acid, and evaporate the solution by heating until copious white fumes begin to evolve. Dissolve the residue in 35 mL of water. The limit is 0.001%.

Lead—Place 1.0 g in a small Kjeldahl flask, add 5 mL of sulfuric acid, and insert a small funnel into the flask. Gently rotate the flask until the sulfuric acid has completely wetted the Gentian Violet, then heat gently until complete carbonization has taken place. Allow to cool, and add, in small quantities, 5 mL of nitric acid. Again heat gently until copious white fumes are evolved. Allow to cool, add another 5 mL of nitric acid, and again heat until white fumes are evolved. Allow to cool, cautiously add about 25 mL of water, and boil for a few minutes. After cooling, neutralize to litmus paper with ammonium hydroxide, and add 5 mL of nitric acid. Transfer the solution to a 100-mL volumetric flask, dilute with water to volume, and mix. Use 20 mL of this solution for the limit test for *Lead* ⟨251⟩. Perform a blank determination, and make any necessary correction. The limit is 0.003%.

Zinc—

Zinc stock standard solution—Transfer about 1 g of zinc, accurately weighed, to a 1000-mL volumetric flask, add 50 mL of nitric acid, and mix to dissolve. Dilute with water to volume, and mix.

Standard preparation—Dilute the *Zinc stock standard solution* with water to obtain a *Standard preparation* containing 0.50 µg of zinc per mL.

Test preparation—Weigh accurately 0.50 g of Gentian Violet in a suitable tared crucible. Place in a low-temperature plasma ashing apparatus, and ash until a constant weight is attained. Pipet 10 mL of 6 *N* nitric acid into the crucible, and heat to dissolve the ash. Transfer the solution to a 500-mL volumetric flask, dilute with water to volume, and mix. Prepare a reagent blank.

Procedure—Concomitantly determine the absorbances of the *Standard preparation*, the *Test preparation*, and the reagent blank at the zinc emission line at 213.9 nm, with a suitable atomic absorption spectrophotometer (see *Spectrophotometry and Light-scattering* ⟨851⟩) equipped with a zinc lamp and an air-acetylene flame, using water as the blank. The absorbance of the *Test preparation*, corrected for that of the reagent blank, is not greater than the absorbance of the *Standard preparation*, similarly corrected (0.05%).

Chromatographic purity—Dissolve 10 mg in 10 mL of methanol to obtain the *Test solution*. Transfer 1.0 mL of *Test solution* to a 100-mL volumetric flask, dilute with methanol to volume, and mix (*Diluted test solution*). On a suitable thin-layer chromatographic plate (see *Chromatography* ⟨621⟩), coated with a 0.25-mm layer of octadecylsilanized chromatographic silica gel, apply 5 µL each of the *Test solution* and the *Diluted test solution*. Allow the spots to dry, and develop the chromatogram in a suitable chromatographic chamber with a solvent system consisting of the upper layer separated from a well-shaken mixture of water, butyl alcohol, and glacial acetic acid (100:80:20), until the solvent front has moved about three-fourths of the length of the plate. Remove the plate from the chamber, allow the solvent to evaporate, and visually locate the spots on the plate: the *Test solution* exhibits a principal spot and not more than one secondary spot which, if present in the chromatogram from the *Test solution*, is not more intense than the principal spot obtained from the *Diluted test solution* (1.0%).

Assay—Transfer about 400 mg of Gentian Violet, accurately weighed, to a 300-mL conical flask, add 25 mL of water and 10 mL of hydrochloric acid, displace the air in the flask with carbon dioxide, and pass a stream of carbon dioxide through the flask during the assay. Add 50.0 mL of 0.1 *N* titanium trichloride VS, heat to boiling, and boil gently for 10 minutes, swirling the liquid occasionally. Cool the solution, add 5 mL of ammonium thiocyanate solution (1 in 10), and titrate with 0.1 *N* ferric ammonium sulfate VS until a faint red color is produced. Perform a blank determination (see *Residual Titrations* ⟨541⟩). Each mL of 0.1 *N* titanium trichloride is equivalent to 20.40 mg of $C_{25}H_{30}ClN_3$.

Gentian Violet Cream

» Gentian Violet Cream is Gentian Violet in a suitable cream base. It contains, in each 100 g, not less than 1.20 g and not more than 1.60 g of gentian violet, calculated as hexamethylpararosaniline chloride ($C_{25}H_{30}ClN_3$).

Packaging and storage—Preserve in collapsible tubes, or in other tight containers, and avoid exposure to excessive heat.

Reference standard—*USP Gentian Violet Reference Standard*—Do not dry; determine the water content at the time of use.

Identification—

A: Dissolve about 0.1 g in 5 mL of sulfuric acid: an orange or brown-red color is produced.

B: The visible absorption spectrum of a portion of the *Assay preparation* employed for measurement of absorbance in the *Assay* exhibits maxima at the same wavelengths as that of a similar solution of USP Gentian Violet RS, concomitantly measured.

Minimum fill ⟨755⟩: meets the requirements.

Assay—

Standard preparation—Transfer an accurately weighed quantity of USP Gentian Violet RS, equivalent to about 13.5 mg of anhydrous gentian violet, to a 100-mL volumetric flask. Add 50 mL of hydrochloric acid, and mix until dissolved. Dilute with hydrochloric acid to volume, and mix. Pipet 5 mL of the solution into a 50-mL volumetric flask, add hydrochloric acid to volume, and mix.

Assay preparation—Transfer about 500 mg of Gentian Violet Cream, accurately weighed, to a 100-mL volumetric flask. Add 15 mL of hydrochloric acid, warm in a water bath to disperse the cream thoroughly, cool, add hydrochloric acid to volume, and mix. Filter the mixture through a fine-porosity, sintered-glass filter. Pipet 2 mL of the solution into a 10-mL volumetric flask, add hydrochloric acid to volume, and mix.

Procedure—Concomitantly determine the absorbances of the *Standard preparation* and the *Assay preparation* in 1-cm cells at the wavelength of maximum absorbance at about 435 nm, with a suitable spectrophotometer, using hydrochloric acid as the blank. [NOTE—Use quartz cells having tightly fitting covers to prevent corrosion caused by the hydrochloric acid.] Calculate the quantity, in mg, of $C_{25}H_{30}ClN_3$ in the portion of Cream taken by the formula:

$$0.5C(A_U/A_S),$$

in which C is the concentration, in µg per mL, of USP Gentian Violet RS, calculated on the anhydrous basis, in the *Standard preparation*, and A_U and A_S are the absorbances of the *Assay preparation* and the *Standard preparation*, respectively.

Gentian Violet Topical Solution

» Gentian Violet Topical Solution contains, in each 100 mL, not less than 0.95 g and not more than 1.05 g of gentian violet, calculated as hexamethylpararosaniline chloride ($C_{25}H_{30}ClN_3$).

Packaging and storage—Preserve in tight containers.

Identification—A 1 in 5 dilution responds to *Identification tests B* and *C* under *Gentian Violet*.

Solution of residue in alcohol—Evaporate 10 mL on a steam bath to dryness: the residue dissolves completely in 10 mL of alcohol.

Alcohol content ⟨611⟩: between 8.0% and 10.0% of C_2H_5OH.

Assay—Pipet 25 mL of Gentian Violet Topical Solution into a 300-mL conical flask, and proceed as directed in the *Assay* under *Gentian Violet*, beginning with "add 25 mL of water."

Gentisic Acid Ethanolamide—*see* Gentisic Acid Ethanolamide NF

Glacial Acetic Acid—*see* Acetic Acid, Glacial

Glaze, Pharmaceutical—*see* Glaze, Pharmaceutical NF

Globulin—*see complete list in index*

Immune Globulin

» Immune Globulin conforms to the regulations of the federal Food and Drug Administration concerning biologics (640.100 to 640.104) (see *Biologics* ⟨1041⟩). It is a sterile, nonpyrogenic solution of globulins that contains many antibodies normally present in adult human blood, prepared by pooling approximately equal amounts of material (source blood, plasma, serum, or placentas) from not less than 1,000 donors. It contains not less than 15 g and not more than 18 g of protein per 100 mL, not less than 90.0 percent of which is gamma globulin. It contains 0.3 *M* glycine as a stabilizing agent and contains a suitable preservative. It has a potency of component antibodies of diphtheria antitoxin based on the U. S. Standard Diphtheria Antitoxin and a diphtheria test toxin, tested in guinea pigs (not less than 2 antitoxin units per mL), and antibodies for measles and poliovirus. It meets the requirements of the tests for heat stability in absence of gelation on heating, and for pH.

Packaging and storage—Preserve at a temperature between 2° and 8°.

Expiration date—The expiration date is not later than 3 years after date of issue from manufacturer's cold storage (5°, 3 years).

Labeling—Label it to state that passive immunization with Immune Globulin modifies hepatitis A, prevents or modifies measles, and provides replacement therapy in persons having hypo- or agammaglobulinemia, that it is not standardized with respect to antibody titers against hepatitis B surface antigen and that it should be used for prophylaxis of viral hepatitis type B only when the specific Immune Globulin is not available, that it may be of benefit in women who have been exposed to rubella in the first trimester of pregnancy but who would not consider a therapeutic abortion, and that it may be used in immunosuppressed patients for passive immunization against varicella if the specific Immune Globulin is not available. Label it also to state that it is not indicated for routine prophylaxis or treatment of rubella, poliomyelitis or mumps, or for allergy or asthma in patients who have normal levels of immunoglobulin, that the plasma units from which it has been derived have been tested and found non-reactive for hepatitis B surface antigen, and that it should not be administered intravenously but be given intramuscularly, preferably in the gluteal region.

Rh_o (D) Immune Globulin

» Rh_o (D) Immune Globulin conforms to the regulations of the federal Food and Drug Administration concerning biologics (see *Biologics* ⟨1041⟩). It is a sterile, nonpyrogenic solution of globulins derived from human blood plasma containing antibody to the erythrocyte factor Rh_o (D). It contains not less than 10 g and not more than 18 g of protein per 100 mL, not less than 90.0 percent of which is gamma globulin. It has a potency, determined by a suitable method, not less than that of the U. S. Reference Rh_o (D) Immune Globulin. It contains 0.3 *M* glycine as a stabilizing agent and contains a suitable preservative.

Packaging and storage—Preserve at a temperature between 2° and 8°.

Expiration date—The expiration date is not later than 6 months from the date of issue from manufacturer's cold storage, or not later than 1 year from the date of manufacture, as indicated on the label.

Anti–Human Globulin Serum

» Anti–Human Globulin Serum conforms to the regulations of the federal Food and Drug Administration concerning biologics (see *Biologics* ⟨1041⟩). It is a sterile, liquid preparation of serum produced by immunizing lower animals such as rabbits or goats with human serum or plasma, or with selected human plasma proteins. It is free from agglutinins and from hemolysins to non-sensitized human red cells of all blood groups. It contains a suitable antimicrobial preservative. Three varieties of Anti–Human Globulin Serum are recognized: (1) a general-purpose polyspecific reagent which, as a minimum, contains antibodies specific for immunoglobulin IgG, and at least the C3d component of human complement (for use in the direct antiglobulin test, it contains this Anti-C3d and Anti-IgG activity) and which may be artificially colored green; (2) a reagent containing antibodies only against immunoglobulin IgG (not heavy chain specific) intended for use in the indirect antiglobulin test, and which may be artificially colored green; and (3) reagents containing antibodies specific for individual or selected components of human complement, such as Anti-C3, and Anti-C3b-C3d-C4, or a single class of immunoglobulins, such as Anti-IgG (heavy chain specific), used only to identify plasma components coated on the surface of red blood cells. Anti–Human Globulin Serums containing Anti-IgG meet the requirements of the test for potency, in parallel with the U. S. Reference Anti–Human Globulin (Anti-IgG) Serum (at a 1:4 dilution) when tested with red cells suspended in isotonic saline sensitized with decreasing amounts of non-agglutinating Anti-D (Anti-Rh_0) serum, and with cells sensitized in the same manner with an immunoglobulin IgG Anti-Fy^a serum of similar potency. Anti–Human Globulin Serum containing one or more Anti-complement components meets the requirements of the tests for potency in giving a 2+ agglutination reaction (i.e., agglutinated cells dislodged into many small clumps of equal size) by the low-ionic sucrose or sucrose-trypsin procedures when tested as recommended in the labeling. Anti–Human Globulin Serum containing Anti-3Cd activity meets the requirements for stability, by potency testing of representative lots every 3 months during the dating period.

Packaging and storage—Preserve at a temperature between 2° and 8°.

Expiration date—Its expiration date is not later than 1 year after the date of issue from manufacturer's cold storage (5°, 1 year; or 0°, 2 years).

Labeling—Label it to state the animal source of the product. Label it also to state the specific antibody activities present; to state the application for which the reagent is intended; to include a cautionary statement that it does not contain antibodies to immunoglobulins or that it does not contain antibodies to complement components, wherever and whichever is applicable; and to state that it is for in-vitro diagnostic use. [NOTE—The lettering on the label of the general-purpose polyspecific reagent is black on a white background. The label of all other Anti–Human Globulin Serum containers is in white lettering on a black background.]

Glucagon

His - Ser - Gln - Gly - Thr - Phe - Thr - Ser - Asp - Tyr - Ser - Lys - Tyr - Leu - Asp - Ser -
1 2 3 4 5 6 7 8 9 10 11 12 13 14 15 16

Arg - Arg - Ala - Gln - Asp - Phe - Val - Gln - Trp - Leu - Met - Asn - Thr
17 18 19 20 21 22 23 24 25 26 27 28 29

$C_{153}H_{225}N_{43}O_{49}S$ 3482.78
Glucagon (pig).
Glucagon [*16941-32-5*].

» Glucagon is a polypeptide hormone, which has the property of increasing the concentration of glucose in the blood. It is obtained from porcine and bovine pancreas glands.

Packaging and storage—Preserve in tight, glass containers, under nitrogen, in a refrigerator.

Water, *Method I* ⟨921⟩: not more than 10.0%.

Residue on ignition ⟨281⟩: not more than 2.5%.

Nitrogen content, *Method II* ⟨461⟩: between 16.0% and 18.5%, calculated on the anhydrous basis.

Zinc content ⟨591⟩: not more than 0.05%.

Glucagon for Injection

» Glucagon for Injection is a mixture of the hydrochloride of Glucagon with one or more suitable, dry diluents. It contains not less than 80.0 percent and not more than 125.0 percent of the labeled amount of glucagon.

Packaging and storage—Preserve in *Containers for Sterile Solids* as described under *Injections* ⟨1⟩. Preserve the accompanying solvent in single-dose or in multiple-dose containers, preferably of Type I glass.

Reference standard—*USP Glucagon Reference Standard*—Store unopened ampuls in a freezer (−20°). For use, dissolve the contents in special diluent, as directed.

Constituted solution—At the time of use, the constituted solution prepared from Glucagon for Injection meets the requirements for *Constituted Solutions* under *Injections* ⟨1⟩.

pH and Clarity of solution—Dissolve it in the solvent and in the concentration recommended in the labeling: the pH of the solution is between 2.5 and 3.0, and the solution is clear.

Other requirements—Both Glucagon for Injection and the accompanying solvent meet the requirements for *Sterility Tests* ⟨71⟩ and *Labeling* under *Injections* ⟨1⟩. Glucagon for Injection meets the requirements for *Uniformity of dosage units* ⟨905⟩.

Assay—

Special diluent—To 1.6 mL of glycerin and 0.2 g of phenol in a 100-mL volumetric flask add recently distilled water to volume, and mix. Adjust to a pH of 2.6 ± 0.1 by the addition of 3 *N* hydrochloric acid.

Standard preparation—Dissolve the contents of 1 ampul of USP Glucagon RS in sufficient *Special diluent* to make 10.0 mL, accurately measured. On the day of the assay, dilute an accurately measured portion of this solution with Water for Injection to obtain the *Standard preparation*, which has a concentration of 0.0005 USP Unit per mL.

Assay preparation—Dilute the contents of not less than 1 vial of Glucagon for Injection with *Special diluent* to a concentration of about 1.0 USP Glucagon Unit per mL. Dilute an accurately measured portion of this solution with Water for Injection to obtain the *Assay preparation*, which has a concentration of 0.0005 USP Unit per mL.

The animal—Select 8 adult domestic cats, each weighing between 2 kg and 4 kg, that are free from gross evidence of disease, and, if female, not in advanced pregnancy. At 16 hours prior to placing the animals under anesthesia, withdraw food, but not water, from each animal, and inject into each, intraperitoneally, 10 mL of a sterile solution of dextrose (1 in 4). On the following day, inject each cat with a suitable dose of a long-acting barbiturate, and when anesthesia has set in, immobilize the animal and expose both of its femoral veins. Cover the animal to maintain its body temperature. Prepare to inject the doses into one femoral vein, and remove the blood samples from the vein on the opposite side.

Procedure—Into each cat inject a dose of the *Standard preparation* equivalent to 0.000025 Unit per kg of body weight. One hour later, withdraw an initial blood sample and assign each animal at random to a different one of the dose regimens in the accompanying table, S_1 and S_2 being 0.00005- and 0.0001-Unit-per-kg doses of the *Standard preparation*, and U_1 and U_2 the corresponding doses of the *Assay preparation*. Inject the doses indicated by the assigned regimen and withdraw blood samples at 12 minutes and 24 minutes, respectively, after each dose. Allow the animal to rest for 66 minutes, and inject the next dose, taking blood samples just prior to and then 12 minutes and 24 minutes following the dose. Repeat the procedure until all four doses have been given and a total of 12 blood samples have been taken. Determine the blood-sugar concentration in each sample as directed in the third and fourth paragraphs under *Procedure* in the *Insulin Assay* ⟨121⟩, beginning with "Determine the dextrose (glucose) content."

Dose No.	*Dose Regimen for Two 4 × 4 Latin Squares*							
	1	2	3	4	5	6	7	8
1	S_2	S_1	U_1	U_2	S_2	S_1	U_2	U_1
2	U_1	U_2	S_2	S_1	U_2	U_1	S_2	S_1
3	S_1	S_2	U_2	U_1	S_1	S_2	U_1	U_2
4	U_2	U_1	S_1	S_2	U_1	U_2	S_1	S_2

Calculation—For each cat, compute the observed maximum increase in blood sugar y following each test dose by subtracting the corresponding initial level from the higher of the two readings that follow the injection. From the four values of y so obtained for each cat, total separately the eight responses y to each of the four doses U_1, U_2, S_1, and S_2. From the totals of the responses (T_t) corresponding to these four doses or treatments, calculate the differences T_a and T_b, where T_a is the difference in the T_t's for $U_1 + U_2 - S_1 - S_2$ and T_b is the difference for $U_2 - U_1 + S_2 - S_1$. The logarithm of the relative potency of the *Assay preparation* is $M' = iT_a/T_b$. The potency of Glucagon for Injection, in USP Glucagon Units per vial, is R(antilog M'), where R is the assumed potency, in USP Glucagon Units, in the vial.

Determine the log confidence interval (L) of the log potency M from this assay with $n' = 2$ Latin squares (see *Glucagon Injection* under *The Confidence Interval and Limits of Potency* in *Design and Analysis of Biological Assays* ⟨111⟩). If the confidence interval (L) exceeds 0.1938, which corresponds at $P = 0.95$ to confidence limits of 80 and 125% of the computed potency, repeat the assay with one or more additional groups of 4 cats each, assigned to the dose regimen 1 to 4 or 5 to 8, and recompute the potency and the confidence interval (L) from the combined data for the $n' = 3$ or more Latin squares until the confidence interval (L) of the combined assay meets this limit.

Glucose, Liquid—*see* Glucose, Liquid NF

Glucose Enzymatic Test Strip

» Glucose Enzymatic Test Strip consists of the enzymes glucose oxidase and horseradish peroxidase, a suitable substrate for the reaction of hydrogen peroxide catalyzed by peroxidase, and other inactive ingredients impregnated and dried on filter paper. When tested in human urine containing known glucose concentrations, it reacts in the specified times to produce colors corresponding to the color chart provided.

Packaging and storage—Preserve in the original container, in a dry place, at controlled room temperature.

Identification—Remove the Test Strips from the container, and test them as directed in the instructions provided by the manufacturer, first in dextrose solution (1 in 50) and then in sucrose solution (1 in 50): color develops from the dextrose solution, but not from the sucrose solution.

Calibration—

Glucose standard solutions—Dissolve anhydrous dextrose in separate portions of freshly voided, normal, glucose-free human urine, previously adjusted to a pH of 6.0 with dilute formic acid (1 in 5), and with sodium hydroxide solution (1 in 2), respectively, to obtain separate solutions of the final concentrations corresponding to the color chart calibrations provided. Allow the standard solutions to stand for 1 hour prior to use.

Procedure—Remove the Test Strips from the container, and test each *Glucose standard solution* as directed in the instructions provided by the manufacturer: the colors formed on each of the Test Strips during the specified times match the colors on the color chart provided.

Glutaral Concentrate

$OCH(CH_2)_3CHO$

$C_5H_8O_2$ 100.12
Pentanedial.
Glutaraldehyde [*111-30-8*].

» Glutaral Concentrate is a solution of glutaraldehyde in Purified Water. It contains not less than 100.0 percent and not more than 104.0 percent of the labeled amount, the labeled amount being 50.0 g of $C_5H_8O_2$ per 100 g of Concentrate.

Packaging and storage—Preserve in tight containers, protected from light, and avoid exposure to excessive heat.

Clarity of solution—Transfer 5.0 mL of Glutaral Concentrate to a glass-stoppered, 100-mL graduated cylinder, add water to obtain 100 mL of mixture, insert the stopper, and mix by inverting the graduated cylinder several times. Allow the bubbles to rise, and view downward through the solution against a dark background: the solution is clear.

Identification—

2,4-Dinitrophenylhydrazine reagent—Add 4 mL of sulfuric acid to 0.8 g of 2,4-dinitrophenylhydrazine, then add 6 mL of water, dropwise, with swirling. When dissolution is essentially complete, add 20 mL of alcohol, mix, and filter. The filtrate is the *2,4-Dinitrophenylhydrazine reagent*.

Procedure—Add 0.4 mL of Glutaral Concentrate to 20 mL of *2,4-Dinitrophenylhydrazine reagent*, mix by swirling, and allow to stand for 5 minutes. Collect the precipitate on a filter, and rinse thoroughly with alcohol. Dissolve the precipitate in 20 mL of hot ethylene dichloride, filter, and cool the filtrate in an ice bath until crystallization occurs. Collect the precipitate on a filter. Redissolve the precipitate by refluxing with 30 mL of acetone, filter, and cool the filtrate in an ice bath until crystallization occurs. Collect the precipitate on a filter: the 2,4-dinitrophenylhydrazine so obtained melts between 185° and 195°, within a 3° range (see *Melting Range or Temperature* ⟨741⟩).

Specific gravity ⟨841⟩: between 1.128 and 1.135 at 20°/20°.

Acidity—Transfer 60.0 g to a conical flask, add phenolphthalein TS, and titrate with 0.10 *N* alcoholic potassium hydroxide to a pink end-point that is permanent for not less than 15 seconds: not more than 40 mL of 0.10 *N* potassium hydroxide is consumed, corresponding to not more than 0.4% (w/w) of acid, calculated as acetic acid.

pH ⟨791⟩: between 3.7 and 4.5.

Heavy metals ⟨231⟩: 0.001%.

Assay—

Hydroxylamine hydrochloride solution—Dissolve 35 g of hydroxylamine hydrochloride in 150 mL of water in a 1000-mL volumetric flask, add isopropyl alcohol to volume, and mix.

Triethanolamine solution—Transfer 65 mL of triethanolamine to a glass-stoppered, 1000-mL volumetric flask, add water to volume, and mix.

Procedure—To 500 mL of *Hydroxylamine hydrochloride solution* add 15 mL of a solution of bromophenol blue in alcohol (1 in 2500), and add *Triethanolamine solution* from a buret to obtain a neutralized solution that appears greenish blue by transmitted light. Transfer 65.0 mL of the neutralized solution to a glass-stoppered, 500-mL conical flask, add 50.0 mL of *Triethanolamine solution*, purge with nitrogen, and insert the stopper. Add about 1.2 g of Glutaral Concentrate, accurately weighed, by means of a suitable weighing pipet, insert the stopper, and allow to stand for 60 minutes, swirling the flask occasionally. Titrate with 0.5 *N* sulfuric acid VS to a greenish blue end-point, and perform a blank determination (see *Residual Titrations* under *Titrimetry* ⟨541⟩). Calculate the percentage, by weight, of $C_5H_8O_2$ in the Glutaral Concentrate taken by the formula:

$$[N(B - A)(0.05006)/W]100,$$

in which *N* is the normality of the sulfuric acid, *B* and *A* are the volumes, in mL, of 0.5 *N* sulfuric acid VS consumed by the blank and specimen solution, respectively, 0.05006 is the milliequivalent weight, in g per milliequivalent, of glutaraldehyde, and *W* is the weight, in g, of Glutaral Concentrate taken.

Glutaral Disinfectant Solution—*see* Glutaral Disinfectant Solution NF

Glutethimide

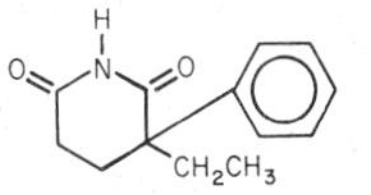

$C_{13}H_{15}NO_2$ 217.27

2,6-Piperidinedione, 3-ethyl-3-phenyl-.

2-Ethyl-2-phenylglutarimide [*77-21-4*].

» Glutethimide, dried over phosphorus pentoxide at 45° to constant weight, contains not less than 98.0 percent and not more than 102.0 percent of $C_{13}H_{15}NO_2$.

Packaging and storage—Preserve in well-closed containers.

Reference standard—*USP Glutethimide Reference Standard*—Dry over phosphorus pentoxide at 45° to constant weight before using.

Identification—

A: The infrared absorption spectrum of a potassium bromide dispersion of it, previously dried, exhibits maxima only at the same wavelengths as that of a similar preparation of USP Glutethimide RS.

B: The ultraviolet absorption spectrum of a 1 in 4000 solution in methanol exhibits maxima and minima at the same wavelengths as that of a similar solution of USP Glutethimide RS, concomitantly measured.

C: The R_f value of the principal spot in the chromatogram of the *Identification preparation* corresponds to that of the *Standard preparation A*, as obtained in the test for *Chromatographic purity*.

Melting range, *Class I* ⟨741⟩: between 86° and 89°.

Loss on drying ⟨731⟩—Dry it over phosphorus pentoxide at 45° to constant weight: it loses not more than 1.0% of its weight.

Residue on ignition ⟨281⟩: not more than 0.1%.

Chromatographic purity—

Standard preparations—Dissolve USP Glutethimide RS in methanol, dilute quantitatively with methanol, and mix to obtain a solution having a known concentration of 1.0 mg per mL. Dilute quantitatively with methanol to obtain *Standard preparations*, designated below by letter, having the following compositions:

Standard preparation	Dilution	Concentration (μg RS per mL)	Percentage (%, for comparison with test specimen)
A	(1 in 2)	500	0.5
B	(2 in 5)	400	0.4
C	(3 in 10)	300	0.3
D	(1 in 5)	200	0.2
E	(1 in 10)	100	0.1

Test preparation—Dissolve an accurately weighed quantity of Glutethimide in methanol to obtain a solution containing 100 mg per mL.

Identification preparation—Dilute a portion of the *Test preparation* quantitatively with methanol to obtain a solution containing 500 μg per mL.

Detection reagent—Prepare (1) a solution of 0.5 g of potassium iodide in 50 mL of water, and (2) a solution of 1.5 g of soluble starch in 50 mL of hot water. Mix 10 mL of each solution with 4 mL of alcohol. [NOTE—The *Detection reagent* so obtained may be used for up to 3 or 4 days.]

Developing solvent systems—Use cyclohexane as *Developing solvent A*. Prepare a mixture of ethyl acetate, methanol, and water (36:2:2) as *Developing solvent B*.

Procedure—On a suitable thin-layer chromatographic plate (see *Chromatography* ⟨621⟩), coated with a 0.25-mm layer of chromatographic silica gel, apply separately 2 μL of the *Test preparation*, 2 μL of each *Standard preparation*, and 2 μL of the *Identification preparation*, and allow the spots to dry. Place *Developing solvent A* and *Developing solvent B* in separate troughs of a double-trough chromatographic chamber, and allow the chamber to equilibrate. Position the plate in *Developing solvent A* in the chromatographic chamber, and develop the chromatograms until the solvent front has moved 12 cm above the line of application. Remove the plate from the chamber, mark the solvent front, and air-dry the plate at room temperature for 10 minutes. Position the plate in *Developing solvent B* in the chromatographic chamber, and develop the chromatograms until the solvent front has moved 12 cm above the line of application. Remove the plate from the chamber, mark the solvent front, and dry the plate at 100° to 110° for 10 minutes. Expose the plate to chlorine gas for 1 minute, and air-dry the plate at room temperature for 2 minutes. Spray the plate with *Detection reagent*. Compare the intensities of any secondary spots observed in the chromatogram of the *Test preparation* with those of the principal spots in the chromatograms of the *Standard preparations:* no secondary spot observed in the chromatogram of the *Test preparation* is larger or more intense than the principal spot in the chromatogram of *Standard preparation E* (0.1%), and the sum of the intensities of all secondary spots obtained from the *Test preparation* corresponds to not more than 0.5%.

Assay—

pH 4.0 acetate buffer—Dissolve 0.82 g of anhydrous sodium acetate in 800 mL of water, adjust with glacial acetic acid to a pH of 4.0, dilute with water to 1000 mL, and mix.

Mobile phase—Prepare a filtered and degassed mixture of *pH 4.0 acetate buffer* and acetonitrile (3:2). Make adjustments if necessary (see *System Suitability* under *Chromatography* ⟨621⟩).

Standard preparation—Dissolve an accurately weighed quantity of USP Glutethimide RS in *Mobile phase*, and dilute quantitatively and stepwise, if necessary, with *Mobile phase* to obtain a solution having a known concentration of about 1 mg per mL.

Assay preparation—Transfer about 100 mg of Glutethimide, accurately weighed, to a 100-mL volumetric flask, dissolve in *Mobile phase*, and dilute with *Mobile phase* to volume, and mix.

Chromatographic system (see *Chromatography* ⟨621⟩)—The liquid chromatograph is equipped with a 254-nm detector and a 4.6-mm × 25-cm column that contains packing L1. The flow rate is about 1 mL per minute. Chromatograph the *Standard preparation*, and record the peak responses as directed under *Procedure:* the column efficiency determined from the analyte peak is not less than 3000 theoretical plates, the tailing factor for the analyte peak is not more than 2.0, and the relative standard deviation for replicate injections is not more than 1.0%.

Procedure—Separately inject equal volumes (about 10 µL) of the *Standard preparation* and the *Assay preparation* into the chromatograph, record the chromatograms, and measure the responses for the major peaks. Calculate the quantity, in mg, of $C_{13}H_{15}NO_2$ in the portion of Glutethimide taken by the formula:

$$100C(r_U/r_S),$$

in which C is the concentration, in mg per mL, of USP Glutethimide RS in the *Standard preparation*, and r_U and r_S are the peak responses obtained from the *Assay preparation* and the *Standard preparation*, respectively.

Glutethimide Capsules

» Glutethimide Capsules contain not less than 95.0 percent and not more than 105.0 percent of the labeled amount of $C_{13}H_{15}NO_2$.

Packaging and storage—Preserve in well-closed containers.

Reference standard—*USP Glutethimide Reference Standard*—Dry over phosphorus pentoxide at 45° to constant weight before using.

Identification—Evaporate a portion of the eluate obtained in the *Assay* to dryness: the residue so obtained, previously dried, responds to *Identification test A* under *Glutethimide.*

Dissolution ⟨711⟩—

Medium: water; 900 mL.

Apparatus 1: 100 rpm.

Time: 45 minutes.

Procedure—Determine the amount of $C_{13}H_{15}NO_2$ dissolved from ultraviolet absorbances at the wavelength of maximum absorbance at about 257 nm of filtered portions of the solution under test, suitably diluted with *Dissolution Medium*, if necessary, in comparison with a Standard solution having a known concentration of USP Glutethimide RS in the same medium.

Tolerances—Not less than 75% (Q) of the labeled amount of $C_{13}H_{15}NO_2$ is dissolved in 45 minutes.

Uniformity of dosage units ⟨905⟩: meet the requirements.

Assay—

Chromatographic column—Proceed as directed for *Column Partition Chromatography* under *Chromatography* ⟨621⟩, packing a chromatographic tube with two segments of packing material. The lower segment is a mixture of 2 g of *Solid Support* and 2.0 mL of 4 *N* sulfuric acid, and the upper segment is a mixture of 2 g of *Solid Support* and 2.0 mL of sodium bicarbonate solution (1 in 12). Wash the column with 50 mL of ether (saturated with water) followed by 25 mL of chloroform (saturated with water).

Procedure—Remove, as completely as possible, the contents of not less than 20 Glutethimide Capsules, and weigh. Mix the combined contents, and transfer an accurately weighed quantity of the powder, equivalent to about 250 mg of glutethimide, to a 100-mL volumetric flask, add chloroform (saturated with water) to volume, and mix. Transfer 10.0 mL of this solution to the column, and elute with 90 mL of chloroform (saturated with water), collecting the eluate in a 100-mL volumetric flask. Dilute with chloroform (saturated with water) to volume, and mix. Concomitantly determine the absorbances of this solution and of a Standard solution of USP Glutethimide RS in the same medium having a known concentration of about 250 µg per mL in 1-cm cells at the wavelength of maximum absorbance at about 257 nm, with a suitable spectrophotometer, using chloroform (saturated with water) as the blank. Calculate the quantity, in mg, of $C_{13}H_{15}NO_2$ in the portion of Capsules taken by the formula:

$$C(A_U/A_S),$$

in which C is the concentration, in µg per mL, of USP Glutethimide RS in the Standard solution, and A_U and A_S are the absorbances of the solution from the Capsules and the Standard solution, respectively.

Glutethimide Tablets

» Glutethimide Tablets contain not less than 90.0 percent and not more than 110.0 percent of the labeled amount of $C_{13}H_{15}NO_2$.

Packaging and storage—Preserve in well-closed containers.

Reference standard—*USP Glutethimide Reference Standard*—Dry over phosphorus pentoxide at 45° to constant weight before using.

Identification—Transfer a quantity of powdered Tablets, equivalent to about 175 mg of glutethimide, to a centrifuge tube containing 10 mL of chloroform, and shake until the powder is suspended. Centrifuge the suspension and withdraw the clear chloroform supernatant liquid: the infrared absorption spectrum of the chloroform solution so obtained, determined in 0.1-mm cells, exhibits maxima only at the same wavelengths as that of a similar preparation of USP Glutethimide RS.

Dissolution ⟨711⟩—

Medium: water; 900 mL.

Apparatus 2: 50 rpm.

Time: 60 minutes.

Procedure—Determine the amount of $C_{13}H_{15}NO_2$ dissolved from ultraviolet absorbances at the wavelength of maximum absorbance at about 257 nm of filtered portions of the solution under test, suitably diluted with *Dissolution Medium*, if necessary, in comparison with a Standard solution having a known concentration of USP Glutethimide RS in the same medium.

Tolerances—Not less than 70% (Q) of the labeled amount of $C_{13}H_{15}NO_2$ is dissolved in 60 minutes.

Uniformity of dosage units ⟨905⟩: meet the requirements.

Assay—

pH 4.0 acetate buffer, Mobile phase, Standard preparation, and *Chromatographic system*—Prepare as directed in the *Assay* under *Glutethimide.*

Assay preparation—Weigh and finely powder not less than 20 Glutethimide Tablets. Transfer an accurately weighed portion of the powder, equivalent to about 100 mg of glutethimide, to a 100-mL volumetric flask. Add about 90 mL of *Mobile phase*, and shake by mechanical means for 5 minutes. Dilute with *Mobile phase* to volume, mix, and filter, discarding the first 10 mL of the filtrate.

Procedure—Proceed with Glutethimide Tablets as directed for *Procedure* in the *Assay* under *Glutethimide.* Calculate the quantity, in mg, of $C_{13}H_{15}NO_2$ in the portion of Tablets taken by the formula:

$$100C(r_U/r_S),$$

in which C is the concentration, in mg per mL, of USP Glutethimide RS in the *Standard preparation*, and r_U and r_S are the peak responses obtained from the *Assay preparation* and the *Standard preparation*, respectively.

Glycerin

$CH_2OH . CHOH . CH_2OH$

$C_3H_8O_3$ 92.09
1,2,3-Propanetriol.
Glycerol [*56-81-5*].

» Glycerin contains not less than 95.0 percent and not more than 101.0 percent of $C_3H_8O_3$.

Packaging and storage—Preserve in tight containers.

Identification—The infrared absorption spectrum of a thin film of it exhibits a very strong, broad band at 2.7 µm to 3.3 µm, a strong doublet at about 3.4 µm, a maximum at about 6.1 µm, a strong region of absorption between 6.7 µm and 8.3 µm, having maxima at about 7.1 µm, 7.6 µm, and 8.2 µm, and a very strong region of bands at about 9.0 µm, 9.6 µm, 10.1 µm, 10.9 µm, and 11.8 µm. [NOTE—Glycerin containing low content of water may not exhibit a maximum at about 6.1 µm.]

Specific gravity ⟨841⟩: not less than 1.249.

Color—Its color, when viewed downward against a white surface in a 50-mL color-comparison tube, is not darker than the color of a standard made by diluting 0.40 mL of ferric chloride CS with water to 50 mL and similarly viewed in a color-comparison tube of approximately the same diameter and color as that containing the Glycerin.

Residue on ignition ⟨281⟩—Heat 50 g in an open, shallow, 100-mL porcelain dish until it ignites, and allow it to burn without further application of heat in a place free from drafts. Cool, moisten the residue with 0.5 mL of sulfuric acid, and ignite to constant weight: the weight of the residue does not exceed 5 mg (0.01%).

Chloride ⟨221⟩—A 7.0-g portion shows no more chloride than corresponds to 0.10 mL of 0.020 *N* hydrochloric acid (0.001%).

Sulfate ⟨221⟩—A 10-g portion shows no more sulfate than corresponds to 0.20 mL of 0.020 *N* sulfuric acid (about 0.002%).

Arsenic, *Method I* ⟨211⟩: 1.5 ppm.

Heavy metals ⟨231⟩—Mix 4.0 g with 2 mL of 0.1 *N* hydrochloric acid, and dilute with water to 25 mL: the limit is 5 ppm.

Chlorinated compounds—Weigh accurately 5 g into a dry, round-bottom, 100-mL flask, add 15 mL of morpholine, and connect the flask by a ground joint to a reflux condenser. Reflux gently for 3 hours. Rinse the condenser with 10 mL of water, receiving the washing in the flask, and cautiously acidify with nitric acid. Transfer the solution to a suitable comparison tube, add 0.50 mL of 0.10 *N* silver nitrate, dilute with water to 50.0 mL, and mix: the turbidity is not greater than that of a blank to which 0.20 mL of 0.020 *N* hydrochloric acid has been added, the refluxing being omitted (0.003% of Cl).

Fatty acids and esters—Mix 50 g with 50 mL of freshly boiled water and 5 mL of 0.5 *N* sodium hydroxide VS, boil the mixture for 5 minutes, cool, add phenolphthalein TS, and titrate the excess alkali with 0.5 *N* hydrochloric acid VS. Perform a blank determination (see *Residual Titrations* under *Titrimetry* ⟨541⟩): not more than 1 mL of 0.5 *N* sodium hydroxide is consumed.

Assay—

Sodium periodate solution—Dissolve 60 g of sodium metaperiodate in sufficient water containing 120 mL of 0.1 *N* sulfuric acid to make 1000 mL. Do not heat to dissolve the periodate. If the solution is not clear, filter through a sintered-glass filter. Store the solution in a glass-stoppered, light-resistant container. Test the suitability of this solution as follows: Pipet 10 mL into a 250-mL volumetric flask, dilute with water to volume, and mix. To about 550 mg of Glycerin dissolved in 50 mL of water add 50 mL of the diluted periodate solution with a pipet. For a blank, pipet 50 mL of the solution into a flask containing 50 mL of water. Allow the solutions to stand for 30 minutes, then to each add 5 mL of hydrochloric acid and 10 mL of potassium iodide TS, and rotate to mix. Allow to stand for 5 minutes, add 100 mL of water, and titrate with 0.1 *N* sodium thiosulfate, shaking continuously and adding 3 mL of starch TS as the end-point is approached. The ratio of the volume of 0.1 *N* sodium thiosulfate required for the glycerin-periodate mixture to that required for the blank should be between 0.750 and 0.765.

Procedure—Transfer about 400 mg of Glycerin, accurately weighed, to a 600-mL beaker, dilute with 50 mL of water, add bromothymol blue TS, and acidify with 0.2 *N* sulfuric acid to a definite green or greenish yellow color. Neutralize with 0.05 *N* sodium hydroxide to a definite blue end-point, free from green color. Prepare a blank containing 50 mL of water, and neutralize in the same manner. Pipet 50 mL of the *Sodium periodate solution* into each beaker, mix by swirling gently, cover with a watch glass, and allow to stand for 30 minutes at room temperature (not exceeding 35°) in the dark or in subdued light. Add 10 mL of a mixture of equal volumes of ethylene glycol and water, and allow to stand for 20 minutes. Dilute each solution with water to about 300 mL, and titrate with 0.1 *N* sodium hydroxide VS to a pH of 8.1 ± 0.1 for the specimen under assay and 6.5 ± 0.1 for the blank, using a pH meter. Each mL of 0.1 *N* sodium hydroxide, after correction for the blank, is equivalent to 9.210 mg of $C_3H_8O_3$.

Glycerin Ophthalmic Solution

» Glycerin Ophthalmic Solution is a sterile, anhydrous solution of Glycerin, containing not less than 98.5 percent of $C_3H_8O_3$. It may contain one or more suitable antimicrobial preservatives. [NOTE—In the preparation of this Ophthalmic Solution, use Glycerin that has a low water content, in order that the Ophthalmic Solution may comply with the *Water* limit. This may be ensured by using Glycerin having a specific gravity of not less than 1.2607, corresponding to a concentration of 99.5%.]

Note—Do not use the Ophthalmic Solution if it contains crystals, or if it is cloudy, discolored, or contains a precipitate.

Packaging and storage—Preserve in tight containers of glass or plastic, containing not more than 15 mL, protected from light. The container or individual carton is sealed and tamper-proof so that sterility is assured at time of first use.

Identification—It responds to the *Identification test* under *Glycerin.*

Sterility—It meets the requirements under *Sterility Tests* ⟨71⟩.

pH ⟨791⟩: between 4.5 and 7.5, determined potentiometrically in a solution prepared by the addition of 5 mL of Sodium Chloride Injection to 5 mL of Glycerin Ophthalmic Solution.

Water, *Method I* ⟨921⟩: not more than 1.0%.

Assay—Transfer an accurately measured volume of Glycerin Solution, equivalent to about 3 g of glycerin, to a 500-mL volumetric flask, dilute with water to volume, and mix. Transfer a 3-mL portion to a conical flask, add 100.0 mL of a solution of potassium periodate (prepared by dissolving 3 g of potassium periodate in about 500 mL of warm water, cooling to room temperature, and then diluting with water to 1000 mL), swirl, and allow to stand at room temperature for 10 minutes. Add 4 g of sodium bicarbonate and 2 g of potassium iodide, and titrate immediately with 0.1 *N* potassium arsenite VS, adding 3 mL of starch TS as the end-point is approached. Perform a blank determination, using water in place of the Ophthalmic Solution, and note the difference in volumes required. Each mL of 0.1 *N* potassium arsenite is equivalent to 2.303 mg of $C_3H_8O_3$.

Glycerin Oral Solution

» Glycerin Oral Solution contains not less than 95.0 percent and not more than 105.0 percent of the labeled amount of $C_3H_8O_3$.

Packaging and storage—Preserve in tight containers.

Identification—Heat a few drops with about 500 mg of potassium bisulfate in a test tube: pungent vapors of acrolein are evolved.

pH ⟨791⟩: between 5.5 and 7.5.

Assay—Transfer an accurately measured volume of Glycerin Oral Solution, equivalent to about 3 g of glycerin, to a 500-mL volumetric flask, dilute with water to volume, and mix. Transfer a 3-mL portion to a conical flask, add 100.0 mL of a solution of potassium periodate (prepared by dissolving 3 g of potassium periodate in about 500 mL of warm water, cooling to room temperature, and then diluting with water to 1000 mL), swirl, and allow to stand at room temperature for 10 minutes. Add 4 g of sodium bicarbonate and 2 g of potassium iodide, and titrate immediately with 0.1 *N* potassium arsenite VS, adding 3 mL of starch TS as the end-point is approached. Perform a blank determination, using water in place of the Oral Solution, and note the difference in volumes required. Each mL of 0.1 *N* potassium arsenite is equivalent to 2.303 mg of $C_3H_8O_3$.

Glycerin Otic Solution—*see* Antipyrine and Benzocaine Otic Solution

Glycerin Suppositories

» Glycerin Suppositories contain Glycerin solidified with Sodium Stearate. [NOTE—If preferred, the Sodium Stearate for Glycerin Suppositories may be prepared during the making of the Suppositories by the direct reaction between Stearic Acid and Sodium Bicarbonate, Sodium Carbonate, or Sodium Hydroxide, these being taken in the correct proportion.] Glycerin Suppositories contain not less than 75.0 percent and not more than 90.0 percent, by weight, of $C_3H_8O_3$.

Packaging and storage—Preserve in well-closed containers.

Reference standard—*USP Stearic Acid Reference Standard*—Do not dry before using.

Identification—

A: Dissolve 1 g of sodium borate in 100 mL of water, add 25 drops of phenolphthalein TS, and mix. To a test tube containing 0.5 mL of this solution add 2 drops of 1 Suppository that has been melted: the pink solution becomes colorless, and when it is heated the pink color reappears.

B: Disperse 12 Suppositories in about 125 mL of water in a 250-mL beaker on a hot plate. Cool, add 1.5 mL of hydrochloric acid, and pour the mixture into a 250-mL separator. Extract with 75 mL of hexanes, discarding the lower aqueous layer and collecting the organic layer in a beaker. Evaporate with the aid of a steam bath to near dryness: the infrared absorption spectrum of a mineral oil dispersion of the residue so obtained exhibits maxima only at the same wavelengths as that of a mineral oil dispersion of USP Stearic Acid RS.

Water, *Method I* ⟨921⟩: not more than 15.0%.

Assay—Transfer an accurately weighed quantity of Glycerin Suppositories, equivalent to about 250 mg of glycerin, to a 250-mL volumetric flask. Dissolve in water, dilute with water to volume, and mix. Pipet 5 mL of this solution into a 250-mL conical flask, and add 50.0 mL of a reagent prepared by mixing 40 mL of dilute sulfuric acid (1 in 20) with 60 mL of potassium periodate solution (1 in 1000) acidified with 3 to 5 drops of sulfuric acid. Heat the solution on a steam bath for 15 minutes, cool to room temperature, and add 1 g of potassium iodide. Allow the flask to stand for 5 minutes, and titrate with 0.02 *N* sodium thiosulfate VS, adding 3 mL of starch TS as the end-point is approached. Perform a blank determination, using water in place of Glycerin Suppositories, and note the difference in volumes required. Each mL of the difference in volume of 0.02 *N* sodium thiosulfate consumed is equivalent to 0.4604 mg of $C_3H_8O_3$.

Glyceryl Behenate—*see* Glyceryl Behenate NF

Glyceryl Monostearate—*see* Glyceryl Monostearate NF

Glycine

NH_2CH_2COOH

$C_2H_5NO_2$ 75.07

Glycine.

Glycine [*56-40-6*].

» Glycine contains not less than 98.5 percent and not more than 101.5 percent of $C_2H_5NO_2$, calculated on the dried basis.

Packaging and storage—Preserve in well-closed containers.

Reference standard—*USP Glycine Reference Standard*—Dry at 105° for 2 hours before using.

Identification—The infrared absorption spectrum of a mineral oil dispersion of it, previously dried, exhibits maxima only at the same wavelengths as that of a similar preparation of USP Glycine RS. If a difference appears, dissolve portions of both the test specimen and the Reference Standard in water, evaporate the solutions to dryness, and repeat the test on the residues.

Loss on drying ⟨731⟩—Dry it at 105° for 2 hours: it loses not more than 0.2% of its weight.

Residue on ignition ⟨281⟩: not more than 0.1%.

Chloride ⟨221⟩—A solution of 1.0 g in water shows no more chloride than corresponds to 0.10 mL of 0.020 *N* hydrochloric acid (0.007%).

Sulfate ⟨221⟩—A solution of 3.0 g in water shows no more sulfate than corresponds to 0.20 mL of 0.020 *N* sulfuric acid (0.0065%).

Heavy metals ⟨231⟩: 0.002%.

Readily carbonizable substances ⟨271⟩—Dissolve 0.5 g in 5 mL of sulfuric acid TS: the solution is colorless.

Hydrolyzable substances—Boil 10 mL of a 1 in 10 solution of it for 1 minute, and set aside for 2 hours: the solution appears as clear and as mobile as 10 mL of the same solution that has not been boiled.

Assay—Transfer about 150 mg of Glycine, accurately weighed, to a 250-mL flask, dissolve in 100 mL of glacial acetic acid, add 1 drop of crystal violet TS, and titrate with 0.1 *N* perchloric acid VS to a green end-point. Perform a blank determination, and make any necessary correction. Each mL of 0.1 *N* perchloric acid is equivalent to 7.507 mg of $C_2H_5NO_2$.

Glycopyrrolate

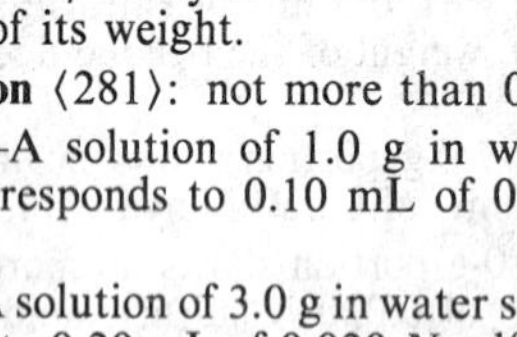

$C_{19}H_{28}BrNO_3$ 398.34

Pyrrolidinium, 3-[(cyclopentylhydroxyphenylacetyl)oxy]-1,1-dimethyl-, bromide.

3-Hydroxy-1,1-dimethylpyrrolidinium bromide α-cyclopentylmandelate [*596-51-0*].

» Glycopyrrolate, dried at 105° for 3 hours, contains not less than 98.0 percent and not more than 100.5 percent of $C_{19}H_{28}BrNO_3$.

Packaging and storage—Preserve in tight containers.

Reference standard—*USP Glycopyrrolate Reference Standard*—Dry at 105° for 3 hours before using.

Identification—

A: The infrared absorption spectrum of a potassium bromide dispersion of it, previously dried, exhibits maxima only at the same wavelengths as that of a similar preparation of USP Glycopyrrolate RS.

B: The ultraviolet absorption spectrum of a solution (1 in 2000) exhibits maxima and minima at the same wavelengths as that of a similar solution of USP Glycopyrrolate RS, concomitantly measured, and the respective absorptivities, calculated on the dried basis, at the wavelength of maximum absorbance at about 258 nm do not differ by more than 3.0%.

C: A solution (1 in 40) responds to the tests for *Bromide* ⟨191⟩.

Melting range, *Class I* ⟨741⟩: between 193° and 198°, but the range between beginning and end of melting does not exceed 2°.

Loss on drying ⟨731⟩—Dry it at 105° for 3 hours: it loses not more than 0.5% of its weight.

Residue on ignition ⟨281⟩: not more than 0.3%.

Assay—Dissolve about 800 mg of Glycopyrrolate, previously dried and accurately weighed, in 50 mL of glacial acetic acid, add 10 mL of mercuric acetate TS and 1 drop of crystal violet TS, and titrate with 0.1 *N* perchloric acid VS. Perform a blank determination, and make any necessary correction. Each mL of 0.1 *N* perchloric acid is equivalent to 39.83 mg of $C_{19}H_{28}BrNO_3$.

Glycopyrrolate Injection

» Glycopyrrolate Injection is a sterile solution of Glycopyrrolate in Water for Injection. It contains not less than 93.0 percent and not more than 107.0 percent of the labeled amount of $C_{19}H_{28}BrNO_3$.

Packaging and storage—Preserve in single-dose or in multiple-dose containers, preferably of Type I glass.

Reference standard—*USP Glycopyrrolate Reference Standard*—Dry at 105° for 3 hours before using.

Identification—

Spray reagent—Dissolve 2 g of bismuth subnitrate in a solution consisting of 100 mL of water and 25 mL of glacial acetic acid (*Solution A*). Dissolve 40 g of potassium iodide in 100 mL of water (*Solution B*). Add 10 mL of *Solution A* and 10 mL of *Solution B* to a solution consisting of 100 mL of water and 20 mL of glacial acetic acid, and mix.

Procedure—Pipet an amount of Injection equivalent to about 1 mg of glycopyrrolate into a 10-mL volumetric flask, dilute with water to volume, and mix to obtain the test solution. Prepare a Standard solution of USP Glycopyrrolate RS in water containing about 0.1 mg of glycopyrrolate per mL. Apply 30 µL of the test solution and 30 µL of the Standard solution to a suitable thin-layer chromatographic plate (see *Chromatography* ⟨621⟩), coated with a 0.25-mm layer of chromatographic silica gel mixture. Allow the spots to dry, and develop the chromatogram in a solvent system consisting of a mixture of butyl alcohol, glacial acetic acid, and water (3:1:1) until the solvent front has moved about three-fourths of the length of the plate. Remove the plate from the developing chamber, and allow to air-dry. Spray the plate with *Spray reagent*, and allow to air-dry: the R_f value and color of the principal spot obtained from the test solution correspond to those obtained from the Standard solution.

pH ⟨791⟩: between 2.0 and 3.0.

Other requirements—It meets the requirements under *Injections* ⟨1⟩.

Assay—

pH 5.3 phosphate buffer—Dissolve 38.0 g of monobasic sodium phosphate and 2.0 g of anhydrous dibasic sodium phosphate in sufficient water to make 1000 mL, and adjust, if necessary, to a pH of 5.3 ± 0.1.

Bromocresol purple solution—Dissolve 400 mg of bromocresol purple in 30 mL of alcohol, add 6.3 mL of 0.1 *N* sodium hydroxide, and dilute with water to 500 mL.

Buffer-dye mixture—On the day of use, mix equal volumes of *pH 5.3 phosphate buffer* and of *Bromocresol purple solution*, and extract with chloroform until the last extract is colorless.

Standard preparation—Transfer about 100 mg of USP Glycopyrrolate RS, accurately weighed, to a 100-mL volumetric flask, add water to volume, and mix. Transfer 10.0 mL of this solution to a second 100-mL volumetric flask, dilute with water to volume, and mix to obtain a *Standard preparation* having a known concentration of about 100 µg per mL.

Assay preparation—Transfer an accurately measured volume of Glycopyrrolate Injection, equivalent to about 5 mg of glycopyrrolate, to a 50-mL volumetric flask, dilute with water to volume, and mix.

Procedure—Transfer 10.0 mL each of the *Standard preparation*, of the *Assay preparation*, and of water to serve as the blank to individual separators, each containing 10.0 mL of *Buffer-dye mixture*. Mix, and extract each solution with four 25-mL portions of chloroform, filtering the extracts from each solution through chloroform-moistened filter paper into separate 100-mL volumetric flasks. Dilute each solution with chloroform to volume, and mix. Concomitantly determine the absorbances of the solutions in 1-cm cells at the wavelength of maximum absorbance at about 410 nm, with a suitable spectrophotometer, using the blank to set the instrument. Calculate the quantity, in µg, of $C_{19}H_{28}BrNO_3$ in each mL of the Injection taken by the formula:

$$(50C/V)(A_U/A_S),$$

in which C is the concentration, in µg per mL, of USP Glycopyrrolate RS in the *Standard preparation*, V is the volume, in mL, of Injection taken, and A_U and A_S are the absorbances of the solutions from the *Assay preparation* and the *Standard preparation*, respectively.

Glycopyrrolate Tablets

» Glycopyrrolate Tablets contain not less than 93.0 percent and not more than 107.0 percent of the labeled amount of $C_{19}H_{28}BrNO_3$.

Packaging and storage—Preserve in tight containers.

Reference standard—*USP Glycopyrrolate Reference Standard*—Dry at 105° for 3 hours before using.

Identification—Blend a portion of finely powdered Tablets, equivalent to about 25 mg of glycopyrrolate, with 50 mL of water in a high-speed blender, and filter through very retentive filter paper: the ultraviolet absorption spectrum of the filtrate exhibits maxima and minima at the same wavelengths as that of a solution of USP Glycopyrrolate RS (1 in 2000).

Dissolution ⟨711⟩—

Medium: water; 500 mL.

Apparatus 1: 100 rpm.

Time: 45 minutes.

Procedure—Determine the amount of $C_{19}H_{28}BrNO_3$ dissolved, employing the procedure set forth in the *Assay*, making any necessary modifications.

Tolerances—Not less than 75% (*Q*) of the labeled amount of $C_{19}H_{28}BrNO_3$ is dissolved in 45 minutes.

Uniformity of dosage units ⟨905⟩: meet the requirements.

Assay—

pH 5.3 phosphate buffer, Bromocresol purple solution, Buffer-dye mixture, and *Standard preparation*—Prepare as directed in the *Assay* under *Glycopyrrolate Injection.*

Assay preparation—Weigh and finely powder not less than 20 Glycopyrrolate Tablets. Transfer an accurately weighed portion of the powder, equivalent to about 10 mg of glycopyrrolate, to a 100-mL volumetric flask, add 50 mL of water, and shake by mechanical means for 10 minutes. Dilute with water to volume, mix, and filter, discarding the first 20 mL of the filtrate. Use the subsequent filtrate as directed in the *Procedure*.

Procedure—Proceed as directed for *Procedure* in the *Assay* under *Glycopyrrolate Injection.* Calculate the quantity, in mg, of $C_{19}H_{28}BrNO_3$ in the portion of Tablets taken by the formula:

$$0.1C(A_U/A_S),$$

in which C is the concentration, in µg per mL, of USP Glycopyrrolate RS in the *Standard preparation.*

Chorionic Gonadotropin

» Chorionic Gonadotropin is a gonad-stimulating polypeptide hormone obtained from the urine of pregnant women. Its potency is not less than 1500 USP Chorionic Gonadotropin Units in each mg, and not less than 80.0 percent and not more than 125.0 percent of the potency stated on the label.

Packaging and storage—Preserve in tight containers, preferably of Type I glass, in a refrigerator.

Reference standard—*USP Chorionic Gonadotropin Reference Standard*—Store in a refrigerator, and do not dry before using. Use a fresh ampul for each group of assays, and discard any unused portion.

Pyrogen—Dissolve a suitable quantity in sterile, pyrogen-free saline TS to obtain a solution containing 500 USP Chorionic Gonadotropin Units per mL. This solution meets the requirements of the *Pyrogen Test* ⟨151⟩, the test dose being 1000 Units per kg.

Acute toxicity—Select five healthy mice, weighing between 18 and 22 g. Prepare a test solution as directed in the test for *Pyrogen*, but containing 2000 USP Chorionic Gonadotropin Units per mL. Inject intravenously a dose of 0.5 mL of the test solution into each of the mice. Observe the animals over the 48 hours following the injection. If, at the end of 48 hours, all of the animals survive and not more than one of the animals shows outward symptoms of a toxic reaction, the requirements of the test are met. If more than one of the animals show outward signs of a toxic reaction or if not more than two of the animals die, repeat the test on ten additional, similar animals: if all of the animals of the repeat test survive for 48 hours and show no symptoms of a toxic reaction, the requirements of the test are met.

Water, *Method I* ⟨921⟩: not more than 5.0%.

Estrogenic activity—Dissolve a suitable quantity in saline TS to obtain a test solution containing the equivalent of 1000 USP Chorionic Gonadotropin Units per mL. Into each of five rats that have been ovariectomized not less than 2 weeks previously, inject subcutaneously 0.25 mL of the test solution in the forenoon and in the afternoon of two successive days. On each of the three following days, take a vaginal smear from each animal: the requirements of the test are met if the cellular elements in the smears consist mainly of leucocytes, and a few nucleated epithelial cells, but no cornified epithelial cells.

Assay—

Standard preparation—Dissolve a suitable quantity of USP Chorionic Gonadotropin RS in a diluent consisting of saline TS, freshly prepared to contain 1 mg per mL of bovine serum albumin and adjusted to a pH between 6.9 and 8.0 with sodium hydroxide TS, to obtain a solution having a known concentration of 10 USP Chorionic Gonadotropin Units in each mL. Using the same diluent, prepare three *Standard solutions* such that the respective concentrations of chorionic gonadotropin constitute a geometric series such as 1:1.2:1.44 or 1:2:4 and such that the activity in each mL lies within the range of 0.1 to 1.0 Unit.

Assay preparation—Following the procedure outlined for the *Standard preparation*, prepare solutions of Chorionic Gonadotropin to give three *Test solutions* corresponding to those of the standard.

The animals—Select 20- to 23-day-old female rats, but restrict the selection so that no rat is more than 30% heavier than the lightest. House the animals under uniform conditions of temperature, lighting, feeding, and watering. Mark the animals for identification, and divide them at random into groups of the same number but not less than 10 animals. Assign one group to each of the three *Standard solutions* and three *Test solutions*, respectively.

Procedure—Inject each rat subcutaneously in the dorsal area with 0.20 mL of the solution to which it was assigned, at approximately the same time on each of three consecutive days. On the afternoon of the fifth day, sacrifice the animals, and excise the uterus from each animal by cutting through the cervix, stripping off the surrounding tissue, and severing at the utero-tubal junction. Gently press out the uterine fluid on moistened absorbent paper, and weigh the uterus to the nearest 0.2 mg, using a suitable balance.

Calculation—Tabulate the observed uterine weight for each rat, designated by the symbol y, for each dosage group of f rats. Proceed as directed in the *Assay* under *Corticotropin Injection*, beginning with "If the data from one or more rats." Compute the log confidence interval L (see *Confidence Intervals for Individual Assay* ⟨111⟩). If the confidence interval is more than 0.1938, which corresponds at $P = 0.95$ to confidence limits of 80% and 125% of the computed potency, repeat the assay until the combined data of two or more assays, redetermined as described under *Combination of Independent Assays* ⟨111⟩, meet this limit.

Chorionic Gonadotropin for Injection

» Chorionic Gonadotropin for Injection is a sterile, dry mixture of Chorionic Gonadotropin with suitable diluents and buffers. Its potency is not less than 80.0 percent and not more than 125.0 percent of the potency stated on the label in USP Chorionic Gonadotropin Units. It may contain an antimicrobial agent.

Packaging and storage—Preserve in *Containers for Sterile Solids* as described under *Injections* ⟨1⟩.

Labeling—Label it to indicate the expiration date.

Constituted solution—At the time of use, the constituted solution prepared from Chorionic Gonadotropin for Injection meets the requirements for *Constituted Solutions* under *Injections* ⟨1⟩.

Pyrogen—The solution prepared for the test for *Estrogenic activity*, adjusted to contain 500 USP Chorionic Gonadotropin Units per mL, meets the requirements of the *Pyrogen Test* ⟨151⟩, the test dose being 1000 Chorionic Gonadotropin Units per kg.

pH ⟨791⟩—The pH of the solution prepared for the test for *Estrogenic activity* is between 6.0 and 8.0.

Estrogenic activity—When constituted as directed in the labeling, it meets the requirements of the test for *Estrogenic activity* under *Chorionic Gonadotropin.*

Other requirements—It meets the requirements for *Sterility Tests* ⟨71⟩, *Uniformity of Dosage Units* ⟨905⟩, and *Labeling* under *Injections* ⟨1⟩.

Assay—Proceed with Chorionic Gonadotropin for Injection as directed in the *Assay* under *Chorionic Gonadotropin*, using an *Assay preparation* obtained by diluting a portion of the solution prepared for the test for *Estrogenic activity* quantitatively and stepwise with the specified diluent.

Gramicidin

Gramicidin.
Gramicidin [*1405-97-6*].

» Gramicidin is an antibacterial substance produced by the growth of *Bacillus brevis* Dubos (Fam. Bacillaceae). It may be obtained from tyrothricin. It has a potency of not less than 900 µg of gramicidin per mg, calculated on the dried basis.

Packaging and storage—Preserve in tight containers.

Reference standard—*USP Gramicidin Reference Standard*—Dry in vacuum at a pressure not exceeding 5 mm of mercury at 60° for 3 hours before using.

Identification—The ultraviolet absorption spectrum of a 1 in 20,000 solution in alcohol exhibits maxima and minima at the same wavelengths as that of a similar solution of USP Gramicidin RS, concomitantly measured. The absorptivities, calculated from the difference in absorbance between the maximum at about 282 nm and the minimum at about 247 nm, do not differ by more than 4.0%, calculated on the dried basis.

Melting temperature, *Class Ia* ⟨741⟩: not lower than 229°, determined after drying.

Crystallinity ⟨695⟩: meets the requirements.

Loss on drying ⟨731⟩—Dry about 100 mg in a capillary-stoppered bottle in vacuum at 60° for 3 hours: it loses not more than 3.0% of its weight.

Residue on ignition ⟨281⟩: not more than 1.0%, the charred residue being moistened with 2 mL of nitric acid and 5 drops of sulfuric acid.

Assay—Proceed with Gramicidin as directed under *Antibiotics—Microbial Assays* ⟨81⟩.

Gramicidin Cream, Neomycin and Polymyxin B Sulfates and—*see* Neomycin and Polymyxin B Sulfates and Gramicidin Cream

Gramicidin Ointment, Neomycin Sulfate and—*see* Neomycin Sulfate and Gramicidin Ointment

Gramicidin Ophthalmic Solution, Neomycin and Polymyxin B Sulfates and—*see* Neomycin and Polymyxin B Sulfates and Gramicidin Ophthalmic Solution

Gramicidin, and Hydrocortisone Acetate Cream, Neomycin and Polymyxin B Sulfates,—*see* Neomycin and Polymyxin B Sulfates, Gramicidin, and Hydrocortisone Acetate Cream

Gramicidin, and Triamcinolone Acetonide Cream, Nystatin, Neomycin Sulfate,—*see* Nystatin, Neomycin Sulfate, Gramicidin, and Triamcinolone Acetonide Cream

Gramicidin, and Triamcinolone Acetonide Ointment, Nystatin, Neomycin Sulfate,—*see* Nystatin, Neomycin Sulfate, Gramicidin, and Triamcinolone Acetonide Ointment

Green Soap

» Green Soap is a potassium soap made by the saponification of suitable vegetable oils, excluding coconut oil and palm kernel oil, without the removal of glycerin.

Green Soap may be prepared as follows:

The Vegetable Oil	380 g
Oleic Acid	20 g
Potassium Hydroxide (total alkali 85 percent)	91.7 g
Glycerin	50 mL
Purified Water, a sufficient quantity, to make about	1000 g

Mix the oil and the Oleic Acid, and heat the mixture to about 80°. Dissolve the Potassium Hydroxide in a mixture of the Glycerin and 100 mL of Purified Water, and add the solution, while it is still hot, to the hot oil. Stir the mixture vigorously until emulsified, then heat while continuing the stirring, until the mixture is homogeneous and a test portion will dissolve to give a clear solution in hot water. Add hot purified water to make the product weigh 1000 g, continuing the stirring until the Soap is homogeneous.

Packaging and storage—Preserve in well-closed containers.

Water—Place about 5 g of Green Soap, quickly weighed to the nearest centigram, in the distilling flask of the apparatus for *Moisture Method by Toluene Distillation* ⟨921⟩. (The Soap is most conveniently weighed in a boat of metal foil, of a size that will just pass through the neck of the flask.) Place 250 mL of toluene and 10 g of anhydrous barium chloride in the flask, connect the flask through a ground-glass joint to the distilling apparatus, fill the receiving tube with toluene, and determine the water as directed, beginning with "Heat the flask gently." The volume of water found corresponds to not more than 52.0% by weight of the Green Soap taken.

Alcohol-insoluble substances—Dissolve about 5 g of Green Soap, rapidly and accurately weighed, in 100 mL of hot neutralized alcohol, collect the residue, if any, on a tared filter, thoroughly wash it with hot neutralized alcohol, and dry at 105° for 1 hour: the weight of the residue so obtained is not more than 3.0% of the weight of the Green Soap taken. Retain the solution for the test for *Free alkali hydroxides*, and retain the residue for the test for *Alkali carbonates*.

Free alkali hydroxides—To the combined filtrate and washings obtained in the test for *Alcohol-insoluble substances* add 0.5 mL of phenolphthalein TS. If a pink color is produced, titrate the solution with 0.1 *N* sulfuric acid VS until the pink color is just discharged. Each mL of 0.1 *N* sulfuric acid is equivalent to 5.611 mg of KOH. The volume of 0.1 *N* sulfuric acid VS consumed corresponds to not more than 0.25% of KOH.

Alkali carbonates—Wash the filter containing the *Alcohol-insoluble substances* with 50 mL of boiling water, cool, add methyl orange TS, and titrate the filtrate with 0.1 *N* sulfuric acid VS. Not more than 0.5 mL of 0.10 *N* sulfuric acid per g of Green Soap originally taken is required (0.35% as K_2CO_3).

Unsaponified matter—A solution of Green Soap in hot water (1 in 20) is nearly clear.

Characteristics of the liberated fatty acids—Dissolve about 30 g of Green Soap in 300 mL of hot water in a beaker, add gradually 60 mL of 2 *N* sulfuric acid, and heat on a steam bath until the liberated acids form a transparent layer. Decant the fatty acids into a separator, and wash them with 50-mL portions of hot water until the last washing, when cool, is neutral to methyl orange TS. Transfer the fatty acids to a dry beaker, and allow them to stand in a warm oven until any water that may be present has separated. Then filter the acids through a dry filter in the warm oven. Determine the *Acid Value* (see *Fats and Fixed Oils* ⟨401⟩) of about 1 g, accurately weighed, of the fatty acids: it is not more than 205. Determine the *Iodine Value* (see *Fats and Fixed Oils* ⟨401⟩) of 150 to 200 mg, accurately weighed, of the fatty acids: it is not less than 85.

Green Soap Tincture

» Prepare Green Soap Tincture as follows:

Green Soap	650 g
Suitable essential oil(s)	
Alcohol	316 mL
Purified Water, a sufficient quantity, to make	1000 mL

Mix the oil(s) and Alcohol, dissolve in this the Green Soap by stirring or by agitation, set the solution aside for 24 hours, filter through paper, and add water to make 1000 mL.

Packaging and storage—Preserve in tight containers.

Identification—Transfer 10 mL to a flask containing 10 mL of water, and carefully add 1 mL of sulfuric acid: fatty material separates.

pH ⟨791⟩: between 9.5 and 11.5.

Alcohol content, *Method II* ⟨611⟩: between 28.0% and 32.0% of C_2H_5OH.

Griseofulvin

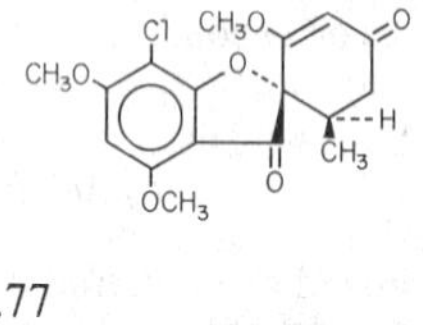

$C_{17}H_{17}ClO_6$ 352.77

Spiro[benzofuran-2(3*H*),1′-[2]cyclohexene]-3,4′-dione, 7-chloro-2′,4,6-trimethoxy-6′-methyl-, (1′*S-trans*)-.

7-Chloro-2′,4,6-trimethoxy-6′β-methylspiro[benzofuran-2(3*H*), 1′-[2]cyclohexene]-3,4′-dione [*126-07-8*].

» Griseofulvin has a potency of not less than 900 μg of $C_{17}H_{17}ClO_6$ per mg.

Packaging and storage—Preserve in tight containers.

Reference standard—*USP Griseofulvin Reference Standard*—Do not dry before using. *USP Griseofulvin Specific Surface Area Reference Standard*—Do not dry before using.

Identification—

A: The infrared absorption spectrum of a mineral oil dispersion of it, previously dried, exhibits maxima only at the same wavelengths as that of a similar preparation of USP Griseofulvin RS.

B: The retention time of the major peak in the chromatogram of the *Assay preparation* corresponds to that in the chromatogram of the *Standard preparation*, both relative to the internal standard, as obtained in the *Assay*.

Melting range ⟨741⟩: between 217° and 224°.

Specific rotation ⟨781⟩: between +348° and +364°, determined in a solution in dimethylformamide containing 10 mg per mL.

Crystallinity ⟨695⟩: meets the requirements.

Loss on drying ⟨731⟩—Dry about 100 mg, accurately weighed, in a capillary-stoppered bottle in vacuum at a pressure not exceeding 5 mm of mercury at 60° for 3 hours: it loses not more than 1.0% of its weight.

Residue on ignition ⟨281⟩: not more than 0.2%.

Heavy metals, *Method II* ⟨231⟩: 0.0025%.

Specific surface area—Determine the apparent particle size in μm by the air-permeation method, using a suitable subsieve sizer. Weigh 1.819 ± 0.001 g of Griseofulvin, and transfer to the compression tube of the apparatus. Compact with moderate pressure so that a uniform porosity is achieved. Pass dry compressed air through the tube, and measure the air pressure with a water manometer. Read the porosity, and calculate the apparent particle size from the instrument equation. Repeat the porosity readings at successively higher degrees of compaction until the apparent particle size reaches a minimum value. Calculate the observed specific surface area, in square meters per g, by the formula:

$$6/(1.455MF),$$

in which M is the minimum apparent particle size, and F is a factor, obtained from the accompanying table, interpolation being used if necessary, to correct the apparent particle size to the true particle size at a given porosity reading.

Porosity Reading	F	Porosity Reading	F
0.80	1.3771	0.56	1.7353
0.76	1.4142	0.52	1.8528
0.72	1.4573	0.48	2.0076
0.68	1.5082	0.44	2.2203
0.64	1.5690	0.40	2.5298
0.60	1.6432		

Concomitantly determine the observed specific surface area of a similar preparation of USP Griseofulvin Specific Surface Area RS. Calculate the specific surface area of the Griseofulvin taken by the formula:

$$O_U(A_S/O_S),$$

in which O_U is the observed specific surface area of the specimen, A_S is the assigned specific surface area of USP Griseofulvin Specific Surface Area RS, and O_S is the observed specific surface area of USP Griseofulvin Specific Surface Area RS: it is between 1.3 and 1.7 square meters per g.

Assay—

Mobile phase—Prepare a suitable filtered mixture of water, acetonitrile, and tetrahydrofuran (60:35:5). Degas for 5 minutes before use, and stir continuously during use. Make adjustments if necessary (see *System Suitability* under *Chromatography* ⟨621⟩).

Internal standard solution—Dissolve a suitable quantity of 3-phenylphenol in methanol to obtain a solution having a concentration of about 1 mg per mL.

Standard preparation—Dissolve an accurately weighed quantity of USP Griseofulvin RS in methanol to obtain a solution having a known concentration of about 1.25 mg per mL. Transfer 5.0 mL of this solution and 4.0 mL of *Internal standard solution* to a 50-mL volumetric flask, dilute with *Mobile phase* to volume, and mix. This solution contains about 0.125 mg of USP Griseofulvin RS in each mL.

Assay preparation—Transfer about 62 mg of Griseofulvin, accurately weighed, to a 50-mL volumetric flask, dissolve in methanol, dilute with methanol to volume, and mix. Transfer 5.0 mL of this solution to a 50-mL volumetric flask, add 4.0 mL of *Internal standard solution*, dilute with *Mobile phase* to volume, and mix.

Chromatographic system (see *Chromatography* ⟨621⟩)—The liquid chromatograph is equipped with a 254-nm detector and a 4.6-mm × 25-cm column that contains packing L10. The flow rate is about 1 mL per minute. The resolution, R, between griseofulvin and 3-phenylphenol is not less than 5.0, and the relative standard deviation of R_S for replicate injections of *Standard preparation* is not more than 2.0%.

Procedure—Separately inject equal volumes (about 20 μL) of the *Standard preparation* and the *Assay preparation* into the chromatograph, and measure the peak responses for the major peaks. The retention time of griseofulvin relative to that of 3-phenylphenol is about 0.8. Calculate the quantity, in μg of $C_{17}H_{17}ClO_6$, in each mg of the Griseofulvin taken by the formula:

$$500(CP/W_U)(R_U/R_S),$$

in which C is the concentration, in mg per mL, of USP Griseofulvin RS in the *Standard preparation*, P is the content, in μg of $C_{17}H_{17}ClO_6$ per mg, of USP Griseofulvin RS, W_U is the quantity, in mg, of Griseofulvin taken, and R_U and R_S are the ratios of the peak response of griseofulvin to that of 3-phenylphenol obtained from the *Assay preparation* and the *Standard preparation*, respectively.

Griseofulvin Capsules

» Griseofulvin Capsules contain not less than 90.0 percent and not more than 115.0 percent of the labeled amount of $C_{17}H_{17}ClO_6$.

Packaging and storage—Preserve in tight containers.

Labeling—The label indicates that the griseofulvin contained is known as griseofulvin (microsize).

Reference standard—*USP Griseofulvin Reference Standard*—Do not dry before using.

Identification—The retention time of the major peak in the chromatogram of the *Assay preparation* corresponds to that in the chromatogram of the *Standard preparation*, both relative to the internal standard, as obtained in the *Assay*.

Dissolution ⟨711⟩—

Medium: water containing 5.4 mg of sodium lauryl sulfate per mL; 1000 mL.

Apparatus 2: 100 rpm.

Time: 30 minutes.

Procedure—Determine the amount of $C_{17}H_{17}ClO_6$ dissolved from ultraviolet absorbances at the wavelength of maximum absorbance at about 291 nm of filtered portions of the solution under test, suitably diluted with a solution of methanol and water (4:1), if necessary, in comparison with a Standard solution having a known concentration of USP Griseofulvin RS in the same medium.

Tolerances—Not less than 80% (Q) of the labeled amount of $C_{17}H_{17}ClO_6$ is dissolved in 30 minutes.

Uniformity of dosage units ⟨905⟩: meet the requirements.

Procedure for content uniformity—Transfer the contents of 1 Capsule to a suitable container, add an accurately measured volume of methanol sufficient to yield a concentration of griseofulvin not greater than 1 mg per mL, shake by mechanical means for 1 hour, or longer if necessary to disperse the specimen completely, and sonicate for 1 minute. Centrifuge a portion of this solution, and quantitatively dilute an accurately measured volume of the clear supernatant liquid to obtain a test solution containing about 10 µg of griseofulvin per mL. Concomitantly determine the absorbances of the test solution and a Standard solution of USP Griseofulvin RS in methanol having a known concentration of about 10 µg per mL at the wavelength of maximum absorbance at about 292 nm, with a suitable spectrophotometer, using methanol as the blank. Calculate the quantity, in mg, of griseofulvin ($C_{17}H_{17}ClO_6$) in the Capsule by the formula:

$$(CL/D)(A_U/A_S),$$

in which C is the concentration, in µg per mL, of USP Griseofulvin RS in the Standard solution, L is the labeled quantity, in mg, of griseofulvin in the Capsule, D is the concentration, in µg per mL, of griseofulvin in the test solution, based on the labeled quantity per Capsule and the extent of dilution, and A_U and A_S are the absorbances of the test solution and the Standard solution, respectively.

Loss on drying ⟨731⟩—Dry about 100 mg, accurately weighed, in a capillary-stoppered bottle in vacuum at a pressure not exceeding 5 mm of mercury at 60° for 3 hours: it loses not more than 1.0% of its weight.

Assay—

Mobile phase, Internal standard solution, Standard preparation, and *Chromatographic system*—Proceed as directed in the *Assay* under *Griseofulvin.*

Assay preparation—Remove, as completely as possible, the contents of not less than 20 Griseofulvin Capsules, and weigh accurately. Mix, and transfer an accurately weighed portion of the powder, equivalent to about 125 mg of griseofulvin, to a 100-mL volumetric flask. Add about 70 mL of methanol, shake by mechanical means for 30 minutes, dilute with methanol to volume, and mix. Filter a portion of this solution, discarding the first 5 mL of the filtrate. Transfer 5.0 mL of the clear filtrate to a 50-mL volumetric flask, add 4.0 mL of *Internal standard solution*, dilute with *Mobile phase* to volume, and mix.

Procedure—Proceed as directed for *Procedure* in the *Assay* under *Griseofulvin.* Calculate the quantity, in mg, of $C_{17}H_{17}ClO_6$ in the portion of Capsules taken by the formula:

$$(PC)(R_U/R_S),$$

in which C is the concentration, in mg per mL, of USP Griseofulvin RS in the *Standard preparation*, and the other terms are as defined therein.

Griseofulvin Oral Suspension

» Griseofulvin Oral Suspension contains not less than 90.0 percent and not more than 115.0 percent of the labeled amount of $C_{17}H_{17}ClO_6$. It contains one or more suitable colors, diluents, flavors, preservatives, and wetting agents.

Packaging and storage—Preserve in tight containers.

Labeling—The label indicates that the griseofulvin contained is known as griseofulvin (microsize).

Reference standard—*USP Griseofulvin Reference Standard*—Do not dry before using.

Identification—The retention time of the major peak in the chromatogram of the *Assay preparation* corresponds to that in the chromatogram of the *Standard preparation*, both relative to the internal standard, as obtained in the *Assay*.

Uniformity of dosage units ⟨905⟩—

FOR SUSPENSION PACKAGED IN SINGLE-UNIT CONTAINERS: meets the requirements.

pH ⟨791⟩: between 5.5 and 7.5.

Assay—

Mobile phase, Internal standard solution, Standard preparation, and *Chromatographic system*—Proceed as directed in the *Assay* under *Griseofulvin.*

Sodium chloride solution—Dissolve a suitable quantity of sodium chloride in water to obtain a solution containing about 0.1 g per mL.

Assay preparation—Transfer an accurately measured volume of Griseofulvin Oral Suspension, freshly mixed and free from air bubbles, equivalent to about 125 mg of griseofulvin, to a glass-stoppered, 50-mL centrifuge tube. Add 20 mL of *Sodium chloride solution* and 20 mL of methylene chloride. Insert the stopper into the tube, and mix by rotating the tube for 10 minutes. Separate the phases by centrifugation, carefully remove the lower methylene chloride layer with a needle and syringe, and filter through methylene chloride–prerinsed anhydrous sodium sulfate into a 100-mL volumetric flask. Repeat the extraction with two additional 20-mL portions of methylene chloride, combining the extracts in the volumetric flask. Dilute with methylene chloride to volume, and mix. Transfer 5.0 mL of the resulting solution to a 50-mL volumetric flask, and evaporate on a steam bath under a stream of nitrogen to dryness. Transfer 4.0 mL of *Internal standard solution* to the flask, swirl to dissolve the residue, dilute with *Mobile phase* to volume, and mix.

Procedure—Proceed as directed for *Procedure* in the *Assay* under *Griseofulvin.* Calculate the quantity, in mg, of $C_{17}H_{17}ClO_6$ in each mL of the Griseofulvin Oral Suspension taken by the formula:

$$(PC/V)(R_U/R_S),$$

in which C is the concentration, in mg per mL, of USP Griseofulvin RS in the *Standard preparation*, V is the volume, in mL, of Oral Suspension taken, and the other terms are as defined therein.

Griseofulvin Tablets

» Griseofulvin Tablets contain not less than 90.0 percent and not more than 115.0 percent of the labeled amount of $C_{17}H_{17}ClO_6$.

Packaging and storage—Preserve in tight containers.

Labeling—The label indicates that the griseofulvin contained is known as griseofulvin (microsize).

Reference standard—*USP Griseofulvin Reference Standard*—Do not dry before using.

Identification—The retention time of the major peak in the chromatogram of the *Assay preparation* corresponds to that in the chromatogram of the *Standard preparation,* both relative to the internal standard, as obtained in the *Assay.*

Dissolution ⟨711⟩—

Medium: water containing 40.0 mg of sodium lauryl sulfate per mL; 1000 mL.

Apparatus 2: 100 rpm.

Time: 60 minutes.

Procedure—Determine the amount of $C_{17}H_{17}ClO_6$ dissolved from ultraviolet absorbances at the wavelength of maximum absorbance at about 291 nm of filtered portions of the solution under test, suitably diluted with a solution of methanol and water (4:1), if necessary, in comparison with a Standard solution having a known concentration of USP Griseofulvin RS in the same medium.

Tolerances—Not less than 70% (*Q*) of the labeled amount of $C_{17}H_{17}ClO_6$ is dissolved in 60 minutes.

Uniformity of dosage units ⟨905⟩: meet the requirements.

Procedure for content uniformity—Transfer 1 Tablet to a suitable container, add an accurately measured volume of methanol sufficient to yield a concentration of griseofulvin not greater than 1 mg per mL, shake by mechanical means for 1 hour, or longer if necessary to disperse the specimen completely, and sonicate for 1 minute. Centrifuge a portion of this solution, and quantitatively dilute an accurately measured volume of the clear supernatant liquid to obtain a test solution containing about 10 µg of griseofulvin per mL. Concomitantly determine the absorbances of the test solution and a Standard solution of USP Griseofulvin RS in methanol having a known concentration of about 10 µg per mL at the wavelength of maximum absorbance at about 292 nm, with a suitable spectrophotometer, using methanol as the blank. Calculate the quantity, in mg, of griseofulvin ($C_{17}H_{17}ClO_6$) in the Tablet by the formula:

$$(CL/D)(A_U/A_S),$$

in which *C* is the concentration, in µg per mL, of USP Griseofulvin RS in the Standard solution, *L* is the labeled quantity, in mg, of griseofulvin in the Tablet, *D* is the concentration, in µg per mL, of griseofulvin in the test solution, based on the labeled quantity per Tablet and the extent of dilution, and A_U and A_S are the absorbances of the test solution and the Standard solution, respectively.

Loss on drying ⟨731⟩—Dry about 100 mg of finely ground Tablets, accurately weighed, in a capillary-stoppered bottle in vacuum at a pressure not exceeding 5 mm of mercury at 60° for 3 hours: it loses not more than 5.0% of its weight.

Assay—

Mobile phase, Internal standard solution, Standard preparation, and *Chromatographic system*—Proceed as directed in the *Assay* under *Griseofulvin.*

Assay preparation—Weigh and finely powder not less than 20 Griseofulvin Tablets, and proceed as directed for *Assay preparation* in the *Assay* under *Griseofulvin Capsules,* beginning with "transfer an accurately weighed portion."

Procedure—Proceed as directed for *Procedure* in the *Assay* under *Griseofulvin.* Calculate the quantity, in mg, of $C_{17}H_{17}ClO_6$ in the portion of Tablets taken by the formula:

$$(PC)(R_U/R_S),$$

in which *C* is the concentration, in mg per mL, of USP Griseofulvin RS in the *Standard preparation,* and the other terms are as defined therein.

Ultramicrosize Griseofulvin Tablets

» Ultramicrosize Griseofulvin Tablets are composed of ultramicrosize crystals of Griseofulvin dispersed in Polyethylene Glycol 6000 or dispersed by other suitable means. They contain not less than 90.0 percent and not more than 115.0 percent of the labeled amount of $C_{17}H_{17}ClO_6$.

Packaging and storage—Preserve in well-closed containers.

Reference standard—*USP Griseofulvin Reference Standard*—Do not dry before using.

Identification—The retention time of the major peak in the chromatogram of the *Assay preparation* corresponds to that in the chromatogram of the *Standard preparation,* both relative to the internal standard, as obtained in the *Assay.*

Dissolution ⟨711⟩—

Medium: water containing 5.4 mg of sodium lauryl sulfate per mL; 1000 mL.

Apparatus 2: 100 rpm.

Time: 60 minutes.

Procedure—Determine the amount of $C_{17}H_{17}ClO_6$ dissolved from ultraviolet absorbances at the wavelength of maximum absorbance at about 291 nm of filtered portions of the solution under test, suitably diluted with a solution of methanol and water (4:1), if necessary, in comparison with a Standard solution having a known concentration of USP Griseofulvin RS in the same medium.

Tolerances—Not less than 85% (*Q*) of the labeled amount of $C_{17}H_{17}ClO_6$ is dissolved in 60 minutes.

Uniformity of dosage units ⟨905⟩: meet the requirements.

Procedure for content uniformity—Proceed as directed for *Procedure for content uniformity* in the test for *Uniformity of dosage units* under *Griseofulvin Tablets.*

Loss on drying ⟨731⟩—Dry about 100 mg of finely ground Tablets, accurately weighed, in a capillary-stoppered bottle in vacuum at a pressure not exceeding 5 mm of mercury at 60° for 3 hours: it loses not more than 5.0% of its weight.

Assay—

Mobile phase, Internal standard solution, Standard preparation, and *Chromatographic system*—Proceed as directed in the *Assay* under *Griseofulvin.*

Assay preparation—Weigh and finely powder not less than 20 Ultramicrosize Griseofulvin Tablets and proceed as directed for *Assay preparation* in the *Assay* under *Griseofulvin Capsules,* beginning with "transfer an accurately weighed portion."

Procedure—Proceed as directed for *Procedure* in the *Assay* under *Griseofulvin.* Calculate the quantity, in mg, of $C_{17}H_{17}ClO_6$ in the portion of Tablets taken by the formula:

$$(PC)(R_U/R_S),$$

in which *C* is the concentration, in mg per mL, of USP Griseofulvin RS in the *Standard preparation,* and the other terms are as defined therein.

Guaifenesin

OH
OCH_2CHCH_2OH
OCH_3

$C_{10}H_{14}O_4$ 198.22

1,2-Propanediol, 3-(2-methoxyphenoxy)-.

3-(*o*-Methoxyphenoxy)-1,2-propanediol [*93-14-1*].

» Guaifenesin contains not less than 98.0 percent and not more than 102.0 percent of $C_{10}H_{14}O_4$, calculated on the dried basis.

Packaging and storage—Preserve in tight containers.

Reference standards—*USP Guaifenesin Reference Standard*—Dry in vacuum, but at a pressure not below 10 mm of mercury, at 60° to constant weight before using. *USP Guaiacol Reference Standard*—After opening ampul, store in tightly closed container.

Identification—

A: The infrared absorption spectrum of a potassium bromide dispersion of it, previously dried, exhibits maxima only at the same wavelengths as that of a similar preparation of USP Guaifenesin RS.

B: The ultraviolet absorption spectrum of a 1 in 25,000 solution in chloroform exhibits maxima and minima at the same wavelengths as that of a similar solution of USP Guaifenesin RS, concomitantly measured.

C: Mix about 5 mg with 1 drop of formaldehyde and a few drops of sulfuric acid: a deep cherry-red to purple color is produced.

Melting range ⟨741⟩: between 78° and 82°, but the range between beginning and end of melting does not exceed 3°.

pH ⟨791⟩: between 5.0 and 7.0, in a solution (1 in 100).

Loss on drying ⟨731⟩—Dry it in vacuum, but at a pressure not below 10 mm of mercury, at 60° to constant weight: it loses not more than 0.5% of its weight.

Heavy metals, *Method I* ⟨231⟩—The limit is 0.0025%.

Free guaiacol—Transfer 1.0 g to a 100-mL volumetric flask, add 25.0 mL of water, warming if necessary to effect solution, add 1.0 mL of potassium ferricyanide TS, and swirl. Add 5 mL of 4-aminoantipyrine solution (1 in 200), and begin timing the reaction with a stop watch. Swirl the flask for 5 seconds, immediately dilute with sodium bicarbonate solution (1 in 1200) to volume, and mix. Fifteen minutes, accurately timed, after the addition of the 4-aminoantipyrine solution, determine the absorbance of the solution, in a 1-cm cell, at the wavelength of maximum absorbance at about 500 nm, with a suitable spectrophotometer, using a reagent blank to set the instrument. The absorbance of the solution is not greater than that of a control similarly prepared with 3.0 mL of a 1 in 10,000 solution of USP Guaiacol RS, the same quantities of the same reagents being used in the same manner (0.03%).

Assay—Transfer about 100 mg of Guaifenesin, accurately weighed, to a 100-mL volumetric flask, add chloroform to volume, and mix. Transfer 4.0 mL of the solution to a second 100-mL volumetric flask, dilute with chloroform to volume, and mix. Concomitantly determine the absorbances of this solution and a Standard solution of USP Guaifenesin RS in the same medium having a known concentration of about 40 µg per mL in 1-cm cells at the wavelength of maximum absorbance at about 276 nm, with a suitable spectrophotometer, using chloroform as the blank. Calculate the quantity, in mg, of $C_{10}H_{14}O_4$ in the Guaifenesin taken by the formula:

$$2.5C(A_U / A_S),$$

in which C is the concentration, in µg per mL, of USP Guaifenesin RS in the Standard solution, and A_U and A_S are the absorbances of the solution from Guaifenesin and the Standard solution, respectively.

Guaifenesin Capsules

» Guaifenesin Capsules contain not less than 90.0 percent and not more than 110.0 percent of the labeled amount of $C_{10}H_{14}O_4$.

Packaging and storage—Preserve in tight containers.

Reference standard—*USP Guaifenesin Reference Standard*—Dry in vacuum, but at a pressure not below 10 mm of mercury, at 60° to constant weight before using.

Identification—

A: Triturate a portion of the contents of Capsules, equivalent to about 100 mg of guaifenesin, with 10 mL of chloroform, filter, and evaporate 1 mL of the filtrate on a watch glass. Mix the residue with 1 drop of formaldehyde and a few drops of sulfuric acid: a deep cherry-red to purple color is produced.

B: The retention time of the guaifenesin peak in the chromatogram of the *Assay preparation* corresponds to that of the guaifenesin peak in the chromatogram of the *Standard preparation*, as obtained in the *Assay*.

Dissolution ⟨711⟩—

Medium: water; 900 mL.

Apparatus 1: 100 rpm.

Time: 45 minutes.

Procedure—Determine the amount of $C_{10}H_{14}O_4$ dissolved, employing the procedure set forth in the *Assay*, making any necessary modifications.

Tolerances—Not less than 75% (Q) of the labeled amount of $C_{10}H_{14}O_4$ is dissolved in 45 minutes.

Uniformity of dosage units ⟨905⟩: meet the requirements.

Assay—

Internal standard solution—Prepare a solution of dextromethorphan hydrobromide in water having a concentration of about 12 mg per mL.

Standard preparation—Dissolve an accurately weighed quantity of USP Guaifenesin RS in water to obtain a solution having a known concentration of about 10 mg per mL. Pipet 10 mL of this solution into a 60-mL separator, add 10.0 mL of *Internal standard solution* and 2 mL of 2.5 *N* sodium hydroxide, and extract with six 20-mL portions of chloroform, filtering each extract through 1 g of anhydrous sodium sulfate supported by a small cotton plug in a funnel into a 150-mL beaker. Wash the sodium sulfate with 10 mL of chloroform, collecting the filtrate in the 150-mL beaker. Evaporate the combined extracts and washing on a steam bath under nitrogen to a volume of about 3 mL, transfer with the aid of chloroform to a 10-mL volumetric flask, dilute with chloroform to volume, and mix.

Assay preparation—Remove, as completely as possible, the contents of not less than 20 Guaifenesin Capsules. Transfer an accurately weighed portion of the powder, equivalent to about 100 mg of guaifenesin, to a 60-mL separator, and proceed as directed under *Standard preparation*, beginning with "add 10.0 mL of *Internal standard solution*."

Chromatographic system—The gas chromatograph is equipped with a flame-ionization detector, and contains a 1.2-m × 4-mm glass column packed with 3 percent phase G19 on support S1AB, conditioned as directed (see *Chromatography* ⟨621⟩). The column is maintained at a temperature of about 185°, and the injection port and detector block are maintained at about 225°. Helium is used as the carrier gas at a flow rate of about 60 mL per minute. Chromatograph the *Standard preparation*, and record the peak responses as directed under *Procedure:* the relative standard deviation for replicate injections of the *Standard preparation* does not exceed 1.5%, the resolution factor between the two peaks is not less than 4.0, and the tailing factor does not exceed 2.0.

Procedure—Inject 1.5-µL portions of the *Assay preparation* and the *Standard preparation* successively into the gas chromatograph. Measure the peak responses for guaifenesin and dextromethorphan in each chromatogram. The retention time for guaifenesin is about 0.6 relative to that of dextromethorphan. Calculate the quantity, in mg, of $C_{10}H_{14}O_4$ in the portion of Capsule contents taken by the formula:

$$10C(R_U / R_S),$$

in which C is the concentration, in mg per mL, of USP Guaifenesin RS in the *Standard preparation*, and R_U and R_S are the ratios of the response of the guaifenesin peak to the response of the dextromethorphan peak obtained from the *Assay preparation* and the *Standard preparation*, respectively.

Guaifenesin Capsules, Theophylline and—*see* Theophylline and Guaifenesin Capsules

Guaifenesin Oral Solution, Theophylline and—*see* Theophylline and Guaifenesin Oral Solution

Guaifenesin Syrup

» Guaifenesin Syrup contains not less than 95.0 percent and not more than 105.0 percent of the labeled amount of $C_{10}H_{14}O_4$.

Packaging and storage—Preserve in tight containers.

Reference standard—*USP Guaifenesin Reference Standard*—Dry in vacuum, but at a pressure not below 10 mm of mercury, at 60° to constant weight before using.

Identification—The retention time of the guaifenesin peak in the chromatogram of the *Assay preparation* corresponds to that of the guaifenesin peak in the chromatogram of the *Standard preparation*, as obtained in the *Assay*.

pH ⟨791⟩: between 2.3 and 3.0.

Alcohol content, *Method I* ⟨611⟩: between 3.0% and 4.0% of C_2H_5OH.

Assay—

Internal standard solution, Standard preparation, and *Chromatographic system*—Prepare as directed in the *Assay* under *Guaifenesin Capsules.*

Assay preparation—Pipet an accurately measured volume of Guaifenesin Syrup, equivalent to about 100 mg of guaifenesin, into a 60-mL separator, and proceed as directed for *Standard preparation* in the *Assay* under *Guaifenesin Capsules*, beginning with "add 10.0 mL of *Internal standard solution.*"

Procedure—Proceed as directed for *Procedure* in the *Assay* under *Guaifenesin Capsules.* Calculate the quantity, in mg, of $C_{10}H_{14}O_4$ in each mL of the Syrup taken by the formula:

$$(10C/V)(R_U/R_S),$$

in which V is the volume, in mL, of Syrup taken, and C, R_U, and R_S are as defined therein.

Guaifenesin Tables

» Guaifenesin Tablets contain not less than 90.0 percent and not more than 110.0 percent of the labeled amount of $C_{10}H_{14}O_4$.

Packaging and storage—Preserve in tight containers.

Reference standard—*USP Guaifenesin Reference Standard*—Dry in vacuum, but at a pressure not below 10 mm of mercury, at 60° to constant weight before using.

Identification—

A: Triturate a quantity of finely powdered Tablets, equivalent to about 100 mg of guaifenesin, with 10 mL of chloroform, filter, and evaporate 1 mL of the filtrate on a watch glass. Mix the residue with 1 drop of formaldehyde and a few drops of sulfuric acid: a deep cherry-red to purple color is produced.

B: The retention time of the guaifenesin peak in the chromatogram of the *Assay preparation* corresponds to that of the guaifenesin peak in the chromatogram of the *Standard preparation,* as obtained in the *Assay.*

Dissolution ⟨711⟩—

Medium: water; 900 mL.

Apparatus 2: 50 rpm.

Time: 45 minutes.

Procedure—Determine the amount of $C_{10}H_{14}O_4$ dissolved in filtered portions of the solution under test from ultraviolet absorbances at the wavelength of maximum absorbance at about 274 nm in comparison with a Standard solution having a known concentration of USP Guaifenesin RS in the same medium.

Tolerances—Not less than 75% (Q) of the labeled amount of $C_{10}H_{14}O_4$ is dissolved in 45 minutes.

Uniformity of dosage units ⟨905⟩: meet the requirements.

Assay—

Mobile phase—Prepare a suitable filtered and degassed mixture of water, methanol, and glacial acetic acid (60:40:1.5). Make adjustments if necessary (see *System Suitability* under *Chromatography* ⟨621⟩).

Internal standard solution—Dissolve benzoic acid in methanol to obtain a solution containing about 2 mg per mL.

Standard preparation—Dissolve an accurately weighed quantity of USP Guaifenesin RS quantitatively in water, with shaking, to obtain a solution having a known concentration of about 2 mg per mL. Transfer 2.0 mL of this solution and 5.0 mL of *Internal standard solution* to a 100-mL volumetric flask; add 40 mL of methanol, dilute with water to volume, and mix to obtain a *Standard preparation* having a known concentration of about 40 µg per mL.

Assay preparation—Weigh and finely powder not less than 20 Guaifenesin Tablets. Transfer an accurately weighed portion of the powder, equivalent to about 200 mg of guaifenesin, to a 100-mL volumetric flask; add about 60 mL of water, and shake for about 15 minutes. Dilute with water to volume, filter if necessary to obtain a clear solution, and mix. Transfer 2.0 mL of this solution and 5.0 mL of *Internal standard solution* to a 100-mL volumetric flask; add 40 mL of methanol, dilute with water to volume, and mix.

Chromatographic system (see *Chromatography* ⟨621⟩)—The liquid chromatograph is equipped with a 276-nm detector and a 4.6-mm × 25-cm column that contains 10-µm packing L1. The flow rate is about 2 mL per minute. Chromatograph the *Standard preparation*, and record the peak responses as directed under *Procedure:* the resolution, R, between the guaifenesin and the benzoic acid peaks is not less than 3, and the relative standard deviation for replicate injections is not more than 2.5%.

Procedure—[Note—Use peak heights where peak responses are indicated.] Separately inject equal volumes (about 20 µL) of the *Standard preparation* and the *Assay preparation* into the chromatograph, record the chromatograms, and measure the responses for the major peaks. The relative retention times are about 0.7 for guaifenesin and 1.0 for benzoic acid. Calculate the quantity, in mg, of $C_{10}H_{14}O_4$ in the portion of Tablets taken by the formula:

$$5C(R_U/R_S),$$

in which C is the concentration, in µg per mL, of USP Guaifenesin RS in the *Standard preparation*, and R_U and R_S are the ratios of the response of the guaifenesin peak to the response of the benzoic acid peak obtained from the *Assay preparation* and the *Standard preparation*, respectively.

Guanabenz Acetate

$C_8H_8Cl_2N_4.C_2H_4O_2$ 291.14

Hydrazinecarboximidamide, 2-[(2,6-dichlorophenyl)methylene]-, monoacetate.

[(2,6-Dichlorobenzylidene)amino]guanidine monoacetate [*23256-50-0*].

» Guanabenz Acetate contains not less than 98.0 percent and not more than 101.5 percent of the labeled amount of $C_{10}H_{12}N_4O_2Cl_2$.

Packaging and storage—Preserve in tight, light-resistant containers.

Reference standard—*USP Guanabenz Acetate Reference Standard*—Dry in vacuum at 60° for 2 hours before using.

Identification—The infrared absorption spectrum of a potassium bromide dispersion of it, previously dried, exhibits maxima only at the same wavelengths as that of a similar preparation of USP Guanabenz Acetate RS.

pH ⟨791⟩: between 5.5 and 7.0, in a solution (7 in 1000).

Loss on drying ⟨731⟩—Dry it in vacuum at 60° for 2 hours: it loses not more than 1.0% of its weight.

Residue on ignition ⟨281⟩: not more than 0.2%.

Limit of 2,6-dichlorobenzaldehyde—

Internal standard solution 1—Dissolve 100 mg of *p*-chlorobenzaldehyde in 100 mL of chloroform, and mix.

Internal standard solution 2—Dilute 1.0 mL of *Internal standard solution 1* to 10.0 mL with chloroform, and mix.

Standard solution—Prepare a solution of 2,6-dichlorobenzaldehyde in chloroform containing 1.0 mg per mL.

Standard preparation—Transfer 4.0 mL of *Standard solution* and 1.0 mL of *Internal standard solution 1* to a 10-mL volumetric flask, dilute with chloroform to volume, and mix.

Test preparation—Transfer 200 mg of Guanabenz Acetate to a 30-mL glass-stoppered centrifuge tube. Add 10 mL of 0.1 *N* hydrochloric acid, shake to dissolve, add 1.0 mL of *Internal standard solution 2*, and shake. Centrifuge, and transfer a portion of the lower layer to a stoppered container. [NOTE—The lower layer must be removed within 10 minutes of adding the acid to the centrifuge tube.]

Chromatographic system (see *Chromatography* ⟨621⟩)—The gas chromatograph is equipped with a flame-ionization detector and a 1.8-m × 3-mm column packed with 20 percent phase G1 on 80- to 100-mesh support S1A. The column is maintained at a temperature of about 190°, the injection port at about 225°, and the detector at about 250°. Nitrogen is used as the carrier gas at a flow rate of about 30 mL per minute.

Procedure—Separately inject 2-µL portions of the *Standard preparation* and the *Test preparation*, successively, into the gas chromatograph. The resolution between 2,6-dichlorobenzaldehyde and *p*-chlorobenzaldehyde is not less than 3.0, and the relative retention time for *p*-chlorobenzaldehyde is 0.5 and for 2,6-dichlorobenzaldehyde is 1.0. The relative peak response ratio obtained from the *Test preparation* does not exceed that obtained from the *Standard preparation* (0.2%).

Chromatographic purity—

Methanolic formic acid—Prepare a mixture of formic acid and methanol (1 in 2000).

Aminoguanidine bicarbonate solution—Transfer 100 mg of aminoguanidine bicarbonate to a test tube, add 0.05 mL of formic acid, and warm gently to effect solution. Quantitatively transfer the contents of the test tube to a 10-mL volumetric flask, dilute with methanol to volume, and mix.

Standard solution A—Transfer 10 mg of *USP Guanabenz Acetate RS* to a 100-mL volumetric flask, and dissolve in 50 mL of *Methanolic formic acid*. Add 1.0 mL of the *Aminoguanidine bicarbonate solution*, dilute with *Methanolic formic acid* to volume, and mix.

Standard solution B—Transfer 5.0 mL of *Standard solution A* to a 10-mL volumetric flask, dilute with *Methanolic formic acid* to volume, and mix.

Standard solution C—Transfer 2.0 mL of *Standard solution A* to a 10-mL volumetric flask, dilute with *Methanolic formic acid* to volume, and mix.

Test solution—Prepare a solution of guanabenz acetate containing 10 mg per mL in *Methanolic formic acid*.

Procedure—Prepare a chromatographic chamber containing a mixture of chloroform, methanol, and ammonium hydroxide (60:40:1) as the developing solvent and allow it to equilibrate for at least 30 minutes before use. Prewash a plate coated with a 0.25-mm layer of chromatographic silica gel mixture (see *Chromatography* ⟨621⟩) by placing it in the chromatographic chamber, allowing the solvent front to rise to the top of the plate, drying it in air and activating it by heating at 105° for 20 minutes. Within 30 minutes after preparation, separately apply 10-µL portions of *Standard solutions A*, *B*, and *C*, the *Test solution*, and *Methanolic formic acid*. Allow the spots to dry and place the plate in the chromatographic chamber. When the solvent has moved about three-fourths of the length of the plate, remove the plate and allow it to air-dry for about 30 minutes. Examine the plate under short-wavelength ultraviolet light. Estimate the amount of any secondary spots (other than any secondary spot with the same R_f as the *Methanolic formic acid*) observed in the chromatogram of the *Test solution* by comparison with the *Standard solutions*. Place the plate in a chamber saturated with iodine vapors for about 10 minutes. Remove and examine the plate. Estimate the amount of any spot in the chromatogram of the *Test solution* that has an R_f corresponding to the R_f of the spot produced by the aminoguanidine bicarbonate by comparison with the *Standard solutions*. No individual secondary spot is greater in size or intensity than the spot produced by *Standard solution B* (0.5%) and the total of any such spots observed is not more than 1%.

Assay—Dissolve about 200 mg of Guanabenz Acetate, accurately weighed, in 50 mL of glacial acetic acid. Titrate with 0.1 *N* perchloric acid VS, determining the end-point potentiometrically. Perform a blank determination (see *Titrimetry* ⟨541⟩), and make any necessary correction. Each mL of 0.1 *N* perchloric acid is equivalent to 29.12 mg of $C_{10}H_{12}N_4O_2Cl_2$.

Guanabenz Acetate Tablets

» Guanabenz Acetate Tablets contain not less than 90.0 percent and not more than 110.0 percent of the labeled amount of guanabenz ($C_8H_8N_4Cl_2$).

Packaging and storage—Preserve in tight, light-resistant containers.

Reference standard—*USP Guanabenz Acetate Reference Standard*—Dry in vacuum at 60° for 2 hours before using.

Identification—Transfer an amount of powdered Tablets, equivalent to about 8 mg of guanabenz, to a 60-mL separator. Add 10 mL of 0.1 *N* hydrochloric acid, and shake to disperse the powder. Shake the mixture with three 10-mL portions of chloroform, discarding the chloroform phase each time. Add 5 mL of 1 *N* sodium hydroxide, and extract with two 25-mL portions of ether, filtering the ether extracts. Evaporate the combined extracts with the aid of a current of air to dryness: the infrared absorption spectrum of a potassium bromide dispersion of the residue so obtained exhibits maxima at the same wavelengths as that of a similar preparation of USP Guanabenz Acetate RS.

Dissolution ⟨711⟩—

Medium: water; 1000 mL.

Apparatus 2: 50 rpm.

Time: 60 minutes.

Procedure—Determine the amount of $C_8H_8N_4Cl_2$ dissolved from ultraviolet absorbances at the wavelength of maximum absorbance at about 272 nm of filtered portions of the solution under test, suitably diluted with *Dissolution Medium*, in comparison with a Standard solution having a known concentration of USP Guanabenz Acetate RS in the same medium.

Tolerances—Not less than 75% (*Q*) of the labeled amount of $C_8H_8N_4Cl_2$ is dissolved in 60 minutes.

Uniformity of dosage units ⟨905⟩: meet the requirements.

Chromatographic purity—

Extracting solvent, *Mobile phase*, *Standard preparation I*, *System suitability solution*, *Chromatographic system*, and *Assay preparation*—Proceed as directed in the *Assay*.

Standard preparation II—Pipet 2 mL of *Standard preparation I* into a 100-mL volumetric flask, dilute with *Extracting solvent* to volume, and mix.

Procedure—Proceed as directed for *Procedure* in the *Assay*, except to substitute *Standard preparation II* for *Standard preparation I*. Calculate the quantity of any impurity observed having a relative retention time corresponding to the component eluting before guanabenz obtained from the *System suitability solution*. The amount of any such impurity observed is not more than 2%.

Assay—

Extracting solvent—Dissolve 8.2 g of sodium acetate in 20 mL of water, add 5.7 mL of glacial acetic acid, dilute to 1 liter with methanol, and mix.

Mobile phase—Prepare a filtered and degassed mixture of water, methanol, and phosphoric acid (57:43:0.3), making adjustments if necessary (see *System Suitability* under *Chromatography* ⟨621⟩).

Standard preparation I—Transfer about 25 mg of USP Guanabenz Acetate RS, accurately weighed, to a 250-mL volumetric flask. Add 25 mL of water, shake to dissolve the solids, dilute with *Extracting solvent* to volume, and mix.

Assay preparation—Transfer 10 Guanabenz Acetate Tablets to a 500-mL volumetric flask. Add 50 mL of water, stir by mechanical means until the solids are well dispersed, add 400 mL of *Extracting solvent*, and stir for 45 minutes. Dilute with *Extracting solvent* to volume, mix, and centrifuge a portion of the mixture until a clear supernatant liquid is obtained. If necessary, dilute a portion of the supernatant liquid quantitatively with a mixture of *Extracting solvent* and water (9:1) to obtain a solution containing about 0.08 mg of guanabenz per mL.

System suitability solution—Transfer about 30 mg of guanabenz acetate to a 100-mL stoppered flask. Add about 50 mL of

0.1 *N* hydrochloric acid, heat on a steam bath for 60 minutes, and allow the solution to cool.

Chromatographic system (see *Chromatography* ⟨621⟩)—The liquid chromatograph is equipped with a 245-nm detector and a 4-mm × 30-cm column that contains packing L1. The flow rate is about 2 mL per minute. Chromatograph replicate injections of *Standard preparation I* and record the peak responses as directed under *Procedure:* the relative standard deviation is not more than 2.0%. Inject a volume (about 20 µL) of the *System suitability solution* into the chromatograph and record the chromatogram: the resolution between guanabenz and the peak eluting before it is not less than 1.6.

Procedure—Separately inject equal volumes (about 20 µL) of *Standard preparation I* and the *Assay preparation* into the chromatograph, record the chromatograms, and measure the responses for the major peaks. Calculate the quantity, in mg, of $C_8H_8N_4Cl_2$ in the portion of Tablets taken by the formula:

$$(231.08/291.14)(CD)(r_U/r_S),$$

in which 231.08 and 291.14 are the molecular weights of guanabenz and guanabenz acetate, respectively, *C* is the concentration, in mg per mL, of USP Guanabenz Acetate RS in *Standard preparation I*, *D* is the *Assay preparation* dilution factor, in mL per Tablet, and r_U and r_S are the peak responses of the *Assay preparation* and the *Standard preparation I*, respectively.

Guanadrel Sulfate

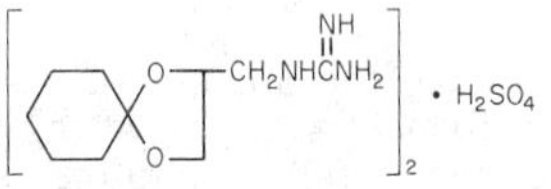

$(C_{10}H_{19}N_3O_2)_2 . H_2SO_4$ 524.63

Guanidine (1,4-dioxaspiro[4.5]dec-2-ylmethyl)-, sulfate (2:1).

(1,4-Dioxaspiro[4.5]dec-2-ylmethyl)guanidine sulfate (2:1) [*22195-34-2*].

» Guanadrel Sulfate contains not less than 97.0 percent and not more than 103.0 percent of $(C_{10}H_{19}N_3O_2)_2 . H_2SO_4$, calculated on the dried basis.

Packaging and storage—Preserve in well-closed containers.

Reference standard—*USP Guanadrel Sulfate Reference Standard*—Dry at room temperature at a pressure not exceeding 5 mm of mercury for 16 hours at time of use for quantitative analyses.

Identification—The infrared absorption spectrum of a mineral oil dispersion of it exhibits maxima only at the same wavelengths as that of a similar preparation of USP Guanadrel Sulfate RS.

Loss on drying ⟨731⟩—Dry it at room temperature and at a pressure not exceeding 5 mm of mercury for 16 hours: it loses not more than 0.5% of its weight.

Residue on ignition ⟨281⟩: not more than 0.5%.

Heavy metals, *Method II* ⟨231⟩: 0.002%.

Assay—

Mobile phase—Prepare a filtered and degassed mixture of 530 mL of water and 470 mL of methanol containing about 6.35 g of *dl*-10-camphorsulfonic acid sodium salt and 0.8 g of ammonium nitrate. Adjust with glacial acetic acid, if necessary, to a pH of between 5.0 and 5.5. Make adjustments if necessary (see *System Suitability* under *Chromatography* ⟨621⟩).

Standard preparation—Dissolve an accurately weighed quantity of USP Guanadrel Sulfate RS in *Mobile phase* to obtain a solution having a known concentration of about 10 mg per mL.

Assay preparation—Transfer about 100 mg of Guanadrel Sulfate, accurately weighed, to a container, add 10.0 mL of *Mobile phase*, and mix.

Resolution solution—Dissolve suitable quantities of USP Guanadrel Sulfate RS and ethylparaben in *Mobile phase* to obtain a solution having known concentrations of about 10 mg and 12 mg, respectively, in each mL.

Chromatographic system (see *Chromatography* ⟨621⟩)—The liquid chromatograph is equipped with a refractive index detector and a 4- to 4.6-mm × 25- to 30-cm stainless steel column containing packing L1. The flow rate is about 1.5 mL per minute. Chromatograph the *Resolution solution*, and record the peak responses as directed under *Procedure:* the resolution, *R*, between the guanadrel and ethylparaben peaks is not less than 1.6 and the relative retention times are about 0.8 for guanadrel and 1.0 for ethylparaben. Chromatograph the *Standard preparation*, and record the peak responses as directed under *Procedure:* the relative standard deviation for replicate injections is not more than 2.5%.

Procedure—[NOTE—Use peak heights where peak responses are indicated.] Separately inject equal volumes (about 25 µL) of the *Standard preparation* and the *Assay preparation* into the chromatograph, record the chromatograms, and measure the responses for the major peaks. Calculate the quantity, in mg, of $(C_{10}H_{19}N_3O_2)_2 . H_2SO_4$ in the portion of Guanadrel Sulfate taken by the formula:

$$10C(r_U/r_S),$$

in which *C* is the concentration, in mg per mL, of USP Guanadrel Sulfate RS in the *Standard preparation*, and r_U and r_S are the peak responses obtained from the *Assay preparation* and the *Standard preparation*, respectively.

Guanadrel Sulfate Tablets

» Guanadrel Sulfate Tablets contain not less than 90.0 percent and not more than 110.0 percent of the labeled amount of $(C_{10}H_{19}N_3O_2)_2 . H_2SO_4$.

Packaging and storage—Preserve in tight, light-resistant containers.

Reference standard—*USP Guanadrel Sulfate Reference Standard*—Dry at room temperature at a pressure not exceeding 5 mm of mercury for 16 hours at time of use for quantitative analyses.

Identification—To a portion of 1 finely powdered Tablet, equivalent to about 4 mg of guanadrel sulfate, add 2 mL of a 1% aqueous alkaline solution of 1-naphthol (containing 6 g of sodium hydroxide and 16 g of sodium carbonate per 100 mL of water) and 1 mL of 2,3-butanedione solution (1 in 2000), and mix. Allow to stand at room temperature: an intense, pinkish red color develops.

Dissolution ⟨711⟩—

Stock buffer solution—Dissolve 245 g of monobasic potassium phosphate and 56.3 g of sodium hydroxide in water, and dilute with water to obtain 2.0 liters of solution. Add, with mixing, phosphoric acid or 1 *N* sodium hydroxide, if necessary, to adjust the solution such that, when this *Stock buffer solution* is diluted 1 in 90 with water, the resulting solution has a pH of 7.40 ± 0.05.

Working buffer solution—Prepare a 1 in 90 dilution of *Stock buffer solution* in water to obtain a *Working buffer solution* having a pH of 7.40 ± 0.05.

Medium: Working buffer solution; 900 mL.

Apparatus 2: 50 rpm.

Time: 20 minutes.

Stock standard solution—Prepare a solution of USP Guanadrel Sulfate RS in *Working buffer solution* having an accurately known final concentration of about 0.5 mg per mL.

Working standard solution—Transfer 5.0 mL of *Stock standard solution* to a 200-mL volumetric flask, dilute with *Working buffer solution* to volume, and mix.

Working color reagent preparation—Transfer 50.0 mL of the *Color reagent preparation*, prepared as directed for *Procedure for content uniformity* under *Uniformity of dosage units*, to a 500-mL volumetric flask, dilute with water to volume, and mix.

Color reagent blank—[NOTE—Prepare this solution concurrently with the preparation of the *Working standard solution* and the solution under test.] Transfer 40.0 mL of *Working buffer solution* to a container, add 8.0 mL of *Working color reagent preparation*, and mix.

Procedure—After 20 minutes, withdraw a portion of the solution under test, and filter immediately. For Tablet potencies of less than 25 mg per Tablet, transfer 20.0 mL of the filtered solution to a container. For Tablet potencies equal to or greater than 25 mg, transfer 10.0 mL of the filtered solution and 10.0 mL of *Working buffer solution* to a container. Transfer 20.0 mL of the *Working standard solution* to another, similar container. Separately add 4.0 mL of *Working color reagent preparation* to the solution of the test specimen and the *Working standard solution*, and mix. Using a suitable spectrophotometer, determine the absorbances of the solutions obtained from the test specimen and the *Working standard solution*, in 5-cm cells, at the wavelength of maximum absorbance at about 494 nm, using the *Color reagent blank* in the reference cell. [NOTE—Once the *Working color reagent preparation* has been added to the filtered dissolution specimens and mixed, determine the absorbances so that none of the solutions stands for less than 20 minutes or more than 80 minutes.] Calculate the amount of $(C_{10}H_{19}N_3O_2)_2 . H_2SO_4$ dissolved by comparison of the absorbances obtained with the solutions obtained from the *Working standard solution* and the solution under test.

Tolerances—Not less than 70% (*Q*) of the labeled amount of $(C_{10}H_{19}N_3O_2)_2 . H_2SO_4$ is dissolved in 20 minutes.

Uniformity of dosage units ⟨905⟩: meet the requirements.

Procedure for content uniformity—

Color reagent preparation—Prepare separately sodium nitroferricyanide solution (1 in 10), potassium ferricyanide solution (1 in 10), and sodium hydroxide solution (1 in 10), and store in separate amber-colored bottles. Mix an equal and sufficient volume of each of these solutions, in the order listed, and allow to stand for about 15 minutes. The solution changes from a deep red-black to a yellow-green color. Prepare a 1 in 10 dilution of the solution in water to obtain the *Color reagent preparation*. [NOTE—Prepare this *Color reagent preparation* on the day of use. The 10% aqueous solutions are stable for about 2 months.]

Standard preparation—Dissolve an accurately weighed quantity of USP Guanadrel Sulfate RS in water to obtain a solution having a known concentration of about 0.1 mg per mL. Transfer 10.0 mL of this solution to a container, and proceed as directed under *Procedure*.

Test preparation—Transfer 1 Tablet to a 100-mL volumetric flask, dilute with water to volume, and shake the flask vigorously for about 4 minutes. Transfer not less than 25 mL of the solution to a vial, and centrifuge for 10 minutes. Transfer an accurately measured volume of this solution, equivalent to 1 mg of guanadrel sulfate, to another container and dilute, if necessary, with an accurately measured volume of water to a volume of 10.0 mL. Mix, and proceed as directed under *Procedure*.

Procedure—Separately add 4.0 mL of the *Color reagent preparation* to the *Standard preparation*, the *Test preparation*, and 10.0 mL of water to provide the blank. Mix the solutions, and allow to stand for 10 minutes. Within 5 minutes, determine the absorbances of the solutions in 1-cm cells at the wavelength of maximum absorbance at about 494 nm, with a suitable spectrophotometer, against the reagent blank. Calculate the quantity, in mg, of $(C_{10}H_{19}N_3O_2)_2 . H_2SO_4$ in the Tablet taken by the formula:

$$(TC/D)(A_U/A_S),$$

in which T is the labeled quantity, in mg, of guanadrel sulfate in the Tablet, C is the concentration, in mg per mL, of USP Guanadrel Sulfate RS in the *Standard preparation*, D is the concentration, in mg per mL, of guanadrel sulfate in the *Test preparation*, based upon the labeled quantity per Tablet and the extent of dilution, and A_U and A_S are the absorbances of the solutions from the solution under test and the *Standard preparation*, respectively.

Assay—

Mobile phase, Standard preparation, Resolution solution, and *Chromatographic system*—Proceed as directed in the *Assay* under *Guanadrel Sulfate*.

Assay preparation—Transfer a number of Tablets, equivalent to about 100 mg of guanadrel sulfate, to a container. Add an accurately measured volume of *Mobile phase* to obtain a final concentration of about 10 mg of guanadrel sulfate per mL of *Mobile phase*, shake by mechanical means for 20 minutes, and centrifuge if necessary.

Procedure—Proceed as directed for *Procedure* in the *Assay* under *Guanadrel Sulfate*. Calculate the quantity, in mg, of $(C_{10}H_{19}N_3O_2)_2 . H_2SO_4$ in the Tablets taken by the formula:

$$VC(r_U/r_S),$$

in which V is the volume, in mL, of *Mobile phase* added to the Tablets, and the other terms are as defined in the *Assay* under *Guanadrel Sulfate*.

Guanethidine Monosulfate

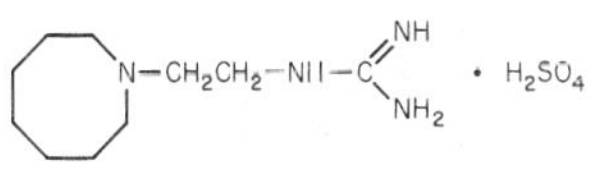

$C_{10}H_{22}N_4 . H_2SO_4$ 296.38

Guanidine, [2-(hexahydro-1(2*H*)-azocinyl)ethyl]-, sulfate (1:1).

[2-(Hexahydro-1(2*H*)-azocinyl)ethyl]guanidine sulfate (1:1) [*645-43-2*].

» Guanethidine Monosulfate contains not less than 97.0 percent and not more than 103.0 percent of $C_{10}H_{22}N_4 . H_2SO_4$, calculated on the dried basis.

Packaging and storage—Preserve in well-closed containers.

Reference standard—*USP Guanethidine Monosulfate Reference Standard*—Dry at 105° to constant weight before using.

Identification—

A: The infrared absorption spectrum of a mineral oil dispersion of it exhibits maxima only at the same wavelengths as that of similar preparation of USP Guanethidine Monosulfate RS.

B: Dissolve 2.5 mg in 10 mL of water. Add 2 mL of a solution prepared by dissolving 500 mg of 1-naphthol, 3 g of sodium hydroxide, and 8 g of sodium carbonate in water to make 50 mL, and 1 mL of a solution of 2,3-butanedione (1 in 2,000). Allow to stand at room temperature: an intense, pinkish red color develops.

pH ⟨791⟩: between 4.7 and 5.7, in a solution containing 20 mg per mL.

Loss on drying ⟨731⟩—Dry it at 105° to constant weight: it loses not more than 0.5% of its weight.

Residue on ignition ⟨281⟩: not more than 0.2%.

Heavy metals, *Method II* ⟨231⟩: 0.001%.

Assay—

Sodium nitroferricyanide–potassium ferricyanide solution—Dissolve 1 g of sodium nitroferricyanide and 1 g of potassium ferricyanide in water to make 100 mL, and mix.

Standard preparation—Dissolve a suitable quantity of USP Guanethidine Monosulfate RS, accurately weighed, in 1 *N* sulfuric acid to obtain a solution having a known concentration of about 1 mg per mL.

Assay preparation—Transfer about 50 mg of Guanethidine Monosulfate, accurately weighed, to a 50-mL volumetric flask, dissolve in 1 *N* sulfuric acid, add 1 *N* sulfuric acid to volume, and mix.

Procedure—Pipet 2 mL each of the *Assay preparation* and the *Standard preparation*, and 2 mL of 1 *N* sulfuric acid to provide the blank, into separate glass-stoppered, 40-mL centrifuge tubes. Add 10.0 mL of water to each tube, and mix. To each tube add 10.0 mL of *Sodium nitroferricyanide–potassium ferricyanide solution*, mix, add 4.0 mL of 1 *N* sodium hydroxide, mix, and allow to stand for 20 minutes, accurately timed. Concomitantly determine the absorbances of both solutions in 1-cm cells against the blank at the wavelength of maximum absorbance at about 500 nm, with a suitable spectrophotometer. Calculate the quantity, in mg, of $C_{10}H_{22}N_4 . H_2SO_4$ in the Guanethidine Monosulfate taken by the formula:

$$50C(A_U/A_S),$$

in which C is the concentration, in mg per mL, of USP Guanethidine Monosulfate RS in the *Standard preparation*, and A_U and A_S are the absorbances of the solutions from the *Assay preparation* and the *Standard preparation*, respectively.

Guanethidine Monosulfate Tablets

» Guanethidine Monosulfate Tablets contain an amount of guanethidine monosulfate ($C_{10}H_{22}N_4 \cdot H_2SO_4$) equivalent to not less than 90.0 percent and not more than 110.0 percent of the labeled amount of guanethidine sulfate [$(C_{10}H_{22}N_4)_2 \cdot H_2SO_4$].

Packaging and storage—Preserve in well-closed containers.

Reference standard—*USP Guanethidine Monosulfate Reference Standard*—Dry at 105° to constant weight.

Identification—Transfer a quantity of powdered Tablets, equivalent to about 120 mg of guanethidine monosulfate, to a glass-stoppered flask, add 20 mL of water, shake by mechanical means for 30 minutes, and filter, discarding the first few mL of the filtrate. Transfer 10 mL of the filtrate to a separator, add 2 mL of 0.1 *N* sodium hydroxide and 2 mL of a saturated solution of picric acid in 0.1 *N* sodium hydroxide, and mix. Extract with 20 mL of chloroform, filter the chloroform extract through cotton, and collect in a beaker. Evaporate the chloroform extract with the aid of a stream of nitrogen to dryness: the infrared absorption spectrum of a mineral oil dispersion of the residue so obtained exhibits maxima only at the same wavelengths as that of a similar preparation of USP Guanethidine Monosulfate RS.

Dissolution ⟨711⟩—

Medium: water; 500 mL.

Apparatus 1: 100 rpm.

Time: 45 minutes.

Procedure—Determine the amount of $(C_{10}H_{22}N_4)_2 \cdot H_2SO_4$ dissolved, employing the procedure set forth in the *Assay*, making any necessary modifications.

Tolerances—Not less than 75% (*Q*) of the labeled amount of $(C_{10}H_{22}N_4)_2 \cdot H_2SO_4$ is dissolved in 45 minutes.

Uniformity of dosage units ⟨905⟩: meet the requirements.

Assay—

Borate solution—Transfer 12.4 g of boric acid to a 1000-mL volumetric flask, dissolve in 100 mL of 1.0 *N* sodium hydroxide, dilute with water to volume, and mix.

Borate buffer—Mix 400 mL of *Borate solution* with 600 mL of 0.1 *N* sodium hydroxide, and adjust with 1.0 *N* sodium hydroxide or 1.0 *N* hydrochloric acid to a pH of 12.3 ± 0.1.

Picrate reagent—Transfer 15 g of picric acid to a 1000-mL volumetric flask, dissolve in 750 mL of 0.1 *N* sodium hydroxide, dilute with water to volume, and mix.

Standard preparation—Prepare a solution in 0.1 *N* sulfuric acid having a known concentration of about 0.3 mg of USP Guanethidine Monosulfate RS per mL.

Assay preparation—Finely powder not less than 20 Guanethidine Monosulfate Tablets. Transfer an accurately weighed portion of the powder, equivalent to about 50 mg of guanethidine sulfate, to a 200-mL volumetric flask, add 150 mL of 0.1 *N* sulfuric acid, shake by mechanical means for 30 minutes, dilute with 0.1 *N* sulfuric acid to volume, and mix. Filter, and discard the first 25 mL of the filtrate.

Procedure—Pipet 5 mL each of the *Standard preparation* and the *Assay preparation* into separate 125-mL separators. To each separator add 20 mL of *Borate buffer* and 20 mL of *Picrate reagent*. Extract with three 20-mL portions of chloroform, filtering the extracts through chloroform-prerinsed cotton, and collect the extracts in low-actinic 100-mL volumetric flasks. Rinse the cotton with 20 mL of chloroform, adding the rinsings to the volumetric flasks, dilute with chloroform to volume, and mix. Concomitantly determine the absorbances of the solutions at the wavelength of maximum absorbance at about 412 nm, with a suitable spectrophotometer, using chloroform as the blank. Calculate the quantity, in mg, of guanethidine sulfate [$(C_{10}H_{22}N_4)_2 \cdot H_2SO_4$] in the portion of Tablets taken by the formula:

$$(247.35/296.38)(200)(C)(A_U/A_S),$$

in which 247.35 is one-half the molecular weight of guanethidine sulfate, 296.38 is the molecular weight of guanethidine monosulfate, C is the concentration, in mg per mL, of USP Guanethidine Monosulfate RS in the *Standard preparation*, and A_U and A_S are the absorbances of the solutions from the *Assay preparation* and the *Standard preparation*, respectively.

Guar Gum—*see* Guar Gum NF

Gutta Percha

» Gutta Percha is the coagulated, dried, purified latex of the trees of the genera *Palaquium* and *Payena* and most commonly *Palaquium gutta* (Hooker) Baillon (Fam. Sapotaceae).

Packaging and storage—Preserve under water in well-closed containers, protected from light.

Residue on ignition ⟨281⟩: not more than 1.7%.

Halazepam

$C_{17}H_{12}ClF_3N_2O$ 352.74

2*H*-1,4-Benzodiazepin-2-one, 7-chloro-1,3-dihydro-5-phenyl-1-(2,2,2-trifluoroethyl)-.

7-Chloro-1,3-dihydro-5-phenyl-1-(2,2,2-trifluoroethyl)-2*H*-1,4-benzodiazepin-2-one [*23092-17-3*].

» Halazepam contains not less than 98.5 percent and not more than 101.0 percent of $C_{17}H_{12}ClF_3N_2O$, calculated on the dried basis.

Packaging and storage—Preserve in well-closed containers.

Reference standards—*USP Halazepam Reference Standard*—Dry at 105° for 3 hours before using. *USP 2-[N-(2,2,2-Trifluoroethyl)amino]-5-chlorobenzophenone Reference Standard*—Do not dry before using.

Identification—

A: The infrared absorption spectrum of a mineral oil dispersion of it, previously dried, exhibits maxima only at the same wavelengths as that of a similar preparation of USP Halazepam RS.

B: The ultraviolet absorption spectrum of a 1 in 100,000 solution of it in a 1 in 120 solution of hydrochloric acid in methanol exhibits maxima and minima at the same wavelengths as that of a similar solution of USP Halazepam RS, concomitantly measured, and the respective absorptivities, calculated on the dried basis, at the wavelength of maximum absorbance at about 284 nm do not differ by more than 3.0%.

C: The principal spot obtained from the chromatogram of the *Test preparation* exhibits an R_f value corresponding to that of *Solution A* in the test for *Related compounds*.

Loss on drying ⟨731⟩—Dry it at 105° for 3 hours: it loses not more than 1.0% of its weight.

Residue on ignition ⟨281⟩: not more than 0.2%.

Heavy metals, *Method II* ⟨231⟩: 0.002%.

Related compounds—

Standard preparations—Dissolve USP Halazepam RS in methanol to obtain a solution having a concentration of 10 mg per mL (*Solution A*). Dilute a portion of *Solution A* quantitatively, and stepwise if necessary, with methanol to obtain a solution containing 0.10 mg per mL (*Solution B*). Pipet 5 mL of *Solution B* into a 10-mL volumetric flask, dilute with methanol to volume, and mix (*Solution C*). Dissolve USP 2-[*N*-(2,2,2-Trifluoroethyl)amino]-5-chlorobenzophenone RS in methanol to obtain a solution having a concentration of 0.10 mg per mL (*Solution D*).

Test preparation—Dissolve Halazepam in methanol to obtain a solution having a concentration of 10 mg per mL.

Procedure—On a suitable thin-layer chromatographic plate (see *Chromatography* ⟨621⟩), coated with a 0.25-mm layer of chromatographic silica gel mixture, apply 50-µL portions of *Solution A* and the *Test preparation*, and similarly apply 10-µL portions of *Solutions B, C,* and *D*. Allow the spots to dry, and develop the chromatogram in a solvent system consisting of toluene, ethyl acetate, and glacial acetic acid (18:3:1) until the solvent front has moved about one-half of the length of the plate. Remove the plate from the developing chamber, mark the solvent front, and dry the plate. Examine the plate under short-wavelength ultraviolet light: the chromatograms from the *Test preparation* and *Solution A* show principal spots at the same R_f value. Estimate the levels of any spots, other than the principal spot, observed in the chromatogram of the *Test preparation* by comparison with the spots in the chromatograms of *Solutions B, C,* and *D:* any spot at the corresponding R_f value produced by the *Test preparation* is not greater in size or intensity than the principal spot produced by *Solution D*, corresponding to not more than 0.2% of 2-[*N*-(2,2,2-Trifluoroethyl)amino]-5-chlorobenzophenone. Not more than 3 additional halazepam-related spots are produced by the *Test preparation*, of which no individual spot exhibits an intensity greater than that of the principal spot produced by *Solution C*, corresponding to not more than 0.1%, and the total of any observed spots does not exceed 0.2%.

Assay—Dissolve about 600 mg of Halazepam, accurately weighed, in 50 mL of a mixture of glacial acetic acid and acetic anhydride (1:1) in a suitable conical flask. Add 2 drops of crystal violet TS, and titrate with 0.1 *N* perchloric acid VS to a green endpoint. Perform a blank determination, and make any necessary correction. Each mL of 0.1 *N* perchloric acid is equivalent to 35.27 mg of $C_{17}H_{12}ClF_3N_2O$.

Halazepam Tablets

» Halazepam Tablets contain not less than 90.0 percent and not more than 110.0 percent of the labeled amount of $C_{17}H_{12}ClF_3N_2O$.

Packaging and storage—Preserve in well-closed containers.

Reference standard—*USP Halazepam Reference Standard*—Dry at 105° for 3 hours before using.

Identification—

A: The retention time of the major peak in the chromatogram of the *Assay preparation* corresponds to that of the *Standard preparation*, both relative to the internal standard, as obtained in the *Assay*.

B: Place a quantity of finely powdered Tablets, equivalent to about 40 mg of halazepam, in a 50-mL centrifuge tube. Add 4.0 mL of methanol, insert a stopper in the tube, and place it in an ultrasonic bath for 10 minutes. Centrifuge the mixture, and use the clear supernatant liquid as the test solution. On a suitable thin-layer chromatographic plate (see *Chromatography* ⟨621⟩), coated with a 0.25-mm layer of chromatographic silica gel mixture, apply separately 25 µL of the test solution and 25 µL of a Standard solution of USP Halazepam RS in methanol containing about 10 mg per mL, and dry the applications with the aid of a stream of nitrogen. Position the plate in a chromatographic chamber, and develop the chromatograms in a solvent system consisting of toluene, ethyl acetate, and glacial acetic acid (18:3:1) until the solvent front has moved about three-fourths of the length of the plate. Remove the plate from the developing chamber, mark the solvent front, and allow the solvent to evaporate. Observe the plate under short-wavelength ultraviolet light: the R_f value of the principal spot obtained from the test solution corresponds to that obtained from the Standard solution.

Dissolution ⟨711⟩—

Medium: 0.1 *N* hydrochloric acid; 900 mL.

Apparatus 1: 100 rpm.

Time: 30 minutes.

Procedure—Determine the amount of $C_{17}H_{12}ClF_3N_2O$ dissolved from ultraviolet absorbances at the wavelength of maximum absorbance at about 284 nm of filtered portions of the solution under test, suitably diluted with *Dissolution Medium*, if necessary, in comparison with a Standard solution having a known concentration of USP Halazepam RS in the same medium.

Tolerances—Not less than 75% (*Q*) of the labeled amount of $C_{17}H_{12}ClF_3N_2O$ is dissolved in 30 minutes.

Uniformity of dosage units ⟨905⟩: meet the requirements.

Assay—

Mobile phase—Prepare a filtered and degassed mixture of water and acetonitrile (3:1). Make adjustments if necessary (see *System Suitability* under *Chromatography* ⟨621⟩).

Internal standard solution—Transfer about 25 mg of diazepam to a 50-mL volumetric flask, add methanol to volume, and mix.

Standard preparation—Dissolve an accurately weighed quantity of USP Halazepam RS in methanol, and dilute quantitatively with methanol to obtain a solution having a known concentration of about 1 mg per mL. Transfer 2.0 mL of this solution to a vial, add 2.0 mL of *Internal standard solution*, and mix to obtain a solution having a known concentration of about 0.5 mg of USP Halazepam RS per mL.

Assay preparation—Weigh and finely powder not less than 10 Halazepam Tablets. Transfer an accurately weighed portion of the powder, equivalent to about 20 mg of halazepam, to a screw-capped, 50-mL centrifuge tube, and add 20.0 mL of methanol. Fasten the cap on the tube, disperse the solid, and rotate the mixture for 15 minutes. [NOTE—Do not use a mechanical shaker.] Centrifuge, transfer 2.0 mL of the supernatant liquid to a vial, add 2.0 mL of *Internal standard solution*, and mix.

Chromatographic system (see *Chromatography* ⟨621⟩)—The liquid chromatograph is equipped with a 254-nm detector and a 2-mm × 50-cm column that contains packing L2. The flow rate is about 0.6 mL per minute. Chromatograph the *Standard preparation*, and record the peak responses as directed under *Procedure:* the resolution, *R*, between the analyte and internal standard peaks is not less than 2, and the relative standard deviation for replicate injections is not more than 2%.

Procedure—Separately inject equal volumes (about 10 µL) of the *Standard preparation* and the *Assay preparation* into the chromatograph, record the chromatograms, and measure the responses for the major peaks. The relative retention times are about 0.6 for diazepam and 1.0 for halazepam. Calculate the quantity, in mg, of $C_{17}H_{12}ClF_3N_2O$ in the portion of Tablets taken by the formula:

$$40C(R_U/R_S),$$

in which *C* is the concentration, in mg per mL, of USP Halazepam RS in the *Standard preparation*, and R_U and R_S are the peak response ratios obtained from the *Assay preparation* and the *Standard preparation*, respectively.

Halazone

HOOC— —SO_2NCl_2

$C_7H_5Cl_2NO_4S$ 270.09

Benzoic acid, 4-[(dichloroamino)sulfonyl]-.

p-(Dichlorosulfamoyl)benzoic acid [*80-13-7*].

» Halazone contains not less than 91.5 percent and not more than 100.5 percent of $C_7H_5Cl_2NO_4S$, calculated on the dried basis.

Packaging and storage—Preserve in tight, light-resistant containers.

Identification—Add about 100 mg to 5 mL of sodium bromide solution (1 in 10): bromine is liberated from the mixture.

Loss on drying ⟨731⟩—Dry it over phosphorus pentoxide for 4 hours: it loses not more than 0.5% of its weight.

Readily carbonizable substances ⟨271⟩—Dissolve 100 mg in 0.5 mL of sulfuric acid TS: no blackening occurs, although some effervescence may take place.

Assay—Add about 150 mg of Halazone accurately weighed, to 10 mL of 2.5 *N* sodium hydroxide in a 250-mL iodine flask, stir well to dissolve, add 75 mL of water, then promptly add 15 mL of potassium iodide solution (1 in 10), and mix. Acidify with 10 mL of 6 *N* acetic acid, and immediately titrate the liberated iodine with 0.1 *N* sodium thiosulfate VS, adding 3 mL of starch TS as the end-point is approached. Perform a blank determination, and make any necessary correction. Each mL of 0.1 *N* sodium thiosulfate is equivalent to 6.752 mg of $C_7H_5Cl_2NO_4S$.

Halazone Tablets for Solution

» Halazone Tablets for Solution contain not less than 90.0 percent and not more than 135.0 percent of the labeled amount of $C_7H_5Cl_2NO_4S$.

Packaging and storage—Preserve in tight, light-resistant containers.

Labeling—Label Tablets to indicate that they are not intended to be swallowed.

Identification—Finely powder a number of Tablets, equivalent to about 150 mg of halazone: a portion of the powder, equivalent to about 100 mg of halazone, responds to the *Identification test* under *Halazone.*

Disintegration ⟨701⟩: 10 minutes.

Uniformity of dosage units ⟨905⟩: meet the requirements, except that if the average value of the dosage units tested is between 100.0 percent and 135.0 percent, *Criterion (B) (3)* applies.

pH ⟨791⟩: not less than 7.0, in a solution of 1 Tablet, containing 4 mg of halazone, in 200 mL of water.

Assay—Transfer a counted number of Halazone Tablets for Solution, equivalent to about 160 mg of halazone, to a suitable container, and proceed as directed in the *Assay* under *Halazone.* Each mL of 0.1 *N* sodium thiosulfate is equivalent to 6.752 mg of $C_7H_5Cl_2NO_4S$.

Halcinonide

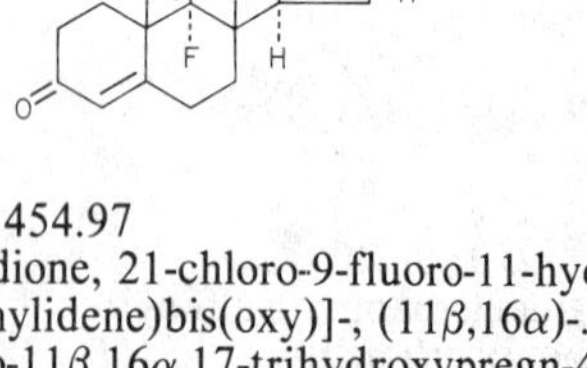

$C_{24}H_{32}ClFO_5$ 454.97

Pregn-4-ene-3,20-dione, 21-chloro-9-fluoro-11-hydroxy-16,17-[(1-methylethylidene)bis(oxy)]-, (11β,16α)-.

21-Chloro-9-fluoro-11β,16α,17-trihydroxypregn-4-ene-3,20-dione cyclic 16,17-acetal with acetone [*3093-35-4*].

» Halcinonide contains not less than 97.0 percent and not more than 102.0 percent of $C_{24}H_{32}ClFO_5$.

Packaging and storage—Preserve in well-closed containers.

Reference standard—*USP Halcinonide Reference Standard*—Dry in vacuum at 100° for 3 hours before using.

Identification—The infrared absorption spectrum of a potassium bromide dispersion of it, prepared by triturating it with potassium bromide and 1 mL of anhydrous methanol, then drying in vacuum at 60° before compressing into a pellet, exhibits maxima only at the same wavelengths as that of a similar preparation of USP Halcinonide RS.

Specific rotation ⟨781⟩: between +150° and +160°, determined in a solution in chloroform containing 20 mg per mL.

Loss on drying ⟨731⟩—Dry it in vacuum at 100° for 3 hours: it loses not more than 1.0% of its weight.

Residue on ignition ⟨281⟩: not more than 0.2%.

Chromatographic purity—Prepare the test solution by dissolving 50 mg of Halcinonide in 5.0 mL of a mixture of chloroform and methanol (1:1). Divide the area of a suitable thin-layer chromatographic plate (see *Chromatography* ⟨621⟩), coated with a 0.25-mm layer of chromatographic silica gel mixture, into three equal sections, the first two sections to be used for the test solution and the third section for the blank. Apply 100 µL of the test solution to appropriate sections of the plate, drying each solution as it is applied with a current of warm air. Using a continuous elution chromatographic chamber, develop the chromatogram in a solvent system consisting of a mixture of chloroform and ethyl acetate (5:1) for about 2 hours. Remove the plate from the developing chamber, dry in an oven at 90° for 15 minutes, and locate the bands by viewing under short-wavelength ultraviolet light. Mark the principal band and any secondary bands. Quantitatively remove the silica gel containing these bands, including a corresponding blank segment, and transfer to separate glass-stoppered, 50-mL centrifuge tubes, combining the impurities if more than one impurity is present. Add 30.0 mL of dehydrated alcohol to the tubes containing the principal band and the corresponding blank, and add 10.0 mL of dehydrated alcohol to the tubes containing the combined impurities and the corresponding blank. Insert stoppers in the tubes, and shake gently on a reciprocating shaker for about 60 minutes. Centrifuge, dilute the principal band eluate and its corresponding blank eluate with an equal volume of dehydrated alcohol, and mix. Determine the absorbances of the clear supernatant eluates in 1-cm cells at the wavelength of maximum absorbance at about 239 nm, with a suitable spectrophotometer, using dehydrated alcohol as the blank. Calculate the percentage of chromatographic impurities by the formula:

$$100A_i/(A_i + 6A_u),$$

in which A_i is the absorbance of the combined impurity bands eluate, corrected for the corresponding blank, and A_u is the absorbance of the principal band eluate, corrected for the corresponding blank. Not more than 3.0% is found.

Assay—

Standard preparation—Dissolve an accurately weighed quantity of USP Halcinonide RS in methanol, and dilute quantitatively and stepwise with methanol to obtain a solution having a known concentration of about 15 µg per mL.

Assay preparation—Weigh accurately about 30 mg of Halcinonide, transfer to a 100-mL volumetric flask, dissolve in methanol, dilute with methanol to volume, and mix. Transfer 5.0 mL of this solution to a second 100-mL volumetric flask, dilute with methanol to volume, and mix.

Procedure—Concomitantly determine the absorbances of the *Assay preparation* and the *Standard preparation* in 1-cm cells at the wavelength of maximum absorbance at about 239 nm, with a suitable spectrophotometer, using methanol as the blank. Calculate the quantity, in mg, of $C_{24}H_{32}ClFO_5$ in the portion of Halcinonide taken by the formula:

$$2C(A_U/A_S),$$

in which *C* is the concentration, in µg per mL, of USP Halcinonide RS in the *Standard preparation*, and A_U and A_S are the absorbances of the *Assay preparation* and the *Standard preparation*, respectively.

Halcinonide Cream

» Halcinonide Cream is Halcinonide in a suitable cream base. It contains not less than 90.0 percent and not more than 110.0 percent of the labeled amount of $C_{24}H_{32}ClFO_5$.

Packaging and storage—Preserve in well-closed containers.

Reference standard—*USP Halcinonide Reference Standard*—Dry in vacuum at 100° for 3 hours before using.

Identification—Transfer a quantity of Cream, equivalent to about 2 mg of halcinonide, to a glass-stoppered, 50-mL centrifuge tube, add 15 mL of warm water, and shake for 2 minutes to disperse. Add 20 mL of chloroform, and shake for 5 minutes. Cool in an ice bath, then centrifuge. Transfer the chloroform layer to a conical flask. Repeat the extraction with an additional 15 mL of chloroform, and combine the chloroform extracts. Evaporate the chloroform extracts on a steam bath under a current of air nearly to dryness, and dissolve the residue in 10.0 mL of chloroform. Apply 20 µL of this solution and 20 µL of a solution of USP Halcinonide RS in chloroform having a concentration of 0.2 mg per mL to a thin-layer chromatographic plate (see *Chromatography* ⟨621⟩) coated with a 0.25-mm layer of chromatographic silica gel mixture. Develop the chromatogram in a solvent system consisting of a mixture of chloroform and ethyl acetate (5:1) until the solvent front has moved about three-fourths of the length of the plate. Remove the plate, air-dry, and view the chromatogram under short-wavelength ultraviolet light: the R_f value of the principal spot obtained from the test solution corresponds to that obtained from the Standard solution.

Microbial limits—It meets the requirements of the tests for absence of *Staphylococcus aureus* and *Pseudomonas aeruginosa* under *Microbial Limit Tests* ⟨61⟩.

Minimum fill ⟨755⟩: meets the requirements.

Assay—

Solvent A—On the day of use, prepare a mixture of acetonitrile and water (2:1).

Mobile phase—Mix approximately equal volumes of acetonitrile and water, adjusting the composition as necessary to achieve acceptable chromatography.

Internal standard solution—Transfer 15 mg of Progesterone to a 100-mL volumetric flask. Dissolve in hexanes-saturated *Solvent A*, dilute with hexanes-saturated *Solvent A* to volume, and mix.

Standard preparation—Transfer about 20 mg of USP Halcinonide RS, accurately weighed, to a 100-mL volumetric flask, dissolve in *Solvent A*, dilute with *Solvent A* to volume, and mix. Transfer 5.0 mL of this solution to a 50-mL volumetric flask. Add 4.0 mL of *Internal standard solution*, dilute with hexanes-saturated *Solvent A* to volume, and mix.

Assay preparation—Transfer an accurately weighed quantity of Halcinonide Cream, equivalent to about 0.5 mg of halcinonide, to a glass-stoppered, 50-mL centrifuge tube, add 12 mL of hexanes-saturated *Solvent A* and 20 mL of hexanes, and shake for 1 minute. Place in a heated ultrasonic bath at 58 ± 2° for 20 minutes, initially shaking for 1 to 2 minutes to ensure dispersion, and at about 5-minute intervals thereafter, on a vibratory mixer. Cool, centrifuge, and transfer the lower layer to a 25-mL volumetric flask. Add 5 mL of hexanes-saturated *Solvent A* to the tube, mix for 1 minute, then centrifuge. Transfer the lower layer to the volumetric flask, and repeat the extraction with an additional 5 mL of hexanes-saturated *Solvent A*, combining the extracts in the flask. Add 2.0 mL of *Internal standard solution* to the flask, dilute with hexanes-saturated *Solvent A* to volume, and mix. If necessary, clarify a portion of the solution by centrifugation.

Chromatographic system (see *Chromatography* ⟨621⟩)—The liquid chromatograph is equipped with a 254-nm detector and a 4-mm × 30-cm column that contains packing L1. The flow rate is about 2 mL per minute. Chromatograph the *Standard preparation*, and record the peak responses as directed under *Procedure:* the resolution, *R*, between the analyte and internal standard peaks is not less than 1.5, and the relative standard deviation for replicate injections is not more than 3.0%.

Procedure—Separately inject equal volumes (about 20 µL) of the *Standard preparation* and the *Assay preparation* into the chromatograph, record the chromatograms, and measure the responses for the major peaks. The relative retention times are about 1.2 for progesterone and 1.0 for halcinonide. Calculate the quantity, in mg, of $C_{24}H_{32}ClFO_5$ in the portion of Cream taken by the formula:

$$25C(R_U/R_S),$$

in which *C* is the concentration, in mg per mL, of USP Halcinonide RS in the *Standard preparation*, and R_U and R_S are the ratios of the peak responses of halcinonide to internal standard obtained from the *Assay preparation* and the *Standard preparation*, respectively.

Halcinonide Ointment

» Halcinonide Ointment is Halcinonide in a suitable ointment base. It contains not less than 90.0 percent and not more than 110.0 percent of the labeled amount of $C_{24}H_{32}ClFO_5$.

Packaging and storage—Preserve in well-closed containers.

Reference standard—*USP Halcinonide Reference Standard*—Dry in vacuum at 100° for 3 hours before using.

Identification—It responds to the *Identification test* under *Halcinonide Cream.*

Microbial limits—It meets the requirements of the tests for absence of *Staphylococcus aureus* and *Pseudomonas aeruginosa* under *Microbial Limit Tests* ⟨61⟩.

Minimum fill ⟨755⟩: meets the requirements.

Assay—

Mobile phase—Mix approximately equal volumes of acetonitrile and water, adjusting the ratio of solvents as necessary to achieve acceptable chromatography.

Internal standard solution—Dissolve Butylparaben in acetonitrile to obtain a solution having a concentration of 6 µg per mL.

Standard preparation—Dissolve an accurately weighed quantity of USP Halcinonide RS in *Internal standard solution* to obtain a solution having a known concentration of about 0.04 mg per mL. Mix 5.0 mL of this solution with 5.0 mL of the *Mobile phase.* Each mL of the *Standard preparation* has a known concentration of about 0.02 mg of USP Halcinonide RS.

Assay preparation—Transfer an accurately weighed quantity of Halcinonide Ointment, equivalent to about 1 mg of halcinonide, to a glass-stoppered, 50-mL centrifuge tube, and add 25.0 mL of *Internal standard solution* and 5.0 mL of hexane. Place in a water bath at 58 ± 2° for 3 minutes, then mix in a vortex mixer for about 1 minute until the specimen is well dispersed. Repeat the above-specified heating and mixing step one more time. Cool in an ice-methanol bath for 15 minutes or until the two phases separate, centrifuging if necessary. Transfer 5.0 mL of the lower layer into a 15-mL centrifuge tube, add 5.0 mL of *Mobile phase*, and mix.

Chromatographic system (see *Chromatography* ⟨621⟩)—The liquid chromatograph is equipped with a 254-nm detector and a 3.9-mm × 30-cm column that contains packing L1. The flow rate is about 2 mL per minute. Chromatograph the *Standard preparation*, and record the peak responses as directed under *Procedure:* the resolution, *R*, between the analyte and internal standard peaks is not less than 2.0, and the relative standard deviation for replicate injections is not more than 3.0%.

Procedure—Separately inject equal volumes (about 20 µL) of the *Standard preparation* and the *Assay preparation* into the chromatograph, record the chromatograms, and measure the responses for the major peaks. The relative retention times are about 0.6 for butylparaben and 1.0 for halcinonide. Calculate the quantity, in mg, of $C_{24}H_{32}ClFO_5$ in the portion of Ointment taken by the formula:

$$50C(R_U/R_S),$$

in which C is the concentration, in mg per mL, of USP Halcinonide RS in the *Standard preparation*, and R_U and R_S are the ratios of the peak responses of halcinonide to internal standard obtained from the *Assay preparation* and the *Standard preparation*, respectively.

Halcinonide Topical Solution

» Halcinonide Topical Solution is Halcinonide in a suitable aqueous vehicle. It contains not less than 90.0 percent and not more than 110.0 percent of the labeled amount of $C_{24}H_{32}ClFO_5$.

Packaging and storage—Preserve in well-closed containers.

Reference standard—*USP Halcinonide Reference Standard*—Dry in vacuum at 100° for 3 hours before using.

Identification—It responds to the *Identification test* under *Halcinonide Cream.*

Microbial limits—It meets the requirements of the tests for absence of *Staphylococcus aureus* and *Pseudomonas aeruginosa* under *Microbial Limit Tests* ⟨61⟩.

Assay—

Mobile phase—Mix approximately equal volumes of acetonitrile and water, adjusting the ratio of solvents as necessary to achieve acceptable chromatography.

Internal standard solution—Transfer 15 mg of Progesterone to a 50-mL volumetric flask. Dissolve in *Mobile phase*, dilute with *Mobile phase* to volume, and mix.

Standard preparation—Transfer about 20 mg of USP Halcinonide RS, accurately weighed, to a 100-mL volumetric flask, dissolve in *Mobile phase*, dilute with *Mobile phase* to volume, and mix. Transfer 5.0 mL of this solution to a 50-mL volumetric flask, add 2.0 mL of *Internal standard solution*, dilute with *Mobile phase* to volume, and mix.

Assay preparation—Transfer an accurately measured quantity of Halcinonide Topical Solution, equivalent to about 1 mg of halcinonide, to a 50-mL volumetric flask, add 2.0 mL of *Internal standard solution*, dilute with *Mobile phase* to volume, and mix.

Chromatographic system and *Procedure*—Proceed as directed in the *Assay* under *Halcinonide Cream.* Calculate the quantity, in mg, of $C_{24}H_{32}ClFO_5$ in the portion of Topical Solution taken by the formula:

$$50C(R_U/R_S),$$

in which the terms are as defined therein.

Haloperidol

Cl–C6H4–C(OH)(piperidine)–N–CH2CH2CH2C(=O)–C6H4–F

$C_{21}H_{23}ClFNO_2$ 375.87

1-Butanone, 4-[4-(4-chlorophenyl)-4-hydroxy-1-piperidinyl]-1-(4-fluorophenyl)-.

4-[4-(*p*-Chlorophenyl)-4-hydroxypiperidino]-4′-fluorobutyrophenone [*52-86-8*].

» Haloperidol contains not less than 98.0 percent and not more than 102.0 percent of $C_{21}H_{23}ClFNO_2$, calculated on the dried basis.

Packaging and storage—Preserve in tight, light-resistant containers.

Reference standards—*USP Haloperidol Reference Standard*—Dry in vacuum at 60° for 3 hours before using. *USP 4,4′-Bis[4-(p-chlorophenyl)-4-hydroxypiperidino]butyrophenone Reference Standard*—Keep container tightly closed and protected from light. Dry in vacuum at 60° for 3 hours before using.

Identification—

A: The infrared absorption spectrum of a potassium bromide dispersion of it, previously dried, exhibits maxima only at the same wavelengths as that of a similar preparation of USP Haloperidol RS.

B: The ultraviolet absorption spectrum of a 1 in 50,000 solution of it in a mixture of 1 volume of dilute hydrochloric acid (1 in 100) and 9 volumes of isopropyl alcohol exhibits maxima and minima at the same wavelengths as that of a similar preparation of USP Haloperidol RS, concomitantly measured, and the respective absorptivities, calculated on the dried basis, at the wavelength of maximum absorbance at about 245 nm do not differ by more than 3.0%.

Melting range ⟨741⟩: between 147° and 152°, determined after drying in vacuum at 60° for 3 hours.

Loss on drying ⟨731⟩—Dry it in vacuum at 60° for 3 hours: it loses not more than 0.5% of its weight.

Residue on ignition ⟨281⟩: not more than 0.1%.

4,4′-Bis[4-(*p*-chlorophenyl)-4-hydroxypiperidino]butyrophenone—

Test solution—Dissolve about 80 mg of Haloperidol, accurately weighed, in 80 mL of isopropyl alcohol in a 100-mL volumetric flask. Add 10 mL of dilute hydrochloric acid (1 in 100), dilute with isopropyl alcohol to volume, and mix.

Standard solution—Prepare a solution containing 800 µg per mL of USP Haloperidol RS and 8 µg per mL of USP 4,4′-Bis[4-(*p*-chlorophenyl)-4-hydroxypiperidino]butyrophenone RS in isopropyl alcohol containing 10 mL of dilute hydrochloric acid (1 in 100) in each 100 mL of solution.

Procedure—Concomitantly determine the absorbances of the *Test solution* and the *Standard solution* at the wavelength of maximum absorbance at about 335 nm, with a suitable spectrophotometer, using isopropyl alcohol containing 10 mL of dilute hydrochloric acid (1 in 100) in each 100 mL of solution as the blank. The absorbance of the *Test solution* is not greater than that of the *Standard solution*, corresponding to not more than 1.0%.

Assay—Dissolve about 125 mg of Haloperidol, accurately weighed, in 25 mL of glacial acetic acid, add 3 drops of *p*-naphtholbenzein TS, and titrate with 0.05 *N* perchloric acid VS. Perform a blank determination, and make any necessary correction. Each mL of 0.05 *N* perchloric acid is equivalent to 18.79 mg of $C_{21}H_{23}ClFNO_2$.

Haloperidol Injection

» Haloperidol Injection is a sterile solution of Haloperidol in Water for Injection, prepared with the aid of Lactic Acid. It may contain a suitable preservative. It contains not less than 90.0 percent and not more than 110.0 percent of the labeled amount of $C_{21}H_{23}ClFNO_2$.

Packaging and storage—Preserve in single-dose or in multiple-dose containers, preferably of Type I glass, protected from light.

Reference standard—*USP Haloperidol Reference Standard*—Dry in vacuum at 60° for 3 hours before using.

Identification—The solution prepared for measurement of absorbance in the *Assay* exhibits a maximum at 245 ± 2 nm.

pH ⟨791⟩: between 3.0 and 3.8.

Other requirements—It meets the requirements under *Injections* ⟨1⟩.

Assay—Transfer an accurately measured volume of Haloperidol Injection, equivalent to about 10 mg of haloperidol, to a separator, and add 20 mL of dilute hydrochloric acid (1 in 20). Extract the solution with four 25-mL portions of ether, and wash the combined ether extracts with four 5-mL portions of dilute hydrochloric acid (1 in 20). Proceed as directed in the *Assay* under *Haloperidol Solution*, beginning with "Discard the ether." Calculate the quantity, in mg, of $C_{21}H_{23}ClFNO_2$ in each mL of Injection taken by the formula:

$$0.5(C/V)(A_U/A_S),$$

in which C is the concentration, in µg per mL, of USP Haloperidol RS in the Standard solution, V is the volume, in mL, of Injection taken, and A_U and A_S are the absorbances of the solution from the Injection and the Standard solution, respectively.

Haloperidol Oral Solution

» Haloperidol Oral Solution is a solution of Haloperidol in Water, prepared with the aid of Lactic Acid. It contains not less than 90.0 percent and not more than 110.0 percent of the labeled amount of $C_{21}H_{23}ClFNO_2$.

Packaging and storage—Preserve in tight, light-resistant containers.

Reference standard—*USP Haloperidol Reference Standard*—Dry in vacuum at 60° for 3 hours before using.

Identification—The solution prepared for measurement of absorbance in the *Assay* exhibits a maximum at 245 ± 2 nm.

pH ⟨791⟩: between 2.75 and 3.75.

Assay—Transfer an accurately measured volume of Haloperidol Oral Solution, equivalent to about 10 mg of haloperidol, to a separator, and add 20 mL of dilute hydrochloric acid (1 in 20). Extract the solution with four 20-mL portions of ether, and wash the combined ether extracts with four 5-mL portions of dilute hydrochloric acid (1 in 20). Discard the ether, and add the acid washings to the aqueous phase. Filter the aqueous phase through a pledget of cotton into a 50-mL volumetric flask, add dilute hydrochloric acid (1 in 20) to volume, and mix. Transfer 10.0 mL of this solution to a 100-mL volumetric flask, dilute with methanol to volume, and mix. Concomitantly determine the absorbances of this solution and of a solution of USP Haloperidol RS in the same medium having a known concentration of about 20 µg per mL, in 1-cm cells at the wavelength of maximum absorbance at about 245 nm, with a suitable spectrophotometer, using a 1 in 10 solution of dilute hydrochloric acid (1 in 20) in methanol as the blank. Calculate the quantity, in mg, of $C_{21}H_{23}ClFNO_2$ in each mL of Oral Solution taken by the formula:

$$0.5(C/V)(A_U/A_S),$$

in which C is the concentration, in µg per mL, of USP Haloperidol RS in the Standard solution, V is the volume, in mL, of Oral Solution taken, and A_U and A_S are the absorbances of the solution from the Oral Solution and the Standard solution, respectively.

Haloperidol Tablets

» Haloperidol Tablets contain not less than 90.0 percent and not more than 110.0 percent of the labeled amount of $C_{21}H_{23}ClFNO_2$.

Packaging and storage—Preserve in tight, light-resistant containers.

Reference standard—*USP Haloperidol Reference Standard*—Dry in vacuum at 60° for 3 hours before using.

Identification—The solution prepared for measurement of absorbance in the *Assay* exhibits a maximum at 245 ± 2 nm.

Dissolution ⟨711⟩—

Medium: simulated gastric fluid TS (without the enzyme); 900 mL.

Apparatus 1: 100 rpm.

Time: 60 minutes.

Mobile phase—Prepare a pH 4.0 degassed and filtered mixture of methanol and 0.05 *M* monobasic potassium phosphate buffer (60:40). Add 1 *N* sodium hydroxide or phosphoric acid, if necessary, to adjust the mixture to a pH of 4.0. Make adjustments if necessary (see *System Suitability* under *Chromatography* ⟨621⟩).

Chromatographic system (see *Chromatography* ⟨621⟩)—The liquid chromatograph is equipped with a 254-nm detector and a 3.9-mm × 25-cm column that contains packing L1. The flow rate is about 1 mL per minute. Chromatograph the Standard solution, and record the peak responses as directed under *Procedure:* the relative standard deviation for replicate injections is not more than 3.0%, and the tailing factor is not more than 2.0.

Procedure—Separately inject equal volumes (about 50 µL) of the Standard solution and filtered portions of the solution under test into the chromatograph, record the chromatograms, and measure the responses for the major peaks. Calculate the quantity of $C_{21}H_{23}ClFNO_2$ dissolved in comparison with a Standard solution having a known concentration of USP Haloperidol RS in the same medium and similarly chromatographed.

Tolerances—Not less than 80% (Q) of the labeled amount of $C_{21}H_{23}ClFNO_2$ is dissolved in 60 minutes.

Uniformity of dosage units ⟨905⟩: meet the requirements.

Procedure for content uniformity—Transfer 1 finely powdered Tablet to a suitable volumetric flask, and prepare a solution containing about 20 µg per mL by adding warm methanol, shaking for 15 minutes, diluting with methanol to volume, and filtering, discarding the first 20 mL of the filtrate. Concomitantly determine the absorbances of this solution and of a solution of USP Haloperidol RS in the same medium, having a known concentration of about 20 µg per mL, in 1-cm cells at the wavelength of maximum absorbance at about 245 nm, with a suitable spectrophotometer, using methanol as the blank. Calculate the quantity, in mg, of $C_{21}H_{23}ClFNO_2$ in the Tablet by the formula:

$$(T/D)C(A_U/A_S),$$

in which T is the labeled quantity, in mg, of haloperidol in the Tablet, D is the concentration, in µg per mL, of the solution from the Tablet, based upon the labeled quantity per Tablet and the extent of dilution, C is the concentration, in µg per mL, of USP Haloperidol RS in the Standard solution, and A_U and A_S are the absorbances of the solution from the Tablet and the Standard solution, respectively.

Assay—[NOTE—Use dilute sulfuric acid (1 in 350) that has been saturated with chloroform.] Weigh and finely powder not less than 20 Haloperidol Tablets. Weigh accurately a portion of the powder, equivalent to about 2.5 mg of haloperidol, and transfer to a glass-stoppered flask. Add 25 mL of sodium hydroxide solution (1 in 250) and 50.0 mL of chloroform, and shake for 30 minutes. Centrifuge the mixture, remove the aqueous layer, and transfer 25.0 mL of the chloroform layer to a glass-stoppered flask. Add 50.0 mL of dilute sulfuric acid (1 in 350), shake for 15 minutes, and centrifuge. Concomitantly determine the absorbances of the acid layer and of a solution of USP Haloperidol RS in dilute sulfuric acid (1 in 350) having a known concentration of about 25 µg per mL, in 1-cm cells at the wavelength of maximum absorbance at about 245 nm, with a suitable spectrophotometer, using dilute sulfuric acid (1 in 350) as the blank. Calculate the quantity, in mg, of $C_{21}H_{23}ClFNO_2$ in the portion of Tablets taken by the formula:

$$0.1C(A_U/A_S),$$

in which C is the concentration, in µg per mL, of USP Haloperidol RS in the Standard solution, and A_U and A_S are the absorbances of the solution from the Tablets and the Standard solution, respectively.

Haloprogin

$C_9H_4Cl_3IO$ 361.39

Benzene, 1,2,4-trichloro-5-[(3-iodo-2-propynyl)oxy]-.

3-Iodo-2-propynyl 2,4,5-trichlorophenyl ether [*777-11-7*].

» Haloprogin contains not less than 95.0 percent and not more than 102.0 percent of $C_9H_4Cl_3IO$, calculated on the dried basis.

Packaging and storage—Preserve in tight, light-resistant containers.

Reference standard—*USP Haloprogin Reference Standard*—Dry in vacuum at 60° for 4 hours before using.

Identification—

A: The infrared absorption spectrum of a potassium bromide dispersion of it, previously dried, exhibits maxima only at the same wavelengths as that of a similar preparation of USP Haloprogin RS.

B: The ultraviolet absorption spectrum of a 1 in 10,000 solution in alcohol exhibits maxima and minima at the same wavelengths as that of a similar solution of USP Haloprogin RS, concomitantly measured.

Melting range ⟨741⟩: between 110° and 114°.

Acidity—Pipet 10 mL of acetone into a 50-mL conical flask, add 2 drops of methyl red TS, and mix. If necessary, add 0.010 *N* sodium hydroxide, dropwise, until the solution just turns yellow. Transfer 1.0 g of Haloprogin, accurately weighed, to the flask, and mix until dissolved. Add 3.0 mL of 0.010 *N* sodium hydroxide to the flask, and mix: the solution remains yellow.

Loss on drying ⟨731⟩—Dry it in vacuum at 60° for 4 hours: it loses not more than 0.5% of its weight.

Residue on ignition ⟨281⟩: not more than 0.1%, a muffle furnace temperature of 600 ± 25° being used.

Heavy metals ⟨231⟩: 0.005%.

Assay—Transfer about 250 mg of Haloprogin, accurately weighed, to a 250-mL conical flask, add 15 mL of methanol, and stir to dissolve. Add 10 mL of a methanolic sodium borohydride solution (1 in 50), slowly, with shaking, cooling the flask under running water. Continue shaking until the foam dissipates, and allow the solution to stand for 15 minutes. Cautiously add 30 mL of water and 40 mL of hydrochloric acid, and cool. Add 7 mL of chloroform, and shake. Titrate, with constant vigorous shaking, with 0.05 *M* Potassium Iodate VS until the color in the chloroform layer is discharged. Each mL of 0.05 *M* Potassium Iodate is equivalent to 36.14 mg of $C_9H_4Cl_3IO$.

Haloprogin Cream

» Haloprogin Cream contains not less than 90.0 percent and not more than 110.0 percent of the labeled amount of $C_9H_4Cl_3IO$.

Packaging and storage—Preserve in tight, light-resistant containers, at controlled room temperature.

Reference standard—*USP Haloprogin Reference Standard*—Dry in vacuum at 60° for 4 hours before using.

Identification—The Assay solution exhibits maxima and minima at the same wavelengths as the Standard solution, prepared as directed in the *Assay*.

Microbial limits—It meets the requirements of the tests under *Microbial Limit Tests* ⟨61⟩ for the absence of *Staphylococcus aureus* and *Pseudomonas aeruginosa* and *Salmonella species* and *Escherichia coli*, and meets the requirements of the test for *Total Aerobic Microbial Count* with a count of not more than 100 microorganisms per g.

Minimum fill ⟨755⟩: meets the requirements.

Water, *Method I* ⟨921⟩: not more than 3.0%.

Assay—Transfer an accurately weighed portion of Haloprogin Cream, equivalent to about 10 mg of haloprogin, to a 100-mL volumetric flask, dissolve in 75 mL of methanol, dilute with methanol to volume, and mix. Concomitantly determine the absorbances of this Assay solution and a Standard solution of USP Haloprogin RS, similarly prepared, having a known concentration of about 100 µg per mL at the wavelength of maximum absorbance at about 298 nm minus the absorbance at 320 nm, with a suitable spectrophotometer, using methanol as the blank. Calculate the quantity, in mg, of $C_9H_4Cl_3IO$ in the portion of Cream taken by the formula:

$$0.1C(A_U/A_S),$$

in which C is the concentration, in µg per mL, of USP Haloprogin RS in the Standard solution, and A_U and A_S are the absorbances of the Assay solution and the Standard solution, respectively.

Haloprogin Topical Solution

» Haloprogin Topical Solution contains not less than 90.0 percent and not more than 110.0 percent of the labeled amount of $C_9H_4Cl_3IO$.

Packaging and storage—Preserve in tight, light-resistant containers, at controlled room temperature.

Reference standard—*USP Haloprogin Reference Standard*—Dry in vacuum at 60° for 4 hours before using.

Identification—It responds to the *Identification test* under *Haloprogin Cream.*

Specific gravity ⟨841⟩: between 0.838 and 0.852.

Alcohol, *Method II* ⟨611⟩: between 95.0% and 105.0% of the labeled amount of C_2H_5OH.

Assay—Pipet a volume of Haloprogin Topical Solution, equivalent to about 10 mg of haloprogin, into a 100-mL volumetric flask, dilute with methanol to volume, and mix. Concomitantly determine the absorbances of the Assay solution and a Standard solution of USP Haloprogin RS, similarly prepared, having a known concentration of about 100 µg per mL at the wavelength of maximum absorbance at about 298 nm minus the absorbance at 320 nm, with a suitable spectrophotometer, using methanol as a blank. Calculate the quantity, in mg, of $C_9H_4Cl_3IO$ in each mL of the Haloprogin Topical Solution taken by the formula:

$$0.1(C/V)(A_U/A_S),$$

in which C is the concentration, in µg per mL, of USP Haloprogin RS in the Standard solution, and A_U and A_S are the absorbances of the Assay solution and the Standard solution, respectively.

Halothane

```
    Br F
    |  |
 H—C—C—F
    |  |
    Cl F
```

$C_2HBrClF_3$ 197.38

Ethane, 2-bromo-2-chloro-1,1,1-trifluoro-.

2-Bromo-2-chloro-1,1,1-trifluoroethane [*151-67-7*].

» Halothane contains not less than 0.008 percent and not more than 0.012 percent of thymol, by weight, as a stabilizer.

Packaging and storage—Preserve in tight, light-resistant containers, preferably of Type NP glass, and avoid exposure to excessive heat. Dispense it only in the original container.

Reference standard—*USP Halothane Reference Standard*—Keep container tightly closed and protected from light.

Identification—The infrared absorption spectrum, determined in a 0.1-mm cell, of a 1 in 25 solution in carbon disulfide exhibits maxima at the same wavelengths as that of a similar preparation of USP Halothane RS.

Specific gravity ⟨841⟩: between 1.872 and 1.877 at 20°.

Distilling range, *Method II* ⟨721⟩—Not less than 95% distils within a 1° range between 49° and 51°, and not less than 100% distils between 49° and 51°, a correction factor of 0.040° per mm being applied as necessary.

Refractive index ⟨831⟩: between 1.369 and 1.371 at 20°.

Acidity or alkalinity—Shake 20 mL with 20 mL of carbon dioxide-free water for 3 minutes, and allow the layers to separate: the aqueous layer requires not more than 0.1 mL of 0.010 *N* sodium hydroxide or not more than 0.6 mL of 0.010 *N* hydrochloric acid for neutralization, bromocresol purple TS being used as the indicator.

Water, *Method I* ⟨921⟩: not more than 0.03%.

Nonvolatile residue—Evaporate 50 mL in a tared dish on a steam bath to dryness, and dry the residue at 105° for 2 hours: the weight of the residue does not exceed 1 mg.

Chloride and bromide—Shake 25 mL with 25 mL of water for 5 minutes, and allow the liquids to separate completely. Draw off the water layer, and to 10 mL add 1 drop of nitric acid and 5 drops of silver nitrate TS: no opalescence is produced.

Thymol content—

Standard thymol solution—Prepare a standard solution of thymol in 0.25 *N* sodium hydroxide containing 0.1 mg of thymol per mL.

Buffer solution—Use pH 8.0 alkaline borate buffer (see under *Solutions* in the section, *Reagents, Indicators, and Solutions*).

Chlorimide solution—Dissolve 100 mg of 2,6-dibromoquinone-chlorimide in 25 mL of dehydrated alcohol. Prepare a fresh solution for each assay.

Standard thymol curve—Pipet into three 100-mL volumetric flasks 1 mL, 3 mL, and 5 mL, respectively, of *Standard thymol solution*, and add 0.25 *N* sodium hydroxide to make the final volume 5.0 mL. Add 5.0 mL of 0.25 *N* sodium hydroxide to a fourth flask in which the blank is to be prepared. To each flask add 10 mL of *Buffer solution*, mix by gentle swirling, and add 1 mL of *Chlorimide solution*. Allow to stand for 15 minutes, accurately timed, add 3 mL of 0.25 *N* sodium hydroxide to each flask, and add water to volume. With a suitable spectrophotometer, measure the absorbances of the thymol-containing solutions relative to the blank at 590 nm. Plot the readings and draw the curve of best fit.

Procedure—Place about 2 mL of Halothane, accurately weighed, in a 100-mL volumetric flask containing 5 mL of 0.25 *N* sodium hydroxide, and mix by gentle swirling. Evaporate the halothane under a stream of nitrogen, and add 10 mL of *Buffer solution* and 1 mL of *Chlorimide solution*. Swirl gently, allow to stand for 15 minutes, accurately timed, add 3 mL of 0.25 *N* sodium hydroxide, and add water to volume. Read the absorbance of the resulting solution, and by reference to the *Standard thymol curve*, calculate the percentage of thymol in the weight of Halothane taken.

Chromatographic purity—

Standard preparation—Add 1.0 µL of 1,1,2-trichloro-1,2,2-trifluoroethane to 20.0 mL of the test specimen.

Chromatographic system—Under typical conditions, the gas chromatograph is equipped with a flame-ionization detector, and contains a 3-m × 2-mm stainless steel column packed with 20 percent G24 on support S1AB. The column is maintained at 60°, the injection port and detector block are maintained at 200°, and nitrogen is used as the carrier gas at a flow rate of about 15 mL per minute. Typical retention times are about 5 minutes for 1,1,2-trichloro-1,2,2-trifluoroethane and about 13 minutes for halothane.

Procedure—Inject separately equal volumes (about 2 µL) of the *Standard preparation* and Halothane into a suitable gas chromatograph, and record the chromatograms. The total area of all peaks (except that of halothane) recorded for the test specimen does not exceed that due to the added 1,1,2-trichloro-1,2,2-trifluoroethane in the *Standard preparation* (0.005%).

Hard Fat—*see* Fat, Hard NF

Helium

He 4.0026
Helium.
Helium [*7440-59-7*].

» Helium contains not less than 99.0 percent, by volume, of He.

Packaging and storage—Preserve in cylinders.

Note—Reduce the container pressure by means of a regulator. Measure the gases with a gas volume meter downstream from the detector tube in order to minimize contamination or change of the specimens.

Identification—The flame of a burning splinter of wood is extinguished when inserted into an inverted test tube filled with Helium. [NOTE—Use caution.] A small balloon filled with Helium shows decided buoyancy.

Odor—Carefully open the container valve to produce a moderate flow of gas. Do not direct the gas stream toward the face, but deflect a portion of the stream toward the nose: no appreciable odor is discernible.

Carbon monoxide—Pass 1050 ± 50 mL through a carbon monoxide detector tube (see under *Reagents* in the section, *Reagents, Indicators, and Solutions*) at the rate specified for the tube: the indicator change corresponds to not more than 0.001%.

Air—Not more than 1.0% of air is present, determined as directed in the *Assay*.

Assay—Introduce a specimen of Helium into a gas chromatograph by means of a gas sampling valve. Select the operating conditions of the gas chromatograph such that the standard peak signal resulting from the following procedure corresponds to not less than 70% of the full-scale reading. Preferably, use an apparatus corresponding to the general type in which the column is 6 m in length and 4 mm in inside diameter and is packed with porous polymer beads, which permits complete separation of nitrogen and oxygen from Helium, although the nitrogen and oxygen may not be separated from each other. Use industrial grade helium (99.99%) as the carrier gas, with a thermal-conductivity detector, and control the column temperature: the peak response produced by the assay specimen exhibits a retention time corresponding to that produced by an air-helium certified standard (see under *Reagents* in the section, *Reagents, Indicators, and Solutions*), and indicates not more than 1.0% of air when compared to the peak response of the air-helium certified standard, and not less than 99.0%, by volume, of He.

Heparin Lock Flush Solution

» Heparin Lock Flush Solution is a sterile preparation of Heparin Sodium Injection with sufficient Sodium Chloride to make it isotonic with blood. Its potency is not less than 90.0 percent and not more than 120.0 percent of the potency stated on the label in terms of USP Heparin Units. It contains not more than 1.00 percent of sodium chloride (NaCl). It may contain a suitable preservative.

Packaging and storage—Preserve in single-dose pre-filled syringes or containers, or in multiple-dose containers, preferably of Type I glass.

Labeling—Label it to indicate the volume of the total contents, and to indicate the potency in terms of USP Heparin Units only per mL, except that single unit-dose containers may be labeled additionally to indicate the single unit-dose volume and the total number of USP Heparin Units in the contents. Where it is labeled with total content, the label states clearly that the entire contents are to be used or, if not, any remaining portion is to be discarded. Label it to indicate the organ and species from which the heparin sodium is derived. The label states also that the Solution is intended for maintenance of patency of intravenous injection devices only, and that it is not to be used for anticoagulant therapy. The label states also that in the case of Solution having a concentration of 10 USP Heparin Units per mL, it may alter, and that in the case of higher concentrations it will alter, the results of blood coagulation tests.

Reference standard—*USP Heparin Sodium Reference Standard*—Store in a cool place and do not freeze.

pH ⟨791⟩: between 5.0 and 7.5.

Particulate matter ⟨788⟩: meets the requirements under *Small-volume Injections.*

Other requirements—It meets the requirements of the *Pyrogen Test* ⟨151⟩, the test dose being 10 mL per kg in the case of Solution containing 100 USP Heparin Units or less in each mL, if it does not contain a stabilizer, and 5 mL per kg in the case of Solution containing more than 100 USP Heparin Units in each mL or where it contains a preservative. It meets also the requirements under *Injections* ⟨1⟩.

Assay for heparin sodium—Proceed as directed in the *Assay* under *Heparin Sodium Injection,* substituting Heparin Lock Flush Solution for the Injection.

Assay for sodium chloride—Pipet 10 mL of Heparin Lock Flush Solution into a suitable container, dilute with water to about 150 mL, add 1.5 mL of potassium chromate TS, and titrate with 0.1 *N* silver nitrate. Each mL of 0.1 *N* silver nitrate is equivalent to 5.844 mg of NaCl.

Heparin Solution, Anticoagulant—*see*
Anticoagulant Heparin Solution

Heparin Calcium

» Heparin Calcium is the calcium salt of forms of a sulfated glycosaminoglycan of mixed mucopolysaccharide nature varying in molecular weights. It is present in mammalian tissues and is usually obtained from the intestinal mucosa or other suitable tissues of domestic mammals used for food by man. It is composed of polymers of alternating derivatives of D-glycosamine (N-sulfated or N-acetylated) and uronic acid (L-iduronic acid or D-glucuronic acid) joined by glycosidic linkages, the components being liberated in varying proportions on complete hydrolysis. It is a mixture of active principles some of which have the property of prolonging the clotting time of blood mainly through the formation of a complex with the plasma protein Anti-thrombin to potentiate the inactivation of Thrombin and to inhibit other coagulation proteases such as Activated Factor X in the clotting sequence. The potency of heparin calcium, calculated on the dried basis, is not less than 140 USP Heparin Units in each mg, and not less than 90.0 percent and not more than 110.0 percent of the potency stated on the label. Heparin Calcium is essentially free from sodium.

NOTE—USP Heparin Units are consistently established on the basis of the *Assay* set forth herein, independently of International Units, and the respective units are not equivalent (see *General Notices*). The USP Units for Anti-factor X_a activity are defined by the USP Heparin Sodium Reference Standard.

Packaging and storage—Preserve in tight containers.

Labeling—Label it to indicate the organ and species from which it is derived.

Reference standards—*USP Heparin Sodium Reference Standard*—Store in a cool place, and do not freeze. *USP Anti-thrombin III Reference Standard. USP Factor X_a Reference Standard.*

Identification—It responds to the test for *Calcium* ⟨191⟩.

Pyrogen—It meets the requirements of the *Pyrogen Test* ⟨151⟩, the test dose being 2 mL per kg, of a solution in pyrogen-free saline TS, containing 1000 USP Heparin Units in each mL.

pH ⟨791⟩: between 5.0 and 7.5, in a solution (1 in 100).

Loss on drying ⟨731⟩—Dry it in vacuum at 60° for 3 hours: it loses not more than 5% of its weight.

Residue on ignition ⟨281⟩: between 28.0% and 41.0%.

Nitrogen content—Determine the nitrogen content as directed in *Method I* (see *Nitrogen Determination* ⟨461⟩): between 1.3% and 2.5% of N is found, calculated on the dried basis.

Protein—To 1 mL of a solution (1 in 100) add 5 drops of trichloroacetic acid solution (1 in 5): no precipitate or turbidity occurs.

Heavy metals, *Method II* ⟨231⟩: 0.003%.

Anti-factor X_a activity—Proceed as directed in the test for *Anti-factor X_a activity* under *Heparin Sodium,* substituting Heparin Calcium for Heparin Sodium for the preparation as directed in the *Test preparation.* In the calculation of the percentage of Anti-factor X_a activity relative to heparin potency, from the formula, P_* is the potency in USP Heparin Units per mg of Heparin Calcium determined by the *Assay* and *C* is the concentration of heparin calcium in mg per mL in the *Test preparation.* The amount of Anti-factor X_a activity, per mg in USP Units for Anti-factor X_a activity, is not less than 80 percent and not more than 120 percent of the potency of heparin in USP Heparin Units per mg as determined by the *Assay.*

Assay—Proceed as directed in the *Assay* under *Heparin Sodium,* substituting Heparin Calcium solution for the solution of heparin sodium prepared as directed under *Assay preparation.* Under *Calculation,* v_U in the equation for *R* is the amount in mg of Heparin Calcium per mL of the *Assay preparation.* The potency of Heparin Calcium in USP Heparin Units per mg is P_* = antilog $\bar{M}$.

Heparin Calcium Injection

» Heparin Calcium Injection is a sterile solution of Heparin Calcium in Water for Injection. It exhibits a potency not less than 90.0 percent and not more than 110.0 percent of the potency stated on the label in terms of USP Heparin Units.

NOTE—USP Heparin Units are consistently established on the basis of the *Assay* set forth herein, independently of International Units, and the respective units are not equivalent (see *General Notices*). The USP Units for Anti-factor X_a activity are defined by the USP Heparin Sodium Reference Standard.

Packaging and storage—Preserve in single-dose or in multiple-dose containers, preferably of Type I glass.

Labeling—Label it to indicate the volume of the total contents and the potency in terms of USP Heparin Units only per mL, except that single unit-dose containers may be labeled additionally to indicate the single unit-dose volume and the total number of USP Heparin Units in the contents. Where it is labeled with total content, the label states also that the entire contents are to be used or, if not, any remaining portion is to be discarded. Label it to indicate also the organ and the species from which it is derived.

Reference standards—*USP Heparin Sodium Reference Standard*—Store in a cool place and do not freeze. *USP Anti-thrombin III Reference Standard. USP Factor X_a Reference Standard.*

Pyrogen—When diluted, if necessary, with pyrogen-free saline TS to a concentration of 1000 USP Heparin Units in each mL, it meets the requirements of the *Pyrogen Test* ⟨151⟩, the test dose being 2 mL per kg.

pH ⟨791⟩: between 5.0 and 7.5.

Particulate matter ⟨788⟩: meets the requirements under *Small-volume Injections.*

Other requirements—It meets the requirements under *Injections* ⟨1⟩.

Anti-factor X_a activity—Proceed as directed in the test for *Anti-factor X_a activity* under *Heparin Sodium Injection*, substituting Heparin Calcium Injection for Heparin Sodium Injection, treated as directed therein. The amount of Anti-factor X_a activity, per mL in USP Units for Anti-factor X_a activity, is not less than 80 percent and not more than 120 percent of the potency of Heparin Calcium Injection in USP Heparin Units per mL as determined by the *Assay*.

Assay—Proceed as directed in the *Assay* under *Heparin Sodium Injection*, substituting Heparin Calcium Injection for the heparin sodium injection, treated as directed therein. The potency of Heparin Calcium Injection in USP Heparin Units per mL is P_* = antilog $\overline{M}$.

Heparin Sodium

» Heparin Sodium is the sodium salt of forms of a sulfated glycosaminoglycan of mixed mucopolysaccharide nature varying in molecular weights. It is present in mammalian tissues and is usually obtained from the intestinal mucosa or other suitable tissues of domestic mammals used for food by man. It is composed of polymers of alternating derivatives of D-glycosamine (N-sulfated or N-acetylated) and uronic acid (L-iduronic acid or D-glucuronic acid) joined by glycosidic linkages, the components being liberated in varying proportions on complete hydrolysis. It is a mixture of active principles some of which have the property of prolonging the clotting time of blood mainly through the formation of a complex with the plasma protein Anti-thrombin to potentiate the inactivation of Thrombin and to inhibit other coagulation proteases such as Activated Factor X in the clotting sequence. The potency of heparin sodium, calculated on the dried basis, is not less than 140 USP Heparin Units in each mg, and not less than 90.0 percent and not more than 110.0 percent of the potency stated on the label.

NOTE—USP Heparin Units are consistently established on the basis of the *Assay* set forth herein, independently of International Units, and the respective units are not equivalent (see *General Notices*). The USP Units for Anti-factor X_a activity are defined by the USP Heparin Sodium Reference Standard.

Packaging and storage—Preserve in tight containers.

Labeling—Label it to indicate the organ and species from which it is derived.

Reference standards—*USP Heparin Sodium Reference Standard*—Store in a cool place and do not freeze. *USP Anti-thrombin III Reference Standard. USP Factor X_a Reference Standard.*

Identification—It responds to the tests for Sodium ⟨191⟩.

Pyrogen—It meets the requirements of the Pyrogen Test ⟨151⟩, the test dose being 2 mL per kg, of a solution in pyrogen-free saline TS, containing 1000 USP Heparin Units in each mL.

pH ⟨791⟩: between 5.0 and 7.5, in a solution (1 in 100).

Loss on drying ⟨731⟩—Dry it in vacuum at 60° for 3 hours: it loses not more than 5% of its weight.

Residue on ignition ⟨281⟩: between 28.0% and 41.0%.

Nitrogen content—Determine the nitrogen content as directed in *Method I* ⟨461⟩: between 1.3% and 2.5% of N is found, calculated on the dried basis.

Protein—To 1 mL of a solution (1 in 100) add 5 drops of trichloroacetic acid solution (1 in 5): no precipitate or turbidity occurs.

Heavy metals, *Method II* ⟨231⟩: 0.003%.

Anti-factor X_a activity—

pH 8.4 buffer—Dissolve amounts of tris(hydroxymethyl)-aminomethane, edetic acid, and sodium chloride in water containing 0.1% of polyethyleneglycol 6000 to obtain a solution having concentrations of 0.050 *M*, 0.0075 *M*, and 0.175 *M*, respectively. Adjust if necessary, with hydrochloric acid or sodium hydroxide solution to the specified pH at 25°.

Antithrombin-III solution—Dissolve bovine Antithrombin-III for Anti-Factor X_a test in *pH 8.4 buffer* to obtain a concentration of 0.5 USP Unit per mL.

Factor X_a solution—Dissolve Factor X_a for Anti-Factor X_a test in Purified Water to obtain a solution giving a constant blank absorbance value, i.e., using saline instead of Heparin solutions (see under *Procedure*) of 0.650 to 0.700A at 405 nm and 37° (approximately 0.4 USP Factor X_a Unit per mL in the final 1-mL reaction mixture).

Chromophore substrate—Dissolve methane sulfonyl-D-Leu-Gly-Arg-pNA substrate in Sterile Water for Injection to obtain a solution having a concentration based on the m.w. stated on the label, of 2.5 to 3.0 μM per mL.

Standard preparation—Prepare 4 solutions of USP Heparin Sodium RS in *pH 8.4 buffer* to obtain concentrations of 0.1, 0.05, 0.025, and 0.0125 USP Heparin Unit per mL.

Test preparation—Dissolve sufficient Heparin Sodium, accurately weighed, in a known volume of *pH 8.4 buffer* to obtain an approximate concentration of 0.025 to 0.05 USP Heparin Unit per mL.

Procedure—For each *Standard preparation* prepare 2 standard solutions each consisting of 100 μL of *Antithrombin-III* solution, 100 μL of a *Standard preparation*, and 600 μL of *pH 8.4 buffer*, mix, and incubate at 37° for 120 seconds, accurately timed. Similarly prepare and process four replicated solutions of the *Test preparation.* Immediately add to each 100 μL of *Factor X_a solution*, mix, and incubate at 37° for 120 seconds, accurately timed. To each tube immediately add 100 μL of *Chromophore substrate*, mix, and record the absorbance at 405 nm and 37°, with a suitable spectrophotometer, continuously, for not less than 60 seconds. [NOTE—Establish a suitable sequence of adding the reagents to the tubes and incubating to enable the timing to be strictly followed.] Calculate from the recordings the mean rate of absorbance change for 60 seconds for each concentration of the *Standard preparation.* Similarly determine the mean rate of absorbance change for the four solutions of the *Test preparation.* Compute the regression of the rate of absorbance change on the log concentration of Heparin Sodium for the final solutions of the *Standard preparation.* Determine the mean of the Anti-Factor X_a content in the final solutions of the *Test preparation* from the regression line. Calculate the percentage of the Anti-Factor X_a activity relative to Heparin potency from the formula:

$$5000U_x/(P_*)(C),$$

in which U_x is the mean of the Anti-Factor X_a content of the final test preparation solutions in USP units per mL, P_* is the potency in USP Units per mg of Heparin Sodium determined by *Heparin Assay*, and *C* is the concentration of Heparin Sodium in mg per mL in the *Test preparation.* The amount of Anti-Factor X_a activity, per mg in USP Units for Anti-Factor X_a, is not less than 80 percent and not more than 120 percent of the potency of heparin in USP Heparin Units per mg as determined by the *Assay.*

Assay—

Standard preparation—Determine by preliminary trial, if necessary, approximately the minimum quantity of USP Heparin Sodium RS which, when added in 0.8 mL of saline TS, maintains fluidity in 1 mL of prepared plasma for 1 hour after the addition of 0.2 mL of calcium chloride solution (1 in 100). This quantity is usually between 1 and 3 USP Heparin Units. On the day of the assay prepare a *Standard preparation* such that it contains, in each 0.8 mL of saline TS, the above-determined quantity of the Reference Standard.

Assay preparation—Dissolve about 25 mg of Heparin Sodium, accurately weighed, in sufficient saline TS to give a concentration of 1 mg per mL, and dilute quantitatively to a concentration estimated to correspond to that of the *Standard preparation.*

Preparation of plasma—Collect blood from sheep directly into a vessel containing 8 percent sodium citrate solution in the proportion of one volume to each 19 volumes of blood to be collected. Mix immediately by gentle agitation and inversion of the vessel. Promptly centrifuge the blood, and pool the separated plasma. To a 1-mL portion of the pooled plasma in a clean test tube add 0.2 mL of calcium chloride solution (1 in 100), and mix. Consider the plasma suitable for use if a solid clot forms within 5 minutes. To store plasma for future use, subdivide the pooled lot into portions not exceeding 100 mL in volume, and store in the frozen state, preventing even partial thawing prior to use. For use in the assay, thaw the frozen plasma in a water bath at a temperature not exceeding 37°. Remove particulate matter by straining the thawed plasma through a coarse filter.

Procedure—To meticulously clean 13-mm × 100-mm test tubes add graded amounts of the *Standard preparation,* selecting the amounts so that the largest does not exceed 0.8 mL and so that they correspond roughly to a geometric series in which each step is approximately 5% greater than the next lower. To each tube so prepared add sufficient saline TS to make the total volume 0.8 mL. Add 1.0 mL of prepared plasma to each tube. Then add 0.2 mL of calcium chloride solution (1 in 100), note the time, immediately insert a suitable stopper in each tube, and mix the contents by inverting three times in such a way that the entire inner surface of the tube is wet.

In the same manner set up a series using the *Assay preparation,* completing the entire process of preparing and mixing the tubes of both the *Standard preparation* and the *Assay preparation* within 20 minutes after the addition of the prepared plasma. One hour, accurately timed, after the addition of the calcium chloride, determine the extent of clotting in each tube, recognizing three grades (0.25, 0.50, and 0.75) between zero and full clotting (1.0). If the series does not contain 2 tubes graded more than 0.5 and 2 tubes graded less than 0.5, repeat the assay, using appropriately modified *Standard* and *Assay preparations.*

Calculation—Convert to logarithms the volumes of *Standard preparation* used in the successive 5 or 6 tubes that bracket a grade of clotting of 0.5, including at least 2 tubes with a larger and 2 tubes with a smaller grade than 0.5. Number and list the tubes serially, and tabulate for each the grade of clotting observed in each tube. From the log-volumes, x, and separately from their corresponding grades of clotting, y, compute the paired averages x_i and y_i of Tubes 1, 2, and 3, of Tubes 2, 3, and 4, of Tubes 3, 4, and 5 and, where the series consists of 6 tubes, of Tubes 4, 5, and 6, respectively. If for one of these paired averages the average grade, y_i, is exactly 0.50, the corresponding x_i is the median log-volume of the *Standard preparation,* x_S. Otherwise, interpolate x_S from the paired values of y_i, x_i and y_{i+1}, x_{i+1} that fall immediately below and above grade 0.5 as

$$x_S = x_i + (y_i - 0.5)(x_{i+1} - x_i)/(y_i - y_{i+1}).$$

From the paired data on the tubes of the *Assay preparation,* compute similarly its median log-volume x_U.

The log potency of the *Assay preparation* is:

$$M = x_S - x_U + \log R,$$

where $R = v_S/v_U$ is the ratio of the USP Heparin Units (v_S) per mL of the *Standard preparation* to the mg (v_U) of Heparin Sodium per mL of the *Assay preparation.*

Repeat the assay independently, and average the two or more values of M to obtain $\bar{M}$. If the second determination of M differs by more than 0.05 from the first determination, continue the assay until the log confidence interval computed as directed under *Confidence Intervals for Individual Assays* in *Design and Analysis of Biological Assays* ⟨111⟩ does not exceed 0.20. The potency of Heparin Sodium in USP Heparin Units per mg is $P_* =$ antilog $\bar{M}$.

Heparin Sodium Injection

» Heparin Sodium Injection is a sterile solution of Heparin Sodium in Water for Injection. It exhibits a potency not less than 90.0 percent and not more than 110.0 percent of the potency stated on the label in terms of USP Heparin Units.

NOTE—USP Heparin Units are consistently established on the basis of the *Assay* set forth herein, independently of International Units, and the respective units are not equivalent (see *General Notices*). The USP Units for Anti-factor X_a activity are defined by the USP Heparin Sodium Reference Standard.

Packaging and storage—Preserve in single-dose or in multiple-dose containers, preferably of Type I glass.

Labeling—Label it to indicate the volume of the total contents and the potency in terms of USP Heparin Units only per mL, except that single unit-dose containers may be labeled additionally to indicate the single unit-dose volume and the total number of USP Heparin Units in the contents. Where it is labeled with total content, the label states also that the entire contents are to be used or, if not, any remaining portion is to be discarded. Label it to indicate also the organ and the species from which it is derived.

Reference standards—*USP Heparin Sodium Reference Standard*—Store in a cool place and do not freeze. *USP Anti-thrombin III Reference Standard. USP Factor X_a Reference Standard.*

Pyrogen—When diluted, if necessary, with pyrogen-free saline TS to a concentration of 1000 USP Heparin Units in each mL, it meets the requirements of the *Pyrogen Test* ⟨151⟩, the test dose being 2 mL per kg.

pH ⟨791⟩: between 5.0 and 7.5.

Particulate matter ⟨788⟩: meets the requirements under *Small-volume Injections.*

Anti-factor X_a activity—Proceed as directed in the test for *Anti-factor X_a activity* under *Heparin Sodium,* substituting the Injection for the solution of Heparin Sodium prepared as directed under *Test preparation.* The amount of Anti-factor X_a activity, per mL in USP Units for Anti-factor X_a activity, is not less than 80 percent and not more than 120 percent of the potency of *Heparin Sodium Injection* in USP Heparin Units per mL as determined by the *Assay.*

Other requirements—It meets the requirements under *Injections* ⟨1⟩.

Assay—Proceed as directed in the *Assay* under *Heparin Sodium,* substituting the Injection for the solution of heparin sodium prepared as directed under *Assay preparation.* Under *Calculation,* v_U in the equation for R is the volume, in mL, of the Injection per mL of the *Assay preparation.* The potency of Heparin Sodium Injection in USP Heparin Units per mL is $P_* =$ antilog $\bar{M}$.

Heparin Sodium, and Lidocaine Hydrochloride Injection, Dihydroergotamine Mesylate,—*see* Dihydroergotamine Mesylate, Heparin Sodium, and Lidocaine Hydrochloride Injection

Hepatitis B Immune Globulin

» Hepatitis B Immune Globulin conforms to the regulations of the federal Food and Drug Administration concerning biologics (see *Biologics* ⟨1041⟩). It is a sterile, nonpyrogenic solution free from turbidity, consisting of globulins derived from the blood plasma of human donors who have high titers of antibodies against hepatitis B surface antigen. It contains not less than 10.0 g and not more than 18.0 g of protein per 100 mL, of which not less than 80 percent is monomeric immunoglobulin G, having no ultracentrifugally detectable fragments, nor aggregates having a sedimentation coefficient greater than 12S. It contains 0.3 *M* glycine as a stabilizing agent, and it contains a suitable preservative. It has a potency per mL not less than that of the U. S. Reference Hepatitis B Immune Globulin tested by an approved radioimmunoassay for the detection and measurement of antibody to hepatitis B surface antigen. It has a pH between 6.4 and 7.2, measured in a solution diluted to contain 1 percent of protein with 0.15 *M* sodium chloride. It meets the requirements of the test for heat stability.

Packaging and storage—Preserve at a temperature between 2° and 8°.

Expiration date—Its minimum expiration date is not later than 1 year after the date of manufacture, such date being that of the first valid potency test of the product.

Labeling—Label it to state that it is not for intravenous injection.

Hepatitis B Virus Vaccine Inactivated

» Hepatitis B Virus Vaccine Inactivated is a sterile preparation consisting of a suspension of particles of Hepatitis B surface antigen (HBsAg) isolated from the plasma of HBsAg carriers; treated with pepsin at pH 2, 8 *M* urea, and 1:4,000 formalin so as to inactivate any hepatitis B virus and any representative viruses from all known virus groups that may be present; purified by ultracentrifugation and biochemical procedures and standardized to a concentration of 35 µg to 55 µg of Lowry (HBsAg) protein per mL. The preparation is adsorbed on aluminum hydroxide and diluted to a concentration of 20 µg Lowry protein per mL or other appropriate concentration, depending on the intended use. The Vaccine meets the requirements for potency in animal tests using mice and by a quantitative parallel line radioimmunoassay. It contains not more than 0.62 mg of aluminum per mL and not more than 0.02% of residual free formaldehyde. It contains thimerosal as a preservative. It meets the requirements of the tests for pyrogen and for general safety (see *Safety Tests—General* under *Biological Reactivity Tests, In-vivo* ⟨88⟩).

Packaging and storage—Preserve at a temperature between 2° and 8°.

Expiration date—The expiration date is 3 years from the date of manufacture, the date of manufacture being the date on which the last valid potency test was initiated.

Labeling—Label it to state the content of HBsAg protein per recommended dose. Label it also to state that it is to be shaken before use and that it is not to be frozen.

Hetacillin

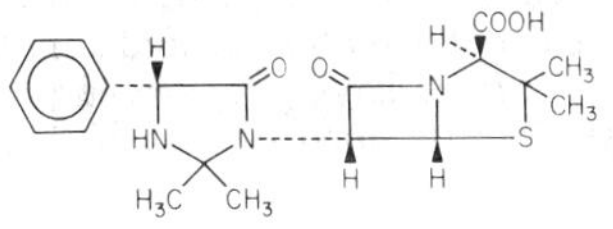

$C_{19}H_{23}N_3O_4S$ 389.47

4-Thia-1-azabicyclo[3.2.0]heptane-2-carboxylic acid, 6-(2,2-dimethyl-5-oxo-4-phenyl-1-imidazolidinyl)-3,3-dimethyl-7-oxo-, [2*S*-[2α,5α,6β(S*)]]-.

6-(2,2-Dimethyl-5-oxo-4-phenyl-1-imidazolidinyl)-3,3-dimethyl-7-oxo-4-thia-1-azabicyclo[3.2.0]heptane-2-carboxylic acid [*3511-16-8*].

» Hetacillin has a potency equivalent to not less than 810 µg of ampicillin ($C_{16}H_{19}N_3O_4S$) per mg.

Packaging and storage—Preserve in tight containers.

Reference standards—*USP Ampicillin Reference Standard*—Do not dry before using. *USP Hetacillin Reference Standard*—Do not dry before using.

Identification—The infrared absorption spectrum of a potassium bromide dispersion of it exhibits maxima only at the same wavelengths as that of a similar preparation of USP Hetacillin RS.

Crystallinity ⟨695⟩: meets the requirements.

pH ⟨791⟩: between 2.5 and 5.5, in a suspension containing 10 mg per mL.

Water, *Method I* ⟨921⟩: not more than 1.0%.

Hetacillin content—

Standard stock solution—Transfer about 100 mg of USP Hetacillin RS, accurately weighed, to a 200-mL volumetric flask, add 150 mL of cold water and 20 mL of 1 *N* hydrochloric acid, and swirl to dissolve. Dilute with water to volume, and mix.

Test preparation—Transfer about 100 mg of Hetacillin, accurately weighed, to a 200-mL volumetric flask, add 150 mL of cold water and 20 mL of hydrochloric acid, and swirl to dissolve. Dilute with water to volume, and mix. Transfer 1.0 mL of this solution to a 25-mL volumetric flask, and add 1.0 mL of hydrochloric acid: the *Test preparation* contains about 0.02 mg of hetacillin per mL.

Standard preparations—Transfer 0.5 mL, 1.0 mL, and 2.0 mL, respectively, of the *Standard stock solution* to separate 25-mL volumetric flasks. Add 1.5 mL and 1.0 mL, respectively, of 0.1 *N* hydrochloric acid to the first two flasks to bring the volume in each flask to 2.0 mL. These *Standard preparations* contain, respectively, about 0.01 mg, 0.02 mg, and 0.04 mg of USP Hetacillin RS per mL.

Procedure—Transfer 2.0 mL of 0.1 *N* hydrochloric acid to a 25-mL volumetric flask (*Blank preparation*). Add 15.0 mL of a freshly prepared mixture of acetone and hydrochloric acid (350:3) to each of the flasks containing the *Standard preparations*, the *Test preparation*, and the *Blank preparation*, and mix. To each flask, add 3.0 mL of a freshly prepared solution of *p*-dimethylaminocinnamaldehyde (1 in 200) in the mixture of acetone and hydrochloric acid, and mix. To each flask, add 3.0 mL of 0.1 *N* hydrochloric acid, dilute with the mixture of acetone and hydrochloric acid to volume, mix, and allow to stand for 30 minutes, accurately timed. Filter if turbid. Concomitantly determine the absorbances of the solutions from the *Standard preparations*, and the *Test preparation*, at the wavelength of maximum absorbance at about 515 nm, with a suitable spectrophotometer, against the solution from the *Blank preparation*. Plot the absorbances of the solutions from the *Standard preparations* versus concentration, in mg per mL, of USP Hetacillin RS, and draw the straight line best fitting the three plotted points. From the graph so obtained, determine the concentration, in mg per mL, of hetacillin in the *Test preparation*. Calculate the percentage of hetacillin in the portion of Hetacillin taken by the formula:

$$500{,}000C/W,$$

in which *C* is the concentration, in mg per mL, of hetacillin in the *Test preparation*, and *W* is the weight, in mg, of Hetacillin taken: between 90.0% and 105.0% is found.

Assay—Proceed as directed for ampicillin under *Antibiotics—Microbial Assays* ⟨81⟩, using an accurately weighed quantity of Hetacillin dissolved in *Buffer No. 3* and diluted quantitatively to obtain a *Test Dilution* having a concentration of ampicillin assumed to be equal to the median dose level of the Standard.

Hetacillin for Oral Suspension

» Hetacillin for Oral Suspension contains the equivalent of not less than 90.0 percent and not more than 120.0 percent of the labeled amount of ampicillin ($C_{16}H_{19}N_3O_4S$). It contains one or more suitable colors, flavors, preservatives, sweeteners, and suspending agents.

Packaging and storage—Preserve in tight containers.

Reference standards—*USP Ampicillin Reference Standard*—Do not dry before using. *USP Hetacillin Reference Standard*—Do not dry before using.

Identification—Dissolve a portion, equivalent to about 125 mg of ampicillin, in 25 mL of a solvent mixture of acetone and 0.1 *N* hydrochloric acid (4:1). Prepare a Standard solution of USP Hetacillin RS to contain 5 mg per mL in the same solvent mixture. Apply separately 2 µL of each solution on a thin-layer chromatographic plate coated with a 0.25-mm layer of chromatographic silica gel mixture (see *Chromatography* ⟨621⟩). Place the plate in a suitable chromatographic chamber, and develop the chromatogram with a mixture of acetone, water, toluene, and glacial acetic acid (650:100:100:25). When the solvent front has moved about three-fourths of the length of the plate, remove the plate from the chamber, and allow to air-dry. Locate the spots on the plate by spraying lightly with a solution of ninhydrin in alcohol containing 3 mg per mL, and dry at 90° for 15 minutes: the R_f value of the principal spot obtained from the test solution corresponds to that obtained from the Standard solution.

Uniformity of dosage units ⟨905⟩—

FOR SOLID PACKAGED IN SINGLE-UNIT CONTAINERS: meets the requirements.

pH ⟨791⟩: between 2.0 and 5.0, in the suspension constituted as directed in the labeling.

Water, *Method I* ⟨921⟩: not more than 2.0%.

Assay—Proceed as directed for ampicillin under *Antibiotics—Microbial Assays* ⟨81⟩, using an accurately measured volume of Hetacillin for Oral Suspension, freshly mixed and free from air bubbles, constituted as directed in the labeling and diluted quantitatively with *Buffer No. 3* to yield a *Test Dilution* having a concentration of ampicillin assumed to be equal to the median dose level of the Standard.

Hetacillin Tablets

» Hetacillin Tablets contain the equivalent of not less than 90.0 percent and not more than 120.0 percent of the labeled amount of ampicillin ($C_{16}H_{19}N_3O_4S$).

Packaging and storage—Preserve in tight containers.

Labeling—Label Tablets to indicate that they are to be chewed before swallowing.

Reference standards—*USP Ampicillin Reference Standard*—Do not dry before using. *USP Hetacillin Reference Standard*—Do not dry before using.

Identification—Powder 1 or more Hetacillin Tablets, and prepare a solution containing the equivalent of 5 mg of ampicillin per mL in a mixture of acetone and 0.1 *N* hydrochloric acid (4:1): the solution so obtained responds to the *Identification test* under *Hetacillin for Oral Suspension.*

Dissolution ⟨711⟩—

Medium: 0.05 *M*, pH 7.6 phosphate buffer (see under *Solutions* in the section, *Reagents, Indicators, and Solutions*); 900 mL.

Apparatus 2: 50 rpm.

Time: 45 minutes.

Procedure—Determine the amount of ampicillin ($C_{16}H_{19}N_3O_4S$) equivalent dissolved, using a suitable validated spectrophotometric procedure, of filtered portions of the solution under test, suitably diluted with *Dissolution Medium*, if necessary, in comparison with a solution of USP Ampicillin RS in the same medium.

Tolerances—Not less than 75% (*Q*) of the labeled amount of $C_{16}H_{19}N_3O_4S$ equivalent is dissolved in 45 minutes.

Uniformity of dosage units ⟨905⟩: meet the requirements.

Water, *Method I* ⟨921⟩: not more than 2.0%.

Assay—Proceed as directed for ampicillin under *Antibiotics—Microbial Assays* ⟨81⟩, using not less than 5 Hetacillin Tablets blended for 4 ± 1 minutes in a high-speed glass blender jar containing an accurately measured volume of *Buffer No. 3.* Dilute an accurately measured volume of this stock solution quantitatively with *Buffer No. 3* to obtain a *Test Dilution* having a concentration of ampicillin assumed to be equal to the median dose level of the Standard.

Hetacillin Potassium

$C_{19}H_{22}KN_3O_4S$ 427.56

4-Thia-1-azabicyclo[3.2.0]heptane-2-carboxylic acid, 6-(2,2-dimethyl-5-oxo-4-phenyl-1-imidazolidinyl)-3,3-dimethyl-7-oxo-, monopotassium salt, [2*S*-[2α,5α,6β(S*)]]-.

Potassium 6-(2,2-dimethyl-5-oxo-4-phenyl-1-imidazolidinyl)-3,3-dimethyl-7-oxo-4-thia-1-azabicyclo[3.2.0]heptane-2-carboxylate [*5321-32-4*].

» Hetacillin Potassium has a potency equivalent to not less than 735 µg of ampicillin ($C_{16}H_{19}N_3O_4S$) per mg.

Packaging and storage—Preserve in tight containers.

Reference standards—*USP Ampicillin Reference Standard*—Do not dry before using. *USP Hetacillin Reference Standard*—Do not dry before using. *USP Hetacillin Potassium Reference Standard*—Do not dry before using.

Identification—The infrared absorption spectrum of a potassium bromide dispersion of it exhibits maxima only at the same wavelengths as that of a similar preparation of USP Hetacillin Potassium RS.

Crystallinity ⟨695⟩: meets the requirements.

pH ⟨791⟩: between 7.0 and 9.0, in a solution containing 10 mg per mL.

Water, *Method I* ⟨921⟩: not more than 1.0%.

Hetacillin content—Proceed as directed for *Hetacillin content* under *Hetacillin*, except to use about 110 mg of Hetacillin Potassium, accurately weighed, in the *Assay preparation.* Between 82.0% and 95.5% of hetacillin ($C_{19}H_{23}N_3O_4S$) is found.

Assay—Proceed as directed for ampicillin under *Antibiotics—Microbial Assays* ⟨81⟩, using an accurately weighed quantity of Hetacillin Potassium dissolved in *Buffer No. 3* to obtain a *Test Dilution* having a concentration of ampicillin assumed to be equal to the median dose level of the Standard.

Hetacillin Potassium Capsules

» Hetacillin Potassium Capsules contain the equivalent of not less than 90.0 percent and not more than

120.0 percent of the labeled amount of ampicillin ($C_{16}H_{19}N_3O_4S$).

Packaging and storage—Preserve in tight containers.

Reference standards—*USP Ampicillin Reference Standard*—Do not dry before using. *USP Hetacillin Reference Standard*—Do not dry before using.

Identification—Prepare a solution containing the equivalent of 5 mg of ampicillin per mL by dissolving the powder from Hetacillin Potassium Capsules in a mixture of acetone and 0.1 *N* hydrochloric acid (4:1): the solution so obtained responds to the *Identification test* under *Hetacillin for Oral Suspension.*

Dissolution ⟨711⟩—

Medium: water; 900 mL.

Apparatus 1: 100 rpm.

Time: 45 minutes.

Procedure—Filter the solution under test, dilute with *Buffer No. 3*, and proceed as directed for ampicillin under *Antibiotics—Microbial Assays* ⟨81⟩.

Tolerances—Not less than 75% (*Q*) of the labeled amount of ampicillin ($C_{16}H_{19}N_3O_4S$) is dissolved in 45 minutes.

Uniformity of dosage units ⟨905⟩: meet the requirements.

Water, *Method I* ⟨921⟩: not more than 3.0%.

Assay—Proceed as directed for ampicillin under *Antibiotics—Microbial Assays* ⟨81⟩, using not less than 5 Hetacillin Potassium Capsules blended for 4 ± 1 minutes in an accurately measured volume of *Buffer No. 3* in a high-speed glass blender jar. Dilute an accurately measured volume of this stock solution with *Buffer No. 3* to obtain a *Test Dilution* having a concentration of ampicillin assumed to be equal to the median dose level of the Standard.

Hetacillin Potassium Intramammary Infusion

» Hetacillin Potassium Intramammary Infusion is a suspension of Hetacillin Potassium in a Peanut Oil vehicle with a suitable dispersing agent. It contains the equivalent of not less than 90.0 percent and not more than 120.0 percent of the labeled amount of ampicillin ($C_{16}H_{19}N_3O_4S$).

Packaging and storage—Preserve in suitable, well-closed, disposable syringes.

Labeling—Label it to indicate that it is for veterinary use only.

Reference standards—*USP Ampicillin Reference Standard*—Do not dry before using. *USP Hetacillin Reference Standard*—Do not dry before using.

Identification—Prepare a solution containing the equivalent of 0.125 mg of ampicillin per mL by dissolving a portion in a mixture of acetone and 0.1 *N* hydrochloric acid (4:1): the solution so obtained responds to the *Identification test* under *Hetacillin for Oral Suspension.*

Water, *Method I* ⟨921⟩: not more than 1.0%, 20 mL of a mixture of carbon tetrachloride, chloroform, and methanol (2:2:1) being used in place of methanol in the titration vessel.

Assay—Proceed as directed for ampicillin under *Antibiotics—Microbial Assays* ⟨81⟩, using the contents of 1 syringe of Hetacillin Potassium Intramammary Infusion blended for 3 to 5 minutes in a high-speed glass blender jar containing 1.0 mL of polysorbate 80 and an accurately measured volume of *Buffer No. 3* to yield a stock solution having a convenient concentration. Allow to stand for 10 minutes, and dilute an accurately measured volume of the aqueous phase quantitatively and stepwise with *Buffer No. 3* to obtain a *Test Dilution* having a concentration of ampicillin assumed to be equal to the median dose level of the Standard.

Sterile Hetacillin Potassium

» Sterile Hetacillin Potassium is Hetacillin Potassium suitable for parenteral use. It contains the equivalent of not less than 90.0 percent and not more than 120.0 percent of the labeled amount of ampicillin ($C_{16}H_{19}N_3O_4S$).

Packaging and storage—Preserve in *Containers for Sterile Solids* as described under *Injections* ⟨1⟩.

Reference standards—*USP Ampicillin Reference Standard*—Do not dry before using. *USP Hetacillin Reference Standard*—Do not dry before using. *USP Hetacillin Potassium Reference Standard*—Do not dry before using.

Pyrogen—When diluted with pyrogen-free saline TS to a concentration of 18 mg of ampicillin per mL, it meets the requirements of the *Pyrogen Test* ⟨151⟩, the test dose being 1 mL per kg.

Sterility—It meets the requirements under *Sterility Tests* ⟨71⟩, when tested as directed in the section, *Test Procedures Using Membrane Filtration.*

pH ⟨791⟩: between 7.0 and 9.0, in the solution constituted as directed in the labeling.

Other requirements—It conforms to the definition, responds to the *Identification test*, and meets the requirements for *pH*, *Water*, *Crystallinity*, and *Hetacillin content* under *Hetacillin Potassium.* It meets also the requirements for *Uniformity of Dosage Units* ⟨905⟩ and *Labeling* under *Injections* ⟨1⟩.

Assay—Proceed as directed for ampicillin under *Antibiotics—Microbial Assays* ⟨81⟩, using the contents of 1 container of Sterile Hetacillin Potassium.

Hetacillin Potassium Oral Suspension

» Hetacillin Potassium Oral Suspension is Hetacillin Potassium suspended in a suitable nonaqueous vehicle. It contains one or more suitable colors, flavors, and gelling agents. It contains the equivalent of not less than 90.0 percent and not more than 120.0 percent of the labeled amount of ampicillin ($C_{16}H_{19}N_3O_4S$).

Packaging and storage—Preserve in tight containers.

Labeling—Label it to indicate that it is for veterinary use only.

Reference standards—*USP Ampicillin Reference Standard*—Do not dry before using. *USP Hetacillin Reference Standard*—Do not dry before using.

Identification—Prepare a solution containing the equivalent of 1 mg of ampicillin per mL by dissolving a portion in a mixture of acetone and 0.1 *N* hydrochloric acid (4:1): the solution so obtained responds to the *Identification test* under *Hetacillin for Oral Suspension.*

Uniformity of dosage units ⟨905⟩—

FOR SUSPENSION PACKAGED IN SINGLE-UNIT CONTAINERS: meets the requirements.

pH ⟨791⟩—Add 10 mL of toluene to 5 mL of Oral Suspension, shake for 3 minutes, and centrifuge. Decant and discard the toluene extract, and add 5 mL of water to the residue: the pH of the resulting solution is between 7.0 and 9.0.

Water, *Method I* ⟨921⟩: not more than 1.0%, 20 mL of a mixture of carbon tetrachloride, chloroform, and methanol (2:2:1) being used in place of methanol in the titration vessel.

Assay—Proceed as directed for ampicillin under *Antibiotics—Microbial Assays* ⟨81⟩, using an accurately measured volume of Hetacillin Potassium Oral Suspension, freshly mixed and free from air bubbles, blended for 3 to 5 minutes in a high-speed glass blender jar containing an accurately measured volume of *Buffer No. 3* to yield a stock solution having a convenient concentration. Dilute an accurately measured volume of this stock solution quantitatively and stepwise with *Buffer No. 3* to obtain a *Test Dilution* having a concentration of ampicillin assumed to be equal to the median dose level of the Standard.

Hetacillin Potassium Tablets

» Hetacillin Potassium Tablets contain the equivalent of not less than 90.0 percent and not more than 120.0 percent of the labeled amount of ampicillin ($C_{16}H_{19}N_3O_4S$).

Packaging and storage—Preserve in tight containers.

Labeling—Label Tablets to indicate that they are for veterinary use only.

Reference standards—*USP Ampicillin Reference Standard*—Do not dry before using. *USP Hetacillin Reference Standard*—Do not dry before using.

Identification—Powder 1 or more Hetacillin Potassium Tablets, and prepare a solution containing the equivalent of 5 mg of ampicillin per mL in a mixture of acetone and 0.1 *N* hydrochloric acid (4:1): the solution so obtained responds to the *Identification test* under *Hetacillin for Oral Suspension.*

Disintegration ⟨701⟩: 30 minutes.

Water, *Method I* ⟨921⟩: not more than 5.0%.

Assay—Proceed as directed for ampicillin under *Antibiotics—Microbial Assays* ⟨81⟩, using not less than 5 Hetacillin Potassium Tablets blended for 4 ± 1 minutes in a high-speed glass blender jar containing an accurately measured volume of *Buffer No. 3.* Dilute an accurately measured volume of this stock solution quantitatively with *Buffer No. 3* to obtain a *Test Dilution* having a concentration of ampicillin assumed to be equal to the median dose level of the Standard.

Hexachlorophene

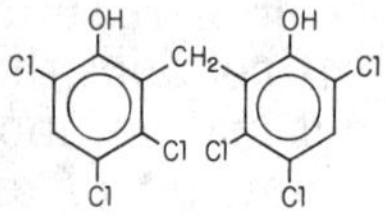

$C_{13}H_6Cl_6O_2$ 406.91

Phenol, 2,2′-methylenebis[3,4,6-trichloro-.

2,2′-Methylenebis[3,4,6-trichlorophenol] [*70-30-4*].

» Hexachlorophene contains not less than 98.0 percent and not more than 100.5 percent of $C_{13}H_6Cl_6O_2$, calculated on the dried basis.

Packaging and storage—Preserve in tight, light-resistant containers.

Reference standard—*USP Hexachlorophene Reference Standard*—Dry at 105° for 4 hours before using.

Identification—

A: The infrared absorption spectrum of a potassium bromide dispersion of it, previously dried, exhibits maxima only at the same wavelengths as that of a similar preparation of USP Hexachlorophene RS.

B: To a solution of about 5 mg in 5 mL of alcohol add 1 drop of ferric chloride TS: a transient purple color is produced immediately.

Melting range ⟨741⟩: between 161° and 167°.

Loss on drying ⟨731⟩—Dry it at 105° for 4 hours: it loses not more than 1.0% of its weight.

Residue on ignition ⟨281⟩: not more than 0.1%.

2,3,7,8-Tetrachlorodibenzo-*p*-dioxin—[*Caution—Since 2,3,7,8-tetrachlorodibenzo-p-dioxin is an extremely toxic substance, exercise all necessary precautions in the conduct of this procedure.*]—Dissolve 10.0 g of Hexachlorophene in 50 mL of methanol, transfer to a 1-liter separator with the aid of 25 mL of methanol, add 25 mL of 2.5 *N* lithium hydroxide and 225 mL of water, and extract with two 200-mL portions of freshly distilled *n*-hexane. Dry the combined *n*-hexane extracts over anhydrous sodium sulfate, filter, and evaporate to a volume of about 15 mL on a rotary evaporator at a bath temperature not exceeding 40°. Transfer this solution in portions to a 12-mL centrifuge tube, concentrating each time to a volume of 1 mL in a gentle stream of nitrogen in a warm water bath. Rinse the flask with 15 mL of *n*-hexane, and evaporate similarly. Wash down the walls of the tube with 10 mL of *n*-hexane, and again evaporate to a volume of 1.0 mL. Cool, and transfer to a micro-column that has been prepared in the following manner: Place a small plug of glass wool in a 5- × 15-mm pipet, add a small amount of sand and 1.0 g of basic alumina, tap several times to pack down the alumina, and heat in a vacuum oven at 110° for 3 hours. Store under vacuum.

Elute the column with 10 mL of a mixture of *n*-hexane and methylene chloride (9:1), using a portion to rinse the tube. Collect the eluate in a 12-mL graduated centrifuge tube, and concentrate in a gentle stream of nitrogen in a warm water bath to a volume of 1.0 mL.

Inject 2.0 µL of the concentrated eluate into a suitable gas chromatograph connected to a mass spectrograph equipped with a multiple-ion detector. The gas chromatograph is fitted with a 2-mm × 1-m glass column containing liquid phase G1 on support S1. The carrier gas is helium, flowing at the rate of 40 mL per minute. The column temperature is maintained at 250° and the injection port is maintained at 300°. Similarly inject 2.0 µL of a Standard solution of 2,3,7,8-tetrachlorodibenzo-*p*-dioxin containing 0.5 µg per mL.* The sum of the peak heights at mass values of 320, 322, and 324 obtained from the solution under test is not greater than the sum of the peak heights at the same mass values obtained from the Standard solution. The limit is 50 ppb.

Assay—Weigh accurately about 1.5 g of Hexachlorophene, dissolve in 25 mL of alcohol, and titrate with 0.1 *N* sodium hydroxide VS, determining the end-point potentiometrically. Perform a blank determination, and make any necessary correction. Each mL of 0.1 *N* sodium hydroxide is equivalent to 40.69 mg of $C_{13}H_6Cl_6O_2$.

Hexachlorophene Cleansing Emulsion

» Hexachlorophene Cleansing Emulsion is Hexachlorophene in a suitable aqueous vehicle. It contains not less than 90.0 percent and not more than 110.0 percent of the labeled amount of $C_{13}H_6Cl_6O_2$. It contains no coloring agents.

Packaging and storage—Preserve in tight, light-resistant, nonmetallic containers.

Reference standard—*USP Hexachlorophene Reference Standard*—Dry at 105° for 4 hours before using.

Identification—Place a volume of Cleansing Emulsion, equivalent to about 150 mg of hexachlorophene, in a glass-stoppered, 25-mL graduated cylinder, dilute with a mixture of equal volumes of chloroform and methanol to volume, mix, and allow to stand for about 5 minutes. On a suitable thin-layer chromatographic plate (see *Chromatography* ⟨621⟩), coated with a 0.25-mm layer of silica gel, apply 10 µL of this solution and 10 µL of a solution of USP Hexachlorophene RS in the same chloroform and methanol mixture containing 6 mg per mL. Develop the chromatogram in a solvent system consisting of a mixture of benzene and glacial acetic acid (9:1) until the solvent moves to about 10 cm above the point of application. Remove the plate, mark the solvent front and evaporate the solvent in a current of warm air. Spray the plate with dilute nitric acid (1 in 5), and warm on a hot plate until yellow spots appear: the R_f value of the principal spot obtained from the solution under test corresponds to that obtained from the Standard solution.

* A solution in anisole is available commercially from KOR Isotopes, Div. of ECO, Inc., 56 Rogers St., Cambridge, Mass. 02142. This solution may be diluted with a mixture of *n*-hexane and methylene chloride (9:1) to the required concentration.

Microbial limits—It meets the requirements of the tests for absence of *Staphylococcus aureus* and *Pseudomonas aeruginosa* under *Microbial Limit Tests* ⟨61⟩.

pH ⟨791⟩—Place 20 mL of the well-shaken Cleansing Emulsion and 10 mL of water in a glass-stoppered, 50-mL graduated cylinder, mix, and determine the pH in a suitable pH meter, using a glass electrode and preferably a sleeve-type calomel electrode: the pH is between 5.0 and 6.0.

Assay—

Standard preparation—Weigh accurately about 50 mg of USP Hexachlorophene RS into a 50-mL volumetric flask, add 10 mL of methanol, shake until dissolved, add methanol to volume, and mix. Pipet 3 mL of this solution into a 100-mL volumetric flask, add 1 mL of dilute hydrochloric acid (1 in 10), add methanol to volume, and mix.

Assay preparation—Transfer an accurately weighed portion of Hexachlorophene Cleansing Emulsion, equivalent to about 30 mg of hexachlorophene, to a 100-mL volumetric flask, add methanol to volume, and mix. Filter the solution through paper, taking adequate precautions to prevent evaporation. Pipet a 10-mL aliquot of the filtrate into a 100-mL volumetric flask, add 1 mL of dilute hydrochloric acid (1 in 10), and add methanol to volume.

Procedure—Concomitantly determine the absorbances of the *Assay preparation* and the *Standard preparation* in 1-cm cells at the wavelength of maximum absorbance at about 299 nm, with a suitable spectrophotometer, using a mixture of 99 volumes of methanol and 1 volume of hydrochloric acid as the blank. Calculate the quantity, in mg, of $C_{13}H_6Cl_6O_2$ in the portion of Cleansing Emulsion taken by the formula:

$$C(A_U/A_S),$$

in which C is the concentration, in µg per mL, of USP Hexachlorophene RS in the *Standard preparation*, and A_U and A_S are the absorbances of the *Assay preparation* and the *Standard preparation*, respectively.

Hexachlorophene Liquid Soap

» Hexachlorophene Liquid Soap is a solution of Hexachlorophene in a 10.0 to 13.0 percent solution of a potassium soap. It contains, in each 100 g, not less than 225 mg and not more than 260 mg of $C_{13}H_6Cl_6O_2$. It may contain suitable water hardness controls.

Note—The inclusion of nonionic detergents in Hexachlorophene Liquid Soap in amounts greater than 8 percent on a total weight basis may decrease the bacteriostatic activity of the Soap.

Packaging and storage—Preserve in tight, light-resistant containers.

Labeling—Solutions of higher concentrations of hexachlorophene and potassium soap, in which the ratios of these components are consistent with the official limits, may be labeled "For the preparation of Hexachlorophene Liquid Soap, USP," provided that the label indicates also that the soap is a concentrate, and provided that directions are given for dilution to the official strength.

Reference standard—*USP Hexachlorophene Reference Standard*—Dry at 105° for 4 hours before using.

Identification—

A: Pour about 2 g into a beaker, and add, with stirring, dilute hydrochloric acid (1 in 100) until the mixture is just acid to litmus. To 10 mL of the mixture, in a beaker, add 10 mL of chloroform, and mix. Add 3 or 4 drops of ferric chloride TS, mix, and allow to stand: the chloroform layer becomes purple.

B: To 2 mL of the mixture prepared in *Identification test A*, in a test tube, add 2 mL of acetone, and mix. Add 1 mL of titanium trichloride solution (1 in 5), and shake vigorously: a yellow oil separates.

Microbial limits—It meets the requirements of the tests for absence of *Staphylococcus aureus* and *Pseudomonas aeruginosa* under *Microbial Limit Tests* ⟨61⟩.

Water—Place about 5 g, quickly weighed to the nearest centigram, in the distilling flask of the apparatus for *Water Determination—Azeotropic Method* ⟨921⟩. (The Soap is most conveniently weighed in a boat of metal foil, of a size that will just pass through the neck of the flask.) Place 250 mL of toluene and 10 g of anhydrous barium chloride in the flask, connect the flask through a ground-glass joint to the distilling apparatus, fill the receiving tube with toluene, and determine the water as directed, beginning with "Heat the flask gently." The volume of water found corresponds to between 86.5% and 90.0% by weight of the portion of Soap taken.

Alcohol-insoluble substances—Dissolve about 5 g, rapidly and accurately weighed, in 100 mL of hot neutralized alcohol, collect the residue, if any, on a tared filter, thoroughly wash it with hot neutralized alcohol, and dry at 105° for 1 hour: the weight of the residue obtained does not exceed 3.0% of the weight of Soap taken. Retain the solution and the residue.

Free alkali hydroxides—To the combined filtrate and washings obtained in the test for *Alcohol-insoluble substances,* add 0.5 mL of phenolphthalein TS. If a pink color is produced, titrate the solution with 0.1 *N* sulfuric acid VS until the pink color is just discharged. Each mL of 0.1 *N* sulfuric acid is equivalent to 5.61 mg of KOH. The volume of 0.1 *N* sulfuric acid consumed corresponds to not more than 0.05% of KOH.

Alkali carbonates—Wash the filter containing the *Alcohol-insoluble substances* with 50 mL of boiling water, cool, add methyl orange TS, and titrate the filtrate with 0.1 *N* sulfuric acid VS. Not more than 0.5 mL of 0.1 *N* sulfuric acid per g of Soap originally taken is required (0.35% calculated as K_2CO_3).

Assay for hexachlorophene—

Alkaline buffer—Dissolve 6.07 g of tris(hydroxymethyl)-aminomethane in 900 mL of methanol. Add 25.0 mL of dilute hydrochloric acid (1 in 10), dilute with water to 1 liter, and mix.

Standard preparation—Place about 50 mg of USP Hexachlorophene RS, accurately weighed, in a 100-mL volumetric flask, dissolve in 10 mL of alcohol, and dilute with *Alkaline buffer* to volume. Preserve in a tight container.

Standard hexachlorophene graph—To 50-mL volumetric flasks add, by pipet and in duplicate, 2-, 3-, 4-, 5-, 6-, and 7-mL portions of the *Standard preparation.* To one flask of each pair of duplicates, add to volume acidified 90 percent methanol containing, in each 100 mL, 5 mL of acetic acid and 0.3 mL of hydrochloric acid, and mix. To the second flask of each pair, add *Alkaline buffer* to volume.

Arrange the two series of standard hexachlorophene solutions in pairs according to their hexachlorophene content, and determine the absorbances of the alkaline solutions at 312 nm, with a suitable spectrophotometer, using the corresponding acid solution as the blank. Plot the observed absorbance on the ordinate scale against the corresponding concentration of hexachlorophene, in mg per 100 mL, on the abscissa scale.

Procedure—Weigh accurately a portion of Hexachlorophene Liquid Soap, containing the equivalent of about 100 mg of hexachlorophene, and transfer to a 100-mL volumetric flask. Add alcohol to volume, and mix. Transfer 25.0 mL of this solution to a 100-mL volumetric flask, add 90 percent methanol to volume, mix, and filter if necessary. Add 10.0 mL of this solution to each of two 50-mL volumetric flasks, and fill one flask to volume with 0.3 *M* acetic acid in 90 percent methanol containing 0.1% of hydrochloric acid, and mix. Fill the other flask to volume with *Alkaline buffer*, and mix. Determine the absorbance of the alkaline solution at 312 nm with the same spectrophotometer used in preparing the *Standard hexachlorophene graph*, using the control as the blank. From the observed absorbance, calculate the weight of hexachlorophene in the Hexachlorophene Liquid Soap taken.

Hexylcaine Hydrochloride

COOCHCH$_2$NH · HCl
CH$_3$

$C_{16}H_{23}NO_2.HCl$ 297.82

2-Propanol, 1-(cyclohexylamino)-, benzoate (ester), hydrochloride.
1-(Cyclohexylamino)-2-propanol benzoate (ester) hydrochloride [*532-76-3*].

» Hexylcaine Hydrochloride, dried in vacuum over phosphorus pentoxide for 4 hours, contains not less than 98.0 percent and not more than 102.0 percent of $C_{16}H_{23}NO_2 . HCl$.

Packaging and storage—Preserve in tight containers.

Reference standard—*USP Hexylcaine Hydrochloride Reference Standard*—Dry in vacuum over phosphorus pentoxide for 4 hours before using.

Identification—

A: The infrared absorption spectrum of a mineral oil dispersion of it, previously dried, exhibits maxima only at the same wavelengths as that of a similar preparation of USP Hexylcaine Hydrochloride RS.

B: The ultraviolet absorption spectrum of a 1 in 100,000 solution in dilute hydrochloric acid (1 in 120) exhibits maxima and minima at the same wavelengths as that of a similar solution of USP Hexylcaine Hydrochloride RS, concomitantly measured, and the respective absorptivities, calculated on the dried basis, at the wavelength of maximum absorbance at about 232 nm do not differ by more than 3.0%.

C: Dissolve about 50 mg in 5 mL of water, add 2 mL of 6 *N* ammonium hydroxide, and filter. Neutralize the clear filtrate with nitric acid, and add 2 drops in excess: the solution responds to the tests for *Chloride* ⟨191⟩.

Melting range, *Class I* ⟨741⟩: between 182° and 184°.

pH ⟨791⟩: between 4.0 and 6.0, in a solution (1 in 20).

Acidity—Dissolve 1.0 g in 25 mL of water, add methyl red TS, and titrate with 0.020 *N* sodium hydroxide: not more than 0.50 mL is required for neutralization.

Loss on drying ⟨731⟩—Dry it in vacuum over phosphorus pentoxide for 4 hours: it loses not more than 0.2% of its weight.

Residue on ignition ⟨281⟩: not more than 0.1%.

Heavy metals, *Method I* ⟨231⟩—Dissolve 0.67 g in 15 mL of water, add 1 mL of dilute hydrochloric acid (1 in 12), and dilute with water to 25 mL: the limit is 0.003%.

Assay—Transfer about 100 mg of Hexylcaine Hydrochloride, previously dried and accurately weighed, to a 100-mL volumetric flask, dissolve in dilute hydrochloric acid (1 in 120), add dilute hydrochloric acid (1 in 120) to volume, and mix. Transfer 10.0 mL of this solution to a second 100-mL volumetric flask, add dilute hydrochloric acid (1 in 120) to volume, and mix. Transfer 10.0 mL of this last solution to a third 100-mL volumetric flask, add dilute hydrochloric acid (1 in 120) to volume, and mix. Concomitantly determine the absorbances of this solution and of a Standard solution of USP Hexylcaine Hydrochloride RS in the same medium having a known concentration of about 10 µg per mL in 1-cm cells at the wavelength of maximum absorbance at about 232 nm, with a suitable spectrophotometer, using dilute hydrochloric acid (1 in 120) as the blank. Calculate the quantity, in mg, of $C_{16}H_{23}NO_2 . HCl$ in the Hexylcaine Hydrochloride taken by the formula:

$$10C(A_U/A_S),$$

in which *C* is the concentration, in µg per mL, of USP Hexylcaine Hydrochloride RS in the Standard solution, and A_U and A_S are the absorbances of the solution from Hexylcaine Hydrochloride and the Standard solution, respectively.

Hexylcaine Hydrochloride Topical Solution

» Hexylcaine Hydrochloride Topical Solution contains not less than 93.0 percent and not more than 107.0 percent of the labeled amount of $C_{16}H_{23}NO_2 . HCl$.

Packaging and storage—Preserve in tight containers.

Reference standard—*USP Hexylcaine Hydrochloride Reference Standard*—Dry in vacuum over phosphorus pentoxide for 4 hours before using.

Identification—It responds to *Identification test A* under *Hexylcaine Hydrochloride Injection.*

pH ⟨791⟩: between 3.0 and 5.0.

Assay—

0.006 M Octylammonium phosphate buffer—Mix 1.0 mL of octylamine with water to obtain 1000 mL of mixture, adjust with phosphoric acid to a pH of 2.9, and filter.

Mobile phase—Prepare a suitable degassed and filtered mixture of water, *0.006 M Octylammonium phosphate buffer*, and acetonitrile (65:20:15).

Standard preparation—Dissolve an accurately weighed quantity of USP Hexylcaine Hydrochloride RS in water to obtain a solution having a known concentration of about 1 mg per mL.

Assay preparation—Transfer an accurately measured volume of Hexylcaine Hydrochloride Topical Solution, equivalent to about 100 mg of hexylcaine hydrochloride, to a 100-mL volumetric flask, dilute with water to volume, and mix.

Chromatographic system (see *Chromatography* ⟨621⟩)—The liquid chromatograph is equipped with a 254-nm detector and a 3.9-mm × 30-cm column that contains 10-µm packing L10. The flow rate is about 1.4 mL per minute. Chromatograph the *Standard preparation*, and record the peak response as directed under *Procedure:* the column efficiency is not less than 900 theoretical plates, and the relative standard deviation for replicate injections is not more than 2.0%.

Procedure—Separately inject equal volumes (about 10 µL) of the *Standard preparation* and the *Assay preparation* into the chromatograph, record the chromatograms, and measure the responses for the major peaks. Calculate the quantity, in mg, of $C_{16}H_{23}NO_2 . HCl$ in each mL of the Topical Solution taken by the formula:

$$100(C/V)(r_U/r_S),$$

in which *C* is the concentration, in mg per mL, of USP Hexylcaine Hydrochloride RS in the *Standard preparation*, *V* is the volume, in mL, of Topical Solution taken, and r_U and r_S are the peak responses obtained for hexylcaine hydrochloride from the *Assay preparation* and the *Standard preparation*, respectively.

Hexylene Glycol—*see* Hexylene Glycol NF

Histamine Phosphate

HN N · $2H_3PO_4$ $CH_2CH_2NH_2$

$C_5H_9N_3 . 2H_3PO_4$ 307.14

1*H*-Imidazole-4-ethanamine, phosphate (1:2).
Histamine phosphate (1:2) [*51-74-1*].

» Histamine Phosphate contains not less than 98.0 percent and not more than 101.0 percent of $C_5H_9N_3 . 2H_3PO_4$, calculated on the dried basis.

Packaging and storage—Preserve in tight, light-resistant containers.

Identification—

A: Dissolve 0.10 g in a mixture of 7 mL of water and 3 mL of 1 *N* sodium hydroxide, and add the solution to a mixture of 50 mg of sulfanilic acid, 10 mL of water, 2 drops of hydrochloric acid, and 2 drops of sodium nitrite solution (1 in 10): a deep red color is produced.

B: Dissolve 50 mg in 5 mL of hot water, add a hot solution of 50 mg of picrolonic acid in 10 mL of alcohol, and allow to crystallize. Filter the crystals with suction, wash with a small amount of ice-cold water, and dry at 105° for 1 hour: the crystals so obtained melt between 250° and 254°, with decomposition.

C: A solution (1 in 10) responds to the tests for *Phosphate* ⟨191⟩.

Loss on drying ⟨731⟩—Dry it at 105° for 2 hours: it loses not more than 3.0% of its weight.

Assay—Dissolve about 150 mg, accurately weighed, of Histamine Phosphate in 10 mL of water. Add 5 mL of chloroform and 25 mL of alcohol, then add 10 drops of thymolphthalein TS, and titrate with 0.2 *N* sodium hydroxide VS. Each mL of 0.2 *N* sodium hydroxide is equivalent to 15.36 mg of $C_5H_9N_3 . 2H_3PO_4$.

Histamine Phosphate Injection

» Histamine Phosphate Injection is a sterile solution of Histamine Phosphate in Water for Injection. It contains not less than 90.0 percent and not more than 110.0 percent of the labeled amount of $C_5H_9N_3 . 2H_3PO_4$.

Packaging and storage—Preserve in single-dose or in multiple-dose containers, preferably of Type I glass, protected from light.

Reference standard—*USP Histamine Dihydrochloride Reference Standard*—Dry over silica gel for 2 hours before using.

Identification—

A: Evaporate a volume of Injection, equivalent to about 2 mg of histamine phosphate, on a steam bath to dryness, dissolve the residue in 0.5 mL of water, and add 0.5 mL of 1 *N* sodium hydroxide. Add 2 drops of sodium nitrite solution (1 in 10), and add 1 mL of a solution prepared by mixing 50 mg of sulfanilic acid with 10 mL of water containing 2 drops of hydrochloric acid: an orange-red color is produced.

B: To 1 mL of Injection, equivalent to not less than 1 mg of histamine phosphate (concentrate a larger volume by evaporation, if necessary), add ammonium molybdate TS dropwise: a yellow precipitate, which is soluble in ammonia TS, is formed.

pH ⟨791⟩: between 3.0 and 6.0.

Other requirements—It meets the requirements under *Injections* ⟨1⟩.

Assay—

Standard preparation—Transfer 33.0 mg of USP Histamine Dihydrochloride RS to a 100-mL volumetric flask, add water to volume, and mix. Prepare fresh daily. Dilute the stock solution with water in the ratio of 1 in 10. The *Standard preparation* thus obtained contains 20 µg of histamine per mL, equivalent to 55.2 µg of histamine phosphate.

Assay preparation—Dilute an accurately measured volume of Histamine Phosphate Injection, equivalent to about 2.75 mg of histamine phosphate, with water in a 50-mL volumetric flask to volume, and mix.

If phenol is present, prepare the *Assay preparation* as follows: Dilute an accurately measured volume of Histamine Phosphate Injection, equivalent to about 2.75 mg of histamine phosphate, with water to about 25 mL. Heat the solution on a steam bath until the odor of phenol no longer is perceptible, adding water as required to maintain a volume of about 15 mL. Transfer to a 50-mL volumetric flask, cool, dilute with water to volume, and mix.

Procedure—Pipet 5 mL each of the *Standard preparation* and the *Assay preparation* into separate, 10-mL volumetric flasks, to each add 1 mL of sodium borate solution (1 in 100), followed by 1 mL of a freshly prepared solution of 50 mg of β-naphthoquinone-4-sodium sulfonate in 10 mL of water. Place the flasks in boiling water for 10 minutes, then immerse them for 5 minutes in water maintained between 5° and 10°. To each flask, add 1 mL of acid-formaldehyde (made by adding 0.5 mL of formaldehyde TS to a mixture of 45 mL of 1 *N* hydrochloric acid and 10 mL of glacial acetic acid and diluting with water to 80 mL), mix, add 1 mL of 0.1 *N* sodium thiosulfate, then dilute with water to volume, and mix. Concomitantly and immediately determine the absorbances of both solutions at the wavelength of maximum absorbance at about 460 nm, with a suitable spectrophotometer, against a reagent blank. Calculate the quantity, in mg, of $C_5H_9N_3 . 2H_3PO_4$ in each mL of the Injection taken by the formula:

$$(2.76 / V)(A_U / A_S),$$

in which V is the volume, in mL, of Injection taken, and A_U and A_S are the absorbances of the solutions from the *Assay preparation* and *Standard preparation*, respectively.

Histidine

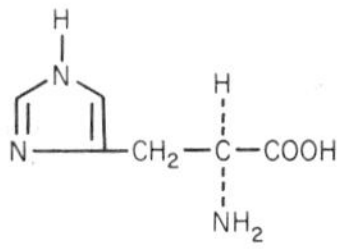

$C_6H_9N_3O_2$ 155.16
L-Histidine.
L-Histidine [*71-00-1*].

» Histidine contains not less than 98.5 percent and not more than 101.5 percent of $C_6H_9N_3O_2$, as L-histidine, calculated on the dried basis.

Packaging and storage—Preserve in well-closed containers.

Reference standard—*USP L-Histidine Reference Standard*—Dry at 105° for 3 hours before using.

Identification—The infrared absorption spectrum of a potassium bromide dispersion of it, previously recrystallized from 80% alcohol and dried, exhibits maxima only at the same wavelengths as that of a similar preparation of USP L-Histidine RS.

Specific rotation ⟨781⟩: between +12.6° and +14.0°, calculated on the dried basis, determined in a solution in 6 *N* hydrochloric acid containing 1.10 g in each 10.0 mL.

pH ⟨791⟩: between 7.0 and 8.5, in a solution (1 in 50).

Loss on drying ⟨731⟩—Dry it at 105° for 3 hours: it loses not more than 0.2% of its weight.

Residue on ignition ⟨281⟩: not more than 0.4%.

Chloride ⟨221⟩—A 0.73-g portion shows no more chloride than corresponds to 0.50 mL of 0.020 *N* hydrochloric acid (0.05%).

Sulfate ⟨221⟩—A 0.33-g portion shows no more sulfate than corresponds to 0.10 mL of 0.020 *N* sulfuric acid (0.03%).

Arsenic ⟨211⟩: 1.5 ppm.

Iron ⟨241⟩: 0.003%.

Heavy metals, *Method I* ⟨231⟩: 0.0015%.

Assay—Transfer about 150 mg of Histidine, accurately weighed, to a 125-mL flask, dissolve in a mixture of 3 mL of formic acid and 50 mL of glacial acetic acid, and titrate with 0.1 *N* perchloric acid VS, determining the end-point potentiometrically. Perform a blank determination, and make any necessary correction. Each mL of 0.1 *N* perchloric acid is equivalent to 15.52 mg of $C_6H_9N_3O_2$.

Histoplasmin

» Histoplasmin conforms to the regulations of the federal Food and Drug Administration concerning biologics (see *Biologics* ⟨1041⟩). It is a clear, colorless, sterile solution containing standardized culture filtrates of *Histoplasma capsulatum* grown on liquid synthetic medium. It has a potency of the 1:100 dilution equivalent to and determined in terms of the Histoplasmin Reference diluted 1:100 tested in guinea pigs. It may contain a suitable antimicrobial agent.

Packaging and storage—Preserve at a temperature between 2° and 8°.

Expiration date—The expiration date is not later than 2 years after date of issue from manufacturer's cold storage (5°, 1 year).

Labeling—Label it to state that only the diluent supplied is to be used for making dilutions, and that it is not to be injected other than intradermally. Label it also to state that a separate syringe and needle shall be used for each individual injection.

Homatropine Hydrobromide

CH3 N H H O OH C O–C–CH– · HBr H

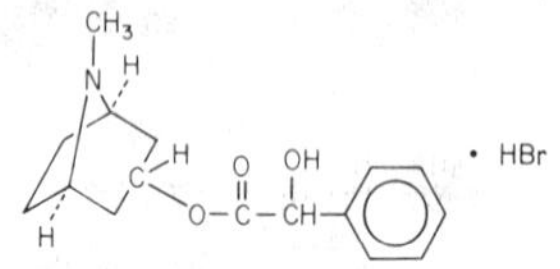

$C_{16}H_{21}NO_3.HBr$ 356.26

Benzeneacetic acid, α-hydroxy-, 8-methyl-8-azabicyclo-[3.2.1]oct-3-yl ester, hydrobromide, *endo*-(±)-.

1αH,5αH-Tropan-3α-ol mandelate (ester) hydrobromide [*51-56-9*].

» Homatropine Hydrobromide contains not less than 98.5 percent and not more than 100.5 percent of $C_{16}H_{21}NO_3.HBr$, calculated on the dried basis.

Packaging and storage—Preserve in tight, light-resistant containers.

Reference standard—*USP Homatropine Hydrobromide Reference Standard*—Dry at 105° for 2 hours before using.

Identification—

A: The infrared absorption spectrum of a potassium bromide dispersion of it exhibits maxima only at the same wavelengths as that of a similar preparation of USP Homatropine Hydrobromide RS.

B: It responds to the tests for *Bromide* ⟨191⟩.

Melting range ⟨741⟩: between 214° and 217°, with slight decomposition.

pH ⟨791⟩: between 5.7 and 7.0, in a solution (1 in 50).

Loss on drying ⟨731⟩—Dry it at 105° for 2 hours: it loses not more than 1.5% of its weight.

Residue on ignition ⟨281⟩: not more than 0.25%.

Assay—Dissolve about 400 mg of Homatropine Hydrobromide, accurately weighed, in water to make 50.0 mL, and mix. Transfer 10.0 mL of this solution to a beaker, add 5 mL of 1 *N* sodium hydroxide, and heat the solution just to boiling. Add 10 mL of 1 *N* nitric acid, add water to make 50 mL, and cool in an ice bath. Concomitantly add 5 mL of 1 *N* nitric acid to a second 10.0-mL portion of the solution of Homatropine Hydrobromide, add water to make 50 mL, and cool in an ice bath. Add 1 drop of nitrophenanthroline TS to each solution, and, while keeping the solutions cold, titrate with 0.05 *N* ceric ammonium nitrate VS until the pink color is discharged. Each mL of the difference in volumes of 0.05 *N* ceric ammonium nitrate required is equivalent to 8.907 mg of $C_{16}H_{21}NO_3.HBr$.

Homatropine Hydrobromide Ophthalmic Solution

» Homatropine Hydrobromide Ophthalmic Solution is a sterile, buffered, aqueous solution of Homatropine Hydrobromide. It contains not less than 95.0 percent and not more than 105.0 percent of the labeled amount of $C_{16}H_{21}NO_3.HBr$. It may contain suitable antimicrobial agents.

Packaging and storage—Preserve in tight containers.

Reference standard—*USP Homatropine Hydrobromide Reference Standard*—Dry at 105° for 2 hours before using.

Identification—

A: Proceed with Ophthalmic Solution as directed under *Identification—Organic Nitrogenous Bases* ⟨181⟩. The specified results are obtained.

B: It responds to the tests for *Bromide* ⟨191⟩.

Sterility—It meets the requirements under *Sterility Tests* ⟨71⟩.

pH ⟨791⟩: between 2.5 and 5.0.

Assay—

Standard preparation—Weigh accurately about 50 mg of USP Homatropine Hydrobromide RS, dissolve in water, and dilute with water in a volumetric flask to 100 mL. Dilute 10.0 mL of this solution with water to 50.0 mL to obtain a solution having a known concentration of about 100 µg per mL. Prepare this solution fresh.

Assay preparation—Transfer a portion of Homatropine Hydrobromide Ophthalmic Solution, equivalent to 50 mg of homatropine hydrobromide, to a 100-mL volumetric flask, and dilute with water to volume. Dilute 10.0 mL of this solution with water to 50.0 mL.

Procedure—Transfer duplicate 2-mL portions of the *Standard preparation* and of the *Assay preparation* to separate glass-stoppered, 40-mL centrifuge tubes. To one set of two tubes add 3 mL of water and 1 mL of sodium hydroxide solution (1 in 100). Heat these tubes in a boiling water bath for 20 minutes, and allow to cool to room temperature. To the remaining set of tubes, which serve as blanks for the *Standard preparation* and the *Assay preparation*, respectively, add 4 mL of water. To each tube, add 2 mL of approximately 0.2 *M* ceric sulfate in diluted sulfuric acid (prepared by dissolving 12.6 g of ceric ammonium sulfate in 50 mL of water and 3 mL of sulfuric acid, and diluting with water to 100 mL) and 20.0 mL of isooctane. Shake by mechanical means for 15 minutes, allow the layers to separate, and remove the isooctane from each tube. Concomitantly determine the absorbances of the isooctane solutions from the hydrolyzed aliquots in 1-cm cells at the wavelength of maximum absorbance at about 242 nm, with a suitable spectrophotometer, against the respective blanks. Calculate the quantity, in mg, of $C_{16}H_{21}NO_3.HBr$ in the portion of Ophthalmic Solution taken by the formula:

$$0.5C(A_U/A_S),$$

in which C is the concentration, in µg per mL, of USP Homatropine Hydrobromide RS in the *Standard preparation*, and A_U and A_S are the absorbances of the solutions from the *Assay preparation* and the *Standard preparation*, respectively.

Homatropine Methylbromide

H3C CH3 N+ H H O OH C O–C–CH– Br⁻ H

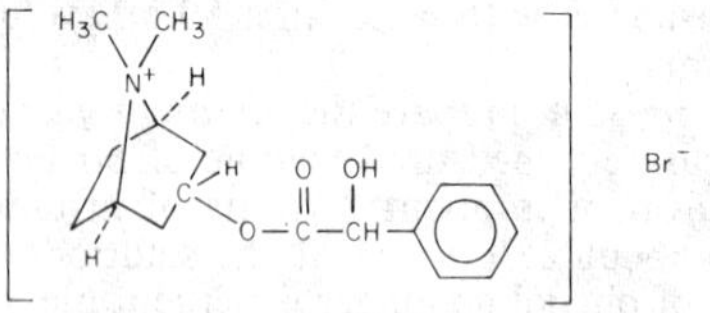

$C_{17}H_{24}BrNO_3$ 370.29

8-Azoniabicyclo[3.2.1]octane, 3-[(hydroxyphenylacetyl)oxy]-8,8-dimethyl-, bromide, *endo*-.

3α-Hydroxy-8-methyl-1αH,5αH-tropanium bromide mandelate [*80-49-9*].

» Homatropine Methylbromide contains not less than 98.5 percent and not more than 100.5 percent of $C_{17}H_{24}BrNO_3$, calculated on the dried basis.

Packaging and storage—Preserve in tight, light-resistant containers.

Reference standard—*USP Homatropine Methylbromide Reference Standard*—Dry at 105° for 3 hours before using.

Identification—

A: The infrared absorption spectrum of a potassium bromide dispersion of it, previously dried, exhibits maxima only at the same wavelengths as that of a similar preparation of USP Homatropine Methylbromide RS. [NOTE—If differences are observed, dissolve the specimen and the Reference Standard separately in methanol, and recrystallize by adding dioxane to each solution.]

B: The ultraviolet absorption spectrum of 1 in 1000 solution in alcohol exhibits maxima and minima at the same wavelengths as that of a similar solution of USP Homatropine Methylbromide RS, concomitantly measured, and the respective absorptivities, calculated on the dried basis, at the wavelength of maximum absorbance at about 258 nm, do not differ by more than 3.0%.

C: Mercuric–potassium iodide TS produces in a solution (1 in 50) a white or slightly yellowish precipitate, but no precipitation is caused by solutions of alkali hydroxides or carbonates, even in concentrated solutions of the substance (*distinction from most alkaloids*).

D: To a solution (1 in 50), add ammonium reineckate TS: a red precipitate is formed.

E: A solution (1 in 20) responds to the tests for *Bromide* ⟨191⟩.

pH ⟨791⟩: between 4.5 and 6.5, in a solution (1 in 100).

Loss on drying ⟨731⟩—Dry it at 105° for 3 hours: it loses not more than 0.5% of its weight.

Residue on ignition ⟨281⟩: not more than 0.2%.

Homatropine, atropine, and other solanaceous alkaloids—To 1 mL of a solution of it (1 in 50), add a few drops of 6 *N* ammonium hydroxide, shake the solution with 5 mL of chloroform, and evaporate the separated chloroform layer on a steam bath to dryness. Warm the residue with 1.5 mL of a solution made by dissolving 500 mg of mercuric chloride in 25 mL of a mixture of 5 volumes of alcohol and 3 volumes of water: no yellow or red color develops.

Assay—Dissolve about 700 mg of Homatropine Methylbromide, accurately weighed, in a mixture of 50 mL of glacial acetic acid and 10 mL of mercuric acetate TS. Add 1 drop of crystal violet TS, and titrate with 0.1 *N* perchloric acid VS to a blue-green end-point. Perform a blank determination, and make any necessary correction. Each mL of 0.1 *N* perchloric acid is equivalent to 37.03 mg of $C_{17}H_{24}BrNO_3$.

Homatropine Methylbromide Tablets

» Homatropine Methylbromide Tablets contain not less than 90.0 percent and not more than 110.0 percent of the labeled amount of $C_{17}H_{24}BrNO_3$.

Packaging and storage—Preserve in tight, light-resistant containers.

Reference standard—*USP Homatropine Methylbromide Reference Standard*—Dry at 105° for 3 hours before using.

Identification—Shake a quantity of finely powdered Tablets, equivalent to about 10 mg of homatropine methylbromide, with 15 mL of a mixture of equal volumes of methanol and water for 10 minutes, and filter. Evaporate the filtrate on a steam bath to dryness, and dry at 105° for 1 hour. The residue of homatropine methylbromide so obtained melts between 190° and 198° (see *Class I* under *Melting Range or Temperature* ⟨741⟩), the temperature at which distinct liquefaction of the specimen is first observed being taken as the beginning of melting.

Dissolution ⟨711⟩—

Medium: water; 900 mL.

Apparatus 2: 50 rpm.

Time: 45 minutes.

Procedure—Determine the amount of $C_{17}H_{24}BrNO_3$ dissolved from ultraviolet absorbances at the wavelength of maximum absorbance at about 258 nm of filtered portions of the solution under test, suitably diluted with *Dissolution Medium*, if necessary, in comparison with a Standard solution having a known concentration of USP Homatropine Methylbromide RS in the same medium.

Tolerances—Not less than 75% (*Q*) of the labeled amount of $C_{17}H_{24}BrNO_3$ is dissolved in 45 minutes.

Uniformity of dosage units ⟨905⟩: meet the requirements.

Procedure for content uniformity—

Standard preparation—Transfer about 25 mg of USP Homatropine Methylbromide RS, accurately weighed, to a 50-mL volumetric flask, add water to volume, and mix. Transfer 10.0 mL of this solution to a second 50-mL volumetric flask, dilute with water to volume, and mix. The concentration of USP Homatropine Methylbromide RS in the *Standard preparation* is about 100 µg per mL.

Test preparation—Transfer 1 finely powdered Tablet to a volumetric flask, suitably sized such that when the specimen is diluted to volume, the concentration is equivalent to about 100 µg of homatropine methylbromide per mL. Add water to about one-half of the volume of the flask, shake for 10 minutes, dilute with water to volume, mix, and filter, discarding the first 10 mL of filtrate. Use the subsequent filtrate as directed in the *Procedure*.

Procedure—Transfer 2.0 mL each of the *Standard preparation* and the *Test preparation* to separate glass-stoppered, 50-mL flasks. To each flask, add 0.1 mL of sodium hydroxide solution (1 in 10), and heat in a water bath at 80° for 15 minutes. Cool to room temperature, add 2.0 mL of 0.2 *M* ceric ammonium sulfate in 1 *N* sulfuric acid, and mix. To each flask, add 20.0 mL of *n*-hexane, and shake for 15 minutes. Decant the hexane layers into separate 1-cm cells, and concomitantly determine the absorbances of the solutions in 1-cm cells at the wavelength of maximum absorbance at about 242 nm, with a suitable spectrophotometer, using *n*-hexane as the blank. Calculate the quantity, in mg, of $C_{17}H_{24}BrNO_3$ in the Tablet by the formula:

$$(TC/D)(A_U/A_S),$$

in which *T* is the labeled quantity, in mg, of homatropine methylbromide in the Tablet, *C* is the concentration, in µg per mL, of USP Homatropine Methylbromide RS in the *Standard preparation*, *D* is the concentration, in µg per mL, of the *Test preparation*, based upon the labeled quantity per Tablet and the extent of dilution, and A_U and A_S are the absorbances of the solutions from the *Test preparation* and the *Standard preparation*, respectively.

Assay—

Standard preparation—Transfer about 25 mg of USP Homatropine Methylbromide RS, accurately weighed, to a 50-mL volumetric flask, dissolve in water, dilute with water to volume, and mix.

Assay preparation—Weigh and finely powder not less than 20 Homatropine Methylbromide Tablets. Weigh accurately a portion of the powder, equivalent to about 12.5 mg of homatropine methylbromide, and shake with 10 mL of water at frequent intervals during 30 minutes. Filter under reduced pressure through a sintered-glass crucible into a test tube placed in the suction flask under the filtering funnel, and wash under suction with several small portions of water. Transfer the contents of the test tube to a 25-mL volumetric flask, dilute with water to volume, and mix.

Procedure—Transfer 10.0 mL each of the *Standard preparation* and the *Assay preparation* to separate test tubes, to each add 1 mL of 5 *N* sulfuric acid and 2 mL of ammonium reineckate TS, shake gently but well, and allow to stand for 1 hour. Filter through a sintered-glass crucible with suction, using portions of the filtrate to transfer the precipitate completely to the filter, and wash it with three 2-mL portions of ice-cold water. Completely dissolve the precipitate by pouring over it 1-mL portions of acetone with the application of suction, receiving the solution in a 10-mL volumetric flask, add acetone to volume, and mix. Concomitantly determine the absorbances of the solutions in 1-cm cells at the wavelength of maximum absorbance at about 525 nm, with a suitable spectrophotometer, using acetone as the blank. Calculate the quantity, in mg, of $C_{17}H_{24}BrNO_3$ in the portion of Tablets taken by the formula:

$$0.025C(A_U/A_S),$$

in which *C* is the concentration, in µg per mL, of USP Homatropine Methylbromide RS in the *Standard preparation*, and A_U and A_S are the absorbances of the solutions from the *Assay preparation* and the *Standard preparation*, respectively.

Human, Albumin—*see* Albumin Human
Human, Insulin—*see* Insulin Human
Human Injection, Insulin—*see* Insulin Human Injection

Hyaluronidase Injection

» Hyaluronidase Injection is a sterile solution of dry, soluble enzyme product, prepared from mammalian testes and capable of hydrolyzing mucopolysaccharides of the type of hyaluronic acid, in Water for Injection. It contains not less than 90.0 percent of the labeled amount of USP Hyaluronidase Units. Hyaluronidase Injection contains not more than 0.25 µg of tyrosine for each USP Hyaluronidase Unit. It may contain suitable stabilizers.

Packaging and storage—Preserve in single-dose or in multiple-dose containers, preferably of Type I glass, in a refrigerator.

Reference standards—*USP Hyaluronidase Reference Standard*—Keep container tightly closed, and store in a cool, dry place, preferably under refrigeration in a desiccator. Do not dry before using. *USP Tyrosine Reference Standard*—Keep container tightly closed. Dry at 105° for 3 hours before using.

Tyrosine—Transfer an accurately measured volume, equivalent to about 120 USP Hyaluronidase Units, to a 15-mL centrifuge tube calibrated at 6 mL, and evaporate at 105° to dryness. Proceed as directed in the test for *Tyrosine* under *Hyaluronidase for Injection*, beginning with "Add 200 µL of 6 *N* sodium hydroxide," except to read Hyaluronidase Injection for Hyaluronidase for Injection in the second and third lines of the second paragraph. Calculate the quantity, in µg, of tyrosine in the volume of Injection taken by the formula:

$$45(A_U/A_S),$$

in which A_U and A_S are the absorbances of the solution from the Injection and the Standard solution, respectively: not more than 0.25 µg of tyrosine is found for each USP Hyaluronidase Unit.

Pyrogen—It meets the requirements of the *Pyrogen Test* ⟨151⟩, the test dose being 1 mL per kg of a solution containing 75 USP Hyaluronidase Units in each mL.

pH ⟨791⟩: between 6.4 and 7.4.

Other requirements—It meets the requirements under *Injections* ⟨1⟩.

Assay—

Acetate buffer solution, Phosphate buffer solution, Hydrolyzed gelatin, Diluent for hyaluronidase solutions, Serum stock solution, Serum solution, Potassium hyaluronate stock solution, Hyaluronate solution, and *Standard solution*—Prepare as directed in the *Assay* under *Hyaluronidase for Injection.*

Assay preparation—Dilute an accurately measured volume of Hyaluronidase Injection quantitatively with cold *Diluent for hyaluronidase solutions*, on the basis of trial or experience, so that the observed absorbances with the three dilutions of the Injection fall on the upper, linear part of the standard curve prepared as directed in the *Procedure.*

Procedure—Proceed as directed for *Procedure* in the *Assay* under *Hyaluronidase for Injection.* The potency, in USP Hyaluronidase Units, of the portion of Injection taken is the average of the six activity values read from the standard curve.

Hyaluronidase for Injection

» Hyaluronidase for Injection is a sterile, dry, soluble, enzyme product prepared from mammalian testes and capable of hydrolyzing mucopolysaccharides of the type of hyaluronic acid. Its potency, in USP Hyaluronidase Units, is not less than the labeled potency. Hyaluronidase for Injection contains not more than 0.25 µg of tyrosine for each USP Hyaluronidase Unit. It may contain a suitable stabilizer.

Packaging and storage—Preserve in *Containers for Sterile Solids* as described under *Injections* ⟨1⟩, preferably of Type I or Type III glass, at controlled room temperature.

Reference standards—*USP Tyrosine Reference Standard*—Keep container tightly closed. Dry at 105° for 3 hours before using. *USP Hyaluronidase Reference Standard*—Keep container tightly closed, and store in a cool, dry place, preferably under refrigeration in a desiccator. Do not dry before using.

Tyrosine—Dissolve the entire contents of 1 or more containers of Hyaluronidase for Injection in sufficient water, accurately measured, to give a concentration of about 60 USP Hyaluronidase Units per mL. Transfer 2.0 mL of the solution to a 15-mL centrifuge tube calibrated at 6 mL, and evaporate at 105° to dryness. Add 200 µL of 6 *N* sodium hydroxide, and heat with steam under pressure at 121° for 3 hours. Add 300 µL of 7 *N* sulfuric acid, then add 1.5 mL of water and 1.5 mL of a 15 in 100 solution of mercuric sulfate in 5 *N* sulfuric acid. Heat on a steam bath for 10 minutes, and cool to room temperature. Add 1 mL of 7 *N* sulfuric acid and 1 mL of sodium nitrite solution (1 in 500), with shaking. Add water to make 6 mL, mix, centrifuge, and decant the supernatant liquid. Twenty minutes after diluting to 6 mL, determine the absorbance of the supernatant liquid at 540 nm, with a suitable spectrophotometer.

Repeat the preceding test, using the same quantities of the same reagents and in the same manner but omitting the Hyaluronidase for Injection and replacing the 1.5 mL of water with 1.5 mL of a solution of USP Tyrosine RS in 0.4 *N* sulfuric acid containing 30 µg in each mL.

Calculate the quantity, in µg, of tyrosine in the 2-mL aliquot of the solution of Hyaluronidase for Injection by the formula:

$$45(A_U/A_S),$$

in which A_U and A_S are the absorbances of the solution from Hyaluronidase for Injection and the Standard solution, respectively: not more than 0.25 µg of tyrosine is found for each USP Hyaluronidase Unit.

Pyrogen—It meets the requirements of the *Pyrogen Test* ⟨151⟩, the test dose being 1.0 mL per kg of a solution containing 75 USP Hyaluronidase Units in each mL.

Sterility—It meets the requirements under *Sterility Tests* ⟨71⟩.

Assay—

Acetate buffer solution—Dissolve 14 g of potassium acetate and 20.5 mL of glacial acetic acid in water to make 1,000 mL.

Phosphate buffer solution—Dissolve 2.5 g of monobasic sodium phosphate, 1.0 g of anhydrous dibasic sodium phosphate, and 8.2 g of sodium chloride in water to make 1000 mL.

Hydrolyzed gelatin—Dissolve 50 g of bacteriological gelatin in 1,000 mL of water, heat in an autoclave at 121° for 90 minutes, and freeze-dry the solution.

Diluent for hyaluronidase solutions—Mix 250 mL of *Phosphate buffer solution* with 250 mL of water and, within 2 hours before use, dissolve 330 mg of *Hydrolyzed gelatin* in the mixture.

Serum stock solution—Constitute dried horse serum with water to its original volume, and dilute it with 9 volumes of *Acetate buffer solution.* Adjust with 4 *N* hydrochloric acid to a pH of 3.1, and allow the solution to stand at room temperature for 18 to 24 hours. Store the solution at 0° to 4°, and use within 30 days.

Serum solution—On the day of the assay, dilute 1 volume of *Serum stock solution* with 3 volumes of *Acetate buffer solution*, and adjust to room temperature.

Potassium hyaluronate stock solution—Prepare a stock solution to contain, in each mL, 500 µg of potassium hyaluronate, previously dried in vacuum over phosphorus pentoxide for 48 hours. Do not keep the hyaluronate over phosphorus pentoxide indefinitely. The use of a weighing bottle is desirable. Store the solution at a temperature not exceeding 5°, and use within 30 days.

Hyaluronate solution—On the day of the assay, dilute 1 volume of *Potassium hyaluronate stock solution* with 1 volume of *Phosphate buffer solution.*

Standard solution—Dissolve a suitable quantity of USP Hyaluronidase RS, accurately weighed, in cold *Diluent for hyaluronidase solutions* to obtain a solution having a known concentration of about 1.5 USP Hyaluronidase Units in each mL. Prepare this solution immediately before use in the assay.

Assay solution—Dissolve the contents of 1 container of Hyaluronidase for Injection by adding cold *Diluent for hyaluronidase solutions* directly to the container. On the basis of trial or experience, dilute the solution so prepared with cold *Diluent for hyaluronidase solutions* so that the observed absorbances with the three dilutions of Hyaluronidase for Injection fall on the upper, linear part of the standard curve prepared as directed below.

Procedure—Prepare a standard concentration–response curve by adding to each of twelve 16- × 100-mm test tubes 500 µL of *Hyaluronate solution.* To each of two of the tubes add, respectively, 500, 400, 300, 200, 100, and 0 µL, of *Diluent for hyaluronidase solutions.* If quantities of *Standard solution* other than those indicated below are used, change the above-specified quantities of *Diluent for hyaluronidase solutions* so that the final volume of solution in each tube, after the addition of the *Standard solution*, is 1.00 mL.

At 30-second intervals, accurately timed, add to each of two tubes 0, 100, 200, 300, 400, and 500 µL, respectively, of *Standard solution.* Mix the contents by gentle shaking, and place each tube in a water bath maintained at 37 ± 0.2°. After 30 ± 0.25 minutes, remove each tube, in order, from the water bath at 30-second intervals, and immediately add 4.0 mL of *Serum solution.* Shake the tube, and allow to stand at room temperature for 30 ± 2 minutes. Again shake the tube, and determine the absorbance at 640 nm, with a suitable spectrophotometer. Perform a blank determination but omit the hyaluronate, and make any necessary correction. Plot the average absorbance value for each level against the hyaluronidase activity expressed in Units, and draw the smooth curve that best fits the plotted points.

Concurrently, to six test tubes add 500 µL of *Hyaluronate solution* and sufficient *Diluent for hyaluronidase solutions* so that the final volume, after the addition of the *Assay solution*, is 1.00 mL. Add, at 30-second intervals, sufficient *Assay solution* so that duplicate tubes contain about 0.30, 0.50, and 0.70 Unit, respectively. Shake each tube gently, and treat as directed in the preceding paragraph for the *Standard solution*, beginning with "place each tube in a water bath." Measure the absorbances in the spectrophotometer, and make any necessary correction for the blank. The potency of Hyaluronidase for Injection is the average of the six activity values read from the standard curve.

Hydralazine Hydrochloride

NHNH$_2$ · HCl

$C_8H_8N_4 . HCl$ 196.64

Phthalazine, 1-hydrazino-, monohydrochloride.

1-Hydrazinophthalazine monohydrochloride [*304-20-1*].

» Hydralazine Hydrochloride contains not less than 98.0 percent and not more than 102.0 percent of $C_8H_8N_4 . HCl$, calculated on the dried basis.

Packaging and storage—Preserve in tight containers.

Reference standard—*USP Hydralazine Hydrochloride Reference Standard*—Dry in vacuum over phosphorus pentoxide for 8 hours before using.

Identification—

A: The infrared absorption spectrum of a mineral oil dispersion of it, previously dried, exhibits maxima only at the same wavelengths as that of a similar preparation of USP Hydralazine Hydrochloride RS.

B: The ultraviolet absorption spectrum of a solution (1 in 100,000) exhibits maxima and minima at the same wavelengths as that of a similar solution of USP Hydralazine Hydrochloride RS, concomitantly measured, and the respective absorptivities, calculated on the dried basis, at the wavelength of maximum absorbance at about 260 nm do not differ by more than 3.0%.

C: A solution (1 in 4000) responds to the tests for *Chloride* ⟨191⟩.

pH ⟨791⟩: between 3.5 and 4.2, in a solution (1 in 50).

Loss on drying ⟨731⟩—Dry it at 110° for 15 hours: it loses not more than 0.5% of its weight.

Residue on ignition ⟨281⟩: not more than 0.1%.

Water-insoluble substances—Transfer 2.0 g to a 250-mL conical flask, add 100 mL of water, and shake by mechanical means for about 30 minutes. Filter the solution through a tared sintered-glass crucible, and wash into the crucible any undissolved residue remaining in the flask. Wash the residue with three 10-mL portions of water, dry at 105° for 3 hours, cool, and weigh: the weight of the residue does not exceed 10 mg (0.5%).

Heavy metals, *Method II* ⟨231⟩: 0.002%.

Assay—

Tetramethylammonium nitrate solution—Dissolve about 1.36 g of tetramethylammonium nitrate in 950 mL of water, add 2 mL of glacial acetic acid, and mix.

Mobile phase—Prepare a filtered and degassed mixture of *Tetramethylammonium nitrate solution* and methanol (95:5). Make adjustments if necessary (see *System Suitability* under *Chromatography* ⟨621⟩).

Standard preparation—Dissolve an accurately weighed quantity of USP Hydralazine Hydrochloride RS in 0.1 *N* hydrochloric acid, dilute quantitatively, and stepwise if necessary, with 0.1 *N* hydrochloric acid to obtain a solution having a known concentration of about 40 µg per mL, and filter.

Assay preparation—Transfer about 100 mg of Hydralazine Hydrochloride, accurately weighed, to a 250-mL volumetric flask, dissolve in 0.1 *N* hydrochloric acid, dilute with the same solvent to volume, and mix. Pipet 10 mL of the resulting solution into a 100-mL volumetric flask, dilute with 0.1 *N* hydrochloric acid to volume, mix, and filter.

Chromatographic system (see *Chromatography* ⟨621⟩)—The liquid chromatograph is equipped with a 254-nm detector and a 3.9-mm × 30-cm column that contains packing L1. The flow rate is about 1 mL per minute. Chromatograph the *Standard preparation*, and record the peak responses as directed under *Procedure:* the tailing factor is not more than 2.5, and the relative standard deviation for replicate injections is not more than 1.0%.

Procedure—Separately inject equal volumes (about 25 µL) of the *Standard preparation* and the *Assay preparation* into the chromatograph, record the chromatograms, and measure the responses for the major peak. Calculate the quantity, in mg, of $C_8H_8N_4 . HCl$ in the portion of Hydralazine Hydrochloride taken by the formula:

$$2.5C(r_U / r_S),$$

in which C is the concentration, in µg per mL, of USP Hydralazine Hydrochloride RS in the *Standard preparation*, and r_U and r_S are the peak responses obtained from the *Assay preparation* and the *Standard preparation*, respectively.

Hydralazine Hydrochloride, and Hydrochlorothiazide Tablets, Reserpine,—*see* Reserpine, Hydralazine Hydrochloride, and Hydrochlorothiazide Tablets

Hydralazine Hydrochloride Injection

» Hydralazine Hydrochloride Injection is a sterile solution of Hydralazine Hydrochloride in Water for Injection. It contains not less than 95.0 percent and not more than 105.0 percent of the labeled amount of $C_8H_8N_4 . HCl$.

Packaging and storage—Preserve in single-dose or in multiple-dose containers, preferably of Type I glass.

Reference standard—*USP Hydralazine Hydrochloride Reference Standard*—Dry in vacuum over phosphorus pentoxide for 8 hours before using.

Identification—Mix a volume of Injection, equivalent to about 60 mg of hydralazine hydrochloride, with an amount of 1 *N* hydrochloric acid sufficient to prepare 25 mL of solution. Place 20 mL of this solution in a separator, wash with 10 mL of methylene chloride, and discard the methylene chloride washing. Mix the aqueous solution in the separator with 2 mL of sodium nitrite solution (14 in 1000), add 10 mL of methylene chloride, shake by mechanical means for 5 minutes, and allow the layers to separate. Pass the methylene chloride layer through a filter of anhydrous sodium sulfate that previously has been washed with methylene chloride, and collect the solution in a 50-mL beaker. Evaporate with the aid of gentle heat and a stream of dry nitrogen to dryness: the infrared absorption spectrum of a potassium bromide dispersion of the residue so obtained exhibits maxima only at the same wavelengths as that of USP Hydralazine Hydrochloride RS similarly treated and prepared.

pH ⟨791⟩: between 3.4 and 4.4.

Particulate matter ⟨788⟩: meets the requirements under *Small-volume Injections*.

Other requirements—It meets the requirements under *Injections* ⟨1⟩.

Assay—Transfer to a 250-mL iodine flask an accurately measured volume of Hydralazine Hydrochloride Injection, equivalent to about 100 mg of hydralazine hydrochloride. Add 20 mL of hydrochloric acid, cool to room temperature, add 5 mL of chloroform, and titrate with 0.02 *M* potassium iodate VS until the purple color of iodine disappears from the chloroform, adding the last portion of the potassium iodate solution dropwise and agitating the mixture vigorously and continuously. Each mL of 0.02 *M* potassium iodate is equivalent to 3.933 mg of $C_8H_8N_4 \cdot$HCl.

Hydralazine Hydrochloride Tablets

» Hydralazine Hydrochloride Tablets contain not less than 90.0 percent and not more than 110.0 percent of the labeled amount of $C_8H_8N_4 \cdot$HCl.

Packaging and storage—Preserve in tight, light-resistant containers.

Reference standard—*USP Hydralazine Hydrochloride Reference Standard*—Dry in vacuum over phosphorus pentoxide for 8 hours before using.

Identification—

A: Transfer a quantity of finely powdered Tablets, equivalent to about 100 mg of hydralazine hydrochloride, to a glass-stoppered flask. Add 40 mL of 1 *N* hydrochloric acid, shake by mechanical means for 5 minutes, and filter, discarding the first few mL of the filtrate. Place 20 mL of the filtrate in a separator, wash with 10 mL of methylene chloride, and discard the methylene chloride washing. Mix the aqueous solution in the separator with 2 mL of sodium nitrite solution (14 in 1000), add 10 mL of methylene chloride, shake by mechanical means for 5 minutes, and allow the layers to separate. Pass the methylene chloride layer through a filter of anhydrous sodium sulfate that previously has been washed with methylene chloride, and collect the solution in a 50-mL beaker. Evaporate with the aid of gentle heat and a stream of dry nitrogen to dryness: the infrared absorption spectrum of a potassium bromide dispersion of the residue so obtained exhibits maxima only at the same wavelengths as that of USP Hydralazine Hydrochloride RS similarly treated and prepared.

B: The retention time of the major peak in the chromatogram of the *Assay preparation* corresponds to that of the *Standard preparation*, as obtained in the *Assay*.

Dissolution ⟨711⟩—

Medium: 0.1 *N* hydrochloric acid; 900 mL.

Apparatus 1: 100 rpm.

Time: 30 minutes.

Procedure—Determine the amount of $C_8H_8N_4 \cdot$HCl dissolved from ultraviolet absorbances at the wavelength of maximum absorbance at about 260 nm of filtered portions of the solution under test, suitably diluted with 0.1 *N* hydrochloric acid, in comparison with a Standard solution having a known concentration of USP Hydralazine Hydrochloride RS in the same medium.

Tolerances—Not less than 60% (*Q*) of the labeled amount of $C_8H_8N_4 \cdot$HCl is dissolved in 30 minutes.

Uniformity of dosage units ⟨905⟩: meet the requirements.

Assay—

Tetramethylammonium nitrate solution, Mobile phase, and *Chromatographic system*—Proceed as directed in the *Assay* under *Hydralazine Hydrochloride.*

Standard preparation—Dissolve an accurately weighed quantity of USP Hydralazine Hydrochloride RS in 0.1 *N* hydrochloric acid, dilute quantitatively, and stepwise if necessary, with 0.1 *N* hydrochloric acid to obtain a solution having a known concentration of about 20 µg per mL, and filter.

Assay preparation—Weigh and finely powder not less than 20 Tablets. Transfer an accurately weighed portion of the powder, equivalent to about 50 mg of hydralazine hydrochloride, to a 250-mL volumetric flask, dissolve in 0.1 *N* hydrochloric acid, dilute with the same solvent to volume, and centrifuge. Pipet 10 mL of the resulting solution into a 100-mL volumetric flask, dilute with 0.1 *N* hydrochloric acid to volume, and filter.

Procedure—Proceed as directed for *Procedure* in the *Assay* under *Hydralazine Hydrochloride.* Calculate the quantity, in mg, of $C_8H_8N_4 \cdot$HCl in the portion of Tablets taken by the formula:

$$2.5C(r_U / r_S).$$

Hydrochloric Acid—*see* Hydrochloric Acid NF

Hydrochloric Acid, Diluted—*see* Hydrochloric Acid, Diluted NF

Hydrochlorothiazide

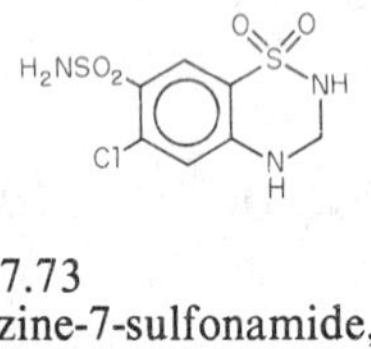

$C_7H_8ClN_3O_4S_2$ 297.73

2*H*-1,2,4-Benzothiadiazine-7-sulfonamide, 6-chloro-3,4-dihydro-, 1,1-dioxide.

6-Chloro-3,4-dihydro-2*H*-1,2,4-benzothiadiazine-7-sulfonamide 1,1-dioxide [*58-93-5*].

» Hydrochlorothiazide contains not less than 98.0 percent and not more than 102.0 percent of $C_7H_8ClN_3O_4S_2$, calculated on the dried basis.

Packaging and storage—Preserve in well-closed containers.

Reference standards—*USP Hydrochlorothiazide Reference Standard*—Dry at 105° for 1 hour before using. *USP 4-Amino-6-chloro-1,3-benzenedisulfonamide Reference Standard*—Keep container tightly closed and protected from light. Dry over silica gel for 4 hours before using.

Identification—

A: The infrared absorption spectrum of a potassium bromide dispersion of it, the potassium bromide–hydrochlorothiazide mixture previously being heated at 105° for 2 hours, exhibits maxima only at the same wavelengths as that of a similar preparation of USP Hydrochlorothiazide RS.

B: The ultraviolet absorption spectrum of a 1 in 100,000 solution in methanol exhibits maxima and minima at the same wavelengths as that of a similar solution of USP Hydrochlorothiazide RS, concomitantly measured.

Loss on drying ⟨731⟩—Dry it at 105° for 1 hour: it loses not more than 1.0% of its weight.

Residue on ignition ⟨281⟩: not more than 0.1%.

Chloride ⟨221⟩—Shake 0.50 g with 40 mL of water for 5 minutes, and filter: the filtrate shows no more chloride than corresponds to 0.50 mL of 0.020 *N* hydrochloric acid (0.07%).

Selenium ⟨291⟩: 0.003%, a 200-mg test specimen being used.

Heavy metals, *Method II* ⟨231⟩: 0.001%.

4-Amino-6-chloro-1,3-benzenedisulfonamide—

Mobile phase, System suitability solution, and *Chromatographic system*—Proceed as directed in the *Assay.*

Test preparation—Proceed as directed for *Assay preparation* in the *Assay.*

Standard preparation—[NOTE—A volume of acetonitrile not exceeding 10% of the total volume of the solution may be used to dissolve the Reference Standard.] Dissolve an accurately weighed quantity of USP 4-Amino-6-chloro-1,3-benzenedisulfonamide RS in *Mobile phase* to obtain a solution having a known concentration of about 1.5 µg per mL.

Procedure—Separately inject equal volumes (about 20 µL) of the *Standard preparation* and the *Test preparation* into the chromatograph, record the chromatograms, and measure the responses for the major peaks. Calculate the quantity, in mg, of 4-amino-6-chloro-1,3-benzenedisulfonamide in the portion of Hydrochlorothiazide taken by the formula:

$$0.2C(r_U/r_S),$$

in which C is the concentration, in µg per mL, of USP 4-Amino-6-chloro-1,3-benzenedisulfonamide RS in the *Standard preparation,* and r_U and r_S are the peak responses of 4-amino-6-chloro-1,3-benzenedisulfonamide obtained from the *Test preparation* and the *Standard preparation,* respectively: not more than 1.0% is present.

Assay—

Mobile phase—Prepare a degassed mixture of 0.1 *M* monobasic sodium phosphate and acetonitrile (9:1), adjust with phosphoric acid to a pH of 3.0 ± 0.1, and filter. Make adjustments if necessary (see *System Suitability* under *Chromatography* ⟨621⟩).

System suitability solution—[NOTE—A volume of acetonitrile not exceeding 10% of the total volume of solution may be used to dissolve the Reference Standards.] Dissolve accurately weighed quantities of USP Hydrochlorothiazide RS and Chlorothiazide in *Mobile phase* to obtain a solution having known concentrations of about 0.15 mg per mL each of USP Hydrochlorothiazide RS and Chlorothiazide.

Standard preparation—[NOTE—A volume of acetonitrile not exceeding 10% of the total volume of the solution may be used to dissolve the Reference Standard.] Dissolve an accurately weighed quantity of USP Hydrochlorothiazide RS in *Mobile phase* to obtain a solution having a known concentration of about 0.15 mg of USP Hydrochlorothiazide RS per mL.

Assay preparation—Transfer about 30 mg of Hydrochlorothiazide, accurately weighed, to a 200-mL volumetric flask, dissolve in a small volume of acetonitrile, not exceeding 10% of the total volume of the solution, dilute with *Mobile phase* to volume, and mix.

Chromatographic system (see *Chromatography* ⟨621⟩)—The liquid chromatograph is equipped with a 254-nm detector and a 4.6-mm × 25-cm column that contains packing L1. The flow rate is about 2.0 mL per minute. Chromatograph replicate injections of the *System suitability solution,* and record the peak responses as directed under *Procedure:* the relative standard deviation is not more than 1.5%, the relative retention times are about 0.8 for chlorothiazide and 1.0 for hydrochlorothiazide, and the resolution, R, between chlorothiazide and hydrochlorothiazide is not less than 2.0.

Procedure—Separately inject equal volumes (about 20 µL) of the *Standard preparation* and the *Assay preparation* into the chromatograph, record the chromatograms, and measure the responses for the major peaks. Calculate the quantity, in mg, of $C_7H_8ClN_3O_4S_2$ in the portion of Hydrochlorothiazide taken by the formula:

$$200C(r_U/r_S),$$

in which C is the concentration, in mg per mL, of USP Hydrochlorothiazide RS in the *Standard preparation,* and r_U and r_S are the peak responses of hydrochlorothiazide obtained from the *Assay preparation* and the *Standard preparation,* respectively.

Hydrochlorothiazide Tablets

» Hydrochlorothiazide Tablets contain not less than 90.0 percent and not more than 110.0 percent of the labeled amount of $C_7H_8ClN_3O_4S_2$.

Packaging and storage—Preserve in well-closed containers.

Reference standards—*USP Hydrochlorothiazide Reference Standard*—Dry at 105° for 1 hour before using. *USP 4-Amino-6-chloro-1,3-benzenedisulfonamide Reference Standard*—Keep container tightly closed and protected from light. Dry over silica gel for 4 hours before using.

Identification—

A: Transfer a portion of finely powdered Tablets, equivalent to about 50 mg of hydrochlorothiazide, to a 50-mL volumetric flask. Add about 20 mL of sodium hydroxide solution (1 in 125), and shake vigorously for 15 minutes. Dilute with the same solvent to volume, mix, and filter, discarding the first few mL of the filtrate. Transfer 5 mL of the filtrate to a 125-mL separator, and add 5 mL of dilute hydrochloric acid (1 in 10). Extract with 50 mL of ether, filter the ether extract through a small, dry, folded filter paper, and evaporate to dryness. Add 5 mL of alcohol, and again evaporate to dryness: the infrared absorption spectrum of a potassium bromide dispersion of the residue so obtained exhibits maxima only at the same wavelengths as that of a similar preparation of USP Hydrochlorothiazide RS previously dissolved in alcohol and recovered by evaporating the solution to dryness.

B: The retention time of the major peak in the chromatogram of the *Assay preparation* corresponds to that of the *Standard preparation,* obtained as directed in the *Assay.*

Dissolution ⟨711⟩—

Medium: 0.1 *N* hydrochloric acid; 900 mL.

Apparatus 1: 100 rpm.

Time: 60 minutes.

Procedure—Determine the amount of $C_7H_8ClN_3O_4S_2$ dissolved from ultraviolet absorbances at the wavelength of maximum absorbance at about 272 nm of filtered portions of the solution under test, suitably diluted with *Dissolution Medium,* if necessary, in comparison with a Standard solution having a known concentration of USP Hydrochlorothiazide RS in the same medium.

Tolerances—Not less than 60% (Q) of the labeled amount of $C_7H_8ClN_3O_4S_2$ is dissolved in 60 minutes.

Uniformity of dosage units ⟨905⟩: meet the requirements.

4-Amino-6-chloro-1,3-benzenedisulfonamide—

Mobile phase, System suitability solution, Standard preparation, Chromatographic system, and *Procedure*—Proceed as directed in the test for *4-Amino-6-chloro-1,3-benzenedisulfonamide* under *Hydrochlorothiazide.*

Test preparation—Proceed as directed for *Assay preparation* in the *Assay.*

Assay—

Mobile phase, System suitability solution, Standard preparation, and *Chromatographic system*—Prepare as directed in the *Assay* under *Hydrochlorothiazide.*

Assay preparation—Weigh and finely powder not less than 20 Hydrochlorothiazide Tablets. Transfer an accurately weighed portion of the powder, equivalent to about 30 mg of hydrochlorothiazide, to a 200-mL volumetric flask. Add about 20 mL of *Mobile phase,* sonicate for 5 minutes, and add about 20 mL of acetonitrile. Sonicate for 5 minutes, add about 50 mL of *Mobile phase,* and shake by mechanical means for 10 minutes. Dilute with *Mobile phase* to volume, mix, and filter, discarding the first 10 mL of the filtrate.

Procedure—Proceed as directed for *Procedure* in the *Assay* under *Hydrochlorothiazide.* Calculate the quantity, in mg, of $C_7H_8ClN_3O_4S_2$ in the portion of Tablets taken by the formula:

$$200C(r_U / r_S),$$

in which C is the concentration, in mg per mL, of USP Hydrochlorothiazide RS in the *Standard preparation*, and r_U and r_S are the peak responses obtained from the *Assay preparation* and the *Standard preparation*, respectively.

Hydrochlorothiazide Tablets, Amiloride Hydrochloride and—*see* Amiloride Hydrochloride and Hydrochlorothiazide Tablets

Hydrochlorothiazide Tablets, Methyldopa and—*see* Methyldopa and Hydrochlorothiazide Tablets

Hydrochlorothiazide Tablets, Metoprolol Tartrate and—*see* Metoprolol Tartrate and Hydrochlorothiazide Tablets

Hydrochlorothiazide Tablets, Propranolol Hydrochloride and—*see* Propranolol Hydrochloride and Hydrochlorothiazide Tablets

Hydrochlorothiazide Tablets, Reserpine and—*see* Reserpine and Hydrochlorothiazide Tablets

Hydrochlorothiazide Tablets, Reserpine, Hydralazine Hydrochloride, and—*see* Reserpine, Hydralazine Hydrochloride, and Hydrochlorothiazide Tablets

Hydrochlorothiazide Tablets, Timolol Maleate and—*see* Timolol Maleate and Hydrochlorothiazide Tablets

Hydrocodone Bitartrate

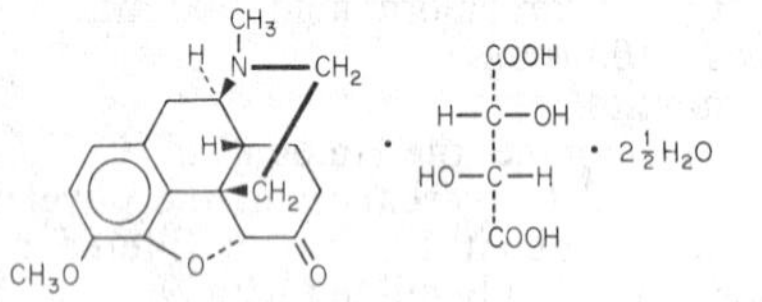

$C_{18}H_{21}NO_3 . C_4H_6O_6 . 2\frac{1}{2}H_2O$ 494.50

Morphinan-6-one, 4,5-epoxy-3-methoxy-17-methyl-, (5α)-, [*R*-(*R**,*R**)]-2,3-dihydroxybutanedioate (1:1), hydrate (2:5).

4,5α-Epoxy-3-methoxy-17-methylmorphinan-6-one tartrate (1:1) hydrate (2:5) [*34195-34-1; 6190-38-1*].

Anhydrous 449.46 [*143-71-5*].

» Hydrocodone Bitartrate contains not less than 98.0 percent and not more than 102.0 percent of $C_{18}H_{21}NO_3 . C_4H_6O_6$, calculated on the anhydrous basis.

Packaging and storage—Preserve in tight, light-resistant containers.

Reference standard—*USP Hydrocodone Bitartrate Reference Standard*—Dry at 105° for 4 hours before using.

Identification—

A: The infrared absorption spectrum of a mineral oil dispersion of it, previously dried, exhibits maxima only at the same wavelengths as that of a similar preparation of USP Hydrocodone Bitartrate RS.

B: The ultraviolet absorption spectrum of a 1 in 10,000 solution in 0.1 *N* sulfuric acid exhibits maxima and minima at the same wavelengths as that of a similar solution of USP Hydrocodone Bitartrate RS, concomitantly measured, and the respective absorptivities, calculated on the dried basis, at the wavelength of maximum absorbance at about 280 nm do not differ by more than 3.0%.

Specific rotation ⟨781⟩: between −79° and −84°, determined in a solution in water containing 200 mg in each 10 mL.

pH ⟨791⟩: between 3.2 and 3.8, in a solution (1 in 50).

Water, *Method I* ⟨921⟩: between 7.5% and 12.0%.

Residue on ignition ⟨281⟩: not more than 0.1%.

Chloride—To 10 mL of a solution (1 in 100), acidified with nitric acid, add a few drops of silver nitrate TS: no opalescence is produced immediately.

Ordinary impurities ⟨466⟩—

Test solution: a mixture of methanol and water (1:1).

Standard solution: a mixture of methanol and water (1:1).

Eluant: a mixture of hexanes, acetone, methanol, and ammonium hydroxide (60:40:20:1.5).

Visualization: 21.

Assay—Transfer about 150 mg of Hydrocodone Bitartrate, accurately weighed, to a 250-mL separator with the aid of 10 mL of water. Add a small piece of red litmus paper, then add, dropwise, 6 *N* ammonium hydroxide until the litmus paper turns blue (about 3 drops). Extract with 25-, 25-, 20-, 20-, 15-, and 15-mL portions of chloroform, and filter the chloroform extracts through a small pledget of cotton into a 250-mL conical flask. Evaporate the combined chloroform extracts almost to dryness, remove the flask from the steam bath, and evaporate the remainder of the chloroform with the aid of a current of air. Dissolve the residue in 80 mL of glacial acetic acid, warming, if necessary. Cool, and titrate with 0.02 *N* perchloric acid VS, determining the end-point potentiometrically. Perform a blank determination, and make any necessary correction. Each mL of 0.02 *N* perchloric acid is equivalent to 8.989 mg of $C_{18}H_{21}NO_3 . C_4H_6O_6$.

Hydrocodone Bitartrate Tablets

» Hydrocodone Bitartrate Tablets contain not less than 90.0 percent and not more than 110.0 percent of the labeled amount of $C_{18}H_{21}NO_3 . C_4H_6O_6 . 2\frac{1}{2}H_2O$.

Packaging and storage—Preserve in tight, light-resistant containers.

Reference standard—*USP Hydrocodone Bitartrate Reference Standard*—Dry at 105° for 4 hours before using.

Identification—Tablets meet the requirements under *Identification—Organic Nitrogenous Bases* ⟨181⟩.

Dissolution ⟨711⟩—

Medium: water; 500 mL.

Apparatus 2: 50 rpm.

Time: 45 minutes.

Phosphate buffer–bromothymol blue solution—Dissolve 3.40 g of monobasic sodium phosphate in about 50 mL of water, add 18 mL of 0.1 *N* sodium hydroxide, dilute with water to 500 mL, and mix. Adjust the solution, if necessary, with 1 *N* sodium hydroxide or 1 *N* phosphoric acid (*Buffer solution*) to a pH of 5.8 ± 0.1. Mix 31.2 mg of bromothymol blue with 1.0 mL of 0.1 *N* sodium hydroxide, add *Buffer solution* to obtain 500 mL of solution, and mix.

Standard preparation—Dissolve an accurately weighed quantity of USP Hydrocodone Bitartrate RS in water, and dilute quantitatively, and stepwise if necessary, with water to obtain a solution having a known concentration of about 10 µg per mL.

Test preparation—Pipet into a 125-mL separator a volume of a filtered portion of the solution under test that is estimated to contain about 0.4 mg of hydrocodone bitartrate. Add 2 drops of 6 *N* ammonium hydroxide, and extract with three 25-mL portions of chloroform. Filter the extracts through about 3 g of anhydrous sodium sulfate supported on filter paper, and combine the filtered extracts in a 100-mL volumetric flask. Add chloroform through the filter to volume, and mix.

Procedure—Pipet 10 mL of the *Standard preparation* into a 125-mL separator, and add 25.0 mL of chloroform. Pipet 25 mL of the *Test preparation* into a second 125-mL separator, and add

10.0 mL of water. Pipet 25 mL of chloroform and 10 mL of water into a third 125-mL separator to provide a blank. Treat each mixture as follows: Add 30 mL of *Phosphate buffer–bromothymol blue solution*, and shake vigorously for not less than 15 minutes. Allow the layers to separate, and filter the chloroform layer through filter paper, discarding the first 10 mL of the filtrate. Determine the absorbance of the clear filtrate at the wavelength of maximum absorbance at about 415 nm. Calculate the amount of hydrocodone bitartrate dissolved in the solution under test from the absorbances of the solutions obtained from the *Test preparation* and the *Standard preparation*, using the solution from the blank to set the spectrophotometer. [NOTE—Use 3-cm spectrophotometer cells for the absorbance measurements.]

Tolerances—Not less than 75% (*Q*) of the labeled amount of $C_{18}H_{21}NO_3 . C_4H_6O_6 . 2\frac{1}{2}H_2O$ is dissolved in 45 minutes.

Uniformity of dosage units ⟨905⟩: meet the requirements.

Procedure for content uniformity—Transfer 1 finely powdered Tablet to a 50-mL volumetric flask, add 0.1 *N* sulfuric acid to volume, and mix. Filter if necessary, discarding the first 20 mL of the filtrate. Concomitantly determine the absorbances of this solution and a solution of USP Hydrocodone Bitartrate RS in the same medium, having a known concentration of about 100 µg per mL, in 1-cm cells at the wavelength of maximum absorbance at about 280 nm, with a suitable spectrophotometer, using 0.1 *N* sulfuric acid as the blank. Calculate the quantity, in mg, of $C_{18}H_{21}NO_3 . C_4H_6O_6 . 2\frac{1}{2}H_2O$ in the Tablet by the formula:

$$(494.50/449.46)(TC/D)(A_U/A_S),$$

in which 494.50 and 449.46 are the molecular weights of hydrated and anhydrous forms, respectively, of hydrocodone bitartrate, *T* is the labeled quantity, in mg, of hydrocodone bitartrate in the Tablet, *C* is the concentration, in µg per mL, of USP Hydrocodone Bitartrate RS in the Standard solution, *D* is the concentration, in µg per mL, of the solution from the Tablet, based upon the labeled quantity per Tablet and the extent of dilution, and A_U and A_S are the absorbances of the solution from the Tablet and the Standard solution, respectively.

Assay—Weigh and finely powder not less than 20 Hydrocodone Bitartrate Tablets. Transfer an accurately weighed portion of the powder, equivalent to about 100 mg of hydrocodone bitartrate, to a 250-mL separator with the aid of 10 mL of water, and proceed as directed in the *Assay* under *Hydrocodone Bitartrate*, beginning with "Add a small piece of red litmus paper." Each mL of 0.02 *N* perchloric acid is equivalent to 9.890 mg of $C_{18}H_{21}NO_3 . C_4H_6O_6 . 2\frac{1}{2}H_2O$.

Hydrocortisone

$C_{21}H_{30}O_5$ 362.47

Pregn-4-ene-3,20-dione, 11,17,21-trihydroxy-, (11β)-.

Cortisol [*50-23-7*].

» Hydrocortisone contains not less than 97.0 percent and not more than 102.0 percent of $C_{21}H_{30}O_5$, calculated on the dried basis.

Packaging and storage—Preserve in well-closed containers.

Reference standard—*USP Hydrocortisone Reference Standard*—Dry at 105° for 3 hours before using.

Identification—

A: The infrared absorption spectrum of a mineral oil dispersion of it, previously dried at 105° for 3 hours, exhibits maxima only at the same wavelengths as that of a similar preparation of USP Hydrocortisone RS.

B: The ultraviolet absorption spectrum of a 1 in 100,000 solution in methanol exhibits maxima and minima at the same wavelengths as that of a similar solution of USP Hydrocortisone RS, concomitantly measured, and the respective absorptivities, calculated on the dried basis, at the wavelength of maximum absorbance at about 242 nm do not differ by more than 2.5%.

Specific rotation ⟨781⟩: between +150° and +156°, calculated on the dried basis, determined in a solution in dioxane containing 100 mg in each 10 mL.

Loss on drying ⟨731⟩—Dry it at 105° for 3 hours: it loses not more than 1.0% of its weight.

Residue on ignition ⟨281⟩: negligible, from 100 mg.

Ordinary impurities ⟨466⟩—

Test solution: alcohol.

Standard solution: alcohol.

Eluant: a mixture of toluene, ethyl acetate, and alcohol (60:20:20), in a nonequilibrated chamber.

Visualization: 5, followed by heating at 120° for 5 minutes and viewing under long-wavelength ultraviolet light.

Assay—

Mobile solvent—Prepare a solution containing butyl chloride, water-saturated butyl chloride, tetrahydrofuran, methanol, and glacial acetic acid (95:95:14:7:6).

Internal standard solution—Prepare a solution of prednisone in water-saturated chloroform containing 0.06 mg per mL.

Standard preparation—Dissolve an accurately weighed quantity of USP Hydrocortisone RS in *Internal standard solution* to obtain a solution having a known concentration of about 0.1 mg per mL.

Assay preparation—Transfer about 10 mg of Hydrocortisone, accurately weighed, to a 100-mL volumetric flask, add *Internal standard solution* to volume, and mix.

Procedure—Inject equal volumes of the *Standard preparation* and the *Assay preparation* into a suitable high-pressure liquid chromatograph (see *Chromatography* ⟨621⟩) equipped with a 254-nm detector and a suitable recorder and capable of providing column pressure up to about 1000 psi. The chromatograph is fitted with a 4-mm × 30-cm stainless steel column containing packing L3. In a suitable chromatogram, the resolution factor, *R* (see *Chromatography* ⟨621⟩), is not less than 3.0 between the peaks for hydrocortisone and the internal standard. Four replicate injections of the *Standard preparation* show a relative standard deviation of not more than 2.0%. Calculate the quantity, in mg, of $C_{21}H_{30}O_5$ in the portion of Hydrocortisone taken by the formula:

$$100C(R_U/R_S),$$

in which *C* is the concentration, in mg per mL, of USP Hydrocortisone RS in the *Standard preparation*, and R_U and R_S are the peak response ratios of the hydrocortisone peak and the internal standard peak obtained from the *Assay preparation* and the *Standard preparation*, respectively.

Hydrocortisone Cream

» Hydrocortisone Cream is Hydrocortisone in a suitable cream base. It contains not less than 90.0 percent and not more than 110.0 percent of the labeled amount of $C_{21}H_{30}O_5$.

Packaging and storage—Preserve in tight containers.

Reference standard—*USP Hydrocortisone Reference Standard*—Dry at 105° for 3 hours before using.

Identification—Transfer a quantity of Cream, equivalent to about 5 mg of hydrocortisone, to a flask, add 5 mL of alcohol, and heat on a steam bath for 5 minutes, with frequent shaking. Cool, and filter. Using the filtrate as the test solution, proceed as directed under *Thin-layer Chromatographic Identification Test* ⟨201⟩.

Microbial limits—It meets the requirements of the tests for absence of *Staphylococcus aureus* and *Pseudomonas aeruginosa* under *Microbial Limit Tests* ⟨61⟩.

Assay—

Mobile phase—Prepare a suitable degassed and filtered solution of water and acetonitrile (about 75:25), such that the retention time of hydrocortisone is about 10 minutes.

Standard preparation—Dissolve an accurately weighed quantity of USP Hydrocortisone RS in methanol to obtain a solution having a known concentration of about 500 µg per mL. Quantitatively dilute 1 volume of this solution with 9 volumes of dilute methanol (1 in 2). The *Standard preparation* has a final known concentration of about 50 µg per mL. [NOTE—If methanol is used in the final dilution of the *Assay preparation*, similarly use methanol instead of aqueous methanol in the final dilution of the *Standard preparation*.]

Assay preparation—Transfer an accurately weighed quantity of Hydrocortisone Cream, equivalent to about 10 mg of hydrocortisone, to a 150-mL beaker. Add 40 mL of methanol, and heat on a steam bath while stirring to melt and disperse the cream. Cool to room temperature, and filter through glass wool into a 100-mL volumetric flask. Repeat the extraction with two 20-mL portions of methanol, combining the filtrates in the 100-mL volumetric flask. Add methanol to volume, and mix. Quantitatively dilute 1 volume of this solution with an equal volume of water, and filter through a 5-µm membrane filter. If precipitation occurs on dilution with water, and the solution is still cloudy after filtration, dilute the initial test solution with methanol instead of water. Filter this solution through a 5-µm membrane filter.

Chromatographic system (see *Chromatography* ⟨621⟩)—The liquid chromatograph is equipped with a 254-nm detector and a 3.9-mm × 30-cm column that contains packing L1. Adjust operating parameters such that the peak obtained from the *Standard preparation* is about 0.6 full-scale. Chromatograph five replicate injections of the *Standard preparation*, and record the peak responses as directed under *Procedure:* the relative standard deviation is not more than 3.0%.

Procedure—Separately inject equal volumes (about 10 to 25 µL) of the *Standard preparation* and the *Assay preparation* into the chromatograph by means of a sampling valve, record the chromatograms, and measure the responses for the major peaks at equivalent retention times. Calculate the quantity, in mg, of $C_{21}H_{30}O_5$ in the portion of Cream taken by the formula:

$$0.2C(r_U/r_S),$$

in which C is the concentration, in µg per mL, of USP Hydrocortisone RS in the *Standard preparation*, and r_U and r_S are the peak responses obtained from the *Assay preparation* and the *Standard preparation*, respectively.

Hydrocortisone Cream, Neomycin Sulfate and—*see* Neomycin Sulfate and Hydrocortisone Cream

Hydrocortisone Enema

» Hydrocortisone Enema contains not less than 90.0 percent and not more than 110.0 percent of the labeled amount of $C_{21}H_{30}O_5$.

Packaging and storage—Preserve in tight containers.

Reference standard—*USP Hydrocortisone Reference Standard*—Dry at 105° for 3 hours before using.

Identification—Prepare a test solution as directed for *Assay preparation* in the *Assay*, except to omit addition of the *Internal standard solution*, and proceed as directed under *Thin-layer Chromatographic Identification Test* ⟨201⟩.

pH ⟨791⟩: between 5.5 and 7.0.

Assay—

Mobile phase—Mix 55 mL of a 5 in 100, water in methanol solution with 1.0 mL of glacial acetic acid, dilute with water-washed 1,2-dichloroethane to 1000 mL, and mix. Degas before using. Make adjustments if necessary (see *System Suitability* under *Chromatography* ⟨621⟩).

Internal standard solution—Dissolve 200 mg of acetaminophen in 4 mL of methanol, dilute with water-washed 1,2-dichloroethane to 200 mL, and mix. Keep the solution tightly stoppered and protected from light.

Standard preparation—Accurately weigh about 8 mg of USP Hydrocortisone RS, add 4 mL of methanol and 4.0 mL of *Internal standard solution*, dilute with chloroform to 100.0 mL, and mix to obtain a solution having a known concentration of about 0.08 mg of USP Hydrocortisone RS per mL.

Assay preparation—Transfer an accurately weighed quantity of Hydrocortisone Enema, equivalent to about 8 mg of hydrocortisone, to a separator. Extract with four 20-mL portions of chloroform, filtering each portion through chloroform-washed cotton into a 100-mL volumetric flask. Add 4 mL of methanol and 4.0 mL of *Internal standard solution*, dilute with chloroform to volume, and mix. Filter the extract through a 0.5-µm porosity polytef membrane filter, discarding the first 20 mL of the filtrate.

Chromatographic system (see *Chromatography* ⟨621⟩)—The liquid chromatograph is equipped with a 254-nm detector and a 4.6-mm × 25-cm column that contains 5-µm packing L3. The flow rate is about 1.5 mL per minute. Chromatograph the *Standard preparation*, and record the peak responses as directed under *Procedure:* the column efficiency determined from the analyte peak is not less than 5000 theoretical plates, the resolution, R, between the analyte and internal standard peaks is not less than 2.5, and the relative standard deviation for replicate injections is not more than 1.0%.

Procedure—Separately inject equal volumes (about 10 µL) of the *Standard preparation* and the *Assay preparation* into the chromatograph, record the chromatograms, and measure the responses for the major peaks. The relative retention times are about 1.3 for acetaminophen and 1.0 for hydrocortisone. Calculate the quantity, in mg, of $C_{21}H_{30}O_5$ in the portion of Hydrocortisone Enema taken by the formula:

$$100C(R_U/R_S),$$

in which C is the concentration, in mg per mL, of USP Hydrocortisone RS in the *Standard preparation*, and R_U and R_S are the peak response ratios obtained from the *Assay preparation* and the *Standard preparation*, respectively.

Hydrocortisone Gel

» Hydrocortisone Gel is Hydrocortisone in a suitable hydroalcoholic gel base. It contains not less than 90.0 percent and not more than 110.0 percent of the labeled amount of $C_{21}H_{30}O_5$.

Packaging and storage—Preserve in tight containers.

Reference standard—*USP Hydrocortisone Reference Standard*—Dry at 105° for 3 hours before using.

Identification—Proceed with Hydrocortisone Gel as directed under *Hydrocortisone Cream*. The specified result is obtained.

Assay—

Mobile solvent—Prepare a mixture of 2 volumes of methanol, 2 volumes of acetonitrile, and 6 volumes of water.

Standard preparation—Dissolve a suitable quantity of USP Hydrocortisone RS, accurately weighed, in alcohol and dichloromethane (75:25) to obtain a solution having a known concentration of about 0.1 mg per mL. Dilute this solution with alcohol to a concentration of 10 µg per mL.

Assay preparation—Weigh accurately an aliquot of Hydrocortisone Gel, equivalent to about 10 mg of hydrocortisone, and dilute with alcohol and dichloromethane (75:25) to obtain a solution having a concentration of about 0.1 mg per mL. Dilute this solution with alcohol to a concentration of 10 µg per mL.

Procedure—Introduce equal volumes (between 5 µL and 15 µL) of the *Assay preparation* and the *Standard preparation* into a high-pressure liquid chromatograph (see *Chromatography* ⟨621⟩) operated at room temperature, by means of a suitable microsyringe or sampling valve, adjusting the specimen size and other operating parameters such that the peak obtained from the *Standard preparation* is about 0.6 full-scale. Typically, the apparatus is fitted with a 4-mm × 30-cm column that contains packing L1 and is equipped with an ultraviolet detector capable of monitoring absorption at 254 nm, and a suitable recorder, and is capable of operating at a column pressure of up to 6000 psi. In a suitable chromatogram, the coefficient of variation for five replicate injections of the *Standard preparation* is not more than 3.0%. Determine the ratios of the peak heights, at equivalent retention times, obtained from the *Assay preparation* and the *Standard preparation*, and calculate the quantity, in mg, of $C_{21}H_{30}O_5$ in the portion of Hydrocortisone Gel taken by the formula:

$$C(H_U/H_S),$$

in which C is the concentration, in µg per mL, of USP Hydrocortisone RS in the *Standard preparation*, and H_U and H_S are the peak heights of the *Assay preparation* and the *Standard preparation*, respectively.

Hydrocortisone Lotion

» Hydrocortisone Lotion is Hydrocortisone in a suitable aqueous vehicle. It contains not less than 90.0 percent and not more than 110.0 percent of the labeled amount of $C_{21}H_{30}O_5$.

Packaging and storage—Preserve in tight containers.

Reference standard—*USP Hydrocortisone Reference Standard*—Dry at 105° for 3 hours before using.

Identification—Transfer a quantity of Lotion, equivalent to about 5 mg of hydrocortisone, to a separator containing 10 mL of methylene chloride, shake for 1 minute, and allow the layers to separate. Filter the methylene chloride extract onto a suitable chromatographic column that has been packed with 2 g of activated magnesium silicate. Wash the column with 25 mL of methylene chloride with the aid of slight air pressure, discarding the washings, and elute the hydrocortisone with 10 mL of methanol. Using USP Hydrocortisone RS to prepare a Standard solution having a concentration of 500 µg per mL, and using as the solvent system a mixture of 180 volumes of chloroform, 15 volumes of methanol, and 1 volume of water, proceed as directed under *Thin-layer Chromatographic Identification Test* ⟨201⟩.

Microbial limits—It meets the requirements of the tests for absence of *Staphylococcus aureus* and *Pseudomonas aeruginosa* under *Microbial Limit Tests* ⟨61⟩.

Assay—

Standard preparation—Prepare as directed for *Standard Preparation* under *Assay for Steroids* ⟨351⟩, using USP Hydrocortisone RS.

Assay preparation—In a tared, 100-mL volumetric flask, weigh 100 mL of Hydrocortisone Lotion that previously has been shaken to ensure homogeneity, allowed to stand until the entrapped air arises, and finally inverted carefully just prior to transfer to the volumetric flask. Transfer to a separator an accurately weighed quantity of Hydrocortisone Lotion, equivalent to about 20 mg of hydrocortisone, and add 15 mL of water. Extract with four 25-mL portions of chloroform, filtering each portion through chloroform-washed cotton into a 200-mL volumetric flask. Add chloroform to volume, and mix. Pipet 20 mL of this solution into a 100-mL volumetric flask, add chloroform to volume, and mix. Pipet 10 mL of the resulting solution into a glass-stoppered, 50-mL conical flask, evaporate on a steam bath just to dryness, cool, and dissolve the residue in 20.0 mL of alcohol.

Procedure—Proceed as directed for *Procedure* under *Assay for Steroids* ⟨351⟩. Calculate the quantity, in mg, of $C_{21}H_{30}O_5$ in the portion of Lotion taken by the formula:

$$2C(A_U/A_S).$$

From the observed weight of 100 mL of the Lotion, calculate the quantity of $C_{21}H_{30}O_5$ in each 100 mL.

Hydrocortisone Ointment

» Hydrocortisone Ointment is Hydrocortisone in a suitable ointment base. It contains not less than 90.0 percent and not more than 110.0 percent of the labeled amount of $C_{21}H_{30}O_5$.

Packaging and storage—Preserve in well-closed containers.

Reference standard—*USP Hydrocortisone Reference Standard*—Dry at 105° for 3 hours before using.

Identification—Transfer a quantity of Ointment, equivalent to about 5 mg of hydrocortisone, to a flask, add 10 mL of methanol, and heat on a steam bath for 5 minutes with frequent shaking. Cool to solidify the ointment base, and filter. Using the filtrate as the test solution, proceed as directed under *Thin-layer Chromatographic Identification Test* ⟨201⟩.

Microbial limits—It meets the requirements of the tests for absence of *Staphylococcus aureus* and *Pseudomonas aeruginosa* under *Microbial Limit Tests* ⟨61⟩.

Assay—Proceed with Hydrocortisone Ointment as directed in the *Assay* under *Hydrocortisone Cream*, except to read Ointment instead of Cream, and to use alcohol instead of methanol.

Hydrocortisone Ointment, Neomycin and Polymyxin B Sulfates, Bacitracin Zinc, and—*see* Neomycin and Polymyxin B Sulfates, Bacitracin Zinc, and Hydrocortisone Ointment

Hydrocortisone Ointment, Neomycin Sulfate and—*see* Neomycin Sulfate and Hydrocortisone Ointment

Hydrocortisone Ointment, Oxytetracycline Hydrochloride and—*see* Oxytetracycline Hydrochloride and Hydrocortisone Ointment

Hydrocortisone Ophthalmic Ointment, Neomycin and Polymyxin B Sulfates, Bacitracin Zinc, and—*see* Neomycin and Polymyxin B Sulfates, Bacitracin Zinc, and Hydrocortisone Ophthalmic Ointment

Hydrocortisone Otic Solution, Polymyxin B Sulfate and—*see* Polymyxin B Sulfate and Hydrocortisone Otic Solution

Sterile Hydrocortisone Suspension

» Sterile Hydrocortisone Suspension is a sterile suspension of Hydrocortisone in Water for Injection. It contains not less than 90.0 percent and not more than 110.0 percent of the labeled amount of $C_{21}H_{30}O_5$.

Packaging and storage—Preserve in single-dose or in multiple-dose containers, preferably of Type I glass.

Reference standard—*USP Hydrocortisone Reference Standard*—Dry at 105° for 3 hours before using.

Identification—It responds to the *Identification test* under *Hydrocortisone Lotion*.

pH ⟨791⟩: between 5.0 and 7.0.

Other requirements—It meets the requirements under *Injections* ⟨1⟩.

Assay—

Standard preparation—Prepare as directed for *Standard Preparation* under *Assay for Steroids* ⟨351⟩, using USP Hydrocortisone RS.

Assay preparation—Transfer to a separator an accurately weighed quantity of Sterile Hydrocortisone Suspension, equivalent to about 50 mg of hydrocortisone, using a total of 25 mL of water to effect the transfer. Extract with four 40-mL portions of chloroform, filtering each portion through chloroform-washed cotton into a 200-mL volumetric flask. Add chloroform to volume, and mix. Pipet 20 mL of this solution into a 100-mL volumetric flask, add chloroform to volume, and mix. Pipet 10 mL of the resulting solution into a glass-stoppered, 100-mL conical flask, evaporate on a steam bath just to dryness, cool, and dissolve the residue in 50.0 mL of alcohol.

Procedure—Proceed as directed for *Procedure* under *Assay for Steroids* ⟨351⟩. Calculate the quantity, in mg, of $C_{21}H_{30}O_5$ in the portion of Suspension taken by the formula:

$$5C(A_U/A_S).$$

Hydrocortisone Ophthalmic Suspension, Neomycin and Polymyxin B Sulfates and—*see* Neomycin and Polymyxin B Sulfates and Hydrocortisone Ophthalmic Suspension

Hydrocortisone Otic Suspension, Neomycin Sulfate and—*see* Neomycin Sulfate and Hydrocortisone Otic Suspension

Hydrocortisone Tablets

» Hydrocortisone Tablets contain not less than 90.0 percent and not more than 110.0 percent of the labeled amount of $C_{21}H_{30}O_5$.

Packaging and storage—Preserve in well-closed containers.

Reference standard—*USP Hydrocortisone Reference Standard*—Dry at 105° for 3 hours before using.

Identification—Powder a number of Tablets, equivalent to about 50 mg of hydrocortisone, and digest with 15 mL of solvent hexane for 15 minutes. Decant the solvent hexane as completely as possible, and extract the residue first with 10 mL of solvent hexane, then with 10 mL of peroxide-free ether in the same manner as before, and discard the extracts. Digest the final residue with 25 mL of dehydrated alcohol for 15 minutes with frequent agitation, filter, and evaporate the alcohol extract on a steam bath to dryness: the residue so obtained responds to *Identification test A* under *Hydrocortisone*.

Dissolution ⟨711⟩—

Medium: water; 900 mL.

Apparatus 2: 50 rpm.

Time: 30 minutes.

Procedure—Determine the amount of $C_{21}H_{30}O_5$ dissolved from ultraviolet absorbances at the wavelength of maximum absorbance at about 248 nm of filtered portions of the solution under test, suitably diluted with *Dissolution Medium*, if necessary, in comparison with a Standard solution having a known concentration of USP Hydrocortisone RS in the same medium.

Tolerances—Not less than 70% (*Q*) of the labeled amount of $C_{21}H_{30}O_5$ is dissolved in 30 minutes.

Uniformity of dosage units ⟨905⟩: meet the requirements.

Procedure for content uniformity—

Mobile solvent, Internal standard solution, and *Standard preparation*—Prepare as directed in the *Assay* under *Hydrocortisone.*

Test preparation—Transfer 1 Tablet to a suitable container and add about 0.3 mL of water directly on the Tablet. Allow the Tablet to stand for about 5 minutes. Shake the container to break up the Tablet and sonicate briefly to ensure complete disintegration. Add a few small glass beads and 50.0 mL of *Internal standard solution* to the container. Shake the container for about 30 minutes. Dilute an accurately measured volume of the clear supernatant liquid with a known, accurately measured volume of *Internal standard solution* to obtain a final concentration of 0.1 mg per mL. Shake the contents of the container to mix, and analyze the clear solution as directed under *Procedure*.

Procedure—Proceed as directed for *Procedure* in the *Assay* under *Hydrocortisone.* Calculate the quantity, in mg, of $C_{21}H_{30}O_5$ in the Tablet by the formula:

$$50(F_2/F_1)C(R_U/R_S),$$

in which F_1 is the volume, in mL, of the supernatant aliquot of the solution from the Tablet taken for dilution, and F_2 is the final volume, in mL, of the *Test preparation*, and the other terms are as defined for *Procedure* in the *Assay* under *Hydrocortisone.*

Assay—

Mobile solvent, Internal standard solution, and *Standard preparation*—Prepare as directed in the *Assay* under *Hydrocortisone.*

Assay preparation—Weigh and finely powder not less than 10 Hydrocortisone Tablets. Weigh a portion of the powder, equivalent to about 5 mg of hydrocortisone, and transfer to a suitable container. Add 50.0 mL of *Internal standard solution.* Shake vigorously for 30 minutes, and centrifuge a portion of this mixture. Use the clear supernatant solution.

Procedure—Proceed as directed for *Procedure* in the *Assay* under *Hydrocortisone.* Calculate the quantity, in mg, of $C_{21}H_{30}O_5$ in the portion of Tablets taken by the formula:

$$50C(R_U/R_S),$$

in which the terms are as defined therein.

Hydrocortisone and Acetic Acid Otic Solution

» Hydrocortisone and Acetic Acid Otic Solution is a solution of Hydrocortisone and Glacial Acetic Acid in a suitable nonaqueous solvent. It contains not less than 90.0 percent and not more than 120.0 percent of the labeled amount of hydrocortisone ($C_{21}H_{30}O_5$), and not less than 85.0 percent and not more than 130.0 percent of the labeled amount of acetic acid ($C_2H_4O_2$).

Packaging and storage—Preserve in tight, light-resistant containers.

Reference standard—*USP Hydrocortisone Reference Standard*—Dry at 105° for 3 hours before using.

Identification—

A: Dilute 5 mL with 10 mL of water, and adjust with 1 *N* sodium hydroxide to a pH of about 7. Add ferric chloride TS: a deep red color is produced, and it is destroyed by the addition of hydrochloric acid.

B: Warm it with sulfuric acid and alcohol: ethyl acetate, recognizable by its characteristic odor, is evolved.

C: The retention time of the major peak in the chromatogram of the *Assay preparation* corresponds to that of the *Standard preparation*, both relative to the internal standard, obtained as directed in the *Assay for acetic acid.*

D: The retention time of the major peak in the chromatogram of the *Assay preparation* corresponds to that of the *Standard preparation*, obtained as directed in the *Assay for hydrocortisone.*

pH ⟨791⟩: between 2.0 and 4.0, when diluted with an equal volume of water.

Assay for acetic acid—

Internal standard solution—Mix 2.0 mL of anisole with methanol to obtain 100 mL of solution.

Standard preparation—Dilute quantitatively an accurately weighed quantity of glacial acetic acid with methanol to obtain a solution having a known concentration of about 20 mg per mL. Transfer 5.0 mL of the resulting solution to a 10-mL volumetric flask, add 2.0 mL of *Internal standard solution*, dilute with methanol to volume, and mix to obtain a *Standard preparation* having a known concentration of about 10 mg of glacial acetic acid per mL.

Assay preparation—Transfer an accurately measured volume of Hydrocortisone and Acetic Acid Otic Solution, equivalent to about 100 mg of acetic acid, to a 10-mL volumetric flask, add 2.0 mL of *Internal standard solution*, dilute with methanol to volume, and mix.

Chromatographic system—The gas chromatograph is equipped with a flame-ionization detector and a 2-mm × 1.8-m glass column packed with 20 percent liquid phase G35 on support S1A. The column is maintained at a temperature of 115° for 12 minutes, programmed to rise at a rate of 35° per minute to a temperature of 190°, and maintained at 190° for 3 minutes, nitrogen being used as the carrier gas at a flow rate of about 25 mL per minute. The injection port and detector are maintained isothermally at temperatures of about 180° and 220°, respectively. Chromatograph the *Standard preparation*, record the chromatogram, and measure the peak responses as directed under *Procedure:* the resolution, *R*, between anisole and acetic acid is not less than 1.5, and the relative standard deviation for replicate injections is not more than 2.0%.

Procedure—Separately inject equal volumes (about 4 µL) of the *Standard preparation* and the *Assay preparation* into the chromatograph, record the chromatograms, and measure the responses of the major peaks. The retention time of acetic acid is about 1.5 relative to that of anisole. Calculate the quantity, in mg, of $C_2H_4O_2$ in each mL of the Otic Solution taken by the formula:

$$10(C/V)(R_U/R_S),$$

in which *C* is the concentration, in mg per mL, of glacial acetic acid in the *Standard preparation*, *V* is the volume, in mL, of Otic Solution taken, and R_U and R_S are the peak response ratios obtained from the *Assay preparation* and the *Standard preparation*, respectively.

Assay for hydrocortisone—

Mobile phase—Prepare a solution containing water and acetonitrile (70:30).

Standard preparation—Using an accurately weighed quantity of USP Hydrocortisone RS, prepare a solution in dilute alcohol (1 in 2) having a known concentration of about 0.5 mg per mL.

Assay preparation—Transfer an accurately measured volume of Hydrocortisone and Acetic Acid Otic Solution, equivalent to about 100 mg of hydrocortisone, to a 200-mL volumetric flask. Add dilute alcohol (1 in 2) to volume, and mix.

Chromatographic system (see *Chromatography* ⟨621⟩)—The liquid chromatograph is equipped with a 254-nm detector and a 4-mm × 30-cm column that contains packing L1. The flow rate is about 2 mL per minute. Chromatograph four replicate injections of the *Standard preparation*, and record the peak responses as directed under *Procedure:* the relative standard deviation is not more than 2.0%.

Procedure—Separately inject equal volumes (about 20 µL) of the *Standard preparation* and the *Assay preparation* into the chromatograph by means of a suitable sampling valve, record the chromatograms, and measure the responses for the major peak. Calculate the quantity, in mg, of $C_{21}H_{30}O_5$ in each mL of the Otic Solution taken by the formula:

$$(200C/V)(r_U/r_S),$$

in which *C* is the concentration, in mg per mL, of USP Hydrocortisone RS in the *Standard preparation*, *V* is the volume, in mL, of Otic Solution taken, and r_U and r_S are the peak responses obtained from the *Assay preparation* and the *Standard preparation*, respectively.

Hydrocortisone Acetate

$C_{23}H_{32}O_6$ 404.50

Pregn-4-ene-3,20-dione, 21-(acetyloxy)-11,17-dihydroxy-, (11β)-.

Cortisol 21-acetate [*50-03-3*].

» Hydrocortisone Acetate contains not less than 97.0 percent and not more than 102.0 percent of $C_{23}H_{32}O_6$, calculated on the dried basis.

Packaging and storage—Preserve in well-closed containers.

Reference standard—*USP Hydrocortisone Acetate Reference Standard*—Dry in vacuum at 60° for 3 hours before using.

Identification—

A: The infrared absorption spectrum of a mineral oil dispersion of it, previously dried, exhibits maxima only at the same wavelengths as that of a similar preparation of USP Hydrocortisone Acetate RS.

B: The ultraviolet absorption spectrum of a 1 in 100,000 solution in methanol exhibits maxima and minima at the same wavelengths as that of a similar solution of USP Hydrocortisone Acetate RS, concomitantly measured, and the respective absorptivities, calculated on the dried basis, at the wavelength of maximum absorbance at about 242 nm do not differ by more than 2.5%.

Specific rotation ⟨781⟩: Between +158° and +165°, calculated on the dried basis, determined in a solution in dioxane containing 100 mg in each 10 mL.

Loss on drying ⟨731⟩—Dry it in vacuum at 60° for 3 hours: it loses not more than 1.0% of its weight.

Residue on ignition ⟨281⟩: negligible, from 100 mg.

Ordinary impurities ⟨466⟩—

Test solution: a mixture of methanol and dimethylformamide (1:1).

Standard solution: a mixture of methanol and dimethylformamide (1:1).

Eluant: a mixture of chloroform, methanol, and water (180:15:1).

Visualization: 5.

Assay—

Internal standard solution—Dissolve fluoxymesterone in water-saturated chloroform to obtain a solution containing about 0.18 mg per mL.

Mobile solvent—Prepare a solution containing butyl chloride, water-saturated butyl chloride, tetrahydrofuran, methanol, and glacial acetic acid (475:475:70:35:30).

Standard preparation—Dissolve an accurately weighed quantity of USP Hydrocortisone Acetate RS in *Internal standard solution* to obtain a solution having a known concentration of about 0.10 mg per mL.

Assay preparation—Transfer about 10 mg of Hydrocortisone Acetate, accurately weighed, to a 100-mL volumetric flask. Dilute with *Internal standard solution* to volume, and mix.

Procedure—Introduce equal volumes of the *Assay preparation* and the *Standard preparation* into a chromatograph fitted with a 254-nm detector. Use a 4-mm (ID), 30-cm long column packed with 10-µm diameter packing L3, and operated at room temperature. In a suitable system, the value of *R* (see *Chromatography* ⟨621⟩) is not less than 2.0 between peaks from hydrocortisone acetate and the internal standard. Six replicate injections of the *Standard preparation* show a relative standard deviation that is not more than 2.0%. Calculate the quantity, in mg, of $C_{23}H_{32}O_6$ in the portion of Hydrocortisone Acetate taken by the formula:

$$100C(R_U/R_S),$$

in which *C* is the concentration, in mg per mL, of USP Hydrocortisone Acetate RS in the *Standard preparation*, and R_U and R_S are the ratios of the peak responses of the hydrocortisone acetate peak and the internal standard peak from the *Assay preparation* and the *Standard preparation*, respectively.

Hydrocortisone Acetate Cream

» Hydrocortisone Acetate Cream is Hydrocortisone Acetate in a suitable cream base. It contains not less than 90.0 percent and not more than 110.0 percent of the labeled amount of $C_{23}H_{32}O_6$.

Packaging and storage—Preserve in well-closed containers.

Reference standard—*USP Hydrocortisone Acetate Reference Standard*—Keep container tightly closed. Dry in vacuum at 60° for 3 hours before using.

Identification—It responds to the *Identification test* under *Hydrocortisone Acetate Ointment.*

Microbial limits—It meets the requirements of the tests for absence of *Staphylococcus aureus* and *Pseudomonas aeruginosa* under *Microbial Limit Tests* ⟨61⟩.

Minimum fill ⟨755⟩: meets the requirements.

Assay—

Internal standard solution—Dissolve fluoxymesterone in tetrahydrofuran to obtain a solution containing about 0.45 mg per mL.

Mobile phase—Prepare a solution containing butyl chloride, water-saturated butyl chloride, tetrahydrofuran, methanol, and glacial acetic acid (475:475:70:35:30).

Standard preparation—Dissolve an accurately weighed quantity of USP Hydrocortisone Acetate RS in *Internal standard solution* to obtain a solution having a known concentration of about 0.25 mg per mL. Transfer 10.0 mL to a suitable container, and mix with 15.0 mL of *Mobile phase* to obtain a *Standard preparation* having a known concentration of about 0.1 mg of USP Hydrocortisone Acetate RS per mL.

Assay preparation—Transfer an accurately weighed quantity of Hydrocortisone Acetate Cream, equivalent to about 25 mg of hydrocortisone acetate, to a suitable container. Add 100.0 mL of *Internal standard solution.* Shake until the cream dissolves. Transfer 10.0 mL of the resulting solution to a suitable container, and mix with 15.0 mL of *Mobile phase.*

Procedure—Introduce equal volumes of the *Assay preparation* and the *Standard preparation* into a chromatograph fitted with a 254-nm detector. Use a 4-mm (ID), 30-cm long column containing packing L3 and operated at room temperature. In a suitable system, the value of *R* (see *Chromatography* ⟨621⟩) is not less than 2.0 between peaks from hydrocortisone acetate and the internal standard. Six replicate injections of the *Standard preparation* show a relative standard deviation of not more than 2.0%. Calculate the quantity, in mg, of $C_{23}H_{32}O_6$ in the portion of Cream taken by the formula:

$$250C(R_U/R_S),$$

in which C is the concentration, in mg per mL, of USP Hydrocortisone Acetate RS in the *Standard preparation,* and R_U and R_S are the ratios of the peak responses of the hydrocortisone acetate peak and the internal standard peak from the *Assay preparation* and the *Standard preparation,* respectively.

Hydrocortisone Acetate Cream, Neomycin and Polymyxin B Sulfates and—*see* Neomycin and Polymyxin B Sulfates and Hydrocortisone Acetate Cream

Hydrocortisone Acetate Cream, Neomycin Sulfate and—*see* Neomycin Sulfate and Hydrocortisone Acetate Cream

Hydrocortisone Acetate Lotion

» Hydrocortisone Acetate Lotion is Hydrocortisone Acetate in a suitable aqueous vehicle. It contains not less than 90.0 percent and not more than 110.0 percent of the labeled amount of $C_{23}H_{32}O_6$.

Packaging and storage—Preserve in tight containers.

Reference standard—*USP Hydrocortisone Acetate Reference Standard*—Dry in vacuum at 60° for 3 hours before using.

Identification—It responds to the *Identification test* under *Hydrocortisone Acetate Ointment.*

Assay—

Internal standard solution—Dissolve fluoxymesterone in water-saturated chloroform to obtain a solution containing about 0.18 mg per mL.

Mobile phase—Prepare a solution containing butyl chloride, water-saturated butyl chloride, tetrahydrofuran, methanol, and glacial acetic acid (475:475:70:35:30).

Standard preparation—Dissolve an accurately weighed quantity of USP Hydrocortisone Acetate RS in *Internal standard solution* to obtain a solution having a known concentration of about 0.10 mg per mL.

Assay preparation—Transfer an accurately weighed quantity of Hydrocortisone Acetate Lotion, equivalent to about 2.5 mg of hydrocortisone acetate, to a closable container. Add 25.0 mL of *Internal standard solution* and about 10 glass beads. Securely close the container, and shake vigorously for approximately 15 minutes. Centrifuge, and use the clear, lower chloroform layer.

Procedure—Introduce equal volumes of the *Assay preparation* and the *Standard preparation* into a high-pressure liquid chromatograph fitted with a 254-nm detector. Use a 4-mm (ID) × 30-cm long column containing packing L3, and operated at room temperature. In a suitable system, the value of *R* (see *Chromatography* ⟨621⟩) is not less than 2.0 between peaks for hydrocortisone acetate and the internal standard. Six replicate injections of the *Standard preparation* show a relative standard deviation of not more than 2.0%. Calculate the quantity, in mg, of $C_{23}H_{32}O_6$ in the portion of Lotion taken by the formula:

$$25C(R_U/R_S),$$

in which C is the concentration, in mg per mL, of USP Hydrocortisone Acetate RS in the *Standard preparation,* and R_U and R_S are the ratios of the peak responses of the hydrocortisone acetate peak and the internal standard peak from the *Assay preparation* and the *Standard preparation,* respectively.

Hydrocortisone Acetate Lotion, Neomycin Sulfate and—*see* Neomycin Sulfate and Hydrocortisone Acetate Lotion

Hydrocortisone Acetate Ointment

» Hydrocortisone Acetate Ointment is Hydrocortisone Acetate in a suitable ointment base. It contains not less than 90.0 percent and not more than 110.0 percent of the labeled amount of $C_{23}H_{32}O_6$.

Packaging and storage—Preserve in well-closed containers.

Reference standard—*USP Hydrocortisone Acetate Reference Standard*—Dry in vacuum at 60° for 3 hours before using.

Identification—Transfer a quantity of Ointment, equivalent to about 5 mg of hydrocortisone acetate, to a flask, add 5 mL of methanol, and heat on a steam bath for 5 minutes with frequent mixing. Cool to solidify the ointment base, and filter. Using the filtrate as the test solution, proceed as directed under *Thin-layer Chromatographic Identification Test* ⟨201⟩. Locate the spots by spraying the dried plate with a 70% methanolic sulfuric acid solution. Heat the plate for 20 to 30 minutes at 90°, allow to cool, and view under long-wavelength ultraviolet light: the R_f value and fluorescence of the principal spot obtained from the test solution correspond to those obtained from the Standard solution.

Microbial limits—It meets the requirements of the tests for absence of *Staphylococcus aureus* and *Pseudomonas aeruginosa* under *Microbial Limit Tests* ⟨61⟩.

Minimum fill ⟨755⟩: meets the requirements.

Assay—Proceed with Hydrocortisone Acetate Ointment as directed in the *Assay* under *Hydrocortisone Acetate Lotion.*

Hydrocortisone Acetate Ointment, Neomycin and Polymyxin B Sulfates, Bacitracin, and—*see* Neomycin and Polymyxin B Sulfates,

Bacitracin, and Hydrocortisone Acetate Ointment

Hydrocortisone Acetate Ointment, Neomycin Sulfate and—*see* Neomycin Sulfate and Hydrocortisone Acetate Ointment

Hydrocortisone Acetate Ophthalmic Ointment

» Hydrocortisone Acetate Ophthalmic Ointment is Hydrocortisone Acetate in a suitable ophthalmic ointment base. It contains not less than 90.0 percent and not more than 110.0 percent of the labeled amount of total steroids, calculated as $C_{23}H_{32}O_6$. It is sterile.

Packaging and storage—Preserve in collapsible ophthalmic ointment tubes.

Reference standard—*USP Hydrocortisone Acetate Reference Standard*—Dry in vacuum at 60° for 3 hours before using.

Identification—It responds to the *Identification test* under *Hydrocortisone Acetate Ointment.*

Sterility—It meets the requirements for *Ophthalmic Ointments* under *Sterility Tests* ⟨71⟩.

Minimum fill ⟨755⟩: meets the requirements.

Particulate matter—It meets the requirements of the test for *Metal Particles in Ophthalmic Ointments* ⟨751⟩.

Assay—

Standard preparation—Prepare as directed for *Standard Preparation* under *Assay for Steroids* ⟨351⟩, using USP Hydrocortisone Acetate RS.

Assay preparation—Transfer to a suitable flask an accurately weighed quantity of Hydrocortisone Acetate Ophthalmic Ointment, equivalent to about 10 mg of hydrocortisone acetate, and add 30 mL of alcohol. Heat on a steam bath to melt the ointment base, and mix. Cool to solidify the ointment base, and filter the alcohol solution into a 100-mL volumetric flask. Repeat the extraction with three 20-mL portions of alcohol, add alcohol to volume, and mix. Pipet 10 mL of this solution into a 100-mL volumetric flask, add alcohol to volume, and mix. Pipet 20 mL of the resulting solution into a glass-stoppered, 50-mL conical flask.

Procedure—Proceed as directed for *Procedure* under *Assay for Steroids* ⟨351⟩. Calculate the quantity, in mg, of $C_{23}H_{32}O_6$ in the portion of Ointment taken by the formula:

$$C(A_U/A_S),$$

in which the terms are as defined therein.

Hydrocortisone Acetate Ophthalmic Ointment, Chloramphenicol, Polymyxin B Sulfate, and—*see* Chloramphenicol, Polymyxin B Sulfate, and Hydrocortisone Acetate Ophthalmic Ointment

Hydrocortisone Acetate Ophthalmic Ointment, Neomycin and Polymyxin B Sulfates, Bacitracin, and—*see* Neomycin and Polymyxin B Sulfates, Bacitracin, and Hydrocortisone Acetate Ophthalmic Ointment

Hydrocortisone Acetate Ophthalmic Ointment, Neomycin and Polymyxin B Sulfates, Bacitracin Zinc, and—*see* Neomycin and Polymyxin B Sulfates, Bacitracin Zinc, and Hydrocortisone Acetate Ophthalmic Ointment

Hydrocortisone Acetate Ophthalmic Ointment, Neomycin Sulfate and—*see* Neomycin Sulfate and Hydrocortisone Acetate Ophthalmic Ointment

Hydrocortisone Acetate Ophthalmic Suspension

» Hydrocortisone Acetate Ophthalmic Suspension is a sterile suspension of Hydrocortisone Acetate in an aqueous medium containing a suitable antimicrobial agent. It contains not less than 90.0 percent and not more than 110.0 percent of the labeled amount of total steroids, calculated as $C_{23}H_{32}O_6$. It may contain suitable buffers and suspending agents.

Packaging and storage—Preserve in tight containers.

Reference standard *USP Hydrocortisone Acetate Reference Standard*—Dry in vacuum at 60° for 3 hours before using.

Identification—Evaporate 50 mL of the *Assay preparation* prepared as directed in the *Assay* on a steam bath just to dryness, dissolve the residue in 1 mL of chloroform, and proceed as directed under *Thin-layer Chromatographic Identification Test* ⟨201⟩.

Sterility—It meets the requirements under *Sterility Tests* ⟨71⟩.

pH ⟨791⟩: between 6.0 and 8.0.

Assay—Proceed with Hydrocortisone Acetate Ophthalmic Suspension as directed in the *Assay* under *Sterile Hydrocortisone Acetate Suspension.*

Hydrocortisone Acetate Ophthalmic Suspension, Neomycin and Polymyxin B Sulfates and—*see* Neomycin and Polymyxin B Sulfates and Hydrocortisone Acetate Ophthalmic Suspension

Hydrocortisone Acetate Ophthalmic Suspension, Neomycin Sulfate and—*see* Neomycin Sulfate and Hydrocortisone Acetate Ophthalmic Suspension

Hydrocortisone Acetate Ophthalmic Suspension, Oxytetracycline Hydrochloride and—*see* Oxytetracycline Hydrochloride and Hydrocortisone Acetate Ophthalmic Suspension

Hydrocortisone Acetate for Ophthalmic Suspension, Chloramphenicol and—*see* Chloramphenicol and Hydrocortisone Acetate for Ophthalmic Suspension

Hydrocortisone Acetate Otic Suspension, Colistin and Neomycin Sulfates and—*see* Colistin and Neomycin Sulfates and Hydrocortisone Acetate Otic Suspension

Hydrocortisone Acetate Topical Suspension, Penicillin G Procaine, Neomycin and Polymyxin B Sulfates, and—*see* Penicillin G Procaine, Neomycin and Polymyxin B Sulfates, and Hydrocortisone Acetate Topical Suspension

Sterile Hydrocortisone Acetate Suspension

» Sterile Hydrocortisone Acetate Suspension is a sterile suspension of Hydrocortisone Acetate in a suitable aqueous medium. It contains not less than 90.0 percent and not more than 110.0 percent of the labeled amount of total steroids, calculated as $C_{23}H_{32}O_6$. It may contain suitable buffers and suspending agents.

Packaging and storage—Preserve in single-dose or in multiple-dose containers, preferably of Type I glass.

Reference standard—*USP Hydrocortisone Acetate Reference Standard*—Dry in vacuum at 60° for 3 hours before using.

Identification—Extract a volume of Suspension, equivalent to about 50 mg of hydrocortisone acetate, with two 10-mL portions of peroxide-free ether, and discard the ether extracts. Filter with suction, wash with small portions of water, and dry at 105° for 1 hour: the hydrocortisone acetate so obtained responds to *Identification test A* under *Hydrocortisone Acetate.*

pH ⟨791⟩: between 5.0 and 7.0.

Other requirements—It meets the requirements under *Injections* ⟨1⟩.

Assay—

Standard preparation—Prepare as directed for *Standard Preparation* under *Assay for Steroids* ⟨351⟩, using USP Hydrocortisone Acetate RS.

Assay preparation—Transfer to a separator an accurately measured volume of Sterile Hydrocortisone Acetate Suspension, equivalent to about 50 mg of hydrocortisone acetate, and dilute with water to about 15 mL. Extract with four 25-mL portions of chloroform, filtering each portion through chloroform-washed cotton into a 250-mL volumetric flask. Add chloroform to volume, and mix. Pipet 10 mL of this solution into a 100-mL volumetric flask, add chloroform to volume, and mix. Pipet 10 mL of the resulting solution into a glass-stoppered, 50-mL conical flask, evaporate the chloroform on a steam bath just to dryness, cool, and dissolve the residue in 20.0 mL of alcohol.

Procedure—Proceed as directed for *Procedure* under *Assay for Steroids* ⟨351⟩. Calculate the quantity, in mg, of $C_{23}H_{32}O_6$ in each mL of the Suspension taken by the formula:

$$5(C/V)(A_U/A_S),$$

in which V is the volume, in mL, of Suspension taken and the other terms are as defined therein.

Hydrocortisone Butyrate

$C_{25}H_{36}O_6$ 432.56

Pregn-4-ene-3,20-dione, 11,21-dihydroxy-17-(1-oxobutoxy)-, (11β)-.

Cortisol 17-butyrate.

11β,17,21-Trihydroxy-pregn-4-ene-3,20-dione 17-butyrate [*13609-67-1*].

» Hydrocortisone Butyrate contains not less than 97.0 percent and not more than 102.0 percent of $C_{25}H_{36}O_6$, calculated on the dried basis.

Packaging and storage—Preserve in well-closed containers.

Reference standard—*USP Hydrocortisone Butyrate Reference Standard*—Dry in vacuum at 78° for 3 hours before using.

Clarity of solution ⟨641⟩—Dissolve 1 g in 10 mL of dichloromethane: the solution is clear.

Identification—

A: The infrared absorption spectrum of a potassium bromide dispersion of it, previously dried, exhibits maxima only at the same wavelengths as that of a similar preparation of USP Hydrocortisone Butyrate RS.

B: The ultraviolet absorption spectrum of a 1 in 100,000 solution in methanol exhibits maxima and minima at the same wavelengths as that of a similar solution of USP Hydrocortisone Butyrate RS, concomitantly measured, and the respective absorptivities, calculated on the dried basis, at the wavelength of maximum absorbance at about 242 nm do not differ by more than 3.0%.

Melting range, *Class Ia* ⟨741⟩: between 197° and 208°, with decomposition, but the range between beginning and end of melting does not exceed 4°.

Specific rotation ⟨781⟩: between +47° and +54°, calculated on the dried basis, determined at 20° in a solution in chloroform containing 10 mg per mL.

Loss on drying ⟨731⟩—Dry it in vacuum at 78° for 3 hours: it loses not more than 1.0% of its weight.

Hydrocortisone 21-butyrate and other related impurities—Proceed as directed in the *Assay*, obtaining the chromatogram over a period twice the retention time of the main component: the sum of the responses of all extraneous peaks, excluding solvent peaks, is not greater than 2.0% of the main peak, with no single peak being greater than 1.0%.

Assay—

Mobile phase—Prepare a solution of water, acetonitrile, and glacial acetic acid (550:450:5). Make adjustments if necessary (see *System Suitability* under *Chromatography* ⟨621⟩).

Standard preparation—Using an accurately weighed quantity of Hydrocortisone Butyrate RS, prepare a solution in a mixture of methanol and glacial acetic acid (1000:1) to obtain a *Standard preparation* having a known concentration of about 40 µg of USP Hydrocortisone Butyrate RS per mL.

Assay preparation—Transfer about 50 mg of Hydrocortisone Butyrate, accurately weighed, to a 50-mL volumetric flask. Add a mixture of methanol and glacial acetic acid (1000:1) to volume, and mix. Pipet 2 mL of this solution into a 50-mL volumetric flask, dilute with the same solvent mixture to volume, and mix.

Chromatographic system (see *Chromatography* ⟨621⟩)—The high-pressure liquid chromatograph is equipped with a 254-nm detector and a 3.9-mm × 30-cm column that contains packing L1. The flow rate is about 2 mL per minute. Chromatograph the *Standard preparation*, and record the peak responses as directed under *Procedure:* the column efficiency determined from the analyte peak is not less than 3000 theoretical plates, and the relative standard deviation for replicate injections is not more than 2.0%.

Procedure—Introduce equal volumes (between 15 µL and 30 µL) of the *Assay preparation* and the *Standard preparation* into the chromatograph, operated at room temperature, adjusting the operating parameters such that the peak obtained from the *Standard preparation* is about 60% full-scale. Determine the peak responses at equivalent retention times, obtained from the *Assay preparation* and the *Standard preparation.* Calculate the quantity, in mg, of $C_{25}H_{36}O_6$ in the portion of Hydrocortisone Butyrate taken by the formula:

$$1.25C(r_U/r_S),$$

in which C is the concentration, in µg per mL, of USP Hydrocortisone Butyrate RS in the *Standard preparation*, and r_U and r_S are the peak responses obtained from the *Assay preparation* and the *Standard preparation*, respectively.

Hydrocortisone Butyrate Cream

» Hydrocortisone Butyrate Cream is Hydrocortisone Butyrate in a suitable cream base. It contains not less than 90.0 percent and not more than 110.0 percent of the labeled amount of $C_{25}H_{36}O_6$.

Packaging and storage—Preserve in well-closed containers.

Reference standard—*USP Hydrocortisone Butyrate Reference Standard*—Dry in vacuum at 78° for 3 hours before using.

Identification—The retention time of the major peak in the chromatogram of the *Assay preparation* corresponds to that of the *Standard preparation*, obtained as directed in the *Assay.*

pH ⟨791⟩: between 3.5 and 4.5.

Microbial limits—It meets the requirements of the tests for absence of *Staphylococcus aureus* and *Pseudomonas aeruginosa* under *Microbial Limit Tests* ⟨61⟩.

Assay—

Mobile phase—Prepare a suitably filtered and degassed solution of water, acetonitrile, and glacial acetic acid (595:400:5). Make adjustments if necessary (see *System Suitability* under *Chromatography* ⟨621⟩).

Internal standard solution—Dissolve propylparaben in methanol to obtain a solution having a concentration of about 200 µg per mL.

Standard preparation—Transfer about 25 mg of USP Hydrocortisone Butyrate RS, accurately weighed, to a 250-mL volumetric flask, dilute with methanol to volume, and mix. Transfer 10.0 mL of this solution and 2.0 mL of *Internal standard solution* to a 50-mL volumetric flask, and dilute with methanol to volume to obtain a solution having a known concentration of about 20 µg of USP Hydrocortisone Butyrate RS per mL.

Assay preparation—Transfer to a 50-mL volumetric flask an accurately weighed quantity of Hydrocortisone Butyrate Cream, equivalent to about 1 mg of hydrocortisone butyrate. Add 2.0 mL of *Internal standard solution* and 20 mL of methanol, and heat the mixture on a steam bath for 5 minutes, with frequent mixing, to disperse the substance completely. Cool to room temperature, dilute with methanol to volume, and mix.

Chromatographic system (see *Chromatography* ⟨621⟩)—The liquid chromatograph is equipped with a 254-nm detector and a column that contains packing L1. The flow rate is about 2 mL per minute. Chromatograph the *Standard preparation*, and record the peak responses as directed under *Procedure:* the resolution, *R*, between the peaks for hydrocortisone butyrate and the internal standard is not less than 2.0, and the relative standard deviation for replicate injections is not more than 3.0%.

Procedure—Separately introduce equal volumes (between 10 µL and 20 µL) of the *Assay preparation* and the *Standard preparation* into the chromatograph, record the chromatograms, and measure the peak responses at equivalent retention times obtained from the *Assay preparation* and the *Standard preparation*. The relative retention times are about 0.7 for propylparaben and 1.0 for hydrocortisone butyrate. Calculate the quantity, in µg, of $C_{25}H_{36}O_6$ in the portion of Cream taken by the formula:

$$50C(R_U/R_S),$$

in which *C* is the concentration, in µg per mL, of USP Hydrocortisone Butyrate RS in the *Standard preparation*, and R_U and R_S are the peak response ratios obtained from the *Assay preparation* and the *Standard preparation*, respectively.

Hydrocortisone Cypionate

$C_{29}H_{42}O_6$ 486.65

Pregn-4-ene-3,20-dione, 21-(3-cyclopentyl-1-oxopropoxy)-11,17-dihydroxy-, (11β)-.

Cortisol 21-cyclopentanepropionate [*508-99-6*].

» Hydrocortisone Cypionate contains not less than 97.0 percent and not more than 103.0 percent of $C_{29}H_{42}O_6$, calculated on the dried basis.

Packaging and storage—Preserve in tight containers, and store in a cold place, protected from light.

Reference standards—*USP Hydrocortisone Cypionate Reference Standard*—Dry in vacuum at 105° for 4 hours before using. *USP Medroxyprogesterone Acetate Reference Standard*—Dry at 105° for 3 hours before using.

Identification—

A: The infrared absorption spectrum of a mineral oil dispersion of it, previously dried, exhibits maxima only at the same wavelengths as that of a similar preparation of USP Hydrocortisone Cypionate RS.

B: The ultraviolet absorption spectrum of a 1 in 50,000 solution in alcohol exhibits maxima and minima at the same wavelengths as that of a similar solution of USP Hydrocortisone Cypionate RS, concomitantly measured, and the respective absorptivities, calculated on the dried basis, at the wavelength of maximum absorbance at about 242 nm do not differ by more than 3.0%.

C: Its retention time, determined as directed for *Assay preparation* in the *Assay*, is the same as that of USP Hydrocortisone Cypionate RS, determined as directed for *Standard preparation* in the *Assay*.

Specific rotation ⟨781⟩: between +142° and +152°, calculated on the dried basis, determined in a solution in chloroform containing 100 mg in each 10 mL.

Loss on drying ⟨731⟩—Dry it in vacuum at 105° for 4 hours: it loses not more than 1.0% of its weight.

Residue on ignition ⟨281⟩: not more than 0.2%.

Ordinary impurities ⟨466⟩—

Test solution: methanol.

Standard solution: methanol.

Eluant: a mixture of chloroform, methanol, and water (180:15:1).

Visualization: 5.

Assay—

Internal standard solution—Prepare a solution of USP Medroxyprogesterone Acetate RS in acetonitrile containing 400 µg per mL.

Mobile solvent—Prepare a suitable degassed solution of acetonitrile in water (about 1 in 2) such that the retention times of the internal standard and hydrocortisone cypionate are about 9 and 12 minutes, respectively.

Standard preparation—Dissolve an accurately weighed quantity of USP Hydrocortisone Cypionate RS in *Internal standard solution* to obtain a solution having a known concentration of about 1 mg per mL.

Assay preparation—Transfer about 25 mg of Hydrocortisone Cypionate, accurately weighed, to a 25-mL volumetric flask, dissolve in *Internal standard solution*, dilute with *Internal standard solution* to volume, and mix.

Procedure—Using a suitable microsyringe or sampling valve, inject equal volumes (about 25 µL) of the *Standard preparation* and the *Assay preparation* into a suitable high-pressure liquid chromatograph (see *Chromatography* ⟨621⟩) of the general type equipped with a detector capable of monitoring absorption at about 254 nm, equipped with a suitable recorder, and capable of providing column pressure up to about 2500 psi. The instrument contains a 4-mm × 30-cm stainless column that contains packing L1. In a suitable chromatogram, the resolution factor, *R* (see *Chromatography* ⟨621⟩), is not less than 3.0 between peaks from the hydrocortisone cypionate and the internal standard. Five replicate injections of the *Standard preparation* show a relative standard deviation of not more than 2.0% for the peak height ratios. Calculate the quantity, in mg, of $C_{29}H_{42}O_6$ in the portion of Hydrocortisone Cypionate taken by the formula:

$$25C(R_U/R_S),$$

in which *C* is the concentration, in mg per mL, of USP Hydrocortisone Cypionate RS in the *Standard preparation*, and R_U and R_S are the peak height ratios of the hydrocortisone cypionate peak and the internal standard peak obtained from the *Assay preparation* and the *Standard preparation*, respectively.

Hydrocortisone Cypionate Oral Suspension

» Hydrocortisone Cypionate Oral Suspension contains an amount of hydrocortisone cypionate ($C_{29}H_{42}O_6$) equivalent to not less than 90.0 percent and not more than 110.0 percent of the labeled amount of hydrocortisone ($C_{21}H_{30}O_5$).

Packaging and storage—Preserve in tight, light-resistant containers.

Reference standards—*USP Hydrocortisone Cypionate Reference Standard*—Dry in vacuum at 105° for 4 hours before using. *USP Medroxyprogesterone Acetate Reference Standard*—Dry at 105° for 3 hours before using.

Identification—Its retention time, determined as directed for *Assay preparation* in the *Assay*, is the same as that for USP Hydrocortisone Cypionate RS, determined as directed for *Standard preparation* in the *Assay*.

pH ⟨791⟩: between 2.8 and 3.2.

Assay—

Internal standard solution, *Mobile solvent*, and *Standard preparation*—Proceed as directed in the *Assay* under *Hydrocortisone Cypionate*.

Assay preparation—Mix the contents of 1 container of Hydrocortisone Cypionate Oral Suspension. Withdraw an air-free portion of the homogeneous suspension, using a tuberculin syringe fitted with a suitable needle (about 15 gauge), and transfer an accurately measured volume, equivalent to about 10 mg of hydrocortisone, to a suitable vial containing 10.0 mL of *Internal standard solution*. Insert the stopper, shake vigorously for 90 minutes, and allow the layers to separate. Use the upper, acetonitrile layer in the chromatography.

Chromatographic system—Proceed as directed in the *Assay* under *Hydrocortisone Cypionate*.

Procedure—Proceed as directed for *Procedure* in the *Assay* under *Hydrocortisone Cypionate*. Calculate the quantity, in mg, of hydrocortisone ($C_{21}H_{30}O_5$) in each mL of the Oral Suspension taken by the formula:

$$(362.47/486.65)(15C/V)(R_U/R_S),$$

in which 362.47 and 486.65 are the molecular weights of hydrocortisone and hydrocortisone cypionate, respectively, V is the volume, in mL, of the Oral Suspension taken, and C, R_U, and R_S are as defined therein.

Hydrocortisone Hemisuccinate

$C_{25}H_{34}O_8 \cdot H_2O$ 480.56

Pregn-4-ene-3,20-dione, 21-(3-carboxy-1-oxopropoxy)-11,17-dihydroxy-, (11β)-, monohydrate.

Cortisol 21-(hydrogen succinate) monohydrate [*83784-20-7*].

Anhydrous 462.54 [*2203-97-6*].

» Hydrocortisone Hemisuccinate contains not less than 97.0 percent and not more than 103.0 percent of $C_{25}H_{34}O_8$, calculated on the dried basis. It contains one molecule of water of hydration or is anhydrous.

Packaging and storage—Preserve in tight containers.

Labeling—Label it to indicate whether it is hydrous or anhydrous.

Reference standards—*USP Hydrocortisone Hemisuccinate Reference Standard*—Dry at 105° for 3 hours before using. *USP Fluorometholone Reference Standard*—Dry in vacuum at 60° for 3 hours before using.

Identification—

A: The infrared absorption spectrum of a mineral oil dispersion of it, previously dried at 105° for 3 hours, exhibits maxima only at the same wavelengths as that of a similar preparation of USP Hydrocortisone Hemisuccinate RS.

B: The ultraviolet absorption spectrum of a 1 in 50,000 solution in alcohol exhibits maxima and minima at the same wavelengths as that of a similar solution of USP Hydrocortisone Hemisuccinate RS, concomitantly measured, and the respective absorptivities, calculated on the dried basis, at the wavelength of maximum absorbance at about 242 nm do not differ by more than 3.0%.

Specific rotation ⟨781⟩: between +124° and +134°, calculated on the dried basis, determined in a solution of acetone containing 100 mg in each 10 mL.

Loss on drying ⟨731⟩—Dry it at 105° for 3 hours: the anhydrous form loses not more than 1.0% of its weight, and the hydrous form loses not more than 4.0% of its weight.

Residue on ignition ⟨281⟩: not more than 0.1%.

Assay—

Internal standard solution—Dissolve USP Fluorometholone RS in tetrahydrofuran to obtain a solution containing about 3 mg per mL.

Mobile solvent—Prepare a solution containing butyl chloride, water-saturated butyl chloride, tetrahydrofuran, methanol, and glacial acetic acid (95:95:14:7:6).

Standard preparation—Transfer about 30 mg of USP Hydrocortisone Hemisuccinate RS, accurately weighed, to a 50-mL volumetric flask. Add 5.0 mL of *Internal standard solution*. Dilute with chloroform containing 3% glacial acetic acid to volume, and mix to dissolve the powder.

Assay preparation—Using about 30 mg of Hydrocortisone Hemisuccinate, accurately weighed, prepare as directed under *Standard preparation*.

Procedure—Chromatograph equal volumes, between 4 and 8 µL, of the *Standard preparation* and the *Assay preparation* into a suitable high-pressure liquid chromatograph (see *Chromatography* ⟨621⟩) of the general type equipped with a detector for monitoring ultraviolet light absorption at about 254 nm, equipped with a suitable recorder, and capable of providing column pressure up to about 1000 psi and fitted with a 4-mm × 30-cm stainless steel column that contains packing L3. In a suitable chromatogram, the resolution factor, R (see *Chromatography* ⟨621⟩) is not less than 2.0 between peaks for hydrocortisone hemisuccinate and the internal standard, and six replicate injections of the *Standard preparation* show a coefficient of variation of not more than 2.0%. Calculate the quantity, in mg, of $C_{25}H_{34}O_8$ in the portion of Hydrocortisone Hemisuccinate taken by the formula:

$$50C(R_U/R_S),$$

in which C is the concentration, in mg per mL, of USP Hydrocortisone Hemisuccinate RS in the *Standard preparation*, and R_U and R_S are the peak area ratios of the hydrocortisone hemisuccinate peak and the internal standard peak obtained from the *Assay preparation* and the *Standard preparation*, respectively.

Hydrocortisone Sodium Phosphate

$C_{21}H_{29}Na_2O_8P$ 486.41

Pregn-4-ene-3,20-dione, 11,17-dihydroxy-21-(phosphonooxy)-, disodium salt, (11β)-.

Cortisol 21-(disodium phosphate) [*6000-74-4*].

» Hydrocortisone Sodium Phosphate contains not less than 96.0 percent and not more than 102.0 percent of $C_{21}H_{29}Na_2O_8P$, calculated on the dried basis.

Packaging and storage—Preserve in tight containers.

Reference standards—*USP Hydrocortisone Reference Standard*—Dry at 105° for 3 hours before using. *USP Hydrocortisone Phosphate Triethylamine Reference Standard*—Dry in vacuum at 60° for 3 hours before using.

Identification—

A: Evaporate 15 mL of a methylene chloride solution of it, prepared as directed under *Procedure* in the *Assay*, on a steam bath to dryness, and dissolve the residue in 1 mL of methylene chloride. Proceed as directed in *Identification test B* under *Hydrocortisone Sodium Phosphate Injection*, beginning with "Apply 5 µL of this solution."

B: The residue from the ignition of about 20 mg of it responds to the tests for *Phosphate* ⟨191⟩ and for *Sodium* ⟨191⟩.

Phosphate ions—

Standard phosphate solution—Dissolve 143.3 mg of dried monobasic potassium phosphate, KH_2PO_4, in water to make 1000.0 mL. This solution contains the equivalent of 0.10 mg of phosphate (PO_4) in each mL.

Phosphate reagent A—Dissolve 5 g of ammonium molybdate in 1 *N* sulfuric acid to make 100 mL.

Phosphate reagent B—Dissolve 350 mg of *p*-methylaminophenol sulfate in 50 mL of water, add 20 g of sodium bisulfite, mix to dissolve, and dilute with water to 100 mL.

Procedure—Dissolve about 50 mg of Hydrocortisone Sodium Phosphate, accurately weighed, in a mixture of 10 mL of water and 5 mL of 2 *N* sulfuric acid contained in a 25-mL volumetric flask, by warming if necessary. Add 1 mL each of *Phosphate reagent A* and *Phosphate reagent B*, dilute with water to 25 mL, mix, and allow to stand at room temperature for 30 minutes. Similarly and concomitantly, prepare a standard solution, using 5.0 mL of *Standard phosphate solution* instead of the 50 mg of the substance under test. Concomitantly determine the absorbances of both solutions in 1-cm cells at 730 nm, with a suitable spectrophotometer, using water as the blank. The absorbance of the test solution is not more than that of the standard solution. The limit is 1.0% of phosphate (PO_4).

Chloride (as NaCl)—Dissolve about 3 g, accurately weighed, in 75 mL of water, add 1 mL of nitric acid, and titrate with 0.1 *N* silver nitrate VS, determining the end-point potentiometrically, using a glass–silver–silver chloride electrode system. Each mL of 0.1 *N* silver nitrate is equivalent to 5.844 mg of NaCl. Not more than 1.00% of NaCl is found.

Specific rotation, pH, and Free hydrocortisone—Place about 2.5 g in a tared 50-mL flask, and weigh accurately (W_U). Add 25 mL of carbon dioxide-free water, and again weigh (W_S). Calculate the quantity, in mg, of anhydrous hydrocortisone sodium phosphate in each g of solution by the formula:

$$1000W_U(1 - L/100)/W_S,$$

in which W_U is the weight of Hydrocortisone Sodium Phosphate taken, L is the average percentage of *Loss on drying*, and W_S is the weight of the solution in carbon dioxide–free water. Use this as the *Test preparation* for the following tests.

Specific rotation ⟨781⟩: between +121° and +129°, calculated on the dried basis, determined in a solution prepared by weighing accurately 5.0 mL of *Test preparation* and diluting with pH 7.0 phosphate buffer (see *Buffer Solutions* in the section *Reagents, Indicators, and Solutions*) to 50.0 mL.

pH ⟨791⟩: between 7.5 and 10.5, in a solution prepared by diluting a portion of *Test preparation* with 9 volumes of carbon dioxide–free water.

Free hydrocortisone—Dilute 1 mL of *Test preparation* with carbon dioxide–free water to 100 mL. Pipet 5 mL of this solution into a glass-stoppered, 50-mL tube, add 25.0 mL of methylene chloride, insert the stopper, and mix by gentle shaking. Prepare a 1 in 500,000 solution of USP Hydrocortisone RS in methylene chloride. Similarly, shake 25 mL of this solution with 5 mL of water. Allow to stand until the methylene chloride layers are clear (about 5 minutes). Determine the absorbances of the methylene chloride solutions in 1-cm cells at 239 nm, with a suitable spectrophotometer, using methylene chloride as the blank. The absorbance of the *Test preparation* does not exceed that of the Standard solution (1.0%).

Loss on drying ⟨731⟩—Dry it in vacuum at 80° for 5 hours: the average percentage weight loss for two determinations (L) does not exceed 5.0%.

Heavy metals, *Method II* ⟨231⟩: 0.004%.

Assay—

pH 9 buffer with magnesium—Mix 3.1 g of boric acid and 500 mL of water in a 1-liter volumetric flask, add 21 mL of 1 *N* sodium hydroxide and 10 mL of 0.1 *M* magnesium chloride, dilute with water to volume, and mix.

Alkaline phosphatase solution—Transfer 250 mg of alkaline phosphatase enzyme to a 25-mL volumetric flask, and dissolve by diluting with *pH 9 buffer with magnesium* to volume. Prepare this solution fresh daily.

Standard preparation—Dissolve about 50 mg of USP Hydrocortisone Phosphate Triethylamine RS, accurately weighed, in carbon dioxide–free water to make 25.0 mL.

Assay preparation—Weigh accurately, in g, 2.0 mL of the *Test preparation*, prepared as directed under *Specific rotation, pH, and Free hydrocortisone*, into a tared 100-mL volumetric flask, and dilute with carbon dioxide–free water that has been saturated with methylene chloride to volume. Pipet 10 mL of this solution into a 125-mL separator, and extract with two 25-mL portions of water-washed methylene chloride, discarding the washings.

Procedure—Weigh accurately 1.0 mL each of the *Standard preparation* (W_S) and the *Assay preparation* (W_A) into separate tared 100-mL volumetric flasks. To each flask, and to a similar flask containing 1.0 mL of water to provide a blank, add 1.0 mL of *Alkaline phosphatase solution* and then 50 mL of methylene chloride, and insert the stopper. Allow the flasks to stand at room temperature (not below 25°) for 2 hours with gentle mixing about every 15 minutes. Add 1 mL of dilute hydrochloric acid (1 in 10) to each flask, and mix gently. Add methylene chloride to each flask until the interfaces are at the 100-mL marks, and mix gently. Remove the aqueous layers by aspiration. Determine the absorbances of the methylene chloride solutions obtained from the *Standard preparation* and the *Assay preparation* at 239 nm, with a suitable spectrophotometer, using the methylene chloride solution blank to set the instrument. Calculate the percentage of $C_{21}H_{29}Na_2O_8P$, on the dried basis, by the formula:

$$100(A_U/A_S)(C_S/C_A)0.895(W_S/W_A),$$

in which A_U and A_S are the absorbances of the solutions from the *Assay preparation* and the *Standard preparation*, respectively, C_A and C_S are the corresponding concentrations, in mg per mL, of those preparations, and 0.895 is the ratio of the molecular weight of hydrocortisone sodium phosphate to that of hydrocortisone phosphate triethylamine.

Hydrocortisone Sodium Phosphate Injection

» Hydrocortisone Sodium Phosphate Injection is a sterile, buffered solution of Hydrocortisone Sodium Phosphate in Water for Injection. It contains the equivalent of not less than 90.0 percent and not more than 115.0 percent of the labeled amount of hydrocortisone ($C_{21}H_{30}O_5$).

Packaging and storage—Preserve in single-dose or in multiple-dose containers, preferably of Type I glass.

Reference standards—*USP Hydrocortisone Reference Standard*—Dry at 105° for 3 hours before using. *USP Hydrocortisone Phosphate Triethylamine Reference Standard*—Dry in vacuum at 60° for 3 hours before using.

Identification—

A: Dilute a portion of Injection with water to obtain a 1 in 50,000 solution of hydrocortisone sodium phosphate: the ultraviolet absorption spectrum of this solution exhibits a maximum at the same wavelength as that of a similar solution of USP Hydrocortisone Phosphate Triethylamine RS, concomitantly measured.

B: Place 5 mL of *Assay preparation*, obtained as directed in the *Assay*, in a glass-stoppered, 50-mL tube, and add 5 mL of a solution prepared by dissolving 50 mg of alkaline phosphatase enzyme in 50 mL of *pH 9 buffer with magnesium* prepared as directed in the *Assay* under *Hydrocortisone Sodium Phosphate.* Allow to stand at room temperature for 2 hours, with occasional mixing, and extract with 25 mL of methylene chloride. Evaporate 15 mL of the methylene chloride extract on a steam bath to dryness, and dissolve the residue in 0.5 mL of methylene chloride. Apply 5 µL of this solution and 5 µL of a solution of USP Hydrocortisone RS in methylene chloride containing 300 µg per mL on a thin-layer chromatographic plate (see *Chromatography* ⟨621⟩) coated with a 0.25-mm layer of chromatographic silica gel mixture. Allow the spots to dry, and develop the chromatogram in a tank completely lined with filter paper, using a solvent system consisting of a mixture of 50 parts of chloroform, 50 parts of acetone, and 1 part of water, until the solvent front has moved about three-fourths of the length of the plate. Remove the plate, mark the solvent front, and dry. Spray the plate with dilute sulfuric acid (1 in 2), and heat at 105° until brown or black spots appear. The R_f value of the principal spot obtained from the solution under test corresponds to that obtained from the Standard solution.

pH ⟨791⟩: between 7.5 and 8.5.

Particulate matter ⟨788⟩: meets the requirements under *Small-volume Injections.*

Other requirements—It meets the requirements under *Injections* ⟨1⟩.

Assay—

Phenylhydrazine hydrochloride solution—Dissolve 65 mg of phenylhydrazine hydrochloride in 100 mL of dilute sulfuric acid (3 in 5), add 50 mL of isopropyl alcohol, and mix. Prepare this solution fresh daily.

Standard preparation—Dissolve a suitable quantity of USP Hydrocortisone Phosphate Triethylamine RS, accurately weighed, in water, and dilute quantitatively and stepwise with water to obtain a solution having a known concentration of about 110 µg per mL.

Assay preparation—Pipet a volume of Hydrocortisone Sodium Phosphate Injection, equivalent to about 100 mg of hydrocortisone sodium phosphate, into a 100-mL volumetric flask, and dilute with water to volume. Pipet 10 mL of this solution into a separator, wash the solution with two 25-mL portions of methylene chloride, and discard the washings. Transfer the aqueous layer to a 100-mL volumetric flask, dilute with water to volume, and mix.

Procedure—Pipet 2 mL each of the *Standard preparation* and the *Assay preparation* into separate glass-stoppered, 50-mL conical flasks. To each flask, and to a similar flask containing 2.0 mL of water to provide a blank, add 10.0 mL of *Phenylhydrazine hydrochloride solution*, and mix. Place the flasks in a water bath maintained at a temperature of 60° for 2 hours, then cool the solutions to room temperature. Concomitantly determine the absorbances of the solutions from the *Assay preparation* and the *Standard preparation* at the wavelength of maximum absorbance at about 410 nm, with a suitable spectrophotometer, using the blank to set the instrument. Calculate the quantity, in mg, of $C_{21}H_{30}O_5$ in each mL of the Injection taken by the formula:

$$0.667(C/V)(A_U/A_S),$$

in which 0.667 is the ratio of the molecular weight of hydrocortisone to that of hydrocortisone phosphate triethylamine, C is the concentration, in µg per mL, of USP Hydrocortisone Phosphate Triethylamine RS in the *Standard preparation*, V is the volume, in mL, of Injection taken, and A_U and A_S are the absorbances of the solutions from the *Assay preparation* and the *Standard preparation*, respectively.

Hydrocortisone Sodium Succinate

$C_{25}H_{33}NaO_8$ 484.52

Pregn-4-ene-3,20-dione, 21-(3-carboxy-1-oxopropoxy)-11,17-dihydroxy-, monosodium salt, (11β)-.

Cortisol 21-(sodium succinate) [*125-04-2*].

» Hydrocortisone Sodium Succinate contains not less than 97.0 percent and not more than 102.0 percent of total steroids, calculated as $C_{25}H_{33}NaO_8$, on the dried basis.

Packaging and storage—Preserve in tight, light-resistant containers.

Reference standard—*USP Hydrocortisone Hemisuccinate Reference Standard*—Dry at 105° for 3 hours before using.

Identification—

A: Transfer about 100 mg to a suitable container, and dissolve in about 10 mL of water. In rapid succession, add 1 mL of 3 *N* hydrochloric acid, shake briefly, immediately decant the aqueous layer, and wash the precipitate with two additional 10-mL portions of water, each time removing the water by decanting. Remove as much of the water as possible, spread the precipitate in a suitable container, and dry in vacuum at about 60° for 3 hours: the infrared absorption spectrum of a mineral oil dispersion of the precipitate so obtained exhibits maxima only at the same wavelengths as that of a similar preparation of USP Hydrocortisone Hemisuccinate RS.

B: The ultraviolet absorption spectrum of a 1 in 50,000 solution in methanol exhibits maxima and minima at the same wavelengths as that of a similar solution of USP Hydrocortisone Hemisuccinate RS, concomitantly measured, and the absorptivity value of the Hydrocortisone Sodium Succinate, calculated on the dried basis, at the wavelength of maximum absorbance at about 242 nm, multiplied by 1.048, does not differ from that of the Reference Standard by more than 3.0%.

C: It responds to the flame test for *Sodium* ⟨191⟩.

Specific rotation ⟨781⟩: between +135° and +145°, calculated on the dried basis, determined in a solution in alcohol containing 100 mg in each 10 mL.

Loss on drying ⟨731⟩—Dry it at 105° for 3 hours: it loses not more than 2.0% of its weight.

Sodium content—Dissolve, with gentle heating, about 1 g, accurately weighed, in 75 mL of glacial acetic acid. Add 20 mL of dioxane, then add crystal violet TS, and titrate with 0.1 *N* perchloric acid VS. Each mL of 0.1 *N* perchloric acid is equivalent to 2.299 mg of Na. Between 4.60% and 4.84% of Na is found, calculated on the dried basis.

Assay—

Standard preparation—Prepare as directed for *Standard Preparation* under *Assay for Steroids* ⟨351⟩, using USP Hydrocortisone Hemisuccinate RS, but dilute the solution with alcohol to a concentration of about 12.5 µg per mL.

Assay preparation—Weigh accurately about 100 mg of Hydrocortisone Sodium Succinate, dissolve in sufficient alcohol to make 200.0 mL, and mix. Pipet 5 mL of this solution into a 200-mL volumetric flask, add alcohol to volume, and mix. Pipet 20 mL of the resulting solution into a glass-stoppered, 50-mL conical flask.

Procedure—To each of the two flasks containing the *Assay preparation* and the *Standard preparation*, respectively, and to a similar flask containing 20.0 mL of alcohol to provide the blank, add 2.0 mL of a solution prepared by dissolving 50 mg of blue tetrazolium in 10 mL of alcohol, and mix. Then to each flask add 4.0 mL of a 1 in 10 solution of tetramethylammonium hydroxide TS in alcohol, mix, allow to stand in the dark for 90 minutes, add 1.0 mL of glacial acetic acid, mix, and proceed as directed for *Procedure* under *Assay for Steroids* ⟨351⟩, beginning with "Concomitantly determine the absorbances." Calculate the quantity, in mg, of $C_{25}H_{33}NaO_8$ in the Hydrocortisone Sodium Succinate taken by the formula:

$$8.38C(A_U/A_S).$$

Hydrocortisone Sodium Succinate for Injection

» Hydrocortisone Sodium Succinate for Injection is a sterile mixture of Hydrocortisone Sodium Succinate and suitable buffers. It may be prepared from Hydrocortisone Sodium Succinate, or from Hydrocortisone Hemisuccinate with the aid of Sodium Hydroxide or Sodium Carbonate. It contains the equivalent of not less than 90.0 percent and not more than 110.0 percent of the labeled amount of hydrocortisone ($C_{21}H_{30}O_5$) in single-compartment containers, or in the volume of solution designated on the label of containers that are constructed to hold in separate compartments the Hydrocortisone Sodium Succinate for Injection and a solvent.

Packaging and storage—Preserve in *Containers for Sterile Solids* as described under *Injections* ⟨1⟩.

Labeling—Label it to indicate that the constituted solution prepared from Hydrocortisone Sodium Succinate for Injection is suitable for use only if it is clear, and that the solution is to be discarded after 3 days. Label it to indicate that it was prepared by freeze-drying, having been filled into its container in the form of a true solution.

Reference standards—*USP Hydrocortisone Hemisuccinate Reference Standard*—Dry at 105° for 3 hours before using. *USP*

Fluorometholone Reference Standard—Dry in vacuum at 60° for 3 hours before using. *USP Hydrocortisone Reference Standard*—Dry at 105° for 3 hours before using.

Constituted solution—At the time of use, the constituted solution prepared from Hydrocortisone Sodium Succinate for Injection meets the requirements for *Constituted Solutions* under *Injections* ⟨1⟩.

Identification—It responds to *Identification test A* under *Hydrocortisone Sodium Succinate.*

pH ⟨791⟩: between 7.0 and 8.0, in a solution containing the equivalent of 50 mg of hydrocortisone per mL.

Loss on drying ⟨731⟩—Dry it at 105° for 3 hours: it loses not more than 2.0% of its weight.

Particulate matter ⟨788⟩: meets the requirements under *Small-volume Injections.*

Free hydrocortisone—Using the chromatograms obtained in the *Assay*, measure the areas of the peaks from the Internal standard and free hydrocortisone. Calculate the ratio of the area of the free hydrocortisone peak to that of the Internal standard in the chromatogram obtained from the *Standard preparation*, S_S, and the same ratio in the chromatogram obtained from the *Assay preparation*, S_U. Calculate the quantity, in mg, of free hydrocortisone in the *Assay preparation* taken by the formula:

$$(100C)(S_U / S_S),$$

in which C is the concentration, in mg per mL, of USP Hydrocortisone RS in the *Standard preparation*, and S_U and S_S are the ratios as defined above. The amount of free hydrocortisone is not more than 6.7% of the labeled amount of hydrocortisone.

Other requirements—It meets the requirements for *Sterility Tests* ⟨71⟩, *Uniformity of Dosage Units* ⟨905⟩, and *Labeling* under *Injections* ⟨1⟩.

Assay—

Internal standard solution and *Mobile solvent*—Prepare as directed in the *Assay* under *Hydrocortisone Hemisuccinate.*

Standard preparation—Weigh accurately about 32.5 mg of USP Hydrocortisone Hemisuccinate RS and transfer it to a 50-mL volumetric flask. Add by pipet 5.0 mL of *Internal standard solution* and 5.0 mL of a solution of glacial acetic acid in chloroform (3 in 100) containing in each mL an accurately known quantity of about 0.30 mg of USP Hydrocortisone RS. Dilute with glacial acetic acid in chloroform (3 in 100) to volume, and mix.

Assay preparation—Mix the constituted solutions prepared from the contents of 10 vials of Hydrocortisone Sodium Succinate for Injection. Transfer an accurately measured volume of the resulting constituted solution, equivalent to about 50 mg of hydrocortisone, to a suitable flask containing 10.0 mL of *Internal standard solution* and dilute with glacial acetic acid in chloroform (3 in 100) to 100.0 mL. Shake thoroughly for 5 minutes, then allow the phases to separate, discarding the upper phase.

Procedure—Proceed as directed for *Procedure* in the *Assay* under *Hydrocortisone Hemisuccinate.* The order of elution of peaks is that from the Internal standard, hydrocortisone hemisuccinate, and successive smaller peaks representing free hydrocortisone and hydrocortisone 17-hemisuccinate, whose relative retention times are about 1.0, 1.5, 2.0, and 2.5, respectively. Measure the areas of the peaks from the Internal standard, hydrocortisone hemisuccinate, and hydrocortisone 17-hemisuccinate. Calculate the ratio of the summation of the areas of the hydrocortisone hemisuccinate and hydrocortisone 17-hemisuccinate peaks to that of the Internal standard in the chromatogram obtained from the *Standard preparation*, R_S, and the same ratio in the chromatogram obtained from the *Assay preparation*, R_U. Calculate the quantity, in mg, of hydrocortisone ($C_{21}H_{30}O_5$) in the portion of constituted solution taken by the formula:

$$0.784(100C)(R_U / R_S),$$

in which 0.784 is the ratio of the molecular weight of hydrocortisone to that of hydrocortisone hemisuccinate, C is the concentration, in mg per mL, of USP Hydrocortisone Hemisuccinate RS in the *Standard preparation*, and R_U and R_S are the ratios as defined above, and to this quantity add the amount, in mg, of free hydrocortisone found in the test for *Free hydrocortisone.*

Hydrocortisone Valerate

$C_{26}H_{38}O_6$ 446.58

Pregn-4-ene-3,20-dione, 11,21-dihydroxy-17-[(1-oxopentyl)oxy]-, (11β)-.

Cortisol 17-valerate.

11β,17,21-Trihydroxypregn-4-ene-3,20-dione 17-valerate [*57524-89-7*].

» Hydrocortisone Valerate contains not less than 97.0 percent and not more than 102.0 percent of $C_{26}H_{38}O_6$, calculated on the dried basis.

Packaging and storage—Preserve in well-closed containers.

Reference standard—*USP Hydrocortisone Valerate Reference Standard*—Dry at 105° for 3 hours before using.

Identification—The infrared absorption spectrum of a potassium bromide dispersion of it exhibits maxima only at the same wavelengths as that of a similar preparation of USP Hydrocortisone Valerate RS.

Specific rotation ⟨781⟩: between 37° and 43°, calculated on the dried basis, determined in a solution in dioxane containing 10 mg per mL.

Loss on drying ⟨731⟩—Dry it at 105° for 3 hours: it loses not more than 1.0% of its weight.

Assay—

Mobile phase—Prepare a suitable solution of acetonitrile in water (45 in 100) such that the retention time of hydrocortisone valerate is approximately 9 minutes.

Internal standard solution—Prepare a solution of ethyl benzoate in methanol containing about 2.0 mg per mL.

Standard preparation—Immediately prior to use, dissolve an accurately weighed quantity of USP Hydrocortisone Valerate RS in methanol to obtain a solution having a known concentration of about 0.5 mg per mL. Pipet 2 mL of this solution and 2 mL of *Internal standard solution* into a 10-mL volumetric flask. Dilute with methanol to volume, and mix to obtain a *Standard preparation* having a known concentration of about 100 µg of hydrocortisone valerate per mL.

Assay preparation—Transfer about 100 mg of Hydrocortisone Valerate, accurately weighed, to a 100-mL volumetric flask. Dissolve in methanol, dilute with methanol to volume, and mix. Pipet 1 mL of this solution and 2 mL of *Internal standard solution* into a 10-mL volumetric flask. Dilute with methanol to volume, and mix.

Chromatographic system (see *Chromatography* ⟨621⟩)—The liquid chromatograph is equipped with a 254-nm detector and a 2.6-mm × 50-cm column that contains packing L2. The flow rate is about 1 mL per minute. Chromatograph three replicate injections of the *Standard preparation*, and record the peak responses as directed under *Procedure:* the relative standard deviation is not more than 2.0%, and the resolution factor between hydrocortisone valerate and ethyl benzoate is not less than 3.0.

Procedure—Separately inject equal volumes (about 10 µL) of the *Standard preparation* and the *Assay preparation* into the chromatograph, record the chromatograms, and measure the responses for the major peaks. Calculate the quantity, in mg, of $C_{26}H_{38}O_6$, in the portion of Hydrocortisone Valerate taken by the formula:

$$C(R_U / R_S),$$

in which C is the concentration, in µg per mL, of USP Hydrocortisone Valerate RS in the *Standard preparation*, and R_U and R_S are the peak response ratios of the hydrocortisone valerate peak and the internal standard peak obtained from the *Assay preparation* and the *Standard preparation*, respectively.

Hydrocortisone Valerate Cream

» Hydrocortisone Valerate Cream is Hydrocortisone Valerate in a suitable cream base. It contains not less than 90.0 percent and not more than 110.0 percent of the labeled amount of $C_{26}H_{38}O_6$.

Packaging and storage—Preserve in well-closed containers.

Reference standard—*USP Hydrocortisone Valerate Reference Standard*—Dry at 105° for 3 hours before using.

Identification—Using the *Assay preparation* and the *Standard preparation* obtained as directed in the *Assay*, proceed as directed under *Thin-layer Chromatographic Identification Test* ⟨201⟩.

Minimum fill ⟨755⟩: meets the requirements.

Assay—

Mobile phase, Internal standard preparation, Standard preparation, and *Chromatographic system*—Prepare as directed in the *Assay* under *Hydrocortisone Valerate.*

Assay preparation—Transfer an accurately weighed quantity of Hydrocortisone Valerate Cream, equivalent to about 1 mg of hydrocortisone valerate, to a screw-capped tube. Add 8.0 mL of a mixture of methanol and water (3:1), and swirl to disperse. Heat at 80° for 1 minute, swirl again, and allow to cool to room temperature. Add 2.0 mL of *Internal standard solution*, and mix. Centrifuge for 5 minutes, and filter, if necessary, to obtain a clear supernatant solution.

Procedure—Proceed as directed for *Procedure* in the *Assay* under *Hydrocortisone Valerate.* Calculate the quantity, in mg, of $C_{26}H_{38}O_6$, in the portion of Cream taken by the formula:

$$0.01C(R_U/R_S),$$

in which the terms are as defined therein.

Hydroflumethiazide

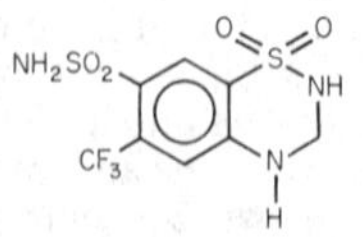

$C_8H_8F_3N_3O_4S_2$ 331.28

2*H*-1,2,4-Benzothiadiazine-7-sulfonamide, 3,4-dihydro-6-(trifluoromethyl)-, 1,1-dioxide.

3,4-Dihydro-6-(trifluoromethyl)-2*H*-1,2,4-benzothiadiazine-7-sulfonamide 1,1-dioxide [*135-09-1*].

» Hydroflumethiazide contains not less than 98.0 percent and not more than 102.0 percent of $C_8H_8F_3N_3O_4S_2$, calculated on the anhydrous basis.

Packaging and storage—Preserve in tight containers.

Reference standards—*USP Hydroflumethiazide Reference Standard*—Dry over silica gel for 4 hours before using. *USP 2,4-Disulfamyl-5-trifluoromethylaniline Reference Standard*—Keep container tightly closed and protected from light. Dry in vacuum over silica gel for 4 hours before using.

Identification—

A: The infrared absorption spectrum of a potassium bromide dispersion of it, previously dried over silica gel for 4 hours, exhibits maxima only at the same wavelengths as that of a similar preparation of USP Hydroflumethiazide RS.

B: The ultraviolet absorption spectrum of a 1 in 100,000 solution in methanol exhibits maxima and minima at the same wavelengths as that of a similar solution of USP Hydroflumethiazide RS, concomitantly measured.

Melting range, *Class I* ⟨741⟩: between 270° and 275°.

pH ⟨791⟩: between 4.5 and 7.5, in a 1 in 100 dispersion in water.

Water, *Method I* ⟨921⟩: not more than 1.0%.

Residue on ignition ⟨281⟩: not more than 1.0%.

Heavy metals, *Method II* ⟨231⟩: 0.002%.

Selenium ⟨291⟩: 0.003%.

Diazotizable substances—

Standard preparation—Transfer 10.0 mg of USP 2,4-Disulfamyl-5-trifluoromethylaniline RS to a 50-mL volumetric flask, dissolve in acetone, dilute with acetone to volume, and mix. Transfer 10.0 mL of this solution to a 100-mL volumetric flask, dilute with water to volume, and mix.

Test preparation—Transfer 100 mg to a 50-mL volumetric flask, dissolve in 5 mL of acetone, dilute with water to volume, and mix.

Procedure—Transfer 5.0 mL each of the *Standard preparation*, the *Test preparation*, and a solution of acetone in water (1 in 10) to provide the blank, to separate 25-mL volumetric flasks. To each flask add 2.0 mL of dilute hydrochloric acid (1 in 5), and immediately add 1 mL of freshly prepared sodium nitrite solution (1 in 100). Mix, and allow to stand for 5 minutes. Add 1 mL of freshly prepared ammonium sulfamate solution (1 in 10) to each flask, mix, and allow to stand for 1 minute, with frequent swirling. Add 1 mL of a freshly prepared solution of *N*-(1-naphthyl)ethylenediamine dihydrochloride (1 in 1000), and mix. After 1 minute, dilute with water to volume, and mix. Concomitantly, and within 5 minutes after mixing, taking care to establish the same elapsed time for each solution, determine the absorbances of the solutions in 1-cm cells at 518 nm, with a suitable spectrophotometer, using the blank to set the instrument: the absorbance of the solution from the *Test preparation* does not exceed that of the solution from the *Standard preparation*, corresponding to not more than 1.0% of diazotizable substances.

Assay—Transfer about 50 mg of Hydroflumethiazide, accurately weighed, to a 100-mL volumetric flask, add methanol to volume, and mix. Transfer 2.0 mL of this solution to a second 100-mL volumetric flask, dilute with methanol to volume, and mix. Concomitantly determine the absorbances of this solution and a Standard solution of USP Hydroflumethiazide RS in the same medium having a known concentration of about 10 µg per mL, in 1-cm cells at the wavelength of maximum absorbance at about 273 nm, with a suitable spectrophotometer, using methanol as the blank. Calculate the quantity, in mg, of $C_8H_8F_3N_3O_4S_2$ in the Hydroflumethiazide taken by the formula:

$$5C(A_U/A_S),$$

in which C is the concentration, in µg per mL, of USP Hydroflumethiazide RS in the Standard solution, and A_U and A_S are the absorbances of the solution of Hydroflumethiazide and the Standard solution, respectively.

Hydroflumethiazide Tablets

» Hydroflumethiazide Tablets contain not less than 95.0 percent and not more than 105.0 percent of the labeled amount of $C_8H_8F_3N_3O_4S_2$.

Packaging and storage—Preserve in tight containers.

Reference standard—*USP Hydroflumethiazide Reference Standard*—Dry over silica gel for 4 hours before using.

Identification—Finely powder a number of Tablets, equivalent to about 100 mg of hydroflumethiazide, and place the powder in a 35-mL, screw-capped centrifuge tube. Add 30 mL of acetone, cap the tube, and allow it to stand for 30 minutes, with occasional shaking. Centrifuge, and decant the supernatant liquid into a 100-mL beaker. Evaporate on a steam bath to dryness, add 10 mL of sodium hydroxide solution (1 in 250) to the residue, and mix. Transfer the liquid to a 125-mL separator. Rinse the beaker with 5 mL of water, and add the rinsing to the main portion. Add 50 mL of anhydrous ethyl ether to the separator, insert the stopper, shake vigorously for 2 minutes, releasing pressure as necessary, and allow the phases to separate. Draw off the lower phase, retaining any emulsion in the separator, and filter it through a membrane filter of 0.2- to 2-µm pore size. Add dilute hydrochloric acid (1 in 10) dropwise to the filtrate in a 50-mL beaker, stirring well and checking the pH with wide-range test paper after each drop. [NOTE—Crystallization begins at about pH 5. Rubbing the bottom of the beaker with a glass stirring rod helps to initiate crystallization.] When precipitation is complete, decant and discard the supernatant liquid, and wash the precipitate with 5 mL of water. Decant and discard the wash water, and dry the precipitate at 105° for 30 minutes. The infrared spectrum of a potassium bromide dispersion of the dried material exhibits maxima only at the same wavelengths as that of a similar preparation of USP Hydroflumethiazide RS.

Dissolution ⟨711⟩—

Medium: dilute hydrochloric acid (1 in 100); 900 mL.

Apparatus 2: 50 rpm.

Time: 60 minutes.

Procedure—Determine the amount of $C_8H_8F_3N_3O_4S_2$ dissolved from ultraviolet absorbances at the wavelength of maximum absorbance at about 273 nm of filtered portions of the solution under test, suitably diluted with *Dissolution Medium*, if necessary, in comparison with a Standard solution having a known concentration of USP Hydroflumethiazide RS in the same medium.

Tolerances—Not less than 80% (Q) of the labeled amount of $C_8H_8F_3N_3O_4S_2$ is dissolved in 60 minutes.

Uniformity of dosage units ⟨905⟩: meet the requirements.

Procedure for content uniformity—Transfer 1 Tablet to a 100-mL volumetric flask, add 50 mL of methanol, and shake the flask until the Tablet is completely disintegrated. Dilute with methanol to volume, mix, and filter, discarding the first 20 mL of the filtrate. Dilute a portion of the subsequent filtrate with methanol to obtain a solution containing approximately 10 µg of hydroflumethiazide per mL. Concomitantly determine the absorbances of this solution and of a Standard solution of USP Hydroflumethiazide RS, in the same medium having a known concentration of about 10 µg per mL in 1-cm cells at the wavelength of maximum absorbance at about 273 nm, with a suitable spectrophotometer, using methanol as the blank. Calculate the quantity, in mg, of $C_8H_8F_3N_3O_4S_2$ in the Tablet by the formula:

$$(TC/D)(A_U/A_S),$$

in which T is the labeled quantity, in mg, of hydroflumethiazide in the Tablet, C is the concentration, in µg per mL, of USP Hydroflumethiazide RS in the Standard solution, D is the concentration, in µg per mL, of hydroflumethiazide in the test solution, based upon the labeled quantity per Tablet and the extent of dilution, and A_U and A_S are the absorbances of the solution from the Tablet and the Standard solution, respectively.

Assay—

Standard preparation—Transfer about 30 mg of USP Hydroflumethiazide RS, accurately weighed, to a 100-mL volumetric flask, add sodium hydroxide solution (1 in 100) to volume, and mix. Transfer 5.0 mL of this solution to a second 100-mL volumetric flask, dilute with sodium hydroxide solution (1 in 100) to volume, and mix. The concentration of USP Hydroflumethiazide RS in the *Standard preparation* is about 15 µg per mL.

Chromatographic column—Proceed as directed for *Column Partition Chromatography* under *Chromatography* ⟨621⟩, packing a chromatographic tube with two segments of packing material. The lower segment is a mixture of 1 g of *Solid Support* and 1 mL of sodium hydroxide solution (1 in 100), and the upper segment is a mixture prepared as directed under *Assay preparation.*

Assay preparation—Weigh and finely powder not less than 20 Hydroflumethiazide Tablets. Transfer an accurately weighed portion of the powder, equivalent to about 75 mg of hydroflumethiazide, to a 50-mL volumetric flask, add about 35 mL of sodium hydroxide solution (1 in 100), shake vigorously, dilute with sodium hydroxide solution (1 in 100) to volume, and mix. Mix 2.0 mL of this solution with 3 g of *Solid Support* as directed under *Chromatographic column*, and transfer to the column. Wash the column with 50 mL of water-saturated chloroform, then with 50 mL of water-saturated ether, and discard the eluates. Elute the hydroflumethiazide from the column with 100 mL of glacial acetic acid in ether (1 in 1000), collecting the eluate in a 250-mL separator. Add 100 mL of a 1 in 1000 solution of glacial acetic acid in ether to a second 250-mL separator to provide a blank, and treat each as follows: Add 60 mL of isooctane to each separator, mix, and extract the resulting solution with three 50-mL portions of sodium hydroxide solution (1 in 100), collecting the extracts in a 200-mL volumetric flask. Dilute with sodium hydroxide solution (1 in 100) to volume, and mix.

Procedure—Concomitantly determine the absorbances of the solutions in 1-cm cells at the wavelength of maximum absorbance at about 273 nm, with a suitable spectrophotometer, using the blank. Calculate the quantity, in mg, of $C_8H_8F_3N_3O_4S_2$ in the portion of Tablets taken by the formula:

$$5C(A_U/A_S),$$

in which C is the concentration, in µg per mL, of USP Hydroflumethiazide RS in the *Standard preparation*, and A_U and A_S are the absorbances of the *Assay preparation* and the *Standard preparation*, respectively.

Hydrogen Peroxide Concentrate

H_2O_2 34.01

Hydrogen peroxide.

Hydrogen peroxide [*7722-84-1*].

» Hydrogen Peroxide Concentrate contains not less than 29.0 percent and not more than 32.0 percent, by weight, of H_2O_2. It contains not more than 0.05 percent of a suitable preservative or preservatives. [*Caution—Hydrogen Peroxide Concentrate is a strong oxidant.*]

Packaging and storage—Preserve in partially-filled containers having a small vent in the closure, and store in a cool place.

Labeling—Label it to indicate the name and amount of any added preservative.

Acidity—Dilute 25 g with water to 250 mL, add phenolphthalein TS, and titrate with 0.10 *N* sodium hydroxide: not more than 2.5 mL is required for neutralization.

Chloride ⟨221⟩—1.5 g diluted with water to 25 mL shows no more chloride than 0.10 mL of 0.020 *N* hydrochloric acid (0.005%).

Other requirements—It responds to the *Identification test* and meets the requirements of the tests for *Nonvolatile residue, Heavy metals*, and *Limit of preservative* (90 mL of it being used) under *Hydrogen Peroxide Topical Solution.*

Assay—Weigh accurately about 1 mL of Hydrogen Peroxide Concentrate in a tared 100-mL volumetric flask, dilute with water to volume, and mix. To 20.0 mL of this solution add 20 mL of 2 *N* sulfuric acid, and titrate with 0.1 *N* potassium permanganate VS. Each mL of 0.1 *N* potassium permanganate is equivalent to 1.701 mg of H_2O_2.

Hydrogen Peroxide Topical Solution

H_2O_2 34.01

Hydrogen peroxide.

Hydrogen peroxide [*7722-84-1*].

» Hydrogen Peroxide Topical Solution contains, in each 100 mL, not less than 2.5 g and not more than 3.5 g of H_2O_2. It contains not more than 0.05 percent of a suitable preservative or preservatives.

Packaging and storage—Preserve in tight, light-resistant containers, at controlled room temperature.

Identification—Shake 1 mL with 10 mL of water containing 1 drop of 2 *N* sulfuric acid, and add 2 mL of ether: the subsequent addition of a drop of potassium dichromate TS produces an evanescent blue color in the water layer which upon agitation and standing passes into the ether layer.

Acidity—To 25 mL add phenolphthalein TS, and titrate with 0.10 *N* sodium hydroxide: not more than 2.5 mL is required for neutralization.

Nonvolatile residue—Evaporate 20 mL, previously shaken, on a steam bath to dryness, and dry the residue at 105° for 1 hour: the weight of the residue does not exceed 30 mg.

Barium—To 10 mL add two drops of 2 *N* sulfuric acid: no turbidity or precipitate is produced within 10 minutes.

Heavy metals ⟨231⟩—Dilute 4 mL, previously shaken, with 20 mL of water, add 2 mL of 6 *N* ammonium hydroxide, and gently boil the solution until the volume is reduced to about 5 mL. Dilute with water to 25 mL: the limit is 5 ppm.

Limit of preservative—Extract 100 mL of well-mixed Topical Solution in a separator with a mixture of 3 volumes of chloroform and 2 volumes of ether, using 50 mL, 25 mL, and 25 mL, respectively. Evaporate the combined extracts at room temperature in a tared glass dish to dryness, and dry over silica gel for 2 hours: the residue, if any, weighs not more than 50 mg (0.05%).

Assay—Pipet 2 mL of Hydrogen Peroxide Topical Solution into a suitable flask containing 20 mL of water. Add 20 mL of 2 *N* sulfuric acid, and titrate with 0.1 *N* potassium permanganate VS. Each mL of 0.1 *N* potassium permanganate is equivalent to 1.701 mg of H_2O_2.

Hydrogenated Castor Oil—*see* Castor Oil, Hydrogenated NF

Hydrogenated Vegetable Oil—*see* Vegetable Oil, Hydrogenated NF

Hydromorphone Hydrochloride

CH3 / H / N—CH2 / H / CH2 / HO / O / O · HCl

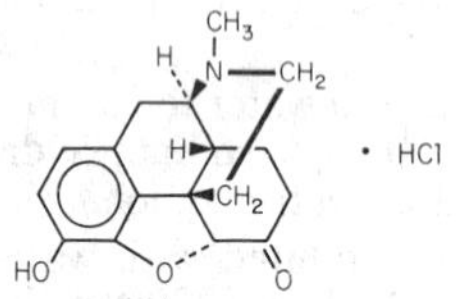

$C_{17}H_{19}NO_3.HCl$ 321.80

Morphinan-6-one, 4,5-epoxy-3-hydroxy-17-methyl-, hydrochloride, (5α)-.

4,5α-Epoxy-3-hydroxy-17-methylmorphinan-6-one hydrochloride [*71-68-1*].

» Hydromorphone Hydrochloride, dried at 105° for 2 hours, contains not less than 98.0 percent and not more than 101.0 percent of $C_{17}H_{19}NO_3.HCl$.

Packaging and storage—Preserve in tight, light-resistant containers.

Reference standard—*USP Hydromorphone Hydrochloride Reference Standard*—Dry at 105° for 2 hours before using.

Identification—

A: The infrared absorption spectrum of a potassium bromide dispersion of it, previously dried, exhibits maxima only at the same wavelengths as that of a similar preparation of USP Hydromorphone Hydrochloride RS.

B: The ultraviolet absorption spectrum of a solution (1 in 10,000) exhibits maxima and minima at the same wavelengths as that of a similar solution of USP Hydromorphone Hydrochloride RS, concomitantly measured, and the respective absorptivities, calculated on the dried basis, at the wavelength of maximum absorbance at about 280 nm do not differ by more than 3.0%.

C: A solution (1 in 20) responds to the tests for *Chloride* ⟨191⟩.

Specific rotation ⟨781⟩: between −136° and −139°, calculated on the dried basis, determined in a solution containing 500 mg in each 10 mL.

Acidity—Dissolve 300 mg in 10 mL of water, add 1 drop of methyl red TS, and titrate with 0.020 *N* sodium hydroxide VS: not more than 0.30 mL is required to produce a yellow color.

Loss on drying ⟨731⟩—Dry it at 105° for 2 hours: it loses not more than 1.5% of its weight.

Residue on ignition ⟨281⟩: not more than 0.3%.

Sulfate—To a solution of 100 mg in 5 mL of water add 0.5 mL of 3 *N* hydrochloric acid and 1 mL of barium chloride TS: no turbidity is produced.

Ordinary impurities ⟨466⟩—

Test solution: water.

Standard solution: water.

Eluant: a mixture of methylene chloride, methanol, and ammonium hydroxide (90:10:1).

Visualization: 3.

Assay—Transfer about 225 mg of Hydromorphone Hydrochloride, previously dried and accurately weighed, to a 250-mL conical flask, and dissolve in 80 mL of glacial acetic acid, warming, if necessary. Cool, and add 5 mL of acetic anhydride and 10 mL of mercuric acetate TS. Add 1 drop of crystal violet TS, and titrate with 0.1 *N* perchloric acid VS to a blue end-point. Perform a blank determination, and make any necessary correction. Each mL of 0.1 *N* perchloric acid is equivalent to 32.18 mg of $C_{17}H_{19}NO_3.HCl$.

Hydromorphone Hydrochloride Injection

» Hydromorphone Hydrochloride Injection is a sterile solution of Hydromorphone Hydrochloride in Water for Injection. It contains not less than 95.0 percent and not more than 105.0 percent of the labeled amount of $C_{17}H_{19}NO_3.HCl$.

Packaging and storage—Preserve in single-dose or in multiple-dose containers, preferably of Type I glass, protected from light.

Reference standard—*USP Hydromorphone Hydrochloride Reference Standard*—Dry at 105° for 2 hours before using.

Identification—Place a volume of Injection, equivalent to about 10 mg of hydromorphone hydrochloride, in a separator. Extract with four 10-mL portions of chloroform, and discard the extracts. Add 1 mL of sodium carbonate TS, and extract with three 10-mL portions of chloroform. Filter the chloroform extracts into a glass-stoppered, 50-mL flask, and evaporate on a steam bath with the aid of a current of air to dryness. Dissolve the residue in 1 mL of chloroform: the infrared absorption spectrum of the solution so obtained exhibits maxima only at the same wavelengths as that of a similar preparation of USP Hydromorphone Hydrochloride RS.

pH ⟨791⟩: between 3.5 and 5.5.

Other requirements—It meets the requirements under *Injections* ⟨1⟩.

Assay—

Standard preparation—Using an accurately weighed quantity of USP Hydromorphone Hydrochloride RS, prepare a solution in water having a known concentration of about 1 mg per mL.

Assay preparation—Transfer an accurately measured volume of Hydromorphone Hydrochloride Injection, equivalent to about 25 mg of hydromorphone hydrochloride, to a 25-mL volumetric flask, dilute with water to volume, and mix.

Procedure—Transfer 4.0 mL each of the *Standard preparation* and the *Assay preparation* to separate 50-mL volumetric flasks. To each flask add, with mixing, 10 mL of 1 *N* hydrochloric acid and 1.0 mL of sodium nitrite solution (1 in 20). Insert the stoppers, allow to stand for 40 to 45 minutes, with occasional swirling, then add 2 mL of ammonium hydroxide, and mix. Allow to stand for 1 minute, then dilute with water to volume, and mix. Concomitantly determine the absorbances of the solutions in 1-cm cells at the wavelength of maximum absorbance at about 440 nm, with a suitable spectrophotometer, using water as the blank. Calculate the quantity, in mg, of $C_{17}H_{19}NO_3.HCl$ in each mL of the Injection taken by the formula:

$$(0.025C/V)(A_U/A_S),$$

in which C is the concentration, in µg per mL, of USP Hydromorphone Hydrochloride RS in the *Standard preparation*, V is the volume, in mL, of Injection taken, and A_U and A_S are the absorbances of the solutions from the *Assay preparation* and the *Standard preparation*, respectively.

Hydromorphone Hydrochloride Tablets

» Hydromorphone Hydrochloride Tablets contain not less than 90.0 percent and not more than 110.0 percent of the labeled amount of $C_{17}H_{19}NO_3.HCl$.

Packaging and storage—Preserve in tight, light-resistant containers.

Reference standard—*USP Hydromorphone Hydrochloride Reference Standard*—Dry at 105° for 2 hours before using.

Identification—Place a quantity of finely powdered Tablets, equivalent to about 10 mg of hydromorphone hydrochloride, in a separator, and proceed as directed in the *Identification test* under *Hydromorphone Hydrochloride Injection*, beginning with "Extract with four 30-mL portions of chloroform."

Dissolution ⟨711⟩—

Medium: water; 500 mL.

Apparatus 2: 50 rpm.

Time: 45 minutes.

Standard preparation—Dissolve an accurately weighed quantity of USP Hydromorphone Hydrochloride RS in *Dissolution Medium*, and dilute quantitatively with *Dissolution Medium* to obtain a solution having a known concentration of about 0.10 mg per mL.

Procedure—Pipet into each of two separate containers a volume of a filtered portion of the solution under test that is estimated to contain about 0.1 mg of hydromorphone hydrochloride and add water, if necessary, to bring each to a volume of 50.0 mL. Label one container as the Test solution and one as the Test blank. Pipet 1 mL of the *Standard preparation* into each of two separate containers, and add water to bring each to a volume of 50.0 mL. Label one container as the Standard solution and one as the Standard blank. Add 4.0 mL of 1.0 *N* hydrochloric acid to all four containers, and add 1.0 mL of sodium nitrite solution (1 in 20) to the Test solution and the Standard solution. Mix, and allow to stand for 15 minutes, accurately timed. Add 2.0 mL of ammonium hydroxide to each of the four containers, mix, and determine the absorbances of the solutions at the wavelength of maximum absorbance at about 440 nm. Calculate the amount of hydromorphone hydrochloride dissolved by comparison of the absorbances obtained from the Test solution and the Standard solution, using the respective blank solutions to correct the absorbances.

Tolerances—Not less than 75% (*Q*) of the labeled amount of $C_{17}H_{19}NO_3.HCl$ is dissolved in 45 minutes.

Uniformity of dosage units ⟨905⟩: meet the requirements.

Assay—

Standard preparation—Prepare as directed in the *Assay* under *Hydromorphone Hydrochloride Injection.*

Assay preparation—Weigh and finely powder not less than 20 Hydromorphone Hydrochloride Tablets. Transfer an accurately weighed portion of the powder, equivalent to about 50 mg of hydromorphone hydrochloride, to a 50-mL volumetric flask, add water to volume, mix, and filter.

Procedure—Proceed as directed for *Procedure* in the *Assay* under *Hydromorphone Hydrochloride Injection.* Calculate the quantity, in mg, of $C_{17}H_{19}NO_3.HCl$ in the portion of Tablets taken by the formula:

$$0.05C(A_U/A_S),$$

in which C is the concentration, in µg per mL, of USP Hydromorphone Hydrochloride RS in the *Standard preparation*, and A_U and A_S are the absorbances of the solutions from the *Assay preparation* and the *Standard preparation*, respectively.

Hydrophilic Ointment—*see* Ointment, Hydrophilic

Hydrophilic Petrolatum—*see* Petrolatum, Hydrophilic

Hydroquinone

$C_6H_6O_2$ 110.11
1,4-Benzenediol.
Hydroquinone [*123-31-9*].

» Hydroquinone contains not less than 99.0 percent and not more than 100.5 percent of $C_6H_6O_2$, calculated on the anhydrous basis.

Packaging and storage—Preserve in tight, light-resistant containers.

Reference standard—*USP Hydroquinone Reference Standard*—Do not dry; determine the water content by the titrimetric method before using for quantitative analyses.

Identification—

A: The infrared absorption spectrum of a potassium bromide dispersion of it exhibits maxima only at the same wavelengths as that of a similar preparation of USP Hydroquinone RS.

B: Prepare a solution of it in methanol containing approximately 1 mg per mL, and prepare a similar solution of USP Hydroquinone RS. Apply 5 µL of each solution on a suitable thin-layer chromatographic plate (see *Chromatography* ⟨621⟩) coated with a 0.25-mm layer of chromatographic silica gel. Allow the spots to dry, and develop the chromatogram in a solvent system consisting of equal volumes of methanol and chloroform until the solvent front has moved about three-fourths of the length of the plate. Remove the plate from the developing chamber, mark the solvent front, and allow the solvent to evaporate. Heat on a hot plate or under a lamp until spots appear: the R_f value of the principal spot obtained from the solution under test corresponds to that obtained from the Standard solution.

C: A 1 in 40,000 solution in methanol exhibits an absorbance maximum at 293 ± 2 nm.

Melting range ⟨741⟩: between 172° and 174°.

Water, *Method I* ⟨921⟩: not more than 0.5%.

Residue on ignition ⟨281⟩: not more than 0.5%.

Assay—Dissolve about 250 mg of Hydroquinone, accurately weighed, in a mixture of 100 mL of water and 10 mL of 0.1 *N* sulfuric acid, add 3 drops of diphenylamine TS, and titrate with 0.1 *N* ceric sulfate VS until a red-violet end-point is reached. Perform a blank determination, and make any necessary correction. Each mL of 0.1 *N* ceric sulfate is equivalent to 5.506 mg of $C_6H_6O_2$.

Hydroquinone Cream

» Hydroquinone Cream contains not less than 94.0 percent and not more than 106.0 percent of the labeled amount of $C_6H_6O_2$.

Packaging and storage—Preserve in well-closed, light-resistant containers.

Reference standard—*USP Hydroquinone Reference Standard*—Do not dry; determine the water content by the titrimetric method before using for quantitative analyses.

Identification—Dissolve a portion of Cream equivalent to about 50 mg of hydroquinone, in a mixture of equal volumes of methanol and chloroform to make 50 mL: a 5-µL portion of this solution responds to *Identification test B* under *Hydroquinone.*

Minimum fill ⟨755⟩: meets the requirements.

Assay—

Standard preparation—Dissolve a suitable quantity of USP Hydroquinone RS in methanol, and dilute quantitatively and stepwise with methanol to obtain a solution having a known concentration of about 10 µg per mL.

Assay preparation—Transfer an accurately weighed portion of Hydroquinone Cream, equivalent to about 20 mg of hydroquinone, to a 100-mL beaker. Triturate the Cream with 50 mL of methanol, and filter the liquid through folded filter paper, previously washed with methanol, into a 500-mL volumetric flask. Repeat the trituration and filtration. Dilute, by washing the contents of the filter paper with methanol through the paper into the volumetric flask, to volume, and mix. Pipet 25 mL of this solution into a 100-mL volumetric flask, add methanol to volume, and mix.

Procedure—Concomitantly determine the absorbances of the *Standard preparation* and the *Assay preparation* in 1-cm cells at the wavelength of maximum absorbance at about 293 nm, with a suitable spectrophotometer, using methanol as the blank. Calculate the quantity, in mg, of $C_6H_6O_2$ in each g of the Cream taken by the formula:

$$2(C/W)(A_U/A_S),$$

in which *C* is the concentration, in µg per mL, of USP Hydroquinone RS in the *Standard preparation*, *W* is the weight, in g, of Cream taken, and A_U and A_S are the absorbances of the *Assay preparation* and the *Standard preparation*, respectively.

Hydroquinone Topical Solution

» Hydroquinone Topical Solution contains not less than 95.0 percent and not more than 110.0 percent of the labeled amount of $C_6H_6O_2$.

Packaging and storage—Preserve in tight, light-resistant containers.

Reference standard—*USP Hydroquinone Reference Standard*—Do not dry; determine the water content by the titrimetric method before using for quantitative analyses.

Identification—The retention time of the major peak in the chromatogram of the *Assay preparation* corresponds to that of the *Standard preparation*, as obtained in the *Assay*.

pH ⟨791⟩: between 3.0 and 4.2.

Assay—

Mobile phase—Mix 55 volumes of methanol and 45 volumes of water.

Standard preparation—Transfer about 250 mg of USP Hydroquinone RS, accurately weighed, to a 25-mL volumetric flask, dilute with *Mobile phase* to volume, and mix. Transfer 3.0 mL of this solution to a 100-mL volumetric flask, dilute with *Mobile phase* to volume, and mix.

Assay preparation—Transfer an accurately measured volume of Hydroquinone Topical Solution, equivalent to about 30 mg of hydroquinone, to a 100-mL volumetric flask, dilute with *Mobile phase* to volume, and mix.

Chromatographic system (see *Chromatography* ⟨621⟩)—The liquid chromatograph is equipped with a 280-nm detector and a 4-mm × 30-cm column that contains packing L1. The flow rate is about 0.8 mL per minute. Chromatograph three replicate injections of the *Standard preparation*, and record the peak responses as directed under *Procedure:* the relative standard deviation is not more than 3.0%.

Procedure—Separately inject equal volumes (about 10 µL) of the *Standard preparation* and the *Assay preparation* into the chromatograph by means of a suitable microsyringe or sampling valve, record the chromatograms, and measure the responses for the major peaks. The retention time is about 4 minutes for hydroquinone. Calculate the quantity, in mg, of $C_6H_6O_2$ in each mL of the Topical Solution taken by the formula:

$$100(C/V)(r_U/r_S),$$

in which *C* is the concentration, in mg per mL, of USP Hydroquinone RS in the *Standard preparation*, *V* is the volume, in mL, of Topical Solution taken, and r_U and r_S are the peak responses of hydroquinone obtained from the *Assay preparation* and the *Standard preparation*, respectively.

Hydrous Benzoyl Peroxide—*see* Benzoyl Peroxide, Hydrous

Hydroxocobalamin

$C_{62}H_{89}CoN_{13}O_{15}P$ 1346.37

Cobinamide, dihydroxide, dihydrogen phosphate (ester), mono(inner salt), 3′-ester with 5,6-dimethyl-1-α-D-ribofuranosyl-1*H*-benzimidazole.

Cobinamide dihydroxide dihydrogen phosphate (ester), mono(inner salt), 3′-ester with 5,6-dimethyl-1-α-D-ribofuranosylbenzimidazole [*13422-51-0*].

» Hydroxocobalamin contains not less than 95.0 percent and not more than 102.0 percent of $C_{62}H_{89}CoN_{13}O_{15}P$, calculated on the dried basis.

Packaging and storage—Preserve in tight, light-resistant containers, and store in a cool place.

Reference standard—*USP Cyanocobalamin Reference Standard*—Dry over silica gel for 4 hours before using.

Identification—

A: The visible absorption spectrum of the solution, prepared for measurement of the absorption as directed under *pH-dependent cobalamins*, exhibits maxima at 426 ± 2 nm, 516 ± 2 nm, and 550 ± 2 nm.

B: Fuse a mixture of about 1 mg of Hydroxocobalamin and about 50 mg of potassium pyrosulfate in a porcelain crucible. Cool, break up the mass with a glass rod, add 3 mL of water, and boil until dissolved. Add 1 drop of phenolphthalein TS, and add 2 *N* sodium hydroxide dropwise until a pink color appears. Add 0.5 g of sodium acetate, 0.5 mL of 1 *N* acetic acid, and 0.5 mL of nitroso R salt solution (1 in 100): a red or orange-red color appears immediately. Add 0.5 mL of hydrochloric acid, and boil for 1 minute: the red or orange-red color persists.

pH ⟨791⟩: between 8.0 and 10.0, in a solution (2 in 100).

Loss on drying ⟨731⟩—Dry it at a pressure below 5 mm of mercury at 100° for 2 hours: it loses between 14.0% and 18.0% of its weight.

pH-dependent cobalamins—

pH 4.0 buffer—Dissolve 2.61 g of sodium acetate and 20.5 g of sodium chloride in 5.25 mL of glacial acetic acid and sufficient water to make 1500 mL of solution, and mix.

pH 9.3 buffer—Dissolve 23.8 g of sodium borate and 402 mg of boric acid in sufficient water to make 1500 mL of solution, and mix.

Procedure—[NOTE—Perform the following test in subdued light.] Transfer about 40 mg of Hydroxocobalamin, accurately weighed, to a 25-mL volumetric flask, dissolve in carbon dioxide–free water, dilute with carbon dioxide–free water to volume, and mix. Transfer 1.0-mL portions of this solution to each of two glass-stoppered test tubes. To one of the tubes, designated *B*, add 3.0 mL of *pH 4.0 buffer*, and mix. To the other tube, designated *U*, add 3.0 mL of *pH 9.3 buffer*, and mix. Determine the absorbance of solution *U*, in a 1-cm cell, at the wavelength of maximum absorbance at about 550 nm, with a suitable spectrophotometer, using solution *B* as the blank. Calculate the percentage of pH-dependent cobalamins, as hydroxocobalamin, by the formula:

$$(100{,}000A)/(19.66W),$$

in which A is the absorbance of solution U, and W is the weight, in mg, of Hydroxocobalamin taken: the content, calculated on the dried basis, is between 95.0% and 102.0%.

Cyanocobalamin—

Cyanocobalamin tracer reagent, Cresol–carbon tetrachloride solution, Butanol–benzalkonium chloride solution, and *Alumina-resin column*—Prepare as directed under *Cobalamin Radiotracer Assay* ⟨371⟩.

Procedure—Transfer about 50 mg of Hydroxocobalamin, accurately weighed, to a 25-mL volumetric flask, dissolve in water, dilute with water to volume, and mix. Transfer 5.0 mL of this solution to a glass-stoppered, 50-mL centrifuge tube, and add 5.0 mL of *Cyanocobalamin tracer reagent* and 15 mL of *Cresol–carbon tetrachloride solution.* Insert the stopper, shake gently, centrifuge, carefully remove the upper, aqueous layer by aspiration, and discard the aspirated liquid. Add 25 mL of 5 N sulfuric acid, insert the stopper, shake gently, centrifuge, and remove and discard the upper, aqueous layer. Repeat the washing with additional 25-mL portions of the 5 N sulfuric acid until the acid wash is colorless (six to eight washings), and discard the acid washings. Add *Cresol–carbon tetrachloride solution* as necessary during the acid washings to maintain the volume of this phase at not less than 10 mL. Wash this solution successively with two 10-mL portions of saturated dibasic sodium phosphate solution and one 10-mL portion of water, and discard all of the aqueous washings. Proceed as directed for *Procedure* under *Cobalamin Radiotracer Assay* ⟨371⟩, beginning with "To the washed extract add 30 mL of a mixture of 2 volumes of *Butanol–Benzalkonium Chloride Solution* and 1 volume of carbon tetrachloride."

Calculation—Calculate the cyanocobalamin content, in µg, of the Hydroxocobalamin taken by the formula:

$$R(C_S/C_U)(A_U/A_S),$$

in which R is the quantity, in µg, of cyanocobalamin in the portion of the Standard solution taken, C_S and C_U are the corrected average radioactivity values, expressed in counts per minute per mL, of the Standard solution and test solution, respectively, and A_U and A_S are the absorbances, determined at 361 nm, of the test solution and the Standard solution, respectively: the limit, calculated on the dried basis, is 5.0%.

Assay—

Cyanocobalamin tracer reagent, Cresol–carbon tetrachloride solution, Phosphate-cyanide solution, Butanol–benzalkonium chloride solution, and *Alumina-resin column*—Prepare as directed under *Cobalamin Radiotracer Assay* ⟨371⟩.

Assay preparation—Transfer about 40 mg of Hydroxocobalamin, accurately weighed, to a 2000-mL volumetric flask, dissolve in water, dilute with water to volume, and mix. Transfer 25.0 mL of this solution to a beaker, add 5.0 mL of *Cyanocobalamin tracer reagent,* and proceed as directed for *Assay Preparation* under *Cobalamin Radiotracer Assay* ⟨371⟩, beginning with "Add, while working *under a hood,* 5 mg of sodium nitrite."

Procedure—Proceed as directed for *Procedure* under *Cobalamin Radiotracer Assay* ⟨371⟩. Calculate the quantity, in µg, of $C_{62}H_{89}CoN_{13}O_{15}P$ in the Hydroxocobalamin taken by the formula:

$$(1346.37/1355.38)(R)(C_S/C_U)(A_U/A_S),$$

in which 1346.37 and 1355.38 are the molecular weights of hydroxocobalamin and cyanocobalamin, respectively, R is the quantity, in µg, of cyanocobalamin in the portion of the Standard solution taken, C_S and C_U are the corrected average radioactivity values, expressed in counts per minute per mL, of the Standard solution and test solution, respectively, and A_U and A_S are the absorbances, determined at 361 nm, of the test solution and the Standard solution, respectively.

Hydroxocobalamin Injection

» Hydroxocobalamin Injection is a sterile solution of Hydroxocobalamin in Water for Injection. It contains not less than 95.0 percent and not more than 115.0 percent of the labeled amount of $C_{62}H_{89}CoN_{13}O_{15}P$.

Packaging and storage—Preserve in single-dose or in multiple-dose containers, preferably of Type I glass, protected from light.

Reference standard—*USP Cyanocobalamin Reference Standard*—Dry over silica gel for 4 hours before using.

Identification—Dilute 3.0 mL of Injection to 100 mL with pH 4.0 buffer (prepared by dissolving 2.61 g of sodium acetate and 20.5 g of sodium chloride in 5.25 mL of glacial acetic acid and sufficient water to make 1500 mL of solution): the ultraviolet-visible absorption spectrum of this solution exhibits maxima at 352 ± 1 nm and 528 ± 2 nm. The ratio A_{352}/A_{528} is between 2.7 and 3.3.

pH ⟨791⟩: between 3.5 and 5.0.

Other requirements—It meets the requirements under *Injections* ⟨1⟩.

Assay—

pH 9.3 buffer—Dissolve 23.8 g of sodium borate and 402 mg of boric acid in sufficient water to make 1500 mL of solution, and mix.

Standard preparation—Dissolve in *pH 9.3 buffer* a suitable quantity of USP Cyanocobalamin RS, accurately weighed, and dilute quantitatively and stepwise, if necessary, to obtain a solution having a known concentration of about 30 µg per mL.

Assay preparation—Transfer an accurately measured volume of Hydroxocobalamin Injection, equivalent to about 5 mg of hydroxocobalamin, to a 50-mL volumetric flask containing about 25 mL of *pH 9.3 buffer.* Add 5.0 mL of potassium cyanide solution (1 in 10,000), allow to stand at room temperature for 30 minutes, dilute with *pH 9.3 buffer* to volume, and mix. Transfer 15.0 mL of this solution to a second 50-mL volumetric flask, dilute with *pH 9.3 buffer* to volume, and mix.

Procedure—Concomitantly determine the absorbances of the solutions in 1-cm cells at the wavelength of maximum absorbance at about 361 nm, with a suitable spectrophotometer, using *pH 9.3 buffer* as the blank. Calculate the quantity, in mg, of $C_{62}H_{89}CoN_{13}O_{15}P$ in each mL of the Injection taken by the formula:

$$(1346.37/1355.38)(0.1667C/V)(A_U/A_S),$$

in which 1346.37 and 1355.38 are the molecular weights of hydroxocobalamin and cyanocobalamin, respectively, C is the concentration, in µg per mL, of USP Cyanocobalamin RS in the *Standard preparation,* V is the volume, in mL, of Injection taken, and A_U and A_S are the absorbances of the *Assay preparation* and the *Standard preparation,* respectively.

Hydroxyamphetamine Hydrobromide

$C_9H_{13}NO.HBr$ 232.12

Phenol, 4-(2-aminopropyl)-, hydrobromide.

(±)-*p*-(2-Aminopropyl)phenol hydrobromide [*306-21-8*].

» Hydroxyamphetamine Hydrobromide contains not less than 98.0 percent and not more than 101.5 percent of $C_9H_{13}NO.HBr$, calculated on the dried basis.

Packaging and storage—Preserve in well-closed, light-resistant containers.

Reference standard—*USP Hydroxyamphetamine Hydrobromide Reference Standard*—Dry at 105° for 2 hours before using.

Identification—

A: The infrared absorption spectrum of a potassium bromide dispersion of it, previously dried, exhibits maxima only at the same wavelengths as that of a similar preparation of USP Hydroxyamphetamine Hydrobromide RS.

B: Dissolve about 500 mg of ammonium molybdate in 10 mL of sulfuric acid, and add to this solution about 2 mg of Hydroxyamphetamine Hydrobromide: an intense blue color is produced (*distinction from similar amino compounds such as amphetamine and methamphetamine, which, lacking a phenolic hydroxyl, do not undergo this reaction*).

C: Dissolve about 200 mg in 2 mL of water, and add a solution of 500 mg of potassium carbonate in 2 mL of water. Extract with two 10-mL portions of ether, allow the clear ether solution to evaporate to dryness, and dry at about 80°: the hydroxyamphetamine so obtained melts between 124° and 127° (see *Class I* under *Melting Range or Temperature* ⟨741⟩).

D: To a solution of about 10 mg of it in 10 mL of water add 1 mL of 2 *N* nitric acid, then add silver nitrate TS: a pale yellow precipitate is formed, and it is slightly soluble in 6 *N* ammonium hydroxide.

Melting range ⟨741⟩: between 189° and 192°.

Loss on drying ⟨731⟩—Dry it at 105° for 2 hours: it loses not more than 0.5% of its weight.

Residue on ignition ⟨281⟩: not more than 0.1%.

Bromide content—Weigh accurately about 400 mg, and dissolve in 50 mL of water. Add 50 mL of methanol and 10 mL of glacial acetic acid, then add eosin Y TS, and titrate with 0.1 *N* silver nitrate VS. Each mL of 0.1 *N* silver nitrate is equivalent to 7.990 mg of Br: the content of Br, calculated on the dried basis, is between 33.6% and 35.2%.

Ordinary impurities ⟨466⟩—

Test solution: methanol.

Standard solution: methanol.

Eluant: a mixture of toluene, methanol, and ammonium hydroxide (10:4:0.25).

Visualization: 1.

Assay—Dissolve about 400 mg of Hydroxyamphetamine Hydrobromide, accurately weighed, in a mixture of 10 mL of glacial acetic acid and 10 mL of mercuric acetate TS, warming slightly, if necessary, to effect solution. Add crystal violet TS, and titrate with 0.1 *N* perchloric acid VS. Perform a blank determination, and make any necessary correction. Each mL of 0.1 *N* perchloric acid is equivalent to 23.21 mg of $C_9H_{13}NO.HBr$.

Hydroxyamphetamine Hydrobromide Ophthalmic Solution

» Hydroxyamphetamine Hydrobromide Ophthalmic Solution is a sterile, buffered, aqueous solution of Hydroxyamphetamine Hydrobromide. It contains not less than 95.0 percent and not more than 105.0 percent of the labeled amount of $C_9H_{13}NO.HBr$. It contains a suitable antimicrobial agent.

Packaging and storage—Preserve in tight, light-resistant containers.

Reference standard—*USP Hydroxyamphetamine Hydrobromide Reference Standard*—Dry at 105° for 2 hours before using.

Identification—

A: Dissolve about 500 mg of ammonium molybdate in 10 mL of sulfuric acid, and add 0.2 mL of Ophthalmic Solution: an intense blue color is produced (*distinction from similar amino compounds such as amphetamine and methamphetamine, which, lacking a phenolic hydroxyl, do not undergo this reaction*).

B: The dried diacetylhydroxyamphetamine obtained in the *Assay* melts between 96° and 100° (see *Class I* under *Melting Range or Temperature* ⟨741⟩), but the range between beginning and end of melting does not exceed 2.0°.

C: It responds to *Identification test D* under *Hydroxyamphetamine Hydrobromide.*

D: Dilute a volume of Ophthalmic Solution, equivalent to about 50 mg of hydroxyamphetamine hydrobromide, with 0.01 *N* hydrochloric acid to 25 mL, and proceed as directed under *Identification—Organic Nitrogenous Bases* ⟨181⟩, using sodium carbonate TS in place of 1 *N* sodium hydroxide, beginning with "Transfer the liquid to a separator": the Ophthalmic Solution meets the requirements of the test.

Sterility—It meets the requirements under *Sterility Tests* ⟨71⟩.

pH ⟨791⟩: between 4.2 and 6.0.

Assay—Transfer an accurately measured volume of Hydroxyamphetamine Hydrobromide Ophthalmic Solution, equivalent to about 100 mg of hydroxyamphetamine hydrobromide, to a 125-mL separator. Wash the solution with 15 mL of chloroform, and discard the washing. Rinse the stopper and the mouth of the separator with a few drops of water. Add 1.05 g of sodium bicarbonate, preventing it from coming in contact with the mouth of the separator, and swirl until most of the bicarbonate has dissolved. By means of a 1-mL syringe, rapidly inject 0.5 mL of acetic anhydride directly into the contents of the separator. Immediately insert the stopper in the separator, and shake vigorously until the evolution of carbon dioxide has ceased (7 to 10 minutes), releasing the pressure as necessary through the stopcock. Allow to stand for 5 minutes, and extract the solution with five 10-mL portions of chloroform, filtering each extract through a pledget of cotton, previously washed with chloroform, into a tared 100-mL beaker. Evaporate the combined chloroform extracts on a steam bath in a current of air or stream of nitrogen to dryness. Dry the residue at 80° for 90 minutes, cool in a desiccator, and weigh. The weight of the diacetylhydroxyamphetamine so obtained, multiplied by 0.9866, represents the weight of $C_9H_{13}NO.HBr$ in the volume of Ophthalmic Solution taken.

Hydroxychloroquine Sulfate

$C_{18}H_{26}ClN_3O.H_2SO_4$ 433.95

Ethanol, 2-[[4-[(7-chloro-4-quinolinyl)amino]pentyl]-ethylamino]-, sulfate (1:1) salt.

2-[[4-[(7-Chloro-4-quinolyl)amino]pentyl]ethylamino]ethanol sulfate (1:1) (salt) [*747-36-4*].

» Hydroxychloroquine Sulfate contains not less than 98.0 percent and not more than 102.0 percent of $C_{18}H_{26}ClN_3O.H_2SO_4$, calculated on the dried basis.

Packaging and storage—Preserve in well-closed, light-resistant containers.

Reference standard—*USP Hydroxychloroquine Sulfate Reference Standard*—Dry at 105° for 2 hours before using.

Identification—

A: The ultraviolet absorption spectrum of a 1 in 100,000 solution in dilute hydrochloric acid (1 in 100) exhibits maxima and minima at the same wavelengths as that of a similar solution of USP Hydroxychloroquine Sulfate RS, concomitantly measured.

B: It meets the requirements under *Identification—Organic Nitrogenous Bases* ⟨181⟩.

C: A solution (1 in 100) responds to the tests for *Sulfate* ⟨191⟩.

Loss on drying ⟨731⟩—Dry it at 105° for 2 hours: it loses not more than 2.0% of its weight.

Ordinary impurities ⟨466⟩—

Test solution: 10% water in methanol.

Standard solution: 10% water in methanol.

Eluant: a mixture of alcohol, water, and ammonium hydroxide (80:16:4).

Visualization: 1.

Assay—Dissolve about 100 mg of Hydroxychloroquine Sulfate, accurately weighed, in about 5 mL of water, and dilute quantitatively and stepwise with dilute hydrochloric acid (1 in 100) to obtain a solution containing about 10 µg per mL. Similarly prepare a Standard solution of USP Hydroxychloroquine Sulfate RS. Concomitantly determine the absorbances of both solutions in 1-cm cells at the wavelength of maximum absorbance at about

343 nm, with a suitable spectrophotometer, using dilute hydrochloric acid (1 in 100) as the blank. Calculate the quantity, in mg, of $C_{18}H_{26}ClN_3O.H_2SO_4$ in the portion of Hydroxychloroquine Sulfate taken by the formula:

$$10C(A_U/A_S),$$

in which C is the concentration, in µg per mL, of USP Hydroxychloroquine Sulfate RS in the Standard solution, and A_U and A_S are the absorbances of the solution of Hydroxychloroquine Sulfate and the Standard solution, respectively.

Hydroxychloroquine Sulfate Tablets

» Hydroxychloroquine Sulfate Tablets contain not less than 93.0 percent and not more than 107.0 percent of the labeled amount of $C_{18}H_{26}ClN_3O.H_2SO_4$.

Packaging and storage—Preserve in well-closed containers.

Reference standard—*USP Hydroxychloroquine Sulfate Reference Standard*—Dry at 105° for 2 hours before using.

Identification—

A: Triturate a quantity of finely powdered Tablets, equivalent to about 1 g of hydroxychloroquine sulfate, with 50 mL of water, and filter: the clear filtrate so obtained responds to *Identification tests B* and *C* under *Hydroxychloroquine Sulfate*.

B: The solution employed for measurement of absorbance in the *Assay* exhibits an absorbance maximum at 343 ± 2 nm.

Dissolution ⟨711⟩—

Medium: water; 900 mL.

Apparatus 2: 100 rpm.

Time: 60 minutes.

Procedure—Determine the amount of $C_{18}H_{26}ClN_3O.H_2SO_4$ dissolved from ultraviolet absorbances at the wavelength of maximum absorbance at about 343 nm of filtered portions of the solution under test, suitably diluted with *Dissolution Medium*, if necessary, in comparison with a Standard solution having a known concentration of USP Hydroxychloroquine Sulfate RS in the same medium.

Tolerances—Not less than 70% (Q) of the labeled amount of $C_{18}H_{26}ClN_3O.H_2SO_4$ is dissolved in 60 minutes.

Uniformity of dosage units ⟨905⟩: meet the requirements.

Assay—Weigh and finely powder not less than 20 Hydroxychloroquine Sulfate Tablets. Weigh accurately a portion of the powder, equivalent to about 800 mg of hydroxychloroquine sulfate, transfer to a 200-mL volumetric flask, add about 100 mL of water, and shake by mechanical means for about 20 minutes. Add water to volume, mix, and filter, discarding the first 50 mL of filtrate. Pipet 50 mL of the clear filtrate into a 250-mL separator, and proceed as directed in the *Assay* under *Chloroquine Phosphate Tablets*, beginning with "add 5 mL of 6 *N* ammonium hydroxide," but use USP Hydroxychloroquine Sulfate RS in preparing the Standard solution. Calculate the quantity, in mg, of $C_{18}H_{26}ClN_3O.H_2SO_4$ in the portion of Tablets taken by the formula:

$$80C(A_U/A_S),$$

in which C is the concentration, in µg per mL, of USP Hydroxychloroquine Sulfate RS in the Standard solution, and A_U and A_S are the absorbances of the solution from the Tablets and the Standard solution, respectively.

Hydroxyethyl Cellulose—*see* Hydroxyethyl Cellulose NF

Hydroxyprogesterone Caproate

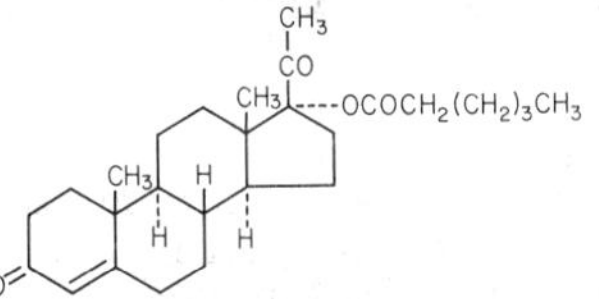

$C_{27}H_{40}O_4$ 428.61

Pregn-4-ene-3,20-dione, 17-[(1-oxohexyl)oxy]-.

17-Hydroxypregn-4-ene-3,20-dione hexanoate [*630-56-8*].

» Hydroxyprogesterone Caproate contains not less than 97.0 percent and not more than 103.0 percent of $C_{27}H_{40}O_4$, calculated on the anhydrous basis.

Packaging and storage—Preserve in well-closed, light-resistant containers.

Reference standard—*USP Hydroxyprogesterone Caproate Reference Standard*—Dry in vacuum over silica gel for 4 hours before using.

Identification—The infrared absorption spectrum of a potassium bromide dispersion of it, the test specimen previously having been dried, exhibits maxima only at the same wavelengths as that of a similar preparation of USP Hydroxyprogesterone Caproate RS.

Melting range ⟨741⟩: between 120° and 124°.

Specific rotation ⟨781⟩: between +58° and +64°, calculated on the anhydrous basis, determined in a solution in chloroform containing 100 mg in each 10 mL.

Water, *Method I* ⟨921⟩: not more than 0.1%.

Free *n*-caproic acid—Dissolve 0.20 g in 25 mL of alcohol that previously has been neutralized to a faint pink color following the addition of 2 or 3 drops of phenolphthalein TS. Promptly titrate with 0.020 *N* sodium hydroxide: not more than 0.50 mL of 0.020 *N* sodium hydroxide is required (0.58%).

Ordinary impurities ⟨466⟩—

Test solution: chloroform.

Standard solution: chloroform.

Eluant: a mixture of chloroform and ethyl acetate (3:1).

Visualization: 5; then view under long-wavelength ultraviolet light.

Assay—Transfer about 50 mg of Hydroxyprogesterone Caproate, accurately weighed, to a 100-mL volumetric flask, add alcohol to volume, and mix. Dilute 2.0 mL of this solution with alcohol to volume in a second 100-mL volumetric flask, and mix. Dissolve in alcohol a suitable quantity of USP Hydroxyprogesterone Caproate RS, and dilute quantitatively and stepwise with alcohol to obtain a Standard solution having a known concentration of about 10 µg per mL. Concomitantly determine the absorbances of both solutions in 1-cm cells at the wavelength of maximum absorbance at about 240 nm, using alcohol as the blank. Calculate the quantity, in mg, of $C_{27}H_{40}O_4$ in the Hydroxyprogesterone Caproate taken by the formula:

$$5C(A_U/A_S),$$

in which C is the concentration, in µg per mL, of USP Hydroxyprogesterone Caproate RS in the Standard solution, and A_U and A_S are the absorbances of the solution of Hydroxyprogesterone Caproate and the Standard solution, respectively.

Hydroxyprogesterone Caproate Injection

» Hydroxyprogesterone Caproate Injection is a sterile solution of Hydroxyprogesterone Caproate in a suitable vegetable oil. It contains not less than 90.0 percent and not more than 110.0 percent of the labeled amount of $C_{27}H_{40}O_4$.

Packaging and storage—Preserve in single-dose or in multiple-dose containers, preferably of Type I or Type III glass.

Reference standard—*USP Hydroxyprogesterone Caproate Reference Standard*—Dry in vacuum over silica gel for 4 hours before using.

Identification—

A: Transfer a volume of Injection, equivalent to 125 mg of hydroxyprogesterone caproate, to a 60-mL separator containing 10 mL of solvent hexane, 8 mL of methanol, and 2 mL of water. Insert the stopper, shake for 2 minutes, and allow the phases to separate. To 3 mL of the lower layer add sulfuric acid dropwise until a color develops, then add 3 mL of methanol: a purple color develops, and the solution, when viewed under long-wavelength ultraviolet light, exhibits a pale yellow fluorescence.

B: Evaporate 4 mL of the *Assay preparation*, obtained as directed in the *Assay*, on a water bath to dryness, and dissolve the residue in 0.5 mL of chloroform. Apply 10 μL of this solution and 10 μL of a solution of USP Hydroxyprogesterone Caproate RS in chloroform, containing 400 μg per mL, on a thin-layer chromatographic plate (see *Chromatography* ⟨621⟩) coated with a 0.25-mm layer of chromatographic silica gel mixture, on a line about 2.5 cm from the bottom edge and about 2 cm apart. Place the plate in a developing chamber that contains and that has been equilibrated with a mixture of 3 volumes of chloroform and 1 volume of ethyl acetate. Develop the plate until the solvent front has moved to about 10 cm above the points of application. Remove the plate, mark the solvent front, and dry. Spray the plate with a mixture of 1 volume of sulfuric acid and 3 volumes of alcohol, and heat in an oven at 105° for 5 minutes: the R_f value of the principal yellowish green spot obtained from the solution under test corresponds to that obtained from the Standard solution.

Water, *Method I* ⟨921⟩: not more than 0.2%.

Other requirements—It meets the requirements under *Injections* ⟨1⟩.

Assay—

Isoniazid reagent—Dissolve 375 mg of isoniazid and 0.47 mL of hydrochloric acid in 500 mL of methanol.

Standard preparation—Dissolve a suitable quantity of USP Hydroxyprogesterone Caproate RS, accurately weighed, in methanol, and dilute quantitatively and stepwise with methanol to obtain a solution having a known concentration of about 50 μg per mL.

Assay preparation—Transfer to a 250-mL volumetric flask an accurately measured volume of Hydroxyprogesterone Caproate Injection, equivalent to about 250 mg of hydroxyprogesterone caproate, add methanol to volume, and mix. Pipet 5 mL of this solution into a 100-mL volumetric flask, add methanol to volume, and mix.

Procedure—Pipet 5 mL of *Assay preparation* into a glass-stoppered, 50-mL conical flask. Pipet 5 mL of *Standard preparation* into a similar flask. To each flask, add 10.0 mL of *Isoniazid reagent*, mix, and allow to stand in a water bath at 30° for about 45 minutes. Concomitantly determine the absorbances of both solutions at the wavelength of maximum absorbance at about 380 nm, with a suitable spectrophotometer, using as a blank a mixture of 5 mL of methanol and 10 mL of *Isoniazid reagent*. Calculate the quantity, in mg, of $C_{27}H_{40}O_4$ in each mL of the Injection taken by the formula:

$$5(C/V)(A_U/A_S),$$

in which C is the concentration, in μg per mL, of USP Hydroxyprogesterone Caproate RS in the *Standard preparation*, V is the volume, in mL, of Injection taken, and A_U and A_S are the absorbances of the solutions from the *Assay preparation* and the *Standard preparation*, respectively.

Hydroxypropyl Cellulose Ocular System

» Hydroxypropyl Cellulose Ocular System contains not less than 85.0 percent and not more than 115.0 percent of the labeled amount of Hydroxypropyl Cellulose. It contains no other substance. It is sterile.

Packaging and storage—Preserve in single-dose containers, at a temperature not exceeding 30°.

Reference standard—*USP Hydroxypropyl Cellulose Reference Standard*—Dry at 105° for 3 hours before using. Keep container tightly closed.

Identification—Prepare a 1 in 100 solution in methanol, based on the labeled amount of Hydroxypropyl Cellulose. Evaporate 2 drops of the solution on a silver chloride plate so that it forms a thin film: the infrared absorption spectrum of the film so obtained exhibits maxima only at the same wavelengths as that of a similar preparation of USP Hydroxypropyl Cellulose RS.

Sterility—It meets the requirements under *Sterility Tests* ⟨71⟩.

Weight variation—Determine the weight of each of a sufficient number of Systems. Not more than 1 out of 20 varies more than 25% from the average or, failing that, not more than 6 out of 60 (including the original 20) vary more than 25% (but none more than 35%) from the average weight.

Assay—

Standard preparation—Dissolve with agitation an accurately weighed quantity of USP Hydroxypropyl Cellulose RS in water to obtain a solution having a known concentration of about 0.05 mg per mL.

Assay preparation—Transfer a sufficient number of Ocular Systems, to provide about 25 mg of hydroxypropyl cellulose, to a 500-mL volumetric flask, add about 250 mL of water, and dissolve with agitation on a mechanical shaker. Dilute with water to volume, and mix.

Procedure—Separately pipet 2 mL of the *Standard preparation*, the *Assay preparation*, and water, to provide a blank, into individual 50-mL centrifuge tubes. Add to each tube 6.0 mL of a 1 in 2000 solution of anthrone in sulfuric acid, and mix on a vortex mixer. After 40 minutes, again mix, and concomitantly determine the absorbances of the solutions obtained from the *Standard preparation* and the *Assay preparation* at 620 nm, with a suitable spectrophotometer, against the solution from the blank. Calculate the quantity, in mg, of hydroxypropyl cellulose in each Ocular System taken by the formula:

$$(500)(C/N)(A_U/A_S),$$

in which C is the concentration, in mg per mL, of USP Hydroxypropyl Cellulose RS in the *Standard preparation*, N is the number of Ocular Systems taken for the *Assay*, and A_U and A_S are the absorbances of the solutions from the *Assay preparation* and the *Standard preparation*, respectively.

Hydroxypropyl Methylcellulose

Cellulose, 2-hydroxypropyl methyl ether.
Cellulose hydroxypropyl methyl ether [*9004-65-3*].

» Hydroxypropyl Methylcellulose is a propylene glycol ether of methylcellulose. When dried at 105° for 2 hours, it contains methoxy ($-OCH_3$) and hydroxypropoxy ($-OCH_2CHOHCH_3$) groups conforming to the limits for the types of Hydroxypropyl Methylcellulose set forth in the accompanying table.

Packaging and storage—Preserve in well-closed containers.

Labeling—Label it to indicate its substitution type and its viscosity type [viscosity of a solution (1 in 50)].

Identification—

A: Gently add 1 g to the top of 100 mL of water in a beaker, and allow to disperse over the surface, tapping the top of the

Substitution Type	Methoxy (percent)		Hydroxypropoxy (percent)	
	Min.	Max.	Min.	Max.
1828	16.5	20.0	23.0	32.0
2208	19.0	24.0	4.0	12.0
2906	27.0	30.0	4.0	7.5
2910	28.0	30.0	7.0	12.0

container to ensure an even dispersion of the substance. Allow the beaker to stand until the substance becomes transparent and mucilaginous (about 5 hours), and swirl the beaker to wet the remaining substance, add a stirring bar, and stir until solution is complete: the mixture remains stable when an equal volume of 1 *N* sodium hydroxide or 1 *N* hydrochloric acid is added.

B: Add 1 g to 100 mL of boiling water, and stir the mixture: a slurry is formed, but the powdered material does not dissolve. Cool the slurry to 20°, and stir: the resulting liquid is a clear or opalescent mucilaginous colloidal mixture.

C: Pour a few mL of the mixture prepared for *Identification test B* onto a glass plate, and allow the water to evaporate: a thin, self-sustaining film results.

Apparent viscosity—Place a quantity, accurately weighed and equivalent to 2 g of solids on the dried basis in a tared, wide-mouth, 250-mL centrifuge bottle, and add 98 g of water previously heated to 80° to 90°. Stir with a propeller-type stirrer for 10 minutes, place the bottle in an ice bath, continue the stirring, and allow to remain in the ice bath for 40 minutes to ensure that hydration and solution are complete. Adjust the weight of the solution to 100 g, if necessary, and centrifuge the solution to expel any entrapped air. Adjust the temperature of the solution to 20 ± 0.1°, and determine the viscosity in a suitable viscosimeter of the Ubbelohde type as directed for *Procedure for Cellulose Derivatives* under *Viscosity* ⟨911⟩. Its viscosity is not less than 80.0% and not more than 120.0% of that stated on the label for viscosity types of 100 centipoises or less, and not less than 75.0% and not more than 140.0% of that stated on the label for viscosity types higher than 100 centipoises.

Loss on drying ⟨731⟩—Dry it at 105° for 2 hours: it loses not more than 5.0% of its weight.

Residue on ignition ⟨281⟩: not more than 1.5% for Hydroxypropyl Methylcellulose having a labeled viscosity of greater than 50 centipoises, not more than 3% for Hydroxypropyl Methylcellulose having a labeled viscosity of 50 centipoises or less, and not more than 5% for Hydroxypropyl Methylcellulose 1828 of all labeled viscosities.

Arsenic, *Method II* ⟨211⟩: 3 ppm.

Heavy metals, *Method II* ⟨231⟩: 0.001%, 1 mL of hydroxylamine hydrochloride solution (1 in 5) being added to the solution of the residue.

Assay—[*Caution—Hydriodic acid and its reaction byproducts are highly toxic. Perform all steps of the Assay preparation and the Standard preparation in a properly functioning hood. Specific safety practices to be followed are to be identified to the analyst performing this test.*]

Internal standard solution—Transfer about 2.5 g of toluene, accurately weighed, to a 100-mL volumetric flask containing 10 mL of *o*-xylene, dilute with *o*-xylene to volume, and mix.

Standard preparation—Into a suitable serum vial weigh about 135 mg of adipic acid and 4.0 mL of hydriodic acid, pipet 4 mL of *Internal standard solution* into the vial, and close the vial securely with a suitable septum stopper. Weigh the vial and contents accurately, add 30 µL of isopropyl iodide through the septum with a syringe, again weigh, and calculate the weight of isopropyl iodide added, by difference. Add 90 µL of methyl iodide similarly, again weigh, and calculate the weight of methyl iodide added, by difference. Shake well, and allow the layers to separate.

Assay preparation—Transfer about 0.065 g of dried Hydroxypropyl Methylcellulose, accurately weighed, to a 5-mL thick-walled reaction vial equipped with a pressure-tight septum-type closure, add an amount of adipic acid equal to the weight of the test specimen, and pipet 2 mL of *Internal standard solution* into the vial. Cautiously pipet 2 mL of hydriodic acid into the mixture, immediately cap the vial tightly, and weigh accurately. Mix the contents of the vial continuously while heating at 150°, for 60 minutes. Allow the vial to cool for about 45 minutes, and again weigh. If the weight loss is greater than 10 mg, discard the mixture, and prepare another *Assay preparation*.

Chromatographic system—Use a gas chromatograph equipped with a thermal conductivity detector. Under typical conditions, the instrument contains a 1.8-m × 4-mm glass column packed with 20 percent liquid phase G28 on 100- to 120-mesh support S1C that is not silanized, the column is maintained at 130°, and helium is used as the carrier gas. In a suitable system, the resolution, *R* (see *Chromatography* ⟨621⟩), between the toluene and isopropyl iodide peaks is not less than 2.0.

Calibration—Inject about 2 µL of the upper layer of the *Standard preparation* into the gas chromatograph, and record the chromatogram. Under the conditions described above, the relative retention times of methyl iodide, isopropyl iodide, toluene, and *o*-xylene are approximately 1.0, 2.2, 3.6, and 8.0, respectively. Calculate the relative response factor, F_{mi}, of equal weights of toluene and methyl iodide by the formula:

$$Q_{smi}/A_{smi},$$

in which Q_{smi} is the quantity ratio of methyl iodide to toluene in the *Standard preparation*, and A_{smi} is the peak area ratio of methyl iodide to toluene obtained from the *Standard preparation*. Similarly, calculate the relative response factor, F_{ii}, of equal weights of toluene and isopropyl iodide by the formula:

$$Q_{sii}/A_{sii},$$

in which Q_{sii} is the quantity ratio of isopropyl iodide to toluene in the *Standard preparation*, and A_{sii} is the peak area ratio of isopropyl iodide to toluene obtained from the *Standard preparation*.

Procedure—Inject about 2 µL of the upper layer of the *Assay preparation* into the gas chromatograph, and record the chromatogram. Calculate the percentage of methoxy in the Hydroxypropyl Methylcellulose by the formula:

$$2(31/142)F_{mi}A_{umi}(W_t/W_u),$$

in which 31/142 is the ratio of the formula weights of methoxy and methyl iodide, F_{mi} is defined under *Calibration*, A_{umi} is the ratio of the area of the methyl iodide peak to that of the toluene peak obtained from the *Assay preparation*, W_t is the weight, in g, of toluene in the *Internal standard solution*, and W_u is the weight, in g, of Hydroxypropyl Methylcellulose taken for the *Assay*. Similarly, calculate the percentage of hydroxypropoxy in the Hydroxypropyl Methylcellulose by the formula:

$$2(75/170)F_{ii}A_{uii}(W_t/W_u),$$

in which 75/170 is the ratio of the formula weights of hydroxypropoxy and isopropyl iodide, F_{ii} is defined under *Calibration*, A_{uii} is the ratio of the area of the isopropyl iodide peak to that of the toluene peak obtained from the *Assay preparation*, W_t is the weight, in g, of toluene in the *Internal standard solution*, and W_u is the weight, in g, of Hydroxypropyl Methylcellulose taken for the *Assay*.

Hydroxypropyl Methylcellulose 2208

Where this monograph is specified or referred to in this Pharmacopeia, see the monograph, *Hydroxypropyl Methylcellulose.*

Hydroxypropyl Methylcellulose 2906

Where this monograph is specified or referred to in this Pharmacopeia, see the monograph, *Hydroxypropyl Methylcellulose.*

Hydroxypropyl Methylcellulose 2910

Where this monograph is specified or referred to in this Pharmacopeia, see the monograph, *Hydroxypropyl Methylcellulose.*

Hydroxypropyl Methylcellulose Ophthalmic Solution

» Hydroxypropyl Methylcellulose Ophthalmic Solution is a sterile solution of Hydroxypropyl Methylcellulose. It contains not less than 85.0 percent and not more than 115.0 percent of the labeled amount of Hydroxypropyl Methylcellulose. It may contain suitable antimicrobial, buffering, and stabilizing agents.

Packaging and storage—Preserve in tight containers.

Reference standard—*USP Hydroxypropyl Methylcellulose Reference Standard*—Dry at 105° for 2 hours before using.

Identification—

A: It responds to *Identification test C* under *Hydroxypropyl Methylcellulose*.

B: Heat 5 mL of Ophthalmic Solution in a test tube over a low flame: the warm solution turns cloudy but clears upon chilling.

Sterility—It meets the requirements under *Sterility Tests* ⟨71⟩.

pH ⟨791⟩: between 6.0 and 7.8.

Assay—

Standard preparation—Dissolve a suitable quantity of USP Hydroxypropyl Methylcellulose RS, accurately weighed, in water, and dilute quantitatively with water to obtain a solution having a known concentration of about 100 µg per mL.

Assay preparation—Dilute quantitatively an accurately measured volume of Hydroxypropyl Methylcellulose Ophthalmic Solution with water to obtain a solution having an equivalent concentration of approximately 100 µg of hydroxypropyl methylcellulose per mL.

Procedure—Pipet 2 mL each of the *Standard preparation*, the *Assay preparation*, and water to provide a blank, into separate, glass-stoppered test tubes. To each tube add 5.0 mL of diphenylamine solution (prepared by dissolving 3.75 g of colorless diphenylamine crystals in 150 mL of glacial acetic acid and diluting the solution with 90 mL of hydrochloric acid), mix, and immediately insert the tubes into an oil bath at 105° to 110° for 30 minutes, the temperature being kept uniform within 0.1° during heating. Remove the tubes, and place them in an ice-water bath for 10 minutes or until thoroughly cool. Concomitantly determine, at room temperature, the absorbances of the solutions from the *Standard preparation* and the *Assay preparation* at 635 nm, with a suitable spectrophotometer, using the water solution as the blank. Calculate the quantity, in mg, of hydroxypropyl methylcellulose in each mL of the Ophthalmic Solution by the formula:

$$0.001C(d/V)(A_U/A_S),$$

in which C is the concentration, in µg per mL, of USP Hydroxypropyl Methylcellulose RS in the Standard solution, d is the dilution fold of V used to obtain the *Assay preparation*, V is the volume, in mL, of Ophthalmic Solution taken, and A_U and A_S are the absorbances of the solutions from the *Assay preparation* and the *Standard preparation*, respectively.

Hydroxypropyl Methylcellulose Phthalate—*see* Hydroxypropyl Methylcellulose Phthalate NF

Hydroxypropyl Methylcellulose Phthalate 200731—*see* Hydroxypropyl Methylcellulose Phthalate 200731 NF

Hydroxypropyl Methylcellulose Phthalate 220824—*see* Hydroxypropyl Methylcellulose Phthalate 220824 NF

Hydroxystilbamidine Isethionate

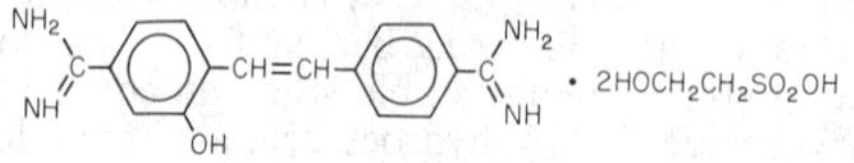

$C_{16}H_{16}N_4O.2C_2H_6O_4S$ 532.58

Benzenecarboximidamide, 4-[2-[4-(aminoiminomethyl)phenyl]-ethenyl]-3-hydroxy-, bis(2-hydroxyethanesulfonate) (salt).

2-Hydroxy-4,4′-stilbenedicarboxamidine bis(2-hydroxyethanesulfonate) (salt) [*533-22-2*].

» Hydroxystilbamidine Isethionate contains not less than 95.0 percent and not more than 105.0 percent of $C_{16}H_{16}N_4O.2C_2H_6O_4S$, calculated on the dried basis.

Packaging and storage—Preserve in tight, light-resistant containers.

Reference standard—*USP Hydroxystilbamidine Isethionate Reference Standard*—Dry in vacuum at 60° for 3 hours before using.

Identification—

A: The infrared absorption spectrum of a potassium bromide dispersion of it, previously dried, exhibits maxima only at the same wavelengths as that of a similar preparation of USP Hydroxystilbamidine Isethionate RS.

B: The ultraviolet absorption spectrum of a 1 in 100,000 solution in 0.01 *N* hydrochloric acid exhibits maxima and minima at the same wavelengths as that of a similar solution of USP Hydroxystilbamidine Isethionate RS, concomitantly measured.

pH ⟨791⟩: between 4.0 and 5.5, in a solution (1 in 100).

Loss on drying ⟨731⟩—Dry about 200 mg, accurately weighed, in vacuum at 60° for 3 hours: it loses not more than 1.0% of its weight.

Residue on ignition ⟨281⟩: not more than 0.1%, 2 g being used for the test.

Selenium ⟨291⟩: 0.003%, a 200-mg test specimen being used.

Heavy metals, *Method II* ⟨231⟩: 0.001%.

Assay—Transfer about 100 mg of Hydroxystilbamidine Isethionate, accurately weighed, to a titration vessel, dissolve in 25 mL of water, and add 100 mL of hydrochloric acid. Titrate the solution with 0.1 *N* bromine VS, determining the end-point potentiometrically, using platinum-calomel electrodes. Each mL of 0.1 *N* bromine is equivalent to 26.63 mg of $C_{16}H_{16}N_4O.2C_2H_6O_4S$.

Sterile Hydroxystilbamidine Isethionate

» Sterile Hydroxystilbamidine Isethionate is Hydroxystilbamidine Isethionate suitable for parenteral use.

Packaging and storage—Preserve in light-resistant *Containers for Sterile Solids* as described under *Injections* ⟨1⟩.

Reference standard—*USP Hydroxystilbamidine Isethionate Reference Standard*—Dry in vacuum at 60° for 3 hours before using.

Completeness of solution ⟨641⟩—The contents of 1 container dissolve in 10 mL of carbon dioxide–free water to yield a clear solution.

Constituted solution—At the time of use, the constituted solution prepared from Sterile Hydroxystilbamidine Isethionate meets the requirements for *Constituted Solutions* under *Injections* ⟨1⟩.

Other requirements—It conforms to the definition, responds to the *Identification tests*, and meets the requirements for *pH, Loss on drying, Residue on ignition, Selenium, Heavy metals*, and *Assay* under *Hydroxystilbamidine Isethionate.* It meets also the requirements for *Sterility Tests* ⟨71⟩, *Uniformity of Dosage Units* ⟨905⟩, and *Labeling* under *Injections* ⟨1⟩.

Hydroxyurea

$H_2NCONHOH$

$CH_4N_2O_2$ 76.05
Urea, hydroxy-.
Hydroxyurea [*127-07-1*].

» Hydroxyurea contains not less than 97.0 percent and not more than 103.0 percent of $CH_4N_2O_2$, calculated on the dried basis.

Packaging and storage—Preserve in tight containers, in a dry atmosphere.

Reference standard—*USP Hydroxyurea Reference Standard*—[*Caution—Hygroscopic; decomposes in the presence of moisture.*] Dry in vacuum at 60° for 3 hours before using.

Identification—The infrared absorption spectrum of a potassium bromide dispersion of it exhibits maxima only at the same wavelengths as that of a similar preparation of USP Hydroxyurea RS.

Loss on drying ⟨731⟩—Dry it in vacuum at 60° for 3 hours: it loses not more than 1.0% of its weight.

Residue on ignition ⟨281⟩: not more than 0.50%.

Heavy metals ⟨231⟩: not more than 0.003%.

Urea and related compounds—

Developing solvent—Shake equal volumes of isobutyl alcohol and water in a separator and allow the layers to separate. Use the upper layer as the *Mobile phase* and the lower layer as the *Stationary phase.*

p-Dimethylaminobenzaldehyde solution, 1 percent—Dissolve 1.0 g of *p*-dimethylaminobenzaldehyde in 50 mL of alcohol, add 2 mL of hydrochloric acid, and dilute with alcohol to 100.0 mL.

pH 6.5 buffer solution—Mix 700 mL of 0.2 *M* dibasic sodium phosphate and 300 mL of 0.1 *M* citric acid.

Standard preparation—Prepare a solution of urea in water, containing 0.1 mg per mL.

Test preparation—Dissolve 10.0 mg of Hydroxyurea in 1.0 mL of water.

Procedure—Treat a suitable chromatographic paper strip (Whatman No. 1 or equivalent) by dipping it in *pH 6.5 buffer solution.* Dry the paper strip, and apply 100 µL of the *Test preparation* and 50 µL of the *Standard preparation.* Place the strip in a chromatographic chamber for descending chromatography containing the *Stationary phase* in the bottom of the chamber and the *Mobile phase* in the trough. Develop for 24 hours, remove the strip from the chamber, air-dry, and develop again for 24 hours. Remove the strip, air-dry, spray with *p-Dimethylaminobenzaldehyde solution, 1 percent*, and heat at 90° for 1 to 2 minutes. Not more than two spots, other than the major component, are present in the *Test preparation*, and their intensities are not greater than the intensity of the spot from the *Standard preparation* (0.5% of each impurity). The R_r values relative to hydroxyurea, the principal spot, are 0.65 and 1.26 (urea).

Assay—

Mobile phase—Use degassed water.

Internal standard solution—Dissolve uracil in water to obtain a solution having a concentration of about 0.12 mg per mL.

Standard preparation—Transfer about 50 mg of USP Hydroxyurea RS, accurately weighed, to a 50-mL volumetric flask, add 10 mL of *Internal standard solution*, dilute with water to volume, and mix.

Resolution solution—Dissolve suitable quantities of USP Hydroxyurea RS and hydroxylamine hydrochloride in water to obtain a solution containing about 1 mg and 4 mg per mL, respectively.

Assay preparation—Transfer about 50 mg of Hydroxyurea, accurately weighed, to a 50-mL volumetric flask, add 10 mL of *Internal standard solution*, dilute with water to volume, and mix.

Chromatographic system (see *Chromatography* ⟨621⟩)—The liquid chromatograph is equipped with a 214-nm detector and a 4.6-mm × 25-cm column that contains 5-µm packing L1. The flow rate is about 0.5 mL per minute. Chromatograph the *Resolution solution*, and record the peak responses as directed under *Procedure:* the resolution, *R*, between the hydroxylamine and hydroxyurea peaks is not less than 1.0. Chromatograph the *Standard preparation*, and record the peak responses as directed under *Procedure:* the relative standard deviation for replicate injections is not more than 2.0%.

Procedure—Separately inject equal volumes (about 20 µL) of the *Standard preparation* and the *Assay preparation* into the chromatograph, record the chromatograms, and measure the responses for the major peaks. The relative retention times are about 0.5 for hydroxyurea and 1.0 for uracil. Calculate the quantity, in mg, of $CH_4N_2O_2$ in the portion of Hydroxyurea taken by the formula:

$$50C(R_U / R_S),$$

in which *C* is the concentration, in mg per mL, of USP Hydroxyurea RS in the *Standard preparation*, and R_U and R_S are the response ratios of the hydroxyurea peak to the internal standard peak obtained from the *Assay preparation* and the *Standard preparation*, respectively.

Hydroxyurea Capsules

» Hydroxyurea Capsules contain not less than 90.0 percent and not more than 110.0 percent of the labeled amount of $CH_4N_2O_2$.

Packaging and storage—Preserve in tight containers, in a dry atmosphere.

Reference standard—*USP Hydroxyurea Reference Standard*—[*Caution—Hygroscopic; decomposes in the presence of moisture.*] Dry in vacuum at 60° for 3 hours before using.

Identification—Transfer a portion of the Capsule contents, equivalent to about 30 mg of hydroxyurea, to a suitable centrifuge tube, and add 10 mL of anhydrous methanol. Mix, and centrifuge for 3 minutes. Transfer 1.0 mL of the clear supernatant liquid to a mortar containing 500 mg of potassium bromide, triturate to a homogeneous blend, dry in a vacuum desiccator at 60° for 3 hours, and prepare a suitable disk: the infrared absorption spectrum exhibits maxima only at the same wavelengths as that of a similar preparation of USP Hydroxyurea RS.

Uniformity of dosage units ⟨905⟩: meet the requirements.

Assay—

Mobile phase, Internal standard solution, Standard preparation, Resolution solution, and *Chromatographic system*—Prepare as directed in the *Assay* under *Hydroxyurea.*

Assay preparation—Remove, as completely as possible, the contents of not less than 20 Hydroxyurea Capsules, weigh, and place in a glass mortar. Grind to a fine powder, and transfer an accurately weighed portion of the powder, equivalent to about 2000 mg of hydroxyurea, to a 1000-mL volumetric flask. Add about 900 mL of water, sonicate for 5 minutes, stir with the aid of a magnetic stirrer for 30 minutes, dilute with water to volume, mix, and sonicate for an additional 5 minutes. Filter a portion of the resulting solution, discarding the first 10 mL of the filtrate. Transfer 25.0 mL of the clear filtrate to a 50-mL volumetric flask, add 10 mL of *Internal standard solution*, dilute with water to volume, and mix.

Procedure—Proceed as directed for *Procedure* in the *Assay* under *Hydroxyurea.* Calculate the quantity, in mg, of $CH_4N_2O_2$ in the portion of Capsules taken by the formula:

$$2000C(R_U / R_S),$$

in which *C* is the concentration, in mg per mL, of USP Hydroxyurea RS in the *Standard preparation*, and R_U and R_S are the response ratios of the hydroxyurea peak to the internal standard peak obtained from the *Assay preparation* and the *Standard preparation*, respectively.

Hydroxyzine Hydrochloride

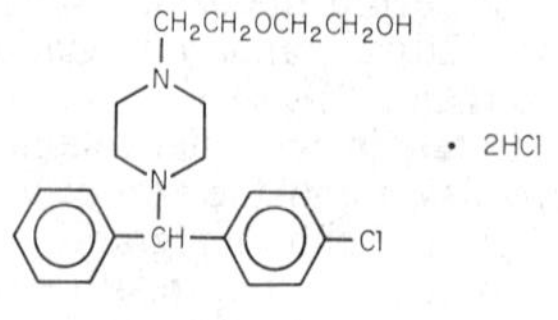

$C_{21}H_{27}ClN_2O_2 . 2HCl$ 447.83

Ethanol, 2-[2-[4-[(4-chlorophenyl)phenylmethyl]-1-piperazinyl]ethoxy]-, dihydrochloride.

2-[2-[4-(*p*-Chloro-α-phenylbenzyl)-1-piperazinyl]ethoxy]ethanol dihydrochloride [*2192-20-3*].

» Hydroxyzine Hydrochloride, dried at 105° for two hours, contains not less than 98.0 percent and not more than 100.5 percent of $C_{21}H_{27}ClN_2O_2 . 2HCl$.

Packaging and storage—Preserve in tight containers.

Reference standards—*USP Hydroxyzine Hydrochloride Reference Standard*—Dry at 105° for 2 hours before using. *USP p-Chlorobenzhydrylpiperazine Reference Standard.*

Identification—

A: The infrared absorption spectrum of a potassium bromide dispersion of it, previously dried, exhibits maxima only at the same wavelengths as that of a similar preparation of USP Hydroxyzine Hydrochloride RS.

B: The ultraviolet absorption spectrum of a 1 in 100,000 solution in alcohol exhibits maxima and minima at the same wavelengths as that of a similar solution of USP Hydroxyzine Hydrochloride RS, concomitantly measured, and the respective absorptivities, calculated on the dried basis, at the wavelength of maximum absorbance at about 230 nm do not differ by more than 3.0%.

C: To 10 mL of a solution (1 in 400) add 2 drops of nitric acid and 1 mL of silver nitrate TS: a curdy, white precipitate, insoluble in 2 *N* nitric acid, but soluble in 6 *N* ammonium hydroxide, separates (*presence of chloride*).

Loss on drying ⟨731⟩—Dry it at 105° for 2 hours: it loses not more than 5.0% of its weight.

Residue on ignition ⟨281⟩: not more than 0.5%.

Heavy metals, *Method II* ⟨231⟩: 0.002%.

Chromatographic purity—

Mobile phase—Prepare a filtered and degassed mixture of acetonitrile and 0.12 *N* sulfuric acid (90:10). Make adjustments if necessary (see *System Suitability* under *Chromatography* ⟨621⟩).

Diluting solvent—Prepare a mixture of acetonitrile and water (90:10).

Standard preparation—Dissolve an accurately weighed quantity of USP Hydroxyzine Hydrochloride RS quantitatively in *Diluting solvent* to obtain a solution having a known concentration of about 1.8 µg per mL.

Resolution solution—Dissolve suitable quantities of USP Hydroxyzine Hydrochloride RS and USP *p*-Chlorobenzhydrylpiperazine RS in *Diluting solvent* to obtain a solution containing 3.6 µg of each per mL.

Test preparation—Transfer an accurately weighed quantity of Hydroxyzine Hydrochloride to a suitable volumetric flask, dissolve in *Diluting solvent*, dilute with *Diluting solvent* to volume to obtain a solution containing a known concentration of about 0.6 mg of specimen per mL, and mix.

Chromatographic system (see *Chromatography* ⟨621⟩)—The liquid chromatograph is equipped with a 230-nm detector, a 5-mm × 6-mm (ID) guard column that contains packing L3, and two series-coupled 3-mm × 10-cm columns that contain packing L3. The flow rate is about 0.4 mL per minute. Chromatograph the *Resolution solution* and the *Standard preparation*, and record the peak responses as directed under *Procedure:* the resolution, *R*, between the *p*-chlorobenzhydrylpiperazine and hydroxyzine peaks is not less than 1.2, and the relative standard deviation for replicate injections of the *Standard preparation* is not more than 2.0%.

Procedure—Separately inject equal volumes (about 20 µL) of the *Standard preparation* and the *Test preparation* into the chromatograph, record the chromatograms for a total time of not less than 1.8 times the retention time of the hydroxyzine peak, and measure the response for each peak, except for the main hydroxyzine peak in the chromatogram obtained from the *Test preparation.* Calculate the apparent percentage of each impurity in the specimen taken by the formula:

$$0.1(C_S/C_U)(r_U/r_S),$$

in which C_S is the concentration, in µg per mL, of USP Hydroxyzine Hydrochloride RS in the *Standard preparation*, C_U is the concentration, in mg per mL, of specimen in the *Test preparation*, r_U is the peak response of a given impurity in the chromatogram obtained from the *Test preparation*, and r_S is the peak response of hydroxyzine in the chromatogram obtained from the *Standard preparation:* not more than 0.3% of any impurity is found, and the sum of all impurities found is not greater than 1.5%.

Assay—Dissolve about 150 mg of Hydroxyzine Hydrochloride, previously dried and accurately weighed, in 10 mL of chloroform. Add 50 mL of glacial acetic acid, 5 mL of mercuric acetate TS, and quinaldine red TS, and titrate with 0.1 *N* perchloric acid VS. Perform a blank determination, and make any necessary correction. Each mL of 0.1 *N* perchloric acid is equivalent to 22.39 mg of $C_{21}H_{27}ClN_2O_2 . 2HCl$.

Hydroxyzine Hydrochloride Injection

» Hydroxyzine Hydrochloride Injection is a sterile solution of Hydroxyzine Hydrochloride in Water for Injection. It contains not less than 90.0 percent and not more than 110.0 percent of the labeled amount of $C_{21}H_{27}ClN_2O_2 . 2HCl$.

Packaging and storage—Preserve in single-dose or in multiple-dose containers, protected from light.

Reference standard—*USP Hydroxyzine Hydrochloride Reference Standard*—Dry at 105° for 2 hours before using.

Identification—Dilute a volume of Injection with 0.1 *N* hydrochloric acid to obtain a solution having a concentration of about 20 µg of hydroxyzine hydrochloride per mL: the ultraviolet absorption spectrum of this solution exhibits maxima and minima at the same wavelengths as that of a 1 in 50,000 solution of USP Hydroxyzine Hydrochloride RS in 0.1 *N* hydrochloric acid, concomitantly measured.

pH ⟨791⟩: between 3.5 and 6.0.

Other requirements—It meets the requirements under *Injections* ⟨1⟩.

Assay and limit of 4-chlorobenzophenone—

Mobile phase—Adjust about 1000 mL of *Buffer No. 1* (see *Phosphate Buffers and Other Solutions* in the section, *Media and Diluents*, under *Antibiotics—Microbial Assays* ⟨81⟩) with 10 *N* potassium hydroxide to a pH of 6.6. To about 35 volumes of this solution add about 65 volumes of methanol, mix, and degas. Make adjustments if necessary (see *System Suitability* under *Chromatography* ⟨621⟩).

Standard preparation—Dissolve accurately weighed quantities of USP Hydroxyzine Hydrochloride RS and 4-chlorobenzophenone in *Mobile phase*, and dilute quantitatively with *Mobile phase* to obtain a solution having known concentrations of about 250 µg of USP Hydroxyzine Hydrochloride RS and 0.5 µg of 4-chlorobenzophenone per mL. Protect this solution from light.

Assay preparation—Transfer an accurately measured volume of Hydroxyzine Hydrochloride Injection, equivalent to about 125 mg of hydroxyzine hydrochloride, to a 50-mL volumetric flask, dilute with *Mobile phase* to volume, and mix. Pipet 10 mL of this solution into a 100-mL volumetric flask, dilute with *Mobile phase* to volume, and mix. Protect this solution from light.

Chromatographic system (see *Chromatography* ⟨621⟩)—The liquid chromatograph is equipped with a 254-nm detector and a 4-mm × 30-cm column that contains packing L1. The flow rate is about 2 mL per minute. Chromatograph the *Standard preparation*, and record the peak responses as directed under *Procedure:* the tailing factors for the 4-chlorobenzophenone and hydroxyzine peaks are not more than 2.5, the resolution, *R*, between the 4-chlorobenzophenone and hydroxyzine peaks is not less than 2.0, and the relative standard deviation for replicate injections of the *Standard preparation* is not more than 2.0%. The relative retention times are about 0.75 for 4-chlorobenzophenone and 1.0 for hydroxyzine.

Procedure—Separately inject equal volumes (about 20 µL) of the *Standard preparation* and the *Assay preparation* into the chromatograph, record the chromatograms, and measure the responses for the major peaks. Calculate the quantity, in mg, of hydroxyzine hydrochloride ($C_{21}H_{27}ClN_2O_2 . HCl$) in each mL of the Injection taken by the formula:

$$0.5(C/V)(r_U/r_S),$$

in which *C* is the concentration, in µg per mL, of USP Hydroxyzine Hydrochloride RS in the *Standard preparation*, *V* is the volume, in mL, of Injection taken, and r_U and r_S are the hydroxyzine peak responses obtained from the *Assay preparation* and the *Standard preparation*, respectively. The ratio of the response of the 4-chlorobenzophenone peak to that of the hydroxyzine peak obtained from the *Assay preparation* does not exceed the corresponding ratio of peak responses obtained from the *Standard preparation* (0.2%).

Hydroxyzine Hydrochloride Syrup

» Hydroxyzine Hydrochloride Syrup contains not less than 90.0 percent and not more than 110.0 percent of the labeled amount of $C_{21}H_{27}ClN_2O_2 . 2HCl$.

Packaging and storage—Preserve in tight, light-resistant containers.

Reference standard—*USP Hydroxyzine Hydrochloride Reference Standard*—Dry at 105° for 2 hours before using.

Identification—Dilute a volume of Syrup, equivalent to about 20 mg of hydroxyzine hydrochloride, with 50 mL of methanol, and mix. On a suitable thin-layer chromatographic plate (see *Chromatography* ⟨621⟩), coated with a 0.25-mm layer of chromatographic silica gel and dried in air for 30 minutes followed by drying in vacuum at 140° for 30 minutes, apply 100 µL of this solution and 100 µL of a solution in the same medium containing about 350 µg of USP Hydroxyzine Hydrochloride RS per mL. Allow the spots to dry, and develop the chromatogram in a solvent system consisting of a mixture of toluene, alcohol, and ammonium hydroxide (150:95:1) until the solvent front has moved about three-fourths of the length of the plate. Remove the plate from the developing chamber, mark the solvent front, and allow the solvent to evaporate. Locate the spots by lightly spraying with potassium iodoplatinate TS: the R_f value of the principal spot obtained from the test solution corresponds to that obtained from the Standard solution.

Assay—

Mobile phase and *Chromatographic system*—Proceed as directed in the *Assay* under *Hydroxyzine Hydrochloride Tablets*.

Standard preparation—Dissolve a suitable quantity of USP Hydroxyzine Hydrochloride RS, accurately weighed, in water to obtain a solution having a known concentration of about 100 µg per mL.

Assay preparation—Transfer an accurately measured volume of Hydroxyzine Hydrochloride Syrup, equivalent to about 20 mg of hydroxyzine hydrochloride, to a 200-mL volumetric flask, dilute with water to volume, mix, and filter a portion through a polytef membrane filter (5-µm or finer porosity).

Procedure—Proceed as directed for *Procedure* in the *Assay* under *Hydroxyzine Hydrochloride Tablets*. Calculate the quantity, in mg, of $C_{21}H_{27}ClN_2O_2 . 2HCl$ in each mL of the Syrup taken by the formula:

$$0.2(C/V)(r_U/r_S),$$

in which *C* is the concentration, in µg per mL, of USP Hydroxyzine Hydrochloride RS in the *Standard preparation*, *V* is the volume, in mL, of Syrup taken, and r_U and r_S are the peak responses obtained from the *Assay preparation* and the *Standard preparation*, respectively.

Hydroxyzine Hydrochloride Tablets

» Hydroxyzine Hydrochloride Tablets contain not less than 90.0 percent and not more than 110.0 percent of the labeled amount of $C_{21}H_{27}ClN_2O_2 . 2HCl$.

Packaging and storage—Preserve in tight containers.

Reference standard—*USP Hydroxyzine Hydrochloride Reference Standard*—Dry at 105° for 2 hours before using.

Identification—Triturate a quantity of finely powdered Tablets, equivalent to about 100 mg of hydroxyzine hydrochloride, with 50 mL of methanol, and filter. On a suitable thin-layer chromatographic plate (see *Chromatography* ⟨621⟩), coated with a 0.25-mm layer of chromatographic silica gel and dried in air for 30 minutes followed by drying in vacuum at 140° for 30 minutes, apply 100 µL of this solution and 100 µL of a solution, in the same medium containing 2 mg of USP Hydroxyzine Hydrochloride RS per mL. Proceed as directed in the *Identification test* under *Hydroxyzine Hydrochloride Syrup*, beginning with "Allow the spots to dry."

Dissolution ⟨711⟩—

Medium: water; 800 mL.

Apparatus—Proceed as directed for *Uncoated Tablets* under *Disintegration* ⟨701⟩, beginning with "Place 1 Tablet in each of the six tubes of the basket," with these exceptions: (a) the disks are not used; (b) the apparatus is adjusted so that the bottom of the basket-rack assembly descends to 1.0 ± 0.1 cm from the inside bottom surface of the vessel on the downward stroke; (c) the 10-mesh, stainless-steel cloth in the basket-rack assembly is replaced with 40-mesh, stainless-steel cloth; and (d) 40-mesh, stainless-steel cloth is fitted to the top of the basket-rack assembly if necessary to prevent any dosage unit from floating out of the tubes of the assembly.

Time: 45 minutes.

Procedure—Determine the amount of $C_{21}H_{27}ClN_2O_2 . 2HCl$ dissolved from ultraviolet absorbances at the wavelength of maximum absorbance at about 230 nm of filtered portions of the solution under test, suitably diluted with water, if necessary, in comparison with a Standard solution having a known concentration of USP Hydroxyzine Hydrochloride RS in the same medium. Calculate the amount of $C_{21}H_{27}ClN_2O_2 . 2HCl$ dissolved per Tablet.

Tolerances—Not less than 75% of the labeled amount of $C_{21}H_{27}ClN_2O_2 . 2HCl$ is dissolved in 45 minutes.

Uniformity of dosage units ⟨905⟩: meet the requirements.

Assay—

Mobile phase—Dissolve 6.8 g of monobasic potassium phosphate in 1000 mL of water, add 1000 mL of methanol, and mix. Filter the solution through a polytef membrane filter (5-µm or finer porosity), and degas.

Standard preparation—Dissolve a suitable quantity of USP Hydroxyzine Hydrochloride RS, accurately weighed, in methanol to obtain a solution having a known concentration of about 100 µg per mL.

Assay preparation—Place 20 Hydroxyzine Hydrochloride Tablets in a high-speed blender jar containing 400.0 mL of methanol, and blend for 5 minutes. The Tablets are completely disintegrated. Allow to settle, and filter a portion of the supernatant liquid through a polytef membrane filter (1-µm or finer porosity). Dilute an accurately measured volume of the filtrate so obtained quantitatively with methanol to obtain an *Assay preparation* having a concentration of about 100 µg of hydroxyzine hydrochloride per mL.

Chromatographic system (see *Chromatography* ⟨621⟩)—The liquid chromatograph is equipped with a 232-nm detector and a 4.6-mm × 25-cm column that contains packing L9. The flow rate is about 2.0 mL per minute. Chromatograph the *Standard preparation,* and record the peak responses as directed under *Procedure:* the relative standard deviation is not more than 2.5%.

Procedure—Separately inject equal volumes (about 20 µL) of the *Standard preparation* and the *Assay preparation* into the chromatograph, record the chromatograms, and measure the responses for the major peaks. Calculate the quantity, in mg, of $C_{21}H_{27}ClN_2O_2.2HCl$ in each Tablet taken by the formula:

$$(L/D)(C)(r_U/r_S),$$

in which L is the labeled quantity, in mg, of hydroxyzine hydrochloride in each Tablet, D is the concentration, in µg per mL, of hydroxyzine hydrochloride in the *Assay preparation* on the basis of the labeled quantity in each Tablet and the extent of dilution, C is the concentration in µg per mL, of USP Hydroxyzine Hydrochloride RS in the *Standard preparation*, and r_U and r_S are the peak responses obtained from the *Assay preparation* and the *Standard preparation*, respectively.

Hydroxyzine Pamoate

$C_{21}H_{27}ClN_2O_2.C_{23}H_{16}O_6$ 763.29

Ethanol, 2-[2-[4-[(4-chlorophenyl)phenylmethyl]-1-piperazinyl]-ethoxy]-, compd. with 4,4′-methylenebis[3-hydroxy-2-naphthalenecarboxylic acid] (1:1).

2-[2-[4-(*p*-Chloro-α-phenylbenzyl)-1-piperazinyl]ethoxy]ethanol 4,4′-methylenebis[3-hydroxy-2-naphthoate] (1:1) [*10246-75-0*].

» Hydroxyzine Pamoate contains not less than 97.0 percent and not more than 102.0 percent of $C_{21}H_{27}ClN_2O_2.C_{23}H_{16}O_6$, calculated on the anhydrous basis.

Packaging and storage—Preserve in tight containers.

Reference standards—*USP Hydroxyzine Pamoate Reference Standard*—Dry over silica gel for 4 hours before using. *USP Hydroxyzine Hydrochloride Reference Standard*—Dry at 105° for 2 hours before using. *USP Pamoic Acid Reference Standard*—Dry in vacuum at 100° for 3 hours before using.

Identification—

A: Transfer about 85 mg (equivalent to about 50 mg of hydroxyzine hydrochloride) to a separator containing 25 mL of 1 *N* sodium hydroxide, and extract with three 20-mL portions of chloroform. Evaporate the combined chloroform extracts on a steam bath with the aid of a current of air to dryness, and dissolve the residue in 0.1 *N* hydrochloric acid to make 100 mL. Dilute a 1-mL portion of this solution with 0.1 *N* hydrochloric acid to 50 mL: the ultraviolet absorption spectrum of the solution so obtained exhibits maxima and minima at the same wavelengths as that of a 1 in 100,000 solution of USP Hydroxyzine Hydrochloride RS in 0.1 *N* hydrochloric acid, concomitantly measured.

B: Prepare a solution of it in a mixture of equal volumes of 0.1 *N* sodium hydroxide and acetone containing 2 mg per mL. On a suitable thin-layer chromatographic plate (see *Chromatography* ⟨621⟩), coated with a 0.50-mm layer of chromatographic silica gel, apply 10 µL of this solution and 10 µL of a solution of USP Hydroxyzine Pamoate RS in the same medium, having a concentration of 2 mg per mL. Allow the spots to dry, and develop the chromatogram in 0.1 *N* hydrochloric acid until the solvent front has moved about three-fourths of the length of the plate. Remove the plate from the developing chamber, mark the solvent front, and locate the spots on the plate by viewing under long-wavelength ultraviolet light: the R_f value of the principal spot, representing the pamoate moiety, obtained from the test solution corresponds to that obtained from the Standard solution. Spray the plate lightly with potassium iodoplatinate TS: the R_f value of the principal spot, representing the hydroxyzine moiety, obtained from the test solution corresponds to that obtained from the Standard solution.

Water, *Method I* ⟨921⟩: not more than 5.0%.

Residue on ignition ⟨281⟩: not more than 0.5%.

Heavy metals, *Method II* ⟨231⟩: 0.005%.

Pamoic acid content—Dissolve an accurately weighed quantity of USP Pamoic Acid RS in dimethylformamide to obtain a solution having a known concentration of about 0.45 mg per mL. Transfer 2.0 mL of the resulting solution to a 50-mL volumetric flask, dilute with *Mobile phase*, prepared as directed in the *Assay*, to volume, and mix. Filter through a membrane filter of 0.5-µm or finer porosity to obtain the *Standard preparation*. Chromatograph this *Standard preparation* as directed in the *Assay*. From the chromatogram of the *Assay preparation* obtained in the *Assay*, calculate the quantity, in mg, of pamoic acid ($C_{23}H_{16}O_6$) in the portion of Hydroxyzine Pamoate taken by the formula:

$$2500C(r_U/r_S),$$

in which C is the concentration, in mg per mL, of USP Pamoic Acid RS in the *Standard preparation*, and r_U and r_S are the pamoic acid peak responses obtained from the *Assay preparation* and the *Standard preparation*, respectively: the content of pamoic acid is between 49.4% and 51.9%.

Assay—

Mobile phase—Dissolve 8.65 g of sodium 1-octanesulfonate in about 1000 mL of water in a 2000-mL volumetric flask, add 4.0 mL of phosphoric acid, dilute with water to volume, mix, and filter through a membrane filter of 0.5-µm or finer porosity. Prepare a suitable mixture of this solution and acetonitrile (45:55), making any adjustments if necessary (see *System Suitability* under *Chromatography* ⟨621⟩).

Standard preparation—Dissolve an accurately weighed quantity of USP Hydroxyzine Hydrochloride RS in dimethylformamide to obtain a solution having a known concentration of about 1 mg per mL. Transfer 2.0 mL of the resulting solution to a 100-mL volumetric flask, dilute with *Mobile phase* to volume, mix, and filter through a membrane filter of 0.5-µm or finer porosity.

Assay preparation—Transfer about 90 mg of Hydroxyzine Pamoate, accurately weighed, to a 50-mL volumetric flask, dissolve in dimethylformamide, dilute with dimethylformamide to volume, and mix. Transfer 2.0 mL of the resulting solution to a 100-mL volumetric flask, dilute with *Mobile phase* to volume, mix, and filter through a membrane filter of 0.5-µm or finer porosity, discarding the first 5 mL of the filtrate.

Chromatographic system (see *Chromatography* ⟨621⟩)—The liquid chromatograph is equipped with a 230-nm detector and a 3.9-mm × 30-cm column that contains packing L1. The flow rate is about 1.5 mL per minute. Chromatograph the *Standard preparation*, and record the peak responses as directed under *Procedure:* the column efficiency, calculated from the analyte peak, is not less than 2000 theoretical plates, and the relative standard deviation for replicate injections is not more than 2%. Chromatograph the *Assay preparation*, and record the peak responses as directed under *Procedure:* the resolution, R, between hydroxyzine and pamoic acid is not less than 1.5.

Procedure—Separately inject equal volumes (about 50 µL) of the *Standard preparation* and the *Assay preparation* into the chromatograph, record the chromatograms, and measure the responses for the major peaks. The relative retention times are about 0.5 for hydroxyzine and 1.0 for pamoic acid. Calculate the quantity, in mg, of $C_{21}H_{27}ClN_2O_2.C_{23}H_{16}O_6$ in the portion of Hydroxyzine Pamoate taken by the formula:

$$2500(763.29/447.83)C(r_U/r_S),$$

in which 763.29 and 447.83 are the molecular weights of hydroxyzine pamoate and hydroxyzine hydrochloride, respectively, C is the concentration, in mg per mL, of USP Hydroxyzine Hydrochloride RS in the *Standard preparation*, and r_U and r_S are the hydroxyzine peak responses obtained from the *Assay preparation* and the *Standard preparation*, respectively.

Hydroxyzine Pamoate Capsules

» Hydroxyzine Pamoate Capsules contain hydroxyzine pamoate ($C_{21}H_{27}ClN_2O_2.C_{23}H_{16}O_6$) equivalent to not less than 90.0 percent and not more than 110.0 percent of the labeled amount of hydroxyzine hydrochloride ($C_{21}H_{27}ClN_2O_2.2HCl$).

Packaging and storage—Preserve in well-closed containers.

Reference standards—*USP Hydroxyzine Pamoate Reference Standard*—Dry over silica gel for 4 hours before using. *USP Hydroxyzine Hydrochloride Reference Standard*—Dry at 105° for 2 hours before using.

Identification—Dissolve a portion of the contents of Capsules, equivalent to about 100 mg of hydroxyzine pamoate, in a mixture of 25 mL of 0.1 *N* sodium hydroxide and 25 mL of acetone, and filter: a 10-µL portion of the filtrate responds to *Identification test B* under *Hydroxyzine Pamoate.*

Dissolution ⟨711⟩—

Medium: 0.1 *N* hydrochloric acid; 900 mL.

Apparatus 2: 50 rpm.

Time: 60 minutes.

Mobile phase—Prepare a suitable degassed solution of methanol and 0.05 *M* monobasic sodium phosphate (6:4).

Chromatographic system (see *Chromatography* ⟨621⟩)—The liquid chromatograph is equipped with a 232-nm detector and a 4.6-mm × 25-cm column that contains packing L9. The flow rate is about 1.9 mL per minute. Chromatograph replicate injections of a Standard solution, and record the peak responses as directed under *Procedure:* the relative standard deviation is not more than 2.0%.

Procedure—Inject alternately 50 µL of a filtered portion of the solution under test and a Standard solution, having a known concentration of USP Hydroxyzine Hydrochloride RS in the same medium, into the chromatograph, record the chromatogram, and measure the response for the major peak. Determine the amount of $C_{21}H_{27}ClN_2O_2.2HCl$ dissolved from the peak response obtained in comparison with the peak response obtained from the Standard solution.

Tolerances—Not less than 75% (*Q*) of the labeled amount of $C_{21}H_{27}ClN_2O_2.2HCl$ is dissolved in 60 minutes.

Uniformity of dosage units ⟨905⟩: meet the requirements.

Assay—

Mobile phase—Dissolve 7.0 g of monobasic sodium phosphate in 1000 mL of water, and adjust with phosphoric acid to a pH of 4.4. Filter the solution through a 5-µm porosity polytef membrane filter. Mix 900 mL of the filtrate with 900 mL of methanol, and degas by stirring under vacuum prior to use.

Standard preparation—Dissolve an accurately weighed quantity of USP Hydroxyzine Pamoate RS in methanol to obtain a solution having a known concentration of about 0.18 mg per mL (equivalent to about 0.1 mg of hydroxyzine hydrochloride per mL).

Assay preparation—Transfer, as completely as possible, the contents of not less than 20 Hydroxyzine Pamoate Capsules to a tared beaker, and determine the average weight per capsule. Mix the combined contents, and transfer an accurately weighed portion, equivalent to about 25 mg of hydroxyzine hydrochloride, to a 250-mL volumetric flask. Add 200 mL of methanol to the flask, sonicate for 5 minutes, shake by mechanical means for 30 minutes, and sonicate for 2 minutes. Dilute with methanol to volume, and mix. Filter the solution through a 5-µm porosity polytef membrane filter equipped with a glass fiber prefilter.

Chromatographic system (see *Chromatography* ⟨621⟩)—The liquid chromatograph is equipped with a 232-nm detector and a 4.6-mm × 25-cm column that contains packing L9. The flow rate is about 2.5 mL per minute. Chromatograph four replicate injections of the *Standard preparation,* and record the peak responses as directed under *Procedure:* the relative standard deviation is not more than 2.0%.

Procedure—Separately inject equal volumes (about 25 µL) of the *Standard preparation* and the *Assay preparation* into the chromatograph by means of a suitable microsyringe or sampling valve, record the chromatograms, and measure the responses for the major peaks. The retention time is about 6 minutes for hydroxyzine. Calculate the equivalent quantity, in mg, of hydroxyzine hydrochloride ($C_{21}H_{27}ClN_2O_2.2HCl$) in the portion of Capsules taken by the formula:

$$(447.83/763.29)(250C)(r_U/r_S),$$

in which 447.83 and 763.29 are the molecular weights of hydroxyzine hydrochloride and hydroxyzine pamoate, respectively, *C* is the concentration, in mg per mL, of USP Hydroxyzine Pamoate RS in the *Standard preparation,* and r_U and r_S are the peak responses obtained from the *Assay preparation* and the *Standard preparation,* respectively.

Hydroxyzine Pamoate Oral Suspension

» Hydroxyzine Pamoate Oral Suspension contains hydroxyzine pamoate ($C_{21}H_{27}ClN_2O_2.C_{23}H_{16}O_6$) equivalent to not less than 90.0 percent and not more than 110.0 percent of the labeled amount of hydroxyzine hydrochloride ($C_{21}H_{27}ClN_2O_2.2HCl$).

Packaging and storage—Preserve in tight, light-resistant containers.

Reference standards—*USP Hydroxyzine Pamoate Reference Standard*—Dry over silica gel for 4 hours before using. *USP Hydroxyzine Hydrochloride Reference Standard*—Dry at 105° for 2 hours before using.

Identification—Dissolve a portion of Oral Suspension, equivalent to about 100 mg of hydroxyzine pamoate, in a mixture of 25 mL of 0.1 *N* sodium hydroxide and 25 mL of acetone, and filter: a 10-µL portion of the filtrate responds to *Identification test B* under *Hydroxyzine Pamoate.*

pH ⟨791⟩: between 4.5 and 7.0.

Assay—

Mobile phase, Standard preparation, and *Chromatographic system*—Prepare as directed in the *Assay* under *Hydroxyzine Pamoate Capsules.*

Assay preparation—Transfer an accurately measured volume of Hydroxyzine Pamoate Oral Suspension, freshly mixed and free from air bubbles, equivalent to about 25 mg of hydroxyzine hydrochloride, to a 250-mL volumetric flask. Dissolve in methanol, dilute with methanol to volume, and mix. Filter this solution through a polytef membrane filter (1-µm or finer porosity).

Procedure—Proceed as directed for *Procedure* in the *Assay* under *Hydroxyzine Pamoate Capsules.* Calculate the equivalent quantity, in mg, of hydroxyzine hydrochloride ($C_{21}H_{27}ClN_2O_2.2HCl$) in each mL of the Hydroxyzine Pamoate Oral Suspension taken by the formula:

$$(447.83/763.29)(250C/V)(r_U/r_S),$$

in which *V* is the volume, in mL, of Oral Suspension taken, and the other terms are as defined therein.

Hyoscyamine

$C_{17}H_{23}NO_3$ 289.37

Benzeneacetic acid, α-(hydroxymethyl)-, 8-methyl-8-azabicyclo[3.2.1]oct-3-yl ester, [3(*S*)-*endo*]-.

1αH,5αH-Tropan-3α-ol (−)-tropate (ester) [*101-31-5*].

» Hyoscyamine contains not less than 98.0 percent and not more than 101.0 percent of $C_{17}H_{23}NO_3$, calculated on the dried basis.

Caution—Handle Hyoscyamine with exceptional care, since it is highly potent.

Packaging and storage—Preserve in tight, light-resistant containers.

Reference standard—*USP Hyoscyamine Sulfate Reference Standard*—Dry in vacuum at 105° for 16 hours before using.

Identification—

A: Transfer 30 mg of Hyoscyamine and 36 mg of USP Hyoscyamine Sulfate RS to individual 60-mL separators with the aid of 5-mL portions of water. To each separator add 1.5 mL of 1 *N* sodium hydroxide and 10 mL of chloroform. Shake for 1 minute, allow the layers to separate, and filter the chloroform extracts through separate filters of about 2 g of anhydrous granular sodium sulfate supported on pledgets of glass wool. Extract each aqueous layer with two additional 10-mL portions of chloroform, filtering and combining with the respective main extracts. Evaporate the chloroform solutions under reduced pressure to dryness, and dissolve each residue in 10 mL of carbon disulfide. The infrared absorption spectrum, determined in a 1-mm cell, of the solution obtained from the test specimen exhibits maxima only at the same wavelengths as that of the solution obtained from the Reference Standard.

B: Dissolve 60 mg in 1 mL of 0.2 *N* hydrochloric acid, and add gold chloride TS, dropwise with shaking, until a definite precipitate separates. Add a small amount of 3 *N* hydrochloric acid, dissolve the precipitate with the aid of heat, and then allow to cool: lustrous golden yellow scales are formed (*distinction from atropine and scopolamine*).

Melting range ⟨741⟩: between 106° and 109°.

Specific rotation ⟨781⟩: between −20° and −23°, calculated on the dried basis, determined in a solution in dilute alcohol (1 in 2) containing 100 mg of Hyoscyamine in each 10 mL.

Loss on drying ⟨731⟩—Dry it in vacuum over silica gel to constant weight: it loses not more than 0.2% of its weight.

Residue on ignition ⟨281⟩: not more than 0.1%.

Foreign alkaloids and other impurities—Prepare a solution of it in methanol containing 20 mg per mL, and by quantitative dilution of a portion of this solution with methanol, prepare a second solution of Hyoscyamine containing 1 mg per mL. On a suitable thin-layer chromatographic plate (see *Chromatography* ⟨621⟩), coated with a 0.5-mm layer of chromatographic silica gel, apply 25 µL of the first (20 mg per mL) Hyoscyamine solution, 1 µL of the second (1 mg per mL) Hyoscyamine solution, and 5 µL of a methanol solution of USP Hyoscyamine Sulfate RS containing 24 mg per mL. Allow the spots to dry, and develop the chromatogram in a solvent system consisting of a mixture of chloroform, acetone, and diethylamine (5:4:1) until the solvent front has moved about three-fourths of the length of the plate. Remove the plate from the developing chamber, mark the solvent front, and allow the solvent to evaporate. Locate the spots on the plate by spraying with potassium iodoplatinate TS. The R_f value of the principal spot obtained from each test solution corresponds to that obtained from the Standard solution, and no secondary spot obtained from the first Hyoscyamine solution exhibits intensity equal to or greater than the principal spot obtained from the second Hyoscyamine solution (0.2%).

Assay—Dissolve about 500 mg of Hyoscyamine, accurately weighed, in 50 mL of glacial acetic acid, add 1 drop of crystal violet TS, and titrate with 0.1 *N* perchloric acid VS to a green end-point. Perform a blank determination, and make any necessary correction. Each mL of 0.1 *N* perchloric acid is equivalent to 28.94 mg of $C_{17}H_{23}NO_3$.

Hyoscyamine Tablets

» Hyoscyamine Tablets contain not less than 90.0 percent and not more than 110.0 percent of the labeled amount of $C_{17}H_{23}NO_3$.

Packaging and storage—Preserve in well-closed, light-resistant containers.

Reference standard—*USP Hyoscyamine Sulfate Reference Standard*—Dry in vacuum at 105° for 16 hours before using.

Identification—Macerate a quantity of powdered Tablets, equivalent to about 5 mg of hyoscyamine, with 20 mL of water, and transfer to a separator. Render the solution alkaline with 6 *N* ammonium hydroxide, and extract the alkaloid with 50 mL of chloroform. Filter the chloroform layer, divide it into two equal portions, and evaporate each to dryness. Perform tests *A* and *B* on the residues.

A: To one portion of the dry residue add 2 drops of nitric acid, evaporate on a steam bath to dryness, and add a few drops of alcoholic potassium hydroxide TS: a violet color is produced.

B: Dissolve the other portion of the residue in 1 mL of 0.1 *N* hydrochloric acid, and add gold chloride TS, dropwise with shaking, until a definite precipitate separates. Slowly heat until the precipitate dissolves, and allow the solution to cool: lustrous golden yellow scales are formed.

Disintegration ⟨701⟩: 30 minutes, the use of disks being omitted.

Uniformity of dosage units ⟨905⟩: meet the requirements.

Assay—

pH 9.0 buffer—Dissolve 34.8 g of dibasic potassium phosphate in 900 mL of water, and adjust to a pH of 9.0, determined electrometrically, by the addition of 3 *N* hydrochloric acid or 1 *N* sodium hydroxide, as necessary, with mixing.

Internal standard solution—Dissolve about 25 mg of homatropine hydrobromide, accurately weighed, in water contained in a 50-mL volumetric flask, add water to volume, and mix. Prepare fresh daily.

Standard preparation—Dissolve about 10 mg of USP Hyoscyamine Sulfate RS, accurately weighed, in water contained in a 100-mL volumetric flask, add water to volume, and mix. Prepare fresh daily. Pipet 10.0 mL of this solution into a separator, add 2.0 mL of *Internal standard solution* and 5.0 mL of *pH 9.0 buffer*, and adjust with 1 *N* sodium hydroxide to a pH of 9.0. Extract with two 10-mL portions of methylene chloride, filter the methylene chloride extracts through 1 g of anhydrous sodium sulfate supported by a small cotton plug in a funnel into a 50-mL beaker, and evaporate under nitrogen to dryness. Dissolve the residue in 2.0 mL of methylene chloride.

Assay preparation—Weigh and finely powder not less than 20 Hyoscyamine Tablets. Transfer an accurately weighed portion of the powder, equivalent to about 0.43 mg of hyoscyamine, to a separator containing 5 mL of *pH 9.0 buffer*, and add, by pipet, 2.0 mL of *Internal standard solution*. Proceed as directed under *Standard preparation*, beginning with "adjust with 1 *N* sodium hydroxide to a pH of 9.0."

Chromatographic system—Under typical conditions, the instrument contains a 1.8-m × 2-mm glass column packed with 3 percent liquid phase G2 on support S1AB, cured as directed (see *Gas Chromatography* ⟨621⟩). Maintain the column at 225°, and use nitrogen as the carrier gas at a flow rate of 25 mL per minute.

System suitability—Chromatograph six to ten injections of the *Standard preparation*, and record peak areas as directed under *Procedure*. The analytical system is suitable for conducting this assay if the relative standard deviation for the ratio of the peak areas does not exceed 2.0%, the resolution factor is not less than 5, and the tailing factor does not exceed 2.0.

Procedure—Inject 1-µL portions of the *Assay preparation* and the *Standard preparation* successively into the gas chromatograph. Measure the areas under the peaks for hyoscyamine and homatropine in each chromatogram. Calculate the ratio, A_U, of the area of the hyoscyamine peak to the area of the internal standard peak in the chromatogram from the *Assay preparation*, and similarly calculate the ratio, A_S, in the chromatogram from the *Standard preparation*. Calculate the quantity, in mg, of $C_{17}H_{23}NO_3$ in the portion of Tablets taken by the formula:

$$(289.37/676.82)(W/10)(A_U/A_S),$$

in which 289.37 and 676.82 are the molecular weights of hyoscyamine and anhydrous hyoscyamine sulfate, respectively, and W is the weight, in mg, of USP Hyoscyamine Sulfate RS taken for the *Standard preparation*.

Hyoscyamine Hydrobromide

$C_{17}H_{23}NO_3.HBr$ 370.29
Benzeneacetic acid, α-(hydroxymethyl)-, 8-methyl-8-azabicyclo[3.2.1]oct-3-yl ester, hydrobromide [3(*S*)-*endo*]-.
1αH,5αH-Tropan-3α-ol (−)-tropate (ester) hydrobromide [*306-03-6*].

» Hyoscyamine Hydrobromide contains not less than 98.5 percent and not more than 100.5 percent of $C_{17}H_{23}NO_3.HBr$, calculated on the dried basis.
Caution—Handle Hyoscyamine Hydrobromide with exceptional care, since it is highly potent.

Packaging and storage—Preserve in tight, light-resistant containers.

Reference standard—*USP Hyoscyamine Sulfate Reference Standard*—Dry in vacuum at 105° for 16 hours before using.

Identification—

A: Transfer 30 mg of Hyoscyamine Hydrobromide and 36 mg of USP Hyoscyamine Sulfate RS to individual 60-mL separators with the aid of 5-mL portions of water. To each separator add 1.5 mL of 1 *N* sodium hydroxide and 10 mL of chloroform. Shake for 1 minute, allow the layers to separate, and filter the chloroform extracts through separate filters of about 2 g of anhydrous granular sodium sulfate supported on pledgets of glass wool. Extract each aqueous layer with two additional 10-mL portions of chloroform, filtering and combining with the respective main extracts. Evaporate the chloroform solutions under reduced pressure to dryness, and dissolve each residue in 10 mL of carbon disulfide. The infrared absorption spectrum, determined in a 1-mm cell, of the solution obtained from the test specimen exhibits maxima only at the same wavelengths as that of the solution obtained from the Reference Standard.

B: To about 1 mL of a solution (1 in 20) add gold chloride TS, dropwise with shaking, until a definite precipitate separates. Add a small amount of 3 *N* hydrochloric acid, dissolve the precipitate with the aid of heat, and allow the solution to cool: lustrous reddish brown scales that may be accompanied by reddish brown needles are formed (*distinction from atropine and scopolamine*).

C: To an aqueous solution (1 in 20) add silver nitrate TS: a yellowish white precipitate is formed, and it is insoluble in nitric acid.

Melting temperature ⟨741⟩: not less than 149°.

Specific rotation ⟨781⟩: not less than −24°, calculated on the dried basis, determined in a solution in water containing 500 mg of Hyoscyamine Hydrobromide in each 10 mL.

Loss on drying ⟨731⟩ Dry it at 105° for 2 hours: it loses not more than 1.0% of its weight.

Residue on ignition ⟨281⟩: not more than 0.2%.

Other alkaloids—Dissolve 250 mg in 1 mL of 0.1 *N* hydrochloric acid, and dilute with water to 15 mL. To 5 mL of the solution add a few drops of platinic chloride TS: no precipitate is formed immediately. To another 5-mL portion add 2 mL of 6 *N* ammonium hydroxide: the mixture may develop a slight opalescence, but no turbidity or precipitate is formed immediately.

Assay—Dissolve about 700 mg of Hyoscyamine Hydrobromide, accurately weighed, in a mixture of 50 mL of glacial acetic acid and 10 mL of mercuric acetate TS. Add 1 drop of crystal violet TS, and titrate with 0.1 *N* perchloric acid VS to a blue-green end-point. Perform a blank determination, and make any necessary correction. Each mL of 0.1 *N* perchloric acid is equivalent to 37.03 mg of $C_{17}H_{23}NO_3.HBr$.

Hyoscyamine Sulfate

$(C_{17}H_{23}NO_3)_2.H_2SO_4.2H_2O$ 712.85
Benzeneacetic acid, α-(hydroxymethyl)-, 8-methyl-8-azabicyclo[3.2.1]oct-3-yl ester, [3(*S*)-*endo*]-, sulfate (2:1), dihydrate.
1αH,5αH-Tropan-3α-ol (−)-tropate (ester) sulfate (2:1) (salt) dihydrate [*6835-16-1*].
Anhydrous 676.82 [*620-61-1*].

» Hyoscyamine Sulfate contains not less than 98.5 percent and not more than 100.5 percent of $(C_{17}H_{23}NO_3)_2.H_2SO_4$, calculated on the dried basis.
Caution—Handle Hyoscyamine Sulfate with exceptional care, since it is highly potent.

Packaging and storage—Preserve in tight, light-resistant containers.

Reference standard—*USP Hyoscyamine Sulfate Reference Standard*—Dry in vacuum at 105° for 16 hours before using.

Identification—

A: The infrared absorption spectrum of a potassium bromide dispersion of it, previously dried, exhibits maxima only at the same wavelengths as that of a similar preparation of USP Hyoscyamine Sulfate RS.

B: To about 1 mL of a solution (1 in 20) add gold chloride TS, dropwise with shaking, until a definite precipitate separates. Add a small amount of 3 *N* hydrochloric acid, dissolve the precipitate with the aid of heat, and allow the solution to cool: lustrous golden yellow scales are formed (*distinction from atropine and scopolamine*).

C: A solution (1 in 20) responds to the tests for *Sulfate* ⟨191⟩.

Melting temperature ⟨741⟩: not less than 200°.

Specific rotation ⟨781⟩: not less than −24°, calculated on the dried basis, determined in a solution containing 500 mg of Hyoscyamine Sulfate in each 10 mL.

Loss on drying ⟨731⟩—Dry it in vacuum at 105° for 16 hours: it loses between 2.0% and 5.5% of its weight.

Residue on ignition ⟨281⟩: not more than 0.2%.

Readily carbonizable substances ⟨271⟩—Dissolve 200 mg in 5 mL of sulfuric acid TS: the solution has no more color than Matching Fluid A.

Other alkaloids—Dissolve 250 mg in 1 mL of 0.1 *N* hydrochloric acid, and dilute with water to 15 mL. To 5 mL of the solution add a few drops of platinic chloride TS: no precipitate is formed. To another 5-mL portion add 2 mL of 6 *N* ammonium hydroxide: the mixture may develop a slight opalescence, but no turbidity of precipitate is formed immediately.

Assay—Dissolve about 1 g of Hyoscyamine Sulfate, accurately weighed, in 50 mL of glacial acetic acid, and titrate with 0.1 *N* perchloric acid VS, determining the end-point potentiometrically. Perform a blank determination, and make any necessary correction. Each mL of 0.1 *N* perchloric acid is equivalent to 67.68 mg of $(C_{17}H_{23}NO_3)_2.H_2SO_4$.

Hyoscyamine Sulfate Elixir

» Hyoscyamine Sulfate Elixir contains not less than 90.0 percent and not more than 110.0 percent of the labeled amount of $(C_{17}H_{23}NO_3)_2.H_2SO_4.2H_2O$.

Packaging and storage—Preserve in tight, light-resistant containers, at controlled room temperature.

Reference standard—*USP Hyoscyamine Sulfate Reference Standard*—[*Caution—Highly potent.*] Dry in vacuum at 105° for 16 hours before using.

Identification—

A: Transfer a quantity of Elixir, equivalent to about 1.25 mg of hyoscyamine sulfate, to a separator. Render alkaline with 6 *N* ammonium hydroxide, and extract with 50 mL of methylene chloride, filtering the extract into a beaker. Evaporate to dryness. Add 2 drops of nitric acid to the dry residue, and evaporate on a steam bath to dryness. Add a few drops of alcoholic potassium hydroxide TS: a violet color appears.

B: Transfer a quantity of Elixir, equivalent to about 2.5 mg of hyoscyamine sulfate, to a separator. Render acidic with 3 *N* sulfuric acid, and extract with methylene chloride. Discard this extract. Render the solution alkaline with 6 *N* ammonium hydroxide, and extract with methylene chloride, filtering the extract into a beaker. Evaporate to dryness. Dissolve the residue in a small amount of 0.1 *N* hydrochloric acid and add gold chloride TS, with shaking, until a definite precipitate separates. Slowly heat until the precipitate dissolves, and allow the solution to cool: lustrous golden yellow scales are formed.

pH ⟨791⟩: between 3.0 and 6.5.

Alcohol content ⟨611⟩: between 90.0% and 110.0% of the labeled amount of C_2H_5OH.

Assay—

Internal standard solution—Dissolve about 25 mg of homatropine hydrobromide in water contained in a 50-mL volumetric flask, add water to volume, and mix. Prepare fresh daily.

Standard preparation—Dissolve about 10 mg of USP Hyoscyamine Sulfate RS, accurately weighed, in water contained in a 100-mL volumetric flask, add water to volume, and mix. Prepare fresh daily. Pipet 5 mL of this solution into a separator, add 2.0 mL of *Internal standard solution*, and adjust with 6 *N* ammonium hydroxide to a pH of 9. Extract with three 10-mL portions of methylene chloride, filter the methylene chloride extracts through 1 g of anhydrous sodium sulfate supported by a small cotton plug in a funnel into a suitable container, and evaporate on a steam bath with the aid of a current of air to dryness. Do not heat past dryness. Dissolve the residue in 2.0 mL of methylene chloride.

Assay preparation—Transfer an accurately measured volume of Hyoscyamine Sulfate Elixir, equivalent to about 0.5 mg of hyoscyamine sulfate, to a separator containing 5 mL of water, add 5 mL of 1 *N* sulfuric acid, and extract with a 25-mL portion of methylene chloride, discarding the extract. Add 2.0 mL of *Internal standard solution.* Proceed as directed under *Standard preparation*, beginning with "adjust with 6 *N* ammonium hydroxide to a pH of 9."

Chromatographic system—Under typical conditions, the gas chromatograph contains a 1.8-m × 2-mm glass column packed with 3 percent liquid phase G3 on support S1AB, conditioned as directed (see *Chromatography* ⟨621⟩). Maintain the column at 225°, and use nitrogen as the carrier gas.

System suitability—Chromatograph a sufficient number of injections of the *Standard preparation*, and record peak areas as directed under *Procedure.* The analytical system is suitable for conducting this assay if the relative standard deviation for the ratio of the peak areas does not exceed 2.0%, the resolution factor is not less than 4.0, and the tailing factor does not exceed 2.0.

Procedure—Inject appropriate portions of the *Assay preparation* and the *Standard preparation* successively into the gas chromatograph. Measure the areas under the peaks for hyoscyamine and homatropine in each chromatogram. Calculate the ratio, R_U, of the area of the hyoscyamine peak to the area of the internal standard peak in the chromatogram from the *Assay preparation*, and similarly calculate the ratio R_S, in the chromatogram from the *Standard preparation.* Calculate the quantity, in mg, of $(C_{17}H_{23}NO_3)_2 \cdot H_2SO_4 \cdot 2H_2O$ in each mL of the Elixir taken by the formula:

$$0.05(1.053)(W/V)(R_U/R_S),$$

in which 1.053 is the ratio of the molecular weight of hydrated hyoscyamine sulfate to that of anhydrous hyoscyamine sulfate, *W* is the weight, in mg, of USP Hyoscyamine Sulfate RS taken for the *Standard preparation*, and *V* is the volume, in mL, of Elixir taken.

Hyoscyamine Sulfate Injection

» Hyoscyamine Sulfate Injection is a sterile solution of Hyoscyamine Sulfate in Water for Injection. It contains not less than 93.0 percent and not more than 107.0 percent of the labeled amount of $(C_{17}H_{23}NO_3)_2 \cdot H_2SO_4 \cdot 2H_2O$.

Packaging and storage—Preserve in single-dose or in multiple-dose containers, preferably of Type I glass, at controlled room temperature.

Reference standard—*USP Hyoscyamine Sulfate Reference Standard*—[*Caution*—*Highly potent*]. Dry in vacuum at 105° for 16 hours before using.

Identification—

A: Transfer a volume of Injection, equivalent to about 2.5 mg of hyoscyamine sulfate, to a separator. Render alkaline with 6 *N* ammonium hydroxide, and extract with 25 mL of methylene chloride, filtering the extract into a beaker. Evaporate to dryness. Add 2 drops of nitric acid to the dry residue, and evaporate on a steam bath to dryness. Add a few drops of alcoholic potassium hydroxide TS: a violet color is produced.

B: Transfer a volume of Injection, equivalent to about 2.5 mg of hyoscyamine sulfate, to a separator. Render the solution acidic with 3 *N* sulfuric acid, and extract with 30 mL of methylene chloride. Discard the extract. Render the solution alkaline with 6 *N* ammonium hydroxide, and extract with 30 mL of methylene chloride, filtering the extract into a beaker. Evaporate to dryness. Dissolve the residue in a small amount of 0.1 *N* hydrochloric acid, and add gold chloride TS, with shaking, until a definite precipitate separates. Heat until the precipitate dissolves, and allow the solution to cool: lustrous golden yellow scales are formed.

C: After evaporation to dryness, or appropriate adjustment of concentration, it responds to the tests for *Sulfate* ⟨191⟩.

D: The angular rotation of the Injection is levorotatory.

pH ⟨791⟩: between 3.0 and 6.5.

Other requirements—It meets the requirements under *Injections* ⟨1⟩.

Assay—

Internal standard solution—Dissolve about 25 mg of homatropine hydrobromide in water contained in a 50-mL volumetric flask, add water to volume, and mix. Prepare fresh daily.

Standard preparation—Dissolve about 10 mg of USP Hyoscyamine Sulfate RS, accurately weighed, in water contained in a 100-mL volumetric flask, add water to volume, and mix. Prepare fresh daily. Pipet 10 mL of this solution into a separator, add 2.0 mL of *Internal standard solution*, and adjust with 6 *N* ammonium hydroxide to a pH of 9. Extract with three 10-mL portions of methylene chloride, filter the methylene chloride extracts through 1 g of anhydrous sodium sulfate supported by a small cotton plug in a funnel into a suitable container, and evaporate on a steam bath with the aid of a current of air to dryness. Do not heat past dryness. Dissolve the residue in 2.0 mL of methylene chloride.

Assay preparation—Transfer an accurately measured volume of Hyoscyamine Sulfate Injection, equivalent to about 1.0 mg of hyoscyamine sulfate, to a separator containing 5 mL of water, and add 2.0 mL of *Internal standard solution.* Proceed as directed under *Standard preparation*, beginning with "adjust with 6 *N* ammonium hydroxide to a pH of 9."

Chromatographic system—Under typical conditions, the gas chromatograph contains a 1.8-m × 2-mm glass column packed with 3 percent liquid phase G3 on support S1AB, conditioned as directed (see *Chromatography* ⟨621⟩). Maintain the column at 225°, and use nitrogen as the carrier gas.

System suitability—Chromatograph a sufficient number of injections of the *Standard preparation*, and record peak areas as directed under *Procedure.* The analytical system is suitable for conducting this assay if the relative standard deviation for the ratio of the peak areas does not exceed 2.0%, the resolution factor is not less than 4.0, and the tailing factor does not exceed 2.0.

Procedure—Inject appropriate portions of the *Assay preparation* and the *Standard preparation* successively into the gas chromatograph. Measure the areas under the peaks for hyoscyamine and homatropine in each chromatogram. Calculate the ratio, R_U, of the area of the hyoscyamine peak to the area of the internal standard peak in the chromatogram from the *Assay preparation*, and similarly calculate the ratio R_S, in the chromatogram from the *Standard preparation.* Calculate the quantity, in mg, of $(C_{17}H_{23}NO_3)_2 \cdot H_2SO_4 \cdot 2H_2O$ in each mL of the Injection taken by the formula:

$$0.1(1.053)(W/V)(R_U/R_S),$$

in which 1.053 is the ratio of the molecular weight of hydrated hyoscyamine sulfate to that of anhydrous hyoscyamine sulfate, *W* is the weight, in mg, of USP Hyoscyamine Sulfate RS taken for the *Standard preparation*, and *V* is the volume, in mL, of Injection taken.

Hyoscyamine Sulfate Oral Solution

» Hyoscyamine Sulfate Oral Solution contains not less than 90.0 percent and not more than 110.0 percent of the labeled amount of $(C_{17}H_{23}NO_3)_2 . H_2SO_4 . 2H_2O$.

Packaging and storage—Preserve in tight, light-resistant containers, at controlled room temperature.

Reference standard—*USP Hyoscyamine Sulfate Reference Standard*—[*Caution—Highly potent.*] Dry in vacuum at 105° for 16 hours before using.

Identification—

A: Transfer a quantity of Oral Solution, equivalent to about 1.25 mg of hyoscyamine sulfate, to a separator. Render alkaline with 6 *N* ammonium hydroxide, and extract with 10 mL of methylene chloride, filtering the extract into a beaker. Evaporate to dryness. Add 2 drops of nitric acid to the dry residue, and evaporate on a steam bath to dryness. Add a few drops of alcoholic potassium hydroxide TS: a violet color is produced.

B: Transfer a quantity of Oral Solution, equivalent to about 2.5 mg of hyoscyamine sulfate to a separator. Render the solution acidic with 3 *N* sulfuric acid, and extract with methylene chloride. Discard the extract. Render the solution alkaline with 6 *N* ammonium hydroxide, and extract with methylene chloride, filtering the extract into a beaker. Evaporate to dryness. Dissolve the residue in a small amount of 0.1 *N* hydrochloric acid, and add gold chloride TS, with shaking, until a definite precipitate separates. Slowly heat until the precipitate dissolves, and allow the solution to cool: lustrous golden yellow scales are formed.

pH ⟨791⟩: between 3.0 and 6.5.

Assay—

Internal standard solution—Dissolve about 25 mg of homatropine hydrobromide in water contained in a 50-mL volumetric flask, add water to volume, and mix. Prepare fresh daily.

Standard preparation—Dissolve about 10 mg of USP Hyoscyamine Sulfate RS, accurately weighed, in water contained in a 100-mL volumetric flask, add water to volume, and mix. Prepare fresh daily. Pipet 10 mL of this solution into a separator, add 2.0 mL of *Internal standard solution*, and adjust with 6 *N* ammonium hydroxide to a pH of 9. Extract with three 15-mL portions of methylene chloride, filter the methylene chloride extracts through 1 g of anhydrous sodium sulfate supported by a small cotton plug in a funnel into a suitable container, and evaporate on a steam bath with the aid of a current of air to dryness. Do not heat past dryness. Dissolve the residue in 2.0 mL of methylene chloride.

Assay preparation—Transfer an accurately measured volume of Hyoscyamine Sulfate Oral Solution, equivalent to about 1.0 mg of hyoscyamine sulfate, to a separator containing 5 mL of water, add 5 mL of 1 *N* sulfuric acid, and extract with a 25-mL portion of methylene chloride, discarding the extract. Add 2.0 mL of *Internal standard solution*. Proceed as directed under *Standard preparation*, beginning with "adjust with 6 *N* ammonium hydroxide to a pH of 9."

Chromatographic system—Under typical conditions, the gas chromatograph contains a 1.8-m × 2-mm glass column packed with 3 percent liquid phase G3 on support S1AB, conditioned as directed (see *Chromatography* ⟨621⟩). Maintain the column at 225°, and use nitrogen as the carrier gas.

System suitability—Chromatograph a sufficient number of injections of the *Standard preparation*, and record peak areas as directed under *Procedure*. The analytical system is suitable for conducting this assay if the relative standard deviation for the ratio of the peak areas does not exceed 2.0%, the resolution factor is not less than 4.0, and the tailing factor does not exceed 2.0.

Procedure—Inject appropriate portions of the *Assay preparation* and the *Standard preparation* successively into the gas chromatograph. Measure the areas under the peaks for hyoscyamine and homatropine in each chromatogram. Calculate the ratio, R_U, of the area of the hyoscyamine peak to the area of the internal standard peak in the chromatogram from the *Assay preparation*, and similarly calculate the ratio, R_S, in the chromatogram from the *Standard preparation*. Calculate the quantity, in mg, of $(C_{17}H_{23}NO_3)_2 . H_2SO_4 . 2H_2O$ in each mL of the Oral Solution taken by the formula:

$$0.1(1.053)(W/V)(R_U/R_S),$$

in which 1.053 is the ratio of the molecular weight of hydrated hyoscyamine sulfate to that of anhydrous hyoscyamine sulfate, *W* is the weight, in mg, of USP Hyoscyamine Sulfate RS taken for the *Standard preparation*, and *V* is the volume, in mL, of Oral Solution taken.

Hyoscyamine Sulfate Tablets

» Hyoscyamine Sulfate Tablets contain not less than 90.0 percent and not more than 110.0 percent of the labeled amount of $(C_{17}H_{23}NO_3)_2 . H_2SO_4 . 2H_2O$.

Packaging and storage—Preserve in tight, light-resistant containers.

Reference standard—*USP Hyoscyamine Sulfate Reference Standard*—Dry in vacuum at 105° for 16 hours before using.

Identification—Macerate a quantity of powdered Tablets, equivalent to about 5 mg of hyoscyamine sulfate, with 20 mL of water, and transfer to a separator. Make the solution alkaline with 6 *N* ammonium hydroxide, and extract the alkaloid with 50 mL of chloroform. Filter the chloroform layer, divide it into two equal portions, and evaporate each to dryness. Perform tests *A* and *B* on the residues.

A: To one portion of the dry residue add 2 drops of nitric acid, evaporate on a steam bath to dryness, and add a few drops of alcoholic potassium hydroxide TS: a violet color is produced.

B: Dissolve the other portion of the residue in 1 mL of 0.1 *N* hydrochloric acid, and add gold chloride TS, dropwise with shaking, until a definite precipitate separates. Slowly heat until the precipitate dissolves, and allow the solution to cool: lustrous golden yellow scales are formed.

C: A filtered solution of Tablets responds to the tests for *Sulfate* ⟨191⟩.

Disintegration ⟨701⟩: 15 minutes.

Uniformity of dosage units ⟨905⟩: meet the requirements.

Assay—

pH 9.0 buffer—Dissolve 34.8 g of dibasic potassium phosphate in 900 mL of water, and adjust to a pH of 9.0, determined electrometrically, by the addition of 3 *N* hydrochloric acid or 1 *N* sodium hydroxide, as necessary, with mixing.

Internal standard solution—Dissolve about 25 mg of homatropine hydrobromide, accurately weighed, in water contained in a 50-mL volumetric flask, add water to volume, and mix. Prepare fresh daily.

Standard preparation—Dissolve about 10 mg of USP Hyoscyamine Sulfate RS, accurately weighed, in water contained in a 100-mL volumetric flask, add water to volume, and mix. Prepare fresh daily. Pipet 10.0 mL of this solution into a separator, add 2.0 mL of *Internal standard solution* and 5.0 mL of *pH 9.0 buffer*, and adjust with 1 *N* sodium hydroxide to a pH of 9.0. Extract with two 10-mL portions of methylene chloride, filter the methylene chloride extracts through 1 g of anhydrous sodium sulfate supported by a small cotton plug in a funnel into a 50-mL beaker, and evaporate under nitrogen to dryness. Dissolve the residue in 2.0 mL of methylene chloride.

Assay preparation—Weigh and finely powder not less than 20 Hyoscyamine Sulfate Tablets. Transfer an accurately weighed portion of the powder, equivalent to about 1.0 mg of hyoscyamine sulfate, to a separator containing 5 mL of *pH 9.0 buffer*, and add, by pipet, 2.0 mL of *Internal standard solution*. Proceed as directed under *Standard preparation*, beginning with "adjust with 1 *N* sodium hydroxide to a pH of 9.0."

Chromatographic system—Under typical conditions, the instrument contains a 1.8-m × 2-mm glass column packed with 3 percent G3 on S1AB, conditioned as directed (see *Chromatography* ⟨621⟩). Maintain the column at 225°, and use nitrogen as the carrier gas at a flow rate of 25 mL per minute.

System suitability—Chromatograph six to ten injections of the *Standard preparation*, and record peak areas as directed under *Procedure*. The analytical system is suitable for conducting this assay if the relative standard deviation for the ratio of the peak areas does not exceed 2.0%, the resolution factor is not less than 4.0, and the tailing factor does not exceed 2.0.

Procedure—Inject 1-µL portions of the *Assay preparation* and the *Standard preparation* successively into the gas chromatograph. Measure the areas under the peaks for hyoscyamine sulfate and homatropine hydrobromide in each chromatogram. Calculate the ratio, A_U, of the area of the hyoscyamine sulfate peak to the area of the internal standard peak in the chromatogram from the *Assay preparation*, and similarly calculate the ratio, A_S, in the chromatogram from the *Standard preparation*. Calculate the quantity, in mg, of $(C_{17}H_{23}NO_3)_2 . H_2SO_4 . 2H_2O$ in the portion of Tablets taken by the formula:

$$1.053(W/10)(A_U/A_S),$$

in which 1.053 is the ratio of the molecular weight of hydrated hyoscyamine sulfate to that of anhydrous hyoscyamine sulfate, and W is the weight, in mg, of USP Hyoscyamine Sulfate RS taken for the *Standard preparation*.

Hypophosphorous Acid—*see* Hypophosphorous Acid NF

Ibuprofen

$C_{13}H_{18}O_2$ 206.28

Benzeneacetic acid, α-methyl-4-(2-methylpropyl), (±)-.

(±)-*p*-Isobutylhydratropic acid.

(±)-2-(*p*-Isobutylphenyl)propionic acid [*15687-27-1*]; (±) mixture [*58560-75-1*].

» Ibuprofen contains not less than 97.0 percent and not more than 103.0 percent of $C_{13}H_{18}O_2$, calculated on the anhydrous basis.

Packaging and storage—Preserve in tight containers.

Reference standard—*USP Ibuprofen Reference Standard*—Do not dry.

Identification—

A: The infrared absorption spectrum of a mineral oil dispersion of it exhibits maxima only at the same wavelengths as that of a similar preparation of USP Ibuprofen RS.

B: The ultraviolet absorption spectrum of a 1 in 4000 solution in 0.1 *N* sodium hydroxide exhibits maxima and minima at the same wavelengths as that of a similar solution of USP Ibuprofen RS, concomitantly measured, and the respective absorptivities, calculated on the anhydrous basis, at the wavelengths of maximum absorbance at about 264 nm and 273 nm do not differ by more than 3.0%.

C: The chromatogram of the *Assay preparation* obtained as directed in the *Assay* exhibits a major peak for ibuprofen the retention time of which, relative to that of the internal standard, corresponds to that exhibited in the chromatogram of the *Standard preparation* obtained as directed in the *Assay*.

Water, *Method I* ⟨921⟩: not more than 1.0%.

Residue on ignition ⟨281⟩: not more than 0.5%.

Heavy metals, *Method II* ⟨231⟩: 0.002%.

Chromatographic purity—

Diazomethane solution—Add 35 mL of 2-(2-ethoxyethoxy)-ethanol and 20 mL of ether to 10 mL of 10 *N* potassium hydroxide in a 100-mL distilling flask fitted with a dropping funnel and an efficient condenser. Heat in a water bath at 70°, stir vigorously, and add through the dropping funnel a solution of 21.5 g of *N*-methyl-*N*-nitroso-*p*-toluenesulfonamide in 200 mL of ether at about the same rate as that of the distillation. When the dropping funnel is empty, add another 40 mL of ether through the dropping funnel at about the same rate as that of the distillation, and continue the distillation until the ether being distilled is colorless. Store the *Diazomethane solution* thus collected in a refrigerator (for up to 7 days) or in a freezer (for up to 6 weeks). [*Caution—Diazomethane is a yellow, toxic gas that may explode if heated. The distillation apparatus employed is not to contain ground glass joints or sharp glass edges. A refrigerator used for storage is to be explosion-proof or otherwise suitable for storage of ether.*]

Standard preparation—Transfer about 20 mg of USP Ibuprofen RS to a suitable vial, and slowly add *Diazomethane solution*, previously allowed to warm to room temperature, until the solution remains yellow (usually about 3 mL). Cap the vial, and allow to stand for 30 minutes. Remove the cap, and evaporate the solution with the aid of a gentle stream of nitrogen. Add 2 mL of methylene chloride to the residue, and swirl.

Test preparation—Proceed as directed for *Standard preparation*, using Ibuprofen instead of the Reference Standard.

Blank preparation—Proceed as directed for *Standard preparation*, except to omit USP Ibuprofen RS.

Chromatographic system—Proceed as directed for *Chromatographic system* in the *Assay*, but maintain the column at about 165°. The column efficiency determined from the analyte peak is not less than 1400 theoretical plates.

Procedure—Inject about 2.0 µL of the *Standard preparation* into the gas chromatograph, record the chromatogram, and locate the major ibuprofen peak. Similarly inject 2.0 µL of the *Test preparation* and 2.0 µL of the *Blank preparation*, and record the chromatograms: the total area of all peaks, other than that of the ibuprofen peak, recorded for the *Test preparation* over a time span of 0.4 to 4 times the retention time for ibuprofen, relative to the total area of all of the peaks recorded during the same time span, corrected for the areas of any peaks recorded for the *Blank preparation*, does not exceed 1.5%.

Assay—

Mobile phase—Dissolve 4.0 g of chloroacetic acid in 400 mL of water, and adjust with ammonium hydroxide to a pH of 3.0. Add 600 mL of acetonitrile, filter, and degas. Make adjustments if necessary (see *System Suitability* under *Chromatography* ⟨621⟩).

Internal standard solution—Prepare a solution of valerophenone in *Mobile phase* having a concentration of about 0.35 mg per mL.

Standard preparation—Dissolve an accurately weighed quantity of USP Ibuprofen RS in *Internal standard solution* to obtain a solution having a known concentration of about 12 mg per mL.

Assay preparation—Transfer about 1200 mg of Ibuprofen, accurately weighed, to a container, add 100.0 mL of *Internal standard solution*, and mix.

Chromatographic system (see *Chromatography* ⟨621⟩)—The liquid chromatograph is equipped with a 254-nm detector and a 4.6-mm × 25-cm column that contains packing L1. The flow rate is about 2 mL per minute. Chromatograph the *Standard preparation*, and record the peak responses as directed under *Procedure:* the resolution, R, between the analyte and internal standard peaks is not less than 2.5, and the relative standard deviation for replicate injections is not more than 2.0%.

Procedure—Separately inject equal volumes (about 5 µL) of the *Standard preparation* and the *Assay preparation* into the chromatograph, record the chromatograms, and measure the responses for the major peaks. The relative retention times are about 1.4 for the internal standard and 1.0 for ibuprofen. Calculate the quantity, in mg, of $C_{13}H_{18}O_2$ in the portion of Ibuprofen taken by the formula:

$$100C(R_U/R_S),$$

in which C is the concentration, in mg per mL, of USP Ibuprofen RS in the *Standard preparation*, and R_U and R_S are the peak response ratios obtained from the *Assay preparation* and the *Standard preparation*, respectively.

Ibuprofen Tablets

» Ibuprofen Tablets contain not less than 90.0 percent and not more than 110.0 percent of the labeled amount of $C_{13}H_{18}O_2$.

Packaging and storage—Preserve in well-closed containers.

Reference standard—*USP Ibuprofen Reference Standard*—Do not dry.

Identification—

A: Grind 1 Tablet to a fine powder in a mortar, add about 5 mL of chloroform, and swirl. Filter the mixture, and evaporate the filtrate with the aid of a stream of nitrogen to dryness: the infrared absorption spectrum of a mineral oil dispersion of the residue so obtained exhibits maxima only at the same wavelengths as that of a similar preparation of USP Ibuprofen RS.

B: Its retention time, relative to that of the internal standard, determined as directed in the *Assay*, corresponds to that of USP Ibuprofen RS.

Dissolution ⟨711⟩—

Medium: pH 7.2 phosphate buffer (see under *Buffers* in the section, *Reagents, Indicators, and Solutions*); 900 mL.

Apparatus 1: 150 rpm.

Time: 30 minutes.

Procedure—Determine the amount of $C_{13}H_{18}O_2$ dissolved from ultraviolet absorbances at the wavelength of maximum absorbance at about 221 nm of filtered portions of the solution under test, suitably diluted with *Dissolution Medium*, if necessary, in comparison with a Standard solution having a known concentration of USP Ibuprofen RS in the same medium.

Tolerances—Not less than 70% (*Q*) of the labeled amount of $C_{13}H_{18}O_2$ is dissolved in 30 minutes.

Uniformity of dosage units ⟨905⟩: meet the requirements.

Water, *Method I* ⟨921⟩: not more than 5.0%.

Assay—

Mobile phase, Internal standard solution, Standard preparation, and *Chromatographic system*—Prepare as directed in the *Assay* under *Ibuprofen.*

Assay preparation—Weigh and finely powder not less than 20 Ibuprofen Tablets. Transfer an accurately weighed portion of the powder, equivalent to about 1200 mg of ibuprofen, to a suitable container, add 100.0 mL of *Internal standard solution*, and shake for 10 minutes. Centrifuge a portion of the suspension so obtained and use the clear supernatant solution as the *Assay preparation.*

Procedure—Proceed as directed for *Procedure* in the *Assay* under *Ibuprofen.* Calculate the quantity, in mg, of $C_{13}H_{18}O_2$ in the portion of Tablets taken by the formula:

$$(CV/N)100C(R_U/R_S),$$

in which the terms are as defined therein.

Ichthammol

Ichthammol.
Ichthammol [*8029-68-3*].

» Ichthammol is obtained by the destructive distillation of certain bituminous schists, sulfonation of the distillate, and neutralization of the product with ammonia. Ichthammol yields not less than 2.5 percent of ammonia (NH_3) and not less than 10.0 percent of total sulfur (S).

Packaging and storage—Preserve in well-closed containers.

Identification—

A: Dilute 10 mL with 90 mL of water, and stir for 5 minutes with a magnetic stirrer. Add 25 mL of hydrochloric acid, and mix: a heavy, resinous precipitate is formed. Remove the liquid by decantation, and wash the precipitate with 2 *N* hydrochloric acid until the last washing is nearly colorless. Transfer the precipitate to absorbent paper, allow it to stand for 10 minutes, and then transfer 10 mg of the precipitate to a 250-mL conical flask. To the flask add 100 mL of ether, attach an air condenser to the flask, and stir for 30 minutes with a magnetic stirrer: the precipitate does not dissolve completely.

B: To a solution (1 in 10) add 1 *N* sodium hydroxide, and heat to the boiling point: ammonia is evolved.

Loss on drying ⟨731⟩—Dry it at 80° for 8 hours, and continue the drying at 100° for constant weight: it loses not more than 50.0% of its weight.

Residue on ignition ⟨281⟩: not more than 0.5%.

Limit for ammonium sulfate—Accurately weigh about 1 g, transfer to a 100-mL beaker, and add 25 mL of alcohol. Stir, filter, and wash the filter with a mixture of equal volumes of ether and alcohol until the last washing is clear and colorless. Air-dry the filter and residue, and pass 200 mL of warm water, slightly acidified with hydrochloric acid, through the residue on the filter. Heat the filtrate to boiling, add barium chloride TS in excess, and heat for 1 hour on a steam bath. Collect the precipitate of barium sulfate on a filter, wash it well, dry, and ignite to constant weight. Each g of barium sulfate is equivalent to 566.1 mg of $(NH_4)_2SO_4$. Ichthammol contains not more than 8.0% of ammonium sulfate.

Assay for ammonia—Dissolve about 5 g of Ichthammol, accurately weighed, in 100 mL of water, transfer the solution to a distillation flask, add 3 g of paraffin, and add 20 mL of sodium hydroxide solution (4 in 10). Connect the flask to a condenser by means of a spray trap, and immerse the lower outlet tube of the condenser in 30.0 mL of 0.5 *N* sulfuric acid VS. Distil slowly, collect about 50 mL of distillate, and then titrate the excess acid with 0.5 *N* sodium hydroxide VS, using methyl red TS as the indicator. Perform a blank determination, and make any necessary correction. Each mL of 0.5 *N* sulfuric acid is equivalent to 8.515 mg of NH_3.

Assay for total sulfur—Transfer from 500 mg to 800 mg of Ichthammol, accurately weighed, to a Kjeldahl flask with the aid of 20 mL of water. Add 3 g of potassium chlorate, then add slowly 30 mL of nitric acid, and evaporate the mixture on a hot plate to about 5 mL. Cool, repeat the oxidation with 3 g of potassium chlorate and 30 mL of nitric acid, and evaporate to about 5 mL. Add 25 mL of hydrochloric acid, and again evaporate to about 5 mL. Add 100 mL of water, heat to boiling, filter, and wash well. To the hot filtrate add 25 mL of barium chloride TS, and heat on a steam bath for 1 hour. Collect the barium sulfate on a previously ignited and tared filtering crucible, wash, dry, and ignite, then cool, and weigh. Each g of barium sulfate is equivalent to 137.4 mg of S.

Ichthammol Ointment

» Ichthammol Ointment contains an amount of Ichthammol equivalent to not less than 0.25 percent of ammonia (NH_3).

Ichthammol	100 g
Anhydrous Lanolin	100 g
Petrolatum	800 g
To make	1000 g

Thoroughly incorporate the Ichthammol with the Anhydrous Lanolin, and combine this mixture with the Petrolatum.

Packaging and storage—Preserve in collapsible tubes or in tight containers, and avoid prolonged exposure to temperatures exceeding 30°.

Assay—

Assay preparation—Transfer an accurately weighed portion of Ichthammol Ointment, equivalent to about 2 g of ichthammol, to a 250-mL beaker, and add about 70 mL of boiling water. Mix with a glass rod, heat on a steam bath, with frequent agitation, for 10 minutes, cover with a watch glass without removing the stirring rod, and allow to stand at room temperature for 15 to 20 minutes. Place in a refrigerator to cause the upper layer to congeal, form a second opening through the congealed layer with the glass rod, and transfer the dark-colored aqueous extract to a funnel containing a pledget of cotton, collecting the filtrate in a 500-mL volumetric flask. Repeat the extraction of the portion of the Ointment several times in the same manner until the aqueous extract is practically colorless, passing each extract through the same cotton filter into the flask containing the main extract. Dilute with water to volume, and mix.

Procedure for ammonia—Transfer 100.0 mL of the *Assay preparation* to a suitable distillation flask, add 3 g of paraffin, and add 20 mL of sodium hydroxide solution (4 in 10). Connect the flask to a condenser by means of a spray trap, and immerse the lower outlet tube of the condenser in 30.0 mL of 0.05 *N* sulfuric acid VS. Distil slowly, collect about 50 mL of distillate, and then titrate the excess acid with 0.05 *N* sodium hydroxide VS, using methyl red TS as the indicator. Perform a blank determination, and make any necessary correction. Each mL of 0.05 *N* sulfuric acid is equivalent to 0.8515 mg of NH_3.

Idoxuridine

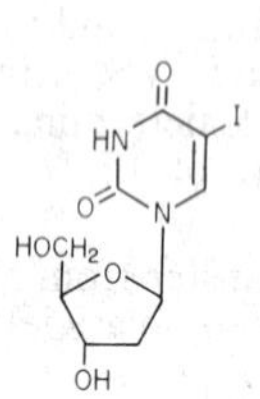

$C_9H_{11}IN_2O_5$ 354.10

Uridine, 2′-deoxy-5-iodo-.

2′-Deoxy-5-iodouridine [*54-42-2*].

» Idoxuridine contains not less than 98.0 percent and not more than 101.0 percent of $C_9H_{11}IN_2O_5$, calculated on the dried basis.

Packaging and storage—Preserve in tight, light-resistant containers.

Reference standard—*USP Idoxuridine Reference Standard*—Dry in vacuum at 60° for 2 hours before using.

Identification—

A: The infrared absorption spectrum of a mineral oil dispersion of it exhibits maxima only at the same wavelengths as that of a similar preparation of USP Idoxuridine RS.

B: The ultraviolet absorption spectrum of a 1 in 30,000 solution in a pH 12.0 buffer (prepared from 7.46 g of potassium chloride and 24 mL of 1 *N* sodium hydroxide dissolved in 2000 mL of water) exhibits maxima and minima at the same wavelengths as that of a similar solution of USP Idoxuridine RS, concomitantly measured, and the respective absorptivities, calculated on the dried basis, at the wavelength of maximum absorbance at about 279 nm do not differ by more than 2.0%.

Loss on drying ⟨731⟩—Dry about 500 mg, accurately weighed, in vacuum at 60° for 2 hours: it loses not more than 1.0% of its weight.

Assay—Dissolve about 250 mg of Idoxuridine, accurately weighed, in 20 mL of dimethylformamide that previously has been neutralized with 0.1 *N* sodium methoxide in benzene VS, a solution of 300 mg of thymol blue in 100 mL of methanol being used as the indicator. Titrate with 0.1 *N* sodium methoxide in benzene VS to a blue end-point, taking precautions against absorption of atmospheric carbon dioxide. Perform a blank determination, and make any necessary correction. Each mL of 0.1 *N* sodium methoxide is equivalent to 35.41 mg of $C_9H_{11}IN_2O_5$.

Idoxuridine Ophthalmic Ointment

» Idoxuridine Ophthalmic Ointment is Idoxuridine in a Petrolatum base. It contains not less than 0.45 percent and not more than 0.55 percent of $C_9H_{11}IN_2O_5$. It is sterile.

Packaging and storage—Preserve in collapsible ophthalmic ointment tubes in a cool place.

Reference standard—*USP Idoxuridine Reference Standard*—Dry in vacuum at 60° for 2 hours before using.

Identification—The ultraviolet absorption spectrum of the solution from the Ophthalmic Ointment employed for measurement of absorbance in the *Assay* exhibits maxima and minima at the same wavelengths as that of the *Standard preparation* prepared for the *Assay*.

Sterility—It meets the requirements for *Ophthalmic Ointments* under *Sterility Tests* ⟨71⟩.

Metal particles—It meets the requirements of the test for *Metal Particles in Ophthalmic Ointments* ⟨751⟩.

Assay—

Chromatographic column—Mix 4 g of chromatographic siliceous earth with 4 mL of 0.1 *N* hydrochloric acid in a glass mortar until the mixture is fluffy. Transfer to a 19- × 250-mm chromatographic tube (see *Chromatography* ⟨621⟩) that contains a pledget of glass wool and is fitted with a stopcock at the bottom. Tamp gently to compress to a uniform mass.

Standard preparation—Transfer about 25 mg of USP Idoxuridine RS, accurately weighed, to a 50-mL volumetric flask, add methanol to volume, and mix. Dilute 5.0 mL of this solution with a mixture of 1 volume of butyl alcohol and 5 volumes of chloroform to 100.0 mL, and mix.

Assay preparation—Mix 4 g of chromatographic siliceous earth with 2 mL of 0.1 *N* hydrochloric acid in a glass mortar until the mixture is fluffy. Add a quantity of Idoxuridine Ophthalmic Ointment, equivalent to about 5 mg of idoxuridine and accurately weighed, to the mixture, and mix.

Procedure—Transfer the *Assay preparation* to the prepared *Chromatographic column*. Transfer 2 g of chromatographic siliceous earth and 2 mL of 0.1 *N* hydrochloric acid to the glass mortar, and mix until fluffy, using this material to rinse the mortar and pick up any remaining Ophthalmic Ointment. Transfer about half of this mixture to the tube, and tamp gently until the column appears uniform. Transfer the remaining portion to the *Chromatographic column*, and tamp as before. Wipe the walls of the mortar with a small pledget of glass wool, and insert the pledget in the top of the column. Pass 50 mL of chloroform through the column at a flow rate of approximately 1 mL per minute, and discard the chloroform. Elute with about 200 mL of a mixture of 1 volume of butyl alcohol and 5 volumes of chloroform at the same flow rate, discarding the first 20 mL of eluate. Collect the remainder of the eluate in a 200-mL volumetric flask, dilute with the eluting solvent to volume, and mix. Concomitantly determine the absorbances of this solution and the *Standard preparation* in 1-cm cells at 320 nm and at the wavelength of maximum absorbance at about 283 nm, with a suitable spectrophotometer, using the butyl alcohol–chloroform mixture as the blank. Calculate the quantity, in mg, of $C_9H_{11}IN_2O_5$ in the Ophthalmic Ointment taken by the formula:

$$0.2C(A_{283} - A_{320})_U/(A_{283} - A_{320})_S,$$

in which C is the concentration, in µg per mL, of USP Idoxuridine RS in the *Standard preparation*, and the parenthetic expressions are the differences in the absorbances of the two solutions at the wavelengths indicated by the subscripts, for the solution from the Ophthalmic Ointment ($_U$) and the *Standard preparation* ($_S$), respectively.

Idoxuridine Ophthalmic Solution

» Idoxuridine Ophthalmic Solution is a sterile, aqueous solution of Idoxuridine. It contains not less than 0.09 percent and not more than 0.11 percent of $C_9H_{11}IN_2O_5$. It may contain suitable buffers, stabilizers, and antimicrobial agents.

Packaging and storage—Preserve in tight, light-resistant containers in a cold place.

Reference standard—*USP Idoxuridine Reference Standard*—Dry in vacuum at 60° for 2 hours before using.

Identification—The ultraviolet absorption spectrum of the solution employed for measurement of absorbance in the *Assay* exhibits maxima and minima at the same wavelengths as that of the *Standard preparation* prepared for the *Assay*.

Sterility—It meets the requirements under *Sterility Tests* ⟨71⟩.

pH ⟨791⟩: between 4.5 and 7.0.

Assay—

Chromatographic column and *Standard preparation*—Prepare as directed in the *Assay* under *Idoxuridine Ophthalmic Ointment*.

Assay preparation—Mix an accurately measured volume of Idoxuridine Ophthalmic Solution, equivalent to about 5 mg of idoxuridine, with 3 g of chromatographic siliceous earth in a glass mortar until the mixture is fluffy.

Procedure—Proceed as directed for *Procedure* in the *Assay* under *Idoxuridine Ophthalmic Ointment*, omitting the treatment of the column with 50 mL of chloroform. Calculate the quantity, in mg, of $C_9H_{11}IN_2O_5$ in each mL of the Ophthalmic Solution taken by the formula:

$$0.2C(A_{283} - A_{320})_U/V(A_{283} - A_{320})_S,$$

in which C is the concentration, in µg per mL, of USP Idoxuridine RS in the *Standard preparation*, V is the volume, in mL, of Ophthalmic Solution taken, and the parenthetic expressions are the differences in the absorbances of the two solutions at the wavelengths indicated by the subscripts, for the Solution ($_U$) and the *Standard preparation* ($_S$), respectively.

Imidurea—*see* Imidurea NF

Imipramine Hydrochloride

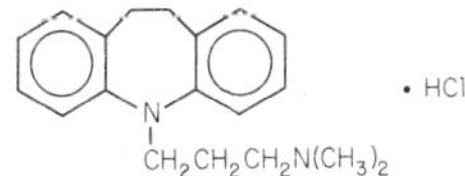

$C_{19}H_{24}N_2 . HCl$ 316.87

5*H*-Dibenz[*b,f*]azepine-5-propanamine, 10,11-dihydro-*N,N*-dimethyl-, monohydrochloride.

5-[3-(Dimethylamino)propyl]-10,11-dihydro-5*H*-dibenz[*b,f*]-azepine monohydrochloride   [*113-52-0*].

» Imipramine Hydrochloride contains not less than 98.0 percent and not more than 102.0 percent of $C_{19}H_{24}N_2 . HCl$, calculated on the dried basis.

Packaging and storage—Preserve in tight containers.

Reference standard—*USP Imipramine Hydrochloride Reference Standard*—Dry at 105° for 2 hours before using.

Identification—

A: The infrared absorption spectrum of a potassium bromide dispersion of it, previously dried, exhibits maxima only at the same wavelengths as that of a similar preparation of USP Imipramine Hydrochloride RS.

B: The ultraviolet absorption spectrum of a 1 in 50,000 solution in 0.1 *N* hydrochloric acid exhibits maxima and minima at the same wavelengths as that of a similar solution of USP Imipramine Hydrochloride RS, concomitantly measured, and the respective absorptivities, calculated on the dried basis, at the wavelength of maximum absorbance at about 250 nm do not differ by more than 3.0%.

C: Dissolve 0.10 g in 2 mL of alcohol, and add 1 mL of 2 *N* nitric acid and 3 drops of silver nitrate TS: a white precipitate is formed, and it dissolves on the dropwise addition of ammonium hydroxide.

Melting range ⟨741⟩: between 170° and 174°.

Loss on drying ⟨731⟩—Dry it at 105° for 2 hours: it loses not more than 0.5% of its weight.

Residue on ignition ⟨281⟩: not more than 0.1%.

Heavy metals, *Method II* ⟨231⟩: 0.001%.

Iminodibenzyl—

Standard preparation—Dissolve an accurately weighed quantity of USP Iminodibenzyl RS in alcohol, and dilute quantitatively and stepwise with alcohol to obtain a solution having a concentration of 50 µg per mL. Transfer 1.0 mL of this solution to a low-actinic, 25-mL volumetric flask, add 10 mL of a mixture of equal volumes of hydrochloric acid and alcohol, and mix.

Test preparation Transfer 50 mg of Imipramine Hydrochloride to a low-actinic 25-mL volumetric flask, add 10 mL of a mixture of equal volumes of hydrochloric acid and alcohol, and mix.

Procedure—To the two flasks containing the *Standard preparation* and the *Test preparation*, and to a third 25-mL volumetric flask containing 10 mL of a mixture of equal volumes of hydrochloric acid and alcohol to provide the blank, add slowly 5 mL of a 0.4% (v/v) solution of furfural in alcohol, mix, then add 5 mL of hydrochloric acid, and allow the flasks to stand in a constant-temperature bath at 25° for 3 hours. Dilute each flask to volume with a mixture of equal volumes of hydrochloric acid and alcohol, and mix. Concomitantly determine the absorbances of the solutions, in 1-cm cells, at the wavelength of maximum absorbance at about 565 nm, with a suitable spectrophotometer, using the blank to set the instrument: the absorbance of the solution from the *Test preparation* is not greater than that from the *Standard preparation* (0.1%).

Assay—Dissolve about 0.3 g of Imipramine Hydrochloride, accurately weighed, in 80 mL of glacial acetic acid. Add 10 mL of mercuric acetate TS and 1 drop of crystal violet TS, and titrate with 0.1 *N* perchloric acid VS to a blue end-point. Perform a blank determination, and make any necessary correction. Each mL of 0.1 *N* perchloric acid is equivalent to 31.69 mg of $C_{19}H_{24}N_2 . HCl$.

Imipramine Hydrochloride Injection

» Imipramine Hydrochloride Injection is a sterile solution of Imipramine Hydrochloride in Water for Injection. It contains, in each mL, not less than 11.5 mg and not more than 13.5 mg of $C_{19}H_{24}N_2 . HCl$.

Packaging and storage—Preserve in single-dose containers, preferably of Type I glass.

Reference standard—*USP Imipramine Hydrochloride Reference Standard*—Dry at 105° for 2 hours before using.

Identification—Transfer 10 mL of Injection to a separator, add 2 mL of 2 *N* hydrochloric acid, extract with 10 mL of chloroform, filter, and evaporate the chloroform solution to about 2 mL. Carefully add ether until the liquid becomes turbid, heat on a steam bath to produce a clear solution, then cool, and allow to stand. Filter the crystalline precipitate, wash with ether, and dry in vacuum at 105° for 30 minutes: the precipitate so obtained responds to *Identification test A* under *Imipramine Hydrochloride*.

pH ⟨791⟩: between 4.0 and 5.0.

Other requirements—It meets the requirements under *Injections* ⟨1⟩.

Assay—Transfer an accurately measured volume of Imipramine Hydrochloride Injection, equivalent to about 25 mg of imipramine hydrochloride, to a 100-mL volumetric flask, add 0.5 *N* hydrochloric acid to volume, and mix. Pipet 10 mL of this solution into a separator, add 10 mL of 1 *N* sodium hydroxide, and extract with four 20-mL portions of ether, shaking each portion for 2 minutes and collecting the extracts in a second separator. Extract the combined ether extracts with four 20-mL portions of 0.5 *N* hydrochloric acid, and combine the extracts in a 250-mL beaker. Aerate this solution with nitrogen to remove residual ether, then transfer to a 100-mL volumetric flask, and rinse the beaker with 0.5 *N* hydrochloric acid, collecting the rinsings in the flask. Add the 0.5 *N* acid to volume, and mix. Dissolve an accurately weighed quantity of USP Imipramine Hydrochloride RS in 0.5 *N* hydrochloric acid, and dilute quantitatively and stepwise with the same solvent to obtain a Standard solution having a known concentration of about 25 µg per mL. Concomitantly determine the absorbances of both solutions in 1-cm cells at the wavelength of maximum absorbance at about 250 nm, with a suitable spectrophotometer, using 0.5 *N* hydrochloric acid as the blank. Calculate the quantity, in mg, of $C_{19}H_{24}N_2 . HCl$ in each mL of the Injection taken by the formula:

$$(C/V)(A_U/A_S),$$

in which *C* is the concentration, in µg per mL, of USP Imipramine Hydrochloride RS in the Standard solution, *V* is the volume, in mL, of Injection taken, and A_U and A_S are the absorbances of the solution from the Injection and the Standard solution, respectively.

Imipramine Hydrochloride Tablets

» Imipramine Hydrochloride Tablets contain not less than 93.0 percent and not more than 107.0 percent of the labeled amount of $C_{19}H_{24}N_2 . HCl$.

Packaging and storage—Preserve in tight containers.

Reference standard—*USP Imipramine Hydrochloride Reference Standard*—Dry at 105° for 2 hours before using.

Identification—Powder a suitable number of Tablets, equivalent to 100 mg of imipramine hydrochloride, and macerate the powder with 10 mL of chloroform. Filter the chloroform extract through paper into a wide-mouth test tube, and evaporate the filtrate to about 3 mL. Carefully add ether until the liquid becomes turbid, heat on a steam bath to produce a clear solution, then cool, and allow to stand. The precipitate that is formed may be recrystallized from acetone. Filter the crystalline precipitate, wash with ether, and dry in vacuum at 105° for 30 minutes: the precipitate so obtained responds to *Identification test A* under *Imipramine Hydrochloride.*

Dissolution ⟨711⟩—

Medium: 0.1 *N* hydrochloric acid; 900 mL.

Apparatus 1: 100 rpm.

Time: 45 minutes.

Procedure—Determine the amount of $C_{19}H_{24}N_2 . HCl$ dissolved from ultraviolet absorbances at the wavelength of maximum absorbance at about 250 nm of filtered portions of the solution under test, suitably diluted with *Dissolution Medium,* in comparison with a Standard solution having a known concentration of USP Imipramine Hydrochloride RS in the same medium.

Tolerances—Not less than 75% (*Q*) of the labeled amount of $C_{19}H_{24}N_2 . HCl$ is dissolved in 45 minutes.

Uniformity of dosage units ⟨905⟩: meet the requirements.

Procedure for content uniformity—Transfer 1 finely powdered Tablet to a 100-mL volumetric flask with the aid of 70 mL of dilute hydrochloric acid (1 in 100), and shake by mechanical means for 30 minutes. Add dilute hydrochloric acid (1 in 100) to volume, mix, and filter, if necessary, discarding the first 20 mL of the filtrate. Transfer an aliquot of the filtrate, equivalent to about 2.5 mg of imipramine hydrochloride, to a 100-mL volumetric flask, add dilute hydrochloric acid (1 in 100) to volume, and mix. Dissolve an accurately weighed quantity of USP Imipramine Hydrochloride RS in dilute hydrochloric acid (1 in 100), and dilute quantitatively and stepwise with the same solvent to obtain a Standard solution having a known concentration of about 25 µg per mL. Concomitantly determine the absorbances of both solutions in 1-cm cells at the wavelength of maximum absorbance at about 250 nm, with a suitable spectrophotometer, using dilute hydrochloric acid (1 in 100) as the blank. Calculate the quantity, in mg, of $C_{19}H_{24}N_2 . HCl$ in the Tablet by the formula:

$$10(C/V)(A_U/A_S),$$

in which *C* is the concentration, in µg per mL, of USP Imipramine Hydrochloride RS in the Standard solution, *V* is the volume, in mL, of the aliquot taken of the solution from the Tablet, and A_U and A_S are the absorbances of the solution from the Tablet and the Standard solution, respectively.

Assay—Weigh and finely powder not less than 20 Imipramine Hydrochloride Tablets. Transfer an accurately weighed portion of the powder, equivalent to about 100 mg of imipramine hydrochloride, to a 200-mL volumetric flask, add about 100 mL of dilute hydrochloric acid (1 in 25), and shake vigorously by mechanical means for 1 hour. Add the dilute acid to volume, mix, and filter, discarding the first 20 mL of the filtrate. Pipet 5 mL of the filtrate into a separator, and proceed as directed in the *Assay* under *Imipramine Hydrochloride Injection,* beginning with "add 10 mL of 1 *N* sodium hydroxide." Calculate the quantity, in mg, of $C_{19}H_{24}N_2 . HCl$ in the portion of Tablets taken by the formula:

$$4C(A_U/A_S),$$

in which *C* is the concentration, in µg per mL, of USP Imipramine Hydrochloride RS in the Standard solution, and A_U and A_S are the absorbances of the solution from the Tablets and the Standard solution, respectively.

Immune Globulin—*see* Globulin, Immune

Immune Globulin, Varicella-Zoster—*see* Varicella-Zoster Immune Globulin

Indigotindisulfonate Sodium

$C_{16}H_8N_2Na_2O_8S_2$ 466.35

1*H*-Indole-5-sulfonic acid, 2-(1,3-dihydro-3-oxo-5-sulfo-2*H*-indol-2-ylidene)-2,3-dihydro-3-oxo-, disodium salt.

Disodium 3,3′-dioxo[$\Delta^{2,2'}$-biindoline]-5,5′-disulfonate [*860-22-0*].

» Indigotindisulfonate Sodium contains not less than 96.0 percent and not more than 102.0 percent of sodium indigotinsulfonates, calculated on the dried basis as $C_{16}H_8N_2Na_2O_8S_2$.

Packaging and storage—Preserve in tight, light-resistant containers.

Reference standard—*USP Indigotindisulfonate Sodium Reference Standard*—Dry at 105° for 3 hours before using.

Identification—

A: Incinerate a portion of it: the residue responds to the tests for *Sodium* ⟨191⟩ and for *Sulfate* ⟨191⟩.

B: The addition of hydrochloric acid to a solution of it changes the color to bluish violet, and further dilution with water restores the original color.

C: The addition of 1 *N* sodium hydroxide to a solution of it changes the color to yellow or olive-brown.

D: The addition of sodium chloride to a solution of it produces a blue precipitate.

Loss on drying ⟨731⟩—Dry it at 105° for 3 hours: it loses not more than 5.0% of its weight.

Water-insoluble substances—Dissolve 1.0 g in 100 mL of water, filter through a tared filtering crucible, wash with water until the filtrate is practically colorless, and dry the residue at 105° for 1 hour: the weight of the residue does not exceed 5 mg.

Arsenic, *Method II* ⟨211⟩: 8 ppm.

Lead—Place 4.0 g in a Kjeldahl flask, moisten with water, and add 10 mL of sulfuric acid and 5 mL of nitric acid. As soon as the first violent reaction subsides, heat until most of the brown fumes are expelled. Repeat the addition of nitric acid, 1 to 3 mL at a time, and heat until the Indigotindisulfonate Sodium is practically decomposed and most of the organic matter is in solution. Then add, *cautiously and in small portions*, 5 mL of perchloric acid. When the violent reaction subsides, continue the addition of small amounts of nitric acid, and heat as before until a colorless solution is obtained. (If the solution fails to become clear in 10 to 20 minutes after the addition of the perchloric acid, add 1 to 3 mL more of this acid, and continue the nitric acid treatment until the solution is colorless.) Boil for 10 to 15 minutes, cool, and neutralize with 1 *N* sodium hydroxide. Transfer to a 100-mL volumetric flask, and dilute with water to volume. Five mL of this solution contains not more than 2 µg of lead (corresponding to not more than 0.001%) when tested according to the limit test for *Lead* ⟨251⟩, 3 mL of *Ammonium Citrate Solution*, 1 mL of *Potassium Cyanide Solution*, and 0.5 mL of *Hydroxylamine Hydrochloride Solution* being used.

Sulfur content—Place about 25 mg, accurately weighed, in halide-free filter paper measuring about 4 cm square, and fold the paper to enclose it. Proceed as directed under *Oxygen Flask Combustion* ⟨471⟩, using a 1-liter flask and using a mixture of 25 mL of water and 5 mL of hydrogen peroxide TS as the absorbing liquid. When the combustion is complete, place a few mL of water in the cup, loosen the stopper, then rinse the stopper, the specimen holder, and the sides of the flask with about 20 mL of water. Add 2 mL of hydrochloric acid, dilute with water to 250 mL, heat to boiling, and slowly add 10 mL of barium chloride TS. Heat the mixture on a steam bath for 1 hour, collect the precipitate of barium sulfate on a filter, wash it until free from chloride, dry, ignite, and weigh. Each g of residue is equivalent to 137.4 mg of sulfur (S). Between 13.0% and 14.0%, calculated on the dried basis, of S is found.

Assay—Dissolve about 500 mg of Indigotindisulfonate Sodium, accurately weighed, in dilute hydrochloric acid (1 in 100), and dilute quantitatively and stepwise with the dilute acid to obtain a solution containing about 10 µg per mL. Concomitantly determine the absorbances of this solution and a Standard solution of USP Indigotindisulfonate Sodium RS in the same medium having a known concentration of about 10 µg per mL, in 1-cm cells at the wavelength of maximum absorbance at about 610 nm, with a suitable spectrophotometer, using dilute hydrochloric acid (1 in 100) as the blank. Calculate the quantity, in mg, of $C_{16}H_8N_2Na_2O_8S_2$ in the portion of Indigotindisulfonate Sodium taken by the formula:

$$50C(A_U/A_S),$$

in which C is the concentration, in µg per mL, of USP Indigotindisulfonate Sodium RS in the Standard solution, and A_U and A_S are the absorbances of the solution of Indigotindisulfonate Sodium and the Standard solution, respectively.

Indigotindisulfonate Sodium Injection

» Indigotindisulfonate Sodium Injection is a sterile solution of Indigotindisulfonate Sodium in Water for Injection. It contains not less than 90.0 percent and not more than 105.0 percent of the labeled amount of $C_{16}H_8N_2Na_2O_8S_2$.

Packaging and storage—Preserve in single-dose, light-resistant containers, preferably of Type I glass.

Reference standard—*USP Indigotindisulfonate Sodium Reference Standard*—Dry at 105° for 3 hours before using.

Identification—It responds to *Identification tests B, C,* and *D* under *Indigotindisulfonate Sodium*.

Pyrogen—It meets the requirements of the *Pyrogen Test* ⟨151⟩, the test dose being 1 mL per kg.

pH ⟨791⟩: between 3.0 and 6.5.

Other requirements—It meets the requirements under *Injections* ⟨1⟩.

Assay—Quantitatively dilute a portion of Indigotindisulfonate Sodium Injection, equivalent to about 40 mg of indigotindisulfonate sodium, with dilute hydrochloric acid (1 in 100) to obtain a solution having a known concentration of about 10 µg of indigotindisulfonate sodium per mL. Proceed as directed in the *Assay* under *Indigotindisulfonate Sodium*, beginning with "Concomitantly determine the absorbances." Calculate the quantity, in mg, of $C_{16}H_8N_2Na_2O_8S_2$ in each mL of the Injection taken by the formula:

$$4(C/V)(A_U/A_S),$$

in which C is the concentration, in µg per mL, of USP Indigotindisulfonate Sodium RS in the Standard solution, V is the volume, in mL, of Injection taken, and A_U and A_S are the absorbances of the solution from the Injection and the Standard solution, respectively.

Indium In 111 Oxyquinoline Solution

» Indium In 111 Oxyquinoline Solution is a sterile, nonpyrogenic, isotonic aqueous solution suitable for the radiolabeling of blood cells, especially leukocytes and platelets, containing radioactive indium (^{111}In) in the form of a complex with 8-hydroxyquinoline, the latter being present in excess. It contains not less than 90.0 percent and not more than 110.0 percent of the labeled amount of ^{111}In as the 8-hydroxyquinoline complex expressed as megabecquerels (millicuries) per mL at the time indicated in the labeling. It may contain sodium chloride, surfactants, and buffers. Other chemical forms of radioactivity do not exceed 10.0 percent of the total radioactivity.

Specific activity: not less than 1.85 gigabecquerels (50 millicuries) per µg of indium.

Packaging and storage—Preserve in single-unit containers at a temperature between 15° and 25°.

Labeling—Label it to contain the following, in addition to the information specified for *Labeling* under *Injections* ⟨1⟩: the time and date of calibration; the amount of ^{111}In as the 8-hydroxyquinoline complex expressed as total megabecquerels (millicuries) and concentration as megabecquerels (millicuries) per mL on the date and time of calibration; the expiration date; the statement, "Not for direct administration. Use only after radiolabeling of blood cells by intravenous injection;" and the statement, "Caution—Radioactive Material." The labeling indicates that in making dosage calculations, correction is to be made for radioactive decay, and also indicates that the radioactive half-life of ^{111}In is 67.9 hours.

Pyrogen—It meets the requirements of the *Pyrogen Test* ⟨151⟩.

pH ⟨791⟩: between 6.5 and 7.5.

Radionuclide identification (see *Radioactivity* ⟨821⟩)—Its gamma-ray spectrum is identical to that of a specimen of ^{111}In that exhibits major photopeaks having energies of 0.171 and 0.245 MeV.

Radiochemical purity—Place a suitable volume, about 100 µL, of Solution, dilute with 3 mL of 0.9 percent sodium chloride solution in a separator, and extract with 6 mL of *n*-octanol by vigorous shaking. Allow the phases to separate and then drain the lower, aqueous layer into a suitable stoppered counting tube. Drain the residual, organic layer into a similar counting tube. Rinse the separator with 1 mL of *n*-octanol and drain this rinse into the counting tube containing the organic layer. Rinse the separator with 5 mL of 2 *N* hydrochloric acid and drain this rinse into a third counting tube. Insert the stopper and measure the radioactivity in each of the three tubes in a suitable gamma counter or ionization chamber calibrated for ^{111}In. The radiochemical purity is calculated by the formula:

$$(A/B),$$

where A is the radioactivity measured in the organic layer and B is the sum of the radioactivity measured in the organic, aqueous, and acid solutions. The radioactivity of the 8-hydroxyquinoline complex is not less than 90.0% of the total radioactivity and is found in the organic layer.

Radionuclidic purity—Using a suitable counting assembly (see *Selection of a Counting Assembly* under *Radioactivity* ⟨821⟩), determine the radioactivity of each radionuclidic impurity, in kBq per MBq (μCi per mCi) of ^{111}In, in the Solution by use of a calibrated system as directed under *Radioactivity* ⟨821⟩.

INDIUM 114m—The limit of ^{114m}In is 3 kBq per MBq (3 μCi per mCi) of ^{111}In. The presence of ^{114m}In in the Solution is demonstrated by a characteristic gamma-ray spectrum with prominent photopeaks having energies of 0.192, 0.558, and 0.725 MeV. The determination is made using a beta-liquid scintillation counter with a high-energy channel set to discriminate against all counts arising from ^{111}In.

ZINC 65—The limit of ^{65}Zn is 3 kBq per MBq (3 μCi per mCi) of ^{111}In. The presence of ^{65}Zn in the Solution is demonstrated by a characteristic gamma-ray spectrum with a prominent photopeak at 1.116 MeV. ^{65}Zn decays with a radioactive half-life of 243.9 days.

Assay for radioactivity—Using a suitable counting assembly (see *Selection of a Counting Assembly* under *Radioactivity* ⟨821⟩), determine the radioactivity, in MBq (mCi) per mL, of Indium In 111 Oxyquinoline Solution by the use of a calibrated system as directed under *Radioactivity* ⟨821⟩.

Indium In 111 Pentetate Injection

» Indium In 111 Pentetate Injection is a sterile, isotonic solution suitable for intrathecal administration, containing radioactive indium (^{111}In) in the form of a chelate of pentetic acid. It contains not less than 90.0 percent and not more than 110.0 percent of the labeled amount of ^{111}In as the pentetic acid complex expressed in megabecquerels (microcuries or millicuries) per mL at the time indicated in the labeling. It may contain sodium chloride and buffers. Other chemical forms of radioactivity do not exceed 10.0 percent of the total radioactivity.

Packaging and storage—Preserve in single-dose containers.

Labeling—Label it to include the following, in addition to the information specified for *Labeling* under *Injections* ⟨1⟩: the time and date of calibration; the amount of ^{111}In as labeled pentetic acid complex expressed as total megabecquerels (millicuries or microcuries) and concentration as megabecquerels (microcuries or millicuries) per mL on the date and time of calibration; the expiration date; and the statement, "Caution—Radioactive Material." The labeling indicates that in making dosage calculations, correction is to be made for radioactive decay, and also indicates that the radioactive half-life of ^{111}In is 2.83 days.

Reference standard—*USP Endotoxin Reference Standard.*

Bacterial endotoxins—It meets the requirements under *Bacterial Endotoxins Test* ⟨85⟩, the limit of endotoxin content being not more than $14/V$ USP Endotoxin Unit per mL of the Injection, when compared with the USP Endotoxin RS, in which V is the maximum recommended total dose, in mL, at the expiration date or time.

pH ⟨791⟩: between 7.0 and 8.0.

Radionuclide identification (see *Radioactivity* ⟨821⟩)—Its gamma-ray spectrum is identical to that of a specimen of ^{111}In that exhibits major photopeaks having energies of 0.173 and 0.247 MeV.

Radiochemical purity—Place 2 μL to 5 μL of Injection about 17 mm from one end of a 65- × 97-mm piece of silica gel–impregnated glass microfiber sheet (see under *Reagents* in the section, *Reagents, Indicators, and Solutions*) (see also *Chromatography* ⟨621⟩), and allow to dry. Repeat applications may be made to obtain a suitable count rate. Develop the chromatogram over a suitable period of time by ascending chromatography, using dilute methanol (8.5 in 10), and dry in an oven at 105 ± 5° for 5 minutes. Determine the radioactivity distribution by scanning the chromatogram with a suitable collimated radiation detector. The radioactivity of the indium pentetic acid complex band is not less than 90.0% of the total radioactivity, and the R_f value is between 0.8 and 1.0.

Radionuclidic purity—Using a suitable counting assembly (see *Selection of a Counting Assembly* under *Radioactivity* ⟨821⟩), determine the radioactivity of each radionuclidic impurity, in kBq per MBq (μCi per mCi) of ^{111}In, in the Injection by use of a calibrated system as directed under *Radioactivity* ⟨821⟩.

INDIUM 114m—The presence of ^{114m}In in the Injection is demonstrated by a characteristic gamma-ray spectrum with prominent photopeaks having energies of 0.192, 0.558, and 0.724 MeV. ^{114m}In decays with a radioactive half-life of 50 days. The amount of ^{114m}In is not greater than 3 kBq per MBq (3 μCi per mCi) of ^{111}In.

ZINC 65—The presence of ^{65}Zn in the Injection is demonstrated by a characteristic gamma-ray spectrum with a prominent photopeak at 1.115 MeV. ^{65}Zn decays with a radioactive half-life of 243.9 days. The amount of ^{65}Zn is not greater than 3 kBq per MBq (3 μCi per mCi) of ^{111}In.

Other requirements—It meets the requirements under *Injections* ⟨1⟩, except that the Injection may be distributed or dispensed prior to the completion of the test for *Sterility*, the latter test being started on the day of final manufacture, and except that it is not subject to the recommendation on *Volume in Container.*

Assay for radioactivity—Using a suitable counting assembly (see *Selection of a Counting Assembly* under *Radioactivity* ⟨821⟩), determine the radioactivity, in MBq per mL, of Indium In 111 Pentetate Injection by use of a calibrated system as directed under *Radioactivity* ⟨821⟩.

Indocyanine Green

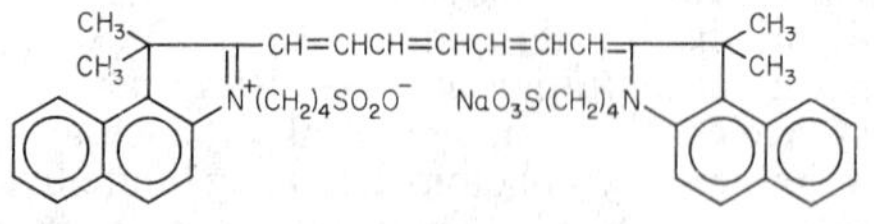

$C_{43}H_{47}N_2NaO_6S_2$ 774.96

1*H*-Benz[*e*]indolium, 2-[7-[1,3-dihydro-1,1-dimethyl-3-(4-sulfobutyl)-2*H*-benz[*e*]indol-2-ylidene]-1,3,5-heptatrienyl]-1,1-dimethyl-3-(4-sulfobutyl)-, hydroxide, inner salt, sodium salt.

2-[7-[1,1-Dimethyl-3-(4-sulfobutyl)benz[*e*]indolin-2-ylidene]-1,3,5-heptatrienyl]-1,1-dimethyl-3-(4-sulfobutyl)-1*H*-benz[*e*]indolium hydroxide, inner salt, sodium salt [*3599-32-4*].

» Indocyanine Green contains not less than 94.0 percent and not more than 105.0 percent of $C_{43}H_{47}N_2NaO_6S_2$, calculated on the dried basis. It contains not more than 5.0 percent of sodium iodide, calculated on the dried basis.

Packaging and storage—Preserve in well-closed containers.

Reference standard—*USP Indocyanine Green Reference Standard*—Dry in vacuum at 50° for 3 hours before using.

Identification—

A: Incinerate a portion of it: the residue responds to the tests for *Sodium* ⟨191⟩, and for *Sulfate* ⟨191⟩.

B: To a solution (1 in 20,000) add 10 drops of 1 *N* sodium hydroxide, and heat to about 60°. Add 10 drops of hydrogen peroxide TS, and mix: a red color develops within about 4 minutes and, on standing, fades to a pale orange.

Loss on drying ⟨731⟩—Dry it in vacuum at 50° for 3 hours: it loses not more than 6.0% of its weight.

Arsenic, *Method II* ⟨211⟩: 8 ppm.

Lead—A 5-mL portion of the solution prepared for the test for *Arsenic* contains not more than 2 µg of lead (corresponding to not more than 0.001%) when tested by the limit test for *Lead* ⟨251⟩, 3 mL of *Ammonium Citrate Solution*, 1 mL of *Potassium Cyanide Solution*, and 0.5 mL of *Hydroxylamine Hydrochloride Solution* being used.

Sodium iodide—Dissolve about 200 mg, accurately weighed, in 100 mL of water, add 1 mL of nitric acid, mix, and titrate with 0.01 *N* silver nitrate VS, determining the end-point potentiometrically, using silver and glass electrodes. Each mL of 0.01 *N* silver nitrate is equivalent to 1.499 mg of sodium iodide. Not more than 5.0% of sodium iodide, calculated on the dried basis, is found.

Assay—Dissolve a quantity of Indocyanine Green, equivalent to about 100 mg of dried indocyanine green and accurately weighed, in methanol, and dilute quantitatively and stepwise with methanol to obtain a solution containing about 2 µg per mL. Dissolve a quantity of USP Indocyanine Green RS, accurately weighed, in methanol, and dilute quantitatively and stepwise with methanol to obtain a Standard solution having a known concentration of about 2 µg per mL. Concomitantly determine the absorbances of both solutions in 1-cm cells at the wavelength of maximum absorbance at about 785 nm, with a suitable spectrophotometer, using methanol as the blank. Calculate the quantity, in mg, of $C_{43}H_{47}N_2NaO_6S_2$ in the Indocyanine Green taken by the formula:

$$50C(A_U/A_S),$$

in which *C* is the concentration, in µg per mL, of USP Indocyanine Green RS in the Standard solution, and A_U and A_S are the absorbances of the solution of Indocyanine Green and the Standard solution, respectively.

Sterile Indocyanine Green

» Sterile Indocyanine Green is Indocyanine Green suitable for parenteral use. It contains not less than 90.0 percent and not more than 110.0 percent of the labeled amount of $C_{43}H_{47}N_2NaO_6S_2$.

Packaging and storage—Preserve in *Containers for Sterile Solids* as described under *Injections* ⟨1⟩.

Reference standard—*USP Indocyanine Green Reference Standard*—Dry in vacuum at 50° for 3 hours before using.

Constituted solution—At the time of use, the constituted solution prepared from Sterile Indocyanine Green meets the requirements for *Constituted Solutions* under *Injections* ⟨1⟩.

pH ⟨791⟩: between 5.5 and 6.5, in a solution (1 in 200).

Content variation—Transfer the contents of each of 5 containers individually to separate 100-mL volumetric flasks with the aid of methanol. To each flask add methanol to volume. Dilute the solutions quantitatively and stepwise with methanol, to obtain a concentration of about 2.5 µg per mL. Proceed as directed in the *Assay* under *Indocyanine Green*, beginning with "Concomitantly determine the absorbances." The requirements are met if the content of each of not less than 4 of the containers tested is within the limits specified under *Uniformity of Dosage Units* ⟨905⟩.

Other requirements—It responds to the *Identification tests*, and meets the requirements for *Arsenic*, *Lead*, *Sodium iodide*, and *Assay* under *Indocyanine Green*. It meets also the requirements for *Sterility Tests* ⟨71⟩ and for *Labeling* under *Injections* ⟨1⟩.

Indomethacin

CO—C6H4—Cl; N; CH3; CH3O; CH2COOH

$C_{19}H_{16}ClNO_4$ 357.79

1*H*-Indole-3-acetic acid, 1-(4-chlorobenzoyl)-5-methoxy-2-methyl-.

1-(*p*-Chlorobenzoyl)-5-methoxy-2-methylindole-3-acetic acid [*53-86-1*].

» Indomethacin contains not less than 98.0 percent and not more than 101.0 percent of $C_{19}H_{16}ClNO_4$, calculated on the dried basis.

Packaging and storage—Preserve in well-closed, light-resistant containers.

Reference standard—*USP Indomethacin Reference Standard*—Dry at a pressure below 5 mm of mercury at 100° for 2 hours before using.

Identification—

A: The infrared absorption spectrum of a mineral oil dispersion of it, previously dried, exhibits maxima only at the same wavelengths as that of a similar preparation of USP Indomethacin RS.

B: The ultraviolet absorption spectrum of a 1 in 40,000 solution in 0.1 *N* methanolic hydrochloric acid exhibits maxima and minima at the same wavelengths as that of a similar solution of USP Indomethacin RS, concomitantly measured, and the respective absorptivities, calculated on the dried basis, at the wavelength of maximum absorbance at about 318 nm do not differ by more than 3.0%.

C: Its X-ray diffraction pattern (see *X-ray Diffraction* ⟨941⟩) conforms to that of USP Indomethacin RS.

Loss on drying ⟨731⟩—Dry it at a pressure below 5 mm of mercury at 100° for 2 hours: it loses not more than 0.5% of its weight.

Residue on ignition ⟨281⟩: not more than 0.2%.

Heavy metals, *Method II* ⟨231⟩: 0.002%.

Assay—

Mobile phase—Prepare a suitable solution of 0.01 *M* monobasic sodium phosphate and 0.01 *M* dibasic sodium phosphate in acetonitrile and water (approximately 1:1).

Standard preparation—Dissolve an accurately weighed quantity of USP Indomethacin RS in *Mobile phase* to obtain a solution having a known concentration of about 0.1 mg per mL.

Assay preparation—Weigh accurately about 100 mg of Indomethacin, and transfer to a 100-mL volumetric flask. Dissolve in *Mobile phase*, dilute with *Mobile phase* to volume, and mix. Pipet 10 mL of this solution into a 100-mL volumetric flask, dilute with *Mobile phase* to volume, and mix.

Chromatographic system (see *Chromatography* ⟨621⟩)—The liquid chromatograph is equipped with a 254-nm detector and a 4-mm × 30-cm column that contains 10-µm packing L1. The flow rate is about 1 mL per minute. Chromatograph the *Standard preparation*, and record the peak responses as directed under *Procedure:* the column efficiency determined from the analyte peak is not less than 500 theoretical plates, and the relative standard deviation for replicate injections is not more than 1.0%.

Procedure—Separately inject equal volumes (about 20 µL) of the *Standard preparation* and the *Assay preparation* into the chromatograph, record the chromatograms, and measure the responses for the major peaks. Calculate the quantity, in mg, of $C_{19}H_{16}ClNO_4$ in the portion of Indomethacin taken by the formula:

$$1000C(r_U/r_S),$$

in which *C* is the concentration, in mg per mL, of USP Indomethacin RS in the *Standard preparation*, and r_U and r_S are the peak responses obtained at equivalent retention times from the *Assay preparation* and the *Standard preparation*, respectively.

Indomethacin Capsules

» Indomethacin Capsules contain not less than 90.0 percent and not more than 110.0 percent of the labeled amount of $C_{19}H_{16}ClNO_4$.

Packaging and storage—Preserve in well-closed containers.

Reference standard—*USP Indomethacin Reference Standard*—Dry at a pressure below 5 mm of mercury at 100° for 2 hours before using.

Identification—

A: Shake a portion of the contents of Capsules, equivalent to about 50 mg of indomethacin, with 10 mL of acetone for about 2 minutes, and filter. Transfer 5 mL of the filtrate to a stoppered flask, add 20 mL of water, and shake for about 2 minutes until a precipitate forms and crystallizes. Filter, and collect the crystals. Dry the crystals in air, then dry at a pressure below 5 mm of mercury at 100° for 2 hours: the infrared absorption spectrum of a potassium bromide dispersion of the dried residue so obtained exhibits maxima only at the same wavelengths as that of a similar preparation of USP Indomethacin RS that has been similarly recrystallized from a solution of 25 mg in 5 mL of acetone.

B: Shake a portion of the contents of Capsules, equivalent to about 25 mg of indomethacin, with 25 mL of methanol, and filter. Separately apply 2 µL of the filtrate so obtained (test solution) and 2 µL of a Standard solution in methanol containing 1 mg of USP Indomethacin RS per mL on a suitable thin-layer chromatographic plate (see *Chromatography* ⟨621⟩), coated with a 0.25-mm layer of chromatographic silica gel mixture, and dry the spots with the aid of a current of air. Develop the chromatogram in a solvent system consisting of a mixture of chloroform and methanol (4:1) until the solvent front has moved about three-fourths of the length of the plate. Remove the plate from the developing chamber, mark the solvent front, allow it to dry, and locate the spots under short-wavelength ultraviolet light: the intensity and R_f value of the principal spot obtained from the test solution corresponds to that obtained from the Standard solution.

Dissolution ⟨711⟩—

Medium: 1 volume of pH 7.2 phosphate buffer (see *Buffer Solutions* in the section, *Reagents, Indicators, and Solutions*) mixed with 4 volumes of water; 750 mL.

Apparatus 1: 100 rpm.

Time: 20 minutes.

Procedure—Determine the amount of $C_{19}H_{16}ClNO_4$ dissolved from ultraviolet absorbances at the wavelength of maximum absorbance at about 318 nm of filtered portions of the solution under test, suitably diluted with *Dissolution Medium,* if necessary, in comparison with a Standard solution having a known concentration of USP Indomethacin RS in the same medium.

Tolerances—Not less than 80% (Q) of the labeled amount of $C_{19}H_{16}ClNO_4$ is dissolved in 20 minutes.

Uniformity of dosage units ⟨905⟩: meet the requirements.

Procedure for content uniformity—Transfer the contents of 1 Capsule to a 100-mL volumetric flask, add 10 mL of water, and allow to stand for 10 minutes, swirling occasionally. Add 60 mL of methanol, shake for 10 minutes, dilute with methanol to volume, mix, and centrifuge. Dilute a portion of the clear solution quantitatively and stepwise, if necessary, with a mixture of equal volumes of methanol and pH 7.0 phosphate buffer (see *Buffer Solutions* in the section, *Reagents, Indicators, and Solutions*) to obtain a solution containing about 25 µg of indomethacin per mL. Concomitantly determine the absorbances of this solution and a Standard solution of USP Indomethacin RS, in the methanol and pH 7.0 phosphate buffer mixture (1:1) having a known concentration of about 25 µg per mL, in 1-cm cells at the wavelength of maximum absorbance at about 318 nm, with a suitable spectrophotometer, using the methanol and pH 7.0 phosphate buffer mixture as the blank. Calculate the quantity, in mg, of $C_{19}H_{16}ClNO_4$ in the Capsule by the formula:

$$(TC/D)(A_U/A_S),$$

in which T is the labeled quantity, in mg, of indomethacin in the Capsule, C is the concentration, in µg per mL, of USP Indomethacin RS in the Standard solution, D is the concentration, in µg per mL, of indomethacin in the test solution, based upon the labeled quantity per Capsule and the extent of dilution, and A_U and A_S are the absorbances of the solution from the Capsule and the Standard solution, respectively.

Assay—

Standard preparation—Transfer about 25 mg of USP Indomethacin RS, accurately weighed, to a 200-mL volumetric flask, dissolve in 2 mL of methanol, dilute with pH 7.2 phosphate buffer (see *Buffer Solutions* in the section, *Reagents, Indicators, and Solutions*) to volume, and mix. Transfer 25.0 mL of this solution to a separator, and extract with three 25-mL portions of methylene chloride. Filter the extracts through a pledget of cotton into a 100-mL volumetric flask, rinse the filter with methylene chloride, dilute with methylene chloride to volume, and mix, to obtain a *Standard preparation* having a known concentration of about 31 µg per mL.

Assay preparation—Transfer, as completely as possible, the contents of not less than 20 Indomethacin Capsules to a suitable tared container, and determine the average content weight per Capsule. Mix the combined contents, and transfer an accurately weighed portion, equivalent to about 25 mg of indomethacin, to a 200-mL volumetric flask, add 2 mL of methanol, shake for 10 minutes, dilute with pH 7.2 phosphate buffer to volume, and mix. Transfer about 50 mL to a centrifuge tube, and centrifuge for 15 minutes. Transfer 25.0 mL of the supernatant liquid to a 125-mL separator, and extract with three 25-mL portions of methylene chloride. Filter the extracts through a pledget of cotton into a 100-mL volumetric flask, rinse the filter with methylene chloride, dilute with methylene chloride to volume, and mix.

Procedure—Concomitantly determine the absorbances of the solutions in 1-cm cells at the wavelength of maximum absorbance at about 318 nm, with a suitable spectrophotometer, using methylene chloride as the blank. Calculate the quantity, in mg, of $C_{19}H_{16}ClNO_4$ in the portion of Capsules taken by the formula:

$$0.8C(A_U/A_S),$$

in which C is the concentration, in µg per mL, of USP Indomethacin RS in the *Standard preparation,* and A_U and A_S are the absorbances of the *Assay preparation* and the *Standard preparation,* respectively.

Indomethacin Extended-release Capsules

» Indomethacin Extended-release Capsules contain not less than 90.0 percent and not more than 110.0 percent of the labeled amount of $C_{19}H_{16}ClNO_4$.

Packaging and storage—Preserve in well-closed containers.

Reference standard—*USP Indomethacin Reference Standard*—Dry at a pressure below 5 mm of mercury at 100° for 2 hours before using.

Identification—

A: The contents of Extended-release Capsules respond to the *Identification tests* under *Indomethacin Capsules.*

B: Transfer a quantity of finely powdered Capsule contents, equivalent to about 100 mg of indomethacin, to a 250-mL flask, add about 100 mL of sodium hydroxide solution (1 in 2500), shake for 5 minutes, and filter. To 1 mL of the clear filtrate add 1 mL of sodium nitrite solution (1 in 1000), mix, and allow to stand for 5 minutes. Add 0.5 mL of sulfuric acid: a golden yellow color develops.

Drug release ⟨724⟩—

Medium: pH 6.2 phosphate buffer (see *Buffer Solutions* in the section, *Reagents, Indicators, and Solutions*); 900 mL.

Apparatus 1: 75 rpm.

Times: 0.083D hours, 0.167D hours, 0.333D hours, 1.000D hours, 2.000D hours.

Procedure—Determine the amount of $C_{19}H_{16}ClNO_4$ dissolved from ultraviolet absorbances at the wavelength of maximum absorbance at about 318 nm of filtered portions of the solution under test, suitably diluted with *Dissolution Medium,* if necessary, in comparison with a Standard solution having a known concentration of USP Indomethacin RS in the same medium.

Tolerances—The percentages of the labeled amount of $C_{19}H_{16}ClNO_4$ dissolved at the times specified conform to *Acceptance Table 1.*

Time (hours)	Amount Dissolved
0.083*D*	between 10% and 32%
0.167*D*	between 20% and 52%
0.333*D*	between 35% and 80%
1.000*D*	not less than 60%
2.000*D*	not less than 80%

Uniformity of dosage units ⟨905⟩: meet the requirements.

Procedure for content uniformity—Transfer the contents of 1 Capsule to a 200-mL volumetric flask, and add 100 mL of a mixture of equal volumes of methanol and pH 7.5 0.1 *M* dibasic potassium phosphate buffer. Sonicate until the contents are dispersed, dilute with the methanol and pH 7.5 phosphate buffer mixture (1:1) to volume, mix, and centrifuge. Dilute a portion of the clear solution quantitatively and stepwise, if necessary, with the methanol and pH 7.5 phosphate buffer mixture (1:1) to obtain a solution containing about 25 µg of indomethacin per mL. Concomitantly determine the absorbances of this solution and a Standard solution of USP Indomethacin RS, in the methanol and pH 7.5 phosphate buffer mixture (1:1) having a known concentration of about 25 µg per mL, in 1-cm cells at the wavelength of maximum absorbance at about 318 nm, with a suitable spectrophotometer, using the methanol and pH 7.5 phosphate buffer mixture as the blank. Calculate the quantity, in mg, of $C_{19}H_{16}ClNO_4$ in the Capsule by the formula:

$$(TC/D)(A_U/A_S),$$

in which *T* is the labeled quantity, in mg, of indomethacin in the Capsule, *C* is the concentration, in µg per mL, of USP Indomethacin RS in the Standard solution, *D* is the concentration, in µg per mL, of indomethacin in the test solution, based upon the labeled quantity per Capsule and the extent of dilution, and A_U and A_S are the absorbances of the solution from the Capsule contents and the Standard solution, respectively.

Assay and limit of 4-chlorobenzoic acid—

Mobile phase—Prepare a suitable mixture of methanol, water, and phosphoric acid (600:400:0.8), and filter through a membrane filter of 0.5-µm or finer porosity. Make adjustments if necessary (see *System Suitability* under *Chromatography* ⟨621⟩).

Diluted phosphoric acid—Dilute 10 mL of phosphoric acid with water to make 1000 mL of solution.

Standard indomethacin preparation—Transfer about 40 mg of USP Indomethacin RS, accurately weighed, to a 50-mL volumetric flask, and dissolve in 30 mL of acetonitrile. Dilute with *Diluted phosphoric acid* to volume, and mix.

Standard 4-chlorobenxoic acid preparation—Dissolve a suitable quantity of 4-chlorobenzoic acid, accurately weighed, in acetonitrile to obtain a solution having a known concentration of about 0.18 mg per mL. Transfer 1.0 mL of this solution to a 50-mL volumetric flask, dilute with *Diluted phosphoric acid* to volume, and mix. This solution contains about 3.6 µg of 4-chlorobenzoic acid per mL.

Assay preparation—Weigh and finely powder the contents of not less than 20 Indomethacin Extended-release Capsules. Transfer an accurately weighed portion of the powder, equivalent to about 75 mg of indomethacin, to a 100-mL volumetric flask, add 40 mL of *Diluted phosphoric acid*, and shake for 1 hour. Sonicate for 15 minutes, add 40 mL of acetonitrile, mix, sonicate for 15 minutes, dilute with acetonitrile to volume, and mix. Centrifuge a portion of this solution, and filter the supernatant liquid through a filter having a porosity of 0.5-µm or finer. Use the filtrate as the *Assay preparation.*

Chromatographic system (see *Chromatography* ⟨621⟩)—The liquid chromatograph is equipped with a 240-nm detector and a 3.9-mm × 30-cm column that contains packing L1. The flow rate is about 2 mL per minute. Chromatograph the *Standard indomethacin preparation*, and record the peak responses as directed under *Procedure:* the column efficiency determined from the indomethacin peak is not less than 1000 theoretical plates, *k′* for the indomethacin peak is not less than 4.0, the tailing factor for the indomethacin peak is not more than 2.0, and the relative standard deviation for replicate injections is not more than 2.0%. Chromatograph the *Standard 4-chlorobenxoic acid preparation*, and record the peak responses as directed under *Procedure: k′* for the 4-chlorobenzoic acid peak not less than 0.9.

Procedure—Separately inject equal volumes (about 20 µL) of the *Standard indomethacin preparation*, the *Standard 4-chlorobenxoic acid preparation*, and the *Assay preparation* into the chromatograph, record the chromatograms, and measure the responses for the major peaks. Calculate the quantity, C_a, in mg, of indomethacin in the portion of Extended-release Capsules taken by the formula:

$$100C(r_U/r_S),$$

in which *C* is the concentration, in mg per mL, of USP Indomethacin RS in the *Standard indomethacin preparation*, and r_U and r_S are the indomethacin peak responses obtained from the *Assay preparation* and the *Standard preparation*, respectively. Calculate the percentage of 4-chlorobenzoic acid ($C_7H_5ClO_2$) in the portion of Extended-release Capsules taken by the formula:

$$10(C_4/C_a)(r_U/r_S),$$

in which C_4 is the concentration, in µg per mL, of 4-chlorobenzoic acid in the *Standard 4-chlorobenxoic acid preparation*, C_a is the quantity, in mg, of indomethacin ($C_{19}H_{16}ClNO_4$) in the portion of Capsule contents taken, determined as directed herein, and r_U and r_S are the 4-chlorobenzoic acid peak responses obtained from the *Assay preparation* and the *Standard 4-chlorobenxoic acid preparation*, respectively: not more than 0.44% is found.

Indomethacin Suppositories

» Indomethacin Suppositories contain not less than 90.0 percent and not more than 110.0 percent of the labeled amount of $C_{19}H_{16}ClNO_4$.

Packaging and storage—Preserve in well-closed containers, at controlled room temperature.

Reference standard—*USP Indomethacin Reference Standard*—Dry in vacuum at a pressure below 5 mm of mercury at 100° for 2 hours before using.

Identification—

Standard preparation—Prepare a solution, containing about 125 µg of USP Indomethacin RS per mL, by first dissolving the Reference Standard in a volume of methanol that is one one-hundredth of the volume of the solution to be prepared, then adding ether to volume, and mixing.

Test preparation—Use the ether extract contained in the 200-mL volumetric flask obtained as directed under *Assay preparation* in the *Assay.*

Procedure—Separately apply 10 µL each of the *Test preparation* and the *Standard preparation* to a thin-layer chromatographic plate (see *Chromatography* ⟨621⟩), coated with a 0.25-mm layer of chromatographic silica gel mixture. Develop the chromatogram in a solvent system consisting of a mixture of chloroform and glacial acetic acid (19:1) until the solvent front has moved about three-fourths of the length of the plate. Remove the plate from the chamber, air-dry, and examine under short-wavelength ultraviolet light: the R_f value of the principal spot in the chromatogram of the *Test preparation* corresponds to that obtained from the *Standard preparation.*

Dissolution ⟨711⟩—

Medium: 0.1 *M*, pH 7.2 phosphate buffer (see *Buffer Solutions* in the section, *Reagents, Indicators, and Solutions*); 900 mL.

Apparatus 2: 50 rpm.

Time: 60 minutes.

Procedure—Determine the amount of $C_{19}H_{16}ClNO_4$ dissolved from ultraviolet absorbances at the wavelength of maximum absorbance at about 320 nm of filtered portions of the solution under test, suitably diluted with *Dissolution Medium*, if necessary, in comparison with a Standard solution having a known concentration of USP Indomethacin RS in the same medium.

Tolerances—Not less than 75% (*Q*) of the labeled amount of $C_{19}H_{16}ClNO_4$ is dissolved in 60 minutes.

Uniformity of dosage units ⟨905⟩: meet the requirements.

Procedure for content uniformity—Place 1 Suppository into a 100-mL volumetric flask containing 80 mL of a solution of methanol and glacial acetic acid (199:1), shake by mechanical means until the Suppository is dissolved, dilute with the methanol–glacial acetic acid solution to volume, and mix. Filter a portion of the solution, discarding the first 15 mL of the filtrate, and dilute an accurately measured volume of the clear filtrate quantitatively and stepwise, if necessary, with the methanol–glacial acetic acid solution to obtain a solution containing about 25 µg of indomethacin per mL. Concomitantly determine the absorbances of this solution and of a Standard solution of USP Indomethacin RS in the same medium having a known concentration of about 25 µg per mL at the wavelength of maximum absorbance at about 320 nm, with a suitable spectrophotometer, using the methanol–glacial acetic acid solution as the blank. Calculate the quantity, in mg, of $C_{19}H_{16}ClNO_4$ in the Suppository by the formula:

$$(TC/D)(A_U/A_S),$$

in which T is the labeled quantity, in mg, of indomethacin in the Suppository, C is the concentration, in µg per mL, of USP Indomethacin RS in the Standard solution, D is the concentration, in µg per mL, of indomethacin in the solution from the Suppository, on the basis of the labeled quantity per Suppository and the extent of dilution, and A_U and A_S are the absorbances of the solution from the Suppository and the Standard solution, respectively.

Assay—

Solvent mixture—Prepare a solution of methanol and glacial acetic acid (199:1).

Standard preparation—Prepare a solution, having a known concentration of about 165 µg of USP Indomethacin RS per mL, by first dissolving an accurately weighed quantity of the Reference Standard in a volume of methanol that is one one-hundredth of the nominal volume of the volumetric flask being used, then adding ether to volume, and mixing. Transfer 15.0 mL of the resulting solution to a 100-mL volumetric flask, dilute with *Solvent mixture* to volume, and mix to obtain a *Standard preparation* having a known concentration of about 25 µg of USP Indomethacin RS per mL.

Assay preparation—Weigh, mash, and then mix not less than 10 Indomethacin Suppositories. Transfer an accurately weighed portion of the mass, equivalent to about 25 mg of indomethacin, to a 125-mL separator, add 15 mL of water and 50 mL of ether, and shake until the mass is dissolved. Transfer the ether layer to a 200-mL volumetric flask, extract the aqueous layer with two additional 50-mL portions of ether, and combine the ether extracts in the 200-mL volumetric flask. Discard the aqueous layer. Dilute with *Solvent mixture* to volume, and mix. Pipet 10 mL of this solution into a 50-mL volumetric flask, dilute with *Solvent mixture* to volume, and mix.

Procedure—Concomitantly determine the absorbances of the *Assay preparation* and the *Standard preparation* at the wavelength of maximum absorbance at about 320 nm, with a suitable spectrophotometer, using *Solvent mixture* as a blank. Calculate the quantity, in mg, of $C_{19}H_{16}ClNO_4$ in the portion of Suppositories taken by the formula:

$$C(A_U/A_S),$$

in which C is the concentration, in µg per mL, of USP Indomethacin RS in the *Standard preparation*, and A_U and A_S are the absorbances of the *Assay preparation* and the *Standard preparation*, respectively.

Influenza Virus Vaccine

» Influenza Virus Vaccine conforms to the regulations of the federal Food and Drug Administration concerning biologics (see *Biologics* ⟨1041⟩). It is a sterile, aqueous suspension of suitably inactivated influenza virus types A and B, either individually or combined, or virus sub-units prepared from the extra-embryonic fluid of influenza virus–infected chicken embryo. The strains of influenza virus used in the preparation of this Vaccine are those designated by the U. S. Government's Expert Committee on Influenza and recommended by the Surgeon General of the U. S. Public Health Service. Influenza Virus Vaccine has a composition of such strains and a content of virus antigen of each, designated for the particular season, of not less than the specified weight (in micrograms) of influenza virus hemagglutinin determined in specific radial-immunodiffusion tests relative to the U. S. Reference Influenza Virus Vaccine. It may contain a suitable antimicrobial agent. If formalin is used for inactivation, it contains not more than 0.02 percent of residual free formaldehyde.

Packaging and storage—Preserve at a temperature between 2° and 8°.

Expiration date—The expiration date is not later than 18 months after date of issue from manufacturer's cold storage (5°, 1 year).

Labeling—Label it to state that it is to be shaken before use and that it is not to be frozen. Label it also to state that it was prepared in embryonated chicken eggs.

Inhalants—*see complete list in index*
Inhalations—*see complete list in index*
Inhalation Solutions—*see complete list in index*
Injections—*see complete list in index*

Insulin

Pork Insulin
Chain A
Glu - Val - Ile - Gly - H (4 3 2 1)
Gln - Cys - Cys - Thr - Ser - Ile - Cys - Ser - Leu - Tyr - Gln - Leu - Glu - Asn - Tyr - Cys - Asn - OH (5 6 7 8 9 10 11 12 13 14 15 16 17 18 19 20 21)

Chain B
Gln - His - Leu - Cys - Gly - Ser - His - Leu - Val - Glu - Ala - Leu - Tyr - Leu - Val - Cys - Gly - Glu (4 5 6 7 8 9 10 11 12 13 14 15 16 17 18 19 20 21)
Asn - Val - Phe - H (3 2 1)
HO - Ala - Lys - Pro - Thr - Tyr - Phe - Phe - Gly - Arg (30 29 28 27 26 25 24 23 22)

$C_{256}H_{381}N_{65}O_{76}S_6$ 5777.59
Insulin (pig) [*12584-58-6*].

Beef Insulin
Chain A
Glu - Val - Ile - Gly - H (4 3 2 1)
Gln - Cys - Cys - Ala - Ser - Val - Cys - Ser - Leu - Tyr - Gln - Leu - Glu - Asn - Tyr - Cys - Asn - OH (5 6 7 8 9 10 11 12 13 14 15 16 17 18 19 20 21)

Chain B
Gln - His - Leu - Cys - Gly - Ser - His - Leu - Val - Glu - Ala - Leu - Tyr - Leu - Val - Cys - Gly - Glu (4 5 6 7 8 9 10 11 12 13 14 15 16 17 18 19 20 21)
Asn - Val - Phe - H (3 2 1)
HO - Ala - Lys - Pro - Thr - Tyr - Phe - Phe - Gly - Arg (30 29 28 27 26 25 24 23 22)

$C_{254}H_{377}N_{65}O_{75}S_6$ 5733.54
Insulin (ox) [*11070-73-8*].

» Insulin is a protein, obtained from the pancreas of healthy bovine and porcine animals used for food by man, that affects the metabolism of glucose. Its biological potency, determined by *Assay A* and calculated on the dried basis, is not less than 26.0 USP Insulin Units in each mg.

Packaging and storage—Preserve in tight containers, protected from light, in a cold place.

Labeling—Label it to indicate the one or more animal species to which it is related, as porcine, as bovine, or as a mixture of porcine and bovine. Where it is highly purified, label it as such.

Reference standards—*USP Insulin Reference Standard*—Preserve in a refrigerator, and after opening the ampul, store in a tight container. *USP Insulin (Beef) Reference Standard. USP Insulin (Pork) Reference Standard. USP Proinsulin (Beef) Reference Standard. USP Proinsulin (Pork) Reference Standard. USP Endotoxin Reference Standard.*

Identification—Prepare a 1 in 500 slurry in water, and add dropwise 3 *N* hydrochloric acid until a clear solution is obtained. The solution responds to the *Identification tests* under *Insulin Injection.*

Microbial limits ⟨61⟩—The total bacterial count does not exceed 300 per g, the test being made on a portion of about 0.2 g, accurately weighed.

Bacterial endotoxins—When tested as directed under *Bacterial Endotoxins Test* ⟨85⟩, the USP Endotoxin RS being used, it contains not more than 20 USP Endotoxin Units in each mg.

Loss on drying ⟨731⟩—Dry about 200 mg, accurately weighed, at 105° for 16 hours: it loses not more than 10.0% of its weight.

Residue on ignition—Accurately weigh about 400 mg in a suitable tared crucible, burn to ash, add 2 drops of nitric acid, and apply additional heat until a gray ash appears. Place in a muffle furnace at 800° to 900° for 30 minutes, then cool in a desiccator, and weigh: the weight of the residue is not more than 2.5%, calculated on the dried basis.

Nitrogen content, *Method II* ⟨461⟩—Determine the nitrogen content of about 10 mg of it, accurately weighed: not less than 14.5% and not more than 16.5% of N is found, calculated on the dried basis.

Zinc content ⟨591⟩—Determine the zinc content of about 10 mg of it, accurately weighed: not more than 1.08% is found, calculated on the dried basis.

Where highly purified, apply both *Assays A* and *B.*

Assay A—Proceed with Insulin as directed under *Insulin Assay* ⟨121⟩, treating about 85 mg of Insulin, accurately weighed, as directed under *Standard Solution* and *Standard Dilutions* to obtain the *Sample Dilutions.*

Assay B—

Mobile phase—Prepare separate, suitable filtered and degassed solutions consisting of a 1 in 1000 solution of trifluoroacetic acid in a mixture of water and acetonitrile (70:30) (*Solution A*), and trifluoroacetic acid in water (1 in 1000) (*Solution B*).

Standard preparation—Dissolve an accurately weighed quantity of the appropriate Reference Standard [USP Insulin (Beef) RS or USP Insulin (Pork) RS] in 0.01 *N* hydrochloric acid to obtain a solution having a known concentration corresponding to that of the *Assay preparation.*

Assay preparation—Dissolve an accurately weighed quantity of Insulin in 0.01 *N* hydrochloric acid to obtain a solution having a concentration of about 40 USP Insulin Units per mL.

Chromatographic system (see *Chromatography* ⟨621⟩)—The liquid chromatograph is equipped with a 280-nm detector and a 30-cm × 3.9-mm column that contains 5- to 10-μm packing L1. The flow rate is about 2 mL per minute, and the *Mobile phase* is *Solution B*, initially, programmed for gradient elution to produce a mixture of *Solution A* and *Solution B* (90:10) in 30 minutes. Chromatograph replicate injections of the *Standard preparation*, and record the peak responses as directed under *Procedure:* the relative standard deviation is not more than 1.5%, and the tailing factor for the insulin peak is not more than 2.5.

Procedure—Separately inject equal volumes (about 20 μL) of the *Standard preparation* and the *Assay preparation* into the chromatograph, record the chromatograms, and measure the responses for the major peaks.

Potency—Calculate the quantity, in USP Insulin Units per mL, of insulin in the portion of the Insulin taken by the formula:

$$C(r_U / r_S),$$

in which C is the concentration, in USP Insulin Units per mL, of the USP Reference Standard in the *Standard preparation*, and r_U and r_S are the peak responses obtained from the *Assay preparation* and the *Standard preparation*, respectively. Its potency, calculated on the dried basis, is not less than 26.0 USP Insulin Units in each mg.

Insulin that is highly purified meets the requirements of the following tests.

Proinsulin content—

REAGENTS—

USP Proinsulin (Porcine) Reference Standard *or* USP Proinsulin (Bovine) Reference Standard.

Proinsulin Specific (Porcine) Antiserum (Guinea pig) *or* Proinsulin Specific (Bovine) Antiserum (Guinea pig).

Labeled (I^{125} *Tyr-C-peptide*) *Antigen.* It has a specific radioactivity of about 75 mCi per mg.

Buffers—

	Buffer A	Buffer B
Monobasic Sodium phosphate	1.05 g	1.05 g
Sodium phosphate dihydrate	5.77 g	5.77 g
Sodium chloride	—	6.0 g
Thimerosal	0.24 g	0.24 g
Albumin Human	1.0 g	10.0 g
Water, a sufficient quantity, to make	1000 mL	1000 mL

Adjust each buffer, if necessary, with 1 *N* hydrochloric acid or 1 *N* sodium hydroxide to a pH of 7.4 ± 0.1.

Alcohol buffer—Mix 18 mL of *Buffer A*, 162 mL of water, and 960 mL of alcohol (96% v/v of C_2H_5OH).

Standard preparation—Dissolve the appropriate Reference Standard [USP Proinsulin (Beef) RS or USP Proinsulin (Pork) RS] in *Buffer B* to obtain a solution having a known concentration of 1.0 μg per mL. Dilute this solution quantitatively with *Buffer B* to obtain a solution having a known concentration of 10 ng per mL. Shake gently, and allow to stand for not less than 15 minutes.

Standard dilutions—In suitable tubes of about 10-mL capacity make dilutions of the *Standard preparation* by taking the specified volume of *Buffer B* in each tube and adding the specified volume of the *Standard preparation* as shown.

Standard dilution (ng per mL)	Buffer B (mL)	Standard preparation (mL)
—	10.0	—
1.0	9.0	1.0
2.5	7.5	2.5
5.0	5.0	5.0
7.5	2.5	7.5
10.0	—	10.0

Shake the tubes gently to mix, and allow to stand for not less than 15 minutes. The standard dilutions can be preserved frozen at −20° and thawed for use several times.

Diluted Specific Antiserum—Dilute the appropriate Specific Antiserum with *Buffer A* to ensure 35% to 50% binding of the diluted *Labeled Antigen.*

Diluted Labeled Antigen—Dilute the *Labeled Antigen* with *Buffer A* until the concentration of the I^{125} C-peptide is 2 ng per mL. Decant the *Diluted Labeled Antigen* into a vial coated with Albumin Human, avoiding transfer of any foam. [NOTE—Vials may be prepared by rinsing the insides with 5 percent (w/v) Albumin Human in *Buffer B*, turning the vials upside down, and allowing them to dry.]

Assay preparation—Dissolve an accurately weighed quantity of the specimen under test, in 0.01 *N* hydrochloric acid, and dilute quantitatively with 0.01 *N* hydrochloric acid to obtain a solution having a concentration corresponding to 10 mg per mL.

Assay dilutions—Make quantitative dilutions of the *Assay preparation*, with *Buffer B*, of two suitable concentrations, e.g., 1 in 10 and 1 in 50, and ensure that the pH of each final dilution is 7.4 ± 0.1.

Procedure—Make three sets of dilution tubes, 100 mm × 10.5 mm, so that there are three replicates at each dilution level. Place 100 μL of each dilution of the *Assay dilutions* and 100 μL of each dilution of the *Standard dilutions* at each set of three tubes. Add to each of the tubes 100 μL of the *Diluted Specific Antiserum.* Mix each by vortexing, and allow to stand at 4°. After 16 to 20 hours, add to each of the tubes 100 μL of *Diluted Labeled Antigen*, and allow to stand at 4° for an additional 16 to 72 hours. After the second incubation, add to each tube 1.6 mL of alcohol (96% v/v of C_2H_5OH). Mix by vortexing, and centrifuge the tubes at about 15° for 15 minutes at about 1720G. Decant the supernatant fluid from each tube, and wash the precipitate (containing bound Labeled Antigen) with 2 mL of *Alcohol buffer.*

Centrifuge the tubes similarly for 10 minutes, and decant the supernatant fluid from each. Dissolve the precipitate in each tube in 0.6 mL of 0.05 *N* sodium hydroxide. Count the radioactivity in a gamma counter appropriately set. Plot the concentration of the *Standard dilutions* on the abscissa, and the mean radioactivity of the precipitate on the ordinate in each case as a percentage of the added activity, and construct a standard curve of best fit visually. Read the mean concentration for each *Assay dilution* from the standard curve, and calculate the concentration, in ppm, of proinsulin activity in the specimen under test. Not more than 10 ppm is found.

High molecular weight protein—

Adsorbant—Use fine gel chromatographic packing that has been allowed to swell for not less than 3 hours in 1 *M* acetic acid, after gentle mixing with a broad spatula or by gentle shaking. Remove the supernatant gel particles by careful decanting or suction, repeating this process to complete the removal of such particles. Store the *Adsorbant*, if desired, after deaeration under vacuum.

Chromatographic tube—Select a chromatographic tube about 100 cm long and of about 2.5 cm internal diameter that is constricted to an outlet at the lower end. Insert at the constriction a small pledget of glass wool, previously washed with chloroform and air-dried.

Chromatographic column—Pack the *Adsorbant* in the *Chromatographic tube* to a height of 40 to 45 cm under a flow of 1 *M* acetic acid at a rate of 18 to 20 mL per hour, pouring all of the *Adsorbant* into the column in one operation. Adjust the height of the *Adsorbant*, if too high in the column, by gently resuspending the surplus and removing it by means of a pipet. Equilibrate the column for not less than 15 hours with the same flow of 1 *M* acetic acid.

Sample preparation—Place a portion of the Insulin corresponding to not less than 2,600 USP Insulin Units (100 mg) in a suitable test tube, and dissolve in a mixture of 1.5 mL of 1 *M* acetic acid and 0.2 mL of glacial acetic acid.

Procedure—Allow the column to drain until the liquid level reaches the surface of the bed. Immediately apply the dissolved sample to the top of the column, and allow it to sink into the bed. Wash the sample tube and the inner surface of the *Chromatographic tube* above the column twice with 1 mL of 1 *M* acetic acid. [NOTE—In all of these operations, take care not to disturb the gel surface of the column.] Continue to elute the column with 1 *M* acetic acid at the same rate as before, using acetic acid from the same batch both to equilibrate the column and to make the elutions. Collect between 160 and 200 fractions of about constant size, each of about 1.8 mL to 2.0 mL. Determine the absorbance of each fraction collected in a 1-cm cell at the wavelength of maximum absorbance, about 276 nm, correcting the spectrophotometer for background absorbance by adjusting it to zero absorbance during the measurement of the void volume, i.e., of the fractions prior to the appearance of any peak. The background absorbance thus compensated for is not more than 0.005 at the specified wavelength. Obtain the sum of the quantities of the high molecular weight fractions, the first peak, and the quantities of the other peaks, taken together. The high molecular weight fractions correspond to not more than 1.0% of the total eluted material.

Insulin Injection

» Insulin Injection is a sterile, acidified or neutral solution of Insulin. It has a biological potency, determined by *Assay A*, of not less than 95.0 percent and not more than 105.0 percent of the potency stated on the label, expressed in USP Insulin Units, the potency being 40, 100, or 500 USP Insulin Units in each mL.

Packaging and storage—Preserve in a refrigerator. Avoid freezing. Dispense it in the unopened, multiple-dose container in which it was placed by the manufacturer. The container for Insulin Injection, up to 100 USP Units in each mL, is of approximately 10-mL capacity and contains not less than 10 mL of the Injection, and the container for Insulin Injection, 500 USP Units per mL, is of approximately 20-mL capacity and contains not less than 20 mL of the Injection.

Labeling—The Injection container label and package label state the potency in USP Insulin Units in each mL, based on the results of *Assay A*, and the expiration date, which is not later than 24 months after the immediate container was filled. If the Injection is prepared from neutral solution, the word "neutral" appears on the label. Label it to indicate the one or more animal species to which it is related, as porcine, as bovine, or as a mixture of porcine and bovine. Where it is highly purified, label it as such. Label it to state that it is to be stored in a refrigerator and that freezing is to be avoided.

Reference standards—*USP Insulin Reference Standard*—Preserve in a refrigerator, and after opening the ampul, store in a tight container. *USP Insulin (Beef) Reference Standard. USP Insulin (Pork) Reference Standard. USP Proinsulin (Beef) Reference Standard. USP Proinsulin (Pork) Reference Standard. USP Endotoxin Reference Standard.*

Identification—

A: Adjust the Injection to a pH between 5.1 and 5.3: a precipitate is formed, and it dissolves when acidified to a pH between 2.5 and 3.5. Insulin Injection having a concentration of not more than 100 Units per mL shows, at most, only a slight haze when adjusted to a pH between 8.0 and 8.5.

B: Where the Injection is treated as directed under *Assay B*, in the chromatographic system, the response(s) for the major peak(s) is (are) identical with the response(s) obtained with the *USP Insulin Reference Standard(s)* from the same animal source(s) for the related species of animal.

Bacterial endotoxins—When tested as directed under *Bacterial Endotoxins Test* ⟨85⟩, the USP Endotoxin RS being used, it contains not more than 80 USP Endotoxin Units for each 100 USP Insulin Units.

Sterility—It meets the requirements under *Sterility Tests* ⟨71⟩, the *Membrane Filtration* procedure being used.

pH ⟨791⟩: between 2.5 and 3.5, determined potentiometrically, for acidified Insulin Injection; between 7.0 and 7.8, determined potentiometrically, for neutral Insulin Injection.

Particulate matter ⟨788⟩: meets the requirements under *Small-volume Injections.*

Residue on ignition—Dry slowly in a suitable tared crucible a volume of Injection, accurately measured and equivalent to not less than 500 USP Insulin Units. To the dry residue add 2 drops of nitric acid, and apply heat, at first very gently, then gradually increase the heat until the carbon that may appear is completely oxidized. Place in a muffle furnace at dull red to medium red heat for 15 minutes. Cool in a desiccator, and weigh: the weight of the residue for Injection prepared from acidified solution is not more than 0.10 mg, and that for Injection prepared from neutral solution is not more than 0.50 mg, per 100 USP Insulin Units.

Nitrogen content, *Method II* ⟨461⟩—Determine the nitrogen content of a quantity of Injection representing not less than 200 USP Insulin Units: not more than 0.7 mg of N is found for each 100 USP Insulin Units.

Zinc content ⟨591⟩: between 10 µg and 40 µg for each 100 USP Insulin Units.

Other requirements—It meets the requirements under *Injections* ⟨1⟩, and where highly purified, it meets also the requirements for *Proinsulin content* and *High molecular weight protein.*

Where the Insulin is highly purified, apply both *Assays A* and *B*.

Assay A—Proceed with Insulin Injection as directed under *Insulin Assay* ⟨121⟩ and as directed in *Assay B* under *Insulin*, where applicable.

Assay B—

Mobile phase—Prepare separate, suitable filtered and degassed solutions consisting of a 1 in 1000 solution of trifluoroacetic acid in a mixture of water and acetonitrile (70:30) (*Solution A*), and trifluoroacetic acid in water (1 in 1000) (*Solution B*).

Standard preparation—Dissolve an accurately weighed quantity of the appropriate Reference Standard [USP Insulin (Beef) RS or USP Insulin (Pork) RS] in 0.01 *N* hydrochloric acid to obtain a solution having a known concentration corresponding to that of the *Assay preparation.*

Assay preparation—Use solutions of Insulin without further mixing or dilution with other solvents. Mix suspensions of Insulin quantitatively with 0.01 *N* hydrochloric acid to obtain clear solutions just before injection into the chromatograph.

Chromatographic system (see *Chromatography* ⟨621⟩)—The liquid chromatograph is equipped with a 280-nm detector and a 30-cm × 3.9-mm column that contains 5 µm to 10 µm of packing L1. The flow rate is about 2 mL per minute, and the *Mobile phase* is *Solution B*, initially, programmed for gradient elution to produce a mixture of *Solution A* and *Solution B* (90:10) in 30 minutes. Chromatograph replicate injections of the *Standard preparation*, and record the peak responses as directed under *Procedure:* the relative standard deviation is not more than 1.5%, and the tailing factor for the insulin peak is not more than 2.5.

Procedure—Separately inject equal volumes (about 20 µL) of the *Standard preparation* and the *Assay preparation* into the chromatograph, record the chromatograms, and measure the responses for the major peaks.

Potency—Calculate the quantity, in USP Insulin Units per mL, of insulin in each mL of the Injection taken by the formula:

$$(C/DV)(r_U/r_S),$$

in which C is the concentration, in USP Insulin Units per mL, of the USP Reference Standard in the *Standard preparation, D* is the concentration, in USP Insulin Units per mL, of insulin in the *Assay preparation*, based upon the labeled quantity in the Injection and the extent of dilution, V is the volume, in mL, of Injection taken, and r_U and r_S are the peak responses obtained from the *Assay preparation* and the *Standard preparation*, respectively. Its potency is not less than 95.0 percent and not more than 105.0 percent of the potency stated on the label, expressed in USP Insulin Units per mL.

Insulin Human

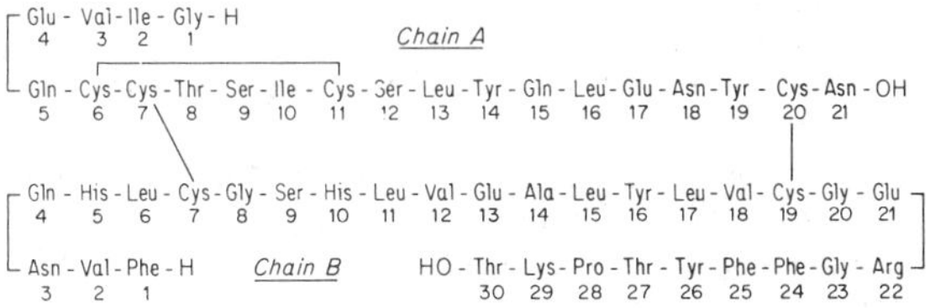

Glu - Val - Ile - Gly - H
4 3 2 1
Chain A
Gln - Cys - Cys - Thr - Ser - Ile - Cys - Ser - Leu - Tyr - Gln - Leu - Glu - Asn - Tyr - Cys - Asn - OH
5 6 7 8 9 10 11 12 13 14 15 16 17 18 19 20 21

Gln - His - Leu - Cys - Gly - Ser - His - Leu - Val - Glu - Ala - Leu - Tyr - Leu - Val - Cys - Gly - Glu
4 5 6 7 8 9 10 11 12 13 14 15 16 17 18 19 20 21

Asn - Val - Phe - H *Chain B* HO - Thr - Lys - Pro - Thr - Tyr - Phe - Phe - Gly - Arg
3 2 1 30 29 28 27 26 25 24 23 22

$C_{257}H_{383}N_{65}O_{77}S_6$ 5807.62
Insulin (human) [*11061-68-0*].

» Insulin Human is a protein corresponding to the active principle elaborated in the human pancreas that affects the metabolism of carbohydrate (particularly glucose), fat, and protein. It is derived by enzymatic modification of insulin from pork pancreas in order to change its amino acid sequence appropriately, or produced by microbial synthesis. Its potency, determined chromatographically and calculated on the dried basis, is not less than 27.5 USP Insulin Human Units in each mg.

Packaging and storage—Preserve in tight containers, in a cold place.

Reference standards—*USP Insulin Human Reference Standard*—Preserve in a freezer at −20° to −18°, and after opening the vial, store in a tight container. After opening the vial, without delay transfer accurately weighed portions to clean, dry volumetric flasks. Keep the flasks tightly closed, and store in a freezer. Do not dry before use for tests and assays. *USP Insulin (Pork) Reference Standard. USP Proinsulin (Pork) Reference Standard. USP Pancreatic Polypeptide (Pork) Reference Standard.*

Identification—

A: Prepare a 1 in 500 slurry in water, and add dropwise 3 *N* hydrochloric acid until a clear solution is obtained. The solution so obtained responds to *Identification test A* under *Insulin Injection.*

B: The retention time of the major peak in the chromatogram of the *Assay preparation* corresponds to that of the *Standard preparation*, obtained as directed in the *Assay.*

Microbial limits ⟨61⟩—The total bacterial count does not exceed 300 per g, the test being performed on a portion of about 0.2 g, accurately weighed.

Biological potency—Proceed with Insulin Human as directed under *Insulin Assay* ⟨121⟩, treating about 85 mg of it, accurately weighed, as directed under *Standard Solution* and *Standard Dilutions* to obtain the *Sample Dilutions*, and determine its biological potency, using USP Insulin Human RS. Its biological potency, calculated on the dried basis, is not less than 26.0 USP Insulin Human Units in each mg.

Loss on drying ⟨731⟩—Dry about 200 mg, accurately weighed, at 105° for 16 hours: it loses not more than 10.0% of its weight.

Residue on ignition ⟨281⟩: not more than 2.5%, calculated on the dried basis, the test being performed on a portion of about 0.4 g, accurately weighed.

Nitrogen content—Determine the nitrogen content as directed under *Method II* ⟨461⟩, using about 10 mg of it, accurately weighed: not less than 14.5% and not more than 16.5% of N, calculated on the dried basis, is found.

Zinc content ⟨591⟩—Determine the zinc content of about 10 mg of it, accurately weighed: not more than 1.08%, calculated on the dried basis, is found.

Proinsulin content—Where derived from pork pancreas, determine the proinsulin pork content by the procedure described for *Proinsulin content* under *Insulin*, but using the corresponding USP Proinsulin Pork RS. Not more than 10 ppm is found.

Pancreatic polypeptide content (where derived from pork pancreas, RIA procedure)—Not more than 1 ppm is found.

High molecular weight protein—Determine the amount of high molecular weight protein by the procedure described therefor under *Insulin*, but of a portion of Insulin Human: the high molecular weight fraction corresponds to not more than 10 parts per thousand of the total eluted material.

Content of desamido insulin and other insulin-related substances—

Mobile phase—Prepare separate, suitable filtered and degassed components consisting of 0.1 *M* monobasic sodium phosphate, adjusted with phosphoric acid to a pH of 2.0 (*Component A*), and acetonitrile (*Component B*).

Test preparations—Dissolve an accurately weighed quantity of Insulin Human in 0.01 *N* hydrochloric acid, and dilute quantitatively with 0.01 *N* hydrochloric acid to obtain a solution having a known concentration of about 10 mg per mL (*Test preparation A*). Dilute an accurately measured volume of *Test preparation A* quantitatively with 0.01 *N* hydrochloric acid to obtain a solution having a known concentration of about 0.10 mg per mL (*Test preparation B*), i.e., 0.01 of the concentration of *Test preparation A.* Store these solutions in a refrigerator.

System suitability preparation—Dilute an accurately measured volume of *Test preparation A* quantitatively with 0.01 *N* hydrochloric acid to obtain a solution having a known concentration of about 1 mg per mL, and allow it to stand at room temperature for 16 to 24 hours to enable desamido insulin to be produced.

Chromatographic system (see *Chromatography* ⟨621⟩)—The liquid chromatograph is equipped with a 214-nm detector and a 25-cm × 4.6-mm column that contains packing L13. The column temperature is maintained at 40°, the flow rate is about 1 mL per minute, and the *Mobile phase* is a mixture of *Component A* and *Component B* (74:26), initially. To determine the suitability of the chromatographic system, chromatograph the *System suitability preparation*, and record the peak responses as directed under *Procedure:* the tailing factor for the insulin peak is not more than 1.8, the resolution, R, between the insulin and desamido insulin peaks is not less than 1.8, and the relative standard deviation for replicate injections is not more than 1.6%. For use in the following *Procedure*, 25 minutes after the column has been subjected to the initial *Mobile phase*, program the *Mobile phase* for gradient elution by increasing the proportion of *Component B* at a rate of 1 volume per minute to produce a mixture of *Component A* and *Component B* (60:40), and maintain that mixture for an additional 6 minutes. Correct for a reagent blank that contains any added substances.

Procedure—Separately inject equal volumes (between 10 µL and 20 µL) of *Test preparation A* and *Test preparation B* into the chromatograph, record the chromatograms, and measure the responses for the major peaks: the peak for insulin from *Test preparation B*, and the peaks for desamido insulin and for the other insulin-related substances, if any, from *Test preparation A*. Calculate separately the amount of desamido insulin from the responses thereto, and the amount of the other insulin-related substances, if any, from the sum of their responses, as a percentage of the amount of insulin human found in the chromatogram. The relative amount of desamido insulin and the relative sum of the amounts of the other insulin-related substances each is not more than 3% of the amount of insulin human found. [NOTE—For use of this result for desamido insulin in the calculation for the *Assay* (see below), calculate the amount of desamido insulin (in equivalents of USP Insulin Human Units) in each mg of Insulin Human on the basis of the results of the chromatographic assay performed as directed below, under *Assay*.] Verify the identity of all peaks of interest by injecting suitable amounts of solutions having similar concentrations of the appropriate Reference Standards in the same medium.

Assay—

Mobile phase—Prepare a filtered and degassed mixture consisting of 74 volumes of 0.1 *M* monobasic sodium phosphate, adjusted with phosphoric acid to a pH of 2.0, and about 26 volumes of acetonitrile. [NOTE—Proportions of the *Mobile phase* components may be adjusted to achieve satisfactory resolution and retention times.]

Standard preparation—Dissolve accurately weighed quantities of USP Insulin Human RS in a solvent made by dissolving 0.6 mg per mL of tetrasodium ethylenediaminetetraacetate in water and adding sufficient 0.01 *N* hydrochloric acid to obtain three solutions, having a pH between 8.5 and 9.0, and having known concentrations between about 34 and about 45 USP Insulin Human Units per mL.

Assay preparation—Dissolve an accurately weighed quantity of Insulin Human in the same solvent used for making the *Standard preparation*, and dilute quantitatively with the same solvent to obtain a solution having a concentration of about 40 USP Insulin Human Units per mL.

Chromatographic system (see *Chromatography* ⟨621⟩)—The liquid chromatograph is equipped with a 214-nm detector and a 25-cm × 4.6-mm column that contains packing L13. The column temperature is maintained at 40°, and the flow rate is about 1 mL per minute. Chromatograph replicate injections of the *Standard preparation* in the three concentrations made, and record the peak responses as directed under *Procedure:* the relative standard deviation of the standard curve is not more than 1.6%, the tailing factor for the insulin peak corresponding to the solution containing 40 USP Insulin Human Units per mL is not more than 1.8, and the resolution, *R*, between insulin and desamido insulin is not less than 1.8.

Procedure—Separately inject equal volumes (between 10 µL and 20 µL) of the three solutions of the *Standard preparation* and the *Assay preparation* into the chromatograph, record the chromatograms, and measure the responses for the major peaks. Plot the mean peak responses for each concentration of the *Standard preparation* against the amounts of Insulin Human (in USP Insulin Human Units per mL), or determine the relationship mathematically by a straight-line fitting procedure using a least-squares method and a test for linearity (see *Calculation of Potency from a Standard Curve* under *Design and Analysis of Biological Assays* ⟨111⟩). Read off or compute the concentration of Insulin Human found, and calculate the amount in each mg of Insulin Human taken (in USP Insulin Human Units).

Potency—Compute the total potency of the portion of Insulin Human taken, as the sum of the amount of Insulin Human found in each mg and 0.9 times the amount of desamido insulin found (see under *Content of desamido insulin and other insulin-related substances*) in each mg in USP Insulin Human Units.

Insulin Human Injection

» Insulin Human Injection is a sterile solution of Insulin Human in Water for Injection. It has a potency, determined chromatographically, of not less than 95.0 percent and not more than 105.0 percent of the potency stated on the label, expressed in USP Insulin Human Units in each mL.

Packaging and storage—Preserve in a refrigerator, and avoid freezing. The container for Insulin Human Injection, 40 or 100 USP Units in each mL, is of approximately 10-mL capacity and contains not less than 10 mL of the Injection, and the container for Insulin Human Injection, 500 USP Units per mL, is of approximately 20-mL capacity and contains not less than 20 mL of the Injection.

Labeling—The Injection container label and package label state the potency in USP Insulin Human Units in each mL on the basis of the results of the *Assay*, and the expiration date, which is not later than 24 months after the immediate container was filled. The labeling states also that it has been prepared either with Insulin Human derived by enzyme modification of pork pancreas Insulin or with Insulin Human obtained from microbial synthesis, whichever is applicable. Label it to state that it is to be stored in a refrigerator and that freezing is to be avoided.

Reference standards—*USP Insulin Human Reference Standard*—Preserve in a freezer at −20° to −18°, and after opening the vial, store in a tight container. After opening the vial, without delay transfer accurately weighed portions to clean, dry volumetric flasks. Keep the flasks tightly closed, and store in a freezer. Do not dry before use for tests and assays. *USP Insulin (Pork) Reference Standard. USP Proinsulin (Pork) Reference Standard. USP Pancreatic Polypeptide (Pork) Reference Standard. USP Endotoxin Reference Standard.*

Identification—

A: It responds to *Identification test A* under *Insulin Injection*.

B: The retention time of the major peak in the chromatogram of the *Assay preparation* corresponds to that of the *Standard preparation*, obtained as directed in the *Assay*.

Bacterial endotoxins—When tested as directed under *Bacterial Endotoxins Test* ⟨85⟩, the USP Endotoxin RS being used, it contains not more than 80 USP Endotoxin Units for each 100 USP Insulin Human Units.

Sterility—It meets the requirements under *Sterility Tests* ⟨71⟩, the *Membrane Filtration* procedure being used.

Biological potency—Proceed with Insulin Human Injection as directed under *Insulin Assay* ⟨121⟩, except to use USP Insulin Human RS, and determine its biological potency. Its biological potency is not less than 95.0 percent and not more than 105.0 percent of the potency stated on the label.

pH ⟨791⟩: between 7.0 and 7.8, determined potentiometrically.

Particulate matter ⟨788⟩: meets the requirements under *Small-volume Injections*.

Nitrogen content, *Method II* ⟨461⟩—Determine the nitrogen content of a quantity of Injection representing not less than 200 USP Insulin Human Units: not more than 0.7 mg of N is found for each 100 USP Insulin Human Units.

Zinc content ⟨591⟩: between 10 µg and 40 µg for each 100 USP Insulin Human Units.

Other requirements—It meets the requirements under *Injections* ⟨1⟩, and meets the requirements for *High molecular weight protein*, and, where derived from pork pancreas insulin, for *Proinsulin content*, and *Pancreatic polypeptide content* under *Insulin Human*.

Assay and content of desamido insulin—

Mobile phase—Prepare a filtered and degassed mixture consisting of 74 volumes of 0.1 *M* monobasic sodium phosphate, adjusted with phosphoric acid to a pH of 2.0, and about 26 volumes of acetonitrile. [NOTE—Proportions of the *Mobile phase* components may be adjusted to achieve satisfactory resolution and retention times.]

be shaken carefully before use. Its container label and its package label state the potency in USP Insulin Units in each mL, and the expiration date, which is not later than 24 months after the immediate container was filled. Label it to state that it is to be stored in a refrigerator and that freezing is to be avoided.

Identification—

A: Acidify the Suspension to a pH of between 2.5 and 3.5: the precipitate dissolves, producing a clear, colorless liquid, which responds to the *Identification tests* under *Insulin Injection.*

B: Add a few drops of ferric chloride TS to the Suspension: a reddish brown color develops, and disappears upon the addition of a mineral acid.

Sterility—It meets the requirements of the test for *Sterility* under *Isophane Insulin Suspension.*

pH ⟨791⟩: between 7.0 and 7.8, determined potentiometrically.

Nitrogen content, *Method II* ⟨461⟩—Determine the nitrogen content of a quantity of Suspension representing not less than 200 USP Insulin Units: not more than 0.70 mg of N is found for each 100 USP Insulin Units.

Zinc content ⟨591⟩: between 0.12 mg and 0.25 mg for each 100 USP Insulin Units.

Zinc in the supernatant liquid—Centrifuge a portion of Suspension sufficient for the test, and determine the zinc content of the clear, supernatant liquid as directed under *Zinc Determination* ⟨591⟩: the zinc concentration, in mg per mL, is between 20% and 65% of the zinc concentration of the Suspension.

Insulin not extracted by buffered acetone solution—Centrifuge 15 mL (40-Unit), 8 mL (80-Unit), or 6 mL (100-Unit) of Suspension, and discard the supernatant liquid. Suspend the residue in 8.4 mL of water, quickly add 16.6 mL of buffered acetone TS, shake or stir vigorously, and centrifuge within 3 minutes after the addition of the buffered acetone TS. Discard the supernatant liquid, repeat the treatment with water and buffered acetone TS, centrifuge, and discard the supernatant liquid: no crystalline residue remains.

Protamine Zinc Insulin Suspension

Insulin protamine zinc.
Insulin protamine zinc [*9004-17-5*].

» Protamine Zinc Insulin Suspension is a sterile suspension of Insulin in buffered Water for Injection, modified by the addition of Zinc Chloride and Protamine Sulfate. The Protamine Sulfate is prepared from the sperm or from the mature testes of fish belonging to the genus *Oncorhynchus* Suckley, or *Salmo* Linné (Fam. Salmonidae), and conforms to the regulations of the federal Food and Drug Administration concerning certification of drugs composed wholly or partly of insulin (21 CFR 429.25, Standards of quality and purity for protamine) (see *Sulfate* and *Nitrogen content* under *Protamine Sulfate*).

In the preparation of Protamine Zinc Insulin Suspension, the amount of insulin used is sufficient to provide 40, 80, or 100 USP Insulin Units for each mL of the Suspension.

Packaging and storage—Preserve in a refrigerator. Avoid freezing. Dispense it in the unopened, multiple-dose container in which it was placed by the manufacturer. The container is of approximately 10-mL capacity and contains not less than 10 mL of the Suspension.

Labeling—Label it to indicate the one or more animal species to which it is related, as porcine, as bovine, or as a mixture of porcine and bovine. Where it is highly purified, label it as such. The Suspension container label states that the Suspension is to be shaken carefully before use. The container label and the package label state the potency in USP Insulin Units in each mL, and the expiration date, which is not later than 24 months after the immediate container was filled. Label it to state that it is to be stored in a refrigerator and that freezing is to be avoided.

Reference standards—*USP Protamine Reference Standard*—Keep container tightly closed, and preserve in a refrigerator. *USP Insulin Reference Standard*—Preserve in a refrigerator, and after opening the ampul, store in a tight container.

Identification—

A: Acidify it to a pH of between 2.5 and 3.5: the suspended solid dissolves, producing a clear, colorless liquid.

B: Acidified as directed in the preceding test, it responds to *Identification test A* under *Insulin Injection.*

Sterility—It meets the requirements of the test for *Sterility* under *Isophane Insulin Suspension.*

pH ⟨791⟩: between 7.1 and 7.4, determined potentiometrically.

Nitrogen content, *Method II* ⟨461⟩—Determine the nitrogen content of a quantity of Suspension representing not less than 200 USP Insulin Units: not more than 1.25 mg of N is found for each 100 USP Insulin Units.

Zinc content ⟨591⟩: between 0.15 mg and 0.25 mg for each 100 USP Insulin Units.

Biological reaction—

Solution 1—Dissolve 183 mg of zinc oxide in 60 mL of 0.1 *N* hydrochloric acid. Add 16 g of glycerin, 2.5 g of phenol (or 2 g of cresol), and sufficient water to make 1000 mL.

Solution 2—Dissolve sufficient dibasic sodium phosphate to provide 4 g of Na_2HPO_4, 16 g of glycerin, and 2.5 g of phenol (or 2 g of cresol) in sufficient water to make 1000 mL.

Solution 3—Dissolve not less than 100 mg of USP Protamine RS in *Solution 1*, in the proportion of 1 mg for each mL. Preserve in a refrigerator, and use within 6 months.

Solution 4—Dissolve a suitable quantity of USP Insulin RS, accurately weighed, in a sufficient quantity of *Solution 3* to make a solution containing 80 USP Insulin Units in each mL. If necessary, add 1 drop of 3 *N* hydrochloric acid to effect complete solution. Preserve in a refrigerator, and use within 6 months.

Standard preparation of protamine zinc insulin, 40 USP Insulin Units per mL—To a suitable volume of *Solution 4*, accurately measured, add an equal volume of *Solution 2*, with gentle shaking, test the reaction, and if the pH is not between 7.1 and 7.4, discard, and prepare a new mixture, using a freshly prepared portion of *Solution 2* in which the pH has been suitably adjusted. Prepare fresh for each test, but do not use before 2 days nor after 6 months.

Standard preparation of protamine zinc insulin, 80 USP Insulin Units per mL—Prepare the standard to contain 80 USP Insulin Units in each mL according to the method under *Standard preparation of protamine zinc insulin, 40 USP Insulin Units per mL*, with zinc and protamine in the same relative proportions per USP Insulin Unit.

Standard preparation of protamine zinc insulin, 100 USP Insulin Units per mL—Prepare the standard to contain 100 USP Insulin Units in each mL according to the method under *Standard preparation of protamine zinc insulin, 40 USP Insulin Units per mL*, with zinc and protamine in the same relative proportions per USP Insulin Unit.

The animals—Use animals such as are described under *Insulin Assay* ⟨121⟩.

Dose of the standard preparation of protamine zinc insulin to be injected—The dose of the *Standard preparation of protamine zinc insulin* to be injected is such that following its injection (without dilution) the average blood sugar level at the time of the final bleeding is between 70% and 95% of the average initial value, but the dose is not so great as to cause convulsions in more than 25% of the animals.

Dose of the Protamine Zinc Insulin Suspension to be injected—For each animal, the dose of Protamine Zinc Insulin Suspension to be injected is the same as the dose of the *Standard preparation of protamine zinc insulin.*

Procedure—Divide the rabbits into two similar groups of approximately equal numbers. Place the rabbits in individual cages, and withhold all food, but not water, for approximately 24 hours before the test. During the test, withhold both food and water until the final sample of blood has been taken. Handle the rabbits with care in order to avoid undue excitement. For the initial determination of blood sugar concentration, as directed under

Insulin Assay ⟨121⟩, obtain slightly more than 1 mL of blood from a small incision in the marginal vein of the ear, collecting the blood in a suitable vessel containing about 3 mg of sodium oxalate. After obtaining this sample of blood, inject subcutaneously, without dilution, into the rabbits of one group, the appropriate volume (determined as directed above) of the *Standard preparation of protamine zinc insulin* of the same strength as the potency declared on the label of the Protamine Zinc Insulin Suspension, and into the rabbits of the other group the appropriate volume of the Protamine Zinc Insulin Suspension. In the same manner as previously, obtain one sample of blood from each rabbit at 1½-hour intervals up to the third hour, and then at 2-hour intervals up to and including the ninth hour after the injection, and determine the blood sugar concentration in each. About 1 week later, inject the Protamine Zinc Insulin Suspension into each rabbit of the group that previously received the *Standard preparation of protamine zinc insulin;* in a similar manner, inject the *Standard preparation of protamine zinc insulin* into each rabbit of the group that previously received the Protamine Zinc Insulin Suspension. Obtain samples of blood in the same manner as previously described, and determine the concentration of blood sugar in each. Complete the test on a total of not less than 30 rabbits.

Interpretation—Subtract the average blood sugar concentration at each bleeding time for those rabbits injected with the Protamine Zinc Insulin Suspension from the average blood sugar concentration at the comparable bleeding time for those rabbits injected with the *Standard preparation of protamine zinc insulin.* The average blood sugar concentrations at each bleeding time do not differ by more than 5 mg per 100 mL, except that at the final bleeding time the average blood sugar concentrations may differ by as much as 8 mg per 100 mL, and the average of the differences for all bleeding times after the injection, the sign of each difference being taken into account, is between −3.0 and 3.0.

Intramammary Infusions—*see complete list in index*

Inulin

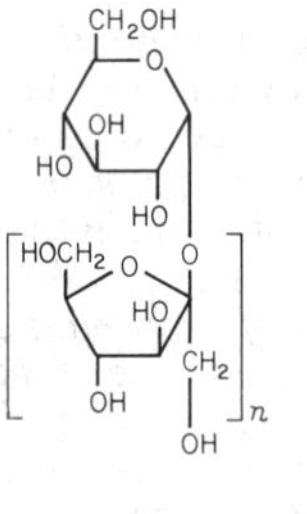

$C_6H_{11}O_5(C_6H_{10}O_5)_nOH$
Inulin.
Inulin [*9005-80-5*].

» Inulin is a polysaccharide which, on hydrolysis, yields mainly fructose. It contains not less than 94.0 percent and not more than 102.0 percent of $C_6H_{11}O_5(C_6H_{10}O_5)_nOH$, calculated on the dried basis.

Packaging and storage—Preserve in well-closed containers.

Reference standards—*USP Fructose Reference Standard*—Dry in vacuum at 70° for 4 hours before using. *USP Dextrose Reference Standard*—Dry at 105° for 16 hours before using.

Completeness of solution—Dissolve 10 g in 20 mL of boiling water in a 200-mL volumetric flask, add 150 mL of water, allow to cool, dilute with water to volume, and mix: the solution is clear.

Specific rotation ⟨781⟩: between −32.0° and −40.0°, calculated on the dried basis, determined in a solution containing 10 g of Inulin and 0.2 mL of 6 *N* ammonium hydroxide in each 100 mL.

Microbial limits ⟨61⟩—It meets the requirements of the tests for absence of *Salmonella* species and *Escherichia coli* and for absence of *Staphylococcus aureus* and *Pseudomonas aeruginosa;* the total aerobic microbial count is less than 1000 per g.

Loss on drying ⟨731⟩: not more than 10.0% after it has been dried at 105° for 2 hours, 2 g of the finely ground powder being used for the test.

Residue on ignition ⟨281⟩—Multiply the percentage of *Calcium* found by 3.4. The residue on ignition does not exceed this percentage by more than 0.05%.

Calcium—Dissolve 10.0 g in 100 mL of water, using heat to facilitate solution. Cool to room temperature, add 15 mL of 1 *N* sodium hydroxide and 300 mg of hydroxy naphthol blue trituration, and titrate with 0.05 *M* disodium ethylenediaminetetraacetate VS to a blue end-point: not more than 5.0 mL is required, corresponding to a limit of 0.10% calcium.

pH, Chloride, Sulfate, Iron, and Reducing sugars—Dissolve 10.0 g in 20 mL of boiling water in a 100-mL volumetric flask, allow to cool, dilute with water to volume, and mix. Use the solution for the following tests.

pH ⟨791⟩—The pH of the solution is between 4.5 and 7.0.

Chloride ⟨221⟩—A 10-mL portion of the solution shows no more chloride than corresponds to 0.20 mL of 0.020 *N* hydrochloric acid (0.014%).

Sulfate—To 10 mL of the solution add 1 mL of barium chloride TS: no turbidity is produced.

Iron—To 10 mL of the solution add 0.5 mL of hydrochloric acid and 3 drops of potassium ferrocyanide TS: the solution does not become blue within 1 minute.

Reducing sugars—To 2 mL of the solution add 5 mL of alkaline cupric tartrate TS: no reduction occurs at room temperature, and only slight reduction occurs after one minute of boiling.

Heavy metals ⟨231⟩—Dissolve 4.0 g in 20 mL of boiling water, allow to cool, and dilute with water to 25 mL: the limit is 5 ppm.

Free fructose—

Blue tetrazolium solution—Dissolve 50 mg of blue tetrazolium in 10 mL of alcohol, and mix.

Tetramethylammonium hydroxide solution—Prepare a mixture of 1 volume of tetramethylammonium hydroxide TS and 9 volumes of alcohol.

Standard stock solution—Transfer about 25 mg of USP Fructose RS, accurately weighed, to a 100-mL volumetric peak, dissolve in water, dilute with water to volume, and mix. Store at about 4°. The solution contains about 250 µg per mL.

Standard preparation—On the day of use, dilute a portion of the *Standard stock solution* quantitatively with alcohol to obtain a solution having a known concentration of about 2.5 µg per mL. Store at about 4°.

Test preparation—Transfer about 2.5 g of Inulin, accurately weighed, to a 100-mL volumetric flask, add about 75 mL of water, heat on a steam bath until solution is complete, cool to room temperature, dilute with water to volume, and mix. Pipet 1 mL into a 100-mL volumetric flask, dilute with alcohol to volume, and mix. If the solution is turbid, filter through fine-porosity filter paper.

Procedure—Pipet 10 mL of the *Test preparation* and 10 mL of the *Standard preparation* into separate glass-stoppered centrifuge tubes. Into each of the tubes, and into a similar tube containing 10.0 mL of alcohol to provide the blank, pipet 1 mL of *Blue tetrazolium solution*, and mix. Then into each tube pipet 1 mL of *Tetramethylammonium hydroxide solution*, mix, and allow to stand in the dark for 60 minutes. Without delay, concomitantly determine the absorbances of the solutions from the *Test preparation* and the *Standard preparation* at 530 nm, with a suitable spectrophotometer, against the blank. Calculate the percentage, *F*, of free fructose in the Inulin taken by the formula:

$$F = (C/W)(A_U/A_S),$$

in which *C* is the concentration, in µg per mL, of USP Fructose RS in the *Standard preparation*, *W* is the quantity, in g, of Inulin taken, and A_U and A_S are the absorbances of the solutions from the *Test preparation* and the *Standard preparation*, respectively. The limit is 2.0%, calculated on the dried basis.

Assay for combined glucose—

Standard stock solution—Transfer about 50 mg of USP Dextrose RS, accurately weighed, to a 100-mL volumetric flask, dissolve in a solution of benzoic acid (1.7 in 1000), dilute with the same solution to volume, and mix. Allow to stand at room temperature for not less than 3 hours before using. This solution is stable for 1 month at about 4°.

Standard preparation—Pipet 7 mL of *Standard stock solution* into a 100-mL volumetric flask, dilute with water to volume, mix, and use at once.

Assay preparation—Transfer about 0.5 g of Inulin, accurately weighed, to a 100-mL volumetric flask, add 5.0 mL of water, dissolve by heating on a steam bath, cool to room temperature, add 0.5 mL of 8 *N* hydrochloric acid, and mix. Place the flask in a boiling water bath for 5 minutes, cool, dilute with water to volume, and mix. Pipet 2 mL of this solution into a 10-mL volumetric flask, dilute with water to volume, and mix. [NOTE—This solution is used also for preparing the *Assay preparation* in the *Assay for inulin.*]

Procedure—Pipet 3-mL portions of glucose oxidase–chromogen TS into 3 separate test tubes, and bring to a temperature of 37 ± 0.5° in a water bath. Pipet 2 mL of *Standard preparation* into one of the tubes, pipet 2 mL of *Assay preparation* into another, and pipet 2 mL of water into the third tube to provide the blank. Maintain at 37 ± 0.5° for an additional 10 minutes, then remove the tubes, and allow them to cool. Determine the absorbances of the solutions from the *Assay preparation* and the *Standard preparation* at about 505 nm, with a suitable spectrophotometer, using the reagent blank as a reference. Calculate the percentage of combined glucose, *G*, in the Inulin taken by the formula:

$$G = 10(C/W)(A_U/A_S),$$

in which *C* is the concentration, in mg per mL, of USP Dextrose RS in the *Standard preparation*, *W* is the amount, in g, of Inulin taken, and A_U and A_S are the absorbances of the solutions from the *Assay preparation* and the *Standard preparation*, respectively. Not less than 2.0% and not more than 5.0%, calculated on the dried basis, is found.

Assay for inulin—

Thiobarbituric acid solution—Dissolve 250 mg of thiobarbituric acid in 100 mL of 8 *N* hydrochloric acid, and mix. This solution is stable for 2 weeks at a temperature of about 4°.

Standard stock solution—Transfer about 100 mg of USP Fructose RS, accurately weighed, to a 100-mL volumetric flask, dissolve in a solution of benzoic acid (1.7 in 1000), dilute with the same solution to volume, and mix. This solution is stable for 1 month at about 4°.

Standard preparation—Dilute the *Standard stock solution* quantitatively with water to one-fiftieth of its concentration. Use immediately.

Assay preparation—Pipet 4 mL of the *Assay preparation* from the *Assay for combined glucose* into a 200-mL volumetric flask, add water to volume, and mix.

Procedure—Pipet 1-mL portions of the *Standard preparation* and the *Assay preparation* into separate glass-stoppered tubes. Pipet 1 mL of water into a third tube to provide the blank. Pipet 5-mL portions of *Thiobarbituric acid solution* into each tube, and mix. Place all of the tubes simultaneously in a water bath maintained at a temperature of about 83°, and allow them to stay immersed for 5 minutes, accurately timed. Remove the tubes simultaneously, and allow them to cool in a dark place for 30 minutes. Determine the absorbances of the solutions from the *Assay preparation* and the *Standard preparation* at about 435 nm, with a suitable spectrophotometer, using the reagent blank as a reference. Calculate the percentage of $C_6H_{11}O_5(C_6H_{10}O_5)_nOH$ in the Inulin taken by the formula:

$$0.900[2.5(C/W)(A_U/A_S) - F] + G,$$

in which 0.900 is the ratio of the formula weight of an anhydrofructose unit of inulin to that of fructose, *C* is the concentration, in µg per mL, of USP Fructose RS in the *Standard preparation*, *W* is the quantity, in g, of Inulin weighed for the *Assay for combined glucose*, A_U and A_S are the absorbances of the solutions from the *Assay preparation* and the *Standard preparation*, respectively, *F* is the percentage of free fructose, and *G* is the percentage of combined glucose.

Inulin in Sodium Chloride Injection

» Inulin in Sodium Chloride Injection is a sterile solution, which may be supersaturated, of Inulin and Sodium Chloride in Water for Injection. It may require heating before use if crystallization has occurred. It contains not less than 90.0 percent and not more than 110.0 percent of the labeled amount of $(C_6H_{12}O_6)_n$ and not less than 95.0 percent and not more than 105.0 percent of the labeled amount of NaCl. It contains no antimicrobial agents.

Packaging and storage—Preserve in single-dose containers, preferably of Type I or Type II glass.

Reference standards—*USP Fructose Reference Standard*—Dry in vacuum for 4 hours at 70° before using. *USP Dextrose Reference Standard*—Dry at 105° for 16 hours before using.

Clarity—If particulate matter is present, heat at 100° for 30 minutes: the resulting solution is free from turbidity and particulate matter.

Note—Before applying the following tests, dissolve any solid matter by heating, and cool to room temperature.

Pyrogen—It meets the requirements of the Pyrogen Test ⟨151⟩, the test dose being 10 mL per kg of a solution containing 28 mg of Inulin in each mL of pyrogen-free saline TS.

pH ⟨791⟩: between 5.0 and 7.0.

Free fructose—

Blue tetrazolium solution, Tetramethylammonium hydroxide solution, Standard stock solution, and *Standard preparation*—Proceed as directed in the test for *Free fructose* under *Inulin.*

Test preparation—Transfer an accurately measured volume of Inulin in Sodium Chloride Injection, equivalent to about 2.5 g of inulin, to a 100-mL volumetric flask, add water to volume, and mix. Just prior to use, pipet 1 mL of this solution into a 100-mL volumetric flask, dilute with alcohol to volume, mix, and, if the solution is turbid, filter through fine-porosity filter paper.

Procedure—Proceed as directed for *Procedure* in the test for *Free fructose* under Inulin. Calculate the quantity, *F*, in mg, of free fructose in each mL of the Injection taken by the formula:

$$10(C/V)(A_U/A_S),$$

in which *C* is the concentration, in µg per mL, of USP Fructose RS in the *Standard preparation*, *V* is the volume, in mL, of Injection taken, and A_U and A_S are the absorbances of the solutions from the *Test preparation* and the *Standard preparation*, respectively. The limit is 2.2 mg per mL.

Other requirements—It meets the other requirements under *Injections* ⟨1⟩.

Assay for inulin—

Assay for combined glucose—Proceed as directed in the *Assay for combined glucose* under *Inulin*, but in preparing the *Assay preparation* use, instead of 0.5 g of inulin, an accurately measured volume of the Inulin in Sodium Chloride Injection equivalent to about 0.5 g of inulin. Calculate the quantity, in mg, of combined glucose, *G*, in each mL of the Injection taken by the formula:

$$100(C/V)(A_U/A_S),$$

in which *C* is the concentration, in mg per mL, of USP Dextrose RS in the *Standard preparation*, *V* is the volume, in mL, of Injection taken, and A_U and A_S are the absorbances of the solutions from the *Assay preparation* and the *Standard preparation*, respectively.

Procedure—Proceed as directed in the *Assay for inulin* under *Inulin.* Calculate the quantity, in mg, of $(C_6H_{12}O_6)_n$ in each mL of the Injection taken by the formula:

$$0.900[25(C/V)(A_U/A_S) - F] + G,$$

in which 0.900 is the ratio of the formula weight of an anhydrofructose unit of inulin to that of fructose, C is the concentration, in µg per mL, of USP Fructose RS in the *Standard preparation*, V is the volume, in mL, of Injection taken, A_U and A_S are the absorbances of the solutions from the *Assay preparation* and the *Standard preparation*, respectively, and F is the quantity, in mg, of free fructose in each mL of Injection.

Assay for sodium chloride—Pipet a volume of Inulin in Sodium Chloride Injection, equivalent to about 90 mg of sodium chloride, into a porcelain casserole, and add 140 mL of water and 1 mL of dichlorofluorescein TS. Mix, and titrate with 0.1 *N* silver nitrate VS until the silver chloride flocculates and the mixture acquires a faint pink color. Each mL of 0.1 *N* silver nitrate is equivalent to 5.844 mg of NaCl.

Invert Sugar Injection—*see* Sugar Injection, Invert

Iocetamic Acid

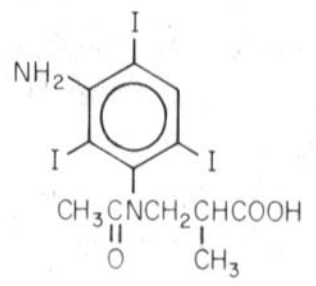

$C_{12}H_{13}I_3N_2O_3$ 613.96

Propanoic acid, 3-[acetyl(3-amino-2,4,6-triiodophenyl)amino]-2-methyl-.

N-Acetyl-*N*-(3-amino-2,4,6-triiodophenyl)-2-methyl-β-alanine [*16034-77-8*].

» Iocetamic Acid contains not less than 98.0 percent and not more than 102.0 percent of $C_{12}H_{13}I_3N_2O_3$, calculated on the dried basis.

Packaging and storage—Preserve in well-closed containers.

Reference standard—*USP Iocetamic Acid Reference Standard*—Dry at 105° for 1 hour before using.

Identification—The infrared absorption spectrum of a potassium bromide dispersion of it exhibits maxima only at the same wavelengths as that of a similar preparation of USP Iocetamic Acid RS.

Loss on drying ⟨731⟩—Dry it at 105° for 4 hours: it loses not more than 1.0% of its weight.

Residue on ignition ⟨281⟩: not more than 0.1%.

Iodide—

Test preparation—Mix 0.80 g with the minimum amount of 0.2 *N* sodium hydroxide necessary to dissolve the test specimen, dilute with water to 10 mL, and mix. Add 2 *N* nitric acid, dropwise, until precipitation of the iocetamic acid is complete, then add 3 mL of 2 *N* nitric acid, and mix. Filter, and wash the precipitate with 5 mL of water, collecting the filtrate and washings in a suitable test tube.

Procedure—In a test tube similar to that used for the *Test preparation*, mix 2.0 mL of potassium iodide solution (1 in 4000) and 3 mL of 2 *N* nitric acid with sufficient water to equal the volume of the *Test preparation*. To each tube add 1 mL of 30 percent hydrogen peroxide and 1 mL of chloroform, and shake each mixture: any violet color in the chloroform layer from the *Test preparation* is not darker than that obtained from the potassium iodide solution (0.005%).

Heavy metals ⟨231⟩—To 1.0 g in a 50-mL beaker add 1 mL of 1 *N* sodium hydroxide and 1 mL of water, and stir. Add another 1 mL of 1 *N* sodium hydroxide, and continue stirring until solution is complete. Add 5 mL of 3 *N* hydrochloric acid, and stir as precipitation occurs. Filter, and wash the filter with 5 to 10 mL of water. Adjust the filtrate by the dropwise addition of 1 *N* sodium hydroxide to a pH of 8.5, transfer to a 50-mL color comparison tube, and dilute with water to about 35 mL. Into another 50-mL color-comparison tube pipet 2 mL of *Standard Lead Solution* (20 µg of Pb), dilute with 10 mL of water, adjust the solution with 1 *N* sodium hydroxide to a pH of 8.5, and dilute with water to about 35 mL. To each tube add 10 mL of sodium sulfide TS, dilute with water to volume, and mix. Any color produced by the solution under test is not greater than that produced by the standard solution (0.002%).

Assay—Transfer about 400 mg of Iocetamic Acid, accurately weighed, to a glass-stoppered, 125-mL conical flask. Add 12 mL of sodium hydroxide solution (1 in 5), 20 mL of water, and 1 g of powdered zinc, connect the flask to a reflux condenser, and reflux for 30 minutes. Cool the flask to room temperature, rinse the condenser with 20 mL of water, disconnect the flask from the condenser, and filter the mixture. Rinse the flask and the filter thoroughly, adding the rinsings to the filtrate. Add 40 mL of 2 *N* sulfuric acid, and titrate immediately with 0.05 *N* silver nitrate VS, determining the end-point potentiometrically, using silver-calomel electrodes and an agar–potassium nitrate salt bridge. Each mL of 0.05 *N* silver nitrate is equivalent to 10.23 mg of $C_{12}H_{13}I_3N_2O_3$.

Iocetamic Acid Tablets

» Iocetamic Acid Tablets contain not less than 90.0 percent and not more than 110.0 percent of the labeled amount of $C_{12}H_{13}I_3N_2O_3$.

Packaging and storage—Preserve in tight containers.

Reference standard—*USP Iocetamic Acid Reference Standard*—Dry at 105° for 1 hour before using.

Identification—A portion of finely powdered Tablets responds to the *Identification test* under *Iocetamic Acid*.

Dissolution ⟨711⟩—

Medium: simulated intestinal fluid TS, prepared without pancreatin; 900 mL.

Apparatus 1: 150 rpm.

Time: 30 minutes and 60 minutes.

Procedure—Filter 15 mL of the solution under test, and transfer 10.0 mL of the filtrate to a 100-mL volumetric flask. Dilute with water to volume, and mix. Determine the amount of $C_{12}H_{13}I_3N_2O_3$ dissolved from ultraviolet absorbances at the wavelength of maximum absorbance at about 315 nm of this solution, using water as the blank, in comparison with a Standard solution having a known concentration of USP Iocetamic Acid RS in the same medium.

Tolerances—Not less than 35% (*Q*) of the labeled amount of $C_{12}H_{13}I_3N_2O_3$ is dissolved in 30 minutes, and not less than 50% (*Q*) is dissolved in 60 minutes.

Uniformity of dosage units ⟨905⟩: meet the requirements.

Assay—Weigh and finely powder not less than 20 Iocetamic Acid Tablets. Weigh accurately a portion of the powder, equivalent to about 400 mg of iocetamic acid, transfer to a 125-mL conical flask, and proceed as directed in the *Assay* under *Iocetamic Acid*, beginning with "Add 12 mL of sodium hydroxide solution (1 in 5)."

Iodine

I (At. wt.) 126.90

Iodine.

Iodine [*7553-56-2*].

» Iodine contains not less than 99.8 percent and not more than 100.5 percent of I.

Packaging and storage—Preserve in tight containers.

Identification—

A: Solutions (1 in 1000) in chloroform, in carbon tetrachloride, and in carbon disulfide have a violet color.

B: To a saturated solution add starch–potassium iodide TS: a blue color is produced. When the mixture is boiled, the color vanishes but reappears as the mixture cools, unless it has been subjected to prolonged boiling.

Nonvolatile residue—Place 5.0 g in a tared porcelain dish, heat on a steam bath until the iodine has been driven off, and dry at 105° for 1 hour: not more than 0.05% of residue remains.

Chloride or bromide—Triturate 250 mg of finely powdered Iodine with 10 mL of water, and filter the solution. Add, dropwise, sulfurous acid (free from chloride), previously diluted with several volumes of water, until the iodine color just disappears. Add 5 mL of 6 *N* ammonium hydroxide, followed by 5 mL of silver nitrate TS in small portions. Filter, and acidify the filtrate with nitric acid: the resulting liquid is not more turbid than a control made with the same quantities of the same reagents to which 0.10 mL of 0.020 *N* hydrochloric acid has been added, the sulfurous acid being omitted (0.028% as chloride).

Assay—Place about 500 mg of powdered Iodine in a tared, glass-stoppered flask, insert the stopper, weigh accurately, and add 1 g of potassium iodide dissolved in 5 mL of water. Dilute with water to about 50 mL, add 1 mL of 3 *N* hydrochloric acid, and titrate with 0.1 *N* sodium thiosulfate VS, adding 3 mL of starch TS as the end-point is approached. Each mL of 0.1 *N* sodium thiosulfate is equivalent to 12.69 mg of I.

Iodine Topical Solution

» Iodine Topical Solution contains, in each 100 mL, not less than 1.8 g and not more than 2.2 g of iodine (I), and not less than 2.1 g and not more than 2.6 g of sodium iodide (NaI).

Iodine	20 g
Sodium Iodide	24 g
Purified Water, a sufficient quantity, to make	1000 mL

Dissolve the Iodine and the Sodium Iodide in 50 mL of Purified Water, then add Purified Water to make 1000 mL.

Packaging and storage—Preserve in tight, light-resistant containers, at a temperature not exceeding 35°.

Identification—

A: Add 1 drop to a mixture of 1 mL of starch TS and 9 mL of water: a deep blue color is produced.

B: Evaporate a few mL on a steam bath to dryness: the residue responds to the flame test for *Sodium* ⟨191⟩, and to the tests for *Iodide* ⟨191⟩.

Assay for iodine—Pipet 5 mL of Iodine Topical Solution into a glass-stoppered, 500-mL flask, and dilute with 10 mL of water. Titrate with 0.1 *N* potassium arsenite VS, using 3 mL of starch TS as the indicator. Each mL of 0.1 *N* potassium arsenite is equivalent to 12.69 mg of I.

Assay for sodium iodide—To the titrated solution obtained in the *Assay for iodine* add 25 mL of hydrochloric acid, cool to room temperature, add 5 mL of chloroform, and titrate with 0.05 *M* potassium iodate VS until the purple color of iodine disappears from the chloroform. Add the last portions of the iodate solution dropwise, while shaking the mixture vigorously and continuously. Allow the mixture to stand for 5 minutes. If the chloroform develops a purple color, titrate the mixture further with the iodate solution. The difference between the number of mL of 0.05 *M* potassium iodate used and the number of mL of 0.1 *N* potassium arsenite used, multiplied by 14.99, represents the number of mg of NaI in the volume of Topical Solution taken.

Strong Iodine Solution

» Strong Iodine Solution contains, in each 100 mL, not less than 4.5 g and not more than 5.5 g of iodine (I), and not less than 9.5 g and not more than 10.5 g of potassium iodide (KI).

Strong Iodine Solution may be prepared by dissolving 50 g of Iodine and 100 g of Potassium Iodide in 100 mL of Purified Water, then adding Purified Water to make the product measure 1000 mL.

Packaging and storage—Preserve in tight containers, preferably at a temperature not exceeding 35°.

Identification—

A: Dilute 1 drop with 10 mL of water, and add 1 mL of starch TS: a deep blue color is produced.

B: Evaporate a few mL on a steam bath to dryness, and ignite gently to volatilize any free iodine: the residue responds to the tests for *Potassium* ⟨191⟩ and for *Iodide* ⟨191⟩.

Assay for iodine—Pipet 5 mL of Strong Iodine Solution into a glass-stoppered, 500-mL flask, add 10 mL of water, and titrate with 0.1 *N* potassium arsenite VS, adding 3 mL of starch TS as the end-point is approached. Each mL of 0.1 *N* potassium arsenite is equivalent to 12.69 mg of I.

Assay for potassium iodide—To the titrated solution obtained in the *Assay for iodine* add 25 mL of hydrochloric acid, cool to room temperature, add 5 mL of chloroform, and titrate with 0.05 *M* potassium iodate VS until the purple color of iodine disappears from the chloroform. Add the last portions of the iodate solution dropwise, while shaking the mixture vigorously and continuously. Allow the mixture to stand for 5 minutes. If the chloroform develops a purple color, titrate the mixture further with the iodate solution. The difference between the number of mL of 0.05 *M* potassium iodate used and the number of mL of 0.1 *N* potassium arsenite used, multiplied by 16.60, represents the number of mg of KI in the volume of the Solution taken.

Iodine Tincture

» Iodine Tincture contains, in each 100 mL, not less than 1.8 g and not more than 2.2 g of iodine (I), and not less than 2.1 g and not more than 2.6 g of sodium iodide (NaI).

Iodine Tincture may be prepared by dissolving 20 g of Iodine and 24 g of Sodium Iodide in 500 mL of Alcohol and then adding Purified Water to make the product measure 1000 mL.

Packaging and storage—Preserve in tight containers.

Identification—

A: Add 1 drop to a mixture of 1 mL of starch TS and 9 mL of water: a deep blue color is produced.

B: Evaporate a few mL on a steam bath to dryness: the residue responds to the flame test for *Sodium* ⟨191⟩, and to the tests for *Iodide* ⟨191⟩.

Alcohol content ⟨611⟩: between 44.0% and 50.0% of C_2H_5OH.

Assay for iodine—Proceed as directed in the *Assay for iodine* under *Strong Iodine Solution*, using 10 mL of Iodine Tincture.

Assay for sodium iodide—Proceed as directed in the *Assay for potassium iodide* under *Strong Iodine Solution*. The difference between the number of mL of 0.05 *M* potassium iodate used and the number of mL of 0.1 *N* potassium arsenite used, multiplied by 14.99, represents the number of mg of NaI in the volume of Iodine Tincture taken.

Strong Iodine Tincture

» Strong Iodine Tincture contains, in each 100 mL, not less than 6.8 g and not more than 7.5 g of iodine (I), and not less than 4.7 g and not more than 5.5 g of potassium iodide (KI).

Strong Iodine Tincture may be prepared by dissolving 50 g of Potassium Iodide in 50 mL of Purified Water, adding 70 g of Iodine, and agitating until solution is effected, and then adding Alcohol to make the product measure 1000 mL.

Packaging and storage—Preserve in tight, light-resistant containers.

Identification—

A: Add 1 drop to a mixture of 1 mL of starch TS and 9 mL of water: a deep blue color is produced.

B: Evaporate a few mL on a steam bath to dryness: the residue responds to the flame test for *Potassium* ⟨191⟩ and to the tests for *Iodide* ⟨191⟩.

Alcohol content ⟨611⟩: between 82.5% and 88.5% of C_2H_5OH.

Assay for iodine—Proceed as directed in the *Assay for iodine* under *Strong Iodine Solution*, using Strong Iodine Tincture.

Assay for potassium iodide—Proceed as directed in the *Assay for potassium iodide* under *Strong Iodine Solution*. The difference between the number of mL of 0.05 *M* potassium iodate used and the number of mL of 0.1 *N* potassium arsenite used, multiplied by 16.60 represents the number of mg of KI in the volume of Strong Iodine Tincture taken.

Iodohippurate Sodium I 123 Injection

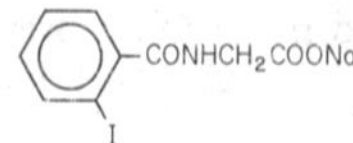

$C_9H_7{}^{123}INNaO_3$
Glycine, *N*-[2-(iodo-^{123}*I*)benzoyl]-, monosodium salt.
Sodium *o*-iodo-^{123}*I*-hippurate [*56254-07-0*].

» Iodohippurate Sodium I 123 Injection is a sterile, aqueous solution containing *o*-iodohippurate sodium in which a portion of the molecules contain radioactive iodine (^{123}I) in the molecular structure. It may contain a preservative or stabilizer.

Iodohippurate Sodium I 123 Injection contains not less than 90.0 percent and not more than 110.0 percent of the labeled amount of I 123 as iodohippurate sodium expressed in megabecquerels (microcuries or millicuries) per mL at the time indicated in the labeling. It contains not less than 90.0 percent and not more than 110.0 percent of the labeled amount of *o*-iodohippuric acid. Other chemical forms of radioactivity do not exceed 3.0 percent of total radioactivity.

Packaging and storage—Preserve in single-dose or in multiple-dose containers that are adequately shielded.

Labeling—Label it to include the following, in addition to the information specified for *Labeling* under *Injections* ⟨1⟩: the time and date of calibration; the amount of I 123 as iodohippurate sodium expressed as total megabecquerels (microcuries or millicuries) per mL at the time of calibration; the name and quantity of any added preservative or stabilizer; the expiration time; and the statement, "Caution—Radioactive Material." The labeling indicates that in making dosage calculations, correction is to be made for radioactive decay, and also indicates that the radioactive half-life of I 123 is 13.2 hours.

Reference standards—*USP Endotoxin Reference Standard. USP o-Iodohippuric Acid RS*—Do not dry before using. Keep container tightly closed and store in a dessicator. Protect from light.

Radionuclidic identification (see *Radioactivity* ⟨821⟩)—Its gamma-ray spectrum is identical to that of a specimen of I 123 of known purity that exhibits a major photopeak having an energy of 159 keV.

Bacterial endotoxins—It meets the requirements of the *Bacterial Endotoxins Test* ⟨85⟩, the limit of endotoxin content being not more than 175/*V* USP Endotoxin Unit per mL of the Injection, when compared with the USP Endotoxin RS, in which *V* is the maximum recommended total dose, in mL, at the expiration time.

pH ⟨791⟩: between 7.0 and 8.5.

Radionuclidic purity—Using a suitable counting assembly (see *Selection of a Counting Assembly* under *Radioactivity* ⟨821⟩), determine the radionuclidic purity of the Injection: not less than 85% of the total radioactivity is present as I 123.

Assay for *o*-iodohippuric acid—

Mobile phase—Prepare a filtered and degassed mixture of water, methanol, and glacial acetic acid (75:25:1). Make adjustments if necessary (see *System Suitability* under *Chromatography* ⟨621⟩).

Standard preparation—Using an accurately weighed quantity of USP *o*-Iodohippuric Acid RS, prepare a solution in water having a known concentration of about 2 mg per mL.

Assay preparation—Use Iodohippurate Sodium I 123 Injection.

Chromatographic system (see *Chromatography* ⟨621⟩)—The liquid chromatograph is equipped with a 265-nm detector and a 8-mm × 10-cm column that contains packing L11. The flow rate is about 5 mL per minute. Chromatograph the *Standard preparation*, and record the peak responses as directed under *Procedure:* the column efficiency determined from the analyte peak is not less than 500 theoretical plates, the tailing factor for the analyte peak is not more than 1.5, and the relative standard deviation for replicate injections is not more than 1.5%.

Procedure—Separately inject equal volumes (about 50 µL) of the *Standard preparation* and the *Assay preparation* into the chromatograph, record the chromatogram, and measure the responses for the major peaks. Calculate the quantity, in mg, of *o*-iodohippuric acid in each mL of Iodohippurate Sodium I 123 Injection taken by the formula:

$$C(r_U/r_S),$$

in which C is the concentration, in mg per mL, of USP *o*-Iodohippuric Acid RS in the *Standard preparation*, and r_U and r_S are the *o*-iodohippuric acid peak responses obtained from the *Assay preparation* and the *Standard preparation*, respectively.

Radiochemical purity—

Mobile phase, Standard preparation, and *Chromatographic system*—Proceed as directed in the *Assay for o-iodohippuric acid*, except to provide that the liquid chromatograph is also equipped with a radioactivity detector (see *Radioactivity* ⟨821⟩).

Procedure—Inject about 50 µL equivalent to 1.8 to 3.7 MBq (50 to 100 µCi) of Iodohippurate Sodium I 123 Injection into the chromatograph, record the chromatogram, and measure the area of all radioactivity peaks. The radioactivity under the *o*-iodohippuric acid I 123 peak is not less than 97.0% of the total area of all peaks observed and its retention time is within ±10% of the value obtained for USP *o*-Iodohippuric Acid RS.

Biological distribution—Inject intravenously between 0.75 and 22 MBq (20 and 600 µCi) of the Injection, in a volume of 0.10 mL to 0.15 mL, into the caudal vein of each of three 125-g to 225-g anesthetized rats. Clamp the opening of the urethra with a hemostat. Sacrifice the animals 1 hour after administration, and carefully remove the intact bladder and contents and thyroid by dissection. Place each organ and remaining carcass (excluding the tail) in separate, suitable counting containers, and determine the radioactivity, in counts per minute, in each container with an appropriate detector using the same counting geometry. Determine the percentage of radioactivity in each organ: not less than 75% of the administered dose is found in the bladder and not more than 3% of the administered dose is found in the thyroid in two of the rats.

Other requirements—It meets the requirements under *Injections* ⟨1⟩, except that the Injection may be distributed or dispensed prior to the completion of the test for *Sterility*, the latter test being started on the day of final manufacture, and except that it is not subject to the recommendation on *Volume in Container.*

Assay for radioactivity—Using a suitable counting assembly (see *Selection of a Counting Assembly* under *Radioactivity* ⟨821⟩), determine the radioactivity in MBq (μCi or mCi) per mL of Iodohippurate Sodium I 123 Injection by use of a calibrated system as directed under *Radioactivity* ⟨821⟩.

Sodium Iodide I 123 Capsules

Sodium iodide ($Na^{123}I$).
Sodium iodide ($Na^{123}I$) [*41927-88-2*].

» Sodium Iodide I 123 Capsules contain radioactive iodine (^{123}I) processed in the form of Sodium Iodide obtained from the bombardment of enriched tellurium 124 with protons or of enriched tellurium 122 with deuterons or by the decay of xenon 123 in such manner that it is carrier-free. Capsules contain not less than 90.0 percent and not more than 110.0 percent of the labeled amount of ^{123}I as iodide expressed in megabecquerels (microcuries or in millicuries) at the time indicated in the labeling. Other chemical forms of radioactivity do not exceed 5 percent of the total radioactivity. The Capsules may contain a stabilizer.

Packaging and storage—Preserve in well-closed containers that are adequately shielded.

Labeling—Label Capsules to include the following: the name of the Capsules; the name, address, and batch or lot number of the manufacturer; the time and date of calibration; the amount of ^{123}I as iodide expressed in megabecquerels (microcuries or in millicuries) per Capsule at the time of calibration; the name and quantity of any added preservative or stabilizer; a statement indicating that the Capsules are for oral use only; the expiration date and time; and the statement, "Caution—Radioactive Material." The labeling indicates that in making dosage calculations, correction is to be made for radioactive decay, and also indicates that the radioactive half-life of ^{123}I is 13.2 hours.

Radionuclide identification—A solution or suspension of 1 or more Capsules in water responds to the test for *Radionuclide identification* under *Sodium Iodide I 123 Solution.*

Uniformity of dosage units ⟨905⟩: meet the requirements.

Procedure for content uniformity—Determine the instrument response of each of 20 Capsules by measurement in a suitable counting assembly and under identical geometric conditions. Calculate the mean radioactivity value per Capsule: the requirements of the test are met if not less than 19 of the Capsules are within the limits of 96.5% and 103.5% of the mean radioactivity value.

Radionuclidic purity—A solution or suspension of 1 or more Capsules in water responds to the test for *Radionuclidic purity* under *Sodium Iodide I 123 Solution.*

Radiochemical purity—Homogenize 1 Capsule in 3 mL of water, add 3 mL of methanol, and centrifuge: the supernatant solution so obtained meets the requirements of the test for *Radiochemical purity* under *Sodium Iodide I 123 Solution.*

Other requirements—A solution or suspension prepared by homogenizing 1 or more Capsules in water to yield a concentration of not less than 1 MBq (25 μCi) per mL meets the requirements of the *Assay for radioactivity* under *Sodium Iodide I 123 Solution.*

Sodium Iodide I 123 Solution

Sodium iodide ($Na^{123}I$).
Sodium iodide ($Na^{123}I$) [*41927-88-2*].

» Sodium Iodide I 123 Solution is a solution, suitable for oral or for intravenous administration, containing radioactive iodine (^{123}I) processed in the form of Sodium Iodide, obtained from the bombardment of enriched tellurium 124 with protons or of enriched tellurium 122 with deuterons, or by the decay of xenon 123 in such manner that it is carrier-free.

Sodium Iodide I 123 Solution contains not less than 90.0 percent and not more than 110.0 percent of the labeled amount of ^{123}I as iodide expressed in megabecquerels (microcuries or in millicuries) per mL at the time indicated in the labeling. Other chemical forms of radioactivity do not exceed 5 percent of the total radioactivity. The Solution may contain a preservative or stabilizer.

Packaging and storage—Preserve in single-dose or in multiple-dose containers that previously have been treated to prevent adsorption, if necessary.

Labeling—Label it to include the following: the time and date of calibration; the amount of ^{123}I as iodide expressed as total megabecquerels (microcuries or millicuries) per mL at the time of calibration; the name and quantity of any added preservative or stabilizer; a statement to indicate whether the contents are intended for oral or for intravenous use; the expiration date and time; and the statement, "Caution—Radioactive Material." The labeling indicates that in making dosage calculations, correction is to be made for radioactive decay, and also indicates that the radioactive half-life of ^{123}I is 13.2 hours.

Reference standard—*USP Endotoxin Reference Standard.*

Radionuclide identification (see *Radioactivity* ⟨821⟩)—Its gamma-ray spectrum is identical to that of a specimen of ^{123}I of known purity that exhibits a major photoelectric peak having an energy of 0.159 MeV.

Radionuclidic purity—Using a suitable counting assembly (see *Selection of a Counting Assembly* under *Radioactivity* ⟨821⟩), determine the radionuclidic purity of the Solution: not less than 85.0% of the total radioactivity is present as I 123.

Bacterial endotoxins—It meets the requirements of the *Bacterial Endotoxins Test* ⟨85⟩, the limit of endotoxin content being not more than 175/V USP Endotoxin Unit per mL of the Injection, when compared with the USP Endotoxin RS, in which V is the maximum recommended total dose, in mL, at the expiration date or time.

pH ⟨791⟩: between 7.5 and 9.0.

Radiochemical purity—Place a measured volume of a solution, containing 100 mg of potassium iodide, 200 mg of potassium iodate, and 1 g of sodium bicarbonate in each 100 mL, 25 mm from one end of a 25- × 300-mm strip of chromatographic paper (see *Chromatography* ⟨621⟩), and allow to dry. To the same area add a similar volume of appropriately diluted Solution such that it provides a count rate of about 20,000 counts per minute, and allow to dry. Develop the chromatogram over a period of about 4 hours by ascending chromatography, using dilute methanol (7 in 10). Dry the chromatogram in air, and determine the radioactivity distribution by scanning with a suitable collimated radiation detector: the radioactivity of the iodide ^{123}I band is not less than 95.0% of the total radioactivity and its R_f value falls within ±5.0% of the value found for sodium iodide when determined under similar conditions. Confirmation of the identity of the iodide band is made by the addition to the suspected iodide band of 6 drops of acidified hydrogen peroxide solution (prepared by adding 6 drops of 1 N hydrochloric acid to 10 mL of hydrogen peroxide solution) followed by the dropwise addition of starch TS: the development of a blue color indicates the presence of iodide.

Other requirements—Solution intended for intravenous use meets the requirements under *Injections* ⟨1⟩, except that it may be distributed or dispensed prior to completion of the test for *Sterility*, the latter test being started on the day of final manufacture, and except that it is not subject to the recommendation on *Volume in Container*.

Assay for radioactivity—Using a suitable counting assembly (see *Selection of a Counting Assembly* under *Radioactivity* ⟨821⟩), determine the radioactivity, in MBq (μCi) per mL, of Sodium Iodide I 123 Solution by use of a calibrated system as directed under *Radioactivity* ⟨821⟩.

Iodinated I 125 Albumin Injection

Albumin labeled with iodine-125.

» Iodinated I 125 Albumin Injection is a sterile, buffered, isotonic solution containing normal human albumin adjusted to provide not more than 37 MBq (1 mCi) of radioactivity per mL. It is derived by mild iodination of normal human albumin with the use of radioactive iodine (^{125}I) to introduce not more than one gram-atom of iodine for each gram-molecule (60,000 g) of albumin.

Iodinated I 125 Albumin Injection contains not less than 95.0 percent and not more than 105.0 percent of the labeled amount of ^{125}I as iodinated albumin, expressed in megabecquerels (microcuries or in millicuries) per mL at the time indicated in the labeling. Other forms of radioactivity do not exceed 3 percent of the total radioactivity. Its production and distribution are subject to federal regulations (see *Biologics* ⟨1041⟩ and *Radioactivity* ⟨821⟩).

Packaging and storage—Preserve in single-dose or multiple-dose containers, at a temperature between 2° and 8°.

Labeling—Label it to include the following, in addition to the information specified for *Labeling* under *Injections* ⟨1⟩: the date of calibration; the amount of ^{125}I as iodinated albumin, expressed as total megabecquerels (microcuries or millicuries), and concentration as megabecquerels (microcuries or millicuries) per mL on the date of calibration; the expiration date; and the statement, "Caution—Radioactive Material." The labeling indicates that in making dosage calculations, correction is to be made for radioactive decay, and also indicates that the radioactive half-life of ^{125}I is 60 days.

Reference standard—*USP Endotoxin Reference Standard*.

Radionuclide identification (see *Radioactivity* ⟨821⟩)—Its gamma-ray spectrum is identical to that of a specimen of ^{125}I of known purity that exhibits a major photopeak having an energy of 0.0355 MeV.

Bacterial endotoxins—It meets the requirements of the *Bacterial Endotoxins Test* ⟨85⟩, the limit of endotoxin content being not more than 175/V USP Endotoxin Unit per mL of the Injection, when compared with the USP Endotoxin RS, in which V is the maximum recommended total dose, in mL, at the expiration date or time.

pH ⟨791⟩: between 7.0 and 8.5.

Radiochemical purity—Place a measured volume, diluted with a suitable diluent so that it provides a count rate of about 20,000 counts per minute, about 25 mm from one end of a 25- × 300-mm strip of chromatographic paper (see *Chromatography* ⟨621⟩), and allow to dry. Develop the chromatogram over a period of about 4 hours by ascending chromatography, using dilute methanol (7 in 10), and air-dry. Determine the radioactivity distribution by scanning the chromatogram with a suitable collimated radiation detector: not less than 97.0% of the total activity is found as albumin (at the point of application).

Other requirements—It meets the requirements under *Biologics* ⟨1041⟩ and under *Injections* ⟨1⟩, except that it is not subject to the recommendation on *Volume in Container*. It meets all other applicable requirements of the federal FDA.

Assay for radioactivity—Using a suitable counting assembly (see *Selection of a Counting Assembly* under *Radioactivity* ⟨821⟩), determine the radioactivity, in MBq (μCi) per mL, of Iodinated I 125 Albumin Injection by use of a calibrated system as directed under *Radioactivity* ⟨821⟩.

Sodium Iodide I 125 Capsules

Sodium iodide ($Na^{125}I$).
Sodium iodide ($Na^{125}I$) [*24359-64-6*].

» Sodium Iodide I 125 Capsules contain radioactive iodine (^{125}I) processed in the form of Sodium Iodide in such manner that it is carrier-free. Capsules contain not less than 90.0 percent and not more than 110.0 percent of the labeled amount of ^{125}I as iodide expressed in megabecquerels (microcuries or in millicuries) at the time indicated in the labeling. Other chemical forms of radioactivity do not exceed 5 percent of the total radioactivity. The Capsules may contain a preservative or stabilizer.

Packaging and storage—Preserve in well-closed containers that are adequately shielded.

Labeling—Label Capsules to include the following: the date of calibration; the amount of ^{125}I as iodide expressed as total megabecquerels (microcuries or millicuries) at the time of calibration; the name and quantity of any added preservative or stabilizer; a statement indicating that the Capsules are for oral use only; the expiration date; and the statement, "Caution—Radioactive Material." The labeling indicates that in making dosage calculations, correction is to be made for radioactive decay, and also indicates that the radioactive half-life of ^{125}I is 60 days.

Radionuclide identification—A solution or suspension of 1 or more Capsules in water responds to the test for *Radionuclide identification* under *Sodium Iodide I 125 Solution*.

Uniformity of dosage units ⟨905⟩: meet the requirements.

Procedure for content uniformity—Determine the instrument response of each of 20 Capsules or their individual contents by measurement in a suitable counting assembly and under identical geometric conditions. Calculate the mean radioactivity value per Capsule: the requirements of the test are met if not less than 19 of the Capsules are within the limits of 96.5% and 103.5% of the mean radioactivity value.

Radiochemical purity—Homogenize 1 Capsule in 3 mL of water, add 3 mL of methanol, and centrifuge: the supernatant solution so obtained meets the requirements of the test for *Radiochemical purity* under *Sodium Iodide I 125 Solution*.

Other requirements—A solution or suspension prepared by homogenizing 1 or more Capsules in water to yield a concentration of not less than 1 MBq (25 μCi) per mL meets the requirements of the *Assay for radioactivity* under *Sodium Iodide I 125 Solution*.

Sodium Iodide I 125 Solution

Sodium iodide ($Na^{125}I$).
Sodium iodide ($Na^{125}I$) [*24359-64-6*].

» Sodium Iodide I 125 Solution is a solution suitable for either oral or intravenous administration, containing radioactive iodine (^{125}I) processed in the form of sodium iodide from the neutron bombardment of xenon gas in such a manner that it is essentially carrier-free. Sodium Iodide I 125 Solution contains not less than 85.0 percent and not more than 115.0 percent of the labeled amount of ^{125}I as iodide expressed in megabecquerels (microcuries or in millicuries) per mL at the time indicated in the labeling. Other chemical forms of radioactivity do not exceed 5 percent of the total radioactivity.

Packaging and storage—Preserve in single-dose or in multiple-dose containers that previously have been treated to prevent adsorption.

Labeling—Label it to include the following: the name of the Solution; the name, address, and batch or lot number of the manufacturer; the date of calibration; the amount of ^{125}I as iodide expressed as total megabecquerels (microcuries or millicuries) and as megabecquerels (microcuries or millicuries) per mL at the time of calibration; the name and quantity of any added preservative or stabilizer; a statement of the intended use, whether oral or intravenous; a statement of whether diagnostic or therapeutic; the expiration date; and the statement, "Caution—Radioactive Material." The labeling indicates that in making dosage calculations, correction is to be made for radioactive decay, and also indicates that the radioactive half-life of ^{125}I is 60 days.

Reference standard—*USP Endotoxin Reference Standard.*

Radionuclide identification (see *Radioactivity* ⟨821⟩)—Its gamma-ray scintillation spectrum is identical to that of a specimen of ^{125}I of known purity that exhibits a major photoelectric peak, having an energy of 0.0355 MeV.

Bacterial endotoxins—It meets the requirements of the *Bacterial Endotoxins Test* ⟨85⟩, the limit of endotoxin content being not more than 175/*V* USP Endotoxin Unit per mL of the Injection, when compared with the USP Endotoxin RS, in which *V* is the maximum recommended total dose, in mL, at the expiration date or time.

pH ⟨791⟩: between 7.5 and 9.0.

Radiochemical purity—Place a measured volume of a solution containing 100 mg of potassium iodide, 200 mg of potassium iodate, and 1 g of sodium bicarbonate in each 100 mL, 25 mm from one end of a 25- × 300-mm strip of chromatographic paper (see *Chromatography* ⟨621⟩), and allow to dry. To the same area add a similar volume of appropriately diluted Solution representing a radioactivity of about 20,000 counts per minute, and allow to dry. Develop the chromatogram over a period of about 4 hours by ascending chromatography, using dilute methanol (7 in 10). Dry the chromatogram in air, and determine the radioactivity distribution by scanning with a suitable collimated radiation detector: the radioactivity of the iodide ^{125}I band is not less than 95.0% of the total radioactivity, and its R_f value falls within ±5% of the value found for sodium iodide when determined under similar conditions. Confirmation of the identity of the iodide band is made by the addition to the suspected iodide band of 6 drops of acidified hydrogen peroxide solution (prepared by adding 6 drops of 1 *N* hydrochloric acid to 10 mL of hydrogen peroxide solution) followed by the dropwise addition of starch TS: the development of a blue color indicates the presence of iodide.

Other requirements—Solution intended for intravenous use meets the requirements under *Injections* ⟨1⟩, except that it is not subject to the recommendation on *Volume in Container.*

Assay for radioactivity—Using a suitable counting assembly (see *Selection of a Counting Assembly* under *Radioactivity* ⟨821⟩), determine the radioactivity, in MBq (μCi) per mL, of Sodium Iodide I 125 Solution by use of a calibrated system as directed under *Radioactivity* ⟨821⟩.

Iodinated I 131 Albumin Injection

Albumin labeled with iodine-131.

» Iodinated I 131 Albumin Injection is a sterile, buffered, isotonic solution containing normal human albumin adjusted to provide not more than 37 MBq (1 mCi) of radioactivity per mL. It is derived by mild iodination of normal human albumin with the use of radioactive iodine (^{131}I) to introduce not more than one gram-atom of iodine for each gram-molecule (60,000 g) of albumin.

Iodinated I 131 Albumin Injection contains not less than 95.0 percent and not more than 105.0 percent of the labeled amount of ^{131}I as iodinated albumin, expressed in megabecquerels (microcuries or in millicuries) per mL at the time indicated in the labeling. Other forms of radioactivity do not exceed 3 percent of the total radioactivity. Its production and distribution are subject to federal regulations (see *Biologics* ⟨1041⟩ and *Radioactivity* ⟨821⟩).

Labeling—Label it to include the following, in addition to the information specified for *Labeling* under *Injections* ⟨1⟩: the date of calibration; the amount of ^{131}I as iodinated albumin, expressed as total megabecquerels (millicuries or microcuries), and concentration as megabecquerels (microcuries or millicuries) per mL on the date of calibration; the expiration date; and the statement, "Caution—Radioactive Material." The labeling indicates that in making dosage calculations, correction is to be made for radioactive decay, and also indicates that the radioactive half-life of ^{131}I is 8.08 days.

Reference standard—*USP Endotoxin Reference Standard.*

Radionuclide identification (see *Radioactivity* ⟨821⟩)—Its gamma-ray spectrum is identical to that of a specimen of ^{131}I of known purity that exhibits a major photopeak having an energy of 0.364 MeV.

Other requirements—It meets the requirements for *Packaging and storage, Bacterial endotoxins, pH, Radiochemical purity,* and *Assay for radioactivity* under *Iodinated I 125 Albumin Injection.* It meets also the requirements under *Biologics* ⟨1041⟩, and the requirements under *Injections* ⟨1⟩, except that it is not subject to the recommendation on *Volume in Container.* It meets all other applicable requirements of the federal FDA.

Iodinated I 131 Albumin Aggregated Injection

Albumin labeled with iodine-131.

» Iodinated I 131 Albumin Aggregated Injection is a sterile aqueous suspension of Albumin Human that has been iodinated with ^{131}I and denatured to produce aggregates of controlled particle size. Each mL of the suspension contains not less than 300 μg and not more than 3.0 mg of aggregated albumin with a specific activity of not less than 7.4 megabecquerels (200 microcuries) per mg and not more than 44.4 megabecquerels (1.2 millicuries) per mg of aggregated albumin. Iodinated I 131 Albumin Aggregated Injection contains not less than 95.0 percent and not more than 105.0 percent of the labeled amount of ^{131}I, as aggregated albumin, expressed in megabecquerels (microcuries) per mL or megabecquerels (millicuries) per mL at the time indicated in the labeling. Other chemical forms of radioactivity do not exceed 6 percent of the total radioactivity. Its production and distribution are subject to federal regulations (see *Biologics* ⟨1041⟩ and *Radioactivity* ⟨821⟩).

Packaging and storage—Preserve in single-dose or in multiple-dose containers, at a temperature between 2° and 8°.

Labeling—Label it to include the following, in addition to the information specified for *Labeling* under *Injections* ⟨1⟩: the time and date of calibration; the amount of ^{131}I as aggregated albumin expressed as total megabecquerels (microcuries or millicuries) and as aggregated albumin in mg per mL on the date of calibration; the expiration date; and the statement, "Caution—Radioactive Material." The labeling indicates that in making dosage calculations, correction is to be made for radioactive decay, and also indicates that the radioactive half-life of ^{131}I is 8.08 days; in addition, the labeling states that it is not to be used if clumping of the albumin is observed and directs that the container be agitated before the contents are withdrawn into a syringe.

Reference standard—*USP Endotoxin Reference Standard.*

Radionuclide identification—Its gamma-ray spectrum is identical to that of a specimen of ^{131}I of known purity that exhibits a major photopeak having an energy of 0.364 MeV.

pH ⟨791⟩: between 5.0 and 6.0.

Other requirements—It meets the requirements under *Biologics* ⟨1041⟩ and under *Injections* ⟨1⟩, except that it is not subject to the recommendation on *Volume in Container.* It meets also the requirements for *Particle size, Bacterial endotoxins,* and *Radiochemical purity* under *Technetium Tc 99m Albumin Aggregated Injection,* except that in the test for *Radiochemical purity,* not more than 6% of the radioactivity is found in the supernatant liquid following centrifugation.

Assay for radioactivity—Using a suitable counting assembly (see *Selection of a Counting Assembly* under *Radioactivity* ⟨821⟩), determine the radioactivity, in MBq (μCi) per mL, of Iodinated I 131 Albumin Aggregated Injection by use of a calibrated system as directed under *Radioactivity* ⟨821⟩.

Iodohippurate Sodium I 131 Injection

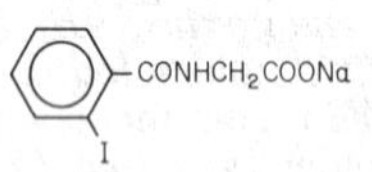

$C_9H_7{}^{131}INNaO_3$
Glycine, *N*-(2-iodo-^{131}*I*-benzoyl)-, monosodium salt.
Monosodium *o*-iodo-^{131}*I*-hippurate [*881-17-4*].

» Iodohippurate Sodium I 131 Injection is a sterile solution containing *o*-iodohippurate sodium in which a portion of the molecules contain radioactive iodine (^{131}I) in the molecular structure.

Iodohippurate Sodium I 131 Injection contains not less than 90.0 percent and not more than 110.0 percent of the labeled amount of ^{131}I as iodohippurate sodium expressed in megabecquerels (microcuries or in millicuries) per mL at the time indicated in the labeling. Other chemical forms of radioactivity do not exceed 3.0 percent of the total radioactivity.

Packaging and storage—Preserve in single-dose or in multiple-dose containers.

Labeling—Label it to include the following, in addition to the information specified for *Labeling* under *Injections* ⟨1⟩: the time and date of calibration; the amount of ^{131}I as iodohippurate sodium expressed as total megabecquerels (microcuries or millicuries) and as megabecquerels (microcuries or millicuries) per mL at the time of calibration; the expiration date; and the statement, "Caution—Radioactive Material." The labeling indicates that in making dosage calculations, correction is to be made for radioactive decay, and also indicates that the radioactive half-life of ^{131}I is 8.08 days.

Reference standard—*USP Endotoxin Reference Standard.*

Radionuclide identification (see *Radioactivity* ⟨821⟩)—Its gamma-ray spectrum is identical to that of a specimen of iodine 131 of known purity that exhibits a major photopeak having an energy of 0.364 MeV.

Bacterial endotoxins—It meets the requirements of the *Bacterial Endotoxins Test* ⟨85⟩, the limit of endotoxin content being not more than 175/*V* USP Endotoxin Unit per mL of the Injection, when compared with the USP Endotoxin RS, in which *V* is the maximum recommended total dose, in mL, at the expiration date or time.

pH ⟨791⟩: between 7.0 and 8.5.

Radiochemical purity—Place 1 drop of a solution containing 0.2 percent potassium iodide, 1.0 percent sodium bicarbonate, and 1.0 percent sodium thiosulfate, about 45 mm from one end of each of two 25- × 300-mm strips of chromatographic paper (see *Chromatography* ⟨621⟩), and allow to dry. Superimpose one of the spots with a measured volume of Injection, appropriately diluted, such that it provides a count rate of about 20,000 counts per minute, on this point of application, and allow to air-dry. Superimpose the second spot with 100 μL of a solution prepared by dissolving 50 mg of non-radioactive iodohippurate sodium in 10 mL of alcohol, and allow to air-dry. Develop the chromatogram over a period of about 2½ hours by descending chromatography, using the upper layer obtained by shaking together 2 volumes of benzene, 2 volumes of glacial acetic acid, and 1 volume of water. Use the aqueous layer to equilibrate the apparatus prior to the start of development. Dry the chromatogram in air, and determine the radioactivity distribution by scanning the chromatogram with a suitable collimated radiation detector. Locate the position of the non-radioactive spot by viewing the chromatogram under short-wavelength ultraviolet light. The radioactivity under the iodohippuric acid band is not less than 97.0% of the total radioactivity and its R_f value is within ±10% of that of the non-radioactive spot.

Other requirements—It meets the requirements under *Injections* ⟨1⟩, except that the Injection may be distributed or dispensed prior to the completion of the test for *Sterility,* the latter test being started on the day of final manufacture and except that it is not subject to the recommendation on *Volume in Container.*

Assay for radioactivity—Using a suitable counting assembly (see *Selection of a Counting Assembly* under *Radioactivity* ⟨821⟩), determine the radioactivity, in MBq (μCi) per mL, of Iodohippurate Sodium I 131 Injection by use of a calibrated system as directed under *Radioactivity* ⟨821⟩.

Rose Bengal Sodium I 131 Injection

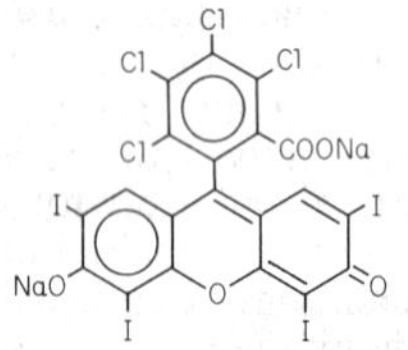

$C_{20}H_2Cl_4{}^{131}I_4Na_2O_5$
Spiro[isobenzofuran-1(3*H*),9′-[9*H*]-xanthene]-3-one, 4,5,6,7-tetrachloro-3′,6′-dihydroxy-2′,4′,5′,7′-tetraiodo-, disodium salt, labeled with iodine-131.
4,5,6,7-Tetrachloro-2′,4′,5′,7′-tetraiodofluorescein disodium salt-^{131}I [*24916-55-0; 50291-21-9; 15251-14-6*].

» Rose Bengal Sodium I 131 Injection is a sterile solution containing rose bengal sodium in which a portion of the molecules contain radioactive iodine (^{131}I) in the molecular structure. It may contain a suitable buffer.

Rose Bengal Sodium I 131 Injection contains not less than 90.0 percent and not more than 110.0 percent of the labeled amount of ^{131}I as rose bengal sodium expressed in megabecquerels (microcuries or in millicuries) per mL at the time indicated in the labeling. The rose bengal sodium content is not less than 90.0 percent and not more than 110.0 percent of the labeled amount. Other chemical forms of radioactivity do not exceed 10.0 percent of the total radioactivity.

Packaging and storage—Preserve in single-dose or in multiple-dose containers.

Labeling—Label it to include the following, in addition to the information specified for *Labeling* under *Injections* ⟨1⟩: the time and date of calibration; the amount of ^{131}I as rose bengal sodium expressed as total megabecquerels (microcuries or millicuries) and as megabecquerels (microcuries or millicuries) per mL on the date of calibration; the expiration date; and the statement, "Caution—Radioactive Material." The labeling indicates that in making dosage calculations, correction is to be made for radioactive decay, and also indicates that the radioactive half-life of ^{131}I is 8.08 days.

Reference standard—*USP Endotoxin Reference Standard.*

Radionuclide identification (see *Radioactivity* ⟨821⟩)—Its gamma-ray spectrum is identical to that of a specimen of ^{131}I of known purity that exhibits a major photopeak having an energy of 0.364 MeV.

Bacterial endotoxins—It meets the requirements of the *Bacterial Endotoxins Test* ⟨85⟩, the limit of endotoxin content being not more than $175/V$ USP Endotoxin Unit per mL of the Injection, when compared with the USP Endotoxin RS, in which V is the maximum recommended total dose, in mL, at the expiration date or time.

pH ⟨791⟩: between 7.0 and 8.5.

Radiochemical purity—Place a measured volume of a solution containing 100 mg of potassium iodide, 200 mg of potassium iodate, and 1 g of sodium bicarbonate in each 100 mL about 25 mm from one end of a 25- × 300-mm strip of chromatographic paper (see *Chromatography* ⟨621⟩), and allow to dry. To the same area add a similar volume of appropriately diluted Injection such that it provides a count rate of about 20,000 counts per minute, and allow to dry. Develop the chromatogram over a period of about 2 hours by ascending chromatography, using 1 *N* acetic acid. Dry the chromatogram in air, and determine the radioactivity distribution by scanning the chromatogram with a suitable collimated radiation detector. The radioactivity under the rose bengal band is not less than 90.0% of the total radioactivity. The rose bengal band is at the point of application.

Other requirements—It meets the requirements under *Injections* ⟨1⟩, except that the Injection may be distributed or dispensed prior to the completion of the test for *Sterility*, the latter test being started on the day of final manufacture, and except that it is not subject to the recommendation on *Volume in Container*.

Assay for rose bengal sodium—Determine the absorbance of Rose Bengal Sodium I 131 Injection, appropriately diluted, in a 1-cm cell at 550 nm, with a suitable spectrophotometer, using a sodium bicarbonate solution adjusted to a pH of 8.0 as the blank. Calculate the quantity, in mg, of rose bengal sodium per mL of the Injection taken by the formula:

$$0.004D(A_U/A_S),$$

in which D is the dilution factor, A_U is the absorbance of the solution, and A_S is the absorbance, similarly determined, of a solution of rose bengal sodium adjusted to a pH of 8.0 by the addition of sodium bicarbonate and containing 4 µg of rose bengal sodium per mL.

Assay for radioactivity—Using a suitable counting assembly (see *Selection of a Counting Assembly* under *Radioactivity* ⟨821⟩), determine the radioactivity, in MBq (µCi) per mL, of Rose Bengal Sodium I 131 Injection by use of a calibrated system as directed under *Radioactivity* ⟨821⟩.

Sodium Iodide I 131 Capsules

Sodium iodide ($Na^{131}I$).
Sodium iodide ($Na^{131}I$) [*7790-26-3*].

» Sodium Iodide I 131 Capsules contain radioactive iodine (^{131}I) processed in the form of Sodium Iodide from products of uranium fission or the neutron bombardment of tellurium in such a manner that it is essentially carrier-free and contains only minute amounts of naturally occurring iodine 127. Capsules contain not less than 90.0 percent and not more than 110.0 percent of the labeled amount of ^{131}I as iodide expressed in megabecquerels (microcuries or in millicuries) at the time indicated in the labeling. Other chemical forms of radioactivity do not exceed 5 percent of the total radioactivity. The Capsules may contain a stabilizing agent.

Packaging and storage—Preserve in well-closed containers.

Labeling—Label Capsules to include the following: the date of calibration; the amount of ^{131}I as iodide expressed in megabecquerels (microcuries or in millicuries) per Capsule at the time of calibration; a statement of whether the contents are intended for diagnostic or therapeutic use; the expiration date; and the statement, "Caution—Radioactive Material." The labeling indicates that in making dosage calculations, correction is to be made for radioactive decay, and also indicates that the radioactive half-life of ^{131}I is 8.08 days.

Radionuclide identification—A solution or suspension of 1 or more Capsules in water responds to the test for *Radionuclide identification* under *Sodium Iodide I 131 Solution.*

Uniformity of dosage units ⟨905⟩: meet the requirements.

Procedure for content uniformity—Determine the instrument response of each of 20 Capsules by measurement in a suitable counting assembly and under identical geometric conditions. Calculate the mean radioactivity value per Capsule. The requirements are met if not less than 19 of the Capsules are within the limits of 96.5% and 103.5% of the mean radioactivity value.

Radiochemical purity—Homogenize 1 Capsule in 3 mL of water, add 3 mL of methanol, and centrifuge. The supernatant solution meets the requirements of the test for *Radiochemical purity* under *Sodium Iodide I 131 Solution.*

Other requirements—A solution or suspension prepared by homogenizing 1 or more Capsules in sufficient water to yield a concentration of not less than 1 MBq (25 µCi) per mL meets the requirements of the *Assay for radioactivity* under *Sodium Iodide I 131 Solution.*

Sodium Iodide I 131 Solution

Sodium iodide ($Na^{131}I$).
Sodium iodide ($Na^{131}I$) [*7790-26-3*].

» Sodium Iodide I 131 Solution is a solution suitable for either oral or intravenous administration, containing radioactive iodine (^{131}I) processed in the form of Sodium Iodide from the products of uranium fission or the neutron bombardment of tellurium in such a manner that it is essentially carrier-free and contains only minute amounts of naturally occurring iodine 127.

Sodium Iodide I 131 Solution contains not less than 90.0 percent and not more than 110.0 percent of the labeled amount of ^{131}I as iodide expressed in megabecquerels (microcuries or in millicuries) per mL at the time indicated in the labeling. Other chemical forms of radioactivity do not exceed 5 percent of the total radioactivity. The Solution may contain a preservative or stabilizer.

Packaging and storage—Preserve in single-dose or in multiple-dose containers that previously have been treated to prevent adsorption.

Labeling—Label it to include the following: the time and date of calibration; the amount of ^{131}I as iodide expressed as total megabecquerels (microcuries or millicuries) and as megabecquerels (microcuries or millicuries) per mL at the time of calibration; the name and quantity of any added preservative or stabilizer; a statement of the intended use, whether oral or intravenous; a statement of whether the contents are intended for diagnostic or therapeutic use; the expiration date; and the statement, "Caution—Radioactive Material." The labeling indicates that in making dosage calculations, correction is to be made for radioactive decay, and also indicates that the radioactive half-life of ^{131}I is 8.08 days.

Reference standard—*USP Endotoxin Reference Standard.*

Radionuclide identification (see *Radioactivity* ⟨821⟩)—Its gamma-ray spectrum is identical to that of a specimen of iodine 131 of known purity that exhibits a major photopeak having an energy of 0.364 MeV.

Bacterial endotoxins—Solution intended for intravenous use meets the requirements of the *Bacterial Endotoxins Test* ⟨85⟩, the limit of endotoxin content being not more than 175/V USP Endotoxin Unit per mL of the Solution, when compared with the USP Endotoxin RS, in which V is the maximum recommended total dose, in mL, at the expiration date or time.

pH ⟨791⟩: between 7.5 and 9.0.

Radiochemical purity—Place a measured volume of a solution containing 100 mg of potassium iodide, 200 mg of potassium iodate, and 1 g of sodium bicarbonate in each 100 mL, 25 mm from one end of a 25- × 300-mm strip of chromatographic paper (see *Chromatography* ⟨621⟩), and allow to dry. To the same area add a similar volume of appropriately diluted Sodium Iodide I 131 Solution such that it provides a count rate of about 20,000 counts per minute, and allow to dry. Develop the chromatogram over a period of about 4 hours by ascending chromatography, using dilute methanol (7.0 in 10). Dry the chromatogram in air, and determine the radioactivity distribution by scanning with a suitable collimated radiation detector: the radioactivity of the iodide ^{131}I band is not less than 95% of the total radioactivity and its R_f value falls within ±5% of the value found for sodium iodide when determined under parallel conditions. Confirmation of the identity of the iodide band is made by the addition to the suspected iodide band of 6 drops of acidified hydrogen peroxide solution (prepared by adding 6 drops of 1 N hydrochloric acid to 10 mL of hydrogen peroxide solution) followed by the dropwise addition of starch TS; the development of a blue color indicates presence of iodide.

Other requirements—Solution intended for intravenous use meets the requirements under *Injections* ⟨1⟩, except that the Solution may be distributed or dispensed prior to completion of the test for *Sterility*, the latter test being started on the day of final manufacture, and except that it is not subject to the recommendation on *Volume in Container.*

Assay for radioactivity—Using a suitable counting assembly (see *Selection of a Counting Assembly* under *Radioactivity* ⟨821⟩), determine the radioactivity, in MBq (μCi) per mL, of Sodium Iodide I 131 Solution by use of a calibrated system as directed under *Radioactivity* ⟨821⟩.

Iodipamide

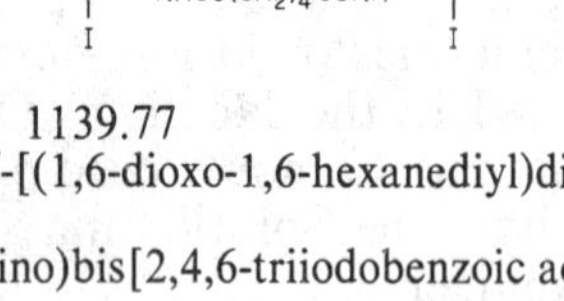

$C_{20}H_{14}I_6N_2O_6$ 1139.77

Benzoic acid, 3,3′-[(1,6-dioxo-1,6-hexanediyl)diimino]bis[2,4,6-triiodo-.

3,3′-(Adipoyldiimino)bis[2,4,6-triiodobenzoic acid] [*606-17-7*].

» Iodipamide contains not less than 98.0 percent and not more than 102.0 percent of $C_{20}H_{14}I_6N_2O_6$, calculated on the anhydrous basis.

Packaging and storage—Preserve in well-closed containers.

Reference standards—*USP 3-Amino-2,4,6-triiodobenzoic Acid Reference Standard*—Dry at 105° for 4 hours before using. Keep container tightly closed and protected from light. *USP Iodipamide Reference Standard*—Dry at 105° for 4 hours before using.

Identification—

A: It responds to the *Thin-layer Chromatographic Identification Test* ⟨201⟩, the test solution and Standard solution being prepared at a concentration of 1 mg per mL in an 0.8 in 1000 solution of sodium hydroxide in methanol, the solvent mixture being a mixture of chloroform, methanol, and ammonium hydroxide (20:10:2), and short-wavelength ultraviolet light being used to locate the spots.

B: Heat about 500 mg in a suitable crucible: violet vapors are evolved.

Water, *Method I* ⟨921⟩: not more than 1.0%.

Residue on ignition ⟨281⟩: not more than 0.1%.

Free aromatic amine—Transfer 1.0 g to a 50-mL volumetric flask, and add 12.5 mL of water and 2.5 mL of 1 N sodium hydroxide. To a second 50-mL volumetric flask transfer 4 mL of water, 10 mL of 0.1 N sodium hydroxide, and 1.0 mL of a Standard solution prepared by dissolving a suitable quantity of USP 3-Amino-2,4,6-triiodobenzoic Acid RS in 0.1 N sodium hydroxide (use 0.2 mL of the 0.1 N sodium hydroxide for each 5.0 mg of the Reference Standard) and by diluting with water to obtain a solution having a known concentration of 500 μg per mL. Proceed as directed in the test for *Free aromatic amine* under *Diatrizoate Meglumine*, beginning with "To a third 50-mL volumetric flask add 5 mL of water."

Other requirements—It meets the requirements of the tests for *Iodine and iodide* and *Heavy metals* under *Diatrizoic Acid.*

Assay—Transfer about 300 mg of Iodipamide, accurately weighed, to a glass-stoppered, 125-mL conical flask, add 30 mL of 1.25 N sodium hydroxide and 500 mg of powdered zinc, connect the flask to a reflux condenser, and reflux the mixture for 30 minutes. Cool the flask to room temperature, rinse the condenser with 20 mL of water, disconnect the flask from the condenser, and filter the mixture. Rinse the filter and the flask thoroughly, adding the rinsings to the filtrate. Add 5 mL of glacial acetic acid and 1 mL of tetrabromophenolphthalein ethyl ester TS, and titrate with 0.05 N silver nitrate VS until the yellow precipitate just turns green. Each mL of 0.05 N silver nitrate is equivalent to 9.498 mg of $C_{20}H_{14}I_6N_2O_6$.

Iodipamide Meglumine Injection

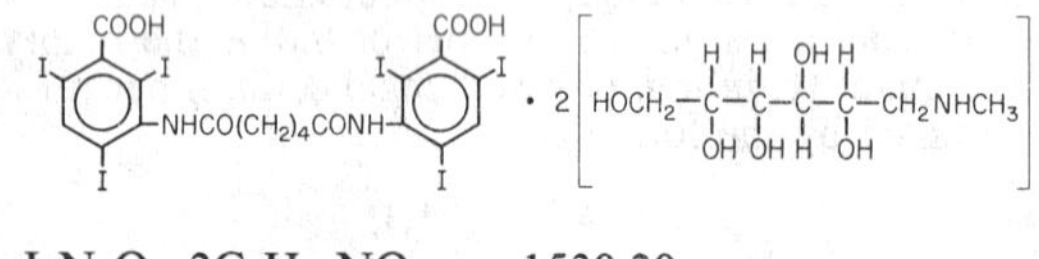

$C_{20}H_{14}I_6N_2O_6 \cdot 2C_7H_{17}NO_5$ 1530.20

Benzoic acid, 3,3′-[(1,6-dioxo-1,6-hexanediyl)diimino]bis[2,4,6-triiodo-, compd. with 1-deoxy-1-(methylamino)-D-glucitol (1:2).

1-Deoxy-1-(methylamino)-D-glucitol 3,3′-(adipoyldiimino)bis[2,4,6-triiodobenzoate] (2:1) (salt) [*3521-84-4*].

» Iodipamide Meglumine Injection is a sterile solution of Iodipamide in Water for Injection, prepared with the aid of Meglumine. It contains not less than 95.0 percent and not more than 105.0 percent of the labeled amount of iodipamide meglumine ($C_{20}H_{14}I_6N_2O_6 \cdot 2C_7H_{17}NO_5$). It may contain small amounts of suitable buffers and of Edetate Calcium Disodium or Edetate Disodium as a stabilizer. Iodipamide Meglumine Injection intended for intravascular use contains no antimicrobial agents.

Packaging and storage—Preserve in single-dose containers, preferably of Type I or Type III glass, protected from light.

Labeling—Label containers of Injection intended for intravascular injection to direct the user to discard any unused portion remaining in the container. Label containers of Injection intended for other than intravascular injection to show that the contents are not intended for intravascular injection.

Reference standards—*USP 3-Amino-2,4,6-triiodobenzoic Acid Reference Standard*—Dry at 105° for 4 hours before using. Keep container tightly closed and protected from light. *USP Iodipamide Reference Standard*—Dry at 105° for 4 hours before using.

Identification—

A: Dilute a volume of the Injection, if necessary, with an 0.8 in 1000 solution of sodium hydroxide in methanol to obtain a test solution having a concentration of 1 mg per mL. The test solution responds to the *Thin-layer Chromatographic Identification Test* ⟨201⟩, the Standard solution being prepared at a concentration of 1 mg of USP Iodipamide RS per mL in an 0.8 in 1000 solution of sodium hydroxide in methanol, the solvent mixture being a mixture of chloroform, methanol, and ammonium hydroxide (20:10:2), and short-wavelength ultraviolet light being used to locate the spots.

B: Evaporate a volume of Injection, equivalent to about 500 mg of iodipamide, to dryness, and heat the residue so obtained in a suitable crucible: violet vapors are evolved.

Pyrogen—It meets the requirements of the *Pyrogen Test* ⟨151⟩, the test dose being the equivalent of 1.25 g of iodipamide meglumine per kg, except that where the Injection is less concentrated than 125 mg per mL, the test dose is 10 mL per kg of body weight of rabbit.

pH ⟨791⟩: between 6.5 and 7.7.

Free aromatic amine—Transfer an accurately measured volume of Injection, equivalent to 1 g of iodipamide meglumine, to a 50-mL volumetric flask, dilute with water to 5 mL, and add 10 mL of 0.1 *N* sodium hydroxide. To a second 50-mL volumetric flask add 10 mL of 0.1 *N* sodium hydroxide, 4 mL of water, and 1.0 mL of a Standard solution prepared by dissolving a suitable quantity of USP 3-Amino-2,4,6-triiodobenzoic Acid RS in 0.1 *N* sodium hydroxide (use 0.2 mL of the 0.1 *N* sodium hydroxide for each 5.0 mg of the Reference Standard) and diluting with water to obtain a solution having a known concentration of 500 µg per mL. Proceed as directed in the test for *Free aromatic amine* under *Diatrizoate Meglumine*, beginning with "To a third 50-mL volumetric flask add 5 mL of water."

Meglumine content—Proceed as directed in the test for *Meglumine content* under *Diatrizoate Meglumine Injection.* The meglumine content is not less than 24.2% and not more than 26.8% of the labeled amount of iodipamide meglumine.

Other requirements—It meets the requirements of the tests for *Iodine and iodide* and *Heavy metals* under *Diatrizoate Meglumine Injection.* It meets also the requirements under *Injections* ⟨1⟩.

Assay—Pipet a volume of Iodipamide Meglumine Injection, equivalent to about 5 g of iodipamide meglumine, into a 100-mL volumetric flask, add 1.25 *N* sodium hydroxide to volume, and mix. Pipet 10 mL of the solution into a glass-stoppered, 250-mL flask, add 20 mL of the sodium hydroxide solution and 500 mg of powdered zinc, and proceed as directed in the *Assay* under *Diatrizoate Meglumine Injection*, beginning with "connect the flask to a reflux condenser." Each mL of 0.05 *N* silver nitrate is equivalent to 12.75 mg of $C_{20}H_{14}I_6N_2O_6 . 2C_7H_{17}NO_5$.

Iodochlorhydroxyquin and its dosage forms—*see* Clioquinol and its dosage forms

Iodoquinol

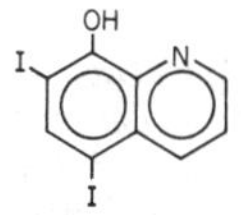

$C_9H_5I_2NO$ 396.95

8-Quinolinol, 5,7-diiodo-.

5,7-Diiodo-8-quinolinol [*83-73-8*].

» Iodoquinol contains not less than 96.0 percent and not more than 100.5 percent of $C_9H_5I_2NO$, calculated on the dried basis.

Packaging and storage—Preserve in well-closed containers.

Reference standard—*USP Iodoquinol Reference Standard*—Dry over silica gel for 4 hours before using.

Identification—

A: Prepare a 1 in 200 solution in carbon disulfide, warming slightly, if necessary, to effect complete solution: the infrared absorption spectrum of this solution, in a 3-mm sodium chloride cell, carbon disulfide being used as the blank, in the region from 7 µm to 11 µm exhibits absorption maxima and minima only at the same wavelengths as that of a similar solution of USP Iodoquinol RS, concomitantly measured.

B: Warm a small quantity of it with 1 mL of sulfuric acid: violet vapors of iodine are evolved.

Loss on drying ⟨731⟩—Dry it over silica gel for 4 hours: it loses not more than 0.5% of its weight.

Residue on ignition ⟨281⟩: not more than 0.5%.

Free iodine and iodide—Shake 1.0 g with 20 mL of water for 30 seconds, allow to stand for 5 minutes, and filter. To 10 mL of the filtrate add 1 mL of 2 *N* sulfuric acid, then add 2 mL of chloroform, and shake: no violet color appears in the chloroform (*free iodine*). To the mixture add 5 mL of 2 *N* sulfuric acid and 1 mL of potassium dichromate TS, and shake for 15 seconds: the color in the chloroform layer is not deeper than that produced in a control test made in the following manner. Dilute 2 mL of potassium iodide solution (1 in 6000) with water to 10 mL, add 6 mL of 2 *N* sulfuric acid, 1 mL of potassium dichromate TS, and 2 mL of chloroform, and shake for 15 seconds (0.05% of *iodide*).

Assay—Using about 14 mg of Iodoquinol, accurately weighed, proceed as directed under *Oxygen Flask Combustion* ⟨471⟩, using a mixture of 10 mL of sodium hydroxide solution (1 in 100) and 1 mL of freshly prepared sodium bisulfite solution (1 in 100) as the absorbing liquid. When the combustion is complete, place a few mL of water around the stopper of the flask, loosen the stopper, then rinse the stopper, the specimen holder, and the sides of the flask with about 20 mL of water, added in small portions. Add 1 mL of an oxidizing solution prepared by adding 5 mL of bromine to 100 mL of a 1 in 10 solution of sodium acetate in glacial acetic acid. Insert the stopper in the flask, and shake vigorously for 1 minute. Add 0.5 mL of formic acid, replace the stopper, and shake vigorously for 1 minute. Remove the stopper, and rinse the stopper, the specimen holder, and the sides of the flask with several small portions of water. Bubble nitrogen through the flask to remove the oxygen and excess bromine, add 500 mg of potassium iodide, swirl to dissolve, add 3 mL of 2 *N* sulfuric acid, swirl to mix, and allow to stand for 2 minutes. Titrate with 0.02 *N* sodium thiosulfate VS, adding 3 mL of starch TS as the end-point is approached. Each mL of 0.02 *N* sodium thiosulfate is equivalent to 0.6616 mg of $C_9H_5I_2NO$.

Iodoquinol Tablets

» Iodoquinol Tablets contain not less than 95.0 percent and not more than 105.0 percent of the labeled amount of $C_9H_5I_2NO$.

Packaging and storage—Preserve in well-closed containers.

Reference standard—*USP Iodoquinol Reference Standard*—Dry over silica gel for 4 hours before using.

Identification—

A: Shake a portion of finely powdered Tablets, equivalent to about 5 mg of iodoquinol, with 10 mL of carbon disulfide, and filter: the filtrate responds to *Identification test A* under *Iodoquinol.*

B: Place a portion of the powdered Tablets prepared for the *Assay*, equivalent to about 50 mg of iodoquinol, in a dry test tube, add 1 mL of sulfuric acid, and warm gently: violet vapors of iodine are evolved.

Disintegration ⟨701⟩: 1 hour.

Uniformity of dosage units ⟨905⟩: meet the requirements.

Soluble iodides—Digest a quantity of powdered Tablets, equivalent to 100 mg of iodoquinol, with 5 mL of water for 10 minutes, cool, and filter. To the filtrate add 1 mL of 3 *N* hydrochloric acid, 2 drops of ferric chloride TS, and 2 mL of chloroform, shake gently, and allow to separate: any violet color in the chloroform is not more intense than that in a blank to which 0.2 mg of potassium iodide has been added.

Assay—Weigh and finely powder not less than 20 Iodoquinol Tablets. Using a portion of the powder, accurately weighed and equivalent to about 14 mg of iodoquinol, proceed as directed in the *Assay* under *Iodoquinol*. Each mL of 0.02 *N* sodium thiosulfate is equivalent to 0.6616 mg of $C_9H_5I_2NO$.

Iopamidol

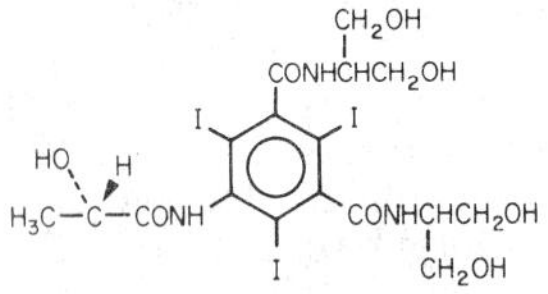

$C_{17}H_{22}I_3N_3O_8$ 777.09

1,3-Benzenedicarboxamide, *N,N′*-bis[2-hydroxy-1-(hydroxymethyl)ethyl]-5-[(2-hydroxy-1-oxopropyl)amino]-2,4,6-triiodo-, (*S*)-.

(*S*)-*N,N′*-bis[2-Hydroxy-1-(hydroxymethyl)ethyl]-2,4,6-triiodo-5-lactamidoisophthalamide [*60166-93-0*].

» Iopamidol contains not less than 98.0 percent and not more than 101.0 percent of iopamidol, calculated on the dried basis.

Packaging and storage—Preserve in well-closed, light-resistant containers.

Reference standards—*USP Iopamidol Reference Standard*—Dry at 105° for 4 hours before using. *USP N,N′-bis-(1,3-dihydroxy-2-propyl)-5-amino-2,4,6-triiodoisophthalamide Reference Standard*—Do not dry before using. Keep container tightly closed and protected from light.

Identification—

A: The infrared absorption spectrum of a mineral oil dispersion of it, previously dried, exhibits maxima only at the same wavelengths as that of a similar preparation of USP Iopamidol RS.

B: Heat about 500 mg in a suitable crucible: violet vapors are evolved.

C: It responds to the *Thin-layer Chromatographic Identification Test* ⟨201⟩, the test solution and the Standard solution being prepared at a concentration of 0.5 mg per mL in a mixture of methanol and water (9:1), the solvent mixture being chloroform, methanol, ammonium hydroxide, and water (60:30:9:1), and short-wavelength ultraviolet light being used to locate the spots.

Specific rotation ⟨781⟩: between −4.6° and −5.2°, calculated on the dried basis, determined at 20° and 436 nm (mercury line), in a solution prepared by dissolving 20 g in 50 mL of water, heating on a water bath, if necessary, to effect solution, then filtering through a suitable membrane filter (3 µm or less).

Loss on drying ⟨731⟩—Dry it at 105° for 4 hours: it loses not more than 0.5% of its weight.

Residue on ignition ⟨281⟩: not more than 0.1%.

Free aromatic amine—Transfer 500 mg to a 25-mL volumetric flask, and add 20 mL of water, heating on a water bath, if necessary, to effect solution. To a second 25-mL volumetric flask transfer 16 mL of water and 4.0 mL of a Standard solution prepared by dissolving a suitable quantity of USP *N,N′*-bis-(1,3-dihydroxy-2-propyl)-5-amino-2,4,6-triiodoisophthalamide RS in water and diluting with water to obtain a solution having a concentration of 62.5 µg per mL. To a third 25-mL volumetric flask add 20 mL of water to provide a blank. Treat each flask as follows: Place the flasks in an ice bath, protected from light, for 5 minutes. [NOTE—In conducting the following steps, keep the flasks in the ice bath and protected from light as much as possible until all of the reagents have been added.] Add slowly 1 mL of hydrochloric acid, mix, and allow to stand for 5 minutes. Add 1 mL of sodium nitrite solution (1 in 50), mix, and allow to stand for 5 minutes. Add 1 mL of ammonium sulfamate solution (3 in 25), shake, and allow to stand for 5 minutes. [*Caution—Considerable pressure is produced.*] Add 1 mL of *N*-(1-naphthyl)ethylenediamine dihydrochloride solution (1 in 1000), and mix. Remove the flasks from the ice bath, and allow to stand in a water bath at about 25° for 10 minutes. Dilute with water to volume, mix, and without delay (about 5 minutes from final dilution), concomitantly determine the absorbances of the solution from the substance under test and the Standard solution in 1-cm cells at the wavelength of maximum absorbance at about 500 nm, with a suitable spectrophotometer, against the prepared blank. The absorbance of the solution from the Iopamidol is not greater than that of the Standard solution (0.05%).

Free iodine—Transfer 2.0 g to a stoppered, 50-mL centrifuge tube, add sufficient water to dissolve, heating on a water bath, if necessary, to effect solution, and dilute with water to 25 mL. Add 5 mL of toluene and 5 mL of 2 *N* sulfuric acid, shake well, and centrifuge: the toluene layer shows no red color.

Free iodide—Transfer 5.0 g to a 125-mL conical flask, and dissolve in 25 mL of water, heating on a water bath, if necessary, to effect solution. Cool, and adjust the solution with glacial acetic acid to a pH of 4.5 to 5.5. Add 2 mL of sodium chloride solution (0.58 in 100), and titrate with 0.001 *N* silver nitrate VS, determining the end-point potentiometrically. Not more than 1.2 mL of 0.001 *N* silver nitrate is required (0.003% *iodide*).

Free acid or alkali—Dissolve 10.0 g in 100 mL of freshly boiled and cooled water. Using a pH meter and a glass-calomel electrode system, determine the volume of 0.01 *N* hydrochloric acid VS or 0.01 *N* sodium hydroxide VS to bring the pH of the test solution to 7.0: not more than 1.37 mL of 0.01 *N* sodium hydroxide, equivalent to a free acid content of 5 mg of hydrochloric acid per 100 g, or not more than 0.75 mL of 0.01 *N* hydrochloric acid, equivalent to a free alkali content of 3 mg of sodium hydroxide per 100 g, is required.

Heavy metals, *Method II* ⟨231⟩: not more than 0.001%.

Chromatographic purity—Dissolve 1.0 g in 1 mL of water in a 10-mL volumetric flask, heating on a water bath, if necessary, to effect solution. Cool, and dilute with methanol to volume to obtain the test solution. Dissolve an accurately weighed quantity of USP Iopamidol RS in water, heating on a water bath, if necessary, to effect solution. Cool, and dilute with methanol, the ratio of methanol to water being 9:1, and mix to obtain a solution having a known concentration of 0.5 mg per mL. On a suitable thin-layer chromatographic plate (see *Chromatography* ⟨621⟩), coated with a 0.25-mm layer of chromatographic silica gel mixture, apply separately 10-µL portions of the test solution and the Standard solution. Position the plate in a chromatographic chamber, and develop the chromatograms in a solvent system consisting of a mixture of chloroform, methanol, ammonium hydroxide, and water (60:30:9:1) until the solvent front has moved about 15 cm from the origin. Remove the plate from the developing chamber, mark the solvent front, air-dry, and view under short-wavelength ultraviolet light: the R_f value of any spot other than the principal spot obtained from the test solution does not exceed, in size or intensity, the principal spot obtained from the Standard solution (0.5%).

Assay—Transfer about 300 mg of Iopamidol, accurately weighed, to a glass-stoppered, 125-mL conical flask, add 40 mL of 1.25 *N* sodium hydroxide and 1 g of powdered zinc, connect the flask to a reflux condenser, and reflux the mixture for 30 minutes. Cool the flask to room temperature, rinse the condenser with 20 mL of water, disconnect the flask from the condenser, and filter the mixture. Rinse the flask and the filter thoroughly, adding the rinsings to the filtrate. Add 5 mL of glacial acetic acid, and titrate with 0.1 *N* silver nitrate VS, determining the end-point potentiometrically. Each mL of 0.1 *N* silver nitrate is equivalent to 25.90 mg of $C_{17}H_{22}I_3N_3O_8$.

Iopamidol Injection

» Iopamidol Injection is a sterile solution of Iopamidol in Water for Injection. It contains not less than 95.0 percent and not more than 105.0 percent of the labeled amount of Iopamidol ($C_{17}H_{22}I_3N_3O_8$). It may contain small amounts of suitable buffers and of Edetate Calcium Disodium as a stabilizer. Iopamidol Injection intended for intravascular or intrathecal use contains no antimicrobial agents.

Packaging and storage—Preserve Injection intended for intravascular or intrathecal use in single-dose containers, preferably of Type I glass, and protected from light.

Labeling—Label containers of Injection to direct the user to discard any unused portion remaining in the container and to check for the presence of particulate matter before using.

Reference standards—*USP Iopamidol Reference Standard*—Dry at 105° for 4 hours before using. *USP N,N'-bis-(1,3-dihydroxy-2-propyl)-5-amino-2,4,6-triiodoisophthalamide Reference Standard*—Do not dry before using. Keep container tightly closed and protected from light.

Identification—

A: Evaporate a volume of Injection, equivalent to about 500 mg of iopamidol, to dryness, and heat the residue so obtained in a suitable crucible: violet vapors are evolved.

B: It responds to the *Thin-layer Chromatographic Identification Test* ⟨201⟩, the test solution and the Standard solution being prepared at a concentration of 0.5 mg per mL in a mixture of methanol and water (9:1), the solvent mixture being chloroform, methanol, ammonium hydroxide, and water (60:30:9:1), and short-wavelength ultraviolet light being used to locate the spots.

Pyrogen—It meets the requirements of the *Pyrogen Test* ⟨151⟩, the test dose being 5 mL of Injection per kg.

pH ⟨791⟩: between 6.5 and 7.5.

Free aromatic amine—Transfer an accurately measured volume of Injection, equivalent to about 500 mg of iopamidol, to a 25-mL volumetric flask, dilute with water to 20 mL, and mix. Proceed as directed in the test for *Free aromatic amine* under *Iopamidol*, beginning with "To a second 25-mL volumetric flask transfer 16 mL of water."

Free iodine—Transfer a volume of Injection, equivalent to 2.0 g of iopamidol, to a glass-stoppered test tube. Add 2 mL of 2 *N* sulfuric acid and 1.0 mL of toluene, shake, and allow the layers to separate: the toluene layer shows no red color.

Free iodide—Transfer 10.0 mL of Injection to a beaker, add 50 mL of water, 2 mL of glacial acetic acid, and 2 mL of sodium chloride solution (0.58 in 100), and mix. Adjust the solution with sodium hydroxide solution (4 in 100) to a pH of 4.5, using pH indicator paper. Titrate with 0.001 *N* silver nitrate VS, determining the end-point potentiometrically. Not more than 3.1 mL of 0.001 *N* silver nitrate is required (0.04 mg of *iodide* per mL).

Other requirements—It meets the requirements under *Injections* ⟨1⟩.

Assay—Dilute an accurately measured volume of Iopamidol Injection, equivalent to about 800 mg of iopamidol, quantitatively and stepwise with water to obtain a solution having a known concentration of about 8 µg per mL. Concomitantly determine the absorbances of this solution and a Standard solution of USP Iopamidol RS in water having a known concentration of about 8 µg per mL, in 1-cm cells at the wavelength of maximum absorbance at about 240 nm, with a suitable spectrophotometer, using water as the blank. Calculate the quantity, in mg, of $C_{17}H_{22}I_3N_3O_8$ in each mL of the Injection taken by the formula:

$$100(C_S/V)(A_U/A_S),$$

in which C_S is the concentration, in µg per mL, of USP Iopamidol RS in the Standard preparation, V is the volume, in mL, of Injection taken, and A_U and A_S are the absorbances of the solution from the Injection and the Standard solution, respectively.

Iopanoic Acid

$C_{11}H_{12}I_3NO_2$ 570.93

Benzenepropanoic acid, 3-amino-α-ethyl-2,4,6-triiodo-.

3-Amino-α-ethyl-2,4,6-triiodohydrocinnamic acid [*96-83-3*].

» Iopanoic Acid contains an amount of iodine equivalent to not less than 97.0 percent and not more than 101.0 percent of $C_{11}H_{12}I_3NO_2$, calculated on the dried basis.

Packaging and storage—Preserve in tight, light-resistant containers.

Identification—Mix about 100 mg with 500 mg of sodium carbonate in a crucible, and heat until thoroughly charred. Cool, add 5 mL of hot water, heat on a steam bath for 5 minutes, and filter: the solution responds to the tests for *Iodide* ⟨191⟩.

Melting range ⟨741⟩: between 152° and 158°, with decomposition.

Loss on drying ⟨731⟩—Dry it at 105° for 1 hour: it loses not more than 1.0% of its weight.

Residue on ignition ⟨281⟩: not more than 0.1%.

Free iodine—Shake about 200 mg with 2 mL of water and 2 mL of chloroform for 1 minute: the chloroform layer shows no violet color.

Halide ions—Place about 500 mg in a glass-stoppered, 50-mL cylinder, add 10 mL of 2 *N* nitric acid and 15 mL of water, shake for 5 minutes, and filter through paper: 10 mL of the filtrate shows no greater turbidity than corresponds to 0.05 mL of 0.020 *N* hydrochloric acid (see *Chloride and Sulfate* ⟨221⟩).

Heavy metals, *Method II* ⟨231⟩: 0.002%.

Assay—Transfer about 250 mg of Iopanoic Acid, accurately weighed, to a glass-stoppered, 250-mL conical flask. Add 30 mL of 1.25 *N* sodium hydroxide and 500 mg of powdered zinc, and reflux the mixture for 30 minutes. Cool to room temperature, wash the condenser with 20 mL of water, and filter the mixture. Wash the flask and the filter with small portions of water, adding the washings to the filtrate. Add to the filtrate 5 mL of glacial acetic acid and 1 mL of tetrabromophenolphthalein ethyl ester TS, and titrate with 0.05 *N* silver nitrate VS until the color of the yellow precipitate just changes to green. Each mL of 0.05 *N* silver nitrate is equivalent to 9.516 mg of $C_{11}H_{12}I_3NO_2$.

Iopanoic Acid Tablets

» Iopanoic Acid Tablets contain not less than 95.0 percent and not more than 105.0 percent of the labeled amount of $C_{11}H_{12}I_3NO_2$.

Packaging and storage—Preserve in tight, light-resistant containers.

Identification—Triturate a quantity of finely powdered Tablets, equivalent to about 1 g of iopanoic acid, with two 10-mL portions of solvent hexane, and decant and discard the liquid. Allow the residue to dry spontaneously, triturate with 15 mL of acetone, and filter. Repeat the trituration with another 15-mL portion of acetone, evaporate the combined filtrates on a steam bath to a volume of not more than 1 mL, add, with constant stirring, 20 mL of water, filter, wash the precipitate with two 5-mL portions of water, and dry at 105° for 2 hours: the iopanoic acid so obtained melts between 150° and 158°, with decomposition, and responds to the *Identification test* under *Iopanoic Acid.*

Disintegration ⟨701⟩: 30 minutes.

Uniformity of dosage units ⟨905⟩: meet the requirements.

Halide ions—A portion of the powdered Tablets prepared for the *Assay*, equivalent to about 500 mg of iopanoic acid, meets the requirements of the test for *Halide ions* under *Iopanoic Acid.*

Assay—Weigh and finely powder not less than 20 Iopanoic Acid Tablets. Weigh accurately a portion of the powder, equivalent to about 1 g of iopanoic acid, and triturate with 10 mL of solvent hexane. Allow the mixture to settle, decant the hexane through a small filter, repeat the trituration with 10 mL of solvent hexane, filter through the same filter, and discard the filtrates. Warm the residue with 10 mL of neutralized alcohol at 70°, filter through the same filter, and wash the undissolved residue with four 10-mL portions of neutralized alcohol at 70°, passing the washings through the same filter. Cool the combined filtrate and washings to room temperature, add 3 to 5 drops of thymol blue TS, and titrate with 0.1 *N* sodium hydroxide VS. Each mL of 0.1 *N* sodium hydroxide is equivalent to 57.09 mg of $C_{11}H_{12}I_3NO_2$.

Iophendylate

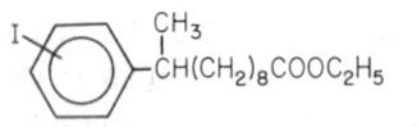

$C_{19}H_{29}IO_2$ 416.34
Benzenedecanoic acid, iodo-ι-methyl-, ethyl ester.
Ethyl 10-(iodophenyl)undecanoate [*1320-11-2*].

» Iophendylate is a mixture of isomers of ethyl iodophenylundecanoate, consisting chiefly of ethyl 10-(iodophenyl)undecanoate. It contains not less than 98.0 percent and not more than 102.0 percent of $C_{19}H_{29}IO_2$.

Packaging and storage—Preserve in tight, light-resistant containers.

Identification—Place about 1 mL of Iophendylate, 15 mL of water, and 7 g of potassium dichromate in a round-bottom, 50-mL flask. Carefully add 10 mL of sulfuric acid, moderating the ensuing vigorous reaction by cooling the flask with tap water. When the reaction has subsided, reflux the mixture for 2 hours. Pour the cooled contents of the flask into 25 mL of water, filter the mixture with suction, and wash the precipitate with a small quantity of cold water. Crystallize the precipitate from 10 mL of diluted alcohol, and sublime the solid so obtained: the sublimate of *p*-iodobenzoic acid melts between 268° and 272°.

Specific gravity ⟨841⟩: between 1.248 and 1.257.

Refractive index ⟨831⟩: between 1.524 and 1.526.

Residue on ignition ⟨281⟩: not more than 0.1%.

Free acids—Transfer about 4 g, accurately weighed, to a small flask, and add 20 mL of alcohol. Swirl to dissolve the test specimen, add 5 drops of phenolphthalein TS, and titrate with 0.050 *N* alcoholic potassium hydroxide to a pink color that persists for 30 seconds: not more than 0.60 mL of 0.050 *N* alcoholic potassium hydroxide is required for neutralization, correction being made for the amount of 0.050 *N* alcoholic potassium hydroxide consumed by a blank (0.3% as iodophenylundecanoic acid).

Free iodine—Determine its absorbance in a 4-cm cell, at 485 nm, with a suitable spectrophotometer, using water as the blank: the absorbance is not greater than 0.16 (7.5 ppm).

Saponification value—Transfer about 1 g, accurately weighed, to a 250-mL flask, add 25.0 mL of 0.5 *N* alcoholic potassium hydroxide VS, and reflux the mixture on a steam bath for 1 hour. Cool, add 25 mL of water and 0.7 mL of phenolphthalein TS, and titrate with 0.5 *N* hydrochloric acid VS. The saponification value (see *Fats and Fixed Oils* ⟨401⟩) is between 132 and 142.

Assay—Dissolve about 50 mg of Iophendylate, accurately weighed, in 5 mL of toluene contained in a 125-mL separator fitted with a suitable, inert plastic stopcock. Add 15 mL of sodium biphenyl, and shake vigorously for 2 minutes. Extract gently with three 10-mL portions of 5 *M* phosphoric acid, combining the lower phases in a 125-mL iodine flask. Add 1 *N* sodium hypochlorite dropwise to the combined extracts until the solution turns brown, and then add an additional 0.5 mL. Shake intermittently for 3 minutes, add 5 mL of freshly prepared, saturated phenol solution, mix, and allow to stand for 1 minute, accurately timed. Add 1 g of potassium iodide, shake for 30 seconds, and titrate rapidly with 0.1 *N* sodium thiosulfate VS, adding 3 mL of starch TS as the end-point is approached. Each mL of 0.1 *N* sodium thiosulfate is equivalent to 6.939 mg of $C_{19}H_{29}IO_2$.

Iophendylate Injection

» Iophendylate Injection is sterile Iophendylate.

Packaging and storage—Preserve in single-dose containers, preferably of Type I glass, protected from light.

Other requirements—It conforms to the definition, responds to the *Identification test*, and meets the requirements for *Specific gravity*, *Refractive index*, *Residue on ignition*, *Free acids*, *Free iodine*, *Saponification value*, and *Assay* under *Iophendylate.* It meets also the requirements under *Injections* ⟨1⟩.

Iothalamate Meglumine Injection

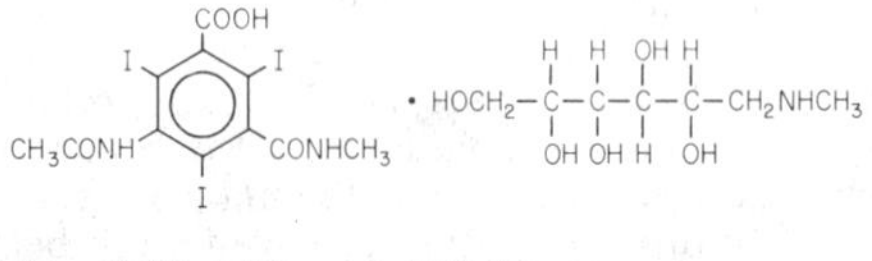

$C_{11}H_9I_3N_2O_4.C_7H_{17}NO_5$ 809.13
Benzoic acid, 3-(acetylamino)-2,4,6-triiodo-5-[(methylamino)carbonyl]-, compd. with 1-deoxy-1-(methylamino)-D-glucitol (1:1).
1-Deoxy-1-(methylamino)-D-glucitol 5-acetamido-2,4,6-triiodo-*N*-methylisophthalamate (salt) [*13087-53-1*].

» Iothalamate Meglumine Injection is a sterile solution of Iothalamic Acid in Water for Injection, prepared with the aid of Meglumine. It contains not less than 95.0 percent and not more than 105.0 percent of the labeled amount of iothalamate meglumine ($C_{11}H_9I_3N_2O_4.C_7H_{17}NO_5$). It may contain small amounts of suitable buffers and of Edetate Calcium Disodium or Edetate Disodium as a stabilizer. Iothalamate Meglumine Injection intended for intravascular use contains no antimicrobial agents.

Packaging and storage—Preserve in single-dose containers, preferably of Type I glass, protected from light.

Labeling—Label containers of Injection intended for intravascular injection to direct the user to discard any unused portion remaining in the container. Label containers of Injection intended for other than intravascular injection to show that the contents are not intended for intravascular injection.

Reference standards—*USP 5-Amino-2,4,6-triiodo-N-methylisophthalamic Acid Reference Standard*—Dry at 105° for 4 hours before using. Keep container tightly closed and protected from light. *USP Iothalamic Acid Reference Standard*—Dry at 105° for 4 hours before using.

Identification—Dilute 3 mL of Injection with water to 100 mL, add an excess of 3 *N* hydrochloric acid, and filter. Wash the precipitated iothalamic acid on the filter with four 10-mL portions of water, and dry at 105° for 4 hours: the dried iothalamic acid responds to the following tests.

A: The infrared absorption spectrum of a 0.5 percent potassium bromide dispersion of the dried acid exhibits maxima only at the same wavelengths as that of a similar preparation of USP Iothalamic Acid RS.

B: Heat about 500 mg of the dried acid in a suitable crucible: violet vapors are evolved.

Pyrogen—It meets the requirements of the *Pyrogen Test* ⟨151⟩, the test dose being the equivalent of 2.5 g of iothalamate meglumine per kg.

pH ⟨791⟩: between 6.5 and 7.7.

Free aromatic amine—Dilute a suitable volume of Injection with water to yield a solution containing 100 mg of iothalamate meglumine per mL. Proceed as directed in the test for *Free aromatic amine* under *Iothalamic Acid*, beginning with "Pipet 5 mL of this solution into a 50-mL volumetric flask."

Iodine and iodide—Dilute a volume of Injection, equivalent to 2 g of iothalamate meglumine, with 20 mL of water in a 50-mL beaker, add 5 mL of 2 *N* sulfuric acid, stir, and filter into a glass-stoppered, 50-mL cylinder. Proceed as directed for *Procedure* in the test for *Iodine and iodide* under *Iothalamic Acid*, beginning with "To the filtrate add 5 mL of toluene."

Heavy metals ⟨231⟩—In a 50-mL color-comparison tube, mix a volume of Injection, equivalent to 1.0 g of iothalamate meglumine, with 5 mL of 1 *N* sodium hydroxide, dilute with water to 40 mL, and mix. Using this as the *Test preparation*, proceed as directed in the test for *Heavy metals* under *Diatrizoate Meglumine:* the limit is 0.002%.

Meglumine content—Proceed as directed in the test for *Meglumine content* under *Diatrizoate Meglumine Injection.* The meglumine content is not less than 22.9% and not more than 25.3% of the labeled amount of iothalamate meglumine.

Other requirements—It meets the requirements under *Injections* ⟨1⟩.

Assay—Pipet a volume of Iothalamate Meglumine Injection, equivalent to about 4 g of iothalamate meglumine, into a 250-mL volumetric flask, dilute with water to volume, and mix. Pipet 25 mL of this solution into a glass-stoppered, 125-mL conical flask, add 12 mL of 5 *N* sodium hydroxide and 1 g of powdered zinc, connect the flask to a reflux condenser, and reflux for 30 minutes. Cool to room temperature, rinse the condenser with 20 mL of water, disconnect the flask from the condenser, and filter the mixture. Rinse the filter and the flask thoroughly, adding the rinsings to the filtrate. Add 40 mL of 2 *N* sulfuric acid, and titrate immediately with 0.05 *N* silver nitrate VS, determining the end-point potentiometrically, using silver-calomel electrodes and an agar–potassium nitrate salt bridge. Each mL of 0.05 *N* silver nitrate is equivalent to 13.49 mg of $C_{11}H_9I_3N_2O_4 \cdot C_7H_{17}NO_5$.

Iothalamate Meglumine and Iothalamate Sodium Injection

» Iothalamate Meglumine and Iothalamate Sodium Injection is a sterile solution of Iothalamic Acid in Water for Injection, prepared with the aid of Meglumine and Sodium Hydroxide. It contains not less than 95.0 percent and not more than 105.0 percent of the labeled amounts of iothalamate meglumine ($C_{11}H_9I_3N_2O_4 \cdot C_7H_{17}NO_5$) and iothalamate sodium ($C_{11}H_8I_3N_2NaO_4$). It may contain small amounts of suitable buffers and of Edetate Calcium Disodium or Edetate Disodium as a stabilizer. Iothalamate Meglumine and Iothalamate Sodium Injection intended for intravascular use contains no antimicrobial agents.

Packaging and storage—Preserve in single-dose containers, preferably of Type I glass, protected from light.

Labeling—Label containers of Injection intended for intravascular injection to direct the user to discard any unused portion remaining in the container. Label containers of Injection intended for other than intravascular injection to show that the contents are not intended for intravascular injection.

Reference standards—*USP 5-Amino-2,4,6-triiodo-N-methylisophthalamic Acid Reference Standard*—Dry at 105° for 4 hours before using. Keep container tightly closed and protected from light. *USP Iothalamic Acid Reference Standard*—Dry at 105° for 4 hours before using.

Identification—Dilute 3 mL of Injection with water to 100 mL, add an excess of 3 *N* hydrochloric acid, mix, and filter. Wash the precipitate of iothalamic acid so obtained with four 10-mL portions of water, and dry at 105° for 4 hours: the dried iothalamic acid so obtained responds to the following tests.

A: The infrared absorption spectrum of a potassium bromide dispersion of it exhibits maxima only at the same wavelengths as that of a similar preparation of USP Iothalamic Acid RS.

B: Heat about 500 mg in a suitable crucible: violet vapors are evolved.

Pyrogen—It meets the requirements of the *Pyrogen Test* ⟨151⟩, the test dose being the equivalent of 2.5 g of the total of iothalamate meglumine and iothalamate sodium per kg.

pH ⟨791⟩: between 6.5 and 7.7.

Free aromatic amine—Dilute a suitable volume of Injection with water to yield a solution containing 100 mg of the total of iothalamate meglumine and iothalamate sodium per mL. Pipet 5 mL of this solution into a 50-mL volumetric flask, and add 10 mL of water. Proceed as directed in the test for *Free aromatic amine* under *Iothalamic Acid*, beginning with "In another flask place 15 mL of water." The absorbance of the solution from the Injection is not greater than that of the Standard solution (0.05%).

Iodine and iodide—Dilute a volume of Injection, equivalent to 2 g of the total of iothalamate meglumine and iothalamate sodium, with 20 mL of water in a 50-mL beaker, and proceed as directed for *Procedure* in the test for *Iodine and iodide* under *Iothalamic Acid*, beginning with "add 5 mL of 2 *N* sulfuric acid." The limit of *Iodine and iodide* is 0.02% of iodide.

Heavy metals ⟨231⟩—In a 50-mL color-comparison tube, mix a volume of Injection, equivalent to 1.0 g of the total of iothalamate meglumine and iothalamate sodium, with 5 mL of 1 *N* sodium hydroxide, dilute with water to 40 mL, and mix. Using this as the *Test preparation*, proceed as directed for *Heavy metals* under *Diatrizoate Meglumine:* the limit is 0.002%.

Other requirements—It meets the requirements under *Injections* ⟨1⟩.

Assay for iothalamate meglumine—Pipet 5 mL of Iothalamate Meglumine and Iothalamate Sodium Injection into a 10-mL volumetric flask, add water to volume, and mix. Determine the angular rotation (see *Optical Rotation* ⟨781⟩) of the diluted Injection, using a 10-cm cell and a suitable polarimeter. Calculate the quantity, in mg per mL, of iothalamate meglumine in the Injection by the formula:

$$2000a/6.01,$$

in which a is the observed angular rotation, in degrees, corrected for the blank, and the factor 6.01 is the specific rotation, in degrees, of iothalamate meglumine.

Assay for iothalamate sodium—Transfer an accurately measured volume of Iothalamate Meglumine and Iothalamate Sodium Injection, equivalent to about 4 g of iothalamate meglumine and iothalamate sodium, to a 250-mL volumetric flask, dilute with water to volume, and mix. Pipet 25 mL of this solution into a glass-stoppered, 125-mL conical flask, add 12 mL of 5 *N* sodium hydroxide and 1 g of powdered zinc, connect the flask to a reflux condenser, and reflux the mixture for 30 minutes. Cool the flask to room temperature, rinse the condenser with 20 mL of water, disconnect the flask from the condenser, and filter the mixture. Rinse the flask and the filter thoroughly, adding the rinsings to the filtrate. Add 40 mL of 2 *N* sulfuric acid, and titrate immediately with 0.05 *N* silver nitrate VS, determining the end-point potentiometrically, using silver-calomel electrodes and an agar–potassium nitrate salt bridge. Calculate the volume, in mL, consumed by the iothalamate meglumine in the portion of solution taken, using the value found in the *Assay for iothalamate meglumine.* Each mL of 0.05 *N* silver nitrate is equivalent to 13.49 mg of $C_{11}H_9I_3N_2O_4 \cdot C_7H_{17}NO_5$. Subtract this volume from the total volume of 0.05 *N* silver nitrate consumed. Use the resulting volume to calculate the amount, in mg per mL, of iothalamate sodium in the Injection. Each mL of 0.05 *N* silver nitrate is equivalent to 10.60 mg of $C_{11}H_8I_3N_2NaO_4$.

Iothalamate Sodium Injection

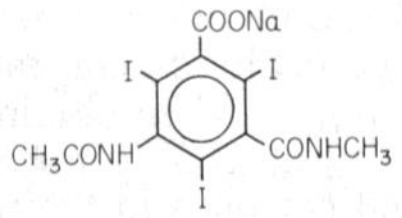

$C_{11}H_8I_3N_2NaO_4$ 635.90

Benzoic acid, 3-(acetylamino)-2,4,6-triiodo-5-[(methylamino)carbonyl]-, monosodium salt.

Monosodium 5-acetamido-2,4,6-triiodo-*N*-methylisophthalamate [*1225-20-3*].

» Iothalamate Sodium Injection is a sterile solution of Iothalamic Acid in Water for Injection prepared with the aid of Sodium Hydroxide. It contains not less than 95.0 percent and not more than 105.0 percent of the labeled amount of iothalamate sodium ($C_{11}H_8I_3N_2NaO_4$). It may contain small amounts of suitable buffers and of Edetate Calcium Disodium or Edetate Disodium as a stabilizer. Iothalamate Sodium Injection intended for intravascular use contains no antimicrobial agents.

Packaging and storage—Preserve in single-dose containers, preferably of Type I glass, protected from light.

Labeling—Label containers of the Injection intended for intravascular injection to direct the user to discard any unused portion remaining in the container. Label containers of the Injection intended for other than intravascular injection to show that the contents are not intended for intravascular injection.

Reference standards—*USP 5-Amino-2,4,6-triiodo-N-methylisophthalamic Acid Reference Standard*—Dry at 105° for 4 hours before using. Keep container tightly closed and protected from light. *USP Iothalamic Acid Reference Standard*—Dry at 105° for 4 hours before using.

Identification—

A: Dilute 3 mL of Injection with water to 100 mL, add an excess of 3 *N* hydrochloric acid, and filter. Wash the precipitated iothalamic acid with four 10-mL portions of water, and dry at 105° for 4 hours: the dried iothalamic acid responds to *Identification tests A* and *B* under *Iothalamate Meglumine Injection.*

B: It responds to the flame test for *Sodium* ⟨191⟩.

Pyrogen—It meets the requirements of the *Pyrogen Test* ⟨151⟩, the test dose being the equivalent of 2.5 g of iothalamate sodium per kg.

pH ⟨791⟩: between 6.5 and 7.7.

Free aromatic amine—Dilute a suitable volume of Injection with water to yield a solution containing 100 mg of iothalamate sodium per mL. Proceed as directed in the test for *Free aromatic amine* under *Iothalamic Acid*, beginning with "Pipet 5 mL of this solution into a 50-mL volumetric flask."

Iodine and iodide—Dilute a volume of Injection, equivalent to about 2 g of iothalamate sodium, with 20 mL of water in a 50-mL beaker, add 5 mL of 2 *N* sulfuric acid, stir, and filter into a glass-stoppered, 50-mL cylinder. Proceed as directed for *Procedure* in the test for *Iodine and iodide* under *Iothalamic Acid*, beginning with "To the filtrate add 5 mL of toluene."

Heavy metals ⟨231⟩—In a 50-mL color-comparison tube, mix a volume of Injection, equivalent to 1.0 g of iothalamate sodium, with 5 mL of 1 *N* sodium hydroxide, dilute with water to 40 mL, and mix. Using this as the *Test preparation*, proceed as directed in the test for *Heavy metals* under *Diatrizoate Meglumine:* the limit is 0.002%.

Other requirements—It meets the requirements under *Injections* ⟨1⟩.

Assay—Proceed with Iothalamate Sodium Injection as directed in the *Assay* under *Iothalamate Meglumine Injection.* Each mL of 0.05 *N* silver nitrate is equivalent to 10.60 mg of $C_{11}H_8I_3N_2NaO_4$.

Iothalamate Sodium Injection, Iothalamate Meglumine and—*see* Iothalamate Meglumine and Iothalamate Sodium Injection

Iothalamic Acid

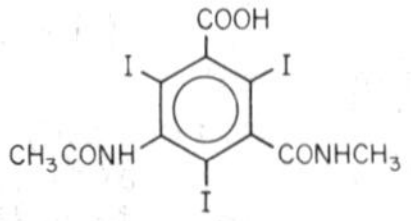

$C_{11}H_9I_3N_2O_4$ 613.92

Benzoic acid, 3-(acetylamino)-2,4,6-triiodo-5-[(methylamino)carbonyl]-.

5-Acetamido-2,4,6-triiodo-*N*-methylisophthalamic acid [*2276-90-6*].

» Iothalamic Acid contains not less than 98.0 percent and not more than 102.0 percent of $C_{11}H_9I_3N_2O_4$, calculated on the anhydrous basis.

Packaging and storage—Preserve in well-closed containers.

Reference standards—*USP 5-Amino-2,4,6-triiodo-N-methylisophthalamic Acid Reference Standard*—Dry at 105° for 4 hours before using. Keep container tightly closed and protected from light. *USP Iothalamic Acid Reference Standard*—Dry at 105° for 4 hours before using.

Identification—

A: The infrared absorption spectrum of a potassium bromide dispersion of it, previously dried at 105° for 4 hours, exhibits maxima only at the same wavelengths as that of a similar preparation of USP Iothalamic Acid RS.

B: Heat about 500 mg in a suitable crucible: violet vapors are evolved.

Water, *Method I* ⟨921⟩: not more than 1.0%.

Residue on ignition ⟨281⟩: not more than 0.1%.

Free aromatic amine—Dissolve 10.0 g of Iothalamic Acid in a minimal amount of 1 *N* sodium hydroxide in a 150-mL beaker, add 75 mL of water, and adjust with 1 *N* sulfuric acid to a pH of 7 ± 0.1. Transfer the solution to a 100-mL cylinder, dilute with water to 100 mL, and mix. Pipet 5 mL of this solution into a 50-mL volumetric flask, and add 10 mL of water. In another flask place 15 mL of water to provide a blank, and to a third flask add 12.5 mL of water and 2.5 mL of a Standard solution prepared as follows: Dissolve 25.0 mg of USP 5-Amino-2,4,6-triiodo-*N*-methylisophthalamic Acid RS, accurately weighed, in a mixture of 0.5 mL of 1 *N* sodium hydroxide and 2.5 mL of water in a 250-mL beaker, swirling to effect solution, then add 225 mL of water, mix, adjust with 1 *N* sulfuric acid to a pH of 7 ± 0.1, transfer to a 250-mL volumetric flask, add water to volume, and mix. Place the three flasks containing the solutions from the substance under test, the Standard solution, and the blank, respectively, in an ice bath. [NOTE—In conducting the following steps, keep the flasks in the ice bath and in the dark as much as possible, until all of the reagents have been added. Chill all reagents and the diluting water to about 5° prior to addition.] Treat each flask as follows: Add 5 mL of freshly prepared sodium nitrite solution (1 in 200), immediately add 10 mL of 1 *N* hydrochloric acid, and swirl gently to mix. [NOTE—Disregard any precipitate that may be formed at this point.] Allow to stand for 2 minutes, accurately timed. Add 10 mL of ammonium sulfamate solution (1 in 50), and shake frequently during 5 minutes. Five minutes after the addition of the ammonium sulfamate solution, add 3 drops of a 1 in 10 solution of 1-naphthol in alcohol. Mix, and allow to stand for 1 minute. Add 3.5 mL of a pH 10 buffer (made by dissolving 67.5 g of ammonium chloride in 300 mL of water, adding 570 mL of ammonium hydroxide, and diluting with water to 1 liter). Mix, remove from the ice bath, and immediately dilute, with water that has been chilled to 5°, to volume. Within 20 minutes of diluting the contents of all three flasks to 50 mL, concomitantly determine the absorbances of the test solution and the Standard solution in 1-cm cells at the wavelength of maximum absorbance at about 485 nm, with a suitable spectrophotometer, versus the prepared blank. The absorbance of the solution from the Iothalamic Acid is not greater than that of the Standard solution (0.05%).

Iodine and iodide—

Test preparation—To 10.0 g in a 50-mL beaker add 16 mL of 1 *N* sodium hydroxide, and stir until solution is complete. Dilute with water to about 35 mL, and adjust the solution to a pH of between 7.0 and 7.5 with 0.1 *N* sodium hydroxide or 0.1 *N* hydrochloric acid. Dilute with water to 50 mL.

Procedure—Dilute 10 mL of *Test preparation* with 20 mL of water in a 50-mL beaker, add 5 mL of 2 *N* sulfuric acid, stir, and filter into a glass-stoppered, 50-mL cylinder. To the filtrate add 5 mL of toluene, and shake well: the toluene layer shows no red color. Add 1 mL of sodium nitrite solution (1 in 50), and shake well: any red color in the toluene layer is not darker than that obtained when a mixture of 2 mL of potassium iodide solution (1 in 4000) and 22 mL of water is substituted for the solution under test (0.02% of iodide).

Heavy metals ⟨231⟩—To a 50-mL color-comparison tube transfer 5.0 mL of solution prepared as directed for *Test preparation* in the test for *Iodine and iodide*, add 5 mL of 1 *N* sodium hydroxide, dilute with water to 40 mL, and mix. Using this as the *Test preparation*, proceed as directed in the test for *Heavy metals* under *Diatrizoate Meglumine:* the limit is 0.002%.

Assay—Transfer about 400 mg of Iothalamic Acid, accurately weighed, to a glass-stoppered, 125-mL conical flask, add 12 mL of 5 *N* sodium hydroxide, 20 mL of water, and 1 g of powdered zinc, connect the flask to a reflux condenser, and reflux for 30 minutes. Cool the flask to room temperature, rinse the condenser with 20 mL of water, disconnect the flask from the condenser, and filter the mixture. Rinse the flask and the filter thoroughly, adding the rinsings to the filtrate. Add 40 mL of 2 *N* sulfuric acid, and titrate immediately with 0.05 *N* silver nitrate VS, determining the end-point potentiometrically, using silver-calomel electrodes and an agar–potassium nitrate salt bridge. Each mL of 0.05 *N* silver nitrate is equivalent to 10.23 mg of $C_{11}H_9I_3N_2O_4$.

Ipecac

» Ipecac consists of the dried rhizome and roots of *Cephaëlis acuminata* Karsten, or of *Cephaëlis ipecacuanha* (Brotero) A. Richard (Fam. Rubiaceae).

Ipecac yields not less than 2.0 percent of the total ether-soluble alkaloids of ipecac. Its content of emetine ($C_{29}H_{40}N_2O_4$) and cephaeline ($C_{28}H_{38}N_2O_4$) together is not less than 90.0 percent of the amount of the total ether-soluble alkaloids. The content of cephaeline varies from an amount equal to, to an amount not more than 2.5 times, the content of emetine.

Reference standard—*USP Emetine Hydrochloride Reference Standard*—Dry small portions only, well exposed, at 105° to constant weight before using.

Botanic characteristics—A mixture of segments of the roots and rhizomes. The root segments are mostly curved and flexuous, occasionally branched, up to 15 cm in length and usually from 3 mm to 6.5 mm in diameter, but may be up to 9 mm in diameter, grayish, grayish brown, or reddish brown, the reddish brown type often having light-colored abrasions, transverse ridges about 0.5 mm to 1.0 mm wide that extend about halfway around the circumference of the root and fade at their tapering extremities into the general surface, with from 1 to 6 of these ridges per cm, and annulations sometimes seen at irregular intervals. The rhizomes are cylindrical, about 2 mm thick, finely longitudinally wrinkled, with a few elliptical scars. The odor is distinctive; the dust is sternutatory. The taste is bitter, nauseating, and acrid.

Histology—At the center of the root is a well-defined primary xylem but no pith. Surrounding this is a dense wood of secondary xylem crossed by medullary rays. These elements are all lignified. External to the wood is a narrow band of secondary phloem and a wide parenchymatous phelloderm surrounded by a narrow layer of cork a few cells thick. The secondary xylem consists of narrow, bordered-pitted tracheidal vessels and tracheids in combination with xylem parenchyma. The latter have simple pits and contain starch grains. Starch is present also in the medullary rays. The phloem occurs as small groups of sieve tissue embedded in parenchyma. The wide phelloderm consists of round-celled cellulose parenchyma filled with starch grains and a few idioblasts, each of which contains a bundle of acicular raphides of calcium oxalate crystals about 30 µm to 80 µm long. The starch grains are rarely single but usually occur as 2 to 4 and sometimes 8 in a clump. Individual grains measure up to 22 µm in diameter.

The rhizome differs from the root in having a ring of xylem around a large pith. The pericycle contains characteristic sclerenchymatous cells. Spiral vessels are found in the protoxylem. The pith is composed of pitted parenchyma, which is somewhat lignified.

Overground stems—The proportion of overground stems does not exceed 5%.

Foreign organic matter ⟨561⟩—The proportion of foreign organic matter does not exceed 2.0%.

Assay for total ether-soluble alkaloids—[NOTE—It is important that the ether used in this assay shall have been shown by test to be free from peroxides within 24 hours prior to use.] The alkaloids may be extracted by either of the methods given in the following two paragraphs.

I—To 10 g of finely powdered Ipecac, in a suitable container, add 100 mL of ether, accurately measured at 25°, insert the stopper in the container tightly, shake the mixture thoroughly, and allow it to stand for 5 minutes. Then add 10 mL of 6 *N* ammonium hydroxide, close again tightly, shake it for 1 hour in a mechanical shaker or intermittently during 2 hours, and allow to stand overnight at a temperature not exceeding 25°. Again shake the mixture intermittently during 30 minutes, and allow the drug to settle at 25°. Transfer to a separator a 50.0-mL aliquot of the clear, supernatant liquid, representing 5 g of Ipecac.

II—Place 5 g of the finely powdered Ipecac in a continuous-extraction thimble. Add enough ether to cover the powder, and allow to stand for 10 minutes with occasional stirring. Add 3 mL of ammonium hydroxide, mix, and allow to stand overnight. Cover the drug with a pledget of cotton, pack well, and extract with ether for 5 hours. Transfer the ether extract to a separator.

Extract the alkaloids from the ether with 2 *N* sulfuric acid, using at first 15 mL, or more, if necessary, to ensure an acid reaction, then successive 10-mL portions until extraction is complete, and filtering all extracts through the same filter into a second separator. To the combined acid solutions add about an equal volume of ether, render the mixture distinctly alkaline with 6 *N* ammonium hydroxide (at least pH 10, by test paper), and extract with successive portions of ether until the last extract shows not more than a slight turbidity when treated as follows: Evaporate 1 mL of the last extraction, dissolve the residue in 0.5 mL of 0.5 *N* hydrochloric acid, and add 1 drop of mercuric iodide TS.

Filter each portion of the ether extract into a flask or beaker, and carefully evaporate the combined ether extracts on a steam bath almost to dryness. Add 5 mL of ether and 10.0 mL of 0.1 *N* sulfuric acid VS, and heat on a steam bath to effect complete solution of the alkaloids, and to remove all the ether. Cool, add 15 mL of water, then add methyl red TS, and titrate the excess acid with 0.1 *N* sodium hydroxide VS. Each mL of 0.1 *N* sulfuric acid is equivalent to 24.0 mg of the total ether-soluble alkaloids of ipecac, calculated as emetine ($C_{29}H_{40}N_2O_4$).

Assay for emetine and cephaeline—

Standard preparation—Weigh accurately a suitable quantity of USP Emetine Hydrochloride RS, and dissolve in 0.5 *N* sulfuric acid. Dilute quantitatively and stepwise with the same dilute sulfuric acid to obtain a solution having a known concentration equivalent to about 50 µg of emetine per mL.

Assay preparation—Prepare a *test sample* as directed under *Vegetable Drugs—Methods of Analysis* ⟨561⟩. Transfer to a 150-mL beaker about 200 mg, accurately weighed, of the fine powder. Add 2 mL of methyl sulfoxide, mix with a flattened stirring rod to assure complete wetting of the powder, and allow to stand for about 30 minutes. Add 2 mL of water and about 1 g of sodium bicarbonate, and mix.

Phosphate buffer—Prepare approximately 0.5 *M* solutions of monobasic potassium phosphate (containing 5.1 g per 75 mL) and dibasic potassium phosphate (containing 2.2 g per 25 mL). Mix 3 volumes of 0.5 *M* monobasic potassium phosphate with 1 volume of 0.5 *M* dibasic potassium phosphate, and adjust by the addition of one or the other of the solutions to a pH of 6.0 ± 0.05. Dissolve 7.5 g of potassium chloride in 100 mL of the resulting solution.

Citric acid buffer—Prepare approximately 0.5 *M* solutions of sodium citrate (containing 6.5 g per 50 mL) and citric acid (containing 4.8 g per 50 mL). Mix equal volumes of these solutions, and adjust by addition of one or the other of the solutions to a pH of 4.0 ± 0.05.

Chromatographic columns—For each column, pack a pledget of fine glass wool in the base of a chromatographic tube (25- × 200-mm test tube to which is fused a 5-cm length of 7-mm tubing) with the aid of a tamping rod having a disk with a diameter about 1 mm less than that of the tube.

Prepare *Column I* as follows: To the *Assay preparation* add 6 g of purified siliceous earth, mix, transfer the mixture to the column, scrub the beaker with about 1 g of the purified siliceous earth, transfer this to the top of the column, and tamp. Prepare *Column II* using 3 g of the purified siliceous earth and 2 mL of *Phosphate buffer;* prepare *Column III* using 2 mL of *Citric acid buffer* and 3 g of the purified siliceous earth; and prepare *Column IV* using 2 mL of sodium hydroxide solution (1 in 50) and 3 g of the purified siliceous earth. Pack a pledget of glass wool on the top of each column.

Procedure—[NOTES—Use water-saturated solvents throughout this procedure. Rinse the tips of the chromatographic columns before discarding them.] Mount *Columns I* and *II* so that the effluent from *Column I* flows onto *Column II*. Pass three 50-mL portions of ether through the columns, and discard *Column I* and the eluate. Mount *Column III* below *Column II* and pass three 50-mL portions of a mixture of 1 volume of ether and 3 volumes of chloroform through the columns. Discard *Column II* and the eluate. Wash *Column III* with 25 mL of the ether-chloroform mixture, followed by 25 mL of a mixture of equal volumes of ether and isooctane, and discard the washings. Wash *Column IV* with 20 mL of a 1 in 50 solution of triethylamine in the ether-isooctane mixture, and discard the washing. Mount *Column IV* below *Column III*, and place as a receiver under *Column IV* a 125-mL separator containing 15 mL of 4 *N* sulfuric acid. Pass through the columns 10 mL of a 1 in 5 solution of triethylamine in the ether-isooctane mixture, followed by three 10-mL portions of a 1 in 50 solution of triethylamine in the ether-isooctane mixture. Discard *Column III*, and pass through *Column IV* 20 mL of the 1 in 50 solution of triethylamine in the ether-isooctane mixture. Shake the separator, allow the phases to separate, and transfer the aqueous extract to a 50-mL volumetric flask. Extract with two additional 10-mL portions of 0.5 *N* sulfuric acid, combining the extracts in the volumetric flask. Add 0.5 *N* sulfuric acid to volume, and mix (*emetine solution*).

Elute *Column IV* with 75 mL of chloroform, collecting the eluate in a 250-mL separator containing 150 mL of ether. Discard *Column IV*. Extract with one 20-mL, and then with two 10-mL, portions of 0.5 *N* sulfuric acid, collecting the extracts in a 50-mL volumetric flask. Rinse the stem of the separator, add the acid to volume, and mix (*cephaeline solution*).

Concomitantly determine the absorbances of the *emetine solution*, the *cephaeline solution*, and the *Standard preparation* in 1-cm cells at the wavelength of maximum absorbance at about 283 nm and at 350 nm, with a suitable spectrophotometer, using 0.5 *N* sulfuric acid as the blank.

Calculate the quantity, in mg, of emetine in the portion of Ipecac taken by the formula:

$$0.05C(A_{283} - A_{350})_U/(A_{283} - A_{350})_S,$$

in which *C* is the concentration, in µg per mL, of emetine in the *Standard preparation*, and the parenthetic expressions are the differences in the absorbances of the solution of *emetine* from the *Assay preparation* ($_U$) and the *Standard preparation* ($_S$), respectively, at the wavelengths indicated by the subscripts.

Calculate the quantity, in mg, of cephaeline in the portion of Ipecac taken by the formula:

$$0.971(0.05C)(A_{283} - A_{350})_U/(A_{283} - A_{350})_S,$$

in which 0.971 is the ratio of the molecular weight of cephaeline to that of emetine, *C* is as defined in the preceding paragraph, and the parenthetic expressions are the differences in the absorbances of the solution of *cephaeline* from the *Assay preparation* ($_U$) and the *Standard preparation* ($_S$), respectively, at the wavelengths indicated by the subscripts.

Powdered Ipecac

» Powdered Ipecac is Ipecac reduced to a fine or a very fine powder and adjusted to a potency of not less than 1.9 percent and not more than 2.1 percent of the total ether-soluble alkaloids of ipecac, by the addition of exhausted marc of ipecac or of other suitable inert diluent or by the addition of powdered ipecac of either a lower or a higher potency.

The content of emetine ($C_{29}H_{40}N_2O_4$) and cephaeline ($C_{28}H_{38}N_2O_4$) together is not less than 90.0 percent of the total amount of the ether-soluble alkaloids. The content of cephaeline varies from an amount equal to, to an amount not more than 2.5 times, the content of emetine.

Packaging and storage—Preserve in tight containers.

Reference standard—*USP Emetine Hydrochloride Reference Standard*—Dry small portions only, well exposed, at 105° to constant weight before using.

Botanic characteristics—Thin-walled, fairly small cork cells, the starch grains rarely simple and usually 2- to 8-compound, the single grains up to 22 µm in diameter; raphides of calcium oxalate 30 to 80 µm in length; tracheids and tracheidal vessels found in groups having very numerous, small, bordered pits; parenchyma of phelloderm filled with starch or acicular crystals of calcium oxalate, having cells thin-walled, oval with intercellular spaces; parenchyma of the xylem composed of small rectangular and longitudinally elongated cells with moderately thick walls and scattered bordered or simple pits; rhizome parenchyma cells larger than root parenchyma cells, with slightly thicker walls and lignified with fairly numerous simple pits; sclereids from the rhizome large, rectangular, with uneven walls and large, conspicuous pits.

Assay for total ether-soluble alkaloids—Proceed with Powdered Ipecac as directed in the *Assay for total ether-soluble alkaloids* under *Ipecac*.

Assay for emetine and cephaeline—

Standard preparation, Phosphate buffer, Citric acid buffer, and *Chromatographic columns*—Prepare as directed in the *Assay for emetine and cephaeline* under *Ipecac*.

Assay preparation—Transfer to a 150-mL beaker about 200 mg, accurately weighed, of Powdered Ipecac. Add 2 mL of methyl sulfoxide, mix with a flattened stirring rod to assure complete wetting of the powder, and allow to stand for about 30 minutes. Add 2 mL of water and about 1 g of sodium bicarbonate, and mix.

Procedure—Proceed as directed for *Procedure* in the *Assay for emetine and cephaeline* under *Ipecac*.

Calculate the quantity, in mg, of emetine in the portion of Powdered Ipecac taken by the formula:

$$0.05C(A_{283} - A_{350})_U/(A_{283} - A_{350})_S,$$

in which the parenthetic expressions are the differences in the absorbances of the solution of emetine from the *Assay preparation* ($_U$) and the *Standard preparation* ($_S$), respectively, at the wavelengths indicated by the subscripts, and *C* is as defined in the *Procedure*.

Calculate the quantity, in mg, of cephaeline in the portion of Powdered Ipecac taken by the formula:

$$0.971(0.05C)(A_{283} - A_{350})_U/(A_{283} - A_{350})_S,$$

in which 0.971 is the ratio of the molecular weight of cephaeline to that of emetine, the parenthetic expressions are the differences in the absorbances of the solution of cephaeline from the *Assay preparation* (U) and the *Standard preparation* (S), respectively, at the wavelengths indicated by the subscripts, and C is as defined in the *Procedure*.

Ipecac Syrup

» Ipecac Syrup yields, from each 100 mL, not less than 123 mg and not more than 157 mg of the total ether-soluble alkaloids of ipecac.

The content of emetine ($C_{29}H_{40}N_2O_4$) and cephaeline ($C_{28}H_{38}N_2O_4$) together is not less than 90.0 percent of the amount of the total ether-soluble alkaloids. The content of cephaeline varies from an amount equal to, to an amount not more than 2.5 times, the content of emetine.

Powdered Ipecac	70 g
Glycerin	100 mL
Syrup, a sufficient quantity, to make	1000 mL

Exhaust the powdered Ipecac by percolation, using a mixture of 3 volumes of alcohol and 1 volume of water as the menstruum, macerating for 72 hours, and percolating slowly. Reduce the entire percolate to a volume of 70 mL by evaporation at a temperature not exceeding 60° and preferably in vacuum, and add 140 mL of water. Allow the mixture to stand overnight, filter, and wash the residue on the filter with water. Evaporate the filtrate and washings to 40 mL, and to this add 2.5 mL of hydrochloric acid and 20 mL of alcohol, mix, and filter. Wash the filter with a mixture of 30 volumes of alcohol, 3.5 volumes of hydrochloric acid, and 66.5 volumes of water, using a volume sufficient to produce 70 mL of the filtrate. Add 100 mL of Glycerin and enough Syrup to make the product measure 1000 mL, and mix.

Packaging and storage—Preserve in tight containers, preferably at a temperature not exceeding 25°. Containers intended for sale to the public without prescription contain not more than 30 mL of Ipecac Syrup.

Reference standard—*USP Emetine Hydrochloride Reference Standard*—Dry small portions only, well exposed, at 105° to constant weight before using.

Microbial limits—It meets the requirements of the tests for absence of *Escherichia coli* under *Microbial Limit Tests* ⟨61⟩.

Alcohol content ⟨611⟩: between 1.0% and 2.5% of C_2H_5OH.

Assay for total ether-soluble alkaloids—[NOTE—It is important that the ether used in this assay shall have been shown by test to be free from peroxides within 24 hours prior to use.] Transfer about 50 mL, accurately measured, of Ipecac Syrup to a liquid-liquid automatic extractor, add water, if necessary, to reduce the viscosity, render the liquid distinctly alkaline with ammonium hydroxide, and extract with ether for at least 4 hours or until the extraction is complete. Use a water bath to boil the ether. Frequently disconnect the extractor from the condenser, and agitate the lower layer by raising and lowering the center tube or by other suitable manipulation. At the conclusion of the extraction period, transfer the ether extract to a separator, and rinse the extraction flask with 2 or more small volumes of ether, adding the rinsings to the separator. Complete the assay as directed in the *Assay for total ether-soluble alkaloids* under *Ipecac*, beginning with "Extract the alkaloids from the ether."

Assay for emetine and cephaeline—

Standard preparation, *Phosphate buffer*, and *Citric acid buffer*—Prepare as directed in the *Assay for emetine and cephaeline* under *Ipecac*.

Assay preparation—Pipet 10 mL of water into a 25-mL volumetric flask. With the aid of a 20-mL pipet, add Ipecac Syrup to volume, taking care to prevent contact of the Syrup with the neck of the flask above the graduation line. Insert the stopper, and mix.

Chromatographic columns—Pack a pledget of fine glass wool in the base of a chromatographic tube (25-mm × 200-mm test tube to which is fused a 5-cm length of 7-mm tubing) with the aid of a tamping rod having a disk with a diameter about 1 mm less than that of the tube.

To prepare *Column I*, transfer 4.0 mL of *Assay preparation* to a 150-mL beaker, add about 1 g of sodium bicarbonate, and mix. Then proceed as directed for *Chromatographic columns* in the *Assay for emetine and cephaeline* under *Ipecac*, beginning with "add 6 g of purified siliceous earth," and prepare *Columns II*, *III*, and *IV* as directed therein.

Procedure—Proceed as directed for *Procedure* in the *Assay for emetine and cephaeline* under *Ipecac*.

Calculate the quantity, in mg, of emetine in each 100 mL of Ipecac Syrup by the formula:

$$2.08C(A_{283} - A_{350})_U/(A_{283} - A_{350})_S,$$

in which the parenthetic expressions are the differences in the absorbances of the solution of emetine from the *Assay preparation* (U) and the *Standard preparation* (S), respectively, at the wavelengths indicated by the subscripts, and C is as defined in the *Procedure*.

Calculate the quantity, in mg, of cephaeline in each 100 mL of Ipecac Syrup by the formula:

$$0.971(2.08C)(A_{283} - A_{350})_U/(A_{283} - A_{350})_S,$$

in which 0.971 is the ratio of the molecular weight of cephaeline to that of emetine, the parenthetic expressions are the differences in the absorbances of the solution of cephaeline from the *Assay preparation* (U) and the *Standard preparation* (S), respectively, at the wavelengths indicated by the subscripts, and C is as defined in the *Procedure*.

Ipodate Calcium

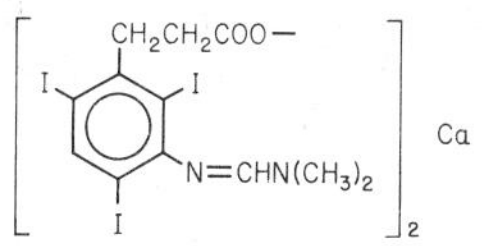

$C_{24}H_{24}CaI_6N_4O_4$ 1233.98

Benzenepropanoic acid, 3-[[(dimethylamino)methylene]amino]-2,4,6-triiodo-, calcium salt.

Calcium 3-[[(dimethylamino)methylene]amino]-2,4,6-triiodohydrocinnamate [*1151-11-7*].

» Ipodate Calcium contains not less than 97.5 percent and not more than 102.5 percent of $C_{24}H_{24}CaI_6N_4O_4$, calculated on the anhydrous basis.

Packaging and storage—Preserve in tight containers.

Reference standard—*USP Ipodate Calcium Reference Standard*—Dry in vacuum at 60° for 4 hours before using.

Identification—

A: The infrared absorption spectrum of a potassium bromide dispersion of it, previously dried in vacuum at 60° for 4 hours, exhibits maxima only at the same wavelengths as that of a similar preparation of USP Ipodate Calcium RS.

B: The ultraviolet absorption spectrum of a 1 in 100,000 solution in methanol exhibits maxima and minima at the same wavelengths as that of a similar preparation of USP Ipodate Calcium RS, concomitantly measured, and the respective absorptivities, calculated on the anhydrous basis, at the wavelength of maximum absorbance at about 235 nm do not differ by more than 3.0%.

C: Heat about 500 mg in a porcelain crucible over a flame: violet vapors of iodine are evolved.

D: Dissolve about 200 mg in 10 mL of 6 *N* acetic acid, and add 2 mL of ammonium oxalate TS: a white precipitate, which is soluble in 3 *N* hydrochloric acid, is formed.

Water, *Method I* ⟨921⟩: not more than 3.5%.

Iodide or iodine—Dissolve about 200 mg in 10 mL of 6 *N* acetic acid, add 1 mL of chloroform, and shake vigorously. Allow the layers to separate: the chloroform layer shows no violet color (absence of free iodine). Add 1 mL of 0.1 *N* potassium iodate, shake vigorously, and allow the layers to separate: the chloroform layer shows at most a slight trace of violet color.

Heavy metals, *Method II* ⟨231⟩: 0.003%.

Assay—Transfer about 300 mg of Ipodate Calcium, accurately weighed, to a glass-stoppered, 250-mL flask, add 30 mL of 1.25 *N* sodium hydroxide and 0.5 g of powdered zinc, and reflux the mixture for 60 minutes. Cool, wash the condenser with 20 mL of water, and filter the mixture. Wash the flask and the filter with small portions of water, adding the washings to the filtrate. Add to the filtrate 5 mL of glacial acetic acid and 3 drops of eosin Y TS, and titrate with 0.05 *N* silver nitrate VS until the entire mixture changes to a permanent pink color. Each mL of 0.05 *N* silver nitrate is equivalent to 10.28 mg of $C_{24}H_{24}CaI_6N_4O_4$.

Ipodate Calcium for Oral Suspension

» Ipodate Calcium for Oral Suspension is a dry mixture of Ipodate Calcium and one or more suitable suspending, dispersing, and flavoring agents. It contains not less than 85.0 percent and not more than 115.0 percent of the labeled amount of $C_{24}H_{24}CaI_6N_4O_4$.

Packaging and storage—Preserve in well-closed containers.

Reference standard—*USP Ipodate Calcium Reference Standard*—Dry in vacuum at 60° for 4 hours before using.

Identification—Stir an amount of it, equivalent to about 2 g of ipodate calcium, with 100 mL of water for several minutes, and filter the suspension with the aid of suction. Wash the filter with several small portions of water, and dry the residue in vacuum at 60° for 4 hours: the ipodate calcium so obtained responds to *Identification tests A, C,* and *D* under *Ipodate Calcium.*

Minimum fill ⟨755⟩: meets the requirements.

Assay—Weigh accurately the contents of not less than 10 containers of Ipodate Calcium for Oral Suspension, and mix. Transfer an accurately weighed portion of the granules, equivalent to about 300 mg of ipodate calcium, to a glass-stoppered, 250-mL flask, and proceed as directed in the *Assay* under *Ipodate Calcium,* beginning with "add 30 mL of 1.25 *N* sodium hydroxide."

Ipodate Sodium

CH_2CH_2COONa I I $N{=}CHN(CH_3)_2$ I

$C_{12}H_{12}I_3N_2NaO_2$ 619.94

Benzenepropanoic acid, 3-[[(dimethylamino)methylene]amino]-2,4,6-triiodo-, sodium salt.

Sodium 3-[[(dimethylamino)methylene]amino]-2,4,6-triiodohydrocinnamate [*1221-56-3*].

» Ipodate Sodium contains not less than 97.5 percent and not more than 102.5 percent of $C_{12}H_{12}I_3N_2NaO_2$, calculated on the dried basis.

Packaging and storage—Preserve in tight containers.

Reference standard—*USP Ipodate Sodium Reference Standard*—Dry in vacuum at 60° for 3 hours before using.

Identification—

A: The infrared absorption spectrum of a potassium bromide dispersion of it, previously dried, exhibits maxima only at the same wavelengths as that of a similar preparation of USP Ipodate Sodium RS.

B: The ultraviolet absorption spectrum of a 1 in 100,000 solution in methanol exhibits maxima and minima at the same wavelengths as that of a similar preparation of USP Ipodate Sodium RS, concomitantly measured, and the respective absorptivities, calculated on the dried basis, at the wavelength of maximum absorbance at about 235 nm do not differ by more than 3.0%.

C: Heat about 500 mg in a porcelain crucible over a free flame: violet vapors of iodine are evolved.

D: It responds to the flame test for *Sodium* ⟨191⟩.

Loss on drying ⟨731⟩—Dry it in vacuum at 60° for 3 hours: it loses not more than 0.5% of its weight.

Iodide or iodine—Dissolve about 200 mg in 10 mL of 6 *N* acetic acid, add 2 mL of 1 *N* sulfuric acid and 15 mL of chloroform, and shake vigorously. Allow the layers to separate: the chloroform layer shows not more than a faint violet color. Add 1 mL of 0.1 *N* potassium iodate, shake vigorously, and allow the layers to separate: the chloroform layer shows at most a slight trace of violet color.

Heavy metals, *Method II* ⟨231⟩: 0.003%.

Assay—Transfer about 300 mg of Ipodate Sodium, accurately weighed, to a 250-mL flask, add 30 mL of 1.25 *N* sodium hydroxide and 0.5 g of powdered zinc, and reflux the mixture for 60 minutes. Cool, wash the condenser with 20 mL of water, and filter the mixture. Wash the flask and the filter with small portions of water, adding the washings to the filtrate. Add to the filtrate 5 mL of glacial acetic acid and 3 mL of a mixture of 2 drops of nitric acid in 5 mL of water, then add 3 drops of eosin Y TS, and titrate with 0.05 *N* silver nitrate VS until the entire mixture changes to a permanent pink color. Each mL of 0.05 *N* silver nitrate is equivalent to 10.33 mg of $C_{12}H_{12}I_3N_2NaO_2$.

Ipodate Sodium Capsules

» Ipodate Sodium Capsules contain not less than 90.0 percent and not more than 110.0 percent of the labeled amount of $C_{12}H_{12}I_3N_2NaO_2$.

Packaging and storage—Preserve in tight containers.

Identification—

A: Transfer a portion of the contents of Capsules, equivalent to about 2 g of ipodate sodium, to a 250-mL separator, add 100 mL of water and 50 mL of solvent hexane, and shake. Transfer the aqueous layer to a beaker, add 5 mL of 3 *N* hydrochloric acid, and mix. Filter (retain the filtrate), and wash the precipitate with several portions of water. Dry the precipitate in vacuum at 60° for 4 hours. A 1 in 100,000 solution of the residue so obtained, in a 1 in 100 mixture of 2 *N* hydrochloric acid in methanol, exhibits an ultraviolet absorbance maximum at 242 ± 2 nm.

B: The residue obtained in *Identification test A* responds to *Identification test C* under *Ipodate Sodium.*

C: The filtrate obtained in *Identification test A* responds to the flame test for *Sodium* ⟨191⟩.

Uniformity of dosage units ⟨905⟩: meet the requirements.

Assay—Place a number of Ipodate Sodium Capsules, equivalent to about 5 g of ipodate sodium, in a 400-mL beaker, add 200 mL of 1 *N* sodium hydroxide and 50 mL of solvent hexane, and stir by mechanical means until the capsules have completely disintegrated. Transfer the mixture to a 500-mL separator, wash the beaker with a total of 25 mL of 1 *N* sodium hydroxide in divided portions, and add the washings to the separator. Allow the layers to separate, and transfer the aqueous layer to a 500-mL volumetric flask. Wash the solvent hexane layer with two 50-mL portions of 1 *N* sodium hydroxide, add the washings to the volumetric flask, dilute with 1 *N* sodium hydroxide to volume, and mix. Pipet 25 mL of the solution, which may be milky in appearance, into a 250-mL conical flask, add 500 mg of powdered zinc, and proceed as directed in the *Assay* under *Ipodate Sodium*, beginning with "and reflux the mixture for 60 minutes."

Ferrous Citrate Fe 59 Injection

$$\left[\begin{array}{c} CH_2COO- \\ | \\ HO-C-COO- \\ | \\ CH_2COO- \end{array}\right]_2 Fe_3$$

$C_{12}H_{10}{}^{59}Fe_3O_{14}$
1,2,3-Propanetricarboxylic acid, 2-hydroxy-, iron(2+)-^{59}*Fe* salt (2:3).
Iron(2+)-^{59}*Fe* citrate (3:2) [*64521-35-3*].

» Ferrous Citrate Fe 59 Injection is a sterile solution of radioactive iron (^{59}Fe) in the ferrous state and complexed with citrate ion in Water for Injection. It may contain added Sodium Chloride in an amount sufficient to render the solution isotonic and may contain bacteriostatic agents. Iron 59 is produced by the neutron bombardment of iron 58.

Ferrous Citrate Fe 59 Injection contains not less than 90.0 percent and not more than 110.0 percent of the labeled amount of ^{59}Fe expressed in megabecquerels (microcuries or in millicuries) per mL at the time indicated in the labeling. Its specific activity is not less than 185 megabecquerels (5 millicuries) per mg of ferrous citrate on the date of manufacture.

Packaging and storage—Preserve in single-dose or in multiple-dose containers.

Labeling—Label it to include the following, in addition to the information specified for *Labeling* under *Injections* ⟨1⟩: the date of calibration; the amount of ferrous citrate expressed as µg of Fe per mL; the amount of ^{59}Fe as ferrous citrate expressed as total megabecquerels (millicuries) and concentration as megabecquerels (millicuries) per mL on the date of calibration; the expiration date; and the statement, "Caution—Radioactive Material." The labeling indicates that correction is to be made for radioactive decay, and also indicates that the radioactive half-life of ^{59}Fe is 44.6 days.

Reference standard—*USP Endotoxin Reference Standard.*

Radionuclide identification (see *Radioactivity* ⟨821⟩)—Its gamma-ray spectrum is identical to that of a specimen of ^{59}Fe of known purity that exhibits major photopeaks having energies of 1.095 MeV and 1.292 MeV.

Bacterial endotoxins—It meets the requirements of the *Bacterial Endotoxins Test* ⟨85⟩, the limit of endotoxin content being not more than 175/*V* USP Endotoxin Unit per mL of the Injection, when compared with the USP Endotoxin RS, in which *V* is the maximum recommended total dose, in mL, at the expiration date or time.

pH ⟨791⟩: between 5.0 and 7.0.

Other requirements—It meets the requirements under *Injections* ⟨1⟩, except that it is not subject to the recommendation on *Volume in Container.*

Assay for radioactivity—Using a suitable counting assembly (see *Selection of a Counting Assembly* under *Radioactivity* ⟨821⟩), determine the radioactivity, in MBq (µCi) per mL, of Ferrous Citrate Fe 59 Injection by use of a calibrated system as directed under *Radioactivity* ⟨821⟩.

Iron Ascorbate Pentetic Acid Complex Injection, Technetium Tc 99m—*see* Technetium Tc 99m Ferpentetate Injection

Iron Dextran Injection

» Iron Dextran Injection is a sterile, colloidal solution of ferric hydroxide in complex with partially hydrolyzed Dextran of low molecular weight, in Water for Injection. It contains not less than 95.0 percent and not more than 105.0 percent of the labeled amount of iron. It may contain not more than 0.5 percent of phenol as a preservative.

Packaging and storage—Preserve in single-dose or in multiple-dose containers, preferably of Type I or Type II glass.

Identification—To 1 mL of Injection on a watch glass add 2 drops of ammonium hydroxide: no precipitate is formed. Add 2 mL of hydrochloric acid, mix, and add 2 mL of ammonium hydroxide: a brown precipitate is formed.

Pyrogen—It meets the requirements of the *Pyrogen Test* ⟨151⟩, the test dose being 5 mL per kg of a solution prepared by dilution of the Injection with pyrogen-free saline TS to obtain a solution containing 5 mg of iron per mL.

Acute toxicity—Select five mice each weighing between 18 and 25 g, maintained on an adequately balanced diet. Inject a dose of Iron Dextran Injection, equivalent to 200 mg of iron per kg of body weight, into a tail vein at a rate not exceeding 0.1 mL per second. Keep the mice under observation for 48 hours after the injection. If none of the mice show outward symptoms of toxicity, the requirements of the test are met. If any of the mice die within the observation period, select four groups of ten mice each weighing between 18 and 25 g. Inject, intravenously, all mice of one group with one of the following doses of Iron Dextran Injection: 375, 500, 750, or 1000 mg of iron per kg of body weight. Observe the mice for 7 days, and record the number of deaths in each group. If more than 16 mice die, calculate the LD_{50} as directed under *Design and Analysis of Biological Assays* ⟨111⟩: the LD_{50} is not less than 500 mg of iron per kg of body weight.

Absorption from injection site—Prepare a site over the semitendinosus muscle of one leg of each of two rabbits by clipping the fur and disinfecting the exposed skin. Inject each site with a dose of 0.4 mL per kg of body weight in the following manner: place the needle in the distal end of the semitendinosus muscle at an angle such as to ensure that the full length of the needle is used, then pass it through the sartorius and vastus medialis muscles. House the rabbits separately. Seven days after the injection, sacrifice the rabbits and dissect the treated legs to examine the muscles: no heavy black deposit of unabsorbed iron compounds is observed, and the tissue is only lightly colored.

pH ⟨791⟩: between 5.2 and 6.5.

Nonvolatile residue—Using a "to contain" pipet, transfer 1.0 mL onto 3 to 5 g of sand spread in a shallow layer in a stainless steel dish, the dish and sand having been previously dried and weighed. Rinse the pipet, with several small portions of water, onto the sand. Evaporate on a steam bath to dryness, continue the drying in an oven at 105° for 15 hours, and weigh: the weight of the residue for Injection labeled to contain 50 mg of iron per mL is not less than 28.0% and not more than 32.0%, that for Injection labeled to contain 75 mg of iron per mL is not less than 35.0% and not more than 40.0%, and that for Injection labeled to contain 100 mg of iron per mL is not less than 37.0% and not more than 43.0%.

Chloride content—Using a "to contain" pipet, transfer 10.0 mL of Injection into a 150-mL beaker, rinsing the pipet into the beaker with several small portions of water. Add 50 mL of water and 2 mL of nitric acid, mix, and titrate with 0.1 *N* silver nitrate VS, determining the end-point potentiometrically with silver-glass electrodes. Each mL of 0.1 *N* silver nitrate consumed is equivalent to 3.545 mg of chloride (Cl). The chloride content of Injection labeled to contain 50 mg of iron per mL is not less than 0.48% and not more than 0.68%, and that of Injection labeled to contain either 75 mg or 100 mg of iron per mL is not less than 0.8% and not more than 1.1%.

Phenol content—Proceed as directed for *Phenol* under *Antimicrobial Agents—Content* ⟨341⟩: not more than 0.5% is found.

Other requirements—It meets the requirements under *Injections* ⟨1⟩.

Assay for iron—

Iron stock solution—Transfer an accurately weighed quantity of about 350 mg of ferrous ammonium sulfate hexahydrate to a 1000-mL volumetric flask, add water to dissolve, dilute with water to volume, and mix to obtain a solution having a concentration of 50 µg of iron per mL.

Calcium chloride solution—Transfer 2.64 g of calcium chloride dihydrate to a 1000-mL volumetric flask, add 500 mL of water, and swirl to dissolve. Add 5.0 mL of hydrochloric acid, and dilute with water to volume.

Standard preparations—To separate 100-mL volumetric flasks transfer 2.0, 4.0, 6.0, 8.0, and 10.0 mL, respectively, of *Iron stock solution.* Dilute each flask with *Calcium chloride solution* to volume, and mix to obtain *Standard preparations* having known concentrations of 1.0, 2.0, 3.0, 4.0 and 5.0 µg of iron per mL.

Assay preparation—Using a "to contain" pipet, transfer an accurately measured volume of Iron Dextran Injection, equivalent to about 100 mg of iron, to a 200-mL volumetric flask. Dilute with *Calcium chloride solution* to volume, and mix. Pipet 2.0 mL of this solution into a 250-mL volumetric flask, dilute with *Calcium chloride solution* to volume, and mix.

Procedure—Concomitantly determine the absorbances of the *Standard preparations* and the *Assay preparation* at the iron emission line of 248.3-nm with a suitable atomic absorption spectrophotometer (see *Spectrophotometry and Light-scattering* ⟨851⟩) equipped with an iron hollow-cathode lamp and an air-acetylene flame, using the *Calcium chloride solution* as the blank. Plot the absorbance of each *Standard preparation* versus concentration, in µg per mL, of iron, and draw the straight line best fitting the five plotted points. From the graph so obtained, determine the concentration, in µg per mL, of iron in the *Assay preparation.* Calculate the quantity, in mg, of iron in each mL of the Injection taken by the formula:

$$25C/V,$$

in which C is the concentration, in µg per mL, of iron in the *Assay preparation* and V is the volume of Injection taken.

Iron Sorbitex Injection

» Iron Sorbitex Injection is a sterile solution of a complex of iron, Sorbitol, and Citric Acid that is stabilized with the aid of Dextrin and an excess of Sorbitol. It contains not less than 94.0 percent and not more than 104.0 percent of the labeled amount of iron.

Packaging and storage—Preserve in single-dose containers, preferably of Type I glass.

Labeling—Label it to indicate its expiration date, which is not more than 24 months after date of manufacture.

Identification—

A: To 1 mL of Injection add 5 mL of water and 1 mL of ammonium hydroxide: no precipitate is formed.

B: To 1 mL of Injection add 0.1 mL of 3 *N* hydrochloric acid and 0.5 mL of potassium ferrocyanide TS: a dark blue precipitate is formed.

C: To 1 mL of Injection in a separator add 4 mL of water and 10 mL of hydrochloric acid, and extract with three 15-mL portions of isopropyl ether. Dilute the aqueous layer with water to 25 mL, and mix. To 0.5 mL of this solution add 10 mL of water, 2 mL of 2 *N* sulfuric acid, and 10 mg of potassium periodate, allow to stand for 10 minutes, and then add 10 mL of 0.1 *N* sodium arsenite. When the solution is colorless, add 10 mL of a 1 in 250 solution of phenylhydrazine hydrochloride in 0.5 *N* hydrochloric acid. Allow to stand for 10 minutes, add 1 mL of potassium ferricyanide solution (1 in 20), allow to stand for 15 minutes, and then add 3 mL of hydrochloric acid: a wine-red color is produced (*presence of sorbitol*).

D: Dilute 1 mL of Injection with 50 mL of water, and to 4 mL of this solution add 1 mL of 6 *N* sulfuric acid and 0.5 mL of phosphoric acid, mix, and allow to stand for about 5 minutes or until decolorized. Add 1 mL of potassium bromide solution (1 in 10) and 1 mL of potassium permanganate solution (1 in 20). After 10 minutes, add hydrogen peroxide TS, dropwise, to discharge the pink color. Transfer the solution to a small separator, shake with 20 mL of solvent hexane, discard the water layer, and wash the hexane layer with 20 mL of water. To the washed hexane solution add 5 mL of a 1 in 25 solution of thiourea in sodium borate solution (1 in 50), and shake the mixture: the aqueous layer that separates shows a yellow color (*presence of citric acid*).

Specific gravity ⟨841⟩: between 1.17 and 1.19 at 20°.

Viscosity ⟨911⟩: between 8 and 13 centipoises, determined at 20° with a capillary tube viscometer.

Pyrogen—It meets the requirements of the *Pyrogen Test* ⟨151⟩, the test dose being 0.5 mL per kg.

pH ⟨791⟩: between 7.2 and 7.9.

Ferrous iron—Using a "to contain" pipet, transfer 10 mL of Injection to a glass-stoppered, 125-mL flask, rinsing the pipet into the flask with several small portions of water. Add 5 mL of sulfuric acid, shake vigorously until decolorization is effected, add 2 drops of orthophenanthroline TS, and immediately titrate with 0.1 *N* ceric sulfate VS. Each mL of 0.1 *N* ceric sulfate is equivalent to 5.585 mg of ferrous Fe. The ferrous iron content is not more than 8.5 mg per mL of Injection.

Other requirements—It meets the requirements under *Injections* ⟨1⟩.

Assay for iron—

Standard preparation—Transfer an accurately weighed portion of ferrous ammonium sulfate, equivalent to about 100 mg of iron (Fe), to a 300-mL Kjeldahl flask, and proceed as directed under *Assay preparation*, beginning with "Add 10 mL of nitric acid," to obtain a *Standard preparation* having a known concentration of about 2 µg of iron per mL.

Assay preparation—Using a "to contain" pipet, transfer an accurately measured volume of Iron Sorbitex Injection, equivalent to about 100 mg of iron, to a 300-mL Kjeldahl flask, rinsing the pipet into the flask with several small portions of water. Add 10 mL of nitric acid, 10 mL of sulfuric acid, and a few glass beads, and boil the solution gently until fumes of sulfur trioxide appear. Cool, add 3 mL of nitric acid, and heat gently again until fumes of sulfur trioxide appear. Continue the addition of nitric acid, followed by gentle boiling, until the solution is clear and light yellow or green in color, and then boil for an additional 30 minutes. Cool, cautiously add 100 mL of water, and boil gently until solution is complete. Cool, transfer the solution to a 500-mL volumetric flask, rinse the Kjeldahl flask with several small portions of water, dilute with water to volume, and mix. Transfer 5.0 mL of the solution to a second 500-mL volumetric flask, add 100 mL of water and 1 g of ascorbic acid, dilute with water to volume, and mix.

Procedure—Transfer 5.0-mL portions of the *Standard preparation*, of the *Assay preparation*, and of water to serve as the blank to separate, clean, dry test tubes, and to each add 3.0 mL of a 1 in 1500 solution of 2,2′-bipyridine in 0.6 *N* glacial acetic acid, mix, and allow to stand for 15 minutes. Concomitantly determine the absorbances of the solutions in 1-cm cells at the wavelength of maximum absorbance at about 510 nm, with a suitable spectrophotometer, using the blank to set the instrument. Calculate the quantity, in mg, of iron in each mL of the Injection taken by the formula:

$$(50C/V)(A_U/A_S),$$

in which C is the concentration, in µg per mL, of the *Standard preparation*, V is the volume, in mL, of Injection taken, and A_U and A_S are the absorbances of the solutions from the *Assay preparation* and *Standard preparation*, respectively.

Irrigation—*see complete list in index*

Isobutane—*see* Isobutane NF

Isocarboxazid

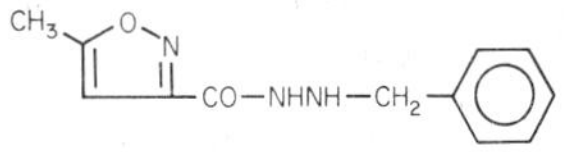

$C_{12}H_{13}N_3O_2$ 231.25

3-Isoxazolecarboxylic acid, 5-methyl-, 2-(phenylmethyl)-hydrazide.

5-Methyl-3-isoxazolecarboxylic acid 2-benzylhydrazide [59-63-2].

» Isocarboxazid contains not less than 98.5 percent and not more than 100.5 percent of $C_{12}H_{13}N_3O_2$, calculated on the dried basis.

Packaging and storage—Preserve in well-closed containers.

Reference standards—*USP Isocarboxazid Reference Standard*—Dry in vacuum over phosphorus pentoxide at 60° for 4 hours before using. *USP Methyl 5-Methyl-3-isoxazolecarboxylate Reference Standard*—Keep container tightly closed and protected from light. Dry over silica gel for 4 hours before using. *USP 1-Benzyl-3-methyl-5-aminopyrazole Hydrochloride Reference Standard*—Keep container tightly closed and protected from light. Dry over silica gel for 4 hours before using.

Identification—

A: The infrared absorption spectrum of a potassium bromide dispersion of it, previously dried, exhibits maxima only at the same wavelengths as that of a similar preparation of USP Isocarboxazid RS.

B: Dissolve about 10 mg in 10 mL of acetone, add 0.2 mL of water, 0.2 mL of ammonium molybdate solution (prepared by dissolving 100 mg of ammonium molybdate in 10 mL of 3 *N* hydrochloric acid), and mix: an orange color develops.

C: Dissolve about 15 mg in alcohol, and add 1 mL of a *p*-dimethylaminobenzaldehyde solution (prepared by dissolving 100 mg of *p*-dimethylaminobenzaldehyde in 1 mL of hydrochloric acid and diluting to 100 mL with alcohol): a yellow color appears.

Melting range, *Class I* ⟨741⟩: between 105° and 108°.

Loss on drying ⟨731⟩—Dry it in vacuum over phosphorus pentoxide at 60° for 4 hours: it loses not more than 0.3% of its weight.

Residue on ignition ⟨281⟩: not more than 0.1%.

Chloride ⟨221⟩—Boil 100 mg with 3 mL of 30 percent hydrogen peroxide, 5 mL of 2 *N* sodium hydroxide, and 7 mL of water for 2 minutes, cool, add water to make a total volume of 30 to 40 mL, and neutralize the solution to litmus with nitric acid: the solution shows no more chloride than corresponds to 0.30 mL of 0.020 *N* hydrochloric acid (0.02%).

Limit of methyl 5-methyl-3-isoxazolecarboxylate and 1-benzyl-3-methyl-5-aminopyrazole—

Standard methyl 5-methyl-3-isoxazolecarboxylate preparation—Transfer 12.5 mg of USP Methyl 5-Methyl-3-isoxazolecarboxylate RS to a 50-mL volumetric flask, add methanol to volume, and mix.

Standard 1-benzyl-3-methyl-5-aminopyrazole preparation—Dissolve 12.5 mg of USP 1-Benzyl-3-methyl-5-aminopyrazole Hydrochloride RS in 50.0 mL of methanol, add 1 g of sodium carbonate, shake for 2 minutes, and filter. Use the filtrate as directed in the *Procedure*.

Standard isocarboxazid preparation—Prepare a solution of USP Isocarboxazid RS, accurately weighed, in methanol containing approximately 50 mg per mL.

Test preparation—Dissolve 500.0 mg of Isocarboxazid in methanol to make 10.0 mL, and mix.

Procedure—Apply 20 µL each of the *Standard isocarboxazid preparation*, the *Standard methyl 5-methyl-3-isoxazolecarboxylate preparation*, the *Standard 1-benzyl-3-methyl-5-aminopyrazole preparation*, and the *Test preparation* at separate locations on a suitable thin-layer chromatographic plate (see *Chromatography* ⟨621⟩), coated with a 250-µm layer of chromatographic silica gel mixture. Develop the chromatogram in a solvent system consisting of ethyl acetate and *n*-heptane (3:2) until the solvent front has moved about three-fourths of the length of the plate. Remove the plate from the developing chamber, and examine the chromatogram under short-wavelength ultraviolet light: any spot produced by the *Test preparation* at about R_f 0.85 does not exceed in magnitude or intensity the spot produced by the *Standard methyl 5-methyl-3-isoxazolecarboxylate preparation* at the same R_f value (0.5%). Spray the plate with a freshly prepared mixture consisting of equal volumes of ferric chloride solution (1 in 10) and potassium ferricyanide solution (1 in 5): any spot produced by the *Test preparation* at about R_f 0.25 does not exceed in magnitude or intensity the spot produced by the *Standard 1-benzyl-3-methyl-5-aminopyrazole preparation* at the same R_f value (0.5%). The *Standard isocarboxazid preparation* produces a spot at about R_f 0.6.

Assay—Dissolve about 700 mg of Isocarboxazid, accurately weighed, in 20 mL of glacial acetic acid, add 20 mL of hydrochloric acid and 40 mL of water, and cool to room temperature. Titrate with 0.1 *M* sodium nitrite VS, determining the end-point potentiometrically, using suitable electrodes. Each mL of 0.1 *M* sodium nitrite is equivalent to 23.13 mg of $C_{12}H_{13}N_3O_2$.

Isocarboxazid Tablets

» Isocarboxazid Tablets contain not less than 95.0 percent and not more than 105.0 percent of the labeled amount of $C_{12}H_{13}N_3O_2$.

Packaging and storage—Preserve in well-closed, light-resistant containers.

Reference standard—*USP Isocarboxazid Reference Standard*—Dry in vacuum over phosphorus pentoxide at 60° for 4 hours before using.

Identification—Proceed as directed for *Identification—Organic Nitrogenous Bases* ⟨181⟩, chloroform being used for extraction instead of carbon disulfide. Instead of determining the spectra of the solutions, evaporate portions of the chloroform solutions under an efficient fume hood: the infrared absorption spectrum of a mineral oil dispersion of the residue obtained from the test specimen exhibits maxima only at the same wavelengths as that of a similar preparation of the residue obtained from the Reference Standard.

Dissolution ⟨711⟩—

Medium: pH 7.5 phosphate buffer (see *Buffer Solutions* in the section, *Reagents, Indicators, and Solutions*); 900 mL.

Apparatus 2: 100 rpm.

Time: 45 minutes.

Procedure—Determine the amount of $C_{12}H_{13}N_3O_2$ dissolved from ultraviolet absorbances at the wavelength of maximum absorbance at about 232 nm of filtered portions of the solution under test, suitably diluted with pH 7.5 phosphate buffer, in comparison with a Standard solution having a known concentration of USP Isocarboxazid RS in the same medium.

Tolerances—Not less than 75% (*Q*) of the labeled amount of $C_{12}H_{13}N_3O_2$ is dissolved in 45 minutes.

Uniformity of dosage units ⟨905⟩: meet the requirements.

Assay—

Ammonium molybdate reagent—Dissolve 1 g of ammonium molybdate in 100 mL of 3 *N* hydrochloric acid.

Standard preparation—Dissolve a suitable quantity of USP Isocarboxazid RS, accurately weighed, in acetone, and dilute quantitatively with acetone to obtain a solution having a known concentration of about 200 µg per mL.

Assay preparation—Weigh and finely powder not less than 20 Isocarboxazid Tablets. Transfer an accurately weighed portion of the powder, equivalent to about 10 mg of isocarboxazid, to a 50-mL volumetric flask, add acetone to volume, and mix. Shake well, centrifuge to clarify, and use the clear supernatant liquid as directed in the *Procedure*.

Procedure—Transfer to separate 100-mL volumetric flasks 5.0 mL each of the *Standard preparation*, the *Assay preparation*, and acetone to provide the blank. To each flask add 1.0 mL of water, 50.0 mL of acetone, and 1.0 mL of *Ammonium molybdate reagent*. Insert the stoppers in the flasks, mix, and allow to stand for 30 minutes with occasional swirling. Concomitantly determine the absorbances of the solutions in 1-cm cells at the wavelength of maximum absorbance at about 420 nm, with a suitable spectrophotometer, using the blank to set the instrument. Calculate the quantity, in mg, of $C_{12}H_{13}N_3O_2$ in the portion of Tablets taken by the formula:

$$0.05C(A_U/A_S),$$

in which C is the concentration, in µg per mL, of USP Isocarboxazid RS in the *Standard preparation*, and A_U and A_S are the absorbances of the solutions from the *Assay preparation* and the *Standard preparation*, respectively.

Isoetharine Inhalation Solution

» Isoetharine Inhalation Solution is a solution of Isoetharine Hydrochloride in Purified Water. It may contain Sodium Chloride. It contains not less than 92.0 percent and not more than 108.0 percent of the labeled amount of $C_{13}H_{21}NO_3.HCl$.

Packaging and storage—Preserve in small, well-filled, tight containers, protected from light.

Reference standard—*USP Isoetharine Hydrochloride Reference Standard*—Dry at 100° for 4 hours before using.

Identification—Dilute the Inhalation Solution with water to obtain a solution containing about 2.5 mg of isoetharine hydrochloride per mL. Apply 10-µL portions of this solution and a solution of USP Isoetharine Hydrochloride RS containing 2.5 mg per mL on a thin-layer chromatographic plate coated with silica gel mixture. Develop the plate in *n*-butyl alcohol, water, and formic acid (70:20:10) to a height of 12 to 14 cm above the point of application. Remove the plate, and evaporate the solvents with the aid of warm, circulating air. Examine under short-wavelength ultraviolet light. Spray the plate with Folin-Ciocalteu phenol TS, and then expose to ammonia vapor until the isoetharine spots develop an intense blue color: the R_f value and color of the principal spot obtained from the solution under test correspond to those obtained from the Standard solution.

pH ⟨791⟩: between 2.5 and 5.5.

Assay—

Standard preparation—Prepare as directed in the *Assay* under *Isoetharine Hydrochloride.*

Assay preparation—Transfer an accurately measured volume of Isoetharine Inhalation Solution, equivalent to about 50 mg of isoetharine hydrochloride, to a 100-mL volumetric flask, dilute with 0.17 *N* acetic acid solution to volume, and mix.

Chromatographic system—Proceed as directed in the *Assay* under *Isoetharine Hydrochloride.*

Procedure—Proceed as directed for *Procedure* in the *Assay* under *Isoetharine Hydrochloride.* Calculate the quantity, in mg, of $C_{13}H_{21}NO_3.HCl$ in each mL of the Inhalation Solution taken by the formula:

$$0.1(C/V)(h_U/h_S),$$

in which V is the volume, in mL, of Inhalation Solution taken, and C, h_U, and h_S are as defined therein.

Isoetharine Hydrochloride

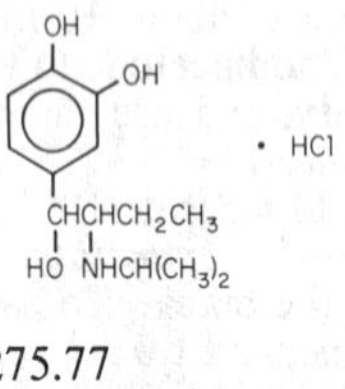

$C_{13}H_{21}NO_3.HCl$ 275.77

1,2-Benzenediol, 4-[1-hydroxy-2-[(1-methylethyl)amino]butyl]-, hydrochloride.

3,4-Dihydroxy-α-[1-(isopropylamino)propyl]benzyl alcohol hydrochloride [*2576-92-3*].

» Isoetharine Hydrochloride contains not less than 97.0 percent and not more than 102.0 percent of $C_{13}H_{21}NO_3.HCl$, calculated on the dried basis.

Packaging and storage—Preserve in tight containers.

Reference standard—*USP Isoetharine Hydrochloride Reference Standard*—Dry at 100° for 4 hours before using.

Identification—

A: The infrared absorption spectrum of a potassium bromide dispersion of it, previously dried, exhibits maxima only at the same wavelengths as that of a similar preparation of USP Isoetharine Hydrochloride RS.

B: A solution (1 in 100) responds to the tests for *Chloride* ⟨191⟩.

pH ⟨791⟩: between 4.0 and 5.6, in a solution (1 in 100).

Loss on drying ⟨731⟩—Dry it at 100° for 4 hours: it loses not more than 1.0% of its weight.

Aromatic ketones—Its absorptivity (see *Spectrophotometry and Light-scattering* ⟨851⟩) at 312 nm, determined in a solution in 0.01 *N* hydrochloric acid containing 2.0 mg per mL, is not more than 0.20.

Assay—

Standard preparation—Dissolve an accurately weighed quantity of USP Isoetharine Hydrochloride RS in freshly prepared sodium bisulfite solution (3 in 1000) to obtain a solution having a concentration of about 5 mg per mL. Transfer 5.0 mL of this solution to a 50-mL volumetric flask, dilute with 0.17 *N* acetic acid to volume, and mix to obtain a solution having a known concentration of about 500 µg per mL.

Assay preparation—Transfer about 125 mg of Isoetharine Hydrochloride, accurately weighed, to a 25-mL volumetric flask, dissolve in sodium bisulfite solution (3 in 1000), dilute with sodium bisulfite solution to volume, and mix. Transfer 5.0 mL of this solution to a 50-mL volumetric flask, dilute with 0.17 *N* acetic acid to volume, and mix.

Chromatographic system (see *Chromatography* ⟨621⟩)—The liquid chromatograph is equipped with a 278-nm detector and a 4-mm × 30-cm column that contains packing L1. The mobile phase is 0.17 *N* acetic acid, having a flow rate of about 2.2 mL per minute. Chromatograph five replicate injections of the *Standard preparation*, and record the peak responses as directed under *Procedure:* the relative standard deviation is not more than 3.0%.

Procedure—Using a microsyringe or sampling valve, chromatograph 10 µL of the *Standard preparation*, and adjust the specimen size and other operating parameters, if necessary, until satisfactory chromatography and peak responses are obtained. Chromatograph equal volumes of the *Standard preparation* and the *Assay preparation*, record the chromatograms, and measure the peak responses. Calculate the quantity, in mg, of $C_{13}H_{21}NO_3.HCl$ in the portion of Isoetharine Hydrochloride taken by the formula:

$$0.25C(h_U/h_S),$$

in which C is the concentration, in µg per mL, of USP Isoetharine Hydrochloride RS in the *Standard preparation*, and h_U and h_S are the peak responses obtained from the *Assay preparation* and the *Standard preparation*, respectively.

Isoetharine Hydrochloride Inhalation—*see* Isoetharine Inhalation Solution

Isoetharine Mesylate

$C_{13}H_{21}NO_3.CH_4O_3S$ 335.41

1,2-Benzenediol, 4-[1-hydroxy-2-[(1-methylethyl)amino]butyl]-, methanesulfonate (salt).

3,4-Dihydroxy-α-[1-(isopropylamino)propyl]benzyl alcohol methanesulfonate (salt) [*7279-75-6*].

» Isoetharine Mesylate contains not less than 97.0 percent and not more than 102.0 percent of $C_{13}H_{21}NO_3 . CH_4O_3S$, calculated on the dried basis.

Packaging and storage—Preserve in tight containers.

Reference standard—*USP Isoetharine Hydrochloride Reference Standard*—Dry at 100° for 4 hours before using.

Identification—

A: It responds to the *Thin-layer Chromatographic Identification Test* ⟨201⟩, the test solution and the Standard solution of USP Isoetharine Hydrochloride RS being prepared at a concentration of 2.5 mg per mL in methanol, the solvent mixture being *n*-butanol, water, and formic acid (64:25:11), and the spots being located by spraying with sodium hydroxide solution (1 in 10).

B: Mix about 50 mg with about 200 mg of powdered sodium hydroxide, transfer the mixture to a small test tube, heat in a small flame to fusion, and continue the heating for a few minutes longer. Cool, add about 0.5 mL of water, then add a moderate excess of hydrochloric acid, and warm: sulfur dioxide, recognizable by its odor, is evolved, and starch iodate paper placed over the mouth of the test tube turns blue.

Melting range ⟨741⟩: between 162° and 168°.

pH ⟨791⟩: between 4.5 and 5.5, in a solution (1 in 100).

Loss on drying ⟨731⟩—Dry it at 80° under vacuum at a pressure of not more than 5 mm of mercury for 4 hours: it loses not more than 1.0% of its weight.

Keto precursor—Its absorptivity (see *Spectrophotometry and Light-scattering* ⟨851⟩) at 312 nm, determined in a solution in 0.01 *N* hydrochloric acid containing 2.0 mg per mL, is not more than 0.20.

Assay—

Standard preparation—Transfer about 60 mg of USP Isoetharine Hydrochloride RS, accurately weighed, to a 25-mL volumetric flask, add 4.0 mL of alcohol, and mix. Add 3 drops of 1 *N* hydrochloric acid, dilute with water to volume, and mix.

Assay preparation—Transfer about 75 mg of Isoetharine Mesylate, accurately weighed, to a 25-mL volumetric flask, add 4.0 mL of alcohol, and mix. Add 3 drops of 1 *N* hydrochloric acid, dilute with water to volume, and mix.

Chromatographic system (see *Chromatography* ⟨621⟩)—The liquid chromatograph is equipped with a 254-nm detector and a 25-cm × 4.6-mm stainless steel column that contains packing L9. The mobile phase consists of 8.0 mL of glacial acetic acid, 100 mL of 1 *M* sodium sulfate, and water to make 1000 mL, and has a flow rate of about 1.5 mL per minute. Chromatograph five replicate injections of the *Standard preparation*, and record the peak responses as directed under *Procedure:* the relative standard deviation is not more than 3.0%.

Procedure—Using a microsyringe or sampling valve, chromatograph 20 µL of the *Standard preparation*, and adjust the specimen size and other operating parameters, if necessary, until satisfactory chromatography and peak responses are obtained. Chromatograph equal volumes of the *Standard preparation* and the *Assay preparation*, and measure the peak responses. Calculate the quantity, in mg, of $C_{13}H_{21}NO_3 . CH_4O_3S$ in the portion of Isoetharine Mesylate taken by the formula:

$$0.025C(335.41/275.77)(h_U/h_S),$$

in which C is the concentration, in µg per mL, of USP Isoetharine Hydrochloride RS in the *Standard preparation*, 335.41 and 275.77 are the molecular weights of isoetharine mesylate and isoetharine hydrochloride, respectively, and h_U and h_S are the peak responses obtained from the *Assay preparation* and the *Standard preparation*, respectively.

Isoetharine Mesylate Inhalation Aerosol

» Isoetharine Mesylate Inhalation Aerosol is a solution of Isoetharine Mesylate in Alcohol in an inert propellant base. It contains not less than 90.0 percent and not more than 110.0 percent of the labeled amount of $C_{13}H_{21}NO_3 . CH_4O_3S$. The quantity of $C_{13}H_{21}NO_3 . CH_4O_3S$ delivered through an oral inhalation actuator is not less than 75.0 percent and not more than 125.0 percent of the labeled dose per inhalation.

Packaging and storage—Preserve in small, nonreactive, light-resistant, aerosol containers equipped with metered-dose valves and provided with oral inhalation actuators.

Reference standard—*USP Isoetharine Hydrochloride Reference Standard*—Dry at 100° for 4 hours before using.

Identification—

A: Expel a quantity of Inhalation Aerosol, equivalent to about 12 mg of isoetharine mesylate, into 2 mL of methanol, dilute with methanol to 5 mL, and mix: this solution responds to *Identification test A* under *Isoetharine Hydrochloride Inhalation.*

B: Expel a quantity of Inhalation Aerosol, equivalent to about 12 mg of isoetharine mesylate, into a test tube, evaporate on a steam bath just to dryness, and add 50 mg of powdered sodium hydroxide. Heat in a small flame to fusion, and continue heating for a few seconds longer. Cool, add about 0.5 mL of water, then add a moderate excess of 3 *N* hydrochloric acid: sulfur dioxide, recognizable by its odor, is evolved, and starch iodate paper placed over the mouth of the test tube turns blue.

Alcohol content ⟨611⟩—Weigh accurately a filled Isoetharine Mesylate Inhalation Aerosol container, and record the weight. Invert the container, and place the outlet tip against the bottom of a 50-mL beaker containing 5 mL of water. Slowly actuate the valve 10 times. Raise the unit above the contents of the beaker, and wash the outlet with 1 mL of water. Collect the washings in the beaker. Dip the outlet stem in alcohol, shake to remove the solvent completely, air-dry the valve, and again weigh the Inhalation Aerosol container. Record the weight of the expelled specimen. Transfer the contents of the beaker, with the aid of 4 mL of water, to a glass-stoppered graduated cylinder. Determine the alcohol content of the *Test solution* thus prepared by the gas-liquid chromatographic procedure, 2 mL of dilute isopropyl alcohol (15 in 100) being used as the internal standard. Calculate the alcohol content of the Inhalation Aerosol taken by the formula:

$$SV/W,$$

in which S is the percentage (w/v) of alcohol in the *Test solution*, V is the total volume, in mL, of the *Test solution*, and W is the weight, in g, of the expelled specimen taken: between 25.9% and 35.0% (w/w) of C_2H_5OH is found.

Unit spray content—

Unit spray-sampling apparatus—Prepare as directed under *Aerosols* ⟨601⟩.

Ferro-citrate solution and *Buffer solution*—Prepare as directed under *Epinephrine Assay* ⟨391⟩.

Standard preparation—Transfer about 15 mg of USP Isoetharine Hydrochloride RS, accurately weighed, to a 100-mL volumetric flask containing 100 mg of sodium bisulfite and about 50 mL of water. Dilute with water to volume, and mix, to obtain a solution having a concentration of about 150 µg per mL.

Test preparation—Transfer 30.0 mL of freshly prepared sodium bisulfite solution (1 in 500) to the collection chamber of the apparatus, and clamp the intake tube in a horizontal position. Fit the aerosol container with its oral inhalation actuator, and prime the unit by alternately shaking and firing it 10 times. Couple the unit to the intake tube of the apparatus, and adjust the airflow to a rate of 10 to 15 liters per minute. Remove the container, leaving the actuator in place. Shake the container, seat it in the actuator, and immediately deliver a single spray into the system. Deliver in this manner a number of sprays, equivalent to 5 mg of isoetharine mesylate. Remove the container and actuator from the assembly, clamp the intake tube in a vertical position, and scavenge the drug remaining in the intake system as follows: Disconnect the collection chamber, transfer the test solution to a small beaker, and wash the inside of the intake tube with the test solution, collecting the washing in the collection chamber placed a few cm below its normal position. Connect the collection chamber into the system again, and draw the test solution into it through the bubbler and intake tube. Recycle the test solution twice in this manner, then add 15 mL of chloroform to the collection chamber, insert the stopper, and shake vigorously for 5 minutes. Use the clear supernatant solution as directed in the *Procedure*.

Procedure—Into three separate 25-mL volumetric flasks pipet 10 mL each of the *Test preparation*, the *Standard preparation*, and sodium bisulfite solution (1 in 1000) to provide the blank, respectively. To each add 0.5 mL of *Ferrocitrate solution* followed by 5 mL of *Buffer solution*. Dilute with sodium bisulfite solution (1 in 1000) to volume, and allow the color to develop for 10 minutes. Determine the absorbances of the *Test preparation* and the *Standard preparation* relative to the blank at the wavelength of maximum absorbance at about 530 nm, with a suitable spectrophotometer. Calculate the quantity, in µg, of $C_{13}H_{21}NO_3 . CH_4O_3S$ delivered per spray taken by the formula:

$$(30C/N)(335.41/275.77)(A_U/A_S),$$

in which C is the concentration, in µg per mL, of USP Isoetharine Hydrochloride RS in the *Standard preparation*, N is the total number of sprays delivered, 335.41 and 275.77 are the molecular weights of isoetharine mesylate and isoetharine hydrochloride, respectively, and A_U and A_S are the absorbances of the solutions from the *Test preparation* and the *Standard preparation*, respectively.

Other requirements—It meets the requirements for *Leak Testing* under *Aerosols* ⟨601⟩.

Assay—

Standard preparation—Prepare as directed for *Standard preparation* in the *Assay* under *Isoetharine Mesylate*.

Assay preparation—Weigh 1 Isoetharine Mesylate Inhalation Aerosol container with contents. Invert the container, and place the outlet tip against the bottom of a 50-mL beaker containing 2.5 mL of 0.01 *N* hydrochloric acid. Slowly actuate the valve about 90 times (the weight of the assay specimen is approximately 5 g). Raise the unit above the contents of the beaker, and wash the outlet with a few mL of water. Collect the washings in the beaker. Dip the outlet stem in alcohol, shake to remove the solvent completely, air-dry the valve, and then again weigh the Inhalation Aerosol container. Record the weight of the expelled specimen. Transfer the contents of the beaker to a 100-mL volumetric flask with the aid of water, dilute with water to volume, and mix.

Procedure—Proceed as directed for *Procedure* in the *Assay* under *Isoetharine Mesylate*. Calculate the percentage of $C_{13}H_{21}NO_3 . CH_4O_3S$ in the portion of Inhalation Aerosol taken by the formula:

$$0.01(335.41/275.77)(C/W)(h_U/h_S),$$

in which W is the weight, in g, of Inhalation Aerosol taken, and the other terms are as defined therein.

Isoflurane

CF_3—CHCl—O—CHF_2

$C_3H_2ClF_5O$ 184.49

Ethane, 2-chloro-2-(difluoromethoxy)-1,1,1-trifluoro-.

1-Chloro-2,2,2-trifluoroethyl difluoromethyl ether [*26675-46-7*].

» Isoflurane contains not less than 99.0 percent and not more than 101.0 percent of $C_3H_2ClF_5O$, calculated on the anhydrous basis.

Packaging and storage—Preserve in tight, light-resistant containers.

Reference standard—*USP Isoflurane Reference Standard*—Do not dry; determine the water content titrimetrically at the time of use for quantitative analyses.

Identification—The infrared absorption spectrum of it obtained using a gas cell exhibits maxima only at the same wavelengths as that of a similar preparation of USP Isoflurane RS.

Refractive index ⟨831⟩: between 1.2990 and 1.3005, at 20°.

Water, *Method I* ⟨921⟩: not more than 0.14%.

Chloride—Pipet 10 mL into a suitable vessel containing 60 mL of isopropyl alcohol and 4 drops of dilute nitric acid (1:1), and stir to dissolve. Titrate potentiometrically with 0.0020 *N* silver nitrate: not more than 2.11 mL is required (0.001%).

Nonvolatile residue—Transfer 10.0 mL to a suitable weighed evaporating dish, evaporate with the aid of a stream of air to dryness, and dry the residue at 50° for 2 hours: the weight of the residue does not exceed 2.0 mg (0.02%).

Fluoride—Transfer 10 mL to a suitable separator, add 10 mL of ammonium hydroxide solution (1 in 20), shake for 1 minute, withdraw the aqueous ammonia layer, and transfer it to a suitable color-comparison tube. Repeat the extraction with two 10-mL portions of ammonium hydroxide solution (1 in 20), and combine the aqueous ammonia layers. Add 30 mL of ammonium hydroxide solution (1 in 20) to a second color-comparison tube to provide a blank, and treat both tubes as follows: Add 50 mL of water, 6 drops of sodium alizarinsulfonate solution (1 in 2000), and slowly add 3 *N* hydrochloric acid until the violet color changes to clear yellow. Adjust the solutions to a pH of 3.0 ± 0.1, using ammonium hydroxide solution (1 in 20) or 0.1 *N* hydrochloric acid.

Titrate the preparation under test with a solution of thorium nitrate, prepared by dissolving 100 mg of thorium nitrate in 1 liter of water, to a tannish pink end-point, and dilute with water to 100 mL. Add the same volume of thorium nitrate solution to the blank preparation, and dilute with water to 100 mL: the pink color of the blank preparation is not darker than that of the test preparation.

If the blank preparation exhibits a darker color than the test preparation, proceed as follows: Titrate the blank preparation with a solution of sodium fluoride having a known concentration of 20 µg per mL of fluoride until the color of the blank preparation corresponds to the color of the test preparation: not more than 15 mL of sodium fluoride is required (0.001%).

Assay—

Chromatographic system (see *Chromatography* ⟨621⟩)—The gas chromatograph is equipped with a thermal conductivity detector and a 2.4-mm (ID) × 3.7-m stainless steel column packed with 10 percent phase G31 and 15 percent phase G18 on support S1C. The column temperature is programmed from 50° to 125° at a rate of 4° per minute, and helium is used as the carrier gas at a flow rate of about 50 mL per minute. The injection port is maintained at about 50° and the detector is maintained at about 215°. Chromatograph five replicate injections of the *Standard preparation*, and record the peak responses as directed under *Procedure:* the relative standard deviation is not more than 2.0% and the tailing factor is not more than 1.5.

Procedure—Separately inject equal volumes (about 20 µL) of USP Isoflurane RS (*Standard preparation*) and the specimen under test (*Assay preparation*) into the chromatograph, record the chromatograms, and measure the responses of the major peaks. Calculate the quantity, in µL, of Isoflurane in the portion taken by the formula:

$$20(r_U/r_S),$$

in which r_U and r_S are the peak responses obtained from the *Assay preparation* and the *Standard preparation*, respectively.

Isoflurophate

$(CH_3)_2CHO{-}P(F)(=O){-}OCH(CH_3)_2$

$C_6H_{14}FO_3P$ 184.15

Phosphorofluoridic acid, bis(1-methylethyl) ester.

Diisopropyl phosphorofluoridate [*55-91-4*].

» Isoflurophate contains not less than 95.0 percent of $C_6H_{14}FO_3P$.

Packaging and storage—Preserve in glass, fuse-sealed containers, or in other suitable sealed containers, in a cool place.

Labeling—Label it to indicate that in the handling of Isoflurophate in open containers, the eyes, nose, and mouth are to be protected with a suitable mask, and contact with the skin is to be avoided.

Identification—

A: *Under a hood with a good draft* place a few drops of Isoflurophate in a small platinum crucible, quickly add 2 mL of sulfuric acid, and immediately cover the crucible with a small, clear watch glass. Allow to stand for 10 minutes, then heat on a steam bath for 5 to 10 minutes: the side of the watch glass exposed to the mixture is visibly etched.

B: Slowly heat *under the hood* the crucible and contents from *Identification test A* until copious white fumes are evolved. Cool, place the crucible in a beaker, and add a sufficient quantity of water nearly to cover the crucible. After a few minutes remove the crucible from the beaker with the aid of a glass rod, add 3 mL of nitric acid, and boil for a few minutes. Cool, cautiously add 6 *N* ammonium hydroxide with stirring until a slight odor of ammonia persists, then add 1 mL of nitric acid. Filter the liquid if not clear, warm to about 40°, and add about 10 mL of ammonium molybdate TS: a yellow precipitate, which is soluble in 6 *N* ammonium hydroxide, is formed.

Acidity—

Mixed indicator—Mix 3 volumes of a 1 in 1000 solution of bromocresol green in alcohol and 1 volume of a 1 in 500 solution of methyl red in alcohol.

0.1 N Sodium hydroxide in dehydrated alcohol—Dissolve about 0.4 g of sodium hydroxide in 100 mL of dehydrated alcohol, and when solution is complete, standardize the alcoholic solution as follows: Pipet 25.0 mL of 0.1 *N* hydrochloric acid VS into a suitable container. Dilute with 50 mL of water, add 2 drops of *Mixed indicator*, and titrate with the alcoholic sodium hydroxide solution to the first appearance of a green color. Calculate the normality. [NOTE—Store in tightly stoppered bottles.]

Test preparation—Prepare as directed in the *Assay*, except to draw 1.0 mL instead of 0.5 mL of Isoflurophate into the bulb.

Procedure—Carefully place the bulb containing the *Test preparation* in a 250-mL flask containing 20 mL of dehydrated alcohol. Break the bulb as directed in the *Procedure* under *Ionic fluorine*. Rinse the glass tube with 30 mL of dehydrated alcohol, and titrate immediately with *0.1 N Sodium hydroxide in dehydrated alcohol*, using the *Mixed indicator*, to the appearance of the first green color. Each mL of *0.1 N Sodium hydroxide in dehydrated alcohol* is equivalent to 100.8 µg of hydrogen ion (*acidity*). The limit is 0.01%.

Ionic fluorine—

Sodium methoxide solution—Dissolve 10 g of sodium methoxide in dehydrated alcohol to make 500 mL, and mix.

Buffer solution—Dissolve 9.55 g of monochloroacetic acid and 2 g of sodium hydroxide in water to make 100 mL. If necessary, adjust by the addition, of either reagent to obtain a solution having a pH of 3.0.

Standard fluoride solution—Dissolve 2.2105 g of sodium fluoride in water to make 1000.0 mL. Each mL is equivalent to 1.00 mg of fluoride ion.

Thorium nitrate solution—Dissolve 9 g of thorium nitrate in water to make 1000.0 mL, and mix. Standardize as directed under *Standard curve*.

Standard curve—Into each of four 180-mL beakers pipet 50.0 mL of *Sodium methoxide solution* and 0.25, 0.50, 1.0, and 2.0 mL, respectively, of *Standard fluoride solution*. Treat each beaker in the same manner, as follows: Add 2 drops of phenolphthalein TS, and render just acid with 6 *N* hydrochloric acid. Add 1.0 mL of sodium alizarinsulfonate solution (1 in 2000), and add 6 *N* hydrochloric acid dropwise until the pink color is discharged. Dilute with water to 100 mL, and add *Buffer solution* (approximately 4 mL) until the pH is 3.1. Titrate with *Thorium nitrate solution*, while stirring constantly and rapidly, to a permanent pink color. During the titration, maintain the solution at a pH between 2.9 and 3.1 by adding small volumes of *Buffer solution*, if necessary, but not more than a total of 10 mL. Plot the mg of fluoride ion versus the mL of *Thorium nitrate solution* consumed.

Test preparation—Prepare as directed in the *Assay*, except to draw 1.0 mL instead of 0.5 mL of Isoflurophate into the bulb.

Procedure—Place 50.0 mL of *Sodium methoxide solution* in a 180-mL beaker, and add 2 drops of phenolphthalein TS. Acidify, dropwise, with 6 *N* hydrochloric acid. Add 1.0 mL of sodium alizarinsulfonate solution (1 in 2000), then add 6 *N* hydrochloric acid dropwise until the pink color disappears. Add 0.5 *N* sodium hydroxide until a faint pink color appears, then add 0.05 *N* hydrochloric acid until the pink color just disappears. Add 4 mL of the *Buffer solution*. Carefully place the *Test preparation* in the beaker with the stem of the bulb inserted in a suitable length of glass tubing. Break the bulb by pressing down on the glass tubing, making sure that the bulb is beneath the surface of the liquid and that the bottom of the beaker is properly supported so that it will not break when the bulb is broken. Wash down the glass tube with water, and dilute with water to 100 mL. Titrate immediately with the *Thorium nitrate solution*. Determine the mg of ionic fluorine present in the *Test preparation* directly from the thorium nitrate standardization curve. Not more than 0.15% of ionic fluorine is found.

Assay—

Solvent—Use dry carbon disulfide, chromatographic grade.

Internal standard solution—Pipet 1.0 mL of chromatographic grade cyclohexanone into a 100-mL volumetric flask, dilute with *Solvent* to volume, and mix. Pipet 3.0 mL of the resulting solution into a 100-mL volumetric flask, dilute with *Solvent* to volume, and mix. Each mL of the *Internal standard solution* contains 0.30 µL of cyclohexanone.

Standard preparation—Dissolve a suitable quantity of Isoflurophate, previously subjected to the *Assay*, in peanut oil, and dilute quantitatively and stepwise with peanut oil to obtain a solution having a known concentration of about 0.8 mg of isoflurophate per g of solution. Transfer about 1.2 g of this Isoflurophate solution in peanut oil, accurately weighed, to a 10-mL volumetric flask, pipet 1.0 mL of *Internal standard solution* into the flask, dilute with *Solvent* to volume, and mix.

Assay preparation—Tare an unsealed, thin-walled glass bulb with a thin, long stem, having a capacity of 1 to 2 mL. *Under a hood*, open the Isoflurophate container and place it in a firmly based container in a suitable vacuum-filtration flask. Insert the stem of the bulb under the surface of the liquid, and insert the stopper in the filtration flask. Allow about 0.5 mL of liquid to be drawn up into the bulb, and release the vacuum. Remove the bulb from the container, wipe the stem clean, fire-seal it without loss of any glass, cool, and again weigh.

Place the glass bulb in a 125-mL conical flask, add about 70 mL of *Solvent*, and break the bulb with the aid of a glass rod by pressing down on the glass tubing over the neck of the bulb. Take care to assure that the bulb is beneath the surface of the liquid in the flask and that the bottom of the flask is properly supported so that it will not break when the bulb is broken. Remove the rod, transfer the solution to a 100-mL volumetric flask, dilute with *Solvent* to volume, and mix (*Solution A*). Pipet 2.0 mL of *Solution A* into a 10-mL volumetric flask, dilute with *Solvent* to volume, and mix (*Solution B*). Pipet 1.0 mL of *Solution B* into another 10-mL volumetric flask, pipet 1.0 mL of *Internal standard solution* into the flask, dilute with *Solvent* to volume, and mix.

Chromatographic system (see *Chromatography* ⟨621⟩)—Under typical conditions, the gas chromatograph is equipped with a flame-ionization detector, and contains a 1.8-m × 4-mm glass column packed with 5 percent phase G33 on 80- to 100-mesh support S1AB, utilizing either a glass-lined sample introduction system or on-column injection. The column is maintained isothermally at a temperature between 75° and 80°, and the injector port and detector block are maintained at 200° and 250°, respectively; dry, oxygen-free helium is used as the carrier gas at a flow rate adjusted to obtain a cyclohexanone peak about 6 minutes after sample introduction.

Procedure—Inject 6 µL of the *Standard preparation* into a suitable gas chromatograph, and record the chromatogram. Measure the areas under the first (cyclohexanone) and second (isoflurophate) peaks, and record the values as A_D and A_S, respectively. Calculate the factor *F* by the formula:

$$(A_D/A_S)(W_S/10)(C/1000),$$

in which W_S is the weight, in mg, of Isoflurophate solution in peanut oil in the *Standard preparation*, and *C* is the weight, in mg, of isoflurophate per g of Isoflurophate solution in peanut oil. Similarly inject 6 µL of the *Assay preparation*, and record the chromatogram. Measure the areas under the first (cyclohexanone) and second (isoflurophate) peaks, and record the values as a_D and a_U, respectively. Calculate the percentage of $C_6H_{14}FO_3P$ in the portion of Isoflurophate taken by the formula:

$$(F)(a_U/a_D)(100/W_U)(5000),$$

in which W_U is the weight, in mg, of Isoflurophate in the *Assay preparation*, and F is the factor as determined above.

Isoflurophate Ophthalmic Ointment

» Isoflurophate Ophthalmic Ointment contains not less than 0.0225 percent and not more than 0.0275 percent of $C_6H_{14}FO_3P$, in a suitable anhydrous ointment base. It is sterile.

Packaging and storage—Preserve in collapsible ophthalmic ointment tubes.

Labeling—Label it to indicate the expiration date, which is not later than 2 years after date of manufacture.

Identification—Place about 100 mg of Ointment in one eye of each of 3 rabbits, and examine the eyes 18 to 20 hours later: the average diameter of the pupils of the treated eyes is not less than 2 mm smaller than the average diameter of the pupils of the untreated eyes.

Irritation—The conjunctivas of the eyes treated as directed in the *Identification test*, as compared with those of the untreated eyes, after 1 hour, show not more than a slight reddening, which practically disappears in 4 hours.

Sterility—It meets the requirements under *Sterility Tests* ⟨71⟩.

Minimum fill ⟨755⟩: meets the requirements.

Water, *Method I* ⟨921⟩—Dissolve about 10 g, accurately weighed, in a mixture of 25 mL each of methanol and toluene. Not more than 0.03% is found.

Metal particles—It meets the requirements of the test for *Metal Particles in Ophthalmic Ointments* ⟨751⟩.

Assay—

Solvent, Internal standard solution, Standard preparation, and *Chromatographic system*—Prepare as directed in the *Assay* under *Isoflurophate.*

Assay preparation—Transfer about 3.5 g of Isoflurophate Ophthalmic Ointment, accurately weighed, to a 50-mL centrifuge tube. Add 9 mL of *Solvent* and 1.0 mL of *Internal standard solution*, shake, and centrifuge. The bottom layer is the *Assay preparation.*

Procedure—Proceed as directed for *Procedure* in the *Assay* under *Isoflurophate.*

Isoleucine

H H
CH_3CH_2—C—C—COOH
CH_3 NH_2

$C_6H_{13}NO_2$ 131.17
L-Isoleucine.
L-Isoleucine [*73-32-5*].

» Isoleucine contains not less than 98.5 percent and not more than 101.5 percent of $C_6H_{13}NO_2$, as L-isoleucine, calculated on the dried basis.

Packaging and storage—Preserve in well-closed containers.

Reference standard—*USP L-Isoleucine Reference Standard*—Dry at 105° for 3 hours before using.

Identification—The infrared absorption spectrum of a potassium bromide dispersion of it, previously dried, exhibits maxima only at the same wavelengths as that of a similar preparation of USP L-Isoleucine RS.

Specific rotation ⟨781⟩: between +38.9° and +41.8°, calculated on the dried basis, determined in a solution in 6 *N* hydrochloric acid containing 0.40 g in each 10.0 mL.

pH ⟨791⟩: between 5.5 and 7.0, in a solution (1 in 100).

Loss on drying ⟨731⟩—Dry it at 105° for 3 hours: it loses not more than 0.3% of its weight.

Residue on ignition ⟨281⟩: not more than 0.3%.

Chloride ⟨221⟩—A 0.73-g portion shows no more chloride than corresponds to 0.50 mL of 0.020 *N* hydrochloric acid (0.05%).

Sulfate ⟨221⟩—A 0.33-g portion shows no more sulfate than corresponds to 0.10 mL of 0.020 *N* sulfuric acid (0.03%).

Arsenic ⟨211⟩: 1.5 ppm.

Iron ⟨241⟩: 0.003%.

Heavy metals, *Method I* ⟨231⟩: 0.0015%.

Assay—Transfer about 130 mg of Isoleucine, accurately weighed, to a 125-mL flask, dissolve in a mixture of 3 mL of formic acid and 50 mL of glacial acetic acid, and titrate with 0.1 *N* perchloric acid VS, determining the end-point potentiometrically. Perform a blank determination, and make any necessary correction. Each mL of 0.1 *N* perchloric acid is equivalent to 13.12 mg of $C_6H_{13}NO_2$.

Isoniazid

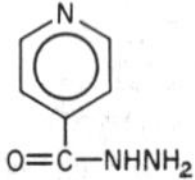

$C_6H_7N_3O$ 137.14
4-Pyridinecarboxylic acid, hydrazide.
Isonicotinic acid hydrazide [*54-85-3*].

» Isoniazid contains not less than 98.0 percent and not more than 102.0 percent of $C_6H_7N_3O$, calculated on the dried basis.

Packaging and storage—Preserve in tight, light-resistant containers.

Reference standard—*USP Isoniazid Reference Standard*—Dry at 105° for 4 hours before using.

Identification—

A: The infrared absorption spectrum of a potassium bromide dispersion of it exhibits maxima only at the same wavelengths as that of a similar preparation of USP Isoniazid RS.

B: Transfer about 50 mg of it to a 500-mL volumetric flask, add water to volume, and mix. Transfer 10.0 mL of the resulting solution to a 100-mL volumetric flask, add 2.0 mL of 0.1 *N* hydrochloric acid, dilute with water to volume, and mix to obtain a 1 in 100,000 solution: the ultraviolet absorption spectrum of the solution so obtained exhibits maxima and minima only at the same wavelengths as that of a similar solution of USP Isoniazid RS, concomitantly measured.

Melting range ⟨741⟩: between 170° and 173°.

pH ⟨791⟩: between 6.0 and 7.5, in a solution (1 in 10).

Loss on drying ⟨731⟩—Dry it at 105° for 4 hours: it loses not more than 1.0% of its weight.

Residue on ignition ⟨281⟩: not more than 0.2%.

Heavy metals, *Method II* ⟨231⟩: 0.002%.

Assay—

Mobile phase—Dissolve 4.4 g of docusate sodium in 600 mL of methanol, add 400 mL of water, adjust with 2 *N* sulfuric acid to a pH of 2.5, and mix. Make adjustments if necessary (see *System Suitability* under *Chromatography* ⟨621⟩).

Standard preparation—Dissolve an accurately weighed quantity of USP Isoniazid RS in *Mobile phase*, and dilute quantitatively with *Mobile phase* to obtain a solution having a known concentration of about 0.32 mg per mL.

Assay preparation—Transfer about 16 mg of Isoniazid, accurately weighed, to a 50-mL volumetric flask, dissolve in *Mobile phase*, dilute with *Mobile phase* to volume, and mix.

Chromatographic system (see *Chromatography* ⟨621⟩)—The liquid chromatograph is equipped with a 254-nm detector and a 4.6-mm × 25-cm column that contains packing L1. The flow rate is about 1.5 mL per minute. Chromatograph the *Standard preparation*, and record the peak responses as directed under *Procedure:* the column efficiency determined from the isoniazid peak is not less than 1800 theoretical plates, the tailing factor for the isoniazid peak is not more than 2.0, and the relative standard deviation for replicate injections is not more than 2.0%.

Procedure—Separately inject equal volumes (about 10 µL) of the *Standard preparation* and the *Assay preparation* into the chromatograph, record the chromatograms, and measure the responses for the major peaks. Calculate the quantity, in mg, of $C_6H_7N_3O$ in the portion of Isoniazid taken by the formula:

$$50C(r_U/r_S),$$

in which C is the concentration, in mg per mL, of USP Isoniazid RS in the *Standard preparation*, and r_U and r_S are the peak responses of isoniazid obtained from the *Assay preparation* and the *Standard preparation*, respectively.

Isoniazid Capsules, Rifampin and—*see* Rifampin and Isoniazid Capsules

Isoniazid Injection

» Isoniazid Injection is a sterile solution of Isoniazid in Water for Injection. It contains not less than 90.0 percent and not more than 110.0 percent of the labeled amount of $C_6H_7N_3O$.

Packaging and storage—Preserve in single-dose or in multiple-dose containers, preferably of Type I glass, protected from light.

Labeling—Its package label states that if crystallization has occurred, the Injection should be warmed to redissolve the crystals prior to use.

Reference standard—*USP Isoniazid Reference Standard*—Dry at 105° for 4 hours before using.

Identification—

A: The retention time exhibited by isoniazid in the chromatogram of the *Assay preparation* corresponds to that of isoniazid in the chromatogram of the *Standard preparation*, as obtained in the *Assay*.

B: A volume of Injection, equivalent to about 50 mg of isoniazid, responds to *Identification test B* under *Isoniazid*.

pH ⟨791⟩: between 6.0 and 7.0.

Other requirements—It meets the requirements under *Injections* ⟨1⟩.

Assay—

Mobile phase, *Standard preparation*, and *Chromatographic system*—Proceed as directed in the *Assay* under *Isoniazid*.

Assay preparation—Transfer an accurately measured volume of Isoniazid Injection, equivalent to about 100 mg of isoniazid, to a 50-mL volumetric flask. Dilute with *Mobile phase* to volume, and mix. Transfer 8.0 mL of this solution to a 50-mL volumetric flask, dilute with *Mobile phase* to volume, and mix.

Procedure—Proceed as directed for **Procedure** in the *Assay* under *Isoniazid*. Calculate the quantity, in mg, of $C_6H_7N_3O$ in each mL of the Injection taken by the formula:

$$312.5(C/V)(r_U/r_S),$$

in which C is the concentration, in mg per mL, of USP Isoniazid RS in the *Standard preparation*, V is the volume, in mL, of Injection taken, and r_U and r_S are the peak responses of isoniazid obtained from the *Assay preparation* and the *Standard preparation*, respectively.

Isoniazid Syrup

» Isoniazid Syrup contains, in each 100 mL, not less than 0.93 g and not more than 1.10 g of $C_6H_7N_3O$.

Packaging and storage—Preserve in tight, light-resistant containers.

Reference standard—*USP Isoniazid Reference Standard*—Dry at 105° for 4 hours before using.

Identification—A volume of Syrup equivalent to about 50 mg of isoniazid responds to *Identification test B* under *Isoniazid*.

Assay—Transfer an accurately measured volume of Isoniazid Syrup, equivalent to about 100 mg of isoniazid, to a 100-mL beaker. Add 50 mL of a mixture of 1 part of potassium bromide in 10 parts of dilute hydrochloric acid (1 in 6), and proceed as directed under *Nitrite Titration* ⟨451⟩, beginning with "cool to 15°." Each mL of 0.1 *M* sodium nitrite is equivalent to 13.71 mg of $C_6H_7N_3O$.

Isoniazid Tablets

» Isoniazid Tablets contain not less than 90.0 percent and not more than 110.0 percent of the labeled amount of $C_6H_7N_3O$.

Packaging and storage—Preserve in well-closed, light-resistant containers.

Reference standard—*USP Isoniazid Reference Standard*—Dry at 105° for 4 hours before using.

Identification—

A: The retention time exhibited by isoniazid in the chromatogram of the *Assay preparation* corresponds to that of isoniazid in the chromatogram of the *Standard preparation*, as obtained in the *Assay*.

B: Transfer a portion of finely powdered Tablets, equivalent to about 50 mg of isoniazid, to a 500-mL volumetric flask. Add water to volume, mix, and filter a portion of the mixture. Proceed as directed in *Identification test B* under *Isoniazid*, beginning with "Transfer 10.0 mL of the resulting solution to a 100-mL volumetric flask."

Dissolution ⟨711⟩—

Medium: 0.1 *N* hydrochloric acid; 900 mL.

Apparatus 1: 100 rpm.

Time: 45 minutes.

Procedure—Determine the amount of $C_6H_7N_3O$ dissolved from ultraviolet absorbances at the wavelength of maximum absorbance at about 263 nm of filtered portions of the solution under test, suitably diluted with water, in comparison with a Standard solution having a known concentration of USP Isoniazid RS in the same medium.

Tolerances—Not less than 80% (Q) of the labeled amount of $C_6H_7N_3O$ is dissolved in 45 minutes.

Uniformity of dosage units ⟨905⟩: meet the requirements.

Procedure for content uniformity—Transfer 1 finely powdered Tablet to a 500-mL volumetric flask with the aid of 200 mL of water. Shake by mechanical means for 30 minutes, add water to volume, and mix. Filter, and discard the first 20 mL of the filtrate. Dilute a portion of the filtrate quantitatively and stepwise, if necessary, with a 3 in 100 mixture of 0.1 *N* hydrochloric acid and water to obtain a solution containing about 10 µg per mL. Dissolve an accurately weighed quantity of USP Isoniazid RS in a volume of water corresponding to that used to dissolve a similar amount of isoniazid from the Tablet, and dilute quantitatively and stepwise, if necessary, with a 3 in 100 mixture of 0.1 *N* hydrochloric acid and water to obtain a Standard solution having a known concentration of about 10 µg per mL. Concomitantly determine the absorbances of both solutions in 1-cm cells at the wavelength of maximum absorbance at about 263 nm, with a suitable spectrophotometer, using water as the blank. Calculate the quantity, in mg, of $C_6H_7N_3O$ in the Tablet by the formula:

$$(TC/D)(A_U/A_S),$$

in which T is the labeled quantity, in mg, of isoniazid in the Tablet, C is the concentration, in µg per mL, of USP Isoniazid RS in the Standard solution, D is the concentration, in µg per mL, of isoniazid in the solution from the Tablet, based on the labeled quantity per Tablet and the extent of dilution, and A_U and A_S are the absorbances of the solution from the Tablet and the Standard solution, respectively.

Assay—

Mobile phase, *Standard preparation*, and *Chromatographic system*—Proceed as directed in the *Assay* under *Isoniazid*.

Assay preparation—Weigh and finely powder not less than 20 Isoniazid Tablets. Weigh accurately a portion of the powder, equivalent to about 80 mg of isoniazid, and transfer with the aid of about 80 mL of *Mobile phase* to a 100-mL volumetric flask. Shake by mechanical means for 30 minutes, dilute with *Mobile phase* to volume, mix, and filter, discarding the first 10 mL of the filtrate. Pipet 10 mL of the filtrate into a 25-mL volumetric flask, dilute with *Mobile phase* to volume, and mix.

Procedure—Proceed as directed for *Procedure* in the *Assay* under *Isoniazid*. Calculate the quantity, in mg, of $C_6H_7N_3O$ in the portion of Tablets taken by the formula:

$$250C(r_U/r_S),$$

in which C is the concentration, in mg per mL, of USP Isoniazid RS in the *Standard preparation*, and r_U and r_S are the peak responses of isoniazid obtained from the *Assay preparation* and the *Standard preparation*, respectively.

Isophane Insulin Suspension—*see* Insulin Suspension, Isophane

Isopropamide Iodide

$(CH_3)_2CH$, CH_3—N^+—$CH_2CH_2CCONH_2$ (with two phenyl rings), $(CH_3)_2CH$; I^-

$C_{23}H_{33}IN_2O$ 480.43

Benzenepropanaminium, γ-(aminocarbonyl)-*N*-methyl-*N*,*N*-bis(1-methylethyl)-γ-phenyl-, iodide.

(3-Carbamoyl-3,3-diphenylpropyl)diisopropylmethylammonium iodide [*71-81-8*].

» Isopropamide Iodide, dried in vacuum at 60° for 2 hours, contains not less than 98.0 percent and not more than 101.0 percent of $C_{23}H_{33}IN_2O$.

Packaging and storage—Preserve in well-closed, light-resistant containers.

Reference standard—*USP Isopropamide Iodide Reference Standard*—Dry in vacuum at 60° for 2 hours before using.

Identification—

A: The infrared absorption spectrum of a 2 in 25 solution of it, previously dried, in chloroform, determined in a 0.1-mm cell, exhibits maxima only at the same wavelengths as that of a similar solution of USP Isopropamide Iodide RS.

B: To 5 mL of a solution (1 in 1000) add 5 mL of sodium carbonate solution (1 in 100), 0.5 mL of bromophenol blue TS, and 10 mL of chloroform, and shake for several minutes: the chloroform layer becomes an intense blue in color.

C: A solution (1 in 1000) responds to the tests for *Iodide* ⟨191⟩.

Loss on drying ⟨731⟩—Dry it in vacuum at 60° for 2 hours: it loses not more than 1.0% of its weight.

Residue on ignition ⟨281⟩: not more than 0.5%, after ignition at 550 ± 25° for 4 hours.

Heavy metals, *Method II* ⟨231⟩: 0.002%.

Ordinary impurities ⟨466⟩—

Test solution: methanol.

Standard solution: methanol.

Eluant: a mixture of methanol, glacial acetic acid, and water (8:1:1).

Visualization: 2.

Assay—Dissolve about 750 mg of Isopropamide Iodide, previously dried and accurately weighed, in 60 mL of glacial acetic acid, add 15 mL of mercuric acetate TS and crystal violet TS, and titrate with 0.1 *N* perchloric acid VS to a blue end-point. Perform a blank determination, and make any necessary correction. Each mL of 0.1 *N* perchloric acid is equivalent to 48.04 mg of $C_{23}H_{33}IN_2O$.

Isopropamide Iodide Tablets

» Isopropamide Iodide Tablets contain an amount of isopropamide iodide ($C_{23}H_{33}IN_2O$) equivalent to not less than 93.0 percent and not more than 107.0 percent of the labeled amount of isopropamide ($C_{23}H_{33}N_2O$).

Packaging and storage—Preserve in well-closed containers.

Reference standard—*USP Isopropamide Iodide Reference Standard*—Dry in vacuum at 60° for 2 hours before using.

Identification—Triturate a portion of powdered Tablets, equivalent to about 10 mg of isopropamide, with 10 mL of water, and filter: the filtrate responds to *Identification tests B* and *C* under *Isopropamide Iodide*.

Dissolution ⟨711⟩—

Medium: water; 500 mL.

Apparatus 2: 100 rpm.

Time: 60 minutes.

Procedure—Determine the amount of isopropamide ($C_{23}H_{33}N_2O$) dissolved from ultraviolet absorbances at the wavelength of maximum absorbance at about 258 nm of filtered portions of the solution under test, suitably diluted with *Dissolution Medium*, if necessary, in comparison with a Standard solution having a known concentration of USP Isopropamide Iodide RS in the same medium.

Tolerances—Not less than 70% (*Q*) of the labeled amount of $C_{23}H_{33}N_2O$ is dissolved in 60 minutes.

Uniformity of dosage units ⟨905⟩: meet the requirements.

Procedure for content uniformity—Crush and transfer 1 Tablet to a 100-mL volumetric flask with the aid of about 50 mL of water, and shake by mechanical means for 30 minutes. Dilute with water to volume, mix, and filter, discarding the first 20 mL of the filtrate. Concomitantly determine the absorbances of this solution and a Standard solution of USP Isopropamide Iodide RS in the same medium having a known concentration of about 70 µg per mL, in 5-cm cells at 280 nm and at the wavelength of maximum absorbance at about 258 nm, with a suitable spectrophotometer, using water as the blank. Calculate the quantity, in mg, of isopropamide ($C_{23}H_{33}N_2O$) in the Tablet by the formula:

$$(353.53/480.43)(TC/D)[(A_{U258} - A_{U280})/(A_{S258} - A_{S280})],$$

in which 353.53 and 480.43 are the molecular weights of isopropamide and isopropamide iodide, respectively, T is the labeled quantity, in mg, of isopropamide in the Tablet, C is the concentration, in µg per mL, of USP Isopropamide Iodide RS in the Standard solution, D is the concentration, in µg per mL, of isopropamide in the test solution, based upon the labeled quantity per Tablet and the extent of dilution, and A_U and A_S are the absorbances of the test solution and the Standard solution, respectively, at the wavelengths indicated by the subscripts.

Assay—

Ion-exchange column—Insert a small pledget of glass wool in the bottom of a 6-mm × 240-mm glass tube fitted with a stopcock, fill the tube with water, and add an aqueous slurry of a suitable anion-exchange resin, chloride form (soaked in water for not less than 24 hours prior to use), until a height of about 200 mm is reached. Wash the column with 50 mL of water, and use without delay.

Standard preparation—Transfer about 135 mg of USP Isopropamide Iodide RS, accurately weighed, to a 250-mL volumetric flask, and dissolve in 150 mL of water. Add 5 mL of aluminum chloride solution (1 in 10) and 2 mL of ammonium hydroxide, then dilute with water to volume, mix, and filter, discarding the first 15 mL of the filtrate. Use the subsequent filtrate as directed in the *Procedure*.

Assay preparation—Weigh and finely powder not less than 25 Isopropamide Iodide Tablets. Transfer an accurately weighed portion of the powder, equivalent to about 100 mg of isopropamide, to a 250-mL volumetric flask, add 150 mL of water, and shake by mechanical means for 60 minutes. Add 5 mL of aluminum chloride solution (1 in 10) and 2 mL of ammonium hydroxide, dilute with water to volume, mix, and filter, discarding the first 15 mL of the filtrate. Use the subsequent filtrate as directed in the *Procedure*.

Procedure—Pipet 50.0 mL each of the *Standard preparation* and the *Assay preparation*, respectively, onto separate ion-exchange columns, and collect the eluates in separate 200-mL volumetric flasks. Regulate the flow of effluent so that it does not exceed 40 drops per minute, and when the liquid level reaches the top of each column, add successively two 5-mL portions and one 10-mL portion of water to each column, allowing each portion just to enter the column before adding the next portion of water. After the eluates have been collected, dilute the contents of each flask with water to volume, and mix. Concomitantly determine the absorbances of the solutions in 5-cm cells at 280 nm and at the maximum at about 258 nm, with a suitable spectrophotometer, using water as the blank. Calculate the quantity, in mg, of isopropamide ($C_{23}H_{33}N_2O$) in the portion of Tablets taken by the formula:

$$(353.53/480.43)(0.25C)[(A_{U258} - A_{U280})/(A_{S258} - A_{S280})],$$

in which 353.53 and 480.43 are the molecular weights of isopropamide and isopropamide iodide, respectively, C is the concentration, in µg per mL, of USP Isopropamide Iodide RS in the *Standard preparation*, and A_U and A_S are the absorbances of the solutions from the *Assay preparation* and the *Standard preparation*, respectively, at the wavelengths indicated by the subscripts.

Isopropyl Alcohol

CH_3CHCH_3 with OH on the central carbon

C_3H_8O 60.10

2-Propanol.

Isopropyl alcohol [*67-63-0*].

» Isopropyl Alcohol contains not less than 99.0 percent of C_3H_8O.

Packaging and storage—Preserve in tight containers, remote from heat.

Identification—Determine the infrared absorption spectrum, in a 0.1-mm cell, of a 1 in 20 solution of it in carbon tetrachloride, with carbon tetrachloride in the reference beam. It exhibits a region of strong absorption between 6.7 µm and 8.0 µm (the most prominent features being the peaks at about 6.8 µm, 7.2 µm, 7.4 µm, 7.7 µm, and 8.0 µm), a region of strong absorption between 8.3 µm and 9.5 µm (the most prominent features being the peaks at about 8.6 µm and 8.7 µm), and a strong peak at about 10.5 µm.

Specific gravity ⟨841⟩: between 0.783 and 0.787.

Refractive index ⟨831⟩: between 1.376 and 1.378 at 20°.

Acidity—To 50 mL in a suitable flask add 100 mL of carbon dioxide–free water. Add 2 drops of phenolphthalein TS, and titrate with 0.020 *N* sodium hydroxide to a pink color that persists for 30 seconds: not more than 0.70 mL of 0.020 *N* sodium hydroxide is required for neutralization.

Nonvolatile residue—Evaporate 50 mL in a tared porcelain dish on a steam bath to dryness, and heat at 105° for 1 hour: the weight of the residue does not exceed 2.5 mg (0.005%).

Assay—Inject about 5 µL of Isopropyl Alcohol into a suitable gas chromatograph, equipped with a thermal conductivity detector. Under typical conditions, the gas chromatograph contains a 1.8-m × 6.4-mm (OD) stainless steel column packed with 10 percent liquid phase G20 on support S1A, the column is maintained at 55°, and helium is used as the carrier gas at a flow rate of 45 mL per minute. Relative retention times of some of the possible components, when present, are: air at 0.09, diethyl ether at 0.14, diisopropyl ether at 0.17, acetone at 0.37, isopropyl alcohol at 1.00, 2-butanol at 1.64, *n*-propyl alcohol at 1.86, and water at 3.14. Calculate the percentage of C_3H_8O in the Isopropyl Alcohol by dividing the area under the isopropyl alcohol peak by the sum of the areas under all of the peaks observed, and multiplying by 100.

Isopropyl Rubbing Alcohol

» Isopropyl Rubbing Alcohol contains not less than 68.0 percent and not more than 72.0 percent of isopropyl alcohol, by volume, the remainder consisting of water, with or without suitable stabilizers, perfume oils, and color additives certified by the FDA for use in drugs.

Packaging and storage—Preserve in tight containers, remote from heat.

Labeling—Label it to indicate that it is flammable.

Specific gravity ⟨841⟩: between 0.872 and 0.883 at 20°.

Acidity—Transfer 50 mL to a suitable flask, and add about 75 mL of carbon dioxide–free water. Titrate potentiometrically to a pH of 8.5: not more than 1.0 mL of 0.020 *N* sodium hydroxide is required for neutralization.

Nonvolatile residue—Evaporate 50 mL to dryness in a tared porcelain dish on a steam bath, and dry at 105° for 1 hour: the weight of the residue does not exceed 5 mg (0.01%).

Assay—Transfer 50.0 mL of Isopropyl Rubbing Alcohol to a 250-mL distilling flask, and add 100 mL of water. Arrange the flask for distillation, distil, and collect 95 mL of distillate in a 100-mL volumetric flask. Dilute to volume with water, mix, and determine the specific gravity of the distillate at 25° (see *Specific Gravity* ⟨841⟩). The specific gravity is between 0.955 and 0.950, corresponding to between 68.0% and 72.0% of isopropyl alcohol in the specimen taken.

Azeotropic Isopropyl Alcohol

» Azeotropic Isopropyl Alcohol contains not less than 91.0 percent and not more than 93.0 percent of isopropyl alcohol, by volume, the remainder consisting of water.

Packaging and storage—Preserve in tight containers, remote from heat.

Identification—The infrared absorption spectrum of a thin film of it exhibits a strong broad band at 3.0 µm; a strong region of absorption between 3.35 µm and 3.5 µm, with its highest peak at 3.36 µm, and others at 3.41 µm and 3.47 µm; many weak peaks between 3.6 µm and 6.0 µm, among the most noticeable being those at 3.68 µm, 3.77 µm, 3.97 µm, 4.17 µm, and 5.26 µm; a broad band at about 6.2 µm; a strong region of absorption between 6.7 µm and 7.8 µm, the most prominent features being the peaks at 6.80 µm, 7.09 µm, 7.25 µm (the highest), 7.46 µm, and 7.63 µm; a strong region of absorption between 8.5 µm and 9.2 µm, peaking at 8.6 µm, 8.85 µm, and 9.0 µm; and strong peaks at 10.5 µm and at 12.3 µm.

Specific gravity ⟨841⟩: between 0.815 and 0.810, indicating between 91.0% and 93.0% by volume of C_3H_8O.

Refractive index ⟨831⟩: between 1.376 and 1.378 at 20°.

Acidity—To 50 mL in a suitable flask add 100 mL of carbon dioxide-free water. Add 2 drops of phenolphthalein TS, and titrate with 0.020 *N* sodium hydroxide to a pink color that persists for 30 seconds: not more than 0.70 mL of 0.020 *N* sodium hydroxide is required for neutralization.

Nonvolatile residue—Evaporate 50 mL of Isopropyl Alcohol in a tared porcelain dish on a steam bath to dryness, and heat at 105° for 1 hour: the weight of the residue does not exceed 2.5 mg (0.005%).

Volatile impurities—Inject about 5 µL into a suitable gas chromatograph, equipped with a thermal conductivity detector. Under typical conditions, the gas chromatograph contains a 1.8-m × 6.4-mm (OD) stainless steel column packed with 10 percent liquid phase G20 on support S1A, the column is maintained at 55°, and helium is used as the carrier gas at a flow rate of 45 mL per minute. Relative retention times of some of the possible components, when present, are: air at 0.09, diethyl ether at 0.14, diisopropyl ether at 0.17, acetone at 0.37, isopropyl alcohol at 1.00, 2-butanol at 1.64, *n*-propyl alcohol at 1.86, and water at 3.14. Divide the area under the isopropyl alcohol peak by the sum of the areas under all of the peaks observed, excluding the water peak: the quotient is not less than 0.99.

Isopropyl Myristate—*see* Isopropyl Myristate NF

Isopropyl Palmitate—*see* Isopropyl Palmitate NF

Isoproterenol Inhalation Solution

» Isoproterenol Inhalation Solution is a solution of Isoproterenol Hydrochloride in Purified Water. It may contain Sodium Chloride. It contains not less than 90.0 percent and not more than 115.0 percent of the labeled amount of $C_{11}H_{17}NO_3.HCl$.

Packaging and storage—Preserve in small, well-filled, tight containers, protected from light.

Labeling—Label it to indicate that the Inhalation Solution is not to be used if it is pinkish to brownish in color or contains a precipitate.

Reference standard—*USP Isoproterenol Hydrochloride Reference Standard*—Dry in vacuum over phosphorus pentoxide for 4 hours before using.

Identification—The retention time of the major peak in the chromatogram of the *Assay preparation* corresponds to that of the *Standard preparation*, as obtained in the *Assay*.

pH ⟨791⟩: between 2.5 and 5.5.

Assay—

Standard preparation—Prepare as directed in the *Assay* under *Isoproterenol Hydrochloride*.

Assay preparation—Transfer an accurately measured volume of Isoproterenol Inhalation Solution, equivalent to about 25 mg of isoproterenol hydrochloride, to a 100-mL volumetric flask, dilute with 0.17 *N* acetic acid solution to volume, and mix.

Chromatographic system—Proceed as directed in the *Assay* under *Isoproterenol Hydrochloride*.

Procedure—Proceed as directed for *Procedure* in the *Assay* under *Isoproterenol Hydrochloride*. Calculate the quantity, in mg, of $C_{11}H_{17}NO_3.HCl$ in each mL of the Inhalation Solution taken by the formula:

$$0.1(C/V)(h_U/h_S),$$

in which V is the volume, in mL, of Inhalation Solution taken, and C, h_U, and h_S are as defined therein.

Isoproterenol Hydrochloride

$C_{11}H_{17}NO_3.HCl$ 247.72

1,2-Benzenediol, 4-[1-hydroxy-2-[(1-methylethyl)amino]ethyl]-, hydrochloride.

3,4-Dihydroxy-α-[(isopropylamino)methyl]benzyl alcohol hydrochloride [*51-30-9*].

» Isoproterenol Hydrochloride contains not less than 97.0 percent and not more than 101.5 percent of $C_{11}H_{17}NO_3.HCl$, calculated on the dried basis.

Packaging and storage—Preserve in tight, light-resistant containers.

Reference standard—*USP Isoproterenol Hydrochloride Reference Standard*—Dry in vacuum over phosphorus pentoxide for 4 hours before using.

Identification—

A: The infrared absorption spectrum of a potassium bromide dispersion of it, previously dried, exhibits maxima only at the same wavelengths as that of a similar preparation of USP Isoproterenol Hydrochloride RS.

B: The ultraviolet absorption spectrum of a solution (1 in 20,000) exhibits maxima and minima at the same wavelengths as that of a similar solution of USP Isoproterenol Hydrochloride RS, concomitantly measured.

Melting range ⟨741⟩: between 165° and 170°.

Loss on drying ⟨731⟩—Dry about 1 g, accurately weighed, in vacuum over phosphorus pentoxide for 4 hours: it loses not more than 1.0% of its weight.

Residue on ignition ⟨281⟩: not more than 0.2%.

Sulfate ⟨221⟩—A 0.10-g portion shows no more sulfate than corresponds to 0.20 mL of 0.020 *N* sulfuric acid (0.2%).

Isoproterenone—Its absorptivity (see *Spectrophotometry and Light-scattering* ⟨851⟩) at 310 nm, determined in a solution containing 2 mg per mL, is not more than 0.2.

Chloride content—Dissolve about 500 mg, accurately weighed, in 5 mL of water. Add 5 mL of glacial acetic acid and 40 mL of methanol. Add eosin Y TS, and titrate with 0.1 *N* silver nitrate VS. Each mL of 0.1 *N* silver nitrate is equivalent to 3.545 mg of Cl. Between 13.9% and 14.6% of Cl is found, calculated on the dried basis.

Assay—

Standard preparation—Dissolve an accurately weighed quantity of USP Isoproterenol Hydrochloride RS in freshly prepared sodium bisulfite solution (3 in 1000) to obtain a solution having a concentration of about 2.5 mg per mL. Transfer 5.0 mL of this solution to a 50-mL volumetric flask, dilute with 0.17 *N* acetic acid to volume, and mix to obtain a solution having a known concentration of about 250 µg per mL.

Assay preparation—Transfer about 125 mg of Isoproterenol Hydrochloride, accurately weighed, to a 25-mL volumetric flask, dissolve in sodium bisulfite solution (3 in 1000), dilute with sodium bisulfite solution to volume, and mix. Transfer 5.0 mL of this solution to a 100-mL volumetric flask, dilute with 0.17 *N* acetic acid to volume, and mix.

Chromatographic system (see *Chromatography* ⟨621⟩)—The liquid chromatograph is equipped with a 278-nm detector and a 30-cm × 4-mm stainless steel column that contains packing L1. The mobile phase is 0.17 *N* acetic acid having a flow rate of about 1.5 mL per minute. Chromatograph five replicate injections of the *Standard preparation*, and record the peak responses as directed under *Procedure:* the relative standard deviation is not more than 3.0%.

Procedure—Using a microsyringe or sampling valve, chromatograph 10 µL of the *Standard preparation*, and adjust the specimen size and other operating parameters, if necessary, until satisfactory chromatography and peak responses are obtained. Chromatograph equal volumes of the *Standard preparation* and the *Assay preparation*, and measure the peak responses. Calculate the quantity, in mg, of $C_{11}H_{17}NO_3 . HCl$ in the portion of Isoproterenol Hydrochloride taken by the formula:

$$0.5C(h_U/h_S),$$

in which C is the concentration, in µg per mL of USP Isoproterenol Hydrochloride RS in the *Standard preparation*, and h_U and h_S are the peak responses obtained from the *Assay preparation* and the *Standard preparation*, respectively.

Isoproterenol Hydrochloride Inhalation Aerosol

» Isoproterenol Hydrochloride Inhalation Aerosol is a solution of Isoproterenol Hydrochloride in Alcohol in an inert propellant base. It contains not less than 90.0 percent and not more than 115.0 percent of the labeled amount of $C_{11}H_{17}NO_3 . HCl$. The quantity of $C_{11}H_{17}NO_3 . HCl$ delivered through an oral inhalation actuator is not less than 75.0 percent and not more than 125.0 percent of the labeled dose per inhalation.

Packaging and storage—Preserve in small, nonreactive, light-resistant aerosol containers equipped with metered-dose valves and provided with oral inhalation actuators.

Reference standard—*USP Isoproterenol Hydrochloride Reference Standard*—Keep the container tightly closed and protected from light. Dry in vacuum over phosphorus pentoxide for 4 hours before using.

Identification—

A: Place 10 mL of water in a small beaker, and deliver 10 sprays from Isoproterenol Hydrochloride Inhalation Aerosol under the surface of the water, actuating the valve by pressing the stem tip against the bottom of the beaker. Filter, place 5 mL of the filtrate in a test tube, and retain the remainder of the solution for *Identification test B*. Add 1 drop of 0.2 *N* sulfuric acid and 0.5 mL of 0.1 *N* iodine, allow to stand for 5 minutes, and add 1 mL of 0.1 *N* sodium thiosulfate: a salmon-pink color is produced.

B: Dilute the remainder of the filtrate obtained in *Identification test A* with an equal volume of water. Add a few drops of 6 *N* ammonium hydroxide, filter, acidify the filtrate with nitric acid, and divide into two equal portions. To each portion add a few drops of silver nitrate TS: white precipitates are formed. To one portion add a slight excess of nitric acid: the white precipitate remains. To the other portion add a slight excess of 6 *N* ammonium hydroxide, and shake: the precipitate dissolves.

Alcohol content ⟨611⟩—Weigh accurately a filled Isoproterenol Hydrochloride Inhalation Aerosol container, and record the weight. Place the container in a dry ice–alcohol bath, and cool for 60 minutes. Remove the container from the bath, and carefully remove the valve with wire cutters, taking precautions to save all pieces of the valve and cap. With the aid of three 5-mL portions of water, transfer the contents of the container to a beaker previously chilled in the bath. Dry the rinsed empty container and all of its parts in an oven at 105° for 2 hours, cool, and weigh. Calculate the weight of the container contents. Add a few boiling chips to the beaker, and carefully stir to help evaporate the propellant. After the bulk of the propellant has evaporated, transfer the contents of the beaker, with the aid of several mL of water, to a glass-stoppered graduated cylinder, measure the volume, and determine the alcohol content of the *Test solution* thus prepared by the gas-liquid chromatographic procedure, methyl ethyl ketone being used as the internal standard. Calculate the alcohol content of the Inhalation Aerosol taken by the formula:

$$SV/W,$$

in which S is the percentage of alcohol (w/v) in the *Test solution*, V is the volume, in mL, of the *Test solution*, and W is the weight, in g, of the container contents: between 28.5% and 38.5% (w/w) of C_2H_5OH is found.

Unit spray content—

Unit spray sampling apparatus—Prepare as directed under *Aerosols* ⟨601⟩.

Ferro-citrate Solution and *Buffer Solution*—Prepare as directed under *Epinephrine Assay* ⟨391⟩.

Standard preparation—Transfer about 50 mg of USP Isoproterenol Hydrochloride RS, accurately weighed, to a 100-mL volumetric flask, add freshly prepared sodium bisulfite solution (1 in 500) to volume, and mix. Transfer 10.0 mL of this solution to a second 100-mL volumetric flask, dilute with the same sodium bisulfite solution to volume, and mix, to obtain a solution having a concentration of about 50 µg per mL.

Test preparation—Transfer 20.0 mL of freshly prepared sodium bisulfite solution (1 in 500) to the collection chamber of the sampling apparatus, and clamp the intake tube in a horizontal position. Fit the aerosol container with its oral inhalation actuator, and prime the unit by firing it 10 times. Couple the unit to the intake tube of the apparatus, and adjust the airflow to a rate of 10 to 15 liters per minute. Deliver into the system a number of sprays equivalent to 1 mg of isoproterenol hydrochloride. Remove the container and actuator from the assembly, clamp the intake tube in a vertical position and scavenge the drug remaining in the intake system as follows: Disconnect the collection chamber, transfer the test solution to a small beaker, and wash the inside of the intake tube with the test solution, collecting the washing in the collection chamber placed a few cm below its normal position. Connect the collection chamber into the system again, and draw the test solution into it through the intake tube and bubbler. Recycle the test solution twice in this manner, then add 10 mL of chloroform to the collection chamber, insert the stopper, and shake vigorously for 1 minute. Pour the mixture into a centrifuge tube, and centrifuge. Use the clear supernatant liquid as directed in the *Procedure*.

Procedure—Proceed as directed for *Procedure* in the *Assay*. Calculate the quantity, in µg, of $C_{11}H_{17}NO_3 . HCl$ delivered per spray by the formula:

$$(20C/N)(A_U/A_S),$$

in which C is the concentration, in µg per mL, of USP Isoproterenol Hydrochloride RS in the *Standard preparation*, N is the total number of sprays delivered, and the other terms are as defined therein.

Other requirements—It meets the requirements for *Leak Testing* under *Aerosols* ⟨601⟩.

Assay—

Ferro-citrate Solution, *Buffer Solution*, and *Standard preparation*—Prepare as directed under *Unit spray content*.

Assay preparation—Fit the Isoproterenol Hydrochloride Inhalation Aerosol container with its inhalation actuator, and prime the unit by firing it 10 times. Remove the actuator, and accurately weigh the container with its remaining contents. Invert the container, place the valve stem tip against the bottom of a 100-mL beaker containing 20 mL of chloroform, and deliver under the surface of the chloroform a number of sprays equivalent to about 500 µg of isoproterenol hydrochloride. Raise the unit above the contents of the beaker, and wash the valve stem with about 2 mL of chloroform. Collect the washings in the beaker. Allow the valve stem to dry, and again accurately weigh the Inhalation Aerosol container with its remaining contents. Record the weight of the expelled specimen. Transfer the contents of the beaker to a centrifuge tube with the aid of two 3-mL portions of chloroform, add 10.0 mL of freshly prepared sodium bisulfite solution (1 in 500), and shake vigorously for 1 minute. Centrifuge, and use the clear supernatant liquid as directed in the *Procedure*.

Procedure—Into three separate test tubes, pipet 5 mL each of the *Assay preparation*, *Standard preparation*, and water to provide the blank, respectively. To each tube add 100 µL of *Ferro-citrate Solution*, followed by 1.0 mL of *Buffer Solution*, and mix. Determine the absorbances of the *Assay preparation* and the *Standard preparation* at the wavelength of maximum absorbance at about 530 nm, against the blank. Calculate the percentage (w/w) of $C_{11}H_{17}NO_3 . HCl$ in the portion of Inhalation Aerosol taken by the formula:

$$0.001(C/W)(A_U/A_S),$$

in which C is the concentration, in µg per mL, of USP Isoproterenol Hydrochloride RS in the *Standard preparation*, W is the weight, in g, of Inhalation Aerosol taken, and A_U and A_S are the absorbances of the solutions from the *Assay preparation* and the *Standard preparation*, respectively.

Isoproterenol Hydrochloride Inhalation—*see* Isoproterenol Inhalation Solution

Isoproterenol Hydrochloride Injection

» Isoproterenol Hydrochloride Injection is a sterile solution of Isoproterenol Hydrochloride in Water for Injection. It contains not less than 90.0 percent and not more than 115.0 percent of the labeled amount of $C_{11}H_{17}NO_3.HCl$.

Packaging and storage—Preserve in single-dose containers, preferably of Type I glass, protected from light.

Labeling—Label it to indicate that the Injection is not to be used if it is pinkish to brownish in color or contains a precipitate.

Reference standard—*USP Isoproterenol Hydrochloride Reference Standard*—Dry in vacuum over phosphorus pentoxide for 4 hours before using.

Identification—

A: Its ultraviolet absorption spectrum exhibits maxima and minima at the same wavelengths as that of a similar solution of USP Isoproterenol Hydrochloride RS, concomitantly measured.

B: The retention time of the major peak in the chromatogram of the *Assay preparation* corresponds to that of the *Standard preparation*, as obtained in the *Assay*.

pH ⟨791⟩: between 2.5 and 4.5.

Particulate matter ⟨788⟩: meets the requirements under *Small-volume Injections*.

Other requirements—It meets the requirements under *Injections* ⟨1⟩.

Assay—

Mobile phase—Dissolve 1.76 g of sodium 1-heptanesulfonate in 800 mL of water. Add 200 mL of methanol, and adjust with 1 *M* phosphoric acid to a pH of 3.0 ± 0.1. Filter the solution through a membrane filter (1-µm or finer porosity).

Standard preparation—Dissolve an accurately weighed quantity of USP Isoproterenol Hydrochloride RS in freshly prepared sodium bisulfite solution (1 in 1000) to obtain a solution having a known concentration of about 20 µg per mL.

Resolution solution—Prepare a solution of epinephrine bitartrate in freshly prepared *Mobile phase* containing 1.0% of sodium bisulfite, having a concentration of about 200 µg per mL. Mix 2.0 mL of this solution and 18.0 mL of *Standard preparation*.

Assay preparation—Dilute an accurately measured volume of Isoproterenol Hydrochloride Injection quantitatively with freshly prepared sodium bisulfite solution (1 in 1000) to obtain a solution having a concentration of about 20 µg per mL.

Chromatographic system (see *Chromatography* ⟨621⟩)—The liquid chromatograph is equipped with a 280-nm detector and a 30-cm × 4-mm column that contains packing L1. The flow rate is about 2 mL per minute. Chromatograph the *Standard preparation*, and record the peak responses as directed under *Procedure*: the relative standard deviation for replicate injections is not more than 1.5%. Chromatograph the *Resolution solution*: the resolution, R, for the epinephrine and isoproterenol peaks is not less than 3.5, and the tailing factors for the epinephrine and isoproterenol peaks are not more than 2.5. The relative retention times are about 0.55 for epinephrine and 1.0 for isoproterenol.

Procedure—Separately inject equal volumes (about 100 µL) of the *Standard preparation* and the *Assay preparation*, record the chromatograms, and measure the peak responses. Calculate the quantity, in mg, of $C_{11}H_{17}NO_3.HCl$ in each mL of Isoproterenol Hydrochloride Injection taken by the formula:

$$C(L/D)(r_U/r_S),$$

in which C is the concentration, in µg per mL, of USP Isoproterenol Hydrochloride RS in the *Standard preparation*, L is the labeled quantity, in µg per mL, of isoproterenol hydrochloride in the Injection, and D is the concentration, in µg per mL, of isoproterenol hydrochloride in the *Assay preparation*, on the basis of the labeled quantity in each mL and the extent of dilution, and r_U and r_S are the peak responses obtained from the *Assay preparation* and the *Standard preparation*, respectively.

Isoproterenol Hydrochloride Inhalation Solution, Acetylcysteine and—*see* Acetylcysteine and Isoproterenol Hydrochloride Inhalation Solution

Isoproterenol Hydrochloride Tablets

» Isoproterenol Hydrochloride Tablets contain not less than 93.0 percent and not more than 107.0 percent of the labeled amount of $C_{11}H_{17}NO_3.HCl$.

Packaging and storage—Preserve in well-closed, light-resistant containers.

Reference standard—*USP Isoproterenol Hydrochloride Reference Standard*—Dry in vacuum over phosphorus pentoxide for 4 hours before using.

Identification—Powder a number of Tablets, equivalent to about 50 mg of isoproterenol hydrochloride, digest with 15 mL of hot dehydrated alcohol for 20 minutes, cool, filter, and evaporate the filtrate on a steam bath to dryness: the residue responds to the *Identification tests* under *Isoproterenol Hydrochloride*.

Dissolution ⟨711⟩—

Medium: water; 900 mL.

Apparatus 2: 50 rpm.

Time: 45 minutes.

Procedure—Determine the amount of $C_{11}H_{17}NO_3.HCl$ dissolved from ultraviolet absorbances at the wavelength of maximum absorbance at about 279 nm of filtered portions of the solution under test, suitably diluted with *Dissolution Medium*, if necessary, in comparison with a Standard solution having a known concentration of USP Isoproterenol Hydrochloride RS in the same medium.

Tolerances—Not less than 75% (Q) of the labeled amount of $C_{11}H_{17}NO_3.HCl$ is dissolved in 45 minutes.

Uniformity of dosage units ⟨905⟩: meet the requirements.

Procedure for content uniformity—Crush 1 Tablet, and transfer it with the aid of 25 mL of water to a 50-mL volumetric flask. Shake gently until no more dissolves, add water to volume, and mix. Filter through a dry filter into a dry flask, rejecting the first 20 mL of the filtrate. Transfer a portion of the filtrate, equivalent to about 2.5 mg of isoproterenol hydrochloride and accurately measured, to a 50-mL volumetric flask, dilute with water to volume, and mix. Dissolve an accurately weighed quantity of USP Isoproterenol Hydrochloride RS in water, and dilute quantitatively and stepwise with water to obtain a Standard solution having a known concentration of about 50 µg per mL. Concomitantly and without delay, determine the absorbances of both solutions in 1-cm cells at the wavelength of maximum absorbance at about 279 nm, with a suitable spectrophotometer, using water as the blank. Calculate the quantity, in mg, of $C_{11}H_{17}NO_3.HCl$ in the Tablet by the formula:

$$2.5(C/V)(A_U/A_S),$$

in which C is the concentration, in µg per mL, of USP Isoproterenol Hydrochloride RS in the Standard solution, V is the volume, in mL, of the filtrate taken, and A_U and A_S are the absorbances of the solution from the Tablet and the Standard solution, respectively.

Assay—

Mobile phase— Adjust 0.1 *M* sodium sulfate with phosphoric acid to a pH of 3.0. Mix 90 parts of this solution with 10 parts of methanol.

Standard preparation—Dissolve an accurately weighed quantity of USP Isoproterenol Hydrochloride RS in 0.1 *N* sulfuric acid, and dilute quantitatively and stepwise with the same solvent to obtain a solution having a known concentration of about 0.1 mg per mL.

Assay preparation—Weigh and finely powder not less than 20 Isoproterenol Hydrochloride Tablets. Weigh accurately a portion of the powder, equivalent to about 10 mg of isoproterenol hydrochloride, and transfer with the aid of 50 mL of 0.1 *N* sulfuric acid to a 100-mL volumetric flask. Shake gently until no more dissolves, dilute with the same solvent to volume, and mix. Filter through a dry filter into a dry flask, rejecting the first 20 mL of the filtrate.

Chromatographic system (see *Chromatography* ⟨621⟩)—The liquid chromatograph is equipped with a 280-nm detector and a 3.9-mm × 30-cm column that contains packing L1. The flow rate is about 1 mL per minute. Chromatograph three replicate injections of the *Standard preparation*, and record the peak responses as directed under *Procedure:* the relative standard deviation is not more than 2.0%.

Procedure—Separately inject equal volumes (about 20 µL) of the *Standard preparation* and the *Assay preparation* into the chromatograph by means of a suitable microsyringe or sampling valve, record the chromatograms, and measure the responses for the major peaks. The retention time is about 3.5 minutes for isoproterenol. Calculate the quantity, in mg, of $C_{11}H_{17}NO_3 . HCl$ in the portion of Tablets taken by the formula:

$$100C(r_U / r_S),$$

in which C is the concentration, in mg per mL, of USP Isoproterenol Hydrochloride RS in the *Standard preparation*, and r_U and r_S are the peak responses obtained from the *Assay preparation* and the *Standard preparation*, respectively.

Isoproterenol Hydrochloride and Phenylephrine Bitartrate Inhalation Aerosol

» Isoproterenol Hydrochloride and Phenylephrine Bitartrate Inhalation Aerosol is a suspension of microfine Isoproterenol Hydrochloride and Phenylephrine Bitartrate in suitable propellants in a pressurized container. It contains not less than 90.0 percent and not more than 110.0 percent of the labeled amounts of isoproterenol hydrochloride ($C_{11}H_{17}NO_3 . HCl$) and phenylephrine bitartrate ($C_9H_{13}NO_2 . C_4H_6O_6$), and it delivers not less than 75.0 percent and not more than 125.0 percent of the labeled dose per inhalation of $C_{11}H_{17}NO_3 . HCl$ and $C_9H_{13}NO_2 . C_4H_6O_6$ through an oral inhalation actuator.

Packaging and storage—Preserve in small, nonreactive, light-resistant aerosol containers equipped with metered-dose valves and provided with oral inhalation actuators.

Reference standards—*USP Isoproterenol Hydrochloride Reference Standard*—Dry in vacuum over phosphorus pentoxide for 4 hours before using. *USP Phenylephrine Hydrochloride Reference Standard*—Dry at 105° for 2 hours before using.

Identification—

A: Place 10 mL of water in a small beaker, and deliver 20 sprays from Isoproterenol Hydrochloride and Phenylephrine Bitartrate Inhalation Aerosol under the surface of the water, actuating the valve by pressing the tip against the bottom of the beaker. To 5 mL of the solution add 1 drop of dilute sulfuric acid (1 in 200), add 0.5 mL of 0.1 *N* iodine, allow to stand for 5 minutes, and add 1 mL of 0.1 *N* sodium thiosulfate: a red-brown color is produced.

B: To the balance of the solution obtained in *Identification test A* add 3 mL of *Mercuric sulfate solution*, prepared as directed in the test for *Unit spray content*. Heat on a steam bath for 5 minutes, cool, and add 3 mL of sodium nitrite solution (1 in 80): a deep red color is produced.

Unit spray content—

Unit spray sampling apparatus—Prepare as directed under *Aerosols* ⟨601⟩.

Ferro-citrate solution and *Buffer solution*—Prepare as directed under *Epinephrine Assay* ⟨391⟩.

Mercuric sulfate solution—While stirring a mixture of 15 g of yellow mercuric oxide and 80 mL of water, slowly add 20 mL of sulfuric acid, and stir until completely dissolved.

Isoproterenol hydrochloride standard preparation—Transfer about 50 mg of USP Isoproterenol Hydrochloride RS, accurately weighed, to a 100-mL volumetric flask, add dilute sulfuric acid (1 in 1000) to volume, and mix. Transfer 10.0 mL of this solution to a 50-mL volumetric flask, add dilute sulfuric acid (1 in 1000) to volume, and mix. The concentration of USP Isoproterenol Hydrochloride RS in the *Isoproterenol hydrochloride standard preparation* is about 100 µg per mL.

Phenylephrine hydrochloride standard preparation—Transfer about 50 mg of USP Phenylephrine Hydrochloride RS, accurately weighed, to a 100-mL volumetric flask, add dilute sulfuric acid (1 in 1000) to volume, and mix. Transfer 10.0 mL of this solution to a 50-mL volumetric flask, add dilute sulfuric acid (1 in 1000) to volume, and mix. The concentration of USP Phenylephrine Hydrochloride RS in the *Phenylephrine hydrochloride standard preparation* is about 100 µg per mL.

Test preparation—Transfer 20.0 mL of dilute sulfuric acid (1 in 1000) to the collection chamber of the sampling apparatus, and clamp the intake tube in a horizontal position. Prime the valve of Isoproterenol Hydrochloride and Phenylephrine Bitartrate Inhalation Aerosol by alternately shaking and firing it 10 times through its oral inhalation actuator. Couple the aerosol container and actuator to the intake tube of the apparatus, turn on the vacuum source, and set the flow regulator for an airflow rate of 12 ± 1 liters per minute. Turn off the vacuum, and remove the aerosol container, leaving the actuator in place. Shake the container, seat it in the actuator, turn on the vacuum, immediately deliver a single spray into the system, and turn off the vacuum. Deliver a total of 12 sprays in this manner. Uncouple the aerosol container and actuator from the assembly, clamp the intake tube in a vertical position, and scavenge the isoproterenol hydrochloride and phenylephrine bitartrate remaining in the intake system as follows: Disconnect the collection chamber, transfer the test solution to a small beaker, then wash the inside of the intake tube with the test solution, collecting the washings in the collection chamber placed a few cm below its normal position. Recycle the test solution twice in this manner, then add 10 mL of chloroform to the collection chamber, insert the stopper, and shake vigorously for 1 minute. Transfer the mixture to a centrifuge tube, centrifuge for 20 minutes, and use the clear supernatant liquid as directed in the *Procedure*.

Procedure—Proceed as directed for *Procedure* in the *Assay*. Calculate the quantity, in mg, of isoproterenol hydrochloride ($C_{11}H_{17}NO_3 . HCl$) delivered per spray by the formula:

$$(0.02C / 12)(A_U / A_S),$$

in which C is the concentration, in µg per mL, of USP Isoproterenol Hydrochloride RS in the *Isoproterenol hydrochloride standard preparation*. Calculate the quantity, in mg, of phenylephrine bitartrate ($C_9H_{13}NO_2 . C_4H_6O_6$) delivered per spray by the formula:

$$(317.29 / 203.67)(0.02C / 12)(A_U / A_S),$$

in which 317.29 and 203.67 are the molecular weights of phenylephrine bitartrate and phenylephrine hydrochloride, respectively, and C is the concentration, in µg per mL, of USP Phenylephrine Hydrochloride RS in the *Phenylephrine hydrochloride standard preparation*.

Particle size—Prime the valve of Isoproterenol Hydrochloride and Phenylephrine Bitartrate Inhalation Aerosol by alternately shaking and firing it several times through its oral inhalation actuator, and then actuate one measured spray onto a clean, dry microscope slide held 5 cm from the end of the actuator, perpendicular to the direction of spray. Carefully rinse the slide with about 2 mL of carbon tetrachloride, and allow to dry. Examine the slide under a microscope, equipped with a calibrated ocular micrometer, using 450× magnification. Focus on the particles of 25 fields of view near the center of the sample pattern, and note the size of the great majority of individual particles: they are less than 5 µm in diameter. Record the size of all individual crystalline particles (not agglomerates) more than 10 µm in length measured along the longest axis: not more than 10 such particles are observed.

Other requirements—It meets the requirements for *Leak Testing* under *Aerosols* ⟨601⟩.

Assay—

Ferro-citrate solution, Buffer solution, Mercuric sulfate solution, and *Standard preparations*—Prepare as directed in the test for *Unit spray content.*

Assay preparation—[NOTE—Valve actuation during sampling is to be performed in a manner that will deliver freely the spray into the solvent in the bottom of the sampling beaker, but with minimal lateral pressure on the valve stem to avoid possible side-of-stem leakage. Any devices and techniques designed to accomplish these objectives may be used (e.g., actuating the valve by pressing the tip into an indentation in the bottom of the beaker or using a special adapter to spray at right angles to the valve held at the bottom of the beaker).] Place 20 mL of chloroform in a suitable 100-mL beaker. Prime the valve of Isoproterenol Hydrochloride and Phenylephrine Bitartrate Inhalation Aerosol by alternately shaking and firing it 10 times through its oral inhalation actuator. Accurately weigh the Inhalation Aerosol, shake it, and immediately deliver a single spray under the surface of the chloroform. Raise the Inhalation Aerosol above the surface of the chloroform, and shake it gently preparatory to delivering another spray similarly under the surface of the chloroform. Deliver a total of 6 sprays in this manner. Rinse the valve stem and ferrule with about 2 mL of chloroform, collecting the rinsing with the sample in the beaker. Allow the Inhalation Aerosol to dry, weigh it, and determine the total weight of the 6 sprays. Transfer the solution to a centrifuge tube with the aid of two 3-mL portions of chloroform, and add 10.0 mL of dilute sulfuric acid (1 in 1000). Insert the stopper, shake vigorously for 1 minute, centrifuge for 20 minutes, and use the clear supernatant liquid as directed in the *Procedure.*

Procedure for isoproterenol hydrochloride—Transfer 3.0 mL each of the *Isoproterenol hydrochloride standard preparation* and the *Assay preparation* to separate test tubes. To each tube add 0.10 mL of *Ferro-citrate solution* and 1.0 mL of *Buffer solution*, and mix. Concomitantly determine the absorbances of the solutions in 1-cm cells at the wavelength of maximum absorbance at about 530 nm, with a suitable spectrophotometer, using water as the blank. Calculate the quantity, in mg, of $C_{11}H_{17}NO_3 . HCl$ in each mL of the Inhalation Aerosol taken by the formula:

$$(0.01Cd/W)(A_U/A_S),$$

in which C is the concentration, in µg per mL, of USP Isoproterenol Hydrochloride RS in the *Isoproterenol hydrochloride standard preparation, d* is the density, in g per mL, of Inhalation Aerosol taken, W is the weight, in g, of the sample taken, and A_U and A_S are the absorbances of the solutions from the *Assay preparation* and the *Isoproterenol hydrochloride standard preparation,* respectively. [The density, d, is determined as follows: Weigh a known volume (v) of the Inhalation Aerosol in a suitable 5-mL gas-tight syringe equipped with a linear valve. Calibrate the volume of the syringe by filling to the 5-mL mark with dichlorotetrafluoroethane withdrawn from a plastic-coated glass vial sealed with a neoprene multiple-dose rubber stopper and an aluminum seal, using 1.456 g per mL as the density of the calibrating liquid. Maintain the dichlorotetrafluoroethane, the Inhalation Aerosol sample, and the syringe (protected from becoming wet) at 25° in a water bath. Obtain the sample, equivalent to the same volume as that obtained during the sampling procedure, from the Inhalation Aerosol by means of a sampling device consisting of a replaceable rubber septum engaged in the plate threads at one end of a threaded fitting, the opposite end of which contains a sharpened tube capable of puncturing the aerosol container, and a rubber gasket around the tube to prevent leakage of the container contents after puncture.* Calculate the density by the formula:

$$w/v,$$

in which w is the weight of the volume, v, of Inhalation Aerosol taken.]

Procedure for phenylephrine bitartrate—Transfer 5.0 mL each of the *Phenylephrine hydrochloride standard preparation,* the *Assay preparation,* and a blank consisting of dilute sulfuric acid (1 in 1000), to separate test tubes. To each tube add 3.0 mL of *Mercuric sulfate solution,* and mix. Immerse in a water bath maintained between 35° and 40° for 10 minutes, remove, and allow to stand at room temperature for 30 minutes. Add 0.25 mL of sodium nitrite solution (1 in 80), mix, and allow to stand, with occasional swirling, for 30 minutes. Concomitantly determine the absorbances of the solutions against the blank in 1-cm cells at the wavelength of maximum absorbance at about 495 nm, with a suitable spectrophotometer. Calculate the quantity, in mg, of $C_9H_{13}NO_2 . C_4H_6O_6$ in each mL of the Inhalation Aerosol taken by the formula:

$$(317.29/203.67)(0.01Cd/W)(A_U/A_S),$$

in which 317.29 and 203.67 are the molecular weights of phenylephrine bitartrate and phenylephrine hydrochloride, respectively, C is the concentration, in µg per mL, of USP Phenylephrine Hydrochloride RS in the *Phenylephrine hydrochloride standard preparation, d* is the density, in g per mL, of Inhalation Aerosol taken, W is the weight, in g, of the sample taken, and A_U and A_S are the absorbances of the solutions from the *Assay preparation* and the *Phenylephrine hydrochloride standard preparation,* respectively.

Isoproterenol Sulfate

$(C_{11}H_{17}NO_3)_2 . H_2SO_4 . 2H_2O$ 556.62

1,2-Benzenediol, 4-[1-hydroxy-2-[(1-methylethyl)amino]ethyl]-, sulfate (2:1) (salt), dihydrate.

3,4-Dihydroxy-α-[(isopropylamino)methyl]benzyl alcohol sulfate (2:1) (salt) dihydrate [*6700-39-6*].

Anhydrous 520.59 [*299-95-6*].

» Isoproterenol Sulfate contains not less than 97.0 percent and not more than 103.0 percent of $(C_{11}H_{17}NO_3)_2 . H_2SO_4$, calculated on the anhydrous basis.

Packaging and storage—Preserve in tight, light-resistant containers.

Reference standard—*USP Isoproterenol Hydrochloride Reference Standard*—Dry in vacuum over phosphorus pentoxide for 4 hours before using.

Identification—

A: The ultraviolet absorption spectrum of a 1 in 20,000 solution in 0.1 *N* hydrochloric acid exhibits maxima and minima at the same wavelengths as that of a similar preparation of USP Isoproterenol Hydrochloride RS.

B: To a solution of 10 mg in 5 mL of water add 1 drop of ferric chloride TS: an intense green color is produced, and it becomes olive-green on standing.

C: To a solution of 10 mg in 1 mL of water add 1 drop of phosphotungstic acid TS: a white precipitate is formed immediately and it becomes brown on standing (*distinction from epinephrine, which forms no precipitate*).

D: Dilute 1.0 mL of a solution (1 in 1000) with water to 10 mL, add 0.1 mL of dilute hydrochloric acid (1 in 120), then add 1.0 mL of 0.10 *N* iodine. Allow to stand for 5 minutes, and add 2.0 mL of 0.10 *N* sodium thiosulfate: a salmon pink color is produced (*distinction from norepinephrine, which, at the same*

* A suitable sampling system is available from Alltech Associates, 2051 Waukegan Rd., Deerfield, IL 60015.

pH, about 3, produces no color or at most only a slight pink color).

E: It responds to the tests for *Sulfate* ⟨191⟩.

Water, *Method I* ⟨921⟩: not more than 7.0%.

Residue on ignition ⟨281⟩: not more than 0.2%.

Chloride ⟨221⟩—A 0.10-g portion shows no more chloride than corresponds to 0.20 mL of 0.020 *N* hydrochloric acid (0.14%).

Isoproterenone—It meets the requirements of the test for *Isoproterenone* under *Isoproterenol Hydrochloride.*

Assay—

Standard preparation—Prepare as directed in the *Assay* under *Isoproterenol Hydrochloride.*

Assay preparation—Transfer about 125 mg of Isoproterenol Sulfate, accurately weighed, to a 25-mL volumetric flask, dissolve in sodium bisulfite solution (3 in 1000), dilute with sodium bisulfite solution to volume, and mix. Transfer 5.0 mL of this solution to a 100-mL volumetric flask, dilute with 0.17 *N* acetic acid to volume, and mix.

Chromatographic system—Proceed as directed for *Procedure* in the *Assay* under *Isoproterenol Hydrochloride.* Calculate the quantity, in mg, of $(C_{11}H_{17}NO_3)_2 \cdot H_2SO_4$ in the portion of Isoproterenol Sulfate taken by the formula:

$$(260.30/247.72)(0.5C)(h_U/h_S),$$

in which 260.30 is one-half of the molecular weight of anhydrous isoproterenol sulfate, 247.72 is the molecular weight of isoproterenol hydrochloride, and C, h_U, and h_S are as defined therein.

Isoproterenol Sulfate Inhalation Aerosol

» Isoproterenol Sulfate Inhalation Aerosol is a suspension of microfine Isoproterenol Sulfate in fluorochlorohydrocarbon propellants in a pressurized container. It contains not less than 90.0 percent and not more than 110.0 percent of the labeled amount of $(C_{11}H_{17}NO_3)_2 \cdot H_2SO_4$, and it delivers not less than 75.0 percent and not more than 125.0 percent of the labeled amount of $(C_{11}H_{17}NO_3)_2 \cdot H_2SO_4$ per inhalation through an oral inhalation actuator.

Packaging and storage—Preserve in small, nonreactive, light-resistant aerosol containers equipped with metered-dose valves and provided with oral inhalation actuators.

Reference standard—*USP Isoproterenol Hydrochloride Reference Standard*—Dry in vacuum over phosphorus pentoxide for 4 hours before using.

Identification—

A: Place 10 mL of water in a small beaker, and deliver 10 sprays from Isoproterenol Sulfate Inhalation Aerosol under the surface of the water, actuating the valve by pressing the tip against the bottom of the beaker. Filter, and to 5 mL of the filtrate add 1 drop of dilute hydrochloric acid (1 in 120). Add 0.50 mL of 0.10 *N* iodine, allow to stand for 5 minutes, and add 1.0 mL of 0.10 *N* sodium thiosulfate: a red-brown color is produced.

B: A portion of the filtrate obtained in *Identification test A* responds to the tests for *Sulfate* ⟨191⟩.

Microbial limits—It meets the requirements of the tests for absence of *Staphylococcus aureus* and *Pseudomonas aeruginosa* under *Microbial Limit Tests* ⟨61⟩.

Unit spray content—

Unit spray sampling apparatus—Prepare as directed under *Aerosols* ⟨601⟩.

Ferro-citrate solution and *Buffer solution*—Prepare as directed under *Epinephrine Assay* ⟨391⟩.

Standard preparation—Prepare as directed in the *Assay.*

Test preparation—Transfer 20.0 mL of a freshly prepared solution of sodium bisulfite (1 in 500) to the collection chamber of the sampling apparatus, and clamp the intake tube in a horizontal position. Prime the valve of Aerosol by alternately shaking and firing it 10 times through its oral inhalation actuator. Couple the aerosol container and actuator to the intake tube of the apparatus, turn on the vacuum source, and set the flow regulator for an airflow rate of 12 ± 1 liters per minute. Turn off the vacuum, and remove the aerosol container, leaving the actuator in place. Shake the container, seat it in the actuator, turn on the vacuum, immediately deliver a single spray into the system, and turn off the vacuum. Deliver a total of 10 sprays in this manner. Uncouple the aerosol container and actuator from the assembly, clamp the intake tube in a vertical position, and scavenge the isoproterenol sulfate remaining in the intake system as follows: Disconnect the collection chamber, transfer the test solution to a small beaker, then wash the inside of the intake tube with the test solution, collecting the washings in the collection chamber placed a few cm below its normal position. Recycle the test solution twice in this manner, then add 10 mL of chloroform to the collection chamber, insert the stopper, and shake vigorously for 1 minute. Transfer the mixture to a centrifuge tube, centrifuge for 5 minutes, and use the clear supernatant liquid as directed in the *Procedure.*

Procedure—Proceed as directed for *Procedure* in the *Assay.* Calculate the quantity, in mg, of $(C_{11}H_{17}NO_3)_2 \cdot H_2SO_4$ delivered per spray by the formula:

$$(260.30/247.72)(0.02C/10)(A_U/A_S),$$

in which 260.30 is one-half of the molecular weight of isoproterenol sulfate (anhydrous), and 247.72 is the molecular weight of isoproterenol hydrochloride, C is the concentration, in µg per mL, of USP Isoproterenol Hydrochloride RS in the *Standard preparation,* and A_U and A_S are the absorbances of the solutions from the *Assay preparation* and the *Standard preparation,* respectively.

Particle size—Prime the valve of Isoproterenol Sulfate Inhalation Aerosol by alternately shaking and firing it several times through its oral inhalation actuator, and then actuate one measured spray onto a clean, dry microscope slide held 5 cm from the end of the actuator, perpendicular to the direction of the spray. Carefully rinse the slide with about 2 mL of carbon tetrachloride, and allow to dry. Examine the slide under a microscope, equipped with a calibrated ocular micrometer, using 450× magnification. Focus on the particles of 25 fields of view near the center of the test specimen pattern, and note the size of the great majority of individual particles: they are less than 5 µm in diameter. Record the number and size of all individual crystalline particles (not agglomerates) more than 10 µm in length measured along the longest axis: not more than 10 such particles are observed.

Other requirements—It meets the requirements for *Leak Testing* under *Aerosols* ⟨601⟩.

Assay—

Ferro-citrate solution and *Buffer solution*—Prepare as directed under *Epinephrine Assay* ⟨391⟩.

Standard preparation—Transfer about 50 mg of USP Isoproterenol Hydrochloride RS, accurately weighed, to a 100-mL volumetric flask, add a freshly prepared solution of sodium bisulfite (1 in 500) to volume, and mix. Transfer 10.0 mL of this solution to a second 100-mL volumetric flask, dilute with the sodium bisulfite solution to volume, and mix. The concentration of USP Isoproterenol Hydrochloride RS in the *Standard preparation* is about 50 µg per mL.

Assay preparation—[NOTE—A suitable sampling beaker is one having a small indentation formed in its inside bottom surface having dimensions adequate to accept the aerosol valve stem during actuation, thereby preventing particle entrapment and side-of-stem leakage during the sample delivery.] Place 20 mL of chloroform in a suitable 100-mL beaker. Prime the valve of Isoproterenol Sulfate Inhalation Aerosol by alternately shaking and firing it 10 times through its oral inhalation actuator. Accurately weigh the Inhalation Aerosol, shake it, and immediately deliver a single spray under the surface of the chloroform, actuating the valve by pressing the tip into the indentation in the bottom of the beaker. Raise the Inhalation Aerosol above the surface of the chloroform, and shake it gently preparatory to delivering another spray similarly under the surface of the chloroform. Deliver a total of 5 sprays in this manner. Rinse the valve stem and ferrule with about 2 mL of chloroform, collecting the rinsing with the sample in the beaker. Allow the Inhalation Aerosol to dry, weigh it and determine the total weight of the 5 sprays. Transfer the solution to a centrifuge tube with the aid of two 3-mL portions of chloroform, and add 10.0 mL of freshly prepared sodium bisulfite solution (1 in 500). Insert the stopper, shake vigorously for 1 minute, centrifuge for 5 minutes, and use the clear supernatant liquid as directed in the *Procedure*.

Procedure—Transfer 5.0 mL each of the *Standard preparation* and the *Assay preparation* to separate test tubes. To each tube add 100 µL of *Ferro-citrate solution* and 1.0 mL of *Buffer solution*, and mix. Concomitantly determine the absorbances of the solutions in 1-cm cells at the wavelength of maximum absorbance at about 530 nm, with a suitable spectrophotometer, using water as the blank. Calculate the quantity, in mg, of $(C_{11}H_{17}NO_3)_2 \cdot H_2SO_4$ in each mL of the Inhalation Aerosol taken by the formula:

$$(260.30/247.72)(0.01Cd/W)(A_U/A_S),$$

in which 260.30 is one-half of the molecular weight of isoproterenol sulfate (anhydrous), and 247.72 is the molecular weight of isoproterenol hydrochloride, *C* is the concentration, in µg per mL, of USP Isoproterenol Hydrochloride RS in the *Standard preparation*, *d* is the density, in g per mL, of the Inhalation Aerosol, *W* is the weight, in g, of the portion of Inhalation Aerosol taken, and A_U and A_S are the absorbances of the solutions from the *Assay preparation* and the *Standard preparation*, respectively. [The density, *d*, is determined as follows: Weigh a known volume (*v*) of the Inhalation Aerosol in a suitable, gas-tight, 5-mL syringe equipped with a linear valve. Calibrate the volume of the syringe by filling to the 5-mL mark with dichlorotetrafluoroethane withdrawn from a plastic-coated glass vial sealed with a neoprene multiple-dose rubber stopper and an aluminum seal, using 1.456 g per mL as the density of the calibrating liquid. Maintain the dichlorotetrafluoroethane, the Inhalation Aerosol assay specimen, and the syringe (protected from becoming wet) in a water bath at 25°. Obtain the specimen equivalent to the same volume as that obtained during the sampling procedure, from the Inhalation Aerosol by means of a sampling device consisting of a replaceable rubber septum engaged in the plate threads at one end of a threaded fitting, the opposite end of which contains a sharpened tube capable of puncturing the aerosol container, and a rubber gasket around the tube to prevent leakage of the container contents after puncture.* Record the weight of the volume (*v*) of the Inhalation Aerosol as *w*, and calculate the density by the formula *w/v*.]

Isoproterenol Sulfate Inhalation Solution

» Isoproterenol Sulfate Inhalation Solution is a solution of Isoproterenol Sulfate in Purified Water. It may contain Sodium Chloride. It contains not less than 90.0 percent and not more than 115.0 percent of the labeled amount of $(C_{11}H_{17}NO_3)_2 \cdot H_2SO_4$.

Packaging and storage—Store in small, well-filled, tight containers, protected from light.

Labeling—Label it to indicate that the Inhalation Solution is not to be used if it is brown or contains a precipitate.

Reference standard—*USP Isoproterenol Hydrochloride Reference Standard*—Dry in vacuum over phosphorus pentoxide for 4 hours before using.

Identification—It responds to *Identification tests C, D*, and *E* under *Isoproterenol Sulfate*.

Assay—

Standard preparation—Prepare as directed in the *Assay* under *Isoproterenol Hydrochloride*.

Assay preparation—Transfer an accurately measured volume of Isoproterenol Sulfate Inhalation Solution, equivalent to about 25 mg of isoproterenol sulfate, to a 100-mL volumetric flask, add 50.0 mL of 0.30 *N* acetic acid, dilute with water to volume, and mix.

Chromatographic system—Proceed as directed in the *Assay* under *Isoproterenol Hydrochloride*.

Procedure—Proceed as directed for *Procedure* in the *Assay* under *Isoproterenol Hydrochloride*. Calculate the quantity, in mg, of $(C_{11}H_{17}NO_3)_2 \cdot H_2SO_4$ in each mL of the Inhalation Solution taken by the formula:

$$(260.30/247.72)(0.1)(C/V)(h_U/h_S),$$

in which 260.30 is one-half of the molecular weight of anhydrous isoproterenol sulfate, 247.72 is the molecular weight of isoproterenol hydrochloride, *V* is the volume, in mL, of Inhalation Solution taken, and *C*, h_U, and h_S are as defined therein.

Isosorbide Concentrate

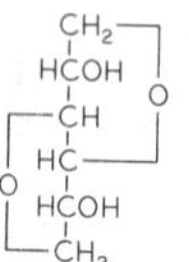

$C_6H_{10}O_4$ 146.14

D-Glucitol, 1,4:3,6-dianhydro-.

1,4:3,6-Dianhydro-D-glucitol [*652-67-5*].

» Isosorbide Concentrate is an aqueous solution containing, in each 100 g, not less than 70.0 g and not more than 80.0 g of $C_6H_{10}O_4$.

Packaging and storage—Preserve in tight, light-resistant containers.

Reference standard—*USP Isosorbide Reference Standard*—After opening the ampul, store in a tightly closed container.

Identification—[NOTE—Isosorbide is hygroscopic. Take precautions to protect isolated isosorbide crystals from atmospheric moisture.]

A: Dry a portion of it in an evaporating dish over phosphorus pentoxide at 70° and at a pressure of 50 mm of mercury for 48 hours, changing the phosphorus pentoxide after 24 hours. Scratch the bottom of the dish with a glass rod or seed with a crystal of isosorbide, if necessary, to initiate crystallization: the crystals so obtained melt between 60° and 63° when tested by the procedure for *Class I* substances (see *Melting Range or Temperature* ⟨741⟩).

B: The infrared absorption spectrum of a potassium bromide dispersion of the crystals obtained as directed in *Identification test A* exhibits maxima and minima only at the same wavelengths as that of a similar preparation of USP Isosorbide RS.

Specific rotation ⟨781⟩: between +44.5° and +47.0°, calculated on the anhydrous basis, determined in a solution containing 800 mg of isosorbide in each 10 mL.

Water, *Method I* ⟨921⟩: between 24.0% and 26.0%.

Residue on ignition ⟨281⟩: not more than 0.01%.

Arsenic ⟨211⟩: not more than 1 ppm, calculated on the anhydrous basis.

Heavy metals, *Method II* ⟨231⟩: not more than 5 ppm, calculated on the anhydrous basis.

Periodate consumption—Dilute about 15 g, accurately weighed, with 25 mL of water, and add 50.0 mL of a solution prepared by dissolving 5.4 g of periodic acid in 100 mL of water and adding 1900 mL of glacial acetic acid. Allow to stand for 1 hour. Add 20 mL of potassium iodide TS, and titrate with 0.1 *N* sodium thiosulfate VS to the disappearance of the brown color. Add 3 mL of starch TS, and complete the titration. Perform a blank determination, and note the difference in volumes required. If the volume required for the specimen is less than 0.8 of that required for the blank, repeat the procedure with a smaller specimen. The difference in volume corresponds to not more than 0.20 mL of 0.1 *N* sodium thiosulfate for each g of Isosorbide Concentrate taken.

Acid value—Dilute about 15 g, accurately weighed, with 50 mL of water, and titrate with 0.02 *N* potassium hydroxide VS to a phenolphthalein end-point. Perform a blank determination, and make any necessary correction. Calculate the acid value by the formula:

* A suitable sampling system is available from Alltech Associates, 2051 Waukegan Rd., Deerfield, IL 60015.

$$56.11(AN/W),$$

in which A is the number of mL of potassium hydroxide VS consumed and N is its normality, and W is the weight, in g, of Isosorbide Concentrate taken. The limit is 0.5, calculated on the anhydrous basis.

Methyl ethyl ketone—

Internal standard solution—Prepare a solution in water containing about 1 mg per mL of methyl isobutyl ketone.

Standard preparation—Prepare a solution in water containing an accurately known concentration of methyl ethyl ketone equivalent to about 1 mg per mL. Pipet 5 mL of this solution into a 100-mL volumetric flask, add 5.0 mL of *Internal standard solution*, add water to volume, and mix.

Test preparation—Pipet 5 mL of *Internal standard solution* into a 100-mL volumetric flask, add Isosorbide Concentrate to volume, and mix.

Support—Place about 90 g of unsilanized support S1A in a crystallizing dish, and cover it with chloroform. Stir the mixture thoroughly, and carefully remove the supernatant chloroform with an aspirator. Spread the moist support on a clean surface, and allow it to air-dry. Place the dried support in the crystallizing dish, and cover it with 0.5 *N* alcoholic potassium hydroxide TS. Allow it to stand for one-half hour with occasional stirring. Carefully pour off the supernatant alcoholic potassium hydroxide solution, and wash the moist support with water until the washing is neutral to phenolphthalein indicator. Spread the moist support on a clean surface, and allow it to air-dry.

Chromatographic system—The gas chromatograph is equipped with a flame-ionization detector and a 0.6-m × 2-mm column packed with 25 percent liquid phase G16 on unsilanized acid- and base-washed *Support* which has been washed with chloroform, and conditioned as directed (see *Chromatography* ⟨621⟩). The column is maintained at 70°, and nitrogen is used as the carrier gas at a flow rate of about 30 mL per minute. In a suitable chromatogram, the relative standard deviation of five replicate injections is not more than 3.0%, and the resolution is not less than 2.0.

Procedure—Inject about 3 µL of the *Standard preparation* into the gas chromatograph, record the chromatogram, and measure the peak response of each component. [NOTE—Clean the syringe after each injection with pentane. Do not use acetone.] Similarly inject about 3 µL of the *Test preparation*, record the chromatogram, and measure the peak response of each component. Calculate the quantity, in mg, of methyl ethyl ketone in each mL of the Isosorbide Concentrate taken by the formula:

$$(1/0.95)C(R_U/R_S),$$

in which C is the concentration, in mg per mL, of methyl ethyl ketone in the *Standard preparation*, and R_U and R_S are the ratios of the response of the methyl ethyl ketone to the response of the internal standard obtained from the *Test preparation* and the *Standard preparation*, respectively. The limit is 0.05 mg per mL.

Assay—

Internal standard solution—Dissolve a suitable quantity of triethylene glycol in water to obtain a solution containing about 15 mg per mL.

Standard solution—Prepare a solution of USP Isosorbide RS in water containing an accurately known concentration equivalent to about 25 mg of $C_6H_{10}O_4$ per mL.

Standard preparations—Pipet 2-, 3-, 4-, and 5-mL quantities of *Standard solution* into separate 50-mL volumetric flasks, add 5.0 mL of *Internal standard solution* to each, add water to volume, and mix.

Assay preparation—Transfer about 200 mg of Isosorbide Concentrate, accurately weighed, to a 100-mL volumetric flask, add 10.0 mL of *Internal standard solution*, add water to volume, and mix.

Chromatographic system—The gas chromatograph is equipped with a flame-ionization detector and a 0.6-m × 3-mm glass column packed with support S9. The column is maintained at 230°, and nitrogen is used as the carrier gas. The retention time of the isosorbide peak is about 1.5, relative to that of triethylene glycol.

System suitability and standard curve—Inject 1-µL portions of each *Standard preparation*, and record each peak response. Plot the ratio of the peak response of isosorbide to that of triethylene glycol versus the concentration, in mg per mL, of isosorbide in the respective *Standard preparation*. The analytical system is suitable for conducting the assay if the correlation coefficient for the Standard curve is greater than 0.980, the resolution, *R*, is not less than 1.5, and neither tailing factor exceeds 2.0.

Procedure—Inject a 1-µL portion of the *Assay preparation*, record the peak responses for the two major peaks, calculate the ratio of the peak responses, and determine the concentration, *C*, in mg per mL, of isosorbide in the *Assay preparation* by reference to the *Standard curve*. Calculate the quantity, in mg, of $C_6H_{10}O_4$ in the Isosorbide Concentrate taken by the formula:

$$100C.$$

Isosorbide Oral Solution

» Isosorbide Oral Solution contains not less than 90.0 percent and not more than 110.0 percent of the labeled amount of $C_6H_{10}O_4$.

Packaging and storage—Preserve in tight containers.

Reference standard—*USP Isosorbide Reference Standard*—After opening the ampul, store in a tightly closed container.

Identification—The retention time of the major peak obtained in the *Assay* is the same as that obtained from the *Standard preparations.*

pH ⟨791⟩: between 3.2 and 3.8.

Assay—

Internal standard solution, Standard solution, Standard preparations, Chromatographic system, and *System suitability and standard curve*—Proceed as directed in the *Assay* under *Isosorbide Concentrate.*

Assay preparation—Transfer an accurately measured volume of Isosorbide Oral Solution, equivalent to about 450 mg of isosorbide, to a 250-mL volumetric flask, add 25.0 mL of *Internal standard solution*, then add water to volume, and mix.

Procedure—Proceed as directed for *Procedure* in the *Assay* under *Isosorbide Concentrate.* Calculate the quantity, in mg, of $C_6H_{10}O_4$ in each mL of the Oral Solution taken by the formula:

$$250(C/V),$$

in which C is the concentration, in mg per mL, of isosorbide in the *Assay preparation* found by reference to the *Standard curve*, and V is the volume, in mL, of Isosorbide Oral Solution taken.

Diluted Isosorbide Dinitrate

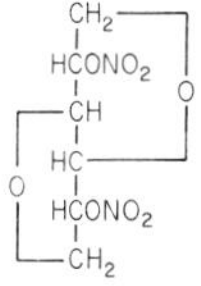

$C_6H_8N_2O_8$ 236.14

D-Glucitol, 1,4:3,6-dianhydro-, dinitrate.

1,4:3,6-Dianhydro-D-glucitol dinitrate [*87-33-2*].

» Diluted Isosorbide Dinitrate is a dry mixture of isosorbide dinitrate ($C_6H_8N_2O_8$) with Lactose, Mannitol, or suitable inert excipients to permit safe handling. It may contain up to 1.0 percent of a suitable stabilizer, such as Ammonium Phosphate. It contains not less than 95.0 percent and not more than 105.0 percent of the labeled amount of $C_6H_8N_2O_8$. It usually contains approximately 25 percent of isosorbide dinitrate.

Caution—Exercise proper precautions in handling undiluted isosorbide dinitrate, which is a powerful explosive and can be exploded by percussion or excessive heat. Only exceedingly small amounts should be isolated.

Packaging and storage—Preserve in tight containers.

Reference standard—*USP Diluted Isosorbide Dinitrate Reference Standard*—This is a mixture containing 25% of isosorbide dinitrate in mannitol. Do not dry before using.

Identification—Transfer to a medium-porosity, sintered-glass filtering crucible a quantity of it, equivalent to about 50 mg of isosorbide dinitrate, and pass three 5-mL portions of acetone through it. Evaporate the combined extracts at a temperature not exceeding 35°, with the aid of a gentle current of air, and dry the residue in vacuum over calcium chloride at room temperature for 16 hours: the infrared absorption spectrum of a 1 in 40 solution of the residue so obtained, in chloroform, determined in a 0.1-mm cell, exhibits maxima only at the same wavelengths as that of a similar preparation from the residue obtained from USP Diluted Isosorbide Dinitrate RS.

Loss on drying ⟨731⟩—Dry it in vacuum over calcium chloride at room temperature for 16 hours: it loses not more than 1.0% of its weight.

Heavy metals, *Method II* ⟨231⟩: 0.001%.

Assay—

Buffer solution—Dissolve 15.4 g of ammonium acetate in water, add 11.5 mL of glacial acetic acid, dilute with water to 1000 mL, and mix to obtain a solution having a pH of about 4.7.

Mobile phase—Mix 350 mL of water, 100 mL of *Buffer solution*, and 550 mL of methanol. Cool to room temperature, dilute with water to 1000 mL, mix, degas, and filter. Make adjustments if necessary (see *System Suitability* under *Chromatography* ⟨621⟩).

Internal standard solution—Transfer about 6.0 g of diluted nitroglycerin (10% in lactose) to a 200-mL flask, add about 120 mL of methanol, sonicate for 5 minutes, shake for 30 minutes, dilute with methanol to 200 mL, and mix. Allow the undissolved lactose to settle, filter, and store the filtrate in an air-tight container.

Standard preparation—Transfer about 125 mg of recently mixed USP Diluted Isosorbide Dinitrate RS, accurately weighed, to a 50-mL volumetric flask, add about 30 mL of *Mobile phase*, shake for 30 minutes, dilute with *Mobile phase* to volume, and mix. Pipet 10 mL of the resulting solution into a 25-mL volumetric flask, and add 4.0 mL of *Internal standard solution* and 4 mL of dilute *Buffer solution* (1 in 10). Cool to room temperature, dilute with *Mobile phase* to volume, and mix to obtain a solution having a known concentration of about 0.25 mg of isosorbide dinitrate per mL, based on the quantity of USP Diluted Isosorbide Dinitrate RS weighed and the labeled content of isosorbide dinitrate. Filter a portion of this solution through a 0.45-μm filter.

Assay preparation—Transfer an accurately weighed quantity of recently mixed Diluted Isosorbide Dinitrate, equivalent to about 30 mg of isosorbide dinitrate, to a 50-mL volumetric flask. Proceed as directed for *Standard preparation*, beginning with "add about 30 mL of *Mobile phase*."

Chromatographic system (see *Chromatography* ⟨621⟩)—The liquid chromatograph is equipped with a 220-nm detector and a 4-mm × 25-cm column that contains packing L1. The flow rate is about 1 mL per minute. Chromatograph the *Standard preparation*, and record the peak responses as directed under *Procedure:* the resolution, *R*, between the isosorbide dinitrate and nitroglycerin peaks is not less than 2.0, and the relative standard deviation of the ratio of the peak responses for replicate injections is not more than 2%.

Procedure—Separately inject equal volumes (about 20 μL) of the *Standard preparation* and the *Assay preparation* into the chromatograph, record the chromatograms, and measure the responses for the major peaks. The relative retention times are about 0.75 for isosorbide dinitrate and 1.0 for nitroglycerin. The relative retention times of isosorbide mononitrates, if present, are about 0.38. Calculate the quantity, in mg, of $C_6H_8N_2O_8$ in the portion of Diluted Isosorbide Dinitrate taken by the formula:

$$125C(R_U/R_S),$$

in which C is the concentration, in mg per mL, of isosorbide dinitrate from the USP Diluted Isosorbide Dinitrate RS taken for the *Standard preparation*, and R_U and R_S are the ratios of the peak responses obtained from the *Assay preparation* and the *Standard preparation*, respectively.

Isosorbide Dinitrate Extended-release Capsules

» Isosorbide Dinitrate Extended-release Capsules contain not less than 90.0 percent and not more than 110.0 percent of the labeled amount of $C_6H_8N_2O_8$.

Packaging and storage—Preserve in well-closed containers.

Reference standard—*USP Diluted Isosorbide Dinitrate Reference Standard*—This is a mixture containing 25% of isosorbide dinitrate in mannitol. Do not dry before using.

Identification—The finely powdered contents of Extended-release Capsules respond to the *Identification test* under *Isosorbide Dinitrate Tablets*. If separation of interferences is required, transfer a quantity of the finely powdered contents of the Capsules, equivalent to about 20 mg of isosorbide dinitrate, to a glass-stoppered centrifuge tube, add 10 mL of sodium hydroxide solution (1 in 250), shake to wet the powder, add 15 mL of solvent hexane, and again shake. Centrifuge the mixture, and transfer the upper phase to a beaker. Place in a freezer, at a temperature of about −14°, the beaker and a short-stem funnel fitted with a cotton plug that previously has been chloroform-washed and dried. After 30 minutes, filter the solution while still in the freezer. Evaporate the solvent, and dry the residue in vacuum over calcium chloride for 16 hours: the infrared absorption spectrum of the residue so obtained, dissolved in 0.4 mL of chloroform and determined with the use of matched 0.1-mm cells, shows all of the significant absorption bands present in the spectrum obtained for a similar preparation from the residue obtained from USP Diluted Isosorbide Dinitrate RS. The major peaks are at about 1650 cm^{-1}, 1284 cm^{-1} and 1275 cm^{-1} (a doublet), 1106 cm^{-1}, and 844 cm^{-1}.

Uniformity of dosage units ⟨905⟩: meet the requirements.

Assay—

Buffer solution, Mobile phase, Internal standard solution, Standard preparation, and *Chromatographic system*—Prepare as directed in the *Assay* under *Diluted Isosorbide Dinitrate*.

Assay preparation—Weigh and finely powder the contents of not less than 20 Isosorbide Dinitrate Extended-release Capsules. Transfer an accurately weighed portion of the powder, equivalent to about 12.5 mg of isosorbide dinitrate, to a dry, 50-mL volumetric flask, add about 30 mL of *Mobile phase*, and shake the mixture by hand immediately, to prevent clumping. If clumping persists, disperse with the aid of sonication, or break the aggregates with a stirring rod, or warm on a steam bath while keeping the flask stoppered, or allow the flask to stand until the clumps dissipate. [NOTE—If clumping still continues, discard the mixture, and instead disperse an accurately weighed test portion in 15 mL of a 1 in 10 dilution of *Buffer solution* in water by heating on a steam bath for 1 hour with frequent shaking, then add 15 mL of methanol.] Shake for 30 minutes. Add 8.0 mL of *Internal standard solution*, cool to room temperature, add 8 mL of a 1 in 10 dilution of *Buffer solution* in water, dilute with *Mobile phase* to volume, and mix. Filter a portion through a microporous membrane filter.

Procedure—Proceed as directed for *Procedure* in the *Assay* under *Diluted Isosorbide Dinitrate*. Calculate the quantity, in mg, of $C_6H_8N_2O_8$ in the portion of Extended-release Capsules taken by the formula:

$$50C(R_U/R_S),$$

in which C is the concentration, in mg per mL, of isosorbide dinitrate from the USP Isosorbide Dinitrate RS taken for the *Standard preparation*, and R_U and R_S are the ratios of the peak

responses obtained from the *Assay preparation* and the *Standard preparation*, respectively.

Isosorbide Dinitrate Tablets

» Isosorbide Dinitrate Tablets contain not less than 90.0 percent and not more than 110.0 percent of the labeled amount of $C_6H_8N_2O_8$.

Packaging and storage—Preserve in well-closed containers.

Reference standard—*USP Diluted Isosorbide Dinitrate Reference Standard*—This is a mixture containing 25% of isosorbide dinitrate in mannitol. Do not dry before using.

Identification—Transfer a suitable quantity of finely powdered Tablets to a glass-stoppered centrifuge tube. Add 10 mL of sodium hydroxide solution (1 in 250), shake to wet the powder, then add 15 mL of solvent hexane, and again shake. Centrifuge the mixture, and transfer the upper phase to a beaker. Evaporate the solvent, and dry the residue in vacuum over anhydrous calcium chloride at room temperature for 16 hours: the infrared absorption spectrum of a suitable solution in chloroform of the residue so obtained exhibits maxima only at the same wavelengths as that of a similar preparation from the residue obtained from USP Diluted Isosorbide Dinitrate RS.

Dissolution ⟨711⟩—

Medium: water; 1000 mL.

Apparatus 2: 75 rpm.

Time: 45 minutes.

Mobile phase—Prepare a suitable degassed and filtered mixture of pH 3.0, 0.1 *M* ammonium sulfate and methanol (50:50). Make adjustments if necessary (see *System Suitability* under *Chromatography* ⟨621⟩), using sulfuric acid for any necessary pH adjustment.

Chromatographic system (see *Chromatography* ⟨621⟩)—The liquid chromatograph is equipped with a 220-nm detector and a 4.6-mm × 5-cm column that contains packing L1. The flow rate is about 1.0 mL per minute. Chromatograph replicate injections of the Standard solution, and record the peak responses as directed under *Procedure:* the relative standard deviation is not more than 2.0%, and the tailing factor is not more than 1.5.

Procedure—Separately inject equal volumes (about 20 µL) of the Standard solution and a filtered aliquot of the solution under test into the chromatograph, record the chromatograms, and measure the responses for the major peaks. Calculate the amount of $C_6H_8N_2O_8$ dissolved in comparison with a Standard solution having a known concentration of USP Isosorbide Dinitrate RS, similarly prepared and chromatographed.

Tolerances—Not less than 70% (*Q*) of the labeled amount of $C_6H_8N_2O_8$ is dissolved in 45 minutes.

Uniformity of dosage units ⟨905⟩: meet the requirements.

Assay—

Buffer solution, Mobile phase, Internal standard solution, Standard preparation, and *Chromatographic system*—Prepare as directed in the *Assay* under *Diluted Isosorbide Dinitrate.*

Assay preparation—Weigh and finely powder not less than 20 Isosorbide Dinitrate Tablets. Transfer an accurately weighed portion of the powder, equivalent to about 12.5 mg of isosorbide dinitrate, to a dry, 50-mL volumetric flask, add about 30 mL of *Mobile phase*, and shake the mixture by hand immediately, to prevent clumping. If clumping persists, disperse with the aid of sonication, or break the aggregates with a stirring rod. Shake for 30 minutes. Add 8.0 mL of *Internal standard solution*, cool to room temperature, add 8 mL of a 1 in 10 dilution of *Buffer solution* in water, dilute with *Mobile phase* to volume, and mix. Filter a portion through a 0.45-µm filter.

Procedure—Proceed as directed for *Procedure* in the *Assay* under *Diluted Isosorbide Dinitrate.* Calculate the quantity, in mg, of $C_6H_8N_2O_8$ in the portion of Tablets taken by the formula:

$$50C(R_U / R_S),$$

in which C is the concentration, in mg per mL, of isosorbide dinitrate from the USP Isosorbide Dinitrate RS taken for the *Standard preparation*, and R_U and R_S are the ratios of the peak responses obtained from the *Assay preparation* and the *Standard preparation*, respectively.

Isosorbide Dinitrate Chewable Tablets

» Isosorbide Dinitrate Chewable Tablets contain not less than 90.0 percent and not more than 110.0 percent of the labeled amount of $C_6H_8N_2O_8$.

Packaging and storage—Preserve in well-closed containers.

Reference standard—*USP Diluted Isosorbide Dinitrate Reference Standard*—This is a mixture containing 25% of isosorbide dinitrate in mannitol. Do not dry before using.

Identification—Chewable Tablets respond to the *Identification test* under *Isosorbide Dinitrate Tablets.* Where separation of interferences is required, use the technique given under the *Identification test* for *Isosorbide Dinitrate Extended-release Capsules.*

Uniformity of dosage units ⟨905⟩: meet the requirements.

Assay—

Buffer solution, Mobile phase, Internal standard solution, Standard preparation, and *Chromatographic system*—Prepare as directed in the *Assay* under *Diluted Isosorbide Dinitrate.*

Assay preparation—Weigh and finely powder not less than 20 Isosorbide Dinitrate Chewable Tablets. Transfer an accurately weighed portion of the powder, equivalent to about 12.5 mg of isosorbide dinitrate, to a dry, 50-mL volumetric flask, add about 30 mL of *Mobile phase*, and shake the mixture by hand immediately, to prevent clumping. If clumping persists, disperse with the aid of sonication, or break the aggregates with a stirring rod, or warm on a steam bath while keeping the flask stoppered, or allow the flask to stand until the clumps dissipate. [NOTE—If clumping still continues, discard the mixture, and instead disperse an accurately weighed test portion in 15 mL of a 1 in 10 dilution of *Buffer solution* in water by heating on a steam bath for 1 hour with frequent shaking, then add 15 mL of methanol.] Shake for 30 minutes. Add 8.0 mL of *Internal standard solution*, cool to room temperature, add 8 mL of a 1 in 10 dilution of *Buffer solution* in water, dilute with *Mobile phase* to volume, and mix. Filter a portion through a microporous membrane filter.

Procedure—Proceed as directed for *Procedure* in the *Assay* under *Diluted Isosorbide Dinitrate.* Calculate the quantity, in mg, of $C_6H_8N_2O_8$ in the portion of Chewable Tablets taken by the formula:

$$50C(R_U / R_S),$$

in which C is the concentration, in mg per mL, of isosorbide dinitrate from the USP Isosorbide Dinitrate RS taken for the *Standard preparation*, and R_U and R_S are the ratios of the peak responses obtained from the *Assay preparation* and the *Standard preparation*, respectively.

Isosorbide Dinitrate Extended-release Tablets

» Isosorbide Dinitrate Extended-release Tablets contain not less than 90.0 percent and not more than 110.0 percent of the labeled amount of $C_6H_8N_2O_8$.

Packaging and storage—Preserve in well-closed containers.

Reference standard—*USP Diluted Isosorbide Dinitrate Reference Standard*—This is a mixture containing 25% of isosorbide dinitrate in mannitol. Do not dry before using.

Identification—Extended-release Tablets respond to the *Identification test* under *Isosorbide Dinitrate Tablets.* Where separation of interferences is required, use the technique given under the *Identification test* for *Isosorbide Dinitrate Extended-release Capsules.*

Uniformity of dosage units ⟨905⟩: meet the requirements.

Assay—

Buffer solution, Mobile phase, Internal standard solution, Standard preparation, and *Chromatographic system*—Prepare as directed in the *Assay* under *Diluted Isosorbide Dinitrate.*

Assay preparation—Weigh and finely powder not less than 20 Isosorbide Dinitrate Extended-release Tablets. Transfer an accurately weighed portion of the powder, equivalent to about 12.5 mg of isosorbide dinitrate, to a dry, 50-mL volumetric flask, add about 30 mL of *Mobile phase,* and shake the mixture by hand immediately, to prevent clumping. If clumping persists, disperse with the aid of sonication, or break the aggregates with a stirring rod, or warm on a steam bath while keeping the flask stoppered, or allow the flask to stand until the clumps dissipate. [NOTE—If clumping still continues, discard the mixture, and instead disperse an accurately weighed test portion in 15 mL of a 1 in 10 dilution of *Buffer solution* in water by heating on a steam bath for 1 hour with frequent shaking, then add 15 mL of methanol.] Shake for 30 minutes. Add 8.0 mL of *Internal standard solution,* cool to room temperature, add 8 mL of a 1 in 10 dilution of *Buffer solution* in water, dilute with *Mobile phase* to volume, and mix. Filter a portion through a microporous membrane filter.

Procedure—Proceed as directed for *Procedure* in the *Assay* under *Diluted Isosorbide Dinitrate.* Calculate the quantity, in mg, of $C_6H_8N_2O_8$ in the portion of Extended-release Tablets taken by the formula:

$$50C(R_U/R_S),$$

in which C is the concentration, in mg per mL, of isosorbide dinitrate from the USP Isosorbide Dinitrate RS taken for the *Standard preparation,* and R_U and R_S are the ratios of the peak responses obtained from the *Assay preparation* and the *Standard preparation,* respectively.

Isosorbide Dinitrate Sublingual Tablets

» Isosorbide Dinitrate Sublingual Tablets contain not less than 90.0 percent and not more than 110.0 percent of the labeled amount of $C_6H_8N_2O_8$.

Packaging and storage—Preserve in well-closed containers.

Reference standard—*USP Diluted Isosorbide Dinitrate Reference Standard*—This is a mixture containing 25% of isosorbide dinitrate in mannitol. Do not dry before using.

Identification—Sublingual Tablets respond to the *Identification test* under *Isosorbide Dinitrate Tablets.*

Disintegration ⟨701⟩: 2 minutes, determined as directed for *Sublingual Tablets.*

Dissolution ⟨711⟩—

Medium: water; 900 mL.

Apparatus 2: 50 rpm.

Times: 15 minutes; 30 minutes.

Mobile phase—Prepare a suitable degassed and filtered mixture of pH 3.0, 0.1 *M* ammonium sulfate and methanol (50:50).

Chromatographic system (see *Chromatography* ⟨621⟩)—The liquid chromatograph is equipped with a 220-nm detector and a 4.6-mm × 5-cm column that contains packing L1. The flow rate is about 1.0 mL per minute. Chromatograph three replicate injections of the *Standard solution,* and record the peak responses as directed under *Procedure:* the relative standard deviation is not more than 2.0%, and the tailing factor is not more than 1.5.

Procedure—Separately inject equal volumes (about 20 µL) of the *Standard solution* and a filtered aliquot of the solution under test into the chromatograph, record the chromatograms, and measure the responses for the major peaks. Calculate the amount of $C_6H_8N_2O_8$ dissolved in comparison with a *Standard solution* having a known concentration of USP Isosorbide Dinitrate RS, similarly prepared and chromatographed.

Tolerances—Not less than 50% (*Q*) of the labeled amount of $C_6H_8N_2O_8$ is dissolved in 15 minutes, and not less than 70% (*Q*) of the labeled amount of $C_6H_8N_2O_8$ is dissolved in 30 minutes.

Uniformity of dosage units ⟨905⟩: meet the requirements.

Assay—

Buffer solution, Mobile phase, Internal standard solution, Standard preparation, and *Chromatographic system*—Prepare as directed in the *Assay* under *Diluted Isosorbide Dinitrate.*

Assay preparation—Weigh and finely powder not less than 20 Isosorbide Dinitrate Sublingual Tablets. Transfer an accurately weighed portion of the powder, equivalent to about 12.5 mg of isosorbide dinitrate, to a 50-mL volumetric flask, add about 30 mL of *Mobile phase,* and shake for 30 minutes. Add 8.0 mL of *Internal standard solution,* cool to room temperature, add 8 mL of a 1 in 10 dilution of *Buffer solution* in water, dilute with *Mobile phase* to volume, and mix. Filter a portion through a 0.45-µm filter.

Procedure—Proceed as directed for *Procedure* in the *Assay* under *Diluted Isosorbide Dinitrate.* Calculate the quantity, in mg, of $C_6H_8N_2O_8$ in the portion of Sublingual Tablets taken by the formula:

$$50C(R_U/R_S),$$

in which C is the concentration, in mg per mL, of isosorbide dinitrate from the USP Isosorbide Dinitrate RS taken for the *Standard preparation,* and R_U and R_S are the ratios of the peak responses obtained from the *Assay preparation* and the *Standard preparation,* respectively.

Isotretinoin

$C_{20}H_{28}O_2$ 300.44

Retinoic acid, 13-*cis*-.

3,7-Dimethyl-9-(2,6,6-trimethyl-1-cyclohexen-1-yl)2-*cis*-4-*trans*-6-*trans*-8-*trans*-nonatetraenoic acid [*4759-48-2*].

» Isotretinoin contains not less than 98.0 percent and not more than 102.0 percent of $C_{20}H_{28}O_2$, calculated on the dried basis.

Packaging and storage—Preserve in tight containers, under an atmosphere of an inert gas, protected from light.

Reference standards—*USP Isotretinoin Reference Standard*—Store ampuls at a temperature below 0°, allow to reach room temperature before opening, and use the contents promptly after opening ampuls. *USP Tretinoin Reference Standard*—Store ampuls at a temperature below 0°, allow to reach room temperature before opening, and use the contents promptly after opening ampuls.

NOTE—Avoid exposure to strong light, and use low-actinic glassware in the performance of the following procedures.

Identification—

A: The infrared absorption spectrum of a mineral oil dispersion of it exhibits maxima only at the same wavelengths as that of a similar preparation of USP Isotretinoin RS.

B: The ultraviolet absorption spectrum of a 1 in 250,000 solution in acidified isopropyl alcohol, prepared by diluting 1 mL of 0.01 *N* hydrochloric acid with isopropyl alcohol to 1000 mL, exhibits maxima and minima at the same wavelengths as that of a similar solution of USP Isotretinoin RS, concomitantly measured, and the respective absorptivities, calculated on the dried basis, at the wavelength of maximum absorbance at about 354 nm do not differ by more than 3.0%.

Loss on drying ⟨731⟩—Dry it at 105° for 3 hours: it loses not more than 0.5% of its weight.

Residue on ignition ⟨281⟩: not more than 0.1%.

Heavy metals, *Method II* ⟨231⟩: 0.002%.

Limit of tretinoin—

Mobile phase—Prepare a suitable filtered and degassed mixture of isooctane, isopropyl alcohol, and glacial acetic acid (99.65:0.25:0.1).

Standard solution—Dissolve an accurately weighed quantity of USP Tretinoin RS in a minimum quantity of methylene chloride, add an amount of isooctane to obtain a solution having a known concentration of about 250 µg per mL, and mix.

Standard preparation—Pipet 1 mL of *Standard solution* into a 100-mL volumetric flask, add isooctane to volume, and mix.

Test preparation—Transfer about 25 mg of Isotretinoin, accurately weighed, to a 100-mL volumetric flask, dissolve in a minimum quantity of methylene chloride, add isooctane to volume, and mix.

System suitability solution—Dissolve a quantity of USP Isotretinoin RS in a minimum amount of methylene chloride, add an amount of isooctane to obtain an isotretinoin concentration of about 250 µg per mL, and mix.

System suitability preparation—Pipet 1 mL of *Standard solution* into a 100-mL volumetric flask, add *System suitability solution* to volume, and mix.

Chromatographic system—The liquid chromatograph (see *Chromatography* ⟨621⟩) is equipped with a 352-nm detector and a 4.0-mm × 25-cm column containing 5-µm packing L3. The flow rate is about 1 mL per minute. Chromatograph about 20 µL of *System suitability preparation*, and record the peak responses. The relative retention times for isotretinoin and tretinoin are about 0.84 and 1.00, respectively. The relative standard deviation of the tretinoin peak response in replicate injections is not more than 2.0%. The resolution, *R*, of isotretinoin and tretinoin is not less than 2.0.

Procedure—Separately inject equal volumes (about 20 µL) of the *Standard preparation* and the *Test preparation* into the chromatograph, record the chromatograms, and measure the responses for the major peaks. Calculate the percentage of tretinoin by the formula:

$$10(C/W)(r_U/r_S),$$

in which *C* is the concentration, in µg per mL, of USP Tretinoin RS in the *Standard preparation*, *W* is the weight, in mg, of Isotretinoin taken, and r_U and r_S are the peak responses of the tretinoin peaks obtained from the *Test preparation* and the *Standard preparation*, respectively. The content of tretinoin is not more than 1.0%.

Assay—Dissolve about 240 mg of Isotretinoin, accurately weighed, in 50 mL of dimethylformamide, add 3 drops of a 1 in 100 solution of thymol blue in dimethylformamide, and titrate with 0.1 *N* sodium methoxide VS to a greenish end-point. Perform a blank determination, and make any necessary correction. Each mL of 0.1 *N* sodium methoxide is equivalent to 30.04 mg of $C_{20}H_{28}O_2$.

Isoxsuprine Hydrochloride

$C_{18}H_{23}NO_3 \cdot HCl$ 337.85

Benzenemethanol, 4-hydroxy-α-[1-[(1-methyl-2-phenoxyethyl)amino]ethyl]-, hydrochloride, stereoisomer.

p-Hydroxy-α-[1-[(1-methyl-2-phenoxyethyl)amino]ethyl]benzyl alcohol hydrochloride.

(±)-(α*R**)-*p*-Hydroxy-α-[(1*S**)-1-[[(1*S**)-1-methyl-2-phenoxyethyl]amino]ethyl]benzyl alcohol hydrochloride
[579-56-6; 34331-89-0].

» Isoxsuprine Hydrochloride contains not less than 97.0 percent and not more than 103.0 percent of $C_{18}H_{23}NO_3 \cdot HCl$, calculated on the dried basis.

Packaging and storage—Preserve in tight containers.

Reference standard—*USP Isoxsuprine Hydrochloride Reference Standard*—Dry at 105° for 1 hour before using.

Identification—

A: The infrared absorption spectrum of a potassium bromide dispersion of it, previously dried, exhibits maxima only at the same wavelengths as that of a similar preparation of USP Isoxsuprine Hydrochloride RS.

B: The ultraviolet absorption spectrum of a solution (1 in 20,000) exhibits maxima and minima at the same wavelengths as that of a similar solution of USP Isoxsuprine Hydrochloride RS, concomitantly measured.

C: To 1 mL of a solution (1 in 100), obtained by heating as necessary, add 3 mL of a 1 in 15 solution of sodium nitrite in 2 *N* sulfuric acid. Add ammonium hydroxide dropwise: a yellow precipitate is formed and it dissolves upon the addition of sodium hydroxide solution (1 in 5).

D: To 1 mL of a solution (1 in 100) add 1 mL of phosphomolybdic acid solution (1 in 100): a pale yellow to white precipitate is formed.

pH ⟨791⟩: between 4.5 and 6.0, in a solution (1 in 100).

Loss on drying ⟨731⟩—Dry it at 105° for 1 hour: it loses not more than 0.5% of its weight.

Residue on ignition ⟨281⟩: not more than 0.2%.

Heavy metals, *Method II* ⟨231⟩: 0.002%.

Related compounds—To 10 mg, accurately weighed in a suitable vial, add 1 mL of *N*-trimethylsilylimidazole, and heat at 65° for 10 minutes. Add 5 mL of isooctane, wash with one 3-mL portion of water, and allow the layers to separate. Inject a 2-µL portion of the isooctane solution into a gas chromatograph equipped with a 2.0-m × 0.3-cm glass column packed with packing S1A containing 3 percent liquid phase G2 and a flame-ionization detector. The column temperature is maintained at 215°, and the injection port and detector are maintained at 250°. The carrier gas is nitrogen, flowing at the rate of 25 mL per minute. Adjust the instrument to provide full-scale response for the major component. Inject a second 2-µL portion of the isooctane solution with the attenuator adjusted to an 8-fold increase in sensitivity, and record the chromatogram from 0.5 to 1.5 relative to the retention time of the major peak. Measure the area of all minor peaks, and correct for differences in sensitivity settings. Calculate the percentage of related compounds present taken by the formula:

$$100A/B,$$

in which *A* is the sum of the corrected area peaks for all minor peaks, and *B* is the sum of the corrected area peaks for the major and minor peaks. Not more than 2.0% is found.

Assay—Transfer about 50 mg of Isoxsuprine Hydrochloride, accurately weighed, to a 1000-mL volumetric flask, add water to volume, and mix. Concomitantly determine the absorbances of this solution and of a Standard solution of USP Isoxsuprine Hydrochloride RS in the same medium having a known concentration of about 50 µg per mL in 1-cm cells at the wavelengths of maximum absorbance at about 269 and 300 nm, with a suitable spectrophotometer, using water as the blank. Calculate the quantity, in mg, of $C_{18}H_{23}NO_3 \cdot HCl$ in the Isoxsuprine Hydrochloride taken by the formula:

$$C(A_{U269} - A_{U300})/(A_{S269} - A_{S300}),$$

in which *C* is the concentration, in µg per mL, of USP Isoxsuprine Hydrochloride RS in the Standard solution, and the parenthetic expressions are the differences in the absorbances of the two solutions at the wavelengths indicated by the subscripts, for the assay solution ($_U$) and the Standard solution ($_S$), respectively.

Isoxsuprine Hydrochloride Injection

» Isoxsuprine Hydrochloride Injection is a sterile solution of Isoxsuprine Hydrochloride in Water for Injection. It contains not less than 95.0 percent and not more than 105.0 percent of the labeled amount of $C_{18}H_{23}NO_3 \cdot HCl$.

Packaging and storage—Preserve in single-dose or in multiple-dose containers, preferably of Type I glass.

Reference standard—*USP Isoxsuprine Hydrochloride Reference Standard*—Dry at 105° for 1 hour before using.

Identification—To a 60-mL separator transfer 10 mL of pH 9.0 buffer (prepared by mixing equal volumes of 0.1 *M* monobasic potassium phosphate and 0.1 *N* sodium hydroxide and, using a pH meter, adjusting to a pH of 9.0 by adding, as necessary, more of either solution) add 1 mL of Injection, and mix. Add 2 mL of chloroform, shake vigorously for 1 minute, filter the chloroform extract through a pledget of cotton, and mix the filtrate with 500 mg of potassium bromide. Evaporate the chloroform, carefully removing the last trace of solvent in a small vacuum flask. The infrared absorption spectrum of a potassium bromide dispersion of the isoxsuprine so obtained exhibits maxima only at the same wavelengths as that of a similar preparation of USP Isoxsuprine Hydrochloride RS that has been treated in the same manner.

Pyrogen—It meets the requirements of the *Pyrogen Test* ⟨151⟩, the test dose being 5 mg per kg, of a solution containing 5 mg in each mL.

pH ⟨791⟩: between 4.9 and 6.0.

Other requirements—It meets the requirements under *Injections* ⟨1⟩.

Assay—

pH 4.0 citrate buffer—Mix equal volumes of 0.5 *M* citric acid and 0.5 *M* sodium citrate, and adjust the pH of the solution to 4.0 ± 0.2 by the addition of either solution as necessary.

Mixed solvent—Shake 40 mL of ether, 160 mL of isooctane, and 10 mL of water in a separator, remove and discard the water phase, and pass the solvent phase through a large pledget of cotton to remove excess water.

Standard preparation—Transfer about 40 mg of USP Isoxsuprine Hydrochloride RS, accurately weighed, to a 50-mL volumetric flask, add 2 *N* sulfuric acid to volume, and mix. Transfer 10.0 mL of this solution to a 100-mL volumetric flask, dilute with 2 *N* sulfuric acid to volume, and mix. The concentration of USP Isoxsuprine Hydrochloride RS in the *Standard preparation* is about 80 µg per mL.

Chromatographic column—Proceed as directed under *Column Partition Chromatography* (see *Chromatography* ⟨621⟩), packing a chromatographic tube with two segments of packing material. The lower segment is a mixture of 2 g of *Solid Support* and 1 mL of *pH 4.0 citrate buffer*, and the upper segment is a mixture prepared as directed under *Assay preparation.*

Assay preparation—Transfer an accurately measured volume of Isoxsuprine Hydrochloride Injection, equivalent to about 4 mg of isoxsuprine hydrochloride, to a 100-mL beaker, add 1 mL of dimethyl sulfoxide, and allow to stand for about 10 minutes, with occasional swirling. Add 1 mL of *pH 4.0 citrate buffer* and 3 g of *Solid Support*, mix as directed under *Chromatographic column*, and transfer to the column. Pass 75 mL of *Mixed solvent* through the column, and discard the eluate. Elute the column with a solution prepared by mixing 0.2 mL of bis(2-ethylhexyl)phosphoric acid with 75 mL of *Mixed solvent*, and collect the eluate in a 125-mL separator. Extract the eluate with two 20-mL portions of 2 *N* sulfuric acid. Transfer the extracts to a 50-mL volumetric flask, dilute with 2 *N* sulfuric acid to volume, and mix.

Procedure—Concomitantly determine the absorbances of the *Assay preparation* and the *Standard preparation* in 1-cm cells at the wavelength of maximum absorbance at about 275 nm, with a suitable spectrophotometer, using a column blank, prepared with *Mixed solvent*, to set the instrument. Calculate the quantity, in mg, of $C_{18}H_{23}NO_3 . HCl$ in the portion of Injection taken by the formula:

$$0.05C(A_U/A_S),$$

in which *C* is the concentration, in µg per mL, of USP Isoxsuprine Hydrochloride RS in the *Standard preparation*, and A_U and A_S are the absorbances of the *Assay preparation* and the *Standard preparation*, respectively.

Isoxsuprine Hydrochloride Tablets

» Isoxsuprine Hydrochloride Tablets contain not less than 93.0 percent and not more than 107.0 percent of the labeled amount of $C_{18}H_{23}NO_3 . HCl$.

Packaging and storage—Preserve in tight containers.

Reference standard—*USP Isoxsuprine Hydrochloride Reference Standard*—Dry at 105° for 1 hour before using.

Identification—Transfer a portion of finely powdered Tablets, equivalent to about 10 mg of isoxsuprine hydrochloride, to a 60-mL beaker, add about 20 mL of water, mix, and filter. Transfer the clear filtrate to a 60-mL separator, add 10 mL of pH 9.0 alkaline borate buffer (see *Buffer Solutions* in the section, *Reagents, Indicators, and Solutions*), and shake vigorously to mix. Extract with 2 mL of chloroform, filter the extract through a pledget of cotton, and mix the filtrate with 500 mg of potassium bromide. Evaporate the chloroform, carefully removing the last trace of solvent in a small vacuum flask. The infrared absorption spectrum of a potassium bromide dispersion of the isoxsuprine so obtained exhibits maxima only at the same wavelengths as that of a similar preparation of USP Isoxsuprine Hydrochloride RS that has been treated in the same manner.

Dissolution ⟨711⟩—

Medium: water; 900 mL.

Apparatus 1: 100 rpm.

Time: 45 minutes.

Procedure—Determine the amount of $C_{18}H_{23}NO_3 . HCl$ dissolved from ultraviolet absorbances at the wavelength of maximum absorbance at about 269 nm of filtered portions of the solution under test, suitably diluted with *Dissolution Medium*, if necessary, in comparison with a Standard solution having a known concentration of USP Isoxsuprine Hydrochloride RS in the same medium.

Tolerances—Not less than 75% (*Q*) of the labeled amount of $C_{18}H_{23}NO_3 . HCl$ is dissolved in 45 minutes.

Uniformity of dosage units ⟨905⟩: meet the requirements.

Assay—

pH 4.0 citrate buffer, Mixed solvent, Standard preparation, and *Chromatographic column*—Prepare as directed in the *Assay* under *Isoxsuprine Hydrochloride Injection.*

Assay preparation—Weigh and finely powder not less than 20 Isoxsuprine Hydrochloride Tablets. Transfer an accurately weighed portion of the powder, equivalent to about 4 mg of isoxsuprine hydrochloride, to a 100-mL beaker, and proceed as directed for *Assay preparation* in the *Assay* under *Isoxsuprine Hydrochloride Injection*, beginning with "add 1 mL of dimethyl sulfoxide."

Procedure—Proceed as directed for *Procedure* in the *Assay* under *Isoxsuprine Hydrochloride Injection.* Calculate the quantity, in mg, of $C_{18}H_{23}NO_3 . HCl$ in the portion of Tablets taken by the formula:

$$0.05C(A_U/A_S),$$

in which the terms are as defined therein.

Jelly, Lidocaine Hydrochloride—*see* Lidocaine Hydrochloride Jelly

Jelly, Pramoxine Hydrochloride—*see* Pramoxine Hydrochloride Jelly

Juniper Tar

» Juniper Tar is the empyreumatic volatile oil obtained from the woody portions of *Juniperus oxycedrus* Linné (Fam. Pinaceae).

Packaging and storage—Preserve in tight, light-resistant containers, and avoid exposure to excessive heat.

Identification—Shake 1 volume with 20 volumes of warm water, filter, and use the filtrate for the following tests.

A: To 5 mL of the cold filtrate add a few drops of silver–ammonium nitrate TS: a black color is produced.

B: To 5 mL of the filtrate add a few drops of alkaline cupric tartrate TS, and heat the solution to boiling: a red precipitate is formed.

Specific gravity ⟨841⟩: between 0.950 and 1.055.

Reaction—The filtrate prepared for the *Identification tests* is acid to litmus.

Rosin or rosin oils—Triturate 1 mL of it with 15 mL of solvent hexane, filter the solution, add an equal volume of cupric acetate solution (1 in 100), shake vigorously, and allow the liquids to separate. Decant the solvent hexane layer into a test tube, and add an equal volume of ether: the liquid does not become dark green or blackish.

Kanamycin Sulfate

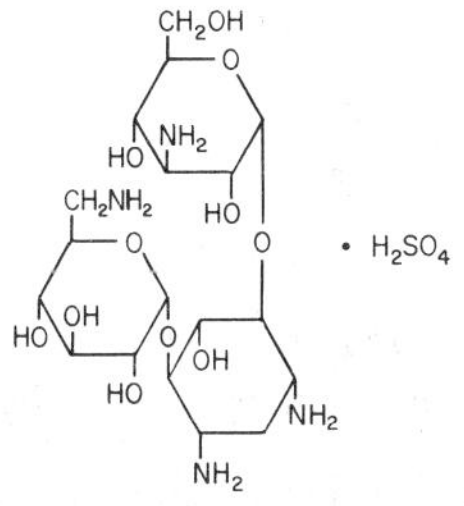

$C_{18}H_{36}N_4O_{11} \cdot H_2SO_4$ 582.58

D-Streptamine, *O*-3-amino-3-deoxy-α-D-glucopyranosyl-(1→6)-*O*-[6-amino-6-deoxy-α-D-glucopyranosyl-(1→4)]-2-deoxy-, sulfate (1:1) (salt).

Kanamycin sulfate (1:1) (salt) [*133-92-6; 25389-94-0*].

» Kanamycin Sulfate has a potency equivalent to not less than 750 µg of kanamycin ($C_{18}H_{36}N_4O_{11}$) per mg, calculated on the dried basis.

Packaging and storage—Preserve in tight containers.

Reference standard—*USP Kanamycin Sulfate Reference Standard*—Do not dry before using.

Identification—Dissolve about 10 mg in 1 mL of water, and add 1 mL of a 1 in 500 solution of ninhydrin in normal butyl alcohol and 0.5 mL of pyridine. Heat in a steam bath for 5 minutes, and add 10 mL of water: a deep-purple color is produced.

Crystallinity ⟨695⟩: meets the requirements.

pH ⟨791⟩: between 6.5 and 8.5, in a solution (1 in 100).

Loss on drying ⟨731⟩—Dry about 100 mg, accurately weighed, in a capillary-stoppered bottle in vacuum at a pressure not exceeding 5 mm of mercury at 60° for 3 hours: it loses not more than 4.0% of its weight.

Residue on ignition ⟨281⟩: not more than 1.0%, the charred residue being moistened with 2 mL of nitric acid and 5 drops of sulfuric acid.

Chromatographic purity—Dissolve a quantity of Kanamycin Sulfate in water to obtain a test solution having a concentration of 30 mg per mL. Dissolve a suitable quantity of USP Kanamycin Sulfate RS in water to obtain a Standard solution having a known concentration of 30 mg per mL. Dilute a portion of this solution quantitatively with water to obtain a *Diluted standard solution* having a concentration of 0.90 mg per mL. Apply separate 10-µL portions of the three solutions on the starting line of a suitable thin-layer chromatographic plate (see *Chromatography* ⟨621⟩), coated with a 0.25-mm layer of chromatographic silica gel and heated at 110° for 1 hour and cooled immediately before use. Allow the spots to dry, and develop the chromatogram in a suitable chamber, previously equilibrated for 18 hours with a developing solvent of monobasic potassium phosphate solution (15 in 100), until the solvent front has moved about three-fourths of the length of the plate. Remove the plate from the chamber, and air-dry. Spray the plate with a 1 in 100 solution of ninhydrin in butyl alcohol. Dry the plate at 110° for 10 minutes, and examine the chromatograms: the chromatograms show principal spots at about the same R_f value, and no secondary spot, if present in the chromatogram from the test solution, is more intense than the principal spot obtained from the *Diluted standard solution*.

Assay—Proceed with Kanamycin Sulfate as directed under *Antibiotics—Microbial Assay* ⟨81⟩.

Kanamycin Sulfate Capsules

» Kanamycin Sulfate Capsules contain the equivalent of not less than 90.0 percent and not more than 115.0 percent of the labeled amount of kanamycin ($C_{18}H_{36}N_4O_{11}$).

Packaging and storage—Preserve in tight containers.

Reference standard—*USP Kanamycin Sulfate Reference Standard*—Do not dry before using.

Identification—Dissolve a suitable quantity of Capsule contents in water to obtain a test solution having a concentration of about 1 mg of kanamycin per mL. Dissolve a suitable quantity of USP Kanamycin Sulfate RS in water to obtain a Standard solution having a concentration of about 1 mg per mL. Apply separate 10-µL portions of these solutions on the starting line of a suitable thin-layer chromatographic plate (see *Chromatography* ⟨621⟩), coated with a 0.25-mm layer of chromatographic silica gel and heated at 110° for 1 hour and cooled immediately before use. Allow the spots to dry, and develop the chromatogram in a suitable chamber, previously equilibrated for 18 hours with a developing solvent of monobasic potassium phosphate solution (15 in 100), until the solvent front has moved about three-fourths of the length of the plate. Remove the plate from the chamber, and air-dry. Spray the plate with a 1 in 100 solution of ninhydrin in butyl alcohol. Dry the plate at 110° for 10 minutes, and examine the chromatograms: the principal spot obtained from the test solution corresponds in R_f value to that obtained from the Standard solution.

Dissolution ⟨711⟩—

Medium: 0.1 *N* hydrochloric acid; 900 mL.

Apparatus 1: 100 rpm.

Time: 45 minutes.

Procedure—Determine the amount of kanamycin ($C_{18}H_{36}N_4O_{11}$) dissolved, employing the procedure set forth in the *Assay*, making any necessary modifications.

Tolerances—Not less than 75% (*Q*) of the labeled amount of $C_{18}H_{36}N_4O_{11}$ is dissolved in 45 minutes.

Loss on drying ⟨731⟩—Dry about 100 mg in a capillary-stoppered bottle in vacuum at 60° for 3 hours: it loses not more than 4.0% of its weight.

Assay—Place not less than 5 Kanamycin Sulfate Capsules in a high-speed glass blender jar containing an accurately measured volume of water, and blend for 4 ± 1 minutes. Proceed as directed under *Antibiotics—Microbial Assays* ⟨81⟩, using an accurately measured volume of this stock test solution diluted quantitatively with water to yield a *Test Dilution* having a concentration assumed to be equal to the median dose level of the Standard.

Kanamycin Sulfate Injection

» Kanamycin Sulfate Injection contains the equivalent of not less than 90.0 percent and not more than 115.0 percent of the labeled amount of kanamycin ($C_{18}H_{36}N_4O_{11}$). It contains suitable buffers and preservatives.

Packaging and storage—Preserve in single-dose or in multiple-dose containers, preferably of Type I or Type III glass.

Reference standard—*USP Kanamycin Sulfate Reference Standard*—Do not dry before using.

Identification—Dilute a suitable volume of Injection with water to obtain a test solution having a concentration of about 1 mg of kanamycin per mL. This solution responds to the *Identification test* under *Kanamycin Sulfate Capsules*.

Pyrogen—It meets the requirements of the *Pyrogen Test* ⟨151⟩, the test dose being 1.0 mL per kg of a solution prepared by diluting the Injection, if necessary, with Water for Injection to a concentration of 10 mg of kanamycin per mL.

Sterility—It meets the requirements under *Sterility Tests* ⟨71⟩, when tested as directed in the section, *Test Procedures Using Membrane Filtration.*

pH ⟨791⟩: between 3.5 and 5.0.

Particulate matter ⟨788⟩: meets the requirements under *Small-volume Injections.*

Assay—Proceed with Kanamycin Sulfate Injection as directed under *Antibiotics—Microbial Assays* ⟨81⟩, using an accurately measured volume of Injection diluted quantitatively with water to obtain a *Test Dilution* having a concentration assumed to be equal to the median dose level of the Standard.

Sterile Kanamycin Sulfate

» Sterile Kanamycin Sulfate is Kanamycin Sulfate suitable for parenteral use. It has a potency equivalent to not less than 750 µg of kanamycin ($C_{18}H_{36}N_4O_{11}$) per mg, calculated on the dried basis.

Packaging and storage—Preserve in *Containers for Sterile Solids* as described under *Injections* ⟨1⟩.

Reference standard—*USP Kanamycin Sulfate Reference Standard*—Do not dry before using.

Pyrogen—It meets the requirements of the *Pyrogen Test* ⟨151⟩, the test dose being 1.0 mL per kg of a solution in Sterile Water for Injection containing 10 mg of kanamycin per mL.

Sterility—It meets the requirements under *Sterility Tests* ⟨71⟩, when tested as directed in the section, *Test Procedures Using Membrane Filtration*, 6 g of specimen aseptically dissolved in 200 mL of *Fluid A* being used.

Other requirements—It responds to the *Identification test* and meets the requirements for *pH, Loss on drying, Residue on ignition, Crystallinity, Kanamycin B content,* and *Assay* under *Kanamycin Sulfate.*

Kaolin

» Kaolin is a native hydrated aluminum silicate, powdered and freed from gritty particles by elutriation.

Packaging and storage—Preserve in well-closed containers.

Identification—Mix 1 g with 10 mL of water and 5 mL of sulfuric acid in a porcelain dish. Evaporate the mixture until the excess water is removed, and continue heating the residue until dense, white fumes of sulfur trioxide appear. Cool, cautiously add 20 mL of water, boil for a few minutes, and filter: there remains on the filter a gray residue (*impure silica*). The filtrate responds to the tests for *Aluminum* ⟨191⟩.

Microbial limit—It meets the requirements of the test for absence of *Escherichia coli* under *Microbial Limit Tests* ⟨61⟩.

Loss on ignition ⟨733⟩—Ignite between 550° and 600°: it loses not more than 15.0% of its weight.

Acid-soluble substances—Digest 1.0 g with 20 mL of 3 *N* hydrochloric acid for 15 minutes, and filter: 10 mL of the filtrate, evaporated to dryness and ignited, leaves not more than 10 mg of residue (2.0%).

Carbonate—Mix 1.0 g with 10 mL of water and 5 mL of sulfuric acid: no effervescence occurs.

Iron—Triturate 2.0 g in a mortar with 10 mL of water, and add 0.50 g of sodium salicylate: the mixture does not acquire more than a slight reddish tint.

Lead—To 1.0 g contained in a centrifuge tube add 10 mL of 1 *N* nitric acid, and digest for 1 hour in a boiling water bath. Centrifuge until the solids are completely separated, and pour the supernatant liquid into a 100-mL volumetric flask. Add 5 mL of 1 *N* nitric acid to the Kaolin, mix, and digest for 15 minutes in a boiling water bath. Centrifuge, and add the supernatant liquid to the previous extract in the volumetric flask. Dilute with water to volume. A 50-mL portion of this solution contains not more than 5 µg of lead (corresponding to not more than 0.001% of Pb) when tested as directed in the test for *Lead* ⟨251⟩, 3 mL of *Ammonium Citrate Solution*, 1 mL of *Potassium Cyanide Solution*, and 500 µL of *Hydroxylamine Hydrochloride Solution* being used.

Ketamine Hydrochloride

$C_{13}H_{16}ClNO.HCl$ 274.19

Cyclohexanone, 2-(2-chlorophenyl)-2-(methylamino)-, hydrochloride.

(±)-2-(*o*-Chlorophenyl)-2-(methylamino)cyclohexanone hydrochloride [*1867-66-9*].

» Ketamine Hydrochloride contains not less than 98.5 percent and not more than 101.0 percent of $C_{13}H_{16}ClNO.HCl$.

Packaging and storage—Preserve in well-closed containers.

Reference standard—*USP Ketamine Hydrochloride Reference Standard*—Do not dry before using.

Clarity and color of solution—Dissolve 1 g in 5 mL of water: the solution is clear and colorless.

Identification—

A: The infrared absorption spectrum of a potassium bromide dispersion of it exhibits maxima only at the same wavelengths as that of a similar preparation of USP Ketamine Hydrochloride RS.

B: *Acid solvent*—The ultraviolet absorption spectrum of a 1 in 3000 solution in 0.1 *N* hydrochloric acid exhibits maxima and minima at the same wavelengths as that of a similar solution of USP Ketamine Hydrochloride RS, concomitantly measured, and the respective absorptivities, at the wavelengths of maximum absorbance at about 269 and 276 nm do not differ by more than 3.0%.

Basic solvent—The ultraviolet absorption spectrum of a 1 in 1250 solution in 0.01 *N* sodium hydroxide, in a 1 in 20 mixture of water and methanol, exhibits maxima and minima at the same wavelengths as that of a similar solution of USP Ketamine Hydrochloride RS, concomitantly measured, and the respective absorptivities, at the wavelength of maximum absorbance at about 302 nm do not differ by more than 3.0%.

Melting range, *Class Ia* ⟨741⟩: between 258° and 261°.

pH ⟨791⟩: between 3.5 and 4.1, in a solution (1 in 10).

Residue on ignition ⟨281⟩: not more than 0.1%.

Heavy metals, *Method I* ⟨231⟩: 0.002%.

Foreign amines—

Modified Dragendorff reagent—Dissolve 1.7 g of bismuth subnitrate in 80 mL of water and 20 mL of glacial acetic acid, warming, if necessary. Cool, add 100 mL of potassium iodide solution (1 in 2), and mix. Refrigerate this stock solution for prolonged storage. For use, dilute 10 mL of this stock solution with water to 100 mL, add 10 mL of glacial acetic acid, and mix. Then add 120 mg of iodine crystals, and shake until the iodine has completely dissolved. Store refrigerated, and discard after 2 weeks.

Procedure—Dissolve 750 mg of Ketamine Hydrochloride in 5 mL of methanol, and prepare a second solution by diluting the first solution 1 in 200 with methanol. Apply 4 µL of each solution at points 2.5 cm from the bottom edge of a suitable thin-layer chromatographic plate (see *Chromatography* ⟨621⟩), coated with a 0.25-mm layer of chromatographic silica gel, and allow the spots to dry. Develop the chromatogram in a solvent system consisting of a mixture of benzene, methanol, and ammonium hydroxide (80:20:1) until the solvent front has moved about three-fourths of the length of the plate. Remove the plate from the developing chamber, mark the solvent front, and allow the solvent to evaporate. Spray the plate with *Modified Dragendorff reagent.* The two chromatograms show principal spots at about the same R_f value, one much weaker than the other, and no secondary spot, if present in the chromatogram from the more concentrated solution, is more intense than the principal spot obtained from the weaker solution.

Chromatographic purity—

Standard preparations—Dissolve USP Ketamine Hydrochloride RS in methanol, and mix to obtain *Standard preparation A* having a known concentration of 0.5 mg per mL. Dilute quantitatively with methanol to obtain *Standard preparations*, designated below by letter, having the following compositions:

Dilution	Concentration (µg RS per mL)	Percentage (%, for comparison with test specimen)
A (undiluted)	500	1.0
B (1 in 2)	250	0.5
C (1 in 5)	100	0.2
D (1 in 10)	50	0.1

Test preparation—Dissolve an accurately weighed quantity of Ketamine Hydrochloride in methanol to obtain a solution containing 50 mg per mL.

Procedure—On a suitable thin-layer chromatographic plate (see *Chromatography* ⟨621⟩), coated with a 0.25-mm layer of chromatographic silica gel mixture, apply separately 10 µL of the *Test preparation* and 10 µL of each *Standard preparation*, and allow to dry. Position the plate in an unsaturated chromatographic chamber, and develop the chromatograms in a solvent system consisting of a mixture of toluene, isopropyl alcohol, and ammonium hydroxide (80:19.5:0.5) until the solvent front has moved about three-fourths of the length of the plate. Remove the plate from the developing chamber, mark the solvent front, allow the solvent to evaporate, and place the plate in an iodine vapor chamber for 1 hour. Compare the intensities of any secondary spots observed in the chromatogram of the *Test preparation* with those of the principal spots in the chromatograms of the *Standard preparations:* the sum of the intensities of all secondary spots obtained from the *Test preparation* corresponds to not more than 1.0% of related compounds, with no single impurity corresponding to more than 0.5%.

Assay—Dissolve about 500 mg of Ketamine Hydrochloride, accurately weighed, in 1 mL of formic acid, and add 50 mL of glacial acetic acid. Add 10 mL of mercuric acetate TS and 1 drop of crystal violet TS, and titrate with 0.1 *N* perchloric acid VS to a blue-green end-point. Perform a blank determination, and make any necessary correction. Each mL of 0.1 *N* perchloric acid is equivalent to 27.42 mg of $C_{13}H_{16}ClNO.HCl$.

Ketamine Hydrochloride Injection

» Ketamine Hydrochloride Injection is a sterile solution of Ketamine Hydrochloride in Water for Injection. It contains an amount of ketamine hydrochloride ($C_{13}H_{16}ClNO.HCl$) equivalent to not less than 95.0 percent and not more than 105.0 percent of the labeled amount of ketamine ($C_{13}H_{16}ClNO$).

Packaging and storage—Preserve in single-dose or in multiple-dose containers, preferably of Type I glass, protected from light and heat.

Reference standard—*USP Ketamine Hydrochloride Reference Standard*—Do not dry before using.

Identification—

A: The ultraviolet absorption spectrum, measured in the region between 250 and 350 nm, of a dilution of Injection in 0.01 *N* methanolic sodium hydroxide containing ketamine hydrochloride equivalent to about 800 µg of ketamine per mL, exhibits maxima and minima at the same wavelengths as that of a similar preparation of USP Ketamine Hydrochloride RS, concomitantly measured.

B: The ultraviolet absorption spectrum of the solution employed for measurement of absorbance of the assay solution, prepared as directed in the *Assay*, exhibits maxima and minima at the same wavelengths as that of the Standard solution, prepared as directed in the *Assay*.

pH ⟨791⟩: between 3.0 and 5.0.

Other requirements—It meets the requirements under *Injections* ⟨1⟩.

Assay—Transfer an accurately measured volume of Ketamine Hydrochloride Injection, equivalent to about 500 mg of ketamine hydrochloride, to a 200-mL volumetric flask, dilute with water to volume, and mix. Transfer 20.0 mL of this solution to a 125-mL separator, add 3 mL of 0.1 *N* sodium hydroxide, and extract with three 15-mL portions of chloroform. Collect the chloroform extracts in a second 125-mL separator, and extract with three 30-mL portions of 0.1 *N* sulfuric acid, collecting the acid extracts in a 200-mL volumetric flask. Dilute with 0.1 *N* sulfuric acid (saturated with chloroform) to volume, and mix. Concomitantly determine the absorbances of this solution and a Standard solution of USP Ketamine Hydrochloride RS in the same medium having a known concentration of about 250 µg per mL, in 1-cm cells at the wavelength of maximum absorbance at about 269 nm, with a suitable spectrophotometer, using 0.1 *N* sulfuric acid (saturated with chloroform) as the blank. Calculate the quantity, in mg, of ketamine ($C_{13}H_{16}ClNO$) in each mL of the Injection taken by the formula:

$$(237.73/274.19)(2C/V)(A_U/A_S),$$

in which 237.73 and 274.19 are the molecular weights of ketamine and ketamine hydrochloride, respectively, C is the concentration, in µg per mL, of USP Ketamine Hydrochloride RS in the Standard solution, V is the volume, in mL, of Injection taken, and A_U and A_S are the absorbances of the solution from the Injection and the Standard solution, respectively.

Ketoconazole

$C_{26}H_{28}Cl_2N_4O_4$ 531.44

Piperazine, 1-acetyl-4-[4-[[2-(2,4-dichlorophenyl)-2-(1*H*-imidazol-1-ylmethyl)-1,3-dioxolan-4-yl]methoxy]phenyl]-, *cis*-.

(±)-*cis*-1-Acetyl-4-[*p*-[[2-(2,4-dichlorophenyl)-2-(imidazol-1-ylmethyl)-1,3-dioxolan-4-yl]methoxy]phenyl]piperazine [*65277-42-1*].

» Ketoconazole contains not less than 98.0 percent and not more than 102.0 percent of $C_{26}H_{28}Cl_2N_4O_4$, calculated on the dried basis.

Packaging and storage—Preserve in well-closed containers.

Reference standard—*USP Ketoconazole Reference Standard*—Dry in vacuum at 80° for 4 hours before using.

Identification—The infrared absorption spectrum of a potassium bromide dispersion of it, previously dried, exhibits maxima only at the same wavelengths as that of a similar preparation of USP Ketoconazole RS.

Melting range ⟨741⟩: between 148° and 152°.

Specific rotation ⟨781⟩: between −1° and +1°, at 20°, calculated on the dried basis, determined in a solution in methanol containing 400 mg in each 10 mL.

Loss on drying ⟨731⟩—Dry it in vacuum at 80° for 4 hours: it loses not more than 0.5% of its weight.

Residue on ignition ⟨281⟩: not more than 0.1% from 2 g.

Heavy metals, *Method II* ⟨231⟩: 0.002%.

Chromatographic purity—Dissolve 30 mg in 3.0 mL of chloroform (*Test solution*). Dissolve a suitable quantity of USP Ketoconazole RS in chloroform to obtain a *Standard solution* having a known concentration of 10 mg per mL. Dilute a portion of this solution quantitatively with chloroform to obtain a *Diluted standard solution* having a concentration of 1.0 mg per mL. Apply separate 10-µL portions of the *Test solution* and the *Standard solution* and a 2-µL portion of the *Diluted standard solution* on the starting line of a suitable thin-layer chromatographic plate (see *Chromatography* ⟨621⟩), coated with a 0.25-mm layer of chromatographic silica gel mixture. Allow the spots to dry, and develop the chromatogram in a suitable unsaturated chamber with a solvent system consisting of a mixture of *n*-hexane, ethyl acetate, methanol, water, and glacial acetic acid (42:40:15:2:1) until the solvent front has moved about three-fourths of the length of the plate. Remove the plate from the chamber, and air-dry. Expose the plate to iodine vapors in a closed chamber, and locate the spots: the principal spot obtained from the *Test solution* has about the same size and R_f value as that obtained from the *Standard solution*, and the sum of the intensities of any secondary spots obtained from the *Test solution* does not exceed the intensity of the principal spot obtained from the *Diluted standard solution.*

Assay—Dissolve about 200 mg of Ketoconazole, accurately weighed, in 40 mL of glacial acetic acid. Titrate with 0.1 *N* perchloric acid VS, determining the end-point potentiometrically. Perform a blank determination, and make any necessary correction. Each mL of 0.1 *N* perchloric acid is equivalent to 26.57 mg of $C_{26}H_{28}Cl_2N_4O_4$.

Ketoconazole Tablets

» Ketoconazole Tablets contain not less than 90.0 percent and not more than 110.0 percent of the labeled amount of $C_{26}H_{28}Cl_2N_4O_4$.

Packaging and storage—Preserve in well-closed containers.

Reference standards—*USP Ketoconazole Reference Standard*—Dry in vacuum at 80° for 4 hours before using. *USP Terconazole Reference Standard.*

Identification—Transfer a quantity of finely powdered Tablets, equivalent to about 50 mg of ketoconazole, to a suitable flask, add 50 mL of chloroform, shake for about 2 minutes, and filter. Apply separate 10-µL portions of this solution and of a Standard solution of USP Ketoconazole RS in chloroform containing 1 mg per mL on the starting line of a thin-layer chromatographic plate (see *Chromatography* ⟨621⟩), coated with a 0.25-mm layer of chromatographic silica gel mixture. Allow the spots to dry, and develop the chromatogram in an unsaturated chamber with a solvent system consisting of a mixture of *n*-hexane, ethyl acetate, methanol, water, and glacial acetic acid (42:40:15:2:1) until the solvent front has moved about three-fourths of the length of the plate. Remove the plate from the chamber, air-dry, and view under short-wavelength ultraviolet light: the R_f value of the principal spot obtained from the test solution corresponds to that obtained from the Standard solution.

Disintegration ⟨701⟩: 10 minutes.

Uniformity of dosage units ⟨905⟩: meet the requirements.

Assay—

Methanol–methylene chloride—Mix equal volumes of methanol and methylene chloride.

Mobile phase—Prepare a suitable 7:3 mixture of a 1 in 500 solution of diisopropylamine in methanol and ammonium acetate solution (1 in 200).

Internal standard solution—Dissolve USP Terconazole RS in *Methanol–methylene chloride* to obtain a solution containing about 5 mg per mL.

Standard preparation—Transfer about 20 mg of USP Ketoconazole RS, accurately weighed, to a 50-mL volumetric flask, add 5.0 mL of *Internal standard solution*, dilute with *Methanol–methylene chloride* to volume, and mix.

Assay preparation—Weigh and finely powder not less than 20 Ketoconazole Tablets. Transfer an accurately weighed portion of the powder, equivalent to about 200 mg of ketoconazole, to a suitable screw-capped bottle, add 50.0 mL of *Methanol–methylene chloride*, shake by mechanical means for 30 minutes, and centrifuge. Transfer 5.0 mL of the clear supernatant solution so obtained to a 50-mL volumetric flask, add 5.0 mL of *Internal standard solution*, dilute with *Methanol–methylene chloride* to volume, and mix.

Chromatographic system (see *Chromatography* ⟨621⟩)—The liquid chromatograph is equipped with a 225-nm detector and a 3.9-mm × 30-cm column that contains packing L1. The flow rate is about 3 mL per minute. Chromatograph the *Standard preparation*, and record the peak responses as directed under *Procedure:* the relative standard deviation is not more than 2.0%, and the resolution, *R*, between ketoconazole and terconazole is not less than 2.0.

Procedure—Separately inject equal volumes (about 20 µL) of the *Standard preparation* and the *Assay preparation* into the chromatograph, record the chromatograms, and measure the responses for the major peaks. The relative retention times are about 0.6 for ketoconazole and 1.0 for terconazole. Calculate the quantity, in mg, of $C_{26}H_{28}Cl_2N_4O_4$ in the portion of Tablets taken by the formula:

$$10W_S(R_U/R_S),$$

in which W_S is the weight, in mg, of USP Ketoconazole RS taken, and R_U and R_S are the ratios of the peak responses of ketoconazole to those of terconazole from the *Assay preparation* and the *Standard preparation*, respectively.

Krypton Kr 81m

Kr 81m
Krypton, isotope of mass 81 (metastable).
Krypton, isotope of mass 81 (metastable) [*15678-91-8*].

» Krypton Kr 81m is a gas suitable only for inhalation in diagnostic studies, and is obtained from a generator that contains rubidium 81 adsorbed on an immobilized suitable column support. Rubidium 81 decays with a half-life of 4.58 hours and forms its radioactive daughter ^{81m}Kr, which is eluted from the generator by passage of humidified oxygen or air through the column. Rubidium 81 is produced in an accelerator by proton bombardment of Kr 82. Other radioisotopes of rubidium are produced and are present on the generator column. These other radioisotopes do not decay to ^{81m}Kr. The column contains not less than 90.0 percent and not more than 110.0 percent of the labeled amount of Rb 81 at the date and time indicated in the labeling, and on elution yields not less than 80.0 percent of ^{81m}Kr.

Packaging and storage—The generator column is enclosed in a lead container. The unit is stored at room temperature.

Labeling—The labeling indicates the name and address of the manufacturer, the name of the generator, the quantity of ^{81}Rb at the date and time of calibration, and the statement, "Caution—Radioactive Material." The labeling indicates that in making dosage calculations, correction is to be made for radioactive decay, and also indicates that the radioactive half-life of ^{81m}Kr is 13.1 seconds.

[NOTE—Perform the following tests and *Assay* quickly, because of the rapid decay of the ^{81m}Kr.]

Radionuclide identification (see *Radioactivity* ⟨821⟩)—The gamma-ray spectrum of eluted ^{81m}Kr exhibits a monoenergetic gamma ray at a mean energy of 191 KeV.

Radionuclidic purity—Using a suitable counting assembly (see *Selection of a Counting Assembly* under *Radioactivity* ⟨821⟩), determine the radioactivity of each radionuclide present in a specimen of Kr 81m gas obtained from eluting the generator by use of a calibrated system as directed under *Radioactivity* ⟨821⟩. Not less than 99.9% of the radioactivity in the specimen eluted from the generator is present as ^{81m}Kr.

Assay for radioactivity—Using a suitable counting assembly (see *Radioactivity* ⟨821⟩), determine the quantity, in MBq (mCi), of Kr 81m contained in an elution of the generator. Decay correct the result to the time of generator elution, and calculate the quantity of ^{81}Rb present in the column at the time of elution. The quantity of ^{81m}Kr eluted is not less than 80.0 percent of the labeled MBq (mCi) of ^{81}Rb present on the column at time of elution.

Labetalol Hydrochloride

CONH₂ OH · HCl
CH₂CH₂CHNHCH₂CH
CH₃ OH

$C_{19}H_{24}N_2O_3 . HCl$ 364.87

Benzamide, 2-hydroxy-5-[1-hydroxy-2-[(1-methyl-3-phenylpropyl)amino]ethyl], monohydrochloride.

5-[1-Hydroxy-2-[(1-methyl-3-phenylpropyl)amino]-ethyl]-salicylamide monohydrochloride [*32780-64-6*].

» Labetalol Hydrochloride contains not less than 97.5 percent and not more than 101.0 percent of $C_{19}H_{24}N_2O_3 . HCl$, calculated on the dried basis.

Packaging and storage—Preserve in tight, light-resistant containers.

Reference standard—*USP Labetalol Hydrochloride Reference Standard*—Dry in vacuum at 105° for 4 hours before using.

Identification—

A: The infrared absorption spectrum of a mineral oil dispersion of it exhibits maxima only at the same wavelengths as that of a similar preparation of USP Labetalol Hydrochloride RS.

B: It responds to the tests for *Chloride* ⟨191⟩.

pH ⟨791⟩: between 4.0 and 5.0, in a solution (1 in 100).

Loss on drying ⟨731⟩: Dry it in a vacuum at 105° for 4 hours: it loses not more than 1.0% of its weight.

Residue on ignition ⟨281⟩: not more than 0.1%.

Heavy metals, *Method II* ⟨231⟩: 0.002%.

Chromatographic purity—

Detection reagent—Transfer 2.5 g of cadmium acetate to a 500-mL volumetric flask, add 10 mL of glacial acetic acid, dilute with alcohol to volume, and mix. Just prior to use, prepare a 0.2 in 100 solution of ninhydrin in the cadmium acetate solution for use as the *Detection reagent*.

Solvent mixture—Prepare a solution of methanol and water (4:1), and mix.

Ammonium chloride reference solution—Dissolve 60 mg of ammonium chloride in 10.0 mL of water, and mix.

Stock standard solution—Dissolve USP Labetalol Hydrochloride RS in *Solvent mixture*, and mix to obtain a solution having a known concentration of 40 mg per mL.

Standard solution 1—Dilute a portion of the *Stock standard solution* quantitatively with *Solvent mixture* to obtain a solution having a known concentration of 0.2 mg per mL.

Standard solution 2—Dilute a portion of the *Standard solution 1* quantitatively with *Solvent mixture* to obtain a solution having a known concentration of 0.1 mg per mL.

Test solution—Dissolve 200 mg of Labetalol Hydrochloride in 5.0 mL of *Solvent mixture*, and mix.

PROCEDURE I—On a suitable thin-layer chromatographic plate (see *Chromatography* ⟨621⟩), coated with a 0.25-mm layer of chromatographic silica gel mixture, apply separately 5-µL portions of the *Stock standard solution, Standard solution 1, Standard solution 2*, and the *Test solution*.

Allow the spots to dry, and develop the chromatograms in a solvent system consisting of a mixture of dichloromethane, methanol, and ammonium hydroxide (15:5:1) until the solvent front has moved about three-fourths of the length of the plate. Remove the plate from the developing chamber, mark the solvent front, and allow the solvent to evaporate. Examine the plate under short-wavelength ultraviolet light: the R_f value of the principal spot from the *Test solution* corresponds to that of the *Stock standard solution*.

Spray the plate with *Detection reagent*, heat the plate at 105° for 15 minutes, cool to room temperature, and examine the chromatogram: no individual secondary spot observed in the chromatogram of the *Test solution* is greater in size or intensity than the principal spot observed in the chromatogram of *Standard solution 1* (0.5% each). [NOTE—The spots appear as dark orange spots on a light orange to yellow background. A "negative image" spot (white) near the origin may be observed in the chromatogram of the *Test solution*. This is due to the formation of ammonium chloride during the chromatographic procedure and may be ignored.]

PROCEDURE II—On a suitable thin-layer chromatographic plate (see *Chromatography* ⟨621⟩), coated with a 0.25-mm layer of chromatographic silica gel mixture, apply separately 10-µL portions of the *Ammonium chloride reference solution*, the *Stock standard solution, Standard solution 1, Standard solution 2*, and the *Test solution*.

Allow the spots to dry, and develop the chromatograms in a solvent system consisting of a mixture of ethyl acetate, isopropyl alcohol, water, and ammonium hydroxide (25:15:8:2) until the solvent front has moved about three-fourths of the length of the plate. Remove the plate from the developing chamber, mark the solvent front, and allow the solvent to evaporate. Examine the plate under short-wavelength ultraviolet light: no individual secondary spot (other than that due to ammonium chloride) observed in the chromatogram of the *Test solution* is greater in size or intensity than the principal spot observed in the chromatogram of *Standard solution 1* (0.5% each).

Total related compounds—The sum of the intensities of all secondary spots (other than those due to ammonium chloride) observed in the chromatograms of the *Test solution* from both *Procedure I* and *Procedure II* does not exceed 1.0%.

Diastereoisomer ratio—

Methylboron dihydroxide solution—Dissolve methylboron dihydroxide acid in pyridine, previously dried over a suitable molecular sieve, and mix to obtain a solution having a known concentration of 12 mg per mL.

System suitability solution—Dissolve an accurately weighed quantity of USP Labetalol Hydrochloride RS in *Methylboron dihydroxide solution*, and dilute quantitatively and stepwise with *Methylboron dihydroxide solution* to obtain a solution having a known concentration of about 1.4 mg of USP Labetalol Hydrochloride RS per mL. Allow the solution to stand at room temperature for 20 minutes before using.

Test solution—Transfer about 1 mg of Labetalol Hydrochloride to a 1-mL reaction vial, add 0.7 mL of *Methylboron dihydroxide solution*, and mix until the labetalol hydrochloride is completely dissolved. Allow the solution to stand at room temperature for 20 minutes before using.

Chromatographic system (see *Chromatography* ⟨621⟩)—The gas chromatograph is equipped with a flame-ionization detector and a 1.2-m × 2-mm glass column packed with 3 percent phase G3 on 100- to 120-mesh support S1AB. The column is maintained at about 265°, and the injection port and the detector block are maintained at about 300°. Nitrogen is used as the carrier gas at the flow rate of about 30 mL per minute. Chromatograph the *System suitability solution*, and record the peak responses as directed under *Procedure:* the resolution, *R*, between the diastereoisomer A methylboron dihydroxide derivative and diastereoisomer B methylboron dihydroxide derivative peaks is not less than 1.5, and the relative standard deviation for replicate injections is not more than 2.0%.

Procedure—Inject about 2 µL of the *Test solution* into the chromatograph, record the chromatograms, and measure the responses for the major peaks. The relative retention times are about 0.8 for the diastereoisomer B methylboron dihydroxide derivative and 1.0 for the diastereoisomer A methylboron dihydroxide derivative. Calculate the diastereoisomer A content, in percentage, by the formula:

$$100r_A/(r_A + r_B),$$

in which r_A is the peak area of the diastereoisomer A methylboron dihydroxide derivative peak and r_B is the peak area of the diastereoisomer B methylboron dihydroxide derivative peak. The diastereoisomer A content is not less than 45.0% and not more than 55.0%.

Assay—

Mobile phase—Prepare a suitable filtered and degassed mixture of 0.1 *M* monobasic sodium phosphate and methanol (65:35). Make adjustments if necessary (see *System Suitability* under *Chromatography* ⟨621⟩).

Standard preparation—Dissolve an accurately weighed quantity of USP Labetalol Hydrochloride RS in *Mobile phase* to obtain a solution having a known concentration of about 0.4 mg per mL.

Assay preparation—Transfer about 40 mg of Labetalol Hydrochloride, accurately weighed, to a 100-mL volumetric flask, dilute with *Mobile phase* to volume, and mix.

Chromatographic system (see *Chromatography* ⟨621⟩)—The liquid chromatograph is equipped with a 230-nm detector and a 4.6-mm × 20-cm column that contains packing L1 and is maintained at 60 ± 1°. The flow rate is about 1.5 mL per minute. Chromatograph the *Standard preparation*, and record the peak responses as directed under *Procedure:* the column efficiency determined from the analyte peak is not less than 700 theoretical plates, the tailing factor for the analyte peak is not more than 2.0, and the relative standard deviation for replicate injections is not more than 1.5%.

Procedure—[NOTE—Use peak areas where peak responses are indicated.] Separately inject equal volumes (about 5 µL) of the *Standard preparation* and the *Assay preparation* into the chromatograph, record the chromatograms, and measure the responses for the major peaks. Calculate the quantity, in mg, of labetalol hydrochloride ($C_{19}H_{24}N_2O_3.HCl$) in the portion of Labetalol Hydrochloride taken by the formula:

$$100C(r_U/r_S),$$

in which *C* is the concentration, in mg per mL, of USP Labetalol Hydrochloride RS in the *Standard preparation*, and r_U and r_S are the peak responses obtained from the *Assay preparation* and the *Standard preparation*, respectively.

Labetalol Hydrochloride Injection

» Labetalol Hydrochloride Injection is a sterile solution of Labetalol Hydrochloride in Water for Injection. It contains not less than 90.0 percent and not more than 110.0 percent of the labeled amount of $C_{19}H_{24}N_2O_3.HCl$.

Packaging and storage—Preserve in single-dose containers, or in multiple-dose containers not exceeding 60 mL in volume, preferably of Type I glass, at a temperature between 2° and 30°. Avoid freezing and exposure to light.

Reference standard—*USP Labetalol Hydrochloride Reference Standard*—Dry in vacuum at 105° for 4 hours before using.

Identification—The retention time of the major peak in the chromatogram of the *Assay preparation* corresponds to that of the *Standard preparation*, as obtained in the *Assay*.

pH ⟨791⟩: between 3.0 and 4.5.

Other requirements—It meets the requirements under *Injections* ⟨1⟩.

Assay—

Mobile phase—Prepare a suitable filtered and degassed mixture of 0.1 *M* monobasic sodium phosphate and methanol (65:35). Make adjustments if necessary (see *System Suitability* under *Chromatography* ⟨621⟩).

Standard preparation—Dissolve an accurately weighed quantity of USP Labetalol Hydrochloride RS in *Mobile phase* to obtain a solution having a known concentration of about 0.5 mg per mL.

Resolution solution—Dissolve a quantity of methylparaben in the *Standard preparation* to obtain a solution containing about 0.08 mg per mL.

Assay preparation—Transfer an accurately measured volume of Labetalol Hydrochloride Injection, equivalent to about 50 mg of labetalol hydrochloride, to a 100-mL volumetric flask, dilute with *Mobile phase* to volume, and mix.

Chromatographic system (see *Chromatography* ⟨621⟩)—The liquid chromatograph is equipped with a 254-nm detector and a 4.6-mm × 20-cm column that contains packing L1 and is maintained at 60 ± 1°. The flow rate is about 1.5 mL per minute. Chromatograph the *Standard preparation*, and record the peak responses as directed under *Procedure:* the column efficiency determined from the analyte peak is not less than 700 theoretical plates, the tailing factor for the analyte peak is not more than 2.0, and the relative standard deviation for replicate injections is not more than 1.5%. Chromatograph the *Resolution solution*, and record the peak responses as directed under *Procedure:* the relative retention times are about 0.6 for methylparaben and 1.0 for labetalol, and the resolution, *R*, between the methylparaben peak and the labetalol peak is not less than 2.0.

Procedure—[NOTE—Use peak areas where peak responses are indicated.] Separately inject equal volumes (about 5 µL) of the *Standard preparation* and the *Assay preparation* into the chromatograph, record the chromatograms, and measure the responses for the major peaks. Calculate the quantity, in mg, of labetalol hydrochloride ($C_{19}H_{24}N_2O_3.HCl$) in each mL of the Injection taken by the formula:

$$100(C/V)(r_U/r_S),$$

in which *C* is the concentration, in mg per mL, of USP Labetalol Hydrochloride RS in the *Standard preparation*, *V* is the volume, in mL, of Injection taken, and r_U and r_S are the peak responses obtained from the *Assay preparation* and the *Standard preparation*, respectively.

Labetalol Hydrochloride Tablets

» Labetalol Hydrochloride Tablets contain not less than 90.0 percent and not more than 110.0 percent of the labeled amount of $C_{19}H_{24}N_2O_3.HCl$.

Packaging and storage—Preserve in tight, light-resistant containers, at a temperature between 2° and 30°.

Reference standard—*USP Labetalol Hydrochloride Reference Standard*—Dry in vacuum at 105° for 4 hours before using.

Identification—The retention time of the major peak in the chromatogram of the *Assay preparation* corresponds to that of the *Standard preparation*, as obtained in the *Assay*.

Dissolution ⟨711⟩—

Medium: water; 900 mL.

Apparatus 2: 50 rpm.

Time: 45 minutes.

Procedure—Determine the amount of $C_{19}H_{24}N_2O_3.HCl$ dissolved from ultraviolet absorbances at the wavelength of maximum absorbance at about 302 nm of filtered portions of the solution under test, suitably diluted with water, if necessary, in comparison with a Standard solution having a known concentration of USP Labetalol Hydrochloride RS in the same medium.

Tolerances—Not less than 80% (*Q*) of the labeled amount of $C_{19}H_{24}N_2O_3.HCl$ is dissolved in 45 minutes.

Uniformity of dosage units ⟨905⟩: meet the requirements.

Assay—

Mobile phase, Standard preparation, and *Chromatographic system*—Proceed as directed in the *Assay* under *Labetalol Hydrochloride.*

Assay preparation—Transfer an accurately counted number of Labetalol Hydrochloride Tablets, equivalent to about 2000 mg of labetalol hydrochloride, to a 500-mL volumetric flask, add 200 mL of water, and shake by mechanical means for 60 minutes. Dilute with water to volume, and mix. Filter the solution through a filter of 0.5-μm or finer porosity, discarding the first few mL of the filtrate. Transfer 10.0 mL of the filtrate to a 100-mL volumetric flask, dilute with *Mobile phase* to volume, and mix.

Procedure—[NOTE—Use peak areas where peak responses are indicated.] Separately inject equal volumes (about 5 μL) of the *Standard preparation* and the *Assay preparation* into the chromatograph, record the chromatograms, and measure the responses for the major peaks. Calculate the quantity, in mg, of labetalol hydrochloride ($C_{19}H_{24}N_2O_3 . HCl$) in each Tablet taken by the formula:

$$5000(C/N)(r_U/r_S),$$

in which C is the concentration, in mg per mL, of USP Labetalol Hydrochloride RS in the *Standard preparation*, N is the number of Tablets taken, and r_U and r_S are the peak responses obtained from the *Assay preparation* and the *Standard preparation*, respectively.

Lactated Ringer's Injection—*see* Ringer's Injection, Lactated

Lactic Acid

Propanoic acid, 2-hydroxy-.
Lactic acid [*50-21-5*].

» Lactic Acid is a mixture of lactic acid ($C_3H_6O_3$) and lactic acid lactate ($C_6H_{10}O_5$) equivalent to a total of not less than 85.0 percent and not more than 90.0 percent, by weight, of $C_3H_6O_3$. It is obtained by the lactic fermentation of sugars or prepared synthetically. Lactic Acid obtained by fermentation of sugars is levorotatory, while that prepared synthetically is racemic. [NOTE—Lactic Acid prepared by fermentation becomes dextrorotatory on dilution, which hydrolyzes L($-$) lactic acid lactate to L($+$) lactic acid.]

Packaging and storage—Preserve in tight containers.

Labeling—Label it to indicate whether it is levorotatory or racemic.

Identification—It responds to the test for *Lactate* ⟨191⟩.

Specific rotation ⟨781⟩: between $-0.05°$ and $+0.05°$, for racemic Lactic Acid.

Residue on ignition ⟨281⟩: not more than 3 mg, from a 5-mL portion (0.05%).

Sugars—To 10 mL of hot alkaline cupric tartrate TS add 5 drops of Lactic Acid: no red precipitate is formed.

Chloride—To 10 mL of a solution (1 in 100) acidified with nitric acid add a few drops of silver nitrate TS: no opalescence is produced immediately.

Citric, oxalic, phosphoric, or tartaric acid—To 10 mL of a solution (1 in 10) add 40 mL of calcium hydroxide TS, and boil for 2 minutes: no turbidity is produced.

Sulfate—To 10 mL of a solution (1 in 100) add 2 drops of hydrochloric acid and 1 mL of barium chloride TS: no turbidity is produced.

Heavy metals, *Method II* ⟨231⟩: 0.001%.

Readily carbonizable substances—Rinse a test tube with sulfuric acid TS, and allow to drain for 10 minutes. Add 5 mL of sulfuric acid TS to the test tube, carefully overlay it with 5 mL of Lactic Acid, and maintain the tube at a temperature of 15°: no dark color develops at the interface of the two acids within 15 minutes.

Assay—To about 2.5 mL of Lactic Acid, accurately weighed in a tared 250-mL flask, add 50.0 mL of 1 N sodium hydroxide VS, and boil the mixture for 20 minutes. Add phenolphthalein TS, and titrate the excess alkali in the hot solution with 1 N sulfuric acid VS. Perform a blank determination (see *Residual Titrations* under *Titrimetry* ⟨541⟩). Each mL of 1 N sodium hydroxide is equivalent to 90.08 mg of $C_3H_6O_3$.

Lactose—*see* Lactose NF

Lactulose Concentrate

$C_{12}H_{22}O_{11}$ 342.30
D-Fructose, 4-*O*-β-D-galactopyranosyl-.
Lactulose.
4-*O*-β-D-Galactopyranosyl-D-fructofuranose [*4618-18-2*].

» Lactulose Concentrate is a solution of sugars prepared from Lactose. It consists principally of lactulose together with minor quantities of lactose and galactose, and traces of other related sugars and water. It contains not less than 95.0 percent and not more than 105.0 percent of the labeled amount of lactulose ($C_{12}H_{22}O_{11}$). It contains no added substances.

Packaging and storage—Preserve in tight containers, preferably at a temperature between 2° and 30°. Avoid subfreezing temperatures.

Reference standards—*USP Lactulose Reference Standard*—Dry at 70° for 4 hours before using. *USP Galactose Reference Standard. USP Lactose Reference Standard. USP Fructose Reference Standard*—Dry at 70° for 4 hours before using.

Identification—

A: The principal spot from the *Test preparation* found in the test for *Related substances* corresponds in R_f value to that obtained from the *Standard lactulose preparation.*

B: Add a few drops of a solution (1 in 20) to 5 mL of hot alkaline cupric tartrate TS: a red precipitate of cuprous oxide is formed.

Refractive index ⟨831⟩: not less than 1.451, at 20°.

Residue on ignition ⟨281⟩: not more than 0.1%.

Related substances—

Standard lactulose preparation—Transfer 50 mg of USP Lactulose RS to a 10-mL volumetric flask, add water to volume, and mix.

Standard galactose preparation—Transfer 11 mg of USP Galactose RS to a 10-mL volumetric flask, add water to volume, and mix.

Standard lactose preparation—Transfer 6 mg of USP Lactose RS to a 10-mL volumetric flask, add water to volume, and mix.

Standard fructose preparation—Transfer 5 mg of USP Fructose RS to a 100-mL volumetric flask, add water to volume, and mix.

Standard lactose preparation for epilactose—Transfer 4 mg of USP Lactose RS to a 10-mL volumetric flask, add water to volume, and mix.

Test preparation—Prepare a solution of Lactulose Concentrate containing 5 mg of lactulose per mL.

Procedure—Apply 2 µL each of the *Standard lactulose preparation*, the *Standard galactose preparation*, the *Standard lactose preparation*, the *Standard lactose preparation for epilactose*, the *Standard fructose preparation*, and the *Test preparation* at separate locations on a suitable thin-layer chromatographic plate (see *Chromatography* ⟨621⟩), coated with a 0.25-mm layer of silica gel mixture. Develop the plate in a solvent system consisting of a mixture of ethyl acetate, methanol, a 1 in 20 aqueous solution of boric acid, and glacial acetic acid (55:20:15:10) until the solvent front has moved about three-fourths of the length of the plate. Remove the plate from the developing chamber, and evaporate the solvent by heating at 110° for 5 minutes. Spray the plate with a 1 in 1000 solution of naphthoresorcinol dissolved in a mixture of alcohol and sulfuric acid (90:10). Heat the plate at 110° for 5 minutes: any spot produced by the *Test preparation* at about R_f 0.35 does not exceed in magnitude or intensity the spot produced by the *Standard galactose preparation* at the same R_f value; any spot produced by the *Test preparation* at about R_f 0.23 does not exceed in magnitude or intensity the spot produced by the *Standard lactose preparation* at the same R_f value; any spot produced by the *Test preparation* at about R_f 0.25 does not exceed in magnitude or intensity the spot produced by the *Standard lactose preparation for epilactose* at about R_f 0.23; any spot produced by the *Test preparation* at about R_f 0.29 does not exceed in magnitude or intensity the spot produced by the *Standard fructose preparation*. The principal spot produced by the *Standard lactulose preparation* occurs at about R_f 0.17.

Assay—

Phosphate buffer—Transfer 13.6 g of monobasic potassium phosphate to a 1000-mL volumetric flask, and dissolve in 700 mL of water. Add 2 *N* potassium hydroxide to adjust to a pH of 7.3, dilute with water to volume, and mix.

Standard preparation—Transfer about 125 mg of USP Lactulose RS, accurately weighed, to a 25-mL volumetric flask, add water to volume, and mix. Pipet 5 mL of this solution into a 50-mL volumetric flask, dilute with *Phosphate buffer* to volume, and mix.

Assay preparation—Transfer an accurately weighed quantity of Lactulose Concentrate containing about 2.5 g of lactulose to a 500-mL volumetric flask, add water to volume, and mix. Pipet 10 mL of this solution into a 100-mL volumetric flask, dilute with *Phosphate buffer* to volume, and mix.

Procedure—Pipet 0.1 mL each of the *Standard preparation* and the *Assay preparation* into separate 1-cm cells. Add 0.03 mL of β-galactosidase suspension, and mix. Allow to stand for 2 hours at 37 ± 0.5°. Add 3.0 mL of nicotinamide adenine dinucleotide phosphate-adenosine-5′-triphosphate mixture and 0.02 mL of hexokinase and glucose-6-phosphate dehydrogenase suspension, and mix. Allow to stand for 10 minutes, then determine the absorbances of the solutions at 340 nm, with a suitable spectrophotometer, using air as the blank. Add to each cell 0.02 mL of phosphoglucose isomerase, mix, allow to stand for 10 minutes, and measure the absorbances at 340 nm as before. Calculate the quantity, in mg, of $C_{12}H_{22}O_{11}$ in the portion of Lactulose Concentrate taken by the formula:

$$5C(A_U/A_S),$$

in which *C* is the concentration, in µg per mL, of USP Lactulose RS in the *Standard preparation*, and A_U and A_S are the differences in the absorbances before and after the addition of phosphoglucose isomerase for the *Assay preparation* and the *Standard preparation*, respectively.

Lactulose Syrup

» Lactulose Syrup is prepared from Lactulose Concentrate. It is a mixture of sugars consisting principally of lactulose together with minor quantities of lactose and galactose, and traces of other related sugars in water. It contains not less than 90.0 percent and not more than 110.0 percent of the labeled amount of lactulose ($C_{12}H_{22}O_{11}$).

Packaging and storage—Preserve in tight containers, preferably at a temperature between 2° and 30°. Avoid subfreezing temperatures.

Reference standards—*USP Lactulose Reference Standard*—Dry at 70° for 4 hours before using. *USP Galactose Reference Standard. USP Lactose Reference Standard. USP Fructose Reference Standard*—Dry at 70° for 4 hours before using.

Microbial limits ⟨61⟩—The total bacterial count does not exceed 100 per g, and the tests for *Salmonella* species and *Escherichia coli* are negative.

pH ⟨791⟩: between 3.0 and 7.0, after 15 minutes of contact with the electrodes.

Other requirements—It responds to the *Identification tests* and meets the requirements for *Related substances* and *Assay* under *Lactulose Concentrate.*

Lanolin

» Lanolin is the purified, fat-like substance from the wool of sheep, *Ovis aries* Linné (Fam. Bovidae), containing not less than 25.0 percent and not more than 30.0 percent of water.

Packaging and storage—Preserve in tight, preferably rust-proof, containers, preferably at controlled room temperature.

Acidity ⟨401⟩—The free acids in 12.5 g require for neutralization not more than 2.0 mL of 0.10 *N* sodium hydroxide.

Alkalinity—Dissolve 2.5 g in 10 mL of ether, and add 2 drops of phenolphthalein TS: the liquid is not colored red.

Water content—Dissolve about 25 g, accurately weighed, in 75 mL of a *Mixed solvent* consisting of 3 parts of chloroform and 2 parts of methanol, and dilute with *Mixed solvent* to 250.0 mL. Determine the water content of a 5.0-mL portion as directed under *Water Determination, Method I* ⟨921⟩. Perform a blank determination on 5.0 mL of the *Mixed solvent*, and make any necessary correction. Between 25.0% and 30.0% is found.

Water-soluble acids and alkalies—Warm 12.5 g with 50 mL of water on a steam bath, constantly stirring the mixture until the Lanolin is melted: the fat separates completely on cooling, leaving the water layer nearly clear and neutral to litmus. Retain the water layer.

Chloride ⟨221⟩—Boil 20 mL of alcohol with 1.0 g of Lanolin under a reflux condenser, cool, add 1 mL of 2 *N* nitric acid, filter, and to the filtrate add 5 drops of a 1 in 50 solution of silver nitrate in alcohol: any turbidity produced does not exceed that of a blank to which 0.50 mL of 0.020 *N* hydrochloric acid has been added (0.035%).

Ammonia—Add 1 mL of 1 *N* sodium hydroxide to a 10-mL portion of the solution from the test for *Water-soluble acids and alkalies*, and boil: the vapors do not turn red litmus blue.

Petrolatum—Heat about 3 g, accurately weighed, on a steam bath, with frequent stirring, until it loses not less than 25.0% of its weight. Boil 40 mL of dehydrated alcohol with 500 mg of the dried lanolin so obtained: the solution is clear or not more than opalescent.

Iodine value ⟨401⟩: between 18 and 36, determined on a portion of 780 mg to 820 mg of the dried material obtained in the test for *Petrolatum*.

Lanolin Alcohols—*see* Lanolin Alcohols NF

Anhydrous Lanolin

» Anhydrous Lanolin is the purified, fat-like substance from the wool of sheep, *Ovis aries* Linné (Fam.

Bovidae), containing not more than 0.25 percent of water.

Packaging and storage—Preserve in well-closed containers, preferably at controlled room temperature.

Melting range, *Class II* ⟨741⟩: between 38° and 44°, determined on a test specimen previously cooled to between 8° and 10°.

Acidity ⟨401⟩—The free acids in 10.0 g require for neutralization not more than 2.0 mL of 0.10 *N* sodium hydroxide.

Alkalinity—Dissolve 2.0 g in 10 mL of ether, and add 2 drops of phenolphthalein TS: the liquid is not colored red.

Water—Dissolve about 25 g, accurately weighed, in 75 mL of a *Mixed solvent* consisting of 3 parts of chloroform and 2 parts of methanol, and dilute with *Mixed solvent* to 100.0 mL. Determine the water content of a 10.0-mL portion as directed under *Water Determination, Method I* ⟨921⟩. Perform a blank determination on 10.0 mL of *Mixed solvent*, and make any necessary correction. Not more than 0.25% is found.

Residue on ignition ⟨281⟩: not more than 0.1%.

Water-soluble acids and alkalies—Warm 10.0 g with 50 mL of water on a steam bath, constantly stirring the mixture until the Lanolin is melted: the fat separates completely on cooling, leaving the water layer nearly clear and neutral to litmus. Retain the water layer.

Water-soluble oxidizable substances—A 10-mL portion of the solution from the test for *Water-soluble acids and alkalies* does not completely decolorize 50 µL of 0.10 *N* potassium permanganate within 10 minutes.

Chloride ⟨221⟩—Boil 20 mL of alcohol with 1.0 g of Anhydrous Lanolin under a reflux condenser, cool, add 1 mL of 2 *N* nitric acid, filter, and to the filtrate add 5 drops of a 1 in 50 solution of silver nitrate in alcohol: any turbidity produced does not exceed that of a blank to which 0.50 mL of 0.020 *N* hydrochloric acid has been added (0.035%).

Ammonia—Add 1 mL of 1 *N* sodium hydroxide to a 10-mL portion of the solution from the test for *Water-soluble acids and alkalies*, and boil: the vapors do not turn red litmus blue.

Iodine value ⟨401⟩: between 18 and 36, determined on a specimen of 780 mg to 820 mg.

Petrolatum—Boil 40 mL of dehydrated alcohol with 500 mg: the solution is clear or not more than opalescent.

Lecithin—*see* Lecithin NF

Leucine

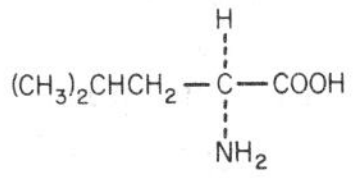

$C_6H_{13}NO_2$ 131.17

L-Leucine.

L-Leucine [*61-90-5*].

» Leucine contains not less than 98.5 percent and not more than 101.5 percent of $C_6H_{13}NO_2$, as L-leucine, calculated on the dried basis.

Packaging and storage—Preserve in well-closed containers.

Reference standard—*USP* L-*Leucine Reference Standard*—Dry at 105° for 3 hours before using.

Identification—The infrared absorption spectrum of a potassium bromide dispersion of it, previously dried, exhibits maxima only at the same wavelengths as that of a similar preparation of USP L-Leucine RS.

Specific rotation ⟨781⟩: between +14.9° and +17.3°, calculated on the dried basis, determined in a solution in 6 *N* hydrochloric acid containing 0.40 g in each 10.0 mL.

pH ⟨791⟩: between 5.5 and 7.0, in a solution (1 in 100).

Loss on drying ⟨731⟩—Dry it at 105° for 3 hours: it loses not more than 0.2% of its weight.

Residue on ignition ⟨281⟩: not more than 0.4%.

Chloride ⟨221⟩—A 0.73-g portion shows no more chloride than corresponds to 0.50 mL of 0.020 *N* hydrochloric acid (0.05%).

Sulfate ⟨221⟩—A 0.33-g portion shows no more sulfate than corresponds to 0.10 mL of 0.020 *N* sulfuric acid (0.03%).

Arsenic ⟨211⟩: 1.5 ppm.

Iron ⟨241⟩: 0.003%.

Heavy metals, *Method II* ⟨231⟩: 0.0015%.

Assay—Transfer about 130 mg of Leucine, accurately weighed, to a 125-mL flask, dissolve in a mixture of 3 mL of formic acid and 50 mL of glacial acetic acid, and titrate with 0.1 *N* perchloric acid VS, determining the end-point potentiometrically. Perform a blank determination, and make any necessary correction. Each mL of 0.1 *N* perchloric acid is equivalent to 13.12 mg of $C_6H_{13}NO_2$.

Leucovorin Calcium

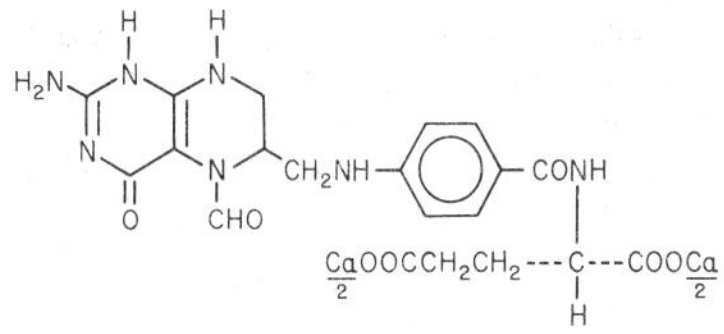

$C_{20}H_{21}CaN_7O_7$ 511.51

L-Glutamic acid, *N*-[4-[[(2-amino-5-formyl-1,4,5,6,7,8-hexahydro-4-oxo-6-pteridinyl)methyl]amino]benzoyl]-, calcium salt (1:1).

Calcium *N*-[*p*-[[[(6*RS*)-2-amino-5-formyl-5,6,7,8-tetrahydro-4-hydroxy-6-pteridinyl]methyl]amino]benzoyl]-L-glutamate (1:1) [*1492-18-8*].

» Leucovorin Calcium contains not less than 95.0 percent and not more than 105.0 percent of $C_{20}H_{21}CaN_7O_7$, calculated on the anhydrous basis.

Packaging and storage—Preserve in well-closed, light-resistant containers.

Reference standard—*USP Leucovorin Calcium Reference Standard*—Do not dry; determine the water content at time of use for quantitative analyses.

Identification—The infrared absorption spectrum of a potassium bromide dispersion of it exhibits maxima only at the same wavelengths as that of a similar preparation of USP Leucovorin Calcium RS.

Water, *Method I* ⟨921⟩: not more than 17.0%.

Heavy metals, *Method II* ⟨231⟩: 0.005%.

Assay—[Note—Use only freshly deionized water wherever water is specified throughout this procedure. Use low-actinic glassware for solutions containing leucovorin calcium and otherwise protect the solutions from unnecessary exposure to light. Complete the assay without prolonged interruption.]

Tetrabutylammonium hydroxide solution—Dissolve tetrabutylammonium hydroxide in methanol to obtain a solution containing 0.25 g per mL.

2 N Monobasic sodium phosphate solution—Dissolve monobasic sodium phosphate monohydrate in water to obtain a solution containing 276 mg per mL.

Mobile phase—Mix 15 mL of *Tetrabutylammonium hydroxide solution* with 835 mL of water. Add 125 mL of acetonitrile, adjust with *2 N Monobasic sodium phosphate solution* to an apparent pH of 7.5 ± 0.1, mix, dilute with water to 1000 mL, and filter. Adjust the concentration of acetonitrile, if necessary (see *System Suitability* under *Chromatography* ⟨621⟩).

Diluting solution—Mix 15 mL of *Tetrabutylammonium hydroxide solution* with 900 mL of water and adjust with *2 N*

Monobasic sodium phosphate solution to a pH of 7.5 ± 0.1. Dilute with water to 1000 mL, and mix.

Standard preparation—Dissolve an accurately weighed quantity of USP Leucovorin Calcium RS in *Diluting solution*, and dilute quantitatively with *Diluting solution* to obtain a solution having a known concentration of about 175 µg of anhydrous USP Leucovorin Calcium RS per mL.

Assay preparation—Dissolve about 20 mg of Leucovorin Calcium, accurately weighed, in *Diluting solution* in a 100-mL volumetric flask. Dilute with *Diluting solution* to volume, and mix.

System suitability preparation—Dissolve folic acid in *Diluting solution* to obtain a solution containing about 175 µg per mL. Mix 1 part of this solution with 4 parts of the *Standard preparation*.

Chromatographic system (see *Chromatography* ⟨621⟩)—The liquid chromatograph is equipped with a 254-nm detector and a 4-mm × 30-cm column that contains packing L1. The flow rate is 1 to 2 mL per minute. Chromatograph the *System suitability preparation*, and record the peak responses as directed under *Procedure:* the resolution, *R*, between the leucovorin calcium and folic acid peaks is not less than 3.6, and the relative standard deviation for replicate injections is not more than 2.0%. The relative retention times for leucovorin and folic acid are 1.0 and about 1.6, respectively.

Procedure—Separately inject equal volumes (about 15 µL) of the *Standard preparation* and the *Assay preparation* into the chromatograph, record the chromatograms, and measure the responses for the major peaks appearing at corresponding retention times in the chromatograms. Calculate the quantity, in mg, of $C_{20}H_{21}CaN_7O_7$ in the portion of Leucovorin Calcium taken by the formula:

$$0.1C(r_U/r_S),$$

in which *C* is the concentration, in µg per mL, of anhydrous USP Leucovorin Calcium RS in the *Standard preparation*, and r_U and r_S are the peak responses obtained from the *Assay preparation* and the *Standard preparation*, respectively.

Leucovorin Calcium Injection

» Leucovorin Calcium Injection is a sterile solution of Leucovorin Calcium in Water for Injection. It contains not less than 90.0 percent and not more than 120.0 percent of the labeled amount of leucovorin ($C_{20}H_{23}N_7O_7$).

Packaging and storage—Preserve in single-dose, light-resistant containers, preferably of Type I glass.

Reference standard—*USP Leucovorin Calcium Reference Standard*—Do not dry; determine the water content at time of use for quantitative analyses.

Identification—Transfer a volume of Injection, equivalent to about 6 mg of leucovorin calcium, to a glass-stoppered, 50-mL centrifuge tube, add about 40 mL of acetone, mix, centrifuge for a few minutes, and decant and discard the liquid phase. Repeat the washing process with an additional 40 mL of acetone. Dry the precipitate so obtained with a stream of dry nitrogen: the residue responds to the *Identification test* under *Leucovorin Calcium*.

pH ⟨791⟩: between 6.5 and 8.5.

Other requirements—It meets the requirements under *Injections* ⟨1⟩.

Assay—[NOTE—Use only freshly deionized water wherever water is specified throughout this procedure. Use low-actinic glassware for solutions containing leucovorin calcium, and otherwise protect the solutions from unnecessary exposure to light. Complete the assay without prolonged interruption.]

Tetrabutylammonium hydroxide solution, 2 N Monobasic sodium phosphate solution, Mobile phase, Diluting solution, Standard preparation, System suitability preparation, and *Chromatographic system*—Prepare as directed in the *Assay* under *Leucovorin Calcium*.

Assay preparation—Transfer an accurately measured volume of Leucovorin Calcium Injection, equivalent to about 9 mg of leucovorin, to a 50-mL volumetric flask, dilute with *Diluting solution* to volume, and mix. Pipet 25 mL of this solution into a 60-mL separator, add 25 mL of methylene chloride, shake the mixture, allow the layers to separate, and discard the methylene chloride extract. Repeat the extraction with two more 25-mL portions of methylene chloride, discarding the methylene chloride extracts. Filter the aqueous layer, discarding the first 5 mL of the filtrate, and collect the remaining filtrate in a glass-stoppered conical flask.

Procedure—Proceed as directed for *Procedure* in the *Assay* under *Leucovorin Calcium*. Calculate the quantity, in mg, of $C_{20}H_{23}N_7O_7$ in each mL of the Injection taken by the formula:

$$(473.44/511.51)(50C/V)(r_U/r_S),$$

in which 473.44 and 511.51 are the molecular weights of leucovorin and anhydrous leucovorin calcium, respectively, *C* is the concentration, in mg per mL, of anhydrous USP Leucovorin Calcium RS in the *Standard preparation*, *V* is the volume, in mL, of Injection taken, and r_U and r_S are the peak responses obtained from the *Assay preparation* and the *Standard preparation*, respectively.

Leukocyte Typing Serum—*see* Blood Groupings

Levobunolol Hydrochloride

$C_{17}H_{25}NO_3 \cdot HCl$ 327.85

1(2*H*)-Naphthalenone, 5-[3-[(1,1-dimethylethyl)-amino]-2-hydroxypropyl]-3,4-dihydro-, hydrochloride.

(−)-5-[3-(*tert*-Butylamino)-2-hydroxylpropoxy]-3,4-dihydro-1(2*H*)-naphthalenone hydrochloride [*27912-14-7*].

» Levobunolol Hydrochloride contains not less than 98.5 percent and not more than 101.0 percent of $C_{17}H_{25}NO_3 \cdot HCl$, calculated on the dried basis.

Packaging and storage—Preserve in well-closed containers.

Reference standard—*USP Levobunolol Hydrochloride Reference Standard*—Dry in vacuum over phosphorus pentoxide in a suitable drying tube at 110° for 4 hours before using. Keep container tightly closed and protected from light.

Identification—

A: The infrared absorption spectrum of a mineral oil dispersion of it, previously dried, exhibits maxima only at the same wavelengths as that of a similar preparation of USP Levobunolol Hydrochloride RS.

B: The ultraviolet absorption spectrum of a 1 in 1,000,000 solution of it in alcohol exhibits maxima and minima at the same wavelengths as that of a similar preparation of USP Levobunolol Hydrochloride RS, concomitantly measured.

Specific rotation ⟨781⟩: between −19° and −20°, calculated on the dried basis, determined in a solution in methanol containing 300 mg in each 10 mL.

Melting range ⟨741⟩: between 206° and 211°, within a range of 3°, determined after drying.

pH ⟨791⟩: between 4.5 and 6.5, in a solution (1 in 20).

Loss on drying ⟨731⟩—Dry about 1 g of it in vacuum over phosphorus pentoxide in a suitable drying tube at 110° for 4 hours: it loses not more than 0.5% of its weight.

Residue on ignition ⟨281⟩: not more than 0.1%.

Assay—

Mobile phase—Dissolve 990 mg of sodium 1-heptanesulfonate in 890 mL of water, add 10 mL of glacial acetic acid and 1100 mL of methanol, mix, filter through a suitable filter having a porosity of 1-μm or less, and degas. Make adjustments if necessary (see *System Suitability* under *Chromatography* ⟨621⟩).

Standard preparation—Dissolve an accurately weighed quantity of USP Levobunolol Hydrochloride RS in *Mobile phase* to obtain a solution having a known concentration of about 1.0 mg per mL.

Assay preparation—Transfer about 100 mg of Levobunolol Hydrochloride, accurately weighed, to a 100-mL volumetric flask, dissolve in *Mobile phase*, dilute with *Mobile phase* to volume, and mix.

Chromatographic system (see *Chromatography* ⟨621⟩)—The liquid chromatograph is equipped with a 254-nm detector and a 4-mm × 30-cm column that contains packing L1. The flow rate is about 1.5 mL per minute. Chromatograph the *Standard preparation*, and record the peak responses as directed under *Procedure:* the column efficiency determined from the analyte peak is not less than 1000 theoretical plates, the capacity factor, k, for levobunolol is between 6 and 10, the tailing factor for the analyte peak is not more than 2.5, and the relative standard deviation for replicate injections is not more than 0.5%.

Procedure—[NOTE—Use peak areas where peak responses are indicated.] Separately inject equal volumes (about 20 μL) of the *Standard preparation* and the *Assay preparation* into the chromatograph, record the chromatograms, and measure the responses for the major peaks. Calculate the quantity, in mg, of $C_{17}H_{25}NO_3.HCl$ in the portion of Levobunolol Hydrochloride taken by the formula:

$$100C(r_U/r_S),$$

in which C is the concentration, in mg per mL, of USP Levobunolol Hydrochloride RS in the *Standard preparation*, and r_U and r_S are the peak responses obtained from the *Assay preparation* and the *Standard preparation*, respectively.

Levobunolol Hydrochloride Ophthalmic Solution

» Levobunolol Hydrochloride Ophthalmic Solution contains not less than 90.0 percent and not more than 110.0 percent of the labeled amount of $C_{17}H_{25}NO_3.HCl$.

Packaging and storage—Preserve in tight containers.

Reference standard—*USP Levobunolol Hydrochloride Reference Standard*—Dry in vacuum over phosphorus pentoxide in a suitable drying tube at 110° for 4 hours before using. Keep container tightly closed and protected from light.

Identification—The retention time of the major peak in the chromatogram of the *Assay preparation* corresponds to that of the major peak in the chromatogram of the *Standard preparation* as obtained in the *Assay*.

pH ⟨791⟩: between 5.5 and 7.5.

Antimicrobial preservatives—Effectiveness ⟨51⟩: meets the requirements.

Sterility—It meets the requirements under *Sterility Tests* ⟨71⟩, when tested as directed in the section, *Test Procedures Using Membrane Filtration*.

Assay—

Mobile phase—Dissolve 990 mg of sodium 1-heptanesulfonate in 890 mL of water, add 10 mL of glacial acetic acid and 1100 mL of methanol, mix, filter through a suitable filter having a porosity of 1 μm or less, and degas. Make adjustments if necessary (see *System Suitability* under *Chromatography* ⟨621⟩).

Standard preparation—Dissolve an accurately weighed quantity of USP Levobunolol Hydrochloride RS in *Mobile phase* to obtain a solution having a known concentration of about 0.1 mg per mL.

Assay preparation—Dilute an accurately measured volume of Levobunolol Hydrochloride Ophthalmic Solution quantitatively, and stepwise if necessary, with *Mobile phase* to obtain a solution containing about 0.1 mg of levobunolol hydrochloride per mL.

Chromatographic system (see *Chromatography* ⟨621⟩)—The liquid chromatograph is equipped with a 254-nm detector and a 4-mm × 30-cm column that contains packing L1. The flow rate is about 1.5 mL per minute. Chromatograph the *Standard preparation*, and record the peak responses as directed under *Procedure:* the column efficiency determined from the analyte peak is not less than 1000 theoretical plates, the capacity factor, k', for levobunolol is between 3 and 6, the tailing factor for the analyte peak is not more than 2.6, and the relative standard deviation for replicate injections is not more than 1.5%.

Procedure—[NOTE—Use peak areas where peak responses are indicated.] Separately inject equal volumes (about 30 μL) of the *Standard preparation* and the *Assay preparation* into the chromatograph, record the chromatograms, and measure the responses for the major peaks. Calculate the quantity, in mg, of $C_{17}H_{25}NO_3.HCl$ in each mL of the Levobunolol Hydrochloride Ophthalmic Solution taken by the formula:

$$(L/D)(C)(r_U/r_S),$$

in which L is the labeled amount, in mg, of levobunolol hydrochloride in each mL of the Ophthalmic Solution, D is the concentration, in mg per mL, of levobunolol hydrochloride in the *Assay preparation*, based on the labeled quantity per mL and the extent of dilution, C is the concentration, in mg per mL, of USP Levobunolol Hydrochloride RS in the *Standard preparation*, and r_U and r_S are the peak responses obtained from the *Assay preparation* and the *Standard preparation*, respectively.

Levocarnitine

HO H
C
$(CH_3)_3N^+—CH_2$ CH_2COO^-

$C_7H_{15}NO_3$ 161.20

(*R*)-3-carboxy-2-hydroxy-*N*,*N*,*N*-trimethyl-1-propanaminium hydroxide, inner salt.

(*R*)-(3-carboxy-2-hydroxypropyl)trimethylammonium hydroxide, inner salt [*541-15-1*].

» Levocarnitine contains not less than 97.0 percent and not more than 103.0 percent of $C_7H_{15}NO_3$, calculated on the anhydrous basis.

Packaging and storage—Preserve in tight containers.

Reference standard—*USP Levocarnitine Reference Standard*—For quantitative use, determine the water content titrimetrically at the time of use.

Identification—The infrared absorption spectrum of a potassium bromide dispersion of it, previously dried in vacuum at 50° for 5 hours, exhibits maxima only at the same wavelengths as that of a similar preparation of USP Levocarnitine RS.

Specific rotation ⟨781⟩: between −29° and −32° calculated on the anhydrous basis, determined in a solution containing 10 g in each 100 mL.

pH ⟨791⟩: between 5.5 and 9.5 in a solution (1 in 20).

Water content ⟨921⟩: not more than 4.0%.

Residue on ignition ⟨281⟩: not more than 0.5%.

Chloride ⟨221⟩: not more than 0.4%.

Arsenic ⟨211⟩: not more than 2 ppm.

Potassium ⟨216⟩: not more than 0.2%.

Sodium ⟨216⟩: not more than 0.1%.

Heavy metals ⟨231⟩: not more than 0.002%.

Assay—Transfer about 100 mg of Levocarnitine, accurately weighed, to a 250-mL flask, dissolve in a mixture of 3 mL of formic acid and 50 mL of glacial acetic acid. Add 2 drops of crystal violet TS, and titrate with 0.1 *N* perchloric acid VS to an emerald green end-point. Perform a blank determination, and make any necessary correction. Each mL of 0.1 *N* perchloric acid is equivalent to 16.12 mg of $C_7H_{15}NO_3$.

Levocarnitine Oral Solution

» Levocarnitine Oral Solution is a solution of Levocarnitine in water and it contains suitable antimicrobial agents. It may contain a suitable flavor. It contains not less than 90.0 percent and not more than 110.0 percent of the labeled amount of $C_7H_{15}NO_3$.

Packaging and storage—Preserve in tight containers.

Reference standard—*USP Levocarnitine Reference Standard*—For quantitative use, determine the water content titrimetrically at the time of use.

Identification—The retention time of the major peak in the chromatogram of the *Assay preparation* corresponds to that in the chromatogram of the *Standard preparation*, both relative to the internal standard, as obtained in the *Assay*.

pH ⟨791⟩: between 4.0 and 6.0.

Assay—

Phosphate buffer 0.05 M—Dissolve 11.5 mL of phosphoric acid in 1900 mL of water, adjust with about 100 mL of 1 *N* sodium hydroxide to a pH of 2.4, and mix.

Mobile phase—Dissolve 555 mg of sodium 1-heptanesulfonate in 980 mL of *0.05 M Phosphate buffer* with stirring, add 20 mL of methanol, mix, degas, and filter. Make adjustments if necessary (see *System Suitability* under *Chromatography* ⟨621⟩).

Internal standard solution—Transfer about 10 mg of *p*-aminobenzoic acid to a 100-mL volumetric flask, dissolve in water, dilute with water to volume, and mix. Transfer 5.0 mL of the resulting solution to a 25-mL volumetric flask, dilute with water to volume, and mix to obtain the *Internal standard solution*.

Standard preparation—Transfer an accurately weighed quantity of about 10 mg of USP Levocarnitine RS to a 5-mL volumetric flask, add 1.0 mL of *Internal standard solution*, dilute with water to volume, and mix.

Assay preparation—Transfer an accurately measured volume of Levocarnitine Oral Solution, equivalent to about 500 mg of levocarnitine to a 50-mL volumetric flask, dilute with water to volume, and mix. Wash a 10-mm × 4-cm disposable column containing 500 mg of packing L1, in the order, with two column volumes of methylene chloride, two column volumes of methanol, and three column volumes of water. Pipet 5.0 mL of the solution prepared above into the washed disposable column, and rinse the column twice with 6.0 mL portions of water. Collect the filtrate and washings in a 25-mL volumetric flask, add 5.0 mL of *Internal standard solution*, dilute with water to volume, and mix to obtain the *Assay preparation*.

Chromatographic system (see *Chromatography* ⟨621⟩)—The liquid chromatograph is equipped with a 225-nm detector and a 3.9-mm × 30-cm column that contains 10-μm packing L1. The flow rate is about 1.5 mL per minute. Chromatograph the *Standard preparation*, and record the peak responses as directed under *Procedure:* the resolution, *R*, between the levocarnitine and internal standard peaks is not less than 1.5, and the relative standard deviation for replicate injections is not more than 2.0%.

Procedure—Separately inject equal volumes (about 40 μL) of the *Standard preparation* and the *Assay preparation* into the chromatograph, record the chromatograms, and measure the responses for the major peaks. The relative retention times are about 0.56 for levocarnitine and 1.0 for *p*-aminobenzoic acid. Calculate the quantity, in mg, of $C_7H_{15}NO_3$ in each mL of the Oral Solution taken by the formula:

$$250C/V(R_U/R_S),$$

in which *C* is the concentration, in mg per mL, of USP Levocarnitine RS in the *Standard preparation*, *V* is the volume, in mL, of Oral Solution taken, and R_U and R_S are the peak response ratios obtained from the *Assay preparation* and the *Standard preparation*, respectively.

Levodopa

$C_9H_{11}NO_4$ 197.19

L-Tyrosine, 3-hydroxy-.

(−)-3-(3,4-Dihydroxyphenyl)-L-alanine [*59-92-7*].

» Levodopa contains not less than 99.0 percent and not more than 100.5 percent of $C_9H_{11}NO_4$, calculated on the dried basis.

Packaging and storage—Preserve in tight, light-resistant containers, in a dry place, and prevent exposure to excessive heat.

Reference standards—*USP Levodopa Reference Standard*—Dry at 105° for 4 hours before using. *USP 3-(3,4,6-Trihydroxyphenyl)alanine Reference Standard*—Use without drying. Keep container tightly closed, and store in a cool, dry place. *USP 3-Methoxytyrosine Reference Standard*—Use without drying. Keep container tightly closed, and store in a cool, dry place.

Identification—

A: The infrared absorption spectrum of a mineral oil dispersion of it exhibits maxima only at the same wavelengths as that of a similar preparation of USP Levodopa RS.

B: The ultraviolet absorption spectrum of a 1 in 25,000 solution in 0.1 *N* hydrochloric acid exhibits maxima and minima at the same wavelengths as that of a similar solution of USP Levodopa RS, concomitantly measured, and the respective absorptivities, calculated on the dried basis, at the wavelength of maximum absorbance do not differ by more than 3.0%.

C: In the test for *Related compounds*, the R_f value of the principal spot obtained from the *Test preparation* corresponds to the R_f value of the principal spot obtained from *Reference preparation A* (at about R_f 0.4).

Specific rotation ⟨781⟩: between −160° and −167°, determined in a solution prepared as follows: Place about 500 mg of Levodopa, accurately weighed, in a 25-mL volumetric flask, add 10 mL of 1 *N* hydrochloric acid to dissolve the solid, then add 5 g of methenamine, swirl the contents to dissolve the methenamine, add 1 *N* hydrochloric acid to volume, and mix. Allow to stand in the dark at 25° for 3 hours, and measure the rotation.

Loss on drying ⟨731⟩—Dry it at 105° for 4 hours: it loses not more than 0.5% of its weight.

Residue on ignition ⟨281⟩: not more than 0.1%.

Heavy metals, *Method II* ⟨231⟩: 0.002%.

Related compounds—

Ferric chloride–potassium ferricyanide solution—Just prior to use, mix 2 volumes of ferric chloride solution (1 in 10) with 1 volume of potassium ferricyanide solution (1 in 20) to obtain about 100 mL of solution.

Sodium metabisulfite solution—Dissolve 100 mg of sodium metabisulfite in 10 mL of 1.2 *N* hydrochloric acid, and dilute with acetone to 100 mL.

Reference preparation A—[NOTE—Use low-actinic glassware.] Dissolve 2.5 mg of USP 3-(3,4,6-Trihydroxyphenyl)-alanine RS in *Sodium metabisulfite solution* to make 25.0 mL. Pipet 1 mL of this solution into a 10-mL volumetric flask containing 100 mg of USP Levodopa RS, and dilute with *Sodium metabisulfite solution* to volume. Mix this solution just prior to application.

Reference preparation B—[NOTE—Use low-actinic glassware.] Dissolve 2.5 mg of USP 3-Methoxytyrosine RS in *Sodium metabisulfite solution* to make 25.0 mL. Pipet 5 mL of this solution into a 10-mL volumetric flask, and dilute with *Sodium metabisulfite solution* to volume. Mix this solution just prior to application.

Test preparation—[NOTE—Use low-actinic glassware.] Just prior to application, dissolve 100 mg of Levodopa in 10.0 mL of *Sodium metabisulfite solution*.

Developing solvent—Prepare a mixture of 150 mL of butyl alcohol, 75 mL of glacial acetic acid, 75 mL of water, and 15 mL of methanol.

Chromatographic plates—Use suitable thin-layer chromatographic plates (see *Chromatography* ⟨621⟩) coated with a 0.25-mm layer of microcrystalline cellulose. Pre-develop a plate in the *Developing solvent* until the solvent front has moved not less than 18 cm from the origin. Remove the plate from the chamber, and dry in a current of air for about 10 minutes. [NOTE—The plate may be developed overnight; solvent overflow during pre-development is of no consequence.]

Procedure—Apply 10 µL each of the *Test preparation* and *Reference preparations A* and *B* to separate points about 3 cm from the bottom of the plate. Dry the spots in a stream of nitrogen, and develop the chromatogram in a suitable low-actinic chamber equilibrated for 5 minutes with a freshly mixed portion of *Developing solvent*, until the solvent front has moved about 15 cm from the line of application. Remove the plate from the chamber, mark the solvent front, and dry in a current of air for about 10 minutes. Spray the plate with *Ferric chloride–potassium ferricyanide solution.* The 3-(3,4,6-trihydroxyphenyl)-alanine produces a spot at about R_f 0.25, and the 3-methoxytyrosine produces a spot at about R_f 0.5. Any spots from the *Test preparation*, occurring at R_f 0.25 and R_f 0.5, are not greater in size or intensity than the spots at the respective R_f values produced by the *Reference preparations*, corresponding to not more than 0.1% of 3-(3,4,6-trihydroxyphenyl)alanine, and not more than 0.5% of 3-methoxytyrosine. (A bleached spot may appear at about R_f 0.6, which is an artifact resulting from the development of the *Sodium metabisulfite solution.*)

Assay—Dissolve about 600 mg of Levodopa, accurately weighed, in 10 mL of 96 percent formic acid, and add 80 mL of glacial acetic acid. Titrate with 0.1 *N* perchloric acid VS, determining the end-point potentiometrically, using a glass-calomel electrode system. Perform a blank determination, and make any necessary correction. Each mL of 0.1 *N* perchloric acid is equivalent to 19.72 mg of $C_9H_{11}NO_4$.

Levodopa Capsules

» Levodopa Capsules contain not less than 90.0 percent and not more than 110.0 percent of the labeled amount of $C_9H_{11}NO_4$.

Packaging and storage—Preserve in tight, light-resistant containers, in a dry place, and prevent exposure to excessive heat.

Reference standards—*USP Levodopa Reference Standard*—Dry at 105° for 4 hours before using. *USP 3-(3,4,6-Trihydroxyphenyl)alanine Reference Standard*—Use without drying. Keep container tightly closed, and store in a cool, dry place. *USP 3-Methoxytyrosine Reference Standard*—Use without drying. Keep container tightly closed, and store in a cool, dry place.

Identification—Shake a quantity of the contents of the Capsules, equivalent to about 500 mg of levodopa, with 25 mL of 3 *N* hydrochloric acid, and filter. Adjust the acidity of the filtrate to pH 3 with 6 *N* ammonium hydroxide, added dropwise with stirring, and allow to stand, protected from light, for several hours. Filter, wash the precipitate with water, and dry at 105°: the residue responds to *Identification test A* under *Levodopa.*

Dissolution ⟨711⟩—

Medium: 0.1 *N* hydrochloric acid; 900 mL.

Apparatus 1: 100 rpm.

Time: 30 minutes.

Procedure—Determine the amount of $C_9H_{11}NO_4$ dissolved from ultraviolet absorbances at the wavelength of maximum absorbance at about 280 nm of filtered portions of the solution under test, suitably diluted with 0.1 *N* hydrochloric acid, in comparison with a Standard solution having a known concentration of USP Levodopa RS in the same medium.

Tolerances—Not less than 75% (*Q*) of the labeled amount of $C_9H_{11}NO_4$ is dissolved in 30 minutes.

Uniformity of dosage units ⟨905⟩: meet the requirements.

Related compounds—The residue obtained in the *Identification test* responds to the test for *Related compounds* under *Levodopa.*

Assay—Weigh the contents of not less than 20 Levodopa Capsules, mix, and transfer an accurately weighed portion of the powder, equivalent to about 175 mg of levodopa, to a 100-mL volumetric flask. Add 50 mL of 0.1 *N* hydrochloric acid, and shake the mixture by mechanical means for 5 minutes. Add 0.1 *N* hydrochloric acid to volume, mix, and filter, discarding the first 20 mL of the filtrate. Dilute 2.0 mL of the subsequent filtrate with 0.1 *N* hydrochloric acid to 100 mL. Concomitantly determine the absorbances of this solution and a similarly prepared Standard solution of USP Levodopa RS, having a known concentration of about 35 µg per mL, in 1-cm cells at the wavelength of maximum absorbance at about 280 nm with a suitable spectrophotometer, using 0.1 *N* hydrochloric acid as the blank. Calculate the quantity, in mg, of $C_9H_{11}NO_4$ in the portion of Capsules contents taken by the formula:

$$5C(A_U / A_S),$$

in which *C* is the concentration, in µg per mL, of USP Levodopa RS in the Standard solution, and A_U and A_S are the absorbances of the solution from the Capsule contents and the Standard solution, respectively.

Levodopa Tablets

» Levodopa Tablets contain not less than 90.0 percent and not more than 110.0 percent of the labeled amount of $C_9H_{11}NO_4$.

Packaging and storage—Preserve in tight, light-resistant containers, in a dry place, and prevent exposure to excessive heat.

Reference standards—*USP Levodopa Reference Standard*—Dry at 105° for 4 hours before using. *USP 3-(3,4,6-Trihydroxyphenyl)alanine Reference Standard*—Use without drying. Keep container tightly closed, and store in a cool, dry place. *USP 3-Methoxytyrosine Reference Standard*—Use without drying. Keep container tightly closed, and store in a cool, dry place.

Identification—Shake a quantity of powdered Tablets, equivalent to about 500 mg of levodopa, with 25 mL of 3 *N* hydrochloric acid, and filter. Adjust the acidity of the filtrate with 6 *N* ammonium hydroxide, added dropwise with stirring, and allow to stand, protected from light, for several hours. Filter, wash the precipitate with water, and dry at 105°: the residue responds to *Identification test A* under *Levodopa.*

Dissolution ⟨711⟩—

Medium: 0.1 *N* hydrochloric acid; 900 mL.

Apparatus 1: 100 rpm.

Time: 30 minutes.

Procedure—Determine the amount of $C_9H_{11}NO_4$ dissolved from ultraviolet absorbances at the wavelength of maximum absorbance at about 280 nm of filtered portions of the solution under test, suitably diluted with 0.1 *N* hydrochloric acid, in comparison with a Standard solution having a known concentration of USP Levodopa RS in the same medium.

Tolerances—Not less than 75% (*Q*) of the labeled amount of $C_9H_{11}NO_4$ is dissolved in 30 minutes.

Uniformity of dosage units ⟨905⟩: meet the requirements.

Related compounds—The residue obtained in the *Identification test* responds to the test for *Related compounds* under *Levodopa.*

Assay—Weigh and finely powder not less than 20 Levodopa Tablets. Weigh accurately a portion of the powder, equivalent to about 175 mg of levodopa, and transfer to a 100-mL volumetric flask. Proceed as directed in the *Assay* under *Levodopa Capsules*, beginning with "Add 50 mL of 0.1 *N* hydrochloric acid."

Levodopa Tablets, Carbidopa and—*see* Carbidopa and Levodopa Tablets

Levonordefrin

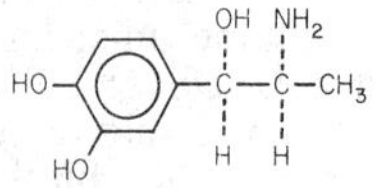

$C_9H_{13}NO_3$ 183.21
1,2-Benzenediol, 4-(2-amino-1-hydroxypropyl)-, [*R*-(*R**,*S**)]-.
(–)-α-(1-Aminoethyl)-3,4-dihydroxybenzyl alcohol
[*18829-78-2; 829-74-3*].

» Levonordefrin, dried in vacuum at 60° for 15 hours, contains not less than 98.0 percent and not more than 102.0 percent of $C_9H_{13}NO_3$.

Packaging and storage—Preserve in well-closed containers.

Reference standard—*USP Levonordefrin Reference Standard*—Dry in vacuum at 60° for 15 hours before using.

Identification—

A: The infrared absorption spectrum of a potassium bromide dispersion of it, previously dried, exhibits maxima only at the same wavelengths as that of a similar preparation of USP Levonordefrin RS.

B: The ultraviolet absorption spectrum of a solution in 0.1 *N* hydrochloric acid, containing about 25 µg per mL, exhibits maxima and minima only at the same wavelengths as that of a similar solution of USP Levonordefrin RS, concomitantly measured.

Specific rotation ⟨781⟩: between −28° and −31°, determined in a solution prepared by dissolving 500 mg of Levonordefrin, previously dried, in 3 mL of dilute hydrochloric acid (1 in 12) and diluting with water to 10.0 mL.

Loss on drying ⟨731⟩—Dry it in vacuum at 60° for 15 hours: it loses not more than 1.0% of its weight.

Residue on ignition ⟨281⟩: not more than 0.2%.

Chromatographic impurities—

Standard preparations—Dissolve an accurately weighed quantity of USP Levonordefrin RS in a mixture of methanol and glacial acetic acid (96:4) to obtain a Standard stock solution having a known concentration of 5 mg per mL. Dilute this solution quantitatively with the mixture of methanol and glacial acetic acid (96:4) to obtain *Standard preparations*, designated below by letter, having the following compositions:

Dilution	Concentration (µg RS per mL)	Percentage (%, for comparison with test specimen)
A (1 in 10)	500	1.0
B (1 in 20)	250	0.5
C (1 in 50)	100	0.2
D (1 in 100)	50	0.1

Test preparation—Dissolve an accurately weighed quantity of Levonordefrin in a mixture of methanol and glacial acetic acid (96:4) to obtain a solution containing 50 mg per mL.

Procedure—On a suitable thin-layer chromatographic plate (see *Chromatography* ⟨621⟩), coated with 0.25-mm layer of chromatographic silica gel mixture, apply separately 5 µL of the *Test preparation* and 5 µL of each *Standard preparation*. Position the plate in a chromatographic chamber, and develop the chromatograms in a solvent system consisting of a mixture of *n*-butyl alcohol, water, and glacial acetic acid (70:20:10) until the solvent front has moved about three-fourths of the length of the plate. Remove the plate from the developing chamber, mark the solvent front, and allow the solvent to evaporate in warm, circulating air. Examine the plate under short-wavelength ultraviolet light. Expose the plate to iodine vapors, and examine again. Compare the intensities, observed by both visualizations, of any secondary spots observed in the chromatogram of the *Test preparation* with those of the principal spots in the chromatograms of the *Standard preparations:* the sum of the intensities of secondary spots obtained from the *Test preparation* corresponds to not more than 1.0% of related compounds, with no single impurity corresponding to more than 0.5%.

Assay—Transfer about 350 mg of Levonordefrin, previously dried and accurately weighed, to a small flask, dissolve in 50 mL of glacial acetic acid, heating, if necessary, add 1 drop of crystal violet TS, and titrate with 0.1 *N* perchloric acid VS to a green end-point. Perform a blank determination, and make any necessary correction. Each mL of 0.1 *N* perchloric acid is equivalent to 18.32 mg of $C_9H_{13}NO_3$.

Levonordefrin Injection, Mepivacaine Hydrochloride and—*see* Mepivacaine Hydrochloride and Levonordefrin Injection

Levonordefrin Injection, Procaine and Tetracaine Hydrochlorides and—*see* Procaine and Tetracaine Hydrochlorides and Levonordefrin Injection

Levonordefrin Injection, Propoxycaine and Procaine Hydrochlorides and—*see* Propoxycaine and Procaine Hydrochlorides and Levonordefrin Injection

Levonorgestrel

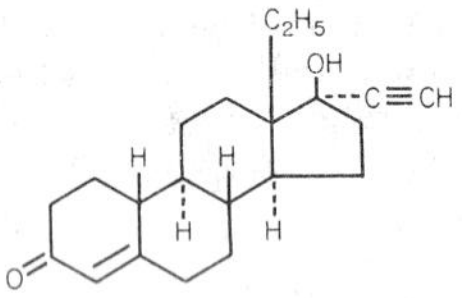

$C_{21}H_{28}O_2$ 312.45
18,19-Dinorpregn-4-en-20-yn-3-one, 13-ethyl-17-hydroxy-, (17α)-(−)-.
(−)-13-Ethyl-17-hydroxy-18,19-dinor-17α-pregn-4-en-20-yn-3-one [*797-63-7*].

» Levonorgestrel contains not less than 98.0 percent and not more than 102.0 percent of $C_{21}H_{28}O_2$, calculated on the dried basis.

Packaging and storage—Preserve in well-closed, light-resistant containers.

Reference standard—*USP Norgestrel Reference Standard*—Dry at 105° for 3 hours before using.

Identification—

A: The infrared absorption spectrum of a potassium bromide dispersion of it, previously dried, exhibits maxima only at the same wavelengths as that of a similar preparation of USP Norgestrel RS.

B: Meeting the requirements of the tests for *Specific rotation* and *Melting range* provides identification distinguishing it from norgestrel.

Melting range ⟨741⟩: between 232° and 239°, but the range between beginning and end of melting does not exceed 4°.

Specific rotation ⟨781⟩: between −30° and −35°, calculated on the dried basis, determined in a solution containing 200 mg of it in each 10 mL of chloroform.

Loss on drying ⟨731⟩—Dry it at 105° for 5 hours: it loses not more than 0.5% of its weight.

Residue on ignition ⟨281⟩: not more than 0.3%.

Ethynyl group—Dissolve 200 mg in about 40 mL of tetrahydrofuran. Add 10 mL of silver nitrate solution (1 in 10), and titrate with 0.1 *N* sodium hydroxide VS, using glass and calomel electrodes, the latter being of standard fiber type but containing potassium nitrate solution as the electrolyte. Perform a blank determination, and make any necessary correction. Each mL of 0.1 *N* sodium hydroxide is equivalent to 2.503 mg of ethynyl group (–C≡CH). Not less than 7.81% and not more than 8.18% of ethynyl group is found.

Chromatographic impurities—Proceed as directed in the test for *Chromatographic impurities* under *Norgestrel*. The requirements of the test are met if the sum of the impurities in the *Test preparation* does not exceed 2.0% and no single impurity is greater than 0.5%.

Assay—Using USP Norgestrel RS, proceed with Levonorgestrel as directed in the *Assay* under *Norgestrel*, except to read "Levonorgestrel" in place of "Norgestrel."

Levonorgestrel and Ethinyl Estradiol Tablets

» Levonorgestrel and Ethinyl Estradiol Tablets contain not less than 90.0 percent and not more than 110.0 percent of the labeled amount of levonorgestrel ($C_{21}H_{28}O_2$) and not less than 90.0 percent and not more than 110.0 percent of the labeled amount of ethinyl estradiol ($C_{20}H_{24}O_2$).

Packaging and storage—Preserve in well-closed containers.

Reference standards—*USP Norgestrel Reference Standard*—Dry at 105° for 3 hours before using. *USP Ethinyl Estradiol Reference Standard*—Dry in vacuum over silica gel for 4 hours before using.

Identification—

A: The retention times of the two major peaks in the chromatogram of the *Assay preparation* correspond to those of norgestrel and ethinyl estradiol in the *Standard preparation*, as obtained in the *Assay*.

B: Finely powder 20 Tablets and transfer a portion of the powder, equivalent to about 4 mg of levonorgestrel, to a suitable container. Add 250 mL of a solvent mixture consisting of isooctane and chloroform (3:1). Sonicate the mixture for 3 minutes, and then stir it by mechanical means for 30 minutes. Filter the mixture and evaporate the filtrate to dryness in a rotating vacuum evaporator. Dissolve the residue in 3 mL of chloroform, and transfer with a pipet to a 60-mL separator containing 18 mL of isooctane. Rinse the evaporator flask with an additional 3-mL portion of chloroform, and add the rinsing to the separator. Add 10 mL of 1 *N* sodium hydroxide, shake vigorously, and allow the layers to separate. Discard the lower aqueous phase, and filter the organic phase through about 3 g of anhydrous sodium sulfate on filter paper into a 50-mL beaker. Rinse the filter with several small portions of the mixture of isooctane and chloroform (3:1), adding the filtered rinsings to the filtrate, and evaporate under nitrogen on a steam bath to dryness. Dissolve the residue in 1 to 2 mL of hot toluene and transfer to a small glass vial with a pipet. Reduce the volume of the solution to about 0.1 mL under nitrogen with warming. [NOTE—During this step, any crystals that deposit on the vial wall should be transferred to the bottom and allowed to redissolve.] Store the vial containing the clear toluene solution at 4° overnight to allow crystallization to occur. Remove and discard the mother liquor with a pipet, rinse the crystals with two 0.5-mL portions of anhydrous ether, and discard the rinsings. Dry the vial containing the rinsed crystals in a vacuum desiccator at 60° for 4 hours: the melting point (see *Melting Range or Temperature* ⟨741⟩) of the dried crystals of levonorgestrel so obtained, using the *Class I* method, is not lower than 220°.

Dissolution ⟨711⟩—

Medium: polysorbate 80 (5 ppm) in water; 500 mL.

Apparatus 2: 75 rpm.

Time—

UNCOATED TABLETS: 30 minutes; 60 minutes.

SUGAR-COATED TABLETS: 60 minutes.

Standard preparation—[NOTE—A volume of alcohol not exceeding 2% of the final total volume of solution may be used to aid in dissolving the Reference Standards.] Prepare a solution of USP Norgestrel RS and USP Ethinyl Estradiol RS in *Dissolution Medium* having accurately known concentrations corresponding approximately to the concentrations that would be obtained by dissolving 1 Tablet in 500 mL of solvent.

Test preparation—

UNCOATED TABLETS—After operating the apparatus for 30 minutes, and for 60 minutes, withdraw 15-mL portions of liquid from each vessel, and filter through a polyvinylidene filter, discarding the first 10 mL of the filtrate. Replace the portion withdrawn at the 30-minute sampling, without delay, by adding 15 mL of *Dissolution Medium* to each vessel to maintain the volume constant at 500 mL.

SUGAR-COATED TABLETS—After operating the apparatus for 60 minutes, withdraw 15-mL portions of liquid from each vessel, and filter through a polyvinylidene filter, discarding the first 10 mL of the filtrate.

Mobile phase—Prepare a filtered and degassed solution of acetonitrile and water (60:40).

Chromatographic system (see *Chromatography* ⟨621⟩)—The liquid chromatograph is equipped with a 247-nm detector (for levonorgestrel analysis), a spectrofluorometric detector (for ethinyl estradiol analysis) with an excitation wavelength of 285 nm and an emission wavelength of 310 nm, and a 4-mm × 15-cm column that contains packing L7. The flow rate is about 1 mL per minute. Chromatograph replicate injections of the *Standard preparation*, and record the peak responses as directed under *Procedure:* the relative standard deviation is not more than 3.0%.

Procedure—Separately inject equal volumes (about 100 µL) of the *Standard preparation* and the *Test preparation* into the chromatograph, record the chromatograms, and measure the responses for the major peaks. The relative retention times are about 0.7 for ethinyl estradiol and 1.0 for levonorgestrel.

Calculations—

UNCOATED TABLETS—Calculate the percentage of each drug dissolved at the 30-minute interval by the formula:

$$(0.5C)(100/K)(r_U/r_S),$$

and at the 60-minute interval by the formula:

$$[(0.5C)(r_U/r_S) + (15/500)(Y)](100/K),$$

in which C is the concentration, in µg per mL, of the appropriate drug in the *Standard preparation*, K is the labeled amount, in mg per Tablet, of the appropriate drug, r_U and r_S are the peak responses obtained from the *Test preparation* and the *Standard preparation*, respectively, and Y is the amount, in mg, of the appropriate drug calculated at the 30-minute interval.

SUGAR-COATED TABLETS—Calculate the percentage of each drug dissolved by the formula:

$$(0.5C)(100/K)(r_U/r_S).$$

Tolerances—

UNCOATED TABLETS—Not less than 60% (*Q*), and not less than 75% (*Q*), of the labeled amounts of levonorgestrel ($C_{12}H_{28}O_2$) and ethinyl estradiol ($C_{20}H_{24}O_2$) are dissolved in 30 minutes, and 60 minutes, respectively.

SUGAR-COATED TABLETS—Not less than 60% (*Q*) of the labeled amounts of levonorgestrel ($C_{21}H_{28}O_2$) and ethinyl estradiol ($C_{20}H_{24}O_2$) are dissolved in 60 minutes.

Uniformity of dosage units ⟨905⟩: meet the requirements.

Procedure for content uniformity—Transfer 1 Tablet to a 40-mL centrifuge tube. Add 10.0 mL of *Mobile phase* and proceed as directed in the *Assay*.

Assay—

Mobile phase—Prepare a deaerated mixture containing 350 mL of acetonitrile, 150 mL of methanol, and 450 mL of water.

Standard preparation—Transfer 15.0 mL of a solution of USP Norgestrel RS in *Mobile phase* and 3.0 mL of a solution of USP Ethinyl Estradiol RS in *Mobile phase*, each solution having a concentration of about 0.1 mg per mL, into a 100-mL volumetric flask. Dilute with *Mobile phase* to volume, and mix. Each mL of this *Standard preparation* has a known concentration of about 15 µg and 3 µg of USP Norgestrel RS and USP Ethinyl Estradiol RS per mL, respectively.

Assay preparation—Transfer a number of Tablets, equivalent to about 3 mg of levonorgestrel, to a 200-mL volumetric flask. Dilute with *Mobile phase* to volume, sonicate to disintegrate the Tablets, then shake by mechanical means for 20 minutes. Centrifuge and use the clear, supernatant solution.

Chromatographic system (see *Chromatography* ⟨621⟩)—The liquid chromatograph is equipped with a 215-nm detector and a

4.6-mm × 15-cm column that contains 5- to 7-µm packing L7. The flow rate is about 1 mL per minute. Chromatograph the *Standard preparation*, and record the peak response as directed under *Procedure:* the resolution, *R*, between the two major peaks, is not less than 2.5, and the relative standard deviation for replicate injections is not more than 2.0.

Procedure—Separately inject equal volumes (about 50 µL) of the *Standard preparation* and the *Assay preparation* into the chromatograph, record the chromatograms, and measure the responses for the major peaks. The relative retention times are about 0.7 for ethinyl estradiol and 1.0 for norgestrel. Calculate the quantities, in mg, of levonorgestrel ($C_{21}H_{28}O_2$) and ethinyl estradiol ($C_{20}H_{24}O_2$), respectively, in the portion of Tablets taken for the *Assay preparation* by the same formula:

$$0.2C(r_U/r_S),$$

in which *C* is the concentration, in µg per mL, of the appropriate USP Reference Standard in the *Standard preparation*, and r_U and r_S are the peak responses of the corresponding analyte obtained from the *Assay preparation* and the *Standard preparation*, respectively.

Levopropoxyphene Napsylate

$C_{22}H_{29}NO_2 . C_{10}H_8O_3S . H_2O$ 565.72

Benzeneethanol, α-[2-(dimethylamino)-1-methylethyl]-α-phenyl-, propanoate (ester), [*R*-(*R**,*S**)]-, compd. with 2-naphthalenesulfonic acid (1:1), monohydrate.

2-Naphthalenesulfonic acid compound with (−)-α-[2-(dimethylamino)-1-methylethyl]-α-phenylphenethyl propionate (1:1) monohydrate [*5667-69-6; 55557-30-7*].

Anhydrous 547.71 [*5714-90-9*].

» Levopropoxyphene Napsylate contains not less than 97.0 percent and not more than 103.0 percent of $C_{22}H_{29}NO_2 . C_{10}H_8O_3S$, calculated on the anhydrous basis.

Packaging and storage—Preserve in tight containers.

Reference standards—*USP Levopropoxyphene Napsylate Reference Standard*—Do not dry; determine the *Water* content by *Method I* before using. *USP α-d-2-Acetoxy-4-(dimethylamino)-1,2-diphenyl-3-methylbutane Reference Standard*—Keep container tightly closed. Do not dry before using. *USP α-d-4-(Dimethylamino)-1,2-diphenyl-3-methyl-2-butanol Hydrochloride Reference Standard*—Keep container tightly closed. Do not dry before using.

Identification—

A: The infrared absorption spectrum of a potassium bromide dispersion of it, undried, exhibits maxima only at the same wavelengths as that of a similar preparation of USP Levopropoxyphene Napsylate RS.

B: The ultraviolet absorption spectrum of a 1 in 25,000 solution in methanol exhibits maxima and minima at the same wavelengths as that of a similar solution of USP Levopropoxyphene Napsylate RS, concomitantly measured, and the respective absorptivities, calculated on the dried basis, at the wavelength of maximum absorbance at about 275 nm do not differ by more than 3.0%.

Melting range, *Class I* ⟨741⟩: between 158° and 165°, but the range between beginning and end of melting does not exceed 4°, determined after drying at 105° for 3 hours.

Specific rotation ⟨781⟩: between −35° and −43°, determined in a solution in chloroform containing 100 mg in each 10 mL.

Water, *Method I* ⟨921⟩: not more than 5.0%.

Residue on ignition ⟨281⟩: not more than 0.5%.

Heavy metals, *Method II* ⟨231⟩: 0.003%.

Related compounds—

Internal standard solution—Dissolve a quantity of *n*-tricosane in chloroform to obtain a solution containing about 1 mg per mL.

Solution A—Weigh accurately 12.5 mg of the acetoxy analog standard, USP α-*d*-2-Acetoxy-4-(dimethylamino)-1,2-diphenyl-3-methylbutane RS, and 12.5 mg of the carbinol hydrochloride standard, USP α-*d*-4-(Dimethylamino)-1,2-diphenyl-3-methyl-2-butanol Hydrochloride RS, and transfer to the same 50-mL volumetric flask. Dilute with chloroform to volume, and mix.

Standard preparation—Transfer about 150 mg of USP Levopropoxyphene Napsylate RS, accurately weighed, to a 125-mL separator, add 15 mL of water, and mix. Add 5.0 mL of sodium hydroxide solution (1 in 10), 2.0 mL of *Solution A*, and 15 mL of chloroform to the contents of the separator, shake for 1 minute, and allow the phases to separate. Drain the chloroform extract through a layer of about 2 g of anhydrous granular sodium sulfate, supported on glass wool and previously washed with chloroform, into a 50-mL volumetric flask containing 1.0 mL of *Internal standard solution.* Extract the aqueous phase in the separator with two additional 15-mL portions of chloroform, draining each chloroform extract through the same sodium sulfate filter into the same volumetric flask. Rinse the sodium sulfate filter with 1 to 2 mL of chloroform, collect the rinsing with the combined chloroform extracts, dilute with chloroform to volume, and mix. Evaporate 10.0 mL of this chloroform solution under a stream of nitrogen to about 2 mL.

Test preparation—Prepare as directed for *Standard preparation*, except to use about 150 mg of Levopropoxyphene Napsylate and to omit the addition of the 2.0 mL of *Solution A.*

Chromatographic system (see *Chromatography* ⟨621⟩)—Under typical conditions, the gas chromatograph is equipped with a flame-ionization detector and contains a 0.6-m × 3-mm glass column packed with 3 percent phase G2 on support S1AB. [NOTE—Frequent resilylation of the column packing may be necessary to prevent on-column decomposition of levopropoxyphene.] The temperatures of the column and injector port are maintained at about 160°, and the temperature of the detector block is maintained at about 190°. [NOTE—Do not allow the temperature to exceed 200°.] Dry helium is used as the carrier gas at a flow rate of about 75 mL per minute.

System suitability—Chromatograph five injections of the *Standard preparation*, and record peak heights as directed under *Procedure.* The development time for one chromatogram is about 10 minutes, and retention times for the carbinol precursor and acetoxy analog are, respectively, about 0.3 and 0.4 relative to the internal standard: the analytical system is suitable for conducting the test if the resolution factor, *R*, between the carbinol and the acetoxy peaks is not less than 1.5, the coefficients of variation for the ratios R_A and R_C do not exceed 6.0%; and there is no evidence of spurious peaks resulting from the decomposition of levopropoxyphene in the system.

Procedure—Inject separately a suitable volume of the *Standard preparation* and an identical volume of the *Test preparation* into a suitable gas chromatograph, and record the chromatogram. Calculate the peak heights of the carbinol precursor, acetoxy analog, and *n*-tricosane peaks in each chromatogram.

Calculate the percentage of carbinol napsylate precursor in the portion of Levopropoxyphene Napsylate taken by the formula:

$$(509.66/319.87)(200C_C/W)(r_C/R_C),$$

in which 509.66 and 319.87 are the molecular weights of carbinol napsylate monohydrate precursor and anhydrous carbinol hydrochloride precursor, respectively, C_C is the concentration, in mg per mL, of USP α-*d*-4-(Dimethylamino)-1,2-diphenyl-3-methyl-2-butanol Hydrochloride RS in *Solution A*, *W* is the weight, in mg, of Levopropoxyphene Napsylate taken, and r_C and R_C are the ratios of the peak heights of the carbinol precursor to those of *n*-tricosane in the *Test preparation* and the *Standard preparation*, respectively. The limit is 0.50%.

Calculate the percentage of acetoxy analog napsylate in the Levopropoxyphene Napsylate taken by the formula:

$$(551.70/325.45)(200C_A/W)(r_A/R_A),$$

in which 551.70 and 325.45 are the molecular weights of acetoxy analog napsylate monohydrate and anhydrous acetoxy analog, respectively, C_A is the concentration, in mg per mL, of USP α-*d*-2-Acetoxy-4-(dimethylamino)-1,2-diphenyl-3-methylbutane RS in *Solution A*, *W* is the weight, in mg, of the Levopropoxyphene Napsylate taken, and r_A and R_A are the ratios of the peak heights of the acetoxy analog to those of *n*-tricosane in the *Test preparation* and the *Standard preparation*, respectively. The limit is 0.60%.

Assay—

pH 12.5 borate buffer—Dissolve 6.18 g of boric acid and 7.5 g of potassium chloride in water to make 500 mL. Dissolve 17.5 g of sodium hydroxide in water to make 500 mL. Mix the two solutions, dilute with water to 2000 mL, and mix.

Procedure—Transfer about 375 mg of Levopropoxyphene Napsylate, accurately weighed, to a 250-mL separator, and dissolve in 40 mL of chloroform. Add 25 mL of *pH 12.5 borate buffer*, shake for 2 minutes, and allow the phases to separate. Filter the chloroform extract through a layer of about 3 g of anhydrous granular sodium sulfate supported on a pledget of cotton into a 250-mL beaker. Extract the aqueous phase in the separator with three additional 40-mL portions of chloroform, pass the extracts through the same sodium sulfate filter, and combine the chloroform extracts in the same beaker. Evaporate the chloroform from the solution on a steam bath with the aid of a current of air nearly to dryness, and continue the evaporation, with gentle heat and the aid of a current of air, to remove all traces of chloroform. Add a magnetic stirring bar and 40 mL of glacial acetic acid, and stir on a magnetic stirrer until the residue is dissolved. Add crystal violet TS, and titrate with 0.1 *N* perchloric acid VS. Perform a blank determination, and make any necessary correction. Each mL of 0.1 *N* perchloric acid is equivalent to 54.77 mg of $C_{22}H_{29}NO_2 \cdot C_{10}H_8O_3S$.

Levopropoxyphene Napsylate Capsules

» Levopropoxyphene Napsylate Capsules contain an amount of levopropoxyphene napsylate ($C_{22}H_{29}NO_2 \cdot C_{10}H_8O_3S \cdot H_2O$) equivalent to not less than 90.0 percent and not more than 110.0 percent of the labeled amount of levopropoxyphene ($C_{22}H_{29}NO_2$).

Packaging and storage—Preserve in tight containers.

Reference standards—*USP Levopropoxyphene Napsylate Reference Standard*—Do not dry; determine the *Water* content by *Method I* before using. *USP α-d-2-Acetoxy-4-(dimethylamino)-1,2-diphenyl-3-methylbutane Reference Standard*—Keep container tightly closed. Do not dry before using. *USP α-d-4-(Dimethylamino)-1,2-diphenyl-3-methyl-2-butanol Hydrochloride Reference Standard*—Keep container tightly closed. Do not dry before using.

Identification—Transfer the contents of Capsules, equivalent to about 100 mg of levopropoxyphene, to a small flask, mix with 10 mL of chloroform, and filter: the solution is levorotatory (see *Optical Rotation* ⟨781⟩).

Dissolution ⟨711⟩—

Medium: pH 4.5 acetate buffer, prepared as directed in the test for *Dissolution* under *Propoxyphene Hydrochloride, Aspirin, and Caffeine Capsules;* 500 mL.

Apparatus 1: 100 rpm.

Time: 60 minutes.

Standard preparation—Transfer 20 mg of USP Levopropoxyphene Napsylate RS, accurately weighed, to a 100-mL volumetric flask. Dissolve in *pH 4.5 acetate buffer*, dilute with the same solvent to volume, and mix.

Procedure—Proceed as directed for *Procedure for propoxyphene hydrochloride* in the test for *Dissolution* under *Propoxyphene Hydrochloride, Aspirin, and Caffeine Capsules.*

Tolerances—Not less than 75% (*Q*) of the labeled amount of $C_{22}H_{29}NO_2 \cdot C_{10}H_8O_3S \cdot H_2O$ is dissolved in 60 minutes.

Uniformity of dosage units ⟨905⟩: meet the requirements.

Procedure for content uniformity—Transfer, as completely as possible, the contents of 1 Capsule to a 100-mL volumetric flask. Rinse the empty capsule with methanol, and add the rinsing to the flask. Add 50 mL of methanol, and shake by mechanical means for 15 minutes. Dilute with methanol to volume, mix, and filter, discarding the first 20 mL of the filtrate. Dilute a portion of the subsequent filtrate quantitatively and stepwise, if necessary, with methanol to obtain a solution containing approximately 25 µg of levopropoxyphene per mL. Concomitantly determine the absorbances of this solution and of a Standard solution of USP Levopropoxyphene Napsylate RS in the same medium having a known concentration of about 40 µg per mL in 1-cm cells at the wavelength of maximum absorbance at about 275 nm, with a suitable spectrophotometer, using methanol as the blank. Calculate the quantity, in mg, of $C_{22}H_{29}NO_2$ in the Capsule by the formula:

$$(339.48/547.71)(TC/D)(A_U/A_S),$$

in which 339.48 and 547.71 are the molecular weights of levopropoxyphene and anhydrous levopropoxyphene napsylate, respectively, *T* is the labeled quantity, in mg, of levopropoxyphene in the Capsule, *C* is the concentration, in µg per mL, of USP Levopropoxyphene Napsylate RS in the Standard solution, *D* is the concentration, in µg per mL, of levopropoxyphene in the solution from the Capsule, based upon the labeled content per Capsule and the extent of dilution, and A_U and A_S are the absorbances of the solution from the Capsule and the Standard solution, respectively.

Related compounds—

Internal standard solution—Dissolve a quantity of *n*-tricosane in chloroform to obtain a solution containing about 20 µg per mL.

Solution A—Prepare as directed in the test for *Related compounds* under *Levopropoxyphene Napsylate*, except to use methanol in place of chloroform for the dilution.

Standard preparation—Transfer about 100 mg of USP Levopropoxyphene Napsylate RS, accurately weighed, to a 50-mL volumetric flask. Add 40 mL of methanol and 2.0 mL of *Solution A*, mix, and sonicate for about 15 minutes. Dilute with methanol to volume, and mix. Transfer 10.0 mL of this solution to a 125-mL separator, and add 10 mL of sodium carbonate solution (1 in 20). Add 10.0 mL of *Internal standard solution* to the contents of the separator, shake for 5 minutes, and allow the phases to separate. Carefully drain the chloroform layer, discarding the first mL, and collect the remaining chloroform extract in a small, glass-stoppered conical flask. Evaporate this solution under a stream of nitrogen to 2 mL, add 1 g of anhydrous granular sodium sulfate, and insert the stopper.

Test preparation—Transfer an accurately weighed portion of the contents of Capsules, equivalent to about 500 mg of levopropoxyphene napsylate, to a 250-mL volumetric flask. Add 200 mL of methanol, mix, and sonicate for about 15 minutes. Dilute with methanol to volume, and mix. Proceed as directed under *Standard preparation*, beginning with "Transfer 10.0 mL of this solution."

Chromatographic system, System suitability, and *Procedure*—Proceed as directed in the test for *Related compounds* under *Levopropoxyphene Napsylate.* Calculate the quantity, in mg, of carbinol napsylate precursor in each g of the Capsules taken by the formula:

$$(509.66/319.87)(10C_C/W)(r_C/R_C),$$

in which 509.66 and 319.87 are the molecular weights of carbinol napsylate monohydrate precursor and anhydrous carbinol hydrochloride precursor, respectively, C_C is the concentration, in mg per mL, of USP α-*d*-4-Dimethylamino-1,2-diphenyl-3-methyl-2-butanol Hydrochloride RS in *Solution A*, *W* is the weight, in g, of the Capsules taken, and r_C and R_C are the ratios of the peak heights of the carbinol precursor to those of *n*-tricosane in the *Test preparation* and the *Standard preparation*, respectively. The limit is 0.5%, based upon the levopropoxyphene napsylate content of the Capsules taken.

Calculate the quantity, in mg, of acetoxy analog napsylate in each g of the Capsules taken by the formula:

$$(551.70/325.45)(10C_A/W)(r_A/R_A),$$

in which 551.70 and 325.45 are the molecular weights of acetoxy analog napsylate monohydrate and anhydrous acetoxy analog, respectively, C_A is the concentration, in mg per mL, of USP α-

d-2-Acetoxy-4-dimethylamino-1,2-diphenyl-3-methylbutane RS in *Solution A*, *W* is the weight, in g, of the Capsules taken, and r_A and R_A are the ratios of the peak heights of the acetoxy analog to those of *n*-tricosane in the *Test preparation* and the *Standard preparation*, respectively. The limit is 0.6%, based upon the levopropoxyphene napsylate content of the Capsules taken.

Assay—

Internal standard solution—Dissolve a quantity of *n*-tricosane in chloroform to obtain a solution containing about 1 mg per mL.

Standard preparation—Transfer about 100 mg of USP Levopropoxyphene Napsylate RS, accurately weighed, to a 50-mL volumetric flask. Add 40 mL of methanol, mix, sonicate for about 15 minutes, dilute with methanol to volume, and mix. Transfer 5.0 mL of this solution to a 125-mL separator, and add 5 mL of sodium carbonate solution (1 in 20). Add 5.0 mL of *Internal standard solution* to the contents of the separator, shake for 5 minutes, and allow the phases to separate. Carefully drain the chloroform layer, discarding the first mL, and collect the next 2 mL of chloroform extract in a small, glass-stoppered conical flask containing about 100 mg of anhydrous granular sodium sulfate, and insert the stopper.

Assay preparation—Remove, as completely as possible, the contents of not less than 20 Levopropoxyphene Napsylate Capsules, and weigh. Transfer an accurately weighed portion of the contents, equivalent to about 500 mg of levopropoxyphene napsylate, to a 250-mL volumetric flask. Add 200 mL of methanol, mix, and sonicate for about 15 minutes. Dilute with methanol to volume, and mix. Proceed as directed under *Standard preparation*, beginning with "Transfer 5.0 mL of this solution."

Chromatographic system (see *Chromatography* ⟨621⟩)—Proceed as directed for *Chromatographic system* in the test for *Related compounds* under *Levopropoxyphene Napsylate*.

System suitability—Chromatograph five injections of the *Standard preparation*, and record peak heights as directed under *Procedure*. The retention time for levopropoxyphene is about 0.5 relative to the internal standard: the analytical system is suitable for conducting the assay if the coefficient of variation for the ratio R_S does not exceed 2%, the resolution factor between the peak heights of levopropoxyphene and *n*-tricosane is not less than 3.0, and the tailing factor (the sum of the distances from peak center to the leading edge and to the tailing edge divided by twice the distance from peak center to the leading edge), measured at 5% of peak heights, does not exceed 1.5 for each component.

Procedure—Inject separately a suitable volume of the *Standard preparation* and an identical volume of the *Assay preparation* into a suitable gas chromatograph, and record the chromatogram. Calculate the peak heights of the levopropoxyphene and *n*-tricosane peaks in each chromatogram. Calculate the quantity, in mg, of $C_{22}H_{29}NO_2$ in the portion of Capsules taken by the formula:

$$(339.48/547.71)(5W_S)(R_U/R_S),$$

in which 339.48 and 547.71 are the molecular weights of levopropoxyphene and anhydrous levopropoxyphene napsylate, respectively, W_S is the weight, in mg, of (anhydrous) USP Levopropoxyphene Napsylate RS taken, and R_U and R_S are the ratios of the peak heights of levopropoxyphene to those of *n*-tricosane in the *Test preparation* and the *Standard preparation*, respectively.

Levopropoxyphene Napsylate Oral Suspension

» Levopropoxyphene Napsylate Oral Suspension contains an amount of levopropoxyphene napsylate ($C_{22}H_{29}NO_2.C_{10}H_8O_3S.H_2O$) equivalent to not less than 90.0 percent and not more than 110.0 percent of the labeled amount of levopropoxyphene ($C_{22}H_{29}NO_2$).

Packaging and storage—Preserve in tight containers, protected from light. Avoid freezing.

Reference standards—*USP Levopropoxyphene Napsylate Reference Standard*—Do not dry; determine the *Water* content by *Method I* before using. *USP α-d-2-Acetoxy-4-(dimethylamino)-1,2-diphenyl-3-methylbutane Reference Standard*—Keep container tightly closed. Do not dry before using. *USP α-d-4-(Dimethylamino)-1,2-diphenyl-3-methyl-2-butanol Hydrochloride Reference Standard*—Keep container tightly closed. Do not dry before using.

Identification—Transfer a volume of Oral Suspension, equivalent to about 100 mg of levopropoxyphene, to a small flask, mix with 10 mL of chloroform, and filter: the chloroform solution is levorotatory (see *Optical Rotation* ⟨781⟩).

Alcohol content, *Method II* ⟨611⟩: between 0.5% and 1.5% of C_2H_5OH.

Related compounds—

Internal standard solution—Dissolve a quantity of *n*-tricosane in ethyl acetate to obtain a solution containing about 20 µg per mL.

Solution A—Weigh accurately 12.5 mg of the acetoxy analog standard, USP α-*d*-2-Acetoxy-4-dimethylamino-1,2-diphenyl-3-methylbutane RS, and 12.5 mg of the carbinol hydrochloride standard, USP α-*d*-4-Dimethylamino-1,2-diphenyl-3-methyl-2-butanol Hydrochloride RS, and transfer to the same 50-mL volumetric flask. Add 5 mL of acetone, and mix (sonicate, if necessary) to dissolve. Dilute with water to volume, and mix.

Standard preparation—Transfer about 50 mg of USP Levopropoxyphene Napsylate RS, accurately weighed, to a 100-mL volumetric flask, add 10 mL of acetone, and mix. Add 1.0 mL of *Solution A*, mix, and sonicate for about 15 minutes. Dilute with water to volume, and mix. Transfer 40.0 mL of this solution to a 125-mL separator, and add 10 mL of sodium carbonate solution (1 in 5). Add 10.0 mL of *Internal standard solution* to the contents of the separator, shake for 5 minutes, and allow the phases to separate. Discard the lower, aqueous phase, and transfer the remaining liquid to a small centrifuge tube. Centrifuge, transfer the upper layer to a small flask containing about 2 g of anhydrous granular sodium sulfate, and mix. Transfer 5.0 mL of this solution to a small flask, and evaporate under a stream of nitrogen to about 1 mL.

Test preparation—Transfer an accurately weighed quantity of well-mixed Oral Suspension, equivalent to about 50 mg of levopropoxyphene napsylate, to a 100-mL volumetric flask, add 10 mL of acetone, and swirl gently (do not shake) to mix. Sonicate for about 15 minutes, dilute with water to volume, and mix. Proceed as directed under *Standard preparation*, beginning with "Transfer 40.0 mL of this solution."

Chromatographic system, *System suitability*, and *Procedure*—Proceed as directed in the test for *Related compounds* under *Levopropoxyphene Napsylate*. Calculate the quantity, in mg, of carbinol napsylate precursor in each mL of the Oral Suspension taken by the formula:

$$(509.66/319.87)(C_C/W)(D)(r_C/R_C),$$

in which 509.66 and 319.87 are the molecular weights of carbinol napsylate monohydrate precursor and anhydrous carbinol hydrochloride precursor, respectively, C_C is the concentration, in mg per mL, of USP α-*d*-4-Dimethylamino-1,2-diphenyl-3-methyl-2-butanol Hydrochloride RS in *Solution A*, *W* is the weight, in g, of the Oral Suspension taken, *D* is the density, in g per mL, of the Oral Suspension, and r_C and R_C are the ratios of the peak heights of the carbinol precursor to those of *n*-tricosane in the *Test preparation* and the *Standard preparation*, respectively. The limit is 0.5%, based upon the levopropoxyphene napsylate content of the Oral Suspension taken.

Calculate the quantity, in mg, of acetoxy analog napsylate in each mL of the Oral Suspension taken by the formula:

$$(551.70/325.45)(C_A/W)(D)(r_A/R_A),$$

in which 551.70 and 325.45 are the molecular weights of acetoxy analog napsylate monohydrate and anhydrous acetoxy analog, respectively, C_A is the concentration, in mg per mL, of USP α-*d*-2-Acetoxy-4-dimethylamino-1,2-diphenyl-3-methylbutane RS in *Solution A*, *W* is the weight, in g, of the Oral Suspension taken, *D* is the density, in g per mL, of the Oral Suspension, and r_A and R_A are the ratios of the peak heights of the acetoxy analog to those of *n*-tricosane in the *Test preparation* and the *Standard preparation*, respectively. The limit is 0.6%, based upon the levopropoxyphene napsylate content of the Oral Suspension taken.

Assay—

Internal standard solution—Dissolve a quantity of *n*-tricosane in ethyl acetate to obtain a solution containing about 1 mg per mL.

Standard preparation—Transfer about 50 mg of USP Levopropoxyphene Napsylate RS, accurately weighed, to a 100-mL volumetric flask, add 10 mL of acetone, mix, and sonicate for about 15 minutes. Dilute with water to volume, and mix. Transfer 10.0 mL of this solution to a 125-mL separator, and add 5 mL of sodium carbonate solution (1 in 5). Add 5.0 mL of *Internal standard solution* to the contents of the separator, shake for 5 minutes, and allow the phases to separate. Discard the lower, aqueous phase, and transfer the remaining liquid to a small centrifuge tube. Centrifuge, transfer the upper layer to a small flask containing about 1 g of anhydrous granular sodium sulfate, and mix.

Assay preparation—Transfer an accurately weighed quantity of well-mixed Levopropoxyphene Napsylate Oral Suspension, equivalent to about 50 mg of levopropoxyphene napsylate, to a 100-mL volumetric flask, add 10 mL of acetone, and swirl gently (do not shake) to mix. Sonicate for about 15 minutes, dilute with water to volume, and mix. Transfer 10.0 mL of this solution to a 125-mL separator, and add 5 mL of sodium carbonate solution (1 in 5). Proceed as directed under *Standard preparation*, beginning with "Add 5.0 mL of *Internal standard solution*."

Chromatographic system, *System suitability*, and *Procedure*—Proceed as directed in the *Assay* under *Levopropoxyphene Napsylate Capsules*. Calculate the quantity, in mg, of $C_{22}H_{29}NO_2$ in each mL of the Oral Suspension taken by the formula:

$$(339.48/547.71)(W_S/W_U)(D)(R_U/R_S),$$

in which 339.48 and 547.71 are the molecular weights of levopropoxyphene and anhydrous levopropoxyphene napsylate, respectively, W_S is the weight, in mg, of USP Levopropoxyphene Napsylate RS taken, W_U is the weight, in g, of the Oral Suspension taken, D is the density, in g per mL, of the Oral Suspension, and R_U and R_S are the ratios of the peak heights of levopropoxyphene to those of *n*-tricosane in the *Test preparation* and the *Standard preparation*, respectively.

Levorphanol Tartrate

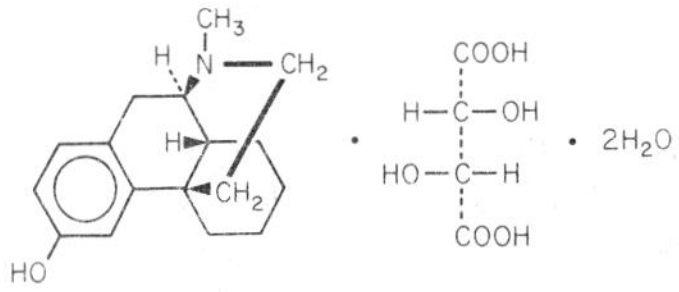

$C_{17}H_{23}NO.C_4H_6O_6.2H_2O$ 443.49

Morphinan-3-ol, 17-methyl-, [*R*-(*R**,*R**)]-2,3-dihydroxybutanedioate (1:1) (salt), dihydrate.

17-Methylmorphinan-3-ol, tartrate (1:1) (salt) dihydrate [*5985-38-6*].

Anhydrous 407.46 [*125-72-4*].

» Levorphanol Tartrate contains not less than 99.0 percent and not more than 101.0 percent of $C_{17}H_{23}NO.C_4H_6O_6$, calculated on the anhydrous basis.

Packaging and storage—Preserve in well-closed containers.

Reference standard—*USP Levorphanol Tartrate Reference Standard*—This is the dihydrate form. Do not dry; determine the *Water* content by *Method I* before using.

Identification—

A: Dissolve 50 mg in 25 mL of water in a 125-mL separator. Add 2 mL of 6 *N* ammonium hydroxide, extract with 25 mL of chloroform, and filter the chloroform extract through a layer of 4 g of granular anhydrous sodium sulfate supported on glass wool into a 125-mL conical flask. Evaporate the chloroform extract on a steam bath with the aid of a stream of nitrogen to dryness. Dissolve the residue in 1 mL of acetone, and evaporate to dryness. Dry in vacuum at 90° for 1 hour: the infrared absorption spectrum of a potassium bromide dispersion of the dried levorphanol so obtained exhibits maxima only at the same wavelengths as that of a similar preparation of USP Levorphanol Tartrate RS.

B: The ultraviolet absorption spectrum of a 1 in 7500 solution in 0.1 *N* hydrochloric acid exhibits maxima and minima at the same wavelengths as that of a similar solution of USP Levorphanol Tartrate RS, concomitantly measured, and the respective absorptivities, calculated on the anhydrous basis, at the wavelength of maximum absorbance at about 279 nm do not differ by more than 3.0%.

Specific rotation ⟨781⟩: between −14.7° and −16.3°, calculated on the anhydrous basis, determined in a solution containing 300 mg in each 10 mL, prepared as follows: Transfer about 750 mg of Levorphanol Tartrate, accurately weighed, to a 25-mL volumetric flask, add about 20 mL of water, and heat on a steam bath, or sonicate, to dissolve. Cool, add water to volume, and mix.

Water, *Method I* ⟨921⟩: between 7.0% and 9.0%.

Residue on ignition ⟨281⟩: not more than 0.1%.

Ordinary impurities ⟨466⟩—

Test solution: water.

Standard solution: water.

Eluant: a mixture of hexanes, dehydrated alcohol, and ammonium hydroxide (80:25:1).

Visualization: 17; then view immediately under short-wavelength ultraviolet light.

Assay—Transfer about 250 mg of Levorphanol Tartrate, accurately weighed, to a 250-mL flask, add 25 mL of dehydrated alcohol, and warm slightly to dissolve. Add 100 mL of chloroform, then add 4 drops of methanolic methyl red TS, and titrate with 0.02 *N* perchloric acid in dioxane VS to a red end-point. Perform a blank determination, and make any necessary correction. Each mL of 0.02 *N* perchloric acid is equivalent to 8.149 mg of $C_{17}H_{23}NO.C_4H_6O_6$.

Levorphanol Tartrate Injection

» Levorphanol Tartrate Injection is a sterile solution of Levorphanol Tartrate in Water for Injection. It contains not less than 93.0 percent and not more than 107.0 percent of the labeled amount of $C_{17}H_{23}NO.C_4H_6O_6.2H_2O$.

Packaging and storage—Preserve in single-dose or in multiple-dose containers, preferably of Type I glass.

Identification—

A: To 1 mL of Injection add 1 drop of 3 *N* hydrochloric acid and 2 drops of ferric chloride TS. Heat to boiling, and add 1 mL of potassium ferricyanide solution (1 in 200): a blue-green color develops.

B: The angular rotation of the Injection is levorotatory (see *Optical Rotation* ⟨781⟩).

Pyrogen—Levorphanol Tartrate Injection, diluted with Sodium Chloride Injection to a concentration of 200 µg of levorphanol tartrate per mL, meets the requirements of the *Pyrogen Test* ⟨151⟩, the test dose being 1 mL per kg.

pH ⟨791⟩: between 4.1 and 4.5.

Other requirements—It meets the requirements under *Injections* ⟨1⟩.

Assay—Transfer an accurately measured volume of Levorphanol Tartrate Injection, equivalent to about 40 mg of levorphanol tartrate, to a 125-mL separator. Add 5 g of sodium chloride and sufficient sodium bicarbonate to render the solution alkaline to litmus, add an additional 100 mg of sodium bicarbonate, and extract the levorphanol with five 20-mL portions of a mixture of 3 volumes of ether and 1 volume of chloroform. Pass the combined extracts through a layer of about 10 g of granular anhydrous sodium sulfate into a 500-mL conical flask, and evaporate to a volume of about 30 mL. Add about 50 mL of chloroform and 1 drop of methanolic methyl red TS, and titrate with 0.01 *N* perchloric acid in dioxane VS to a red end-point. Perform a blank determination, and make any necessary correction. Each mL of 0.01 *N* perchloric acid is equivalent to 4.435 mg of $C_{17}H_{23}NO.C_4H_6O_6.2H_2O$.

Levorphanol Tartrate Tablets

» Levorphanol Tartrate Tablets contain not less than 93.0 percent and not more than 107.0 percent of the labeled amount of $C_{17}H_{23}NO.C_4H_6O_6.2H_2O$.

Packaging and storage—Preserve in well-closed containers.

Reference standard—*USP Levorphanol Tartrate Reference Standard*—This is the dihydrate form. Do not dry; determine the *Water* content by *Method I* before using.

Identification—

A: Powder finely a number of Tablets. To a portion of the powder, equivalent to about 1 mg of levorphanol tartrate, add 1 mL of water, 1 drop of 3 *N* hydrochloric acid, and 2 drops of ferric chloride TS, and heat to boiling. To the hot solution add 1 mL of potassium ferricyanide solution (1 in 200): a bluish color develops.

B: Powder a number of Tablets, equivalent to about 60 mg of levorphanol tartrate, and transfer the mixture to a small separator. Add 10 mL of water, dissolve as much of the powder as possible, add about 400 mg of sodium bicarbonate, and extract with a 50-mL portion of chloroform. Evaporate the filtered chloroform extract on a steam bath to a small volume, dilute with chloroform to 10 mL, and determine the angular rotation: the solution is levorotatory (see *Optical Rotation* ⟨781⟩).

Dissolution ⟨711⟩—

Medium: water; 500 mL.

Apparatus 2: 50 rpm.

Time: 30 minutes.

Procedure—Determine the amount of $C_{17}H_{23}NO.C_4H_6O_6.2H_2O$ dissolved from ultraviolet absorbances at the wavelength of maximum absorbance at about 279 nm of filtered portions of the solution under test, suitably diluted with water, in comparison with a Standard solution having a known concentration of USP Levorphanol Tartrate RS in the same medium.

Tolerances—Not less than 75% (*Q*) of the labeled amount of $C_{17}H_{23}NO.C_4H_6O_6.2H_2O$ is dissolved in 30 minutes.

Uniformity of dosage units ⟨905⟩: meet the requirements.

Procedure for content uniformity—Transfer 1 Tablet to a glass-stoppered flask, add 25.0 mL of 0.1 *N* hydrochloric acid, and allow the Tablet to disintegrate. Shake well, and filter through a small filter paper, discarding the first portion of the filtrate. Dilute a portion of the filtrate quantitatively and stepwise, if necessary, to provide a solution containing about 80 µg of levorphanol tartrate per mL. Concomitantly determine the absorbances of this solution and of a solution of USP Levorphanol Tartrate RS in the same medium having a known concentration of about 80 µg of anhydrous levorphanol tartrate per mL, in 1-cm cells at the wavelength of maximum absorbance at about 279 nm, with a suitable spectrophotometer, using 0.1 *N* hydrochloric acid as the blank. Calculate the quantity, in mg, of $C_{17}H_{23}NO.C_4H_6O_6.2H_2O$ in the Tablet by the formula:

$$(443.49/407.46)(TC/D)(A_U/A_S),$$

in which 443.49 and 407.46 are the molecular weights of the hydrated and anhydrous forms of levorphanol tartrate, respectively, *T* is the labeled quantity, in mg, of levorphanol tartrate in the Tablet, *C* is the concentration, in µg per mL, of USP Levorphanol Tartrate RS, on the anhydrous basis, in the Standard solution, *D* is the concentration, in µg per mL, of levorphanol tartrate in the solution from the Tablet, based on the labeled quantity per Tablet and the extent of dilution, and A_U and A_S are the absorbances of the solution from the Tablet and the Standard solution, respectively.

Assay—Weigh and finely powder not less than 20 Levorphanol Tartrate Tablets. Weigh accurately a portion of the powder, equivalent to about 40 mg of levorphanol tartrate, transfer to a 125-mL separator, add 20 mL of water and sufficient sodium bicarbonate to render the suspension alkaline to litmus, and proceed as directed in the *Assay* under *Levorphanol Tartrate Injection*, beginning with "add an additional 100 mg of sodium bicarbonate."

Levothyroxine Sodium

$$HO-C_6H_2I_2-O-C_6H_2I_2-CH_2-CH(NH_2)-COONa \cdot xH_2O$$

$C_{15}H_{10}I_4NNaO_4.xH_2O$ (anhydrous) 798.86

L-Tyrosine, *O*-(4-hydroxy-3,5-diiodophenyl)-3,5-diiodo-, monosodium salt, hydrate.

Monosodium L-thyroxine hydrate [*25416-65-3*].

Anhydrous [*55-03-8*].

» Levothyroxine Sodium is the sodium salt of the levo isomer of thyroxine, an active physiological principle obtained from the thyroid gland of domesticated animals used for food by man or prepared synthetically. It contains not less than 97.0 percent and not more than 103.0 percent of $C_{15}H_{10}I_4NNaO_4$, calculated on the anhydrous basis.

Packaging and storage—Preserve in tight containers, protected from light.

Reference standards—*USP Liothyronine Reference Standard*—Keep container tightly closed and protected from light. Use without drying; correct for moisture, determined by drying a separate portion in vacuum at 60° for 3 hours. *USP Levothyroxine Reference Standard*—Keep container tightly closed and protected from light. Use without drying; correct for moisture, determined by drying a separate portion in vacuum at 60° for 4 hours.

Identification—

A: Ignite about 50 mg in a platinum dish over a flame: it decomposes and liberates iodine vapors.

B: To about 0.5 mg add 7.5 mL of acid sodium chloride solution (prepared by mixing 300 mL of water, 250 mL of alcohol, 100 mL of 1 *N* sodium hydroxide, and 100 mL of hydrochloric acid) and 1 mL of sodium nitrite solution (1 in 100). Allow to stand in the dark for 20 minutes, and add 1.25 mL of ammonium hydroxide: a pink color is produced.

Specific rotation ⟨781⟩: between −5° and −6°, determined in a mixture of 1 part of 1 *N* sodium hydroxide and 2 parts of alcohol containing the equivalent of 300 mg of anhydrous Levothyroxine Sodium in each 10 mL.

Water, *Method III* ⟨921⟩—Dry about 500 mg, accurately weighed, over phosphorus pentoxide at 60° and at a pressure not exceeding 10 mm of mercury for 4 hours: it loses not more than 11.0% of its weight.

Soluble halides—Shake 10 mg with 10 mL of water containing 1 drop of 2 *N* nitric acid for 5 minutes, and filter. Dilute the filtrate with water to 10 mL, and add 3 drops of silver nitrate TS: any turbidity produced is not greater than that of a control containing 0.10 mL of 0.020 *N* hydrochloric acid (0.7% as chloride).

Liothyronine sodium—

Mobile phase, 0.01 M Methanolic sodium hydroxide, and *Chromatographic system*—Prepare as directed in the *Assay.*

Standard preparation—Dissolve an accurately weighed quantity of USP Liothyronine RS in *0.01 M Methanolic sodium hydroxide,* and dilute quantitatively and stepwise with *0.01 M Methanolic sodium hydroxide* to obtain a solution having a known concentration of 4.0 µg of liothyronine, as the sodium salt, per mL.

Test preparation—Transfer 20 mg of Levothyroxine Sodium, accurately weighed, to a 100-mL volumetric flask, add *0.01 M Methanolic sodium hydroxide* to volume, and mix.

System suitability preparation—Mix equal volumes of the *Standard preparation* and the *Standard preparation* prepared for the *Assay*. Chromatograph this mixture, and record the peak responses as directed under *Procedure:* the resolution, *R*, between the liothyronine and levothyroxine is not less than 4.0.

Procedure—Separately inject equal volumes (about 50 µL) of the *Standard preparation* and the *Test preparation* into the chromatograph, record the chromatograms, and measure the responses for the liothyronine peaks: the peak response obtained from the *Test preparation* is not greater than that obtained from the *Standard preparation*, corresponding to not more than 2.0% of liothyronine sodium.

Assay—

Mobile phase—Prepare a suitable degassed and filtered mixture of water and acetonitrile (65:35) that contains 1 mL of phosphoric acid in each 1000 mL of solution.

0.01 M Methanolic sodium hydroxide—Dissolve 400 mg of sodium hydroxide in 500 mL of water. Cool, add 500 mL of methanol, and mix.

Standard preparation—Dissolve an accurately weighed quantity of USP Levothyroxine RS in *0.01 M Methanolic sodium hydroxide*, and dilute quantitatively and stepwise with *0.01 M Methanolic sodium hydroxide* to obtain a solution having a known concentration of about 4 µg per mL.

Assay preparation—Transfer about 10 mg of Levothyroxine Sodium, accurately weighed, to a 25-mL volumetric flask, dissolve in *0.01 M Methanolic sodium hydroxide*, dilute with the same solvent to volume, and mix. Pipet 1 mL of this solution to a 100-mL volumetric flask, dilute with *0.01 M Methanolic sodium hydroxide* to volume, and mix.

Chromatographic system (see *Chromatography* ⟨621⟩)—The liquid chromatograph is equipped with a 225-nm detector and a 25- to 30-cm column that contains packing L10. The flow rate is about 1 mL per minute. Chromatograph five replicate injections of the *Standard preparation*, and record the peak responses as directed under *Procedure:* the relative standard deviation is not more than 2.0%, and the tailing factor is not more than 1.8.

Procedure—Separately inject equal volumes (about 50 µL) of the *Standard preparation* and the *Assay preparation* into the chromatograph, record the chromatograms, and measure the responses for the major peaks. Calculate the quantity, in mg, of $C_{15}H_{10}I_4NNaO_4$ in the portion of Levothyroxine Sodium taken by the formula:

$$(798.86/776.87)(2.5C)(r_U/r_S),$$

in which 798.86 and 776.87 are the molecular weights of levothyroxine sodium and levothyroxine, respectively, C is the concentration, in µg per mL, of USP Levothyroxine RS in the *Standard preparation*, and r_U and r_S are the peak responses obtained from the *Assay preparation* and the *Standard preparation*, respectively.

Levothyroxine Sodium Tablets

» Levothyroxine Sodium Tablets contain not less than 90.0 percent and not more than 110.0 percent of the labeled amount of $C_{15}H_{10}I_4NNaO_4$.

Packaging and storage—Preserve in tight, light-resistant containers.

Reference standards—*USP Liothyronine Reference Standard*—Keep container tightly closed and protected from light. Use without drying; correct for moisture, determined by drying a separate portion in vacuum at 60° for 3 hours. *USP Levothyroxine Reference Standard*—Keep container tightly closed and protected from light. Use without drying; correct for moisture, determined by drying a separate portion in vacuum at 60° for 4 hours.

Identification—

Solvent system—Mix 5 volumes of *tert*-amyl alcohol, 4 volumes of water, and 1 volume of ammonium hydroxide, shake, and allow to stand. Transfer the upper phase to a suitable chromatographic chamber, arranged for thin-layer chromatography, pouring it over the paper lining, cover the chamber, and allow to stand for 1 hour.

Detection reagent—Add 65 mL of 2 *N* hydrochloric acid to 50 mL of a 1 in 10 solution of sodium arsenite in 1 *N* sodium hydroxide, with vigorous stirring. Mix 1 volume of this solution with 5 volumes of a 27 in 1000 solution of ferric chloride in 2 *N* hydrochloric acid and 5 volumes of freshly prepared potassium ferricyanide solution (35 in 1000).

Standard preparation—Prepare a solution of about 15 mg of USP Levothyroxine RS, accurately weighed, in 100 mL of a mixture of 19 volumes of methanol and 1 volume of ammonium hydroxide. Dilute 10.0 mL of this solution with the same solvent to 50.0 mL, and mix.

Test preparation—Shake an amount of powdered Levothyroxine Sodium Tablets, equivalent to about 60 µg of levothyroxine sodium, with 2 mL of a mixture of 19 volumes of methanol and 1 volume of ammonium hydroxide in a centrifuge tube for 10 minutes, and centrifuge.

Procedure—Apply 10-µL volumes of the *Test preparation* and of the *Standard preparation*, respectively, to a thin-layer chromatographic plate coated with a 0.1-mm layer of cellulose. Develop the plate in the *Solvent system* until the solvent front has moved not less than 10 cm beyond the point of application of the *Test preparation*, air-dry, and spray the plate with *Detection reagent:* the chromatogram of the *Test preparation* shows a blue spot corresponding in R_f value to the chromatogram from the levothyroxine *Standard preparation*.

Disintegration ⟨701⟩: 30 minutes.

Dissolution ⟨711⟩—[NOTE—All containers that are in contact with solutions containing levothyroxine sodium are to be made of glass.]

Medium: 0.05 *M*, pH 7.4 phosphate buffer (see *Buffer Solutions* in the section, *Reagents, Indicators, and Solutions*); 500 mL.

Apparatus 2: 100 rpm.

Time: 80 minutes.

Procedure—

0.01 M Methanolic sodium hydroxide—Add 1 mL of 10 *N* sodium hydroxide to 750 mL of methanol in a 1-liter volumetric flask, and mix. Dilute with water to volume, and mix.

Mobile phase—Prepare a filtered and degassed mixture of methanol and 0.1% phosphoric acid (60:40).

Standard solutions—Prepare a solution of USP Levothyroxine RS in *0.01 M Methanolic sodium hydroxide* having an accurately known concentration of about 0.5 mg per mL. Dilute aliquots of this solution with 0.05 *M*, pH 7.4 phosphate buffer containing 5% of a 2% methanolic–ammonium hydroxide solution to obtain Standard solutions having concentrations similar to those expected in the Test solutions.

Test solution—[NOTE—Prior to use, check the filters for absorptive loss of drug.] Transfer 20.0 mL of a filtered portion of the solution under test to a flask, add 1.0 mL of a 2% methanolic–ammonium hydroxide solution, and mix.

Chromatographic system (see *Chromatography* ⟨621⟩)—The liquid chromatograph is equipped with a 225-nm detector and a 4.6-mm × 25-cm column that contains packing L1. The flow rate is about 2 mL per minute. Chromatograph replicate injections of the *Standard solutions*, and record the peak responses as directed under *Procedure:* the relative standard deviation is not more than 4.0%, and the tailing factor is not more than 1.5.

Procedure—Separately inject equal volumes (about 800 µL) of the *Standard solution* and the *Test solution* into the chromatograph, record the chromatograms, and measure the responses for the major peaks. Calculate the amount of $C_{15}H_{10}I_4NNaO_4$ dissolved.

Tolerances—Not less than 55% (*Q*) of the labeled amount of $C_{15}H_{10}I_4NNaO_4$ is dissolved in 80 minutes.

Uniformity of dosage units ⟨905⟩: meet the requirements.

Soluble halides—Place a portion of finely powdered Tablets, equivalent to about 2.5 mg of anhydrous levothyroxine sodium, in a large test tube, add 1 g of chloride-free activated charcoal and 25 mL of water, insert the stopper in the tube, heat to about 40°, and shake for 5 minutes. Add 3 drops of dilute nitric acid (2 in 5), and filter. To the filtrate add 8 drops of silver nitrate TS: any turbidity produced does not exceed that in a control containing 0.25 mL of 0.020 *N* hydrochloric acid (7.1%).

Liothyronine sodium—

Mobile phase, 0.01 M Methanolic sodium hydroxide, and *Chromatographic system*—Prepare as directed in the *Assay.*

Standard preparation—Dissolve an accurately weighed quantity of USP Liothyronine RS in *0.01 M Methanolic sodium hydroxide,* and dilute quantitatively and stepwise with *0.01 M Methanolic sodium hydroxide* to obtain a solution having a known concentration of 4.0 µg of liothyronine, as the sodium salt, per mL.

Test preparation—Transfer an accurately weighed quantity of finely powdered Levothyroxine Tablets, equivalent to about 2 mg of levothyroxine sodium, to a 10-mL volumetric flask, add *0.01 M Methanolic sodium hydroxide* to volume, mix, and filter. The clear filtrate is the *Test preparation.*

System suitability preparation—Mix equal volumes of the *Standard preparation* and the *Standard preparation* prepared for the *Assay.* Chromatograph this mixture, and record the peak responses as directed under *Procedure:* the resolution, *R*, between the liothyronine and levothyroxine peaks is not less than 4.0.

Procedure—Separately inject equal volumes (about 40 µL) of the *Standard preparation* and the *Test preparation* into the chromatograph, record the chromatograms, and measure the responses for the liothyronine peaks: the peak response obtained from the *Test preparation* is not greater than that obtained from the *Standard preparation,* corresponding to not more than 2.0% of liothyronine sodium.

Assay—

Mobile phase, 0.01 M Methanolic sodium hydroxide, Standard preparation, and *Chromatographic system*—Prepare as directed in the *Assay* under *Levothyroxine Sodium.*

Assay preparation—Weigh and finely powder not less than 20 Levothyroxine Sodium Tablets. Transfer an accurately weighed portion of the powder, equivalent to about 400 µg of levothyroxine sodium, to a 100-mL volumetric flask, add about 50 mL of *0.01 M Methanolic sodium hydroxide,* and place in an ultrasonic bath for 1 minute. Shake the mixture for 5 minutes, dilute with *0.01 M Methanolic sodium hydroxide* to volume, mix, and filter.

Procedure—Proceed as directed for *Procedure* in the *Assay* under *Levothyroxine Sodium,* and calculate the quantity, in mg, of $C_{15}H_{10}I_4NNaO_4$ in the portion of Tablets taken by the formula:

$$(798.86/776.87)(0.1C)(r_U/r_S),$$

in which the terms are as defined therein.

Lidocaine

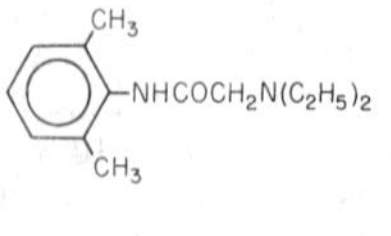

$C_{14}H_{22}N_2O$ 234.34

Acetamide, 2-(diethylamino)-*N*-(2,6-dimethylphenyl)-.

2-(Diethylamino)-2′,6′-acetoxylidide [*137-58-6*].

» Lidocaine contains not less than 97.5 percent and not more than 102.5 percent of $C_{14}H_{22}N_2O$.

Packaging and storage—Preserve in well-closed containers.

Reference standard—*USP Lidocaine Reference Standard*—Dry in vacuum over silica gel for 24 hours before using. Keep the container tightly closed.

Identification—

A: The infrared absorption spectrum of a potassium bromide dispersion of it, previously dried in vacuum over silica gel for 24 hours, exhibits maxima only at the same wavelengths as that of a similar preparation of USP Lidocaine RS.

B: To about 100 mg dissolved in 1 mL of alcohol add 10 drops of cobaltous chloride TS, and shake the solution for about 2 minutes: a bright green color develops, and a fine precipitate is formed.

Melting range ⟨741⟩: between 66° and 69°.

Residue on ignition ⟨281⟩: not more than 0.1%.

Sulfate—Dissolve about 200 mg in a mixture of 2 mL of 2 *N* nitric acid and 20 mL of water, and filter if necessary. To one-half of the filtrate add 1 mL of barium chloride TS: no more turbidity is produced than is present in the remaining portion of the filtrate to which nothing has been added.

Chloride ⟨221⟩—Dissolve 1.0 g in a mixture of 3 mL of 2 *N* nitric acid and 12 mL of water, and add 1 mL of silver nitrate TS: the turbidity does not exceed that produced by 50 µL of 0.020 *N* hydrochloric acid (0.0035%).

Heavy metals, *Method I* ⟨231⟩—Dissolve 1.0 g in a mixture of 2 mL of 3 *N* hydrochloric acid and 10 mL of water, evaporate on a steam bath to dryness, and dissolve the residue in 25 mL of water: the limit is 0.002%.

Assay—

Mobile phase, Standard preparation, Resolution preparation, and *Chromatographic system*—Prepare as directed in the *Assay* under *Lidocaine Hydrochloride.*

Assay preparation—Dissolve about 85 mg of Lidocaine, accurately weighed, with warming if necessary, in 0.5 mL of 1 *N* hydrochloric acid in a 50-mL volumetric flask, dilute with *Mobile phase* to volume, and mix.

Procedure—Proceed as directed for *Procedure* in the *Assay* under *Lidocaine Hydrochloride.* Calculate the quantity, in mg, of $C_{14}H_{22}N_2O$ in the portion of Lidocaine taken by the formula:

$$50C(r_U/r_S),$$

in which *C* is the concentration, in mg per mL, of USP Lidocaine RS in the *Standard preparation,* and r_U and r_S are the lidocaine peak responses obtained from the *Assay preparation* and the *Standard preparation,* respectively.

Lidocaine Topical Aerosol

» Lidocaine Topical Aerosol is a solution of Lidocaine in a suitable flavored vehicle with suitable propellants in a pressurized container equipped with a metering valve. It contains not less than 90.0 percent and not more than 110.0 percent of the labeled amount of $C_{14}H_{22}N_2O$, and it delivers not less than 85.0 percent and not more than 115.0 percent of the labeled amount of $C_{14}H_{22}N_2O$ per actuation.

Packaging and storage—Preserve in non-reactive aerosol containers equipped with metered-dose valves.

Reference standard—*USP Lidocaine Reference Standard*—Dry in vacuum over silica gel for 24 hours before using.

Identification—

A: To about 5 mL of Topical Aerosol spray, collected in a separator, add about 10 mL of water and 3 mL of dilute hydrochloric acid (1 in 2), wash with two 15-mL portions of chloroform, and discard the chloroform washings. Render the solution in the separator alkaline with 5 to 6 mL of ammonium hydroxide, and extract with three 20-mL portions of chloroform, filtering the chloroform extracts through a pledget of cotton previously moistened with chloroform. Evaporate the combined chloroform extracts with the aid of gentle heat to dryness, and dry the residue in vacuum over silica gel for 24 hours: a potassium bromide dispersion of the lidocaine so obtained exhibits maxima only at the same wavelengths as that of a similar preparation of USP Lidocaine RS.

B: To about 2 mL of Topical Aerosol spray, collected in a test tube, add 10 to 15 drops of cobaltous chloride TS, and shake for about 2 minutes: a bright green color develops, and a fine precipitate is formed (lidocaine).

C: To about 2 mL of Topical Aerosol spray, collected in a test tube, add 5 mL of water, 1 mL of 2 *N* nitric acid, and 3 mL of mercuric nitrate TS: a light yellow color develops (lidocaine).

Microbial limits—It meets the requirements of the tests for absence of *Staphylococcus aureus* and *Pseudomonas aeruginosa* under *Microbial Limit Tests* ⟨61⟩.

Other requirements—It meets the requirements for *Leak Testing* under *Aerosols* ⟨601⟩.

Assay—Accurately weigh 1 Lidocaine Topical Aerosol container and actuator. Transfer a counted number of not less than 10 doses to a 125-mL conical flask by carefully discharging the doses in a manner such as to avoid loss of material, and take precautions to protect the specimen from absorption of atmospheric moisture. Accurately weigh the container and actuator to obtain the specimen weight. To the specimen add 20 mL of chloroform, mix, and add 10 mL of dioxane and 2 drops of crystal violet TS. Titrate with 0.1 *N* perchloric acid in dioxane VS to a blue end-point. Perform a blank determination, and make any necessary correction. Each mL of 0.1 *N* perchloric acid is equivalent to 23.43 mg of $C_{14}H_{22}N_2O$.

Lidocaine Ointment

» Lidocaine Ointment is Lidocaine in a suitable hydrophilic ointment base. It contains not less than 95.0 percent and not more than 105.0 percent of the labeled amount of $C_{14}H_{22}N_2O$.

Packaging and storage—Preserve in tight containers.

Reference standard—*USP Lidocaine Reference Standard*—Dry in vacuum over silica gel for 24 hours before using. Keep the container tightly closed.

Identification—Stir a quantity of Ointment, equivalent to about 300 mg of lidocaine, with 20 mL of water, transfer to a separator, and extract with two 30-mL portions of solvent hexane. Wash the combined hexane extracts with 10 mL of water, evaporate with the aid of a current of warm air, and dry the residue in vacuum over silica gel for 24 hours: the crystalline precipitate so obtained responds to the *Identification test A* under *Lidocaine*.

Microbial limits—It meets the requirements of the tests for absence of *Staphylococcus aureus* and *Pseudomonas aeruginosa* under *Microbial Limit Tests* ⟨61⟩.

Minimum fill ⟨755⟩: meets the requirements.

Assay—Proceed with Lidocaine Ointment as directed in the *Assay* under *Lidocaine Hydrochloride Jelly*. Each mL of 0.01 *N* sulfuric acid is equivalent to 2.343 mg of $C_{14}H_{22}N_2O$.

Lidocaine Ointment, Neomycin and Polymyxin B Sulfates, Bacitracin Zinc, and—*see* Neomycin and Polymyxin B Sulfates, Bacitracin Zinc, and Lidocaine Ointment

Lidocaine Oral Topical Solution

» Lidocaine Oral Topical Solution contains not less than 95.0 percent and not more than 105.0 percent of the labeled amount of $C_{14}H_{22}N_2O$. It contains a suitable flavor.

Packaging and storage—Preserve in tight containers.

Reference standard—*USP Lidocaine Reference Standard*—Dry in vacuum over silica gel for 24 hours.

Identification—Transfer a quantity of Oral Topical Solution, equivalent to about 250 mg of lidocaine, to a separator with 20 mL of water, and extract with 20 mL of chloroform. Wash the chloroform extract with 20 mL of water, and evaporate the chloroform extract with the aid of a current of warm air. Dissolve the residue in hexane, evaporate with the aid of a current of warm air, and dry the residue in vacuum over silica gel for 24 hours: the crystalline precipitate so obtained responds to *Identification test A* under *Lidocaine*.

Assay—Transfer an accurately measured volume of Lidocaine Oral Topical Solution, equivalent to about 150 mg of lidocaine, to a 125-mL conical flask, and protect from atmospheric moisture with a stopper fitted with a tube containing silica gel. Add 20 mL of glacial acetic acid and 2 drops of crystal violet TS. Titrate immediately with 0.1 *N* perchloric acid VS to a blue end-point. Perform a blank determination, and make any necessary correction. Each mL of 0.1 *N* perchloric acid is equivalent to 23.43 mg of $C_{14}H_{22}N_2O$.

Lidocaine and Epinephrine Injection

» Lidocaine and Epinephrine Injection is a sterile solution prepared from Lidocaine Hydrochloride and Epinephrine with the aid of Hydrochloric Acid in Water for Injection, or a sterile solution prepared from Lidocaine and Epinephrine with the aid of Hydrochloric Acid in Water for Injection, or a sterile solution of Lidocaine Hydrochloride and Epinephrine Bitartrate in Water for Injection. The content of epinephrine does not exceed 0.002 percent (1 in 50,000). Lidocaine and Epinephrine Injection contains the equivalent of not less than 95.0 percent and not more than 105.0 percent of the labeled amount of lidocaine hydrochloride ($C_{14}H_{22}N_2O \cdot HCl$) and the equivalent of not less than 90.0 percent and not more than 115.0 percent of the labeled amount of epinephrine ($C_9H_{13}NO_3$).

Packaging and storage—Preserve in single-dose or in multiple-dose, light-resistant containers, preferably of Type I glass.

Reference standards—*USP Lidocaine Reference Standard*—Dry in vacuum over silica gel for 24 hours before using. *USP Epinephrine Bitartrate Reference Standard*—Dry in vacuum over silica gel for 3 hours before using.

pH ⟨791⟩: between 3.3 and 5.5.

Other requirements—It responds to the *Identification test* under *Lidocaine Hydrochloride Injection*. It meets also the requirements under *Injections* ⟨1⟩.

Assay for lidocaine hydrochloride—

Mobile phase—Mix 50 mL of glacial acetic acid and 930 mL of water, and adjust with 1 *N* sodium hydroxide to a pH of 3.40. Mix about 4 volumes of this solution with 1 volume of acetonitrile, such that the retention time of lidocaine is about 4 to 6 minutes. Filter through a membrane filter (1-µm or finer porosity), and degas. Make adjustments if necessary (see *System Suitability* under *Chromatography* ⟨621⟩).

Standard preparation—Dissolve about 85 mg of USP Lidocaine RS, accurately weighed, with warming if necessary, in 0.5 mL of 1 *N* hydrochloric acid in a 50-mL volumetric flask, dilute with *Mobile phase* to volume, and mix to obtain a *Standard preparation* having a known concentration of about 1.7 mg of lidocaine per mL.

Assay preparation—Transfer an accurately measured volume of Lidocaine and Epinephrine Injection, equivalent to about 100 mg of lidocaine hydrochloride, to a 50-mL volumetric flask, dilute with *Mobile phase* to volume, and mix.

Resolution preparation—Prepare a solution of methylparaben in *Mobile phase* containing about 220 µg per mL. Mix 2 mL of this solution and 20 mL of *Standard preparation*.

Chromatographic system (see *Chromatography* ⟨621⟩)—The liquid chromatograph is equipped with a 254-nm detector and a 30-cm × 3.9-mm column that contains packing L1, and is operated at a temperature between 20° and 25° maintained at ± 1.0° of the selected temperature. The flow rate is about 1.5 mL per minute. Chromatograph the *Standard preparation*, and record the peak responses as directed under *Procedure:* the relative standard deviation for replicate injections is not more than 1.5%. Chromatograph about 20 µL of the *Resolution preparation*, and record the peak responses as directed under *Procedure:* the res-

olution, *R*, between the lidocaine and methylparaben peaks is not less than 3.0.

Procedure—Separately inject equal volumes (about 20 µL) of the *Assay preparation* and the *Standard preparation* into the chromatograph. Record the chromatograms, and measure the responses for the major peaks. Calculate the quantity, in mg, of $C_{14}H_{22}N_2O.HCl$ in each mL of the Injection taken by the formula:

$$(270.80/234.34)(50)(C/V)(r_U/r_S),$$

in which 270.80 and 234.34 are the molecular weights of lidocaine hydrochloride and lidocaine, respectively, *C* is the concentration, in mg per mL, of USP Lidocaine RS in the *Standard preparation*, *V* is the volume, in mL, of Injection taken, and r_U and r_S are the lidocaine peak responses obtained from the *Assay preparation* and the *Standard preparation*, respectively.

Assay for epinephrine—

Mobile phase—Mix 50 mL of glacial acetic acid and 930 mL of water, and adjust with 1 *N* sodium hydroxide to a pH of 3.40. Dissolve 1.1 g of sodium 1-heptanesulfonate in this solution, add 1.0 mL of 0.1 *M* disodium ethylenediaminetetraacetate, and mix. Mix about 9 volumes of this solution with 1 volume of methanol, such that the retention time of epinephrine is about 4 to 6 minutes. Filter through a membrane filter (1-µm or finer porosity), and degas.

Standard preparation—Dissolve an accurately weighed quantity of USP Epinephrine Bitartrate RS in *Mobile phase* to obtain a solution having a known concentration of about 9 µg of epinephrine bitartrate per mL. Pipet 10 mL of this solution into a 50-mL volumetric flask, dilute with *Mobile phase* to volume, and mix to obtain a *Standard preparation* having a known concentration of about 1.8 µg of epinephrine bitartrate per mL.

Assay preparation—Transfer an accurately measured volume of Lidocaine and Epinephrine Injection, equivalent to about 50 µg of epinephrine, to a 50-mL volumetric flask, dilute with *Mobile phase* to volume, and mix.

Chromatographic system (see *Chromatography* ⟨621⟩)—The liquid chromatograph is fitted with a 30-cm × 3.9-mm stainless steel column packed with packing L1, and is equipped with an electrochemical detector held at a potential of +650 mV, a controller capable of regulating the background current, and a suitable recorder, and it is operated at a temperature between 20° and 25° maintained at ±1.0° of the selected temperature. The flow rate is about 1 mL per minute. Chromatograph the *Standard preparation* as directed under *Procedure.* The relative standard deviation of the peak responses of successive injections of the *Standard preparation* is not more than 1.5%.

Procedure—Separately inject equal volumes (about 20 µL) of the *Assay preparation* and the *Standard preparation* into the chromatograph by means of a suitable microsyringe or sampling valve, adjusting the specimen size and other operating parameters such that satisfactory chromatography and peak responses are obtained. Record the chromatograms, and measure the responses for the major peaks. Calculate the quantity, in µg, of epinephrine ($C_9H_{13}NO_3$) in each mL of the Injection taken by the formula:

$$(183.21/333.29)(50)(C/V)(r_U/r_S),$$

in which 183.21 and 333.29 are the molecular weights of epinephrine and epinephrine bitartrate, respectively, *C* is the concentration, in µg per mL, of USP Epinephrine Bitartrate RS in the *Standard preparation*, *V* is the volume, in mL, of Injection taken, and r_U and r_S are the peak responses obtained from the *Assay preparation* and the *Standard preparation*, respectively.

Lidocaine Hydrochloride

$C_{14}H_{22}N_2O.HCl.H_2O$ 288.82

Acetamide, 2-(diethylamino)-*N*-(2,6-dimethylphenyl)-, monohydrochloride, monohydrate.

2-(Diethylamino)-2′,6′-acetoxylidide monohydrochloride monohydrate [*6108-05-0*].

Anhydrous 270.80 [*73-78-9*].

» Lidocaine Hydrochloride contains not less than 97.5 percent and not more than 102.5 percent of $C_{14}H_{22}N_2O.HCl$, calculated on the anhydrous basis.

Packaging and storage—Preserve in well-closed containers.

Reference standard—*USP Lidocaine Reference Standard*—Dry in vacuum over silica gel for 24 hours before using. Keep the container tightly closed.

Identification—

A: Dissolve about 300 mg in 5 to 10 mL of water in a separator, add 4 mL of 6 *N* ammonium hydroxide, and extract with four 15-mL portions of chloroform. Combine the chloroform extracts, evaporate with the aid of a current of warm air, and dry the residue in vacuum over silica gel for 24 hours: the crystalline precipitate so obtained responds to the *Identification tests* under *Lidocaine.*

B: Its solutions respond to the tests for *Chloride* ⟨191⟩.

Melting range ⟨741⟩: between 74° and 79°.

Water, *Method I* ⟨921⟩: between 5.0% and 7.0%.

Residue on ignition ⟨281⟩: not more than 0.1%.

Sulfate—Dissolve about 200 mg in 20 mL of water, add 2 mL of 3 *N* hydrochloric acid, mix, and divide into two parts. To one part of the solution add 1 mL of barium chloride TS: no more turbidity is produced than is present in the remaining portion of the solution to which nothing has been added.

Heavy metals, *Method I* ⟨231⟩: 0.002%.

Assay—

Mobile phase—Mix 50 mL of glacial acetic acid and 930 mL of water, and adjust with 1 *N* sodium hydroxide to a pH of 3.40. Mix about 4 volumes of this solution with 1 volume of acetonitrile, such that the retention time of lidocaine is about 4 to 6 minutes. Filter through a membrane filter (1-µm or finer porosity), and degas. Make adjustments if necessary (see *System Suitability* under *Chromatography* ⟨621⟩).

Standard preparation—Dissolve about 85 mg of USP Lidocaine RS, accurately weighed, with warming if necessary, in 0.5 mL of 1 *N* hydrochloric acid in a 50-mL volumetric flask, dilute with *Mobile phase* to volume, and mix to obtain a *Standard preparation* having a known concentration of about 1.7 mg of lidocaine per mL.

Assay preparation—Transfer about 100 mg of Lidocaine Hydrochloride, accurately weighed, to a 50-mL volumetric flask, dilute with *Mobile phase* to volume, and mix.

Resolution preparation—Prepare a solution of methylparaben in *Mobile phase* containing about 220 µg per mL. Mix 2 mL of this solution and 20 mL of *Standard preparation.*

Chromatographic system (see *Chromatography* ⟨621⟩)—The liquid chromatograph is equipped with a 254-nm detector and a 30-cm × 3.9-mm column that contains packing L1, and is operated at a temperature between 20° and 25° maintained at ±1.0° of the selected temperature. The flow rate is about 1.5 mL per minute. Chromatograph the *Standard preparation*, and record the peak responses as directed under *Procedure:* the relative standard deviation for replicate injections is not more than 1.5%. Chromatograph about 20 µL of the *Resolution preparation*, and record the peak responses as directed under *Procedure:* the resolution, *R*, between the lidocaine and methylparaben peaks is not less than 3.0.

Procedure—Separately inject equal volumes (about 20 µL) of the *Assay preparation* and the *Standard preparation* into the chromatograph. Record the chromatograms, and measure the responses for the major peaks. Calculate the quantity, in mg, of $C_{14}H_{22}N_2O.HCl$ in the portion of Lidocaine Hydrochloride taken by the formula:

$$(270.80/234.34)(50C)(r_U/r_S),$$

in which 270.80 and 234.34 are the molecular weights of lidocaine hydrochloride and lidocaine, respectively, *C* is the concentration, in mg per mL, of USP Lidocaine RS in the *Standard preparation*, and r_U and r_S are the lidocaine peak responses obtained from the *Assay preparation* and the *Standard preparation*, respectively.

Lidocaine Hydrochloride Injection

» Lidocaine Hydrochloride Injection is a sterile solution of Lidocaine Hydrochloride in Water for Injection, or a sterile solution prepared from Lidocaine with the aid of Hydrochloric Acid in Water for Injection. It contains not less than 95.0 percent and not more than 105.0 percent of the labeled amount of $C_{14}H_{22}N_2O.HCl$.

Packaging and storage—Preserve in single-dose or in multiple-dose containers, preferably of Type I glass. Injection may be packaged in 50-mL multiple-dose containers.

Labeling—Injections that are of such concentration that they are not intended for direct injection into tissues are labeled to indicate that they are to be diluted prior to administration.

Reference standard—*USP Lidocaine Reference Standard*—Dry in vacuum over silica gel for 24 hours before using.

Identification—Place in a separator a volume of Injection equivalent to about 300 mg of lidocaine hydrochloride, and extract with four 15-mL portions of chloroform, discarding the chloroform extracts. Add 2 mL of 2 *N* sodium hydroxide to the aqueous solution remaining in the separator, and extract with four 15-mL portions of chloroform. Combine the chloroform extracts, and evaporate with the aid of a current of warm air to dryness. Dissolve the crystals so obtained in solvent hexane, evaporate with the aid of warm air, and dry the residue in vacuum over silica gel for 24 hours: the residue so obtained responds to *Identification test A* under *Lidocaine.*

pH ⟨791⟩: between 5.0 and 7.0.

Particulate matter ⟨788⟩: meets the requirements under *Small-volume Injections.*

Other requirements—It meets the requirements under *Injections* ⟨1⟩.

Assay—Proceed with Lidocaine Hydrochloride Injection as directed in the *Assay for lidocaine hydrochloride* under *Lidocaine and Epinephrine Injection.*

Lidocaine Hydrochloride Injection, Dihydroergotamine Mesylate, Heparin Sodium, and—*see* Dihydroergotamine Mesylate, Heparin Sodium, and Lidocaine Hydrochloride Injection

Lidocaine Hydrochloride Jelly

» Lidocaine Hydrochloride Jelly is Lidocaine Hydrochloride in a suitable, water-soluble, sterile, viscous base. It contains not less than 95.0 percent and not more than 105.0 percent of the labeled amount of $C_{14}H_{22}N_2O.HCl$.

Packaging and storage—Preserve in tight containers.

Reference standard—*USP Lidocaine Reference Standard*—Dry in vacuum over silica gel for 24 hours before using.

Identification—Place in a separator containing 10 to 15 mL of water a quantity of Jelly, equivalent to about 300 mg of lidocaine hydrochloride, mix to assure thorough dilution of the Jelly, add 4 mL of 6 *N* ammonium hydroxide, and extract with four 15-mL portions of chloroform. Combine the chloroform extracts, and evaporate with the aid of a current of warm air to dryness. Redissolve the crystals in solvent hexane, evaporate with the aid of warm air, and dry the residue in vacuum over silica gel for 24 hours: the lidocaine so obtained responds to *Identification test A* under *Lidocaine.*

Sterility—It meets the requirements under *Sterility Tests* ⟨71⟩.

Minimum fill ⟨755⟩: meets the requirements.

pH ⟨791⟩: between 6.0 and 7.0.

Assay—Accurately weigh into a separator containing 10 to 15 mL of water a quantity of Lidocaine Hydrochloride Jelly, equivalent to 20 to 30 mg of lidocaine hydrochloride, mix to assure thorough dilution of the Jelly, add 1 mL of 6 *N* ammonium hydroxide, and extract by shaking with four 20-mL portions of chloroform. Combine the chloroform extracts, and evaporate with the aid of a current of warm air, adding 25.0 mL of 0.01 *N* sulfuric acid VS just before the last trace of chloroform is expelled. Complete the evaporation of the chloroform, and titrate the excess acid with 0.01 *N* sodium hydroxide VS, determining the end-point potentiometrically (see *Residual Titrations* under *Titrimetry* ⟨541⟩). Each mL of 0.01 *N* sulfuric acid is equivalent to 2.708 mg of $C_{14}H_{22}N_2O.HCl$.

Lidocaine Hydrochloride Oral Topical Solution

» Lidocaine Hydrochloride Oral Topical Solution contains not less than 95.0 percent and not more than 105.0 percent of the labeled amount of $C_{14}H_{22}N_2O.HCl$. It contains a suitable flavor.

Packaging and storage—Preserve in tight containers.

Reference standard—*USP Lidocaine Reference Standard*—Dry in vacuum over silica gel for 24 hours.

Identification—Place in a separator a volume of Oral Topical Solution, equivalent to about 300 mg of lidocaine hydrochloride, and extract with four 15-mL portions of chloroform, discarding the chloroform extracts. Add 2 mL of 2 *N* sodium hydroxide to the aqueous solution remaining in the separator, and extract with four 15-mL portions of chloroform. Combine the chloroform extracts, and evaporate with the aid of a current of warm air to dryness. Dissolve the crystals so obtained in solvent hexane, evaporate with the aid of warm air, and dry the residue in vacuum over silica gel for 24 hours: the residue so obtained responds to *Identification test A* under *Lidocaine.*

pH ⟨791⟩: between 5.0 and 7.0.

Assay—

Mobile phase, Standard preparation, and *Resolution preparation*—Prepare as directed in the *Assay for lidocaine hydrochloride* under *Lidocaine and Epinephrine Injection.*

Assay preparation—Transfer an accurately measured volume of Lidocaine Hydrochloride Oral Topical Solution, equivalent to about 100 mg of lidocaine hydrochloride, to a 50-mL volumetric flask, dilute with *Mobile phase* to volume, and mix.

Procedure—Proceed as directed for *Procedure* in the *Assay for lidocaine hydrochloride* under *Lidocaine and Epinephrine Injection.* Calculate the quantity, in mg, of $C_{14}H_{22}N_2O.HCl$ in each mL of the Oral Topical Solution taken by the formula:

$$(270.80/234.34)(50)(C/V)(r_U/r_S),$$

in which V is the volume, in mL, of Oral Topical Solution taken, and the other terms are as defined therein.

Lidocaine Hydrochloride Topical Solution

» Lidocaine Hydrochloride Topical Solution contains not less than 95.0 percent and not more than 105.0 percent of the labeled amount of $C_{14}H_{22}N_2O.HCl$.

Packaging and storage—Preserve in tight containers.

Reference standard—*USP Lidocaine Reference Standard*—Dry in vacuum over silica gel for 24 hours.

Identification—A portion of Topical Solution equivalent to about 200 mg of lidocaine hydrochloride responds to the *Identification test* under *Lidocaine Hydrochloride Injection.*

pH ⟨791⟩: between 5.0 and 7.0.

Assay—

Mobile phase, Standard preparation, and *Resolution preparation*—Prepare as directed in the *Assay for lidocaine hydrochloride* under *Lidocaine and Epinephrine Injection.*

Assay preparation—Transfer an accurately measured volume of Lidocaine Hydrochloride Topical Solution, equivalent to about 100 mg of lidocaine hydrochloride, to a 50-mL volumetric flask, dilute with *Mobile phase* to volume, and mix.

Procedure—Proceed as directed for *Procedure* in the *Assay for lidocaine hydrochloride* under *Lidocaine and Epinephrine Injection.* Calculate the quantity, in mg, of $C_{14}H_{22}N_2O.HCl$ in each mL of the Topical Solution taken by the formula:

$$(270.80/234.34)(50)(C/V)(r_U/r_S),$$

in which V is the volume, in mL, of Topical Solution taken, and the other terms are as defined therein.

Sterile Lidocaine Hydrochloride

» Sterile Lidocaine Hydrochloride is Lidocaine Hydrochloride suitable for parenteral use. It contains not less than 95.0 percent and not more than 105.0 percent of the labeled amount of $C_{14}H_{22}N_2O.HCl$.

Packaging and storage—Preserve in *Containers for Sterile Solids* as described under *Injections* ⟨1⟩.

Reference standard—*USP Lidocaine Reference Standard*—Dry in vacuum over silica gel for 24 hours before using.

Other requirements—It responds to the *Identification test* and meets the requirements for *Melting range, Water, Residue on ignition, Sulfate, Heavy metals,* and *Assay* under *Lidocaine Hydrochloride.* It meets also the requirements for *Sterility Tests* ⟨71⟩, *Uniformity of Dosage Units* ⟨905⟩, and *Constituted Solutions* and *Labeling* under *Injections* ⟨1⟩.

Lidocaine Hydrochloride and Dextrose Injection

» Lidocaine Hydrochloride and Dextrose Injection is a sterile solution of Lidocaine Hydrochloride and Dextrose in Water for Injection. It contains not less than 95.0 percent and not more than 105.0 percent of the labeled amounts of lidocaine hydrochloride ($C_{14}H_{22}N_2O.HCl$) and dextrose ($C_6H_{12}O_6.H_2O$).

Packaging and storage—Preserve in single-dose containers of Type I or Type II glass, or of a suitable plastic material.

Reference standard—*USP Lidocaine Reference Standard*—Dry in vacuum over silica gel for 24 hours before using.

Identification—

A: Place in a separator a volume of Injection equivalent to about 300 mg of lidocaine hydrochloride, add 2 mL of 2 *N* sodium hydroxide, and extract with four 15-mL portions of chloroform. Combine the chloroform extracts, and evaporate with the aid of a current of warm air to dryness. Dissolve the crystals so obtained in solvent hexane, evaporate with the aid of warm air, and dry the residue in vacuum over silica gel for 24 hours: the residue so obtained responds to *Identification test A* under *Lidocaine.*

B: It responds to the *Identification test* under *Dextrose.*

pH ⟨791⟩: between 3.0 and 7.0.

Other requirements—It meets the requirements under *Injections* ⟨1⟩.

Assay for lidocaine hydrochloride—Proceed with Lidocaine Hydrochloride and Dextrose Injection as directed in the *Assay for lidocaine hydrochloride* under *Lidocaine and Epinephrine Injection.*

Assay for dextrose—Determine the angular rotation of Lidocaine Hydrochloride and Dextrose Injection in a suitable polarimeter tube at 25° (see *Optical Rotation* ⟨781⟩). The observed rotation in degrees, multiplied by 1.0425A, in which A is the ratio 200 divided by the length, in mm, of the polarimeter tube employed, represents the weight, in g, of $C_6H_{12}O_6.H_2O$ in 100 mL of the Injection.

Light Mineral Oil—*see* Mineral Oil, Light NF

Lime

CaO 56.08
Calcium oxide [*1305-78-8*].

» Lime, when freshly ignited to constant weight, contains not less than 95.0 percent of CaO.

Packaging and storage—Preserve in tight containers.

Identification—

A: Moisten it with water: heat is generated and a white powder is obtained (calcium hydroxide or slaked lime). Mix the powder with 3 or 4 times its weight of water: a smooth magma of lime forms that is alkaline to litmus.

B: Slake 1 g with 20 mL of water, and add 6 *N* acetic acid until the lime is dissolved: the resulting solution responds to the test for *Calcium* ⟨191⟩.

Loss on ignition—Ignite a portion to constant weight in a tared platinum crucible at 1100 ± 50°: it loses not more than 10.0% of its weight.

Insoluble substances—Slake 5.0 g, then mix with 100 mL of water, followed by hydrochloric acid, dropwise, with agitation, until solution takes place: the resulting solution after boiling and cooling is acid, and when filtered through a tared crucible, washed with water until free of chlorides, and dried at 105° for 1 hour yields not more than 50 mg of insoluble substances (1.0%).

Carbonate—Slake 1 g, mix with 50 mL of water, and decant the greater portion of the milky liquid: the addition of an excess of 3 *N* hydrochloric acid to the residue does not cause more than a slight effervescence.

Magnesium and alkali salts—Dissolve 500 mg in 30 mL of water and 15 mL of 3 *N* hydrochloric acid. Neutralize the solution with 6 *N* ammonium hydroxide, heat to boiling, and add ammonium oxalate TS to precipitate the calcium completely. Heat the mixture on a steam bath for 1 hour, cool, dilute with water to 100 mL, mix, and filter. To 50 mL of the filtrate add 0.5 mL of sulfuric acid, evaporate to dryness, and ignite in a tared platinum crucible to constant weight. The weight of the residue does not exceed 9 mg.

Assay—Ignite about 1 g of Lime in a muffle furnace to constant weight, cool, weigh accurately, and dissolve in 20 mL of 3 *N* hydrochloric acid. Cool the solution, transfer to a 500-mL volumetric flask with the aid of water, dilute with water to volume, and mix. Transfer 50.0 mL to a suitable container, add 100 mL of water, 15 mL of 1 *N* sodium hydroxide, and 300 mg of hydroxy naphthol blue, and titrate with 0.05 *M* disodium ethylenediaminetetraacetate VS until the solution is deep blue in color. Each mL of 0.05 *M* disodium ethylenediaminetetraacetate is equivalent to 2.804 mg of CaO.

Lincomycin Hydrochloride

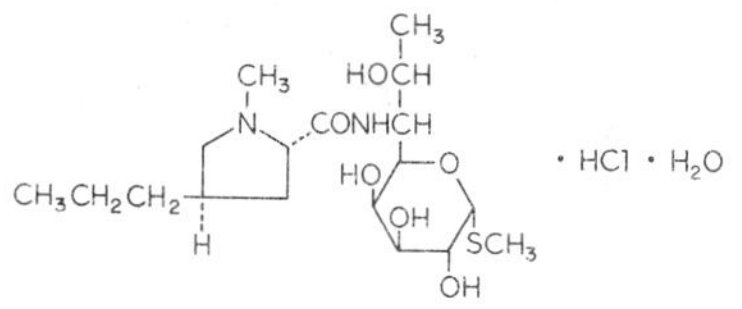

$C_{18}H_{34}N_2O_6S.HCl.H_2O$ 461.01
D-*erythro*-α-D-*galacto*-Octopyranoside, methyl 6,8-dideoxy-6-[[(1-methyl-4-propyl-2-pyrrolidinyl)carbonyl]amino]-1-thio-, monohydrochloride, monohydrate, (2*S*-*trans*)-.
Methyl 6,8-dideoxy-6-(1-methyl-*trans*-4-propyl-L-2-pyrrolidinecarboxamido)-1-thio-D-*erythro*-α-D-*galacto*-octopyranoside monohydrochloride monohydrate [*7179-49-9*].
Anhydrous 443.00 [*859-18-7*].

» Lincomycin Hydrochloride has a potency equivalent to not less than 790 μg of lincomycin ($C_{18}H_{34}N_2O_6S$) per mg.

Packaging and storage—Preserve in tight containers.

Reference standard—*USP Lincomycin Hydrochloride Reference Standard*—Do not dry before using.

Identification—The infrared absorption spectrum of a mineral oil dispersion of it exhibits maxima only at the same wavelengths as that of a similar preparation of USP Lincomycin Hydrochloride RS.

Specific rotation ⟨781⟩: between +135° and +150°, calculated on the anhydrous basis, in a solution containing 20 mg per mL.

Crystallinity ⟨695⟩: meets the requirements.

pH ⟨791⟩: between 3.0 and 5.5, in a solution (1 in 10).

Water, *Method I* ⟨921⟩: between 3.0% and 6.0%.

Lincomycin B—Use the chromatogram obtained in the *Assay:* the area of the lincomycin B peak is not greater than 5.0% of the sum of the areas of the lincomycin B peak and the lincomycin peak.

Other requirements—Lincomycin Hydrochloride intended for use in making Lincomycin Hydrochloride Injection complies also with the requirements for *Depressor substances* and *Pyrogen* under *Sterile Lincomycin Hydrochloride.*

Assay—

Chloroform-imidazole solution—Prepare a chloroform solution containing 20 mg of imidazole per mL.

Internal standard solution—Dissolve *n*-dotriacontane in chloroform to obtain a solution having a concentration of about 8.5 mg per mL.

Standard preparation—Transfer about 110 mg of USP Lincomycin Hydrochloride RS, accurately weighed, to a suitable flask, add 10.0 mL of *Internal standard solution* and 90 mL of *Chloroform-imidazole solution* to the flask, and shake vigorously until completely dissolved.

Assay preparation—Using Lincomycin Hydrochloride, prepare as directed under *Standard preparation.*

Chromatographic system—The gas chromatograph is equipped with a flame-ionization detector and contains a 3-mm × 122-cm glass column packed with 3 percent liquid phase G3 on support S1AB and is cured as directed (see *Chromatography* ⟨621⟩). The column is maintained at a temperature of about 255°, the detector is maintained at about 270°, and dry helium is used as the carrier gas flowing at a rate of about 45 mL per minute.

System suitability—Under typical conditions, the retention time for lincomycin B, lincomycin, and *n*-dotriacontane are about 3, 4, and 6 minutes, respectively. Chromatograph the solution from the *Standard preparation,* and record the peak responses as directed under *Procedure:* the relative standard deviation for replicate injections is not greater than 2.5%, and the resolution, *R,* between lincomycin and *n*-dotriacontane is not less than 3.0.

Procedure—Transfer equal volumes (about 4 mL) of the *Standard preparation* and the *Assay preparation* to separate glass-stoppered, 15-mL centrifuge tubes. To each tube add 1.0 mL of a 1 in 100 solution of trimethylchlorosilane in bis(trimethylsilyl)acetamide, and swirl gently. Position the glass stoppers loosely in the tubes, and heat at 65° for 30 minutes. Allow to cool, and mix. Inject separate suitable portions (0.5 μL to 1.5 μL) of the solutions from the *Standard preparation* and the *Assay preparation* into the gas chromatograph. Record the chromatograms, and measure the responses for the major peaks. Calculate the quantity, in μg, of lincomycin ($C_{18}H_{34}N_2O_6S$) in the portion of Lincomycin Hydrochloride taken by the formula:

$$WP(R_U/R_S),$$

in which W is the weight, in mg, of USP Lincomycin Hydrochloride RS taken, P is the potency, in μg of lincomycin ($C_{18}H_{34}N_2O_6S$) per mg, of USP Lincomycin Hydrochloride RS, and R_U and R_S are the peak response ratios of the lincomycin peak to the *n*-dotriacontane peak obtained with the solutions from the *Assay preparation* and the *Standard preparation,* respectively.

Lincomycin Hydrochloride Capsules

» Lincomycin Hydrochloride Capsules contain an amount of $C_{18}H_{34}N_2O_6S.HCl.H_2O$ equivalent to not less than 90.0 percent and not more than 120.0 percent of the labeled amount of lincomycin ($C_{18}H_{34}N_2O_6S$).

Packaging and storage—Preserve in tight containers.

Reference standard—*USP Lincomycin Hydrochloride Reference Standard*—Do not dry before using.

Dissolution ⟨711⟩—

Medium: water; 500 mL.
Apparatus 1: 100 rpm.
Time: 45 minutes.

Procedure—Filter a portion of about 20 mL of the solution under test. Transfer about 5 mL of the eluent into a small test tube and add 250 μL of 0.01 *M* sodium sulfate internal standard solution. Evaporate until dry using a vacuum centrifuge. Add 10.0 μL of water to the precipitate and vortex the tube until all solid material is dissolved. Transfer this solution to a capillary tube, place it in a Raman spectrometer, and obtain the Raman spectrum using suitable instrumental conditions (see *Spectrophotometry and Light-scattering* ⟨851⟩). Integrate the Raman intensity, applying baseline corrections, between 660 cm^{-1} and 720 cm^{-1}. Divide this result by the integrated intensity between 966 cm^{-1} and 994 cm^{-1}. Determine the amount of $C_{18}H_{34}N_2O_6S$ dissolved in comparison with an aqueous Standard solution having a known concentration of USP Lincomycin Hydrochloride RS.

Tolerances—Not less than 75% (*Q*) of the labeled amount of $C_{18}H_{34}N_2O_6S$ is dissolved in 45 minutes.

Uniformity of dosage units ⟨905⟩: meet the requirements.

Water, *Method I* ⟨921⟩: not more than 7.0%.

Assay—

Chloroform-imidazole solution, Internal standard solution, and *Standard preparation*—Proceed as directed in the *Assay* under *Lincomycin Hydrochloride.*

Assay preparation—Remove, as completely as possible, the contents of not less than 20 Lincomycin Hydrochloride Capsules. Transfer an accurately weighed portion of the powder, equivalent to about 100 mg of lincomycin, to a 250-mL conical flask, add 10.0 mL of *Internal standard solution* and 90 mL of *Chloroform-imidazole solution,* shake by mechanical means for 15 minutes, and filter.

Chromatographic system, System suitability, and *Procedure*—Proceed as directed in the *Assay* under *Lincomycin Hydrochloride.* Calculate the quantity, in mg, of lincomycin ($C_{18}H_{34}N_2O_6S$) in the portion of Capsules taken by the formula:

$$(WP/1000)(R_U/R_S),$$

in which the terms are as defined therein.

Lincomycin Hydrochloride Injection

» Lincomycin Hydrochloride Injection is a sterile solution of Lincomycin Hydrochloride in Water for Injection. It contains the equivalent of not less than 90.0 percent and not more than 120.0 percent of the labeled amount of lincomycin ($C_{18}H_{34}N_2O_6S$). It contains benzyl alcohol as a preservative.

Packaging and storage—Preserve in single-dose or in multiple-dose containers, preferably of Type I glass.

Reference standard—*USP Lincomycin Hydrochloride Reference Standard*—Do not dry before using.

Depressor substances—When diluted with sterile saline TS to a concentration of 3.0 mg of lincomycin ($C_{18}H_{34}N_2O_6S$) per mL, it meets the requirements of the *Depressor Substances Test* ⟨101⟩, the test dose being 1.0 mL per kg.

Pyrogen—When diluted with sterile pyrogen-free saline TS to a concentration of 0.5 mg of lincomycin ($C_{18}H_{34}N_2O_6S$) per mL, it meets the requirements of the *Pyrogen Test* ⟨151⟩, the test dose being 1.0 mL per kg.

Sterility—It meets the requirements under *Sterility Tests* ⟨71⟩, when tested as directed in the section, *Test Procedures Using Membrane Filtration.*

pH ⟨791⟩: between 3.0 and 5.5.

Particulate matter ⟨788⟩: meets the requirements under *Small-volume Injections.*

Other requirements—It meets the requirements under *Injections* ⟨1⟩.

Assay—

Chloroform-imidazole solution, Internal standard solution, and *Standard preparation*—Proceed as directed in the *Assay* under *Lincomycin Hydrochloride.*

Assay preparation—Transfer an accurately measured volume of Lincomycin Hydrochloride Injection, equivalent to about 100 mg of lincomycin, to a suitable container, add an equal volume of methanol, and evaporate to dryness. Add 10.0 mL of *Internal standard solution* and 90 mL of *Chloroform-imidazole solution,* and shake by mechanical means until the residue dissolves.

Chromatographic system, System suitability, and *Procedure*—Proceed as directed in the *Assay* under *Lincomycin Hydrochloride.* Calculate the quantity, in mg, of lincomycin ($C_{18}H_{34}N_2O_6S$) in each mL of the Injection taken by the formula:

$$(WP/1000V)(R_U/R_S),$$

in which V is the volume, in mL, of Injection taken, and the other terms are as defined therein.

Sterile Lincomycin Hydrochloride

» Sterile Lincomycin Hydrochloride is Lincomycin Hydrochloride suitable for parenteral use. It has a potency equivalent to not less than 790 μg of lincomycin ($C_{18}H_{34}N_2O_6S$) per mg.

Packaging and storage—Preserve in *Containers for Sterile Solids* as described under *Injections* ⟨1⟩.

Reference standard—*USP Lincomycin Hydrochloride Reference Standard*—Do not dry before using.

Depressor substances—It meets the requirements of the *Depressor Substances Test* ⟨101⟩, the test dose being 1.0 mL per kg of a solution prepared to contain 3.0 mg of lincomycin ($C_{18}H_{34}N_2O_6S$) per mL in sterile saline TS.

Pyrogen—It meets the requirements of the *Pyrogen Test* ⟨151⟩, the test dose being 1.0 mL per kg of a solution prepared to contain 0.5 mg of lincomycin ($C_{18}H_{34}N_2O_6S$) per mL in pyrogen-free saline TS.

Sterility—It meets the requirements under *Sterility Tests* ⟨71⟩, when tested as directed in the section, *Test Procedures Using Membrane Filtration,* 6 g of specimen aseptically dissolved in 200 mL of *Fluid A* being used.

Other requirements—It conforms to the definition, responds to the *Identification test,* and meets the requirements for *Specific rotation, pH, Water, Crystallinity, Lincomycin B,* and *Assay* under *Lincomycin Hydrochloride.*

Lincomycin Hydrochloride Syrup

» Lincomycin Hydrochloride Syrup contains an amount of $C_{18}H_{34}N_2O_6S \cdot HCl \cdot H_2O$ equivalent to not less than 90.0 percent and not more than 120.0 percent of the labeled amount of lincomycin ($C_{18}H_{34}N_2O_6S$), and one or more suitable colors, flavors, preservatives, and sweeteners in water.

Packaging and storage—Preserve in tight containers.

Reference standard—*USP Lincomycin Hydrochloride Reference Standard*—Do not dry before using.

Uniformity of dosage units ⟨905⟩—

FOR SYRUP PACKAGED IN SINGLE-UNIT CONTAINERS: meets the requirements.

pH ⟨791⟩: between 3 and 5.5.

Assay—

Chloroform-imidazole solution, Internal standard solution, and *Standard preparation*—Proceed as directed in the *Assay* under *Lincomycin Hydrochloride.*

Assay preparation—Transfer an accurately measured volume of Lincomycin Hydrochloride Syrup, equivalent to about 250 mg of lincomycin, to a 125-mL separator, add 75 mL of *Chloroform-imidazole solution,* and shake for 5 minutes. Filter the chloroform layer through a pledget of cotton into a suitable flask containing 25.0 mL of *Internal standard solution.* Repeat the extraction with three 50-mL portions of *Chloroform-imidazole solution.* Combine the extracts, and mix.

Chromatographic system, System suitability, and *Procedure*—Proceed as directed in the *Assay* under *Lincomycin Hydrochloride.* Calculate the quantity, in mg, of lincomycin ($C_{18}H_{34}N_2O_6S$) in each mL of the Syrup taken by the formula:

$$(WP/1000V)(R_U/R_S),$$

in which V is the volume, in mL, of Syrup taken, and the other terms are as defined therein.

Lindane

$C_6H_6Cl_6$ 290.83

Cyclohexane, 1,2,3,4,5,6-hexachloro-, (1α,2α,3β,4α,5α,6β)-.

γ-1,2,3,4,5,6-Hexachlorocyclohexane [*58-89-9*].

» Lindane is the gamma isomer of hexachlorocyclohexane. It contains not less than 99.0 percent and not more than 100.5 percent of $C_6H_6Cl_6$.

Packaging and storage—Preserve in well-closed containers.

Reference standard—*USP Lindane Reference Standard*—Do not dry. Keep container tightly closed.

Identification—The infrared absorption spectrum of a potassium bromide dispersion of it exhibits maxima only at the same wavelengths as that of a similar preparation of USP Lindane RS.

Congealing temperature ⟨651⟩: not less than 112.0°.

Water, *Method I* ⟨921⟩: not more than 0.5%.

Chloride ion—Place about 100 mg in a test tube with 10 mL of water, shake well, and filter. Add 1 mL of nitric acid and 3 mL of silver nitrate TS to the filtrate: no turbidity develops.

Assay—Transfer about 400 mg of Lindane, accurately weighed, to a wide-mouth, glass-stoppered, 250-mL conical flask, add 20 mL of alcohol, and warm on a steam bath until solution is complete. Cool, add 20 mL of a 1 in 20 solution of potassium hydroxide in alcohol, swirl gently, and allow to stand for 10 minutes. Dilute with water to about 100 mL, neutralize with 2 *N* nitric acid, and add 5 mL in excess. Pipet into the solution 50 mL of 0.1 *N* silver nitrate VS, then add 5 mL of nitrobenzene, and shake vigorously. Add ferric ammonium sulfate TS, and titrate the excess silver nitrate with 0.1 *N* ammonium thiocyanate VS. Perform a blank determination (see *Residual Titrations* under *Titrimetry* ⟨541⟩). Each mL of 0.1 *N* silver nitrate is equivalent to 9.694 mg of $C_6H_6Cl_6$.

Lindane Cream

» Lindane Cream is Lindane in a suitable cream base. It contains not less than 90.0 percent and not more than 110.0 percent of the labeled amount of $C_6H_6Cl_6$.

Packaging and storage—Preserve in tight containers.

Reference standard—*USP Lindane Reference Standard*—Do not dry. Keep container tightly closed.

Identification—Wind a strip of 20-mesh copper gauze 1.5 cm wide and 5 cm long around the end of a copper wire. Heat the gauze in the nonluminous flame of a Bunsen burner until it glows without coloring the flame green. Allow the gauze to cool, and repeat the heating and cooling step several times until a thorough coating of oxide is formed. Apply a small amount of Cream to the cooled gauze, ignite, and allow to burn freely in the air. Hold the gauze in the outer edge of the burner flame at a height of about 4 cm: a bright green color is imparted to the flame.

pH ⟨791⟩: between 8.0 and 9.0, in a 1 in 5 dilution.

Assay—

Internal standard solution—Prepare a solution in methylene chloride containing 1 mg of *n*-docosane in each mL.

Standard preparation—Dissolve an accurately weighed quantity of USP Lindane RS in methylene chloride to obtain a solution having a known concentration of about 2 mg per mL. Transfer 5.0 mL of this solution to a graduated centrifuge tube, add 5.0 mL of *Internal standard solution*, mix, and evaporate with the aid of gentle heat and a current of dry air to 3 mL. [NOTE—Avoid evaporating to dryness. If the mixture is inadvertently evaporated to dryness, discard it, and begin another *Standard preparation*.]

Solid support—Use 60- to 100-mesh magnesium silicate that has been previously heated at 300° for 2 hours.

Mobile phase—Mix 18 mL of anhydrous ethyl ether with 280 mL of chromatographic solvent hexane.

Assay preparation—Place a pledget of cotton on a removable porous plate at the base of a 25-mm × 200-mm chromatographic tube that is fitted with a polytef stopcock. Add 50 mL of *Mobile phase* and 10 g of *Solid support*, and stir the mixture to expel air bubbles. Add 1.5 g of anhydrous sodium sulfate to the column, and elute until the surface of the liquid is about 4 cm above the *Solid support*, discarding the eluate. Transfer an accurately weighed portion of Lindane Cream, corresponding to about 10 mg of lindane, to a 150-mL beaker, and add 10 g of *Solid support*. Mix with a spatula, adding chromatographic solvent hexane as necessary to produce a homogeneous mixture, and continue stirring until a free-flowing powder is produced. Transfer this mixture to the chromatographic column with the aid of three 5-mL portions of *Mobile phase*, and elute the column with 225 mL of the *Mobile phase* at a flow rate of 2 mL to 3 mL per minute, collecting the eluate in a 250-mL beaker. Remove the chromatographic column, add 5.0 mL of *Internal standard solution* to the eluate, and evaporate with the aid of gentle heat and a current of dry air to about 5 mL. Transfer this solution to a graduated centrifuge tube with the aid of 1 mL of methylene chloride, and evaporate with the aid of gentle heat and a current of dry air to about 3 mL. [See *Note* under *Standard preparation*.]

Chromatographic system (see *Chromatography* ⟨621⟩)—The gas chromatograph is equipped with a flame-ionization detector and contains a 1.8-m × 2-mm glass column packed with 3 percent liquid phase G3 on support S1A. Maintain the column at 195°, and maintain the injection port and the detector at 250°. Use dry nitrogen as the carrier gas at a flow rate of about 40 mL per minute. Chromatograph six to ten replicate injections of the *Standard preparation*, and record the peak responses as directed under *Procedure:* the relative standard deviation is not more than 3.0%, the tailing factor is not more than 2.0, and the resolution factor between lindane and *n*-docosane is not less than 5.

Procedure—Separately inject equal volumes (about 1 µL) of the *Standard preparation* and the *Assay preparation* into the chromatograph, record the chromatograms, and measure the responses for the major peaks. Calculate the percentage of $C_6H_6Cl_6$ in the Cream taken by the formula:

$$500(C/W)(R_U/R_S),$$

in which C is the concentration, in mg per mL, of USP Lindane RS in the solution prepared, prior to the addition of *Internal standard solution*, for the *Standard preparation*, W is the weight, in mg, of Cream taken, and R_U and R_S are the ratios of the peak responses of lindane to those of *n*-docosane from the *Assay preparation* and the *Standard preparation*, respectively.

Lindane Lotion

» Lindane Lotion is Lindane in a suitable aqueous vehicle. It contains not less than 90.0 percent and not more than 110.0 percent of the labeled amount of $C_6H_6Cl_6$.

Packaging and storage—Preserve in tight containers.

Reference standard—*USP Lindane Reference Standard*—Do not dry. Keep container tightly closed.

Identification—It responds to the *Identification test* under *Lindane Cream*.

pH ⟨791⟩: between 6.5 and 8.5.

Assay—Proceed as directed in the *Assay* under *Lindane Cream*, substituting "Lotion" for "Cream" throughout.

Lindane Shampoo

» Lindane Shampoo is Lindane in a suitable vehicle. It contains not less than 90.0 percent and not more than 110.0 percent of the labeled amount of $C_6H_6Cl_6$.

Packaging and storage—Preserve in tight containers.

Reference standard—*USP Lindane Reference Standard*—Do not dry. Keep container tightly closed.

Identification—It responds to the *Identification test* under *Lindane Cream*.

pH ⟨791⟩: between 6.2 and 7.0.

Assay—Proceed with Lindane Shampoo as directed in the *Assay* under *Lindane Cream*, substituting "Shampoo" for "Cream" throughout.

Liothyronine Sodium

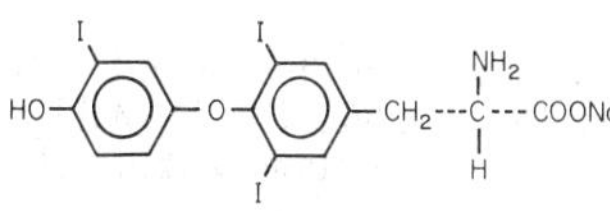

$C_{15}H_{11}I_3NNaO_4$ 672.96

L-Tyrosine, *O*-(4-hydroxy-3-iodophenyl)-3,5-diiodo-, monosodium salt.
Monosodium L-3-[4-(4-hydroxy-3-iodophenoxy)-3,5-diiodophenyl]alanine [*55-06-1*].

» Liothyronine Sodium is the sodium salt of L-3,3′,5-triiodothyronine. It contains not less than 95.0 percent and not more than 101.0 percent of $C_{15}H_{11}I_3NNaO_4$, calculated on the dried basis.

Packaging and storage—Preserve in tight containers.

Reference standards—*USP Liothyronine Reference Standard*—Keep container tightly closed and protected from light. Use without drying; correct for moisture, determined by drying a separate portion in vacuum at 60° for 3 hours. *USP Levothyroxine Reference Standard*—Keep container tightly closed and protected from light. Use without drying; correct for moisture, determined by drying a separate portion in vacuum at 60° for 4 hours.

Identification—

A: The ultraviolet absorption spectrum of a 1 in 10,000 solution in dilute hydrochloric acid (1 in 50) in 80 percent alcohol exhibits maxima at the same wavelengths as that of a similar solution of USP Liothyronine RS, concomitantly measured, and the respective absorptivities, both calculated on the dried basis in terms of the acid, at the wavelength of maximum absorbance at about 297 nm do not differ by more than 5.0%.

B: Heat about 50 mg with a few drops of sulfuric acid in a porcelain crucible: violet vapors of iodine are evolved.

C: The residue from the ignition of it responds to the tests for *Sodium* ⟨191⟩.

Specific rotation ⟨781⟩: between +18° and +22°, calculated on the dried basis, determined in a solution containing 20 mg in each mL of a solvent consisting of a mixture of 4 volumes of alcohol and 1 volume of dilute hydrochloric acid (1 in 10).

Loss on drying ⟨731⟩—Dry it at 105° for 2 hours: it loses not more than 4.0% of its weight.

Inorganic iodide—Weigh accurately 100 mg, and dissolve in 25 mL of water contained in a 125-mL separator. Add 5 mL of 2 *N* sulfuric acid, 10 mL of chloroform, and 3 drops of 0.05 *M* potassium iodate, shake for 30 seconds, and allow the layers to separate. Filter the chloroform layer through paper into a color-comparison tube: any pink color produced is not darker than that produced when 25 mL of a freshly prepared solution containing 111 µg of potassium iodide is treated and extracted similarly (0.08%).

Chloride content—Weigh accurately 100 mg, previously dried, and transfer to a platinum dish. Ignite over a low flame, protecting the dish from air currents during the ignition. When carbonization is complete, cool the dish, add 2 drops of water, and break up the charred mass thoroughly with a stirring rod. Add 10 mL of water and 5 mL of ammonium hydroxide, and mix. Transfer the slurry to a glass-stoppered, 50-mL flask, and wash the platinum dish and the stirring rod with water, adding the washings to the flask, until the volume of the solution is about 25 mL. Add 10 mL of silver nitrate solution (1 in 20), shake thoroughly, and filter through a retentive paper into a 50-mL color-comparison tube. Wash the flask and the filter paper with 10 mL of water, and add the washings to the tube. Acidify the combined filtrate and washings to litmus with nitric acid, and dilute with water to 50 mL. Prepare a control by mixing 5 mL of ammonium hydroxide, 20 mL of water, and 10 mL of silver nitrate solution (1 in 20), filtering the mixture through a retentive paper into a 50-mL color-comparison tube, then washing the filter paper with 10 mL of water into the tube, acidifying the contents of the tube to litmus with nitric acid, diluting with water to 50 mL, and adding sodium chloride solution (1 in 1000) in 0.1-mL increments until the turbidity of the control matches that of the test solution. Not more than 2.0 mL of sodium chloride is required (1.2%).

Sodium content—Weigh accurately about 100 mg, previously dried, and transfer to a platinum dish. Add 8 to 10 drops of sulfuric acid, and ignite to constant weight, taking care to avoid spattering. Each mg of residue is equivalent to 0.324 mg of Na. Correct the result for the amount of sodium equivalent to the NaCl found in the test for *Chloride content:* not less than 2.9% and not more than 4.0% is found.

Levothyroxine sodium—

Mobile phase, 0.01 M Methanolic sodium hydroxide, and *Chromatographic system*—Prepare as directed in the *Assay*.

Standard preparation—Dissolve an accurately weighed quantity of USP Levothyroxine RS in *0.01 M Methanolic sodium hydroxide*, and dilute quantitatively and stepwise with *0.01 M Methanolic sodium hydroxide* to obtain a solution having a known concentration of 4.0 µg of levothyroxine, as the sodium salt, per mL.

Test preparation—Transfer 8.0 mg of Liothyronine Sodium, accurately weighed, to a 100-mL volumetric flask, add *0.01 M Methanolic sodium hydroxide* to volume, and mix.

System suitability preparation—Mix equal volumes of the *Standard preparation* and the *Standard preparation* prepared for the *Assay*. Chromatograph this mixture, and record the peak responses as directed under *Procedure:* the resolution, *R*, between the liothyronine and levothyroxine peaks is not less than 4.0.

Procedure—Separately inject equal volumes (about 50 µL) of the *Standard preparation* and the *Test preparation* into the chromatograph, record the chromatograms, and measure the responses for the levothyroxine peaks: the peak response obtained from the *Test preparation* is not greater than that obtained from the *Standard preparation*, corresponding to not more than 5.0% of levothyroxine sodium.

Assay—

Mobile phase—Prepare a suitable degassed and filtered mixture of water and acetonitrile (7:3) that contains 5 mL of phosphoric acid in each 1000 mL of solution.

0.01 M Methanolic sodium hydroxide—Dissolve 400 mg of sodium hydroxide in 500 mL of water. Cool, add 500 mL of methanol, and mix.

Standard preparation—Dissolve an accurately weighed quantity of USP Liothyronine RS in *0.01 M Methanolic sodium hydroxide*, and dilute quantitatively and stepwise with *0.01 M Methanolic sodium hydroxide* to obtain a solution having a known concentration of about 5 µg per mL.

Assay preparation—Transfer about 12.5 mg of Liothyronine Sodium, accurately weighed, to a 25-mL volumetric flask, dissolve in *0.01 M Methanolic sodium hydroxide*, dilute with the same solvent to volume, and mix. Pipet 1 mL of this solution into a 100-mL volumetric flask, dilute with *0.01 M Methanolic sodium hydroxide* to volume, and mix.

Chromatographic system (see *Chromatography* ⟨621⟩)—The liquid chromatograph is equipped with a 225-nm detector and a 3.9-mm × 30-cm column that contains packing L10. The flow rate is about 1 mL per minute. Chromatograph five replicate injections of the *Standard preparation*, and record the peak responses as directed under *Procedure:* the relative standard deviation is not more than 2.0%, and the tailing factor is not more than 1.8.

Procedure—Separately inject equal volumes (about 50 µL) of the *Standard preparation* and the *Assay preparation* into the chromatograph, record the chromatograms, and measure the responses for the major peaks. Calculate the quantity, in mg, of $C_{15}H_{11}I_3NNaO_4$ in the portion of Liothyronine Sodium taken by the formula:

$$(672.96/650.98)(2.5C)(r_U/r_S),$$

in which 672.96 and 650.98 are the molecular weights of liothyronine sodium and liothyronine, respectively, *C* is the concentration, in µg per mL, of USP Liothyronine RS in the *Standard preparation*, and r_U and r_S are the peak responses obtained from the *Assay preparation* and the *Standard preparation*, respectively.

Liothyronine Sodium Tablets

» Liothyronine Sodium Tablets contain an amount of $C_{15}H_{11}I_3NNaO_4$ equivalent to not less than 90.0

percent and not more than 110.0 percent of the labeled amount of liothyronine ($C_{15}H_{12}I_3NO_4$).

Packaging and storage—Preserve in tight containers.

Reference standards—*USP Liothyronine Reference Standard*—Keep container tightly closed and protected from light. Use without drying; correct for moisture, determined by drying a separate portion in vacuum at 60° for 3 hours. *USP Levothyroxine Reference Standard*—Keep container tightly closed and protected from light. Use without drying; correct for moisture, determined by drying a separate portion in vacuum at 60° for 4 hours.

Identification—

A: Mix a portion of finely powdered Tablets, equivalent to about 0.1 mg of liothyronine, with an equal amount of anhydrous potassium carbonate, and transfer to a crucible. Ignite the crucible at 700° for 5 minutes, and cool. Add 5 mL of water to the crucible, heat on a steam bath for 5 minutes, cool, and filter. To the filtrate add 1 mL of chloroform, 1 mL of dilute phosphoric acid (1 in 2), and 1 mL of sodium nitrite solution (1 in 100), shake vigorously, and allow to separate: the chloroform layer is colored purple.

B: Shake a portion of finely powdered Tablets, equivalent to 0.1 mg of liothyronine, with 15 mL of water for 1 minute. Add 2 drops of hydrochloric acid and 10 mL of butyl alcohol, and shake for 1 minute. Centrifuge the mixture for 5 minutes. Remove as much as possible of the clear, upper layer by means of a pipet, and evaporate it on a steam bath until the odor of butyl alcohol is no longer present. Add 3 drops of methanol to the residue, and rotate the container to wet the contents thoroughly. Transfer, with the aid of a capillary tube, as much of the methanol as possible to a small area on a filter paper. When the filter paper is dry, spray it with diazotized sulfanilamide, prepared by mixing 5 mL of a 1 in 100 solution of sulfanilamide in dilute hydrochloric acid (1 in 10) with 5 mL of sodium nitrite solution (1 in 20) for 1 minute, adding butyl alcohol to make 50 mL, shaking for 1 minute, allowing to stand for 4 minutes, and decanting the butyl alcohol layer to be used as the spraying solution. Dry the filter paper in a stream of air, and spray it with sodium carbonate solution (1 in 10): pink color is produced on the paper in the area where the test specimen was applied.

Disintegration ⟨701⟩: 30 minutes.

Uniformity of dosage units ⟨905⟩: meet the requirements.

Assay—

Mobile phase, 0.01 M Methanolic sodium hydroxide, Standard preparation, and *Chromatographic system*—Prepare as directed in the *Assay* under *Liothyronine Sodium*.

Assay preparation—Weigh and finely powder not less than 20 Liothyronine Sodium Tablets. Transfer an accurately weighed portion of the powder, equivalent to about 125 µg of liothyronine sodium, to a 25-mL volumetric flask, add about 15 mL of *0.01 M Methanolic sodium hydroxide*, and place in an ultrasonic bath for 1 minute. Shake the mixture for 5 minutes, dilute with *0.01 M Methanolic sodium hydroxide* to volume, mix, and filter.

Procedure—Proceed as directed for *Procedure* in the *Assay* under *Liothyronine Sodium*, and calculate the quantity, in mg, of liothyronine ($C_{15}H_{12}I_3NO_4$) in the portion of Tablets taken by the formula:

$$(0.1C)(r_U/r_S),$$

in which the terms are as defined therein.

Liotrix Tablets

» Liotrix Tablets contain not less than 90.0 percent and not more than 110.0 percent of the labeled amounts of levothyroxine sodium ($C_{15}H_{10}I_4NNaO_4$) and liothyronine sodium ($C_{15}H_{11}I_3NNaO_4$).

Packaging and storage—Preserve in tight containers.

Reference standards—*USP Liothyronine Reference Standard*—Keep container tightly closed and protected from light. Use without drying; correct for moisture, determined by drying a separate portion in vacuum at 60° for 3 hours. *USP Levothyroxine Reference Standard*—Keep container tightly closed and protected from light. Use without drying; correct for moisture, determined by drying a separate portion in vacuum at 60° for 4 hours.

Identification—

Solvent system—Place a suitable volume of tertiary amyl alcohol in a separator, add an equal volume of 3 *N* ammonium hydroxide, and shake to equilibrate. Discard the lower layer, and transfer the upper layer to the developing chamber, cover the chamber, and allow to stand for 1 hour before using.

Detection reagent—Add 65 mL of 2 *N* hydrochloric acid to 50 mL of a 1 in 10 solution of sodium arsenite in 1 *N* sodium hydroxide, with vigorous stirring. Mix 1 volume of this solution with 5 volumes of a 27 in 1000 solution of ferric chloride in 2 *N* hydrochloric acid and 5 volumes of freshly prepared potassium ferricyanide solution (35 in 1000).

Standard preparations—Prepare a solution of about 15 mg of USP Levothyroxine RS, accurately weighed, in 100 mL of a mixture of equal volumes of methanol and 3 *N* ammonium hydroxide. Prepare a solution of about 4 mg of USP Liothyronine RS, accurately weighed, in the same solvent mixture. Dilute 20.0 mL of each solution with the same solvent to 100.0 mL.

Test preparation—Shake an amount of powdered Liotrix Tablets, equivalent to about 60 µg of levothyroxine sodium and 15 µg of liothyronine sodium, with 2 mL of a mixture of equal volumes of methanol and 3 *N* ammonium hydroxide in a centrifuge tube for 10 minutes, and centrifuge.

Procedure—Apply 10-µL volumes of the *Test preparation* and each of the *Standard preparations*, respectively, to a thin-layer chromatographic plate coated with a 0.25-mm layer of chromatographic microcrystalline cellulose containing a fluorescent indicator. Develop the plate until the solvent front has moved not less than 10 cm beyond the point of application, air-dry, and spray the plate with *Detection reagent:* the chromatogram of the *Test preparation* shows blue spots corresponding in R_f value to the chromatograms from the levothyroxine and liothyronine *Standard preparations*, respectively.

Disintegration ⟨701⟩: 30 minutes.

Uniformity of dosage units ⟨905⟩: meet the requirements.

Assay—

Mobile phase—Prepare a suitable degassed and filtered mixture of water and acetonitrile (7:3) that contains 5 mL of phosphoric acid in each 1000 mL of solution.

0.01 M Methanolic sodium hydroxide—Dissolve 400 mg of sodium hydroxide in 500 mL of water. Cool, add 500 mL of methanol, and mix.

Standard preparation—Dissolve accurately weighed quantities of USP Levothyroxine RS and USP Liothyronine RS in *0.01 M Methanolic sodium hydroxide*, and dilute quantitatively and stepwise with *0.01 M Methanolic sodium hydroxide* to obtain a solution having known concentrations of about 10 µg of liothyronine per mL and 40 µg of levothyroxine per mL, respectively.

Assay preparation—Weigh and finely powder not less than 20 Liotrix Tablets. Transfer an accurately weighed portion of the powder, equivalent to about 100 µg of levothyroxine sodium, to a 25-mL volumetric flask, add about 15 mL of *0.01 M Methanolic sodium hydroxide*, and place in an ultrasonic bath for 1 minute. Shake the mixture for 5 minutes, dilute with *0.01 M Methanolic sodium hydroxide* to volume, mix, and filter.

Chromatographic system (see *Chromatography* ⟨621⟩)—The liquid chromatograph is equipped with a 225-nm detector and a 25- to 30-cm column that contains packing L10. The flow rate is about 1 mL per minute. Chromatograph five replicate injections of the *Standard preparation*, and record the peak responses as directed under *Procedure:* the relative standard deviation is not more than 2.0%, and the tailing factor is not more than 1.8 for both major peaks. The resolution, *R*, between the peaks for levothyroxine and liothyronine is not less than 4.0.

Procedure—Separately inject equal volumes (about 50 µL) of the *Standard preparation* and the *Assay preparation* into the chromatograph, record the chromatograms, and measure the responses for the major peaks. The relative retention times are about 1.0 for levothyroxine and 0.5 for liothyronine. Calculate the quantity, in mg, of levothyroxine sodium ($C_{15}H_{10}I_4NNaO_4$) in the portion of Tablets taken by the formula:

$$(798.86/776.87)(0.025C)(r_U/r_S),$$

in which 798.86 and 776.87 are the molecular weights of levothyroxine sodium and levothyroxine, respectively, C is the concentration, in µg per mL, of USP Levothyroxine RS in the *Standard preparation*, and r_U and r_S are the peak responses obtained for levothyroxine from the *Assay preparation* and the *Standard preparation*, respectively. Calculate the quantity, in mg, of liothyronine sodium ($C_{15}H_{11}I_3NNaO_4$) in the portion of Tablets taken by the formula:

$$(672.96/650.98)(0.025C)(r_U/r_S),$$

in which 672.96 and 650.98 are the molecular weights of liothyronine sodium and liothyronine, respectively, C is the concentration, in µg per mL, of USP Liothyronine RS in the *Standard preparation*, and r_U and r_S are the liothyronine peak responses obtained from the *Assay preparation* and the *Standard preparation*, respectively.

Liquefied Phenol—*see* Phenol, Liquefied
Liquid Glucose—*see* Glucose, Liquid NF

Lithium Carbonate

Li_2CO_3 73.89
Carbonic acid, dilithium salt.
Dilithium carbonate [*554-13-2*].

» Lithium Carbonate contains not less than 99.0 percent of Li_2CO_3, calculated on the dried basis.

Packaging and storage—Preserve in well-closed containers.

Identification—

A: It effervesces upon the addition of an acid, yielding a colorless gas which, when passed into calcium hydroxide TS, immediately causes a white precipitate to form.

B: When moistened with hydrochloric acid, it imparts an intense crimson color to a nonluminous flame.

Reaction—A saturated solution is alkaline to litmus.

Loss on drying ⟨731⟩—Dry it at 200° for 4 hours: it loses not more than 1.0% of its weight.

Insoluble substances—Transfer 10 g to a 250-mL beaker, add 50 mL of water, then add slowly 50 mL of 6 *N* hydrochloric acid. Cover with a watch glass, and boil the solution for 1 hour. Filter the solution through a dried, tared filtering crucible fitted with a glass-fiber filter disk, using suction. Wash the filter with hot water until the last washing is free from chloride when tested with silver nitrate TS. Dry the crucible in an oven at 110° for 1 hour: the weight of the residue is not more than 0.02% of the weight of Lithium Carbonate taken.

Chloride—To 500 mg of it add 1.2 mL of nitric acid, dilute with water to 50 mL, and add 1 mL of silver nitrate TS. Prepare a standard solution of equal volume containing 1.2 mL of nitric acid, 0.50 mL of 0.020 *N* hydrochloric acid, and 1.0 mL of silver nitrate TS. The turbidity produced in the test solution is not greater than that produced in the standard solution (0.07%).

Sulfate—Dissolve 1.0 g in 10 mL of 3 *N* hydrochloric acid, dilute with water to 40 mL, and add 1 mL of barium chloride TS. Prepare a standard solution of equal volume containing 1.0 mL of 0.020 *N* sulfuric acid, 1 mL of 3 *N* hydrochloric acid, and 1 mL of barium chloride TS. The turbidity produced in the test solution, after 3 minutes, is not greater than that produced in the standard solution (0.1%).

Aluminum and iron—Dissolve 500 mg in 10 mL of water by the dropwise addition, with agitation, of hydrochloric acid. Boil the solution, then cool it, and to 5 mL of the solution add 6 *N* ammonium hydroxide until the reaction is alkaline: no turbidity or precipitate is observed.

Arsenic, *Method I* ⟨211⟩: 8 ppm.

Calcium—Suspend 5.0 g in 50 mL of water, and add a slight excess of 3 *N* hydrochloric acid. Boil the clear solution to expel carbon dioxide, add 5 mL of ammonium oxalate TS, render alkaline with 6 *N* ammonium hydroxide, and allow to stand for 4 hours. Filter through a filtering crucible, and wash with warm water until the last washing yields no turbidity with calcium chloride TS. Place the crucible in a beaker, cover it with water, add 3 mL of sulfuric acid, heat to 70°, and titrate with 0.10 *N* potassium permanganate to a pale pink color that persists for 30 seconds. Not more than 3.76 mL of 0.10 *N* potassium permanganate is consumed (0.15%).

Sodium—

Standard preparation—Dissolve 1.271 g of sodium chloride, previously dried at 130° to constant weight, in water in a 1000-mL volumetric flask, dilute with water to volume, and mix. Each mL contains 500 µg of Na.

Stock solution—Suspend 20.0 g of Lithium Carbonate in 100 mL of water, cautiously add 50.0 mL of hydrochloric acid, transfer to a 200-mL volumetric flask, dilute with water to volume, and mix.

Test preparation—Pipet 5 mL of *Stock solution* into a 100-mL volumetric flask, add water to volume, and mix.

Control solution—Pipet 5 mL of *Stock solution* and 1 mL of *Standard preparation* into a 100-mL volumetric flask, add water to volume, and mix.

Procedure—Set a suitable flame photometer for maximum emission at about 589 nm, using the *Control solution.* Measure the emission intensities of the *Test preparation* at 580 nm and 589 nm. The difference between the intensities observed at 580 nm and 589 nm for the *Test preparation* does not exceed the difference between the intensities observed at 589 nm for the *Test preparation* and the *Control solution*, respectively. The sodium limit is 0.1%.

Heavy metals ⟨231⟩—Dissolve 1 g in 10 mL of 3 *N* hydrochloric acid, and dilute with water to 25 mL: the limit is 0.002%.

Assay—Dissolve about 1 g of Lithium Carbonate, accurately weighed, in 50.0 mL of 1 *N* sulfuric acid VS, add methyl orange TS, and titrate the excess acid with 1 *N* sodium hydroxide VS. Perform a blank determination (see *Residual Titrations* under *Titrimetry* ⟨541⟩). Each mL of 1 *N* sulfuric acid is equivalent to 36.95 mg of Li_2CO_3.

Lithium Carbonate Capsules

» Lithium Carbonate Capsules contain not less than 95.0 percent and not more than 105.0 percent of the labeled amount of Li_2CO_3.

Packaging and storage—Preserve in well-closed containers.

Reference standard—*USP Lithium Carbonate Reference Standard*—Dry at 200° for 4 hours before using.

Identification—A portion of the Capsule contents responds to the *Identification tests* under *Lithium Carbonate.*

Dissolution ⟨711⟩—

Medium: water; 900 mL.

Apparatus 1: 100 rpm.

Time: 30 minutes.

Procedure—Determine the amount of Li_2CO_3 dissolved in filtered portions of the solution under test with a suitable flame photometer, as directed in the *Assay.*

Tolerances—Not less than 60% (*Q*) of the labeled amount of Li_2CO_3 is dissolved in 30 minutes.

Uniformity of dosage units ⟨905⟩: meet the requirements.

Assay—

Standard preparation—Transfer to a 100-mL volumetric flask about 30 mg of USP Lithium Carbonate RS, accurately weighed.

Add about 20 mL of water and 0.5 mL of hydrochloric acid, shake until dissolved, dilute with water to volume, and mix. Pipet 20 mL of the resulting solution into a 1000-mL volumetric flask, add about 800 mL of water and 20 mL of a suitable surfactant solution, appropriately diluted, dilute with water to volume, and mix.

Assay preparation—Empty as completely as possible the contents of not less than 20 Lithium Carbonate Capsules. Weigh accurately a portion of the powder, equivalent to about 600 mg of lithium carbonate, and transfer to a 1000-mL volumetric flask. Add 40 mL of water and 5 mL of hydrochloric acid, shake until the sample is well disintegrated, dilute with water to volume, and mix. Pipet 10 mL of the resulting solution into a 1000-mL volumetric flask, add about 800 mL of water and 20 mL of the surfactant solution, dilute with water to volume, and mix.

Procedure—Employ a suitable flame photometer, and adjust the instrument with the surfactant solution. Aspirate into the photometer the *Standard preparation* and the *Assay preparation*, and measure the emission at about 671 nm. Calculate the quantity, in mg, of Li_2CO_3 in the portion of Capsules taken by the formula:

$$100C(A/S),$$

in which *C* is the concentration, in µg per mL, of USP Lithium Carbonate RS in the *Standard preparation*, and *A* and *S* are the photometer readings of the *Assay preparation* and the *Standard preparation*, respectively.

Lithium Carbonate Tablets

» Lithium Carbonate Tablets contain not less than 95.0 percent and not more than 105.0 percent of the labeled amount of Li_2CO_3.

Packaging and storage—Preserve in well-closed containers.

Reference standard—*USP Lithium Carbonate Reference Standard*—Dry at 200° for 4 hours before using.

Identification—A portion of the powdered Tablets responds to the *Identification tests* under *Lithium Carbonate.*

Dissolution ⟨711⟩—

Medium: water; 900 mL.

Apparatus 1: 100 rpm.

Time: 30 minutes.

Procedure—Determine the amount in solution by flame photometry, the test specimen being prepared by diluting the 900 mL of the solution under test to 1000 mL and a 20.0-mL filtered aliquot being transferred to a 1000-mL volumetric flask and treated with 500 mL of water, 1 drop of hydrochloric acid, and 20 mL of a suitable surfactant solution, appropriately diluted. Add water to volume, mix, and read on a suitable photometer together with the *Standard preparation* as directed in the *Assay* under *Lithium Carbonate Capsules.*

Tolerances—Not less than 60% (*Q*) of the labeled amount of Li_2CO_3 is dissolved in 30 minutes.

Uniformity of dosage units ⟨905⟩: meet the requirements.

Assay—

Standard preparation—Prepare as directed in the *Assay* under *Lithium Carbonate Capsules.*

Assay preparation—Weigh and finely powder not less than 20 Lithium Carbonate Tablets. Weigh accurately a portion of the powder, equivalent to about 600 mg of lithium carbonate, and transfer to a 1000-mL volumetric flask. Add 400 mL of water and 5 mL of hydrochloric acid, shake until the solid is well disintegrated, dilute with water to volume, and mix. Pipet 10 mL of the resulting solution into a 1000-mL volumetric flask, add about 800 mL of water and 20 mL of the surfactant solution, dilute with water to volume, and mix.

Procedure—Proceed as directed for *Procedure* in the *Assay* under *Lithium Carbonate Capsules.* Calculate the quantity, in mg, of Li_2CO_3 in the portion of Tablets taken by the formula:

$$100C(A/S),$$

in which *C* is the concentration, in µg per mL, of USP Lithium Carbonate RS in the *Standard preparation*, and *A* and *S* are the photometer readings of the *Assay preparation* and the *Standard preparation*, respectively.

Lithium Citrate

$CH_2(COOLi)C(OH)(COOLi)CH_2COOLi . 4H_2O$

$C_6H_5Li_3O_7 . 4H_2O$ 282.00

1,2,3-Propanetricarboxylic acid, 2-hydroxy-trilithium salt tetrahydrate.

Trilithium citrate tetrahydrate [*6080-58-6*].

Anhydrous 209.92 [*919-16-4*].

» Lithium Citrate contains not less than 98.0 percent and not more than 102.0 percent of $C_6H_5Li_3O_7$, calculated on the anhydrous basis.

Packaging and storage—Preserve in tight containers.

Reference standard—*USP Lithium Carbonate Reference Standard*—Dry at 200° for 4 hours before using.

Identification—

A: When moistened with hydrochloric acid, it imparts an intense crimson color to a nonluminous flame.

B: It responds to the test for *Citrate* ⟨191⟩.

pH ⟨791⟩: between 7.0 and 10.0, in a solution (1 in 20).

Water, *Method III* ⟨921⟩—Dry it at 150° for 3 hours: it loses between 24.0% and 28.0% of its weight.

Carbonate—Add about 0.5 g to 5 mL of 6 *N* acetic acid: not more than a slight effervescence is produced.

Arsenic, *Method I* ⟨211⟩: 5 ppm.

Heavy metals ⟨231⟩—Dissolve 2.0 g in 2 mL of 0.1 *N* hydrochloric acid, and dilute with water to 25 mL: the limit is 0.001%.

Assay—

Standard preparation—Transfer to a 100-mL volumetric flask about 30 mg of USP Lithium Carbonate RS, accurately weighed. Add about 20 mL of water and 0.5 mL of hydrochloric acid, shake until dissolved, dilute with water to volume, and mix. Pipet 20 mL of the resulting solution into a 1000-mL volumetric flask, add about 800 mL of water and 20 mL of a suitable surfactant solution, appropriately diluted, dilute with water to volume, and mix.

Assay preparation—Transfer about 800 mg of Lithium Citrate, accurately weighed, to a 1000-mL volumetric flask. Dissolve in water, add 0.5 mL of hydrochloric acid, dilute with water to volume, and mix. Pipet 20 mL of this solution into a 1000-mL volumetric flask, add about 800 mL of water and 20 mL of the surfactant solution, dilute with water to volume, and mix.

Procedure—Employ a suitable flame photometer, and adjust the instrument with the surfactant solution. Aspirate into the instrument the *Standard preparation* and the *Assay preparation*, and measure the emission at about 671 nm. Calculate the quantity, in mg, of $C_6H_5Li_3O_7$ in the portion of Lithium Citrate taken by the formula:

$$(419.84/221.67)(50C)(A/S),$$

in which 419.84 is twice the molecular weight of anhydrous lithium citrate, 221.67 is three times the molecular weight of lithium carbonate, *C* is the concentration, in µg per mL, of USP Lithium Carbonate RS in the *Standard preparation*, and *A* and *S* are the photometer readings of the *Assay preparation* and the *Standard preparation*, respectively.

Lithium Citrate Syrup

» Lithium Citrate Syrup is prepared from Lithium Citrate or Lithium Hydroxide to which an excess of Citric Acid has been added. It contains not less than 90.0 percent and not more than 110.0 percent of the labeled amount of lithium (Li).

Packaging and storage—Preserve in tight containers.

Reference standard—*USP Lithium Carbonate Reference Standard*—Dry at 200° for 4 hours before using.

Identification—

A: When diluted with an equal volume of 3 *N* hydrochloric acid, it imparts an intense crimson color to a nonluminous flame.

B: It responds to the test for *Citrate* ⟨191⟩.

pH ⟨791⟩: between 4.0 and 5.0.

Assay—

Standard preparation—Prepare as directed in the *Assay* under *Lithium Citrate.* Determine its pH.

Assay preparation—Transfer an accurately measured volume of Lithium Citrate Syrup, equivalent to about 60 mg of lithium, to a 1000-mL volumetric flask, dilute with water to volume, and mix. Pipet 20 mL of the resulting solution into a 1000-mL volumetric flask, add about 950 mL of water, 2 mL of 1 *N* hydrochloric acid, and 20 mL of a surfactant solution, and mix. Adjust with 1 *N* hydrochloric acid or 1 *N* sodium hydroxide to the same pH (±0.1 pH unit) as that of the *Standard preparation,* dilute with water to volume, and mix.

Procedure—Employ a suitable flame photometer, and adjust the instrument with the surfactant solution. Aspirate into the photometer the *Standard preparation* and the *Assay preparation,* and measure the emission at about 671 nm. Calculate the quantity, in mg, of lithium in each mL of the Syrup taken by the formula:

$$(13.88/73.89)(50C/V)(A/S),$$

in which 13.88 is twice the atomic weight of lithium, 73.89 is the molecular weight of lithium carbonate, C is the concentration, in µg per mL, of USP Lithium Carbonate RS in the *Standard preparation,* V is the volume, in mL, of Syrup taken, and A and S are the photometer readings of the *Assay preparation* and the *Standard preparation,* respectively.

Lithium Hydroxide

$LiOH.H_2O$ 41.97
Lithium hydroxide monohydrate.
Lithium hydroxide monohydrate [*1310-66-3*].
Anhydrous 23.95 [*1310-65-2*].

» Lithium Hydroxide contains not less than 98.0 percent and not more than 102.0 percent of LiOH, calculated on the anhydrous basis.

Caution—Exercise great care in handling Lithium Hydroxide, as it rapidly destroys tissues.

Packaging and storage—Preserve in tight containers.

Reference standard—*USP Lithium Carbonate Reference Standard*—Dry at 200° for 4 hours before using.

Identification—When moistened with hydrochloric acid, it imparts an intense crimson color to a nonluminous flame.

Water, *Method III* ⟨921⟩—Dry it at 135° at a pressure not exceeding 5 mm of mercury for 1 hour: it loses between 41.0% and 43.5% of its weight.

Carbonate—[NOTE—While pipeting and during the subsequent titrations, keep the contents of the flasks blanketed with a stream of carbon dioxide–free air.] To the flask containing the completed *Final titration* obtained in the *Assay* add 1 drop of methyl orange TS, and titrate with 0.1 *N* hydrochloric acid VS until a persistent orange color is produced and no undissolved barium carbonate remains. Perform a blank titration to determine the volume of 0.1 *N* hydrochloric acid consumed in going from the phenolphthalein end-point to the methyl orange end-point. To 100 mL of carbon dioxide–free water in a 250-mL conical flask add 3 drops of the *Assay preparation,* 20 mL of 1 *N* barium chloride, and 3 drops of phenolphthalein TS, and allow to stand for 2 minutes. Titrate this solution with 0.1 *N* hydrochloric acid. At the discharge of the pink color of the indicator, add 1 drop of methyl orange TS, and titrate with 0.1 *N* hydrochloric acid VS until a persistent orange color is produced. The titration shows no more CO_2 than corresponds to 1.5 mL of 0.10 *N* hydrochloric acid (0.7%).

Sulfate ⟨221⟩—A 2.0-g portion shows no more sulfate than corresponds to 1.0 mL of 0.020 *N* sulfuric acid (0.05%).

Arsenic, *Method I* ⟨211⟩: 5 ppm.

Calcium—Dissolve 3.33 g in 50 mL of 3 *N* hydrochloric acid, and proceed as directed in the test for *Calcium* under *Lithium Carbonate,* beginning with "Boil the clear solution." Not more than 3.34 mL of 0.10 *N* potassium permanganate is consumed (0.20%).

Heavy metals, *Method I* ⟨231⟩—Dissolve 1.0 g in 15 mL of 3 *N* hydrochloric acid, and dilute with water to 25 mL: the limit is 0.002%.

Lithium content—

Standard preparation—Prepare as directed in the *Assay* under *Lithium Citrate.* Determine its pH.

Test preparation—Transfer about 400 mg of Lithium Hydroxide, accurately weighed, to a 1000-mL volumetric flask. Dissolve in water, dilute with water to volume, and mix. Pipet 20 mL of this solution into a 1000-mL volumetric flask, add about 950 mL of water, 2 mL of 1 *N* hydrochloric acid, and 20 mL of a surfactant solution, and mix. Adjust with 1 *N* hydrochloric acid or 1 *N* sodium hydroxide to the same pH (±0.1 pH unit) as that of the *Standard preparation,* dilute with water to volume, and mix.

Procedure—Employ a suitable flame photometer, and adjust the instrument with the surfactant solution. Aspirate into the instrument the *Standard preparation* and the *Assay preparation,* and measure the emission at about 671 nm. Calculate the quantity, in mg, of Li in the portion of Lithium Hydroxide taken by the formula:

$$(13.88/73.89)(50C)(A/S),$$

in which 13.88 is twice the atomic weight of lithium, 73.89 is the molecular weight of lithium carbonate, C is the concentration, in µg per mL, of USP Lithium Carbonate RS in the *Standard preparation,* and A and S are the photometric readings of the *Assay preparation* and the *Standard preparation,* respectively. The lithium content is between 28.4% and 29.1%, calculated on the anhydrous basis.

Assay—

Assay preparation—Transfer an accurately weighed quantity of Lithium Hydroxide, accurately weighed, equivalent to about 10 g of anhydrous lithium hydroxide, to a 1000-mL volumetric flask, dissolve in carbon dioxide–free water, dilute with carbon dioxide–free water to volume, and mix.

Preliminary titration—Pipet 50 mL of the *Assay preparation* into a 250-mL conical flask. Start the titration by adding 35 mL of 0.5 *N* hydrochloric acid VS with continuous vigorous stirring. Add 20 mL of 1 *N* barium chloride and 3 drops of phenolphthalein TS, mix, and allow to stand for 2 minutes. Continue the titration with 0.5 *N* hydrochloric acid VS, and at the discharge of the pink color of the indicator, record the volume of acid solution consumed.

Final titration—Pipet 50 mL of the *Assay preparation* into a 250-mL conical flask. [NOTE—While pipeting and during the subsequent titrations, keep the contents of the flask blanketed with a stream of carbon dioxide–free air.] Start the titration by adding with continuous vigorous swirling a volume of 0.5 *N* hydrochloric acid VS that is 0.50 mL less than that consumed in the preliminary titration. Add 20 mL of 1 *N* barium chloride and 3 drops of phenolphthalein TS, mix, and allow to stand for

2 minutes. Rinse the sides of the flask with carbon dioxide-free water, and continue the titration with 0.1 *N* hydrochloric acid VS. At the discharge of the pink color of the indicator, record the volume of acid solution consumed. Each mL of 0.5 *N* and 0.1 *N* hydrochloric acid is equivalent to 11.975 and 2.395 mg of total alkali, respectively, calculated as LiOH.

Loperamide Hydrochloride

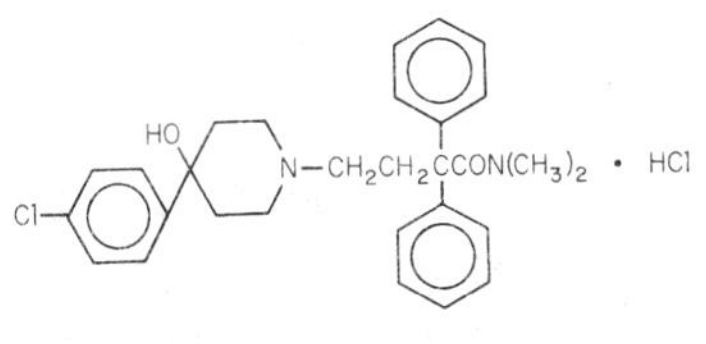

$C_{29}H_{33}ClN_2O_2.HCl$ 513.51

1-Piperidinebutanamide, 4-(4-chlorophenyl)-4-hydroxy-*N,N*-dimethyl-*α,α*-diphenyl-, monohydrochloride.

4-(*p*-Chlorophenyl)-4-hydroxy-*N,N*-dimethyl-*α,α*-diphenyl-1-piperidinebutyramide monohydrochloride [*34552-83-5*].

» Loperamide Hydrochloride contains not less than 98.0 percent and not more than 102.0 percent of $C_{29}H_{33}ClN_2O_2.HCl$, calculated on the dried basis.

Packaging and storage—Preserve in well-closed containers.

Reference standard—*USP Loperamide Hydrochloride Reference Standard*—Dry in vacuum at 80° for 4 hours before using.

Identification—

A: The infrared absorption spectrum of a potassium bromide dispersion of it, previously dried, exhibits maxima only at the same wavelengths as that of a similar preparation of USP Loperamide Hydrochloride RS.

B: Transfer about 40 mg, accurately weighed, to a 100-mL volumetric flask, dissolve in about 50 mL of isopropyl alcohol, add 10 mL of 0.1 *N* hydrochloric acid, dilute with isopropyl alcohol to volume, and mix: the ultraviolet absorption spectrum between 250 nm and 300 nm of this solution exhibits maxima and minima at the same wavelengths as that of a similar solution of USP Loperamide Hydrochloride RS, concomitantly measured.

Loss on drying ⟨731⟩—Dry it in vacuum at 80° for 4 hours: it loses not more than 0.5% of its weight.

Residue on ignition ⟨281⟩: not more than 0.2%.

Chloride content—Using about 13 mg, accurately weighed, proceed as directed under *Oxygen Flask Combustion* ⟨471⟩, using a mixture of 10 mL of 0.02 *N* sodium hydroxide and 2 drops of 30 percent hydrogen peroxide as the absorbing liquid. When combustion is complete and the combustion gases absorbed, rinse the stopper, sample holder, and inner walls of the flask with 50 mL of isopropyl alcohol. Add 4 mL of 0.1 *N* nitric acid, and titrate with 0.01 *N* mercuric nitrate VS, using diphenylcarbazone TS as the indicator. Each mL of 0.01 *N* mercuric nitrate is equivalent to 0.3545 mg of chlorine: between 13.52% and 14.20% is found.

Heavy metals, *Method II* ⟨231⟩: 0.002%.

Chromatographic purity—Prepare a test solution in chloroform containing 10 mg per mL. On a thin-layer chromatographic plate (see *Chromatography* ⟨621⟩), coated with a 0.25-mm layer of chromatographic silica gel mixture, apply 10 µL of this solution and 10 µL of a Standard solution of USP Loperamide Hydrochloride RS in chloroform containing 10 mg per mL. Allow the spots to dry, and develop the chromatogram in a solvent system consisting of a mixture of chloroform, methanol, and formic acid (85:10:5) until the solvent front has moved about three-fourths of the length of the plate. Remove the plate from the developing chamber, mark the solvent front, and allow the plate to air-dry. Locate the spots on the plate by exposing it to fumes of iodine: the spot obtained from the test solution corresponds in R_f value, color, and intensity to that obtained from the Standard solution, and no secondary spots are observed.

Assay—

Neutralized acetic acid—Dissolve 10 mg of *α*-naphtholbenzein in 100 mL of glacial acetic acid, and titrate with 0.1 *N* perchloric acid to a green end-point, disregarding the amount of titrant consumed.

Procedure—Dissolve about 375 mg, accurately weighed, of Loperamide Hydrochloride in 25 mL of *Neutralized acetic acid.* Add 10 mL of mercuric acetate solution (prepared by dissolving 1 g of mercuric acetate in 33 mL of *Neutralized acetic acid*) and titrate with 0.1 *N* perchloric acid VS to the original green color of the *Neutralized acetic acid.* Each mL of 0.1 *N* perchloric acid is equivalent to 51.35 mg of $C_{29}H_{33}ClN_2O_2.HCl$.

Loperamide Hydrochloride Capsules

» Loperamide Hydrochloride Capsules contain not less than 90.0 percent and not more than 110.0 percent of the labeled amount of $C_{29}H_{33}ClN_2O_2.HCl$.

Packaging and storage—Preserve in well-closed containers.

Reference standard—*USP Loperamide Hydrochloride Reference Standard*—Dry in vacuum at 80° for 4 hours before using.

Identification—

A:

Buffer solution—To 30 mL of water, add 10 mL of 1 *N* acetic acid, and adjust with 1 *N* sodium hydroxide to a pH of 4.7. Dilute with water to 50 mL, and mix.

Modified Dragendorff spray reagent—*Solution A*—Dissolve 1.7 g of bismuth subnitrate and 20 g of tartaric acid in 80 mL of water. *Solution B*—Dissolve 16 g of potassium iodide in 40 mL of water. Mix equal parts of *Solutions A* and *B*, and mix 5 mL of the mixture with 50 mL of tartaric acid solution (1 in 5).

Procedure—Transfer a quantity of the contents of the Capsules, equivalent to about 10 mg of loperamide hydrochloride, to a 37-mL stoppered vial, add 10 mL of methanol, shake for 5 minutes, and filter. On a thin-layer chromatographic plate (see *Chromatography* ⟨621⟩), coated with a 0.25-mm layer of chromatographic silica gel mixture, apply 10 µL of this test solution and 1 µL of a Standard solution of USP Loperamide Hydrochloride RS in methanol containing 10 mg per mL. Allow the spots to dry, and develop the chromatogram in a solvent system consisting of a mixture of methanol and *Buffer solution* (95:5) until the solvent front has moved about three-fourths of the length of the plate. Remove the plate from the developing chamber, mark the solvent front, and dry the plate at 40° to 60°. Visualize the spots by spraying with *Modified Dragendorff spray reagent:* the R_f value of the spot obtained from the test solution corresponds to that obtained from the *Standard solution.*

B: The retention time of the major peak in the chromatogram of the *Assay preparation* corresponds to that of the *Standard preparation*, as obtained in the *Assay.*

Dissolution ⟨711⟩—

Medium: pH 4.7 acetate buffer, prepared by mixing 200 mL of 1 *N* acetic acid with 600 mL of water, adjusting with 1 *N* sodium hydroxide to a pH of 4.70 ± 0.05, diluting with water to 1000 mL, and mixing; 500 mL.

Apparatus 1: 100 rpm.

Time: 30 minutes.

Determine the amount of loperamide hydrochloride dissolved using the following method.

Mobile phase and *Chromatographic system*—Proceed as directed in the *Assay.*

Procedure—Inject a volume (about 50 µL) of a filtered portion of the solution under test into the chromatograph, record the chromatogram, and measure the response for the major peak. Calculate the quantity of $C_{29}H_{33}ClN_2O_2.HCl$ dissolved in comparison with a Standard solution having a known concentration of USP Loperamide Hydrochloride RS in the same medium and similarly chromatographed.

Tolerances—Not less than 70% (*Q*) of the labeled amount of $C_{29}H_{33}ClN_2O_2.HCl$ is dissolved in 30 minutes.

Uniformity of dosage units ⟨905⟩: meet the requirements.

Assay—

Mobile phase—Transfer 500 mL of acetonitrile to a 1000-mL volumetric flask. Dilute with water to volume, add 20 drops of Phosphoric acid, mix, and filter. Make adjustments if necessary (see *System Suitability* under *Chromatography* ⟨621⟩).

Standard preparation—Dissolve an accurately weighed quantity of USP Loperamide Hydrochloride RS in a mixture of acetonitrile and water (1:1) to obtain a solution having a known concentration of about 0.2 mg per mL. Transfer 5.0 mL of this solution to a 100-mL volumetric flask, dilute with a mixture of acetonitrile and water (1:1) to volume, and mix to obtain a solution having a known concentration of about 10 µg per mL.

Assay preparation—Transfer, as completely as possible, the contents of not less than 20 Loperamide Hydrochloride Capsules to a suitable tared container, and determine the average weight per capsule. Mix the combined contents, and transfer an accurately weighed portion of the powder, equivalent to about 20 mg of loperamide hydrochloride, to a 100-mL volumetric flask. Add about 70 mL of a mixture of acetonitrile and water (1:1), and sonicate for 15 minutes. Dilute with a mixture of acetonitrile and water (1:1) to volume, mix, and filter. Transfer 5.0 mL of this solution to a 100-mL volumetric flask, dilute with a mixture of acetonitrile and water (1:1) to volume, and mix.

Chromatographic system (see *Chromatography* ⟨621⟩)—The liquid chromatograph is equipped with a 220-nm detector and a 4-mm × 25-cm column that contains 10-µm packing L10. The flow rate is about 2 mL per minute. Chromatograph the *Standard preparation*, and record the peak responses as directed under *Procedure:* the column efficiency, *N*, determined from the analyte peak is not less than 1900 theoretical plates, the capacity factor, K^1, is not less than 3.5, and the relative standard deviation for replicate injections is not more than 2.0%.

Procedure—Separately inject equal volumes (about 50 µL) of the *Standard preparation* and the *Assay preparation* into the chromatograph, record the chromatograms, and measure the responses for the major peaks. Calculate the quantity, in mg, of $C_{29}H_{33}ClN_2O_2 \cdot HCl$ in the portion of Capsules taken by the formula:

$$2000C(r_U/r_S),$$

in which *C* is the concentration, in mg per mL, of USP Loperamide Hydrochloride RS in the *Standard preparation*, and r_U and r_S are the peak responses obtained from the *Assay preparation* and the *Standard preparation*, respectively.

Lorazepam

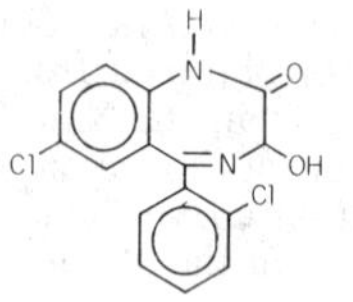

$C_{15}H_{10}Cl_2N_2O_2$ 321.16

2*H*-1,4-Benzodiazepin-2-one, 7-chloro-5-(2-chlorophenyl)-1,3-dihydro-3-hydroxy-.

7-Chloro-5-(*o*-chlorophenyl)-1,3-dihydro-3-hydroxy-2*H*-1,4-benzodiazepin-2-one [*846-49-1*].

» Lorazepam contains not less than 98.0 percent and not more than 102.0 percent of $C_{15}H_{10}Cl_2N_2O_2$, calculated on the dried basis.

Packaging and storage—Preserve in tight, light-resistant containers.

Reference standards—*USP Lorazepam Reference Standard*—Dry in vacuum at 105° for 3 hours before using. *USP 7-Chloro-5-(o-chlorophenyl)-1, 3-dihydro-3-acetoxy-2H-1,4-benzodiazepin-2-one Reference Standard*—Keep container tightly closed and protected from light. Do not dry before using. *USP 2-Amino-2′,5-dichlorobenzophenone Reference Standard*—Keep container tightly closed and protected from light. Do not dry before using.

Identification—

A: The infrared absorption spectrum of a potassium bromide dispersion of it, previously dried in vacuum at 105° for 3 hours, exhibits maxima only at the same wavelengths as that of a similar preparation of USP Lorazepam RS.

B: The R_f value of the principal spot observed in the chromatogram of the *Test preparation* obtained as directed in *Related compounds test A* corresponds to that obtained from the *Identification preparation.*

Loss on drying ⟨731⟩—Dry it in vacuum at 105° for 3 hours: it loses not more than 0.5% of its weight.

Residue on ignition ⟨281⟩: not more than 0.3%.

Heavy metals, *Method II* ⟨231⟩: not more than 0.002%.

Related compounds—

A: Dissolve Lorazepam in chloroform to obtain a *Test preparation* containing 2 mg per mL. Dissolve USP Lorazepam RS in chloroform to obtain an *Identification preparation* having a known concentration of 2 mg per mL. Dissolve USP 7-Chloro-5-(*o*-chlorophenyl)-1, 3-dihydro-3-acetoxy-2*H*-1,4-benzodiazepin-2-one RS in chloroform to obtain a *Standard preparation* having a known concentration of 20 µg per mL. Dilute portions of this *Standard preparation* quantitatively with chloroform to obtain solutions having concentrations of 10 µg per mL (*Diluted standard preparation A*) and 4 µg per mL (*Diluted standard preparation B*), respectively. Within 30 minutes after preparation, apply separately 50 µL of the *Test preparation*, the *Identification preparation*, the *Standard preparation*, the *Diluted standard preparation A*, and the *Diluted standard preparation B* to a suitable thin-layer chromatographic plate (see *Chromatography* ⟨621⟩), coated with a 0.25-mm layer of chromatographic silica gel mixture and previously washed with a mixture of chloroform, ethyl acetate, and methanol (2:1:1) and dried in air. Allow the spots to dry, and develop the chromatograms in a solvent system consisting of a mixture of chloroform, dioxane, and glacial acetic acid (91:5:4) until the solvent front has moved to within 2 cm to 3 cm from the top of the plate. Remove the plate from the developing chamber, mark the solvent front, and allow to air-dry for about 30 minutes. Examine the plate under short-wavelength ultraviolet light. Compare the intensities of any secondary spots observed in the chromatogram of the *Test preparation* with those of the principal spots in the chromatograms of the *Standard preparation* and the *Diluted standard preparations:* the sum of the intensities of all secondary spots obtained from the *Test preparation* corresponds to not more than 1.0%.

B: Transfer 50.0 mg of Lorazepam to a 10-mL conical flask, add 2.5 mL of acetone, and shake. Allow any undissolved particles to settle, and use the supernatant solution as the *Test preparation.* Dissolve USP 2-Amino-2′,5-dichlorobenzophenone RS in acetone to obtain a *Standard preparation* having a known concentration of 10 µg per mL. Apply separately 50 µL of the *Test preparation* and 10 µL of the *Standard preparation* to a suitable thin-layer chromatographic plate (see test *A* under *Related compounds*). Allow the spots to dry, and develop the chromatograms in a solvent system consisting of a mixture of chloroform, dioxane, and glacial acetic acid (91:5:4) until the solvent front has moved not less than 10 cm from the origin. Remove the plate from the developing chamber, mark the solvent front, and allow the solvent to evaporate. Lightly spray the plate with 2 *N* sulfuric acid, dry at 105° for 15 minutes, and spray successively with sodium nitrite solution (1 in 1000), ammonium sulfamate solution (1 in 200), and *N*-(1-naphthyl)ethylenediamine dihydrochloride solution (1 in 1000), drying the plate with a current of air after each spraying. Observe the plate under visible light: the spot produced by the *Test preparation* is not greater in size or intensity than the principal spot produced at the corresponding R_f value by the *Standard preparation*, corresponding to not more than 0.01% of 2-amino-2′,5-dichlorobenzophenone.

Assay—Dissolve about 400 mg of Lorazepam, accurately weighed, in 50 mL of N,N-dimethylformamide. Titrate the solution with 0.1 *N* tetrabutylammonium hydroxide VS, taking precautions against the absorption of atmospheric carbon dioxide, determining the end-point potentiometrically, using a glass electrode and a calomel electrode containing a saturated solution of potassium chloride in methanol (see *Titrimetry* ⟨541⟩). Perform a blank determination, and make any necessary correction. Each mL of

0.1 *N* tetrabutylammonium hydroxide is equivalent to 32.12 mg of $C_{15}H_{10}Cl_2N_2O_2$.

Lorazepam Injection

» Lorazepam Injection is a sterile solution of Lorazepam in a suitable medium. It contains not less than 90.0 percent and not more than 110.0 percent of the labeled amount of $C_{15}H_{10}Cl_2N_2O_2$.

Packaging and storage—Preserve in a single-dose or in multiple-dose containers, preferably of Type I glass, protected from light.

Reference standards—*USP Loraxepam Reference Standard*—Dry in vacuum at 105° for 3 hours before using. *USP 2-Amino-2′,5-dichlorobenxophenone Reference Standard*—Keep container tightly closed and protected from light. Do not dry before using. *USP 6-Chloro-4-(o-chlorophenyl)-2-quinaxolinecarboxaldehyde Reference Standard*—Do not dry before using. *USP 6-Chloro-4-(o-chlorophenyl)-2-quinaxolinecarboxylic acid Reference Standard*—Do not dry before using.

Identification—

A: The retention time of the major peak in the chromatogram of the *Assay preparation* corresponds to that in the chromatogram of the *Standard preparation* obtained in the *Assay*.

B: Dissolve USP Lorazepam RS in alcohol to obtain a solution having a concentration of 1 mg per mL. Transfer 10 mL of this solution to a suitable container. Transfer a volume of Injection, equivalent to about 10 mg of lorazepam, to a second container. Separately add 5 mL of hydrochloric acid to each container, heat each solution on a steam bath for 20 minutes, and cool. Transfer the solutions to separators, and add 8 mL of 10 *N* sodium hydroxide to each separator. Extract each solution with two 25-mL portions of ether, filtering the ether extracts through cotton plugs into suitable containers. Evaporate both ether extracts to about 2 mL, and add 8 mL of methanol to each. Apply separately 10 µL of the test solution and the Standard solution to a suitable thin-layer chromatographic plate (see *Chromatography* ⟨621⟩) coated with a 0.25-mm layer of chromatographic silica gel. Allow the spots to dry, and develop the chromatograms in toluene until the solvent front has moved about 15 cm. Remove the plate from the developing chamber, mark the solvent front, and allow the solvent to evaporate. Spray the plate with a freshly prepared 1 in 80 solution of sodium nitrite in 0.5 *N* hydrochloric acid. Heat the plate at 100° for 5 minutes, allow to cool, and spray with a 1 in 1000 solution of *N*-(1-naphthyl)ethylenediamine dihydrochloride in alcohol: the R_f value of the principal spot obtained from the test solution corresponds to that obtained from the Standard solution.

Pyrogen—When diluted with 0.9 percent Sodium Chloride Injection to a concentration of 1.0 mg of lorazepam per mL, it meets the requirements of the *Pyrogen Test* ⟨151⟩, the test dose being 0.2 mL of the *Injection* per kg.

Related compounds—

A: *Mobile phase* and *Chromatographic system*—Prepare as directed in the *Assay*.

Standard preparation—Prepare a solution in *Mobile phase* having known concentrations of about 3.2 µg each of USP 6-Chloro-4-(*o*-chlorophenyl)-2-quinazolinecarboxaldehyde RS and USP 6-Chloro-4-(*o*-chlorophenyl)-2-quinazolinecarboxylic acid RS per mL.

Test preparation—Prepare as directed for *Assay preparation* in the *Assay*.

Procedure—Separately inject equal volumes (about 20 µL) of the *Standard preparation* and the *Test preparation* into the chromatograph, record the chromatograms, and measure the peak responses of any peaks observed other than the lorazepam peak. Do not include as an impurity any peak observed in the chromatogram of the *Test preparation* that has a retention time shorter than that of the 6-chloro-4-(*o*-chlorophenyl)-2-quinazolinecarboxylic acid peak in the *Standard preparation*. Calculate the percentage of 6-chloro-4-(*o*-chlorophenyl)-2-quinazolinecarboxaldehyde and the percentage of 6-chloro-4-(*o*-chlorophenyl)-2-quinazolinecarboxylic acid by the formula:

$$100(C_S/C_U)(r_U/r_S),$$

in which C_S is the concentration, in µg per mL, of the corresponding component in the *Standard preparation*, C_U is the concentration, in µg per mL, of Lorazepam in the *Test preparation*, r_U is the peak response of 6-chloro-4-(*o*-chlorophenyl)-2-quinazolinecarboxaldehyde or 6-chloro-4-(*o*-chlorophenyl)-2-quinazolinecarboxylic acid in the chromatogram obtained from the *Test preparation*, and r_S is the peak response of the corresponding component in the *Standard preparation*. The total of all impurities detected does not exceed 4.0%.

B: Transfer 5.0 mL of Injection to a suitable separator, and add 50 mL of 0.1 *N* sodium hydroxide. Extract with three 10-mL portions of chloroform, and collect the chloroform extracts in a second separator. Wash the chloroform extracts with 10 mL of water, and transfer the chloroform extracts to a centrifuge tube. Evaporate the chloroform extracts with the aid of a current of air to dryness, and dissolve the residue in acetone to obtain a *Test preparation* having a concentration of 10 mg per mL. Dissolve USP 2-Amino-2′,5-dichlorobenzophenone RS in acetone to obtain a *Standard preparation* having a known concentration of 0.1 mg per mL. Apply separately 50 µL of the *Test preparation* and 5 µL of the *Standard preparation* to a suitable thin-layer chromatographic plate (see *Chromatography* ⟨621⟩) coated with a 0.25-mm layer of chromatographic silica gel mixture. Allow the spots to dry, and develop the chromatograms in a solvent system consisting of a mixture of chloroform, *n*-heptane, and alcohol (10:10:1) until the solvent front has moved not less than 10 cm from the origin. Remove the plate from the developing chamber, mark the solvent front, and allow the solvent to evaporate. Lightly spray the plate with 2 *N* sulfuric acid, dry at 105° for 15 minutes, and spray successively with sodium nitrite solution (1 in 1000), ammonium sulfamate solution (1 in 200), and *N*-(1-naphthyl)ethylenediamine dihydrochloride solution (1 in 1000), drying the plate with a current of air after each spraying. Observe the plate under visible light: the spot produced by the *Test preparation* is not greater in size or intensity than the principal spot produced at the corresponding R_f value by the *Standard preparation*, corresponding to not more than 0.1% of 2-amino-2′,5-dichlorobenzophenone.

Other requirements—It meets the requirements under *Injections* ⟨1⟩.

Assay—

Mobile phase—Prepare a mixture of methanol and 0.05 *M* monobasic ammonium phosphate (50:50). Adjust with ammonium hydroxide to a pH of 6.5, filter, and degas. Make adjustments if necessary (see *System Suitability* under *Chromatography* ⟨621⟩).

Standard preparation—Dissolve an accurately weighed quantity of USP Lorazepam RS in methanol to obtain a solution having a known concentration of about 1.0 mg per mL. Transfer 4.0 mL of this solution to a 25-mL volumetric flask, dilute with *Mobile phase* to volume, and mix to obtain a solution having a known concentration of about 0.16 mg per mL.

Assay preparation—Transfer an accurately measured volume of Injection, equivalent to about 4 mg of lorazepam, to a 25-mL volumetric flask, dilute with *Mobile phase* to volume, and mix.

System suitability preparation—Prepare a solution of Lorazepam in *Mobile phase* containing about 0.04 mg of lorazepam per mL and about 0.032 mg each of 6-chloro-4-(*o*-chlorophenyl)-2-quinazolinecarboxaldehyde and 6-chloro-4-(*o*-chlorophenyl)-2-quinazolinecarboxylic acid per mL.

Chromatographic system (see *Chromatography* ⟨621⟩)—The liquid chromatograph is equipped with a 240-nm detector and 4.6-mm × 10- to 15-cm column that contains packing L1. The flow rate is about 2 mL per minute. Chromatograph replicate injections of the *Standard preparation*, and record the peak responses as directed under *Procedure:* the relative standard deviation is not more than 2.0%. Chromatograph the *System suitability preparation*, and record the peak responses as directed under *Procedure:* the resolution, *R*, between any of the major peaks is not less than 1.2. The relative retention times are about 0.7 for 6-chloro-4-(*o*-chlorophenyl)-2-quinazolinecarboxylic acid, 1.0 for lorazepam, and 2.7 for 6-chloro-4-(*o*-chlorophenyl)-2-quinazolinecarboxaldehyde.

Procedure—Separately inject equal volumes (about 20 µL) of the *Standard preparation* and the *Assay preparation* into the

chromatograph, record the chromatograms, and measure the responses for the major peaks. Calculate the quantity, in mg, of $C_{15}H_{10}Cl_2N_2O_2$ in each mL of the Injection taken by the formula:

$$25(C/V)(r_U/r_S),$$

in which C is the concentration, in mg per mL, of USP Lorazepam RS in the *Standard preparation*, V is the volume, in mL, of Injection taken, and r_U and r_S are the peak responses obtained from the *Assay preparation* and the *Standard preparation*, respectively.

Lorazepam Tablets

» Lorazepam Tablets contain not less than 90.0 percent and not more than 110.0 percent of the labeled amount of $C_{15}H_{10}Cl_2N_2O_2$.

Packaging and storage—Preserve in tight, light-resistant containers.

Reference standards—*USP Loraxepam Reference Standard*—Dry in vacuum at 105° for 3 hours before using. *USP 2-Amino-2′,5-dichlorobenxophenone Reference Standard*—Keep container tightly closed and protected from light. Do not dry before using. *USP 6-Chloro-4-(o-chlorophenyl)-2-quinaxolinecarboxaldehyde Reference Standard*—Do not dry before using. *USP 6-Chloro-4-(o-chlorophenyl)-2-quinaxolinecarboxylic acid Reference Standard*—Do not dry before using. *USP 6-Chloro-4-(o-chlorophenyl)-2-quinaxoline methanol Reference Standard*—Do not dry before using.

Identification—

A: The retention time of the major peak in the chromatogram of the *Assay preparation* corresponds to that in the chromatogram of the *Standard preparation* obtained in the *Assay*.

B: Stir a portion of finely powdered Tablets, equivalent to about 15 mg of lorazepam, with 40 mL of acetone for 5 minutes. Filter through very retentive filter paper pre-washed with acetone. Evaporate the filtrate on a steam bath with the aid of a current of air to dryness. Dissolve the residue in 1 mL of acetone, and add 20 mL of isooctane. Heat the solution on a hot plate to a gentle boil, and evaporate to a volume of about 10 mL. Remove the solution from the hot plate, and evaporate with the aid of a current of air to dryness. Dry the residue in vacuum at 60° for 1 hour: the infrared absorption spectrum of a mineral oil dispersion of the residue so obtained exhibits maxima only at the same wavelengths as that of a similar preparation of USP Lorazepam RS.

Dissolution ⟨711⟩—

Medium: water; 500 mL.

Apparatus 1: 100 rpm.

Times: 30 minutes; 60 minutes.

Mobile phase and *Chromatographic system*—Prepare as directed in the *Assay*.

Procedure—Inject an accurately measured volume (about 50 µL) of a filtered portion of the solution under test into the chromatograph, record the chromatogram, and measure the response for the major peak. Calculate the quantity of $C_{15}H_{10}Cl_2N_2O_2$ dissolved by comparison of the peak response obtained from a similarly chromatographed Standard solution having a known concentration of USP Lorazepam RS in water. [NOTE—A volume of alcohol not exceeding 10% of the final volume of the Standard solution is used initially to dissolve USP Lorazepam RS.]

Tolerances—The percentage of the labeled amount of $C_{15}H_{10}Cl_2N_2O_2$ dissolved from the Tablets is not less than 60% (Q) in 30 minutes and not less than 80% (Q) in 60 minutes.

Uniformity of dosage units ⟨905⟩: meet the requirements.

Procedure for content uniformity—Transfer 1 Tablet to a 100-mL volumetric flask, add 15 mL of water, and shake until the Tablet disintegrates. Add about 60 mL of alcohol, and shake for 15 minutes. Dilute with alcohol to volume, mix, and centrifuge. If necessary, dilute an accurately measured volume of the supernatant liquid quantitatively with dilute alcohol (85 in 100) to obtain a *Test preparation* containing about 5 µg of lorazepam per mL. Dissolve an accurately weighed quantity of USP Lorazepam RS in dilute alcohol (85 in 100) to obtain a *Standard preparation* containing about 5 µg of USP Lorazepam RS per mL. Concomitantly determine the absorbances of both solutions in 1-cm cells at the wavelength of maximum absorbance at about 230 nm, with a suitable spectrophotometer, using dilute alcohol (85 in 100) as the blank. Calculate the quantity, in mg, of $C_{15}H_{10}Cl_2N_2O_2$ in the Tablet by the formula:

$$(TC/D)(A_U/A_S),$$

in which T is the labeled quantity, in mg, of lorazepam in the Tablets, C is the concentration, in µg per mL, of USP Lorazepam RS in the *Standard preparation*, D is the concentration, in µg per mL, of lorazepam in the *Test preparation*, based upon the labeled quantity per Tablet and the extent of dilution, and A_U and A_S are the absorbances of the *Test preparation* and the *Standard preparation*, respectively.

Related compounds—

A: *Standard preparations*—Prepare a solution in chloroform having known concentrations of 1.0 mg each of USP 6-Chloro-4-(*o*-chlorophenyl)-2-quinazolinecarboxaldehyde RS, USP 6-Chloro-4-(*o*-chlorophenyl)-2-quinazolinecarboxylic acid RS and USP 6-Chloro-4-(*o*-chlorophenyl)-2-quinazoline methanol RS per mL. Dilute quantitatively with chloroform to obtain *Standard preparations*, designated below by letter, having the following compositions:

Dilution	Concentration (µg of each RS per mL)	Percentage (%, for comparison with test specimen)
A (1 in 25)	40	2.0
B (1 in 50)	20	1.0
C (1 in 100)	10	0.5

Test preparation—Transfer a quantity of finely powdered Tablets, equivalent to 4.0 mg of lorazepam, to a sintered-glass funnel. Extract with two 1-mL portions of chloroform followed by two 1-mL portions of methanol, collecting the filtrate in a centrifuge tube. Evaporate the filtrate with the aid of a stream of nitrogen at room temperature to dryness. Dissolve the residue in 2.0 mL of chloroform, and centrifuge. Use the clear supernatant liquid as the *Test preparation*.

Procedure—Within 30 minutes after preparation, apply separately 50 µL of the *Test preparation* and 50 µL of each *Standard preparation* to a suitable thin-layer chromatographic plate (see *Chromatography* ⟨621⟩) coated with a 0.25-mm layer of chromatographic silica gel mixture and previously washed with a mixture of chloroform, ethyl acetate, and methanol (2:1:1) and dried in air. Allow the spots to dry, and develop the chromatograms in a solvent system consisting of a mixture of chloroform, dioxane, and glacial acetic acid (91:5:4) until the solvent front has moved to within 2 cm to 3 cm from the top of the plate. Remove the plate from the developing chamber, mark the solvent front, and allow to air-dry for about 30 minutes. Examine the plate under short-wavelength ultraviolet light. Compare the intensities of any secondary spots observed in the chromatogram of the *Test preparation* with those of the principal spots in the chromatograms of the *Standard preparations:* [NOTE—The R_f value and the intensity of the spot for the USP 6-Chloro-4-(*o*-chlorophenyl)-2-quinazoline methanol RS in the *Standard preparations* correspond closely, but not necessarily precisely, to those observed for one of the secondary spots observed in the chromatogram of the *Test preparation.*] the sum of the intensities of all secondary spots obtained from the *Test preparation* corresponds to not more than 4.0%.

B: Transfer a quantity of finely powdered Tablets, equivalent to 25.0 mg of lorazepam, to a tapered 15-mL centrifuge tube, add 2.5 mL of acetone, insert a stopper into the tube, mix by mechanical means, and centrifuge. Use the supernatant solution as the *Test preparation*. Dissolve USP 2-Amino-2′,5-dichlorobenzophenone RS in acetone to obtain a *Standard preparation* having a known concentration of 100 µg per mL. Apply separately 50 µL of the *Test preparation* and 5 µL of the *Standard preparation* to a suitable thin-layer chromatographic plate (see *Chromatography* ⟨621⟩) coated with a 0.25-mm layer of chromatographic silica gel mixture and previously washed with a mixture of chloroform, ethyl acetate, and methanol (2:1:1) and dried in air. Proceed as directed in test *B* for *Related compounds* under *Loraxepam*, beginning with "Allow the spots to dry." The spot produced by the *Test preparation* is not greater in size or intensity than the principal spot produced at the corresponding R_f value by the *Standard preparation*, corresponding to not more than 0.1% of 2-amino-2′,5-dichlorobenzophenone.

Assay—

Mobile phase—Prepare a filtered and degassed mixture of water, acetonitrile, and glacial acetic acid (55:45:2). Make adjustments if necessary (see *System Suitability* under *Chromatography* ⟨621⟩).

Standard preparation—Dissolve an accurately weighed quantity of USP Lorazepam RS in methanol, and dilute quantitatively, and stepwise if necessary, with methanol to obtain a solution having a known concentration of about 0.10 mg per mL.

Assay preparation—Transfer 20 Lorazepam Tablets to a 100-mL volumetric flask, add 10 mL of water, and shake the mixture until the Tablets have disintegrated. Add about 30 mL of methanol, shake by mechanical means for about 20 minutes, dilute with methanol to volume, mix, and centrifuge. Dilute an accurately measured volume (V_S mL) of the clear supernatant liquid quantitatively with methanol to obtain a solution (V_A mL) containing about 0.1 mg of lorazepam per mL.

System suitability preparation—Dissolve 10 mg each of Lorazepam and 6-chloro-4-(*o*-chlorophenyl)-2-quinazoline methanol in 100 mL of methanol.

Chromatographic system (see *Chromatography* ⟨621⟩)—The liquid chromatograph is equipped with a 254-nm detector and a 4-mm × 30-cm column that contains packing L1. The flow rate is about 2 mL per minute. Chromatograph replicate injections of the *Standard preparation*, and record the peak responses as directed under *Procedure:* the relative standard deviation is not more than 2.0%. Chromatograph the *System suitability preparation*, and record the peak responses as directed under *Procedure:* the resolution, *R*, between the lorazepam and 6-chloro-4-(*o*-chlorophenyl)-2-quinazoline methanol peaks is not less than 2.0. The relative retention times are about 0.6 for lorazepam and 1.0 for 6-chloro-4-(*o*-chlorophenyl)-2-quinazoline methanol.

Procedure—Separately inject equal volumes (about 20 µL) of the *Standard preparation* and the *Assay preparation* into the chromatograph, record the chromatograms, and measure the responses for the major peaks. Calculate the quantity, in mg, of $C_{15}H_{10}Cl_2N_2O_2$ in each Tablet taken by the formula:

$$100(C/20)(V_A/V_S)(r_U/r_S),$$

in which *C* is the concentration, in mg per mL, of USP Lorazepam RS in the *Standard preparation*, and r_U and r_S are the peak responses obtained from the *Assay preparation* and the *Standard preparation*, respectively.

Lotions—*see complete list in index*

Lozenges, Cetylpyridinium Chloride—*see* Cetylpyridinium Chloride Lozenges

Lypressin Nasal Solution

» Lypressin Nasal Solution is a solution, in a suitable diluent, of the polypeptide hormone, prepared synthetically and free from foreign proteins, which has the properties of causing the contraction of vascular and other smooth muscle and of producing antidiuresis, and which is present in the posterior lobe of the pituitary of healthy pigs. It contains suitable preservatives, and is packaged in a form suitable for nasal administration so that the required dosage can be controlled as required. Each mL of Lypressin Nasal Solution possesses a pressor activity of not less than 85.0 percent and not more than 120.0 percent of that stated on the label in USP Posterior Pituitary Units.

Packaging and storage—Preserve in containers suitable for administering the contents by spraying into the nasal cavities in a controlled individualized dosage.

Reference standard—*USP Posterior Pituitary Reference Standard*—Do not dry before using. Store at a temperature of 0° or below. Each mg represents 2.4 USP Posterior Pituitary Units of oxytocic activity and 2.1 USP Posterior Pituitary Units of vasopressor activity.

Labeling—Label it to indicate that it is for intranasal administration only. Label it also to state that the package insert should be consulted for instructions to regulate the dosage according to symptoms.

Oxytocic activity—Proceed with Lypressin Nasal Solution as directed in the test for *Oxytocic activity* under *Vasopressin Injection.*

pH ⟨791⟩: between 3.0 and 4.3.

Assay—Proceed with Lypressin Nasal Solution as directed in the *Assay* under *Vasopressin Injection.*

Lysine Acetate

$$H_2N(CH_2)_3CH_2-\overset{\displaystyle H}{\underset{\displaystyle NH_2}{C}}-COOH \cdot CH_3COOH$$

$C_6H_{14}N_2O_2 . C_2H_4O_2$ 206.24

L-Lysine monoacetate.

L-Lysine monoacetate [*57282-49-2*].

» Lysine Acetate contains not less than 98.0 percent and not more than 102.0 percent of $C_6H_{14}N_2O_2 . C_2H_4O_2$, as L-lysine acetate, calculated on the dried basis.

Packaging and storage—Preserve in well-closed containers.

Reference standard—*USP L-Lysine Acetate Reference Standard*—Dry at 80° for 3 hours before using.

Identification—The infrared absorption spectrum of a potassium bromide dispersion of it, previously dried, exhibits maxima only at the same wavelengths as that of a similar preparation of USP L-Lysine Acetate RS.

Specific rotation ⟨781⟩: between +8.0° and +10.0°, calculated on the dried basis, determined in a solution containing 1 g in each 10 mL.

Loss on drying ⟨731⟩—Dry it at 80° for 3 hours: it loses not more than 0.2% of its weight.

Residue on ignition ⟨281⟩: not more than 0.4%.

Chloride ⟨221⟩—A 0.73-g portion shows no more chloride than corresponds to 0.50 mL of 0.020 *N* hydrochloric acid (0.05%).

Sulfate ⟨221⟩—A 0.33-g portion shows no more sulfate than corresponds to 0.10 mL of 0.020 *N* sulfuric acid (0.03%).

Arsenic ⟨211⟩: 1.5 ppm.

Iron ⟨241⟩: 0.003%.

Heavy metals, *Method I* ⟨231⟩: 0.0015%.

Assay—Transfer about 100 mg of Lysine Acetate, accurately weighed, to a 125-mL flask, dissolve in a mixture of 3 mL of formic acid and 50 mL of glacial acetic acid, and titrate with 0.1 *N* perchloric acid VS, determining the end-point potentiometrically. Perform a blank determination and make any necessary correction. Each mL of 0.1 *N* perchloric acid is equivalent to 10.31 mg of $C_6H_{14}N_2O_2 . C_2H_4O_2$.

Lysine Hydrochloride

$C_6H_{14}N_2O_2.HCl$ 182.65

» Lysine Hydrochloride contains not less than 98.5 percent and not more than 101.5 percent of $C_6H_{14}N_2O_2.HCl$, as L-lysine hydrochloride, calculated on the dried basis.

Packaging and storage—Preserve in well-closed containers.

Reference standard—*USP* L-*Lysine Hydrochloride Reference Standard*—Dry at 105° for 3 hours before using.

Identification—The infrared absorption spectrum of a potassium bromide dispersion of it, previously dried, exhibits maxima only at the same wavelengths as that of a similar preparation of USP L-Lysine Hydrochloride RS.

Specific rotation ⟨781⟩: between +20.4° and +21.4°, calculated on the dried basis, determined in a solution in 6 *N* hydrochloric acid containing 800 mg in each 10 mL.

Loss on drying ⟨731⟩—Dry it at 105° for 3 hours: it loses not more than 0.4% of its weight.

Residue on ignition ⟨281⟩: not more than 0.1%.

Chloride content ⟨221⟩—Transfer about 350 mg, accurately weighed, to a porcelain casserole, and add 140 mL of water and 1 mL of dichlorofluorescein TS. Mix, and titrate with 0.1 *N* silver nitrate VS until the silver chloride flocculates and the mixture acquires a faint pink color. Each mL of 0.1 *N* silver nitrate is equivalent to 3.545 mg of chloride: between 19.0% and 19.6% is found.

Sulfate ⟨221⟩—A solution containing 0.33 g shows no more sulfate than corresponds to 0.10 mL of 0.020 *N* sulfuric acid (0.03%).

Arsenic ⟨211⟩: 1.5 ppm.

Iron ⟨241⟩: 0.003%.

Heavy metals, *Method I* ⟨231⟩: 0.0015%.

Assay—Transfer about 90 mg of Lysine Hydrochloride, accurately weighed, to a 125-mL flask, and dissolve in a mixture of 3 mL of formic acid and 50 mL of glacial acetic acid. Add 10 mL of mercuric acetate TS, and titrate with 0.1 *N* perchloric acid VS, determining the end-point potentiometrically. Perform a blank determination, and make any necessary correction. Each mL of 0.1 *N* perchloric acid is equivalent to 9.133 mg of $C_6H_{14}N_2O_2.HCl$.

Mafenide Acetate

H_2NCH_2—⟨benzene ring⟩—SO_2NH_2 · $HC_2H_3O_2$

$C_7H_{10}N_2O_2S.C_2H_4O_2$ 246.28

Benzenesulfonamide, 4-(aminomethyl)-, monoacetate.

α-Amino-*p*-toluenesulfonamide monoacetate [*13009-99-9*].

» Mafenide Acetate contains not less than 98.0 percent and not more than 102.0 percent of $C_7H_{10}N_2O_2S.C_2H_4O_2$, calculated on the dried basis.

Packaging and storage—Preserve in tight, light-resistant containers.

Reference standards—*USP Mafenide Acetate Reference Standard*—Dry in vacuum at 60° for 16 hours before using. *USP 4-Formylbenzenesulfonamide Reference Standard*—Preserve in tight, light-resistant containers. Dry in vacuum at 60° for 4 hours before using.

Identification—

A: The infrared absorption spectrum of a potassium bromide dispersion of it, previously dried, exhibits maxima only at the same wavelengths as that of a similar preparation of USP Mafenide Acetate RS.

B: The R_f value of the principal spot in the chromatogram of the *Identification preparation* corresponds to that of *Standard preparation A*, as obtained in the test for *Chromatographic purity*.

Melting range ⟨741⟩: between 162° and 171°, but the range between beginning and end of melting does not exceed 4°.

pH ⟨791⟩: between 6.4 and 6.8, in a solution (1 in 10).

Loss on drying ⟨731⟩—Dry it in vacuum at 60° for 16 hours: it loses not more than 1.0% of its weight.

Residue on ignition ⟨281⟩: not more than 0.2%.

Selenium ⟨291⟩: 0.003%, a 200-mg test specimen being used.

Heavy metals, *Method II* ⟨231⟩: 0.002%.

Chromatographic purity—

Standard preparations—Dissolve USP Mafenide Acetate RS in methanol, mix to obtain *Standard preparation A* having a known concentration of 500 µg per mL, dissolve USP 4-Formylbenzenesulfonamide RS in methanol, and mix to obtain *Standard preparation D* having a known concentration of 500 µg per mL. Dilute portions of these solutions quantitatively with methanol to obtain *Standard preparations* having the following compositions:

Standard preparation	Dilution	Concentration (µg RS per mL)	Percentage (%, for comparison with test specimen)
A	(Undiluted)	500	1.0
B	5 in 10	250	0.5
C	1 in 5	100	0.2
D	(Undiluted)	500	1.0
E	5 in 10	250	0.5
F	1 in 5	100	0.2

Test preparation—Dissolve an accurately weighed quantity of Mafenide Acetate in methanol to obtain a solution containing 50 mg per mL.

Identification preparation—Dilute a portion of the *Test preparation* quantitatively with methanol to obtain a solution containing 500 µg per mL.

Ninhydrin solution—Dissolve 300 mg of triketohydrindene hydrate in 100 mL of butyl alcohol, add 3 mL of glacial acetic acid and mix.

Procedure—On a suitable thin-layer chromatographic plate (see *Chromatography* ⟨621⟩), coated with a 0.25-mm layer of chromatographic silica gel mixture, apply separately 5 µL of the *Test preparation*, 5 µL of the *Identification preparation*, and 5 µL of each *Standard preparation*. Position the plate in a chromatographic chamber, and develop the chromatograms in a solvent system consisting of a mixture of ethyl acetate, methanol, and isopropylamine (77:20:3) until the solvent front has moved about three-fourths of the length of the plate. Remove the plate from the developing chamber, mark the solvent front, and allow the solvent to evaporate in warm, circulating air. Examine the plate under short-wavelength ultraviolet light, and compare the intensities of any secondary spots observed in the chromatogram of the *Test preparation* at the R_f value corresponding to those of the principal spots in the chromatograms of *Standard preparations D*, *E*, and *F*. Spray the plate with the *Ninhydrin solution*, heat the plate at 105° for 5 minutes, and examine the plate. Compare the intensities of any secondary spots observed in the chromatogram of the *Test preparation* to those of the principal spots in the chromatograms of *Standard preparations A*, *B*, and *C*. No secondary spot, observed by both visualizations, from the chromatogram of the *Test preparation* is larger or more intense than the principal spots obtained from *Standard preparation B* (0.5%) and *Standard preparation E* (0.5%), and the sum of the intensities of all secondary spots obtained from the *Test preparation* corresponds to not more than 1.0%.

Assay—Transfer about 100 mg of Mafenide Acetate, accurately weighed, to a 50-mL volumetric flask, dissolve in about 20 mL of water, dilute with water to volume, and mix. Pipet 10 mL of this solution into a 100-mL volumetric flask containing 1 mL of 1 *N* hydrochloric acid, dilute with water to volume, and mix. Dissolve an accurately weighed quantity of USP Mafenide Acetate RS in 0.01 *N* hydrochloric acid, and dilute quantitatively and stepwise with the same solvent to obtain a Standard solution having a known concentration of about 200 µg per mL. Concomitantly determine the absorbance of both solutions in 1-cm cells at the wavelength of maximum absorbance at about 267 nm, with a suitable spectrophotometer, using 0.01 *N* hydrochloric acid as the blank. Calculate the quantity, in mg, of $C_7H_{10}N_2O_2S.C_2H_4O_2$ in the portion of Mafenide Acetate taken by the formula:

$$0.5C(A_U/A_S),$$

in which *C* is the concentration, in µg per mL, of USP Mafenide Acetate RS in the Standard solution, and A_U and A_S are the absorbances of the solution of Mafenide Acetate and the Standard solution, respectively.

Mafenide Acetate Cream

» Mafenide Acetate Cream is Mafenide Acetate in a water-miscible, oil-in-water cream base, containing suitable preservatives. It contains not less than 90.0 percent and not more than 110.0 percent of $C_7H_{10}N_2O_2S.C_2H_4O_2$ in terms of the labeled amount of mafenide ($C_7H_{10}N_2O_2S$).

Packaging and storage—Preserve in tight, light-resistant containers, and avoid exposure to excessive heat.

Reference standard—*USP Mafenide Acetate Reference Standard*—Dry in vacuum at 60° for 16 hours before using.

Identification—

A: The ultraviolet absorption spectrum of the solution from the Cream employed for measurement of absorbance in the *Assay* exhibits maxima and minima at the same wavelengths as that of a similar solution of USP Mafenide Acetate RS, concomitantly measured.

B: Place about 1 g in a beaker, warm to melt the cream, add about 25 mL of water, and mix: the solution responds to the tests for *Acetate* ⟨191⟩.

Assay—

Standard preparation—Dissolve an accurately weighed quantity of USP Mafenide Acetate RS in 0.01 *N* hydrochloric acid, and dilute quantitatively and stepwise with the same solvent to obtain a solution having a known concentration of about 200 µg per mL.

Assay preparation—Transfer a quantity of Mafenide Acetate Cream, equivalent to about 100 mg of mafenide acetate and accurately weighed, to a 60-mL separator, and add 20 mL of chloroform to dissolve it. Add 20 mL of water, shake for 2 minutes, allow the layers to separate completely, and discard the lower, chloroform layer. Repeat this washing with two separate 20-mL portions of chloroform, and discard the chloroform washings. Filter the aqueous phase through a dry filter into a 100-mL volumetric flask. Rinse the separator and the filter with water, passing all rinses through the filter, add water to volume, and mix. Centrifuge about 30 mL of the *Assay preparation*, then pipet 20 mL of the clear, supernatant liquid into a 100-mL volumetric flask, add 1 mL of 1 *N* hydrochloric acid, add water to volume, and mix.

Procedure—Concomitantly determine the absorbances of the *Standard preparation* and the *Assay preparation* in 1-cm cells at the wavelength of maximum absorbance at about 267 nm, with a suitable spectrophotometer, using 0.01 *N* hydrochloric acid as the blank. Calculate the quantity, in mg, of $C_7H_{10}N_2O_2S.C_2H_4O_2$ in the portion of Cream taken by the formula:

$$0.5C(A_U/A_S),$$

in which *C* is the concentration, in µg per mL, of USP Mafenide Acetate RS in the *Standard preparation*, and A_U and A_S are the absorbances of the solutions from the *Assay preparation* and the *Standard preparation*, respectively.

Magaldrate

Aluminum magnesium hydroxide sulfate ($Al_5Mg_{10}(OH)_{31}(SO_4)_2.xH_2O$).
Aluminum magnesium hydroxide sulfate, hydrate [*74978-16-8*].
Anhydrous 1097.38

» Magaldrate is a chemical combination of aluminum and magnesium hydroxides and sulfate, corresponding approximately to the formula:

$$Al_5Mg_{10}(OH)_{31}(SO_4)_2.xH_2O.$$

It contains the equivalent of not less than 90.0 percent and not more than 105.0 percent of $Al_5Mg_{10}(OH)_{31}(SO_4)_2$, calculated on the dried basis.

Packaging and storage—Preserve in well-closed containers.

Reference standard—*USP Magaldrate Reference Standard*—Keep container tightly closed, and store in a cool place. Do not dry before using.

Identification—

A: Dissolve about 600 mg in 20 mL of 3 *N* hydrochloric acid, add 3 drops of methyl red TS and about 30 mL of water, and heat to boiling. Add 6 *N* ammonium hydroxide until the color just changes to yellow, continue boiling for 2 minutes, and filter: the filtrate responds to the tests for *Magnesium* ⟨191⟩.

B: Wash the precipitate obtained in *Identification test A* with 50 mL of hot ammonium chloride solution (1 in 50), then dissolve the precipitate in 15 mL of 3 *N* hydrochloric acid: the solution responds to the tests for *Aluminum* ⟨191⟩.

C: Its X-ray diffraction pattern (see *X-ray Diffraction* ⟨941⟩) in the d-spacings region below 0.257 nm (2.57 angstrom units) conforms to that of USP Magaldrate RS.

Microbial limit—It meets the requirements of the test for absence of *Escherichia coli* under *Microbial Limit Tests* ⟨61⟩.

Loss on drying ⟨731⟩—Dry it at 200° for 4 hours: it loses between 10.0% and 20.0% of its weight.

Soluble chloride—Boil 1 g of it, accurately weighed, with 50.0 mL of water for 5 minutes, cool, add water to restore the original volume, mix, and filter. To 25.0 mL of the filtrate add 0.1 mL of potassium chromate TS, and titrate with 0.10 *N* silver nitrate until a persistent pink color is obtained: not more than 5.0 mL of 0.10 *N* silver nitrate is required (3.5%).

Soluble sulfate ⟨221⟩—A 2.5-mL portion of the filtrate obtained in the test for *Soluble chloride* shows no more sulfate than corresponds to 1.0 mL of 0.020 *N* sulfuric acid (1.9%).

Sodium—Transfer 2 g of it, accurately weighed, to a 100-mL volumetric flask, place in an ice bath, add 5 mL of nitric acid, and swirl to dissolve. Allow to warm to room temperature, dilute with water to volume, and mix. Filter, if necessary, to obtain a clear solution. Dilute 10.0 mL of the filtrate with water to 100.0 mL: the emission intensity of this solution, determined with a suitable flame photometer at 589 nm and corrected for background transmission at 580 nm, is not greater than that produced by a standard containing 2.2 µg of Na per mL, similarly measured (0.11%).

Arsenic, *Method I* ⟨211⟩: 8 ppm.

Heavy metals ⟨231⟩—Dissolve 330 mg in 10 mL of 3 *N* hydrochloric acid, filter if necessary to obtain a clear solution, and dilute with water to 25 mL: the limit is 0.006%.

Magnesium hydroxide content—Dissolve about 100 mg, accurately weighed, in 3 mL of dilute hydrochloric acid (1 in 10), and dilute with water to about 200 mL. Add, with stirring, 1 g of ammonium chloride, 20 mL of triethanolamine, 10 mL of ammonia–ammonium chloride buffer TS, and 0.1 mL of eriochrome black TS, and titrate with 0.05 *M* disodium ethylenediaminetetraacetate VS to a blue color. Perform a blank determination, and make any necessary correction. Each mL of 0.05 *M* disodium ethylenediaminetetraacetate is equivalent to 2.916 mg of $Mg(OH)_2$: between 49.2% and 66.6% of $Mg(OH)_2$ is found, calculated on the dried basis.

Aluminum hydroxide content—

Disodium ethylenediaminetetraacetate titrant—Prepare and standardize as directed in the *Assay* under *Ammonium Alum.*

Procedure—Dissolve about 100 mg of Magaldrate, accurately weighed, in 3 mL of dilute hydrochloric acid (1 in 10), and dilute with water to about 30 mL. Add, with stirring, 25.0 mL of *Disodium ethylenediaminetetraacetate titrant*, mix, and allow to stand for 5 minutes. Then add 20 mL of acetic acid–ammonium acetate buffer TS, 60 mL of alcohol, and 2 mL of dithizone TS, and titrate with 0.05 *M* zinc sulfate to a bright rose-pink color. Perform a blank determination, and make any necessary correction. Each mL of 0.05 *M Disodium ethylenediaminetetraacetate titrant* is equivalent to 3.900 mg of $Al(OH)_3$: between 32.1% and 45.9% of $Al(OH)_3$ is found, calculated on the dried basis.

Sulfate content—

Chromatographic column—Transfer 15 mL of strongly acidic 50- to 100-mesh styrene-divinylbenzene cation-exchange resin to a 1-cm inside diameter glass column. Wash the resin with 30 mL of water.

Indicator solution—Prepare a solution in water containing 2 mg of sodium alizarinsulfonate per mL.

Magnesium acetate solution—Dissolve 26.8 g of magnesium acetate in 500 mL of water.

0.05 M Barium chloride—Dissolve 12.2 g of barium chloride in about 900 mL of water, adjust with 1 *N* hydrochloric acid to a pH of 3.0, dilute with water to 1000 mL, and mix. Standardize this solution as follows: Transfer 10.0 mL of 0.1 *N* sulfuric acid VS to a 125-mL conical flask. Adjust by adding *Magnesium acetate solution* to a pH of 3.0. Add 25 mL of methanol and 3 or 4 drops of *Indicator solution.* Add from a buret an accurately measured volume of 8 to 9 mL of 0.05 *M* barium chloride. Add an additional 4 drops of *Indicator solution* and titrate slowly until the yellow color disappears and a pink tinge is visible. Calculate the molarity of the barium chloride titrant by the formula:

$$5(N/V),$$

in which *N* is the normality of the sulfuric acid, and *V* is the volume, in mL, of titrant consumed.

Test preparation—Transfer about 875 mg of Magaldrate, accurately weighed, to a 25-mL volumetric flask. Dissolve in 10 mL of water and 5 mL of glacial acetic acid, dilute with water to volume, and mix. Transfer 5.0 mL of this solution to the chromatographic column and wash the column with 15 mL of water, collecting the eluate in a 125-mL conical flask (*Test preparation*).

Procedure—Add to the *Test preparation* 5 mL of *Magnesium acetate solution*, 32 mL of methanol, and 3 or 4 drops of *Indicator solution.* Add from a buret an accurately measured volume of 5.0 to 5.5 mL of 0.05 *M* barium chloride. Add an additional 3 drops of *Indicator solution*, and titrate slowly until the yellow color disappears and a pink tinge is visible. Each mL of 0.05 *M* barium chloride is equivalent to 4.803 mg of sulfate (SO_4): between 16.0% and 21.0% of SO_4 is found, calculated on the dried basis.

Assay—Transfer about 3 g of Magaldrate, accurately weighed, to a 250-mL beaker, add 100.0 mL of 1 *N* hydrochloric acid VS, and stir until the solution becomes clear. Titrate the excess acid with 1 *N* sodium hydroxide VS to a pH of 3.0, determined potentiometrically. Perform a blank determination (see *Residual Titrations* under *Titrimetry* ⟨541⟩). Each mL of 1 *N* hydrochloric acid is equivalent to 35.40 mg of $Al_5Mg_{10}(OH)_{31}(SO_4)_2$.

Magaldrate Oral Suspension

» Magaldrate Oral Suspension contains not less than 90.0 percent and not more than 110.0 percent of the labeled amount of magaldrate [$Al_5Mg_{10}(OH)_{31}(SO_4)_2$].

Packaging and storage—Preserve in tight containers.

Reference standard—*USP Magaldrate Reference Standard*—Keep container tightly closed, and store in a cool place. Do not dry before using.

Identification—

A: Dissolve an amount of Oral Suspension, equivalent to about 800 mg of magaldrate, in 20 mL of 3 *N* hydrochloric acid, dilute with water to about 50 mL, add 3 drops of methyl red TS, and proceed as directed in *Identification test A* under *Magaldrate*, beginning with "and heat to boiling."

B: It responds to *Identification test B* under *Magaldrate.*

C: Transfer an amount of Oral Suspension, equivalent to about 1 g of magaldrate, to a 100-mL centrifuge tube. Add about 60 mL of water, cap, and shake for 3 minutes. Centrifuge the suspension, and discard the supernatant liquid. Repeat the washing of the residue with three 60-mL portions of water. Transfer the residue to a 250-mL beaker, and heat on a steam bath to dryness: the x-ray diffraction pattern (see *X-ray Diffraction* ⟨941⟩), in the d-spacings region below 2.57 angstrom units, of the residue so obtained conforms to that of USP Magaldrate RS.

Microbial limits ⟨61⟩—Its total aerobic microbial count does not exceed 100 per mL, and it meets the requirements of the test for absence of *Escherichia coli.*

Acid-neutralizing capacity ⟨301⟩—The acid consumed by the minimum single dose recommended in the labeling is not less than 5 mEq, and not less than the number of mEq calculated by the formula:

$$0.8(0.0282M),$$

in which 0.0282 is the theoretical acid-neutralizing capacity, in mEq per mg, of magaldrate, and *M* is the quantity, in mg, of the labeled amount of magaldrate.

Magnesium hydroxide content—

Test preparation—Transfer an accurately measured quantity of Magaldrate Oral Suspension, equivalent to about 1 g of magaldrate, to a 100-mL volumetric flask, add 30 mL of dilute hydrochloric acid (1 in 10), shake to dissolve, dilute with water to volume, and mix.

Procedure—Transfer 10.0 mL of *Test preparation* to a 400-mL beaker, and proceed as directed in the test for *Magnesium hydroxide content* under *Magaldrate*, beginning with "and dilute with water to about 200 mL." Not less than 492 mg and not more than 666 mg of magnesium hydroxide [$Mg(OH)_2$] per g of the labeled amount of magaldrate is found.

Aluminum hydroxide content—

Disodium ethylenediaminetetraacetate titrant—Prepare and standardize as directed in the *Assay* under *Ammonium Alum.*

Test preparation—Prepare as directed in the test for *Magnesium hydroxide content.*

Procedure—Transfer 10.0 mL of *Test preparation* and 20 mL of water to a 250-mL beaker, and proceed as directed for *Procedure* in the test for *Aluminum hydroxide content* under *Magaldrate*, beginning with "Add, with stirring, 25.0 mL of *Disodium ethylenediaminetetraacetate titrant.*" Not less than 321 mg and not more than 459 mg of aluminum hydroxide [$Al(OH)_3$] per g of the labeled amount of magaldrate is found.

Other requirements—Evaporate a volume of Oral Suspension, equivalent to about 5 g of magaldrate, on a steam bath to dryness: the residue so obtained meets the requirements of the tests for *Arsenic* and *Heavy metals* under *Magaldrate.*

Assay—Transfer an accurately measured quantity of Magaldrate Oral Suspension, equivalent to about 3 g of magaldrate, to a beaker. Add 100.0 mL of 1 *N* hydrochloric acid VS, and mix, using a magnetic stirrer to achieve dissolution. Titrate the excess acid with 1 *N* sodium hydroxide VS to a pH of 3.0, determined potentiometrically. Perform a blank determination (see *Residual Titrations* under *Titrimetry* ⟨541⟩). Each mL of 1 *N* hydrochloric acid is equivalent to 35.40 mg of $Al_5Mg_{10}(OH)_{31}(SO_4)_2$.

Magaldrate Tablets

» Magaldrate Tablets contain not less than 90.0 percent and not more than 110.0 percent of the labeled amount of magaldrate [$Al_5Mg_{10}(OH)_{31}(SO_4)_2$].

Packaging and storage—Preserve in well-closed containers.

Labeling—Label Tablets to indicate whether they are to be swallowed or to be chewed.

Reference standard—*USP Magaldrate Reference Standard*—Keep container tightly closed, and store in a cool place. Do not dry before using.

Identification—Transfer a quantity of powdered Tablets, equivalent to about 2 g of magaldrate, to a 100-mL centrifuge tube. Add about 60 mL of water, cap, and shake for 3 minutes. Centrifuge the suspension, and discard the supernatant liquid. Repeat the washing with three more 60-mL portions of water. Transfer the residue to a 250-mL beaker, and heat on a steam bath to dryness: the residue so obtained meets the requirements of the *Identification tests* under *Magaldrate.*

Microbial limit—It meets the requirements of the test for absence of *Escherichia coli* under *Microbial Limit Tests* ⟨61⟩.

Disintegration ⟨701⟩: 2 minutes, for Magaldrate Tablets labeled to be swallowed.

Uniformity of dosage units ⟨905⟩: meet the requirements.

Acid-neutralizing capacity—Proceed as directed under *Acid-neutralizing Capacity* ⟨301⟩. The acid consumed by the minimum single dose recommended in the labeling is not less than 5 mEq, and not less than the number of mEq calculated by the formula:

$$0.8(0.0282M),$$

in which 0.0282 is the theoretical acid-neutralizing capacity, in mEq per mg, of magaldrate, and M is the quantity, in mg, of the labeled amount of magaldrate.

Magnesium hydroxide content—

Test preparation—Weigh and finely powder not less than 20 Magaldrate Tablets. Transfer an accurately weighed portion of the powder, equivalent to about 1 g of magaldrate, to a 100-mL volumetric flask, add 30 mL of dilute hydrochloric acid (1 in 10), shake for 15 minutes, dilute with water to volume, and mix.

Procedure—Transfer 10.0 mL of *Test preparation* to a 400-mL beaker, and proceed as directed in the test for *Magnesium hydroxide content* under *Magaldrate*, beginning with "and dilute with water to about 200 mL." Not less than 492 mg and not more than 666 mg of magnesium hydroxide [$Mg(OH)_2$] per g of the labeled amount of magaldrate is found.

Aluminum hydroxide content—

Disodium ethylenediaminetetraacetate titrant—Prepare and standardize as directed in the *Assay* under *Ammonium Alum.*

Test preparation—Prepare as directed in the test for *Magnesium hydroxide content.*

Procedure—Transfer 10.0 mL of *Test preparation* and 20 mL of water to a 250-mL beaker, and proceed as directed for *Procedure* in the test for *Aluminum hydroxide content* under *Magaldrate*, beginning with "Add, with stirring, 25.0 mL of *Disodium ethylenediaminetetraacetate titrant.*" Not less than 321 mg and not more than 459 mg of aluminum hydroxide [$Al(OH)_3$] per g of the labeled amount of magaldrate is found.

Assay—Weigh and finely powder not less than 20 Magaldrate Tablets. Transfer an accurately weighed portion of the powder, equivalent to about 6 g of magaldrate, to a 200-mL volumetric flask. Add 100.0 mL of 2 *N* hydrochloric acid VS, and swirl by mechanical means for 30 minutes. Dilute with water to volume, mix, and filter. Transfer 100.0 mL of the filtrate to a beaker. Titrate the excess acid with 1 *N* sodium hydroxide VS to a pH of 3.0, determined potentiometrically. Perform a blank determination (see *Residual Titrations* under *Titrimetry* ⟨541⟩). Each mL of 2 *N* hydrochloric acid is equivalent to 70.80 mg of $Al_5Mg_{10}(OH)_{31}(SO_4)_2$.

Magaldrate and Simethicone Oral Suspension

» Magaldrate and Simethicone Oral Suspension contains not less than 90.0 percent and not more than 110.0 percent of the labeled amount of magaldrate [$Al_5Mg_{10}(OH)_{31}(SO_4)_2$], and an amount of polydimethylsiloxane [$-(CH_3)_2SiO-]_n$ that is not less than 85.0 percent and not more than 115.0 percent of the labeled amount of simethicone.

Packaging and storage—Preserve in tight containers, and keep from freezing.

Reference standards—*USP Magaldrate Reference Standard*—Keep container tightly closed, and store in a cool place. Do not dry before using. *USP Polydimethylsiloxane Reference Standard*—Keep container tightly closed. Do not dry before using.

Identification—

A: Dissolve an amount of Oral Suspension, equivalent to about 800 mg of magaldrate, in 20 mL of 3 *N* hydrochloric acid, dilute with water to about 50 mL, add 3 drops of methyl red TS, and proceed as directed in *Identification test A* under *Magaldrate*, beginning with "and heat to boiling."

B: It responds to *Identification test B* under *Magaldrate.*

C: Transfer an amount of Oral Suspension, equivalent to about 1 g of magaldrate, to a 100-mL centrifuge tube. Add about 60 mL of water, insert the cap, and shake for 3 minutes. Centrifuge the suspension, and discard the supernatant liquid. Repeat the washing of the residue with three 60-mL portions of water. Transfer the residue to a 250-mL beaker, and heat on a steam bath to dryness: the x-ray diffraction pattern (see *X-ray Diffraction* ⟨941⟩), in the d-spacings region below 2.57 angstrom units, of the residue so obtained conforms to that of USP Magaldrate RS.

D: The infrared absorption spectrum, in the 7- to 15-μm region, determined in a 0.1-mm cell, of the *Assay preparation* prepared as directed in the *Assay for polydimethylsiloxane* exhibits maxima only at the same wavelengths as that of the *Standard preparation* prepared as directed in the *Assay for polydimethylsiloxane.*

Acid-neutralizing capacity ⟨301⟩—The acid consumed by the minimum single dose recommended in the labeling is not less than 5 mEq, and not less than the number of mEq calculated by the formula:

$$0.8(0.0282M),$$

in which 0.0282 is the theoretical acid-neutralizing capacity, in mEq per mg, of magaldrate, and M is the quantity, in mg, of the labeled amount of magaldrate.

Defoaming activity—

Foaming solution—Dissolve 5 mg of FD&C Blue No. 1 and 10 g of polyoxyethylene (23) lauryl ether in 1000 mL of water. Warm to 37° before use.

Procedure—[NOTE—For each test use a clean 250-mL cylindrical jar (50-mm internal diameter × 110-mm internal height) with a 50-mm opening fitted with a tight cap with an inert liner.] Transfer a volume of well-mixed Oral Suspension, equivalent to 20 mg of simethicone, to a 250-mL cylindrical jar containing 50 mL of 0.6 *N* hydrochloric acid that has been warmed to 37°. Cap the jar, and clamp it in an upright position on a wrist-action shaker. Employing a radius of 13.3 ± 0.4 cm (measured from the center of shaft to center of bottle), shake for 30 seconds through an arc of 10 degrees at a frequency of 300 ± 30 strokes per minute. Uncap the jar, add 50 mL of *Foaming solution*, recap the jar, and shake for 10 seconds using the conditions specified above. Record the time required for the foam to collapse. The time, in seconds, for foam collapse is determined at the instant the first portion of foam-free liquid surface appears, measured from the end of the shaking period. The defoaming activity time does not exceed 45 seconds.

System suitability tests—[NOTE—For each of the following tests use a separate clean 250-mL cylindrical jar having the dimensions specified under *Procedure.*] Transfer 50 mL of *Foaming solution* to a 250-mL cylindrical jar containing 50 mL of 0.6 *N* hydrochloric acid that has been warmed to 37°. Cap the jar, and shake it for 10 seconds using the conditions specified under *Procedure:* the foam layer remains intact for not less than 5 minutes. Transfer 0.15 mL of simethicone and 50 mL of *Foaming solution* to a second 250-mL cylindrical jar containing 50 mL of 0.6 *N* hydrochloric acid that has been warmed to 37°. Cap the jar, and shake it for 10 seconds using the conditions specified under *Procedure:* the time required for the foam to collapse does not exceed 45 seconds.

Microbial limits ⟨61⟩—Its total aerobic microbial count does not exceed 100 per mL, and it meets the requirements of the test for absence of *Escherichia coli*.

Magnesium oxide hydroxide content—

Test preparation—Transfer an accurately measured quantity of Magaldrate and Simethicone Oral Suspension, equivalent to about 1 g of magaldrate, to a 100-mL volumetric flask, add 30 mL of dilute hydrochloric acid (1 in 10), shake to dissolve, dilute with water to volume, and mix.

Procedure—Transfer 10.0 mL of *Test preparation* to a 400-mL beaker, and proceed as directed in the test for *Magnesium oxide hydroxide content* under *Magaldrate*, beginning with "and dilute with water to about 200 mL." Not less than 492 mg and not more than 666 mg of magnesium hydroxide [$Mg(OH)_2$] per g of the labeled amount of magaldrate is found.

Aluminum hydroxide content—

Disodium ethylenediaminetetraacetate titrant—Prepare and standardize as directed in the *Assay* under *Ammonium Alum*.

Test preparation—Prepare as directed in the test for *Magnesium oxide hydroxide content*.

Procedure—Transfer 10.0 mL of *Test preparation* and 20 mL of water to a 250-mL beaker, and proceed as directed for *Procedure* in the test for *Aluminum hydroxide content* under *Magaldrate*, beginning with "Add, with stirring, 25.0 mL of *Disodium ethylenediaminetetraacetate titrant*." Not less than 321 mg and not more than 459 mg of aluminum hydroxide [$Al(OH)_3$] per g of the labeled amount of magaldrate is found.

Other requirements—Evaporate a volume of Oral Suspension, equivalent to about 5 g of magaldrate, on a steam bath to dryness: the residue so obtained meets the requirements of the tests for *Arsenic* and *Heavy metals* under *Magaldrate*.

Assay for magaldrate—Transfer an accurately measured quantity of Magaldrate and Simethicone Oral Suspension, equivalent to about 3 g of magaldrate, to a beaker. Add 100.0 mL of 1 *N* hydrochloric acid VS, and mix, using a magnetic stirrer to achieve dissolution. Titrate the excess acid with 1 *N* sodium hydroxide VS to a pH of 3.0, determined potentiometrically. Perform a blank determination (see *Residual Titrations* under *Titrimetry* ⟨541⟩). Each mL of 1 *N* hydrochloric acid is equivalent to 35.40 mg of magaldrate [$Al_5Mg_{10}(OH)_{31}(SO_4)_2$].

Assay for polydimethylsiloxane—Transfer an accurately measured quantity of Magaldrate and Simethicone Oral Suspension, equivalent to about 250 mg of simethicone, to a 200-mL centrifuge bottle. Add an equal volume of hydrochloric acid, swirl to dissolve the Oral Suspension, add 25.0 mL of hexanes, and immediately close the bottle securely with a cap having an inert liner. Shake the bottle for 30 minutes, and centrifuge the mixture until a clear supernatant layer is obtained (*Assay preparation*). Prepare a *Standard preparation* of USP Polydimethylsiloxane RS in hexanes having a known concentration of about 10 mg per mL. Concomitantly determine the absorbances of the *Assay preparation* and the *Standard preparation* in 0.1-mm cells at the wavelength of maximum absorbance at about 7.9 µm and at the wavelengths of minimum absorbance at about 7.5 µm and 8.3 µm, with a suitable infrared spectrophotometer, using hexanes as the blank. Draw a linear baseline between the two minima, and determine the absorbances for the *Standard preparation* and the *Assay preparation* with respect to the baseline, making any necessary correction for the blank. Calculate the quantity, in mg, of [$-(CH_3)_2SiO-]_n$ in the portion of Oral Suspension taken by the formula:

$$25C(A_U/A_S),$$

in which C is the concentration, in mg per mL, of USP Polydimethylsiloxane RS in the *Standard preparation*, and A_U and A_S are the absorbances of the *Assay preparation* and the *Standard preparation*, respectively.

Magaldrate and Simethicone Tablets

» Magaldrate and Simethicone Tablets contain not less than 90.0 percent and not more than 110.0 percent of the labeled amount of magaldrate [$Al_5Mg_{10}(OH)_{31}(SO_4)_2$], and an amount of polydimethylsiloxane [$-(CH_3)_2SiO-]_n$ that is not less than 85.0 percent and not more than 115.0 percent of the labeled amount of simethicone.

Packaging and storage—Preserve in well-closed containers.

Labeling—Label Tablets to indicate that they are to be chewed before being swallowed.

Reference standards—*USP Magaldrate Reference Standard*—Keep container tightly closed, and store in a cool place. Do not dry before using. *USP Polydimethylsiloxane Reference Standard*—Keep container tightly closed. Do not dry before using.

Identification—

A: Transfer a quantity of powdered Tablets, equivalent to about 2 g of magaldrate, to a 100-mL centrifuge tube. Add about 60 mL of water, cap, and shake for 3 minutes. Centrifuge the suspension, and discard the supernatant liquid. Repeat the washing with three more 60-mL portions of water. Transfer the residue to a 250-mL beaker, and heat on a steam bath to dryness: the residue so obtained meets the requirements of the *Identification tests* under *Magaldrate*.

B: The infrared absorption spectrum, in the 7- to 11-µm region, determined in a 0.5-mm cell, of the *Assay preparation* prepared as directed in the *Assay for polydimethylsiloxane* exhibits maxima only at the same wavelengths as that of the *Standard preparation* containing about 2 mg of USP Polydimethylsiloxane RS per mL prepared as directed in the *Assay for polydimethylsiloxane*.

Microbial limit—Tablets meet the requirements of the test for absence of *Escherichia coli* under *Microbial Limit Tests* ⟨61⟩.

Uniformity of dosage units ⟨905⟩: meet the requirements for *Weight Variation* with respect to magaldrate.

Acid-neutralizing capacity—Proceed as directed under *Acid-neutralizing Capacity* ⟨301⟩. The acid consumed by the minimum single dose recommended in the labeling is not less than 5 mEq, and not less than the number of mEq calculated by the formula:

$$0.8(0.0282M),$$

in which 0.0282 is the theoretical acid-neutralizing capacity, in mEq per mg, of magaldrate, and M is the quantity, in mg, of the labeled amount of magaldrate.

Magnesium hydroxide content—

Test preparation—Weigh and finely powder not less than 20 Magaldrate and Simethicone Tablets. Transfer an accurately weighed portion of the powder, equivalent to about 1 g of magaldrate, to a 100-mL volumetric flask, add 30 mL of dilute hydrochloric acid (1 in 10), shake for 15 minutes, dilute with water to volume, and mix.

Procedure—Transfer 10.0 mL of *Test preparation* to a 400-mL beaker, and proceed as directed in the test for *Magnesium hydroxide content* under *Magaldrate*, beginning with "and dilute with water to about 200 mL." Not less than 492 mg and not more than 666 mg of magnesium hydroxide [$Mg(OH)_2$] per g of the labeled amount of magaldrate is found.

Aluminum hydroxide content—

Disodium ethylenediaminetetraacetate titrant—Prepare and standardize as directed in the *Assay* under *Ammonium Alum*.

Test preparation—Prepare as directed in the test for *Magnesium hydroxide content*.

Procedure—Transfer 10.0 mL of *Test preparation* and 20 mL of water to a 250-mL beaker, and proceed as directed for *Procedure* in the test for *Aluminum hydroxide content* under *Magaldrate*, beginning with "Add, with stirring, 25.0 mL of *Disodium ethylenediaminetetraacetate titrant*." Not less than 321 mg and not more than 459 mg of aluminum hydroxide [$Al(OH)_3$] per g of the labeled amount of magaldrate is found.

Assay for magaldrate—Weigh and finely powder not less than 20 Magaldrate and Simethicone Tablets. Transfer an accurately weighed portion of the powder, equivalent to about 6 g of magaldrate, to a 200-mL volumetric flask. Add 100.0 mL of 2 *N* hydrochloric acid VS, and swirl by mechanical means for 30 minutes. Dilute with water to volume, mix, and filter. Transfer 100.0 mL of the filtrate to a beaker. Titrate the excess acid with

1 *N* sodium hydroxide VS to a pH of 3.0, determined potentiometrically. Perform a blank determination (see *Residual Titrations* under *Titrimetry* ⟨541⟩). Each mL of 2 *N* hydrochloric acid is equivalent to 70.80 mg of $Al_5Mg_{10}(OH)_{31}(SO_4)_2$.

Assay for polydimethylsiloxane—Weigh and finely powder not less than 20 Magaldrate and Simethicone Tablets. Transfer an accurately weighed portion of the powder, equivalent to about 20 mg of simethicone, to a screw-capped, 50-mL centrifuge tube. Add 10.0 mL of carbon tetrachloride and 25 mL of 6 *N* hydrochloric acid. Immediately close the centrifuge tube securely with a cap having an inert liner, and shake by mechanical means for not less than 30 minutes. Centrifuge until clear layers are obtained. Remove and discard the upper aqueous layer. Add 25 mL of water, cap tightly, and shake for 3 minutes. Centrifuge until clear layers are obtained. Remove and discard the upper aqueous layer. Add 25 mL of 4 *N* sodium hydroxide. Immediately close the centrifuge tube securely with a cap having an inert liner, and shake by mechanical means for 1 hour. Centrifuge the mixture until clear layers are obtained. Draw off and discard as much of the upper aqueous layer as possible. Transfer not less than 5 mL of the lower layer to a test tube containing about 0.5 g of anhydrous sodium sulfate. Insert the stopper in the tube, shake vigorously, and centrifuge, if necessary, to obtain a clear supernatant liquid (*Assay preparation*). Prepare three *Standard preparations* in carbon tetrachloride having known concentrations of about 1.6, 2.0, and 2.4 mg of USP Polydimethylsiloxane RS per mL, respectively. Concomitantly determine the absorbances of the *Assay preparation* and the *Standard preparations* in 0.5-mm cells at the wavelength of maximum absorbance at about 7.9 µm and at the wavelengths of minimum absorbance at about 7.6 µm and 8.1 µm, with a suitable infrared spectrophotometer, using carbon tetrachloride as the blank. Draw a linear baseline between the two minima, and determine the absorbances for the *Standard preparations* and the *Assay preparation* with respect to the baseline, making any necessary correction for the blank. Plot the absorbances for the *Standard preparations* versus concentration, in mg per mL, of USP Polydimethylsiloxane RS, and draw the straight line best fitting the three plotted points. From the graph so obtained, determine the concentration, *C*, in mg per mL, of polydimethylsiloxane in the *Assay preparation*. Calculate the quantity, in mg, of $[-(CH_3)_2SiO-]_n$ in the portion of Tablets taken by multiplying *C* by 10.

Magma, Dihydroxyaluminum Aminoacetate—*see* Dihydroxyaluminum Aminoacetate Magma

Milk of Magnesia

$Mg(OH)_2$ 58.32
Magnesium hydroxide.
Magnesium hydroxide [*1309-42-8*].

» Milk of Magnesia is a suspension of Magnesium Hydroxide. Milk of Magnesia, Double-strength Milk of Magnesia, and Triple-strength Milk of Magnesia contain not less than 90.0 percent and not more than 115.0 percent of the labeled amount of $Mg(OH)_2$, the labeled amount being 80, 160, and 240 mg of $Mg(OH)_2$ per mL, respectively. It may contain not more than 0.05 percent of a volatile oil or a blend of volatile oils, suitable for flavoring purposes.

Packaging and storage—Preserve in tight containers, preferably at a temperature not exceeding 35°. Avoid freezing.

Labeling—Double- or Triple-strength Milk of Magnesia is so labeled, or may be labeled as 2× or 3× Concentrated Milk of Magnesia, respectively.

Identification—A solution of the equivalent of 1 g of regular-strength Milk of Magnesia in 2 mL of 3 *N* hydrochloric acid responds to the tests for *Magnesium* ⟨191⟩.

Microbial limits ⟨61⟩—Its total aerobic microbial count does not exceed 100 per mL, and it meets the requirements of the test for absence of *Escherichia coli*.

Soluble alkalies—Transfer about 25 mL of it to a filter, and reject the first 5 mL of the filtrate. Dilute 5 mL of the clear filtrate with 40 mL of water. Add 1 drop of methyl red TS, and titrate the solution with 0.10 *N* sulfuric acid to the production of a persistent pink color: not more than 1.0 mL of the acid is required. Where the specimen is Double- or Triple-strength Milk of Magnesia, not more than 2.0 mL or 3.0 mL of the acid is required, respectively.

Soluble salts—To 5.0 mL of the clear filtrate obtained in the test for *Soluble alkalies* add 3 drops of sulfuric acid, evaporate on a steam bath to dryness, and ignite gently to constant weight: the residue weighs not more than 12 mg, or 24 mg or 36 mg where the specimen is Double- or Triple-strength Milk of Magnesia, respectively.

Carbonate and acid-insoluble matter—To the equivalent of 1 g of regular-strength Milk of Magnesia add 2 mL of 3 *N* hydrochloric acid: not more than a slight effervescence occurs, and the solution is not more than slightly turbid.

Arsenic, *Method I* ⟨211⟩—Prepare a *Standard Preparation* as directed in the test for *Arsenic* ⟨211⟩, except to contain 2 µg of arsenic instead of 3 µg. Prepare a *Test Preparation* consisting of the equivalent of 3.3 g of regular-strength Milk of Magnesia in 20 mL of 7 *N* sulfuric acid. The limit is 2/*W* ppm, *W* being the weight, in g, of specimen taken.

Calcium ⟨216⟩: not more than 0.07%, 0.70 mL of *Standard Calcium Ion Solution* being used in the *Standard Preparation*, and the equivalent of 2.0 g of regular-strength Milk of Magnesia being used as the *Test preparation*.

Heavy metals ⟨231⟩—To the equivalent of 4.0 g of regular-strength Milk of Magnesia add 6 mL of 3 *N* hydrochloric acid, and evaporate the solution on a steam bath to dryness, with frequent stirring. Dissolve the residue in 20 mL of water, and evaporate to dryness in the same manner as before. Redissolve the residue in 20 mL of water, filter if necessary, and dilute with water to 25 mL: the limit is 20/*W* ppm, *W* being the weight, in g, of specimen taken.

Assay—Transfer an accurately measured quantity of Milk of Magnesia, previously shaken in its original container, equivalent to about 800 mg of magnesium hydroxide, to a 250-mL volumetric flask. Dissolve in 30 mL of 3 *N* hydrochloric acid, dilute with water to volume, and mix. Filter, if necessary, and transfer 25.0 mL of the filtrate to a beaker containing 75 mL of water, and mix. Adjust the reaction of the solution to a pH of 7 (using pH indicator paper; see *Indicator and Test Papers* under *Reagents*, in the section, *Reagents, Indicators, and Solutions*) with 1 *N* sodium hydroxide, add 5 mL of ammonia–ammonium chloride buffer TS and 0.15 mL of eriochrome black TS, and titrate with 0.05 *M* disodium ethylenediaminetetraacetate VS to a blue end-point. Each mL of 0.05 *M* disodium ethylenediaminetetraacetate is equivalent to 2.916 mg of $Mg(OH)_2$.

Magnesia Tablets

» Magnesia Tablets contain not less than 93.0 percent and not more than 107.0 percent of the labeled amount of magnesium hydroxide [$Mg(OH)_2$].

Packaging and storage—Preserve in well-closed containers.

Identification—Crush several Tablets, and dissolve 1 g of the powder in 20 mL of 3 *N* hydrochloric acid: the solution responds to the tests for *Magnesium* ⟨191⟩.

Disintegration ⟨701⟩: 10 minutes, simulated gastric fluid TS being substituted for water in the test.

Uniformity of dosage units ⟨905⟩: meet the requirements.

Assay—Weigh and finely powder not less than 20 Magnesia Tablets. Transfer an accurately weighed portion of the powder, equivalent to about 250 mg of magnesium hydroxide, to a 100-mL volumetric flask, and proceed as directed in the *Assay* under

Milk of Magnesia beginning with "Dissolve in 10 mL of 3 *N* hydrochloric acid."

Magnesia Tablets, Alumina and—*see* Alumina and Magnesia Tablets

Magnesia Tablets, Aspirin, Alumina, and—*see* Aspirin, Alumina, and Magnesia Tablets

Magnesia Tablets, Aspirin, Codeine Phosphate, Alumina, and—*see* Aspirin, Codeine Phosphate, Alumina, and Magnesia Tablets

Magnesia Tablets, Calcium Carbonate and—*see* Calcium Carbonate and Magnesia Tablets

Magnesia and Alumina Oral Suspension

» Magnesia and Alumina Oral Suspension is a mixture containing aluminum hydroxide [$Al(OH)_3$] and Magnesium Hydroxide [$Mg(OH)_2$] in equal amounts, or with $Mg(OH)_2$ predominating. It contains not less than 90.0 percent and not more than 110.0 percent of the labeled amounts of aluminum hydroxide [$Al(OH)_3$] and magnesium hydroxide [$Mg(OH)_2$]. It may contain flavoring agents, and may contain suitable antimicrobial agents.

pH ⟨791⟩: between 7.3 and 8.5.

Other requirements—It responds to the *Identification tests*, and meets the requirements for *Packaging and storage, Labeling, Microbial limits, Chloride, Sulfate,* and *Acid-neutralizing capacity* under *Alumina and Magnesia Oral Suspension.* It meets also the requirements of the tests for *Arsenic* and *Heavy metals* under *Aluminum Hydroxide Gel.*

Assay for aluminum hydroxide—Proceed with Magnesia and Alumina Oral Suspension as directed in the *Assay for aluminum hydroxide* under *Alumina and Magnesia Oral Suspension,* except, under *Procedure,* to omit the heating.

Assay for magnesium hydroxide—Proceed with Magnesia and Alumina Oral Suspension as directed in the *Assay for magnesium hydroxide* under *Alumina and Magnesia Oral Suspension.*

Magnesia and Alumina Tablets

» Magnesia and Alumina Tablets contain not less than 90.0 percent and not more than 110.0 percent of the labeled amounts of magnesium hydroxide [$Mg(OH)_2$] and aluminum hydroxide [$Al(OH)_3$].

Other requirements—Tablets respond to the *Identification tests* and meet the requirements for *Packaging and storage, Labeling, Disintegration, Uniformity of dosage units* ⟨905⟩, and *Acid-neutralizing capacity* under *Alumina and Magnesia Tablets.*

Assay for magnesium hydroxide—Proceed with Magnesia and Alumina Tablets as directed in the *Assay for magnesium hydroxide* under *Alumina and Magnesia Tablets.*

Assay for aluminum hydroxide—

Disodium ethylenediaminetetraacetate titrant—Prepare and standardize as directed in the *Assay* under *Ammonium Alum.*

Assay preparation—Weigh and finely powder not less than 20 Magnesia and Alumina Tablets. Transfer an accurately weighed portion of the powder, equivalent to about 600 mg of aluminum hydroxide, to a 150-mL beaker, add 20 mL of water, stir, and slowly add 30 mL of 3 *N* hydrochloric acid. Proceed as directed for *Assay preparation* in the *Assay for aluminum hydroxide* under *Alumina and Magnesia Oral Suspension,* beginning with "Heat gently, if necessary."

Procedure—Proceed as directed for *Procedure* in the *Assay for aluminum hydroxide* under *Alumina and Magnesia Oral Suspension,* except to omit the heating. Each mL of 0.05 *M Disodium ethylenediaminetetraacetate titrant* is equivalent to 3.900 mg of $Al(OH)_3$.

Magnesia, and Calcium Carbonate Oral Suspension, Alumina,—*see* Alumina, Magnesia, and Calcium Carbonate Oral Suspension

Magnesia, and Calcium Carbonate Tablets, Alumina,—*see* Alumina, Magnesia, and Calcium Carbonate Tablets

Magnesia, and Simethicone Oral Suspension, Alumina,—*see* Alumina, Magnesia, and Simethicone Oral Suspension

Magnesia, and Simethicone Tablets, Alumina,—*see* Alumina, Magnesia, and Simethicone Tablets

Magnesium Aluminum Silicate—*see* Magnesium Aluminum Silicate NF

Magnesium Carbonate

Carbonic acid, magnesium salt, basic; or, Carbonic acid, magnesium salt (1:1), hydrate.
Magnesium carbonate, basic [*39409-82-0*]; or, Magnesium carbonate (1:1) hydrate [*23389-33-5*].
Anhydrous 84.31 [*546-93-0*].

» Magnesium Carbonate is a basic hydrated magnesium carbonate or a normal hydrated magnesium carbonate. It contains the equivalent of not less than 40.0 percent and not more than 43.5 percent of magnesium oxide (MgO).

Packaging and storage—Preserve in well-closed containers.

Identification—When treated with 3 *N* hydrochloric acid, it dissolves with effervescence, and the resulting solution responds to the tests for *Magnesium* ⟨191⟩.

Microbial limit—It meets the requirements of the test for absence of *Escherichia coli* under *Microbial Limit Tests* ⟨61⟩.

Soluble salts—Mix 2.0 g with 100 mL of a mixture of equal volumes of *n*-propyl alcohol and water. Heat the mixture to the boiling point with constant stirring, cool to room temperature, dilute with water to 100 mL, and filter. Evaporate 50 mL of the filtrate on a steam bath to dryness, and dry at 105° for 1 hour: the weight of the residue does not exceed 10 mg (1.0%).

Acid-insoluble substances—Mix 5.0 g with 75 mL of water, add hydrochloric acid in small portions, with agitation, until no more of the magnesium carbonate dissolves, and boil for 5 minutes. If an insoluble residue remains, filter, wash well with water until the last washing is free from chloride, and ignite: the weight of the ignited residue does not exceed 2.5 mg (0.05%).

Arsenic, *Method I* ⟨211⟩—Prepare the *Test Preparation* by dissolving 750 mg in 25 mL of 3 *N* hydrochloric acid. The limit is 4 ppm.

Calcium—

Dilute hydrochloric acid—Dilute 100 mL of hydrochloric acid with water to 1000 mL.

Lanthanum solution—To 58.65 g of lanthanum oxide add 400 mL of water, and add, gradually with stirring, 250 mL of hydrochloric acid. Stir until dissolved, dilute with water to 1000 mL, and mix.

Standard preparations—Transfer 249.7 mg of calcium carbonate, previously dried at 300° for 3 hours and cooled in a desiccator for 2 hours, to a 100-mL volumetric flask, dissolve in a minimum amount of hydrochloric acid, dilute with water to volume, and mix. Transfer 5.0, 10.0, and 15.0 mL of this stock solution to separate 1000-mL volumetric flasks, each containing 20 mL of *Lanthanum solution* and 40 mL of *Dilute hydrochloric acid*, add water to volume, and mix. These *Standard preparations* contain 5.0, 10.0, and 15.0 µg of Ca in each mL, respectively.

Blank solution—Transfer 4 mL of *Lanthanum solution* and 10 mL of *Dilute hydrochloric acid* to a 200-mL volumetric flask, dilute with water to volume, and mix.

Test preparation—Transfer 250 mg of Magnesium Carbonate to a beaker, add 30 mL of *Dilute hydrochloric acid*, and stir until dissolved, heating if necessary. Transfer the solution so obtained to a 200-mL volumetric flask containing 4 mL of *Lanthanum solution*, dilute with water to volume, and mix.

Procedure— Concomitantly determine the absorbances of the *Standard preparations* and the *Test preparation* at the calcium emission line at 422.7 nm with a suitable atomic absorption spectrophotometer (see *Spectrophotometry and Light-scattering* ⟨851⟩) equipped with a calcium hollow-cathode lamp and a nitrous oxide–acetylene flame, using the *Blank solution* as the blank. Plot the absorbances of the *Standard preparations* versus their concentrations of calcium, in µg per mL, and draw a straight line through the three points. From the line so obtained and the absorbance of the *Test preparation*, determine the concentration, in µg per mL, of calcium in the *Test preparation*. Calculate the percentage of Ca in the specimen by multiplying this value by 0.08: the limit is 0.45%.

Heavy metals, *Method I* ⟨231⟩—Dissolve 0.67 g in 10 mL of 3 *N* hydrochloric acid in a suitable crucible, and evaporate the solution on a steam bath to dryness. Ignite at 550 ± 25° until all carbonaceous material is consumed. Dissolve the residue in 15 mL of water and 5 mL of hydrochloric acid, and evaporate to dryness. Toward the end of the evaporation, stir frequently to disintegrate the residue so that finally a dry powder is obtained. Dissolve the residue in 20 mL of water, and evaporate in the same manner as before to dryness. Redissolve the residue in 20 mL of water, filter, if necessary, and add to the filtrate 2 mL of 1 *N* acetic acid and water to make 25 mL: the limit is 0.003%.

Iron ⟨241⟩—Boil 50 mg with 5 mL of 2 *N* nitric acid for 1 minute. Cool, dilute with water to 45 mL, add 2 mL of hydrochloric acid, and mix: the limit is 0.02%.

Assay—Dissolve about 1 g of Magnesium Carbonate, accurately weighed, in 30.0 mL of 1 *N* sulfuric acid VS, add methyl orange TS, and titrate the excess acid with 1 *N* sodium hydroxide VS. From the volume of 1 *N* sulfuric acid consumed, deduct the volume of 1 *N* sulfuric acid corresponding to the content of calcium in the weight of Magnesium Carbonate taken for the assay. The difference is the volume of 1 *N* sulfuric acid equivalent to the magnesium oxide present. Each mL of 1 *N* sulfuric acid is equivalent to 20.15 mg of MgO and to 20.04 mg of Ca.

Magnesium Carbonate, and Magnesium Oxide Tablets, Alumina,—*see* Alumina, Magnesium Carbonate, and Magnesium Oxide Tablets

Magnesium Carbonate Oral Suspension, Alumina and—*see* Alumina and Magnesium Carbonate Oral Suspension

Magnesium Carbonate and Sodium Bicarbonate for Oral Suspension

» Magnesium Carbonate and Sodium Bicarbonate for Oral Suspension contains not less than 90.0 percent and not more than 110.0 percent of the labeled amounts of $MgCO_3$ and $NaHCO_3$. It may contain suitable flavors.

Packaging and storage—Preserve in tight containers.

Identification—

A: Place about 1 g in a flask equipped with a stopper and glass tubing, the tip of which is immersed in calcium hydroxide TS in a test tube. Add 5 mL of 3 *N* hydrochloric acid to the flask, and immediately insert the stopper: gas evolves in the flask and a precipitate is formed in the test tube.

B: The solution remaining in the flask responds to the tests for *Magnesium* ⟨191⟩ and for *Sodium* ⟨191⟩.

Minimum fill ⟨755⟩: meets the requirements.

Assay for magnesium carbonate—Transfer an accurately weighed portion of Magnesium Carbonate and Sodium Bicarbonate for Oral Suspension, equivalent to about 4.2 g of $MgCO_3$, to a 500-mL volumetric flask. Add 200 mL of 1 *N* hydrochloric acid, and mix. When dissolved, dilute with water to volume, and mix. Transfer 10.0 mL of this stock solution to a suitable container, dilute with water to 100 mL, add 10 mL of ammonia–ammonium chloride buffer TS, 5 mL of triethanolamine, and 0.3 mL of eriochrome black TS, and titrate with 0.05 *M* disodium ethylenediaminetetraacetate VS to a blue end-point. Each mL of 0.05 *M* disodium ethylenediaminetetraacetate consumed is equivalent to 4.216 mg of $MgCO_3$.

Assay for sodium bicarbonate—

Standard preparations—Dissolve a suitable quantity of sodium chloride, previously dried at 125° for 30 minutes and accurately weighed, in water, and dilute quantitatively with water to obtain a solution having a known concentration of about 600 µg per mL. On the day of use, further dilute this solution quantitatively with water to obtain three solutions containing 6.0, 12.0, and 18.0 µg of sodium chloride per mL, respectively.

Assay preparation—Transfer an accurately measured volume of the stock solution remaining from the *Assay for magnesium carbonate*, equivalent to about 180 mg of $NaHCO_3$, to a 100-mL volumetric flask, dilute with water to volume, and mix. Transfer 10.0 mL of the resulting solution to a 1000-mL volumetric flask, dilute with water to volume, and mix.

Procedure—Concomitantly determine the absorbances of the *Standard preparations* and the *Assay preparation* at the sodium emission line at about 589.0 nm, with a suitable atomic absorption spectrophotometer (see *Spectrophotometry and Light-scattering* ⟨851⟩) equipped with a sodium hollow-cathode lamp and an air-acetylene flame, using water as the blank. Plot the absorbances of the *Standard preparations* versus concentration, in µg of sodium chloride per mL, and draw the straight line best fitting the three plotted points. From the graph so obtained, determine the concentration, in µg per mL, of sodium chloride equivalent in the *Assay preparation*. Calculate the quantity, in g, of $NaHCO_3$ in the portion of Magnesium Carbonate and Sodium Bicarbonate for Oral Suspension taken by the formula:

$$(84.01/58.44)(5C/V),$$

in which 84.01 and 58.44 are the molecular weights of sodium bicarbonate and sodium chloride, respectively, C is the concentration, in µg per mL, of sodium chloride equivalent in the *Assay preparation*, and V is the volume, in mL, of the stock solution remaining from the *Assay for magnesium carbonate* taken.

Magnesium Carbonate Tablets, Alumina and—*see* Alumina and Magnesium Carbonate Tablets

Magnesium Carbonates Tablets, Calcium and—*see* Calcium and Magnesium Carbonates Tablets

Magnesium Chloride

$MgCl_2 \cdot 6H_2O$ 203.30
Magnesium chloride, hexahydrate.
Magnesium chloride hexahydrate [*7791-18-6*].
Anhydrous 95.21 [*7786-30-3*].

» Magnesium Chloride contains not less than 98.0 percent and not more than 101.0 percent of $MgCl_2 . 6H_2O$.

Packaging and storage—Preserve in tight containers.

Identification—A solution (1 in 20) responds to the tests for *Magnesium* ⟨191⟩ and for *Chloride* ⟨191⟩.

pH ⟨791⟩: between 4.5 and 7.0, in a 1 in 20 solution in carbon dioxide-free water.

Insoluble matter—Dissolve 20 g, accurately weighed, in 200 mL of water, heat to boiling, and digest in a covered beaker on a steam bath for 1 hour. Filter through a tared filtering crucible, wash thoroughly, and dry at 115°: the weight of the residue does not exceed 1 mg (0.005%).

Sulfate ⟨221⟩—A 10-g portion shows no more sulfate than corresponds to 0.50 mL of 0.020 *N* sulfuric acid (0.005%).

Arsenic, *Method I* ⟨211⟩: 3 ppm.

Barium—Dissolve 1 g in 10 mL of water, and add 1 mL of 2 *N* sulfuric acid: no turbidity is produced within 2 hours.

Calcium ⟨216⟩: not more than 0.01%, a 10-g specimen being used and 0.50 mL of *Standard Calcium Ion Solution* being used in the *Standard preparation.*

Potassium—Dissolve 5 g in 5 mL of water, and add 0.2 mL of sodium bitartrate TS: no turbidity is produced within 5 minutes.

Heavy metals ⟨231⟩—Dissolve 2 g in water to make 25 mL: the limit is 0.001%.

Assay—Weigh accurately about 450 mg of Magnesium Chloride, dissolve in 25 mL of water, add 5 mL of ammonia–ammonium chloride buffer TS and 0.1 mL of eriochrome black TS, and titrate with 0.05 *M* disodium ethylenediaminetetraacetate VS to a blue end-point. Each mL of 0.05 *M* disodium ethylenediaminetetraacetate is equivalent to 10.17 mg of $MgCl_2 . 6H_2O$.

Magnesium Citrate Oral Solution

$$\left[\begin{array}{l} CH_2COO- \\ HO-C-COO- \\ CH_2COO- \end{array}\right]_2 Mg_3$$

$C_{12}H_{10}Mg_3O_{14}$ 451.12

1,2,3-Propanetricarboxylic acid, hydroxy-, magnesium salt (2:3).

Magnesium citrate (3:2) [*3344-18-1*].

» Magnesium Citrate Oral Solution is a sterilized or pasteurized solution containing, in each 100 mL, not less than 7.59 g of anhydrous citric acid ($C_6H_8O_7$) and an amount of magnesium citrate equivalent to not less than 1.55 g and not more than 1.9 g of magnesium oxide (MgO).

Magnesium Citrate Oral Solution may be prepared as follows:

Magnesium Carbonate	15 g
Anhydrous Citric Acid	27.4 g
Syrup	60 mL
Talc	5 g
Lemon Oil	0.1 mL
Potassium Bicarbonate	2.5 g
Purified Water, a sufficient quantity, to make	350 mL

Dissolve the anhydrous Citric Acid in 150 mL of hot Purified Water in a suitable dish, slowly add the Magnesium Carbonate, previously mixed with 100 mL of Purified Water, and stir until it is dissolved. Then add the Syrup, heat the mixed liquids to the boiling point, immediately add the Lemon Oil, previously triturated with the Talc, and filter the mixture, while hot, into a strong bottle (previously rinsed with boiling Purified Water) of suitable capacity. Add boiled Purified Water to make the product measure 350 mL. Use Purified Cotton as a stopper for the bottle, allow to cool, add the Potassium Bicarbonate, and immediately insert the stopper in the bottle securely. Finally, shake the solution occasionally until the Potassium Bicarbonate is dissolved, cap the bottle, and sterilize or pasteurize the solution.

NOTE—An amount (30 g) of citric acid containing 1 molecule of water of hydration, equivalent to 27.4 g of anhydrous citric acid, may be used in the foregoing formula. In this process the 2.5 g of potassium bicarbonate may be replaced by 2.1 g of sodium bicarbonate, preferably in tablet form. The Oral Solution may be further carbonated by the use of CO_2 under pressure.

Packaging and storage—Preserve at controlled room temperature or in a cool place, in bottles containing not less than 200 mL.

Identification—

A: It responds to the tests for *Magnesium* ⟨191⟩.

B: To 5 mL of Oral Solution add 1 mL of potassium permanganate TS and 5 mL of mercuric sulfate TS, and heat the solution: a white precipitate is formed.

Chloride ⟨221⟩—A 2.0-mL portion shows no more chloride than corresponds to 0.30 mL of 0.020 *N* hydrochloric acid (0.01%).

Sulfate ⟨221⟩—A 2.0-mL portion shows no more sulfate than corresponds to 0.30 mL of 0.020 *N* sulfuric acid (0.015%).

Tartaric acid—To 10 mL in a test tube add 1 mL of glacial acetic acid and 3 mL of a solution of potassium acetate (1 in 2), shake the mixture vigorously, then gently rub the inner wall of the test tube with a glass rod for a few minutes, and allow to stand for 1 hour: no white, crystalline precipitate soluble in 6 *N* ammonium hydroxide is formed.

Assay for anhydrous citric acid—Measure accurately 10 mL of Magnesium Citrate Oral Solution, which previously has been freed from excessive carbon dioxide by repeated pouring, into a 250-mL beaker, and add 30 mL of water. Then add phenolphthalein TS and just enough 1 *N* sodium hydroxide to give the liquid a persistent pink color, and acidify with 4 drops of 1 *N* hydrochloric acid. Add 20 mL of calcium chloride TS, and concentrate, by boiling, to about 30 mL, stirring constantly with a rubber-tipped glass rod during the boiling. Completely transfer the precipitate from the hot mixture to a filter of from 9 cm to 11 cm in diameter with the aid of small quantities of boiling water, then wash the precipitate five times with boiling water. Collect the filtrate and washings in a 150-mL beaker, and concentrate the solution, by boiling, to about 20 mL. Add sufficient 6 *N* ammonium hydroxide, dropwise, to give the liquid a distinct red color, and then concentrate to about 10 mL. Transfer the precipitate completely from the hot mixture to a filter of from 7 cm to 9 cm in diameter with the aid of small quantities of boiling water, and wash the precipitate six times with 5-mL portions of boiling water.

Dry the two filters with the precipitates, and incinerate them together in a loosely covered platinum crucible, heating first at a low temperature until the precipitates are well charred, and then removing the cover and raising the temperature until the residue is nearly white. If a gas flame is used, prevent its contact with the mass in the crucible. Cool, place the crucible with its contents in a suitable beaker, and add about 30 mL of water and then 50.0 mL of 0.5 *N* hydrochloric acid VS. When the residue has dissolved, remove the crucible, rinsing it well with water into the beaker. Add 100 mL of water, cover the beaker with a watch glass, and boil gently for 10 minutes. Cool, and titrate the excess acid with 0.5 *N* sodium hydroxide VS, using phenolphthalein TS as the indicator. Each mL of 0.5000 *N* hydrochloric acid is equivalent to 32.02 mg of $C_6H_8O_7$.

Assay for magnesium oxide—Transfer to a 100-mL volumetric flask 50.0 mL of Magnesium Citrate Oral Solution that has been previously freed from excessive carbon dioxide by repeated pouring. Dilute to volume with water, and mix. Transfer 5.0 mL of this solution to a beaker containing 150 mL of water heated to 70° to 80°, and add 1 mL of ammonium chloride TS and then 3 mL of ammonium hydroxide. Mix, and add slowly, with stirring, 8 mL of 8-hydroxyquinoline TS. After standing for 30 minutes, filter through a sintered-glass crucible, previously dried and weighed, and wash the precipitate with ten 10-mL portions of water. Dry the crucible and contents at 105° for 3 hours, cool, and weigh. Determine the equivalent of MgO in 100 mL of the Oral Solution by multiplying the weight of $C_{18}H_{12}MgN_2O_2 . 2H_2O$ so obtained by 4.624.

Magnesium Gluconate

$[HOCH_2-CH(OH)-CH(OH)-CH(OH)-CH(OH)-COO-]_2\ Mg \cdot xH_2O$

$C_{12}H_{22}MgO_{14} . xH_2O$ (anhydrous) 414.60
D-Gluconic acid, magnesium salt (2:1), hydrate.
Magnesium D-gluconate (1:2) hydrate.
Magnesium D-gluconate (1:2) dihydrate 450.63 [*59625-89-7*].
Anhydrous [*3632-91-5*].

» Magnesium Gluconate contains not less than 98.0 percent and not more than 102.0 percent of $C_{12}H_{22}MgO_{14}$, calculated on the anhydrous basis.

Packaging and storage—Preserve in well-closed containers.

Reference standard—*USP Potassium Gluconate Reference Standard*—Dry in vacuum at 105° for 4 hours before using.

Identification—

A: A solution (1 in 10) responds to the tests for *Magnesium* ⟨191⟩.

B: It responds to *Identification test B* under *Calcium Gluconate.*

pH ⟨791⟩: between 6.0 and 7.8, in a solution (1 in 20).

Water, *Method Ib* ⟨921⟩: between 3.0% and 12.0%, 30 minutes being allowed for solubilization of the specimen and for the reaction to reach completion, and a blank determination being performed with the same volume of *Reagent* but without the specimen. Calculate the water content of the specimen, in mg, by the formula:

$$F(X_b - X)R,$$

in which X_b is the volume, in mL, of standardized *Water-Methanol Solution* required to neutralize the unconsumed *Reagent* in the blank determination, and the other terms are as defined therein.

Chloride ⟨221⟩—A 1.0-g portion shows no more chloride than corresponds to 0.70 mL of 0.020 *N* hydrochloric acid (0.05%).

Sulfate ⟨221⟩—A 2.0-g portion shows no more sulfate than corresponds to 1.0 mL of 0.020 *N* sulfuric acid (0.05%).

Arsenic, *Method I* ⟨211⟩—Dissolve 1.0 g in 35 mL of water: the limit is 3 ppm.

Heavy metals ⟨231⟩—Dissolve 1.0 g in 10 mL of water, add 6 mL of 3 *N* hydrochloric acid, and dilute with water to 25 mL: the limit is 0.002%.

Reducing substances—Transfer 1.0 g to a 250-mL conical flask, dissolve in 10 mL of water, and add 25 mL of alkaline cupric citrate TS. Cover the flask, boil gently for 5 minutes, accurately timed, and cool rapidly to room temperature. Add 25 mL of 0.6 *N* acetic acid, 10.0 mL of 0.1 *N* iodine VS, and 10 mL of 3 *N* hydrochloric acid, and titrate with 0.1 *N* sodium thiosulfate VS, adding 3 mL of starch TS as the end-point is approached. Perform a blank determination, omitting the specimen, and note the difference in volumes required. Each mL of the difference in volume of 0.1 *N* sodium thiosulfate consumed is equivalent to 2.7 mg of reducing substances (as dextrose): the limit is 1.0%.

Assay—Weigh accurately about 800 mg of Magnesium Gluconate, dissolve in 20 mL of water, add 5 mL of ammonia–ammonium chloride buffer TS and 0.1 mL of eriochrome black TS, and titrate with 0.05 *M* disodium ethylenediaminetetraacetate VS to a blue end-point. Each mL of 0.05 *M* disodium ethylenediaminetetraacetate is equivalent to 20.73 mg of $C_{12}H_{22}MgO_{14}$.

Magnesium Gluconate Tablets

» Magnesium Gluconate Tablets contain not less than 95.0 percent and not more than 105.0 percent of the labeled amount of $C_{12}H_{22}MgO_{14}$.

Packaging and storage—Preserve in well-closed containers.

Reference standard—*USP Potassium Gluconate Reference Standard*—Dry in vacuum at 105° for 4 hours before using.

Identification—A filtered solution of Tablets, equivalent to magnesium gluconate solution (1 in 10), diluted with water where necessary, responds to the *Identification tests* under *Magnesium Gluconate.*

Dissolution ⟨711⟩—

Medium: water; 900 mL.

Apparatus 2: 50 rpm.

Time: 30 minutes.

Procedure—Determine the amount of $C_{12}H_{22}MgO_{14}$ dissolved, employing atomic absorption spectrophotometry at a wavelength of about 285.2 nm, in filtered portions of the solution under test, suitably diluted with water, in comparison with a Standard solution having a known concentration of magnesium in the same medium.

Tolerances—Not less than 80% (*Q*) of the labeled amount of $C_{12}H_{22}MgO_{14}$ is dissolved in 30 minutes.

Uniformity of dosage units ⟨905⟩: meet the requirements.

Assay—Weigh and finely powder not less than 20 Magnesium Gluconate Tablets. Weigh accurately a portion of the powder, equivalent to about 800 mg of magnesium gluconate, transfer to a suitable crucible, and ignite, gently at first, until free from carbon. Cool the crucible, add 25 mL of water and 5 mL of hydrochloric acid, and stir. Heat on a steam bath for 5 minutes, and filter, rinsing the filter with several portions of water. Dilute the combined filtrate and washings with water to about 150 mL. Add ammonia–ammonium chloride buffer TS until the solution is neutral to litmus. Add 5 mL of ammonia–ammonium chloride buffer TS and 0.1 mL of eriochrome black TS, and titrate with 0.05 *M* disodium ethylenediaminetetraacetate VS to a blue end-point. Each mL of 0.05 *M* disodium ethylenediaminetetraacetate is equivalent to 20.73 mg of $C_{12}H_{22}MgO_{14}$.

Magnesium Hydroxide

$Mg(OH)_2$ 58.32
Magnesium hydroxide.
Magnesium hydroxide [*1309-42-8*].

» Magnesium Hydroxide, dried at 105° for 2 hours, contains not less than 95.0 percent and not more than 100.5 percent of $Mg(OH)_2$.

Packaging and storage—Preserve in tight containers.

Identification—A 1 in 20 solution in 3 *N* hydrochloric acid responds to the tests for *Magnesium* ⟨191⟩.

Microbial limit—It meets the requirements of the test for absence of *Escherichia coli* under *Microbial Limit Tests* ⟨61⟩.

Loss on drying ⟨731⟩—Dry it at 105° for 2 hours: it loses not more than 2.0% of its weight.

Loss on ignition ⟨733⟩—Ignite it at 800°, increasing the heat gradually, to constant weight: it loses between 30.0% and 33.0% of its weight.

Soluble salts—Boil 2.0 g with 100 mL of water for 5 minutes in a covered beaker, filter while hot, cool, and dilute the filtrate with water to 100 mL. Titrate 50 mL of the diluted filtrate with 0.10 *N* sulfuric acid, using methyl red TS as the indicator: not more than 2.0 mL of the acid is consumed. Evaporate 25 mL of the diluted filtrate to dryness, and dry at 105° for 3 hours: not more than 10 mg of residue remains.

Carbonate—Boil a mixture of 0.10 g with 5 mL of water, cool, and add 5 mL of 6 *N* acetic acid: not more than a slight effervescence is observed.

Arsenic, *Method I* ⟨211⟩—Prepare the *Test Preparation* by dissolving 1.0 g in 25 mL of 3 *N* hydrochloric acid. The limit is 3 ppm.

Calcium—

Dilute hydrochloric acid, Lanthanum solution, Standard preparations, and *Blank solution*—Prepare as directed in the test for *Calcium* under *Magnesium Carbonate.*

Test preparation—Transfer 250 mg of Magnesium Hydroxide, previously dried, to a beaker, add 30 mL of *Dilute hydrochloric acid*, and stir until dissolved, heating if necessary. Transfer the solution so obtained to a 200-mL volumetric flask containing 4 mL of *Lanthanum solution*, dilute with water to volume, and mix.

Procedure—Proceed as directed in the test for *Calcium* under *Magnesium Carbonate:* the limit is 0.7%.

Heavy metals, *Method I* ⟨231⟩—Dissolve 1.0 g in 15 mL of 3 *N* hydrochloric acid, and evaporate the solution on a steam bath to dryness. Toward the end of the evaporation, stir the residue frequently, disintegrate it so that finally a dry powder is obtained, dissolve the residue in 20 mL of water, and filter. To the filtrate, which should be neutral to litmus, add 2 mL of 1 *N* acetic acid, and dilute with water to 50 mL: the limit is 0.004%.

Lead ⟨251⟩—Prepare a *Test Preparation* by dissolving 1 g of Magnesium Hydroxide in 20 mL of 3 *N* hydrochloric acid. Use 10 mL of *Diluted Standard Lead Solution* (10 µg of Pb) for the test: the limit is 0.001%.

Assay—Transfer about 400 mg of Magnesium Hydroxide, previously dried and accurately weighed, to a conical flask. Add 25.0 mL of 1 *N* sulfuric acid VS, and after solution is complete, add methyl red TS. Titrate the excess acid with 1 *N* sodium hydroxide VS. From the volume of 1 *N* sulfuric acid consumed, deduct the volume of 1 *N* sulfuric acid corresponding to the content of calcium in the Magnesium Hydroxide taken for the assay. The difference is the volume of 1 *N* sulfuric acid equivalent to the $Mg(OH)_2$ in the portion of Magnesium Hydroxide taken. Each mL of 1 *N* sulfuric acid is equivalent to 29.16 mg of $Mg(OH)_2$ and to 20.04 mg of Ca.

Magnesium Hydroxide Paste

» Magnesium Hydroxide Paste is an aqueous paste of magnesium hydroxide, each 100 g of which contains not less than 29.0 g and not more than 33.0 g of $Mg(OH)_2$.

Packaging and storage—Preserve in tight containers.

Identification—One g of Paste dissolved in 10 mL of 3 *N* hydrochloric acid responds to the tests for *Magnesium* ⟨191⟩.

Microbial limits ⟨61⟩—Its total aerobic microbial count does not exceed 400 per g, and it meets the requirements of the test for absence of *Escherichia coli.*

Soluble alkalies—Mix 25.0 g of Paste with 75.0 mL of water, transfer about 25 mL of this diluted Paste to a filter, and reject the first 5 mL of the filtrate. [NOTE—Retain the remaining diluted Paste for the tests for *Carbonate and acid-insoluble matter, Arsenic*, and *Heavy metals.*] Dilute 5 mL of the clear filtrate with 40 mL of water. Add 1 drop of methyl red TS, and titrate the solution with 0.10 *N* sulfuric acid to the production of a persistent pink color: not more than 1.0 mL of the acid is required.

Soluble salts—To 5.0 mL of the clear filtrate obtained in the test for *Soluble alkalies* add 3 drops of sulfuric acid, evaporate on a steam bath to dryness, and ignite gently to constant weight: the residue weighs not more than 12 mg.

Carbonate and acid-insoluble matter—To 1 mL of the diluted Paste obtained in the test for *Soluble alkalies* add 2 mL of 3 *N* hydrochloric acid: not more than a slight effervescence occurs, and the solution is not more than slightly turbid.

Arsenic, *Method I* ⟨211⟩—Prepare a *Standard Preparation* as directed in the test for *Arsenic* ⟨211⟩, except to have it contain 2 µg of arsenic instead of 3 µg. Dissolve 3.3 mL of the diluted Paste obtained in the test for *Soluble alkalies* in 20 mL of 7 *N* sulfuric acid, and add 35 mL of water: the limit is 0.6 ppm, based on the amount of diluted Paste taken.

Calcium—

Potassium nitrate solution and *Standard preparation*—Prepare as directed in the test for *Calcium* under *Magnesium Carbonate.*

Test preparation—Transfer a portion of the Paste, equivalent to 1.0 g of $Mg(OH)_2$, to a 100-mL volumetric flask, add 25 mL of 3 *N* hydrochloric acid, shake until dissolved, dilute with *Potassium nitrate solution* to volume, and mix. Transfer 10.0 mL of this solution to a second 100-mL volumetric flask, dilute with *Potassium nitrate solution* to volume, and mix.

Procedure—Proceed as directed in the test for *Calcium* under *Magnesium Carbonate:* the limit is 0.7%.

Heavy metals, *Method I* ⟨231⟩—To 4.0 mL of the diluted Paste obtained in the test for *Soluble alkalies* add 6 mL of 3 *N* hydrochloric acid, and evaporate the solution on a steam bath to dryness, with frequent stirring. Dissolve the residue in 20 mL of water, and evaporate to dryness in the same manner as before. Redissolve in 20 mL of water, filter if necessary, and dilute with water to 25 mL: the limit is 5 ppm, based on the amount of diluted Paste taken.

Assay—Transfer an accurately weighed portion of Magnesium Hydroxide Paste, equivalent to about 250 mg of magnesium hydroxide, to a 100-mL volumetric flask. Dissolve in 10 mL of 3 *N* hydrochloric acid, dilute with water to volume, and mix. Filter, if necessary, and transfer 25.0 mL of the filtrate to a beaker containing 75 mL of water, and mix. Adjust the reaction of the solution to a pH of 7 (using pH indicator paper; see *Indicator and Test Papers* under *Reagents*, in the section *Reagents, Indicators, and Solutions*) with 1 *N* sodium hydroxide, add 5 mL of ammonia–ammonium chloride buffer TS and 0.15 mL of eriochrome black TS, and titrate with 0.05 *M* disodium ethylenediaminetetraacetate VS to a blue end-point. Each mL of 0.05 *M* disodium ethylenediaminetetraacetate is equivalent to 2.916 mg of $Mg(OH)_2$.

Magnesium Oxide

MgO 40.30
Magnesium oxide.
Magnesium oxide [*1309-48-4*].

» Magnesium Oxide, after ignition, contains not less than 96.0 percent and not more than 100.5 percent of MgO.

Packaging and storage—Preserve in tight containers.

Labeling—Label it to indicate whether it is Light Magnesium Oxide or Heavy Magnesium Oxide.

Identification—A solution in diluted hydrochloric acid responds to the tests for *Magnesium* ⟨191⟩.

Loss on ignition ⟨733⟩—Transfer to a tared platinum crucible about 500 mg, weigh accurately, and ignite at 800 ± 25° to constant weight: it loses not more than 10.0% of its weight.

Free alkali and soluble salts—Boil 2.0 g with 100 mL of water for 5 minutes in a covered beaker, and filter while hot. To 50 mL of the cooled filtrate add methyl red TS, and titrate with

0.10 *N* sulfuric acid: not more than 2.0 mL of the acid is consumed. Evaporate 25 mL of the remaining filtrate to dryness, and dry at 105° for 1 hour: not more than 10 mg of residue remains (2.0%).

Acid-insoluble substances—Mix 2 g with 75 mL of water, add hydrochloric acid in small portions, with agitation, until no more dissolves, and boil for 5 minutes. If an insoluble residue remains, filter, wash well with water until the last washing is free from chloride, and ignite: the weight of the ignited residue does not exceed 2 mg (0.1%).

Arsenic, *Method I* ⟨211⟩—Prepare the *Test Preparation* by dissolving 1.0 g in 30 mL of 7 *N* sulfuric acid, and diluting with water to 55 mL: the resulting solution meets the requirements of the test, the addition of 20 mL of 7 *N* sulfuric acid specified under *Procedure* being omitted. The limit is 3 ppm.

Calcium—

Dilute hydrochloric acid, Lanthanum solution, Standard preparations, and *Blank solution*—Prepare as directed in the test for *Calcium* under *Magnesium Carbonate.*

Test preparation—Transfer 250 mg of Magnesium Oxide, freshly ignited, to a beaker, add 30 mL of *Dilute hydrochloric acid*, and stir until dissolved, heating if necessary. Transfer the solution so obtained to a 200-mL volumetric flask containing 4 mL of *Lanthanum solution*, dilute with water to volume, and mix.

Procedure—Proceed as directed in the test for *Calcium* under *Magnesium Carbonate:* the limit is 1.1%.

Heavy metals ⟨231⟩—Dissolve 1 g in 20 mL of 3 *N* hydrochloric acid, and evaporate the solution on a steam bath to dryness. Toward the end of the evaporation, stir frequently to disintegrate the residue so that finally a dry powder is obtained. Dissolve the residue in 20 mL of water, and evaporate to dryness in the same manner as before. Redissolve the residue in 20 mL of water, filter if necessary, and dilute with water to 40 mL. To 20 mL add water to make 25 mL: the limit is 0.004%.

Iron ⟨241⟩—Boil 40 mg with 5 mL of 2 *N* nitric acid for 1 minute. Cool, dilute with water to 50 mL, and mix. Dilute 25 mL of this solution with water to 45 mL, and add 2 mL of hydrochloric acid: the limit is 0.05%.

Assay—Ignite about 500 mg of Magnesium Oxide to constant weight in a tared platinum crucible, weigh the residue accurately, dissolve it in 30.0 mL of 1 *N* sulfuric acid VS, add methyl orange TS, and titrate the excess acid with 1 *N* sodium hydroxide VS. From the volume of 1 *N* sulfuric acid consumed, deduct the volume of 1 *N* sulfuric acid corresponding to the content of calcium in the Magnesium Oxide taken for the assay. The difference is the volume of 1 *N* sulfuric acid equivalent to the MgO in the portion of Magnesium Oxide taken. Each mL of 1 *N* sulfuric acid is equivalent to 20.15 mg of MgO and to 20.04 mg of Ca.

Magnesium Oxide Capsules

» Magnesium Oxide Capsules contain not less than 90.0 percent and not more than 110.0 percent of the labeled amount of MgO.

Packaging and storage—Preserve in well-closed containers.

Identification—

A: Transfer the contents of 1 Capsule to a beaker, add 10 mL of 3 *N* hydrochloric acid and 5 drops of methyl red TS, heat to boiling, add 6 *N* ammonium hydroxide until the color of the solution changes to deep yellow, then continue boiling for 2 minutes, and filter: the filtrate so obtained responds to the tests for *Magnesium* ⟨191⟩.

B: Wash the solids on the filter obtained in *Identification test A* with hot ammonium chloride solution (1 in 50), add 10 mL of 3 *N* hydrochloric acid, and filter. Transfer the filter paper and contents to a small platinum dish, ignite, cool in a desiccator, and weigh. Moisten the residue with water, and add 6 mL of hydrofluoric acid. Evaporate to dryness, ignite for 5 minutes, cool in a desiccator, and weigh: a loss of more than 10% in relation to the weight of the residue from the initial ignition indicates SiO_2.

Disintegration ⟨701⟩: 10 minutes, simulated gastric fluid TS being substituted for water in the test.

Uniformity of dosage units ⟨905⟩: meet the requirements.

Acid-neutralizing capacity ⟨301⟩: not less than 5 mEq of acid is consumed by the minimum single dose recommended in the labeling, and not less than 85.0% of the expected mEq value calculated from the results of the *Assay* is obtained. Each mg of MgO has an expected acid-neutralizing capacity value of 0.0492 mEq.

Assay—Weigh accurately the contents of not less than 20 Magnesium Oxide Capsules, and mix. Transfer an accurately weighed portion of the powder, equivalent to about 500 mg of magnesium oxide, to a beaker, add 20 mL of water, and slowly add 40 mL of 3 *N* hydrochloric acid, with mixing. Heat the mixture to boiling, cool, and filter into a 200-mL volumetric flask. Wash the beaker with water, adding the washings to the filter. Add water to volume, and mix. Transfer 20.0 mL of this solution to a 400-mL beaker, add 180 mL of water and 20 mL of triethanolamine, and stir. Add 10 mL of ammonia–ammonium chloride buffer TS and 3 drops of an eriochrome black indicator solution prepared by dissolving 200 mg of eriochrome black T in a mixture of 15 mL of triethanolamine and 5 mL of dehydrated alcohol, and mix. Cool the solution to between 3° and 4° by immersion of the beaker in an ice bath, then remove, and titrate with 0.05 *M* disodium ethylenediaminetetraacetate VS to a blue end-point. Perform a blank determination, substituting 20 mL of water for the assay solution, and make any necessary correction. Each mL of 0.05 *M* disodium ethylenediaminetetraacetate consumed is equivalent to 2.015 mg of MgO.

Magnesium Oxide Tablets

» Magnesium Oxide Tablets contain not less than 90.0 percent and not more than 110.0 percent of the labeled amount of MgO.

Packaging and storage—Preserve in well-closed containers.

Disintegration ⟨701⟩: 10 minutes, simulated gastric fluid TS being substituted for water in the test.

Uniformity of dosage units ⟨905⟩: meet the requirements.

Acid-neutralizing capacity ⟨301⟩: not less than 5 mEq of acid is consumed by the minimum single dose recommended in the labeling, and not less than 85.0% of the expected mEq value calculated from the results of the *Assay* is obtained. Each mg of MgO has an expected acid-neutralizing capacity value of 0.0492 mEq.

Other requirements—One powdered Tablet responds to the *Identification tests* under *Magnesium Oxide Capsules.*

Assay—Weigh and finely powder not less than 20 Magnesium Oxide Tablets. Transfer an accurately weighed portion of the powder, equivalent to about 500 mg of magnesium oxide, to a beaker. Proceed as directed in the *Assay* under *Magnesium Oxide Capsules* beginning with "add 20 mL of water." Each mL of 0.05 *M* disodium ethylenediaminetetraacetate consumed is equivalent to 2.015 of mg of MgO.

Magnesium Oxide, and Sodium Carbonate Irrigation, Citric Acid,—*see* Citric Acid, Magnesium Oxide, and Sodium Carbonate Irrigation

Magnesium Oxide Tablets, Alumina, Magnesium Carbonate, and—*see* Alumina, Magnesium Carbonate, and Magnesium Oxide Tablets

Magnesium Phosphate

$Mg_3(PO_4)_2 \cdot 5H_2O$ 352.93
Phosphoric acid, magnesium salt (2:3), pentahydrate.
Magnesium phosphate (3:2) pentahydrate [*10233-87-1*].
Anhydrous 262.86 [*7757-87-1*].

» Magnesium Phosphate, ignited at 425° to constant weight, contains not less than 98.0 percent and not more than 101.5 percent of $Mg_3(PO_4)_2$.

Packaging and storage—Preserve in well-closed containers.

Identification—

A: Dissolve about 200 mg in 10 mL of 2 *N* nitric acid, and add, dropwise, ammonium molybdate TS: a greenish yellow precipitate of ammonium phosphomolybdate is formed and it is soluble in 6 *N* ammonium hydroxide.

B: Dissolve 0.1 g in 0.7 mL of 1 *N* acetic acid and 20 mL of water. Add 1 mL of ferric chloride TS, allow to stand for 5 minutes, and filter: 5 mL of the filtrate responds to the test for *Magnesium* ⟨191⟩.

Microbial limit—It meets the requirements of the test for absence of *Escherichia coli* under *Microbial Limit Tests* ⟨61⟩.

Loss on ignition ⟨733⟩—Ignite it at 425° to constant weight: it loses between 20.0% and 27.0% of its weight.

Acid-insoluble substances—If an insoluble residue remains in the test for *Carbonate*, filter the solution, wash well with hot water until the last washing is free from chloride, and ignite the residue: the weight of the residue does not exceed 4 mg (0.2%).

Soluble substances—Digest 2.0 g with 100 mL of water on a steam bath for 30 minutes, cool, add sufficient water to restore the original volume, mix, and filter. Evaporate 50 mL of the filtrate to dryness, and ignite gently to constant weight: the weight of the residue does not exceed 15 mg (1.5%).

Carbonate—Mix 2.0 g with 20 mL of water, and add hydrochloric acid, dropwise, to effect solution: no effervescence occurs when the acid is added.

Chloride ⟨221⟩—Dissolve 0.50 g in 50 mL of 2 *N* nitric acid, and add 1 mL of silver nitrate TS: the turbidity does not exceed that produced by 1.0 mL of 0.020 *N* hydrochloric acid (0.14%).

Nitrate—Mix 0.20 g with 5 mL of water, and add just sufficient hydrochloric acid to effect solution. Dilute with water to 10 mL, add 0.1 mL of indigo carmine TS, then add, with stirring, 10 mL of sulfuric acid: the blue color persists for not less than 5 minutes.

Sulfate ⟨221⟩—Dissolve 0.50 g in the smallest possible amount of 3 *N* hydrochloric acid, dilute with water to 48 mL, and add 2 mL of barium chloride TS: the turbidity does not exceed that produced by 3.0 mL of 0.020 *N* sulfuric acid (0.6%).

Arsenic, *Method I* ⟨211⟩—Prepare a *Test Preparation* by dissolving 1.0 g in just sufficient 3 *N* hydrochloric acid (about 9 mL) to dissolve the specimen. The limit is 3 ppm.

Barium—Mix 2.0 g with 40 mL of water, heat, add hydrochloric acid, dropwise, to effect solution, and then add 1 mL of acid in excess. Cool, dilute with water to 50 mL, and filter. To 5 mL of the filtrate add 1 mL of potassium sulfate TS: no turbidity is produced within 15 minutes.

Calcium—Mix 0.50 g with 15 mL of water, heat, and add sufficient hydrochloric acid, in small portions, to effect solution. Cool, add 6 *N* ammonium hydroxide, in small portions, to produce a slight permanent precipitate, then add 2 mL of 6 *N* acetic acid. Dilute with water to 25 mL, and filter. To 10 mL of the filtrate add 2 mL of ammonium oxalate TS: not more than a slight turbidity is produced within 5 minutes.

Heavy metals, *Method I* ⟨231⟩—Dissolve 0.67 g in 4.5 mL of 3 *N* hydrochloric acid, and dilute with water to 25 mL: the limit is 0.003%.

Dibasic salt and magnesium oxide—Ignite about 2.5 g to constant weight. Weigh accurately about 2 g of ignited salt, and dissolve it by warming with 50.0 mL of 1 *N* hydrochloric acid VS. Cool, add 1 or 2 drops of methyl orange TS, and slowly titrate the excess 1 *N* hydrochloric acid VS with 1 *N* sodium hydroxide VS to a yellow color, vigorously shaking the mixture during the titration. Between 14.8 and 15.4 mL of 1 *N* hydrochloric acid is consumed for each g of the ignited salt.

Lead ⟨251⟩—Prepare a *Test Preparation* by dissolving 1.0 g in 20 mL of 3 *N* hydrochloric acid, evaporating on a steam bath to about 10 mL, diluting with water to about 20 mL, and cooling. Use 5 mL of *Diluted Standard Lead Solution* (5 µg of Pb) for the test: the limit is 5 ppm.

Assay—Weigh accurately about 200 mg of Magnesium Phosphate, previously ignited at 425° to constant weight, and dissolve in a mixture of 25 mL of water and 10 mL of 2 *N* nitric acid. Filter, if necessary, wash any precipitate, add sufficient 6 *N* ammonium hydroxide to the filtrate to produce a slight precipitate, and then dissolve the precipitate by the addition of 1 mL of 2 *N* nitric acid. Adjust the temperature to about 50°, add 75 mL of ammonium molybdate TS, and maintain the temperature at about 50° for 30 minutes, stirring occasionally. Wash the precipitate once or twice with water by decantation, using from 30 to 40 mL each time and passing the washings through a filter. Transfer the precipitate to the filter, and wash with potassium nitrate solution (1 in 100) until the last washing is not acid to litmus. Transfer the precipitate and filter to the precipitation vessel, add 50 mL of water and 40.0 mL of 1 *N* sodium hydroxide VS, agitate until the precipitate is dissolved, add phenolphthalein TS, and then titrate the excess alkali with 1 *N* sulfuric acid VS. Each mL of 1 *N* sodium hydroxide is equivalent to 5.716 mg of $Mg_3(PO_4)_2$.

Magnesium Salicylate

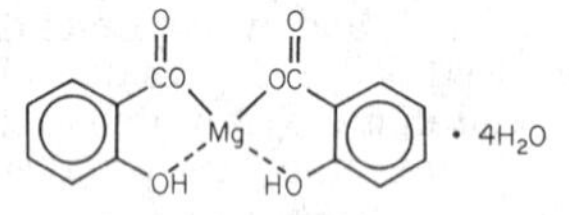

$\cdot 4H_2O$

$C_{14}H_{10}MgO_6 \cdot 4H_2O$ 370.60
Magnesium, bis(2-hydroxybenzoato-O^1,O^2)-, tetrahydrate.
Magnesium salicylate (1:2), tetrahydrate [*18917-95-8*].
Anhydrous 298.53 [*18917-89-0*].

» Magnesium Salicylate contains not less than 98.0 percent and not more than 103.0 percent of $C_{14}H_{10}MgO_6 \cdot 4H_2O$.

Packaging and storage—Store in tight containers.

Reference standards—*USP Magnesium Salicylate Reference Standard*—Dry at 105° for 4 hours before using. *USP Salicylic Acid Reference Standard*—Dry over silica gel for 3 hours before using.

Identification—

A: The infrared absorption spectrum of a potassium bromide dispersion of it, previously dried, exhibits maxima only at the same wavelengths as that of a similar preparation of USP Magnesium Salicylate RS.

B: The ultraviolet absorption spectrum of a solution (1 in 50,000) exhibits maxima and minima at the same wavelengths as that of a similar solution of USP Magnesium Salicylate RS, concomitantly measured.

C: It responds to the test for *Magnesium* ⟨191⟩.

Loss on drying ⟨731⟩—Dry it at 105° for 4 hours: it loses between 17.5% and 20.0% of its weight.

Heavy metals, *Method I* ⟨231⟩: 0.004%.

Magnesium content—Transfer about 800 mg, accurately weighed, to a 200-mL volumetric flask. Dissolve in water, dilute with water to volume, and mix. Stir the resulting solution continuously for about 15 minutes and filter, discarding the first 10 mL of the filtrate, into a flask. Transfer 50.0 mL of the filtrate to a 250-mL conical flask, add 50 mL of water, 5 mL of ammonia–ammonium chloride buffer TS, and 0.15 mL of eriochrome black TS, and titrate with 0.05 *M* disodium ethylenediaminetetraacetate VS to a blue end-point. Each mL of 0.05 *M* disodium ethylenediaminetetraacetate is equivalent to 1.215 mg of magnesium: between 6.3% and 6.7% of magnesium is found.

Assay—Transfer 3.0 mL of the filtrate prepared in the test for *Magnesium content* to a 500-mL volumetric flask, dilute with water to volume, and mix. Concomitantly determine the absorbances of this solution and a solution of USP Salicylic Acid RS in water having a known concentration of about 18 µg per mL, in 1-cm cells at the wavelength of maximum absorbance at about 296 nm, with a suitable spectrophotometer, using water as the blank. Calculate the quantity, in mg, of $C_{14}H_{10}MgO_6 \cdot 4H_2O$ in the Magnesium Salicylate taken in the test for *Magnesium content* by the formula:

$$(370.60/276.24)(33.3C)(A_U/A_S),$$

in which 370.60 is the molecular weight of magnesium salicylate tetrahydrate, 276.24 is twice the molecular weight of salicylic acid, C is the concentration, in µg per mL, of USP Salicylic Acid RS in the Standard solution, and A_U and A_S are the absorbances of the test solution and the Standard solution, respectively.

Magnesium Salicylate Tablets

» Magnesium Salicylate Tablets contain an amount of magnesium salicylate ($C_{14}H_{10}MgO_6 \cdot 4H_2O$) equivalent to not less than 95.0 percent and not more than 105.0 percent of the labeled amount of $C_{14}H_{10}MgO_6$.

Packaging and storage—Preserve in tight containers.

Reference standards—*USP Magnesium Salicylate Reference Standard*—Dry at 105° for 4 hours before using. *USP Salicylic Acid Reference Standard*—Dry over silica gel for 3 hours before using.

Identification—

A: The infrared absorption spectrum of a potassium bromide dispersion of a quantity of finely powdered Tablets exhibits maxima at the same wavelengths as that of a similar preparation of USP Magnesium Salicylate RS.

B: A filtered solution of Tablets, equivalent to magnesium salicylate solution (1:20), responds to the test for *Magnesium* ⟨191⟩.

Dissolution ⟨711⟩—

Medium: water; 900 mL.

Apparatus 2: 50 rpm.

Time: 120 minutes.

Procedure—Determine the amount of $C_{14}H_{10}MgO_6$ dissolved from ultraviolet absorbances at the wavelength of maximum absorbance at about 296 nm of filtered portions of the solution under test, suitably diluted with water, in comparison with a Standard solution having a known concentration of USP Magnesium Salicylate RS in the same medium, using water as the blank.

Tolerances—Not less than 80% (Q) of the labeled amount of $C_{14}H_{10}MgO_6$ is dissolved in 120 minutes.

Uniformity of dosage units ⟨905⟩: meet the requirements.

Assay—Weigh and finely powder not less than 20 Magnesium Salicylate Tablets. Weigh accurately a portion of the powder, equivalent to about 500 mg of magnesium salicylate, and transfer to a 250-mL volumetric flask. Dilute with water to volume, mix, and filter, discarding the first 20 mL of the filtrate. Dilute an accurately measured portion of the filtrate quantitatively and stepwise, if necessary, to obtain a final concentration of about 20 µg per mL. Dissolve an accurately weighed quantity of USP Salicylic Acid RS in water, and dilute quantitatively and stepwise, if necessary, with water to obtain a Standard solution having a known concentration of about 18 µg per mL. Concomitantly determine the absorbances of both solutions in 1-cm cells at the wavelength of maximum absorbance at about 296 nm, with a suitable spectrophotometer, using water as the blank. Calculate the quantity, in mg, of $C_{14}H_{10}MgO_6$ in the portion of Tablets taken by the formula:

$$(298.53/276.24)(L/D)(C)(A_U/A_S),$$

in which 298.53 is the molecular weight of anhydrous magnesium salicylate, 276.24 is twice the molecular weight of salicylic acid, L is the labeled quantity, in mg, of magnesium salicylate in each Tablet, D is the concentration, in mg per mL, of magnesium salicylate in the solution from the Tablets, based on the labeled quantity per Tablet and the extent of dilution, C is the concentration, in µg per mL, of USP Salicylic Acid RS in the Standard solution, and A_U and A_S are the absorbances of the solution from the Tablets and the Standard solution, respectively.

Magnesium Silicate—*see* Magnesium Silicate NF

Magnesium Stearate—*see* Magnesium Stearate NF

Magnesium Sulfate

$MgSO_4 \cdot xH_2O$

Sulfuric acid magnesium salt (1:1), hydrate.

Magnesium sulfate (1:1) monohydrate 138.36

Magnesium sulfate (1:1) heptahydrate 246.47 [*10034-99-8*].

Anhydrous 120.36 [*7487-88-9*].

» Magnesium Sulfate, rendered anhydrous by ignition, contains not less than 99.0 percent and not more than 100.5 percent of $MgSO_4$.

Packaging and storage—Preserve in well-closed containers.

Labeling—The label states whether it is the monohydrate, the dried form, or the heptahydrate.

Identification—A solution (1 in 20) responds to the tests for *Magnesium* ⟨191⟩ and for *Sulfate* ⟨191⟩.

pH ⟨791⟩: between 5.0 and 9.2, in a solution (1 in 20).

Loss on ignition ⟨733⟩—Weigh accurately about 1 g in a crucible, heat at 105° for 2 hours, then ignite in a muffle furnace at 450 ± 25° to constant weight: the monohydrate loses between 13.0% and 16.0% of its weight, the dried form loses between 22.0% and 28.0% of its weight, and the heptahydrate loses between 40.0% and 52.0% of its weight.

Chloride ⟨221⟩—A 1.0-g portion shows no more chloride than corresponds to 0.20 mL of 0.020 *N* hydrochloric acid (0.014%).

Arsenic, *Method I* ⟨211⟩: 3 ppm.

Heavy metals ⟨231⟩—Dissolve 2 g in 25 mL of water: the limit is 0.001%.

Selenium ⟨291⟩—Dissolve 200 mg in 50 mL of 0.25 *N* nitric acid to obtain the *Test Solution.* The limit is 0.003%.

Assay—Weigh accurately about 250 mg of the ignited Magnesium Sulfate obtained in the test for *Loss on ignition*, and dissolve in 100 mL of water and the minimum amount of 3 *N* hydrochloric acid required for a clear solution. Adjust the reaction of the solution (using pH indicator paper; see *Indicator and Test Papers* under *Reagents* in the section, *Reagents, Indicators, and Solutions*) with 1 *N* sodium hydroxide to a pH of 7, add 5 mL of ammonia–ammonium chloride buffer TS and 0.15 mL of eriochrome black TS, and titrate with 0.05 *M* disodium ethylenediaminetetraacetate VS to a blue end-point. Each mL of 0.05 *M* disodium ethylenediaminetetraacetate is equivalent to 6.018 mg of $MgSO_4$.

Magnesium Sulfate Injection

» Magnesium Sulfate Injection is a sterile solution of Magnesium Sulfate in Water for Injection. It contains magnesium sulfate equivalent to not less than 93.0 percent and not more than 107.0 percent of the labeled amount of $MgSO_4 \cdot 7H_2O$.

Packaging and storage—Preserve in single-dose or in multiple-dose containers, preferably of Type I glass.

Labeling—The label states the total osmolar concentration in mOsmol per liter. Where the contents are less than 100 mL, or where the label states that the Injection is not for direct injection but is to be diluted before use, the label alternatively may state the total osmolar concentration in mOsmol per mL.

Identification—It responds to the tests for *Magnesium* ⟨191⟩ and for *Sulfate* ⟨191⟩.

pH ⟨791⟩: between 5.5 and 7.0, when diluted to a concentration of 5% (w/v).

Particulate matter ⟨788⟩: meets the requirements under *Small-volume Injections.*

Other requirements—It meets the requirements under *Injections* ⟨1⟩.

Assay—Transfer to a beaker an accurately measured volume of Magnesium Sulfate Injection, equivalent to about 250 mg of anhydrous magnesium sulfate, and dilute with water to 100 mL. Adjust the reaction of the solution to a pH of 7 (using pH indicator paper; see *Indicator and Test Papers* under *Reagents* in the section, *Reagents, Indicators, and Solutions*) with 1 *N* sodium hydroxide, add 5 mL of ammonia–ammonium chloride buffer TS and 0.15 mL of eriochrome black TS, and titrate with 0.05 *M* disodium ethylenediaminetetraacetate VS to a blue end-point. Each mL of 0.05 *M* disodium ethylenediaminetetraacetate is equivalent to 12.32 mg of $MgSO_4.7H_2O$.

Magnesium Trisilicate

$2MgO.3SiO_2.xH_2O$ (anhydrous) 260.86
Silicic acid ($H_4Si_3O_8$), magnesium salt (1:2), hydrate.
Magnesium silicate hydrate ($Mg_2Si_3O_8.xH_2O$) [*39365-87-2*].
Anhydrous [*14987-04-3*].

» Magnesium Trisilicate is a compound of Magnesium Oxide and silicon dioxide with varying proportions of water. It contains not less than 20.0 percent of magnesium oxide (MgO) and not less than 45.0 percent of silicon dioxide (SiO_2).

Packaging and storage—Preserve in well-closed containers.

Identification—

A: Mix about 500 mg with 10 mL of 3 *N* hydrochloric acid, filter, and neutralize the filtrate to litmus paper with 6 *N* ammonium hydroxide: the neutralized filtrate responds to the tests for *Magnesium* ⟨191⟩.

B: Prepare a bead by fusing a few crystals of sodium ammonium phosphate on a platinum loop in the flame of a Bunsen burner. Place the hot, transparent bead in contact with Magnesium Trisilicate, and again fuse: silica floats about in the bead, producing, upon cooling, an opaque bead with a web-like structure.

Water, *Method III* ⟨921⟩—Weigh accurately about 1 g in a tared platinum crucible provided with a cover. Gradually apply heat to the crucible at first, then strongly ignite to constant weight: it loses between 17.0% and 34.0% of its weight.

Soluble salts—Boil 10.0 g with 150 mL of water for 15 minutes. Cool to room temperature, allow the mixture to stand for 15 minutes, filter with the aid of suction, transfer the filtrate to a 200-mL volumetric flask, dilute with water to volume, and mix. Evaporate 50.0 mL of this solution, representing 2.5 g of the Trisilicate, in a tared platinum dish to dryness, and ignite gently to constant weight: the weight of the residue does not exceed 38.0 mg (1.5%).

Chloride ⟨221⟩—A 20-mL portion of the diluted filtrate prepared in the test for *Soluble salts,* representing 1 g of Magnesium Trisilicate, shows no more chloride than corresponds to 0.75 mL of 0.020 *N* hydrochloric acid (0.055%).

Sulfate—Treat the residue obtained in the test for *Soluble salts* with 2 mL of hydrofluoric acid, and evaporate on a steam bath to dryness. Mix the residue with water, transfer to a filter, and wash, using approximately 50 mL of water for the complete procedure. Heat the filtrate to boiling, and add 0.1 mL of hydrochloric acid and 5 mL of barium chloride TS. Maintain the mixture near its boiling point for 1 hour, filter, wash the precipitate thoroughly with water, dry, and ignite to constant weight: the weight of the residue does not exceed 30 mg (0.5%).

Free alkali—Add 2 drops of phenolphthalein TS to 20 mL of the diluted filtrate prepared in the test for *Soluble salts,* representing 1 g of the Trisilicate: if a pink color is produced, not more than 1.0 mL of 0.10 *N* hydrochloric acid is required to discharge it.

Arsenic, *Method I* ⟨211⟩: 8 ppm.

Heavy metals ⟨231⟩—Boil 2.67 g with a mixture of 50 mL of water and 5 mL of hydrochloric acid for 20 minutes, adding water to maintain the volume during the boiling. Add ammonium hydroxide until the mixture is only slightly acid to litmus paper. Filter with the aid of suction, and wash with 15 to 20 mL of water, combining the washing with the original filtrate. Add 2 drops of phenolphthalein TS, then add a slight excess of 6 *N* ammonium hydroxide. Discharge the pink color with dilute hydrochloric acid (1 in 100), then add 8 mL of dilute hydrochloric acid (1 in 100). Dilute with water to 100 mL, and use 25 mL of the solution for the test: the limit is 0.003%.

Acid-consuming capacity—Weigh accurately about 200 mg into a glass-stoppered, 125-mL conical flask. Add 30.0 mL of 0.1 *N* hydrochloric acid VS and 20.0 mL of water. Place the flask in a bath maintained at 37°, and shake the mixture occasionally during a period of 4 hours but leave the mixture undisturbed during the last 15 minutes of the heating period. Cool to room temperature. To 25.0 mL of the supernatant liquid add methyl red TS, and titrate the excess acid with 0.1 *N* sodium hydroxide VS. One g of Magnesium Trisilicate, calculated on the anhydrous basis, consumes not less than 140 mL and not more than 160 mL of 0.10 *N* hydrochloric acid.

Assay for magnesium oxide—Weigh accurately about 1.5 g, and transfer to a 250-mL conical flask. Add 50.0 mL of 1 *N* sulfuric acid VS, and digest on a steam bath for 1 hour. Cool to room temperature, add methyl orange TS, and titrate the excess acid with 1 *N* sodium hydroxide VS. Each mL of 1 *N* sulfuric acid is equivalent to 20.15 mg of MgO.

Assay for silicon dioxide—Transfer about 700 mg of Magnesium Trisilicate, accurately weighed, to a small platinum dish. Add 10 mL of 1 *N* sulfuric acid, and heat on a steam bath to dryness, leaving the dish uncovered. Treat the residue with 25 mL of water, and digest on a steam bath for 15 minutes. Decant the supernatant liquid through an ashless filter paper, with the aid of suction, and wash the residue, by decantation, three times with hot water, passing the washings through the filter paper. Finally transfer the residue to the filter, and wash thoroughly with hot water. Transfer the filter paper and its contents to the platinum dish previously used. Heat to dryness, incinerate, ignite strongly for 30 minutes, cool, and weigh. Moisten the residue with water, and add 6 mL of hydrofluoric acid and 3 drops of sulfuric acid. Evaporate to dryness, ignite for 5 minutes, cool, and weigh: the loss in weight represents the weight of SiO_2.

Ratio of SiO_2 to MgO—Divide the percentage of SiO_2 obtained in the *Assay for silicon dioxide* by the percentage of MgO obtained in the *Assay for magnesium oxide:* the quotient obtained is between 2.10 and 2.37.

Magnesium Trisilicate Oral Suspension, Alumina and—*see* Alumina and Magnesium Trisilicate Oral Suspension

Magnesium Trisilicate Tablets, Alumina and—*see* Alumina and Magnesium Trisilicate Tablets

Magnesium Trisilicate Tablets

» Magnesium Trisilicate Tablets contain not less than 90.0 percent and not more than 110.0 percent of the labeled amount of $Mg_2Si_3O_8$.

Packaging and storage—Preserve in well-closed containers.

Identification—

A: Powder 1 Tablet, add 10 mL of 3 *N* hydrochloric acid and 5 drops of methyl red TS, heat to boiling, add 6 *N* ammonium hydroxide until the color of the solution changes to deep yellow, then continue boiling for 2 minutes, and filter: the filtrate so obtained responds to the tests for *Magnesium* ⟨191⟩.

B: Wash the solids on the filter obtained in *Identification test A* with hot ammonium chloride solution (1 in 50), add 10 mL of 3 *N* hydrochloric acid, and filter. Transfer the filter paper and contents to a small platinum dish, ignite, cool in a desiccator, and weigh. Moisten the residue with water, and add 6 mL of hydrofluoric acid. Evaporate to dryness, ignite for 5 minutes, cool in a desiccator, and weigh: a loss of more than 10% in relation to the weight of the residue from the initial ignition indicates SiO_2.

Disintegration ⟨701⟩: 10 minutes, simulated gastric fluid TS being substituted for water in the test.

Uniformity of dosage units ⟨905⟩: meet the requirements.

Assay—Weigh and finely powder not less than 20 Magnesium Trisilicate Tablets. Transfer an accurately weighed portion of the powder, equivalent to about 1 g of magnesium trisilicate, to a beaker, add 20 mL of water, and slowly add 40 mL of 3 *N* hydrochloric acid, with mixing. Heat the mixture to boiling, cool, and filter into a 200-mL volumetric flask. Wash the beaker with water, adding the washings to the filter. Add water to volume, and mix. Transfer 20.0 mL of this solution to a 400-mL beaker, add 180 mL of water and 20 mL of triethanolamine, and stir. Add 10 mL of ammonia–ammonium chloride buffer TS and 3 drops of an eriochrome black indicator solution prepared by dissolving 200 mg of eriochrome black T in a mixture of 15 mL of triethanolamine and 5 mL of dehydrated alcohol, and mix. Cool the solution to between 3° and 4° by immersion of the beaker in an ice bath, then remove, and titrate with 0.05 *M* disodium ethylenediaminetetraacetate VS to a blue end-point. Perform a blank determination, substituting 20 mL of water for the assay solution, and make any necessary correction. Each mL of 0.05 *M* disodium ethylenediaminetetraacetate consumed is equivalent to 6.521 mg of $Mg_2Si_3O_8$.

Malathion

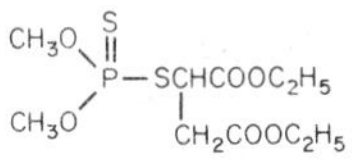

$C_{10}H_{19}O_6PS_2$ 330.35

Butanedioic acid, [(dimethoxyphosphinothioyl)-thio]-, diethyl ester.

Diethyl mercaptosuccinate, *S*-ester with *O,O*-dimethyl phosphorodithioate [*121-75-5*].

» Malathion contains not less than 98.0 percent and not more than 102.0 percent of $C_{10}H_{19}O_6PS_2$.

Packaging and storage—Preserve in tight, light-resistant containers.

Reference standards—*USP Isomalathion Reference Standard—*Keep container tightly closed. Store in a refrigerator. *USP Malathion Reference Standard—*Keep container tightly closed. Store in a refrigerator.

Identification—The infrared absorption spectrum of a thin film of it exhibits maxima only at the same wavelengths as that of a similar preparation of USP Malathion RS.

Specific gravity ⟨841⟩: between 1.220 and 1.240.

Water, *Method I* ⟨921⟩: not more than 0.1%.

Isomalathion—

*Mobile phase—*Prepare a suitable degassed solution of methanol and water (50:30). Make adjustments if necessary (see *System Suitability* under *Chromatography* ⟨621⟩).

*Standard preparation—*Dissolve an accurately weighed quantity of USP Isomalathion RS in methanol to obtain a solution having a known concentration of about 0.1 mg per mL.

*Test preparation—*Dissolve an accurately weighed quantity of Malathion in methanol to obtain a solution having a known concentration of about 20 mg per mL.

Chromatographic system (see *Chromatography* ⟨621⟩)—The liquid chromatograph is equipped with a 210-nm detector and a 4-mm × 30-cm column that contains 10-µm packing L1. The flow rate is about 1 mL per minute. Chromatograph the *Standard preparation,* and record the peak responses as directed under *Procedure:* the relative standard deviation for replicate injections is not more than 2.0%.

*Procedure—*Separately inject equal volumes (about 20 µL) of the *Standard preparation* and the *Test preparation* into the chromatograph, record the chromatograms, and measure the responses of the major peaks. The relative retention times are about 0.5 for isomalathion and 1.0 for malathion. Calculate the percentage of isomalathion in the Malathion taken by the formula:

$$(C_S / C_U)(P)(r_U / r_S),$$

in which C_S is the concentration, in mg per mL, of USP Isomalathion RS in the *Standard preparation,* C_U is the concentration, in mg per mL, of specimen in the *Test preparation, P* is the stated purity, in percentage, of the USP Isomalathion RS, and r_U and r_S are the isomalathion peak responses obtained from the *Test preparation* and the *Standard preparation,* respectively: not more than 0.3% of isomalathion is found.

Assay—

*Mobile phase—*Prepare a suitable degassed solution of methanol and water (50:30). Make adjustments if necessary (see *System Suitability* under *Chromatography* ⟨621⟩).

*Internal standard solution—*Dissolve propylparaben in *Mobile phase* to obtain a solution containing about 0.6 mg per mL.

*Standard preparation—*Transfer about 500 mg of USP Malathion RS, accurately weighed, to a 50-mL volumetric flask, add 5.0 mL of *Internal standard solution,* dilute with methanol to volume, and mix.

*Assay preparation—*Transfer about 500 mg of Malathion, accurately weighed, to a 50-mL volumetric flask, add 5.0 mL of *Internal standard solution,* dilute with methanol to volume, and mix.

Chromatographic system (see *Chromatography* ⟨621⟩)—The liquid chromatograph is equipped with a 254-nm detector and a 4-mm × 30-cm column that contains 10-µm packing L1. The flow rate is about 1 mL per minute. Chromatograph the *Standard preparation,* and record the peak responses as directed under *Procedure:* the resolution, *R,* of the propylparaben and malathion peaks is not less than 4, and the relative standard deviation for replicate injections is not more than 2.0%.

*Procedure—*Separately inject equal volumes (about 20 µL) of the *Standard preparation* and the *Assay preparation* into the chromatograph, record the chromatograms, and measure the responses for the major peaks. The relative retention times are about 0.6 for propylparaben and 1.0 for malathion. Calculate the quantity, in mg, of $C_{10}H_{19}O_6PS_2$ in the portion of Malathion taken by the formula:

$$50C(R_U / R_S),$$

in which *C* is the concentration, in mg per mL, of USP Malathion RS in the *Standard preparation,* and R_U and R_S are the ratios of the peak responses of malathion to propylparaben obtained from the *Assay preparation* and the *Standard preparation,* respectively.

Malathion Lotion

» Malathion Lotion is Malathion in a suitable isopropyl alcohol vehicle. It contains not less than 90.0 percent and not more than 110.0 percent of the labeled amount of $C_{10}H_{19}O_6PS_2$.

Packaging and storage—Preserve in tight, glass containers.

Labeling—The labeling states the percentage (v/v) of isopropyl alcohol in the Lotion.

Reference standard—*USP Malathion Reference Standard*—Keep container tightly closed. Store in a refrigerator.

Identification—The chromatogram of the *Assay preparation* obtained as directed in the *Assay* exhibits a major peak for malathion, the retention time of which corresponds to that exhibited in the chromatogram of the *Standard preparation*, both relative to the internal standard, obtained as directed in the *Assay*.

Isopropyl alcohol content—

Internal standard solution—Mix 4 volumes of ethyl acetate and 1 volume of dehydrated alcohol.

Standard preparation—Transfer 2.0 mL of isopropyl alcohol and 5.0 mL of *Internal standard solution* to a 200-mL volumetric flask, dilute with ethyl acetate to volume, and mix.

Test preparation—Transfer an accurately measured volume of Malathion Lotion, equivalent to about 2.0 mL of isopropyl alcohol, to a 200-mL volumetric flask. Add 5.0 mL of *Internal standard solution*, dilute with ethyl acetate to volume, and mix.

Chromatographic system (see *Chromatography* ⟨621⟩)—The gas chromatograph is equipped with a flame-ionization detector and contains a 1.8-m × 2-mm glass column packed with 110- to 120-mesh support S2. Maintain the column, the injector port, and the detector block at 130°, 200°, and 220°, respectively. Use dry nitrogen as the carrier gas at a flow rate of about 7 mL per minute. Chromatograph the *Standard preparation*, and record the peak responses as directed under *Procedure:* the relative standard deviation of the ratio of the isopropyl alcohol peak response to the internal standard peak response for replicate injections is not more than 2.0%.

Procedure—[NOTE—Use peak areas where peak responses are indicated.] Separately inject equal volumes (about 1 µL) of the *Standard preparation* and the *Test preparation* into the gas chromatograph, record the chromatograms, and measure the responses for the major peaks. Calculate the percentage of isopropyl alcohol (C_3H_8O) in the Lotion by the formula:

$$(200/V)(R_U/R_S),$$

in which V is the volume, in mL, of Lotion taken, and R_U and R_S are the ratios of the peak responses of isopropyl alcohol to internal standard obtained from the *Test preparation* and the *Standard preparation*, respectively: between 90% and 110% of the labeled amount of C_3H_8O is found.

Assay—

Solvent mixture—Mix 4 volumes of methyl ethyl ketone and 1 volume of *n*-hexane.

Internal standard solution—Prepare a solution of parathion in *Solvent mixture* containing about 2 mg per mL.

Standard preparation—Dissolve an accurately weighed quantity of USP Malathion RS in *Solvent mixture* to obtain a solution having a known concentration of about 2 mg per mL. Transfer 5.0 mL of this solution to a 25-mL volumetric flask, add 5.0 mL of *Internal standard solution*, dilute with *Solvent mixture* to volume, and mix.

Assay preparation—Transfer an accurately measured volume of Malathion Lotion, equivalent to about 10 mg of malathion, to a 25-mL volumetric flask, add 5.0 mL of *Internal standard solution*, dilute with *Solvent mixture* to volume, and mix.

Chromatographic system (see *Chromatography* ⟨621⟩)—The gas chromatograph is equipped with a flame-ionization detector and contains a 1.8-m × 2-mm glass column packed with 5 percent G6 liquid phase on 110- to 120-mesh support S1A. Maintain the column, the injector port, and the detector block at 190°, 230°, and 250°, respectively. Use dry nitrogen as the carrier gas at a flow rate of about 15 mL per minute. Chromatograph the *Standard preparation*, and record the peak responses as directed under *Procedure:* the resolution, *R*, between the malathion and parathion peaks is not less than 3.0, and the relative standard deviation for replicate injections is not more than 2.0%.

Procedure—Separately inject equal volumes (about 1 µL) of the *Standard preparation* and the *Assay preparation* into the chromatograph, record the chromatograms, and measure the responses of the major peaks. The relative retention times are 1.0 for malathion and about 1.3 for parathion. Calculate the quantity, in mg, of malathion ($C_{10}H_{19}O_6PS_2$) in each mL of the Malathion Lotion taken by the formula:

$$25(C/V)(R_U/R_S),$$

in which C is the concentration, in mg per mL, of USP Malathion RS in the *Standard preparation*, V is the volume, in mL, of Malathion Lotion taken, and R_U and R_S are the ratios of the peak responses of malathion to parathion obtained from the *Assay preparation* and the *Standard preparation*, respectively.

Malic Acid—*see* Malic Acid NF

Manganese Chloride

$MnCl_2 . 4H_2O$ 197.91

Manganese chloride ($MnCl_2$) tetrahydrate.

Manganese(2+) chloride tetrahydrate [*13446-34-9*].

Anhydrous 125.84 [*7773-01-5*].

» Manganese Chloride contains not less than 98.0 percent and not more than 101.0 percent of $MnCl_2$, calculated on the dried basis.

Packaging and storage—Preserve in tight containers.

Identification—It responds to the tests for *Chloride* ⟨191⟩ and for *Manganese* ⟨191⟩.

Loss on drying ⟨731⟩—Dry it at 50° for 2 hours, then raise the temperature to 150° for 24 hours: it loses between 36.0% and 38.5% of its weight.

pH ⟨791⟩: between 3.5 and 6.0, 10 g dissolved in 200 mL of carbon dioxide– and ammonia–free water being used.

Insoluble matter—Transfer 10 g to a 250-mL beaker, add 150 mL of water, cover the beaker, and heat to boiling. Digest the hot solution on a steam bath for 1 hour, and filter through a tared, fine-porosity filtering crucible. Rinse the beaker with hot water, passing the rinsings through the filter, and finally wash the filter with additional hot water. Dry the filter at 105°: the residue weighs not more than 0.5 mg (0.005%).

Sulfate ⟨221⟩—A 2.0 g portion shows no more sulfate than corresponds to 0.10 mL of 0.020 *N* sulfuric acid (0.005%).

Substances not precipitated by ammonium sulfide—Dissolve 2.0 g in about 90 mL of water, add 5 mL of ammonium hydroxide, and warm the solution to about 80°. Pass a stream of hydrogen sulfide through the solution for 30 minutes. Dilute with water to 100 mL, mix, and allow the precipitate to settle. Decant the supernatant liquid through a fine-porosity filter, and transfer 50.0 mL to an evaporating dish that previously has been ignited and tared. Evaporate the filtrate to dryness, cool, add 0.5 mL of sulfuric acid, heat gently to remove the excess acid, and ignite at 800 ± 25° for 15 minutes: the weight of the residue is not greater than 2.0 mg (0.2% as sulfate).

Heavy metals, *Method I* ⟨231⟩—Dissolve 6.0 g in 30 mL of water. Use 25 mL of this solution in the *Test Preparation*, and use the remaining 5.0 mL in preparing the *Standard Preparation:* the limit is 5 ppm.

Iron ⟨241⟩—Dissolve 2.0 g in 40 mL of water: the limit is 5 ppm.

Zinc—Dissolve 1 g in a mixture of 48 mL of water and 2 mL of sulfuric acid, and add, slowly and with constant agitation, 1 mL of potassium ferrocyanide solution (1 in 100): no turbidity is produced within 5 minutes.

Assay—Transfer about 425 mg of Manganese Chloride, accurately weighed, to a 400-mL beaker, dissolve in about 25 mL of water, add 300 mg of ammonium chloride and 0.5 g of hydroxylamine hydrochloride, and swirl to dissolve. Warm slightly on a hot plate, and dilute with water to 100 mL. Add about 3 mL of triethanolamine, stir the solution, using, preferably, a magnetic stirrer, begin the titration by adding about 25 mL of 0.05 *M* disodium ethylenediaminetetraacetate VS, using a suitable buret, then add 10 mL of ammonia–ammonium chloride buffer TS, and 1 mL of eriochrome black TS, and complete the titration with 0.05 *M* disodium ethylenediaminetetraacetate VS to a blue end-point. Each mL of 0.05 *M* disodium ethylenediaminetetraacetate is equivalent to 6.292 mg of $MnCl_2$.

Manganese Chloride Injection

» Manganese Chloride Injection is a sterile solution of Manganese Chloride in Water for Injection. It contains not less than 95.0 percent and not more than 105.0 percent of the labeled amount of manganese (Mn).

Packaging and storage—Preserve in single-dose or in multiple-dose containers, preferably of Type I or Type II glass.

Labeling—Label the Injection to indicate that it is to be diluted to the appropriate strength with Sterile Water for Injection or other suitable fluid prior to administration.

Identification—The *Assay preparation*, prepared as directed in the *Assay*, exhibits an absorption maximum at about 279 nm when tested as directed for *Procedure* in the *Assay*.

Pyrogen—When diluted with Sodium Chloride Injection to contain 4 µg of manganese per mL it meets the requirements of the *Pyrogen Test* ⟨151⟩.

pH ⟨791⟩: between 1.5 and 2.5.

Particulate matter ⟨788⟩: meets the requirements under *Small-volume Injections*.

Other requirements—It meets the requirements under *Injections* ⟨1⟩.

Assay—

Manganese stock solution—Transfer 1.000 g of manganese to a 1000-mL volumetric flask, dissolve in 20 mL of nitric acid, dilute with 0.1 *N* hydrochloric acid to volume, and mix. This solution contains 1000 µg of manganese per mL. Store in a polyethylene bottle.

Standard preparations—Pipet 10 mL of the *Manganese stock solution* into a 500-mL volumetric flask, dilute with water to volume, and mix. Transfer 5.0, 10.0, 15.0, and 20.0 mL, respectively, of this solution to separate 100-mL volumetric flasks, dilute the contents of each flask with water to volume, and mix. These *Standard preparations* contain, respectively, 1.0, 2.0, 3.0, and 4.0 µg of manganese per mL.

Assay preparation—Transfer an accurately measured volume of Manganese Chloride Injection, equivalent to about 1 mg of manganese, to a 100-mL volumetric flask, dilute with water to volume, and mix. Pipet 10 mL of this solution into a 50-mL volumetric flask, dilute with water to volume, and mix.

Procedure—Concomitantly determine the absorbances of the *Standard preparations* and the *Assay preparation* at the manganese emission line of 279 nm, with a suitable atomic absorption spectrophotometer (see *Spectrophotometry and Light-scattering* ⟨851⟩) equipped with a manganese hollow-cathode lamp and an air-acetylene flame, using water as the blank. Plot the absorbances of the *Standard preparations* versus concentration, in µg per mL, of manganese, and draw the straight line best fitting the four plotted points. From the graph so obtained, determine the concentration, in µg per mL, of manganese in the *Assay preparation*. Calculate the quantity, in mg, of manganese in each mL of the Injection taken by the formula:

$$0.5C/V,$$

in which *C* is the concentration, in µg per mL, of manganese in the *Assay preparation*, and *V* is the volume, in mL, of Injection taken.

Manganese Gluconate

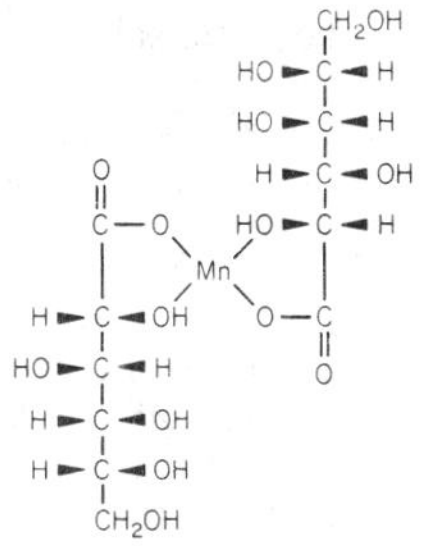

$C_{12}H_{22}MnO_{14}$ (anhydrous) 445.24
Bis(D-gluconato-O^1,O^2) manganese.
Manganese D-gluconate (1:2).
Dihydrate 481.27.

» Manganese Gluconate is dried or contains two molecules of water of hydration. It contains not less than 98.0 percent and not more than 102.0 percent of $C_{12}H_{22}MnO_{14}$, calculated on the anhydrous basis.

Packaging and storage—Preserve in well-closed containers.

Labeling—The label indicates whether it is the dried or the dihydrate form.

Reference standard—*USP Potassium Gluconate Reference Standard*—Dry in vacuum at 105° for 4 hours before using.

Identification—

A: A solution (1 in 20) responds to the tests for *Manganese* ⟨191⟩.

B: It responds to *Identification test B* under *Calcium Gluconate*.

Water, *Method I* ⟨921⟩ (where labeled as the dried form): between 3.0% and 9.0%, the determination being performed by stirring the mixture containing the *Test preparation*, maintained at a temperature of 50°, for 30 minutes before titrating with the *Reagent:* where labeled as the dihydrate it is between 6.0% and 9.0%.

Chloride ⟨221⟩—A 1.0-g portion shows no more chloride than corresponds to 0.70 mL of 0.020 *N* hydrochloric acid (0.05%).

Sulfate ⟨221⟩—A 2.0-g portion shows no more sulfate than corresponds to 4.0 mL of 0.020 *N* sulfuric acid (0.2%).

Arsenic, *Method I* ⟨211⟩—Dissolve 1.0 g in 35 mL of water: the limit is 3 ppm.

Reducing substances—Transfer 1.0 g to a 250-mL conical flask, dissolve in 10 mL of water, and add 25 mL of alkaline cupric citrate TS. Cover the flask, boil gently for 5 minutes, accurately timed, and cool rapidly to room temperature. Add 25 mL of 0.6 *N* acetic acid, 10.0 mL of 0.1 *N* iodine VS, and 10 mL of 3 *N* hydrochloric acid, and titrate with 0.1 *N* sodium thiosulfate VS, adding 3 mL of starch TS as the end-point is approached. Perform a blank determination, omitting the specimen, and note the difference in volumes required. Each mL of the difference in volume of 0.1 *N* sodium thiosulfate consumed is equivalent to 2.7 mg of reducing substances (as dextrose): the limit is 1.0%.

Heavy metals ⟨231⟩—Dissolve 1 g in 10 mL of water, add 6 mL of 3 *N* hydrochloric acid, and dilute with water to 25 mL: the limit is 0.004%.

Lead—[NOTE—For the preparation of all aqueous solutions and for the rinsing of glassware before use, employ water that has been passed through a strong-acid, strong-base, mixed-bed ion-exchange resin before use. Select all reagents to have as low a content of lead as practicable, and store all reagent solutions in containers of borosilicate glass. Cleanse glassware before use by soaking in warm 8 *N* nitric acid for 30 minutes and by rinsing with deionized water.]

Ascorbic acid–sodium iodide solution—Dissolve 20 g of ascorbic acid and 38.5 g of sodium iodide in water in a 200-mL volumetric flask, dilute with water to volume, and mix.

Trioctylphosphine oxide solution—[*Caution—This solution causes irritation. Avoid contact with eyes, skin, and clothing.*

Take special precautions in disposing of unused portions of solutions to which this reagent is added.] Dissolve 5.0 g of trioctylphosphine oxide in 4-methyl-2-pentanone in a 100-mL volumetric flask, dilute with the same solvent to volume, and mix.

Standard preparation and *Blank*—Transfer 5.0 mL of *Lead Nitrate Stock Solution*, prepared as directed in the test for *Heavy Metals* ⟨231⟩, to a 100-mL volumetric flask, dilute with water to volume, and mix. Transfer 2.0 mL of the resulting solution to a 50-mL volumetric flask. To this volumetric flask and to a second, empty 50-mL volumetric flask (*Blank*) add 10 mL of 9 *N* hydrochloric acid and about 10 mL of water. To each flask add 20 mL of *Ascorbic acid–sodium iodide solution* and 5.0 mL of *Trioctylphosphine oxide solution*, shake for 30 seconds, and allow to separate. Add water to bring the organic solvent layer into the neck of each flask, shake again, and allow to separate. The organic solvent layers are the *Blank* and the *Standard preparation*, and they contain 0.0 and 2.0 µg of lead per mL, respectively.

Test preparation—Add 1.0 g of Manganese Gluconate, 10 mL of 9 *N* hydrochloric acid, about 10 mL of water, 20 mL of *Ascorbic acid–sodium iodide solution*, and 5.0 mL of *Trioctylphosphine oxide solution* to a 50-mL volumetric flask, shake for 30 seconds, and allow to separate. Add water to bring the organic solvent layer into the neck of the flask, shake again, and allow to separate. The organic solvent layer is the *Test preparation.*

Procedure—Concomitantly determine the absorbances of the *Blank*, the *Standard preparation*, and the *Test preparation* at the lead emission line at 283.3 nm, with a suitable atomic absorption spectrophotometer (see *Spectrophotometry and Light-scattering* ⟨851⟩) equipped with a lead hollow-cathode lamp and an air-acetylene flame, using 4-methyl-2-pentanone to set the instrument to zero. In a suitable analysis, the absorbance of the *Blank* is not greater than 20% of the difference between the absorbance of the *Standard preparation* and the absorbance of the *Blank:* the absorbance of the *Test preparation* does not exceed that of the *Standard preparation* (0.001%).

Assay—Dissolve about 700 mg of Manganese Gluconate, accurately weighed, in 50 mL of water. Add 1 g of ascorbic acid, 10 mL of ammonia–ammonium chloride buffer TS, and 0.1 mL of eriochrome black TS, and titrate with 0.05 *M* disodium ethylenediaminetetraacetate VS until the solution is deep blue in color. Each mL of 0.05 *M* disodium ethylenediaminetetraacetate is equivalent to 22.26 mg of $C_{12}H_{22}MnO_{14}$.

Manganese Sulfate

$MnSO_4 . H_2O$ 169.01

Sulfuric acid, manganese(2+) salt (1:1) monohydrate.

Manganese(2+) sulfate (1:1) monohydrate [*10034-96-5*].

Anhydrous 151.00 [*7785-87-7*].

» Manganese Sulfate contains not less than 98.0 percent and not more than 102.0 percent of $MnSO_4 . H_2O$.

Packaging and storage—Preserve in tight containers.

Identification—A solution (1 in 10) responds to the tests for *Manganese* ⟨191⟩ and for *Sulfate* ⟨191⟩.

Loss on ignition ⟨733⟩—Ignite it at 450° to constant weight: it loses between 10.0% and 13.0% of its weight.

Substances not precipitated by ammonium sulfide—Dissolve 2.0 g in 90 mL of water, add 5 mL of ammonium hydroxide, warm the solution, and pass hydrogen sulfide through the solution for about 30 minutes. Dilute with water to 100 mL, mix, and allow the precipitate to settle. Decant the supernate through a filter, transfer 50 mL of the clear filtrate to a tared dish, evaporate to dryness, and ignite, gently at first and finally at 800 ± 25°: the weight of the residue does not exceed 5 mg (0.5%).

Assay—Dissolve about 350 mg of Manganese Sulfate, accurately weighed, in 200 mL of water. Add about 10 mg of ascorbic acid, begin the titration by adding about 25 mL of 0.05 *M* disodium ethylenediaminetetraacetate VS, using a suitable buret, then add 10 mL of ammonia–ammonium chloride buffer TS, and about 0.15 mL of eriochrome black TS, and complete the titration with 0.05 *M* disodium ethylenediaminetetraacetate VS to a blue endpoint. Each mL of 0.05 *M* disodium ethylenediaminetetraacetate is equivalent to 8.451 mg of $MnSO_4 . H_2O$.

Manganese Sulfate Injection

» Manganese Sulfate Injection is a sterile solution of Manganese Sulfate in Water for Injection. It contains not less than 95.0 percent and not more than 105.0 percent of the labeled amount of manganese (Mn).

Packaging and storage—Preserve in single-dose or in multiple-dose containers, preferably of Type I or Type II glass.

Labeling—Label the Injection to indicate that it is to be diluted to the appropriate strength with Sterile Water for Injection or other suitable fluid prior to administration.

Identification—The *Assay preparation*, prepared as directed in the *Assay*, exhibits an absorption maximum at about 279 nm when tested as directed for *Procedure* in the *Assay*.

Pyrogen—When diluted with Sodium Chloride Injection to contain 4 µg of manganese per mL, it meets the requirements of the *Pyrogen Test* ⟨151⟩.

pH ⟨791⟩: between 2.0 and 3.5.

Particulate matter ⟨788⟩: meets the requirements under *Small-volume Injections*.

Other requirements—It meets the requirements under *Injections* ⟨1⟩.

Assay—

Manganese stock solution and *Standard preparations*—Prepare as directed in the *Assay* under *Manganese Chloride Injection.*

Assay preparation—Transfer an accurately measured volume of Manganese Sulfate Injection, equivalent to about 1 mg of manganese, to a 100-mL volumetric flask, dilute with water to volume, and mix. Pipet 10 mL of this solution into a 50-mL volumetric flask, dilute with water to volume, and mix.

Procedure—Proceed as directed for *Procedure* in the *Assay* under *Manganese Chloride Injection.*

Mannitol

```
          H   H   OH  OH
          |   |   |   |
HOCH2—C—C—C—C—CH2OH
          |   |   |   |
          OH  OH  H   H
```

$C_6H_{14}O_6$ 182.17

D-Mannitol.

D-Mannitol [*69-65-8*].

» Mannitol contains not less than 96.0 percent and not more than 101.5 percent of $C_6H_{14}O_6$, calculated on the dried basis.

Packaging and storage—Preserve in well-closed containers.

Reference standard—*USP Mannitol Reference Standard.*

Identification—

A: Add 5 drops of a saturated solution of it to 1 mL of ferric chloride TS. Add 5 drops of water to a second tube containing 1 mL of ferric chloride TS. Add 5 drops of 5 *N* sodium hydroxide to each tube: a brown precipitate of ferric hydroxide is formed in the tube containing no Mannitol, and a yellow precipitate is formed in the tube containing Mannitol. Shake the tubes vigorously: a clear solution results in the tube containing Mannitol, but the precipitate remains in the other tube. Additional 5 *N* sodium hydroxide does not cause precipitation in the tube containing Mannitol, but further precipitation takes place in the other.

B: To about 500 mg of it in a test tube add 3 mL of acetic anhydride and 1 mL of pyridine. Heat the mixture in a bath of boiling water for 15 minutes, with frequent shaking, or until solution is complete, and continue heating for 5 minutes. Cool the mixture, add 20 mL of water, mix, allow to stand for 5 minutes, and collect the precipitate on a sintered-glass filter: the mannitol hexaacetate so obtained, after being dried in vacuum at 60° for 1 hour or after recrystallization from ether, melts between 119° and 124°.

Melting range ⟨741⟩: between 165° and 169°, with softening at a lower temperature.

Specific rotation ⟨781⟩—Transfer about 1 g of Mannitol, accurately weighed, to a 100-mL volumetric flask, and add 40 mL of ammonium molybdate solution (1 in 10), which previously has been filtered if necessary. Add 20 mL of 1 *N* sulfuric acid, dilute with water to volume, and mix: the specific rotation is between +137° and +145°.

Acidity—Dissolve 5.0 g in 50 mL of carbon dioxide–free water, add 3 drops of phenolphthalein TS, and titrate with 0.020 *N* sodium hydroxide to a distinct pink end-point: not more than 0.30 mL of 0.020 *N* sodium hydroxide is required for neutralization.

Loss on drying ⟨731⟩—Dry it at 105° for 4 hours: it loses not more than 0.3% of its weight.

Chloride ⟨221⟩—A 2.0-g portion shows no more chloride than corresponds to 0.20 mL of 0.020 *N* hydrochloric acid (0.007%).

Sulfate ⟨221⟩—A 2.0-g portion shows no more sulfate than corresponds to 0.20 mL of 0.020 *N* sulfuric acid (0.01%).

Arsenic, *Method II* ⟨211⟩: 1 ppm.

Reducing sugars—To 5 mL of alkaline cupric citrate TS add 1 mL of a saturated solution of Mannitol (about 200 mg). Heat for 5 minutes in a boiling water bath: not more than a very slight precipitate is formed.

Assay—

Mobile phase—Use degassed water.

Resolution solution—Dissolve sorbitol and USP Mannitol RS in water to obtain a solution having concentrations of about 4.8 mg per mL of each.

Standard preparation—Dissolve an accurately weighed quantity of USP Mannitol RS in water, and dilute quantitatively with water to obtain a solution having a known concentration of about 4.8 mg per mL.

Assay preparation—Transfer about 0.24 g of Mannitol, accurately weighed, to a 50-mL volumetric flask, dissolve in 10 mL of water, dilute with water to volume, and mix.

Chromatographic system (see *Chromatography* ⟨621⟩)—The liquid chromatograph is equipped with a refractive index detector that is maintained at a constant temperature and a 7.8-mm × 30-cm column that contains packing L19. The column temperature is maintained at 30 ± 2° and the flow rate is about 0.2 mL per minute. Chromatograph the *Standard preparation*, and record the peak responses as directed under *Procedure:* the relative standard deviation for replicate injections is not more than 2.0%. In a similar manner, chromatograph the *Resolution solution:* the resolution, *R*, between the sorbitol and mannitol peaks is not less than 2.0.

Procedure—Separately inject equal volumes (about 20 µL) of the *Assay preparation* and the *Standard preparation* into the chromatograph, record the chromatograms, and measure the responses for the major peaks. Calculate the quantity, in mg, of $C_6H_{14}O_6$, in the Mannitol taken by the formula:

$$50C(r_U / r_S),$$

in which *C* is the concentration, in mg per mL, of USP Mannitol RS in the *Standard preparation*, and r_U and r_S are the peak responses obtained from the *Assay preparation* and the *Standard preparation*, respectively.

Mannitol Injection

» Mannitol Injection is a sterile solution, which may be supersaturated, of Mannitol in Water for Injection. It may require warming or autoclaving before use if crystallization has occurred. It contains not less than 95.0 percent and not more than 105.0 percent of the labeled amount of $C_6H_{14}O_6$. It contains no antimicrobial agents.

Packaging and storage—Preserve in single-dose containers, preferably of Type I or Type II glass.

Labeling—The label states the total osmolar concentration in mOsmol per liter. Where the contents are less than 100 mL, or where the label states that the Injection is not for direct injection but is to be diluted before use, the label alternatively may state the total osmolar concentration in mOsmol per mL.

Identification—Evaporate a portion of Injection on a steam bath to dryness, and dry the residue at 105° for 4 hours: the residue responds to the *Identification tests* under *Mannitol.*

Specific rotation ⟨781⟩—Transfer an accurately measured volume of Injection, equivalent to about 1 g of mannitol as determined by the *Assay*, to a 100-mL volumetric flask: it meets the requirements of the test for *Specific rotation* under *Mannitol.*

Pyrogen—It meets the requirements of the *Pyrogen Test* ⟨151⟩, after being diluted, if necessary, with *Water for Injection* to contain not more than 10% of $C_6H_{14}O_6$.

pH ⟨791⟩: between 4.5 and 7.0, determined potentiometrically, on a portion to which 0.30 mL of saturated potassium chloride solution has been added for each 100 mL, and which previously has been diluted with water, if necessary, to a concentration of not more than 5% of mannitol.

Particulate matter ⟨788⟩: meets the requirements under *Small-volume Injections.*

Other requirements—It meets the requirements under *Injections* ⟨1⟩.

Assay—Transfer an accurately measured volume of Mannitol Injection, equivalent to about 1 g of mannitol, to a 1000-mL volumetric flask, add water to volume, and mix. Transfer 4.0 mL of this solution to a 250-mL conical flask, and proceed as directed in the *Assay* under *Mannitol*, beginning with "add 50.0 mL of a reagent." Each mL of the difference in volume of 0.02 *N* sodium thiosulfate consumed is equivalent to 0.3643 mg of $C_6H_{14}O_6$.

Mannitol in Sodium Chloride Injection

» Mannitol in Sodium Chloride Injection is a sterile solution of Mannitol and Sodium Chloride in Water for Injection. It contains not less than 95.0 percent and not more than 105.0 percent of the labeled amounts of $C_6H_{14}O_6$ and NaCl. It contains no antimicrobial agents.

Labeling—The label states the total osmolar concentration in mOsmol per liter. Where the contents are less than 100 mL, or where the label states that the Injection is not for direct injection but is to be diluted before use, the label alternatively may state the total osmolar concentration in mOsmol per mL.

Identification—

A: Evaporate a portion of Injection on a steam bath to dryness, and dry the residue at 105° for 4 hours: the residue responds to the *Identification tests* under *Mannitol.*

B: It responds to the tests for *Sodium* ⟨191⟩, and for *Chloride* ⟨191⟩.

pH ⟨791⟩: between 4.5 and 7.0.

Other requirements—It meets the requirements for *Packaging and storage* and *Pyrogen* under *Mannitol Injection.* It meets also the requirements under *Injections* ⟨1⟩.

Assay for mannitol—Proceed with Mannitol in Sodium Chloride Injection as directed in the *Assay* under *Mannitol Injection.*

Assay for sodium chloride—Proceed with Mannitol in Sodium Chloride Injection as directed in the *Assay* under *Sodium Chloride Injection.*

Maprotiline Hydrochloride

$CH_2CH_2CH_2NHCH_3$ · HCl

$C_{20}H_{23}N.HCl$ 313.87

9,10-Ethanoanthracene-9(10*H*)-propanamine, *N*-methyl-, hydrochloride.

N-Methyl-9,10-ethanoanthracene-9(10*H*)-propylamine hydrochloride [*10347-81-6*].

» Maprotiline Hydrochloride contains not less than 99.0 percent and not more than 101.0 percent of the labeled amount of $C_{20}H_{23}N.HCl$, calculated on the dried basis.

Packaging and storage—Preserve in tight containers.

Reference standard—*USP Maprotiline Hydrochloride Reference Standard*—Dry in vacuum at 80° to constant weight before using.

Identification—

A: The infrared absorption spectrum of a potassium bromide dispersion of it, previously dried, exhibits maxima only at the same wavelengths as that of a similar preparation of USP Maprotiline Hydrochloride RS.

B: The ultraviolet absorption spectrum of a 1 in 10,000 solution in methanol exhibits maxima and minima at the same wavelengths as that of a similar solution of USP Maprotiline Hydrochloride RS, concomitantly measured, and the respective absorptivities, calculated on the dried basis, at the wavelengths of maximum absorbance at about 266 nm and 272 nm, do not differ by more than 3.0%.

C: A solution (1 in 200) responds to the tests for *Chloride* ⟨191⟩, when tested as specified for alkaloidal hydrochlorides.

Loss on drying ⟨731⟩—Dry it in vacuum at 80° to constant weight: it loses not more than 1.0% of its weight.

Residue on ignition ⟨281⟩: not more than 0.1%.

Heavy metals, *Method II* ⟨231⟩: 0.001%.

Chromatographic purity—

Standard preparations—Dissolve USP Maprotiline Hydrochloride RS in methanol, and mix to obtain a stock standard solution having a known concentration of 20 mg per mL. Dilute quantitatively with methanol to obtain *Standard preparation A* having a known concentration of 0.10 mg per mL. Dilute quantitatively with methanol to obtain *Standard preparations*, designated below by letter, having the following compositions:

Dilution	Concentration (µg RS per mL)	Percentage (%, for comparison with test specimen)
A (undiluted)	100	0.5
B (4 in 5)	80	0.4
C (3 in 5)	60	0.3
D (2 in 5)	40	0.2
E (1 in 5)	20	0.1

Test preparation—Dissolve an accurately weighed quantity of Maprotiline Hydrochloride in methanol to obtain a solution containing 20 mg per mL.

Procedure—In a suitable chromatographic chamber (see *Chromatography* ⟨621⟩), place a volume of a solvent system consisting of a mixture of secondary butyl alcohol, ethyl acetate, and 2 *N* ammonium hydroxide (6:3:1) sufficient to develop a chromatogram. Place a beaker containing 25 mL of ammonium hydroxide in the bottom of the chamber, and allow it to equilibrate for 1 hour. On a suitable thin-layer chromatographic plate coated with a 0.25-mm layer of chromatographic silica gel that has been prewashed with chloroform by allowing chloroform to move the full length of the plate, and dried at 100° for 30 minutes, apply 5-µL portions of the *Test preparation*, the stock standard solution, and each of the *Standard preparations*, and allow the spots to dry. Develop the chromatograms until the solvent front has moved about three-fourths of the length of the plate, remove the plate from the developing chamber, mark the solvent front, and allow the solvent to evaporate. Expose the plate to hydrogen chloride vapor for 30 minutes, expose it to a high-intensity ultraviolet light irradiator (1000 to 1600 watts) [*Caution—Ultraviolet irradiators emit ultraviolet radiation that is harmful to eyes and skin*] until the spot in the chromatogram of *Standard preparation E* is clearly visible, and compare the chromatograms under long-wavelength ultraviolet light: the R_f value of the principal spot obtained from the *Test preparation* corresponds to that obtained from the stock standard solution, and the sum of the intensities of all secondary spots obtained from the *Test preparation*, compared with those of the principal spots obtained from the *Standard preparations*, corresponds to not more than 1.0%.

Assay—Dissolve about 60 mg of Maprotiline Hydrochloride, accurately weighed, in 25 mL of mercuric acetate TS, and titrate with 0.1 *N* perchloric acid VS, determining the end-point potentiometrically, using a glass electrode and a calomel electrode containing saturated lithium chloride in glacial acetic acid (see *Titrimetry* ⟨541⟩). Perform a blank determination, and make any necessary correction. Each mL of 0.1 *N* perchloric acid is equivalent to 31.39 mg of $C_{20}H_{23}N.HCl$.

Maprotiline Hydrochloride Tablets

» Maprotiline Hydrochloride Tablets contain not less than 90.0 percent and not more than 110.0 percent of the labeled amount of $C_{20}H_{23}N.HCl$.

Packaging and storage—Preserve in well-closed containers.

Reference standard—*USP Maprotiline Hydrochloride Reference Standard*—Dry in vacuum at 80° to constant weight before using.

Identification—

Standard solution—Dissolve 20 mg of USP Maprotiline Hydrochloride RS in 1.0 mL of methanol.

Test solution—Transfer a portion of powdered Tablets, equivalent to about 100 mg of maprotiline hydrochloride, to a glass-stoppered centrifuge tube. Add 5.0 mL of methanol to the tube, sonicate for 10 minutes, shake by mechanical means for 10 minutes, and centrifuge.

Procedure—In a suitable chromatographic chamber (see *Chromatography* ⟨621⟩), place a volume of a solvent system consisting of a mixture of secondary butyl alcohol, ethyl acetate, and 2 *N* ammonium hydroxide (6:3:1) sufficient to develop a chromatogram. Place a beaker containing 25 mL of ammonium hydroxide in the bottom of the chamber, and allow it to equilibrate for 1 hour. On a suitable thin-layer chromatographic plate, coated with a 0.25-mm layer of chromatographic silica gel that has been prewashed with chloroform by allowing chloroform to travel the full length of the plate and dried at 100° for 30 minutes, apply 5-µL portions of the *Test solution*, the *Standard solution*, and allow the spots to dry. Develop the chromatogram until the solvent front has moved about three-fourths of the length of the plate, remove the plate from the developing chamber, mark the solvent front, and allow the solvent to evaporate. Expose the plate to hydrogen chloride vapor for 30 minutes, expose it to a high-intensity ultraviolet light irradiator (1000 to 1600 watts) [*Caution—Ultraviolet irradiators emit ultraviolet radiation that is harmful to eyes and skin*] for 5 minutes, and compare the chromatograms under long-wavelength ultraviolet light: the R_f value of the principal spot obtained from the *Test solution* corresponds to that obtained from the *Standard solution*.

Dissolution ⟨711⟩—

Medium: dilute hydrochloric acid (7 in 1000); 900 mL.

Apparatus 2: 50 rpm.

Time: 60 minutes.

Procedure—Determine the amount of $C_{20}H_{23}N.HCl$ dissolved from ultraviolet absorbances, using the difference between the absorbance maximum at about 272 nm and the absorbance minimum at about 268 nm, of filtered portions of the solution under test, suitably diluted with dilute hydrochloric acid (7 in 1000), in comparison with a Standard solution having a known concentration of USP Maprotiline Hydrochloride RS in the same medium.

Tolerances—Not less than 75% (*Q*) of the labeled amount of $C_{20}H_{23}N.HCl$ is dissolved in 60 minutes.

Uniformity of dosage units ⟨905⟩: meet the requirements.

Assay—

Standard preparation—Transfer an accurately weighed quantity of USP Maprotiline Hydrochloride RS to a suitable volumetric flask, and prepare a solution having a concentration of about 1 mg per mL in a mixture of alcohol and water (7:3) by dissolving the Reference Standard in the water, adding the alcohol to volume, and mixing. Pipet 10 mL of this solution into a 100-mL volumetric flask, dilute with alcohol to volume, and mix to obtain a *Standard preparation* having a known concentration of about 100 µg per mL.

Assay preparation—Weigh and finely powder not less than 20 Maprotiline Hydrochloride Tablets. Weigh accurately a portion of the powder, equivalent to about 100 mg of maprotiline hydrochloride, and transfer to a 100-mL volumetric flask. Add 30 mL of water and 40 mL of alcohol, sonicate for 10 minutes, shake by mechanical means for 10 minutes, dilute with alcohol to volume, mix, and filter through a medium-porosity, sintered-glass filter. Pipet 10 mL of the filtrate into a 100-mL volumetric flask, dilute with alcohol to volume, and mix.

Procedure—Concomitantly determine the absorbances of the *Standard preparation* and the *Assay preparation*, using the difference between the absorbance maximum, at about 273 nm, and the absorbance at 300 mm, with a suitable spectrophotometer, using alcohol as the blank. Calculate the quantity, in mg, of $C_{20}H_{23}N.HCl$ in the portion of Tablets taken by the formula:

$$C(A_U/A_S),$$

in which *C* is the concentration, in µg per mL, of USP Maprotiline Hydrochloride RS in the *Standard preparation*, and A_U and A_S are the absorbances of the *Assay preparation* and the *Standard preparation*, respectively.

Mazindol

$C_{16}H_{13}ClN_2O$ 284.74

3*H*-Imidazo[2,1-*a*]isoindol-5-ol, 5-(4-chlorophenyl)-2,5-dihydro-.

5-(*p*-Chlorophenyl)-2,5-dihydro-3*H*-imidazo[2,1-*a*]isoindol-5-ol [*22232-71-9*].

» Mazindol contains not less than 98.0 percent and not more than 102.0 percent of $C_{16}H_{13}ClN_2O$, calculated on the dried basis.

Packaging and storage—Preserve in tight containers.

Reference standard—*USP Mazindol Reference Standard*—Dry in vacuum at 60° for 2 hours before using.

Clarity and color of solution—A 1 in 100 solution of Mazindol in a mixture of chloroform and methanol (9:1) is clear and not darker in color than a solution prepared by mixing equal volumes of Matching Fluid C (see *Color and Achromicity* ⟨631⟩) and water.

Identification—

A: The infrared absorption spectrum of a potassium bromide dispersion of it, previously dried, exhibits maxima only at the same wavelengths as that of a similar preparation of USP Mazindol RS.

B: The ultraviolet absorption spectrum of a 1 in 100,000 solution in 0.6 *N* hydrochloric acid exhibits maxima and minima at the same wavelengths as that of a similar preparation of USP Mazindol RS, concomitantly measured, and the respective absorptivities, calculated on the dried basis, at the wavelength of maximum absorption at about 272 nm, do not differ by more than 3.0%.

Loss on drying ⟨731⟩—Dry it in vacuum at 60° for 2 hours: it loses not more than 0.5% of its weight.

Residue on ignition ⟨281⟩: not more than 0.1%.

Heavy metals, *Method II* ⟨231⟩: 0.002%.

Sulfate ⟨221⟩—Triturate a 500-mg portion with 10 mL of water in a mortar. Filter the suspension through a water-washed filter, and rinse the mortar and filter with 30 mL of water, collecting the combined filtrate and washings in a 50-mL color-comparison tube. The filtrate shows no more sulfate than corresponds to 0.20 mL of 0.020 *N* sulfuric acid (0.04%).

Chromatographic purity—Dissolve 10 mg in 2.0 mL of a mixture of chloroform and methanol (9:1) to obtain the *Test solution.* Dissolve a suitable quantity of USP Mazindol RS in a mixture of chloroform and methanol (9:1) to obtain a *Standard solution* having a concentration of 5.0 mg per mL. Dilute portions of this solution quantitatively and stepwise with the mixture of chloroform and methanol (9:1) to obtain a series of *Diluted standard solutions* having concentrations of 0.100, 0.050, 0.025, and 0.0125 mg per mL, respectively. On a suitable thin-layer chromatographic plate (see *Chromatography* ⟨621⟩), coated with a 0.25-mm layer of chromatographic silica gel mixture, separately apply a 20-µL portion of the *Test solution* and 20-µL portions of the *Standard solution* and each *Diluted standard solution.* Allow the spots to dry, and develop the chromatogram in a solvent system consisting of a mixture of chloroform, alcohol, and ammonium hydroxide (80:20:1) until the solvent front has moved about three-fourths of the length of the plate. Remove the plate from the developing chamber, mark the solvent front, and allow the solvent to evaporate. Locate the spots on the plate by examination under short-wavelength ultraviolet light: the chromatograms show principal spots at about the same R_f value. Estimate the concentration of any secondary spots, if present in the chromatogram from the *Test solution,* by comparison with the *Diluted standard solutions:* the principal spots from the 0.100, 0.050, 0.025, and 0.0125 mg per mL dilutions are equivalent to 2.0%, 1.0%, 0.50%, and 0.25% of impurities, respectively. No individual impurity is greater than 1.0%, and the sum of the impurities is not greater than 2.0%.

Assay—Transfer about 230 mg of Mazindol, accurately weighed, to a suitable flask, dissolve in 40 mL of glacial acetic acid, add 3 drops of crystal violet TS, and titrate with 0.1 *N* perchloric acid VS to an emerald-green end-point. Perform a blank determination, and make any necessary correction. Each mL of 0.1 *N* perchloric acid is equivalent to 28.47 mg of $C_{16}H_{13}ClN_2O$.

Mazindol Tablets

» Mazindol Tablets contain not less than 90.0 percent and not more than 110.0 percent of the labeled amount of $C_{16}H_{13}ClN_2O$.

Packaging and storage—Preserve in tight containers, at a temperature not exceeding 25°.

Reference standard—*USP Mazindol Reference Standard*—Dry in vacuum at 60° for 2 hours before using.

Identification—Place a portion of powdered Tablets, equivalent to about 1 mg of mazindol, in a suitable flask. Add 40 mL of methanol, shake by mechanical means for not less than 5 minutes, and heat for several minutes on a steam bath to boiling. Cool, dilute with methanol to about 100 mL, and filter. Separate the filtrate into two approximately equal portions, add 2 drops of hydrochloric acid to one portion, and mix: the ultraviolet absorption spectra of the solutions so obtained exhibit maxima and minima at the same wavelengths as those of similar solutions prepared from USP Mazindol RS, concomitantly measured.

Dissolution ⟨711⟩—

Medium: 0.1 *N* hydrochloric acid; 500 mL.

Apparatus 2: 50 rpm.

Time: 60 minutes.

Mobile phase—Mix 11.50 g of monobasic ammonium phosphate and 1.32 g of dibasic ammonium phosphate with water to obtain 1000 mL of an ammonium phosphate buffer. The *Mobile phase* is a suitably filtered and degassed mixture of the ammonium phosphate buffer and acetonitrile (55:45). Make adjustments if necessary (see *System Suitability* under *Chromatography* ⟨621⟩).

Chromatographic system (see *Chromatography* ⟨621⟩)—The liquid chromatograph is equipped with a 271-nm detector and a 4-mm × 30-cm column that contains packing L7. The flow rate is about 2 mL per minute. Chromatograph three replicate injections of the Standard solution, and record the peak responses as directed under *Procedure:* the relative standard deviation is not more than 3.0%.

Procedure—Inject an appropriate volume (50 µL to 500 µL) of a filtered portion of the solution under test into the chromatograph, record the chromatogram, and measure the response for the major peak. Calculate the quantity of $C_{16}H_{13}ClN_2O$ dissolved in comparison with a Standard solution having a known concentration of USP Mazindol RS in the same medium and similarly chromatographed.

Tolerances—Not less than 60% (*Q*) of the labeled amount of $C_{16}H_{13}ClN_2O$ is dissolved in 60 minutes.

Uniformity of dosage units ⟨905⟩: meet the requirements.

Procedure for content uniformity—

Dye solution—Dissolve 100 mg of bromocresol purple in 1000 mL of 0.33 *N* acetic acid, and mix.

Standard preparation—Dissolve an accurately weighed quantity of USP Mazindol RS in 0.33 *N* acetic acid, and dilute quantitatively and stepwise with 0.33 *N* acetic acid to obtain a solution having a known concentration of about 20 µg per mL.

Test preparation—Mix 1 finely powdered Tablet with an accurately measured volume of 0.33 *N* acetic acid, sufficient to provide a solution having a concentration of about 20 µg of mazindol per mL, shake by mechanical means for 30 minutes, and filter, discarding the first few mL of the filtrate.

Procedure—Transfer 25.0 mL each of the *Standard preparation*, the *Test preparation*, and 0.33 *N* acetic acid to provide the blank, to individual 125-mL separators. Add 30 mL of *Dye solution* and 50.0 mL of chloroform to each, and shake by mechanical means for 15 minutes. Allow the layers to separate, and filter the chloroform layers. Concomitantly determine the absorbances of the filtered solutions obtained from the *Test preparation* and the *Standard preparation* at the wavelength of maximum absorbance at about 420 nm, using the blank to set the instrument. Calculate the quantity, in mg, of $C_{16}H_{13}ClN_2O$ in the Tablet by the formula:

$$(TC/D)(A_U/A_S),$$

in which *T* is the labeled quantity, in mg, of mazindol in the Tablet, *C* is the concentration, in µg per mL, of USP Mazindol RS in the *Standard preparation*, *D* is the concentration, in µg per mL, of mazindol in the *Test preparation*, based on the labeled quantity per Tablet and the extent of dilution, and A_U and A_S are the absorbances of the solutions from the *Test preparation* and the *Standard preparation*, respectively.

Assay—

Internal standard solution—Dissolve 50 mg of Amitriptyline Hydrochloride in 250 mL of methanol, and mix.

Standard preparation—Transfer about 32 mg of USP Mazindol RS, accurately weighed, to a 100-mL volumetric flask, add about 50 mL of *Internal standard solution*, and shake by mechanical means for 30 minutes. Dilute with *Internal standard solution* to volume, and mix.

Assay preparation—Weigh and finely powder not less than 20 Mazindol Tablets. Transfer an accurately weighed portion of the powder, equivalent to about 8 mg of mazindol, to a suitable flask, add 25.0 mL of *Internal standard solution*, and shake by mechanical means for 30 minutes. Filter through a fine-porosity, sintered-glass filter, discarding the first few mL of the filtrate.

Mobile phase—Transfer 200 mL of aqueous 0.01 *M* dibasic ammonium phosphate to a 1000-mL volumetric flask, dilute with methanol to volume, and mix. Filter through a 0.5-µm porosity polytef filter, and degas under vacuum. Protect this solution from light.

Chromatographic system (see *Chromatography* ⟨621⟩)—The liquid chromatograph is equipped with a 254-nm detector and a 4-mm × 30-cm column that contains packing L10. Inject three replicate portions of the *Standard preparation*, and record the peak responses as directed under *Procedure:* the relative standard deviation is not more than 3.0%, and the resolution factor is not less than 2.0.

Procedure—Separately inject equal volumes (about 20 µL) of the *Standard preparation* and the *Assay preparation* into the chromatograph, and record the chromatograms. Measure the peak responses for mazindol and amitriptyline hydrochloride in each chromatogram. Calculate the quantity, in mg, of $C_{16}H_{13}ClN_2O$ in the portion of Tablets taken by the formula:

$$25C(R_U/R_S),$$

in which *C* is the concentration, in mg per mL, of USP Mazindol RS in the *Standard preparation*, and R_U and R_S are the ratios of peak responses of mazindol and amitriptyline hydrochloride obtained from the *Assay preparation* and the *Standard preparation*, respectively.

Measles Virus Vaccine Live

» Measles Virus Vaccine Live conforms to the regulations of the federal Food and Drug Administration concerning biologics (630.30 to 630.37) (see *Biologics* ⟨1041⟩). It is a bacterially sterile preparation of live virus derived from a strain of measles virus tested for neurovirulence in monkeys, for safety, and for immunogenicity, free from all demonstrable viable microbial agents except unavoidable bacteriophage, and found suitable for human immunization. The strain is grown for purposes of vaccine production on chicken embryo primary cell tissue cultures derived from pathogen-free flocks, meets the requirements of the specific safety tests in adult and suckling mice; the requirements of the tests in monkey kidney, chicken embryo and human tissue cell cultures and embryonated eggs; and the requirements of the tests for absence of *Mycobacterium tuberculosis* and of avian leucosis, unless the production cultures were derived from certified avian leucosis-free sources and the control fluids were tested for avian leucosis. The strain cultures are treated to remove all intact tissue cells. The Vaccine meets the requirements of the specific tissue culture test for live virus titer, in a single immunizing dose, of not less than the equivalent of 1000 $TCID_{50}$ (quantity of virus estimated to infect 50% of inoculated cultures × 1000) when tested in parallel with the U. S. Reference Measles Virus, Live Attenuated. It may contain suitable antimicrobial agents.

Packaging and storage—Preserve in single-dose containers, or in light-resistant, multiple-dose containers, at a temperature between 2° and 8°. Multiple-dose containers for 50 doses are adapted for use only in jet injectors, and those for 10 doses for use by jet or syringe injection.

Expiration date—The expiration date is 1 to 2 years, depending on the manufacturer's data, after date of issue from manufacturer's cold storage (−20°, 1 year).

Labeling—Label the Vaccine in multiple-dose containers to indicate that the contents are intended solely for use by jet injector or for use by either jet or syringe injection, whichever is applicable. Label the Vaccine in single-dose containers, if such containers are not light-resistant, to state that it should be protected from sunlight. Label it also to state that constituted Vaccine should be discarded if not used within 8 hours.

Measles and Mumps Virus Vaccine Live

» Measles and Mumps Virus Vaccine Live conforms to the regulations of the federal Food and Drug Administration concerning biologics (see *Biologics* ⟨1041⟩). It is a bacterially sterile preparation of a combination of live measles virus and live mumps virus such that each component is prepared in conformity with and meets the requirements for Measles Virus Vaccine Live and for Mumps Virus Vaccine Live, whichever is applicable. Each component provides an immunizing dose and meets the requirements of the corresponding Virus Vaccine in the total dosage prescribed in the labeling. It may contain suitable antimicrobial agents.

Packaging and storage—Preserve in single-dose containers, or in light-resistant, multiple-dose containers, at a temperature between 2° and 8°. Multiple-dose containers for 50 doses are adapted for use only in jet injectors, and those for 10 doses for use by jet or syringe injection.

Expiration date—The expiration date is 1 to 2 years, depending on the manufacturer's data, after date of issue from manufacturer's cold storage (−20°, 1 year).

Labeling—Label the Vaccine in multiple-dose containers to indicate that the contents are intended solely for use by jet injector or for use by either jet or syringe injection, whichever is applicable. Label the Vaccine in single-dose containers, if such containers are not light-resistant, to state that it should be protected from sunlight. Label it also to state that constituted Vaccine should be discarded if not used within 8 hours.

Measles, Mumps, and Rubella Virus Vaccine Live

» Measles, Mumps, and Rubella Virus Vaccine Live conforms to the regulations of the federal Food and Drug Administration concerning biologics (see *Biologics* ⟨1041⟩). It is a bacterially sterile preparation of a combination of live measles virus, live mumps virus, and live rubella virus such that each component is prepared in conformity with and meets the requirements for Measles Virus Vaccine Live, for Mumps Virus Vaccine Live, and for Rubella Virus Vaccine Live, whichever is applicable. Each component provides an immunizing dose and meets the requirements of the corresponding Virus Vaccine in the total dosage prescribed in the labeling. It may contain suitable antimicrobial agents.

Packaging and storage—Preserve in single-dose containers, or in light-resistant, multiple-dose containers, at a temperature between 2° and 8°. Multiple-dose containers for 50 doses are adapted for use only in jet injectors, and those for 10 doses for use by jet or syringe injection.

Expiration date—The expiration date is 1 to 2 years, depending on the manufacturer's data, after date of issue from manufacturer's cold storage (−20°, 1 year).

Labeling—Label the Vaccine in multiple-dose containers to indicate that the contents are intended solely for use by jet injector or for use by either jet or syringe injection, whichever is applicable. Label the Vaccine in single-dose containers, if such containers are not light-resistant, to state that it should be protected from sunlight. Label it also to state that constituted Vaccine should be discarded if not used within 8 hours.

Measles and Rubella Virus Vaccine Live

» Measles and Rubella Virus Vaccine Live conforms to the regulations of the federal Food and Drug Administration concerning biologics (see *Biologics* ⟨1041⟩). It is a bacterially sterile preparation of a combination of live measles virus and live rubella virus such that each component is prepared in conformity with and meets the requirements for Measles Virus Vaccine Live and for Rubella Virus Vaccine Live, whichever is applicable. Each component provides an immunizing dose and meets the requirements of the corresponding Virus Vaccine in the total dosage prescribed in the labeling. It may contain suitable antimicrobial agents.

Packaging and storage—Preserve in single-dose containers, or in light-resistant, multiple-dose containers, at a temperature between 2° and 8°. Multiple-dose containers for 50 doses are adapted for use only in jet injectors, and those for 10 doses for use by jet or syringe injection.

Expiration date—The expiration date is 1 to 2 years, depending on the manufacturer's data, after date of issue from manufacturer's cold storage (−20°, 1 year).

Labeling—Label the Vaccine in multiple-dose containers to indicate that the contents are intended solely for use by jet injector or for use by either jet or syringe injection, whichever is applicable. Label the Vaccine in single-dose containers, if such containers are not light-resistant, to state that it should be protected from sunlight. Label it also to state that constituted Vaccine should be discarded if not used within 8 hours.

Mebendazole

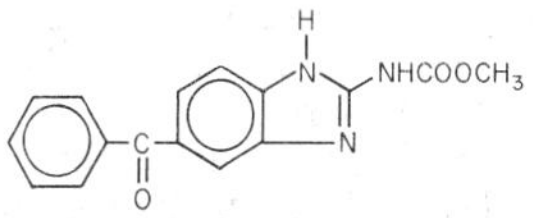

$C_{16}H_{13}N_3O_3$ 295.30

Carbamic acid, (5-benzoyl-1*H*-benzimidazol-2-yl), methyl ester.

Methyl 5-benzoyl-2-benzimidazolecarbamate [*31431-39-7*].

» Mebendazole contains not less than 98.0 percent and not more than 102.0 percent of $C_{16}H_{13}N_3O_3$, calculated on the dried basis.

Packaging and storage—Preserve in well-closed containers.

Reference standard—*USP Mebendazole Reference Standard*—Dry at 105° for 4 hours before using.

Identification—The infrared absorption spectrum of a potassium bromide dispersion of it, previously dried, exhibits maxima only at the same wavelengths as that of a similar preparation of USP Mebendazole RS.

Loss on drying ⟨731⟩—Dry it at 105° for 4 hours: it loses not more than 0.5% of its weight.

Residue on ignition ⟨281⟩: not more than 0.1%.

Heavy metals, *Method II* ⟨231⟩: 0.002%.

Chromatographic purity—Dissolve 50 mg in 1.0 mL of 96 percent formic acid in a 10-mL volumetric flask, add chloroform to volume, and mix. Similarly prepare a solution of USP Mebendazole RS in the same medium having a concentration of 5 mg per mL. Transfer 1.0 mL of this Standard solution to a 200-mL volumetric flask, add a mixture of chloroform and 96 percent formic acid (9:1) to volume, and mix (diluted Standard solution). On a suitable thin-layer chromatographic plate (see *Chromatography* ⟨621⟩), coated with a 0.25-mm layer of chromatographic silica gel mixture, apply 10-µL portions of the test solution, the Standard solution, and the diluted Standard solution. Allow the spots to dry, and develop the chromatogram in a solvent system consisting of a mixture of chloroform, methanol, and 96 percent formic acid (90:5:5) until the solvent front has moved about three-fourths of the length of the plate. Remove the plate from the developing chamber, mark the solvent front, allow the solvent to evaporate, and examine the plate under short-wavelength ultraviolet light: the R_f value of the principal spot obtained from the test solution corresponds to that obtained from the Standard solution, and no spot, other than the principal spot, in the chromatogram of the test solution is larger or more intense than the principal spot obtained from the diluted Standard solution.

Assay—Dissolve about 225 mg of Mebendazole, accurately weighed, in 30 mL of glacial acetic acid. Titrate with 0.1 *N* perchloric acid VS, determining the end-point potentiometrically, using a calomel-glass electrode system. Perform a blank determination, and make any necessary correction. Each mL of 0.1 *N* perchloric acid is equivalent to 29.53 mg of $C_{16}H_{13}N_3O_3$.

Mebendazole Tablets

» Mebendazole Tablets contain not less than 90.0 percent and not more than 110.0 percent of the labeled amount of $C_{16}H_{13}N_3O_3$.

Packaging and storage—Preserve in well-closed containers.

Reference standard—*USP Mebendazole Reference Standard*—Dry at 105° for 4 hours before using.

Identification—Finely powder a quantity of Tablets, equivalent to about 200 mg of mebendazole, and mix the powder with 20 mL of a mixture of chloroform and 96 percent formic acid (19:1). Warm the suspension on a water bath for a few minutes, cool, and filter through a medium-porosity, sintered-glass filter. On a suitable thin-layer chromatographic plate (see *Chromatography* ⟨621⟩), coated with a 0.25-mm layer of chromatographic silica gel mixture, apply 10 µL of this solution and 10 µL of a Standard solution of USP Mebendazole RS in a mixture of chloroform and 96 percent formic acid (19:1) containing 10 mg per mL. Allow the spots to dry, and develop the chromatogram in a solvent system consisting of a mixture of chloroform, methanol, and 96 percent formic acid (90:5:5) until the solvent front has moved about three-fourths of the length of the plate. Remove the plate from the developing chamber, mark the solvent front, allow the solvent to evaporate, and examine the plate under short-wavelength ultraviolet light: the R_f value of the principal spot obtained from the test solution corresponds to that obtained from the Standard solution.

Disintegration ⟨701⟩: 10 minutes.

Uniformity of dosage units ⟨905⟩: meet the requirements.

Procedure for content uniformity—

Standard preparation—Transfer about 20 mg of USP Mebendazole RS, accurately weighed, to a 10-mL volumetric flask, add 4 mL of 96 percent formic acid, and mix to dissolve. Add isopropyl alcohol to volume, and mix. Pipet 0.5 mL of this solution into a 100-mL volumetric flask, dilute with isopropyl alcohol to volume, and mix.

Test preparation—Mix 1 Tablet with 20 mL of 96 percent formic acid in a 100-mL volumetric flask, and heat on a steam bath for 15 minutes. Cool, add isopropyl alcohol to volume, mix, and filter through a medium-porosity, sintered-glass filter. Transfer an accurately measured portion of the filtrate, equivalent to 1 mg of mebendazole, to a 100-mL volumetric flask, dilute with isopropyl alcohol to volume, and mix.

Procedure—Concomitantly determine the absorbances of the *Standard preparation* and the *Test preparation* in 1-cm cells at the wavelength of maximum absorbance at about 310 nm, with a suitable spectrophotometer, using a 1 in 500 solution of 96 percent formic acid in isopropyl alcohol as the blank. Calculate the quantity, in mg, of $C_{16}H_{13}N_3O_3$ in the Tablet by the formula:

$$(TC/D)(A_U/A_S),$$

in which *T* is the labeled quantity, in mg, of mebendazole in the Tablet, *C* is the concentration, in µg per mL, of USP Mebendazole RS in the *Standard preparation*, *D* is the concentration, in µg per mL, of mebendazole in the *Test preparation*, based upon the labeled quantity per Tablet and the extent of dilution, and A_U and A_S are the absorbances of the solutions from the *Test preparation* and the *Standard preparation*, respectively.

Assay—

Standard preparation—Transfer about 10 mg of USP Mebendazole RS, accurately weighed, to a 100-mL volumetric flask, and add 90 mL of chloroform, 7 mL of isopropyl alcohol, and 2 mL of 96 percent formic acid. Agitate until the solid has dissolved, add isopropyl alcohol to volume, and mix. Pipet 5 mL of this solution into a second 100-mL volumetric flask, dilute with isopropyl alcohol to volume, and mix to obtain a solution having a known concentration of about 5 µg per mL.

Assay preparation—Weigh and finely powder not less than 20 Mebendazole Tablets. Transfer an accurately weighed portion of the powder, equivalent to about 100 mg of mebendazole, to a 100-mL volumetric flask with the aid of 50 mL of 96 percent formic acid, and heat in a water bath at a temperature of 50° for 15 minutes. Cool, add water to volume, mix, and filter through a medium-porosity, sintered-glass filter. Pipet 10 mL of the filtrate into a 250-mL separator, and add 50 mL of water and 50 mL of chloroform. Shake for about 2 minutes, allow the phases to separate, and transfer the chloroform layer to a second 250-mL separator. Wash the aqueous layer with two 10-mL portions of chloroform, adding the chloroform washings to the second separator, and discard the aqueous layer. Wash the combined chloroform solutions with a mixture of 4 mL of 1 *N* hydrochloric acid and 50 mL of a 1 in 10 solution of 96 percent formic acid in water, and transfer the chloroform layer to a 100-mL volumetric flask. Extract the aqueous washing with two 10-mL portions of chloroform, add these chloroform extracts to the chloroform solution in the volumetric flask, dilute with isopropyl alcohol to volume, and mix. Pipet 5 mL of this solution into another 100-mL volumetric flask, dilute with isopropyl alcohol to volume, and mix.

Procedure—Mix 90 mL of chloroform with 2 mL of 96 percent formic acid in a 100-mL volumetric flask, add isopropyl alcohol to volume, and mix. Pipet 5 mL of this solution into a second 100-mL volumetric flask, dilute with isopropyl alcohol to volume, and mix to obtain a reagent blank. Concomitantly determine the absorbances of the *Assay preparation* and the *Standard preparation* in 1-cm cells at the wavelength of maximum absorbance at about 247 nm, with a suitable spectrophotometer, using the reagent blank to set the instrument. Calculate the quantity, in mg, of $C_{16}H_{13}N_3O_3$ in the portion of Tablets taken by the formula:

$$20C(A_U/A_S),$$

in which *C* is the concentration, in µg per mL, of USP Mebendazole RS in the *Standard preparation*, and A_U and A_S are the absorbances of the *Assay preparation* and the *Standard preparation*, respectively.

Mechlorethamine Hydrochloride

$CH_3N(CH_2CH_2Cl)_2 \cdot HCl$

$C_5H_{11}Cl_2N.HCl$ 192.52

Ethanamine, 2-chloro-*N*-(2-chloroethyl)-*N*-methyl-, hydrochloride.

2,2′-Dichloro-*N*-methyldiethylamine hydrochloride [*55-86-7*].

» Mechlorethamine Hydrochloride contains not less than 97.5 percent and not more than 100.5 percent of $C_5H_{11}Cl_2N.HCl$, calculated on the anhydrous basis.

Packaging and storage—Preserve in tight, light-resistant containers.

Labeling—The label bears a warning that great care should be taken to prevent inhaling particles of Mechlorethamine Hydrochloride and exposing the skin to it.

Reference standard—*USP Mechlorethamine Hydrochloride Reference Standard*—Do not dry before using.

Identification—

A: The infrared absorption spectrum of a potassium bromide dispersion of it exhibits maxima only at the same wavelengths as that of a similar preparation of USP Mechlorethamine Hydrochloride RS.

B: Transfer 100 mg to a test tube containing 1 mL of sodium thiosulfate solution (prepared by dissolving 1 g of sodium thiosulfate and 100 mg of sodium carbonate in 40 mL of water), shake, allow to stand for 2 hours, then add 1 drop of iodine TS: the color of free iodine remains.

Melting range ⟨741⟩: between 108° and 111°.

pH ⟨791⟩: between 3.0 and 5.0, in a solution (1 in 500).

Water, *Method I* ⟨921⟩: not more than 0.4%.

Ionic chloride content—Dissolve about 30 mg, accurately weighed, in 30 mL of water contained in a beaker. Add 5 mL of nitric acid and 10 mL of gelatin solution (1 in 100). Stir the solution with the rotating platinum electrode of an amperometric titration assembly, the remainder of which consists of a suitable microammeter and an agar–potassium nitrate salt bridge. When the current flow has become constant at about 20 microamperes to 50 microamperes, begin titrating rapidly with 0.01 *N* silver nitrate. When the current flow begins to increase, add the titrant slowly, noting the volumes at three points suitable for determining the end-point graphically from the plotted volume-current flow relationship. Each mL of 0.01 *N* silver nitrate is equivalent to 0.3545 mg of ionic chloride: between 18.0% and 19.3% of ionic chloride is found.

Assay—Transfer about 100 mg of Mechlorethamine Hydrochloride, accurately weighed, to a 125-mL conical flask. Add 100 mg of sodium bicarbonate and 20.0 mL of 0.1 *N* sodium thiosulfate VS. Allow to stand for 2½ hours, add 3 mL of starch TS, and titrate the excess sodium thiosulfate with 0.1 *N* iodine VS. Each mL of 0.1 *N* sodium thiosulfate is equivalent to 9.626 mg of $C_5H_{11}Cl_2N.HCl$.

Mechlorethamine Hydrochloride for Injection

» Mechlorethamine Hydrochloride for Injection is a sterile mixture of Mechlorethamine Hydrochloride with Sodium Chloride or other suitable diluent. It contains not less than 90.0 percent and not more than 110.0 percent of the labeled amount of $C_5H_{11}Cl_2N.HCl$.

Packaging and storage—Preserve in *Containers for Sterile Solids* as described under *Injections* ⟨1⟩.

Labeling—It meets the requirements for *Labeling* under *Injections* ⟨1⟩.

The label bears a warning that great care should be taken to prevent inhaling particles of Mechlorethamine Hydrochloride for Injection and exposing the skin to it.

Reference standard—*USP Mechlorethamine Hydrochloride Reference Standard*—Do not dry before using.

Completeness of solution ⟨641⟩—A 0.10-g portion dissolves in 10 mL of carbon dioxide–free water to yield a clear solution.

Constituted solution—At the time of use, the constituted solution prepared from Mechlorethamine Hydrochloride for Injection meets the requirements for *Constituted Solutions* under *Injections* ⟨1⟩.

Identification—It responds to the *Identification tests* under *Mechlorethamine Hydrochloride.*

pH ⟨791⟩: between 3.0 and 5.0, in a solution (1 in 50).

Water, *Method I* ⟨921⟩: not more than 1.0%.

Particulate matter ⟨788⟩: meets the requirements under *Small-volume Injections.*

Other requirements—It meets the requirements for *Sterility Tests* ⟨71⟩ and *Uniformity of Dosage Units* ⟨905⟩.

Assay—

Assay preparation—Select a counted number of not less than 10 containers of Mechlorethamine Hydrochloride for Injection, equivalent to about 100 mg of mechlorethamine hydrochloride. Dissolve the contents of each container in water, and transfer the resulting solutions to a 250-mL conical flask.

Procedure—Immediately proceed as directed in the *Assay* under *Mechlorethamine Hydrochloride,* beginning with "Add 100 mg of sodium bicarbonate." Calculate the average content, in mg, of $C_5H_{11}Cl_2N.HCl$ per container of Mechlorethamine Hydrochloride for Injection taken by the formula:

$$(9.626)(V/N),$$

in which V is the volume, in mL, of 0.1 *N* sodium thiosulfate consumed, and N is the number of containers selected to prepare the *Assay preparation.*

Meclizine Hydrochloride

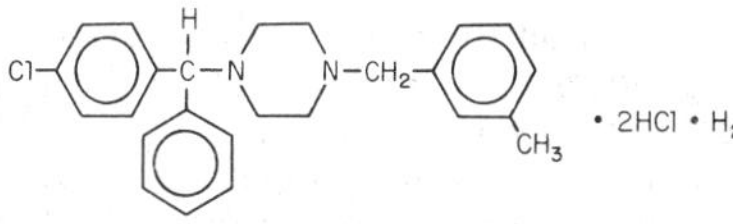

$C_{25}H_{27}ClN_2.2HCl.H_2O$ 481.89

Piperazine, 1-[(4-chlorophenyl)phenylmethyl]-4-[(3-methylphenyl)methyl]-, dihydrochloride, monohydrate.

1-(*p*-Chloro-α-phenylbenzyl)-4-(*m*-methylbenzyl)piperazine dihydrochloride monohydrate [*31884-77-2*].

Anhydrous 463.88 [*1104-22-9*].

» Meclizine Hydrochloride contains not less than 97.0 percent and not more than 100.5 percent of $C_{25}H_{27}ClN_2.2HCl$, calculated on the anhydrous basis.

Packaging and storage—Preserve in tight containers.

Reference standard—*USP Meclizine Hydrochloride Reference Standard*—Do not dry; determine the water content titrimetrically at the time of use.

Identification—

A: The infrared absorption spectrum of a potassium bromide dispersion of it exhibits maxima only at the same wavelengths as that of a similar preparation of USP Meclizine Hydrochloride RS.

B: The ultraviolet absorption spectrum of a 1 in 100,000 solution in dilute hydrochloric acid (1 in 100) exhibits maxima and minima at the same wavelengths as that of a similar solution of USP Meclizine Hydrochloride RS, concomitantly measured.

C: Dissolve 25 mg in a mixture of 3 mL of 2 *N* nitric acid and 5 mL of alcohol: the solution responds to the tests for *Chloride* ⟨191⟩.

Melting range ⟨741⟩: between 217° and 224°, with decomposition, determined after drying the material over silica gel for 24 hours.

Water, *Method I* ⟨921⟩: not more than 5.0%.

Residue on ignition ⟨281⟩: not more than 0.1%.

Chromatographic purity—

Mobile phase—Dissolve 1.5 g of sodium 1-heptanesulfonate in 300 mL of water and mix this solution with 700 mL of acetonitrile. Adjust with 0.1 *N* sulfuric acid to a pH of 4, filter, and degas. Make adjustments if necessary (see *System Suitability* under *Chromatography* ⟨621⟩).

Diluting solution—Prepare a mixture of acetonitrile and water (70:30).

Standard preparations—Dissolve an accurately weighed quantity of USP Meclizine Hydrochloride RS in *Diluting solution,* and dilute quantitatively, and stepwise if necessary, with *Diluting solution* to obtain a solution having a known concentration of about 2.5 µg per mL.

Test preparation—Prepare a solution of Meclizine Hydrochloride in *Diluting solution* having a known concentration of about 0.25 mg per mL.

System suitability solution—Prepare a solution in *Diluting solution* containing about 0.01 mg of USP Meclizine Hydrochloride RS and 0.01 mg of 4-chlorobenzophenone per mL.

Chromatographic system (see *Chromatography* ⟨621⟩)—The liquid chromatograph is equipped with a 230-nm detector and a 4.2-mm × 25-cm column that contains 10-µm packing L1. The flow rate is about 1.3 mL per minute. Chromatograph the *System suitability solution*, and record the peak responses as directed under *Procedure:* the relative retention times are about 0.8 for Meclizine Hydrochloride and 1.0 for 4-chlorobenzophenone, and the resolution, *R*, between the 4-chlorobenzophenone and Meclizine Hydrochloride peaks is not less than 2.0. Chromatograph the *Standard preparation*, and record the peak responses as directed under *Procedure:* the tailing factor for the analyte peak is not more than 1.5, and the column efficiency, *N*, determined from the analyte peak is not less than 1800 theoretical plates. The relative standard deviation for replicate injections is not more than 1.5%.

Procedure—Separately inject equal volumes (about 20 µL) of the *Standard preparation* and the *Test preparation* into the chromatograph. Allow the *Test preparation* to elute for not less than three times the retention time of Meclizine Hydrochloride. Record the chromatograms and measure all of the peak responses: the sum of the peak responses, excluding that of Meclizine Hydrochloride, from the *Test preparation* is not more than two times the Meclizine Hydrochloride response from the *Standard preparation* (2.0%), and no single peak response is greater than that of the Meclizine Hydrochloride response from the *Standard preparation* (1.0%).

Assay—Dissolve about 350 mg of Meclizine Hydrochloride, accurately weighed, in about 50 mL of chloroform. Add 50 mL of glacial acetic acid, 5 mL of acetic anhydride, and 10 mL of mercuric acetate TS, and titrate with 0.1 *N* perchloric acid VS, determining the end-point potentiometrically, using a calomel-glass electrode system (see *Titrimetry* ⟨541⟩). Perform a blank determination, and make any necessary correction. Each mL of 0.1 *N* perchloric acid is equivalent to 23.19 mg of $C_{25}H_{27}ClN_2.2HCl$.

Meclizine Hydrochloride Tablets

» Meclizine Hydrochloride Tablets contain not less than 95.0 percent and not more than 110.0 percent of the labeled amount of $C_{25}H_{27}ClN_2.2HCl$.

Packaging and storage—Preserve in well-closed containers.

Reference standard—*USP Meclizine Hydrochloride Reference Standard*—Do not dry; determine the water content titrimetrically at the time of use.

Identification—

A: Tablets meet the requirements under *Identification—Organic Nitrogenous Bases* ⟨181⟩.

B: Extract a quantity of finely powdered Tablets, equivalent to about 125 mg of meclizine hydrochloride, by shaking for 15 minutes with 50 mL of methanol. Apply 50 µL of this solution and 50 µL of a solution of USP Meclizine Hydrochloride RS in methanol, containing 2.5 mg per mL, on a thin-layer chromatographic plate coated with a 0.5-mm layer of chromatographic silica gel mixture. Place the plate in a developing chamber that contains and that has been equilibrated with a mixture of cyclohexane, toluene, and diethylamine (15:3:2). Develop the chromatogram until the solvent front has moved about three-fourths of the length of the plate. Remove the plate, mark the solvent front, and allow the solvent to evaporate. Locate the spots on the plate by examination under short-wavelength ultraviolet light: the R_f value of the principal spot obtained with the test specimen corresponds to that obtained with the Reference Standard.

Dissolution ⟨711⟩—

Medium: 0.1 *N* hydrochloric acid; 900 mL.

Apparatus 1: 100 rpm.

Time: 45 minutes.

Mobile phase—Prepare a suitable degassed and filtered mixture of water and methanol (55:45) that contains 0.69 g of monobasic sodium phosphate in each 100 mL and is adjusted with phosphoric acid, if necessary, to a pH of 4.0.

Chromatographic system (see *Chromatography* ⟨621⟩)—The liquid chromatograph is equipped with a 230-nm detector, a 4.6-mm × 4-cm precolumn positioned between the pump and the injection valve that contains packing L4, and a 4.6-mm × 25-cm analytical column that contains packing L9. The flow rate is about 2 mL per minute. Chromatograph replicate injections of the Standard solution, and record the peak responses as directed under *Procedure:* the relative standard deviation is not more than 2.0%.

Procedure—Inject about 100 µL of a filtered portion of the solution under test, suitably diluted with *Mobile phase*, if necessary, into the chromatograph, record the chromatogram, and measure the response for the major peak. Determine the amount of $C_{25}H_{27}ClN_2.2HCl$ dissolved from the peak response obtained in comparison with the peak response obtained from a Standard solution having a known concentration of USP Meclizine Hydrochloride RS in a mixture of *Dissolution Medium* and *Mobile phase* (1:1), similarly chromatographed. An amount of alcohol not to exceed 1% of the total volume of the Standard solution may be used to dissolve USP Meclizine Hydrochloride RS prior to dilution.

Tolerances—Not less than 75% (*Q*) of the labeled amount of $C_{25}H_{27}ClN_2.2HCl$ is dissolved in 45 minutes.

Uniformity of dosage units ⟨905⟩: meet the requirements.

Procedure for content uniformity—Place 1 Tablet in a 100-mL volumetric flask, add 50 mL of dilute hydrochloric acid (1 in 100), shake by mechanical means for 30 minutes, add the dilute acid to volume, and filter, discarding the first 20 mL of the filtrate. Dilute quantitatively and stepwise with the same acid to obtain a solution having a concentration of about 15 µg of meclizine hydrochloride per mL. Similarly, prepare a Standard solution of USP Meclizine Hydrochloride RS in dilute hydrochloric acid (1 in 100) having a known concentration of about 15 µg per mL. Concomitantly determine the absorbances of the solution from the Tablet and the Standard solution in 1-cm cells at the wavelength of maximum absorbance at about 232 nm, with a suitable spectrophotometer, using dilute hydrochloric acid (1 in 100) as the blank. Calculate the quantity, in mg, of $C_{25}H_{27}ClN_2.2HCl$ in the Tablet by the formula:

$$(T/D)C(A_U/A_S),$$

in which *T* is the quantity, in mg, of meclizine hydrochloride in the Tablet, *D* is the concentration, in µg per mL, of meclizine hydrochloride in the solution from the Tablet, on the basis of the labeled quantity per Tablet and the extent of dilution, *C* is the concentration, in µg per mL, of USP Meclizine Hydrochloride RS in the Standard solution, and A_U and A_S are the absorbances of the solution from the Tablet and the Standard solution, respectively.

Assay—Weigh and finely powder not less than 20 Meclizine Hydrochloride Tablets. Weigh accurately a portion of the powder, equivalent to about 350 mg of meclizine hydrochloride, and transfer to a suitable vessel. Extract with three 50-mL portions of chloroform, each time stirring the mixture for 30 minutes, then allowing the undissolved matter to settle and decanting the supernatant liquid onto a fine, sintered-glass filter. Finally, transfer the residue to the filter with the aid of chloroform, and wash it, the vessel, and the filter with a 20-mL portion of chloroform. Evaporate the combined extracts on a steam bath to about 10 mL. Cool, and add 50 mL of glacial acetic acid, 5 mL of acetic anhydride, and 10 mL of mercuric acetate TS. Then titrate with 0.1 *N* perchloric acid VS, determining the end-point potentiometrically, using a calomel-glass electrode system (see *Titrimetry* ⟨541⟩). Perform a blank determination, and make any necessary correction. Each mL of 0.1 *N* perchloric acid is equivalent to 23.19 mg of $C_{25}H_{27}ClN_2.2HCl$.

Meclocycline Sulfosalicylate

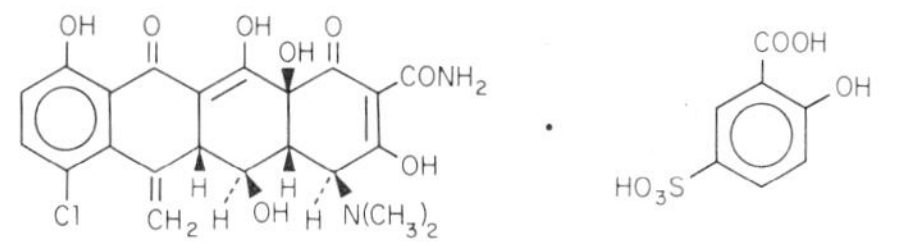

$C_{22}H_{21}ClN_2O_8 . C_7H_6O_6S$ 695.05

2-Naphthacenecarboxamide, 7-chloro-4-(dimethylamino)-1,4,4a,5,5a,6,11,12a-octahydro-3,5,10,12,12a-pentahydroxy-6-methylene-1,11-dioxo-, [4*S*-(4α,4aα,5α,-5aα,12aα)]- mono(2-hydroxy-5-sulfobenzoate) (salt).

(4*S*,4a*R*,5*S*,5a*R*,12a*S*)-7-Chloro-4-(dimethylamino)-1,4,4a,5,5a,6,11,12a-octahydro-3,5,10,12,12a-pentahydroxy-6-methylene-1,11-dioxo-2-naphthacenecarboxamide mono(5-sulfosalicylate) (salt) [*73816-42-9*].

» Meclocycline Sulfosalicylate has a potency equivalent to not less than 620 μg of meclocycline ($C_{22}H_{21}ClN_2O_8$) per mg.

Packaging and storage—Preserve in tight containers, protected from light.

Reference standard—*USP Meclocycline Reference Standard*—Dry in vacuum at 60° and at a pressure not exceeding 5 mm of mercury for 3 hours before using.

Crystallinity ⟨695⟩: meets the requirements.

pH ⟨791⟩: between 2.5 and 3.5, in a solution containing 10 mg per mL.

Water, *Method I* ⟨921⟩: not more than 4.0%, 20 mL of a mixture of carbon tetrachloride, chloroform, and methanol (2:2:1) being used in place of methanol in the titration vessel.

Assay—

0.001 M Ammonium edetate—Transfer 293 mg of edetic acid, accurately weighed, to a 1000-mL volumetric flask, add 1 mL of methanol and 7 mL of ammonium hydroxide, and shake to dissolve the edetic acid. Add 900 mL of water, adjust with glacial acetic acid to a pH of 6.6, dilute with water to volume, and mix.

Mobile phase—Prepare a solution of *0.001 M Ammonium edetate* and tetrahydrofuran (85:15). Filter and degas the solution before use.

Standard preparation—Transfer 25 mg of USP Meclocycline RS, accurately weighed, to a 50-mL volumetric flask, dilute with methanol to volume, and mix. Transfer 3.0 mL of this solution to a 25-mL volumetric flask, dilute with *Mobile phase* to volume, and mix to obtain a solution having a known concentration of about 60 μg of meclocycline per mL.

Assay preparation—Using 36 mg of meclocycline sulfosalicylate, accurately weighed, prepare as directed for *Standard preparation.*

Procedure—Introduce equal volumes (about 10 μL) of the *Assay preparation* and the *Standard preparation* into a high-pressure liquid chromatograph (see *Chromatography* ⟨621⟩), operated at room temperature, by means of a suitable microsyringe or sampling valve, adjusting the specimen size and other operating parameters such that the peak obtained from the *Standard preparation* is about 0.6 full-scale. Typically, the apparatus is fitted with a 25-cm × 4-mm column packed with packing L1 and is equipped with an ultraviolet detector capable of monitoring absorption at 340 nm, and a suitable recorder. In a suitable chromatogram the coefficient of variation for five replicate injections of the *Standard preparation* is not more than 3.0%. Measure the peak responses at equivalent retention times, obtained from the *Assay preparation* and the *Standard preparation*, and calculate the quantity, in mg, of $C_{22}H_{21}ClN_2O_8$ in the portion of Meclocycline Sulfosalicylate taken by the formula:

$$(1.25/3)(C)(R_U/R_S),$$

in which *C* is the equivalent, in μg per mL, of meclocycline from the USP Meclocycline RS in the *Standard preparation*, and R_U and R_S are the peak responses obtained from the *Assay preparation* and the *Standard preparation*, respectively.

Meclocycline Sulfosalicylate Cream

» Meclocycline Sulfosalicylate Cream contains the equivalent of not less than 90.0 percent and not more than 125.0 percent of the labeled amount of meclocycline ($C_{22}H_{21}ClN_2O_8$).

Packaging and storage—Preserve in tight containers, protected from light.

Reference standard—*USP Meclocycline Reference Standard*—Dry in vacuum at a pressure not exceeding 5 mm of mercury at 60° for 3 hours before using.

Minimum fill ⟨755⟩: meets the requirements.

Assay—

Mobile phase—Prepare as directed under *Meclocycline Sulfosalicylate.*

Standard preparation—Transfer 25 mg of USP Meclocycline RS, accurately weighed, to a 50-mL volumetric flask, dilute with methanol to volume, and mix. Transfer 2.0 mL of this solution to a 100-mL volumetric flask, dilute with *Mobile phase* to volume, and mix to obtain a solution having a known concentration of about 10 μg of meclocycline per mL.

Assay preparation—Transfer an accurately weighed quantity of Meclocycline Sulfosalicylate Cream, equivalent to about 5 mg of meclocycline, to a glass-stoppered, 50-mL centrifuge tube, add 20 mL of methanol and 20 mL of 0.025 *N* sulfuric acid, and shake vigorously for 15 minutes. Transfer the solution to a 50-mL volumetric flask, rinsing the centrifuge tube with two 5-mL portions of methanol and adding the rinsings to the flask, dilute with methanol to volume, and mix. Centrifuge a portion of this solution for 5 minutes, transfer 5 mL of the supernatant liquid to a 50-mL volumetric flask, dilute with *Mobile phase* to volume, mix, and filter.

Procedure—Proceed as directed in the *Assay* under *Meclocycline Sulfosalicylate.* Calculate the quantity, in mg, of $C_{22}H_{21}ClN_2O_8$ in the portion of Cream taken by the formula:

$$0.5C(R_U/R_S),$$

in which *C* is the concentration, in μg per mL, of meclocycline in the *Standard preparation*, and R_U and R_S are the peak responses obtained from the *Assay preparation* and the *Standard preparation*, respectively.

Meclofenamate Sodium

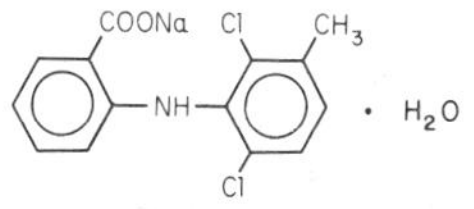

$C_{14}H_{10}Cl_2NNaO_2 . H_2O$ 336.15

Benzoic acid, 2-[(2,6-dichloro-3-methylphenyl)amino]-, monosodium salt, monohydrate.

Monosodium *N*-(2,6-dichloro-*m*-tolyl)anthranilate monohydrate [*6385-02-0*].

Anhydrous 318.13

» Meclofenamate Sodium contains not less than 97.0 percent and not more than 103.0 percent of $C_{14}H_{10}Cl_2NNaO_2$, calculated on the anhydrous basis.

Packaging and storage—Preserve in tight, light-resistant containers.

Reference standard—*USP Meclofenamate Sodium Reference Standard*—Do not dry; determine the *Water* content by *Method I* ⟨921⟩ before using.

Identification—

A: The infrared absorption spectrum of a potassium bromide dispersion of it exhibits maxima only at the same wavelengths as that of a similar preparation of undried USP Meclofenamate Sodium RS.

B: The ultraviolet absorption spectrum of a 1 in 40,000 solution in 0.01 *N* hydrochloric acid in methanol exhibits maxima and minima at the same wavelengths as that of a similar preparation of USP Meclofenamate Sodium RS, concomitantly measured, and the respective absorptivities, calculated on the anhydrous basis, at the wavelengths of maximum absorbance at about 242 nm, at about 279 nm, and at about 336 nm do not differ by more than 3.0%: the ultraviolet absorption spectrum of a 1 in 40,000 solution in 0.1 *N* sodium hydroxide exhibits maxima and minima at the same wavelengths as that of a similar preparation of USP Meclofenamate Sodium RS, concomitantly measured, and the respective absorptivities, calculated on the anhydrous basis, at the wavelengths of maximum absorbance at about 279 nm and at about 317 nm do not differ by more than 3.0%.

Water, *Method I* ⟨921⟩: between 4.8% and 5.8%.

Copper—

Standard copper solution—Dissolve 1000 mg of copper wire in 6 mL of nitric acid in a 1-liter volumetric flask. Add 8 mL of hydrochloric acid, dilute with water to volume, and mix. Dilute this solution quantitatively and stepwise with water to obtain a *Standard copper solution* having a known concentration of 0.6 µg per mL.

Test solution—Transfer 2 g of Meclofenamate Sodium, accurately weighed, to a 100-mL volumetric flask, and add 1 drop of ammonium hydroxide. Dissolve in water, dilute with water to volume, and mix.

Procedure—Concomitantly determine the absorbances of the *Standard copper solution* and the *Test solution* at the copper emission line at about 325 nm, with a suitable atomic absorption spectrophotometer (see *Spectrophotometry and Light-scattering* ⟨851⟩) equipped with a copper hollow–cathode lamp, using water as the blank. Adjust the operating conditions to obtain about 70% full-scale detector response with the *Standard copper solution.* The detector response obtained with the *Test solution* is not greater than that obtained with the *Standard copper solution* (0.003%).

Chromatographic purity—

Standard preparations—Dissolve an accurately weighed quantity of USP Meclofenamate Sodium RS in methanol to obtain a solution containing 20 mg per mL (*Solution A*). Dilute 1.0 mL of *Solution A* with sufficient methanol to obtain 200 mL of solution (*Solution B*).

Test preparation—Dissolve 200 mg of Meclofenamate Sodium in 10.0 mL of methanol.

Procedure—On a suitable thin-layer chromatographic plate (see *Chromatography* ⟨621⟩), coated with a 0.25-mm layer of chromatographic silica gel mixture, apply 10-µL portions of *Solution A*, *Solution B*, and the *Test preparation.* Allow the spots to dry, and develop the chromatogram in a solvent system consisting of a mixture of methylene chloride, methyl ethyl ketone, and glacial acetic acid (50:48:2) until the solvent front has moved about eight-tenths of the length of the plate. Remove the plate from the developing chamber, mark the solvent front, and allow the solvent to evaporate. Examine the plate under short-wavelength ultraviolet light: the chromatograms show a principal spot at about the same R_f value, and any secondary spot, if present in the chromatogram from the *Test preparation* is not more intense than the principal spot obtained from *Solution B* (0.5%).

Assay—Transfer about 350 mg of Meclofenamate Sodium, accurately weighed, to a 125-mL separator, add 10 mL of water, and mix to dissolve. To this solution add 3 mL of 3 *N* hydrochloric acid, shake, and extract with three 30-mL portions of chloroform, collecting the chloroform extracts in an evaporating flask. Evaporate the chloroform extracts to dryness. Dissolve the residue in 5 mL of dimethyl sulfoxide and 25 mL of methanol. Mix, add 5 drops of phenolphthalein TS, and titrate the mixture with 0.1 *N* sodium hydroxide VS. Each mL of 0.1 *N* sodium hydroxide is equivalent to 31.81 mg of $C_{14}H_{10}Cl_2NNaO_2$.

Meclofenamate Sodium Capsules

» Meclofenamate Sodium Capsules contain an amount of $C_{14}H_{10}Cl_2NNaO_2$ equivalent to not less than 90.0 percent and not more than 110.0 percent of the labeled amount of meclofenamic acid ($C_{14}H_{11}Cl_2NO_2$).

Packaging and storage—Preserve in tight, light-resistant containers.

Reference standard—*USP Meclofenamate Sodium Reference Standard*—Do not dry; determine the *Water* content by *Method I* ⟨921⟩ before using.

Identification—Prepare a solution of Capsule contents in methanol containing 20 mg per mL, and filter. The clear filtrate so obtained meets the requirements of the *Thin-layer Chromatographic Identification Test* ⟨201⟩, the solvent mixture consisting of methylene chloride, methyl ethyl ketone, and glacial acetic acid (50:48:2).

Dissolution ⟨711⟩—

Medium: 0.05 *M* pH 8.0 phosphate buffer (see under *Buffer solutions* in the section, *Reagents, Indicators, and Solutions*); 900 mL.

Apparatus 2: 50 rpm.

Time: 45 minutes.

Procedure—Determine the amount of meclofenamic acid ($C_{14}H_{11}Cl_2NO_2$) dissolved from ultraviolet absorbances at the wavelength of maximum absorbance at about 279 nm of filtered portions of the solution under test, suitably diluted with *Dissolution Medium*, if necessary, in comparison with a Standard solution having a known concentration of USP Meclofenamate Sodium RS in the same medium.

Tolerances—Not less than 75% (*Q*) of the labeled amount of $C_{14}H_{11}Cl_2NO_2$ is dissolved in 45 minutes.

Uniformity of dosage units ⟨905⟩: meet the requirements.

Assay—Remove, as completely as possible, the contents of not less than 20 Meclofenamate Sodium Capsules, and weigh accurately. Mix the combined contents, and transfer an accurately weighed quantity of the powder, equivalent to about 50 mg of meclofenamic acid, to a 200-mL volumetric flask. Add 0.01 *N* hydrochloric acid in methanol to volume, and mix. Filter, discarding the first 20 mL of the filtrate. Transfer 10.0 mL of the filtrate to a 100-mL volumetric flask, add 0.01 *N* hydrochloric acid in methanol to volume, and mix. Dissolve an accurately weighed quantity of USP Meclofenamate Sodium RS in 0.01 *N* hydrochloric acid in methanol to obtain a solution having a known concentration of about 27 µg per mL. Concomitantly determine the absorbances of both solutions in 1-cm cells at the wavelength of maximum absorbance at about 336 nm, with a suitable spectrophotometer, using 0.01 *N* hydrochloric acid in methanol as the blank. Calculate the quantity, in mg, of $C_{14}H_{11}Cl_2NO_2$ in the portion of Capsule contents taken by the formula:

$$2C(296.15/318.13)(A_U/A_S),$$

in which C is the concentration, in µg per mL, of USP Meclofenamate Sodium RS in the Standard solution, 296.15 and 318.13 are the molecular weights of meclofenamic acid and meclofenamate sodium, respectively, and A_U and A_S are the absorbances of the solution from the Capsule contents and the Standard solution, respectively.

Medroxyprogesterone Acetate

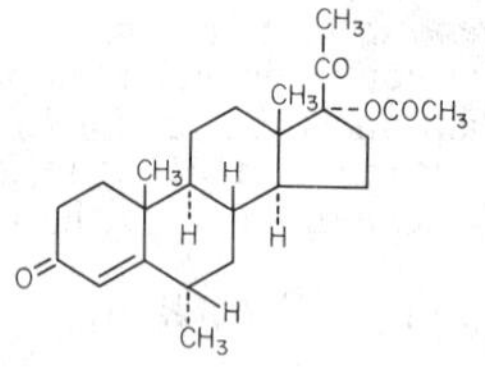

$C_{24}H_{34}O_4$ 386.53

Pregn-4-ene-3,20-dione, 17-(acetyloxy)-6-methyl-, (6α)-.

17-Hydroxy-6α-methylpregn-4-ene-3,20-dione acetate [*71-58-9*].

» Medroxyprogesterone Acetate contains not less than 97.0 percent and not more than 103.0 percent of $C_{24}H_{34}O_4$, calculated on the dried basis.

Packaging and storage—Preserve in tight, light-resistant containers.

Reference standards—*USP Medroxyprogesterone Acetate Reference Standard*—Dry at 105° for 3 hours before using. *USP Progesterone Reference Standard*—Dry in vacuum over silica gel for 4 hours before using.

Identification—

A: The infrared absorption spectrum of a potassium bromide dispersion of it, previously dried, exhibits maxima only at the same wavelengths as that of a similar preparation of USP Medroxyprogesterone Acetate RS.

B: The ultraviolet absorption spectrum of a 1 in 100,000 solution in alcohol exhibits maxima and minima at the same wavelengths as that of a similar solution of USP Medroxyprogesterone Acetate RS, concomitantly measured, and the respective absorptivities, calculated on the dried basis, at the wavelength of maximum absorbance at about 241 nm do not differ by more than 2.0%.

Specific rotation ⟨781⟩: between +45° and +51°, calculated on the dried basis, determined in a solution in dioxane containing 10 mg per mL.

Loss on drying ⟨731⟩—Dry it at 105° for 3 hours: it loses not more than 1.0% of its weight.

Assay—

Mobile phase—Prepare a filtered and degassed mixture of acetonitrile and water (650:350). Make adjustments if necessary (see *System Suitability* under *Chromatography* ⟨621⟩).

Standard preparation—Dissolve an accurately weighed quantity of USP Medroxyprogesterone Acetate RS in acetonitrile to obtain a solution having a known concentration of about 1 mg per mL.

Chromatographic system (see *Chromatography* ⟨621⟩)—The liquid chromatograph is equipped with a 254-nm detector and a 4-mm × 30-cm column that contains packing L1. The flow rate is about 2 mL per minute. Chromatograph the *Standard preparation*, and record the peak responses as directed under *Procedure:* the tailing factor is not more than 2, and the relative standard deviation of the peak responses for replicate injections is not more than 2.0%.

Procedure—Separately inject equal volumes (about 10 µL) of the *Standard preparation* and the *Assay preparation* into the chromatograph, record the chromatograms, and measure the responses for the major peaks. Calculate the quantity, in mg, of $C_{24}H_{34}O_4$ in the portion of Medroxyprogesterone Acetate taken by the formula:

$$25C(r_U/r_S),$$

in which C is the concentration, in mg per mL, of USP Medroxyprogesterone Acetate RS in the *Standard preparation*, and r_U and r_S are the peak responses obtained from the *Assay preparation* and the *Standard preparation*, respectively.

Sterile Medroxyprogesterone Acetate Suspension

» Sterile Medroxyprogesterone Acetate Suspension is a sterile suspension of Medroxyprogesterone Acetate in a suitable aqueous medium. It contains not less than 90.0 percent and not more than 110.0 percent of the labeled amount of $C_{24}H_{34}O_4$.

Packaging and storage—Preserve in single-dose or in multiple-dose containers, preferably of Type I glass.

Reference standards—*USP Medroxyprogesterone Acetate Reference Standard*—Dry at 105° for 3 hours before using. *USP Progesterone Reference Standard*—Dry in vacuum over silica gel for 4 hours before using.

Identification—Transfer a volume of Suspension, equivalent to about 50 mg of medroxyprogesterone acetate, to a centrifuge tube, centrifuge, decant the supernatant liquid, and wash the solids with two 15-mL portions of water, discarding the water washings. Dissolve the solids in 10 mL of chloroform, transfer to a small beaker, evaporate the chloroform on a steam bath, and dry at 105° for 3 hours: the residue so obtained responds to *Identification test A* under *Medroxyprogesterone Acetate.*

pH ⟨791⟩: between 3.0 and 7.0.

Other requirements—It meets the requirements under *Injections* ⟨1⟩.

Assay—

Mobile phase—Mix 700 mL of butyl chloride, 300 mL of hexane, both previously saturated with water, and 80 mL of acetonitrile. The acetonitrile concentration may be varied to meet system suitability requirements and to provide elution times of about 12 and 15 minutes for progesterone and medroxyprogesterone acetate, respectively. Filter the solution through a membrane filter (1 µm or finer porosity).

Internal standard solution—Prepare a solution of USP Progesterone RS in *Mobile phase* containing 0.25 mg per mL.

Standard preparation—Dissolve about 8 mg of USP Medroxyprogesterone Acetate RS, accurately weighed, in 20.0 mL of *Internal standard solution.*

Assay preparation—Transfer to a suitable container an accurately measured volume of Sterile Medroxyprogesterone Acetate Suspension, equivalent to about 50 mg of medroxyprogesterone acetate. Pipet 25 mL of chloroform into the container, shake for about 20 minutes, and centrifuge. Pipet 4 mL of the chloroform layer into a suitable container, and evaporate to dryness. Pipet 20 mL of *Internal standard solution* into the container to dissolve the residue.

Procedure—Proceed as directed in the *Assay* under *Medroxyprogesterone Acetate.* Calculate the quantity, in mg, of $C_{24}H_{34}O_4$ in each mL of the Suspension taken by the formula:

$$125(C/V)(R_U/R_S),$$

in which C is the concentration, in mg, of USP Medroxyprogesterone Acetate RS in the *Standard preparation*, V is the volume, in mL, of Suspension taken, and R_U and R_S are the ratios of peak areas of medroxyprogesterone acetate peak to internal standard peak obtained from the *Assay preparation* and the *Standard preparation*, respectively.

Medroxyprogesterone Acetate Tablets

» Medroxyprogesterone Acetate Tablets contain not less than 93.0 percent and not more than 107.0 percent of the labeled amount of $C_{24}H_{34}O_4$.

Packaging and storage—Preserve in well-closed containers.

Reference standards—*USP Medroxyprogesterone Acetate Reference Standard*—Dry at 105° for 3 hours before using. *USP Progesterone Reference Standard*—Dry in vacuum over silica gel for 4 hours before using.

Identification—Triturate a number of Tablets, equivalent to about 25 mg of medroxyprogesterone acetate, with 15 mL of chloroform, filter, evaporate the chloroform on a steam bath, and dry the residue at 105° for 3 hours: the residue so obtained responds to *Identification test A* under *Medroxyprogesterone Acetate.*

Disintegration ⟨701⟩: 30 minutes.

Uniformity of dosage units ⟨905⟩: meet the requirements.

Procedure for content uniformity—Dissolve an accurately weighed portion of USP Medroxyprogesterone Acetate RS in a mixture of alcohol and water (3:1) to obtain a solution having a known concentration of about 15 µg per mL. Transfer 1 Tablet to a volumetric flask, add a mixture of alcohol and water (3:1) to volume, and shake for about 15 minutes. Filter, and quantitatively dilute a portion of the filtrate as required to obtain a final solution containing about 15 µg per mL. Concomitantly determine the absorbances of this solution and the Standard solution in 1-cm cells at the wavelength of maximum absorbance at about 242 nm. Calculate the quantity, in mg, of $C_{24}H_{34}O_4$ in the Tablet by the formula:

$$(T/D)C(A_U/A_S),$$

in which T is the labeled quantity, in mg, of medroxyprogesterone acetate in the Tablet, D is the concentration, in µg per mL, of medroxyprogesterone acetate in the solution from the Tablet, C is the concentration, in µg per mL, of USP Medroxyprogesterone Acetate RS in the Standard solution, and A_U and A_S are the absorbances of the solution from the Tablet and the Standard solution, respectively.

Assay—

Mobile phase, *Standard preparation*, and *Chromatographic system*—Prepare as directed in the *Assay* under *Medroxyprogesterone Acetate.*

Assay preparation—Weigh and finely powder not less than 20 Medroxyprogesterone Acetate Tablets. Weigh accurately a portion of the powder, equivalent to about 25 mg of medroxyprogesterone acetate, into a 50-mL glass centrifuge tube. Pipet 25 mL of acetonitrile into the tube, sonicate for about 5 minutes, and centrifuge. Use the clear supernatant liquid as the *Assay preparation.*

Procedure—Proceed as directed for *Procedure* in the *Assay* under *Medroxyprogesterone Acetate.* Calculate the quantity, in mg, of $C_{24}H_{34}O_4$ in the portion of Tablets taken by the formula:

$$25C(r_U/r_S),$$

in which C is the concentration, in mg per mL, of USP Medroxyprogesterone Acetate RS in the *Standard preparation,* and r_U and r_S are the peak responses obtained from the *Assay preparation* and the *Standard preparation,* respectively.

Medrysone

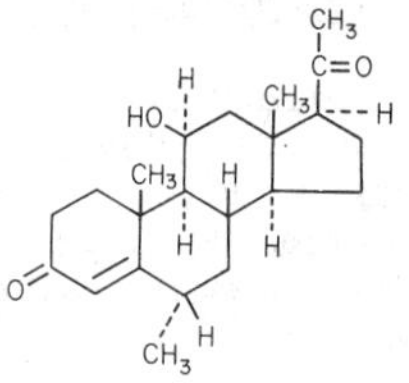

$C_{22}H_{32}O_3$ 344.49

Pregn-4-ene-3,20-dione, 11-hydroxy-6-methyl-, (6α,11β)-.

11β-Hydroxy-6α-methylpregn-4-ene-3,20-dione [*2668-66-8*].

» Medrysone contains not less than 97.0 percent and not more than 103.0 percent of $C_{22}H_{32}O_3$, calculated on the dried basis.

Packaging and storage—Preserve in well-closed containers.

Reference standard—*USP Medrysone Reference Standard*—Dry in vacuum at 50° for 3 hours before using.

Identification—

A: The infrared absorption spectrum of a potassium bromide dispersion of it, previously dried, exhibits maxima only at the same wavelengths as that of a similar preparation of USP Medrysone RS.

B: It responds to the *Thin-layer Chromatographic Identification Test* ⟨201⟩, a solution containing about 1 mg per mL of Medrysone in chloroform being used as the test solution, the thin-layer chromatographic plate being activated at 105° for 15 minutes prior to the test, and a solvent mixture of cyclohexane, chloroform, and glacial acetic acid (7:2:1) being used.

Specific rotation ⟨781⟩: between +186° and +194°, calculated on the dried basis, determined in a solution in chloroform containing 10 mg per mL.

Loss on drying ⟨731⟩—Dry it in vacuum at 50° for 3 hours: it loses not more than 3.0% of its weight.

Ordinary impurities ⟨466⟩—

Test solution: methanol.

Standard solution: methanol.

Application volume: 10 µL.

Eluant: a mixture of toluene and isopropyl alcohol (90:10), in a nonequilibrated chamber.

Visualization: 5.

Assay—Dissolve about 100 mg of Medrysone, accurately weighed, in alcohol, and dilute quantitatively and stepwise with alcohol to obtain a solution having a concentration of about 10 µg per mL. Dissolve an accurately weighed quantity of USP Medrysone RS in alcohol, and dilute quantitatively and stepwise with alcohol to obtain a Standard solution having a known concentration of about 10 µg per mL. Concomitantly determine the absorbances of both solutions in 1-cm cells at the wavelength of maximum absorbance at about 242 nm, with a suitable spectrophotometer, using alcohol as the blank. Calculate the quantity, in mg, of $C_{22}H_{32}O_3$ in the Medrysone taken by the formula:

$$10C(A_U/A_S),$$

in which C is the concentration, in µg per mL, of USP Medrysone RS in the Standard solution, and A_U and A_S are the absorbances of the solution of Medrysone and the Standard solution, respectively.

Medrysone Ophthalmic Suspension

» Medrysone Ophthalmic Suspension is a sterile suspension of Medrysone in a buffered aqueous medium containing a suitable antimicrobial agent and preservative. It contains not less than 90.0 percent and not more than 115.0 percent of the labeled amount of $C_{22}H_{32}O_3$.

Packaging and storage—Preserve in tight, light-resistant containers.

Reference standard—*USP Medrysone Reference Standard*—Dry in vacuum at 50° for 3 hours before using.

Identification—Centrifuge about 2 mL of Ophthalmic Suspension, and decant the supernatant liquid. Suspend the residue in 10 mL of water, by vigorous shaking, centrifuge, and decant the supernatant liquid. Repeat the washing with two additional 10-mL portions of water: the residue, previously dried in vacuum at 50°, responds to *Identification test A* under *Medrysone.*

Sterility—It meets the requirements under *Sterility Tests* ⟨71⟩.

pH ⟨791⟩: between 6.2 and 7.5.

Assay—

Standard preparation—Dissolve about 50 mg of USP Medrysone RS, accurately weighed, in alcohol contained in a 100-mL volumetric flask, dilute with alcohol to volume, and mix. Pipet 2 mL of this solution into a second 100-mL volumetric flask, add diluted alcohol to volume, and mix.

Assay preparation—Transfer a volume of Medrysone Ophthalmic Suspension, previously mixed and accurately measured, equivalent to about 10 mg of medrysone, to a 100-mL volumetric flask, add diluted alcohol to volume, and mix. Pipet 10 mL of this solution into a second 100-mL volumetric flask, add diluted alcohol to volume, and mix.

Procedure—Concomitantly determine the absorbances of the *Standard preparation* and the *Assay preparation* in 1-cm cells at the wavelength of maximum absorbance at about 242 nm, with a suitable spectrophotometer, using diluted alcohol as the blank. Calculate the quantity, in mg, of $C_{22}H_{32}O_3$ in each mL of the Ophthalmic Suspension taken by the formula:

$$(C/V)(A_U/A_S),$$

in which C is the concentration, in µg per mL, of USP Medrysone RS in the *Standard preparation*, V is the volume, in mL, of Ophthalmic Suspension taken, and A_U and A_S are the absorbances of the solutions from the *Assay preparation* and the *Standard preparation*, respectively.

Megestrol Acetate

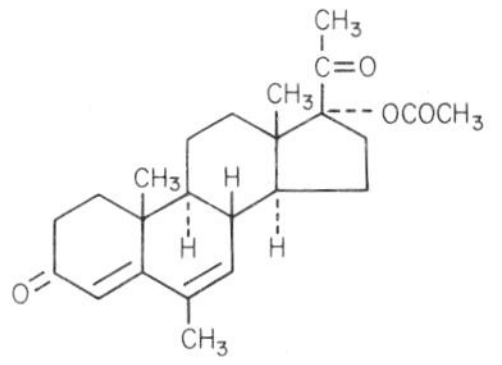

$C_{24}H_{32}O_4$ 384.51

Pregna-4,6-diene-3,20-dione, 17-(acetyloxy)-6-methyl-.

17-Hydroxy-6-methylpregna-4,6-diene-3,20-dione acetate [595-33-5].

» Megestrol Acetate contains not less than 97.0 percent and not more than 103.0 percent of $C_{24}H_{32}O_4$, calculated on the anhydrous basis.

Packaging and storage—Preserve in well-closed containers, protected from light.

Reference standard—*USP Megestrol Acetate Reference Standard*—Do not dry; determine the *Water* content by *Method I* ⟨921⟩ before using.

Completeness of solution ⟨641⟩: meets the requirements, 500 mg being dissolved in 10 mL of acetone.

Identification—The infrared absorption spectrum of a potassium bromide dispersion of it exhibits maxima only at the same wavelengths as that of a similar preparation of USP Megestrol Acetate RS.

Melting range ⟨741⟩: between 213° and 220°, but the range between beginning and end of melting does not exceed 3°.

Specific rotation ⟨781⟩: between +8.8° and +12.0°, calculated on the anhydrous basis, determined at 20° in a solution in chloroform containing 200 mg in each 10 mL.

Water, *Method I* ⟨921⟩: not more than 0.5%.

Residue on ignition ⟨281⟩: not more than 0.2%, a platinum dish being used, with ignition at 600 ± 25°.

Heavy metals, *Method II* ⟨231⟩: not more than 0.002%.

Assay—

Mobile phase—Prepare a solution of acetonitrile and water (55:45), mix, and degas. The acetonitrile concentration may be varied slightly to meet system suitability test requirements and to provide a suitable elution time.

Solvent mixture—Mix 60 volumes of water and 40 volumes of acetonitrile.

Internal standard solution—Transfer about 80 mg of propylparaben to a 100-mL volumetric flask, dissolve in acetonitrile, add acetonitrile to volume, and mix.

Standard preparation—Using an accurately weighed quantity of USP Megestrol Acetate RS, prepare a solution in acetonitrile containing about 1 mg per mL. Transfer 4.0 mL of this solution and 5.0 mL of *Internal standard solution* to a 50-mL volumetric flask, dilute with *Solvent mixture* to volume, and mix. The *Standard preparation* has a known concentration of about 80 µg of megestrol acetate per mL.

Assay preparation—Transfer about 100 mg of Megestrol Acetate, accurately weighed, to a 100-mL volumetric flask. Dissolve in acetonitrile, add acetonitrile to volume, and mix. Transfer 4.0 mL of this solution and 5.0 mL of *Internal standard solution* to a 50-mL volumetric flask, dilute with *Solvent mixture* to volume, and mix.

Chromatographic system—The liquid chromatograph is equipped with an ultraviolet detector that monitors absorption at 280 nm and 3.9-mm × 30-cm column containing packing L1. The flow rate is about 1 mL per minute. Chromatograph the *Standard preparation*, and record the peak responses as directed under *Procedure:* the relative standard deviation of the peak response ratio, R_S, for replicate injections is not more than 2.0%, and the resolution factor, R (see *Chromatography* ⟨621⟩), between the peaks for propylparaben and megestrol acetate is not less than 8.0.

Procedure—Separately inject equal volumes (about 25 µL) of the *Standard preparation* and the *Assay preparation* into the chromatograph, record the chromatograms, and measure the peak responses of the major peaks. The relative retention times are about 0.4 for propylparaben and 1.0 for megestrol acetate. Calculate the quantity, in mg, of $C_{24}H_{32}O_4$ in the portion of Megestrol Acetate taken by the formula:

$$1.25C(R_U/R_S),$$

in which C is the concentration, in µg per mL, of USP Megestrol Acetate RS in the *Standard preparation*, and R_U and R_S are the ratios of the peak responses of megestrol acetate and propylparaben obtained from the *Assay preparation* and the *Standard preparation*, respectively.

Megestrol Acetate Tablets

» Megestrol Acetate Tablets contain not less than 93.0 percent and not more than 107.0 percent of $C_{24}H_{32}O_4$.

Packaging and storage—Preserve in well-closed containers.

Reference standard—*USP Megestrol Acetate Reference Standard*—Do not dry; determine the *Water* content by *Method I* ⟨921⟩ before using.

Identification—Grind a suitable number of Tablets in a known volume of chloroform, not less than 10 mL, to obtain a solution containing about 4 mg of megestrol acetate per mL. Filter into a beaker. Introduce 0.6 mL of the filtrate via a transfer pipet into a stainless steel grinding vial containing 500 mg of potassium bromide, dry with a current of air, grind, pellet, and record the infrared spectrum: the infrared absorption spectrum of the potassium bromide dispersion so obtained exhibits maxima only at the same wavelengths as that of a similar preparation of USP Megestrol Acetate RS.

Dissolution ⟨711⟩—

Medium: 1% sodium lauryl sulfate; 900 mL.

Apparatus 2: 100 rpm.

Time: 60 minutes.

Procedure—Determine the amount of $C_{24}H_{32}O_4$ dissolved from ultraviolet absorbances at the wavelength of maximum absorbance at about 292 nm of filtered portions of the solution under test, suitably diluted with 1% sodium lauryl sulfate, if necessary, in comparison with a Standard solution having a known concentration of USP Megestrol Acetate RS in the same medium.

Tolerances—Not less than 75% (Q) of the labeled amount of $C_{24}H_{32}O_4$ is dissolved in 60 minutes.

Uniformity of dosage units ⟨905⟩: meet the requirements.

Procedure for content uniformity—Place 1 Tablet in a volumetric flask of suitable size so that the final expected solution concentration is between about 0.2 and 1.0 mg of megestrol acetate per mL. Add 1 mL of water, and gently shake until the Tablet has disintegrated. Fill the flask to three-quarters of its nominal capacity with methanol, and shake by mechanical means for 20 minutes. Dilute with methanol to volume, mix, and filter, discarding the first 15 mL of the filtrate. Dilute 5.0 mL of the subsequent filtrate quantitatively with methanol to obtain a solution containing about 10 µg of megestrol acetate per mL. Prepare a Standard solution of USP Megestrol Acetate RS in the same medium having a known concentration of about 10 µg per mL. Record the absorbances of the solutions in 1-cm cells, against a blank of methanol, scanning from 350 nm to 260 nm. Measure the absorbances at the wavelength of maximum absorbance at about 288 nm. Calculate the quantity, in mg, of $C_{24}H_{32}O_4$ in the Tablet taken by the formula:

$$(TC/D)(A_U/A_S),$$

in which T is the labeled quantity, in mg, of megestrol acetate in the Tablet, C is the concentration, in µg per mL of USP Megestrol Acetate RS in the Standard solution, D is the concentration, in µg per mL, of the solution from the Tablet, based upon the labeled quantity per Tablet and the extent of dilution, and A_U and A_S are the absorbances of the solution from the Tablet and the Standard solution, respectively.

Assay—

Mobile phase, Solvent mixture, Internal standard solution, Standard preparation, and *Chromatographic system*—Prepare as directed in the *Assay* under *Megestrol Acetate.*

Assay preparation—Weigh and finely powder not less than 20 Megestrol Acetate Tablets. Transfer an accurately weighed portion of the powder, equivalent to about 80 mg of megestrol acetate, to a 100-mL volumetric flask. Add about 10 mL of water, and shake for 10 minutes. Add 75 mL of acetonitrile, and shake for 30 minutes, then dilute with acetonitrile to volume, and mix. Place a 25-mL aliquot in a glass-stoppered, 35-mL centrifuge tube, insert the stopper, and centrifuge for 10 minutes. Transfer 5.0 mL of the supernatant solution and 5.0 mL of *Internal standard solution* to a 50-mL volumetric flask, dilute with *Solvent mixture* to volume, and mix.

Procedure—Proceed as directed for *Procedure* in the *Assay* under *Megestrol Acetate.* Calculate the quantity, in mg, of $C_{24}H_{32}O_4$ in the portion of Tablets taken by the formula:

$$C(R_U/R_S),$$

in which the terms are as defined therein.

Meglumine

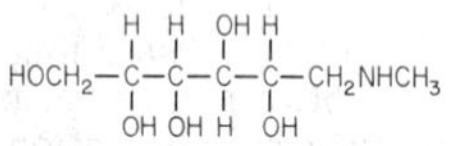

$C_7H_{17}NO_5$ 195.21

D-Glucitol, 1-deoxy-1-(methylamino)-.

1-Deoxy-1-(methylamino)-D-glucitol [*6284-40-8*].

» Meglumine contains not less than 99.0 percent and not more than 100.5 percent of $C_7H_{17}NO_5$, calculated on the dried basis.

Packaging and storage—Preserve in well-closed containers.

Completeness and color of solution—A solution (1 in 5) is clear, and its absorbance, determined in a 1-cm cell at 420 nm, with a suitable spectrophotometer, water being used as the blank, is not greater than 0.030.

Identification—Transfer about 250 mg to a dry, 50-mL centrifuge tube, add 500 mg of sodium metaperiodate, then add 5 mL of water rapidly in one portion. Allow to stand undisturbed: the solution instantly turns yellow and heat is produced. The color then changes from deep yellow to orange-brown (rust), and after 20 minutes, the rust-colored solution is cloudy. Then add 2 mL of 2.5 *N* sodium hydroxide: the mixture turns bright yellow and becomes clear.

Melting range ⟨741⟩: between 128° and 132°.

Specific rotation ⟨781⟩: between −15.7° and −17.3°, determined in a solution containing 1.000 g in each 10 mL.

Loss on drying ⟨731⟩—Dry about 1 g at 105° to constant weight: it loses not more than 1.0% of its weight.

Residue on ignition ⟨281⟩: not more than 0.1%.

Absence of reducing substances—To 5 mL of a solution (1 in 20) add 5 mL of alkaline cupric tartrate TS, and heat to boiling: the color of the solution does not change.

Heavy metals ⟨231⟩—Dissolve 1 g in 20 mL of water, add phenolphthalein TS, neutralize with 3 *N* hydrochloric acid, and dilute with water to 25 mL. The limit is 0.002%.

Assay—Transfer about 500 mg of Meglumine, accurately weighed, to a conical flask, dissolve in about 40 mL of water, add 2 drops of methyl red TS, and titrate with 0.1 *N* hydrochloric acid VS. Each mL of 0.1 *N* hydrochloric acid is equivalent to 19.52 mg of $C_7H_{17}NO_5$.

Melphalan

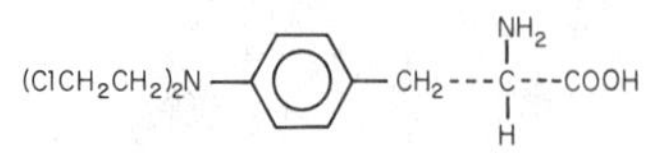

$C_{13}H_{18}Cl_2N_2O_2$ 305.20

L-Phenylalanine, 4-[bis(2-chloroethyl)amino]-.

L-3-[*p*-[Bis(2-chloroethyl)amino]phenyl]alanine [*148-82-3*].

» Melphalan contains not less than 93.0 percent and not more than 100.5 percent of $C_{13}H_{18}Cl_2N_2O_2$, calculated on the dried and ionizable chlorine-free basis.

Caution—Handle Melphalan with exceptional care since it is a highly potent agent.

Packaging and storage—Preserve in tight, light-resistant, glass containers.

Reference standard—*USP Melphalan Hydrochloride Reference Standard*—[Caution—Avoid contact.] Do not dry before using. Dry a portion separate from the analytical specimen at 105° to constant weight, and apply a correction for *Loss on drying* for quantitative analyses. Keep container tightly closed and protected from light.

Identification—

A: The ultraviolet absorption spectrum of a 1 in 200,000 solution in methanol exhibits maxima and minima at the same wavelengths as that of a similar solution of USP Melphalan Hydrochloride RS, concomitantly measured.

B: To 1 mL of 1 in 10,000 solution in alcohol in a glass-stoppered test tube add 1 mL of pH 4.0 acid phthalate buffer (see under *Solutions* in the section, *Reagents, Indicators, and Solutions*), 1 mL of a 1 in 20 solution of 4-(*p*-nitrobenzyl)pyridine in acetone, and 1 mL of saline TS. Heat on a water bath at 80° for 20 minutes, and cool quickly. Add 10 mL of alcohol and 1 mL of 1 *N* potassium hydroxide: a violet to red-violet color is produced.

C: Heat 100 mg with 10 mL of 0.1 *N* sodium hydroxide on a water bath for 10 minutes: the resulting solution, after acidification with 2 *N* nitric acid, responds to the tests for *Chloride* ⟨191⟩.

Specific rotation ⟨781⟩: between −30° and −36°, calculated on the dried basis, determined in a solution in methanol containing 70 mg in each 10 mL, prepared with the aid of gentle heating.

Loss on drying ⟨731⟩: Dry it in vacuum at 105° to constant weight: it loses not more than 7.0% of its weight.

Residue on ignition ⟨281⟩: not more than 0.3%.

Ionizable chlorine—Dissolve about 500 mg of Melphalan, accurately weighed, in a mixture of 75 mL of water and 2 mL of nitric acid, allow to stand for 2 minutes, and titrate with 0.1 *N* silver nitrate VS, determining the end-point potentiometrically: not more than 1.0 mL of 0.1 *N* silver nitrate is required for each 500 mg of test specimen.

Nitrogen content ⟨461⟩—Determine the nitrogen content as directed under *Method II*, using about 325 mg of Melphalan, accurately weighed, and 0.1 *N* sulfuric acid VS for the titration: not less than 8.90% and not more than 9.45% of N is found, calculated on the dried basis.

Assay—Transfer to a beaker about 200 mg of Melphalan, accurately weighed, and dissolve in 20 mL of 0.5 *N* sodium hydroxide. Cover the beaker with a watch glass, and boil the solution for 30 minutes, adding water as necessary to maintain the volume. Cool, neutralize to phenolphthalein TS with acetic acid, and add 1 mL of acetic acid in excess. Titrate with 0.1 *N* silver nitrate VS, determining the end-point potentiometrically, using silver and calomel electrodes, the latter modified to contain saturated potassium sulfate solution. From the results obtained in the test for *Ionixable chlorine*, calculate the volume, in mL, of 0.1 *N* silver nitrate that is equivalent to the ionizable chlorine in the quantity of Melphalan taken for the *Assay*, and subtract it from the *Assay* titration volume. Each mL of 0.1 *N* silver nitrate is equivalent to 15.26 mg of $C_{13}H_{18}Cl_2N_2O_2$.

Melphalan Tables

» Melphalan Tablets contain not less than 90.0 percent and not more than 110.0 percent of the labeled amount of $C_{13}H_{18}Cl_2N_2O_2$.

Packaging and storage—Preserve in well-closed, light-resistant, glass containers.

Reference standard—*USP Melphalan Hydrochloride Reference Standard*—[Caution—Avoid contact.] Do not dry before using. Dry a portion separate from the analytical specimen at 105° to constant weight, and apply a correction for *Loss on drying* for quantitative analyses. Keep container tightly closed and protected from light.

Identification—

A: The retention time of the major peak in the chromatogram of the *Assay preparation* corresponds to that of the *Standard preparation*, obtained as directed in the *Assay*.

B: Shake a portion of finely powdered Tablets, equivalent to about 2 mg of melphalan, with 20 mL of alcohol, and filter: a 1-mL portion of the solution so obtained responds to *Identification test B* under *Melphalan.*

Disintegration ⟨701⟩: 15 minutes, the use of disks being omitted.

Uniformity of dosage units ⟨905⟩: meet the requirements.

Procedure for content uniformity—Place 1 Tablet in a 200-mL volumetric flask, add 10 mL of water and 10 mL of alcohol, sonicate to dissolve the soluble components in the mixture, dilute with alcohol to volume, mix, and filter to obtain a clear solution. Dissolve an accurately weighed quantity of USP Melphalan Hydrochloride RS in alcohol to obtain a Standard solution having a known concentration of about 10 µg per mL. Concomitantly determine the absorbances of the solution from the Tablet and the Standard solution in 1-cm cells at the wavelength of maximum absorbance at about 260 nm, with a suitable spectrophotometer, using alcohol as the blank. Calculate the quantity, in mg, of $C_{13}H_{18}Cl_2N_2O_2$ in the Tablet by the formula:

$$(305.20/341.67)(T/D)C(A_U/A_S),$$

in which 305.20 and 341.67 are the molecular weights of melphalan and melphalan hydrochloride, respectively, T is the labeled quantity, in mg, of melphalan in the Tablet, D is the concentration, in µg per mL, of melphalan in the solution from the Tablet, on the basis of the labeled quantity per Tablet and the extent of dilution, C is the concentration, in µg per mL, of USP Melphalan Hydrochloride RS in the Standard solution, and A_U and A_S are the absorbances of the solution from the Tablet and the Standard solution, respectively.

Assay—

Mobile phase—Prepare a solution of 0.025 *M* diethylamine in a mixture of methanol and water (1:1), adjust with 3.5 *N* hydrochloric acid to a pH of 5.5, filter, and degas. Make adjustments if necessary (see *System Suitability* under *Chromatography* ⟨621⟩).

Standard preparation—Dissolve an accurately weighed quantity of USP Melphalan Hydrochloride RS in alcohol, and dilute quantitatively with alcohol to obtain a solution having a known concentration equivalent to about 0.9 mg of anhydrous melphalan hydrochloride per mL. Pipet 10 mL of this solution into a 100-mL volumetric flask containing 75 mL of alcohol and 2.0 mL of glacial acetic acid, dilute with alcohol to volume, and mix to obtain a *Standard preparation* having a known concentration of about 90 µg of anhydrous USP Melphalan Hydrochloride RS per mL (equivalent to about 80 µg of anhydrous melphalan per mL).

Assay preparation—Weigh and finely powder not less than 20 Melphalan Tablets. Transfer an accurately weighed portion of the powder, equivalent to 8 mg of anhydrous melphalan, to a 100-mL volumetric flask. Add about 75 mL of alcohol and 2.0 mL of glacial acetic acid to the flask, and sonicate for 15 minutes. Cool, dilute with alcohol to volume, and mix. Filter through a medium-porosity, sintered-glass funnel, discarding the first few mL of the filtrate, and use the remainder of the filtrate as the *Assay preparation.*

Chromatographic system (see *Chromatography* ⟨621⟩)—The liquid chromatograph is equipped with a 254-nm detector and a 4.2-mm × 25-cm column that contains packing L7. The flow rate is about 1 mL per minute. Chromatograph the *Standard preparation*, and record the peak responses as directed under *Procedure:* the tailing factor for the analyte peak is not more than 2.0, and the relative standard deviation for replicate injections is not more than 2.0%.

Procedure—Separately inject equal volumes (between 10 and 20 µL) of the *Standard preparation* and the *Assay preparation* into the chromatograph, record the chromatograms, and measure the responses for the major peaks. Calculate the quantity, in mg, of $C_{13}H_{18}Cl_2N_2O_2$ in the portion of Tablets taken by the formula:

$$(305.20/341.67)(0.1C)(r_U/r_S),$$

in which 305.20 and 341.67 are the molecular weights of melphalan and melphalan hydrochloride, respectively, C is the concentration, in µg per mL, of anhydrous melphalan hydrochloride in the *Standard preparation*, and r_U and r_S are the peak responses obtained from the *Assay preparation* and the *Standard preparation*, respectively.

Menadiol Sodium Diphosphate

OPO_3Na_2 CH_3 · $6H_2O$ OPO_3Na_2

$C_{11}H_8Na_4O_8P_2 \cdot 6H_2O$ 530.18

1,4-Naphthalenediol, 2-methyl-, bis(dihydrogen phosphate), tetrasodium salt, hexahydrate.

2-Methyl-1,4-naphthalenediol bis(dihydrogen phosphate) tetrasodium salt, hexahydrate [*6700-42-1*].

Anhydrous 422.09 [*131-13-5*].

» Menadiol Sodium Diphosphate contains not less than 97.5 percent and not more than 102.0 percent of $C_{11}H_8Na_4O_8P_2$, calculated on the anhydrous basis.

Packaging and storage—Preserve in tight, light-resistant containers, and store in a cold place.

Identification—

A: Dissolve about 200 mg of Menadiol Sodium Diphosphate in 10 mL of water, add 10 mL of 2 *N* sulfuric acid, 10 mL of 0.1 *N* ceric sulfate, and 1 mL of 30 percent hydrogen peroxide previously diluted with 5 mL of water, and extract the solution with two 10-mL portions of chloroform. Gently evaporate the clear chloroform solution on a steam bath to dryness, and dry the residue at 80° for 1 hour: the menadione so obtained melts between 104° and 107°.

B: To 50 mg of the dried residue obtained in *Identification test A* add 5 mL of water, then add 75 mg of sodium bisulfite, and heat on a steam bath, shaking vigorously until the substance is dissolved and the solution is practically colorless. Dilute with water to 50 mL, and mix. To 2 mL of the solution add 2 mL of alcoholic ammonia (prepared by mixing equal volumes of alcohol and ammonium hydroxide), shake, and add 3 drops of ethyl cyanoacetate: a deep purplish blue color is produced, and on the addition of 1 mL of sodium hydroxide solution (1 in 3), it changes to green and then to yellow.

C: To about 20 mg contained in a small beaker add 1 mL of water, 2 drops of nitric acid, and 1 mL of sulfuric acid, and heat slowly to the evolution of white fumes. Cool, cautiously dilute with water to about 10 mL, and filter if not clear. Render the filtrate slightly alkaline to litmus with 6 *N* ammonium hydroxide, then render it acid with nitric acid, and add to the warm solution 3 mL of ammonium molybdate TS: a yellow precipitate is formed within a few minutes.

Water, *Method I* ⟨921⟩: between 19.0% and 21.5%.

Assay—Dissolve about 100 mg of Menadiol Sodium Diphosphate, accurately weighed, in 25 mL of water, and add 25 mL of glacial acetic acid and 25 mL of 3 *N* hydrochloric acid. Titrate the solution with 0.02 *N* ceric sulfate VS, determining the endpoint potentiometrically using a calomel–platinum electrode system. Each mL of 0.02 *N* ceric sulfate is equivalent to 4.221 mg of $C_{11}H_8Na_4O_8P_2$.

Menadiol Sodium Diphosphate Injection

» Menadiol Sodium Diphosphate Injection is a sterile solution of Menadiol Sodium Diphosphate in Water for Injection. It contains not less than 95.0 percent and not more than 110.0 percent of the labeled amount of $C_{11}H_8Na_4O_8P_2.6H_2O$.

Packaging and storage—Preserve in single-dose, light-resistant containers, preferably of Type I glass.

Reference standard—*USP Menadione Reference Standard*—[*Caution—Avoid contact; avoid exposure to light.*] Dry over silica gel for 4 hours before using.

Identification—

A: Transfer a volume of Injection, equivalent to about 100 mg of menadiol sodium diphosphate, to a separator, add 10 mL of 2 *N* sulfuric acid, and extract with six 25-mL portions of ether, discarding the ether extracts. To the aqueous solution add 1 mL of 0.5 *N* ceric sulfate and 1 mL of 30 percent hydrogen peroxide, and extract with two 10-mL portions of chloroform. Evaporate the combined chloroform extracts on a steam bath just to dryness, then dry at 80° for 1 hour: the infrared absorption spectrum of a potassium bromide dispersion of the menadione so obtained exhibits maxima at the same wavelengths as that of a similar preparation of USP Menadione RS. The solid also responds to *Identification test B* under *Menadiol Sodium Diphosphate.*

B: Adjust, if necessary, a volume of Injection, equivalent to about 20 mg of menadiol sodium diphosphate, by evaporation or dilution with water, as required, to 2 mL: the solution responds to *Identification test C* under *Menadiol Sodium Diphosphate.*

pH ⟨791⟩: between 7.5 and 8.5.

Other requirements—It meets the requirements under *Injections* ⟨1⟩.

Assay—Transfer an accurately measured volume of Menadiol Sodium Diphosphate Injection, equivalent to about 50 mg of menadiol sodium diphosphate, to a 125-mL separator, and extract with three 25-mL portions of chloroform, discarding the chloroform extracts. Transfer the aqueous solution to a 250-mL beaker, add 25 mL of glacial acetic acid and 25 mL of 3 *N* hydrochloric acid, vigorously bubble nitrogen through this solution for not less than 15 minutes, and titrate with 0.01 *N* ceric sulfate VS, determining the end-point potentiometrically using a calomel-platinum electrode system. Each mL of 0.01 *N* ceric sulfate is equivalent to 2.651 mg of $C_{11}H_8Na_4O_8P_2.6H_2O$.

Menadiol Sodium Diphosphate Tablets

» Menadiol Sodium Diphosphate Tablets contain not less than 95.0 percent and not more than 110.0 percent of the labeled amount of $C_{11}H_8Na_4O_8P_2.6H_2O$.

Packaging and storage—Preserve in well-closed, light-resistant containers.

Reference standard—*USP Menadione Reference Standard*—[Caution—Avoid contact; avoid exposure to light.] Dry over silica gel for 4 hours before using.

Identification—

A: Triturate a quantity of powdered Tablets, equivalent to about 100 mg of menadiol sodium diphosphate, with a mixture of 10 mL of water and 10 mL of 2 *N* sulfuric acid, centrifuge the mixture, and filter the supernatant liquid. To the filtrate add 1 mL of 0.5 *N* ceric sulfate, mix, extract with 10 mL of chloroform, and centrifuge. Evaporate the chloroform extract on a steam bath just to dryness, then dry at 80° for 1 hour: the infrared absorption spectrum of a potassium bromide dispersion of the menadione so obtained exhibits maxima at the same wavelengths as that of a similar preparation of USP Menadione RS.

B: To 50 mg of the menadione obtained in *Identification test A* add 5 mL of water, then add 75 mg of sodium bisulfite, and heat on a steam bath, shaking vigorously until the substance is dissolved and the solution is almost colorless. Add water to make 50 mL, and mix. To 2 mL of the solution add 2 mL of alcoholic ammonia (prepared by mixing equal volumes of alcohol and ammonium hydroxide), shake, and add 3 drops of ethyl cyanoacetate: a deep purplish blue color is produced, and, on the addition of 1 mL of sodium hydroxide solution (1 in 3), it changes to green and then to yellow.

C: Triturate a quantity of powdered Tablets, equivalent to about 20 mg of menadiol sodium diphosphate, with 10 mL of water, centrifuge the mixture, filter the supernatant liquid, and evaporate to a volume of about 2 mL. Add 2 drops of nitric acid and 1 mL of sulfuric acid, and heat slowly to the evolution of white fumes. Cool, cautiously dilute with water to about 10 mL, and filter if not clear. Render the filtrate slightly alkaline to litmus with 6 *N* ammonium hydroxide, then render it acid with nitric acid, and add to the warm solution 3 mL of ammonium molybdate TS: a yellow precipitate is formed within a few minutes.

Dissolution ⟨711⟩—

Medium: 0.1 *N* hydrochloric acid; 900 mL.

Apparatus 1: 100 rpm.

Time: 30 minutes.

Procedure—Determine the amount of $C_{11}H_8Na_4O_8P_2.6H_2O$ dissolved from ultraviolet absorbances at the wavelength of maximum absorbance at about 227 nm of filtered portions of the solution under test, suitably diluted with *Dissolution Medium*, in comparison with a Standard solution having a known concentration of Menadiol Sodium Diphosphate, previously subjected to the *Assay* and dried in vacuum over phosphorus pentoxide for 4 hours, in the same medium.

Tolerances—Not less than 75% (*Q*) of the labeled amount of $C_{11}H_8Na_4O_8P_2.6H_2O$ is dissolved in 30 minutes.

Uniformity of dosage units ⟨905⟩: meet the requirements.

Procedure for content uniformity—[NOTE—Use low-actinic glassware.] Transfer 1 finely powdered Tablet to a glass-stoppered centrifuge tube, add 25 mL of pH 8.0 phosphate buffer (see under *Solutions* in the section, *Reagents, Indicators, and Solutions*), and shake vigorously for several minutes. Filter into a 50-mL volumetric flask, rinse the centrifuge tube, and filter with three 5-mL portions of pH 8.0 phosphate buffer, adding the rinsings to the volumetric flask, dilute with pH 8.0 phosphate buffer to volume, and mix. Dilute a portion of this solution quantitatively and stepwise, if necessary, with pH 8.0 phosphate buffer to provide a solution containing approximately 40 µg of menadiol sodium diphosphate per mL. Concomitantly determine the absorbances of this solution and of a solution of Menadiol Sodium Diphosphate, previously dried in vacuum over phosphorus pentoxide for 4 hours, in the same medium having a known concentration of about 40 µg per mL, at the wavelength of maximum absorbance at about 297 nm, with a suitable spectrophotometer, using pH 8.0 phosphate buffer as the blank. Calculate the quantity, in mg, of $C_{11}N_8Na_4O_8P_2.6H_2O$ in the Tablet, taken by the formula:

$$(TC/D)(A_U/A_S),$$

in which *T* is the labeled quantity, in mg, of menadiol sodium diphosphate in the Tablet, *C* is the concentration, in µg per mL, of $C_{11}H_8Na_4O_8P_2.6H_2O$ in the standard solution, *D* is the concentration, in µg per mL, of menadiol sodium diphosphate in the test solution, based upon the labeled quantity per Tablet and the extent of dilution, and A_U and A_S are the absorbances of the solution from the Tablet and the standard solution, respectively.

Assay—Weigh and finely powder not less than 20 Menadiol Sodium Diphosphate Tablets. Transfer an accurately weighed portion of the powder, equivalent to about 50 mg of menadiol sodium diphosphate, to a 250-mL beaker. Moisten the powder with a few mL of glacial acetic acid, and then add sufficient of the acid to make 25 mL. Add 25 mL of 3 *N* hydrochloric acid and 25 mL of water, mix, and titrate with 0.01 *N* ceric sulfate VS, determining the end-point potentiometrically using a calomel-platinum electrode system. Each mL of 0.01 *N* ceric sulfate is equivalent to 2.651 mg of $C_{11}H_8Na_4O_8P_2 . 6H_2O$.

Menadione

$C_{11}H_8O_2$ 172.18
1,4-Naphthalenedione, 2-methyl-.
2-Methyl-1,4-naphthoquinone [*58-27-5*].

» Menadione contains not less than 98.5 percent and not more than 101.0 percent of $C_{11}H_8O_2$, calculated on the dried basis.

Caution—Menadione powder is irritating to the respiratory tract and to the skin, and a solution of it in alcohol is a vesicant.

Packaging and storage—Preserve in well-closed, light-resistant containers.

Reference standard—*USP Menadione Reference Standard*—[*Caution—Avoid contact; avoid exposure to light.*] Dry over silica gel for 4 hours before using.

Identification—

A: The infrared absorption spectrum of a potassium bromide dispersion of it, previously dried, exhibits maxima only at the same wavelengths as that of a similar preparation of USP Menadione RS.

B: The ultraviolet absorption spectrum of a 1 in 200,000 solution in alcohol exhibits maxima and minima at the same wavelengths as that of a similar solution of USP Menadione RS, concomitantly measured, and the respective absorptivities, calculated on the dried basis, at the wavelength of maximum absorbance at about 250 nm do not differ by more than 3.0%.

Melting range, *Class I* ⟨741⟩: between 105° and 107°.

Loss on drying ⟨731⟩—Dry it over silica gel for 4 hours: it loses not more than 0.3% of its weight.

Residue on ignition ⟨281⟩: not more than 0.1%.

Ordinary impurities ⟨466⟩—
Test solution: methanol.
Standard solution: methanol.
Eluant: chloroform.
Visualization: 1.

Assay—Weigh accurately about 150 mg of Menadione, and transfer to a 150-mL flask. Add 15 mL of glacial acetic acid and 15 mL of 3 *N* hydrochloric acid, and rotate the flask until the Menadione is dissolved. Then add about 3 g of zinc dust, close the flask with a stopper bearing a Bunsen valve, shake, and allow to stand in the dark for 1 hour, with frequent shaking. Rapidly decant the solution through a pledget of cotton into another flask, immediately wash the reduction flask with three 10-mL portions of freshly boiled and cooled water, add 0.1 mL of orthophenanthroline TS, and immediately titrate the combined filtrate and washings with 0.1 *N* ceric sulfate VS. Perform a blank determination, and make any necessary correction. Each mL of 0.1 *N* ceric sulfate is equivalent to 8.609 mg of $C_{11}H_8O_2$.

Menadione Injection

» Menadione Injection is a sterile solution of Menadione in oil. It contains not less than 90.0 percent and not more than 120.0 percent of the labeled amount of $C_{11}H_8O_2$.

Packaging and storage—Preserve in single-dose or in multiple-dose containers, preferably of Type I glass.

Reference standard—*USP Menadione Reference Standard*—[*Caution—Avoid contact; avoid exposure to light.*] Dry over silica gel for 4 hours before using.

Other requirements—It meets the requirements under *Injections* ⟨1⟩.

Assay—[NOTE—Avoid exposing Menadione and its solutions to light throughout the *Assay*.]

Standard preparation—Transfer about 25 mg of USP Menadione RS, accurately weighed, to a 100-mL volumetric flask, dissolve in a mixture of equal volumes of alcohol and ether, dilute with the same mixture to volume, and mix. Keep the solution tightly closed in a dark, cool place, and use it within 7 days.

Assay preparation—Transfer an accurately measured volume of Menadione Injection, equivalent to about 25 mg of menadione, to a 100-mL volumetric flask, dilute with a mixture of equal volumes of ether and alcohol to volume, and mix.

Procedure—Transfer 1.0 mL each of the *Standard preparation* and the *Assay preparation* to separate 50-mL volumetric flasks, add to each 4 mL of alcohol, and mix. Then to each flask add 1.0 mL of a solution prepared by dissolving 50 mg of 2,4-dinitrophenylhydrazine in 20 mL of a mixture of 2 volumes of 3 *N* hydrochloric acid and 1 volume of water. Place the flasks in a bath maintained at 70° to 75° for 15 minutes, shaking vigorously every 2 to 3 minutes. Immediately after the heating, cool the flasks to about 25°; then add to each 5 mL of alcoholic ammonia, prepared by mixing equal volumes of alcohol and ammonium hydroxide. Shake the flasks thoroughly, add alcohol to make 50.0 mL, mix, allow to stand for 15 minutes, and decant from any separated oil. Determine the absorbances of the solutions, in 1-cm cells at the wavelength of maximum absorbance at about 635 nm, with a suitable spectrophotometer, using a reagent blank to set the instrument. Calculate the quantity, in mg, of $C_{11}H_8O_2$ in each mL of the Injection taken by the formula:

$$(0.1C/V)(A_U/A_S),$$

in which C is the concentration, in µg per mL, of USP Menadione RS in the *Standard preparation*, V is the volume, in mL, of Injection taken, and A_U and A_S are the absorbances of the solutions from the *Assay preparation* and the *Standard preparation*, respectively.

Meningococcal Polysaccharide Vaccine Group A

» Meningococcal Polysaccharide Vaccine Group A conforms to the regulations of the federal Food and Drug Administration concerning biologics (see *Biologics* ⟨1041⟩). It is a sterile preparation of the group-specific polysaccharide antigen from *Neisseria meningitidis*, Group A, consisting of a polymer of *N*-acetyl mannosamine phosphate. The Vaccine contains 50 µg of isolated product and 2.5 mg to 5 mg of lactose as a stabilizer per 0.5-mL dose, when constituted as directed. The constituting fluid is Bacteriostatic Sodium Chloride Injection in which the antimicrobial agent is Thimerosal in a suitable concentration. It meets the requirements of the tests for potency by immunization of not less than 25 healthy adult human subjects, in whom the antibody titers of the sera from not less than 90 percent of the subjects show a fourfold or greater rise after immunization, and by gel-permeation chromatography determination.

Packaging and storage—Preserve in multiple-dose containers for subcutaneous or jet injection at a temperature between 2° and 8°. [NOTE—Use the constituted vaccine immediately after its constitution, or if stored in a refrigerator within 8 hours after constitution.]

Expiration date—The expiration date is not later than 18 months after date of issue from manufacturer's cold storage (−20°, 6 months).

Meningococcal Polysaccharide Vaccine Group C

» Meningococcal Polysaccharide Vaccine Group C conforms to the regulations of the federal Food and Drug Administration concerning biologics (see *Biologics* ⟨1041⟩). It is a sterile preparation of the group-specific polysaccharide antigen from *Neisseria meningitidis*, Group C, consisting of a polymer of sialic acid. The Vaccine contains 50 µg of isolated product and 2.5 mg to 5 mg of lactose as a stabilizer per 0.5-mL dose, when constituted as directed. The constituting fluid is Bacteriostatic Sodium Chloride Injection in which the antimicrobial agent is Thimerosal in a suitable concentration. It meets the requirements of the tests for potency by immunization of not less than 25 healthy adult human subjects, in whom the antibody titers of the sera from not less than 90 percent of the subjects show a fourfold or greater rise after immunization, and by gel-permeation chromatography determination.

Packaging and storage—Preserve in multiple-dose containers for subcutaneous or jet injection at a temperature between 2° and 8°. [NOTE—Use the constituted vaccine immediately after its constitution, or if stored in a refrigerator within 8 hours after constitution.]

Expiration date—The expiration date is not later than 18 months after date of issue from manufacturer's cold storage (−20°, 6 months).

Menotropins

» Menotropins is an extract of human female urine containing both follicle-stimulating hormone activity and luteinizing hormone activity, having the property in females of stimulating growth and maturation of ovarian follicles and the properties in males of maintaining and stimulating testicular interstitial cells (Leydig tissue) related to testosterone production and of being responsible for the full development and maturation of spermatozoa in the seminiferous tubules. It has a potency of not less than 40 USP Follicle-stimulating Hormone Units and not less than 40 USP Luteinizing Hormone Units per mg, and it contains not less than 80 percent and not more than 125 percent of each of the hormone potencies stated on the label. The ratio of units of Follicle-stimulating Hormone to units of Luteinizing Hormone is approximately 1.

Packaging and storage—Preserve in tight containers, preferably of Type I glass, in a refrigerator.

Reference standards—*USP Menotropins Reference Standard. USP Chorionic Gonadotropin Reference Standard*—Store in a refrigerator, and do not dry before using. Use a fresh ampul for each group of assays, and discard any unused portion.

Pyrogen—A solution of it in Sodium Chloride Injection containing 1 USP Follicle-stimulating Hormone Unit per mL meets the requirements of the *Pyrogen Test* ⟨151⟩, the test dose being 2 USP Follicle-stimulating Hormone Units per kg.

Safety—Prepare a test solution of it in Sodium Chloride Injection containing 75 USP Follicle-stimulating Hormone Units per mL. Select five healthy mice, each weighing between 18 and 22 g. Inject intravenously one dose of 1.0 mL of the test solution into each of the mice. Observe the animals over the 48 hours following the injection. If, at the end of 48 hours, not more than 1 of the animals shows outward symptoms of a toxic reaction, the requirements of the test are met. If 1 or 2 of the animals die, repeat the test on 10 additional, similar animals; if all of the animals survive for 48 hours and show no symptoms of a toxic reaction, the requirements of the test are met.

Water, *Method I* ⟨921⟩: not more than 5.0%.

Assay—

Standard preparation—Dissolve a suitable quantity of USP Menotropins RS in saline TS freshly prepared to contain 1 mg per mL of bovine serum albumin and adjusted with 1 *N* sodium hydroxide (*Diluent A*) to a pH of 7.2 ± 0.2, to obtain a solution having a known concentration of USP Follicle-stimulating Hormone Units or Luteinizing Hormone Units, depending on the hormone activity to be measured.

Assay preparation—Following the procedure given under *Standard preparation*, prepare solutions of Menotropins to give *Test solutions* corresponding to those of the standard.

Control solution—Use the same injection medium of saline TS for dose determination trials. Store all solutions at 5 ± 3° for the duration of the assay; properly dispose of any unused portions.

Test animals—For *Procedure for follicle-stimulating hormone activity*, select 20 to 21-day old female rats with weights within 5 g of each other. For *Procedure for luteinizing hormone activity*, select 20 to 21-day old male rats with weights within 10 g of each other. House the animals under uniform conditions of temperature, lighting, feeding, and watering. Mark the animals for identification, and divide them at random into 7 groups of the same number but not less than 6 animals per group. Assign one group to each of three *Standard solutions* and three *Test solutions*, and a seventh group to provide a control.

Dose determination trials—Use the methods described under *Procedures* below to determine a 3-dose range in which the lowest dose produces a definite response in some of the rats in the low dose group (as compared with the control group) and the highest dose produces a sub-maximal to maximal response in the high dose group. Doses must be established in a geometric progression. For Follicle-stimulating Hormone determinations, doses of 0.5, 1.0 and 2.0 USP Follicle-stimulating Hormone Units for three days are suggested for an initial trial, but the concentrations may need to be adjusted as rat strain responses may vary significantly. For Luteinizing Hormone determinations, doses of 7, 14, and 28 USP Luteinizing Hormone Units are suggested for an initial trial, but adjustment may be required depending on sensitivity of the rat strain.

PROCEDURE FOR FOLLICLE-STIMULATING HORMONE ACTIVITY—To *Diluent A*, add USP Chorionic Gonadotropin RS to a known concentration; adjust with 1 *N* sodium hydroxide (*Diluent B*) to a pH of 7.2 ± 0.2. Dilute the *Standard preparation* and *Test solutions* using *Diluent B A*, after adding a calculated volume of *Diluent B* so that the daily dose for each rat in each group may be administered in 0.2 mL to 0.5 mL, and contains not less than 14 USP units of Chorionic Gonadotropin. Store the solution at 5 ± 3° during the test. Inject each rat of the six groups assigned to the doses of *Standard solutions* and *Test solutions* subcutaneously in the dorsal area with a suitable volume between 0.2 and 0.5 mL of the solution to which it was assigned. For *Dose determination trials* only, similarly inject each rat in the seventh control group with 0.2 mL of *Diluent B*. Repeat these injections at approximately the same time of day after 24 hours and 48 hours. Twenty-four hours after the last injection weigh each rat, sacrifice the animals, and carefully dissect out the ovaries of each rat, removing any fat and fibrous tissue. Thoroughly dry the ovaries by pressing against absorbent paper, avoiding damage to follicles on the ovary surface, and immediately weigh them to the nearest 0.2 mg, using a suitable balance.

Calculation—Tabulate the observed ovarian pair weight for each rat designated by the symbol y, for each dosage group of f rats. For apparently outlying ovarian weight gain, an attempt may be made to correct the organ mass relative to the mass of the rat from which it was taken. For the y-value in question, calculate for each of the f rats in the appropriate group the ratio of ovarian weight to the total body weight. Reject the y-value if its corresponding ratio differs from the rest of the group by more than 1.5 standard deviations. If the data from one or more rats are missing, adjust to groups of equal size by suitable means (see *Replacement of Missing Values* ⟨111⟩). Total the values of y in each group, and designate each total as T, using subscripts 1 to 3 for the three successive dosage levels and subscripts S and U for the Standard and the material under test, respectively. Test both the agreement in slope of the dosage-response lines for the Standard and for the material under test, and the lack of curvature as directed for a 3-dose balanced assay (see *Tests of Assay Validity* ⟨111⟩). If the combined discrepancy as measured by F_3 exceeds its tabular value in Table 9 (see *Combination of Independent Assays* ⟨111⟩), regard these data as preliminary, and repeat the assay.

Determine the logarithm of Follicle-stimulating Hormone potency of the Menotropins by the formula:

$$M = (4iT_a/3T_b) + \log R,$$

in which $T_a = \Sigma(T_U - T_S)$, $T_b = \Sigma(T_3 - T_1)$, i is the interval between successive log doses of both the *Standard preparation* and *Assay preparation*, and $R = v_S/v_U$ is the ratio of the high dose of the Standard in USP Units (v_S) to the high dose of the Menotropins in mg (v_U). Compute the log confidence interval (see *Design and Analysis of Biological Assays* ⟨111⟩). Repeat the entire determination at least once. Test the agreement among the two or more independent determinations, and compute the weights/mean log-potency M and its confidence interval, L_C (see *Confidence Intervals for Individual Assays* ⟨111⟩). If this exceeds 0.18, repeat the assay until the confidence interval of the combined results is 0.18 or less.

The potency P_* is satisfactory if P_* = antilog M is not less than 80% and not more than 125% of the labeled potency and if the log confidence interval does not exceed 0.18.

PROCEDURE FOR LUTEINIZING HORMONE ACTIVITY—Further dilute the 35 Luteinizing Hormone *Standard solution* and *Test solutions* with *Diluent A* so that the daily dose for each rat in each group may be administered in 0.2 mL. Store the solution at 5 ± 3° during the test.

Inject each rat of each group assigned to the *Standard solution* and *Test solutions* subcutaneously in the dorsal area with 0.2 mL of the assigned solution. For Dose determination trials only, similarly inject each rat of the seventh (control) group with 0.2 mL of *Diluent A*. Repeat these injections at approximately the same time of day after 24, 48 and 72 hours.

Twenty-four hours after the last injection, weigh, sacrifice and carefully dissect out the seminal vessicle of each rat, removing any fat and fibrous tissue. Thoroughly dry the vessicles by pressing against absorbant paper and immediately weigh them to the nearest 0.2 mg.

Calculation—Tabulate the observed seminal vessicle weights (y) for each dosage group of f rats. For apparently outlying seminal vessicle weights, an attempt may be made to correct the organ mass relative to the mass of the rat from which it was taken. For the y-value in question, calculate for each of the f rats in the appropriate group the ratio of seminal vessicle weight to total body weight. Reject the y-value if its corresponding ratio differs from the rest of the group by more than 1.5 standard deviations.

If the data from one or more rats are missing, adjust to groups of equal size by suitable means (see *Replacement of Missing Values* ⟨111⟩). Total the values of y each group, and designate each total as T, using subscripts 1 to 3 for the three successive dosage levels and subscripts S and U for the Standard and the material under test, respectively. Test both the agreement in slope of the dosage-response lines for the Standard and for the material under test, and the lack of curvature as directed for a 3-dose balanced assay (see *Tests of Assay Validity* ⟨111⟩). If the combined discrepancy as measured by F_3 exceeds its tabular value in Table 9, repeat the assay.

Determine the logarithm of Luteinizing Hormone potency of the Menotropins by the formula:

$$M = (4iT_a/3T_b) + \log R,$$

in which $T_a = \Sigma(T_u - T_s)$, $T_b = \Sigma(T_3 - T_1)$, i is the interval between successive log doses of both the *Standard preparation* and the *Assay preparation*, and $R = v_s/v_u$ is the ratio of the high dose of the Standard in USP Luteinizing Hormone Units (v_s) to the high dose of the Menotropins in mg (v_u). Compute the log confidence interval (see *Design and Analysis of Biological Assays* ⟨111⟩).

Replication—Repeat the entire determination at least once. Test the agreement among the two or more independent determinations, and compute the weight for each (see *Combination of Independent Assays* ⟨111⟩). Calculate the weighted mean log-potency M and its confidence interval, L_c (see *Confidence Intervals for Individual Assays* ⟨111⟩). The potency, P_*, is satisfactory if P_* = antilog M is not less than 80% and not more than 125% of the labeled potency and if the confidence interval does not exceed 0.18.

Menotropins for Injection

» Menotropins for Injection is a sterile, freeze-dried mixture of menotropins and suitable excipients. Its potency is not less than 80 percent and not more than 125 percent of each of the Follicle-stimulating Hormone and Luteinizing Hormone potencies stated on the label. It may contain an antimicrobial agent.

Packaging and storage—Preserve in *Containers for Sterile Solids* as described under *Injections* ⟨1⟩.

Reference standards—*USP Menotropins Reference Standard. USP Chorionic Gonadotropin Reference Standard*—Store in a refrigerator, and do not dry before using. Use a fresh ampul for each group of assays, and discard any unused portion.

Constituted solution—At the time of use, the constituted solution prepared from Menotropins for Injection meets the requirements for *Constituted Solutions* under *Injections* ⟨1⟩.

Pyrogen—The solution constituted as directed in the labeling and adjusted to contain 1 USP Follicle-stimulating Hormone Unit per mL meets the requirements of the *Pyrogen Test* ⟨151⟩, the test dose being 2 USP Follicle-stimulating Hormone Units per kg.

pH ⟨791⟩: between 6.0 and 7.0, in the solution constituted as directed in the labeling.

Other requirements—It meets the requirements for *Sterility Tests* ⟨71⟩, *Uniformity of Dosage Units* ⟨905⟩, and *Labeling* under *Injections* ⟨1⟩.

Assay—Using an *Assay preparation* obtained by diluting a portion of the solution constituted as directed in the labeling, proceed with Menotropins for Injection as directed in the *Assay* under *Menotropins*.

Menthol

CH3
OH
CH3—CH—CH3

$C_{10}H_{20}O$ 156.27

Cyclohexanol, 5-methyl-2-(1-methylethyl)-.

» Menthol is an alcohol obtained from diverse mint oils or prepared synthetically. Menthol may be levorotatory (*l*-Menthol), from natural or synthetic sources, or racemic (*dl*-Menthol).

Packaging and storage—Preserve in tight containers, preferably at controlled room temperature.

Labeling—Label it to indicate whether it is levorotatory or racemic.

Identification—When it is triturated with about an equal weight of camphor, of chloral hydrate, or of phenol, the mixture liquefies.

Melting range of *l*-Menthol ⟨741⟩: between 41° and 44°.

Congealing range of *dl*-Menthol ⟨651⟩—*Perform this test preferably in a room having a temperature below 30° and a relative humidity below 50%.* Place about 10 g of racemic Menthol, previously dried in a desiccator over silica gel for 24 hours, in a dry test tube of from 18- to 20-mm internal diameter, and melt the contents at a temperature of about 40°. Suspend the test tube in water having a temperature of 23° to 25°, and stir the contents of the tube continually with a thermometer, keeping the bulb of the thermometer immersed in the liquid. Racemic Menthol congeals at a temperature between 27° and 28°. Shortly after the temperature has stabilized at the congealing point, add a few mg of dried racemic Menthol to the congealed mass, and continue the stirring: after a few minutes the temperature of the mass quickly rises to 30.5° to 32.0°.

Specific rotation ⟨781⟩: between −45° and −51° for *l*-Menthol, determined in a solution in alcohol containing 1 g of *l*-Menthol in each 10 mL; between −2° and +2° for *dl*-Menthol, determined in the same manner.

Nonvolatile residue—Volatilize 2 g, accurately weighed, in a tared open porcelain dish on a steam bath, and dry the residue at 105° for 1 hour: the residue weighs not more than 1 mg (0.05%).

Readily oxidizable substances in *dl*-Menthol—Place 500 mg of *dl*-Menthol in a clean, dry test tube, add 10 mL of a solution of potassium permanganate, prepared by diluting 3 mL of 0.1 *N* potassium permanganate with water to 100 mL, and place the test tube in a beaker of water at a temperature between 45° and 50°. Remove the tube from the bath at intervals of 30 seconds, and mix quickly by shaking: the purple color of potassium permanganate is still apparent after 5 minutes.

Chromatographic purity—

System suitability preparation—Dissolve decanol and Menthol in ether to obtain a solution having concentrations of about 0.05 mg of each per mL.

Test preparation—Dissolve 10 mg of Menthol in 50 mL of ether, and mix. Dilute 25 mL of this solution with ether to 100 mL, and mix.

Chromatographic system—The gas chromatograph is equipped with a flame-ionization detector and contains a 1.8-m × 2-mm column packed with 10 percent phase G16 on support S1AB. The column is maintained at about 170°, the injection port is maintained at about 260°, and the detector block is maintained at about 240°. Dry helium is used as the carrier gas at a flow rate of about 50 mL per minute. Chromatograph the *System suitability preparation*, and record the peak responses as directed under *Procedure*. The retention time of menthol is about 0.7 relative to decanol. In a suitable chromatogram, the resolution, *R*, of the 2 peaks is not less than 2.5, and the relative standard deviation of the ratio of the peak response obtained with menthol to that obtained with decanol is not more than 2%.

Procedure—Inject about 2 µL of the *Test preparation* into the gas chromatograph, and measure the peak responses. The peak response due to menthol is not less than 97 percent of the sum of all the peak responses, excluding any due to the ether.

Menthol Ointment, Tetracaine and—*see* Tetracaine and Menthol Ointment

Meperidine Hydrochloride

CH3 N • HCl COOC2H5

$C_{15}H_{21}NO_2 . HCl$ 283.80

4-Piperidinecarboxylic acid, 1-methyl-4-phenyl-, ethyl ester, hydrochloride.

Ethyl 1-methyl-4-phenylisonipecotate hydrochloride [*50-13-5*].

» Meperidine Hydrochloride contains not less than 98.0 percent and not more than 101.0 percent of $C_{15}H_{21}NO_2 . HCl$, calculated on the dried basis.

Packaging and storage—Preserve in well-closed, light-resistant containers.

Reference standard—*USP Meperidine Hydrochloride Reference Standard*—Dry in vacuum at 80° for 4 hours before using.

Identification—

A: It meets the requirements under *Identification—Organic Nitrogenous Bases* ⟨181⟩.

B: A solution (1 in 100) responds to the tests for *Chloride* ⟨191⟩.

Melting range ⟨741⟩: between 186° and 189°, determined after drying in vacuum at 80° for 4 hours.

Loss on drying ⟨731⟩—Dry it in vacuum at 80° for 4 hours: it loses not more than 1.0% of its weight.

Residue on ignition ⟨281⟩: not more than 0.1%.

Chloride content—Weigh accurately about 500 mg, previously dried, and transfer to a 250-mL conical flask. Add 15 mL of water, 5 mL of glacial acetic acid, 50 mL of methanol, and 0.2 mL of eosin Y TS, and titrate with 0.1 *N* silver nitrate VS to a rose-colored end-point. Each mL of 0.1 *N* silver nitrate is equivalent to 3.545 mg of Cl. Not less than 12.2% and not more than 12.7% of Cl is found.

Chromatographic purity—Dissolve it in water to obtain a solution containing about 10 mg per mL. Inject 2.0 µL of the solution into a suitable gas chromatograph equipped with a flame-ionization detector. Under typical conditions, the instrument contains a 2-m × 2-mm glass column packed with 10 percent phase G3 on support S1A. Maintain the temperature of the column at 190°, the injection port at 255°, and the detector at 280°. Use oxygen-free nitrogen as the carrier gas at a flow rate of 28 mL per minute. Calculate the area percentage of each peak observed in the chromatogram. No peak, other than the principal peak, constitutes more than 1.0% of the total area.

Assay—Transfer about 500 mg of Meperidine Hydrochloride, accurately weighed, to a small flask, and dissolve in a mixture of 10 mL of glacial acetic acid and 10 mL of mercuric acetate TS, warming slightly if necessary to effect solution. Add 2 drops of crystal violet TS, and titrate with 0.1 *N* perchloric acid VS. Perform a blank determination, and make any necessary correction. Each mL of 0.1 *N* perchloric acid is equivalent to 28.38 mg of $C_{15}H_{21}NO_2 . HCl$.

Meperidine Hydrochloride Injection

» Meperidine Hydrochloride Injection is a sterile solution of Meperidine Hydrochloride in Water for Injection. It contains not less than 95.0 percent and not more than 105.0 percent of the labeled amount of $C_{15}H_{21}NO_2 . HCl$.

Packaging and storage—Preserve in single-dose or in multiple-dose containers, preferably of Type I glass.

Reference standard—*USP Meperidine Hydrochloride Reference Standard*—Dry in vacuum at 80° for 4 hours before using.

Identification—It meets the requirements under *Identification—Organic Nitrogenous Bases* ⟨181⟩.

pH ⟨791⟩: between 3.5 and 6.0.

Other requirements—It meets the requirements under *Injections* ⟨1⟩.

Assay—Transfer to a separator an accurately measured volume of Meperidine Hydrochloride Injection, equivalent to about 400 mg of meperidine hydrochloride, add water, if necessary, to make about 20 mL, and saturate with sodium chloride. Render the solution strongly alkaline by the addition of 3 mL of 1 *N* sodium hydroxide, and extract the precipitated base with five 10-mL portions of chloroform. Filter the extracts through a pledget of cotton, wash the cotton with 5 mL of chloroform, and collect the combined filtrates in a 250-mL conical flask. Add 10 mL of glacial acetic acid and 2 drops of crystal violet TS, and titrate with 0.1 *N* perchloric acid VS. Perform a blank determination, and make any necessary correction. Each mL of 0.1 *N* perchloric acid is equivalent to 28.38 mg of $C_{15}H_{21}NO_2 . HCl$.

Meperidine Hydrochloride Syrup

» Meperidine Hydrochloride Syrup contains not less than 95.0 percent and not more than 105.0 percent of the labeled amount of $C_{15}H_{21}NO_2 . HCl$.

Packaging and storage—Preserve in tight, light-resistant containers.

Reference standard—*USP Meperidine Hydrochloride Reference Standard*—Dry in vacuum at 80° for 4 hours before using.

Identification—Transfer a volume of Syrup, equivalent to about 100 mg of meperidine hydrochloride, to a 125-mL separator, add 40 mL of water and 3 mL of 1 *N* sodium hydroxide, and extract with three 25-mL portions of *n*-hexane. Wash the combined extracts with two 20-mL portions of water, discard the water, and then extract with three 25-mL portions of 0.1 *N* hydrochloric acid. Transfer the extracts to a 100-mL volumetric flask, dilute with 0.1 *N* hydrochloric acid to volume, and mix: the ultraviolet absorption spectrum of this solution exhibits maxima and minima at the same wavelengths as that of a similar preparation of USP Meperidine Hydrochloride RS, concomitantly measured.

pH ⟨791⟩: between 3.5 and 3.9.

Assay—Transfer an accurately measured volume of Meperidine Hydrochloride Syrup, equivalent to about 250 mg of meperidine hydrochloride, to a separator, and add 3 mL of 1 *N* sodium hydroxide. Extract with five 20-mL portions of chloroform, and filter the extracts through a pledget of cotton into a 250-mL conical flask. Wash the cotton with 5 mL of chloroform, and add the washing to the combined filtrates. Add 10 mL of glacial acetic acid and 2 drops of crystal violet TS, and titrate with 0.1 *N* perchloric acid VS to a blue end-point. Perform a blank determination, and make any necessary correction. Each mL of 0.1 *N* perchloric acid is equivalent to 28.38 mg of $C_{15}H_{21}NO_2 . HCl$.

Meperidine Hydrochloride Tablets

» Meperidine Hydrochloride Tablets contain not less than 95.0 percent and not more than 105.0 percent of the labeled amount of $C_{15}H_{21}NO_2 . HCl$.

Packaging and storage—Preserve in well-closed, light-resistant containers.

Reference standard—*USP Meperidine Hydrochloride Reference Standard*—Dry in vacuum at 80° for 4 hours before using.

Identification—Transfer a quantity of powdered Tablets, equivalent to about 50 mg of meperidine hydrochloride, to a separator, add 10 mL of water, and shake. Add 5 mL of saturated sodium chloride solution and 1 mL of sodium hydroxide solution (1 in 25). Extract with three 20-mL portions of chloroform, filtering the extracts through cotton overlaid with anhydrous sodium sulfate. Evaporate the chloroform on a steam bath, and dissolve the residue in 4 mL of carbon disulfide. In a second separator, prepare a similar solution, using 50 mg of USP Meperidine Hydrochloride RS. Proceed as directed under *Identification—Organic Nitrogenous Bases* ⟨181⟩, beginning with "Determine the absorption spectra."

Dissolution ⟨711⟩—

Medium: water; 500 mL.

Apparatus 1: 100 rpm.

Time: 45 minutes.

Procedure—[Note—Perform baseline correction in determining the absorbance by extrapolating the baseline through the absorbance minima at about 350 nm and 280 nm and beyond 257 nm.] Determine the amount of $C_{15}H_{21}NO_2 . HCl$ dissolved from ultraviolet absorbances at the wavelength of maximum absorbance at about 257 nm of filtered portions of the solution under test, suitably diluted with *Dissolution Medium*, if necessary, in comparison with a Standard solution having a known concentration of USP Meperidine Hydrochloride RS in the same medium.

Tolerances—Not less than 75% (*Q*) of the labeled amount of $C_{15}H_{21}NO_2 . HCl$ is dissolved in 45 minutes.

Uniformity of dosage units ⟨905⟩—meet the requirements.

Procedure for content uniformity—Place 1 Tablet in a suitable flask, add about 15 mL of water, and allow the Tablet to disintegrate, crushing if necessary with the flat end of a glass rod. Rinse the rod with water, and collect the rinse water in the flask. Dilute this solution quantitatively and stepwise with water to obtain a solution containing about 2 mg of meperidine hydrochloride per mL. Filter if necessary. Transfer 20.0 mL of the solution to a separator, add 3.0 mL of sodium hydroxide solution (1 in 250), and shake. Extract with six 8-mL portions of chloroform. Filter the combined chloroform extracts through a glass wool plug wetted with chloroform into a 50-mL volumetric flask, dilute with chloroform to volume, and mix. Transfer 20.0 mL of the chloroform extract to a separator, and extract with six 8-mL portions of 0.1 *N* hydrochloric acid. Add 0.1 *N* hydrochloric acid to make 50.0 mL, and mix, to obtain the test solution. Dissolve a suitable quantity of USP Meperidine Hydrochloride RS in 0.1 *N* hydrochloric acid to obtain a solution having a known concentration of about 0.4 mg per mL. Concomitantly determine the absorbances of the test solution and the Standard solution in 1-cm cells at the wavelength of maximum absorbance at about 257 nm, with a suitable spectrophotometer, using 0.1 *N* hydrochloric acid as the blank. Calculate the quantity, in mg, of $C_{15}H_{21}NO_2 . HCl$ in the Tablet by the formula:

$$(TC/D)(A_U/A_S),$$

in which T is the labeled quantity, in mg, of meperidine hydrochloride in the Tablet, C is the concentration, in µg per mL, of USP Meperidine Hydrochloride RS in the Standard solution, D is the concentration, in µg per mL, of meperidine hydrochloride in the test solution, based on the labeled quantity per Tablet and the extent of dilution, and A_U and A_S are the absorbances of the test solution and the Standard solution, respectively.

Assay—Weigh and finely powder not less than 20 Meperidine Hydrochloride Tablets. Transfer an accurately weighed portion of the powder, equivalent to about 1 g of meperidine hydrochloride, to a 100-mL flask. Add 50 mL of chloroform, gently heat on a steam bath for 10 minutes, cool, add 40 mL of chloroform, mix, filter through a sintered-glass filter, and transfer to a 100-mL volumetric flask. Add chloroform to volume, and mix. Transfer 50.0 mL of the filtrate to a 250-mL conical flask, evaporate the chloroform, and proceed with the residue as directed in the *Assay* under *Meperidine Hydrochloride*, beginning with "dissolve in a mixture."

Mephentermine Sulfate

$(C_{11}H_{17}N)_2 \cdot H_2SO_4$ 424.60
Benzeneethanamine, *N*,α,α-trimethyl-, sulfate (2:1).
N,α,α-Trimethylphenethylamine sulfate (2:1) [*1212-72-2*].
Dihydrate 460.63 [*6190-60-9*].

» Mephentermine Sulfate is anhydrous or contains two molecules of water of hydration. It contains not less than 98.0 percent and not more than 102.0 percent of $(C_{11}H_{17}N)_2 \cdot H_2SO_4$, calculated on the anhydrous basis.

Packaging and storage—Preserve in well-closed, light-resistant containers.

Labeling—Label it to indicate whether it is anhydrous or hydrous.

Reference standard—*USP Mephentermine Sulfate Reference Standard*—Dry at 105° to constant weight before using.

Identification—

A: The infrared absorption spectrum of a mineral oil dispersion of it, previously dried at 105° to constant weight, exhibits maxima only at the same wavelengths as that of a similar preparation of USP Mephentermine Sulfate RS.

B: The ultraviolet absorption spectrum of a 1 in 2000 solution in 0.1 *N* sulfuric acid exhibits maxima and minima at the same wavelengths as that of a similar preparation of USP Mephentermine Sulfate RS, concomitantly measured, and the respective absorptivities, calculated on the anhydrous basis, at the wavelength of maximum absorbance at about 257 nm do not differ by more than 3.0%.

C: A solution (1 in 500) yields a dark brown precipitate when shaken with an equal volume of iodine TS and a white precipitate when shaken with an equal volume of mercuric–potassium iodide TS.

D: Dissolve about 100 mg in 5 mL water, and add, in small portions and with stirring, 10 mL of trinitrophenol TS. Allow to stand for 30 minutes, filter, and wash the precipitate with small portions of cold water until the last washing is practically colorless: the picrate so obtained, after drying at 105°, melts between 154° and 158°. (*Caution—Picrates may be explosive.*)

E: It responds to the tests for *Sulfate* ⟨191⟩.

Water, *Method I* ⟨921⟩: not more than 0.2% in the anhydrous form, and between 6.8% and 8.8% in the hydrated form.

Residue on ignition ⟨281⟩: not more than 0.1%.

Chromatographic purity—

Standard preparations—Dissolve an accurately weighed quantity of USP Mephentermine Sulfate RS in dehydrated alcohol to obtain *Standard preparation A* having a known concentration of 50 mg per mL. Use sonication, if necessary, to aid solution. Dilute *Standard preparation A* quantitatively with dehydrated alcohol to obtain Standard preparations having the following compositions:

Standard preparation	Dilution	Concentration (μg RS per mL)	Percentage (%, for comparison with test specimen)
B	(1 in 50)	1000	2.0
C	(1 in 200)	250	0.5
D	(1 in 500)	100	0.2

Test preparation—Dissolve an accurately weighed quantity of Mephentermine Sulfate in dehydrated alcohol to obtain a solution containing 50 mg per mL.

Iodine spray reagent—Dissolve 1 g of potassium iodide and 1 g of iodine in 100 mL of dehydrated alcohol.

Procedure—On a suitable thin-layer chromatographic plate (see *Chromatography* ⟨621⟩), coated with a 0.25-mm layer of chromatographic silica gel mixture, apply separately 5 μL of the *Test preparation* and 5 μL of each *Standard preparation.* Position the plate in a chromatographic chamber, and develop the chromatograms in a solvent system consisting of a mixture of *n*-butyl alcohol, ammonium hydroxide, water, and alcohol (60:15:15:10) until the solvent front has moved about three-fourths of the length of the plate. Remove the plate from the developing chamber, mark the solvent front, and allow the solvent to evaporate. Spray the plate evenly with *Iodine spray reagent*, followed by 3 *N* hydrochloric acid. Examine under light, and compare the intensities of any secondary spots observed in the chromatogram of the *Test preparation* with those of the principal spots in the chromatograms of the *Standard preparations:* the sum of the intensities of secondary spots obtained from the *Test preparation* corresponds to not more than 2.0% of related compounds, with no single impurity corresponding to more than 0.5%.

Assay—Dissolve about 300 mg of Mephentermine Sulfate, accurately weighed, in 50 mL of glacial acetic acid, add 4 drops of *p*-naphtholbenzein TS, and titrate with 0.1 *N* perchloric acid VS to a green end-point. Perform a blank determination, and make any necessary correction. Each mL of 0.1 *N* perchloric acid is equivalent to 42.46 mg of $(C_{11}H_{17}N)_2 \cdot H_2SO_4$.

Mephentermine Sulfate Injection

» Mephentermine Sulfate Injection is a sterile solution of Mephentermine Sulfate in Water for Injection. It contains an amount of mephentermine sulfate $[(C_{11}H_{17}N)_2 \cdot H_2SO_4]$ equivalent to not less than 95.0 percent and not more than 105.0 percent of the labeled amount of mephentermine ($C_{11}H_{17}N$).

Packaging and storage—Preserve in single-dose or in multiple-dose containers, preferably of Type I glass.

Reference standard—*USP Mephentermine Sulfate Reference Standard*—Dry at 105° to constant weight before using.

Identification—

A: It responds to *Identification tests C* and *E* under *Mephentermine Sulfate.*

B: Transfer a volume of Injection, equivalent to about 60 mg of mephentermine, to a 60-mL separator, and add water to obtain a total volume of about 10 mL. In a second 60-mL separator dissolve about 85 mg of USP Mephentermine Sulfate RS in about 10 mL of water. Treat each solution similarly and concomitantly as follows: Add 1 mL of 1 *N* sodium hydroxide, shake, and without delay extract with two 10-mL portions and one 5-mL portion of carbon disulfide, collecting the extracts in a 50-mL beaker. Evaporate with the aid of a current of air, but without heat, to a volume of about 5 mL, and filter through a dry filter. Proceed as directed under *Identification—Organic Nitrogenous Bases* ⟨181⟩, beginning with "Without delay determine the absorption spectra": the Injection meets the requirements of the test.

pH ⟨791⟩: between 4.0 and 6.5.

Particulate matter ⟨788⟩: meets the requirements under *Small-volume Injections.*

Other requirements—It meets the requirements under *Injections* ⟨1⟩.

Assay—Transfer to a coarse-porosity, sintered-glass filtering crucible 5 g of purified siliceous earth and 500 mg of chromatographic magnesium oxide, mix with the aid of a glass rod, add an accurately measured volume of Mephentermine Sulfate Injection, equivalent to about 60 mg of mephentermine, and again mix. Add 15 mL of hot chloroform, mix, and with gentle suction draw off the chloroform into 40 mL of glacial acetic acid. Repeat the extraction with four 10-mL portions of hot chloroform, successively and similarly applied, collecting each portion in the glacial acetic acid. Add 6 drops of *p*-naphtholbenzein TS, and titrate with 0.1 *N* perchloric acid VS to a green end-point. Perform a blank determination, and make any necessary correction. Each mL of 0.1 *N* perchloric acid is equivalent to 16.33 mg of $C_{11}H_{17}N$.

Mephenytoin

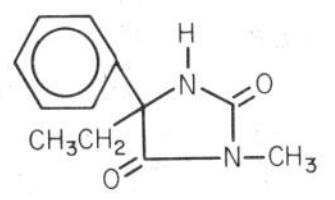

$C_{12}H_{14}N_2O_2$ 218.25
2,4-Imidazolidinedione, 5-ethyl-3-methyl-5-phenyl-.
5-Ethyl-3-methyl-5-phenylhydantoin [*50-12-4*].

» Mephenytoin contains not less than 98.0 percent and not more than 102.0 percent of $C_{12}H_{14}N_2O_2$, calculated on the dried basis.

Packaging and storage—Preserve in well-closed containers.

Reference standard—*USP Mephenytoin Reference Standard*—Dry at 105° for 4 hours before using.

Identification—

A: The infrared absorption spectrum of a potassium bromide dispersion of it, previously dried, exhibits maxima only at the same wavelengths as that of a similar preparation of USP Mephenytoin RS.

B: The ultraviolet absorption spectrum of a 1 in 1000 solution in methanol exhibits maxima and minima at the same wavelengths as that of a similar solution of USP Mephenytoin RS, concomitantly measured.

Melting range ⟨741⟩: between 136° and 140°.

Loss on drying ⟨731⟩—Dry it at 105° for 4 hours: it loses not more than 1.0% of its weight.

Residue on ignition ⟨281⟩: not more than 0.1%.

Heavy metals, *Method II* ⟨231⟩: 0.002%.

Chromatographic purity—Prepare a *Test solution* in chloroform containing 200 mg per mL. Dissolve an accurately weighed quantity of USP Mephenytoin RS in chloroform, and dilute quantitatively and stepwise with chloroform to obtain solutions containing 3 mg per mL (*Standard solution A*), 2 mg per mL (*Standard solution B*), and 1 mg per mL (*Standard solution C*). Apply 5 µL each of the *Test solution, Standard solution A, Standard solution B*, and *Standard solution C* to separate points on a suitable thin-layer chromatographic plate (see *Chromatography* ⟨621⟩) coated with a 0.25-mm layer of chromatographic silica gel mixture. Allow the spots to dry, and develop the chromatogram in a chromatographic chamber containing a solvent system consisting of a mixture of chloroform and alcohol (98:2) until the solvent front has moved about three-fourths of the length of the plate. Remove the plate from the developing chamber, mark the solvent front, and allow the solvent to evaporate. Locate the spots on the plate by examination under short-wavelength ultraviolet light: the chromatograms show principal spots at about the same R_f value; any spot from the *Test solution*, other than the principal spot, is not greater in size or intensity than the spot obtained from *Standard solution B* corresponding to not more than 1.0% of any individual impurity, and the sum of the sizes and intensities of all spots, other than the principal spot, is not greater than that obtained from *Standard solution A*, corresponding to not more than 1.5% of total impurities.

Ordinary impurities ⟨466⟩—

Test solution: 200 mg of mephenytoin per mL in chloroform.
Standard solution: chloroform.
Application volume: 5 µL.
Eluant: a mixture of chloroform and alcohol (49:1).
Visualization: 1.

Assay—Transfer about 50 mg of Mephenytoin, accurately weighed, to a 100-mL volumetric flask, dissolve in methanol, dilute with methanol to volume, and mix. Concomitantly determine the absorbances of this solution and a Standard solution of Mephenytoin RS in the same medium having a known concentration of about 0.5 mg per mL, in 1-cm cells at the wavelength of maximum absorbance at about 257 nm, with a suitable spectrophotometer, using methanol as the blank. Calculate the quantity, in mg, of $C_{12}H_{14}N_2O_2$ in the Mephenytoin taken by the formula:

$$100C(A_U / A_S),$$

in which C is the concentration, in mg per mL, of USP Mephenytoin RS in the Standard solution, and A_U and A_S are the absorbances of the solution of Mephenytoin and the Standard solution, respectively.

Mephenytoin Tablets

» Mephenytoin Tablets contain not less than 90.0 percent and not more than 110.0 percent of the labeled amount of $C_{12}H_{14}N_2O_2$.

Packaging and storage—Preserve in well-closed containers.

Reference standard—*USP Mephenytoin Reference Standard*—Dry at 105° for 4 hours before using.

Dissolution ⟨711⟩—

Medium: water; 500 mL.
Apparatus 2: 75 rpm.
Time: 60 minutes.

Procedure—Determine the amount of $C_{12}H_{14}N_2O_2$ dissolved from ultraviolet absorbances at the wavelength of maximum absorbance at about 257 nm of filtered portions of the solution under test, suitably diluted with *Dissolution Medium*, if necessary, in comparison with a Standard solution having a known concentration of USP Mephenytoin RS in the same medium.

Tolerances—Not less than 70% (*Q*) of the labeled amount of $C_{12}H_{14}N_2O_2$ is dissolved in 60 minutes.

Uniformity of dosage units ⟨905⟩: meet the requirements.

Assay—

Mobile phase—[NOTE—Mix the acetonitrile and methanol, then add the water, and mix.] Prepare a filtered and degassed mixture of water, methanol, and acetonitrile (10:9:1), making adjustments if necessary (see *System Suitability* under *Chromatography* ⟨621⟩).

Standard preparation—Dissolve an accurately weighed quantity of USP Mephenytoin RS in methanol. Sonicate until the solution is clear, and mix. Dilute quantitatively, and stepwise if necessary, with methanol to obtain a solution having a known concentration of about 5 mg per mL.

Assay preparation—Weigh and finely powder not less than 20 Mephenytoin Tablets. Transfer an accurately weighed portion of the powder, equivalent to about 500 mg of mephenytoin, to a 100-mL volumetric flask. Add about 50 mL of methanol, and shake by mechanical means for 30 minutes. Dilute with methanol to volume, mix, and filter, discarding the first portion of the filtrate.

Chromatographic system (see *Chromatography* ⟨621⟩)—The liquid chromatograph is equipped with a 257-nm detector and a 4.6-mm × 25-cm column that contains packing L1. The flow rate is about 2 mL per minute. Chromatograph replicate injections of the *Standard preparation*, and record the peak responses as directed under *Procedure:* the column efficiency determined from the mephenytoin peak is not less than 300 theoretical plates, the tailing factor for the mephenytoin peak is not more than 2.0, and the relative standard deviation is not more than 2.0%.

Procedure—Separately inject equal volumes (about 10 µL) of the *Standard preparation* and the *Assay preparation* into the chromatograph, record the chromatograms, and measure the responses for the major peaks. Calculate the quantity, in mg, of $C_{12}H_{14}N_2O_2$ in the portion of Tablets taken by the formula:

$$100C(r_U / r_S),$$

in which C is the concentration, in mg per mL, of USP Mephenytoin RS in the *Standard preparation*, and r_U and r_S are the peak responses obtained from the *Assay preparation* and the *Standard preparation*, respectively.

Mephobarbital

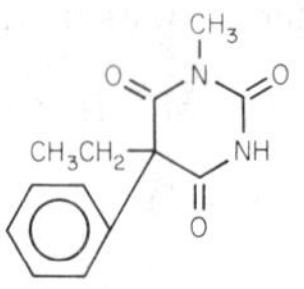

$C_{13}H_{14}N_2O_3$ 246.27
2,4,6(1*H*,3*H*,5*H*)-Pyrimidinetrione, 5-ethyl-1-methyl-5-phenyl-.
5-Ethyl-1-methyl-5-phenylbarbituric acid [*115-38-8*].

» Mephobarbital contains not less than 98.0 percent and not more than 100.5 percent of $C_{13}H_{14}N_2O_3$, calculated on the dried basis.

Packaging and storage—Preserve in well-closed containers.

Reference standard—*USP Mephobarbital Reference Standard*—Dry at 105° for 4 hours before using.

Identification—

A: The infrared absorption spectrum of a mineral oil dispersion of it, previously dried, exhibits maxima only at the same wavelengths as that of a similar preparation of USP Mephobarbital RS.

B: Boil about 200 mg with 10 mL of 1 *N* sodium hydroxide: ammonia is evolved.

C: Shake about 60 mg with 5 mL of sodium hydroxide solution (1 in 500), and filter. To a 1-mL portion of the filtrate add 3 drops of mercuric nitrate TS: a white precipitate is formed, and it is soluble in 6 *N* ammonium hydroxide. To another 1-mL portion of the filtrate add silver nitrate TS, dropwise: a white precipitate is formed, and it dissolves readily in 6 *N* ammonium hydroxide.

Melting range, *Class I* ⟨741⟩: between 176° and 181°.

Loss on drying ⟨731⟩—Dry it at 105° for 4 hours: it loses not more than 1.0% of its weight.

Residue on ignition ⟨281⟩: not more than 0.1%.

Assay—Dissolve about 500 mg of Mephobarbital, accurately weighed, in 50 mL of dimethylformamide in a 200-mL flask. Add 4 drops of thymolphthalein TS, and titrate with 0.1 *N* lithium methoxide VS, using a magnetic stirrer and a cover for the flask to protect against atmospheric carbon dioxide. Perform a blank determination, and make any necessary correction. Each mL of 0.1 *N* lithium methoxide is equivalent to 24.63 mg of $C_{13}H_{14}N_2O_3$.

Mephobarbital Tablets

» Mephobarbital Tablets contain not less than 95.0 percent and not more than 110.0 percent of the labeled amount of $C_{13}H_{14}N_2O_3$.

Packaging and storage—Preserve in well-closed containers.

Reference standard—*USP Mephobarbital Reference Standard*—Dry at 105° for 4 hours before using.

Identification—The mephobarbital obtained in the *Assay* melts between 174° and 181° (see *Melting Range or Temperature* ⟨741⟩), and responds to *Identification test A* under *Mephobarbital.*

Dissolution ⟨711⟩—

Medium: pH 10.0 alkaline borate buffer (see *Buffer Solutions* in the section, *Reagents, Indicators, and Solutions*); 900 mL.

Apparatus 2: 75 rpm.

Time: 75 minutes.

Procedure—Determine the amount of $C_{13}H_{14}N_2O_3$ dissolved from ultraviolet absorbances at the wavelength of maximum absorbance at about 242 nm of filtered portions of the solution under test, suitably diluted with *pH 10.0 alkaline borate buffer*, in comparison with a Standard solution having a known concentration of USP Mephobarbital RS in the same medium.

Tolerances—Not less than 70% (*Q*) of the labeled amount of $C_{13}H_{14}N_2O_3$ is dissolved in 75 minutes.

Uniformity of dosage units ⟨905⟩: meet the requirements.

Procedure for content uniformity—

Alkaline borate solution—Dissolve 6.2 g of boric acid and 7.45 g of potassium chloride in 500 mL of water, add 210 mL of sodium hydroxide solution (1 in 25), and mix. Add water to make 2000 mL of solution, and mix.

Standard preparation—[NOTE—Prepare the *Standard preparation* and the *Test preparation* concomitantly.] Dissolve an accurately weighed quantity of USP Mephobarbital RS in *Alkaline borate solution* to obtain a solution having a concentration of about 10 mg per mL, and dilute quantitatively with water to obtain a solution having a known concentration of about 1.5 mg per mL.

Test preparation—Transfer 1 Tablet to a glass-stoppered centrifuge tube, crush the tablet, and add 25.0 mL of *Alkaline borate solution.* Insert the stopper, shake for 10 minutes, and centrifuge until clear, filtering the supernatant liquid, if necessary. Dilute a portion of the subsequent liquid quantitatively and stepwise, if necessary, with water to obtain a solution having a concentration of about 1.5 mg of mephobarbital per mL.

Procedure—Transfer 3.0 mL each of the *Standard preparation* and the *Test preparation* to separate 200-mL volumetric flasks, dilute each with a 1 in 3 solution of *Alkaline borate solution* in water to volume, and mix. Concomitantly determine the absorbances of the solutions in 1-cm cells at the wavelength of maximum absorbance at about 245 nm, with a suitable spectrophotometer, using a 1 in 3 solution of *Alkaline borate solution* in water as the blank. Calculate the quantity, in mg, of $C_{13}H_{14}N_2O_3$ in the Tablet taken by the formula:

$$(TC/D)(A_U/A_S),$$

in which *T* is the labeled quantity, in mg, of mephobarbital in the Tablet, *C* is the concentration, in mg per mL, of USP Mephobarbital RS in the *Standard preparation, D* is the concentration, in mg per mL, of mephobarbital in the *Test preparation,* based upon the labeled quantity per Tablet and the extent of dilution, and A_U and A_S are the absorbances of the solutions from the *Assay preparation* and the *Standard preparation,* respectively.

Assay—Weigh and finely powder not less than 20 Mephobarbital Tablets. Transfer an accurately weighed portion of the powder, equivalent to about 300 mg of mephobarbital, to an extraction thimble. Extract with 15 mL of solvent hexane, allow the thimble to drain, transfer to a continuous-extraction apparatus provided with a tared flask, and extract the mephobarbital with chloroform for 2 hours. Evaporate the chloroform on a steam bath with the aid of a current of air, cool, dissolve the residue in about 10 mL of alcohol, evaporate, dry the residue at 105° for 1 hour, cool, and weigh. The weight of the residue represents the weight of $C_{13}H_{14}N_2O_3$ in the portion of Tablets taken.

Mepivacaine Hydrochloride

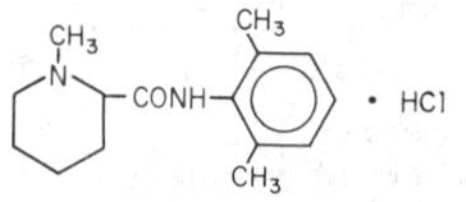

$C_{15}H_{22}N_2O.HCl$ 282.81
2-Piperidinecarboxamide, *N*-(2,6-dimethylphenyl)-1-methyl-, monohydrochloride.
1-Methyl-2′,6′-pipecoloxylidide monohydrochloride [*1722-62-9*].

» Mepivacaine Hydrochloride contains not less than 98.0 percent and not more than 102.0 percent of $C_{15}H_{22}N_2O.HCl$, calculated on the dried basis.

Packaging and storage—Preserve in well-closed containers.

Reference standard—*USP Mepivacaine Hydrochloride Reference Standard*—Dry at 105° for 4 hours before using.

Identification—

A: The infrared absorption spectrum of a potassium bromide dispersion of it, previously dried, exhibits maxima only at the same wavelengths as that of a similar preparation of USP Mepivacaine Hydrochloride RS.

B: Dissolve about 250 mg in 10 mL of water, render slightly alkaline with sodium carbonate TS, extract the precipitate with ether, evaporate the ether extract on a steam bath to dryness, and dry the residue in vacuum at 60° for 1 hour: the mepivacaine so obtained melts between 149° and 153°.

C: A solution (1 in 100) responds to the tests for *Chloride* ⟨191⟩.

Loss on drying ⟨731⟩—Dry it at 105° for 4 hours: it loses not more than 1.0% of its weight.

Residue on ignition ⟨281⟩: not more than 0.1%.

Chromatographic purity—Mix about 500 mg with 10 mL of water to obtain a test solution. Inject about 1 µL of this solution into a suitable gas chromatograph (see *Chromatography* ⟨621⟩), equipped with a flame-ionization detector. Under typical conditions, the instrument contains a 1.8-m × 4-mm glass column packed with 3 percent phase G19 on packing S1A. The column is maintained at about 230°, and dry helium, flowing at the rate of about 40 mL per minute, is used as the carrier gas. Similarly inject about 1 µL of a Standard solution of USP Mepivacaine Hydrochloride RS in water having a concentration of about 5 mg per mL. Locate the mepivacaine peak in each chromatogram. The total area of all extraneous peaks (except that of the solvent peak) recorded for the test solution, over a time span of not less than 1.3 × the retention time for mepivacaine, relative to the total area of all peaks does not exceed 0.4%.

Assay—Transfer about 550 mg of Mepivacaine Hydrochloride, accurately weighed, to a 200-mL beaker, and dissolve in 50 mL of glacial acetic acid, heating if necessary. Add 10 mL of mercuric acetate TS and 1 drop of crystal violet TS, and titrate with 0.1 *N* perchloric acid VS to a green end-point. Perform a blank determination, and make any necessary correction. Each mL of 0.1 *N* perchloric acid is equivalent to 28.28 mg of $C_{15}H_{22}N_2O.HCl$.

Mepivacaine Hydrochloride Injection

» Mepivacaine Hydrochloride Injection is a sterile solution of Mepivacaine Hydrochloride in Water for Injection. It contains not less than 95.0 percent and not more than 105.0 percent of the labeled amount of $C_{15}H_{22}N_2O.HCl$.

Packaging and storage—Preserve in single-dose or in multiple-dose containers, preferably of Type I glass. Injection labeled to contain 2% or less of mepivacaine hydrochloride may be packaged in 50-mL multiple-dose containers.

Reference standard—*USP Mepivacaine Hydrochloride Reference Standard*—Dry at 105° for 4 hours before using.

Identification—

A: It meets the requirements under *Identification—Organic Nitrogenous Bases* ⟨181⟩.

B: Extract a volume of Injection, equivalent to about 200 mg of mepivacaine, with two 10-mL portions of ether, and discard the ether extracts: the remaining solution responds to *Identification test B* under *Mepivacaine Hydrochloride*.

pH ⟨791⟩: between 4.5 and 6.8.

Other requirements—It meets the requirements under *Injections* ⟨1⟩.

Assay—Transfer to a 60-mL separator an accurately measured volume of Injection, equivalent to about 40 mg of mepivacaine hydrochloride. Dilute with water to 10 mL, add 1 mL of 1 *N* sodium hydroxide, and completely extract the mepivacaine with three 10-mL portions of chloroform. Wash the combined chloroform extracts with 10 mL of water, filter the chloroform through a pledget of cotton, wash the filter with chloroform, add 2 drops of methyl red TS to the combined chloroform filtrate and washings, and titrate with 0.01 *N* perchloric acid in dioxane VS to a red end-point. Perform a blank determination, and make any necessary correction. Each mL of 0.01 *N* perchloric acid is equivalent to 2.828 mg of $C_{15}H_{22}N_2O.HCl$.

Mepivacaine Hydrochloride and Levonordefrin Injection

» Mepivacaine Hydrochloride and Levonordefrin Injection is a sterile solution of Mepivacaine Hydrochloride and Levonordefrin in Water for Injection. It contains not less than 95.0 percent and not more than 105.0 percent of the labeled amount of mepivacaine hydrochloride ($C_{15}H_{22}N_2O.HCl$) and not less than 90.0 percent and not more than 110.0 percent of the labeled amount of levonordefrin ($C_9H_{13}NO_3$).

Packaging and storage—Preserve in single-dose or in multiple-dose containers, preferably of Type I glass.

Reference standard—*USP Epinephrine Bitartrate Reference Standard*—Keep the container tightly closed and protected from light. Dry in vacuum over silica gel for 18 hours before using.

Identification—

A: Extract a volume of Injection, equivalent to about 200 mg of mepivacaine, with two 10-mL portions of ether, and discard the ether extracts. Render slightly alkaline with sodium carbonate TS, extract the precipitate with ether, evaporate the ether extract on a steam bath to dryness, and dry the residue in vacuum at 60° for 1 hour: the mepivacaine so obtained melts between 149° and 153°.

B: It responds to the tests for *Chloride* ⟨191⟩.

pH ⟨791⟩: between 3.3 and 5.5.

Other requirements—It meets the requirements under *Injections* ⟨1⟩.

Assay for mepivacaine hydrochloride—Transfer to a 60-mL separator an accurately measured volume of Mepivacaine Hydrochloride and Levonordefrin Injection, equivalent to about 40 mg of mepivacaine hydrochloride. Dilute with water to 10 mL, add 1 mL of 1 *N* sodium hydroxide, and completely extract the mepivacaine with three 10-mL portions of chloroform. Wash the combined chloroform extracts with 10 mL of water, filter the chloroform through a pledget of cotton, wash the filter with chloroform, add 2 drops of methyl red TS to the combined chloroform filtrate and washings, and titrate with 0.01 *N* perchloric acid in dioxane VS to a red end-point. Perform a blank determination, and make any necessary correction. Each mL of 0.01 *N* perchloric acid is equivalent to 2.828 mg of $C_{15}H_{22}N_2O.HCl$.

Assay for levonordefrin—

Ferro-citrate solution, Buffer solution, and *Standard preparation*—Prepare as directed under *Epinephrine Assay* ⟨391⟩.

Assay preparation—Using Mepivacaine Hydrochloride and Levonordefrin Injection, prepare as directed for *Assay Preparation* under *Epinephrine Assay* ⟨391⟩, except to read levonordefrin wherever epinephrine is called for.

Procedure—Proceed as directed under *Epinephrine Assay* ⟨391⟩, except to read levonordefrin wherever epinephrine (base) is called for. When the *Ferro-citrate solution* and the *Buffer solution* are mixed with the *Assay preparation*, a fine precipitate may be formed. Remove this precipitate by centrifugation or by filtration through dry filter paper before the colorimetric measurements are taken. Calculate the quantity, in mg, of levonordefrin ($C_9H_{13}NO_3$) in each mL of the Injection taken by the formula:

$$(183.21/333.29)(0.05C/V)(A_U/A_S),$$

in which 183.21 and 333.29 are the molecular weights of levonordefrin and epinephrine bitartrate, respectively, *C* is the concentration, in µg per mL, of USP Epinephrine Bitartrate RS in the *Standard preparation*, and *V* is the volume, in mL, of Injection taken.

Meprednisone

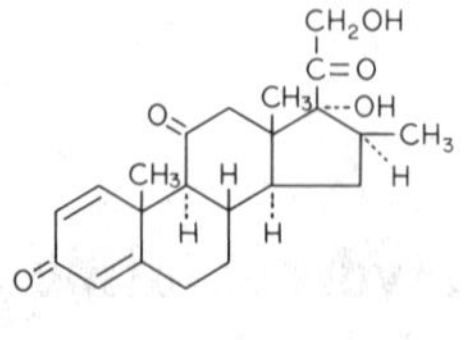

$C_{22}H_{28}O_5$ 372.46
Pregna-1,4-diene-3,11,20-trione, 17,21-dihydroxy-16-methyl-, (16β)-.
17,21-Dihydroxy-16β-methylpregna-1,4-diene-3,11,20-trione [*1247-42-3*].

» Meprednisone contains not less than 97.5 percent and not more than 102.5 percent of $C_{22}H_{28}O_5$, calculated on the dried basis.

Packaging and storage—Preserve in tight, light-resistant containers, and avoid exposure to excessive heat.

Reference standard—*USP Meprednisone Reference Standard*—Dry at 105° for 3 hours before using.

Identification—

A: The infrared absorption spectrum of a mineral oil dispersion of it, previously dried, exhibits maxima only at the same wavelengths as that of a similar preparation of USP Meprednisone RS.

B: The ultraviolet absorption spectrum of a 1 in 100,000 solution in methanol exhibits maxima and minima at the same wavelengths as that of a similar solution of USP Meprednisone RS, concomitantly measured, and the respective absorptivities, calculated on the dried basis, at the wavelength of maximum absorbance at about 238 nm do not differ by more than 3.0%.

C: Prepare a solution of it in a mixture of toluene and alcohol (1:1) containing 20 mg per mL. On a suitable thin-layer chromatographic plate (see *Chromatography* ⟨621⟩), coated with a 0.25-mm layer of chromatographic silica gel, apply 10 µL of this solution and 10 µL of a solution of USP Meprednisone RS in a mixture of toluene and alcohol (1:1) containing 20 mg per mL. Allow the spots to dry, and develop the chromatogram in a solvent system consisting of a mixture of chloroform, methanol, and water (180:15:1) until the solvent front has moved about three-fourths of the length of the plate. Remove the plate from the developing chamber, mark the solvent front, and allow the solvent to evaporate. Locate the spots on the plate by spraying with a mixture of sulfuric acid, methanol, and nitric acid (10:10:1) and heating at 105° for 10 minutes: the R_f value of the principal spot obtained from the test solution corresponds to that obtained from the Standard solution.

Specific rotation ⟨781⟩: between +180° and +188°, calculated on the dried basis, determined in a solution in dioxane containing 100 mg in each 10 mL.

Loss on drying ⟨731⟩—Dry it at 105° for 3 hours: it loses not more than 1.0% of its weight.

Residue on ignition ⟨281⟩: not more than 0.1%.

Assay—

Standard preparation—Prepare as directed under *Single-steroid Assay* ⟨511⟩, using USP Meprednisone RS.

Assay preparation—Weigh accurately about 20 mg of Meprednisone, previously dried, dissolve it in a sufficient quantity of a mixture of equal volumes of alcohol and chloroform to make 10.0 mL, and mix.

Procedure—Proceed as directed for *Procedure* under *Single-steroid Assay* ⟨511⟩, using a solvent system consisting of chloroform, methanol, and water (180:15:1), through the fourth sentence of the second paragraph under *Procedure.* Then centrifuge the tubes for 5 minutes, and determine the absorbances of the supernatant solutions in 1-cm cells at the wavelength of maximum absorbance at about 238 nm, with a suitable spectrophotometer, against the blank. Calculate the quantity, in mg, of $C_{22}H_{28}O_5$ in the portion of Meprednisone taken by the formula:

$$10C(A_U/A_S),$$

in which C is the concentration, in mg per mL, of USP Meprednisone RS in the *Standard preparation,* and A_U and A_S are the absorbances of the solutions from the *Assay preparation* and the *Standard preparation,* respectively.

Meprobamate

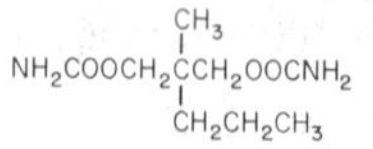

$C_9H_{18}N_2O_4$ 218.25
1,3-Propanediol, 2-methyl-2-propyl-, dicarbamate.
2-Methyl-2-propyl-1,3-propanediol dicarbamate [*57-53-4*].

» Meprobamate contains not less than 97.0 percent and not more than 101.0 percent of $C_9H_{18}N_2O_4$, calculated on the dried basis.

Packaging and storage—Preserve in tight containers.

Reference standard—*USP Meprobamate Reference Standard*—Dry in vacuum at 60° for 3 hours before using.

Identification—

A: The infrared absorption spectrum of a potassium bromide dispersion of it (about 1 mg in 200 mg), previously dried, exhibits maxima only at the same wavelengths as that of a similar preparation of USP Meprobamate RS. If a difference appears, dissolve portions of both the test specimen and the Reference Standard in acetone at a concentration of 8 mg per mL. Dilute 0.1-mL portions of the acetone solutions with 1 mL of *n*-heptane, and remove the solvents by evaporation under nitrogen at a temperature of about 30°. Dry the residues in vacuum at room temperature for 30 minutes, and repeat the test on the residues.

B: The R_f value of the principal spot in the chromatogram of the *Identification preparation* corresponds to that of *Standard preparation A*, as obtained in the test for *Chromatographic purity.*

Melting range ⟨741⟩: between 103° and 107°, but the range between beginning and end of melting does not exceed 2°.

Loss on drying ⟨731⟩—Dry it in vacuum at 60° for 3 hours: it loses not more than 0.5% of its weight.

Chromatographic purity—

Standard preparations—Dissolve USP Meprobamate RS in alcohol, and mix to obtain *Standard preparation A* having a known concentration of 1.0 mg per mL. Dilute quantitatively with alcohol to obtain *Standard preparations*, designated below by letter, having the following compositions:

Dilution	Concentration (mg RS per mL)	Percentage (%, for comparison with test specimen)
A (undiluted)	1.0	1.0
B (4 in 5)	0.8	0.8
C (3 in 5)	0.6	0.6
D (2 in 5)	0.4	0.4
E (1 in 5)	0.2	0.2

Test preparation—Dissolve an accurately weighed quantity of Meprobamate in alcohol to obtain a solution containing 100 mg per mL.

Identification preparation—Dilute a portion of the *Test preparation* quantitatively with alcohol to obtain a solution containing 1.0 mg per mL.

Procedure—On a suitable thin-layer chromatographic plate (see *Chromatography* ⟨621⟩), coated with a 0.25-mm layer of chromatographic silica gel, apply separately 2 µL of the *Test preparation,* 2 µL of the *Identification preparation,* and 2 µL of each *Standard preparation.* Position the plate in a chromatographic chamber, and develop the chromatograms in a solvent system consisting of a mixture of hexane, acetone, and pyridine (7:3:1) until the solvent front has moved about three-fourths of the length of the plate. Remove the plate from the developing chamber, mark the solvent front, and air-dry the plate for 15 minutes. Heat the plate at 100° for 15 minutes, cool, and spray with a solution prepared by dissolving 1 g of vanillin in a cooled mixture of sulfuric acid and alcohol (160:40). Heat the plate at 110° for 15 to 20 minutes, cool, and allow the plate to develop blue-purple spots at room temperature. [NOTE—Color development requires approximately 30 to 60 minutes.] Examine the plate, and compare the intensities of any secondary spots observed in the chromatogram of the *Test preparation* with those of the principal spots in the chromatograms of the *Standard preparations.* No secondary spot from the chromatogram of the *Test preparation* is larger or more intense than the principal spot obtained from *Standard preparation A* (1.0%), and the sum of the intensities of all secondary spots obtained from the *Test preparation* corresponds to not more than 2.0%.

Methyl carbamate—

Mobile phase—Use filtered and degassed water.

Standard preparation—Dissolve an accurately weighed quantity of methyl carbamate in water to obtain a solution having a known concentration of 1.0 mg per mL.

Test preparation—Transfer 1.0 g of finely powdered Meprobamate, accurately weighed, to a beaker, add 5.0 mL of water, and stir to wet the powder completely. Filter the slurry through a small plug of glass wool in the stem of a glass funnel. Use the clear filtrate as the *Test preparation.*

Chromatographic system (see *Chromatography* ⟨621⟩)—The liquid chromatograph is equipped with a 200-nm detector and a 3.9- to 4.6-mm × 25- to 30-cm column that contains packing L1. The flow rate is about 1 mL per minute. Chromatograph replicate injections of the *Standard preparation*, and record the peak responses as directed under *Procedure:* the relative standard deviation is not more than 2.0%.

Procedure—Separately inject equal volumes (about 50 µL) of the *Standard preparation* and the *Test preparation* into the chromatograph, record the chromatograms, and measure the responses for the methyl carbamate peaks: the peak response obtained from the *Test preparation* is not greater than that obtained from the *Standard preparation*, corresponding to not more than 0.5% of methyl carbamate.

Assay—Transfer about 400 mg of Meprobamate, accurately weighed, to a conical flask, add 40 mL of hydrochloric acid and several boiling chips, and reflux for 90 minutes. Remove the condenser, and continue boiling until the volume is reduced to between 5 and 10 mL. Cool the flask to room temperature, add 50 mL of water and 1 drop of methyl red TS, and, while cooling the flask continuously, cautiously neutralize the acid with 10 *N* sodium hydroxide until the indicator begins to change color. If necessary, add 1 *N* hydrochloric acid to restore the pink color, and carefully neutralize with 0.1 *N* sodium hydroxide VS. Add a mixture of 15 mL of formaldehyde TS and 15 mL of water, which previously has been neutralized with 0.1 *N* sodium hydroxide VS to phenolphthalein TS, and titrate with 0.1 *N* sodium hydroxide VS to a yellow end-point. Add 0.2 mL of phenolphthalein TS, and continue the titration with 0.1 *N* sodium hydroxide VS to a distinct pink color. Perform a blank determination, and make any necessary correction. Each mL of the total volume of 0.1 *N* sodium hydroxide consumed after the addition of formaldehyde TS is equivalent to 10.91 mg of $C_9H_{18}N_2O_4$.

Meprobamate Oral Suspension

» Meprobamate Oral Suspension contains not less than 95.0 percent and not more than 110.0 percent of the labeled amount of $C_9H_{18}N_2O_4$.

Packaging and storage—Preserve in tight containers.

Identification—Mix 2 mL of Oral Suspension with 2 mL of acetone and 2 mL of a 1 in 100 solution of furfural in glacial acetic acid, add 5 mL of hydrochloric acid, and shake: a purple color is produced, and, on standing, it changes to blue, then to blue-black, and finally to black-brown.

Assay—Transfer an accurately measured volume of Meprobamate Oral Suspension, equivalent to about 400 mg of meprobamate, to a separator, and completely extract the meprobamate with 20-mL portions of chloroform, filtering the extracts through a pledget of cotton enclosed in glass wool that previously has been moistened with chloroform. Collect the filtrate in a conical flask, add several glass beads to the flask, and evaporate on a steam bath to dryness. To the residue add 20 mL of water, heat on a steam bath for several minutes, then add 40 mL of hydrochloric acid, and reflux for 90 minutes. Remove the condenser, and continue boiling until the volume is reduced to about 20 mL. Cool to room temperature, add 50 mL of water, and cool in an ice bath. Add 1 drop of methyl red TS, and, while cooling continuously, cautiously neutralize with sodium hydroxide solution (4 in 10) until the indicator begins to change color. Add hydrochloric acid, if necessary, to restore the pink color, then carefully neutralize with 0.1 *N* sodium hydroxide VS. Add 30 mL of neutral formaldehyde solution (18% w/w), and titrate with 0.1 *N* sodium hydroxide VS until the solution becomes yellow. Add 0.2 mL of phenolphthalein TS, and continue the titration with 0.1 *N* sodium hydroxide VS to a distinct pink color. Perform a blank determination, and make any necessary correction. Each mL of the total volume of 0.1 *N* sodium hydroxide consumed after the addition of the formaldehyde solution is equivalent to 10.91 mg of $C_9H_{18}N_2O_4$.

Meprobamate Tablets

» Meprobamate Tablets contain not less than 95.0 percent and not more than 105.0 percent of the labeled amount of $C_9H_{18}N_2O_4$.

Packaging and storage—Preserve in well-closed containers.

Reference standard—*USP Meprobamate Reference Standard*—Dry in vacuum at 60° for 3 hours before using.

Identification—

A: To a portion of finely powdered Tablets, equivalent to about 800 mg of meprobamate, add 5 mL of dehydrated alcohol, and heat just below the boiling temperature for about 5 minutes, with occasional swirling. Cool, filter into 15 mL of solvent hexane, and mix. With the aid of suction, filter the crystals that form, and dry at 60°: the crystals of meprobamate so obtained melt between 103° and 107°, but the range between beginning and end of melting does not exceed 2°.

B: A portion of the crystals obtained in *Identification test A* responds to *Identification test A* under *Meprobamate.*

Dissolution ⟨711⟩—

Medium: deaerated water; 900 mL.

Apparatus 1: 100 rpm.

Time: 30 minutes.

Procedure—Determine the amount in solution using the reagents described in the *Assay* and using, as the *Test preparation*, an aliquot of the filtered *Dissolution Medium* estimated to contain between 4 mg and 6 mg of meprobamate, diluted with water to 25.0 mL. Prepare the *Standard preparation* by dissolving an accurately weighed quantity of USP Meprobamate RS in water to obtain a solution having a known concentration of about 200 µg per mL.

Pipet 1 mL of *Test preparation* into a 50-mL volumetric flask. Into a similar flask pipet 1 mL of water to provide a reagent blank, and into a third 50-mL volumetric flask pipet 1 mL of *Standard preparation.* Treat each flask as follows: Add by pipet 2 mL of *Chlorinating solution*, taking care to deliver the solution into the bottom of the flask, gently swirl to mix, and insert the stopper. Twenty minutes, accurately timed, after the addition of the *Chlorinating solution*, add 2.0 mL of *Acidified phenol*, washing down the neck of the flask, swirl, insert the stopper, and allow to stand for 5 minutes, accurately timed, then add 1.0 mL of *Potassium iodide solution*, swirl, immediately add dehydrated alcohol to volume, and mix. [NOTE—Dilute immediately with dehydrated alcohol to volume after the introduction of the *Potassium iodide solution.*] After 30 minutes, accurately timed, concomitantly determine the absorbances of the treated *Standard preparation* and *Test preparation* in 1-cm cells at the wavelength of maximum absorbance at about 358 nm, with a suitable spectrophotometer, against the reagent blank. Calculate the percentage of dissolution of the Tablet by the formula:

$$(22.5 \times 10^5)(C/WV)(A_U/A_S),$$

in which C is the concentration, in mg per mL, of USP Meprobamate RS in the *Standard preparation*, W is the labeled quantity, in mg, of meprobamate in the Tablet, V is the volume, in mL, of the aliquot of *Dissolution Medium* used in the *Test preparation*, and A_U and A_S are the absorbances of the solutions from the *Test preparation* and the *Standard preparation*, respectively.

Tolerances—Not less than 60% (Q) of the labeled amount of $C_9H_{18}N_2O_4$ is dissolved in 30 minutes.

Uniformity of dosage units ⟨905⟩: meet the requirements.

Assay—

Borate buffer—Dissolve 3.1 g of boric acid and 3.7 g of potassium chloride in approximately 900 mL of water, adjust with 2 N sodium hydroxide to a pH of 10.5 ± 0.1, transfer to a 1000-mL volumetric flask, dilute with water to volume, and mix.

Chlorinating solution—Pipet 3 mL of sodium hypochlorite TS into a 100-mL volumetric flask, dilute with *Borate buffer* to volume, and mix. Prepare this solution fresh, and store in light-resistant containers.

Acidified phenol—Place 5.0 g of phenol in a 1000-mL volumetric flask, dilute with 0.1 N hydrochloric acid to volume, and mix.

Potassium iodide solution—Place 1.0 g of potassium iodide in a 100-mL volumetric flask, dilute with water to volume, and mix. Prepare this solution fresh, and store in light-resistant containers.

Standard preparation—Dissolve about 50 mg, accurately weighed, of USP Meprobamate RS in 50 mL of alcohol in a 250-mL volumetric flask, dilute with water to volume, and mix.

Assay preparation—Weigh and finely powder not less than 20 Meprobamate Tablets. Weigh accurately a portion of the powder, equivalent to about 50 mg of meprobamate, transfer to a 250-mL volumetric flask, add 50 mL of alcohol, stir for 1 minute, dilute with water to volume, mix, and filter.

Procedure—Pipet 1 mL of *Assay preparation* into a 50-mL volumetric flask. Into a similar flask pipet 1 mL of a 1 in 5 solution of alcohol in water to provide a reagent blank, and into a third 50-mL volumetric flask pipet 1 mL of *Standard preparation*. Treat each flask as follows: Add by pipet 2 mL of *Chlorinating solution*, taking care to deliver the solution into the bottom of the flask, gently swirl to mix, and insert the stopper. Twenty minutes, accurately timed, after the addition of the *Chlorinating solution*, add 2.0 mL of *Acidified phenol*, washing down the neck of the flask, swirl, insert the stopper, and allow to stand for 5 minutes, accurately timed, then add 1.0 mL of *Potassium iodide solution*, swirl, immediately add dehydrated alcohol to volume, and mix. [NOTE—Dilute immediately with dehydrated alcohol to volume after the introduction of the *Potassium iodide solution*.] After 30 minutes, accurately timed, concomitantly determine the absorbances of the treated *Standard preparation* and *Assay preparation* in 1-cm cells at the wavelength of maximum absorbance at about 358 nm, with a suitable spectrophotometer, against the reagent blank. Calculate the quantity, in mg, of $C_9H_{18}N_2O_4$ in the portion of Tablets taken by the formula:

$$W(A_U/A_S),$$

in which W is the weight, in mg, of USP Meprobamate RS used in the *Standard preparation*, and A_U and A_S are the absorbances of the solutions from the *Assay preparation* and *Standard preparation*, respectively.

Meprylcaine Hydrochloride

CH3 · HCl
COOCH2CNHCH2CH2CH3
CH3

$C_{14}H_{21}NO_2.HCl$ 271.79

1-Propanol, 2-methyl-2-(propylamino)-, benzoate (ester), hydrochloride.

2-Methyl-2-(propylamino)-1-propanol benzoate (ester) hydrochloride [*956-03-6*].

» Meprylcaine Hydrochloride contains not less than 98.5 percent and not more than 101.5 percent of $C_{14}H_{21}NO_2.HCl$, calculated on the dried basis.

Packaging and storage—Preserve in well-closed containers.

Reference standard—*USP Meprylcaine Hydrochloride Reference Standard*—Dry in a suitable vacuum drying tube, using phosphorus pentoxide as the desiccant, for 6 hours before using.

Identification—

A: The infrared absorption spectrum of a potassium bromide dispersion of it, previously dried, exhibits maxima only at the same wavelengths as that of a similar preparation of USP Meprylcaine Hydrochloride RS.

B: The ultraviolet absorption spectrum of a 1 in 10,000 solution in 0.1 N hydrochloric acid exhibits maxima and minima at the same wavelengths as that of a similar solution of USP Meprylcaine Hydrochloride RS, concomitantly measured, and the respective absorptivities, calculated on the dried basis, at the wavelength of maximum absorbance at about 274 nm do not differ by more than 3.0%.

C: A solution (1 in 100) responds to the tests for *Chloride* ⟨191⟩.

D: Dissolve about 100 mg in 10 mL of water, heat almost to boiling, and add, with stirring, 1 mL of a saturated solution of trinitrophenol in 20 percent alcohol. Cool slowly, collect the precipitate on a filter, wash with a few small portions of water, and dry in vacuum over phosphorus pentoxide for 18 hours: the picrate so obtained melts between 196° and 200° with decomposition (see *Melting Range or Temperature* ⟨741⟩). [Caution—Picrates may explode.]

Melting range ⟨741⟩: between 150° and 153°.

Loss on drying ⟨731⟩—Dry it in a suitable vacuum drying tube, using phosphorus pentoxide as the desiccant, for 6 hours: it loses not more than 0.5% of its weight.

Residue on ignition ⟨281⟩: not more than 0.1%.

Assay—Transfer about 150 mg of Meprylcaine Hydrochloride, accurately weighed, to a separator, dissolve in 20 mL of water, add 3 mL of 6 N ammonium hydroxide, and completely extract the meprylcaine with six 20-mL portions of isooctane. Filter the isooctane extracts into a beaker, wash the filter with isooctane, and evaporate the combined filtrate and washing in a current of warm air nearly to dryness. To the residue add 5 mL of neutralized alcohol and 40.0 mL of 0.02 N sulfuric acid VS, and heat on a steam bath until the odor of isooctane no longer is perceptible. Cool, add 2 drops of methyl red TS, and titrate with 0.02 N sodium hydroxide VS (see *Residual Titrations* under *Titrimetry* ⟨541⟩). Each mL of 0.02 N sulfuric acid is equivalent to 5.436 mg of $C_{14}H_{21}NO_2.HCl$.

Meprylcaine Hydrochloride and Epinephrine Injection

» Meprylcaine Hydrochloride and Epinephrine Injection is a sterile solution of Meprylcaine Hydrochloride and Epinephrine in Water for Injection. It contains not less than 90.0 percent and not more than 110.0 percent of the labeled amount of meprylcaine hydrochloride ($C_{14}H_{21}NO_2.HCl$) and not less than 95.0 percent and not more than 110.0 percent of the labeled amount of epinephrine ($C_9H_{13}NO_3$).

Packaging and storage—Preserve in single-dose containers, preferably of Type I glass, protected from light.

Reference standard—*USP Epinephrine Bitartrate Reference Standard*—Dry in vacuum over silica gel for 3 hours before using.

Identification—It responds to *Identification test D* under *Meprylcaine Hydrochloride*.

Other requirements—It meets the requirements under *Injections* ⟨1⟩.

Assay for meprylcaine hydrochloride—Transfer to a separator an accurately measured volume of Meprylcaine Hydrochloride and Epinephrine Injection, equivalent to about 150 mg of meprylcaine hydrochloride, and proceed as directed in the *Assay* under *Meprylcaine Hydrochloride*, beginning with "add 3 mL of 6 *N* ammonium hydroxide."

Assay for epinephrine—Proceed with Meprylcaine Hydrochloride and Epinephrine Injection as directed under *Epinephrine Assay* ⟨391⟩.

Mercaptopurine

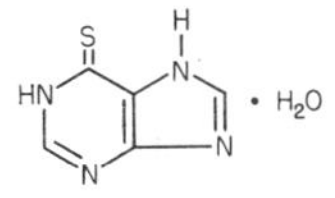

$C_5H_4N_4S.H_2O$ 170.19
6*H*-Purine-6-thione, 1,7-dihydro-, monohydrate.
Purine-6-thiol monohydrate [*6112-76-1*].
Anhydrous 152.17 [*50-44-2*].

» Mercaptopurine contains not less than 97.0 percent and not more than 102.0 percent of $C_5H_4N_4S$, calculated on the anhydrous basis.

Packaging and storage—Preserve in well-closed containers.

Reference standard—*USP Mercaptopurine Reference Standard*—Do not dry; determine the *Water* content by *Method I* ⟨921⟩ before using.

Identification—

A: The ultraviolet absorption spectrum of a 1 in 200,000 solution in 0.1 *N* hydrochloric acid exhibits maxima at the same wavelengths as that of a similar solution of USP Mercaptopurine RS, concomitantly measured, and the respective absorptivities, calculated on the anhydrous basis, at the wavelength of maximum absorbance at about 325 nm, do not differ by more than 3.0%. The ratio A_{255}/A_{325} does not exceed 0.06.

B: To a solution of 600 mg in 6 mL of sodium hydroxide solution (1 in 33) add slowly, with shaking, 0.5 mL of methyl iodide. Allow the mixture to stand for 2 hours at room temperature, cool in an ice bath, and adjust with acetic acid to a pH of about 5. Collect the crystals, and recrystallize from hot water: the methylmercaptopurine trihydrate so obtained, dried at 120° for 30 minutes, melts between 218° and 222°, with decomposition.

Water, *Method I* ⟨921⟩: not more than 12.0%.

Phosphorus—Digest 200 mg with 2 mL of 15 *N* sulfuric acid in a large test tube, periodically adding nitric acid, dropwise and with caution. Continue heating until practically all of the liquid has evaporated and the residue is colorless. Transfer the residue, with the aid of small portions of water, to a 25-mL volumetric flask, add 1 mL of 15 *N* sulfuric acid, 0.5 mL of nitric acid, 0.75 mL of ammonium molybdate TS, and 1 mL of aminonaphtholsulfonic acid TS, then dilute with water to volume, and mix. Allow to stand for 5 minutes, and determine the absorbance of this solution at 750 nm, with a suitable spectrophotometer, using as a blank a solution of the same quantities of the same reagents, prepared in the same manner. The absorbance of this solution is not greater than that of a solution obtained by treating as directed above, beginning with "add 1 mL of 15 *N* sulfuric acid," a 2-mL portion of a standard phosphate solution prepared to contain, in each mL, 43.96 µg of dried monobasic potassium phosphate, equivalent to 10 µg of P (0.010%).

Assay—Dissolve about 300 mg of Mercaptopurine, accurately weighed, in 80 mL of dimethylformamide. Add 5 drops of a 1 in 100 solution of thymol blue in dimethylformamide, and titrate with 0.1 *N* sodium methoxide VS, using a magnetic stirrer and taking precautions against absorption of atmospheric carbon dioxide. Perform a blank determination, and make any necessary correction. Each mL of 0.1 *N* sodium methoxide is equivalent to 15.22 mg of $C_5H_4N_4S$.

Mercaptopurine Tablets

» Mercaptopurine Tablets contain not less than 93.0 percent and not more than 110.0 percent of the labeled amount of $C_5H_4N_4S.H_2O$.

Packaging and storage—Preserve in well-closed containers.

Reference standard—*USP Mercaptopurine Reference Standard*—Do not dry; determine the *Water* content by *Method I* ⟨921⟩ before using.

Identification—

A: The ultraviolet absorption spectrum of the solution of Tablets employed for measurement of absorbance in the *Assay* exhibits a maximum at 325 ± 2 nm, and the ratio A_{255}/A_{325} does not exceed 0.09.

B: Triturate a quantity of finely powdered Tablets, equivalent to about 600 mg of mercaptopurine, with three 25-mL portions of hot alcohol. Filter the hot alcohol extracts, and evaporate the filtrate on a steam bath to dryness. Add to the residue 5 mL of sodium hydroxide solution (1 in 33), agitate well, and filter: the clear filtrate so obtained responds to *Identification test B* under *Mercaptopurine*.

Disintegration ⟨701⟩: 30 minutes.

Uniformity of dosage units ⟨905⟩: meet the requirements.

Assay—Weigh and finely powder not less than 20 Mercaptopurine Tablets. Weigh accurately a portion of the powder, equivalent to about 50 mg of mercaptopurine, and transfer to a 100-mL volumetric flask. Add 20 mL of water and 1.5 mL of 1 *N* sodium hydroxide, and swirl for not more than 5 minutes. Dilute with water to volume, mix, and filter, discarding the first 20 mL of the filtrate. Dilute an accurately measured portion of the filtrate quantitatively and stepwise with 0.1 *N* hydrochloric acid to give a final concentration of about 5 µg per mL. Dissolve an accurately weighed quantity of USP Mercaptopurine RS in a mixture of 10 mL of water and 1 mL of 1 *N* sodium hydroxide contained in a 100-mL volumetric flask, dilute with water to volume, and mix. Dilute an aliquot of this solution quantitatively and stepwise with 0.1 *N* hydrochloric acid to obtain a Standard solution having a known concentration of about 5 µg per mL. Concomitantly determine the absorbances of both solutions in 1-cm cells at the wavelength of maximum absorbance at about 325 nm, with a suitable spectrophotometer, using 0.1 *N* hydrochloric acid as the blank. Calculate the quantity, in mg, of $C_5H_4N_4S.H_2O$ in the portion of Tablets taken by the formula:

$$(170.19/152.17)10C(A_U/A_S),$$

in which 170.19 and 152.17 are the molecular weights of mercaptopurine monohydrate and anhydrous mercaptopurine, respectively, *C* is the concentration, in µg per mL, of USP Mercaptopurine RS in the Standard solution, and A_U and A_S are the absorbances of the solution from Mercaptopurine Tablets and the Standard solution, respectively.

Ammoniated Mercury

$Hg(NH_2)Cl$ 252.07
Mercury amide chloride.
Mercury amide chloride [*10124-48-8*].

» Ammoniated Mercury contains not less than 98.0 percent and not more than 100.5 percent of $Hg(NH_2)Cl$.

Packaging and storage—Preserve in well-closed, light-resistant containers.

Identification—

A: A 0.1-g portion is soluble, with the evolution of ammonia, in a cold solution of 1 g of sodium thiosulfate in 2 mL of water. When this solution is heated gently, a rust-colored mixture is formed, from which a red precipitate is obtained on centrifugation. If the solution is strongly heated, a black mixture forms.

B: When heated with 1 *N* sodium hydroxide, it becomes yellow, and ammonia is evolved.

C: A solution in warm acetic acid yields with potassium iodide TS a red precipitate, which is soluble in an excess of the reagent. The solution yields a white precipitate with silver nitrate TS.

Residue on ignition ⟨281⟩: not more than 0.2%.

Mercurous compounds—Dissolve 2.5 g in 25 mL of warm hydrochloric acid, filter through a tared filtering crucible, wash with water, and dry at 60° to constant weight: the weight of the residue does not exceed 5 mg (0.2%).

Assay—Mix about 0.25 g of Ammoniated Mercury, accurately weighed, with about 10 mL of water. Add 3 g of potassium iodide, mix occasionally until dissolved, and add about 40 mL of water. Add methyl red TS, and titrate with 0.1 *N* hydrochloric acid VS. Perform a blank determination, and make any necessary correction. Each mL of 0.1 *N* hydrochloric acid is equivalent to 12.60 mg of $Hg(NH_2)Cl$.

Ammoniated Mercury Ointment

» Ammoniated Mercury Ointment contains not less than 90.0 percent and not more than 110.0 percent of the labeled amount of $Hg(NH_2)Cl$ in a suitable oleaginous ointment base.

Packaging and storage—Preserve in collapsible tubes or in well-closed, light-resistant containers.

Identification—Transfer an amount of Ointment, equivalent to about 250 mg of ammoniated mercury, to a fine-porosity filtering crucible, and warm to melt the ointment base. Add 20 mL of chloroform, stir to dissolve the ointment base, and draw the chloroform solution through the filter with suction. Wash the residue on the filter successively with two 5-mL portions of chloroform, draw each chloroform washing through the filter, and discard the chloroform solution and washings: the residue responds to the *Identification tests* under *Ammoniated Mercury.*

Minimum fill ⟨755⟩: meets the requirements.

Assay—Transfer to a separator an accurately weighed amount of Ammoniated Mercury Ointment, equivalent to about 150 mg of ammoniated mercury. Warm it slightly to soften the Ointment, and while rotating, add 75 mL of ether, and then shake the mixture until the ointment base is dissolved. Add 10 mL of dilute hydrochloric acid (1 in 2), and shake vigorously until all of the ammoniated mercury has dissolved. Filter into a 250-mL beaker the water layer that separates, and wash the remaining ether solution with several 10-mL portions of dilute hydrochloric acid (1 in 10), adding the washings to the main extract. Dilute with water to about 150 mL, add 5 mL of hydrochloric acid, and pass hydrogen sulfide gas through the solution. Collect the precipitate in a tared filtering crucible, and wash it successively with water and two 10-mL portions of alcohol. Then, without suction, wash with two 10-mL portions of carbon tetrachloride. Dry the crucible and contents at 105° for 2 hours. The weight of mercuric sulfide so obtained, multiplied by 1.083, represents the weight of $Hg(NH_2)Cl$ in the portion of Ointment taken.

Ammoniated Mercury Ophthalmic Ointment

» Ammoniated Mercury Ophthalmic Ointment is a sterile ointment containing not less than 90.0 percent and not more than 110.0 percent of the labeled amount of $Hg(NH_2)Cl$ in a suitable oleaginous ointment base.

Packaging and storage—Preserve in collapsible ophthalmic ointment tubes.

Sterility—It meets the requirements under *Sterility Tests* ⟨71⟩.

Metal particles—It meets the requirements of the test for *Metal Particles in Ophthalmic Ointments* ⟨751⟩.

Other requirements—It responds to the *Identification tests*, and meets the requirements for *Minimum fill* ⟨755⟩ and *Assay* under *Ammoniated Mercury Ointment.*

Mesoridazine Besylate

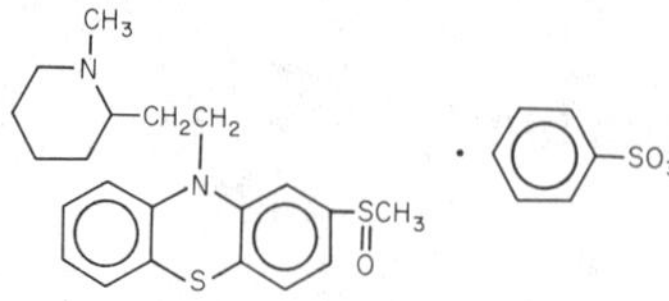

$C_{21}H_{26}N_2OS_2 . C_6H_6O_3S$ 544.74

10*H*-Phenothiazine, 10-[2-(1-methyl-2-piperidinyl)ethyl]-2-(methylsulfinyl)-, monobenzenesulfonate.

10-[2-(1-Methyl-2-piperidyl)ethyl]-2-(methylsulfinyl)-phenothiazine monobenzenesulfonate [*32672-69-8*].

» Mesoridazine Besylate contains not less than 98.0 percent and not more than 102.0 percent of $C_{21}H_{26}N_2OS_2 . C_6H_6O_3S$, calculated on the dried basis.

Packaging and storage—Preserve in tight, light-resistant containers.

Reference standard—*USP Mesoridazine Besylate Reference Standard*—Dry at 105° for 4 hours before using.

NOTE—Throughout the following procedures, protect test or assay specimens, the Reference Standard, and solutions containing them, by conducting the procedures without delay, under subdued light, or using low-actinic glassware.

Identification—

A: The infrared absorption spectrum of a mineral oil dispersion of it, previously dried, exhibits maxima only at the same wavelengths as that of a similar preparation of USP Mesoridazine Besylate RS.

B: The ultraviolet absorption spectrum of a 1 in 100,000 solution in methanol exhibits maxima and minima at the same wavelengths as that of a similar solution of USP Mesoridazine Besylate RS, concomitantly measured, and the respective absorptivities, calculated on the dried basis, at the wavelength of maximum absorbance at about 263 nm do not differ by more than 3.0%.

pH ⟨791⟩: between 4.2 and 5.7, in a freshly prepared solution (1 in 100).

Loss on drying ⟨731⟩—Dry it at 105° for 4 hours: it loses not more than 0.5% of its weight.

Residue on ignition ⟨281⟩: not more than 0.2%.

Heavy metals, *Method II* ⟨231⟩: 0.002%.

Selenium ⟨291⟩—The absorbance of the solution from the *Test Solution,* prepared with 100 mg of Mesoridazine Besylate and 100 mg of magnesium oxide, is not greater than one-half that from the *Standard Solution* (0.003%).

Assay—Dissolve about 150 mg of Mesoridazine Besylate, accurately weighed, in 70 mL of acetic anhydride, and titrate with 0.1 *N* perchloric acid VS, determining the end-point potentiometrically. Perform a blank determination, and make any necessary correction. Each mL of 0.1 *N* perchloric acid is equivalent to 27.24 mg of $C_{21}H_{26}N_2OS_2 . C_6H_6O_3S$.

Mesoridazine Besylate Injection

» Mesoridazine Besylate Injection is a sterile solution of Mesoridazine Besylate in Water for Injection. It contains mesoridazine besylate ($C_{21}H_{26}N_2OS_2 . C_6H_6O_3S$) equivalent to not less than 90.0 percent and not more than 110.0 percent of the labeled amount of mesoridazine ($C_{21}H_{26}N_2OS_2$).

Packaging and storage—Preserve in single-dose containers, preferably of Type I glass, protected from light.

Reference standard—*USP Mesoridazine Besylate Reference Standard*—Dry at 105° for 4 hours before using.

NOTE—Throughout the following procedures, protect test or assay specimens, the Reference Standard, and solutions containing them, by conducting the procedures without delay, under subdued light, or using low-actinic glassware.

Identification—Dilute a volume of Injection, equivalent to about 50 mg of mesoridazine besylate, with 0.01 *N* hydrochloric acid to 25 mL, and proceed as directed under *Identification—Organic Nitrogenous Bases* ⟨181⟩, beginning with "Transfer the liquid to a separator": the Injection meets the requirements of the test.

pH ⟨791⟩: between 4.0 and 5.0.

Other requirements—It meets the requirements under *Injections* ⟨1⟩.

Assay—[NOTE—Conduct this procedure with minimum exposure to light.] Proceed with Mesoridazine Besylate Injection as directed under *Salts of Organic Nitrogenous Bases* ⟨501⟩, except to use 1.0 mL each of the *Standard Preparation* and the *Assay Preparation* in the *Procedure*, and determine the absorbances at the wavelength of maximum absorbance at about 262 nm. Calculate the quantity, in mg, of $C_{21}H_{26}N_2OS_2$ in each mL of the Injection taken by the formula:

$$(386.57/544.74)(0.05C/V)(A_U/A_S),$$

in which 386.57 and 544.74 are the molecular weights of mesoridazine and mesoridazine besylate, respectively, C is the concentration, in µg per mL, of USP Mesoridazine Besylate RS in the *Standard Preparation*, and V is the volume, in mL, of Injection taken.

Mesoridazine Besylate Oral Solution

» Mesoridazine Besylate Oral Solution contains not less than 90.0 percent and not more than 110.0 percent of the labeled amount of mesoridazine ($C_{21}H_{26}N_2OS_2$).

Packaging and storage—Preserve in tight, light-resistant containers at a temperature not exceeding 25°.

Labeling—Label it to indicate that it is to be diluted to the appropriate strength with water or other suitable fluid prior to administration.

Reference standard—*USP Mesoridazine Besylate Reference Standard*—Dry at 105° for 4 hours before using.

NOTE—Throughout the following procedures, protect test or assay specimens, the Reference Standard, and solutions containing them, by conducting the procedures without delay, under subdued light, or using low-actinic glassware.

Identification—[NOTE—Conduct this test without exposure to daylight and with the minimum necessary exposure to artificial light.]

Standard preparation—Prepare a solution of USP Mesoridazine Besylate RS in methanol to contain 14 mg per mL.

Test preparation—Transfer 4.0 mL of the Oral Solution into a separator, add 6 mL of 1 *N* sodium hydroxide and 10 mL of chloroform, shake for 2 minutes, and filter the chloroform layer through anhydrous sodium sulfate into a small, glass-stoppered conical flask.

Developing solvent—To a separator add benzene, alcohol, and ammonium hydroxide (10:2:1), shake, and allow the layers to separate. Use the upper layer.

Procedure—Into a suitable chromatographic chamber arranged for thin-layer chromatography, place a volume of *Developing solvent* sufficient to develop the chromatogram, and allow to equilibrate. On a suitable thin-layer chromatographic plate coated with a 0.25-mm layer of chromatographic silica gel mixture, apply separate 10-µL portions of the *Test preparation* and the *Standard preparation*. Allow the spots to dry, and develop the chromatogram until the solvent front has moved about three-fourths of the length of the plate. Remove the plate from the developing chamber, mark the solvent front, and allow the solvent to evaporate in a fume hood. Spray the plate with a solution prepared by diluting 15 mL of perchloric acid with water to 100 mL, and heat at 80° for 2 minutes: the principal spot obtained from the *Test preparation* corresponds in R_f value and color to that of the *Standard preparation*.

Alcohol content, *Method I* ⟨611⟩: between 0.25% and 1.0% of C_2H_5OH.

Assay—[NOTE—Conduct this procedure with the minimum necessary exposure to light.]

Standard preparation—Transfer about 14 mg of USP Mesoridazine Besylate RS, accurately weighed, to a 125-mL separator containing 30 mL of water. Render the solution alkaline with 10 mL of 1 *N* sodium hydroxide, and extract with three 30-mL portions of chloroform. Filter the extracts through anhydrous sodium sulfate into a 100-mL volumetric flask. Rinse the filter with small portions of chloroform, collecting the rinsings in the volumetric flask, dilute with chloroform to volume, and mix. Dilute 10.0 mL of this solution with chloroform to 100.0 mL, and mix.

Assay preparation—Pipet a volume of Mesoridazine Besylate Oral Solution, equivalent to about 100 mg of mesoridazine, into a separator containing 30 mL of water. Proceed as directed under *Standard preparation*, beginning with "Render the solution alkaline." Pipet 10.0 mL of this solution into a third 100-mL volumetric flask, dilute with chloroform to volume, and mix.

Procedure—Concomitantly determine the absorbances of the *Standard preparation* and the *Assay preparation* in 1-cm cells at the wavelength of maximum absorbance at about 267 nm, with a suitable spectrophotometer, using chloroform as the blank. Calculate the quantity, in mg, of $C_{21}H_{26}N_2OS_2$ in each mL of the Oral Solution taken by the formula:

$$(386.57/544.74)(10C/V)(A_U/A_S),$$

in which 386.57 and 544.74 are the molecular weights of mesoridazine and mesoridazine besylate, respectively, C is the concentration, in µg per mL, of USP Mesoridazine Besylate RS in the *Standard preparation*, V is the volume, in mL, of Oral Solution taken, and A_U and A_S are the absorbances of the *Assay preparation* and the *Standard preparation*, respectively.

Mesoridazine Besylate Tablets

» Mesoridazine Besylate Tablets contain mesoridazine besylate ($C_{21}H_{26}N_2OS_2 . C_6H_6O_3S$) equivalent to not less than 90.0 percent and not more than 110.0 percent of the labeled amount of mesoridazine ($C_{21}H_{26}N_2OS_2$).

Packaging and storage—Preserve in well-closed, light-resistant containers. Preserve Tablets having an opaque coating in well-closed containers.

[NOTE—Throughout the following procedures, protect test or assay specimens, the Reference Standard, and solutions containing them, by conducting the procedures without delay, under subdued light, or using low-actinic glassware.

Reference standard—*USP Mesoridazine Besylate Reference Standard*—Dry at 105° for 4 hours before using.

Identification—Tablets meet the requirements under *Identification—Organic Nitrogenous Bases* ⟨181⟩.

Dissolution ⟨711⟩—

Medium: 0.1 *N* hydrochloric acid; 1000 mL.

Apparatus 2: 100 rpm.

Time: 60 minutes.

Procedure—Determine the amount of mesoridazine ($C_{21}H_{26}N_2OS_2$) dissolved from ultraviolet absorbances at the wavelength of maximum absorbance at about 261 nm of filtered portions of the solution under test, suitably diluted with *Dissolution*

Medium, if necessary, in comparison with a Standard solution having a known concentration of USP Mesoridazine Besylate RS in the same medium. [NOTE—A volume of methanol not exceeding 1% of the final total volume may be used to prepare the Standard solution.]

Tolerances—Not less than 80% (*Q*) of the labeled amount of $C_{21}H_{26}N_2OS_2$ is dissolved in 60 minutes.

Uniformity of dosage units ⟨905⟩: meet the requirements.

Assay—

Mobile phase—Prepare a filtered and degassed mixture of acetonitrile, water, and triethylamine (850:150:1). Make adjustments if necessary (see *System Suitability* under *Chromatography* ⟨621⟩).

Standard preparation—Dissolve an accurately weighed quantity of USP Mesoridazine Besylate RS in methanol, and dilute quantitatively, and stepwise if necessary, with methanol to obtain a solution having a known concentration of about 0.35 mg per mL.

Assay preparation—Weigh and finely powder not less than 20 Mesoridazine Besylate Tablets. Transfer an accurately weighed portion of the powder, equivalent to about 50 mg of mesoridazine, to a 200-mL volumetric flask. Add about 150 mL of methanol, shake by mechanical means for about 15 minutes, dilute with methanol to volume, and mix. Sonicate for 30 minutes, and allow dispersed material to settle. Filter through a 0.25-μm disk, discarding the first 20 mL of the filtrate.

System suitability preparation—Dissolve a suitable quantity of Thioridazine Hydrochloride in a portion of the *Standard preparation,* and mix to obtain a solution containing 0.025 mg of thioridazine hydrochloride per mL.

Chromatographic system (see *Chromatography* ⟨621⟩)—The liquid chromatograph is equipped with a 265-nm detector and a 4.6-mm × 25-cm column that contains packing L1. The flow rate is about 2.5 mL per minute. Chromatograph the *Standard preparation* and the *System suitability preparation,* and record the peak areas as directed under *Procedure:* the column efficiency determined from the analyte peak is not less than 750 theoretical plates, the resolution, *R,* between the mesoridazine besylate and thioridazine hydrochloride peaks is not less than 1.0, and the relative standard deviation for replicate injections of the *Standard preparation* is not more than 2.0%.

Procedure—Separately inject equal volumes (about 10 μL) of the *Standard preparation* and the *Assay preparation* into the chromatograph, record the chromatograms, and measure the areas for the major peaks. Calculate the quantity, in mg, of mesoridazine ($C_{21}H_{26}N_2OS_2$) in the portion of Tablets taken by the formula:

$$(386.57/544.74)(200C)(r_U/r_S),$$

in which 386.57 and 544.74 are the molecular weights of mesoridazine and mesoridazine besylate, respectively, *C* is the concentration, in mg per mL, of USP Mesoridazine Besylate RS in the *Standard preparation,* and r_U and r_S are the peak areas obtained from the *Assay preparation* and the *Standard preparation,* respectively.

Mestranol

$C_{21}H_{26}O_2$ 310.44

19-Norpregna-1,3,5(10)-trien-20-yn-17-ol, 3-methoxy-, (17α)-.
3-Methoxy-19-nor-17α-pregna-1,3,5(10)-trien-20-yn-17-ol [*72-33-3*].

» Mestranol contains not less than 97.0 percent and not more than 102.0 percent of $C_{21}H_{26}O_2$, calculated on the dried basis.

Packaging and storage—Preserve in well-closed, light-resistant containers.

Reference standard—*USP Mestranol Reference Standard*—Dry at 105° for 3 hours before using.

Identification—

A: The infrared absorption spectrum of a potassium bromide dispersion of it, previously dried, exhibits maxima only at the same wavelengths as that of a similar preparation of USP Mestranol RS.

B: The ultraviolet absorption spectrum of a 1 in 10,000 solution in methanol exhibits maxima and minima at the same wavelengths as that of a similar solution of USP Mestranol RS, concomitantly measured.

C: Prepare a solution in chloroform to contain 1 mg of mestranol per mL. Apply 10 μL of this solution and 10 μL of a solution of USP Mestranol RS in chloroform containing 1 mg per mL on a line parallel to and about 2.5 cm from the bottom edge of a thin-layer chromatographic plate (see *Chromatography* ⟨621⟩) coated with a 0.25-mm layer of chromatographic silica gel mixture. Place the plate in a developing chamber containing and equilibrated with a mixture of 29 volumes of chloroform and 1 volume of dehydrated alcohol. Develop the chromatogram until the solvent front has moved about 15 cm above the line of application. Remove the plate, allow the solvent to evaporate, and spray with *Methanol–sulfuric acid* prepared as described in the *Assay.* Heat the plate in an oven at 105° for 5 minutes, and observe under long-wavelength ultraviolet light: the R_f value of the principal spot obtained from the solution under test corresponds to that obtained with the Standard solution.

Melting range ⟨741⟩: between 146° and 154°, but the range between beginning and end of melting does not exceed 4°.

Specific rotation ⟨781⟩: between +2° and +8°, determined in a solution in dioxane containing 200 mg in each 10 mL, the test specimen previously having been dried.

Loss on drying ⟨731⟩—Dry it at 105° for 3 hours: it loses not more than 1.0% of its weight.

Assay—

Methanol–sulfuric acid—Pipet 30 mL of methanol into a 100-mL volumetric flask contained in an ice bath. Add slowly and cautiously, and with continuous stirring, about 65 mL of sulfuric acid, taking care that the temperature remains below 15°. Allow the solution to warm to room temperature, and dilute with sulfuric acid to 100 mL.

Standard preparation—Dissolve a suitable quantity of USP Mestranol RS, accurately weighed, in chloroform, and dilute quantitatively and stepwise with chloroform to obtain a solution having a known concentration of about 5 μg per mL.

Assay preparation—Weigh accurately about 20 mg of Mestranol, dissolve in chloroform to make 200.0 mL, and mix. Pipet 5 mL of this solution into a 100-mL volumetric flask, add chloroform to volume, and mix.

Procedure—Pipet 4 mL each of the *Standard preparation* and the *Assay preparation* into separate glass-stoppered, 25-mL conical flasks. Evaporate the solutions under a slow stream of air, without the aid of heat, to dryness. Dissolve the residue in 0.3 mL of methanol. Place the flasks in a water bath maintained at a temperature of 25°, and pipet into each, with constant swirling, 10 mL of *Methanol–sulfuric acid.* Insert the stoppers in the flasks. At 6 minutes, accurately timed, after the addition of the *Methanol–sulfuric acid,* concomitantly determine the absorbances of the solutions obtained from the *Assay preparation* and the *Standard preparation* at the wavelength of maximum absorbance at about 545 nm, with a suitable spectrophotometer, using *Methanol–sulfuric acid* as the blank. Calculate the quantity, in mg, of $C_{21}H_{26}O_2$ in the Mestranol taken by the formula:

$$4C(A_U/A_S),$$

in which *C* is the concentration, in μg per mL, of USP Mestranol RS in the *Standard preparation,* and A_U and A_S are the absorbances of the solutions from the *Assay preparation* and the *Standard preparation,* respectively.

Mestranol Tablets, Ethynodiol Diacetate and—*see* Ethynodiol Diacetate and Mestranol Tablets

Mestranol Tablets, Norethindrone and—*see* Norethindrone and Mestranol Tablets

Methacrylic Acid Copolymer—*see* Methacrylic Acid Copolymer NF

Metaproterenol Sulfate

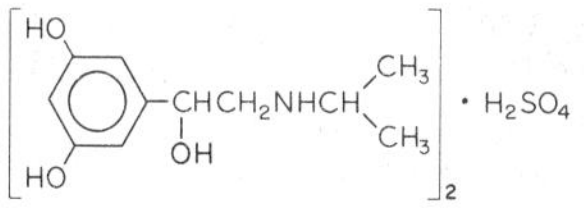

$(C_{11}H_{17}NO_3)_2 \cdot H_2SO_4$ 520.59

1,3-Benzenediol, 5-[1-hydroxy-2-[(1-methylethyl)amino]ethyl]-, sulfate (2:1) (salt).

3,5-Dihydroxy-α-[(isopropylamino)methyl]benzyl alcohol sulfate (2:1) [*5874-97-5*].

» Metaproterenol Sulfate contains not less than 98.0 percent and not more than 102.0 percent of $(C_{11}H_{17}NO_3)_2 \cdot H_2SO_4$, calculated on the anhydrous, isopropyl alcohol–free, and methanol-free basis.

Packaging and storage—Preserve in tight, light-resistant containers.

Reference standard—*USP Metaproterenol Sulfate Reference Standard*—Dry at 105° for 1 hour before using.

Identification—

A: The infrared absorption spectrum of a potassium bromide dispersion of it exhibits maxima and minima only at the same wavelengths as that of a similar preparation of USP Metaproterenol Sulfate RS.

B: To a solution of 10 mg in 1 mL of water add 1 drop of ferric chloride TS: a violet color is produced.

C: It responds to the tests for *Sulfate* ⟨191⟩.

D: The chromatogram of the *Assay preparation* obtained as directed in the *Assay* exhibits a major peak for metaproterenol, the retention time of which corresponds with that exhibited in the chromatogram of the *Standard preparation* obtained as directed in the *Assay*.

pH ⟨791⟩: between 4.0 and 5.5, in a solution containing 100 mg per mL.

Water, *Method I* ⟨921⟩: not more than 2.0%.

Residue on ignition ⟨281⟩: not more than 0.1%.

Heavy metals, *Method II* ⟨231⟩: 0.001%.

Iron ⟨241⟩—Dissolve 2.0 g in 45 mL of water, add 2 mL of hydrochloric acid, and mix: the limit is 5 ppm.

Metaproteronone—Its absorptivity (see *Spectrophotometry and Light-scattering* ⟨851⟩) at 328 nm, determined in a 0.01 *N* hydrochloric acid solution containing 9.0 mg per mL, is not more than 0.0079 (0.1%).

Isopropyl alcohol and methanol—

Isopropyl alcohol standard solution—Transfer about 0.3 g of isopropyl alcohol, accurately weighed, to a 100-mL volumetric flask containing about 10 mL of water, dilute with water to volume, and mix. Pipet 10 mL of the resulting solution into a 100-mL volumetric flask, add about 85 mL of pyridine, mix, and allow to stand for 1 hour. Dilute with pyridine to volume, and mix. Pipet 5 mL of this solution to a 50-mL volumetric flask, dilute with pyridine to volume, and mix. The solution so obtained contains about 30 µg of isopropyl alcohol per mL.

Methanol standard solution—Prepare as directed for *Isopropyl alcohol standard solution*, using about 0.1 g of methanol, accurately weighed. The resulting solution contains about 10 µg of methanol per mL.

Test preparation—Transfer about 1 g of Metaproterenol Sulfate, accurately weighed, to a 100-mL volumetric flask, dissolve in about 2 mL of water, dilute with pyridine to volume, and mix.

Chromatographic system—The gas chromatograph is equipped with a flame-ionization detector and contains a 2-m × 2-mm column packed with 0.1 percent liquid phase G25 on 80- to 100-mesh support S7. The injection port is maintained at a temperature of about 150°; the column is programmed for 2 minutes at 40°, to increase at a rate of about 15° per minute to 200°, and for 10 minutes at 200°; the detector is maintained at about 250°; and helium is used as the carrier gas at a flow rate of about 15 mL per minute.

Procedure—Inject equal volumes (about 2 µL) of the *Test preparation*, the *Isopropyl alcohol standard solution*, and the *Methanol standard solution* successively into the gas chromatograph. Measure the responses of the isopropyl alcohol peak and the methanol peak in each chromatogram. Determine the quantities, in mg, of isopropyl alcohol and methanol in the portion of Metaproterenol Sulfate taken by the formula:

$$0.1C(r_U / r_S),$$

in which C is the concentration, in µg per mL, of isopropyl alcohol or methanol in the *Isopropyl alcohol standard solution* or the *Methanol standard solution*, and r_U and r_S are the responses of the respective analytes in the *Test preparation* and of the corresponding *Isopropyl alcohol standard solution* or *Methanol standard solution:* not more than 0.3% of isopropyl alcohol and not more than 0.1% of methanol are found.

Assay—

Mobile phase—Dissolve 11.9 g of anhydrous dibasic sodium phosphate in water to make 1000 mL of solution, and mix (*Solution A*). Dissolve 9.1 g of monobasic potassium phosphate in water to make 1000 mL of solution, and mix (*Solution B*). Mix 735 mL of *Solution A* and 140 mL of *Solution B*, add 125 mL of methanol, and mix. Filter and degas this solution before use.

Standard preparation—Dissolve an accurately weighed quantity of USP Metaproterenol Sulfate RS in 0.01 *N* hydrochloric acid to obtain a solution having a known concentration of about 2 mg per mL.

Assay preparation—Transfer about 100 mg of Metaproterenol Sulfate, accurately weighed, to a 50-mL volumetric flask, dilute with 0.01 *N* hydrochloric acid to volume, and mix.

Chromatographic system (see *Chromatography* ⟨621⟩)—The liquid chromatograph is equipped with a 278-nm detector and a 4.6-mm × 5-cm guard column that contains packing L7 and a 4.6-mm × 25-cm analytical column that contains 10-µm packing L1. The flow rate is about 2 mL per minute. Chromatograph the *Standard preparation*, and record the peak responses as directed under *Procedure:* the column efficiency determined from the analyte peak is not less than 500 theoretical plates, the tailing factor for the analyte peak is not more than 3.0, and the relative standard deviation for replicate injections is not more than 2.0%.

Procedure—Separately inject equal volumes (about 10 µL) of the *Standard preparation* and the *Assay preparation* into the chromatograph, record the chromatograms, and measure the responses for the major peaks. Calculate the quantity, in mg, of $(C_{11}H_{17}NO_3)_2 \cdot H_2SO_4$ in the portion of Metaproterenol Sulfate taken by the formula:

$$50C(r_U / r_S),$$

in which C is the concentration, in mg per mL, of USP Metaproterenol Sulfate RS in the *Standard preparation*, and r_U and r_S are the peak responses from the *Assay preparation* and the *Standard preparation*, respectively.

Metaproterenol Sulfate Inhalation Aerosol

» Metaproterenol Sulfate Inhalation Aerosol is a suspension of microfine Metaproterenol Sulfate in fluorochlorohydrocarbon propellants in a pressurized container. It contains not less than 90.0 percent and not more than 110.0 percent of the labeled amount of $(C_{11}H_{17}NO_3)_2 \cdot H_2SO_4$, and it delivers not less than

75.0 percent and not more than 125.0 percent of the labeled amount of $(C_{11}H_{17}NO_3)_2.H_2SO_4$ per inhalation through an oral inhalation actuator.

Packaging and storage—Preserve in small, nonreactive, light-resistant aerosol containers equipped with metered-dose valves and provided with oral inhalation actuators.

Reference standard—*USP Metaproterenol Sulfate Reference Standard*—Dry at 105° for 1 hour before using.

Identification—The ultraviolet absorption spectrum of the solution from the *Assay preparation*, obtained as directed in the *Assay*, exhibits maxima and minima at the same wavelengths as that of the *Standard preparation* prepared as directed in the *Assay*.

Unit spray content—

Unit spray sampling apparatus—Prepare as directed under *Aerosols* ⟨601⟩.

Standard preparation—Using a suitable quantity of USP Metaproterenol Sulfate RS, accurately weighed, prepare a solution in 0.01 *N* hydrochloric acid to obtain a solution having a known concentration of 0.05 mg per mL.

Test preparation—Transfer 35.0 mL of 0.01 *N* hydrochloric acid to the collection chamber of the apparatus, and clamp the intake tube in a horizontal position. Fit a Metaproterenol Sulfate Inhalation Aerosol container with its oral inhalation actuator, and prime the unit by alternately shaking and firing it 10 times. Couple the unit to the intake tube of the apparatus, and adjust the air flow to a rate of 10 to 15 liters per minute. Remove the container, leaving the actuator in place. Shake the container, seat it in the actuator, and immediately deliver a single spray into the system. Deliver a total of 10 sprays in this manner. Remove the container and actuator from the assembly, clamp the intake tube in a vertical position, and scavenge the drug remaining in the intake system as follows: Disconnect the collection chamber, transfer the sample solution quantitatively to a 100-mL volumetric flask, and then wash the inside of the intake tube with 0.01 *N* hydrochloric acid, collecting the washings in the collection chamber placed a few cm below its normal position. Connect the collection chamber into the system again, and draw additional 0.01 *N* hydrochloric acid into it through the bubbler and intake tube. Recycle 0.01 *N* hydrochloric acid twice in this manner collecting the rinsings in the 100-mL volumetric flask, dilute with 0.01 *N* hydrochloric acid to volume, and mix.

Procedure—Transfer 20.0 mL portions of *Standard preparation*, *Test preparation*, and 0.01 *N* hydrochloric acid to serve as a blank to separate centrifuge tubes. Add 10.0 mL of chloroform to each, shake by mechanical means for 5 minutes, and separate the layers by centrifuging for 5 minutes. Determine the absorbances of the respective aqueous layers in 1-cm cells, at the wavelength of maximum absorbance at about 276 nm, with a suitable spectrophotometer, against the blank. Calculate the quantity, in mg, of $(C_{11}H_{17}NO_3)_2.H_2SO_4$ delivered per spray by the formula:

$$10C(A_U/A_S),$$

in which C is the concentration, in mg per mL, of USP Metaproterenol Sulfate RS in the *Standard preparation*, and A_U and A_S are the absorbances of the solutions from the *Test preparation* and the *Standard preparation*, respectively.

Particle size—Prime the valve of a Metaproterenol Sulfate Inhalation Aerosol container by alternately shaking and firing it several times and then actuate one measured spray onto a clean, dry microscope slide held 5 cm from the end of the oral inhalation actuator, perpendicular to the direction of the spray. Carefully rinse the slide with about 2 mL of chloroform, and allow to dry. Examine the slide under a microscope equipped with a calibrated ocular micrometer, using 450× magnification. Focus on the particles of 25 fields of view near the center of the test specimen pattern, and note the size of the great majority of individual particles: they are less than 5 µm along the longest axis. Record the number and size of all individual crystalline particles (not agglomerates) more than 10 µm in length measured along the longest axis: not more than 10 such particles are observed.

Water—Transfer the contents of a weighed container to the titration vessel by attaching the valve stem to an inlet tube. Weigh the empty container and determine the weight of the specimen taken. The water content, determined by *Method I* ⟨921⟩, is not more than 0.075%.

Other requirements—It meets the requirements for *Leak Testing* under *Aerosols* ⟨601⟩.

Assay—Cool an accurately weighed Metaproterenol Sulfate Inhalation Aerosol container for 10 minutes in a bath consisting of a mixture of acetone and solid carbon dioxide. Cut the valve from the aerosol container and allow the container to warm to room temperature. When most of the propellants have evaporated, transfer the residue in the container to a 250-mL separator with the aid of 30 mL of chloroform and 50 mL of 0.01 *N* hydrochloric acid. Reserve the valve and the empty container. Shake the separator for 1 minute and allow the phases to separate. Transfer the chloroform phase to a second 250-mL separator and the aqueous phase to a 250-mL volumetric flask. Wash the chloroform phase with two 50-mL portions of 0.01 *N* hydrochloric acid, add the washings to the 250-mL volumetric flask, dilute with 0.01 *N* hydrochloric acid to volume, and mix. Transfer an accurately measured volume of this stock solution, equivalent to about 10 mg of metaproterenol sulfate, to a 100-mL volumetric flask, dilute with 0.01 *N* hydrochloric acid to volume, and mix. Dissolve an accurately weighed quantity of USP Metaproterenol Sulfate RS in 0.01 *N* hydrochloric acid, and dilute quantitatively and stepwise with the same solvent to obtain a Standard solution having a known concentration of about 100 µg per mL. Concomitantly determine the absorbances of both solutions at the wavelength of maximum absorbance at about 276 nm, with a suitable spectrophotometer, using 0.01 *N* hydrochloric acid as the blank. Rinse the empty aerosol container and the valve with water and dry them at 105° for 10 minutes, allow to cool, and weigh. Subtract the weight thus obtained from the original weight of the Inhalation Aerosol container to obtain the weight of the Inhalation Aerosol taken. Calculate the quantity, in mg, of $(C_{11}H_{17}NO_3)_2.H_2SO_4$ in each mL of the Inhalation Aerosol taken by the formula:

$$25(C/V)(d/W)(A_U/A_S),$$

in which C is the concentration, in µg per mL, of USP Metaproterenol Sulfate RS in the Standard solution, V is the volume, in mL, of stock solution taken, W is the weight, in g, of the Inhalation Aerosol taken, and A_U and A_S are the absorbances of the solution from the Inhalation Aerosol and the Standard solution, respectively. [The density, d, is determined as follows: Weigh a known volume (v) of the Inhalation Aerosol in a suitable 5-mL gas-tight syringe equipped with a linear valve. Calibrate the volume of the syringe by filling to the 5-mL mark with dichlorotetrafluoroethane withdrawn from a plastic-coated glass vial sealed with a neoprene multiple-dose rubber stopper and an aluminum seal, using 1.456 g per mL as the density of the calibrating liquid. Maintain the dichlorotetrafluoroethane, the Inhalation Aerosol sample, and the syringe (protected from becoming wet) at 25° in a water bath. Obtain the sample, equivalent to the same volume as that obtained during the sampling procedure, from the Inhalation Aerosol by means of a sampling device consisting of a replaceable rubber septum engaged in the plate threads at one end of a threaded fitting, the opposite end of which contains a sharpened tube capable of puncturing the aerosol container, and a rubber gasket around the tube to prevent leakage of the container contents after puncture.* Calculate the density by the formula:

$$w/v,$$

in which w is the weight of the volume, v, of the Inhalation Aerosol taken.]

Metaproterenol Sulfate Inhalation Solution

» Metaproterenol Sulfate Inhalation Solution is a solution of Metaproterenol Sulfate in Purified Water.

* A suitable sampling system is available from Alltek Associates, P. O. Box 498, Arlington Heights, IL 60006.

It may contain Sodium Chloride. It contains not less than 90.0 percent and not more than 110.0 percent of the labeled amount of $(C_{11}H_{17}NO_3)_2 . H_2SO_4$.

Packaging and storage—Store in small, well-filled, tight containers, protected from light.

Labeling—Label it to indicate that the Inhalation Solution is not to be used if it is brown or contains a precipitate.

Reference standard—*USP Metaproterenol Sulfate Reference Standard*—Dry at 105° for 1 hour before using.

Identification—

A: On a suitable thin-layer chromatographic plate (see *Chromatography* ⟨621⟩), coated with a 0.25-mm layer of chromatographic silica gel mixture, apply 4 µL of the Inhalation Solution and 4 µL of an aqueous solution of USP Metaproterenol Sulfate RS containing about 50 mg per mL. Allow the spots to dry, and develop the chromatogram in a solvent system consisting of the upper layer of a freshly prepared mixture of butyl alcohol, water, and formic acid (50:25:7) until the solvent front has moved about three-fourths of the length of the plate. Remove the plate from the developing chamber, mark the solvent front, and allow the solvent to evaporate. Locate the spots on the plate by examination under short-wavelength ultraviolet light: the R_f value of the principal spot obtained from the Inhalation Solution corresponds to that obtained from the Standard solution.

B: The chromatogram of the *Assay preparation* obtained as directed in the *Assay* exhibits a major peak for metaproterenol, the retention time of which corresponds to that exhibited in the chromatogram of the *Standard preparation* obtained as directed in the *Assay*.

pH ⟨791⟩: between 2.8 and 4.0.

Assay—

Mobile phase, Standard preparation, and *Chromatographic system*—Prepare as directed in the *Assay* under *Metaproterenol Sulfate.*

Assay preparation—Transfer an accurately measured volume of Metaproterenol Sulfate Inhalation Solution, equivalent to about 200 mg of metaproterenol sulfate, to a 100-mL volumetric flask, dilute with 0.01 *N* hydrochloric acid to volume, and mix.

Procedure—Proceed as directed for *Procedure* in the *Assay* under *Metaproterenol Sulfate.* Calculate the quantity, in mg, of $(C_{11}H_{17}NO_3)_2 . H_2SO_4$ in each mL of the Inhalation Solution taken by the formula:

$$100(C/V)(r_U/r_S),$$

in which V is the volume, in mL, of Inhalation Solution taken, and C, r_U, and r_S are as defined therein.

Metaproterenol Sulfate Syrup

» Metaproterenol Sulfate Syrup contains not less than 90.0 percent and not more than 110.0 percent of the labeled amount of $(C_{11}H_{17}NO_3)_2 . H_2SO_4$.

Packaging and storage—Preserve in tight, light-resistant containers.

Reference standard—*USP Metaproterenol Sulfate Reference Standard*—Dry at 105° for 1 hour before using.

Identification—

A: Transfer a portion of Syrup, equivalent to about 10 mg of metaproterenol sulfate, to a separator, and extract with four 30-mL portions of ether, discarding the ether extracts. On the lower right corner of a suitable thin-layer chromatographic plate (see *Chromatography* ⟨621⟩), coated with a 0.25-mm layer of chromatographic silica gel mixture, apply 10 µL of the extracted portion of Syrup and allow to dry. Develop the chromatogram in a solvent system consisting of the lower layer of a well-shaken mixture of dioxane, methylene chloride, alcohol, and ammonium hydroxide (4:4:1:1). Allow the solvent front to move about three-fourths of the length of the plate. Remove the plate from the developing chamber, mark the solvent front, and dry in vacuum at 35° to 40° for 30 minutes. Rotate the plate 90°. At a point about four-fifths of the distance between the initial application of the Syrup extract and the solvent front, apply 10 µL of a Standard solution of USP Metaproterenol Sulfate RS in water containing about 2 mg per mL. Proceed as directed in *Identification test A* under *Metaproterenol Sulfate Inhalation Solution*, beginning with "Allow the spots to dry:" the R_f value of the principal spot obtained from the Syrup corresponds to that obtained from the Standard solution.

B: The chromatogram of the *Assay preparation* obtained as directed in the *Assay* exhibits a major peak for metaproterenol, the retention time of which corresponds to that exhibited in the chromatogram of the *Standard preparation* obtained as directed in the *Assay*.

pH ⟨791⟩: between 2.5 and 4.0, in a solution obtained by mixing 1 volume of Syrup and 4 volumes of water.

Assay—

Mobile phase—Mix 10 mL of formic acid and water to make 1000 mL of solution. Filter and degas this solution before use. Make adjustments if necessary (see *System Suitability* under *Chromatography* ⟨621⟩).

Standard preparation—Dissolve an accurately weighed quantity of USP Metaproterenol Sulfate RS in water to obtain a solution having a known concentration of about 0.2 mg per mL.

Assay preparation—Transfer an accurately measured volume of Metaproterenol Sulfate Syrup, equivalent to about 20 mg of metaproterenol sulfate, to a 100-mL volumetric flask, dilute with water to volume, and mix.

Chromatographic system (see *Chromatography* ⟨621⟩)—The liquid chromatograph is equipped with a 278-nm detector, a 4.6-mm × 5-cm guard column that contains packing L2, and a 3.9-mm × 30-cm analytical column that contains packing L1. [NOTE—After use, rinse the analytical column with water and store with water in it.] The flow rate is about 2 mL per minute. Chromatograph the *Standard preparation*, and record the peak responses as directed under *Procedure:* the tailing factor for the analyte peak is not more than 3.0, and the relative standard deviation for replicate injections is not more than 2.0%.

Procedure—Separately inject equal volumes (about 100 µL) of the *Standard preparation* and the *Assay preparation* into the chromatograph, record the chromatograms, and measure the responses for the major peaks. Calculate the quantity, in mg, of $(C_{11}H_{17}NO_3)_2 . H_2SO_4$ in each mL of the Syrup taken by the formula:

$$100(C/V)(r_U/r_S),$$

in which C is the concentration, in mg per mL, of USP Metaproterenol Sulfate RS in the *Standard preparation*, V is the volume, in mL, of Syrup taken, and r_U and r_S are the peak responses from the *Assay preparation* and the *Standard preparation*, respectively.

Metaproterenol Sulfate Tablets

» Metaproterenol Sulfate Tablets contain not less than 92.0 percent and not more than 108.0 percent of the labeled amount of $(C_{11}H_{17}NO_3)_2 . H_2SO_4$.

Packaging and storage—Preserve in well-closed, light-resistant containers.

Reference standard—*USP Metaproterenol Sulfate Reference Standard*—Dry at 105° for 1 hour before using.

Identification—

A: Powder a number of Tablets, equivalent to about 100 mg of metaproterenol sulfate, add 10 mL of water, stir for about 3 minutes, and centrifuge. Use the clear solution so obtained as the *Test solution.* Dissolve a suitable quantity of USP Metaproterenol Sulfate RS in water to obtain a Standard solution having a concentration of 10 mg per mL. Apply separate 10-µL portions of the *Test solution* and the Standard solution on a thin-layer chromatographic plate (see *Chromatography* ⟨621⟩), coated with a 0.25-mm layer of chromatographic silica gel mixture. Proceed as directed in *Identification test A* under *Metaproterenol Sulfate Inhalation Solution*, beginning with "Allow the spots to dry:" the R_f value of the principal spot obtained from the *Test solution* corresponds to that obtained from the Standard solution.

B: Mix a quantity of powdered Tablets, equivalent to about 20 mg of metaproterenol sulfate, with 5 mL of water, and filter: the filtrate responds to the tests for *Sulfate* ⟨191⟩.

C: The chromatogram of the *Assay preparation* obtained as directed in the *Assay* exhibits a major peak for metaproterenol, the retention time of which corresponds to that exhibited in the chromatogram of the *Standard preparation* obtained as directed in the *Assay*.

Dissolution ⟨711⟩—

Medium: water; 500 mL.

Apparatus 2: 50 rpm.

Time: 30 minutes.

Procedure—Determine the amount of $(C_{11}H_{17}NO_3)_2 \cdot H_2SO_4$ dissolved from ultraviolet absorbances at the wavelength of maximum absorbance at about 276 nm of filtered portions of the solution under test, suitably diluted with *Dissolution Medium*, if necessary, in comparison with a Standard solution having a known concentration of USP Metaproterenol Sulfate RS in the same medium.

Tolerances—Not less than 70% (*Q*) of the labeled amount of $(C_{11}H_{17}NO_3)_2 \cdot H_2SO_4$ is dissolved in 30 minutes.

Uniformity of dosage units ⟨905⟩: meet the requirements.

Assay—

Mobile phase, *Standard preparation*, and *Chromatographic system*—Prepare as directed in the *Assay* under *Metaproterenol Sulfate*.

Assay preparation—Transfer 20 Metaproterenol Sulfate Tablets to a 500-mL conical flask. Add an accurately measured volume of 0.01 *N* hydrochloric acid sufficient to yield a solution containing about 2 mg of metaproterenol sulfate per mL, shake by mechanical means for 30 minutes, and filter. Use the filtrate so obtained as the *Assay preparation*.

Procedure—Proceed as directed for *Procedure* in the *Assay* under *Metaproterenol Sulfate*. Calculate the quantity, in mg, of $(C_{11}H_{17}NO_3)_2 \cdot H_2SO_4$ in each Tablet taken by the formula:

$$(CV/20)(r_U/r_S),$$

in which V is the volume, in mL, of 0.01 *N* hydrochloric acid added, and C, r_U, and r_S are as defined therein.

Methacholine Chloride

$$\left[CH_3COOCHCH_2N^+(CH_3)_3 \atop \quad CH_3 \right] Cl^-$$

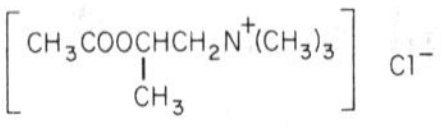

$C_8H_{18}ClNO_2$ 195.69

1-Propanaminium, 2-(acetyloxy)-*N*,*N*,*N*-trimethyl-, chloride.

(2-Hydroxypropyl)trimethylammonium chloride acetate [*62-51-1*].

» Methacholine Chloride, dried at 105° for 4 hours, contains not less than 98.0 percent and not more than 101.0 percent of $C_8H_{18}ClNO_2$.

Packaging and storage—Preserve in tight containers.

Identification—

A: Dissolve about 100 mg in about 2 mL of water on a watch glass, and add 3 mL of platinic chloride TS: small rhombohedric plates are formed, which melt between 220° and 225° (see *Melting Range or Temperature* ⟨741⟩) (*distinction from acetylcholine chloride, which forms needles radiating from a central point, and from choline chloride, which forms no crystals*).

B: To 1 mL of a solution (1 in 10) add 1 mL of alcohol and 1 mL of sulfuric acid, and heat gently: the odor of ethyl acetate is perceptible.

C: To 5 mL of a solution (1 in 10) add 2 g of potassium hydroxide and heat gently: the odor of trimethylamine is perceptible.

D: A solution (1 in 50) responds to the tests for *Chloride* ⟨191⟩.

Melting range ⟨741⟩—Dissolve about 100 mg in 2 to 3 mL of chloroform in a small beaker. Heat at 110° for 1 hour. While the test specimen is still hot, quickly powder the dry residue with a glass rod, and transfer to a melting point tube in the usual manner. Determine the melting range without delay. It melts between 170° and 173°.

Loss on drying ⟨731⟩—Dry it at 105° for 4 hours: it loses not more than 1.5% of its weight.

Residue on ignition ⟨281⟩: not more than 0.1%.

Acetylcholine chloride—To 2 mL of a solution (1 in 10) add 3 mL of a solution of sodium perchlorate (1 in 5), shake, and immerse in ice water for 5 minutes: no precipitate is formed.

Heavy metals, *Method II* ⟨231⟩: 0.002%.

Assay—Transfer to a conical flask about 400 mg of Methacholine Chloride, previously dried and accurately weighed (because it is very hygroscopic, store the dried material in a vacuum desiccator), dissolve it in 50 mL of glacial acetic acid, add 10 mL of mercuric acetate TS and 1 drop of crystal violet TS, and titrate with 0.1 *N* perchloric acid VS to a blue-green end-point. Perform a blank determination, and make any necessary correction. Each mL of 0.1 *N* perchloric acid is equivalent to 19.57 mg of $C_8H_{18}ClNO_2$.

Methacycline Hydrochloride

OH O OH O OH CONH2 · HCl OH CH2 H OH H N(CH3)2

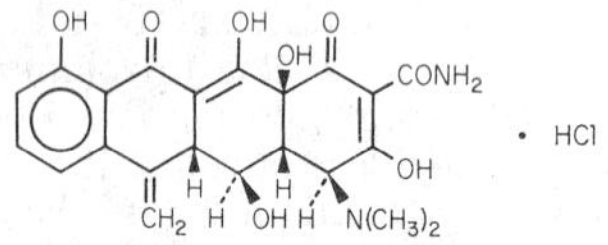

$C_{22}H_{22}N_2O_8 \cdot HCl$ 478.89

2-Naphthacenecarboxamide, 4-(dimethylamino)-1,4,4a,5,5a,6,11,12a-octahydro-3,5,10,12,12a-pentahydroxy-6-methylene-1,11-dioxo-, monohydrochloride, [4*S*-(4α,4aα,5α,5aα,12aα)]-.

4-(Dimethylamino)-1,4,4a,5,5a,6,11,12a-octahydro-3,5,10,12,12a-pentahydroxy-6-methylene-1,11-dioxo-2-naphthacenecarboxamide monohydrochloride [*3963-95-9*].

» Methacycline Hydrochloride has a potency equivalent to not less than 832 μg of methacycline ($C_{22}H_{22}N_2O_8$) per mg.

Packaging and storage—Preserve in tight, light-resistant containers.

Reference standard—*USP Methacycline Hydrochloride Reference Standard*—Dry in vacuum at a pressure not exceeding 5 mm of mercury at 60° for 3 hours before using.

Identification—The ultraviolet absorption spectrum of a 1 in 50,000 solution in a 1 in 1200 mixture of hydrochloric acid and methanol exhibits maxima and minima at the same wavelengths as that of a similar solution of USP Methacycline Hydrochloride RS, concomitantly measured, and the absorptivity, calculated on the dried basis, at the wavelength of maximum absorbance at about 345 nm is between 88.4% and 96.4% of that of the USP Methacycline Hydrochloride RS, the potency of the Reference Standard being taken into account.

Crystallinity ⟨695⟩: meets the requirements.

pH ⟨791⟩: between 2.0 and 3.0, in a solution containing 10 mg of methacycline per mL.

Water, *Method I* ⟨921⟩: not more than 2.0%.

Assay—Proceed with Methacycline Hydrochloride as directed under *Antibiotics—Microbial Assays* ⟨81⟩.

Methacycline Hydrochloride Capsules

» Methacycline Hydrochloride Capsules contain the equivalent of not less than 90.0 percent and not more than 120.0 percent of the labeled amount of methacycline ($C_{22}H_{22}N_2O_8$).

Packaging and storage—Preserve in tight, light-resistant containers.

Reference standard—*USP Methacycline Hydrochloride Reference Standard*—Dry in vacuum at a pressure not exceeding 5 mm of mercury at 60° for 3 hours before using.

Dissolution ⟨711⟩—

Medium: water; 900 mL.

Apparatus 1: 100 rpm.

Time: 60 minutes.

Procedure—Determine the amount of $C_{22}H_{22}N_2O_8.HCl$ dissolved from ultraviolet absorbances at the wavelength of maximum absorbance at about 345 nm of filtered portions of the solution under test, suitably diluted with water, in comparison with a Standard solution having a known concentration of USP Methacycline Hydrochloride RS in the same medium.

Tolerances—Not less than 70% (*Q*) of the labeled amount of $C_{22}H_{22}N_2O_8.HCl$ is dissolved in 60 minutes.

Uniformity of dosage units ⟨905⟩: meet the requirements.

Water, *Method I* ⟨921⟩: not more than 7.5%.

Assay—Place not less than 5 Methacycline Hydrochloride Capsules into a high-speed glass blender jar containing an accurately measured volume of water, and blend for 3 to 5 minutes to obtain a stock solution having a convenient concentration of methacycline. Proceed as directed under *Antibiotics—Microbial Assays* ⟨81⟩, using an accurately measured volume of this stock solution diluted quantitatively and stepwise with water to yield a *Test Dilution* having a concentration assumed to be equal to the median dose level of the Standard.

Methacycline Hydrochloride Oral Suspension

» Methacycline Hydrochloride Oral Suspension contains the equivalent of not less than 90.0 percent and not more than 125.0 percent of the labeled amount of methacycline ($C_{22}H_{22}N_2O_8$). It contains one or more suitable and harmless buffers, colors, diluents, dispersants, flavors, and preservatives.

Packaging and storage—Preserve in tight, light-resistant containers.

Reference standard—*USP Methacycline Hydrochloride Reference Standard*—Dry in vacuum at a pressure not exceeding 5 mm of mercury at 60° for 3 hours before using.

Uniformity of dosage units ⟨905⟩—

FOR SUSPENSION PACKAGED IN SINGLE-UNIT CONTAINERS: meets the requirements.

pH ⟨791⟩: between 6.5 and 8.0.

Assay—Transfer an accurately measured quantity of Methacycline Hydrochloride Oral Suspension, freshly mixed and free from air bubbles, to a suitable volumetric flask, dilute with water to volume, and mix. Proceed as directed under *Antibiotics—Microbial Assays* ⟨81⟩, using an accurately measured volume of this stock solution diluted quantitatively and stepwise with water to yield a *Test Dilution* having a concentration assumed to be equal to the median dose level of the Standard.

Methadone Hydrochloride

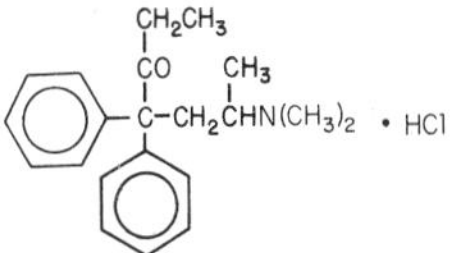

$C_{21}H_{27}NO.HCl$ 345.91

3-Heptanone, 6-(dimethylamino)-4,4-diphenyl-, hydrochloride.

6-(Dimethylamino)-4,4-diphenyl-3-heptanone hydrochloride [*1095-90-5*].

» Methadone Hydrochloride contains not less than 98.5 percent and not more than 100.5 percent of $C_{21}H_{27}NO.HCl$, calculated on the dried basis.

Packaging and storage—Preserve in tight, light-resistant containers.

Reference standard—*USP Methadone Hydrochloride Reference Standard*—Dry at 105° for 1 hour before using.

Identification—

A: The infrared absorption spectrum of a potassium bromide dispersion of it, previously dried, exhibits absorption maxima only at the same wavelengths as that of a similar preparation of USP Methadone Hydrochloride RS.

B: A solution of it responds to the tests for *Chloride* ⟨191⟩.

pH ⟨791⟩: between 4.5 and 6.5, in a solution (1 in 100).

Loss on drying ⟨731⟩—Dry about 500 mg, accurately weighed, at 105° for 1 hour: it loses not more than 0.3% of its weight.

Residue on ignition ⟨281⟩: not more than 0.1%.

Ordinary impurities ⟨466⟩—

Test solution: alcohol.

Standard solution: alcohol.

Eluant: a mixture of methanol and ammonium hydroxide (100:1.5).

Visualization: 3.

Limits—The sum of the intensities of all secondary spots obtained from the *Test solution* corresponds to not more than 1.0%.

Assay—Dissolve about 500 mg of Methadone Hydrochloride, accurately weighed, in a mixture of 10 mL of glacial acetic acid and 10 mL of mercuric acetate TS, warming slightly if necessary to effect solution. Cool the solution to room temperature, add 10 mL of dioxane, then add crystal violet TS, and titrate rapidly with 0.1 *N* perchloric acid VS. Perform a blank determination, and make any necessary correction. Each mL of 0.1 *N* perchloric acid is equivalent to 34.59 mg of $C_{21}H_{27}NO.HCl$.

Methadone Hydrochloride Oral Concentrate

» Methadone Hydrochloride Oral Concentrate contains, in each mL, not less than 9.0 mg and not more than 11.0 mg of $C_{21}H_{27}NO.HCl$. It contains a suitable preservative and may contain suitable coloring, flavoring, and surface-active agents.

Packaging and storage—Preserve in tight containers, protected from light, at controlled room temperature.

Labeling—Label it to indicate that it is to be diluted with water or other liquid to 30 mL or more prior to administration.

Reference standard—*USP Methadone Hydrochloride Reference Standard*—Dry at 105° for 1 hour before using.

Identification—

A: Shake a volume of Oral Concentrate, equivalent to about 5 mg of methadone hydrochloride, with 5 mL of sodium carbonate TS, and extract with 5 mL of chloroform: the extract so obtained responds to the *Thin-layer Chromatographic Identification Test* ⟨201⟩, a solvent mixture of alcohol, glacial acetic acid, and water

(5:3:2) being used for development and iodoplatinate TS being used to visualize the spots.

B: It responds to the tests for *Chloride* ⟨191⟩.

pH ⟨791⟩: between 3.0 and 6.0.

Assay—

Mobile phases—

A: To 100 mL of freshly distilled formic acid and 1000 mL of water in a 2000-mL volumetric flask add 2.5 mL of ammonium hydroxide, dilute with water to volume, and mix.

B: Spectrophotometric acetonitrile.

Standard preparation—On the day of use, dissolve an accurately weighed quantity of USP Methadone Hydrochloride RS in water to obtain a solution having a known concentration of about 2.5 mg per mL.

Assay preparation—Transfer 75.0 mL of water to a dry 100-mL volumetric flask. Taking precautions to keep the Oral Concentrate off the neck of the flask, dilute with Methadone Hydrochloride Oral Concentrate to volume, and mix.

Procedure—By means of a sampling valve, introduce separately equal volumes (about 50 µL) of the *Assay preparation* and the *Standard preparation* into a gradient elution high-pressure liquid chromatograph (see *Chromatography* ⟨621⟩), operated at room temperature. Adjust the operating parameters so that the peak obtained from the *Standard preparation* is about 0.6 full-scale. Typically, the apparatus is fitted with a 3.2- × 250-mm chloride-resistant stainless steel column, slurry-packed with packing L1, equipped with a 280-nm, double-beam, photometric detector and recorder, and capable of providing column pressures up to about 1500 psi. The gradient elution is programmed so that the mobile phase changes from 100% *Mobile phase A* to 80% *Mobile phase B* in *Mobile phase A* in about 10 to 20 minutes. In a suitable chromatogram, five replicate injections of a single test specimen show a relative standard deviation of not more than 2.0%. Measure the area of the peaks obtained from the *Assay preparation* and the *Standard preparation*, at identical retention times, and calculate the quantity, in mg, of $C_{21}H_{27}NO.HCl$ in each mL of the Oral Concentrate taken by the formula:

$$4C(A_U/A_S),$$

in which C is the concentration, in mg per mL, of USP Methadone Hydrochloride RS in the *Standard preparation*, and A_U and A_S are the peak areas obtained from the *Assay preparation* and *Standard preparation*, respectively.

Methadone Hydrochloride Injection

» Methadone Hydrochloride Injection is a sterile solution of Methadone Hydrochloride in Water for Injection. It contains, in each mL, not less than 9.5 mg and not more than 10.5 mg of $C_{21}H_{27}NO.HCl$.

Packaging and storage—Preserve in single-dose or in multiple-dose, light-resistant containers, preferably of Type I glass.

Reference standard—*USP Methadone Hydrochloride Reference Standard*—Dry at 105° for 1 hour before using.

Identification—It meets the requirements under *Identification—Organic Nitrogenous Bases* ⟨181⟩.

pH ⟨791⟩: between 3.0 and 6.5.

Other requirements—It meets the requirements under *Injections* ⟨1⟩.

Assay—

Internal standard solution—Weigh about 100 mg of procaine, and dissolve in 20 mL of methylene chloride.

Standard preparation—Weigh accurately about 10 mg of USP Methadone Hydrochloride RS, transfer to a 60-mL separator, add 1 mL of water and 2 mL of 0.5 *N* sodium hydroxide, and proceed as directed for *Assay preparation*, beginning with "extract with three 10-mL portions of chromatographic grade methylene chloride."

Assay preparation—Transfer 1.0 mL of Methadone Hydrochloride Injection, equivalent to 10 mg of methadone hydrochloride, to a 60-mL separator, add 2 mL of 0.5 *N* sodium hydroxide, and extract with three 10-mL portions of chromatographic grade methylene chloride, combining the extracts in a vessel containing about 3 g of anhydrous sodium sulfate. Transfer 2.0 mL of *Internal standard solution* to the vessel containing the extracts, insert the stopper, and mix. Decant about 15 mL of the methylene chloride solution to a test tube, and evaporate to a volume of 2 to 3 mL, using vacuum or a stream of nitrogen.

Procedure—Use a suitable gas chromatograph equipped with a flame-ionization detector and a glass column 1.2-m long and 4-mm in diameter, packed with 3 percent phase G2 on 100- to 200-mesh support S1A. Maintain the column temperature at 170°, the injection port at 225°, and the detector at 240°. Use dry helium as the carrier gas, at a flow rate of about 55 mL per minute. In a suitable chromatogram, six replicate injections of the *Standard preparation* show a coefficient of variation of not more than 1 percent in the ratios of the peak areas of methadone to the peak area of procaine, and the resolution factor is not less than 5. Inject, separately, suitable volumes of the *Assay preparation*, containing about 5 µg of methadone, and of the *Standard preparation*. Calculate the quantity, in mg, of $C_{21}H_{27}NO.HCl$ in each mL of the Injection taken by the formula:

$$W(R_U/R_S),$$

in which W is the weight, in mg, of USP Methadone Hydrochloride RS in the *Standard preparation*, and R_U and R_S are the ratios of the peak areas of methadone to the peak area of procaine in the *Assay preparation* and the *Standard preparation*, respectively.

Methadone Hydrochloride Oral Solution

» Methadone Hydrochloride Oral Solution contains not less than 90.0 percent and not more than 110.0 percent of the labeled amount of $C_{21}H_{27}NO.HCl$.

Packaging and storage—Preserve in tight containers, protected from light, at controlled room temperature.

Reference standard—*USP Methadone Hydrochloride Reference Standard*—Dry at 105° for 1 hour before using.

Identification—

A: Shake a volume of Oral Solution, equivalent to about 5 mg of methadone hydrochloride, with 5 mL of sodium carbonate TS, and extract with 5 mL of chloroform: the extract so obtained responds to the *Thin-layer Chromatographic Identification Test* ⟨201⟩, a solvent mixture of alcohol, glacial acetic acid, and water (5:3:2) being used for development and iodoplatinate TS being used to visualize the spots.

B: It responds to the tests for *Chloride* ⟨191⟩.

pH ⟨791⟩: between 1.0 and 4.0.

Alcohol ⟨611⟩: between 7.0% and 9.0% of C_2H_5OH.

Assay—

Mobile phase—Prepare a solution containing about 40 volumes of acetonitrile and 60 volumes of 0.033 *M* monobasic potassium phosphate adjusted, dropwise, with phosphoric acid to a pH of 4.0.

Internal standard solution—Prepare a solution of pyrilamine maleate in water containing 250 µg per mL.

Standard preparation—Transfer about 20 mg of USP Methadone Hydrochloride RS, accurately weighed, to a 25-mL volumetric flask, add 2.0 mL of *Internal standard solution*, dilute with water to volume, and mix.

Assay preparation—Transfer an accurately measured volume of Methadone Hydrochloride Oral Solution, equivalent to about 20 mg of methadone hydrochloride, to a 125-mL separator. Extract the specimen with two 50-mL portions of ether, collecting the ether extracts in a second separator. Wash the combined ether extracts with 2 mL of water, and discard the ether extract. Transfer the aqueous wash and the aqueous specimen to a 25-mL volumetric flask, add 2.0 mL of *Internal standard solution*, dilute with water to volume, and mix. Filter the solution through a 5-µm filter.

Chromatographic system (see *Chromatography* ⟨621⟩)—The liquid chromatograph is equipped with a 254-nm detector and a

3.9-mm × 30-cm column that contains packing L11. The flow rate is about 1.3 mL per minute. Chromatograph five replicate injections of the *Standard preparation*, and record the peak responses as directed under *Procedure:* the relative standard deviation is not more than 2.0%.

Procedure—Separately inject equal volumes (about 10 µL) of the *Standard preparation* and the *Assay preparation* into the chromatograph by means of a suitable microsyringe or sampling valve, record the chromatograms, and measure the responses for the major peaks. The relative retention times are about 5.5 minutes for the internal standard and 9 minutes for methadone hydrochloride. Calculate the quantity, in mg, of $C_{21}H_{27}NO.HCl$ in each mL of the Oral Solution taken by the formula:

$$25(C/V)(R_U/R_S),$$

in which C is the concentration, in mg per mL, of USP Methadone Hydrochloride RS in the *Standard preparation*, V is the volume, in mL, of Oral Solution taken, and R_U and R_S are the peak response ratios of the internal standard and methadone hydrochloride peaks obtained from the *Assay preparation* and the *Standard preparation*, respectively.

Methadone Hydrochloride Tablets

» Methadone Hydrochloride Tablets contain not less than 93.0 percent and not more than 107.0 percent of the labeled amount of $C_{21}H_{27}NO.HCl$.

Packaging and storage—Preserve in well-closed containers.

Labeling—Label Tablets that are not intended for oral administration as intact Tablets to state that they are dispersible tablets or to indicate that they are intended for dispersion in a liquid prior to oral administration of the prescribed dose.

Reference standard—*USP Methadone Hydrochloride Reference Standard*—Dry at 105° for 1 hour before using.

Identification—A quantity of powdered Tablets, equivalent to about 5 mg of methadone hydrochloride, responds to *Identification test A* under *Methadone Hydrochloride Oral Solution.*

Disintegration ⟨701⟩: for dispersible Tablets, 15 minutes, the use of disks being omitted.

Dissolution ⟨711⟩—

FOR TABLETS INTENDED TO BE SWALLOWED—

Medium: water; 500 mL.

Apparatus 1: 100 rpm.

Time: 45 minutes.

Procedure—Filter a portion of the solution under test, and pipet a volume of the filtrate, equivalent to about 400 µg of methadone hydrochloride, into a suitable separator. Add 1 mL of glacial acetic acid and 20 mL of a solution of bromocresol purple, prepared by dissolving 200 mg of bromocresol purple in 1000 mL of dilute glacial acetic acid (1 in 50), mix, and extract with 20.0 mL of chloroform. Determine the amount of $C_{21}H_{27}NO.HCl$ dissolved from ultraviolet absorbances at the wavelength of maximum absorbance at about 405 nm of the chloroform extract so obtained in comparison with the chloroform extract similarly prepared from a Standard solution having a known concentration of USP Methadone Hydrochloride RS in water.

Tolerances—Not less than 75% (Q) of the labeled amount of $C_{21}H_{27}NO.HCl$ is dissolved in 45 minutes.

Uniformity of dosage units ⟨905⟩: meet the requirements.

Assay—

Internal standard solution and *Standard preparation*—Prepare as directed in the *Assay* under *Methadone Hydrochloride Injection.*

Assay preparation—Weigh and finely powder not less than 20 Methadone Hydrochloride Tablets. Transfer an accurately weighed portion of the powder, equivalent to about 10 mg of methadone hydrochloride, to a suitable vessel, add 20 mL of 0.06 *N* acetic acid, and shake by mechanical means for 15 minutes. Using gentle suction, filter through a medium-porosity, sintered-glass funnel. Wash the solids in the funnel with three 7-mL portions of 0.06 *N* acetic acid, relieving suction during washing, and mixing the solids with a glass rod. Transfer the filtrate and washings to a 125-mL separator with the aid of about 10 mL of water, add 15 drops of sodium hydroxide solution (1 in 2) to the solution in the separator, and proceed as directed for *Assay preparation* in the *Assay* under *Methadone Hydrochloride Injection*, beginning with "extract with three 10-mL portions of chromatographic grade methylene chloride."

Procedure—Proceed as directed for *Procedure* in the *Assay* under *Methadone Hydrochloride Injection.* Calculate the quantity, in mg, of $C_{21}H_{27}NO.HCl$ in the portion of Tablets taken by the formula:

$$W(R_U/R_S),$$

in which W is the weight, in mg, of USP Methadone Hydrochloride RS in the *Standard preparation*, and R_U and R_S are the ratios of the peak areas of methadone to the peak area of procaine in the *Assay preparation* and the *Standard preparation*, respectively.

Methantheline Bromide

$C_{21}H_{26}BrNO_3$ 420.35

Ethanaminium, *N,N*-diethyl-*N*-methyl-2-[(9*H*-xanthen-9-ylcarbonyl)oxy]-, bromide.

Diethyl(2-hydroxyethyl)methylammonium bromide xanthene-9-carboxylate [*53-46-3*].

» Methantheline Bromide contains not less than 98.0 percent and not more than 102.0 percent of $C_{21}H_{26}BrNO_3$, calculated on the dried basis.

Packaging and storage—Preserve in well-closed containers.

Reference standards—*USP Methantheline Bromide Reference Standard*—Dry at 105° for 4 hours before using. *USP Xanthanoic Acid Reference Standard*—Dry at 105° for 1 hour before using.

Identification—

A: The infrared absorption spectrum of a potassium bromide dispersion of it, previously dried, exhibits maxima only at the same wavelengths as that of a similar preparation of USP Methantheline Bromide RS.

B: The ultraviolet absorption spectrum of a 1 in 20,000 solution in methanol exhibits maxima and minima at the same wavelengths as that of a similar solution of USP Methantheline Bromide RS, concomitantly measured, and the respective absorptivities, calculated on the dried basis, at the wavelength of maximum absorbance at about 282 nm do not differ by more than 3.0%.

C: Dissolve about 250 mg in 10 mL of water. To 2 mL of the solution add 5 mL of water, 2 mL of 2 *N* nitric acid, and 2 mL of silver nitrate TS: a light yellow, curdy precipitate forms which is only slightly soluble in 6 *N* ammonium hydroxide. To the remainder of the solution add 10 mL of 1 *N* sodium hydroxide, and heat to boiling. Keep the mixture hot for about 2 minutes, cool slightly, and acidify with about 5 mL of 2 *N* hydrochloric acid. Cool, filter, wash the precipitate thoroughly with water, recrystallize from diluted alcohol, and dry at 105° for 1 hour: the infrared absorption spectrum of a potassium bromide dispersion of the residue so obtained exhibits maxima only at the same wavelengths as that of a similar preparation of USP Xanthanoic Acid RS.

Melting range, *Class I* ⟨741⟩: between 171° and 177°.

Loss on drying ⟨731⟩—Dry it at 105° for 4 hours: it loses not more than 0.5% of its weight.

Residue on ignition ⟨281⟩: not more than 0.2%.

Assay—Dissolve about 800 mg of Methantheline Bromide, accurately weighed, in 50 mL of glacial acetic acid. Add 10 mL of mercuric acetate TS and 1 drop of crystal violet TS, and titrate with 0.1 *N* perchloric acid VS to a blue-green end-point. Perform a blank determination, and make any necessary correction. Each mL of 0.1 *N* perchloric acid is equivalent to 42.04 mg of $C_{21}H_{26}BrNO_3$.

Sterile Methantheline Bromide

» Sterile Methantheline Bromide contains not less than 98.0 percent and not more than 102.0 percent of $C_{21}H_{26}BrNO_3$, calculated on the dried basis.

Packaging and storage—Preserve in *Containers for Sterile Solids* as described under *Injections* ⟨1⟩.

Reference standards—*USP Methantheline Bromide Reference Standard*—Dry at 105° for 4 hours before using. *USP Xanthanoic Acid Reference Standard*—Dry at 105° for 1 hour before using.

Completeness of solution ⟨641⟩—The contents of 1 container dissolve in 1 mL of water to yield a clear solution.

Constituted solution—At the time of use, the constituted solution prepared from Sterile Methantheline Bromide meets the requirements for *Constituted Solutions* under *Injections* ⟨1⟩.

Pyrogen—It meets the requirements of the *Pyrogen Test* ⟨151⟩.

Uniformity of dosage units ⟨905⟩: meets the requirements.

Other requirements—It responds to the *Identification tests* and meets the requirements for *Melting range, Loss on drying*, and *Residue on ignition* under *Methantheline Bromide.* It meets also the requirements for *Sterility Tests* ⟨71⟩ and for *Labeling* under *Injections* ⟨1⟩.

Assay—Proceed with *Sterile Methantheline Bromide* as directed in the *Assay* under *Methantheline Bromide.*

Methantheline Bromide Tablets

» Methantheline Bromide Tablets contain not less than 95.0 percent and not more than 105.0 percent of the labeled amount of $C_{21}H_{26}BrNO_3$.

Packaging and storage—Preserve in well-closed containers.

Reference standards—*USP Methantheline Bromide Reference Standard*—Dry at 105° for 4 hours before using. *USP Xanthanoic Acid Reference Standard*—Dry at 105° for 1 hour before using.

Identification—Triturate a quantity of powdered Tablets, equivalent to about 500 mg of methantheline bromide, with 10 mL of chloroform. Filter the mixture, and evaporate the filtrate with the aid of a current of air to a thick oil. Stir the residue with about 5 mL of ether: a white powder is obtained which responds to the *Identification tests* under *Methantheline Bromide.*

Dissolution ⟨711⟩—

Medium: water; 900 mL.

Apparatus 2: 50 rpm.

Time: 45 minutes.

Procedure—Determine the amount of $C_{21}H_{26}BrNO_3$ dissolved from ultraviolet absorbances at the wavelength of maximum absorbance at about 282 nm of filtered portions of the solution under test, suitably diluted with water, if necessary, in comparison with a Standard solution having a known concentration of USP Methantheline Bromide RS in the same medium.

Tolerances—Not less than 75% (*Q*) of the labeled amount of $C_{21}H_{26}BrNO_3$ is dissolved in 45 minutes.

Uniformity of dosage units ⟨905⟩: meet the requirements.

Procedure for content uniformity—Triturate 1 Tablet with methanol in a glass mortar, and transfer the mixture with the aid of methanol to a 100-mL volumetric flask. Dilute with methanol to volume, mix, and filter, discarding the first 25 mL of the filtrate. Dilute a portion of the subsequent filtrate quantitatively and stepwise, if necessary, with methanol to provide a solution containing approximately 50 µg of methantheline bromide per mL. Concomitantly determine the absorbances of this solution and a Standard solution of USP Methantheline Bromide RS in the same medium having a known concentration of about 50 µg per mL, in 1-cm cells at the wavelength of maximum absorbance at about 282 nm, with a suitable spectrophotometer, using methanol as the blank. Calculate the quantity, in mg, of $C_{21}H_{26}BrNO_3$ in the Tablet by the formula:

$$(TC/D)(A_U/A_S),$$

in which *T* is the labeled quantity, in mg, of methantheline bromide in the Tablet, *C* is the concentration, in µg per mL, of USP Methantheline Bromide RS in the Standard solution, *D* is the concentration, in µg per mL, of methantheline bromide in the solution from the Tablet, based upon the labeled quantity per Tablet and the extent of dilution, and A_U and A_S are the absorbances of the solution from the Tablet and the Standard solution, respectively.

Assay—Weigh and finely powder not less than 20 Methantheline Bromide Tablets. Weigh accurately a portion of the powder, equivalent to about 800 mg of methantheline bromide, and transfer to a medium-porosity, sintered-glass funnel that is fitted into a small suction flask. Wash the powder with 10 mL of ether, stir with a small rod, and remove the ether by suction. Wash with two additional 10-mL portions of ether, and discard the ether washings. In the same manner, extract the powder with four 10-mL portions of chloroform. Transfer the combined chloroform extracts to a beaker, and evaporate carefully on a hot plate until only a heavy oil remains, avoiding local overheating. Cool, and dry the residue in vacuum for 15 minutes. Add 10 mL of ether to the oily residue, and stir until the residue becomes a white, crystalline powder. Allow the solids to settle, and decant the clear liquid, avoiding loss of solid. Dry the solids in air until the odor of ether is no longer perceptible, heat at 105° for 30 minutes, and cool. Add 50 mL of glacial acetic acid to dissolve, and proceed as directed in the *Assay* under *Methantheline Bromide,* beginning with "Add 10 mL of mercuric acetate TS."

Metharbital

CH3
O N O
CH3CH2
CH3CH2 NH
O

$C_9H_{14}N_2O_3$ 198.22

2,4,6(1*H*,3*H*,5*H*)-Pyrimidinetrione, 5,5-diethyl-1-methyl-.

5,5-Diethyl-1-methylbarbituric acid [*50-11-3*].

» Metharbital, dried at 105° for 4 hours, contains not less than 98.0 percent and not more than 100.5 percent of $C_9H_{14}N_2O_3$.

Packaging and storage—Preserve in tight containers.

Reference standard—*USP Metharbital Reference Standard*—Dry at 105° for 4 hours before using.

Identification—

A: The infrared absorption spectrum of a potassium bromide dispersion of it, previously dried at 105° for 4 hours, exhibits maxima only at the same wavelengths as that of a similar preparation of USP Metharbital RS.

B: The ultraviolet absorption spectrum of a 1 in 100,000 solution in sodium hydroxide solution (1 in 250) exhibits maxima and minima at the same wavelengths as that of a similar solution of USP Metharbital RS, concomitantly measured.

C: Shake about 300 mg with 5 mL of sodium hydroxide solution (1 in 125) for 2 minutes, filter, and to 1 mL of the filtrate add several drops of silver nitrate TS: a white precipitate is formed, and it is soluble in 6 *N* ammonium hydroxide. To a second 1-mL portion of the filtrate add 1 drop of mercuric chloride TS: a white precipitate is formed, and it is soluble in 6 *N* ammonium hydroxide.

Melting range, *Class I* ⟨741⟩: between 151° and 155°.

Water, *Method I* ⟨921⟩: not more than 1.0%.

Residue on ignition ⟨281⟩: not more than 0.1%.

Assay—Transfer to a 100-mL volumetric flask about 100 mg of Metharbital, previously dried and accurately weighed, dissolve in sodium hydroxide solution (1 in 250), dilute with sodium hydroxide solution (1 in 250) to volume, and mix. Transfer 10.0 mL of this solution to a second 100-mL volumetric flask, dilute with the same sodium hydroxide solution to volume, and mix. Transfer 10.0 mL of the resulting solution to a third 100-mL volumetric flask, dilute with the sodium hydroxide solution to volume, and mix. Concomitantly determine the absorbances of this solution and a Standard solution of USP Metharbital RS in the same medium having a known concentration of about 10 µg per mL in 1-cm cells at the wavelength of maximum absorbance at about 244 nm, with a suitable spectrophotometer, using sodium hydroxide solution (1 in 250) as the blank. Calculate the quantity, in mg, of $C_9H_{14}N_2O_3$ in the Metharbital taken by the formula:

$$10C(A_U/A_S),$$

in which *C* is the concentration, in µg per mL, of USP Metharbital RS in the Standard solution, and A_U and A_S are the absorbances of the solution of Metharbital and the Standard solution, respectively.

Metharbital Tablets

» Metharbital Tablets contain not less than 95.0 percent and not more than 105.0 percent of the labeled amount of $C_9H_{14}N_2O_3$.

Packaging and storage—Preserve in tight containers.

Reference standard—*USP Metharbital Reference Standard*—Dry at 105° for 4 hours before using.

Identification—Extract a portion of powdered Tablets, equivalent to about 500 mg of metharbital, with solvent hexane in a continuous-extraction apparatus. Dry the residue in the thimble, extract with alcohol in the apparatus, evaporate the alcohol extract to dryness, and dry at 105° for 2 hours: a portion of the residue of metharbital so obtained responds to *Identification test A* under *Metharbital.*

Dissolution ⟨711⟩—

Medium: water; 900 mL.

Apparatus 1: 100 rpm.

Time: 45 minutes.

Procedure—Determine the amount of $C_9H_{14}N_2O_3$ dissolved from ultraviolet absorbances at the wavelength of maximum absorbance at about 244 nm of filtered portions of the solution under test, suitably diluted with freshly prepared dilute ammonium hydroxide (1 in 40), in comparison with a Standard solution having a known concentration of USP Metharbital RS in the same medium.

Tolerances—Not less than 75% (*Q*) of the labeled amount of $C_9H_{14}N_2O_3$ is dissolved in 45 minutes.

Uniformity of dosage units ⟨905⟩: meet the requirements.

Assay—Weigh and finely powder not less than 20 Metharbital Tablets. Weigh accurately a portion of the powder, equivalent to about 100 mg of metharbital, and transfer to a 100-mL volumetric flask. Add 50 mL of sodium hydroxide solution (1 in 250), and shake by mechanical means for 15 minutes. Dilute with the sodium hydroxide solution to volume, mix, and filter. Transfer 10.0 mL of the filtrate to a second 100-mL volumetric flask, dilute with the sodium hydroxide solution to volume, and mix. Transfer 10.0 mL of this last solution to a third 100-mL volumetric flask, dilute with the sodium hydroxide solution to volume, and mix. Concomitantly determine the absorbances of this solution and of a Standard solution of USP Metharbital RS in the same medium having a known concentration of about 10 µg per mL in 1-cm cells at the wavelength of maximum absorbance at about 244 nm, with a suitable spectrophotometer, using sodium hydroxide solution (1 in 250) as the blank. Calculate the quantity, in mg, of $C_9H_{14}N_2O_3$ in the portion of Tablets taken by the formula:

$$10C(A_U/A_S),$$

in which *C* is the concentration, in µg per mL, of USP Metharbital RS in the Standard solution, and A_U and A_S are the absorbances of the solution from the Tablets and the Standard solution, respectively.

Methazolamide

CH_3CON … S … SO_2NH_2 / CH_3-N——N

$C_5H_8N_4O_3S_2$ 236.26

Acetamide, *N*-[5-(aminosulfonyl)-3-methyl-1,3,4-thiadiazol-2(3*H*)-ylidene]-.

N-(4-Methyl-2-sulfamoyl-Δ^2-1,3,4-thiadiazolin-5-ylidene)-acetamide [*554-57-4*].

» Methazolamide contains not less than 96.0 percent and not more than 100.5 percent of $C_5H_8N_4O_3S_2$, calculated on the dried basis.

Packaging and storage—Preserve in well-closed, light-resistant containers.

Reference standard—*USP Methazolamide Reference Standard*—Dry at 105° for 2 hours before using.

Identification—

A: The infrared absorption spectrum of a potassium bromide dispersion of it exhibits maxima only at the same wavelengths as that of a similar preparation of USP Methazolamide RS.

B: The ultraviolet absorption spectrum of a 1 in 100,000 solution of it in sodium hydroxide solution (1 in 250) exhibits maxima at the same wavelengths as that of a similar solution of USP Methazolamide RS, concomitantly measured.

Loss on drying ⟨731⟩—Dry it in vacuum at 105° for 2 hours: it loses not more than 0.5% of its weight.

Residue on ignition ⟨281⟩: not more than 0.1%.

Selenium ⟨291⟩: 0.003%, a 200-mg specimen being used.

Heavy metals, *Method II* ⟨231⟩: 0.002%.

Assay—Transfer about 50 mg of Methazolamide, accurately weighed, to a 100-mL volumetric flask, add 50 mL of pH 8.4 alkaline borate buffer (see *Buffer Solutions* in the section, *Reagents, Indicators, and Solutions*), and shake the flask until solution is complete. Dilute with pH 8.4 alkaline borate buffer to volume, and mix. Pipet 2 mL of this solution into a 100-mL volumetric flask, dilute with pH 8.4 alkaline borate buffer to volume, and mix. Dissolve an accurately weighed quantity of USP Methazolamide RS in pH 8.4 alkaline borate buffer, and dilute quantitatively and stepwise with the borate buffer to obtain a Standard solution having a known concentration of about 10 µg per mL. Concomitantly determine the absorbances of both solutions in 1-cm cells at the wavelength of maximum absorbance at about 288 nm, with a suitable spectrophotometer, using pH 8.4 alkaline borate buffer as the blank. Calculate the quantity, in mg, of $C_5H_8N_4O_3S_2$ in the portion of Methazolamide taken by the formula:

$$5C(A_U/A_S),$$

in which C is the concentration, in μg per mL, of USP Methazolamide RS in the Standard solution, and A_U and A_S are the absorbances of the solution of Methazolamide and the Standard solution, respectively.

Methazolamide Tablets

» Methazolamide Tablets contain not less than 95.0 percent and not more than 105.0 percent of the labeled amount of $C_5H_8N_4O_3S_2$.

Packaging and storage—Preserve in well-closed containers.

Reference standard—*USP Methazolamide Reference Standard*—Dry at 105° for 2 hours before using.

Identification—

A: Extract a quantity of finely powdered Tablets, equivalent to about 250 mg of methazolamide, with about 50 mL of acetone. Filter, and add solvent hexane until a heavy white precipitate is formed. Collect the solid on a filter, and dry: the infrared absorption spectrum of a potassium bromide dispersion of the methazolamide so obtained exhibits maxima only at the same wavelengths as that of a similar preparation of USP Methazolamide RS.

B: Dissolve about 100 mg of the dried solid obtained in *Identification test A* in 5 mL of 1 *N* sodium hydroxide, and add 5 mL of a mixture of 1 g of hydroxylamine hydrochloride and 500 mg of cupric sulfate in 100 mL of water. Heat the solution on a steam bath for 15 minutes: the solution turns dark amber, then a black precipitate is formed.

C: The polarogram of the solution employed for measurement in the *Assay* exhibits a half-wave potential ($E_{1/2}$) within 1% of that of USP Methazolamide RS, similarly measured ($E_{1/2}$ is about −1.3 volts against the saturated calomel electrode).

Dissolution ⟨711⟩—

Medium: pH 4.5 acetate buffer, prepared by mixing 2.99 g of sodium acetate trihydrate and 1.66 mL of glacial acetic acid with water to obtain 1000 mL of a solution having a pH of 4.5; 900 mL.

Apparatus 2: 100 rpm.

Time: 45 minutes.

Procedure—Determine the amount of $C_5H_8N_4O_3S_2$ dissolved from ultraviolet absorbances at the wavelength of maximum absorbance at about 252 nm of filtered portions of the solution under test, suitably diluted with pH 4.5 acetate buffer, in comparison with a Standard solution having a known concentration of USP Methazolamide RS in the same medium.

Tolerances—Not less than 75% (Q) of the labeled amount of $C_5H_8N_4O_3S_2$ is dissolved in 45 minutes.

Uniformity of dosage units ⟨905⟩: meet the requirements.

Assay—Weigh and finely powder not less than 20 Methazolamide Tablets. Weigh accurately a portion of the powder, equivalent to about 50 mg of methazolamide, and transfer to a 200-mL volumetric flask. Add methanol to volume, and mix. Allow any undissolved matter to settle, pipet 10 mL of the clear, supernatant liquid into a 50-mL volumetric flask, and dilute with 0.1 *M* ammonium chloride to volume. Transfer a portion of this solution to a polarographic cell that is immersed in a water bath regulated at 24.5° to 25.5°. Deaerate by bubbling purified nitrogen through the assay solution for 15 minutes. Insert the dropping mercury electrode of a suitable polarograph, and record the polarogram from −0.8 volt to −1.80 volts, using a saturated calomel electrode as the reference electrode. Determine the height of the diffusion current at −1.62 ± 0.03 volts. Calculate the quantity, in mg, of $C_5H_8N_4O_3S_2$ in the portion of Tablets taken by the formula:

$$C[(i_d)_U/(i_d)_S],$$

in which $(i_d)_U$ is the observed diffusion current of the assay solution and $(i_d)_S$ is that determined similarly in a solution of USP Methazolamide RS, the concentration of which is C μg per mL (about 50 μg per mL).

Methdilazine

$C_{18}H_{20}N_2S$ 296.43

10*H*-Phenothiazine, 10-[(1-methyl-3-pyrrolidinyl)methyl]-.

10-[(1-Methyl-3-pyrrolidinyl)methyl]phenothiazine [*1982-37-2*].

» Methdilazine contains not less than 97.0 percent and not more than 103.0 percent of $C_{18}H_{20}N_2S$, calculated on the dried basis.

Packaging and storage—Preserve in tight, light-resistant containers.

Reference standard—*USP Methdilazine Reference Standard*—Dry in vacuum at 70° for 4 hours before using.

NOTE—Throughout the following procedures, protect test or assay specimens, the Reference Standard, and solutions containing them, by conducting the procedures without delay, under subdued light, or using low-actinic glassware.

Identification—

A: The infrared absorption spectrum of a potassium bromide dispersion of it, previously dried, exhibits maxima only at the same wavelengths as that of a similar preparation of USP Methdilazine RS.

B: Dissolve 25.0 mg in 25 mL of 1 *N* hydrochloric acid, and dilute quantitatively and stepwise with water to a concentration of 5 μg per mL: the ultraviolet absorption spectrum of this solution exhibits maxima and minima at the same wavelengths as that of a 1 in 200,000 solution of USP Methdilazine RS in the same medium, concomitantly measured.

Melting range, *Class I* ⟨741⟩: between 83° and 88°, but the range between beginning and end of melting does not exceed 2°.

Loss on drying ⟨731⟩—Dry it in vacuum at 70° for 4 hours: it loses not more than 1.0% of its weight.

Residue on ignition ⟨281⟩: not more than 0.5%.

Selenium ⟨291⟩—The absorbance of the solution from the *Test Solution*, prepared with 100 mg of Methdilazine and 100 mg of magnesium oxide, is not greater than one-half that from the *Standard Solution* (0.003%).

Heavy metals, *Method II* ⟨231⟩: 0.002%.

Assay—Transfer about 100 mg of Methdilazine, accurately weighed, to a 1000-mL volumetric flask, dissolve in 100 mL of 1 *N* hydrochloric acid, dilute with water to volume, and mix. Transfer 5.0 mL of this solution to a 100-mL volumetric flask, dilute with water to volume, and mix. Concomitantly determine the absorbances of this solution and of a Standard solution of USP Methdilazine RS in the same medium having a known concentration of about 5 μg per mL, in 1-cm cells, at 252 and 275 nm, with a suitable spectrophotometer, using water as the blank. Calculate the quantity, in mg, of $C_{18}H_{20}N_2S$ in the Methdilazine taken by the formula:

$$20C(A_{252} - A_{275})_U/(A_{252} - A_{275})_S,$$

in which C is the concentration, in μg per mL, of USP Methdilazine RS in the Standard solution, and the parenthetic expressions are the differences in the absorbances of the two solutions at the wavelengths indicated by the subscripts, for the solution of Methdilazine (U) and the Standard solution (S), respectively.

Methdilazine Tablets

» Methdilazine Tablets contain not less than 93.0 percent and not more than 107.0 percent of the labeled amount of $C_{18}H_{20}N_2S$.

Packaging and storage—Preserve in tight, light-resistant containers.

Reference standard—*USP Methdilazine Reference Standard*—Dry in vacuum at 70° for 4 hours before using.

NOTE—Throughout the following procedures, protect test or assay specimens, the Reference Standard, and solutions containing them, by conducting the procedures without delay, under subdued light, or using low-actinic glassware.

Identification—Place in a mortar a number of Tablets, equivalent to about 20 mg of methdilazine, add 10 mL of ether, and grind thoroughly. Cool the mixture by evaporating a small amount of the ether in a current of air, and filter. Again cool the filtrate by evaporating a small amount of the ether. Filter if a precipitate forms. Evaporate the filtrate on a steam bath to dryness, dissolve the methdilazine so obtained in about 3 mL of chloroform, and transfer to a suitable mortar containing 500 mg of potassium bromide. Grind thoroughly, dry under vacuum, repeat if necessary to remove the last trace of solvent, and prepare a pressed disk from the triturate: the infrared absorption spectrum of a potassium bromide dispersion of the methdilazine so obtained exhibits maxima only at the same wavelengths as that of a similar preparation of USP Methdilazine RS in the 6- to 15-μm region.

Disintegration ⟨701⟩: 30 minutes.

Uniformity of dosage units ⟨905⟩: meet the requirements.

Assay—

Standard preparation—Dissolve a suitable quantity of USP Methdilazine RS, accurately weighed, in chloroform, and dilute quantitatively with chloroform to obtain a solution having a known concentration of about 400 μg per mL.

Assay preparation—Weigh and finely powder not less than 20 Methdilazine Tablets, and transfer an accurately weighed portion of the powder, equivalent to about 40 mg of methdilazine, to a 100-mL volumetric flask. Add 60 mL of chloroform, shake for 20 minutes, dilute with chloroform to volume, and mix. Filter, discarding the first 15 mL of the filtrate. Use the subsequent filtrate as directed in the *Procedure*.

Procedure—Into three separate 100-mL volumetric flasks transfer 10.0 mL each of the *Standard preparation*, the *Assay preparation*, and chloroform to provide the blank. To each flask add 20 mL of chloroform and 4.0 mL of buffered palladium chloride TS, dilute with alcohol to volume, and mix. Concomitantly determine the absorbances of the solutions in 1-cm cells at the wavelength of maximum absorbance at about 460 nm, with a suitable spectrophotometer, using the blank to set the instrument. Calculate the quantity, in mg, of $C_{18}H_{20}N_2S$ in the portion of Tablets taken by the formula:

$$0.1C(A_U/A_S),$$

in which C is the concentration, in μg per mL, of USP Methdilazine RS in the *Standard preparation*, and A_U and A_S are the absorbances of the solutions from the *Assay preparation* and the *Standard preparation*, respectively.

Methdilazine Hydrochloride

$C_{18}H_{20}N_2S \cdot HCl$ 332.89

10*H*-Phenothiazine, 10-[(1-methyl-3-pyrrolidinyl)methyl]-, monohydrochloride.

10-[(1-Methyl-3-pyrrolidinyl)methyl]phenothiazine monohydrochloride [*1229-35-2*].

» Methdilazine Hydrochloride contains not less than 97.0 percent and not more than 103.0 percent of $C_{18}H_{20}N_2S \cdot HCl$, calculated on the dried basis.

Packaging and storage—Preserve in tight, light-resistant containers.

Reference standard—*USP Methdilazine Hydrochloride Reference Standard*—Dry in vacuum at 65° for 16 hours before using.

NOTE—Throughout the following procedures, protect test or assay specimens, the Reference Standard, and solutions containing them, by conducting the procedures without delay, under subdued light, or using low-actinic glassware.

Identification—

A: The infrared absorption spectrum of a potassium bromide dispersion of it, previously dried, exhibits maxima only at the same wavelengths as that of a similar preparation of USP Methdilazine Hydrochloride RS.

B: The ultraviolet absorption spectrum of a solution (1 in 200,000) exhibits maxima and minima at the same wavelengths as that of a similar solution of USP Methdilazine Hydrochloride RS, concomitantly measured.

C: A solution of it responds to the tests for *Chloride* ⟨191⟩.

Melting range, *Class I* ⟨741⟩: between 184° and 190°.

pH ⟨791⟩: between 4.8 and 6.0, in a solution (1 in 100).

Loss on drying ⟨731⟩—Dry it in vacuum at 65° for 16 hours: it loses not more than 1.0% of its weight.

Residue on ignition ⟨281⟩: not more than 0.5%.

Heavy metals, *Method II* ⟨231⟩: 0.002%.

Selenium ⟨291⟩—The absorbance of the solution from the *Test Solution*, prepared with 100 mg of Methdilazine Hydrochloride and 100 mg of magnesium oxide, is not greater than one-half that from the *Standard Solution* (0.003%).

Ordinary impurities ⟨466⟩—

Test solution: methanol.

Standard solution: methanol.

Application volume: 10 μL.

Eluant: a mixture of toluene, isopropyl alcohol, and ammonium hydroxide (70:29:1), in a nonequilibrated chamber.

Visualization: 5.

Assay—Transfer about 100 mg of Methdilazine Hydrochloride, accurately weighed, to a 1000-mL volumetric flask, add water to volume, and mix. Transfer 5.0 mL of this solution to a 100-mL volumetric flask, dilute with water to volume, and mix. Concomitantly determine the absorbances of this solution and of a Standard solution of USP Methdilazine Hydrochloride RS in the same medium having a known concentration of about 5 μg per mL, in 1-cm cells at 252 and 275 nm, with a suitable spectrophotometer, using water as the blank. Calculate the quantity, in mg, of $C_{18}H_{20}N_2S \cdot HCl$ in the portion of Methdilazine Hydrochloride taken by the formula:

$$20C(A_{252} - A_{275})_U/(A_{252} - A_{275})_S,$$

in which C is the concentration, in μg per mL, of USP Methdilazine Hydrochloride RS in the Standard solution, and the parenthetic expressions are the differences in the absorbances of the two solutions at the wavelengths indicated by the subscripts, for the solution of Methdilazine Hydrochloride (U) and the Standard solution (S), respectively.

Methdilazine Hydrochloride Syrup

» Methdilazine Hydrochloride Syrup contains not less than 93.0 percent and not more than 107.0 percent of the labeled amount of $C_{18}H_{20}N_2S \cdot HCl$.

Packaging and storage—Preserve in tight, light-resistant containers.

Reference standard—*USP Methdilazine Hydrochloride Reference Standard*—Dry in vacuum at 65° for 16 hours before using.

NOTE—Throughout the following procedures, protect test or assay specimens, the Reference Standard, and solutions containing them, by conducting the procedures without delay, under subdued light, or using low-actinic glassware.

Identification—Transfer a volume of Syrup, equivalent to about 4 mg of methdilazine hydrochloride, to a 60-mL separator, add 5 mL of 0.1 *N* hydrochloric acid, and extract with 10 mL of ether, discarding the extract. Add 10 mL of sodium bicarbonate solution (1 in 10) to the separator, and extract with 3 mL of chloroform. Filter the extract through a pledget of cotton. Evaporate the chloroform, carefully removing the last trace of solvent in a small vacuum flask: the infrared absorption spectrum of a potassium bromide dispersion of the methdilazine so obtained exhibits maxima only at the same wavelengths as that of a similar preparation of USP Methdilazine Hydrochloride RS that has been treated in the same manner.

pH ⟨791⟩: between 3.3 and 4.1.

Alcohol content, *Method II* ⟨611⟩: between 6.5% and 7.5% of C_2H_5OH.

Assay—

Standard preparation—Dissolve a suitable quantity of USP Methdilazine Hydrochloride, accurately weighed, in chloroform, and dilute quantitatively with chloroform to obtain a solution having a known concentration of about 400 µg per mL.

Assay preparation—Transfer a volume of Methdilazine Hydrochloride Syrup, equivalent to about 4 mg of methdilazine hydrochloride, to a 60-mL separator, add 10 mL of a saturated solution of sodium chloride, and extract with three 10-mL portions of chloroform, transferring the extracts to a 100-mL volumetric flask.

Procedure—Transfer 10.0 mL of *Standard preparation* to a 100-mL volumetric flask, and add 20 mL of chloroform. To this flask and to the flask containing the *Assay preparation* add 4.0 mL of buffered palladium chloride TS, dilute with alcohol to volume, and mix. Concomitantly determine the absorbances of the solutions in 1-cm cells at the wavelength of maximum absorbance at about 460 nm, with a suitable spectrophotometer, using a mixture of 30 mL of chloroform, 4 mL of palladium chloride TS, and 66 mL of alcohol as the blank. Calculate the quantity, in mg, of $C_{18}H_{20}N_2S.HCl$ in each mL of the Syrup taken by the formula:

$$(0.01C/V)(A_U/A_S),$$

in which C is the concentration, in µg per mL, of USP Methdilazine Hydrochloride RS in the *Standard preparation*, V is the volume, in mL, of Syrup taken, and A_U and A_S are the absorbances of the solutions from the *Assay preparation* and the *Standard preparation*, respectively.

Methdilazine Hydrochloride Tablets

» Methdilazine Hydrochloride Tablets contain not less than 93.0 percent and not more than 107.0 percent of the labeled amount of $C_{18}H_{20}N_2S.HCl$.

Packaging and storage—Preserve in tight, light-resistant containers.

Reference standard—*USP Methdilazine Hydrochloride Reference Standard*—Dry in vacuum at 65° for 16 hours before using.

NOTE—Throughout the following procedures, protect test or assay specimens, the Reference Standard, and solutions containing them, by conducting the procedures without delay, under subdued light, or using low-actinic glassware.

Identification—Transfer a portion of finely powdered Tablets, equivalent to about 8 mg of methdilazine hydrochloride, to a 60-mL separator, add 10 mL of sodium bicarbonate solution (1 in 10), and extract with 3 mL of chloroform. Filter the extract through a pledget of cotton. Evaporate the chloroform, carefully removing the last trace of solvent in a small vacuum flask: the infrared absorption spectrum of a potassium bromide dispersion of the methdilazine so obtained exhibits maxima only at the same wavelengths as that of a similar preparation of USP Methdilazine Hydrochloride RS, similarly treated and measured.

Dissolution ⟨711⟩—

Medium: water; 900 mL.

Apparatus 1: 100 rpm.

Time: 45 minutes.

Procedure—Determine the amount of $C_{18}H_{20}N_2S.HCl$ dissolved from ultraviolet absorbances at the wavelength of maximum absorbance at about 252 nm of filtered portions of the solution under test, suitably diluted with *Dissolution Medium*, if necessary, in comparison with a Standard solution having a known concentration of USP Methdilazine Hydrochloride RS in the same medium.

Tolerances—Not less than 75% (*Q*) of the labeled amount of $C_{18}H_{20}N_2S.HCl$ is dissolved in 45 minutes.

Uniformity of dosage units ⟨905⟩: meet the requirements.

Assay—

Standard preparation—Dissolve a suitable quantity of USP Methdilazine Hydrochloride RS, accurately weighed, in chloroform, and dilute quantitatively with chloroform to obtain a solution having a known concentration of about 400 µg per mL.

Assay preparation—Weigh and finely powder not less than 20 Methdilazine Hydrochloride Tablets, and transfer an accurately weighed portion of the powder, equivalent to about 80 mg of methdilazine hydrochloride, to a 200-mL volumetric flask. Add 60 mL of chloroform, shake for 20 minutes, dilute with chloroform to volume, and mix. Filter, discarding the first 15 mL of the filtrate. Use the subsequent filtrate as directed in the *Procedure*.

Procedure—Into three separate 100-mL volumetric flasks transfer 10.0 mL each of the *Standard preparation*, the *Assay preparation*, and chloroform to provide the blank. To each flask add 20 mL of chloroform and 4.0 mL of buffered palladium chloride TS, dilute with alcohol to volume, and mix. Concomitantly determine the absorbances of the solutions in 1-cm cells at the wavelength of maximum absorbance at about 460 nm, with a suitable spectrophotometer, using the blank to set the instrument. Calculate the quantity, in mg, of $C_{18}H_{20}N_2S.HCl$ in the portion of Tablets taken by the formula:

$$0.2C(A_U/A_S),$$

in which C is the concentration, in µg per mL, of USP Methdilazine Hydrochloride RS in the *Standard preparation*, and A_U and A_S are the absorbances of the solutions from the *Assay preparation* and the *Standard preparation*, respectively.

Methenamine

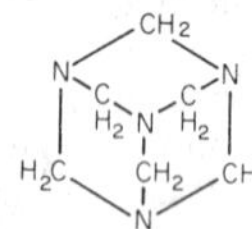

$C_6H_{12}N_4$ 140.19

1,3,5,7-Tetraazatricyclo[3.3.1.1^{3,7}]decane.

Hexamethylenetetramine [*100-97-0*].

» Methenamine, dried over phosphorus pentoxide for 4 hours, contains not less than 99.0 percent and not more than 100.5 percent of $C_6H_{12}N_4$.

Packaging and storage—Preserve in well-closed containers.

Reference standard—*USP Methenamine Reference Standard*—Dry over phosphorus pentoxide for 4 hours before using.

Identification—

A: The infrared absorption spectrum of a potassium bromide dispersion of it, previously dried, exhibits maxima only at the same wavelengths as that of a similar preparation of USP Methenamine RS.

B: Heat a solution (1 in 10) with 2 *N* sulfuric acid: formaldehyde is liberated, recognizable by its odor and by its darkening of paper moistened with silver ammonium nitrate TS. On the subsequent addition of an excess of 1 *N* sodium hydroxide to the solution, ammonia is evolved.

Loss on drying ⟨731⟩—Dry it over phosphorus pentoxide for 4 hours: it loses not more than 2.0% of its weight.

Residue on ignition ⟨281⟩: not more than 0.1%.

Chloride ⟨221⟩—A 1.0-g portion shows no more chloride than corresponds to 0.20 mL of 0.020 *N* hydrochloric acid (0.014%).

Sulfate—To 10 mL of a solution (1 in 50), acidified with 5 drops of hydrochloric acid, add 5 drops of barium chloride TS: no turbidity is produced within 1 minute.

Ammonium salts—To 10 mL of a solution (1 in 20) add 1 mL of alkaline mercuric–potassium iodide TS: the mixture is not darker in color than a mixture of 1 mL of the reagent and 10 mL of water.

Heavy metals, *Method I* ⟨231⟩—Dissolve 2 g in 10 mL of water, add 2 mL of 3 *N* hydrochloric acid, and dilute with water to 25 mL. Proceed as directed, except to use glacial acetic acid to adjust the pH: the limit is 0.001%.

Assay—

Chromotropic acid spot test solution—Suspend 100 mg of chromotropic acid in 2 mL of water, and *cautiously* add 3 mL of sulfuric acid. Allow to cool. Add 25 mL of sulfuric acid, and mix. [NOTE—If excessive heat generated during mixing causes a violet color to appear in the solution, discard the solution and prepare another, taking precautions to avoid excessive heat.]

Procedure—Transfer about 1 g of Methenamine, previously dried and accurately weighed, to a beaker. Add 40.0 mL of 1 *N* sulfuric acid VS, and heat to a gentle boil, adding water from time to time, if necessary, until the formaldehyde has been expelled. Test for the absence of formaldehyde by adding a drop of the assay solution to a glass fiber filter disk, on a watch glass, on which has previously been placed 3 or 4 drops of *Chromotropic acid spot test solution.* Formaldehyde produces a violet color with this reagent; repeat the test until no violet color is obtained on the warmed test filter disk upon comparison with a blank filter disk to which no assay specimen is added. Cool, add 20 mL of water, then add methyl red TS, and titrate the excess acid with 1 *N* sodium hydroxide VS. Perform a blank determination (see *Residual Titrations* under *Titrimetry* ⟨541⟩). Each mL of 1 *N* sulfuric acid is equivalent to 35.05 mg of $C_6H_{12}N_4$.

Methenamine Elixir

» Methenamine Elixir contains not less than 90.0 percent and not more than 110.0 percent of the labeled amount of $C_6H_{12}N_4$.

Packaging and storage—Preserve in tight containers.

Reference standard—*USP Methenamine Reference Standard*—Dry over phosphorus pentoxide for 4 hours before using.

Identification—Heat a volume of Elixir, equivalent to about 1 g of methenamine, with 10 mL of 2 *N* sulfuric acid: formaldehyde is liberated, recognizable by its odor and by its darkening of paper moistened with silver ammonium nitrate TS. On the subsequent addition of an excess of 1 *N* sodium hydroxide to the solution, ammonia is evolved.

Alcohol content, *Method I* ⟨611⟩: between 19.0% and 21.0% of C_2H_5OH.

Assay—

Chromotropic acid solution—Mix 100 mg of chromotropic acid with 50 mL of water in a 100-mL volumetric flask. Cool in an ice bath and, while cooling, cautiously and slowly add 50 mL of sulfuric acid, and mix. Allow the solution to reach room temperature, and add dilute sulfuric acid (1 in 2) to volume. [NOTE—If excessive heat generated during mixing causes a violet color to appear in the solution, discard the solution, and prepare another, taking precautions to avoid excessive heat.]

Standard preparation—Transfer about 50 mg of USP Methenamine RS, accurately weighed, to a 1000-mL volumetric flask, dissolve in water, dilute with water to volume, and mix. Proceed as directed under *Assay preparation*, beginning with "Transfer a 2.0-mL portion of this stock solution to a 100-mL volumetric flask." The concentration of USP Methenamine RS in the *Standard preparation* is about 1 µg per mL.

Assay preparation—Transfer an accurately measured volume of Methenamine Elixir, equivalent to about 1.5 g of methenamine, to a 500-mL volumetric flask, dissolve in water, dilute with water to volume, and mix. Transfer 2.0 mL of this solution to a 100-mL volumetric flask, dilute with water to volume, and mix to obtain the stock solution. Transfer a 2.0-mL portion of this stock solution to a 100-mL volumetric flask, add 25 mL of *Chromotropic acid solution* and 50 mL of dilute sulfuric acid (1 in 2), and mix. Transfer another 2.0-mL portion of the stock solution to a second 100-mL volumetric flask to provide a blank, add 75 mL of dilute sulfuric acid (1 in 2), and mix. Place the two 100-mL flasks in a boiling water bath for 30 minutes, accurately timed, then remove them from the bath, cool immediately to room temperature, add dilute sulfuric acid (1 in 2) to volume, and mix.

Procedure—Concomitantly determine the absorbances of the solutions in 1-cm cells at the wavelength of maximum absorbance at about 570 nm, with a suitable spectrophotometer, using dilute sulfuric acid (1 in 2) to set the instrument. Calculate the quantity, in mg, of $C_6H_{12}N_4$ in each mL of the Elixir taken by the formula:

$$(1250C/V)[(A_U - B_U)/(A_S - B_S)],$$

in which C is the concentration, in µg per mL, of USP Methenamine RS in the *Standard preparation*, V is the volume, in mL, of Elixir taken, A_U and A_S are the absorbances of the *Assay preparation* and the *Standard preparation*, respectively, and B_U and B_S are the absorbances of the blanks from the *Assay preparation* and the *Standard preparation*, respectively.

Methenamine Tablets

» Methenamine Tablets contain not less than 95.0 percent and not more than 105.0 percent of the labeled amount of $C_6H_{12}N_4$.

Packaging and storage—Preserve in well-closed containers.

Reference standard—*USP Methenamine Reference Standard*—Dry over phosphorus pentoxide for 4 hours before using.

Identification—Dissolve about 500 mg of powdered Tablets in 10 mL of water, add 10 mL of 2 *N* sulfuric acid, and heat: formaldehyde is liberated, recognizable by its odor and by its darkening of paper moistened with silver ammonium nitrate TS. On the subsequent addition of an excess of 1 *N* sodium hydroxide to the solution, ammonia is evolved.

Dissolution ⟨711⟩—

Medium: water; 900 mL.

Apparatus 1: 100 rpm.

Time: 45 minutes.

Procedure—Determine the amount of $C_6H_{12}N_4$ dissolved, employing the procedure set forth in the *Assay*, making any necessary modifications.

Tolerances—Not less than 75% (Q) of the labeled amount of $C_6H_{12}N_4$ is dissolved in 45 minutes.

Uniformity of dosage units ⟨905⟩: meet the requirements.

Assay—

Chromotropic acid solution and *Standard preparation*—Prepare as directed in the *Assay* under *Methenamine Elixir.*

Assay preparation—Weigh and finely powder not less than 20 Methenamine Tablets. Transfer an accurately weighed portion of the powder, equivalent to about 500 mg of methenamine, to a 250-mL volumetric flask, dilute with water to volume, mix, and filter, discarding the first 20 mL of the filtrate. Transfer 25.0 mL of the subsequent filtrate to a 1000-mL volumetric flask, dilute with water to volume, and mix. Proceed as directed for *Assay preparation* in the *Assay* under *Methenamine Elixir*, beginning with "Transfer a 2.0-mL portion of this stock solution."

Procedure—Proceed as directed for *Procedure* in the *Assay* under *Methenamine Elixir.* Calculate the quantity, in mg, of $C_6H_{12}N_4$ in the portion of Tablets taken by the formula:

$$500C[(A_U - B_U)/(A_S - B_S)],$$

in which C is the concentration, in µg per mL, of USP Methenamine RS in the *Standard preparation*, and A_U, A_S, B_U, and B_S are as defined therein.

Methenamine and Monobasic Sodium Phosphate Tablets

» Methenamine and Monobasic Sodium Phosphate Tablets contain not less than 92.5 percent and not more than 107.5 percent of the labeled amount of methenamine ($C_6H_{12}N_4$) and of monobasic sodium phosphate ($NaH_2PO_4 . H_2O$).

Packaging and storage—Preserve in tight containers.

Reference standard—*USP Methenamine Reference Standard*—Dry over phosphorus pentoxide for 4 hours before using.

Identification—

A: Powder several Tablets, and heat a filtered solution of the powder, equivalent to a methenamine solution (1 in 10), with 2 *N* sulfuric acid: it decomposes with the liberation of formaldehyde, recognizable by its odor and by its darkening of paper moistened with silver ammonium nitrate TS. On the subsequent addition of an excess of 1 *N* sodium hydroxide to the solution, ammonia is evolved.

B: Add 1 mL of alkaline mercuric-potassium iodide TS to a filtered solution of Tablets, equivalent to a methenamine solution (1 in 20): a yellow precipitate is formed, and after acidification with hydrochloric acid, the precipitate becomes soluble upon the addition of 6 *N* ammonium hydroxide.

C: A filtered solution of Tablets, equivalent to a monobasic sodium phosphate solution (1 in 20) responds to the tests for *Sodium* ⟨191⟩ and for *Phosphate* ⟨191⟩.

Dissolution ⟨711⟩—

Medium: water; 900 mL.

Apparatus 1: 100 rpm.

Time: 45 minutes.

Procedure—Determine the amount of $C_6H_{12}N_4$ dissolved, employing the procedure set forth in the *Assay for methenamine*, making any necessary modifications.

Tolerances—Not less than 75% (*Q*) of the labeled amount of $C_6H_{12}N_4$ is dissolved in 45 minutes.

Uniformity of dosage units ⟨905⟩: meet the requirements.

Ammonium salts—Dissolve 500 mg of powdered Tablets in 10 mL of water, add 5 mL of 1 *N* sodium hydroxide, and then add 1 mL of alkaline mercuric-potassium iodide TS: the color produced in 15 seconds may be yellow, but not orange-yellow or darker.

Assay for methenamine—

Chromotropic acid solution and *Standard preparation*—Prepare as directed in the *Assay* under *Methenamine Elixir.*

Assay preparation—Weigh and finely powder not less than 20 Methenamine and Monobasic Sodium Phosphate Tablets. Transfer an accurately weighed portion of the powder, equivalent to about 500 mg of methenamine, to a 250-mL volumetric flask, add water to volume, and shake thoroughly to dissolve. Filter, if necessary, discarding the first 20 mL of the filtrate, and dilute 25.0 mL of the subsequent filtrate to 1000.0 mL with water to obtain the stock solution. Proceed as directed for *Assay preparation* in the *Assay* under *Methenamine Elixir*, beginning with "Transfer a 2.0-mL portion of this stock solution."

Procedure—Proceed as directed for *Procedure* in the *Assay* under *Methenamine Elixir.* Calculate the quantity, in mg, of $C_6H_{12}N_4$ in the portion of Tablets taken by the formula:

$$500C[(A_U - B_U)/(A_S - B_S)],$$

in which *C* is the concentration, in µg per mL, of USP Methenamine RS in the *Standard preparation*, and A_U, A_S, B_U, and B_S are as defined therein.

Assay for monobasic sodium phosphate—Weigh and finely powder not less than 10 Methenamine and Monobasic Sodium Phosphate Tablets. Transfer an accurately weighed portion of the powder, equivalent to about 60 mg of monobasic sodium phosphate, to a 600-mL beaker, and add 25 mL of nitric acid and 10 mL of sulfuric acid. Cover with a watch glass, and boil gently, adding additional nitric acid, if necessary, until all organic matter is destroyed. Increase the heat, and digest until dense, white fumes are evolved. Cool, dilute with water to about 100 mL, and neutralize the solution with ammonium hydroxide. Cool, and make the solution just acidic with nitric acid. Add 10 g of ammonium nitrate, warm to 35° to 40°, add 75 mL of ammonium molybdate TS with stirring, and continue to stir the mixture at room temperature for 15 minutes. Allow the mixture to stand for 15 minutes, and decant the supernatant liquid through a filter crucible. Wash the precipitate in the beaker, by decantation, with two 30-mL portions of water, stirring thoroughly and allowing to settle before passing the supernatant liquid through the filter crucible. Transfer the precipitate to the filter, wash with cold water until the washings are neutral to litmus, and place the crucible and contents in the beaker used for the precipitation. Dissolve the yellow precipitate in the smallest possible excess of 0.5 *N* sodium hydroxide VS, accurately measuring the volume of 0.5 *N* sodium hydroxide VS added. Add 2.0 mL more of 0.5 *N* sodium hydroxide VS, and boil for 20 minutes. While the solution is still warm, add phenolphthalein TS, titrate the excess alkali with 0.5 *N* sulfuric acid VS just to discharge the pink color, and add 2.0 mL of 0.5 *N* sulfuric acid VS in excess. Boil the solution for 15 minutes, cool, add phenolphthalein TS, and titrate with 0.5 *N* sodium hydroxide VS to the first definite pink color obtained. Subtract the total volume of 0.5 *N* sulfuric acid used from the total volume of 0.5 *N* sodium hydroxide used to determine the amount of base consumed. Each mL of 0.5 *N* sodium hydroxide is equivalent to 2.654 mg of monobasic sodium phosphate ($NaH_2PO_4 . H_2O$).

Methenamine Hippurate

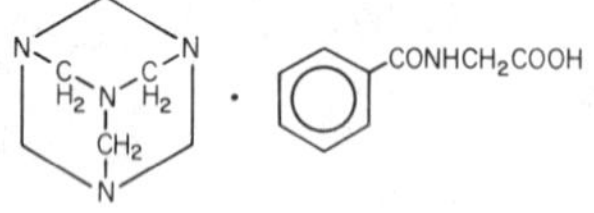

$C_6H_{12}N_4 . C_9H_9NO_3$ 319.36

Glycine, *N*-benzoyl, compd. with 1,3,5,7-tetraazatricyclo[3.3.1.1^{3,7}]decane (1:1).

Hexamethylenetetramine monohippurate [*5714-73-8*].

» Methenamine Hippurate, dried in vacuum at 60° for 1 hour, contains not less than 95.5 percent and not more than 102.0 percent of $C_6H_{12}N_4 . C_9H_9NO_3$, and contains not less than 54.0 percent and not more than 58.0 percent of hippuric acid ($C_9H_9NO_3$).

Packaging and storage—Preserve in well-closed containers.

Reference standard—*USP Methenamine Hippurate Reference Standard*—Dry in vacuum at 60° for 1 hour before using.

Identification—The infrared absorption spectrum of a mineral oil dispersion of it, previously dried, exhibits maxima only at the same wavelengths as that of a similar preparation of USP Methenamine Hippurate RS.

Loss on drying ⟨731⟩—Dry it in vacuum at 60° for 1 hour: it loses not more than 1.0% of its weight.

Residue on ignition ⟨281⟩: not more than 0.1%.

Sulfate—Dissolve 200 mg in 10 mL of water, and add 5 drops of 3 *N* hydrochloric acid and 5 drops of barium chloride TS: no turbidity appears within 1 minute.

Heavy metals, *Method II* ⟨231⟩: 0.0015%.

Hippuric acid content—Transfer about 1 g, accurately weighed, to a 250-mL conical flask, and add 50 mL of water. When solution is complete, add phenolphthalein TS, and titrate with 0.1 *N* sodium hydroxide VS. Each mL of 0.1 *N* sodium hydroxide is equivalent to 17.92 of $C_9H_9NO_3$.

Assay—Dissolve about 700 mg of Methenamine Hippurate, accurately weighed, in 50 mL of glacial acetic acid, and titrate with 0.1 *N* perchloric acid VS, determining the end-point potentiometrically. Perform a blank determination, and make any necessary correction. Each mL of 0.1 *N* perchloric acid is equivalent to 31.94 mg of $C_6H_{12}N_4 . C_9H_9NO_3$.

Methenamine Hippurate Tablets

» Methenamine Hippurate Tablets contain not less than 95.0 percent and not more than 105.0 percent of the labeled amount of $C_6H_{12}N_4 \cdot C_9H_9NO_3$.

Packaging and storage—Preserve in well-closed containers.

Reference standard—*USP Methenamine Hippurate Reference Standard*—Dry in vacuum at 60° for 1 hour before using.

Identification—A portion of finely powdered Tablets responds to the *Identification test* under *Methenamine Hippurate.*

Disintegration ⟨701⟩: 30 minutes.

Uniformity of dosage units ⟨905⟩: meet the requirements.

Assay—Weigh and finely powder not less than 20 Methenamine Hippurate Tablets. Transfer an accurately weighed portion of the powder, equivalent to about 700 mg of methenamine hippurate, to a 250-mL conical flask. Add 50 mL of alcohol, then add thymolphthalein TS, and titrate with 0.1 *N* sodium hydroxide VS. Perform a blank determination on a mixture of 50 mL of alcohol and 20 mL of water, and make any necessary correction. Each mL of 0.1 *N* sodium hydroxide is equivalent to 31.94 mg of $C_6H_{12}N_4 \cdot C_9H_9NO_3$.

Methenamine Mandelate

$C_6H_{12}N_4 \cdot C_8H_8O_3$ 292.34

Benzeneacetic acid, α-hydroxy-, compd. with 1,3,5,7-tetraazatricyclo[3.3.1.1^{3,7}]decane (1:1).

Hexamethylenetetramine monomandelate [*587-23-5*].

» Methenamine Mandelate contains not less than 95.5 percent and not more than 102.0 percent of $C_6H_{12}N_4 \cdot C_8H_8O_3$, and contains not less than 50.0 percent and not more than 53.0 percent of mandelic acid ($C_8H_8O_3$), calculated on the dried basis.

Packaging and storage—Preserve in well-closed containers.

Reference standard—*USP Methenamine Mandelate Reference Standard*—Dry over silica gel for 18 hours before using.

Identification—The infrared absorption spectrum of a potassium bromide dispersion of it, previously dried, exhibits maxima only at the same wavelengths as that of a similar preparation of USP Methenamine Mandelate RS.

Loss on drying ⟨731⟩—Dry it over silica gel for 18 hours: it loses not more than 1.5% of its weight.

Residue on ignition ⟨281⟩: not more than 0.1%.

Chloride ⟨221⟩—Dissolve 1.0 g in 10 mL of water, and add gradually 500 mg of anhydrous sodium carbonate. Evaporate to dryness, and ignite the residue at a dull-red heat. Add 20 mL of 2 *N* nitric acid, stir gently, and filter: the filtrate shows no more chloride than corresponds to 0.15 mL of 0.020 *N* hydrochloric acid (0.01%).

Sulfate—Dissolve 0.20 g in 10 mL of water, and add 5 drops of 3 *N* hydrochloric acid and 5 drops of barium chloride TS: no turbidity appears within 1 minute.

Heavy metals ⟨231⟩—Dissolve 1.3 g in 10 mL of water, add 2 mL of 3 *N* hydrochloric acid, and dilute with water to 25 mL: the limit is 0.0015%.

Mandelic acid content—Transfer about 90 mg, accurately weighed, to a 250-mL conical flask containing 50 mL of water. When solution is complete, titrate the magnetically stirred solution with 0.05 *N* ceric ammonium nitrate VS, determining the end-point potentiometrically. Each mL of 0.05 *N* ceric ammonium nitrate is equivalent to 3.804 mg of $C_8H_8O_3$.

Assay—

0.05 N Silver nitrate in dehydrated alcohol—Dissolve, by stirring, about 8.5 g of silver nitrate in 1000 mL of dehydrated alcohol. Transfer about 100 mg, accurately weighed, of sodium chloride, previously dried at 110° for 2 hours, to a 100-mL beaker, and dissolve in 50 mL of water. Titrate with the silver nitrate solution to the potentiometric end-point, using a silver billet indicator electrode and a silver–silver chloride double-junction reference electrode containing a potassium nitrate salt bridge. Calculate the normality of the titrant.

Procedure—Transfer about 60 mg of Methenamine Mandelate, accurately weighed, to a 250-mL conical flask. Add 15 mL of dehydrated alcohol, stir to dissolve, and add 40 mL of chloroform. Titrate with *0.05 N Silver nitrate in dehydrated alcohol*, determining the end-point potentiometrically, using a silver billet indicator electrode and a silver–silver chloride double-junction reference electrode containing a potassium nitrate salt bridge. Each mL of 0.05 *N* silver nitrate is equivalent to 7.308 mg of $C_6H_{12}N_4 \cdot C_8H_8O_3$.

Methenamine Mandelate for Oral Solution

» Methenamine Mandelate for Oral Solution contains not less than 90.0 percent and not more than 110.0 percent of the labeled amount of $C_6H_{12}N_4 \cdot C_8H_8O_3$.

Packaging and storage—Preserve in well-closed containers.

Labeling—Label Methenamine Mandelate for Oral Solution that contains insoluble ingredients to indicate that the aqueous constituted Oral Solution contains dissolved methenamine mandelate but may remain turbid because of the presence of added substances.

Reference standard—*USP Methenamine Mandelate Reference Standard*—Dry over silica gel for 18 hours before using.

Identification—A finely powdered portion, equivalent to about 100 mg of methenamine mandelate, responds to the *Identification test* under *Methenamine Mandelate Oral Suspension.*

pH ⟨791⟩: between 4.0 and 4.5, in a mixture of 1 g with 30 mL of water.

Water, *Method I* ⟨921⟩: not more than 0.5%.

Assay—Weigh accurately the contents of not less than 10 containers of Methenamine Mandelate for Oral Solution, and reduce to a fine powder. Transfer an accurately weighed portion of the powder, equivalent to about 60 mg of methenamine mandelate, to a 150-mL beaker. Add 15 mL of dehydrated alcohol, stir to dissolve, and proceed as directed in the *Assay* under *Methenamine Mandelate*, beginning with "add 40 mL of chloroform."

Methenamine Mandelate Oral Suspension

» Methenamine Mandelate Oral Suspension is Methenamine Mandelate suspended in vegetable oil. It contains not less than 90.0 percent and not more than 110.0 percent of the labeled amount of $C_6H_{12}N_4 \cdot C_8H_8O_3$.

Packaging and storage—Preserve in tight containers.

Reference standard—*USP Methenamine Mandelate Reference Standard*—Dry over silica gel for 18 hours before using.

Identification—Triturate a quantity, equivalent to about 100 mg of methenamine mandelate, with 10 mL of chloroform, and filter through a 0.45-μm membrane filter. Evaporate the solvent, wash the residue with 5 small portions of ether, and allow it to air-dry: the infrared absorption spectrum of a potassium bromide dispersion of the residue so obtained exhibits maxima only at the same wavelengths as that of a potassium bromide dispersion of USP Methenamine Mandelate RS.

Water, *Method I* ⟨921⟩: not more than 0.1%.

Assay—Shake the Methenamine Mandelate Oral Suspension, then pipet, using a "to contain" pipet, an amount equivalent to 1 g of methenamine mandelate, into a 100-mL volumetric flask. Add 5.0 mL of dehydrated alcohol, mix, and add methylene chloride to volume. Pipet 5 mL of this solution into a 250-mL conical flask. Add 15 mL of dehydrated alcohol, and proceed as directed in the *Assay* under *Methenamine Mandelate*, beginning with "add 40 mL of chloroform."

Methenamine Mandelate Tablets

» Methenamine Mandelate Tablets contain not less than 95.0 percent and not more than 105.0 percent of the labeled amount of $C_6H_{12}N_4 . C_8H_8O_3$.

Packaging and storage—Preserve in well-closed containers.

Reference standard—*USP Methenamine Mandelate Reference Standard*—Dry over silica gel for 18 hours before using.

Identification—Triturate a quantity of finely powdered Tablets, equivalent to about 5.0 mg of methenamine mandelate, with 5 mL of chloroform, and filter through a 0.45-μm membrane filter. Evaporate the solvent, and allow it to air-dry: the infrared absorption spectrum of a potassium bromide dispersion of the residue so obtained exhibits maxima only at the same wavelengths as that of a potassium bromide dispersion of USP Methenamine Mandelate RS.

Disintegration ⟨701⟩: FOR ENTERIC-COATED TABLETS—2 hours and 30 minutes, determined as directed under *Enteric-coated Tablets.*

Dissolution ⟨711⟩—

FOR UNCOATED OR PLAIN COATED TABLETS—

Medium: water; 900 mL.

Apparatus 1: 100 rpm.

Time: 45 minutes.

Procedure—Determine the amount of $C_6H_{12}N_4 . C_8H_8O_3$ dissolved from ultraviolet absorbances at the wavelength of maximum absorbance at about 257 nm of portions of the solution under test, filtered through a 0.45-μm filter and suitably diluted with *Dissolution Medium*, if necessary, in comparison with a Standard solution having a known concentration of USP Methenamine Mandelate RS in the same medium.

Tolerances—Not less than 75% (*Q*) of the labeled amount of $C_6H_{12}N_4 . C_8H_8O_3$ is dissolved in 45 minutes.

Uniformity of dosage units ⟨905⟩: meet the requirements.

Assay—Weigh and finely powder not less than 20 Methenamine Mandelate Tablets. Weigh accurately a portion of the powder, equivalent to about 60 mg of methenamine mandelate, and transfer to a 250-mL conical flask. Proceed as directed in the *Assay* under *Methenamine Mandelate*, beginning with "Add 15 mL of dehydrated alcohol."

Methicillin Sodium for Injection

» Methicillin Sodium for Injection is a sterile mixture of Sterile Methicillin Sodium and Sodium Citrate. It may contain one or more suitable preservatives. It contains not less than 90.0 percent and not more than 115.0 percent of the labeled amount of methicillin ($C_{17}H_{20}N_2O_6S$).

Packaging and storage—Preserve in *Containers for Sterile Solids* as described under *Injections* ⟨1⟩, at controlled room temperature.

Reference standard—*USP Methicillin Sodium Reference Standard*—Do not dry before using.

Constituted solution—At the time of use, the constituted solution prepared from Methicillin Sodium for Injection meets the requirements for *Constituted Solutions* under *Injections* ⟨1⟩.

Pyrogen—It meets the requirements of the *Pyrogen Test* ⟨151⟩, the test dose being 1.0 mL per kg of a solution in pyrogen-free saline TS containing 60 mg of methicillin per mL.

Sterility—It meets the requirements under *Sterility Tests* ⟨71⟩, when tested as directed in the section, *Test Procedures Using Membrane Filtration*, using suitable sampling techniques.

pH ⟨791⟩: between 6.0 and 8.5, in a solution containing 10 mg per mL.

Water, *Method I* ⟨921⟩: not more than 6.0%.

Particulate matter ⟨788⟩: meets the requirements under *Small-volume Injections.*

Other requirements—It responds to the *Identification test* under *Sterile Methicillin Sodium* and meets the requirements for *Uniformity of Dosage Units* ⟨905⟩ and *Labeling* under *Injections* ⟨1⟩.

Assay—Proceed as directed under *Antibiotics—Microbial Assays* ⟨81⟩, using Methicillin Sodium for Injection constituted as directed in the labeling. Remove an accurately measured portion, and dilute quantitatively with *Buffer No. 1* to obtain a *Test Dilution* having a concentration assumed to be equal to the median dose level of the Standard.

Sterile Methicillin Sodium

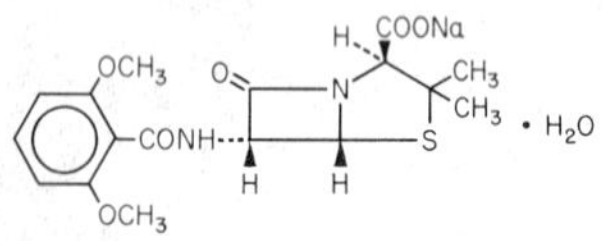

$C_{17}H_{19}N_2NaO_6S . H_2O$ 420.41

4-Thia-1-azabicyclo[3.2.0]heptane-2-carboxylic acid, 6-[(2,6-dimethoxybenzoyl)amino]-3,3-dimethyl-7-oxo-, monosodium salt, monohydrate, [2*S*-(2α,5α,6β)]-.

Monosodium (2*S*,5*R*,6*R*)-6-(2,6-dimethoxybenzamido)-3,3-dimethyl-7-oxo-4-thia-1-azabicyclo[3.2.0]heptane-2-carboxylate monohydrate   [*7246-14-2*].

Anhydrous   402.40   [*132-92-3*].

» Sterile Methicillin Sodium is Methicillin Sodium suitable for parenteral use. It has a potency equivalent to not less than 815 μg of methicillin ($C_{17}H_{20}N_2O_6S$) per mg.

Packaging and storage—Preserve in *Containers for Sterile Solids* as described under *Injections* ⟨1⟩, at controlled room temperature.

Reference standard—*USP Methicillin Sodium Reference Standard*—Do not dry before using.

Identification—Dissolve an accurately weighed portion in water to obtain a solution having a concentration of about 0.2 mg per mL. Using a suitable spectrophotometer equipped with a 1-cm cell and water as the blank, determine the absorbances at the wavelength of maximum absorbance at about 280 nm and of minimum absorbance at about 264 nm: the ratio A_{280}/A_{264} for Sterile Methicillin Sodium is between 1.30 and 1.45.

Crystallinity ⟨695⟩: meets the requirements.

Pyrogen—It meets the requirements of the *Pyrogen Test* ⟨151⟩, the test dose being 1.0 mL per kg of a solution in pyrogen-free saline TS containing 60 mg of methicillin per mL.

Sterility—It meets the requirements under *Sterility Tests* ⟨71⟩, when tested as directed in the section, *Test Procedures Using Membrane Filtration*, using 6 g of specimen aseptically dissolved in 200 mL of *Fluid A*.

pH ⟨791⟩: between 5.0 and 7.5, in a solution containing 10 mg per mL.

Water, *Method I* ⟨921⟩: between 3.0% and 6.0%.

Methicillin content—Transfer about 20 mg of Sterile Methicillin Sodium, accurately weighed, to a 100-mL volumetric flask, add water to volume, and mix. Concomitantly determine the absorbances of this solution and of a similar solution of USP Methicillin Sodium RS at the wavelength of maximum absorbance at about 280 nm, with a suitable spectrophotometer, using water as the blank. Calculate the percentage of methicillin ($C_{17}H_{20}N_2O_6S$) by the formula:

$$P(a_U / a_S),$$

in which P is the percentage content of methicillin in USP Methicillin Sodium RS, and a_U and a_S are the absorptivities of the specimen and the Standard, respectively: not less than 81.5% is found.

Assay—Proceed with Sterile Methicillin Sodium as directed under *Antibiotics—Microbial Assays* ⟨81⟩.

Methimazole

CH3 N S NH

$C_4H_6N_2S$ 114.16
2*H*-Imidazole-2-thione, 1,3-dihydro-1-methyl-.
1-Methylimidazole-2-thiol [*60-56-0*].

» Methimazole contains not less than 98.0 percent and not more than 101.0 percent of $C_4H_6N_2S$, calculated on the dried basis.

Packaging and storage—Preserve in well-closed, light-resistant containers.

Reference standard—*USP Methimazole Reference Standard*—Dry at 105° for 2 hours before using.

Identification—

A: The infrared absorption spectrum of a potassium bromide dispersion of it exhibits maxima only at the same wavelengths as that of a similar preparation of USP Methimazole RS.

B: Mercuric chloride TS produces in a solution (1 in 200) a white precipitate, but no precipitation is produced by trinitrophenol TS. The solution is colored intensely blue by molybdophosphotungstate TS.

Melting range ⟨741⟩: between 143° and 146°.

Loss on drying ⟨731⟩—Dry it at 105° for 2 hours: it loses not more than 0.5% of its weight.

Residue on ignition ⟨281⟩: not more than 0.1%.

Selenium ⟨291⟩: 0.003%, a 200-mg specimen being used.

Ordinary impurities ⟨466⟩—

Test solution: ethyl acetate.

Standard solution: ethyl acetate.

Eluant: a mixture of toluene, isopropyl alcohol, and ammonium hydroxide (70:29:1), in a nonequilibrated chamber.

Visualization: 2.

Assay—Dissolve about 250 mg of Methimazole, accurately weighed, in 75 mL of water. Add from a buret 15 mL of 0.1 *N* sodium hydroxide VS, mix, and add, with agitation, about 30 mL of 0.1 *N* silver nitrate. Add 1 mL of bromothymol blue TS, and continue the titration with the 0.1 *N* sodium hydroxide VS until a permanent, blue-green color is produced. Each mL of 0.1 *N* sodium hydroxide is equivalent to 11.42 mg of $C_4H_6N_2S$.

Methimazole Tablets

» Methimazole Tablets contain not less than 94.0 percent and not more than 106.0 percent of the labeled amount of $C_4H_6N_2S$.

Packaging and storage—Preserve in well-closed, light-resistant containers.

Reference standard—*USP Methimazole Reference Standard*—Dry at 105° for 2 hours before using.

Identification—Digest a quantity of powdered Tablets, equivalent to about 10 mg of methimazole, with 10 mL of warm chloroform for 20 minutes, filter, and evaporate the filtrate on a steam bath to dryness: the residue responds to the *Identification tests* under *Methimazole.*

Dissolution ⟨711⟩—

Medium: water; 500 mL.

Apparatus 1: 100 rpm.

Time: 30 minutes.

Procedure—Determine the amount of $C_4H_6N_2S$ dissolved from ultraviolet absorbances at the wavelength of maximum absorbance at about 252 nm of filtered portions of the solution under test, suitably diluted with *Dissolution Medium,* if necessary, in comparison with a Standard solution having a known concentration of USP Methimazole RS in the same medium.

Tolerances—Not less than 80% (*Q*) of the labeled amount of $C_4H_6N_2S$ is dissolved in 30 minutes.

Uniformity of dosage units ⟨905⟩: meet the requirements.

Procedure for content uniformity—Place 1 Tablet, previously crushed or finely powdered, in a 100-mL volumetric flask, add 50 mL of water, and shake by mechanical means for 30 minutes. Dilute with water to volume, mix, and filter, discarding the first 20 mL of filtrate. Dilute a portion of the subsequent filtrate quantitatively and stepwise with water so that the assumed concentration of methimazole is about 5 µg per mL. Dissolve an accurately weighed quantity of USP Methimazole RS in water, and dilute quantitatively and stepwise with water to obtain a Standard solution having a known concentration of about 5 µg per mL. Concomitantly determine the absorbances of both solutions in 1-cm cells at the wavelength of maximum absorbance at about 252 nm, with a suitable spectrophotometer, using water as the blank. Calculate the quantity, in mg, of $C_4H_6N_2S$ in the Tablet by the formula:

$$(T/5)C(A_U / A_S),$$

in which T is the labeled quantity, in mg, of methimazole in the Tablet, C is the concentration, in µg per mL, of USP Methimazole RS in the Standard solution, and A_U and A_S are the absorbances of the solution from the Tablet and the Standard solution, respectively.

Assay—Weigh and finely powder not less than 20 Methimazole Tablets. Weigh accurately a portion of the powder, equivalent to about 120 mg of methimazole, and place in a 100-mL volumetric flask. Add about 80 mL of water, insert the stopper, and shake by mechanical means or occasionally by hand during 30 minutes. Dilute with water to volume, and mix. Filter, and transfer 50.0 mL of the filtrate to a 125-mL conical flask. Add from a buret 3.5 mL of 0.1 *N* sodium hydroxide VS, mix, and add, with agitation, about 7 mL of 0.1 *N* silver nitrate. Add 1 mL of bromothymol blue TS, and continue the titration with the 0.1 *N* sodium hydroxide VS until a permanent, blue-green color is produced. Each mL of 0.1 *N* sodium hydroxide is equivalent to 11.42 mg of $C_4H_6N_2S$.

Methionine

H
$CH_3SCH_2CH_2-C-COOH$
NH_2

$C_5H_{11}NO_2S$ 149.21
L-Methionine.
L-Methionine [*63-68-3*].

» Methionine contains not less than 98.5 percent and not more than 101.5 percent of $C_5H_{11}NO_2S$, as L-methionine, calculated on the dried basis.

Packaging and storage—Preserve in well-closed containers.

Reference standard—*USP L-Methionine Reference Standard*—Dry at 105° for 3 hours before using.

Identification—The infrared absorption spectrum of a potassium bromide dispersion of it, previously dried, exhibits maxima only at the same wavelengths as that of a similar preparation of USP L-Methionine RS.

Specific rotation ⟨781⟩: between +21.9° and +24.1°, calculated on the dried basis, determined in a solution in 6 *N* hydrochloric acid containing 200 mg in each 10 mL.

pH ⟨791⟩: between 5.6 and 6.1 in a solution (1 in 100).

Loss on drying ⟨731⟩—Dry it at 105° for 3 hours: it loses not more than 0.3% of its weight.

Residue on ignition ⟨281⟩: not more than 0.4%.

Chloride ⟨221⟩—A 0.73-g portion shows no more chloride than corresponds to 0.50 mL of 0.020 *N* hydrochloric acid (0.05%).

Sulfate ⟨221⟩—A 0.33-g portion shows no more sulfate than corresponds to 0.10 mL of 0.020 *N* sulfuric acid (0.03%).

Arsenic ⟨211⟩: 1.5 ppm.

Iron ⟨241⟩: 0.003%.

Heavy metals, *Method I* ⟨231⟩: 0.0015%.

Assay—Transfer about 140 mg of Methionine, accurately weighed, to a 125-mL flask, dissolve in a mixture of 3 mL of formic acid and 50 mL of glacial acetic acid, and titrate with 0.1 *N* perchloric acid VS, determining the end-point potentiometrically. Perform a blank determination, and make any necessary correction. Each mL of 0.1 *N* perchloric acid is equivalent to 14.92 mg of $C_5H_{11}NO_2S$.

Methionine—*see* Racemethionine

Methionine Capsules—*see* Racemethionine Capsules

Methionine Tablets—*see* Racemethionine Tablets

Methocarbamol

OH | OCH$_2$CHCH$_2$OOCNH$_2$ | OCH$_3$

$C_{11}H_{15}NO_5$ 241.24

1,2-Propanediol, 3-(2-methoxyphenoxy)-, 1-carbamate.

3-(*o*-Methoxyphenoxy)-1,2-propanediol 1-carbamate [*532-03-6*].

» Methocarbamol contains not less than 98.5 percent and not more than 101.5 percent of $C_{11}H_{15}NO_5$, calculated on the dried basis.

Packaging and storage—Preserve in tight containers.

Reference standard—*USP Methocarbamol Reference Standard*—Dry at 60° for 2 hours before using.

Identification—

A: The infrared absorption spectrum of a potassium bromide dispersion of it, previously dried, exhibits maxima only at the same wavelengths as that of a similar preparation of USP Methocarbamol RS.

B: The ultraviolet absorption spectrum of a 1 in 25,000 solution in alcohol exhibits maxima and minima at the same wavelengths as that of a similar solution of USP Methocarbamol RS, concomitantly measured.

Loss on drying ⟨731⟩—Dry it at 60° for 2 hours: it loses not more than 0.5% of its weight.

Residue on ignition ⟨281⟩: not more than 0.1%.

Heavy metals, *Method I* ⟨231⟩—Dissolve 1.0 g in a mixture of 7 mL of methanol and 3 mL of 1 *N* acetic acid, and dilute with water to 25 mL. The limit is 0.002%.

Related impurities—

pH 4.5 buffer solution—Dissolve 6.8 g of monobasic potassium phosphate in 1000 mL of water. Adjust with 18 *N* phosphoric acid or 10 *N* potassium hydroxide to a pH of 4.5 ± .05.

Mobile phase—Prepare a suitably filtered and degassed solution of *pH 4.5 buffer solution* and methanol (about 75:25) (see *System suitability*).

Guaifenesin standard solution—Transfer 20.0 mg of USP Guaifenesin RS to a 50-mL volumetric flask. Dissolve in methanol, dilute with methanol to volume, and mix.

Standard preparation—Transfer 98.0 mg of USP Methocarbamol RS to a 50-mL volumetric flask. Add 5.0 mL of *Guaifenesin standard solution* and 8.0 mL of methanol to dissolve the methocarbamol. Dilute with *pH 4.5 buffer solution* to volume, and mix. Use this solution within 24 hours.

Test preparation—Transfer about 100 mg of Methocarbamol, accurately weighed, to a 50-mL volumetric flask, add 13 mL of methanol to dissolve, dilute with *pH 4.5 buffer solution* to volume, and mix. Use this solution within 24 hours.

Chromatographic system (see *Chromatography* ⟨621⟩)—The liquid chromatograph is equipped with a 274-nm detector and a 4-mm × 25-cm column that contains packing L1. Adjust the operating conditions so that *System suitability* requirements are met.

System suitability—Chromatograph three replicate 20-µL portions of the *Standard preparation* as directed under *Procedure*. The analytical system is suitable for use if the peak area percent of guaifenesin is 2.4 ± 1.0, the relative standard deviation for the peak area percent is not greater than 4.0%, and the resolution factor between guaifenesin and methocarbamol is not less than 2.0.

Procedure—By means of a suitable microsyringe or sampling valve, inject about 20 µL of the *Test preparation* into the chromatograph. Determine the peak areas of the methocarbamol peak and all extraneous peaks having a retention time greater than 0.5 of the retention time of methocarbamol. The relative retention times are about 0.8 for guaifenesin and 1.0 for methocarbamol. Calculate the percentage of related impurities by the formula:

$$100(2.4/G)(P_E/P_T),$$

in which G is the peak area percent of guaifenesin in the *Standard preparation* determined under *System suitability*, P_E is the peak area of all extraneous peaks, and P_T is the total of the peak areas of all extraneous peaks and the methocarbamol. The limit is 2.0%.

Assay—Transfer about 100 mg of Methocarbamol, accurately weighed, to a 100-mL volumetric flask, add methanol to volume, and mix. Transfer 4.0 mL of this solution to a second 100-mL volumetric flask, dilute with methanol to volume, and mix. Concomitantly determine the absorbances of this solution and a Standard solution of USP Methocarbamol RS in methanol having a known concentration of about 40 µg per mL at the wavelength of maximum absorbance at about 274 nm in 1-cm cells, using methanol as the blank. Calculate the quantity, in mg, of $C_{11}H_{15}NO_5$ in the portion of Methocarbamol taken by the formula:

$$2.5C(A_U/A_S),$$

in which C is the concentration, in µg per mL, of USP Methocarbamol RS in the Standard solution, and A_U and A_S are the absorbances of the assay solution and the Standard solution, respectively.

Methocarbamol Injection

» Methocarbamol Injection is a sterile solution of Methocarbamol in an aqueous solution of Polyethylene Glycol 300. It contains not less than 95.0 percent and not more than 105.0 percent of the labeled amount of $C_{11}H_{15}NO_5$.

Packaging and storage—Preserve in single-dose containers, preferably of Type I glass.

Reference standard—*USP Methocarbamol Reference Standard*—Dry at 60° for 2 hours before using.

Identification—Mix a volume of Injection, equivalent to about 500 mg of methocarbamol, with 40 mL of water in a small separator. Extract with 10 mL of ethyl acetate, and dry the ethyl acetate layer over anhydrous sodium sulfate. Evaporate the ethyl acetate with the use of a water bath maintained at 40° under a stream of nitrogen to dryness: the methocarbamol so obtained responds to *Identification test A* under *Methocarbamol.*

pH ⟨791⟩: between 3.5 and 6.0.

Particulate matter ⟨788⟩: meets the requirements under *Small-volume Injections.*

Aldehydes—Transfer to a 25-mL volumetric flask an accurately measured volume of Injection, equivalent to 400 mg of methocarbamol, add 2.0 mL of a filtered 1 in 100 solution of phenylhydrazine hydrochloride in dilute alcohol (1 in 5), and allow to stand for 10 minutes. Add 1 mL of potassium ferricyanide solution (1 in 100), and allow to stand for 5 minutes. Add 4 mL of hydrochloric acid, dilute with alcohol to volume, and mix. Determine the absorbance of this solution in a 1-cm cell at the wavelength of maximum absorbance at about 515 nm, with a suitable spectrophotometer, using a reagent blank to set the instrument: the absorbance is not greater than that produced by 4 mL of formaldehyde solution (1 in 100,000), treated in the same manner as the portion of Injection taken and similarly measured, corresponding to not more than 0.01%, as formaldehyde, based upon the content of $C_{11}H_{15}NO_5$ as determined in the *Assay.*

Other requirements—It meets the requirements under *Injections* ⟨1⟩.

Assay—

pH 4.5 buffer solution—Dissolve 6.8 g of monobasic potassium phosphate in 1000 mL of water. Adjust with 18 *N* phosphoric acid or 10 *N* potassium hydroxide to a pH of 4.5 ± 0.05.

Mobile phase—Prepare a suitable, filtered and degassed solution of *pH 4.5 buffer solution* and methanol (about 70:30) (see *System Suitability* under *Chromatography* ⟨621⟩).

Internal standard solution—Prepare a solution of caffeine in methanol containing about 3 mg per mL.

Standard preparation—Transfer about 100 mg of USP Methocarbamol RS, accurately weighed, to a 100-mL volumetric flask. Dissolve in about 50 mL of *pH 4.5 buffer solution* and 25 mL of methanol. Add 5.0 mL of *Internal standard solution*, dilute with *pH 4.5 buffer solution* to volume, and mix.

Assay preparation—Accurately transfer a volume of Methocarbamol Injection, equivalent to 100 mg of methocarbamol, to a 100-mL volumetric flask. Add *pH 4.5 buffer solution*, 25 mL of methanol, and 5.0 mL of *Internal standard solution*, and mix. Dilute with *pH 4.5 buffer solution* to volume, and mix.

Chromatographic system (see *Chromatography* ⟨621⟩)—The liquid chromatograph is equipped with a 274-nm detector and a 4.6-mm × 10-cm column that contains 3- or 5-μm packing L1, operated at 30°. The flow rate is 1 mL per minute. Chromatograph the *Standard preparation*, and record the peak responses as directed under *Procedure:* the resolution, *R*, between the methocarbamol and internal standard peaks is not less than 9.5, and the relative standard deviation for replicate injections is not more than 2.0%.

Procedure—Separately inject equal volumes (about 20 μL) of the *Standard preparation* and the *Assay preparation* into the chromatograph, record the chromatograms, and measure the responses for the major peaks. The relative retention times are about 0.5 for caffeine and 1.0 for methocarbamol. Calculate the quantity, in mg, of $C_{11}H_{15}NO_5$ in each mL of the Injection taken by the formula:

$$(100C/V)(R_U/R_S),$$

in which *C* is the concentration, in mg per mL, of USP Methocarbamol RS in the *Standard preparation*, *V* is the volume, in mL, of Injection taken, and R_U and R_S are the peak response ratios of methocarbamol and caffeine obtained from the *Assay preparation* and the *Standard preparation*, respectively.

Methocarbamol Tablets

» Methocarbamol Tablets contain not less than 95.0 percent and not more than 105.0 percent of the labeled amount of $C_{11}H_{15}NO_5$.

Packaging and storage—Preserve in tight containers.

Reference standard—*USP Methocarbamol Reference Standard*—Dry at 60° for 2 hours before using.

Identification—Mix a portion of finely powdered Tablets, equivalent to about 1 g of methocarbamol, with 25 mL of water in a separator, and extract with 25 mL of chloroform. Filter the extract, and evaporate to dryness: the residue of methocarbamol so obtained responds to *Identification test A* under *Methocarbamol.*

Dissolution ⟨711⟩—

Medium: water; 900 mL.

Apparatus 2: 50 rpm.

Time: 45 minutes.

Procedure—Determine the amount of $C_{11}H_{15}NO_5$ dissolved, employing the procedure set forth in the *Assay*, making any necessary modifications.

Tolerances—Not less than 75% (*Q*) of the labeled amount of $C_{11}H_{15}NO_5$ is dissolved in 45 minutes.

Uniformity of dosage units ⟨905⟩: meet the requirements.

Assay—

pH 4.5 buffer solution, Mobile phase, Internal standard solution, Standard preparation, and *Chromatographic system*—Proceed as directed in the *Assay* under *Methocarbamol Injection.*

Assay preparation—Weigh and powder not less than 10 Methocarbamol Tablets. Transfer an accurately weighed portion of the powder, equivalent to 100 mg of methocarbamol, to a 100-mL volumetric flask. Add about 50 mL of *pH 4.5 buffer solution*, 25 mL of methanol, and 5.0 mL of *Internal standard solution.* Shake vigorously for 10 minutes, dilute with *pH 4.5 buffer solution* to volume, and mix. Filter, discarding the first 20 mL of the filtrate.

Procedure—Proceed as directed for *Procedure* in the *Assay* under *Methocarbamol Injection.* Calculate the quantity, in mg, of $C_{11}H_{15}NO_5$ in the portion of Tablets taken by the formula:

$$100C(R_U/R_S),$$

in which the terms are as defined therein.

Methohexital

$C_{14}H_{18}N_2O_3$ 262.31

2,4,6(1*H*,3*H*,5*H*)-Pyrimidinetrione, 1-methyl-5-(1-methyl-2-pentynyl)-5-(2-propenyl)-, (±)-.

(±)-5-Allyl-1-methyl-5-(1-methyl-2-pentynyl)barbituric acid [*18652-93-2*].

» Methohexital contains not less than 98.0 percent and not more than 101.0 percent of $C_{14}H_{18}N_2O_3$, calculated on the anhydrous basis.

Packaging and storage—Preserve in well-closed containers.

Reference standard—*USP Methohexital Reference Standard*—Dry over silica gel to constant weight before using.

Identification—The infrared absorption spectrum of a 1 in 100 solution in chloroform, determined in a 0.1-mm cell, exhibits maxima only at the same wavelengths as that of a similar solution of USP Methohexital RS.

Melting range ⟨741⟩: between 92° and 96°, but the range between beginning and end of melting does not exceed 3°.

Water, *Method I* ⟨921⟩: not more than 2.0%.

Chloride ⟨221⟩—Dissolve 200 mg in a mixture of 75 mL of ether and 25 mL of water, agitate, and allow to separate: the water solution shows no more chloride than corresponds to 0.17 mL of 0.010 *N* hydrochloric acid (0.03%).

Heavy metals, *Method II* ⟨231⟩: 0.001%.

Ordinary impurities ⟨466⟩—

Test solution: methanol.

Standard solution: methanol.

Eluant: a mixture of chloroform and acetone (7:3).

Visualization: 1.

Assay—Dissolve about 100 mg of Methohexital, accurately weighed, in chloroform, and dilute quantitatively and stepwise with chloroform to obtain a solution having a concentration of about 10 mg per mL. Dissolve an accurately weighed quantity of USP Methohexital RS in chloroform, and dilute quantitatively and stepwise with chloroform to obtain a Standard solution having a known concentration of about 10 mg per mL. Concomitantly determine the absorbances of both solutions in 0.1-mm cells at the wavelength of maximum absorbance at about 5.93 µm, with a suitable spectrophotometer, using chloroform as the blank. Calculate the quantity, in mg, of $C_{14}H_{18}N_2O_3$ in the portion of Methohexital taken by the formula:

$$10C(A_U/A_S),$$

in which *C* is the concentration, in mg per mL, of USP Methohexital RS in the Standard solution, and A_U and A_S are the absorbances of the solution of Methohexital and the Standard solution, respectively.

Methohexital Sodium for Injection

$C_{14}H_{17}N_2NaO_3$ 284.29

2,4,6(1*H*,3*H*,5*H*)-Pyrimidinetrione, 1-methyl-5-(1-methyl-2-pentynyl)-5-(2-propenyl)-, (±)-, monosodium salt.

Sodium 5-allyl-1-methyl-5-(1-methyl-2-pentynyl)barbiturate [*309-36-4; 22151-68-4*].

» Methohexital Sodium for Injection is a freeze-dried, sterile mixture of methohexital sodium and anhydrous Sodium Carbonate as a buffer, prepared from an aqueous solution of Methohexital, Sodium Hydroxide, and Sodium Carbonate. It contains not less than 90.0 percent and not more than 110.0 percent of the labeled amount of $C_{14}H_{17}N_2NaO_3$.

Packaging and storage—Preserve in *Containers for Sterile Solids* as described under *Injections* ⟨1⟩.

Reference standard—*USP Methohexital Reference Standard*—Dry over silica gel to constant weight before using.

Completeness of solution—Mix 1 g with 20 mL of carbon dioxide–free water: after 1 minute, the solution is clear and free from undissolved solid.

Constituted solution—At the time of use, the constituted solution prepared from Methohexital Sodium for Injection meets the requirements for *Constituted Solutions* under *Injections* ⟨1⟩.

Identification—

A: Dissolve about 500 mg in 10 mL of water in a separator, add 10 mL of 3 *N* hydrochloric acid, and extract the liberated methohexital with two 25-mL portions of chloroform. Evaporate the combined chloroform extracts to dryness, add 10 mL of ether, evaporate again, and dry the residue in vacuum at 80° for 4 hours. Dissolve 50 mg of the residue so obtained in 5 mL of chloroform: the solution exhibits infrared absorption maxima at the same wavelengths as that of a similar preparation of USP Methohexital RS.

B: The methohexital obtained and dried as directed for *Identification test A* melts between 92° and 96°.

Uniformity of dosage units ⟨905⟩: meets the requirements.

Procedure for content uniformity—

Standard solution—Transfer about 23 mg of USP Methohexital RS, accurately weighed, to a 250-mL volumetric flask, add 50 mL of water, 0.5 mL of sodium hydroxide solution (1 in 10), and 1.5 mL of sodium carbonate solution (1 in 1000), and mix. Dilute with water to volume, and mix. Transfer 20.0 mL of this solution to a 100-mL volumetric flask, dilute with water to volume, and mix.

Test solution—Transfer the contents of 1 vial of Methohexital Sodium for Injection with the aid of water to a 1000-mL volumetric flask, dilute with water to volume, and mix. Transfer an accurately measured volume of this solution, equivalent to about 100 mg of methohexital sodium, to a 1000-mL volumetric flask, add about 200 mL of water and 2.0 mL of sodium hydroxide solution (1 in 10), mix, dilute with water to volume, and again mix. Transfer 20.0 mL of the resulting solution to a 100-mL volumetric flask, dilute with water to volume, and mix.

Procedure—Concomitantly determine the absorbances of the *Standard solution* and the *Test solution* in 1-cm cells at the wavelength of maximum absorbance at about 247 nm, with a suitable spectrophotometer, using water as the blank. Calculate the quantity, in mg, of $C_{14}H_{17}N_2NaO_3$ in the Methohexital Sodium for Injection taken by the formula:

$$(284.29/262.31)(TC/D)(A_U/A_S),$$

in which 284.29 and 262.31 are the molecular weights of methohexital sodium and methohexital, respectively, *T* is the labeled quantity, in mg, of methohexital sodium in the Methohexital Sodium for Injection, *C* is the concentration, in µg per mL, of USP Methohexital RS in the *Standard solution, D* is the concentration, in µg per mL, of methohexital sodium in the *Test solution* based on the labeled quantity per container and the extent of dilution, and A_U and A_S are the absorbances of the *Test solution* and the *Standard solution*, respectively.

pH ⟨791⟩: between 10.6 and 11.6 in the solution prepared in the test for *Completeness of solution.*

Loss on drying ⟨731⟩—Dry it at 105° for 4 hours: it loses not more than 2.0% of its weight.

Heavy metals, *Method II* ⟨231⟩: 0.001%.

Other requirements—It meets the requirements under *Injections* ⟨1⟩.

Assay—

Internal standard solution—Dissolve aprobarbital in chloroform to obtain a solution having a concentration of about 1.35 mg per mL.

Standard preparation—Dissolve an accurately weighed quantity of USP Methohexital RS in chloroform to obtain a solution having a known concentration of about 0.46 mg per mL. Transfer 5.0 mL of the resulting solution to a 10-mL volumetric flask, add 2.0 mL of *Internal standard solution*, dilute with chloroform to volume, and mix to obtain a *Standard preparation* having a known concentration of about 230 µg per mL.

Assay preparation—Combine and mix the constituted solutions prepared from the contents of 5 vials of Methohexital Sodium for Injection. Transfer an accurately measured volume of the resulting solution, equivalent to about 50 mg of methohexital sodium, to a 125-mL separator containing 25 mL of water, and mix. Add 0.2 mL of dilute hydrochloric acid (1 in 2), and mix. Extract with three 25-mL portions of chloroform, shaking each extraction for 2 minutes and filtering the extracts through about 15 g of anhydrous sodium sulfate, that previously has been washed with about 5 mL of chloroform, into a 100-mL volumetric flask. Wash the sodium sulfate with several small portions of chloroform, collecting the washings in the 100-mL volumetric flask. Dilute with chloroform to volume, and mix. Transfer 5.0 mL of this solution to a 10-mL volumetric flask, add 2.0 mL of *Internal standard solution*, dilute with chloroform to volume, and mix.

Chromatographic system (see *Chromatography* ⟨621⟩)—The gas chromatograph is equipped with a flame-ionization detector and contains a 1.2-m × 4-mm column packed with 3 percent phase G10 on support S1AB. The column is maintained at about 230°, the injection port at about 265°, and the detector block at about 265°. Dry helium is used as the carrier gas at a flow rate of about 60 mL per minute. Chromatograph replicate injections of the *Standard preparation*, and record the peak responses as directed under *Procedure*. The resolution, R, between methohexital and aprobarbital is not less than 4.0, and the relative standard deviation is not more than 2.0%.

Procedure—Separately inject equal volumes (about 2 µL) of the *Assay preparation* and the *Standard preparation* into the gas chromatograph, and measure the peak responses for the major peak. The relative retention times are about 0.6 for methohexital and 1.0 for aprobarbital. Calculate the quantity, in mg, of $C_{14}H_{17}N_2NaO_3$ in the portion of Methohexital Sodium for Injection taken by the formula:

$$(284.29/262.31)(0.2C)(R_U/R_S),$$

in which 284.29 and 262.31 are the molecular weights of methohexital sodium and methohexital, respectively, C is the concentration, in µg per mL, of USP Methohexital RS in the *Standard preparation*, and R_U and R_S are the peak response ratios obtained from the *Assay preparation* and the *Standard preparation*, respectively.

Methotrexate

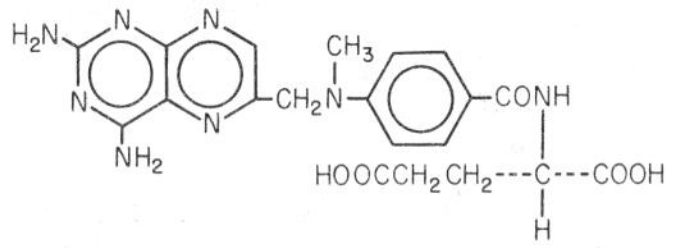

$C_{20}H_{22}N_8O_5$ 454.44

L-Glutamic acid, *N*-[4-[[(2,4-diamino-6-pteridinyl)-methyl]methylamino]benzoyl]-.

L-(+)-*N*-[*p*-[[(2,4-Diamino-6-pteridinyl)methyl]methyl-aminobenzoyl]glutamic acid [*59-05-2*].

» Methotrexate is a mixture of 4-amino-10-methylfolic acid and closely related compounds, and it contains not less than 94.0 percent of $C_{20}H_{22}N_8O_5$, calculated on the anhydrous basis.

Caution—Great care should be taken to prevent inhaling particles of Methotrexate and exposing the skin to it.

Packaging and storage—Preserve in tight, light-resistant containers.

Reference standard—*USP Methotrexate Reference Standard*—Do not dry; determine the *Water* content by *Method I* ⟨921⟩ before using.

Identification—

A: The infrared absorption spectrum of a potassium bromide dispersion of it exhibits maxima only at the same wavelengths as that of a similar preparation of USP Methotrexate RS.

B: The ultraviolet absorption spectrum of a 1 in 100,000 solution in 0.1 *N* hydrochloric acid exhibits maxima and minima at the same wavelengths as that of a similar solution of USP Methotrexate RS, concomitantly measured.

Water, *Method I* ⟨921⟩: not more than 12.0%.

Residue on ignition ⟨281⟩: not more than 0.1%.

Assay—

pH 6.0 buffer solution—Mix 370 mL of 0.1 *M* citric acid solution with 630 mL of 0.2 *M* dibasic sodium phosphate solution, and adjust to a pH of 6.0 by adding small volumes of either solution as necessary.

Mobile phase—Prepare a suitable degassed solution of *pH 6.0 buffer solution* and acetonitrile in the approximate ratio of 90:10 such that the retention time of methotrexate is approximately 7 to 10 minutes. Filter the solution through a 0.45-µm membrane filter before using.

Standard preparation—Dissolve a suitable quantity of USP Methotrexate RS, accurately weighed, in *Mobile phase* to obtain a final solution having a known concentration of about 100 µg per mL.

Assay preparation—Transfer about 25 mg of Methotrexate, accurately weighed, to a 250-mL volumetric flask, dissolve in *Mobile phase*, dilute with *Mobile phase* to volume, and mix.

System suitability preparation—Prepare a solution containing about 0.1 mg per mL each of USP Methotrexate RS and folic acid in *Mobile phase*.

System suitability test—Inject 10 µL of the *System suitability preparation* into the instrument described under *Procedure*. Six replicate injections show a resolution factor between methotrexate and folic acid of not less than 8.0 and a relative standard deviation for the methotrexate peak of not more than 2.5%.

Procedure—By means of a sampling valve introduce equal volumes (about 10 µL) of the *Assay preparation* and the *Standard preparation* into a high-pressure liquid chromatograph operated at room temperature and a flow rate of about 1.2 mL per minute. Typically, the apparatus is fitted with a 25-cm × 4.6-mm column that contains packing L1 and is equipped with an ultraviolet detector capable of monitoring absorption at 302 nm and a suitable recorder. Measure the peak responses at identical retention times obtained with the *Assay preparation* and the *Standard preparation*, and calculate the quantity, in mg, of $C_{20}H_{22}N_8O_5$ in the portion of Methotrexate taken by the formula:

$$(0.25C)(P_U/P_S),$$

in which C is the concentration, in µg per mL, of USP Methotrexate RS in the Standard preparation, and P_U and P_S are the peak responses obtained from the *Assay preparation* and the *Standard preparation*, respectively.

Methotrexate Tablets

» Methotrexate Tablets contain not less than 90.0 percent and not more than 110.0 percent of the labeled amount of $C_{20}H_{22}N_8O_5$.

Packaging and storage—Preserve in well-closed containers.

Reference standard—*USP Methotrexate Reference Standard*—Do not dry; determine the *Water* content by *Method I* ⟨921⟩ before using.

Identification—Dissolve 1 Tablet in 100 mL of dilute hydrochloric acid (1 in 100), and filter the solution: the ultraviolet absorption spectrum of the filtrate exhibits maxima and minima at the same wavelengths as that of a solution containing about 2.5 mg of USP Methotrexate RS in 100 mL of dilute hydrochloric acid (1 in 100).

Dissolution ⟨711⟩—

Medium: 0.1 *N* hydrochloric acid; 900 mL.

Apparatus 2: 50 rpm.

Time: 45 minutes.

Procedure—Determine the amount of $C_{20}H_{22}N_8O_5$ dissolved from ultraviolet absorbances at the wavelength of maximum absorbance at about 306 nm of filtered portions of the solution under test, suitably diluted with *Dissolution Medium*, in comparison with a Standard solution having a known concentration of USP Methotrexate RS in the same medium.

Tolerances—Not less than 75% (Q) of the labeled amount of $C_{20}H_{22}N_8O_5$ is dissolved in 45 minutes.

Uniformity of dosage units ⟨905⟩: meet the requirements.

Assay—

pH 6.0 buffer solution, Mobile phase, System suitability preparation, System suitability test, and *Standard preparation*—Proceed as directed in the *Assay* under *Methotrexate*.

Assay preparation—Weigh and finely powder not less than 20 Methotrexate Tablets. Weigh accurately a portion of the powder, equivalent to about 25 mg of methotrexate, and transfer to a 250-mL volumetric flask. Add about 200 mL of *Mobile phase*, and dissolve the methotrexate using a mechanical shaker or ultrasonic bath. Dilute with *Mobile phase* to volume, and mix.

Procedure—Proceed as directed for *Procedure* in the *Assay* under *Methotrexate*. Calculate the quantity, in mg, of $C_{20}H_{22}N_8O_5$ in the portion of Tablets taken by the formula:

$$250C(P_U/P_S),$$

in which C is the concentration, in mg per mL, of USP Methotrexate RS in the *Standard preparation*, and P_U and P_S are the peak responses obtained from the *Assay preparation* and the *Standard preparation*, respectively.

Methotrexate Sodium Injection

» Methotrexate Sodium Injection is a sterile solution of Methotrexate in Water for Injection prepared with the aid of Sodium Hydroxide. It contains not less than 95.0 percent and not more than 115.0 percent of the labeled amount of methotrexate ($C_{20}H_{22}N_8O_5$).

Packaging and storage—Preserve in single-dose or in multiple-dose containers, preferably of Type I glass, protected from light.

Reference standard—*USP Methotrexate Reference Standard*—Do not dry; determine the *Water* content by *Method I* ⟨921⟩ before using.

Identification—Dilute, if necessary, a volume of Injection, equivalent to about 25 mg of methotrexate, with water to obtain a solution having a concentration of about 2.5 mg per mL. Adjust with 0.1 *N* hydrochloric acid to a pH of 4.0. Place the slurry in a 50-mL centrifuge tube, and centrifuge. Decant the supernatant liquid, add 25 mL of acetone, shake, and filter through a solvent-resistant, membrane filter of 0.45-µm pore size. Air-dry the filtered precipitate: the methotrexate so obtained responds to *Identification test A* under *Methotrexate*.

Pyrogen—It meets the requirements of the *Pyrogen Test* ⟨151⟩, the test dose being 0.5 mg per kg.

pH ⟨791⟩: between 7.5 and 9.0.

Other requirements—It meets the requirements under *Injections* ⟨1⟩.

Assay—

pH 6.0 buffer solution, Mobile phase, System suitability preparation, System suitability test, and *Standard preparation*—Proceed as directed in the *Assay* under *Methotrexate*.

Assay preparation—Transfer an accurately measured volume of Methotrexate Sodium Injection, equivalent to about 25 mg of methotrexate, to a 250-mL volumetric flask, dilute with *Mobile phase* to volume, and mix.

Procedure—Proceed as directed for *Procedure* in the *Assay* under *Methotrexate*. Calculate the quantity, in mg, of $C_{20}H_{22}N_8O_5$ in each mL of the Injection taken by the formula:

$$250(C/V)(P_U/P_S),$$

in which C is the concentration, in mg per mL, of USP Methotrexate RS in the *Standard preparation*, V is the volume, in mL, of Injection taken, and P_U and P_S are the peak responses obtained from the *Assay preparation* and the *Standard preparation*, respectively.

Methotrimeprazine

$C_{19}H_{24}N_2OS$ 328.47

10*H*-Phenothiazine-10-propanamine, 2-methoxy-*N*,*N*,β-trimethyl-, (−)-.

(−)-10-[3-(Dimethylamino)-2-methylpropyl]-2-methoxyphenothiazine [*60-99-1*].

» Methotrimeprazine contains not less than 98.0 percent and not more than 101.0 percent of $C_{19}H_{24}N_2OS$, calculated on the dried basis.

Packaging and storage—Preserve in well-closed, light-resistant containers.

Reference standard—*USP Methotrimeprazine Reference Standard*—Dry at 100° for 3 hours before using.

NOTE—Throughout the following procedures, protect test or assay specimens, the Reference Standard, and solutions containing them, by conducting the procedures without delay, under subdued light, or using low-actinic glassware.

Identification—

A: The infrared absorption spectrum of a potassium bromide dispersion of it, previously dried, exhibits maxima only at the same wavelengths as that of a similar preparation of USP Methotrimeprazine RS.

B: The ultraviolet absorption spectrum of a 1 in 150,000 solution in alcohol exhibits maxima and minima at the same wavelengths as that of a similar solution of USP Methotrimeprazine RS, concomitantly measured, and the respective absorptivities, calculated on the dried basis, at the wavelength of maximum absorbance at about 255 nm do not differ by more than 3.0%.

Specific rotation ⟨781⟩: between −15° and −18°, calculated on the dried basis, determined in a solution in chloroform containing 500 mg in each 10 mL.

Loss on drying ⟨731⟩—Dry it at 100° for 3 hours: it loses not more than 0.5% of its weight.

Selenium ⟨291⟩—The absorbance of the solution from the *Test Solution*, prepared with 100 mg of Methotrimeprazine and 100 mg of magnesium oxide, is not greater than one-half that from the *Standard Solution* (0.003%).

Assay—Dissolve about 700 mg of Methotrimeprazine, accurately weighed, in 100 mL of chloroform, add 1 drop of a 1 in 500 solution of crystal violet in chloroform, and titrate with 0.1 *N* perchloric acid VS to the first disappearance of the violet tinge. Perform a blank determination, and make any necessary correction. Each mL of 0.1 *N* perchloric acid is equivalent to 32.85 mg of $C_{19}H_{24}N_2OS$.

Methotrimeprazine Injection

» Methotrimeprazine Injection is a sterile solution of Methotrimeprazine in Water for Injection, prepared with the aid of hydrochloric acid. It contains not less than 90.0 percent and not more than 110.0 percent of the labeled amount of $C_{19}H_{24}N_2OS$, as the hydrochloride.

Packaging and storage—Preserve in single-dose or in multiple-dose containers, preferably of Type I glass, protected from light.

Reference standard—*USP Methotrimeprazine Reference Standard*—Dry at 100° for 3 hours before using.

NOTE—Throughout the following procedures, protect test or assay specimens, the Reference Standard, and solutions containing them, by conducting the procedures without delay, under subdued light, or using low-actinic glassware.

Identification—Place 1 mL of Injection in a 125-mL separator, and add 1 *N* sodium hydroxide dropwise until the solution becomes opaque white. Extract with 50 mL of ether, wash the ether extract with 25 mL of water, and discard the washing. Filter the ether extract through a layer of anhydrous sodium sulfate into a beaker, evaporate the filtrate by means of a stream of nitrogen to complete dryness, and dry at 100° for 3 hours: the methotrimeprazine so obtained responds to *Identification test A* under *Methotrimeprazine*.

pH ⟨791⟩: between 3.0 and 5.0.

Other requirements—It meets the requirements under *Injections* ⟨1⟩.

Assay—[NOTE—Use low-actinic glassware.]

pH 5.3 phosphate buffer—Dissolve 38.0 g of monobasic sodium phosphate and 2.0 g of anhydrous dibasic sodium phosphate in 900 mL of water, adjust to a pH of 5.3 ± 0.1, if necessary, dilute with water to 1000 mL, and mix.

Bromocresol green solution—Dissolve 200 mg of bromocresol green in a mixture of 6.4 mL of 0.1 *N* sodium hydroxide and 100 mL of water. Dilute with water to 250 mL, extract with five 25-mL portions of chloroform, discard the extracts, and filter the aqueous phase.

Standard preparation—Dissolve about 20 mg of USP Methotrimeprazine RS, accurately weighed, in 1 mL of 0.01 *N* hydrochloric acid in a 10-mL volumetric flask, dilute with water to volume, and mix. Transfer 4.0 mL of this solution to a 100-mL volumetric flask, dilute with water to volume, and mix. The concentration of USP Methotrimeprazine RS in the *Standard preparation* is about 80 µg per mL.

Assay preparation—Transfer an accurately measured volume of Methotrimeprazine Injection, equivalent to about 40 mg of methotrimeprazine, to a 500-mL volumetric flask, dilute with water to volume, and mix.

Procedure—Transfer 6.0 mL each of the *Assay preparation* and the *Standard preparation* to individual 125-mL separators. To each separator add 5 mL of *Bromocresol green solution* and 10 mL of *pH 5.3 phosphate buffer*. Extract with three 25-mL portions of chloroform, filtering the extracts from each separator into separate 100-mL volumetric flasks. Dilute the filtrate in each flask with chloroform to volume, and mix. Concomitantly determine the absorbances of the solutions in 1-cm cells at the wavelength of maximum absorbance at about 415 nm, with a suitable spectrophotometer, using chloroform as the blank. Calculate the quantity, in mg, of $C_{19}H_{24}N_2OS$ in each mL of the Injection taken by the formula:

$$(0.5C/V)(A_U/A_S),$$

in which *C* is the concentration, in µg per mL, of USP Methotrimeprazine RS in the *Standard preparation*, *V* is the volume, in mL, of Injection taken, and A_U and A_S are the absorbances of the solutions from the *Assay preparation* and the *Standard preparation*, respectively.

Methoxamine Hydrochloride

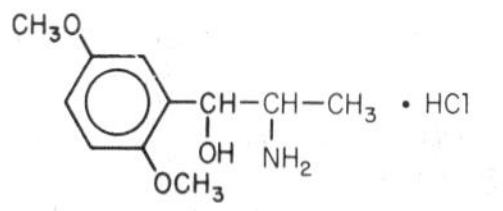

$C_{11}H_{17}NO_3 \cdot HCl$ 247.72

Benzenemethanol, α-(1-aminoethyl)-2,5-dimethoxy-, hydrochloride.

(±)-α-(1-Aminoethyl)-2,5-dimethoxybenzyl alcohol hydrochloride [*61-16-5*].

» Methoxamine Hydrochloride contains not less than 98.0 percent and not more than 100.5 percent of $C_{11}H_{17}NO_3 \cdot HCl$, calculated on the dried basis.

Packaging and storage—Preserve in well-closed, light-resistant containers.

Reference standard—*USP Methoxamine Hydrochloride Reference Standard*—Dry at 105° for 2 hours before using.

Identification—

A: The infrared absorption spectrum of a potassium bromide dispersion of it exhibits maxima only at the same wavelengths as that of a similar preparation of USP Methoxamine Hydrochloride RS.

B: To about 1 mg in a porcelain dish add 3 drops of a mixture of 3 mL of sulfuric acid and 2 drops of formaldehyde TS: a purple color is produced immediately, and it changes to brown and then to green.

C: It responds to the tests for *Chloride* ⟨191⟩.

Melting range ⟨741⟩: between 214° and 219°.

Loss on drying ⟨731⟩—Dry it at 105° for 2 hours: it loses not more than 1.0% of its weight.

Residue on ignition ⟨281⟩: not more than 0.2%.

Chloride content—Dissolve about 400 mg, accurately weighed, in 50 mL of water. Add 50 mL of methanol and 10 mL of glacial acetic acid. Then add eosin Y TS, and titrate with 0.1 *N* silver nitrate VS. Each mL of 0.1 *N* silver nitrate is equivalent to 3.545 mg of Cl. Not less than 14.1% and not more than 14.5% of Cl, calculated on the dried basis, is found.

Assay—Dissolve about 400 mg of Methoxamine Hydrochloride, accurately weighed, in a mixture of 10 mL of glacial acetic acid and 10 mL of mercuric acetate TS, warming slightly if necessary to effect solution. Add crystal violet TS, and titrate with 0.1 *N* perchloric acid VS. Perform a blank determination, and make any necessary correction. Each mL of 0.1 *N* perchloric acid is equivalent to 24.77 mg of $C_{11}H_{17}NO_3 \cdot HCl$.

Methoxamine Hydrochloride Injection

» Methoxamine Hydrochloride Injection is a sterile solution of Methoxamine Hydrochloride in Water for Injection. It contains not less than 93.0 percent and not more than 107.0 percent of the labeled amount of $C_{11}H_{17}NO_3 \cdot HCl$.

Packaging and storage—Preserve in single-dose or in multiple-dose containers, preferably of Type I glass, and protect from light.

Reference standard—*USP Methoxamine Hydrochloride Reference Standard*—Dry at 105° for 2 hours before using.

Identification—

A: Extract 2 mL of Injection with 10 mL of ether, and evaporate the water layer to dryness: the residue so obtained, previously dried at 105° for 2 hours, responds to *Identification test A* under *Methoxamine Hydrochloride*.

B: Evaporate 2 or 3 drops in a porcelain dish to dryness, and cool: the residue responds to *Identification test B* under *Methoxamine Hydrochloride*.

C: It responds to the tests for *Chloride* ⟨191⟩.

pH ⟨791⟩: between 3.0 and 5.0.

Other requirements—It meets the requirements under *Injections* ⟨1⟩.

Assay—

Mobile phase—Transfer 2.20 g of sodium 1-heptanesulfonate, accurately weighed, to a 2000-mL volumetric flask. Add 500 mL of water and 500 mL of methanol to the flask, dilute with water to volume, and mix. Adjust the solution with 0.1 *N* hydrochloric acid to a pH of 3.5, degas, and filter through a 0.8-µm alcohol-resistant membrane. Make adjustments if necessary (see *System Suitability* under *Chromatography* ⟨621⟩).

Standard preparation—Dissolve an accurately weighed quantity of USP Methoxamine Hydrochloride RS in water to obtain a stock solution having a known concentration of about 1 mg per mL. Pipet 10 mL of this stock solution into a 25-mL volumetric flask, dilute with 0.1 *N* hydrochloric acid to volume, and mix to obtain a solution having a known concentration of about 0.4 mg per mL.

Assay preparation—Pipet a volume of Methoxamine Hydrochloride Injection, equivalent to about 10 mg of methoxamine hydrochloride, into a 25-mL volumetric flask. Dilute with 0.1 *N* hydrochloric acid to volume, and mix.

System suitability preparation—Dissolve 2,5-dimethoxybenzaldehyde in 0.1 *N* hydrochloric acid to obtain a solution having a concentration of about 14 µg per mL. Mix 3 volumes of this solution with 2 volumes of the stock solution prepared as directed for the *Standard preparation*.

Chromatographic system (see *Chromatography* ⟨621⟩)—The liquid chromatograph is equipped with a 290-nm detector and a 3.9-mm × 30-cm column that contains packing L1. The flow rate is about 3 mL per minute. Chromatograph the *Standard preparation*, and the *System suitability preparation*, and record the peak responses as directed under *Procedure:* the tailing factor for the analyte peak is not more than 2.0, the resolution, *R*, between the methoxamine and 2,5-dimethoxybenzaldehyde peaks is not less than 2.0, and the relative standard deviation for replicate injections of the *Standard preparation* is not more than 2.0%. The relative retention times are about 1.0 for methoxamine and 0.75 for 2,5-dimethoxybenzaldehyde.

Procedure—Separately inject equal volumes (between 20 µL and 30 µL) of the *Standard preparation* and the *Assay preparation* into the chromatograph, record the chromatograms, and measure the responses for the major peaks. Calculate the quantity, in mg, of $C_{11}H_{17}NO_3HCl$ in each mL of the Injection taken by the formula:

$$25(C/V)(r_U/r_S),$$

in which C is the concentration, in mg per mL, of USP Methoxamine Hydrochloride RS in the *Standard preparation*, V is the volume, in mL, of Injection taken, and r_U and r_S are the peak responses obtained from the *Assay preparation* and the *Standard preparation*, respectively.

Methoxsalen

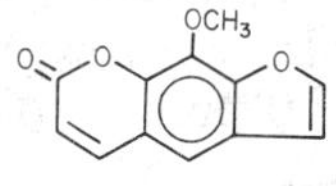

$C_{12}H_8O_4$ 216.19

7*H*-Furo[3,2-*g*][1]benzopyran-7-one, 9-methoxy-.

9-Methoxy-7*H*-furo[3,2-*g*][1]benzopyran-7-one [*298-81-7*].

» Methoxsalen contains not less than 98.0 percent and not more than 102.0 percent of $C_{12}H_8O_4$, calculated on the anhydrous basis.

Caution—Avoid contact with the skin.

Packaging and storage—Preserve in well-closed, light-resistant containers.

Reference standard—*USP Methoxsalen Reference Standard*—Do not dry; determine the *Water* content by *Method I* ⟨921⟩ before using.

Identification—The infrared absorption spectrum of a potassium bromide dispersion of it exhibits maxima and minima only at the same wavelengths as that of a similar preparation of USP Methoxsalen RS.

Melting range, *Class I* ⟨741⟩: between 143° and 148°.

Water, *Method I* ⟨921⟩: not more than 0.5%.

Residue on ignition ⟨281⟩: not more than 0.1%, a 1-g specimen being used.

Heavy metals, *Method II* ⟨231⟩: 0.002%.

Chromatographic impurities—Prepare a solution of it in chloroform containing about 20 mg per mL (*Solution A*). Dilute 1.0 mL of it with chloroform to 100.0 mL (*Solution B*). Apply 5-µL portions of both solutions at points along a line about 2.5 cm from one edge of a thin-layer chromatographic plate coated with a 0.25-mm layer of chromatographic silica gel mixture and previously dried at 105° for 30 minutes. Develop the plate in a suitable chamber, without previous equilibration, using a mixture of 9 volumes of benzene and 1 volume of ethyl acetate, until the solvent front has moved to about 15 cm above the line of application. Remove the plate from the chamber, air-dry, and observe under long-wavelength ultraviolet light: any spot in the chromatogram from *Solution A*, other than the principal spot, is not more intense than the spot from *Solution B* (1.0%).

Assay—

Mobile phase—Prepare a solution of acetonitrile in water (35 in 100). Make adjustments if necessary (see *System Suitability* under *Chromatography* ⟨621⟩).

Internal standard preparation—Dissolve trioxsalen in alcohol to obtain a solution containing about 0.2 mg per mL.

Standard preparation—Using an accurately weighed quantity of USP Methoxsalen RS, prepare a solution in alcohol having a known concentration of about 0.2 mg per mL. Transfer 2.0 mL of this solution to a 100-mL volumetric flask, add 2.0 mL of *Internal standard preparation*, dilute with *Mobile phase* to volume, and mix to obtain a *Standard preparation* having a known concentration of about 4 µg of USP Methoxsalen RS per mL. Filter through a 0.45-µm disk before using.

Assay preparation—Using 20 mg of methoxsalen, accurately weighed, proceed as directed for *Standard preparation*.

Chromatographic system (see *Chromatography* ⟨621⟩)—The liquid chromatograph is equipped with a 254-nm detector and a 4-mm × 30-cm column that contains packing L1. The flow rate is about 1.5 mL per minute. Chromatograph the *Standard preparation*, and record the peak responses as directed under *Procedure:* the resolution, *R*, between the analyte and internal standard peaks is not less than 4.0, and the relative standard deviation for replicate injections is not more than 2.0%.

Procedure—Separately inject equal volumes (about 20 µL) of the *Standard preparation* and the *Assay preparation* into the chromatograph, record the chromatograms, and measure the responses for the major peaks. The relative retention times are about 2.1 for trioxsalen and 1.0 for methoxsalen. Calculate the quantity, in mg, of $C_{12}H_8O_4$, in the portion of Methoxsalen taken by the formula:

$$5C(R_U/R_S),$$

in which C is the concentration, in µg per mL, of USP Methoxsalen RS in the *Standard preparation*, and R_U and R_S are the ratios of the peak responses of methoxsalen to the internal standard obtained from the *Assay preparation* and the *Standard preparation*, respectively.

Methoxsalen Capsules

» Methoxsalen Capsules contain not less than 90.0 percent and not more than 110.0 percent of the labeled amount of $C_{12}H_8O_4$.

Packaging and storage—Preserve in tight, light-resistant containers.

Reference standard—*USP Methoxsalen Reference Standard*—Do not dry; determine the *Water* content by *Method I* ⟨921⟩ before using.

Identification—

A: The retention time exhibited by methoxsalen in the chromatogram of the *Assay preparation* corresponds to that of methoxsalen in the chromatogram of the *Standard preparation* as obtained in the *Assay*.

B: Place one capsule in 50 mL of alcohol contained in a high-speed glass blender jar and blend thoroughly until the shell is completely dispersed. Dilute a portion quantitatively with alcohol to obtain a solution having a concentration of about 4 µg per mL: the ultraviolet absorption spectrum of the solution so obtained exhibits maxima and minima at the same wavelengths as that of a similar solution of USP Methoxsalen RS, concomitantly measured.

Dissolution ⟨711⟩—

Medium: water; 900 mL.

Apparatus 2: 50 rpm.

Time: 45 minutes.

Procedure—Determine the amount of $C_{12}H_8O_4$ dissolved from ultraviolet absorbances at the wavelength of maximum absorbance at about 300 nm using filtered portions of the solution under test, suitably diluted with water, if necessary, in comparison with a Standard solution having a known concentration of USP Methoxsalen RS in the same medium. [NOTE—An amount of alcohol not to exceed 1% of the total volume of the Standard solution may be used to bring the Reference Standard into solution prior to dilution with *Dissolution Medium.*]

Tolerances—Not less than 75% (*Q*) of the labeled amount of $C_{12}H_8O_4$ is dissolved in 45 minutes.

Uniformity of dosage units ⟨905⟩: meet the requirements.

Assay—

Mobile phase—Prepare a filtered and degassed mixture of acetonitrile and water (65:35). Make adjustments if necessary (see *System Suitability* under *Chromatography* ⟨621⟩).

Internal standard solution—Prepare a solution of trioxsalen in alcohol having a known concentration of 0.2 mg per mL.

Standard preparation—Prepare a solution in alcohol having an accurately known concentration of 0.2 mg of USP Methoxsalen RS per mL. Pipet 2.0 mL of this solution into a 100-mL volumetric flask containing 2.0 mL of *Internal standard solution*, dilute with *Mobile phase* to volume, and mix.

Assay preparation—Place not less than 10 Methoxsalen Capsules in a high-speed glass blender jar containing 100.0 mL of alcohol, and blend thoroughly. Transfer an accurately measured volume of the aliquot from the blender jar, equivalent to about 2 mg of Methoxsalen, to a 50-mL volumetric flask containing 10.0 mL of *Internal standard solution*, dilute with alcohol to volume, mix, and filter. Transfer 5.0 mL of this solution to a 50-mL volumetric flask, dilute with *Mobile phase* to volume, mix, and filter.

Chromatographic system (see *Chromatography* ⟨621⟩)—The liquid chromatograph is equipped with a 254-nm detector and a 4-mm × 30-cm column that contains packing L1. The flow rate is about 1.5 mL per minute. Chromatograph the *Standard preparation*, and record the peak responses as directed under *Procedure:* the resolution, *R*, between the analyte peak and internal standard peak is not less than 4.0, and the relative standard deviation for replicate injections is not more than 2.0%.

Procedure—Separately inject equal volumes (about 20 µL) of the *Standard preparation* and the *Assay preparation* into the chromatograph, record the chromatograms, and measure the responses for the major peaks. The relative retention times are about 0.5 for Methoxsalen and 1.0 for Trioxsalen. Calculate the quantity, in mg, per Capsule taken by the formula:

$$500(C/V)(R_U/R_S),$$

in which C is the concentration, in mg per mL, of USP Methoxsalen in the *Standard preparation*, V is the volume, in mL, of *Assay preparation* taken, and R_U and R_S are the peak response ratios obtained from the *Assay preparation* and the *Standard preparation*, respectively.

Methoxsalen Topical Solution

» Methoxsalen Topical Solution is a solution of Methoxsalen in a suitable vehicle. It contains not less than 9.2 mg and not more than 10.8 mg of $C_{12}H_8O_4$ per mL.

Packaging and storage—Preserve in tight, light-resistant containers.

Reference standard—*USP Methoxsalen Reference Standard*—Do not dry; determine the *Water* content by *Method I* ⟨921⟩ before using.

Identification—Transfer a volume of Topical Solution to a 100-mL volumetric flask, and dilute quantitatively and stepwise with alcohol to obtain a concentration of about 8 µg per mL: the ultraviolet absorption spectrum of this solution exhibits maxima and minima at the same wavelengths as that of a similar solution of USP Methoxsalen RS, concomitantly measured.

Alcohol content ⟨611⟩: between 66.5% and 77.0% of C_2H_5OH.

Assay—

Mobile phase, Internal standard preparation, Standard preparation, and *Chromatographic system*—Prepare as directed in the *Assay* under *Methoxsalen.*

Assay preparation—Transfer an accurately measured volume of Methoxsalen Topical Solution, equivalent to about 20 mg of methoxsalen, to a 100-mL volumetric flask. Dilute with alcohol to volume, and mix. Transfer 2.0 mL of this solution and 2.0 mL of *Internal standard preparation* to a 100-mL volumetric flask. Dilute with *Mobile phase* to volume, and mix. Filter through a 0.45-µm disk before using.

Procedure—Proceed as directed for **Procedure** in the *Assay* under *Methoxsalen.* Calculate the quantity, in mg, of $C_{12}H_8O_4$ in each mL of the Topical Solution taken by the formula:

$$5(C/V)(R_U/R_S),$$

in which V is the volume, in mL, of Topical Solution taken, and the other terms are as defined therein.

Methoxyflurane

$CHCl_2CF_2—O—CH_3$

$C_3H_4Cl_2F_2O$ 164.97
Ethane, 2,2-dichloro-1,1-difluoro-1-methoxy-.
2,2-Dichloro-1,1-difluoroethyl methyl ether [*76-38-0*].

» Methoxyflurane contains not less than 99.9 percent and not more than 100.0 percent of $C_3H_4Cl_2F_2O$, calculated on the anhydrous basis. It may contain a suitable stabilizer.

Packaging and storage—Preserve in tight, light-resistant containers, and avoid exposure to excessive heat.

Reference standard—*USP Methoxyflurane Reference Standard*—Keep container tightly closed and protected from light.

Identification—

A: The infrared absorption spectrum of a 1 in 20 solution in chloroform exhibits maxima only at the same wavelengths as that of a similar solution of USP Methoxyflurane RS.

B: To 1 mL in a test tube add 1 mL of sulfuric acid: the specimen forms a layer over the acid (*distinction from halothane*).

C: Cautiously heat the contents of the test tube from *Identification test B* with agitation: the interface disappears, and hydrofluoric acid is evolved (*distinction from chloroform, trichloroethylene, and halothane*).

Specific gravity ⟨841⟩: between 1.420 and 1.425.

Nonvolatile residue—Allow 50 mL to evaporate at room temperature in a tared evaporating dish, and dry the residue at 105° for 1 hour: the weight of the residue does not exceed 1 mg.

Acidity—Shake 25 mL with 25 mL of carbon dioxide–free water for 2 minutes, and allow the layers to separate. Add 1 drop of methyl red TS to the water extract, boil for 1 minute, and titrate with 0.010 *N* sodium hydroxide: not more than 0.50 mL of 0.010 *N* sodium hydroxide is required to produce a distinct yellow color.

Water, *Method I* ⟨921⟩: not more than 0.1%.

Foreign odor—Place 10 mL on a dry watch glass, and allow it to evaporate spontaneously to about 1 mL: no foreign odor is perceptible during the evaporation. To the remaining test specimen add a piece of odorless filter paper, and allow the paper to dry, checking it for foreign odor every few seconds during the drying period. As the last traces of odor leave the paper, a faint, characteristic odor may be detected for a few seconds, but no residual foreign odor is detected.

Assay—Inject a volume of Methoxyflurane of suitable size, but not more than 30 µL, into a suitable gas chromatograph (see *Chromatography* ⟨621⟩) equipped with a thermal conductivity detector. Under typical conditions, the instrument contains a 3-m × 4-mm stainless steel column packed with liquid phase G11 on support S1A, the column is maintained at a temperature of 100° to 110°, the injection port is maintained at about 150°, and dry helium is used as the carrier gas at a flow rate of about 60 mL per minute. From the area under the curve, calculate the percentage (a/a) of $C_3H_4Cl_2F_2O$ in the portion of Methoxyflurane taken.

Methscopolamine Bromide

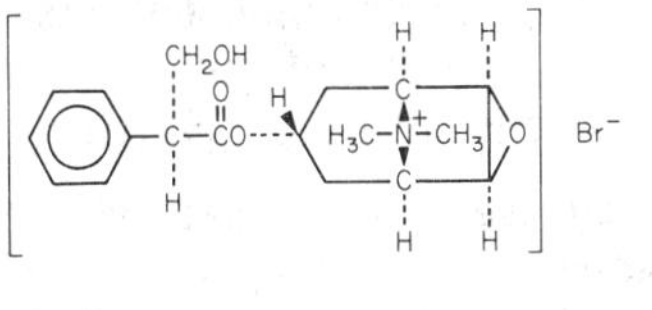

$C_{18}H_{24}BrNO_4$ 398.30
3-Oxa-9-azoniatricyclo[3.3.1.0^{2,4}]nonane, 7-(3-hydroxy-1-oxo-2-phenylpropoxy)-9,9-dimethyl-, bromide, [7(*S*)-(1α,2β,4β,5α,7β)]-.

6β,7β-Epoxy-3α-hydroxy-8-methyl-1αH,5αH-tropanium bromide (−)-tropate [*155-41-9*].

» Methscopolamine Bromide contains not less than 97.0 percent and not more than 103.0 percent of $C_{18}H_{24}BrNO_4$, calculated on the dried basis.

Packaging and storage—Preserve in tight, light-resistant containers.

Reference standard—*USP Methscopolamine Bromide Reference Standard*—Dry at 105° for 3 hours before using.

Identification—

A: The infrared absorption spectrum of a potassium bromide dispersion of it, previously dried, exhibits maxima only at the same wavelengths as that of a similar preparation of USP Methscopolamine Bromide RS.

B: A solution (1 in 20) responds to the tests for *Bromide* ⟨191⟩.

Specific rotation ⟨781⟩: between −21° and −25°, calculated on the dried basis, determined in a solution containing 500 mg in each 10 mL.

Loss on drying ⟨731⟩—Dry it at 105° for 3 hours: it loses not more than 2.0% of its weight.

Residue on ignition ⟨281⟩: not more than 0.1%.

Assay—

Mobile phase—Prepare a solution containing 2.6 g of decyl sodium sulfate, 0.8 g of ammonium nitrate, 600 mL of methanol, 400 mL of water, and 10 mL of glacial acetic acid.

Internal standard solution—Dissolve benzphetamine hydrochloride in water to obtain a solution containing about 0.1 mg per mL.

Standard preparation—Transfer about 20 mg of USP Methscopolamine Bromide RS, accurately weighed, to a 100-mL volumetric flask. Add *Internal standard solution* to volume, and mix.

Assay preparation—Prepare as directed under *Standard preparation*, using about 20 mg of Methscopolamine Bromide, accurately weighed.

Procedure—Inject 50-µL aliquots of the *Assay preparation* and the *Standard preparation* into a chromatograph of the general type fitted with a 30-cm × 3.9-mm column containing packing L1 (see *Chromatography* ⟨621⟩), which is operated at room temperature and is equipped with a 254-nm detector. In a suitable system, the retention times for the methscopolamine bromide and benzphetamine hydrochloride peaks are about 9 and 17 minutes, respectively. Six replicate injections of the *Standard preparation* show a relative standard deviation for the peak response ratio that is not greater than 2.0%. Calculate the quantity, in mg, of $C_{18}H_{24}BrNO_4$ in the Methscopolamine Bromide taken by the formula:

$$100C(R_U/R_S),$$

in which C is the concentration, in mg per mL, of USP Methscopolamine Bromide RS in the *Standard preparation*, and R_U and R_S are the ratios of the peak responses of the methscopolamine bromide and benzphetamine hydrochloride obtained from the *Assay preparation* and the *Standard preparation*, respectively.

Methscopolamine Bromide Tablets

» Methscopolamine Bromide Tablets contain not less than 93.0 percent and not more than 107.0 percent of the labeled amount of $C_{18}H_{24}BrNO_4$.

Packaging and storage—Preserve in tight containers.

Reference standard—*USP Methscopolamine Bromide Reference Standard*—Dry at 105° for 3 hours before using.

Identification—

A: *pH 7.3 dye-buffer solution*—Prepare a solution containing, in each 500 mL, 200 mg of bromothymol blue, 3.2 mL of 0.1 *N* sodium hydroxide, 577.5 mg of citric acid monohydrate, and 6.3 mg of sodium phosphate, dibasic, anhydrous.

Procedure—Finely powder 1 Tablet, and transfer an amount equivalent to about 0.5 mg of methscopolamine bromide to a suitable container. Add 20 mL of water, heat for 5 minutes on a steam bath with frequent agitation, and centrifuge to obtain a clear supernate. Transfer 10 mL of the supernate to a vessel containing 10 mL of chloroform and 10 mL of *pH 7.3 dye-buffer solution*. Shake vigorously for 3 minutes, centrifuge, and transfer 8 mL of the chloroform layer to a suitable container. Evaporate to dryness, and dissolve the residue in 1 mL of chloroform to obtain the test solution. Prepare a 0.025-mg-per-mL Standard solution of USP Methscopolamine Bromide RS in water, and treat as directed above, beginning with "Transfer 10 mL of the supernate." On a thin-layer chromatographic plate (see *Chromatography* ⟨621⟩), coated with a 0.25-mm layer of chromatographic silica gel mixture, apply 50-µL volumes of the Standard solution and test solution. Allow the spots to dry, and develop the plate in a solvent system prepared as follows: In a suitable container, mix water, butyl alcohol, and glacial acetic acid (5:4:1), then transfer a measured volume of the upper, organic layer to a suitable container, and mix with a volume of alcohol equivalent to 20% of the volume of the organic layer. Allow the solvent front to move about three-fourths of the length of the plate, and remove the plate from the developing chamber, mark the solvent front, and dry the plate under a current of air for 30 minutes. Spray the plate evenly with potassium-bismuth iodide TS: the chromatogram of the test solution shows a bright orange spot on a yellow background corresponding in R_f value (about 0.25) to the chromatogram from the Standard solution. Bromothymol blue produces a dark yellow spot at an R_f value of about 0.8.

B: Powder a number of Tablets, equivalent to about 5 mg of methscopolamine bromide, digest with 5 mL of water for 10 minutes, and filter: a portion of the clear solution so obtained responds to *Identification test B* under *Methscopolamine Bromide*.

Disintegration ⟨701⟩: 15 minutes, the use of disks being omitted.

Uniformity of dosage units ⟨905⟩: meet the requirements.

Assay—

Mobile phase, Internal standard solution, and *Standard preparation*—Prepare as directed in the *Assay* under *Methscopolamine Bromide*.

Assay preparation—Weigh and finely powder not less than 10 Methscopolamine Bromide Tablets. Weigh accurately a portion of the powder, equivalent to about 5 mg of methscopolamine bromide, and transfer completely to a suitable vial. Add 25.0 mL of *Internal standard solution*. Shake for 30 minutes. Centrifuge, and use the clear, supernatant solution.

Procedure—Proceed as directed for *Procedure* in the *Assay* under *Methscopolamine Bromide*. Calculate the quantity, in mg, of $C_{18}H_{24}BrNO_4$ in the portion of Tablets taken by the formula:

$$25C(R_U/R_S),$$

in which the terms are as defined therein.

Methsuximide

$C_{12}H_{13}NO_2$ 203.24

2,5-Pyrrolidinedione, 1,3-dimethyl-3-phenyl-.

N,2-Dimethyl-2-phenylsuccinimide [*77-41-8*].

» Methsuximide contains not less than 97.0 percent and not more than 103.0 percent of $C_{12}H_{13}NO_2$, calculated on the dried basis.

Packaging and storage—Preserve in tight containers.

Reference standard—*USP Methsuximide Reference Standard*—Dry over phosphorus pentoxide for 16 hours before using.

Identification—

A: The infrared absorption spectrum of a potassium bromide dispersion of it, previously dried, exhibits maxima only at the same wavelengths as that of a similar preparation of USP Methsuximide RS.

B: The ultraviolet absorption spectrum of a 1 in 3000 solution in alcohol exhibits maxima and minima at the same wavelengths as that of a similar solution of USP Methsuximide RS, concomitantly measured.

Melting range ⟨741⟩: between 50° and 56°, determined by a *Class I* procedure, except that the test specimen is inserted into the bath at about room temperature.

Loss on drying ⟨731⟩—Dry it over phosphorus pentoxide for 16 hours: it loses not more than 0.5% of its weight.

Residue on ignition ⟨281⟩: not more than 0.2%.

Cyanide—Dissolve 1.0 g in 10 mL of alcohol, and add 3 drops of ferrous sulfate TS, 1 mL of 1 *N* sodium hydroxide, and a few drops of ferric chloride TS. Warm gently, and finally acidify with 2 *N* sulfuric acid: no blue precipitate or blue color develops within 15 minutes.

Assay—Transfer about 150 mg of Methsuximide, accurately weighed, to a 50-mL volumetric flask, dissolve in 40 mL of alcohol, then dilute with alcohol to volume, and mix. Transfer 5.0 mL of this solution to a second 50-mL volumetric flask, dilute with alcohol to volume, and mix. Concomitantly determine the absorbances of this solution and a Standard solution of USP Methsuximide RS in the same medium having a known concentration of about 300 µg per mL, in 1-cm cells at the wavelength of maximum absorbance at about 247 nm, with a suitable spectrophotometer, using alcohol as the blank. Calculate the quantity, in mg, of $C_{12}H_{13}NO_2$ in the Methsuximide taken by the formula:

$$0.5C(A_U/A_S),$$

in which *C* is the concentration, in µg per mL, of USP Methsuximide RS in the Standard solution, and A_U and A_S are the absorbances of the solution of Methsuximide and the Standard solution, respectively.

Methsuximide Capsules

» Methsuximide Capsules contain not less than 92.0 percent and not more than 108.0 percent of the labeled amount of $C_{12}H_{13}NO_2$.

Packaging and storage—Preserve in tight containers, and avoid exposure to excessive heat.

Reference standard—*USP Methsuximide Reference Standard*—Dry over phosphorus pentoxide for 16 hours before using.

Identification—Mix a portion of the contents of Capsules, equivalent to about 200 mg of methsuximide, with 25 mL of water in a separator, extract with 50 mL of ether, and discard the aqueous layer. Wash the ether extract with 25 mL of water, and discard the water. Filter the extract, evaporate with the aid of a current of warm air to dryness, and dry the methsuximide over phosphorus pentoxide for 16 hours: the residue so obtained responds to *Identification test A* under *Methsuximide.*

Dissolution ⟨711⟩—

Medium: water; 900 mL.

Apparatus 1: 100 rpm.

Time: 120 minutes.

Procedure—Determine the amount of $C_{12}H_{13}NO_2$ dissolved, employing the procedure set forth in the *Assay*, making any necessary modifications.

Tolerances—Not less than 75% (*Q*) of the labeled amount of $C_{12}H_{13}NO_2$ is dissolved in 120 minutes.

Uniformity of dosage units ⟨905⟩: meet the requirements.

Assay—Transfer, as completely as possible, the contents of not less than 20 Methsuximide Capsules to a suitable tared container, and weigh. Mix with a glass rod. Weigh accurately a portion of the powder, equivalent to about 300 mg of methsuximide, and transfer to a 125-mL separator. Add 20 mL of water, and extract with three 40-mL portions of chloroform, filtering each successive chloroform extract through a chloroform-washed pledget of cotton into a 150-mL extraction flask. Evaporate the combined chloroform extracts on a steam bath to about 3 mL, then allow the remaining chloroform to evaporate at room temperature. Dissolve the residue of methsuximide in 30 mL of alcohol, transfer to a 100-mL volumetric flask, dilute with alcohol to volume, and mix. Pipet 5.0 mL of the solution into a 50-mL volumetric flask, dilute with alcohol to volume, and mix. Concomitantly determine the absorbances of this solution and a Standard solution of USP Methsuximide RS in the same medium having a known concentration of about 300 µg per mL, in 1-cm cells at the wavelength of maximum absorbance at about 247 nm, with a suitable spectrophotometer, using alcohol as the blank. Calculate the quantity, in mg, of $C_{12}H_{13}NO_2$ in the portion of Capsules taken by the formula:

$$C(A_U/A_S),$$

in which *C* is the concentration, in µg per mL, of USP Methsuximide RS in the Standard solution, and A_U and A_S are the absorbances of the solution from the Capsules and the Standard solution, respectively.

Methyclothiazide

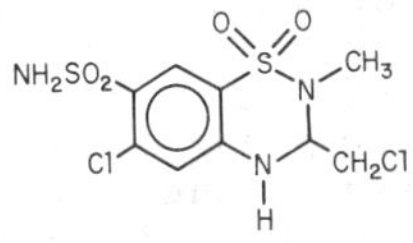

$C_9H_{11}Cl_2N_3O_4S_2$ 360.23

2*H*-1,2,4-Benzothiadiazine-7-sulfonamide, 6-chloro-3-(chloromethyl)-3,4-dihydro-2-methyl-, 1,1-dioxide.

6-Chloro-3-(chloromethyl)-3,4-dihydro-2-methyl-2*H*-1,2,4-benzothiadiazine-7-sulfonamide 1,1-dioxide [*135-07-9*].

» Methyclothiazide contains not less than 97.0 percent and not more than 102.0 percent of $C_9H_{11}Cl_2N_3O_4S_2$, calculated on the dried basis.

Packaging and storage—Preserve in well-closed containers.

Reference standards—*USP Methyclothiazide Reference Standard*—Dry at 105° for 4 hours before using. *USP 4-Amino-6-chloro-N³-methyl-m-benzenedisulfonamide Reference Standard*—Keep container tightly closed and protected from light. Dry over silica gel for 4 hours before using.

Identification—

A: The infrared absorption spectrum of a potassium bromide dispersion of it, previously dried, exhibits maxima only at the same wavelengths as that of a similar preparation of USP Methyclothiazide RS.

B: The ultraviolet absorption spectrum of a 1 in 50,000 solution in methanol exhibits maxima and minima at the same wavelengths as that of a similar solution of USP Methyclothiazide RS, concomitantly measured.

C: Fuse about 100 mg of it with a pellet of sodium hydroxide: the ammonia fumes produced turn moistened red litmus paper blue. The fused mixture responds to the test for Sulfite ⟨191⟩.

Loss on drying ⟨731⟩—Dry it at 105° for 4 hours: it loses not more than 0.5% of its weight.

Residue on ignition ⟨281⟩: not more than 0.2%.

Chloride ⟨221⟩—Shake 750 mg with 25 mL of water for 2 minutes, filter through a suitable membrane filter, and add 4 or 5 drops of 2 *N* nitric acid: the acidified filtrate shows no more chloride than corresponds to 0.20 mL of 0.020 *N* hydrochloric acid (0.02%).

Selenium ⟨291⟩: 0.003%.

Heavy metals, *Method II* ⟨231⟩: 0.002%.

Diazotizable substances—

Standard preparation—Transfer about 10 mg of USP 4-Amino-6-chloro-N^3-methyl-*m*-benzenedisulfonamide RS, accurately weighed, into a 50-mL volumetric flask, dilute with acetonitrile to volume, and mix. Pipet 25 mL of the solution into a 100-mL volumetric flask, dilute with acetonitrile to volume, and mix. Each mL of *Standard preparation* contains about 50 µg of the Reference Standard.

Test preparation—Weigh accurately about 500 mg of Methyclothiazide, transfer to a 100-mL volumetric flask, dissolve in acetonitrile, dilute with acetonitrile to volume, and mix.

Procedure—Pipet 2 mL each of the *Standard preparation* and the *Test preparation* into separate 50-mL volumetric flasks. Pipet 2 mL of acetonitrile into a third 50-mL flask to provide the blank. To each flask add 4 mL of 0.1 *N* hydrochloric acid, and mix. Add 3.0 mL of sodium nitrite solution (1 in 200) to each flask, mix, place the flasks in an ice bath for 5 minutes, shaking occasionally. Add to each flask 3.0 mL of ammonium sulfamate solution (1 in 50), mix, and allow the flasks to remain in the ice bath for 1 additional minute. Remove the flasks from the ice bath, add 1.0 mL of *N*-1-naphthylethylenediamine dihydrochloride solution (1 in 1000), and mix. Allow the flasks to stand at room temperature for 1 minute, then dilute with water to volume, and mix. Concomitantly determine the absorbances of the solutions obtained from the *Standard preparation* and the *Test preparation* in 1-cm cells at 525 nm, with a suitable spectrophotometer, using the reagent blank to set the instrument. The absorbance of the solution from the *Test preparation* does not exceed that of the solution from the *Standard preparation*, corresponding to not more than 1.0% of diazotizable substances.

Assay—Transfer about 350 mg of Methyclothiazide, accurately weighed, to a 250-mL conical flask, add 40 mL of a 1 in 20 solution of potassium hydroxide in methanol, and reflux at full boil for 1 hour. Cool, rinse the inner walls of the condenser with 20 mL of water and two 20-mL portions of methanol, add 10 mL of glacial acetic acid and 2 drops of eosin Y TS, and titrate with 0.1 *N* silver nitrate VS to the first appearance of a definite pink color. Each mL of 0.1 *N* silver nitrate is equivalent to 36.02 mg of $C_9H_{11}Cl_2N_3O_4S_2$.

Methyclothiazide Tablets

» Methyclothiazide Tablets contain not less than 90.0 percent and not more than 110.0 percent of the labeled amount of $C_9H_{11}Cl_2N_3O_4S_2$.

Packaging and storage—Preserve in well-closed containers.

Reference standard—*USP Methyclothiazide Reference Standard*—Dry at 105° for 4 hours before using.

Identification—Powder a number of Tablets, equivalent to about 50 mg of methyclothiazide, and transfer to a 100-mL volumetric flask with the aid of methanol. Add about 60 mL of methanol, and shake the flask for 1 hour. Dilute with methanol to volume, mix, and centrifuge a portion of the solution. Pipet 2 mL of the clear supernatant liquid into a second 100-mL volumetric flask, dilute with methanol to volume, and mix: the ultraviolet absorption spectrum of this solution exhibits maxima and minima only at the same wavelengths as that of a similar solution of USP Methyclothiazide RS.

Dissolution ⟨711⟩—

Medium: 0.1 *N* hydrochloric acid; 900 mL.
Apparatus 2: 50 rpm.
Time: 60 minutes.

Procedure—Determine the amount of $C_9H_{11}Cl_2N_3O_4S_2$ dissolved from ultraviolet absorbances at the wavelength of maximum absorbance at about 270 nm of filtered portions of the solution under test, suitably diluted with water, if necessary, in comparison with a Standard solution having a known concentration of USP Methyclothiazide RS in the same medium. An amount of alcohol not to exceed 1% of the total volume of the Standard solution may be used to dissolve USP Methyclothiazide RS prior to dilution with water.

Tolerances—Not less than 70% (*Q*) of the labeled amount of $C_9H_{11}Cl_2N_3O_4S_2$ is dissolved in 60 minutes.

Uniformity of dosage units ⟨905⟩: meet the requirements.

Procedure for content uniformity—Transfer 1 finely powdered Tablet to a 50-mL volumetric flask, add about 30 mL of methanol, and shake by mechanical means for 1 hour. Dilute with methanol to volume, mix, and centrifuge a portion of the mixture. Dilute quantitatively with methanol to obtain a solution containing approximately 10 µg of methyclothiazide per mL. Concomitantly determine the absorbances of this solution and a Standard solution of USP Methyclothiazide RS in the same medium, having a known concentration of about 10 µg per mL, in 1-cm cells at the wavelength of maximum absorbance at about 267 nm, with a suitable spectrophotometer, using methanol as the blank. Calculate the quantity, in mg, of $C_9H_{11}Cl_2N_3O_4S_2$ in the Tablet by the formula:

$$(TC/D)(A_U/A_S),$$

in which *T* is the labeled quantity, in mg, of methyclothiazide in the Tablet, *C* is the concentration, in µg per mL, of USP Methyclothiazide RS in the Standard solution, *D* is the concentration, in µg per mL, of methyclothiazide in the solution from the Tablet, based upon the labeled quantity per Tablet and the extent of dilution, and A_U and A_S are the absorbances of the solution from the Tablet and the Standard solution, respectively.

Assay—

Standard preparation—Transfer about 20 mg of USP Methyclothiazide RS, accurately weighed, to a 100-mL volumetric flask, add methanol to volume, and mix. Transfer 10.0 mL of this solution to a 200-mL volumetric flask, add chloroform to volume, and mix.

Assay preparation—Weigh and finely powder not less than 20 Methyclothiazide Tablets. Transfer an accurately weighed portion of the powder, equivalent to about 2 mg of methyclothiazide, to a 150-mL beaker, add 2.0 mL of methanol, mix, allow the mixture to stand for 30 minutes while taking precautions against loss of solvent, add 2.0 mL of 0.1 *M* sodium bicarbonate, and mix.

Procedure—[NOTE—Use water-saturated solvents throughout this procedure.] Mix about 3 g of chromatographic siliceous earth with 2.0 mL of 0.1 *M* sodium bicarbonate in a 150-mL beaker. Pack the mixture into a 25-mm × 200-mm chromatographic column. Add 4 g of chromatographic siliceous earth to the *Assay preparation*, mix, transfer the mixture to the column, and pack. Dry-wash the beaker with 1 g of the siliceous earth mixed with 3 drops of water, and transfer to the column. Place a small pad of glass wool above the column packing, pass 75 mL of a mixture of isooctane and ether (9:1) through the column, and discard the eluate. Using a 200-mL volumetric flask as a receiver, pass 100 mL of chloroform through the column, wash the tip of column with ether, add 10.0 mL of methanol, dilute with chloroform to volume, and mix. Concomitantly determine the absorbances of this solution and the *Standard preparation* at the wavelength of maximum absorbance at about 268 nm, with a suitable spectrophotometer, using chloroform as the blank. Calculate the quantity, in mg, of $C_9H_{11}Cl_2N_3O_4S_2$ in the portion of Tablets taken by the formula:

$$0.2C(A_U/A_S),$$

in which *C* is the concentration, in µg per mL, of USP Methyclothiazide RS in the *Standard preparation*, and A_U and A_S are the absorbances of the solution from the *Assay preparation* and the *Standard preparation*, respectively.

Methyl Alcohol—*see* Methyl Alcohol NF

Methyl Isobutyl Ketone—*see* Methyl Isobutyl Ketone NF

Methyl Salicylate—*see* Methyl Salicylate NF

Methylbenzethonium Chloride

$[(CH_3)_3CCH_2C(CH_3)_2-C_6H_3(CH_3)-OCH_2CH_2OCH_2CH_2N^+(CH_3)_2CH_2C_6H_5]\ Cl^- \cdot H_2O$

$C_{28}H_{44}ClNO_2.H_2O$ 480.13

Benzenemethanaminium, *N,N*-dimethyl-*N*-[2-[2-[methyl-4-(1,1,3,3-tetramethylbutyl)phenoxy]ethoxy]ethyl]-, chloride, monohydrate.

Benzyldimethyl[2-[2-[[4-(1,1,3,3-tetramethylbutyl)tolyl]oxy]-ethoxy]ethyl]ammonium chloride monohydrate [*1320-44-1*].

Anhydrous 462.11 [*25155-18-4*].

» Methylbenzethonium Chloride contains not less than 97.0 percent and not more than 103.0 percent of $C_{28}H_{44}ClNO_2$, calculated on the dried basis.

Packaging and storage—Preserve in tight containers.

Identification—

A: To 1 mL of a solution (1 in 100) add 2 mL of alcohol, 0.5 mL of 2 *N* nitric acid, and 1 mL of silver nitrate TS: a white precipitate, which is insoluble in 2 *N* nitric acid and soluble in 6 *N* ammonium hydroxide, is formed.

B: Treat separate portions of a solution (1 in 100) with 2 *N* nitric acid and with mercuric chloride TS, respectively: precipitates are formed that dissolve upon the addition of alcohol.

C: To 10 mL of a solution (1 in 20,000) add 0.1 g of sodium carbonate, 1 mL of bromophenol blue TS, and 10 mL of chloroform, and shake the mixture: the chloroform layer is blue.

D: Dissolve 100 mg in 1 mL of sulfuric acid, add 1.0 g of potassium nitrate, and heat on a steam bath for 3 minutes. Cautiously dilute the solution with water to 10 mL, add 0.5 g of granulated zinc, and warm the mixture for 10 minutes. Cool, add 0.2 g of sodium nitrite to 1 mL of the clear liquid, and add this mixture to 20 mg of naphthol potassium disulfonate in 1 mL of ammonium hydroxide: the solution turns orange-red, and a brown precipitate may be formed.

Melting range ⟨741⟩: between 159° and 163°, the specimen having been previously dried.

Loss on drying ⟨731⟩—Dry it at 105° for 4 hours: it loses not more than 5.0% of its weight.

Residue on ignition ⟨281⟩: not more than 0.1%.

Ammonia—To 5 mL of a solution (1 in 50) add 3 mL of 1 *N* sodium hydroxide, and heat to boiling: the odor of ammonia is not perceptible.

Assay—Weigh accurately a quantity of Methylbenzethonium Chloride, equivalent to about 500 mg of dried methylbenzethonium chloride, and transfer, with the aid of 35 mL of water, to a glass-stoppered, 250-mL conical separator containing 25 mL of chloroform. Add 10.0 mL of freshly prepared potassium iodide solution (1 in 20), insert the stopper in the separator, shake, allow the layers to separate, and discard the chloroform layer. Wash the aqueous layer with three 10-mL portions of chloroform, and discard the washings. Transfer the aqueous layer to a glass-stoppered, 250-mL conical flask, and rinse the separator with three 5-mL portions of water, adding the washings to the flask. Add 40 mL of cold hydrochloric acid to the flask, mix, and titrate with 0.05 *M* potassium iodate VS until the solution becomes light brown in color. Add 5 mL of chloroform, insert the stopper in the flask, and shake vigorously. Continue the titration, dropwise, with shaking after each addition, until the chloroform layer becomes colorless and the aqueous layer is clear yellow. Perform a blank determination, using 20 mL of water as the test specimen (see *Residual Titrations* under *Titrimetry* ⟨541⟩). Each mL of 0.05 *M* potassium iodate is equivalent to 46.21 mg of $C_{28}H_{44}ClNO_2$.

Methylbenzethonium Chloride Lotion

» Methylbenzethonium Chloride Lotion is an emulsion containing not less than 90.0 percent and not more than 110.0 percent of the labeled amount of $C_{28}H_{44}ClNO_2.H_2O$.

Packaging and storage—Preserve in tight containers.

Reference standard—*USP Docusate Sodium Reference Standard*—Do not dry; determine the water content at the time of use.

Identification—Suspend about 0.5 mL of Lotion in 20 mL of water, add 0.1 g of sodium carbonate, 1 mL of bromophenol blue TS, and 10 mL of chloroform, and shake the mixture: the chloroform layer is blue.

pH ⟨791⟩: between 5.2 and 6.0.

Assay—

0.0001 N Docusate sodium—Dissolve an accurately weighed quantity of USP Docusate Sodium RS in isopropyl alcohol, and dilute quantitatively with isopropyl alcohol to obtain a solution having a concentration of 4.446 mg of anhydrous docusate sodium per mL. Store this solution in a tightly stoppered glass container. On the day of use, pipet 10 mL of this solution into a 1000-mL volumetric flask, add water to volume, and mix to obtain a 0.0001 *N* solution.

Procedure—Transfer an accurately weighed portion of Methylbenzethonium Chloride Lotion, equivalent to about 0.5 mg of methylbenzethonium chloride, to a glass-stoppered, 50-mL cylinder. Add 5 mL of chloroform (freshly purified by shaking 100 mL with 10 g of silica gel, allowing to settle, and withdrawing the supernatant liquid), 5 mL of phosphoric acid solution (1 in 10), and 1 mL of safranin O solution (1 in 20,000). Titrate with *0.0001 N Docusate sodium* until about 1 mL from the end-point, then shake the stoppered tube vigorously for about 2 minutes, and continue the titration in 0.1-mL increments, shaking vigorously after each addition, until a pink color appears in the chloroform layer. Perform a blank determination, and make any necessary correction. Each mL of *0.0001 N Docusate sodium* is equivalent to 48.01 µg of $C_{28}H_{44}ClNO_2.H_2O$.

Methylbenzethonium Chloride Ointment

» Methylbenzethonium Chloride Ointment contains not less than 90.0 percent and not more than 110.0 percent of the labeled amount of $C_{28}H_{44}ClNO_2.H_2O$.

Packaging and storage—Preserve in collapsible tubes or in tight containers.

Reference standard—*USP Docusate Sodium Reference Standard*—Do not dry; determine the water content at the time of use.

Identification—Suspend about 0.5 g of Ointment in 10 mL of water, add 0.1 g of sodium carbonate, 1 mL of bromophenol blue TS, and 10 mL of chloroform, and shake: the chloroform layer is blue.

pH ⟨791⟩: between 5.0 and 7.0, in a dispersion of it in carbon dioxide–free water (1 in 100).

Assay—

0.0001 N Docusate sodium—Prepare as directed for *0.0001 N Docusate sodium* in the *Assay* under *Methylbenzethonium Chloride Lotion.*

Procedure—Transfer an accurately weighed portion of Methylbenzethonium Chloride Ointment, equivalent to about 0.5 mg of methylbenzethonium chloride, to a glass-stoppered, 50-mL cylinder, and proceed as directed in the *Assay* under *Methylbenzethonium Chloride Lotion*, beginning with "Add 5 mL of chloroform."

Methylbenzethonium Chloride Powder

» Methylbenzethonium Chloride Powder contains not less than 85.0 percent and not more than 115.0 percent of the labeled amount of $C_{28}H_{44}ClNO_2.H_2O$ in a suitable fine powder base, free from grittiness.

Packaging and storage—Preserve in well-closed containers.

Reference standard—*USP Docusate Sodium Reference Standard*—Do not dry; determine the water content at the time of use.

Identification—Suspend about 0.1 g in 10 mL of water, add 0.1 g of sodium carbonate, 1 mL of bromophenol blue TS, and 10 mL of chloroform, and shake the mixture: the chloroform layer is blue.

pH ⟨791⟩: between 9.0 and 10.5, in a dispersion of it in carbon dioxide-free water (1 in 100).

Other tests—Not less than 99% of it passes through a No. 200 sieve (see *Powder Fineness* ⟨811⟩).

Assay—

0.0001 N Docusate sodium—Prepare as directed for *0.0001 N Docusate sodium* in the *Assay* under *Methylbenzethonium Chloride Lotion.*

Procedure—Transfer an accurately weighed portion of Methylbenzethonium Chloride Powder, equivalent to about 0.5 mg of methylbenzethonium chloride, to a glass-stoppered, 50-mL cylinder, and proceed as directed in the *Assay* under *Methylbenzethonium Chloride Lotion*, beginning with "Add 5 mL of chloroform."

Methylcellulose

Cellulose, methyl ether.
Cellulose methyl ether [*9004-67-5*].

» Methylcellulose is a methyl ether of cellulose. When dried at 105° for 2 hours, it contains not less than 27.5 percent and not more than 31.5 percent of methoxy (OCH_3) groups.

Packaging and storage—Preserve in well-closed containers.

Labeling—Label it to indicate its viscosity type [viscosity of a solution (1 in 50)].

Identification—

A: Gently add 1 g to the top of 100 mL of water in a beaker, and allow to disperse over the surface, tapping the top of the container to ensure an even dispersion of the test specimen. Allow the beaker to stand until the specimen becomes transparent and mucilaginous (about 5 hours), and swirl the beaker to wet the remaining substance, add a stirring bar, and stir until solution is complete: the mixture remains stable when an equal volume of 1 *N* sodium hydroxide or 1 *N* hydrochloric acid is added.

B: Heat a few mL of the solution prepared for *Identification test A:* the solution becomes cloudy and a flaky precipitate, which redissolves as the solution cools, appears.

C: Pour a few mL of the solution prepared for *Identification test A* onto a glass plate, and allow the water to evaporate: a thin, self-sustaining film results.

Apparent viscosity—Place a quantity, accurately weighed and equivalent to 2 g of solids on the dried basis in a tared, wide-mouth, 250-mL centrifuge bottle, and add 98 g of water previously heated to 80° to 90°. Stir with a propeller-type stirrer for 10 minutes, place the bottle in an ice bath, continue the stirring, and allow to remain in the ice bath for 40 minutes to ensure that hydration and solution are complete. Adjust the weight of the solution to 100 g if necessary, and centrifuge the solution to expel any entrapped air. Adjust the temperature of the solution to 20 ± 0.1°, and determine the viscosity in a suitable viscosimeter of the Ubbelohde type as directed for *Procedure for Cellulose Derivatives* under *Viscosity* ⟨911⟩. Its viscosity is not less than 80.0% and not more than 120.0% of that stated on the label for viscosity types of 100 centipoises or less, and not less than 75.0% and not more than 140.0% of that stated on the label for viscosity types higher than 100 centipoises.

Loss on drying ⟨731⟩—Dry it at 105° for 2 hours: it loses not more than 5.0% of its weight.

Residue on ignition ⟨281⟩: not more than 1.5%.

Arsenic, *Method II* ⟨211⟩: 3 ppm.

Heavy metals, *Method II* ⟨231⟩: 0.001%, adding 1 mL of hydroxylamine hydrochloride solution (1 in 5) to the solution of the residue.

Assay—[*Caution—Perform all steps involving hydriodic acid carefully, in a well-ventilated hood. Use goggles, acid-resistant gloves, and other appropriate safety equipment. Be exceedingly careful when handling the hot vials, since they are under pressure. In the event of hydriodic acid exposure, wash with copious amounts of water, and seek medical attention at once.*]

Internal standard solution—Transfer about 2.5 g of toluene, accurately weighed, to a 100-mL volumetric flask containing 10 mL of *o*-xylene, dilute with *o*-xylene to volume, and mix.

Standard preparation—Into a suitable serum vial weigh about 135 mg of adipic acid, add 4.0 mL of hydriodic acid, then pipet 4 mL of *Internal standard solution* into the vial, and close the vial securely with a suitable septum stopper. Weigh the vial and contents accurately, add 90 µL of methyl iodide with a syringe through the septum, again weigh, and calculate the weight of methyl iodide added, by difference. Shake, and allow the layers to separate.

Assay preparation—Transfer about 0.065 g of dried Methylcellulose, accurately weighed, to a 5-mL thick-walled reaction vial equipped with a pressure-tight septum closure, add an amount of adipic acid equal to the weight of the test specimen, and pipet 2 mL of *Internal standard solution* into the vial. Cautiously pipet 2 mL of hydriodic acid into the mixture, immediately secure the closure, and weigh it accurately. Shake the vial for 30 seconds, heat at 150° for 20 minutes, using a heating block or a protective wrapping, remove it from the source of heat, shake it again, using extreme caution, and heat it as before at 150° for an additional 40 minutes. Allow the vial to cool for about 45 minutes, and again weigh it. If the weight loss is greater than 10 mg, discard the mixture, and prepare another *Assay preparation.*

Chromatographic system—Use a gas chromatograph equipped with a thermal conductivity detector. Under typical conditions, the instrument contains a 1.8-m × 4-mm glass column packed with 10 percent liquid phase G1 on 100- to 120-mesh support S1A, the column is maintained at 100°, the injection port and the detector are maintained at 200°, and helium is used as the carrier gas at a flow rate of 20 mL per minute.

Calibration—Inject about 2 µL of the upper layer of the *Standard preparation* into the gas chromatograph, and record the chromatogram. The retention times for methyl iodide, toluene, and *o*-xylene are approximately 3, 7, and 13 minutes, respectively. Calculate the relative response factor, F_{mi}, of equal weights of toluene and methyl iodide taken by the formula:

$$Q_{smi}/A_{smi},$$

in which Q_{smi} is the quantity ratio of methyl iodide to toluene in the *Standard preparation*, and A_{smi} is the peak area ratio of the methyl iodide to toluene obtained from the *Standard preparation.*

Procedure—Inject about 2 µL of the upper layer of the *Assay preparation* into the gas chromatograph, and record the chromatogram. Calculate the percentage of methoxy in the Methylcellulose taken by the formula:

$$2(31/142)F_{mi}A_{umi}(W_t/W_u),$$

in which 31/142 is the ratio of the formula weights of methoxy and methyl iodide, F_{mi} is defined under *Calibration*, A_{umi} is the ratio of the area of the methyl iodide peak to that of the toluene peak obtained from the *Assay preparation*, W_t is the weight, in g, of toluene in the *Internal standard solution*, and W_u is the weight, in g, of Methylcellulose taken for the *Assay*.

Methylcellulose Ophthalmic Solution

» Methylcellulose Ophthalmic Solution is a sterile solution of Methylcellulose. It contains not less than 85.0 percent and not more than 115.0 percent of the labeled amount of methylcellulose. It may contain suitable antimicrobial, buffering, and stabilizing agents.

Packaging and storage—Preserve in tight containers.

Identification—It responds to *Identification tests B* and *C* under *Methylcellulose.*

Sterility—It meets the requirements under *Sterility Tests* ⟨71⟩.

pH ⟨791⟩: between 6.0 and 7.8.

Assay—To boiling flask *A*, as described under *Methoxy Determination* ⟨431⟩, pipet a quantity of Methylcellulose Ophthalmic Solution, equivalent to 50 mg of methylcellulose. Evaporate on a steam bath to dryness, cool the flask in an ice bath, add the specified amount of hydriodic acid, and proceed as directed under *Methoxy Determination* ⟨431⟩. Each mL of 0.1 *N* sodium thiosulfate is equivalent to 1.753 mg of methylcellulose.

Methylcellulose Oral Solution

» Methylcellulose Oral Solution is a flavored solution of Methylcellulose. It contains not less than 85.0 percent and not more than 115.0 percent of the labeled amount of methylcellulose.

Packaging and storage—Preserve in tight, light-resistant containers, and avoid exposure to direct sunlight and to excessive heat. Avoid freezing.

Identification—It responds to *Identification tests B* and *C* under *Methylcellulose.*

Microbial limits ⟨61⟩—Its total aerobic microbial count does not exceed 100 per mL, and it meets the requirements of the test for the absence of *Escherichia coli.*

Alcohol content, *Method II* ⟨611⟩: between 3.5% and 6.5% of C_2H_5OH.

Assay—To boiling flask *A*, as described under *Methoxy Determination* ⟨431⟩, transfer an accurately measured volume of Methylcellulose Oral Solution, equivalent to 50 mg of methylcellulose. Evaporate on a steam bath to dryness, cool the flask in an ice bath, add the specified amount of hydriodic acid, and proceed as directed under *Methoxy Determination* ⟨431⟩. Each mL of 0.1 *N* sodium thiosulfate is equivalent to 1.753 mg of methylcellulose.

Methylcellulose Tablets

» Methylcellulose Tablets contain not less than 90.0 percent and not more than 110.0 percent of the labeled amount of methylcellulose.

Packaging and storage—Preserve in well-closed containers.

Identification—Add the residue obtained in the *Assay* to 50 mL of water: the solution responds to the *Identification tests* under *Methylcellulose.*

Disintegration ⟨701⟩: 30 minutes.

Uniformity of dosage units ⟨905⟩: meet the requirements.

Assay—Weigh and finely powder not less than 20 Methylcellulose Tablets. Weigh accurately a portion of the powder, equivalent to about 500 mg of methylcellulose, and transfer to a tared, fine fritted-glass, low-form, 30-mL crucible having a fitted crucible lid. Add 20 mL of alcohol, and macerate the solid for about 5 minutes, mixing intermittently with a glass stirring rod. Repeat the extraction with ten consecutive 10-mL portions of alcohol. Test for completeness of extraction by evaporating the last alcohol extract on a steam bath to dryness, taking up the residue in about 1 mL of water, and adding this to 5 mL of hot alkaline cupric tartrate TS (no red precipitate of cuprous oxide is formed within 5 minutes). If a precipitate is formed, continue with the alcohol extractions until the test is negative. Wash the completely extracted residue with a 10-mL portion of ether, using suction to drain off the liquid. Dry the residue in the crucible in a drying oven at 105° to constant weight. Weigh the crucible with the crucible lid in place. The weight of residue is the weight of methylcellulose present in the portion of powdered Tablets taken.

Methyldopa

HO, HO, NH_2, CH_2---C---COOH · $1\frac{1}{2}H_2O$, CH_3

$C_{10}H_{13}NO_4 . 1\frac{1}{2}H_2O$ 238.24

L-Tyrosine, 3-hydroxy-α-methyl-, sesquihydrate.

L-3-(3,4-Dihydroxyphenyl)-2-methylalanine sesquihydrate [*41372-08-1*].

Anhydrous 211.22 [*555-30-6*].

» Methyldopa contains not less than 98.0 percent and not more than 101.0 percent of $C_{10}H_{13}NO_4$, calculated on the anhydrous basis.

Packaging and storage—Preserve in well-closed, light-resistant containers.

Reference standards—*USP Methyldopa Reference Standard*—Do not dry; determine the *Water* content by *Method I* before using. *USP 3-O-Methylmethyldopa Reference Standard*—Do not dry; use as is.

Identification—

A: The infrared absorption spectrum of a mineral oil dispersion of it exhibits maxima only at the same wavelengths as that of a similar preparation of USP Methyldopa RS.

B: The ultraviolet absorption spectrum of a 1 in 25,000 solution in 0.1 *N* hydrochloric acid exhibits maxima and minima at the same wavelengths as that of a similar solution of USP Methyldopa RS, concomitantly measured, and the respective absorptivities, calculated on the anhydrous basis, at the wavelength of maximum absorbance at about 280 nm do not differ by more than 3.0%.

C: To 10 mg add 0.15 mL of a 1 in 250 solution of triketohydrindene hydrate in sulfuric acid: a dark purple color is produced within 5 to 10 minutes. Add 0.15 mL of water: the color changes to pale brownish yellow.

Specific rotation ⟨781⟩: between −25° and −28°, calculated on the anhydrous basis, determined in a solution containing 440 mg in each 10 mL, the solvent being a 2 in 3 solution, in water, of aluminum chloride that previously has been treated with activated charcoal, filtered, and adjusted with sodium hydroxide solution (1 in 100) to a pH of 1.5.

Acidity—Dissolve 1.0 g in carbon dioxide–free water with the aid of heat, add 1 drop of methyl red TS, and titrate with 0.10 *N* sodium hydroxide to a yellow end-point: not more than 0.50 mL is required.

Water, *Method I* ⟨921⟩: between 10.0% and 13.0%.

Residue on ignition ⟨281⟩: not more than 0.1%.

Heavy metals, *Method II* ⟨231⟩: 0.001%.

3-*O*-Methylmethyldopa—

Developing solvent—Mix 65 parts by volume of butyl alcohol, 15 parts by volume of glacial acetic acid, and 25 parts by volume of water. Prepare this mixture fresh.

Chromatographic plate—Prepare a thin-layer chromatographic plate with a suitable grade of cellulose, 250 µm thick, pre-washed with the *Developing solvent.* Wash the plate by placing it in a tank containing the solvent system and allowing the solvent to rise to the top of the plate. Dry with the aid of a stream of dry air.

Spray solution 1—Dissolve 300 mg of *p*-nitroaniline in 100 mL of 10 *N* hydrochloric acid (*Solution A*). Dissolve 2.5 g of sodium nitrite in 50 mL of water (*Solution B*). Mix 90 mL of *Solution A* and 10 mL of *Solution B* (*Spray solution 1*). Prepare all solutions fresh, just before spraying.

Spray solution 2—Dissolve 25 g of sodium carbonate in 100 mL of water, and mix.

Test solution—Dissolve 100 mg of Methyldopa in methanol, and dilute with methanol to 10.0 mL.

Standard solution—Dissolve 5.0 mg of USP 3-*O*-Methylmethyldopa RS in methanol, and dilute with methanol to 50.0 mL to obtain a Standard solution having a known concentration of 100 µg per mL.

Procedure—Apply 20 µL of *Test solution* in two 10-µL increments and 10 µL of *Standard solution* to the *Chromatographic plate,* so that the spots are not larger than 0.5 cm in diameter. Develop the chromatogram using the *Developing solvent* until the solvent front has moved about 10 cm from the origin. Remove the plate from the chamber, and dry with the aid of a stream of dry air until no odor of acetic acid is perceptible. Place the plate in a vertical position, and evenly spray with *Spray solution 1* until the adsorbent layer is uniformly soaked down to the glass (do not overspray). Place the plate in a horizontal position, and dry as completely as possible with the aid of a stream of warm dry air (no odor of hydrochloric acid is perceptible). Place the plate in a vertical position, and evenly spray with *Spray solution 2* until the plate is uniformly wet (do not overspray). The major methyldopa spot is black on a pale pink or orange background at an R_f value of about 0.50, and the 3-*O*-methylmethyldopa spot is dark on a similar background at an R_f value of about 0.65. The area and intensity of any 3-*O*-methylmethyldopa spot from the *Test solution* are not greater than those from the *Standard solution* (0.5%).

Assay—Dissolve about 200 mg of Methyldopa, accurately weighed, in 25 mL of glacial acetic acid, with the aid of heat. Cool to room temperature, and add 0.1 mL of crystal violet TS and 50 mL of acetonitrile. Titrate with 0.1 *N* perchloric acid VS to a blue end-point. Perform a blank determination, and make any necessary correction. Each mL of 0.1 *N* perchloric acid is equivalent to 21.12 mg of $C_{10}H_{13}NO_4$.

Methyldopa Oral Suspension

» Methyldopa Oral Suspension is an aqueous suspension of Methyldopa. It contains one or more suitable flavors, wetting agents, and preservatives, and it may contain Sucrose. It contains not less than 90.0 percent and not more than 110.0 percent of the labeled amount of $C_{10}H_{13}NO_4$.

Packaging and storage—Preserve in tight, light-resistant containers at a temperature not exceeding 26°.

Reference standards—*USP Methyldopa Reference Standard*—Do not dry; determine the *Water* content by *Method I* ⟨921⟩ before using. *USP Methyldopa-glucose Reaction Product Reference Standard*—Use as is.

Identification—On a suitable thin-layer chromatographic plate (see *Chromatography* ⟨621⟩), coated with a 0.25-mm layer of chromatographic silica gel mixture, apply 10-µL portions of the *Assay preparation* and the *Standard preparation* prepared as directed in the *Assay.* Allow to dry, develop the chromatogram in a solvent system consisting of equal volumes of acetone, butyl alcohol, glacial acetic acid, toluene, and water until the solvent front has moved about three-fourths of the length of the plate. Remove the plate from the developing chamber, mark the solvent front, and allow the solvent to evaporate. Locate the spots by staining the plate with iodine vapor for about 50 minutes, then view the plate under short-wavelength ultraviolet light: the R_f value of the principal spot obtained from the *Assay preparation* corresponds to that obtained from the *Standard preparation.*

pH ⟨791⟩: between 3.0 and 5.0; between 3.2 and 3.8 if sucrose is present.

Limit of methyldopa-glucose reaction product [TO BE DETERMINED IF SUCROSE IS PRESENT]—

Mobile phase—Prepare as directed in the *Assay.*

Solution A—Dissolve a suitable, accurately weighed quantity of USP Methyldopa-glucose Reaction Product RS in 0.1 *N* sulfuric acid to obtain a solution having a known concentration of about 0.45 mg per mL.

Standard preparation—Transfer about 25 mg of USP Methyldopa RS, accurately weighed, to a 25-mL volumetric flask, add 1.0 mL of *Solution A,* dilute with 0.1 *N* sulfuric acid to volume, and mix. The *Standard preparation* has a known concentration of about 18 µg of USP Methyldopa-glucose Reaction Product RS per mL.

Test preparation—Prepare as directed for *Assay preparation* in the *Assay.*

Chromatographic system—Use the system described under *Chromatographic system* in the *Assay.* The relative retention times for methyldopa and methyldopa-glucose reaction product are about 1.0 and 0.8, respectively. Chromatograph three replicate injections of the *Standard preparation:* the resolution factor, *R,* between methyldopa and methyldopa-glucose reaction product is not less than 2.0. The relative standard deviations for three replicate injections of the *Standard preparation* are not more than 2.0% and 3.0% for methyldopa and methyldopa-glucose reaction product, respectively.

Procedure—Proceed as directed for *Procedure* in the *Assay.* Calculate the quantity, in µg, of methyldopa equivalent to the methyldopa-glucose reaction product in each mL of the Oral Suspension taken by the formula:

$$(211.22/373.35)(250)(CD/W)(r_U/r_S),$$

in which 211.22 and 373.35 are the molecular weights of anhydrous methyldopa and methyldopa-glucose reaction product, respectively, *C* is the concentration, in µg per mL, of USP Methyldopa-glucose Reaction Product RS in the *Standard preparation,* r_U and r_S are the peak responses of the methyldopa-glucose reaction product obtained from the *Test preparation* and the *Standard preparation,* respectively, and the other terms are as defined therein. The limit is 10.0%, based on the methyldopa content of the Oral Suspension as determined in the *Assay.*

Assay—

Mobile phase—To 6.8 g of monobasic potassium phosphate add 750 mL of water, and stir until solution is complete. Adjust with 1 *M* phosphoric acid to a pH of 3.5, dilute with water to 1000 mL, mix, and filter through a membrane filter (10-µm or finer porosity).

Standard preparation—Dissolve an accurately weighed quantity of USP Methyldopa RS in 0.1 *N* sulfuric acid to obtain a solution having a known concentration of about 1 mg of anhydrous methyldopa per mL.

Assay preparation—Transfer an accurately measured volume of Methyldopa Oral Suspension, freshly mixed, equivalent to about 250 mg of methyldopa, to a 250-mL volumetric flask, dilute with 0.1 *N* sulfuric acid to volume, and mix to dissolve the methyldopa. Filter the solution through a 0.45-µm membrane filter before using.

Chromatographic system (see *Chromatography* ⟨621⟩)—The liquid chromatograph is equipped with a 280-mm detector and a 3.9-mm × 30-cm column that contains packing L1. The flow rate is about 1 mL per minute. Chromatograph three replicate injections of the *Standard preparation,* and record the peak responses as directed under *Procedure:* the relative standard deviation is not more than 2.0%.

Procedure—Separately inject equal volumes (about 50 µL) of the *Standard preparation* and the *Assay preparation* into the chromatograph by means of a suitable microsyringe or sampling valve, record the chromatograms, and measure the responses for the major peaks. Calculate the quantity, in mg, of $C_{10}H_{13}NO_4$ in each mL of the Oral Suspension taken by the formula:

$$250(C/V)(r_U/r_S),$$

in which C is the concentration, in mg per mL, of USP Methyldopa RS in the *Standard preparation*, V is the volume, in mL, of Oral Suspension taken, and r_U and r_S are the peak responses obtained from the *Assay preparation* and the *Standard preparation*, respectively.

Methyldopa Tablets

» Methyldopa Tablets contain not less than 90.0 percent and not more than 110.0 percent of the labeled amount of $C_{10}H_{13}NO_4$.

Packaging and storage—Preserve in well-closed containers.

Reference standard—*USP Methyldopa Reference Standard*—Do not dry; determine the *Water* content by *Method I* before using.

Identification—

A: To about 10 mg of finely ground Tablets add 3 drops of a 1 in 250 solution of triketohydrindene hydrate in sulfuric acid: a dark purple color is produced within 5 to 10 minutes. Add 3 drops of water: the color changes to pale brownish yellow.

B: To about 10 mg of finely ground Tablets add 2 mL of 0.1 *N* sulfuric acid and 2 mL of *Ferrous tartrate solution* prepared as directed in the *Assay*, then add 0.25 mL of 6 *N* ammonium hydroxide, and mix: a dark purple color is produced immediately.

Dissolution ⟨711⟩—

Medium: 0.1 *N* hydrochloric acid; 900 mL.

Apparatus 2: 50 rpm.

Time: 20 minutes.

Procedure—Determine the amount of $C_{10}H_{13}NO_4$ dissolved from ultraviolet absorbances at the wavelength of maximum absorbance at about 280 nm of filtered portions of the solution under test, suitably diluted with *Dissolution Medium*, if necessary, in comparison with a Standard solution having a known concentration of USP Methyldopa RS in the same medium.

Tolerances—Not less than 80% (Q) of the labeled amount of $C_{10}H_{13}NO_4$ is dissolved in 20 minutes.

Uniformity of dosage units ⟨905⟩: meet the requirements.

Assay—

Ferrous tartrate solution—Dissolve 1 g of ferrous sulfate, 2 g of potassium sodium tartrate, and 100 mg of sodium bisulfite in water to make 100 mL. Prepare this solution fresh.

Buffer solution—Dissolve 50 g of ammonium acetate in 1000 mL of 20 percent alcohol. Adjust with 6 *N* ammonium hydroxide to a pH of 8.5.

Standard preparation—Dissolve a suitable quantity of USP Methyldopa RS in 0.1 *N* sulfuric acid to obtain a solution having a known concentration of about 1 mg of anhydrous methyldopa per mL.

Assay preparation—Weigh and finely powder not less than 20 Methyldopa Tablets. Transfer an accurately weighed portion of the powder, equivalent to about 100 mg of methyldopa, to a 100-mL volumetric flask, add 50 mL of 0.1 *N* sulfuric acid, agitate by mechanical means for 15 minutes, add the dilute acid to volume, and mix. Filter the solution, rejecting the first 20 mL of the filtrate.

Procedure—Pipet 5 mL each of the *Assay preparation* and the *Standard preparation* into separate 100-mL volumetric flasks. To a third 100-mL volumetric flask add 5 mL of water to provide a blank. Add to each flask 5 mL of *Ferrous tartrate solution*, and dilute with *Buffer solution* to volume. Determine the absorbances of both solutions at the wavelength of maximum absorbance at about 520 nm, with a suitable spectrophotometer, using the blank in the reference cell. Calculate the quantity, in mg, of $C_{10}H_{13}NO_4$ in the portion of Tablets taken by the formula:

$$100C(A_U/A_S),$$

in which C is the concentration, in mg per mL, of USP Methyldopa RS in the Standard solution, and A_U and A_S are the absorbances of the solutions from the *Assay preparation* and the *Standard preparation*, respectively.

Methyldopa and Chlorothiazide Tablets

» Methyldopa and Chlorothiazide Tablets contain not less than 90.0 percent and not more than 110.0 percent of the labeled amounts of methyldopa ($C_{10}H_{13}NO_4$) and chlorothiazide ($C_7H_6ClN_3O_4S_2$).

Packaging and storage—Preserve in well-closed containers.

Reference standards—*USP Methyldopa Reference Standard*—Do not dry; determine the *Water* content by *Method I* ⟨921⟩ before using. *USP Chlorothiazide Reference Standard*—Dry at 105° for one hour before using.

Identification—Transfer the finely ground contents of 1 Tablet to a test tube, add 10 mL of dilute alcohol (1 in 2), shake for 5 minutes, and centrifuge. Use the clear supernatant liquid as the Test solution. Prepare a solution of alcohol and 0.1 *N* sodium hydroxide (1:1) containing in each mL about 10 mg of USP Methyldopa RS and 10 mg of USP Chlorothiazide RS. Apply 20 µL of the Test solution on a line parallel to and about 2 cm from the bottom edge of a 20-cm × 10-cm thin-layer chromatographic plate (see *Chromatography* ⟨621⟩) coated with chromatographic silica gel mixture, and apply 20 µL of the Standard solution separately on the starting line. Allow the spots to dry, develop the chromatogram in a solvent system consisting of equal volumes of glacial acetic acid, acetone, butyl alcohol, toluene, and water until the solvent front has moved about three-fourths of the length of the plate. Remove the plate from the developing tank, and allow the solvent to evaporate. View the plate under short-wavelength ultraviolet light: the solution under test exhibits two major spots having R_f values corresponding to those of the two major spots obtained with the Standard solution.

Dissolution ⟨711⟩—

PROCEDURE FOR METHYLDOPA—

Medium: 0.1 *N* hydrochloric acid; 900 mL.

Apparatus 2: 75 rpm.

Time: 30 minutes.

Standard preparation—Dissolve an accurately weighed quantity of USP Methyldopa RS in *Dissolution Medium*, and dilute quantitatively with the same solvent to obtain a solution having a known concentration of about 275 µg of anhydrous methyldopa per mL.

Ferrous tartrate solution—Dissolve 1 g of ferrous sulfate, 2 g of potassium sodium tartrate, and 100 mg of sodium bisulfite in water to make 100 mL, and mix. Use a freshly prepared solution.

Buffer solution—Dissolve 50 g of ammonium acetate in 1000 mL of dilute alcohol (1 in 5). Adjust with 6 *N* ammonium hydroxide to a pH of 8.5.

Procedure—Filter 35 mL of the solution under test, and transfer an aliquot estimated to contain between 2 mg and 3 mg of methyldopa to a 100-mL volumetric flask. Adjust the final volume, if necessary, with *Dissolution Medium* to 10 mL. To a second 100-mL volumetric flask add 10.0 mL of *Standard preparation*, and to a third 100-mL volumetric flask add 10.0 mL of *Dissolution Medium* to provide a blank. Pipet 5.0 mL of *Ferrous tartrate solution* into each flask, dilute with *Buffer solution* to volume, and mix. Concomitantly determine the absorbances of the treated *Standard preparation* and test solution at the wavelength of maximum absorbance at about 520 nm, with a suitable spectrophotometer, against the reagent blank. Calculate the amount of $C_{10}H_{13}NO_4$ dissolved, in mg, by the formula:

$$9(C/V)(A_U/A_S),$$

in which C is the concentration, in µg of anhydrous methyldopa per mL, of USP Methyldopa RS in the *Standard preparation*, V is the volume, in mL, of the aliquot of test solution used, and A_U and A_S are the absorbances of the solutions from the test solution and the *Standard preparation*, respectively.

Tolerances—Not less than 80% (Q) of the labeled amount of $C_{10}H_{13}NO_4$ is dissolved in 30 minutes.

PROCEDURE FOR CHLOROTHIAZIDE—

Medium: 0.05 *M*, pH 8.0 phosphate buffer (see *Buffer Solutions* in the section, *Reagents, Indicators, and Solutions*) containing sodium sulfite (1 in 5000); 900 mL.

Apparatus 2: 75 rpm.

Time: 60 minutes.

Procedure—Determine the amount of $C_7H_6ClN_3O_4S_2$ dissolved from ultraviolet absorbances of the solution under test, suitably diluted with *Dissolution Medium*, if necessary, at the wavelength of maximum absorbance at about 317 nm in comparison with a Standard solution having a known concentration of USP Chlorothiazide RS in the same medium.

Tolerances—Not less than 75% (Q) of the labeled amount of $C_7H_6ClN_3O_4S_2$ is dissolved in 60 minutes.

Uniformity of dosage units ⟨905⟩: meet the requirements with respect to methyldopa and to chlorothiazide.

Assay—

Mobile phase—Prepare a filtered and degassed mixture of 0.08 *M* monobasic sodium phosphate and methanol (95:5). Adjust by the addition of phosphoric acid to a pH of 2.8. Make adjustments if necessary (see *System Suitability* under *Chromatography* ⟨621⟩).

Standard preparation—Transfer to a 100-mL volumetric flask accurately weighed quantities of USP Methyldopa RS and USP Chlorothiazide RS, equivalent to one-fifth of their labeled amounts, in mg, per Tablet. Add 15 mL of water and 5 mL of 1 *N* hydrochloric acid, and sonicate for about 3 minutes. Add 10 mL of acetonitrile, and sonicate for 2 minutes. Dilute with water to volume, and mix.

Assay preparation—Weigh and finely powder not less than 10 Methyldopa and Chlorothiazide Tablets. Transfer an accurately weighed portion of the powder, equivalent to the weight of 1 Tablet, to a 500-mL volumetric flask. Add 75 mL of water and 25 mL of 1 *N* hydrochloric acid, and sonicate for about 5 minutes. Add 50 mL of acetonitrile, and sonicate for 10 minutes. Dilute with water to volume, and mix. Filter through a 0.45- to 2.0-µm membrane filter, discarding the first 10 mL.

Chromatographic system (see *Chromatography* ⟨621⟩)—The liquid chromatograph is equipped with a 280-nm detector and a 3.9-mm × 30-cm column that contains packing L1. The flow rate is about 2 mL per minute. Chromatograph the *Standard preparation*, and record the peak responses as directed under *Procedure:* the column efficiency determined from the chlorothiazide peak is not less than 1300 theoretical plates, the tailing factor for chlorothiazide peak is not more than 2, the resolution, *R*, between the chlorothiazide and methyldopa peaks is not less than 7, and the relative standard deviation for replicate injections is not more than 2.0%.

Procedure—Separately inject equal volumes (about 10 µL) of the *Standard preparation* and the *Assay preparation* into the chromatograph, record the chromatograms, and measure the responses for the major peaks. The relative retention times are about 1.0 for methyldopa and 2.5 for chlorothiazide. Calculate the quantity, in mg, of chlorothiazide ($C_6H_6ClN_3O_4S_2$) in the portion of Tablets taken by the formula:

$$500C(r_U/r_S),$$

in which C is the concentration, in mg per mL, of USP Chlorothiazide RS in the *Standard preparation*, and r_U and r_S are the peak responses of the chlorothiazide peak obtained from the *Assay preparation* and the *Standard preparation*, respectively. Calculate the quantity, in mg, of methyldopa ($C_{10}H_{13}NO_4$) taken by the same formula, reading "methyldopa" instead of "chlorothiazide."

Methyldopa and Hydrochlorothiazide Tablets

» Methyldopa and Hydrochlorothiazide Tablets contain not less than 90.0 percent and not more than 110.0 percent of the labeled amounts of methyldopa ($C_{10}H_{13}NO_4$) and hydrochlorothiazide ($C_7H_8ClN_3O_4S_2$).

Packaging and storage—Preserve in well-closed containers.

Reference standards—*USP Methyldopa Reference Standard*—Do not dry; determine the *Water* content by *Method I* ⟨921⟩ before using. *USP Hydrochlorothiazide Reference Standard*—Dry at 105° for 1 hour before using.

Identification—

A: The retention times of the 2 major peaks in the chromatogram of the *Assay preparation* correspond to those in the chromatogram of the *Standard preparation* as obtained in the *Assay*.

B: A portion of crushed Tablets, equivalent to about 10 mg of methyldopa, responds to *Identification test C* under *Methyldopa*.

Dissolution ⟨711⟩—

Medium: 0.1 *N* hydrochloric acid; 900 mL.

Apparatus 2: 50 rpm.

Times: 30 minutes; 60 minutes.

PROCEDURE FOR METHYLDOPA—

Standard preparation—Dissolve an accurately weighed quantity of USP Methyldopa RS in *Dissolution Medium*, and dilute quantitatively with the same solvent to obtain a solution having a known concentration of about 275 µg of anhydrous methyldopa per mL.

Ferrous tartrate solution—Dissolve 1 g of ferrous sulfate, 2 g of potassium sodium tartrate, and 100 mg of sodium bisulfite in water to make 100 mL. Use a freshly prepared solution.

Buffer solution—Dissolve 50 g of ammonium acetate in 1000 mL of dilute alcohol (1 in 5). Adjust with 6 *N* ammonium hydroxide to a pH of 8.5.

Procedure—Filter 35 mL of the solution under test through paper, and transfer an aliquot estimated to contain between 2 mg and 3 mg of methyldopa into a 100-mL volumetric flask. Adjust the final volume, if necessary, with *Dissolution Medium* to 10 mL. To a second 100-mL volumetric flask add 10.0 mL of *Standard preparation*, and to a third 100-mL volumetric flask add 10.0 mL of *Dissolution Medium* to provide a blank. Treat each flask as follows: Add by pipet 5 mL of *Ferrous tartrate solution* and, dilute with *Buffer solution* to volume. Concomitantly determine the absorbances of the treated *Standard preparation* and test solution in 1-cm cells at the wavelength of maximum absorbance at about 520 nm, with a suitable spectrophotometer, against the reagent blank. Calculate the amount of $C_{10}H_{13}NO_4$ dissolved, in mg, by the formula:

$$9(C/V)(A_U/A_S),$$

in which C is the concentration, in µg of anhydrous methyldopa per mL, of USP Methyldopa RS in the *Standard preparation*, V is the volume, in mL, of the aliquot of test solution used, and A_U and A_S are the absorbances of the solutions from the test solution and the *Standard preparation*, respectively.

PROCEDURE FOR HYDROCHLOROTHIAZIDE—Determine the amount of $C_7H_8ClN_3O_4S_2$ dissolved from ultraviolet absorbances at the wavelength of maximum absorbance at about 317 nm in 1-cm cells, of filtered portions of the solution under test, suitably diluted with *Dissolution Medium*, in comparison with a Standard solution having a known concentration of USP Hydrochlorothiazide RS in the same medium.

Tolerances—Not less than 80% (Q) of the labeled amount of methyldopa ($C_{10}H_{13}NO_4$) is dissolved in 30 minutes, and not less than 80% (Q) of the labeled amount of hydrochlorothiazide ($C_7H_8ClN_3O_4S_2$) is dissolved in 60 minutes.

Uniformity of dosage units ⟨905⟩: meet the requirements.

Assay—

pH 2.8 sodium phosphate solution—Dissolve 11.04 g of monobasic sodium phosphate in 950 mL of water. Adjust this solution with phosphoric acid to a pH of 2.8. Transfer the solution to a 1-liter volumetric flask, add water to volume, and mix. Filter through a membrane filter.

Mobile phase—Prepare a solution containing 95 volumes of *pH 2.8 sodium phosphate solution* and 5 volumes of methanol.

Standard preparation—Transfer a suitable quantity of USP Methyldopa RS to a 100-mL volumetric flask to obtain a solution having a known concentration of about 1 mg of anhydrous methyldopa per mL. Add an accurately weighed quantity of USP Hydrochlorothiazide RS that corresponds to the ratio of hydrochlorothiazide to methyldopa in the Tablets. Dissolve in 10 mL of water, 10 mL of acetonitrile, and 5 mL of 1 *N* hydrochloric acid. Dilute with water to volume, and mix.

Assay preparation—Weigh and finely powder not less than 20 Methyldopa and Hydrochlorothiazide Tablets. Transfer an accurately weighed portion of the powder, equivalent to about 250 mg of methyldopa, to a 250-mL volumetric flask, and add 50 mL of water, 25 mL of acetonitrile, and 13 mL of 1 *N* hydrochloric acid. Shake the flask for 5 minutes, dilute with water to volume, and mix.

Chromatographic system (see *Chromatography* ⟨621⟩)—The liquid chromatograph is equipped with a 270-nm detector and a 3.9-mm × 30-cm column that contains packing L1. The flow rate, about 2 mL per minute, is adjusted until the relative retention times for methyldopa and hydrochlorothiazide are about 0.38 and 1.0, respectively. Chromatograph five replicate injections of the *Standard preparation*, and record the peak responses as directed under *Procedure:* the relative standard deviation is not more than 2.0% and the resolution factor between methyldopa and hydrochlorothiazide is not less than 6.

Procedure—Separately inject equal volumes (about 10 µL) of the *Standard preparation* and the *Assay preparation* into the chromatograph by means of a suitable microsyringe or sampling valve. Record the chromatograms, and measure the responses for the major peaks. Calculate the quantity, in mg, of methyldopa ($C_{10}H_{13}NO_4$) in the portion of Tablets taken by the formula:

$$(250C)(r_U / r_S),$$

in which C is the concentration, in mg per mL, of USP Methyldopa RS in the *Standard preparation*, and r_U and r_S are the responses of the methyldopa peak obtained from the *Assay preparation* and the *Standard preparation*, respectively. Calculate the quantity, in mg, of hydrochlorothiazide ($C_7H_8ClN_3O_4S_2$) in the portion of Tablets taken by the same formula, reading "hydrochlorothiazide" instead of "methyldopa."

Methyldopate Hydrochloride

HO, HO–C₆H₃–CH₂–C(NH₂)(CH₃)–COOC₂H₅ · HCl

$C_{12}H_{17}NO_4.HCl$ 275.73

L-Tyrosine, 3-hydroxy-α-methyl-, ethyl ester, hydrochloride.

L-3-(3,4-Dihydroxyphenyl)-2-methylalanine ethyl ester hydrochloride [*5123-53-5; 2508-79-4*].

» Methyldopate Hydrochloride contains not less than 98.0 percent and not more than 101.0 percent of $C_{12}H_{17}NO_4.HCl$, calculated on the dried basis.

Packaging and storage—Preserve in well-closed containers.

Reference standard—*USP Methyldopate Hydrochloride Reference Standard*—Dry at a pressure not exceeding 5 mm of mercury at 100° for 2 hours before using.

Identification—

A: The infrared absorption spectrum of a mineral oil dispersion of it exhibits maxima only at the same wavelengths as that of a similar preparation of USP Methyldopate Hydrochloride RS.

B: The ultraviolet absorption spectrum of a 1 in 20,000 solution in 0.1 *N* hydrochloric acid exhibits maxima and minima at the same wavelengths as that of a similar solution of USP Methyldopate Hydrochloride RS, concomitantly measured, and the respective absorptivities, calculated on the dried basis, at the wavelength of maximum absorbance at about 280 nm do not differ by more than 3.0%.

C: It responds to *Identification test C* under *Methyldopa.*

D: It responds to the tests for *Chloride* ⟨191⟩, except that on the addition of the slight excess of 6 *N* ammonium hydroxide a brown precipitate is formed.

Specific rotation ⟨781⟩: between −13.5° and −14.9° measured at 405 nm, calculated on the dried basis, determined in a solution in 0.1 *N* hydrochloric acid containing 400 mg in each 10 mL.

pH ⟨791⟩: between 3.0 and 5.0, in a solution (1 in 100).

Loss on drying ⟨731⟩—Dry it at 100° and at a pressure not exceeding 5 mm of mercury for 2 hours: it loses not more than 0.5% of its weight.

Residue on ignition ⟨281⟩: not more than 0.1%.

Heavy metals, *Method II* ⟨231⟩: 0.001%.

Assay—

Mobile solvent—Prepare a suitable solution of 0.02 *M* monobasic sodium phosphate and 0.015 *M* phosphoric acid in a water and methanol solution (approximately 15.5:4.5) such that the retention time of methyldopate hydrochloride is approximately 6.5 minutes.

Standard preparation—Dissolve an accurately weighed quantity of USP Methyldopate Hydrochloride RS in the *Mobile solvent* to obtain a solution containing about 1 mg per mL.

Assay preparation—Transfer about 50 mg of Methyldopate Hydrochloride, accurately weighed, to a 50-mL volumetric flask, dissolve in *Mobile solvent*, and dilute with *Mobile solvent* to volume.

Procedure—Introduce separately 20-µL portions of the *Assay preparation* and the *Standard preparation* into a high-pressure liquid chromatograph (see *Chromatography* ⟨621⟩) operated at 25°, by means of a suitable microsyringe or sampling valve, adjusting the operating parameters such that the peak obtained with the *Standard preparation* is 100% full-scale. Typically, the apparatus is fitted with a 4-mm × 30-cm column that contains packing L1, is equipped with an ultraviolet detector capable of monitoring absorption at 280 nm and a suitable recorder, and is capable of operating at a column pressure between 700 and 1700 psi. In a suitable chromatogram, three replicate injections of the *Standard preparation* show a relative standard deviation of not more than 1.5%. Determine the peak areas, at equivalent retention times, obtained with the *Assay preparation* and the *Standard preparation*, and calculate the quantity, in mg, of $C_{12}H_{17}NO_4$.HCl in the portion of Methyldopate Hydrochloride taken by the formula:

$$50C(A_U / A_S),$$

in which C is the concentration, in mg per mL, of USP Methyldopate Hydrochloride RS in the *Standard preparation*, and A_U and A_S are the peak areas obtained from the *Assay preparation* and the *Standard preparation*, respectively.

Methyldopate Hydrochloride Injection

» Methyldopate Hydrochloride Injection is a sterile solution of Methyldopate Hydrochloride in Water for Injection. It contains not less than 90.0 percent and not more than 110.0 percent of the labeled amount of $C_{12}H_{17}NO_4.HCl$.

Packaging and storage—Preserve in single-dose containers, preferably of Type I glass.

Reference standard—*USP Methyldopate Hydrochloride Reference Standard*—Dry at a pressure not exceeding 5 mm of mercury at 100° for 2 hours before using.

Identification—

A: Dilute a volume of Injection with a mixture of chloroform and methanol (1:1), if necessary, to obtain a solution containing about 5 mg of methyldopate hydrochloride per mL. On a suitable thin-layer chromatographic plate (see *Chromatography* ⟨621⟩), coated with a 0.25-mm layer of chromatographic silica gel, apply separately 10 µL of this solution and 10 µL of a Standard solution of USP Methyldopate Hydrochloride RS in a solvent mixture of chloroform and methanol (1:1) containing 5 mg per mL. Allow the spots to dry, and develop the chromatogram in a saturated chamber with a solvent system consisting of a mixture of butyl alcohol, water, and formic acid (7:2:1), until the front has moved about three-fourths of the length of the plate. Remove the plate from the developing chamber, mark the solvent front, and allow the solvent to evaporate. Locate the spots on the plate by lightly spraying with Folin-Ciocalteu Phenol TS followed by spraying with sodium carbonate solution (1 in 10): the R_f value of the principal spot obtained from the test solution corresponds to that obtained from the Standard solution.

B: It responds to the tests for *Chloride* ⟨191⟩.

pH ⟨791⟩: between 3.0 and 4.2.

Particulate matter ⟨788⟩: meets the requirements under *Small-volume Injections*.

Other requirements—It meets the requirements under *Injections* ⟨1⟩.

Assay—

Buffer solution—To 214 g of monobasic potassium phosphate add 700 mL of water, and stir. Cautiously add 75 mL of sodium hydroxide solution (1 in 2), and stir until solution is complete. Adjust with sodium hydroxide solution (1 in 2) to a pH of 8.0, and dilute with water to 1000.0 mL.

Water-saturated tributyl phosphate—Shake 800 mL of tributyl phosphate with 100 mL of water, and discard the lower, aqueous phase. Filter the upper phase.

Standard preparation—Transfer about 25 mg of USP Methyldopate Hydrochloride RS, accurately weighed, to a 25-mL volumetric flask, add water to volume, and mix. Transfer 5 mL of this solution to a 100-mL volumetric flask, add 0.1 *N* sulfuric acid to volume, and mix. Use a freshly prepared solution. The *Standard preparation* contains about 50 µg per mL.

Assay preparation—Transfer to a 50-mL volumetric flask an accurately measured volume of Methyldopate Hydrochloride Injection, equivalent to about 50 mg of methyldopate hydrochloride, add water to volume, and mix. Transfer a 5.0-mL aliquot of the solution to a 60-mL separator, add 15 mL of *Buffer solution* and 10 mL of *Water-saturated tributyl phosphate*, and shake for about 1 minute. Allow the phases to separate, and transfer the lower, aqueous phase to a second 60-mL separator. To this separator add a second 10-mL portion of *Water-saturated tributyl phosphate*, shake for about 1 minute, allow the phases to separate, discard the lower, aqueous phase, and add the upper tributyl phosphate phase to the phase retained in the first separator. Rinse the second separator with about 2 mL of *Water-saturated tributyl phosphate*, and add the rinsing to the first separator. Extract the phase contained in the first separator with two 25-mL portions of 0.1 *N* sulfuric acid. Collect the acid extracts in a 100-mL volumetric flask, add 0.1 *N* sulfuric acid to volume, and mix. Filter, if necessary, to obtain a clear solution.

Procedure—Concomitantly determine the absorbances of the *Assay preparation* and the *Standard preparation* in 1-cm cells at the wavelength of maximum absorbance at about 283 nm, with a suitable spectrophotometer, using 0.1 *N* sulfuric acid as the blank. Calculate the quantity, in mg, of $C_{12}H_{17}NO_4 \cdot HCl$ in each mL of the Injection taken by the formula:

$$(C/V)(A_U/A_S),$$

in which C is the concentration, in µg per mL, of USP Methyldopate Hydrochloride RS in the *Standard preparation*, V is the volume, in mL, of Injection taken, and A_U and A_S are the absorbances of the *Assay preparation* and the *Standard preparation*, respectively.

Methylene Blue

$C_{16}H_{18}ClN_3S \cdot 3H_2O$ 373.90

Phenothiazin-5-ium, 3,7-bis(dimethylamino)-, chloride, trihydrate.

C.I. Basic Blue 9 trihydrate [*7220-79-3*].

Anhydrous 319.85 [*61-73-4*].

» Methylene Blue contains not less than 98.0 percent and not more than 103.0 percent of $C_{16}H_{18}ClN_3S$, calculated on the dried basis.

Packaging and storage—Preserve in well-closed containers.

Reference standard—*USP Methylene Blue Reference Standard*—Dry at a pressure not exceeding 5 mm of mercury at 75° for 4 hours before using.

Identification—The infrared absorption spectrum of a potassium bromide dispersion of it, previously dried at 75° and at a pressure not exceeding 5 mm of mercury for 4 hours, exhibits maxima only at the same wavelengths as that of a similar preparation of USP Methylene Blue RS.

Loss on drying ⟨731⟩—Dry it at 75° and at a pressure not exceeding 5 mm of mercury for 4 hours: it loses between 8.0% and 18.0% of its weight.

Residue on ignition ⟨281⟩: not more than 1.2%.

Arsenic, *Method I* ⟨211⟩—Prepare the *Test Preparation* by mixing 0.375 g with 10 mL of water in the arsine generator flask. Add 15 mL of nitric acid and 5 mL of perchloric acid, mix, and heat cautiously to the production of strong fumes of perchloric acid. Cool, wash down the sides of the flask with water, and again heat to strong fumes. Again cool, wash down the sides of the flask, and heat to fumes. Cool, dilute with water to 52 mL, and add 3 mL of hydrochloric acid: the resulting solution meets the requirements of the test, the addition of 20 mL of 7 *N* sulfuric acid specified under *Procedure* being omitted. The limit is 8 ppm.

Copper or zinc—Ignite 1.0 g in a porcelain crucible, using as low a temperature as practicable, until all of the carbon is oxidized. Cool the residue, add 15 mL of 2 *N* nitric acid, and boil for 5 minutes. Filter the cooled solution, and wash any residue with 10 mL of water. To the combined filtrate and washing add an excess of 6 *N* ammonium hydroxide, and filter the solution into a 50-mL volumetric flask. Wash the precipitate with small portions of water, adding the washings to the filtrate, dilute the solution with water to volume, and mix. To 25 mL of the solution add 10 mL of hydrogen sulfide TS: no turbidity is produced within 5 minutes (*absence of zinc*). Any dark color produced does not exceed that of a control prepared by boiling a quantity of cupric sulfate, equivalent to 200 µg of copper, with 15 mL of 2 *N* nitric acid for 5 minutes, and treating this solution as directed above, beginning with "Filter the cooled solution" (0.02% of copper).

Chromatographic purity—Dissolve an accurately weighed quantity of Methylene Blue quantitatively in methanol to obtain a *Test solution* containing 1.0 mg per mL. Dissolve a suitable quantity of USP Methylene Blue RS in methanol to obtain a *Standard solution* having a concentration of 100 µg per mL. Dilute a portion of this solution quantitatively with methanol to obtain a *Diluted standard solution* having a concentration of 10 µg per mL. On a suitable thin-layer chromatographic plate (see *Chromatography* ⟨621⟩), coated with a 0.25-mm layer of octadecylsilanized chromatographic silica gel, apply 5 µL each of the *Test solution*, the *Standard solution*, and the *Diluted standard solution*. Allow the spots to dry, and develop the chromatogram in a chromatographic chamber with a solvent system consisting of the upper layer separated from a well-shaken mixture of water, *n*-butanol, and glacial acetic acid (100:80:20), until the solvent front has moved about three-fourths of the length of the plate. Remove the plate from the chamber, allow the solvent to evaporate, and visually locate the spots on the plate: the R_f value of the principal spot in the chromatogram from the *Test solution* corresponds to that from the *Standard solution*, and other spots, if present in the chromatogram from the *Test solution*, consist of a secondary spot that does not exceed in size or intensity, the principal spot obtained from the *Standard solution* (10%), and not more than two additional spots, neither of which exceeds in size or intensity the principal spot from the *Diluted standard solution* (1%).

Assay—Dissolve in diluted alcohol about 100 mg of Methylene Blue, accurately weighed, and dilute quantitatively and stepwise with diluted alcohol to obtain a solution containing about 2 µg per mL. Dissolve an accurately weighed quantity of USP Methylene Blue RS in diluted alcohol, and dilute quantitatively and stepwise with diluted alcohol to obtain a Standard solution having a known concentration of about 2 µg per mL. Concomitantly determine the absorbances of both solutions in 1-cm cells at the wavelength of maximum absorbance at about 663 nm, with a suitable spectrophotometer, using diluted alcohol as the blank. Calculate the quantity, in mg, of $C_{16}H_{18}ClN_3S$ in the Methylene Blue taken by the formula:

$$50C(A_U/A_S),$$

in which C is the concentration, in µg per mL, of anhydrous methylene blue in the Standard solution, and A_U and A_S are the absorbances of the solution of Methylene Blue and the Standard solution, respectively.

Methylene Blue Injection

» Methylene Blue Injection is a sterile solution of Methylene Blue in Water for Injection. It contains, in each mL, not less than 9.5 mg and not more than 10.5 mg of $C_{16}H_{18}ClN_3S.3H_2O$.

Packaging and storage—Preserve in single-dose containers, preferably of Type I glass.

Reference standard—*USP Methylene Blue Reference Standard*—Dry at a pressure not exceeding 5 mm of mercury at 75° for 4 hours before using.

Identification—

A: The visible absorption spectrum of the solution employed for measurement of absorbance in the *Assay* exhibits maxima and minima at the same wavelengths as that of the Standard solution employed in the *Assay*, concomitantly measured.

B: Dilute a portion of the Injection with an equal volume of methanol. Dissolve 5 mg of USP Methylene Blue RS in 1 mL of a mixture of equal volumes of methanol and water. On a thin-layer chromatographic plate, coated with a 0.25-mm layer of chromatographic silica gel, apply 1 µL of each solution, allow the spots to dry, and develop the chromatogram, using a mixture of water, ethanol, and acetic acid (4:3:3) as the solvent system, until the solvent front has moved about 10 cm above the line of application. Remove the plate from the developing chamber, and allow the solvent to evaporate: the R_f value of the principal spot obtained from the Methylene Blue corresponds to that obtained from the Reference Standard.

pH ⟨791⟩: between 3.0 and 4.5.

Other requirements—It meets the requirements under *Injections* ⟨1⟩.

Assay—Dilute an accurately measured volume of Methylene Blue Injection, equivalent to about 100 mg of methylene blue, quantitatively and stepwise with diluted alcohol to a concentration of about 2 µg per mL. Prepare a Standard solution of USP Methylene Blue RS as directed in the *Assay* under *Methylene Blue*. Concomitantly determine the absorbance of both solutions in 1-cm cells at the wavelength of maximum absorbance at about 663 nm, with a suitable spectrophotometer, using diluted alcohol as the blank. Calculate the quantity, in mg, of $C_{16}H_{18}ClN_3S.3H_2O$ in each mL of the Injection taken by the formula:

$$58.45(C/V)(A_U/A_S),$$

in which C is the concentration, in µg per mL, of USP Methylene Blue RS in the Standard solution, V is the volume, in mL, of Injection taken, and A_U and A_S are the absorbances of the solution from the Injection and the Standard solution, respectively.

Methylene Chloride—*see* Methylene Chloride NF

Methylergonovine Maleate

H CONH—C—CH2CH3 CH2OH N—CH3 H HN · HC—COOH ‖ HC—COOH

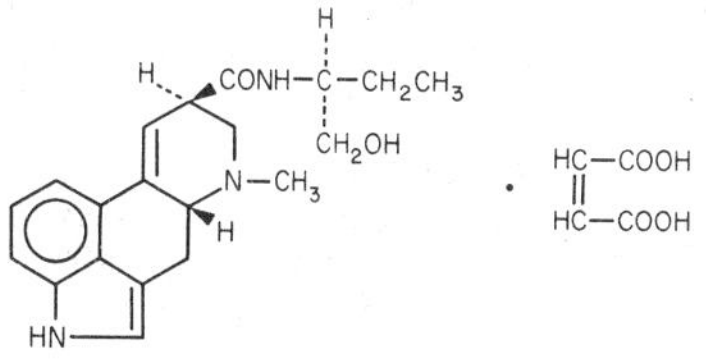

$C_{20}H_{25}N_3O_2.C_4H_4O_4$ 455.51

Ergoline-8-carboxamide, 9,10-didehydro-*N*-[1-(hydroxymethyl)propyl]-6-methyl-, [8β(*S*)]-, (*Z*)-2-butenedioate (1:1) (salt).

9,10-Didehydro-*N*-[(*S*)-1-(hydroxymethyl)propyl]-6-methylergoline-8β-carboxamide maleate (1:1) (salt) [*7054-07-1*].

» Methylergonovine Maleate contains not less than 97.0 percent and not more than 103.0 percent of $C_{20}H_{25}N_3O_2.C_4H_4O_4$, calculated on the dried basis.

Packaging and storage—Preserve in tight, light-resistant containers, and store in a cold place.

Reference standard—*USP Methylergonovine Maleate Reference Standard*—Dry in vacuum at 80° to constant weight before using.

Identification—

A: The infrared absorption spectrum of a potassium bromide dispersion of it, previously dried, exhibits maxima only at the same wavelengths as that of a similar preparation of USP Methylergonovine Maleate RS.

B: The ultraviolet absorption spectrum of a 1 in 50,000 solution in dilute sulfuric acid (1 in 90) exhibits maxima and minima at the same wavelengths as that of a similar solution of USP Methylergonovine Maleate RS, concomitantly measured, and the respective absorptivities, calculated on the dried basis, at the wavelength of maximum absorbance at about 311 nm do not differ by more than 3.0%.

C: The R_f values of the principal fluorescent spot and the principal blue spot obtained from the *Test preparation* correspond to those obtained from *Standard preparation A* in the chromatogram prepared as directed in the test for *Related alkaloids*.

D: Dissolve about 2 mg in 3 mL of water: the solution exhibits a bluish fluorescence under ultraviolet light. To the solution add 2 mL of a 1 in 2 mixture of glacial acetic acid in ethyl acetate, and stratify 2 mL of sulfuric acid under the mixture: a bluish purple ring appears at the interface of the two liquids.

Specific rotation ⟨781⟩: between +44° and +50°, calculated on the dried basis, determined in a solution in water containing 50 mg in each 10 mL.

pH ⟨791⟩: between 4.4 and 5.2, in a solution (1 in 5000).

Loss on drying ⟨731⟩—Dry it in vacuum at 80° to constant weight: it loses not more than 2.0% of its weight.

Residue on ignition ⟨281⟩: not more than 0.1%.

Related alkaloids—Proceed as directed for *Related alkaloids* under *Ergonovine Maleate*, using Methylergonovine Maleate in place of Ergonovine Maleate.

Assay—

Standard preparation—Dissolve, with warming, if necessary, a suitable quantity of USP Methylergonovine Maleate RS in water, and dilute quantitatively with water to obtain a solution having a known concentration of about 50 µg per mL.

Assay preparation—Transfer about 25 mg of Methylergonovine Maleate, accurately weighed, to a 500-mL volumetric flask, dissolve in 125 mL of water, warming, if necessary, dilute with water to volume, and mix.

Procedure—Transfer 4.0 mL each of the *Standard preparation*, the *Assay preparation*, and water to provide the blank, to separate, glass-stoppered, 50-mL conical flasks. Place in an ice bath, swirl continuously while adding to each flask 10.0 mL of *p*-dimethylaminobenzaldehyde TS, mix, and allow to stand in the dark for 1 hour. Concomitantly determine the absorbances of the solutions in 1-cm cells at the wavelength of maximum absorbance at about 555 nm, with a suitable spectrophotometer, using the blank to set the instrument. Calculate the quantity, in mg, of $C_{20}H_{25}N_3O_2 . C_4H_4O_4$ in the Methylergonovine Maleate taken by the formula:

$$0.5C(A_U/A_S),$$

in which C is the concentration, in µg per mL, of USP Methylergonovine Maleate RS in the *Standard preparation*, and A_U and A_S are the absorbances of the solutions from the *Assay preparation* and the *Standard preparation*, respectively.

Methylergonovine Maleate Injection

» Methylergonovine Maleate Injection is a sterile solution of Methylergonovine Maleate in Water for Injection. It contains, in each mL, not less than 90.0 percent and not more than 110.0 percent of the labeled amount of $C_{20}H_{25}N_3O_2 . C_4H_4O_4$.

Packaging and storage—Preserve in single-dose, light-resistant containers, preferably of Type I glass.

Reference standard—*USP Methylergonovine Maleate Reference Standard*—Dry in vacuum at 80° to constant weight before using.

Identification—It responds to *Identification tests C* and *D* under *Methylergonovine Maleate.*

pH ⟨791⟩: between 2.7 and 3.5.

Related alkaloids—[NOTE—Conduct this test without exposure to daylight and with minimum exposure to artificial light.]

Solvent mixture—Mix 9 volumes of alcohol with 1 volume of ammonium hydroxide.

Test preparation—Transfer a volume of Injection, equivalent to about 5 mg of methylergonovine maleate, to a separator, and extract with three 5-mL portions of chloroform. Discard the chloroform extracts. Render alkaline to litmus with 6 *N* ammonium hydroxide, and extract with three 5-mL portions of chloroform. Evaporate the combined extracts with the aid of a current of air, but without heat, to dryness. Dissolve the residue so obtained in 0.5 mL of *Solvent mixture.*

Standard preparation and *Standard dilutions*—Prepare a solution of USP Methylergonovine Maleate RS in *Solvent mixture* to contain 10 mg per mL (*Standard preparation*). Prepare a series of dilutions of the *Standard preparation* in *Solvent mixture* to contain 0.50 mg, 0.20 mg, 0.10 mg, and 0.05 mg per mL (*Standard dilutions*).

Procedure—In a suitable chromatographic chamber arranged for thin-layer chromatography place a volume of a solvent system consisting of a mixture of chloroform, methanol, and water (75:25:3) sufficient to develop the chromatogram, cover, and allow to equilibrate for 30 minutes. On a suitable thin-layer chromatographic plate (see *Chromatography* ⟨621⟩), coated with a 0.25-mm layer of chromatographic silica gel, apply 5-µL portions of *Test preparation, Standard preparation*, and each of the three *Standard dilutions.* Allow the spots to dry, and develop the chromatogram until the solvent front has moved about three-fourths of the length of the plate. Remove the plate from the developing chamber, mark the solvent front, and allow the solvent to evaporate. Locate the spots on the plate by spraying thoroughly and evenly with a solution prepared by dissolving 1 g of *p*-dimethylaminobenzaldehyde in a cooled mixture of 50 mL of alcohol and 50 mL of hydrochloric acid: the R_f value of the principal spot obtained from the *Test preparation* corresponds to that obtained from the *Standard preparation.* Estimate the concentration of any other spots observed in the lane for the *Test preparation* by comparison with the *Standard dilutions:* the spots from the 0.50-, 0.20-, 0.10-, and 0.05-mg-per-mL dilutions are equivalent to 5.0%, 2.0%, 1.0%, and 0.50% of impurities, respectively. The sum of the impurities is not greater than 5.0%.

Other requirements—It meets the requirements under *Injections* ⟨1⟩.

Assay—

Standard preparation—Prepare as directed in the *Assay* under *Methylergonovine Maleate.*

Assay preparation—Transfer an accurately measured volume of Methylergonovine Maleate Injection, equivalent to about 2.5 mg of methylergonovine maleate, to a 50-mL volumetric flask, dilute with water to volume, and mix.

Procedure—Proceed as directed for *Procedure* in the *Assay* under *Methylergonovine Maleate.* Calculate the quantity, in mg, of $C_{20}H_{25}N_3O_2 . C_4H_4O_4$ in each mL of the Injection taken by the formula:

$$(0.05C/V)(A_U/A_S),$$

in which V is the volume, in mL, of Injection taken, and C, A_U, and A_S are as defined therein.

Methylergonovine Maleate Tablets

» Methylergonovine Maleate Tablets contain not less than 90.0 percent and not more than 110.0 percent of the labeled amount of $C_{20}H_{25}N_3O_2 . C_4H_4O_4$.

Packaging and storage—Preserve in tight, light-resistant containers.

Reference standard—*USP Methylergonovine Maleate Reference Standard*—Dry in vacuum at 80° to constant weight before using.

Identification—

A: Tablets respond to *Identification test C* under *Methylergonovine Maleate.*

B: Transfer a quantity of powdered Tablets, equivalent to about 4 mg of methylergonovine maleate, to a separator, add 20 mL of water, and render alkaline to litmus with sodium carbonate solution (1 in 10). Extract with three 20-mL portions of chloroform, filter the combined chloroform extracts into a small evaporating dish, and evaporate on a steam bath to dryness. Dissolve the residue in a mixture of 6 mL of water and 0.3 mL of hydrochloric acid, and filter, if necessary: the solution so obtained responds to *Identification test D* under *Methylergonovine Maleate.*

Dissolution ⟨711⟩—

Medium: tartaric acid solution (1 in 200); 900 mL.

Apparatus 2: 100 rpm.

Time: 30 minutes.

Procedure—Filter a portion of the solution under test into a flask. Concomitantly determine the fluorescence intensity of this solution in comparison with a Standard solution of USP Methylergonovine Maleate RS in the same medium having a known concentration of about 0.22 µg per mL in a fluorometer at an excitation wavelength of about 327 nm and an emission wavelength of about 428 nm, using tartaric acid solution (1 in 200) as the blank.

Tolerances—Not less than 70% (Q) of the labeled amount of $C_{20}H_{25}N_3O_2 . C_4H_4O_4$ is dissolved in 30 minutes.

Uniformity of dosage units ⟨905⟩: meet the requirements.

Related alkaloids—[NOTE—Conduct this test without exposure to daylight and with the minimum exposure to artificial light.]

Solvent mixture—Mix 75 volumes of chloroform, 25 volumes of methanol, and 1 volume of ammonium hydroxide.

Detecting reagent—Cautiously dissolve 800 mg of *p*-dimethylaminobenzaldehyde in a mixture of alcohol and sulfuric acid (101:11).

Test preparation—Transfer a quantity of finely powdered Tablets, equivalent to 5.0 mg of methylergonovine maleate, to a suitable container, add 50 mL of *Solvent mixture*, and stir with the aid of a magnetic stirrer for 40 minutes. Filter, rinsing the container with two 10-mL portions of *Solvent mixture.* Evaporate the combined filtrates in vacuum at 25° to 30°, and dissolve the residue in 2.0 mL of *Solvent mixture.*

Standard stock solution—Transfer 25 mg of USP Methylergonovine Maleate RS to a 10-mL volumetric flask, add *Solvent mixture* to volume, and mix to obtain a solution having a known concentration of 2.5 mg per mL.

Standard preparations A, B, C, and D—Dilute accurately measured volumes of *Standard stock solution* quantitatively with *Solvent mixture* (designated below as parts by volume of *Standard stock solution* in total parts by volume of the finished *Stan-*

dard preparation) to obtain *Standard preparations*, designated below by letter, having the following concentrations and percentage assignments:

A—(1 in 20); 125 µg per mL (5.0%).
B—(1 in 33); 75 µg per mL (3.0%).
C—(1 in 100); 25 µg per mL (1.0%).
D—(1 in 200); 12.5 µg per mL (0.5%).

Procedure—On a suitable thin-layer chromatographic plate (see *Chromatography* ⟨621⟩), coated with a 0.25-mm layer of chromatographic silica gel mixture, apply separately 20 µL of the *Test preparation* and 20 µL of each *Standard preparation.* Dry the plate with the aid of a stream of cool air. Position the plate in a chromatographic chamber, and develop the chromatograms in *Solvent mixture* until the solvent front has moved about three-fourths of the length of the plate. Remove the plate from the developing chamber, mark the solvent front, and allow the solvent to evaporate in a stream of cool air. Examine the plate under long-wavelength ultraviolet light. Mark the principal and any secondary fluorescent spots. Spray the plate with *Detecting reagent*, and mark the principal and secondary blue spots. Compare the intensities of any secondary spots observed in the chromatogram of the *Test preparation* with those of the principal spots in the chromatograms of the *Standard preparations:* the sum of the intensities of secondary spots obtained from the *Test preparation* corresponds to not more than 5.0% of related compounds.

Assay—[NOTE—Conduct this procedure with a minimum exposure to light.]

Mobile phase—Prepare a filtered and degassed mixture of 800 mL of monobasic potassium phosphate solution (1 in 500) and 200 mL of acetonitrile. Make adjustments if necessary (see *System Suitability* under *Chromatography* ⟨621⟩).

Internal standard solution—Transfer 5 g of tartaric acid to a 1000-mL volumetric flask. Add 750 mL of a mixture of methanol and water (1:1), and shake to dissolve. Add 10 mg of ergonovine maleate, and shake. Add a mixture of methanol and water (1:1) to volume, and mix with shaking if necessary to effect solution.

Standard preparation—Using an accurately weighed quantity of USP Methylergonovine Maleate RS, prepare a solution in *Internal standard solution* having a known concentration of about 10 µg per mL.

Assay preparation—Transfer a number of Tablets, equivalent to about 2 mg of methylergonovine maleate, to a 200-mL volumetric flask. Add about 150 mL of *Internal standard solution*, and shake by mechanical means for 30 minutes. Dilute with *Internal standard solution* to volume, and mix. Filter a portion through a 0.45-µm membrane, discarding the first 5 mL of the filtrate.

Chromatographic system (see *Chromatography* ⟨621⟩)—The liquid chromatograph is equipped with a fluorometer operating at an excitation wavelength of 326 nm and an emission wavelength of 428 nm, and a 4-mm × 25-cm column that contains packing L7. The flow rate is about 2 mL per minute. Chromatograph the *Standard preparation*, and record the peak responses as directed under *Procedure:* the column efficiency determined from the analyte peak is not less than 1200 theoretical plates, the resolution, *R*, between the analyte and internal standard peaks is not less than 1.2, and the relative standard deviation for replicate injections is not more than 2.0%.

Procedure—Separately inject equal volumes (about 10 µL) of the *Standard preparation* and the *Assay preparation* into the chromatograph, record the chromatograms, and measure the responses for the major peaks. The relative retention times are about 0.7 for the internal standard and 1.0 for methylergonovine. Calculate the quantity, in mg, of $C_{20}H_{25}N_3O_2 . C_4H_4O_4$ in the Tablets taken by the formula:

$$0.2C(R_U/R_S),$$

in which C is the concentration, in µg per mL, of USP Methylergonovine Maleate RS in the *Standard preparation*, and R_U and R_S are the peak response ratios obtained from the *Assay preparation* and the *Standard preparation*, respectively.

Methylparaben—*see* Methylparaben NF

Methylparaben Sodium—*see* Methylparaben Sodium NF

Methylphenidate Hydrochloride

[Structure: phenyl–CH(COOCH$_3$)–CH (piperidine ring, HN) · HCl]

$C_{14}H_{19}NO_2 . HCl$ 269.77

2-Piperidineacetic acid, α-phenyl-, methyl ester, hydrochloride, (R*,R*)-(±)-.

Methyl α-phenyl-2-piperidineacetate hydrochloride [*298-59-9*].

» Methylphenidate Hydrochloride contains not less than 98.0 percent and not more than 100.5 percent of $C_{14}H_{19}NO_2 . HCl$, calculated on the dried basis.

Packaging and storage—Preserve in well-closed containers.

Reference standards—*USP Methylphenidate Hydrochloride Reference Standard*—Dry in vacuum at 60° for 4 hours before using. *USP Methylphenidate Hydrochloride Erythro Isomer Reference Standard*—Do not dry before using. *USP α-Phenyl-2-piperidineacetic Acid Hydrochloride Reference Standard*—Do not dry; use as is.

Identification—

A: The infrared absorption spectrum of a mineral oil dispersion of it, previously dried, exhibits maxima only at the same wavelengths as that of a similar preparation of USP Methylphenidate Hydrochloride RS.

B: It responds to the tests for *Chloride* ⟨191⟩.

Loss on drying ⟨731⟩—Dry it in vacuum at 60° for 4 hours: it loses not more than 0.5% of its weight.

Residue on ignition ⟨281⟩: not more than 0.1%.

Heavy metals, *Method II* ⟨231⟩: 0.001%.

Limit of erythro [(R*,S*)] isomer—

Mobile solvent—Mix 190 volumes of chloroform with 10 volumes of methanol and 1 volume of ammonium hydroxide.

Detecting reagent—Dissolve 0.7 g of bismuth subnitrate in 40 mL of a mixture of 1 volume of glacial acetic acid and 4 volumes of water. Add 40 mL of potassium iodide solution (2 in 5), then add 120 mL of glacial acetic acid and 250 mL of water.

Procedure—Prepare methanol solutions of Methylphenidate Hydrochloride and of USP Methylphenidate Hydrochloride Erythro Isomer RS containing 50 mg per mL and 0.5 mg per mL, respectively. Apply 20-µL portions of each solution on a suitable thin-layer chromatographic plate (see *Chromatography* ⟨621⟩), coated with a 0.25-mm layer of chromatographic silica gel. Allow the spots to dry, and develop the chromatogram, using the *Mobile solvent*, in a suitable chamber, lined with absorbent paper and previously equilibrated with the *Mobile solvent*, until the solvent front has moved about three-fourths of the length of the plate. Remove the plate from the developing chamber, and allow the solvent to evaporate. Locate the spots on the plate by spraying first with the *Detecting reagent* and then with 1 *N* sulfuric acid. Any spot in the lane from the methylphenidate hydrochloride at the same R_f as the erythro isomer is not larger or more intense than that produced by USP Methylphenidate Hydrochloride Erythro Isomer RS, when viewed under ordinary lighting (1%).

Limit of α-phenyl-2-piperidineacetic acid hydrochloride—

Mobile solvent—Mix 65 volumes of chloroform with 25 volumes of methanol and 5 volumes of acetic acid.

Sodium hydroxide–methanol—Prepare a 1 in 2500 solution of sodium hydroxide in methanol.

Standard preparation—Dissolve a suitable quantity of USP α-Phenyl-2-piperidineacetic Acid Hydrochloride RS in *Sodium hydroxide–methanol* to obtain a solution having a known concentration of about 240 µg per mL.

Test preparation—Dissolve 400 mg of Methylphenidate Hydrochloride, accurately weighed, in *Sodium hydroxide–methanol* to make 10.0 mL. Use immediately after preparation.

Procedure—On a suitable thin-layer chromatographic plate (see *Chromatography* ⟨621⟩), coated with a 0.25-mm layer of chromatographic silica gel, apply 10-µL portions of the *Test preparation* and the *Standard preparation*. Allow the spots to dry, and develop the chromatogram, using the *Mobile solvent*, in a suitable chamber, lined with absorbent paper and previously equilibrated with *Mobile solvent*, until the solvent front has moved about three-fourths of the length of the plate. Remove the plate from the developing chamber, and allow the plate to dry for 30 minutes. Expose the plate to high-intensity, ultraviolet radiation for 10 minutes. View the plate under long-wavelength ultraviolet light: any spot in the lane from the *Test preparation* having the same R_f value as the principal spot from the *Standard preparation* is not larger or more intense than that produced by the *Standard preparation* (0.6%).

Assay—Dissolve about 225 mg of Methylphenidate Hydrochloride, accurately weighed, in 50 mL of glacial acetic acid in a 125-mL conical flask. Add 15 mL of mercuric acetate TS and 5 drops of *p*-naphtholbenzein TS, and titrate with 0.1 *N* perchloric acid VS to a green end-point. Perform a blank determination, and make any necessary correction. Each mL of 0.1 *N* perchloric acid is equivalent to 26.98 mg of $C_{14}H_{19}NO_2.HCl$.

Methylphenidate Hydrochloride Tablets

» Methylphenidate Hydrochloride Tablets contain not less than 93.0 percent and not more than 107.0 percent of the labeled amount of $C_{14}H_{19}NO_2.HCl$.

Packaging and storage—Preserve in tight containers.

Reference standard—*USP Methylphenidate Hydrochloride Reference Standard*—Dry in vacuum at 60° for 4 hours before using.

Identification—Place a portion of powdered Tablets, equivalent to about 50 mg of methylphenidate hydrochloride, in a 40-mL centrifuge tube, add 10 mL of chloroform, shake, and centrifuge. Filter the clear extract through a medium-sized sintered-glass funnel into a beaker, and repeat the extraction with an additional 10-mL portion of chloroform. Evaporate the combined chloroform extracts on a steam bath to dryness. Agitate the dried residue with 2 mL of acetonitrile, and filter the mixture through a small sintered-glass funnel. Wash the crystals with an additional 2 mL of acetonitrile, and dry them with the aid of suction: the infrared absorption spectrum of a mineral oil dispersion of the residue exhibits maxima only at the same wavelengths as that of a similar preparation of USP Methylphenidate Hydrochloride RS.

Dissolution ⟨711⟩—

Medium: water; 900 mL.

Apparatus 1: 100 rpm.

Time: 45 minutes.

Standard preparation—Dissolve a suitable quantity of USP Methylphenidate Hydrochloride RS, accurately weighed, in water to obtain a solution having a known concentration of about 5 µg per mL.

Procedure—Pipet a volume of a filtered portion of the solution under test that contains about 0.5 mg of methylphenidate hydrochloride and 100 mL of the *Standard preparation* into individual separators. To each separator add 25.0 mL of pH 5 buffer (prepared by mixing 48.5 mL of 0.1 *M* citric acid with 51.5 mL of 0.2 *M* dibasic sodium phosphate, adjusting with either solution if necessary to a pH of 5.0), and 1.0 mL of a 1 in 100 solution of picric acid in 0.1 *N* sodium hydroxide. Extract with three 25-mL portions of chloroform, filtering each portion through chloroform-moistened cotton into a 100-mL volumetric flask, dilute with chloroform to volume, and mix. Concomitantly determine the absorbances of both solutions in 5-cm cells at the wavelength of maximum absorbance at about 405 nm, with a suitable spectrophotometer, using chloroform as the blank, and determine the amount of $C_{14}H_{19}NO_2.HCl$ dissolved.

Tolerances—Not less than 75% (*Q*) of the labeled amount of $C_{14}H_{19}NO_2.HCl$ is dissolved in 45 minutes.

Uniformity of dosage units ⟨905⟩: meet the requirements.

Assay—

Standard preparation—Dissolve a suitable quantity of USP Methylphenidate Hydrochloride RS, accurately weighed, in 0.1 *N* sulfuric acid to obtain a solution having a known concentration of about 200 µg per mL.

Assay preparation—Weigh and finely powder not less than 20 Methylphenidate Hydrochloride Tablets. Weigh accurately a portion of the powder, equivalent to 50 mg of methylphenidate hydrochloride, into a 250-mL volumetric flask, add about 150 mL of 0.1 *N* sulfuric acid, and shake by mechanical means for 45 minutes. Add 0.1 *N* sulfuric acid to volume, and filter, discarding the first few mL of the filtrate. The subsequent clear filtrate is the *Assay preparation*.

Procedure—Transfer 10.0 mL each of the *Assay preparation* and the *Standard preparation* to separate 125-mL separators. To each separator add 10.0 mL of 0.1 *N* sodium hydroxide, 10.0 mL of pH 5 buffer (prepared by mixing 48.5 mL of 0.1 *M* citric acid with 51.5 mL of 0.2 *M* dibasic sodium phosphate, adjusting with either solution if necessary to a pH of 5.0), and 1.0 mL of a 1 in 100 solution of picric acid in 0.1 *N* sodium hydroxide. Extract with three 25-mL portions of chloroform, filtering each portion through chloroform-moistened cotton into a 100-mL volumetric flask, dilute with chloroform to volume, and mix. Concomitantly determine the absorbances of both solutions in 1-cm cells at the wavelength of maximum absorbance at about 405 nm, with a suitable spectrophotometer, using chloroform as the blank. Calculate the quantity, in mg, of $C_{14}H_{19}NO_2.HCl$ in the portion of Tablets taken by the formula:

$$0.25C(A_U/A_S),$$

in which C is the concentration, in µg per mL, of USP Methylphenidate Hydrochloride RS in the *Standard preparation*, and A_U and A_S are the absorbances of the solutions from the *Assay preparation* and the *Standard preparation*, respectively.

Methylphenidate Hydrochloride Extended-release Tablets

» Methylphenidate Hydrochloride Extended-release Tablets contain not less than 90.0 percent and not more than 110.0 percent of the labeled amount of $C_{14}H_{19}NO_2.HCl$.

Packaging and storage—Preserve in tight containers.

Reference standard—*USP Methylphenidate Hydrochloride Reference Standard*—Dry in vacuum at 60° for 4 hours before using.

Identification—Place a portion of powdered Tablets, equivalent to about 100 mg of methylphenidate hydrochloride, in a 100-mL beaker. Add 20 mL of chloroform, stir for 5 minutes, and filter, collecting the filtrate. Evaporate the filtrate to about 5 mL. Add ethyl ether slowly, with stirring, until crystals form. Filter the crystals, wash with ethyl ether, and dry at 80° for 30 minutes: the infrared absorption spectrum of a mineral oil dispersion of the crystals so obtained exhibits maxima only at the same wavelengths as that of a similar preparation of USP Methylphenidate Hydrochloride RS.

Drug release ⟨724⟩—

Medium: water; 500 mL.

Apparatus 2: 50 rpm.

Times: 0.125D hours; 0.250D hours; 0.438D hours; 0.625D hours; 0.875D hours.

Determine the amount of Methylphenidate Hydrochloride dissolved, using the following method.

Picric acid solution (1 in 100)—Dissolve 1.0 g of picric acid in 0.1 *N* sodium hydroxide to make 100 mL.

pH 5.0 buffer solution—Mix 48.5 mL of 0.1 *M* citric acid with 51.5 mL of 0.2 *M* dibasic sodium phosphate. Adjust with either solution to a pH of 5.0, if necessary.

Standard solution—Accurately weigh about 20 mg of USP Methylphenidate Hydrochloride RS into a 500-mL volumetric flask, dissolve in water, dilute with water to volume, and mix.

Procedure—Pipet 20 mL of a filtered portion of the solution under test and 20 mL of the *Standard solution* into individual 125-mL separators. To each add 10.0 mL of *pH 5.0 buffer solution* and 1.0 mL of *Picric acid solution*, and extract with three 25-mL portions of chloroform, filtering the chloroform extracts through cotton (previously rinsed with chloroform) into a 100-mL volumetric flask. Dilute with chloroform to volume, and mix. Determine the absorbance of the test solution and the *Standard solution* against chloroform as the blank at the wavelength of maximum absorbance at about 405 nm, using a suitable spectrophotometer.

Calculate the amount of $C_{14}H_{19}NO_2 . HCl$ dissolved.

Tolerances—The percentages of the labeled amount of $C_{14}H_{19}NO_2 . HCl$ dissolved at the times specified conform to *Acceptance Table 1*.

Time (hours)	Amount dissolved (%)
0.125*D*	between 20% and 50%
0.250*D*	between 35% and 70%
0.438*D*	between 53% and 83%
0.625*D*	between 70% and 95%
0.875*D*	not less than 80%

Uniformity of dosage units ⟨905⟩: meet the requirements.

Assay—

Standard preparation, Assay preparation, and *Procedure*—Proceed as directed in the *Assay* under *Methylphenidate Hydrochloride Tablets*.

Methylprednisolone

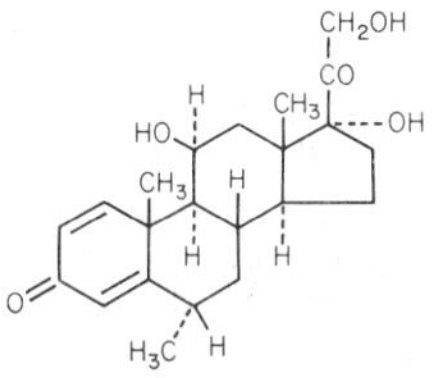

$C_{22}H_{30}O_5$ 374.48

Pregna-1,4-diene-3,20-dione, 11,17,21-trihydroxy-6-methyl-, (6α,11β)-.

11β,17,21-Trihydroxy-6α-methylpregna-1,4-diene-3,20-dione [*83-43-2*].

» Methylprednisolone contains not less than 97.0 percent and not more than 103.0 percent of $C_{22}H_{30}O_5$, calculated on the dried basis.

Packaging and storage—Preserve in tight, light-resistant containers.

Reference standard—*USP Methylprednisolone Reference Standard*—Dry at 105° for 3 hours before using.

Identification—

A: The infrared absorption spectrum of a potassium bromide dispersion of it, previously dried, exhibits maxima only at the same wavelengths as that of a similar preparation of USP Methylprednisolone RS.

B: The ultraviolet absorption spectrum of a 1 in 100,000 solution in alcohol exhibits maxima and minima at the same wavelengths as that of a similar solution of USP Methylprednisolone RS, concomitantly measured, and the respective absorptivities, calculated on the dried basis, at the wavelength of maximum absorbance at about 243 nm do not differ by more than 3.0%.

C: Dissolve about 5 mg in 2 mL of sulfuric acid: a red color is produced.

Specific rotation ⟨781⟩: between +79° and +86°, calculated on the dried basis, determined in a solution in dioxane containing 100 mg in each 10 mL.

Loss on drying ⟨731⟩—Dry it at 105° for 3 hours: it loses not more than 1.0% of its weight.

Residue on ignition ⟨281⟩: not more than 0.2%.

Ordinary impurities ⟨466⟩—

Test solution: methanol.

Standard solution: methanol.

Application volume: 10 µL.

Eluant: a mixture of acetone, toluene, and 2 *N* formic acid (65:30:5), in a nonequilibrated chamber.

Visualization: 5.

Assay—

Mobile phase—Prepare a solution containing a mixture of butyl chloride, water-saturated butyl chloride, tetrahydrofuran, methanol, and glacial acetic acid (475:475:70:35:30).

Internal standard solution—Dissolve prednisone in a 3 in 100 solution of glacial acetic acid in chloroform to obtain a solution having a concentration of about 0.2 mg per mL.

Standard preparation—Dissolve an accurately weighed quantity of USP Methylprednisolone RS in *Internal standard solution* to obtain a solution having a known concentration of about 0.2 mg per mL.

Assay preparation—Using about 10 mg of Methylprednisolone, accurately weighed, proceed as directed for *Standard preparation*.

Chromatographic system (see *Chromatography* ⟨621⟩)—The liquid chromatograph is equipped with a 254-nm detector and a 4-mm × 25-cm column that contains packing L3. The flow rate is about 1 mL per minute. Chromatograph the *Standard preparation*, and record the peak responses as directed under *Procedure:* the resolution, *R*, between the methylprednisolone and internal standard peaks is not less than 4.0, and the relative standard deviation for replicate injections is not more than 2.0%.

Procedure—Separately inject equal volumes (about 10 µL) of the *Standard preparation* and the *Assay preparation* into the chromatograph, record the chromatograms, and measure the responses for the major peaks. The relative retention times are about 0.7 for prednisone and 1.0 for methylprednisolone. Calculate the quantity, in mg, of $C_{22}H_{30}O_5$ in the portion of Prednisolone taken by the formula:

$$50C(R_U/R_S),$$

in which *C* is the concentration, in mg per mL, of USP Methylprednisolone RS in the *Standard preparation*, and R_U and R_S are the ratios of the peak responses for the methylprednisolone peak and the internal standard peak obtained from the *Assay preparation* and the *Standard preparation*, respectively.

Methylprednisolone Tablets

» Methylprednisolone Tablets contain not less than 92.5 percent and not more than 107.5 percent of the labeled amount of $C_{22}H_{30}O_5$.

Packaging and storage—Preserve in tight containers.

Reference standard—*USP Methylprednisolone Reference Standard*—Dry at 105° for 3 hours before using.

Identification—Powder a number of Tablets, equivalent to about 40 mg of methylprednisolone, and digest with 25 mL of solvent hexane for 15 minutes. Filter, and discard the filtrate. Digest the residue with 25 mL of chloroform for 15 minutes. Filter, evaporate the filtrate to dryness, and dry at 105° for 2 hours: the residue so obtained responds to *Identification tests A* and *C* under *Methylprednisolone*.

Dissolution ⟨711⟩—

Medium: water; 900 mL.

Apparatus 1: 100 rpm.

Time: 30 minutes.

Procedure—Measure the ultraviolet absorption of filtered aliquots removed from the *Dissolution Medium* and suitably diluted, if necessary, in 1-cm cells at 246 nm, with a suitable spectrophotometer, using water as the blank and utilizing a standard curve, representing the absorbance versus concentration of USP Methylprednisolone RS. [NOTE—Dissolve about 20 mg of USP

Methylprednisolone RS, accurately weighed, in 1 mL of alcohol, dilute in a 1000-mL volumetric flask with water to volume, and mix. Prepare quantitative dilutions of this solution for the development of a standard curve.]

Tolerances—Not less than 50% (*Q*) of the labeled amount of $C_{22}H_{30}O_5$ is dissolved.

Uniformity of dosage units ⟨905⟩: meet the requirements.

Procedure for content uniformity—

Mobile phase, Internal standard solution, Standard preparation, and *Chromatographic system*—Proceed as directed in the *Assay* under *Methylprednisolone.*

Test preparation—Place 1 Tablet in a suitable container. For tablet labeled strengths of 10 mg or less, add 0.5 mL of water. For tablet labeled strengths greater than 10 mg, add 1.0 mL of water. Allow the tablet to stand for about 2 minutes, then swirl the container to disperse the tablet. Add 5.0 mL of *Internal standard solution* for each mg of labeled tablet strength, shake for 15 minutes, and filter or centrifuge a portion of the test specimen. Analyze the clear solution as directed under *Procedure.*

Procedure—Proceed as directed for *Procedure* in the *Assay* under *Methylprednisolone.* Calculate the quantity, in mg, of $C_{22}H_{30}O_5$ in the Tablet by the formula:

$$(FW_S)(R_U/R_S),$$

in which *F* is the ratio of the volume of *Internal standard preparation*, in mL, in the *Test preparation* to the volume, in mL, of the *Internal standard preparation* in the *Standard preparation*, W_S is the weight, in mg, of USP Methylprednisolone RS taken for the *Standard preparation*, and the other terms are as defined for *Procedure* in the *Assay* under *Methylprednisolone.*

Assay—

Mobile phase, Internal standard solution, Standard preparation, and *Chromatographic system*—Proceed as directed in the *Assay* under *Methylprednisolone.*

Assay preparation—Accurately weigh 20 Tablets, and grind to a fine powder in a mortar and pestle. Accurately weigh a portion of the powder, equivalent to about 10 mg of methylprednisolone, and transfer to a suitable container. Add 2.5 mL of water to the ground tablet material and swirl to form a fine slurry. Add 50.0 mL of *Internal standard solution*, and shake for 15 minutes. Filter or centrifuge a portion of the liquid so obtained, if necessary, and analyze the clear solution as directed under *Procedure.*

Procedure—Proceed as directed for *Procedure* in the *Assay* under *Methylprednisolone.* Calculate the quantity, in mg, of $C_{22}H_{30}O_5$ in the portion of Tablets taken by the formula:

$$50C(R_U/R_S),$$

in which the terms are as defined therein.

Methylprednisolone Acetate

$C_{24}H_{32}O_6$ 416.51

Pregna-1,4-diene-3,20-dione, 21-(acetyloxy)-11,17-dihydroxy-6-methyl-, (6α,11β)-.

11β,17,21-Trihydroxy-6α-methylpregna-1,4-diene-3,20-dione 21-acetate [*53-36-1*].

» Methylprednisolone Acetate contains not less than 97.0 percent and not more than 103.0 percent of $C_{24}H_{32}O_6$, calculated on the dried basis.

Packaging and storage—Preserve in tight, light-resistant containers.

Reference standard—*USP Methylprednisolone Acetate Reference Standard*—Dry at 105° for 3 hours before using.

Identification—

A: The infrared absorption spectrum of a potassium bromide dispersion of it, previously dried at 105° for 3 hours, exhibits maxima only at the same wavelengths as that of a similar preparation of USP Methylprednisolone Acetate RS.

B: The ultraviolet absorption spectrum of a 1 in 100,000 solution in alcohol exhibits maxima and minima at the same wavelengths as that of a similar preparation of USP Methylprednisolone Acetate RS, concomitantly measured, and the respective absorptivities, calculated on the dried basis, at the wavelength of maximum absorbance at about 243 nm do not differ by more than 3.0%.

C: Dissolve about 5 mg in 2 mL of sulfuric acid: a wine-red color is produced.

D: Place about 5 mg in a test tube, add 2 mL of alcoholic potassium hydroxide TS, and heat in a boiling water bath for 5 minutes. Cool, add 2 mL of dilute sulfuric acid (1 in 3.5), and boil gently for about 1 minute: the odor of ethyl acetate is perceptible.

Specific rotation ⟨781⟩: between +97° and +105°, calculated on the dried basis, determined in a solution in dioxane containing 100 mg in each 10 mL.

Loss on drying ⟨731⟩—Dry it at 105° for 3 hours: it loses not more than 1.0% of its weight.

Residue on ignition ⟨281⟩: not more than 0.2%.

Assay—

Mobile phase—Prepare a solvent mixture of *n*-butyl chloride, water-saturated *n*-butyl chloride, tetrahydrofuran, methanol, and glacial acetic acid (475:475:70:35:30).

Internal standard solution—Prepare a solution of prednisone containing about 6 mg per mL in a mixture of chloroform and glacial acetic acid (97:3), by first adding the entire amount of glacial acetic acid to the prednisone contained in a 100-mL volumetric flask, followed by sonication. Then slowly add the chloroform, using sonication and shaking to dissolve the material. Dilute with chloroform to volume, and mix.

Standard preparation—Transfer about 20 mg of USP Methylprednisolone Acetate RS, accurately weighed, and 5.0 mL of the *Internal standard solution* to a 100-mL volumetric flask. Dilute with chloroform to volume, and shake to dissolve the specimen.

Assay preparation—Prepare a solution of Methylprednisolone Acetate as directed under *Standard preparation.*

Chromatographic system (see *Chromatography* ⟨621⟩)—The liquid chromatograph is equipped with a 254-nm detector and a 4-mm × 25-cm column that contains packing L3. The flow rate is about 1 mL per minute. Chromatograph the *Standard preparation*, and record the peak responses as directed under *Procedure:* the resolution, *R*, between the analyte and internal standard peaks is not less than 2.5, and the relative standard deviation for replicate injections is not more than 2.0%.

Procedure—Separately inject equal volumes (about 10 µL) of the *Standard preparation* and the *Assay preparation* into the chromatograph, record the chromatograms, and measure the responses for the major peaks. The relative retention times are about 1.3 for prednisone and 1.0 for methylprednisolone acetate. Calculate the quantity, in mg, of $C_{24}H_{32}O_6$ in the portion of Methylprednisolone Acetate taken by the formula:

$$100C(R_U/R_S),$$

in which *C* is the concentration, in mg per mL, of USP Methylprednisolone Acetate RS in the *Standard preparation*, and R_U and R_S are the peak height response ratios of the methylprednisolone acetate peak and the internal standard peak obtained from the *Assay preparation* and the *Standard preparation*, respectively.

Methylprednisolone Acetate Cream

» Methylprednisolone Acetate Cream contains not less than 90.0 percent and not more than 110.0 percent of the labeled amount of $C_{24}H_{32}O_6$.

Packaging and storage—Preserve in collapsible tubes or in tight containers, protected from light.

Reference standard—*USP Methylprednisolone Acetate Reference Standard*—Dry at 105° for 3 hours before using.

Identification—In the thin-layer chromatogram, prepared as directed in the *Assay*, the R_f value of the principal spot obtained from the *Assay preparation* corresponds to that obtained from the *Standard preparation*, prepared as directed in the *Assay*.

Minimum fill ⟨755⟩: meets the requirements.

Assay—

Standard preparation—Dissolve an accurately weighed quantity of USP Methylprednisolone Acetate RS in a mixture of equal volumes of alcohol and chloroform, and dilute quantitatively with the same solvent to obtain a solution having a known concentration of about 500 µg per mL.

Assay preparation—Transfer an accurately weighed quantity of Methylprednisolone Acetate Cream, equivalent to about 5 mg of methylprednisolone acetate, to a 125-mL separator, add 50 mL of solvent hexane, and mix. Extract with three 10-mL portions of acetonitrile, and evaporate the combined extracts on a steam bath with the aid of a current of air nearly to dryness. Transfer the residue to a 10-mL volumetric flask with the aid of one 5-mL portion and two 2-mL portions of a mixture of equal volumes of alcohol and chloroform, dilute with the same solvent to volume, and mix.

Procedure—Divide the area of a suitable thin-layer chromatographic plate (see *Chromatography* ⟨621⟩), coated with a 0.5-mm layer of chromatographic silica gel mixture, into three equal sections, the left and right sections to be used for the *Assay preparation* and the *Standard preparation*, respectively, and the center section for the blank. Apply 250 µL each of the *Assay preparation* and of the *Standard preparation* as streaks 2.5 cm from the bottom of the designated section of the plate, and dry the streaks with the aid of a current of air. Develop the chromatogram in a solvent system consisting of a mixture of ethyl acetate and chloroform (7:5) until the solvent front has moved about three-fourths of the length of the plate. Remove the plate from the developing chamber, mark the solvent front, and allow the solvent to evaporate. Locate the principal bands occupied by the *Standard preparation* and the *Assay preparation* (see also the *Identification test*) by viewing under short-wavelength ultraviolet light. Mark these bands and the corresponding band in the section of the plate representing the blank. Quantitatively remove the silica gel containing these bands, and transfer to separate glass-stoppered, 50-mL centrifuge tubes. Add 25.0 mL of alcohol to each tube, shake for 2 minutes, and centrifuge at about 1500 rpm for 5 minutes. Transfer 20.0 mL of each supernatant solution to separate glass-stoppered, 50-mL conical flasks, add 2.0 mL of blue tetrazolium TS to each solution, mix, and to each flask add 2.0 mL of a mixture of 1 volume of tetramethylammonium hydroxide TS and 9 volumes of alcohol. Mix, and allow the solutions to stand in the dark for 90 minutes. Concomitantly determine the absorbances of the solutions in 1-cm cells at the wavelength of maximum absorbance at about 525 nm, with a suitable spectrophotometer, against the blank. Calculate the quantity, in mg, of $C_{24}H_{32}O_6$ in the portion of Cream taken by the formula:

$$0.01C(A_U/A_S),$$

in which C is the concentration, in µg per mL, of USP Methylprednisolone Acetate RS in the *Standard preparation*, and A_U and A_S are the absorbances of the solutions from the *Assay preparation* and the *Standard preparation*, respectively.

Methylprednisolone Acetate Cream, Neomycin Sulfate and—*see* Neomycin Sulfate and Methylprednisolone Acetate Cream

Methylprednisolone Acetate for Enema

» Methylprednisolone Acetate for Enema is a dry mixture of Methylprednisolone Acetate with one or more suitable excipients. It contains not less than 90.0 percent and not more than 110.0 percent of the labeled amount of $C_{24}H_{32}O_6$.

Packaging and storage—Preserve in well-closed containers.

Reference standard—*USP Methylprednisolone Acetate Reference Standard*—Dry at 105° for 3 hours before using.

Identification—Transfer a quantity of it, equivalent to about 40 mg of methylprednisolone acetate, to a suitable container, add 20 mL of chloroform, and digest for about 15 minutes. Filter the mixture, evaporate the clear filtrate on a steam bath to a volume of about 2 mL, then evaporate at room temperature with the aid of a current of air to dryness. Dry the residue at 105° for 3 hours: the infrared absorption spectrum of a mineral oil dispersion of the dried residue so obtained exhibits maxima only at the same wavelengths as that of a similar preparation of USP Methylprednisolone Acetate RS.

Uniformity of dosage units ⟨905⟩: meets the requirements.

Procedure for content uniformity—

Mobile phase, *Internal standard solution*, and *Chromatographic system*—Proceed as directed in the *Assay* under *Methylprednisolone Acetate*.

Standard preparation—Proceed as directed for *Standard preparation* in the *Assay*.

Test preparation—Transfer the contents of 1 container to a volumetric flask of appropriate size so that the concentration of methylprednisolone acetate after dilution to volume is about 0.20 mg per mL. Rinse the container not less than three times with approximately equal portions of a solvent mixture of chloroform and glacial acetic acid (97:3). Add the rinsings to the volumetric flask. Add an accurately measured volume of *Internal standard solution*, equal to 5 percent of the nominal volume of the volumetric flask, and dilute with the mixture of chloroform and glacial acetic acid (97:3) to volume. Shake for 15 minutes to dissolve the specimen, and centrifuge a portion to obtain a clear solution.

Procedure—Proceed as directed for *Procedure* in the *Assay* under *Methylprednisolone Acetate*. Calculate the quantity, in mg, of $C_{24}H_{32}O_6$ in the container taken by the formula:

$$V_U C(R_U/R_S),$$

in which V_U is the volume, in mL, of the volumetric flask to which the contents of the container have been added, and the other terms are as defined therein.

Assay—

Mobile phase, *Internal standard solution*, and *Chromatographic system*—Proceed as directed in the *Assay* under *Methylprednisolone Acetate*.

Standard preparation—Transfer about 40 mg of USP Methylprednisolone Acetate RS, accurately weighed, and 10.0 mL of *Internal standard solution* to a 200-mL volumetric flask. Dilute with a mixture of chloroform and glacial acetic acid (97:3) to volume, and shake well to dissolve.

Assay preparation—Transfer an accurately weighed quantity of Methylprednisolone Acetate for Enema, equivalent to about 40 mg of methylprednisolone acetate, to a 200-mL volumetric flask. Add 10.0 mL of *Internal standard solution* and dilute with a mixture of chloroform and glacial acetic acid (97:3) to volume. Shake for about 15 minutes to dissolve the specimen. Centrifuge a suitable portion to obtain a clear solution.

Procedure—Proceed as directed for *Procedure* in the *Assay* under *Methylprednisolone Acetate*. Calculate the quantity, in mg, of $C_{24}H_{32}O_6$ in the portion of Methylprednisolone Acetate for Enema taken by the formula:

$$200C(R_U/R_S),$$

in which the terms are as defined therein.

Sterile Methylprednisolone Acetate Suspension

» Sterile Methylprednisolone Acetate Suspension is a sterile suspension of Methylprednisolone Acetate in a suitable aqueous medium. It contains not less than 90.0 percent and not more than 110.0 percent of the labeled amount of $C_{24}H_{32}O_6$.

Packaging and storage—Preserve in single-dose or in multiple-dose containers, preferably of Type I glass.

Reference standard—*USP Methylprednisolone Acetate Reference Standard*—Dry at 105° for 3 hours before using.

Identification—Filter a volume of Suspension, equivalent to about 100 mg of methylprednisolone acetate, through paper. Wash the residue with several 5-mL portions of water, and dry at 105° for 3 hours: the infrared absorption spectrum of a potassium bromide dispersion of the residue so obtained exhibits maxima only at the same wavelengths as that of a similar preparation of USP Methylprednisolone Acetate RS.

Uniformity of dosage units ⟨905⟩: meets the requirements.

pH ⟨791⟩: between 3.5 and 7.0.

Particle size—Transfer 1 drop to a microscope slide, and spread evenly, diluting with water, if necessary, to decrease the density of the field. Examine the slide under a microscope, equipped with a calibrated ocular micrometer, using about 400× magnification. Scan the entire slide, and note the size of the individual particles: not less than 99% of the particles are less than 20 μm in length when measured along the longest axis, and not less than 75% of the particles are less than 10 μm.

Other requirements—It meets the requirements under *Injections* ⟨1⟩.

Assay—

Mobile phase, Internal standard solution, Standard preparation, and *Chromatographic system*—Proceed as directed in the *Assay* under *Methylprednisolone Acetate.*

Assay preparation—Swirl the Suspension to ensure uniformity prior to analysis. Transfer an accurately measured volume of Sterile Methylprednisolone Acetate Suspension, equivalent to about 40 mg of methylprednisolone acetate, to a 25-mL volumetric flask, add 10.0 mL of *Internal standard solution*, dilute with chloroform to volume, and shake for about 15 minutes or until the aqueous layer is clear. Transfer 4.0 mL of the chloroform layer to a suitable vial, add 30 mL of chloroform and a small quantity (about 400 mg) of anhydrous sodium sulfate, shake for 5 minutes, and use the clear solution.

Procedure—Proceed as directed for *Procedure* in the *Assay* under *Methylprednisolone Acetate.* Calculate the quantity, in mg, of $C_{24}H_{32}O_6$ in each mL of the Suspension taken by the formula:

$$(200C/V)(R_U/R_S),$$

in which V is the volume, in mL, of Sterile Suspension taken, and the other terms are as defined therein.

Methylprednisolone Hemisuccinate

$C_{26}H_{34}O_8$ 474.55

Pregna-1,4-diene-3,20-dione, 21-(3-carboxy-1-oxopropoxy)-11,17-dihydroxy-6-methyl-, (6α,11β)-.

11β,17,21-Trihydroxy-6α-methylpregna-1,4-diene-3,20-dione 21-(hydrogen succinate) [*2921-57-5*].

» Methylprednisolone Hemisuccinate contains not less than 97.0 percent and not more than 103.0 percent of $C_{26}H_{34}O_8$, calculated on the dried basis.

Packaging and storage—Preserve in tight containers.

Reference standards—*USP Methylprednisolone Hemisuccinate Reference Standard*—Dry in vacuum at 100° for 2 hours before using. *USP Fluorometholone Reference Standard*—Dry in vacuum at 60° for 3 hours before using.

Identification—

A: The infrared absorption spectrum of a mineral oil dispersion of it, previously dried at 105° for 3 hours, exhibits maxima only at the same wavelengths as that of a similar preparation of USP Methylprednisolone Hemisuccinate RS.

B: The ultraviolet absorption spectrum of a 1 in 50,000 solution in alcohol exhibits maxima and minima at the same wavelengths as that of a similar solution of USP Methylprednisolone Hemisuccinate RS, concomitantly measured, and the respective absorptivities, calculated on the dried basis, at the wavelength of maximum absorbance at about 243 nm do not differ by more than 3.0%.

Specific rotation ⟨781⟩: between +87° and +95°, calculated on the dried basis, determined in a solution in dioxane containing 100 mg in each 10 mL.

Loss on drying ⟨731⟩—Dry it at 105° for 3 hours: it loses not more than 1.0% of its weight.

Residue on ignition ⟨281⟩: not more than 0.2%.

Assay—

Internal standard solution—Dissolve USP Fluorometholone RS in tetrahydrofuran to obtain a solution containing about 6 mg per mL.

Mobile solvent—Prepare a solution containing a mixture of butyl chloride, water-saturated butyl chloride, tetrahydrofuran, methanol, and glacial acetic acid (475:475:70:35:30).

Standard preparation—Transfer about 40 mg of USP Methylprednisolone Hemisuccinate RS, accurately weighed, to a 100-mL volumetric flask. Add 5.0 mL of *Internal standard solution.* Dilute with chloroform containing 3 percent glacial acetic acid to volume, and mix to dissolve the powder.

Assay preparation—Using about 40 mg of Methylprednisolone Hemisuccinate, accurately weighed, prepare as directed under *Standard preparation.*

Procedure—Using a suitable microsyringe or sampling valve, inject separately suitable portions, between 4 and 8 μL, of the *Standard preparation* and the *Assay preparation* into a suitable high-pressure liquid chromatograph (see *Chromatography* ⟨621⟩) of the general type equipped with a detector for monitoring ultraviolet absorption at about 254 nm, equipped with a suitable recorder, capable of providing column pressure up to about 1000 psi and fitted with a 4-mm × 30-cm stainless steel column that contains packing L3. In a suitable chromatogram, the resolution factor R (see *Chromatography* ⟨621⟩) is not less than 2.0 between peaks for methylprednisolone hemisuccinate and the internal standard, and six replicate injections of the *Standard preparation* show a coefficient of variation of not more than 2.0%. Calculate the quantity, in mg, of $C_{26}H_{34}O_8$ in the portion of Methylprednisolone Hemisuccinate taken by the formula:

$$100C(R_U/R_S),$$

in which C is the concentration, in mg per mL, of USP Methylprednisolone Hemisuccinate RS in the *Standard preparation,* and R_U and R_S are the peak area ratios of the methylprednisolone hemisuccinate peak and the internal standard peak obtained from the *Assay preparation* and the *Standard preparation,* respectively.

Methylprednisolone Sodium Succinate

$C_{26}H_{33}NaO_8$ 496.53

Pregna-1,4-diene-3,20-dione, 21-(3-carboxy-1-oxopropoxy)-11,17-dihydroxy-6-methyl-, monosodium salt, (6α,11β)-.

11β,17,21-Trihydroxy-6α-methylpregna-1,4-diene-3,20-dione 21-(sodium succinate) [*2375-03-3*].

» Methylprednisolone Sodium Succinate contains not less than 97.0 percent and not more than 103.0 percent of $C_{26}H_{33}NaO_8$, calculated on the dried basis.

Packaging and storage—Preserve in tight, light-resistant containers.

Reference standard—*USP Methylprednisolone Hemisuccinate Reference Standard*—Dry in vacuum at 100° for 2 hours before using.

Identification—

A: Transfer about 100 mg to a separator, dissolve in 10 mL of water, add 1 mL of 3 *N* hydrochloric acid, and extract immediately with 50 mL of chloroform. Filter the chloroform extract through cotton, evaporate on a steam bath to dryness, and dry in vacuum at 60° for 3 hours: the infrared absorption spectrum of a mineral oil dispersion of the residue so obtained exhibits maxima only at the same wavelengths as that of a similar preparation of USP Methylprednisolone Hemisuccinate RS.

B: The ultraviolet absorption spectrum of a 1 in 50,000 solution in methanol exhibits maxima and minima at the same wavelengths as that of a similar solution of USP Methylprednisolone Hemisuccinate RS, concomitantly measured, and the absorptivity of the Methylprednisolone Sodium Succinate, calculated on the dried basis, at the wavelength of maximum absorbance at about 243 nm, multiplied by 1.046, does not differ from that of the Reference Standard by more than 3.0%.

C: It responds to the flame test for *Sodium* ⟨191⟩.

Specific rotation ⟨781⟩: between +96° and +104°, calculated on the dried basis, determined in a solution in alcohol containing 100 mg in each 10 mL.

Loss on drying ⟨731⟩—Dry it at 105° for 3 hours: it loses not more than 3.0% of its weight.

Sodium content—Dissolve, with gentle heating, about 1 g of it, accurately weighed, in 75 mL of glacial acetic acid. Add 20 mL of dioxane, then add 1 drop of crystal violet TS, and titrate with 0.1 *N* perchloric acid VS to a blue-green end-point. Perform a blank determination, and make any necessary correction. Each mL of 0.1 *N* perchloric acid is equivalent to 2.299 mg of Na. Not less than 4.49% and not more than 4.77%, calculated on the dried basis, is found.

Assay—

Standard preparation—Proceed as directed under *Assay for Steroids* ⟨351⟩, using USP Methylprednisolone Hemisuccinate RS to obtain a solution having a known concentration of about 12.5 µg per mL.

Assay preparation—Weigh accurately about 100 mg of Methylprednisolone Sodium Succinate, dissolve it in alcohol to make 200.0 mL, and mix. Pipet 5 mL of this solution into a 200-mL volumetric flask, add alcohol to volume, and mix. Pipet 20 mL of the resulting solution into a glass-stoppered, 50-mL conical flask.

Procedure—To each of the flasks containing the *Assay preparation* and the *Standard preparation*, and to a similar flask containing 20.0 mL of alcohol, to provide the blank, add 2.0 mL of a solution prepared by dissolving 50 mg of blue tetrazolium in 10 mL of alcohol, and mix. Then to each flask add 4.0 mL of a mixture of 1 volume of tetramethylammonium hydroxide TS and 9 volumes of alcohol. Mix, allow to stand in the dark for 90 minutes, add 1.0 mL of glacial acetic acid, mix, and proceed as directed for *Procedure* under *Assay for Steroids* ⟨351⟩, beginning with "Concomitantly determine the absorbances." Calculate the quantity, in mg, of $C_{26}H_{33}NaO_8$ in the portion of Methylprednisolone Sodium Succinate taken by the formula:

$$8.37C(A_U/A_S).$$

Methylprednisolone Sodium Succinate for Injection

» Methylprednisolone Sodium Succinate for Injection is a sterile mixture of Methylprednisolone Sodium Succinate with suitable buffers. It may be prepared from Methylprednisolone Sodium Succinate or from Methylprednisolone Hemisuccinate with the aid of Sodium Hydroxide or Sodium Carbonate. It contains the equivalent of not less than 90.0 percent and not more than 110.0 percent of the labeled amount of methylprednisolone ($C_{22}H_{30}O_5$) in the volume of constituted solution designated on the label.

Packaging and storage—Preserve in *Containers for Sterile Solids* as described under *Injections* ⟨1⟩.

Reference standards—*USP Methylprednisolone Hemisuccinate Reference Standard*—Dry in vacuum at 100° for 2 hours before using. *USP Fluorometholone Reference Standard*—Dry in vacuum at 60° for 3 hours before using. *USP Methylprednisolone Reference Standard*—Dry at 105° for 3 hours before using.

Constituted solution—At the time of use, the constituted solution prepared from Methylprednisolone Sodium Succinate for Injection meets the requirements for *Constituted Solutions* under *Injections* ⟨1⟩.

Identification—It responds to *Identification test A* under *Methylprednisolone Sodium Succinate.*

pH ⟨791⟩: between 7.0 and 8.0, in a solution containing about 50 mg of methylprednisolone sodium succinate per mL.

Loss on drying ⟨731⟩—Dry it at 105° for 3 hours: it loses not more than 2.0% of its weight.

Particulate matter ⟨788⟩: meets the requirements under *Small-volume Injections.*

Free methylprednisolone—Using the chromatograms obtained in the *Assay*, measure the areas of the peaks from the Internal standard and free methylprednisolone. Calculate the ratio of the area of the free methylprednisolone peak to that of the Internal standard in the chromatogram obtained from the *Standard preparation*, S_S, and the same ratio in the chromatogram obtained from the *Assay preparation*, S_U. Calculate the quantity, in mg, of free methylprednisolone in the *Assay preparation* by the formula:

$$(100C)(S_U/S_S),$$

in which C is the concentration, in mg per mL, of USP Methylprednisolone RS in the *Standard preparation*, and S_U and S_S are the ratios as defined above. The amount of free methylprednisolone is not more than 6.6% of the labeled amount of methylprednisolone.

Other requirements—It meets the requirements for *Sterility Tests* ⟨71⟩, *Uniformity of Dosage Units* ⟨905⟩, and *Labeling* under *Injections* ⟨1⟩.

Assay—

Internal standard solution and *Mobile solvent*—Prepare as directed in the *Assay* under *Hydrocortisone Hemisuccinate.*

Standard preparation—Weigh accurately about 32.5 mg of USP Methylprednisolone Hemisuccinate RS and transfer it to a 50-mL volumetric flask. Add by pipet 5.0 mL of *Internal standard solution* and 5.0 mL of a solution of glacial acetic acid in chloroform (3 in 100) containing in each mL an accurately known quantity of about 0.30 mg of USP Methylprednisolone RS. Dilute with glacial acetic acid in chloroform (3 in 100) to volume, and mix.

Assay preparation—Mix the constituted solutions prepared from the contents of 10 vials of Methylprednisolone Sodium Succinate for Injection. Transfer an accurately measured volume of the resulting constituted solution, equivalent to about 50 mg of methylprednisolone, to a suitable flask containing 10.0 mL of *Internal standard solution*, and dilute with glacial acetic acid in chloroform (3 in 100) to 100.0 mL. Shake thoroughly for 5 minutes, then allow the phases to separate, discarding the upper phase.

Procedure—Proceed as directed for *Procedure* in the *Assay* under *Hydrocortisone Hemisuccinate.* The order of elution of peaks is that from the Internal standard, methylprednisolone hemisuccinate, and successive smaller peaks representing free methylprednisolone and methylprednisolone 17-hemisuccinate. Measure the areas of the peaks from the Internal standard, methylprednisolone hemisuccinate, and methylprednisolone 17-hemisuccinate. Calculate the ratio of the summation of the areas of the methylprednisolone hemisuccinate and methylprednisolone 17-hemisuccinate peaks to that of the Internal standard in the chromatogram obtained from the *Standard preparation*, R_S, and the same ratio in the chromatogram obtained from the *Assay preparation*, R_U. Calculate the quantity, in mg, of methylprednisolone ($C_{22}H_{30}O_5$) in the portion of constituted solution taken by the formula:

$$0.789(100C)(R_U/R_S),$$

in which 0.789 is the ratio of the molecular weight of methylprednisolone to that of methylprednisolone hemisuccinate, C is the concentration, in mg per mL, of USP Methylprednisolone Hemisuccinate RS in the *Standard preparation*, and R_U and R_S are the ratios as defined above, and to this quantity add the amount, in mg, of free methylprednisolone found in the test for *Free methylprednisolone.*

Methyltestosterone

$C_{20}H_{30}O_2$ 302.46
Androst-4-en-3-one, 17-hydroxy-17-methyl-, (17β)-.
17β-Hydroxy-17-methylandrost-4-en-3-one [58-18-4].

» Methyltestosterone contains not less than 97.0 percent and not more than 103.0 percent of $C_{20}H_{30}O_2$, calculated on the dried basis.

Packaging and storage—Preserve in well-closed, light-resistant containers.

Reference standard—*USP Methyltestosterone Reference Standard*—Dry at 105° for 4 hours before using.

Identification—

A: The infrared absorption spectrum of a potassium bromide dispersion of it, previously dried, exhibits maxima only at the same wavelengths as that of a similar preparation of USP Methyltestosterone RS.

B: The ultraviolet absorption spectrum of a 1 in 100,000 solution in alcohol exhibits maxima and minima at the same wavelengths as that of a similar preparation of USP Methyltestosterone RS, concomitantly measured.

Melting range ⟨741⟩: between 162° and 167°.

Specific rotation ⟨781⟩: between +79° and +85°, calculated on the dried basis, determined in a solution in alcohol containing 100 mg in each 10 mL.

Loss on drying ⟨731⟩—Dry it at 105° for 4 hours: it loses not more than 2.0% of its weight.

Chromatographic purity—

Standard preparations—Dissolve an accurately weighed quantity of USP Methyltestosterone RS in a mixture of chloroform and methanol (9:1) to obtain *Standard preparation A* having a known concentration of 20 mg per mL. Dilute quantitatively with the mixture of chloroform and methanol (9:1) to obtain *Standard preparations* having the following compositions:

Standard preparation	Dilution	Concentration (mg RS per mL)	Percentage (%, for comparison with test specimen)
A	(undiluted)	20	100
B	(1 in 100)	0.20	1.0
C	(1 in 200)	0.10	0.5

System suitability preparation—Dissolve 10 mg of USP Testosterone RS in 0.5 mL of *Standard preparation A*, dilute with the mixture of chloroform and methanol (9:1) to 10.0 mL, and mix.

Test preparation—Dissolve an accurately weighed quantity of methyltestosterone in the mixture of chloroform and methanol (9:1) to obtain a solution containing 20 mg per mL.

Procedure—On a suitable thin-layer chromatographic plate (see *Chromatography* ⟨621⟩), coated with a 0.25-mm layer of chromatographic silica gel mixture, apply separately 5 µL of the *Test preparation* and 5 µL of each *Standard preparation*, and the *System suitability preparation.* Position the plate in a chromatographic chamber, and develop the chromatograms in a solvent system consisting of a mixture of butyl acetate, hexanes, and glacial acetic acid (70:30:1) until the solvent front has moved about three-fourths of the length of the plate. Remove the plate from the developing chamber, mark the solvent front, and allow the solvent to evaporate in air. Examine the plate under short-wavelength ultraviolet light. The test is valid if the chromatogram obtained from the *System suitability preparation* shows two clearly separated spots. Compare the intensities of any secondary spots observed in the chromatogram of the *Test preparation* with those of the principal spots in the chromatograms of the *Standard preparations:* the sum of the intensities of secondary spots obtained from the *Test preparation* corresponds to not more than 1.0% of related compounds, with no single impurity corresponding to more than 0.5%.

Assay—

Mobile phase—Prepare a filtered and degassed mixture of acetonitrile and water (55:45). Make adjustments if necessary (see *System Suitability* under *Chromatography* ⟨621⟩).

Standard preparation—Transfer about 25 mg of USP Methyltestosterone RS, accurately weighed, to a 100-mL volumetric flask, dilute with methanol to volume, and mix. Pipet 8 mL of this solution into a 100-mL volumetric flask, dilute with *Mobile phase* to volume, and mix to obtain the *Standard preparation* having a known concentration of about 20 µg per mL.

Assay preparation—Transfer about 50 mg of Methyltestosterone, accurately weighed, to a 100-mL volumetric flask, dilute with methanol to volume, and mix. Pipet 8 mL of this solution to a 200-mL volumetric flask, dilute with *Mobile phase* to volume, and mix.

System suitability preparation—Prepare a solution of testosterone in methanol containing about 250 µg per mL. Dilute 4 mL of this solution with the *Standard preparation* to 50 mL, and mix.

Chromatographic system (see *Chromatography* ⟨621⟩)—The liquid chromatograph is equipped with a 241-nm detector and a 4-mm × 25-cm column that contains packing L1. The flow rate is about 1 mL per minute. Chromatograph the *Standard preparation* and the *System suitability preparation*, and record the peak responses as directed under *Procedure:* the column efficiency determined from the analyte peak is not less than 2000 theoretical plates, the tailing factor for the analyte peak is not more than 2.7, the resolution, *R*, between the testosterone and methyltestosterone peaks is not less than 2.0, and the relative standard deviation for replicate injections is not more than 2.0%. The relative retention times are about 0.8 for testosterone and 1.0 for methyltestosterone.

Procedure—Separately inject equal volumes (about 50 µL) of the *Standard preparation* and the *Assay preparation* into the chromatograph, record the chromatograms, and measure the responses for the major peaks. Calculate the quantity, in mg, of $C_{20}H_{30}O_2$ in the portion of methyltestosterone taken by the formula:

$$2500C(r_U/r_S),$$

in which C is the concentration, in mg per mL, of USP Methyltestosterone RS in the *Standard preparation*, and r_U and r_S are the peak responses obtained from the *Assay preparation* and the *Standard preparation*, respectively.

Methyltestosterone Capsules

» Methyltestosterone Capsules contain not less than 90.0 percent and not more than 110.0 percent of the labeled amount of $C_{20}H_{30}O_2$.

Packaging and storage—Preserve in well-closed containers.

Reference standard—*USP Methyltestosterone Reference Standard*—Dry in vacuum over silica gel for 4 hours before using.

Identification—Evaporate 25 mL of the first alcohol solution of methyltestosterone obtained in the *Assay* to dryness: the infrared absorption spectrum of a potassium bromide dispersion of the residue so obtained exhibits maxima at the same wavelengths as that of a similar preparation of USP Methyltestosterone RS.

Dissolution ⟨711⟩—

Medium: water; 900 mL.

Apparatus 1: 100 rpm.

Time: 30 minutes.

Procedure—Determine the amount in solution on filtered portions of the *Dissolution Medium*, suitably diluted quantitatively and stepwise with water to obtain a solution containing about 10 µg of methyltestosterone per mL at the wavelength of maximum absorbance at about 248 nm, with a suitable spectrophotometer, in comparison with a solution of USP Methyltestosterone RS in the same medium at the same concentration, water being used as the blank.

Tolerances—Not less than 60% (Q) of the labeled amount of $C_{20}H_{30}O_2$ is dissolved in 30 minutes.

Uniformity of dosage units ⟨905⟩: meet the requirements.

Procedure for content uniformity—Transfer the contents of 1 Capsule to a 100-mL volumetric flask, add 50 mL of methanol, and shake by mechanical means for 60 minutes. Dilute with methanol to volume, and filter, discarding the first 20 mL of the filtrate. Dilute a portion of the subsequent filtrate quantitatively and stepwise, if necessary, with methanol to obtain a solution containing about 10 µg of methyltestosterone per mL. Concomitantly determine the absorbances of this solution and a Standard solution of USP Methyltestosterone RS in the same medium having a known concentration of about 10 µg per mL, in 1-cm cells at the wavelength of maximum absorbance at about 241 nm, with a suitable spectrophotometer, using methanol as the blank. Calculate the quantity, in mg, of $C_{20}H_{30}O_2$ in the Capsule by the formula:

$$(T/D)C(A_U/A_S),$$

in which T is the labeled quantity, in mg, of methyltestosterone in the Capsule, D is the concentration, in µg per mL, of methyltestosterone in the solution from the Capsule, based on the labeled quantity per Capsule and the extent of dilution, C is the concentration, in µg per mL, of USP Methyltestosterone RS in the Standard solution, and A_U and A_S are the absorbances of the solution from the Capsule and the Standard solution, respectively.

Assay—Remove as completely as possible and weigh the contents of not less than 20 Methyltestosterone Capsules, and mix. Transfer an accurately weighed portion of the powder, equivalent to about 10 mg of methyltestosterone, to a 125-mL separator with the aid of about 5 mL of water. Extract with four 20-mL portions of chloroform, filtering each through chloroform-washed cotton. Evaporate the combined extracts on a steam bath, with the aid of a current of air, to dryness. Dissolve the residue in alcohol, transfer to a 50-mL volumetric flask, add alcohol to volume, and mix. Pipet a 5-mL aliquot into a 100-mL volumetric flask, add alcohol to volume, and mix. Concomitantly determine the absorbances of this solution and a Standard solution of USP Methyltestosterone RS in the same medium having a known concentration of about 10 µg per mL, in 1-cm cells at the wavelength of maximum absorbance at about 241 nm, with a suitable spectrophotometer, using alcohol as the blank. Calculate the quantity, in mg, of $C_{20}H_{30}O_2$ in the portion of Capsules contents taken by the formula:

$$C(A_U/A_S),$$

in which C is the concentration, in µg per mL, of USP Methyltestosterone RS in the Standard solution, and A_U and A_S are the absorbances of the solution from the Capsules and the Standard solution, respectively.

Methyltestosterone Tablets

» Methyltestosterone Tablets contain not less than 90.0 percent and not more than 110.0 percent of the labeled amount of $C_{20}H_{30}O_2$.

Packaging and storage—Preserve in well-closed containers.

Reference standard—*USP Methyltestosterone Reference Standard*—Dry at 105° for 4 hours before using.

Identification—Evaporate to dryness 25 mL of the first alcohol solution of methyltestosterone obtained in the *Assay:* the infrared absorption spectrum of a potassium bromide dispersion of the residue so obtained exhibits maxima at the same wavelengths as that of a similar preparation of USP Methyltestosterone RS.

Disintegration ⟨701⟩: 30 minutes. Tablets intended for buccal administration meet the requirements for *Buccal Tablets* ⟨701⟩.

Uniformity of dosage units ⟨905⟩: meet the requirements.

Procedure for content uniformity—Transfer 1 finely powdered Tablet to a 100-mL volumetric flask, add 50 mL of methanol, and shake by mechanical means for 60 minutes. Dilute with methanol to volume, and filter, discarding the first 20 mL of the filtrate. Dilute a portion of the subsequent filtrate quantitatively and stepwise, if necessary, with methanol to provide a solution containing about 10 µg of methyltestosterone per mL. Concomitantly determine the absorbances of this solution and a solution of USP Methyltestosterone RS in the same medium having a known concentration of about 10 µg per mL, in 1-cm cells at the wavelength of maximum absorbance at about 241 nm, with a suitable spectrophotometer, using methanol as the blank. Calculate the quantity, in mg, of $C_{20}H_{30}O_2$ in the Tablet by the formula:

$$(T/D)C(A_U/A_S),$$

in which T is the labeled quantity, in mg, of methyltestosterone in the Tablet, D is the concentration, in µg per mL, of methyltestosterone in the solution from the Tablet, based on the labeled quantity per Tablet and the extent of dilution, C is the concentration, in µg per mL, of USP Methyltestosterone RS in the Standard solution, and A_U and A_S are the absorbances of the solution from the Tablet and the Standard solution, respectively.

Assay—Weigh and finely powder not less than 20 Methyltestosterone Tablets. Weigh accurately a portion of the powder, equivalent to about 10 mg of methyltestosterone, and transfer to a 125-mL separator with the aid of about 5 mL of water. Extract with four 25-mL portions of chloroform, filtering each through chloroform-washed cotton. Evaporate the combined extracts on a steam bath, with the aid of a current of air, to dryness. Dissolve the residue in alcohol, transfer to a 50-mL volumetric flask, add alcohol to volume, and mix. Pipet a 5-mL aliquot into a 100-mL volumetric flask, add alcohol to volume, and mix. Concomitantly determine the absorbances of this solution and a solution of USP Methyltestosterone RS in the same medium having a known concentration of about 10 µg per mL, in 1-cm cells at the wavelength of maximum absorbance at about 241 nm, with a suitable spectrophotometer, using alcohol as the blank. Calculate the quantity, in mg, of $C_{20}H_{30}O_2$ in the portion of Tablets taken by the formula:

$$C(A_U/A_S),$$

in which C is the concentration, in µg per mL, of USP Methyltestosterone RS in the Standard solution, and A_U and A_S are the absorbances of the solution from the Tablets and the Standard solution, respectively.

Methyprylon

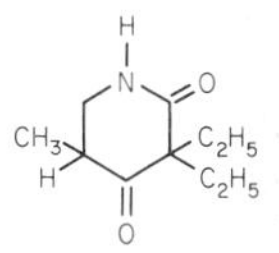

$C_{10}H_{17}NO_2$ 183.25

2,4-Piperidinedione, 3,3-diethyl-5-methyl-.

3,3-Diethyl-5-methyl-2,4-piperidinedione [*125-64-4*].

» Methyprylon contains not less than 98.0 percent and not more than 101.0 percent of $C_{10}H_{17}NO_2$, calculated on the dried basis.

Packaging and storage—Preserve in well-closed, light-resistant containers.

Reference standard—*USP Methyprylon Reference Standard*—Dry at a pressure between 150 mm and 200 mm of mercury at 56° for 3 hours before using.

Identification—

A: The infrared absorption spectrum of a potassium bromide dispersion of it, previously dried, exhibits maxima only at the same wavelengths as that of a similar preparation of USP Methyprylon RS.

B: The ultraviolet absorption spectrum of a 1 in 333 solution in isopropyl alcohol exhibits maxima and minima at the same wavelengths as that of a similar solution of USP Methyprylon RS, concomitantly measured. The absorptivity, calculated on the dried basis, at the wavelength of maximum absorbance at about 295 nm is not more than 0.25.

C: Add 50 mg of it to 7 mL of 0.3 *N* sodium hydroxide, mix, and add 0.5 mL of potassium ferricyanide TS: the solution shows a green fluorescence under ultraviolet light.

Melting range, *Class I* ⟨741⟩: between 74.0° and 77.5°.

Loss on drying ⟨731⟩—Dry it at 56° and at a pressure between 150 mm and 200 mm of mercury for 3 hours: it loses not more than 1.0% of its weight.

Residue on ignition ⟨281⟩: not more than 0.1%.

Assay—

0.1 N Potassium ferricyanide—Transfer about 33 g of potassium ferricyanide, accurately weighed, and 12.4 g of monohydrated sodium carbonate to a 1000-mL volumetric flask. Dissolve in water, and dilute with water to volume. Determine the normality of the solution as directed for *Procedure*, using USP Methyprylon RS as the primary standard. Store the solution in amber-colored bottles, and standardize just before use.

Procedure—Transfer about 250 mg of Methyprylon, accurately weighed, to a 150-mL beaker, and add 60 mL of water and 10 mL of 1 *N* potassium hydroxide. Calculate the expected end-point, and titrate, determining the end-point potentiometrically, using a calomel-platinum electrode system. Add the *0.1 N Potassium ferricyanide*, rapidly at first, to within about 3 mL of the calculated end-point and then in 0.2-mL increments until the final deflection of the instrument. Each mL of *0.1 N Potassium ferricyanide* is equivalent to 9.162 mg of $C_{10}H_{17}NO_2$.

Methyprylon Capsules

» Methyprylon Capsules contain not less than 90.0 percent and not more than 110.0 percent of the labeled amount of $C_{10}H_{17}NO_2$.

Packaging and storage—Preserve in tight, light-resistant containers.

Reference standard—*USP Methyprylon Reference Standard*—Dry at a pressure between 150 mm and 200 mm of mercury at 56° for 3 hours before using.

Identification—Extract the contents of 1 Capsule, equivalent to about 300 mg of methyprylon, with 10 mL of ether, filter the extract into a beaker, and evaporate on a steam bath to dryness: the residue so obtained responds to *Identification tests A* and *C* under *Methyprylon.*

Dissolution ⟨711⟩—

Medium: water; 500 mL.

Apparatus 1: 100 rpm.

Time: 45 minutes.

Mobile phase—Prepare a suitable and degassed solution in a mixture of methanol and water (60:40).

Internal standard solution—Prepare a solution of ethylparaben in *Mobile phase* having a final concentration of about 24 µg of ethylparaben per mL.

Standard preparation—Mix an accurately weighed quantity of USP Methyprylon RS with an accurately measured volume of *Internal standard solution*, and dilute quantitatively with *Mobile phase* to obtain a solution having a known concentration of about 360 µg of methyprylon per mL and a concentration of about 2 µg of ethylparaben per mL.

Test preparation—Mix an accurately measured, filtered portion of the solution under test with an accurately measured volume of *Internal standard solution*, and dilute with *Mobile phase* to obtain a solution having a concentration of about 360 µg of methyprylon per mL and a concentration of about 2 µg of ethylparaben per mL.

Chromatographic system (see *Chromatography* ⟨621⟩)—The liquid chromatograph is equipped with a 280-nm detector and a 4.6-mm × 25-cm column that contains packing L1. The flow rate is about 0.7 mL per minute. Chromatograph five replicate injections of the *Standard preparation*, and record the peak responses as directed under *Procedure:* the relative standard deviation is not more than 2.0%, and the resolution factor between methyprylon and ethylparaben is not less than 2.2.

Procedure—Separately inject equal volumes (about 20 µL) of the *Standard preparation* and the *Assay preparation* into the chromatograph, record the chromatograms, measure the responses for the major peaks, and determine the peak response ratios for methyprylon to ethylparaben, R_U and R_S, obtained from the *Test preparation* and the *Standard preparation*, respectively. The relative retention times are about 0.81 for methyprylon and 1.0 for ethylparaben. Determine the amount of $C_{10}H_{17}NO_2$ dissolved from the peak response ratios obtained.

Tolerances—Not less than 75% (*Q*) of the labeled amount of $C_{10}H_{17}NO_2$ is dissolved in 45 minutes.

Uniformity of dosage units ⟨905⟩: meet the requirements.

Assay—

0.1 N Potassium ferricyanide—Prepare and standardize as directed in the *Assay* under *Methyprylon.*

Procedure—Transfer, as completely as possible, the contents of not less than 20 Methyprylon Capsules to a suitable tared container, and determine the average weight per capsule. Mix the combined contents, and transfer an accurately weighed portion, equivalent to about 250 mg of methyprylon, to a 150-mL beaker. Add 70 mL of water and 10 mL of 1 *N* potassium hydroxide, and proceed as directed for *Procedure* in the *Assay* under *Methyprylon*, beginning with "Calculate the expected end-point."

Methyprylon Tablets

» Methyprylon Tablets contain not less than 95.0 percent and not more than 105.0 percent of the labeled amount of $C_{10}H_{17}NO_2$.

Packaging and storage—Preserve in tight, light-resistant containers.

Reference standard—*USP Methyprylon Reference Standard*—Dry at a pressure between 150 mm and 200 mm of mercury at 56° for 3 hours before using.

Identification—Extract a quantity of finely powdered Tablets, equivalent to about 200 mg of methyprylon, with 10 mL of ether. Filter the ether extract into a beaker, and evaporate on a steam bath to dryness: the residue so obtained responds to *Identification tests A* and *C* under *Methyprylon.*

Dissolution ⟨711⟩—

Medium: water; 500 mL.

Apparatus 1: 100 rpm.

Time: 45 minutes.

Mobile phase—Prepare a suitable and degassed solution in a mixture of methanol and water (60:40).

Internal standard solution—Prepare a solution of ethylparaben in *Mobile phase* having a final concentration of about 120 µg of ethylparaben per mL.

Standard preparation—Mix an accurately weighed quantity of USP Methyprylon RS with an accurately measured volume of *Internal standard solution*, and dilute quantitatively with *Mobile phase* to obtain a solution having a known concentration of about 80 µg of methyprylon per mL and a concentration of about 0.5 µg of ethylparaben per mL.

Test preparation—Mix an accurately measured, filtered portion of the solution under test with an accurately measured volume of *Internal standard solution*, and dilute with *Mobile phase* to obtain a solution having a concentration of about 80 µg of methyprylon per mL and a concentration of about 0.5 µg of ethylparaben per mL.

Chromatographic system (see *Chromatography* ⟨621⟩)—The liquid chromatograph is equipped with a 280-nm detector and a 4.6-mm × 25-cm column that contains packing L1. The flow rate is about 0.7 mL per minute. Chromatograph five replicate injections of the *Standard preparation*, and record the peak responses as directed under *Procedure:* the relative standard deviation is not more than 2.0%, and the resolution factor between methyprylon and ethylparaben is not less than 2.2.

Procedure—Separately inject equal volumes (about 70 µL) of the *Standard preparation* and the *Assay preparation* into the chromatograph, record the chromatograms, measure the responses for the major peaks, and determine the peak response ratios for methyprylon to ethylparaben, R_U and R_S, obtained from the *Test preparation* and the *Standard preparation*, respectively: the relative retention times are about 0.81 for methyprylon and 1.0 for ethylparaben. Determine the amount of $C_{10}H_{17}NO_2$ dissolved from the peak response ratios obtained.

Tolerances—Not less than 75% (*Q*) of the labeled amount of $C_{10}H_{17}NO_2$ is dissolved in 45 minutes.

Uniformity of dosage units ⟨905⟩: meet the requirements.

Assay—

0.1 N Potassium ferricyanide—Prepare and standardize as directed in the *Assay* under *Methyprylon*.

Procedure—Weigh and finely powder not less than 20 Methyprylon Tablets. Weigh accurately a portion of the powder, equivalent to about 250 mg of methyprylon, and transfer to a 150-mL beaker. Add 70 mL of water and 10 mL of 1 *N* potassium hydroxide, and proceed as directed for *Procedure* in the *Assay* under *Methyprylon*, beginning with "Calculate the expected end-point."

Methysergide Maleate

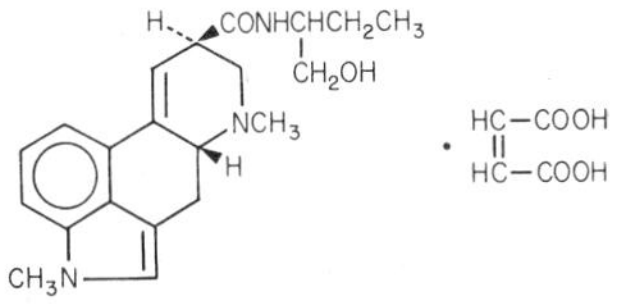

$C_{21}H_{27}N_3O_2 . C_4H_4O_4$ 469.54

Ergoline-8-carboxamide, 9,10-didehydro-*N*-[1-(hydroxymethyl)propyl]-1,6-dimethyl-, (8β)-, (*Z*)-2-butenedioate (1:1) (salt).

9,10-Didehydro-*N*-[1-(hydroxymethyl)propyl]-1,6-dimethylergoline-8β-carboxamide maleate (1:1) (salt) [*129-49-7*].

» Methysergide Maleate contains not less than 97.0 percent and not more than 103.0 percent of $C_{21}H_{27}N_3O_2 . C_4H_4O_4$, calculated on the dried basis.

Packaging and storage—Preserve in tight, light-resistant containers, in a cold place.

Reference standard—*USP Methysergide Maleate Reference Standard*—Dry in vacuum at 120° for 2 hours before using.

Identification—

A: The infrared absorption spectrum of a potassium bromide dispersion of it, previously dried, exhibits maxima only at the same wavelengths as that of a similar preparation of USP Methysergide Maleate RS.

B: [NOTE—Conduct this test without exposure to daylight, and with the minimum exposure to artificial light.] In a suitable chromatographic chamber, arranged for thin-layer chromatography (see *Chromatography* ⟨621⟩), place a volume of a solvent system consisting of a mixture of chloroform and methanol (20:1) sufficient to develop the chromatogram. Place a beaker containing 25 mL of ammonium hydroxide in the chamber, cover, and allow to equilibrate for 30 minutes. Prepare a test solution of Methysergide Maleate in methanol containing 5 mg per mL. On a suitable thin-layer chromatographic plate (see *Chromatography* ⟨621⟩), coated with a 0.25-mm layer of chromatographic silica gel, apply 25 µL of this solution and 25 µL of a Standard solution of USP Methysergide Maleate RS in methanol containing 5 mg per mL. Develop the chromatogram until the solvent front has moved about three-fourths of the length of the plate. Remove the plate from the developing chamber, mark the solvent front, and allow the solvent to evaporate. Locate the spots on the plate by lightly spraying with a solution prepared by dissolving 800 mg of *p*-dimethylaminobenzaldehyde in a cooled mixture of 80 mL of alcohol and 20 mL of sulfuric acid, allow the plate to dry, then expose it briefly to fumes of a mixture of nitric and hydrochloric acids: the R_f value of the principal spot obtained from the test solution corresponds to that obtained from the Standard solution.

Specific rotation ⟨781⟩: between +35° and +45°, calculated on the dried basis, determined in a solution containing 25 mg in each 10 mL.

pH ⟨791⟩: between 3.7 and 4.7, in a 1 in 500 solution in carbon dioxide-free water.

Loss on drying ⟨731⟩—Dry it in vacuum at 120° for 2 hours: it loses not more than 7.0% of its weight.

Ordinary impurities ⟨466⟩—

Test solution: methanol.

Standard solution: methanol.

Eluant: prepare as directed under *Identification test B*.

Visualization: 1.

Assay—Dissolve about 200 mg of Methysergide Maleate, accurately weighed, in 30 mL of glacial acetic acid, add 1 drop of crystal violet TS, and titrate with 0.1 *N* perchloric acid VS to a blue end-point. Perform a blank determination, and make any necessary correction. Each mL of 0.1 *N* perchloric acid is equivalent to 46.95 mg of $C_{21}H_{27}N_3O_2 . C_4H_4O_4$.

Methysergide Maleate Tablets

» Methysergide Maleate Tablets contain not less than 90.0 percent and not more than 110.0 percent of the labeled amount of $C_{21}H_{27}N_3O_2 . C_4H_4O_4$.

Packaging and storage—Preserve in tight containers.

Reference standard—*USP Methysergide Maleate Reference Standard*—Dry in vacuum at 120° for 2 hours before using.

Identification—To a portion of finely powdered Tablets, equivalent to about 10 mg of methysergide maleate, add 50 mL of tartaric acid solution (1 in 100) and 4 drops of benzalkonium chloride solution (1 in 10), and shake vigorously by mechanical means for 30 minutes. Filter the mixture. To 5 mL of the filtrate add 10 mL of *p*-dimethylaminobenzaldehyde TS: a violet-blue color develops.

Dissolution ⟨711⟩—

Medium: tartaric acid solution (1 in 200); 900 mL.

Apparatus 2: 100 rpm.

Time: 30 minutes.

Procedure—Filter a portion of the solution under test into a flask. Concomitantly determine the fluorescence intensity of this solution in comparison with a Standard solution of USP Methysergide Maleate RS in the same medium having a known concentration of about 2.2 µg per mL in a fluorometer at an excitation wavelength of about 327 nm and an emission wavelength of about 428 nm, using tartaric acid solution (1 in 200) as the blank.

Tolerances—Not less than 70% (*Q*) of the labeled amount of $C_{21}H_{27}N_3O_2 . C_4H_4O_4$ is dissolved in 30 minutes.

Uniformity of dosage units ⟨905⟩: meet the requirements.

Assay—[NOTE—Conduct this procedure with a minimum exposure to light.]

Mobile phase—Dissolve 6.8 g of monobasic potassium phosphate in 700 mL of water, add 300 mL of acetonitrile, and mix. Filter through a 0.45-µm membrane, and degas under vacuum. Make adjustments if necessary (see *System Suitability* under *Chromatography* ⟨621⟩).

Solvent mixture—Dissolve 10 g of tartaric acid in 1 liter of water, add 1 liter of methanol, and mix.

Standard preparation—Dissolve an accurately weighed quantity of USP Methysergide Maleate RS in *Solvent mixture* with the help of sonication, and dilute quantitatively, and stepwise if necessary, with the same solvent to obtain a solution having a known concentration of about 0.1 mg per mL.

Assay preparation—Weigh and finely powder not less than 20 Methysergide Maleate Tablets. Transfer an accurately weighed portion of the powder, equivalent to about 10 mg of methysergide maleate, to a 100-mL volumetric flask. Add 75 mL of *Solvent mixture*, and shake by mechanical means for 60 minutes. Add *Solvent mixture* to volume, mix, and filter through a 0.45-µm membrane, discarding the first 20 mL of the filtrate.

Chromatographic system (see *Chromatography* ⟨621⟩)—The liquid chromatograph is equipped with a 318-nm detector and a 4.6-mm × 15-cm column that contains packing L7. The flow rate is about 2 mL per minute. Chromatograph the *Standard preparation*, and record the peak responses as directed under *Procedure:* the column efficiency determined from the analyte peak is not less than 1000 theoretical plates, the tailing factor for the analyte peak is not more than 2.5, the resolution, *R*, between the analyte and the closest adjacent peak is not less than 1.0, and the relative standard deviation for replicate injections is not more than 2.0%.

Procedure—Separately inject equal volumes (about 20 µL) of the *Standard preparation* and the *Assay preparation* into the chromatograph, record the chromatograms, and measure the responses for the major peaks. Calculate the quantity, in mg, of $C_{21}H_{37}N_3O_2 . C_4H_4O_4$ in the portion of Tablets taken by the formula:

$$(100C)(r_U / r_S),$$

in which *C* is the concentration, in mg per mL, of USP Methysergide Maleate RS in the *Standard preparation*, and r_U and r_S are the peak responses obtained from the *Assay preparation* and the *Standard preparation*, respectively.

Metoclopramide Hydrochloride

[Structural formula: Cl, $CONHCH_2CH_2N(C_2H_5)_2$, H_2N, OCH_3 · HCl · H_2O]

$C_{14}H_{22}ClN_3O_2 . HCl . H_2O$ 354.28

Benzamide, 4-amino-5-chloro-*N*-[2-(diethylamino)ethyl]-2-methoxy-, monohydrochloride, monohydrate.

4-Amino-5-chloro-*N*-[2-(diethylamino)ethyl]-*o*-anisamide monohydrochloride monohydrate [*54143-57-6*].

» Metoclopramide Hydrochloride contains not less than 98.0 percent and not more than 101.0 percent of $C_{14}H_{22}ClN_3O_2 . HCl$, calculated on the anhydrous basis.

Packaging and storage—Preserve in tight, light-resistant containers.

Reference standard—*USP Metoclopramide Hydrochloride Reference Standard*—Do not dry; determine the water content titrimetrically at time of use for quantitative analyses.

Identification—

A: The infrared absorption spectrum of a potassium bromide dispersion of it, previously dried, exhibits maxima only at the same wavelengths as that of a similar preparation of USP Metoclopramide Hydrochloride RS.

B: Dissolve 50 mg in 5 mL of water, and add 5 mL of a 1 in 100 solution of *p*-dimethylaminobenzaldehyde in 1 *N* hydrochloric acid: a yellow-orange color is produced.

C: The R_f value of the principal spot in the chromatogram of the *Identification preparation* corresponds to that of the *Standard preparation A*, as obtained in the test for *Chromatographic purity*.

Water, *Method I* ⟨921⟩: between 4.5% and 6.0%.

Residue on ignition ⟨281⟩: not more than 0.1%.

Chromatographic purity—

Standard preparations—Dissolve USP Metoclopramide Hydrochloride RS in methanol, and mix to obtain a solution having a known concentration of 1 mg per mL. Dilute quantitatively with methanol to obtain three *Standard preparations*, designated below by letters, having the following compositions:

Standard preparation	Dilution	Concentration (µg RS per mL)	Percentage (%, for comparison with test specimen)
A	(1 in 4)	250	0.5
B	(3 in 20)	150	0.3
C	(1 in 20)	50	0.1

Test preparation—Dissolve an accurately weighed quantity of Metoclopramide Hydrochloride in methanol to obtain a solution containing 50 mg per mL.

Identification preparation—Dilute a portion of the *Test preparation* quantitatively with methanol to obtain a solution containing 500 µg per mL.

Procedure—On a suitable thin-layer chromatographic plate (see *Chromatography* ⟨621⟩), coated with a 0.25-mm layer of chromatographic silica gel mixture, apply separately 10 µL of the *Test preparation*, 10 µL of the *Identification preparation*, and 10 µL of each *Standard preparation*. Allow the spots to dry, position the plate in a chromatographic chamber, and develop the chromatograms in a solvent system consisting of a mixture of chloroform, methanol, toluene, and ammonium hydroxide (140:60:20:1) until the solvent front has moved about three-fourths of the length of the plate. Remove the plate from the developing chamber, mark the solvent front, and allow the solvent to evaporate. Examine the plate under short-wavelength ultraviolet light, and compare the intensities of any secondary spots observed in the chromatogram of the *Test preparation* with those of the principal spots in the chromatograms of the *Standard preparations*. [NOTE—Disregard any spots observed at the origins of the chromatograms.] No secondary spot from the chromatogram of the *Test preparation* is larger or more intense than the principal spot obtained from *Standard preparation A* (0.5%), and the sum of the intensities of all secondary spots obtained from the *Test preparation* corresponds to not more than 1.0%.

Assay—Transfer about 300 mg of Metoclopramide Hydrochloride, accurately weighed, to a stoppered, 125-mL flask, add 10 mL of mercuric acetate TS and 2 mL of acetic anhydride, and allow to stand for 3 hours. Add 80 mL of glacial acetic acid, and titrate with 0.1 *N* perchloric acid VS, determining the endpoint potentiometrically (see *Titrimetry* ⟨541⟩). Perform a blank determination, and make any necessary correction. Each mL of 0.1 *N* perchloric acid is equivalent to 33.63 mg of $C_{14}H_{22}ClN_3O_2 . HCl$.

Metoclopramide Injection

» Metoclopramide Injection is a sterile solution of Metoclopramide Hydrochloride in Water for Injection. It contains the equivalent of not less than 90.0 percent and not more than 110.0 percent of the labeled amount of metoclopramide ($C_{14}H_{22}ClN_3O_2$).

Packaging and storage—Preserve in single-dose or in multiple-dose containers, preferably of Type I glass, protected from light. [NOTE—Injection containing an antioxidant agent does not require protection from light.]

Reference standard—*USP Metoclopramide Hydrochloride Reference Standard*—Do not dry; determine the water content titrimetrically at time of use for quantitative analyses.

Identification—

A: The retention time of the major peak in the chromatogram of the *Assay preparation* corresponds to that of the *Standard preparation*, as obtained in the *Assay*.

B: Mix a volume of Injection, equivalent to about 50 mg of metoclopramide, with 5 mL of water and 5 mL of a 1 in 100 solution of *p*-dimethylaminobenzaldehyde in 1 *N* hydrochloric acid: a yellow-orange color is produced.

Pyrogen—It meets the requirements of the *Pyrogen Test* ⟨151⟩.

pH ⟨791⟩: between 2.5 and 6.5.

Particulate matter ⟨788⟩: meets the requirements under *Small-volume Injections*.

Other requirements—It meets the requirements under *Injections* ⟨1⟩.

Assay—

Mobile phase—Dissolve 2.7 g of sodium acetate in 500 mL of water, add 500 mL of acetonitrile, 1 mL of glacial acetic acid, and 2 mL of tetramethylammonium hydroxide solution in methanol (1 in 5), mix, filter, and degas. Make adjustments if necessary (see *System Suitability* under *Chromatography* ⟨621⟩).

Standard preparation—Dissolve an accurately weighed quantity of USP Metoclopramide Hydrochloride RS in 0.01 *M* phosphoric acid to obtain a stock solution having a known concentration of about 9 mg of anhydrous metoclopramide hydrochloride per mL. Transfer 5.0 mL of this stock solution to a 100-mL volumetric flask, dilute with 0.01 *M* phosphoric acid to volume, and mix to obtain a *Standard preparation* having a known concentration of about 450 µg of USP Metoclopramide Hydrochloride RS, on the anhydrous basis, per mL (equivalent to about 400 µg of anhydrous metoclopramide per mL).

System suitability solution—Transfer about 125 mg of benzenesulfonamide to a 25-mL volumetric flask, add 15 mL of methanol, and shake to dissolve. Dilute with 0.01 *M* phosphoric acid to volume, and mix. Pipet 5 mL of this solution and 5 mL of the stock solution used to prepare the *Standard preparation* into a 100-mL volumetric flask, dilute with 0.01 *M* phosphoric acid to volume, and mix.

Assay preparation—Transfer an accurately measured volume of Metoclopramide Injection, equivalent to about 40 mg of metoclopramide, to a 100-mL volumetric flask, dilute with 0.01 *M* phosphoric acid to volume, and mix.

Chromatographic system (see *Chromatography* ⟨621⟩)—The liquid chromatograph is equipped with a 270-nm detector and a 4.6-mm × 25-cm column that contains packing L1. The flow rate is about 1.5 mL per minute. Chromatograph the *System suitability solution*, and record the peak responses as directed under *Procedure:* the relative retention times are about 0.7 for benzenesulfonamide and 1.0 for metoclopramide, and the resolution, *R*, between the benzenesulfonamide peak and the metoclopramide peak is not less than 2.0. Chromatograph the *Standard preparation*, and record the peak responses as directed under *Procedure:* the tailing factor for the metoclopramide peak is not more than 2.0, and the relative standard deviation for replicate injections is not more than 2.0%.

Procedure—Separately inject equal volumes (about 20 µL) of the *Standard preparation* and the *Assay preparation* into the chromatograph, record the chromatograms, and measure the responses for the major peaks. Calculate the quantity, in mg, of $C_{14}H_{22}ClN_3O_2$ in each mL of the Injection taken by the formula:

$$(299.80/336.26)(0.1C/V)(r_U/r_S),$$

in which 299.80 and 336.26 are the molecular weights of metoclopramide and anhydrous metoclopramide hydrochloride, respectively, *C* is the concentration, in µg per mL, of USP Metoclopramide Hydrochloride RS, on the anhydrous basis, in the *Standard preparation*, *V* is the volume, in mL, of Injection taken, and r_U and r_S are the peak responses of metoclopramide obtained from the *Assay preparation* and the *Standard preparation*, respectively.

Metoclopramide Oral Solution

» Metoclopramide Oral Solution contains an amount of Metoclopramide Hydrochloride ($C_{14}H_{22}ClN_3O_2$.-$HCl.H_2O$) equivalent to not less than 90.0 percent and not more than 110.0 percent of the labeled amount of metoclopramide ($C_{14}H_{22}ClN_3O_2$).

Packaging and storage—Store in tight, light-resistant containers in a cool place. Protect from freezing.

Reference standard—*USP Metoclopramide Hydrochloride Reference Standard*—Do not dry; determine the water content titrimetrically at time of use for quantitative analysis.

Identification—The retention time of the major peak in the chromatogram of the *Assay preparation* corresponds to that of the *Standard preparation* as obtained in the *Assay*.

pH ⟨791⟩: between 2.0 and 4.0.

Assay—

Mobile phase—Dissolve 2.7 g of sodium acetate in 600 mL of water, add 400 mL of acetonitrile, and 4 mL of Tetramethylammonium hydroxide solution in methanol (25%), and mix. Adjust with glacial acetic acid to a pH of 6.5, filter, and degas. Make adjustments if necessary (see *System Suitability* under *Chromatography* ⟨621⟩).

Standard preparation—Dissolve an accurately weighed quantity of USP Metoclopramide Hydrochloride RS in 0.01 *M* phosphoric acid to obtain a stock solution having a known concentration of about 9 mg of anhydrous metoclopramide hydrochloride per mL. Transfer 5.0 mL of this stock solution to a 250-mL volumetric flask, dilute with 0.01 *M* phosphoric acid to volume, and mix to obtain a *Standard preparation* having a known concentration of about 180 µg of USP Metoclopramide Hydrochloride RS, on the anhydrous basis, per mL (equivalent to about 160 µg of anhydrous metoclopramide per mL).

System suitability solution—Transfer about 125 mg of benzenesulfonamide to a 25-mL volumetric flask, add 15 mL of methanol, and shake to dissolve. Dilute with 0.01 *M* phosphoric acid to volume, and mix. Pipet 15 mL of this solution and 5 mL of the stock solution used to prepare the *Standard preparation* into a 250-mL volumetric flask, dilute with 0.01 *M* phosphoric acid to volume, and mix.

Assay preparation—Transfer an accurately measured volume of Metoclopramide Oral Solution, equivalent to about 4 mg of metoclopramide, to a 25-mL volumetric flask, dilute with 0.01 *M* phosphoric acid to volume, and mix.

Chromatographic system (see *Chromatography* ⟨621⟩)—Prepare as directed in the *Assay* under *Metoclopramide Injection.* The relative retention times are about 0.2 for benzenesulfonamide and 1.0 for metoclopramide.

Procedure—Separately inject equal volumes (about 20 µL) of the *Standard preparation* and the *Assay preparation* into the chromatograph, record the chromatograms, and measure the responses for the major peaks. Calculate the quantity, in mg, of $C_{14}H_{22}ClN_3O_2$ in each mL of the Oral Solution taken by the formula:

$$(299.80/336.26)(25C/V)(r_U/r_S),$$

in which 299.80 and 336.26 are the molecular weights of metoclopramide and anhydrous metoclopramide hydrochloride, respectively, *C* is the concentration, in mg per mL, of USP Metoclopramide Hydrochloride RS on the anhydrous basis, in the *Standard preparation*, *V* is the volume, in mL, of Oral Solution taken, and r_U and r_S are the peak responses of metoclopramide obtained from the *Assay preparation* and the *Standard preparation*, respectively.

Metoclopramide Tablets

» Metoclopramide Tablets contain an amount of metoclopramide hydrochloride ($C_{14}H_{22}ClN_3O_2$.HCl.-H_2O) equivalent to not less than 90.0 percent and not more than 110.0 percent of the labeled amount of metoclopramide ($C_{14}H_{22}ClN_3O_2$).

Packaging and storage—Preserve in tight, light-resistant containers.

Reference standard—*USP Metoclopramide Hydrochloride Reference Standard*—Do not dry; determine the water content titrimetrically at time of use for quantitative analyses.

Identification—

A: The retention time of the major peak in the chromatogram of the *Assay preparation* corresponds to that of the *Standard preparation*, as obtained in the *Assay*.

B: Transfer a quantity of finely ground Tablets, equivalent to about 50 mg of metoclopramide, to a suitable flask, add 5 mL of water, shake by mechanical means, and filter. Add to the filtrate 5 mL of a 1 in 100 solution of *p*-dimethylaminobenzaldehyde in 1 *N* hydrochloric acid: a yellow-orange color is produced.

Dissolution ⟨711⟩—
Medium: water; 900 mL.
Apparatus 1: 50 rpm.
Time: 30 minutes.
Procedure—Determine the amount of $C_{14}H_{22}ClN_3O_2$ dissolved from ultraviolet absorbances at the wavelength of maximum absorbance at about 309 nm of filtered portions of the solution under test, suitably diluted with water, if necessary, in comparison with a Standard solution having a known concentration of USP Metoclopramide Hydrochloride RS in the same medium.
Tolerances—Not less than 75% (*Q*) of the labeled amount of $C_{14}H_{22}ClN_3O_2$ is dissolved in 30 minutes.

Uniformity of dosage units ⟨905⟩: meet the requirements.

Assay—
Mobile phase, Standard preparation, System suitability solution, and *Chromatographic system*—Prepare as directed in the *Assay* under *Metoclopramide Injection.*
Assay preparation—Weigh and finely powder not less than 20 Metoclopramide Tablets. Transfer an accurately weighed portion of the powder, equivalent to about 40 mg of metoclopramide, to a 100-mL volumetric flask, add about 70 mL of 0.01 *M* phosphoric acid, and sonicate for 5 minutes. Cool to room temperature, dilute with 0.01 *M* phosphoric acid to volume, and mix. Filter the solution through a 0.45-μm filter, discarding the first portion of the filtrate.
Procedure—Proceed as directed for *Procedure* in the *Assay* under *Metoclopramide Injection.* Calculate the quantity, in mg, of $C_{14}H_{22}ClN_3O_2$ in the portion of Tablets taken by the formula:

$$(299.80/336.26)(0.1C)(r_U/r_S),$$

in which 299.80 and 336.26 are the molecular weights of metoclopramide and anhydrous metoclopramide hydrochloride, respectively, *C* is the concentration, in μg per mL, of USP Metoclopramide Hydrochloride RS, on the anhydrous basis, in the *Standard preparation,* and r_U and r_S are the peak responses of metoclopramide obtained from the *Assay preparation* and the *Standard preparation,* respectively.

Metocurine Iodide

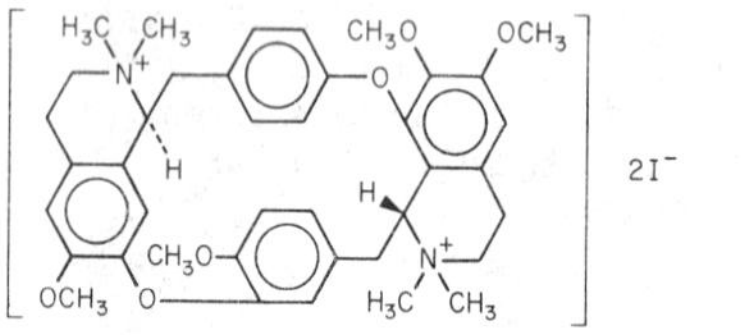

$C_{40}H_{48}I_2N_2O_6$ 906.64
Tubocuraranium, 6,6′,7′,12′-tetramethoxy-2,2,2′,2′-tetramethyl-, diiodide.
(+)-*O,O*′-Dimethylchondrocurarine diiodide [*7601-55-0*].

» Metocurine Iodide contains not less than 95.0 percent and not more than 105.0 percent of $C_{40}H_{48}I_2N_2O_6$, calculated on the anhydrous basis.

Caution—Handle Metocurine Iodide with exceptional care since it is a highly potent skeletal muscle relaxant.

Packaging and storage—Preserve in tight containers.

Reference standard—*USP Metocurine Iodide Reference Standard*—Do not dry; determine the water content by the titrimetric method before using for quantitative analyses.

Identification—
A: The infrared absorption spectrum of a potassium bromide dispersion of it exhibits maxima only at the same wavelengths as that of a similar preparation of USP Metocurine Iodide RS.
B: The ultraviolet absorption spectrum of a 1 in 20,000 solution in alcohol exhibits maxima and minima at the same wavelengths as that of a similar solution of USP Metocurine Iodide RS, concomitantly measured.
C: The chromatogram of the *Assay preparation* obtained as directed in the *Assay* exhibits a peak for metocurine iodide, the retention time of which corresponds to that exhibited by the *Standard preparation.*

Specific rotation ⟨781⟩: between +148° and +158°, calculated on the anhydrous basis, determined in a solution in water containing 25 mg in each 10 mL.

Water, *Method I* ⟨921⟩: not more than 7.0%, a diluted Karl Fischer reagent being used.

Related substances—In the chromatogram obtained from the *Assay preparation* in the *Assay,* the sum of the response of any peaks detected, other than the two peaks due to metocurine and iodide, is not more than 5.0% of the total of all the peak responses.

Assay—
Mobile phase—Mix 3 volumes of acetonitrile and 2 volumes of methanol, and cool to room temperature. To 270 mL of this solution in a 1-liter graduated cylinder add 20.0 mL of 25% tetramethylammonium hydroxide solution in methanol, and add water to 1 liter. Adjust with phosphoric acid to a pH of 4.0, filter, and degas.
Standard preparation—Dissolve an accurately weighed quantity of USP Metocurine Iodide RS in *Mobile phase* to obtain a solution having a known concentration of about 0.2 mg per mL.
Assay preparation—Transfer about 10 mg of Metocurine Iodide, accurately weighed, to a 50-mL volumetric flask. Dissolve in *Mobile phase,* dilute with *Mobile phase* to volume, and mix.
System suitability preparation—Dissolve suitable quantities of metocurine iodide, tubocurarine chloride, and phenol in *Mobile phase* to obtain a solution containing about 0.2 mg, 0.30 mg, and 0.5 mg per mL, respectively.
Chromatographic system (see *Chromatography* ⟨621⟩)—The liquid chromatograph is equipped with a 220-nm detector, and a 4-mm × 25-cm column that contains packing L1. The flow rate is about 1 mL per minute. Chromatograph the *System suitability preparation,* and record the peak responses as directed under *Procedure:* the resolution, *R,* between any two major peaks is not less than 2.0 and the tailing factor, *T,* for tubocurarine chloride is not more than 2.0. The relative standard deviation for replicate injections of the *Standard preparation* is not more than 2.0%. The relative retention times are about 0.18, 0.35, 0.5, and 1.0 for iodide, tubocurarine, phenol, and metocurine, respectively.
Procedure—Separately inject equal volumes (about 10 μL) of the *Standard preparation* and the *Assay preparation* into the chromatograph, record the chromatograms, and measure the responses for the major peaks. Calculate the quantity, in mg, of $C_{40}H_{48}I_2N_2O_6$ in the portion of Metocurine Iodide taken by the formula:

$$50C(r_U/r_S),$$

in which *C* is the concentration, in mg per mL, of USP Metocurine Iodide RS in the *Standard preparation,* and r_U and r_S are the peak responses obtained from the *Assay preparation* and the *Standard preparation,* respectively.

Metocurine Iodide Injection

» Metocurine Iodide Injection is a sterile solution of Metocurine Iodide in isotonic sodium chloride solution. It contains not less than 93.0 percent and not more than 107.0 percent of the labeled amount of $C_{40}H_{48}I_2N_2O_6$.

Packaging and storage—Preserve in single-dose or in multiple-dose containers, preferably of Type I glass. Phenol, 0.5%, or some other suitable bacteriostatic substance, is added to the Injection in multiple-dose containers.

Reference standard—*USP Metocurine Iodide Reference Standard*—Do not dry; determine the water content by the titrimetric method before using for quantitative analyses.

Identification—

A: Evaporate a quantity of Injection, equivalent to about 25 mg of metocurine iodide, on a steam bath to dryness. Add 5 mL of water, and again evaporate to dryness. Repeat the process with another 5 mL of water: the residue so obtained responds to *Identification test A* under *Metocurine Iodide*.

B: The chromatogram of the *Assay preparation* obtained as directed in the *Assay* exhibits a peak for Metocurine Iodide, the retention time of which corresponds to that exhibited by the *Standard preparation*.

Other requirements—It meets the requirements under *Injections* ⟨1⟩.

Assay—

Mobile phase, Standard preparation, System suitability preparation, and *Chromatographic system*—Prepare as directed in the *Assay* under *Metocurine Iodide*.

Assay preparation—Transfer an accurately measured volume of Metocurine Iodide Injection, equivalent to about 10 mg of metocurine iodide, to a 50-mL volumetric flask, dilute with *Mobile phase* to volume, and mix.

Procedure—Separately inject equal volumes (about 10 µL) of the *Standard preparation* and the *Assay preparation* into the chromatograph, record the chromatograms, and measure the responses for the major peaks. Calculate the quantity, in mg, of $C_{40}H_{48}I_2N_6O_6$ in each mL of the Metocurine Iodide Injection taken by the formula:

$$50(C/V)(r_U/r_S),$$

in which C is the concentration, in mg per mL, of USP Metocurine Iodide RS in the *Standard preparation*, V is the volume, in mL, of Injection taken, and r_U and r_S are the peak responses obtained from the *Assay preparation* and the *Standard preparation*, respectively.

Metoprolol Tartrate

$[CH_3OCH_2CH_2-C_6H_4-OCH_2CH(OH)CH_2NHCH(CH_3)_2]_2 \cdot HOOC-CH(OH)-CH(OH)-COOH$

$(C_{15}H_{25}NO_3)_2 . C_4H_6O_6$ 684.82

2-Propanol, 1-[4-(2-methoxyethyl)phenoxy]-3-[(1-methylethyl)amino]-, (±)-, [*R*-(*R**,*R**)]-2,3-dihydroxybutanedioate (2:1) (salt).

(±)-1-(Isopropylamino)-3-[*p*-(2-methoxyethyl)phenoxy]-2-propanol L-(+)-tartrate (2:1) (salt).

1-(Isopropylamino)-3-[*p*-(2-methoxyethyl)phenoxy]-2-propanol (2:1) *dextro*-tartrate salt [*56392-17-7*].

» Metoprolol Tartrate contains not less than 99.0 percent and not more than 101.0 percent of $(C_{15}H_{25}NO_3)_2 . C_4H_6O_6$, calculated on the dried basis.

Packaging and storage—Preserve in tight, light-resistant containers.

Reference standard—*USP Metoprolol Tartrate Reference Standard*—Dry in vacuum at 60° for 4 hours before using.

Identification—The infrared absorption spectrum of a mineral oil dispersion of it exhibits maxima only at the same wavelengths as that of a similar preparation of USP Metoprolol Tartrate RS.

Specific rotation ⟨781⟩: between +6.5° and +10.5°, determined at 20° in a solution having a known concentration of 200 mg in each 10 mL.

pH ⟨791⟩: between 6.0 and 7.0, in a solution (1 in 10).

Loss on drying ⟨731⟩—Dry it in vacuum at 60° for 4 hours: it loses not more than 0.5% of its weight.

Residue on ignition ⟨281⟩: not more than 0.1%.

Heavy metals, *Method I* ⟨231⟩: 0.001%.

Chromatographic purity—

Standard solution and *Standard dilutions*—Dissolve a suitable quantity of USP Metoprolol Tartrate RS, accurately weighed, in chloroform to obtain a solution containing 100 mg per mL (*Standard solution*). Dilute this solution quantitatively and stepwise with chloroform to obtain solutions having known concentrations of 1.0, 0.5, 0.2, and 0.1 mg per mL, respectively (*Standard dilutions*).

Test solution—Dissolve a quantity of Metoprolol Tartrate in chloroform to obtain a solution containing 100 mg per mL.

Chromatographic chamber—Line a suitable chamber (see *Chromatography* ⟨621⟩) with absorbent paper, and pour into the chamber 200 mL of chloroform. Place several beakers, each containing 45 mL of ammonium hydroxide, on the bottom of the chamber. Saturate the chamber for 1.5 hours before using.

Detecting reagent—Prepare separate solutions of potassium iodide (1 in 100) and soluble starch (prepared by triturating 3 g in 10 mL of cold water and adding the mixture to 90 mL of boiling water with constant stirring). Just prior to use, mix 10 mL of each solution with 3 mL of alcohol.

Procedure—On a suitable thin-layer chromatographic plate (see *Chromatography* ⟨621⟩), coated with a 0.25-mm layer of chromatographic silica gel mixture, separately apply 5-µL portions of the *Test solution* and each of the *Standard dilutions*. Place the plate in the *Chromatographic chamber*, seal the chamber, and allow the chromatogram to develop until the solvent front has moved about three-fourths of the length of the plate. Remove the plate, and dry in a current of warm air until the odor of ammonia is no longer perceptible (about 45 minutes). Place a beaker containing 0.5 g of potassium permanganate in a chamber. Add 5 mL of 6 *N* hydrochloric acid to the beaker and allow to equilibrate for 5 minutes. Place the plate in the chamber for 5 minutes. Remove the plate from the chamber, allow to stand in a current of cool air for 1 hour, and spray with *Detecting reagent*. If spots other than the principal spot are observed in the lane of the *Test solution*, estimate the concentration of each by comparison with the *Standard dilutions:* the spots from the 1.0, 0.5, 0.2, and 0.1 mg per mL *Standard dilutions* correspond to 1.0, 0.5, 0.2, and 0.1% of impurities, respectively. The sum of any observed impurities in the *Test solution* is not greater than 1.0%.

Assay—Dissolve about 280 mg of Metoprolol Tartrate, accurately weighed, in 20 mL of glacial acetic acid, and titrate with 0.1 *N* perchloric acid VS, determining the end-point potentiometrically, using a glass electrode and a calomel electrode containing glacial acetic acid saturated with lithium chloride (see *Titrimetry* ⟨541⟩). Perform a blank determination, and make any necessary correction. Each mL of 0.1 *N* perchloric acid is equivalent to 34.24 mg of $(C_{15}H_{25}NO_3)_2 . C_4H_6O_6$.

Metoprolol Tartrate Injection

» Metoprolol Tartrate Injection is a sterile solution of Metoprolol Tartrate in Water for Injection. It contains Sodium Chloride as a tonicity-adjusting agent. It contains not less than 90.0 percent and not more than 110.0 percent of the labeled amount of $(C_{15}H_{25}NO_3)_2 . C_4H_6O_6$.

Packaging and storage—Preserve in single-dose light-resistant containers, preferably of Type I or Type II glass.

Reference standards—*USP Metoprolol Tartrate Reference Standard*—Dry in vacuum at 60° for 4 hours before using. *USP Oxprenolol Hydrochloride Reference Standard*—Dry in vacuum at 60° for 6 hours before using.

Identification—Place a volume of Injection, equivalent to about 40 mg of metoprolol tartrate, in a separator, add 4 mL of dilute ammonium hydroxide (1 in 3), and extract with 20 mL of chloroform, filtering the chloroform extract through chloroform-prerinsed anhydrous sodium sulfate. Evaporate the chloroform to dryness, and place in a freezer to congeal the residue: the infrared absorption spectrum of a potassium bromide dispersion of the residue so obtained exhibits maxima only at the same wavelengths as that of a similar preparation of USP Metoprolol Tartrate RS.

pH ⟨791⟩: between 5.0 and 8.0.

Sterility—It meets the requirements under *Sterility Tests* ⟨71⟩, when tested as directed in the section, *Test Procedures Using Membrane Filtration.*

Other requirements—It meets the requirements under *Injections* ⟨1⟩.

Assay—

Mobile phase, Internal standard solution, and *Chromatographic system*—Prepare as directed in the *Assay* under *Metoprolol Tablets.*

Sodium chloride solution—Dissolve 9.0 g of sodium chloride in water to make 1000 mL.

Standard preparation—Dissolve an accurately weighed quantity of USP Metoprolol Tartrate RS in *Sodium chloride solution* to obtain a stock solution having a known concentration of about 1000 µg per mL. Mix equal volumes, accurately measured, of this stock solution and of *Internal standard solution.*

Assay preparation—Dilute an accurately measured volume of Metoprolol Tartrate Injection, if necessary, quantitatively with *Sodium chloride solution* to obtain a stock solution having a concentration of about 1000 µg per mL. Mix equal volumes, accurately measured, of this stock solution and of *Internal standard solution.*

Procedure—Separately inject equal volumes (about 10 µL) of the *Standard preparation* and the *Assay preparation* into the chromatograph, record the chromatograms, and measure the responses for the major peaks. The relative retention times are about 0.8 for metoprolol tartrate and 1.0 for oxprenolol hydrochloride. Calculate the quantity, in mg, of $(C_{15}H_{25}NO_3)_2 \cdot C_4H_6O_6$ in each mL of the Injection taken by the formula:

$$(L/D)(C)(R_U/R_S),$$

in which L is the labeled quantity, in mg, of metoprolol tartrate in the Injection, D is the concentration, in µg per mL, of metoprolol tartrate in the *Assay preparation,* on the basis of the labeled quantity in each mL of Injection taken and the extent of dilution, C is the concentration, in µg per mL, of USP Metoprolol Tartrate RS in the *Standard preparation,* and R_U and R_S are the peak response ratios of metoprolol tartrate to oxprenolol hydrochloride obtained from the *Assay preparation* and the *Standard preparation,* respectively.

Metoprolol Tartrate Tablets

» Metoprolol Tartrate Tablets contain not less than 90.0 percent and not more than 110.0 percent of the labeled amount of $(C_{15}H_{25}NO_3)_2 \cdot C_4H_6O_6$.

Packaging and storage—Preserve in tight, light-resistant containers.

Reference standards—*USP Metoprolol Tartrate Reference Standard*—Dry in vacuum at 60° for 4 hours before using. *USP Oxprenolol Hydrochloride Reference Standard*—Dry in vacuum at 60° for 6 hours before using.

Identification—Place a quantity of finely powdered Tablets, equivalent to about 40 mg of metoprolol tartrate, in a separator, add 25 mL of water and 4 mL of dilute ammonium hydroxide (1 in 3), and extract with 20 mL of chloroform, filtering the chloroform extract through chloroform-prerinsed anhydrous sodium sulfate. Evaporate the chloroform to dryness, and place in a freezer to congeal the residue: the infrared absorption spectrum of a potassium bromide dispersion of the residue so obtained exhibits maxima only at the same wavelengths as that of a similar preparation of USP Metoprolol Tartrate RS.

Dissolution ⟨711⟩—

Medium: simulated gastric fluid TS (without enzyme); 900 mL.

Apparatus 1: 100 rpm.

Time: 30 minutes.

Procedure—Determine the amount of $(C_{15}H_{25}NO_3)_2 \cdot C_4H_6O_6$ dissolved in filtered portions of the solution under test from ultraviolet absorbances at the wavelength of maximum absorbance at about 275 nm in comparison with a Standard solution having a known concentration of USP Metoprolol Tartrate RS in the same medium.

Tolerances—Not less than 75% (Q) of the labeled amount of $(C_{15}H_{25}NO_3)_2 \cdot C_4H_6O_6$ is dissolved in 30 minutes.

Uniformity of dosage units ⟨905⟩: meet the requirements.

Procedure for content uniformity—Transfer 1 Tablet to a suitable volumetric flask, add 0.1 *N* hydrochloric acid to fill to about 60% of the nominal volume, sonicate for 15 minutes, and shake by mechanical means for 30 minutes. Dilute with 0.1 *N* hydrochloric acid to volume, to obtain a final solution having a concentration of approximately 1000 µg per mL. Mix, and filter, discarding the first 20 mL of the filtrate. Pipet 10 mL of filtrate, 10 mL of a Standard solution of USP Metoprolol Tartrate RS in the same medium having a known concentration of about 1000 µg per mL, and 10 mL of 0.1 *N* hydrochloric acid to provide a blank into individual separators. To each separator add 2.0 mL of 2.5 *N* sodium hydroxide, and extract with three 25-mL portions of chloroform, passing the chloroform extracts through chloroform-prerinsed glass wool into individual 100-mL volumetric flasks. Dilute with chloroform to volume, and mix. Determine the absorbances of the solutions in 1-cm cells at the wavelength of maximum absorbance at about 276 nm, with a suitable spectrophotometer, using the blank to set the instrument. Calculate the quantity, in mg, of $(C_{15}H_{25}NO_3)_2 \cdot C_4H_6O_6$ in the Tablet by the formula:

$$(T/1000)C(A_U/A_S),$$

in which T is the labeled quantity, in mg, of metoprolol tartrate in the Tablet, C is the concentration, in µg per mL, of USP Metoprolol Tartrate RS in the Standard solution, and A_U and A_S are the absorbances of the solution from the Tablet and the Standard solution, respectively.

Assay—

Mobile phase—Prepare a suitable and degassed solution by dissolving 961 mg of 1-pentanesulfonic acid sodium salt (monohydrate) and 82 mg of anhydrous sodium acetate in a mixture of 550 mL of methanol and 470 mL of water and adding 0.57 mL of glacial acetic acid.

Internal standard solution—Dissolve USP Oxprenolol Hydrochloride RS in freshly prepared *Mobile phase* to obtain a solution containing about 720 µg per mL.

Standard preparation—Dissolve an accurately weighed quantity of USP Metoprolol Tartrate RS in freshly prepared *Internal standard solution* to obtain a solution having a known concentration of about 1000 µg per mL. Transfer 50.0 mL of this solution to a 100-mL volumetric flask, dilute with methanol to volume, and mix.

Assay preparation—Weigh and finely powder not less than 20 Metoprolol Tartrate Tablets. Transfer an accurately weighed portion of the powder, equivalent to about 50 mg of metoprolol tartrate, to a 50-mL volumetric flask, add 30 mL of methanol, heat on a steam bath, and sonicate for 10 minutes. Allow the solution to cool to room temperature, dilute with methanol to volume, and mix. Centrifuge a portion of the solution, and transfer 25.0 mL of the supernatant liquid to a suitable flask, add 25.0 mL of *Internal standard solution,* mix, and allow to come to room temperature. Filter a portion of this solution through a membrane filter (0.5 µm) discarding the first few mL of the filtrate.

Chromatographic system (see *Chromatography* ⟨621⟩)—The liquid chromatograph is equipped with a 254-nm detector and a 3.9-mm × 30-cm column that contains packing L1. The flow rate is about 1 mL per minute. Chromatograph three replicate injections of the *Standard preparation,* and record the peak responses as directed under *Procedure:* the relative standard deviation is not more than 2.0%, and the resolution factor between metoprolol tartrate and oxprenolol hydrochloride is not less than 2.0.

Procedure—Separately inject equal volumes (about 10 µL) of the *Standard preparation* and the *Assay preparation* into the chromatograph, record the chromatograms, and measure the responses for the major peaks. The relative retention times are about 0.8 for metoprolol tartrate and 1.0 for oxprenolol hydrochloride. Calculate the quantity, in mg, of $(C_{15}H_{25}NO_3)_2 \cdot C_4H_6O_6$ in the portion of Tablets taken by the formula:

$$0.1C(R_U/R_S),$$

in which C is the concentration, in µg per mL, of USP Metoprolol Tartrate RS in the *Standard preparation*, and R_U and R_S are the peak response ratios of metoprolol tartrate to oxprenolol hydrochloride obtained from the *Assay preparation* and the *Standard preparation*, respectively.

Metoprolol Tartrate and Hydrochlorothiazide Tablets

» Metoprolol Tartrate and Hydrochlorothiazide Tablets contain not less than 90.0 percent and not more than 110.0 percent of the labeled amounts of metoprolol tartrate $[(C_{15}H_{25}NO_3)_2 \cdot C_4H_6O_6]$ and hydrochlorothiazide $(C_7H_8ClN_3O_4S_2)$.

Packaging and storage—Preserve in tight, light-resistant containers.

Reference standards—*USP Metoprolol Tartrate Reference Standard*—Dry in vacuum at 60° for 4 hours before using. *USP Oxprenolol Hydrochloride Reference Standard*—Dry in vacuum at 60° for 6 hours before using. *USP Hydrochlorothiazide Reference Standard*—Dry at 105° for 1 hour before using. *USP 4-Amino-6-chloro-1,3-benzenedisulfonamide Reference Standard*—Keep container tightly closed and protected from light. Dry over silica gel for 4 hours before using.

Identification—

A: Place a quantity of finely powdered Tablets, equivalent to about 100 mg of metoprolol tartrate, in a 50-mL volumetric flask. Add about 30 mL of 0.1 *N* sodium hydroxide, shake for 20 minutes, dilute with 0.1 *N* sodium hydroxide to volume, and mix. Filter a portion of this mixture, discarding the first 10 mL of the filtrate. Place 25 mL of the filtrate into a separator, and extract with three 15-mL portions of chloroform, filtering the chloroform extracts through chloroform-prerinsed anhydrous sodium sulfate, and combining the extracts in a suitable container. [NOTE—Retain the aqueous layer remaining after extraction for *Identification test B*.] Evaporate the chloroform to dryness, and place in a freezer to congeal the residue: the infrared absorption spectrum of a potassium bromide dispersion of the residue so obtained exhibits maxima only at the same wavelengths as that of a similar preparation of USP Metoprolol Tartrate RS.

B: Pass the aqueous layer from *Identification test A* through 0.1 *N* sodium hydroxide–prerinsed cotton. Dilute a portion of the filtrate quantitatively and stepwise with 0.1 *N* sodium hydroxide to obtain a solution having a concentration of about 0.01 mg of hydrochlorothiazide per mL: the ultraviolet absorption spectrum of this solution exhibits maxima and minima at the same wavelengths as a Standard solution prepared as follows: Dissolve 25 mg of USP Hydrochlorothiazide RS in 50 mL of 0.1 *N* sodium hydroxide in a separator, and extract with three 15-mL portions of chloroform. Discard the chloroform extracts, and pass the aqueous solution through 0.1 *N* sodium hydroxide–prerinsed cotton. Pipet 2 mL of the filtrate into a 100-mL volumetric flask, dilute with 0.1 *N* sodium hydroxide to volume, and mix.

Dissolution ⟨711⟩—

Medium: simulated gastric fluid TS (without enzyme); 900 mL.

Apparatus 1: 100 rpm.

Time: 30 minutes.

Determination of dissolved metoprolol tartrate—Remove about 125 mL of the solution under test, allow to cool to room temperature, and filter, discarding the first 25 mL of the filtrate. [NOTE—Retain about 30 mL of the remaining filtrate of the solution under test for the *Determination of dissolved hydrochlorothiazide*.] If necessary, quantitatively dilute a portion of the filtrate with fresh *Dissolution Medium* to obtain a solution having a concentration of about 0.05 mg of metoprolol tartrate per mL. Transfer to separate separators 50.0 mL of the filtrate, 50.0 mL of a Standard solution in *Dissolution Medium* having a known concentration of about 0.05 mg of USP Metoprolol Tartrate RS per mL, and 50.0 mL of *Dissolution Medium* to provide the blank. Add 10 mL of 2.5 *N* sodium hydroxide to each separator, and extract each with three 15-mL portions of chloroform, filtering the chloroform extracts through pledgets of chloroform-prerinsed glass wool into individual 50-mL volumetric flasks. Dilute the contents of each flask with chloroform to volume, and mix. Determine the absorbances of the solutions from the filtrate and the Standard solution in 2-cm cells at the wavelength of maximum absorbance at about 276 nm, with a suitable spectrophotometer, using the solution from the blank to set the instrument. Calculate the quantity, in mg, of $(C_{15}H_{25}NO_3)_2 \cdot C_4H_6O_6$ dissolved by the formula:

$$900Cf(A_U/A_S),$$

in which C is the concentration, in mg per mL, of USP Metoprolol Tartrate RS in the Standard solution, f is the dilution factor for the solution from the filtrate, and A_U and A_S are the absorbances of the solution from the filtrate and of the Standard solution, respectively.

Determination of dissolved hydrochlorothiazide—Filter a portion of the filtrate retained from the *Determination of dissolved metoprolol tartrate* through a filter of 0.8-µm or finer porosity, discarding the first 5 mL of the filtrate. If necessary, quantitatively dilute a portion of the filtrate with fresh *Dissolution Medium* to obtain a solution having a concentration of about 0.03 mg of hydrochlorothiazide per mL. Prepare a Standard solution in *Dissolution Medium* having a known concentration of about 0.03 mg of USP Hydrochlorothiazide RS per mL. Determine the absorbances of these solutions in 2-cm cells at the wavelength of maximum absorbance at about 316 nm, using *Dissolution Medium* as the blank. Calculate the quantity, in mg, of $C_7H_8ClN_3O_4S_2$ dissolved by the formula:

$$900Cf(A_U/A_S),$$

in which C is the concentration, in mg per mL, of USP Hydrochlorothiazide RS in the Standard solution, f is the dilution factor for the solution from the filtrate, and A_U and A_S are the absorbances of the solution from the filtrate and of the Standard solution, respectively.

Tolerances—Not less than 75% (*Q*) of the labeled amount of metoprolol tartrate $[(C_{15}H_{25}NO_3)_2 \cdot C_4H_6O_6]$ and not less than 60% (*Q*) of the labeled amount of hydrochlorothiazide $(C_7H_8ClN_3O_4S_2)$ are dissolved in 30 minutes.

Uniformity of dosage units ⟨905⟩: meet the requirements for *Content Uniformity* with respect to metoprolol tartrate and to hydrochlorothiazide.

Procedure for content uniformity of metoprolol tartrate—Using 1 Metoprolol Tartrate and Hydrochlorothiazide Tablet, proceed as directed for *Procedure for content uniformity* in the test for *Uniformity of dosage units* under *Metoprolol Tartrate Tablets*.

Procedure for content uniformity of hydrochlorothiazide—Transfer 1 Tablet to a 100-mL volumetric flask containing 15 mL of water, and shake by mechanical means for 15 minutes. Add 60 mL of methanol, sonicate for 5 minutes, and shake by mechanical means for 30 minutes. Dilute with methanol to volume, and mix. Centrifuge 40 mL of this suspension. Dilute an accurately measured portion of the supernatant liquid quantitatively with methanol to obtain a solution having a concentration of about 0.05 mg of hydrochlorothiazide per mL. Filter a portion of this solution through a 0.5-µm porosity filter, discarding the first few mL of the filtrate. Use the filtrate as the test solution. Transfer about 25 mg of USP Hydrochlorothiazide RS, accurately weighed, to a 100-mL volumetric flask containing 15 mL of water, dilute with methanol to volume, and mix. Transfer 10.0 mL of this solution to a 50-mL volumetric flask, dilute with methanol to volume, and mix (Standard solution). Concomitantly determine the absorbances of the test solution and the Standard solution at the wavelength of maximum absorbance at about 316 nm, with a suitable spectrophotometer, using methanol as the blank. Calculate the quantity, in mg, of hydrochlorothiazide $(C_7H_8ClN_3O_4S_2)$ in the Tablet by the formula:

$$(LC/D)(A_U/A_S),$$

in which L is the labeled quantity, in mg, of hydrochlorothiazide in the Tablet, C is the concentration, in mg per mL, of USP Hydrochlorothiazide RS in the Standard solution, D is the concentration, in mg per mL, of hydrochlorothiazide in the test solution based on the labeled quantity in the Tablet and the extent of dilution, and A_U and A_S are the absorbances of the test solution and the Standard solution, respectively.

Diazotizable substances—

Standard preparation—Weigh accurately 5 mg of USP 4-Amino-6-chloro-1,3-benzenedisulfonamide RS, and dissolve in 2 mL of methanol contained in a 50-mL volumetric flask. Dilute with water to volume, and mix. Pipet 5 mL of the resulting solution into a 100-mL volumetric flask, add 20 mL of methanol, dilute with water to volume, and mix. Each mL of *Standard preparation* contains 5 µg of the reference standard.

Test preparation—Transfer a portion of the powdered Tablets prepared for the *Assay*, accurately weighed and equivalent to 50 mg of hydrochlorothiazide, to a 100-mL volumetric flask containing a mixture of 20 mL of methanol and 20 mL of water. Shake continuously for 5 to 10 minutes, dilute with water to volume, mix, and filter. Use the filtrate as the *Test preparation* immediately after preparation.

Procedure—Pipet 5 mL each of the *Standard preparation* and the *Test preparation* into separate, 50-mL volumetric flasks. Pipet 5 mL of water into a third 50-mL volumetric flask to provide a blank. To each flask add 1 mL of freshly prepared sodium nitrite solution (1 in 100) and 5 mL of dilute hydrochloric acid (1 in 10), and allow to stand for 5 minutes. Add 2 mL of ammonium sulfamate solution (1 in 50), allow to stand for 5 minutes with frequent swirling, then add 2 mL of freshly prepared disodium chromotropate solution (1 in 100) and 10 mL of sodium acetate TS. Dilute with water to volume, and mix. Concomitantly determine the absorbances of the solutions obtained from the *Standard preparation* and the *Test preparation* at 500 nm, with a suitable spectrophotometer, against the blank. The absorbance of the solution from the *Test preparation* does not exceed that of the solution from the *Standard preparation*, corresponding to not more than 1.0% of diazotizable substances.

Assay for metoprolol tartrate—

Mobile phase and *Chromatographic system*—Proceed as directed in the *Assay* under *Metoprolol Tartrate Tablets*.

Internal standard solution—Dissolve USP Oxprenolol Hydrochloride RS in freshly prepared *Mobile phase* to obtain a solution containing about 360 µg per mL.

Standard preparation—Dissolve an accurately weighed quantity of USP Metoprolol Tartrate RS in 0.1 *N* hydrochloric acid to obtain a stock solution having a known concentration of about 1000 µg per mL. Transfer 10.0 mL of this stock solution to a suitable separator, add 2.0 mL of 2.5 *N* sodium hydroxide, and extract with three 25-mL portions of chloroform. Pass the chloroform extracts through chloroform-prerinsed glass wool into a round-bottom flask, and evaporate on a rotary evaporator under vacuum to dryness. Add 20.0 mL of *Internal standard solution* to the flask, sonicate for 3 minutes, and gently swirl to dissolve the residue in the flask.

Assay preparation—Weigh and finely powder not less than 20 Metoprolol Tartrate and Hydrochlorothiazide Tablets. Transfer an accurately weighed portion of the powder, equivalent to about 100 mg of metoprolol tartrate, to a 100-mL volumetric flask, add 60-mL of 0.1 *N* hydrochloric acid, heat on a steam bath for 3 minutes, and sonicate for 5 minutes. Shake for 30 minutes. Allow the solution to cool to room temperature, dilute with 0.1 *N* hydrochloric acid to volume, and mix. Filter a portion of this solution, discarding the first 20 mL of the filtrate. Transfer 10.0 mL of the filtrate to a separator, add 2.0 mL of 2.5 *N* sodium hydroxide, and extract with three 25-mL portions of chloroform. Pass the chloroform extracts through chloroform-prerinsed glass wool into a round-bottom flask, and evaporate on a rotary evaporator under vacuum to dryness. Add 20.0 mL of *Internal standard solution* to the flask, sonicate for 3 minutes, and gently swirl to dissolve the residue in the flask. Filter a portion of this solution through a filter of 0.5-µm or finer porosity, discarding the first few mL of the filtrate. Use the filtrate as the *Assay preparation*.

Procedure—Proceed as directed for *Procedure* in the *Assay* under *Metoprolol Tartrate Tablets*. Calculate the quantity, in mg, of metoprolol tartrate $[(C_{15}H_{25}NO_3)_2 . C_4H_6O_6]$ in the portion of Tablets taken by the formula:

$$0.1C(R_U/R_S),$$

in which C is the concentration, in µg per mL, of USP Metoprolol Tartrate RS in the stock solution used to prepare the *Standard preparation*, and R_U and R_S are the peak response ratios of metoprolol tartrate to oxprenolol hydrochloride obtained from the *Assay preparation* and the *Standard preparation*, respectively.

Assay for hydrochlorothiazide—

Mobile phase—[NOTE—Filter the methanol and water through 0.5-µm porosity filters before use.] Dissolve 1.38 g of monobasic sodium phosphate in 780 mL of water, add 220 mL of methanol, and mix. Degas before use. Make adjustments if necessary (see *System Suitability* under *Chromatography* ⟨621⟩).

Internal standard solution—Dissolve a quantity of sulfanilamide in methanol to obtain a solution containing about 0.4 mg per mL.

System suitability solution—Dissolve a quantity of USP 4-Amino-6-chloro-1,3-benzenedisulfonamide RS in *Internal standard solution* to obtain a solution containing about 1 mg per mL. Mix 1 mL of this solution and 4 mL of methanol.

Standard preparation—Transfer about 50 mg of USP Hydrochlorothiazide RS, accurately weighed, to a 100-mL volumetric flask, add 20.0 mL of *Internal standard solution*, dilute with methanol to volume, and mix.

Assay preparation—Weigh and finely powder not less than 20 Metoprolol Tartrate and Hydrochlorothiazide Tablets. Transfer an accurately weighed portion of the powder, equivalent to about 25 mg of hydrochlorothiazide, to a 50-mL volumetric flask. Add 10.0 mL of *Internal standard solution* and 20 mL of methanol, and sonicate for 5 minutes. Shake by mechanical means for 30 minutes, dilute with methanol to volume, and mix. Centrifuge a portion of this solution, and filter a portion of the supernatant liquid through a 0.5-µm porosity filter, discarding the first few mL of the filtrate. Use the filtrate as the *Assay preparation*.

Chromatographic system (see *Chromatography* ⟨621⟩)—The liquid chromatograph is equipped with a 254-nm detector and a 3.9-mm × 30-cm column that contains packing L1. The flow rate is about 0.6 mL per minute. Chromatograph replicate injections of the *System suitability solution*, and record the peak responses as directed under *Procedure:* the relative retention times are about 0.7 for sulfanilamide and 1.0 for 4-amino-6-chloro-1,3-benzenedisulfonamide, and the resolution *R*, between the peaks is not less than 2.0. Chromatograph the *Standard preparation*, and record the peak responses as directed under *Procedure:* the relative standard deviation for replicate injections is not more than 2.0%.

Procedure—Separately inject equal volumes (about 4 µL) of the *Standard preparation* and the *Assay preparation* into the chromatograph, record the chromatograms, and measure the responses for the major peaks. The relative retention times are about 0.6 for sulfanilamide and 1.0 for hydrochlorothiazide. Calculate the quantity, in mg, of hydrochlorothiazide ($C_7H_8ClN_3O_4S_2$) in the portion of Tablets taken by the formula:

$$50C(R_U/R_S),$$

in which C is the concentration, in mg per mL, of USP Hydrochlorothiazide RS in the *Standard preparation*, and R_U and R_S are the peak response ratios of hydrochlorothiazide to sulfanilamide obtained from the *Assay preparation* and the *Standard preparation*, respectively.

Metronidazole

$C_6H_9N_3O_3$ 171.16

1*H*-Imidazole-1-ethanol, 2-methyl-5-nitro-.

2-Methyl-5-nitroimidazole-1-ethanol [*443-48-1*].

» Metronidazole contains not less than 99.0 percent and not more than 101.0 percent of $C_6H_9N_3O_3$, calculated on the dried basis.

Packaging and storage—Preserve in well-closed, light-resistant containers.

Reference standard—*USP Metronidazole Reference Standard*—Dry at 105° for 2 hours before using.

Identification—

A: The infrared absorption spectrum of a potassium bromide dispersion of it exhibits maxima only at the same wavelengths as that of a similar preparation of USP Metronidazole RS.

B: The ultraviolet absorption spectrum of a 1 in 50,000 solution of it in a 1 in 350 solution of sulfuric acid in methanol exhibits maxima and minima at the same wavelengths as that of a similar solution of USP Metronidazole RS, concomitantly measured.

Melting range ⟨741⟩: between 159° and 163°.

Loss on drying ⟨731⟩—Dry it at 105° for 2 hours: it loses not more than 0.5% of its weight.

Residue on ignition ⟨281⟩: not more than 0.1%.

Heavy metals, *Method II* ⟨231⟩: 0.005%.

Non-basic substances—A 1-g portion dissolves completely in 10 mL of dilute hydrochloric acid (1 in 2).

Chromatographic purity—Dissolve 100 mg of Metronidazole in 10.0 mL of acetone. Similarly prepare a Standard solution of USP Metronidazole RS in acetone having a concentration of 3.0 mg per mL. Dilute an aliquot of this Standard solution quantitatively and stepwise with acetone to obtain a solution having a concentration of 30 µg per mL (diluted Standard solution). On a suitable thin-layer chromatographic plate (see *Chromatography* ⟨621⟩), coated with a 0.25-mm layer of chromatographic silica gel, apply separately 20-µL portions of the test solution, the Standard solution, and the diluted Standard solution. Allow the spots to dry, and develop the chromatogram in a solvent system consisting of a mixture of chloroform, dehydrated alcohol, diethylamine, and water (80:10:10:1) until the solvent front has moved about three-fourths of the length of the plate. Remove the plate from the developing chamber, mark the solvent front, and allow the solvent to evaporate. Spray the plate with titanium trichloride (20% solution), heat at 110° until the blue-gray color begins to disappear, and cool the plate. Spray (*Caution—Use fast blue salt B spray with adequate ventilation, avoid inhalation of vapors, and avoid contact with the skin*) the plate with fast blue salt B solution (1 in 100), allow to stand for 3 minutes, and spray with a mixture of alcohol, water, and ammonium hydroxide (50:30:20): the R_f value of the principal spot obtained from the test solution corresponds to that obtained from the Standard solution, and no spot, other than the principal spot, obtained from the test solution is larger or more intense than the principal spot obtained from the diluted Standard solution.

Assay—Dissolve about 100 mg of Metronidazole, accurately weighed, in 20 mL of acetic anhydride, warming slightly to effect solution. Cool, add 1 drop of malachite green TS, and titrate with 0.1 *N* perchloric acid VS from a 10-mL microburet to a yellow-green end-point. Perform a blank determination, and make any necessary correction. Each mL of 0.1 *N* perchloric acid is equivalent to 17.12 mg of $C_6H_9N_3O_3$.

Metronidazole Injection

» Metronidazole Injection is a sterile, isotonic, buffered solution of Metronidazole in Water for Injection. It contains not less than 90.0 percent and not more than 110.0 percent of the labeled amount of $C_6H_9N_3O_3$.

Packaging and storage—Preserve in single-dose containers of Type I or Type II glass, or in suitable plastic containers, protected from light.

Reference standard—*USP Metronidazole Reference Standard*—Dry at 105° for 2 hours before using.

Identification—

A: On a suitable thin-layer chromatographic plate (see *Chromatography* ⟨621⟩), coated with a 0.25-mm layer of chromatographic silica gel mixture, apply a measured volume of Injection containing 0.025 mg of metronidazole and a measured volume of a solution of USP Metronidazole RS containing 0.025 mg of metronidazole. Allow the spots to dry, and develop the chromatogram in a solvent system consisting of a mixture of chloroform, methanol, water, and ammonium hydroxide (70:28:4:2) until the solvent front has moved about three-fourths of the length of the plate. Remove the plate from the developing chamber, mark the solvent front, and allow the solvent to evaporate. Locate the spots on the plate by viewing under short-wavelength ultraviolet light: the R_f value of the principal spot obtained from the test solution corresponds to that obtained from the Standard solution.

B: The retention time of the major peak obtained from the *Assay preparation* corresponds to that obtained from the *Standard preparation*, obtained as directed in the *Assay*.

Pyrogen—It meets the requirements of the *Pyrogen Test* ⟨151⟩, the test dose being the equivalent of 15 mg of metronidazole per kg.

pH ⟨791⟩: between 4.5 and 7.0.

Particulate matter ⟨788⟩: meets the requirements under *Small-volume Injections*.

Other requirements—It meets the requirements under *Injections* ⟨1⟩.

Assay—

Mobile phase—Prepare a suitable filtered and degassed mixture of monobasic potassium phosphate, prepared by dissolving 0.68 g of monobasic potassium phosphate in 930 mL of water, and methanol (930:70), and adjust with 1 *M* phosphoric acid to a pH of 4.0 ± 0.5.

Standard preparation—Transfer about 25 mg of USP Metronidazole RS, accurately weighed, to a 25-mL volumetric flask, dissolve in methanol, dilute with methanol to volume, and mix. Pipet 2 mL of this solution into a 10-mL volumetric flask containing 2 mL of water, dilute with *Mobile phase* to volume, and mix.

Assay preparation—Transfer an accurately measured volume of Metronidazole Injection, equivalent to about 25 mg of metronidazole, to a 25-mL volumetric flask, dilute with water to volume, and mix. Pipet 2 mL of this solution into a 10-mL volumetric flask containing 2 mL of methanol, dilute with *Mobile phase* to volume, and mix.

Chromatographic system (see *Chromatography* ⟨621⟩)—The liquid chromatograph is equipped with a 320-nm detector and a 4.6-mm × 25-cm column that contains packing L1. The flow rate is about 2.0 mL per minute. Chromatograph five replicate injections of the *Standard preparation*, and record the peak responses as directed under *Procedure:* the relative standard deviation is not more than 2.0%, and the tailing factor is not greater than 2.0.

Procedure—Separately inject equal volumes (about 20 µL) of the *Standard preparation* and the *Assay preparation* into the chromatograph, record the chromatograms, and measure the responses for the major peaks. Calculate the quantity, in mg, of $C_6H_9N_3O_3$ in each mL of the Injection taken by the formula:

$$125C/V(r_U/r_S),$$

in which C is the concentration, in mg per mL, of USP Metronidazole RS in the *Standard preparation*, V is the volume, in mL, of Injection taken, and r_U and r_S are the peak responses for metronidazole obtained from the *Assay preparation* and the *Standard preparation*, respectively.

Metronidazole Tablets

» Metronidazole Tablets contain not less than 90.0 percent and not more than 110.0 percent of the labeled amount of $C_6H_9N_3O_3$.

Packaging and storage—Preserve in well-closed, light-resistant containers.

Reference standard—*USP Metronidazole Reference Standard*—Dry at 105° for 2 hours before using.

Identification—

A: To a portion of powdered Tablets, equivalent to about 300 mg of metronidazole, add 20 mL of dilute hydrochloric acid (1 in 100), shake for several minutes, and filter: suitable aliquots of the filtrate respond to *Identification test B* under *Metronidazole*.

B: The retention time of the major peak in the chromatogram of the *Assay preparation* corresponds to that of the *Standard preparation*, as obtained in the *Assay*.

Dissolution ⟨711⟩—

Medium: 0.1 *N* hydrochloric acid; 900 mL.

Apparatus 1: 100 rpm.

Time: 60 minutes.

Procedure—Determine the amount of $C_6H_9N_3O_3$ dissolved from ultraviolet absorbances at the wavelength of maximum absorbance at about 274 nm of filtered portions of the solution under test, suitably diluted with 0.1 *N* hydrochloric acid, in comparison with a Standard solution having a known concentration of USP Metronidazole RS in the same medium.

Tolerances—Not less than 85% (*Q*) of the labeled amount of $C_6H_9N_3O_3$ is dissolved in 60 minutes.

Uniformity of dosage units ⟨905⟩: meet the requirements.

Procedure for content uniformity—Using a mortar and pestle, triturate a Tablet with 40 mL of methanol. Transfer the mixture, with the aid of methanol, to a 250-mL volumetric flask. Add methanol until there is approximately 100 mL in the flask. Shake for 15 minutes, dilute with methanol to volume, and mix. Filter, discarding the first 15 mL of the filtrate. Dilute the filtrate quantitatively with methanol, to obtain a solution having a concentration of about 0.2 mg of metronidazole per mL. Pipet 10 mL of this solution into a 100-mL volumetric flask, add 1 mL of 1 *N* hydrochloric acid, dilute with methanol to volume, and mix. Concomitantly determine the absorbance of this Test solution and that of a similarly prepared Standard solution of USP Metronidazole RS, having a known concentration of about 20 µg per mL, in 1-cm matched cells, at the wavelength of maximum absorbance at about 274 nm, with a suitable spectrophotometer, using methanol as the blank. Calculate the quantity, in mg, of $C_6H_9N_3O_3$ in the Tablet by the formula:

$$(TC/D)(A_U/A_S),$$

in which *T* is the labeled quantity, in mg, of the metronidazole in the Tablet; *C* is the concentration, in µg per mL, of USP Metronidazole RS in the Standard solution; *D* is the concentration, in µg per mL, of metronidazole in the Test solution, on the basis of the labeled quantity per Tablet and the extent of dilution; and A_U and A_S are the absorbances of the Test solution and the Standard solution, respectively.

Assay—

Mobile phase—Prepare a filtered and degassed mixture of water and methanol (80:20), making adjustments if necessary (see *System Suitability* under *Chromatography* ⟨621⟩).

Standard preparation—Dissolve an accurately weighed quantity of USP Metronidazole RS in *Mobile phase* to obtain a solution having a known concentration of about 0.5 mg per mL.

Assay preparation—Weigh and finely powder not less than 10 Metronidazole Tablets. Transfer an accurately weighed portion of the powder, equivalent to about 1 g of metronidazole, to a 100-mL volumetric flask. Add about 50 mL of methanol, and sonicate to dissolve. Dilute with methanol to volume, and mix. Allow the solution to stand until the insoluble material has settled. Pipet 5.0 mL of the clear supernatant liquid into a 100-mL volumetric flask, dilute with *Mobile phase* to volume, and mix. Filter the solution.

Chromatographic system (see *Chromatography* ⟨621⟩)—The liquid chromatograph is equipped with a 254-nm detector and a 4.6-mm × 15-cm column that contains packing L7. The flow rate is about 1 mL per minute. Chromatograph the *Standard preparation*, and record the peak response as directed under *Procedure:* the tailing factor is not more than 5 and the relative standard deviation for replicate injections is not more than 2.0%.

Procedure—Separately inject equal volumes (about 20 µL) of the *Standard preparation* and the *Assay preparation* into the chromatograph, record the chromatograms, and measure the peak responses. Calculate the quantity, in mg, of $C_6H_9N_3O_3$ in the portion of Tablets taken by the formula:

$$2000C(r_U/r_S),$$

in which *C* is the concentration, in mg per mL, of USP Metronidazole RS in the *Standard preparation*, and r_U and r_S are the metronidazole peak responses obtained from the *Assay preparation* and the *Standard preparation*, respectively.

Metyrapone

$C_{14}H_{14}N_2O$ 226.28

1-Propanone, 2-methyl-1,2-di-3-pyridinyl-.

2-Methyl-1,2-di-3-pyridyl-1-propanone [*54-36-4*].

» Metyrapone contains not less than 98.0 percent and not more than 102.0 percent of $C_{14}H_{14}N_2O$, calculated on the dried basis.

Packaging and storage—Preserve in tight containers, protected from heat and light.

Reference standard—*USP Metyrapone Reference Standard*—Dry in vacuum at room temperature for 6 hours before using.

Identification—

A: The infrared absorption spectrum of a mineral oil dispersion of it, previously dried, exhibits maxima only at the same wavelengths as that of a similar preparation of USP Metyrapone RS.

B: The ultraviolet absorption spectrum of a 1 in 100,000 solution in 1 *N* sulfuric acid exhibits maxima and minima at the same wavelengths as that of a similar solution of USP Metyrapone RS, concomitantly measured.

Loss on drying ⟨731⟩—Dry it in vacuum at room temperature for 6 hours: it loses not more than 0.5% of its weight.

Heavy metals, *Method II* ⟨231⟩: 0.001%.

Residue on ignition ⟨281⟩: not more than 0.1%.

Chromatographic purity—

Standard preparations—Dissolve USP Metyrapone RS in methanol, and mix to obtain a solution having a known concentration of 0.2 mg per mL. Dilute quantitatively with methanol to obtain *Standard preparation A*, containing 40 µg of the Reference Standard per mL, and *Standard preparation B*, containing 20 µg of the Reference Standard per mL.

Test preparation—Dissolve an accurately weighed quantity of Metyrapone in methanol to obtain a solution containing 20 mg per mL.

Procedure—On a suitable thin-layer chromatographic plate (see *Chromatography* ⟨621⟩), coated with a 0.25-mm layer of chromatographic silica gel mixture, apply separately 5 µL of the *Test preparation* and 5 µL of each *Standard preparation.* Position the plate in a chromatographic chamber, and develop the chromatograms in a solvent system consisting of a mixture of chloroform and methanol (48:3) until the solvent front has moved about three-fourths of the length of the plate. Remove the plate from the developing chamber, mark the solvent front, and dry under a current of nitrogen for about 10 minutes. Position the dried plate once again in the same chromatographic chamber, and again develop the chromatograms, until the solvent front has moved about three-fourths of the length of the plate. Remove the plate from the developing chamber, mark the solvent front, and dry under a current of warm air for about 15 minutes. Examine the plate under short-wavelength ultraviolet light, and compare the intensities of any secondary spots observed in the chromatogram of the *Test preparation* with those of the principal spots in the chromatograms of the *Standard preparations:* no secondary spot from the chromatogram of the *Test preparation* is larger or more intense than the principal spot obtained from *Standard preparation A* (0.2%), and the sum of the intensities of the secondary spots obtained from the *Test preparation* corresponds to not more than 1.0%.

Assay—Transfer about 50 mg of Metyrapone, accurately weighed, to a 100-mL volumetric flask. Dissolve in 1 *N* sulfuric acid, dilute with the same solvent to volume, and mix. Pipet 2 mL of this solution into a 100-mL volumetric flask, add 1 *N* sulfuric acid to volume, and mix. Dissolve an accurately weighed quantity of USP Metyrapone RS in 1 *N* sulfuric acid, and dilute quantitatively and stepwise with 1 *N* sulfuric acid to obtain a Standard solution having a known concentration of about 10 µg per mL. Concomitantly determine the absorbances of both solutions in 1-cm cells at the wavelength of maximum absorbance at about 260 nm, with a suitable spectrophotometer, using 1 *N* sulfuric acid as the blank. Calculate the quantity, in mg, of $C_{14}H_{14}N_2O$ in the portion of Metyrapone taken by the formula:

$$5C(A_U/A_S),$$

in which *C* is the concentration, in µg per mL, of USP Metyrapone RS in the Standard solution, and A_U and A_S are the absorbances of the solution of Metyrapone and the Standard solution, respectively.

Metyrapone Tablets

» Metyrapone Tablets contain not less than 95.0 percent and not more than 105.0 percent of the labeled amount of $C_{14}H_{14}N_2O$.

Packaging and storage—Preserve in tight, light-resistant containers, and avoid exposure to excessive heat.

Reference standard—*USP Metyrapone Reference Standard*—Dry in vacuum at room temperature for 6 hours before using.

Identification—

A: Transfer to a centrifuge tube a quantity of powdered Tablets, equivalent to about 500 mg of metyrapone, add 10 mL of 1 *N* sodium hydroxide, and mix. Extract with 10 mL of chloroform, centrifuge, and filter: the infrared absorption spectrum of this solution, determined in a 0.5-mm cell against chloroform, exhibits maxima only at the same wavelengths as that of a similar solution of USP Metyrapone RS.

B: Transfer to a centrifuge tube 1 mL of the filtrate obtained in *Identification test A*, add 20 mL of chloroform, and extract with 30 mL of 1 *N* sulfuric acid, centrifuging and filtering the sulfuric acid layer through a pledget of cotton. Dilute 1 mL of this solution with 1 *N* sulfuric acid to 100 mL: the ultraviolet absorption spectrum of the resulting solution exhibits maxima and minima at the same wavelengths as that of a similar solution of USP Metyrapone RS, concomitantly measured.

Dissolution ⟨711⟩—

Medium: 0.1 *N* hydrochloric acid; 900 mL.

Apparatus 1: 100 rpm.

Time: 45 minutes.

Procedure—Determine the amount of $C_{14}H_{14}N_2O$ dissolved from ultraviolet absorbances at the wavelength of maximum absorbance at about 259 nm of filtered portions of the solution under test, suitably diluted with 0.1 *N* hydrochloric acid, in comparison with a Standard solution having a known concentration of USP Metyrapone RS in the same medium.

Tolerances—Not less than 60% (*Q*) of the labeled amount of $C_{14}H_{14}N_2O$ is dissolved in 45 minutes.

Uniformity of dosage units ⟨905⟩: meet the requirements.

Assay—

2,4-Dinitrophenylhydrazine in methanol—Shake by mechanical means about 1 g of 2,4-dinitrophenylhydrazine with 75 mL of methanol in a 100-mL flask for about 15 minutes, and filter through paper. Prepare fresh daily.

2,4-Dinitrophenylhydrazine hydrochloride—Mix 2 mL of hydrochloric acid with 23 mL of *2,4-Dinitrophenylhydrazine in methanol.*

Potassium hydroxide in methanol—Dissolve 5 g of potassium hydroxide in 100 mL of methanol, and filter.

Standard preparation—Dissolve an accurately weighed quantity of USP Metyrapone RS in chloroform, and dilute quantitatively and stepwise with chloroform to obtain a solution having a known concentration of about 100 µg per mL.

Assay preparation—Weigh and finely powder not less than 20 Metyrapone Tablets. Transfer an accurately weighed portion of the powder, equivalent to about 25 mg of metyrapone, with the aid of 10 mL of 1 *N* sodium hydroxide, to a centrifuge tube. Shake gently, and extract with three 15-mL portions of chloroform. Centrifuge each extract, filtering through a pledget of cotton, previously washed with chloroform, into a 50-mL volumetric flask. Add chloroform to volume, and mix. Pipet 5 mL of this solution into a 25-mL volumetric flask, add chloroform to volume, and mix.

Procedure—Pipet 3 mL each of the *Standard preparation*, the *Assay preparation*, and chloroform to provide a blank, into separate 50-mL volumetric flasks. To each flask add 1 mL of *2,4-Dinitrophenylhydrazine hydrochloride*, and shake lightly. Evaporate the solutions on a steam bath to near dryness. Wash down the sides of the flask with 1 mL of chloroform and methanol (1:1), and again evaporate the solutions to near dryness. Heat the flasks in an oven maintained at 110° to 120° for 30 minutes. Remove the flasks, pipet 10 mL of *Potassium hydroxide in methanol* into each flask, and heat again in the boiling water bath for 1 minute. Allow to cool to room temperature, insert the stoppers, shake by mechanical means for 5 minutes, add methanol to volume, and mix. Concomitantly determine the absorbances of the solutions in 1-cm cells relative to the blank, at the wavelength of maximum absorbance at about 450 nm, with a suitable spectrophotometer. Calculate the quantity, in mg, of $C_{14}H_{14}N_2O$ in the portion of Tablets taken by the formula:

$$250C(A_U/A_S),$$

in which *C* is the concentration, in mg per mL, of USP Metyrapone RS in the *Standard preparation*, and A_U and A_S are the absorbances of the solutions from the *Assay preparation* and the *Standard preparation*, respectively.

Metyrosine

CH3
HO—⟨ring⟩—CH2—C—COOH
NH2

$C_{10}H_{13}NO_3$ 195.22

L-Tyrosine, α-methyl-, (−)-.

(−)-α-Methyl-L-tyrosine [*672-87-7*].

» Metyrosine contains not less than 98.6 percent and not more than 101.0 percent of $C_{10}H_{13}NO_3$, calculated on the dried basis.

Packaging and storage—Preserve in well-closed containers.

Reference standard—*USP Metyrosine Reference Standard*—Dry at a pressure not exceeding 5 mm of mercury at 100° for two hours before using.

Identification—

A: The infrared absorption spectrum of a mineral oil dispersion of it, previously dried, exhibits maxima only at the same wavelengths as that of a similar preparation of USP Metyrosine RS.

B: The ultraviolet absorption spectrum of a 1 in 60,000 solution in 0.1 *N* hydrochloric acid exhibits maxima and minima at the same wavelengths as that of a similar solution of USP Metyrosine RS concomitantly measured, and the respective absorptivities, calculated on the dried basis, at the wavelength of maximum absorbance at about 224 nm do not differ by more than 3.0%.

Specific rotation ⟨781⟩—

Solvent A—Transfer 20.0 g of anhydrous sodium acetate to a 250-mL volumetric flask, dissolve in about 150 mL of water, add 50.0 mL of glacial acetic acid, dilute with water to volume, and mix.

Solvent B—Transfer 62.5 g of cupric sulfate to a 200-mL volumetric flask, dissolve in water, dilute with water to volume, and mix.

Solvent C—Mix *Solvent A* and *Solvent B* in a 1000-mL volumetric flask, dilute with water to volume, and mix.

Procedure—Prepare a solution of Metyrosine in *Solvent C* containing 125 mg in each 25 mL. [NOTE—Sonication may be necessary to dissolve the metyrosine.] Measure the specific rotation of this solution at 546 nm at 30° in a 0.5-dm tube: the specific rotation of Metyrosine is between +185° and +195°, calculated on the dried basis.

Loss on drying ⟨731⟩—Dry it at a pressure not exceeding 5 mm of mercury at 100° for two hours: it loses not more than 1.0% of its weight.

Residue on ignition ⟨281⟩: not more than 0.1%.

Heavy metals, *Method II* ⟨231⟩: 0.003%.

Chromatographic purity—

Standard preparation—Dissolve USP Metyrosine RS in a solvent mixture of methanol and ammonium hydroxide (7:3) to obtain a solution having a concentration of 10 mg per mL (*Solution A*). Pipet 1 mL of *Solution A* into a 100-mL volumetric flask, dilute with the same solvent mixture to volume, and mix (*Solution B*). Pipet 5 mL of *Solution B* into a 10-mL volumetric flask, dilute with the same solvent mixture to volume, and mix (*Solution C*). Pipet 5 mL of *Solution C* into a 10-mL volumetric flask, dilute with the same solvent mixture to volume, and mix (*Solution D*).

Test preparation—Dissolve Metyrosine in the solvent mixture of methanol and ammonium hydroxide (7:3) to obtain a solution having a concentration of 10 mg per mL.

Procedure—On a suitable thin-layer chromatographic plate (see *Chromatography* ⟨621⟩), coated with a 0.25-mm layer of chromatographic silica gel mixture and previously washed with methanol, apply 10-µL portions of *Solutions A, B, C,* and *D* and the *Test preparation.* Allow the spots to dry, and develop the chromatogram in a solvent system consisting of a mixture of *n*-propyl alcohol and ammonium hydroxide (7:3) until the solvent front has moved about three-fourths of the length of the plate. Remove the plate from the developing chamber, mark the solvent front, and dry the plate. Expose the plate to iodine vapors, and examine under short-wavelength ultraviolet light: the chromatogram shows principal spots at about the same R_f value. Estimate the levels of any additional spots observed in the chromatogram of the *Test preparation* by comparison with the spots in the chromatograms of *Solutions B, C,* and *D:* the sum of the intensities of any spots observed is not greater than that of the principal spot obtained from *Solution B*, corresponding to not more than 1%.

Assay—Dissolve about 300 mg of Metyrosine, accurately weighed, in about 100 mL of glacial acetic acid, sonicate for about 5 minutes, and titrate with 0.1 *N* perchloric acid VS, determining the end-point potentiometrically, using a platinum ring electrode and a sleeve-type calomel electrode containing 0.1 *N* lithium perchlorate in glacial acetic acid (see *Titrimetry* ⟨541⟩). Perform a blank determination, and make any necessary correction. Each mL of 0.1 *N* perchloric acid is equivalent to 19.52 mg of $C_{10}H_{13}NO_3$.

Metyrosine Capsules

» Metyrosine Capsules contain not less than 90.0 percent and not more than 110.0 percent of the labeled amount of $C_{10}H_{13}NO_3$.

Packaging and storage—Preserve in well-closed containers.

Reference standard—*USP Metyrosine Reference Standard*—Dry at a pressure not exceeding 5 mm of mercury at 100° for 2 hours before using.

Identification—The ultraviolet absorption spectrum of a 1 in 10,000 solution of the Capsule contents in dilute hydrochloric acid (1 in 100) exhibits maxima and minima at the same wavelengths as that of a similar solution of USP Metyrosine RS, concomitantly measured.

Dissolution ⟨711⟩—

Medium: 0.1 *N* hydrochloric acid; 750 mL.

Apparatus 1: 100 rpm.

Time: 60 minutes.

Procedure—Determine the amount of $C_{10}H_{13}NO_3$ dissolved from ultraviolet absorbances at the wavelength of maximum absorbance at about 274 nm of filtered portions of the solution under test, suitably diluted with *Dissolution Medium*, if necessary, in comparison with a Standard solution having a known concentration of USP Metyrosine RS in the same medium.

Tolerances—Not less than 75% (*Q*) of the labeled amount of $C_{10}H_{13}NO_3$ is dissolved in 60 minutes.

Uniformity of dosage units ⟨905⟩: meet the requirements.

Assay—

Standard preparation—Dissolve a suitable quantity of USP Metyrosine RS, accurately weighed, in dilute hydrochloric acid (1 in 100) to obtain a solution having a known concentration of about 100 µg per mL.

Assay preparation—Combine the contents of not less than 20 Metyrosine Capsules, and transfer an accurately weighed portion of the combined contents, equivalent to about 100 mg of metyrosine, to a 100-mL volumetric flask. Add 50 mL of dilute hydrochloric acid (1 in 100), shake by mechanical means for 45 minutes, dilute with dilute hydrochloric acid (1 in 100) to volume, mix, and filter. Transfer 10.0 mL of the filtrate to a 100-mL volumetric flask, dilute with dilute hydrochloric acid solution (1 in 100) to volume, and mix. Concomitantly determine the absorbances of this solution and the *Standard preparation* at the wavelength of maximum absorbance at about 274 nm, with a suitable spectrophotometer, using dilute hydrochloric acid solution (1 in 100) as the blank. Calculate the quantity, in mg, of $C_{10}H_{13}NO_3$ in the portion of Capsules taken by the formula:

$$C(A_U/A_S),$$

in which *C* is the concentration, in µg per mL, of USP Metyrosine RS in the *Standard preparation*, and A_U and A_S are the absorbances of the solutions obtained from the *Assay preparation* and the *Standard preparation*, respectively.

Mexiletine Hydrochloride

NH2
OCH2CHCH3
H3C CH3 · HCl

$C_{11}H_{17}NO.HCl$ 215.72

2-Propanamine, 1-(2,6-dimethylphenoxy)-, hydrochloride.

1-Methyl-2-(2,6-xylyloxy)ethylamine hydrochloride [*5370-01-04*].

» Mexiletine Hydrochloride contains not less than 98.0 percent and not more than 102.0 percent of $C_{11}H_{17}NO.HCl$, calculated on the dried basis.

Packaging and storage—Preserve in tight containers.

Reference standard—*USP Mexiletine Hydrochloride Reference Standard*—Dry at 105° for 2 hours before using.

Identification—

A: The infrared absorption spectrum of a mineral oil dispersion of it, previously dried, exhibits maxima only at the same wavelengths as that of a similar preparation of USP Mexiletine Hydrochloride RS.

B: Prepare a test solution by dissolving a suitable quantity of it in methanol to obtain a concentration of about 10 mg per mL. Similarly prepare a Standard solution, using USP Mexiletine Hydrochloride RS. Separately apply 5-µL portions of the test solution and the Standard solution on a suitable thin-layer chromatographic plate (see *Chromatography* ⟨621⟩) coated with a 0.25-mm layer of chromatographic silica gel. Develop the chromatogram in a suitable chromatographic chamber half saturated with solvent vapor, using a solvent system consisting of a mixture of chloroform, methanol, and ammonium hydroxide (425:70:5) until the solvent front has moved about three-fourths of the length of the plate. Remove the plate from the developing chamber, mark the solvent front, and allow the solvent to evaporate. Spray the plate with a 1 in 500 solution of fast blue salt BB in methanol, and dry the plate at 105° for 15 minutes. Locate the spots on the plate by spraying it with a 1 in 5 solution of potassium hydroxide in methanol: the R_f value of the principal spot obtained from the test solution corresponds to that obtained from the Standard solution.

C: To 3 mL of a solution (1 in 60) add 1 mL of 6 *N* ammonium hydroxide, filter, and acidify the filtrate with 2 mL of nitric acid. Then add 1 mL of silver nitrate TS: a curdy, white precipitate is formed, and it is soluble in an excess of 6 *N* ammonium hydroxide (presence of chloride).

Chromatographic purity—

Mobile phase, Standard preparation, and *Resolution solution*—Prepare as directed in the *Assay*.

Standard solution—Transfer 10.0 mL of the *Standard preparation* prepared as directed in the *Assay* to a 1000-mL volumetric flask, dilute with *Mobile phase* to volume, and mix. This solution contains about 20 µg of USP Mexiletine Hydrochloride RS per mL.

Test solution—Transfer about 500 mg of Mexiletine Hydrochloride, accurately weighed, to a 5-mL volumetric flask, dilute with *Mobile phase* to volume, and mix.

Chromatographic system—Prepare as directed in the *Assay*, except that the relative standard deviation of replicate injections of the *Standard solution* is not more than 3.0%.

Procedure—Proceed as directed for *Procedure* in the *Assay*. Calculate the percentage of each impurity observed by the formula:

$$5(C/W)(r_U/r_S),$$

in which *C* is the concentration, in mg per mL, of USP Mexiletine Hydrochloride RS in the *Standard solution*, *W* is the weight, in mg, of the Mexiletine Hydrochloride taken to prepare the *Test solution*, r_U is the peak response obtained from an individual impurity observed in the chromatogram of the *Test solution*, and r_S is the mexiletine peak response obtained from the *Standard solution:* not more than 1% of any individual impurity is found, and the total of all observed impurities is not more than 1.5%.

pH ⟨791⟩: between 3.5 and 5.5, in a solution (1 in 10).

Loss on drying ⟨731⟩—Dry it at 105° for 2 hours: it loses not more than 0.5% of its weight.

Residue on ignition ⟨281⟩: not more than 0.1%.

Heavy metals, *Method II* ⟨231⟩: 0.001%.

Assay—

Sodium acetate buffer solution—Dissolve 11.5 g of anhydrous sodium acetate in 500 mL of water, add 3.2 mL of glacial acetic acid, mix, and allow to cool. Adjust with hydrochloric acid to a pH of 4.8 ± 0.1, dilute with water to 1000 mL, and mix.

Mobile phase—Prepare a suitable filtered and degassed mixture of methanol and *Sodium acetate buffer solution* (600:400). Make adjustments if necessary (see *System Suitability* under *Chromatography* ⟨621⟩).

Standard preparation—Dissolve an accurately weighed quantity of USP Mexiletine Hydrochloride RS in *Mobile phase* to obtain a solution having a known concentration of about 2 mg per mL.

Resolution solution—Prepare a solution of 2-phenylethylamine hydrochloride in *Standard preparation* containing about 1 mg per mL.

Assay preparation—Transfer about 100 mg of Mexiletine Hydrochloride, accurately weighed, to a 50-mL volumetric flask, dissolve in *Mobile phase*, dilute with *Mobile phase* to volume, and mix.

Chromatographic system (see *Chromatography* ⟨621⟩)—The liquid chromatograph is equipped with a 254-nm detector, a guard column containing packing L1, and a 3.9-mm × 30-cm column that contains 10-µm packing L1. The flow rate is about 1 mL per minute. Chromatograph about 20 µL of the *Resolution preparation*, and record the peak responses as directed under *Procedure:* the resolution, *R*, between the 2-phenylethylamine and mexiletine peaks is not less than 3.0. Chromatograph the *Standard preparation*, and record the peak responses as directed under *Procedure:* the relative standard deviation for replicate injections is not more than 2.0%.

Procedure—[NOTE—Use peak areas where peak responses are indicated.] Separately inject equal volumes (about 20 µL) of the *Assay preparation* and the *Standard preparation* into the chromatograph, record the chromatograms, and measure the responses for the major peaks. The relative retention times are about 0.7 for 2-phenylethylamine and 1.0 for mexiletine. Calculate the quantity, in mg, of $C_{11}H_{17}NO.HCl$ in the portion of Mexiletine Hydrochloride taken by the formula:

$$50C(r_U/r_S),$$

in which *C* is the concentration, in mg per mL, of USP Mexiletine Hydrochloride RS in the *Standard preparation*, and r_U and r_S are the mexiletine peak responses obtained from the *Assay preparation* and the *Standard preparation*, respectively.

Mexiletine Hydrochloride Capsules

» Mexiletine Hydrochloride Capsules contain not less than 90.0 percent and not more than 110.0 percent of the labeled amount of $C_{11}H_{17}NO.HCl$.

Packaging and storage—Preserve in tight containers.

Reference standard—*USP Mexiletine Hydrochloride Reference Standard*—Dry at 105° for 2 hours before using.

Identification—

A: Transfer a quantity of Capsule contents, equivalent to about 250 mg of mexiletine hydrochloride, to a suitable test tube, add 10 mL of methanol, and vortex for 1 minute. Filter the mixture, evaporate the filtrate under a stream of nitrogen to dryness, and dry the residue in vacuum at 60° for 1 hour: the infrared absorption spectrum of a mineral oil dispersion of the dried residue so obtained exhibits maxima only at the same wavelengths as that of a similar preparation of USP Mexiletine Hydrochloride RS.

B: The retention time of the major peak in the chromatogram of the *Assay preparation* corresponds to that of the *Standard preparation*, obtained as directed in the *Assay*.

Dissolution ⟨711⟩—

Medium: water; 900 mL.

Apparatus 2: 50 rpm.

Time: 30 minutes.

Procedure—Determine the amount of $C_{11}H_{17}NO.HCl$ dissolved from fluorescence intensities, using the maximum excitation wavelength at about 265 nm and the maximum emission wavelength at about 295 nm, of filtered portions of the solution under test, suitably diluted with *Dissolution Medium*, if necessary, in comparison with a Standard solution having a known concentration of USP Mexiletine Hydrochloride RS in the same medium.

Tolerances—Not less than 80% (*Q*) of the labeled amount of $C_{11}H_{17}NO.HCl$ is dissolved in 30 minutes.

Uniformity of dosage units ⟨905⟩: meet the requirements.

Related impurities—

Mobile phase, Standard preparation, and *Resolution solution*—Prepare as directed in the *Assay* under *Mexiletine Hydrochloride.*

Standard solution—Transfer 10.0 mL of the *Standard preparation* prepared as directed in the *Assay* under *Mexiletine Hydrochloride* to a 1000-mL volumetric flask, dilute with *Mobile phase* to volume, and mix. This solution contains about 20 µg of USP Mexiletine Hydrochloride RS per mL.

Test solution—Use the *Assay preparation* prepared as directed in the *Assay*.

Chromatographic system—Prepare as directed in the *Assay* under *Mexiletine Hydrochloride*, except that the relative standard deviation of replicate injections of the *Standard solution* is not more than 3.0%.

Procedure—[NOTE—Use peak areas where peak responses are indicated.] Separately inject equal volumes (about 20 µL) of the *Standard solution* and the *Test solution* into the chromatograph, record the chromatograms using a high sensitivity setting for the recorder, and measure the responses for the peaks. Calculate the percentage of each impurity observed by the formula:

$$0.1(C/L)(r_U/r_S),$$

in which C is the concentration, in mg per mL, of USP Mexiletine Hydrochloride RS in the *Standard solution*, L is the quantity, in mg, of mexiletine hydrochloride in each mL of the *Test solution*, based on the labeled amount in the portion of Capsule contents used to prepare the *Assay preparation* and the extent of dilution, r_U is the peak response obtained from an individual impurity observed in the chromatogram of the *Test solution*, and r_S is the mexiletine peak response obtained from the *Standard solution:* not more than 1% of any individual impurity is found, and the total of all observed impurities is not more than 1.5%.

Assay—

Mobile phase, Standard preparation, Resolution solution, and *Chromatographic system*—Prepare as directed in the *Assay* under *Mexiletine Hydrochloride*.

Assay preparation—Weigh the contents of not less than 20 Mexiletine Hydrochloride Capsules, and calculate the average weight per Capsule. Mix the combined contents of the Capsules, and transfer an accurately weighed portion, equivalent to about 50 mg of mexiletine hydrochloride, to a stoppered, 50-mL centrifuge tube. Add 25.0 mL of *Mobile phase*, insert the stopper, and shake by mechanical means for 15 minutes. Centrifuge, and use the clear supernatant solution as the *Assay preparation*. [NOTE—Reserve a portion of this solution for use as the *Test solution* in the test for *Related impurities*.]

Procedure—Proceed as directed for *Procedure* in the *Assay* under *Mexiletine Hydrochloride*. Calculate the quantity, in mg, of $C_{11}H_{17}NO.HCl$ in the portion of Capsule contents taken by the formula:

$$25C(r_U/r_S),$$

in which C is the concentration, in mg per mL, of USP Mexiletine Hydrochloride RS in the *Standard preparation*, and r_U and r_S are the mexiletine peak responses obtained from the *Assay preparation* and the *Standard preparation*, respectively.

Sterile Mezlocillin Sodium

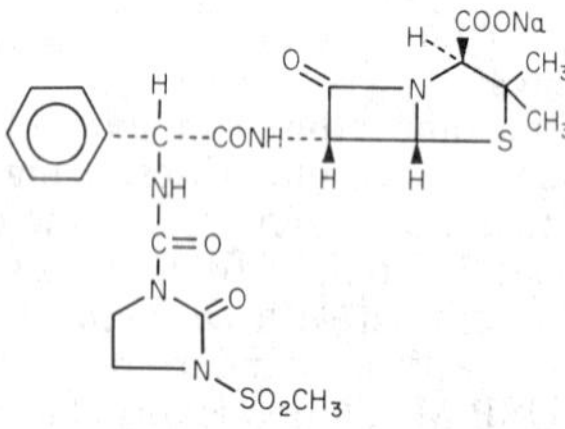

$C_{21}H_{24}NaN_5O_8S_2$ 561.56

4-Thia-1-azabicyclo[3.2.0]heptane-2-carboxylic acid, 3,3-dimethyl-6-[[[[[3-(methylsulfonyl)-2-oxo-1-imidazolidinyl]carbonyl]amino]phenylacetyl]amino]-7-oxo-, monosodium salt, [2*S*[2α,5α,6β(*S**)]]-.

Sodium (2*S*,5*R*,6*R*)-3,3-dimethyl-6-[(*R*)-2-[3-(methylsulfonyl)-2-oxo-1-imidazolidinecarboxamido]-2-phenylacetamido]-7-oxo-4-thia-1-azabicyclo[3.2.0]heptane-2-carboxylate [*59798-30-0*].

Monohydrate 579.58

» Sterile Mezlocillin Sodium is mezlocillin sodium suitable for parenteral use. It has a potency equivalent to not less than 838 µg and not more than 978 µg of mezlocillin ($C_{21}H_{25}N_5O_8S_2$) per mg, calculated on the anhydrous basis. In addition, where packaged for dispensing, it contains the equivalent of not less than 90.0 percent and not more than 115.0 percent of the labeled amount of mezlocillin ($C_{21}H_{25}N_5O_8S_2$).

Packaging and storage—Preserve in *Containers for Sterile Solids* as described under *Injections* ⟨1⟩.

Reference standard—*USP Mezlocillin Sodium Reference Standard*—Do not dry before using.

Constituted solution—At the time of use, the constituted solution prepared from Sterile Mezlocillin Sodium meets the requirements for *Constituted Solutions* under *Injections* ⟨1⟩.

Identification—Prepare a test solution containing the equivalent of 4 mg of mezlocillin per mL. Prepare a Standard solution of USP Mezlocillin Sodium RS containing 4 mg per mL. Use within 10 minutes after preparation. Apply separately 5 µL of each solution on a thin-layer chromatographic plate coated with a 0.25-mm layer of chromatographic silica gel mixture (see *Chromatography* ⟨621⟩). Place the plate in a suitable chromatographic chamber, and develop the chromatogram with a solvent system consisting of a mixture of methanol, chloroform, water, and pyridine (90:80:30:10) until the solvent front has moved about three-fourths of the length of the plate. Remove the plate from the chamber, and dry with a current of warm air for 10 minutes. Locate the spots on the plate by exposing it to iodine vapors in a closed chamber for about 30 seconds: the R_f value of the principal spot obtained from the test solution corresponds to that obtained from the Standard solution.

Specific rotation ⟨781⟩: between +175° and +195°, determined in a solution containing 10 mg per mL.

Pyrogen—It meets the requirements of the *Pyrogen Test* ⟨151⟩, the test dose being 1.0 mL per kg of a solution prepared by diluting Sterile Mezlocillin Sodium with Sterile Water for Injection to a concentration of 100 mg of mezlocillin per mL.

Sterility—It meets the requirements under *Sterility Tests* ⟨71⟩, when tested as directed in the section, *Test Procedures Using Membrane Filtration*.

pH ⟨791⟩: between 4.5 and 8.0, in a solution (1 in 10).

Water, *Method I* ⟨921⟩: not more than 6.0%.

Particulate matter ⟨788⟩: meets the requirements under *Small-volume Injections*.

Other requirements—It meets the requirements for *Uniformity of Dosage Units* ⟨905⟩ and *Labeling* under *Injections* ⟨1⟩.

Assay—

Buffer solution—Dissolve 200 g of tromethamine in water to make 1000 mL. Filter before using.

Standard preparation—Dissolve a suitable quantity of USP Mezlocillin Sodium RS, accurately weighed, in water to obtain a solution having a known concentration of about 2 mg of mezlocillin per mL.

Assay preparation 1—Using a suitable quantity of Sterile Mezlocillin Sodium, accurately weighed, proceed as directed under *Standard preparation*.

Assay preparation 2 (where it is represented as being in a single-dose container)—Constitute Sterile Mezlocillin Sodium in a volume of water, accurately measured, corresponding to the volume of solvent specified in the labeling. Withdraw all of the withdrawable contents, using a suitable hypodermic needle and syringe, and dilute quantitatively with water to obtain a solution containing about 2 mg of mezlocillin per mL.

Assay preparation 3 (where the label states the quantity of mezlocillin in a given volume of constituted solution)—Constitute Sterile Mezlocillin Sodium in a volume of water, accurately measured, corresponding to the volume of solvent specified in the labeling. Dilute an accurately measured portion of the constituted solution quantitatively with water to obtain a solution containing about 2 mg of mezlocillin per mL.

Procedure—Proceed as directed for *Procedure* in the section, *Antibiotics—Hydroxylamine Assay*, under *Automated Methods of Analysis* ⟨16⟩, except to use *Buffer solution* instead of *Acetate buffer*. Calculate the potency, in µg of mezlocillin ($C_{21}H_{25}N_5O_8S_2$) per mg, of the Sterile Mezlocillin Sodium taken by the formula:

$$(CP/W)(A_U/A_S),$$

in which W is the weight, in mg, of Sterile Mezlocillin Sodium taken in each mL of *Assay preparation 1*, and the other terms are as defined therein. Calculate the quantity, in mg, of mezlocillin in the container, or in the portion of constituted solution taken by the formula:

$$(L/D)(CP/1000)(A_U/A_S),$$

in which L is the labeled quantity, in mg, of mezlocillin in the container, or in the volume of constituted solution taken, and D is the concentration, in mg per mL, of mezlocillin in *Assay preparation 2* or in *Assay preparation 3*, on the basis of the labeled quantity in the container, or in the portion of constituted solution taken, respectively, and the extent of dilution.

Miconazole

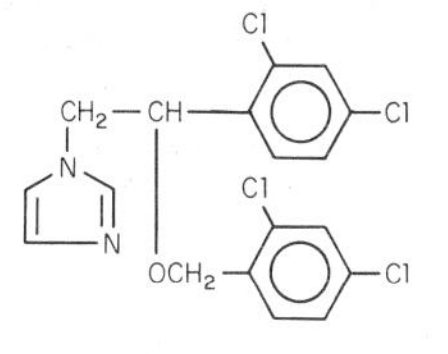

$C_{18}H_{14}Cl_4N_2O$ 416.14

1*H*-Imidazole, 1-[2-(2,4-dichlorophenyl)-2-[(2,4-dichlorophenyl)methoxy]ethyl]-.

1-[2,4-Dichloro-β-[(2,4-dichlorobenzyl)oxy]phenethyl]imidazole [*22916-47-8*].

» Miconazole contains not less than 98.0 percent and not more than 102.0 percent of $C_{18}H_{14}Cl_4N_2O$, calculated on the dried basis.

Packaging and storage—Preserve in well-closed containers, protected from light.

Reference standard—*USP Miconazole Reference Standard*—Dry in vacuum at 60° for 4 hours before using.

Identification—

A: The infrared absorption spectrum of a potassium bromide dispersion of it, previously dried, exhibits maxima only at the same wavelengths as that of a similar preparation of USP Miconazole RS.

B: Transfer 40 mg to a 100-mL volumetric flask, dissolve in 50 mL of isopropyl alcohol, add 10 mL of 0.1 *N* hydrochloric acid, dilute with isopropyl alcohol to volume, and mix: the ultraviolet absorption spectrum of this solution exhibits maxima and minima at the same wavelengths as that of a similar solution of USP Miconazole RS, concomitantly measured.

Loss on drying ⟨731⟩—Dry it in vacuum at 60° for 4 hours: it loses not more than 0.5% of its weight.

Residue on ignition ⟨281⟩: not more than 0.2%.

Chromatographic purity—Dissolve 30 mg in 3.0 mL of chloroform to obtain the *Test preparation*. Dissolve a suitable quantity of USP Miconazole RS in chloroform to obtain a *Standard solution* having a concentration of 10.0 mg per mL. Dilute a portion of this solution quantitatively with chloroform to obtain a *Diluted standard solution* having a concentration of 100 µg per mL. Apply separate 5-µL portions of the three solutions on the starting line of a suitable thin-layer chromatographic plate (see *Chromatography* ⟨621⟩), coated with a 0.25-mm layer of chromatographic silica gel mixture. Develop the chromatogram in a suitable chamber with a freshly prepared solvent system consisting of a mixture of *n*-hexane, chloroform, methanol, and ammonium hydroxide (60:30:10:1) until the solvent front has moved about three-fourths of the length of the plate. Remove the plate from the chamber, and allow the solvent to evaporate. Expose the plate to iodine vapors in a closed chamber for about 30 minutes, and locate the spots: the R_f value of the principal spot obtained from the *Test solution* corresponds to that obtained from the *Standard solution*, and any other spot obtained from the *Test solution* does not exceed, in size or intensity, the principal spot obtained from the *Diluted standard solution* (1.0%).

Assay—Dissolve about 300 mg of Miconazole, accurately weighed, in 40 mL of glacial acetic acid, add 4 drops of *p*-naphtholbenzein TS, and titrate with 0.1 *N* perchloric acid VS to a green endpoint. Perform a blank determination, and make any necessary correction. Each mL of 0.1 *N* perchloric acid is equivalent to 41.61 mg of $C_{18}H_{14}Cl_4N_2O$.

Miconazole Injection

» Miconazole Injection is a sterile solution of Miconazole in Water for Injection. It contains not less than 90.0 percent and not more than 110.0 percent of the labeled amount of $C_{18}H_{14}Cl_4N_2O$.

Packaging and storage—Preserve in single-dose containers, preferably of Type I glass, at controlled room temperature.

Reference standard—*USP Miconazole Reference Standard*—Dry in vacuum at 60° for 4 hours before using.

Identification—

Dragendorff's reagent—Dissolve 0.85 g of bismuth subnitrate in a mixture of 40 mL of water and 10 mL of glacial acetic acid (*Solution A*). Dissolve 8 g of potassium iodide in 20 mL of water (*Solution B*). Transfer 5 mL of *Solution A*, 5 mL of *Solution B*, and 20 mL of glacial acetic acid to a 100-mL volumetric flask, dilute with water to volume, and mix.

Procedure—Transfer a volume of Injection, equivalent to about 50 mg of miconazole, to a 10-mL volumetric flask, dilute with methanol to volume, and mix. Dissolve a suitable quantity of USP Miconazole RS in methanol to obtain a Standard solution having a known concentration of about 5 mg per mL. Apply separate 5-µL portions of the two solutions on the starting line of a suitable thin-layer chromatographic plate (see *Chromatography* ⟨621⟩), coated with a 0.25-mm layer of chromatographic silica gel mixture. Develop the chromatogram in a suitable chamber with a freshly prepared solvent system consisting of a mixture of *n*-hexane, chloroform, methanol, and ammonium hydroxide (60:30:10:1) until the solvent front has moved about three-fourths of the length of the plate. Remove the plate from the chamber, and allow the solvent to evaporate. Locate the spots on the plate by spraying with *Dragendorff's reagent:* the R_f value of one of the principal spots obtained from the test solution corresponds to that obtained from the Standard solution.

Pyrogen—It meets requirements of the *Pyrogen Test* ⟨151⟩.

pH ⟨791⟩: between 3.7 and 5.7.

Particulate matter ⟨788⟩: meets the requirements under *Small-volume Injections*.

Other requirements—It meets the requirements under *Injections* ⟨1⟩.

Assay—

Color reagent—Transfer 100 mg of tropaeolin OO to a 100-mL volumetric flask containing 80 mL of water, and dissolve by warming. Cool, dilute with water to volume, mix, and filter.

pH 3 buffer—Dissolve 1.67 g of citric acid monohydrate and 1.10 g of dibasic sodium phosphate heptahydrate in 100 mL of water.

Acidified methanol—Carefully add 1 mL of sulfuric acid to 60 mL of methanol, dilute with methanol to 100 mL, and mix.

Standard preparation—Dissolve a suitable quantity of USP Miconazole RS, accurately weighed, in chloroform to obtain a solution having a known concentration of about 2 mg per mL.

Assay preparation—Transfer an accurately measured volume of Injection, equivalent to about 100 mg of miconazole, to a 50-mL volumetric flask, dilute with methanol to volume, and mix.

Chromatographic plate—Prepare a suitable thin-layer chromatographic plate (see *Chromatography* ⟨621⟩) coated with a 0.25-mm layer of chromatographic silica gel mixture treated with 0.5 *N* sodium hydroxide. Heat the plates at 105° for 30 minutes, and cool.

Procedure—Apply 500-µL portions, accurately measured, of the *Assay preparation* and the *Standard preparation* in 6-cm streaks along the starting line of the *Chromatographic plate*. Allow the streaks to dry, and develop the chromatograms in a suitable developing chamber lined with a paper wick, using a freshly prepared solvent system consisting of a mixture of *n*-hexane, chloroform, and methanol (60:30:10) until the solvent front has moved about three-fourths of the length of the plate. Remove the plate from the chamber, and allow the solvent to evaporate. Locate the spots of miconazole under short-wavelength ultraviolet light, and mark the miconazole bands to include similar zones of silica gel for the *Assay preparation* and the *Standard preparation*. Scrape off these zones, and transfer them to separate stoppered, 50-mL centrifuge tubes. Add 15 mL of chloroform to each tube, shake for 1 minute, centrifuge, and filter the supernatant liquids through small pledgets of cotton into separate 50-mL volumetric flasks. Repeat the washing of the silica gel scrapings with three 10-mL portions of chloroform, combining the washings in the respective 50-mL volumetric flasks. Dilute with chloroform to volume, and mix. Transfer 10.0-mL portions of these solutions to separate 125-mL separators containing 3.0 mL of *Color reagent* and 10.0 mL of *pH 3 buffer*. Shake for 1 minute, and allow the layers to separate. Transfer the chloroform extracts to separate 50-mL volumetric flasks. Repeat the extractions with three 10-mL portions of chloroform, combining the extracts in the respective 50-mL volumetric flasks. To each flask add 3.0 mL of *Acidified methanol*, dilute with chloroform to volume, and mix. Concomitantly determine the absorbances of the solutions in 1-cm cells at the wavelength of maximum absorbance at about 540 nm, with a suitable spectrophotometer, using chloroform as the blank. Calculate the quantity, in mg, of $C_{18}H_{14}Cl_4N_2O$ in each mL of the Injection taken by the formula:

$$50(C/V)(A_U/A_S),$$

in which C is the concentration, in mg per mL, of USP Miconazole RS in the *Standard preparation*, V is the volume, in mL, of Injection taken, and A_U and A_S are the absorbances of the solutions from the *Assay preparation* and the *Standard preparation*, respectively.

Miconazole Nitrate

$C_{18}H_{14}Cl_4N_2O.HNO_3$ 479.15

1*H*-Imidazole, 1-[2-(2,4-dichlorophenyl)-2-[(2,4-dichlorophenyl)methoxy]ethyl]-, mononitrate.

1-[2,4-Dichloro-β-[(2,4-dichlorobenzyl)oxy]phenethyl]imidazole mononitrate [*22832-87-7*].

» Miconazole Nitrate contains not less than 98.0 percent and not more than 102.0 percent of $C_{18}H_{14}Cl_4N_2O.HNO_3$, calculated on the dried basis.

Packaging and storage—Preserve in well-closed containers, protected from light.

Reference standard—*USP Miconazole Nitrate Reference Standard*—Dry at 105° for 2 hours before using.

Identification—

A: The infrared absorption spectrum of a potassium bromide dispersion of it, previously dried, exhibits maxima only at the same wavelengths as that of a similar preparation of USP Miconazole Nitrate RS.

B: The ultraviolet absorption spectrum of a 1 in 2500 solution in a 1 in 10 solution of 0.1 *N* hydrochloric acid in isopropyl alcohol exhibits maxima and minima at the same wavelengths as that of a similar preparation of USP Miconazole Nitrate RS, concomitantly measured.

Loss on drying ⟨731⟩—Dry it at 105° for 2 hours: it loses not more than 0.5% of its weight.

Residue on ignition ⟨281⟩: not more than 0.2%.

Chromatographic purity—Dissolve 100 mg in a solvent consisting of a mixture of chloroform and methanol (1:1), and dilute with the same solvent to 10.0 mL to obtain the *Test solution*. Prepare a *Standard solution* of USP Miconazole Nitrate RS in a solvent consisting of a mixture of chloroform and methanol (1:1) to contain 10 mg per mL. Dilute a portion of this solution quantitatively and stepwise with the same solvent to obtain a *Diluted standard solution* having a concentration of 25 µg per mL. Apply separate 50-µL portions of the three solutions on the starting line of a suitable thin-layer chromatographic plate (see *Chromatography* ⟨621⟩), coated with a 0.25-mm layer of chromatographic silica gel. Develop the chromatogram in a suitable chamber with a freshly prepared solvent system consisting of a mixture of *n*-hexane, chloroform, methanol, and ammonium hydroxide (60:30:10:1) until the solvent front has moved about three-fourths of the length of the plate. Remove the plate from the chamber, air-dry, spray with a solution of iodine in chloroform (1 in 20), and examine the chromatogram: the R_f value of the principal spot from the *Test solution* corresponds to that of the *Standard solution*, and any other spot obtained from the *Test solution* does not exceed, in size or intensity, the principal spot obtained from the *Diluted standard solution* (0.25%).

Ordinary impurities ⟨466⟩—

Test solution: methanol.

Standard solution: methanol.

Eluant: a mixture of toluene, isopropyl alcohol, and ammonium hydroxide (70:29:1), in a nonequilibrated chamber.

Visualization: 1.

Assay—Dissolve about 350 mg of Miconazole Nitrate, accurately weighed, in 50 mL of glacial acetic acid, and titrate with 0.1 *N* perchloric acid VS, determining the end-point potentiometrically, using a glass-calomel electrode system. Perform a blank determination, and make any necessary correction. Each mL of 0.1 *N* perchloric acid is equivalent to 47.92 mg of $C_{18}H_{14}Cl_4N_2O.HNO_3$.

Miconazole Nitrate Cream

» Miconazole Nitrate Cream contains not less than 90.0 percent and not more than 110.0 percent of the labeled amount of $C_{18}H_{14}Cl_4N_2O.HNO_3$.

Packaging and storage—Preserve in collapsible tubes or in tight containers.

Reference standard—*USP Miconazole Nitrate Reference Standard*—Dry at 105° for 2 hours before using.

Identification—Place about 25 mL of the stock solution, prepared as directed in the *Assay*, in a 50-mL beaker, and evaporate on a steam bath with the aid of a current of filtered air to dryness. Dry the residue at 105° for 10 minutes: the infrared absorption spectrum of a potassium bromide dispersion of it so obtained exhibits maxima only at the same wavelengths as that of a similar preparation of USP Miconazole Nitrate RS.

Minimum fill ⟨755⟩: meets the requirements.

Assay—

Internal standard solution—Dissolve a suitable quantity of cholestane in a mixture of chloroform and methanol (1:1) to obtain a solution having a concentration of about 1 mg per mL.

Standard preparation—Dissolve an accurately weighed quantity of USP Miconazole Nitrate RS in methanol to obtain a solution having a known concentration of about 500 µg per mL. Transfer 10.0 mL of this solution to a test tube, and evaporate on a steam bath with the aid of a current of filtered air to dryness. Dissolve the residue in 2.0 mL of *Internal standard solution*. This *Standard preparation* has a concentration of about 2500 µg per mL.

Assay preparation—Transfer an accurately weighed portion of Cream, equivalent to about 100 mg of miconazole nitrate, to a 100-mL volumetric flask. Dissolve in isopropyl alcohol and chloroform (1:1) solvent mixture, dilute with the same solvent to volume, and mix. Pipet 25 mL of this solution into a 150-mL beaker and evaporate on a steam bath with the aid of a stream of nitrogen to dryness. Add 10 mL of chloroform to the residue and heat on a steam bath just to boiling. Remove the beaker from the steam bath and stir to effect solution. [NOTE—Avoid excessive evaporation of chloroform.] Add 50 mL of pentane in small portions with continuous stirring. Allow to crystallize for 10 to 15 minutes. Filter through a medium-porosity sintered-glass filter funnel with the aid of air applied to the surface through a one-hole stopper fitted onto the funnel. Wash the beaker with four 5-mL portions of pentane and add the washings to the filter funnel. Wash the funnel and precipitate with four 5-mL portions of pentane. Dry the precipitate on the filter by allowing filtered air to pass through the funnel for several minutes. Dissolve the precipitate by washing the beaker and funnel with small portions of methanol, and collect the filtrate in a 50-mL volumetric flask using filtered air applied to the top of the funnel to aid in filtration. Dilute with methanol to volume, and mix. Transfer 10.0 mL of this stock solution to a test tube, and evaporate on a steam bath with the aid of a current of filtered air to dryness. Dissolve the residue in 2.0 mL of *Internal standard solution.*

Chromatographic system (see *Chromatography* ⟨621⟩)—The gas chromatograph is equipped with a flame-ionization detector and contains a 1.2-m × 2-mm column packed with 3 percent phase G32 on support S1A. The injection port, detector, and column are maintained at temperatures of about 250°, 300°, and 250°, respectively, and helium is used as the carrier gas, flowing at a rate of about 50 mL per minute. The relative retention times for cholestane and miconazole nitrate are about 0.44 and 1.0, respectively. Chromatograph the *Standard preparation,* and record the peak responses as directed under *Procedure:* the relative standard deviation of replicate injections is not more than 3.0%, and the resolution between the cholestane and miconazole nitrate peaks is not less than 2.0.

Procedure—Separately inject equal volumes (about 5 µL) of the *Assay preparation* and the *Standard preparation* into the gas chromatograph, record the chromatograms, and measure the peak responses. Calculate the quantity, in mg, of $C_{18}H_{14}Cl_4N_2O \cdot HNO_3$ in the portion of Cream taken by the formula:

$$0.04C(R_U/R_S),$$

in which C is the concentration, in µg per mL, of USP Miconazole Nitrate RS in the *Standard preparation,* and R_U and R_S are the peak response ratios of miconazole nitrate to cholestane obtained from the *Assay preparation* and the *Standard preparation,* respectively.

Miconazole Nitrate Vaginal Suppositories

» Miconazole Nitrate Vaginal Suppositories contain not less than 90.0 percent and not more than 110.0 percent of the labeled amount of $C_{18}H_{14}Cl_4N_2O \cdot HNO_3$.

Packaging and storage—Preserve in tight containers, at controlled room temperature.

Reference standard—*USP Miconazole Nitrate Reference Standard*—Dry at 105° for 2 hours before using.

Identification—Place a portion of the stock solution, prepared as directed in the *Assay,* containing about 25 mg of miconazole nitrate, in a 50-mL beaker, and evaporate on a steam bath with the aid of a current of filtered air to dryness. Dry the residue at 105° for 10 minutes: the infrared absorption spectrum of a potassium bromide dispersion of it so obtained exhibits maxima only at the same wavelengths as that of a similar preparation of USP Miconazole Nitrate RS.

Assay—

Internal standard solution, Standard preparation, and *Chromatographic system*—Prepare as directed in the *Assay* under *Miconazole Nitrate Cream.*

Assay preparation—Transfer 1 Miconazole Nitrate Vaginal Suppository to a stoppered, 50-mL centrifuge tube. Add 30 mL of pentane, and shake by mechanical means for 20 minutes to dissolve the suppository base and to disperse the miconazole nitrate. Centrifuge to obtain a clear supernatant liquid. Aspirate and discard the clear liquid. Wash the residue with three 20-mL portions of pentane, shaking, centrifuging, and aspirating in the same manner. Discard the pentane washings. Evaporate the residual pentane from the residue with the aid of a current of filtered air. Using small portions of methanol, transfer the residue to a 100-mL volumetric flask. Dissolve in methanol, dilute with methanol to volume, and mix. Transfer an accurately measured volume of this stock solution, equivalent to about 5 mg of miconazole nitrate, to a suitable container, and evaporate on a steam bath with the aid of a current of filtered air to dryness. Dissolve the residue in 2.0 mL of *Internal standard solution.*

Procedure—Proceed as directed for *Procedure* in the *Assay* under *Miconazole Nitrate Cream.* Calculate the quantity, in mg, of $C_{18}H_{14}Cl_4N_2O \cdot HNO_3$ in the Suppository taken by the formula:

$$(0.2C/V)(R_U/R_S),$$

in which V is the volume, in mL, of stock solution used to prepare the *Assay preparation,* and the other terms are as defined therein.

Microcrystalline Cellulose—*see* Cellulose, Microcrystalline NF

Microcrystalline Wax—*see* Wax, Microcrystalline NF

Milk of Bismuth—*see* Bismuth, Milk of

Milk of Magnesia—*see* Magnesia, Milk of

Mineral Oil

» Mineral Oil is a mixture of liquid hydrocarbons obtained from petroleum. It may contain a suitable stabilizer.

Packaging and storage—Preserve in tight containers.

Labeling—Label it to indicate the name of any substance added as a stabilizer.

Specific gravity ⟨841⟩: between 0.845 and 0.905.

Viscosity ⟨911⟩—It has a kinematic viscosity of not less than 34.5 centistokes at 40.0°.

Neutrality—Boil 10 mL with an equal volume of alcohol: the alcohol remains neutral to moistened litmus paper.

Readily carbonizable substances—Place 5 mL in a glass-stoppered test tube that previously has been rinsed with chromic acid cleansing mixture (see *Cleaning Glass Apparatus* ⟨1051⟩), then rinsed with water, and dried. Add 5 mL of sulfuric acid containing from 94.5% to 94.9% of H_2SO_4, and heat in a boiling water bath for 10 minutes. After the test tube has been in the bath for 30 seconds, remove it quickly, and, while holding the stopper in place, give three vigorous, vertical shakes over an amplitude of about 5 inches. Repeat every 30 seconds. Do not keep the test tube out of the bath longer than 3 seconds for each shaking period. At the end of 10 minutes from the time when first placed in the water bath, remove the test tube: the Oil may turn hazy, but it remains colorless, or shows a slight pink or yellow color, and the acid does not become darker than the standard color produced by mixing in a similar test tube 3 mL of ferric chloride CS, 1.5 mL of cobaltous chloride CS, and 0.5 mL of cupric sulfate CS, this mixture being overlaid with 5 mL of Mineral Oil (see *Readily Carbonizable Substances Test* ⟨271⟩).

Limit of polynuclear compounds—

Methyl sulfoxide—Use methyl sulfoxide that has an absorbance, compared to that of water, in a 1-cm cell, not greater than 1.0 at 264 nm, and that shows no extraneous impurity peaks in the wavelength region up to 350 nm.

Standard solution—Dissolve a suitable quantity of naphthalene, accurately weighed, in isooctane, and dilute quantitatively and stepwise with isooctane to obtain a solution having a concentration of 7.0 µg per mL. Determine the absorbance of this solution in a 1-cm cell at the maximum at about 275 nm, with a suitable spectrophotometer, using isooctane as the blank.

Procedure—Transfer 25.0 mL of Mineral Oil and 25 mL of *n*-hexane to a 125-mL separator, and mix. [NOTE—Use only *n*-hexane that previously has been washed by being shaken twice with one-fifth its volume of *Methyl sulfoxide.* Use no lubricants other than water on the stopcock, or use a separator equipped with a suitable polymeric stopcock.] Add 5.0 mL of *Methyl sulfoxide,* and shake the mixture vigorously for 1 minute. Allow to stand until the lower layer is clear, transfer the lower layer to another 125-mL separator, add 2 mL of *n*-hexane, and shake vigorously. Separate the lower layer, and determine its absorbance in a 1-cm cell, in the range of 260 nm to 350 nm, with a suitable spectrophotometer, using as the blank *Methyl sulfoxide* that previously has been shaken vigorously for 1 minute with *n*-hexane in the ratio of 5 mL of *Methyl sulfoxide* and 25 mL of *n*-hexane. The absorbance at any wavelength in the specified range is not greater than one-third of the absorbance, at 275 nm, of the *Standard solution.*

Solid paraffin—Fill a tall, cylindrical, standard oil-sample bottle of colorless glass of about 120-mL capacity with Mineral Oil that has been dried previously in a beaker at 105° for 2 hours and cooled to room temperature in a desiccator over silica gel. Insert the stopper, and immerse the bottle in a mixture of ice and water for 4 hours: the Oil is sufficiently clear that a black line 0.5 mm in width, on a white background, held vertically behind the bottle, is clearly visible.

Mineral Oil Emulsion

» Prepare Mineral Oil Emulsion as follows:

Mineral Oil	500 mL
Acacia, in very fine powder	125 g
Syrup	100 mL
Vanillin	40 mg
Alcohol	60 mL
Purified Water, a sufficient quantity, to make	1000 mL

Mix the Mineral Oil with the Powdered Acacia in a dry mortar, add 250 mL of Purified Water all at once, and emulsify the mixture. Then add, in divided portions, triturating after each addition, a mixture of the Syrup, 50 mL of Purified Water, and the Vanillin dissolved in the alcohol. Finally add Purified Water to make the product measure 1000 mL, and mix.

The Vanillin may be replaced by not more than 1 percent of any other official flavoring substance or any mixture of official flavoring substances. Sixty mL of sweet orange peel tincture or 2 g of benzoic acid may be used as a preservative in place of the Alcohol.

For other permissible modifications, see *Emulsions* ⟨1151⟩.

Packaging and storage—Preserve in tight containers.

Alcohol content, *Method I* ⟨611⟩: between 4.0% and 6.0% of C_2H_5OH.

Mineral Oil Enema

» Mineral Oil Enema is Mineral Oil that has been suitably packaged.

Packaging and storage—Preserve in tight, single-unit containers.

Specific gravity ⟨841⟩: between 0.845 and 0.905.

Viscosity ⟨911⟩—It has a kinematic viscosity of not less than 34.5 centistokes at 40.0°.

Neutrality—Boil 10 mL with an equal volume of alcohol: the alcohol remains neutral to moistened litmus paper.

Mineral Oil, Light—*see* Mineral Oil, Light NF

Topical Light Mineral Oil

» Topical Light Mineral Oil is Light Mineral Oil that has been suitably packaged.

Packaging and storage—Preserve in tight containers.

Labeling—Label it to indicate the name of any substance added as a stabilizer, and label packages intended for direct use by the public to indicate that it is not intended for internal use.

Specific gravity ⟨841⟩: between 0.818 and 0.880.

Viscosity ⟨911⟩—It has a kinematic viscosity of not more than 33.5 centistokes at 40°.

Neutrality, Readily carbonizable substances, and Solid paraffin—It meets the requirements of the tests for *Neutrality, Readily carbonizable substances,* and *Solid paraffin* under *Mineral Oil.*

Minocycline Hydrochloride

$C_{23}H_{27}N_3O_7 . HCl$ 493.94

2-Naphthacenecarboxamide, 4,7-bis(dimethylamino)-1,4,4a,5,-5a,6,11,12a-octahydro-3,10,12,12a-tetrahydroxy-1,11-dioxo-, monohydrochloride, [4*S*-(4α,4aα,5aα,12aα)]-.

4,7-Bis(dimethylamino)-1,4,4a,5,5a,6,11,12a-octahydro-3,10,-12,12a-tetrahydroxy-1,11-dioxo-2-naphthacenecarboxamide monohydrochloride [*13614-98-7*].

» Minocycline Hydrochloride contains the equivalent of not less than 875 µg and not more than 950 µg of minocycline ($C_{23}H_{27}N_3O_7$) per mg, calculated on the anhydrous basis.

Packaging and storage—Preserve in tight containers, protected from light.

Reference standard—*USP Minocycline Hydrochloride Reference Standard*—Do not dry before using.

Identification—The infrared absorption spectrum of a potassium chloride dispersion of it, previously dried at 100° for 2 hours, exhibits maxima only at the same wavelengths as that of a similar preparation of USP Minocycline Hydrochloride RS.

Crystallinity ⟨695⟩: meets the requirements.

pH ⟨791⟩: between 3.5 and 4.5, in a solution containing the equivalent of 10 mg of minocycline per mL.

Water, *Method I* ⟨921⟩: between 4.3% and 8.0%.

Heavy metals, *Method II* ⟨231⟩: 0.005%.

Chromatographic purity—Calculate the area percentage of each peak (except the solvent peak) observed in the chromatogram of the *Assay preparation* obtained in the *Assay:* epi-minocycline, if present, has a retention time of about 0.86 relative to that of minocycline, and constitutes not more than 2.0% of the total area, and the sum of any other extraneous peaks constitutes not more than 2.0% of the total area.

Assay—

Mobile phase—Prepare a suitable mixture of 0.2 *M* ammonium oxalate, dimethylformamide, and 0.1 *M* disodium ethylenediaminetetraacetate (550:250:200). Adjust with 0.4 *M* tetrabutylammonium hydroxide to a pH of 6.2, and filter through a membrane filter (0.5-µm or finer porosity).

Standard preparation—Dissolve an accurately weighed quantity of USP Minocycline Hydrochloride RS in *Mobile phase* to obtain a solution having a known concentration of about 500 µg of minocycline ($C_{23}H_{27}N_3O_7$) per mL.

Resolution solution—Prepare a solution in *Mobile phase* containing, in each mL, 0.24 mg of USP Minocycline Hydrochloride RS and 0.24 mg of chlortetracycline hydrochloride.

Assay preparation—Transfer an accurately weighed quantity of Minocycline Hydrochloride, equivalent to about 50 mg of minocycline ($C_{23}H_{27}N_3O_7$), to a 100-mL volumetric flask, dilute with *Mobile phase* to volume, and mix.

Chromatographic system (see *Chromatography* ⟨621⟩)—The liquid chromatograph is equipped with a 280-nm detector, a 4.6-mm × 3-cm guard column that contains 10-μm packing L7, and a 4.6-mm × 15-cm analytical column that contains 5-μm packing L7. The flow rate is about 2 mL per minute. Chromatograph the *Standard preparation*, and record the peak responses as directed under *Procedure:* the relative standard deviation for replicate injections is not more than 2.0%. Chromatograph the *Resolution solution:* the resolution, *R*, between the minocycline and chlortetracycline peaks is not less than 3.0.

Procedure—Separately inject equal volumes (about 10 μL) of the *Standard preparation* and the *Assay preparation* into the chromatograph, record the chromatograms, and measure the responses for the major peaks. Calculate the content of minocycline ($C_{23}H_{27}N_3O_7$), in μg per mg, taken by the formula:

$$100(C/W)(r_U/r_S),$$

in which *C* is the concentration, in μg per mL, of minocycline ($C_{23}H_{27}N_3O_7$) in the *Standard preparation*, *W* is the weight, in mg, of Minocycline Hydrochloride taken, and r_U and r_S are the peak responses of the major peaks obtained from the *Assay preparation* and the *Standard preparation*, respectively.

Minocycline Hydrochloride Capsules

» Minocycline Hydrochloride Capsules contain the equivalent of not less than 90.0 percent and not more than 115.0 percent of the labeled amount of minocycline ($C_{23}H_{27}N_3O_7$).

Packaging and storage—Preserve in tight, light-resistant containers.

Reference standard—*USP Minocycline Hydrochloride Reference Standard*—Do not dry before using.

Identification—The retention time of the major peak in the chromatogram of the *Assay preparation* corresponds to that of the *Standard preparation*, obtained as directed in the *Assay*.

Dissolution ⟨711⟩—

Medium: water; 900 mL.

Apparatus 2: 50 rpm.

Time: 45 minutes.

Procedure—Determine the amount of $C_{23}H_{27}N_3O_7$ dissolved from ultraviolet absorbances at the wavelength of maximum absorbance at about 358 nm of filtered portions of the solution under test, suitably diluted with *Dissolution Medium*, if necessary, in comparison with a Standard solution having a known concentration of USP Minocycline Hydrochloride RS in the same medium.

Tolerances—Not less than 75% (*Q*) of the labeled amount of $C_{23}H_{27}N_3O_7$ is dissolved in 45 minutes.

Uniformity of dosage units ⟨905⟩: meet the requirements.

Water, *Method I* ⟨921⟩: not more than 12.0%.

Assay—

Mobile phase, Standard preparation, Resolution solution, and *Chromatographic system*—Proceed as directed in the *Assay* under *Minocycline Hydrochloride.*

Assay preparation—Weigh the contents of not less than 20 Minocycline Hydrochloride Capsules, and calculate the average weight per Capsule. Mix the combined contents of the Capsules, transfer an accurately weighed portion, equivalent to about 50 mg of minocycline ($C_{23}H_{27}N_3O_7$), to a 100-mL volumetric flask, add about 50 mL of *Mobile phase*, and shake to dissolve. Dilute with *Mobile phase* to volume, mix, and filter.

Procedure—Proceed as directed for *Procedure* in the *Assay* under *Minocycline Hydrochloride.* Calculate the quantity, in mg, of $C_{23}H_{27}N_3O_7$ in the portion of Capsules taken by the formula:

$$0.1C(r_U/r_S).$$

Sterile Minocycline Hydrochloride

» Sterile Minocycline Hydrochloride is sterile, freeze-dried Minocycline Hydrochloride suitable for parenteral use. It contains the equivalent of not less than 90.0 percent and not more than 120.0 percent of the labeled amount of minocycline ($C_{23}H_{27}N_3O_7$).

Packaging and storage—Preserve in *Containers for Sterile Solids* as described under *Injections* ⟨1⟩, protected from light.

Reference standard—*USP Minocycline Hydrochloride Reference Standard*—Do not dry before using.

Constituted solution—At the time of use, the constituted solution prepared from Sterile Minocycline Hydrochloride meets the requirements for *Constituted Solutions* under *Injections* ⟨1⟩.

Identification—The retention time of the major peak in the chromatogram of the *Assay preparation* corresponds to that of the *Standard preparation*, obtained as directed in the *Assay*.

Depressor substances—When diluted with sterile saline TS to a concentration of 5.0 mg of minocycline ($C_{23}H_{27}N_3O_7$) per mL, it meets the requirements of the *Depressor Substances Test* ⟨101⟩, the test dose being 0.6 mL per kg.

Pyrogen—When diluted with Sterile Water for Injection to a concentration of 5.0 mg of minocycline ($C_{23}H_{27}N_3O_7$) per mL, it meets the requirements of the *Pyrogen Test* ⟨151⟩, the test dose being 1.0 mL per kg.

pH ⟨791⟩: between 2.0 and 3.5, in a solution containing the equivalent of 10 mg of minocycline per mL.

Water, *Method I* ⟨921⟩: not more than 3.0%, the *Test preparation* being prepared as directed for a hygroscopic specimen.

Particulate matter ⟨788⟩: meets the requirements under *Small-volume Injections.*

Other requirements—It conforms to the definition, and meets the requirements for *Heavy metals, Chromatographic purity*, and *Assay* under *Minocycline Hydrochloride.* It meets also the requirements for *Sterility Tests* ⟨71⟩, *Uniformity of Dosage Units* ⟨905⟩, and *Labeling* under *Injections* ⟨1⟩.

Assay—

Mobile phase, Standard preparation, Resolution solution, and *Chromatographic system*—Proceed as directed in the *Assay* under *Minocycline Hydrochloride.*

Assay preparation 1 (where it is represented as being in a single-dose container)—Constitute Sterile Minocycline Hydrochloride in a volume of water, accurately measured, corresponding to the volume of solvent specified in the labeling. Withdraw all of the withdrawable contents, using a hypodermic needle and syringe, and dilute quantitatively with *Mobile phase* to obtain a solution containing the equivalent of about 0.5 mg of minocycline ($C_{23}H_{27}N_3O_7$) per mL.

Assay preparation 2 (where the label states the quantity of minocycline in a given volume of constituted solution)—Constitute Sterile Minocycline Hydrochloride in a volume of water, accurately measured, corresponding to the volume of solvent specified in the labeling. Dilute an accurately measured portion of the constituted solution quantitatively with *Mobile phase* to obtain a solution containing the equivalent of about 0.5 mg of minocycline ($C_{23}H_{27}N_3O_7$) per mL.

Procedure—Proceed as directed for *Procedure* in the *Assay* under *Minocycline Hydrochloride.* Calculate the quantity, in mg, of minocycline ($C_{23}H_{27}N_3O_7$) in the container, or in the portion of constituted solution taken by the formula:

$$0.001C(L/D)(r_U/r_S),$$

in which *L* is the labeled quantity, in mg, of minocycline in the container, or in the volume of constituted solution taken, *D* is the concentration, in mg per mL, of minocycline in *Assay preparation 1* or in *Assay preparation 2*, on the basis of the labeled quantity in the container, or in the portion of constituted solution taken, respectively, and the extent of dilution, and the other terms are as defined therein.

Minocycline Hydrochloride Oral Suspension

» Minocycline Hydrochloride Oral Suspension contains the equivalent of not less than 90.0 percent and not more than 130.0 percent of the labeled amount of minocycline ($C_{23}H_{27}N_3O_7$), and one or more suitable diluents, flavors, preservatives, and wetting agents in an aqueous vehicle.

Packaging and storage—Preserve in tight, light-resistant containers.

Reference standard—*USP Minocycline Hydrochloride Reference Standard*—Do not dry before using.

Identification—The retention time of the major peak in the chromatogram of the *Assay preparation* corresponds to that of the *Standard preparation*, obtained as directed in the *Assay*.

Uniformity of dosage units ⟨905⟩—
FOR SUSPENSION PACKAGED IN SINGLE-UNIT CONTAINERS: meets the requirements.

pH ⟨791⟩: between 7.0 and 9.0.

Assay—
Mobile phase, Standard preparation, Resolution solution, and *Chromatographic system*—Proceed as directed in the *Assay* under *Minocycline Hydrochloride.*
Assay preparation—Transfer an accurately measured quantity of Minocycline Hydrochloride Oral Suspension, freshly mixed and free from air bubbles, equivalent to about 50 mg of $C_{23}H_{27}N_3O_7$, to a 100-mL volumetric flask, dilute with *Mobile phase* to volume, mix, and filter.
Procedure—Proceed as directed for *Procedure* in the *Assay* under *Minocycline Hydrochloride.* Calculate the quantity, in mg, of $C_{23}H_{27}N_3O_7$ in each mL of the Oral Suspension taken by the formula:

$$0.1(C/V)(r_U/r_S),$$

in which V is the volume, in mL, of Oral Suspension taken, and the other terms are as defined therein.

Minocycline Hydrochloride Tablets

» Minocycline Hydrochloride Tablets contain the equivalent of not less than 90.0 percent and not more than 115.0 percent of the labeled amount of minocycline ($C_{23}H_{27}N_3O_7$).

Packaging and storage—Preserve in tight, light-resistant containers.

Reference standard—*USP Minocycline Hydrochloride Reference Standard*—Do not dry before using.

Identification—The retention time of the major peak in the chromatogram of the *Assay preparation* corresponds to that of the *Standard preparation*, obtained as directed in the *Assay*.

Dissolution ⟨711⟩—
Medium: water; 900 mL.
Apparatus 2: 50 rpm.
Time: 45 minutes.
Procedure—Determine the amount of $C_{23}H_{27}N_3O_7$ dissolved from ultraviolet absorbances at the wavelength of maximum absorbance at about 358 nm of filtered portions of the solution under test, suitably diluted with *Dissolution Medium*, if necessary, in comparison with a Standard solution having a known concentration of USP Minocycline Hydrochloride RS in the same medium.
Tolerances—Not less than 75% (Q) of the labeled amount of $C_{23}H_{27}N_3O_7$ is dissolved in 45 minutes.

Uniformity of dosage units ⟨905⟩: meet the requirements.

Water, *Method I* ⟨921⟩: not more than 12.0%.

Assay—
Mobile phase, Standard preparation, Resolution solution, and *Chromatographic system*—Proceed as directed in the *Assay* under *Minocycline Hydrochloride.*
Assay preparation—Weigh and finely powder not less than 20 Minocycline Hydrochloride Tablets. Transfer an accurately weighed portion of the powder, equivalent to about 50 mg of minocycline ($C_{23}H_{27}N_3O_7$), to a 100-mL volumetric flask, add about 50 mL of *Mobile phase*, and shake for about 1 minute. Dilute with *Mobile phase* to volume, mix, and filter.
Procedure—Proceed as directed for *Procedure* in the *Assay* under *Minocycline Hydrochloride.* Calculate the quantity, in mg, of $C_{23}H_{27}N_3O_7$ in the portion of Tablets taken by the formula:

$$0.1C(r_U/r_S).$$

Minoxidil

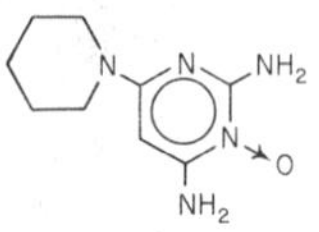

$C_9H_{15}N_5O$ 209.25
2,4-Pyrimidinediamine, 6-(1-piperidinyl)-, 3-oxide.
2,4-Diamino-6-piperidinopyrimidine 3-oxide [*38304-91-5*].

» Minoxidil contains not less than 97.0 percent and not more than 103.0 percent of $C_9H_{15}N_5O$, calculated on the dried basis.

Packaging and storage—Preserve in well-closed containers.

Reference standard—*USP Minoxidil Reference Standard*—Do not dry before using.

Identification—The infrared absorption spectrum of a mineral oil dispersion of it exhibits maxima only at the same wavelengths as that of a similar preparation of USP Minoxidil RS.

Loss on drying ⟨731⟩—Dry it at 50° and at a pressure not exceeding 5 mm of mercury for 3 hours: it loses not more than 0.5% of its weight.

Residue on ignition ⟨281⟩: not more than 0.5%.

Heavy metals, *Method II* ⟨231⟩: 0.002%.

Chromatographic purity—
Mobile phase and *Chromatographic system*—Prepare as directed in the *Assay*.
Test solution—Prepare a solution of Minoxidil in *Mobile phase* having a concentration of about 0.25 mg per mL.
Procedure—Inject about 10 µL of *Test solution* into the chromatograph, record the chromatogram, and measure the peak response for each component. Calculate the total percentage of impurities by the formula:

$$100S/(S + A),$$

in which S is the sum of the areas of the minor component peaks detected, and A is the area of the major component. The total of any impurities detected is not more than 1.5%.

Assay—
Mobile phase—Prepare a solution consisting of a mixture of methanol, water, and glacial acetic acid (700:300:10), add 3.0 g of docusate sodium per liter of solution, and mix. Adjust with perchloric acid to a pH of 3.0, filter, and degas.
Internal standard solution—Prepare a solution of medroxyprogesterone acetate in *Mobile phase* having a concentration of about 0.2 mg per mL.
Standard preparation—Dissolve an accurately weighed quantity of USP Minoxidil RS in *Internal standard solution* to obtain a solution having a known concentration of about 0.25 mg per mL.
Assay preparation—Transfer about 5 mg of Minoxidil, accurately weighed, to a container, add 20.0 mL of *Internal standard solution*, and mix.

Chromatographic system (see *Chromatography* ⟨621⟩)—The liquid chromatograph is equipped with a 254-nm detector and a 4-mm × 25-cm column that contains packing L1. The flow rate is about 1 mL per minute. Chromatograph not less than four replicate injections of the *Standard preparation*, and record the peak responses as directed under *Procedure:* the relative standard deviation is not more than 2.0%, and the resolution, *R*, between the internal standard and minoxidil is not less than 2.0.

Procedure—Separately inject equal volumes (about 10 µL) of the *Standard preparation* and the *Assay preparation* into the chromatograph, record the chromatograms, and measure the responses for the major peaks. The relative retention times are about 0.8 for the internal standard and 1.0 for minoxidil. Calculate the quantity, in mg, of $C_9H_{15}N_5O$ in the portion of Minoxidil taken by the formula:

$$20C(R_U/R_S),$$

in which *C* is the concentration, in mg per mL, of USP Minoxidil RS in the *Standard preparation*, and R_U and R_S are the peak response ratios obtained from the *Assay preparation* and the *Standard preparation*, respectively.

Minoxidil Tablets

» Minoxidil Tablets contain not less than 90.0 percent and not more than 110.0 percent of the labeled amount of $C_9H_{15}N_5O$.

Packaging and storage—Preserve in tight containers.

Reference standard—*USP Minoxidil Reference Standard*—Do not dry before using.

Identification—Transfer a portion of finely powdered Tablets, equivalent to about 10 mg of minoxidil, to a separator. Add 25 mL of water, and extract with three 15-mL portions of chloroform. Combine the chloroform extracts, and evaporate with the aid of a stream of nitrogen. Wash the inside of the container with about 5 mL of alcohol, add 300 mg of potassium bromide, and evaporate under vacuum at 50° until dry: the infrared absorption spectrum of a potassium bromide dispersion prepared from the residue so obtained exhibits maxima at the same wavelengths as that of a similar preparation of USP Minoxidil RS.

Dissolution ⟨711⟩—

Medium: pH 7.2 phosphate buffer (see under *Buffer Solutions* in the section, *Reagents, Indicators, and Solutions*); 900 mL.

Apparatus 1: 75 rpm.

Time: 15 minutes.

Procedure—Determine the amount of $C_9H_{15}N_5O$ dissolved from ultraviolet absorbances at the wavelength of maximum absorbance of filtered portions of the solution under test, suitably diluted with *Dissolution Medium*, if necessary, in comparison with a Standard solution having a known concentration of USP Minoxidil RS in the same medium. For Tablets containing up to 10 mg of minoxidil, measurement is made at about 231 nm; for Tablets containing more than 10 mg, the wavelength used is about 287 nm.

Tolerances—Not less than 50% (*Q*) of the labeled amount of $C_9H_{15}N_5O$ is dissolved in 15 minutes.

Uniformity of dosage units ⟨905⟩: meet the requirements.

Assay—

Mobile phase, Internal standard solution, Standard preparation, and *Chromatographic system*—Proceed as directed in the *Assay* under *Minoxidil.*

Assay preparation—Weigh and finely powder not less than 10 Minoxidil Tablets. To an accurately weighed portion of the powder, equivalent to about 5 mg of minoxidil, add 20.0 mL of *Internal standard solution*, shake for 5 minutes, and centrifuge.

Procedure—Proceed as directed for *Procedure* in the *Assay* under *Minoxidil.* Calculate the quantity, in mg, of $C_9H_{15}N_5O$ in the portion of Tablets taken by the formula:

$$20C(R_U/R_S),$$

in which the terms are as defined therein.

Mithramycin—*see* Plicamycin

Mithramycin for Injection—*see* Plicamycin for Injection

Mitomycin

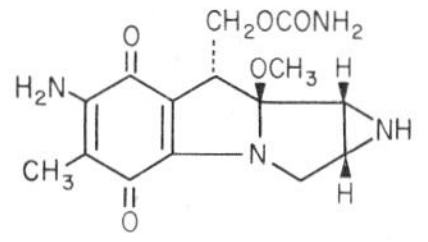

$C_{15}H_{18}N_4O_5$ 334.33

Azirino[2′,3′:3,4]pyrrolo[1,2-*a*]indole-4,7-dione, 6-amino-8-[[(aminocarbonyl)oxy]methyl]-1,1a,2,8,8a,8b-hexahydro-8a-methoxy-5-methyl-, [1a*R*-(1aα,8β,8aα,8bα)]-.

6-Amino-1,1a,2,8,8a,8b-hexahydro-8-(hydroxymethyl)-8a-methoxy-5-methylazirino[2′,3′:3,4]pyrrolo[1,2-*a*]indole-4,7-dione carbamate (ester).

Mitomycin C [*50-07-7*].

» Mitomycin has a potency of not less than 900 µg of $C_{15}H_{18}N_4O_5$ per mg.

Packaging and storage—Preserve in tight, light-resistant containers.

Reference standard—*USP Mitomycin Reference Standard*—Do not dry before using.

Identification—

A: The infrared absorption spectrum of a mineral oil dispersion of it exhibits maxima only at the same wavelengths as that of a similar preparation of USP Mitomycin RS.

B: Transfer about 25 mg, accurately weighed, to a 50-mL volumetric flask, dilute with methanol to volume, and mix. Dilute a portion of this solution with methanol to obtain a solution having a concentration of about 0.005 mg of mitomycin per mL: the ultraviolet absorption spectrum of the solution so obtained exhibits maxima and minima at the same wavelengths as that of a similar solution having a known concentration of USP Mitomycin RS, concomitantly measured, and the absorptivity, calculated on the anhydrous basis, at the wavelength of maximum absorbance at about 357 nm is not less than 95.0% and not more than 105.0% of that of the USP Mitomycin RS.

Crystallinity ⟨695⟩: meets the requirements.

pH ⟨791⟩: between 6.0 and 8.0, in a solution containing 5 mg per mL.

Water, *Method I* ⟨921⟩: not more than 5.0%, a mixture of carbon tetrachloride, chloroform, and methanol (2:2:1) being used in place of methanol in the titration vessel.

Assay—

Mobile phase—Dissolve 1.54 g of ammonium acetate in 250 mL of methanol, add 5.0 mL of 0.83 *N* acetic acid and water to make 1000 mL, and mix. Filter through a 0.5-µm or finer porosity filter, and degas. Make adjustments if necessary (see *System Suitability* under *Chromatography* ⟨621⟩).

Standard preparation—Dissolve an accurately weighed quantity of USP Mitomycin RS quantitatively in *N,N*-dimethylacetamide to obtain a solution having a known concentration of about 0.5 mg per mL.

Resolution solution—Dissolve suitable quantities of USP Mitomycin RS and 3-ethoxy-4-hydroxybenzaldehyde in *N,N*-dimethylacetamide to obtain a solution containing about 0.5 mg and 7.5 mg per mL, respectively.

Assay preparation—Transfer about 25 mg of Mitomycin, accurately weighed, to a 50-mL volumetric flask, add *N,N*-dimethylacetamide to volume, and mix to dissolve.

Chromatographic system (see *Chromatography* ⟨621⟩)—The liquid chromatograph is equipped with a 365-nm detector and a 4-mm × 30-cm column that contains packing L11. The flow rate is about 2 mL per minute. Chromatograph the *Resolution solution*, and record the peak responses as directed under *Procedure:* the resolution, *R*, between the mitomycin and the 3-ethoxy-4-hydroxybenzaldehyde peaks is not less than 1.8. The relative retention times are about 1.0 for mitomycin and 1.4 for 3-ethoxy-4-hydroxybenzaldehyde. Chromatograph the *Standard preparation*, and record the peak responses as directed under *Procedure:* the tailing factor for the mitomycin peak is not more than 1.3, and the relative standard deviation for replicate injections is not more than 2.0%.

Procedure—[NOTE—Use peak areas where peak responses are indicated.] Separately inject equal volumes (about 10 µL) of the *Standard preparation* and the *Assay preparation* into the chromatograph, record the chromatograms, and measure the responses for the major peaks. Calculate the quantity, in mg, of $C_{15}H_{18}N_4O_5$ in the portion of Mitomycin taken by the formula:

$$50C(r_U/r_S),$$

in which C is the concentration, in mg per mL, of USP Mitomycin RS in the *Standard preparation*, and r_U and r_S are the peak responses obtained from the *Assay preparation* and the *Standard preparation*, respectively.

Mitomycin for Injection

» Mitomycin for Injection is a dry mixture of Mitomycin and Mannitol. It contains not less than 90.0 percent and not more than 120.0 percent of the labeled amount of $C_{15}H_{18}N_4O_5$.

Packaging and storage—Preserve in *Containers for Sterile Solids* as described under *Injections* ⟨1⟩, protected from light.

Reference standard—*USP Mitomycin Reference Standard*—Do not dry before using.

Constituted solution—At the time of use, the constituted solution prepared from Mitomycin for Injection meets the requirements for *Constituted Solutions* under *Injections* ⟨1⟩.

Identification—Dissolve a quantity in water, and dilute with water to obtain a solution having a concentration of about 1 mg of mitomycin per mL. On a suitable thin-layer chromatographic plate (see *Chromatography* ⟨621⟩), coated with a 0.25-mm layer of chromatographic silica gel mixture, apply 2 µL of this solution and 2 µL of a Standard solution of USP Mitomycin RS, similarly prepared. Allow the spots to dry, and develop the chromatograms in a solvent system consisting of a mixture of butyl alcohol, glacial acetic acid, and water (4:2:1). Remove the plate from the developing chamber, mark the solvent front, and allow the solvent to evaporate. Spray the plate with a 1 in 100 solution of ninhydrin in alcohol. Heat the plate in an oven at 110° for 15 minutes. Mitomycin appears as a pink spot: the R_f value of the principal spot obtained from the solution under test corresponds to that obtained from the Standard solution.

Depressor substances—When diluted with sterile saline TS to a concentration of 0.05 mg of mitomycin ($C_{15}H_{18}N_4O_5$) per mL, it meets the requirements of the *Depressor Substances Test* ⟨101⟩, the test dose being 1.0 mL per kg.

Pyrogen—When diluted with sterile, pyrogen-free saline TS to a concentration of 0.5 mg of mitomycin ($C_{15}H_{18}N_4O_5$) per mL, it meets the requirements of the *Pyrogen Test* ⟨151⟩, the test dose being 1.0 mL per kg.

Sterility—It meets the requirements under *Sterility Tests* ⟨71⟩, when tested as directed in the section, *Test Procedures Using Membrane Filtration.*

pH ⟨791⟩: between 6.0 and 8.0, in the solution constituted as directed in the labeling.

Water, *Method I* ⟨921⟩: not more than 5.0%.

Other requirements—It meets the requirements under *Injections* ⟨1⟩.

Assay—

Mobile phase, Standard preparation, Resolution solution, and *Chromatographic system*—Prepare as directed in the *Assay* under *Mitomycin.*

Assay preparation—Add an accurately measured volume of *N,N*-dimethylacetamide to 1 container of Mitomycin for Injection to obtain a solution having a concentration of about 0.5 mg of mitomycin per mL.

Procedure—[NOTE—Use peak areas where peak responses are indicated.] Separately inject equal volumes (about 10 µL) of the *Standard preparation* and the *Assay preparation* into the chromatograph, record the chromatograms, and measure the responses for the major peaks. Calculate the quantity, in mg, of $C_{15}H_{18}N_4O_5$ in the container of Mitomycin for Injection taken by the formula:

$$C(L/D)(r_U/r_S),$$

in which C is the concentration, in mg per mL, of USP Mitomycin RS in the *Standard preparation*, L is the labeled quantity, in mg, of mitomycin in the container, D is the concentration, in mg per mL, of mitomycin in the *Assay preparation*, on the basis of the labeled quantity and the extent of dilution, and r_U and r_S are the peak responses obtained from the *Assay preparation* and the *Standard preparation*, respectively.

Mitotane

Cl, CH, CHCl2, Cl

$C_{14}H_{10}Cl_4$ 320.04

Benzene, 1-chloro-2-[2,2-dichloro-1-(4-chlorophenyl)ethyl]-.

1,1-Dichloro-2-(*o*-chlorophenyl)-2-(*p*-chlorophenyl)ethane [*53-19-0*].

» Mitotane contains not less than 97.0 percent and not more than 103.0 percent of $C_{14}H_{10}Cl_4$, calculated on the dried basis.

Caution—Handle Mitotane with exceptional care, since it is a highly potent agent.

Packaging and storage—Preserve in tight, light-resistant containers.

Reference standard—*USP Mitotane Reference Standard*—Dry in vacuum at 60° for 2 hours before using.

Identification—

A: The infrared absorption spectrum of a mineral oil dispersion of it exhibits maxima only at the same wavelengths as that of a similar preparation of USP Mitotane RS.

B: The ultraviolet absorption spectrum of a 1 in 5000 solution in methanol exhibits maxima and minima at the same wavelengths as that of a similar solution of USP Mitotane RS, concomitantly measured.

Melting range ⟨741⟩: between 75° and 81°.

Loss on drying ⟨731⟩—Dry it in vacuum at 60° for 2 hours: it loses not more than 0.5% of its weight.

Residue on ignition ⟨281⟩: not more than 0.5%.

Assay—Dissolve about 0.1 g of Mitotane, accurately weighed, in methanol, and dilute quantitatively and stepwise with methanol to obtain a solution having a concentration of about 200 µg per mL. Similarly prepare a solution in methanol of USP Mitotane RS, diluted quantitatively and stepwise with methanol to yield a Standard solution having a known concentration of about 200 µg per mL. Concomitantly determine the absorbances of both solutions in 1-cm cells at the wavelength of maximum absorbance at about 268 nm, with a suitable spectrophotometer, using methanol as the blank. Calculate the quantity, in mg, of $C_{14}H_{10}Cl_4$ in the portion of Mitotane taken by the formula:

$$0.5C(A_U/A_S),$$

in which C is the concentration, in µg per mL, of USP Mitotane RS in the Standard solution, and A_U and A_S are the absorbances of the solution of Mitotane and the Standard solution, respectively.

Mitotane Tablets

» Mitotane Tablets contain not less than 90.0 percent and not more than 110.0 percent of the labeled amount of $C_{14}H_{10}Cl_4$.

Packaging and storage—Preserve in tight, light-resistant containers.

Reference standard—*USP Mitotane Reference Standard*—Dry in vacuum at 60° for 2 hours before using.

Identification—Triturate a quantity of finely powdered Tablets, equivalent to about 500 mg of mitotane, with 10 mL of water, filter on a sintered-glass filter funnel, and wash the residue with two 5-mL portions of water. Transfer the residue to a small beaker, add 4 mL of alcohol, heat to boiling, and filter immediately. Allow the filtrate to cool, filter the crystals of mitotane, wash once with 2 mL of alcohol, and dry in vacuum at 60° for 2 hours: the infrared absorption spectrum of a mineral oil dispersion of the mitotane so obtained exhibits maxima only at the same wavelengths as that of a similar preparation of USP Mitotane RS.

Disintegration ⟨701⟩: 15 minutes, the use of disks being omitted.

Uniformity of dosage units ⟨905⟩: meet the requirements.

Assay—

Standard preparation—Dissolve about 50 mg of USP Mitotane RS, accurately weighed, in methanol, and dilute quantitatively and stepwise with methanol to obtain a solution having a known concentration of about 200 µg per mL.

Assay preparation—Weigh and finely powder not less than 10 Mitotane Tablets. Transfer an accurately weighed portion of the powder, equivalent to about 100 mg of mitotane, to a 250-mL volumetric flask, add 100 mL of methanol, and shake occasionally for 5 minutes, then dilute with methanol to volume, and mix. Filter, rejecting the first portion of the filtrate, transfer 25.0 mL of the filtrate to a 50-mL volumetric flask, dilute with methanol to volume, and mix.

Procedure—Proceed as directed in the *Assay* under *Mitotane*, beginning with "Concomitantly determine the absorbances of both solutions."

Monobasic Potassium Phosphate—*see* Potassium Phosphate, Monobasic NF

Monobasic Sodium Phosphate—*see* Sodium Phosphate, Monobasic

Monobenzone

$C_{13}H_{12}O_2$ 200.24
Phenol, 4-(phenylmethoxy)-.
p-(Benzyloxy)phenol [*103-16-2*].

» Monobenzone, dried at 105° for 3 hours, contains not less than 98.0 percent and not more than 102.0 percent of $C_{13}H_{12}O_2$.

Packaging and storage—Preserve in tight, light-resistant containers, and avoid exposure to temperatures above 30°.

Reference standard—*USP Monobenzone Reference Standard*—Dry at 105° for 3 hours before using.

Identification—

A: The infrared absorption spectrum of a potassium bromide dispersion of it, previously dried, exhibits maxima only at the same wavelengths as that of a similar preparation of USP Monobenzone RS.

B: The ultraviolet absorption spectrum of a 1 in 100,000 solution of it in methanol exhibits maxima at the same wavelengths as that of a similar solution of USP Monobenzone RS, concomitantly measured, and the respective absorptivities, calculated on the dried basis, at the wavelength of maximum absorbance at about 292 nm, do not differ by more than 3.0%.

C: Transfer about 500 mg of Monobenzone, previously dried, to a 150-mL flask fitted with a reflux condenser, employing a suitable glass joint. Add 5 mL of pyridine and 3 mL of acetic anhydride, reflux for 10 minutes, and cool. Add 100 mL of water and 6 mL of acetone to the flask, and insert a stopper. Cool the contents of the flask in a refrigerator for 1 hour, collect the precipitate in a sintered-glass crucible, and wash the precipitate with water until no odor of pyridine remains. Dry the precipitate for 16 hours in a vacuum desiccator over phosphorus pentoxide. The monobenzone acetate so obtained melts between 110° and 113° when determined as directed for *Class I* (see *Melting Range or Temperature* ⟨741⟩).

Melting range, *Class I* ⟨741⟩: between 117° and 120°.

Loss on drying ⟨731⟩—Dry it at 105° for 3 hours: it loses not more than 1.0% of its weight.

Residue on ignition ⟨281⟩: not more than 0.5%.

Assay—

Standard preparation—Dissolve an accurately weighed quantity of USP Monobenzone RS in methanol, and dilute quantitatively, and stepwise if necessary, with methanol to obtain a solution having a known final concentration of about 40 µg per mL.

Assay preparation—Transfer about 100 mg of Monobenzone, accurately weighed, to a 100-mL volumetric flask, dissolve in methanol, dilute with methanol to volume, and mix. Pipet 4 mL of this solution into a 100-mL volumetric flask, dilute with methanol to volume, and mix.

Procedure—Concomitantly determine the absorbances of the *Standard preparation* and the *Assay preparation* at the wavelength of maximum absorbance at about 292 nm, with a suitable spectrophotometer, using methanol as a blank. Calculate the quantity, in mg, of $C_{13}H_{12}O_2$ in the portion of Monobenzone taken by the formula:

$$2500C(A_U/A_S),$$

in which C is the concentration, in mg per mL, of USP Monobenzone RS in the *Standard preparation*, and A_U and A_S are the absorbances obtained from the *Assay preparation* and the *Standard preparation*, respectively.

Monobenzone Cream

» Monobenzone Cream contains not less than 94.0 percent and not more than 106.0 percent of the labeled amount of $C_{13}H_{12}O_2$.

Packaging and storage—Preserve in tight containers, and avoid exposure to temperatures above 30°.

Reference standard—*USP Monobenzone Reference Standard*—Dry at 105° for 3 hours before using.

Identification—Transfer a quantity of Monobenzone Cream, equivalent to about 500 mg of monobenzone, to a centrifuge bottle, add 100 mL of water, and shake until the cream is completely dispersed. Centrifuge the suspension, decant the supernatant liquid, wash the monobenzone with water, again centrifuge, and decant the water. Transfer the monobenzone to a separator with the aid of water, and adjust the volume to about 100 mL. Extract with four 25-mL portions of chloroform, filtering the extracts through a pledget of cotton into a 150-mL flask. Evaporate the chloroform in a current of warm air, and add 5 mL of pyridine and 3 mL of acetic anhydride to the dry residue. Connect the flask to a reflux condenser, reflux for 10 minutes, cool, and proceed as directed in *Identification test C* under *Monobenzone*, beginning with "Add 100 mL of water." It responds to *Identification test C* under *Monobenzone*.

Assay—

Standard preparation—Prepare as directed in the *Assay* under *Monobenzone*.

Assay preparation—Transfer an accurately weighed portion of Monobenzone Cream, equivalent to about 200 mg of monobenzone, to a suitable container, add 100 mL of methanol, and shake for about 30 minutes. Transfer the mixture to a 200-mL volumetric flask. Rinse the container with two 25-mL portions of methanol, and add the rinsings to the 200-mL volumetric flask. Dilute with methanol to volume, mix, and filter, discarding the first 20 mL of the filtrate. Pipet 4 mL of this solution into a 100-mL volumetric flask, dilute with methanol to volume, and mix.

Procedure—Proceed as directed in the *Assay* under *Monobenzone*, except to calculate by the formula:

$$5000C(A_U/A_S).$$

Mono- and Di-glycerides—*see* Mono- and Diglycerides NF

Mono- and Di-acetylated Monoglycerides—*see* Mono- and Di-acetylated Monoglycerides NF

Monoethanolamine—*see* Monoethanolamine NF

Monosodium Glutamate—*see* Monosodium Glutamate NF

Monothioglycerol—*see* Monothioglycerol NF

Morphine Sulfate

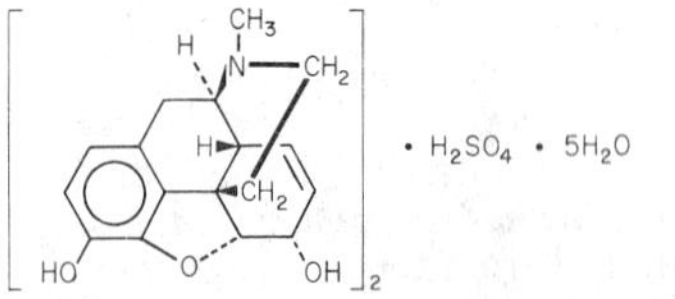

$\cdot H_2SO_4 \cdot 5H_2O$

$(C_{17}H_{19}NO_3)_2.H_2SO_4.5H_2O$ 758.83

Morphinan-3,6-diol, 7,8-didehydro-4,5-epoxy-17-methyl-, (5α,6α)-, sulfate (2:1) (salt), pentahydrate.

7,8-Didehydro-4,5α-epoxy-17-methylmorphinan-3,6α-diol sulfate (2:1) (salt) pentahydrate [*6211-15-0*].

Anhydrous 668.76 [*64-31-3*].

» Morphine Sulfate contains not less than 98.0 percent and not more than 102.0 percent of $(C_{17}H_{19}NO_3)_2.H_2SO_4$, calculated on the anhydrous basis.

Packaging and storage—Preserve in tight, light-resistant containers.

Reference standard—*USP Morphine Sulfate Reference Standard*—This is the pentahydrate form. Do not dry before using, except where directed in this monograph. Determine the water content titrimetrically at time of use for quantitative analyses.

Identification—

A: The infrared absorption spectrum of a potassium bromide dispersion of it, previously dried at 145° for 1 hour, exhibits maxima only at the same wavelengths as that of a similar preparation of USP Morphine Sulfate RS.

B: To 1 mg in a porcelain crucible or small dish add 0.5 mL of sulfuric acid containing, in each mL, 1 drop of formaldehyde TS: an intense purple color is produced at once, and quickly changes to deep blue-violet (distinction from *codeine* which gives at once an intense violet-blue color, and from *hydromorphone* which gives at first a yellow to brown color, changing to pink and then to purplish red).

C: To a solution of 5 mg in 5 mL of sulfuric acid in a test tube add 1 drop of ferric chloride TS, mix, and heat in boiling water for 2 minutes: a blue color is produced, and, when 1 drop of nitric acid is added, it changes to dark red-brown (*codeine* and *ethylmorphine* give the same color reactions, but *hydromorphone* and *papaverine* do not produce this color change).

D: A solution (1 in 50) responds to the tests for *Sulfate* ⟨191⟩.

Specific rotation ⟨781⟩: between −107° and −109.5°, calculated on the anhydrous basis, determined in a solution containing the equivalent of 200 mg in each 10 mL.

Acidity—Dissolve 500 mg in 15 mL of water, add 1 drop of methyl red TS, and titrate with 0.020 *N* sodium hydroxide: not more than 0.50 mL is required to produce a yellow color.

Water, *Method I* ⟨921⟩: between 10.4% and 13.4% is found.

Residue on ignition ⟨281⟩: not more than 0.1%, from 500 mg.

Chloride—To 10 mL of a solution (1 in 100) add 1 mL of 2 *N* nitric acid and 1 mL of silver nitrate TS: no precipitate or turbidity is produced immediately.

Ammonium salts—Heat 200 mg with 5 mL of 1 *N* sodium hydroxide on a steam bath for 1 minute: no odor of ammonia is perceptible.

Foreign alkaloids—Dissolve 1.00 g in 10 mL of 1 *N* sodium hydroxide in a separator, and shake the solution with three successive portions of 15, 10, and 10 mL of chloroform, passing the chloroform solutions through a small filter previously moistened with chloroform. Shake the combined chloroform solutions with 5 mL of water, separate the chloroform layer, and carefully evaporate on a steam bath to dryness. To the residue add 10.0 mL of 0.020 *N* sulfuric acid, and heat gently until dissolved. Cool, add 2 drops of methyl red TS, and titrate the excess acid with 0.020 *N* sodium hydroxide: not less than 7.5 mL is required (1.5%).

Assay—

Mobile phase—Dissolve 0.73 g of sodium 1-heptanesulfonate in 720 mL of water, add 280 mL of methanol and 10 mL of glacial acetic acid, mix, filter, and degas. Make adjustments if necessary (see *System Suitability* under *Chromatography* ⟨621⟩).

Standard preparation—Dissolve an accurately weighed quantity of USP Morphine Sulfate RS in *Mobile phase*, and dilute quantitatively, and stepwise if necessary, with *Mobile phase* to obtain a solution having a known concentration of about 0.24 mg per mL. Prepare a fresh solution daily.

Assay preparation—Transfer about 24 mg of Morphine Sulfate, accurately weighed, to a 100-mL volumetric flask, dissolve in *Mobile phase*, dilute with *Mobile phase* to volume, and mix.

Chromatographic system (see *Chromatography* ⟨621⟩)—The liquid chromatograph is equipped with a 284-nm detector and a 4.6-mm × 30-cm column that contains packing L1. The flow rate is about 1.5 mL per minute. Chromatograph the *Standard preparation*, and record the peak responses as directed under *Procedure:* the tailing factor for the morphine sulfate peak is not more than 2.0, and the relative standard deviation for replicate injections is not more than 2.0%.

Procedure—Separately inject equal volumes (about 25 µL) of the *Standard preparation* and the *Assay preparation* into the chromatograph, record the chromatograms, and measure the responses for the major peaks. Calculate the quantity, in mg, of $(C_{17}H_{19}NO_3)_2.H_2SO_4$ in the portion of Morphine Sulfate taken by the formula:

$$100C(r_U/r_S),$$

in which C is the concentration, in mg per mL, of anhydrous morphine sulfate in the *Standard preparation*, as determined from the concentration of USP Morphine Sulfate RS corrected for moisture content by a titrimetric water determination, and r_U and r_S are the peak responses obtained from the *Assay preparation* and the *Standard preparation*, respectively.

Morphine Sulfate Injection

» Morphine Sulfate Injection is a sterile solution of Morphine Sulfate in Water for Injection. It contains not less than 90.0 percent and not more than 110.0 percent of the labeled amount of $(C_{17}H_{19}NO_3)_2$.-

$H_2SO_4 . 5H_2O$. Injection intended for intramuscular or intravenous administration may contain sodium chloride as a tonicity-adjusting agent, and suitable antioxidants and antimicrobial agents. Injection intended for intrathecal or epidural use may contain sodium chloride as a tonicity-adjusting agent, but contains no other added substances.

Packaging and storage—Preserve in single-dose or in multiple-dose containers, preferably of Type I glass, protected from light. Preserve Injection labeled "Preservative-free" in single-dose containers.

Labeling—It meets the requirements for *Labeling* under *Injections* ⟨1⟩. Label it also to state that the Injection is not to be used if its color is darker than pale yellow, if it is discolored in any other way, or if it contains a precipitate. Injection containing no antioxidant or antimicrobial agents prominently bears on its label the words "Preservative-free," and includes, in its labeling, its routes of administration and the statement that it is not to be heat-sterilized. Injection containing antioxidant or antimicrobial agents includes in its labeling its routes of administration and the statement that it is not for intrathecal or epidural use.

Reference standards—*USP Morphine Sulfate Reference Standard*—This is the pentahydrate form. Do not dry before using. Determine the water content titrimetrically at time of use for quantitative analyses. *USP Endotoxin Reference Standard.*

Identification—

A: Dilute with methanol, if necessary, a volume of the Injection to obtain a solution containing 500 µg per mL. On a suitable thin-layer chromatographic plate (see *Chromatography* ⟨621⟩), coated with a 250-µm layer of chromatographic silica gel mixture, apply 20 µL of this solution and 20 µL of a solution of USP Morphine Sulfate RS in a mixture of methanol and water (1:1) containing 500 µg per mL. Allow the spots to dry, and develop the chromatogram in a solvent system consisting of a mixture of acetone, methanol, and ammonium hydroxide (50:50:1) until the solvent front has moved about three-fourths of the length of the plate. Remove the plate from the developing chamber, mark the solvent front, and allow the solvent to evaporate. Locate the spots on the plate by examination under short-wavelength ultraviolet light: the R_f value of the principal spot obtained from the Injection corresponds to that obtained from the Standard solution.

B: It responds to the tests for *Sulfate* ⟨191⟩.

Bacterial endotoxins—When tested as directed under *Bacterial Endotoxins Test* ⟨85⟩, the USP Endotoxin RS being used, it contains not more than 17.0 USP Endotoxin Units per mg of morphine sulfate; if labeled for intrathecal use, it contains not more than 3.3 USP Endotoxin Units per mg of morphine sulfate.

pH ⟨791⟩: between 2.5 and 6.5.

Particulate matter ⟨788⟩: meets the requirements under *Small-volume Injections.*

Other requirements—It meets the requirements under *Injections* ⟨1⟩.

Assay—*Mobile phase* and *Standard preparation*—Prepare as directed in the *Assay* under *Morphine Sulfate.*

System suitability preparation—Dissolve suitable quantities of USP Morphine Sulfate RS and phenol in *Mobile phase* to obtain a solution containing about 0.24 mg and 0.15 mg per mL, respectively.

Assay preparation—Transfer an accurately measured volume of Morphine Sulfate Injection, equivalent to about 24 mg of morphine sulfate, to a 100-mL volumetric flask, dilute with *Mobile phase* to volume, and mix.

Chromatographic system (see *Chromatography* ⟨621⟩)—Use the system as directed in the *Assay* under *Morphine Sulfate.* Chromatograph the *Standard preparation* and the *System suitability preparation*, and record the peak responses as directed under *Procedure:* the tailing factor for the morphine sulfate peak is not more than 2.0, the resolution, *R*, between the phenol and morphine sulfate peaks is not less than 2.0, and the relative standard deviation for replicate injections of the *Standard preparation* is not more than 2.0%. The relative retention times are about 0.7 for phenol and 1.0 for morphine sulfate.

Procedure—Proceed as directed for *Procedure* in the *Assay* under *Morphine Sulfate.* Calculate the quantity, in mg, of $(C_{17}H_{19}NO_3)_2 . H_2SO_4 . 5H_2O$ in each mL of the Injection taken by the formula:

$$(758.83/668.76)(100C/V)(r_U/r_S),$$

in which 758.83 and 668.76 are the molecular weights of morphine sulfate pentahydrate and anhydrous morphine sulfate, respectively, C is the concentration, in mg per mL, of anhydrous morphine sulfate in the *Standard preparation*, as determined from the concentration of USP Morphine Sulfate RS corrected for moisture content by a titrimetric water determination, V is the volume, in mL, of Injection taken, and r_U and r_S are the peak responses obtained from the *Assay preparation* and the *Standard preparation*, respectively.

Morrhuate Sodium Injection

» Morrhuate Sodium Injection is a sterile solution of the sodium salts of the fatty acids of Cod Liver Oil. It contains, in each mL, not less than 46.5 mg and not more than 53.5 mg of morrhuate sodium. A suitable antimicrobial agent, not to exceed 0.5 percent, and ethyl alcohol or benzyl alcohol, not to exceed 3.0 percent, may be added.

NOTE—Morrhuate Sodium Injection may show a separation of solid matter on standing. Do not use the material if such solid does not dissolve completely upon warming.

Packaging and storage—Preserve in single-dose or in multiple-dose containers, preferably of Type I glass. It may be packaged in 50-mL multiple-dose containers.

Identification—Evaporate about 5 mL of the chloroform solution of the fatty acids obtained in the test for *Iodine value of the fatty acids* on a steam bath nearly to dryness, dissolve the residue in 1 mL of chloroform, and add 1 drop of sulfuric acid: a transient red color is produced, and it changes to brown-red.

Acidity and alkalinity—To 5 mL of Injection add 5 mL of alcohol and 2 drops of phenolphthalein TS. If no red color is produced, not more than 0.50 mL of 0.10 *N* sodium hydroxide is required to impart a distinct red color. If a red color is produced, not more than 0.30 mL of 0.10 *N* acid is required to discharge it. For concentrations of morrhuate sodium other than 5%, no larger than proportional volumes of alkali and acid are required.

Iodine value of the fatty acids—Transfer to a tared, 125-mL conical flask the solvent hexane solution of the fatty acids obtained in the *Assay*. Evaporate at about 60° to dryness, dry the residue in vacuum over silica gel for 18 hours, and weigh. Dissolve the residue in chloroform to make 100.0 mL of solution, and determine the iodine value (see *Fats and Fixed Oils* ⟨401⟩) on a 25.0-mL aliquot of the solution: the iodine value is not less than 130.

Other requirements—It meets the requirements under *Injections* ⟨1⟩, except that at times it may show a slight turbidity or precipitate.

Assay—Transfer an accurately measured volume of Morrhuate Sodium Injection, equivalent to about 500 mg of morrhuate sodium, to a small separator containing 30.0 mL of 0.1 *N* sulfuric acid VS, add 25 mL of solvent hexane, shake gently, and allow to separate. Withdraw the aqueous layer into a beaker or flask, and wash the solvent hexane layer with two 10-mL portions of water, adding the washings to the main aqueous solution. Retain the hexane solution for the test for *Iodine value of the fatty acids*. Add methyl orange TS, and titrate the excess acid in the aqueous solution with 0.1 *N* sodium hydroxide VS. Each mL of 0.1 *N* sulfuric acid is equivalent to 32.4 mg of morrhuate sodium.

Moxalactam Disodium for Injection

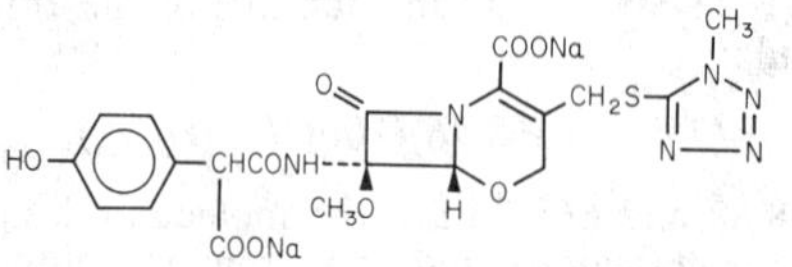

$C_{20}H_{18}N_6Na_2O_9S$ 564.44

5-Oxa-1-azabicyclo[4.2.0]oct-2-ene-2-carboxylic acid, 7-[[carboxy(4-hydroxyphenyl)acetyl]amino]-7-methoxy-3-[[(1-methyl-1*H*-tetrazol-5-yl)thio]methyl]-8-oxo-, disodium salt.

N-[(6*R*,7*R*)-2-Carboxy-7-methoxy-3-[[(1-methyl-1*H*-tetrazol-5-yl)thio]methyl]-8-oxo-5-oxa-1-azabicyclo[4.2.0]oct-2-en-7-yl]-2-(*p*-hydroxyphenyl)malonamic acid disodium salt [*64953-12-4*].

» Moxalactam Disodium for Injection is a sterile mixture of moxalactam disodium and Mannitol. The mixture has a potency equivalent to not less than 722 µg of moxalactam ($C_{20}H_{20}N_6O_9S$) per mg. It contains an amount of moxalactam disodium equivalent to not less than 90.0 percent and not more than 120.0 percent of the labeled amount of moxalactam ($C_{20}H_{20}N_6O_9S$).

Packaging and storage—Preserve in *Containers for Sterile Solids* as described under *Injections* ⟨1⟩.

Reference standard—*USP Moxalactam Disodium Reference Standard*—Do not dry; determine the water content titrimetrically after equilibrium at 42% relative humidity for 16 hours. Store equilibrated material in a tightly closed container, and use within 36 hours.

Constituted solution—At the time of use, the constituted solution prepared from Moxalactam Disodium for Injection meets the requirements for *Constituted Solutions* under *Injections* ⟨1⟩.

Identification—Prepare a test solution in water containing 10 mg of moxalactam per mL. Prepare a Standard solution of USP Moxalactam Disodium RS in water containing 10 mg of moxalactam per mL. Apply separately 10 µL of each solution on a thin-layer chromatographic plate coated with a 0.25-mm layer of chromatographic silica gel mixture (see *Chromatography* ⟨621⟩). Place the plate in a suitable chromatographic chamber, and develop the chromatogram in a solvent system consisting of a mixture of ethyl acetate, water, acetonitrile, and glacial acetic acid (21:9:7:7) until the solvent front has moved about three-fourths of the length of the plate. Remove the plate from the chamber, and dry with the aid of a current of warm air for 10 minutes. Locate the spots on the plate by exposing it to iodine vapors in a closed chamber for about 40 minutes: the intensity and R_f value of the principal spot obtained from the test solution correspond to that obtained from the Standard solution.

Pyrogen—It meets the requirements of the *Pyrogen Test* ⟨151⟩, the test dose being 1.0 mL per kg of a solution prepared by diluting Moxalactam Disodium for Injection with Sterile Water for Injection to a concentration of 50 mg of moxalactam per mL.

Sterility—It meets the requirements under *Sterility Tests* ⟨71⟩, when tested as directed in the section, *Test Procedures Using Membrane Filtration.*

pH ⟨791⟩: between 4.5 and 7.0, in a solution (1 in 10).

Water, *Method I* ⟨921⟩: not more than 3.0%.

Particulate matter ⟨788⟩: meets the requirements under *Small-volume Injections.*

Isomer ratio—Use the chromatogram of *Assay preparation 1* obtained as directed in the *Assay:* the ratio of the response of the R-isomer to the response of the S-isomer is between 0.8 and 1.4.

Other requirements—It meets the requirements for *Uniformity of Dosage Units* ⟨905⟩ and *Labeling* under *Injections* ⟨1⟩.

Assay—

Mobile phase—Prepare a mixture of 0.01 *M* ammonium acetate and methanol (19:1). Filter through a membrane filter (1-µm or finer porosity), and degas.

Standard preparation—Dissolve a suitable quantity of USP Moxalactam Disodium RS, accurately weighed, in water to obtain a solution having a known concentration of about 0.5 mg of moxalactam per mL.

Assay preparation 1—Using a suitable quantity of Moxalactam Disodium for Injection, accurately weighed, proceed as directed under *Standard preparation.*

Assay preparation 2 (where it is represented as being in a single-dose container)—Constitute Moxalactam Disodium for Injection in a volume of water, accurately measured, corresponding to the volume of solvent specified in the labeling. Withdraw all of the withdrawable contents, using a suitable hypodermic needle and syringe, and dilute quantitatively with water to obtain a solution containing about 0.5 mg of moxalactam per mL.

Assay preparation 3 (where the label states the quantity of moxalactam in a given volume of constituted solution)—Constitute Moxalactam Disodium for Injection in a volume of water, accurately measured, corresponding to the volume of solvent specified in the labeling. Dilute an accurately measured volume of the constituted solution quantitatively with water to obtain a solution containing about 0.5 mg of moxalactam per mL.

Chromatographic system (see *Chromatography* ⟨621⟩)—The liquid chromatograph is equipped with a 254-nm detector and a 4.0-mm × 30-cm column that contains packing L1. The flow rate is about 0.5 mL per minute. Chromatograph three replicate injections of the *Standard preparation,* and record the peak responses as directed under *Procedure:* the relative standard deviation is not more than 2.0%, and the resolution (*R*) between the R-isomer and the S-isomer is not less than 4.0.

Procedure—Separately inject equal volumes (about 5 µL) of the *Standard preparation* and the *Assay preparation* into the chromatograph by means of a suitable microsyringe or sampling valve, record the chromatograms, and measure the responses for the major peaks. The retention times are about 15.5 minutes for the R-isomer and 21.5 minutes for the S-isomer. Calculate the quantity, in µg of moxalactam ($C_{20}H_{20}N_6O_9S$) per mg of the Moxalactam Disodium for Injection taken by the formula:

$$1000(C/M)(r_U/r_S),$$

in which *C* is the concentration, in mg of moxalactam ($C_{20}H_{20}N_6O_9S$) per mL, of the *Standard preparation, M* is the concentration, in mg per mL, of the *Assay preparation* based on the weight of Moxalactam Disodium for Injection taken and the extent of dilution, and r_U and r_S are the sums of the responses of the R-isomer and S-isomer peaks of the *Assay preparation* and the *Standard preparation,* respectively. Calculate the quantity, in mg of moxalactam ($C_{20}H_{20}N_6O_9S$) withdrawn from the container, or in the portion of constituted solution taken, by the formula:

$$(L/D)(C)(r_U/r_S),$$

in which *L* is the labeled quantity, in mg of moxalactam ($C_{20}H_{20}N_6O_9S$), in the container, or in the volume of constituted solution taken, and *D* is the concentration, in mg of moxalactam ($C_{20}H_{20}N_6O_9S$) per mL, of *Assay preparation 2* or *Assay preparation 3,* based on the labeled quantity in the container or in the portion of constituted solution taken, respectively, and the extent of dilution.

Mumps Skin Test Antigen

» Mumps Skin Test Antigen conforms to the regulations of the federal Food and Drug Administration concerning biologics (see *Biologics* ⟨1041⟩). It is a sterile suspension of formaldehyde-inactivated mumps virus prepared from the extra-embryonic fluids of the mumps virus–infected chicken embryo, concentrated and purified by differential centrifugation, and diluted with isotonic sodium chloride solution. It contains not less than 20 complement-fixing units in each mL. It contains approximately 0.006 *M* glycine as a stabilizing agent, and it contains a preservative.

Packaging and storage—Preserve at a temperature between 2° and 8°.

Expiration date—The expiration date is not later than 18 months after date of manufacture or date of issue from manufacturer's cold storage (5°, 1 year).

Labeling—Label it to state that it was prepared in embryonated chicken eggs and that a separate syringe and needle are to be used for each individual injection.

Mumps Virus Vaccine Live

» Mumps Virus Vaccine Live conforms to the regulations of the federal Food and Drug Administration concerning biologics (630.50 to 630.57) (see *Biologics* ⟨1041⟩). It is a bacterially sterile preparation of live virus derived from a strain of mumps virus tested for neurovirulence in monkeys, and for immunogenicity, free from all demonstrable viable microbial agents except unavoidable bacteriophage, and found suitable for human immunization. The strain is grown for the purpose of vaccine production on chicken embryo primary cell tissue cultures derived from pathogen-free flocks, meets the requirements of the specific safety tests in adult and suckling mice; the requirements of the tests in monkey kidney, chicken embryo and human tissue cell cultures and embryonated eggs; and the requirements of the tests for absence of *Mycobacterium tuberculosis* and of avian leucosis, unless the production cultures were derived from certified avian leucosis-free sources and the control fluids were tested for avian leucosis. The strain cultures are treated to remove all intact tissue cells. The Vaccine meets the requirements of the specific tissue culture test for live virus titer, in a single immunizing dose, of not less than the equivalent of 5000 $TCID_{50}$ (quantity of virus estimated to infect 50% of inoculated cultures × 5000) when tested in parallel with the U.S. Reference Mumps Virus, Live. It may contain suitable antimicrobial agents.

Packaging and storage—Preserve in single-dose containers, or in light-resistant, multiple-dose containers, at a temperature between 2° and 8°. Multiple-dose containers for 50 doses are adapted for use only in jet injectors, and those for 10 doses for use by jet or syringe injection.

Expiration date—The expiration date is 1 to 2 years, depending on the manufacturer's data, after date of issue from manufacturer's cold storage (−20°, 1 year).

Labeling—Label the Vaccine in multiple-dose containers to indicate that the contents are intended solely for use by jet injector or for use by either jet or syringe injection, whichever is applicable. Label the Vaccine in single-dose containers, if such containers are not light-resistant, to state that it should be protected from sunlight. Label it also to state that constituted Vaccine should be discarded if not used within 8 hours.

Mumps, and Rubella Virus Vaccine Live, Measles,—*see* Measles, Mumps, and Rubella Virus Vaccine Live

Mumps Virus Vaccine Live, Measles and—*see* Measles and Mumps Virus Vaccine Live

Mumps Virus Vaccine Live, Rubella and—*see* Rubella and Mumps Virus Vaccine Live

Myristyl Alcohol—*see* Myristyl Alcohol NF

Nadolol

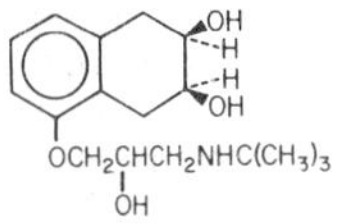

$C_{17}H_{27}NO_4$ 309.41

2,3-Naphthalenediol, 5-[3-[(1,1-dimethylethyl)amino]-2-hydroxypropoxy]-1,2,3,4-tetrahydro-, *cis*-.

1-(*tert*-Butylamino)-3-[(5,6,7,8-tetrahydro-*cis*-6,7-dihydroxy-1-naphthyl)oxy]-2-propanol [*42200-33-9*].

» Nadolol contains not less than 98.0 percent and not more than 101.5 percent of $C_{17}H_{27}NO_4$, calculated on the dried basis.

Packaging and storage—Preserve in well-closed containers.

Reference standard—*USP Nadolol Reference Standard*—Dry in vacuum at 60° for 3 hours before using.

Identification—The infrared absorption spectrum of a mineral oil dispersion of it, previously dried, exhibits maxima only at the same wavelengths as that of a similar preparation of USP Nadolol RS.

Loss on drying ⟨731⟩—Dry it in vacuum at 60° for 3 hours: it loses not more than 2.0% of its weight.

Residue on ignition ⟨281⟩: not more than 0.1%.

Heavy metals, *Method II* ⟨231⟩: 0.003%.

Racemate composition—Prepare a mineral oil dispersion of Nadolol, previously dried, adjusting the thickness of the mull to give an absorbance reading of 0.6 ± 0.1 at 6.3 µm. Record the spectrum from 6 µm to 9 µm, using mineral oil in the reference beam. Calculate the percentage of racemate A by the formula:

$$(50/0.9)(A_a/A_b),$$

in which 0.9 is the average value of (A_a/A_b) in a 1:1 mixture of racemates A and B, A_a is the uncorrected absorbance at the wavelength of a maximum absorbance at about 7.90 µm, corresponding to racemate A, and A_b is the uncorrected absorbance at the wavelength of a maximum absorbance at about 8.05 µm, corresponding to racemate B: the content of racemate A is between 40% and 60%.

Chromatographic purity—Prepare the test solution by dissolving about 500 mg of Nadolol in 10.0 mL of a mixture of methanol and chloroform (1:1). Prepare a solution of USP Nadolol RS in a mixture of methanol and chloroform (1:1) containing about 50 mg per mL. [NOTE—This Standard solution is used only to identify the nadolol zone.] Divide a thin-layer chromatographic plate (see *Chromatography* ⟨621⟩), coated with a 0.25-mm layer of chromatographic silica gel mixture, into four equal sections, the first section to be used for the Standard solution, the next two sections to be used for the test solution, and the last section for the blank. Apply, as streaks, 100 µL of the Standard solution, two 100-µL portions of the test solution, and 100 µL of the mixture of methanol and chloroform (1:1) to provide the blank to appropriate sections of the plate, drying each solution as it is applied with a current of cool air. Develop the chromatogram in a solvent system consisting of a mixture of acetone, chloroform, and 2 *M* ammonium hydroxide (8:1:1) until the solvent front has moved about three-fourths of the length of the plate. Remove the plate from the developing chamber, air-dry, and locate the bands by viewing under short-wavelength ultraviolet light. Identify the nadolol zones by comparison of the chromatograms of the test solution with the chromatogram of the Standard solution. Mark the nadolol zones and the separated impurity zones in the chromatograms of the test solution, and the corresponding zones in the blank section of the plate. Remove the silica gel from the nadolol zone in each chromatogram of the test solution, and transfer to separate 50-mL centrifuge tubes, and similarly transfer the silica gel removed from the corresponding area of the blank chromatogram to a third 50-mL centrifuge tube. Remove the silica gel from the combined impurity zones in each chromatogram of the test solution, and transfer to separate 50-mL centrifuge tubes, and similarly transfer the silica gel removed from the corresponding area of the blank chromatogram to a sixth 50-mL centrifuge tube. Add 30.0 mL of alcohol to each of the 2 tubes containing the mixtures from the nadolol zones and to the third tube containing the corresponding portion of the blank mixture, and add 10.0 mL of alcohol to each of the 2 tubes containing the mixtures from the impurity zones and to the sixth tube containing the corresponding portion of the blank mixture. Shake for 60 minutes, and centrifuge to clarify. Concomitantly determine the absorbances of the supernatant solutions at the wavelength of maximum absorbance at about 278 nm, with a suitable spectrophotometer, using alcohol as the spectrophotometer blank. Calculate the percentage of impurities by the formula:

$$100A_i/(A_i + 3A_U),$$

in which A_i is the average absorbance of the impurity zone eluates corrected for the corresponding blank, and A_U is the average absorbance of the nadolol zone eluates corrected for the corresponding blank: not more than 2.0% is found.

Assay—

Perchloric acid titrant—Mix 8.5 mL of perchloric acid with 500 mL of glacial acetic acid, cool, dilute with glacial acetic acid to 1000 mL, and mix. Standardize this solution as directed for *Perchloric Acid, Tenth-Normal (0.1 N) (in Glacial Acetic Acid)* in the section, *Volumetric Solutions*, under *Reagents, Indicators, and Solutions*.

Procedure—Transfer about 280 mg of Nadolol, accurately weighed, to a 250-mL conical flask. Add 100 mL of glacial acetic acid, and place in an ultrasonic bath until solution is complete. Add 2 drops of crystal violet TS, and titrate with *Perchloric acid titrant* to an emerald-green end-point. Perform a blank determination, and make any necessary correction. Each mL of 0.1 *N* perchloric acid is equivalent to 30.94 mg of $C_{17}H_{27}NO_4$.

Nadolol Tablets

» Nadolol Tablets contain not less than 90.0 percent and not more than 110.0 percent of the labeled amount of $C_{17}H_{27}NO_4$.

Packaging and storage—Preserve in tight containers.

Reference standard—*USP Nadolol Reference Standard*—Dry in vacuum at 60° for 3 hours before using.

Identification—Transfer a quantity of powdered Tablets, equivalent to about 50 mg of nadolol, to a conical flask. Add 10 mL of 0.1 *N* hydrochloric acid, stir for 30 minutes, using a magnetic stirrer, and place in an ultrasonic bath for an additional 30 minutes. Centrifuge, and use the supernatant solution for the test solution. On a thin-layer chromatographic plate (see *Chromatography* ⟨621⟩) coated with a 0.25-mm layer of chromatographic silica gel mixture, apply as streaks, 100 µL of the test solution and 100 µL of a Standard solution of USP Nadolol RS in 0.1 *N* hydrochloric acid having a concentration of about 5 mg per mL. Develop the chromatogram in a solvent system consisting of acetone, chloroform, and 2 *N* ammonium hydroxide (8:1:1) until the solvent front has moved about three-fourths of the length of the plate. Remove the plate from the developing chamber, allow the solvent to evaporate, and examine the chromatogram under short-wavelength ultraviolet light: the R_f value of the principal spot obtained from the test solution corresponds to that obtained from the Standard solution.

Dissolution ⟨711⟩—

Medium: 0.1 *N* hydrochloric acid; 900 mL.

Apparatus 1: 100 rpm.

Time: 50 minutes.

Procedure—Determine the amount of $C_{17}H_{27}NO_4$ dissolved, employing the procedure set forth in the *Assay*, except to prepare the *Mobile phase* using 560 mL of methanol and 1440 mL of water instead of 700 mL and 1300 mL, respectively, using filtered portions of the solution under test, suitably diluted with *Dissolution Medium*, if necessary, in comparison with a Standard solution having a known concentration of USP Nadolol RS in the same medium.

Tolerances—Not less than 80% (*Q*) of the labeled amount of $C_{17}H_{27}NO_4$ is dissolved in 50 minutes.

Uniformity of dosage units ⟨905⟩: meet the requirements.

Assay—

Mobile phase—Prepare a filtered and degassed mixture of 700 mL of methanol and 1300 mL of water containing 5.84 g of sodium chloride and 1.0 mL of 0.1 *N* hydrochloric acid.

Standard preparation—Dissolve an accurately weighed quantity of USP Nadolol RS in *Mobile phase* to obtain a solution having a known concentration of about 0.2 mg per mL.

Assay preparation—Weigh and finely powder not less than 20 Nadolol Tablets. Weigh accurately a portion of the powder, equivalent to about 20 mg of nadolol, and transfer to a 100-mL volumetric flask. Add about 75 mL of *Mobile phase*, place in an ultrasonic bath for 15 minutes, shaking intermittently, add *Mobile phase* to volume, and mix. Clarify the solution by filtration or centrifugation.

Chromatographic system (see *Chromatography* ⟨621⟩)—The liquid chromatograph is equipped with a 220-nm detector and a 4.6-mm × 25-cm column that contains packing L16. The flow rate is about 1 mL per minute. Chromatograph replicate injections of the *Standard preparation*, and record the peak responses as directed under *Procedure:* the relative standard deviation is not more than 2.0%, and the tailing factor for the nadolol peak is not more than 3.

Procedure—Separately inject equal volumes (about 20 µL) of the *Standard preparation* and the *Assay preparation* into the chromatograph, record the chromatograms, and measure the responses for the major peaks. Calculate the quantity, in mg, of $C_{17}H_{27}NO_4$ in the portion of Tablets taken by the formula:

$$100C(r_U/r_S),$$

in which *C* is the concentration, in mg per mL, of USP Nadolol RS in the *Standard preparation*, and r_U and r_S are the peak responses obtained from the *Assay preparation* and the *Standard preparation*, respectively.

Nafcillin Sodium

$C_{21}H_{21}N_2NaO_5S.H_2O$ 454.47

4-Thia-1-azabicyclo[3.2.0]heptane-2-carboxylic acid, 6-[[(2-ethoxy-1-naphthalenyl)carbonyl]amino]-3,3-dimethyl-7-oxo-, monosodium salt, monohydrate, [2*S*-(2α,5α,6β)].

Monosodium (2*S*,5*R*,6*R*)-6-(2-ethoxy-1-naphthamido)-3,3-dimethyl-7-oxo-4-thia-1-azabicyclo[3.2.0]heptane-2-carboxylate monohydrate [*7177-50-6*].

Anhydrous 436.46 [*985-16-0*].

» Nafcillin Sodium has a potency equivalent to not less than 820 µg of nafcillin ($C_{21}H_{22}N_2O_5S$) per mg.

Packaging and storage—Preserve in tight containers.

Reference standard—*USP Nafcillin Sodium Reference Standard*—Do not dry before using.

Identification—The ultraviolet absorption spectrum of a solution (1 in 20,000) exhibits maxima and minima at the same wavelengths as that of a similar preparation of USP Nafcillin Sodium RS, concomitantly measured.

Crystallinity ⟨695⟩: meets the requirements.

pH ⟨791⟩: between 5.0 and 7.0, in a solution containing 30 mg per mL.

Water, *Method I* ⟨921⟩: between 3.5% and 5.3%.

Nafcillin content—Transfer about 25 mg of Nafcillin Sodium, accurately weighed, to a 100-mL volumetric flask, add water to volume, and mix. Pipet 10 mL of this solution into a 50-mL volumetric flask, dilute with water to volume, and mix. Concomitantly determine the absorbances of this solution and a similarly prepared Standard solution of USP Nafcillin Sodium RS at the wavelength of maximum absorbance at about 280 nm. Calculate the percentage of nafcillin by the formula:

$$P(a_U / a_S),$$

in which P is the percentage content of nafcillin in the USP Nafcillin Sodium RS, and a_U and a_S are the absorptivities of the specimen and the Standard, respectively: not less than 82.0% is found.

Assay—Proceed with Nafcillin Sodium as directed under *Antibiotics—Microbial Assays* ⟨81⟩.

Nafcillin Sodium Capsules

» Nafcillin Sodium Capsules contain not less than 90.0 percent and not more than 120.0 percent of the labeled amount of nafcillin ($C_{21}H_{22}N_2O_5S$).

Packaging and storage—Preserve in tight containers.

Reference standard—*USP Nafcillin Sodium Reference Standard*—Do not dry before using.

Dissolution ⟨711⟩—

Medium: water; 900 mL.

Apparatus 1: 100 rpm.

Time: 45 minutes.

Procedure—Determine the amount of nafcillin ($C_{21}H_{22}N_2O_5S$) by a suitable validated spectrophotometric analysis of a filtered portion of the solution under test, suitably diluted with *Dissolution Medium*, if necessary, in comparison with a Standard solution having a known concentration of USP Nafcillin RS in the same medium.

Tolerances—Not less than 75% (Q) of the labeled amount of $C_{21}H_{22}N_2O_5S$ is dissolved in 45 minutes.

Uniformity of dosage units ⟨905⟩: meet the requirements.

Water, *Method I* ⟨921⟩: not more than 5.0%.

Assay—Proceed as directed under *Antibiotics—Microbial Assays* ⟨81⟩, using not less than 5 Nafcillin Sodium Capsules blended for 4 ± 1 minutes in a high-speed glass blender jar containing an accurately measured volume of *Buffer No. 1*. Dilute an accurately measured volume of this stock solution quantitatively with *Buffer No. 1* to obtain a *Test Dilution* having a concentration assumed to be equal to the median dose level of the Standard.

Nafcillin Sodium for Injection

» Nafcillin Sodium for Injection is a sterile, dry mixture of Nafcillin Sodium and a suitable buffer. It contains not less than 90.0 percent and not more than 120.0 percent of the labeled amount of nafcillin ($C_{21}H_{22}N_2O_5S$).

Packaging and storage—Preserve in *Containers for Sterile Solids* as described under *Injections* ⟨1⟩.

Reference standard—*USP Nafcillin Sodium Reference Standard*—Do not dry before using.

Constituted solution—At the time of use, the constituted solution prepared from Nafcillin Sodium for Injection meets the requirements for *Constituted Solutions* under *Injections* ⟨1⟩.

Pyrogen—It meets the requirements of the *Pyrogen Test* ⟨151⟩, the test dose being 1 mL per kg of a solution in pyrogen-free saline TS containing 80 mg of nafcillin per mL.

Sterility—It meets the requirements under *Sterility Tests* ⟨71⟩, when tested as directed in the section, *Test Procedures Using Membrane Filtration.*

pH ⟨791⟩: between 6.0 and 8.5, in the solution constituted as directed in the labeling.

Water, *Method I* ⟨921⟩: between 3.5% and 5.3%.

Particulate matter ⟨788⟩: meets the requirements under *Small-volume Injections.*

Other requirements—It responds to the *Identification test* under Nafcillin Sodium and meets the requirements for *Uniformity of Dosage Units* ⟨905⟩ and *Labeling* under *Injections* ⟨1⟩.

Assay—Proceed as directed under *Antibiotics—Microbial Assays* ⟨81⟩, using Nafcillin Sodium for Injection constituted as directed in the labeling. Remove an accurately measured portion, and dilute quantitatively with *Buffer No. 1* to obtain a *Test Dilution* having a concentration assumed to be equal to the median dose level of the Standard.

Nafcillin Sodium for Oral Solution

» Nafcillin Sodium for Oral Solution contains not less than 90.0 percent and not more than 120.0 percent of the labeled amount of nafcillin ($C_{21}H_{22}N_2O_5S$). It contains one or more suitable buffers, colors, diluents, dispersants, flavors, and preservatives.

Packaging and storage—Preserve in tight containers.

Reference standard—*USP Nafcillin Sodium Reference Standard*—Do not dry before using.

Uniformity of dosage units ⟨905⟩—

FOR SOLID PACKAGED IN SINGLE-UNIT CONTAINERS: meets the requirements.

pH ⟨791⟩: between 5.5 and 7.5, in the solution constituted as directed in the labeling.

Water, *Method I* ⟨921⟩: not more than 5.0%.

Assay—Proceed as directed under *Antibiotics—Microbial Assays* ⟨81⟩, using Nafcillin Sodium for Oral Solution constituted as directed in the labeling. Dilute an accurately measured volume of the solution quantitatively with *Buffer No. 1* to obtain a *Test Dilution* having a concentration assumed to be equal to the median dose level of the Standard.

Sterile Nafcillin Sodium

» Sterile Nafcillin Sodium is Nafcillin Sodium suitable for parenteral use. It has a potency equivalent to not less than 820 μg of nafcillin ($C_{21}H_{22}N_2O_5S$) per mg.

Packaging and storage—Preserve in *Containers for Sterile Solids* as described under *Injections* ⟨1⟩.

Reference standard—*USP Nafcillin Sodium Reference Standard*—Do not dry before using.

Pyrogen—It meets the requirements of the *Pyrogen Test* ⟨151⟩, the test dose being 1 mL per kg of a solution in pyrogen-free saline TS containing 80 mg of nafcillin per mL.

Sterility—It meets the requirements under *Sterility Tests* ⟨71⟩, when tested as directed in the section, *Test Procedures Using Membrane Filtration*, using 6 g of specimen aseptically dissolved in 200 mL of *Fluid A.*

Other requirements—It conforms to the definition, responds to the *Identification test*, and meets the requirements for *pH, Water, Crystallinity*, and *Nafcillin content* under *Nafcillin Sodium.*

Assay—Proceed with Sterile Nafcillin Sodium as directed under *Antibiotics—Microbial Assays* ⟨81⟩.

Nafcillin Sodium Tablets

» Nafcillin Sodium Tablets contain not less than 90.0 percent and not more than 120.0 percent of the labeled amount of nafcillin ($C_{21}H_{22}N_2O_5S$).

Packaging and storage—Preserve in tight, light-resistant containers.

Reference standard—*USP Nafcillin Sodium Reference Standard*—Do not dry before using.

Dissolution ⟨711⟩—

pH 4.0 buffer—Transfer 10.94 g of anhydrous dibasic sodium phosphate and 12.92 g of citric acid monohydrate to a 1-liter volumetric flask, dissolve in water, dilute with water to volume, and mix.

Medium: pH 4.0 buffer; 900 mL.

Apparatus 2: 50 rpm.

Time: 45 minutes.

Procedure—Determine the amount of nafcillin ($C_{21}H_{22}N_2O_5S$) dissolved from ultraviolet absorbances, at the wavelength of maximum absorbance at about 280 nm of filtered portions of the solution under test, suitably diluted with *Dissolution Medium*, if necessary, in comparison with a Standard solution having a known concentration of USP Nafcillin Sodium RS in the same medium.

Tolerances—Not less than 75% (*Q*) of the labeled amount of nafcillin ($C_{21}H_{22}N_2O_5S$) is dissolved in 45 minutes.

Uniformity of dosage units ⟨905⟩: meet the requirements.

Water, *Method I* ⟨921⟩: not more than 5.0%.

Assay—Proceed as directed under *Antibiotics—Microbial Assays* ⟨81⟩, using not less than 5 Nafcillin Sodium Tablets blended for 4 ± 1 minutes in a high-speed glass blender jar containing an accurately measured volume of *Buffer No. 1.* Dilute an accurately measured volume of this stock solution quantitatively with *Buffer No. 1* to obtain a *Test Dilution* having a concentration assumed to be equal to the median dose level of the Standard.

Nalidixic Acid

$C_{12}H_{12}N_2O_3$ 232.24

1,8-Naphthyridine-3-carboxylic acid, 1-ethyl-1,4-dihydro-7-methyl-4-oxo-.

1-Ethyl-1,4-dihydro-7-methyl-4-oxo-1,8-naphthyridine-3-carboxylic acid [*389-08-2*].

» Nalidixic Acid contains not less than 98.0 percent and not more than 102.0 percent of $C_{12}H_{12}N_2O_3$, calculated on the dried basis.

Packaging and storage—Preserve in tight containers.

Reference standard—*USP Nalidixic Acid Reference Standard*—Dry at 105° for 2 hours before using.

Identification—

A: The infrared absorption spectrum of a potassium bromide dispersion of it, previously dried, exhibits maxima only at the same wavelengths as that of a similar preparation of USP Nalidixic Acid RS.

B: The ultraviolet absorption spectrum of a 1 in 200,000 solution in 0.01 *N* sodium hydroxide exhibits maxima and minima at the same wavelengths as that of a similar solution of USP Nalidixic Acid RS, concomitantly measured, and the respective absorptivities, calculated on the dried basis, at the wavelength of maximum absorbance at about 258 nm do not differ by more than 3.0%.

Melting range ⟨741⟩: between 225° and 231°.

Loss on drying ⟨731⟩—Dry it at 105° for 2 hours: it loses not more than 0.5% of its weight.

Residue on ignition ⟨281⟩: not more than 0.2%.

Heavy metals, *Method II* ⟨231⟩: 0.002%.

Chromatographic purity—

Standard preparation—Prepare a solution of USP Nalidixic Acid RS in chloroform containing 1.0 mg per mL. Dilute quantitatively with chloroform to obtain Standard preparations having the following composition:

Standard preparation	Dilution	Concentration (µg RS per mL)	Percentage (%, for comparison with test specimen)
A	1 in 10	0.1	0.5
B	1 in 25	0.04	0.2
C	1 in 50	0.02	0.1

Test preparation—Dissolve an accurately weighed quantity of Nalidixic Acid in chloroform to obtain a solution containing 20 mg per mL.

Procedure—On a suitable thin-layer chromatographic plate (see *Chromatography* ⟨621⟩), coated with a 0.25-mm layer of chromatographic silica gel mixture, apply separately 10 µL of the *Test preparation* and 10 µL of each *Standard preparation.* Position the plate in a chromatographic chamber, and develop the chromatograms in a solvent system consisting of a mixture of alcohol, chloroform, and 5 *M* ammonium hydroxide (70:20:10) until the solvent front has moved about three-fourths of the length of the plate. Remove the plate from the developing chamber, mark the solvent front, and allow the solvent to evaporate, with the aid of warm circulating air. Examine the plate under short-wavelength ultraviolet light. Compare the intensities of any secondary spots observed in the chromatogram of the *Test preparation* with those of the principal spots in the chromatograms of the *Standard preparations:* no secondary spot is more intense than the principal spot obtained from *Standard preparation A* (0.5%), and the sum of the intensities of all secondary spots obtained from the *Test preparation* does not exceed 1.0%.

Assay—Dissolve about 250 mg of Nalidixic Acid, accurately weighed, in 30 mL of dimethylformamide that previously has been neutralized to thymolphthalein TS, and titrate with 0.1 *N* lithium methoxide VS, using a magnetic stirrer and taking precautions against absorption of atmospheric carbon dioxide. Each mL of 0.1 *N* lithium methoxide is equivalent to 23.22 mg of $C_{12}H_{12}N_2O_3$.

Nalidixic Acid Oral Suspension

» Nalidixic Acid Oral Suspension contains not less than 95.0 percent and not more than 105.0 percent of the labeled amount of $C_{12}H_{12}N_2O_3$ in a suitable aqueous vehicle.

Packaging and storage—Preserve in tight containers.

Reference standard—*USP Nalidixic Acid Reference Standard*—Dry at 105° for 2 hours before using.

Identification—

A: Place a portion of Oral Suspension, freshly mixed, equivalent to about 100 mg of nalidixic acid, in a separator, and extract with 50 mL of chloroform. Evaporate the chloroform extract on a steam bath to dryness, and dry the residue at 105° for 2 hours: the infrared absorption spectrum of a potassium bromide dispersion of the residue so obtained exhibits maxima only at the same wavelengths as that of a similar preparation of USP Nalidixic Acid RS.

B: The ultraviolet absorption spectrum of the solution from the Oral Suspension, prepared as directed in the *Assay*, exhibits maxima and minima at the same wavelengths as that of the Standard solution, prepared as directed in the *Assay*, concomitantly measured.

Assay—Transfer an accurately measured volume of Nalidixic Acid Oral Suspension, freshly mixed, equivalent to about 100 mg of nalidixic acid, to a 100-mL volumetric flask, add 1 mL of 1 *N* sodium hydroxide, and mix to dissolve the nalidixic acid. Add water to volume, mix, and filter if necessary. Pipet 2 mL of the clear solution into a 250-mL volumetric flask, dilute with 0.01 *N* sodium hydroxide to volume, and mix. Concomitantly determine the absorbances of this solution and a Standard solution of USP Nalidixic Acid RS in 0.01 *N* sodium hydroxide having a known concentration of about 8 µg per mL, in 1-cm cells at the wavelength of maximum absorbance at about 258 nm, with a suitable spectrophotometer, using 0.01 *N* sodium hydroxide as the blank. Calculate the quantity, in mg, of $C_{12}H_{12}N_2O_3$ in each mL of the Oral Suspension taken by the formula:

$$12.5(C/V)(A_U/A_S),$$

in which *C* is the concentration, in µg per mL, of USP Nalidixic Acid RS in the Standard solution, *V* is the volume, in mL, of Oral Suspension taken, and A_U and A_S are the absorbances of the solution from the Oral Suspension and the Standard solution, respectively.

Nalidixic Acid Tablets

» Nalidixic Acid Tablets contain not less than 93.0 percent and not more than 107.0 percent of the labeled amount of $C_{12}H_{12}N_2O_3$.

Packaging and storage—Preserve in tight containers.

Reference standard—*USP Nalidixic Acid Reference Standard*—Dry at 105° for 2 hours before using.

Identification—

A: The ultraviolet absorption spectrum of the solution from the Tablets, prepared as directed in the *Assay*, exhibits maxima and minima at the same wavelengths as that of the Standard solution, prepared as directed in the *Assay*, concomitantly measured.

B: Dissolve a portion of finely powdered Tablets, equivalent to about 100 mg of nalidixic acid, in 50 mL of chloroform, shake for 15 minutes, and filter. Evaporate the filtrate on a steam bath to dryness, and dry the residue at 105° for 2 hours: the infrared absorption spectrum of a potassium bromide dispersion of the residue so obtained exhibits maxima only at the same wavelengths as that of a similar preparation of USP Nalidixic Acid RS.

Dissolution ⟨711⟩—

Medium: pH 8.60 buffer, prepared by mixing 2.3 volumes of 0.2 *M* sodium hydroxide with 2.5 volumes of 0.2 *M* monobasic potassium phosphate and 2.0 volumes of methanol, cooling, mixing with water to obtain 10 volumes of solution, and adjusting, if necessary, by the addition of 1 *N* sodium hydroxide to a pH of 8.60 ± 0.05. The initial volume for the test is 900 mL.

Apparatus 2: 60 rpm.

Time: 30 minutes.

Procedure—Determine the amount of $C_{12}H_{12}N_2O_3$ dissolved from ultraviolet absorbances at the wavelength of maximum absorbance at about 258 nm of filtered portions of the solution under test, suitably diluted with 0.01 *N* sodium hydroxide, if necessary, in comparison with a Standard solution of known concentration of USP Nalidixic Acid RS in 0.01 *N* sodium hydroxide, using as the blank a mixture of *Dissolution Medium* and 0.01 *N* sodium hydroxide in the same proportions as present in the test solution.

Tolerances—Not less than 80% (*Q*) of the labeled amount of $C_{12}H_{12}N_2O_3$ is dissolved in 30 minutes.

Uniformity of dosage units ⟨905⟩: meet the requirements.

Assay—Weigh and finely powder not less than 20 Nalidixic Acid Tablets. Transfer an accurately weighed portion of the powder, equivalent to about 150 mg of nalidixic acid, to a separator, add 100 mL of chloroform, and shake for 5 minutes. Extract the mixture with five 20-mL portions of 1 *N* sodium hydroxide, collecting each portion in a 200-mL volumetric flask. Dilute with 1 *N* sodium hydroxide to volume, mix, and filter, discarding the first 20 mL of the filtrate. Transfer 10.0 mL of the subsequent filtrate to a 100-mL volumetric flask, dilute with water to volume, and mix. Finally transfer 10.0 mL of this solution to a second 100-mL volumetric flask, dilute with water to volume, and mix. Concomitantly determine the absorbances of this solution and a Standard solution of USP Nalidixic Acid RS in the same medium having a known concentration of about 7.5 µg per mL, in 1-cm cells at the wavelength of maximum absorbance at about 258 nm, with a suitable spectrophotometer, using water as the blank. Calculate the quantity, in mg, of $C_{12}H_{12}N_2O_3$ in the portion of Tablets taken by the formula:

$$20C(A_U/A_S),$$

in which *C* is the concentration, in µg per mL, of USP Nalidixic Acid RS in the Standard solution, and A_U and A_S are the absorbances of the solution from the Tablets and the Standard solution, respectively.

Nalorphine Hydrochloride

$C_{19}H_{21}NO_3.HCl$ 347.84

Morphinan-3,6-diol, 7,8-didehydro-4,5-epoxy-17-(2-propenyl)-(5α,6α)-, hydrochloride.

17-Allyl-7,8-didehydro-4,5α-epoxymorphinan-3,6α-diol hydrochloride [*57-29-4*].

» Nalorphine Hydrochloride contains not less than 97.0 percent and not more than 103.0 percent of $C_{19}H_{21}NO_3.HCl$, calculated on the dried basis.

Packaging and storage—Preserve in tight, light-resistant containers.

Reference standard—*USP Nalorphine Hydrochloride Reference Standard*—Dry in vacuum at 100° for 2 hours before using.

Identification—

A: The infrared absorption spectrum of a potassium bromide dispersion of it, previously dried, exhibits maxima only at the same wavelengths as that of a similar preparation of USP Nalorphine Hydrochloride RS.

B: The ultraviolet absorption spectrum of a solution (1 in 10,000) exhibits maxima and minima at the same wavelengths as that of a similar solution of USP Nalorphine Hydrochloride RS, concomitantly measured.

C: A solution of it responds to the tests for *Chloride* ⟨191⟩.

Specific rotation ⟨781⟩: between −122° and −125°, calculated on the dried basis, determined in a solution in water containing 200 mg in each 10 mL.

Loss on drying ⟨731⟩—Dry it in vacuum at 100° for 2 hours: it loses not more than 0.5% of its weight.

Residue on ignition ⟨281⟩: not more than 0.1%.

Assay—Transfer about 25 mg of Nalorphine Hydrochloride, accurately weighed, to a 250-mL volumetric flask, dissolve in water, dilute with water to volume, and mix. Concomitantly determine the absorbances of this solution and of a Standard solution of USP Nalorphine Hydrochloride RS in the same medium having a known concentration of about 100 µg per mL in 1-cm cells at the wavelength of maximum absorbance at about 285 nm, with a suitable spectrophotometer, using water as the blank. Calculate the quantity, in mg, of $C_{19}H_{21}NO_3.HCl$ in the Nalorphine Hydrochloride taken by the formula:

$$0.25C(A_U/A_S),$$

in which *C* is the concentration, in µg per mL, of USP Nalorphine

Hydrochloride RS in the Standard solution, and A_U and A_S are the absorbances of the solution of Nalorphine Hydrochloride and the Standard solution, respectively.

Nalorphine Hydrochloride Injection

» Nalorphine Hydrochloride Injection is a suitably buffered, sterile solution of Nalorphine Hydrochloride in Water for Injection. It contains not less than 90.0 percent and not more than 110.0 percent of the labeled amount of $C_{19}H_{21}NO_3.HCl$.

Packaging and storage—Preserve in single-dose or in multiple-dose containers, preferably of Type I glass.

Reference standard—*USP Nalorphine Hydrochloride Reference Standard*—Dry in vacuum at 100° for 2 hours before using.

Identification—On a suitable thin-layer chromatographic plate (see *Chromatography* ⟨621⟩), coated with a 0.25-mm layer of chromatographic silica gel mixture, apply 15 µL of Nalorphine Hydrochloride Injection and 15 µL of a Standard solution of USP Nalorphine Hydrochloride RS in methanol containing 5 mg per mL. Allow the applications to dry, and develop the chromatogram in an equilibrated chamber containing methanol until the solvent front has moved about three-fourths of the length of the plate. Remove the plate from the developing chamber, mark the solvent front, and allow the solvent to evaporate. Observe the plate under short-wavelength and long-wavelength ultraviolet light: the R_f value of the principal spot obtained from the Injection corresponds to that obtained from the Standard solution.

pH ⟨791⟩: between 6.0 and 7.5.

Other requirements—It meets the requirements under *Injections* ⟨1⟩.

Assay—Transfer an accurately measured volume of Nalorphine Hydrochloride Injection, equivalent to about 10 mg of nalorphine hydrochloride, to a 25-mL centrifuge separator, add 1 mL of 3 *N* hydrochloric acid, and dilute with water to about 10 mL. Extract with five 5-mL portions of chloroform, separating the layers by centrifugation before drawing off each chloroform extract, and discard the chloroform extracts. Transfer the aqueous layer to a 100-mL volumetric flask with the aid of water, dilute with water to volume, and mix. Proceed as directed in the *Assay* under *Nalorphine Hydrochloride*, beginning with "Concomitantly determine the absorbances." Calculate the quantity, in mg, of $C_{19}H_{21}NO_3.HCl$ in each mL of the Injection taken by the formula:

$$(0.1C/V)(A_U/A_S),$$

in which V is the volume, in mL, of Injection taken, and C, A_U, and A_S are as defined therein.

Naloxone Hydrochloride

$C_{19}H_{21}NO_4.HCl$ 363.84

Morphinan-6-one, 4,5-epoxy-3,14-dihydroxy-17-(2-propenyl)-, hydrochloride, (5α)-.

17-Allyl-4,5α-epoxy-3,14-dihydroxymorphinan-6-one hydrochloride [*357-08-4*].

Dihydrate 399.87 [*51481-60-8*].

» Naloxone Hydrochloride is anhydrous or contains two molecules of water of hydration. It contains not less than 98.0 percent and not more than 102.0 percent of $C_{19}H_{21}NO_4.HCl$, calculated on the dried basis.

Packaging and storage—Preserve in tight, light-resistant containers.

Reference standards—*USP Naloxone Reference Standard*—Dry at 105° to constant weight before using. *USP Noroxymorphone Hydrochloride Reference Standard*—Do not dry before using.

Identification—Dissolve about 150 mg in 25 mL of water in a small separator, add a few drops of 6 *N* ammonium hydroxide, extract with three 5-mL portions of chloroform, and filter the extracts through a dry filter, collecting the filtrate in a small flask. Evaporate the filtrate on a steam bath to dryness, and dry at 105° for 1 hour: the infrared absorption spectrum of a 1 in 50 solution in chloroform of the residue so obtained, determined in a 0.5-mm cell, exhibits maxima only at the same wavelengths as that of a similar solution of USP Naloxone RS.

Specific rotation ⟨781⟩: between −170° and −181°, calculated on the dried basis, determined in a solution containing 25 mg per mL.

Loss on drying ⟨731⟩—Dry it at 105° to constant weight: the anhydrous form loses not more than 0.5% of its weight, and the hydrous form loses not more than 11.0% of its weight.

Noroxymorphone hydrochloride and other impurities—Transfer about 40 mg, accurately weighed, to a 5-mL volumetric flask, dissolve completely in 2.0 mL of water, add methanol to volume, and mix, to obtain the test solution. Prepare a solution of USP Naloxone RS in chloroform containing about 7.6 mg per mL. Prepare a solution of USP Noroxymorphone Hydrochloride [(−)-4,5α-epoxy-3,14-dihydroxymorphinan-6-one hydrochloride] RS in dilute methanol (3 in 5) containing 0.084 mg per mL. Apply 5 µL each of the test solution and the two Standard solutions on a thin-layer chromatographic plate (see *Chromatography* ⟨621⟩) coated with a 0.25-mm layer of chromatographic silica gel that previously has been activated by heating for 15 minutes at 105°. Immediately place the plate in a suitable chromatographic chamber containing a 1 in 20 solution of methanol in ammoniacal butanol previously prepared by shaking 100 mL of butyl alcohol with 60 mL of ammonium hydroxide solution (1 in 100) and discarding the lower layer. Develop the chromatogram, protected from light, until the solvent front has moved about 10 cm from the point of application. Remove the plate, dry thoroughly, and spray with ferric chloride–potassium ferricyanide reagent prepared, immediately prior to use, by dissolving 100 mg of potassium ferricyanide in 20 mL of ferric chloride solution (1 in 10). Other than the principal spot corresponding in R_f value to that of USP Naloxone RS and the spot at the origin (ammonium chloride), no other spot is more intense than the spot corresponding to that of USP Noroxymorphone Hydrochloride RS (1.0%).

Chloride content—Dissolve about 300 mg, accurately weighed, in 50 mL of methanol contained in a 125-mL conical flask, add 5 mL of glacial acetic acid and 2 drops of eosin Y TS, and titrate with 0.1 *N* silver nitrate VS to a pink end-point. Each mL of 0.1 *N* silver nitrate is equivalent to 3.545 mg of chloride. Not less than 9.54% and not more than 9.94%, calculated on the dried basis, is found.

Assay—Dissolve about 300 mg of Naloxone Hydrochloride, accurately weighed, in a mixture of 40 mL of glacial acetic acid and 10 mL of acetic anhydride, add 10 mL of mercuric acetate TS and 1 drop of methyl violet TS, and titrate with 0.1 *N* perchloric acid VS. Perform a blank determination, and make any necessary correction. Each mL of 0.1 *N* perchloric acid is equivalent to 36.38 mg of $C_{19}H_{21}NO_4.HCl$.

Naloxone Hydrochloride Injection

» Naloxone Hydrochloride Injection is a sterile, isotonic solution of Naloxone Hydrochloride in Water for Injection. It contains not less than 90.0 percent and not more than 110.0 percent of the labeled amount of $C_{19}H_{21}NO_4.HCl$. It may contain suitable preservatives.

Packaging and storage—Preserve in single-dose or in multiple-dose containers of Type I glass, protected from light.

Reference standards—*USP Naloxone Reference Standard*—Dry at 105° to constant weight before using. *USP Endotoxin Reference Standard.*

Identification—The retention time of the major peak in the chromatogram of the *Assay preparation* corresponds to that of the *Standard preparation* as obtained in the *Assay*.

Bacterial endotoxin—When tested as directed under *Bacterial Endotoxins Test* ⟨85⟩, the USP Endotoxin RS being used, it contains not more than 500 USP Endotoxin Units per mg of Naloxone Hydrochloride.

pH ⟨791⟩: between 3.0 and 4.5.

Other requirements—It meets the requirements under *Injections* ⟨1⟩.

Assay—

Mobile phase—Prepare a filtered and degassed mixture of 0.68 g of sodium 1-octanesulfonate, 1.0 g of sodium chloride, 550 mL of water, 450 mL of methanol, and 1.0 mL of phosphoric acid. Make adjustments if necessary (see *System Suitability* under *Chromatography* ⟨621⟩).

Diluting solvent—Prepare a filtered and degassed mixture of water, methanol, and phosphoric acid (550:450:1).

Standard preparation—Dissolve an accurately weighed quantity of USP Naloxone RS in *Diluting solvent*, and dilute quantitatively, and stepwise if necessary, with *Diluting solvent* to obtain a solution having a known concentration of about 10 µg per mL.

Assay preparation FOR INJECTION LABELED TO CONTAIN NOT MORE THAN 100 µg OF NALOXONE HYDROCHLORIDE PER mL—Transfer an accurately measured volume of Naloxone Hydrochloride Injection, equivalent to about 100 µg of naloxone hydrochloride, to a 10-mL volumetric flask, add *Diluting solvent* to volume, and mix.

Assay preparation FOR INJECTION LABELED TO CONTAIN MORE THAN 100 µg OF NALOXONE HYDROCHLORIDE PER mL—Transfer an accurately measured volume of Naloxone Hydrochloride Injection, equivalent to about 2000 µg of naloxone hydrochloride, to a 200-mL volumetric flask, add *Diluting solvent* to volume, and mix.

System suitability preparation—Prepare a solution in *Diluting solvent* containing about 20 µg of USP Naloxone RS and about 9 µg of acetaminophen per mL.

Chromatographic system (see *Chromatography* ⟨621⟩)—The liquid chromatograph is equipped with a 229-nm detector and a 4.6-mm × 25-cm column that contains packing L1. The flow rate is about 1 mL per minute. Chromatograph the *Standard preparation* and the *System suitability preparation*, and record the peak responses as directed under *Procedure:* the resolution, *R*, between the acetaminophen and naloxone peaks is not less than 1.5, and the relative standard deviation for replicate injections of the *Standard preparation* is not more than 1.5%.

Procedure—[NOTE—Use peak areas where peak responses are indicated.] Separately inject equal volumes (about 20 µL) of the *Standard preparation* and the appropriate *Assay preparation* into the chromatograph, record the chromatograms, and measure the responses for the major peaks. The relative retention times are about 0.6 for acetaminophen and 1.0 for naloxone. Calculate the quantity, in µg, of $C_{19}H_{21}NO_4 . HCl$ in each mL of the Injection taken by the formula:

$$(363.84/327.38)V_a(C/V)(r_U/r_S),$$

in which 363.84 and 327.38 are the molecular weights of anhydrous naloxone hydrochloride and naloxone, respectively, V_a is the volume, in mL, of the *Assay preparation*, *C* is the concentration, in µg per mL, of USP Naloxone RS in the *Standard preparation*, *V* is the volume, in mL, of Injection taken, and r_U and r_S are the peak responses obtained from the *Assay preparation* and the *Standard preparation*, respectively.

Naloxone Hydrochlorides Tablets, Pentazocine and—*see* Pentazocine and Naloxone Hydrochlorides Tablets

Nandrolone Decanoate

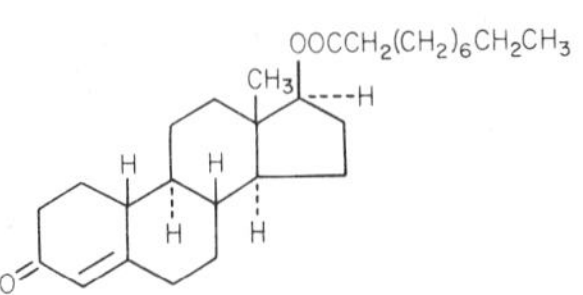

$C_{28}H_{44}O_3$ 428.65
Estr-4-en-3-one, 17-[(1-oxodecyl)oxy]-, (17β)-.
17β-Hydroxyestr-4-en-3-one decanoate [*360-70-3*].

» Nandrolone Decanoate contains not less than 97.0 percent and not more than 103.0 percent of $C_{28}H_{44}O_3$, calculated on the dried basis.

Packaging and storage—Preserve in tight, light-resistant containers, and store in a refrigerator.

Reference standard—*USP Nandrolone Decanoate Reference Standard*—Dry in vacuum over silica gel for 4 hours before using.

Completeness and clarity of solution—A 1 in 50 solution in dioxane is clear.

Identification—

A: The infrared absorption spectrum of a potassium bromide dispersion of it, previously dried, exhibits maxima only at the same wavelengths as that of a similar preparation of USP Nandrolone Decanoate RS.

B: The ultraviolet absorption spectrum of a 1 in 100,000 solution in alcohol exhibits maxima and minima at the same wavelengths as that of a similar solution of USP Nandrolone Decanoate RS, concomitantly measured, and the respective absorptivities, calculated on the dried basis, at the wavelength of maximum absorbance at about 239 nm, do not differ by more than 3.0%.

C: Prepare a solution in acetone containing 5 mg per mL. On a suitable thin-layer chromatographic plate (see *Chromatography* ⟨621⟩), coated with a 0.25-mm layer of chromatographic silica gel, apply 10 µL of this solution and 10 µL of a solution of USP Nandrolone Decanoate RS in acetone containing 5 mg per mL. Allow the spots to dry, and develop the chromatogram in a solvent system consisting of a mixture of *n*-heptane and acetone (3:1) until the solvent front has moved about three-fourths of the length of the plate. Remove the plate from the developing chamber, mark the solvent front, and allow the solvent to evaporate. Locate the spots on the plate by lightly spraying with a 1 in 50 mixture of sulfuric acid in alcohol and heating in an oven at 110° for 15 minutes: the R_f value of the principal spot obtained from the test solution corresponds to that obtained from the Standard solution.

Melting range ⟨741⟩: between 33° and 37°.

Specific rotation ⟨781⟩: between +32° and +36°, determined in a solution in dioxane containing 100 mg, previously dried, in 10 mL.

Loss on drying ⟨731⟩—Dry it in vacuum over silica gel for 4 hours: it loses not more than 0.5% of its weight.

Assay—

Standard preparation—Prepare as directed under *Single-steroid Assay* ⟨511⟩, using USP Nandrolone Decanoate RS.

Assay preparation—Weigh accurately about 20 mg of Nandrolone Decanoate, previously dried, dissolve in a sufficient quantity of a mixture of equal volumes of alcohol and chloroform to make 10.0 mL, and mix.

Procedure—Proceed as directed for *Procedure* under *Single-steroid Assay* ⟨511⟩, using a solvent system consisting of a mixture of *n*-heptane and acetone (3:1), through the fourth sentence of the second paragraph under *Procedure*. Then centrifuge the tubes for 5 minutes, and determine the absorbances of the supernatant solutions in 1-cm cells at the wavelength of maximum absorbance at about 239 nm, with a suitable spectrophotometer, against the blank. Calculate the quantity, in mg, of $C_{28}H_{44}O_3$ in the portion of Nandrolone Decanoate taken by the formula:

$$10C(A_U/A_S),$$

in which *C* is the concentration, in mg per mL, of USP Nandrolone Decanoate RS in the *Standard preparation*, and A_U and A_S are the absorbances of the solutions from the *Assay preparation* and the *Standard preparation*, respectively.

Nandrolone Decanoate Injection

» Nandrolone Decanoate Injection is a sterile solution of Nandrolone Decanoate in Sesame Oil, with a suitable preservative. It contains not less than 90.0 percent and not more than 110.0 percent of the labeled amount of $C_{28}H_{44}O_3$.

Packaging and storage—Preserve in single-dose or in multiple-dose containers, preferably of Type I glass, protected from light.

Reference standards—*USP Nandrolone Decanoate Reference Standard*—Dry in vacuum over silica gel for 4 hours before using. *USP Nandrolone Reference Standard*—Keep container tightly closed and protected from light, and store in a refrigerator. Do not dry before using.

Identification—Dilute a volume of Injection with acetone to provide a solution containing approximately 5 mg of nandrolone decanoate per mL. This solution responds to *Identification test C* under *Nandrolone Decanoate*, 5-µL portions of the test solution and the Standard solution being used.

Other requirements—It meets the requirements under *Injections* ⟨1⟩.

Nandrolone—

Standard preparation—Dissolve 25.0 mg of USP Nandrolone RS in 50.0 mL of acetone. Dilute 5.0 mL of this solution with acetone to 50.0 mL, and mix.

Test preparation—Transfer an accurately measured volume of Injection, equivalent to about 50 mg of nandrolone decanoate, to a 10-mL volumetric flask, dilute with acetone to volume, and mix.

Procedure—On a suitable thin-layer chromatographic plate (see *Chromatography* ⟨621⟩), coated with a 0.25-mm layer of chromatographic silica gel mixture, apply 10 µL each of the *Standard preparation* and of the *Test preparation*. Allow the spots to dry, and develop the chromatogram in a solvent system consisting of a mixture of *n*-heptane and acetone (3:1) until the solvent front has moved about three-fourths of the length of the plate. Remove the plate from the developing chamber, mark the solvent front, and allow the solvent to evaporate. Return the dry plate to the developing chamber containing the same solvent system, and again develop the chromatogram until the solvent front has moved the same distance from the origin. Remove the plate from the developing chamber, and allow the solvent to evaporate. Locate the spots on the plate by lightly spraying with a 4 in 10 solution of sulfuric acid in methanol and heating at about 100° for 10 minutes. Cool, and examine under long-wavelength ultraviolet light: any yellow fluorescent spot from the *Test preparation* at an R_f value of about 0.2 is not greater in size or intensity than that produced by the *Standard preparation* at the same R_f value, corresponding to not more than 1.0% of nandrolone.

Assay—

Isoniazid reagent—Dissolve 500 mg of isoniazid in about 250 mL of methanol, add 0.63 mL of hydrochloric acid, dilute with methanol to 500.0 mL, and mix.

Standard preparation—Transfer about 25 mg of USP Nandrolone Decanoate RS, accurately weighed, to a 100-mL volumetric flask, dissolve in chloroform, dilute with chloroform to volume, and mix. Transfer 5.0 mL of this solution to a 50-mL volumetric flask, dilute with chloroform to volume, and mix.

Assay preparation—Transfer to a 200-mL volumetric flask an accurately measured volume of Nandrolone Decanoate Injection, equivalent to about 50 mg of nandrolone decanoate, add chloroform to volume, and mix. Transfer 10.0 mL of this solution to a 100-mL volumetric flask, dilute with chloroform to volume, and mix.

Procedure—Transfer 5.0 mL each of the *Standard preparation*, the *Assay preparation*, and chloroform to provide the blank, to separate 10-mL volumetric flasks, dilute each with *Isoniazid reagent* to volume, and mix. Allow the flasks to stand for 1 hour with occasional shaking. Concomitantly determine the absorbances of the solutions in 1-cm cells at the wavelength of maximum absorbance at about 380 nm, with a suitable spectrophotometer, using the blank to set the instrument. Calculate the quantity, in mg, of $C_{28}H_{44}O_3$ in each mL of the Injection taken by the formula:

$$(2C/V)(A_U/A_S),$$

in which C is the concentration, in µg per mL, of USP Nandrolone Decanoate RS in the *Standard preparation*, V is the volume, in mL, of Injection taken, and A_U and A_S are the absorbances of the solutions from the *Assay preparation* and the *Standard preparation*, respectively.

Naphazoline Hydrochloride

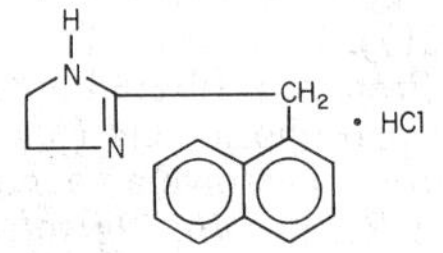

$C_{14}H_{14}N_2 \cdot HCl$ 246.74

1*H*-Imidazole, 4,5-dihydro-2-(1-naphthalenylmethyl)-, monohydrochloride.

2-(1-Naphthylmethyl)-2-imidazoline monohydrochloride [*550-99-2*].

» Naphazoline Hydrochloride contains not less than 98.0 percent and not more than 100.5 percent of $C_{14}H_{14}N_2 \cdot HCl$, calculated on the dried basis.

Packaging and storage—Preserve in tight, light-resistant containers.

Reference standard—*USP Naphazoline Hydrochloride Reference Standard*—Dry at 105° for 2 hours before using.

Identification—

A: The infrared absorption spectrum of a potassium bromide dispersion of it, previously dried, exhibits maxima only at the same wavelengths as that of a similar preparation of USP Naphazoline Hydrochloride RS.

B: The ultraviolet absorption spectrum of a 1 in 50,000 solution in methanol exhibits maxima and minima at the same wavelengths as that of a similar solution of USP Naphazoline Hydrochloride RS, concomitantly measured, and the respective absorptivities, calculated on the dried basis, at the wavelength of maximum absorbance at about 280 nm, do not differ by more than 3.0%.

C: A solution (1 in 100) responds to the tests for *Chloride* ⟨191⟩.

pH ⟨791⟩: between 5.0 and 6.6, in a 1 in 100 solution in carbon dioxide-free water, and the solution is clear and colorless.

Loss on drying ⟨731⟩—Dry it at 105° for 2 hours: it loses not more than 0.5% of its weight.

Residue on ignition ⟨281⟩: not more than 0.2%.

Ordinary impurities ⟨466⟩—

Test solution: methanol.

Standard solution: methanol.

Eluant: a mixture of methanol, glacial acetic acid, and water (8:1:1).

Visualization: 2.

Assay—Dissolve about 300 mg of Naphazoline Hydrochloride, accurately weighed, in 50 mL of glacial acetic acid, and add 10 mL of mercuric acetate TS and 1 drop of crystal violet TS. Titrate (see *Titrimetry* ⟨541⟩) with 0.1 *N* perchloric acid VS to a blue-green end-point. Perform a blank determination, and make any necessary correction. Each mL of 0.1 *N* perchloric acid is equivalent to 24.67 mg of $C_{14}H_{14}N_2 \cdot HCl$.

Naphazoline Hydrochloride Nasal Solution

» Naphazoline Hydrochloride Nasal Solution is a solution of Naphazoline Hydrochloride in water adjusted to a suitable pH and tonicity. It contains not less than 90.0 percent and not more than 110.0 percent of the labeled amount of $C_{14}H_{14}N_2 . HCl$.

Packaging and storage—Preserve in tight, light-resistant containers.

Reference standard—*USP Naphaxoline Hydrochloride Reference Standard*—Dry at 105° for 2 hours before using.

Identification—The retention time of the major peak in the chromatogram of the *Assay preparation* corresponds to that of the *Standard preparation* as obtained in the *Assay*.

Assay—

Mobile phase—Dissolve 1.1 g of sodium 1-heptanesulfonate in about 400 mL of water. Add 250 mL of acetonitrile and 10 mL of glacial acetic acid, dilute with water to 1000 mL, and mix. Sonicate for 10 minutes, filter, and degas to obtain a solution having a pH of about 3.5. Make adjustments if necessary (see *System Suitability* under *Chromatography* ⟨621⟩).

Standard preparation—Dissolve an accurately weighed quantity of USP Naphazoline Hydrochloride RS in water, and dilute quantitatively, and stepwise if necessary, with water to obtain a solution having a known concentration of about 250 µg per mL.

Assay preparation—Pipet a volume of Nasal Solution, equivalent to about 25 mg of naphazoline hydrochloride, into a 100-mL volumetric flask, dilute with water to volume, and mix.

Chromatographic system (see *Chromatography* ⟨621⟩)—The liquid chromatograph is equipped with a 280-nm detector and a 4-mm × 30-cm column that contains packing L11. The flow rate is about 2 mL per minute. Chromatograph the *Standard preparation*, and record the peak responses as directed under *Procedure:* the tailing factor for the naphazoline hydrochloride peak is not more than 2.0, and the relative standard deviation for replicate injections is not more than 1.5%.

Procedure—Separately inject equal volumes (about 15 µL) of the *Standard preparation* and the *Assay preparation* into the chromatograph, record the chromatograms, and measure the responses for the major peaks. Calculate the quantity, in mg, of $C_{14}H_{14}N_2.HCl$ in each mL of the Nasal Solution taken by the formula:

$$0.1(C/V)(r_U/r_S),$$

in which C is the concentration, in µg per mL, of USP Naphazoline Hydrochloride RS in the *Standard preparation*, V is the volume, in mL, of Nasal Solution taken, and r_U and r_S are the peak responses obtained from the *Assay preparation* and the *Standard preparation*, respectively.

Naphazoline Hydrochloride Ophthalmic Solution

» Naphazoline Hydrochloride Ophthalmic Solution is a sterile, buffered solution of Naphazoline Hydrochloride in water adjusted to a suitable tonicity. It contains not less than 90.0 percent and not more than 115.0 percent of the labeled amount of $C_{14}H_{14}N_2 . HCl$. It contains a suitable preservative.

Packaging and storage—Preserve in tight containers.

Reference standard—*USP Naphazoline Hydrochloride Reference Standard*—Dry at 105° for 2 hours before using.

Identification—Place in a separator a volume of Ophthalmic Solution, equivalent to about 25 mg of naphazoline hydrochloride, add 5 mL of 1 *N* sodium hydroxide, saturate with sodium chloride, and extract with two 25-mL portions of ether. Wash the ether solution with 5 mL of water, filter the ether through a small paper filter, evaporate the filtrate to about 5 mL, transfer the residual solution to a 10- to 15-mL beaker, allow to evaporate spontaneously, and dry the residue at 80° for 1 hour: the naphazoline so obtained melts between 115° and 120° when determined as directed for *Class Ia* under *Melting Range or Temperature* ⟨741⟩.

Sterility—It meets the requirements under *Sterility Tests* ⟨71⟩.

pH ⟨791⟩: between 5.5 and 7.0.

Assay—Transfer to a 100-mL volumetric flask an accurately measured volume of Naphazoline Hydrochloride Ophthalmic Solution, equivalent to about 2 mg of naphazoline hydrochloride, dilute with methanol to volume, and mix. Concomitantly determine the absorbances of this solution and of a Standard solution of USP Naphazoline Hydrochloride RS in the same medium, having a known concentration of about 20 µg per mL, in 1-cm cells at the wavelength of maximum absorbance at about 280 nm, with a suitable spectrophotometer, using methanol as the blank. Calculate the quantity, in mg, of $C_{14}H_{14}N_2 . HCl$ in each mL of the Ophthalmic Solution taken by the formula:

$$0.1(C/V)(A_U/A_S),$$

in which C is the concentration, in µg per mL, of USP Naphazoline Hydrochloride RS in the Standard solution, V is the volume, in mL, of Ophthalmic Solution taken, and A_U and A_S are the absorbances of the solution from the Ophthalmic Solution and the Standard solution, respectively.

Naproxen

CH_3 CHCOOH CH_3O

$C_{14}H_{14}O_3$ 230.26

2-Naphthaleneacetic acid, 6-methoxy-α-methyl-, (+)-.

(+)-6-Methoxy-α-methyl-2-naphthaleneacetic acid [*22204-53-1*].

» Naproxen contains not less than 98.5 percent and not more than 101.5 percent of $C_{14}H_{14}O_3$, calculated on the dried basis.

Packaging and storage—Preserve in tight containers.

Reference standard—*USP Naproxen Reference Standard*—Dry at 105° for 3 hours before using.

Identification—

A: The infrared absorption spectrum of a potassium bromide dispersion of it exhibits maxima only at the same wavelengths as that of a similar preparation of USP Naproxen RS.

B: The ultraviolet absorption spectrum of a 1 in 40,000 solution in methanol exhibits maxima and minima at the same wavelengths as that of a similar solution of USP Naproxen RS, concomitantly measured, and the respective absorptivities, calculated on the dried basis, at the wavelength of maximum absorbance at about 271 nm do not differ by more than 3%.

Specific rotation ⟨781⟩: between +63.0° and +68.5°, calculated on the dried basis, determined in a solution in chloroform containing 100 mg in each 10 mL.

Loss on drying ⟨731⟩—Dry it at 105° for 3 hours: it loses not more than 0.5% of its weight.

Heavy metals, *Method II* ⟨231⟩: 0.002%.

Chromatographic purity—Dissolve 100 mg of Naproxen in methanol, and dilute with methanol to 5.0 mL to obtain the *Test solution*. Dissolve a suitable quantity of USP Naproxen RS in methanol to obtain a *Standard solution* having a known concentration of about 20 mg per mL. Dilute a portion of this solution quantitatively and stepwise with methanol to obtain three *Comparison solutions* having concentrations of 20, 60, and 100 µg per mL (0.1%, 0.3%, and 0.5% of the *Standard solution*), respectively. Apply separate 10-µL portions of the five solutions on the starting line of a suitable thin-layer chromatographic plate (see *Chromatography* ⟨621⟩), coated with a 0.25-mm layer of chromatographic silica gel mixture. Develop the chromatogram in a solvent system consisting of a mixture of toluene, tetrahydrofuran, and glacial acetic acid (30:3:1) until the solvent front has moved about three-fourths of the length of the plate. Remove the plate from the chamber, mark the solvent front, air-dry, and view under short- and long-wavelength ultraviolet light: the R_f value of the principal spot in the chromatogram of the *Test solution* corresponds to that of the *Standard solution*, and any other spot obtained from the *Test solution* does not exceed, in size or intensity, the principal spot obtained from the 100-µg-per-mL *Comparison solution* (0.5%), and the sum of the intensities of any secondary spots, similarly compared, does not exceed 2.0%.

Assay—Dissolve about 500 mg of Naproxen, accurately weighed, in a mixture of 75 mL of methanol and 25 mL of water that has been previously neutralized to the phenolphthalein end-point with 0.1 *N* NaOH. Dissolve by gentle warming, if necessary, add phenolphthalein TS, and titrate with 0.1 *N* sodium hydroxide VS. Each mL of 0.1 *N* sodium hydroxide is equivalent to 23.03 mg of $C_{14}H_{14}O_3$.

Naproxen Tablets

» Naproxen Tablets contain not less than 90.0 percent and not more than 110.0 percent of the labeled amount of $C_{14}H_{14}O_3$.

Packaging and storage—Preserve in well-closed containers.

Reference standard—*USP Naproxen Reference Standard*—Dry at 105° for 3 hours before using.

Identification—Prepare a mixture of the *Standard preparation* and the *Assay preparation* (1:1), prepared as directed in the *Assay*, and chromatograph as directed in the *Assay:* the chromatogram so obtained exhibits two main peaks, corresponding to naproxen and the internal standard.

Dissolution ⟨711⟩—

0.1 M, pH 7.4 phosphate buffer—Dissolve 2.62 g of monobasic sodium phosphate and 11.50 g of anhydrous dibasic sodium phosphate in 1000 mL of water, and mix.

Medium: 0.1 M, pH 7.4 phosphate buffer; 900 mL.

Apparatus 2: 50 rpm.

Time: 45 minutes.

Procedure—Determine the amount of $C_{14}H_{14}O_3$ dissolved from ultraviolet absorbances at the wavelength of maximum absorbance at about 332 nm of filtered portions of the solution under test, suitably diluted with *0.1 M, pH 7.4 phosphate buffer*, in comparison with a Standard solution having a known concentration of USP Naproxen RS in the same medium.

Tolerances—Not less than 70% (*Q*) of the labeled amount of $C_{14}H_{14}O_3$ is dissolved in 45 minutes.

Uniformity of dosage units ⟨905⟩: meet the requirements.

Assay—

Mobile phase—Prepare a suitable mixture of acetonitrile, water, and glacial acetic acid (50:49:1). Make adjustments if necessary (see *System Suitability* under *Chromatography* ⟨621⟩). Increased resolution may be achieved by increasing the proportion of water in the *Mobile phase.*

Solvent mixture—Prepare a suitable mixture of acetonitrile and water (90:10).

Internal standard solution—Dilute 5 mL of butyrophenone with acetonitrile to make 100 mL. Dilute 1 mL of the resulting solution with acetonitrile to make 100 mL. Each mL of this solution contains about 0.5 µL of butyrophenone.

Standard preparation—Dissolve an accurately weighed quantity of USP Naproxen RS in *Solvent mixture* to obtain a solution having a known concentration of about 2.5 mg per mL. Transfer 1.0 mL of the resulting solution and 2.0 mL of *Internal standard solution* to a 100-mL volumetric flask, dilute with *Mobile phase* to volume, and mix. This solution contains about 25 µg of USP Naproxen RS per mL.

Assay preparation—Weigh and finely powder not less than 20 Naproxen Tablets. Transfer an accurately weighed quantity of the powder, equivalent to about 250 mg of naproxen, to a 100-mL volumetric flask. Add 10 mL of water, and shake until the material is completely dispersed. Dilute with acetonitrile to volume, and mix. Allow any insoluble matter to settle, then transfer 1.0 mL of the clear supernatant solution to a 100-mL volumetric flask, add 2.0 mL of *Internal standard solution*, dilute with *Mobile phase* to volume, and mix.

Chromatographic system (see *Chromatography* ⟨621⟩)—The liquid chromatograph is equipped with a 254-nm detector and a 4.6-mm × 15-cm column that contains 5-µm packing L1. The flow rate is about 1.2 mL per minute. Chromatograph the *Standard preparation*, and record the peak responses as directed under *Procedure:* the column efficiency, determined from the analyte peak, is not less than 4000 theoretical plates when calculated by the formula:

$$5.545(t/W_{h/2})^2,$$

the resolution between the analyte and internal standard peaks is not less than 11.5 when calculated by the formula:

$$2(t_2 - t_1)/[1.699(W_{1h/2} + W_{2h/2})],$$

and the relative standard deviation of replicate injections is not more than 1.5%.

Procedure—Separately inject equal volumes (about 20 µL) of the *Standard preparation* and the *Assay preparation* into the chromatograph, record the chromatograms, and measure the responses for the major peaks. The relative retention times are about 0.6 for naproxen and 1.0 for the internal standard. Calculate the quantity, in mg, of $C_{14}H_{14}O_3$ in the portion of Tablets taken by the formula:

$$10C(R_U/R_S),$$

in which *C* is the concentration, in µg per mL, of USP Naproxen RS in the *Standard preparation*, and R_U and R_S are the ratios of the response of the naproxen peak to the response of the internal standard peak obtained from the *Assay preparation* and the *Standard preparation*, respectively.

Naproxen Sodium

$C_{14}H_{13}NaO_3$ 252.24

2-Naphthaleneacetic acid, 6-methoxy-α-methyl-, sodium salt, (−)-.

(−)-Sodium 6-methoxy-α-methyl-2-naphthaleneacetate [*26159-34-2*].

» Naproxen Sodium contains not less than 98.0 percent and not more than 102.0 percent of $C_{14}H_{13}NaO_3$, calculated on the dried basis.

Packaging and storage—Preserve in tight containers.

Reference standard—*USP Naproxen Sodium Reference Standard*—Dry in vacuum at 105° for 3 hours before using.

Identification—

A: The infrared absorption spectrum of a potassium bromide dispersion of it, previously dried, exhibits maxima only at the same wavelengths as that of a similar preparation of USP Naproxen Sodium RS.

B: The ultraviolet absorption spectrum of a 1 in 40,000 solution in methanol exhibits maxima and minima at the same wavelengths as that of a similar solution of USP Naproxen Sodium RS, concomitantly measured, and the respective absorptivities, calculated on the dried basis, at the wavelength of maximum absorbance at about 272 nm do not differ by more than 3%.

Specific rotation ⟨781⟩: between −17.0° and −15.3°, calculated on the dried basis, determined in 0.1 *N* sodium hydroxide containing 500 mg in each 10 mL.

Loss on drying ⟨731⟩—Dry it in vacuum at 105° for 3 hours: it loses not more than 1.0% of its weight.

Heavy metals, *Method I* ⟨231⟩—Dissolve 1.0 g in 20 mL of water in a separator, add 5 mL of 1 *N* hydrochloric acid, and extract with successive 20-mL, 20-mL, and 10-mL portions of methylene chloride. Discard the methylene chloride extracts, and use the aqueous layer for the test: the limit is 0.002%.

Chromatographic purity—Dissolve 100 mg in 5 mL of methanol. Dissolve a suitable quantity of USP Naproxen Sodium RS in methanol to obtain a *Standard solution* having a known concentration of about 20 mg per mL. Dilute a portion of this solution quantitatively with methanol to obtain three *Comparison solutions* having concentrations of 20, 60, and 100 µg per mL (0.1%, 0.3%, and 0.5% of the *Standard solution*), respectively. Apply separate 10-µL portions of the five solutions on the starting line of a suitable thin-layer chromatographic plate (see *Chromatography* ⟨621⟩) coated with a 0.25-mm layer of chromatographic silica gel mixture. Develop the chromatogram in a solvent system consisting of a mixture of toluene, tetrahydrofuran, and glacial acetic acid (30:3:1) until the solvent front has moved about three-fourths of the length of the plate. Remove the plate from the chamber, mark the solvent front, air-dry, and view under short- and long-wavelength ultraviolet light: the R_f value of the principal spot in the chromatogram of the solution under test corresponds to that of the *Standard solution*, the intensity of any individual secondary spot does not exceed that of the 100-µg-per-mL *Comparison solution* (0.5%), and the sum of the intensities of any secondary spots, similarly compared, does not exceed 2.0%.

Free naproxen—Dissolve about 5.0 g in 25 mL of water in a separator, and extract the solution with three 15-mL portions of chloroform. Evaporate the combined extracts on a steam bath to dryness. Dissolve the residue in 10 mL of a mixture of methanol and water (3:1) previously neutralized with 0.1 *N* sodium hydroxide to the phenolphthalein end-point. Add phenolphthalein TS, and titrate with 0.10 *N* sodium hydroxide: not more than 2.2 mL is consumed (1.0%).

Assay—Dissolve about 200 mg of Naproxen Sodium, accurately weighed, in 50 mL of glacial acetic acid containing 2 drops of 1-naphtholbenzein previously neutralized with 0.1 *N* perchloric acid if necessary. Titrate with 0.1 *N* perchloric acid VS. Each mL of 0.1 *N* perchloric acid is equivalent to 25.22 mg of $C_{14}H_{13}NaO_3$.

Naproxen Sodium Tablets

» Naproxen Sodium Tablets contain not less than 90.0 percent and not more than 110.0 percent of the labeled amount of $C_{14}H_{13}NaO_3$.

Packaging and storage—Preserve in well-closed containers.

Reference standard—*USP Naproxen Sodium Reference Standard*—Dry in vacuum at 105° for 3 hours before using.

Identification—

A: Transfer a quantity of finely powdered Tablets, equivalent to about 250 mg of naproxen sodium, to a centrifuge tube, and add 12 mL of water and 1 mL of hydrochloric acid: a dense white precipitate is formed. Centrifuge the mixture: the clear, supernatant solution responds to the identification test for *Sodium* ⟨191⟩.

B: Prepare a mixture of the *Standard preparation* and the *Assay preparation* (1:1), prepared as directed in the *Assay*, and chromatograph as directed in the *Assay:* the chromatogram thus obtained exhibits two main peaks, corresponding to naproxen and the internal standard.

Dissolution ⟨711⟩—

Medium: 0.1 *M* phosphate buffer (pH 7.4), prepared by dissolving 2.62 g of monobasic sodium phosphate and 11.50 g of anhydrous dibasic sodium phosphate in water to make 1000 mL; 900 mL.

Apparatus 2: 50 rpm.

Time: 45 minutes.

Standard preparation—Dissolve an accurately weighed portion of USP Naproxen Sodium RS in *Dissolution Medium* to [illegible]in a solution having a known concentration of about 50 µg

[illegible]*re*—Dilute a filtered portion of the solution under test [illegible]y with *Dissolution Medium* as necessary to obtain [illegible]aving a concentration of about 50 µg per mL of $C_{14}H_{13}NaO_3$. Determine the amount of $C_{14}H_{13}NaO_3$ dissolved from ultraviolet absorbances at the wavelength of maximum absorbance at about 332 nm of this solution in comparison with the *Standard preparation.*

Tolerances—Not less than 70% (*Q*) of the labeled amount of $C_{14}H_{13}NaO_3$ is dissolved in 45 minutes.

Uniformity of dosage units ⟨905⟩: meet the requirements.

Assay—

Mobile phase, Solvent mixture, Internal standard solution, and *Chromatographic system*—Prepare as directed in the *Assay* under *Naproxen Tablets.*

Standard preparation—Dissolve an accurately weighed quantity of USP Naproxen Sodium RS in *Solvent mixture* to obtain a solution having a known concentration of about 2.75 mg per mL. Transfer 1.0 mL of the resulting solution and 2.0 mL of *Internal standard solution* to a 100-mL volumetric flask, dilute with *Mobile phase* to volume, and mix. This solution contains about 27.5 µg of USP Naproxen Sodium RS per mL.

Assay preparation—Weigh and finely powder not less than 20 Naproxen Sodium Tablets. Transfer an accurately weighed quantity of the powder, equivalent to about 275 mg of naproxen sodium, to a 100-mL volumetric flask. Add 10 mL of water, and shake until the material is completely dispersed. Dilute with acetonitrile to volume, and mix. Allow any insoluble matter to settle, then transfer 1.0 mL of the clear supernatant solution to a 100-mL volumetric flask, add 2.0 mL of *Internal standard solution*, dilute with *Mobile phase* to volume, and mix.

Procedure—Proceed as directed for *Procedure* in the *Assay* under *Naproxen Tablets.* Calculate the quantity, in mg, of $C_{14}H_{13}NaO_3$ in the portion of Tablets taken by the formula:

$$10C(R_U/R_S),$$

in which *C* is the concentration, in µg per mL, of USP Naproxen Sodium RS in the *Standard preparation*, and R_U and R_S are the ratios of the response of the naproxen peak to the response of the internal standard peak obtained from the *Assay preparation* and the *Standard preparation*, respectively.

Nasal Solutions—*see complete list in index*

Natamycin

$C_{33}H_{47}NO_{13}$ 665.73

Stereoisomer of 22-[(3-amino-3,6-dideoxy-β-D-mannopyranosyl)-oxy]-1,3,26-trihydroxy-12-methyl-10-oxo-6,11,28-trioxatricyclo[22.3.1.0^{5,7}]octacosa-8,14,16,18,20-pentaene-25-carboxylic acid [*7681-93-8*].

» Natamycin contains not less than 90.0 percent and not more than 102.0 percent of $C_{33}H_{47}NO_{13}$, calculated on the anhydrous basis.

Packaging and storage—Preserve in tight, light-resistant containers.

Reference standard—*USP Natamycin Reference Standard*—Do not dry before using.

Identification—Transfer 50 mg, accurately weighed, to a 200-mL volumetric flask, add 5.0 mL of water, and moisten the specimen. Add 100 mL of a 1 in 1000 solution of glacial acetic acid in methanol, and shake by mechanical means in the dark until dissolved. Dilute with the acetic acid–methanol solution to volume, and mix. Transfer 2.0 mL of this solution to a 100-mL volumetric flask, dilute with the acetic acid–methanol solution to volume, and mix: the ultraviolet absorption spectrum of the solution so obtained exhibits maxima and minima at the same wavelengths as that of a similar solution of USP Natamycin RS, concomitantly measured.

Crystallinity ⟨695⟩: meets the requirements.

pH ⟨791⟩: between 5.0 and 7.5, in an aqueous suspension containing 10 mg per mL.

Water, *Method I* ⟨921⟩: between 6.0% and 9.0%.

Assay—[NOTE—Throughout the *Assay*, protect from direct light all solutions containing natamycin.]

Mobile phase—Dissolve 3.0 g of ammonium acetate and 1.0 g of ammonium chloride in 760 mL of water, and mix. Add 5.0 mL of tetrahydrofuran and 240 mL of acetonitrile, and mix. Filter this solution through a 0.5-µm or finer porosity filter. Make adjustments if necessary (see *System Suitability* under *Chromatography* ⟨621⟩).

Standard preparation—Transfer about 20 mg of USP Natamycin RS, accurately weighed, to a 100-mL volumetric flask. Add 5.0 mL of tetrahydrofuran, and sonicate for 10 minutes. Add 60 mL of methanol, and swirl to dissolve. Add 25 mL of water, and mix. Allow to cool to room temperature. Dilute with water to volume, and mix. Filter this solution through a suitable membrane filter of 0.5-µm or finer porosity.

Resolution solution—Dissolve 20 mg of Natamycin in 100 mL of a 1 in 24,000 solution of hydrochloric acid in methanol. Allow to stand for 2 hours at 45°, and cool to room temperature.

Assay preparation—Transfer about 20 mg of Natamycin, accurately weighed, to a 100-mL volumetric flask. Proceed as directed under *Standard preparation*, beginning with "add 5.0 mL of tetrahydrofuran."

Chromatographic system (see *Chromatography* ⟨621⟩)—The liquid chromatograph is equipped with a 313-nm detector and a 4.6-mm × 25-cm analytical column that contains packing L1. [NOTE—A 3.9-mm × 20-mm precolumn may be used to extend the useful life of the analytical column.] The flow rate is about 3 mL per minute. Chromatograph the *Standard preparation*, and record the peak responses as directed under *Procedure:* the column efficiency determined from the analyte peak is not less than 3000 theoretical plates, the tailing factor is not less than 0.8 and not more than 1.3, and the relative standard deviation for replicate injections is not more than 1.0%. Chromatograph the *Resolution solution*, and record the peak responses as directed under *Procedure:* the resolution, R, between natamycin and its acid degradation product is not less than 2.5. The relative retention times are about 0.7 for natamycin and 1.0 for its acid degradation product.

Procedure—[NOTE—Use peak areas where peak responses are indicated.] Separately inject equal volumes (about 20 µL) of the *Standard preparation* and the *Assay preparation* into the chromatograph, record the chromatograms, and measure the responses for the major peaks. Calculate the percentage of natamycin ($C_{33}H_{47}NO_{13}$) in the portion of Natamycin taken by the formula:

$$0.1(W_S P_S / W_U)(r_U / r_S),$$

in which W_S is the weight, in mg, of USP Natamycin RS taken to prepare the *Standard preparation*, P_S is the stated content, in µg per mg, of USP Natamycin RS, W_U is the weight, in mg, of Natamycin taken to prepare the *Assay preparation*, and r_U and r_S are the peak responses obtained from the *Assay preparation* and the *Standard preparation*, respectively.

Natamycin Ophthalmic Suspension

» Natamycin Ophthalmic Suspension is a sterile suspension of Natamycin in a suitable aqueous vehicle. It contains not less than 90.0 percent and not more than 125.0 percent of the labeled amount of $C_{33}H_{47}NO_{13}$. It contains one or more suitable preservatives.

Packaging and storage—Preserve in tight, light-resistant containers. The containers or individual cartons are sealed and tamper-proof so that sterility is assured at time of first use.

Reference standard—*USP Natamycin Reference Standard*—Do not dry before using.

Identification—Transfer a volume of Ophthalmic Suspension, equivalent to about 50 mg of natamycin, to a 200-mL volumetric flask, and add water to make a volume of 5 mL. Proceed as directed in the *Identification test* under *Natamycin*, beginning with "Add 100 mL of a 1 in 1000 solution of glacial acetic acid in methanol:" the specified result is obtained.

Sterility—It meets the requirements under *Sterility Tests* ⟨71⟩, when tested as directed in the section, *Test Procedures for Direct Transfer to Test Media*, using 0.25 mL of the Ophthalmic Suspension taken from each container.

pH ⟨791⟩: between 5.0 and 7.5.

Assay—[NOTE—Throughout the *Assay* protect from direct light all solutions containing natamycin.]

Mobile phase, Standard preparation, Resolution solution, and *Chromatographic system*—Proceed as directed in the *Assay* under *Natamycin.*

Assay preparation—Transfer an accurately measured volume of Natamycin Ophthalmic Suspension, equivalent to about 50 mg of natamycin, to a 250-mL volumetric flask. Add 12.5 mL of tetrahydrofuran, and sonicate for 10 minutes. Add 150 mL of methanol, and swirl to dissolve. Add 60 mL of water, and mix. Allow to cool to room temperature. Dilute with water to volume, and mix. Filter this solution through a suitable membrane filter of 0.5-µm or finer porosity.

Procedure—Proceed as directed in the *Assay* under *Natamycin.* Calculate the quantity, in mg, of $C_{33}H_{47}NO_{13}$ in each mL of the Ophthalmic Suspension taken by the formula:

$$0.25C(P_S / V)(r_U / r_S),$$

in which C is the concentration, in mg per mL, of USP Natamycin RS in the *Standard preparation*, P_S is the stated content, in µg per mg, of USP Natamycin RS, V is the volume, in mL, of Ophthalmic Suspension taken, and r_U and r_S are the peak responses obtained from the *Assay preparation* and the *Standard preparation*, respectively.

Neomycin Sulfate

Neomycin sulfate.
Neomycins sulfate [*1405-10-3*].

» Neomycin Sulfate is the sulfate salt of a kind of neomycin, an antibacterial substance produced by the growth of *Streptomyces fradiae* Waksman (Fam. Streptomycetaceae), or a mixture of two or more such salts. It has a potency equivalent to not less than 600 µg of neomycin per mg, calculated on the dried basis.

Packaging and storage—Preserve in tight, light-resistant containers.

Reference standard—*USP Neomycin Sulfate Reference Standard*—Dry in vacuum at a pressure not exceeding 5 mm of mercury at 60° for 3 hours before using.

Identification—

A: On a suitable thin-layer chromatographic plate (see *Chromatography* ⟨621⟩), coated with a 0.25-mm layer of chromatographic silica gel, apply separately 1 µL of a solution containing 20 mg of specimen per mL and 1 µL of a Standard solution containing 20 mg of USP Neomycin Sulfate RS per mL. Allow the spots to dry, place the plate in a suitable chromatographic chamber, and develop the chromatogram with a freshly prepared solvent system consisting of a mixture of water, ammonium hydroxide, and acetone (71.5:8.5:20) until the solvent front has moved about three-fourths of the length of the plate. Remove the plate from the developing chamber, air-dry, and heat the plate at 105° for 1 hour. Spray the plate with a 1 in 100 solution of ninhydrin in butanol, heat at 105° for 5 minutes, and examin the chromatograms: the R_f value of the principal red spot obtained from the solution under test corresponds to that obtained from the Standard solution.

B: Dissolve about 10 mg in 1 mL of water, add 5 mL of 15 *N* sulfuric acid, and heat at 100° for 100 minutes. Allow to cool, add 10 mL of xylene, and shake for 10 minutes. Allow to separate, and decant the xylene layer. To the xylene layer add 10 mL of *p*-bromoaniline TS, and shake: a vivid pink-red color develops upon standing.

C: A solution (1 in 20) responds to the tests for *Sulfate* ⟨191⟩.

pH ⟨791⟩: between 5.0 and 7.5, in a solution containing 33 mg of neomycin per mL.

Loss on drying ⟨731⟩—Dry about 100 mg in vacuum at a pressure not exceeding 5 mm of mercury at 60° for 3 hours: it loses not more than 8.0% of its weight.

Other requirements—Neomycin Sulfate intended for use in making ophthalmic ointments meets the requirements under *Sterility Tests* ⟨71⟩, when tested as directed in the section, *Test Procedures Using Membrane Filtration*.

Assay—Proceed with Neomycin Sulfate as directed under *Antibiotics—Microbial Assays* ⟨81⟩.

Neomycin Sulfate, Gramicidin, and Triamcinolone Acetonide Cream, Nystatin,—*see* Nystatin, Neomycin Sulfate, Gramicidin, and Triamcinolone Acetonide Cream

Neomycin Sulfate, Gramicidin, and Triamcinolone Acetonide Ointment, Nystatin,—*see* Nystatin, Neomycin Sulfate, Gramicidin, and Triamcinolone Acetonide Ointment

Neomycin Sulfates and Hydrocortisone Acetate Otic Suspension, Colistin and—*see* Colistin and Neomycin Sulfates and Hydrocortisone Acetate Otic Suspension

Neomycin Sulfate Cream

» Neomycin Sulfate Cream contains the equivalent of not less than 90.0 percent and not more than 135.0 percent of the labeled amount of neomycin.

Packaging and storage—Preserve in well-closed containers, preferably at controlled room temperature.

Reference standard—*USP Neomycin Sulfate Reference Standard*—Dry in vacuum at a pressure not exceeding 5 mm of mercury at 60° for 3 hours before using.

Identification—Shake a quantity of Cream, equivalent to about 70 mg of neomycin, with 100 mL of chloroform, add 5 mL of water, and shake. Allow to separate, filter the aqueous layer, and use the filtrate as the test solution. On a suitable thin-layer chromatographic plate coated with a 0.25-mm layer of chromatographic silica gel (see *Chromatography* ⟨621⟩), apply separately 5 µL of the test solution and 5 µL of a Standard solution containing 20 mg of USP Neomycin Sulfate RS per mL. Allow the spots to dry, place the plate in a suitable chromatographic chamber, and develop the chromatogram with a solvent system consisting of a mixture of methanol, ammonium hydroxide, chloroform, and water (6:3:2:1) until the solvent front has moved about three-fourths of the length of the plate. Remove the plate from the chamber, and air-dry. Spray the plate with a 1 in 100 solution of ninhydrin in butyl alcohol, and heat the plate at 105° for 2 minutes: the R_f value of the principal red spot in the chromatogram obtained from the test solution corresponds to that obtained from the Standard solution.

[Mini]mum fill ⟨755⟩: meets the requirements.

[Assay—]Proceed as directed under *Antibiotics—Microbial As[says* ⟨81⟩,] using an accurately weighed portion of Neomycin [Sulfate Cre]am, equivalent to about 1.75 mg of neomycin, shaken in a separator with about 50 mL of ether, and extracted with four 20-mL portions of *Buffer No. 3*. Combine the aqueous extracts, and dilute with *Buffer No. 3* to an appropriate volume to obtain a stock solution of convenient concentration. Dilute this stock solution quantitatively and stepwise with *Buffer No. 3* to obtain a *Test Dilution* having a concentration assumed to be equal to the median dose level of the Standard.

Neomycin Sulfate Ointment

» Neomycin Sulfate Ointment contains the equivalent of not less than 90.0 percent and not more than 135.0 percent of the labeled amount of neomycin.

Packaging and storage—Preserve in well-closed containers, preferably at controlled room temperature.

Reference standard—*USP Neomycin Sulfate Reference Standard*—Dry in vacuum at a pressure not exceeding 5 mm of mercury at 60° for 3 hours before using.

Identification—It responds to the *Identification test* under *Neomycin Sulfate Cream.*

Minimum fill ⟨755⟩: meets the requirements.

Water, *Method I* ⟨921⟩: not more than 1.0%, 20 mL of a mixture of carbon tetrachloride, chloroform, and methanol (2:2:1) being used in place of methanol in the titration vessel.

Assay—Proceed as directed under *Antibiotics—Microbial Assays* ⟨81⟩, using an accurately weighed portion of Neomycin Sulfate Ointment, equivalent to about 3.5 mg of neomycin, shaken in a separator with about 50 mL of ether, and extracted with four 20-mL portions of *Buffer No. 3.* Combine the aqueous extracts, and dilute with *Buffer No. 3* to an appropriate volume to obtain a stock solution of convenient concentration. Dilute this stock solution quantitatively and stepwise with *Buffer No. 3* to obtain a *Test Dilution* having a concentration assumed to be equal to the median dose level of the Standard.

Neomycin Sulfate Ophthalmic Ointment

» Neomycin Sulfate Ophthalmic Ointment is a sterile preparation of Neomycin Sulfate in a suitable ointment base. It contains the equivalent of not less than 90.0 percent and not more than 135.0 percent of the labeled amount of neomycin.

Packaging and storage—Preserve in collapsible ophthalmic ointment tubes.

Reference standard—*USP Neomycin Sulfate Reference Standard*—Dry in vacuum at a pressure not exceeding 5 mm of mercury at 60° for 3 hours before using.

Identification—It responds to the *Identification test* under *Neomycin Sulfate Cream.*

Sterility—It meets the requirements under *Sterility Tests* ⟨71⟩.

Minimum fill ⟨755⟩: meets the requirements.

Water, *Method I* ⟨921⟩: not more than 1.0%, 20 mL of a mixture of carbon tetrachloride, chloroform, and methanol (2:2:1) being used in place of methanol in the titration vessel.

Metal particles—It meets the requirements of the test for *Metal Particles in Ophthalmic Ointments* ⟨751⟩.

Assay—Proceed with Neomycin Sulfate Ophthalmic Ointment as directed in the *Assay* under *Neomycin Sulfate Ointment.*

Neomycin Sulfate Oral Solution

» Neomycin Sulfate Oral Solution contains the equivalent of not less than 90.0 percent and not more than 125.0 percent of the labeled amount of neomycin. It may contain one or more suitable colors, flavors, and preservatives.

Packaging and storage—Preserve in tight, light-resistant containers, preferably at controlled room temperature.

Reference standard—*USP Neomycin Sulfate Reference Standard*—Dry in vacuum at a pressure not exceeding 5 mm of mercury at 60° for 3 hours before using.

Identification—Dilute a portion of Oral Solution with water to obtain a test solution containing the equivalent of 14 mg of neomycin per mL. Proceed as directed in the *Identification test* under *Neomycin Sulfate Cream*, beginning with "On a suitable thin-layer chromatographic plate:" the specified result is observed.

pH ⟨791⟩: between 5.0 and 7.5.

Assay—Proceed as directed under *Antibiotics—Microbial Assays* ⟨81⟩, using an accurately measured volume of Neomycin Sulfate Oral Solution diluted quantitatively with *Buffer No. 3* to yield a solution having a convenient concentration of neomycin. Dilute this stock solution quantitatively with *Buffer No. 3* to obtain a *Test Dilution* having a concentration assumed to be equal to the median dose level of the Standard.

Sterile Neomycin Sulfate

» Sterile Neomycin Sulfate is Neomycin Sulfate suitable for parenteral use. It has a potency equivalent to not less than 600 µg of neomycin per mg, calculated on the dried basis. In addition, where packaged for dispensing, it contains not less than 90.0 percent and not more than 120.0 percent of the labeled amount of neomycin.

Packaging and storage—Preserve in *Containers for Sterile Solids* as described under *Injections* ⟨1⟩.

Reference standard—*USP Neomycin Sulfate Reference Standard*—Dry in vacuum at a pressure not exceeding 5 mm of mercury at 60° for 3 hours before using.

Pyrogen—It meets the requirements of the *Pyrogen Test* ⟨151⟩, the test dose being 1.0 mL per kg of a solution in pyrogen-free saline TS containing 10 mg of neomycin per mL.

Sterility—It meets the requirements under *Sterility Tests* ⟨71⟩, when tested as directed in the section, *Test Procedures Using Membrane Filtration.*

Other requirements—It conforms to the definition, responds to the *Identification test*, and meets the requirements for *pH* and *Loss on drying* under *Neomycin Sulfate.* Where packaged for dispensing, it meets also the requirements for *Uniformity of Dosage Units* ⟨905⟩ and *Labeling* under *Injections* ⟨1⟩. Where intended for use in preparing sterile ophthalmic dosage forms, it is exempt from the requirements for *Pyrogen.*

Assay—

Assay preparation 1—Dissolve a suitable quantity of Sterile Neomycin Sulfate, accurately weighed, in *Buffer No. 3*, and dilute quantitatively with *Buffer No. 3* to obtain a solution having a convenient concentration.

Assay preparation 2 (where it is packaged for dispensing)—Constitute Sterile Neomycin Sulfate as directed in the labeling. Withdraw all of the withdrawable contents, using a suitable hypodermic needle and syringe, and dilute quantitatively with *Buffer No. 3* to obtain a solution having a convenient concentration.

Assay preparation 3 (where it is packaged for dispensing and where the labeling states the quantity of neomycin in a given volume of constituted solution)—Constitute Sterile Neomycin Sulfate as directed in the labeling. Dilute an accurately measured volume of the constituted solution quantitatively with *Buffer No. 3* to obtain a solution having a convenient concentration.

Procedure—Proceed as directed under *Antibiotics—Microbial Assays* ⟨81⟩, using an accurately measured volume of *Assay preparation* diluted quantitatively and stepwise with *Buffer No. 3* to yield a *Test Dilution* having a concentration assumed to be equal to the median dose level of the Standard.

Neomycin Sulfate Tablets

» Neomycin Sulfate Tablets contain the equivalent of not less than 90.0 percent and not more than 125.0 percent of the labeled amount of neomycin.

Packaging and storage—Preserve in tight containers.

Reference standard—*USP Neomycin Sulfate Reference Standard*—Dry in vacuum at a pressure not exceeding 5 mm of mercury at 60° for 3 hours before using.

Identification—Shake a portion of ground Tablet powder, equivalent to about 70 mg of neomycin, with 5 mL of water, and filter. Using the filtrate as the test solution, proceed as directed in the *Identification test* under *Neomycin Sulfate Cream*, beginning with "On a suitable thin-layer chromatographic plate:" the specified result is observed.

Disintegration ⟨701⟩: 60 minutes.

Uniformity of dosage units ⟨905⟩: meet the requirements.

Loss on drying ⟨731⟩—Dry about 100 mg of powdered Tablets, accurately weighed, in a capillary-stoppered bottle in vacuum at a pressure not exceeding 5 mm of mercury at 60° for 3 hours: it loses not more than 10.0% of its weight.

Assay—Proceed as directed under *Antibiotics—Microbial Assays* ⟨81⟩, using not less than 5 Neomycin Sulfate Tablets blended at high-speed in a blender jar for 3 to 5 minutes with a sufficient accurately measured volume of *Buffer No. 3* to obtain a stock solution having a convenient concentration. Dilute this stock solution quantitatively and stepwise with *Buffer No. 3* to obtain a *Test Dilution* having a concentration assumed to be equal to the median dose level of the Standard.

Neomycin Sulfate and Bacitracin Ointment

» Neomycin Sulfate and Bacitracin Ointment contains the equivalent of not less than 90.0 percent and not more than 130.0 percent of the labeled amounts of neomycin and bacitracin.

Packaging and storage—Preserve in tight, light-resistant containers, preferably at controlled room temperature.

Reference standards—*USP Neomycin Sulfate Reference Standard*—Dry in vacuum at a pressure not exceeding 5 mm of mercury at 60° for 3 hours before using. *USP Bacitracin Zinc Reference Standard*—Dry in vacuum at a pressure not exceeding 5 mm of mercury at 60° for 3 hours before using.

Identification—

A: It responds to the *Identification test* under *Neomycin Sulfate Cream.*

B: It responds to the *Identification test* under *Bacitracin Ointment.*

Minimum fill ⟨755⟩: meets the requirements.

Water, *Method I* ⟨921⟩: not more than 0.5%, 20 mL of a mixture of carbon tetrachloride, chloroform, and methanol (2:2:1) being used in place of methanol in the titration vessel.

Assay for neomycin—Proceed with Neomycin Sulfate and Bacitracin Ointment as directed in the *Assay* under *Neomycin Sulfate Ointment.*

Assay for bacitracin—Proceed with Neomycin Sulfate and Bacitracin Ointment as directed in the *Assay* under *Bacitracin Ointment.*

Neomycin Sulfate and Bacitracin Zinc Ointment

» Neomycin Sulfate and Bacitracin Zinc Ointment contains the equivalent of not less than 90.0 percent and not more than 130.0 percent of the labeled amounts of neomycin and bacitracin.

Packaging and storage—Preserve in collapsible tubes or in well-closed containers.

Reference standards—*USP Neomycin Sulfate Reference Standard*—Dry in vacuum at a pressure not exceeding 5 mm of mercury at 60° for 3 hours before using. *USP Bacitracin Zinc Reference Standard*—Dry in vacuum at a pressure not exceeding 5 mm of mercury at 60° for 3 hours before using.

Identification—

A: It responds to the *Identification test* under *Neomycin Sulfate Cream.*

B: It responds to the *Identification test* under *Bacitracin Ointment.*

Minimum fill ⟨755⟩: meets the requirements.

Water, *Method I* ⟨921⟩: not more than 0.5%, 20 mL of a mixture of carbon tetrachloride, chloroform, and methanol (2:2:1) being used in place of methanol in the titration vessel.

Assay for neomycin and Assay for bacitracin—Proceed with Neomycin Sulfate and Bacitracin Zinc Ointment as directed in the *Assay for neomycin* and in the *Assay for bacitracin* under *Neomycin and Polymyxin B Sulfates and Bacitracin Zinc Ophthalmic Ointment.*

Neomycin Sulfate and Dexamethasone Sodium Phosphate Cream

» Neomycin Sulfate and Dexamethasone Sodium Phosphate Cream contains the equivalent of not less than 90.0 percent and not more than 135.0 percent of the labeled amount of neomycin, and the equivalent of not less than 90.0 percent and not more than 110.0 percent of the labeled amount of dexamethasone phosphate ($C_{22}H_{30}FO_8P$).

Packaging and storage—Preserve in collapsible tubes or in tight containers.

Reference standards—*USP Neomycin Sulfate Reference Standard*—Dry in vacuum at a pressure not exceeding 5 mm of mercury at 60° for 3 hours before using. *USP Dexamethasone Reference Standard*—Dry at 105° for 3 hours before using. *USP Dexamethasone Phosphate Reference Standard*—Dry in vacuum at a pressure of 5 mm of mercury at 40° to constant weight before using.

Identification—

A: It responds to the *Identification test* under *Neomycin Sulfate Cream.*

B: The *Assay preparation,* prepared as directed in the *Assay for dexamethasone phosphate,* responds to the *Identification test* under *Dexamethasone Sodium Phosphate Cream.*

Minimum fill ⟨755⟩: meets the requirements.

Assay for neomycin—Proceed with Neomycin Sulfate and Dexamethasone Sodium Phosphate Cream as directed in the *Assay* under *Neomycin Sulfate Cream.*

Assay for dexamethasone phosphate—

Alcohol–aqueous phosphate buffer, 0.05 M Phosphate buffer, [illegible]*bile phase, Standard preparation,* and *Chromatographic sys-* [illegible] Prepare as directed in the *Assay* under *Dexamethasone* [illegible] *Phosphate Cream.*

[illegible]*paration*—Using an accurately weighed portion of [illegible]lfate and Dexamethasone Sodium Phosphate Cream, prepare as directed in the *Assay* under *Dexamethasone Sodium Phosphate Cream.*

Procedure—Proceed as directed for *Procedure* in the *Assay* under *Dexamethasone Sodium Phosphate Cream.* Calculate the quantity, in mg, of dexamethasone phosphate ($C_{22}H_{30}FO_8P$) in the portion of Cream taken by the formula:

$$0.1C(r_U / r_S).$$

Neomycin Sulfate and Dexamethasone Sodium Phosphate Ophthalmic Ointment

» Neomycin Sulfate and Dexamethasone Sodium Phosphate Ophthalmic Ointment is a sterile ointment containing Neomycin Sulfate and Dexamethasone Sodium Phosphate. It contains the equivalent of not less than 90.0 percent and not more than 135.0 percent of the labeled amount of neomycin, and the equivalent of not less than 90.0 percent and not more than 110.0 percent of the labeled amount of dexamethasone phosphate ($C_{22}H_{30}FO_8P$).

NOTE—Where Neomycin Sulfate and Dexamethasone Sodium Phosphate Ophthalmic Ointment is prescribed without reference to the quantity of neomycin or dexamethasone phosphate contained therein, a product containing 3.5 mg of neomycin and 0.5 mg of dexamethasone phosphate per g shall be dispensed.

Packaging and storage—Preserve in collapsible ophthalmic ointment tubes.

Reference standards—*USP Neomycin Sulfate Reference Standard*—Dry in vacuum at a pressure not exceeding 5 mm of mercury at 60° for 3 hours before using. *USP Dexamethasone Reference Standard*—Dry at 105° for 3 hours before using. *USP Dexamethasone Phosphate Reference Standard*—Dry at a pressure of 5 mm of mercury at 40° to constant weight before using.

Identification—

A: It responds to the *Identification test* under *Neomycin Sulfate Cream.*

B: The *Assay preparation,* prepared as directed in the *Assay for dexamethasone phosphate,* responds to the *Identification test* under *Dexamethasone Sodium Phosphate Cream.*

Sterility—It meets the requirements under *Sterility Tests* ⟨71⟩.

Minimum fill ⟨755⟩: meets the requirements.

Water, *Method I* ⟨921⟩: not more than 1.0%, 20 mL of a mixture of carbon tetrachloride, chloroform, and methanol (2:2:1) being used in place of methanol in the titration vessel.

Metal particles—It meets the requirements of the test for *Metal Particles in Ophthalmic Ointments* ⟨751⟩.

Assay for neomycin—Proceed as directed under *Antibiotics—Microbial Assays* ⟨81⟩, using an accurately weighed portion of Neomycin Sulfate and Dexamethasone Sodium Phosphate Ophthalmic Ointment shaken in a separator with about 50 mL of ether, and extracted with four 20-mL portions of *Buffer No. 3.* Combine the aqueous extracts, and dilute with *Buffer No. 3* to an appropriate volume to obtain a stock solution. Dilute this stock solution quantitatively and stepwise with *Buffer No. 3* to obtain a *Test Dilution* having a concentration assumed to be equal to the median dose level of the Standard.

Assay for dexamethasone phosphate—

Alcohol–aqueous phosphate buffer, 0.05 M Phosphate buffer, Mobile phase, Standard preparation, and *Chromatographic system*—Prepare as directed in the *Assay* under *Dexamethasone Sodium Phosphate Cream.*

Assay preparation—Using an accurately weighed portion of Neomycin Sulfate and Dexamethasone Sodium Phosphate Ophthalmic Ointment, prepare as directed in the *Assay* under *Dexamethasone Sodium Phosphate Cream.*

Procedure—Proceed as directed for *Procedure* in the *Assay* under *Dexamethasone Sodium Phosphate Cream.* Calculate the quantity, in mg, of dexamethasone phosphate ($C_{22}H_{30}FO_8P$) in the portion of Ophthalmic Ointment taken by the formula:

$$0.1C(r_U/r_S).$$

Neomycin Sulfate and Dexamethasone Sodium Phosphate Ophthalmic Solution

» Neomycin Sulfate and Dexamethasone Sodium Phosphate Ophthalmic Solution is a sterile, aqueous solution of Neomycin Sulfate and Dexamethasone Sodium Phosphate. It contains the equivalent of not less than 90.0 percent and not more than 130.0 percent of the labeled amount of neomycin, and the equivalent of not less than 90.0 percent and not more than 115.0 percent of the labeled amount of dexamethasone phosphate ($C_{22}H_{30}FO_8P$). It may contain one or more suitable buffers, dispersants, and preservatives.

NOTE—Where Neomycin Sulfate and Dexamethasone Sodium Phosphate Ophthalmic Solution is prescribed, without reference to the amount of neomycin or dexamethasone phosphate contained therein, a product containing 3.5 mg of neomycin and 1.0 mg of dexamethasone phosphate per mL shall be dispensed.

Packaging and storage—Preserve in tight, light-resistant containers, and avoid exposure to excessive heat.

Reference standards—*USP Neomycin Sulfate Reference Standard*—Dry in vacuum at a pressure not exceeding 5 mm of mercury at 60° for 3 hours before using. *USP Dexamethasone Reference Standard*—Dry at 105° for 3 hours before using. *USP Dexamethasone Phosphate Reference Standard*—Dry at a pressure of 5 mm of mercury at 40° to constant weight before using.

Identification—The *Assay preparation*, prepared as directed in the *Assay for dexamethasone phosphate*, responds to the *Identification test* under *Dexamethasone Sodium Phosphate Cream.*

Sterility—It meets the requirements under *Sterility Tests* ⟨71⟩.

pH ⟨791⟩: between 6.0 and 8.0.

Assay for neomycin—Proceed as directed under *Antibiotics—Microbial Assays* ⟨81⟩, using an accurately measured volume of Neomycin Sulfate and Dexamethasone Sodium Phosphate Ophthalmic Solution diluted quantitatively and stepwise with *Buffer No. 3* to yield a *Test Dilution* having a concentration assumed to be equal to the median dose level of the Standard (1.0 µg of neomycin per mL).

Assay for dexamethasone phosphate—

Mobile phase, Standard preparation, and *Chromatographic system*—Prepare as directed in the *Assay* under *Dexamethasone Sodium Phosphate Injection.*

Assay preparation—Transfer accurately measured volume of Neomycin Sulfate and Dexamethasone Sodium Phosphate Ophthalmic Solution, equivalent to about 8 mg of dexamethasone phosphate, to a 100-mL volumetric flask, dilute with *Mobile phase* to volume, and mix.

Procedure—Proceed as directed for *Procedure* in the *Assay* under *Dexamethasone Sodium Phosphate Injection.* Calculate the quantity, in mg, of dexamethasone phosphate ($C_{22}H_{30}FO_8P$) in each mL of the Ophthalmic Solution taken by the formula:

$$0.1(C/V)(r_U/r_S),$$

in which V is the volume, in mL, of Ophthalmic Solution taken.

Neomycin Sulfate and Fluocinolone Acetonide Cream

» Neomycin Sulfate and Fluocinolone Acetonide Cream contains the equivalent of not less than 90.0 percent and not more than 135.0 percent of the labeled amount of neomycin, and the equivalent of not less than 90.0 percent and not more than 110.0 percent of the labeled amount of fluocinolone acetonide ($C_{24}H_{30}F_2O_6$).

Packaging and storage—Preserve in collapsible tubes or in tight containers.

Reference standards—*USP Neomycin Sulfate Reference Standard*—Dry in vacuum at a pressure not exceeding 5 mm of mercury at 60° for 3 hours before using. *USP Fluocinolone Acetonide Reference Standard*—Dry at 105° for 3 hours before using.

Identification—

A: It responds to the *Identification test* under *Neomycin Sulfate Cream.*

B: It responds to the *Identification test* under *Fluocinolone Acetonide Cream.*

Minimum fill ⟨755⟩: meets the requirements.

Assay for neomycin—Proceed with Neomycin Sulfate and Fluocinolone Acetonide Cream as directed in the *Assay* under *Neomycin Sulfate Cream.*

Assay for fluocinolone acetonide—Proceed with Neomycin Sulfate and Fluocinolone Acetonide Cream as directed in the *Assay* under *Fluocinolone Acetonide Cream.*

Neomycin Sulfate and Fluorometholone Ointment

» Neomycin Sulfate and Fluorometholone Ointment contains the equivalent of not less than 90.0 percent and not more than 135.0 percent of the labeled amount of neomycin, and not less than 90.0 percent and not more than 110.0 percent of the labeled amount of fluorometholone ($C_{22}H_{29}FO_4$).

Packaging and storage—Preserve in collapsible tubes or in well-closed containers.

Reference standards—*USP Neomycin Sulfate Reference Standard*—Dry in vacuum at a pressure not exceeding 5 mm of mercury at 60° for 3 hours before using. *USP Fluorometholone Reference Standard*—Dry in vacuum at 60° for 3 hours before using. *USP Fluoxymesterone Reference Standard*—Dry at 105° for 3 hours before using.

Identification—

A: It responds to the *Identification test* under *Neomycin Sulfate Cream.*

B: The ratios of the retention time of the main peak to that of the internal standard peak obtained from the *Standard preparation* and the *Assay preparation* as directed in the *Assay for fluorometholone* do not differ by more than 2.0%.

Minimum fill ⟨755⟩: meets the requirements.

Water, *Method I* ⟨921⟩: not more than 1.0%, 20 mL of a mixture of carbon tetrachloride, chloroform, and methanol (2:2:1) being used in place of methanol in the titration vessel.

Assay for neomycin—Proceed with Neomycin Sulfate and Fluorometholone Ointment as directed in the *Assay* under *Neomycin Sulfate Ointment.*

Assay for fluorometholone—

Internal standard solution, Mobile solvent, and *Standard preparation*—Prepare as directed in the *Assay* under *Fluorometholone Cream.*

Assay preparation—Transfer an accurately weighed quantit of Neomycin Sulfate and Fluorometholone Ointment, equival to about 1 mg of fluorometholone, to a suitable container, add 20.0 mL of *Internal standard solution*, and mix.

Procedure—Treat 20.0 mL each of the *Standard preparation* and the *Assay preparation* in the following manner. To each add 10.0 mL of hexane, shake for about 15 minutes, then allow the layers to separate, and centrifuge, if necessary. Using the lower (acetonitrile) layer, proceed as directed for *Procedure* in the *Assay* under *Fluorometholone Cream*, beginning with "Using a suitable microsyringe." Calculate the quantity, in mg, of $C_{22}H_{29}FO_4$ in the portion of Ointment taken by the formula:

$$20C(R_U/R_S).$$

Neomycin Sulfate and Flurandrenolide Cream

» Neomycin Sulfate and Flurandrenolide Cream contains the equivalent of not less than 90.0 percent and not more than 135.0 percent of the labeled amount of neomycin, and not less than 90.0 percent and not more than 110.0 percent of the labeled amount of flurandrenolide ($C_{24}H_{33}FO_6$).

Packaging and storage—Preserve in collapsible tubes or in tight containers, protected from light.

Reference standards—*USP Neomycin Sulfate Reference Standard*—Dry in vacuum at a pressure not exceeding 5 mm of mercury at 60° for 3 hours before using. *USP Flurandrenolide Reference Standard*—Dry in vacuum at 105° for 4 hours before using.

Identification—

A: It responds to the *Identification test* under *Neomycin Sulfate Cream.*

B: It responds to the *Identification test* under *Flurandrenolide Cream.*

Minimum fill ⟨755⟩: meets the requirements.

Assay for neomycin—Proceed with Neomycin Sulfate and Flurandrenolide Cream as directed in the *Assay* under *Neomycin Sulfate Ointment.*

Assay for flurandrenolide—Proceed with Neomycin Sulfate and Flurandrenolide Cream as directed in the *Assay* under *Flurandrenolide Cream.* Calculate the quantity, in mg, of $C_{24}H_{33}FO_6$ in the portion of Cream taken by the formula:

$$10C(r_U/r_S),$$

in which C is the concentration, in mg per mL, of USP Flurandrenolide RS in the *Standard preparation*, and r_U and r_S are the peak responses obtained from the *Assay preparation* and the *Standard preparation*, respectively.

Neomycin Sulfate and Flurandrenolide Lotion

» Neomycin Sulfate and Flurandrenolide Lotion contains the equivalent of not less than 90.0 percent and not more than 130.0 percent of the labeled amount of neomycin, and not less than 90.0 percent and not more than 110.0 percent of the labeled amount of flurandrenolide ($C_{24}H_{33}FO_6$).

Packaging and storage—Preserve in tight containers, protected from light.

Reference standards—*USP Neomycin Sulfate Reference Standard*—Dry in vacuum at a pressure not exceeding 5 mm of mercury at 60° for 3 hours before using. *USP Flurandrenolide Reference Standard*—Dry in vacuum at 105° for 4 hours before using.

Identification—

A: It responds to the *Identification test* under *Neomycin Sulfate Cream.*

B: It responds to the *Identification test* under *Flurandrenolide Cream.*

Microbial limits—It meets the requirements of the tests for absence of *Staphylococcus aureus* and *Pseudomonas aeruginosa* under *Microbial Limit Tests* ⟨61⟩.

Minimum fill ⟨755⟩: meets the requirements.

Assay for neomycin—Proceed as directed for neomycin under *Antibiotics—Microbial Assays* ⟨81⟩, using an accurately weighed portion of Neomycin Sulfate and Flurandrenolide Lotion, equivalent to about 3.5 mg of neomycin, blended for 3 to 5 minutes in a high-speed glass blender jar containing an accurately measured volume of *Buffer No. 3* sufficient to obtain a stock solution having a convenient concentration of neomycin. Dilute an accurately measured volume of this stock solution quantitatively with *Buffer No. 3* to obtain a *Test Dilution* having a concentration of neomycin assumed to be equal to the median dose level of the Standard.

Assay for flurandrenolide—Proceed with Neomycin Sulfate and Flurandrenolide Lotion as directed in the *Assay* under *Flurandrenolide Cream.* Calculate the quantity, in mg, of $C_{24}H_{33}FO_6$ in the portion of Lotion taken by the formula:

$$10C(r_U/r_S),$$

in which C is the concentration, in mg per mL, of USP Flurandrenolide RS in the *Standard preparation*, and r_U and r_S are the peak responses obtained from the *Assay preparation* and the *Standard preparation*, respectively.

Neomycin Sulfate and Flurandrenolide Ointment

» Neomycin Sulfate and Flurandrenolide Ointment contains the equivalent of not less than 90.0 percent and not more than 135.0 percent of the labeled amount of neomycin, and not less than 90.0 percent and not more than 110.0 percent of the labeled amount of flurandrenolide ($C_{24}H_{33}FO_6$).

Packaging and storage—Preserve in collapsible tubes or in tight containers, protected from light.

Reference standards—*USP Neomycin Sulfate Reference Standard*—Dry in vacuum at a pressure not exceeding 5 mm of mercury at 60° for 3 hours before using. *USP Flurandrenolide Reference Standard*—Dry in vacuum at 105° for 4 hours before using.

Identification—

A: It responds to the *Identification test* under *Neomycin Sulfate Cream.*

B: It responds to the *Identification test* under *Flurandrenolide Cream.*

Minimum fill ⟨755⟩: meets the requirements.

Water, *Method I* ⟨921⟩: not more than 1.0%, 20 mL of a mixture of carbon tetrachloride, chloroform, and methanol (2:2:1) being used in place of methanol in the titration vessel.

Assay for neomycin—Proceed with Neomycin Sulfate and Flurandrenolide Ointment as directed in the *Assay* under *Neomycin Sulfate Ointment.*

Assay for flurandrenolide—Proceed with Neomycin Sulfate and Flurandrenolide Ointment as directed in the *Assay* under *Flurandrenolide Cream.* Calculate the quantity, in mg, of $C_{24}H_{33}FO_6$ in the portion of Ointment taken by the formula:

$$10C(r_U/r_S),$$

in which C is the concentration, in mg per mL, of USP Flurandrenolide RS in the *Standard preparation*, and r_U and r_S are the peak responses obtained from the *Assay preparation* and the *Standard preparation*, respectively.

Neomycin Sulfate and Gramicidin Ointment

» Neomycin Sulfate and Gramicidin Ointment contains the equivalent of not less than 90.0 percent and not more than 140.0 percent of the labeled amounts of neomycin and gramicidin.

Packaging and storage—Preserve in collapsible tubes or in well-closed containers.

Reference standards—*USP Neomycin Sulfate Reference Standard*—Dry in vacuum at a pressure not exceeding 5 mm of mercury at 60° for 3 hours before using. *USP Gramicidin Reference Standard*—Dry in vacuum at a pressure not exceeding 5 mm of mercury at 60° for 3 hours before using.

Identification—It responds to the *Identification test* under *Neomycin Sulfate Cream.*

Minimum fill ⟨755⟩: meets the requirements.

Water, *Method I* ⟨921⟩: not more than 1.0%, 20 mL of a mixture of carbon tetrachloride, chloroform, and methanol (2:2:1) being used in place of methanol in the titration vessel.

Assay for neomycin—Proceed with Neomycin Sulfate and Gramicidin Ointment as directed in the *Assay* under *Neomycin Sulfate Ointment.*

Assay for gramicidin—Proceed as directed for gramicidin under *Antibiotics—Microbial Assays* ⟨81⟩, using an accurately weighed portion of Neomycin Sulfate and Gramicidin Ointment dissolved in 50 mL of hexanes in a separator, and extracted with four 20-mL portions of 80 percent alcohol. Combine the extracts in a suitable volumetric flask, dilute with alcohol to volume, and mix. Dilute this solution quantitatively and stepwise with alcohol to obtain a *Test Dilution* having a concentration of gramicidin assumed to be equal to the median dose level of the Standard.

Neomycin Sulfate and Hydrocortisone Cream

» Neomycin Sulfate and Hydrocortisone Cream contains the equivalent of not less than 90.0 percent and not more than 135.0 percent of the labeled amount of neomycin, and not less than 90.0 percent and not more than 110.0 percent of the labeled amount of hydrocortisone ($C_{21}H_{30}O_5$).

Packaging and storage—Preserve in collapsible tubes or in well-closed containers.

Reference standards—*USP Neomycin Sulfate Reference Standard*—Dry in vacuum at a pressure not exceeding 5 mm of mercury at 60° for 3 hours before using. *USP Hydrocortisone Reference Standard*—Dry at 105° for 3 hours before using.

Identification—

A: It responds to the *Identification test* under *Neomycin Sulfate Cream.*

B: The chromatogram of the *Assay preparation* obtained as directed in the *Assay for hydrocortisone* exhibits a major peak for hydrocortisone, the retention time of which corresponds to that exhibited in the chromatogram of the *Standard preparation* obtained as directed in the *Assay for hydrocortisone.*

Minimum fill ⟨755⟩: meets the requirements.

Assay for neomycin—Proceed with Neomycin Sulfate and Hydrocortisone Cream as directed in the *Assay* under *Neomycin Sulfate Ointment.*

Assay for hydrocortisone—Proceed with Neomycin Sulfate and Hydrocortisone Cream as directed in the *Assay for hydrocortisone* under *Neomycin and Polymyxin B Sulfates, Bacitracin Zinc, and Hydrocortisone Ophthalmic Ointment.*

Neomycin Sulfate and Hydrocortisone Ointment

» Neomycin Sulfate and Hydrocortisone Ointment contains the equivalent of not less than 90.0 percent and not more than 135.0 percent of the labeled amount of neomycin, and not less than 90.0 percent and not more than 110.0 percent of the labeled amount of hydrocortisone ($C_{21}H_{30}O_5$).

Packaging and storage—Preserve in collapsible tubes or in well-closed containers.

Reference standards—*USP Neomycin Sulfate Reference Standard*—Dry in vacuum at a pressure not exceeding 5 mm of mercury at 60° for 3 hours before using. *USP Hydrocortisone Reference Standard*—Dry at 105° for 3 hours before using.

Identification—

A: It responds to the *Identification test* under *Neomycin Sulfate Cream.*

B: The chromatogram of the *Assay preparation* obtained as directed in the *Assay for hydrocortisone* exhibits a major peak for hydrocortisone, the retention time of which corresponds to that exhibited in the chromatogram of the *Standard preparation* obtained as directed in the *Assay for hydrocortisone.*

Minimum fill ⟨755⟩: meets the requirements.

Water, *Method I* ⟨921⟩: not more than 1.0%, 20 mL of a mixture of carbon tetrachloride, chloroform, and methanol (2:2:1) being used in place of methanol in the titration vessel.

Assay for neomycin—Proceed with Neomycin Sulfate and Hydrocortisone Ointment as directed in the *Assay* under *Neomycin Sulfate Ointment.*

Assay for hydrocortisone—Proceed with Neomycin Sulfate and Hydrocortisone Ointment as directed in the *Assay for hydrocortisone* under *Neomycin and Polymyxin B Sulfates, Bacitracin Zinc, and Hydrocortisone Ophthalmic Ointment.*

Neomycin Sulfate and Hydrocortisone Otic Suspension

» Neomycin Sulfate and Hydrocortisone Otic Suspension is a sterile suspension containing not less than 90.0 percent and not more than 130.0 percent of the labeled amount of neomycin, and not less than 90.0 percent and not more than 110.0 percent of the labeled amount of hydrocortisone. It contains Acetic Acid, and may contain one or more suitable buffers, dispersants, and preservatives.

NOTE—Where Neomycin Sulfate and Hydrocortisone Otic Suspension is prescribed, without reference to the quantity of neomycin or hydrocortisone contained therein, a product containing 3.5 mg of neomycin and 10 mg of hydrocortisone per mL shall be dispensed.

Packaging and storage—Preserve in tight, light-resistant containers.

Reference standards—*USP Neomycin Sulfate Reference Standard*—Dry in vacuum at a pressure not exceeding 5 mm of mercury at 60° for 3 hours before using. *USP Hydrocortisone Reference Standard*—Dry at 105° for 3 hours before using.

Sterility—It meets the requirements under *Sterility Tests* ⟨71⟩.

pH ⟨791⟩: between 4.5 and 6.0.

Assay for neomycin—Using an accurately measured volume of Neomycin Sulfate and Hydrocortisone Otic Suspension, freshly mixed and free from entrapped air, proceed as directed in the

Assay for neomycin under *Neomycin and Polymyxin B Sulfates and Hydrocortisone Otic Solution.*

Assay for hydrocortisone—

Mobile phase and *Standard preparation*—Prepare as directed in the *Assay for hydrocortisone content* under *Neomycin and Polymyxin B Sulfates, Bacitracin Zinc, and Hydrocortisone Ophthalmic Ointment.*

Assay preparation—Transfer 3.0 mL of Neomycin Sulfate and Hydrocortisone Otic Suspension, freshly mixed and free from entrapped air, to a 200-mL volumetric flask, dilute with a mixture of methanol and water (1:1) to volume, and mix. Filter the solution, rejecting the first 10 mL of the filtrate.

Procedure—Proceed as directed for *Procedure* in the *Assay for hydrocortisone content* under *Neomycin and Polymyxin B Sulfates, Bacitracin Zinc, and Hydrocortisone Ophthalmic Ointment.* Calculate the quantity, in mg, of $C_{21}H_{30}O_5$ in each mL of the Otic Suspension taken by the formula:

$$(66.67C)(r_U / r_S),$$

in which C is the concentration, in mg per mL, of USP Hydrocortisone RS in the *Standard preparation,* and r_U and r_S are the peak responses obtained from the *Assay preparation* and the *Standard preparation,* respectively.

Neomycin Sulfate and Hydrocortisone Acetate Cream

» Neomycin Sulfate and Hydrocortisone Acetate Cream contains the equivalent of not less than 90.0 percent and not more than 135.0 percent of the labeled amount of neomycin, and not less than 90.0 percent and not more than 110.0 percent of the labeled amount of hydrocortisone acetate ($C_{23}H_{32}O_6$).

Packaging and storage—Preserve in well-closed containers.

Reference standards—*USP Neomycin Sulfate Reference Standard*—Dry in vacuum at a pressure not exceeding 5 mm of mercury at 60° for 3 hours before using. *USP Hydrocortisone Acetate Reference Standard*—Dry in vacuum at 60° for 3 hours before using.

Identification—

A: It responds to the *Identification test* under *Neomycin Sulfate Cream.*

B: The chromatogram of the *Assay preparation* obtained as directed in the *Assay for hydrocortisone acetate* exhibits a major peak for hydrocortisone acetate, the retention time of which corresponds to that exhibited in the chromatogram of the *Standard preparation* obtained as directed in the *Assay for hydrocortisone acetate.*

Minimum fill ⟨755⟩: meets the requirements.

Assay for neomycin—Proceed with Neomycin Sulfate and Hydrocortisone Acetate Cream as directed in the *Assay* under *Neomycin Sulfate Ointment.*

Assay for hydrocortisone acetate—Proceed with Neomycin Sulfate and Hydrocortisone Acetate Cream as directed in the *Assay* under *Hydrocortisone Acetate Lotion.*

Neomycin Sulfate and Hydrocortisone Acetate Lotion

» Neomycin Sulfate and Hydrocortisone Acetate Lotion contains the equivalent of not less than 90.0 percent and not more than 130.0 percent of the labeled amount of neomycin, and not less than 90.0 percent and not more than 110.0 percent of the labeled amount of hydrocortisone acetate ($C_{23}H_{32}O_6$).

Packaging and storage—Preserve in well-closed containers.

Reference standards—*USP Neomycin Sulfate Reference Standard*—Dry in vacuum at a pressure not exceeding 5 mm of mercury at 60° for 3 hours before using. *USP Hydrocortisone Acetate Reference Standard*—Dry in vacuum at 60° for 3 hours before using.

Identification—

A: It responds to the *Identification test* under *Neomycin Sulfate Cream.*

B: The chromatogram of the *Assay preparation* obtained as directed in the *Assay for hydrocortisone acetate* exhibits a major peak for hydrocortisone acetate, the retention time of which corresponds to that exhibited in the chromatogram of the *Standard preparation* obtained as directed in the *Assay for hydrocortisone acetate.*

Minimum fill ⟨755⟩: meets the requirements.

Assay for neomycin—Proceed as directed for neomycin under *Antibiotics—Microbial Assays* ⟨81⟩, blending an accurately measured volume of Neomycin Sulfate and Hydrocortisone Acetate Lotion for 3 to 5 minutes in a high-speed glass blender jar containing an accurately measured volume of *Buffer No. 3.* Dilute an accurately measured volume of the solution so obtained quantitatively and stepwise with *Buffer No. 3* to obtain a *Test Dilution* having a concentration of neomycin assumed to be equal to the median dose level of the Standard.

Assay for hydrocortisone acetate—Proceed with Neomycin Sulfate and Hydrocortisone Acetate Lotion as directed in the *Assay* under *Hydrocortisone Acetate Lotion.*

Neomycin Sulfate and Hydrocortisone Acetate Ointment

» Neomycin Sulfate and Hydrocortisone Acetate Ointment contains the equivalent of not less than 90.0 percent and not more than 135.0 percent of the labeled amount of neomycin, and not less than 90.0 percent and not more than 110.0 percent of the labeled amount of hydrocortisone acetate ($C_{23}H_{32}O_6$).

Packaging and storage—Preserve in collapsible tubes or in well-closed containers.

Reference standards—*USP Neomycin Sulfate Reference Standard*—Dry in vacuum at a pressure not exceeding 5 mm of mercury at 60° for 3 hours before using. *USP Hydrocortisone Acetate Reference Standard*—Dry in vacuum at 60° for 3 hours before using.

Identification—

A: It responds to the *Identification test* under *Neomycin Sulfate Cream.*

B: The chromatogram of the *Assay preparation* obtained as directed in the *Assay for hydrocortisone acetate* exhibits a major peak for hydrocortisone acetate, the retention time of which corresponds to that exhibited in the chromatogram of the *Standard preparation* obtained as directed in the *Assay for hydrocortisone acetate.*

Minimum fill ⟨755⟩: meets the requirements.

Water, *Method I* ⟨921⟩: not more than 1.0%, 20 mL of a mixture of carbon tetrachloride, chloroform, and methanol (2:2:1) being used in place of methanol in the titration vessel.

Assay for neomycin—Proceed with Neomycin Sulfate and Hydrocortisone Acetate Ointment as directed in the *Assay* under *Neomycin Sulfate Ointment.*

Assay for hydrocortisone acetate—Proceed with Neomycin Sulfate and Hydrocortisone Acetate Ointment as directed in the *Assay* under *Hydrocortisone Acetate Lotion.*

Neomycin Sulfate and Hydrocortisone Acetate Ophthalmic Ointment

» Neomycin Sulfate and Hydrocortisone Acetate Ophthalmic Ointment contains the equivalent of not less than 90.0 percent and not more than 135.0 percent of the labeled amount of neomycin, and not less than 90.0 percent and not more than 110.0 percent of the labeled amount of hydrocortisone acetate ($C_{23}H_{32}O_6$).

Packaging and storage—Preserve in collapsible ophthalmic ointment tubes.

Reference standards—*USP Neomycin Sulfate Reference Standard*—Dry in vacuum at a pressure not exceeding 5 mm of mercury at 60° for 3 hours before using. *USP Hydrocortisone Acetate Reference Standard*—Dry in vacuum at 60° for 3 hours before using.

Identification—

A: It responds to the *Identification test* under *Neomycin Sulfate Cream.*

B: The chromatogram of the *Assay preparation* obtained as directed in the *Assay for hydrocortisone acetate* exhibits a major peak for hydrocortisone acetate, the retention time of which corresponds to that exhibited in the chromatogram of the *Standard preparation* obtained as directed in the *Assay for hydrocortisone acetate.*

Sterility—It meets the requirements under *Sterility Tests* ⟨71⟩, when tested as directed in the section, *Test Procedures Using Membrane Filtration.*

Minimum fill ⟨755⟩: meets the requirements.

Water, *Method I* ⟨921⟩: not more than 1.0%, 20 mL of a mixture of carbon tetrachloride, chloroform, and methanol (2:2:1) being used in place of methanol in the titration vessel.

Metal particles—It meets the requirements of the test for *Metal Particles in Ophthalmic Ointments* ⟨751⟩.

Assay for neomycin—Proceed with Neomycin Sulfate and Hydrocortisone Acetate Ophthalmic Ointment as directed in the *Assay* under *Neomycin Sulfate Ointment.*

Assay for hydrocortisone acetate—Proceed with Neomycin Sulfate and Hydrocortisone Acetate Ophthalmic Ointment as directed in the *Assay* under *Hydrocortisone Acetate Lotion.*

Neomycin Sulfate and Hydrocortisone Acetate Ophthalmic Suspension

» Neomycin Sulfate and Hydrocortisone Acetate Ophthalmic Suspension is a sterile, aqueous suspension containing the equivalent of not less than 90.0 percent and not more than 130.0 percent of the labeled amount of neomycin, and not less than 90.0 percent and not more than 110.0 percent of the labeled amount of hydrocortisone acetate ($C_{23}H_{32}O_6$).

Packaging and storage—Preserve in tight containers. The containers or individual cartons are sealed and tamper-proof so that sterility is assured at time of first use.

Reference standards—*USP Neomycin Sulfate Reference Standard*—Dry in vacuum at a pressure not exceeding 5 mm of mercury at 60° for 3 hours before using. *USP Hydrocortisone Acetate Reference Standard*—Dry in vacuum at 60° for 3 hours before using.

Identification—The chromatogram of the *Assay preparation* obtained as directed in the *Assay for hydrocortisone acetate* exhibits a major peak for hydrocortisone acetate, the retention time of which corresponds with that exhibited in the chromatogram of the *Standard preparation* obtained as directed in the *Assay for hydrocortisone acetate.*

Sterility—It meets the requirements under *Sterility Tests* ⟨71⟩.

pH ⟨791⟩: between 5.5 and 7.5.

Assay for neomycin—Proceed as directed for neomycin under *Antibiotics—Microbial Assays* ⟨81⟩, using an accurately measured volume of Neomycin Sulfate and Hydrocortisone Acetate Ophthalmic Suspension, freshly mixed and free from air bubbles, diluted quantitatively and stepwise with *Buffer No. 3* to obtain a *Test Dilution* having a concentration assumed to be equal to the median dose level of the Standard.

Assay for hydrocortisone acetate—

Mobile phase—Prepare a solution containing *n*-butyl chloride, water-saturated *n*-butyl chloride, tetrahydrofuran, methanol, and glacial acetic acid (95:95:14:7:6).

Internal standard solution—Prepare a solution of fluoxymesterone in chloroform containing 0.8 mg per mL.

Standard preparation—Dissolve about 10 mg of USP Hydrocortisone Acetate RS, accurately weighed, in 10.0 mL of *Internal standard solution*, dilute with about 40 mL of chloroform, and mix.

Assay preparation—Transfer an accurately measured volume of Neomycin Sulfate and Hydrocortisone Acetate Ophthalmic Suspension, freshly mixed and free from air bubbles, equivalent to about 10 mg of hydrocortisone acetate, to a suitable container. Add 10.0 mL of *Internal standard solution* and about 40 mL of chloroform, shake vigorously for about 5 minutes, and allow the phases to separate. Use the clear chloroform layer as the *Assay preparation.*

Chromatographic system (see *Chromatography* ⟨621⟩)—The liquid chromatograph is equipped with a 254-nm detector and a 4-mm × 30-cm column that contains packing L3. The flow rate is about 1 mL per minute. Chromatograph the *Standard preparation*, and record the peak responses as directed under *Procedure:* the resolution, *R*, between the analyte and internal standard peaks is not less than 3.0, and the relative standard deviation for replicate injections is not more than 2.0%.

Procedure—Separately inject equal volumes (about 15 µL) of the *Standard preparation* and the *Assay preparation* into the chromatograph, record the chromatograms, and measure the responses for the major peaks. The relative retention times are about 0.7 and 1.0 for hydrocortisone acetate and fluoxymesterone, respectively. Calculate the quantity, in mg, of hydrocortisone acetate ($C_{23}H_{32}O_6$) in each mL of the Ophthalmic Suspension taken by the formula:

$$(W/V)(R_U/R_S),$$

in which *W* is the quantity, in mg, of USP Hydrocortisone Acetate RS taken to prepare the *Standard preparation*, *V* is the volume, in mL, of Ophthalmic Suspension taken, and R_U and R_S are the peak response ratios of the hydrocortisone acetate peak to the internal standard peak obtained from the *Assay preparation* and the *Standard preparation*, respectively.

Neomycin Sulfate and Methylprednisolone Acetate Cream

» Neomycin Sulfate and Methylprednisolone Acetate Cream contains the equivalent of not less than 90.0 percent and not more than 135.0 percent of the labeled amount of neomycin, and not less than 90.0 percent and not more than 110.0 percent of the labeled amount of methylprednisolone acetate ($C_{24}H_{32}O_6$).

Packaging and storage—Preserve in collapsible tubes or in tight containers, protected from light.

Reference standards—*USP Neomycin Sulfate Reference Standard*—Dry in vacuum at a pressure not exceeding 5 mm of mercury at 60° for 3 hours before using. *USP Methylprednisolone Acetate Reference Standard*—Dry at 105° for 3 hours before using.

Identification—

A: It responds to the *Identification test* under *Neomycin Sulfate Cream.*

B: In the thin-layer chromatogram prepared as directed in the *Assay for methylprednisolone acetate*, the R_f value of the principal spot obtained from the *Assay preparation* corresponds to that obtained from the *Standard preparation*, prepared as directed in the *Assay.*

Minimum fill ⟨755⟩: meets the requirements.

Assay for neomycin—Proceed with Neomycin Sulfate and Methylprednisolone Acetate Cream as directed in the *Assay* under *Neomycin Sulfate Ointment.*

Assay for methylprednisolone acetate—Proceed with Neomycin Sulfate and Methylprednisolone Acetate Cream as directed in the *Assay* under *Methylprednisolone Acetate Cream.*

Neomycin and Polymyxin B Sulfates Cream

» Neomycin and Polymyxin B Sulfates Cream contains the equivalent of not less than 90.0 percent and not more than 130.0 percent of the labeled amounts of neomycin and polymyxin B.

Packaging and storage—Preserve in well-closed containers, preferably at controlled room temperature.

Reference standards—*USP Neomycin Sulfate Reference Standard*—Dry in vacuum at a pressure not exceeding 5 mm of mercury at 60° for 3 hours before using. *USP Polymyxin B Sulfate Reference Standard*—Dry in vacuum at a pressure not exceeding 5 mm of mercury at 60° for 3 hours before using.

Identification—

A: It responds to the *Identification test* under *Neomycin Sulfate Cream.*

B: Shake a quantity of Cream, equivalent to about 50,000 USP Polymyxin B Units, with 20 mL of chloroform, add 5 mL of 0.1 *N* hydrochloric acid, shake vigorously, centrifuge, and use the clear supernatant liquid as the test solution. On a suitable thin-layer chromatographic plate coated with a 0.25-mm layer of chromatographic silica gel (see *Chromatography* ⟨621⟩) apply separately 10 µL of the test solution and 10 µL of a Standard solution of USP Polymyxin B Sulfate RS in 0.1 *N* hydrochloric acid containing 10,000 USP Polymyxin B Units per mL. Place the plate in a suitable chromatographic chamber, and develop the chromatogram with a solvent system consisting of a mixture of isopropyl alcohol, water, and ammonium hydroxide (24:17:3) until the solvent front has moved about three-fourths of the length of the plate. Remove the plate from the chamber, and dry at 105° for 5 minutes. Spray the plate with a 1 in 200 solution of ninhydrin in butyl alcohol, and heat the plate at 105° for 15 minutes: the R_f value of the principal spot in the chromatogram obtained from the test solution corresponds to that of the principal spot in the chromatogram obtained from the Standard solution.

Minimum fill ⟨755⟩: meets the requirements.

Assay for neomycin—Proceed as directed under *Antibiotics—Microbial Assays* ⟨81⟩, using an accurately weighed portion of Neomycin and Polymyxin B Sulfates Cream, equivalent to about 1.75 mg of neomycin, shaken in a separator with about 50 mL of ether, and extracted with four 20-mL portions of *Buffer No. 3.* Combine the aqueous extracts, and dilute with *Buffer No. 3* to an appropriate volume to obtain a stock solution of convenient concentration. Dilute this stock solution quantitatively and stepwise with *Buffer No. 3* to obtain a *Test Dilution* having a concentration assumed to be equal to the median dose level of the Standard.

Assay for polymyxin B—Proceed as directed under *Antibiotics—Microbial Assays* ⟨81⟩, using an accurately weighed portion of Neomycin and Polymyxin B Sulfates Cream shaken with about 50 mL of ether in a separator, and extracted with four 25-mL portions of *Buffer No. 6.* Combine the aqueous extracts, and dilute with *Buffer No. 6* to an appropriate volume to obtain a stock solution. Dilute this stock solution quantitatively and stepwise with *Buffer No. 6* to obtain a *Test Dilution* having a concentration assumed to be equal to the median dose level of the Standard (10 Polymyxin B Units per mL). Add to each test dilution of the Standard a quantity of USP Neomycin Sulfate RS, dissolved in *Buffer No. 6*, to obtain the same concentration of neomycin present in the *Test Dilution.*

Neomycin and Polymyxin B Sulfates Solution for Irrigation

» Neomycin and Polymyxin B Sulfates Solution for Irrigation is a sterile, aqueous solution containing the equivalent of not less than 90.0 percent and not more than 130.0 percent of the labeled amounts of neomycin and of polymyxin B. It may contain a suitable preservative.

Packaging and storage—Preserve in tight containers.

Labeling—Label it to indicate that it is to be diluted for use in a urinary bladder irrigation and is not intended for injection.

Reference standards—*USP Neomycin Sulfate Reference Standard*—Dry in vacuum at a pressure not exceeding 5 mm of mercury at 60° for 3 hours before using. *USP Polymyxin B Sulfate Reference Standard*—Dry in vacuum at a pressure not exceeding 5 mm of mercury at 60° for 3 hours before using.

Sterility—It meets the requirements under *Sterility Tests* ⟨71⟩, when tested as directed in the section, *Test Procedures Using Membrane Filtration.*

pH ⟨791⟩: between 4.5 and 6.0.

Assay for neomycin and Assay for polymyxin B—Proceed with Neomycin and Polymyxin B Sulfates Solution for Irrigation as directed in the *Assay for neomycin* and in the *Assay for polymyxin B* under *Neomycin and Polymyxin B Sulfates and Hydrocortisone Otic Solution.*

Neomycin and Polymyxin B Sulfates Ophthalmic Ointment

» Neomycin and Polymyxin B Sulfates Ophthalmic Ointment is a sterile ointment containing Neomycin Sulfate and Polymyxin B Sulfate. It contains the equivalent of not less than 90.0 percent and not more than 125.0 percent of the labeled amounts of neomycin and polymyxin B.

Packaging and storage—Preserve in collapsible ophthalmic ointment tubes.

Reference standards—*USP Neomycin Sulfate Reference Standard*—Dry in vacuum at a pressure not exceeding 5 mm of mercury at 60° for 3 hours before using. *USP Polymyxin B Sulfate Reference Standard*—Dry in vacuum at a pressure not exceeding 5 mm of mercury at 60° for 3 hours before using.

Identification—

A: It responds to the *Identification test* under *Neomycin Sulfate Cream.*

B: It responds to *Identification test B* under *Neomycin and Polymyxin B Sulfates Cream.*

Sterility—It meets the requirements under *Sterility Tests* ⟨71⟩, when tested as directed in the section, *Test Procedures Using Membrane Filtration.*

Minimum fill ⟨755⟩: meets the requirements.

Water, *Method I* ⟨921⟩: not more than 0.5%, 20 mL of a mixture of carbon tetrachloride, chloroform, and methanol (2:2:1) being used in place of methanol in the titration vessel.

Metal particles—It meets the requirements of the test for *Metal Particles in Ophthalmic Ointments* ⟨751⟩.

Assay for neomycin and Assay for polymyxin B—Proceed with Neomycin and Polymyxin B Sulfates Ophthalmic Ointment as directed in the *Assay for neomycin* and in the *Assay for polymyxin B* under *Neomycin and Polymyxin B Sulfates and Bacitracin Zinc Ophthalmic Ointment.*

Neomycin and Polymyxin B Sulfates Ophthalmic Solution

» Neomycin and Polymyxin B Sulfates Ophthalmic Solution contains the equivalent of not less than 90.0 percent and not more than 125.0 percent of the labeled amounts of neomycin and of polymyxin B. It may contain one or more suitable buffers, dispersants, irrigants, and preservatives.

Packaging and storage—Preserve in tight containers, and avoid exposure to excessive heat.

Reference standards—*USP Neomycin Sulfate Reference Standard*—Dry in vacuum at a pressure not exceeding 5 mm of mercury at 60° for 3 hours before using. *USP Polymyxin B Sulfate Reference Standard*—Dry in vacuum at a pressure not exceeding 5 mm of mercury at 60° for 3 hours before using.

Sterility—It meets the requirements under *Sterility Tests* ⟨71⟩, when tested as directed in the section, *Test Procedures Using Membrane Filtration.*

pH ⟨791⟩: between 5.0 and 7.0.

Assay for neomycin and Assay for polymyxin B—Proceed with Neomycin and Polymyxin B Sulfates Ophthalmic Solution as directed in the *Assay for neomycin* and in the *Assay for polymyxin B* under *Neomycin and Polymyxin B Sulfates and Hydrocortisone Otic Solution.*

Neomycin and Polymyxin B Sulfates and Bacitracin Ointment

» Neomycin and Polymyxin B Sulfates and Bacitracin Ointment contains the equivalent of not less than 90.0 percent and not more than 130.0 percent of the labeled amounts of neomycin, polymyxin B, and bacitracin.

Packaging and storage—Preserve in tight, light-resistant containers, preferably at controlled room temperature.

Reference standards—*USP Neomycin Sulfate Reference Standard*—Dry in vacuum at a pressure not exceeding 5 mm of mercury at 60° for 3 hours before using. *USP Polymyxin B Sulfate Reference Standard*—Dry in vacuum at a pressure not exceeding 5 mm of mercury at 60° for 3 hours before using. *USP Bacitracin Zinc Reference Standard*—Dry in vacuum at a pressure not exceeding 5 mm of mercury at 60° for 3 hours before using.

Identification—

A: It responds to the *Identification test* under *Neomycin Sulfate Cream.*

B: Shake a quantity of Ointment, equivalent to about 2500 USP Bacitracin Units, with 20 mL of chloroform, add 5 mL of 0.1 *N* hydrochloric acid, shake vigorously, centrifuge, and use the clear supernatant liquid as the test solution. On a suitable thin-layer chromatographic plate coated with a 0.25-mm layer of chromatographic silica gel (see *Chromatography* ⟨621⟩) apply separately 10 µL of the test solution, 10 µL of a Standard bacitracin solution of USP Bacitracin Zinc RS in 0.1 *N* hydrochloric acid containing 500 USP Bacitracin Units per mL, and 10 µL of a Standard polymyxin B solution of USP Polymyxin B Sulfate RS in 0.1 *N* hydrochloric acid containing 500*J* USP Polymyxin B Units per mL, *J* being the ratio of the labeled amount of USP Polymyxin B Units to the labeled amount of USP Bacitracin Units in each g of the specimen. Place the plate in a suitable chromatographic chamber, and develop the chromatogram with a solvent system consisting of a mixture of isopropyl alcohol, water, and ammonium hydroxide (24:17:3) until the solvent front has moved about three-fourths of the length of the plate. Remove the plate from the chamber, and dry at 105° for 5 minutes. Spray the plate with a 1 in 200 solution of ninhydrin in butyl alcohol, and heat the plate at 105° for 15 minutes: the R_f values of the two principal spots in the chromatogram obtained from the test solution correspond to those of the single principal spots in the chromatograms obtained from the Standard bacitracin solution and the Standard polymyxin solution, respectively.

Minimum fill ⟨755⟩: meets the requirements.

Water, *Method I* ⟨921⟩: not more than 0.5%, 20 mL of a mixture of carbon tetrachloride, chloroform, and methanol (2:2:1) being used in place of methanol in the titration vessel.

Assay for neomycin and Assay for polymyxin B—Proceed with Neomycin and Polymyxin B Sulfates and Bacitracin Ointment as directed in the *Assay for neomycin* and in the *Assay for polymyxin B* under *Neomycin and Polymyxin B Sulfates and Bacitracin Zinc Ophthalmic Ointment.*

Assay for bacitracin—Proceed with Neomycin and Polymyxin B Sulfates and Bacitracin Ointment as directed in the *Assay* under *Bacitracin Ointment.*

Neomycin and Polymyxin B Sulfates and Bacitracin Ophthalmic Ointment

» Neomycin and Polymyxin B Sulfates and Bacitracin Ophthalmic Ointment is a sterile ointment containing Neomycin Sulfate, Polymyxin B Sulfate, and Bacitracin. It contains the equivalent of not less than 90.0 percent and not more than 140.0 percent of the labeled amounts of neomycin, polymyxin B, and bacitracin.

Packaging and storage—Preserve in collapsible ophthalmic ointment tubes.

Reference standards—*USP Neomycin Sulfate Reference Standard*—Dry in vacuum at a pressure not exceeding 5 mm of mercury at 60° for 3 hours before using. *USP Polymyxin B Sulfate Reference Standard*—Dry in vacuum at a pressure not exceeding 5 mm of mercury at 60° for 3 hours before using. *USP Bacitracin Zinc Reference Standard*—Dry in vacuum at a pressure not exceeding 5 mm of mercury at 60° for 3 hours before using.

Identification—

A: It responds to the *Identification test* under *Neomycin Sulfate Cream.*

B: It responds to *Identification test B* under *Neomycin and Polymyxin B Sulfates and Bacitracin Ointment.*

Sterility—It meets the requirements under *Sterility Tests* ⟨71⟩, when tested as directed in the section, *Test Procedures Using Membrane Filtration.*

Minimum fill ⟨755⟩: meets the requirements.

Water, *Method I* ⟨921⟩: not more than 0.5%, 20 mL of a mixture of carbon tetrachloride, chloroform, and methanol (2:2:1) being used in place of methanol in the titration vessel.

Metal particles—It meets the requirements of the test for *Metal Particles in Ophthalmic Ointments* ⟨751⟩.

Assay for neomycin and Assay for polymyxin B—Proceed with Neomycin and Polymyxin B Sulfates and Bacitracin Ophthalmic Ointment as directed in the *Assay for neomycin* and in the *Assay for polymyxin B* under *Neomycin and Polymyxin B Sulfates and Bacitracin Zinc Ophthalmic Ointment.*

Assay for bacitracin—Proceed with Neomycin and Polymyxin B Sulfates and Bacitracin Ophthalmic Ointment as directed in the *Assay for bacitracin* under *Bacitracin Ointment.*

Neomycin and Polymyxin B Sulfates, Bacitracin, and Hydrocortisone Acetate Ointment

» Neomycin and Polymyxin B Sulfates, Bacitracin, and Hydrocortisone Acetate Ointment contains the equivalent of not less than 90.0 percent and not more than 130.0 percent of the labeled amounts of neomycin, polymyxin B, and bacitracin, and not less than 90.0 percent and not more than 110.0 percent of the labeled amount of hydrocortisone acetate in a suitable ointment base.

Packaging and storage—Preserve in collapsible tubes or in well-closed containers.

Reference standards—*USP Neomycin Sulfate Reference Standard*—Dry in vacuum at a pressure not exceeding 5 mm of mercury at 60° for 3 hours before using. *USP Polymyxin B Sulfate Reference Standard*—Dry in vacuum at a pressure not exceeding 5 mm of mercury at 60° for 3 hours before using. *USP Bacitracin Zinc Reference Standard*—Dry in vacuum at a pressure not exceeding 5 mm of mercury at 60° for 3 hours before using. *USP Hydrocortisone Acetate Reference Standard*—Dry in vacuum at 60° for 3 hours before using.

Identification—

A: It responds to the *Identification test* under *Neomycin Sulfate Cream.*

B: It responds to *Identification test B* under *Neomycin and Polymyxin B Sulfates and Bacitracin Ointment.*

C: The chromatogram of the *Assay preparation* obtained as directed in the *Assay for hydrocortisone acetate* exhibits a major peak for hydrocortisone acetate, the retention time of which corresponds to that exhibited in the chromatogram of the *Standard preparation* obtained as directed in the *Assay for hydrocortisone acetate.*

Minimum fill ⟨755⟩: meets the requirements.

Water, *Method I* ⟨921⟩: not more than 0.5%, 20 mL of a mixture of carbon tetrachloride, chloroform, and methanol (2:2:1) being used in place of methanol in the titration vessel.

Assay for neomycin and Assay for polymyxin B—Proceed with Neomycin and Polymyxin B Sulfates, Bacitracin, and Hydrocortisone Acetate Ointment as directed in the *Assay for neomycin* and in the *Assay for polymyxin B* under *Neomycin and Polymyxin B Sulfates and Bacitracin Zinc Ophthalmic Ointment.*

Assay for bacitracin—Proceed with Neomycin and Polymyxin B Sulfates, Bacitracin, and Hydrocortisone Acetate Ointment as directed in the *Assay* under *Bacitracin Ointment.*

Assay for hydrocortisone acetate—Proceed with Neomycin and Polymyxin B Sulfates, Bacitracin, and Hydrocortisone Acetate Ointment as directed in the *Assay* under *Hydrocortisone Acetate Lotion.*

Neomycin and Polymyxin B Sulfates, Bacitracin, and Hydrocortisone Acetate Ophthalmic Ointment

» Neomycin and Polymyxin B Sulfates, Bacitracin, and Hydrocortisone Acetate Ophthalmic Ointment contains the equivalent of not less than 90.0 percent and not more than 140.0 percent of the labeled amounts of neomycin, polymyxin B, and bacitracin, and not less than 90.0 percent and not more than 110.0 percent of the labeled amount of hydrocortisone acetate, in a suitable ointment base.

Packaging and storage—Preserve in collapsible ophthalmic ointment tubes.

Reference standards—*USP Neomycin Sulfate Reference Standard*—Dry in vacuum at a pressure not exceeding 5 mm of mercury at 60° for 3 hours before using. *USP Polymyxin B Sulfate Reference Standard*—Dry in vacuum at a pressure not exceeding 5 mm of mercury at 60° for 3 hours before using. *USP Bacitracin Zinc Reference Standard*—Dry in vacuum at a pressure not exceeding 5 mm of mercury at 60° for 3 hours before using. *USP Hydrocortisone Acetate Reference Standard*—Dry in vacuum at 60° for 3 hours before using.

Identification—

A: It responds to the *Identification test* under *Neomycin Sulfate Cream.*

B: It responds to *Identification test B* under *Neomycin and Polymyxin B Sulfates and Bacitracin Ointment.*

C: The chromatogram of the *Assay preparation* obtained as directed in the *Assay for hydrocortisone acetate* exhibits a major peak for hydrocortisone acetate, the retention time of which corresponds to that exhibited in the chromatogram of the *Standard preparation* obtained as directed in the *Assay for hydrocortisone acetate.*

Sterility—It meets the requirements under *Sterility Tests* ⟨71⟩, when tested as directed in the section, *Test Procedures Using Membrane Filtration.*

Minimum fill ⟨755⟩: meets the requirements.

Water, *Method I* ⟨921⟩: not more than 0.5%, 20 mL of a mixture of carbon tetrachloride, chloroform, and methanol (2:2:1) being used in place of methanol in the titration vessel.

Metal particles—It meets the requirements of the test for *Metal Particles in Ophthalmic Ointments* ⟨751⟩.

Assay for neomycin and Assay for polymyxin B—Proceed with Neomycin and Polymyxin B Sulfates, Bacitracin, and Hydrocortisone Acetate Ophthalmic Ointment as directed in the *Assay for neomycin* and in the *Assay for polymyxin B* under *Neomycin and Polymyxin B Sulfates and Bacitracin Zinc Ophthalmic Ointment.*

Assay for bacitracin—Proceed with Neomycin and Polymyxin B Sulfates, Bacitracin, and Hydrocortisone Acetate Ophthalmic Ointment as directed in the *Assay* under *Bacitracin Ointment.*

Assay for hydrocortisone acetate—Proceed with Neomycin and Polymyxin B Sulfates, Bacitracin, and Hydrocortisone Acetate Ophthalmic Ointment as directed in the *Assay* under *Hydrocortisone Acetate Lotion.*

Neomycin and Polymyxin B Sulfates, Bacitracin, and Lidocaine Ointment

» Neomycin and Polymyxin B Sulfates, Bacitracin, and Lidocaine Ointment contains the equivalent of not less than 90.0 percent and not more than 130.0 percent of the labeled amounts of neomycin, polymyxin B, and bacitracin, and not less than 90.0 percent and not more than 110.0 percent of the labeled amount of lidocaine ($C_{14}H_{22}N_2O$).

Packaging and storage—Preserve in well-closed containers, preferably at controlled room temperature.

Reference standards—*USP Neomycin Sulfate Reference Standard*—Dry in vacuum at a pressure not exceeding 5 mm of mercury at 60° for 3 hours before using. *USP Polymyxin B Sulfate Reference Standard*—Dry in vacuum at a pressure not exceeding 5 mm of mercury at 60° for 3 hours before using. *USP Bacitracin Zinc Reference Standard*—Dry in vacuum at a pressure not exceeding 5 mm of mercury at 60° for 3 hours before using. *USP Lidocaine Reference Standard*—Dry in vacuum over silica gel for 24 hours before using.

Identification—

A: It responds to the *Identification test* under *Neomycin Sulfate Cream.*

B: It responds to *Identification test B* under *Neomycin and Polymyxin B Sulfates and Bacitracin Ointment.*

C: The chromatogram of the *Assay preparation* obtained as directed in the *Assay for lidocaine* exhibits a major peak for lidocaine, the retention time of which corresponds to that exhibited in the chromatogram of the *Standard preparation* obtained as directed in the *Assay for lidocaine.*

Minimum fill ⟨755⟩: meets the requirements.

Water, *Method I* ⟨921⟩: not more than 0.5%, 20 mL of a mixture of carbon tetrachloride, chloroform, and methanol (2:2:1) being used in place of methanol in the titration vessel.

Assay for neomycin and Assay for polymyxin B—Proceed with Neomycin and Polymyxin B Sulfates, Bacitracin, and Lidocaine Ointment as directed in the *Assay for neomycin* and in the *Assay for polymyxin B* under *Neomycin and Polymyxin B Sulfates and Bacitracin Zinc Ophthalmic Ointment.*

Assay for bacitracin—Proceed with Neomycin and Polymyxin B Sulfates, Bacitracin, and Lidocaine Ointment as directed in the *Assay* under *Bacitracin Ointment.*

Assay for lidocaine—

Mobile phase—Dissolve 4.44 g of docusate sodium in 1000 mL of a mixture of methanol and water (4:1), add 1 mL of 0.1 *N* sulfuric acid, and mix. Make adjustments if necessary (see *System Suitability* under *Chromatography* ⟨621⟩).

Standard preparation—Dissolve a suitable quantity of USP Lidocaine RS, accurately weighed, in *Mobile phase* to obtain a solution having a known concentration of about 0.4 mg per mL.

Assay preparation—Transfer an accurately weighed quantity of Neomycin and Polymyxin B Sulfates, Bacitracin, and Lidocaine Ointment, equivalent to about 40 mg of lidocaine, to a separator, add 50 mL of *n*-hexane, and shake until the specimen is in solution. Add 30 mL of *Mobile phase,* shake for 1 minute, and allow the layers to separate. Drain the lower layer into a 100-mL volumetric flask, and extract the *n*-hexane layer remaining in the separator with two 30-mL portions of *Mobile phase,* combining the lower layers in the volumetric flask. Dilute the combined extracts in the 100-mL volumetric flask with *Mobile phase* to volume, and mix.

Chromatographic system (see *Chromatography* ⟨621⟩)—The liquid chromatograph is equipped with a 230-nm detector and a 4-mm × 25-cm column that contains packing L1. The flow rate is about 1 mL per minute. Chromatograph the *Standard preparation,* and record the peak response as directed under *Procedure:* the column efficiency determined from the analyte peak is not less than 500 theoretical plates, and the relative standard deviation for replicate injections is not more than 2.0%.

Procedure—Separately inject equal volumes (about 20 µL) of the *Standard preparation* and the *Assay preparation* into the chromatograph, record the chromatograms, and measure the responses for the major peaks. Calculate the quantity, in mg, of lidocaine ($C_{14}H_{22}N_2O$) in the portion of Ointment taken by the formula:

$$100C(r_U / r_S),$$

in which *C* is the concentration, in mg per mL, of USP Lidocaine RS in the *Standard preparation,* and r_U and r_S are the peak responses obtained from the *Assay preparation* and the *Standard preparation,* respectively.

Neomycin and Polymyxin B Sulfates and Bacitracin Zinc Ointment

» Neomycin and Polymyxin B Sulfates and Bacitracin Zinc Ointment contains the equivalent of not less than 90.0 percent and not more than 130.0 percent of the labeled amounts of neomycin, polymyxin B, and bacitracin.

Packaging and storage—Preserve in well-closed containers, preferably at controlled room temperature.

Reference standards—*USP Neomycin Sulfate Reference Standard*—Dry in vacuum at a pressure not exceeding 5 mm of mercury at 60° for 3 hours before using. *USP Polymyxin B Sulfate Reference Standard*—Dry in vacuum at a pressure not exceeding 5 mm of mercury at 60° for 3 hours before using. *USP Bacitracin Zinc Reference Standard*—Dry in vacuum at a pressure not exceeding 5 mm of mercury at 60° for 3 hours before using.

Identification—

A: It responds to the *Identification test* under *Neomycin Sulfate Cream.*

B: It responds to *Identification test B* under *Neomycin and Polymyxin B Sulfates and Bacitracin Ointment.*

Minimum fill ⟨755⟩: meets the requirements.

Water, *Method I* ⟨921⟩: not more than 0.5%, 20 mL of a mixture of carbon tetrachloride, chloroform, and methanol (2:2:1) being used in place of methanol in the titration vessel.

Assay for neomycin, Assay for polymyxin B, and Assay for bacitracin—Proceed with Neomycin and Polymyxin B Sulfates and Bacitracin Zinc Ointment as directed in the *Assay for neomycin,* in the *Assay for polymyxin B,* and in the *Assay for bacitracin* under *Neomycin and Polymyxin B Sulfates and Bacitracin Zinc Ophthalmic Ointment.*

Neomycin and Polymyxin B Sulfates and Bacitracin Zinc Ophthalmic Ointment

» Neomycin and Polymyxin B Sulfates and Bacitracin Zinc Ophthalmic Ointment contains the equivalent of not less than 90.0 percent and not more than 140.0 percent of the labeled amounts of neomycin, polymyxin B, and bacitracin.

Packaging and storage—Preserve in collapsible ophthalmic ointment tubes.

Reference standards—*USP Neomycin Sulfate Reference Standard*—Dry in vacuum at a pressure not exceeding 5 mm of mercury at 60° for 3 hours before using. *USP Polymyxin B Sulfate Reference Standard*—Dry in vacuum at a pressure not exceeding 5 mm of mercury at 60° for 3 hours before using. *USP Bacitracin Zinc Reference Standard*—Dry in vacuum at a pressure not exceeding 5 mm of mercury at 60° for 3 hours before using.

Identification—

A: It responds to the *Identification test* under *Neomycin Sulfate Cream.*

B: It responds to *Identification test B* under *Neomycin and Polymyxin B Sulfates and Bacitracin Ointment.*

Sterility—It meets the requirements under *Sterility Tests* ⟨71⟩.

Minimum fill ⟨755⟩: meets the requirements.

Water, *Method I* ⟨921⟩: not more than 0.5%, 20 mL of a mixture of carbon tetrachloride, chloroform, and methanol (2:2:1) being used in place of methanol in the titration vessel.

Metal particles—It meets the requirements of the test for *Metal Particles in Ophthalmic Ointments* ⟨751⟩.

Assay for neomycin—Proceed as directed under *Antibiotics—Microbial Assays* ⟨81⟩, using an accurately weighed portion of Neomycin and Polymyxin B Sulfates and Bacitracin Zinc Ophthalmic Ointment shaken in a separator with about 50 mL of ether, and extracted with four 20-mL portions of *Buffer No. 3.* Combine the aqueous extracts, and dilute with *Buffer No. 3* to an appropriate volume to obtain a stock solution. Dilute this stock solution quantitatively and stepwise with *Buffer No. 3* to obtain a *Test Dilution* having a concentration assumed to be equal to the median dose level of the Standard.

Assay for polymyxin B—Proceed as directed under *Antibiotics—Microbial Assays* ⟨81⟩, using an accurately weighed portion of Neomycin and Polymyxin B Sulfates and Bacitracin Zinc Ophthalmic Ointment shaken with about 50 mL of ether in a separator, and extracted with four 25-mL portions of *Buffer No. 6.* Combine the aqueous extracts, and dilute with *Buffer No. 6* to an appropriate volume to obtain a stock solution. Dilute this stock solution quantitatively and stepwise with *Buffer No. 6* to obtain a *Test Dilution* having a concentration assumed to be equal to the median dose level of the Standard (10 Polymyxin B Units per mL). Add to each test dilution of the Standard a quantity of Neomycin Standard, dissolved in *Buffer No. 6*, to obtain the same concentration of neomycin present in the *Test Dilution.*

Assay for bacitracin—Proceed as directed under *Antibiotics—Microbial Assays* ⟨81⟩, using an accurately weighed portion of Neomycin and Polymyxin B Sulfates and Bacitracin Zinc Ophthalmic Ointment shaken with about 50 mL of ether in a separator, and extracted with four 25-mL portions of 0.01 *N* hydrochloric acid. Combine the acid extracts, and dilute with 0.01 *N* hydrochloric acid to an appropriate volume to obtain a stock solution. Dilute this stock solution quantitatively and stepwise with *Buffer No. 1* to obtain a *Test Dilution* having a concentration assumed to be equal to the median dose level of the Standard (1.0 Bacitracin Unit per mL). [NOTE—If the stock solution has a concentration of less than 100 Bacitracin Units per mL, add additional hydrochloric acid to each test dilution of the Standard to obtain the same concentration of hydrochloric acid as the *Test Dilution.*]

Neomycin and Polymyxin B Sulfates, Bacitracin Zinc, and Hydrocortisone Ointment

» Neomycin and Polymyxin B Sulfates, Bacitracin Zinc, and Hydrocortisone Ointment contains the equivalent of not less than 90.0 percent and not more than 130.0 percent of the labeled amounts of neomycin, polymyxin B, and bacitracin, and not less than 90.0 percent and not more than 110.0 percent of the labeled amount of hydrocortisone.

Packaging and storage—Preserve in well-closed containers, preferably at controlled room temperature.

Reference standards—*USP Neomycin Sulfate Reference Standard*—Dry in vacuum at a pressure not exceeding 5 mm of mercury at 60° for 3 hours before using. *USP Polymyxin B Sulfate Reference Standard*—Dry in vacuum at a pressure not exceeding 5 mm of mercury at 60° for 3 hours before using. *USP Bacitracin Zinc Reference Standard*—Dry in vacuum at a pressure not exceeding 5 mm of mercury at 60° for 3 hours before using. *USP Hydrocortisone Reference Standard*—Dry at 105° for 3 hours before using.

Identification—

A: It responds to the *Identification test* under *Neomycin Sulfate Cream.*

B: It responds to *Identification test B* under *Neomycin and Polymyxin B Sulfates and Bacitracin Ointment.*

C: The chromatogram of the *Assay preparation* obtained as directed in the *Assay for hydrocortisone* exhibits a major peak for hydrocortisone, the retention time of which corresponds to that exhibited in the chromatogram of the *Standard preparation* obtained as directed in the *Assay for hydrocortisone.*

Minimum fill ⟨755⟩: meets the requirements.

Water, *Method I* ⟨921⟩: not more than 0.5%, 20 mL of a mixture of carbon tetrachloride, chloroform, and methanol (2:2:1) being used in place of methanol in the titration vessel.

Assay for neomycin, Assay for polymyxin B, and Assay for bacitracin—Proceed with Neomycin and Polymyxin B Sulfates, Bacitracin Zinc, and Hydrocortisone Ointment as directed in the *Assay for neomycin,* in the *Assay for polymyxin B,* and in the *Assay for bacitracin* under *Neomycin and Polymyxin B Sulfates and Bacitracin Zinc Ophthalmic Ointment.*

Assay for hydrocortisone—Proceed with Neomycin and Polymyxin B Sulfates, Bacitracin Zinc, and Hydrocortisone Ointment as directed in the *Assay for hydrocortisone* under *Neomycin and Polymyxin B Sulfates, Bacitracin Zinc, and Hydrocortisone Ophthalmic Ointment.*

Neomycin and Polymyxin B Sulfates, Bacitracin Zinc, and Hydrocortisone Ophthalmic Ointment

» Neomycin and Polymyxin B Sulfates, Bacitracin Zinc, and Hydrocortisone Ophthalmic Ointment is a sterile ointment containing Neomycin Sulfate, Polymyxin B Sulfate, Bacitracin Zinc, and Hydrocortisone. It contains the equivalent of not less than 90.0 percent and not more than 140.0 percent of the labeled amounts of neomycin, polymyxin B, and bacitracin, and not less than 90.0 percent and not more than 110.0 percent of the labeled amount of hydrocortisone.

Packaging and storage—Preserve in collapsible ophthalmic ointment tubes.

Reference standards—*USP Neomycin Sulfate Reference Standard*—Dry in vacuum at a pressure not exceeding 5 mm of mercury at 60° for 3 hours before using. *USP Polymyxin B Reference Standard*—Dry in vacuum at a pressure not exceeding 5 mm of mercury at 60° for 3 hours before using. *USP Bacitracin Zinc Reference Standard*—Dry in vacuum at a pressure not exceeding 5 mm of mercury at 60° before using. *USP Hydrocortisone Reference Standard*—Dry at 105° for 3 hours before using.

Identification—

A: It responds to the *Identification test* under *Neomycin Sulfate Cream.*

B: It responds to *Identification test B* under *Neomycin and Polymyxin B Sulfates and Bacitracin Ointment.*

C: The chromatogram of the *Assay preparation* obtained as directed in the *Assay for hydrocortisone* exhibits a major peak for hydrocortisone, the retention time of which corresponds to that exhibited in the chromatogram of the *Standard preparation* obtained as directed in the *Assay for hydrocortisone.*

Sterility—It meets the requirements under *Sterility Tests* ⟨71⟩.

Minimum fill ⟨755⟩: meets the requirements.

Water, *Method I* ⟨921⟩: not more than 0.5%, 20 mL of a mixture of carbon tetrachloride, chloroform, and methanol (2:2:1) being used in place of methanol in the titration vessel.

Metal particles—It meets the requirements of the test for *Metal Particles in Ophthalmic Ointments* ⟨751⟩.

Assay for neomycin, Assay for polymyxin B, and Assay for bacitracin—Proceed as directed in the *Assay for neomycin,* the *Assay for polymyxin B,* and the *Assay for bacitracin* under *Neomycin and Polymyxin B Sulfates and Bacitracin Zinc Ophthalmic Ointment.*

Assay for hydrocortisone—

Mobile phase—Prepare a suitable solution of about 500 volumes of methanol, 500 volumes of water, and 1 volume of glacial acetic acid, such that the retention time of hydrocortisone is between 6 and 10 minutes.

Standard preparation—Dissolve a suitable quantity of USP Hydrocortisone RS, accurately weighed, in methanol and water (1:1) to obtain a solution having a known concentration of about 0.15 mg per mL.

Assay preparation—Transfer to a separator about 1.5 g, accurately weighed, of Neomycin and Polymyxin B Sulfates, Bacitracin Zinc, and Hydrocortisone Ophthalmic Ointment. Add 3 mL of *n*-hexane, and warm gently on a steam bath with mild agitation until dissolved. Add 7 mL of *n*-hexane, mix by swirling, and extract with four 15-mL portions of methanol and water (1:1). Collect the extracts in a 100-mL volumetric flask, dilute with methanol and water (1:1) to volume, and mix. Filter the solution, rejecting the first 10 mL of the filtrate.

Chromatographic system (see *Chromatography* ⟨621⟩)—The chromatograph is equipped with a 254-nm detector and a 4-mm × 30-cm column that contains packing L1. The flow rate is about 2 mL per minute. Chromatograph five replicate injections of the *Standard preparation* and record the peak responses as directed under *Procedure:* the relative standard deviation is not more than 2.0%.

Procedure—Separately inject equal volumes (about 10 µL) of the *Standard preparation* and the *Assay preparation* into the chromatograph by means of a suitable microsyringe or sampling valve, adjusting the specimen size and other operating parameters such that the peak obtained from the *Standard preparation* is about 0.6 full scale. Record the chromatograms, and measure the responses for the major peaks. Calculate the quantity, in mg per g, of $C_{21}H_{30}O_5$ in the Ophthalmic Ointment taken by the formula:

$$(100C/W)(r_U/r_S),$$

in which C is the concentration, in mg per mL, of USP Hydrocortisone RS in the *Standard preparation*, W is the weight, in g, of the portion of Ophthalmic Ointment taken, and r_U and r_S are the peak responses obtained from the *Assay preparation* and the *Standard preparation*, respectively.

Neomycin and Polymyxin B Sulfates, Bacitracin Zinc, and Hydrocortisone Acetate Ophthalmic Ointment

» Neomycin and Polymyxin B Sulfates, Bacitracin Zinc, and Hydrocortisone Acetate Ophthalmic Ointment is a sterile ointment containing Neomycin Sulfate, Polymyxin B Sulfate, Bacitracin Zinc, and Hydrocortisone Acetate. It contains the equivalent of not less than 90.0 percent and not more than 140.0 percent of the labeled amounts of neomycin, polymyxin B, and bacitracin, and not less than 90.0 percent and not more than 110.0 percent of the labeled amount of hydrocortisone acetate.

Packaging and storage—Preserve in collapsible ophthalmic ointment tubes.

Reference standards—*USP Neomycin Sulfate Reference Standard*—Dry in vacuum at a pressure not exceeding 5 mm of mercury at 60° for 3 hours before using. *USP Polymyxin B Sulfate Reference Standard*—Dry in vacuum at a pressure not exceeding 5 mm of mercury at 60° for 3 hours before using. *USP Bacitracin Zinc Reference Standard*—Dry in vacuum at a pressure not exceeding 5 mm of mercury at 60° for 3 hours before using. *USP Hydrocortisone Acetate Reference Standard*—Dry in vacuum at 60° for 3 hours before using.

Identification—

A: It responds to the *Identification test* under *Neomycin Sulfate Cream.*

B: It responds to *Identification test B* under *Neomycin and Polymyxin B Sulfates and Bacitracin Ointment.*

C: The chromatogram of the *Assay preparation* obtained as directed in the *Assay for hydrocortisone acetate* exhibits a major peak for hydrocortisone acetate, the retention time of which corresponds to that exhibited in the chromatogram of the *Standard preparation* obtained as directed in the *Assay for hydrocortisone acetate.*

Sterility—It meets the requirements under *Sterility Tests* ⟨71⟩, when tested as directed in the section, *Test Procedures Using Membrane Filtration.*

Minimum fill ⟨755⟩: meets the requirements.

Water, *Method I* ⟨921⟩: not more than 0.5%, 20 mL of a mixture of carbon tetrachloride, chloroform, and methanol (2:2:1) being used in place of methanol in the titration vessel.

Metal particles—It meets the requirements of the test for *Metal Particles in Ophthalmic Ointments* ⟨751⟩.

Assay for neomycin, Assay for polymyxin B, and Assay for bacitracin—Proceed with Neomycin and Polymyxin B Sulfates, Bacitracin Zinc, and Hydrocortisone Acetate Ophthalmic Ointment as directed in the *Assay for neomycin,* in the *Assay for polymyxin B,* and in the *Assay for bacitracin* under *Neomycin and Polymyxin B Sulfates and Bacitracin Zinc Ophthalmic Ointment.*

Assay for hydrocortisone acetate—Proceed with Neomycin and Polymyxin B Sulfates, Bacitracin Zinc, and Hydrocortisone Acetate Ophthalmic Ointment as directed in the *Assay* under *Hydrocortisone Acetate Lotion.*

Neomycin and Polymyxin B Sulfates and Dexamethasone Ophthalmic Ointment

» Neomycin and Polymyxin B Sulfates and Dexamethasone Ophthalmic Ointment contains the equivalent of not less than 90.0 percent and not more than 125.0 percent of the labeled amounts of neomycin and polymyxin B, and not less than 90.0 percent and not more than 110.0 percent of the labeled amount of dexamethasone.

Packaging and storage—Preserve in collapsible ophthalmic ointment tubes.

Reference standards—*USP Neomycin Sulfate Reference Standard*—Dry in vacuum at a pressure not exceeding 5 mm of mercury at 60° for 3 hours before using. *USP Polymyxin B Sulfate Reference Standard*—Dry in vacuum at a pressure not exceeding 5 mm of mercury at 60° for 3 hours before using. *USP Dexamethasone Reference Standard*—Dry at 105° for 3 hours before using.

Identification—

A: It responds to the *Identification test* under *Neomycin Sulfate Cream.*

B: It responds to *Identification test B* under *Neomycin and Polymyxin B Sulfates Cream.*

C: The chromatogram of the *Assay preparation* obtained as directed in the *Assay for dexamethasone* exhibits a major peak for dexamethasone, the retention time of which corresponds to that exhibited in the chromatogram of the *Standard preparation* obtained as directed in the *Assay for dexamethasone.*

Sterility—It meets the requirements under *Sterility Tests* ⟨71⟩, when tested as directed in the section, *Test Procedures Using Membrane Filtration.*

Minimum fill ⟨755⟩: meets the requirements.

Water, *Method Ib* ⟨921⟩: not more than 0.5%, 20 mL of a mixture of carbon tetrachloride, chloroform, and methanol (2:2:1) being used in place of methanol in the titration vessel.

Metal particles—It meets the requirements of the test for *Metal Particles in Ophthalmic Ointments* ⟨751⟩.

Assay for neomycin and Assay for polymyxin B—Proceed with Neomycin and Polymyxin B Sulfates and Dexamethasone Ophthalmic Ointment as directed in the *Assay for neomycin* and in the *Assay for polymyxin B* under *Neomycin and Polymyxin B Sulfates and Bacitracin Zinc Ophthalmic Ointment.*

Assay for dexamethasone—

Mobile phase—Prepare a suitable aqueous solution of acetonitrile, approximately 1 in 3, such that the retention time of dexamethasone is about 5 minutes.

Standard preparation—Dissolve an accurately weighed quantity of USP Dexamethasone RS in a mixture of acetonitrile and methanol (1:1) to obtain a solution having a known concentration of about 60 µg per mL.

Assay preparation—Transfer an accurately weighed portion of Neomycin and Polymyxin B Sulfates and Dexamethasone Ophthalmic Ointment, equivalent to about 3 mg of dexamethasone, to a suitable test tube, add about 15 mL of cyclohexane, and heat in a water bath at 75 ± 5° for 10 minutes. [NOTE—If the ointment is not fully dissolved, heat on a steam bath for about 30 seconds, place a cap on the test tube, and vortex.] Filter with suction through a medium-porosity sintered-glass filter. Rinse the test tube twice with 10-mL portions of cyclohexane, filtering the rinsings through the filter, and discard the filtrates. Wash the filter with about 10 mL of a mixture of acetonitrile and methanol (1:1), and collect the filtrate in a 50-mL beaker. Wash the test tube and the filter with several 10-mL portions of the same solvent, and combine the washings in the 50-mL beaker. Transfer the contents of the beaker to a 50-mL volumetric flask, with the aid of a mixture of acetonitrile and methanol (1:1), dilute with the same solvent to volume, and mix.

Chromatographic system (see *Chromatography* ⟨621⟩)—The liquid chromatograph is equipped with a 254-nm detector and a 4.6-mm × 25-cm column that contains 5- to 10-μm packing L1. The flow rate is about 2 mL per minute. Chromatograph the *Standard preparation*, and record the peak response as directed under *Procedure:* the column efficiency is not less than 4000 theoretical plates, and the relative standard deviation for replicate injections is not more than 1.5%.

Procedure—Separately inject equal volumes (about 10 μL) of the *Standard preparation* and the *Assay preparation* into the chromatograph, record the chromatograms, and measure the responses for the major peaks. Calculate the quantity, in mg, of $C_{22}H_{29}FO_5$ in the portion of Ophthalmic Ointment taken by the formula:

$$50C(r_U/r_S),$$

in which C is the concentration, in μg per mL, of USP Dexamethasone RS in the *Standard preparation*, and r_U and r_S are the peak responses of the *Assay preparation* and the *Standard preparation*, respectively.

Neomycin and Polymyxin B Sulfates and Dexamethasone Ophthalmic Suspension

» Neomycin and Polymyxin B Sulfates and Dexamethasone Ophthalmic Suspension contains the equivalent of not less than 90.0 percent and not more than 125.0 percent of the labeled amounts of neomycin and polymyxin B, and not less than 90.0 percent and not more than 110.0 percent of the labeled amount of dexamethasone. It may contain one or more suitable buffers, stabilizers, preservatives, and suspending agents.

Packaging and storage—Preserve in tight, light-resistant containers in a cool place or at controlled room temperature. The containers or individual cartons are sealed and tamper-proof so that sterility is assured at time of first use.

Reference standards—*USP Neomycin Sulfate Reference Standard*—Dry in vacuum at a pressure not exceeding 5 mm of mercury at 60° for 3 hours before using. *USP Polymyxin B Sulfate Reference Standard*—Dry in vacuum at a pressure not exceeding 5 mm of mercury at 60° for 3 hours before using. *USP Dexamethasone Reference Standard*—Dry at 105° for 3 hours before using.

Identification—Transfer a quantity of Ophthalmic Suspension, equivalent to about 2.5 mg of dexamethasone, to a suitable test tube, add 5 mL of chloroform, mix, and centrifuge. On a suitable thin-layer chromatographic plate (see *Chromatography* ⟨621⟩), coated with a 0.25-mm layer of chromatographic silica gel, apply 25 μL of the lower chloroform layer and 25 μL of a Standard solution of USP Dexamethasone RS in chloroform containing 500 μg per mL. Allow the spots to dry, and develop the chromatogram in a solvent system consisting of a mixture of chloroform and diethylamine (2:1) until the solvent front has moved about three-fourths of the length of the plate. Remove the plate from the developing chamber, mark the solvent front, and allow the solvent to evaporate. Locate the spots on the plate by examination under short-wavelength ultraviolet light: the R_f value of the principal spot obtained from the test solution corresponds to that obtained from the Standard solution.

Sterility—It meets the requirements under *Sterility Tests* ⟨71⟩, when tested as directed in the section, *Test Procedures Using Membrane Filtration.*

pH ⟨791⟩: between 3.5 and 6.0.

Assay for neomycin—Proceed as directed for neomycin under *Antibiotics—Microbial Assays* ⟨81⟩, using an accurately measured volume of Neomycin and Polymyxin B Sulfates and Dexamethasone Ophthalmic Suspension, freshly mixed and free from air bubbles, diluted quantitatively and stepwise with *Buffer No. 3* to yield a *Test Dilution* having a concentration assumed to be equal to the median dose level of the Standard.

Assay for polymyxin B—Proceed as directed for polymyxin B under *Antibiotics—Microbial Assays* ⟨81⟩, using an accurately measured volume of Neomycin and Polymyxin B Sulfates and Dexamethasone Ophthalmic Suspension, freshly mixed and free from air bubbles, diluted quantitatively and stepwise with *Buffer No. 6* to yield a *Test Dilution* having a concentration assumed to be equal to the median dose level of the Standard. Add to each test dilution of the Standard a quantity of USP Neomycin Sulfate RS, dissolved in *Buffer No. 6*, to obtain the same concentration of neomycin as is present in the *Test Dilution.*

Assay for dexamethasone—

Mobile phase and *Chromatographic system*—Proceed as directed in the *Assay for dexamethasone* under *Neomycin and Polymyxin B Sulfates and Dexamethasone Ophthalmic Ointment.*

Standard preparation—Dissolve an accurately weighed quantity of USP Dexamethasone RS in *Mobile phase* to obtain a solution having a known concentration of about 0.12 mg per mL.

Assay preparation—Dilute an accurately measured volume of freshly mixed Neomycin and Polymyxin B Sulfates and Dexamethasone Ophthalmic Suspension quantitatively with *Mobile phase* to obtain a solution containing about 0.12 mg of dexamethasone per mL.

Procedure—Proceed as directed for *Procedure* in the *Assay for dexamethasone* under *Neomycin and Polymyxin B Sulfates and Dexamethasone Ophthalmic Ointment.* Calculate the quantity, in mg per mL, of $C_{22}H_{29}FO_5$ in the Ophthalmic Suspension taken by the formula:

$$(CL/D)(r_U/r_S),$$

in which L is the labeled quantity, in mg per mL, of dexamethasone in the Ophthalmic Suspension, D is the concentration, in mg per mL, of dexamethasone in the *Assay preparation* based on the labeled quantity in the Ophthalmic Suspension and the extent of dilution, and the other terms are as defined therein.

Neomycin and Polymyxin B Sulfates and Gramicidin Cream

» Neomycin and Polymyxin B Sulfates and Gramicidin Cream contains the equivalent of not less than 90.0 percent and not more than 120.0 percent of the labeled amounts of neomycin, polymyxin B, and gramicidin.

Packaging and storage—Preserve in collapsible tubes or in well-closed containers.

Reference standards—*USP Neomycin Sulfate Reference Standard*—Dry in vacuum at a pressure not exceeding 5 mm of mercury at 60° for 3 hours before using. *USP Polymyxin B Sulfate Reference Standard*—Dry in vacuum at a pressure not exceeding 5 mm of mercury at 60° for 3 hours before using. *USP Gramicidin Reference Standard*—Dry in vacuum at a pressure not exceeding 5 mm of mercury at 60° for 3 hours before using.

Minimum fill ⟨755⟩: meets the requirements.

Assay for neomycin and Assay for polymyxin B—Proceed with Neomycin and Polymyxin B Sulfates and Gramicidin Cream as directed in the *Assay for neomycin* and in the *Assay for polymyxin B* under *Neomycin and Polymyxin B Sulfates and Bacitracin Zinc Ophthalmic Ointment.*

Assay for gramicidin—Proceed with Neomycin and Polymyxin B Sulfates and Gramicidin Cream as directed in the *Assay for gramicidin* under *Neomycin Sulfate and Gramicidin Ointment.*

Neomycin and Polymyxin B Sulfates and Gramicidin Ophthalmic Solution

» Neomycin and Polymyxin B Sulfates and Gramicidin Ophthalmic Solution is a sterile, isotonic aqueous solution of Neomycin Sulfate, Polymyxin B Sulfate, and Gramicidin. It contains the equivalent of not less than 90.0 percent and not more than 130.0 percent of the labeled amounts of neomycin, polymyxin B, and gramicidin.

Packaging and storage—Preserve in tight containers. The containers or individual cartons are sealed and tamper-proof so that sterility is assured at time of first use.

Reference standards—*USP Neomycin Sulfate Reference Standard*—Dry in vacuum at a pressure not exceeding 5 mm of mercury at 60° for 3 hours before using. *USP Polymyxin B Sulfate Reference Standard*—Dry in vacuum at a pressure not exceeding 5 mm of mercury at 60° for 3 hours before using. *USP Gramicidin Reference Standard*—Dry in vacuum at a pressure not exceeding 5 mm of mercury at 60° for 3 hours before using.

Sterility—It meets the requirements under *Sterility Tests* ⟨71⟩, when tested as directed in the section, *Test Procedures Using Membrane Filtration.*

pH ⟨791⟩: between 4.7 and 6.0.

Assay for neomycin—Proceed as directed for neomycin under *Antibiotics—Microbial Assays* ⟨81⟩, using an accurately measured volume of Neomycin and Polymyxin B Sulfates and Gramicidin Ophthalmic Solution diluted quantitatively and stepwise with *Buffer No. 3* to yield a *Test Dilution* having a concentration assumed to be equal to the median dose level of the Standard.

Assay for polymyxin B—Proceed as directed for polymyxin B under *Antibiotics—Microbial Assays* ⟨81⟩, using an accurately measured volume of Neomycin and Polymyxin B Sulfates and Gramicidin Ophthalmic Solution diluted quantitatively and stepwise with *Buffer No. 6* to yield a *Test Dilution* having a concentration assumed to be equal to the median dose level of the Standard. Add to each test dilution of the Standard a quantity of USP Neomycin Sulfate RS, dissolved in *Buffer No. 6*, to obtain the same concentration of neomycin as is present in the *Test Dilution.*

Assay for gramicidin—Proceed as directed for gramicidin under *Antibiotics—Microbial Assays* ⟨81⟩, using an accurately measured volume of Neomycin and Polymyxin B Sulfates and Gramicidin Ophthalmic Solution diluted quantitatively and stepwise with alcohol to yield a *Test Dilution* having a concentration assumed to be equal to the median dose level of the Standard.

Neomycin and Polymyxin B Sulfates, Gramicidin, and Hydrocortisone Acetate Cream

» Neomycin and Polymyxin B Sulfates, Gramicidin, and Hydrocortisone Acetate Cream contains the equivalent of not less than 90.0 percent and not more than 120.0 percent of the labeled amounts of neomycin, polymyxin B, and gramicidin, and not less than 90.0 percent and not more than 110.0 percent of the labeled amount of hydrocortisone acetate ($C_{23}H_{32}O_6$).

Packaging and storage—Preserve in well-closed containers.

Reference standards—*USP Neomycin Sulfate Reference Standard*—Dry in vacuum at a pressure not exceeding 5 mm of mercury at 60° for 3 hours before using. *USP Polymyxin B Sulfate Reference Standard*—Dry in vacuum at a pressure not exceeding 5 mm of mercury at 60° for 3 hours before using. *USP Gramicidin Reference Standard*—Dry in vacuum at a pressure not exceeding 5 mm of mercury at 60° for 3 hours before using. *USP Hydrocortisone Acetate Reference Standard*—Dry in vacuum at 60° for 3 hours before using.

Minimum fill ⟨755⟩: meets the requirements.

Assay for neomycin and Assay for polymyxin B—Proceed with Neomycin and Polymyxin B Sulfates, Gramicidin, and Hydrocortisone Acetate Cream as directed in the *Assay for neomycin* and in the *Assay for polymyxin B* under *Neomycin and Polymyxin B Sulfates and Bacitracin Zinc Ophthalmic Ointment.*

Assay for gramicidin—Proceed with Neomycin and Polymyxin B Sulfates, Gramicidin, and Hydrocortisone Cream as directed in the *Assay for gramicidin* under *Neomycin Sulfate and Gramicidin Ointment.*

Assay for hydrocortisone acetate—Proceed with Neomycin and Polymyxin B Sulfates, Gramicidin, and Hydrocortisone Acetate Cream as directed in the *Assay* under *Hydrocortisone Acetate Lotion.*

Neomycin and Polymyxin B Sulfates and Hydrocortisone Otic Solution

» Neomycin and Polymyxin B Sulfates and Hydrocortisone Otic Solution is a sterile solution containing the equivalent of not less than 90.0 percent and not more than 130.0 percent of the labeled amounts of neomycin and polymyxin B. It contains not less than 90.0 percent and not more than 110.0 percent of the labeled amount of hydrocortisone. It may contain one or more suitable buffers, dispersants, and solvents.

Packaging and storage—Preserve in tight, light-resistant containers. The containers or individual cartons are sealed and tamper-proof so that sterility is assured at time of first use.

Reference standards—*USP Neomycin Sulfate Reference Standard*—Dry in vacuum at a pressure not exceeding 5 mm of mercury at 60° for 3 hours before using. *USP Polymyxin B Sulfate Reference Standard*—Dry in vacuum at a pressure not exceeding 5 mm of mercury at 60° for 3 hours before using. *USP Hydrocortisone Reference Standard*—Dry at 105° for 3 hours before using.

Sterility—It meets the requirements under *Sterility Tests* ⟨71⟩.

pH ⟨791⟩: between 2.0 and 4.5.

Assay for neomycin—Proceed as directed under *Antibiotics—Microbial Assays* ⟨81⟩, using an accurately measured volume of Neomycin and Polymyxin B Sulfates and Hydrocortisone Otic Solution diluted quantitatively and stepwise with *Buffer No. 3* to yield a *Test Dilution* having a concentration assumed to be equal to the median dose level of the Standard (1.0 µg of neomycin per mL).

Assay for polymyxin B—Proceed as directed under *Antibiotics—Microbial Assays* ⟨81⟩, using an accurately measured volume of Neomycin and Polymyxin B Sulfates and Hydrocortisone Otic Solution diluted quantitatively and stepwise with *Buffer No. 6* to yield a *Test Dilution* having a concentration assumed to be equal to the median dose level of the Standard (10 Polymyxin B Units per mL). Add to each test dilution of the Standard a quantity of Neomycin Standard, dissolved in *Buffer No. 6*, to obtain the same concentration of neomycin present in the *Test Dilution.*

Assay for hydrocortisone—

Mobile phase, Standard preparation, and *Chromatographic system*—Prepare as directed in the *Assay for hydrocortisone* under *Neomycin and Polymyxin B Sulfates, Bacitracin Zinc, and Hydrocortisone Ophthalmic Ointment.*

Assay preparation—Transfer 3.0 mL of Neomycin and Polymyxin B Sulfates and Hydrocortisone Otic Solution to a 200-mL volumetric flask, dilute with a mixture of methanol and water (1:1) to volume, and mix.

Procedure—Proceed as directed for *Procedure* in the *Assay for hydrocortisone* under *Neomycin and Polymyxin B Sulfates, Bacitracin Zinc, and Hydrocortisone Ophthalmic Ointment.* Calculate the quantity, in mg, of $C_{21}H_{30}O_5$ in each mL of the Otic Solution taken by the formula:

$$(66.67C)(r_U / r_S),$$

in which C is the concentration, in mg per mL, of USP Hydrocortisone RS in the *Standard preparation,* and r_U and r_S are the peak responses obtained from the *Assay preparation* and the *Standard preparation,* respectively.

Neomycin and Polymyxin B Sulfates and Hydrocortisone Ophthalmic Suspension

» Neomycin and Polymyxin B Sulfates and Hydrocortisone Ophthalmic Suspension is a sterile, aqueous suspension containing the equivalent of not less than 90.0 percent and not more than 130.0 percent of the labeled amounts of neomycin and of polymyxin B. It contains not less than 90.0 percent and not more than 110.0 percent of the labeled amount of hydrocortisone.

Packaging and storage—Preserve in tight containers. The containers or individual cartons are sealed and tamper-proof so that sterility is assured at time of first use.

Reference standards—*USP Neomycin Sulfate Reference Standard*—Dry in vacuum at a pressure not exceeding 5 mm of mercury at 60° for 3 hours before using. *USP Polymyxin B Sulfate Reference Standard*—Dry in vacuum at a pressure not exceeding 5 mm of mercury at 60° for 3 hours before using. *USP Hydrocortisone Reference Standard*—Dry at 105° for 3 hours before using.

Sterility—It meets the requirements under *Sterility Tests* ⟨71⟩.

pH ⟨791⟩: between 4.1 and 7.0.

Assay for neomycin—Proceed as directed for neomycin under *Antibiotics—Microbial Assays* ⟨81⟩, using an accurately measured volume of Neomycin and Polymyxin B Sulfates and Hydrocortisone Ophthalmic Suspension, freshly mixed and free from air bubbles, diluted quantitatively and stepwise with *Buffer No. 3* to yield a *Test Dilution* having a concentration assumed to be equal to the median dose level of the Standard.

Assay for polymyxin B—Proceed as directed for polymyxin B under *Antibiotics—Microbial Assays* ⟨81⟩, using an accurately measured volume of Neomycin and Polymyxin B Sulfates and Hydrocortisone Ophthalmic Suspension, freshly mixed and free from air bubbles, diluted quantitatively and stepwise with *Buffer No. 6* to yield a *Test Dilution* having a concentration assumed to be equal to the median dose level of the Standard. Add to each test dilution of the Standard a quantity of Neomycin Sulfate RS, dissolved in *Buffer No. 6,* to yield the same concentration of neomycin as is present in the *Test Dilution.*

Assay for hydrocortisone—

Mobile phase, Standard preparation, and *Chromatographic system*—Prepare as directed in the *Assay for hydrocortisone* under *Neomycin and Polymyxin B Sulfates, Bacitracin Zinc, and Hydrocortisone Ophthalmic Ointment.*

Assay preparation—Transfer an accurately measured volume of Neomycin and Polymyxin B Sulfates and Hydrocortisone Ophthalmic Suspension, freshly mixed and free from air bubbles, equivalent to about 30 mg of hydrocortisone, to a 200-mL volumetric flask, dilute with a mixture of methanol and water (1:1) to volume, and mix. Filter the solution, rejecting the first 10 mL of the filtrate.

Procedure—Proceed as directed for *Procedure* in the *Assay for hydrocortisone* under *Neomycin and Polymyxin B Sulfates, Bacitracin Zinc, and Hydrocortisone Ophthalmic Ointment.* Calculate the quantity, in mg, of $C_{21}H_{30}O_5$ in each mL of the Ophthalmic Suspension taken by the formula:

$$200(C / V)(r_U / r_S),$$

in which C is the concentration, in mg per mL, of USP Hydrocortisone RS in the *Standard preparation,* V is the volume, in mL, of Ophthalmic Suspension taken, and r_U and r_S are the peak responses obtained from the *Assay preparation* and the *Standard preparation,* respectively.

Neomycin and Polymyxin B Sulfates and Hydrocortisone Otic Suspension

» Neomycin and Polymyxin B Sulfates and Hydrocortisone Otic Suspension is a sterile suspension containing the equivalent of not less than 90.0 percent and not more than 130.0 percent of the labeled amounts of neomycin and of polymyxin B. It contains not less than 90.0 percent and not more than 110.0 percent of the labeled amount of hydrocortisone. It may contain one or more suitable buffers, dispersants, and preservatives.

Packaging and storage—Preserve in tight, light-resistant containers. The containers or individual cartons are sealed and tamper-proof so that sterility is assured at time of first use.

Reference standards—*USP Neomycin Sulfate Reference Standard*—Dry in vacuum at a pressure not exceeding 5 mm of mercury at 60° for 3 hours before using. *USP Polymyxin B Sulfate Reference Standard*—Dry in vacuum at a pressure not exceeding 5 mm of mercury at 60° for 3 hours before using. *USP Hydrocortisone Reference Standard*—Dry at 105° for 3 hours before using.

Sterility—It meets the requirements under *Sterility Tests* ⟨71⟩.

pH ⟨791⟩: between 3.0 and 5.5.

Assay for neomycin and Assay for polymyxin B—Using an accurately measured volume of Neomycin and Polymyxin B Sulfates and Hydrocortisone Otic Suspension, freshly mixed and free from air bubbles, proceed as directed in the *Assay for neomycin* and the *Assay for polymyxin B* under *Neomycin and Polymyxin B Sulfates and Hydrocortisone Otic Solution.*

Assay for hydrocortisone—

Mobile phase, Standard preparation, and *Chromatographic system*—Prepare as directed in the *Assay for hydrocortisone* under *Neomycin and Polymyxin B Sulfates, Bacitracin Zinc, and Hydrocortisone Ophthalmic Ointment.*

Assay preparation—Transfer 3.0 mL of Neomycin and Polymyxin B Sulfates and Hydrocortisone Otic Suspension, freshly mixed and free from air bubbles, to a 200-mL volumetric flask, dilute with a mixture of methanol and water (1:1) to volume, and mix. Filter the solution, rejecting the first 10 mL of the filtrate.

Procedure—Proceed as directed for *Procedure* in the *Assay for hydrocortisone* under *Neomycin and Polymyxin B Sulfates, Bacitracin Zinc, and Hydrocortisone Ophthalmic Ointment.* Calculate the quantity, in mg, of $C_{21}H_{30}O_5$ in each mL of the Otic Suspension taken by the formula:

$$(66.67C)(r_U / r_S),$$

in which C is the concentration, in mg per mL, of USP Hydrocortisone RS in the *Standard preparation,* and r_U and r_S are the peak responses obtained from the *Assay preparation* and the *Standard preparation,* respectively.

Neomycin and Polymyxin B Sulfates and Hydrocortisone Acetate Cream

» Neomycin and Polymyxin B Sulfates and Hydrocortisone Acetate Cream contains the equivalent of not less than 90.0 percent and not more than 130.0 percent of the labeled amounts of neomycin and polymyxin B, and not less than 90.0 percent and not more than 110.0 percent of the labeled amount of hydrocortisone acetate ($C_{23}H_{32}O_6$).

Packaging and storage—Preserve in well-closed containers.

Reference standards—*USP Neomycin Sulfate Reference Standard*—Dry in vacuum at a pressure not exceeding 5 mm of mercury at 60° for 3 hours before using. *USP Polymyxin B Sulfate Reference Standard*—Dry in vacuum at a pressure not exceeding 5 mm of mercury at 60° for 3 hours before using. *USP Hydrocortisone Acetate Reference Standard*—Dry in vacuum at 60° for 3 hours before using.

Identification—

A: It responds to the *Identification test* under *Neomycin Sulfate Cream.*

B: It responds to *Identification test B* under *Neomycin and Polymyxin B Sulfates Cream.*

C: The chromatogram of the *Assay preparation* obtained as directed in the *Assay for hydrocortisone acetate* exhibits a major peak for hydrocortisone acetate, the retention time of which corresponds to that exhibited in the chromatogram of the *Standard preparation* obtained as directed in the *Assay for hydrocortisone acetate.*

Minimum fill ⟨755⟩: meets the requirements.

Assay for neomycin and Assay for polymyxin B—Proceed with Neomycin and Polymyxin B Sulfates and Hydrocortisone Acetate Cream as directed in the *Assay for neomycin* and in the *Assay for polymyxin B* under *Neomycin and Polymyxin B Sulfates Cream.*

Assay for hydrocortisone acetate—Proceed with Neomycin and Polymyxin B Sulfates and Hydrocortisone Acetate Cream as directed in the *Assay* under *Hydrocortisone Acetate Lotion.*

Neomycin and Polymyxin B Sulfates and Hydrocortisone Acetate Ophthalmic Suspension

» Neomycin and Polymyxin B Sulfates and Hydrocortisone Acetate Ophthalmic Suspension is a sterile suspension of Hydrocortisone Acetate in an aqueous solution of Neomycin Sulfate and Polymyxin B Sulfate. It contains the equivalent of not less than 90.0 percent and not more than 125.0 percent of the labeled amounts of neomycin and polymyxin B, and not less than 90.0 percent and not more than 110.0 percent of the labeled amount of hydrocortisone acetate ($C_{23}H_{32}O_6$). It may contain suitable buffers, preservatives, and suspending agents.

Packaging and storage—Preserve in tight containers. The containers or individual cartons are sealed and tamper-proof so that sterility is assured at time of first use.

Reference standards—*USP Neomycin Sulfate Reference Standard*—Dry in vacuum at a pressure not exceeding 5 mm of mercury at 60° for 3 hours before using. *USP Polymyxin B Sulfate Reference Standard*—Dry in vacuum at a pressure not exceeding 5 mm of mercury at 60° for 3 hours before using. *USP Hydrocortisone Acetate Reference Standard*—Dry in vacuum at 60° for 3 hours before using.

Sterility—It meets the requirements under *Sterility Tests* ⟨71⟩, when tested as directed in the section, *Test Procedures Using Membrane Filtration.*

pH ⟨791⟩: between 5.0 and 7.0.

Assay for neomycin and Assay for polymyxin B—Proceed with Neomycin and Polymyxin B Sulfates and Hydrocortisone Acetate Ophthalmic Suspension as directed in the *Assay for neomycin* and in the *Assay for polymyxin B* under *Neomycin and Polymyxin B Sulfates and Prednisolone Acetate Ophthalmic Suspension.*

Assay for hydrocortisone acetate—Proceed with Neomycin and Polymyxin B Sulfates and Hydrocortisone Acetate Ophthalmic Suspension as directed in the *Assay* under *Sterile Hydrocortisone Acetate Suspension.*

Neomycin and Polymyxin B Sulfates, and Hydrocortisone Acetate Topical Suspension, Penicillin G Procaine,—*see* Penicillin G Procaine, Neomycin and Polymyxin B Sulfates, and Hydrocortisone Acetate Topical Suspension

Neomycin and Polymyxin B Sulfates and Prednisolone Acetate Ophthalmic Suspension

» Neomycin and Polymyxin B Sulfates and Prednisolone Acetate Ophthalmic Suspension is a sterile suspension of Prednisolone Acetate in an aqueous solution of Neomycin Sulfate and Polymyxin B Sulfate. It contains the equivalent of not less than 90.0 percent and not more than 125.0 percent of the labeled amounts of neomycin and polymyxin B, and not less than 90.0 percent and not more than 110.0 percent of the labeled amount of prednisolone acetate ($C_{23}H_{30}O_6$). It may contain suitable buffers, preservatives, and suspending agents.

Packaging and storage—Preserve in tight containers. The containers or individual cartons are sealed and tamper-proof so that sterility is assured at time of first use.

Reference standards—*USP Neomycin Sulfate Reference Standard*—Dry in vacuum at a pressure not exceeding 5 mm of mercury at 60° for 3 hours before using. *USP Polymyxin B Sulfate Reference Standard*—Dry in vacuum at a pressure not exceeding 5 mm of mercury at 60° for 3 hours before using. *USP Prednisolone Acetate Reference Standard*—Dry at 105° for 3 hours before using.

Identification—The chromatogram of the *Assay preparation* obtained as directed in the *Assay for prednisolone acetate* exhibits a major peak for prednisolone acetate, the retention time of which corresponds to that exhibited in the chromatogram of the *Standard preparation* obtained as directed in the *Assay for prednisolone acetate.*

Sterility—It meets the requirements under *Sterility Tests* ⟨71⟩.

pH ⟨791⟩: between 5.0 and 7.0.

Assay for neomycin—Proceed as directed for neomycin under *Antibiotics—Microbial Assays* ⟨81⟩, using an accurately measured volume of Neomycin and Polymyxin B Sulfates and Prednisolone Acetate Ophthalmic Suspension, freshly mixed and free from air bubbles, diluted quantitatively and stepwise with *Buffer No. 3* to yield a *Test Dilution* having a concentration assumed to be equal to the median dose level of the Standard.

Assay for polymyxin B—Proceed as directed for polymyxin B under *Antibiotics—Microbial Assays* ⟨81⟩, using an accurately measured volume of Neomycin and Polymyxin B Sulfates and Prednisolone Acetate Ophthalmic Suspension, freshly mixed and free from air bubbles, diluted quantitatively and stepwise with *Buffer No. 6* to yield a *Test Dilution* having a concentration assumed to be equal to the median dose level of the Standard. Add to each test dilution of the Standard a quantity of Neomycin Sulfate RS, dissolved in *Buffer No. 6*, to obtain the same concentration of neomycin as is present in the *Test Dilution.*

Assay for prednisolone acetate—

Mobile phase, Internal standard solution, Standard preparation, and *Chromatographic system*—Prepare as directed in the *Assay for prednisolone acetate* under *Neomycin Sulfate and Prednisolone Acetate Ophthalmic Suspension.*

Assay preparation—Transfer an accurately measured volume of Neomycin and Polymyxin B Sulfates and Prednisolone Acetate Ophthalmic Suspension, freshly mixed and free from air bubbles, equivalent to about 2.5 mg of prednisolone acetate, to a suitable container, add 5.0 mL of *Internal standard solution* and about 100 mL of water-saturated chloroform, and shake by mechanical means for about 15 minutes. Allow to separate for about 15 minutes, and use the clear chloroform layer as the *Assay preparation.*

Procedure—Proceed as directed in the *Assay for prednisolone acetate* under *Neomycin Sulfate and Prednisolone Acetate Ophthalmic Suspension.* Calculate the quantity, in mg, of prednisolone acetate ($C_{23}H_{30}O_6$) in each mL of the Ophthalmic Suspension taken by the formula:

$$0.1(C/V)(R_U/R_S),$$

in which C is the concentration, in µg per mL, of USP Prednisolone Acetate RS in the *Standard preparation,* V is the volume, in mL, of Ophthalmic Suspension taken, and R_U and R_S are the peak response ratios of prednisolone acetate to betamethasone obtained from the *Assay preparation* and the *Standard preparation,* respectively.

Neomycin Sulfate and Prednisolone Acetate Ointment

» Neomycin Sulfate and Prednisolone Acetate Ointment contains the equivalent of not less than 90.0 percent and not more than 135.0 percent of the labeled amount of neomycin, and not less than 90.0 percent and not more than 110.0 percent of the labeled amount of prednisolone acetate ($C_{23}H_{30}O_6$).

Packaging and storage—Preserve in collapsible tubes or in tight containers, protected from light.

Reference standards—*USP Neomycin Sulfate Reference Standard*—Dry in vacuum at a pressure not exceeding 5 mm of mercury at 60° for 3 hours before using. *USP Prednisolone Acetate Reference Standard*—Dry at 105° for 3 hours before using.

Identification—

A: It responds to the *Identification test* under *Neomycin Sulfate Cream.*

B: The chromatogram of the *Assay preparation* obtained as directed in the *Assay for prednisolone acetate* exhibits a major peak for prednisolone acetate, the retention time of which corresponds to that exhibited in the chromatogram of the *Standard preparation* obtained as directed in the *Assay for prednisolone acetate.*

Minimum fill ⟨755⟩: meets the requirements.

Water, *Method I* ⟨921⟩: not more than 1.0%, 20 mL of a mixture of carbon tetrachloride, chloroform, and methanol (2:2:1) being used in place of methanol in the titration vessel.

Assay for neomycin—Proceed with Neomycin Sulfate and Prednisolone Acetate Ointment as directed in the *Assay* under *Neomycin Sulfate Ointment.*

Assay for prednisolone acetate—

Mobile phase, Internal standard solution, Standard preparation, and *Chromatographic system*—Prepare as directed in the *Assay for prednisolone acetate* under *Neomycin Sulfate and Prednisolone Acetate Ophthalmic Suspension.*

Assay preparation—Transfer an accurately weighed portion of Neomycin Sulfate and Prednisolone Acetate Ointment, equivalent to about 1 mg of prednisolone acetate, to a suitable container, add 2.0 mL of *Internal standard solution,* dilute with water-saturated chloroform to about 35 mL, and shake to dissolve the ointment. Transfer about 5 mL of this solution to a suitable container, and evaporate to dryness. Add about 5 mL of water-saturated chloroform, and sonicate for 5 minutes. Filter, and use the clear solution as the *Assay preparation.*

Procedure—Proceed as directed for *Procedure* in the *Assay for prednisolone acetate* under *Neomycin Sulfate and Prednisolone Acetate Ophthalmic Suspension.* Calculate the quantity, in mg, of prednisolone acetate ($C_{23}H_{30}O_6$) in the portion of Ointment taken by the formula:

$$0.04C(R_U/R_S),$$

in which C is the concentration, in µg per mL, of USP Prednisolone Acetate RS in the *Standard preparation,* and R_U and R_S are the peak response ratios of prednisolone acetate to betamethasone obtained from the *Assay preparation* and the *Standard preparation,* respectively.

Neomycin Sulfate and Prednisolone Acetate Ophthalmic Ointment

» Neomycin Sulfate and Prednisolone Acetate Ophthalmic Ointment is a sterile ointment containing Neomycin Sulfate and Prednisolone Acetate. It contains the equivalent of not less than 90.0 percent and not more than 135.0 percent of the labeled amount of neomycin, and not less than 90.0 percent and not more than 110.0 percent of the labeled amount of prednisolone acetate ($C_{23}H_{30}O_6$).

Packaging and storage—Preserve in collapsible ophthalmic ointment tubes.

Reference standards—*USP Neomycin Sulfate Reference Standard*—Dry in vacuum at a pressure not exceeding 5 mm of mercury at 60° for 3 hours before using. *USP Prednisolone Acetate Reference Standard*—Dry at 105° for 3 hours before using.

Identification—

A: It responds to the *Identification test* under *Neomycin Sulfate Cream.*

B: The chromatogram of the *Assay preparation* obtained as directed in the *Assay for prednisolone acetate* exhibits a major peak for prednisolone acetate, the retention time of which corresponds to that exhibited in the chromatogram of the *Standard preparation* obtained as directed in the *Assay for prednisolone acetate.*

Sterility—It meets the requirements under *Sterility Tests* ⟨71⟩, when tested as directed in the section, *Test Procedures Using Membrane Filtration.*

Minimum fill ⟨755⟩: meets the requirements.

Water, *Method I* ⟨921⟩: not more than 1.0%, 20 mL of a mixture of carbon tetrachloride, chloroform, and methanol (2:2:1) being used in place of methanol in the titration vessel.

Metal particles—It meets the requirements of the test for *Metal Particles in Ophthalmic Ointments* ⟨751⟩.

Assay for neomycin—Proceed with Neomycin Sulfate and Prednisolone Acetate Ophthalmic Ointment as directed in the *Assay* under *Neomycin Sulfate Ointment.*

Assay for prednisolone acetate—

Mobile phase, Internal standard solution, Standard preparation, and *Chromatographic system*—Prepare as directed in the *Assay for prednisolone acetate* under *Neomycin Sulfate and Prednisolone Acetate Ophthalmic Suspension.*

Assay preparation—Using Neomycin Sulfate and Prednisolone Acetate Ophthalmic Ointment, proceed as directed for *Assay preparation* in the *Assay for prednisolone acetate* under *Neomycin Sulfate and Prednisolone Acetate Ointment.*

Procedure—Proceed as directed for *Procedure* in the *Assay for prednisolone acetate* under *Neomycin Sulfate and Prednisolone Acetate Ophthalmic Suspension.* Calculate the quantity, in mg, of prednisolone acetate ($C_{23}H_{30}O_6$) in the portion of Ophthalmic Ointment taken by the formula:

$$0.04C(R_U/R_S),$$

in which C is the concentration, in µg per mL, of USP Prednisolone Acetate RS in the *Standard preparation*, and R_U and R_S are the peak response ratios of prednisolone acetate to betamethasone obtained from the *Assay preparation* and the *Standard preparation*, respectively.

Neomycin Sulfate and Prednisolone Acetate Ophthalmic Suspension

» Neomycin Sulfate and Prednisolone Acetate Ophthalmic Suspension contains the equivalent of not less than 90.0 percent and not more than 130.0 percent of the labeled amount of neomycin, and not less than 90.0 percent and not more than 110.0 percent of the labeled amount of prednisolone acetate ($C_{23}H_{30}O_6$).

Packaging and storage—Preserve in tight containers. The containers or individual cartons are sealed and tamper-proof so that sterility is assured at time of first use.

Reference standards—*USP Neomycin Sulfate Reference Standard*—Dry in vacuum at a pressure not exceeding 5 mm of mercury at 60° for 3 hours before using. *USP Prednisolone Acetate Reference Standard*—Dry at 105° for 3 hours before using.

Identification—

A: Filter a portion of Ophthalmic Suspension, freshly mixed but free from air bubbles, equivalent to about 60 mg of prednisolone acetate, discarding the filtrate. Wash the filter with about 10 mL of water, and dry at 105° for 3 hours: the infrared absorption spectrum of a potassium bromide dispersion of the dried residue on the filter so obtained exhibits maxima only at the same wavelengths as that of a similar preparation of USP Prednisolone Acetate RS.

B: The chromatogram of the *Assay preparation* obtained as directed in the *Assay for prednisolone acetate* exhibits a major peak for prednisolone acetate, the retention time of which corresponds to that exhibited in the chromatogram of the *Standard preparation* obtained as directed in the *Assay for prednisolone acetate*.

Sterility—It meets the requirements under *Sterility Tests* ⟨71⟩, when tested as directed in the section, *Test Procedures Using Membrane Filtration*.

pH ⟨791⟩: between 5.5 and 7.5.

Assay for neomycin—Proceed with Neomycin Sulfate and Prednisolone Acetate Ophthalmic Suspension as directed in the *Assay for neomycin* under *Neomycin and Polymyxin B Sulfates and Prednisolone Acetate Ophthalmic Suspension*.

Assay for prednisolone acetate—

Mobile phase—Prepare a solution containing *n*-butyl chloride, water-saturated *n*-butyl chloride, tetrahydrofuran, methanol, and glacial acetic acid (95:95:14:7:6).

Internal standard solution—Prepare a solution of betamethasone in tetrahydrofuran containing 10 mg per mL. Dilute this solution with water-saturated chloroform, and mix to obtain a solution having a concentration of about 1 mg per mL.

Standard preparation—Dissolve about 5 mg of USP Prednisolone Acetate RS, accurately weighed, in 10.0 mL of *Internal standard solution*. Use sonication, if necessary, dilute with water-saturated chloroform to 200.0 mL, and mix to obtain a solution having a known concentration of about 25 µg per mL.

Assay preparation—Transfer an accurately measured volume of Neomycin Sulfate and Prednisolone Acetate Ophthalmic Suspension, freshly mixed and free from air bubbles, equivalent to about 2.5 mg of prednisolone acetate, to a suitable container, add 5.0 mL of *Internal standard solution* and about 100 mL of water-saturated chloroform, and shake by mechanical means for about 15 minutes. Allow to separate for about 15 minutes, and use the clear chloroform layer as the *Assay preparation*.

Chromatographic system (see *Chromatography* ⟨621⟩)—The liquid chromatograph is equipped with a 254-nm detector and a 4-mm × 30-cm column that contains packing L3. The flow rate is about 1 mL per minute. Chromatograph the *Standard preparation*, and record the peak responses as directed under *Procedure:* the resolution, R, between the analyte and internal standard peaks is not less than 3.0, and the relative standard deviation for replicate injections is not more than 2.0%.

Procedure—Separately inject equal volumes (about 10 µL) of the *Standard preparation* and the *Assay preparation* into the chromatograph, record the chromatograms, and measure the responses for the major peaks. The relative retention times are about 1.6 for betamethasone and 1.0 for prednisolone acetate. Calculate the quantity, in mg, of prednisolone acetate ($C_{23}H_{30}O_6$) in each mL of the Ophthalmic Suspension taken by the formula:

$$0.1(C/V)(R_U/R_S),$$

in which C is the concentration, in µg per mL, of USP Prednisolone Acetate RS in the *Standard preparation*, V is the volume, in mL, of Ophthalmic Suspension taken, and R_U and R_S are the peak response ratios of prednisolone acetate to betamethasone obtained from the *Assay preparation* and the *Standard preparation*, respectively.

Neomycin Sulfate and Prednisolone Sodium Phosphate Ophthalmic Ointment

» Neomycin Sulfate and Prednisolone Sodium Phosphate Ophthalmic Ointment is a sterile ointment containing Neomycin Sulfate and Prednisolone Sodium Phosphate. It contains the equivalent of not less than 90.0 percent and not more than 135.0 percent of the labeled amount of neomycin, and the equivalent of not less than 90.0 percent and not more than 115.0 percent of the labeled amount of prednisolone phosphate ($C_{21}H_{29}O_8P$).

NOTE—Where Neomycin Sulfate and Prednisolone Sodium Phosphate Ophthalmic Ointment is prescribed without reference to the quantity of neomycin or prednisolone phosphate contained therein, a product containing 3.5 mg of neomycin and 2.5 mg of prednisolone phosphate per g shall be dispensed.

Packaging and storage—Preserve in collapsible ophthalmic ointment tubes.

Reference standards—*USP Neomycin Reference Standard*—Dry at a pressure not exceeding 5 mm of mercury at 60° for 3 hours before using. *USP Prednisolone Reference Standard*—Dry in vacuum at 105° for 3 hours before using.

Identification—

A: It responds to the *Identification test* under *Neomycin Sulfate Cream*.

B: Shake a quantity of Ophthalmic Ointment, equivalent to about 40 mg of prednisolone phosphate, with 25 mL of sodium chloride solution (1 in 20) and 25 mL of methylene chloride, for 2 minutes. Transfer the methylene chloride layer to a second separator containing 15 mL of sodium chloride (1 in 20). Shake for 1 minute, and discard the methylene chloride layer. Repeat the operation with a second portion of 25 mL of methylene chloride. Combine the aqueous phase from the second separator with the aqueous phase of the first separator. Add 10 mL of *Alkaline phosphatase solution*, prepared as directed in the *Assay for prednisolone phosphate*, and add 50 mL of methylene chloride. Insert the stopper, and allow to stand, with occasional gentle inversion (about once every 15 minutes), for 2 hours. Filter the methylene chloride layer through a dry paper, and evaporate 25 mL of the filtrate to dryness: the residue so obtained responds to *Identification test A* under *Prednisolone*.

Sterility—It meets the requirements under *Sterility Tests* ⟨71⟩.

Minimum fill ⟨755⟩: meets the requirements.

Water, *Method I* ⟨921⟩: not more than 1.0%, 20 mL of a mixture of carbon tetrachloride, chloroform, and methanol (2:2:1) being used in place of methanol in the titration vessel.

Metal particles—It meets the requirements of the test for *Metal Particles in Ophthalmic Ointments* ⟨751⟩.

Assay for neomycin—Proceed as directed under *Antibiotics—Microbial Assays* ⟨81⟩, using an accurately weighed portion of Neomycin Sulfate and Prednisolone Sodium Phosphate Ophthalmic Ointment shaken in a separator with about 50 mL of ether, and extracted with four 20-mL portions of *Buffer No. 3*. Combine the aqueous extracts, and dilute with *Buffer No. 3* to an appropriate volume to obtain a stock solution. Dilute this stock solution quantitatively and stepwise with *Buffer No. 3* to obtain a *Test Dilution* having a concentration assumed to be equal to the median dose level of the Standard.

Assay for prednisolone phosphate—

pH 9 buffer with magnesium—Prepare as directed in the *Assay* under *Dexamethasone Sodium Phosphate*.

Alkaline phosphatase solution—Prepare as directed in the *Assay* under *Dexamethasone Sodium Phosphate Injection*.

Standard preparation—Prepare as directed for *Standard Preparation* under *Assay for Steroids* ⟨351⟩, using USP Prednisolone RS.

Assay preparation—Transfer an accurately weighed portion of Neomycin Sulfate and Prednisolone Sodium Phosphate Ophthalmic Ointment, equivalent to about 3 mg of prednisolone phosphate, to a 125-mL separator. Add 25 mL of sodium chloride solution (1 in 20) and 25 mL of methylene chloride, and shake for not less than 2 minutes to disperse the assay specimen. Transfer the methylene chloride layer to a second separator containing 15 mL of sodium chloride solution (1 in 20). Shake for 1 minute, and discard the methylene chloride layer. Repeat the operation with a second portion of 25 mL of methylene chloride. Transfer the aqueous phases from both separators to a 50-mL volumetric flask, rinsing the first separator with the aqueous phase of the second separator. Rinse both separators with the same 5 mL of sodium chloride solution (1 in 20), and add the rinsing to the volumetric flask. Add sodium chloride solution (1 in 20) to volume, and mix.

Pipet 5 mL of the resulting solution into a 125-mL separator, add 8.0 mL of *Alkaline phosphatase solution*, mix, and allow to stand for 2 hours. Extract the solution with two 25-mL portions of methylene chloride, filtering the extracts through methylene chloride–washed cotton into a small beaker. Evaporate the methylene chloride on a steam bath nearly to dryness, then evaporate with the aid of a current of air to dryness. Dissolve the residue in 25.0 mL of alcohol.

Prepare a blank by evaporating 50 mL of methylene chloride to dryness and dissolving the residue in 25 mL of alcohol.

Procedure—Pipet 20 mL each of the *Assay preparation*, the *Standard preparation*, and the blank solution into separate glass-stoppered flasks, and proceed as directed for *Procedure* under *Assay for Steroids* ⟨351⟩, beginning with "add 2.0 mL of a solution prepared by dissolving 50 mg of blue tetrazolium." Calculate the quantity, in mg, of prednisolone phosphate ($C_{21}H_{29}O_8P$) in the portion of Ophthalmic Ointment taken by the formula:

$$0.25C(A_U/A_S)(440.43/360.45),$$

in which C is the concentration, in µg per mL, of USP Prednisolone RS in the *Standard preparation*, A_U and A_S are the absorbances of the solutions from the *Assay preparation* and the *Standard preparation*, respectively, and 440.43 and 360.45 are the molecular weights of prednisolone phosphate and prednisolone, respectively.

Neomycin Sulfate, Sulfacetamide Sodium, and Prednisolone Acetate Ophthalmic Ointment

» Neomycin Sulfate, Sulfacetamide Sodium, and Prednisolone Acetate Ophthalmic Ointment contains the equivalent of not less than 90.0 percent and not more than 135.0 percent of the labeled amount of neomycin, and not less than 90.0 percent and not more than 110.0 percent of the labeled amounts of sulfacetamide sodium ($C_8H_9N_2NaO_3S.H_2O$) and prednisolone acetate ($C_{23}H_{30}O_6$).

Packaging and storage—Preserve in collapsible ophthalmic ointment tubes.

Reference standards—*USP Neomycin Sulfate Reference Standard*—Dry in vacuum at a pressure not exceeding 5 mm of mercury at 60° for 3 hours before using. *USP Prednisolone Acetate Reference Standard*—Dry at 105° for 3 hours before using.

Identification—

A: Dissolve a quantity of Ophthalmic Ointment, equivalent to about 1 g of sulfacetamide sodium, in 100 mL of ether in a separator, and extract the mixture with 25 mL of water. Wash the extract with 25 mL of ether, and warm the water extract on a steam bath to remove the last traces of ether. Adjust with 6 *N* acetic acid to a pH of between 4 and 5, and filter. Wash the precipitate with water, and dry at 105° for 2 hours: the sulfacetamide so obtained melts between 180° and 184°, and responds to *Identification tests B, D,* and *E* under *Sulfacetamide Sodium*.

B: To a quantity of Ophthalmic Ointment, equivalent to about 25 mg of prednisolone acetate, add 15 mL of water, extract with two 10-mL portions of peroxide-free ether, discard the ether extracts, and extract with two 10-mL portions of chloroform. Evaporate the combined, clear chloroform extracts, with the aid of a current of air, to dryness: the residue so obtained responds to *Identification test A* under *Prednisolone Acetate*.

Sterility—It meets the requirements for *Ophthalmic Ointments* under *Sterility Tests* ⟨71⟩, when tested as directed in the section, *Test Procedures Using Membrane Filtration*.

Minimum fill ⟨755⟩: meets the requirements.

Metal particles—It meets the requirements of the test for *Metal Particles in Ophthalmic Ointments* ⟨751⟩.

Assay for neomycin—Proceed with Neomycin Sulfate, Sulfacetamide Sodium, and Prednisolone Acetate Ophthalmic Ointment as directed in the *Assay* under *Neomycin Sulfate Ointment*.

Assay for sulfacetamide sodium—Weigh accurately a quantity of Neomycin Sulfate, Sulfacetamide Sodium, and Prednisolone Acetate Ophthalmic Ointment, equivalent to about 500 mg of sulfacetamide sodium, and transfer to a 125-mL separator. Dissolve the ointment in 50 mL of ether, and extract the mixture with six 25-mL portions of water. Warm the combined extracts on a steam bath to remove the last traces of ether, add 20 mL of hydrochloric acid, and proceed as directed under *Nitrite Titration* ⟨451⟩, beginning with "cool to 15°." Each mL of 0.1 *M* sodium nitrite is equivalent to 25.42 mg of $C_8H_9N_2NaO_3S.H_2O$.

Assay for prednisolone acetate—

Standard preparation—Prepare as directed for *Standard Preparation* under *Assay for Steroids* ⟨351⟩, using USP Prednisolone Acetate RS.

Assay preparation—Transfer to a suitable flask an accurately weighed quantity of Neomycin Sulfate, Sulfacetamide Sodium, and Prednisolone Acetate Ophthalmic Ointment, equivalent to about 10 mg of prednisolone acetate, and add 30 mL of alcohol. Heat on a steam bath to melt the ointment base, and mix. Cool to solidify the ointment base, and filter the alcohol solution into a 100-mL volumetric flask. Repeat the extraction with three 20-mL portions of alcohol, add alcohol to volume, and mix. Pipet 10 mL of this solution into a 100-mL volumetric flask, add alcohol to volume, and mix. Pipet 20 mL of the resulting solution into a glass-stoppered, 50-mL conical flask.

Procedure—Proceed as directed for *Procedure* under *Assay for Steroids* ⟨351⟩. Calculate the quantity, in mg, of $C_{23}H_{30}O_6$ in the portion of Ophthalmic Ointment taken by the formula:

$$C(A_U/A_S).$$

Neomycin Sulfate and Triamcinolone Acetonide Cream

» Neomycin Sulfate and Triamcinolone Acetonide Cream contains the equivalent of not less than 90.0 percent and not more than 135.0 percent of the labeled amount of neomycin, and not less than 90.0 percent and not more than 110.0 percent of the labeled amount of triamcinolone acetonide ($C_{24}H_{34}FO_6$).

Packaging and storage—Preserve in collapsible tubes or in tight containers.

Reference standards—*USP Neomycin Sulfate Reference Standard*—Dry in vacuum at a pressure not exceeding 5 mm of mercury at 60° for 3 hours before using. *USP Triamcinolone Acetonide Reference Standard*—Do not dry; use as is.

Identification—

A: It responds to the *Identification test* under *Neomycin Sulfate Cream.*

B: Place 2 g of Cream in a conical flask, add 5.0 mL of chloroform, and shake for 10 minutes. Add 15 mL of alcohol, and shake for an additional 10 minutes. Filter the solution into a centrifuge tube, and evaporate the filtrate to dryness. Dissolve the residue in alcohol to obtain a solution containing about 250 µg of triamcinolone acetonide per mL. Proceed as directed in the *Identification test* under *Triamcinolone Acetonide Cream,* beginning with "Apply 10 µL of this solution:" the specified result is observed.

Minimum fill ⟨755⟩: meets the requirements.

Assay for neomycin—Proceed with Neomycin Sulfate and Triamcinolone Acetonide Cream as directed in the *Assay* under *Neomycin Sulfate Cream.*

Assay for triamcinolone acetonide—Proceed with Neomycin Sulfate and Triamcinolone Acetonide Cream as directed in the *Assay* under *Triamcinolone Acetonide Cream.*

Neomycin Sulfate and Triamcinolone Acetonide Ophthalmic Ointment

» Neomycin Sulfate and Triamcinolone Acetonide Ophthalmic Ointment contains the equivalent of not less than 90.0 percent and not more than 135.0 percent of the labeled amount of neomycin, and not less than 90.0 percent and not more than 110.0 percent of the labeled amount of triamcinolone acetonide ($C_{24}H_{31}FO_6$).

Packaging and storage—Preserve in collapsible ophthalmic ointment tubes.

Reference standards—*USP Neomycin Sulfate Reference Standard*—Dry in vacuum at a pressure not exceeding 5 mm of mercury at 60° for 3 hours before using. *USP Triamcinolone Acetonide Reference Standard*—Do not dry; use as is.

Identification—

A: It responds to the *Identification test* under *Neomycin Sulfate Cream.*

B: Place 2 g of Ophthalmic Ointment in a conical flask, add 5.0 mL of chloroform, and shake for 10 minutes. Add 15 mL of alcohol, and shake for an additional 10 minutes. Filter the solution into a centrifuge tube, and evaporate the filtrate to dryness. Dissolve the residue in alcohol to obtain a solution containing about 250 µg of triamcinolone acetonide per mL. Proceed as directed in the *Identification test* under *Triamcinolone Acetonide Cream,* beginning with "Apply 10 µL of this solution:" the specified result is observed.

Sterility—It meets the requirements for *Ophthalmic Ointments* under *Sterility Tests* ⟨71⟩, when tested as directed in the section, *Test Procedures Using Membrane Filtration.*

Minimum fill ⟨755⟩: meets the requirements.

Water, *Method I* ⟨921⟩: not more than 1.0%, 20 mL of a mixture of carbon tetrachloride, chloroform, and methanol (2:2:1) being used in place of methanol in the titration vessel.

Metal particles—It meets the requirements of the test for *Metal Particles in Ophthalmic Ointments* ⟨751⟩.

Assay for neomycin—Proceed with Neomycin Sulfate and Triamcinolone Ophthalmic Ointment as directed in the *Assay* under *Neomycin Sulfate Ointment.*

Assay for triamcinolone acetonide—Proceed with Neomycin Sulfate and Triamcinolone Acetonide Ophthalmic Ointment as directed in the *Assay* under *Triamcinolone Acetonide Cream,* except to read "Ophthalmic Ointment" in place of "Cream" throughout.

Netilmicin Sulfate

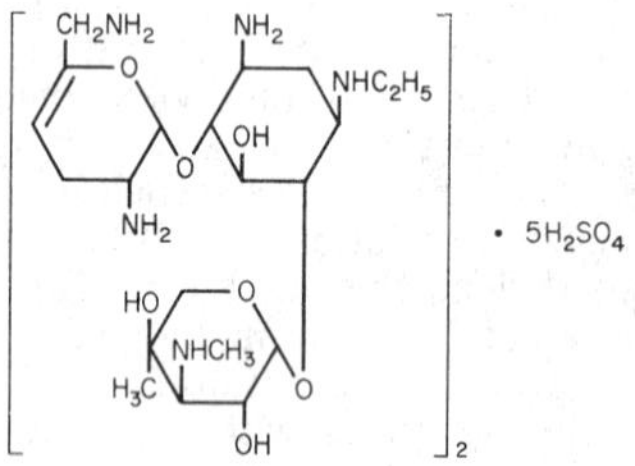

CH2NH2 NH2 NHC2H5 O OH O NH2 HO O NHCH3 H3C O OH 2 · 5H2SO4

$(C_{21}H_{41}N_5O_7)_2 \cdot 5H_2SO_4$ 1441.54

D-Streptamine, *O*-3-deoxy-4-*C*-methyl-3-(methylamino)-β-L-arabinopyranosyl-(1→6)-*O*-[2,6-diamino-2,3,4,6-tetradeoxy-α-D-*glycero*-hex-4-enopyranosyl-(1→4)]-2-deoxy-N^1-ethyl-, sulfate (2:5) (salt).

O-3-Deoxy-4-*C*-methyl-3-(methylamino)-β-L-arabinopyranosyl-(1→4)-*O*-[2,6-diamino-2,3,4,6-tetradeoxy-α-D-*glycero*-hex-4-enopyranosyl-(1→6)]-2-deoxy-N^3-ethyl-L-streptamine sulfate (2:5) (salt) [*56391-57-2*].

» Netilmicin Sulfate has a potency equivalent to not less than 595 µg of netilmicin ($C_{21}H_{41}N_5O_7$) per mg, calculated on the dried basis.

Packaging and storage—Preserve in tight containers.

Reference standard—*USP Netilmicin Sulfate Reference Standard*—Do not dry before using; calculate its weight on the dried basis.

Identification—Prepare a solution containing 10 mg of netilmicin per mL. On a suitable thin-layer chromatographic plate (see *Chromatography* ⟨621⟩), coated with a 0.25-mm layer of chromatographic silica gel mixture, apply 5 µL of this solution, 5 µL of a Standard solution of USP Netilmicin Sulfate RS containing 10 mg of netilmicin per mL, and 5 µL of a 1:1 mixture of the two solutions. Allow the spots to dry, place the plate in a developing chamber fitted for continuous-flow, thin-layer chromatography, develop the chromatogram in a solvent system consisting of a mixture of methanol, ammonium hydroxide, and chloroform (60:30:25) for 1.5 hours. Remove the plate from the developing chamber, allow the solvent to evaporate, and heat the plate at 110° for 15 minutes. Spray the plate with a 1 in 100 solution of ninhydrin in butanol to which 1 mL of pyridine has been added: netilmicin appears as a brown spot, and the spots obtained from the test solution and from the mixture of test solution and Standard solution, respectively, correspond in distance from the origin to that of the spot from the Standard solution.

Specific rotation ⟨781⟩: between +88° and +96°, calculated on the dried basis, in a solution containing 30 mg per mL.

pH ⟨791⟩: between 3.5 and 5.5, in a solution containing 40 mg of netilmicin per mL.

Loss on drying ⟨731⟩—Dry about 100 mg in vacuum at a pressure not exceeding 5 mm of mercury at 110° for 3 hours: it loses not more than 15.0% of its weight.

Residue on ignition ⟨281⟩: not more than 1.0%, the charred residue being moistened with 2 mL of nitric acid and 5 drops of sulfuric acid.

Assay—Proceed as directed under *Antibiotics—Microbial Assays* ⟨81⟩.

Netilmicin Sulfate Injection

» Netilmicin Sulfate Injection is a sterile solution of Netilmicin Sulfate in Water for Injection. It contains the equivalent of not less than 90.0 percent and not more than 115.0 percent of the labeled amount of netilmicin ($C_{21}H_{41}N_5O_7$). It may contain one or more suitable buffers, chelating agents, and preservatives.

Packaging and storage—Preserve in single-dose or in multiple-dose containers, preferably of Type I glass.

Reference standard—*USP Netilmicin Sulfate Reference Standard*—Do not dry before using; calculate its weight on the dried basis.

Identification—It responds to the *Identification test* under *Netilmicin Sulfate.*

Pyrogen—When diluted, if necessary, with pyrogen-free saline TS to a concentration of 10 mg of netilmicin per mL, it meets the requirements of the *Pyrogen Test* ⟨151⟩, the test dose being 1 mL per kg.

Sterility—It meets the requirements under *Sterility Tests* ⟨71⟩, when tested as directed in the section, *Test Procedures Using Membrane Filtration.*

pH ⟨791⟩: between 3.5 and 6.0.

Particulate matter ⟨788⟩: meets the requirements under *Small-volume Injections.*

Other requirements—It meets the requirements under *Injections* ⟨1⟩.

Assay—Proceed as directed under *Antibiotics—Microbial Assays* ⟨81⟩, using an accurately measured volume of Netilmicin Sulfate Injection diluted quantitatively and stepwise with *Buffer No. 3* to obtain a *Test Dilution* having a concentration assumed to be equal to the median dose level of the Standard.

Niacin

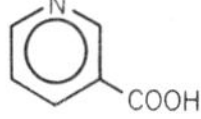

$C_6H_5NO_2$ 123.11
3-Pyridinecarboxylic acid.
Nicotinic acid [*59-67-6*].

» Niacin contains not less than 99.0 percent and not more than 101.0 percent of $C_6H_5NO_2$, calculated on the dried basis.

Packaging and storage—Preserve in well-closed containers.

Reference standard—*USP Niacin Reference Standard*—Dry at 105° for 1 hour before using.

Identification—

A: The infrared absorption spectrum of a mineral oil dispersion of it, previously dried, exhibits maxima only at the same wavelengths as that of a similar preparation of USP Niacin RS.

B: The ultraviolet absorption spectrum of a solution (1 in 50,000) exhibits maxima and minima at the same wavelengths as that of a similar solution of USP Niacin RS, concomitantly measured, and the ratio A_{237}/A_{262} is 0.37 ± 0.02.

Loss on drying ⟨731⟩—Dry it at 105° for 1 hour: it loses not more than 1.0% of its weight.

Residue on ignition ⟨281⟩: not more than 0.1%.

Chloride ⟨221⟩—A 0.50-g portion shows no more chloride than corresponds to 0.15 mL of 0.020 *N* hydrochloric acid (0.02%).

Sulfate ⟨221⟩—A 0.50-g portion shows no more sulfate than corresponds to 0.10 mL of 0.020 *N* sulfuric acid (0.02%).

Heavy metals, *Method I* ⟨231⟩—Mix 1 g with 4 mL of 1 *N* acetic acid, dilute with water to 25 mL, heat gently until solution is complete, and cool: the limit is 0.002%.

Assay—Transfer about 200 mg of Niacin, accurately weighed, to a 500-mL volumetric flask, add water to dissolve, dilute with water to volume, and mix. Transfer 5.0 mL of this solution to a 100-mL volumetric flask, dilute with water to volume, and mix. Concomitantly determine the absorbances of this solution and a solution of USP Niacin RS in the same medium, at a concentration of about 20 µg per mL, in 1-cm cells at the wavelength of maximum absorbance at about 262 nm, with a suitable spectrophotometer, using water as the blank. Calculate the quantity, in mg, of $C_6H_5NO_2$ in the Niacin taken by the formula:

$$10C(A_U/A_S),$$

in which C is the concentration, in µg per mL, of USP Niacin RS in the Standard solution, and A_U and A_S are the absorbances of the solution of Niacin and the Standard solution, respectively.

Niacin Injection

» Niacin Injection is a sterile solution of Niacin and niacin sodium in Water for Injection, made with the aid of Sodium Carbonate or Sodium Hydroxide. It contains not less than 95.0 percent and not more than 110.0 percent of the labeled amount of $C_6H_5NO_2$.

Packaging and storage—Preserve in single-dose or in multiple-dose containers, preferably of Type I glass.

Reference standard—*USP Niacin Reference Standard*—Dry at 105° for 1 hour before using.

Identification—To a volume of Injection, equivalent to about 100 mg of niacin, add 0.3 mL of 3 *N* hydrochloric acid, evaporate, if necessary, on a steam bath to about 2 mL, and allow to stand for 1 hour in a cool place. Filter by suction, wash with small volumes of ice-cold water until the last washing does not give a reaction for chloride, and dry at 105° for 1 hour: the niacin so obtained responds to *Identification tests A* and *B* under *Niacin.*

pH ⟨791⟩: between 4.0 and 6.0.

Other requirements—It meets the requirements under *Injections* ⟨1⟩.

Assay—Proceed with Niacin Injection as directed for *Chemical Method* under *Niacin or Niacinamide Assay* ⟨441⟩, using *Standard Niacin Preparation* as the *Standard Preparation* in the *Assay Procedure*, and the following as the *Assay Preparation:* Transfer an accurately measured volume of Niacin Injection, equivalent to about 50 mg of niacin, to a 500-mL volumetric flask, dilute with water to volume, and mix. Transfer 10.0 mL of this solution to a 200-mL volumetric flask, dilute with water to volume, and mix. Calculate the quantity, in mg, of $C_6H_5NO_2$ in each mL of the Injection taken by the formula:

$$(50/V)(A_U/A_S),$$

in which V is the volume, in mL, of Injection taken.

Niacin Tablets

» Niacin Tablets contain not less than 90.0 percent and not more than 110.0 percent of the labeled amount of $C_6H_5NO_2$.

Packaging and storage—Preserve in well-closed containers.

Reference standard—*USP Niacin Reference Standard*—Dry at 105° for 1 hour before using.

Identification—Heat a quantity of finely powdered Tablets, equivalent to about 500 mg of niacin, with 25 mL of alcohol on a steam bath for a few minutes, filter, and wash the residue with a few mL of hot alcohol. To the filtrate add 30 mL of water, and evaporate to about 25 mL on the steam bath. Cool, filter if insoluble matter separates, and evaporate the filtrate to about 10 mL. Cool, and place in a refrigerator for 1 hour. Filter the separated niacin with suction, wash it with a few mL of cold alcohol, and dry at 105° for 1 hour: the niacin so obtained responds to *Identification tests A* and *B* under *Niacin.*

Disintegration ⟨701⟩: 30 minutes.

Uniformity of dosage units ⟨905⟩: meet the requirements.

Assay—Proceed with Niacin Tablets as directed for *Chemical Method* under *Niacin or Niacinamide Assay* ⟨441⟩, using *Standard Niacin Preparation* as the *Standard Preparation* in the *Assay Procedure*, and the following as the *Assay Preparation:* Weigh and finely powder not less than 20 Niacin Tablets. Weigh accurately a portion of the powder, equivalent to about 25 mg of niacin, and transfer with the aid of about 50 mL of water to a 250-mL volumetric flask. Heat, if necessary, until no more dissolves, cool, dilute with water to volume, and mix. Transfer 10.0 mL of this solution to a 200-mL volumetric flask, dilute with water to volume, and mix. Calculate the quantity, in mg, of $C_6H_5NO_2$ in the portion of Tablets taken by the formula:

$$25(A_U/A_S).$$

Niacinamide

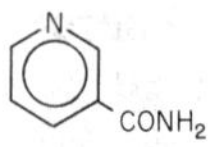

$C_6H_6N_2O$ 122.13
3-Pyridinecarboxamide.
Nicotinamide [*98-92-0*].

» Niacinamide contains not less than 98.5 percent and not more than 101.5 percent of $C_6H_6N_2O$, calculated on the dried basis.

Packaging and storage—Preserve in tight containers.

Reference standard—*USP Niacinamide Reference Standard*—Dry over silica gel for 4 hours before using.

Identification—

A: The infrared absorption spectrum of a potassium bromide dispersion of it exhibits maxima only at the same wavelengths as that of a similar preparation of USP Niacinamide RS.

B: The ultraviolet absorption spectrum of a solution (1 in 50,000) exhibits maxima and minima at the same wavelengths as that of a similar solution of USP Niacinamide RS, concomitantly measured, and the ratio A_{245}/A_{262} is between 0.63 and 0.67.

Melting range ⟨741⟩: between 128° and 131°.

Loss on drying ⟨731⟩—Dry it over silica gel for 4 hours: it loses not more than 0.5% of its weight.

Residue on ignition ⟨281⟩: not more than 0.1%.

Heavy metals, *Method II* ⟨231⟩: 0.003%.

Readily carbonizable substances ⟨271⟩—Dissolve 200 mg in 5 mL of sulfuric acid TS: the solution has no more color than *Matching Fluid A*.

Assay—

Mobile phase—Prepare a filtered and degassed solution containing 0.005 *M* Sodium 1-heptanesulfonate and methanol (70: 30).

Standard preparation—Transfer about 50 mg of USP Niacinamide RS, accurately weighed, to a 100-mL volumetric flask, dissolve in about 3 mL of water, dilute with *Mobile phase* to volume, and mix. Dilute 4.0 mL of the resulting solution with *Mobile phase* to 50.0 mL, and mix.

Assay preparation—Prepare as directed under *Standard preparation*, using Niacinamide instead of the Reference Standard.

Chromatographic system (see *Chromatography* ⟨621⟩)—The liquid chromatograph is equipped with a 254-nm detector and a 3.9-mm × 30-cm column containing packing L1. The flow rate is about 2 mL per minute. Chromatograph a solution prepared by combining equal volumes of *Standard preparation* and of a niacin solution prepared similarly at the same concentration. The resolution of the two peaks is not less than 3.0. Chromatograph replicate injections of the *Standard preparation*, and record the peak responses as directed under *Procedure:* the relative standard deviation is not more than 2.0%.

Procedure—Separately inject equal volumes (about 20 µL) of the *Standard preparation* and the *Assay preparation* into the chromatograph, record the chromatograms, and measure the responses for the major peaks. Calculate the quantity, in mg, of $C_6H_6N_2O$ in the portion of Niacinamide taken by the formula:

$$1250C(r_U/r_S),$$

in which *C* is the concentration, in mg per mL, of USP Niacinamide RS in the *Standard preparation*, and r_U and r_S are the peak responses for the *Assay preparation* and the *Standard preparation*, respectively.

Niacinamide Injection

» Niacinamide Injection is a sterile solution of Niacinamide in Water for Injection. It contains not less than 95.0 percent and not more than 110.0 percent of the labeled amount of $C_6H_6N_2O$.

Packaging and storage—Preserve in single-dose or in multiple-dose containers, preferably of Type I glass.

Reference standard—*USP Niacinamide Reference Standard*—Dry over silica gel for 4 hours before using.

Identification—Dilute a quantity of the Injection, equivalent to about 200 mg of niacinamide, with water to about 10 mL. Add 1 mL of 2.5 *N* sodium hydroxide, evaporate on a steam bath to dryness, add 5 mL of water, and similarly evaporate to about 1 mL: during the initial evaporation, the odor of ammonia is perceptible. Neutralize to litmus paper with 3 *N* hydrochloric acid, add 1 mL of the acid in excess, and place the solution in a refrigerator for 2 hours. Then filter, wash the precipitated niacin with small portions of ice-cold water until free from chloride, and dry at 105° for 1 hour: the infrared absorption spectrum of a potassium bromide dispersion of the residue so obtained exhibits maxima only at the same wavelengths as that of a similar preparation of USP Niacinamide RS.

pH ⟨791⟩: between 5.0 and 7.0.

Other requirements—It meets the requirements under *Injections* ⟨1⟩.

Assay—Proceed with Niacinamide Injection as directed for *Chemical Method* under *Niacin or Niacinamide Assay* ⟨441⟩, using *Standard Niacinamide Preparation* as the *Standard Preparation* in the *Assay Procedure*, and the following as the *Assay Preparation:* Dilute an accurately measured volume of Niacinamide Injection, equivalent to about 50 mg of niacinamide, with water to 500 mL in a volumetric flask, and mix. Pipet 10 mL of the solution into a 100-mL volumetric flask, dilute with water to volume, and mix. Calculate the quantity, in mg, of $C_6H_6N_2O$ in each mL of the Injection taken by the formula:

$$(50/V)(A_U/A_S),$$

in which V is the volume, in mL, of Injection taken.

Niacinamide Tablets

» Niacinamide Tablets contain not less than 90.0 percent and not more than 110.0 percent of the labeled amount of $C_6H_6N_2O$.

Packaging and storage—Preserve in tight containers.

Reference standard—*USP Niacinamide Reference Standard*—Dry over silica gel for 4 hours before using.

Identification—

A: Extract a quantity of powdered Tablets, equivalent to about 500 mg of niacinamide, with two 10-mL portions of alcohol, evaporate the filtered alcohol extracts on a steam bath, and dry at 80° for 2 hours: the infrared absorption spectrum of a potassium bromide dispersion of the residue so obtained exhibits maxima only at the same wavelengths as that of a similar preparation of USP Niacinamide RS.

B: From the niacinamide obtained in *Identification test A*, prepare a solution (1 in 50,000), and determine the absorbance of this solution in a 1-cm cell at 245 nm and at 262 nm, with a suitable spectrophotometer, using water as the blank: the ratio A_{245}/A_{262} is between 0.63 and 0.67.

Disintegration ⟨701⟩: 30 minutes.

Uniformity of dosage units ⟨905⟩: meet the requirements.

Assay—Proceed with Niacinamide Tablets as directed for *Chemical Method* under *Niacin or Niacinamide Assay* ⟨441⟩, using *Standard Niacinamide Preparation* as the *Standard Preparation* in the *Assay Procedure*, and the following as the *Assay Preparation:* Weigh and finely powder not less than 10 Niacinamide Tablets. Weigh accurately a quantity of the powder, equivalent to about 25 mg of niacinamide, and transfer with the aid of about 50 mL of water to a 250-mL volumetric flask. Heat, if necessary, until no more dissolves, cool, dilute with water to volume, and mix. Pipet 10 mL of the solution into a 100-mL volumetric flask, dilute with water to volume, and mix. Calculate the quantity, in mg, of $C_6H_6N_2O$ in the portion of Tablets taken by the formula:

$$25(A_U/A_S).$$

Nifedipine

$C_{17}H_{18}N_2O_6$ 346.34

3,5-Pyridinedicarboxylic acid, 1,4-dihydro-2,6-dimethyl-4-(2-nitrophenyl)-, dimethyl ester.

Dimethyl 1,4-dihydro-2,6-dimethyl-4-(*o*-nitrophenyl)-3,5-pyridinedicarboxylate [*21829-25-4*].

» Nifedipine contains not less than 98.0 percent and not more than 102.0 percent of $C_{17}H_{18}N_2O_6$, calculated on the dried basis.

Packaging and storage—Preserve in tight, light-resistant containers.

Reference standards—*USP Nifedipine Reference Standard*—Keep container tightly closed and protected from light. Do not dry before using. *USP Nifedipine Nitrophenylpyridine Analog Reference Standard. USP Nifedipine Nitrosophenylpyridine Analog Reference Standard.*

NOTE—Nifedipine, when exposed to daylight and certain wavelengths of artificial light, readily converts to a nitrosophenylpyridine derivative. Exposure to ultraviolet light leads to the formation of a nitrophenylpyridine derivative. Perform assays and tests in the dark or under golden fluorescent or other low-actinic light. Use low-actinic glassware.

Identification—

A: The infrared absorption spectrum of a potassium bromide dispersion of it exhibits maxima only at the same wavelengths as that of a similar preparation of USP Nifedipine RS.

B: To a 50-mL volumetric flask containing 70 mg of Nifedipine add 5.0 mL of chloroform, dilute with methanol to volume, and mix. Pipet a 1.0-mL aliquot of the solution into a 100-mL volumetric flask, dilute with methanol to volume, and mix to obtain the test solution. Using a suitable recording spectrophotometer, record the absorption spectrum of the test solution in a 1-cm cell from 450 nm to 220 nm, with methanol in the reference cell. The spectrum of the test solution exhibits maxima and minima at the same wavelengths as that of a similar preparation of USP Nifedipine RS.

C: The retention time of the major peak in the chromatogram of the *Assay preparation* corresponds to that of the *Standard preparation*, as obtained in the *Assay*.

Melting range, *Class Ia* ⟨741⟩: between 171° and 175°.

Loss on drying ⟨731⟩—Dry it at 105° to constant weight: it loses not more than 0.5% of its weight.

Residue on ignition ⟨281⟩: not more than 0.1%, an ignition temperature of 600° being used.

Heavy metals, *Method II* ⟨231⟩: 0.001%.

Perchloric acid titration—Transfer about 4 g, accurately weighed, to a 250-mL conical flask, and dissolve in 160 mL of glacial acetic acid with the aid of an ultrasonic bath. Add 3 drops of *p*-naphtholbenzein TS, and titrate to a green end-point with 0.1 *N* perchloric acid VS: not more than 0.12 mL of 0.1 *N* perchloric acid is consumed for each g of Nifedipine.

Chloride and Sulfate—To 5.0 g in a 140-mL beaker add 4.0 mL of 6 *N* acetic acid and 46 mL of water. Bring carefully to a boil on a hot plate, cool, and filter through paper free of chloride and sulfate. Use this Nifedipine filtrate for the following tests.

Chloride—Pipet 2.5 mL of the Nifedipine filtrate into a 50-mL color-comparison tube, and add 12.5 mL of water. Into a matched color-comparison tube pipet 10 mL of a Standard solution containing 8.2 µg of sodium chloride per mL corresponding to 5 µg of chloride per mL, add 5.0 mL of water, and mix. To each tube add 0.15 mL of 0.3 *M* nitric acid and 0.3 mL of silver nitrite TS, and mix. The opalescence exhibited by the Nifedipine filtrate does not exceed that of the Standard solution (0.02%).

Sulfate—Pipet into each of two 50-mL matched color-comparison tubes 1.5 mL of sulfate solution consisting of sufficient potassium sulfate dissolved in water to yield a sulfate concentration of 10 µg per mL. To each tube add, successively and with continuous shaking, 0.75 mL of alcohol, 0.5 mL of a 6.1% aqueous solution of barium chloride, and 0.25 mL of 6 *N* acetic acid. Shake for an additional 30 seconds. Pipet into one tube, designated the Standard tube, 15 mL of the sulfate solution. Pipet into the other tube, designated the Specimen tube, 3 mL of the Nifedipine filtrate and 12 mL of water: the turbidity exhibited by the Specimen tube does not exceed that of the Standard tube (0.05%).

Related compounds—[NOTE—Protect the *Standard nifedipine solution* and the *Test preparation* from actinic light. Conduct this test promptly after preparation of the *Standard nifedipine solution* and the *Test preparation.*]

Mobile phase—Prepare as directed in the *Assay*.

Standard nifedipine solution—Dissolve an accurately weighed quantity of USP Nifedipine RS in methanol (about 1 mg per mL), and dilute quantitatively with *Mobile phase* to obtain a solution having a known concentration of about 0.3 mg per mL.

Reference solution A—Dissolve an accurately weighed quantity of USP Nifedipine Nitrophenylpyridine Analog RS in methanol (about 1 mg per mL), and dilute quantitatively with *Mobile phase* to obtain a solution having a known concentration of about 0.6 µg per mL.

Reference solution B—Dissolve an accurately weighed quantity of USP Nifedipine Nitrosophenylpyridine Analog RS in methanol (about 1 mg per mL), and dilute quantitatively with *Mobile phase* to obtain a solution having a known concentration of about 0.6 µg per mL.

Standard preparation—Transfer 5.0 mL of each of the two *Reference solutions* to a container, add 5.0 mL of *Mobile phase*, and mix.

Test preparation—Prepare as directed for the *Assay preparation* in the *Assay*.

System suitability preparation—Mix equal volumes of the *Standard nifedipine solution* and of each of the two *Reference solutions*.

Chromatographic system—Prepare as directed in the *Assay*. Chromatograph the *System suitability preparation*, and record the peak responses as directed under *Procedure:* the resolution, *R*, between the nitrophenylpyridine analog and nitrosophenylpyridine analog peaks is not less than 1.5, the resolution, *R*, between the nitrosophenylpyridine analog and nifedipine peaks is not less than 1.0, and the relative standard deviation of the response for each analog in replicate injections is not more than 10%. The relative retention times are about 0.8 for the nitrophenylpyridine analog, about 0.9 for the nitrosophenylpyridine analog, and 1.0 for nifedipine.

Procedure—Separately inject equal volumes (about 25 µL) of the *Standard preparation* and the *Test preparation* into the chromatograph, record the chromatograms, and measure the responses for the major peaks. Calculate the quantity, in mg, of each related compound in the portion of Nifedipine taken by the formula:

$$250C(r_U/r_S),$$

in which *C* is the concentration, in mg per mL, of the appropriate USP Nifedipine Analog RS, in the *Standard preparation*, and r_U and r_S are the peak responses of the corresponding related compound obtained from the *Test preparation* and the *Standard preparation*, respectively. Not more than 0.2% of each of dimethyl 4-(2-nitrophenyl)-2,6-dimethylpyridine-3,5-dicarboxylate and dimethyl-4-(2-nitrosophenyl)-2,6-dimethylpyridine-3,5-dicarboxylate, corresponding to Nifedipine Nitrophenylpyridine Analog and Nifedipine Nitrosophenylpyridine Analog, respectively, is found.

Assay—[NOTE—Protect the *Standard preparation* and the *Assay preparation* from actinic light. Conduct the *Assay* promptly after preparation of the *Standard preparation* and the *Assay preparation.*]

Mobile phase—Prepare a suitable mixture of water, acetonitrile, and methanol (50:25:25), and degas. Make adjustments if necessary (see *System Suitability* under *Chromatography* ⟨621⟩).

Standard preparation—Dissolve an accurately weighed quantity of USP Nifedipine RS in methanol (about 1 mg per mL), and dilute quantitatively with *Mobile phase* to obtain a solution having a known concentration of about 0.1 mg per mL.

Assay preparation—Transfer about 25 mg of Nifedipine, accurately weighed, to a 250-mL volumetric flask. Dissolve in 25 mL of methanol, dilute with *Mobile phase* to volume, and mix to obtain a solution having a concentration of about 0.1 mg per mL.

Chromatographic system (see *Chromatography* ⟨621⟩)—The liquid chromatograph is equipped with a 235-nm detector and a 25-cm × 4.6-mm column that contains 5-µm packing L1. The flow rate is about 1.0 mL per minute. Chromatograph the *Standard preparation*, and record the peak responses as directed under *Procedure:* the column efficiency is not less than 16000 theoretical plates per meter, the tailing factor is not more than 1.5, and the relative standard deviation of the responses of the main peak is not more than 1.0%.

Procedure—Separately inject equal volumes (about 25 µL) of the *Standard preparation* and the *Assay preparation* into the chromatograph, record the chromatograms, and measure the responses for the major peaks. Calculate the quantity, in mg, of $C_{17}H_{18}N_2O_6$ in the portion of Nifedipine taken by the formula:

$$250C(r_U/r_S),$$

in which *C* is the concentration, in mg per mL, of USP Nifedipine RS in the *Standard preparation*, and r_U and r_S are the peak responses obtained from the *Assay preparation* and the *Standard preparation*, respectively.

Nifedipine Capsules

» Nifedipine Capsules contain not less than 90.0 percent and not more than 110.0 percent of the labeled amount of $C_{17}H_{18}N_2O_6$.

Packaging and storage—Preserve in tight, light-resistant containers at a temperature between 15° and 25°.

Reference standards—*USP Nifedipine Reference Standard*—Keep container tightly closed and protected from light. Do not dry before using. *USP Nifedipine Nitrophenylpyridine Analog Reference Standard. USP Nifedipine Nitrosophenylpyridine Analog Reference Standard.*

NOTE—Nifedipine, when exposed to daylight and certain wavelengths of artificial light, readily converts to a nitrosophenylpyridine derivative. Exposure to ultraviolet light leads to the formation of a nitrophenylpyridine derivative. Perform assays and tests in the dark or under golden fluorescent or other low-actinic light. Use low-actinic glassware.

Identification—

A: *Visualizing solution*—In a 100-mL volumetric flask dissolve 3 g of bismuth subnitrate and 30 g of potassium iodide with 10 mL of 3 *N* hydrochloric acid. Dilute with water to volume, and mix. Prior to use, transfer 10.0 mL of solution to a 100-mL volumetric flask, add 10 mL of 3 *N* hydrochloric acid, dilute with water to volume, and mix.

Standard solution—Prepare a *Standard solution* of USP Nifedipine RS in methylene chloride containing about 1.2 mg per mL.

Test solution—Using the technique described under *Procedure for content uniformity* in the test for *Uniformity of dosage units*, transfer the contents of 3 Nifedipine Capsules to a centrifuge tube, rinsing the scissors with about 20 mL of 0.1 *N* sodium hydroxide. Pipet 25 mL of methylene chloride into the tube, insert a stopper, invert several times, and carefully release the pressure in the tube. Insert the stopper again tightly, and shake gently for 1 hour. Centrifuge the tube for 10 minutes at 2000 to 2500 rpm. Remove the supernatant aqueous phase by aspiration with a syringe, and transfer 5.0 mL of the clarified lower layer to a suitable vial.

Procedure—Mix equal portions of the *Standard solution* and the *Test solution*. Apply separately 500 µL each of the *Standard solution*, the *Test solution*, and their mixture on a suitable thin-layer chromatographic plate (see *Chromatography* ⟨621⟩) coated with a 0.5-mm layer of chromatographic silica gel mixture. Allow the spots to dry, and develop the chromatogram, protected from light, in a solvent system consisting of a mixture of ethyl acetate and cyclohexane (1:1) until the solvent front has moved about three-fourths of the length of the plate. Remove the plate from the developing chamber, mark the solvent front, and air-dry the plate until no odor is detectable. Immediately view the plate under short-wavelength ultraviolet light: each solution exhibits a dark blue major band at the same R_f value of about 0.3. Spray the plate with *Visualizing solution:* each appears as a compact light orange band on a yellow background.

B: The retention time of the major peak in the chromatogram of the *Assay preparation* corresponds to that of the *Standard preparation*, as obtained in the *Assay*.

Dissolution ⟨711⟩—

Medium: simulated gastric fluid TS (without pepsin); 900 mL.

Apparatus 2: 50 rpm.

Time: 20 minutes.

Determine the amount of nifedipine dissolved using the following method.

Mobile phase and *Chromatographic system*—Prepare as directed in the *Assay*.

Procedure—Separately inject into the chromatograph equal volumes of a filtered portion of the solution under test and a *Standard solution* having a known concentration of USP Nifedipine RS in *Dissolution Medium*. Record the chromatograms and measure the responses for the major peaks. Calculate the amount of $C_{17}H_{18}N_2O_6$ dissolved.

Tolerances—Not less than 80% (Q) of the labeled amount of $C_{17}H_{18}N_2O_6$ is dissolved in 20 minutes.

Uniformity of dosage units ⟨905⟩: meet the requirements.

Procedure for content uniformity—With the point of a pair of sharp scissors, make a small hole at the end of 1 Capsule. Squeeze most of the contents into a 200-mL volumetric flask, cut the capsule in half, and drop it into the flask. Rinse the scissors with about 20 mL of methanol, collecting the rinse quantitatively in the flask. Dilute with methanol to volume, and mix to obtain the test solution. Dissolve an accurately weighed quantity of USP Nifedipine RS in methanol, and dilute quantitatively and stepwise with methanol to obtain a Standard solution having a known concentration of about 50 µg per mL. Concomitantly determine the absorbances of both solutions in 1-cm cells at the wavelength of maximum absorbance at about 350 nm, with a suitable spectrophotometer, using methanol as the blank. Calculate the quantity, in mg, of $C_{17}H_{18}N_2O_6$ in the Capsule by the formula:

$$(T/D)C(A_U/A_S),$$

in which T is the labeled quantity, in mg, of nifedipine in the Capsule, D is the concentration, in µg per mL, of nifedipine in the solution from the Capsule, on the basis of the labeled quantity per Capsule and the extent of dilution, C is the concentration, in µg per mL, of USP Nifedipine RS in the Standard solution, and A_U and A_S are the absorbances of the solution from the Capsule and the Standard solution, respectively.

Related compounds—[NOTE—Protect the *Standard nifedipine solution* and the *Test preparation* from actinic light. Conduct this test promptly after preparation of the *Standard nifedipine solution* and the *Test preparation.*]

Mobile phase—Prepare as directed in the *Assay* under *Nifedipine.*

Standard nifedipine solution—Prepare as directed for *Standard nifedipine solution* in the test for *Related compounds* under *Nifedipine.*

Reference solution A—Prepare as directed for *Reference solution A* in the test for *Related compounds* under *Nifedipine*, except to make the final known concentration of about 6 µg per mL.

Reference solution B—Prepare as directed for *Reference solution B* in the test for *Related compounds* under *Nifedipine*, except to make the final known concentration of about 1.5 µg per mL.

Standard preparation—Transfer 5.0 mL of each of the two *Reference solutions* to a container, add 5.0 mL of *Mobile phase*, and mix.

Test preparation—Prepare as directed for the *Assay preparation* in the *Assay*.

System suitability preparation—Mix equal volumes of the *Standard nifedipine solution* and of each of the two *Reference solutions*.

Chromatographic system—Prepare as directed in the *Assay*. Chromatograph the *System suitability preparation*, and record the peak responses as directed under *Procedure:* the resolution, R, between the nitrophenylpyridine analog and nitrosophenylpyridine analog peaks is not less than 1.5, the resolution, R, between the nitrosophenylpyridine analog and nifedipine peaks is not less than 1.0, and the relative standard deviation of the response for each analog in replicate injections is not more than 10%. The relative retention times are about 0.8 for the nitrophenylpyridine analog, about 0.9 for the nitrosophenylpyridine analog, and 1.0 for nifedipine.

Procedure—Separately inject equal volumes (about 25 µL) of the *Standard preparation* and the *Test preparation* into the chromatograph, record the chromatograms, and measure the responses for the major peaks. Calculate the quantity, in mg, of each related compound in the portion of Capsules taken by the formula:

$$(V/5)C(r_U/r_S),$$

in which V is the volume, in mL, of the *Test preparation*, C is the concentration, in mg per mL, of the appropriate USP Nifedipine Analog RS in the *Standard preparation*, and r_U and r_S are the peak responses of the corresponding related compound obtained from the *Test preparation* and the *Standard preparation*, respectively: not more than 2.0% of dimethyl 4-(2-nitrophenyl)-2,6-dimethylpyridine-3,5-dicarboxylate, corresponding to Nifedipine Nitrophenylpyridine Analog, and not more than 0.5% of dimethyl 4-(2-nitrosophenyl)-2,6-dimethylpyridine-3,5-dicarboxylate, corresponding to Nifedipine Nitrosophenylpyridine Analog, both relative to the nifedipine content, are found.

Assay—[NOTE—Protect the *Standard preparation* and the *Assay preparation* from actinic light. Conduct the *Assay* promptly after preparation of the *Standard preparation* and the *Assay preparation.*]

Mobile phase and *Standard preparation*—Prepare as directed in the *Assay* under *Nifedipine.*

Assay preparation—Transfer the contents of 5 Nifedipine Capsules with the aid of a small amount of methanol to a volumetric flask, dilute quantitatively with *Mobile phase* to obtain a total volume, V mL, of solution having a concentration of about 0.1 mg of nifedipine per mL. Filter through a solvent-resistant filter.

Chromatographic system (see *Chromatography* ⟨621⟩)—The liquid chromatograph is equipped with a 265-nm detector, a 25-cm × 4.6-mm column that contains 5-µm packing L1, and a guard column that contains packing L1. The flow rate is about 1.0 mL per minute. Chromatograph the *Standard preparation*, and record the peak responses as directed under *Procedure:* the column efficiency is not less than 16000 theoretical plates per meter, the tailing factor is not more than 1.5, and the relative standard deviation for replicate injections is not more than 1.0%.

Procedure—Separately inject equal volumes (about 25 µL) of the *Standard preparation* and the *Assay preparation* into the chromatograph, record the chromatograms, and measure the responses for the major peaks. Calculate the quantity, in mg, of $C_{17}H_{18}N_2O_6$ in the portion of Capsule contents taken by the formula:

$$(V/5)C(r_U/r_S),$$

in which V is the volume, in mL, of the *Assay preparation*, C is the concentration, in mg per mL, of USP Nifedipine RS in the *Standard preparation*, and r_U and r_S are the peak responses obtained from the *Assay preparation* and the *Standard preparation*, respectively.

Nitric Acid—*see* Nitric Acid NF

Nitrofurantoin

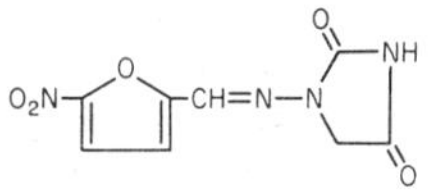

$C_8H_6N_4O_5$ (anhydrous) 238.16
2,4-Imidazolidinedione, 1-[[(5-nitro-2-furanyl)methylene]amino]-.
1-[(5-Nitrofurfurylidene)amino]hydantoin [*67-20-9*].
Monohydrate 256.17 [*17140-81-7*].

» Nitrofurantoin is anhydrous or contains one molecule of water of hydration. It contains not less than 98.0 percent and not more than 102.0 percent of $C_8H_6N_4O_5$, calculated on the anhydrous basis.

Caution—Nitrofurantoin and solutions of it are discolored by alkali and by exposure to light, and are decomposed upon contact with metals other than stainless steel and aluminum.

Packaging and storage—Preserve in tight, light-resistant containers.

Labeling—Label it to indicate whether it is anhydrous or hydrous.

Reference standards—*USP Nitrofurantoin Reference Standard*—Dry at 140° for 30 minutes before using. *USP Nitrofurazone Reference Standard*—Dry at 105° for 1 hour before using. *USP Nitrofurfural Diacetate Reference Standard*—Keep container tightly closed and protected from light. Dry at 105° for 1 hour before using.

Identification—

A: The infrared absorption spectrum of a mineral oil dispersion of it, previously dried at 105° for 1 hour, exhibits maxima only at the same wavelengths as that of a similar preparation of USP Nitrofurantoin RS.

B: The chromatogram of the *Assay preparation* obtained as directed in the *Assay* exhibits a major peak, the retention time of which corresponds to that exhibited in the chromatogram of the *Standard preparation.*

Water, *Method III* ⟨921⟩—Dry it at 140° for 30 minutes: the anhydrous form loses not more than 1.0%, and the hydrous form between 6.5% and 7.5%, of its weight.

Nitrofurfural diacetate—In a 10-mL volumetric flask dissolve 100 mg of Nitrofurantoin in 1 mL of dimethylformamide, add acetone to volume, and mix. On a suitable thin-layer chromatographic plate (see *Chromatography* ⟨621⟩), coated with a 0.25-mm layer of chromatographic silica gel mixture, apply 10 µL of this solution and 10 µL of a Standard solution of USP Nitrofurfural Diacetate RS in a 1 in 10 mixture of dimethylformamide in acetone containing 100 µg per mL. Allow the spots to dry, and develop the chromatogram in a solvent system consisting of a mixture of chloroform and methanol (9:1) until the solvent front has moved about three-fourths of the length of the plate. Remove the plate from the developing chamber, mark the solvent front, allow it to air-dry for 5 minutes, and heat the plate at 105° for 5 minutes. Remove the plate from the oven and, while it is still warm, locate the spots by spraying the plate with a solution prepared by dissolving 0.75 g of phenylhydrazine hydrochloride in 50 mL of water, decolorizing with activated charcoal, adding 25 mL of hydrochloric acid, and mixing with water to produce 200 mL: any spot produced by the test specimen, at an R_f value of about 0.7, is not greater in size or intensity than that produced by the Standard solution at the same R_f value (1.0% of nitrofurfural diacetate).

Nitrofurazone—

pH 7.0 phosphate buffer—Dissolve 6.8 g of monobasic potassium phosphate in about 500 mL of water. Add a volume of 1.0 *N* sodium hydroxide (about 30 mL) sufficient to adjust to a pH of 7.0, dilute with water to 1 liter, and mix.

Mobile phase—Prepare a filtered and degassed solution containing *pH 7.0 phosphate buffer* and tetrahydrofuran (9:1).

Standard preparation—Prepare a solution of USP Nitrofurazone RS in dimethylformamide containing 5.0 µg per mL. Pipet 2 mL of this solution into a glass-stoppered flask, add 20.0 mL of water, and mix.

Test preparation—Dissolve 100 mg of Nitrofurantoin in 2.0 mL of dimethylformamide in a glass-stoppered, 25-mL flask. Add 20.0 mL of water, mix, and allow to stand for about 15 minutes to allow precipitate to form. Filter a portion of the solution through a 0.45-µm pore size nylon filter, and use the clear filtrate.

Chromatographic system (see *Chromatography* ⟨621⟩)—The liquid chromatograph is equipped with a 375-nm detector and a 3.9-mm × 30-cm column containing packing L1. The flow rate is about 1.6 mL per minute. Chromatograph the *Standard preparation,* adjusting the operating parameters so that the nitrofurazone peak has a retention time of about 10.5 minutes and its height is about 0.1 full-scale. The relative standard deviation of the peak height in replicate injections is not more than 2.0%. Prepare a solution containing about 5.0 µg each of nitrofurazone and nitrofurantoin per mL in dimethylformamide. Dilute this solution 1:10 with *Mobile phase,* and inject 60 µL to 100 µL: the resolution, *R,* of the two peaks is not less than 4.0.

Procedure—Inject equal volumes (60 µL to 100 µL) of the *Standard preparation* and the *Test preparation* separately into the chromatograph, and record the chromatograms. The height of any peak appearing in the chromatogram of the *Test preparation* at a retention time corresponding to that of the main peak from the *Standard preparation* is not greater than the height of the main peak from the *Standard preparation* (0.01%).

Assay—

pH 7.0 phosphate buffer—Prepare as directed in the test for *Nitrofurazone.*

Mobile phase—Prepare a filtered and degassed solution containing *pH 7.0 phosphate buffer* and acetonitrile (88:12).

Internal standard solution—Prepare a solution containing about 1 mg of acetanilide per mL in water, and mix.

Standard preparation—Dissolve about 50 mg of USP Nitrofurantoin RS, accurately weighed, in 40.0 mL of dimethylformamide in a glass-stoppered flask, add 50.0 mL of *Internal standard solution,* and mix.

Assay preparation—Using about 50 mg of Nitrofurantoin, accurately weighed, proceed as directed for *Standard preparation.*

Chromatographic system (see *Chromatography* ⟨621⟩)—The liquid chromatograph is equipped with a 254-nm detector and a 3.9-mm × 30-cm column that contains packing L1. Chromatograph the *Standard preparation,* adjusting the operating parameters so that the retention time of the nitrofurantoin peak is about 8 minutes and the peak heights are about half full-scale. The relative standard deviation of the ratio of the peak responses in replicate injections is not more than 2.0%. The resolution, *R,* of the acetanilide and nitrofurantoin peaks is not less than 3.0.

Procedure—Inject equal volumes (5 µL to 10 µL) of the *Standard preparation* and the *Assay preparation* separately into the chromatograph, and record the chromatograms. Measure the responses for the major peaks. Calculate the quantity, in mg, of $C_8H_6N_4O_5$ in the portion of Nitrofurantoin taken by the formula:

$$W(R_U/R_S),$$

in which *W* is the weight, in mg, of USP Nitrofurantoin RS in the *Standard preparation,* and R_U and R_S are the peak response ratios obtained from the *Assay preparation* and the *Standard preparation,* respectively.

Nitrofurantoin Capsules

» Nitrofurantoin Capsules contain not less than 90.0 percent and not more than 110.0 percent of the labeled amount of $C_8H_6N_4O_5$.

Packaging and storage—Preserve in tight, light-resistant containers.

Reference standards—*USP Nitrofurantoin Reference Standard*—Dry at 140° for 30 minutes before using. *USP Nitrofurazone Reference Standard*—Dry at 105° for 1 hour before using.

Identification—

A: Add 10 mL of 6 *N* acetic acid to a quantity of the contents of Capsules, equivalent to about 100 mg of nitrofurantoin, boil for a few minutes, and filter while hot. Cool to room temperature, collect the precipitate of nitrofurantoin, and dry at 105° for 1 hour: the infrared absorption spectrum of a mineral oil dispersion of the precipitate so obtained exhibits maxima only at the same wavelength as that of a similar preparation of USP Nitrofurantoin RS.

B: It responds to *Identification test B* under *Nitrofurantoin.*

Uniformity of dosage units ⟨905⟩: meet the requirements.

Procedure for content uniformity—Proceed as directed in the *Assay,* using the following *Test preparation* instead of the *Assay preparation.*

Test preparation for content uniformity—Transfer the contents of 1 Capsule to a suitable flask, and add a volume of dimethylformamide to obtain a solution having a concentration of about 1.2 mg of nitrofurantoin per mL. Shake the flask for 15 minutes. Proceed as directed for *Assay preparation* in the *Assay,* beginning with "add 50.0 mL of *Internal standard solution.*"

Nitrofurazone—

pH 7.0 phosphate buffer, Mobile phase, Standard preparation, Chromatographic system, and *Procedure*—Proceed as directed in the test for *Nitrofurazone* under *Nitrofurantoin.*

Test preparation—Into a glass-stoppered, 25-mL flask weigh a portion of contents of Capsules equivalent to 100 mg of nitrofurantoin. Add 2.0 mL of dimethylformamide, and shake for 5 minutes. Add 20.0 mL of water, mix, and allow to stand for 15 minutes. Filter a portion of the mixture through a 0.45-μm pore size nylon filter.

Assay—

pH 7.0 phosphate buffer, Mobile phase, Internal standard solution, Standard preparation, and *Chromatographic system*—Proceed as directed in the *Assay* under *Nitrofurantoin.*

Assay preparation—Transfer, as completely as possible, the contents of 20 Nitrofurantoin Capsules to a 125-mL flask. Place the emptied capsules in a beaker, add 25 mL of dimethylformamide, and agitate for 1 minute. Decant into the flask containing the Capsule contents. Rinse the emptied capsules with another 20-mL portion of dimethylformamide, and decant into the flask. Insert the stopper in the flask, and shake by mechanical means for 15 minutes. Filter through a medium-porosity, sintered-glass filter into a 100-mL volumetric flask. Rinse the flask and the filter with several portions of dimethylformamide, and dilute with dimethylformamide to volume. Transfer an aliquot, equivalent to about 50 mg of nitrofurantoin, to a flask. Add an accurately measured volume of dimethylformamide to bring the volume in the flask to 40.0 mL. To the flask add 50.0 mL of *Internal standard solution*, mix, and cool to room temperature. Filter a portion of the solution through a 0.45-μm pore size nylon filter, discarding the first few mL of the filtrate.

Procedure—Proceed as directed for *Procedure* in the *Assay* under *Nitrofurantoin.* Calculate the quantity, in mg, of $C_8H_6N_4O_5$ in the portion of the powder included in the sample aliquot by the formula:

$$W(R_U / R_S),$$

in which W is the weight, in mg, of USP Nitrofurantoin RS in the *Standard preparation*, and R_U and R_S are the ratios of the peak responses of the nitrofurantoin to the internal standard obtained from the *Assay preparation* and the *Standard preparation*, respectively.

Nitrofurantoin Oral Suspension

» Nitrofurantoin Oral Suspension is a suspension of Nitrofurantoin in a suitable, aqueous vehicle. It contains, in each 100 mL, not less than 460 mg and not more than 540 mg of $C_8H_6N_4O_5$.

Packaging and storage—Preserve in tight, light-resistant containers.

Reference standards—*USP Nitrofurantoin Reference Standard*—Dry at 140° for 30 minutes before using. *USP N-(aminocarbonyl)-N-[([5-nitro-2-furanyl]methylene)amino]glycine Reference Standard*—Do not dry before using.

Identification—

A: To 10 mL of Oral Suspension add 15 mL of acetone, and warm to 50°, with stirring, to coagulate the excipients. Filter, evaporate the acetone with the aid of a warm air blast nearly to dryness, add 10 mL of acetic acid, heat to boiling, and filter while hot. Cool the filtrate to room temperature. Filter the precipitated nitrofurantoin, and dry at 105° for 1 hour: the infrared absorption spectrum of a mineral oil dispersion of the precipitate so obtained exhibits maxima only at the same wavelengths as that of a similar preparation of USP Nitrofurantoin RS.

B: It responds to *Identification test B* under *Nitrofurantoin.*

pH ⟨791⟩: between 4.5 and 6.5.

Limit of N-(aminocarbonyl)-N-[([5-nitro-2-furanyl]methylene)-amino]glycine (NF 250) and Assay—

pH 7.0 phosphate buffer and *Mobile phase*—Prepare as directed in the *Assay* under *Nitrofurantoin.*

Internal standard solution—Dissolve about 13 mg of acetanilide in *Mobile phase*, dilute with *Mobile phase* to 200 mL, and mix.

Standard NF 250 preparation—Prepare a solution of USP N-(aminocarbonyl)-N-[([5-nitro-2-furanyl]methylene)amino]glycine RS in *Mobile phase* to contain 125 μg per mL. Dilute 2.0 mL of this solution with *Mobile phase* to 100.0 mL, and mix.

Standard nitrofurantoin preparation—Transfer about 25 mg of USP Nitrofurantoin RS, accurately weighed, to a 100-mL volumetric flask with the aid of about 50 mL of dimethylformamide. Add 20 mL of water, cool to room temperature, and dilute with dimethylformamide to volume to obtain a *Standard solution.* Transfer a 4.0-mL aliquot of this *Standard solution* to a glass-stoppered flask, add 15.0 mL of *Internal standard solution*, and mix.

Assay preparation—Transfer an accurately measured volume of freshly mixed Nitrofurantoin Oral Suspension, equivalent to about 25 mg of nitrofurantoin, to a 100-mL volumetric flask, add 20 mL of water to the flask, and mix. Add about 50 mL of dimethylformamide, and shake the flask for about 20 minutes. Cool to room temperature, and dilute with dimethylformamide to volume. Centrifuge a portion of the solution, and transfer a 4.0-mL aliquot of the supernatant liquid to a glass-stoppered flask. Add 15.0 mL of *Internal standard solution*, and mix. Filter a portion of the solution through a 5-μm pore size polytef filter, discarding the first few mL of the filtrate.

Test preparation—Transfer an accurately measured volume of the freshly mixed Oral Suspension, equivalent to 5 mg of nitrofurantoin, to a 100-mL volumetric flask. Dilute with *Mobile phase* to volume, and mix. Centrifuge a portion of this solution. Filter a portion of the supernatant liquid through a 5-μm pore size polytef filter, discarding the first few mL of the filtrate.

Chromatographic system (see *Chromatography* ⟨621⟩)—The liquid chromatograph is equipped with both a 254-nm detector and a 375-nm detector and a 3.9-mm × 30-cm column that contains packing L1. For the *Assay*, chromatograph the *Standard nitrofurantoin preparation*, adjusting the operating parameters so that the retention time of the nitrofurantoin peak is about 8 minutes and its peak height is about half full-scale: the relative standard deviation of the ratio of the peak responses in replicate injections is not more than 2.0%, and the resolution, R, of the acetanilide and nitrofurantoin peaks is not less than 3.5. The flow rate is about 1.2 mL per minute. For the NF 250 test, adjust the operating parameters so that the NF 250 peak has a retention time of between 3 and 6 minutes and its height is about 0.1 full-scale. The flow rate is about 1.2 mL per minute.

Procedure for limit of N-(aminocarbonyl)-N-[([5-nitro-2-furanyl]methylene)amino]glycine—Inject separately equal volumes (30 μL to 60 μL) of *Standard NF 250 preparation* and the *Test preparation* into the chromatograph, and record the peak responses with the 375-nm detector: the height of any peak appearing in the chromatogram of the *Test preparation* at a retention time corresponding to that of the main peak in the *Standard NF 250 preparation* is not greater than the height of the latter (5.0%).

Procedure for assay—Inject equal volumes (about 15 μL) of *Standard nitrofurantoin preparation* and the *Assay preparation* separately into the chromatograph, and record the peak responses with the 254-nm detector. Calculate the quantity, in mg, of $C_8H_6N_4O_5$ in each mL of the Oral Suspension taken by the formula:

$$0.1(C/V)(R_U / R_S),$$

in which C is the concentration, in μg per mL, of USP Nitrofurantoin RS in the *Standard solution*, V is the volume, in mL, of Oral Suspension taken, and R_U and R_S are the ratios of the peak responses of the nitrofurantoin to the internal standard obtained from the *Assay preparation* and the *Standard nitrofurantoin preparation*, respectively.

Nitrofurantoin Tablets

» Nitrofurantoin Tablets contain not less than 90.0 percent and not more than 110.0 percent of the labeled amount of $C_8H_6N_4O_5$.

Packaging and storage—Preserve in tight, light-resistant containers.

Reference standards—*USP Nitrofurantoin Reference Standard*—Dry at 140° for 30 minutes before using. *USP Nitrofurazone Reference Standard*—Dry at 105° for 1 hour before using.

Identification—

A: Add 10 mL of 6 *N* acetic acid to a quantity of powdered Tablets, equivalent to about 100 mg of nitrofurantoin, boil for a few minutes, and filter while hot. Cool to room temperature, collect the precipitate of nitrofurantoin, and dry at 105° for 1 hour: the infrared absorption spectrum of a mineral oil dispersion of the precipitate so obtained exhibits maxima only at the same wavelengths as that of a similar preparation of USP Nitrofurantoin RS.

B: It responds to *Identification test B* under *Nitrofurantoin.*

Dissolution ⟨711⟩—

Medium: pH 7.2 phosphate buffer (see *Buffer Solutions* in the section, *Reagents, Indicators, and Solutions*); 900 mL.

Apparatus 1: 100 rpm.

Times: 60 minutes, 120 minutes.

Standard preparation—Dissolve about 50 mg, accurately weighed, of USP Nitrofurantoin RS in 25 mL of dimethylformamide, dilute with *Dissolution Medium* to 500 mL, mix, and dilute a suitable aliquot of the resulting solution with *Dissolution Medium* to obtain a solution having a known concentration of about 10 µg per mL.

Procedure—Determine the amount of $C_8H_6N_4O_5$ dissolved from absorbances at the wavelength of maximum absorbance at about 375 nm of filtered portions of the solution under test, suitably diluted with *Dissolution Medium*, if necessary, using *Dissolution Medium* as the blank, in comparison with the *Standard preparation.*

Tolerances—Not less than 25% (*Q*) of the labeled amount of $C_8H_6N_4O_5$ is dissolved in 60 minutes, and not less than 85% (*Q*) of the labeled amount of $C_8H_6N_4O_5$ is dissolved in 120 minutes.

Uniformity of dosage units ⟨905⟩: meet the requirements.

Nitrofurazone: meet the requirements of the test for *Nitrofurazone* under *Nitrofurantoin Capsules*, powdered Tablets being used in place of contents of Capsules.

Assay—

pH 7.0 phosphate buffer, Mobile phase, Internal standard solution, Standard preparation, and *Chromatographic system*—Proceed as directed in the *Assay* under *Nitrofurantoin.*

Assay preparation—Weigh and finely powder not less than 20 Nitrofurantoin Tablets. Weigh accurately a portion of the powder, equivalent to about 50 mg of nitrofurantoin, into a glass-stoppered flask. Add 40.0 mL of dimethylformamide, and shake by mechanical means for 15 minutes. Add 50.0 mL of *Internal standard solution*, mix, and cool to room temperature. Filter a portion of the solution through a 0.45-µm pore size nylon filter, discarding the first few mL of the filtrate.

Procedure—Proceed as directed for *Procedure* in the *Assay* under *Nitrofurantoin.* Calculate the quantity, in mg, of $C_8H_6N_4O_5$ in the portion of the powder taken by the formula:

$$W(R_U/R_S),$$

in which *W* is the weight, in mg, of USP Nitrofurantoin RS in the *Standard preparation* and R_U and R_S are the ratios of the peak responses of the nitrofurantoin to the internal standard obtained from the *Assay preparation* and the *Standard preparation*, respectively.

Nitrofurazone

NO₂ – furan – CH=NNHCONH₂

$C_6H_6N_4O_4$ 198.14

Hydrazinecarboxamide, 2-[(5-nitro-2-furanyl)methylene]-.

5-Nitro-2-furaldehyde semicarbazone [*59-87-0*].

» Nitrofurazone, dried at 105° for 1 hour, contains not less than 98.0 percent and not more than 102.0 percent of $C_6H_6N_4O_4$.

NOTE—Avoid exposing solutions of nitrofurazone at all times to direct sunlight, excessive heat, strong fluorescent lighting, and alkaline materials.

Packaging and storage—Preserve in tight, light-resistant containers, and avoid exposure to direct sunlight and to excessive heat.

Reference standards—*USP Nitrofurazone Reference Standard*—Dry at 105° for 1 hour before using. *USP 5-Nitro-2-furfuraldazine Reference Standard*—Keep container tightly closed and protected from light. Dry at 105° for 1 hour before using.

Identification—

A: The infrared absorption spectrum of a potassium bromide dispersion of it, previously dried, exhibits maxima only at the same wavelengths as that of a similar preparation of USP Nitrofurazone RS.

B: The ultraviolet absorption spectrum of a 1 in 125,000 solution of it, prepared as directed in the *Assay*, exhibits an absorbance maximum at 375 ± 2 nm and an absorbance minimum at 306 ± 2 nm. The ratio A_{306}/A_{375} does not exceed 0.25.

C: Dissolve 400 mg of potassium hydroxide in 10 mL of alcohol. Immediately before use dilute this solution with dimethylformamide to 100 mL. To 10 mL of the prepared solution add a few crystals of Nitrofurazone: a purple solution results.

pH ⟨791⟩—Suspend 1 g in 100 mL of water, shake for 15 minutes, allow the suspension to settle, and filter: the pH of the filtrate is between 5.0 and 7.5.

Loss on drying ⟨731⟩—Dry it at 105° for 1 hour: it loses not more than 0.5% of its weight.

Residue on ignition ⟨281⟩: not more than 0.1%.

Ordinary impurities ⟨466⟩—

Test solution: dimethylformamide.

Standard solution: dimethylformamide.

Application volume: 10 µL.

Eluant: a mixture of chloroform, methanol, and ammonium hydroxide (60:24:3), in a nonequilibrated chamber.

Visualization: 1.

Limit of 5-nitro-2-furfuraldazine—

Standard preparation—Transfer 50.0 mg of USP 5-Nitro-2-furfuraldazine RS to a 100-mL volumetric flask, dissolve in dimethylformamide, dilute with dimethylformamide to volume, and mix. Transfer 5.0 mL of this solution to a 25-mL volumetric flask, add 10 mL of dimethylformamide, dilute with acetone to volume, and mix.

Test preparation—Transfer 2.0 g to a 100-mL volumetric flask. Dissolve in 60 mL of dimethylformamide, dilute with acetone to volume, and mix.

Procedure—Apply 5 µL each of the *Standard preparation* and of the *Test preparation* to a suitable thin-layer chromatographic plate (see *Chromatography* ⟨621⟩) coated with a 0.5-mm layer of chromatographic silica gel. Develop the chromatogram in a solvent system consisting of a mixture of benzene and acetone (4:1) until the solvent front has moved about three-fourths of the length of the plate. Scan the spot produced by the *Standard preparation* and any spot from the *Test preparation*, having the same R_f as that produced by the *Standard preparation*, with a suitable densitometer equipped with a filter having its maximum transmittance at about 360 nm. The area and intensity of any spot from the *Test preparation* are not greater than the area and intensity produced by the *Standard* spot (0.5%).

Assay—Transfer about 100 mg of Nitrofurazone, previously dried and accurately weighed, to a 250-mL volumetric flask, dissolve in 50 mL of dimethylformamide, dilute with water to volume, and mix. Transfer 5.0 mL of this solution to a 250-mL volumetric flask, dilute with water to volume, and mix. Concomitantly determine the absorbances of this solution and a Standard solution of USP Nitrofurazone RS in the same medium having a known concentration of about 8 µg per mL, in 1-cm cells at the wavelength of maximum absorbance at about 375 nm, with a suitable spectrophotometer, using water as the blank. Calculate the quantity, in mg, of $C_6H_6N_4O_4$ in the Nitrofurazone taken by the formula:

$$12.5C(A_U/A_S),$$

in which *C* is the concentration, in µg per mL, of USP Nitrofurazone RS in the Standard solution, and A_U and A_S are the absorbances of the solution of Nitrofurazone and the Standard solution, respectively.

Nitrofurazone Cream

» Nitrofurazone Cream is Nitrofurazone in a suitable, emulsified, water-miscible base. It contains not less than 90.0 percent and not more than 110.0 percent of the labeled amount of $C_6H_6N_4O_4$.

NOTE—Avoid exposure at all times to direct sunlight, excessive heat, strong fluorescent lighting, and alkaline materials.

Packaging and storage—Preserve in tight, light-resistant containers. Avoid exposure to direct sunlight, strong fluorescent lighting, and excessive heat.

Reference standard—*USP Nitrofurazone Reference Standard*—Dry at 105° for 1 hour before using.

Identification—Dissolve 400 mg of potassium hydroxide in a mixture of 9.5 mL of alcohol and 0.5 mL of methanol. Immediately before use, dilute with dimethylformamide to 100 mL. To 10 mL of this solution add a quantity of Cream, equivalent to about 10 µg of nitrofurazone, and mix: a purple solution results.

Minimum fill ⟨755⟩: meets the requirements.

Assay—[NOTE—Protect all solutions from light.] Transfer an accurately weighed quantity of Nitrofurazone Cream, equivalent to about 4 mg of nitrofurazone, to a 100-mL beaker, slurry with 25 mL of chloroform, and transfer to a 100-mL volumetric flask with the aid of dimethylformamide. Dilute with dimethylformamide to volume, and mix. Transfer 15.0 mL of this solution to a second 100-mL volumetric flask, dilute with dimethylformamide to volume, and mix. Concomitantly determine the absorbances of this solution and a Standard solution of USP Nitrofurazone RS in the same medium having a known concentration of about 6 µg per mL, in 1-cm cells at the wavelength of maximum absorbance at about 385 nm, with a suitable spectrophotometer, using dimethylformamide containing 4 mL of chloroform per 100 mL as the blank. Calculate the quantity, in mg, of $C_6H_6N_4O_4$ in the portion of Cream taken by the formula:

$$0.667C(A_U/A_S),$$

in which C is the concentration, in µg per mL, of USP Nitrofurazone RS in the Standard solution, and A_U and A_S are the absorbances of the solution from the Cream and the Standard solution, respectively.

Nitrofurazone Ointment

» Nitrofurazone Ointment is Nitrofurazone in a suitable water-miscible base. It contains not less than 90.0 percent and not more than 110.0 percent of the labeled amount of $C_6H_6N_4O_4$.

NOTE—Avoid exposure at all times to direct sunlight, excessive heat, strong fluorescent lighting, and alkaline materials.

Packaging and storage—Preserve in tight, light-resistant containers. Avoid exposure to direct sunlight, strong fluorescent lighting, and excessive heat.

Reference standard—*USP Nitrofurazone Reference Standard*—Dry at 105° for 1 hour before using.

Completeness of solution—One g dissolves in 9 mL of water to form a clear solution.

Identification—It responds to the *Identification test* under *Nitrofurazone Cream.*

Assay—

Chromatographic column—Pack a pledget of fine glass wool at the base of a standard 20- × 220-mm chromatographic tube. Add chromatographic alumina (pH between 9.5 and 10) to the tube to a height of about 80 mm, and pack carefully to avoid air pockets or channels. Wash the column so prepared with 30 mL of dehydrated alcohol, and discard the washings. Do not allow the column to run dry.

[NOTE—Protect all solutions from light.]

Standard preparation—Transfer about 100 mg of USP Nitrofurazone RS, accurately weighed, to a 250-mL volumetric flask, dissolve in 50 mL of dimethylformamide, dilute with water to volume, and mix. Transfer 5.0 mL of this solution to a 250-mL volumetric flask, dilute with water to volume, and mix. The concentration of USP Nitrofurazone RS in the *Standard preparation* is about 8 µg per mL.

Assay preparation—Transfer an accurately weighed portion of Nitrofurazone Ointment, equivalent to about 4 mg of nitrofurazone, to a small beaker, and dissolve in about 30 mL of warm dehydrated alcohol (not exceeding 50°). With the aid of dehydrated alcohol, transfer the clear, yellow solution to a 100-mL volumetric flask, dilute with dehydrated alcohol to volume, and mix.

Procedure—Transfer 20.0 mL of the *Assay preparation* to the top of the column, then after almost all the solution has passed into the column, wash the column with 40 mL of dehydrated alcohol, maintaining the liquid level above the adsorbent column throughout the wash. Discard the washings. Pass 70 mL of water through the column to elute the nitrofurazone. Dilute the eluate with water to 100.0 mL, mix, centrifuge for 15 minutes, and filter through a medium-porosity, sintered-glass filter, discarding the first 20 mL of the filtrate. Concomitantly determine the absorbances of this solution and the *Standard preparation* in 1-cm cells at the wavelength of maximum absorbance at about 375 nm, with a suitable spectrophotometer, using water as the blank. Calculate the quantity, in mg, of $C_6H_6N_4O_4$ in the portion of Ointment taken by the formula:

$$0.5C(A_U/A_S),$$

in which C is the concentration, in µg per mL, of USP Nitrofurazone RS in the *Standard preparation*, and A_U and A_S are the absorbances of the solution from the *Assay preparation* and the *Standard preparation*, respectively.

Nitrofurazone Topical Solution

» Nitrofurazone Topical Solution contains not less than 95.0 percent and not more than 105.0 percent (w/w) of the labeled amount of $C_6H_6N_4O_4$.

NOTE—Avoid exposure at all times to direct sunlight, excessive heat, and alkaline materials.

Packaging and storage—Preserve in tight, light-resistant containers. Avoid exposure to direct sunlight and excessive heat.

Reference standard—*USP Nitrofurazone Reference Standard*—Dry at 105° for 1 hour before using.

Identification—Dissolve 400 mg of potassium hydroxide in a mixture of 9.5 mL of alcohol and 0.5 mL of methanol. Immediately before use dilute with dimethylformamide to 100 mL. To 10 mL of this solution add 1 drop of Topical Solution: a purple solution results.

Assay—

Chromatographic column and *Standard preparation*—Prepare as directed in the *Assay* under *Nitrofurazone Ointment.*

[NOTE—Protect all solutions from light.]

Assay preparation—Transfer an accurately weighed portion of Nitrofurazone Topical Solution, equivalent to about 4 mg of nitrofurazone, to a 100-mL volumetric flask, mix with about 30 mL of dehydrated alcohol, dilute with dehydrated alcohol to volume, and mix.

Procedure—Proceed as directed for *Procedure* in the *Assay* under *Nitrofurazone Ointment.*

Nitrogen—*see* Nitrogen NF

Diluted Nitroglycerin

$CH_2ONO_2 . CHONO_2 . CH_2ONO_2$

$C_3H_5N_3O_9$ 227.09
1,2,3-Propanetriol, trinitrate.
Nitroglycerin [*55-63-0*].

» Diluted Nitroglycerin is a mixture of nitroglycerin ($C_3H_5N_3O_9$) with lactose, dextrose, alcohol, propylene glycol, or other suitable inert excipient to permit safe handling. It contains not less than 90.0 percent and not more than 110.0 percent of the labeled amount of $C_3H_5N_3O_9$. It usually contains approximately 10 percent of nitroglycerin.

Caution—Exercise proper precautions in handling undiluted nitroglycerin, since it is a powerful explosive and can be exploded by percussion or excessive heat. Only exceedingly small amounts should be isolated.

Packaging and storage—Preserve in tight, light-resistant containers, and prevent exposure to excessive heat.

Reference standard—*USP Diluted Nitroglycerin Reference Standard.*

Identification—

A: The R_f value of the principal spot in the chromatogram of the *Identification test preparation* corresponds to that of the *Standard preparation*, as obtained in the test for *Chromatographic purity*.

B: The retention time of the major peak in the chromatogram of the *Assay preparation* corresponds to that obtained from the *Standard preparation* obtained as directed in the *Assay*.

Chromatographic purity—

Standard preparation—Dissolve an accurately weighed quantity of USP Diluted Nitroglycerin RS in methanol, and dilute quantitatively with methanol to obtain a solution having a concentration of 400 µg of nitroglycerin per mL.

Identification test preparation—Prepare a clear solution in methanol containing an amount of Diluted Nitroglycerin equivalent to about 400 µg of nitroglycerin per mL.

Test preparation—Transfer an accurately weighed portion, equivalent to 100 mg of nitroglycerin, to a 5-mL volumetric flask. Dissolve (or suspend) in methanol, dilute with methanol to volume, and mix. Centrifuge a portion, if necessary, to obtain a clear liquid phase.

Procedure—On a suitable thin-layer chromatographic plate (see *Chromatography* ⟨621⟩), coated with a 0.25-mm layer of chromatographic silica gel mixture, apply separately 20 µL of the *Test preparation*, 5 µL, 10 µL, 15 µL, and 20 µL of the *Standard preparation*, and 20 µL of the *Identification test preparation*. Allow the spots to dry, position the plate in a chromatographic chamber, and develop the chromatograms in a solvent system consisting of a mixture of toluene and ethyl acetate (4:1) until the solvent front has moved about three-fourths of the length of the plate. Remove the plate from the developing chamber, mark the solvent front, and allow the solvent to evaporate. Spray the plate with a 1 in 100 solution of diphenylamine in methanol, and irradiate the plate with short- and long-wavelength ultraviolet light for about 15 minutes, and examine the chromatogram: any spot obtained from the *Test preparation*, other than the principal spot, is not more intense than the spot in the chromatogram from the 20-µL application of the *Standard preparation*. Compare the intensities of any secondary spots observed in the chromatogram of the *Test preparation* with those of the principal spots in the chromatograms of the *Standard preparation* (corresponding to 0.5%, 1.0%, 1.5%, and 2.0%, respectively): the sum of the intensities of the secondary spots obtained in the *Test preparation* is not more than 3%. [NOTE—Nitrates of glycerin typically have R_f values of about 0.21, 0.37, and 0.61 for mono-, di-, and trisubstituted glycerins, respectively.]

Assay—

Mobile phase—Prepare a degassed solution containing equal volumes of methanol and water, making adjustments if necessary (see *System Suitability* under *Chromatography* ⟨621⟩).

System suitability preparation—Dissolve diluted pentaerythritol tetranitrate and USP Diluted Nitroglycerin RS in a mixture of methanol and water (65:35) to obtain a solution having concentrations of about 0.1 mg of pentaerythritol tetranitrate per mL and about 0.075 mg of nitroglycerin per mL. [NOTE—In case of incomplete solution due to insolubility of the diluent of the pentaerythritol tetranitrate, use the supernatant solution.]

Standard preparation—Dissolve an accurately weighed quantity of USP Diluted Nitroglycerin RS in *Mobile phase* to obtain a solution having a known concentration of about 0.075 mg of nitroglycerin per mL.

Assay preparation—Transfer an accurately weighed portion of Diluted Nitroglycerin, equivalent to about 7.5 mg of nitroglycerin, to a 100-mL volumetric flask, and dissolve in about 75 mL of *Mobile phase*. If necessary, sonicate for 2 minutes or until the solid is totally dispersed, then shake by mechanical means for 30 minutes. Dilute with *Mobile phase* to volume, and mix. Filter through a 0.7-µm filter paper, if necessary.

Chromatographic system (see *Chromatography* ⟨621⟩)—The liquid chromatograph is equipped with a 220-nm detector and a 3.9-mm × 30-cm column that contains packing L1, and a short precolumn that contains packing L1 is used if necessary. The flow rate is about 1 mL per minute. Chromatograph replicate injections of the *System suitability preparation*, and record the peak responses as directed under *Procedure:* the relative standard deviation is not more than 3.0%, the relative retention of nitroglycerin to pentaerythritol tetranitrate is about 0.7, and the resolution between the nitroglycerin and pentaerythritol tetranitrate peaks is not less than 2.0.

Procedure—Separately inject equal volumes (about 20 µL) of the *Standard preparation* and the *Assay preparation* into the chromatograph, by means of a microsyringe or sampling valve, record the chromatograms, and measure the responses for the major peaks. Calculate the quantity, in mg, of $C_3H_6N_3O_9$ in the portion of Diluted Nitroglycerin taken by the formula:

$$100C(r_U / r_S),$$

in which C is the concentration, in mg per mL, of nitroglycerin in the *Standard preparation*, and r_U and r_S are the peak responses of nitroglycerin obtained from the *Assay preparation* and the *Standard preparation*, respectively.

Nitroglycerin Injection

» Nitroglycerin Injection is a sterile solution prepared from Diluted Nitroglycerin; the solvent may contain Alcohol, Propylene Glycol, and Water for Injection. Nitroglycerin Injection contains not less than 90.0 percent and not more than 110.0 percent of the labeled amount of $C_3H_5N_3O_9$.

Packaging and storage—Preserve in single-dose or in multiple-dose containers, preferably of Type I or Type II glass.

Labeling—Where necessary, label it to indicate that it is to be diluted before use.

Reference standard—*USP Diluted Nitroglycerin Reference Standard.*

Identification—The retention time of the major peak in the chromatogram of the *Assay preparation* corresponds to that of the *Standard preparation*, obtained as directed in the *Assay*.

Pyrogen—It meets the requirements of the *Pyrogen Test* ⟨151⟩, the test dose being 50 µg of nitroglycerin per kg.

pH ⟨791⟩: between 3.0 and 6.5, determined potentiometrically in a solution prepared by adding 5 mL of water and 1 drop of saturated potassium chloride solution to 5 mL of the Injection.

Alcohol content, *Method II* ⟨611⟩: between 90.0% and 110.0% of the labeled amount of C_2H_5OH, isopropyl alcohol being used as the internal standard.

Particulate matter ⟨788⟩: meets the requirements under *Small-volume Injections.*

Other requirements—It meets the requirements under *Injections* ⟨1⟩.

Assay—

Mobile phase, System suitability preparation, Standard preparation, and *Chromatographic system*—Prepare as directed in the *Assay* under *Diluted Nitroglycerin.*

Assay preparation—Transfer an accurately measured volume of Nitroglycerin Injection, equivalent to about 7.5 mg of nitroglycerin, to a 100-mL volumetric flask, dissolve in *Mobile phase,* dilute with *Mobile phase* to volume, and mix.

Procedure—Separately inject equal volumes (about 20 µL) of the *Standard preparation* and the *Assay preparation* into the chromatograph by means of a microsyringe or sampling valve, record the chromatograms, and measure the responses for the major peaks. Calculate the quantity, in mg, of $C_3H_5N_3O_9$ in the portion of Nitroglycerin Injection taken by the formula:

$$100C(r_U/r_S),$$

in which C is the concentration, in mg per mL, of nitroglycerin in the *Standard preparation,* and r_U and r_S are the peak responses of nitroglycerin obtained from the *Assay preparation* and the *Standard preparation,* respectively.

Nitroglycerin Ointment

» Nitroglycerin Ointment is Diluted Nitroglycerin in a suitable ointment base. It contains not less than 90.0 percent and not more than 115.0 percent of the labeled amount of $C_3H_5N_3O_9$.

Packaging and storage—Preserve in tight containers.

Labeling—Label multiple-dose containers with a direction to close tightly, immediately after each use.

Reference standard—*USP Diluted Nitroglycerin Reference Standard.*

Identification—The retention time of the major peak obtained from the *Assay preparation* corresponds to that from the *Standard preparation,* both relative to the internal standard, as obtained in the *Assay.*

Minimum fill ⟨755⟩: meets the requirements.

Homogeneity—In the case of single-dose containers, perform the *Assay* on specimens from each of 10 containers. In the case of multiple-dose containers, perform the *Assay* on one specimen from the top and one from the bottom of each of 5 containers. Each specimen contains not less than 90.0% and not more than 110.0% of the mean value.

Assay—

Mobile phase—Prepare a degassed solution containing equal volumes of methanol and water. The composition of this solution may be varied to meet system suitability requirements.

Internal standard solution—Dissolve diluted pentaerythritol tetranitrate in methanol to obtain a solution having a concentration of about 0.1 mg of pentaerythritol tetranitrate per mL.

Standard preparation—Dissolve an accurately weighed quantity of USP Diluted Nitroglycerin RS in *Internal standard solution* to obtain a solution having a known concentration of about 0.075 mg of nitroglycerin per mL. Filter through 0.7-µm filter paper.

Assay preparation—Transfer an accurately weighed portion of Nitroglycerin Ointment, equivalent to about 2.0 mg of nitroglycerin, to a glass-stoppered, 50-mL conical flask, and add 25.0 mL of *Internal standard solution.* Immerse for 10 minutes in a water bath maintained at a temperature of 50°, shaking intermittently until the specimen becomes dispersed. Remove from the bath, and shake vigorously for 1 minute to obtain a coagulated solid. Repeat the heating and shaking steps one more time. Filter through 0.7-µm filter paper.

Chromatographic system (see *Chromatography* ⟨621⟩)—The liquid chromatograph is equipped with a 220-nm detector and a 3.9-mm × 30-cm column that contains packing L1, and a short precolumn that contains packing L1 is used if necessary. The flow rate is about 1 mL per minute. Chromatograph three replicate injections of the *Standard preparation,* and record the peak responses as directed under *Procedure:* the relative standard deviation is not more than 3.0%, the relative retention of nitroglycerin to internal standard is about 0.7, the resolution between the nitroglycerin and internal standard peaks is not less than 2.0, and the tailing factors for the nitroglycerin and internal standard peaks are not more than 2.0.

Procedure—Separately inject equal volumes of the *Standard preparation* and the *Assay preparation* into the chromatograph, by means of a microsyringe or sampling valve, record the chromatograms, and measure the responses for the major peaks. Calculate the quantity, in mg, of $C_3H_5N_3O_9$ in the portion of Ointment taken by the formula:

$$25C(R_U/R_S),$$

in which C is the concentration, in mg per mL, of nitroglycerin in the *Standard preparation,* and R_U and R_S are the ratios of the peak responses of nitroglycerin to the internal standard obtained from the *Assay preparation* and the *Standard preparation,* respectively.

Nitroglycerin Tablets

» Nitroglycerin Tablets contain not less than 90.0 percent and not more than 110.0 percent of the labeled amount of $C_3H_5N_3O_9$.

Packaging and storage—Preserve in tight containers, preferably of glass, at controlled room temperature. Each container holds not more than 100 Tablets.

Labeling—The labeling indicates that the Tablets are for sublingual use, and the label directs that the Tablets be dispensed in the original, unopened container, labeled with the following statement directed to the patient. "Warning: to prevent loss of potency, keep these tablets in the original container or in a supplemental Nitroglycerin container specifically labeled as being suitable for Nitroglycerin Tablets. Close tightly immediately after each use."

Reference standard—*USP Diluted Nitroglycerin Reference Standard.*

Identification—

A: *Standard preparation*—Mix a portion of USP Diluted Nitroglycerin RS with acetone to obtain a solution having a concentration of about 1 mg per mL.

Test preparation—Transfer an amount of finely powdered Tablets, equivalent to about 1 mg of nitroglycerin, to a glass-stoppered vessel, add 1 mL of acetone, shake by mechanical means for 30 minutes, and filter.

Procedure—On a thin-layer chromatographic plate (see *Chromatography* ⟨621⟩), coated with a 0.25-mm layer of chromatographic silica gel, apply separately 10 µL portions of the *Standard preparation* and the *Test preparation.* Develop the chromatogram in a solvent system consisting of a mixture of toluene, ethyl acetate, and glacial acetic acid (16:4:1) until the solvent front has moved about three-fourths of the length of the plate. Remove the plate from the developing chamber, allow the solvent to evaporate, spray with a 1 in 100 solution of diphenylamine in methanol, and irradiate the plate with short- and long-wavelength ultraviolet light for about 10 minutes: the R_f value of the principal spot obtained from the *Test preparation* corresponds to that obtained from the *Standard preparation.*

B: The retention time of the major peak obtained from the *Assay preparation* corresponds to that from the *Standard preparation* obtained in the *Assay.*

Disintegration ⟨701⟩: 2 minutes, determined as set forth for *Sublingual Tablets.*

Uniformity of dosage units ⟨905⟩: meet the requirements.

Assay—

Mobile phase, System suitability preparation, Standard preparation, and *Chromatographic system*—Prepare as directed in the *Assay* under *Diluted Nitroglycerin.*

Assay preparation—Weigh and finely powder not less than 20 Nitroglycerin Tablets. Transfer an accurately weighed portion of the powder, equivalent to about 7.5 mg of nitroglycerin, to a 100-mL volumetric flask, add about 80 mL of methanol, sonicate for 2 minutes or until the powder is completely dispersed, then shake by mechanical means for 30 minutes. Dilute with methanol to volume, mix, and filter through a 0.7-μm filter paper.

Procedure—Separately inject equal volumes (about 20 μL) of the *Standard preparation* and the *Assay preparation* into the chromatograph, record the chromatograms, and measure the responses for the major peaks. Calculate the quantity, in mg, of $C_3H_5N_3O_9$ in the portion of Tablets taken by the formula:

$$100C(r_U / r_S),$$

in which C is the concentration, in mg per mL, of nitroglycerin in the *Standard preparation*, and r_U and r_S are the peak responses of nitroglycerin obtained from the *Assay preparation* and the *Standard preparation*, respectively.

Nitromersol

```
        NO2
         |
Hg—C=C—CH
 |  |  ||
O—C=C—CH
       |
      CH3
```

$C_7H_5HgNO_3$ 351.71

7-Oxa-8-mercurabicyclo[4.2.0]octa-1,3,5-triene, 5-methyl-2-nitro-.

5-Methyl-2-nitro-7-oxa-8-mercurabicyclo[4.2.0]octa-1,3,5-triene [*133-58-4*].

» Nitromersol, dried at 105° for 2 hours, contains not less than 98.0 percent and not more than 100.5 percent of $C_7H_5HgNO_3$.

Packaging and storage—Preserve in tight, light-resistant containers.

Identification—

A: A solution (1 in 1000) in 1 *N* sodium hydroxide possesses a reddish orange color. The addition of 3 *N* hydrochloric acid to this solution causes the color to disappear and a yellowish, flocculent precipitate to form.

B: To a solution prepared by dissolving 250 mg of Nitromersol in 2.5 mL of 1 *N* sodium hydroxide and diluting with water to 20 mL add about 3 mL of 3 *N* hydrochloric acid: a yellowish precipitate is formed. Upon filtration, the filtrate is nearly colorless or slightly yellow. Retain the filtrate for the test for *Mercury ions.* Dissolve the precipitate in 20 mL of water to which 2.5 mL of 1 *N* sodium hydroxide has been added, add 0.5 g of sodium hydrosulfite, and heat to boiling: a heavy deposit of metallic mercury is formed.

Loss on drying ⟨731⟩—Dry it at 105° for 2 hours: it loses not more than 1.0% of its weight.

Residue on ignition ⟨281⟩: not more than 0.1%.

Mercury ions—To the filtrate obtained in *Identification test B* add an equal volume of hydrogen sulfide TS: no darkening in color is produced, although a small amount of a flocculent, light yellow precipitate may form.

Alkali-insoluble substances—Add 7 mL of 1 *N* sodium hydroxide to 1.0 g of Nitromersol, then dilute with water to 20 mL. The resulting solution, upon standing in a glass-stoppered vessel in the dark for 24 hours, shows no more than a slight amount of insoluble material. Collect the insoluble residue, if any, in a tared filter crucible, wash the residue with warm water, and dry at 105° for 1 hour: the weight of the insoluble material does not exceed 1 mg (0.1%).

Uncombined nitrocresol—Shake 500 mg of Nitromersol with 50 mL of benzene, filter, evaporate the filtrate in a tared dish to dryness, and dry the residue at 80° for 2 hours: the weight of the residue does not exceed 5 mg (1%).

Assay—Weigh accurately about 200 mg of Nitromersol, previously ground to a fine powder and dried, and transfer to a 500-mL Kjeldahl flask. Add 15 mL of sulfuric acid, digest cautiously with occasional swirling over a flame until the solution becomes a clear, light yellowish brown, cool, and add, dropwise, enough 30 percent hydrogen peroxide to decolorize the solution. Digest for 2 to 3 minutes, adding more hydrogen peroxide, if necessary, to produce a colorless solution. Cool, dilute with water to about 100 mL, and add potassium permanganate TS until a permanent pink color persists on heating. Then add hydrogen peroxide TS, dropwise, until the color is completely discharged. Cool, and add 5 mL of nitric acid which has been diluted with 10 mL of water. Add 5 mL of ferric ammonium sulfate TS, and titrate with 0.1 *N* ammonium thiocyanate VS. Each mL of 0.1 *N* ammonium thiocyanate is equivalent to 17.59 mg of $C_7H_5HgNO_3$.

Nitromersol Topical Solution

» Nitromersol Topical Solution yields, from each 100 mL, not less than 180.0 mg and not more than 220.0 mg of $C_7H_5HgNO_3$.

Nitromersol	2	g
Sodium Hydroxide	0.4	g
Sodium Carbonate, monohydrate	4.25	g
Purified Water, a sufficient quantity, to make	1000	mL

Dissolve the Sodium Hydroxide and the monohydrated Sodium Carbonate in 50 mL of Purified Water, add the Nitromersol, and stir until dissolved. Gradually add Purified Water to make 1000 mL.

NOTE—Prepare dilutions of Nitromersol Topical Solution as needed, since they tend to precipitate upon standing.

Packaging and storage—Preserve in tight, light-resistant containers.

Identification—

A: To 100 mL add 3 mL of 3 *N* hydrochloric acid: a yellowish precipitate is formed. Filter, and retain both the filtrate and the precipitate.

B: Add the precipitate from *Identification test A* to 20 mL of water and 2.5 mL of 1 *N* sodium hydroxide. Add 500 mg of sodium hydrosulfite, and heat to boiling: a heavy deposit of metallic mercury is formed.

Specific gravity ⟨841⟩: between 1.005 and 1.010.

Mercury ions—To the filtrate obtained in *Identification test A* add an equal volume of hydrogen sulfide TS: no darkening in color is produced, although a small amount of a flocculent, light yellow precipitate may be formed.

Assay—Transfer 50.0 mL of Nitromersol Topical Solution to a 500-mL Kjeldahl flask, add a few glass beads, and evaporate to about 5 mL. Proceed as directed in the *Assay* under *Nitromersol,* beginning with "Add 15 mL of sulfuric acid."

Nitrous Oxide

N_2O 44.01

Nitrogen oxide (N_2O).

Nitrogen oxide (N_2O) [*10024-97-2*].

» Nitrous Oxide contains not less than 99.0 percent, by volume, of N_2O.

Packaging and storage—Preserve in cylinders.

Note—The following tests are designed to reflect the quality of Nitrous Oxide in both its vapor and liquid phases, which are present in previously unopened cylinders. Reduce the container pressure by means of a regulator. Withdraw the samples for the tests with the least possible release of Nitrous Oxide consistent with proper purging of the sampling apparatus. Measure the gases with a gas volume meter downstream from the detector tubes in order to minimize contamination or change of the specimens. Perform tests in the sequence in which they are listed.

NOTE—The various detector tubes called for in the respective tests are listed under *Reagents* in the section, *Reagents, Indicators, and Solutions.*

Identification—

A: With the container temperatures the same and maintained between 15° and 25°, concomitantly read the pressure of the Nitrous Oxide container and of a container of nitrous oxide certified standard (see under *Reagents* in the section, *Reagents, Indicators, and Solutions*). [NOTE—Do not use the nitrous oxide certified standard if it has been depleted to less than half of its full capacity.] The pressure of the Nitrous Oxide container is within 50 psi of that of the nitrous oxide certified standard.

B: Pass 100 ± 5 mL released from the vapor phase of the contents of the Nitrous Oxide container through a carbon dioxide detector tube at the rate specified for the tube: no color change is observed (*distinction from carbon dioxide*).

Carbon monoxide—Pass 1050 ± 50 mL, released from the vapor phase of the contents of the container, through a carbon monoxide detector tube at the rate specified for the tube: the indicator change corresponds to not more than 0.001%.

Nitric oxide—Pass 550 ± 50 mL, released from the vapor phase of the contents of the container, through a nitric oxide–nitrogen dioxide detector tube at the rate specified for the tube: the indicator change corresponds to not more than 1 ppm.

Nitrogen dioxide—Arrange a container so that when its valve is opened, a portion of the liquid phase of the contents is released through a piece of tubing of sufficient length to allow all of the liquid to vaporize during passage through it, and to prevent frost from reaching the inlet of the detector tube. Release into the tubing a flow of liquid sufficient to provide 550 mL of the vaporized sample plus any excess necessary to assure adequate flushing of air from the system. Pass 550 ± 50 mL of this gas through a nitric oxide–nitrogen dioxide detector tube at the rate specified for the tube: the indicator change corresponds to not more than 1 ppm.

Halogens—Pass 1050 ± 50 mL, released from the vapor phase of the contents of the container, through a bromine detector tube at the rate specified for the tube: the indicator change corresponds to not more than 1 ppm.

Carbon dioxide—Pass 1050 ± 50 mL, released from the vapor phase of the contents of the container, through a carbon dioxide detector tube at the rate specified for the tube: the indicator change corresponds to not more than 0.03%.

Ammonia—Proceed with Nitrous Oxide as directed in the test for *Carbon monoxide*, except to use an ammonia detector tube: the indicator change corresponds to not more than 0.0025%.

Water—It meets the requirements of the test for *Water* under *Carbon Dioxide*.

Air—Not more than 1.0% of air is present, determined as directed in the *Assay*.

Assay—Introduce a specimen of Nitrous Oxide taken from the liquid phase, as directed in the test for *Nitrogen dioxide*, into a gas chromatograph by means of a gas-sampling valve. Select the operating conditions of the gas chromatograph such that the peak response resulting from the following procedure corresponds to not less than 70% of the full-scale reading. Preferably, use an apparatus corresponding to the general type in which the column is 6 m in length and 4 mm in inside diameter and is packed with porous polymer beads, which permits complete separation of N_2 and O_2 from N_2O, although the N_2 and O_2 may not be separated from each other. Use industrial grade helium (99.99%) as the carrier gas, with a thermal-conductivity detector, and control the column temperature: the peak response produced by the assay specimen exhibits a retention time corresponding to that produced by an air-helium certified standard (see under *Reagents* in the section, *Reagents, Indicators, and Solutions*), and is equivalent to not more than 1.0% of air when compared to the peak response of the air-helium certified standard, indicating not less than 99.0%, by volume, of N_2O.

Nonabsorbable Surgical Suture—*see* Suture, Nonabsorbable Surgical

Nonoxynol 9

$$C_9H_{19}-C_6H_4-(OCH_2CH_2)_nOH$$

(n varies from 6 to 16)

α-(*p*-Nonylphenyl)-ω-hydroxynona(oxyethylene) [*26027-38-3*].

» Nonoxynol 9 is an anhydrous liquid mixture consisting chiefly of monononylphenyl ethers of polyethylene glycols corresponding to the formula:

$$C_9H_{19}C_6H_4(OCH_2CH_2)_nOH,$$

in which the average value of n is about 9. It contains not less than 90.0 percent and not more than 110.0 percent of nonoxynol 9.

Packaging and storage—Preserve in tight containers.

Reference standard—*USP Nonoxynol 9 Reference Standard.*

Identification—Its infrared absorption spectrum, obtained by spreading a capillary film of it between sodium chloride plates, exhibits maxima only at the same wavelengths as that of a similar preparation of USP Nonoxynol 9 RS.

Acid value ⟨401⟩: not more than 0.2.

Water, *Method I* ⟨921⟩: not more than 0.5%.

Polyethylene glycol—Transfer about 10 g, accurately weighed, to a 250-mL beaker. Add 100 mL of ethyl acetate, and stir on a magnetic stirrer to effect solution. Transfer, with the aid of 100 mL of 5 *N* sodium chloride, to a pear-shaped, 500-mL separator fitted with a glass stopper. Insert the stopper, and shake vigorously for 1 minute. Remove the stopper carefully to release the pressure. Immerse a thermometer in the mixture, and support the separator so that it is partially immersed in a water bath maintained at 50°. Swirl the separator gently while letting the internal temperature rise to between 40° and 45°, then immediately remove the separator from the bath, dry the outside surface, and drain the salt (lower) layer into another pear-shaped, 500-mL separator. In the same manner, extract the ethyl acetate layer a second time with 100 mL of fresh 5 *N* sodium chloride, combining the two aqueous extracts. Discard the ethyl acetate layer. Wash the combined aqueous layers with 100 mL of ethyl acetate, using the same technique, and drain the salt (lower) layer into a clean pear-shaped, 500-mL separator. Discard the ethyl acetate layer. Extract the aqueous layer with two successive 100-mL portions of chloroform, draining the chloroform (lower) layers through Whatman folded filter paper 2V, and combining them into a 250-mL beaker. Evaporate on a steam bath to dryness, and continue heating until the odor of chloroform is no longer perceptible. Allow the beaker to cool. Add 25 mL of acetone, and dissolve the residue on a magnetic stirrer. Filter through Whatman folded filter paper 2V into a tared 250-mL beaker, rinsing with two 25-mL portions of acetone. Evaporate on a steam bath to dryness. Dry in vacuum at 60° for 1 hour. Allow the beaker to cool, and weigh: not more than 1.6% of polyethylene glycol is found.

Cloud point—Transfer 1.0 g to a 250-mL beaker, add 99 g of water, and mix to dissolve. Pour about 30 mL of the solution into a 70-mL test tube. Support the test tube in a hot water bath, and stir the contents with a thermometer constantly until the solution becomes cloudy, then remove the test tube from the bath immediately, so that the temperature rises not more than 2°

further, and continue stirring. The cloud point is the temperature at which the solution returns to its original clarity: it is between 52° and 56°.

Free ethylene oxide—

Stripped nonoxynol 9—Maintain Nonoxynol 9 at a temperature of 150° with constant stirring in an open vessel until it no longer displays a peak for ethylene oxide when chromatographed as directed below.

Standard solutions—[NOTE—Ethylene oxide is toxic and flammable. Prepare these solutions in a well-ventilated hood, using great care.] Chill all apparatus and reagents used in the preparation of standards in a refrigerator or freezer before use. Fill a chilled pressure bottle with liquid ethylene oxide from a lecture bottle, and store in a freezer when not in use. Use a small piece of polyethylene film to protect the liquid from contact with the rubber gasket. Transfer about 100 mL of chilled isopropyl alcohol to a 500-mL volumetric flask. Using a chilled graduated cylinder, transfer 25 mL of ethylene oxide to the isopropyl alcohol, and swirl gently to mix. Dilute with additional chilled isopropyl alcohol to volume, replace the stopper, and swirl gently to mix. This stock solution contains about 43.6 mg of ethylene oxide per mL. Pipet 25 mL of 0.5 *N* alcoholic hydrochloric acid, prepared by mixing 45 mL of hydrochloric acid with 1 liter of alcohol, into a 500-mL conical flask containing 40 g of magnesium chloride hexahydrate. Shake the mixture to effect saturation. Pipet 10 mL of the ethylene oxide solution into the flask, and add 20 drops of bromocresol green TS. If the solution is not yellow (acid), add an additional volume, accurately measured, of 0.5 *N* alcoholic hydrochloric acid to give an excess of about 10 mL. Record the total volume of 0.5 *N* alcoholic hydrochloric acid added. Insert the stopper in the flask, and allow to stand for 30 minutes. Titrate the excess acid with 0.5 *N* alcoholic potassium hydroxide VS. Perform a blank titration, using 10.0 mL of isopropyl alcohol instead of ethylene oxide solution, adding the same total volume of 0.5 *N* alcoholic hydrochloric acid, and note the difference in volumes required. Each mL of the difference in volumes of 0.5 *N* alcoholic potassium hydroxide consumed is equivalent to 22.02 mg of ethylene oxide. Calculate the concentration, in mg per mL, of ethylene oxide in the stock solution. Standardize daily. Store in a refrigerator. Prepare a 1000-ppm standard by pipeting the calculated volume (about 2 mL) of cold stock solution which, on the basis of the standardization, contains 88.6 mg of ethylene oxide, into a container and adding 87.0 g of *Stripped nonoxynol 9*. Prepare 10-, 5-, and 0.5-ppm standards by quantitatively diluting the 1000-ppm standard with additional *Stripped nonoxynol 9*.

Standard preparations—Transfer 5 ± 0.01 g of each *Standard solution* to suitable serum vials equipped with pressure-tight septum closures designed to relieve any excessive pressure, and seal them.

Test preparation—Transfer 5 ± 0.01 g of Nonoxynol 9 to a serum vial of the same kind as the vials used for the *Standard preparations*.

Chromatographic system (see *Chromatography* ⟨621⟩)—Use a gas chromatograph equipped with a flame-ionization detector. Under typical conditions, the instrument contains a 6.4-m × 2.1-mm (ID) nickel column packed with 60- to 80-mesh support S9, the column is maintained at 100°, the injection port is maintained at 160°, the detector is maintained at 200°, and helium is used as the carrier gas at a flow rate of 30 mL per minute. The resolution, *R*, of ethylene oxide and acetaldehyde, upon chromatographing a solution containing 10 µg per mL of each in *Stripped nonoxynol 9*, is not less than 1.5. None of the points used for constructing the straight line *Calibration* curve deviates from the line by more than 10 percent.

Calibration—Place the vial containing the 10-ppm ethylene oxide *Standard preparation* in an oven, and heat at 90° for 30 minutes. Remove the vial from the oven. Using a gas-tight syringe, immediately inject a 100-µL aliquot of the head space gas into the gas chromatograph. Obtain the area for the ethylene oxide peak (retention time approximately 8 minutes). Raise the temperature of the column to 200° after ethylene oxide elutes to volatilize heavy components. Re-equilibrate the column at 100°. Repeat the foregoing steps, using the vials containing the 5- and 0.5-ppm *Standard preparation*. Plot area units versus ppm ethylene oxide for the standards on linear graph paper, and draw the best straight line through the points.

Procedure—Place the vial containing the *Test preparation* in an oven, and heat at 90° for 30 minutes. Remove the vial from the oven. Immediately inject a 100-µL aliquot of the head space gas into the gas chromatograph, and obtain the area for the ethylene oxide peak. Calculate the concentration of ethylene oxide in the test specimen, in ppm, by the formula:

$$r_U S,$$

in which r_U is the peak area obtained from the *Test preparation*, and *S* is the slope of the standard curve, in ppm per area unit. Not more than 5 ppm is found.

Dioxane—It meets the requirements under *Dioxane* ⟨224⟩.

Assay—Transfer about 50 mg of Nonoxynol 9, accurately weighed, to a 200-mL volumetric flask, add methanol to volume, and mix. Concomitantly determine the absorbances of this solution and of a Standard solution of USP Nonoxynol 9 RS in methanol having a known concentration of about 250 µg per mL at the wavelength of maximum absorbance at about 275 nm in 1-cm cells, using methanol as the blank. Calculate the quantity, in mg, of Nonoxynol 9 in the portion taken by the formula:

$$0.2C(A_U / A_S),$$

in which *C* is the concentration, in µg per mL, of USP Nonoxynol 9 RS in the Standard solution, and A_U and A_S are the absorbances of the assay solution and the Standard solution, respectively.

Nonoxynol 10—*see* Nonoxynol 10 NF

Norepinephrine Bitartrate

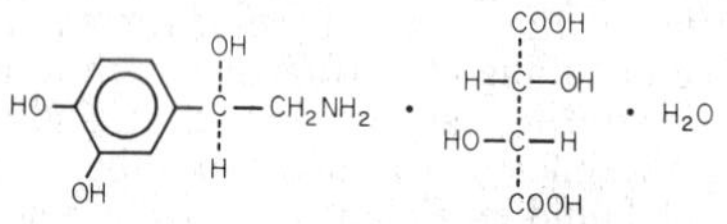

$C_8H_{11}NO_3 . C_4H_6O_6 . H_2O$ 337.28

1,2-Benzenediol, 4-(2-amino-1-hydroxyethyl)-, (*R*)-, [*R*-(*R**,*R**)]-2,3-dihydroxybutanedioate (1:1) (salt), monohydrate.

(−)-α-(Aminomethyl)-3,4-dihydroxybenzyl alcohol tartrate (1:1) (salt), monohydrate [*69815-49-2*].

Anhydrous 319.27 [*51-40-1*].

» Norepinephrine Bitartrate contains not less than 97.0 percent and not more than 102.0 percent of $C_8H_{11}NO_3 . C_4H_6O_6$, calculated on the anhydrous basis.

Packaging and storage—Preserve in tight, light-resistant containers.

Reference standard—*USP Norepinephrine Bitartrate Reference Standard*—This is the monohydrate form of norepinephrine bitartrate. Do not dry before using.

Identification—

A: The infrared absorption spectrum of a potassium bromide dispersion of it exhibits maxima only at the same wavelengths as that of a similar preparation of USP Norepinephrine Bitartrate RS.

B: To a solution of 10 mg in 2 mL of water add 1 drop of ferric chloride TS: an intensely green color develops.

C: To 10 mL of a solution (1 in 10,000) add 1.0 mL of 0.10 *N* iodine. Allow to stand for 5 minutes, and add 2.0 mL of 0.10 *N* sodium thiosulfate: the solution is colorless or has at most a slight pink or slight violet color (*epinephrine* and *isoproterenol* at the same pH, about 3.5, give a strong red-brown or violet color).

Specific rotation ⟨781⟩: between −10° and −12°, calculated on the anhydrous basis, determined in a solution containing 500 mg in each 10 mL.

Water, *Method I* ⟨921⟩: between 4.5% and 5.8%.

Residue on ignition ⟨281⟩: negligible, from 200 mg.

Arterenone—Its absorptivity (see *Spectrophotometry and Light-scattering* ⟨851⟩) at 310 nm, determined in a solution containing 2 mg per mL, is not more than 0.2.

Assay—Dissolve about 500 mg of Norepinephrine Bitartrate, accurately weighed, in 20 mL of glacial acetic acid, warming slightly if necessary to effect solution. Add 2 drops of crystal violet TS, and titrate with 0.1 *N* perchloric acid VS. Perform a blank determination, and make any necessary correction. Each mL of 0.1 *N* perchloric acid is equivalent to 31.93 mg of $C_8H_{11}NO_3 . C_4H_6O_6$.

Norepinephrine Bitartrate Injection

» Norepinephrine Bitartrate Injection is a sterile solution of Norepinephrine Bitartrate in Water for Injection. It contains not less than 90.0 percent and not more than 115.0 percent of the labeled amount of norepinephrine ($C_8H_{11}NO_3$).

Note—Do not use the Injection if it is brown or contains a precipitate.

Packaging and storage—Preserve in single-dose, light-resistant containers, preferably of Type I glass.

Labeling—Label the Injection in terms of mg of norepinephrine per mL, and, where necessary, label it to indicate that it must be diluted prior to use.

Reference standard—*USP Norepinephrine Bitartrate Reference Standard*—This is the monohydrate form of norepinephrine bitartrate. Do not dry before using.

Identification—

A: It responds to *Identification test B* under *Norepinephrine Bitartrate*.

B: Dilute the Injection with water to a concentration of 1 mg in 5 mL. To 10 mL of the dilution add 2.0 mL of 0.10 *N* iodine, allow to stand for 5 minutes, then add 3.0 mL of 0.10 *N* sodium thiosulfate: the solution is colorless or has at most a slight pink or slight violet color (*epinephrine* and *isoproterenol* at the same pH, about 3.5, give a red-brown or violet color).

pH ⟨791⟩: between 3.0 and 4.5.

Particulate matter ⟨788⟩: meets the requirements under *Small-volume Injections*.

Other requirements—It meets the requirements under *Injections* ⟨1⟩.

Assay—

Mobile phase—Dissolve 1.1 g of sodium 1-heptanesulfonate in 800 mL of water. Add 200 mL of methanol, and adjust with 1 *M* phosphoric acid to a pH of 3.0 ± 0.1. Filter through a membrane filter. Make adjustments if necessary (see *System Suitability* under *Chromatography* ⟨621⟩).

Standard preparation—Dissolve an accurately weighed quantity of USP Norepinephrine Bitartrate RS in freshly prepared dilute acetic acid (1 in 25), and dilute quantitatively and stepwise, if necessary, to obtain a solution having a known concentration of about 0.4 mg of norepinephrine bitartrate monohydrate per mL.

Assay preparation—Transfer an accurately measured volume of Norepinephrine Bitartrate Injection, equivalent to about 5 mg of norepinephrine, to a 25-mL volumetric flask, add dilute acetic acid (1 in 25) to volume, and mix.

System suitability preparation—Dissolve a suitable quantity of isoproterenol hydrochloride in the *Standard preparation* to obtain a solution containing, in each mL, 0.4 mg of USP Norepinephrine Bitartrate RS and 0.4 mg of isoproterenol hydrochloride.

Chromatographic system (see *Chromatography* ⟨621⟩)—The liquid chromatograph is equipped with a 280-nm detector and a 4.6-mm × 25-cm column that contains packing L1. The flow rate is about 2 mL per minute. Chromatograph the *Standard preparation* and the *System suitability preparation*, and record the peak responses as directed under *Procedure:* the tailing factor for the analyte peak is not more than 2.5, the resolution, *R*, between the norepinephrine and isoproterenol peaks is not less than 4.0, and the relative standard deviation for replicate injections is not more than 2.0%.

Procedure—Separately inject equal volumes (about 20 µL) of the *Standard preparation* and the *Assay preparation* into the chromatograph, record the chromatograms, and measure the responses for the major peaks. Calculate the quantity, in mg, of norepinephrine ($C_8H_{11}NO_3$) in each mL of the Injection taken by the formula:

$$(169.18/337.28)(25C/V)(r_U/r_S),$$

in which 169.18 and 337.28 are the molecular weights of norepinephrine and norepinephrine bitartrate monohydrate, respectively, *C* is the concentration, in mg per mL, of USP Norepinephrine Bitartrate RS in the *Standard preparation*, *V* is the volume, in mL, of Injection taken, and r_U and r_S are the peak responses obtained from the *Assay preparation* and the *Standard preparation*, respectively.

Norepinephrine Bitartrate Injection, Propoxycaine and Procaine Hydrochlorides and—*see* Propoxycaine and Procaine Hydrochlorides and Norepinephrine Bitartrate Injection

Norethindrone

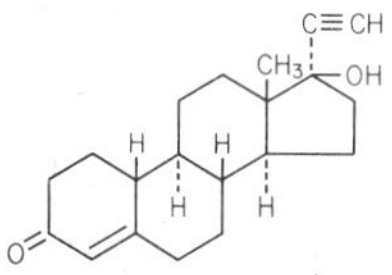

$C_{20}H_{26}O_2$ 298.42

19-Norpregn-4-en-20-yn-3-one, 17-hydroxy-, (17α)-.

17-Hydroxy-19-nor-17α-pregn-4-en-20-yn-3-one [*68-22-4*].

» Norethindrone contains not less than 97.0 percent and not more than 102.0 percent of $C_{20}H_{26}O_2$, calculated on the dried basis.

Packaging and storage—Preserve in well-closed containers.

Reference standard—*USP Norethindrone Reference Standard*—Dry in vacuum at 105° for 3 hours before using.

Completeness of solution—The solution called for in the test for *Specific rotation* is clear and free from undissolved solid.

Identification—The infrared absorption spectrum of a potassium bromide dispersion of it, previously dried, exhibits maxima only at the same wavelengths as that of a similar preparation of USP Norethindrone RS.

Melting range ⟨741⟩: between 202° and 208°.

Specific rotation ⟨781⟩: between −30° and −38°, determined in a solution in dioxane containing 200 mg in each 10 mL.

Loss on drying ⟨731⟩—Dry it in vacuum at 105° for 3 hours: it loses not more than 0.5% of its weight.

Chromatographic impurities—

Test preparation—Prepare a solution of Norethindrone in chloroform to contain 10 mg per mL.

Standard solutions—Prepare a solution of USP Norethindrone RS in chloroform to contain 10 mg per mL (*Standard stock solution*). Dilute accurately measured volumes of *Standard stock solution* with chloroform to obtain *Standard solutions A*, *B*, *C*, and *D* having known concentrations of 150, 50, 30, and 10 µg per mL, respectively.

Procedure—Apply 10-µL volumes, as two 5-µL portions, of the *Test preparation*, and *Standard solutions A*, *B*, *C*, and *D* at equidistant points along a line 2.5 cm from one edge of a 20- × 20-cm thin-layer chromatographic plate (see *Chromatography* ⟨621⟩) coated with a 0.25-mm layer of chromatographic silica gel mixture. Place the plate in a suitable chromatographic chamber previously equilibrated with a mixture of 95 volumes of chloroform and 5 volumes of methanol, seal the chamber, and allow the chromatogram to develop until the solvent front has moved 15 cm above the line of application. Remove the plate, and allow the solvent to evaporate. Spray the plate with a mixture of methanol and sulfuric acid (7:3), then heat the plate at 100° for 5 minutes. The lane of the *Test preparation* exhibits its principal spot at the same R_f as the principal spot of *Standard solution A;* no subsidiary spot is more intense than the spot from *Standard solution B* (0.5%) and the sum of the intensities of all subsidiary spots is not more intense than the spot from *Standard solution A* (1.5%).

Ethynyl group—Dissolve 200 mg in about 40 mL of tetrahydrofuran. Add 10 mL of silver nitrate solution (1 in 10), and titrate with 0.1 *N* sodium hydroxide VS, using glass and calomel electrodes, the latter being of standard fiber type but containing potassium nitrate solution as the electrolyte. Perform a blank determination, and make any necessary correction. Each mL of 0.1 *N* sodium hydroxide is equivalent to 2.503 mg of ethynyl group (–C≡CH). Not less than 8.18% and not more than 8.43% of ethynyl group is found.

Assay—Dissolve about 100 mg of Norethindrone, accurately weighed, in alcohol, and dilute quantitatively and stepwise with alcohol to obtain a solution containing about 10 µg per mL. Dissolve an accurately weighed quantity of USP Norethindrone RS in alcohol, and dilute quantitatively and stepwise with alcohol to obtain a Standard solution having a known concentration of about 10 µg per mL. Concomitantly determine the absorbances of both solutions in 1-cm cells at the wavelength of maximum absorbance at about 240 nm, with a suitable spectrophotometer, using alcohol as the blank. Calculate the quantity, in mg, of $C_{20}H_{26}O_2$ in the portion of Norethindrone taken by the formula:

$$10C(A_U/A_S),$$

in which *C* is the concentration, in µg per mL, of USP Norethindrone RS in the Standard solution, and A_U and A_S are the absorbances of the solution of Norethindrone and the Standard solution, respectively.

Norethindrone Tablets

» Norethindrone Tablets contain not less than 90.0 percent and not more than 110.0 percent of the labeled amount of $C_{20}H_{26}O_2$.

Packaging and storage—Preserve in well-closed containers.

Reference standard—*USP Norethindrone Reference Standard*—Dry in vacuum at 105° for 3 hours before using.

Identification—Mix a portion of powdered Tablets, equivalent to about 50 mg of norethindrone, with 15 mL of solvent hexane, and stir occasionally for 15 minutes. Centrifuge the mixture, then decant and discard the solvent hexane. Extract the residue with two 10-mL portions of solvent hexane, centrifuging and decanting as before, and discard the solvent hexane. Add 25 mL of chloroform to the residue, mix by shaking for 1 to 2 minutes, and filter. Evaporate the filtrate to about 3 mL, add a few mL of solvent hexane to induce crystallization, and evaporate to dryness: the infrared absorption spectrum of a potassium bromide dispersion prepared from the residue so obtained exhibits maxima only at the same wavelengths as that of a similar preparation of USP Norethindrone RS.

Disintegration ⟨701⟩: 15 minutes, the use of disks being omitted.

Uniformity of dosage units ⟨905⟩: meet the requirements.

Assay—

Isoniazid reagent—Dissolve 1.0 g of isoniazid in 1000 mL of anhydrous methanol, add 1.3 mL of hydrochloric acid, and mix.

Procedure—Weigh and finely powder not less than 20 Norethindrone Tablets. Transfer an accurately weighed portion of the powder, equivalent to about 0.7 mg of norethindrone, to a 50-mL volumetric flask, and add anhydrous methanol to volume. Mix, and allow to stand for 10 minutes, with occasional mixing. Filter a portion of the mixture to clarify the solution, and transfer 10.0 mL of the filtrate to a suitable container. Add 2.0 mL of *Isoniazid reagent*, mix, seal, and allow to stand for 30 minutes. This is the *Assay preparation.* Transfer a second 10.0-mL portion of the filtrate to a suitable container, add 2.0 mL of methanol, and mix. This is the *Assay blank preparation.* Transfer 10.0 mL of methanol to a suitable container, add 2.0 mL of *Isoniazid reagent*, mix, seal, and allow to stand for 30 minutes. This is the *Reagent blank preparation.* Prepare a *Standard preparation* by transferring 10.0 mL of a solution of USP Norethindrone RS in methanol having a concentration of about 14 µg per mL to a suitable container. Add 2.0 mL of *Isoniazid reagent*, mix, seal, and allow to stand for 30 minutes. Concomitantly determine the absorbances of these solutions in 1-cm cells, at about 380 nm, with a suitable spectrophotometer, using methanol as the reference for the *Assay blank preparation*, and using the *Reagent blank preparation* as the reference for the *Assay preparation* and the *Standard preparation.* Calculate the quantity, in mg, of $C_{20}H_{26}O_2$ in the portion of Tablets taken by the formula:

$$0.05C(A_U - A_B)/A_S,$$

in which *C* is the concentration, in µg per mL, of the *Standard preparation*, and A_U, A_B, and A_S are the absorbances of the *Assay preparation*, the *Assay blank preparation*, and the *Standard preparation*, respectively.

Norethindrone and Ethinyl Estradiol Tablets

» Norethindrone and Ethinyl Estradiol Tablets contain not less than 90.0 percent and not more than 110.0 percent of the labeled amount of norethindrone ($C_{20}H_{26}O_2$), and not less than 90.0 percent and not more than 110.0 percent of the labeled amount of ethinyl estradiol ($C_{20}H_{24}O_2$).

Packaging and storage—Preserve in well-closed containers.

Reference standards—*USP Norethindrone Reference Standard*—Dry in vacuum at 105° for 3 hours before using. *USP Ethinyl Estradiol Reference Standard*—Dry in vacuum over silica gel for 4 hours before using.

Identification—Crush 1 Tablet in 1 mL of alcohol in a 15-mL conical centrifuge tube, and warm to 50° for 10 minutes with gentle swirling. Cool, and centrifuge to obtain a clear solution. Apply 20 µL of this test solution and 20 µL of an alcohol solution containing, in each mL, about 1 mg of USP Norethindrone RS and about 50 µg of USP Ethinyl Estradiol RS at equidistant points along a line about 2.5 cm from the bottom of a thin-layer chromatographic plate (see *Chromatography* ⟨621⟩) coated with a 0.25-mm layer of chromatographic silica gel and previously activated by heating at 105° for 30 minutes. Develop the chromatogram in a mixture of benzene and ethyl acetate (4:1) in a suitable chamber, previously equilibrated with the solvent mixture, until the solvent front has moved about 10 cm. Remove the plate, and air-dry. Spray the plate with *Sulfuric acid–methanol*, prepared by cautiously adding 70 mL of sulfuric acid in small increments to 30 mL of methanol chilled in an ice-bath, and mixing: the spots from the test solution have the same R_f values as the spots from USP Ethinyl Estradiol RS (about 0.7) and from USP Norethindrone RS (about 0.4).

Dissolution ⟨711⟩—[NOTE—Exercise care in filtering and handling solutions containing ethinyl estradiol to prevent adsorptive loss of the drug. Centrifugation may be used instead of filtration with non-adsorptive membrane filters. Withdraw dissolution aliquots with glass or polytef pipets or syringes that have been checked for adsorptive loss. Use glass dissolution vessels and polytef-coated or solid polytef paddles.]

Medium: 0.1 *N* hydrochloric acid; 600 mL.

Apparatus 2: 75 rpm.

Time: 60 minutes.

Determine the amounts of norethindrone ($C_{20}H_{26}O_2$) and ethinyl estradiol ($C_{20}H_{24}O_2$) dissolved, employing the following method.

Mobile phase—Prepare a degassed and filtered mixture of water and acetonitrile (55:45). Make adjustments if necessary (see *System Suitability* under *Chromatography* ⟨621⟩).

Chromatographic system (see *Chromatography* ⟨621⟩)—The liquid chromatograph is equipped with a 200-nm detector, and a 5-mm × 8.3-cm column that contains 3-μm packing L1. The flow rate is about 1 mL per minute. Chromatograph replicate injections of a filtered portion of a *Standard solution* in 0.1 *N* hydrochloric acid [NOTE—A volume of methanol not exceeding 4% of the total final volume of the *Standard solution* may be used in preparing *Standard solutions*] of USP Norethindrone RS and USP Ethinyl Estradiol RS having concentrations similar to those expected in the *Dissolution Medium*, and record the peak responses as directed under *Procedure:* the relative standard deviation is not more than 3.0%, the minimum number of theoretical plates for the ethinyl estradiol peak is not less than 7000, the resolution, *R*, for norethindrone and ethinyl estradiol is not less than 1.5, and the tailing factor does not exceed 2.0 for either peak.

Procedure—Separately inject equal volumes (about 100 μL) of the *Standard solution* and a filtered portion of the solution under test into the chromatograph, record the chromatograms, and measure the responses for the major peaks. The relative retention times are about 0.9 for norethindrone and 1.0 for ethinyl estradiol. Calculate the quantities of norethindrone ($C_{20}H_{26}O_2$) and ethinyl estradiol ($C_{20}H_{24}O_2$) dissolved by comparison of the corresponding peak responses obtained from the *Standard solution* and the *Test solutions*.

Tolerances—Not less than 50% (*Q*) and 75% (*Q*) of the labeled amounts of $C_{20}H_{26}O_2$ and $C_{20}H_{24}O_2$, respectively, are dissolved in 60 minutes.

Uniformity of dosage units ⟨905⟩: meet the requirements for *Content Uniformity* with respect to norethindrone and to ethinyl estradiol.

Assay—

Mobile phase—Prepare a filtered and degassed mixture of acetonitrile and water (60:40). Make adjustments if necessary (see *System Suitability* under *Chromatography* ⟨621⟩).

Internal standard solution—Transfer about 15 mg of valerophenone into a 250-mL volumetric flask, add 125 mL of acetonitrile, dilute with water to volume, and mix.

Ethinyl estradiol standard stock solution—Dissolve an accurately weighed quantity of USP Ethinyl Estradiol RS in acetonitrile, and dilute quantitatively and stepwise with acetonitrile to obtain a solution having a known concentration of about 0.09 mg per mL.

Norethindrone standard stock solution—Using an accurately weighed quantity of USP Norethindrone RS, prepare a solution in acetonitrile having a known concentration of about 1.25 mg per mL.

Mixed standard preparation—Transfer 5.0 mL of *Internal standard solution* into a 100-mL volumetric flask. Add accurately measured volumes of *Ethinyl estradiol standard stock solution* and *Norethindrone standard stock solution* so that the final known concentrations, in mg per mL, of the Reference Standards correspond numerically to about one-twentieth of the labeled amounts of the corresponding ingredients in the Tablets. Add (26 − *X*) mL of acetonitrile, *X* being the total volume of the *standard stock solution* taken. Dilute with a mixture of acetonitrile and water (45 in 100) to volume, and mix.

Assay preparation—Transfer 10 Norethindrone and Ethinyl Estradiol Tablets to a 100-mL volumetric flask, add 20 mL of water, and shake by mechanical means until the tablets are completely disintegrated. Add 10.0 mL of *Internal standard solution* and 60 mL of acetonitrile, and mix. Sonicate for about 2 minutes. Dilute with acetonitrile to volume, and mix. Allow solid particles to settle, or centrifuge if necessary to obtain a slightly turbid solution. Dilute 5.0 mL of this solution with a mixture of acetonitrile and water (45 in 100) to 10.0 mL, and mix.

Chromatographic system (see *Chromatography* ⟨621⟩)—The liquid chromatograph is equipped with a 200-nm detector and a 4.6-mm × 25-cm column that contains packing L1. The flow rate is about 1.0 mL per minute. Chromatograph the *Standard preparation*, and record the peak responses as directed under *Procedure:* the column efficiency determined from the internal standard peak is not less than 8000 theoretical plates, the resolution, *R*, between the norethindrone and ethinyl estradiol peaks is not less than 2.0, and the relative standard deviation for six replicate injections is not more than 2.0% (both peaks).

Procedure—Separately inject equal volume (about 25 μL) of the *Standard preparation* and the *Assay preparation* into the chromatograph, record the chromatograms, and measure the responses for the major peaks. The relative retention times are about 0.9 for ethinyl estradiol and 1.0 for norethindrone. Calculate the quantities, in mg, of norethindrone ($C_{20}H_{26}O_2$) and ethinyl estradiol ($C_{20}H_{24}O_2$) in each Tablet taken by the formula:

$$20C(R_U/R_S),$$

in which *C* is the concentration, in mg per mL, of the appropriate USP Reference Standard in the *Mixed standard preparation*, and R_U and R_S are the peak response ratios, at corresponding retention times, obtained from the *Assay preparation* and the *Standard preparation*, respectively.

Norethindrone and Mestranol Tablets

» Norethindrone and Mestranol Tablets contain not less than 90.0 percent and not more than 110.0 percent of the labeled amount of norethindrone ($C_{20}H_{26}O_2$), and not less than 90.0 percent and not more than 110.0 percent of the labeled amount of mestranol ($C_{21}H_{26}O_2$).

Packaging and storage—Preserve in well-closed containers.

Reference standards—*USP Norethindrone Reference Standard*—Dry in vacuum at 105° for 3 hours before using. *USP Mestranol Reference Standard*—Dry at 105° for 3 hours before using.

Identification—Crush 1 Tablet in 1 mL of alcohol in a 15-mL conical centrifuge tube, and centrifuge briefly. Apply 10 μL of this test solution and 10 μL each of solutions containing, respectively, about 1 mg per mL of USP Norethindrone RS in alcohol and about 50 μg per mL of USP Mestranol RS in alcohol at equidistant points along a line about 2.5 cm from the bottom of a thin-layer chromatographic plate (see *Chromatography* ⟨621⟩) coated with a 0.25-mm layer of chromatographic silica gel and previously activated by heating at 105° for 30 minutes. Develop the chromatogram in a mixture of equal volumes of ethyl acetate and cyclohexane in a suitable chamber, previously equilibrated with the solvent mixture, until the solvent front has moved about 15 cm. Remove the plate, air-dry, and observe under short-wavelength ultraviolet light: the principal spot from the test solution appears at the same R_f value as the principal spot from USP Norethindrone RS, at about R_f 0.6. Spray the plate with *Sulfuric acid–methanol*, prepared as directed in the *Assay* under *Ethinyl Estradiol*, and heat it at 105° for 10 minutes: the pink spot from the test solution appears at the same R_f value as the pink spot from USP Mestranol RS (about R_f 0.8).

Dissolution ⟨711⟩—[NOTE—Exercise care in filtering solutions containing mestranol to prevent adsorptive loss of the drug. Centrifugation may be used instead of filtration with nonadsorptive membrane filters. Withdraw dissolution aliquots with glass or polytef pipets or syringes that have been checked for adsorptive loss. Use glass dissolution vessels and polytef-coated or solid polytef paddles.]

Medium: 0.1 *N* hydrochloric acid and isopropyl alcohol (97:3); 600 mL.

Apparatus 2: 75 rpm.

Time: 60 minutes.

Determine the amounts of norethindrone ($C_{10}H_{26}O_2$) and mestranol ($C_{21}H_{26}O_2$) dissolved, employing the following method.

Mobile phase—Prepare a degassed and filtered mixture of acetonitrile and water (60:40). Make adjustments if necessary (see *System Suitability* under *Chromatography* ⟨621⟩).

Chromatographic system (see *Chromatography* ⟨621⟩)—The liquid chromatograph is equipped with a 254-nm detector for norethindrone and a 200-nm detector for mestranol, and a 4-mm × 15-cm column that contains packing L1. The flow rate is about 1 mL per minute. Chromatograph replicate injections of a filtered portion of a Standard solution of USP Norethindrone

RS and USP Mestranol RS in *Dissolution Medium* having known concentrations similar to those expected in the solution under test, and record the peak responses as directed under *Procedure:* the relative standard deviation is not more than 2.0%, the minimum number of theoretical plates for the mestranol peak is 4000, and the minimum retention volume for norethindrone is 2.4 mL.

Procedure—Separately inject equal volumes (about 200 µL) of the Standard solution and a filtered portion of the solution under test into the chromatograph, record the chromatograms, and measure the responses for the major peaks. The relative retention times are about 0.4 for norethindrone and 1.0 for mestranol. Calculate the quantities of norethindrone and mestranol dissolved by comparison of the corresponding peak responses obtained from the Standard solution and the test solutions.

Tolerances—Not less than 50% (*Q*) of $C_{20}H_{26}O_2$ and 75% (*Q*) of $C_{21}H_{26}O_2$ are dissolved in 60 minutes.

Uniformity of dosage units ⟨905⟩: meet the requirements for *Content Uniformity* with respect to norethindrone and to mestranol.

Assay—

Mobile phase—Prepare a filtered and degassed mixture of acetonitrile and water (50:50). Make adjustments if necessary (see *System Suitability* under *Chromatography* ⟨621⟩).

Internal standard solution—Transfer about 80 mg of progesterone into a 100-mL volumetric flask, add 50 mL of acetonitrile, dilute with water to volume, and mix.

Mestranol standard stock solution—Dissolve an accurately weighed quantity of USP Mestranol RS in acetonitrile, and dilute quantitatively and stepwise with acetonitrile to obtain a solution having a known concentration of about 0.05 mg per mL.

Norethindrone standard stock solution—Using an accurately weighed quantity of USP Norethindrone RS, prepare a solution in acetonitrile having a known concentration of about 1 mg per mL.

Mixed standard preparation—Transfer 2.0 mL of *Internal standard solution* into a 100-mL volumetric flask. Add accurately measured volumes of *Mestranol standard stock solution* and *Norethindrone standard stock solution* so that the final known concentrations, in mg per mL, of the Reference Standards correspond numerically to about one-fiftieth of the labeled amounts of the corresponding ingredients in the Tablets. Add 50 mL of water, dilute with acetonitrile to volume, and mix.

Assay preparation—Transfer 10 Norethindrone and Mestranol Tablets to a 250-mL volumetric flask, add 50 mL of water, and shake by mechanical means until the tablets are completely disintegrated. Add 10.0 mL of *Internal standard solution* and 165 mL of acetonitrile, and mix. Sonicate for about 2 minutes. Dilute with acetonitrile to volume, and mix. Allow solid particles to settle, or centrifuge if necessary to obtain a slightly turbid solution. Transfer 5.0 mL of this solution to a 10-mL volumetric flask. Add 1.0 mL of acetonitrile, dilute with water to volume, and mix.

Chromatographic system (see *Chromatography* ⟨621⟩)—The liquid chromatograph is equipped with a 200-nm detector and a 4.6-mm × 15-cm column that contains packing L7. The flow rate is about 1.0 mL per minute. Chromatograph the *Standard preparation*, and record the peak responses as directed under *Procedure:* the column efficiency determined from the mestranol peak is not less than 6000 theoretical plates, the resolution, *R*, between the progesterone and mestranol peaks is not less than 5.0, and the relative standard deviation for six replicate injections is not more than 2.0% (both peaks).

Procedure—Separately inject equal volumes (about 25 µL) of the *Standard preparation* and the *Assay preparation* into the chromatograph, record the chromatograms, and measure the responses for the major peaks. The relative retention times are about 2.5 for mestranol and 1.0 for norethindrone. Calculate the quantities, in mg, of norethindrone ($C_{20}H_{26}O_2$) and mestranol ($C_{21}H_{26}O_2$) in each Tablet taken by the formula:

$$50C(R_U/R_S),$$

in which *C* is the concentration, in mg per mL, of the appropriate USP Reference Standard in the *Mixed standard preparation*, and R_U and R_S are the peak response ratios, at corresponding retention times, obtained from the *Assay preparation* and the *Standard preparation*, respectively.

Norethindrone Acetate

$C_{22}H_{28}O_3$ 340.46

19-Norpregn-4-en-20-yn-3-one, 17-(acetyloxy)-, (17α).

17-Hydroxy-19-nor-17α-pregn-4-en-20-yn-3-one acetate [*51-98-9*].

» Norethindrone Acetate contains not less than 97.0 percent and not more than 103.0 percent of $C_{22}H_{28}O_3$, calculated on the dried basis.

Packaging and storage—Preserve in well-closed containers.

Reference standard—*USP Norethindrone Acetate Reference Standard*—Dry in vacuum at 105° for 3 hours before using.

Completeness of solution—The solution prepared for the determination of *Specific rotation* is clear and free from undissolved solid.

Identification—The infrared absorption spectrum of a potassium bromide dispersion of it, previously dried, exhibits maxima only at the same wavelengths as that of a similar preparation of USP Norethindrone Acetate RS.

Specific rotation ⟨781⟩: between −32° and −38°, calculated on the dried basis, determined in a solution in dioxane containing 200 mg in each 10 mL.

Loss on drying ⟨731⟩—Dry it at 105° for 3 hours: it loses not more than 0.5% of its weight.

Chromatographic impurities—

Test preparation—Prepare a solution of Norethindrone Acetate in chloroform to contain 10 mg per mL.

Standard solutions—Prepare a solution of USP Norethindrone Acetate RS in chloroform to contain 10 mg per mL (*Standard stock solution*). Dilute accurately measured volumes of *Standard stock solution* with chloroform to obtain *Standard solutions A, B, C,* and *D* having known concentrations of 150, 50, 30, and 10 µg per mL, respectively.

Procedure—Apply 10-µL volumes, as two 5-µL portions, of the *Test preparation*, and *Standard solutions A, B, C,* and *D* at equidistant points along a line 2.5 cm from one edge of a 20- × 20-cm thin-layer chromatographic plate (see *Chromatography* ⟨621⟩) coated with a 0.25-mm layer of chromatographic silica gel mixture. Place the plate in a suitable chromatographic chamber previously equilibrated with a mixture of toluene and ethyl acetate (1:1), seal the chamber, and allow the chromatogram to develop until the solvent front has moved 15 cm above the line of application. Remove the plate, and allow the solvent to evaporate. Spray the plate with a mixture of methanol and sulfuric acid (7:3), then heat the plate at 100° for 5 minutes. The lane of the *Test preparation* exhibits its principal spot at the same R_f as the principal spot of *Standard solution A*; no subsidiary spot is more intense than the spot from *Standard solution B* (0.5%) and the sum of the intensities of all subsidiary spots is not more intense than the spot from *Standard solution A* (1.5%).

Ethynyl group—Proceed as directed in the test for *Ethynyl group* under *Norethindrone.* Not less than 7.13% and not more than 7.57% of ethynyl group is found.

Assay—Transfer about 100 mg of Norethindrone Acetate, accurately weighed, to a 200-mL volumetric flask, add alcohol to volume, and mix. Transfer 5.0 mL of this solution to a 250-mL volumetric flask, dilute with alcohol to volume, and mix. Dissolve an accurately weighed quantity of USP Norethindrone Acetate RS in alcohol, and dilute quantitatively and stepwise with alcohol to obtain a Standard solution having a known concentration of about 10 µg per mL. Concomitantly determine the absorbances of both solutions in 1-cm cells at the wavelength of maximum absorbance at about 240 nm, with a suitable spectrophotometer, using alcohol as the blank. Calculate the quantity, in mg, of $C_{22}H_{28}O_3$ in the Norethindrone Acetate taken by the formula:

$$10C(A_U/A_S),$$

in which *C* is the concentration, in µg per mL, of USP Norethindrone Acetate RS in the Standard solution, and A_U and A_S are the absorbances of the solution of Norethindrone Acetate and the Standard solution, respectively.

Norethindrone Acetate Tablets

» Norethindrone Acetate Tablets contain not less than 90.0 percent and not more than 110.0 percent of the labeled amount of $C_{22}H_{28}O_3$.

Packaging and storage—Preserve in well-closed containers.

Reference standard—*USP Norethindrone Acetate Reference Standard*—Dry in vacuum at 105° for 3 hours before using.

Identification—It responds to the *Identification test* under *Norethindrone Tablets*, USP Norethindrone Acetate RS being used to prepare the Standard preparation.

Dissolution ⟨711⟩—

Medium: dilute hydrochloric acid (1 in 100) containing 0.02% of sodium lauryl sulfate; 900 mL.

Apparatus 1: 100 rpm.

Time: 60 minutes.

Procedure—Determine the amount of $C_{22}H_{28}O_3$ dissolved from ultraviolet absorbances at the wavelength of maximum absorbance at about 248 nm, measured from a baseline drawn from 350 nm through 310 nm and extending beyond the peak maximum, of filtered portions of the solution under test, suitably diluted with *Dissolution Medium*, in comparison with a Standard solution having a known concentration of USP Norethindrone Acetate RS in the same medium. [NOTE—The Standard solution may be prepared by dissolving the Reference Standard in a volume of methanol, not exceeding 0.5% of the final volume of the solution, and diluting quantitatively with *Dissolution Medium.*]

Tolerances—Not less than 70% (*Q*) of the labeled amount of $C_{22}H_{28}O_3$ is dissolved in 60 minutes.

Uniformity of dosage units ⟨905⟩: meet the requirements.

Procedure for content uniformity—Transfer 1 finely powdered Tablet to a 100-mL volumetric flask with the aid of about 75 mL of alcohol. Heat the alcohol to boiling, and allow the mixture to remain at a temperature just below the boiling point for about 15 minutes, with occasional swirling. Cool to room temperature, dilute with alcohol to volume, mix, and centrifuge a portion of the contents at about 2000 rpm until the solution becomes clear. Dilute a portion of the supernatant solution quantitatively and stepwise with alcohol to obtain a solution containing approximately 10 µg of norethindrone acetate per mL. Concomitantly determine the absorbances of this solution and of a Standard solution of USP Norethindrone Acetate RS in alcohol having a known concentration of about 10 µg per mL in 1-cm cells at the wavelength of maximum absorbance at about 240 nm, with a suitable spectrophotometer, using alcohol as the blank. Calculate the quantity, in mg, of $C_{22}H_{28}O_3$ in the Tablet by the formula:

$$(TC/D)(A_U/A_S),$$

in which *T* is the labeled quantity, in mg, of norethindrone acetate in the Tablet, *C* is the concentration, in µg per mL, of USP Norethindrone Acetate RS in the Standard solution, *D* is the concentration, in µg per mL, of the solution from the Tablet, based upon the labeled quantity per Tablet and the extent of dilution, and A_U and A_S are the absorbances of the solution from the Tablet and the Standard solution, respectively.

Assay—Weigh and finely powder not less than 20 Norethindrone Acetate Tablets. Transfer an accurately weighed portion of the powder, equivalent to about 20 mg of norethindrone acetate, to a separator, add 10 mL of water, and extract with three 25-mL portions of chloroform, filtering each extract through chloroform-washed cotton. Evaporate the combined chloroform extracts on a steam bath to dryness, reducing the heat as dryness is approached. Dissolve the residue in alcohol, transfer the solution to a 100-mL volumetric flask, dilute with alcohol to volume, and mix. Transfer a 5.0-mL aliquot to a 100-mL volumetric flask, dilute with alcohol to volume, and mix. Concomitantly determine the absorbances of this solution and a Standard solution of USP Norethindrone Acetate RS in alcohol having a known concentration of about 10 µg per mL in 1-cm cells at the wavelength of maximum absorbance at about 240 nm, with a suitable spectrophotometer, using alcohol as the blank. Calculate the quantity, in mg, of $C_{22}H_{28}O_3$ in the portion of Tablets taken by the formula:

$$2C(A_U/A_S),$$

in which *C* is the concentration, in µg per mL, of USP Norethindrone Acetate RS in the Standard solution, and A_U and A_S are the absorbances of the solution from the Tablets and the Standard solution, respectively.

Norethindrone Acetate and Ethinyl Estradiol Tablets

» Norethindrone Acetate and Ethinyl Estradiol Tablets contain not less than 90.0 percent and not more than 110.0 percent of the labeled amount of norethindrone acetate ($C_{22}H_{28}O_3$), and not less than 88.0 percent and not more than 112.0 percent of the labeled amount of ethinyl estradiol ($C_{20}H_{24}O_2$).

Packaging and storage—Preserve in well-closed containers.

Reference standards—*USP Norethindrone Acetate Reference Standard*—Dry in vacuum at 105° for 3 hours before using. *USP Ethinyl Estradiol Reference Standard*—Dry in vacuum over silica gel for 4 hours before using.

Identification—

A: Wash the isooctane-benzene solution obtained in the *Assay for ethinyl estradiol* with 5 mL of water, filter, and evaporate to dryness: the infrared absorption spectrum of a potassium bromide dispersion of the residue so obtained exhibits maxima only at the same wavelengths as that of a similar preparation of USP Norethindrone Acetate RS.

B: Crush 1 Tablet in 1 mL of alcohol in a 15-mL conical centrifuge tube, and centrifuge briefly. Apply 5 µL of this solution on a suitable thin-layer chromatographic plate (see *Chromatography* ⟨621⟩), coated with a 0.25-mm layer of chromatographic silica gel mixture, activated at 105° for 60 minutes immediately prior to use. Apply to the same plate 5 µL of an alcohol solution containing in each mL an amount of USP Norethindrone Acetate RS, accurately weighed, corresponding to the labeled quantity of norethindrone acetate per Tablet, and similarly apply 5 µL of an alcohol solution containing 50 µg of USP Ethinyl Estradiol RS, accurately weighed, per mL. Allow the spots to dry, and develop the chromatogram in a solvent system consisting of a mixture of chloroform and glacial acetic acid (95:5) until the solvent front has moved about 10 cm from the origin. Remove the plate from the chromatographic chamber, allow the solvent to evaporate, then heat in an oven at 105° for 10 minutes. Remove the plate from the oven and, while it is still hot, spray lightly with dilute sulfuric acid (3 in 4). Observe the plate under long-wavelength ultraviolet light: any red fluorescent spot produced by the test solution at an R_f value of about 0.4, and any yellow fluorescent spot produced by the test solution at an R_f value of about 0.2, are not greater in size and intensity than the corresponding spots from the solutions from USP Norethindrone Acetate RS and USP Ethinyl Estradiol RS, respectively.

Disintegration ⟨701⟩: 20 minutes, the use of disks being omitted.

Uniformity of dosage units ⟨905⟩: meet the requirements for *Content Uniformity* with respect to norethindrone acetate and to ethinyl estradiol.

Procedure for content uniformity for ethinyl estradiol—Place 1 Tablet in a 125-mL separator, add 5 mL of water, and shake until the Tablet has disintegrated completely. Add 25.0 mL of a mixture of benzene and isooctane (3:2), shake thoroughly, allow to settle, and remove and discard the aqueous phase. Transfer 20.0 mL of the isooctane-benzene solution to a second 125-mL separator, avoiding mechanical transfer of any of the aqueous phase. Transfer 8.0 mL of a solution of USP Ethinyl Estradiol RS in benzene, containing in each mL an amount of ethinyl estradiol equal to one-tenth of the labeled quantity per Tablet, to a third 125-mL separator, and add 12 mL of isooctane. To the second and third separators add 8.0 mL of a 1 in 25 solution of sodium hydroxide in dilute alcohol (1 in 10), shake, allow to settle, and transfer the sodium hydroxide solution from each separator into separate suitable containers. Proceed as directed under *Assay for ethinyl estradiol*, beginning with "Add, dropwise, 5.0 mL of the sodium hydroxide extract," but determine the absorbances of the final solutions using 5-cm cells. Calculate the quantity, in µg, of $C_{20}H_{24}O_2$ in the Tablet by the formula:

$$10C(A_U/A_S),$$

in which C is the concentration, in µg per mL, of the Standard solution in benzene, and A_U and A_S are the absorbances of the test solution and the Standard solution, respectively.

Procedure for content uniformity for norethindrone acetate—Transfer 1 finely powdered Tablet to a 100-mL volumetric flask with the aid of about 75 mL of alcohol, heat to boiling, and allow to remain at a temperature just below the boiling temperature for about 15 minutes, with occasional swirling. Cool to room temperature, dilute with alcohol to volume, and mix. Centrifuge a portion of the contents at about 2000 rpm until the solution becomes clear. If necessary, dilute a portion of the supernatant solution quantitatively with alcohol to provide a solution containing about 10 µg of norethindrone acetate per mL. Concomitantly determine the absorbances of this solution and of a solution of USP Norethindrone Acetate RS in alcohol having a known concentration of about 10 µg per mL in 1-cm cells at the wavelength of maximum absorbance at about 240 nm, with a suitable spectrophotometer, using alcohol as the blank. Calculate the quantity, in mg, of $C_{22}H_{28}O_3$ in the Tablet by the formula:

$$(T/D)C(A_U/A_S),$$

in which T is the labeled quantity, in mg, of norethindrone acetate in the Tablet, D is the concentration, in µg per mL, of norethindrone acetate in the test solution, based on the labeled quantity per Tablet and the extent of dilution, C is the concentration, in µg per mL, of USP Norethindrone Acetate RS in the Standard solution, and A_U and A_S are the absorbances of the solution from the Tablet and the Standard solution, respectively.

Assay for norethindrone acetate—Place 20 Norethindrone Acetate and Ethinyl Estradiol Tablets in a 125-mL separator, add 20 mL of water, and shake until the Tablets have disintegrated completely. Extract with three 30-mL portions of chloroform, filtering each extract through chloroform-moistened cotton into a round-bottom, 250-mL flask. Evaporate the combined extracts under vacuum to dryness, with the aid of gentle heat (not more than 40°). Cool, add 5 mL of water and 100.0 mL of a mixture of isooctane-benzene (3:2), insert the stopper, and shake for 2 to 3 minutes. Transfer an accurately measured volume of the supernatant isooctane-benzene solution, containing about 1.5 mg of norethindrone acetate, to a round-bottom, 250-mL flask, and similarly evaporate under vacuum to dryness, taking care to assure complete removal of residual benzene. [NOTE—Retain the remaining isooctane-benzene solution for the *Assay for ethinyl estradiol*.] Dissolve the residue in 100.0 mL of alcohol, and mix. Concomitantly determine the absorbances of this solution and a solution of USP Norethindrone Acetate RS in the same medium, having a known concentration of about 15 µg per mL, in 1-cm cells at the wavelength of maximum absorbance at about 240 nm, with a suitable spectrophotometer, using alcohol as the blank. Calculate the quantity, in mg, of $C_{22}H_{28}O_3$ in the accurately measured volume of isooctane-benzene solution of the Tablets taken by the formula:

$$0.1C(A_U/A_S),$$

in which C is the concentration, in µg per mL, of USP Norethindrone Acetate RS in the *Standard solution*, and A_U and A_S are the absorbances of the solution from the Tablets and the Standard solution, respectively.

Assay for ethinyl estradiol—Transfer 25.0 mL of the isooctane-benzene solution prepared from the Tablets as directed in the *Assay for norethindrone acetate* to a 125-mL separator, avoiding mechanical transfer of any of the aqueous phase. Transfer 10.0 mL of a benzene solution of USP Ethinyl Estradiol RS, containing in each mL a known amount of ethinyl estradiol equal to one-half of the labeled quantity per Tablet, to another 125-mL separator containing 15.0 mL of isooctane. Add 10.0 mL of a 1 in 25 solution of sodium hydroxide in dilute alcohol (1 in 10) to each separator, and shake gently for 3 minutes. Allow to settle, and transfer the sodium hydroxide solution from each separator into separate suitable containers. [NOTE—Retain the isooctane-benzene solution for *Identification test A*.] Add, dropwise, 5.0 mL of the sodium hydroxide extract from the Tablets to 25.0 mL of dilute sulfuric acid (4 in 5) contained in a 150-mL beaker and previously chilled in an ice bath. Stir the acid solution continuously during the addition with the aid of a magnetic stirrer, and keep it in the ice bath. [NOTE—Stir the acid solution rapidly and introduce the alkaline solution near the perimeter of the rapidly swirling acid solution, rather than near the vortex. Add the alkaline solution slowly, dropwise.] Treat 5.0 mL of the sodium hydroxide solution from the Standard in the same manner, and allow the solutions to reach room temperature. Concomitantly determine the absorbances of both solutions in 1-cm cells at the wavelength of maximum absorbance at about 536 nm, with a suitable spectrophotometer, using water as the blank. [NOTE—Use 2-cm cells for Tablets labeled to contain 30 µg or less of ethinyl estradiol.] Calculate the quantity, in µg, of $C_{20}H_{24}O_2$ in the portion of isooctane-benzene solution taken by the formula:

$$10C(A_U/A_S),$$

in which C is the concentration, in µg per mL, of USP Ethinyl Estradiol RS in the Standard solution, and A_U and A_S are the absorbances of the solutions from the Tablets and the Reference Standard, respectively.

Norethynodrel

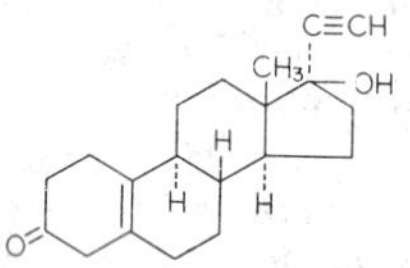

$C_{20}H_{26}O_2$ 298.42

19-Norpregn-5(10)-en-20-yn-3-one, 17-hydroxy-, (17α)-.

17-Hydroxy-19-nor-17α-pregn-5(10)-en-20-yn-3-one [*68-23-5*].

» Norethynodrel contains not less than 97.0 percent and not more than 101.0 percent of $C_{20}H_{26}O_2$.

Packaging and storage—Preserve in well-closed containers.

Reference standards—*USP Norethynodrel Reference Standard*—Do not dry before using. *USP Norethindrone Reference Standard*—Dry in vacuum at 105° for 3 hours before using.

Identification—The infrared absorption spectrum, determined in a 0.1-mm cell, of a 1 in 20 solution in chloroform exhibits maxima only at the same wavelengths as that of a similar solution of USP Norethynodrel RS.

Specific rotation ⟨781⟩: between +119° and +125°, determined in a solution in dioxane containing 100 mg in each 10 mL.

Ethynyl group—Dissolve 200 mg in about 40 mL of tetrahydrofuran. Add 10 mL of silver nitrate solution (1 in 10), and titrate with 0.1 *N* sodium hydroxide VS, using a glass-calomel electrode system in which the calomel electrode is of standard fiber type but contains potassium nitrate solution as the electrolyte. Each mL of 0.1 *N* sodium hydroxide is equivalent to 2.503 mg of ethynyl group (–C≡CH). Not less than 8.18% and not more than 8.43% of ethynyl group is found.

Limit of norethindrone—

Test preparation—Prepare a solution of Norethynodrel in chloroform containing 10 mg per mL.

Standard solution—Prepare a solution of USP Norethindrone RS in chloroform to contain 1 mg per mL. Dilute 2 mL of the solution with chloroform to 10 mL.

Procedure—Apply 10-µL volumes of the *Test preparation* and the *Standard solution* (see *Chromatography* ⟨621⟩) on a thin-layer chromatographic plate coated with a 0.25-mm layer of chromatographic silica gel mixture, and allow not more than 5 minutes between spotting the plate and starting development of the chromatogram. Place the plate in a suitable chromatographic chamber previously equilibrated with a mixture of cyclohexane, ethyl acetate, and methanol (60:40:2), and allow the solvent front to move 15 cm. Spray the plate with dilute sulfuric acid (1 in 2), heat the plate at 105° for 5 minutes, and view under long-wavelength ultraviolet light. Locate any norethindrone impurity in the *Test preparation* by comparison with the R_f value from the *Standard solution*. If present, the norethindrone spot from the *Test preparation* is not larger or more intense than the spot from the *Standard solution* (2.0%).

Ordinary impurities ⟨466⟩—

Test solution: chloroform.

Standard solution: chloroform.

Eluant: ether.

Visualization: 5, followed by viewing under long-wavelength ultraviolet light.

Assay—

Standard preparation—Dissolve a suitable quantity of USP Norethynodrel RS, accurately weighed, in methanol, and dilute quantitatively with methanol to obtain a solution having a known concentration of about 1 mg per mL.

Assay preparation—Dissolve about 100 mg of Norethynodrel, accurately weighed, in methanol to make 100.0 mL, and mix.

Procedure—Transfer 10.0 mL each of the *Standard preparation* and the *Assay preparation* to separate 100-mL volumetric flasks. To each flask add 40 mL of methanol, then add 5 mL of a mixture of 3 volumes of hydrochloric acid and 2 volumes of water, mix quickly, and allow to stand at a temperature of about 25° for 1 hour, accurately timed. Prior to the end of the 1-hour period, prepare blanks as follows: Add 1.0 mL each of the *Standard preparation* and the *Assay preparation* to separate 100-mL volumetric flasks, each containing a mixture of 50 mL of methanol and 2 mL of water, dilute each with methanol to volume, and mix. At the end of the 1-hour reaction period, dilute each of the acid-containing solutions with methanol to volume, and mix. Transfer 10.0 mL of each into separate 100-mL volumetric flasks, add 2 mL of water to each, dilute with methanol to volume, and mix. Concomitantly determine the absorbances of the solutions in 1-cm cells at the wavelength of maximum absorbance at about 240 nm, with a suitable spectrophotometer, relative to the corresponding blanks. Calculate the quantity, in mg, of $C_{20}H_{26}O_2$ in the portion of Norethynodrel taken by the formula:

$$100C(A_U/A_S),$$

in which C is the concentration, in mg per mL, of USP Norethynodrel RS in the *Standard preparation*, and A_U and A_S are the absorbances of the solutions from the *Assay preparation* and the *Standard preparation*, respectively.

Norfloxacin

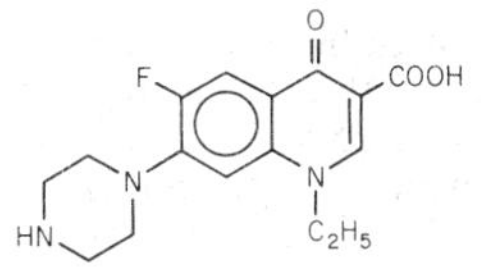

$C_{16}H_{18}FN_3O_3$ 319.34

3-Quinolinecarboxylic acid, 1-ethyl-6-fluoro-1,4-dihydro-4-oxo-7-(1-piperazinyl)-.

1-Ethyl-6-fluoro-1,4-dihydro-4-oxo-7-(1-piperazinyl)-3-quinolinecarboxylic acid [*70458-96-7*].

» Norfloxacin contains not less than 99.0 percent and not more than 101.0 percent of $C_{16}H_{18}FN_3O_3$, calculated on the dried basis.

Packaging and storage—Preserve in tight, light-resistant containers.

Reference standard—*USP Norfloxacin Reference Standard*—Dry in vacuum at a pressure not exceeding 5 mm of mercury at 100° for 2 hours before using.

Identification—

A: The infrared absorption spectrum of a mineral oil dispersion of it, previously dried, exhibits maxima only at the same wavelengths as that of a similar preparation of USP Norfloxacin RS.

B: [NOTE—Use low-actinic glassware in this procedure.] The ultraviolet absorption spectrum of a 1 in 200,000 solution of it, previously dried, in 0.1 *N* sodium hydroxide, exhibits maxima and minima at the same wavelengths as that of a similar preparation of USP Norfloxacin RS, concomitantly measured, and the respective absorptivities, calculated on the dried basis, at the wavelength of maximum absorbance at about 273 nm do not differ by more than 3.0%.

Loss on drying ⟨731⟩—Dry it in vacuum at a pressure not exceeding 5 mm of mercury at 100° to constant weight: it loses not more than 1.0% of its weight.

Residue on ignition ⟨281⟩: not more than 0.1%.

Heavy metals, *Method II* ⟨231⟩: 0.0015%.

Chromatographic purity—Dissolve a quantity of Norfloxacin in a mixture of methanol and methylene chloride (1:1) to obtain a test solution containing 8.0 mg per mL. Dissolve 4.0 mg of USP Norfloxacin RS in 1 mL of glacial acetic acid, add 4 mL of methanol, and mix. To 1 mL of this Standard stock solution add 9 mL of the mixture of methanol and methylene chloride (1:1) to obtain *Comparison solution A*. Dilute a portion of this solution with an equal volume of the mixture of methanol and methylene chloride (1:1) to obtain *Comparison solution B*. On a suitable high-performance thin-layer chromatographic plate (see *Chromatography* ⟨621⟩), coated with a 0.25-mm layer of silica gel mixture, previously washed with methanol and air-dried, separately apply 5 µL of the test solution, 1, 2, and 5 µL of *Comparison solution A*, and 5 µL of *Comparison solution B*. The spots of *Comparison solutions A* and *B* are equivalent to 0.2, 0.4, 1.0, and 0.5% of impurities, respectively. Place the plate in a suitable paper-lined chromatographic chamber previously equilibrated with a solvent system consisting of a mixture of chloroform, methanol, toluene, diethylamine, and water (40:40:20:14:8). Seal the chamber and allow the chromatogram to develop until the solvent front has moved about nine-tenths of the length of the plate. Remove the plate from the chamber, mark the solvent front, allow the solvent to evaporate, and examine the plate under both short- and long-wavelength ultraviolet light. Compare the intensities of any secondary spots observed in the chromatogram of the test solution with those of the principal spots in the chromatograms of *Comparison solutions A* and *B*: the sum of the intensities of secondary spots obtained from the test solution corresponds to not more than 1.0% of impurities.

Assay—Dissolve about 460 mg of Norfloxacin, accurately weighed, in 100 mL of glacial acetic acid. Titrate potentiometrically with 0.1 *N* perchloric acid VS using a suitable anhydrous electrode system (see *Titrimetry* ⟨541⟩). [NOTE—Remove any aqueous solution in the electrode(s), render anhydrous, and fill with 0.1 *N* lithium perchlorate in acetic anhydride.] Perform a blank determination, and make any necessary correction. Each mL of 0.1 *N* perchloric acid is equivalent to 31.93 mg of $C_{16}H_{18}FN_3O_3$.

Norfloxacin Tablets

» Norfloxacin Tablets contain not less than 90.0 percent and not more than 110.0 percent of the labeled amount of $C_{16}H_{18}FN_3O_3$.

Packaging and storage—Preserve in well-closed containers.

Reference standard—*USP Norfloxacin Reference Standard*—Dry in vacuum at a pressure not exceeding 5 mm of mercury at 100° for 2 hours before using.

Identification—

A: The retention time of the major peak in the chromatogram of the *Assay preparation* corresponds to that of the *Standard preparation* obtained as directed in the *Assay*.

B: Shake a quantity of finely powdered Tablets, equivalent to about 75 mg of norfloxacin, with 50 mL of a mixture of acidic methanol (prepared by mixing 1000 mL of methanol and 9 mL of hydrochloric acid) and methylene chloride (1:1). Centrifuge a portion of the suspension thus obtained, and use the clear supernatant solution as the test solution. On a suitable thin-layer chromatographic plate (see *Chromatography* ⟨621⟩), coated with a 0.25-mm layer of chromatographic silica gel mixture, apply 50 µL each of the test solution and a standard solution of USP Norfloxacin RS in the same solvent containing 1.5 mg per mL. Place the plate in a suitable chromatographic chamber that contains and has been equilibrated with a developing system consisting of a mixture of chloroform, methanol, toluene, diethylamine, and water (40:40:20:14:8), and develop the chromatogram until the solvent front has moved about three-fourths of the length of the plate. Remove the plate from the chamber, mark the solvent front, and allow the solvent to evaporate. Locate the spots on the plate by examination under short-wavelength ultraviolet light: the R_f value of the principal spot obtained from the test solution corresponds to that obtained from the Standard solution.

Dissolution ⟨711⟩—

pH 4.0 buffer—To 900 mL of water in a 1000-mL volumetric flask add 2.86 mL of glacial acetic acid and 1.0 mL of a 50% (w/w) solution of sodium hydroxide, dilute with water to volume, and mix. If necessary, adjust with glacial acetic acid or the sodium hydroxide solution to a pH of 4.0.

Medium: pH 4.0 buffer; 750 mL.

Apparatus 2: 50 rpm.

Time: 30 minutes.

Procedure—Determine the amount of $C_{16}H_{18}FN_3O_3$ dissolved from ultraviolet absorbances at the wavelength of maximum absorbance at about 313 nm of filtered portions of the solution under test, suitably diluted with *Dissolution Medium*, if necessary, in comparison with a Standard solution having a known concentration of USP Norfloxacin RS in the same medium.

Tolerances—Not less than 80% (*Q*) of the labeled amount of $C_{16}H_{18}FN_3O_3$ is dissolved in 30 minutes.

Uniformity of dosage units ⟨905⟩: meet the requirements.

Assay—

Mobile phase—Prepare a filtered and degassed mixture of phosphoric acid solution (1 in 1000) and acetonitrile (850:150). Make adjustments if necessary (see *System Suitability* under *Chromatography* ⟨621⟩).

Standard preparation—Dissolve an accurately weighed quantity of USP Norfloxacin RS quantitatively in *Mobile phase*, and dilute quantitatively, and stepwise if necessary, with *Mobile phase* to obtain a solution having a known concentration of about 0.2 mg per mL.

Assay preparation—Weigh and finely powder not less than 20 Norfloxacin Tablets. Transfer an accurately weighed portion of the powder, equivalent to about 100 mg of norfloxacin, to a 200-mL volumetric flask. Add 80 mL of *Mobile phase*, sonicate for 10 minutes, dilute with phosphoric acid solution (1 in 1000) to volume, and mix. Transfer 10.0 mL of this solution to a 25-mL volumetric flask, dilute with *Mobile phase* to volume, mix, and filter through a filter having a porosity of 1 µm or less.

Chromatographic system (see *Chromatography* ⟨621⟩)—The liquid chromatograph is equipped with a 275-nm detector and a 3.9-mm × 30-cm column that contains packing L1, and is operated at 40° ± 1.0°. Precondition the column for several hours with degassed 0.1 *M* monobasic sodium phosphate adjusted with phosphoric acid to a pH of 4.0. The flow rate is about 2 mL per minute. Chromatograph the *Standard preparation*, and record the peak responses as directed under *Procedure:* the tailing factor for the norfloxacin peak is not more than 2.0, and the relative standard deviation for replicate injections is not more than 2.0%.

Procedure—[NOTE—Use peak areas where peak responses are indicated.] Separately inject equal volumes (about 10 µL) of the *Standard preparation* and the *Assay preparation* into the chromatograph, record the chromatograms, and measure the peak responses for the major peaks. Calculate the quantity, in mg, of $C_{16}H_{18}FN_3O_3$ in the portion of Tablets taken by the formula:

$$500C(r_U/r_S),$$

in which *C* is the concentration, in mg per mL, of USP Norfloxacin RS in the *Standard preparation*, and r_U and r_S are the norfloxacin peak responses obtained from the *Assay preparation* and the *Standard preparation*, respectively.

Norgestrel

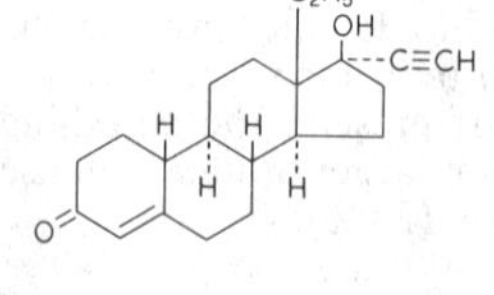

$C_{21}H_{28}O_2$ 312.45

18,19-Dinorpregn-4-en-20-yn-3-one, 13-ethyl-17-hydroxy-, (17α)-(±)-.

(±)-13-Ethyl-17-hydroxy-18,19-dinor-17α-pregn-4-en-20-yn-3-one [*6533-00-2*].

» Norgestrel contains not less than 98.0 percent and not more than 102.0 percent of $C_{21}H_{28}O_2$, calculated on the dried basis.

Packaging and storage—Preserve in well-closed containers.

Reference standard—*USP Norgestrel Reference Standard*—Dry at 105° for 3 hours before using.

Identification—The infrared absorption spectrum of a potassium bromide dispersion of it, previously dried at 105° for 3 hours, exhibits maxima only at the same wavelengths as that of a similar preparation of USP Norgestrel RS. If differences appear, dissolve portions of both the test specimen and the Reference Standard in ethyl acetate, evaporate the solutions on a steam bath to dryness, and repeat the test on the dried, solvent-free residues.

Melting range, *Class I* ⟨741⟩: between 205° and 212°, but the range between beginning and end of melting does not exceed 4°.

Optical rotation ⟨781⟩—A 1 in 20 solution in chloroform of a previously dried specimen exhibits an angular rotation between −0.1° and +0.1°, when determined in a 100-mm tube.

Loss on drying ⟨731⟩—Dry it at 105° for 5 hours: it loses not more than 0.5% of its weight.

Residue on ignition ⟨281⟩: not more than 0.3%.

Chromatographic impurities—

Phosphomolybdic acid reagent—Add 10 g of phosphomolybdic acid to 100 mL of alcohol, and stir the mixture for not less than 30 minutes. Filter before use.

Test preparation—Prepare a solution of Norgestrel in chloroform to contain 10.0 mg per mL.

Standard solution and *Standard dilutions*—Prepare a solution of USP Norgestrel RS in chloroform to contain 10 mg per mL (*Standard solution*). Prepare a series of dilutions of *Standard solution* in chloroform to contain 0.20, 0.10, 0.05, 0.02, and 0.01 mg per mL (*Standard dilutions*).

Procedure—Apply 10-µL volumes of *Standard solution*, the *Test preparation*, and each of the five *Standard dilutions* at equidistant points along a line 2.5 cm from one edge of a 20- × 20-cm thin-layer chromatographic plate (see *Chromatography* ⟨621⟩) coated with a 0.25-mm layer of chromatographic silica gel mixture and previously activated by heating at 100° for 15 minutes. Place the plate in a suitable developing chamber that contains and has been equilibrated with a mixture of 96 volumes of chloroform and 4 volumes of alcohol, seal the chamber, and allow the chromatogram to develop until the solvent front has moved 15 cm above the line of application. Remove the plate, allow the solvent to evaporate, then spray uniformly with *Phosphomolybdic acid reagent*, and heat it at 105° for 10 to 15 minutes. The lane of the *Test preparation* exhibits its principal spot at the same R_f as the principal spot of *Standard solution*. If spots other than the principal spot are observed in the lane of the *Test preparation*, estimate the concentration of each by comparison with the *Standard dilutions*. The spots from the 0.20-, 0.10-, 0.05-, 0.02-, and 0.01-mg per mL dilutions are equivalent to 2.0, 1.0, 0.5, 0.2, and 0.1% of impurities, respectively. The requirements of the test are met if the sum of the impurities in the *Test preparation* does not exceed 2.0%.

Ethynyl group—Proceed as directed in the test for *Ethynyl group* under *Norethindrone*. Not less than 7.81% and not more than 8.18% of ethynyl group is found.

Assay—Dissolve about 100 mg of Norgestrel, accurately weighed, in alcohol, and dilute quantitatively and stepwise with alcohol to obtain a solution containing about 10 µg per mL. Dissolve an accurately weighed quantity of USP Norgestrel RS in alcohol to obtain a Standard solution having a known concentration of about 10 µg per mL. Concomitantly determine the absorbances of both solutions in 1-cm cells at the wavelength of maximum absorbance at about 241 nm, with a suitable spectrophotometer, using alcohol as the blank. Calculate the quantity, in mg, of $C_{21}H_{28}O_2$ in the portion of Norgestrel taken by the formula:

$$10C(A_U/A_S),$$

in which C is the concentration, in µg per mL, of USP Norgestrel RS in the Standard solution, and A_U and A_S are the absorbances of the solution of Norgestrel and the Standard solution, respectively.

Norgestrel Tablets

» Norgestrel Tablets contain not less than 90.0 percent and not more than 110.0 percent of the labeled amount of $C_{21}H_{28}O_2$.

Packaging and storage—Preserve in well-closed containers.

Reference standard—*USP Norgestrel Reference Standard*—Dry at 105° for 3 hours before using.

Identification—Finely powder 20 Tablets, triturate the powder with 5 mL of chloroform, and allow the solids to settle. Apply 60 µL of the extract and 60 µL of a chloroform solution containing about 300 µg of USP Norgestrel RS per mL at points about 3 cm from one edge of a thin-layer chromatographic plate (see *Chromatography* ⟨621⟩), coated with a 0.25-mm layer of chromatographic silica gel mixture. Place the plate in a developing chamber containing a mixture of chloroform and alcohol (96:4) to a depth of 2 cm, the chamber having been previously equilibrated with the solvent mixture. Remove the plate when the solvent has moved about 15 cm from the line of application, dry at room temperature, spray with a mixture of 80 volumes of sulfuric acid and 20 volumes of alcohol, and heat at 105° for several minutes: the spot from the solution under test exhibits an R_f value identical to that of the spot from the Standard solution, and, when viewed under long-wavelength ultraviolet light, exhibits a red fluorescence similar to that from the Standard solution.

Disintegration ⟨701⟩: 15 minutes, the use of disks being omitted.

Uniformity of dosage units ⟨905⟩: meet the requirements.

Assay—

Isoniazid reagent—Dissolve 0.25 g of isoniazid and 0.3 mL of hydrochloric acid in 500 mL of dehydrated alcohol.

Procedure—Weigh and finely powder not less than 20 Norgestrel Tablets. Transfer an accurately weighed portion of the powder, equivalent to about 75 µg of norgestrel, to a 30-mL separator containing 5 mL of water. Extract with three 5-mL portions of chloroform, shaking for about 1 minute each time, and collecting the chloroform extracts through glass wool, previously moistened with chloroform, into a glass-stoppered test tube. Add 1 mL of dilute hydrochloric acid (1 in 12) to the remaining aqueous phase and extract with a fourth 5-mL portion of chloroform, collecting this chloroform extract as before and combining it with the previous three. To another glass-stoppered test tube transfer 20.0 mL of a solution of USP Norgestrel RS, in chloroform, having a known concentration of about 3.75 µg per mL. Evaporate the contents of both tubes in a water bath with the aid of a current of air to dryness. Add 5.0 mL of *Isoniazid reagent* to each tube, insert the stopper in each tube, and swirl occasionally for 1 hour. Concomitantly determine the absorbances of both solutions in 1-cm cells, at the wavelength of maximum absorbance at about 380 nm, using a suitable spectrophotometer, and using *Isoniazid reagent* as the blank. Calculate the quantity, in µg, of $C_{21}H_{28}O_2$ in the portion of Tablets taken by the formula:

$$20C(A_U/A_S),$$

in which C is the concentration, in µg per mL, of USP Norgestrel RS in the Standard solution, and A_U and A_S are the absorbances of the solutions from the Tablets and the Standard solution, respectively.

Norgestrel and Ethinyl Estradiol Tablets

» Norgestrel and Ethinyl Estradiol Tablets contain not less than 90.0 percent and not more than 110.0 percent of the labeled amount of norgestrel ($C_{21}H_{28}O_2$) and not less than 90.0 percent and not more than 110.0 percent of the labeled amount of ethinyl estradiol ($C_{20}H_{24}O_2$).

Packaging and storage—Preserve in well-closed containers.

Reference standards—*USP Norgestrel Reference Standard*—Dry at 105° for 3 hours before using. *USP Ethinyl Estradiol Reference Standard*—Dry in vacuum over silica gel for 4 hours before using.

Identification—Finely powder 20 Tablets, triturate the powder with 15 mL of methylene chloride, and allow the solids to settle. Apply 30 µL of the extract and 30 µL of a methylene chloride solution containing about 670 µg of USP Norgestrel RS and about 67 µg of USP Ethinyl Estradiol RS per mL at points about 3 cm from one edge of a thin-layer chromatographic plate (see *Chromatography* ⟨621⟩), consisting of a 0.25-mm layer of chromatographic silica gel mixture. Place the plate in a developing chamber containing a mixture of 96 volumes of chloroform and 4 volumes of alcohol to a depth of 2 cm, the chamber having been previously equilibrated with the solvent mixture. Remove the plate when the solvent has moved about 15 cm from the line of application, dry it at room temperature, spray it with a mixture of 80 volumes of sulfuric acid and 20 volumes of methanol, and heat it at 105° for several minutes: the spots from the solution under test exhibit R_f values equivalent to those from the Standard solution, and, when viewed under long-wavelength ultraviolet light, fluoresce similarly to those from the Standard solution (red for norgestrel and orange-yellow for ethinyl estradiol).

Disintegration ⟨701⟩: 15 minutes, the use of disks being omitted.

Uniformity of dosage units ⟨905⟩: meet the requirements for *Content Uniformity* with respect to norgestrel and to ethinyl estradiol, 5-cm cells and a *Standard preparation* containing about 1.7 µg per mL being used in the test for ethinyl estradiol.

Assay for norgestrel—Weigh and finely powder not less than 20 Norgestrel and Ethinyl Estradiol Tablets, and transfer a portion of the powder, equivalent to about 500 µg of norgestrel, to a 50-mL volumetric flask. Add 5 mL of water, shake for about 1 minute, then add 30 mL of alcohol, and shake for about 20 minutes. Add alcohol to volume, mix, and centrifuge the mixture. Dissolve an accurately weighed quantity of USP Norgestrel RS in alcohol, each 100 mL of which contains 10 mL of water, and dilute quantitatively and stepwise with this solvent to obtain a Standard solution having a known concentration of about 10 µg per mL. Concomitantly determine the absorbances of both solutions in 1-cm cells at the wavelength of maximum absorbance at about 241 nm, using alcohol, each 100 mL of which contains 10 mL of water, as the blank. Calculate the quantity, in µg, of $C_{21}H_{28}O_2$ in the portion of Tablets taken by the formula:

$$50C(A_U/A_S),$$

in which C is the concentration, in µg per mL, of USP Norgestrel RS in the Standard solution, and A_U and A_S are the absorbances of the solutions from the Tablets and the Standard solution, respectively.

Assay for ethinyl estradiol—[NOTE—Use separators fitted with polytetrafluoroethylene stopcocks, and complete the procedure promptly.]

Sulfuric acid solution—Cautiously add sulfuric acid to 200 mL of chilled water in a 1-liter volumetric flask, in small increments and with mixing. Adjust to room temperature, add sulfuric acid to volume, and mix.

Standard preparation—Transfer about 50 mg of USP Ethinyl Estradiol RS, accurately weighed, to a 100-mL volumetric flask, add chloroform to volume, and mix. Pipet 1 mL of this solution into a 100-mL volumetric flask, add chloroform to volume, and mix.

Assay preparation—Weigh a portion of the powdered Norgestrel and Ethinyl Estradiol Tablets, equivalent to about 150 µg of ethinyl estradiol, and transfer it to a 30-mL separator. Add 4 mL of water, shake the mixture for 1 minute, and allow it to stand for an additional 5 minutes. Add 1 mL of glacial acetic acid, shake vigorously for 2 minutes, and immediately extract the ethinyl estradiol with three 10-mL portions of chloroform, shaking the separator vigorously for 3 minutes each time. Combine the chloroform extracts in a 125-mL separator.

Procedure—Transfer 30 mL of the *Standard preparation* to a 125-mL separator. Reduce the volumes of the *Standard preparation* and *Assay preparation* in the separators, with the aid of currents of air, to about 3 mL each. Add 50 mL of isooctane to each separator, and immediately extract the ethinyl estradiol with three 8-mL portions of 1 *N* sodium hydroxide, shaking the separator vigorously for 3 minutes each time. Break up emulsions, if necessary, by probing with a thin metal spatula. Combine the aqueous extracts from each separator in a 25-mL volumetric flask, add 1 *N* sodium hydroxide to volume, and mix. Transfer 15 mL of *Sulfuric acid solution* to each of two 125-mL conical flasks, and place the flasks in an ice bath. Stir the contents of the flasks rapidly and at the same rate, using magnetic stirring devices, and add a 5.0-mL portion of the combined alkaline aqueous extracts obtained from the *Assay preparation* and the *Standard preparation*, respectively, to each of two separate flasks, introducing the solutions dropwise and away from the vortexes. When the addition is complete, continue stirring for 2 additional minutes before removing the flasks from the ice baths. Within 10 to 15 minutes after acidification of the extracts, concomitantly determine the absorbances of both solutions in 1-cm cells at the wavelength of maximum absorbance at about 536 nm, with a suitable spectrophotometer, using water as the blank. Calculate the quantity, in µg, of $C_{20}H_{24}O_2$ in the portion of Tablets taken by the formula:

$$30C(A_U/A_S),$$

in which C is the concentration, in µg per mL, of USP Ethinyl Estradiol RS in the *Standard preparation*, and A_U and A_S are the absorbances of the solutions from the *Assay preparation* and the *Standard preparation*, respectively.

Nortriptyline Hydrochloride

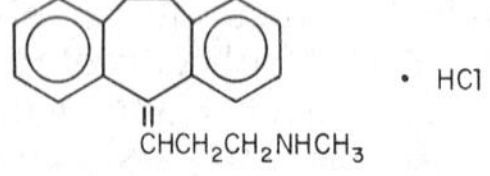

$C_{19}H_{21}N.HCl$ 299.84

1-Propanamine, 3-(10,11-dihydro-5*H*-dibenzo[*a,d*]cyclohepten-5-ylidene)-*N*-methyl-, hydrochloride.

10,11-Dihydro-*N*-methyl-5*H*-dibenzo[*a,d*]cycloheptene-Δ^5,γ-propylamine hydrochloride [*894-71-3*].

» Nortriptyline Hydrochloride contains not less than 97.0 percent and not more than 101.5 percent of $C_{19}H_{21}N.HCl$, calculated on the dried basis.

Packaging and storage—Preserve in tight, light-resistant containers.

Reference standard—*USP Nortriptyline Hydrochloride Reference Standard*—Dry at 105° for 3 hours before using.

Identification—

A: The infrared absorption spectrum of a 1 in 20 solution of it, previously dried, in chloroform exhibits maxima only at the same wavelengths as that of a similar solution of USP Nortriptyline Hydrochloride RS.

B: The ultraviolet absorption spectrum of a 1 in 100,000 solution in methanol exhibits maxima and minima at the same wavelengths as that of a similar solution of USP Nortriptyline Hydrochloride RS, concomitantly measured, and the respective absorptivities, calculated on the dried basis, at the wavelength of maximum absorbance at about 239 nm do not differ by more than 3.0%.

C: It responds to the tests for *Chloride* ⟨191⟩ when tested as specified for alkaloidal hydrochlorides.

Melting range, *Class I* ⟨741⟩: between 215° and 220°, but the range between beginning and end of melting does not exceed 3°.

Loss on drying ⟨731⟩—Dry it at 105° for 3 hours: it loses not more than 0.5% of its weight.

Residue on ignition ⟨281⟩: not more than 0.1%.

Heavy metals, *Method II* ⟨231⟩—The limit is 0.001%.

Assay—Dissolve about 600 mg of Nortriptyline Hydrochloride, accurately weighed, in 50 mL of glacial acetic acid, add 10 mL of mercuric acetate TS, and titrate with 0.1 *N* perchloric acid VS, determining the end-point potentiometrically. Perform a blank determination, and make any necessary correction. Each mL of 0.1 *N* perchloric acid is equivalent to 29.98 mg of $C_{19}H_{21}N.HCl$.

Nortriptyline Hydrochloride Capsules

» Nortriptyline Hydrochloride Capsules contain nortriptyline hydrochloride equivalent to not less than 90.0 percent and not more than 110.0 percent of the labeled amount of nortriptyline ($C_{19}H_{21}N$).

Packaging and storage—Preserve in tight containers.

Reference standard—*USP Nortriptyline Hydrochloride Reference Standard*—Dry at 105° for 3 hours before using.

Identification—

A: Transfer the contents of Capsules, equivalent to about 50 mg of nortriptyline hydrochloride, to a suitable flask. Add 15 mL of chloroform, insert the stopper in the flask, and shake for 15 minutes. Transfer the mixture to a suitable centrifuge tube, and centrifuge at about 2900 rpm for about 5 minutes. Filter through a suitable filter paper containing a small amount of anhydrous sodium sulfate. Evaporate the filtrate to dryness, and dissolve the residue in 0.5 mL of chloroform: the infrared absorption spectrum of this solution exhibits maxima only at the same wavelengths as that of a Standard solution prepared by dissolving 50 mg of USP Nortriptyline Hydrochloride RS in 0.5 mL of chloroform.

B: A filtered solution in water of the contents of Capsules, equivalent to nortriptyline hydrochloride solution (1 in 20), responds to the tests for *Chloride* ⟨191⟩, when tested as specified for alkaloidal hydrochlorides.

Dissolution ⟨711⟩—

Medium: water; 500 mL.

Apparatus 1: 100 rpm.

Time: 30 minutes.

Procedure—Determine the amount of $C_{19}H_{21}N$ dissolved from ultraviolet absorbances at the wavelength of maximum absorbance at about 239 nm of filtered portions of the solution under test, suitably diluted with water, if necessary, in comparison with a Standard solution having a known concentration of USP Nortriptyline Hydrochloride RS in the same medium.

Tolerances—Not less than 70% (*Q*) of the labeled amount of $C_{19}H_{21}N$ is dissolved in 30 minutes.

Uniformity of dosage units ⟨905⟩: meet the requirements.

Assay—

Ammonium carbonate solution—Dissolve 300 mg of ammonium carbonate in 50 mL of water, and mix.

Mobile phase—Prepare a filtered and degassed mixture of acetonitrile, methanol, and *Ammonium carbonate solution* (475:475:50), making adjustments if necessary (see *System Suitability* under *Chromatography* ⟨621⟩).

Standard preparation—Dissolve an accurately weighed quantity of USP Nortriptyline Hydrochloride RS in methanol, and dilute quantitatively, and stepwise if necessary, with methanol to obtain a solution having a known concentration of about 114 µg of USP Nortriptyline Hydrochloride RS per mL (equivalent to about 100 µg of nortriptyline per mL).

Assay preparation—Weigh and mix the contents of not less than 20 Nortriptyline Hydrochloride Capsules. Transfer an accurately weighed portion of Capsule contents, equivalent to about 25 mg of nortriptyline, to a 250-mL volumetric flask. Add about 200 mL of methanol, and shake by mechanical means for 15 minutes. Dilute with methanol to volume, mix, and filter, discarding the first 5 mL of the filtrate.

Chromatographic system (see *Chromatography* ⟨621⟩)—The liquid chromatograph is equipped with a 239-nm detector and a 4.0-mm × 25-cm column that contains packing L16. The flow rate is about 3 mL per minute. Chromatograph replicate injections of the *Standard preparation*, and record the peak responses as directed under *Procedure:* the column efficiency determined from the nortriptyline hydrochloride peak is not less than 500 theoretical plates, the tailing factor for the nortriptyline hydrochloride peak is not more than 3.0, and the relative standard deviation is not more than 2.0%.

Procedure—Separately inject equal volumes (about 20 µL) of the *Standard preparation* and the *Assay preparation* into the chromatograph, record the chromatograms, and measure the responses for the major peaks. Calculate the quantity, in mg, of nortriptyline ($C_{19}H_{21}N$) in the portion of Capsules taken by the formula:

$$(263.38/299.84)(0.25C)(r_U/r_S),$$

in which 263.38 and 299.84 are the molecular weights of nortriptyline and nortriptyline hydrochloride, respectively, *C* is the concentration, in µg per mL, of USP Nortriptyline Hydrochloride RS in the *Standard preparation*, and r_U and r_S are the peak responses obtained from the *Assay preparation* and the *Standard preparation*, respectively.

Nortriptyline Hydrochloride Oral Solution

» Nortriptyline Hydrochloride Oral Solution contains nortriptyline hydrochloride equivalent to not less than 90.0 percent and not more than 110.0 percent of the labeled amount of nortriptyline ($C_{19}H_{21}N$).

Packaging and storage—Preserve in tight, light-resistant containers.

Reference standard—*USP Nortriptyline Hydrochloride Reference Standard*—Dry at 105° for 3 hours before using.

Identification—

A: Transfer a measured volume of Oral Solution, equivalent to about 50 mg of nortriptyline hydrochloride, to a suitable separator, and render the solution distinctly alkaline (to a pH of 11 or above as indicated by pH indicator paper) by the dropwise addition of 1 *N* sodium hydroxide. Extract with 15 mL of chloroform, and filter the chloroform extract through about 2 g of anhydrous sodium sulfate that has been previously washed with chloroform. Evaporate the chloroform extract with the aid of heat and a current of air to dryness, and dissolve the residue in 0.5 mL of chloroform: the infrared absorption spectrum of this solution exhibits maxima only at the same wavelengths as that of a Standard solution obtained by dissolving 50 mg of USP Nortriptyline Hydrochloride RS in 25 mL of water and proceeding as directed for the test specimen.

B: It responds to the tests for *Chloride* ⟨191⟩, when tested as specified for alkaloidal hydrochlorides.

pH ⟨791⟩: between 2.5 and 4.0.

Alcohol content, *Method II* ⟨611⟩: between 3.0% and 5.0% of C_2H_5OH.

Assay—Transfer an accurately measured volume of Nortriptyline Hydrochloride Oral Solution, equivalent to about 10 mg of nortriptyline, to a 125-mL separator. Add 20 mL of water, mix, and render the solution distinctly alkaline (to a pH of 11 or above as indicated by pH indicator paper) by the dropwise addition of sodium hydroxide solution (1 in 2). Extract the nortriptyline with four 25-mL portions of chloroform, filtering each extract into a 250-mL beaker through about 12 g of anhydrous sodium sulfate previously washed with 25 mL of chloroform. Rinse the sodium sulfate with four 5-mL portions of chloroform, and collect the rinsings in the beaker. Evaporate the combined chloroform solution with the aid of heat and a current of air to about 10 mL. Transfer the contents of the beaker with the aid of chloroform to a 200-mL volumetric flask. Evaporate the chloroform with the aid of air alone to dryness. [*Caution—Do not use heat.*] Dissolve the residue in 1.7 mL of hydrochloric acid, dilute with water to volume, and mix. Transfer 10.0 mL of the solution to a 50-mL volumetric flask, dilute with water to volume, and mix to obtain the *Assay preparation.* Concomitantly determine the absorbances of the *Assay preparation* and a Standard solution of USP Nortriptyline Hydrochloride RS in water having a known concentration of about 11.4 µg per mL in 1-cm cells at the wavelength of maximum absorbance at about 239 nm, with a suitable spectrophotometer, using water as the blank. Calculate the quantity, in mg, of $C_{19}H_{21}N$ in the portion of Oral Solution taken by the formula:

$$(263.38/299.84)(C)(A_U/A_S),$$

in which 263.38 and 299.84 are the molecular weights of nortriptyline and nortriptyline hydrochloride, respectively, *C* is the concentration, in µg per mL, of USP Nortriptyline Hydrochloride RS in the Standard solution, and A_U and A_S are the absorbances of the *Assay preparation* and the Standard solution, respectively.

Noscapine

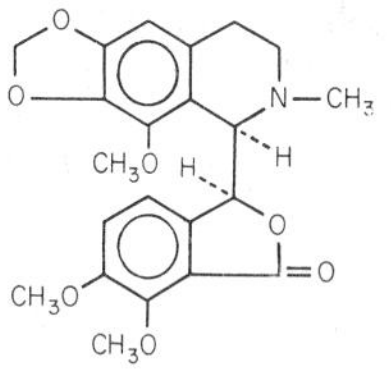

$C_{22}H_{23}NO_7$ 413.43

1(3*H*)-Isobenzofuranone, 6,7-dimethoxy-3-(5,6,7,8-tetrahydro-4-methoxy-6-methyl-1,3-dioxolo[4,5-*g*]isoquinolin-5-yl), [*S*-(*R**,*S**)]-.

Narcotine [*128-62-1*].

» Noscapine contains not less than 99.0 percent and not more than 100.5 percent of $C_{22}H_{23}NO_7$, calculated on the anhydrous basis.

Packaging and storage—Preserve in well-closed containers.

Reference standard—*USP Noscapine Reference Standard*—Do not dry; determine the water content at time of use.

Identification—

A: The infrared absorption spectrum of a potassium bromide dispersion of it exhibits maxima only at the same wavelengths as that of a similar preparation of USP Noscapine RS.

B: The ultraviolet absorption spectrum of a 1 in 16,000 solution in methanol exhibits maxima and minima at the same wavelengths as that of a similar solution of USP Noscapine RS, concomitantly measured.

C: Place about 100 mg in a small porcelain dish, add a few drops of sulfuric acid, and stir: a greenish yellow solution is produced, and on warming it becomes red and then turns violet.

Melting range ⟨741⟩: between 174° and 176°.

Specific rotation ⟨781⟩: between +42° and +48°, calculated on the anhydrous basis, determined in a solution in 0.1 *N* hydrochloric acid containing 200 mg in each 10 mL.

Water, *Method I* ⟨921⟩: not more than 1.0%.

Residue on ignition ⟨281⟩: not more than 0.1%.

Chloride ⟨221⟩—A 700-mg portion shows no more chloride than corresponds to 0.20 mL of 0.020 *N* hydrochloric acid (0.02%).

Morphine—Dissolve 100 mg in 10 mL of 0.1 *N* hydrochloric acid. To 1.0 mL of this solution add 5.0 mL of diluted ferricyanide reagent (prepared by dissolving 0.50 g of potassium ferricyanide in 50 mL of water, adding 0.50 mL of ferric chloride TS, and diluting 5.0 mL of the resulting solution to 25.0 mL): no blue or dark green color develops within 1 minute.

Ordinary impurities ⟨466⟩—
Test solution: chloroform.
Standard solution: chloroform.
Eluant: a mixture of ethyl acetate and ether (80:20).
Visualization: 17; then examine the plate immediately.
Limits—The sum of the intensities of all secondary spots obtained from the *Test solution* corresponds to not more than 1.0%.

Assay—Dissolve about 1.5 g of Noscapine, accurately weighed, in 25 mL of glacial acetic acid. Add 25 mL of dioxane and 5 drops of crystal violet TS, and titrate with 0.1 *N* perchloric acid VS to a blue end-point. Perform a blank determination, and make any necessary correction. Each mL of 0.1 *N* perchloric acid is equivalent to 41.34 mg of $C_{22}H_{23}NO_7$.

Novobiocin Calcium

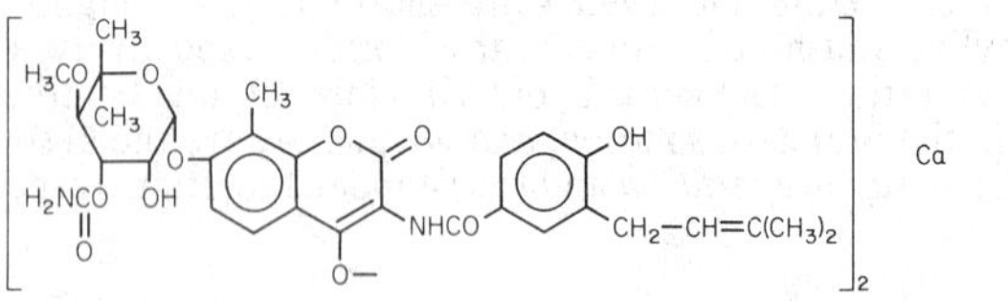

$C_{62}H_{70}CaN_4O_{22}$ 1263.33
Benzamide, *N*-[7-[[3-*O*-(aminocarbonyl)-6-deoxy-5-*C*-methyl-4-*O*-methyl-β-L-*lyxo*-hexopyranosyl]oxy]-4-hydroxy-8-methyl-2-oxo-2*H*-1-benzopyran-3-yl]-4-hydroxy-3-(3-methyl-2-butenyl)-, calcium salt (2:1).
Novobiocin, calcium salt [*4309-70-0*].

» Novobiocin Calcium has a potency equivalent to not less than 840 μg of novobiocin ($C_{31}H_{36}N_2O_{11}$) per mg, calculated on the dried basis.

Packaging and storage—Preserve in tight containers.

Reference standard—*USP Novobiocin Reference Standard*—Dry in vacuum at a pressure not exceeding 5 mm of mercury at 100° for 4 hours before using.

Identification—

A: Prepare a test solution by dissolving a quantity of it in methanol to obtain a concentration of about 1 mg of novobiocin per mL. Similarly prepare a Standard solution, using USP Novobiocin RS. Separately apply 1-μL portions of the test solution and the Standard solution on a suitable thin-layer chromatographic plate (see *Chromatography* ⟨621⟩) coated with a 0.25-mm layer of chromatographic silica gel mixture, and allow the spots to dry. Place the plate in a chromatographic chamber equilibrated with a solvent system consisting of a mixture of chloroform, methanol, and ammonium hydroxide (75:25:1), and develop the chromatogram. When the solvent front has moved about three-fourths of the length of the plate, remove the plate from the chamber, and allow to dry. Locate the spots on the plate by examination under short-wavelength ultraviolet light: the R_f value of the principal spot obtained from the test solution corresponds to that obtained from the Standard solution.

B: The residue obtained by igniting it responds to the tests for *Calcium* ⟨191⟩.

Specific rotation ⟨781⟩: between −50° and −58°, calculated on the dried basis, determined in a solution containing 50 mg in each mL of a solvent consisting of methanol and hydrochloric acid (100:1).

Crystallinity ⟨695⟩: meets the requirements.

pH ⟨791⟩: between 6.5 and 8.5, in a saturated aqueous suspension containing 25 mg per mL.

Loss on drying ⟨731⟩—Dry about 100 mg, accurately weighed, in a capillary-stoppered bottle in vacuum at a pressure not exceeding 5 mm of mercury at 60° for 3 hours: it loses not more than 10.0% of its weight.

Assay—Proceed with Novobiocin Calcium as directed under *Antibiotics—Microbial Assays* ⟨81⟩.

Novobiocin Calcium Oral Suspension

» Novobiocin Calcium Oral Suspension is prepared from Novobiocin Calcium or Novobiocin Sodium reacted with a suitable calcium salt, and it contains one or more suitable buffers, colors, diluents, flavors, and preservatives. It contains the equivalent of not less than 90.0 percent and not more than 120.0 percent of the labeled amount of novobiocin ($C_{31}H_{36}N_2O_{11}$).

Packaging and storage—Preserve in tight, light-resistant containers.

Reference standard—*USP Novobiocin Reference Standard*—Dry in vacuum at a pressure not exceeding 5 mm of mercury at 100° for 4 hours before using.

Identification—Shake a suitable quantity of Oral Suspension with methanol to obtain a solution containing 1 mg of novobiocin per mL, and filter: the filtrate (test solution) responds to *Identification test A* under *Novobiocin Calcium*.

Uniformity of dosage units ⟨905⟩—
FOR SUSPENSION PACKAGED IN SINGLE-UNIT CONTAINERS: meets the requirements.

pH ⟨791⟩: between 6.0 and 7.5.

Assay—Transfer an accurately measured quantity of Novobiocin Calcium Oral Suspension, freshly mixed and free from air bubbles, to a high-speed blender containing an accurately measured volume of alcohol sufficient to give a concentration of about 1000 μg of novobiocin per mL, and blend for about 4 minutes. Proceed as directed under *Antibiotics—Microbial Assays* ⟨81⟩, using an accurately measured volume of this solution diluted quantitatively and stepwise with *Buffer No. 6* to yield a *Test Dilution* having a concentration assumed to be equal to the median dose level of the Standard.

Novobiocin Sodium

$C_{31}H_{35}N_2NaO_{11}$ 634.62
Benzamide, *N*-[7-[[3-*O*-(aminocarbonyl)-6-deoxy-5-*C*-methyl-4-*O*-methyl-β-L-*lyxo*-hexopyranosyl]oxy]-4-hydroxy-8-methyl-2-oxo-2*H*-1-benzopyran-3-yl]-4-hydroxy-3-(3-methyl-2-butenyl)-, monosodium salt.
Novobiocin, monosodium salt [*1476-53-5*].

» Novobiocin Sodium has a potency equivalent to not less than 850 μg of novobiocin ($C_{31}H_{36}N_2O_{11}$) per mg, calculated on the dried basis.

Packaging and storage—Preserve in tight containers.

Reference standard—*USP Novobiocin Reference Standard*—Dry in vacuum at a pressure not exceeding 5 mm of mercury at 100° for 4 hours before using.

Identification—

A: It responds to *Identification test A* under *Novobiocin Calcium*.

B: The residue obtained by igniting it responds to the tests for *Sodium* ⟨191⟩.

Specific rotation ⟨781⟩: between −50° and −58°, calculated on the dried basis, determined in a solution containing 50 mg in each mL of a solvent consisting of a mixture of methanol and hydrochloric acid (100:1).

Crystallinity ⟨695⟩: meets the requirements.

pH ⟨791⟩: between 6.5 and 8.5, in a solution containing 25 mg per mL.

Loss on drying ⟨731⟩—Dry about 100 mg, accurately weighed, in a capillary-stoppered bottle in vacuum at a pressure not exceeding 5 mm of mercury at 60° for 3 hours: it loses not more than 6.0% of its weight.

Residue on ignition ⟨281⟩: between 10.5% and 12.0%, the charred residue being moistened with 2 mL of sulfuric acid and an ignition temperature of 550 ± 50° being used.

Assay—Dissolve a suitable quantity of Novobiocin Sodium, accurately weighed, in an accurately measured volume of *Buffer No. 3* sufficient to obtain a stock solution of convenient concentration. Proceed as directed under *Antibiotics—Microbial Assays* ⟨81⟩, using an accurately measured volume of this stock solution diluted quantitatively and stepwise with *Buffer No. 6* to yield a *Test Dilution* having a concentration assumed to be equal to the median dose level of the Standard.

Novobiocin Sodium Capsules

» Novobiocin Sodium Capsules contain not less than 90.0 percent and not more than 120.0 percent of the labeled amount of novobiocin ($C_{31}H_{36}N_2O_{11}$).

Packaging and storage—Preserve in tight, light-resistant containers.

Reference standard—*USP Novobiocin Reference Standard*—Dry in vacuum at a pressure not exceeding 5 mm of mercury at 100° for 4 hours before using.

Identification—Shake the contents of 1 Capsule with methanol to obtain a solution having a concentration of about 1 mg of novobiocin per mL, and filter: the filtrate (test solution) responds to *Identification test A* under *Novobiocin Calcium.*

Loss on drying ⟨731⟩—Dry about 100 mg, accurately weighed, of the powder obtained from 4 Capsules, in vacuum at a pressure not exceeding 5 mm of mercury at 60° for 3 hours: it loses not more than 6.0% of its weight.

Assay—Place not less than 5 Novobiocin Sodium Capsules in a high-speed glass blender jar containing 1.0 mL of polysorbate 80 and an accurately measured volume of *Buffer No. 3* sufficient to give a stock solution of convenient concentration, and blend for about 4 minutes. Proceed as directed under *Antibiotics—Microbial Assays* ⟨81⟩, using an accurately measured volume of this stock solution diluted quantitatively and stepwise with *Buffer No. 6* to yield a *Test Dilution* having a concentration assumed to be equal to the median dose level of the Standard.

Novobiocin Sodium Capsules, Tetracycline Phosphate Complex and—*see* Tetracycline Phosphate Complex and Novobiocin Sodium Capsules

Novobiocin Sodium Intramammary Infusion, Penicillin G Procaine and—*see* Penicillin G Procaine and Novobiocin Sodium Intramammary Infusion

Novobiocin Sodium, and Prednisolone Tablets, Tetracycline Hydrochloride,—*see* Tetracycline Hydrochloride, Novobiocin Sodium, and Prednisolone Tablets

Novobiocin Sodium Tablets, Tetracycline Hydrochloride and—*see* Tetracycline Hydrochloride and Novobiocin Sodium Tablets

Nylidrin Hydrochloride

HO–C6H4–CH(OH)–CH(CH3)–NHCH(CH3)CH2CH2–C6H5 · HCl

$C_{19}H_{25}NO_2 . HCl$ 335.87

Benzenemethanol, 4-hydroxy-α-[1-[(1-methyl-3-phenylpropyl)amino]ethyl]-, hydrochloride.

p-Hydroxy-α-[1-[(1-methyl-3-phenylpropyl)amino]ethyl]benzyl alcohol hydrochloride [*900-01-6; 849-55-8*].

» Nylidrin Hydrochloride contains not less than 98.0 percent and not more than 102.0 percent of $C_{19}H_{25}NO_2 . HCl$, calculated on the dried basis.

Packaging and storage—Preserve in tight containers.

Reference standard—*USP Nylidrin Hydrochloride Reference Standard*—Dry in vacuum at 60° for 3 hours before using.

Identification—

A: The infrared absorption spectrum of a potassium bromide dispersion of it, previously dried, exhibits maxima only at the same wavelengths as that of a similar preparation of USP Nylidrin Hydrochloride RS.

B: The ultraviolet absorption spectrum of a 1 in 10,000 solution in alcohol exhibits maxima and minima at the same wavelengths as that of a similar solution of USP Nylidrin Hydrochloride RS, concomitantly measured, and the respective absorptivities, calculated on the dried basis, at the wavelength of maximum absorbance at about 277 nm do not differ by more than 3.0%.

C: Dissolve about 50 mg in 5 mL of water, with heating, and cool. Add 2 drops of nitric acid and 1 mL of silver nitrate TS: a white precipitate is formed, and it is insoluble in diluted nitric acid, but soluble in 6 *N* ammonium hydroxide (*presence of chloride*).

pH ⟨791⟩: between 4.5 and 6.5, in a solution (1 in 100).

Loss on drying ⟨731⟩—Dry it in vacuum at 60° for 3 hours: it loses not more than 0.5% of its weight.

Residue on ignition ⟨281⟩: not more than 0.5%.

Assay—

Mobile phase—Prepare a solution of 0.01 *M* dibasic ammonium phosphate, and adjust with phosphoric acid to a pH of 7.5. Mix with methanol (about 1 in 4) such that the retention times for nylidrin hydrochloride and fluorene are approximately 5 and 7 minutes, respectively. Filter through a 0.45-μm porosity membrane filter to degas prior to use.

Internal standard solution—Dissolve fluorene with *Mobile phase* to obtain a solution having a concentration of about 0.5 mg per mL.

Standard preparation—Accurately weigh about 30 mg of USP Nylidrin Hydrochloride RS, and transfer to a 25-mL volumetric flask. Pipet 5 mL of *Internal standard solution* into the volumetric flask, and dilute with *Mobile phase* to volume. Shake vigorously until the standard dissolves. This *Standard preparation* contains about 1.2 mg of nylidrin hydrochloride and about 0.1 mg of fluorene per mL.

Assay preparation—Weigh accurately about 30 mg of Nylidrin Hydrochloride, and prepare as directed under *Standard preparation.*

Chromatographic system (see *Chromatography* ⟨621⟩)—The liquid chromatograph is equipped with an ultraviolet detector capable of monitoring absorption at 276 nm, a suitable recorder, is fitted with a 4-mm × 25-cm stainless steel column that contains packing L1, and is operated at room temperature. The *Mobile solvent* is flowing at a rate of 1.5 mL per minute. In a suitable chromatogram, five replicate injections of the *Standard preparation* show a relative standard deviation of not more than 2.0%, the tailing factors for the nylidrin hydrochloride and fluorene peaks are not more than 2.0, and the resolution factor is not less than 1.5 between the two peaks.

Procedure—Separately inject equal volumes (about 20 μL) of the *Standard preparation* and the *Assay preparation* into the chromatograph by means of a suitable microsyringe or sampling valve, and adjust the operating parameters, if necessary, until satisfactory chromatographic responses are obtained. Calculate the quantity, in mg, of $C_{19}H_{25}NO_2 . HCl$ in the portion of Nylidrin Hydrochloride taken by the formula:

$$25C(R_U/R_S),$$

in which C is the concentration, in mg per mL, of USP Nylidrin Hydrochloride RS in the *Standard preparation*, and R_U and R_S are the ratios of peak areas of nylidrin hydrochloride to the internal standard obtained from the *Assay preparation* and the *Standard preparation*, respectively.

Nylidrin Hydrochloride Injection

» Nylidrin Hydrochloride Injection is a sterile solution of Nylidrin Hydrochloride in Water for Injection. It contains not less than 95.0 percent and not more than 105.0 percent of the labeled amount of $C_{19}H_{25}NO_2 . HCl$.

Packaging and storage—Preserve in single-dose or in multiple-dose containers, preferably of Type I glass.

Reference standard—*USP Nylidrin Hydrochloride Reference Standard*—Dry in vacuum at 60° for 3 hours before using.

Identification—

A: Transfer a volume of Injection, equivalent to about 100 mg of nylidrin hydrochloride, to a 60-mL separator, and add 5 mL of sodium carbonate TS. Extract the liberated base with two 10-mL portions of chloroform. Dry the extract with anhydrous sodium sulfate, allowing the mixture to stand for 10 minutes after an excess of the drying agent has been added. Filter through a dry filter paper, collect the filtrate in a 50-mL beaker, and evaporate on a steam bath with the aid of a current of warm air to dryness. Recrystallize the residue from diluted alcohol, and dry in a vacuum desiccator over phosphorus pentoxide for 30 minutes: the infrared absorption spectrum of the nylidrin so obtained exhibits maxima only at the same wavelengths as that of a similar preparation of USP Nylidrin Hydrochloride RS, similarly treated.

B: It responds to *Identification test C* under *Nylidrin Hydrochloride.*

Other requirements—It meets the requirements under *Injections* ⟨1⟩.

Assay—Transfer an accurately measured volume of Nylidrin Hydrochloride Injection, equivalent to about 40 mg of nylidrin hydrochloride, to a 125-mL separator, and dilute with water to 35 mL. Add 5 mL of sodium carbonate TS, and extract the solution with four 10-mL portions of chloroform. Wash the combined chloroform extracts with 10 mL of water, and extract the wash solution with 5 mL of chloroform. Filter the combined chloroform extracts through a tightly packed, chloroform-washed pledget of cotton, and wash the cotton with 5 mL of chloroform. Add methanolic methyl red TS to the chloroform solution, and titrate with 0.01 *N* perchloric acid in dioxane VS. Perform a blank determination, and make any necessary correction. Each mL of 0.01 *N* perchloric acid is equivalent to 3.359 mg of $C_{19}H_{25}NO_2 . HCl$.

Nylidrin Hydrochloride Tablets

» Nylidrin Hydrochloride Tablets contain not less than 93.0 percent and not more than 107.0 percent of the labeled amount of $C_{19}H_{25}NO_2 . HCl$.

Packaging and storage—Preserve in tight containers.

Reference standard—*USP Nylidrin Hydrochloride Reference Standard*—Dry in vacuum at 60° for 3 hours before using.

Identification—Transfer an amount of powdered Tablets, equivalent to about 100 mg of nylidrin hydrochloride, to a 60-mL separator, add 20 mL of water, and proceed as directed in *Identification test A* under *Nylidrin Hydrochloride Injection*, beginning with "add 5 mL of sodium carbonate TS": the specified result is obtained.

Dissolution ⟨711⟩—

Medium: water; 900 mL.

Apparatus 1: 100 rpm.

Time: 30 minutes.

Procedure—Determine the amount of $C_{19}H_{25}NO_2 . HCl$ dissolved in a 10.0-mL filtered portion of the solution under test, to which is added 1.0 mL of tribasic sodium phosphate solution (1 in 10) from ultraviolet absorbances at the wavelength of maximum absorbance at about 242 nm, using a similarly prepared solution of water and buffer as the blank, in comparison with a Standard solution having a known concentration of USP Nylidrin Hydrochloride RS in the same medium.

Tolerances—Not less than 75% (*Q*) of the labeled amount of $C_{19}H_{25}NO_2 . HCl$ is dissolved in 30 minutes.

Uniformity of dosage units ⟨905⟩: meet the requirements.

Procedure for content uniformity—Transfer 1 finely powdered Tablet to a 100-mL volumetric flask, add tribasic sodium phosphate solution (1 in 50) to volume, mix, and centrifuge a portion of the mixture at high speed. Dilute an accurately measured volume of the supernatant liquid quantitatively and stepwise, if necessary, with the phosphate solution to obtain a solution containing about 10 µg of nylidrin hydrochloride per mL. Concomitantly determine the absorbances of this solution and of a Standard solution of USP Nylidrin Hydrochloride RS in the same medium having a known concentration of about 10 µg per mL in 1-cm cells at the wavelength of maximum absorbance at about 242 nm, with a suitable spectrophotometer, using the phosphate solution as the blank. Calculate the quantity, in mg, of $C_{19}H_{25}NO_2 . HCl$ in the Tablet by the formula:

$$(TC/D)(A_U/A_S),$$

in which T is the labeled quantity, in mg, of nylidrin hydrochloride in the Tablet, C is the concentration, in µg per mL, of USP Nylidrin Hydrochloride RS in the Standard solution, D is the concentration, in µg per mL, of nylidrin hydrochloride in the solution from the Tablet, on the basis of the labeled quantity per Tablet and the extent of dilution, and A_U and A_S are the absorbances of the solution from the Tablet and the Standard solution, respectively.

Assay—

Mobile phase, Internal standard solution, Chromatographic system, and *Standard preparation*—Prepare as directed in the *Assay* under *Nylidrin Hydrochloride.*

Assay preparation—Weigh and finely powder not less than 20 Nylidrin Hydrochloride Tablets. Transfer an accurately weighed portion of the powder, equivalent to about 30 mg of nylidrin hydrochloride, to a 50-mL centrifuge tube. Pipet 5 mL of *Internal standard solution* into the centrifuge tube, and add 20 mL of *Mobile phase.* Sonicate the tube for about 2 minutes, shake by mechanical means for 30 minutes, and centrifuge. Filter a portion of the supernatant solution through a 0.45-µm porosity membrane filter to obtain a clear filtrate.

Procedure—Separately inject equal volumes (about 20 µL) of the *Standard preparation* and the *Assay preparation* into the chromatograph by means of a suitable microsyringe or sampling valve, and adjust the operating parameters, if necessary, until satisfactory chromatographic responses are obtained. Calculate the quantity, in mg, of $C_{19}H_{25}NO_2 . HCl$ in the portion of Tablets taken by the formula:

$$25C(R_U/R_S),$$

in which C is the concentration, in mg per mL, of USP Nylidrin Hydrochloride RS in the *Standard preparation,* and R_U and R_S are the ratios of peak areas of nylidrin hydrochloride to the internal standard obtained from the *Assay preparation* and the *Standard preparation*, respectively.

Nystatin

Nystatin.
Nystatin [*1400-61-9*].

» Nystatin is a substance, or a mixture of two or more substances, produced by the growth of *Streptomyces noursei* Brown et al. (Fam. Streptomycetaceae). It has a potency of not less than 4400 USP Nystatin Units per mg, or, where intended for use in the extemporaneous preparation of oral suspensions, not less than 5000 USP Nystatin Units per mg.

Packaging and storage—Preserve in tight, light-resistant containers.

Labeling—Where packaged for use in the extemporaneous preparation of oral suspensions, the label so states.

Reference standard—*USP Nystatin Reference Standard*—Dry in vacuum at a pressure not exceeding 5 mm of mercury at 40° for 2 hours before using.

Identification—Transfer about 50 mg to a glass-stoppered, 100-mL volumetric flask, add 25 mL of methanol and 5 mL of glacial acetic acid to dissolve the specimen, dilute with methanol to volume, and mix. Pipet 2 mL of this solution into a 100-mL volumetric flask, dilute with methanol to volume, and mix: the ultraviolet absorption spectrum of this solution, determined immediately, exhibits maxima and minima at the same wavelengths as that of a similar solution of USP Nystatin RS, concomitantly measured, a 1 in 1000 solution of glacial acetic acid in methanol being used as the blank, and the ratio of its absorbance at 230 nm to that at 279 nm is between 0.90 and 1.25.

Suspendibility (where packaged for use in the extemporaneous preparation of oral suspensions)—Transfer about 200 mg, accurately weighed, to a 250-mL beaker containing 200.0 mL of water, and disperse by stirring gently with a stirring rod. Allow to stand for 2 minutes, and observe the suspension: the material is in suspension and little or no sediment is present on the bottom of the beaker. If there is any sediment, assay the undisturbed suspension as directed for nystatin under *Antibiotics—Microbial Assays* ⟨81⟩, using an accurately measured volume of it blended in a high-speed blender for 3 to 5 minutes with a sufficient accurately measured volume of dimethylformamide to give a concentration of about 400 USP Nystatin Units per mL. Dilute this stock solution quantitatively with *Buffer No. 6* to obtain a *Test Dilution* having a concentration assumed to be equal to the median dose level of the Standard: not less than 90.0% of the expected number of USP Nystatin Units is found, based on the potency obtained in the *Assay*.

Crystallinity (where packaged for use in the extemporaneous preparation of oral suspensions) ⟨695⟩: meets the requirements.

pH ⟨791⟩: between 6.5 and 8.0, in a 3% aqueous suspension.

Loss on drying ⟨731⟩—Dry about 100 mg, accurately weighed, in a capillary-stoppered bottle in vacuum at a pressure not exceeding 5 mm of mercury at 60° for 3 hours: it loses not more than 5.0% of its weight.

Assay—Proceed with Nystatin as directed under *Antibiotics—Microbial Assays* ⟨81⟩.

Nystatin Capsules, Demeclocycline Hydrochloride and—*see* Demeclocycline Hydrochloride and Nystatin Capsules

Nystatin Capsules, Oxytetracycline and—*see* Oxytetracycline and Nystatin Capsules

Nystatin Capsules, Tetracycline Hydrochloride and—*see* Tetracycline Hydrochloride and Nystatin Capsules

Nystatin Cream

» Nystatin Cream contains not less than 90.0 percent and not more than 130.0 percent of the labeled amount of USP Nystatin Units.

Packaging and storage—Preserve in collapsible tubes, or in other tight containers, and avoid exposure to excessive heat.

Reference standard—*USP Nystatin Reference Standard*—Dry in vacuum at a pressure not exceeding 5 mm of mercury at 40° for 2 hours before using.

Minimum fill ⟨755⟩: meets the requirements.

Assay—Proceed as directed for Nystatin under *Antibiotics—Microbial Assays* ⟨81⟩, blending a suitable accurately weighed portion of Nystatin Cream in a high-speed blender for 3 to 5 minutes with a sufficient accurately measured volume of dimethylformamide to give a concentration of about 400 USP Nystatin Units per mL. Dilute this stock solution quantitatively with *Buffer No. 6* to obtain a *Test Dilution* having a concentration assumed to be equal to the median dose level of the Standard.

Nystatin Lotion

» Nystatin Lotion contains not less than 90.0 percent and not more than 140.0 percent of the labeled amount of USP Nystatin Units.

Packaging and storage—Preserve in tight containers, at controlled room temperature.

Reference standard—*USP Nystatin Reference Standard*—Dry in vacuum at a pressure not exceeding 5 mm of mercury at 40° for 2 hours before using.

pH ⟨791⟩: between 5.5 and 7.5.

Assay—Proceed with Nystatin Lotion as directed in the *Assay* under *Nystatin Cream.*

Nystatin Lozenges

» Nystatin Lozenges contain not less than 90.0 percent and not more than 125.0 percent of the labeled amount of USP Nystatin Units.

Packaging and storage—Preserve in tight, light-resistant containers.

Reference standards—*USP Nystatin Reference Standard*—Dry in vacuum at a pressure not exceeding 5 mm of mercury at 40° for 2 hours before using.

Disintegration ⟨701⟩: 90 minutes, determined as set forth under *Uncoated Tablets.*

pH ⟨791⟩: between 5.0 and 7.5, in a solution prepared by dissolving 1 Lozenge in 100 mL of water at 37° and allowing the solution to cool to room temperature.

Assay—Proceed as directed for Nystatin under *Antibiotics—Microbial Assays* ⟨81⟩, blending not less than 5 Nystatin Lozenges for 18 to 20 minutes in a high-speed blender jar containing 100.0 mL of water. Add 400.0 mL of dimethylformamide and blend for an additional 10 minutes. Dilute an accurately measured volume of this solution quantitatively with a mixture of dimethylformamide and water (4:1) to obtain a stock solution containing about 400 USP Nystatin Units per mL. Dilute an accurately measured volume of this stock solution quantitatively with *Buffer No. 6* to obtain a *Test Dilution* having a nystatin concentration assumed to be equal to the median dose level of the Standard. [NOTE—The *Test Dilution* of the specimen and the test dilutions of the Standard contain the same amount of dimethylformamide (about 4%).]

Nystatin Ointment

» Nystatin Ointment contains not less than 90.0 percent and not more than 130.0 percent of the labeled amount of USP Nystatin Units.

Packaging and storage—Preserve in well-closed containers, preferably at controlled room temperature.

Reference standard—*USP Nystatin Reference Standard*—Dry in vacuum at a pressure not exceeding 5 mm of mercury at 40° for 2 hours before using.

Minimum fill ⟨755⟩: meets the requirements.

Water, *Method I* ⟨921⟩: not more than 0.5%, 20 mL of a mixture of carbon tetrachloride, chloroform, and methanol (2:2:1) being used in place of methanol in the titration vessel.

Assay—Proceed with Nystatin Ointment as directed in the *Assay* under *Nystatin Cream.*

Nystatin Topical Powder

» Nystatin Topical Powder is a dry powder composed of Nystatin and Talc. It contains not less than 90.0 percent and not more than 130.0 percent of the labeled amount of USP Nystatin Units.

Packaging and storage—Preserve in well-closed containers.

Reference standard—*USP Nystatin Reference Standard*—Dry in vacuum at a pressure not exceeding 5 mm of mercury at 40° for 2 hours before using.

Loss on drying ⟨731⟩—Dry about 100 mg, accurately weighed, in a capillary-stoppered bottle in vacuum at a pressure not exceeding 5 mm of mercury at 60° for 3 hours: it loses not more than 2.0% of its weight.

Assay—Proceed with Nystatin Topical Powder as directed in the *Assay* under *Nystatin Cream.*

Nystatin Vaginal Suppositories

» Nystatin Vaginal Suppositories contain not less than 90.0 percent and not more than 130.0 percent of the labeled amount of USP Nystatin Units.

Packaging and storage—Preserve in tight, light-resistant containers, at controlled room temperature.

Reference standard—*USP Nystatin Reference Standard*—Dry in vacuum at a pressure not exceeding 5 mm of mercury at 40° for 2 hours before using.

Water, *Method I* ⟨921⟩: not more than 1.5%.

Assay—Proceed with Nystatin Vaginal Suppositories as directed in the *Assay* under *Nystatin Tablets.*

Nystatin Oral Suspension

» Nystatin Oral Suspension contains not less than 90.0 percent and not more than 130.0 percent of the labeled amount of USP Nystatin Units. It contains suitable dispersants, flavors, preservatives, and suspending agents.

Packaging and storage—Preserve in tight, light-resistant containers.

Reference standard—*USP Nystatin Reference Standard*—Dry in vacuum at a pressure not exceeding 5 mm of mercury at 40° for 2 hours before using.

Uniformity of dosage units ⟨905⟩—

FOR SUSPENSION PACKAGED IN SINGLE-UNIT CONTAINERS: meets the requirements.

Procedure for content uniformity—[NOTE—Use low-actinic glassware.] Transfer the well-shaken contents of 1 container of Oral Suspension to a 100-mL volumetric flask, dissolve in methanol, dilute with methanol to volume, and mix. Dilute an accurately measured volume of this solution quantitatively, and stepwise if necessary, with methanol to obtain a test solution containing about 25 USP Nystatin Units per mL. Similarly, prepare a Standard solution of USP Nystatin RS in methanol having a known concentration of about 25 USP Nystatin Units per mL. Concomitantly determine the absorbances of the test solution and the Standard solution at the wavelength of maximum absorbance at about 304 nm, with a suitable spectrophotometer, using methanol as the blank. Calculate the quantity, in USP Nystatin Units, in the container by the formula:

$$(CL/D)(A_U/A_S),$$

in which C is the concentration, in USP Nystatin Units per mL, of the Standard solution, L is the labeled quantity, in USP Nystatin Units, in the container, D is the concentration, in USP Nystatin Units, in the test solution, on the basis of the labeled quantity in the container and the extent dilution, and A_U and A_S are the absorbances of the test solution and the Standard solution, respectively.

pH ⟨791⟩: between 4.5 and 6.0; or if it contains glycerin, between 6.0 and 7.5.

Assay—Proceed as directed for Nystatin under *Antibiotics—Microbial Assays* ⟨81⟩, blending a suitable accurately measured volume of Nystatin Oral Suspension, freshly mixed and free from air bubbles, for 3 to 5 minutes in a high-speed blender with a sufficient accurately measured volume of dimethylformamide to obtain a solution of convenient concentration. Dilute an accurately measured portion of this solution quantitatively with dimethylformamide to obtain a stock solution containing about 400 USP Nystatin Units per mL. Dilute this stock solution quantitatively with *Buffer No. 6* to obtain a *Test Dilution* having a concentration assumed to be equal to the median dose level of the Standard.

Nystatin for Oral Suspension

» Nystatin for Oral Suspension is a dry mixture of Nystatin with one or more suitable colors, diluents, suspending agents, flavors, and preservatives. It contains the equivalent of not less than 90.0 percent and not more than 140.0 percent of the labeled amount of USP Nystatin Units.

Packaging and storage—Preserve in tight containers.

Reference standard—*USP Nystatin Reference Standard*—Dry in vacuum at a pressure not exceeding 5 mm of mercury at 40° for 2 hours before using.

pH ⟨791⟩: between 4.9 and 5.5, in the suspension constituted as directed in the labeling.

Water, *Method I* ⟨921⟩: not more than 7.0%.

Assay—Constitute Nystatin for Oral Suspension as directed in the labeling, and proceed as directed in the *Assay* under *Nystatin Oral Suspension.*

Nystatin for Oral Suspension, Oxytetracycline and—*see* Oxytetracycline and Nystatin for Oral Suspension

Nystatin Tablets

» Nystatin Tablets contain not less than 90.0 percent and not more than 130.0 percent of the labeled amount of USP Nystatin Units.

Packaging and storage—Preserve in tight, light-resistant containers.

Labeling—Label the Tablets to indicate that they are intended for oral use only (as distinguished from Vaginal Tablets).

Reference standard—*USP Nystatin Reference Standard*—Dry in vacuum at a pressure not exceeding 5 mm of mercury at 40° for 2 hours before using.

Disintegration ⟨701⟩: if plain-coated, 120 minutes.

Loss on drying ⟨731⟩—Dry about 100 mg, accurately weighed, of powdered Tablets in a capillary-stoppered bottle in vacuum at a pressure not exceeding 5 mm of mercury at 60° for 3 hours:

if plain-coated, it loses not more than 5.0% of its weight; if film-coated, it loses not more than 8.0% of its weight.

Assay—Proceed as directed for Nystatin under *Antibiotics—Microbial Assays* ⟨81⟩, blending not less than 5 Nystatin Tablets for 3 to 5 minutes in a high-speed blender with a sufficient accurately measured volume of dimethylformamide to obtain a solution of convenient concentration. Dilute an accurately measured portion of this solution quantitatively with dimethylformamide to obtain a stock solution containing about 400 USP Nystatin Units per mL. Dilute this stock solution quantitatively with *Buffer No. 6* to obtain a *Test Dilution* having a concentration assumed to be equal to the median dose level of the Standard.

Nystatin Tablets, Demeclocycline Hydrochloride and—*see* Demeclocycline Hydrochloride and Nystatin Tablets

Nystatin Vaginal Tablets

» Nystatin Vaginal Tablets are tablets composed of Nystatin with suitable binders, diluents, and lubricants. Tablets contain not less than 90.0 percent and not more than 140.0 percent of the labeled amount of USP Nystatin Units.

Packaging and storage—Preserve in tight, light-resistant containers and, where so specified in the labeling, in a refrigerator.

Reference standard—*USP Nystatin Reference Standard*—Dry in vacuum at a pressure not exceeding 5 mm of mercury at 40° for 2 hours before using.

Disintegration ⟨701⟩: 60 minutes.

Loss on drying ⟨731⟩—Dry about 100 mg, accurately weighed, of powdered Vaginal Tablets in a capillary-stoppered bottle in vacuum at a pressure not exceeding 5 mm of mercury at 60° for 3 hours: it loses not more than 5.0% of its weight.

Assay—Proceed with Nystatin Vaginal Tablets as directed in the *Assay* under *Nystatin Tablets*.

Nystatin and Clioquinol Ointment

» Nystatin and Clioquinol Ointment contains not less than 90.0 percent and not more than 140.0 percent of the labeled amount of USP Nystatin Units, and not less than 90.0 percent and not more than 110.0 percent of the labeled amount of clioquinol (C_9H_5ClINO).

Packaging and storage—Preserve in collapsible tubes or in tight, light-resistant containers, at controlled room temperature.

Reference standards—*USP Nystatin Reference Standard*—Dry in vacuum at a pressure not exceeding 5 mm of mercury at 40° for 2 hours before using. *USP Clioquinol Reference Standard*—Dry over phosphorus pentoxide for 5 hours before using.

Identification—On a suitable thin-layer chromatographic plate (see *Chromatography* ⟨621⟩), coated with a 0.25-mm layer of chromatographic silica gel mixture, separately apply 10-µL each of the stock solution prepared as directed under *Assay preparation* in the *Assay for clioquinol* (*Test solution*) and of a Standard solution of USP Clioquinol RS in acetone containing 500 µg per mL. Allow the spots to dry, and develop the chromatogram in a solvent system consisting of a mixture of benzene and methanol (9:1) until the solvent front has moved about three-fourths of the length of the plate. Remove the plate, mark the solvent front, allow the solvent to evaporate, and examine the plate under short-wavelength ultraviolet light: the R_f value and intensity of the principal spot obtained from the *Test solution* correspond to those obtained from the Standard solution.

Minimum fill ⟨755⟩: meets the requirements.

Water, *Method I* ⟨921⟩: not more than 0.5%, 20 mL of a mixture of carbon tetrachloride, chloroform, and methanol (2:2:1) being used in place of methanol in the titration vessel.

Assay for nystatin—Proceed as directed for nystatin under *Antibiotics—Microbial Assays* ⟨81⟩, blending a suitable, accurately weighed portion of Nystatin and Clioquinol Ointment in a high-speed blender for 3 to 5 minutes with a sufficient accurately measured volume of dimethylformamide to obtain a convenient concentration. Dilute an accurately measured volume of this solution quantitatively with dimethylformamide to obtain a stock solution containing about 400 USP Nystatin Units per mL. Dilute an accurately measured volume of this stock solution quantitatively with *Buffer No. 6* to obtain a *Test Dilution* having a concentration assumed to be equal to the median dose level of the Standard.

Assay for clioquinol—

Ferric chloride solution—Dissolve 1.0 g of ferric chloride in 1000 mL of 0.012 *N* hydrochloric acid.

Standard preparation—Dissolve an accurately weighed quantity of USP Clioquinol RS in methoxyethanol to obtain a solution having a known concentration of about 1 mg per mL. Transfer 5.0 mL of this *Standard preparation* to a 50-mL volumetric flask.

Assay preparation—Transfer an accurately weighed portion of Nystatin and Clioquinol Ointment, equivalent to about 50 mg of clioquinol, to a 125-mL conical flask, add 50 mL of acetone, warm on a steam bath, and shake gently. Allow to cool, and filter through a pledget of glass wool, collecting the filtrate in a 100-mL volumetric flask. Wash the conical flask with two 20-mL portions of acetone, filtering the washings through the pledget of glass wool, and collecting the filtrates in the volumetric flask. Dilute with acetone to volume, and mix. Transfer 10.0 mL of this stock solution to a 50-mL volumetric flask, and evaporate on a steam bath to dryness. Add 20 mL of methoxyethanol, and swirl. (Retain about 5 mL of the stock solution for the *Identification test.*)

Procedure—To each volumetric flask containing the *Assay preparation* and the *Standard preparation*, and to a third 50-mL volumetric flask, to provide a blank, add 2.0 mL of *Ferric chloride solution*, dilute with methoxyethanol to volume, and mix. Concomitantly determine the absorbances of the solutions from the *Assay preparation* and the *Standard preparation* at the wavelength of maximum absorbance at about 650 nm, with a suitable spectrophotometer, against the blank. Calculate the quantity, in mg, of C_9H_5ClINO in the portion of Ointment taken by the formula:

$$50C(A_U/A_S),$$

in which C is the concentration, in mg per mL, of USP Clioquinol RS in the *Standard preparation*, and A_U and A_S are the absorbances of the solutions from the *Assay preparation* and the *Standard preparation*, respectively.

Nystatin, Neomycin Sulfate, Gramicidin, and Triamcinolone Acetonide Cream

» Nystatin, Neomycin Sulfate, Gramicidin, and Triamcinolone Acetonide Cream contains not less than 90.0 percent and not more than 140.0 percent of the labeled amounts of nystatin, neomycin, and gramicidin, and not less than 90.0 percent and not more than 110.0 percent of the labeled amount of triamcinolone acetonide ($C_{24}H_{31}FO_6$).

Packaging and storage—Preserve in tight containers.

Reference standards—*USP Nystatin Reference Standard*—Dry in vacuum at a pressure not exceeding 5 mm of mercury at 40° for 2 hours before using. *USP Neomycin Sulfate Reference Standard*—Dry in vacuum at a pressure not exceeding 5 mm of mercury at 60° for 3 hours before using. *USP Gramicidin Reference Standard*—Dry in vacuum at a pressure not exceeding 5 mm of mercury at 60° for 3 hours before using. *USP Triamcinolone Acetonide Reference Standard*—Do not dry; use as is.

Identification—Place 2 g of Cream in a conical flask, add 5.0 mL of chloroform, and shake for 10 minutes. Add 15 mL of alcohol, and shake for an additional 10 minutes. Filter the solution into a centrifuge tube, and evaporate the filtrate to dryness. Dissolve the residue in alcohol to obtain a solution containing about 250 µg of triamcinolone acetonide per mL. Proceed as directed in the *Identification test* under *Triamcinolone Acetonide Cream*, beginning with "Apply 10 µL of this solution": the specified result is observed.

Minimum fill ⟨755⟩: meets the requirements.

Assay for nystatin—Proceed as directed for nystatin under *Antibiotics—Microbial Assays* ⟨81⟩, blending a suitable, accurately weighed portion of Nystatin, Neomycin Sulfate, Gramicidin, and Triamcinolone Acetonide Cream in a high-speed blender for 3 to 5 minutes with a sufficient, accurately measured volume of dimethylformamide to give a convenient concentration. Dilute an accurately measured volume of the solution so obtained quantitatively with dimethylformamide to obtain a stock solution containing about 400 USP Nystatin Units per mL. Dilute an accurately measured volume of this stock solution quantitatively with *Buffer No. 6* to obtain a *Test Dilution* having a concentration of nystatin assumed to be equal to the median dose level of the Standard.

Assay for neomycin—Proceed with Nystatin, Neomycin Sulfate, Gramicidin, and Triamcinolone Acetonide Cream as directed in the *Assay* under *Neomycin Sulfate Cream.*

Assay for gramicidin—Proceed as directed for gramicidin under *Antibiotics—Microbial Assays* ⟨81⟩, using an accurately weighed portion of Nystatin, Neomycin Sulfate, Gramicidin, and Triamcinolone Acetonide Cream dissolved in 50 mL of hexanes in a separator, and extracted with four 20-mL portions of 80% alcohol. Combine the extracts in a suitable volumetric flask, dilute with alcohol to volume, and mix. Dilute an accurately measured volume of the solution so obtained quantitatively and stepwise with alcohol to obtain a *Test Dilution* having a concentration of gramicidin assumed to be equal to the median dose level of the Standard.

Assay for triamcinolone acetonide—Proceed with Nystatin, Neomycin Sulfate, Gramicidin, and Triamcinolone Acetonide Cream as directed in the *Assay* under *Triamcinolone Acetonide Cream.*

Nystatin, Neomycin Sulfate, Gramicidin, and Triamcinolone Acetonide Ointment

» Nystatin, Neomycin Sulfate, Gramicidin, and Triamcinolone Acetonide Ointment contains not less than 90.0 percent and not more than 140.0 percent of the labeled amounts of nystatin, neomycin, and gramicidin, and not less than 90.0 percent and not more than 110.0 percent of the labeled amount of triamcinolone acetonide ($C_{24}H_{31}FO_6$).

Packaging and storage—Preserve in tight containers.

Reference standards—*USP Nystatin Reference Standard*—Dry in vacuum at a pressure not exceeding 5 mm of mercury at 40° for 2 hours before using. *USP Neomycin Sulfate Reference Standard*—Dry in vacuum at a pressure not exceeding 5 mm of mercury at 60° for 3 hours before using. *USP Gramicidin Reference Standard*—Dry in vacuum at a pressure not exceeding 5 mm of mercury at 60° for 3 hours before using. *USP Triamcinolone Acetonide Reference Standard*—Do not dry; use as is.

Identification—Place 2 g of Ointment in a conical flask, add 5.0 mL of chloroform, and shake for 10 minutes. Add 15 mL of alcohol, and shake for an additional 10 minutes. Filter the solution into a centrifuge tube, and evaporate the filtrate to dryness. Dissolve the residue in alcohol to obtain a solution containing about 250 µg of triamcinolone acetonide per mL. Proceed as directed in the *Identification test* under *Triamcinolone Acetonide Cream*, beginning with "Apply 10 µL of this solution": the specified result is observed.

Minimum fill ⟨755⟩: meets the requirements.

Water, *Method I* ⟨921⟩: not more than 0.5%, 20 mL of a mixture of carbon tetrachloride, chloroform, and methanol (2:2:1) being used in place of methanol in the titration vessel.

Assay for nystatin—Proceed as directed for nystatin under *Antibiotics—Microbial Assays* ⟨81⟩, blending a suitable, accurately weighed portion of Nystatin, Neomycin Sulfate, Gramicidin, and Triamcinolone Acetonide Ointment in a high-speed blender for 3 to 5 minutes with a sufficient, accurately measured volume of dimethylformamide to give a convenient concentration. Dilute an accurately measured volume of the solution so obtained quantitatively with dimethylformamide to obtain a stock solution containing about 400 USP Nystatin Units per mL. Dilute an accurately measured volume of this stock solution quantitatively with *Buffer No. 6* to obtain a *Test Dilution* having a concentration of nystatin assumed to be equal to the median dose level of the Standard.

Assay for neomycin—Proceed with Nystatin, Neomycin Sulfate, Gramicidin, and Triamcinolone Acetonide Ointment as directed in the *Assay* under *Neomycin Sulfate Ointment.*

Assay for gramicidin—Proceed as directed for gramicidin under *Antibiotics—Microbial Assays* ⟨81⟩, using an accurately weighed portion of Nystatin, Neomycin Sulfate, Gramicidin, and Triamcinolone Acetonide Ointment dissolved in 50 mL of hexanes in a separator, and extracted with four 20-mL portions of 80% alcohol. Combine the extracts in a suitable volumetric flask, dilute with alcohol to volume, and mix. Dilute an accurately measured volume of the solution so obtained quantitatively and stepwise with alcohol to obtain a *Test Dilution* having a concentration of gramicidin assumed to be equal to the median dose level of the Standard.

Assay for triamcinolone acetonide—Proceed with Nystatin, Neomycin Sulfate, Gramicidin, and Triamcinolone Acetonide Ointment as directed in the *Assay* under *Triamcinolone Acetonide Cream*, except to read "Ointment" in place of "Cream" throughout.

Nystatin and Triamcinolone Acetonide Cream

» Nystatin and Triamcinolone Acetonide Cream contains not less than 90.0 percent and not more than 140.0 percent of the labeled amount of USP Nystatin Units and not less than 90.0 percent and not more than 110.0 percent of the labeled amount of triamcinolone acetonide ($C_{24}H_{31}FO_6$).

Packaging and storage—Preserve in tight containers.

Reference standards—*USP Nystatin Reference Standard*—Dry in vacuum at a pressure not exceeding 5 mm of mercury at 40° for 2 hours before using. *USP Triamcinolone Acetonide Reference Standard*—Do not dry; use as is.

Identification—Place 2 g of Cream in a conical flask, add 5.0 mL of chloroform, and shake for 10 minutes. Add 15 mL of alcohol, and shake for an additional 10 minutes. Filter the solution into a centrifuge tube, and evaporate the filtrate to dryness. Dissolve the residue in alcohol to obtain a solution containing about 250 µg of triamcinolone acetonide per mL. Proceed as directed in the *Identification test* under *Triamcinolone Acetonide Cream*, beginning with "Apply 10 µL of this solution": the specified result is observed.

Minimum fill ⟨755⟩: meets the requirements.

Assay for nystatin—Proceed as directed for nystatin under *Antibiotics—Microbial Assays* ⟨81⟩, blending a suitable, accurately weighed portion of Nystatin and Triamcinolone Acetonide Cream in a high-speed blender for 3 to 5 minutes with a sufficient, accurately measured volume of dimethylformamide to give a convenient concentration. Dilute an accurately measured volume of the solution so obtained quantitatively with dimethylformamide to obtain a stock solution containing about 400 USP Nystatin Units per mL. Dilute an accurately measured volume of this stock solution quantitatively with *Buffer No. 6* to obtain a *Test Dilution* having a concentration of nystatin assumed to be equal to the median dose level of the Standard.

Assay for triamcinolone acetonide—Proceed with Nystatin and Triamcinolone Acetonide Cream as directed in the *Assay* under *Triamcinolone Acetonide Cream.*

Nystatin and Triamcinolone Acetonide Ointment

» Nystatin and Triamcinolone Acetonide Ointment contains not less than 90.0 percent and not more than 140.0 percent of the labeled amount of USP Nystatin Units and not less than 90.0 percent and not more than 110.0 percent of the labeled amount of triamcinolone acetonide ($C_{24}H_{31}FO_6$).

Packaging and storage—Preserve in tight containers.

Reference standards—*USP Nystatin Reference Standard*—Dry in vacuum at a pressure not exceeding 5 mm of mercury at 40° for 2 hours before using. *USP Triamcinolone Acetonide Reference Standard*—Do not dry; use as is.

Identification—Place 2 g of Ointment in a conical flask, add 5.0 mL of chloroform, and shake for 10 minutes. Add 15 mL of alcohol, and shake for an additional 10 minutes. Filter the solution into a centrifuge tube, and evaporate the filtrate to dryness. Dissolve the residue in alcohol to obtain a solution containing about 250 µg of triamcinolone acetonide per mL. Proceed as directed in the *Identification test* under *Triamcinolone Acetonide Cream*, beginning with "Apply 10 µL of this solution": the specified result is observed.

Minimum fill ⟨755⟩: meets the requirements.

Water, *Method I* ⟨921⟩: not more than 0.5%, 20 mL of a mixture of carbon tetrachloride, chloroform, and methanol (2:2:1) being used in place of methanol in the titration vessel.

Assay for nystatin—Proceed as directed for nystatin under *Antibiotics—Microbial Assays* ⟨81⟩, blending a suitable, accurately weighed portion of Nystatin and Triamcinolone Acetonide Ointment in a high-speed blender for 3 to 5 minutes with a sufficient, accurately measured volume of dimethylformamide to give a convenient concentration. Dilute an accurately measured volume of the solution so obtained quantitatively with dimethylformamide to obtain a stock solution containing about 400 USP Nystatin Units per mL. Dilute an accurately measured volume of this stock solution quantitatively with *Buffer No. 6* to obtain a *Test Dilution* having a concentration of nystatin assumed to be equal to the median dose level of the Standard.

Assay for triamcinolone acetonide—Proceed with Nystatin and Triamcinolone Acetonide Ointment as directed in the *Assay* under *Triamcinolone Acetonide Cream*, except to read "Ointment" in place of "Cream" throughout.

Octoxynol 9—*see* Octoxynol 9 NF
Octyldodecanol—*see* Octyldodecanol NF
Oils—*see complete list in index*

Hydrophilic Ointment

» Prepare Hydrophilic Ointment as follows:

Methylparaben	0.25 g
Propylparaben	0.15 g
Sodium Lauryl Sulfate	10 g
Propylene Glycol	120 g
Stearyl Alcohol	250 g
White Petrolatum	250 g
Purified Water	370 g
To make about	1000 g

Melt the Stearyl Alcohol and the White Petrolatum on a steam bath, and warm to about 75°. Add the other ingredients, previously dissolved in the water and warmed to 75°, and stir the mixture until it congeals.

Packaging and storage—Preserve in tight containers.

White Ointment

» Prepare White Ointment as follows:

White Wax	50 g
White Petrolatum	950 g
To make	1000 g

Melt the White Wax in a suitable dish on a water bath, add the White Petrolatum, warm until liquefied, then discontinue the heating, and stir the mixture until it begins to congeal.

Packaging and storage—Preserve in well-closed containers.

Yellow Ointment

» Prepare Yellow Ointment as follows:

Yellow Wax	50 g
Petrolatum	950 g
To make	1000 g

Melt the Yellow Wax in a suitable dish on a steam bath, add the Petrolatum, warm until liquefied, then discontinue the heating, and stir the mixture until it begins to congeal.

Packaging and storage—Preserve in well-closed containers.

Ointments—*see complete list in index*
Oleic Acid—*see* Oleic Acid NF

Oleovitamin A and D

» Oleovitamin A and D is a solution of vitamin A and vitamin D in fish liver oil or in an edible vegetable oil. The vitamin D is present as ergocalciferol or cholecalciferol obtained by the activation of ergosterol or 7-dehydrocholesterol or from natural sources. Oleovitamin A and D contains not less than 90.0 percent of the labeled amounts of vitamins A and D.

Packaging and storage—Preserve in tight containers, protected from light and air, preferably under an atmosphere of an inert gas. Store in a dry place.

Labeling—Label it to indicate the content of vitamin A in mg per g. The vitamin A content may also be expressed in USP Vitamin A Units per g. Label it to show whether it contains ergocalciferol, cholecalciferol, or vitamin D from a natural source. Label it to indicate also the vitamin D content in µg per g. Its vitamin D content may be expressed also in USP Vitamin D Units per g.

Reference standard—*USP Ergocalciferol Reference Standard* or *USP Cholecalciferol Reference Standard*—Store in a cold place, protected from light. Allow it to attain room temperature before opening ampul. Use the material promptly, and discard the unused portion.

Assay for vitamin A—Proceed as directed under *Vitamin A Assay* ⟨571⟩.

Assay for vitamin D—Accurately measure a portion of Oleovitamin A and D expected to contain about the equivalent of 125 to 250 µg of vitamin D, but not more than 7.5 mg of vitamin A, and proceed as directed for *Chemical Method* under *Vitamin D Assay* ⟨581⟩. If the assay specimen contains less than the equivalent of 2.5 µg of vitamin D per g, or if the ratio of vitamin A to vitamin D exceeds 300 to 1, proceed as directed for *Biological Method* under *Vitamin D Assay* ⟨581⟩.

Oleovitamin A and D Capsules

» Oleovitamin A and D Capsules contain not less than 90.0 percent of the labeled amounts of vitamins A and D. The oil in Oleovitamin A and D Capsules conforms to the definition for *Oleovitamin A and D.*

Packaging and storage—Preserve in tight, light-resistant containers. Store in a dry place.

Labeling—Label the Capsules to indicate the content, in mg, of vitamin A in each capsule. The vitamin A content in each capsule may be expressed also in USP Vitamin A Units. Label the Capsules to show whether they contain ergocalciferol, cholecalciferol, or vitamin D from a natural source. Label the Capsules to indicate also the vitamin D content, in µg, in each capsule. The vitamin D content may be expressed also in USP Vitamin D Units in each capsule.

Reference standard—*USP Ergocalciferol Reference Standard* or *USP Cholecalciferol Reference Standard*—Store in a cold place, protected from light. Allow it to attain room temperature before opening ampul. Use the material promptly, and discard the unused portion.

Assay for vitamin A—Transfer to a saponification flask not less than 5 Oleovitamin A and D Capsules, add 10 to 20 mL of water, and heat for about 10 minutes. Crush any remaining solids with the blunt end of a glass rod, and proceed as directed in the second paragraph under *Procedure* in the *Vitamin A Assay* ⟨571⟩, beginning with "Reflux in an *all-glass* apparatus."

Assay for vitamin D—Transfer to a saponification flask not less than five Oleovitamin A and D Capsules, add 10 to 20 mL of water, and heat for about 10 minutes. Crush any remaining solids with the blunt end of a glass rod, and proceed as directed for *Chemical Method* under *Vitamin D Assay* ⟨581⟩. The *Test Preparation* used in the chromatographic procedures shall contain not more than the equivalent of 7.5 mg of vitamin A or less than 125 µg of vitamin D. If the ratio of vitamin A to vitamin D in the *Test Preparation* exceeds 300 to 1, proceed as directed for *Biological Method* under *Vitamin D Assay* ⟨581⟩.

Oleyl Alcohol—*see* Oleyl Alcohol NF
Olive Oil—*see* Olive Oil NF
Ophthalmic Ointments—*see complete list in index*
Ophthalmic Solutions—*see complete list in index*
Ophthalmic Suspensions—*see complete list in index*

Opium

» Opium is the air-dried milky exudate obtained by incising the unripe capsules of *Papaver somniferum* Linné or its variety *album* De Candolle (Fam. Papaveraceae). It yields not less than 9.5 percent of anhydrous morphine.

Botanic characteristics—More or less rounded, oval, brick-shaped or elongated, somewhat flattened masses usually about 8 cm to 15 cm in diameter and weighing about 300 g to 2 kg each. Externally, it is pale olive-brown or olive-gray, having a coarse surface and being covered with a thin coating consisting of fragments of poppy leaves and, at times, with fruits of a species of *Rumex* adhering from the packing; it is more or less plastic when fresh, becoming hard or tough on storage. Internally, it is reddish brown and coarsely granular.

Assay—

Chromatographic tubes, Citrate buffer, and *Standard preparation*—Prepare as directed in the *Assay* under *Paregoric.*

Assay preparation—Transfer about 2 g of Opium, accurately weighed, to a 250-mL beaker, add 20 mL of methyl sulfoxide, and heat for 20 minutes on a steam bath, intermittently dispersing the substance with a flat-end stirring rod. Allow to stand for 15 minutes to permit undissolved material to settle, and carefully decant the supernatant liquid into a 100-mL volumetric flask. Add another 20 mL of methyl sulfoxide to the residue, rinsing the sides of the beaker with the methyl sulfoxide. Disperse and heat the substance as before, allow to settle, and decant into the volumetric flask. Repeat the dissolution one or two times, until the opium is dissolved (other than for small leaf fragments, sand-like particles, gelatinous materials, etc.). Rinse the beaker, and transfer the residue to the flask with the aid of water. Dilute with water to about 90 mL, and mix. If necessary, add 1 drop of alcohol to dispel any foam. Cool to room temperature, adjust with water to volume, and mix. Filter the resulting solution through a medium-porosity filter paper, discarding the first 20 mL of the filtrate.

Chromatographic columns—Pack a pledget of glass wool at the base of each of the three tubes, and fill with adsorbant prepared as follows, using chromatographic siliceous earth as the base of the adsorbant, and tamp it firmly in place. Pack *Column I* in two layers, the lower layer consisting of 3 g of chromatographic siliceous earth mixed with 2 mL of *Citrate buffer* and the upper layer of 3 g of chromatographic siliceous earth mixed with 2.0 mL of the *Assay preparation* and 0.5 mL of *Citrate buffer*. Dry-rinse the beaker in which the components of the two layers have been mixed with 1 g of chromatographic siliceous earth, and add it also to the top of *Column I*. Pack *Column II* with 3 g of chromatographic siliceous earth mixed with 2 mL of dibasic potassium phosphate solution (1 in 5.75). Pack *Column III* with 3 g of chromatographic siliceous earth mixed with 2 mL of sodium hydroxide solution (1 in 50). Place a small pad of glass wool above each column packing.

Procedure—Proceed as directed in the *Assay* under *Paregoric.* Calculate the percentage of anhydrous morphine in the Opium taken by the formula:

$$0.25(C/W)(A_U/A_S),$$

in which C is the concentration, in µg per mL, of anhydrous morphine in the *Standard preparation,* W is the weight, in g, of Opium taken, and A_U and A_S are the corrected absorbances of the solutions from the *Assay preparation* and the *Standard preparation,* respectively.

Powdered Opium

» Powdered Opium is Opium dried at a temperature not exceeding 70°, and reduced to a very fine powder.

Powdered Opium yields not less than 10.0 percent and not more than 10.5 percent of anhydrous morphine. It may contain any of the diluents, with the exception of starch, permitted for powdered extracts under *Extracts* ⟨1151⟩.

Packaging and storage—Preserve in well-closed containers.

Botanic characteristics—Consists chiefly of yellowish brown to yellow, more or less irregular and granular fragments of latex, varying from 15 µm to 150 µm in diameter; a few fragments of strongly lignified, thick-walled, 4- to 5-sided or narrowly elongated, epidermal cells of the poppy capsule; very few fragments of tissues of poppy leaves, poppy capsules, and, occasionally, *Rumex* fruits. In addition, there will be the microscopic characteristics of the diluent if any has been used in the preparation of the powder.

Assay—Proceed with Powdered Opium as directed in the *Assay* under *Opium*.

Opium Tincture

» Opium Tincture contains, in each 100 mL, not less than 0.90 g and not more than 1.10 g of anhydrous morphine.

Opium Tincture may be prepared as follows.

Place 100 g of granulated or sliced Opium in a suitable vessel. [NOTE—Do not use Powdered Opium.] Add 500 mL of boiling water, and allow to stand, with frequent stirring, for 24 hours. Transfer the mixture to a percolator, allow it to drain, percolate with water as the menstruum to complete extraction, and evaporate the percolate to a volume of 400 mL. Boil actively for not less than 15 minutes, and allow to stand overnight. Heat the mixture to 80°, add 50 g of paraffin, and heat until the paraffin is melted. Beat the mixture thoroughly, and cool.

Remove the paraffin, and filter the concentrate, washing the paraffin and the filter with sufficient water to make the filtrate measure 750 mL. Add 188 mL of alcohol to the filtrate, mix, and assay a 10-mL portion of the resulting solution as directed in the *Assay*. Dilute the remaining solution with a mixture of 1 volume of alcohol and 4 volumes of water to obtain a Tincture containing 1 g of anhydrous morphine in each 100 mL. Mix.

Packaging and storage—Preserve in tight, light-resistant containers, and avoid exposure to direct sunlight and to excessive heat.

Alcohol content ⟨611⟩: between 17.0% and 21.0% of C_2H_5OH, determined by the gas-liquid procedure, acetone being used as the internal standard.

Assay—

Chromatographic tubes, Citrate buffer, Standard preparation, and *Chromatographic columns*—Prepare as directed in the *Assay* under *Paregoric*.

Assay preparation—Transfer 10.0 mL of Opium Tincture to a 50-mL volumetric flask containing 10.0 mL of alcohol, add purified water to volume, and mix. Transfer a 2.0-mL aliquot of the solution, equivalent to about 4 mg of Morphine, to a 50-mL beaker, and add 0.5 mL of *Citrate buffer*.

Procedure—Proceed as directed for *Procedure* in the *Assay* under *Paregoric*. Calculate the weight of anhydrous morphine, in g per 100 mL of the Tincture taken by the formula:

$$0.250W(A_U/A_S),$$

in which W is the weight, in mg, of anhydrous morphine in the 50 mL of *Standard preparation*, and A_U and A_S are the corrected absorbances of the solution from the *Assay preparation* and the *Standard preparation*, respectively.

Oral Rehydration Salts—*see* Rehydration Salts, Oral

Oral Solutions—*see complete list in index*

Oral Suspensions—*see complete list in index*

Orange Flower Oil—*see* Orange Flower Oil NF

Orphenadrine Citrate

$(CH_3)_2NCH_2CH_2OCH$ · HO—C—COOH (CH_2COOH, CH_2COOH); CH_3

$C_{18}H_{23}NO.C_6H_8O_7$ 461.51

Ethanamine, *N,N*-dimethyl-2-[(2-methylphenyl)phenyl methoxy]-, 2-hydroxy-1,2,3-propanetricarboxylate (1:1).

N,N-Dimethyl-2-[(*o*-methyl-α-phenylbenzyl)oxy]ethylamine citrate (1:1) [*4682-36-4*].

» Orphenadrine Citrate contains not less than 98.0 percent and not more than 101.5 percent of $C_{18}H_{23}NO.C_6H_8O_7$, calculated on the dried basis.

Packaging and storage—Preserve in tight, light-resistant containers.

Reference standard—*USP Orphenadrine Citrate Reference Standard*—Dry at 105° for 3 hours before using.

Clarity and color of solution—Mix 1 g of it with 10 mL of a 1 in 28 solution of hydrochloric acid in alcohol: the solution is clear and its absorbance at 436 nm is not greater than 0.050.

Identification—

A: The infrared absorption spectrum of a mineral oil dispersion of it, previously dried, exhibits maxima only at the same wavelengths as that of a similar preparation of USP Orphenadrine Citrate RS.

B: The ultraviolet absorption spectrum of a 1 in 2000 solution in alcohol exhibits maxima and minima at the same wavelengths as that of a similar solution of USP Orphenadrine Citrate RS, concomitantly measured, and the respective absorptivities, calculated on the dried basis, at the wavelength of maximum absorbance at about 264 nm do not differ by more than 3.0%.

Melting range ⟨741⟩: between 134° and 138°.

Loss on drying ⟨731⟩—Dry it at 105° for 3 hours: it loses not more than 0.5% of its weight.

Residue on ignition ⟨281⟩: not more than 0.1%.

Chromatographic purity—

Standard preparation—Transfer 50 mg, accurately weighed, of USP Orphenadrine Citrate RS to a 10-mL volumetric flask, add about 7 mL of methanol, and swirl to dissolve. Dilute with methanol to volume, and mix.

Test preparation—Prepare as directed under *Standard preparation*, using Orphenadrine Citrate instead of the USP Reference Standard.

Procedure—Divide a thin-layer chromatographic plate, coated with chromatographic silica gel mixture, into two equal parts. To one part of the plate apply 20 µL of *Standard preparation* in 5-µL portions, drying after each application, and similarly apply 20 µL of *Test preparation* to the second part of the plate. Allow the spots to dry, and develop the chromatogram in an unsaturated chamber in a solvent system consisting of a mixture of methanol and ammonium hydroxide (100 : 1) until the solvent front has moved at least 10 cm from the origin. Remove the plate from the chamber, and allow the solvent to evaporate. Irradiate the plate under short-wavelength ultraviolet light, and without delay locate the spots by visual inspection: the R_f value of the principal spot obtained from the *Test preparation* corresponds to that obtained from the *Standard preparation*, and no spots are observed other than the principal spot and a spot at the origin representing residual citrate.

Isomer content—

Solvent—Use carbon tetrachloride.

NMR reference—Use tetramethylsilane.

Test preparation—Place about 1 g of Orphenadrine Citrate and 10 mL of water in a 60-mL separator, slowly add about 20 drops of sodium hydroxide solution (1 in 2), with swirling, to obtain a solution having a pH of about 10, and extract with three 15-mL portions of ether. Combine the ether extracts in a beaker, discarding the aqueous phase, and evaporate to about one-half the volume by warming on a steam bath under a stream of nitrogen. Transfer to a 60-mL separator, wash with three 20-mL portions of water, and dry the ether solution with about 15 g of anhydrous sodium sulfate in a 125-mL conical flask for 1 hour, with intermittent swirling. Decant the dried ether solution through a pledget of glass wool into a small beaker. Rinse the sodium sulfate with two 10-mL portions of ether, and add the rinsings to the beaker. Evaporate most of the ether by warming under a stream of nitrogen, and remove the last traces of ether by drying at a pressure not exceeding 2 mm of mercury at 60°. Transfer 400 mg of the orphenadrine so obtained to a small weighing bottle, add 0.5 mL of carbon tetrachloride and 1 drop of tetramethylsilane, and swirl to dissolve.

Procedure—Proceed as directed for *Relative Method of Quantitation* under *Nuclear Magnetic Resonance* ⟨761⟩, using the calculation formula given therein, in which A_1 is the sum of the average areas of the combined methine peaks associated with the *meta-* and *para-*methylbenzyl isomers, appearing at about 5.23 ppm, and A_2 is the area of the methine peak associated with the *ortho-*methylbenzyl isomer, appearing at about 5.47 ppm, with reference to the tetramethylsilane singlet at 0 ppm, and both n_1 and n_2 are equal to 1: the limit of combined *meta-* and *para-*methylbenzyl isomers is 3.0%.

Assay—Dissolve about 1 g of Orphenadrine Citrate, accurately weighed, in 50 mL of glacial acetic acid, add 1 drop of crystal violet TS, and titrate with 0.1 *N* perchloric acid VS to a blue end-point. Perform a blank determination, and make any necessary correction. Each mL of 0.1 *N* perchloric acid is equivalent to 46.15 mg of $C_{18}H_{23}NO.C_6H_8O_7$.

Orphenadrine Citrate Injection

» Orphenadrine Citrate Injection is a sterile solution of Orphenadrine Citrate in Water for Injection, prepared with the aid of Sodium Hydroxide. It contains not less than 93.0 percent and not more than 107.0 percent of the labeled amount of $C_{18}H_{23}NO.C_6H_8O_7$.

Packaging and storage—Preserve in single-dose or in multiple-dose containers, preferably of Type I glass, protected from light.

Reference standard—*USP Orphenadrine Citrate Reference Standard*—Dry at 105° for 3 hours before using.

Identification—

A: Dilute a volume of Injection, equivalent to about 30 mg of orphenadrine citrate, with water to 100 mL: the ultraviolet absorption spectrum of the solution so obtained, measured between 240 nm and 300 nm, exhibits maxima and minima at the same wavelengths as that of USP Orphenadrine Citrate RS solution (1 in 3300), concomitantly measured.

B: A few drops of Injection respond to the test for *Citrate* ⟨191⟩.

pH ⟨791⟩: between 5.0 and 6.0.

Other requirements—It meets the requirements under *Injections* ⟨1⟩.

Assay—

Trinitrophenol reagent—Dissolve 200 mg of trinitrophenol in 1000 mL of toluene. [NOTE—Use toluene that has been dried over anhydrous sodium sulfate for 16 hours and filtered, if necessary, until clear.]

Standard preparation—Transfer about 30 mg of USP Orphenadrine Citrate RS, accurately weighed, to a 50-mL volumetric flask, dissolve in 0.1 *N* hydrochloric acid, dilute with 0.1 *N* hydrochloric acid to volume, and mix.

Assay preparation—Transfer an accurately measured volume of Orphenadrine Citrate Injection, equivalent to about 60 mg of orphenadrine citrate, to a 100-mL volumetric flask, dilute with 0.1 *N* hydrochloric acid to volume, and mix.

Procedure—Transfer 1.0 mL each of the *Standard preparation*, the *Assay preparation*, and 0.1 *N* hydrochloric acid to provide the blank, to separate glass-stoppered centrifuge tubes. To each tube add 10.0 mL of toluene and 1.0 mL of 1 *N* sodium hydroxide. Insert the stopper, shake by mechanical means for 15 minutes, and centrifuge at 1500 rpm for 10 minutes. Transfer 5.0 mL of each clear toluene layer to three separate, small glass-stoppered flasks, each containing 5.0 mL of *Trinitrophenol reagent*, and mix. Concomitantly determine the absorbances of the solutions in 1-cm cells at the wavelength of maximum absorbance at about 410 nm, with a suitable spectrophotometer, using the blank to set the instrument. Calculate the quantity, in mg, of $C_{18}H_{23}NO.C_6H_8O_7$ in each mL of the Injection taken by the formula:

$$(0.1C/V)(A_U/A_S),$$

in which C is the concentration, in µg per mL, of USP Orphenadrine Citrate RS in the *Standard preparation*, V is the volume, in mL, of Injection taken, and A_U and A_S are the absorbances of the solutions from the *Assay preparation* and the *Standard preparation*, respectively.

Otic Solutions—*see complete list in index*

Oxacillin Sodium

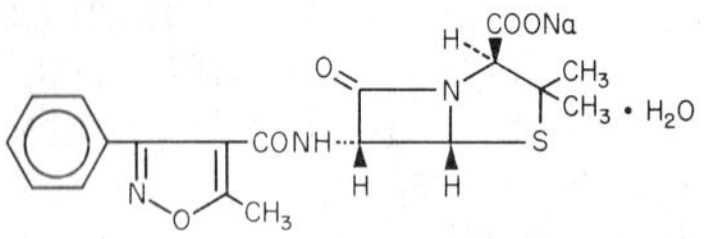

$C_{19}H_{18}N_3NaO_5S.H_2O$ 441.43

4-Thia-1-azabicyclo[3.2.0]heptane-2-carboxylic acid, 3,3-dimethyl-6-[[(5-methyl-3-phenyl-4-isoxazolyl)carbonyl]amino]-7-oxo-, monosodium salt, monohydrate, [2*S*-(2α,5α,6β)]-.

Monosodium (2*S*,5*R*,6*R*)-3,3-dimethyl-6-(5-methyl-3-phenyl-4-isoxazolecarboxamido)-7-oxo-4-thia-1-azabicyclo[3.2.0]-heptane-2-carboxylate monohydrate [*7240-38-2*].

Anhydrous 423.42 [*1173-88-2*].

» Oxacillin Sodium has a potency equivalent to not less than 815 µg and not more than 950 µg of oxacillin ($C_{19}H_{19}N_3O_5S$) per mg.

Packaging and storage—Preserve in tight containers, at controlled room temperature.

Reference standard—*USP Oxacillin Sodium Reference Standard*—Do not dry before using.

Identification—The ultraviolet absorption spectrum, between 230 nm and 260 nm, of the test solution prepared as directed in the test for *Oxacillin content*, exhibits maxima and minima at the same wavelengths as that of the Standard solution prepared as directed in the test for *Oxacillin content*, concomitantly measured.

Crystallinity ⟨695⟩: meets the requirements.

pH ⟨791⟩: between 4.5 and 7.5, in a solution containing 30 mg per mL.

Water, *Method I* ⟨921⟩: between 3.5% and 5.0%.

Oxacillin content—Transfer about 60 mg, accurately weighed, to a 100-mL volumetric flask. Dissolve in water, dilute with water to volume, and mix. Transfer 5.0 mL of this solution to a 22- × 200-mm test tube, add 5.0 mL of 10 *N* sodium hydroxide, and mix. Heat the tube in a water bath for 60 minutes. Cool the tube, carefully add 10 mL of 6 *N* hydrochloric acid, and mix. Place the tube in the water bath for 10 minutes, positioning the tube so that the level of liquid in it is the same as the water level in the bath. Remove the tube from the bath, carefully agitate it, and cool to room temperature. Transfer the contents of the tube to a 250-mL volumetric flask, add 200 mL of freshly boiled and cooled water, then add 4.0 mL of 7.5 *N* ammonium hydroxide, dilute with freshly boiled and cooled water to volume, and mix. Concomitantly determine the absorbances of this test solution and a similarly prepared Standard solution of USP Oxacillin Sodium RS, at the wavelength of maximum absorbance at about 235 nm, using a reagent blank. Calculate the percentage of oxacillin ($C_{19}H_{19}N_3O_5S$) by the formula:

$$P(W_S/W_U)(A_U/A_S),$$

in which P is the percentage content of oxacillin in the USP Oxacillin Sodium RS, W_S and W_U are the amounts of USP Oxacillin Sodium RS and of Oxacillin Sodium taken, respectively, and A_U and A_S are the absorbances of the test solution and the Standard solution, respectively: between 81.5% and 95.0% is found.

Assay—Proceed with Oxacillin Sodium as directed under *Antibiotics—Microbial Assays* ⟨81⟩.

Oxacillin Sodium Capsules

» Oxacillin Sodium Capsules contain the equivalent of not less than 90.0 percent and not more than 120.0 percent of the labeled amount of oxacillin ($C_{19}H_{19}N_3O_5S$).

Packaging and storage—Preserve in tight containers, at controlled room temperature.

Reference standard—*USP Oxacillin Sodium Reference Standard*—Do not dry before using.

Dissolution ⟨711⟩—

Medium: water; 900 mL.

Apparatus 1: 100 rpm.

Time: 45 minutes.

Procedure—Determine the amount of oxacillin ($C_{19}H_{19}N_3O_5S$) by a suitable validated spectrophotometric analysis of a filtered portion of the solution under test, suitably diluted with *Dissolution Medium*, if necessary, in comparison with a Standard solution having a known concentration of USP Oxacillin Sodium RS in the same medium.

Tolerances—Not less than 75% (Q) of the labeled amount of $C_{19}H_{19}N_3O_5S$ is dissolved in 45 minutes.

Uniformity of dosage units ⟨905⟩: meet the requirements.

Water, *Method I* ⟨921⟩: not more than 6.0%.

Assay—Proceed as directed under *Antibiotics—Microbial Assays* ⟨81⟩, using not less than 5 Oxacillin Sodium Capsules blended for 4 ± 1 minutes in a high-speed blender jar containing an accurately measured volume of *Buffer No. 1*. Dilute an accurately measured volume of this stock solution quantitatively with *Buffer No. 1* to obtain a *Test Dilution* having a concentration assumed to be equal to the median dose level of the Standard.

Oxacillin Sodium for Injection

» Oxacillin Sodium for Injection is a sterile, dry mixture of Oxacillin Sodium and one or more suitable buffers. It contains the equivalent of not less than 90.0 percent and not more than 115.0 percent of the labeled amount of oxacillin ($C_{19}H_{19}N_3O_5S$).

Packaging and storage—Preserve in *Containers for Sterile Solids* as described under *Injections* ⟨1⟩, at controlled room temperature.

Reference standard—*USP Oxacillin Sodium Reference Standard*—Do not dry before using.

Constituted solution—At the time of use, the constituted solution prepared from Oxacillin Sodium for Injection meets the requirements for *Constituted Solutions* under *Injections* ⟨1⟩.

Pyrogen—It meets the requirements of the *Pyrogen Test* ⟨151⟩, the test dose being 1.0 mL per kg of a solution in pyrogen-free saline TS containing 20 mg of oxacillin per mL.

Sterility—It meets the requirements under *Sterility Tests* ⟨71⟩, when tested as directed in the section, *Test Procedures Using Membrane Filtration.*

Uniformity of dosage units ⟨905⟩: meets the requirements.

pH ⟨791⟩: between 6.0 and 8.5, in a solution containing 30 mg per mL.

Water, *Method I* ⟨921⟩: not more than 6.0%.

Particulate matter ⟨788⟩: meets the requirements under *Small-volume Injections.*

Assay—Proceed as directed under *Antibiotics—Microbial Assays* ⟨81⟩, using Oxacillin Sodium for Injection constituted as directed in the labeling. Remove an accurately measured portion, and dilute quantitatively with *Buffer No. 1* to obtain a *Test Dilution* having a concentration assumed to be equal to the median dose level of the Standard.

Oxacillin Sodium for Oral Solution

» Oxacillin Sodium for Oral Solution contains the equivalent of not less than 90.0 percent and not more than 120.0 percent of the labeled amount of oxacillin ($C_{19}H_{19}N_3O_5S$). It contains one or more suitable buffers, colors, flavors, preservatives, and stabilizers.

Packaging and storage—Preserve in tight containers, at controlled room temperature.

Reference standard—*USP Oxacillin Sodium Reference Standard*—Do not dry before using.

Uniformity of dosage units ⟨905⟩—

FOR SOLID PACKAGED IN SINGLE-UNIT CONTAINERS: meets the requirements.

pH ⟨791⟩: between 5.0 and 7.5, in the solution constituted as directed in the labeling.

Water, *Method I* ⟨921⟩: not more than 1.0%.

Assay—Proceed as directed under *Antibiotics—Microbial Assays* ⟨81⟩, using Oxacillin Sodium for Oral Solution constituted as directed in the labeling. Dilute an accurately measured volume of the solution quantitatively with *Buffer No. 1* to obtain a *Test Dilution* having a concentration assumed to be equal to the median dose level of the Standard.

Sterile Oxacillin Sodium

» Sterile Oxacillin Sodium is Oxacillin Sodium suitable for parenteral use. It has a potency equivalent to not less than 815 µg and not more than 950 µg of oxacillin ($C_{19}H_{19}N_3O_5S$) per mg.

Packaging and storage—Preserve in *Containers for Sterile Solids* as described under *Injections* ⟨1⟩.

Reference standard—*USP Oxacillin Sodium Reference Standard*—Do not dry before using.

Pyrogen—It meets the requirements of the *Pyrogen Test* ⟨151⟩, the test dose being 1 mL per kg of a solution in sterile saline TS containing 20 mg of oxacillin per mL.

Sterility—It meets the requirements under *Sterility Tests* ⟨71⟩, when tested as directed in the section, *Test Procedures Using Membrane Filtration*, 6 g being aseptically dissolved in 200 mL of *Fluid A*.

Other requirements—It conforms to the definition, responds to the *Identification test*, and meets the requirements for *pH*, *Water*, *Crystallinity*, and *Oxacillin content* under *Oxacillin Sodium*.

Assay—Proceed with Sterile Oxacillin Sodium as directed under *Antibiotics—Microbial Assays* ⟨81⟩.

Oxamniquine

H
O_2N $CH_2NHCH(CH_3)_2$
$HOCH_2$

$C_{14}H_{21}N_3O_3$ 279.34
6-Quinolinemethanol, 1,2,3,4-tetrahydro-2-[[(1-methylethyl)-amino]methyl]-7-nitro-.
1,2,3,4-Tetrahydro-2-[(isopropylamino)methyl]-7-nitro-6-quinolinemethanol [*21738-42-1*].

» Oxamniquine contains not less than 97.0 percent and not more than 103.0 percent of $C_{14}H_{21}N_3O_3$, calculated on the anhydrous basis.

Packaging and storage—Preserve in well-closed containers.

Reference standards—*USP Oxamniquine Reference Standard*—Do not dry before using. *USP Related Compound A Reference Standard*—Do not dry before using. *USP Related Compound B Reference Standard*—Do not dry before using.

Identification—

A: The infrared absorption spectrum of a mineral oil dispersion of it exhibits maxima only at the same wavelengths as that of a similar preparation of USP Oxamniquine RS.

B: Prepare a solution of 25 mg in 1 mL of warm chloroform. On a suitable thin-layer chromatographic plate (see *Chromatography* ⟨621⟩), coated with a 0.25-mm layer of chromatographic silica gel, apply 20 µL of this solution and 20 µL of a Standard solution of USP Oxamniquine RS in the same solvent having a concentration of 25 mg per mL. Allow the spots to dry, and develop the chromatogram in a solvent system consisting of a mixture of chloroform, *n*-hexane, isopropyl alcohol, and isopropylamine (200:100:20:5) until the solvent front has moved about three-fourths of the length of the plate. Remove the plate from the developing chamber, mark the solvent front, and locate the spots on the plate by viewing under short- and long-wavelength ultraviolet light: the R_f value of the principal spot obtained from the test solution corresponds to that obtained from the Standard solution.

Melting range, *Class I* ⟨741⟩: between 145° and 152°.

Specific rotation ⟨781⟩: between −4° and +4°, determined in a solution containing 20 mg per mL, tetrahydrofuran being used as the solvent.

pH ⟨791⟩: between 8.0 and 10.0 in a suspension (1 in 100).

Water, *Method I* ⟨921⟩: not more than 1.0%.

Residue on ignition ⟨281⟩: not more than 0.2%, determined on a 2.0-g test specimen.

Iron ⟨241⟩—Dissolve the residue obtained in the test for *Residue on ignition* in 4 mL of hydrochloric acid with the aid of gentle heating, dilute with water to 100 mL, and mix. Dilute 10 mL of this solution with water to 47 mL: the limit is 0.005%.

Alcohol—

Standard preparation—By stepwise quantitative dilution of an accurately measured quantity of alcohol, with the use of methanol as the diluent, prepare a solution containing 0.75 mg of C_2H_5OH per mL.

Test preparation—Prepare a solution of Oxamniquine in methanol containing 100 mg per mL.

Chromatographic system (see *Chromatography* ⟨621⟩)—The gas chromatograph is equipped with a flame-ionization detector and a 1-m × 2-mm column that contains support S3. The column temperature is maintained at about 100°, and nitrogen is used as the carrier gas. The alcohol peak elutes at about 8 minutes. The oven temperature is adjusted, if necessary, to yield an alcohol peak that is just baseline separated from the methanol peak.

Procedure—Separately inject equal volumes (about 2 µL) of the *Standard preparation* and the *Test preparation* into the chromatograph, record the chromatograms, and measure the alcohol peaks. The response obtained from the *Test preparation* is not greater than that from the *Standard preparation* (0.75%).

Methyl isobutyl ketone—

Standard preparation—By stepwise quantitative dilution of an accurately measured quantity of methyl isobutyl ketone, with the use of methanol as the diluent, prepare a solution containing 0.10 mg per mL.

Test preparation—Prepare a solution of Oxamniquine in methanol containing 100 mg per mL.

Chromatographic system (see *Chromatography* ⟨621⟩)—The gas chromatograph is equipped with a flame-ionization detector and a 1-m × 2-mm column that contains support S3. The column temperature is maintained at about 220°, and nitrogen is used as the carrier gas. The methyl isobutyl ketone peak elutes at about 5 minutes. The oven temperature is adjusted, if necessary, to yield a methyl isobutyl ketone peak that is just baseline separated from the methanol peak.

Procedure—Separately inject equal volumes (about 2 µL) of the *Standard preparation* and the *Test preparation* into the chromatograph, record the chromatograms, and measure the methyl isobutyl ketone peaks. The response obtained from the *Test preparation* is not greater than that from the *Standard preparation* (0.10%).

Heavy metals, *Method II* ⟨231⟩: 0.005%.

Related compounds—Prepare, by warming, a solution in chloroform containing 25.0 mg of Oxamniquine per mL. Apply 20 µL of this solution to a suitable thin-layer chromatographic plate (see *Chromatography* ⟨621⟩) coated with a 0.25-mm layer of chromatographic silica gel. Apply to the same plate 20 µL of a methanol solution containing 175 µg per mL of USP Related Compound A RS and 20 µL of a methanol solution containing 125 µg per mL of USP Related Compound B RS. Allow the spots to dry, and develop the chromatogram in a solvent system consisting of a mixture of chloroform, *n*-hexane, isopropyl alcohol, and isopropylamine (200:100:20:5) until the solvent front has moved about three-fourths of the length of the plate. Remove the plate from the developing chamber, mark the solvent front, and locate the spots on the plate by viewing under short- and long-wavelength ultraviolet light: any spots from the Oxamniquine are not greater in size or intensity than the spots, occurring at the respective R_f values, produced by the Standard solutions corresponding to not more than 0.5% of 1,2,3,4-Tetrahydro-2-[(isopropylamino)methyl]-7-nitro-6-methylquinoline and not more than 0.5% of 1,2,3,4-Tetrahydro-2-[(isopropylamino)methyl]-5-nitro-6-quino-linemethanol, corresponding to Related Compounds A and B, respectively. [NOTE—USP Related Compound A RS is the methanesulfonate salt of the related compound.]

Assay—Transfer about 100 mg of Oxamniquine, accurately weighed, to a 200-mL volumetric flask, dissolve in methanol, add methanol to volume, and mix. Pipet 5 mL of this solution into a 500-mL volumetric flask, dilute with methanol to volume, and mix. Concomitantly determine the absorbances of this assay solution and of a Standard solution containing an accurately known concentration of about 5 µg per mL of USP Oxamniquine RS in methanol, using 1-cm cells, at the wavelength of maximum absorbance at about 251 nm, against a methanol blank. Calculate the quantity, in mg, of $C_{14}H_{21}N_3O_3$ in the portion of Oxamniquine taken by the formula:

$$20C(A_U/A_S),$$

in which *C* is the concentration, in µg per mL, of USP Oxamniquine RS in the Standard solution, and A_U and A_S are the absorbances of the assay solution and the Standard solution, respectively.

Oxamniquine Capsules

» Oxamniquine Capsules contain not less than 90.0 percent and not more than 110.0 percent of the labeled amount of $C_{14}H_{21}N_3O_3$.

Packaging and storage—Preserve in tight containers.

Reference standard—*USP Oxamniquine Reference Standard*—Do not dry before using.

Identification—Dissolve a portion of the contents of Capsules in a solvent consisting of equal volumes of tetrahydrofuran and methanol to obtain a solution containing 20 mg of oxamniquine per mL. Shake by mechanical means for 10 minutes, clarify a portion of the mixture by centrifugation, filter, if necessary, and use the clear supernatant solution or filtrate for the test. On a suitable thin-layer chromatographic plate (see *Chromatography* ⟨621⟩), coated with a 0.25-mm layer of chromatographic silica gel mixture, apply 20 µL of this test solution, 20 µL of a Standard solution containing 20 mg of USP Oxamniquine RS per mL in the same medium, and 20 µL of a mixture of equal volumes of the test solution and the Standard solution. Allow the spots to dry, and develop the chromatogram in a solvent system consisting of a mixture of toluene, methanol, glacial acetic acid, and acetone (80:20:5:5) until the solvent front has moved to about 2.5 cm from the top edge of the plate. Remove the plate from the developing chamber, mark the solvent front, and air-dry for 15 minutes. Locate the spots on the plate by viewing under short-wavelength ultraviolet light: the R_f value of the principal spot obtained from the test solution corresponds to those obtained from the Standard solution and the mixed test-Standard solution.

Dissolution ⟨711⟩—

Medium: 0.1 *N* hydrochloric acid; 900 mL.

Apparatus 2: 50 rpm.

Time: 60 minutes.

Procedure—Determine the amount of $C_{14}H_{21}N_3O_3$ dissolved from ultraviolet absorbances at the wavelength of maximum absorbance at about 247 nm of the filtered portion of the solution under test, suitably diluted with *Dissolution Medium*, if necessary, in comparison with a Standard solution having a known concentration of USP Oxamniquine RS in the same medium.

Tolerances—Not less than 70% (*Q*) of the labeled amount of $C_{14}H_{21}N_3O_3$ is dissolved in 60 minutes.

Uniformity of dosage units ⟨905⟩: meet the requirements.

Assay—Transfer, as completely as possible, the contents of not less than 20 Oxamniquine Capsules to a tared beaker, and weigh. Mix the combined contents, and transfer an accurately weighed portion, equivalent to about 250 mg of Oxamniquine, to a 500-mL volumetric flask. Add about 400 mL of methanol, shake by mechanical means for 30 minutes, dilute with methanol to volume, and filter through filter paper, discarding the first 20 mL of the filtrate. Transfer 10 mL of the clear filtrate to a 100-mL volumetric flask, dilute with methanol to volume, and mix. Pipet 10 mL of this solution into a 100-mL volumetric flask, dilute with methanol to volume, and mix. Dissolve an accurately weighed quantity of USP Oxamniquine RS in methanol, and dilute quantitatively and stepwise with methanol to obtain a Standard solution having a known concentration of about 5 µg per mL. Concomitantly determine the absorbances of both solutions in 1-cm cells at the wavelength of maximum absorbance at about 251 nm, with a suitable spectrophotometer, using methanol as the blank. Calculate the quantity, in mg, of $C_{14}H_{31}N_3O_3$ in the portion of Capsules taken by the formula:

$$50C(A_U/A_S),$$

in which *C* is the concentration, in µg per mL, of USP Oxamniquine RS in the Standard solution, and A_U and A_S are the absorbances of the solution from the Capsules and the Standard solution, respectively.

Oxandrolone

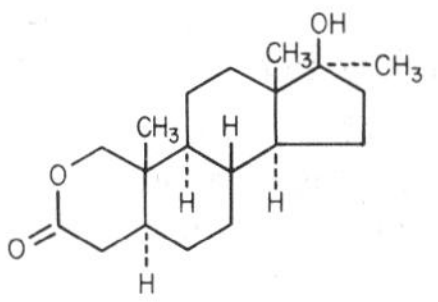

$C_{19}H_{30}O_3$ 306.44

2-Oxaandrostan-3-one, 17-hydroxy-17-methyl-, (5α,17β)-.

17β-Hydroxy-17-methyl-2-oxa-5α-androstan-3-one [*53-39-4*].

» Oxandrolone contains not less than 97.0 percent and not more than 100.5 percent of $C_{19}H_{30}O_3$, calculated on the dried basis.

Packaging and storage—Preserve in well-closed, light-resistant containers.

Reference standard—*USP Oxandrolone Reference Standard*—Dry at 105° for 3 hours before using.

Identification—

A: The infrared absorption spectrum of a potassium bromide dispersion of it, previously dried, exhibits maxima only at the same wavelengths as that of a similar preparation of USP Oxandrolone RS.

B: Prepare a solution in chloroform containing 5 mg per mL. On a suitable thin-layer chromatographic plate (see *Chromatography* ⟨621⟩), coated with a 0.25-mm layer of chromatographic silica gel, apply 10 µL each of this solution and a solution of USP Oxandrolone RS in chloroform, containing 5 mg per mL. Allow the spots to dry, and develop the chromatogram in a solvent system consisting of a mixture of chloroform and methanol (19:1) until the solvent front has moved about three-fourths of the length of the plate. Remove the plate from the developing chamber, mark the solvent front, and allow the solvent to evaporate. Locate the spots on the plate by lightly spraying with dilute sulfuric acid (1 in 2) and heating on a hot plate or under a lamp until spots appear: the R_f value of the principal spot obtained from the test solution corresponds to that obtained from the Standard solution.

Specific rotation ⟨781⟩: between −18° and −24°, calculated on the dried basis, determined in a solution in chloroform containing 100 mg in each 10 mL.

Loss on drying ⟨731⟩—Dry it at 105° for 3 hours: it loses not more than 1.0% of its weight.

Residue on ignition ⟨281⟩: not more than 0.2%.

Ordinary impurities ⟨466⟩—

Test solution: methanol.

Standard solution: methanol.

Application volume: 10 µL.

Eluant: a mixture of toluene and isopropyl alcohol (90:10), in a nonequilibrated chamber.

Visualization: 5.

Assay—Transfer about 500 mg of Oxandrolone, accurately weighed, to a 250-mL conical flask, and add 25.0 mL of 0.1 *N* alcoholic potassium hydroxide VS. Insert into the neck of the flask, by means of a perforated stopper, an air condenser consisting of a glass tube 70 to 80 cm in length and 5 to 8 mm in diameter, and heat the flask on a steam bath for 30 minutes, frequently rotating the contents. Cool, add 1 mL of phenolphthalein TS, and titrate the excess alkali with 0.1 *N* hydrochloric acid VS. Perform a blank determination, and make any necessary correction. Each mL of 0.1 *N* alcoholic potassium hydroxide is equivalent to 30.64 mg of $C_{19}H_{30}O_3$.

Oxandrolone Tablets

» Oxandrolone Tablets contain not less than 92.0 percent and not more than 108.0 percent of the labeled amount of $C_{19}H_{30}O_3$.

Packaging and storage—Preserve in tight, light-resistant containers.

Reference standard—*USP Oxandrolone Reference Standard*—Dry at 105° for 3 hours before using.

Identification—Transfer a portion of finely powdered Tablets, equivalent to about 20 mg of oxandrolone, to a glass-stoppered, 50-mL centrifuge tube, add 4 mL of chloroform, shake by mechanical means for 10 minutes, and filter a portion of the chloroform layer through a fine-porosity, sintered-glass filter with suction: the filtrate responds to *Identification test B* under *Oxandrolone*, beginning with "On a suitable thin-layer chromatographic plate."

Disintegration ⟨701⟩: 15 minutes.

Uniformity of dosage units ⟨905⟩: meet the requirements.

Assay—

Internal standard solution—Dissolve 100 mg of *n*-octacosane in 200 mL of chloroform.

Standard preparation—Transfer about 50 mg of USP Oxandrolone RS, accurately weighed, to a 100-mL volumetric flask, dissolve in *Internal standard solution*, dilute with *Internal standard solution* to volume, and mix.

Assay preparation—Weigh and finely powder not less than 20 Oxandrolone Tablets. Transfer an accurately weighed portion of the powder, equivalent to about 5 mg of oxandrolone, to a suitable container, add 10.0 mL of *Internal standard solution*, and shake by mechanical means for 30 minutes. Filter through Whatman No. 2 filter paper, or equivalent, discarding the first 5 mL of the filtrate.

Procedure—Inject separately 2-μL portions of the *Standard preparation* and the *Assay preparation* into a suitable gas chromatograph equipped with a flame-ionization detector (see *Chromatography* ⟨621⟩) and a 2-m × 4-mm glass column, packed with 3 percent methylsilicone oil on 80- to 100-mesh acid-, base-, and water-washed, flux-calcined, silanized siliceous earth. The column is maintained at about 250°, the injection port is maintained at about 290°, and the detector block is maintained at about 300°; helium is used as the carrier gas at a flow rate of about 60 mL per minute. Calculate the quantity, in mg, of $C_{19}H_{30}O_3$ in the portion of Tablets taken by the formula:

$$10C(R_U / R_S),$$

in which C is the concentration, in mg per mL, of USP Oxandrolone RS in the *Standard preparation*, and R_U and R_S are the ratios of the peak heights of the oxandrolone peak and the internal standard peak from the *Assay preparation* and the *Standard preparation*, respectively.

Oxazepam

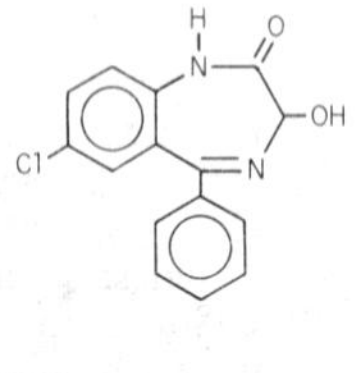

$C_{15}H_{11}ClN_2O_2$ 286.72

2*H*-1,4-Benzodiazepin-2-one, 7-chloro-1,3-dihydro-3-hydroxy-5-phenyl-.

7-Chloro-1,3-dihydro-3-hydroxy-5-phenyl-2*H*-1,4-benzodiazepin-2-one [*604-75-1*].

» Oxazepam contains not less than 98.0 percent and not more than 102.0 percent of $C_{15}H_{11}ClN_2O_2$, calculated on the dried basis.

Packaging and storage—Preserve in well-closed containers.

Reference standard—*USP Oxazepam Reference Standard*—Dry at a pressure below 5 mm of mercury at 105° for 3 hours before using.

Identification—

A: The infrared absorption spectrum of a potassium bromide dispersion of it, previously dried, exhibits maxima only at the same wavelengths as that of a similar preparation of USP Oxazepam RS.

B: The ultraviolet absorption spectrum of a 1 in 250,000 solution in alcohol exhibits maxima and minima at the same wavelengths as that of a similar solution of USP Oxazepam RS, concomitantly measured, and the respective absorptivities, calculated on the dried basis, at the wavelength of maximum absorbance at about 229 nm do not differ by more than 3.0%.

pH ⟨791⟩: between 4.8 and 7.0, in an aqueous suspension (1 in 50).

Loss on drying ⟨731⟩—Dry it at a pressure below 5 mm of mercury at 105° for 3 hours: it loses not more than 2.0% of its weight.

Residue on ignition ⟨281⟩: not more than 0.3%.

Assay—Transfer about 400 mg of Oxazepam, accurately weighed, to a beaker, dissolve in 100 mL of dimethylformamide, and titrate with 0.1 *N* tetrabutylammonium hydroxide VS, determining the end-point potentiometrically using a calomel-glass electrode system and taking precautions against absorption of atmospheric carbon dioxide. Perform a blank determination, and make any necessary correction. Each mL of 0.1 *N* tetrabutylammonium hydroxide is equivalent to 28.67 mg of $C_{15}H_{11}ClN_2O_2$.

Oxazepam Capsules

» Oxazepam Capsules contain not less than 90.0 percent and not more than 110.0 percent of the labeled amount of $C_{15}H_{11}ClN_2O_2$.

Packaging and storage—Preserve in well-closed containers.

Reference standard—*USP Oxazepam Reference Standard*—Dry at a pressure below 5 mm of mercury at 105° for 3 hours before using.

Identification—The solution prepared for measurement of absorbance in the *Assay* exhibits a maximum at 229 ± 2 nm.

Dissolution ⟨711⟩—

Medium: 0.1 *N* hydrochloric acid; 1000 mL.

Apparatus 2: 75 rpm.

Time: 60 minutes.

Chromatographic system (see *Chromatography* ⟨621⟩)—The liquid chromatograph is fitted with a 232-nm detector and a 4-mm × 15-cm column that contains packing L7. The mobile phase is a degassed mixture of methanol, water, and glacial acetic acid (60:40:1), and the flow rate is about 2 mL per minute. Chromatograph the Standard solution, and record the peak response as directed under *Procedure:* the tailing factor for the oxazepam peak is not more than 1.5, and the relative standard deviation observed following replicate injections is not more than 3.0%. Make adjustments if necessary (see *System Suitability* under *Chromatography* ⟨621⟩).

Procedure—Inject an accurately measured volume (about 20 μL) of a filtered portion of the solution under test into the chromatograph, record the chromatogram, and measure the response for the major peak. Calculate the quantity of $C_{15}H_{11}ClN_2O_2$ dissolved in comparison with a Standard solution having a known concentration of USP Oxazepam RS in 0.1 *N* hydrochloric acid (NOTE—a volume of methanol not to exceed 10% of the final total volume may be used to dissolve the Oxazepam reference standard) and similarly chromatographed.

Tolerances—Not less than 75% (*Q*) of the labeled amount of $C_{15}H_{11}ClN_2O_2$ is dissolved in 60 minutes.

Uniformity of dosage units ⟨905⟩: meet the requirements.

Assay—Remove, as completely as possible, the contents of not less than 20 Oxazepam Capsules, and weigh. Transfer an accurately weighed portion of the mixed powder, equivalent to about 50 mg of oxazepam, to a medium-porosity, sintered-glass funnel that is fitted into a small suction flask. Add 25 mL of alcohol, mix with the aid of a stirring rod, and after about 5 minutes apply gentle suction to remove the extract. Repeat the extraction with four additional 25-mL portions of alcohol, transfer the extracts to a 250-mL volumetric flask, dilute with alcohol to volume, and mix. Transfer 2.0 mL of this solution to a 100-mL volumetric flask, dilute with alcohol to volume, and mix. Concomitantly determine the absorbances of this solution and of a Standard solution of USP Oxazepam RS in the same medium having a known concentration of about 4 µg per mL in 1-cm cells at the wavelength of maximum absorbance at about 229 nm, with a suitable spectrophotometer, using alcohol as the blank. Calculate the quantity, in mg, of $C_{15}H_{11}ClN_2O_2$ in the portion of Capsules taken by the formula:

$$12.5C(A_U/A_S),$$

in which C is the concentration, in µg per mL, of USP Oxazepam RS in the Standard solution, and A_U and A_S are the absorbances of the solution from the Capsules and the Standard solution, respectively.

Oxazepam Tablets

» Oxazepam Tablets contain not less than 90.0 percent and not more than 110.0 percent of the labeled amount of $C_{15}H_{11}ClN_2O_2$.

Packaging and storage—Preserve in well-closed containers.

Reference standard—*USP Oxazepam Reference Standard*—Dry at a pressure below 5 mm of mercury at 105° for 3 hours before using.

Identification—The solution prepared for measurement of absorbance in the *Assay* exhibits a maximum at 229 ± 2 nm.

Dissolution ⟨711⟩—

Medium: 0.1 *N* hydrochloric acid; 1000 mL.

Apparatus 2: 50 rpm.

Time: 60 minutes.

Chromatographic system—The liquid chromatograph is fitted with a 232-nm detector and a 4-mm × 30-cm column that contains packing L7. The mobile phase is a degassed mixture of methanol, water, and glacial acetic acid (60:40:1), and the flow rate is about 2 mL per minute. In a suitable system, the relative standard deviation observed following replicate injections is not more than 3.0%.

Procedure—Inject an accurately measured volume (about 10 µL) of a filtered portion of the solution under test into the chromatograph by means of a microsyringe or a sampling valve, record the chromatogram, and measure the response for the major peak. Calculate the quantity of $C_{15}H_{11}ClN_2O_2$ dissolved in comparison with a similarly chromatographed Standard solution having a known concentration of USP Oxazepam RS in 0.1 *N* hydrochloric acid. [NOTE—A volume of alcohol not exceeding 10% of the final total volume of the Standard solution may be used to dissolve the Reference Standard.]

Tolerances—Not less than 80% (*Q*) of the labeled amount of $C_{15}H_{11}ClN_2O_2$ is dissolved in 60 minutes.

Uniformity of dosage units ⟨905⟩: meet the requirements.

Assay—Weigh and finely powder not less than 20 Oxazepam Tablets. Transfer an accurately weighed portion of the powder, equivalent to about 50 mg of oxazepam, to a medium-porosity, sintered-glass funnel that is fitted into a small suction flask, and proceed as directed in the *Assay* under *Oxazepam Capsules*, beginning with "Add 25 mL of alcohol." Calculate the quantity, in mg, of $C_{15}H_{11}ClN_2O_2$ in the portion of Tablets taken by the formula:

$$12.5C(A_U/A_S),$$

in which C is the concentration, in µg per mL, of USP Oxazepam RS in the Standard solution, and A_U and A_S are the absorbances of the solution from the Tablets and the Standard solution, respectively.

Oxidized Cellulose—*see* Cellulose, Oxidized

Oxprenolol Hydrochloride

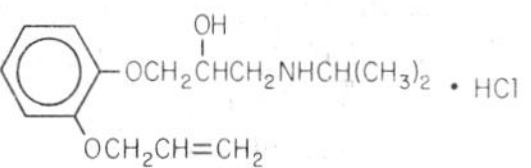

$C_{15}H_{23}NO_3 \cdot HCl$ 301.81

2-Propanol, 1-(*o*-allyloxyphenoxy)-3-isopropylamino-, hydrochloride.

1-(*o*-Allyloxyphenoxy)-3-isopropylamino-2-propanol hydrochloride [*6452-73-9*].

» Oxprenolol Hydrochloride contains not less than 98.5 percent and not more than 101.0 percent of $C_{15}H_{23}NO_3 \cdot HCl$, calculated on the dried basis.

Packaging and storage—Preserve in well-closed containers.

Reference standard—*USP Oxprenolol Hydrochloride Reference Standard*—Dry in vacuum at 60° for 6 hours before using. Keep container tightly closed and protected from light.

Clarity of solution—Dissolve 1 g in 10 mL of water: the solution is clear.

Identification—

A: The infrared absorption spectrum of a potassium bromide dispersion of it, previously dried, exhibits maxima only at the same wavelengths as that of a similar preparation of USP Oxprenolol Hydrochloride RS.

B: A solution of it responds to the tests for *Chloride* ⟨191⟩.

pH ⟨791⟩: between 4.0 and 6.0, in a solution (1 in 10).

Loss on drying ⟨731⟩—Dry about 3 g of it in vacuum at 60° for 6 hours: it loses not more than 0.5% of its weight.

Residue on ignition ⟨281⟩: not more than 0.1%.

Heavy metals, *Method II* ⟨231⟩: 0.001%.

Chromatographic purity—

Diluting solvent—Prepare a mixture of chloroform and dehydrated alcohol (1:1).

Standard solution A—Prepare a solution of USP Oxprenolol Hydrochloride RS in *Diluting solvent* containing 20 mg per mL.

Standard solution B—Dilute an accurately measured volume of *Standard solution A* quantitatively, and stepwise if necessary, with *Diluting solvent* to obtain a solution containing 0.04 mg per mL.

Test solution—Transfer 200 mg of it to a 10-mL volumetric flask, dissolve in *Diluting solvent*, dilute with *Diluting solvent* to volume, and mix.

Procedure—Apply separate 5-µL portions of the *Test solution* and the *Standard solutions* on the starting line of a suitable thin-layer chromatographic plate (see *Chromatography* ⟨621⟩), coated with a 0.25-mm layer of chromatographic silica gel mixture, previously washed with methanol until the solvent front reaches the top of the plate, dried first in air and then at 100° for 20 minutes, and cooled in a desiccator. Allow the spots to dry. Line a suitable chromatographic chamber with filter paper, saturate the paper with 100 mL of a solvent system consisting of a mixture of chloroform and methanol (90:10), then place two 20-mL beakers of ammonium hydroxide in the chamber and allow to stand for 1 hour. Place the plate in the chamber, and develop the chromatogram until the solvent front has moved about three-fourths of the length of the plate. Remove the plate from the chamber and dry at 100° for 30 minutes. Spray the plate uniformly with a detection reagent consisting of a freshly prepared mixture of equal volumes of potassium ferricyanide solution (1 in 100) and ferric chloride solution (1 in 5). Dry the plate in a current of warm air for about 5 minutes or until a spot from *Standard solution B* is visible. Examine the chromatograms in ordinary light: the R_f value of the principal spot from the *Test solution* corresponds to that obtained from *Standard solution A*. No spot other than the principal spot obtained from the *Test solution* exceeds in size or intensity the principal spot obtained from *Standard solution B* (0.2%, corresponding to 0.4% of related compounds, the response factors for which are about double that of oxprenolol hydrochloride).

Assay—Dissolve about 450 mg of Oxprenolol Hydrochloride in 100 mL of glacial acetic acid. Add 10 mL of mercuric acetate

TS, and titrate with 0.1 *N* perchloric acid VS, determining the end-point potentiometrically, using a glass-calomel electrode system (with a salt bridge of a saturated solution of lithium chloride in glacial acetic acid). Perform a blank determination, and make any necessary correction. Each mL of 0.1 *N* perchloric acid is equivalent to 30.18 mg of $C_{15}H_{23}NO_3 . HCl$.

Oxprenolol Tablets

» Oxprenolol Tablets contain not less than 90.0 percent and not more than 110.0 percent of the labeled amount of oxprenolol hydrochloride ($C_{15}H_{23}NO_3 . HCl$).

Packaging and storage—Preserve in well-closed, light-resistant containers.

Labeling—Label Tablets to state both the content of the active moiety and the content of the salt used in formulating the article.

Reference standard—*USP Oxprenolol Hydrochloride Reference Standard*—Dry in vacuum at 60° for 6 hours before using. Keep container tightly closed and protected from light.

Identification—Transfer a portion of powdered Tablets, equivalent to about 100 mg of oxprenolol hydrochloride, to a suitable test tube containing 5 mL of water. Shake this mixture for about 1 minute and allow it to settle. Use the clear supernatant liquid as the test solution. Prepare a Standard solution containing 20 mg of USP Oxprenolol Hydrochloride RS per mL. Apply separate 1-µL portions of the test solution and of the Standard solution on a suitable thin-layer chromatographic plate (see *Chromatography* ⟨621⟩), coated with a 0.25-mm layer of chromatographic silica gel mixture. Line a suitable chromatographic chamber with filter paper, saturate the paper with the developing solvent consisting of a mixture of chloroform, methanol, and ammonium hydroxide (44:6:1), and allow to stand for about 30 minutes. Place the plate in the chamber, and develop the chromatogram until the solvent front has moved about one-half of the length of the plate. Remove the plate from the chamber and dry in a current of warm air. Spray the plate uniformly with a detection reagent prepared by mixing, in the following order, 0.5 mL of 4-methoxybenzaldehyde, 10 mL of glacial acetic acid, 85 mL of methanol, and 5 mL of sulfuric acid. Examine the chromatograms in ordinary light: the R_f value of the principal spot obtained from the test solution corresponds to that obtained from the Standard solution.

Dissolution ⟨711⟩—

Medium: 0.1 *N* hydrochloric acid; 900 mL.

Apparatus 1: 100 rpm.

Time: 30 minutes.

Procedure—Determine the amount of $C_{15}H_{23}NO_3 . HCl$ dissolved from ultraviolet absorbances at the wavelength of maximum absorbance at about 272 nm of filtered portions of the solution under test, suitably diluted with *Dissolution Medium*, if necessary, in comparison with a Standard solution having a known concentration of USP Oxprenolol Hydrochloride RS in the same medium.

Tolerances—Not less than 80% (*Q*) of the labeled amount of $C_{15}H_{23}NO_3 . HCl$ is dissolved in 30 minutes.

Uniformity of dosage units ⟨905⟩: meet the requirements.

Assay—Weigh and finely powder not less than 20 Oxprenolol Tablets. Transfer an accurately weighed portion of the powder, equivalent to about 80 mg of oxprenolol hydrochloride, to a 100-mL volumetric flask. Add 80 mL of a *Diluting solvent*, which consists of a mixture of methanol and 0.1 *N* hydrochloric acid (9:1), and shake by mechanical means for 1 hour. Dilute with *Diluting solvent* to volume, and mix. Filter, discarding the first 10 mL of the filtrate, transfer 25.0 mL of the clear filtrate to a 200-mL volumetric flask, dilute with *Diluting solvent* to volume, and mix. Concomitantly determine the absorbances of this assay solution and a Standard solution of USP Oxprenolol Hydrochloride RS in the same solvent having a known concentration of about 0.1 mg per mL at the wavelength of maximum absorbance at about 274 nm and, in addition, at 300 nm, using *Diluting solvent* as the blank. Calculate the quantity, in mg, of $C_{15}H_{23}NO_3 . HCl$ in the portion of Tablets taken by the formula:

$$800C(A_U / A_S),$$

in which *C* is the concentration, in mg per mL, of USP Oxprenolol Hydrochloride RS in the Standard solution, and A_U and A_S are the differences between the absorbances at 274 nm and 300 nm of the assay solution and the Standard solution, respectively.

Oxtriphylline

[(CH₃)₃N⁺CH₂CH₂OH] [theophylline anion structure: O, CH₃, N, N⁻, N, O, N, CH₃]

$C_{12}H_{21}N_5O_3$ 283.33

Ethanaminium, 2-hydroxy-*N*,*N*,*N*-trimethyl-, salt with 3,7-dihydro-1,3-dimethyl-1*H*-purine-2,6-dione.

Choline salt with theophylline (1:1) [*4499-40-5; 13930-27-3*].

» Oxtriphylline contains not less than 61.7 percent and not more than 65.5 percent of anhydrous theophylline ($C_7H_8N_4O_2$), calculated on the dried basis.

Packaging and storage—Preserve in tight containers.

Identification—Dissolve about 1 g in 20 mL of water, and use the solution for the following tests.

A: To 10 mL of the test solution add 5 mL of mercuric–potassium iodide TS: a pale yellow precipitate is formed (*presence of choline*).

B: To 10 mL of the test solution add 5 drops of 6 *N* ammonium hydroxide and 5 mL of silver nitrate TS: a gelatinous precipitate is formed, and it coagulates on heating (*presence of theophylline*).

Melting range, *Class I* ⟨741⟩: between 185° and 189°.

Loss on drying ⟨731⟩—Dry it at 80° for 4 hours: it loses not more than 1.0% of its weight.

Residue on ignition ⟨281⟩: not more than 0.3%.

Chloride ⟨221⟩—A 0.50-g portion shows no more chloride than corresponds to 0.15 mL of 0.020 *N* hydrochloric acid (0.02%).

Ordinary impurities ⟨466⟩—

Test solution: a mixture of chloroform, alcohol, and formic acid (88:10:2).

Standard solution: a mixture of chloroform, alcohol, and formic acid (88:10:2).

Eluant: a mixture of chloroform, alcohol, and formic acid (88:10:2).

Visualization: 1.

Choline content—Dissolve about 900 mg of Oxtriphylline, accurately weighed, in 50 mL of water, and add 4 drops of a solution prepared by dissolving 30 mg of methyl red in 100 mL of methanol, adding 15 mL of methylene blue solution (1 in 1000), and mixing. Mix, and titrate with 0.1 *N* sulfuric acid VS to a purple end-point. Each mL of 0.1 *N* sulfuric acid is equivalent to 12.12 mg of $C_5H_{15}NO_2$. The content of choline ($C_5H_{15}NO_2$) is between 652 mg and 693 mg per g of $C_7H_8N_4O_2$ found in the *Assay*. Retain the final solution for the *Assay for theophylline*.

Assay for theophylline—To the solution retained in the *Choline content* add 35 mL of silver nitrate TS, swirl gently to promote complete precipitation, and titrate with 0.1 *N* sodium hydroxide VS to a green end-point. Each mL of 0.1 *N* sodium hydroxide is equivalent to 18.02 mg of $C_7H_8N_4O_2$.

Oxtriphylline Elixir

» Oxtriphylline Elixir contains an amount of oxtriphylline equivalent to not less than 90.0 percent and not more than 110.0 percent of the labeled amount of anhydrous theophylline ($C_7H_8N_4O_2$).

Packaging and storage—Preserve in tight containers.

Labeling—Label Elixir to state both the content of oxtriphylline and the content of anhydrous theophylline.

Reference standard—*USP Oxtriphylline Reference Standard*—Dry at 80° for 4 hours before using.

Identification—Place a volume of Elixir, equivalent to about 100 mg of oxtriphylline, in a 60-mL separator containing 1 mL of glacial acetic acid and 40 mL of chloroform. Shake for 1 minute, allow the phases to separate, and filter the lower phase through dry cotton into a 100-mL beaker. Transfer a portion of the chloroform solution, equivalent to about 10 mg of oxtriphylline, to a porcelain dish, and evaporate on a steam bath with the aid of a current of dry air to dryness. Add 1 mL of hydrochloric acid and 100 mg of potassium chlorate, evaporate on a steam bath to dryness, and invert the dish over a vessel containing a few drops of 6 *N* ammonium hydroxide: the residue acquires a purple color.

pH ⟨791⟩: between 8.0 and 9.0.

Alcohol content, *Method I* ⟨611⟩: between 18.0% and 22.0% of C_2H_5OH.

Assay—

Standard preparation—Transfer about 50 mg of USP Oxtriphylline RS, accurately weighed, to a 250-mL volumetric flask, and dissolve in about 150 mL of water. Add 25 mL of sodium hydroxide solution (1 in 250), dilute with water to volume, and mix. Transfer 5.0 mL of this solution to a 100-mL volumetric flask, dilute with sodium hydroxide solution (1 in 2500) to volume, and mix. The concentration of USP Oxtriphylline RS in the *Standard preparation* is about 10 µg per mL.

Assay preparation—Transfer an accurately measured volume of Oxtriphylline Elixir, equivalent to about 100 mg of oxtriphylline, to a 1000-mL volumetric flask. Add about 400 mL of water and 100 mL of sodium hydroxide solution (1 in 250), dilute with water to volume, and mix. Transfer 10.0 mL of this solution to a 125-mL separator, and add 10 mL of sodium hydroxide solution (1 in 250). Extract with three 30-mL portions of chloroform, combine the extracts in a second separator, wash this chloroform solution with 10 mL of sodium hydroxide solution (1 in 2500), and discard the chloroform phase. Combine the aqueous phases in a 100-mL beaker, and heat on a steam bath for 15 minutes, using a current of air to expel any chloroform that may be present. Cool, transfer to a 100-mL volumetric flask, dilute with water to volume, and mix.

Procedure—Concomitantly determine the absorbances of the solutions in 1-cm cells at the wavelength of maximum absorbance at about 275 nm, with a suitable spectrophotometer, using sodium hydroxide solution (1 in 2500) as the blank. Calculate the quantity, in mg, of $C_7H_8N_4O_2$ in each mL of the Elixir taken by the formula:

$$(180.17/283.33)(10C/V)(A_U/A_S),$$

in which 180.17 and 283.33 are the molecular weights of anhydrous theophylline and oxtriphylline, respectively, *C* is the concentration, in µg per mL, of USP Oxtriphylline RS in the *Standard preparation*, *V* is the volume, in mL, of Elixir taken, and A_U and A_S are the absorbances of the *Assay preparation* and the *Standard preparation*, respectively.

Oxtriphylline Delayed-release Tablets

» Oxtriphylline Delayed-release Tablets contain an amount of oxtriphylline equivalent to not less than 90.0 percent and not more than 110.0 percent of the labeled amount of anhydrous theophylline ($C_7H_8N_4O_2$).

Packaging and storage—Preserve in tight containers.

Labeling—Label Tablets to state both the content of oxtriphylline and the content of anhydrous theophylline.

Reference standard—*USP Oxtriphylline Reference Standard*—Dry at 80° for 4 hours before using.

Identification—Remove the coating from several Oxtriphylline Delayed-release Tablets by washing with water. Dry the tablet cores in air before finely powdering them. Extract a portion of the powder, equivalent to about 300 mg of oxtriphylline, with three 15-mL portions of alcohol. Filter the combined extracts through Whatman No. 42, or equivalent, filter paper, evaporate the filtrate to dryness, and dissolve the residue in 10 mL of water. Add 5 mL of mercuric–potassium iodide TS to the solution so obtained: a yellow precipitate is formed.

Disintegration ⟨701⟩—Test Oxtriphylline Delayed-release Tablets as directed for *Enteric-coated Tablets* (see ⟨701⟩): the tablets do not disintegrate after 30 minutes of agitation in simulated gastric fluid TS; continue agitation in simulated gastric fluid TS for an additional 30 minutes: the tablets may disintegrate during this period; if all of the tablets have not disintegrated, place the basket in simulated intestinal fluid TS, and operate the apparatus: all of the tablets disintegrate within 90 minutes (2.5 hours total disintegration time).

Uniformity of dosage units ⟨905⟩: meet the requirements.

Assay—[NOTE—In this assay, use chloroform recently saturated with water.]

Standard preparation—Transfer about 75 mg of USP Oxtriphylline RS, accurately weighed, to a 200-mL volumetric flask, add a 1 in 100 solution of glacial acetic acid in chloroform to volume, and mix. Transfer 3.0 mL of this solution to a 100-mL volumetric flask, dilute with a 1 in 100 solution of glacial acetic acid in chloroform to volume, and mix.

Chromatographic column—Proceed as directed under *Column Partition Chromatography* (see *Chromatography* ⟨621⟩), packing a chromatographic tube with two segments of packing material. The lower segment is a mixture of 500 mg of *Solid Support* and 0.3 mL of tribasic sodium phosphate solution (1 in 4), and the upper segment is a mixture prepared as directed under *Assay preparation.*

Assay preparation—Weigh and finely powder not less than 20 Oxtriphylline Delayed-release Tablets. Transfer an accurately weighed portion of the powder, equivalent to about 75 mg of oxtriphylline, to a 100-mL beaker, mix with 2.0 mL of tribasic sodium phosphate solution (1 in 4) and 3 g of *Solid Support* as directed under *Chromatographic column*, and transfer quantitatively to the column in three portions. Wash the column with three 25-mL portions of chloroform, allowing each portion to enter the column before adding the next, and discard the washings. Elute the theophylline with 5 mL of a 1 in 10 solution of glacial acetic acid in chloroform, followed by 150 mL of a 1 in 100 solution of glacial acetic acid in chloroform, collecting the eluates in a 200-mL volumetric flask. Dilute with a 1 in 100 solution of glacial acetic acid in chloroform to volume, and mix. Transfer 3.0 mL of this solution to a 100-mL volumetric flask, dilute with a 1 in 100 solution of glacial acetic acid in chloroform to volume, and mix.

Procedure—Concomitantly determine the absorbances of the solutions in 1-cm cells at the wavelength of maximum absorbance at about 277 nm, with a suitable spectrophotometer, using a 1 in 100 solution of glacial acetic acid in chloroform as the blank. Calculate the quantity, in mg, of $C_7H_8N_4O_2$ in the portion of Tablets taken by the formula:

$$(6.67)(180.17/283.33)C(A_U/A_S),$$

in which 180.17 and 283.33 are the molecular weights of anhydrous theophylline and oxtriphylline, respectively, *C* is the concentration, in µg per mL, of USP Oxtriphylline RS in the *Standard preparation*, and A_U and A_S are the absorbances of the *Assay preparation* and the *Standard preparation*, respectively.

Oxybenzone

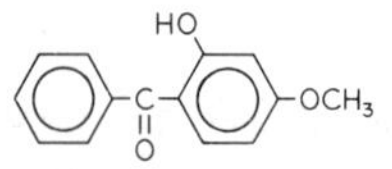

$C_{14}H_{12}O_3$ 228.25
Methanone, (2-hydroxy-4-methoxyphenyl)phenyl-.
2-Hydroxy-4-methoxybenzophenone [*131-57-7*].

» Oxybenzone contains not less than 97.0 percent and not more than 103.0 percent of $C_{14}H_{12}O_3$, calculated on the dried basis.

Packaging and storage—Preserve in tight, light-resistant containers.

Reference standard—*USP Oxybenzone Reference Standard*—Dry in vacuum at 40° for 2 hours before using.

Identification—

A: The infrared absorption spectrum of a potassium bromide dispersion of it, previously dried, exhibits maxima only at the same wavelengths as that of a similar preparation of USP Oxybenzone RS.

B: The ultraviolet absorption spectrum of a 1 in 100,000 solution in methanol exhibits maxima and minima at the same wavelengths as that of a similar preparation of USP Oxybenzone RS, concomitantly measured, and the respective absorptivities, calculated on the dried basis, at the wavelength of maximum absorbance at about 290 nm, do not differ by more than 3.0%.

Congealing temperature ⟨651⟩: not lower than 62.0°.

Loss on drying ⟨731⟩—Dry it in vacuum at 40° for 2 hours: it loses not more than 2.0% of its weight.

Assay—Dissolve about 100 mg of Oxybenzone, accurately weighed, in methanol in a 100-mL volumetric flask, dilute with methanol to volume, and mix. Agitate gently, if necessary, to hasten solution. Pipet 1 mL of this solution into a second 100-mL volumetric flask, add methanol to volume, and mix. Similarly, prepare a Standard solution of USP Oxybenzone RS, accurately weighed, having a known concentration of about 10 µg per mL. Concomitantly determine the absorbances of both solutions in 1-cm cells at the wavelength of maximum absorbance at about 285 nm, with a suitable spectrophotometer, using methanol as the blank. Calculate the quantity, in mg, of $C_{14}H_{12}O_3$ in the portion of Oxybenzone taken by the formula:

$$10C(A_U/A_S),$$

in which C is the concentration, in µg per mL, of USP Oxybenzone RS in the Standard solution, and A_U and A_S are the absorbances of the solution of Oxybenzone and the Standard solution, respectively.

Oxybenzone Cream, Dioxybenzone and—*see* Dioxybenzone and Oxybenzone Cream

Oxybutynin Chloride

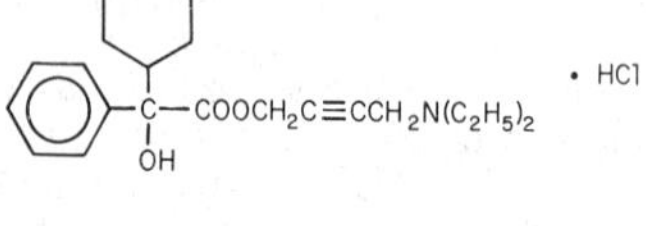

$C_{22}H_{31}NO_3.HCl$ 393.95

Benzeneacetic acid, α-cyclohexyl-α-hydroxy-, 4-(diethylamino)-2-butynyl ester hydrochloride.
4-(Diethylamino)-2-butynyl α-phenylcyclohexaneglycolate hydrochloride [*1508-65-2*].

» Oxybutynin Chloride contains not less than 97.0 percent and not more than 101.0 percent of $C_{22}H_{31}NO_3.HCl$, calculated on the dried basis.

Packaging and storage—Preserve in well-closed containers.

Reference standards—*USP Oxybutynin Chloride Reference Standard*—Dry in vacuum at 60° for 24 hours before using. *USP Phenylcyclohexylglycolic Acid Reference Standard.*

Identification—The infrared absorption spectrum of a potassium bromide dispersion of it exhibits maxima only at the same wavelengths as that of a similar preparation of USP Oxybutynin Chloride RS.

Melting range ⟨741⟩: between 124° and 129°.

Loss on drying ⟨731⟩: Dry it at 105° for two hours: it loses not more than 3% of its weight.

Residue on ignition ⟨281⟩: not more than 0.1%.

Heavy metals, *Method I* ⟨231⟩: 0.002%.

Chromatographic purity—

Triethanolamine phosphate solution—Dissolve 2 mL of triethanolamine in about 900 mL of water in a 1000-mL volumetric flask, adjust with phosphoric acid to a pH of 3.5, dilute with water to volume, and mix.

Mobile phase—Prepare a degassed and filtered mixture of acetonitrile and *Triethanolamine phosphate solution* (65:35). Make adjustments if necessary (see *System Suitability* under *Chromatography* ⟨621⟩).

Standard preparation—Prepare a solution containing 2.0 mg of USP Oxybutynin Chloride RS per mL and 10 µg of USP Phenylcyclohexylglycolic Acid RS per mL in acetonitrile. Filter through a 0.45-µm filter.

Test preparation—Transfer 200 mg of Oxybutynin Chloride, accurately weighed, to a 100-mL volumetric flask, dissolve in acetonitrile, dilute with acetonitrile to volume, and mix.

Chromatographic system (see *Chromatography* ⟨621⟩)—The liquid chromatograph is equipped with a 200-nm detector and a 4.6-mm × 25-cm column that contains packing L7, and a short precolumn that contains packing L7 is used if necessary. The column is maintained at a temperature of 45°. The flow rate is about 3 mL per minute. Chromatograph a solution in acetonitrile containing 40 µg per mL each of USP Oxybutynin Chloride RS and USP Phenylcyclohexylglycolic Acid RS, and record the peak responses as directed under *Procedure:* the relative standard deviation of the responses for replicate injections is not more than 3.0%, and the resolution, R, of the main peaks is not less than 5.0.

Procedure—Separately inject equal volumes (about 20 µL) of the *Standard preparation* and the *Test preparation* into the chromatograph, and record the chromatograms. The relative retention times are about 0.2 for phenylcyclohexylglycolic acid and 1.0 for oxybutynin. In the chromatogram of the *Test preparation*, any peak corresponding in retention time to phenylcyclohexylglycolic acid has a response not greater than that in the chromatogram of the *Standard preparation*, and the sum of the responses of all peaks, excluding the solvent front and oxybutynin, is not greater than twice that value.

Chloride content—Dissolve about 600 mg of oxybutynin chloride, previously dried and accurately weighed, in 100 mL of water, and add 5 mL of nitric acid. Titrate (see *Titrimetry* ⟨541⟩) with 0.1 N silver nitrate VS, determining the end-point potentiometrically, using a platinum–silver chloride electrode system. Each mL of 0.1 N silver nitrate is equivalent to 3.545 mg of Cl: the content is between 8% and 10%.

Assay—Dissolve about 500 mg of oxybutynin chloride, accurately weighed, in 50 mL of glacial acetic acid, and add 10 mL of mercuric acetate TS. Add five drops of crystal violet TS, and titrate (see *Titrimetry* ⟨541⟩) with 0.1 N perchloric acid VS to an emerald-green end-point. Perform a blank determination, and make any necessary correction. Each mL of 0.1 N perchloric acid is equivalent to 39.40 mg of $C_{22}H_{31}NO_3.HCl$.

Oxybutynin Chloride Syrup

» Oxybutynin Chloride Syrup contains not less than 90.0 percent and not more than 110.0 percent of the labeled amount of $C_{22}H_{31}NO_3.HCl$.

Packaging and storage—Preserve in tight, light-resistant containers.

Reference standard—*USP Oxybutynin Chloride Reference Standard*—Dry in vacuum at 60° for 24 hours before using.

Identification—Place a volume of Syrup, equivalent to about 50 mg of oxybutynin chloride, in a separator, and extract with 10 mL of chloroform. The extract so obtained responds to the *Thin-layer Chromatographic Identification Test* ⟨201⟩, methanol being used as the developing solvent, and iodine vapor being used to visualize the spots.

Assay—

pH 4 phosphate buffer—Place 38 mL of 0.2 *M* dibasic sodium phosphate in a 100-mL volumetric flask. Dilute with 0.1 *M* citric acid to volume, and mix. Adjust the pH, if necessary, with either the dibasic sodium phosphate solution or the citric acid solution.

pH 5.6 phosphate buffer—Place 58 mL of 0.2 *M* dibasic sodium phosphate in a 100-mL volumetric flask. Dilute with 0.1 *M* citric acid to volume, and mix. Adjust the pH, if necessary, with either the dibasic sodium phosphate solution or the citric acid solution.

Bromocresol green solution—Transfer 125 mg of bromocresol green to a 25-mL volumetric flask, dissolve in 3.5 mL of 0.05 *N* sodium hydroxide, dilute with water to volume, and mix.

Standard preparation—Dissolve an accurately weighed quantity of USP Oxybutynin Chloride RS in 0.05 *N* sulfuric acid to obtain a solution having a known concentration of about 100 µg per mL.

Assay preparation—Transfer an accurately measured volume of Oxybutynin Chloride Syrup, equivalent to about 10 mg of oxybutynin chloride, to a 100-mL volumetric flask, dilute with water to volume, and mix.

Procedure—Separately transfer 10.0 mL of the *Standard preparation* and the *Assay preparation* to separate 125-mL separators. Add 20 mL of *pH 4 phosphate buffer* to each separator, and extract each solution with a 25-mL portion of chloroform. [NOTE—Allow at least 10 minutes for the layers to separate.] Collect the chloroform extracts in respective 125-mL separators, each containing a mixture of 2 mL of *pH 5.6 phosphate buffer* and 1 mL of *Bromocresol green solution.* Shake the separators, and filter the chloroform extracts through rayon pledgets, collecting the extracts in respective 100-mL volumetric flasks. Repeat the double extractions with 25-mL portions of chloroform. Wash the rayon pledgets with chloroform, collecting the washings in the respective 100-mL volumetric flasks. Dilute both solutions with chloroform to volume, and mix. Concomitantly determine the absorbances of both solutions at the wavelength of maximum absorbance at about 415 nm, with a suitable spectrophotometer, against a blank prepared using 10 mL of 0.05 *N* sulfuric acid treated in the same manner as the *Standard preparation* and the *Assay preparation.* Calculate the quantity, in mg, of $C_{22}H_{31}NO_3.HCl$ in each mL of Syrup taken by the formula:

$$(0.1C/V)(A_U/A_S),$$

in which C is the concentration, in µg per mL, of USP Oxybutynin Chloride RS in the *Standard preparation*, V is the volume, in mL, of Syrup taken, and A_U and A_S are the absorbances of the solutions from the *Assay preparation* and the *Standard preparation*, respectively.

Oxybutynin Chloride Tablets

» Oxybutynin Chloride Tablets contain not less than 90.0 percent and not more than 110.0 percent of the labeled amount of $C_{22}H_{31}NO_3.HCl$.

Packaging and storage—Preserve in tight, light-resistant containers.

Reference standard—*USP Oxybutynin Chloride Reference Standard*—Dry in vacuum at 60° for 24 hours before using.

Identification—Add a portion of powdered Tablets, equivalent to about 50 mg of oxybutynin chloride, to 10 mL of chloroform. Mix for two minutes, and centrifuge. The supernatant layer of the solution so obtained responds to the *Thin-layer Chromatographic Identification Test* ⟨201⟩, methanol being used as the developing solvent, and iodine vapor being used to visualize the spots.

Dissolution ⟨711⟩—

Medium: water; 900 mL.

Apparatus 2: 50 rpm.

Time: 30 minutes.

Determine the amount of oxybutynin chloride dissolved using the following method.

Bromocresol green solution—Dissolve 0.2 g of bromocresol green in about 50 mL of 0.05 *N* sodium hydroxide. Add 200 mL of pH 2.8 phosphate buffer (0.05 *M*) and adjust to a pH of 2.8 by adding a few drops of phosphoric acid, if necessary. Filter to obtain a clear solution.

Standard preparation—Dissolve an accurately weighed quantity of USP Oxybutynin Chloride RS in water, and dilute quantitatively and stepwise, with water to obtain a solution having a known concentration of about 5.6 µg per mL.

Procedure—Filter the solutions under test with a filter that has been shown to not absorb the drug by testing the *Standard preparation.* Separately pipet 20.0 mL of the *Standard preparation* and the specimens under test to a 50-mL stoppered or screw-topped centrifuge tube. Add 2.0 mL of *Bromocresol green solution* and 10.0 mL of chloroform. Similarly prepare a blank replacing the specimen under test with 20.0 mL of water. Shake the tubes vigorously for at least 30 seconds and centrifuge to separate the layers. Concomitantly determine the absorbances of the chloroform layers at the wavelength of maximum absorbance at about 415 nm, with a suitable spectrophotometer, against the blank. Calculate the quantity, in mg, of $C_{22}H_{31}NO_3.HCl$ dissolved from each specimen by the formula:

$$0.9CA_U/A_S,$$

in which C is the concentration, in µg per mL, of USP Oxybutynin Chloride RS in the solution from the *Standard preparation*, and A_U and A_S are the absorbances of the solution from the specimen under test and the *Standard preparation*, respectively.

Tolerances—Not less than 80% (*Q*) of the labeled amount of $C_{22}H_{31}NO_3.HCl$ is dissolved in 30 minutes.

Uniformity of dosage units ⟨905⟩—meet the requirements.

Assay—

pH 4 phosphate buffer, *pH 5.6 phosphate buffer*, and *Bromocresol green solution*—Prepare as directed in the *Assay* under *Oxybutynin Chloride Syrup.*

Standard preparation—Dissolve an accurately weighed quantity of USP Oxybutynin Chloride RS in 0.01 *N* sulfuric acid to obtain a solution having a known concentration of about 100 µg per mL.

Assay preparation—Weigh and finely powder not less than 20 Oxybutynin Chloride Tablets. Transfer an accurately weighed portion of the powder, equivalent to about 50 mg of oxybutynin chloride, to a 500-mL volumetric flask. Add about 300 mL of water and 5 mL of 1.0 *N* sulfuric acid, and shake for 30 minutes. Dilute with water to volume, mix, and filter, discarding the first portion of the filtrate.

Procedure—Proceed as directed in the *Assay* under *Oxybutynin Chloride Syrup*, except to determine the absorbances against a blank prepared using 10 mL of 0.01 *N* sulfuric acid and 28 mL of *pH 4 phosphate buffer* treated in the same manner as the *Standard preparation* and the *Assay preparation.* Calculate the quantity, in mg, of $C_{22}H_{31}NO_3.HCl$ in the portion of Tablets taken by the formula:

$$0.5C(A_U/A_S),$$

in which C is the concentration, in µg per mL, of USP Oxybutynin Chloride RS in the *Standard preparation*, and A_U and A_S are the absorbances of the solutions from the *Assay preparation* and the *Standard preparation*, respectively.

Oxycodone Tablets

» Oxycodone Tablets contain not less than 90.0 percent and not more than 110.0 percent of the labeled amount of oxycodone hydrochloride ($C_{18}H_{21}NO_4 \cdot HCl$).

Packaging and storage—Preserve in tight, light-resistant containers.

Labeling—Label Tablets to state both the content of the active moiety and the content of the salt used in formulating the article.

Reference standard—*USP Oxycodone Hydrochloride Reference Standard*—Dry at 105° for 2 hours before using.

Identification—

A: The ultraviolet absorption spectrum of the *Assay preparation* exhibits maxima and minima at the same wavelengths as that of the *Standard preparation*, as obtained in the *Assay*.

B: Transfer a portion of powdered Tablets, equivalent to about 15 mg of oxycodone hydrochloride, to a suitable screw-capped tube, add 5 mL of chloroform, sonicate for about 30 seconds, shake for several minutes, and centrifuge. Use the clear supernatant liquid as the test solution. Prepare a Standard solution in chloroform containing 1 mg of USP Oxycodone Hydrochloride RS per mL. Apply separate 20-µL portions of the test solution and the Standard solution on a suitable thin-layer chromatographic plate (see *Chromatography* ⟨621⟩), coated with a 0.25-mm layer of chromatographic silica gel mixture. Allow the spots to dry, and develop the chromatogram in a solvent system consisting of a mixture of acetone, toluene, ether, and ammonium hydroxide (6:4:1:0.3) until the solvent front has moved about three-fourths of the length of the plate. Remove the plate from the chamber, mark the solvent front, allow the solvent to evaporate, and spray with iodoplatinate TS: the principal spot obtained from the test solution corresponds in color, size, and R_f value to that obtained from the Standard solution, and no other spots are observed.

Dissolution ⟨711⟩—

Medium: water; 500 mL.

Apparatus 2: 50 rpm.

Time: 45 minutes.

Procedure—Determine the amount of $C_{18}H_{21}NO_4 \cdot HCl$ dissolved from ultraviolet absorbances at the wavelength of maximum absorbance at about 225 nm of filtered portions of the solution under test, suitably diluted with *Dissolution Medium*, if necessary, in comparison with a Standard solution having a known concentration of USP Oxycodone Hydrochloride RS in the same medium.

Tolerances—Not less than 70% (*Q*) of the labeled amount of $C_{18}H_{21}NO_4 \cdot HCl$ is dissolved in 45 minutes.

Uniformity of dosage units ⟨905⟩: meet the requirements.

Assay—

Standard preparation—Dissolve an accurately weighed quantity of USP Oxycodone Hydrochloride RS quantitatively in 0.01 *N* hydrochloric acid to obtain a solution having a known concentration of about 0.1 mg per mL.

Assay preparation—Weigh and finely powder not less than 20 Oxycodone Tablets. Transfer an accurately weighed portion of the powder, equivalent to about 20 mg of oxycodone hydrochloride, to a 200-mL volumetric flask containing about 180 mL of 0.01 *N* hydrochloric acid, sonicate for about 1 minute, and shake by mechanical means for 30 minutes. Dilute with 0.01 *N* hydrochloric acid to volume, mix, and filter. Use the clear filtrate as the *Assay preparation*.

Procedure—Concomitantly determine the absorbances of the *Standard preparation* and the *Assay preparation* in 1-cm cells at the wavelength of maximum absorbance at about 281 nm, with a suitable spectrophotometer, using 0.01 *N* hydrochloric acid as the blank. Calculate the quantity, in mg, of $C_{18}H_{21}NO_4 \cdot HCl$ in the portion of Tablets taken by the formula:

$$200C(A_U/A_S),$$

in which *C* is the concentration, in mg per mL, of USP Oxycodone Hydrochloride RS in the *Standard preparation*, and A_U and A_S are the absorbances of the *Assay preparation* and the *Standard preparation*, respectively.

Oxycodone and Acetaminophen Capsules

» Oxycodone and Acetaminophen Capsules contain Oxycodone Hydrochloride and Acetaminophen, or Oxycodone Hydrochloride, Oxycodone Terephthalate, and Acetaminophen. Capsules contain not less than 90.0 percent and not more than 110.0 percent of the labeled amounts of oxycodone hydrochloride or oxycodone hydrochloride and oxycodone terephthalate, calculated as total oxycodone ($C_{18}H_{21}NO_4$), and not less than 90.0 percent and not more than 110.0 percent of the labeled amount of acetaminophen ($C_8H_9NO_2$).

Packaging and storage—Preserve in tight, light-resistant containers.

Labeling—Label Capsules to indicate whether they contain Oxycodone Hydrochloride or Oxycodone Hydrochloride and Oxycodone Terephthalate. Capsules may be labeled to indicate the total content of oxycodone ($C_{18}H_{21}NO_4$) equivalent. Each mg of oxycodone hydrochloride or oxycodone terephthalate is equivalent to 0.8963 mg or 0.7915 mg of oxycodone ($C_{18}H_{21}NO_4$), respectively.

Reference standards—*USP Acetaminophen Reference Standard*—Dry over silica gel for 18 hours before using. *USP Oxycodone Hydrochloride Reference Standard*—Dry at 105° for 2 hours before using.

Identification—The retention times of the major peaks in the chromatogram of the *Assay preparation* correspond to those of the *Standard preparation*, as obtained in the *Assay*.

Dissolution ⟨711⟩—

Medium: 0.1 *N* hydrochloric acid; 900 mL.

Apparatus 2: 50 rpm.

Time: 45 minutes.

Procedure—Determine the amounts of oxycodone ($C_{18}H_{21}NO_4$) and acetaminophen ($C_8H_9NO_2$) dissolved, employing the procedure set forth in the *Assay*, making any necessary volumetric adjustments, including adjusting the solution under test to a pH of about 5.5 before injecting.

Tolerances—Not less than 75% (*Q*) of the labeled amounts of $C_{18}H_{21}NO_4$ and $C_8H_9NO_2$ are dissolved in 45 minutes.

Uniformity of dosage units ⟨905⟩: meet the requirements for *Content Uniformity* with respect to oxycodone and acetaminophen.

Assay—

Solvent mixture—Prepare a suitable mixture of 0.05 *M* dibasic potassium phosphate and methanol (9:1), and adjust with phosphoric acid to a pH of 4.0. Make adjustments if necessary (see *System Suitability* under *Chromatography* ⟨621⟩).

Mobile phase—Add 950 mg of monobasic potassium phosphate to 1000 mL of water. Add 1 mL of phosphoric acid, and stir until dissolved. Add 1 mL of *n*-nonylamine with stirring, and stir until a clear solution is obtained. Adjust with potassium hydroxide solution (1 in 2) to a pH of 4.9 ± 0.1. Mix 9 volumes of this solution with 1 volume of methanol. Make adjustments if necessary (see *System Suitability* under *Chromatography* ⟨621⟩).

Oxycodone hydrochloride standard stock solution—Dissolve an accurately weighed quantity of USP Oxycodone Hydrochloride RS in *Solvent mixture* to obtain a solution having a known concentration of about 0.085 mg per mL.

Standard preparation—Transfer about 0.85*J* mg of USP Acetaminophen RS, accurately weighed, *J* being the ratio of the labeled amount, in mg, of acetaminophen to that of oxycodone equivalent, to a 25-mL volumetric flask, add about 10 mL of *Solvent mixture*, and mix to dissolve. Add 10.0 mL of *Oxycodone hydrochloride standard stock solution*, dilute with *Solvent mixture* to volume, and mix. Transfer 5.0 mL of the resulting solution to a 50-mL volumetric flask, dilute with *Mobile phase* to volume, and mix. This solution contains about 0.0034 mg of USP Oxycodone Hydrochloride RS and 0.0034*J* mg of USP Acetaminophen RS per mL.

Assay preparation—Weigh the contents of not less than 20 Oxycodone and Acetaminophen Capsules. Mix the contents, and transfer an accurately weighed portion of the powder, equivalent to about 4.5 mg of oxycodone ($C_{18}H_{21}NO_4$), to a suitable container. Add 150.0 mL of *Solvent mixture*, and shake by mechanical means for 1 hour. Transfer 5.0 mL of this solution to a 50-mL volumetric flask, dilute with *Mobile phase* to volume, and mix. Filter the resulting solution through a membrane filter of 0.5-µm or finer porosity, discarding the first 10 mL of the filtrate. Use the filtrate as the *Assay preparation.*

Chromatographic system (see *Chromatography* ⟨621⟩)—The liquid chromatograph is equipped with a 214-nm detector and a 4.6-mm × 25-cm column that contains 5 µm of packing L1. The column is maintained at a temperature of 40°. The flow rate is about 2 mL per minute. Chromatograph the *Standard preparation*, and record the peak responses as directed under *Procedure:* the resolution, *R*, between acetaminophen and oxycodone is not less than 2.4, and the relative standard deviation for replicate injections is not more than 2.0%.

Procedure—Separately inject equal volumes (about 20 µL) of the *Standard preparation* and the *Assay preparation* into the chromatograph, record the chromatograms, and measure the responses for the major peaks. The relative retention times are about 0.6 for oxycodone and 1.0 for acetaminophen. Calculate the quantity, in mg, of oxycodone ($C_{18}N_{21}NO_4$) in the portion of Capsules taken by the formula:

$$(315.37/351.83)(1500C)(r_U/r_S),$$

in which 315.37 and 351.83 are the molecular weights of oxycodone and oxycodone hydrochloride, respectively, *C* is the concentration, in mg per mL, of USP Oxycodone Hydrochloride RS in the *Standard preparation*, and r_U and r_S are the peak oxycodone responses obtained from the *Assay preparation* and the *Standard preparation*, respectively. Calculate the quantity, in mg, of acetaminophen ($C_8H_9NO_2$) in the portion of Capsules taken by the formula:

$$1500C(r_U/r_S),$$

in which *C* is the concentration, in mg per mL, of USP Acetaminophen RS in the *Standard preparation*, and r_U and r_S are the peak acetaminophen responses obtained from the *Assay preparation* and the *Standard preparation*, respectively.

Oxycodone and Acetaminophen Tablets

» Oxycodone and Acetaminophen Tablets contain Oxycodone Hydrochloride and Acetaminophen. Tablets contain the equivalent of not less than 90.0 percent and not more than 110.0 percent of the labeled amount of oxycodone ($C_{18}H_{21}NO_4$), and not less than 90.0 percent and not more than 110.0 percent of the labeled amount of acetaminophen ($C_8H_9NO_2$).

Packaging and storage—Preserve in tight, light-resistant containers.

Labeling—Tablets may be labeled to indicate the content of oxycodone hydrochloride ($C_{18}H_{21}NO_4.HCl$) equivalent. Each mg of oxycodone is equivalent to 1.116 mg of oxycodone hydrochloride.

Reference standards—*USP Acetaminophen Reference Standard*—Dry over silica gel for 18 hours before using. *USP Oxycodone Hydrochloride Reference Standard*—Dry at 105° for 2 hours before using.

Identification—The retention times of the major peaks in the chromatogram of the *Assay preparation* correspond to those of the *Standard preparation*, as obtained in the *Assay*.

Dissolution ⟨711⟩—

Medium: 0.1 *N* hydrochloric acid; 900 mL.

Apparatus 2: 50 rpm.

Time: 45 minutes.

Procedure—Determine the amounts of oxycodone ($C_{18}H_{21}NO_4$) and acetaminophen ($C_8H_9NO_2$) dissolved, employing the procedure set forth in the *Assay*, making any necessary volumetric adjustments, including adjusting the pH of the solution under test to about 5.5 before injecting.

Tolerances—Not less than 75% (*Q*) of the labeled amounts of $C_{18}H_{21}NO_4$ and $C_8H_9NO_2$ are dissolved in 45 minutes.

Uniformity of dosage units ⟨905⟩: meet the requirements for *Content Uniformity* with respect to oxycodone and to acetaminophen.

Assay—

Solvent mixture, Mobile phase, Oxycodone hydrochloride standard stock solution, Standard preparation, and *Chromatographic system*—Proceed as directed in the *Assay* under *Oxycodone and Acetaminophen Capsules.*

Assay preparation—Weigh and finely powder not less than 20 Oxycodone and Acetaminophen Tablets. Transfer an accurately weighed portion of powder, equivalent to about 4.5 mg of oxycodone ($C_{18}H_{21}NO_4$), to a suitable container. Add 150.0 mL of *Solvent mixture*, and shake by mechanical means for 1 hour. Transfer 5.0 mL of this solution to a 50-mL volumetric flask, dilute with *Mobile phase* to volume, and mix. Filter the resulting solution through a membrane filter of 0.5-µm or finer porosity, discarding the first 10 mL of the filtrate. Use the filtrate as the *Assay preparation.*

Procedure—Proceed as directed for *Procedure* in the *Assay* under *Oxycodone and Acetaminophen Capsules.* Calculate the quantity, in mg, of oxycodone ($C_{18}N_{21}NO_4$) in the portion of Tablets taken by the formula:

$$(315.37/351.83)(1500C)(r_U/r_S),$$

in which 315.37 and 351.83 are the molecular weights of oxycodone and oxycodone hydrochloride, respectively, *C* is the concentration, in mg per mL, of USP Oxycodone Hydrochloride RS in the *Standard preparation*, and r_U and r_S are the peak oxycodone responses obtained from the *Assay preparation* and the *Standard preparation*, respectively. Calculate the quantity, in mg, of acetaminophen ($C_8H_9NO_2$) in the portion of Tablets taken by the formula:

$$1500C(r_U/r_S),$$

in which *C* is the concentration, in mg per mL, of USP Acetaminophen RS in the *Standard preparation*, and r_U and r_S are the peak acetaminophen responses obtained from the *Assay preparation* and the *Standard preparation*, respectively.

Oxycodone and Aspirin Tablets

» Oxycodone and Aspirin Tablets contain Oxycodone Hydrochloride and Aspirin, or Oxycodone Hydrochloride, Oxycodone Terephthalate, and Aspirin. Tablets contain not less than 93.0 percent and not more than 107.0 percent of the labeled amount of oxycodone ($C_{18}H_{21}NO_4$), and not less than 90.0 percent and not more than 110.0 percent of the labeled amount of aspirin ($C_9H_8O_4$).

Packaging and storage—Preserve in tight, light-resistant containers.

Reference standards—*USP Oxycodone Hydrochloride Reference Standard*—Dry at 105° for 2 hours before using. *USP Aspirin Reference Standard*—Dry over silica gel for 5 hours before using. *USP Salicylic Acid Reference Standard*—Dry over silica gel for 3 hours before using.

Identification—The retention times of the oxycodone peak and the aspirin peak in the chromatograms of the respective *Assay preparations* correspond to those of the corresponding analytes of the respective *Standard preparations*, as obtained in the *Assay for oxycodone* and the *Assay for aspirin*, respectively.

Uniformity of dosage units ⟨905⟩: meet the requirements.

Salicylic acid—

Mobile phase—Dissolve 2 g of sodium 1-heptanesulfonate in a mixture of 850 mL of water and 150 mL of acetonitrile, and adjust with glacial acetic acid to a pH of 3.4. Make adjustments if necessary (see *System Suitability* under *Chromatography* ⟨621⟩).

Diluting solution—Prepare a mixture of acetonitrile and formic acid (99:1).

Standard preparation—Dissolve an accurately weighed quantity of USP Salicylic Acid RS in *Diluting solution* to obtain a solution having a known concentration of about 0.008 mg per mL.

Test preparation—Weigh and finely powder not less than 20 Tablets. Transfer an accurately weighed quantity of the powder, equivalent to about 380 mg of aspirin, to a 100-mL volumetric flask, add about 20 mL of *Diluting solution*, and sonicate for about 15 minutes. Dilute with *Diluting solution* to volume, and mix. Centrifuge a portion of this mixture, and use the clear supernatant liquid as the *Test preparation*.

Chromatographic system (see *Chromatography* ⟨621⟩)—The liquid chromatograph is equipped with a 299-nm detector and a 3.9-mm × 30-cm column that contains packing L1. The flow rate is about 2 mL per minute. Chromatograph the *Standard preparation*, and record the responses as directed under *Procedure:* the relative standard deviation for replicate injections is not more than 4.0%.

Procedure—Separately inject equal volumes (about 20 µL) of the *Test preparation* and the *Standard preparation* into the chromatograph, record the chromatograms, and measure the responses for the salicylic acid peaks. Calculate the percentage of salicylic acid in the portion of Tablets taken by the formula:

$$100(C/a)(r_U/r_S),$$

in which *C* is the concentration, in mg per mL, of USP Salicylic Acid RS in the *Standard preparation*, *a* is the quantity, in mg, of aspirin in the portion of Tablets taken, as determined in the *Assay for aspirin*, and r_U and r_S are the salicylic acid peak responses obtained from the *Test preparation* and the *Standard preparation*, respectively: not more than 3.0% is found.

Assay for aspirin—

Mobile phase—Prepare a suitable mixture of water, methanol, and glacial acetic acid (594:400:6). Filter through a filter of 0.5-µm or finer porosity. Make adjustments if necessary (see *System Suitability* under *Chromatography* ⟨621⟩).

Standard oxycodone solution—Transfer about 25 mg of USP Oxycodone Hydrochloride RS, accurately weighed, to a 50-mL volumetric flask, add 25 mL of methanol, sonicate for about 5 minutes, and shake by mechanical means for about 5 minutes. Dilute with water to volume, and mix.

Standard preparation—Transfer about 0.9*J* mg of USP Aspirin RS, accurately weighed, to a 10-mL volumetric flask, *J* being the ratio of the labeled amount, in mg, of aspirin to that of oxycodone equivalent. Add 4 mL of methanol, sonicate for about 5 minutes, and shake by mechanical means for 5 minutes. Add 2.0 mL of *Standard oxycodone solution*, dilute with water to volume, and mix. This solution contains about 0.09*J* mg of USP Aspirin RS and 0.1 mg of USP Oxycodone Hydrochloride RS, equivalent to 0.09 mg of oxycodone, per mL.

Assay preparation—Weigh and finely powder not less than 20 Oxycodone and Aspirin Tablets. Transfer an accurately weighed portion of the powder, equivalent to about 4.5 mg of oxycodone, to a 50-mL volumetric flask. Add 5 mL of water, sonicate for about 1 minute, add 25 mL of methanol, and shake by mechanical means for 15 minutes. Dilute with water to volume, and mix. Filter the mixture through a filter of 0.5-µm or finer porosity. Use the clear filtrate as the *Assay preparation*.

Chromatographic system (see *Chromatography* ⟨621⟩)—The liquid chromatograph is equipped with a 254-nm detector and a 3.9-mm × 30-cm column that contains packing L1. The flow rate is about 1.5 mL per minute. Chromatograph the *Standard preparation*, and record the responses as directed under *Procedure:* the relative standard deviation for replicate injections is not more than 2.0%.

Procedure—[NOTE—Use peak areas where peak responses are indicated.] Separately inject equal volumes (about 15 µL) of the *Standard preparation* and the *Assay preparation* into the chromatograph, and record the responses for the major peaks. Calculate the quantity, in mg, of $C_9H_8O_4$ in the portion of Tablets taken by the formula:

$$50C(r_U/r_S),$$

in which *C* is the concentration, in mg per mL, of USP Aspirin RS in the *Standard preparation*, and r_U and r_S are the aspirin peak responses obtained from the *Assay preparation* and the *Standard preparation*, respectively.

Assay for oxycodone—

Mobile phase—Prepare a suitable mixture of 900 mL of methanol, 90 mL of water, and 1.3 mL of dibutylamine, adjust with phosphoric acid to a pH of 3.5 ± 0.1, dilute with water to 1000 mL, and mix. Filter through a filter of 0.5-µm or finer porosity. Make adjustments if necessary (see *System Suitability* under *Chromatography* ⟨621⟩).

Standard preparation and *Assay preparation*—Prepare as directed in the *Assay for aspirin*.

Chromatographic system (see *Chromatography* ⟨621⟩)—The liquid chromatograph is equipped with a 280-nm detector and a 4.6-mm × 25-cm column that contains packing L10. The flow rate is about 1.5 mL per minute. Chromatograph the *Standard preparation*, and record the responses as directed under *Procedure:* the relative standard deviation for replicate injections is not more than 2.0%.

Procedure—[NOTE—Use peak areas where peak responses are indicated.] Separately inject equal volumes (about 50 µL) of the *Standard preparation* and the *Assay preparation* into the chromatograph, and record the responses for the major peaks. Calculate the quantity, in mg, of oxycodone ($C_{18}H_{21}NO_4$) in the portion of Tablets taken by the formula:

$$(315.37/351.83)(50C)(r_U/r_S),$$

in which 315.37 and 351.83 are the molecular weights of oxycodone and oxycodone hydrochloride, respectively, *C* is the concentration, in mg per mL, of USP Oxycodone Hydrochloride RS in the *Standard preparation*, and r_U and r_S are the oxycodone peak responses obtained from the *Assay preparation* and the *Standard preparation*, respectively.

Oxycodone Hydrochloride

$C_{18}H_{21}NO_4.HCl$ 351.83

Morphinan-6-one, 4,5-epoxy-14-hydroxy-3-methoxy-17-methyl-, hydrochloride, (5α)-.

4,5α-Epoxy-14-hydroxy-3-methoxy-17-methylmorphinan-6-one hydrochloride [*124-90-3*].

» Oxycodone Hydrochloride contains not less than 97.0 percent and not more than 103.0 percent of $C_{18}H_{21}NO_4.HCl$, calculated on the dried basis.

Packaging and storage—Preserve in tight containers.

Reference standard—*USP Oxycodone Hydrochloride Reference Standard*—Dry at 105° for 2 hours before using.

Identification—

A: Dissolve 250 mg in 25 mL of water. Render the solution alkaline with 6 *N* ammonium hydroxide. Allow the mixture to stand until a precipitate is formed. Filter, wash the precipitate with 50 mL of cold water, and dry for 2 hours at 105°: the precipitate so obtained melts between 218° and 223°, but the range between beginning and end of melting does not exceed 2° (see *Melting Range or Temperature* ⟨741⟩).

B: The infrared absorption spectrum of a potassium bromide dispersion of a portion of the dried precipitate obtained in *Identification test A* exhibits maxima only at the same wavelengths as that of a similar preparation of USP Oxycodone Hydrochloride RS.

Specific rotation ⟨781⟩: between −137° and −149°, calculated on the dried basis, determined in a solution containing 2.5 g in each 100 mL.

Loss on drying ⟨731⟩—Dry it at 105° for 2 hours: it loses not more than 7.0% of its weight.

Residue on ignition ⟨281⟩: not more than 0.05%.

Chromatographic purity—Dissolve an accurately weighed quantity of it in methanol to obtain a test solution containing 10 mg per mL. Prepare a solution of USP Oxycodone Hydrochloride RS in methanol containing 10 mg per mL (*Standard solution A*). Dilute portions of *Standard solution A* quantitatively with methanol to obtain *Standard solution B* containing 100 µg per mL and *Standard solution C* containing 50 µg per mL. Apply separate 20-µL portions of the four solutions on a suitable thin-layer chromatographic plate (see *Chromatography* ⟨621⟩), coated with a 0.25-mm layer of chromatographic silica gel mixture. Develop the chromatogram in a solvent system consisting of a mixture of butyl alcohol, water, and glacial acetic acid (4:2:1). Remove the plate from the chamber, and air-dry. Expose the plate to iodine vapors in a closed chamber: the R_f value of the principal spot obtained from the test solution corresponds to that obtained from *Standard solution A*. Any other spot obtained from the test solution does not exceed in size or intensity the principal spot obtained from *Standard solution B* (1%), and the sum of the impurities represented by all of the spots other than the principal spot, based on a comparison of the intensities of such spots with the intensities of the principal spots obtained from *Standard solutions B* and *C* does not exceed 2%.

Chloride content—Dissolve about 300 mg, accurately weighed, in 50 mL of methanol, add 5 mL of glacial acetic acid, and titrate with 0.1 *N* silver nitrate VS, determining the end-point potentiometrically. Each mL of 0.1 *N* silver nitrate is equivalent to 3.545 mg of Cl: the content of Cl is between 9.8% and 10.4%, calculated on the dried basis.

Assay—Dissolve about 700 mg of Oxycodone Hydrochloride, accurately weighed, in 50 mL of glacial acetic acid, add 10 mL of mercuric acetate TS, and titrate with 0.1 *N* perchloric acid VS, determining the end-point potentiometrically. Each mL of 0.1 *N* perchloric acid is equivalent to 35.18 mg of $C_{18}H_{21}NO_4 \cdot HCl$.

Oxycodone Hydrochloride Oral Solution

» Oxycodone Hydrochloride Oral Solution contains not less than 90.0 percent and not more than 110.0 percent of the labeled amount of $C_{18}H_{21}NO_4 \cdot HCl$.

Packaging and storage—Preserve in tight, light-resistant containers.

Reference standard—*USP Oxycodone Hydrochloride Reference Standard*—Dry at 105° for 2 hours before using.

Identification—

A: The ultraviolet absorption spectrum of the *Assay preparation*, prepared as directed in the *Assay*, exhibits maxima and minima at the same wavelengths as the *Standard preparation*, prepared as directed in the *Assay*, concomitantly measured.

B: Separately evaporate 5 mL of the *Assay preparation* and 5 mL of the *Standard preparation* just to dryness. Dissolve each residue in 1.0 mL of chloroform. Apply separate 20-µL portions of the test solution from the *Assay preparation* and the Standard solution from the *Standard preparation* on a suitable thin-layer chromatographic plate (see *Chromatography* ⟨621⟩), coated with a 0.25-mm layer of chromatographic silica gel mixture. Allow the spots to dry, and develop the chromatogram in solvent system consisting of a mixture of acetone, toluene, ether, and ammonium hydroxide (6:4:1:0.3) until the solvent front has moved about three-fourths of the length of the plate. Remove the plate from the chamber, mark the solvent front, allow the solvent to evaporate, and spray with iodoplatinate TS: the principal spot obtained from the test solution corresponds in color, size, and R_f value to that obtained from the Standard solution, and no other spots are observed.

pH ⟨791⟩: between 1.4 and 4.0.

Alcohol content, *Method II* ⟨611⟩: between 7.0% and 9.0% of C_2H_5OH, determined by the gas liquid chromatographic method, acetone being used as the internal standard.

Assay—

Standard preparation—Transfer about 15 mg of USP Oxycodone Hydrochloride RS, accurately weighed, to a suitable separator, add 25 mL of 0.01 *N* hydrochloric acid, and swirl to dissolve. Extract with four 40-mL portions of chloroform, collecting the chloroform extracts in a second separator. Wash the combined chloroform extracts with 5 mL of 0.01 *N* hydrochloric acid, and discard the chloroform layer. Combine the acidic wash with the aqueous solution remaining in the first separator, and adjust with 6 *N* ammonium hydroxide to a pH of 9.5 ± 0.5. Extract with one 50-mL and two 20-mL portions of chloroform, and filter the chloroform extracts through chloroform-washed cotton, collecting the filtrates in a 100-mL volumetric flask. Dilute with water-saturated chloroform to volume, and mix.

Assay preparation—Transfer an accurately measured volume of Oxycodone Hydrochloride Oral Solution, equivalent to about 15 mg of oxycodone hydrochloride, to a suitable separator, add 10 mL of 0.01 *N* hydrochloric acid, and proceed as directed for *Standard preparation*, beginning with "Extract with four 40-mL portions of chloroform."

Procedure—Concomitantly determine the absorbances of the *Standard preparation* and the *Assay preparation* in 1-cm cells at the wavelength of maximum absorbance at about 283.5 nm, with a suitable spectrophotometer, using chloroform as the blank. Calculate the quantity, in mg, of $C_{18}H_{21}NO_4 \cdot HCl$ in each mL of the Oral Solution taken by the formula:

$$(W_S / V)(A_U / A_S),$$

in which W_S is the weight, in mg, of USP Oxycodone Hydrochloride RS, used to prepare the *Standard preparation*, V is the volume, in mL, of Oral Solution taken to prepare the *Assay preparation*, and A_U and A_S are the absorbances of the *Assay preparation* and the *Standard preparation*, respectively.

Oxygen

O_2 32.00
Oxygen.
Oxygen [*7782-44-7*].

» Oxygen contains not less than 99.0 percent, by volume, of O_2. [NOTE—Oxygen that is produced by the air-liquefaction process is exempt from the requirements of the tests for *Carbon dioxide* and *Carbon monoxide*.]

Packaging and storage—Preserve in cylinders or in a pressurized storage tank. Containers used for Oxygen must not be treated with any toxic, sleep-inducing, or narcosis-producing compounds, and must not be treated with any compound that will be irritating to the respiratory tract when the Oxygen is used.

NOTE—*Reduce the container pressure by means of a regulator. Measure the gases with a gas volume meter downstream from the detector tube in order to minimize contamination or change of the specimens.*

Labeling—Label it to indicate whether or not it has been produced by the air-liquefaction process. Where it is piped directly from the cylinder or storage tank to the point of use, label each outlet "Oxygen."

NOTE—The various detector tubes called for in the respective tests are listed under *Reagents* in the section, *Reagents, Indicators, and Solutions*.

Identification—

A: When tested as directed in the *Assay*, not more than 1.0 mL of gas remains.

B: Pass 100 ± 5 mL released from the vapor phase of the contents of the Oxygen container through a carbon dioxide detector tube at the rate specified for the tube: no color change is observed (*distinction from carbon dioxide*).

Odor—Carefully open the container valve to produce a moderate flow of gas. Do not direct the gas stream toward the face, but deflect a portion of the stream toward the nose: no appreciable odor is discernible.

Carbon dioxide—Pass 1050 ± 50 mL through a carbon dioxide detector tube at the rate specified for the tube: the indicator change corresponds to not more than 0.03%.

Carbon monoxide—Pass 1050 ± 50 mL through a carbon monoxide detector tube at the rate specified for the tube: the indicator change corresponds to not more than 0.001%.

Assay—Place a sufficient quantity of ammonium chloride–ammonium hydroxide solution, prepared by mixing equal volumes of water and ammonium hydroxide and saturating with ammonium chloride at room temperature, in a test apparatus composed of a calibrated 100-mL buret, provided with a two-way stopcock, a gas absorption pipet, and a leveling bulb, both of suitable capacity and all suitably interconnected. Fill the gas absorption pipet with metallic copper in the form of wire coils, wire mesh, or other suitable configuration. Eliminate all gas bubbles from the liquid in the test apparatus. Activate the test solution by performing two or three tests that are not for record purposes. Fill the calibrated buret, all interconnecting tubing, both stopcock openings, and the intake tube with liquid. Draw 100.0 mL of Oxygen into the buret by lowering the leveling bulb. Open the stopcock to the absorption pipet, and force the Oxygen into the absorption pipet by raising the leveling bulb. Agitate the pipet to provide frequent and intimate contact of the liquid, gas, and copper. Continue agitation until no further diminution in volume occurs. Draw the residual gas back into the calibrated buret, and measure its volume: not more than 1.0 mL of gas remains.

Oxygen 93 Percent

» Oxygen 93 Percent is Oxygen produced from air by the molecular sieve process. It contains not less than 90.0 percent and not more than 96.0 percent, by volume, of O_2, the remainder consisting mostly of argon and nitrogen.

Packaging and storage—Preserve in cylinders or in a low pressure collecting tank. Containers used for Oxygen 93 Percent must not be treated with any toxic, sleep-inducing, or narcosis-producing compounds, and must not be treated with any compound that will be irritating to the respiratory tract when the Oxygen 93 Percent is used.

Labeling—Where it is piped directly from the collecting tank to the point of use, label each outlet "Oxygen 93 Percent."

NOTE—The various detector tubes called for in the respective tests are listed under *Reagents*, in the section, *Reagents, Indicators, and Solutions.*

Where it is preserved in cylinders, reduce the pressure by means of a regulator. Measure the gases with a gas volume meter downstream from the detector tube in order to minimize contamination or change of the specimens.

Identification—

A: When tested as directed in the *Assay*, not more than 10.0 mL and not less than 4.0 mL of gas remains.

B: Pass 100 ± 5 mL released from the vapor phase of the contents of the Oxygen 93 Percent container or from the outlet at the point of use through a carbon dioxide detector tube at the rate specified for the tube: no color change is observed (*distinction from carbon dioxide*).

Odor—Carefully open the container valve or system outlet to produce a moderate flow of gas. Do not direct the gas stream toward the face, but deflect a portion of the stream toward the nose: no appreciable odor is discernible.

Carbon dioxide—Pass 1050 ± 50 mL through a carbon dioxide detector tube at the rate specified for the tube: the indicator change corresponds to not more than 0.03%.

Carbon monoxide—Pass 1050 ± 50 mL through a carbon monoxide detector tube at the rate specified for the tube: the indicator change corresponds to not more than 0.001%.

Assay—Place a sufficient quantity of ammonium chloride–ammonium hydroxide solution, prepared by mixing equal volumes of water and ammonium hydroxide and saturating with ammonium chloride at room temperature, in a test apparatus composed of a calibrated 100-mL buret, provided with a two-way stopcock, a gas absorption pipet, and a leveling bulb, both of suitable capacity and all suitably interconnected. Fill the gas absorption pipet with metallic copper in the form of wire coils, wire mesh, or other suitable configuration. Eliminate all gas bubbles from the liquid in the test apparatus. Activate the test solution by performing two or three tests that are not for record purposes. Fill the calibrated buret, all interconnecting tubing, both stopcock openings, and the intake tube with liquid. Draw 100.0 mL of Oxygen 93 Percent into the buret by lowering the leveling bulb. Open the stopcock to the absorption pipet, and force the Oxygen 93 Percent into the absorption pipet by raising the leveling bulb. Agitate the pipet to provide frequent and intimate contact of the liquid, gas, and copper. Continue agitation until no further diminution in volume occurs. Draw the residual gas back into the calibrated buret, and measure its volume: not more than 10.0 mL and not less than 4.0 mL of gas remains.

Oxymetazoline Hydrochloride

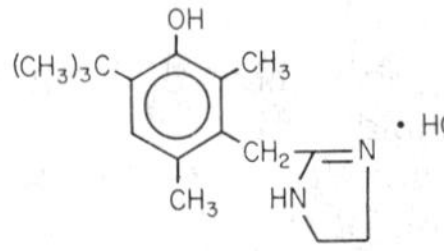

$C_{16}H_{24}N_2O.HCl$ 296.84

Phenol, 3-[(4,5-dihydro-1*H*-imidazol-2-yl)methyl]-6-(1,1-dimethylethyl)-2,4-dimethyl-, monohydrochloride.

6-*tert*-Butyl-3-(2-imidazolin-2-ylmethyl)-2,4-dimethylphenol monohydrochloride [*2315-02-8*].

» Oxymetazoline Hydrochloride contains not less than 98.5 percent and not more than 101.5 percent of $C_{16}H_{24}N_2O.HCl$, calculated on the dried basis.

Packaging and storage—Preserve in tight containers.

Reference standard—*USP Oxymetazoline Hydrochloride Reference Standard*—Dry at 105° for 3 hours before using.

Identification—

A: The infrared absorption spectrum of a mineral oil dispersion of it, previously dried, exhibits maxima only at the same wavelengths as that of a similar preparation of USP Oxymetazoline Hydrochloride RS.

B: The ultraviolet absorption spectrum of a solution (1 in 10,000) exhibits maxima and minima at the same wavelengths as that of a similar preparation of USP Oxymetazoline Hydrochloride RS, concomitantly measured, and the respective absorptivities, calculated on the dried basis, at the wavelength of maximum absorbance at about 279 nm do not differ by more than 3.0%.

C: To a solution of about 50 mg in 3 mL of water add 1 mL of 6 *N* ammonium hydroxide, filter, and acidify the filtrate with nitric acid: the filtrate responds to the tests for *Chloride* ⟨191⟩.

pH ⟨791⟩: between 4.0 and 6.5, in a solution (1 in 20).

Loss on drying ⟨731⟩—Dry it at 105° for 3 hours: it loses not more than 1.0% of its weight.

Residue on ignition ⟨281⟩: not more than 0.1%.

Heavy metals, *Method II* ⟨231⟩: 0.001%.

Assay—

Mobile phase—Prepare a suitable degassed solution of water, methanol, 1 *M* sodium acetate, and glacial acetic acid (46:40:10:4).

Standard preparation—Prepare a solution in *Mobile phase* of USP Oxymetazoline Hydrochloride RS having a known concentration of about 0.5 mg per mL.

Assay preparation—Transfer about 25 mg of Oxymetazoline Hydrochloride, accurately weighed, to a 50-mL volumetric flask, dissolve in *Mobile phase*, dilute with *Mobile phase* to volume, and mix.

Chromatographic system (see *Chromatography* ⟨621⟩)—The liquid chromatograph is equipped with a 280-nm detector and a 4.6-mm × 0.25-m column that contains packing L9. The flow rate is about 1 mL per minute. Chromatograph five replicate injections of the *Standard preparation*, and record the peak responses as directed under *Procedure:* the tailing factor is not more than 2.0 and the relative standard deviation is not more than 2.0%.

Procedure—Separately inject equal volumes (about 20 µL) of the *Standard preparation* and the *Assay preparation* into the chromatograph, record the chromatograms, and measure the response for the major peak. Calculate the quantity, in mg, of $C_{16}H_{24}N_2O \cdot HCl$ in the portion of Oxymetazoline Hydrochloride taken by the formula:

$$50C(r_U / r_S),$$

in which *C* is the concentration, in mg per mL, of USP Oxymetazoline RS in the *Standard preparation*, and r_U and r_S are the peak responses obtained from the *Assay preparation* and the *Standard preparation*, respectively.

Oxymetazoline Hydrochloride Nasal Solution

» Oxymetazoline Hydrochloride Nasal Solution is a solution of Oxymetazoline Hydrochloride in water adjusted to a suitable tonicity. It contains not less than 90.0 percent and not more than 110.0 percent of the labeled amount of $C_{16}H_{24}N_2O \cdot HCl$.

Packaging and storage—Preserve in tight containers.

Reference standard—*USP Oxymetazoline Hydrochloride Reference Standard*—Dry at 105° for 3 hours before using.

Identification—Place a volume of Nasal Solution, equivalent to about 2.5 mg of oxymetazoline hydrochloride, in a 60-mL separator, and add water to make about 10 mL. Add 2 mL of sodium carbonate solution (1 in 10), extract with 10 mL of chloroform, and transfer the chloroform extract to a second 60-mL separator. Extract the chloroform solution with 10 mL of 0.1 *N* hydrochloric acid, allow to separate, and discard the chloroform layer. Transfer 8 mL of the acidic aqueous layer to a test tube, neutralize by the dropwise addition of 1 *N* sodium hydroxide, add 1 drop of 1 *N* sodium hydroxide in excess, and mix. Add a few drops of sodium nitroferricyanide TS and 2 drops of sodium hydroxide solution (15 in 100), mix, and allow to stand for 10 minutes. Add 0.1 *N* hydrochloric acid dropwise until the pH is between 8 and 9, and allow to stand for 10 minutes: a violet color develops.

pH ⟨791⟩: between 4.0 and 6.5.

Assay—

Mobile phase—Prepare as directed in the *Assay* under *Oxymetazoline Hydrochloride.*

Standard preparation—Prepare a solution of USP Oxymetazoline Hydrochloride RS in *Mobile phase*, having a known concentration, approximately equal to the labeled concentration of the Nasal Solution.

Assay preparation—Use Oxymetazoline Hydrochloride Nasal Solution.

Chromatographic system and *Procedure*—Proceed as directed in the *Assay* under *Oxymetazoline Hydrochloride*, except to calculate the quantity, in mg, of $C_{16}H_{24}N_2O \cdot HCl$ in each mL of the Nasal Solution taken by the formula:

$$C(r_U / r_S),$$

in which the terms are as defined therein.

Oxymetazoline Hydrochloride Ophthalmic Solution

» Oxymetazoline Hydrochloride Ophthalmic Solution is a sterile, buffered solution of Oxymetazoline Hydrochloride in water adjusted to a suitable tonicity. It contains not less than 90.0 percent and not more than 110.0 percent of the labeled amount of $C_{16}H_{24}N_2O \cdot HCl$. It contains a suitable preservative.

Packaging and storage—Preserve in tight containers.

Reference standard—*USP Oxymetazoline Hydrochloride Reference Standard*—Dry at 105° for 3 hours before using.

Identification—A volume of Ophthalmic Solution, equivalent to about 2.5 mg of oxymetazoline hydrochloride, responds to the *Identification test* under *Oxymetazoline Hydrochloride Nasal Solution.*

Sterility—It meets the requirements under *Sterility Tests* ⟨71⟩.

pH ⟨791⟩: between 5.8 and 6.8.

Assay—

Mobile phase—Prepare as directed in the *Assay* under *Oxymetazoline Hydrochloride.*

Standard preparation—Prepare a solution of USP Oxymetazoline Hydrochloride RS in *Mobile phase*, having a known concentration approximately equal to the labeled concentration of the Ophthalmic Solution.

Assay preparation—Use Oxymetazoline Hydrochloride Ophthalmic Solution.

Chromatographic system and *Procedure*—Proceed as directed in the *Assay* under *Oxymetazoline Hydrochloride*, except to calculate the quantity, in mg, of $C_{16}H_{24}N_2O \cdot HCl$ in each mL of the Ophthalmic Solution taken by the formula:

$$C(r_U / r_S),$$

in which the terms are as defined therein.

Oxymetholone

$C_{21}H_{32}O_3$ 332.48

Androstan-3-one, 17-hydroxy-2-(hydroxymethylene)-17-methyl-, (5α,17β)-.

17β-Hydroxy-2-(hydroxymethylene)-17-methyl-5α-androstan-3-one [*434-07-1*].

» Oxymetholone contains not less than 97.0 percent and not more than 103.0 percent of $C_{21}H_{32}O_3$, calculated on the dried basis.

Packaging and storage—Preserve in well-closed containers.

Reference standard—*USP Oxymetholone Reference Standard*—Dry in vacuum over phosphorus pentoxide for 4 hours before using.

Completeness of solution—Dissolve 100 mg in 5 mL of dioxane: the solution is clear and free from undissolved solid.

Identification—

A: The infrared absorption spectrum of a potassium bromide dispersion of it, previously dried, exhibits maxima only at the same wavelengths as that of a similar preparation of USP Oxymetholone RS.

B: The ultraviolet absorption spectrum of a 1 in 100,000 solution in 0.01 *N* methanolic sodium hydroxide exhibits maxima and minima at the same wavelengths as that of a similar solution of USP Oxymetholone RS, concomitantly measured.

Melting range ⟨741⟩: between 172° and 180°.

Specific rotation ⟨781⟩: between +34° and +38°, calculated on the dried basis, determined in a solution in dioxane containing 200 mg in each 10 mL.

Loss on drying ⟨731⟩—Dry it in vacuum over phosphorus pentoxide for 4 hours: it loses not more than 1.0% of its weight.

Assay—

Standard preparation—Prepare as directed under *Single-steroid Assay* ⟨511⟩, using USP Oxymetholone RS.

Assay preparation—Weigh accurately about 20 mg of Oxymetholone, previously dried, dissolve in a sufficient quantity of a mixture of equal volumes of alcohol and chloroform to make 10.0 mL, and mix.

Procedure—Proceed as directed for *Procedure* under *Single-steroid Assay* ⟨511⟩, using a solvent system consisting of a mixture of benzene and alcohol (98:2), through the fourth sentence of the second paragraph under *Procedure.* Then centrifuge the tubes for 5 minutes, and determine the absorbances of the supernatant solutions in 1-cm cells at the wavelength of maximum absorbance at about 315 nm, with a suitable spectrophotometer, against the blank. [NOTE—Use 0.01 *N* alcoholic sodium hydroxide, rather than alcohol, to elute the silica gel bands.] Calculate the quantity, in mg, of $C_{21}H_{32}O_3$ in the portion of Oxymetholone taken by the formula:

$$10C(A_U/A_S),$$

in which *C* is the concentration, in mg per mL, of USP Oxymetholone RS in the *Standard preparation*, and A_U and A_S are the absorbances of the solutions from the *Assay preparation* and the *Standard preparation*, respectively.

Oxymetholone Tablets

» Oxymetholone Tablets contain not less than 90.0 percent and not more than 110.0 percent of the labeled amount of $C_{21}H_{32}O_3$.

Packaging and storage—Preserve in well-closed containers.

Reference standard—*USP Oxymetholone Reference Standard*—Dry in vacuum over phosphorus pentoxide for 4 hours before using.

Identification—Mix an amount of powdered Tablets, equivalent to about 50 mg of oxymetholone, with 15 mL of solvent hexane, and stir occasionally for 15 minutes. Centrifuge the mixture, and decant and discard the solvent hexane. Extract the residue with two 10-mL portions of solvent hexane, centrifuging and decanting as before, and discard the solvent hexane. Add 25 mL of chloroform to the residue, mix by shaking for 1 to 2 minutes, and filter. Evaporate the filtrate to about 3 mL, add a few mL of solvent hexane to induce crystallization, and evaporate to dryness: the infrared absorption spectrum of a potassium bromide dispersion prepared from the oxymetholone so obtained, and previously dried, exhibits maxima only at the same wavelengths as that of a similar preparation of USP Oxymetholone RS, crystallized from the same solvent mixture.

Dissolution ⟨711⟩—

Medium: 0.05 *M* pH 8.5 alkaline borate buffer (see under *Solutions* in the section, *Reagents, Indicators, and Solutions*); 900 mL.

Apparatus 1: 100 rpm.

Time: 45 minutes.

Procedure—Determine the amount of $C_{21}H_{32}O_3$ dissolved from ultraviolet absorbances at the wavelength of maximum absorbance at about 313 nm of filtered portions of the solution under test, suitably diluted with *Dissolution Medium* if necessary, in comparison with a Standard solution having a known concentration of USP Oxymetholone RS in the same medium. [NOTE—An amount of acetonitrile not to exceed 5% of the total volume of the Standard solution may be used to bring the Reference Standard into solution prior to dilution with *Dissolution Medium.*]

Tolerances—Not less than 75% (*Q*) of the labeled amount of $C_{21}H_{32}O_3$ is dissolved in 45 minutes.

Uniformity of dosage units ⟨905⟩: meet the requirements.

Procedure for content uniformity—Transfer 1 finely powdered Tablet to a 100-mL volumetric flask with the aid of about 75 mL of methanol. Heat the methanol to boiling, and allow to remain at a temperature just below the boiling point for 15 minutes with occasional swirling. Cool to room temperature, dilute with methanol to volume, and mix. Centrifuge a portion of the mixture at about 2000 rpm until the solution becomes clear. Transfer a portion of the supernatant solution, equivalent to about 1 mg of oxymetholone, to a 100-mL volumetric flask. Add 10 mL of a 1 in 250 solution of sodium hydroxide in methanol, and dilute with methanol to volume. Without delay, concomitantly determine the absorbances of this solution and a freshly prepared Standard solution of USP Oxymetholone RS in the same medium having a known concentration of about 10 µg per mL in 1-cm cells at the wavelength of maximum absorbance at about 315 nm, with a suitable spectrophotometer, using a 1 in 2500 solution of sodium hydroxide in methanol as the blank. Calculate the quantity, in mg, of $C_{21}H_{32}O_3$ in the Tablet by the formula:

$$(TC/D)(A_U/A_S),$$

in which *T* is the labeled quantity, in mg, of oxymetholone in the Tablet, *C* is the concentration, in µg per mL, of USP Oxymetholone RS in the Standard solution, *D* is the concentration, in µg per mL, of oxymetholone in the solution from the Tablet, based upon the labeled quantity per Tablet and the extent of dilution, and A_U and A_S are the absorbances of the solution from the Tablet and the Standard solution, respectively.

Assay—Weigh and finely powder not less than 20 Oxymetholone Tablets. Transfer an accurately weighed portion of the powder, equivalent to about 20 mg of oxymetholone, to a separator, add 10 mL of water, and extract with three 25-mL portions of chloroform, filtering each extract through chloroform-washed cotton. Evaporate the combined chloroform extracts on a steam bath to dryness, reducing the application of heat as dryness is approached. Dissolve the residue in methanol, transfer to a 100-mL volumetric flask, dilute with methanol to volume, and mix. Transfer 5.0 mL of the solution to a 100-mL volumetric flask, add 10 mL of a 1 in 250 solution of sodium hydroxide in methanol, dilute with methanol to volume, and mix. Without delay, concomitantly determine the absorbances of this solution and a freshly prepared Standard solution of USP Oxymetholone RS in the same medium having a known concentration of about 10 µg per mL in 1-cm cells at the wavelength of maximum absorbance at about 315 nm, with a suitable spectrophotometer, using a 1 in 2500 solution of sodium hydroxide in methanol as the blank. Calculate the quantity, in mg, of $C_{21}H_{32}O_3$ in the portion of Tablets taken by the formula:

$$2C(A_U/A_S),$$

in which *C* is the concentration, in µg per mL, of USP Oxymetholone RS in the Standard solution, and A_U and A_S are the absorbances of the solution from the Tablets and the Standard solution, respectively.

Oxymorphone Hydrochloride

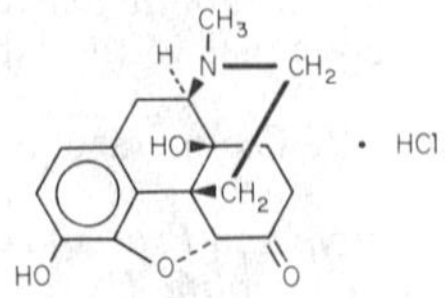

CH₃ H N CH₂ HO CH₂ · HCl HO O O

$C_{17}H_{19}NO_4 \cdot HCl$ 337.80

Morphinan-6-one, 4,5-epoxy-3,14-dihydroxy-17-methyl-, hydrochloride, (5α)-.
4,5α-Epoxy-3,14-dihydroxy-17-methylmorphinan-6-one hydrochloride [*357-07-3*].

» Oxymorphone Hydrochloride contains not less than 97.0 percent and not more than 102.0 percent of $C_{17}H_{19}NO_4 . HCl$, calculated on the dried basis.

Packaging and storage—Preserve in tight, light-resistant containers.

Reference standard—*USP Oxymorphone Reference Standard*—Keep container tightly closed and protected from light. Do not dry before using.

Identification—

A: Dissolve about 250 mg in 25 mL of water, and render the solution alkaline with a saturated solution of sodium bicarbonate. Extract the liberated oxymorphone with two 15-mL portions of chloroform. Reserve the chloroform extracts, combined in a second separator, for *Identification test B*: the aqueous phase, acidified with 2 *N* nitric acid, responds to the tests for *Chloride* ⟨191⟩.

B: Wash the combined chloroform extracts from *Identification test A* with 5 mL of water, and filter. Evaporate the chloroform solution on a steam bath nearly to dryness, then add a few mL of ether, and continue the evaporation with stirring until the solvent is removed: the infrared absorption spectrum of a 1 in 50 solution in alcohol-free chloroform of the oxymorphone so obtained, determined in a 0.5-mm cell, exhibits maxima only at the same wavelengths as that of a similar solution of USP Oxymorphone RS.

C: The ultraviolet absorption spectrum of a 1 in 6500 solution in 0.1 *N* hydrochloric acid exhibits maxima and minima at the same wavelengths as that of a solution of USP Oxymorphone RS, prepared by dissolving about 20 mg of the Reference Standard in 10 mL of 1 *N* hydrochloric acid and diluting with water to 100.0 mL. The ratio A_{281}/A_{264} is 1.75 ± 0.2.

D: Dissolve 10 mg in 1 mL of water, and add a few drops of ferric chloride TS: a blue color is produced immediately.

Specific rotation ⟨781⟩: between −145° and −155°, calculated on the dried basis, determined in a solution containing 1 g in each 10 mL.

Acidity—Dissolve 300 mg in 10 mL of water, add 1 drop of methyl red TS, and titrate with 0.020 *N* sodium hydroxide: not more than 0.30 mL is required to produce a yellow color.

Loss on drying ⟨731⟩—Dry it at 105° for 18 hours: it loses not more than 8.0% of its weight.

Residue on ignition ⟨281⟩: not more than 0.3%.

Ordinary impurities ⟨466⟩—

Test solution: methanol.

Standard solution: methanol.

Eluant: a mixture of dehydrated alcohol, cyclohexane, and ammonium hydroxide (10:5:1).

Visualization: 1.

Chloride content—Dissolve about 300 mg, accurately weighed, in 50 mL of methanol in a glass-stoppered flask, add 5 mL of glacial acetic acid and 3 drops of eosin Y TS, and titrate with 0.1 *N* silver nitrate VS. Each mL of 0.1 *N* silver nitrate is equivalent to 3.545 mg of Cl: the content is between 10.2% and 10.8%, calculated on the dried basis.

Nonphenolic substances—Dissolve 1 g in 15 mL of water, add 5 mL of sodium hydroxide solution (2 in 25), and extract with three 10-mL portions of chloroform. Filter the combined extracts through a small chloroform-moistened filter paper, and wash the filtrate with 5 mL of water. Filter the chloroform layer through chloroform-moistened filter paper into a tared, 50-mL beaker, and evaporate on a steam bath with the aid of a gentle current of filtered air to dryness. Dry the beaker and residue at 105° for 1 hour, and weigh: the residue so obtained does not exceed 15 mg.

Assay—Transfer about 700 mg of Oxymorphone Hydrochloride, accurately weighed, to a glass-stoppered flask containing 50 mL of glacial acetic acid and 10 mL of mercuric acetate TS. Add 3 mL of acetic anhydride and 1 drop of methyl violet TS, and titrate with 0.1 *N* perchloric acid VS to a clear blue color. Perform a blank determination, and make any necessary correction. Each mL of 0.1 *N* perchloric acid is equivalent to 33.78 mg of $C_{17}H_{19}NO_4 . HCl$.

Oxymorphone Hydrochloride Injection

» Oxymorphone Hydrochloride Injection is a sterile solution of Oxymorphone Hydrochloride in Water for Injection. It contains not less than 93.0 percent and not more than 107.0 percent of the labeled amount of $C_{17}H_{19}NO_4 . HCl$.

Packaging and storage—Preserve in single-dose or in multiple-dose containers of Type I glass, protected from light.

Reference standard—*USP Oxymorphone Reference Standard*—Keep container tightly closed and protected from light. Do not dry before using.

Identification—The solution prepared for measurement of absorbance in the *Assay* exhibits maxima and minima at the same wavelengths as the *Standard preparation* prepared as directed in the *Assay*.

pH ⟨791⟩: between 2.7 and 4.5.

Other requirements—It meets the requirements under *Injections* ⟨1⟩.

Assay—

Standard preparation—Transfer about 45 mg of USP Oxymorphone RS, accurately weighed, to a 50-mL volumetric flask, dissolve in about 10 mL of chloroform, dilute with chloroform to volume, and mix. Transfer 15.0 mL of this solution to a 100-mL volumetric flask, dilute with chloroform to volume, and mix. The concentration of USP Oxymorphone RS in the *Standard preparation* is about 135 µg per mL.

Assay preparation—Transfer an accurately measured volume of Oxymorphone Hydrochloride Injection, equivalent to about 15 mg of oxymorphone hydrochloride, to a 125-mL separator, and add water, if necessary, to bring the volume to 15 mL. Adjust to a pH of less than 2 by the addition of hydrochloric acid, extract with five 15-mL portions of chloroform, and discard the chloroform extracts. Adjust the aqueous phase with ammonium hydroxide to a pH of 9.5, and extract with four 20-mL portions of chloroform. Filter the chloroform extracts through a chloroform-moistened pledget of cotton into a 100-mL volumetric flask, dilute with chloroform to volume, and mix.

Procedure—Concomitantly determine the absorbances of the solutions in 1-cm cells at the wavelength of maximum absorbance at about 282 nm, with a suitable spectrophotometer, using chloroform as the blank. Calculate the quantity, in mg, of $C_{17}H_{19}NO_4 . HCl$ in each mL of the Injection taken by the formula:

$$(337.80/301.34)(0.1C/V)(A_U/A_S),$$

in which 337.80 and 301.34 are the molecular weights of oxymorphone hydrochloride and oxymorphone, respectively, C is the concentration, in µg per mL, of USP Oxymorphone RS in the *Standard preparation*, V is the volume, in mL, of Injection taken, and A_U and A_S are the absorbances of the *Assay preparation* and the *Standard preparation*, respectively.

Oxymorphone Hydrochloride Suppositories

» Oxymorphone Hydrochloride Suppositories contain not less than 93.0 percent and not more than 107.0 percent of the labeled amount of $C_{17}H_{19}NO_4 . HCl$.

Packaging and storage—Preserve in well-closed containers, and store in a refrigerator.

Reference standard—*USP Oxymorphone Reference Standard*—Keep container tightly closed and protected from light. Do not dry before using.

Identification—Place a number of Suppositories, equivalent to 5 mg of oxymorphone hydrochloride, in a 125-mL separator. Add 25 mL of 0.1 *N* hydrochloric acid, and shake without heating until the specimen is dissolved. Wash the solution with five 25-mL portions of chloroform, shaking the separator gently to avoid forming emulsions, and discard the chloroform washings. Adjust with 6 *N* ammonium hydroxide to a pH of about 9.5, using short-range pH indicator paper, and extract with three 25-mL portions of chloroform, filtering the extracts through chloroform-moistened glass wool into a 200-mL round-bottom flask. Evaporate the combined extracts to dryness, using a rotary evaporator. Add 25 mL of 0.1 *N* hydrochloric acid, insert the stopper, and swirl to dissolve the residue: the ultraviolet absorption spectrum of the solution so obtained exhibits maxima and minima at the same wavelengths as that of a similar solution of USP Oxymorphone RS, concomitantly measured.

Assay—

Mobile phase—0.05 *M* Sodium borate adjusted to a pH of about 9.1.

Internal standard solution—Prepare a solution of procaine hydrochloride in 0.01 *N* hydrochloric acid having a concentration of about 3 mg per mL.

Standard preparation—Using an accurately weighed quantity of USP Oxymorphone RS, prepare a solution in 0.01 *N* hydrochloric acid having a known concentration of about 4.5 mg per mL, sonicating, if necessary, to effect solution. Transfer 10.0 mL of this solution, 10.0 mL of the *Internal standard solution*, and 5.0 mL of 0.01 *N* hydrochloric acid to a 125-mL separator. Extract with four 25-mL portions of chloroform, discarding the chloroform layer each time. Transfer the aqueous layer to a suitable flask, and bubble filtered air through the solution for 10 minutes to remove final traces of chloroform. The concentration of USP Oxymorphone RS in the *Standard preparation* is about 1.8 mg per mL.

Assay preparation—Transfer a number of Suppositories, accurately counted and equivalent to about 50 mg of oxymorphone hydrochloride, to a 125-mL separator. Add 15.0 mL of 0.01 *N* hydrochloric acid, 10.0 mL of *Internal standard solution*, and 25 mL of chloroform. Shake until the suppositories dissolve. Discard the chloroform layer. Extract the aqueous layer with three 25-mL portions of chloroform, discarding the chloroform each time. Transfer the aqueous layer to a suitable flask and bubble filtered air through the solution for 10 minutes to remove final traces of chloroform.

Chromatographic system (see *Chromatography* ⟨621⟩)—The liquid chromatograph is equipped with a 254-nm detector and a 2.1-mm × 100-cm column that contains packing L12. The flow rate is about 1 mL per minute. Chromatograph five replicate injections of the *Standard preparation*, and record the peak responses as directed under *Procedure:* the relative standard deviation is not more than 2.0%, and the resolution factor between oxymorphone hydrochloride and procaine hydrochloride is not less than 1.5.

Procedure—Separately inject equal volumes (about 15 µL) of the *Standard preparation* and the *Assay preparation* into the chromatograph by means of a suitable microsyringe or sampling valve, record the chromatograms, and measure the responses for the major peaks. The retention times are about 5 and 7.5 minutes for oxymorphone hydrochloride and procaine hydrochloride, respectively. Calculate the quantity, in mg, of $C_{17}H_{19}NO_4 . HCl$ in each Suppository taken by the formula:

$$(337.80/301.34)(25C/N)(R_U/R_S),$$

in which 337.80 and 301.34 are the molecular weights of oxymorphone hydrochloride and oxymorphone, respectively, *C* is the concentration, in mg per mL, of USP Oxymorphone RS in the *Standard preparation*, *N* is the number of Suppositories taken, and R_U and R_S are the ratios of the peak responses of oxymorphone hydrochloride and procaine hydrochloride obtained from the *Assay preparation* and the *Standard preparation*, respectively.

Oxyphenbutazone

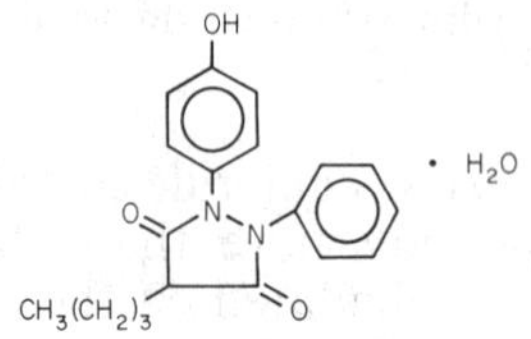

OH · H_2O $CH_3(CH_2)_3$

$C_{19}H_{20}N_2O_3 . H_2O$ 342.39

3,5-Pyrazolidinedione, 4-butyl-1-(4-hydroxyphenyl)-2-phenyl-, monohydrate.

4-Butyl-1-(*p*-hydroxyphenyl)-2-phenyl-3,5-pyrazolidinedione monohydrate [*7081-38-1*].

Anhydrous 324.38 [*129-20-4*].

» Oxyphenbutazone contains not less than 98.0 percent and not more than 100.5 percent of $C_{19}H_{20}N_2O_3$, calculated on the anhydrous basis.

Packaging and storage—Preserve in tight containers.

Reference standard—*USP Oxyphenbutazone Reference Standard*—Do not dry before using.

Identification—

A: The infrared absorption spectrum of a 1 in 50 solution of it, previously dried over a suitable desiccant, in methylene chloride exhibits maxima only at the same wavelengths as that of a similar preparation of USP Oxyphenbutazone RS.

B: The ultraviolet absorption spectrum of a 1 in 100,000 solution in 0.01 *N* sodium hydroxide exhibits maxima and minima at the same wavelengths as that of a similar solution of USP Oxyphenbutazone RS, concomitantly measured, and the respective absorptivities, calculated on the anhydrous basis, at the wavelength of maximum absorbance at about 254 nm do not differ by more than 2.0%.

C: Dissolve about 20 mg in 2 mL of methanol, and add 3 mL of Millon's reagent: a precipitate, which turns cherry-red when the mixture is heated, is formed.

Water, *Method I* ⟨921⟩: between 5.0% and 6.0%.

Residue on ignition ⟨281⟩: not more than 0.1%.

Chloride—Boil 1 g with 20 mL of water for 5 minutes, cool, and filter. If the filtrate is not clear, add a few mg of talc, boil again, cool, and filter. To 1 mL of the filtrate add 1 mL of 2.5 *N* nitric acid and 1 mL of silver nitrate TS: no opalescence is produced.

Chromatographic purity—[NOTE—Conduct this entire test without delay.]

Ascorbic acid solution—Dissolve 1.5 g of ascorbic acid and 20 mg of butylated hydroxytoluene in 100 mL of alcohol, with warming on a steam bath.

Standard preparations—Dissolve USP Oxyphenbutazone RS in *Ascorbic acid solution*, and mix to obtain a solution having a known concentration of 0.80 mg per mL. Dilute quantitatively with methanol to obtain 4 diluted *Standard preparations A*, *B*, *C*, and *D* having the following compositions:

Dilution	Concentration (µg RS per mL)	Percentage (%, for comparison with test specimen)
A (1 in 2)	400	2.0
B (1 in 4)	200	1.0
C (1 in 8)	100	0.5
D (1 in 16)	50	0.25

Test preparation—Mix 100 mg of Oxyphenbutazone with 5.0 mL of *Ascorbic acid solution* in a 10-mL centrifuge tube. Shake by mechanical means for 30 minutes, centrifuge, and use the clear supernatant solution without delay.

Procedure—On a suitable thin-layer chromatographic plate (see *Chromatography* ⟨621⟩), coated with a 0.25-mm layer of chromatographic silica gel mixture, apply separately 5 µL of the *Test preparation* and 5 µL of each *Standard preparation*. Position the plate in a chromatographic chamber, and develop the chromatograms in a solvent system consisting of a mixture of 80 mL of chloroform, 20 mL of glacial acetic acid, and 20 mg of butylated hydroxytoluene until the solvent front has moved about three-fourths of the length of the plate. Remove the plate from the developing chamber, mark the solvent front, and allow the solvent to evaporate. Immediately examine the plate under short-wavelength ultraviolet light, and compare the intensities of any secondary spots observed in the chromatogram of the *Test preparation* with those of the principal spots in the chromatograms of the *Standard preparations:* the sum of the intensities of secondary spots obtained from the *Test preparation* corresponds to not more than 1.0%.

Assay—Dissolve about 500 mg of Oxyphenbutazone, accurately weighed, in 100 mL of methanol, and titrate with 0.1 *N* sodium hydroxide VS, determining the end-point potentiometrically, using a calomel-glass electrode system with a saturated salt bridge of potassium chloride in methanol. Perform a blank determination, and make any necessary correction. Each mL of 0.1 *N* sodium hydroxide is equivalent to 32.44 mg of $C_{19}H_{20}N_2O_3$.

Oxyphenbutazone Tablets

» Oxyphenbutazone Tablets contain not less than 94.0 percent and not more than 106.0 percent of the labeled amount of $C_{19}H_{20}N_2O_3 . H_2O$.

Packaging and storage—Preserve in tight containers.

Reference standard—*USP Oxyphenbutazone Reference Standard*—Do not dry before using.

Identification—Mix a portion of powdered Tablets, equivalent to about 300 mg of oxyphenbutazone, with 20 mL of ether, and boil on a steam bath for 2 to 3 minutes. Filter the hot mixture, wash the residue with two 5-mL portions of hot ether, and concentrate the filtrate on a steam bath to about 12 mL. Add 12 mL of solvent hexane, again concentrate on a steam bath to about 12 mL, and cool. Filter the mixture, wash the crystals with two 5-mL portions of solvent hexane, and dry the crystals in vacuum at 100° for 1 hour: the infrared absorption spectrum of a 1 in 50 solution of the oxyphenbutazone (anhydrous) so obtained, in methylene chloride, exhibits maxima only at the same wavelengths as that of a similar preparation of USP Oxyphenbutazone RS, similarly treated and measured.

Dissolution ⟨711⟩—

Medium: pH 7.5 phosphate buffer (see under *Buffer solutions* in the section, *Reagents, Indicators, and Solutions*); 900 mL.

Apparatus 1: 100 rpm.

Time: 30 minutes.

Procedure—Determine the amount of $C_{19}H_{20}N_2O_3 . H_2O$ dissolved from ultraviolet absorbances at the wavelength of maximum absorbance at about 262 nm of filtered portions of the solution under test, suitably diluted with pH 7.5 phosphate buffer, in comparison with a Standard solution having a known concentration of USP Oxyphenbutazone RS in the same medium.

Tolerances—Not less than 60% (*Q*) of the labeled amount of $C_{19}H_{20}N_2O_3 . H_2O$ is dissolved in 30 minutes.

Uniformity of dosage units ⟨905⟩: meet the requirements.

Assay—Weigh and finely powder not less than 20 Oxyphenbutazone Tablets. Transfer an accurately weighed portion of the powder, equivalent to about 100 mg of oxyphenbutazone, to a glass-stoppered, 250-mL conical flask, add 100 mL of ether, and shake by mechanical means for 15 minutes. Filter, with the aid of vacuum, through a fine-porosity, sintered-glass funnel into a 250-mL separator, rinsing the original flask and funnel with three 15-mL portions of ether and collecting the rinsings in the separator. Extract the ether solution with three 30-mL portions of 0.1 *N* sodium hydroxide, and combine the extracts in a 250-mL volumetric flask. Bubble nitrogen through this solution for 10 minutes to remove any residual ether. Add 10 mL of 0.1 *N* sodium hydroxide, dilute with water to volume, and mix. Filter, discarding the first 25 mL of the filtrate. Transfer 25.0 mL of the subsequent filtrate to a 100-mL volumetric flask, dilute with water to volume, and mix. Finally, transfer 10.0 mL of this last solution to a second 100-mL volumetric flask, dilute with 0.01 *N* sodium hydroxide to volume, and mix, to obtain the *Assay preparation.* Concomitantly determine the absorbances of the *Assay preparation* and a Standard solution of USP Oxyphenbutazone RS in 0.01 *N* sodium hydroxide having a known concentration of about 10 µg per mL, in 1-cm cells at the wavelength of maximum absorbance at about 254 nm, with a suitable spectrophotometer, using 0.01 *N* sodium hydroxide as the blank. Calculate the quantity, in mg, of $C_{19}H_{20}N_2O_3 . H_2O$ in the portion of Tablets taken by the formula:

$$10C(A_U/A_S),$$

in which *C* is the concentration, in µg per mL, of USP Oxyphenbutazone RS in the Standard solution, and A_U and A_S are the absorbances of the *Assay preparation* and the Standard solution, respectively.

Oxyphencyclimine Hydrochloride

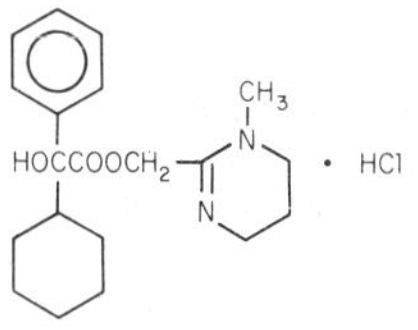

$C_{20}H_{28}N_2O_3 . HCl$ 380.91

Benzeneacetic acid, α-cyclohexyl-α-hydroxy-, (1,4,5,6-tetrahydro-1-methyl-2-pyrimidinyl)methyl ester monohydrochloride.

(1,4,5,6-Tetrahydro-1-methyl-2-pyrimidinyl)methyl α-phenylcyclohexaneglycolate monohydrochloride [*125-52-0*].

» Oxyphencyclimine Hydrochloride, dried in vacuum at 60° for 2.5 hours, contains not less than 98.0 percent and not more than 101.0 percent of $C_{20}H_{28}N_2O_3 . HCl$.

Packaging and storage—Preserve in tight containers.

Reference standard—*USP Oxyphencyclimine Hydrochloride Reference Standard*—Dry in vacuum at 60° for 2.5 hours before using.

Identification—

A: The infrared absorption spectrum of a potassium bromide dispersion of it, previously dried, exhibits maxima only at the same wavelengths as that of a similar preparation of USP Oxyphencyclimine Hydrochloride RS.

B: Prepare a solution of it in 0.01 *N* hydrochloric acid in methanol containing 4 mg per mL. On a suitable thin-layer chromatographic plate (see *Chromatography* ⟨621⟩), coated with a 500-µm layer of chromatographic silica gel mixture, apply 100 µL of this solution and 100 µL of a solution of USP Oxyphencyclimine Hydrochloride RS in the methanolic hydrochloric acid containing 4 mg per mL. Develop the chromatogram in a solvent system consisting of a mixture of chloroform and methanol (65:27) until the solvent front has moved about three-fourths of the length of the plate. Remove the plate from the developing chamber, mark the solvent front, and allow the solvent to evaporate. Locate the spots on the plate by lightly spraying with a mixture of 1 volume of a 3 in 1000 solution of blue tetrazolium in alcohol and 2 volumes of 2.5 *N* sodium hydroxide, then expose the sprayed plate to short-wavelength ultraviolet light for about 10 minutes: the R_f value of the principal spot obtained from the test solution corresponds to that obtained from the Standard solution.

Loss on drying ⟨731⟩—Dry it in vacuum at 60° for 2.5 hours: it loses not more than 2.0% of its weight.

Residue on ignition ⟨281⟩: not more than 0.2%.

Heavy metals, *Method II* ⟨231⟩: 0.005%.

Assay—Dissolve about 300 mg of Oxyphencyclimine Hydrochloride, previously dried and accurately weighed, in 20 mL of chloroform, add 50 mL of glacial acetic acid, 10 mL of mercuric acetate TS, and 2 drops of quinaldine red TS, and titrate with 0.1 *N* perchloric acid VS. Perform a blank determination, and make any necessary correction. Each mL of 0.1 *N* perchloric acid is equivalent to 38.09 mg of $C_{20}H_{28}N_2O_3 . HCl$.

Oxyphencyclimine Hydrochloride Tablets

» Oxyphencyclimine Hydrochloride Tablets contain not less than 90.0 percent and not more than 110.0 percent of the labeled amount of $C_{20}H_{28}N_2O_3 . HCl$.

Packaging and storage—Preserve in tight containers.

Reference standard—*USP Oxyphencyclimine Hydrochloride Reference Standard*—Dry in vacuum at 60° for 2.5 hours before using.

Identification—Transfer an amount of finely powdered Tablets, equivalent to about 100 mg of oxyphencyclimine hydrochloride, to a 40-mL centrifuge tube, add 25 mL of a 1 in 1200 solution of hydrochloric acid in methanol, shake for 10 minutes, then centrifuge for 5 minutes: a 0.1-mL portion of the supernatant liquid so obtained responds to *Identification test B* under *Oxyphencyclimine Hydrochloride.*

Dissolution ⟨711⟩—

Medium: water; 500 mL.

Apparatus 1: 100 rpm.

Time: 45 minutes.

Procedure—Determine the amount of $C_{20}H_{28}N_2O_3 . HCl$ dissolved from ultraviolet absorbances at the wavelength of maximum absorbance at about 218 nm of filtered portions of the solution under test, suitably diluted with *Dissolution Medium*, if necessary, in comparison with a Standard solution having a known concentration of USP Oxyphencyclimine Hydrochloride RS in the same medium.

Tolerances—Not less than 75% (*Q*) of the labeled amount of $C_{20}H_{28}N_2O_3 . HCl$ is dissolved in 45 minutes.

Uniformity of dosage units ⟨905⟩: meet the requirements.

Assay—

Standard preparation—Transfer about 25 mg of USP Oxyphencyclimine Hydrochloride RS, accurately weighed, to a 50-mL volumetric flask, dilute with methanol to volume, and mix. Transfer 10.0 mL of this solution to a 50-mL volumetric flask, dilute with water to volume, and mix to obtain a *Standard preparation* having a known concentration of about 100 µg per mL.

Assay preparation—Weigh and finely powder not less than 20 Oxyphencyclimine Hydrochloride Tablets. Transfer an accurately weighed portion of the powder, equivalent to about 50 mg of oxyphencyclimine hydrochloride, to a 100-mL volumetric flask, add 60 mL of methanol, and shake by mechanical means for 15 minutes. Dilute with methanol to volume, mix, and filter, discarding the first 15 mL of the filtrate. Transfer 10.0 mL of the subsequent filtrate to a 50-mL volumetric flask, dilute with water to volume, and mix.

Procedure—Transfer duplicate 2.0-mL portions of the *Standard preparation* and of the *Assay preparation* to separate, glass-stoppered, 40-mL centrifuge tubes. To one set of two tubes add 3 mL of water and 1 mL of 0.25 *N* sodium hydroxide. Heat these tubes in a boiling water bath for 10 minutes, and allow them to cool to room temperature. To the remaining set of tubes, which provide blanks for the *Standard preparation* and the *Assay preparation*, respectively, add 4 mL of water. To all tubes add 2 mL of approximately 0.2 *M* ceric sulfate (prepared by dissolving 12.6 g of ceric ammonium sulfate in 50 mL of water and 3 mL of sulfuric acid and diluting with water to 100 mL) and 20.0 mL of isooctane. Shake by mechanical means for 15 minutes, allow the layers to separate, and remove the isooctane from each tube. Concomitantly determine the absorbances of the isooctane solutions from the hydrolyzed aliquots in 1-cm cells at the wavelength of maximum absorbance at about 239 nm, with a suitable spectrophotometer, using the respective blanks to set the instrument. Calculate the quantity, in mg, of $C_{20}H_{28}N_2O_3 . HCl$ in the portion of Tablets taken by the formula:

$$0.5C(A_U / A_S),$$

in which *C* is the concentration, in µg per mL, of USP Oxyphencyclimine Hydrochloride RS in the *Standard preparation*, and A_U and A_S are the absorbances of the solutions from the *Assay preparation* and the *Standard preparation*, respectively.

Oxyquinoline Sulfate—*see* Oxyquinoline Sulfate NF

Oxytetracycline

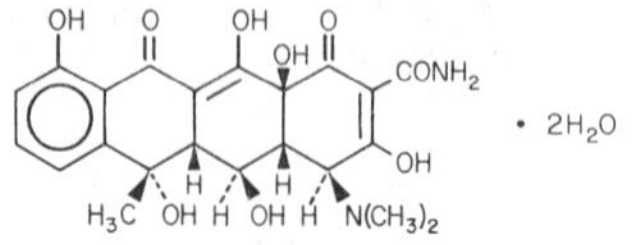

$C_{22}H_{24}N_2O_9 . 2H_2O$ 496.47

2-Naphthacenecarboxamide, 4-(dimethylamino)-1,4,4a,5,5a,-6,11,12a-octahydro-3,5,6,10,12,12a-hexahydroxy-6-methyl-1,11-dioxo-, [4*S*-(4α,4aα,5α,5aα,6β,12aα)]-, dihydrate.

4-(Dimethylamino)-1,4,4a,5,5a,6,11,12a-octahydro-3,5,6,10,-12,12a-hexahydroxy-6-methyl-1,11-dioxo-2-naphthacene-carboxamide dihydrate [*6153-64-6*].

Anhydrous 460.44 [*79-57-2*].

» Oxytetracycline has a potency equivalent to not less than 832 µg of $C_{22}H_{24}N_2O_9$ per mg.

Packaging and storage—Preserve in tight, light-resistant containers.

Reference standard—*USP Oxytetracycline Reference Standard*—Do not dry before using.

Identification—

A: The ultraviolet absorption spectrum of a 1 in 50,000 solution in 0.1 *N* hydrochloric acid exhibits maxima and minima at the same wavelengths as that of a similar solution of USP Oxytetracycline RS, concomitantly measured, and the absorptivity, calculated on the anhydrous basis, at the wavelength of maximum absorbance at about 353 nm is between 96.0% and 104.0% of that of USP Oxytetracycline RS, the potency of the Reference Standard being taken into account.

B: To 1 mg add 2 mL of sulfuric acid: a light red color is produced.

Crystallinity ⟨695⟩: meets the requirements.

pH ⟨791⟩: between 4.5 and 7.0, in an aqueous suspension containing 10 mg per mL.

Water, *Method I* ⟨921⟩: between 6.0% and 9.0%.

Assay—Proceed with Oxytetracycline as directed under *Antibiotics—Microbial Assays* ⟨81⟩.

Oxytetracycline Injection

» Oxytetracycline Injection is a sterile solution of Oxytetracycline with or without one or more suitable anesthetics, antioxidants, buffers, complexing agents, preservatives, and solvents. It contains the equivalent of not less than 90.0 percent and not more than 120.0 percent of the labeled amount of $C_{22}H_{24}N_2O_9$, the labeled amount being 50 or 125 mg per mL.

Packaging and storage—Preserve in single-dose or in multiple-dose containers, protected from light.

Reference standard—*USP Oxytetracycline Reference Standard*—Do not dry before using.

Depressor substances—When diluted with sterile water to a concentration of 5.0 mg of oxytetracycline ($C_{22}H_{24}N_2O_9$) per mL, it meets the requirements of the *Depressor Substances Test* ⟨101⟩, the test dose being 0.6 mL per kg.

Pyrogen—When diluted with Sterile Water for Injection to a concentration of 5.0 mg of oxytetracycline ($C_{22}H_{24}N_2O_9$) per mL, it meets the requirements of the *Pyrogen Test* ⟨151⟩, the test dose being 1.0 mL per kg.

Sterility—It meets the requirements under *Sterility Tests* ⟨71⟩, when tested as directed in the section, *Test Procedures Using Membrane Filtration.*

pH ⟨791⟩: between 8.0 and 9.0.

Assay—Dilute an accurately measured volume of Oxytetracycline Injection quantitatively with 0.1 *N* hydrochloric acid to obtain a stock solution that contains not less than 150 µg of oxytetracycline ($C_{22}H_{24}N_2O_9$) per mL. Proceed as directed under *Antibiotics—Microbial Assays* ⟨81⟩, using an accurately measured volume of this stock solution diluted quantitatively and stepwise with sterile water to yield a *Test Dilution* having a concentration assumed to be equal to the median dose level of the Standard.

Sterile Oxytetracycline

» Sterile Oxytetracycline is Oxytetracycline suitable for parenteral use. It has a potency of not less than 832 µg of oxytetracycline ($C_{22}H_{24}N_2O_9$) per mg.

Packaging and storage—Preserve in *Containers for Sterile Solids* as described under *Injections* ⟨1⟩, protected from light.

Reference standard—*USP Oxytetracycline Reference Standard*—Do not dry before using.

Depressor substances—Dissolve 200 mg of Sterile Oxytetracycline, accurately weighed, in 10.0 mL of 0.1 *N* hydrochloric acid. When this solution is diluted with sterile water to a concentration of 5.0 mg of oxytetracycline ($C_{22}H_{24}N_2O_9$) per mL, it meets the requirements of the *Depressor Substances Test* ⟨101⟩, the test dose being 0.6 mL per kg.

Pyrogen—Dissolve 200 mg of Sterile Oxytetracycline, accurately weighed, in 10.0 mL of 0.1 *N* hydrochloric acid. When this solution is diluted with Sterile Water for Injection to a concentration of 5.0 mg of oxytetracycline ($C_{22}H_{24}N_2O_9$) per mL, it meets the requirements of the *Pyrogen Test* ⟨151⟩, the test dose being 1.0 mL per kg.

Sterility—It meets the requirements under *Sterility Tests* ⟨71⟩, when tested as directed in the section, *Test Procedures Using Membrane Filtration, Fluid D* being used instead of *Fluid A.*

Other requirements—It conforms to the definition, responds to the *Identification tests*, and meets the requirements for *pH, Water, Crystallinity*, and *Assay* under *Oxytetracycline.*

Oxytetracycline Tablets

» Oxytetracycline Tablets contain the equivalent of not less than 90.0 percent and not more than 120.0 percent of the labeled amount of $C_{22}H_{24}N_2O_9$.

Packaging and storage—Preserve in tight, light-resistant containers.

Reference standard—*USP Oxytetracycline Reference Standard*—Do not dry before using.

Dissolution ⟨711⟩—

Medium: 0.1 *N* hydrochloric acid; 900 mL.

Apparatus 1: 100 rpm.

Time: 45 minutes.

Procedure—Determine the amount of $C_{22}H_{24}N_2O_9$ dissolved from ultraviolet absorbances at the wavelength of maximum absorbance at about 353 nm of filtered portions of the solution under test, suitably diluted with *Dissolution Medium*, if necessary, in comparison with a Standard solution having a known concentration of USP Oxytetracycline RS in the same medium.

Tolerances—Not less than 75% (*Q*) of the labeled amount of $C_{22}H_{24}N_2O_9$ is dissolved in 45 minutes.

Uniformity of dosage units ⟨905⟩: meet the requirements.

Water, *Method I* ⟨921⟩: not more than 7.5%.

Assay—Place not less than 5 Oxytetracycline Tablets in a high-speed glass blender jar containing an accurately measured volume of 0.1 *N* hydrochloric acid, so that the stock solution so obtained contains not less than 150 µg of oxytetracycline ($C_{22}H_{24}N_2O_9$) per mL, and blend for 3 to 5 minutes. Proceed as directed under *Antibiotics—Microbial Assays* ⟨81⟩, using an accurately measured volume of this stock solution diluted quantitatively and stepwise with water to yield a *Test Dilution* having a concentration assumed to be equal to the median dose level of the Standard.

Oxytetracycline and Nystatin Capsules

» Oxytetracycline and Nystatin Capsules contain not less than 90.0 percent and not more than 120.0 percent of the labeled amount of oxytetracycline ($C_{22}H_{24}N_2O_9$), and not less than 90.0 percent and not more than 135.0 percent of the labeled amount of USP Nystatin Units.

Packaging and storage—Preserve in tight, light-resistant containers.

Reference standards—*USP Oxytetracycline Reference Standard*—Do not dry before using. *USP Nystatin Reference Standard*—Dry in vacuum at a pressure not exceeding 5 mm of mercury at 40° for 2 hours before using.

Identification—Shake a suitable quantity of Capsule contents with methanol to obtain a solution containing about 1 mg of oxytetracycline per mL, and filter. Using the filtrate as the *Test Solution*, proceed as directed under *Identification—Tetracyclines* ⟨193⟩.

Dissolution ⟨711⟩—

Medium: 0.1 *N* hydrochloric acid; 900 mL.

Apparatus 1: 100 rpm.

Time: 45 minutes.

Procedure—Determine the amount of oxytetracycline ($C_{22}H_{24}N_2O_9$) dissolved from ultraviolet absorbances at the wavelength of maximum absorbance at about 353 nm of filtered portions of the solution under test, suitably diluted with *Dissolution Medium*, if necessary, in comparison with a Standard solution having a known concentration of USP Oxytetracycline RS in the same medium.

Tolerances—Not less than 75% (*Q*) of the labeled amount of $C_{22}H_{24}N_2O_9$ is dissolved in 45 minutes.

Uniformity of dosage units ⟨905⟩: meet the requirements for *Weight Variation* with respect to oxytetracycline.

Water, *Method I* ⟨921⟩: not more than 7.5%.

Assay for oxytetracycline—Place not less than 5 Oxytetracycline and Nystatin Capsules in a high-speed glass blender jar containing an accurately measured volume of 0.1 *N* hydrochloric acid, and blend for 3 to 5 minutes, so that the stock solution so obtained contains not less than 150 µg of oxytetracycline ($C_{22}H_{24}N_2O_9$) per mL. Proceed as directed under *Antibiotics—Microbial Assays* ⟨81⟩, using an accurately measured volume of this stock solution diluted quantitatively and stepwise with water to yield a *Test Dilution* having a concentration of oxytetracycline assumed to be equal to the median dose level of the Standard.

Assay for nystatin—Proceed as directed for Nystatin under *Antibiotics—Microbial Assays* ⟨81⟩, blending not less than 5 Oxytetracycline and Nystatin Capsules for 3 to 5 minutes in a high-speed blender with a sufficient accurately measured volume of dimethylformamide to obtain a solution of convenient concentration. Dilute an accurately measured portion of this solution quantitatively with dimethylformamide to obtain a stock solution containing about 400 USP Nystatin Units per mL. Dilute this stock solution quantitatively with *Buffer No. 6* to obtain a *Test Dilution* having a concentration of nystatin assumed to be equal to the median dose level of the Standard.

Oxytetracycline and Nystatin for Oral Suspension

» Oxytetracycline and Nystatin for Oral Suspension is a dry mixture of Oxytetracycline and Nystatin with one or more suitable buffers, colors, diluents, flavors, suspending agents, and preservatives. When constituted as directed in the labeling, it contains not less than 90.0 percent and not more than 120.0 percent of the labeled amount of oxytetracycline ($C_{22}H_{24}N_2O_9$), and not less than 90.0 percent and not more than 135.0 percent of the labeled amount of USP Nystatin Units.

Packaging and storage—Preserve in tight, light-resistant containers, at controlled room temperature.

Reference standards—*USP Oxytetracycline Reference Standard*—Do not dry before using. *USP Nystatin Reference Standard*—Dry in vacuum at a pressure not exceeding 5 mm of mercury at 40° for 2 hours before using.

Uniformity of dosage units ⟨905⟩—

FOR SOLID PACKAGED IN SINGLE-UNIT CONTAINERS: meets the requirements for *Content Uniformity* with respect to oxytetracycline and nystatin.

pH ⟨791⟩: between 4.5 and 7.5, in the suspension constituted as directed in the labeling.

Water, *Method I* ⟨921⟩: not more than 2.0%.

Assay for oxytetracycline—Constitute Oxytetracycline and Nystatin for Oral Suspension as directed in the labeling. Transfer an accurately measured volume of the suspension so obtained, freshly mixed and free from air bubbles, to a suitable volumetric flask, dilute with 0.1 *N* hydrochloric acid to volume so that the stock solution so obtained contains not less than 150 µg of oxytetracycline per mL, and mix. Proceed as directed for oxytetracycline under *Antibiotics—Microbial Assays* ⟨81⟩, using an accurately measured volume of the stock solution diluted quantitatively and stepwise with water to yield a *Test Dilution* having a concentration assumed to be equal to the median dose level of the Standard.

Assay for nystatin—Constitute Oxytetracycline and Nystatin for Oral Suspension as directed in the labeling. Transfer an accurately measured volume of the suspension so obtained, freshly mixed and free from air bubbles, to a blender jar containing a sufficient, accurately measured volume of dimethylformamide to yield a solution of convenient concentration, and blend at high speed for 3 to 5 minutes. Dilute an accurately measured volume of this solution quantitatively with dimethylformamide to obtain a stock solution containing about 400 USP Nystatin Units per mL. Proceed as directed for nystatin under *Antibiotics—Microbial Assays* ⟨81⟩, using an accurately measured volume of this solution diluted quantitatively with *Buffer No. 6* to yield a *Test Dilution* having a concentration of nystatin assumed to be equal to the median dose level of the Standard.

Oxytetracycline Calcium

$C_{44}H_{46}CaN_4O_{18}$ 958.94

2-Naphthacenecarboxamide, 4-(dimethylamino)-1,4,4a,5,5a,-6,11,12a-octahydro-3,5,6,10,12,12a-hexahydroxy-6-methyl-1,11-dioxo-, calcium salt, [4*S*-(4α,4aα,5α,5aα,-6β,12aα)]-.

4-(Dimethylamino)-1,4,4a,5,5a,6,11,12a-octahydro-3,5,6,10,12,-12a-hexahydroxy-6-methyl-1,11-dioxo-2-naphthacene-carboxamide calcium salt [*15251-48-6*].

» Oxytetracycline Calcium has a potency equivalent to not less than 865 µg of oxytetracycline ($C_{22}H_{24}N_2O_9$) per mg, calculated on the anhydrous basis.

Packaging and storage—Preserve in tight, light-resistant containers, and store in a cool place.

Reference standard—*USP Oxytetracycline Reference Standard*—Do not dry before using.

Identification—Dissolve a suitable quantity in methanol to obtain a *Test Solution* containing 1 mg of oxytetracycline per mL, and proceed as directed under *Identification—Tetracyclines* ⟨193⟩.

Crystallinity ⟨695⟩: meets the requirements.

pH ⟨791⟩: between 6.0 and 8.0, in an aqueous suspension containing 25 mg per mL.

Water, *Method I* ⟨921⟩: between 8.0% and 14.0%.

Calcium content—Proceed as directed under *Residue on Ignition* ⟨281⟩, except to ignite at 550 ± 50° instead of at 800 ± 25°: the weight of residue so obtained, multiplied by 0.2944, gives the equivalent of calcium in the Oxytetracycline Calcium taken. The calcium content is between 3.85% and 4.35%, calculated on the anhydrous basis.

Assay—Dissolve an accurately weighed quantity of Oxytetracycline Calcium in an accurately measured volume of 0.1 *N* hydrochloric acid to obtain a stock solution having a concentration of about 1000 µg of oxytetracycline per mL. Proceed as directed for oxytetracycline under *Antibiotics—Microbial Assays* ⟨81⟩, using an accurately measured volume of the stock solution diluted quantitatively and stepwise with water to yield a *Test Dilution* having a concentration assumed to be equal to the median dose level of the Standard.

Oxytetracycline Calcium Oral Suspension

» Oxytetracycline Calcium Oral Suspension contains the equivalent of not less than 90.0 percent and not more than 120.0 percent of the labeled amount of oxytetracycline ($C_{22}H_{24}N_2O_9$). It contains one or more suitable buffers, colors, flavors, preservatives, stabilizers, and suspending agents. In addition, it may contain *N*-acetylglucosamine.

Packaging and storage—Preserve in tight, light-resistant containers.

Reference standard—*USP Oxytetracycline Reference Standard*—Do not dry before using.

Identification—Shake a suitable quantity of Oral Suspension with methanol to obtain a solution containing 1 mg of oxytetracycline per mL, and filter. Using the filtrate as the *Test Solution,* proceed as directed under *Identification—Tetracyclines* ⟨193⟩.

Uniformity of dosage units ⟨905⟩—

FOR SUSPENSION PACKAGED IN SINGLE-UNIT CONTAINERS: meets the requirements.

pH ⟨791⟩: between 5.0 and 8.0.

Assay—Transfer an accurately measured quantity of Oxytetracycline Calcium Oral Suspension, freshly mixed and free from air bubbles, equivalent to about 150 mg of oxytetracycline, to a 1000-mL volumetric flask, dilute with 0.1 *N* hydrochloric acid to volume, and mix. Proceed as directed for oxytetracycline under *Antibiotics—Microbial Assays* ⟨81⟩, using an accurately measured volume of this stock solution diluted quantitatively and stepwise with water to yield a *Test Dilution* having a concentration assumed to be equal to the median dose level of the Standard.

Oxytetracycline Hydrochloride

$C_{22}H_{24}N_2O_9 \cdot HCl$ 496.90
2-Naphthacenecarboxamide, 4-(dimethylamino)-1,4,4a,5,5a,6,11,12a-octahydro-3,5,6,10,12,12a-hexahydroxy-6-methyl-1,11-dioxo-, monohydrochloride, [4*S*-(4α,4aα,5α,5aα,6β,12aα)]-.
4-(Dimethylamino)-1,4,4a,5,5a,6,11,12a-octahydro-3,5,6,10,12,-12a-hexahydroxy-6-methyl-1,11-dioxo-2-naphthacenecarboxamide monohydrochloride [*2058-46-0*].

» Oxytetracycline Hydrochloride has a potency equivalent to not less than 835 μg of oxytetracycline ($C_{22}H_{24}N_2O_9$) per mg, calculated on the dried basis.

Packaging and storage—Preserve in tight, light-resistant containers.

Reference standard—*USP Oxytetracycline Reference Standard*—Do not dry before using.

Identification—

A: The ultraviolet absorption spectrum of a 1 in 50,000 solution in 0.1 *N* hydrochloric acid exhibits maxima and minima at the same wavelengths as that of a similar solution of USP Oxytetracycline RS, concomitantly measured, and the absorptivity, calculated on the dried basis, at the wavelength of maximum absorbance at about 353 nm is between 88.2% and 96.8% of that of the USP Oxytetracycline RS, the potency of the Reference Standard being taken into account.

B: To 1 mg add 2 mL of sulfuric acid: a light red color is produced.

Crystallinity ⟨695⟩: meets the requirements.

pH ⟨791⟩: between 2.0 and 3.0, in a solution containing 10 mg per mL.

Loss on drying—Dry about 100 mg, accurately weighed, in a capillary-stoppered bottle in vacuum at a pressure not exceeding 5 mm of mercury at 60° for 3 hours: it loses not more than 2.0% of its weight.

Assay—Proceed with Oxytetracycline Hydrochloride as directed under *Antibiotics—Microbial Assays* ⟨81⟩.

Oxytetracycline Hydrochloride Capsules

» Oxytetracycline Hydrochloride Capsules contain the equivalent of not less than 90.0 percent and not more than 120.0 percent of the labeled amount of oxytetracycline ($C_{22}H_{24}N_2O_9$).

Packaging and storage—Preserve in tight, light-resistant containers.

Reference standard—*USP Oxytetracycline Reference Standard*—Do not dry before using.

Identification—Shake a suitable quantity of Capsule contents with methanol to obtain a solution containing 1 mg of oxytetracycline per mL, and filter. Using the filtrate as the *Test Solution*, proceed as directed under *Identification—Tetracyclines* ⟨193⟩.

Dissolution ⟨711⟩—

Medium: water; 900 mL.

Apparatus 2: 75 rpm.

Time: 60 minutes.

Procedure—Determine the amount of $C_{22}H_{24}N_2O_9$ dissolved from ultraviolet absorbances at the wavelength of maximum absorbance at about 273 nm of filtered portions of the solution under test, suitably diluted with water, in comparison with a Standard solution having a known concentration of USP Oxytetracycline RS in the same medium, using 5 mL of 0.1 *N* hydrochloric acid to dissolve the Standard.

Tolerances—Not less than 80% (*Q*) of the labeled amount of $C_{22}H_{24}N_2O_9$ is dissolved in 60 minutes.

Uniformity of dosage units ⟨905⟩: meet the requirements.

Loss on drying ⟨731⟩—Dry about 100 mg of Capsule contents, accurately weighed, in a capillary-stoppered bottle in vacuum at a pressure not exceeding 5 mm of mercury at 60° for 3 hours: it loses not more than 5.0% of its weight.

Assay—Place not less than 5 Oxytetracycline Hydrochloride Capsules in a high-speed blender jar containing an accurately measured volume of 0.1 *N* hydrochloric acid, so that the stock solution thus obtained contains not less than 150 μg of oxytetracycline per mL, and blend for about 3 to 5 minutes. Proceed as directed for oxytetracycline under *Antibiotics—Microbial Assays* ⟨81⟩, using an accurately measured volume of the stock solution diluted quantitatively and stepwise with water to yield a *Test Dilution* having a concentration assumed to be equal to the median dose level of the Standard.

Oxytetracycline Hydrochloride for Injection

» Oxytetracycline Hydrochloride for Injection is a sterile dry mixture of Sterile Oxytetracycline Hydrochloride and a suitable buffer. It contains the equivalent of not less than 90.0 percent and not more than 115.0 percent of the labeled amount of oxytetracycline ($C_{22}H_{24}N_2O_9$).

Packaging and storage—Preserve in *Containers for Sterile Solids* as described under *Injections* ⟨1⟩, protected from light.

Reference standard—*USP Oxytetracycline Reference Standard*—Do not dry before using.

Constituted solution—At the time of use, the constituted solution prepared from Oxytetracycline Hydrochloride for Injection meets the requirements for *Constituted Solutions* under *Injections* ⟨1⟩.

pH ⟨791⟩: between 1.8 and 2.8, in a solution containing 25 mg per mL.

Loss on drying ⟨731⟩—Dry about 100 mg, accurately weighed, in a capillary-stoppered bottle in vacuum at a pressure not exceeding 5 mm of mercury at 60° for 3 hours: it loses not more than 3.0% of its weight.

Particulate matter ⟨788⟩: meets the requirements under *Small-volume Injections*.

Other requirements—It responds to *Identification test B* under *Oxytetracycline Hydrochloride*, and meets the requirements for *Depressor substances*, *Pyrogen*, and *Sterility* under *Sterile Oxytetracycline Hydrochloride*. It also meets the requirements for *Uniformity of Dosage Units* ⟨905⟩ and *Labeling* under *Injections* ⟨1⟩.

Assay—

Assay preparation 1 (where it is represented as being in a single-dose container)—Constitute Oxytetracycline Hydrochloride for Injection in a volume of water, accurately measured, corresponding to the volume of solvent specified in the labeling. Withdraw all of the withdrawable contents, using a suitable hypodermic needle and syringe, and dilute quantitatively with water to obtain a solution having a convenient concentration of oxytetracycline.

Assay preparation 2 (where the label states the quantity of oxytetracycline in a given volume of constituted solution)—Constitute Oxytetracycline Hydrochloride for Injection in a volume of water, accurately measured, corresponding to the volume of solvent specified in the labeling. Dilute an accurately measured volume of the constituted solution quantitatively with water to obtain a solution having a convenient concentration of oxytetracycline.

Procedure—Proceed as directed under *Antibiotics—Microbial Assays* ⟨81⟩, using an accurately measured volume of *Assay preparation* diluted quantitatively with water to yield a *Test Dilution* having a concentration assumed to be equal to the median dose level of the Standard.

Sterile Oxytetracycline Hydrochloride

» Sterile Oxytetracycline Hydrochloride is Oxytetracycline Hydrochloride suitable for parenteral use. It has a potency equivalent to not less than 835 µg of oxytetracycline ($C_{22}H_{24}N_2O_9$) per mg, calculated on the dried basis.

Packaging and storage—Preserve in *Containers for Sterile Solids* as described under *Injections* ⟨1⟩, protected from light.

Reference standard—*USP Oxytetracycline Reference Standard*—Do not dry before using.

Depressor substances—When dissolved in sterile saline TS and the solution is diluted with the same solvent to a concentration of 5.0 mg of oxytetracycline ($C_{22}H_{24}N_2O_9$) per mL, it meets the requirements of the *Depressor Substances Test* ⟨101⟩, the test dose being 0.6 mL per kg.

Pyrogen—When dissolved in Sterile Water for Injection and the solution is diluted with the same solvent to a concentration of 5.0 mg of oxytetracycline ($C_{22}H_{24}N_2O_9$) per mL, it meets the requirements of the *Pyrogen Test* ⟨151⟩, the test dose being 1.0 mL per kg.

Sterility—It meets the requirements under *Sterility Tests* ⟨71⟩, when tested as directed in the section, *Test Procedures Using Membrane Filtration*, *Fluid D* being used instead of *Fluid A*.

Other requirements—It conforms to the definition, responds to the *Identification tests*, and meets the requirements for *pH*, *Loss on drying*, *Crystallinity*, and *Assay* under *Oxytetracycline Hydrochloride*.

Oxytetracycline Hydrochloride and Hydrocortisone Ointment

» Oxytetracycline Hydrochloride and Hydrocortisone Ointment contains the equivalent of not less than 90.0 percent and not more than 115.0 percent of the labeled amount of oxytetracycline ($C_{22}H_{24}N_2O_9$), and not less than 90.0 percent and not more than 110.0 percent of the labeled amount of hydrocortisone.

Packaging and storage—Preserve in collapsible tubes or in well-closed, light-resistant containers.

Reference standards—*USP Oxytetracycline Reference Standard*—Do not dry before using. *USP Hydrocortisone Reference Standard*—Dry at 105° for 3 hours before using.

Minimum fill ⟨755⟩: meets the requirements.

Water, *Method I* ⟨921⟩: not more than 1.0%, 20 mL of a mixture of carbon tetrachloride, chloroform, and methanol (2:2:1) being used in place of methanol in the titration vessel.

Assay for oxytetracycline—Transfer a suitable, accurately weighed quantity of Oxytetracycline Hydrochloride and Hydrocortisone Ointment to a separator, add 50 mL of ether, and shake. Add 20 mL of 0.1 *N* hydrochloric acid, shake, and allow to separate. Collect the acid layer, and repeat the extraction with three additional 20-mL portions of 0.1 *N* hydrochloric acid. Combine the acid extracts in a 100-mL volumetric flask, dilute with 0.1 *N* hydrochloric acid to volume, and mix. Quantitatively dilute a portion of this solution with 0.1 *N* hydrochloric acid so that the solution so obtained contains not less than 150 µg of oxytetracycline per mL. Proceed as directed for oxytetracycline under *Antibiotics—Microbial Assays* ⟨81⟩, using an accurately measured volume of this solution diluted quantitatively and stepwise with water to obtain a *Test Dilution* having a concentration assumed to be equal to the median dose level of the Standard.

Assay for hydrocortisone—Proceed with Oxytetracycline Hydrochloride and Hydrocortisone Ointment as directed in the *Assay for hydrocortisone* under *Neomycin and Polymyxin B Sulfate, Bacitracin Zinc, and Hydrocortisone Ophthalmic Ointment*.

Oxytetracycline Hydrochloride and Hydrocortisone Acetate Ophthalmic Suspension

» Oxytetracycline Hydrochloride and Hydrocortisone Acetate Ophthalmic Suspension is a sterile suspension of Oxytetracycline Hydrochloride and Hydrocortisone Acetate in a suitable oil vehicle with one or more suitable suspending agents. It contains the equivalent of not less than 90.0 percent and not more than 115.0 percent of the labeled amount of oxytetracycline ($C_{22}H_{24}N_2O_9$) and not less than 90.0 percent and not more than 110.0 percent of the labeled amount of hydrocortisone acetate ($C_{23}H_{32}O_6$).

Packaging and storage—Preserve in tight, light-resistant containers. The containers are sealed and tamper-proof so that sterility is assured at time of first use.

Reference standards—*USP Oxytetracycline Reference Standard*—Do not dry before using. *USP Hydrocortisone Acetate Reference Standard*—Dry in vacuum at 105° for 3 hours before using.

Sterility—It meets the requirements under *Sterility Tests* ⟨71⟩, when tested as directed in the section, *Test Procedures for Direct Transfer to Test Media*, except to use 0.25 mL of specimen instead of 1.0 mL.

Water, *Method I* ⟨921⟩: not more than 1.0%, 60 mL of a mixture of methanol and chloroform (2:1) being used instead of methanol in the titration vessel.

Assay for oxytetracycline—Transfer an accurately measured volume of Oxytetracycline Hydrochloride and Hydrocortisone Acetate Ophthalmic Suspension to a separator, add 50 mL of ether, and shake. Add 25 mL of 0.1 *N* hydrochloric acid, shake, and allow to separate. Collect the acid layer, and repeat the extraction with three additional 25-mL portions of 0.1 *N* hydrochloric acid. Combine the acid extracts in a 200-mL volumetric flask, dilute with 0.1 *N* hydrochloric acid to volume, and mix. Proceed as directed for oxytetracycline under *Antibiotics—Microbial Assays* ⟨81⟩, using an accurately measured volume of this solution diluted quantitatively and stepwise with water to obtain a *Test Dilution* having a concentration assumed to be equal to the median dose level of the Standard.

Assay for hydrocortisone acetate—

Mobile phase—Prepare a degassed and filtered mixture of water and methanol (50:50).

Standard preparation—Dissolve an accurately weighed quantity of USP Hydrocortisone Acetate RS in alcohol to obtain a solution having a known concentration of about 0.06 mg per mL.

Assay preparation—Transfer an accurately measured volume of Oxytetracycline Hydrochloride and Hydrocortisone Acetate Ophthalmic Suspension, equivalent to about 30 mg of hydrocortisone acetate, to a separator containing 25 mL of pH 9.0 alkaline borate buffer (see under *Buffer solutions* in the section, *Reagents, Indicators, and Solutions*). Extract with four 25-mL portions of chloroform, filtering each chloroform extract through a thin layer of chloroform-washed anhydrous sodium sulfate into a 250-mL volumetric flask. Rinse the sodium sulfate with chloroform, collecting the filtrate in the volumetric flask, dilute with chloroform to volume, and mix. Transfer 25.0 mL of the resulting solution to a 50-mL conical flask, and evaporate slowly with the aid of mild heat until about 5 mL remains. Add about 15 mL of alcohol, and evaporate slowly until about 5 mL remains. Transfer this solution to a 50-mL volumetric flask, dilute with alcohol to volume, and mix.

Chromatographic system (see *Chromatography* ⟨621⟩)—The liquid chromatograph is equipped with a 254-nm detector and a 3.9-mm × 30-cm column that contains packing L1. The flow rate is about 1 mL per minute. Chromatograph the *Standard preparation*, and record the peak responses as directed under *Procedure:* the column efficiency determined from the analyte peak is not less than 235 theoretical plates, the tailing factor for the analyte peak is not more than 1.7, and the relative standard deviation of replicate injections is not more than 2.0%.

Procedure—Separately inject equal volumes (about 25 µL) of the *Standard preparation* and the *Assay preparation* into the chromatograph, record the chromatograms, and measure the responses for the major peaks. Calculate the quantity, in mg, of $C_{23}H_{32}O_6$ in each mL of the Ophthalmic Suspension taken by the formula:

$$500(C/V)(r_U/r_S),$$

in which C is the concentration, in mg per mL, of USP Hydrocortisone Acetate RS in the *Standard preparation*, V is the volume, in mL, of Ophthalmic Suspension taken, and r_U and r_S are the peak responses obtained from the *Assay preparation* and the *Standard preparation*, respectively.

Oxytetracycline and Phenazopyridine Hydrochlorides and Sulfamethizole Capsules

» Oxytetracycline and Phenazopyridine Hydrochlorides and Sulfamethizole Capsules contain the equivalent of not less than 90.0 percent and not more than 120.0 percent of the labeled amount of oxytetracycline ($C_{22}H_{24}N_2O_9$), not less than 90.0 percent and not more than 115.0 percent of the labeled amount of phenazopyridine hydrochloride ($C_{11}H_{11}N_5 . HCl$), and not less than 90.0 percent and not more than 110.0 percent of the labeled amount of sulfamethizole ($C_9H_{10}N_4O_2S_2$).

Packaging and storage—Preserve in tight, light-resistant containers.

Reference standards—*USP Oxytetracycline Reference Standard*—Do not dry before using. *USP Phenazopyridine Hydrochloride Reference Standard*—Dry at 105° for 4 hours before using. *USP Sulfamethizole Reference Standard*—Dry at 105° for 2 hours before using.

Uniformity of dosage units ⟨905⟩: meet the requirements for *Content Uniformity* with respect to oxytetracycline, to phenazopyridine hydrochloride, and to sulfamethizole.

Loss on drying ⟨731⟩—Dry about 100 mg of Capsule contents, accurately weighed, in a capillary-stoppered bottle in vacuum at a pressure not exceeding 5 mm of mercury at 60° for 3 hours: the substance loses not more than 5.0% of its weight.

Assay for oxytetracycline—Place not less than 5 Oxytetracycline and Phenazopyridine Hydrochlorides and Sulfamethizole Capsules in a high-speed blender jar containing an accurately measured volume of 0.1 *N* hydrochloric acid so that the stock solution so obtained contains not less than 150 µg of oxytetracycline per mL, and blend for about 3 to 5 minutes. Proceed as directed for oxytetracycline under *Antibiotics—Microbial Assays* ⟨81⟩, using an accurately measured volume of the stock solution diluted quantitatively and stepwise with water to obtain a *Test Dilution* having a concentration assumed to be equal to the median dose level of the Standard.

Assay for phenazopyridine hydrochloride—Transfer, as completely as possible, the contents of not less than 20 Oxytetracycline and Phenazopyridine Hydrochlorides and Sulfamethizole Capsules to a tared container, and determine the average weight per capsule. Mix the combined contents, and transfer an accurately weighed portion, equivalent to about 100 mg of phenazopyridine hydrochloride, to a separator. (Reserve the remaining combined contents for the *Assay for sulfamethizole*). Add 50 mL of 1 *N* sodium hydroxide to the separator, and extract with five 40-mL portions of chloroform, shaking each extraction for 2 minutes. Filter the extracts through anhydrous sodium sulfate into a 250-mL volumetric flask, dilute with chloroform to volume, and mix. Transfer 3.0 mL of this solution to a 200-mL volumetric flask, dilute with chloroform to volume, and mix. Similarly prepare a Standard solution, using 100 mg of USP Phenazopyridine Hydrochloride RS. Concomitantly determine the absorbances of the solution from the Capsules and the Standard solution in 1-cm cells at the wavelength of maximum absorbance at about 398 nm, with a suitable spectrophotometer, using chloroform as the blank. Calculate the quantity, in mg, of $C_{11}H_{11}N_5 . HCl$ in the portion of the contents of the Capsules taken by the formula:

$$16.67C(A_U/A_S),$$

in which C is the concentration, in µg per mL, of USP Phenazopyridine Hydrochloride RS in the Standard solution, and A_U and A_S are the absorbances of the solution from the Capsules and the Standard solution, respectively.

Assay for sulfamethizole—Transfer an accurately weighed portion of the combined contents of Oxytetracycline and Phenazopyridine Hydrochlorides and Sulfamethizole Capsules remaining from the *Assay for phenazopyridine hydrochloride*, equivalent to about 375 mg of sulfamethizole, to a 200-mL volumetric flask. Add about 120 mL of 1 *N* sodium hydroxide, shake by mechanical means for 10 minutes, dilute with 1 *N* sodium hydroxide to volume, and mix. Filter a portion of this solution, discarding the first 20 mL of the filtrate. Transfer 4.0 mL of the filtrate to a 250-mL volumetric flask, dilute with 0.1 *N* hydrochloric acid to volume, and mix. Dissolve an accurately weighed quantity of USP Sulfamethizole RS in 0.1 *N* hydrochloric acid, and dilute quantitatively and stepwise with the same medium to obtain a Standard solution having a known concentration of about 30 µg per mL. Transfer 2.0 mL each of the solution from the Capsules, the Standard solution, and 0.1 *N* hydrochloric acid to provide the blank, to separate 25-mL volumetric flasks, add 6.0 mL of a 1 in 50 solution of *p*-dimethylaminobenzaldehyde in alcohol, dilute with 0.1 *N* hydrochloric acid to volume, mix, and allow to stand for 20 minutes. Concomitantly determine the absorbances of the solutions from the Capsules and the Standard solution in 1-cm cells at the wavelength of maximum absorbance at about 450 nm, with a suitable spectrophotometer, using the blank to set the instrument. Calculate the quantity, in mg, of $C_9H_{10}N_4O_2S_2$ in the portion of the combined contents of the Capsules taken by the formula:

$$12.5(C)(A_U/A_S),$$

in which C is the concentration, in µg per mL, of USP Sulfamethizole RS in the Standard solution, and A_U and A_S are the absorbances of the solution from the Capsules and the Standard solution, respectively.

Oxytetracycline Hydrochloride and Polymyxin B Sulfate Ointment

» Oxytetracycline Hydrochloride and Polymyxin B Sulfate Ointment contains the equivalent of not less than 90.0 percent and not more than 120.0 percent of the labeled amount of oxytetracycline ($C_{22}H_{24}N_2O_9$), and not less than 90.0 percent and not more than 125.0 percent of the labeled amount of polymyxin B.

Packaging and storage—Preserve in collapsible tubes, or in well-closed, light-resistant containers.

Reference standards—*USP Oxytetracycline Reference Standard*—Do not dry before using. *USP Polymyxin B Sulfate Reference Standard*—Dry in vacuum at a pressure not exceeding 5 mm of mercury at 60° for 3 hours before using.

Minimum fill ⟨755⟩: meets the requirements.

Water, *Method I* ⟨921⟩: not more than 1.0%, 10 mL of a mixture of carbon tetrachloride, chloroform, and methanol (2:2:1) being used in place of methanol in the titration vessel.

Assay for oxytetracycline—Proceed with Oxytetracycline Hydrochloride and Polymyxin B Sulfate Ointment as directed in the

Assay for oxytetracycline under *Oxytetracycline Hydrochloride and Hydrocortisone Ointment.*

Assay for polymyxin B—Transfer an accurately weighed quantity of Oxytetracycline Hydrochloride and Polymyxin B Sulfate Ointment, equivalent to about 10,000 USP Polymyxin B Units, to a 15-mL centrifuge tube, add 10 mL of ether, stir, and centrifuge for 10 minutes. Decant and discard the clear ether. Wash the residue with 10 mL of ether, centrifuge for 10 minutes, decanting and discarding the clear ether. Wash the residue with several 10-mL portions of acetone, centrifuging, decanting, and discarding each washing until the yellow color is removed from the residue. [NOTE—Take care not to remove any of the residue with the washings.] Add 0.2 mL of polysorbate 80 to the residue, and mix. Transfer the mixture to a 100-mL volumetric flask with the aid of *Buffer No. 6*, dilute with the same solvent to volume, and mix. Proceed as directed for polymyxin B under *Antibiotics—Microbial Assays* ⟨81⟩, using an accurately measured volume of this solution diluted quantitatively with *Buffer No. 6* to yield a *Test Dilution* having a concentration of polymyxin B assumed to be equal to the median dose level of the Standard.

Oxytetracycline Hydrochloride and Polymyxin B Sulfate Ophthalmic Ointment

» Oxytetracycline Hydrochloride and Polymyxin B Sulfate Ophthalmic Ointment is a sterile ointment containing Oxytetracycline Hydrochloride and Polymyxin B Sulfate. It contains the equivalent of not less than 90.0 percent and not more than 120.0 percent of the labeled amount of oxytetracycline, and not less than 90.0 percent and not more than 125.0 percent of the labeled amount of polymyxin B.

Packaging and storage—Preserve in collapsible ophthalmic ointment tubes.

Reference standards—*USP Oxytetracycline Reference Standard*—Do not dry before using. *USP Polymyxin B Sulfate Reference Standard*—Dry in vacuum at a pressure not exceeding 5 mm of mercury at 60° for 3 hours before using.

Sterility—It meets the requirements for *Ophthalmic Ointments* under *Sterility Tests* ⟨71⟩.

Minimum fill ⟨755⟩: meets the requirements.

Water, *Method I* ⟨921⟩: not more than 1.0%, 20 mL of a mixture of carbon tetrachloride, chloroform, and methanol (2:2:1) being used in place of methanol in the titration vessel.

Metal particles—It meets the requirements of the test for *Metal Particles in Ophthalmic Ointments* ⟨751⟩.

Assay for oxytetracycline—Proceed with Oxytetracycline Hydrochloride and Polymyxin B Sulfate Ophthalmic Ointment as directed in the *Assay for oxytetracycline* under *Oxytetracycline Hydrochloride and Hydrocortisone Ointment.*

Assay for polymyxin B—Proceed with Oxytetracycline Hydrochloride and Polymyxin B Sulfate Ophthalmic Ointment as directed in the *Assay for polymyxin B* under *Oxytetracycline Hydrochloride and Polymyxin B Sulfate Ointment.*

Oxytetracycline Hydrochloride and Polymyxin B Sulfate Topical Powder

» Oxytetracycline Hydrochloride and Polymyxin B Sulfate Topical Powder contains the equivalent of not less than 90.0 percent and not more than 120.0 percent of the labeled amounts of oxytetracycline ($C_{22}H_{24}N_2O_9$) and polymyxin B in a suitable, fine powder base.

Packaging and storage—Preserve in well-closed containers.

Reference standards—*USP Oxytetracycline Reference Standard*—Do not dry before using. *USP Polymyxin B Sulfate Reference Standard*—Dry in vacuum at a pressure not exceeding 5 mm of mercury at 60° for 3 hours before using.

Minimum fill ⟨755⟩: meets the requirements.

Loss on drying ⟨731⟩—Dry about 100 mg, accurately weighed, in a capillary-stoppered bottle in vacuum at a pressure not exceeding 5 mm of mercury at 60° for 3 hours: it loses not more than 2.0% of its weight.

Assay for oxytetracycline—Transfer a suitable, accurately weighed quantity of Oxytetracycline Hydrochloride and Polymyxin B Sulfate Topical Powder to a glass blender jar containing a sufficient, accurately measured volume of 0.1 *N* hydrochloric acid to yield a stock solution containing not less than 150 µg of oxytetracycline per mL, and blend at high speed for 3 to 5 minutes. Proceed as directed for oxytetracycline under *Antibiotics—Microbial Assays* ⟨81⟩, using an accurately measured volume of this solution diluted quantitatively and stepwise with water to obtain a *Test Dilution* having a concentration of oxytetracycline assumed to be equal to the median dose level of the Standard.

Assay for polymyxin B—Transfer an accurately weighed quantity of Oxytetracycline Hydrochloride and Polymyxin B Sulfate Topical Powder, equivalent to about 10,000 USP Polymyxin B Units, to a 50-mL centrifuge tube, add 15 mL of acetone and 0.05 mL of hydrochloric acid, and stir. Add 20 mL of acetone, and centrifuge for 10 minutes. Decant and discard the clear liquid. Add 15 mL of acetone and 0.05 mL of hydrochloric acid to the residue, stir, add 20 mL of acetone, and centrifuge for 10 minutes, decanting and discarding the clear liquid. [NOTE—Take care not to discard any of the residue with the clear liquid.] Add 5 mL of *Buffer No. 6* to the residue, and mix. Transfer the mixture to a 100-mL volumetric flask with the aid of *Buffer No. 6*, dilute with the same solvent to volume, and mix. Proceed as directed for polymyxin B under *Antibiotics—Microbial Assays* ⟨81⟩, using an accurately measured volume of this stock solution diluted quantitatively with *Buffer No. 6* to obtain a *Test Dilution* having a concentration of polymyxin B assumed to be equal to the median dose level of the Standard.

Oxytetracycline Hydrochloride and Polymyxin B Sulfate Vaginal Tablets

» Oxytetracycline Hydrochloride and Polymyxin B Sulfate Vaginal Tablets contain the equivalent of not less than 90.0 percent and not more than 120.0 percent of the labeled amounts of oxytetracycline ($C_{22}H_{24}N_2O_9$) and polymyxin B.

Packaging and storage—Preserve in well-closed containers.

Reference standards—*USP Oxytetracycline Reference Standard*—Do not dry before using. *USP Polymyxin B Sulfate Reference Standard*—Dry in vacuum at a pressure not exceeding 5 mm of mercury at 60° for 3 hours before using.

Loss on drying ⟨731⟩—Dry about 100 mg, accurately weighed, of powdered Vaginal Tablets in a capillary-stoppered bottle in vacuum at a pressure not exceeding 5 mm of mercury at 60° for 3 hours: it loses not more than 3.0% of its weight.

Assay for oxytetracycline—Place not less than 5 Oxytetracycline Hydrochloride and Polymyxin B Sulfate Vaginal Tablets in a high-speed blender jar containing an accurately measured volume of 0.1 *N* hydrochloric acid, so that the stock solution obtained after blending for 3 to 5 minutes contains not less than 150 µg of oxytetracycline per mL. Proceed as directed for oxytetracycline under *Antibiotics—Microbial Assays* ⟨81⟩, using an accurately measured volume of this solution diluted quantitatively and stepwise with water to obtain a *Test Dilution* having a concentration of oxytetracycline assumed to be equal to the median dose level of the Standard.

Assay for polymyxin B—Weigh and finely powder not less than 5 Oxytetracycline Hydrochloride and Polymyxin B Sulfate Vaginal Tablets. Transfer an accurately weighed portion of the powder, equivalent to about 100,000 USP Polymyxin B Units, to a filter funnel equipped with a solvent-resistant membrane filter (1-μm or finer porosity). Wash the powder with five 20-mL portions of acetone, applying vacuum, and discarding the accumulated filtrate. [NOTE—If necessary, wash the powder with additional portions of acetone to remove any yellow color.] Carefully transfer the filter and the washed powder to a beaker containing about 400 mL of *Buffer No. 6*, and stir. Transfer the contents of the beaker to a 500-mL volumetric flask with the aid of *Buffer No. 6*, dilute with the same solvent to volume, and mix. Proceed as directed for polymyxin B under *Antibiotics—Microbial Assays* ⟨81⟩, using an accurately measured volume of this solution diluted quantitatively and stepwise with *Buffer No. 6* to yield a *Test Dilution* having a concentration of polymyxin B assumed to be equal to the median dose level of the Standard.

Oxytocin Injection

H-Cys-Tyr-Ile-Glu(NH_2)-Asp(NH_2)-Cys-Pro-Leu-Gly-NH_2
1 2 3 4 5 6 7 8 9

$C_{43}H_{66}N_{12}O_{12}S_2$ 1007.19
Oxytocin.
Oxytocin [*50-56-6*].

» Oxytocin Injection is a sterile solution, in a suitable diluent, of material containing the polypeptide hormone having the property of causing the contraction of uterine, vascular, and other smooth muscle, which is prepared by synthesis or obtained from the posterior lobe of the pituitary of healthy, domestic animals used for food by man. Each mL of Oxytocin Injection possesses an oxytocic activity of not less than 85.0 percent and not more than 120.0 percent of that stated on the label in USP Posterior Pituitary Units.

Packaging and storage—Preserve in single-dose or in multiple-dose containers, preferably of Type I glass. Do not freeze.

Reference standard—*USP Posterior Pituitary Reference Standard*—Do not dry before using. Store at a temperature of 0° or below. Each mg represents 2.4 USP Posterior Pituitary Units of oxytocic activity and 2.1 USP Posterior Pituitary Units of vasopressor activity.

pH ⟨791⟩: between 2.5 and 4.5.

Particulate matter ⟨788⟩: meets the requirements under *Small-volume Injections*.

Pressor activity—

Standard preparation—Prepare a stock solution as directed in the first paragraph under *Standard preparation* in the *Assay*.

On the day of the test, dilute the stock solution by adding sufficient saline TS to a measured volume of it so that the resulting *Standard preparation* contains a known potency of pressor activity of 0.1 USP Posterior Pituitary Unit per mL.

Test preparation—Dilute the Injection to correspond in pressor activity to the *Standard preparation*, on the assumption that the level of pressor activity of the Injection is not greater than 0.01 USP Posterior Pituitary Unit for each USP Unit of oxytocic activity.

The animal—Prepare as directed for *The animal* in the *Assay* under *Vasopressin Injection*.

Procedure—Determine by trial the three doses of the diluted *Standard preparation* that, when injected intravenously at regular intervals of 12 to 15 minutes, will produce consistent and graded blood-pressure elevations of between 20 and 70 mm of mercury. Repeat the intermediate dose: the second response does not differ from the first by more than ±10%. Inject a series of five doses, of which three are of the determined intermediate dose of the diluted *Standard preparation*, including the last repeated dose in the determination above, alternated with two doses of the same volume of the *Test preparation*. Measure the change in blood pressure following each of the five doses. Compute the difference between each response to the dose of the *Test preparation* and the mean of the associated doses of the *Standard preparation* that immediately precede and follow it. The average of the differences is such that the pressor response to the *Test preparation* is not greater than that of the dilution of the *Standard preparation*.

Other requirements—It meets the requirements under *Injections* ⟨1⟩.

Assay—

Standard preparation—Place 50 to 100 mg of USP Posterior Pituitary RS, accurately weighed, in a 250-mL flask, and add 0.5 mL of 0.045 *N* acetic acid for each unit of oxytocic activity represented by the weight of Reference Standard taken. Insert a small, short-stemmed funnel in the neck of the flask. Heat the flask and contents, with continuous, gentle shaking, in a boiling water bath for 5 minutes. Cool the flask under cold running water, and filter. Each mL of the filtrate contains potency, in USP Posterior Pituitary Units, equivalent to 2 USP Units of oxytocic activity and 1.75 USP Units of vasopressin activity. Place this stock solution in hard-glass ampuls, seal, and heat in flowing steam for 20 minutes on each of three successive days. Store the solution in a refrigerator and use within 6 months. On the day of the assay, prepare a primary test dilution by adding sufficient saline TS to a measured volume of the stock solution so that the resulting *Standard preparation* contains a known oxytocic activity not greater than 0.4 USP Posterior Pituitary Unit in each mL, the concentration being indicated by the *Determination of the sensitivity of the animal preparation*.

Assay preparation—To an accurately measured quantity of Oxytocin Injection add sufficient saline TS so that each mL of the resulting *Assay preparation* is expected to have the same potency as the *Standard preparation*.

The animal—Select a young adult domestic chicken weighing between 1 and 2.5 kg. Anesthetize the chicken deeply enough to prevent voluntary muscular movements, using an anesthetic substance that favors the maintenance of a prolonged anesthesia and a uniform blood pressure. Expose an ischiatic artery by careful dissection, and cannulate it after occluding the afferent branches for a distance of about 2 cm. Arrange to obtain a continuous record of the blood pressure. Prepare to inject the test dilutions intravenously by exposing a crural vein or by tying back one of the wings and exposing the brachial vein.

Determination of the sensitivity of the animal preparation—Determine by trial injections the quantity of Standard that elicits a precipitous but evanescent decrease in blood pressure of 20 to 40 mm of mercury. If necessary at the outset or during the assay, alter the concentration of the *Standard preparation* so that a dose of not less than 0.15 mL and not more than 0.5 mL of it elicits a response in the indicated range. If severe tachyphylaxis occurs, as shown by a rapidly decreasing response to successive injections of the Standard, discard the chicken as unsuitable for use in the assay.

Procedure—Inject the selected dose of the *Standard preparation*, and repeat at regular intervals of 3 to 10 minutes thereafter. At the mid-point of the time interval between successive, equal doses of the *Standard preparation*, inject one of two doses of the *Assay preparation* so selected that their geometric mean corresponds approximately to the dose of the *Standard preparation* and the lesser of the two doses is expected to produce a response approximately four-fifths of that of the *Standard preparation*. Inject three or more sets of doses, each set consisting of the low and high levels of the *Assay preparation* and the alternating doses of the *Standard preparation*, randomizing the order of the low and high doses within each set. Keep the dose of the standard constant as long as the recorded blood-pressure decreases are between 20 and 40 mm of mercury, but if necessary, because of changes in sensitivity of the animal, alter the dose appropriately between sets to keep the responses within this range, starting the next set with the changed dose of the standard. Whenever the dose of the *Standard preparation* is altered, adjust both doses of the *Assay preparation* so as to maintain the same proportionate relationship between the low and high levels (i.e., a constant log-interval, *i*) throughout the assay. Measure and record the responses observed to the nearest mm of mercury.

Calculation—Tabulate, in chronological order, the logarithms of the doses of the *Standard preparation* and of the *Assay preparation* and the respective responses thereto. Average the response to the standard that immediately precedes and the response to the standard that immediately follows each response to the *Assay preparation*, and subtract each average from the intervening response. Designate the difference in response as y. Compute similarly the corresponding difference, x, in the respective log-doses and compute the average, $\bar{x}$, of all the values of x. Total the differences, y, to obtain T_1 and T_2 for the low dose and high dose, respectively, of the *Assay preparation*.

The logarithm of the relative potency of the *Assay preparation* is:

$$M' = \{i(T_1 + T_2)/2(T_2 - T_1)\} - \bar{x}.$$

The potency of the Injection, in USP Units of oxytocic activity per mL, is R(antilog M'), where $R = v_S/v_U$, in which v_S is the number of USP Units of oxytocic activity in each mL of the *Standard preparation*, and v_U is the volume, in mL, of the Injection in each mL of the *Assay preparation*.

Compute the approximate log confidence interval, L (see *Design and Analysis of Biological Assays* ⟨111⟩). If L exceeds 0.20, repeat the assay or increase the number of sets of responses until the approximate confidence interval of the combined results is 0.20 or less.

Oxytocin Nasal Solution

» Oxytocin Nasal Solution is a solution, in a suitable diluent, of the polypeptide hormone, prepared synthetically, which has the property of causing the contraction of uterine, vascular, and other smooth muscle, and which is present in the posterior lobe of the pituitary of healthy, domestic animals used for food by man. It contains suitable preservatives, and is packaged in a form suitable for nasal administration. Each mL of Oxytocin Nasal Solution possesses an oxytocic activity of not less than 85.0 percent and not more than 120.0 percent of that stated on the label in USP Posterior Pituitary Units.

Packaging and storage—Preserve in containers suitable for administering the contents by spraying into the nasal cavities with the patient in the upright position, or for instillation in drop form.

Reference standard—*USP Posterior Pituitary Reference Standard*—Do not dry before using. Store at a temperature of 0° or below. Each mg represents 2.4 USP Posterior Pituitary Units of oxytocic activity and 2.1 USP Posterior Pituitary Units of vasopressor activity.

Labeling—Label it to indicate that it is for intranasal administration only.

pH ⟨791⟩: between 3.7 and 4.3.

Pressor activity—Proceed with Oxytocin Nasal Solution as directed in the test for *Pressor activity* under *Oxytocin Injection*.

Assay—Proceed with Oxytocin Nasal Solution as directed in the *Assay* under *Oxytocin Injection*.

Padimate O

$COOCH_2CHCH_2CH_2CH_2CH_3$ / C_2H_5 / $N(CH_3)_2$

$C_{17}H_{27}NO_2$ 277.41

Benzoic acid, 4-(dimethylamino)-, 2-ethylhexyl ester.

2-Ethylhexyl *p*-(dimethylamino)benzoate [*21245-02-3*].

» Padimate O contains not less than 97.0 percent and not more than 103.8 percent of $C_{17}H_{27}NO_2$.

Packaging and storage—Preserve in tight, light-resistant containers.

Reference standard—*USP Padimate O Reference Standard*—Do not dry; determine the water content titrimetrically, for quantitative application, at the time of use.

Identification—

A: The infrared absorption spectrum of a thin film of it exhibits maxima only at the same wavelengths as that of a similar preparation of USP Padimate O RS.

B: The ultraviolet absorption spectrum of a 1 in 200,000 solution in alcohol exhibits maxima and minima at the same wavelengths as that of a similar solution of USP Padimate O RS, concomitantly measured, and the respective absorptivities at the wavelength of maximum absorbance at about 312 nm do not differ by more than 4.0%.

Specific gravity ⟨841⟩: between 0.990 and 1.000.

Refractive index ⟨831⟩: between 1.5390 and 1.5430.

Acid value ⟨401⟩: not more than 1.0.

Saponification value ⟨401⟩: between 195 and 215, reflux temperature being maintained for 4 hours.

Chromatographic purity—

Chromatographic system (see *Chromatography* ⟨621⟩)—The gas chromatograph is equipped with a flame-ionization detector and a 1.8-m × 3-mm stainless steel column packed with 10 percent liquid phase G9 on support S1A. The column temperature is programmed at a rate of 10° per minute from 150° to 250°, then maintained at 250° for 10 minutes, and helium is used as the carrier gas.

Procedure—Chromatograph 2 µL of a 1 in 100 solution of Padimate O in chloroform: the response due to padimate O is not less than 98.0% of the sum of the responses on the chromatogram, exclusive of the chloroform peak.

Assay—Dissolve about 500 mg of Padimate O, accurately weighed, in 75 mL of acetic anhydride, and titrate with 0.1 *N* perchloric acid VS, determining the end-point potentiometrically. Each mL of 0.1 *N* perchloric acid is equivalent to 27.74 mg of $C_{17}H_{27}NO_2$.

Padimate O Lotion

» Padimate O Lotion contains not less than 90.0 percent and not more than 110.0 percent of the labeled amount of $C_{17}H_{27}NO_2$.

Packaging and storage—Preserve in tight, light-resistant containers.

Reference standard—*USP Padimate O Reference Standard*—Do not dry; determine the water content titrimetrically, for quantitative application, at the time of use.

Identification—The retention time of the major peak in the chromatogram of the *Assay preparation* corresponds to that in the chromatogram of the *Standard preparation*, as obtained in the *Assay*.

Assay—

Mobile phase—Prepare a suitable filtered and degassed solution containing methanol, water, and glacial acetic acid (85:15:0.5).

Standard preparation—Dissolve an accurately weighed quantity of USP Padimate O RS in isopropyl alcohol and dilute quantitatively, and stepwise if necessary, with isopropyl alcohol to obtain a solution having a known concentration of about 100 µg per mL.

Assay preparation—Transfer an accurately weighed quantity of Padimate O Lotion, equivalent to about 100 mg of Padimate O, to a 100-mL volumetric flask, and add about 75 mL of isopropyl alcohol. Heat gently with swirling until the specimen is dispersed. Cool to room temperature, dilute with isopropyl alcohol to volume, and mix. Pipet 10.0 mL of this solution into a

100-mL volumetric flask, dilute with isopropyl alcohol to volume, and mix.

Chromatographic system (see *Chromatography* ⟨621⟩)—The liquid chromatograph is equipped with a 308-nm detector and a 4.6-mm × 25-cm column that contains 5-μm *base-deactivated* packing L1. The flow rate is about 1.5 mL per minute. Chromatograph the *Standard preparation*, and record the peak responses as directed under *Procedure:* the tailing factor is not more than 2.5 and the relative standard deviation for replicate injections is not more than 2.0%.

Procedure—Separately inject equal volumes (about 10 μL) of the *Standard preparation* and the *Assay preparation* into the chromatograph, record the chromatograms, and measure the responses for the major peaks. Calculate the quantity, in mg, of $C_{17}H_{27}NO_2$ in the portion of Lotion taken by the formula:

$$C(r_U / r_S),$$

in which C is the concentration, in μg per mL, of USP Padimate O RS in the *Standard preparation*, and r_U and r_S are the peak responses for padimate O obtained from the *Assay preparation* and the *Standard preparation*, respectively.

Pancreatin

Pancreatin.
Pancreatin [*8049-47-6*].

» Pancreatin is a substance containing enzymes, principally amylase, lipase, and protease, obtained from the pancreas of the hog, *Sus scrofa* Linné var. *domesticus* Gray (Fam. Suidae) or of the ox, *Bos taurus* Linné (Fam. Bovidae). Pancreatin contains, in each mg, not less than 25 USP Units of amylase activity, not less than 2.0 USP Units of lipase activity, and not less than 25 USP Units of protease activity. Pancreatin of a higher digestive power may be labeled as a whole-number multiple of the three minimum activities or may be diluted by admixture with lactose, or with sucrose containing not more than 3.25 percent of starch, or with pancreatin of lower digestive power.

NOTE—One USP Unit of amylase activity is contained in the amount of pancreatin that decomposes starch at an initial rate such that one microequivalent of glucosidic linkage is hydrolyzed per minute under the conditions of the *Assay for amylase activity*.

One USP Unit of lipase activity is contained in the amount of pancreatin that liberates 1.0 μEq of acid per minute at a pH of 9.0 and 37° under the conditions of the *Assay for lipase activity*.

One USP Unit of protease activity is contained in the amount of pancreatin that digests 1.0 mg of casein under the conditions of the *Assay for protease activity*.

Packaging and storage—Preserve in tight containers, at a temperature not exceeding 30°.

Reference standards—*USP Sodium Taurocholate Reference Standard*—Keep container tightly closed. [*Caution—Avoid inhaling airborne particles.*] Dry at 105° for 4 hours before using. *USP Pancreatin Reference Standard*—Keep container tightly closed, and store in a refrigerator. Do not open while cold, and do not dry before using.

Microbial limit—It meets the requirements of the test for absence of *Salmonella* species under *Microbial Limit Tests* ⟨61⟩.

Loss on drying ⟨731⟩—Dry it in vacuum at 60° for 4 hours: it loses not more than 5.0% of its weight.

Fat—Place 2.0 g of Pancreatin in a flask of about 50-mL capacity, add 20 mL of ether, insert the stopper, and set it aside for 2 hours, mixing by rotating at frequent intervals. Decant the supernatant ether by means of a guiding rod into a plain filter about 7 cm in diameter, previously moistened with ether, and collect the filtrate in a tared beaker. Repeat the extraction with a 10-mL portion of ether, proceeding as directed before, then with another 10-mL portion of ether, and transfer the ether and the remainder of the Pancreatin to the filter. Allow to drain, evaporate the ether spontaneously, and dry the residue at 105° for 2 hours: the residue of fat obtained from Pancreatin possessing three or more times the three minimum activities weighs not more than 120 mg (6.0%); that obtained from Pancreatin possessing less than three times the three minimum activities weighs not more than 60 mg (3.0%).

Assay for amylase activity (*Starch digestive power*)—

pH 6.8 phosphate buffer—On the day of use, dissolve 13.6 g of monobasic potassium phosphate in water to make 500 mL of solution. Dissolve 14.2 g of anhydrous dibasic sodium phosphate in water to make 500 mL of solution. Mix 51 mL of the monobasic potassium phosphate solution with 49 mL of the dibasic sodium phosphate solution. If necessary, adjust by the dropwise addition of the appropriate solution to a pH of 6.8.

Substrate solution—On the day of use, stir a portion of purified soluble starch equivalent to 2.0 g of dried substance with 10 mL of water, and add this mixture to 160 mL of boiling water. Rinse the beaker with 10 mL of water, add it to the hot solution, and heat to boiling, with continuous mixing. Cool to room temperature, and add water to make 200 mL.

Standard preparation—Weigh accurately about 20 mg of USP Pancreatin RS into a suitable mortar. Add about 30 mL of *pH 6.8 phosphate buffer*, and triturate for 5 to 10 minutes. Transfer the mixture with the aid of *pH 6.8 phosphate buffer* to a 50-mL volumetric flask, dilute with *pH 6.8 phosphate buffer* to volume, and mix. Calculate the activity, in USP Units of amylase activity per mL, of the resulting solution from the declared potency on the label of the Reference Standard.

Assay preparation—For Pancreatin having about the same amylase activity as the USP Pancreatin RS, weigh accurately about 40 mg of Pancreatin into a suitable mortar. [NOTE—For Pancreatin having a different amylase activity, weigh accurately the amount necessary to obtain an *Assay preparation* having amylase activity per mL corresponding approximately to that of the *Standard preparation*.] Add about 3 mL of *pH 6.8 phosphate buffer*, and triturate for 5 to 10 minutes. Transfer the mixture with the aid of *pH 6.8 phosphate buffer* to a 100-mL volumetric flask, dilute with *pH 6.8 phosphate buffer* to volume, and mix.

Procedure—Prepare four stoppered, 250-mL conical flasks, and mark them *S*, *U*, *BS*, and *BU*. Pipet into each flask 25 mL of *Substrate solution*, 10 mL of *pH 6.8 phosphate buffer*, and 1 mL of sodium chloride solution (11.7 in 1000), insert the stoppers, and mix. Place the flasks in a water bath maintained at 25 ± 0.1°, and allow them to equilibrate. To flasks *BU* and *BS* add 2 mL of 1 *N* hydrochloric acid, mix, and return the flasks to the water bath. To flasks *U* and *BU* add 1.0-mL portions of the *Assay preparation*, and to flasks *S* and *BS* add 1.0 mL of the *Standard preparation*. Mix each, and return the flasks to the water bath. After 10 minutes, accurately timed from the addition of the enzyme, add 2-mL portions of 1 *N* hydrochloric acid to flasks *S* and *U*, and mix. To each flask, with continuous stirring, add 10.0 mL of 0.1 *N* iodine VS, and immediately add 45 mL of 0.1 *N* sodium hydroxide. Place the flasks in the dark at a temperature between 15° and 25° for 15 minutes. To each flask add 4 mL of 2 *N* sulfuric acid, and titrate with 0.1 *N* sodium thiosulfate VS to the disappearance of the blue color. Calculate the amylase activity, in USP Units per mg, taken by the formula:

$$100(C_S / W_U)(V_{BU} - V_U)/(V_{BS} - V_S),$$

in which C_S is the amylase activity of the *Standard preparation*, in USP Units per mL, W_U is the amount, in mg, of Pancreatin taken, and V_U, V_S, V_{BU}, and V_{BS} are the volumes, in mL, of 0.1 *N* sodium thiosulfate consumed in the titration of the solutions in flasks *U*, *S*, *BU*, and *BS*, respectively.

Assay for lipase activity (*Fat digestive power*)—

Acacia solution—Centrifuge a solution of acacia (1 in 10) until clear. Use only the clear solution.

Olive oil substrate—Combine 165 mL of *Acacia solution*, 20 mL of olive oil, and 15 g of crushed ice in the cup of an electric blender. Cool the mixture in an ice bath to 5°, and homogenize at high speed for 15 minutes, intermittently cooling in an ice bath to prevent the temperature from exceeding 30°.

Test for suitability of mixing as follows: Place a drop of the homogenate on a microscope slide, and gently press a cover slide in place to spread the liquid. Examine the entire field under high power (43× objective lens and 5× ocular), using an eyepiece equipped with a calibrated micrometer. The substrate is satisfactory if 90% of the particles do not exceed 2 μm in diameter and none exceeds 10 μm in diameter.

Buffer solution—Dissolve 60 mg of tris(hydroxymethyl)aminomethane and 234 mg of sodium chloride in water to make 100 mL.

Sodium taurocholate solution—Prepare a solution to contain 80.0 mg of USP Sodium Taurocholate RS in each mL.

Standard test dilution—Suspend about 200 mg of USP Pancreatin RS, accurately weighed, in about 3 mL of cold water in a mortar, triturate for 10 minutes, and add cold water to a volume necessary to produce a concentration of 8 to 16 USP Units of lipase activity per mL, based upon the declared potency on the label of the Reference Standard. Maintain the suspension at 4°, and mix before using. For each determination withdraw 5 to 10 mL of the cold suspension, and allow the temperature to rise to 20° before pipeting the exact volume.

Assay test dilution—Suspend about 200 mg of Pancreatin, accurately weighed, in about 3 mL of cold water in a mortar, triturate for 10 minutes, and add cold water to a volume necessary to produce a concentration of 8 to 16 USP Units of lipase activity per mL, based upon the estimated potency of the test material. Maintain the suspension at 4°, and mix before using. For each determination withdraw 5 to 10 mL of the cold suspension, and allow the temperature to rise to 20° before pipeting the exact volume.

Procedure—Mix 10.0 mL of *Olive oil substrate*, 8.0 mL of *Buffer solution*, 2.0 mL of *Sodium taurocholate solution*, and 9.0 mL of water in a jacketed glass vessel of about 50-mL capacity, the outer chamber of which is connected to a thermostatically controlled water bath. Cover the mixture, and stir continuously with a mechanical stirring device. With the mixture maintained at a temperature of 37 ± 0.1°, add 0.1 *N* sodium hydroxide VS, from a microburet inserted through an opening in the cover, and adjust to a pH of 9.20 potentiometrically using a calomel-glass electrode system. Add 1.0 mL of the *Assay test dilution*, and then continue adding the 0.1 *N* sodium hydroxide VS for 5 minutes to maintain the pH at 9.0. Determine the volume of 0.1 *N* sodium hydroxide VS added after each minute.

In the same manner, titrate 1.0 mL of *Standard test dilution*.

Calculation of potency—Plot the volume of 0.1 *N* sodium hydroxide VS titrated against time. Using only the points which fall on the straight-line segment of the curve, calculate the mean acidity released per minute by the sample and Standard. Taking into consideration dilution factors, calculate the lipase activity of the assay preparation in USP Units by comparison to the activity of the Standard, using the lipase activity of the USP Pancreatin RS stated on the label.

Assay for protease activity (*Casein digestive power*)—

Casein substrate—Place 1.25 g of finely powdered casein in a 100-mL conical flask containing 5 mL of water, shake to form a suspension, add 10 mL of 0.1 *N* sodium hydroxide, shake for 1 minute, add 50 mL of water, and shake for about 1 hour to dissolve the casein. The resulting solution should have a pH of about 8. If necessary, adjust the pH to about 8, using 1 *N* sodium hydroxide or 1 *N* hydrochloric acid. Transfer the solution to a 100-mL volumetric flask, dilute with water to volume, and mix. Use this substrate on the day it is prepared.

Buffer solution—Dissolve 6.8 g of monobasic potassium phosphate and 1.8 g of sodium hydroxide in 950 mL of water in a 1000-mL volumetric flask, adjust to a pH of 7.5 ± 0.2, using 0.2 *N* sodium hydroxide, dilute with water to volume, and mix. Store this solution in a refrigerator.

Trichloroacetic acid solution—Dissolve 50 g of trichloroacetic acid in 1000 mL of water. Store this solution at room temperature.

Filter paper—Determine the suitability of the filter paper by filtering a 5-mL portion of *Trichloroacetic acid solution* through the paper and measuring the absorbance of the filtrate at 280 nm, using an unfiltered portion of the same *Trichloroacetic acid solution* as the blank: the absorbance is not more than 0.04. If the absorbance is more than 0.04, the filter paper may be washed repeatedly with *Trichloroacetic acid solution* until the absorbance of the filtrate, determined as above, is not more than 0.04.

Standard test dilution—Add about 100 mg of USP Pancreatin RS, accurately weighed, to 100.0 mL of *Buffer solution*, and mix by shaking intermittently at room temperature for about 25 minutes. Dilute quantitatively with *Buffer solution* to produce a concentration of about 2.5 USP Units of protease activity per mL, based upon the potency declared on the label of the Reference Standard.

Assay test dilution—Weigh accurately about 100 mg of Pancreatin into a suitable mortar. Add about 3 mL of *Buffer solution*, and triturate for 5 to 10 minutes. Transfer the mixture with the aid of *Buffer solution* to a 100-mL volumetric flask, dilute with *Buffer solution* to volume, and mix. Dilute quantitatively with *Buffer solution* to obtain a dilution that corresponds in activity to the *Standard test dilution*.

Procedure—Label test tubes in duplicate S_1, S_2, and S_3 for the standard series, and U for the sample. Pipet into tubes S_1 2.0 mL, into S_2 and U 1.5 mL, and into S_3 1.0 mL of *Buffer solution*. Pipet into tubes S_1 1.0 mL, into S_2 1.5 mL, and into S_3 2.0 mL of the *Standard test dilution*. Pipet into tubes U 1.5 mL of the *Assay test dilution*. Pipet into one tube each of S_1, S_2, S_3, and U 5.0 mL of *Trichloroacetic acid solution*, and mix. Designate these tubes as S_{1B}, S_{2B}, S_{3B}, and U_B, respectively. Prepare a blank by mixing 3 mL of *Buffer solution* and 5 mL of *Trichloroacetic acid solution* in a separate test tube marked *B*. Place all the tubes in a 40° water bath, insert a glass stirring rod into each tube, and allow for temperature equilibration. At zero time, add to each tube, at timed intervals, 2.0 mL of the *Casein substrate*, preheated to the bath temperature, and mix. Sixty minutes, accurately timed, after the addition of the *Casein substrate* stop the reaction in tubes S_1, S_2, S_3, and U by adding 5.0 mL of *Trichloroacetic acid solution* at the corresponding time intervals, stir, and remove all the tubes from the bath. Allow to stand for 10 minutes at room temperature for complete protein precipitation, and filter. The filtrates must be free from haze. Determine the absorbances of the filtrates, in 1-cm cells, at 280 nm, with a suitable spectrophotometer, using the filtrate from the blank (tube *B*) to set the instrument.

Calculation of potency—Correct the absorbance values for the filtrates from tubes S_1, S_2, and S_3 by subtracting the absorbance values for the filtrates from tubes S_{1B}, S_{2B}, and S_{3B}, respectively, and plot the corrected absorbance values against the corresponding volumes of the *Standard test dilution* used. From the curve, using the corrected absorbance value ($U - U_B$) for the Pancreatin taken, and taking into consideration the dilution factors, calculate the protease activity, in USP Units, of the Pancreatin taken by comparison with that of the Standard, using the protease activity stated on the label of USP Pancreatin RS.

Pancreatin Capsules

» Pancreatin Capsules contain not less than 90.0 percent of the labeled amount of pancreatin.

Packaging and storage—Preserve in tight containers, preferably at a temperature not exceeding 30°.

Labeling—Label Capsules to indicate minimum pancreatin fat digestive power; i.e., single strength, double strength, triple strength.

Reference standards— *USP Sodium Taurocholate Reference Standard*—Keep container tightly closed. [*Caution—Avoid inhaling airborne particles.*] Dry at 105° for 4 hours before using. *USP Pancreatin Reference Standard*—Keep container tightly closed, and store in a refrigerator. Do not open while cold, and do not dry before using.

Microbial limit—Capsules meet the requirements of the test for absence of *Salmonella* species under *Microbial Limit Tests* ⟨61⟩.

Assay for amylase activity (*Starch digestive power*)—Transfer, as completely as possible, the contents of not less than 20 Pancreatin Capsules to a suitable tared container, and determine the average content weight per capsule. Finely powder and mix the combined contents, avoiding the production of heat during the process. Proceed as directed in the *Assay for amylase activity* under *Pancreatin*, using an accurately weighed portion of the powder, equivalent to 40 mg of pancreatin, as the assay preparation. (Use an inversely proportionate amount of the powder if the Capsules are labeled to contain a whole-number multiple of the minimum requirement for pancreatin digestive activity.)

Assay for lipase activity (*Fat digestive power*)—

Acacia solution, *Olive oil substrate*, *Buffer solution*, *Sodium taurocholate solution*, and *Standard test dilution*—Prepare as directed in the *Assay for lipase activity* under *Pancreatin*.

Assay test dilution—Proceed as directed for *Assay test dilution* in the *Assay for lipase activity* under *Pancreatin*, using as the assay preparation an accurately weighed portion of the powder, prepared as directed in the *Assay for amylase activity*, equivalent to about 200 mg of pancreatin.

Procedure and *Calculation of potency*—Proceed as directed in the *Assay for lipase activity* under *Pancreatin*.

Assay for protease activity (*Casein digestive power*)—

Casein substrate, *Buffer solution*, *Trichloroacetic acid solution*, *Filter paper*, and *Standard test dilution*—Prepare as directed in the *Assay for protease activity* under *Pancreatin*.

Assay test dilution—Add an accurately weighed portion of the powder, prepared as directed in the *Assay for amylase activity*, equivalent to about 100 mg of pancreatin, to 100.0 mL of *Buffer solution*, and mix by shaking intermittently at room temperature for 25 minutes. Dilute quantitatively with *Buffer solution* to obtain a dilution that corresponds in activity to the *Standard test dilution*.

Procedure and *Calculation of potency*—Proceed as directed in the *Assay for protease activity* under *Pancreatin*.

Pancreatin Tablets

» Pancreatin Tablets contain not less than 90.0 percent of the labeled amount of pancreatin.

Packaging and storage—Preserve in tight containers, preferably at a temperature not exceeding 30°.

Labeling—Label Tablets to indicate minimum pancreatin fat digestive power; i.e., single strength, double strength, triple strength.

Reference standards— *USP Sodium Taurocholate Reference Standard*—Keep container tightly closed. [*Caution—Avoid inhaling airborne particles.*] Dry at 105° for 4 hours before using. *USP Pancreatin Reference Standard*—Keep container tightly closed, and store in a refrigerator. Do not open while cold, and do not dry before using.

Microbial limit—Tablets meet the requirements of the test for absence of *Salmonella* species under *Microbial Limit Tests* ⟨61⟩.

Disintegration ⟨701⟩: 60 minutes.

Assay for amylase activity (*Starch digestive power*)—Weigh and finely powder not less than 20 Pancreatin Tablets, avoiding the production of heat during the process. Proceed as directed in the *Assay for amylase activity* under *Pancreatin*, using as the assay preparation an accurately weighed portion of the powder, equivalent to 40 mg of pancreatin. (Use an inversely proportionate amount of the powder if the Tablets are labeled to contain a whole-number multiple of the minimum requirement for pancreatin digestive activity.)

Assay for lipase activity (*Fat digestive power*)—

Acacia solution, *Olive oil substrate*, *Buffer solution*, *Sodium taurocholate solution*, and *Standard test dilution*—Prepare as directed in the *Assay for lipase activity* under *Pancreatin*.

Assay test dilution—Proceed as directed for *Assay test dilution* in the *Assay for lipase activity* under *Pancreatin*, using as the assay preparation an accurately weighed portion of the powder, prepared as directed in the *Assay for amylase activity*, equivalent to about 200 mg of pancreatin.

Procedure and *Calculation of potency*—Proceed as directed in the *Assay for lipase activity* under *Pancreatin*.

Assay for protease activity (*Casein digestive power*)—

Casein substrate, *Buffer solution*, *Trichloroacetic acid solution*, *Filter paper*, and *Standard test dilution*—Prepare as directed in the *Assay for protease activity* under *Pancreatin*.

Assay test dilution—Add an accurately weighed portion of the powder, prepared as directed in the *Assay for amylase activity*, equivalent to about 100 mg of pancreatin, to 100.0 mL of *Buffer solution*, and mix by shaking intermittently at room temperature for 25 minutes. Dilute quantitatively with *Buffer solution* to obtain a dilution that corresponds in activity to the *Standard test dilution*.

Procedure and *Calculation of potency*—Proceed as directed in the *Assay for protease activity* under *Pancreatin*.

Pancrelipase

» Pancrelipase is a substance containing enzymes, principally lipase, with amylase and protease, obtained from the pancreas of the hog, *Sus scrofa* Linné var. *domesticus* Gray (Fam. Suidae). It contains, in each mg, not less than 24 USP Units of lipase activity, not less than 100 USP Units of amylase activity, and not less than 100 USP Units of protease activity.

NOTE—One USP Unit of amylase activity is contained in the amount of pancrelipase that decomposes starch at an initial rate such that one microequivalent of glucosidic linkage is hydrolyzed per minute under the conditions of the *Assay for amylase activity*.

One USP Unit of lipase activity is contained in the amount of pancrelipase that liberates 1.0 μEq of acid per minute at pH 9.0 and 37° under the conditions of the *Assay for lipase activity*.

One USP Unit of protease activity is contained in the amount of pancrelipase that digests 1.0 mg of casein under the conditions of the *Assay for protease activity*.

Packaging and storage—Preserve in tight containers, preferably at a temperature not exceeding 25°.

Labeling—Label it to indicate lipase activity in USP Units.

Reference standards—*USP Sodium Taurocholate Reference Standard*—Keep container tightly closed. [*Caution—Avoid inhaling airborne particles.*] Dry at 105° for 4 hours before using. *USP Pancreatin Reference Standard*—Keep container tightly closed, and store in a refrigerator. Do not open while cold, and do not dry before using.

Microbial limits—It meets the requirements of the tests for absence of *Salmonella* species and *Escherichia coli* under *Microbial Limit Tests* ⟨61⟩.

Loss on drying ⟨731⟩—Dry it in vacuum at 60° for 4 hours: it loses not more than 5.0% of its weight.

Fat—Place 2.0 g of Pancrelipase in a flask of about 50-mL capacity, add 20 mL of ether, insert the stopper, and set aside for 2 hours, mixing by rotating at frequent intervals. Decant the supernatant ether by means of a guiding rod into a plain filter about 7 cm in diameter, previously moistened with ether, and collect the filtrate in a tared beaker. Repeat the extraction with a 10-mL portion of ether, then with another 10-mL portion of ether, transfer the ether and the remainder of the Pancrelipase to the filter. Allow to drain, evaporate the ether spontaneously, and dry the residue at 105° for 2 hours: the residue of fat weighs not more than 100 mg (5.0%).

Assay for amylase activity (*Starch digestive power*)—

pH 6.8 phosphate buffer—On the day of use, dissolve 13.6 g of monobasic potassium phosphate in water to make 500 mL of solution. Dissolve 14.2 g of anhydrous dibasic sodium phosphate in water to make 500 mL of solution. Mix 51 mL of the monobasic potassium phosphate solution with 49 mL of the dibasic sodium phosphate solution. If necessary, adjust by the dropwise addition of the appropriate solution to a pH of 6.8.

Substrate solution—On the day of use, stir a portion of purified soluble starch equivalent to 2.0 g of dried substance with 10 mL of water, and add this mixture to 160 mL of boiling water. Rinse the beaker with 10 mL of water, add it to the hot solution, and heat to boiling, with continuous mixing. Cool to room temperature, and add water to make 200 mL.

Standard preparation—Weigh accurately about 20 mg of USP Pancreatin RS into a suitable mortar. Add about 30 mL of *pH 6.8 phosphate buffer*, and triturate for 5 to 10 minutes. Transfer the mixture with the aid of *pH 6.8 phosphate buffer* to a 50-mL volumetric flask, dilute with *pH 6.8 phosphate buffer* to volume, and mix. Calculate the activity, in USP Units of amylase activity per mL, of the resulting solution from the declared potency on the label of the Reference Standard.

Assay preparation—For Pancrelipase having about 4 times the amylase activity of the USP Pancreatin RS, weigh accurately about 10 mg of Pancrelipase into a suitable mortar. [NOTE—For Pancrelipase having a different amylase activity, weigh accurately the amount necessary to obtain an *Assay preparation* having amylase activity per mL corresponding approximately to that of the *Standard preparation*.] Add about 3 mL of *pH 6.8 phosphate buffer*, and triturate for 5 to 10 minutes. Transfer the mixture with the aid of *pH 6.8 phosphate buffer* to a 100-mL volumetric flask, dilute with *pH 6.8 phosphate buffer* to volume, and mix.

Procedure—Prepare four stoppered, 250-mL conical flasks, and mark them *S*, *U*, *BS*, and *BU*. Pipet into each flask 25 mL of *Substrate solution*, 10 mL of *pH 6.8 phosphate buffer*, and 1 mL of sodium chloride solution (11.7 in 1000), insert the stoppers, and mix. Place the flasks in a water bath maintained at 25 ± 0.1°, and allow them to equilibrate. To flasks *BU* and *BS* add 2 mL of 1 *N* hydrochloric acid, mix, and return the flasks to the water bath. To flasks *U* and *BU* add 1.0-mL portions of *Assay preparation*, and to flasks *S* and *BS* add 1.0 mL of *Standard preparation*. Mix each, and return the flasks to the water bath. After 10 minutes, accurately timed from the addition of the enzyme, add 2-mL portions of 1 *N* hydrochloric acid to flasks *S* and *U*, and mix. To each flask, with continuous stirring, add 10.0 mL of 0.1 *N* iodine VS, and immediately add 45 mL of 0.1 *N* sodium hydroxide. Place the flasks in the dark at a temperature between 15° and 25° for 15 minutes. To each flask add 4 mL of 2 *N* sulfuric acid, and titrate with 0.1 *N* sodium thiosulfate VS to the disappearance of the blue color. Calculate the amylase activity, in USP Units per mg, taken by the formula:

$$100(C_S/W_U)(V_{BU} - V_U)/(V_{BS} - V_S),$$

in which C_S is the amylase activity of the *Standard preparation*, in USP Units per mL, W_U is the amount, in mg, of Pancrelipase taken, and V_U, V_S, V_{BU}, and V_{BS} are the volumes, in mL, of 0.1 *N* sodium thiosulfate consumed in the titration of the solutions in flasks *U*, *S*, *BU*, and *BS*, respectively.

Assay for lipase activity (*Fat digestive power*)—

Acacia solution—Centrifuge a solution of acacia (1 in 10) until clear. Use only the clear solution.

Olive oil substrate—Combine 165 mL of *Acacia solution*, 20 mL of olive oil, and 15 g of crushed ice in the cup of an electric blender. Cool the mixture in an ice bath to 5°, and homogenize at high speed for 15 minutes, intermittently cooling in an ice bath to prevent the temperature from exceeding 30°.

Test for suitability of mixing as follows. Place a drop of the homogenate on a microscope slide, and gently press a cover slide in place to spread the liquid. Examine the entire field under high power (43× objective lens and 5× ocular), using an eyepiece equipped with a calibrated micrometer. The substrate is satisfactory if 90% of the particles do not exceed 2 µm in diameter and none exceeds 10 µm in diameter.

Buffer solution—Dissolve 60 mg of tris(hydroxymethyl)aminomethane and 234 mg of sodium chloride in water to make 100 mL.

Sodium taurocholate solution—Prepare a solution to contain 80.0 mg of USP Sodium Taurocholate RS in each mL.

Standard test dilution—Suspend about 200 mg of USP Pancreatin RS, accurately weighed, in about 3 mL of cold water in a mortar, triturate for 10 minutes, and add cold water to a volume necessary to produce a concentration of 8 to 16 USP Units of lipase activity per mL, based upon the declared potency on the label of the Reference Standard. Maintain the suspension at 4°, and mix before using. For each determination withdraw 5 to 10 mL of the cold suspension, and allow the temperature to rise to 20° before pipeting the exact volume.

Assay test dilution—Suspend about 200 mg of Pancrelipase, accurately weighed, in about 3 mL of cold water in a mortar, triturate for 10 minutes, and add cold water to a volume necessary to produce a concentration of 8 to 16 USP Units of lipase activity per mL, based upon the estimated potency of the test material. Maintain the suspension at 4°, and mix before using. For each determination withdraw 5 to 10 mL of the cold suspension, and allow the temperature to rise to 20° before pipeting the exact volume.

Procedure—Mix 10.0 mL of *Olive oil substrate*, 8.0 mL of *Buffer solution*, 2.0 mL of *Sodium taurocholate solution*, and 9.0 mL of water in a jacketed glass vessel of about 50-mL capacity, the outer chamber of which is connected to a thermostatically controlled water bath. Cover the mixture, and stir continuously with a mechanical stirring device. With the mixture maintained at a temperature of 37 ± 0.1°, add 0.1 *N* sodium hydroxide VS, from a microburet inserted through an opening in the cover, to adjust the pH to 9.20 potentiometrically using a calomel–glass electrode system. Add 1.0 mL of *Assay test dilution*, and then continue adding the 0.1 *N* sodium hydroxide VS for 5 minutes to maintain the pH at 9.0. Determine the volume of 0.1 *N* sodium hydroxide VS added after each minute.

In the same manner titrate 1.0 mL of *Standard test dilution*.

Calculation of potency—Plot the volume of 0.1 *N* sodium hydroxide VS titrated against time. Using only the points which fall on the straight-line segment of the curve, calculate the mean acidity released per minute by the sample and Standard. Taking into consideration dilution factors, calculate the lipase activity of the sample in USP Units by comparison to the activity of the Standard, using the lipase activity of the USP Pancreatin RS stated on the label.

Assay for protease activity (*Casein digestive power*)—

Casein substrate—Place 1.25 g of finely powdered casein in a 100-mL conical flask containing 5 mL of water, shake to form a suspension, add 10 mL of 0.1 *N* sodium hydroxide, shake for 1 minute, add 50 mL of water, and shake for about 1 hour to dissolve the casein. If necessary, adjust to a pH of about 8, using 1 *N* sodium hydroxide or 1 *N* hydrochloric acid. Transfer the solution quantitatively to a 100-mL volumetric flask, dilute with water to volume, and mix. Use this substrate on the day it is prepared.

Buffer solution—Dissolve 6.8 g of monobasic potassium phosphate and 1.8 g of sodium hydroxide in 950 mL of water in a 1000-mL volumetric flask, adjust to a pH of 7.5 ± 0.2, using 0.2 *N* sodium hydroxide, dilute with water to volume, and mix. Store this solution in a refrigerator.

Trichloroacetic acid solution—Dissolve 50 g of trichloroacetic acid in 1000 mL of water. This solution may be stored at room temperature.

Filter paper—Determine the suitability of the filter paper by filtering a 5-mL portion of *Trichloroacetic acid solution* through the paper and measuring the absorbance of the filtrate at 280 nm, using an unfiltered portion of the same *Trichloroacetic acid solution* as the blank: the absorbance is not more than 0.04. If the absorbance is more than 0.04, wash the filter paper repeatedly with *Trichloroacetic acid solution* until the absorbance of the filtrate, determined as above, is not more than 0.04.

Standard test dilution—Add about 100 mg of USP Pancreatin RS, accurately weighed, to 100.0 mL of *Buffer solution*, and mix by shaking intermittently at room temperature for about 25 minutes. Dilute quantitatively with *Buffer solution* to produce a concentration of about 2.5 USP Units of protease activity per mL, based upon the declared potency on the label of the Reference Standard.

Assay test dilution—Weigh accurately about 100 mg of Pancrelipase into a suitable mortar. Add about 3 mL of *Buffer solution*, and triturate for 5 to 10 minutes. Transfer the mixture with the aid of *Buffer solution* to a 100-mL volumetric flask, dilute with *Buffer solution* to volume, and mix. Dilute quantitatively with *Buffer solution* to obtain a dilution that corresponds in activity to the *Standard test dilution.*

Procedure—Label test tubes in duplicate S_1, S_2, and S_3 for the standard series, and U for the sample. Pipet into tubes S_1 2.0 mL, into S_2 and U 1.5 mL, and into S_3 1.0 mL of *Buffer solution.* Pipet into tubes S_1 1.0 mL, into S_2 1.5 mL, and into S_3 2.0 mL of the *Standard test dilution.* Pipet into tubes U 1.5 mL of the *Assay test dilution.* Pipet into one tube each of S_1, S_2, S_3, and U 5.0 mL of *Trichloroacetic acid solution*, and mix. Designate these tubes as S_{1B}, S_{2B}, S_{3B}, and U_B, respectively. Prepare a blank by mixing 3 mL of *Buffer solution* and 5 mL of *Trichloroacetic acid solution* in a separate test tube marked B. Place all the tubes in a 40° water bath, insert a glass stirring rod into each tube, and allow for temperature equilibration. At zero time, add to each tube, at timed intervals, 2.0 mL of the *Casein substrate*, preheated to the bath temperature, and mix. Sixty minutes, accurately timed, after the addition of the *Casein substrate* stop the reaction in tubes S_1, S_2, S_3, and U by adding 5.0 mL of *Trichloroacetic acid solution* at the corresponding time intervals, stir, and remove all the tubes from the bath. Allow to stand at room temperature for 10 minutes for complete protein precipitation, and filter. The filtrates must be free from haze. Determine the absorbances of the filtrates, in 1-cm cells, at 280 nm, with a suitable spectrophotometer, using the filtrate from the blank (tube B) to set the instrument.

Calculation of potency—Correct the absorbance values for the filtrates from tubes S_1, S_2, and S_3 by subtracting the absorbance values for the filtrates from tubes S_{1B}, S_{2B}, and S_{3B}, respectively, and plot the corrected absorbance values against the corresponding volumes of the *Standard test dilution* used. From the curve, using the corrected absorbance value ($U - U_B$) for the Pancrelipase taken, and taking into consideration the dilution factors, calculate the protease activity, in USP Units, of the Pancrelipase taken by comparison with that of the Standard, using the protease activity stated on the label of USP Pancreatin RS.

Pancrelipase Capsules

» Pancrelipase Capsules contain an amount of pancrelipase equivalent to not less than 90.0 percent and not more than 125.0 percent of the labeled lipase activity expressed in USP Units, the labeled activity being not less than 8000 USP Units per capsule. They contain, in each capsule, the pancrelipase equivalent of not less than 30,000 USP Units of amylase activity, and not less than 30,000 USP Units of protease activity.

Packaging and storage—Preserve in tight containers, preferably with a desiccant, at a temperature not exceeding 25°.

Labeling—Label Capsules to indicate lipase activity in USP Units.

Reference standards— *USP Sodium Taurocholate Reference Standard*—Keep container tightly closed. [*Caution—Avoid inhaling airborne particles.*] Dry at 105° for 4 hours before using. *USP Pancreatin Reference Standard*—Keep container tightly closed, and store in a refrigerator. Do not open while cold, and do not dry before using.

Microbial limits—Capsules meet the requirements of the tests for absence of *Salmonella* species and *Escherichia coli* under *Microbial Limit Tests* ⟨61⟩.

Loss on drying ⟨731⟩—Dry the contents of 10 Capsules in vacuum at 60° for 4 hours: it loses not more than 5.0% of its weight.

Assay—Weigh the contents of not less than 10 Pancrelipase Capsules, and determine the average weight per capsule. Mix the combined contents, and proceed as directed for *Assay for amylase activity*, *Assay for lipase activity*, and *Assay for protease activity* under *Pancrelipase.*

Pancrelipase Tablets

» Pancrelipase Tablets contain an amount of pancrelipase equivalent to not less than 90.0 percent and not more than 125.0 percent of the labeled lipase activity expressed in USP Units, the labeled activity being not less than 8000 USP Units per Tablet. They contain, in each Tablet, the pancrelipase equivalent of not less than 30,000 USP Units of amylase activity, and not less than 30,000 USP Units of protease activity.

Packaging and storage—Preserve in tight containers, preferably with a desiccant, at a temperature not exceeding 25°.

Labeling—Label Tablets to indicate the lipase activity in USP Units.

Reference standards— *USP Sodium Taurocholate Reference Standard*—Keep container tightly closed. [*Caution—Avoid inhaling airborne particles.*] Dry at 105° for 4 hours before using. *USP Pancreatin Reference Standard*—Keep container tightly closed, and store in a refrigerator. Do not open while cold, and do not dry before using.

Microbial limits—Tablets meet the requirements of the test for absence of *Salmonella* species and *Escherichia coli* under *Microbial Limit Tests* ⟨61⟩.

Disintegration ⟨701⟩: 75 minutes.

Loss on drying ⟨731⟩—Dry about 5 g, accurately weighed, of finely ground Tablets in vacuum at 60° for 4 hours: it loses not more than 5.0% of its weight.

Assay—Weigh and finely powder not less than 20 Pancrelipase Tablets, avoiding the production of heat during the process, and proceed as directed for *Assay for amylase activity*, *Assay for lipase activity*, and *Assay for protease activity* under *Pancrelipase.*

Panthenol

$$HOCH_2C(CH_3)_2-CH(OH)-CONHCH_2CH_2CH_2OH$$

$C_9H_{19}NO_4$ 205.25

Butanamide, 2,4-dihydroxy-*N*-(3-hydroxypropyl)-3,3-dimethyl-, (±)-.

(±)-2,4-Dihydroxy-*N*-(3-hydroxypropyl)-3,3-dimethylbutyramide.

(±)-Pantothenyl alcohol [*16485-10-2*].

» Panthenol is a racemic mixture of the dextrorotatory and levorotatory isomers of panthenol. It contains not less than 99.0 percent and not more than 102.0 percent of $C_9H_{19}NO_4$, calculated on the dried basis.

Packaging and storage—Preserve in tight containers.

Reference standard—*USP Racemic Panthenol Reference Standard.*

Identification—

A: The infrared absorption spectrum of a mineral oil dispersion of it exhibits maxima only at the same wavelengths as that of a similar preparation of USP Racemic Panthenol RS.

B: To 1 mL of a solution (1 in 10) add 5 mL of 1 N sodium hydroxide and 1 drop of cupric sulfate TS, and shake vigorously: a deep blue color develops.

C: To 1 mL of a solution (1 in 100) add 1 mL of 1 *N* hydrochloric acid, and heat on a steam bath for about 30 minutes. Cool, add 100 mg of hydroxylamine hydrochloride, mix, and add 5 mL of 1 *N* sodium hydroxide. Allow to stand for 5 minutes, then adjust with 1 *N* hydrochloric acid to a pH of between 2.5 and 3.0, and add 1 drop of ferric chloride TS: a purplish red color develops.

Melting range, *Class I* ⟨741⟩: between 64.5° and 68.5°.

Specific rotation ⟨781⟩: between −0.05° and +0.05°, calculated on the dried basis, determined in a solution containing 500 mg in each 10 mL.

Loss on drying ⟨731⟩—Dry it in vacuum over phosphorus pentoxide at 56° for 4 hours: it loses not more than 0.5% of its weight.

Residue on ignition ⟨281⟩: not more than 0.1%.

Aminopropanol—Transfer about 10 g, accurately weighed, to a 50-mL flask, and dissolve in 25 mL of water. Add bromothymol blue TS, and titrate with 0.01 *N* sulfuric acid VS to a yellow end-point. Each mL of 0.01 *N* sulfuric acid is equivalent to 750 µg of aminopropanol. Not more than 0.10% is found.

Assay—

Potassium biphthalate solution—Dissolve 20.42 g of potassium biphthalate in glacial acetic acid contained in a 1000-mL volumetric flask. If necessary, warm the mixture on a steam bath to achieve complete solution, observing precautions against absorption of moisture. Cool to room temperature, dilute with glacial acetic acid to volume, and mix.

Procedure—Transfer about 400 mg of Panthenol, accurately weighed, to a 300-mL flask fitted to a reflux condenser by means of a standard-taper glass joint, add 50.0 mL of 0.1 *N* perchloric acid VS, and reflux for 5 hours. Cool, observing precautions to prevent atmospheric moisture from entering the condenser, and rinse the condenser with glacial acetic acid, collecting the rinsings in the flask. Add 5 drops of crystal violet TS, and titrate with *Potassium biphthalate solution* to a blue-green end-point. Perform a blank determination, and note the difference in volumes required. Each mL of the difference in volumes of 0.1 *N* perchloric acid consumed is equivalent to 20.53 mg of $C_9H_{19}NO_4$.

Papain

Papain.
Papain [*9001-73-4*].

» Papain is a purified proteolytic substance derived from *Carica papaya* Linné (Fam. Caricaceae). Papain, when assayed as directed herein, contains not less than 6000 Units per mg. Papain of a higher digestive power may be reduced to the official standard by admixture with papain of lower activity, lactose, or other suitable diluents.

One USP Unit of Papain activity is the activity that releases the equivalent of 1 µg of tyrosine from a specified casein substrate under the conditions of the *Assay*, using the enzyme concentration that liberates 40 µg of tyrosine per mL of test solution.

Packaging and storage—Preserve in tight, light-resistant containers, in a cool place.

Reference standard—*USP Papain Reference Standard*—Do not dry before using.

pH ⟨791⟩: between 4.8 and 6.2, in a solution (1 in 50).

Loss on drying ⟨731⟩—Dry it in a vacuum oven at 60° for 4 hours: it loses not more than 7.0% of its weight.

Assay (*Casein digestive power*)—

Dibasic sodium phosphate, 0.05 M—Dissolve 7.1 g of anhydrous dibasic sodium phosphate in water to make 1000 mL. Add 1 drop of toluene as a preservative.

Citric acid, 0.05 M—Dissolve 10.5 g of citric acid monohydrate in water to make 1000 mL. Add 1 drop of toluene as a preservative.

Casein substrate—Disperse 1 g of Hammersten-type casein in 50 mL of *0.05 M Dibasic sodium phosphate.* Place in a boiling water bath for 30 minutes with occasional stirring. Cool to room temperature, and add *0.05 M Citric acid* to adjust to a pH of 6.0 ± 0.1. Stir the solution rapidly and continuously during the addition of the *0.05 M Citric acid* to prevent precipitation of the casein. Dilute with water to 100 mL. Prepare fresh daily.

Buffer solution (Phosphate-Cysteine Disodium ethylenediaminetetraacetate Buffer)—Dissolve 3.55 g of anhydrous dibasic phosphate in 400 mL of water in a 500-mL volumetric flask. Add 7.0 g of disodium ethylenediaminetetraacetate and 3.05 g of cysteine hydrochloride monohydrate. Adjust with 1 *N* hydrochloric acid or 1 *N* sodium hydroxide to a pH of 6.0 ± 0.1, dilute with water to volume, and mix. Prepare fresh daily.

Trichloroacetic acid solution—Dissolve 30 g of reagent grade trichloroacetic acid in water, and dilute with water to 100 mL. This solution may be stored at room temperature.

Standard preparation—Weigh accurately 100 mg of USP Papain RS in a 100-mL volumetric flask, and add *Buffer solution* to dissolve. Dilute with *Buffer solution* to volume, and mix. Transfer 2.0 mL of this solution to a 50-mL volumetric flask, dilute with *Buffer solution* to volume, and mix. Use within 30 minutes after preparation.

Assay preparation—Transfer an accurately weighed amount of Papain, equivalent to about 100 mg of USP Papain RS, to a 100-mL volumetric flask, dilute with *Buffer solution* to volume, and mix. Transfer 2.0 mL of this solution to a 50-mL volumetric flask, dilute with *Buffer solution* to volume, and mix.

Procedure—Into each of 12 test tubes (18-mm × 150-mm) pipet 5.0 mL of *Casein substrate.* Place in a water bath at 40°, and allow 10 minutes to reach bath temperature. Into each of two of the tubes (the tests are run in duplicate except for the blanks) labeled S_1, pipet 1.0 mL of the *Standard preparation* and 1.0 mL of the *Buffer solution,* mix by swirling, note zero time, insert the stopper, and replace in the bath. Into each of 2 other tubes, labeled S_2, pipet 1.5 mL of *Standard preparation* and 0.5 mL of *Buffer solution,* and proceed as before. Repeat this procedure for 2 tubes, labeled S_3, to which 2.0 mL of *Standard preparation* is added, and for 2 tubes, labeled U_2, to which 1.5 mL of *Assay preparation* and 0.5 mL of *Buffer solution* are added. After 60 minutes, accurately timed, add to all 12 tubes 3.0 mL of *Trichloroacetic acid solution,* and shake vigorously. With the 4 tubes to which no *Standard preparation* or *Assay preparation* were added, prepare blanks by pipeting, respectively, 1.0 mL of *Standard preparation* and 1.0 mL of *Buffer solution;* 1.5 mL of *Standard preparation* and 0.5 mL of *Buffer solution;* 2.0 mL of *Standard preparation;* and 1.5 mL of *Assay preparation* and 0.5 mL of *Buffer solution.* Replace all tubes in the 40° water bath, for 30 to 40 minutes, to allow to coagulate fully the precipitated protein. Filter through medium-porosity filter paper, discarding the first 3 mL of the filtrate (filtrates used are clear). Read the absorbances, at 280 nm, of the filtrates of all solutions against their respective blanks. Plot the absorbance readings for S_1, S_2, and S_3 against the enzyme concentration of each corresponding level. By interpolation from this curve, taking into consideration dilution factors, calculate the potency in Units, in the weight of Papain taken by the formula:

$$(50{,}000/3)CA,$$

in which 50,000/3 is a factor derived by the expression 100(50/2)(10/1.5), *C* is the concentration, in mg per mL, obtained from the standard curve, and *A* is the activity of the Reference Standard in Units per mg.

Papain Tablets for Topical Solution

» Papain Tablets for Topical Solution contain not less than 100.0 percent of the labeled potency.

Packaging and storage—Preserve in tight, light-resistant containers in a cool place.

Reference standard—*USP Papain Reference Standard*—Do not dry before using.

Completeness of solution ⟨641⟩—Prepare a solution of 50 Tablets in 500.0 mL of water, allow to stand for 4 hours, filter through 2 superimposed, matched-weight, 47-mm diameter, 0.8-μm porosity membrane filters, and wash the residue by rinsing the flask at the sides of the holder with water. Dry both filters in a desiccator under vacuum, over phosphorus pentoxide, for 6 to 18 hours, weigh the filters separately, and subtract the weight of the lower filter from that of the upper filter: the difference in the weights is not more than 50 mg (1 mg per Tablet).

Microbial limits ⟨61⟩—It meets the requirements of the tests for absence of *Staphylococcus aureus* and *Pseudomonas aeruginosa.*

Disintegration ⟨701⟩: not more than 15 minutes at 23 ± 2°.

pH ⟨791⟩: between 6.9 and 8.0, determined in a solution of 1 Tablet in 10 mL.

Assay—

Dibasic sodium phosphate, 0.05 M; Citric acid, 0.05 M; Casein substrate; Buffer solution; Trichloroacetic acid solution; and *Standard preparation*—Prepare as directed in the *Assay* under *Papain.*

Assay preparation—Place a counted number of Papain Tablets for Topical Solution, equivalent to about 600,000 USP Units of Papain, in a 100-mL volumetric flask, dissolve in *Buffer solution,* dilute with *Buffer solution* to volume, and mix. Transfer 2.0 mL of this solution to a 50-mL volumetric flask, dilute with *Buffer solution* to volume, and mix.

Procedure—Proceed as directed for *Procedure* in the *Assay* under *Papain.* By interpolation from the standard curve, calculate the potency, in Units, in the number of Tablets taken by the formula:

$$(50{,}000/3)CA,$$

in which the factors are as defined therein.

Papaverine Hydrochloride

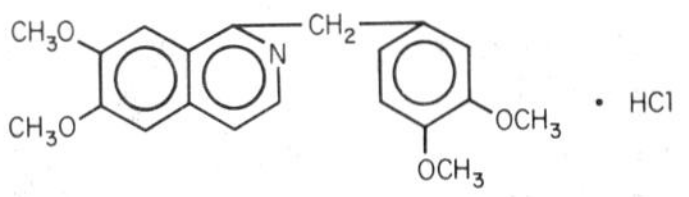

$C_{20}H_{21}NO_4 \cdot HCl$ 375.85

Isoquinoline, 1-[(3,4-dimethoxyphenyl)methyl]-6,7-dimethoxy-, hydrochloride.

6,7-Dimethoxy-1-veratrylisoquinoline hydrochloride [*61-25-6*].

» Papaverine Hydrochloride contains not less than 98.5 percent and not more than 100.5 percent of $C_{20}H_{21}NO_4 \cdot HCl$, calculated on the dried basis.

Packaging and storage—Preserve in tight, light-resistant containers.

Reference standard—*USP Papaverine Hydrochloride Reference Standard*—Dry at 105° for 2 hours before using.

Completeness of solution—A 1 in 15 solution in chloroform is clear and free from undissolved solid.

Identification—

A: The infrared absorption spectrum of a potassium bromide dispersion of it, previously dried, exhibits maxima only at the same wavelengths as that of a similar preparation of USP Papaverine Hydrochloride RS.

B: The ultraviolet absorption spectrum of a 1 in 400,000 solution in 0.1 *N* hydrochloric acid exhibits maxima and minima at the same wavelengths as that of a similar solution of USP Papaverine Hydrochloride RS, concomitantly measured, and the respective absorptivities, calculated on the dried basis, at the wavelength of maximum absorbance at about 251 nm do not differ by more than 3.0%.

C: A solution (1 in 50) responds to the tests for *Chloride* ⟨191⟩.

pH ⟨791⟩: between 3.0 and 4.5, in a solution (1 in 50).

Loss on drying ⟨731⟩—Dry it at 105° for 2 hours: it loses not more than 0.5% of its weight.

Residue on ignition ⟨281⟩: not more than 0.1%.

Cryptopine, thebaine, or other organic impurities—Dissolve 50 mg in 2 mL of sulfuric acid in a small test tube: the resulting solution is not more yellow-brown in color than Matching Fluid S (see *Readily Carbonizable Substances Test* ⟨271⟩), and it is not more pink than a standard prepared, in equal volume, by diluting 3.0 mL of 0.1 *N* potassium permanganate with water to 1000 mL.

Assay—Dissolve about 700 mg of Papaverine Hydrochloride, accurately weighed, in 80 mL of glacial acetic acid, add 10 mL of mercuric acetate TS and 1 drop of crystal violet TS, and titrate with 0.1 *N* perchloric acid VS to a blue-green end-point. Perform a blank determination, and make any necessary correction. Each mL of 0.1 *N* perchloric acid is equivalent to 37.59 mg of $C_{20}H_{21}NO_4 \cdot HCl$.

Papaverine Hydrochloride Injection

» Papaverine Hydrochloride Injection is a sterile solution of Papaverine Hydrochloride in Water for Injection. It contains not less than 95.0 percent and not more than 105.0 percent of the labeled amount of $C_{20}H_{21}NO_4 \cdot HCl$.

Packaging and storage—Preserve in single-dose or in multiple-dose containers, preferably of Type I glass.

Reference standard—*USP Papaverine Hydrochloride Reference Standard*—Dry at 105° for 2 hours before using.

Identification—

A: Add 2 mL of alcohol to 1 mL of Injection, and evaporate on a steam bath, with the aid of a current of nitrogen, to dryness. Dry the residue at 105° for 2 hours: it responds to *Identification test A* under *Papaverine Hydrochloride.*

B: It responds to *Identification test C* under *Papaverine Hydrochloride.*

pH ⟨791⟩: not less than 3.0.

Other requirements—It meets the requirements under *Injections* ⟨1⟩.

Assay—Transfer 1.0 mL of Papaverine Hydrochloride Injection to a 200-mL volumetric flask, and dilute with water to volume. Pipet 3 mL of this solution into a separator, add 10 mL of water, and render alkaline with 6 *N* ammonium hydroxide. Extract the alkaloid with successive 5-mL portions of chloroform, and evaporate the extracts to dryness. Dissolve the residue in 0.1 *N* hydrochloric acid, and dilute with the same medium to 100.0 mL. Concomitantly determine the absorbances of this solution and of a Standard solution of USP Papaverine Hydrochloride RS in 0.1 *N* hydrochloric acid having a known concentration of about 4.5 μg per mL in 1-cm cells at the wavelength of maximum absorbance at about 251 nm, with a suitable spectrophotometer, using 0.1 *N* hydrochloric acid as the blank. Calculate the quantity, in mg, of $C_{20}H_{21}NO_4 \cdot HCl$ in the portion of Injection taken by the formula:

$$6.67C(A_U/A_S),$$

in which *C* is the concentration, in μg per mL, of USP Papaverine Hydrochloride RS in the Standard solution, and A_U and A_S are the absorbances of the solution from the Injection and the Standard solution, respectively.

Papaverine Hydrochloride Tablets

» Papaverine Hydrochloride Tablets contain not less than 93.0 percent and not more than 107.0 percent of the labeled amount of $C_{20}H_{21}NO_4 \cdot HCl$.

Packaging and storage—Preserve in tight containers.

Reference standard—*USP Papaverine Hydrochloride Reference Standard*—Dry at 105° for 2 hours before using.

Identification—Add a portion of powdered Tablets, equivalent to about 30 mg of papaverine hydrochloride, to 10 mL of 0.1 *N* hydrochloric acid in a separator. Extract the mixture with 10 mL of chloroform, filter the chloroform phase through paper, evaporate the solvent on a steam bath, and dry the residue at 105° for 2 hours: it responds to *Identification test A* under *Papaverine Hydrochloride*.

Dissolution ⟨711⟩—

Medium: water; 900 mL.

Apparatus 1: 100 rpm.

Time: 30 minutes.

Procedure—Determine the amount of $C_{20}H_{21}NO_4.HCl$ dissolved from ultraviolet absorbances at the wavelength of maximum absorbance at about 250 nm of filtered portions of the solution under test, suitably diluted with 0.1 *N* hydrochloric acid, in comparison with a Standard solution having a known concentration of USP Papaverine Hydrochloride RS in the same medium.

Tolerances—Not less than 80% (*Q*) of the labeled amount of $C_{20}H_{21}NO_4.HCl$ is dissolved in 30 minutes.

Uniformity of dosage units ⟨905⟩: meet the requirements.

Procedure for content uniformity—Transfer 1 finely powdered Tablet to a 250-mL volumetric flask, add 50 mL of water and 3 mL of hydrochloric acid, mix, and allow to stand for 15 minutes with occasional agitation. Dilute with water to volume, mix, and filter, discarding the first 20 mL of the filtrate. Dilute a portion of the subsequent filtrate quantitatively and stepwise, if necessary, with water to provide a solution containing approximately 2.4 µg of papaverine hydrochloride per mL. Concomitantly determine the absorbances of this solution and a solution of USP Papaverine Hydrochloride RS, in the same medium at a concentration of about 2.4 µg per mL, in 1-cm cells, at the wavelength of maximum absorbance at about 250 nm, with a suitable spectrophotometer, using water as the blank. Calculate the quantity, in mg, of $C_{20}H_{21}NO_4.HCl$ in the Tablet by the formula:

$$(TC/D)(A_U/A_S),$$

in which *T* is the labeled quantity, in mg, of papaverine hydrochloride in the Tablet; *C* is the concentration, in µg per mL, of USP Papaverine Hydrochloride RS in the Standard solution; *D* is the concentration, in µg per mL, of the solution from the Tablet based upon the labeled quantity per Tablet and the extent of dilution; and A_U and A_S are the absorbances of the solution from the Tablet and the Standard solution, respectively.

Assay—Weigh and finely powder not less than 20 Papaverine Hydrochloride Tablets. Transfer an accurately weighed portion of the powder, equivalent to about 30 mg of papaverine hydrochloride, to a glass-stoppered conical flask, add about 100 mL of 0.1 *N* hydrochloric acid, and shake by mechanical means for 15 minutes. Filter the mixture into a 200-mL volumetric flask, and add 0.1 *N* hydrochloric acid to volume. Proceed as directed in the *Assay* under *Papaverine Hydrochloride Injection*, beginning with "Pipet 3 mL of this solution into a separator." Calculate the quantity, in mg, of $C_{20}H_{21}NO_4.HCl$ in the portion of Tablets taken by the formula:

$$6.67C(A_U/A_S),$$

in which *C* is the concentration, in µg per mL, of USP Papaverine Hydrochloride RS in the Standard solution, and A_U and A_S are the absorbances of the solution from the Tablets and the Standard solution, respectively.

Paper Strip, Biological Indicator for Dry-heat Sterilization,—*see* Biological Indicator for Dry-heat Sterilization, Paper Strip

Paper Strip, Biological Indicator for Ethylene Oxide Sterilization,—*see* Biological Indicator for Ethylene Oxide Sterilization, Paper Strip

Paper Strip, Biological Indicator for Steam Sterilization,—*see* Biological Indicator for Steam Sterilization, Paper Strip

Parachlorophenol

Cl–⟨C₆H₄⟩–OH

C_6H_5ClO 128.56

Phenol, 4-chloro-.

p-Chlorophenol [*106-48-9*].

» Parachlorophenol contains not less than 99.0 percent and not more than 100.5 percent of C_6H_5ClO.

Packaging and storage—Preserve in tight, light-resistant containers.

Clarity and reaction of solution—A 1 in 100 solution is clear and is acid to litmus.

Identification—

A: To a 1 in 100 solution of it add bromine TS dropwise: a white precipitate is formed, and at first it redissolves, but then it becomes permanent as an excess of the reagent is added.

B: Add 1 drop of ferric chloride TS to 10 mL of a 1 in 100 solution of it: the solution acquires a violet-blue color.

C: Heat a few crystals, held on a copper wire, in the edge of a nonluminous flame: a green color is imparted to the flame.

D: To a mixture of 1 g of it and 5 mL of sodium hydroxide solution (1 in 3) add 1.5 g of monochloroacetic acid. Shake, and heat on a steam bath for 1 hour. Cool, dilute with 15 mL of water, and acidify with hydrochloric acid. Extract with 50 mL of ether, wash the ether solution with 10 mL of cold water, then extract the ether solution with 25 mL of sodium carbonate solution (1 in 20). Acidify the solution with hydrochloric acid, collect the resulting precipitate on a filter, and recrystallize it from hot water: the resulting parachlorophenoxyacetic acid melts between 154° and 158°.

Congealing temperature ⟨651⟩: between 42° and 44°.

Nonvolatile residue—Heat about 1 g, accurately weighed, in a tared container on a steam bath until it is volatilized, and dry at 105° for 1 hour: not more than 0.1% of residue remains.

Chloride—Acidify 10 mL of a 1 in 100 solution with 2 *N* nitric acid, and add a few drops of silver nitrate TS: no turbidity or opalescence is produced.

Assay—Transfer about 1 g of Parachlorophenol, accurately weighed, to a 500-mL volumetric flask, dissolve in water, dilute with water to volume, and mix. Transfer a 25.0-mL portion of the solution to an iodine flask, cool in an ice bath to about 4°, and add 20.0 mL of 0.1 *N* bromine VS. Add 5 mL of hydrochloric acid, and immediately insert the stopper. Maintain the flask at a temperature of 4° for 30 minutes, shaking at frequent intervals. Allow it to stand for 15 minutes, remove the stopper just sufficiently to introduce quickly 5 mL of potassium iodide solution (1 in 5), taking care that no bromine vapor escapes, and at once insert the stopper in the flask. Shake thoroughly, remove the stopper, and rinse it and the neck of the flask with a small portion of water, allowing the washings to flow into the flask. Shake the mixture, and titrate the liberated iodine with 0.1 *N* sodium thiosulfate VS, using 3 mL of starch TS as the indicator. Perform a blank determination (see *Residual Titrations* under *Titrimetry* ⟨541⟩). Each mL of 0.1 *N* bromine is equivalent to 3.214 mg of C_6H_5ClO.

Camphorated Parachlorophenol

» Camphorated Parachlorophenol is a triturated mixture that contains not less than 33.0 percent and not more than 37.0 percent of parachlorophenol

(C_6H_5ClO) and not less than 63.0 percent and not more than 67.0 percent of camphor ($C_{10}H_{16}O$). The sum of the percentages of parachlorophenol and camphor is not less than 97.0 and not more than 103.0.

Packaging and storage—Preserve in tight, light-resistant containers.

Assay for parachlorophenol—Transfer about 1 g of Camphorated Parachlorophenol, accurately weighed, to a wide-mouth conical flask, and add a few glass beads, 6 mL of sodium hydroxide solution (1 in 2), and 130 mL of water. Heat the solution to boiling, add 70 mL of potassium permanganate solution (3 in 50), and continue to boil for 20 minutes. To the hot solution add 40 mL of 0.1 *N* silver nitrate. Add 50 mL of 18 *N* sulfuric acid, and sodium sulfite crystals, in divided portions and with swirling until the permanganate color is discharged and no manganese dioxide remains. Boil until the vapors are no longer acid to litmus, keeping the volume nearly constant by the addition of water. Add 5 mL of nitric acid, and continue to boil for 5 minutes. Cool, and collect the precipitate on a tared filtering crucible, wash well with water, then with 10 mL of alcohol, dry at 105° for 1 hour, cool, and weigh. Each 1.000 g of the silver chloride so obtained is equivalent to 897.0 mg of C_6H_5ClO.

Assay for camphor—Transfer about 300 mg of Camphorated Parachlorophenol, accurately weighed, to a 200-mL pressure bottle containing 50 mL of freshly prepared dinitrophenylhydrazine TS. Close the pressure bottle, immerse it in a water bath, and maintain it at about 75° for 4 hours. Cool to room temperature, then transfer the contents to a beaker with the aid of 100 mL of 3 *N* sulfuric acid and allow it to stand overnight. Collect the precipitate on a tared filtering crucible, wash with 100 mL of 3 *N* sulfuric acid and then with 75 mL of cold water, in divided portions, to remove the acid. Dry at 80° for 2 hours, cool, and weigh. The weight of the precipitate so obtained, multiplied by 0.4581, represents the weight of $C_{10}H_{16}O$ in the sample taken.

Paraffin—*see* Paraffin NF

Paraldehyde

CH3 O CH3 / O O / CH3

$C_6H_{12}O_3$ 132.16
1,3,5-Trioxane, 2,4,6-trimethyl-.
2,4,6-Trimethyl-*s*-trioxane [*123-63-7*].

» *Note—Paraldehyde is subject to oxidation to form acetic acid. It may contain a suitable stabilizer.*

Packaging and storage—Preserve in well-filled, tight, light-resistant containers, preferably of Type I or Type II glass, holding not more than 30 mL, at a temperature not exceeding 25°. Paraldehyde may be shipped in bulk containers holding a minimum of 22.5 kg (50 lb) to commercial drug repackagers only.

Labeling—The label of all containers of Paraldehyde, including those dispensed by the pharmacist, includes a statement directing the user to discard the unused contents of any container that has been opened for more than 24 hours.

[Note—The label of bulk containers of Paraldehyde directs the commercial drug repackager to demonstrate compliance with the USP purity tests for Paraldehyde immediately prior to repackaging, and not to repackage from a container that has been opened longer than 24 hours.]

Identification—Heat it with a small quantity of 2 *N* sulfuric acid: acetaldehyde, recognizable by its pungent odor, is produced.

Congealing temperature ⟨651⟩: not lower than 11°.

Distilling range, *Method I* ⟨721⟩—It distils completely between 120° and 126°, a correction factor of 0.050° per mm being applied as necessary.

Acidity—To a solution of 6.0 mL in 100 mL of water add 5 drops of phenolphthalein TS, and titrate with 1.0 *N* sodium hydroxide: not more than 0.50 mL is required to produce a pink color (0.5% as acetic acid).

Nonvolatile residue—Heat 5.0 mL in a small, tared evaporating dish on a steam bath: no disagreeable odor is noticeable as the last portions evaporate, and, when dried at 105° for 1 hour, not more than 3 mg of residue remains (0.06%).

Chloride—To 5 mL of a solution (1 in 10) add 1 drop of nitric acid and 3 drops of silver nitrate TS: no opalescence is produced immediately.

Sulfate—To 5 mL of a solution (1 in 10) add 1 drop of hydrochloric acid and 3 drops of barium chloride TS: no turbidity is produced.

Acetaldehyde—Place 100 mL of water in a 250-mL conical flask, add 5.0 mL of Paraldehyde, and shake the mixture gently until solution is complete. Add 5 mL of hydroxylamine hydrochloride solution (3.5 in 100). Shake the mixture gently for 30 seconds, add methyl orange TS, and titrate immediately with 0.50 *N* sodium hydroxide. Perform a blank titration: the difference between the titers does not exceed 1 mL of 0.50 *N* sodium hydroxide (0.4%).

Paramethadione

CH3 O O / CH3CH2 / O N CH3

$C_7H_{11}NO_3$ 157.17
2,4-Oxazolidinedione, 5-ethyl-3,5-dimethyl-.
5-Ethyl-3,5-dimethyl-2,4-oxazolidinedione [*115-67-3*].

» Paramethadione contains not less than 98.0 percent and not more than 100.5 percent of $C_7H_{11}NO_3$.

Packaging and storage—Preserve in tight containers.

Reference standard—*USP Paramethadione Reference Standard*—Allow ampul to attain room temperature prior to opening. Do not dry before using. Discard any unused portion.

Identification—

A: The infrared absorption spectrum of a portion of it between sodium chloride plates exhibits maxima only at the same wavelengths as that of a similar preparation of USP Paramethadione RS.

B: To 5 mL of a solution (1 in 200) add 2 mL of barium hydroxide TS: a precipitate is formed immediately.

Refractive index ⟨831⟩: between 1.449 and 1.501.

Residue on ignition ⟨281⟩: not more than 0.1%.

Urethane—

Standard preparation—Dissolve about 25 mg of urethane, accurately weighed, in about 10 mL of methanol, dilute with methanol to 100.0 mL, and mix to obtain a solution having a known urethane concentration of about 250 µg per mL. Transfer 5.0 µL of this solution to a suitable vessel containing about 250 mg of USP Paramethadione RS, accurately weighed, and mix to obtain a solution having a known urethane concentration of about 5 ppm.

Test preparation—Transfer about 2 g of Paramethadione to a small stoppered flask, and mix.

Chromatographic system (see *Chromatography* ⟨621⟩)—The gas chromatograph is equipped with a flame-ionization detector and a 3-m × 2-mm glass column packed with a liquid phase consisting of 13.3 percent phase G2 plus 1.7 percent phase G16 on support S1A. The column temperature is programmed as follows. Start at 115° and maintain this temperature for 13 minutes after injection, then program to increase the temperature, at the rate of 25° per minute, to 220°, and maintain this temperature for 20 minutes. The injection port and detector are maintained at temperatures of 200° and 250°, respectively. Helium is used as the carrier gas at a flow rate of about 20 mL per minute. [NOTE—The *Standard preparation* and the *Test preparation* must be mixed immediately prior to injection into the chromatograph.] Chromatograph the *Standard preparation* and a 1:5 mixture of the 250 µg urethane per mL methanol solution used in preparing the *Standard preparation* and methanol, to assure correct identification of the urethane peak by its retention time, as directed under *Procedure*: the relative standard deviation of the urethane peak response for replicate injections of the *Standard preparation* is not more than 5.0%. Under typical conditions, the approximate retention time of urethane is 9 minutes.

Procedure—Separately inject equal volumes (about 2 µL) of the *Standard preparation* and the *Test preparation* into the chromatograph, record the chromatograms, and measure the responses for the urethane peak. Calculate the concentration of urethane, in ppm, in the Paramethadione taken by the formula:

$$C(r_U / r_S),$$

in which C is the concentration, in ppm, of urethane in the *Standard preparation*, and r_U and r_S are the urethane peak responses obtained from the *Test preparation* and the *Standard preparation*, respectively. Not more than 5 ppm are found.

Ordinary impurities ⟨466⟩—

Test solution: methanol.

Standard solution: methanol.

Eluant: a mixture of chloroform and acetone (7:1).

Visualization: 16.

Assay—

Internal standard solution—Transfer about 2.5 g of decanol, accurately weighed, to a 500-mL volumetric flask, add dehydrated alcohol to volume, and mix.

Standard preparation—Dissolve an accurately weighed quantity of USP Paramethadione RS in *Internal standard solution* to obtain a solution having a known concentration of about 10 mg per mL.

Assay preparation—Transfer about 500 mg of Paramethadione, accurately weighed, to a 50-mL volumetric flask, dilute with *Internal standard solution* to volume, and mix.

Procedure—Inject 2 µL or other suitable volume of *Standard preparation* into a suitable gas chromatograph equipped with a flame-ionization detector, and record the chromatogram. Under typical conditions, the instrument contains a 75-cm × 3-mm (OD) stainless steel column packed with a suitable 100- to 120-mesh column support, preferably of type S4. The injection port and detector are maintained at 220°, the column temperature is 210°, helium is used as the carrier gas at a flow rate of 45 mL per minute, and in the detector, hydrogen is introduced at a rate of 40 mL per minute and air at a rate of 350 mL per minute. Under these conditions, the paramethadione and the decanol retention times are at about 8 minutes and 25 minutes, respectively. In a suitable chromatogram, the resolution factor, R, is not less than 5.0, between the paramethadione and the internal standard peaks. Measure the areas of the peaks of USP Paramethadione RS and the internal standard. Under the same conditions as described for *Standard preparation*, inject 2 µL or other suitable volume of *Assay preparation* into the chromatograph, record the chromatogram, and measure the areas of the peaks of paramethadione and the internal standard. Calculate the quantity, in mg, of $C_7H_{11}NO_3$ in the Paramethadione taken by the formula:

$$50C(R_U / R_S),$$

in which C is the concentration, in mg per mL, of USP Paramethadione RS in the *Standard preparation*, and R_U and R_S are the ratios of the peak areas of Paramethadione to those of the internal standard of the *Assay preparation* and the *Standard preparation*, respectively.

Paramethadione Capsules

» Paramethadione Capsules contain not less than 90.0 percent and not more than 110.0 percent of the labeled amount of $C_7H_{11}NO_3$.

Packaging and storage—Preserve in tight containers.

Reference standard—*USP Paramethadione Reference Standard*—Allow ampul to attain room temperature prior to opening. Do not dry before using. Discard any unused portion.

Identification—The contents of the Capsules respond to *Identification test B* under *Paramethadione.*

Uniformity of dosage units ⟨905⟩: meet the requirements.

Assay—

Internal standard solution—Dilute 1.0 mL of cyclohexanol with dehydrated alcohol to make 50.0 mL of solution.

Standard preparation—Dissolve an accurately weighed quantity of USP Paramethadione RS in dehydrated alcohol to obtain a solution having a known concentration of about 15 mg per mL.

Assay preparation—Place a counted number of Paramethadione Capsules, equivalent to about 3 g of paramethadione, in a 250-mL tall-form beaker, add about 50 mL of dehydrated alcohol, cool in a solid carbon dioxide–acetone mixture for 10 minutes, and blend with a high-speed blender until all of the solids are reduced to fine particles (do not blend for more than 5 minutes). Transfer the mixture to a 200-mL volumetric flask, dilute with dehydrated alcohol to volume, and mix.

Procedure—Mix 5.0 mL of *Standard preparation* with 2.0 mL of *Internal standard solution*, inject 2 µL or other suitable volume of this mixture into a suitable gas chromatograph equipped with a thermal-conductivity detector, and record the chromatogram. Under typical conditions, the instrument contains a 1.5-m × 3-mm (OD) stainless steel column packed with a suitable 100- to 120-mesh column support, preferably of type S6. The injection port and detector are maintained at 220°, the column temperature is 190°, and helium is used as the carrier gas at a flow rate of 30 mL per minute. Under these conditions, the paramethadione and cyclohexanol retention times are about 11 minutes and 4 minutes, respectively. In a suitable chromatogram, the resolution factor, R, is not less than 5.0 between the paramethadione and the internal standard peaks. Mix 5.0 mL of *Assay preparation* with 2.0 mL of *Internal standard solution*, inject 2 µL or other suitable volume into the chromatograph, and record the chromatogram under the same conditions as described for the *Standard preparation.* Measure the areas of the peaks for Internal standard cyclohexanol, USP Paramethadione RS, and the test specimen. Calculate the quantity, in mg, of $C_7H_{11}NO_3$ in the portion of Paramethadione Capsules taken by the formula:

$$200C(R_U / R_S),$$

in which C is the concentration, in mg per mL, of USP Paramethadione RS in the *Standard preparation*, and R_U and R_S are the ratios of the peak areas of paramethadione to those of the internal standard of the *Assay preparation* and the *Standard preparation*, respectively.

Paramethadione Oral Solution

» Paramethadione Oral Solution is a solution of Paramethadione in dilute Alcohol. It contains, in each mL, not less than 282 mg and not more than 318 mg of $C_7H_{11}NO_3$.

Packaging and storage—Preserve in tight, light-resistant containers.

Reference standard—*USP Paramethadione Reference Standard*—Allow ampul to attain room temperature prior to opening. Do not dry before using. Discard any unused portion.

Identification—It responds to *Identification test B* under *Paramethadione.*

Alcohol content, *Method II* ⟨611⟩: between 62.0% and 68.0% of C_2H_5OH.

Assay—

Internal standard solution—Dilute 1.0 mL of cyclohexanol with dehydrated alcohol to make 50.0 mL of solution.

Standard preparation—Dissolve an accurately weighed quantity of USP Paramethadione RS in dehydrated alcohol to obtain a solution having a known concentration of about 15 mg per mL.

Assay preparation—Pipet 5 mL of Paramethadione Oral Solution into a 100-mL volumetric flask, dilute with dehydrated alcohol to volume, and mix.

Procedure—Proceed as directed for *Procedure* in the *Assay* under *Paramethadione Capsules.* Calculate the quantity, in mg, of $C_7H_{11}NO_3$ in each mL of Oral Solution by the formula:

$$100(C/V)(R_U/R_S),$$

in which C is the concentration, in mg per mL, of USP Paramethadione RS in the *Standard preparation,* V is the volume, in mL, of the Oral Solution taken, and R_U and R_S are the ratios of the peak areas of paramethadione to those of the internal standard of the *Assay preparation* and the *Standard preparation,* respectively.

Paramethasone Acetate

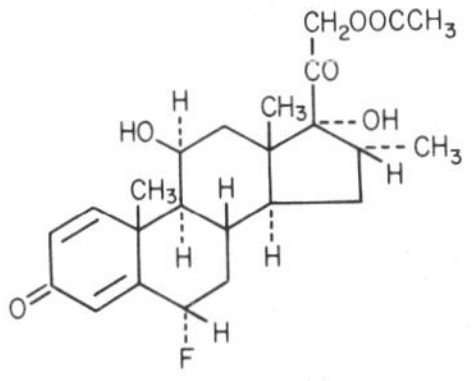

$C_{24}H_{31}FO_6$ 434.50

Pregna-1,4-diene-3,20-dione, 21-(acetyloxy)-6-fluoro-11,17-dihydroxy-16-methyl-, (6α,11β,16α)-.

6α-Fluoro-11β,17,21-trihydroxy-16α-methylpregna-1,4-diene-3,20-dione 21-acetate [*1597-82-6*].

» Paramethasone Acetate contains not less than 95.0 percent and not more than 101.0 percent of $C_{24}H_{31}FO_6$, calculated on the dried basis.

Packaging and storage—Preserve in tight containers.

Reference standard—*USP Paramethasone Acetate Reference Standard*—Dry in vacuum at 105° for 4 hours before using.

Identification—

A: The infrared absorption spectrum of a potassium bromide dispersion of it, previously dried, exhibits maxima only at the same wavelengths as that of a similar preparation of USP Paramethasone Acetate RS.

B: The ultraviolet absorption spectrum of a 1 in 50,000 solution in methanol exhibits maxima and minima at the same wavelengths as that of a similar solution of USP Paramethasone Acetate RS, concomitantly measured, and the respective absorptivities, calculated on the dried basis, at the wavelength of maximum absorbance at about 242 nm do not differ by more than 4.0%.

C: Prepare a solution of Paramethasone Acetate in a mixture of chloroform and methanol (1:1) containing 2 mg per mL. On a suitable thin-layer chromatographic plate (see *Chromatography* ⟨621⟩), coated with a 0.25-mm layer of chromatographic silica gel mixture, apply 10 µL each of this solution and a chloroform-methanol (1:1) solution of USP Paramethasone Acetate RS containing 2 mg per mL. Allow the spots to dry, and develop the chromatogram in a solvent system consisting of a mixturte of methylene chloride, nitromethane, and glacial acetic acid (60:40:1) until the solvent front has moved about three-fourths of the length of the plate. Remove the plate from the developing chamber, and allow to dry in air for 15 minutes. Return the plate to the develop ing chamber, and develop again, in the same solvent system, until the solvent front has moved about three-fourths of the length of the plate. Remove the plate from the developing chamber, mark the solvent front, and allow the solvent to evaporate. Locate the spots on the plate by examination under short-wavelength ultraviolet light: the R_f value of the principal spot obtained from the test solution corresponds to that obtained from the Standard solution.

Specific rotation ⟨781⟩: between +67° and +77°, calculated on the dried basis, determined in a solution in chloroform containing 100 mg of Paramethasone Acetate in each 10 mL.

X-ray diffraction ⟨941⟩—The X-ray diffraction pattern of Paramethasone Acetate conforms to either one or a mixture of the patterns described by the data in the accompanying table.

Form A		Form B	
d	I/I_1	d	I/I_1
12.09	10	11.62	20
8.42	20	7.80	8
7.78	10	7.13	10
6.41	100	6.50	60
5.65	100	5.98	100
5.50	10	5.63	70
5.18	2	5.30	100
4.59	30	4.85	60
4.41	20	4.65	60
4.24	40	4.43	8
3.93	20	4.30	2
3.64	30b	3.93	60
3.48	15	3.72	6
3.27	20	3.58	4
3.12	30	3.45	4
3.03	8	3.26	10
2.90	2	3.09	10
2.82	8	2.96	10
2.70	10	2.88	10
2.61	8	2.81	6b
2.51	10	2.66	8b
2.35	2	2.55	6
2.29	8	2.48	6b
2.24	6	2.38	10
2.11	4	2.30	6
2.08	2	2.26	8
2.04	4	2.19	6
2.00	4	2.11	10
1.95	1	2.04	8
1.92	4	1.99	10
1.87	2		
1.84	2		
1.82	1		

Loss on drying ⟨731⟩—Dry it in vacuum at 105° for 4 hours: it loses not more than 1.0% of its weight.

Assay—

Standard preparation—Prepare as directed under *Single-steroid Assay* ⟨511⟩, using USP Paramethasone Acetate RS to prepare a solution containing approximately 10 mg per mL.

Assay preparation—Weigh accurately about 50 mg of Paramethasone Acetate, previously dried, dissolve it in a sufficient quantity of a mixture of equal volumes of alcohol and chloroform to make 5.0 mL, and mix.

Procedure—Proceed as directed for *Procedure* under *Single-steroid Assay* ⟨511⟩, applying 10 µL each of the *Assay preparation* and the *Standard preparation* to the chromatographic plate, and using a solvent system consisting of methylene chloride, nitromethane, and glacial acetic acid (60:40:1), through the fourth sentence of the second paragraph under *Procedure.* Then centrifuge the tubes for 5 minutes, and determine the absorbances of the supernatant solutions in 1-cm cells at the wavelength of maximum absorbance at about 242 nm, with a suitable spectrophotometer, against the blank. Calculate the quantity, in mg, of $C_{24}H_{31}FO_6$ in the portion of Paramethasone Acetate taken by the formula:

$$5C(A_U/A_S),$$

in which C is the concentration, in mg per mL, of USP Paramethasone Acetate RS in the *Standard preparation,* and A_U and A_S are the absorbances of the solutions from the *Assay preparation* and the *Standard preparation,* respectively.

Paramethasone Acetate Tablets

» Paramethasone Acetate Tablets contain not less than 85.0 percent and not more than 115.0 percent of the labeled amount of $C_{24}H_{31}FO_6$.

Packaging and storage—Preserve in well-closed containers.

Reference standard—*USP Paramethasone Acetate Reference Standard*—Dry in vacuum at 105° for 4 hours before using.

Identification—The infrared absorption spectrum of the *Assay preparation*, prepared as directed in the *Assay*, exhibits maxima only at the same wavelengths as that of the *Standard preparation*, prepared as directed in the *Assay*.

Disintegration ⟨701⟩: 15 minutes, the use of disks being omitted.

Uniformity of dosage units ⟨905⟩: meet the requirements.

Procedure for content uniformity—Transfer 1 finely powdered Tablet to a 50-mL volumetric flask, add 25 mL of chloroform, and shake by mechanical means for 15 minutes. Dilute with chloroform to volume, mix, and filter, discarding the first 20 mL of the filtrate. Dilute a portion of the subsequent filtrate quantitatively and stepwise, if necessary, with chloroform to obtain a solution containing approximately 20 µg of paramethasone acetate per mL. Transfer 10.0 mL each of this solution and of a solution of USP Paramethasone Acetate RS in the same medium having a known concentration of about 20 µg per mL to separate 25-mL volumetric flasks, and transfer 10 mL of chloroform to a third flask to provide the blank. To each flask add 3.0 mL of a 1 in 4000 solution of blue tetrazolium in alcohol and 5.0 mL of a 1 in 20 solution of tetramethylammonium hydroxide TS in alcohol, mixing after each addition. Fifteen minutes, accurately timed, after the addition of the last reagent, add 1 mL of glacial acetic acid to each flask, dilute with chloroform to volume, and mix. Concomitantly determine the absorbances of the solutions in 1-cm cells at the wavelength of maximum absorbance at about 525 nm, with a suitable spectrophotometer, against the blank. Calculate the quantity, in mg, of $C_{24}H_{31}FO_6$ in the Tablet by the formula:

$$(TC/D)(A_U/A_S),$$

in which T is the labeled quantity, in mg, of paramethasone acetate in the Tablet, C is the concentration, in µg per mL, of USP Paramethasone Acetate RS in the Standard solution, D is the concentration, in µg per mL, of the solution from the Tablet, based upon the labeled quantity per Tablet and the extent of dilution, and A_U and A_S are the absorbances of the solution from the Tablet and the Standard solution, respectively.

Assay—

Standard preparation—Transfer about 6 mg of USP Paramethasone Acetate RS, accurately weighed, to a separator containing 15 mL of water, and proceed as directed under *Assay preparation*, beginning with "add 3 drops of hydrochloric acid."

Assay preparation—Weigh and finely powder not less than 20 Paramethasone Acetate Tablets. Transfer an accurately weighed portion of the powder, equivalent to about 6 mg of paramethasone acetate, to a separator containing 15 mL of water, add 3 drops of hydrochloric acid, and heat on a steam bath for 5 minutes, mixing frequently. Cool the separator to room temperature, add 4 drops of sodium hydroxide solution (1 in 2), and immediately extract with four 25-mL portions of chloroform. Filter the extracts through anhydrous sodium sulfate, collecting the extracts in a beaker. [*Caution—Do not allow the filter paper to extend above the top of the funnel.*] Rinse the filter with several small portions of chloroform, add the rinsings to the beaker, and evaporate the chloroform on a steam bath with the aid of a current of air until about 3 mL remains. Transfer the residual liquid, with the aid of several small portions of chloroform, to a glass-stoppered, 10-mL conical flask, and evaporate on a steam bath with the aid of a current of air to dryness. Add 2.0 mL of chloroform to the flask, insert the stopper, and mix to dissolve the residue.

Procedure—Concomitantly determine the absorbances of the solutions in 1-mm cells at the wavelength of maximum absorbance at 6.04 µm, with a suitable infrared spectrophotometer, using chloroform as the blank. Calculate the quantity, in mg, of $C_{24}H_{31}FO_6$ in the portion of Tablets taken by the formula:

$$W(A_U/A_S),$$

in which W is the weight, in mg, of USP Paramethasone Acetate RS used in preparing the *Standard preparation*, and A_U and A_S are the absorbances of the *Assay preparation* and the *Standard preparation*, respectively.

Paregoric

» Paregoric yields, from each 100 mL, not less than 35 mg and not more than 45 mg of anhydrous morphine.

Paregoric may be prepared as follows:

Powdered Opium	4.3 g
Suitable essential oil(s)	
Benzoic Acid	3.8 g
Diluted Alcohol	900 mL
Glycerin	38 mL
To make about	950 mL

Macerate for 5 days the Powdered Opium, Benzoic Acid, and essential oil(s), with occasional agitation, in a mixture of 900 mL of Diluted Alcohol and 38 mL of Glycerin. Then filter, and pass enough Diluted Alcohol through the filter to obtain 950 mL of total filtrate. Assay a portion of this filtrate as directed herein, and dilute the remainder with a sufficient quantity of Diluted Alcohol containing, in each 100 mL, 400 mg of Benzoic Acid, 4 mL of Glycerin, and sufficient essential oil(s) to yield a solution containing, in each 100 mL, 40 mg of anhydrous morphine.

NOTE—Paregoric may be prepared also by using Opium or Opium Tincture instead of Powdered Opium, the anhydrous morphine content being adjusted to 40 mg in each 100 mL and the alcohol content being adjusted to 45%.

Packaging and storage—Preserve in tight, light-resistant containers, and avoid exposure to direct sunlight and to excessive heat.

Alcohol content ⟨611⟩: between 43.0% and 47.0% of C_2H_5OH, determined by the gas-liquid procedure, acetone being used as the internal standard.

Assay—

Chromatographic tubes—Prepare three similar tubes, each about 260 mm long and consisting of about 200 mm of 25-mm tubing and about 6 cm of 6-mm tubing. In each of the tubes, place a pledget of glass wool at a point where the 6-mm tubing is constricted slightly, about 2 cm from the junction.

Citrate buffer—Mix equal volumes of 0.1 *M* sodium citrate and 0.1 *M* citric acid.

Standard preparation—Prepare a solution by dissolving an accurately weighed quantity of USP Morphine Sulfate RS, equivalent to about 40 mg of anhydrous morphine, in 0.5 mL of triethylamine contained in a 100-mL volumetric flask, and add methanol to volume. Pipet 10 mL of this solution into a 50-mL volumetric flask, add 1 mL each of triethylamine and hydrochloric acid, and add water-saturated chloroform to volume.

Assay preparation—Evaporate 10.0 mL of Paregoric (equivalent to about 4 mg of morphine) on a steam bath under a stream of air to about 2 mL, and cool. [NOTE—Avoid reducing the volume to less than 2 mL.] Add 0.5 mL of *Citrate buffer*.

Chromatographic columns—Fill the three tubes with adsorbant prepared as follows, using chromatographic siliceous earth as the base of the adsorbant, and tamp it firmly in place. Pack *Column I* in two layers, the lower layer consisting of 3 g of chromatographic siliceous earth mixed with 2 mL of *Citrate buffer* and the upper layer of 3 g of chromatographic siliceous earth mixed with the *Assay preparation*. Dry-rinse the beaker in which the components of the two layers have been mixed with 1 g of chromatographic siliceous earth, and add it also to the top of *Column I*. Pack *Column II* with 3 g of chromatographic siliceous earth mixed with 2 mL of dibasic potassium phosphate solution (1 in 5.75). Pack *Column III* with 3 g of chromatographic siliceous earth mixed with 2 mL of sodium hydroxide solution (1 in 50). Place a small pad of glass wool above each column packing.

Procedure—[NOTES—(1) Use water-saturated solvents throughout this procedure; (2) prepare eluants fresh daily; and (3) avoid bringing the solutions into contact with metal.] Wash *Column I* with 100 mL of ether, followed by 100 mL of chloroform, rinse the tip of the column with chloroform, and discard the solvents. In the following operations, rinse each column tip before discarding the column or changing receivers. Mount the three columns vertically so that the effluent from *Column I* flows into *Column II*, and the effluent from the latter flows into *Column III*. Pass through the three columns 5 mL of a 1 in 5 solution of triethylamine in chloroform, followed by four 10-mL portions of a 1 in 100 solution of triethylamine in chloroform, allowing each portion to pass through completely before subsequent additions. Discard *Column I*. Pass three 5-mL portions of the 1 in 100 solution of triethylamine in chloroform through the two remaining columns. Discard *Column II*. Wash *Column III* successively with 10 mL of the 1 in 100 solution of triethylamine in chloroform, 50 mL of chloroform, 2 mL of a 1 in 10 solution of glacial acetic acid in chloroform, and 50 mL of a 1 in 100 solution of glacial acetic acid in chloroform. Discard all washings. Arrange to collect eluate from *Column III* in a 50-mL volumetric flask containing 10 mL of methanol and 1 mL of hydrochloric acid. Elute the column with 5 mL of a 1 in 5 solution of triethylamine in chloroform, followed by 33 mL of a 1 in 100 solution of triethylamine in chloroform. Dilute with chloroform to volume, and mix. Concomitantly record the spectra of this solution and the *Standard preparation* in 1-cm cells, with a suitable spectrophotometer, from 255 nm to 360 nm, using chloroform as the blank, and plot the corresponding wavelength-absorbance curves. Correct the absorbance of each solution, at the wavelength of maximum absorbance at about 285 nm, by extrapolating the portion of the base-line curve between 340 nm and 310 nm to this wavelength. Calculate the weight of anhydrous morphine, in mg per 100 mL of Paregoric, by the formula:

$$10W(A_U/A_S),$$

in which W is the weight, in mg, of anhydrous morphine in the 50 mL of *Standard preparation*, and A_U and A_S are the corrected absorbances of the solution from the *Assay preparation* and the *Standard preparation*, respectively.

Pargyline Hydrochloride

(structure: C_6H_5–$CH_2N(CH_3)CH_2C{\equiv}CH$ · HCl)

$C_{11}H_{13}N.HCl$ 195.69

Benzenemethanamine, *N*-methyl-*N*-2-propynyl-, hydrochloride.
N-Methyl-*N*-2-propynylbenzylamine hydrochloride [*306-07-0*].

» Pargyline Hydrochloride contains not less than 98.0 percent and not more than 101.0 percent of $C_{11}H_{13}N.HCl$, calculated on the anhydrous basis.

Packaging and storage—Preserve in tight containers.

Reference standard—*USP Pargyline Hydrochloride Reference Standard*—Dry over silica gel for 4 hours before using.

Identification—

A: The infrared absorption spectrum of a mineral oil dispersion of it, previously dried over silica gel for 4 hours, exhibits maxima only at the same wavelengths as that of a similar preparation of USP Pargyline Hydrochloride RS.

B: The ultraviolet absorption spectrum of a 1 in 2500 solution in 0.1 *N* hydrochloric acid exhibits maxima and minima at the same wavelengths as that of a similar solution of USP Pargyline Hydrochloride RS, concomitantly measured, and the respective absorptivities, calculated on the anhydrous basis, at the wavelength of maximum absorbance at about 262 nm do not differ by more than 3.0%.

C: Dissolve 1.5 g of cupric chloride and 3 g of ammonium chloride in 20 mL of ammonium hydroxide, and dilute with water to 50 mL. Just prior to use, mix 1 volume of this solution with 2 volumes of hydroxylamine hydrochloride solution (1 in 10), and then add 3 drops of this solution to 100 mg of Pargyline Hydrochloride dissolved in 1 mL of water: a copious, pale yellow precipitate is formed.

D: Dissolve about 50 mg in 3 mL of water, add 6 *N* ammonium hydroxide until the solution is alkaline, and filter. Acidify the filtrate with 2 *N* nitric acid, and add 1 mL of silver nitrate TS: a white precipitate, which is insoluble in 2 *N* nitric acid but soluble in 6 *N* ammonium hydroxide, is formed.

Melting range ⟨741⟩: between 158° and 162°.

Water, *Method I* ⟨921⟩: not more than 1.0%.

Residue on ignition ⟨281⟩: not more than 0.2%.

Heavy metals, *Method I* ⟨231⟩—Dissolve 1 g in 20 mL of water, add 2 mL of 1 *N* acetic acid, and dilute with water to 25 mL: the limit is 0.002%.

Ordinary impurities ⟨466⟩—

Test solution: methanol.

Standard solution: methanol.

Eluant: a mixture of butyl alcohol, glacial acetic acid, and water (5:4:1).

Visualization: 1.

Assay—Dissolve about 200 mg of Pargyline Hydrochloride, accurately weighed, in 90 mL of glacial acetic acid, add 10 mL of mercuric acetate TS, and titrate with 0.1 *N* perchloric acid VS, determining the end-point potentiometrically. Perform a blank determination, and make any necessary correction. Each mL of 0.1 *N* perchloric acid is equivalent to 19.57 mg of $C_{11}H_{13}N.HCl$.

Pargyline Hydrochloride Tablets

» Pargyline Hydrochloride Tablets contain not less than 90.0 percent and not more than 110.0 percent of the labeled amount of $C_{11}H_{13}N.HCl$.

Packaging and storage—Preserve in well-closed containers.

Reference standard—*USP Pargyline Hydrochloride Reference Standard*—Dry over silica gel for 4 hours before using.

Identification—Dissolve a quantity of powdered Tablets, equivalent to about 10 mg of pargyline hydrochloride, in 2 mL of water, and filter: the filtrate responds to *Identification test C* under *Pargyline Hydrochloride*.

Dissolution ⟨711⟩—

Medium: pH 6.0 phosphate buffer (see *Buffer Solutions* in the section, *Reagents, Indicators, and Solutions*); 900 mL.

Apparatus 2: 100 rpm.

Time: 60 minutes.

Sodium acetate buffer—Transfer 13.6 g of sodium acetate to a 500-mL volumetric flask, add 400 mL of water, and mix. Adjust with glacial acetic acid to a pH of 5.0 ± 0.1, dilute with water to volume, and mix.

Mobile phase—Prepare a suitable mixture of acetonitrile, water, and *Sodium acetate buffer* (600:360:40). Filter through a membrane filter of 1-µm or finer porosity, and degas. The ratio of acetonitrile to *Sodium acetate buffer* may be varied to meet system suitability requirements.

Standard preparations—Transfer about 150 mg of USP Pargyline Hydrochloride RS, accurately weighed, to a 100-mL volumetric flask. Dissolve in water and dilute with water to volume, and mix. Transfer 10.0 mL of this solution to a 100-mL volumetric flask, dilute with *Dissolution Medium* to volume, and mix. Quantitatively dilute five accurately measured volumes of this stock solution with *Dissolution Medium* to obtain five solutions containing known concentrations of about 6, 12, 18, 30, and 60 µg per mL, respectively. Filter a portion of each of these solutions through a membrane filter of 1-µm or finer porosity to obtain the *Standard preparations*.

Test preparation—Filter about 15 mL of the solution under test through a membrane filter of 0.45-µm porosity, discarding the first 5 mL of the filtrate.

Chromatographic system (see *Chromatography* ⟨621⟩)—The liquid chromatograph is equipped with a 207-nm detector and a 4-mm × 30-cm column that contains 10-µm packing L1. The flow rate is between 0.5 mL and 4.5 mL per minute. Separately inject equal volumes (about 50 µL) of the *Standard preparations* and the *Test preparation* into the chromatograph, record the chromatograms, and measure the responses for the major peaks. Plot the peak responses of the *Standard preparations* versus the concentration, in µg per mL, and draw the straight line best fitting the five plotted points. From the graph so obtained, determine the concentration, in µg per mL, of pargyline hydrochloride in the *Test preparation*, and calculate the amount of $C_{11}H_{13}N.HCl$ dissolved.

Tolerances—Not less than 70% (*Q*) of the labeled amount of $C_{11}H_{13}N.HCl$ is dissolved in 60 minutes.

Uniformity of dosage units ⟨905⟩: meet the requirements.

Procedure for content uniformity—Place 1 Tablet in a 250-mL separator, add about 50 mL of water, and allow the Tablet to disintegrate completely, using a glass rod if necessary to crush it. Make the solution distinctly alkaline (pH 10) with 1 *N* sodium hydroxide, and extract with four 25-mL portions of chloroform, filtering each extract through granular anhydrous sodium sulfate, supported on a pledget of glass wool, into a beaker. Rinse the separator and the anhydrous sodium sulfate with a few small portions of chloroform, collect the rinsings in the beaker, add 25 mL of acetonitrile, and titrate with 0.01 *N* perchloric acid in dioxane VS, determining the end-point potentiometrically. Perform a blank determination, and make any necessary corrections. Each mL of 0.01 *N* perchloric acid is equivalent to 1.957 mg of $C_{11}H_{13}N.HCl$.

Assay—

Sodium phosphate buffer, pH 8.0—Dissolve 2.68 g of dibasic sodium phosphate in about 40 mL of water. Adjust with glacial acetic acid to a pH of 8.0, dilute with water to 50 mL, and mix.

Mobile phase—Prepare a suitable mixture of water, acetonitrile, and *Sodium phosphate buffer, pH 8.0* (580:400:20). Filter through a membrane filter of 1.0-µm or finer porosity, and degas. The acetonitrile concentration may be varied to meet system suitability requirements.

Internal standard solution—Dissolve benzophenone in methanol to obtain a solution having a concentration of about 0.05 mg per mL.

Standard preparation—Transfer about 100 mg of USP Pargyline Hydrochloride RS, accurately weighed, to a 200-mL volumetric flask. Transfer 20.0 mL of *Internal standard solution* and 8.0 mL of methanol to the flask, dilute with *Mobile phase* to volume, and mix.

Assay preparation—Weigh and finely powder not less than 20 Pargyline Hydrochloride Tablets. Transfer an accurately weighed portion of the powder, equivalent to about 50 mg of pargyline hydrochloride, to a 100-mL volumetric flask. Transfer 10.0 mL of *Internal standard solution* and 4.0 mL of methanol to the flask, and dilute with *Mobile phase* to volume. Place a magnetic stirring bar in the flask and stir vigorously for 10 minutes. Filter a portion of the solution through a membrane filter of 1.0-µm or finer porosity to obtain the *Assay preparation*.

Chromatographic system (see *Chromatography* ⟨621⟩)—The liquid chromatograph is equipped with a 262-nm detector and a 4-mm × 30-cm column that contains 10-µm packing L1. The flow rate is between 1 mL and 3.5 mL per minute. Chromatograph the *Standard preparation* and record the peak responses as directed under *Procedure:* the resolution, *R*, between the analyte and internal standard peaks is not less than 1.8, and the relative standard deviation for replicate injections is not more than 2.5%.

Procedure—Separately inject equal volumes (about 50 µL) of the *Standard preparation* and the *Assay preparation* into the chromatograph, record the chromatograms, and measure the responses for the major peaks. Calculate the quantity, in mg, of $C_{11}H_{13}N.HCl$ in the portion of Tablets taken by the formula:

$$100C(R_U / R_S),$$

in which *C* is the concentration, in mg per mL, of USP Pargyline Hydrochloride RS in the *Standard preparation*, and R_U and R_S are the ratios of the peak responses of the pargyline hydrochloride peak to the benzophenone peak obtained from the *Assay preparation* and the *Standard preparation*, respectively.

Paromomycin Sulfate

CH₂OH O HO OH H₂N O H₂N · xH₂SO₄ H₂N OH HOCH₂ O O O CH₂NH₂ HO OH O OH NH₂

$C_{23}H_{45}N_5O_{14}.xH_2SO_4$

D-Streptamine, *O*-2-amino-2-deoxy-α-D-glucopyranosyl-(1→4)-*O*-[*O*-2,6-diamino-2,6-dideoxy-β-L-idopyranosyl-(1→3)-β-D-ribofuranosyl-(1→5)]-2-deoxy-, sulfate (salt).

O-2,6-Diamino-2,6-dideoxy-β-L-idopyranosyl-(1→3)-*O*-β-D-ribofuranosyl-(1→5)-*O*-[2-amino-2-deoxy-α-D-glucopyranosyl-(1→4)]-2-deoxystreptamine sulfate (salt) [*1263-89-4*].

Base 615.63 [*59-04-1; 7542-37-2*].

» Paromomycin Sulfate is the sulfate salt of an antibiotic substance or substances produced by the growth of *Streptomyces rimosus* var. *paromomycinus*, or a mixture of two or more such salts. It has a potency equivalent to not less than 675 µg of paromomycin ($C_{23}H_{45}N_5O_{14}$) per mg, calculated on the dried basis.

Packaging and storage—Preserve in tight containers.

Reference standard—*USP Paromomycin Sulfate Reference Standard*—Dry in vacuum at a pressure not exceeding 5 mm of mercury at 60° for 3 hours before using.

Identification—Prepare a test solution in water containing 20 mg of paromomycin per mL. On a suitable thin-layer chromatographic plate (see *Chromatography* ⟨621⟩), coated with a 0.25-mm layer of chromatographic silica gel, apply 1 µL of this solution and 1 µL of a Standard solution of USP Paromomycin Sulfate RS containing 20 mg of paromomycin per mL. Allow the spots to dry, place the plate in a developing chamber, and develop the chromatogram in a solvent system consisting of freshly prepared ammonium acetate solution (4 in 100) until the solvent front has moved about three-fourths of the length of the plate. Remove the plate from the chamber, mark the solvent front, and allow it to air-dry for 10 minutes. Heat the plate at 105° for 1 hour, allow to cool, and spray with a 1 in 100 solution of ninhydrin in butanol. Heat the plate at 105° for 5 minutes: paromomycin appears as a red spot, and the R_f value of the principal spot obtained from the test solution corresponds to that obtained from the Standard solution.

Specific rotation ⟨781⟩: between +50° and +55°, calculated on the anhydrous basis, determined in a solution containing 50 mg per mL.

pH ⟨791⟩: between 5.0 and 7.5, in a solution (3 in 100).

Loss on drying ⟨731⟩—Dry about 100 mg in a capillary-stoppered bottle in vacuum at a pressure not exceeding 5 mm of mercury at 60° for 3 hours: it loses not more than 5.0% of its weight.

Residue on ignition ⟨281⟩: not more than 2.0%, the charred residue being moistened with 2 mL of nitric acid and 5 drops of sulfuric acid.

Assay—Proceed with Paromomycin Sulfate as directed under *Antibiotics—Microbial Assays* ⟨81⟩.

Paromomycin Sulfate Capsules

» Paromomycin Sulfate Capsules contain the equivalent of not less than 90.0 percent and not more than 125.0 percent of the labeled amount of paromomycin ($C_{23}H_{45}N_5O_{14}$).

Packaging and storage—Preserve in tight containers.

Reference standard—*USP Paromomycin Sulfate Reference Standard*—Dry in vacuum at a pressure not exceeding 5 mm of mercury at 60° for 3 hours before using.

Identification—Mix the contents of 1 Capsule with water to obtain a solution containing the equivalent of about 20 mg of paromomycin per mL. This test solution responds to the *Identification test* under *Paromomycin Sulfate*.

Disintegration ⟨701⟩: 15 minutes, the use of disks being omitted.

Uniformity of dosage units ⟨905⟩: meet the requirements.

Loss on drying ⟨731⟩—Dry about 100 mg in a capillary-stoppered bottle in vacuum at 60° for 3 hours: it loses not more than 7.0% of its weight.

Assay—Proceed with Paromomycin Sulfate Capsules as directed under *Antibiotics—Microbial Assays* ⟨81⟩, blending not less than 5 Capsules for 5 minutes in a high-speed blender with sufficient *Buffer No. 3* to obtain a stock solution of convenient concentration. Dilute this stock solution quantitatively with the same buffer to obtain a *Test Dilution* having a concentration assumed to be equal to the median dose level of the Standard.

Paromomycin Sulfate Syrup

» Paromomycin Sulfate Syrup contains the equivalent of not less than 90.0 percent and not more than 130.0 percent of the labeled amount of paromomycin ($C_{23}H_{45}N_5O_{14}$). It may contain one or more suitable buffers, colors, flavors, preservatives, and solvents.

Packaging and storage—Preserve in tight containers.

Reference standard—*USP Paromomycin Sulfate Reference Standard*—Dry in vacuum at a pressure not exceeding 5 mm of mercury at 60° for 3 hours before using.

Uniformity of dosage units ⟨905⟩—

FOR SYRUP PACKAGED IN SINGLE-UNIT CONTAINERS: meets the requirements.

pH ⟨791⟩: between 7.5 and 8.5.

Assay—Proceed with Paromomycin Syrup as directed under *Antibiotics—Microbial Assays* ⟨81⟩, diluting an accurately measured volume of Syrup with *Buffer No. 3* to yield a stock solution of convenient concentration. Dilute this stock solution quantitatively with the same buffer to obtain a *Test Dilution* having a concentration assumed to be equal to the median dose level of the Standard.

Paste, Magnesium Hydroxide—*see* Magnesium Hydroxide Paste

Paste, Zinc Oxide—*see* Zinc Oxide Paste

Peanut Oil—*see* Peanut Oil NF

Pectin

Pectin.
Pectin [*9000-69-5*].

» Pectin is a purified carbohydrate product obtained from the dilute acid extract of the inner portion of the rind of citrus fruits or from apple pomace. It consists chiefly of partially methoxylated polygalacturonic acids.

Pectin yields not less than 6.7 percent of methoxy groups (–OCH_3) and not less than 74.0 percent of galacturonic acid ($C_6H_{10}O_7$), calculated on the dried basis.

NOTE—Commercial pectin for the production of jellied food products is standardized to the convenient "150 jelly grade" by addition of dextrose or other sugars, and sometimes contains sodium citrate or other buffer salts. This monograph refers to the pure pectin to which no such additions have been made.

Packaging and storage—Preserve in tight containers.

Labeling—Label it to indicate whether it is of apple or of citrus origin.

Identification—

A: Heat 1 g with 9 mL of water on a steam bath until a solution is formed, replacing water lost by evaporation: it forms a stiff gel on cooling.

B: To a solution (1 in 100) add an equal volume of alcohol: a translucent, gelatinous precipitate is formed (*distinction from most gums*).

C: To 10 mL of a solution (1 in 100) add 1 mL of thorium nitrate TS, stir, and allow to stand for 2 minutes: a stable precipitate or gel forms (*distinction from gums*).

D: To 5 mL of a solution (1 in 100) add 1 mL of 2 *N* sodium hydroxide, and allow to stand at room temperature for 15 minutes: a gel or semigel forms (*distinction from tragacanth*).

E: Acidify the gel from the preceding test with 3 *N* hydrochloric acid, and shake: a voluminous, colorless, gelatinous precipitate forms, which upon boiling becomes white and flocculent (*pectic acid*).

Microbial limit—It meets the requirements of the test for absence of *Salmonella* species under *Microbial Limit Tests* ⟨61⟩.

Loss on drying ⟨731⟩—Dry it at 105° for 3 hours: it loses not more than 10.0% of its weight.

Arsenic, *Method II* ⟨211⟩: 3 ppm.

Lead—Add 2.0 g of Pectin to 20 mL of nitric acid in a 250-mL conical flask, mix, and heat the contents carefully until the Pectin is dissolved. Continue the heating until the volume is reduced to about 7 mL. Cool rapidly to room temperature, transfer to a 100-mL volumetric flask, and dilute with water to volume. A 50.0-mL portion of this solution contains not more than 5 µg of lead (corresponding to not more than 0.0005% of Pb) when tested according to the limit test for *Lead* ⟨251⟩, 15 mL of ammonium citrate solution, 3 mL of potassium cyanide solution, and 500 µL of hydroxylamine hydrochloride solution being used. After the first dithizone extractions, wash the combined chloroform layers with 5 mL of water, discarding the water layer and continuing in the usual manner by extracting with 20 mL of dilute nitric acid (1 in 100).

Sugars and organic acids—Place 1 g in a 500-mL flask, moisten it with 3 to 5 mL of alcohol, pour in rapidly 100 mL of water, shake, and allow to stand until solution is complete. To this solution add 100 mL of alcohol containing 0.3 mL of hydrochloric acid, mix, and filter rapidly. Measure 25 mL of the filtrate into a tared dish, evaporate the liquid on a steam bath and dry the residue in a vacuum oven at 50° for 2 hours: the weight of the residue does not exceed 20 mg.

Assay for methoxy groups—Transfer 5.00 g of Pectin to a suitable beaker, and stir for 10 minutes with a mixture of 5 mL of hydrochloric acid and 100 mL of 60 percent alcohol. Transfer to a sintered-glass filter (30- to 60-mL crucible or Büchner type, coarse), and wash with six 15-mL portions of the hydrochloric acid–60 percent alcohol mixture, followed by 60 percent alcohol until the filtrate is free from chlorides. Finally wash with 20 mL of alcohol, dry for 1 hour at 105°, cool, and weigh. Transfer exactly one-tenth of the total net weight of the dried sample (representing 500 mg of the original unwashed sample) to a 250-mL conical flask, and moisten with 2 mL of alcohol. Add 100 mL of carbon dioxide–free water, insert the stopper, and swirl occasionally until the Pectin is completely dissolved. Add 5 drops of phenolphthalein TS, titrate with 0.5 *N* sodium hydroxide VS, and record the results as the *initial titer*. Add 20.0 mL of 0.5 *N* sodium hydroxide VS, insert the stopper, shake vigorously, and allow to stand for 15 minutes. Add 20.0 mL of 0.5 *N* hydrochloric acid VS, and shake until the pink color disappears. Add phenolphthalein TS, and titrate with 0.5 *N* sodium hydroxide VS to a faint pink color that persists after vigorous shaking: record this value as the *saponification titer*. Each mL of 0.5 *N* sodium hydroxide used in the *saponification titer* is equivalent to 15.52 mg of $-OCH_3$.

Assay for galacturonic acid—Each mL of 0.5 *N* sodium hydroxide used in the total titration (the *initial titer* added to the *saponification titer*) in the *Assay for methoxy groups* is equivalent to 97.07 mg of $C_6H_{10}O_7$.

Pellets, Desoxycorticosterone Acetate—*see* Desoxycorticosterone Acetate Pellets

Pellets, Testosterone—*see* Testosterone Pellets

Penicillamine

$$HS-\overset{CH_3}{\underset{CH_3}{C}}-\overset{H}{\underset{NH_2}{C}}-COOH$$

$C_5H_{11}NO_2S$ 149.21

D-Valine, 3-mercapto-.

D-3-Mercaptovaline [*52-67-5*].

» Penicillamine contains not less than 97.0 percent and not more than 100.5 percent of $C_5H_{11}NO_2S$, calculated on the dried basis.

Packaging and storage—Preserve in tight containers.

Reference standards—*USP Penicillamine Reference Standard*—Store in a cool place, and do not dry before using. *USP Penicillin G Potassium Reference Standard*—Do not dry before using.

Identification—

A: The infrared absorption spectrum of a mineral oil dispersion of it (50 mg in 300 mg) exhibits maxima only at the same wavelengths as that of the USP Penicillamine RS, similarly treated.

B: Dissolve 10 mg in 5 mL of water, and add 1 drop of 5 *N* sodium hydroxide and 20 mg of ninhydrin: a blue or violet-blue color is produced immediately.

C: Dissolve 20 mg in 4 mL of water, add 2 mL of phosphotungstic acid solution (1 in 10), and heat nearly to boiling: a deep blue color is produced immediately.

Specific rotation ⟨781⟩: between −60.5° and −64.5°, determined in a 1 in 20 solution in 1.0 *N* sodium hydroxide, calculated on the dried basis.

pH ⟨791⟩: between 4.5 and 5.5, in a solution (1 in 100).

Loss on drying ⟨731⟩—Dry about 100 mg, accurately weighed, in a capillary-stoppered bottle in vacuum at a pressure not exceeding 5 mm of mercury at 60° for 3 hours: it loses not more than 0.5% of its weight.

Residue on ignition ⟨281⟩: not more than 0.1%, the charred residue being moistened with 2 mL of nitric acid and 5 drops of sulfuric acid.

Heavy metals, *Method II* ⟨231⟩: not more than 0.002%.

Limit of penicillin activity—

pH 2.5 buffer—Dissolve 100 g of monobasic potassium phosphate in water, add 0.2 mL of hydrochloric acid, dilute with water to 1000 mL, and mix. Adjust, if necessary, with phosphoric acid or with 10 *N* potassium hydroxide to a pH of 2.5.

Standard preparation—Prepare as directed for Penicillin G in Table 2 under *Antibiotics—Microbial Assays* ⟨81⟩, except to prepare a final stock solution containing 100 Penicillin G Units per mL and six test dilutions ranging from 0.005 Penicillin G Unit per mL to 0.2 Penicillin G Unit per mL, and to use a median dose of the Standard of 0.050 Penicillin G Unit per mL.

Test preparation—Dissolve 1.0 g in water to make 18.0 mL, transfer 9.0 mL of this solution to a separator, add 20 mL of amyl acetate and 1 mL of *pH 2.5 buffer*, and shake. Allow the layers to separate, and draw off the aqueous layer into a second separator, retaining the amyl acetate extract in the first separator. Check the pH of the aqueous layer, and if it is greater than 3.0 adjust it with hydrochloric acid to a pH of 2.5, and extract with 20 mL of amyl acetate. Discard the aqueous layer, and add the amyl acetate extract to the first separator. Wash the combined amyl acetate extracts with 10 mL of diluted *pH 2.5 buffer* (1 in 10), and discard the aqueous layer. Extract the amyl acetate with 10.0 mL of *Buffer No. 1* (see *Phosphate Buffers and Other Solutions* in the section, *Media and Diluents*, under *Antibiotics—Microbial Assays* ⟨81⟩). Use a portion of the buffer extract as *Test solution A*. To a 5-mL portion of the extract add 0.1 mL of penicillinase solution, and incubate at 36° to 37.5° for 60 minutes (*Test solution B*).

Preparation of inoculum—Prepare as directed under *Antibiotics—Microbial Assays* ⟨81⟩, using *Micrococcus luteus* (ATCC 9341) as the test organism, and an inoculum that gives clear sharp zones of inhibition 17 mm to 21 mm in diameter with the median dose level of the Standard.

Procedure—Proceed as directed for the *Cylinder-Plate Method* under *Antibiotics—Microbial Assays* ⟨81⟩, using 10 mL of Medium 1 for the base layer and 4 mL of inoculated Medium 4 for the seed layer, and incubating the plates at 29° to 31°, except on each test plate to fill 2 cylinders with *Test solution A*, 2 cylinders with *Test solution B*, and 2 cylinders with the median dose of the Standard. If *Test solution A* yields a zone of inhibition and *Test solution B* does not, not more than 0.02 Penicillin G Unit is found in each mL of *Test solution A* (0.2 Penicillin G Unit per g).

Mercury—

Dithizone titrant—Use *Standard Dithizone Solution.*

Standard solution—Transfer 135.4 mg of mercuric chloride to a 100-mL volumetric flask, add 0.25 *N* sulfuric acid to volume, and mix. This solution contains the equivalent of 100 mg of Hg in 100 mL.

Diluted standard solution—Pipet 2 mL of *Standard solution* into a 100-mL volumetric flask, add 0.25 *N* sulfuric acid to volume, and mix. Each mL of this solution contains the equivalent of 20 µg of Hg.

Standardization—Pipet 1 mL of *Diluted standard solution* into a 250-mL separator, and add 100 mL of 0.25 *N* sulfuric acid, 90 mL of water, and 10 mL of hydroxylamine hydrochloride solution (1 in 5). Then add 1 mL of disodium ethylenediaminetetraacetate solution (1 in 50), 1 mL of glacial acetic acid, and 5 mL of chloroform, shake for 1 minute, allow to separate, and discard the chloroform layer. To the solution add *Dithizone titrant*, in portions of 0.3 mL to 0.5 mL, from a 10-mL buret. After each addition, shake the mixture 20 times, and allow the chloroform layer to separate and discard it. Continue until an addition of *Dithizone titrant* remains green after the shaking. Calculate the quantity, in µg, of mercury equivalent to 1 mL of *Dithizone titrant* by dividing 20 by the number of mL of *Dithizone titrant* added.

Procedure—[NOTE—Conduct this procedure in subdued light, since mercuric dithizonate is sensitive to light.] Transfer 500 mg of Penicillamine to a 650-mL Kjeldahl flask containing a few glass beads, incline the flask at an angle of about 45°, and add 2.5 mL of nitric acid through a small funnel placed in the mouth of the flask. Allow the mixture to stand at room temperature until nitrous oxide fumes are evolved and vigorous reaction subsides (5 to 30 minutes). Add 2.5 mL of sulfuric acid through the funnel, and heat, gently at first and then to the production of fumes of sulfur trioxide, then cool. Cautiously add 2.5 mL of nitric acid, again heat to the production of sulfur trioxide fumes, and cool. Repeat the treatment with nitric acid and heat, then cool, and cautiously add 50 mL of water, rinsing the funnel and collecting the rinsings in the flask. Remove the funnel, boil the solution down to approximately half its volume (about 25 mL), and cool to room temperature. Transfer to a 250-mL separator with the aid of water, and add water to make about 50 mL. Add 1 mL of disodium ethylenediaminetetraacetate solution (1 in 50) and 1 mL of glacial acetic acid, and extract with small portions of chloroform until the last chloroform extract remains colorless. Discard the chloroform extract, and add 50 mL of 0.25 *N* sulfuric acid, 90 mL of water, and 10 mL of hydroxylamine hydrochloride solution (1 in 5). Add *Dithizone titrant*, in portions of 0.3 mL to 0.5 mL, from a 10-mL buret. After each addition, shake the mixture 20 times, and allow the chloroform layer to separate and discard it. Continue until an addition of *Dithizone titrant* remains green after the shaking. Calculate the amount of mercury present: the limit is 10 µg (0.002%).

Assay—

Mercuric acetate titrant—Dissolve 16 g of mercuric acetate in about 200 mL of water, add 5 mL of glacial acetic acid, dilute with water to make 1000 mL, and standardize the solution as follows. Dissolve about 300 mg of USP Penicillamine RS, accurately weighed, in 200 mL of water, add 10 g of sodium acetate, and render the solution acid to litmus with acetic acid. Add 1 mL of a freshly prepared 1 in 200 solution of diphenylcarbazone in alcohol, and titrate with *Mercuric acetate titrant* until a rose-violet color persists for 2 to 3 minutes. Perform a blank determination, and make any necessary correction. Calculate the molarity of the solution by the formula:

$$W/149.21V,$$

in which W is the weight, in mg, of USP Penicillamine RS taken, and V is the volume, in mL, of *Mercuric acetate titrant* consumed.

Procedure—Dissolve about 300 mg of Penicillamine, accurately weighed, in 200 mL of water, add 10 g of sodium acetate, and render the solution acid to litmus with acetic acid. Add 1 mL of a freshly prepared 1 in 200 solution of diphenylcarbazone in alcohol, and titrate with *Mercuric acetate titrant* until a rose-violet color persists for 2 to 3 minutes. Perform a blank determination, and make any necessary correction. Each mL of 0.05 *M Mercuric acetate titrant* is equivalent to 7.461 mg of $C_5H_{11}NO_2S$.

Penicillamine Capsules

» Penicillamine Capsules contain not less than 90.0 percent and not more than 110.0 percent of the labeled amount of $C_5H_{11}NO_2S$.

Packaging and storage—Preserve in tight containers.

Reference standard—*USP Penicillamine Reference Standard*—Store in a cool place, and do not dry before using.

Identification—The contents of the Capsules respond to *Identification test A* under *Penicillamine Tablets* and to *Identification test C* under *Penicillamine*.

Dissolution ⟨711⟩—

Medium: 0.1 *N* hydrochloric acid; 900 mL.

Apparatus 1: 100 rpm.

Time: 30 minutes.

Dilute hydrochloric acid—Dilute 37 mL of hydrochloric acid with water to one liter.

Ammonium sulfamate reagent—Dissolve 250 mg of ammonium sulfamate in 100 mL of *Dilute hydrochloric acid.*

N-1-Naphthylethylenediamine dihydrochloride reagent—Dissolve 100 mg of *N*-1-naphthylethylenediamine dihydrochloride in 100 mL of *Dilute hydrochloric acid.*

Sulfanilamide–mercuric chloride reagent—Dissolve 100 mg of sulfanilamide and 100 mg of mercuric chloride in 100 mL of *Dilute hydrochloric acid.*

Sodium nitrite reagent—Dissolve 200 mg of sodium nitrite in 100 mL of dilute sulfuric acid (1 in 50). Prepare fresh.

Standard preparation—Dissolve an accurately weighed quantity of USP Penicillamine RS in 0.1 *N* hydrochloric acid to obtain a solution having a known concentration of about 250 µg per mL.

Procedure—Pipet an aliquot of the filtered test solution, estimated to contain about 278 µg of penicillamine, into a 100-mL volumetric flask. Into a similar flask pipet an equivalent volume of 0.1 *N* hydrochloric acid to provide a reagent blank, and into a third 100-mL volumetric flask pipet 1 mL of *Standard preparation.* Treat each flask as follows. Add by pipet 3 mL of *Sodium nitrite reagent*, and mix by swirling occasionally. After 5 minutes, add 10 mL of *Ammonium sulfamate reagent*, swirl, and allow to stand for an additional 5 minutes. Add 5 mL of *Sulfanilamide–mercuric chloride reagent*, swirl, and immediately add 10 mL of *N-1-Naphthylethylenediamine dihydrochloride reagent.* Dilute with water to volume, and mix. Determine the absorbances of both solutions in 1-cm cells at the wavelength of maximum absorbance at about 540 nm, with a suitable spectrophotometer, against the reagent blank. Calculate the percentage dissolution of the Capsule by the formula:

$$90(C/WV)(A_U/A_S),$$

in which C is the concentration, in µg per mL, of USP Penicillamine RS in the *Standard preparation*, W is the labeled quantity, in mg, of penicillamine in the Capsule, V is the volume, in mL, of the aliquot of test solution used, and A_U and A_S are the absorbances of the solutions from the test solution and the *Standard preparation*, respectively.

Tolerances—Not less than 80% (*Q*) of the labeled amount of $C_5H_{11}NO_2S$ is dissolved in 30 minutes.

Uniformity of dosage units ⟨905⟩: meet the requirements.

Water, *Method I* ⟨921⟩: not more than 7.5%.

Assay—

Mercuric acetate titrant—Prepare and standardize as directed in the *Assay* under *Penicillamine.*

Procedure—Empty the contents of not less than 10 Penicillamine Capsules into a tared weighing bottle, weigh, and mix. Calculate the average weight per Capsule, and dissolve an accurately weighed quantity of the powder, equivalent to about 300 mg of penicillamine, in 200 mL of water. Proceed as directed in the *Assay* under *Penicillamine*, beginning with "add 10 g of sodium acetate." Each mL of 0.05 *M Mercuric acetate titrant* is equivalent to 7.461 mg of $C_5H_{11}NO_2S$.

Penicillamine Tablets

» Penicillamine Tablets contain not less than 90.0 percent and not more than 110.0 percent of the labeled amount of $C_5H_{11}NO_2S$.

Packaging and storage—Preserve in tight containers.

Reference standard—*USP Penicillamine Reference Standard*—Store in a cool place, and do not dry before using.

Identification—

A: Transfer a portion of finely powdered Tablets, equivalent to about 100 mg of penicillamine, to a 10-mL volumetric flask, dilute with methanol to volume, add 2 drops of 3 *N* hydrochloric acid, mix, and filter. Use the filtrate as the test solution. Prepare a Standard solution by dissolving 100 mg of USP Penicillamine RS in 10 mL of methanol, adding 2 drops of 3 *N* hydrochloric acid, and mixing. On a suitable thin-layer chromatographic plate (see *Chromatography* ⟨621⟩), coated with a 0.25-mm layer of chromatographic silica gel mixture, heated at 105° for 30 minutes, and allowed to cool before use, separately apply 10-µL portions of the test solution and the Standard solution. Allow the spots to dry, and develop the chromatogram in a solvent system consisting of a mixture of butyl alcohol, glacial acetic acid, and water (8:2:2) until the solvent front has moved about three-fourths of the length of the plate. Remove the plate, mark the solvent front, allow the solvent to evaporate, and place the plate in an atmosphere of iodine vapors. After a few minutes, spray the plate with a 1 in 300 solution of ninhydrin in dehydrated alcohol, heat it at 105° for about 10 minutes, allow it to cool, and examine it: the R_f values, colors, and intensities of the principal spots obtained from the test solution correspond to those obtained from the Standard solution.

B: A portion of powdered Tablets responds to *Identification test C* under *Penicillamine.*

Dissolution ⟨711⟩—

Medium: 0.5 percent disodium ethylenediaminetetraacetate; 900 mL.

Apparatus 1: 150 rpm.

Time: 60 minutes.

Mobile phase—Prepare a filtered and degassed solution of 0.01 *M* dibasic sodium phosphate and 0.01 *M* monobasic potassium phosphate (60:40). If necessary, adjust the solution by the addition of 0.01 *M* dibasic sodium phosphate or 0.01 *M* monobasic potassium phosphate to a pH of 7.0 ± 0.1.

Standard solution—Prepare a solution of USP Penicillamine RS in 0.5 percent disodium ethylenediaminetetraacetate having an accurately known concentration of about 0.28 mg per mL.

Chromatographic system (see *Chromatography* ⟨621⟩)—The liquid chromatograph is equipped with a 254-nm detector and a 3.9-mm × 30-cm column that contains packing L1. The flow rate is about 1 mL per minute. Chromatograph replicate injections of the *Standard solution*, and record the peak responses as directed under *Procedure:* the relative standard deviation is not more than 2.0%, and the resolution factor between the solvent peak and penicillamine is not less than 1.5.

Procedure—Separately inject equal volumes (about 80 µL) of the *Standard solution* and a filtered portion of the solution under test into the chromatograph, record the chromatograms, measure the responses for the major peaks, and calculate the amount of $C_5H_{11}NO_2S$ dissolved per Tablet.

Tolerances—Not less than 60% (*Q*) of the labeled amount of $C_5H_{11}NO_2S$ is dissolved in 60 minutes.

Uniformity of dosage units ⟨905⟩: meet the requirements.

Loss on drying ⟨731⟩—Dry about 100 mg of finely ground Tablets, accurately weighed, in a capillary-stoppered bottle in vacuum at a pressure not exceeding 5 mm of mercury at 60° for 3 hours: it loses not more than 3.0% of its weight.

Assay—

Mercuric acetate titrant—Prepare and standardize as directed in the *Assay* under *Penicillamine.*

Procedure—Weigh and finely powder not less than 10 Penicillamine Tablets. Suspend an accurately weighed portion of the powder, equivalent to about 300 mg of penicillamine, in 200 mL of water, and mix. Filter, washing the filter with about 20 mL of water. Proceed as directed in the *Assay* under *Penicillamine*, beginning with "add 10 g of sodium acetate." Each mL of 0.05 *M Mercuric acetate titrant* is equivalent to 7.461 mg of $C_5H_{11}NO_2S$.

Penicillin G Benzathine

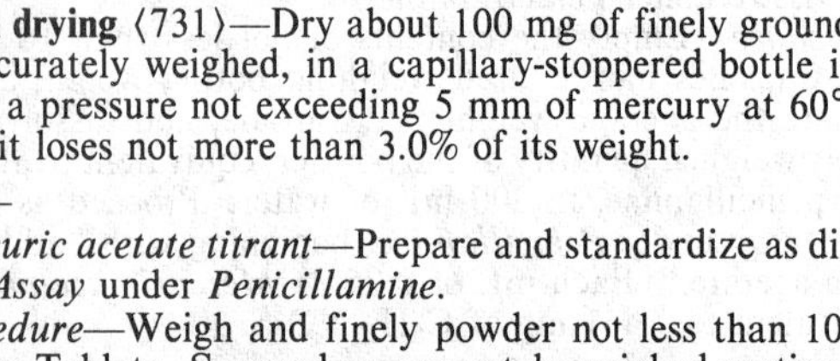

$(C_{16}H_{18}N_2O_4S)_2 \cdot C_{16}H_{20}N_2 \cdot 4H_2O$ 981.19

4-Thia-1-azabicyclo[3.2.0]heptane-2-carboxylic acid, 3,3-dimethyl-7-oxo-6-[(phenylacetyl)amino]-, [2*S*-(2α,5α,6β)]-, compd. with *N,N'*-bis(phenylmethyl)-1,2-ethanediamine (2:1), tetrahydrate.

(2*S*,5*R*,6*R*)-3,3-Dimethyl-7-oxo-6-(2-phenylacetamido)-4-thia-1-azabicyclo[3.2.0]heptane-2-carboxylic acid compound with *N,N'*-dibenzylethylenediamine (2:1), tetrahydrate [*41372-02-5*].

Anhydrous 909.13 [*1538-09-6*].

» Penicillin G Benzathine has a potency of not less than 1090 Penicillin G Units and not more than 1272 Penicillin G Units per mg.

Packaging and storage—Preserve in tight containers.

Reference standards—*USP Penicillin G Potassium Reference Standard*—Do not dry before using. *USP Penicillin G Benzathine Reference Standard*—Do not dry before using.

Identification—Transfer about 50 mg, accurately weighed, to a 100-mL volumetric flask, dissolve in methanol, dilute with methanol to volume, and mix: the ultraviolet absorption spectrum of the solution so obtained exhibits maxima and minima at the same wavelengths as that of a similar solution of USP Penicillin G Benzathine RS, concomitantly measured, and the absorptivity at the wavelength of maximum absorbance at about 263 nm is between 85.0% and 110.0% of that of the USP Penicillin G Benzathine RS.

Crystallinity ⟨695⟩: meets the requirements.

pH ⟨791⟩: between 4.0 and 6.5, in a solution prepared by dissolving 50 mg in 50 mL of dehydrated alcohol, adding 50 mL of water, and mixing.

Water, *Method I* ⟨921⟩: between 5.0% and 8.0%.

Penicillin G content (see *Penicillin G Determination* ⟨475⟩)—Transfer about 40 mg of USP Penicillin G Potassium RS, accurately weighed, to a 50-mL volumetric flask, disperse in 10 mL of acetonitrile, and add 5 mL of methanol to dissolve. Without delay, dilute with *0.05 M Phosphate Buffer, pH 6* to volume, and mix (*Standard Preparation*). Similarly prepare a *Test Preparation*, using 53 mg of Penicillin G Benzathine, accurately weighed: between 61.3% and 71.6% of $C_{16}H_{18}N_2O_4S$ is found.

Assay—

Standard preparation—Using USP Penicillin G Potassium RS, prepare as directed for *Standard Preparation* under *Iodometric Assay—Antibiotics* ⟨425⟩.

Assay preparation—Dissolve an accurately weighed quantity of Penicillin G Benzathine in 1 *N* sodium hydroxide to obtain an *Assay preparation* containing about 2000 Penicillin G Units per mL. Pipet 2.0 mL of this solution into a glass-stoppered, 125-mL conical flask.

Blank preparation—Dilute an accurately weighed quantity of Penicillin G Benzathine quantitatively with *Buffer No. 1* to obtain a suspension containing 2000 Penicillin G Units per mL. Pipet 2.0 mL of this solution into a glass-stoppered, 125-mL conical flask.

Procedure—Proceed as directed for *Procedure* under *Iodometric Assay—Antibiotics* ⟨425⟩, except in performing the *Inactivation and titration* to omit the addition of 1.0 *N* sodium hydroxide to the *Assay preparation*, and in performing the *Blank determination* to use the *Blank preparation* in place of the *Assay preparation.* Calculate the potency, in Penicillin G Units per mg, of the Penicillin G Benzathine taken by the formula:

$$(F/2D)(B - I),$$

in which *D* is the concentration, in mg per mL, of the *Assay preparation*, on the basis of the weight of Penicillin G Benzathine taken and the extent of dilution.

Sterile Penicillin G Benzathine

» Sterile Penicillin G Benzathine is Penicillin G Benzathine suitable for parenteral use. It has a potency of not less than 1090 and not more than 1272 Penicillin G Units per mg.

Packaging and storage—Preserve in *Containers for Sterile Solids* as described under *Injections* ⟨1⟩.

Reference standards—*USP Penicillin G Potassium Reference Standard*—Do not dry before using. *USP Penicillin G Benzathine Reference Standard*—Do not dry before using.

Pyrogen—It meets the requirements of the *Pyrogen Test* ⟨151⟩, the test dose being 0.5 mL per kg of a solution in pyrogen-free saline TS containing 4000 Penicillin G Units per mL.

Sterility—It meets the requirements under *Sterility Tests* ⟨71⟩, when tested as directed in the section, *Test Procedures for Direct Transfer to Test Media*, except to use Fluid Thioglycollate Medium and Soybean-Casein Digest Medium containing polysorbate 80 solution (1 in 200) and an amount of sterile penicillinase sufficient to inactivate the penicillin G in each tube, and to shake the vessels once daily.

Other requirements—It responds to the *Identification test*, and meets the requirements for *pH*, *Water*, *Crystallinity*, *Penicillin G content*, and *Assay* under *Penicillin G Benzathine*.

Penicillin G Benzathine Oral Suspension

» Penicillin G Benzathine Oral Suspension contains not less than 90.0 percent and not more than 120.0 percent of the labeled amount of penicillin G. It contains one or more suitable buffers, colors, dispersants, flavors, and preservatives.

Packaging and storage—Preserve in tight containers.

Reference standards—*USP Penicillin G Potassium Reference Standard*—Do not dry before using. *USP Penicillin G Benzathine Reference Standard*—Do not dry before using.

Identification—Mix a portion of it with methanol to obtain a solution containing about 3000 Penicillin G Units per mL. Apply 20 µL of this test solution and 20 µL of a Standard solution of USP Penicillin G Benzathine RS in methanol containing 2.5 mg per mL to a suitable thin-layer chromatographic plate (see *Chromatography* ⟨621⟩) coated with a 0.25-mm layer of chromatographic silica gel, and allow the spots to dry. Using an unlined developing chamber, develop the chromatogram in a solvent system consisting of a mixture of methanol, acetonitrile, and ammonium hydroxide (70:30:3) until the solvent front has moved about three-fourths of the length of the plate. Remove the plate from the developing chamber, and allow to air-dry. Spray the plate uniformly with a spray reagent prepared as follows. Dissolve 20 g of tartaric acid and 1.7 g of bismuth subnitrate in 80 mL of water. Add 2.5 mL of this solution, 2.5 mL of potassium iodide solution (4 in 10), and 10 g of tartaric acid to 50 mL of water, and mix. Examine the chromatograms: the principal spot obtained from the test solution corresponds in R_f value to that obtained from the Standard solution.

Uniformity of dosage units ⟨905⟩—

FOR SUSPENSION PACKAGED IN SINGLE-UNIT CONTAINERS: meets the requirements.

pH ⟨791⟩: between 6.0 and 7.0.

Assay—

Standard preparation—Using Penicillin G Potassium RS, prepare as directed for *Standard Preparation* under *Iodometric Assay—Antibiotics* ⟨425⟩.

Assay preparation—Dilute an accurately measured volume of Penicillin G Benzathine Oral Suspension, freshly mixed and free from air bubbles, quantitatively with 1 *N* sodium hydroxide to obtain a solution having a concentration of about 2000 Penicillin G Units per mL. Pipet 2 mL of this solution into a glass-stoppered, 125-mL conical flask.

Blank preparation—Dilute an accurately measured volume of Penicillin G Benzathine Oral Suspension, freshly mixed and free from air bubbles, quantitatively with *Buffer No. 1* to obtain a suspension containing about 2000 Penicillin G Units per mL. Pipet 2 mL of this solution into a glass-stoppered, 125-mL conical flask.

Procedure—Proceed as directed for *Procedure* under *Iodometric Assay—Antibiotics* ⟨425⟩, except in performing the *Inactivation and titration* to omit the addition of 1.0 *N* sodium hydroxide to the *Assay preparation*, and in performing the *Blank determination* to use the *Blank preparation* in place of the *Assay preparation*. Calculate the quantity, in Penicillin G Units, in each mL of the Oral Suspension taken by the formula:

$$(L/2D)(F)(B - I),$$

in which L is the labeled quantity, in Penicillin G Units per mL, in the Penicillin G Benzathine Oral Suspension taken, and D is the concentration, in Penicillin G Units per mL, of the *Assay preparation* on the basis of the labeled quantity in the Penicillin G Benzathine Oral Suspension and the extent of dilution.

Sterile Penicillin G Benzathine Suspension

» Sterile Penicillin G Benzathine Suspension is a sterile suspension of Sterile Penicillin G Benzathine in Water for Injection with one or more suitable buffers, dispersants, preservatives, and suspending agents. It contains not less than 90.0 percent and not more than 115.0 percent of the labeled amount of penicillin.

Packaging and storage—Preserve in single-dose or in multiple-dose containers, preferably of Type I or Type II glass, in a refrigerator.

Reference standards—*USP Penicillin G Potassium Reference Standard*—Do not dry before using. *USP Penicillin G Benzathine Reference Standard*—Do not dry before using.

Identification—It responds to the *Identification test* under *Penicillin G Benzathine Oral Suspension.*

Pyrogen—Dilute a portion of the Suspension with pyrogen-free saline TS to a concentration of 4000 Penicillin G Units per mL: it meets the requirements of the *Pyrogen Test* ⟨151⟩, the test dose being 0.5 mL per kg.

Sterility—It meets the requirements under *Sterility Tests* ⟨71⟩, when tested as directed in the section, *Test Procedures for Direct Transfer to Test Media*, except to use Fluid Thioglycollate Medium and Soybean-Casein Digest Medium containing polysorbate 80 solution (1 in 200) and an amount of sterile penicillinase sufficient to inactivate the penicillin G in each vessel, and to shake the vessels once daily.

pH ⟨791⟩: between 5.0 and 7.5.

Other requirements—It meets the requirements under *Injections* ⟨1⟩.

Assay—

Standard preparation—Using Penicillin G Potassium RS, prepare as directed for *Standard Preparation* under *Iodometric Assay—Antibiotics* ⟨425⟩.

Assay preparation—Using a suitable hypodermic needle and syringe, withdraw an accurately measured volume of Sterile Penicillin G Benzathine Suspension, equivalent to about 300,000 Penicillin G Units, and dilute quantitatively with 1 *N* sodium hydroxide to obtain an *Assay preparation* containing about 2000 Penicillin G Units per mL. Pipet 2.0 mL of this solution into a glass-stoppered, 125-mL conical flask.

Blank preparation—Using a suitable hypodermic needle and syringe, withdraw an accurately measured volume of Sterile Penicillin G Benzathine Suspension, equivalent to about 300,000 Penicillin G Units, and dilute quantitatively with *Buffer No. 1* to obtain a suspension containing about 2000 Penicillin G Units per mL. Pipet 2 mL of this solution into a glass-stoppered, 125-mL conical flask.

Procedure—Proceed as directed for *Procedure* under *Iodometric Assay—Antibiotics* ⟨425⟩, except in performing the *Inactivation and titration* to omit the addition of 1.0 *N* sodium hydroxide to the *Assay preparation*, and in performing the *Blank determination* to use the *Blank preparation* in place of the *Assay preparation.* Calculate the quantity, in Penicillin G Units, in each mL of the Suspension taken by the formula:

$$(L/2D)(F)(B - I),$$

in which L is the labeled quantity, in Penicillin G Units per mL, in the Sterile Penicillin G Benzathine Suspension taken, and D is the concentration, in Penicillin G Units per mL, in the *Assay preparation* on the basis of the labeled quantity in the Sterile Penicillin G Benzathine Suspension and the extent of dilution.

Penicillin G Benzathine Tablets

» Penicillin G Benzathine Tablets contain not less than 90.0 percent and not more than 120.0 percent of the labeled amount of penicillin G.

Packaging and storage—Preserve in tight containers.

Reference standards—*USP Penicillin G Potassium Reference Standard*—Do not dry before using. *USP Penicillin G Benzathine Reference Standard*—Do not dry before using.

Identification—Shake a suitable quantity of finely powdered Tablets with methanol to obtain a solution containing about 3000 Penicillin G Units per mL, and filter: the filtrate so obtained responds to the *Identification test* under *Penicillin G Benzathine Oral Suspension.*

Disintegration ⟨701⟩: 60 minutes, simulated gastric fluid TS being used in place of water as the test medium.

Uniformity of dosage units ⟨905⟩: meet the requirements.

Water, *Method I* ⟨921⟩: not more than 8.0%.

Assay—

Standard preparation—Using Penicillin G Potassium RS, prepare as directed for *Standard Preparation* under *Iodometric Assay—Antibiotics* ⟨425⟩.

Assay preparation—Weigh and finely powder not less than 20 Penicillin G Benzathine Tablets. Transfer a portion of the powder, accurately weighed, equivalent to about 200,000 Penicillin G Units, to a 100-mL volumetric flask, add 10 mL of 1.0 *N* sodium hydroxide, and mix. Allow to stand for 15 minutes, add 10 mL of 1.2 *N* hydrochloric acid, dilute with water to volume, and mix. Pipet 2 mL of this solution into a glass-stoppered, 125-mL conical flask.

Blank preparation—Transfer an accurately weighed portion of the powdered Tablets remaining from the preparation of the *Assay preparation*, equivalent to about 200,000 Penicillin G Units, to a 100-mL volumetric flask, dilute with *Buffer No. 1* to volume, and mix. Pipet 2 mL of this solution into a glass-stoppered, 125-mL conical flask.

Procedure—Proceed as directed for *Procedure* under *Iodometric Assay—Antibiotics* ⟨425⟩, except in performing the *Inactivation and titration* to omit the addition of 1.0 *N* sodium hydroxide and 1.2 *N* hydrochloric acid, and in the *Blank determination* to use the *Blank preparation* in place of the *Assay preparation.* Calculate the quantity, in Penicillin G Units, in the portion of Tablets taken by the formula:

$$(T/2D)(F)(B - I),$$

in which T is the labeled quantity, in Penicillin G Units, in each Tablet, and D is the concentration, in Penicillin G Units per mL, of the *Assay preparation* on the basis of the labeled quantity in each Tablet and the extent of dilution.

Sterile Penicillin G Benzathine and Penicillin G Procaine Suspension

» Sterile Penicillin G Benzathine and Penicillin G Procaine Suspension is a sterile suspension of Sterile Penicillin G Benzathine and Sterile Penicillin G Procaine in Water for Injection. It may contain one or more suitable buffers, preservatives, and suspending agents. It contains not less than 90.0 percent and not more than 115.0 percent of the labeled amounts of Penicillin G Benzathine and Penicillin G Procaine.

Packaging and storage—Preserve in single-dose or in multiple-dose containers, preferably of Type I or Type III glass.

Reference standards—*USP Penicillin G Benzathine Reference Standard*—Do not dry before using. *USP Penicillin G Potassium Reference Standard*—Do not dry before using. *USP Procaine Hydrochloride Reference Standard*—Dry over silica gel for 18 hours before using.

Identification—

A: It responds to the *Identification test* under *Penicillin G Benzathine Oral Suspension:* the spot obtained from the test solution, corresponding in R_f value to that obtained from the Standard solution, is completely resolved from a second spot, produced by penicillin G procaine.

B: It responds to the *Identification test* under *Sterile Penicillin G Procaine.*

pH ⟨791⟩: between 5.0 and 7.5.

Other requirements—It meets the requirements for *Pyrogen* and *Sterility* under *Sterile Penicillin G Procaine Suspension.* It meets also the requirements under *Injections* ⟨1⟩.

Assay for penicillin G procaine—

Standard preparations—Dissolve an accurately weighed quantity of USP Procaine Hydrochloride RS in water to obtain a stock solution having a known concentration of about 28 µg per mL. Transfer 1.0, 2.0, 3.0, 4.0, and 5.0 mL, respectively, of this stock solution to each of five 25-mL volumetric flasks. Transfer 4.0, 3.0, 2.0, and 1.0 mL of water to the first four flasks, respectively.

Assay preparation—Where the Suspension is represented as being in a single-dose container, withdraw all of the withdrawable contents, using a suitable hypodermic needle and syringe. Where the label states the quantities of Penicillin G Benzathine and Penicillin G Procaine in a given volume of Suspension, remove an accurately measured volume of the Suspension. For each 300,000 Penicillin G Benzathine Units in the specimen of Suspension taken, add 20 mL of 0.5 *N* sodium hydroxide, and mix. After 15 minutes, add 0.5 mL of 1.2 *N* hydrochloric acid for each mL of 0.5 *N* sodium hydroxide used, and dilute quantitatively with water to obtain a solution containing 36 Penicillin G Procaine Units per mL. Transfer 5.0 mL of this solution to a 50-mL volumetric flask.

Procedure—To each of the flasks containing the *Standard preparations* and the *Assay preparation*, and to a seventh 50-mL volumetric flask containing 5.0 mL of water to provide the blank, add 0.5 mL of 4 *N* hydrochloric acid, 1.0 mL of sodium nitrite solution (1 in 1000), 1.0 mL of ammonium sulfamate solution (1 in 200), and 1.0 mL of *N*-1-naphthylethylenediamine dihydrochloride solution (1 in 1000), mixing and allowing 2 minutes to elapse after each addition. Dilute the contents of each flask with water to volume, and mix. Concomitantly determine the absorbances of the solutions from the *Standard preparations* and the *Assay preparation* at the wavelength of maximum absorbance at about 550 nm, with a suitable spectrophotometer, using the blank to set the instrument at zero. Plot the absorbance values of solutions from the *Standard preparations* versus concentration, in mg per mL, of procaine hydrochloride in the solutions from the *Standard preparations*, and draw the straight line best fitting the five plotted points. From the graph so obtained, determine the concentration (C), in mg per mL, of procaine hydrochloride in the solution from the *Test preparation.* Calculate the quantity, in Penicillin G Procaine Units, in each container, or in each mL of the Suspension taken by the formula:

$$(588.72/272.77)(1009.1)(50CL/D),$$

in which 588.72 and 272.77 are the molecular weights of penicillin G procaine monohydrate and procaine hydrochloride, respectively, 1009.1 is the theoretical potency, in Penicillin G Units, in each mg of penicillin G procaine, L is the labeled amount, in Penicillin G Procaine Units, in each container, or in each mL, of Suspension taken, and D is the concentration, in Penicillin G Procaine Units, per mL, of the solution from the *Assay prepa-*

ration, on the basis of the labeled amount in each container, or in each mL, of Suspension taken and the extent of dilution.

Assay for penicillin G benzathine—

Standard preparation—Using USP Penicillin G Potassium RS, prepare as directed for *Standard Preparation* under *Iodometric Assay—Antibiotics* ⟨425⟩.

Assay preparation—Where the Suspension is represented as being in a single-dose container, withdraw all of the withdrawable contents, using a suitable hypodermic needle and syringe. Where the label states the quantities of Penicillin G Benzathine and Penicillin G Procaine in a given volume of Suspension, remove an accurately measured volume of the Suspension, freshly mixed but free from air bubbles. Dilute the specimen of Suspension taken quantitatively with 1 *N* sodium hydroxide to obtain a solution containing about 2000 Penicillin G Units per mL. Pipet 2 mL of this solution into each of two glass-stoppered, 125-mL conical flasks.

Blank preparation—Prepare as directed for *Assay preparation*, except to use *Buffer No. 1* instead of 1 *N* sodium hydroxide.

Procedure—Proceed as directed for *Procedure* under *Iodometric Assay—Antibiotics* ⟨425⟩, except in performing the *Inactivation and titration* to omit the addition of 1.0 *N* sodium hydroxide to the *Assay preparation*, and in performing the *Blank determination* to use the *Blank preparation* in place of the *Assay preparation*. Calculate the total quantity, *T*, in Penicillin G Units, in each mL of the Suspension taken by the formula:

$$(L/2D)(F)(B - I),$$

in which *L* is the labeled quantity, in Penicillin G Units in each container, or per mL, in the Sterile Penicillin G Benzathine and Penicillin G Procaine Suspension taken, and *D* is the concentration, in Penicillin G Units per mL, in the *Assay preparation* on the basis of the labeled quantity in the Sterile Penicillin G Benzathine and Penicillin G Procaine Suspension and the extent of dilution. Calculate the quantity, in Penicillin G Benzathine Units, in each container, or in each mL, of the Sterile Penicillin G Benzathine and Penicillin G Procaine Suspension taken by the formula:

$$T - P,$$

in which *P* is the quantity, in Penicillin G Procaine Units, in each container, or in each mL, of Suspension taken, as determined in the *Assay for penicillin G procaine*.

Penicillin G Potassium

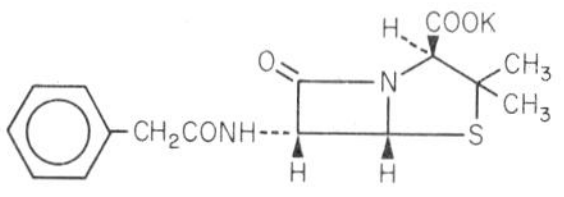

$C_{16}H_{17}KN_2O_4S$ 372.48

4-Thia-1-azabicyclo[3.2.0]heptane-2-carboxylic acid, 3,3-dimethyl-7-oxo-6-[(phenylacetyl)amino]-, monopotassium salt, [2*S*-(2α,5α,6β)]-.

Monopotassium (2*S*,5*R*,6*R*)-3,3-dimethyl-7-oxo-6-(2-phenylacetamido)-4-thia-1-azabicyclo[3.2.0]heptane-2-carboxylate [*113-98-4*].

» Penicillin G Potassium has a potency of not less than 1440 Penicillin G Units and not more than 1680 Penicillin G Units per mg.

Packaging and storage—Preserve in tight containers.

Reference standard—*USP Penicillin G Potassium Reference Standard*—Do not dry before using.

Identification—The infrared absorption spectrum of a potassium bromide dispersion of it, previously dried, exhibits maxima only at the same wavelengths as that of a similar preparation of USP Penicillin G Potassium RS.

Crystallinity ⟨695⟩: meets the requirements.

pH ⟨791⟩: between 5.0 and 7.5, in a solution containing 60 mg per mL.

Loss on drying ⟨731⟩—Dry about 100 mg, accurately weighed, in a capillary-stoppered bottle in vacuum at 60° for 3 hours: it loses not more than 1.5% of its weight.

Penicillin G content ⟨475⟩: between 80.8% and 94.3% of $C_{16}H_{18}N_2O_4S$.

Assay—

Standard preparation—Prepare as directed for *Standard Preparation* under *Iodometric Assay—Antibiotics* ⟨425⟩, using USP Penicillin G Potassium RS.

Assay preparation—Proceed as directed for *Assay Preparation* under *Iodometric Assay—Antibiotics* ⟨425⟩.

Procedure—Proceed as directed for *Procedure* under *Iodometric Assay—Antibiotics* ⟨425⟩. Calculate the potency, in Penicillin G Units per mg, of the Penicillin G Potassium taken by the formula:

$$(F)(B - I)/(2D),$$

in which *D* is the concentration, in mg per mL, of the *Assay preparation*, on the basis of the weight of Penicillin G Potassium taken and the extent of dilution.

Penicillin G Potassium Capsules

» Penicillin G Potassium Capsules contain not less than 90.0 percent and not more than 120.0 percent of the labeled number of Penicillin G Units.

Packaging and storage—Preserve in tight containers.

Reference standard—*USP Penicillin G Potassium Reference Standard*—Do not dry before using.

Identification—Shake a quantity of Capsule contents, equivalent to about 100,000 Penicillin G Units, with 8 mL of a solvent mixture consisting of acetone, 0.1 *M* citric acid, and 0.1 *M* sodium citrate (2:1:1), and filter: the filtrate so obtained responds to the *Identification test* under *Penicillin G Potassium for Injection*.

Dissolution ⟨711⟩—

Medium: Buffer No. 1 (see *Media and Diluents* under *Antibiotics—Microbial Assays* ⟨81⟩); 900 mL.

Apparatus 1: 100 rpm.

Time: 45 minutes.

Standard preparation—Dissolve an accurately weighed quantity of USP Penicillin G Potassium RS in *Buffer No. 1*, and dilute quantitatively with the same solvent to obtain a solution having a known concentration of about 400 Penicillin G Units per mL.

Procedure—Filter the solution under test, dilute with *Buffer No. 1*, if necessary, and proceed as directed for *Procedure* in the section, *Antibiotics—Hydroxylamine Assay*, under *Automated Methods of Analysis* ⟨16⟩. Calculate the number of Penicillin G Units in each mL of the solution taken by the formula:

$$C(A_U / A_S),$$

in which *C* is the concentration, in Penicillin G Units per mL, of the *Standard preparation*.

Tolerances—Not less than 75% (*Q*) of the labeled amount of Penicillin G Units is dissolved in 45 minutes.

Uniformity of dosage units ⟨905⟩: meet the requirements.

Loss on drying ⟨731⟩—Dry about 100 mg, accurately weighed, of powder from Capsules, in a capillary-stoppered bottle in vacuum at 60° for 3 hours: it loses not more than 1.5% of its weight.

Assay—

Standard preparation—Prepare as directed for *Standard Preparation* under *Iodometric Assay—Antibiotics* ⟨425⟩, using USP Penicillin G Potassium RS.

Assay preparation—Place not less than 5 Penicillin G Potassium Capsules in a high-speed glass blender jar containing an accurately measured volume of *Buffer No. 1*, and blend for 4 ± 1 minutes. Dilute an accurately measured volume of this stock solution quantitatively with *Buffer No. 1* to obtain an *Assay Preparation* containing about 2000 Penicillin G Units per mL. Pipet 2 mL of this solution into each of two glass-stoppered, 125-mL conical flasks.

Procedure—Proceed as directed for *Procedure* under *Iodometric Assay—Antibiotics* ⟨425⟩. Calculate the quantity, in Penicillin G Units in each Capsule taken by the formula:

$$(L/2D)(F)(B - I),$$

in which L is the labeled quantity, in Penicillin G Units in each Capsule, and D is the concentration, in Units per mL, of Penicillin G in the *Assay preparation* on the basis of the labeled quantity in each Capsule and the extent of dilution.

Penicillin G Potassium for Injection

» Penicillin G Potassium for Injection is a sterile, dry mixture of Penicillin G Potassium with not less than 4.0 percent and not more than 5.0 percent of Sodium Citrate, of which not more than 0.15 percent may be replaced by Citric Acid. It has a potency of not less than 1355 and not more than 1595 Penicillin G Units per mg. In addition, where packaged for dispensing, it contains not less than 90.0 percent and not more than 120.0 percent of the labeled number of Penicillin G Units.

Packaging and storage—Preserve in *Containers for Sterile Solids* as described under *Injections* ⟨1⟩.

Reference standard—*USP Penicillin G Potassium Reference Standard*—Do not dry before using.

Constituted solution—At the time of use, the constituted solution prepared from Penicillin G Potassium for Injection meets the requirements for *Constituted Solutions* under *Injections* ⟨1⟩.

Identification—Prepare a solution of it containing about 12,000 Penicillin G Units per mL in a solvent mixture consisting of acetone, 0.1 *M* citric acid, and 0.1 *M* sodium citrate (2:1:1). Prepare a Standard solution of USP Penicillin G Potassium RS containing about 12,000 Penicillin G Units per mL in the same solvent mixture. Apply separately 20 µL of each solution on a thin-layer chromatographic plate (see *Chromatography* ⟨621⟩) coated with a 0.25-mm layer of chromatographic silica gel mixture. Place the plate in a suitable chromatographic chamber, and develop the chromatogram in a solvent system consisting of a mixture of toluene, dioxane, and glacial acetic acid (90:25:4) until the solvent front has moved about three-fourths of the length of the plate. Remove the plate from the chamber, mark the solvent front, and allow to air-dry. Spray the plate with starch TS followed by dilute iodine TS (1 in 10). Penicillin G appears as a white spot on a purple background: the R_f value of the penicillin G spot obtained from the test solution corresponds to that obtained from the Standard solution.

Crystallinity ⟨695⟩: meets the requirements.

Pyrogen—It meets the requirements of the *Pyrogen Test* ⟨151⟩, the test dose being 1.0 mL per kg of a solution in Sterile Water for Injection containing 20,000 Penicillin G Units per mL.

Sterility—It meets the requirements under *Sterility Tests* ⟨71⟩, when tested as directed in the section, *Test Procedures Using Membrane Filtration*.

pH ⟨791⟩: between 6.0 and 8.5, in a solution containing 60 mg per mL, or, where packaged for dispensing, in the solution constituted as directed in the labeling.

Loss on drying ⟨731⟩—Dry about 100 mg, accurately weighed, in a capillary-stoppered bottle in vacuum at 60° for 3 hours: it loses not more than 1.5% of its weight.

Particulate matter ⟨788⟩: meets the requirements under *Small-volume Injections*.

Penicillin G content ⟨475⟩: between 76.3% and 89.8% of $C_{16}H_{18}N_2O_4S$.

Other requirements—It meets the requirements for *Uniformity of Dosage Units* ⟨905⟩, and *Labeling* under *Injections* ⟨1⟩.

Assay—Using Penicillin G Potassium for Injection, proceed as directed in the *Assay* under *Sterile Penicillin G Potassium*.

Penicillin G Potassium for Oral Solution

» Penicillin G Potassium for Oral Solution is a dry mixture of Penicillin G Potassium and one or more suitable buffers, colors, diluents, flavors, and preservatives. It contains not less than 90.0 percent and not more than 130.0 percent of the labeled number of Penicillin G Units when constituted as directed in the labeling.

Packaging and storage—Preserve in tight containers.

Reference standard—*USP Penicillin G Potassium Reference Standard*—Do not dry before using.

Identification—Shake a portion of it, equivalent to about 100,000 Penicillin G Units, with 8 mL of a solvent mixture consisting of acetone, 0.1 *M* citric acid, and 0.1 *M* sodium citrate (2:1:1): the solution so obtained responds to the *Identification test* under *Penicillin G Potassium for Injection*.

Uniformity of dosage units ⟨905⟩—

FOR SOLID PACKAGED IN SINGLE-UNIT CONTAINERS: meets the requirements.

pH ⟨791⟩: between 5.5 and 7.5, in the solution constituted as directed in the labeling.

Water, *Method I* ⟨921⟩: not more than 1.0%.

Assay—

Standard preparation—Prepare as directed for *Standard preparation* under *Iodometric Assay—Antibiotics* ⟨425⟩, using USP Penicillin G Potassium RS.

Assay preparation—Dilute an accurately measured volume of solution of Penicillin G Potassium for Oral Solution, constituted as directed in the labeling, quantitatively and stepwise with *Buffer No. 1* (see *Media and Diluents* under *Antibiotics—Microbial Assay* ⟨81⟩) to obtain a solution containing about 2000 Penicillin G Units per mL. Pipet 2 mL of this solution into each of two glass-stoppered, 125-mL conical flasks.

Procedure—Proceed as directed for *Procedure* under *Iodometric Assay—Antibiotics* ⟨425⟩. Calculate the quantity, in Penicillin G Units, in each mL of the constituted solution of Penicillin G Potassium for Oral Solution taken by the formula:

$$(L/2D)(F)(B - I),$$

in which L is the labeled quantity, in Penicillin G Units per mL in the constituted solution of Penicillin G Potassium for Oral Solution, and D is the concentration, in Units per mL, of Penicillin G in the *Assay preparation* on the basis of the labeled quantity in the constituted solution of Penicillin G Potassium for Oral Solution and the extent of dilution.

Sterile Penicillin G Potassium

» Sterile Penicillin G Potassium is Penicillin G Potassium suitable for parenteral use. It has a potency of not less than 1440 Penicillin G Units and not more than 1680 Penicillin G Units per mg and, in addition, where packaged for dispensing, it contains not less than 90.0 percent and not more than 115.0 percent of the labeled number of Penicillin G Units.

Packaging and storage—Preserve in *Containers for Sterile Solids* as described under *Injections* ⟨1⟩.

Reference standard—*USP Penicillin G Potassium Reference Standard*—Do not dry before using.

Constituted solution—At the time of use, the constituted solution prepared from Sterile Penicillin G Potassium meets the requirements for *Constituted Solutions* under *Injections* ⟨1⟩.

Pyrogen—It meets the requirements of the *Pyrogen Test* ⟨151⟩, the test dose being 1.0 mL per kg of a solution in Sterile Water for Injection containing 20,000 Penicillin G Units per mL.

Sterility—It meets the requirements under *Sterility Tests* ⟨71⟩, when tested as directed in the section, *Test Procedures Using Membrane Filtration.*

Particulate matter ⟨788⟩: meets the requirements under *Small-volume Injections.*

Other requirements—It responds to the *Identification test,* and meets the requirements for *pH, Loss on drying, Crystallinity,* and *Penicillin G content* under *Penicillin G Potassium.* It meets also the requirements for *Uniformity of Dosage Units* ⟨905⟩, and *Labeling* under *Injections* ⟨1⟩.

Assay—

Standard preparation—Prepare as directed for *Standard Preparation* under *Iodometric Assay—Antibiotics* ⟨425⟩, using USP Penicillin G Potassium RS.

Assay preparation 1—Dissolve an accurately weighed quantity of Sterile Penicillin G Potassium in *Buffer No. 1* to obtain an *Assay preparation* containing about 2000 Penicillin G Units per mL. Pipet 2 mL of this solution into each of two glass-stoppered, 125-mL conical flasks.

Assay preparation 2 (where it is packaged for dispensing and is represented as being in a single-dose container)—Constitute Sterile Penicillin G Potassium as directed in the labeling. Withdraw all of the withdrawable contents, using a suitable hypodermic needle and syringe, and dilute quantitatively with *Buffer No. 1* to obtain a solution containing about 2000 Penicillin G Units per mL. Pipet 2 mL of this solution into each of two glass-stoppered, 125-mL conical flasks.

Assay preparation 3 (where the label states the quantity of penicillin G potassium in a given volume of constituted solution)—Constitute Sterile Penicillin G Potassium as directed in the labeling. Dilute an accurately measured volume of the constituted solution quantitatively with *Buffer No. 1* to obtain a solution containing about 2000 Penicillin G Units per mL. Pipet 2 mL of this solution into each of two glass-stoppered, 125-mL conical flasks.

Procedure—Proceed as directed for *Procedure* under *Iodometric Assay—Antibiotics* ⟨425⟩. Calculate the potency, in Penicillin G Units per mg, of the Sterile Penicillin G Potassium taken by the formula:

$$(F)(B - I)/2D_1,$$

in which D_1 is the concentration, in mg per mL, of *Assay preparation 1,* on the basis of the weight of Sterile Penicillin G Potassium taken and the extent of dilution. Calculate the quantity, in Penicillin G Units, in the container, and in the portion of constituted solution taken, by the formula:

$$(L/2D)(F)(B - I),$$

in which L is the labeled quantity, in Penicillin G Units, in the container, or in the volume of constituted solution taken, and D is the concentration, in Penicillin G Units per mL, of *Assay preparation 2,* or of *Assay preparation 3,* on the basis of the labeled quantity in the container, or in the portion of constituted solution taken, respectively, and the extent of dilution.

Penicillin G Potassium Tablets

» Penicillin G Potassium Tablets contain not less than 90.0 percent and not more than 120.0 percent of the labeled number of Penicillin G Units.

Packaging and storage—Preserve in tight containers.

Reference standard—*USP Penicillin G Potassium Reference Standard*—Do not dry before using.

Identification—Shake a quantity of ground Tablet powder, equivalent to about 100,000 Penicillin G Units, with 8 mL of a solvent mixture consisting of acetone, 0.1 *M* citric acid, and 0.1 *M* sodium citrate (2:1:1), and filter: the filtrate so obtained responds to the *Identification test* under *Penicillin G Potassium for Injection.*

Dissolution ⟨711⟩—

Medium: pH 6.0 phosphate buffer (see *Buffer Solutions* in the section, *Reagents, Indicators, and Solutions*); 900 mL.

Apparatus 2: 75 rpm.

Time: 60 minutes.

Standard preparation—Dissolve an accurately weighed quantity of USP Penicillin G Potassium RS in *Dissolution Medium,* and dilute quantitatively with the same solvent to obtain a solution having a known concentration of about 400 Penicillin G Units per mL.

Procedure—Filter the solution under test, dilute with *Dissolution Medium,* if necessary, and proceed as directed for *Procedure* in the section, *Antibiotics—Hydroxylamine Assay,* under *Automated Methods of Analysis* ⟨16⟩. Calculate the number of Penicillin G Units in each mL of the solution taken by the formula:

$$C(A_U/A_S),$$

in which C is the concentration, in Penicillin G Units per mL, of the *Standard preparation.*

Tolerances—Not less than 70% (*Q*) of the labeled amount of Penicillin G Units is dissolved in 60 minutes.

Uniformity of dosage units ⟨905⟩: meet the requirements.

Loss on drying ⟨731⟩—Dry about 100 mg, accurately weighed, of powdered Tablets in a capillary-stoppered bottle in vacuum at 60° for 3 hours: it loses not more than 1.0% of its weight.

Assay—

Standard preparation—Prepare as directed for *Standard Preparation* under *Iodometric Assay—Antibiotics* ⟨425⟩, using USP Penicillin G Potassium RS.

Assay preparation—Place not less than 5 Penicillin G Potassium Tablets in a high-speed glass blender jar containing an accurately measured volume of *Buffer No. 1,* and blend for 4 ± 1 minutes. Dilute an accurately measured volume of this stock solution quantitatively with *Buffer No. 1* to obtain an *Assay preparation* containing about 2000 Penicillin G Units per mL. Pipet 2 mL of this solution into each of two glass-stoppered, 125-mL conical flasks.

Procedure—Proceed as directed for *Procedure* under *Iodometric Assay—Antibiotics* ⟨425⟩. Calculate the quantity, in Penicillin G Units in each Tablet taken by the formula:

$$(L/2D)(F)(B - I),$$

in which L is the labeled quantity, in Penicillin G Units, in each Tablet, and D is the concentration, in Units per mL, of Penicillin G in the *Assay preparation* on the basis of the labeled quantity in each Tablet and the extent of dilution.

Penicillin G Potassium Tablets for Oral Solution

» Penicillin G Potassium Tablets for Oral Solution contain not less than 90.0 percent and not more than 120.0 percent of the labeled number of Penicillin G Units.

Packaging and storage—Preserve in tight containers.

Reference standard—*USP Penicillin G Potassium Reference Standard*—Do not dry before using.

Identification—Dissolve 1 Tablet in a volume of a solvent mixture consisting of acetone, 0.1 *M* citric acid, and 0.1 *M* sodium citrate (2:1:1) to obtain a solution containing about 12,000 Penicillin G Units per mL: this solution responds to the *Identification test* under *Penicillin G Potassium for Injection.*

Loss on drying ⟨731⟩—Dry about 100 mg, accurately weighed, of powdered Tablets in a capillary-stoppered bottle in vacuum at 60° for 3 hours: it loses not more than 1.0% of its weight.

Other requirements—Tablets for Oral Solution meet the requirements for *Uniformity of Dosage Units* ⟨905⟩ and *Assay* under *Penicillin G Potassium Tablets.*

Penicillin G Procaine Intramammary Infusion

» Penicillin G Procaine Intramammary Infusion is a suspension of Penicillin G Procaine in a suitable vegetable oil vehicle. It may contain one or more buffers, dispersants, preservatives, and thickening agents. It contains not less than 90.0 percent and not more than 115.0 percent of the labeled amount of penicillin G.

Packaging and storage—Preserve in well-closed disposable syringes.

Labeling—Label it to indicate that it is for veterinary use only.

Reference standards—*USP Peniciilin G Potassium Reference Standard*—Do not dry before using. *USP Penicillin G Procaine Reference Standard*—Do not dry before using.

Identification—Transfer a portion of it, equivalent to about 100,000 Penicillin G Units, to a test tube, add 25 mL of methanol, and shake. Allow to separate, and use the methanol layer as the test solution. Prepare a Standard solution of USP Penicillin G Procaine RS in methanol containing about 4.5 mg per mL. Apply separately 10 µL of each solution on a thin-layer chromatographic plate (see *Chromatography* ⟨621⟩) coated with a 0.25-mm layer of chromatographic silica gel mixture. Allow the spots to dry, and develop the chromatogram in a solvent system consisting of a mixture of butanol, isopropyl alcohol, acetone, and water (4:4:2:2) until the solvent front has moved about three-fourths of the length of the plate. Remove the plate from the developing chamber, mark the solvent front, and allow the solvent to evaporate. Expose the plate to iodine vapors in a closed chamber for about 15 minutes, and locate the spots: the R_f values and colors of the two principal spots obtained from the test solution correspond to those obtained from the Standard solution.

Water, *Method I* ⟨921⟩: not more than 1.4%, 20 mL of a mixture of carbon tetrachloride, chloroform, and methanol (2:2:1) being used in place of methanol in the titration vessel.

Assay—Proceed as directed under *Antibiotics—Microbial Assays* ⟨81⟩, expelling the contents of 1 syringe of Penicillin G Procaine Intramammary Infusion into a high-speed glass blender jar containing 499.0 mL of *Buffer No. 1* and 1.0 mL of polysorbate 80, and blending for 3 to 5 minutes. Allow to stand for about 10 minutes, and dilute an accurately measured volume of the aqueous phase quantitatively and stepwise with *Buffer No. 1* to obtain a *Test Dilution* having a concentration assumed to be equal to the median dose level of the Standard.

Sterile Penicillin G Procaine

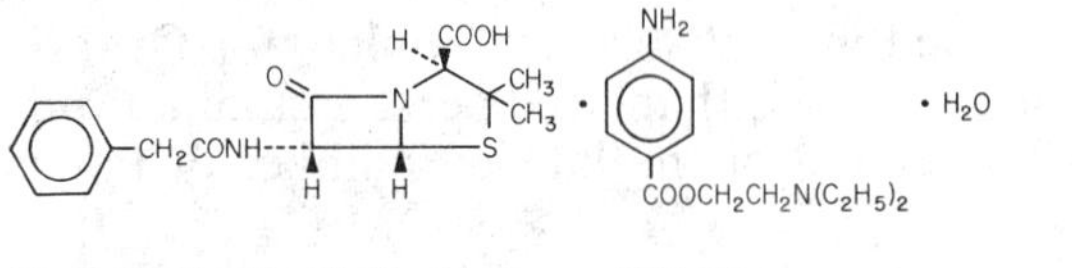

$C_{16}H_{18}N_2O_4S.C_{13}H_{20}N_2O_2.H_2O$ 588.72

4-Thia-1-azabicyclo[3.2.0]heptane-2-carboxylic acid, 3,3-dimethyl-7-oxo-6-[(phenylacetyl)amino]-, [2*S*-(2α,5α,6β)]-, compd. with 2-(diethylamino)ethyl 4-aminobenzoate (1:1) monohydrate.

(2*S*,5*R*,6*R*)-3,3-Dimethyl-7-oxo-6-(2-phenylacetamido)-4-thia-1-azabicyclo[3.2.0]heptane-2-carboxylic acid compound with 2-(diethylamino)ethyl *p*-aminobenzoate (1:1) monohydrate [*6130-64-9*].

Anhydrous 570.70 [*54-35-3*].

» Sterile Penicillin G Procaine is penicillin G procaine suitable for parenteral use. It has a potency of not less than 900 Penicillin G Units and not more than 1050 Penicillin G Units per mg.

Packaging and storage—Preserve in *Containers for Sterile Solids* as described under *Injections* ⟨1⟩.

Reference standards—*USP Penicillin G Potassium Reference Standard*—Do not dry before using. *USP Procaine Hydrochloride Reference Standard*—Dry over silica gel for 18 hours before using.

Identification—Prepare a solution of it containing about 12,000 Penicillin G Units per mL in a solvent mixture consisting of acetone, 0.1 *M* citric acid, and 0.1 *M* sodium citrate (2:1:1). Prepare a Standard solution of USP Penicillin G Potassium RS containing about 12,000 Penicillin G Units per mL in the same solvent mixture (*Standard solution A*). Prepare a Standard solution of USP Procaine Hydrochloride RS containing about 5 mg per mL in the same solvent system (*Standard solution B*). Apply separately 100 µL of each solution on a thin-layer chromatographic plate (see *Chromatography* ⟨621⟩) coated with a 0.25-mm layer of chromatographic silica gel mixture. Place the plate in a suitable chromatographic chamber, and develop the chromatogram in a solvent system consisting of a mixture of toluene, dioxane, and glacial acetic acid (90:25:4) until the solvent front has moved about three-fourths of the length of the plate. Remove the plate from the chamber, mark the solvent front, and allow to air-dry. Examine the plate under short-wavelength and long-wavelength ultraviolet light, noting the positions of the spots. Spray the plate with starch TS followed by dilute iodine TS (1 in 10). Penicillin G appears as a white spot on a purple background: the R_f value of the penicillin G spot obtained from the test solution corresponds to that obtained from *Standard solution A*. Spray the location of the spots visualized with ultraviolet light with a 1 in 20 solution of *p*-dimethylaminobenzaldehyde in methanol. Procaine appears as a bright yellow spot: the R_f value of the procaine spot obtained from the test solution corresponds to that obtained from *Standard solution B*.

Crystallinity ⟨695⟩: meets the requirements.

Pyrogen—It meets the requirements of the *Pyrogen Test* ⟨151⟩, the test dose being 2.0 mL per kg of a solution in pyrogen-free saline TS containing 2000 Penicillin G Units per mL.

Sterility—It meets the requirements under *Sterility Tests* ⟨71⟩, when tested as directed in the section, *Test Procedures Using Membrane Filtration*, except to use *Fluid A* to which has been added sufficient sterile penicillinase to inactivate the penicillin G and to swirl the vessel until solution is complete before filtering.

pH ⟨791⟩: between 5.0 and 7.5, in a (saturated) solution containing about 300 mg per mL.

Water, *Method I* ⟨921⟩: between 2.8% and 4.2%.

Penicillin G and procaine contents—

Mobile phase—Dissolve 14 g of monobasic potassium phosphate and 6.5 g of tetrabutylammonium hydroxide solution (4 in 10) in about 700 mL of water, adjust with 1 *N* potassium hydroxide to a pH of 7.0, dilute with water to 1000 mL, and mix. Mix 500 mL of this solution, 250 mL of acetonitrile, and 250 mL of water. Adjust with 1 *N* potassium hydroxide or dilute phosphoric acid (1 in 10) to a pH of 7.5 ± 0.05, filter through a membrane filter of 5-µm or finer porosity, and degas. Make adjustments if necessary (see *System Suitability* under *Chromatography* ⟨621⟩).

Standard preparation—Using accurately weighed quantities of USP Penicillin G Potassium RS and USP Procaine Hydrochloride RS, prepare a solution in *Mobile phase* having known concentrations of about 0.8 mg per mL and 0.54 mg per mL, respectively.

Test preparation—Transfer about 70 mg of Sterile Penicillin G Procaine, accurately weighed, to a 50-mL volumetric flask, add about 30 mL of *Mobile phase*, sonicate to dissolve, dilute with *Mobile phase* to volume, and mix.

Resolution solution—Prepare a solution of penicillin V potassium in *Mobile phase* containing 2.4 mg per mL. Mix 1 volume of this solution and 3 volumes of *Standard preparation.*

Chromatographic system (see *Chromatography* ⟨621⟩)—The liquid chromatograph is equipped with a 235-nm detector and a 4-mm × 30-cm column that contains 10-µm packing L1. The flow rate is about 1 mL per minute. Chromatograph the *Standard preparation*, and record the peak responses as directed under *Procedure:* the relative standard deviation for replicate injections is not more than 3.0%. Chromatograph about 10 µL of the *Resolution solution*, and record the peak responses as directed under *Procedure:* the resolution, R, between the penicillin G and penicillin V peaks is not less than 2.0.

Procedure—Separately inject equal volumes (about 10 µL) of the *Standard preparation* and the *Test preparation* into the chromatograph, record the chromatograms, and measure the responses for the major peaks. The relative retention times are 1.0 for procaine and about 2.2 for penicillin G. Calculate the percentage of penicillin G ($C_{16}H_{18}N_2O_4S$) in the specimen under test by the formula:

$$50C(G_S/W_U)(r_U/r_S),$$

in which C is the concentration, in mg per mL, of USP Penicillin G Potassium RS in the *Standard preparation*, G_S is the designated penicillin G content, in percentage, of USP Penicillin G Potassium RS, W_U is the amount, in mg, of Sterile Penicillin G Procaine taken, and r_U and r_S are the responses of the penicillin G peaks obtained from the *Test preparation* and the *Standard preparation*, respectively: between 51.0% and 59.6% of $C_{16}H_{18}N_2O_4S$ is found. Calculate the percentage of procaine ($C_{13}H_{20}N_2O_2$) in the specimen under test by the formula:

$$(236.31/272.77)(50C/W_U)(r_U/r_S),$$

in which 236.31 and 272.77 are the molecular weights of procaine and procaine hydrochloride, respectively, C is the concentration, in mg per mL, of USP Procaine Hydrochloride RS in the *Standard preparation*, W_U is the amount, in mg, of Sterile Penicillin G Procaine taken, and r_U and r_S are the responses of the procaine peaks obtained from the *Test preparation* and the *Standard preparation*, respectively: between 37.5% and 43.0% is found.

Assay—

Standard preparation—Using USP Penicillin G Potassium RS, prepare as directed for *Standard Preparation* under *Iodometric Assay—Antibiotics* ⟨425⟩.

Assay preparation—Prepare as directed for *Assay Preparation* under *Iodometric Assay—Antibiotics* ⟨425⟩, except to dissolve about 100 mg of Sterile Penicillin G Procaine, accurately weighed, in 2.0 mL of methanol, and to dilute quantitatively with *Buffer No. 1* to obtain a solution containing about 2000 Penicillin G Units per mL.

Procedure—Proceed as directed for *Procedure* under *Iodometric Assay—Antibiotics* ⟨425⟩. Calculate the potency, in Penicillin G Units per mg, of the Sterile Penicillin G Procaine taken by the formula:

$$(F)(B - I)/(2D),$$

in which D is the concentration, in mg per mL, of the *Assay preparation*, on the basis of the weight of Sterile Penicillin G Procaine taken and the extent of dilution.

Sterile Penicillin G Procaine Suspension

» Sterile Penicillin G Procaine Suspension is a sterile suspension of Sterile Penicillin G Procaine in Water for Injection and contains one or more suitable buffers, dispersants, or suspending agents, and a suitable preservative. It may contain procaine hydrochloride in a concentration not exceeding 2.0 percent, and may contain one or more suitable stabilizers. It contains not less than 90.0 percent and not more than 115.0 percent of the labeled amount of penicillin G, the labeled amount being not less than 300,000 Penicillin G Units per mL or per container.

Packaging and storage—Preserve in single-dose or in multiple-dose containers, preferably of Type I or Type III glass, in a refrigerator.

Reference standards—*USP Penicillin G Potassium Reference Standard*—Do not dry before using. *USP Procaine Hydrochloride Reference Standard*—Dry over silica gel for 18 hours before using.

Identification—It responds to the *Identification test* under *Sterile Penicillin G Procaine.*

Pyrogen—It meets the requirements of the *Pyrogen Test* ⟨151⟩, the test dose being 2.0 mL per kg of a solution in pyrogen-free saline TS containing 2000 Penicillin G Units per mL.

Sterility—It meets the requirements under *Sterility Tests* ⟨71⟩, when tested as directed in the section, *Test Procedures Using Membrane Filtration*, except to use a portion of specimen from each container equivalent to 300,000 Penicillin G Units, instead of the minimum volume specified in the Table of *Quantities for Liquid Articles* in general chapter ⟨71⟩ *Sterility Tests*, and to use *Fluid A* to which has been added sufficient sterile penicillinase to inactivate the penicillin G and to swirl the vessel until solution is complete before filtering. If the Suspension contains lecithin, use *Fluid D.* If it contains carboxymethylcellulose sodium, add sufficient sterile carboxymethylcellulase to *Fluid A* or *Fluid D* to dissolve the carboxymethylcellulose sodium before filtering. If it does not dissolve completely, proceed as directed in the section, *Test Procedures for Direct Transfer to Test Media*, except to use Fluid Thioglycollate Medium and Soybean-Casein Digest Medium containing an amount of sterile penicillinase sufficient to inactivate the penicillin G in each vessel.

pH ⟨791⟩: between 5.0 and 7.5.

Other requirements—It meets the requirements under *Injections* ⟨1⟩.

Assay—

Standard preparation—Using USP Penicillin G Potassium RS, prepare as directed for *Standard Preparation* under *Iodometric Assay—Antibiotics* ⟨425⟩.

Assay preparation 1 (where it is represented as being in a single-dose container)—Withdraw all of the withdrawable contents of Sterile Penicillin G Procaine Suspension, using a suitable hypodermic needle and syringe, and dilute quantitatively with *Buffer No. 1* to obtain a solution containing about 2000 Penicillin G Units per mL. Pipet 2 mL of this solution into each of two glass-stoppered, 125-mL conical flasks.

Assay preparation 2 (where the label states the quantity of penicillin G procaine in a given volume of Suspension)—Dilute an accurately measured volume of Sterile Penicillin G Procaine Suspension quantitatively with *Buffer No. 1* to obtain a solution containing about 2000 Penicillin G Units per mL. Pipet 2.0 mL of this solution into each of two glass-stoppered, 125-mL conical flasks.

Procedure—Proceed as directed for *Procedure* under *Iodometric Assay—Antibiotics* ⟨425⟩. Calculate the quantity, in Penicillin G Units, in the container, or in the portion of Suspension taken, by the formula:

$$(L/2D)(F)(B - I),$$

in which L is the labeled quantity in Penicillin G Units, in the container, or in the volume of Suspension taken, and D is the concentration, in Penicillin G Units per mL, of *Assay preparation 1*, or of *Assay preparation 2*, on the basis of the labeled quantity in the container, or in the portion of Suspension taken, respectively, and the extent of dilution.

Penicillin G Procaine Suspension, Sterile Penicillin G Benzathine and—*see* Penicillin G Benzathine and Penicillin G Procaine Suspension, Sterile

Sterile Penicillin G Procaine for Suspension

» Sterile Penicillin G Procaine for Suspension is a sterile mixture of Sterile Penicillin G Procaine and one or more suitable buffers, dispersants, or suspending agents, and preservatives. It contains not less than 90.0 percent and not more than 115.0 percent of the labeled amount of penicillin G, the labeled amount being not less than 300,000 Penicillin G Units per container or per mL of constituted Suspension.

Packaging and storage—Preserve in single-dose or in multiple-dose containers, preferably of Type I or Type III glass.

Reference standards—*USP Penicillin G Potassium Reference Standard*—Do not dry before using. *USP Procaine Hydrochloride Reference Standard*—Dry over silica gel for 18 hours before using.

Identification—It responds to the *Identification test* under *Sterile Penicillin G Procaine.*

pH ⟨791⟩: between 5.0 and 7.5, when constituted as directed in the labeling.

Water, *Method I* ⟨921⟩: between 2.8% and 4.2%.

Other requirements—It meets the requirements for *Pyrogen* and *Sterility* under *Sterile Penicillin G Procaine Suspension.* It meets also the requirements under *Injections* ⟨1⟩ and *Uniformity of Dosage Units* ⟨905⟩.

Assay—

Standard preparation—Using USP Penicillin G Potassium RS, prepare as directed for *Standard Preparation* under *Iodometric Assay—Antibiotics* ⟨425⟩.

Assay preparation 1 (where it is represented as being in a single-dose container)—Constitute Sterile Penicillin G Procaine for Suspension as directed in the labeling. Withdraw all of the withdrawable contents, using a suitable hypodermic needle and syringe, and dilute quantitatively with *Buffer No. 1* to obtain a solution containing about 2000 Penicillin G Units per mL. Pipet 2 mL of this solution into each of two glass-stoppered, 125-mL conical flasks.

Assay preparation 2 (where the label states the quantity of penicillin G procaine in a given volume of constituted suspension)—Constitute Sterile Penicillin G Procaine for Suspension as directed in the labeling. Dilute an accurately measured volume of the constituted suspension quantitatively with *Buffer No. 1* to obtain a solution containing about 2000 Penicillin G Units per mL. Pipet 2 mL of this solution into each of two glass-stoppered, 125-mL conical flasks.

Procedure—Proceed as directed for *Procedure* under *Iodometric Assay—Antibiotics* ⟨425⟩. Calculate the quantity, in Penicillin G Units, in the container, or in the portion of constituted suspension taken by the formula:

$$(L/2D)(F)(B - I),$$

in which L is the labeled quantity, in Penicillin G Units, in the container, or in the volume of constituted suspension taken, and D is the concentration, in Penicillin G Units per mL, of *Assay preparation 1* or of *Assay preparation 2* on the basis of the labeled quantity in the container or in the portion of constituted suspension taken, respectively, and the extent of dilution.

Sterile Penicillin G Procaine with Aluminum Stearate Suspension

» Sterile Penicillin G Procaine with Aluminum Stearate Suspension is a sterile suspension of Sterile Penicillin G Procaine in a refined vegetable oil with one or more suitable dispersants and hardening agents. It contains not less than 90.0 percent and not more than 115.0 percent of the labeled amount of penicillin G.

Packaging and storage—Preserve in single-dose or in multiple-dose containers, preferably of Type I or Type III glass.

Reference standard—*USP Penicillin G Potassium Reference Standard*—Do not dry before using.

Sterility—It meets the requirements under *Sterility Tests* ⟨71⟩, when tested as directed in the section, *Test Procedures for Direct Transfer to Test Media*, except to use a portion of specimen from each container equivalent to 300,000 Penicillin G Units, instead of the minimum volume specified in the Table of *Quantities for Liquid Articles* in general chapter ⟨71⟩ *Sterility Tests*, and to use Fluid Thioglycollate Medium and Soybean-Casein Digest Medium containing an amount of sterile penicillinase sufficient to inactivate the penicillin G in each vessel.

Water, *Method I* ⟨921⟩: not more than 1.4%, 20 mL of a mixture of carbon tetrachloride, chloroform, and methanol (2:2:1) being used in place of methanol in the titration vessel.

Other requirements—It meets the requirements under *Injections* ⟨1⟩.

Assay—Proceed as directed under *Antibiotics—Microbial Assays* ⟨81⟩, using an accurately measured volume of Sterile Penicillin G Procaine with Aluminum Stearate Suspension or, where it is represented as being in a single-dose container, the entire contents of the container. Transfer the specimen to a 50-mL volumetric flask, add 4 mL of chloroform, mix, dilute with dehydrated alcohol to volume, and mix. Immediately dilute this solution quantitatively and stepwise with *Buffer No. 1* to obtain a *Test Dilution* having a concentration assumed to be equal to the median dose level of the Standard.

Penicillin G Procaine and Dihydrostreptomycin Sulfate Intramammary Infusion

» Penicillin G Procaine and Dihydrostreptomycin Sulfate Intramammary Infusion is a suspension of Penicillin G Procaine and Dihydrostreptomycin Sulfate in a suitable vegetable oil vehicle. It may contain suitable gelling and thickening agents. It contains not less than 90.0 percent and not more than 120.0 percent of the labeled amounts of Penicillin G Units and of dihydrostreptomycin ($C_{21}H_{41}N_7O_{12}$).

Packaging and storage—Preserve in well-closed, disposable syringes.

Labeling—Label it to indicate that it is intended for veterinary use only.

Reference standards—*USP Dihydrostreptomycin Sulfate Reference Standard*—Dry in vacuum at a pressure not exceeding 5 mm of mercury at 100° for 4 hours before using. *USP Penicillin G Potassium Reference Standard*—Do not dry before using. *USP Penicillin G Procaine Reference Standard*—Do not dry before using.

Identification—

A: It responds to the *Identification test* under *Penicillin G Procaine Intramammary Infusion.*

B: Place a portion of it, equivalent to about 100 mg of dihydrostreptomycin, in a separator, add 20 mL of chloroform and 20 mL of water, and shake by mechanical means for 15 minutes. Allow to separate, and discard the lower chloroform layer. Repeat the extraction with a 20-mL portion of chloroform, discarding the chloroform layer. Use the aqueous layer as the test solution. Prepare a Standard solution of USP Dihydrostreptomycin Sulfate RS in water containing 6.5 mg per mL. Apply separately 30 µL of each solution on a thin-layer chromatographic plate (see *Chromatography* ⟨621⟩) coated with a 0.25-mm layer of chromatographic silica gel mixture. Allow the spots to dry, and develop the chromatogram in a solvent system consisting of a mixture of *n*-propyl alcohol, water, pyridine, and glacial acetic acid (15:12:10:2) until the solvent front has moved about three-fourths of the length of the plate. Remove the plate from the developing chamber, mark the solvent front, and allow the solvent to evaporate. Expose the plate to iodine vapors in a closed chamber for about 15 minutes, and locate the spots: the R_f value and color of the principal spot obtained from the test solution correspond to those obtained from the Standard solution.

Water, *Method I* ⟨921⟩: not more than 1.4%, 20 mL of a mixture of carbon tetrachloride, chloroform, and methanol (2:2:1) being used in the titration vessel in place of methanol.

Assay for penicillin G—Proceed as directed for penicillin G under *Antibiotics—Microbial Assays* ⟨81⟩, expelling the contents of 1 syringe of Penicillin G Procaine and Dihydrostreptomycin Sulfate Intramammary Infusion into a high-speed glass blender jar containing 499.0 mL of *Buffer No. 1* and 1.0 mL of polysorbate 80, and blending for 3 to 5 minutes. Allow to stand for about 10 minutes, and dilute an accurately measured volume of the aqueous phase quantitatively and stepwise with *Buffer No. 1* to obtain a *Test Dilution* having a concentration of penicillin G assumed to be equal to the median dose level of the Standard.

Assay for dihydrostreptomycin—Proceed as directed for the cylinder-plate assay for dihydrostreptomycin under *Antibiotics—Microbial Assays* ⟨81⟩, expelling the contents of 1 syringe of Penicillin G Procaine and Dihydrostreptomycin Sulfate Intramammary Infusion into a high-speed glass blender jar containing 499.0 mL of *Buffer No. 3* and 1.0 mL of polysorbate 80, and blending for 3 to 5 minutes. Allow to stand for about 10 minutes, and to an accurately measured volume of the aqueous phase add an accurately measured volume of penicillinase sufficient to inactivate the penicillin G contained therein. Dilute this solution quantitatively with *Buffer No. 3* to obtain a *Test Dilution* having a concentration of dihydrostreptomycin assumed to be equal to the median dose level of the Standard, and store at 37° for 30 minutes before filling the cylinders.

Sterile Penicillin G Procaine and Dihydrostreptomycin Sulfate Suspension

» Sterile Penicillin G Procaine and Dihydrostreptomycin Sulfate Suspension is a sterile suspension of Sterile Penicillin G Procaine in a solution of Dihydrostreptomycin Sulfate in Water for Injection, and contains one or more suitable buffers, preservatives, and dispersing or suspending agents. It may contain Procaine Hydrochloride in a concentration not exceeding 2.0 percent, and it may contain one or more suitable stabilizers. It contains not less than 90.0 percent and not more than 115.0 percent of the labeled amounts of Penicillin G Units and of dihydrostreptomycin ($C_{21}H_{41}N_7O_{12}$).

Packaging and storage—Preserve in single-dose or in multiple-dose, tight containers.

Labeling—Label it to indicate that it is intended for veterinary use only.

Reference standards—*USP Penicillin G Potassium Reference Standard*—Do not dry before using. *USP Dihydrostreptomycin Sulfate Reference Standard*—Dry in vacuum at a pressure not exceeding 5 mm of mercury at 100° for 4 hours before using.

Pyrogen—It meets the requirements of the *Pyrogen Test* ⟨151⟩, the test dose being 1.0 mL per kg of a solution in Sterile Water for Injection containing 10 mg of dihydrostreptomycin per mL.

Sterility—It meets the requirements under *Sterility Tests* ⟨71⟩, when tested as directed in the section, *Test Procedures Using Membrane Filtration*, except to use *Fluid A* to which has been added sufficient sterile penicillinase to inactivate the penicillin G and to swirl the vessel until solution is complete before filtering. If the Suspension contains lecithin, use *Fluid D* to which has been added sufficient penicillinase to inactivate the penicillin G and to swirl the vessel until solution is complete before filtering. If it contains carboxymethylcellulose sodium, add also sufficient sterile carboxymethylcellulase to *Fluid A* or *Fluid D* to dissolve the carboxymethylcellulose sodium before filtering. If it does not dissolve completely, proceed as directed in the section, *Test Procedures for Direct Transfer to Test Media*, except to use Fluid Thioglycollate Medium containing an amount of sterile penicillinase sufficient to inactivate the penicillin G in each vessel.

pH ⟨791⟩: between 5.0 and 8.0.

Assay for penicillin G—

Standard preparation—Using USP Penicillin G Potassium RS, prepare as directed for *Standard preparation* under *Iodometric Assay—Antibiotics* ⟨425⟩.

Assay preparation—Dilute an accurately measured volume of Sterile Penicillin G Procaine and Dihydrostreptomycin Sulfate Suspension quantitatively with *Buffer No. 1* to obtain a solution containing about 2000 Penicillin G Units per mL. Pipet 2 mL of this solution into each of two glass-stoppered, 125-mL conical flasks.

Procedure—Proceed as directed for *Procedure* under *Iodometric Assay—Antibiotics* ⟨425⟩, except in the *Blank determination* to add 0.1 mL of 1.2 *N* hydrochloric acid immediately before the 10.0 mL of 0.01 *N* iodine VS. Calculate the quantity, in Penicillin G Units, in the portion of Suspension taken by the formula:

$$(L/2D)(F)(B - I),$$

in which L is the labeled quantity, in Penicillin G Units, in the volume of Suspension taken, and D is the concentration, in Penicillin G Units per mL, of the *Assay preparation*, on the basis of the labeled quantity in the portion of Suspension taken and the extent of dilution.

Assay for dihydrostreptomycin—Proceed as directed for the turbidimetric assay for dihydrostreptomycin under *Antibiotics—Microbial Assays* ⟨81⟩, using an accurately measured volume of Sterile Penicillin G Procaine and Dihydrostreptomycin Sulfate Suspension diluted quantitatively with water to yield a *Test Dilution* having a concentration assumed to be equal to the median dose level of the Standard.

Sterile Penicillin G Procaine, Dihydrostreptomycin Sulfate, Chlorpheniramine Maleate, and Dexamethasone Suspension

» Sterile Penicillin G Procaine, Dihydrostreptomycin Sulfate, Chlorpheniramine Maleate, and Dexamethasone Suspension is a sterile suspension of Sterile Penicillin G Procaine and Dexamethasone in a solution of Sterile Dihydrostreptomycin Sulfate and Chlorpheniramine Maleate in Water for Injection. It contains one or more suitable buffers, preservatives, and dispersing or suspending agents. It may contain Procaine Hydrochloride in a concentration not exceeding 2.0 percent, and it may contain one or more suitable stabilizers. It contains not less than 90.0 percent and not more than 115.0 percent of the labeled amounts of Penicillin G Units and of dihydrostreptomycin ($C_{21}H_{41}N_7O_{12}$), and not less than 90.0 percent and not more than 110.0 percent of the labeled amounts of chlorpheniramine maleate ($C_{16}H_{19}ClN_2 . C_4H_4O_4$) and of dexamethasone ($C_{22}H_{29}FO_5$).

Packaging and storage—Preserve in single-dose or in multiple-dose, tight containers, in a cool place.

Labeling—Label it to indicate that it is intended for veterinary use only.

Reference standards—*USP Penicillin G Potassium Reference Standard*—Do not dry before using. *USP Penicillin G Procaine Reference Standard*—Do not dry before using. *USP Dihydrostreptomycin Sulfate Reference Standard*—Dry in vacuum at a pressure not exceeding 5 mm of mercury at 100° for 4 hours before using. *USP Chlorpheniramine Maleate Reference Standard*—Dry at 105° for 3 hours before using. *USP Dexamethasone Reference Standard*—Dry at 105° for 3 hours before using.

Identification—

A: Transfer, with the aid of water, a portion of the Suspension, freshly mixed and free from air bubbles, equivalent to about 400,000 Penicillin G Units, to a separator, add 50 mL of chloroform, and shake by mechanical means for 15 minutes. Allow to separate, and filter the lower chloroform layer through about 4 g of anhydrous sodium sulfate supported on a pledget of glass wool, collecting the filtrate in a 100-mL volumetric flask. Repeat the extraction with two 25-mL portions of chloroform, combining the filtrates in the 100-mL volumetric flask. Dilute with chloroform to volume, and mix. [Retain the aqueous phase for *Identification test B.*] Prepare a Standard solution of USP Penicillin G Procaine RS in chloroform containing about 4.5 mg per mL. Apply separately 10 µL of each solution on a thin-layer chromatographic plate (see *Chromatography* ⟨621⟩) coated with a 0.25-mm layer of chromatographic silica gel mixture. Allow the spots to dry, and develop the chromatogram in a solvent system consisting of a mixture of butanol, isopropyl alcohol, acetone, and water (4:4:2:2) until the solvent front has moved about three-fourths of the length of the plate. Remove the plate from the developing chamber, mark the solvent front, and allow the solvent to evaporate. Expose the plate to iodine vapors in a closed chamber for about 15 minutes, and locate the spots: the R_f values and colors of the two principal spots obtained from the test solution correspond to those obtained from the Standard solution.

B: Dilute the aqueous phase retained from *Identification test A* with water to obtain a test solution containing about 5 mg of dihydrostreptomycin per mL. Prepare a Standard solution of USP Dihydrostreptomycin Sulfate RS in water containing 6.5 mg per mL. Apply separately 30 µL of each solution on a thin-layer chromatographic plate (see *Chromatography* ⟨621⟩) coated with a 0.25-mm layer of chromatographic silica gel mixture. Allow the spots to dry, and develop the chromatogram in a solvent system consisting of a mixture of *n*-propyl alcohol, water, pyridine, and glacial acetic acid (15:12:10:2) until the solvent front has moved about three-fourths of the length of the plate. Remove the plate from the developing chamber, mark the solvent front, and allow the solvent to evaporate. Expose the plate to iodine vapors in a closed chamber for about 15 minutes, and locate the spots: the R_f value and color of the principal spot obtained from the test solution correspond to those obtained from the Standard solution.

C: The chromatogram of the *Assay preparation* obtained as directed in the *Assay for chlorpheniramine maleate* exhibits a major peak for chlorpheniramine, the retention time of which corresponds to that exhibited in the chromatogram of the *Standard preparation* similarly determined, both relative to the internal standard.

D: The chromatogram of the *Assay preparation* obtained as directed in the *Assay for dexamethasone* exhibits a major peak for dexamethasone, the retention time of which corresponds to that exhibited in the chromatogram of the *Standard preparation* similarly determined, both relative to the internal standard.

pH ⟨791⟩: between 5.0 and 6.0.

Other requirements—It meets the requirements of the tests for *Pyrogen* and *Sterility* under *Sterile Penicillin G Procaine and Dihydrostreptomycin Sulfate Suspension*, and the requirements under *Injections* ⟨1⟩.

Assay for penicillin G—

Standard preparation—Using USP Penicillin G Potassium RS, prepare as directed for *Standard preparation* under *Iodometric Assay—Antibiotics* ⟨425⟩.

Assay preparation—Dilute an accurately measured volume of Sterile Penicillin G Procaine, Dihydrostreptomycin Sulfate, Chlorpheniramine Maleate, and Dexamethasone Suspension, freshly mixed and free from air bubbles, quantitatively with *Buffer No. 1* to yield a solution containing about 2000 Penicillin G Units per mL. Pipet 2 mL of this solution into each of two glass-stoppered, 125-mL conical flasks.

Procedure—Proceed as directed for *Procedure* under *Iodometric Assay—Antibiotics* ⟨425⟩, except in the *Blank determination* to add 0.1 mL of 1.2 *N* hydrochloric acid immediately before the 10.0 mL of 0.01 *N* iodine VS. Calculate the quantity, in Penicillin G Units, in the portion of Suspension taken by the formula:

$$(L/2D)(F)(B - 1),$$

in which *L* is the labeled quantity, in Penicillin G Units, in the volume of Suspension taken, and *D* is the concentration, in Penicillin G Units per mL, of the *Assay preparation*, on the basis of the labeled quantity in the portion of Suspension taken and the extent of dilution.

Assay for dihydrostreptomycin—Proceed as directed for the turbidimetric assay for dihydrostreptomycin under *Antibiotics—Microbial Assays* ⟨81⟩, using an accurately measured volume of Sterile Penicillin G Procaine, Dihydrostreptomycin Sulfate, Chlorpheniramine Maleate, and Dexamethasone Suspension, freshly mixed and free from air bubbles, diluted quantitatively with water to yield a *Test Dilution* having a concentration of dihydrostreptomycin assumed to be equal to the median dose level of the Standard.

Assay for chlorpheniramine maleate—

Internal standard solution—Prepare a solution of brompheniramine maleate in water having a concentration of about 7 mg per mL.

Standard preparation—Dissolve an accurately weighed quantity of USP Chlorpheniramine Maleate RS in water to obtain a stock solution having a known concentration of about 6 mg per mL. Transfer 5.0 mL of this solution to a 50-mL centrifuge tube. Add 5.0 mL of *Internal standard solution*, and adjust with sodium hydroxide solution (1 in 2) to a pH of about 10. Add 25.0 mL of hexanes, place the cap on the tube, shake for about 2 minutes, and centrifuge. Use the upper hexanes layer as the *Standard preparation.*

Assay preparation—Transfer an accurately measured volume of Sterile Penicillin G Procaine, Dihydrostreptomycin Sulfate, Chlorpheniramine Maleate, and Dexamethasone Suspension, freshly mixed and free from air bubbles, equivalent to about 30 mg of chlorpheniramine maleate, to a 50-mL centrifuge tube. Proceed as directed under *Standard preparation*, beginning with "Add 5.0 mL of *Internal standard solution.*" Use the upper hexanes layer as the *Assay preparation.*

Chromatographic system (see *Chromatography* ⟨621⟩)—The gas chromatograph is equipped with a flame-ionization detector, and contains a 4-mm × 1.8-m glass column packed with 1.2 percent liquid phase G16 and 0.5 percent potassium hydroxide on 100- to 120-mesh support S1A. The column is maintained isothermally at about 180°, and the injection port and the detector block are maintained at about 200°. Dry nitrogen is used as the carrier gas at a flow rate of about 50 mL per minute. Chromatograph the *Standard preparation*, and record the peak responses as directed under *Procedure:* the resolution, *R*, between the analyte and internal standard peaks is not less than 2.0, and the relative standard deviation for replicate injections is not more than 2.0%.

Procedure—Separately inject equal volumes (about 1.5 µL) of the *Standard preparation* and the *Assay preparation* into the chromatograph, record the chromatograms, and measure the responses for the major peaks. The relative retention times are about 0.75 for chlorpheniramine and 1.0 for brompheniramine. Calculate the quantity, in mg, of chlorpheniramine maleate ($C_{16}H_{19}ClN_2 . C_4H_4O_4$) in each mL of the Suspension taken by the formula:

$$5(C/V)(R_U/R_S),$$

in which *C* is the concentration of USP Chlorpheniramine Maleate RS in the stock solution used to prepare the *Standard preparation*, *V* is the volume, in mL, of Suspension taken, and R_U and R_S are the peak response ratios of the chlorpheniramine maleate peak to the internal standard peak obtained from the *Assay preparation* and the *Standard preparation*, respectively.

Assay for dexamethasone—

Mobile phase—Prepare a suitable filtered mixture of water and acetonitrile (2:1). Make adjustments if necessary (see *System Suitability* under *Chromatography* ⟨621⟩).

Internal standard solution—Dissolve about 30 mg of beclomethasone in 2 mL of methanol in a 50-mL volumetric flask, dilute with methylene chloride to volume, and mix.

Standard preparation—Transfer about 25 mg of USP Dexamethasone RS, accurately weighed, to a 50-mL volumetric flask. Add about 1 mL of methanol, swirl to dissolve, dilute with methylene chloride to volume, and mix. Transfer 5.0 mL of this solution to a suitable flask, and add 5.0 mL of *Internal standard solution.* Heat the flask on a steam bath, and evaporate under a stream of nitrogen just to dryness. Add 10.0 mL of methanol to the flask, and swirl to dissolve the residue. This *Standard preparation* contains about 0.25 mg of USP Dexamethasone RS and 0.3 mg of beclomethasone per mL.

Assay preparation—Transfer an accurately measured volume of Sterile Penicillin G Procaine, Dihydrostreptomycin Sulfate, Chlorpheniramine Maleate, and Dexamethasone Suspension, freshly mixed and free from air bubbles, equivalent to about 2.5 mg of dexamethasone, to a separator containing 50 mL of 0.1 *N* hydrochloric acid, add 5.0 mL of *Internal standard solution*, and extract with four 25-mL portions of methylene chloride, combining the extracts in a second separator. Wash the combined extracts with 50 mL of sodium bicarbonate solution (1 in 20), filtering the lower methylene chloride layer through about 4 g of anhydrous sodium sulfate supported on a cotton pledget previously washed with methylene chloride, and collecting the filtrate in a suitable flask. Wash the aqueous layer with 25 mL of methylene chloride, and filter the lower methylene chloride layer through the same filter, collecting the filtrate in the same flask. Heat the flask on a steam bath, and evaporate under a stream of nitrogen just to dryness. Add 10.0 mL of methanol, and swirl to dissolve the residue.

Chromatographic system (see *Chromatography* ⟨621⟩)—The liquid chromatograph is equipped with a 254-nm detector and a 4-mm × 30-cm column that contains packing L1. The flow rate is about 1.2 mL per minute. Chromatograph the *Standard preparation*, and record the peak responses as directed under *Procedure:* the resolution, *R*, of the analyte and the internal standard peaks is not less than 2.0, and the relative standard deviation for replicate injections is not more than 2.0%.

Procedure—Separately inject equal volumes (about 10 µL) of the *Standard preparation* and the *Assay preparation* into the chromatograph, record the chromatograms, and measure the responses for the major peaks. The relative retention times are about 0.8 for dexamethasone and 1.0 for beclomethasone. Calculate the quantity, in mg, of dexamethasone ($C_{22}H_{29}FO_5$) in each mL of the Suspension taken by the formula:

$$10(C/V)(R_U/R_S),$$

in which C is the concentration, in mg per mL, of USP Dexamethasone RS in the *Standard preparation*, V is the volume, in mL, of Suspension taken, and R_U and R_S are the peak response ratios of the dexamethasone peak to the internal standard peak obtained from the *Assay preparation* and the *Standard preparation*, respectively.

Sterile Penicillin G Procaine, Dihydrostreptomycin Sulfate, and Prednisolone Suspension

» Sterile Penicillin G Procaine, Dihydrostreptomycin Sulfate, and Prednisolone Suspension is a sterile suspension of Sterile Penicillin G Procaine and Prednisolone in a solution of Sterile Dihydrostreptomycin Sulfate in Water for Injection. It contains one or more suitable buffers, dispersants, preservatives, and suspending agents. It may contain not more than 2.0 percent of procaine hydrochloride, and one or more suitable stabilizing agents. It contains not less than 90.0 percent and not more than 115.0 percent of the labeled number of Penicillin G Units, not less than 90.0 percent and not more than 115.0 percent of the labeled amount of dihydrostreptomycin ($C_{21}H_{41}N_7O_{12}$), and not less than 90.0 percent and not more than 110.0 percent of the labeled amount of prednisolone ($C_{21}H_{28}O_5$).

Packaging and storage—Preserve in single-dose or in multiple-dose, tight containers.

Reference standards—*USP Penicillin G Potassium Reference Standard*—Do not dry before using. *USP Dihydrostreptomycin Sulfate Reference Standard*—Dry in vacuum at a pressure not exceeding 5 mm of mercury at 100° for 4 hours before using. *USP Prednisolone Reference Standard*—Dry in vacuum at 105° for 3 hours before using.

Labeling—Label it to indicate that it is intended for veterinary use only, and is not to be used in animals to be slaughtered for human consumption.

Other requirements—It meets the requirements of the tests for *Pyrogen*, *Sterility*, and *pH* under *Sterile Penicillin G Procaine and Dihydrostreptomycin Sulfate Suspension.*

Assay for penicillin G—

Standard preparation—Using USP Penicillin G Potassium RS, prepare as directed for *Standard Preparation* under *Iodometric Assay—Antibiotics* ⟨425⟩.

Assay preparation—Dilute an accurately measured volume of Sterile Penicillin G Procaine, Dihydrostreptomycin Sulfate, and Prednisolone Suspension quantitatively with *Buffer No. 1* to obtain a solution containing about 2000 Penicillin G Units per mL. Pipet 2 mL of this solution into each of two glass-stoppered, 125-mL conical flasks.

Procedure—Proceed as directed for *Procedure* under *Iodometric Assay—Antibiotics* ⟨425⟩, except in the *Blank determination* to add 0.1 mL of 1.2 *N* hydrochloric acid immediately before the 10.0 mL of 0.01 *N* iodine VS. Calculate the quantity, in Penicillin G Units, in the portion of Suspension taken by the formula:

$$(L/2D)(F)(B - I),$$

in which L is the labeled quantity, in Penicillin G Units, in the volume of Suspension taken, and D is the concentration, in Penicillin G Units per mL, of the *Assay preparation*, on the basis of the labeled quantity in the portion of Suspension taken and the extent of dilution.

Assay for dihydrostreptomycin—Proceed as directed for the turbidimetric assay for dihydrostreptomycin as directed under *Antibiotics—Microbial Assays* ⟨81⟩, using an accurately measured volume of Sterile Penicillin G Procaine, Dihydrostreptomycin Sulfate, and Prednisolone Suspension diluted quantitatively with water to yield a *Test Dilution* having a concentration assumed to be equal to the median dose level of the Standard.

Assay for prednisolone—

Standard preparation—Prepare as directed for *Standard Preparation* under *Single-steroid Assay* ⟨511⟩, using USP Prednisolone RS.

Assay preparation—Transfer to a separator an accurately measured volume of Sterile Penicillin G Procaine, Dihydrostreptomycin Sulfate, and Prednisolone Suspension, and add 15 mL of water. Extract with three 25-mL portions and finally with one 20-mL portion of chloroform, filtering each portion through chloroform-washed cotton into a 100-mL volumetric flask. Add chloroform to volume, and mix. Pipet 20 mL of this solution into a suitable glass-stoppered flask or tube, evaporate the chloroform on a steam bath just to dryness, cool, and dissolve the residue in 2.0 mL, accurately measured, of a mixture of equal volumes of chloroform and alcohol.

Procedure—Proceed as directed for *Procedure* under *Single-steroid Assay* ⟨511⟩, using *Solvent A* to develop the chromatogram. Calculate the quantity, in mg, of prednisolone ($C_{21}H_{28}O_5$) in each mL of the Suspension taken by the formula:

$$0.01(C/V)(A_U/A_S)$$

in which the terms are as defined therein.

Penicillin G Procaine, Neomycin and Polymyxin B Sulfates, and Hydrocortisone Acetate Topical Suspension

» Penicillin G Procaine, Neomycin and Polymyxin B Sulfates, and Hydrocortisone Acetate Topical Suspension is a suspension of Penicillin G Procaine, Neomycin Sulfate, Polymyxin B Sulfate and Hydrocortisone Acetate in Peanut Oil or Sesame Oil. It may contain one or more suitable dispersing and suspending agents. It contains not less than 90.0 percent and not more than 140.0 percent of the labeled amounts of Penicillin G Units, of neomycin, and of polymyxin B Units, and not less than 90.0 percent and not more than 110.0 percent of the labeled amount of hydrocortisone acetate ($C_{23}H_{32}O_6$).

Packaging and storage—Preserve in well-closed containers.

Labeling—Label it to indicate that it is intended for veterinary use only.

Reference standards—*USP Penicillin G Potassium Reference Standard*—Do not dry before using. *USP Neomycin Sulfate Reference Standard*—Dry in vacuum at a pressure not exceeding 5 mm of mercury at 60° for 3 hours before using. *USP Polymyxin B Sulfate Reference Standard*—Dry in vacuum at a pressure not exceeding 5 mm of mercury at 60° for 3 hours before using. *USP Hydrocortisone Acetate Reference Standard*—Dry in vacuum at 60° for 3 hours before using.

Water, *Method I* ⟨921⟩: not more than 1.0%, 20 mL of a mixture of carbon tetrachloride, chloroform, and methanol (2:2:1) being used in place of methanol in the titration vessel.

Assay for penicillin G—Proceed as directed for penicillin G under *Antibiotics—Microbial Assays* ⟨81⟩, using an accurately measured volume of Penicillin G Procaine, Neomycin and Polymyxin B Sulfates, and Hydrocortisone Acetate Topical Suspension blended for 2 minutes in a high-speed glass blender jar with 499.0 mL of *Buffer No. 1* and 1.0 mL of polysorbate 80. Allow to stand for 10 minutes, and dilute an accurately measured volume of the aqueous phase quantitatively and stepwise with *Buffer No. 1* to obtain a *Test Dilution* having a concentration of penicillin G assumed to be equal to the median dose level of the Standard.

Assay for neomycin—Proceed as directed for neomycin under *Antibiotics—Microbial Assays* ⟨81⟩, using an accurately measured volume of Penicillin G Procaine, Neomycin and Polymyxin B Sulfates, and Hydrocortisone Acetate Topical Suspension shaken in a separator with about 50 mL of ether, and extracted with four 20-mL portions of *Buffer No. 3*. Combine the aqueous extracts, and dilute with *Buffer No. 3* to an appropriate volume to obtain a stock solution. To an accurately measured volume of this stock solution add an accurately measured volume of penicillinase sufficient to inactivate the penicillin G therein, heat at 37° for 30 minutes, and dilute quantitatively and stepwise with *Buffer No. 3* to obtain a *Test Dilution* having a concentration of neomycin assumed to be equal to the median dose level of the Standard.

Assay for polymyxin B—Proceed as directed for polymyxin B under *Antibiotics—Microbial Assays* ⟨81⟩, using an accurately measured volume of Penicillin G Procaine, Neomycin and Polymyxin B Sulfates, and Hydrocortisone Acetate Topical Suspension blended for 2 minutes in a high-speed glass blender jar containing 499.0 mL of *Buffer No. 6* and 1.0 mL of polysorbate 80. Allow to stand for 10 minutes, and to an accurately measured volume of the aqueous phase add an accurately measured volume of penicillinase sufficient to inactivate the penicillin G therein. Heat the solution at 37° for 30 minutes, and dilute quantitatively and stepwise with *Buffer No. 6* to obtain a *Test Dilution* having a concentration of polymyxin assumed to be equal to the median dose level of the Standard. Add to each test dilution of the Standard a quantity of USP Neomycin Sulfate RS dissolved in *Buffer No. 6* to obtain the same concentration of neomycin present in the *Test Dilution*.

Assay for hydrocortisone acetate—Using an accurately measured volume of Penicillin G Procaine, Neomycin and Polymyxin B Sulfates, and Hydrocortisone Acetate Topical Suspension, proceed as directed in the *Assay* under *Hydrocortisone Acetate Lotion*.

Penicillin G Procaine and Novobiocin Sodium Intramammary Infusion

» Penicillin G Procaine and Novobiocin Sodium Intramammary Infusion is a suspension of Penicillin G Procaine and Novobiocin Sodium in a suitable vegetable oil vehicle. It contains a suitable preservative and suspending agent. It contains not less than 90.0 percent and not more than 125.0 percent of the labeled amounts of Penicillin G Units and novobiocin ($C_{31}H_{36}N_2O_{11}$).

Packaging and storage—Preserve in disposable syringes that are well-closed containers.

Labeling—Label it to indicate that it is for veterinary use only.

Reference standards—*USP Penicillin G Potassium Reference Standard*—Do not dry before using. *USP Novobiocin Reference Standard*—Dry in vacuum at a pressure not exceeding 5 mm of mercury at 100° for 4 hours before using.

Water, *Method I* ⟨921⟩: not more than 1.0%, 20 mL of a mixture of carbon tetrachloride, chloroform, and methanol (2:2:1) being used in place of methanol in the titration vessel.

Assay for penicillin G—Proceed as directed for penicillin G under *Antibiotics—Microbial Assays* ⟨81⟩, except to use *Staphylococcus aureus* ATCC No. 12692 as the test organism. Prepare the inoculum by growing the organism at 32° to 35° for 24 hours on Medium 1 to which has been added a solution of novobiocin sodium, containing the equivalent of 2.5 mg of novobiocin per mL that has been filtered through a membrane filter of 0.2-μm porosity, so that the medium contains the equivalent of 10 μg of novobiocin per mL. Use an inoculum composition of about 5 mL of stock suspension in each 100 mL of Medium 1. Expel the contents of a syringe of Penicillin G Procaine and Novobiocin Sodium Intramammary Infusion into a high-speed glass blender jar containing 1.0 mL of polysorbate 80 and 499.0 mL of *Buffer No. 1*, and blend for 3 to 5 minutes. Allow to stand for 10 minutes, and dilute an accurately measured volume of the aqueous phase quantitatively and stepwise with *Buffer No. 1* to obtain a *Test Dilution* having a concentration of penicillin G assumed to be equal to the median dose level of the Standard.

Assay for novobiocin—Proceed as directed for novobiocin under *Antibiotics—Microbial Assays* ⟨81⟩, expelling the contents of a syringe of Penicillin G Procaine and Novobiocin Sodium Intramammary Infusion into a high-speed blender jar containing 1.0 mL of polysorbate 80 and 499.0 mL of *Buffer No. 3*, and blend for 3 to 5 minutes. Allow to stand for 10 minutes. To an accurately measured volume of the aqueous phase add sufficient penicillinase to inactivate the penicillin G therein, and dilute quantitatively and stepwise with *Buffer No. 6* to obtain a *Test Dilution* having a concentration of novobiocin assumed to be equal to the median dose level of the Standard. [NOTE—Store this *Test Dilution* at 37° for 30 minutes and allow to cool before using it to fill the cylinders on the plates.]

Penicillin G Sodium for Injection

» Penicillin G Sodium for Injection is a sterile mixture of penicillin G sodium and not less than 4.0 percent and not more than 5.0 percent of Sodium Citrate, of which not more than 0.15 percent may be replaced by Citric Acid. It has a potency of not less than 1420 Penicillin G Units and not more than 1667 Penicillin G Units per mg. In addition, where packaged for dispensing, it contains not less than 90.0 percent and not more than 120.0 percent of the labeled amount of penicillin G.

NOTE—It contains 2.0 mEq of sodium per million Penicillin G Units.

Packaging and storage—Preserve in *Containers for Sterile Solids* as described under *Injections* ⟨1⟩.

Reference standard—*USP Penicillin G Potassium Reference Standard*—Do not dry before using.

Constituted solution—At the time of use, the constituted solution prepared from Sterile Penicillin G Sodium for Injection meets the requirements for *Constituted Solutions* under *Injections* ⟨1⟩.

Identification—It responds to the *Identification test* under *Penicillin G Potassium for Injection.*

Crystallinity ⟨695⟩: meets the requirements.

pH ⟨791⟩: between 6.0 and 7.5, in a solution containing 60 mg per mL.

Loss on drying ⟨731⟩—Dry about 100 mg, accurately weighed, in a capillary-stoppered bottle in vacuum at a pressure not exceeding 5 mm of mercury at 60° for 3 hours: it loses not more than 1.5% of its weight.

Particulate matter ⟨788⟩: meets the requirements under *Small-volume Injections.*

Penicillin G content ⟨475⟩: between 80.0% and 93.8% of $C_{16}H_{18}N_2O_4S$.

Other requirements—It meets the requirements for *Pyrogen* and *Sterility* under *Sterile Penicillin G Sodium.* It meets also the requirements for *Uniformity of Dosage Units* ⟨905⟩ and *Labeling* under *Injections* ⟨1⟩.

Assay—Using Penicillin G Sodium for Injection, proceed as directed in the *Assay* under *Sterile Penicillin G Sodium.*

Sterile Penicillin G Sodium

$C_{16}H_{17}N_2NaO_4S$ 356.37

4-Thia-1-azabicyclo[3.2.0]heptane-2-carboxylic acid, 3,3-dimethyl-7-oxo-6-[(phenylacetyl)amino]-, [2*S*-(2α,5α,6β)]-, monosodium salt.

Monosodium (2*S*,5*R*,6*R*)-3,3-dimethyl-7-oxo-6-(2-phenylacetamido)-4-thia-1-azabicyclo[3.2.0]heptane-2-carboxylate [*69-57-8*].

» Sterile Penicillin G Sodium is penicillin G sodium suitable for parenteral use. It has a potency of not less than 1500 Penicillin G Units and not more than 1750 Penicillin G Units per mg. In addition, where packaged for dispensing, it contains not less than 90.0 percent and not more than 115.0 percent of the labeled amount of penicillin G.

Packaging and storage—Preserve in *Containers for Sterile Solids* as described under *Injections* ⟨1⟩.

Reference standards—*USP Penicillin G Potassium Reference Standard*—Do not dry before using. *USP Penicillin G Sodium Reference Standard.*

Constituted solution—At the time of use, the constituted solution prepared from Sterile Penicillin G Sodium meets the requirements for *Constituted Solutions* under *Injections* ⟨1⟩.

Identification—The infrared absorption spectrum of a potassium bromide dispersion of it exhibits maxima only at the same wavelengths as that of a similar preparation of USP Penicillin G Sodium RS.

Crystallinity ⟨695⟩: meets the requirements.

Pyrogen—It meets the requirements of the *Pyrogen Test* ⟨151⟩, the test dose being 1.0 mL per kg of a solution in Sterile Water for Injection containing 20,000 Penicillin G Units per mL.

Sterility—It meets the requirements under *Sterility Tests* ⟨71⟩, when tested as directed in the section, *Test Procedures Using Membrane Filtration.*

pH ⟨791⟩: between 5.0 and 7.5, in a solution containing 60 mg per mL.

Loss on drying ⟨731⟩—Dry about 100 mg, accurately weighed, in a capillary-stoppered bottle in vacuum at a pressure not exceeding 5 mm of mercury at 60° for 3 hours: it loses not more than 1.5% of its weight.

Particulate matter ⟨788⟩: meets the requirements under *Small-volume Injections.*

Penicillin G content ⟨475⟩: between 84.5% and 98.5% of $C_{16}H_{18}N_2O_4S$.

Other requirements—It meets the requirements for *Uniformity of Dosage Units* ⟨905⟩ and *Labeling* under *Injections* ⟨1⟩.

Assay—

Standard preparation—Using USP Penicillin G Potassium RS, prepare as directed for *Standard Preparation* under *Iodometric Assay—Antibiotics* ⟨425⟩.

Assay preparation 1—Dissolve an accurately weighed quantity of Sterile Penicillin G Sodium in *Buffer No. 1* to obtain an *Assay preparation* containing about 2000 Penicillin G Units per mL. Pipet 2 mL of this solution into each of two glass-stoppered, 125-mL conical flasks.

Assay preparation 2 (where it is packaged for dispensing and is represented as being in a single-dose container)—Constitute Sterile Penicillin G Sodium as directed in the labeling. Withdraw all of the withdrawable contents, using a suitable hypodermic needle and syringe, and dilute quantitatively with *Buffer No. 1* to obtain a solution containing about 2000 Penicillin G Units per mL. Pipet 2 mL of this solution into each of two glass-stoppered, 125-mL conical flasks.

Assay preparation 3 (where the label states the quantity of penicillin G sodium in a given volume of constituted solution)—Constitute Sterile Penicillin G Sodium as directed in the labeling. Dilute an accurately measured volume of the constituted solution quantitatively with *Buffer No. 1* to obtain a solution containing about 2000 Penicillin G Units per mL. Pipet 2 mL of this solution into each of two glass-stoppered, 125-mL conical flasks.

Procedure—Proceed as directed for *Procedure* under *Iodometric Assay—Antibiotics* ⟨425⟩. Calculate the potency, in Penicillin G Units per mg, of the Sterile Penicillin G Sodium taken by the formula:

$$(F)(B - I)/2D_1,$$

in which D_1 is the concentration, in mg per mL, of *Assay preparation 1*, on the basis of the weight of Sterile Penicillin G Sodium taken and the extent of dilution. Calculate the quantity, in Penicillin G Units, in the container, and in the portion of constituted solution taken by the formula:

$$(L/2D)(F)(B - I),$$

in which L is the labeled quantity, in Penicillin G Units, in the container, or in the volume of constituted solution taken, and D is the concentration, in Penicillin G Units per mL, of *Assay preparation 2*, or of *Assay preparation 3*, on the basis of the labeled quantity in the container, or in the portion of constituted solution taken, respectively, and the extent of dilution.

Penicillin V

$C_{16}H_{18}N_2O_5S$ 350.39

4-Thia-1-azabicyclo[3.2.0]heptane-2-carboxylic acid, 3,3-dimethyl-7-oxo-6-[(phenoxyacetyl)amino]-, [2*S*-(2α,5α,6β)]-.

(2*S*,5*R*,6*R*)-3,3-Dimethyl-7-oxo-6-(2-phenoxyacetamido)-4-thia-1-azabicyclo[3.2.0]heptane-2-carboxylic acid [*87-08-1*].

» Penicillin V has a potency of not less than 1525 and not more than 1780 Penicillin V Units per mg.

Packaging and storage—Preserve in tight containers.

Reference standards—*USP Penicillin V Potassium Reference Standard*—Do not dry before using. *USP Penicillin V Reference Standard*—Do not dry before using.

Labeling—Label it to indicate that it is to be used in the manufacture of nonparenteral drugs only.

Identification—The infrared absorption spectrum of a potassium bromide dispersion of it exhibits maxima only at the same wavelengths as that of a similar preparation of USP Penicillin V RS.

Crystallinity ⟨695⟩: meets the requirements.

pH ⟨791⟩: between 2.5 and 4.0, in a suspension containing 30 mg per mL.

Water, *Method I* ⟨921⟩: not more than 2.0%.

Assay—

Mobile phase—Prepare a suitable filtered and degassed mixture of water, acetonitrile, and glacial acetic acid (650:350:5.75). Make adjustments if necessary (see *System Suitability* under *Chromatography* ⟨621⟩).

Standard preparation—Dissolve an accurately weighed quantity of USP Penicillin V Potassium RS in *Mobile phase* to obtain a solution having a known concentration of about 2.5 mg per mL.

Assay preparation—Transfer about 125 mg of Penicillin V, accurately weighed, to a 50-mL volumetric flask, dilute with *Mobile phase* to volume, and mix.

Resolution solution—Prepare a solution in *Mobile phase* containing about 2.5 mg of penicillin G potassium and 2.5 mg of penicillin V potassium per mL.

Chromatographic system (see *Chromatography* ⟨621⟩)—The liquid chromatograph is equipped with a 254-nm detector and a 4-mm × 30-cm column that contains packing L1. The flow rate is about 1 mL per minute. Chromatograph the *Resolution solution*, and record the peak responses as directed under *Procedure:* the relative retention times are about 0.8 for penicillin G and 1.0 for penicillin V, and column efficiency determined from the penicillin V peak is not less than 1800 theoretical plates, and the resolution, *R*, between the penicillin G and penicillin V peaks is not less than 3.0. Chromatograph the *Standard preparation*, and record the peak responses as directed under *Procedure:* the relative standard deviation for replicate injections is not more than 1.0%.

Procedure—Separately inject equal volumes (about 10 µL) of the *Standard preparation* and the *Assay preparation* into the chromatograph, record the chromatograms, and measure the responses for the major peaks. Calculate the quantity, in USP Penicillin V Units, in each mg of the Penicillin V taken by the formula:

$$25(CP / W_U)(r_U / r_S)$$

in which *C* is the concentration, in mg per mL, of USP Penicillin V Potassium RS in the *Standard preparation*, *P* is the designated potency, in USP Penicillin V Units per mg, of USP Penicillin V Potassium RS, W_U is the weight, in mg, of Penicillin V taken to prepare the *Assay preparation*, and r_U and r_S are the penicillin V peak responses obtained from the *Assay preparation* and the *Standard preparation*, respectively.

Penicillin V for Oral Suspension

» Penicillin V for Oral Suspension is a dry mixture of Penicillin V with or without one or more suitable buffers, colors, flavors, and suspending agents. It contains not less than 90.0 percent and not more than 120.0 percent of the labeled number of Penicillin V Units when constituted as directed.

Packaging and storage—Preserve in tight containers.

Labeling—It may be labeled in terms of the weight of penicillin V contained therein, in addition to or instead of Units, on the basis that 1695 Penicillin V Units are equivalent to 1 mg of penicillin V.

Reference standards—*USP Penicillin V Potassium Reference Standard*—Do not dry before using. *USP Penicillin V Reference Standard*—Do not dry before using.

Identification—To a portion of the powder, equivalent to about 34,000 Penicillin V Units (20 mg of penicillin V), add 20 mL of methanol, mix, add 20 mL of water, and mix to obtain a test solution containing about 0.5 mg of penicillin V per mL. Similarly, prepare a Standard solution of USP Penicillin V RS in the same solvents containing 0.5 mg per mL. Apply separate 5-µL portions of the test solution and the Standard solution on a suitable thin-layer chromatographic plate (see *Chromatography* ⟨621⟩), coated with a 0.25-mm layer of chromatographic silica gel mixture. Develop the chromatogram in a solvent system consisting of a mixture of chloroform and glacial acetic acid (90:20) until the solvent front has moved about three-fourths of the length of the plate. Remove the plate from the chamber, allow it to dry briefly under a current of air, and then at 100° for 3 minutes. Allow to cool, place the plate in a chamber saturated with iodine vapor, and examine after 10 minutes: the R_f value of the principal spot obtained from the test solution corresponds to that obtained from the Standard solution.

Uniformity of dosage units ⟨905⟩—

FOR SOLID PACKAGED IN SINGLE-UNIT CONTAINERS: meets the requirements.

pH ⟨791⟩: between 2.0 and 4.0, in the suspension constituted as directed in the labeling.

Water, *Method I* ⟨921⟩: not more than 1.0%.

Assay—

Mobile phase, Standard preparation, Resolution solution, and *Chromatographic system*—Proceed as directed in the *Assay* under *Penicillin V.*

Assay preparation—Transfer an accurately measured volume of Penicillin V for Oral Suspension, freshly mixed and free from air bubbles, constituted as directed in the labeling, equivalent to about 400,000 USP Penicillin V Units, to a 100-mL volumetric flask, dilute with *Mobile phase* to volume, and mix. Filter a portion of this solution through a suitable filter of 0.5-µm or finer porosity, and use the filtrate as the *Assay preparation.*

Procedure—Proceed as directed for *Procedure* in the *Assay* under *Penicillin V.* Calculate the number of USP Penicillin V Units in each mL of the constituted Penicillin V for Oral Suspension taken by the formula:

$$100(CP / V)(r_U / r_S),$$

in which *V* is the volume, in mL, of constituted Penicillin V for Oral Suspension taken, and the other terms are as defined therein.

Penicillin V Tablets

» Penicillin V Tablets contain not less than 90.0 percent and not more than 120.0 percent of the labeled number of Penicillin V Units.

Packaging and storage—Preserve in tight containers.

Labeling—Tablets may be labeled in terms of the weight of penicillin V contained therein, in addition to or instead of Units, on the basis that 1695 Penicillin V Units are equivalent to 1 mg of penicillin V.

Reference standards—*USP Penicillin V Potassium Reference Standard*—Do not dry before using. *USP Penicillin V Reference Standard*—Do not dry before using.

Identification—A portion of powdered Tablets responds to the *Identification test* under *Penicillin V for Oral Suspension.*

Dissolution ⟨711⟩—

Medium: water; 900 mL.

Apparatus 2: 50 rpm.

Time: 45 minutes.
Procedure—Determine the amount of penicillin V ($C_{16}H_{18}N_2O_5S$) by a suitable validated spectrophotometric analysis of a filtered portion of the solution under test, suitably diluted with *Dissolution Medium*, if necessary, in comparison with a Standard solution having a known concentration of USP Penicillin V Potassium RS in the same medium.
Tolerances—Not less than 75% (*Q*) of the labeled amount of $C_{16}H_{18}N_2O_5S$ is dissolved in 45 minutes.

Uniformity of dosage units ⟨905⟩: meet the requirements.

Water, *Method I* ⟨921⟩: not more than 3.0%.

Assay—
Mobile phase, Standard preparation, Resolution solution, and *Chromatographic system*—Proceed as directed in the *Assay* under *Penicillin V.*
Assay preparation—Weigh and finely powder not less than 20 Penicillin V Tablets. Transfer an accurately weighed portion of the powder, equivalent to about 400,000 USP Penicillin V Units, to a 100-mL volumetric flask, dilute with *Mobile phase* to volume, and shake for about 5 minutes. Filter a portion of this solution through a suitable filter of 0.5-µm or finer porosity, and use the filtrate as the *Assay preparation.*
Procedure—Proceed as directed for *Procedure* in the *Assay* under *Penicillin V.* Calculate the number of USP Penicillin V Units in the portion of Tablets taken by the formula:

$$100CP(r_U/r_S),$$

in which the terms are as defined therein.

Penicillin V Benzathine

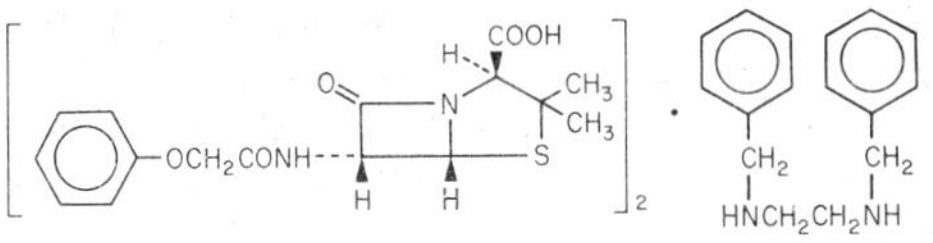

$(C_{16}H_{18}N_2O_5S)_2 . C_{16}H_{20}N_2$ 941.12
4-Thia-1-azabicyclo[3.2.0]heptane-2-carboxylic acid, 3,3-dimethyl-7-oxo-6-[(2-phenoxyacetyl)amino]-, [2*S*-(2α,5α,6β)]-, compd. with *N,N'*-bis(phenylmethyl)-1,2-ethanediamine (2:1).
(2*S*,5*R*,6*R*)-3,3-Dimethyl-7-oxo-6-(2-phenoxyacetamido)-4-thia-1-azabicyclo[3.2.0]heptane-2-carboxylic acid compound with *N,N'*-dibenzylethylenediamine (2:1) [*5928-84-7*].
Tetrahydrate 1013.19 [*63690-57-3*].

» Penicillin V Benzathine has a potency of not less than 1060 and not more than 1240 Penicillin V Units per mg.

Packaging and storage—Preserve in tight containers.

Reference standard—*USP Penicillin V Potassium Reference Standard*—Do not dry before using.

Crystallinity ⟨695⟩: meets the requirements.

pH ⟨791⟩: between 4.0 and 6.5, in a suspension containing about 30 mg per mL.

Water, *Method I* ⟨921⟩: between 5.0% and 8.0%.

Penicillin V content—Transfer about 40 mg, accurately weighed, to a 100-mL volumetric flask, add methanol to volume, and mix. Concomitantly determine the absorbances of this solution and of a similarly prepared Standard solution prepared with about 30 mg of USP Penicillin V Potassium RS at the wavelength of maximum absorbance at about 276 nm. Determine the percentage of penicillin V by the formula:

$$P(a_U/a_S),$$

in which *P* is the percentage content of penicillin V in the USP Penicillin V Potassium RS, and a_U and a_S are the absorptivities of the solution of the specimen and the Standard solution, respectively: between 62.3% and 72.5% is found.

Assay—
Standard preparation—Prepare as directed for *Standard Preparation* under *Iodometric Assay—Antibiotics* ⟨425⟩, using USP Penicillin V Potassium RS.
Assay preparations—Dissolve a quantity of Penicillin V Benzathine in 1.0 *N* sodium hydroxide to obtain a solution containing 2000 Penicillin V Units per mL. Pipet 2 mL of this solution into a glass-stoppered, 125-mL conical flask, and use as the *Assay preparation* for *Inactivation and titration.* Dilute a quantity of Penicillin V Benzathine quantitatively with water to obtain a suspension containing 2000 Penicillin V Units per mL. Pipet 2 mL of this suspension into a glass-stoppered, 125-mL conical flask, and use as the *Assay preparation* for the *Blank determination.*
Procedure—Proceed as directed for *Procedure* under *Iodometric Assay—Antibiotics* ⟨425⟩, except in the *Inactivation and titration* of the specimen to omit the addition of 2.0 mL of 1.0 *N* sodium hydroxide. Calculate the potency, in Penicillin V Units per mg, in the Penicillin V Benzathine taken by the formula:

$$(F)(B - I)/(2D),$$

in which *D* is the concentration, in mg per mL, of the *Assay preparation* for *Inactivation and titration* on the basis of the weight of Penicillin V Benzathine taken and the extent of dilution.

Penicillin V Benzathine Oral Suspension

» Penicillin V Benzathine Oral Suspension contains not less than 90.0 percent and not more than 120.0 percent of the labeled number of Penicillin V Units per mL. It contains one or more suitable buffers, colors, dispersants, flavors, and preservatives.

Packaging and storage—Preserve in tight containers, and store in a refrigerator.

Labeling—It may be labeled in terms of the weight of penicillin V contained therein, in addition to or instead of Units, on the basis that 1695 Penicillin V Units are equivalent to 1 mg of penicillin V.

Reference standard—*USP Penicillin V Potassium Reference Standard*—Do not dry before using.

Uniformity of dosage units ⟨905⟩—
FOR SUSPENSION PACKAGED IN SINGLE-UNIT CONTAINERS: meets the requirements.

pH ⟨791⟩: between 6.0 and 7.0.

Assay—
Standard preparation—Prepare as directed for *Standard Preparation* under *Iodometric Assay—Antibiotics* ⟨425⟩, using USP Penicillin V Potassium RS.
Assay preparations—Dilute an accurately measured volume of Penicillin V Benzathine Oral Suspension, freshly mixed and free from air bubbles, quantitatively with 1.0 *N* sodium hydroxide to obtain a solution containing 2000 Penicillin V Units per mL. Pipet 2 mL of this solution into a glass-stoppered, 125-mL conical flask, and use as the *Assay preparation* for *Inactivation and titration.* Dilute an accurately measured volume of Penicillin V Benzathine Oral Suspension quantitatively with water to obtain a suspension containing 2000 Penicillin V Units per mL. Pipet 2 mL of this suspension into a glass-stoppered, 125-mL conical flask, and use as the *Assay preparation* for the *Blank determination.*
Procedure—Proceed as directed for *Procedure* under *Iodometric Assay—Antibiotics* ⟨425⟩, except in the *Inactivation and titration* to omit the addition of 2.0 mL of 1.0 *N* sodium hydroxide. Calculate the quantity, in Penicillin V Units, in each mL of the Oral Suspension taken by the formula:

$$(T/2D)F(B - I),$$

in which *T* is the labeled quantity, in Penicillin V Units per mL, in the Penicillin V Benzathine Oral Suspension, and *D* is the concentration, in Penicillin V Units per mL, in the *Assay prep-*

aration for *Inactivation and titration* on the basis of the volume of Oral Suspension taken and the extent of dilution.

Penicillin V Potassium

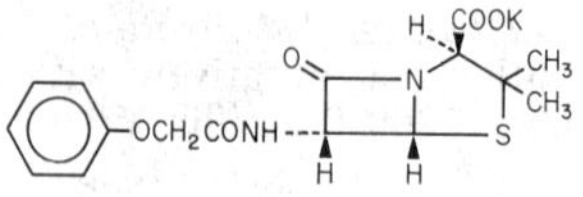

$C_{16}H_{17}KN_2O_5S$ 388.48

4-Thia-1-azabicyclo[3.2.0]heptane-2-carboxylic acid, 3,3-dimethyl-7-oxo-6-[(phenoxyacetyl)amino]-, monopotassium salt, [2*S*-(2α,5α,6β)]-.

Monopotassium (2*S*,5*R*,6*R*)-3,3-dimethyl-7-oxo-6-(2-phenoxyacetamido)-4-thia-1-azabicyclo[3.2.0]heptane-2-carboxylate [*132-98-9*].

» Penicillin V Potassium has a potency of not less than 1380 and not more than 1610 Penicillin V Units per mg.

Packaging and storage—Preserve in tight containers.

Labeling—Label it to indicate that it is to be used in the manufacture of nonparenteral drugs only.

Reference standard—*USP Penicillin V Potassium Reference Standard*—Do not dry before using.

Identification—The infrared absorption spectrum of a potassium bromide dispersion of it exhibits maxima only at the same wavelengths as that of a similar preparation of USP Penicillin V Potassium RS.

Crystallinity ⟨695⟩: meets the requirements.

pH ⟨791⟩: between 4.0 and 7.5, in a solution containing 30 mg per mL.

Loss on drying ⟨731⟩—Dry about 100 mg in a capillary-stoppered bottle in vacuum at 60° for 3 hours: it loses not more than 1.5% of its weight.

Assay—

Mobile phase, Standard preparation, Resolution solution, and *Chromatographic system*—Proceed as directed in the *Assay* under *Penicillin V*.

Assay preparation—Transfer about 125 mg of Penicillin V Potassium, accurately weighed, to a 50-mL volumetric flask, dilute with *Mobile phase* to volume, and mix.

Procedure—Proceed as directed for *Procedure* in the *Assay* under *Penicillin V*. Calculate the quantity, in USP Penicillin V Units, in each mg of the Penicillin V Potassium taken by the formula:

$$25(CP/W_U)(r_U/r_S),$$

in which W_U is the weight, in mg, of Penicillin V Potassium taken to prepare the *Assay preparation*, and the other terms are as defined therein.

Penicillin V Potassium for Oral Solution

» Penicillin V Potassium for Oral Solution is a dry mixture of Penicillin V Potassium with or without one or more suitable buffers, colors, flavors, preservatives, and suspending agents. It contains not less than 90.0 percent and not more than 135.0 percent of the labeled number of Penicillin V Units when constituted as directed.

Packaging and storage—Preserve in tight containers.

Labeling—It may be labeled in terms of the weight of penicillin V contained therein, in addition to or instead of Units, on the basis that 1695 Penicillin V Units are equivalent to 1 mg of penicillin V.

Reference standard—*USP Penicillin V Potassium Reference Standard*—Do not dry before using.

Identification—To a portion of the powder, equivalent to about 34,000 Penicillin V Units (20 mg of penicillin V), add 40 mL of water, and mix to obtain a test solution containing about 0.5 mg of penicillin V per mL. Similarly, prepare a Standard solution of USP Penicillin V Potassium RS in water containing 0.5 mg per mL. Apply separate 5-μL portions of the test solution and the Standard solution on a suitable thin-layer chromatographic plate (see *Chromatography* ⟨621⟩), coated with a 0.25-mm layer of chromatographic silica gel mixture. Develop the chromatogram in a solvent system consisting of a mixture of chloroform and glacial acetic acid (90:20) until the solvent front has moved about three-fourths of the length of the plate. Remove the plate from the chamber, allow it to dry briefly under a current of air, and then at 100° for 3 minutes. Allow to cool, place the plate in a chamber saturated with iodine vapor, and examine after 10 minutes: the R_f value of the principal spot obtained from the test solution corresponds to that obtained from the *Standard solution*.

Uniformity of dosage units ⟨905⟩—

FOR SOLID PACKAGED IN SINGLE-UNIT CONTAINERS: meets the requirements.

pH ⟨791⟩: between 5.0 and 7.5, when constituted as directed in the labeling.

Water, *Method I* ⟨921⟩: not more than 1.0%.

Assay—

Mobile phase, Standard preparation, Resolution solution, and *Chromatographic system*—Proceed as directed in the *Assay* under *Penicillin V*.

Assay preparation—Transfer an accurately measured volume of Penicillin V Potassium for Oral Solution, constituted as directed in the labeling, equivalent to about 400,000 USP Penicillin V Units, to a 100-mL volumetric flask, dilute with *Mobile phase* to volume, and mix. Filter a portion of this solution through a suitable filter of 0.5-μm or finer porosity, and use the filtrate as the *Assay preparation*.

Procedure—Proceed as directed for *Procedure* in the *Assay* under *Penicillin V*. Calculate the number of USP Penicillin V Units in each mL of the constituted Penicillin V Potassium for Oral Solution taken by the formula:

$$100(CP/V)(r_U/r_S),$$

in which V is the volume, in mL, of constituted Penicillin V Potassium for Oral Suspension taken, and the other terms are as defined therein.

Penicillin V Potassium Tablets

» Penicillin V Potassium Tablets contain not less than 90.0 percent and not more than 120.0 percent of the labeled number of Penicillin V Units.

Packaging and storage—Preserve in tight containers.

Labeling—Label Tablets to indicate whether they are to be chewed before swallowing. Tablets may be labeled in terms of the weight of penicillin V contained therein, in addition to or instead of Units, on the basis that 1695 Penicillin V Units are equivalent to 1 mg of penicillin V.

Reference standard—*USP Penicillin V Potassium Reference Standard*—Do not dry before using.

Identification—To a portion of powdered Tablets add sufficient water to obtain a test solution containing about 0.5 mg of penicillin V per mL: the solution responds to the *Identification test* under *Penicillin V Potassium for Oral Solution*.

Dissolution ⟨711⟩—

Medium: pH 6.0 phosphate buffer (see *Buffer Solutions* in the section, *Reagents, Indicators, and Solutions*); 900 mL.

Apparatus 2: 50 rpm.

Time: 45 minutes.

Procedure—Determine the amount of Penicillin V Units by a suitable validated spectrophotometric analysis of a filtered portion of the solution under test, suitably diluted with *Dissolution Medium*, if necessary, in comparison with a Standard solution of USP Penicillin V Potassium RS in the same medium having a known concentration of Penicillin V Units.

Tolerances—Not less than 75% (Q) of the labeled amount of Penicillin V Units is dissolved in 45 minutes.

Uniformity of dosage units ⟨905⟩: meet the requirements.

Loss on drying ⟨731⟩—Dry about 100 mg in a capillary-stoppered bottle in vacuum at 60° for 3 hours: it loses not more than 1.5% of its weight.

Assay—

Mobile phase, Standard preparation, Resolution solution, and *Chromatographic system*—Proceed as directed in the *Assay* under *Penicillin V.*

Assay preparation—Weigh and finely powder not less than 20 Penicillin V Potassium Tablets. Transfer an accurately weighed portion of the powder, equivalent to about 400,000 USP Penicillin V Units, to a 100-mL volumetric flask, dilute with *Mobile phase* to volume, and shake for about 5 minutes. Filter a portion of this solution through a suitable filter of 0.5-µm or finer porosity, and use the filtrate as the *Assay preparation.*

Procedure—Proceed as directed for *Procedure* in the *Assay* under *Penicillin V.* Calculate the number of USP Penicillin V Units in the portion of Tablets taken by the formula:

$$100CP(r_U / r_S),$$

in which the terms are as defined therein.

Diluted Pentaerythritol Tetranitrate

$$\begin{array}{c} CH_2{-}ONO_2 \\ | \\ O_2NO{-}CH_2CCH_2{-}ONO_2 \\ | \\ CH_2{-}ONO_2 \end{array}$$

$C_5H_8N_4O_{12}$ 316.14

1,3-Propanediol, 2,2-bis[(nitrooxy)methyl]-, dinitrate (ester).

2,2-Bis(hydroxymethyl)-1,3-propanediol tetranitrate [*78-11-5*].

» Diluted Pentaerythritol Tetranitrate is a dry mixture of pentaerythritol tetranitrate ($C_5H_8N_4O_{12}$) with Lactose or Mannitol or other suitable inert excipients, to permit safe handling and compliance with federal Interstate Commerce Commission regulations pertaining to interstate shipment. It contains not less than 95.0 percent and not more than 105.0 percent of the labeled amount of $C_5H_8N_4O_{12}$.

Caution—Undiluted pentaerythritol tetranitrate is a powerful explosive; take proper precautions in handling. It can be exploded by percussion or by excessive heat. Only exceedingly small amounts should be isolated.

Packaging and storage—Preserve in tight containers, and prevent exposure to excessive heat.

Identification—Transfer to a medium-porosity, sintered-glass filter an amount equivalent to not more than 10 mg of pentaerythritol tetranitrate, and pass several small portions of dry acetone through the test specimen to extract the pentaerythritol tetranitrate. Evaporate the combined extracts in a beaker at a temperature not exceeding 60°, with the aid of a gentle current of air, and dry the residue at 60° for 4 hours: the crystals of pentaerythritol tetranitrate so obtained melt between 138° and 142°.

Assay—

Standard preparation—Transfer about 130 mg of potassium nitrate, previously dried at 105° for 4 hours and accurately weighed, to a 200-mL volumetric flask, dissolve in 3 mL of water, dilute with glacial acetic acid to volume, and mix.

Assay preparation—Transfer an accurately weighed portion of Diluted Pentaerythritol Tetranitrate, equivalent to about 50 mg of pentaerythritol tetranitrate, to a 100-mL volumetric flask. Add 50 mL of acetone, heat the mixture on a water bath not exceeding 60° to boiling, and boil gently, with occasional swirling, for 5 minutes, maintaining the volume at about 50 mL. Cool, dilute with acetone to volume, and mix. Transfer a portion of the mixture to a glass-stoppered centrifuge tube, and centrifuge at 1500 rpm for 5 minutes. Transfer 1.0 mL of the supernatant solution to a 100-mL volumetric flask, and evaporate at 35° with the aid of a current of air to dryness. To the residue add 1.0 mL of glacial acetic acid, and swirl to dissolve.

Procedure—Transfer 1.0 mL of the *Standard preparation* to a 100-mL volumetric flask. To this flask and to the *Assay preparation*, add 2 mL of phenoldisulfonic acid TS, mix, and allow to stand for 5 minutes. Add to each flask 25 mL of water and 20 mL of 6 N ammonium hydroxide, cool, dilute with water to volume, and mix. Concomitantly determine the absorbances of the solutions in 1-cm cells at thc wavelength of maximum absorbance at about 409 nm, with a suitable spectrophotometer, using water as the blank. Calculate the quantity, in mg, of $C_5H_8N_4O_{12}$ in the Diluted Pentaerythritol Tetranitrate taken by the formula:

$$(316.14 / 101.10)(0.025C)(A_U / A_S),$$

in which 316.14 and 101.10 are the molecular weights of pentaerythritol tetranitrate and potassium nitrate, respectively, C is the concentration, in µg per mL, of potassium nitrate in the *Standard preparation*, and A_U and A_S are the absorbances of the solutions from the *Assay preparation* and the *Standard preparation*, respectively.

Pentaerythritol Tetranitrate Tablets

» Pentaerythritol Tetranitrate Tablets are prepared from Diluted Pentaerythritol Tetranitrate. They contain not less than 93.0 percent and not more than 107.0 percent of the labeled amount of $C_5H_8N_4O_{12}$.

Caution—Undiluted pentaerythritol tetranitrate is a powerful explosive; take proper precautions in handling. It can be exploded by percussion or by excessive heat. Only exceedingly small amounts should be isolated.

Packaging and storage—Preserve in tight containers.

Identification—Extract a portion of finely powdered Tablets, equivalent to about 10 mg of pentaerythritol tetranitrate, with several small portions of solvent hexane, and discard the extracts. Dry the residue in air until the odor of hexane is no longer perceptible, extract the residue with about 10 mL of acetone, and filter. Evaporate the filtrate at a temperature not exceeding 60°, with the aid of a gentle current of air to dryness, then dry the residue at 60° for 4 hours: the crystals of pentaerythritol tetranitrate so obtained melt between 138° and 142°.

Disintegration ⟨701⟩: 10 minutes.

Uniformity of dosage units ⟨905⟩: meet the requirements.

Assay—

Standard preparation—Prepare as directed in the *Assay* under Diluted Pentaerythritol Tetranitrate.

Assay preparation—Weigh and finely powder not less than 20 Pentaerythritol Tetranitrate Tablets. Transfer an accurately weighed portion of the powder, equivalent to about 50 mg of pentaerythritol tetranitrate, to a 100-mL volumetric flask, and proceed as directed for *Assay preparation* in the *Assay* under *Diluted Pentaerythritol Tetranitrate*, beginning with "Add 50 mL of acetone."

Procedure—Proceed as directed for *Procedure* in the *Assay* under *Diluted Pentaerythritol Tetranitrate.* Calculate the quantity, in mg, of $C_5H_8N_4O_{12}$ in the portion of Tablets taken by the formula:

$$(316.14/101.10)(0.025C)(A_U/A_S),$$

in which 316.14 and 101.10 are the molecular weights of pentaerythritol tetranitrate and potassium nitrate, respectively, C is the concentration, in µg per mL, of potassium nitrate in the *Standard preparation,* and A_U and A_S are the absorbances of the solutions from the *Assay preparation* and the *Standard preparation,* respectively.

Pentazocine

$C_{19}H_{27}NO$ 285.43

2,6-Methano-3-benzazocin-8-ol, 1,2,3,4,5,6-hexahydro-6,11-dimethyl-3-(3-methyl-2-butenyl)-, (2α,6α,11*R**)-.

(2*R**,6*R**,11*R**)-1,2,3,4,5,6-Hexahydro-6,11-dimethyl-3-(3-methyl-2-butenyl)-2,6-methano-3-benzazocin-8-ol [*359-83-1*].

» Pentazocine contains not less than 98.0 percent and not more than 101.5 percent of $C_{19}H_{27}NO$, calculated on the dried basis.

Packaging and storage—Preserve in tight, light-resistant containers.

Reference standard—*USP Pentazocine Reference Standard*—Dry at a pressure not exceeding 5 mm of mercury at 60° to constant weight before using.

Identification—

A: The infrared absorption spectrum of a potassium bromide dispersion of it, previously dried, exhibits maxima only at the same wavelengths as that of a similar preparation of USP Pentazocine RS.

B: The ultraviolet absorption spectrum of a 1 in 12,500 solution in 0.01 *N* hydrochloric acid exhibits maxima and minima at the same wavelengths as that of a similar solution of USP Pentazocine RS, concomitantly measured, and the respective absorptivities, calculated on the dried basis, at the wavelength of maximum absorbance at about 278 nm do not differ by more than 3.0%.

Melting range ⟨741⟩: between 147° and 158°, with slight darkening.

Loss on drying ⟨731⟩—Dry it at a pressure not exceeding 5 mm of mercury at 60° to constant weight: it loses not more than 1.0% of its weight.

Residue on ignition ⟨281⟩: not more than 0.2%.

Ordinary impurities ⟨466⟩—

Test solution: methanol.

Standard solution: methanol.

Eluant: a mixture of chloroform, methanol, and isopropylamine (94:3:3).

Visualization—Heat the plate in an oven at 105° for 15 minutes, cool, follow with visualization technique 17, and view under short-wavelength ultraviolet light.

Assay—Dissolve about 500 mg of Pentazocine, accurately weighed, in 50 mL of glacial acetic acid, add 1 drop of crystal violet TS, and titrate with 0.1 *N* perchloric acid VS to a green end-point. Perform a blank determination, and make any necessary correction. Each mL of 0.1 *N* perchloric acid is equivalent to 28.54 mg of $C_{19}H_{27}NO$.

Pentazocine Hydrochloride

$C_{19}H_{27}NO.HCl$ 321.89

2,6-Methano-3-benzazocin-8-ol, 1,2,3,4,5,6-hexahydro-6,11-dimethyl-3-(3-methyl-2-butenyl)-, hydrochloride, (2α,6α,11*R**)-.

(2*R**,6*R**,11*R**)-1,2,3,4,5,6-Hexahydro-6,11-dimethyl-3-(3-methyl-2-butenyl)-2,6-methano-3-benzazocin-8-ol hydrochloride [*64024-15-3*].

» Pentazocine Hydrochloride contains not less than 98.0 percent and not more than 102.0 percent of $C_{19}H_{27}NO.HCl$, calculated on the dried basis.

Packaging and storage—Preserve in tight, light-resistant containers.

Reference standard—*USP Pentazocine Reference Standard*—Dry at a pressure not exceeding 5 mm of mercury at 60° to constant weight before using.

Identification—

A: The ultraviolet absorption spectrum of a solution in 0.01 *N* hydrochloric acid, equivalent to a 1 in 12,500 solution of pentazocine, exhibits maxima and minima at the same wavelengths as that of a similar solution of USP Pentazocine RS, concomitantly measured, and the respective molar absorptivities, calculated on the dried basis, at the wavelength of maximum absorbance at about 278 nm, do not differ by more than 3.0%. [NOTE—The molecular weight of pentazocine ($C_{19}H_{27}NO$) is 285.43.]

B: Dissolve 50 mg of USP Pentazocine RS in 25 mL of 0.01 *N* hydrochloric acid in a separator, and use this in place of the Standard solution specified under *Identification—Organic Nitrogenous Bases* ⟨181⟩: Pentazocine Hydrochloride meets the requirements of the test.

C: A solution (1 in 100) responds to the tests for *Chloride* ⟨191⟩.

Loss on drying ⟨731⟩—Dry it at a pressure not exceeding 5 mm of mercury at 100° to constant weight: it loses not more than 1.0% of its weight.

Residue on ignition ⟨281⟩: not more than 0.2%.

Ordinary impurities ⟨466⟩—

Test solution: methanol.

Standard solution: methanol, USP Pentazocine RS being used.

Eluant: a mixture of chloroform, methanol, and isopropylamine (94:3:3).

Visualization—Heat the plate in an oven at 105° for 15 minutes, cool, follow with visualization technique 17, and view under short-wavelength ultraviolet light.

Limits—The total of any ordinary impurities observed does not exceed 1.0%.

Assay—Dissolve about 650 mg of Pentazocine Hydrochloride, accurately weighed, in 50 mL of glacial acetic acid, and add 10 mL of mercuric acetate TS. Add 1 drop of crystal violet TS, and titrate with 0.1 *N* perchloric acid VS to a green end-point. Perform a blank determination, and make any necessary correction. Each mL of 0.1 *N* perchloric acid is equivalent to 32.19 mg of $C_{19}H_{27}NO.HCl$.

Pentazocine Hydrochloride Tablets

» Pentazocine Hydrochloride Tablets contain an amount of $C_{19}H_{27}NO.HCl$ equivalent to not less than 90.0 percent and not more than 110.0 percent of the labeled amount of pentazocine ($C_{19}H_{27}NO$).

Packaging and storage—Preserve in tight, light-resistant containers.

Reference standard—*USP Pentazocine Reference Standard*—Dry at 60° and at a pressure not exceeding 5 mm of mercury to constant weight before using.

Identification—

A: Place a portion of finely powdered Tablets, equivalent to about 20 mg of pentazocine, in a glass-stoppered conical flask.

Add 10 mL of chloroform, shake for 10 minutes, allow the solids to settle, and filter the chloroform solution. Evaporate 1 mL of the filtrate to dryness: the ultraviolet absorption spectrum of a solution of the residue so obtained in 100 mL of 0.1 *N* sodium hydroxide exhibits maxima and minima at the same wavelengths as that of a 1 in 50,000 solution of USP Pentazocine RS in the same medium, concomitantly measured.

B: Dissolve 50 mg of USP Pentazocine RS in 25 mL of 0.01 *N* hydrochloric acid in a separator, and use this in place of the Standard solution specified under *Identification—Organic Nitrogenous Bases* ⟨181⟩. Tablets meet the requirements of the test.

Dissolution ⟨711⟩—

Medium: water; 900 mL.

Apparatus 2: 50 rpm.

Time: 45 minutes.

Procedure—Determine the amount of pentazocine ($C_{19}H_{27}NO$) dissolved from ultraviolet absorbances at the wavelength of maximum absorbance at about 278 nm of filtered portions of the solution under test, mixed with sufficient hydrochloric acid to provide a concentration of 0.01 *N* hydrochloric acid, and suitably diluted with *Dissolution Medium*, if necessary, in comparison with a Standard solution having a known concentration of USP Pentazocine RS in the same medium.

Tolerances—Not less than 75% (*Q*) of the labeled amount of $C_{19}H_{27}NO$ is dissolved in 45 minutes.

Uniformity of dosage units ⟨905⟩: meet the requirements.

Assay—Proceed with Pentazocine Hydrochloride Tablets as directed under *Salts of Organic Nitrogenous Bases* ⟨501⟩, but for the *Procedure* dilute 20.0 mL of the *Standard Preparation* and of the *Assay Preparation* with 0.5 *N* sulfuric acid to 100.0 mL, and determine the absorbances at the wavelength of maximum absorbance at about 278 nm. Calculate the quantity, in mg, of pentazocine ($C_{19}H_{27}NO$) in the portion of Tablets taken by the formula:

$$0.25C(A_U / A_S),$$

in which *C* is the concentration, in µg per mL, of USP Pentazocine RS in the *Standard Preparation* when diluted as directed and A_U and A_S are as defined therein.

Pentazocine Hydrochloride and Aspirin Tablets

» Pentazocine Hydrochloride and Aspirin Tablets contain not less than 90.0 percent and not more than 110.0 percent of the labeled amount of pentazocine ($C_{19}H_{27}NO$) and not less than 90.0 percent and not more than 110.0 percent of the labeled amount of aspirin ($C_9H_8O_4$).

Packaging and storage—Preserve in tight, light-resistant containers.

Reference standards—*USP Aspirin Reference Standard*—Dry over silica gel for 5 hours before using. *USP Pentazocine Reference Standard*—Dry at a pressure not exceeding 5 mm of mercury at 60° to constant weight before using. *USP Salicylic Acid Reference Standard*—Dry over silica gel for 3 hours before using.

Identification—Prepare a solution of USP Pentazocine RS in a mixture of chloroform and methanol (1:1) containing 2.5 mg per mL. Using the same solvent prepare a solution of USP Aspirin RS containing 65 mg per mL. Shake a quantity of finely powdered Tablets equivalent to about 25 mg of pentazocine and 650 mg of aspirin with 10 mL of the same solvent in an ultrasonic bath for 2 minutes. Allow the solids to settle. Apply 10 µL of this test solution and of each of the Standard solutions on a line parallel to and about 2.5 cm from the bottom edge of a thin-layer chromatographic plate (see *Chromatography* ⟨621⟩) coated with a 0.25-mm layer of chromatographic silica gel mixture. Evaporate the solvents from the spots in warm, circulating air. Place the plate in a developing chamber containing, and equilibrated with, a solvent system consisting of a mixture of ethyl acetate, methanol, and formic acid (90:5:5). Develop the chromatogram until the solvent front has moved 12 to 15 cm above the point of application. Remove the plate from the developing chamber and mark the solvent front. Evaporate the solvents thoroughly in warm, circulating air, and examine the plate under short-wavelength ultraviolet light. Expose the plate to iodine vapor for about 5 minutes and observe. Then spray the plate with an iodoplatinate spray reagent prepared by dissolving 300 mg of platinic chloride in 100 mL of water and adding 100 mL of potassium iodide solution (6 in 100): the chromatogram obtained with the test solution shows two principal spots which correspond in R_f values, size, and intensity of color with the spots obtained with the Standard solutions.

Non-aspirin salicylates—

Ferric chloride–urea reagent—To a mixture of 8 mL of ferric chloride solution (6 in 10) and 42 mL of 0.05 *N* hydrochloric acid add 60 g of urea. Dissolve the urea by swirling and without the aid of heat, and adjust the resulting solution, if necessary, with 6 *N* hydrochloric acid to a pH of 3.2. Prepare on the day of use.

Procedure—Insert a small pledget of glass wool above the stem constriction of a 20- × 2.5-cm chromatographic tube, and uniformly pack with a mixture of about 1 g of chromatographic siliceous earth and 0.5 mL of 5 *M* phosphoric acid. Directly above this layer, pack a similar mixture of about 3 g of chromatographic siliceous earth and 2 mL of *Ferric chloride–urea reagent*. To an accurately weighed quantity of finely powdered Tablets, equivalent to 50 mg of aspirin, add 10 mL of chloroform, stir for 3 minutes, and transfer to the chromatographic adsorption column with the aid of 5 mL of chloroform. Pass 50 mL of chloroform in several portions through the column, rinse the tip of the chromatographic tube with chloroform, and discard the eluate. If the purple zone reaches the bottom of the tube, discard the column, and repeat the test with a smaller quantity of powdered Tablets.

Elute the adsorbed salicylic acid into a 100-mL volumetric flask containing 20 mL of methanol and 4 drops of hydrochloric acid by passing two 10-mL portions of a 1 in 10 solution of glacial acetic acid in water-saturated ether, and then 30 mL of chloroform, through the column, and dilute the eluate with chloroform to volume. Dissolve a suitable, accurately weighed quantity of salicylic acid in chloroform to obtain a Standard solution containing 150 µg of salicylic acid per mL. Pipet 5 mL of this solution into a 50-mL volumetric flask containing 10 mL of methanol, 2 drops of hydrochloric acid, and 10 mL of a 1 in 10 solution of glacial acetic acid in ether. Add chloroform to volume, and mix. Concomitantly determine the absorbances of both solutions in 1-cm cells at the wavelength of maximum absorbance at about 306 nm, using as the blank a solvent mixture of the same composition as that of the Standard solution: the absorbance of the solution from the Tablets does not exceed that of the Standard solution, any necessary adjustment being made for having used a smaller sample (3.0%).

Dissolution ⟨711⟩—

Medium: water; 900 mL.

Apparatus 1: 80 rpm.

Time: 30 minutes.

Strongly basic, anion-exchange resin—Mix a suitable quantity of anion-exchange resin with 10 volumes of dilute glacial acetic acid (1 in 50) and shake for 20 minutes. Allow the resin to settle and decant the supernatant solution. Repeat the acetic acid washing four more times. Wash with water until 5.0 mL of the water wash gives a negligible response when substituted for 5.0 mL of the Test preparation, and carried through the *Determination of dissolved pentazocine* below.

Procedure—To a suitable 50-mL flask, add 0.4 g of *Strongly basic, anion-exchange resin* and 25 mL of the solution under test. Shake by mechanical means for 15 minutes. Allow to settle and use the clear, supernatant solution as the Test preparation in the following determinations.

Determination of dissolved pentazocine—Transfer 5.0 mL portions of the Test preparation, a Standard solution in dilute glacial acetic acid (1 in 50) containing 13 µg of USP Pentazocine RS per mL, and water to serve as the reagent blank, into three separate 125-mL separators. To each separator add 10 mL of a filtered 1 in 4000 solution of bromocresol purple in dilute glacial acetic acid (1 in 50) and 20.0 mL of chloroform. Insert the stopper and shake gently for 1 minute, accurately timed. Allow the layers to separate, and determine the absorbances of the clear chloroform layers from the Standard solution and the Test preparation in 1-cm cells at the wavelength of maximum absorbance at about 408 nm with a suitable spectrophotometer against the chloroform layer from the reagent blank. Determine the amount of pentazocine ($C_{19}H_{27}NO$) in solution by comparison with the Standard solution.

Determination of dissolved aspirin—Transfer 1.0 mL of the Test preparation to a 25-mL volumetric flask containing 1.0 mL of sodium hydroxide solution (1 in 10), and swirl. Allow to stand for 10 minutes. Dilute with water to volume, and mix. Concomitantly determine the absorbances of this solution and of a Standard solution containing 15 µg of USP Salicylic Acid RS per mL of 0.1 *N* sodium hydroxide, in 1-cm cells at the wavelength of maximum absorbance at about 296 nm with a suitable spectrophotometer, using 0.1 *N* sodium hydroxide as the blank. Calculate the quantity, in mg, of aspirin ($C_9H_8O_4$) in solution by comparison with the Standard solution, using the quantity (180.16/138.12), the ratio of the molecular weight of aspirin to that of salicylic acid, to convert the amount of salicylic acid measured to the amount of aspirin in solution.

Tolerances—Not less than 80% (*Q*) of the labeled amount of $C_{19}H_{27}NO$ and not less than 70% (*Q*) of the labeled amount of $C_9H_8O_4$ are dissolved in 30 minutes.

Uniformity of dosage units ⟨905⟩: meet the requirements.

Assay—

Chromatographic column—Use a 200-mm tube consisting of about a 90-mm length of 22-mm tubing fused to about a 100-mm length of 5-mm tubing having a stopcock at the bottom of this section. Place a pledget of glass wool at the bottom of the 5-mm portion just above the stopcock. Transfer a suitable quantity of sulfonic acid cation-exchange resin to a beaker and wash three times with water, discarding the water wash each time by decantation. Cover the resin with a mixture of methanol and 6 *N* hydrochloric acid (1:1), and allow to stand for one hour. Decant the acid wash; if it is colored yellow or orange, repeat this step until the wash is almost colorless. Then wash the resin by repeated 15-minute soakings in a mixture of methanol and water (1:1) followed by decantation until the wash is neutral to wide-range indicator paper. Fill the tube to a height of 100 mm with slurry of the washed resin in a mixture of methanol and water (1:1). Wash the column with 25 mL of methanol and water (1:1).

Procedure—Weigh and finely powder not fewer than 20 Pentazocine Hydrochloride and Aspirin Tablets. Transfer a portion of the freshly powdered Tablets, equivalent to about 25 mg of pentazocine and 650 mg of aspirin, accurately weighed, to a suitable 250-mL flask. Add 100.0 mL of a mixture of methanol and water (1:1), and shake by mechanical means for 20 minutes. Centrifuge a suitable quantity for 5 minutes. Transfer 25.0 mL of the clear supernatant solution to the prepared *Chromatographic column*, followed by five 10-mL portions of the mixture of methanol and water (1:1), collecting the eluate in a 250-mL volumetric flask containing 10.0 mL of 2.5 *N* sodium hydroxide. Dilute with water to volume, and mix. Reserve this *Solution A* for the *Determination of aspirin*. Next pass through the column five 5-mL portions of a mixture of methanol and 6 *N* hydrochloric acid (1:1) followed by 10 mL of water. Collect the eluate in a 100-mL volumetric flask, dilute with water to volume, and use this *Solution P* for the *Determination of pentazocine*.

Determination of aspirin—Pipet 4 mL of *Solution A* into a 100-mL volumetric flask, dilute with 0.1 *N* sodium hydroxide to volume, and mix. Concomitantly determine the absorbances of this solution and a Standard solution of USP Salicylic Acid RS in 0.1 *N* sodium hydroxide having a known concentration of about 18 µg per mL, in 1-cm cells at the wavelength of maximum absorbance at about 296 nm with a suitable spectrophotometer, using 0.1 *N* sodium hydroxide as the blank. Calculate the quantity, in mg, of aspirin ($C_9H_8O_4$), in the portion of Tablets taken by the formula:

$$25C(180.16/138.12)(A_U/A_S),$$

in which *C* is the concentration, in µg per mL, of USP Salicylic Acid RS in the Standard solution, (180.16/138.12) is the ratio of the molecular weight of aspirin to that of salicylic acid, and A_U and A_S are the absorbances of diluted *Solution A* and the Standard solution, respectively.

Determination of pentazocine—Concomitantly determine the absorbances of *Solution P* and a standard solution of USP Pentazocine RS in a mixture of water, methanol, and 6 *N* hydrochloric acid (6:1:1) having a known concentration of about 62.5 µg per mL, in 1-cm cells at the wavelength of maximum absorbance at about 278 nm, with a suitable spectrophotometer, using the solvent for the Standard solution as the blank. Calculate the quantity, in mg, of pentazocine ($C_{19}H_{27}NO$) in the portion of Tablets taken by the formula:

$$0.4C(A_U/A_S),$$

in which *C* is the concentration, in µg per mL, of USP Pentazocine RS in the Standard solution, and A_U and A_S are the absorbances of *Solution P* and the Standard solution, respectively.

Pentazocine and Naloxone Hydrochlorides Tablets

» Pentazocine and Naloxone Hydrochlorides Tablets contain not less than 90.0 percent and not more than 110.0 percent of the labeled amounts of pentazocine ($C_{19}H_{27}NO$) and naloxone ($C_{19}H_{21}NO_4$).

Packaging and storage—Preserve in tight, light-resistant containers.

Reference standards—*USP Pentazocine Reference Standard*—Dry at 60° and at a pressure not exceeding 5 mm of mercury to constant weight before using. *USP Naloxone Reference Standard*—Dry at 105° to constant weight before using.

Identification—Crush 1 Tablet in 10 mL of a mixture of chloroform and methanol (1:1), and mix. Sonicate for about two minutes, and filter (*Solution A*). Evaporate 5 mL of *Solution A* to dryness on a steam bath under a stream of nitrogen. Dissolve the residue in 0.2 mL of the mixture of chloroform and methanol (1:1) (*Solution B*). On a suitable thin-layer chromatographic plate (see *Chromatography* ⟨621⟩), coated with a 0.25-mm layer of chromatographic silica gel mixture, apply 10 µL of *Solution A*, 5 µL of *Solution B*, 10 µL of a standard solution of USP Pentazocine RS in the 1:1 mixture of chloroform and methanol containing 5.0 mg per mL, and 5 µL of a standard solution of USP Naloxone RS in the 1:1 mixture of chloroform and methanol containing 1.3 mg per mL. Develop the chromatograms in a solvent system consisting of a mixture of 1-butanol, water, and glacial acetic acid (70:20:10) until the solvent front has moved about three-fourths of the length of the plate. Remove the plate from the developing chamber, mark the solvent front, and dry under a current of warm air. Spray the plate with Folin-Ciocalteu Phenol TS followed by sodium hydroxide solution (1 in 10). Test *Solution A* and *Solution B* exhibit spots having the same R_f values and approximately the same size and shape as their respective standard solutions.

Dissolution ⟨711⟩—

Medium: water; 900 mL.

Apparatus 2: 50 rpm.

Time: 45 minutes.

Procedure—Determine the amount of $C_{19}H_{27}NO$ dissolved from ultraviolet absorbances at the wavelength of maximum absorbance at about 279 nm (corrected for absorbance at 305 nm) of filtered portions of the solution under test, suitably diluted with *Dissolution Medium*, if necessary, in comparison with a Standard solution having a known concentration of USP Pentazocine RS prepared by dissolving the standard in a minimum volume of 0.1 *N* hydrochloric acid (about 4 mL per 100 mg) and diluting quantitatively and stepwise with water.

Tolerances—Not less than 75% (*Q*) of the labeled amount of pentazocine ($C_{19}H_{27}NO$) is dissolved in 45 minutes.

Uniformity of dosage units ⟨905⟩: meet the requirements.

Procedure for content uniformity for pentazocine and naloxone—

Solvent mixture—Use a mixture containing methanol, water, and phosphoric acid (500:500:1).

Mobile phase—Dissolve 675 mg of sodium 1-octanesulfonate and 426 mg of anhydrous dibasic sodium phosphate in a mixture of 625 mL of water, 375 mL of methanol and 10 mL of phosphoric acid. Make adjustments if necessary (see *System Suitability* under *Chromatography* ⟨621⟩).

Strong anion-exchange resin—Transfer about 30 g of strong anion-exchange resin to a 250-mL beaker. Wash the resin with two 200-mL portions of water, decanting the water after each wash. Then wash with two 200-mL portions of dilute glacial acetic acid (1 in 20), decanting the first wash and finally filtering with the aid of suction.

Naloxone standard preparation—Transfer about 20 mg of USP Naloxone RS, accurately weighed, to a 100-mL volumetric flask. Dissolve in *Solvent mixture* to volume, and mix.

Mixed Naloxone and Pentazocine standard preparation—Transfer about 100 mg of USP Pentazocine RS, accurately weighed, to a 50-mL volumetric flask. Dissolve in about 30 mL of *Solvent mixture*. Add 5.0 mL of the *Naloxone standard preparation*, dilute with *Solvent mixture* to volume, and mix.

Test preparation—Transfer 1 Tablet to a 25-mL glass-stoppered cylinder. Add 25.0 mL of *Solvent mixture*. Sonicate for 10 minutes and shake intermittently for 15 minutes. Filter into a glass-stoppered conical flask. Add about 125 mg of *Strong anion-exchange resin*, and shake for 30 minutes.

Chromatographic system (see *Chromatography* ⟨621⟩)—The liquid chromatograph is equipped with a 229-nm detector and a 4.6-mm × 25-cm column that contains packing L1. The flow rate is about 1.5 mL per minute. Chromatograph the *Mixed Naloxone and Pentazocine standard preparation*, and record the peak responses as directed under *Procedure:* the relative standard deviation is not more than 2.0% for replicate injections and the resolution, *R*, between pentazocine and naloxone is not less than 6.

Procedure—[NOTE—Where peak responses are indicated, use peak areas.] Separately inject equal volumes (about 20 µL) of the *Mixed standard preparation* and the *Test preparation* into the liquid chromatograph, adjusting the operating parameters such that satisfactory chromatography and peak responses are obtained with the *Mixed standard preparation*. Record the chromatograms, and measure the responses for the major peaks. The relative retention times are about 0.3 for naloxone and 1.0 for pentazocine. Calculate the quantities, in mg, of pentazocine ($C_{19}H_{27}NO$) and naloxone ($C_{19}H_{21}NO_4$), in the Tablet by the same formula:

$$25C(r_U/r_S),$$

in which C is the concentration, in mg per mL, of the appropriate reference standard in the *Mixed standard preparation*, and r_U and r_S are the peak responses for the corresponding analyte obtained from the *Test preparation* and the *Mixed standard preparation*, respectively.

Assay for pentazocine and naloxone—Proceed as directed for *Procedure for content uniformity for pentazocine and naloxone* except to use the following *Assay preparation* in place of the *Test preparation*.

Assay preparation—Weigh and finely powder not less than 20 Pentazocine and Naloxone Hydrochlorides Tablets. Transfer an accurately weighed portion of the powder, equivalent to about 100 mg of pentazocine, to a 100-mL volumetric flask, and add 50.0 mL of the *Solvent mixture*. Sonicate for 5 minutes, and shake intermittently for 15 minutes. Filter into a glass-stoppered conical flask. Add about 250 mg of *Strong anion-exchange resin*, and shake for 30 minutes.

Procedure—Proceed as directed for *Procedure* under *Uniformity of dosage units*. Calculate the quantities, in mg, of pentazocine ($C_{19}H_{27}NO$) and naloxone ($C_{19}H_{21}NO_4$), in the portion of the finely powdered tablets taken by the same formula:

$$50C(r_U/r_S),$$

in which C is the concentration, in mg per mL, of the appropriate USP Reference Standard in the *Mixed standard preparation*, and r_U and r_S are the peak responses of the corresponding analyte obtained from the *Assay preparation* and *Mixed standard preparation*, respectively.

Pentazocine Lactate Injection

$C_{19}H_{27}NO.C_3H_6O_3$ 375.51

2,6-Methano-3-benzazocin-8-ol, 1,2,3,4,5,6-hexahydro-6,11-dimethyl-3-(3-methyl-2-butenyl)-, (2α,6α,11*R**)-, compd. with 2-hydroxypropanoic acid (1:1).

(2*R**,6*R**,11*R**)-1,2,3,4,5,6-Hexahydro-6,11-dimethyl-3-(3-methyl-2-butenyl)-2,6-methano-3-benzazocin-8-ol lactate (salt) [*17146-95-1*].

» Pentazocine Lactate Injection is a sterile solution of pentazocine lactate in Water for Injection, prepared from Pentazocine with the aid of Lactic Acid. It contains not less than 95.0 percent and not more than 105.0 percent of the labeled amount of pentazocine ($C_{19}H_{27}NO$).

Packaging and storage—Preserve in single-dose or in multiple-dose containers, preferably of Type I glass.

Reference standard—*USP Pentazocine Reference Standard*—Dry at a pressure not exceeding 5 mm of mercury at 60° to constant weight before using.

Identification—

A: Transfer a volume of Injection, equivalent to about 15 mg of lactic acid, to a 50-mL conical flask, add 1 mL of 2 *N* sulfuric acid, and mix. Add, dropwise, potassium permanganate solution (3.2 in 100), until a slight excess has been added, as evidenced by a violet color. [NOTE—The addition of a large excess of potassium permanganate may result in a false negative test for lactate.] Moisten a piece of filter paper with a color-indicating solution (previously prepared by dissolving 250 mg of sodium nitroferricyanide in water to make 9 mL of solution, adding 1 mL of morpholine, and mixing). Place the moistened filter paper over the conical flask opening, and heat the solution moderately: the acetaldehyde fumes produced turn the moistened filter paper blue.

B: Dissolve 50 mg of USP Pentazocine RS in 25 mL of 0.01 *N* hydrochloric acid in a separator, and use this in place of the Standard solution specified under *Identification—Organic Nitrogenous Bases* ⟨181⟩: a volume of Injection, equivalent to about 50 mg of pentazocine, meets the requirements of the test.

pH ⟨791⟩: between 4.0 and 5.0.

Other requirements—It meets the requirements under *Injections* ⟨1⟩.

Assay—

Ion-exchange column—Place a pledget of glass wool in the base of a 6-mm (ID) tube equipped with a stopcock, and fill the tube to a height of about 25 mm with a styrene-divinylbenzene cation-exchange resin that has been previously soaked in 3 *N* hydrochloric acid for not less than 2 hours. Pass 10 mL of methanol through the column followed by 50 mL of 3 *N* hydrochloric acid, then wash the column with water until the eluate is neutral. [*Caution—Do not permit the column to become dry at any time.*]

Standard preparation—Transfer about 60 mg of USP Pentazocine RS, accurately weighed, to a 50-mL volumetric flas[illegible] dilute with a mixture of equal volumes of methanol [illegible] hydrochloric acid to volume, and mix. Transfer [illegible] solution to a 100-mL volumetric flask, dil[illegible] 6 *N* hydrochloric acid mixture to [illegible] tration of USP Pentazocine R[illegible] about 120 µg per mL.

Assay preparation—Dilute an [illegible] Pentazocine Lactate Injection, eq[illegible] tazocine, with water to 10.0 mL, [illegible] this solution to the column, then p[illegible] the column at a rate of about 2 mL [illegible] eluate. Place a 100-mL volumetric [illegible] pass through the column a mixture of [illegible] and 6 *N* hydrochloric acid until appr[illegible]

has been collected. Remove the flask, dilute with the methanol–6 *N* hydrochloric acid mixture to volume, and mix.

Procedure—Concomitantly determine the absorbances of the *Assay preparation* and the *Standard preparation* in 1-cm cells at the wavelength of maximum absorbance at about 278 nm, with a suitable spectrophotometer, using a mixture of equal volumes of methanol and 6 *N* hydrochloric acid as the blank. Calculate the quantity, in mg, of pentazocine ($C_{19}H_{27}NO$) in each mL of Injection taken by the formula:

$$(0.5C/V)(A_U/A_S),$$

in which C is the concentration, in µg per mL, of USP Pentazocine RS in the *Standard preparation*, V is the volume, in mL, of Injection taken, and A_U and A_S are the absorbances of the *Assay preparation* and the *Standard preparation*, respectively.

Pentetate Sodium Injection, Technetium Tc 99m—
see Technetium Tc 99m Pentetate Injection

Pentobarbital

$C_{11}H_{18}N_2O_3$ 226.27

2,4,6(1*H*,3*H*,5*H*)-Pyrimidinetrione, 5-ethyl-5-(1-methylbutyl)-.
5-Ethyl-5-(1-methylbutyl)barbituric acid [*76-74-4*].

» Pentobarbital contains not less than 98.5 percent and not more than 101.0 percent of $C_{11}H_{18}N_2O_3$, calculated on the dried basis.

Packaging and storage—Preserve in tight containers.

Reference standard—*USP Pentobarbital Reference Standard*—Dry at 105° for 2 hours before using.

Identification—

A: The infrared absorption spectrum, determined in a 0.1-mm cell, of a 7 in 100 solution in chloroform, the specimen having been previously dried, exhibits maxima only at the same wavelengths as that of a similar preparation of USP Pentobarbital RS.

B: The ultraviolet absorption spectrum of a 1 in 62,500 solution in 0.1 *N* sodium hydroxide exhibits maxima and minima at the same wavelengths as that of a similar solution of USP Pentobarbital RS, concomitantly measured, and the respective absorptivities, calculated on the dried basis, at the wavelength of maximum absorbance at about 240 nm do not differ by more than 3.0%.

C: Shake about 300 mg with 5 mL of sodium hydroxide solution (1 in 125) for 2 minutes, filter, and to 1 mL of the filtrate add about 1.2 mL of silver nitrate TS: a white precipitate is formed, and it is soluble in 6 *N* ammonium hydroxide. To a second 1-mL portion of the filtrate add 3 drops of mercuric chloride TS: a white precipitate is formed, and it is soluble in 6 *N* ammonium hydroxide.

Melting range, *Class I* ⟨741⟩: between 127° and 133°.

Loss on drying ⟨731⟩—Dry it at 105° for 2 hours: it loses not more than 1.0% of its weight.

Isomer content—Transfer 300 ± 5 mg to a round-bottom flask equipped with a standard-taper, ground-glass joint, and add 4 mL [illegible] water. Add, dropwise, sodium hydroxide solution (4 in 10) [illegible] to dissolve the specimen (about 3 or 4 drops), then add 10 [illegible] alcohol. Add 300 ± 5 mg of *p*-nitrobenzyl bromide, mix, [illegible] a condenser, and reflux for 30 minutes. Cool, filter under [illegible] pressure, and wash the residue with four 5-mL portions [illegible] Transfer the residue, as completely as possible, to a [illegible] add 25 mL of alcohol, and reflux for 10 minutes: [illegible]olves completely. Cool, and filter under reduced [illegible]-nitrobenzyl derivative so obtained, after being dried at 105° for 30 minutes, melts completely between 136° and 146°, when determined by the procedure for *Class Ia* (see *Melting Range or Temperature* ⟨741⟩).

Residue on ignition ⟨281⟩: not more than 0.1%.

Heavy metals, *Method II* ⟨231⟩: 0.002%.

Assay—

0.1 N Tetrabutylammonium hydroxide in chlorobenzene—Dilute 100 mL of 1 *N* tetrabutylammonium hydroxide VS with chlorobenzene to 1000 mL, and mix.

Standardization of 0.1 N Tetrabutylammonium hydroxide in chlorobenzene—Dissolve about 180 mg, accurately weighed, of primary standard benzoic acid in about 100 mL of acetone, and titrate with *0.1 N Tetrabutylammonium hydroxide in chlorobenzene*, determining the end-point potentiometrically, using a glass electrode and a calomel electrode containing 0.1 *N* methanolic tetrabutylammonium chloride (see *Titrimetry* ⟨541⟩). Each mL of *0.1 N Tetrabutylammonium hydroxide in chlorobenzene* is equivalent to 12.21 mg of benzoic acid.

Procedure—Transfer about 330 mg of Pentobarbital, accurately weighed, to a suitable beaker, and dissolve in 100 mL of acetone. Titrate with *0.1 N Tetrabutylammonium hydroxide in chlorobenzene*, determining the end-point potentiometrically, using a glass electrode and a calomel electrode containing 0.1 *N* methanolic tetrabutylammonium chloride. Each mL of *0.1 N Tetrabutylammonium hydroxide in chlorobenzene* is equivalent to 22.63 mg of $C_{11}H_{18}N_2O_3$.

Pentobarbital Elixir

» Pentobarbital Elixir contains not less than 92.5 percent and not more than 107.5 percent of the labeled amount of $C_{11}H_{18}N_2O_3$.

Packaging and storage—Preserve in tight containers.

Reference standard—*USP Pentobarbital Reference Standard*—Dry at 105° for 2 hours before using.

Identification—Dilute a volume of Elixir with alcohol to a concentration of about 1 mg of pentobarbital per mL. On a suitable thin layer chromatographic plate (see *Chromatography* ⟨621⟩), coated with a 0.25-mm layer of chromatographic silica gel mixture, apply 50 µL of this solution and 50 µL of a Standard solution of USP Pentobarbital RS in alcohol containing 1 mg per mL as streaks about 1 cm in length along the spotting line. Allow the streaks to dry, and develop the chromatogram in a solvent system consisting of a mixture of isopropyl alcohol, chloroform, acetone, and ammonium hydroxide (9:2:2:4) until the solvent front has moved about three-fourths of the length of the plate. Remove the plate from the developing chamber, mark the solvent front, and allow the solvent to evaporate. Locate the spots by viewing the plate under short-wavelength ultraviolet light: the R_f value of the principal spot obtained from the test solution corresponds to that obtained from the Standard solution.

Alcohol content, *Method I* ⟨611⟩: between 16.0% and 20.0% of C_2H_5OH.

Assay—

Internal standard—*n*-Tricosane.

Internal standard solution—Dissolve an accurately weighed quantity of *n*-tricosane in chloroform, and dilute quantitatively with chloroform to obtain a solution having a known concentration of about 0.6 mg per mL.

Standard preparation—Dissolve accurately weighed quantities of USP Pentobarbital RS and *n*-tricosane in chloroform, and dilute quantitatively with chloroform to obtain a solution that contains, in each mL, known amounts of about 1 mg of USP Pentobarbital RS and about 0.4 mg of *n*-tricosane.

Assay preparation—Transfer an accurately measured volume of Pentobarbital Elixir, equivalent to about 20 mg of pentobarbital, to a separator, add 1 mL of dilute hydrochloric acid (1 in 5), and extract with four 10-mL portions of chloroform. Filter the extracts through about 15 g of anhydrous sodium sulfate that is supported on a funnel by a small pledget of glass wool. Collect the combined filtrate in a 50-mL volumetric flask, wash the sodium sulfate with 5 mL of chloroform, dilute with chloroform to volume, and mix. Combine 4.0 mL of this solution with 1.0 mL of *Internal standard solution* in a suitable container, and reduce the volume to about 1.5 mL by evaporation, with the aid of a stream of dry nitrogen, at room temperature.

Chromatographic system and *System suitability*—Proceed as directed for *Chromatographic System* and *System Suitability* under *Barbiturate Assay* ⟨361⟩, the resolution, *R*, between pentobarbital and *n*-tricosane being not less than 2.3. [NOTE—Relative retention times are, approximately, 0.5 for *n*-tricosane and 1.0 for pentobarbital.]

Procedure—Proceed as directed for *Procedure* under *Barbiturate Assay* ⟨361⟩. Calculate the quantity, in mg, of $C_{11}H_{18}N_2O_3$ in each mL of the Elixir taken by the formula:

$$12.5(R_U)(Q_S)(C_i)/V(R_S),$$

in which *V* is the volume, in mL, of Elixir taken and the other terms are as defined therein.

Pentobarbital Sodium

$C_{11}H_{17}N_2NaO_3$ 248.26

2,4,6(1*H*,3*H*,5*H*)-Pyrimidinetrione, 5-ethyl-5-(1-methylbutyl)-, monosodium salt.

Sodium 5-ethyl-5-(1-methylbutyl)barbiturate [*57-33-0*].

» Pentobarbital Sodium contains not less than 98.5 percent and not more than 101.0 percent of $C_{11}H_{17}N_2NaO_3$, calculated on the dried basis.

Packaging and storage—Preserve in tight containers.

Reference standard—*USP Pentobarbital Reference Standard*—Dry at 105° for 2 hours before using.

Completeness of solution—Mix 1.0 g with 10 mL of carbon dioxide-free water: after 1 minute, the solution is clear and free from undissolved solid.

Identification—

A: Recrystallize from hot alcohol the residue obtained in the *Assay*, and dry the recrystallized residue at 105° for 30 minutes: the infrared absorption spectrum of a potassium bromide dispersion of the residue so obtained exhibits maxima only at the same wavelengths as that of a similar preparation of USP Pentobarbital RS.

B: Ignite about 200 mg: the residue effervesces with acids, and responds to the tests for *Sodium* ⟨191⟩.

pH ⟨791⟩: between 9.8 and 11.0, in the solution prepared in the test for *Completeness of solution*.

Loss on drying ⟨731⟩—Dry it at 105° for 6 hours: it loses not more than 3.5% of its weight.

Isomer content—Dissolve 300 ± 5 mg in 5.0 mL of water, and dissolve 300 ± 5 mg of *p*-nitrobenzyl bromide in 10.0 mL of alcohol. Mix the two solutions, reflux for 30 minutes, cool to 25°, and filter by suction. Wash the collected solid with four 5-mL portions of water, transfer as completely as practicable to a small flask, add 25.0 mL of alcohol, and reflux for 10 minutes: the solid dissolves completely. Cool the solution to 25°, and filter by suction: the collected solid, after being dried at 105° for 30 minutes, melts completely between 136° and 146°, when determined by the procedure for *Class Ia* (see *Melting Range or Temperature* ⟨741⟩).

Heavy metals, *Method II* ⟨231⟩: 0.003%.

Assay—Dissolve about 500 mg of Pentobarbital Sodium, accurately weighed, in 15 mL of water in a separator. To the solution add 2 mL of hydrochloric acid, shake, and completely extract the liberated pentobarbital with 25-mL portions of chloroform. Test for completeness of extraction by extracting with an additional 10-mL portion of chloroform and evaporating the solvent: not more than 0.5 mg of residue remains. Filter each extract through a pledget of chloroform-washed cotton, or other suitable filter, into a tared beaker, and finally wash the separator and the filter with several small portions of chloroform. Evaporate the combined filtrate and washings on a steam bath with the aid of a current of air, add 10 mL of ether, again evaporate, dry the residue at 105° for 2 hours, cool, and weigh. The weight of the residue, multiplied by 1.097, represents the weight of $C_{11}H_{17}N_2NaO_3$.

Pentobarbital Sodium Capsules

» Pentobarbital Sodium Capsules contain not less than 92.5 percent and not more than 107.5 percent of the labeled amount of $C_{11}H_{17}N_2NaO_3$.

Packaging and storage—Preserve in tight containers.

Reference standard—*USP Pentobarbital Reference Standard*—Dry at 105° for 2 hours before using.

Identification—Mix a quantity of the contents of Capsules, equivalent to about 100 mg of pentobarbital sodium, with 15 mL of water in a separator. Filter, if necessary, and saturate the solution with sodium chloride. To the solution add 2 mL of hydrochloric acid, shake, and extract the liberated pentobarbital with five 25-mL portions of chloroform. Filter each extract through a pledget of chloroform-washed cotton, or other suitable filter, into a beaker, and finally wash the separator and the filter with several small portions of chloroform. Evaporate the combined filtrate and washings on a steam bath with the aid of a current of air, add 10 mL of ether, again evaporate, recrystallize the residue from hot alcohol, and dry the recrystallized residue at 105° for 30 minutes: the residue so obtained responds to *Identification test A* under *Pentobarbital Sodium*.

Dissolution ⟨711⟩—

Medium: water; 900 mL.

Apparatus 1: 100 rpm.

Time: 45 minutes.

Standard preparation—Dissolve an accurately weighed quantity of USP Pentobarbital RS in freshly prepared dilute ammonium hydroxide (1 in 20) to obtain a solution having a known concentration of about 10 µg of pentobarbital per mL. The concentration of pentobarbital, multiplied by 1.097, represents the equivalent amount of pentobarbital sodium.

Procedure—Determine the amount of $C_{11}H_{17}N_2NaO_3$ dissolved from ultraviolet absorbances at the wavelength of maximum absorbance at about 240 nm of filtered portions of the solution under test, suitably diluted with freshly prepared dilute ammonium hydroxide (1 in 20), in comparison with the *Standard preparation*.

Tolerances—Not less than 75% (*Q*) of the labeled amount of $C_{11}H_{17}N_2NaO_3$ is dissolved in 45 minutes.

Uniformity of dosage units ⟨905⟩: meet the requirements.

Procedure for content uniformity—Transfer the contents of 1 Capsule to a 250-mL volumetric flask, with the aid of about 5 mL of alcohol. Add 10 mL of freshly prepared dilute ammonium hydroxide (1 in 200), and without delay dilute with the same solution to volume. Mix, filter if necessary, and discard the first 20 mL of the filtrate. Dilute a portion of the clear solution with dilute ammonium hydroxide (1 in 200) to obtain a solution having a concentration of about 10 µg of pentobarbital sodium per mL. Dissolve a suitable quantity of USP Pentobarbital RS in dilute ammonium hydroxide (1 in 200) to obtain a Standard solution having a known concentration of about 10 µg per mL. Concomitantly determine the absorbances of both solutions in 1-cm cells at the wavelength of maximum absorbance at about 240 nm, with a suitable spectrophotometer, using dilute ammonium hydroxide (1 in 200) as the blank. Calculate the quantity, in mg, of $C_{11}H_{17}N_2NaO_3$ in the Capsule by the formula:

$$1.097(T/C_U)C_S(A_U/A_S),$$

in which *T* is the labeled quantity, in mg, of pentobarbital sodium in the Capsule, C_U is the concentration, in µg per mL, of pentobarbital sodium in the solution from the Capsule contents, on the basis of the labeled quantity per Capsule and the extent of dilution, C_S is the concentration, in µg per mL, of USP Pentobarbital RS in the Standard solution, and A_U and A_S are the absorbances of the solution from the Capsule contents and the Standard solution, respectively.

Assay—

Internal standard—*n*-Tricosane.

Internal standard solution—Dissolve an accurately weighed quantity of *n*-tricosane in chloroform, and dilute quantitatively with chloroform to obtain a solution having a known concentration of about 0.4 mg per mL.

Standard preparation—Dissolve accurately weighed quantities of USP Pentobarbital RS and *n*-tricosane in chloroform, and dilute quantitatively with chloroform to obtain a solution that contains, in each mL, known amounts of about 0.9 mg of USP Pentobarbital RS and about 0.4 mg of *n*-tricosane.

Assay preparation—Weigh not less than 20 Pentobarbital Sodium Capsules, and transfer the contents as completely as possible to a suitable container. Remove any residual powder from the empty capsules with the aid of a current of air, and weigh the capsule shells, determining the weight of the contents by difference. Mix the contents of the Capsules, transfer an accurately weighed portion of the powder, equivalent to about 50 mg of pentobarbital sodium, to a separator. Add 15 mL of water and 1 mL of hydrochloric acid, and extract with five 25-mL portions of chloroform. Filter the extracts through about 15 g of anhydrous sodium sulfate that is supported on a funnel by a small pledget of glass wool. Collect the combined filtrate in a 100-mL volumetric flask, wash the sodium sulfate with 15 mL of chloroform, collecting the washing with the filtrate, dilute with chloroform to volume, and mix. Combine 2.0 mL of this solution with 1.0 mL of *Internal standard solution* in a suitable container, and reduce the volume to about 1 mL by evaporation, with the aid of a stream of dry nitrogen, at room temperature.

Chromatographic system and *System suitability*—Proceed as directed for *Chromatographic System* and *System Suitability* under *Barbiturate Assay* ⟨361⟩, the resolution, *R*, between pentobarbital and *n*-tricosane being not less than 2.3. [NOTE—Relative retention times are, approximately, 0.5 for *n*-tricosane barbital and 1.0 for pentobarbital.]

Procedure—Proceed as directed for *Procedure* under *Barbiturate Assay* ⟨361⟩. Calculate the quantity, in mg, of $C_{11}H_{17}N_2NaO_3$ in the portion of Capsules taken by the formula:

$$(248.26/226.27)(50)(R_U)(Q_S)(C_i/(R_S),$$

in which 248.26 and 226.27 are the molecular weights of pentobarbital sodium and pentobarbital, respectively.

Pentobarbital Sodium Injection

» Pentobarbital Sodium Injection is a sterile solution of Pentobarbital Sodium in a suitable solvent. Pentobarbital may be substituted for the equivalent amount of Pentobarbital Sodium, for adjustment of the pH. The Injection contains the equivalent of not less than 92.0 percent and not more than 108.0 percent of the labeled amount of $C_{11}H_{17}N_2NaO_3$.

Packaging and storage—Preserve in single-dose or in multiple-dose containers, preferably of Type I glass. The Injection may be packaged in 50-mL containers.

Labeling—The label indicates that the Injection is not to be used if it contains a precipitate.

Reference standard—*USP Pentobarbital Reference Standard*—Dry at 105° for 2 hours before using.

Identification—The residue obtained in the *Assay* responds to *Identification test A* under *Pentobarbital Sodium.*

pH ⟨791⟩: between 9.0 and 10.5.

Other requirements—It meets the requirements under *Injections* ⟨1⟩.

Assay—Pipet a volume of Pentobarbital Sodium Injection, equivalent to about 500 mg of pentobarbital sodium, into a separator, and dilute with water to about 15 mL. Proceed as directed in the *Assay* under *Pentobarbital Sodium*, beginning with "To the solution add 2 mL of hydrochloric acid."

Peppermint—*see* Peppermint NF

Peppermint Oil—*see* Peppermint Oil NF

Peppermint Spirit

» Peppermint Spirit contains, in each 100 mL, not less than 9.0 mL and not more than 11.0 mL of peppermint oil.

Peppermint Oil	100 mL
Peppermint, in coarse powder	10 g
Alcohol, a sufficient quantity, to make	1000 mL

Macerate the peppermint leaves, freed as much as possible from stems and coarsely powdered, for 1 hour in 500 mL of purified water, and then strongly express them. Add the moist, macerated leaves to 900 mL of alcohol, and allow the mixture to stand for 6 hours with frequent agitation. Filter, and to the filtrate add the oil and add alcohol to make the product measure 1000 mL.

Packaging and storage—Preserve in tight containers, protected from light.

Alcohol content, *Method II* ⟨611⟩: between 79.0% and 85.0% of C_2H_5OH.

Assay—Transfer 5.0 mL of Peppermint Spirit to a Babcock bottle, graduated to 8%, add 1.0 mL of kerosene, and mix. Add saturated calcium chloride solution, acidified with hydrochloric acid, almost to fill the bulb of the bottle. Rotate the bottle vigorously to ensure mixing, and then add a sufficient quantity of the calcium chloride solution to bring the separated oil into the neck of the bottle. Centrifuge at about 1500 rpm for 5 minutes, and read the volume of oil in the stem. Subtract five divisions for the kerosene added, and multiply the remaining number of divisions by 4.2 to obtain the volume, in mL, of peppermint oil in 100 mL of the Spirit.

Peppermint Water—*see* Peppermint Water NF

Perphenazine

$CH_2CH_2CH_2$–N(piperazine)N–CH_2CH_2OH; N; Cl; S

$C_{21}H_{26}ClN_3OS$ 403.97

1-Piperazineethanol, 4-[3-(2-chloro-10*H*-phenothiazin-10-yl)propyl]-.

4-[3-(2-Chlorophenothiazin-10-yl)propyl]-1-piperazineethanol [*58-39-9*].

» Perphenazine contains not less than 98.0 percent and not more than 102.0 percent of $C_{21}H_{26}ClN_3OS$, calculated on the dried basis.

Packaging and storage—Preserve in tight, light-resistant containers.

Reference standard—*USP Perphenazine Reference Standard*—Dry in vacuum at 65° for 3 hours before using.

Clarity and color of solution—Dissolve 500 mg in 25 mL of methanol: the solution is clear and not more than light yellow.

NOTE—Throughout the following procedures, protect test or assay specimens, the Reference Standard, and solutions containing them, by conducting the procedures without delay, under subdued light, or using low-actinic glassware.

Identification—

A: The infrared absorption spectrum of a mineral oil dispersion of it, previously dried, exhibits maxima only at the same wavelengths as that of a similar preparation of USP Perphenazine RS.

B: The ultraviolet absorption spectrum of a 1 in 100,000 solution in methanol exhibits maxima and minima at the same wavelengths as that of a similar solution of USP Perphenazine RS, concomitantly measured, and the respective absorptivities, calculated on the dried basis, at the wavelength of maximum absorbance at about 257 nm do not differ by more than 2.5%.

Melting range, *Class I* ⟨741⟩: between 94° and 100°.

Loss on drying ⟨731⟩—Dry it in vacuum at 65° for 3 hours: it loses not more than 0.5% of its weight.

Residue on ignition ⟨281⟩: not more than 0.1%.

Assay—Dissolve about 400 mg of Perphenazine, previously dried and accurately weighed, in 50 mL of glacial acetic acid, warming slightly to effect solution. Cool to room temperature, add 10 mL of acetic anhydride, and allow to stand for 5 minutes. Add 1 drop of crystal violet TS, and titrate with 0.1 *N* perchloric acid VS to a green end-point. Perform a blank determination, and make any necessary correction. Each mL of 0.1 *N* perchloric acid is equivalent to 20.20 mg of $C_{21}H_{26}ClN_3OS$.

Perphenazine Injection

» Perphenazine Injection is a sterile solution of Perphenazine in Water for Injection, prepared with the aid of Citric Acid. It contains not less than 90.0 percent and not more than 110.0 percent of the labeled amount of $C_{21}H_{26}ClN_3OS$, as the citrate.

Packaging and storage—Preserve in single-dose or in multiple-dose containers, preferably of Type I glass, protected from light.

Reference standard—*USP Perphenazine Reference Standard*—Dry in vacuum at 65° for 3 hours before using.

NOTE—Throughout the following procedures, protect test or assay specimens, the Reference Standard, and solutions containing them, by conducting the procedures without delay, under subdued light, or using low-actinic glassware.

Identification—Dilute 1 mL with methanol to 5 mL. Apply 5 µL each of this solution and a solution of USP Perphenazine RS in methanol containing 1 mg per mL on a suitable thin-layer chromatographic plate, coated with a 0.25-mm layer of chromatographic silica gel. Develop the chromatogram in a solvent system consisting of a mixture of acetone and ammonium hydroxide (200:1) until the solvent front has moved about 15 cm. Air-dry the plate, and spray lightly with a solution of iodoplatinic acid prepared by dissolving 100 mg of chloroplatinic acid in 1 mL of 1 *N* hydrochloric acid, adding 25 mL of potassium iodide solution (4 in 100), diluting with water to 100 mL, and adding 0.50 mL of formic acid: the R_f value of the principal spot obtained from the Injection corresponds to that obtained from the Standard solution.

pH ⟨791⟩: between 4.2 and 5.6.

Other requirements—It meets the requirements under *Injections* ⟨1⟩.

Assay—

Acid-alcohol solution—Transfer 10 mL of hydrochloric acid to a 1000-mL flask containing 500 mL of alcohol and 300 mL of water. Dilute with water to volume.

Palladium chloride solution—Dissolve 100 mg of palladium chloride in a mixture of 1 mL of hydrochloric acid and 50 mL of water in a 100-mL volumetric flask, heating on a steam bath to effect solution. Cool, dilute with water to volume, and mix. Store in an amber bottle and use within 30 days. On the day of use, transfer 50 mL to a 500-mL volumetric flask, add 4 mL of hydrochloric acid and 4.1 g of anhydrous sodium acetate, dilute with water to volume, and mix.

Standard preparation—Dissolve an accurately weighed quantity of USP Perphenazine RS in *Acid-alcohol solution* to obtain a solution having a known concentration of about 150 µg per mL.

Assay preparation—Dilute 3.0 mL of Perphenazine Injection with *Acid-alcohol solution* to 100 mL in a volumetric flask.

Procedure—Mix 10.0 mL each of the *Assay preparation* and the *Standard preparation* with 15.0 mL of *Palladium chloride solution*, filter, if necessary, and concomitantly determine the absorbances of these solutions, against a reagent blank, in 1-cm cells at the wavelength of maximum absorbance at about 480 nm, with a suitable spectrophotometer. Calculate the quantity, in mg, of $C_{21}H_{26}ClN_3OS$ in the volume of Injection taken by the formula:

$$0.1C(A_U/A_S),$$

in which C is the concentration, in µg per mL, of USP Perphenazine RS in the *Standard preparation*, and A_U and A_S are the absorbances of the solutions from the *Assay preparation* and the *Standard preparation*, respectively.

Perphenazine Oral Solution

» Perphenazine Oral Solution contains not less than 90.0 percent and not more than 110.0 percent of the labeled amount of $C_{21}H_{26}ClN_3OS$.

Packaging and storage—Preserve in well-closed, light-resistant containers.

Reference standards—*USP Perphenazine Reference Standard*—Dry in vacuum at 65° for 3 hours before using. *USP Perphenazine Sulfoxide Reference Standard*—Use as is.

NOTE—Throughout the following procedures, protect test or assay specimens, the Reference Standard, and solutions containing them, by conducting the procedures without delay, under subdued light, or using low-actinic glassware.

Identification—It responds to the *Identification test* under *Perphenazine Injection.*

Perphenazine sulfoxide—

Mobile phase and *Chromatographic system*—Proceed as directed in the *Assay.*

Standard preparation—Using an accurately weighed quantity of USP Perphenazine Sulfoxide RS, prepare a solution in methanol having a known concentration of about 10 µg per mL.

Test preparation—Using a "to contain" pipet, transfer an accurately measured volume of the Oral Solution, equivalent to 20 mg of perphenazine to a 100-mL volumetric flask. Rinse the pipet with methanol, collecting the methanol in the volumetric flask. Dilute with methanol to volume, and mix.

Procedure—Separately inject equal volumes (about 10 µL) of the *Standard preparation* and the *Test preparation* into the chromatograph, and record the chromatograms. The *Test preparation* may exhibit a minor peak whose retention time corresponds to the peak exhibited by the *Standard preparation;* the peak response of the minor peak from the *Test preparation* is not greater than the peak response of the *Standard preparation* corresponding to not more than 5.0% of perphenazine sulfoxide.

Assay—[NOTE—Conduct this procedure with a minimum exposure to light.]

Mobile phase—Prepare a suitably degassed mixture of methanol and 0.005 *M* ammonium acetate (4:1). Prior to mixing, adjust the 0.005 *M* ammonium acetate with glacial acetic acid if necessary to a pH of 5.25 ± 0.05. Make adjustments if necessary (see *System Suitability* under *Chromatography* ⟨621⟩).

Internal standard solution—Transfer about 40 mg of amitriptyline hydrochloride to a 100-mL volumetric flask, add methanol to volume, and mix.

Standard preparation—Transfer about 20 mg of USP Perphenazine RS, accurately weighed, to a 50-mL volumetric flask, add methanol to volume, and mix. Transfer 2.0 mL of this solution to a 100-mL volumetric flask, add 10.0 mL of *Internal standard solution*, dilute with methanol to volume, and mix to obtain a solution having a known concentration of about 8 µg of USP Perphenazine RS per mL.

Assay preparation—Using a "to contain" pipet, transfer an accurately measured volume of Perphenazine Oral Solution, equivalent to about 16 mg of perphenazine to a 200-mL volumetric flask. Rinse the pipet with methanol, collecting the methanol in the volumetric flask. Dilute with methanol to volume, and mix. Pipet 10 mL of this solution into a 100-mL volumetric flask, add 10.0 mL of *Internal standard solution*, dilute with methanol to volume, and mix.

Chromatographic system (see *Chromatography* ⟨621⟩)—The liquid chromatograph is equipped with a 254-nm detector and a 4.6-mm × 25-cm column that contains packing L10. The flow rate is about 3.5 mL per minute. Chromatograph the *Standard preparation*, and record the peak responses as directed under *Procedure:* the tailing factor for the analyte peak is not more than 3.0, the resolution, *R*, between the analyte and internal standard peaks is not less than 3.0, and the relative standard deviation for replicate injections is not more than 2.0%.

Procedure—Separately inject equal volumes (about 10 µL) of the *Standard preparation* and the *Assay preparation* into the chromatograph, record the chromatograms, and measure the responses for the major peaks. The relative retention times are about 1.6 for amitriptyline and 1.0 for perphenazine. Calculate the quantity, in mg, of $C_{21}H_{26}ClN_3OS$ in each mL of Perphenazine Oral Solution taken by the formula:

$$(2C/V)(R_U/R_S),$$

in which *C* is the concentration, in µg per mL, of USP Perphenazine RS in the *Standard preparation*, *V* is the volume, in mL, of Oral Solution taken for the *Assay preparation*, and R_U and R_S are the peak response ratios obtained from the *Assay preparation* and the *Standard preparation*, respectively.

Perphenazine Syrup

» Perphenazine Syrup contains not less than 90.0 percent and not more than 110.0 percent of the labeled amount of $C_{21}H_{26}ClN_3OS$.

Packaging and storage—Preserve in well-closed, light-resistant containers.

Reference standard—*USP Perphenazine Reference Standard*—Dry in vacuum at 65° for 3 hours before using.

NOTE—Throughout the following procedures, protect test or assay specimens, the Reference Standard, and solutions containing them, by conducting the procedures without delay, under subdued light, or using low-actinic glassware.

Identification—Add 10 mL of water to a volume of Syrup, equivalent to about 4 mg of perphenazine, render alkaline by dropwise addition of sodium hydroxide to a pH of 11 to 12, and extract with four 5-mL portions of chloroform, combining the extracts through a bed of anhydrous sodium sulfate in a funnel into a beaker. Evaporate the extracts on a steam bath nearly to dryness, and dissolve the residue in 4 mL of methanol: the solution so obtained responds to the *Identification test* under *Perphenazine Injection.*

Assay—

Acid-alcohol solution and *Palladium chloride solution*—Prepare as directed in the *Assay* under *Perphenazine Injection.*

Standard preparation—Prepare as directed in the *Assay* under *Perphenazine Solution.*

Assay preparation—Transfer an accurately measured volume of Perphenazine Syrup, equivalent to about 6 mg of perphenazine, to a 25-mL volumetric flask, dilute with water to volume, and mix. Transfer 10 mL to a 125-mL separator, add 25 mL of water, adjust with ammonium hydroxide to a pH of 10 to 11, and extract with four 20-mL portions of chloroform, filtering the extracts through anhydrous sodium sulfate. Evaporate the combined extracts on a steam bath with the aid of a stream of nitrogen to about 5 mL. Complete the evaporation without application of heat, and dissolve the residue in 15.0 mL of *Acid-alcohol solution*, filtering if necessary.

Procedure—Mix 10.0 mL each of the *Assay preparation* and the *Standard preparation* with 15.0 mL of *Palladium chloride solution*, filter if necessary, and concomitantly determine the absorbances of these solutions, against a reagent blank, in 1-cm cells at the wavelength of maximum absorbance at about 480 nm, with a suitable spectrophotometer. Calculate the quantity, in mg, of $C_{21}H_{26}ClN_3OS$ in each mL of the Syrup taken by the formula:

$$0.0375(C/V)(A_U/A_S),$$

in which *C* is the concentration, in µg per mL, of USP Perphenazine RS in the *Standard preparation*, *V* is the volume, in mL, of Syrup taken, and A_U and A_S are the absorbances of the solutions from the *Assay preparation* and the *Standard preparation*, respectively.

Perphenazine Tablets

» Perphenazine Tablets contain not less than 90.0 percent and not more than 110.0 percent of the labeled amount of $C_{21}H_{26}ClN_3OS$.

Packaging and storage—Preserve in tight, light-resistant containers.

Reference standard—*USP Perphenazine Reference Standard*—Dry in vacuum at 65° for 3 hours before using.

NOTE—Throughout the following procedures, protect test or assay specimens, the Reference Standard, and solutions containing them, by conducting the procedures without delay, under subdued light, or using low-actinic glassware.

Identification—Shake a portion of finely powdered Tablets, equivalent to about 5 mg of perphenazine, with about 10 mL of chloroform, filter, evaporate the filtrate on a steam bath nearly to dryness, and dissolve the residue in 5 mL of methanol: the solution so obtained responds to the *Identification test* under *Perphenazine Injection.*

Dissolution ⟨711⟩—

Medium: 0.1 *N* hydrochloric acid; 900 mL.

Apparatus 2: 50 rpm.

Time: 45 minutes.

Procedure—Determine the amount of $C_{21}H_{26}ClN_3OS$ dissolved from ultraviolet absorbances at the wavelength of maximum absorbance at about 257 nm of filtered portions of the solution under test, suitably diluted with *Dissolution Medium*, if necessary, in comparison with a Standard solution having a known concentration of USP Perphenazine RS in the same medium.

Tolerances—Not less than 75% (*Q*) of the labeled amount of $C_{21}H_{26}ClN_3OS$ is dissolved in 45 minutes.

Uniformity of dosage units ⟨905⟩: meet the requirements.

Assay—

Acid-alcohol solution and *Palladium chloride solution*—Prepare as directed in the *Assay* under *Perphenazine Injection.*

Standard preparation—Prepare as directed in the *Assay* under *Perphenazine Solution.*

Assay preparation—Weigh and finely powder not less than 20 Perphenazine Tablets. Transfer a portion of the powder, equivalent to about 4 mg of perphenazine, to a glass-stoppered conical flask, pipet into the flask 25 mL of *Acid-alcohol solution*, shake by mechanical means for 30 minutes, and centrifuge a portion of the mixture. The clear supernatant fluid is the *Assay preparation.*

Procedure—Proceed as directed for *Procedure* in the *Assay* under *Perphenazine Injection.* Calculate the quantity, in mg, of $C_{21}H_{26}ClN_3OS$ in the portion of Tablets taken by the formula:

$$0.025C(A_U/A_S),$$

in which *C* is the concentration, in µg per mL, of USP Perphenazine RS in the *Standard preparation*, and A_U and A_S are the absorbances of the solutions from the *Assay preparation* and the *Standard preparation*, respectively.

Perphenazine and Amitriptyline Hydrochloride Tablets

» Perphenazine and Amitriptyline Hydrochloride Tablets contain not less than 90.0 percent and not more than 110.0 percent of the labeled amounts of perphenazine ($C_{21}H_{26}ClN_3OS$) and amitriptyline hydrochloride ($C_{20}H_{23}N.HCl$).

Packaging and storage—Preserve in well-closed containers.

Reference standards—*USP Perphenazine Reference Standard*—Dry in vacuum at 65° for 3 hours before using. *USP Amitriptyline Hydrochloride Reference Standard*—Dry at a pressure not exceeding 5 mm of mercury at 60° to constant weight before using.

NOTE—Throughout the following procedures, protect test or assay specimens, the Reference Standard, and solutions containing them, by conducting the procedures without delay, under subdued light, or using low-actinic glassware.

Identification—Transfer a portion of powdered Tablets, equivalent to about 40 mg of perphenazine, to a 100-mL volumetric flask containing about 50 mL of alcohol. Agitate for 20 minutes, add alcohol to volume, mix, and filter or centrifuge. Separately prepare two Standard solutions containing 0.4 mg per mL of USP Perphenazine RS and USP Amitriptyline Hydrochloride RS, respectively, in alcohol. Separately apply 5 µL of the test solution and 5 µL of each Standard solution to a thin-layer chromatographic plate (see *Chromatography* ⟨621⟩) coated with a 0.25-mm layer of chromatographic silica gel mixture. Develop the chromatogram using a solvent system consisting of a mixture of cyclohexane, ethyl acetate, and diethylamine (85:25:5) until the solvent front has moved about 15 cm. Remove the plate from the developing chamber, air-dry for 20 minutes, and examine the plate under short-wavelength ultraviolet light: the R_f values of the principal spots obtained from the test solution correspond to those obtained from the Standard solutions.

Dissolution ⟨711⟩—

Medium: 0.1 *N* hydrochloric acid; 900 mL.

Apparatus 2: 50 rpm.

Time: 60 minutes.

Procedure—Determine the amounts of perphenazine and amitriptyline hydrochloride in solution in filtered portions of the solution under test, in comparison with a Standard solution having known concentrations of USP Perphenazine RS and USP Amitriptyline Hydrochloride RS in the same medium, as directed for *Procedure* in the *Assay*.

Tolerances—Not less than 75% (*Q*) of the labeled amounts of perphenazine ($C_{21}H_{26}ClN_3OS$) and amitriptyline hydrochloride ($C_{20}H_{23}N.HCl$) is dissolved in 60 minutes.

Uniformity of dosage units ⟨905⟩: meet the requirements for *Content uniformity* with respect to perphenazine and to amitriptyline hydrochloride.

Assay—

Mobile phase—Prepare a filtered and degassed mixture of water, acetonitrile, methanol, and methanesulfonic acid (490:310:200:2). Make adjustments if necessary (see *System Suitability* under *Chromatography* ⟨621⟩).

Standard preparation—Dissolve an accurately weighed quantity of USP Perphenazine RS in methanol, and dilute quantitatively with methanol to obtain a solution having a known concentration of about 0.8 mg per mL (*Solution P*). Transfer 4*J* mg of USP Amitriptyline Hydrochloride RS to a 50-mL volumetric flask, *J* being the ratio of the labeled amount, in mg, of amitriptyline hydrochloride to the labeled amount, in mg, of perphenazine per Tablet. Add 5.0 mL of *Solution P* and 20 mL of 0.2 *N* acetic acid, shake, and sonicate to dissolve the Reference Standards. Dilute with methanol to volume, and mix. Pipet 25 mL of this solution into a 100-mL volumetric flask, dilute with a mixture of methanol and 0.04 *N* acetic acid (3:2) to volume, and mix to obtain a *Standard preparation* having known concentrations of about 20 µg of USP Perphenazine RS per mL and about 20*J* µg of USP Amitriptyline Hydrochloride RS per mL.

Assay preparation—Transfer 10 Perphenazine and Amitriptyline Hydrochloride Tablets to a 250-mL volumetric flask, add 100 mL of 0.2 *N* acetic acid, and shake the mixture until the Tablets have disintegrated. Add methanol to volume, mix, and filter. Dilute an accurately measured volume (V_F mL) of the clear filtrate quantitatively with a mixture of methanol and 0.04 *N* acetic acid (3:2) to obtain a solution (V_A mL) containing about 20 µg of perphenazine per mL, and filter through a membrane filter.

Chromatographic system (see *Chromatography* ⟨621⟩)—The liquid chromatograph is equipped with a 254-nm detector and a 3.9-mm × 30-cm column that contains packing L1. The flow rate is about 1 mL per minute, and is adjusted until the relative retention times for perphenazine and amitriptyline are about 1 and 1.5, respectively. Chromatograph the *Standard preparation*, and record the peak responses as directed under *Procedure:* the relative standard deviation is not more than 2.0% for replicate injections, and the resolution, *R*, between perphenazine and amitriptyline is not less than 4.

Procedure—Separately inject equal volumes (about 20 µL) of the *Standard preparation* and the *Assay preparation* into the chromatograph, record the chromatograms, and measure the responses for the major peaks. Calculate the quantity, in mg, of perphenazine ($C_{21}H_{26}ClN_3OS$) in each Tablet taken by the formula:

$$0.25(C/10)(V_A/V_F)(r_U/r_S),$$

in which *C* is the concentration, in µg per mL, of USP Perphenazine RS in the *Standard preparation*, V_A is the volume, in mL, of the *Assay preparation*, V_F is the volume, in mL, of the filtrate taken for the *Assay preparation*, and r_U and r_S are the responses of the perphenazine peaks obtained from the *Assay preparation* and the *Standard preparation*, respectively. Calculate the quantity, in mg, of amitriptyline hydrochloride ($C_{20}H_{23}N.HCl$) taken by the same formula, reading amitriptyline hydrochloride instead of perphenazine.

Persic Oil—*see* Persic Oil NF

Pertechnetate, Tc 99m Injection, Sodium—*see under* "technetium"

Pertussis Immune Globulin

» Pertussis Immune Globulin conforms to the regulations of the federal Food and Drug Administration concerning biologics (see *Biologics* ⟨1041⟩). It is a sterile, non-pyrogenic solution of globulins derived from the blood plasma of adult human donors who have been immunized with pertussis vaccine such that each 1.25 mL contains not less than the amount of immune globulin to be equivalent to 25 mL of human hyperimmune serum. It may contain 0.3 *M* glycine as a stabilizing agent, and it contains a suitable preservative.

Packaging and storage—Preserve at a temperature between 2° and 8°.

Expiration date—The expiration date is not later than 3 years after date of issue from manufacturer's cold storage (5°, 3 years).

Labeling—Label it to state that it is not intended for intravenous injection.

Pertussis Vaccine

» Pertussis Vaccine conforms to the regulations of the federal Food and Drug Administration concerning biologics (620.1 to 620.7) (see *Biologics* ⟨1041⟩). It is a sterile bacterial fraction or suspension of killed pertussis bacilli (*Bordetella pertussis*) of a strain or strains selected for high antigenic efficiency. It has a potency determined by the specific mouse potency test based on the U.S. Standard Pertussis Vaccine, and a pertussis challenge, of 12 protective units per total immunizing dose, and, in the case of whole bacterial vaccine, such dose contains not more than 60 opacity units. It meets the requirements of the specific mouse toxicity test. It contains a preservative.

Packaging and storage—Preserve at a temperature between 2° and 8°.

Expiration date—The expiration date is not later than 18 months after date of issue from manufacturer's cold storage (5°, 1 year).

Labeling—Label it to state that it is to be well shaken before use and that it is not to be frozen.

Pertussis Vaccine, Diphtheria and Tetanus Toxoids and—*see* Diphtheria and Tetanus Toxoids and Pertussis Vaccine

Pertussis Vaccine Adsorbed

» Pertussis Vaccine Adsorbed conforms to the regulations of the federal Food and Drug Administration concerning biologics (620.1 to 620.7) (see *Biologics* ⟨1041⟩). It is a sterile bacterial fraction or suspension, in a suitable diluent, of killed pertussis bacilli (*Bordetella pertussis*) of a strain or strains selected for high antigenic efficiency precipitated or adsorbed by the addition of aluminum hydroxide or aluminum phosphate, and re-suspended. It has a potency determined by the specific mouse potency test based on the U.S. Standard Pertussis Vaccine, and a pertussis challenge, of 12 protective units per total immunizing dose, and, in the case of whole bacterial vaccine, such dose contains not more than 48 opacity units. It meets the requirements of the specific mouse toxicity test. It contains a preservative.

Packaging and storage—Preserve at a temperature between 2° and 8°, and avoid freezing.

Expiration date—The expiration date is not later than 18 months after date of issue from manufacturer's cold storage (5°, 1 year).

Labeling—Label it to state that it is to be well shaken before use and that it is not to be frozen.

Pertussis Vaccine Adsorbed, Diphtheria and Tetanus Toxoids and—*see* Diphtheria and Tetanus Toxoids and Pertussis Vaccine Adsorbed

Petrolatum

» Petrolatum is a purified mixture of semisolid hydrocarbons obtained from petroleum. It may contain a suitable stabilizer.

Packaging and storage—Preserve in well-closed containers.

Labeling—Label it to indicate the name and proportion of any added stabilizer.

Specific gravity ⟨841⟩: between 0.815 and 0.880 at 60°.

Melting range, *Class III* ⟨741⟩: between 38° and 60°.

Consistency—

Apparatus—Determine the consistency of Petrolatum by means of a penetrometer fitted with a polished cone-shaped metal plunger weighing 150 g, having a detachable steel tip of the following dimensions: the tip of the cone has an angle of 30°, the point being truncated to a diameter of 0.381 ± 0.025 mm, the base of the tip is 8.38 ± 0.05 mm in diameter, and the length of the tip is 14.94 ± 0.05 mm. The remaining portion of the cone has an angle of 90°, is about 28 mm in height, and has a maximum diameter at the base of about 65 mm. The containers for the test are flat-bottom metal or glass cylinders that are 102 ± 6 mm in diameter and not less than 60 mm in height.

Procedure—Melt a quantity of Petrolatum at 82 ± 2.5°, pour into one or more of the containers, filling to within 6 mm of the rim. Cool to 25 ± 2.5° over a period of not less than 16 hours, protected from drafts. Two hours before the test, place the containers in a water bath at 25 ± 0.5°. If the room temperature is below 23.5° or above 26.5°, adjust the temperature of the cone to 25 ± 0.5° by placing it in the water bath.

Without disturbing the surface of the substance under test, place the container on the penetrometer table, and lower the cone until the tip just touches the top surface of the test substance at a spot 25 mm to 38 mm from the edge of the container. Adjust the zero setting and quickly release the plunger, then hold it free for 5 seconds. Secure the plunger, and read the total penetration from the scale. Make three or more trials, each so spaced that there is no overlapping of the areas of penetration. Where the penetration exceeds 20 mm, use a separate container of the test substance for each trial. Read the penetration to the nearest 0.1 mm. Calculate the average of the three or more readings, and conduct further trials to a total of 10 if the individual results differ from the average by more than ±3%: the final average of the trials is not less than 10.0 mm and not more than 30.0 mm, indicating a consistency value between 100 and 300.

Alkalinity—Introduce 35 g into a suitable beaker, add 100 mL of boiling water, cover, and place on a stirring hot-plate maintained at the boiling point of water. After 5 minutes, allow the phases to separate. Draw off the separated water into a casserole, wash the petrolatum further with two 50-mL portions of boiling water, and add the washings to the casserole. To the pooled washings add 1 drop of phenolphthalein TS, and boil: the solution does not acquire a pink color.

Acidity—If the addition of phenolphthalein TS in the test for *Alkalinity* produces no pink color, add 0.1 mL of methyl orange TS: no red or pink color is produced.

Residue on ignition ⟨281⟩—Heat 2 g in an open porcelain or platinum dish over a Bunsen flame: it volatilizes without emitting an acrid odor, and on ignition yields not more than 0.1% of residue.

Organic acids—Weigh 20.0 g, add 100 mL of a 1 in 2 mixture of neutralized alcohol and water, agitate thoroughly, and heat to boiling. Add 1 mL of phenolphthalein TS, and titrate rapidly with 0.1 *N* sodium hydroxide VS, with vigorous agitation to the production of a sharp pink end-point, noting the color change in the alcohol-water layer: not more than 400 µL of 0.100 *N* sodium hydroxide is required.

Fixed oils, fats, and rosin—Digest 10 g with 50 mL of 5 *N* sodium hydroxide at 100° for 30 minutes. Separate the water layer, and acidify it with 5 *N* sulfuric acid: no oily or solid matter separates.

Color—Melt about 10 g on a steam bath, and pour about 5 mL of the liquid into a clear-glass 15-mm × 150-mm test tube, keeping the petrolatum melted. The petrolatum is not darker than a solution made by mixing 3.8 mL of ferric chloride CS and 1.2 mL of cobaltous chloride CS in a similar tube, the comparison of the two being made in reflected light against a white background, the petrolatum tube being held directly against the background at such an angle that there is no fluorescence.

Petrolatum Gauze—*see* Gauze, Petrolatum

Hydrophilic Petrolatum

» Prepare Hydrophilic Petrolatum as follows:

Cholesterol	30 g
Stearyl Alcohol	30 g
White Wax	80 g
White Petrolatum	860 g
To make	1000 g

Melt the Stearyl Alcohol and White Wax together on a steam bath, then add the Cholesterol, and stir until completely dissolved. Add the White Petrolatum, and mix. Remove from the bath, and stir until the mixture congeals.

White Petrolatum

» White Petrolatum is a purified mixture of semisolid hydrocarbons obtained from petroleum, and wholly or nearly decolorized. It may contain a suitable stabilizer.

Residue on ignition ⟨281⟩—Heat 2 g in an open porcelain or platinum dish over a flame: it volatilizes without emitting an acrid odor, and on ignition yields not more than 0.05% of residue.

Color—Melt about 10 g on a steam bath, and pour 5 mL of the liquid into a clear-glass, 16-mm × 150-mm bacteriological test tube: the warm, melted liquid is not darker than a solution made by mixing 1.6 mL of ferric chloride CS and 3.4 mL of water in a similar tube, the comparison of the two being made in reflected light against a white background, the tubes being held directly against the background at such an angle that there is no fluorescence.

Other requirements—It meets the requirements for *Packaging and storage*, *Labeling*, *Specific gravity*, *Melting range*, *Consistency*, *Alkalinity*, *Acidity*, *Organic acids*, and *Fixed oils, fats, and rosin* under *Petrolatum*.

Pharmaceutical Glaze—*see* Glaze, Pharmaceutical NF

Phenacemide

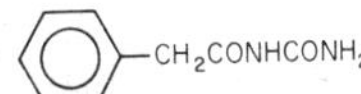

$C_9H_{10}N_2O_2$ 178.19
Benzeneacetamide, *N*-(aminocarbonyl)-.
(Phenylacetyl)urea [*63-98-9*].

» Phenacemide contains not less than 98.0 percent and not more than 100.5 percent of $C_9H_{10}N_2O_2$, calculated on the dried basis.

Packaging and storage—Preserve in tight containers.

Reference standard—*USP Phenacemide Reference Standard*—Dry at 105° for 4 hours before using.

Identification—

A: The infrared absorption spectrum of a potassium bromide dispersion of it, previously dried, exhibits maxima only at the same wavelengths as that of a similar preparation of USP Phenacemide RS.

B: The ultraviolet absorption spectrum of a 3 in 10,000 solution in methanol exhibits maxima and minima at the same wavelengths as that of a similar solution of USP Phenacemide RS, concomitantly measured, and the respective absorptivities, calculated on the dried basis, at the wavelength of maximum absorbance at about 257 nm do not differ by more than 3.0%.

Loss on drying ⟨731⟩—Dry it at 105° for 4 hours: it loses not more than 1.0% of its weight.

Residue on ignition ⟨281⟩: not more than 0.1%.

Heavy metals, *Method II* ⟨231⟩: 0.002%.

Ordinary impurities ⟨466⟩—

Test solution: alcohol.
Standard solution: alcohol.
Application volume: 10 µL.
Eluant: a mixture of toluene, ethyl acetate, and formic acid (50:45:5), in a nonequilibrated chamber.
Visualization: 1.

Assay—Transfer about 400 mg of Phenacemide, accurately weighed, to a round-bottom flask, add 50 mL of water and 8 mL of sulfuric acid, attach a condenser, and reflux gently for 1 hour. Cool, transfer the contents of the flask to a separator, rinse the condenser and flask with three 15-mL portions of chloroform, and add the rinsings to the separator. Shake, allow the layers to separate, and transfer the chloroform layer to a flask. Extract the aqueous layer with five 30-mL portions of chloroform after previously rinsing the condenser and flask with each portion, and add the chloroform extracts to the main extract. Filter the chloroform solution, rinse the filter with a few mL of hot chloroform, add the rinsing to the filtrate, and evaporate without heating, with the aid of a current of air, to dryness. Dissolve the residue in 25 mL of dehydrated alcohol, add phenolphthalein TS, and titrate with 0.1 *N* sodium hydroxide VS. Perform a blank determination, and make any necessary correction. Each mL of 0.1 *N* sodium hydroxide is equivalent to 17.82 mg of $C_9H_{10}N_2O_2$.

Phenacemide Tablets

» Phenacemide Tablets contain not less than 95.0 percent and not more than 105.0 percent of the labeled amount of $C_9H_{10}N_2O_2$.

Packaging and storage—Preserve in well-closed containers.

Reference standard—*USP Phenacemide Reference Standard*—Dry at 105° for 4 hours before using.

Identification—Transfer a quantity of powdered Tablets, equivalent to about 300 mg of phenacemide, to a separator, add 50 mL of water, and extract with five 20-mL portions of chloroform. Filter the chloroform extracts, evaporate on a steam bath with the aid of a gentle current of air to dryness, and dry at 105° for 2 hours: the residue so obtained responds to *Identification test A* under *Phenacemide*.

Dissolution ⟨711⟩—

Medium: 0.1 *N* hydrochloric acid; 900 mL.
Apparatus 2: 100 rpm.
Time: 60 minutes.
Procedure—Determine the amount of $C_9H_{10}N_2O_2$ dissolved from ultraviolet absorbances at the wavelength of maximum absorbance at about 257 nm of filtered portions of the solution under test, suitably diluted with 0.1 *N* hydrochloric acid, in comparison with a Standard solution having a known concentration of USP Phenacemide RS in the same medium.
Tolerances—Not less than 35% (*Q*) of the labeled amount of $C_9H_{10}N_2O_2$ is dissolved in 60 minutes.

Uniformity of dosage units ⟨905⟩: meet the requirements.

Assay—Weigh and finely powder not less than 20 Phenacemide Tablets. Using an accurately weighed portion of the powder, equivalent to about 400 mg of phenacemide, proceed as directed in the *Assay* under *Phenacemide*.

Phenazopyridine Hydrochloride

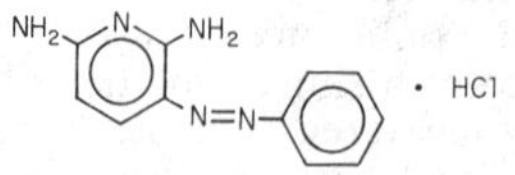

$C_{11}H_{11}N_5 \cdot HCl$ 249.70

2,6-Pyridinediamine, 3-(phenylazo)-, monohydrochloride.

2,6-Diamino-3-(phenylazo)pyridine monohydrochloride [*136-40-3*].

» Phenazopyridine Hydrochloride contains not less than 99.0 percent and not more than 101.0 percent of $C_{11}H_{11}N_5 \cdot HCl$, calculated on the dried basis.

Packaging and storage—Preserve in tight containers.

Reference standard—*USP Phenazopyridine Hydrochloride Reference Standard*—Dry at 105° for 4 hours before using.

Identification—

A: The infrared absorption spectrum of a potassium bromide dispersion of it, previously dried, exhibits maxima only at the same wavelengths as that of a similar preparation of USP Phenazopyridine Hydrochloride RS.

B: The ultraviolet absorption spectrum of a 1 in 200,000 solution of it in dilute alcoholic sulfuric acid (1 in 360) exhibits maxima and minima at the same wavelengths as that of a similar solution of USP Phenazopyridine Hydrochloride RS, concomitantly measured.

C: Prepare a solution of it in alcohol containing about 0.2 mg per mL. Transfer 10 mL of this solution to a glass-stoppered, 100-mL graduated cylinder, add chloroform to volume, and mix. Apply 10 µL of the solution so obtained to a suitable thin-layer chromatographic plate (see *Chromatography* ⟨621⟩), coated with a 0.25-mm layer of chromatographic silica gel. Apply to the same plate 10 µL of a Standard solution of USP Phenazopyridine Hydrochloride RS in the same medium having a known concentration of about 0.02 mg per mL. Develop the chromatogram in a solvent system consisting of a mixture of chloroform, ethyl acetate, and methanol (85:10:5) until the solvent front has moved about three-fourths of the length of the plate. Remove the plate from the chamber and allow it to dry. Locate the spots by spraying the plate lightly with 2 *N* hydrochloric acid: the R_f of the principal spot in the chromatogram of the test solution corresponds to that obtained from the Standard solution.

Loss on drying ⟨731⟩—Dry it at 105° for 4 hours: it loses not more than 1.0% of its weight.

Residue on ignition ⟨281⟩: not more than 0.2%.

Water-insoluble substances—Dissolve about 2 g, accurately weighed, in 200 mL of water, heat to boiling, then heat in a covered container on a steam bath for 1 hour. Filter through a tared, fine-porosity, sintered-glass crucible, wash thoroughly with water, and dry at 105° to constant weight: the weight of the residue does not exceed 0.1% of the weight of Phenazopyridine Hydrochloride taken.

Heavy metals, *Method II* ⟨231⟩: 0.002%.

Ordinary impurities ⟨466⟩—

Test solution—Prepare a solution of it in alcohol having a concentration of 2.0 mg per mL.

Standard solutions—Prepare solutions of USP Phenazopyridine Hydrochloride RS in alcohol containing 0.04, 0.02, and 0.01 mg per mL, respectively.

Eluant: a mixture of chloroform, ethyl acetate, and methanol (85:10:5).

Visualization—Spray the plate with 5 *N* hydrochloric acid.

Assay—Transfer about 100 mg of Phenazopyridine Hydrochloride, accurately weighed, to a 200-mL volumetric flask. Add about 100 mL of a mixture of sulfuric acid and alcohol (1 in 360), heat gently on a steam bath for 10 minutes, shake by mechanical means to dissolve, cool to room temperature, dilute with the alcoholic sulfuric acid to volume, and mix. Transfer 10.0 mL of this solution to a 100-mL volumetric flask, dilute with the alcoholic sulfuric acid to volume, and mix. Transfer 5.0 mL of the resulting solution to a 50-mL volumetric flask, dilute with the alcoholic sulfuric acid to volume, and mix. Concomitantly determine the absorbances of this solution and a Standard solution of USP Phenazopyridine Hydrochloride RS in the same medium having a known concentration of about 5 µg per mL, in 1-cm cells at the wavelength of maximum absorbance at about 390 nm, with a suitable spectrophotometer, using dilute alcoholic sulfuric acid (1 in 360) as the blank. Calculate the quantity, in mg, of $C_{11}H_{11}N_5 \cdot HCl$ in the Phenazopyridine Hydrochloride taken by the formula:

$$20C(A_U / A_S),$$

in which C is the concentration, in µg per mL, of USP Phenazopyridine Hydrochloride RS in the Standard solution, and A_U and A_S are the absorbances of the solution of Phenazopyridine Hydrochloride and the Standard solution, respectively.

Phenazopyridine Hydrochloride Tablets

» Phenazopyridine Hydrochloride Tablets contain not less than 95.0 percent and not more than 105.0 percent of the labeled amount of $C_{11}H_{11}N_5 \cdot HCl$.

Packaging and storage—Preserve in tight containers.

Reference standard—*USP Phenazopyridine Hydrochloride Reference Standard*—Dry at 105° for 4 hours before using.

Identification—Transfer a quantity of finely ground Tablets, equivalent to about 50 mg of phenazopyridine hydrochloride, to a 125-mL separator, add 50 mL of water, 1 mL of 1 *N* hydrochloric acid, and 5 mL of a saturated sodium chloride solution, and shake to dissolve. Extract with two 25-mL portions of chloroform, and discard the chloroform. Add 5 mL of 1 *N* sodium hydroxide to the aqueous solution, and extract with one 50-mL portion of chloroform. Transfer the chloroform layer to a second 125-mL separator, and wash with one 50-mL portion of 0.1 *N* sodium hydroxide. Filter the chloroform layer through a pledget of cotton previously washed with chloroform. Add 5 drops of hydrochloric acid to the filtrate, and evaporate under a current of air on a steam bath to dryness. Add 5 mL of alcohol, and evaporate. Dry the residue at 105° for 4 hours: the infrared absorption spectrum of a potassium bromide dispersion of the dried residue so obtained exhibits maxima only at the same wavelengths as that of a similar preparation of USP Phenazopyridine Hydrochloride RS.

Disintegration ⟨701⟩: 20 minutes.

Uniformity of dosage units ⟨905⟩: meet the requirements.

Assay—

Alcoholic sulfuric acid—Prepare a mixture of sulfuric acid and alcohol (1 in 360).

Mixed solvent—Mix 2 volumes of *Alcoholic sulfuric acid* with 1 volume of 0.1 *N* aqueous sulfuric acid.

Standard preparation—Transfer about 50 mg of USP Phenazopyridine Hydrochloride RS, accurately weighed, to a 100-mL volumetric flask, and add 15.0 mL of *Mixed solvent*. Add about 40 mL of *Alcoholic sulfuric acid*, mix, and sonicate for not less than 15 minutes. Cool, dilute with *Alcoholic sulfuric acid* to volume, and mix.

Assay preparation—Weigh and finely powder not less than 20 Phenazopyridine Hydrochloride Tablets. Transfer an accurately weighed portion of the powder, equivalent to about 50 mg of phenazopyridine hydrochloride, to a 100-mL volumetric flask, and add 15.0 mL of *Mixed solvent*. Add about 40 mL of *Alcoholic sulfuric acid*, mix, and sonicate for not less than 15 minutes. Cool, dilute with *Alcoholic sulfuric acid* to volume, and mix.

Procedure—Transfer 5.0 mL each of the *Standard preparation* and the *Assay preparation*, respectively, to separate 100-mL volumetric flasks [NOTE—Cover the tip of the pipet used to transfer the *Assay preparation* with cotton to serve as a filter], dilute each with *Alcoholic sulfuric acid* to volume, and mix. Transfer 10.0 mL of each solution to separate 50-mL volumetric flasks, dilute with *Alcoholic sulfuric acid* to volume, and mix. Concomitantly determine the absorbances of the solutions in 1-cm cells at the wavelength of maximum absorbance at about 390 nm, with a suitable spectrophotometer, using *Alcoholic sulfuric acid* as the blank. Calculate the quantity, in mg, of $C_{11}H_{11}N_5.HCl$ in the portion of Tablets taken by the formula:

$$0.1C(A_U/A_S),$$

in which C is the concentration, in µg per mL, of USP Phenazopyridine Hydrochloride RS in the *Standard preparation*, and A_U and A_S are the absorbances of the solutions from the *Assay preparation* and the *Standard preparation*, respectively.

Phenazopyridine Hydrochlorides and Sulfamethizole Capsules, Oxytetracycline and—*see* Oxytetracycline and Phenazopyridine Hydrochlorides and Sulfamethizole Capsules

Phendimetrazine Tartrate

$C_{12}H_{17}NO.C_4H_6O_6$ 341.36

Morpholine, 3,4-dimethyl-2-phenyl-, (2*S-trans*)-, [*R*-(*R**,*R**)]-2,3-dihydroxybutanedioate (1:1).

(2*S*,3*S*)-3,4-Dimethyl-2-phenylmorpholine L-(+)-tartrate (1:1) [*50-58-8*].

» Phendimetrazine Tartrate contains not less than 98.0 percent and not more than 102.0 percent of $C_{12}H_{17}NO.C_4H_6O_6$, calculated on the dried basis.

Packaging and storage—Preserve in tight containers.

Reference standard—*USP Phendimetrazine Tartrate Reference Standard*—Dry at 105° to constant weight before using.

Identification—

A: The infrared absorption spectrum of a potassium bromide dispersion of it exhibits maxima only at the same wavelengths as that of a similar preparation of USP Phendimetrazine Tartrate RS.

B: The ultraviolet absorption spectrum of a 1 in 1000 solution in methanol exhibits maxima and minima at the same wavelengths as that of a similar solution of USP Phendimetrazine Tartrate RS, concomitantly measured.

C: It responds to the test for *Tartrate* ⟨191⟩.

Melting range ⟨741⟩: between 182° and 188°, with decomposition, but the range between beginning and end of melting does not exceed 3°.

Specific rotation ⟨781⟩: between 32° and 36°, calculated on the dried basis, determined in a solution containing 100 mg per mL.

pH ⟨791⟩: between 3.0 and 4.0, in a solution (1 in 40).

Loss on drying ⟨731⟩—Dry it to constant weight at 105°: it loses not more than 0.5% of its weight.

Residue on ignition ⟨281⟩: not more than 0.1%.

Chloride ⟨221⟩—A 1.0-g portion shows no more chloride than corresponds to 0.50 mL of 0.020 *N* hydrochloric acid (0.035%).

Sulfate ⟨221⟩—A 1.0-g portion shows no more sulfate than corresponds to 0.10 mL of 0.020 *N* sulfuric acid (0.01%).

Heavy metals ⟨231⟩: 0.001%.

Chromatographic purity—Dissolve 500 mg in water, dilute with water to 5.0 mL, and mix. On the starting line of a suitable thin-layer chromatographic plate (see *Chromatography* ⟨621⟩), coated with a 0.25-mm layer of chromatographic silica gel mixture, apply 10 µL of this preparation and 10 µL of an aqueous solution of USP Phendimetrazine Tartrate RS containing about 100 mg per mL. Develop the chromatogram in a suitable chamber with a solvent system consisting of a mixture of acetone, methanol, and ammonium hydroxide (50:50:1) until the solvent front has moved about three-fourths of the length of the plate. Remove the plate from the chamber, air-dry, view under short-wavelength ultraviolet light, and observe the location of the spots. Expose the plate to iodine vapors in a closed chamber: yellow spots appear at the same locations as the spots observed under ultraviolet light, and the R_f value of the spot obtained from the test preparation corresponds to that obtained from the Standard solution, and no other spot is obtained.

L-*erythro* isomer—Dissolve 3.0 g of Phendimetrazine Tartrate in 25 mL of sodium hydroxide solution (1 in 20) in a suitable separator. Add 25 mL of sodium hydroxide solution (1 in 2), swirl, and allow the phendimetrazine base to separate. Discard the lower, alkaline layer, and collect the upper layer, centrifuging, if necessary, to obtain a clear liquid. Inject 1.0 µL of this liquid into a suitable gas chromatograph equipped with a flame-ionization detector, a 100:1 specimen splitter, and a 25-m × 0.25-mm capillary column, the inside wall of which is coated with a 0.4-µm film of liquid phase G1. The temperatures of the injection port, column, and detector block are 250°, 140°, and 280°, respectively. The carrier gas is helium. Preferably using an electronic integrator, determine the areas of all peaks in the chromatogram. The retention times are about 8.5 minutes for the D-*threo* isomer and 9 minutes for the L-*erythro* isomer. Calculate the percentage of L-*erythro* isomer in the test specimen by the formula:

$$100(r_U/r_S),$$

in which r_U is the peak area response of the L-*erythro* isomer peak and r_S is the sum of the areas of the L-*erythro* isomer peak and the D-*threo* isomer peak: the limit is 0.1%.

Assay—Transfer to a beaker about 500 mg of Phendimetrazine Tartrate, accurately weighed, and dissolve in 50 mL of glacial acetic acid. Add 1 drop of crystal violet TS, and titrate with 0.1 *N* perchloric acid VS to a green end-point. Perform a blank determination, and make any necessary correction. Each mL of 0.1 *N* perchloric acid is equivalent to 34.14 mg of $C_{12}H_{17}NO.C_4H_6O_6$.

Phendimetrazine Tartrate Capsules

» Phendimetrazine Tartrate Capsules contain not less than 95.0 percent and not more than 105.0 percent of the labeled amount of $C_{12}H_{17}NO.C_4H_6O_6$.

Packaging and storage—Preserve in tight containers.

Reference standard—*USP Phendimetrazine Tartrate Reference Standard*—Dry at 105° to constant weight before using.

Identification—

A: Shake a quantity of Capsule contents, equivalent to about 300 mg of phendimetrazine tartrate, with about 50 mL of water, filter, and transfer the filtrate to a 200-mL separator. Add 3 mL of 12.5 *N* sodium hydroxide, and extract with two 50-mL portions of chloroform. Extract the combined chloroform extracts in a 250-mL separator with two 15-mL portions of 0.5 *N* hydrochloric acid, and evaporate the combined aqueous extracts on a steam bath to dryness. Dissolve the residue in 5 mL of acetone, and add 50 mL of anhydrous ether to the solution. On standing, phendimetrazine hydrochloride crystallizes out. Filter the precipitate, wash with anhydrous ether, and dry at 105°: the crystals so obtained melt between 189° and 193°, but the range between beginning and end of melting does not exceed 2°.

B: A portion of Capsule contents responds to the test for *Tartrate* ⟨191⟩.

Uniformity of dosage units ⟨905⟩: meet the requirements.

Assay—

Mobile phase—Dissolve 1.1 g of sodium 1-heptanesulfonate in 575 mL of water, add 400 mL of methanol, 25 mL of dilute acetic acid (14 in 100), and mix. Adjust with glacial acetic acid to a pH of 3.0 ± 0.1, if necessary. Filter through a 0.45-µm membrane filter, and degas. Make adjustments if necessary (see *System Suitability* under *Chromatography* ⟨621⟩).

Diluent—Prepare a mixture of water, methanol, and dilute acetic acid (14 in 100) (57.5:40:2.5).

Internal standard solution—Prepare a solution of salicylamide in *Diluent* having a concentration of about 0.1 mg per mL.

Standard preparation—Dissolve an accurately weighed quantity of USP Phendimetrazine Tartrate RS in *Internal standard solution*, and dilute quantitatively with *Internal standard solution* to obtain a solution having a known concentration of about 0.7 mg of USP Phendimetrazine Tartrate RS per mL.

Assay preparation—Remove, as completely as possible, the contents of not less than 20 Phendimetrazine Tartrate Capsules, and weigh accurately. Mix the combined contents, and transfer an accurately weighed quantity of the powder, equivalent to about 35 mg of phendimetrazine tartrate, to a 50-mL volumetric flask, add 25 mL of *Internal standard solution*, and sonicate for about 15 minutes. Cool the solution to room temperature, dilute with *Internal standard solution* to volume, mix, and filter through a 0.45-μm membrane filter.

Chromatographic system (see *Chromatography* ⟨621⟩)—The liquid chromatograph is equipped with a 256-nm detector and a 3.9-mm × 30-cm column that contains packing L1. The flow rate is about 1 mL per minute. Chromatograph the *Standard preparation*, and record the peak responses as directed under *Procedure:* the resolution, *R*, between the analyte and internal standard peaks is not less than 3.0, and the relative standard deviation for replicate injections is not more than 1.0%.

Procedure—Separately inject equal volumes (about 20 μL) of the *Standard preparation* and the *Assay preparation* into the chromatograph, record the chromatograms, and measure the responses for the major peaks. The relative retention times are about 0.5 for salicylamide and 1.0 for phendimetrazine tartrate. Calculate the quantity, in mg, of $C_{12}H_{17}NO.C_4H_6O_6$ in the portion of Capsules taken by the formula:

$$50C(R_U/R_S),$$

in which C is the concentration, in mg per mL, of USP Phendimetrazine Tartrate RS in the *Standard preparation*, and R_U and R_S are the peak response ratios obtained from the *Assay preparation* and the *Standard preparation*, respectively.

Phendimetrazine Tartrate Tablets

» Phendimetrazine Tartrate Tablets contain not less than 90.0 percent and not more than 110.0 percent of the labeled amount of $C_{12}H_{17}NO.C_4H_6O_6$.

Packaging and storage—Preserve in well-closed containers.

Reference standard—*USP Phendimetrazine Tartrate Reference Standard*—Dry at 105° to constant weight before using.

Identification—A quantity of finely powdered Tablets, equivalent to about 300 mg of phendimetrazine tartrate, responds to the *Identification tests* under *Phendimetrazine Tartrate Capsules*.

Dissolution ⟨711⟩—

Medium: water; 900 mL.

Apparatus 1: 100 rpm.

Time: 45 minutes.

pH 7.5 phosphate buffer—Prepare a solution of 0.025 *M* monobasic potassium phosphate, and adjust to a pH of 7.5 by the addition of 1 *N* potassium hydroxide.

Mobile phase—Prepare a suitable degassed and filtered mixture of acetonitrile and *pH 7.5 phosphate buffer* (65:35).

Chromatographic system (see *Chromatography* ⟨621⟩)—The liquid chromatograph is equipped with a 210-nm detector and a 4-mm × 15-cm column that contains packing L15. The flow rate is about 1.0 mL per minute. Chromatograph three replicate injections of the Standard solution, and record the peak responses as directed under *Procedure:* the relative standard deviation is not more than 3.0%.

Procedure—Separately inject equal volumes (about 50 μL) of the Standard solution and a filtered aliquot of the solution under test into the chromatograph, record the chromatograms, and measure the responses for the major peaks. Calculate the percentage of $C_{12}H_{17}NO.C_4H_6O_6$ dissolved in comparison with a Standard solution of USP Phendimetrazine Tartrate RS, similarly prepared and chromatographed.

Tolerances—Not less than 60% (*Q*) of the labeled amount of $C_{12}H_{17}NO.C_4H_6O_6$ is dissolved in 45 minutes.

Uniformity of dosage units ⟨905⟩: meet the requirements.

Assay—

Mobile phase, Diluent, Internal standard solution, Standard preparation, and *Chromatographic system*—Prepare as directed in the *Assay* under *Phendimetrazine Tartrate Capsules*.

Assay preparation—Weigh and finely powder not less than 20 Phendimetrazine Tartrate Tablets. Transfer an accurately weighed portion of the powder, equivalent to about 35 mg of phendimetrazine tartrate, to a 50-mL volumetric flask, add 25 mL of *Internal standard solution*, and sonicate for about 15 minutes. Cool the solution to room temperature, dilute with *Internal standard solution* to volume, mix, and filter through a 0.45-μm membrane filter.

Procedure—Proceed as directed for *Procedure* in the *Assay* under *Phendimetrazine Tartrate Capsules*. Calculate the quantity, in mg, of $C_{12}H_{17}NO.C_4H_6O_6$ in the portion of Tablets taken by the formula:

$$50C(R_U/R_S),$$

in which C is the concentration, in mg per mL, of USP Phendimetrazine Tartrate RS in the *Standard preparation*, and R_U and R_S are the peak response ratios obtained from the *Assay preparation* and the *Standard preparation*, respectively.

Phenelzine Sulfate

$CH_2CH_2NHNH_2$ · H_2SO_4

$C_8H_{12}N_2.H_2SO_4$ 234.27

Hydrazine, (2-phenylethyl)-, sulfate (1:1).

Phenethylhydrazine sulfate (1:1) [*156-51-4*].

» Phenelzine Sulfate contains not less than 97.0 percent and not more than 100.5 percent of $C_8H_{12}N_2.H_2SO_4$, calculated on the dried basis.

Packaging and storage—Preserve in tight containers, protected from heat and light.

Reference standard—*USP Phenelzine Sulfate Reference Standard*—Dry at a pressure not above 5 mm of mercury over silica gel at 80° for 2 hours before using.

Identification—

A: The infrared absorption spectrum of a potassium bromide dispersion of it, previously dried, exhibits maxima only at the same wavelengths as that of a similar preparation of USP Phenelzine Sulfate RS.

B: Dissolve 100 mg in 5 mL of water, render the solution alkaline with 1 *N* sodium hydroxide, and add 1 mL of alkaline cupric tartrate TS: a red to yellow-red precipitate is formed.

C: A solution (1 in 10) responds to the tests for *Sulfate* ⟨191⟩.

Melting range ⟨741⟩: between 164° and 168°.

pH ⟨791⟩: between 1.4 and 1.9, in a solution (1 in 100).

Loss on drying ⟨731⟩—Dry it at a pressure not above 5 mm of mercury over silica gel at 80° for 2 hours: it loses not more than 1.0% of its weight.

Heavy metals, *Method I* ⟨231⟩: 0.002%.

Ordinary impurities ⟨466⟩—

Test solution: a mixture of methanol and water (1:1).

Standard solution: a mixture of methanol and water (1:1).

Eluant: acetone.

Visualization: 1.

Assay—Dissolve about 235 mg of Phenelzine Sulfate, accurately weighed, in 50 mL of water in a glass-stoppered flask. Dissolve 1.5 g of sodium bicarbonate in the solution, add 50.0 mL of 0.1 *N* iodine VS, insert the stopper, and allow to stand for 90 minutes. Cautiously add 20 mL of 3 *N* hydrochloric acid, and titrate the excess iodine with 0.1 *N* sodium thiosulfate VS, adding 3 mL of starch TS as the end-point is approached. Perform a blank determination (see *Residual Titrations* under *Titrimetry* ⟨541⟩). Each mL of 0.1 *N* iodine is equivalent to 5.857 mg of $C_8H_{12}N_2.H_2SO_4$.

Phenelzine Sulfate Tablets

» Phenelzine Sulfate Tablets contain an amount of phenelzine sulfate ($C_8H_{12}N_2.H_2SO_4$) equivalent to not less than 95.0 percent and not more than 105.0 percent of the labeled amount of phenelzine ($C_8H_{12}N_2$).

Packaging and storage—Preserve in tight containers, protected from heat and light.

Reference standard—*USP Phenelzine Sulfate Reference Standard*—Dry at a pressure not above 5 mm of mercury over silica gel at 80° for 2 hours before using.

Identification—Extract a portion of powdered Tablets, equivalent to about 30 mg of phenelzine, with 10 mL of water, and filter: the filtrate responds to *Identification tests B* and *C* under *Phenelzine Sulfate*.

Disintegration ⟨701⟩: 1 hour.

Uniformity of dosage units ⟨905⟩: meet the requirements.

Assay—Weigh and finely powder not less than 20 Phenelzine Sulfate Tablets. Transfer an accurately weighed portion of the powder, equivalent to about 120 mg of phenelzine, to a glass-stoppered flask. Add 20 mL of water, slowly add 3 *N* hydrochloric acid (about 1 mL) to adjust to a pH of about 1.0, and heat to boiling, with constant agitation. Boil for about 30 seconds, and dilute with water to between 30 mL and 40 mL. Cool to room temperature, adjust the pH with sodium bicarbonate (about 1.5 g) to 7.5 ± 0.5, add 50.0 mL of 0.1 *N* iodine VS, insert the stopper, and allow to stand for 90 minutes. Cautiously add 15 mL of 3 *N* hydrochloric acid, and titrate with 0.1 *N* sodium thiosulfate VS, over a suitable source of light in order to facilitate recognition of the color changes, adding 3 mL of starch TS as the end-point is approached. Perform a blank determination (see *Residual Titrations* under *Titrimetry* ⟨541⟩). Each mL of 0.1 *N* iodine is equivalent to 3.405 mg of $C_8H_{12}N_2$.

Phenindione

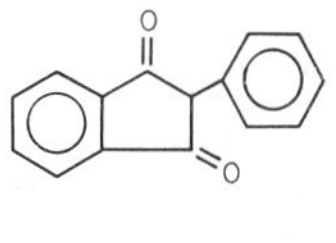

$C_{15}H_{10}O_2$ 222.24
1*H*-Indene-1,3(2*H*)-dione, 2-phenyl-.
2-Phenyl-1,3-indandione [*83-12-5*].

» Phenindione contains not less than 98.0 percent and not more than 100.5 percent of $C_{15}H_{10}O_2$, calculated on the dried basis.

Packaging and storage—Preserve in well-closed containers.

Reference standard—*USP Phenindione Reference Standard*—Dry at 105° for 2 hours before using.

Identification—

A: The infrared absorption spectrum of a potassium bromide dispersion of it, previously dried, exhibits maxima only at the same wavelengths as that of a similar preparation of USP Phenindione RS.

B: Prepare a 1 in 500 solution in alcohol, heating, if necessary, to effect solution. Dilute quantitatively and stepwise with sodium hydroxide solution (1 in 250) to obtain a 1 in 250,000 solution of Phenindione: the ultraviolet absorption spectrum of this solution exhibits maxima and minima at the same wavelengths as that of a similar solution of USP Phenindione RS, concomitantly measured, and the respective absorptivities, calculated on the dried basis, at the wavelength of maximum absorbance at about 278 nm do not differ by more than 3.0%.

C: Place a few crystals in a test tube, and add 1 mL of sulfuric acid: a deep blue to violet color is produced. Add this solution to 1 mL of water in a second test tube: the mixture becomes colorless, and a white precipitate is formed.

Melting range, *Class I* ⟨741⟩: between 148° and 151°.

Loss on drying ⟨731⟩—Dry it at 105° for 2 hours: it loses not more than 1.0% of its weight.

Residue on ignition ⟨281⟩: not more than 0.2%.

Heavy metals, *Method II* ⟨231⟩: 0.002%.

Ordinary impurities ⟨466⟩—

Test solution: chloroform.

Standard solution: chloroform.

Eluant: a mixture of chloroform, methanol, and glacial acetic acid (90:10:1).

Visualization: 1.

Assay—Transfer about 300 mg of Phenindione, accurately weighed, to a glass-stoppered conical flask, add 25 mL of glacial acetic acid, and warm to effect solution. Cool, add 0.5 mL of bromine, and allow to stand for 10 minutes with occasional swirling. Add 10 mL of water, mix, then add 1 g to 2 g of phenol, and shake until the color of bromine is completely discharged and the solution becomes colorless. Cool in an ice bath, and add 10 mL of potassium iodide TS and 5 mL of hydrochloric acid. Titrate the liberated iodine with 0.1 *N* sodium thiosulfate VS, adding 3 mL of starch TS as the end-point is approached. Each mL of 0.1 *N* sodium thiosulfate is equivalent to 11.11 mg of $C_{15}H_{10}O_2$.

Phenindione Tablets

» Phenindione Tablets contain not less than 90.0 percent and not more than 110.0 percent of the labeled amount of $C_{15}H_{10}O_2$.

Packaging and storage—Preserve in well-closed containers.

Reference standard—*USP Phenindione Reference Standard*—Dry at 105° for 2 hours before using.

Identification—Shake a portion of finely powdered Tablets, equivalent to about 50 mg of phenindione, with about 5 mL of ether, filter, and evaporate the filtrate with the aid of a current of air to dryness: the phenindione so obtained responds to *Identification test A* under *Phenindione*.

Dissolution ⟨711⟩—

Medium: pH 8.0 phosphate buffer (see *Buffer Solutions* under *Reagents, Indicators, and Solutions*); 900 mL.

Apparatus 1: 100 rpm.

Time: 45 minutes.

Procedure—Determine the amount of $C_{15}H_{10}O_2$ dissolved from ultraviolet absorbances at the wavelength of maximum absorbance at about 460 nm of filtered portions of the solution under test, suitably diluted with 0.1 *N* sodium hydroxide, in comparison with a Standard solution having a known concentration of USP Phenindione RS in the same medium.

Tolerances—Not less than 85% (*Q*) of the labeled amount of $C_{15}H_{10}O_2$ is dissolved in 45 minutes.

Uniformity of dosage units ⟨905⟩: meet the requirements.

Procedure for content uniformity—Transfer 1 finely powdered Tablet to a 100-mL volumetric flask containing about 50 mL of sodium hydroxide solution (1 in 125). Shake frequently for 15 minutes, dilute with sodium hydroxide solution (1 in 125) to volume, and mix. Allow the insoluble material to settle, and filter, discarding the first 20 mL of the filtrate. Dilute a portion of the subsequent filtrate quantitatively and stepwise, if necessary, with sodium hydroxide solution (1 in 125) to provide a solution containing approximately 70 µg of phenindione per mL. Concomitantly determine the absorbances of this solution and of a solution of USP Phenindione RS in the same medium having a known concentration of about 70 µg per mL in 1-cm cells at the wavelength of maximum absorbance at about 460 nm, with a suitable spectrophotometer, using water as the blank. Calculate the quantity, in mg, of $C_{15}H_{10}O_2$ in the Tablet by the formula:

$$(TC/D)(A_U/A_S),$$

in which T is the labeled quantity, in mg, of phenindione in the Tablet, C is the concentration, in µg per mL, of USP Phenindione RS in the Standard solution, D is the concentration, in µg per mL of phenindione in the solution from the Tablet based upon the labeled quantity per Tablet and the extent of dilution, and A_U and A_S are the absorbances of the solution from the Tablet and the Standard solution, respectively.

Assay—Weigh and finely powder not less than 20 Phenindione Tablets. Transfer an accurately weighed portion of the powder, equivalent to about 70 mg of phenindione, to a 100-mL volumetric flask containing about 80 mL of sodium hydroxide solution (1 in 125), and shake the mixture frequently for 15 minutes. Dilute with sodium hydroxide solution (1 in 125) to volume, mix, and allow the insoluble material to settle. Transfer 10.0 mL of the supernatant solution to a separator, add 5 mL of 3 *N* hydrochloric acid, and extract with three 50-mL portions of *n*-hexane, collecting the *n*-hexane extracts in a second separator. Extract the combined *n*-hexane solution with three 25-mL portions of sodium hydroxide solution (1 in 125), collecting the extracts in a 100-mL volumetric flask. Expel any residual *n*-hexane by flushing with a stream of nitrogen (3 to 5 minutes), dilute with sodium hydroxide solution (1 in 125) to volume, and mix. Concomitantly determine the absorbances of this solution and of a solution of USP Phenindione RS, in the same medium having a known concentration of about 70 µg per mL, in 1-cm cells at the wavelength of maximum absorbance at about 460 nm, using water as the blank. Calculate the quantity, in mg, of $C_{15}H_{10}O_2$ in the portion of Tablets taken by the formula:

$$C(A_U/A_S),$$

in which C is the concentration, in µg per mL, of USP Phenindione RS in the Standard solution, and A_U and A_S are the absorbances of the solution from the Tablets and the Standard solution, respectively.

Phenmetrazine Hydrochloride

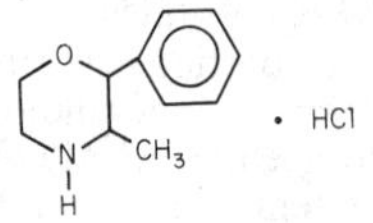

$C_{11}H_{15}NO.HCl$ 213.71

Morpholine, 3-methyl-2-phenyl-, hydrochloride.

3-Methyl-2-phenylmorpholine hydrochloride [*29488-54-8*].

» Phenmetrazine Hydrochloride, dried at 105° for 2 hours, contains not less than 98.0 percent and not more than 102.0 percent of $C_{11}H_{15}NO.HCl$.

Packaging and storage—Preserve in tight containers.

Reference standard—*USP Phenmetrazine Hydrochloride Reference Standard*—Dry at 105° for 2 hours before using.

Identification—

A: The infrared absorption spectrum of a solution of it, previously dried, in chloroform (1 in 20), determined in a 0.1-mm cell, exhibits maxima only at the same wavelengths as that of a similar preparation of USP Phenmetrazine Hydrochloride RS.

B: The ultraviolet absorption spectrum of a 1 in 2000 solution in 0.5 *N* hydrochloric acid exhibits maxima and minima at the same wavelengths as that of a similar solution of USP Phenmetrazine Hydrochloride RS, concomitantly measured.

Melting range, *Class Ia* ⟨741⟩: between 172° and 182°, but the range between beginning and end of melting does not exceed 3°.

pH ⟨791⟩: between 4.5 and 5.5, in a solution (1 in 40).

Loss on drying ⟨731⟩—Dry it at 105° for 2 hours: it loses not more than 0.5% of its weight.

Residue on ignition ⟨281⟩: not more than 0.1%.

Sulfate ⟨221⟩—A 2.0-g portion shows no more sulfate than corresponds to 0.20 mL of 0.020 *N* sulfuric acid (0.01%).

Chloride content—Transfer about 350 mg, previously dried and accurately weighed, to a 250-mL beaker. Add about 125 mL of water and 10 drops of sulfuric acid, and stir for 15 minutes with a magnetic stirrer. Titrate the solution potentiometrically with 0.1 *N* silver nitrate VS, using a silver–mercurous sulfate electrode system with a saturated salt bridge of potassium sulfate. Each mL of 0.1 *N* silver nitrate is equivalent to 3.545 mg of Cl: the content is between 16.3% and 17.0%.

Heavy metals, *Method II* ⟨231⟩: 0.001%.

Ordinary impurities ⟨466⟩—

Test solution: methanol.

Standard solution: methanol.

Eluant: a mixture of chloroform, absolute alcohol, and ammonium hydroxide (80:20:1).

Visualization: 1.

Assay—Transfer to a 200-mL volumetric flask about 100 mg of Phenmetrazine Hydrochloride, previously dried and accurately weighed. Dissolve in 0.5 *N* hydrochloric acid, dilute with 0.5 *N* hydrochloric acid to volume, and mix to obtain the *Assay preparation.* Concomitantly determine the absorbances of the *Assay preparation* and of a Standard solution of USP Phenmetrazine Hydrochloride RS, in the same medium having a known concentration of about 500 µg per mL, in 1-cm cells, at the wavelength of maximum absorbance at about 256 nm, with a suitable spectrophotometer, using 0.5 *N* hydrochloric acid as the blank. Calculate the quantity, in mg, of $C_{11}H_{15}NO.HCl$ in the Phenmetrazine Hydrochloride taken by the formula:

$$0.2C(A_U/A_S),$$

in which C is the concentration, in µg per mL, of USP Phenmetrazine Hydrochloride RS in the Standard solution, and A_U and A_S are the absorbances from the *Assay preparation* and the Standard solution, respectively.

Phenmetrazine Hydrochloride Tablets

» Phenmetrazine Hydrochloride Tablets contain not less than 93.0 percent and not more than 107.0 percent of the labeled amount of $C_{11}H_{15}NO.HCl$.

Packaging and storage—Preserve in tight containers.

Reference standard—*USP Phenmetrazine Hydrochloride Reference Standard*—Dry at 105° for 2 hours before using.

Identification—Dissolve 5 Tablets in 40 mL of water in a 250-mL separator. Add 3 mL of sodium hydroxide solution (1 in 2), and extract with two 50-mL portions of chloroform. Extract the combined chloroform extracts in a 250-mL separator with two 15-mL portions of 0.5 *N* hydrochloric acid, and evaporate the combined aqueous extracts on a steam bath to dryness. Dissolve the residue in 5 mL of acetone, and add 50 mL of anhydrous ether to the solution. On standing, phenmetrazine hydrochloride will crystallize out. Filter the precipitate, wash with anhydrous ether, and dry at 105°: the crystals so obtained melt within a range of 3° between 172° and 182° (see *Melting Range or Temperature* ⟨741⟩).

Dissolution ⟨711⟩—

Medium: water; 900 mL.

Apparatus 2: 50 rpm.

Time: 45 minutes.

Procedure—Determine the amount of $C_{11}H_{15}NO.HCl$ dissolved from ultraviolet absorbances at the wavelength of maximum absorbance at about 256 nm of filtered portions of the solution under test, suitably diluted with *Dissolution Medium,* if necessary, in comparison with a Standard solution having a known concentration of USP Phenmetrazine Hydrochloride RS in the same medium.

Tolerances—Not less than 75% (*Q*) of the labeled amount of $C_{11}H_{15}NO.HCl$ is dissolved in 45 minutes.

Uniformity of dosage units ⟨905⟩: meet the requirements.

Assay—Weigh and finely powder not less than 20 Phenmetrazine Hydrochloride Tablets. Transfer an accurately weighed portion of the powder, equivalent to about 250 mg of phenmetrazine hydrochloride, to a 250-mL volumetric flask, add about 125 mL of 0.5 *N* hydrochloric acid, shake by mechanical means for 1 hour, dilute with 0.5 *N* hydrochloric acid to volume, and mix. Transfer 50.0 mL of the solution to a 250-mL separator, add 5 mL of sodium hydroxide solution (1 in 2), and extract with four 50-mL portions of chloroform, collecting the chloroform extracts in a second 250-mL separator. Extract the combined chloroform extracts with six 15-mL portions of 0.5 *N* hydrochloric acid, collecting the aqueous extracts in a 100-mL volumetric flask, and dilute with 0.5 *N* hydrochloric acid to volume to obtain the *Assay preparation.* Concomitantly determine the absorbances of the *Assay preparation* and of a Standard solution of USP Phenmetrazine Hydrochloride RS in the same medium, having a known concentration of about 500 µg per mL, in 1-cm cells, at the wavelength of maximum absorbance at about 256 nm, with a suitable spectrophotometer, using 0.5 *N* hydrochloric acid as the blank. Calculate the quantity, in mg, of $C_{11}H_{15}NO \cdot HCl$ in the portion of Tablets taken by the formula:

$$0.5C(A_U/A_S),$$

in which *C* is the concentration, in µg per mL, of USP Phenmetrazine Hydrochloride RS in the Standard solution, and A_U and A_S are the absorbances from the *Assay preparation* and the Standard solution, respectively.

Phenobarbital

$C_{12}H_{12}N_2O_3$ 232.24
2,4,6(1*H*,3*H*,5*H*)-Pyrimidinetrione, 5-ethyl-5-phenyl-.
5-Ethyl-5-phenylbarbituric acid [*50-06-6*].

» Phenobarbital contains not less than 98.0 percent and not more than 101.0 percent of $C_{12}H_{12}N_2O_3$, calculated on the dried basis.

Packaging and storage—Preserve in well-closed containers.

Reference standard—*USP Phenobarbital Reference Standard*—Dry at 105° for 2 hours before using.

Identification—The infrared absorption spectrum of a potassium bromide dispersion of it exhibits maxima only at the same wavelengths as that of a similar preparation of USP Phenobarbital RS. If a difference appears, dissolve portions of both the test specimen and the Reference Standard in a suitable solvent, evaporate the solutions to dryness, and repeat the test on the residues.

Melting range ⟨741⟩: between 174° and 178°.

Loss on drying ⟨731⟩—Dry it at 105° for 2 hours: it loses not more than 1.0% of its weight.

Residue on ignition ⟨281⟩: not more than 0.15%.

Phenylbarbituric acid—Reflux 2.0 g with 10 mL of alcohol for 3 minutes: a clear and complete solution is produced.

Assay—Transfer about 500 mg of Phenobarbital, accurately weighed, to a suitable beaker, dissolve in 40 mL of neutralized alcohol, and add 20 mL of water. Titrate with 0.1 *N* sodium hydroxide VS, determining the end-point potentiometrically. Each mL of 0.1 *N* sodium hydroxide is equivalent to 23.22 mg of $C_{12}H_{12}N_2O_3$.

Phenobarbital Capsules, Ephedrine Sulfate and—*see* Ephedrine Sulfate and Phenobarbital Capsules

Phenobarbital Elixir

» Phenobarbital Elixir contains not less than 90.0 percent and not more than 110.0 percent of the labeled amount of $C_{12}H_{12}N_2O_3$.

Packaging and storage—Preserve in tight, light-resistant containers.

Reference standard—*USP Phenobarbital Reference Standard*—Dry at 105° for 2 hours before using.

Identification—

A: Place 10 mL of Elixir in a separator containing 20 mL of water, add 5 mL of 1 *N* sodium hydroxide, and extract with two 10-mL portions of chloroform, discarding the chloroform extracts. Add 5 mL of 3 *N* hydrochloric acid, and extract with two 25-mL portions of chloroform, filtering the extracts through paper into a beaker. Remove the chloroform by evaporation on a steam bath, and dry the residue at 105° for 2 hours: the residue so obtained responds to the *Identification test* under *Phenobarbital.*

B: The retention time of the major peak in the chromatogram of the *Assay preparation* corresponds to that of the *Standard preparation,* both relative to the internal standard, as obtained in the *Assay.*

Alcohol content, *Method II* ⟨611⟩: between 12.0% and 15.0% of C_2H_5OH.

Assay—

pH 4.5 buffer solution, Mobile phase, Chromatographic system, and *System suitability*—Prepare as directed in the *Assay* under *Phenobarbital Tablets.*

Internal standard solution—Dissolve a sufficient quantity of caffeine in a solvent mixture of dichloromethane and methanol (4:1) to obtain a solution having a concentration of about 1 mg per mL.

Standard preparation—Dissolve about 40 mg of USP Phenobarbital RS, accurately weighed, in 4.0 mL of *Internal standard solution,* and evaporate the solvent with the aid of a stream of nitrogen. Dissolve the residue in 20 mL of methanol, add 10 mL of *pH 4.5 buffer solution,* and mix.

Assay preparation—Pipet a quantity of Phenobarbital Elixir, equivalent to about 20 mg of phenobarbital, into a separator. Add 1 mL of hydrochloric acid, and extract with three 10-mL portions of dichloromethane. Filter the extracts through a funnel containing about 15 mg of anhydrous sodium sulfate supported on a small pledget of glass wool. Collect the extracts in a 50-mL volumetric flask containing 2.0 mL of *Internal standard solution,* dilute with dichloromethane to volume, and evaporate about 5 mL of the extract with the aid of a stream of nitrogen. Dissolve the residue in 1 mL of methanol, add 0.5 mL of *pH 4.5 buffer solution,* and mix.

Procedure—Proceed as directed for *Procedure* in the *Assay* under *Phenobarbital Tablets.* Calculate the quantity, in mg, of phenobarbital in each mL of the Elixir taken by the formula:

$$0.5(W/V)(R_U/R_S),$$

in which *W* is the weight, in mg, of USP Phenobarbital RS taken for the *Standard preparation, V* is the volume, in mL, of Elixir taken, and R_U and R_S are the ratios of the peak responses of phenobarbital and caffeine obtained from the *Assay preparation* and the *Standard preparation,* respectively.

Phenobarbital Tablets

» Phenobarbital Tablets contain not less than 90.0 percent and not more than 110.0 percent of the labeled amount of $C_{12}H_{12}N_2O_3$.

Packaging and storage—Preserve in well-closed containers.

Reference standard—*USP Phenobarbital Reference Standard*—Dry at 105° for 2 hours before using.

Identification—

A: Triturate a quantity of finely powdered Tablets, equivalent to about 60 mg of phenobarbital, with 50 mL of chloroform, and filter. Evaporate the clear filtrate to dryness, and dry at 105° for 2 hours: the residue so obtained responds to the *Identification test* under *Phenobarbital.*

B: The retention time of the major peak in the chromatogram of the *Assay preparation* corresponds to that of the *Standard preparation,* both relative to the internal standard, as obtained in the *Assay.*

Dissolution ⟨711⟩—

Medium: water; 900 mL.

Apparatus 2: 50 rpm.

Time: 45 minutes.

Procedure—Determine the amount of $C_{12}H_{12}N_2O_3$ dissolved from ultraviolet absorbances at the wavelength of maximum absorbance at about 240 nm of filtered portions of the solution under test, suitably diluted with pH 9.6 alkaline borate buffer (see *Buffer Solutions* in the section, *Reagents, Indicators, and Solutions*), in comparison with a Standard solution having a known concentration of USP Phenobarbital RS in the same medium.

Tolerances—Not less than 75% (*Q*) of the labeled amount of $C_{12}H_{12}N_2O_3$ is dissolved in 45 minutes.

Uniformity of dosage units ⟨905⟩: meet the requirements.

Procedure for content uniformity—Transfer 1 finely powdered Tablet to a 200-mL volumetric flask with the aid of, first, 10 mL of alcohol, and then about 150 mL of pH 9.6 alkaline borate buffer (see *Buffer Solutions* in the section, *Reagents, Indicators, and Solutions*). Shake the mixture vigorously, dilute with pH 9.6 alkaline borate buffer to volume, mix, and filter. Dilute the filtrate quantitatively with a 1 in 20 solution of alcohol in pH 9.6 alkaline borate buffer to obtain a test solution having a concentration of about 10 µg per mL. Concomitantly determine the absorbances of this solution and of a Standard solution of USP Phenobarbital RS in the same medium, having a known concentration of about 10 µg per mL, in 1-cm cells at the wavelength of maximum absorbance at about 240 nm, with a suitable spectrophotometer, using a 1 in 20 solution of alcohol in pH 9.6 alkaline borate buffer as the blank. Calculate the quantity, in mg, of $C_{12}H_{12}N_2O_3$ in the Tablet by the formula:

$$(TC/D)(A_U/A_S),$$

in which *T* is the labeled quantity, in mg, of phenobarbital in the Tablet, *C* is the concentration, in µg per mL, of USP Phenobarbital RS in the Standard solution, *D* is the concentration, in µg per mL, of phenobarbital in the test solution, on the basis of the labeled quantity per Tablet and the extent of dilution, and A_U and A_S are the absorbances of the test solution and the Standard solution, respectively.

Assay—

pH 4.5 buffer solution—Dissolve about 6.6 g of sodium acetate trihydrate and 3.0 mL of glacial acetic acid in 1000 mL of water, and adjust, if necessary, with glacial acetic acid to a pH of 4.5 ± 0.1.

Mobile phase—Prepare a 2 in 5 mixture of methanol in *pH 4.5 buffer solution,* filter through a 0.5-µm porosity filter, and degas under vacuum.

Internal standard solution—Dissolve a sufficient quantity of caffeine in a mixture of methanol and *pH 4.5 buffer solution* (1:1) to obtain a solution having a final concentration of about 125 µg per mL.

Standard preparation—Dissolve about 20 mg of USP Phenobarbital RS, accurately weighed, in 15.0 mL of *Internal standard solution.* Sonicate if necessary.

Assay preparation—Weigh and finely powder not less than 20 Phenobarbital Tablets. Weigh accurately a portion of the powder, equivalent to about 20 mg of phenobarbital, add 15.0 mL of *Internal standard solution,* mix, and sonicate for 15 minutes. Filter through a membrane filter (0.5-µm) before use.

Chromatographic system (see *Chromatography* ⟨621⟩)—The liquid chromatograph is equipped with a 254-nm detector and a 4-mm × 25-cm column that contains packing L1. The flow rate is about 2 mL per minute. Chromatograph five replicate injections of the *Standard preparation,* and record the peak responses as directed under *Procedure:* the relative standard deviation is not more than 2.0%, the resolution factor between phenobarbital and caffeine is not less than 1.2, and the tailing factor for the two peaks is not greater than 2.0.

Procedure—Separately inject equal volumes (about 10 µL) of the *Standard preparation* and the *Assay preparation* into the chromatograph. Adjust the flow rate and/or the composition of the *Mobile phase* such that system suitability requirements are met, record the chromatograms, and measure the responses for the major peaks. The relative retention times are about 0.6 for caffeine and 1.0 for phenobarbital. Calculate the quantity, in mg, of phenobarbital in the portion of Tablets taken by the formula:

$$(W)(R_U/R_S),$$

in which *W* is the weight, in mg, of USP Phenobarbital RS taken for the *Standard preparation,* and R_U and R_S are the ratios of the peak responses of phenobarbital and caffeine obtained from the *Assay preparation* and the *Standard preparation,* respectively.

Phenobarbital Tablets, Theophylline, Ephedrine Hydrochloride, and—*see* Theophylline, Ephedrine Hydrochloride, and Phenobarbital Tablets

Phenobarbital Sodium

$C_{12}H_{11}N_2NaO_3$ 254.22

2,4,6(1*H*,3*H*,5*H*)-Pyrimidinetrione, 5-ethyl-5-phenyl-, monosodium salt.

Sodium 5-ethyl-5-phenylbarbiturate [*57-30-7*].

» Phenobarbital Sodium contains not less than 98.5 percent and not more than 101.0 percent of $C_{12}H_{11}N_2NaO_3$, calculated on the dried basis.

Packaging and storage—Preserve in tight containers.

Reference standard—*USP Phenobarbital Reference Standard*—Dry at 105° for 2 hours before using.

Completeness of solution—Mix 1.0 g with 10 mL of carbon dioxide–free water: after 1 minute, the solution is clear and free from undissolved solid.

Identification—

A: Evaporate a 50-mL portion of the chloroform solution of phenobarbital obtained in the *Assay* on a steam bath with the aid of a current of air. Add 10 mL of ether, again evaporate, and dry the residue at 105° for 2 hours: the infrared absorption spectrum of a potassium bromide dispersion of the residue so obtained exhibits maxima only at the same wavelengths as that of a similar preparation of USP Phenobarbital RS.

B: Ignite about 200 mg: the residue effervesces with acids, and responds to the tests for *Sodium* ⟨191⟩.

pH ⟨791⟩: between 9.2 and 10.2, in the solution prepared in the test for *Completeness of solution.*

Loss on drying ⟨731⟩—Dry it at 150° for 4 hours: it loses not more than 7.0% of its weight.

Heavy metals ⟨231⟩—Dissolve 2.0 g in 52 mL of water. Add slowly, with vigorous stirring, 8 mL of 1 *N* hydrochloric acid, and filter, rejecting the first 5 mL of the filtrate. Dilute 20 mL of the subsequent filtrate with water to 25 mL: the limit is 0.003%.

Assay—Dissolve about 50 mg of Phenobarbital Sodium, accurately weighed, in 15 mL of water in a separator, add 2 mL of hydrochloric acid, shake, and extract the liberated phenobarbital with four 25-mL portions of chloroform. Filter the combined extracts through a pledget of cotton or other suitable filter into a 250-mL volumetric flask, and wash the separator and the filter with several small portions of chloroform. Add chloroform to volume, and mix. Transfer a 5.0-mL aliquot to a beaker, and evaporate the chloroform on a steam bath just to dryness. Transfer the residue to a 100-mL volumetric flask with the aid, first, of 5 mL of alcohol and then pH 9.6 alkaline borate buffer (see under *Solutions* in the section, *Reagents, Indicators, and Solutions*). Add the buffer to volume, and mix. Dissolve a suitable quantity of USP Phenobarbital RS, accurately weighed, in 5 mL of alcohol contained in a 100-mL volumetric flask, add pH 9.6 alkaline borate buffer to volume, and mix. If necessary, dilute quantitatively and stepwise with pH 9.6 alkaline borate buffer, each 100 mL of which contains 5 mL of added alcohol, to obtain a Standard solution having a known concentration of about 10 µg per mL. Concomitantly determine the absorbances of both solutions in 1-cm cells at the wavelength of maximum absorbance at about 240 nm, with a suitable spectrophotometer, using pH 9.6 alkaline borate buffer, each 100 mL of which contains 5 mL of added alcohol, as the blank. Calculate the quantity, in mg, of $C_{12}H_{11}N_2NaO_3$ in the Phenobarbital Sodium taken by the formula:

$$(5)(1.095)C(A_U/A_S),$$

in which C is the concentration, in µg per mL, of USP Phenobarbital RS in the Standard solution, 1.095 is the ratio of the molecular weight of phenobarbital sodium to that of phenobarbital, and A_U and A_S are the absorbances of the solution of Phenobarbital Sodium and the Standard solution, respectively.

Phenobarbital Sodium Injection

» Phenobarbital Sodium Injection is a sterile solution of Phenobarbital Sodium in a suitable solvent. Phenobarbital may be substituted for the equivalent amount of Phenobarbital Sodium, for adjustment of the pH. The Injection contains the equivalent of not less than 90.0 percent and not more than 105.0 percent of the labeled amount of $C_{12}H_{11}N_2NaO_3$.

Packaging and storage—Preserve in single-dose or in multiple-dose containers, preferably of Type I glass.

Labeling—The label indicates that the Injection is not to be used if it contains a precipitate.

Reference standard—*USP Phenobarbital Reference Standard*—Dry at 105° for 2 hours before using.

Identification—It responds to *Identification test A* under *Phenobarbital Sodium.*

pH ⟨791⟩: between 9.2 and 10.2.

Other requirements—It meets the requirements under *Injections* ⟨1⟩.

Assay—Transfer to a separator an accurately measured volume of Phenobarbital Sodium Injection, equivalent to about 50 mg of phenobarbital sodium, add 15 mL of water, and proceed as directed in the *Assay* under *Phenobarbital Sodium,* beginning with "add 2 mL of hydrochloric acid."

Sterile Phenobarbital Sodium

» Sterile Phenobarbital Sodium is Phenobarbital Sodium suitable for parenteral use.

Packaging and storage—Preserve in *Containers for Sterile Solids* as described under *Injections* ⟨1⟩.

Reference standard—*USP Phenobarbital Reference Standard*—Dry at 105° for 2 hours before using.

Constituted solution—At the time of use, the constituted solution prepared from Sterile Phenobarbital Sodium meets the requirements for *Constituted Solutions* under *Injections* ⟨1⟩.

Other requirements—It conforms to the definition, responds to the *Identification tests,* and meets the requirements for *Completeness of solution, pH, Loss on drying, Heavy metals,* and *Assay* under *Phenobarbital Sodium.* It meets also the requirements for *Sterility Tests* ⟨71⟩, *Uniformity of Dosage Units* ⟨905⟩, and *Labeling* under *Injections* ⟨1⟩.

Phenol

C_6H_5OH

C_6H_6O 94.11
Phenol.
Phenol [*108-95-2*].

» Phenol contains not less than 99.0 percent and not more than 100.5 percent of C_6H_6O, calculated on the anhydrous basis. It may contain a suitable stabilizer.

Caution—Avoid contact with skin, since serious burns may result.

Packaging and storage—Preserve in tight, light-resistant containers.

Labeling—Label it to indicate the name and amount of any substance added as a stabilizer.

Clarity of solution and reaction—A solution (1 in 15) is clear, and is neutral or acid to litmus paper.

Identification—

A: To a solution add bromine TS: a white precipitate is formed, and it dissolves at first but becomes permanent as more of the reagent is added.

B: To 10 mL of a solution (1 in 100) add 1 drop of ferric chloride TS: a violet color is produced.

Congealing temperature ⟨651⟩: not lower than 39°.

Water, *Method I* ⟨921⟩: not more than 0.5%.

Nonvolatile residue—Heat about 5 g, accurately weighed, in a tared porcelain dish on a steam bath until it has evaporated, and dry the residue at 105° for 1 hour: not more than 0.05% of residue remains.

Assay—Place about 2 g of Phenol, accurately weighed, in a 1000-mL volumetric flask, dilute with water to volume, and mix. Pipet 20 mL of the solution into an iodine flask, add 30.0 mL of 0.1 *N* bromine VS, then add 5 mL of hydrochloric acid, and immediately insert the stopper. Shake the flask repeatedly during 30 minutes, allow it to stand for 15 minutes, add quickly 5 mL of potassium iodide solution (1 in 5), taking precautions against the escape of bromine vapor, and at once insert the stopper in the flask. Shake thoroughly, remove the stopper, and rinse it and the neck of the flask with a small quantity of water, so that the washing flows into the flask. Add 1 mL of chloroform, shake the mixture well, and titrate the liberated iodine with 0.1 *N* sodium thiosulfate VS, adding 3 mL of starch TS as the end-point is approached. Perform a blank determination (see *Residual Titrations* under *Titrimetry* ⟨541⟩). Each mL of 0.1 *N* bromine is equivalent to 1.569 mg of C_6H_6O.

Liquefied Phenol

» Liquefied Phenol is Phenol maintained in a liquid condition by the presence of about 10 percent of water. It contains not less than 89.0 percent by weight of C_6H_6O. It may contain a suitable stabilizer.

Caution—Avoid contact with skin, since serious burns may result.

Note—When phenol is to be mixed with a fixed oil, mineral oil, or white petrolatum, use crystalline Phenol, not Liquefied Phenol.

Packaging and storage—Preserve in tight, light-resistant containers.

Labeling—Label it to indicate the name and amount of any substance added as a stabilizer.

Distilling range, *Method I* ⟨721⟩: not higher than 182.5°, an air-cooled condenser being used.

Other requirements—It responds to the *Identification tests*, and meets the requirements of the tests for *Clarity of solution and reaction* and *Nonvolatile residue*, under *Phenol.*

Assay—Proceed with Liquefied Phenol as directed in the *Assay* under *Phenol.*

Phenolated Calamine Lotion—*see* Calamine Lotion, Phenolated

Phenolphthalein

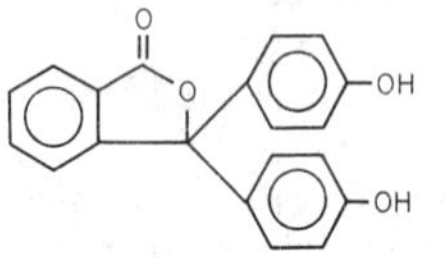

$C_{20}H_{14}O_4$ 318.33

1(3*H*)-Isobenzofuranone, 3,3-bis(4-hydroxyphenyl)-.

3,3-Bis(*p*-hydroxyphenyl)phthalide [*77-09-8*].

» Phenolphthalein contains not less than 98.0 percent and not more than 101.0 percent of $C_{20}H_{14}O_4$, calculated on the dried basis.

Packaging and storage—Preserve in well-closed containers.

Reference standard—*USP Phenolphthalein Reference Standard*—Dry over phosphorus pentoxide for 4 hours before using.

Color of solution—A solution of 500 mg in 30 mL of alcohol has no more than Matching Fluid A (see *Color and Achromicity* ⟨631⟩).

Identification—

A: It is readily dissolved by solutions of alkali hydroxides and by hot solutions of alkali carbonates, yielding red liquids. These solutions are decolorized by the addition of an excess of acid or by high concentrations of alkali hydroxides.

B: The chromatogram of the *Assay preparation* obtained as directed in the *Assay* exhibits a major peak for phenolphthalein, the retention time of which corresponds to that exhibited in the chromatogram of the *Standard preparation* obtained as directed in the *Assay*.

Melting temperature ⟨741⟩: not lower than 258°.

Loss on drying ⟨731⟩—Dry it over phosphorus pentoxide for 4 hours: it loses not more than 1.0% of its weight.

Residue on ignition ⟨281⟩: not more than 0.1%.

Arsenic, *Method II* ⟨211⟩: 8 ppm.

Heavy metals, *Method I* ⟨231⟩—Heat 1.3 g with 25 mL of 1 *N* acetic acid on a steam bath for 5 minutes, filter, and evaporate the filtrate to dryness. To the residue add 1 mL of 0.1 *N* acetic acid, and dilute with water to 25 mL: the limit is 0.0015%.

Fluoran—A 500-mg portion of Phenolphthalein dissolves completely in a mixture of 4 mL of 1 *N* sodium hydroxide and 50 mL of water.

Chromatographic impurities—Proceed as directed in the *Assay*, except to inject 50 µL of *Assay preparation* into the liquid chromatograph. Calculate the area of each peak observed other than the solvent peak: the sum of the areas other than the area of the principal peak is not more than 1.0% of the total area of all peaks observed.

Assay—

Mobile phase—Prepare a filtered and degassed mixture of methanol, water, and glacial acetic acid (50:50:1).

Standard preparation—Transfer about 25 mg of USP Phenolphthalein RS, accurately weighed, to a 50-mL volumetric flask, add 25 mL of methanol, and swirl to dissolve. Add diluted glacial acetic acid (1 in 100) to volume, and mix.

Assay preparation—Transfer about 25 mg of Phenolphthalein, accurately weighed, to a 50-mL volumetric flask, and proceed as directed under *Standard preparation.*

Chromatographic system (see *Chromatography* ⟨621⟩)—The liquid chromatograph is equipped with a 280-nm detector and a 4.6-mm × 25-cm column that contains 10-µm packing L1. The flow rate is about 1.5 mL per minute. Chromatograph the *Standard preparation*, and record the peak responses as directed under *Procedure:* the column efficiency determined from the analyte peak is not less than 900 theoretical plates, the tailing factor for the analyte peak is not more than 2.0, and the relative standard deviation for replicate injections is not more than 2.0%.

Procedure—Separately inject equal volumes (about 10 µL) of the *Standard preparation* and the *Assay preparation* into the chromatograph, record the chromatograms, and measure the responses for the major peaks. Calculate the quantity, in mg, of $C_{20}H_{14}O_4$ in the portion of Phenolphthalein taken by the formula:

$$50C(r_U/r_S),$$

in which C is the concentration, in mg per mL, of USP Phenolphthalein RS in the *Standard preparation*, and r_U and r_S are the peak responses of phenolphthalein obtained from the *Assay preparation* and the *Standard preparation*, respectively.

Phenolphthalein Tablets

» Phenolphthalein Tablets contain not less than 90.0 percent and not more than 110.0 percent of the labeled amount of $C_{20}H_{14}O_4$.

Packaging and storage—Preserve in tight containers.

Labeling—Where Tablets contain Yellow Phenolphthalein, the labeling so indicates.

Reference standard—*USP Phenolphthalein Reference Standard*—Dry over phosphorus pentoxide for 4 hours before using.

Identification—

A: Tablets produce a red liquid when mixed with alkali hydroxide solutions or with hot alkali carbonate solutions. The red liquid is decolorized by the addition of an excess of acid.

B: The chromatogram of the *Assay preparation* obtained as directed in the *Assay* exhibits a major peak for phenolphthalein, the retention time of which corresponds to that exhibited in the chromatogram of the *Standard preparation* obtained as directed in the *Assay*.

Disintegration ⟨701⟩: 30 minutes.

Uniformity of dosage units ⟨905⟩: meet the requirements.

Assay—

Mobile phase and *Chromatographic system*—Proceed as directed in the *Assay* under *Phenolphthalein.*

Standard preparation—Transfer about 60 mg of USP Phenolphthalein RS, accurately weighed, to a 100-mL volumetric flask, add 60 mL of methanol, and swirl to dissolve. Dilute with *Mobile phase* to volume, and mix.

Assay preparation—Weigh and finely powder not less than 20 Phenolphthalein Tablets. Transfer a portion of the powder, equivalent to about 60 mg of phenolphthalein, to a suitable flask with the aid of about 60 mL of methanol. Shake by mechanical means for 20 minutes, and filter into a 100-mL volumetric flask. Wash the extraction flask and the filter with three 10-mL portions of *Mobile phase*, collecting the washings in the volumetric flask, dilute with *Mobile phase* to volume, and mix.

Procedure—Proceed as directed for *Procedure* in the *Assay* under *Phenolphthalein.* Calculate the quantity, in mg, of $C_{20}H_{14}O_4$ in the portion of Tablets taken by the formula:

$$100C(r_U/r_S).$$

Yellow Phenolphthalein

» Yellow Phenolphthalein contains not less than 93.0 percent and not more than 98.0 percent of $C_{20}H_{14}O_4$, calculated on the dried basis.

Packaging and storage—Preserve in well-closed containers.

Reference standard—*USP Phenolphthalein Reference Standard*—Dry over phosphorus pentoxide for 4 hours before using.

Melting temperature ⟨741⟩: not lower than 255°.

Chromatographic impurities—Proceed as directed in the *Assay*, except to inject 50 µL of *Assay preparation* into the liquid chromatograph. Calculate the area of each peak observed other than the solvent peak: the sum of the areas other than the area of the principal peak is not more than 6.0% of the total area of all peaks observed.

Other requirements—It responds to the *Identification tests*, and meets the requirements for *Loss on drying*, *Residue on ignition*, *Arsenic*, and *Heavy metals* under *Phenolphthalein*.

Assay—Proceed with Yellow Phenolphthalein as directed in the *Assay* under *Phenolphthalein.*

Phenprocoumon

$C_{18}H_{16}O_3$ 280.32

2*H*-1-Benzopyran-2-one, 4-hydroxy-3-(1-phenylpropyl)-.

3-(α-Ethylbenzyl)-4-hydroxycoumarin [*435-97-2*].

» Phenprocoumon contains not less than 98.0 percent and not more than 101.0 percent of $C_{18}H_{16}O_3$, calculated on the dried basis.

Packaging and storage—Preserve in well-closed containers.

Reference standard—*USP Phenprocoumon Reference Standard*—Dry in a suitable vacuum drying tube, using phosphorus pentoxide as the desiccant, at 100° for 4 hours before using.

Identification—

A: The infrared absorption spectrum of a potassium bromide dispersion of it, previously dried, exhibits maxima only at the same wavelengths as that of a similar preparation of USP Phenprocoumon RS.

B: The ultraviolet absorption spectrum of a 1 in 100,000 solution in dilute alcoholic hydrochloric acid (1 in 12) exhibits maxima and minima at the same wavelengths as that of a similar solution of USP Phenprocoumon RS, concomitantly measured, and the respective absorptivities, calculated on the dried basis, at the wavelength of maximum absorbance at about 311 nm do not differ by more than 3.0%.

C: Dissolve about 50 mg in 2 mL of chloroform in a test tube, and add a few drops of pyridine and 0.5 mL of a saturated solution of ferric chloride in chloroform, freshly prepared and filtered: an intense, bluish violet color is produced.

Melting range, *Class I* ⟨741⟩: between 177° and 181°.

Acidity—Transfer 1.0 g to a suitable flask, add 20 mL of water, heat on a steam bath with constant stirring for 1 minute, and filter: the filtrate requires for neutralization not more than 0.50 mL of 0.010 *N* sodium hydroxide, methyl red TS being used as the indicator.

Loss on drying ⟨731⟩—Dry it in a suitable vacuum drying tube, using phosphorus pentoxide as the desiccant, at 100° for 4 hours: it loses not more than 0.5% of its weight.

Residue on ignition ⟨281⟩: not more than 0.1%.

Salicylic acid—Add 25 mL of water to 100 mg of Phenprocoumon. Heat on a steam bath for 10 minutes, cool to room temperature, and filter into one of two matched color-comparison tubes, rinsing the filter with water. Adjust the volume with water to 49 mL. To the second tube add 48 mL of water and 1 mL of a standard solution containing 0.10 mg of salicylic acid per mL. To each tube add 1 mL of a freshly prepared solution of ferric ammonium sulfate, prepared by adding 1 mL of dilute hydrochloric acid (1 in 12) to 2 mL of ferric ammonium sulfate TS and diluting to 100 mL. Mix the contents of each tube: the color in the tube containing the test specimen is not more intense than that in the tube containing the standard, indicating not more than 0.1% of salicylic acid.

Assay—

Mobile phase—Prepare a filtered and degassed mixture of water, tetrahydrofuran, methanol, and glacial acetic acid (65:35:10:0.1). Make adjustments if necessary (see *System Suitability* under *Chromatography* ⟨621⟩).

Standard preparation—Dissolve an accurately weighed quantity of USP Phenprocoumon RS in *Mobile phase*, and dilute quantitatively, and stepwise if necessary, with *Mobile phase* to obtain a solution having a known concentration of about 0.12 mg per mL.

Assay preparation—Transfer about 120 mg of Phenprocoumon, accurately weighed, to a 100-mL volumetric flask. Dissolve in *Mobile phase*, dilute with *Mobile phase* to volume, and mix. Dilute 1.0 mL of this solution with *Mobile phase* to 10.0 mL, and mix.

Chromatographic system (see *Chromatography* ⟨621⟩)—The liquid chromatograph is equipped with a 311-nm detector and a 3.9-mm × 30-cm column that contains packing L1. The flow rate is about 1.5 mL per minute. Chromatograph the *Standard preparation*, and record the peak responses as directed under *Procedure:* the tailing factor for the analyte peak is not more than 2.0, and the relative standard deviation for replicate injections is not more than 2.0%.

Procedure—Separately inject equal volumes (about 20 µL of the *Standard preparation* and the *Assay preparation* into the chromatograph, record the chromatograms, and measure the responses for the major peaks. Calculate the quantity, in mg, of $C_{18}H_{16}O_3$ in the portion of Phenprocoumon taken by the formula:

$$1000C(r_U/r_S),$$

in which C is the concentration, in mg per mL, of USP Phenprocoumon RS in the *Standard preparation*, and r_U and r_S are the Phenprocoumon peak responses obtained from the *Assay preparation* and the *Standard preparation*, respectively.

Phenprocoumon Tablets

» Phenprocoumon Tablets contain not less than 90.0 percent and not more than 110.0 percent of the labeled amount of $C_{18}H_{16}O_3$.

Packaging and storage—Preserve in well-closed containers.

Reference standard—*USP Phenprocoumon Reference Standard*—Dry in a suitable vacuum drying tube, using phosphorus pentoxide as the desiccant, at 100° for 4 hours before using.

Identification—

A: Extract a portion of finely powdered Tablets, equivalent to about 30 mg of phenprocoumon, with 50 mL of chloroform, and filter until clear. Evaporate the clear filtrate on a steam bath to dryness, and dry the residue at 105° for 30 minutes: the infrared absorption spectrum of a potassium bromide dispersion of the phenprocoumon so obtained exhibits maxima only at the same wavelengths as that of a similar preparation of USP Phenprocoumon RS.

B: The retention time of the major peak in the chromatogram of the *Assay preparation* corresponds to that in the chromatogram of the *Standard preparation* obtained as directed in the *Assay*.

Dissolution ⟨711⟩—

Medium: water; 900 mL.

Apparatus 2: 50 rpm.

Time: 45 minutes.

Procedure—Determine the amount of $C_{18}H_{16}O_3$ dissolved from ultraviolet absorbances at the wavelength of maximum absorbance at about 311 nm of filtered portions of the solution under test, suitably diluted with *Dissolution Medium*, if necessary, in comparison with a Standard solution having a known concentration of USP Phenprocoumon RS in the same medium.

Tolerances—Not less than 75% (*Q*) of the labeled amount of $C_{18}H_{16}O_3$ is dissolved in 45 minutes.

Uniformity of dosage units ⟨905⟩: meet the requirements.

Assay—

Mobile phase, *Standard preparation*, and *Chromatographic system*—Prepare as directed in the *Assay* under *Phenprocoumon*.

Assay preparation—Weigh and finely powder not less than 20 Phenprocoumon Tablets. Transfer an accurately weighed portion of the powder, equivalent to about 24 mg of Phenprocoumon, to a 200-mL volumetric flask, add about 100 mL of *Mobile phase*, sonicate to dissolve, and cool. Dilute with *Mobile phase* to volume, mix, and filter.

Procedure—Proceed as directed for *Procedure* in the *Assay* under *Phenprocoumon*. Calculate the quantity, in mg, of $C_{18}H_{16}O_3$ in the portion of Phenprocoumon Tablets taken by the formula:

$$200C(r_U/r_S),$$

in which C is the concentration, in mg per mL, of USP Phenprocoumon RS in the *Standard preparation*, and r_U and r_S are the phenprocoumon peak responses obtained from the *Assay preparation* and the *Standard preparation*, respectively.

Phensuximide

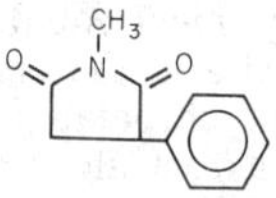

$C_{11}H_{11}NO_2$ 189.21

2,5-Pyrrolidinedione, 1-methyl-3-phenyl-.

N-Methyl-2-phenylsuccinimide [*86-34-0*].

» Phensuximide contains not less than 97.0 percent and not more than 103.0 percent of $C_{11}H_{11}NO_2$, calculated on the anhydrous basis.

Packaging and storage—Preserve in tight containers.

Reference standard—*USP Phensuximide Reference Standard*—Dry in vacuum at 50° for 4 hours before using.

Identification—

A: The infrared absorption spectrum of a potassium bromide dispersion of it, previously dried in vacuum at 50° for 4 hours, exhibits maxima only at the same wavelengths as that of a similar preparation of USP Phensuximide RS.

B: The ultraviolet absorption spectrum of a 1 in 2500 solution in alcohol exhibits maxima and minima at the same wavelengths as that of a similar solution of USP Phensuximide RS, concomitantly measured.

Melting range, *Class I* ⟨741⟩: between 68° and 74°.

Water, *Method I* ⟨921⟩: not more than 1.0%.

Residue on ignition ⟨281⟩: not more than 0.5%.

Cyanide—Dissolve 1.0 g in 10 mL of warm alcohol, and add 3 drops of ferrous sulfate TS, 1 mL of 1 *N* sodium hydroxide, and a few drops of ferric chloride TS. Warm gently, and finally acidify with 2 *N* sulfuric acid: no blue precipitate or blue color is formed within 15 minutes.

Assay—Transfer about 200 mg of Phensuximide, accurately weighed, to a 50-mL volumetric flask. Dissolve in 40 mL of alcohol, dilute with alcohol to volume, and mix. Transfer 5.0 mL of this solution to a 50-mL volumetric flask, dilute with alcohol to volume, and mix. Concomitantly determine the absorbances of this solution and of a Standard solution of USP Phensuximide RS, in the same medium having a known concentration of about 400 µg per mL, in 1-cm cells at the wavelength of maximum absorbance at about 258 nm, with a suitable spectrophotometer, using alcohol as the blank. Calculate the quantity, in mg, of $C_{11}H_{11}NO_2$ in the Phensuximide taken by the formula:

$$0.5C(A_U/A_S),$$

in which C is the concentration, in µg per mL, of USP Phensuximide RS in the Standard solution, and A_U and A_S are the absorbances of the solution from Phensuximide and the Standard solution, respectively.

Phensuximide Capsules

» Phensuximide Capsules contain not less than 93.0 percent and not more than 107.0 percent of the labeled amount of $C_{11}H_{11}NO_2$.

Packaging and storage—Preserve in tight containers.

Reference standard—*USP Phensuximide Reference Standard*—Dry in vacuum at 50° for 4 hours before using.

Identification—The contents of Capsules respond to *Identification test A* under *Phensuximide*.

Dissolution ⟨711⟩—

Medium: water; 900 mL.

Apparatus 1: 100 rpm.

Time: 120 minutes.

Procedure—Determine the amount of $C_{11}H_{11}NO_2$ dissolved, employing the procedure set forth in the *Assay*, making any necessary modifications.

Tolerances—Not less than 75% (*Q*) of the labeled amount of $C_{11}H_{11}NO_2$ is dissolved in 120 minutes.

Uniformity of dosage units ⟨905⟩: meet the requirements.

Assay—Transfer, as completely as possible, the contents of not less than 20 Phensuximide Capsules to a suitable tared container, and weigh. Mix with a glass rod. Weigh accurately a portion of the powder, equivalent to about 200 mg of phensuximide, and transfer to a 125-mL separator. Add 20 mL of water, and extract with three 40-mL portions of chloroform, filtering each successive chloroform extract through a chloroform-washed pledget of cotton into a 150-mL extraction flask. Evaporate the combined chloroform extracts on a steam bath to about 3 mL, and then allow the remaining chloroform to evaporate at room temperature. Dissolve the residue of phensuximide in 20 mL of alcohol, transfer to a 50-mL volumetric flask, and dilute with alcohol to volume. Transfer 5.0 mL of this solution to a 50-mL volumetric flask, dilute with alcohol to volume, and mix. Concomitantly determine the absorbances of this solution and of a Standard solution of USP Phensuximide RS in the same medium having a known concentration of about 400 µg per mL, in 1-cm cells at the wavelength of maximum absorbance at about 258 nm, with a suitable spectrophotometer, using alcohol as the blank. Calculate the quantity, in mg, of $C_{11}H_{11}NO_2$ in the portion of Capsules taken by the formula:

$$0.5C(A_U/A_S),$$

in which C is the concentration, in µg per mL, of USP Phensuximide RS in the Standard solution, and A_U and A_S are the absorbances of the solution from the Tablets and the Standard solution, respectively.

Phentermine Hydrochloride

$C_{10}H_{15}N.HCl$ 185.7

Benzeneethanamine, α,α-dimethyl-, hydrochloride.

α,α-Dimethylphenethylamine hydrochloride [*1197-21-3*].

» Phentermine Hydrochloride contains not less than 98.0 percent and not more than 101.0 percent of $C_{10}H_{15}N.HCl$, calculated on the dried basis.

Packaging and storage—Preserve in tight containers.

Reference standard—*USP Phentermine Hydrochloride Reference Standard*—Dry at 105° for 3 hours before using.

Identification—

A: The infrared absorption spectrum of a potassium bromide dispersion of it, previously dried, exhibits maxima only at the same wavelength as that of a similar preparation of USP Phentermine Hydrochloride RS.

B: The ultraviolet absorption spectrum of a 6 in 10,000 solution in 0.1 *N* hydrochloric acid exhibits maxima and minima at the same wavelengths as that of a similar solution of USP Phentermine Hydrochloride RS, concomitantly measured, and the respective absorptivities, calculated on the dried basis, at the wavelength of maximum absorbance at about 256 nm do not differ by more than 2.0%.

C: It responds to the tests for *Chloride* ⟨191⟩.

Melting range ⟨741⟩: between 202° and 205°.

pH ⟨791⟩: between 5.0 and 6.0, in a solution (1 in 50).

Loss on drying ⟨731⟩—Dry it at 105° for 3 hours: it loses not more than 2.0% of its weight.

Residue on ignition ⟨281⟩: not more than 0.1%.

Chromatographic purity—

Standard preparations—Dissolve an accurately weighed quantity of USP Phentermine Hydrochloride RS in chloroform to obtain *Standard preparation A* having a known concentration of 2 mg per mL. Dilute this solution quantitatively with chloroform to obtain *Standard preparations*, designated below by letter, having the following compositions:

Dilution	Concentration (μg RS per mL)	Percentage (%, for comparison with test specimen)
A (undiluted)	2	1.0
B (1 in 2)	1	0.5
C (1 in 5)	0.4	0.2
D (1 in 10)	0.2	0.1

Test preparation—Dissolve an accurately weighed quantity of Phentermine Hydrochloride in chloroform to obtain a solution containing 200 mg per mL.

Procedure—On a suitable thin-layer chromatographic plate (see *Chromatography* ⟨621⟩), coated with a 0.25-mm layer of chromatographic silica gel mixture, apply separately 10 μL of the *Test preparation* and 10 μL of each *Standard preparation.* Position the plate in a chromatographic chamber, and develop the chromatograms in a solvent system consisting of a mixture of chloroform, cyclohexane, and diethylamine (50:40:10) until the solvent front has moved about three-fourths of the length of the plate. Remove the plate from the developing chamber, mark the solvent front, and allow the solvent to evaporate in air. Examine the plate under short-wavelength ultraviolet light. Compare the intensities of any secondary spots observed in the chromatogram of the *Test preparation* with those of the principal spots in the chromatograms of the *Standard preparations:* the sum of the intensities of secondary spots obtained from the *Test preparation* corresponds to not more than 1.0% of related compounds, with no single impurity corresponding to more than 0.5%.

Assay—Dissolve about 400 mg of Phentermine Hydrochloride, accurately weighed, in 40 mL of glacial acetic acid, and add 10 mL of mercuric acetate TS, warming slightly to effect solution. Cool to room temperature, and titrate with 0.1 *N* perchloric acid VS, determining the end-point potentiometrically. Perform a blank determination, and make any necessary correction. Each mL of 0.1 *N* perchloric acid is equivalent to 18.57 mg of $C_{10}H_{15}N.HCl$.

Phentermine Hydrochloride Capsules

» Phentermine Hydrochloride Capsules contain not less than 90.0 percent and not more than 110.0 percent of the labeled amount of $C_{10}H_{15}N.HCl$.

Packaging and storage—Preserve in tight containers.

Reference standard—*USP Phentermine Hydrochloride Reference Standard*—Dry at 105° for 3 hours before using.

Identification—The retention time of the major peak in the chromatogram of the *Assay preparation* corresponds to that of the *Standard preparation*, as obtained in the *Assay*.

Dissolution ⟨711⟩—

Medium: water; 900 mL. Use 500 mL for Capsules containing 15 mg of phentermine hydrochloride or less.

Apparatus 2: 50 rpm.

Time: 45 minutes.

Procedure—Determine the amount of $C_{10}H_{15}N.HCl$ dissolved, employing the procedure set forth in the *Assay*, making any necessary modifications including concentration of the analyte in the volume of test solution taken.

Tolerances—Not less than 75% (*Q*) of the labeled amount of $C_{10}H_{15}N.HCl$ is dissolved in 45 minutes.

Uniformity of dosage units ⟨905⟩: meet the requirements.

Assay—

Mobile phase—Dissolve 1.1 g of sodium 1-heptanesulfonate in 575 mL of water. Add 25 mL of dilute glacial acetic acid (14 in 100) and 400 mL of methanol. Adjust dropwise, if necessary, with glacial acetic acid to a pH of 3.3 ± 0.1. Filter through a 0.5-μm membrane filter. The volume of methanol may be adjusted to provide a suitable retention time for phentermine hydrochloride (about 8 minutes).

Standard preparation—Using an accurately weighed quantity of USP Phentermine Hydrochloride RS, prepare a solution in 0.04 *M* phosphoric acid having a known concentration of about 0.4 mg per mL.

Assay preparation—Remove, as completely as possible, the contents of not less than 20 Phentermine Hydrochloride Capsules, and weigh. Transfer an accurately weighed portion of the mixed powder, equivalent to about 20 mg of phentermine hydrochloride, to a 50-mL volumetric flask. Add 40 mL of 0.04 *M* phosphoric acid, and sonicate for 15 minutes. Dilute with 0.04 *M* phosphoric acid to volume, and mix. Filter through a 0.5-μm membrane filter, discarding the first few mL of the filtrate.

Chromatographic system (see *Chromatography* ⟨621⟩)—The liquid chromatograph is equipped with a 254-nm detector and a 3.9-mm × 30-cm column that contains packing L1. The flow rate is about 2 mL per minute. Chromatograph three replicate injections of the *Standard preparation*, and record the peak response as directed under *Procedure:* the relative standard deviation is not more than 2.0%.

Procedure—Separately inject equal volumes (about 50 μL) of the *Standard preparation* and the *Assay preparation* into the chromatograph by means of a suitable sampling valve, record the chromatograms, and measure the responses for the major peaks. Calculate the quantity, in mg, of $C_{10}H_{15}N.HCl$ in the portion of Capsules taken by the formula:

$$50C(r_U/r_S),$$

in which C is the concentration, in mg per mL, of USP Phentermine Hydrochloride RS in the *Standard preparation*, and r_U and r_S are the peak responses obtained from the *Assay preparation* and the *Standard preparation*, respectively.

Phentermine Hydrochloride Tablets

» Phentermine Hydrochloride Tablets contain not less than 90.0 percent and not more than 110.0 percent of the labeled amount of $C_{10}H_{15}N.HCl$.

Packaging and storage—Preserve in tight containers.

Reference standard—*USP Phentermine Hydrochloride Reference Standard*—Dry at 105° for 3 hours before using.

Identification—The retention time of the major peak in the chromatogram of the *Assay preparation* corresponds to that of the *Standard preparation*, as obtained in the *Assay*.

Dissolution ⟨711⟩—

Medium: water; 900 mL. Use 500 mL for Tablets containing 15 mg of phentermine hydrochloride or less.

Apparatus 2: 50 rpm.

Time: 45 minutes.

Procedure—Determine the amount of $C_{10}H_{15}N.HCl$ dissolved, employing the procedure set forth in the *Assay*, making any necessary modifications including concentration of the analyte in the volume of test solution taken.

Tolerances—Not less than 75% (*Q*) of the labeled amount of $C_{10}H_{15}N.HCl$ is dissolved in 45 minutes.

Uniformity of dosage units ⟨905⟩: meet the requirements.

Assay—Proceed as directed in the *Assay* under *Phentermine Hydrochloride Capsules*, except to read "Tablets" in place of "Capsules" and to use not less than 20 finely powdered Tablets for the *Assay preparation.*

Phentolamine Mesylate

$C_{17}H_{19}N_3O.CH_4O_3S$ 377.46

Phenol, 3-[[(4,5-dihydro-1*H*-imidazol-2-yl)methyl](4-methylphenyl)amino]-, monomethanesulfonate (salt).

m-[*N*-(2-Imidazolin-2-ylmethyl)-*p*-toluidino]phenol monomethanesulfonate (salt) [*65-28-1*].

» Phentolamine Mesylate contains not less than 98.0 percent and not more than 102.0 percent of $C_{17}H_{19}N_3O.CH_4O_3S$, calculated on the dried basis.

Packaging and storage—Preserve in tight, light-resistant containers.

Reference standard—*USP Phentolamine Mesylate Reference Standard*—Dry in vacuum at 60° for 4 hours before using.

Identification—

A: The infrared absorption spectrum of a mineral oil dispersion of it exhibits maxima only at the same wavelengths as that of a similar preparation of USP Phentolamine Mesylate RS.

B: The ultraviolet absorption spectrum of a solution of it (1 in 50,000) exhibits maxima and minima at the same wavelengths as that of a similar solution of USP Phentolamine Mesylate RS, concomitantly measured.

C: The R_f value of the principal spot in the chromatogram of the *Identification preparation* corresponds to that of *Standard preparation A*, as obtained in the test for *Chromatographic purity.*

Loss on drying ⟨731⟩—Dry it in vacuum at 60° for 4 hours: it loses not more than 0.5% of its weight.

Residue on ignition ⟨281⟩: not more than 0.1%.

Sulfate ⟨221⟩—A 0.10-g portion shows no more sulfate than corresponds to 0.20 mL of 0.020 *N* sulfuric acid (0.2%).

Chromatographic purity—

Standard preparations—Dissolve USP Phentolamine Mesylate RS in methanol, and mix to obtain *Standard preparation A* having a known concentration of 50 µg per mL. Dilute quantitatively with methanol to obtain *Standard preparations*, designated below by letter, having the following compositions:

Standard preparation	Dilution	Concentration (µg RS per mL)	Percentage (%, for comparison with test specimen)
A	(undiluted)	50	0.5
B	(3 in 5)	30	0.3
C	(1 in 5)	10	0.1

Test preparation—Dissolve an accurately weighed quantity of Phentolamine Mesylate in methanol to obtain a solution containing 10 mg per mL.

Identification preparation—Dilute a portion of the *Test preparation* quantitatively with methanol to obtain a solution containing 50 µg per mL.

Detection reagent—Prepare (1) a solution of 1 g of potassium ferricyanide in 20 mL of water, and (2) a solution of 1.9 g of ferric chloride in 20 mL of water. Just prior to use, mix equal volumes of the solutions.

Procedure—On a suitable thin-layer chromatographic plate (see *Chromatography* ⟨621⟩), coated with a 0.25-mm layer of chromatographic silica gel, apply separately 5 µL of the *Test preparation*, 5 µL of the *Identification preparation*, and 5 µL of each *Standard preparation*, and allow to dry. Position the plate in a chromatographic chamber, and develop the chromatograms in a solvent system consisting of a mixture of chloroform, diethylamine, and methanol (15:3:2) until the solvent front has moved about three-fourths of the length of the plate. Remove the plate from the developing chamber, mark the solvent front, and dry the plate at 100° for 1 hour. Spray the plate with *Detection reagent.* Within 15 minutes after spraying, compare the intensities of any secondary spots observed in the chromatogram of the *Test preparation* with those of the principal spots in the chromatograms of the *Standard preparations:* no secondary spot from the chromatogram of the *Test preparation* is larger or more intense than the principal spot obtained from *Standard preparation A* (0.5%), and the sum of the intensities of all secondary spots obtained from the *Test preparation* corresponds to not more than 1.0%.

Assay—

0.1 N Tetrabutylammonium hydroxide in isopropyl alcohol—Dilute with dehydrated isopropyl alcohol a commercially available 25% solution of tetrabutylammonium hydroxide in methanol, and standardize as directed under *Tetrabutylammonium Hydroxide, Tenth-Normal (0.1 N)* (see *Volumetric Solutions* in the section, *Reagents, Indicators, and Solutions*), using dehydrated isopropyl alcohol instead of dimethylformamide.

Procedure—Dissolve with the aid of sonication, if necessary, about 300 mg of Phentolamine Mesylate, accurately weighed, in 100 mL of dehydrated isopropyl alcohol. Titrate in an atmosphere of nitrogen with *0.1 N Tetrabutylammonium hydroxide in isopropyl alcohol*, determining the end-point potentiometrically, using a glass electrode and a calomel electrode containing a saturated solution of tetramethylammonium chloride in dehydrated isopropyl alcohol (see *Titrimetry* ⟨541⟩). Perform a blank determination, and make any necessary correction. Each mL of 0.1 *N* tetrabutylammonium hydroxide is equivalent to 37.75 mg of $C_{17}H_{19}N_3O.CH_4O_3S$.

Phentolamine Mesylate for Injection

» Phentolamine Mesylate for Injection is sterile Phentolamine Mesylate or a sterile mixture of Phentolamine Mesylate with a suitable buffer or suitable diluents. It contains not less than 90.0 percent and not more than 110.0 percent of the labeled amount of $C_{17}H_{19}N_3O.CH_4O_3S$.

Packaging and storage—Preserve in *Containers for Sterile Solids* as described under *Injections* ⟨1⟩.

Reference standard—*USP Phentolamine Mesylate Reference Standard*—Dry in vacuum at 60° for 4 hours before using.

Constituted solution—At the time of use, the constituted solution prepared from Phentolamine Mesylate for Injection meets the requirements for *Constituted Solutions* under *Injections* ⟨1⟩.

Identification—Mix a portion of it, equivalent to about 40 mg of phentolamine mesylate, with about 15 mL of chloroform. Filter into a beaker, and evaporate to dryness, taking precautions against introducing moisture: the residue so obtained responds to *Identification test A* under *Phentolamine Mesylate.*

Uniformity of dosage units ⟨905⟩: meets the requirements.

Procedure for content uniformity—Dissolve the contents of 1 container in water to provide a solution containing about 20 µg of phentolamine mesylate per mL. Concomitantly determine the absorbances of this solution and of a solution of USP Phentolamine Mesylate RS, in the same medium, at a concentration of about 20 µg per mL, in 1-cm cells at the wavelength of maximum absorbance at about 278 nm, with a suitable spectrophotometer, using water as the blank. Calculate the quantity, in mg, of $C_{17}H_{19}N_3O.CH_4O_3S$ in the Phentolamine Mesylate for Injection taken by the formula:

$$(T/D)C(A_U/A_S),$$

in which *T* is the labeled quantity, in mg, of phentolamine mesylate in the Phentolamine Mesylate for Injection, *D* is the concentration, in µg per mL, of phentolamine mesylate in the solution from the Phentolamine Mesylate for Injection, based on the labeled quantity per container and the extent of dilution, *C* is the concentration, in µg per mL, of USP Phentolamine Mesylate RS in the Standard solution, and A_U and A_S are the absorbances of the solution from the Phentolamine Mesylate for Injection and the Standard solution, respectively.

pH ⟨791⟩: between 4.5 and 6.5, in a freshly prepared solution having a concentration of about 1 in 100.

Other requirements—It meets the requirements for *Sterility Tests* ⟨71⟩ and *Labeling* under *Injections* ⟨1⟩.

Assay—

Standard preparation—Transfer about 25 mg of USP Phentolamine Mesylate RS, accurately weighed, to a 50-mL volumetric flask, add water to volume, and mix.

Assay preparation—Dissolve the contents of 10 containers of Phentolamine Mesylate for Injection in a volume of water corresponding to the volume of solvent specified in the labeling. Transfer an aliquot, equivalent to about 25 mg of phentolamine mesylate, to a 50-mL volumetric flask, add water to volume, and mix.

Procedure—Pipet 5-mL portions, respectively, of the *Standard preparation*, *Assay preparation*, and water to provide a blank, into separate 125-mL separators. Into each separator pipet 5-mL portions of 0.1 *N* hydrochloric acid and saturated picric acid solution. Extract with three 25-mL portions of chloroform, filtering each portion through chloroform-washed cotton into a 100-mL volumetric flask. Dilute with chloroform to volume, and mix. Concomitantly determine the absorbances of the solutions from the *Assay preparation* and the *Standard preparation* in 1-cm cells at the wavelength of maximum absorbance at about 410 nm, with a suitable spectrophotometer, against the blank. Calculate the quantity, in mg, of $C_{17}H_{19}N_3O \cdot CH_4O_3S$ in the aliquot of Phentolamine Mesylate for Injection taken by the formula:

$$50C(A_U/A_S),$$

in which *C* is the concentration, in mg per mL, of USP Phentolamine Mesylate RS in the *Standard preparation*, and A_U and A_S are the absorbances of the solutions from the *Assay preparation* and the *Standard preparation*, respectively.

Phenylalanine

$C_9H_{11}NO_2$ 165.19
L-Phenylalanine.
L-Phenylalanine [*63-91-2*].

» Phenylalanine contains not less than 98.5 percent and not more than 101.5 percent of $C_9H_{11}NO_2$, as L-phenylalanine, calculated on the dried basis.

Packaging and storage—Preserve in well-closed containers.

Reference standard—*USP L-Phenylalanine Reference Standard*—Dry at 105° for 3 hours before using.

Identification—The infrared absorption spectrum of a potassium bromide dispersion of it, previously dried, exhibits maxima only at the same wavelengths as that of a similar preparation of USP L-Phenylalanine RS.

Specific rotation ⟨781⟩: between −32.7° and −34.7°, calculated on the dried basis, determined in a solution containing 200 mg in each 10 mL.

pH ⟨791⟩: between 5.4 and 6.0, in a solution (1 in 100).

Loss on drying ⟨731⟩—Dry it at 105° for 3 hours: it loses not more than 0.3% of its weight.

Residue on ignition ⟨281⟩: not more than 0.4%.

Chloride ⟨221⟩—A 0.73-g portion shows no more chloride than corresponds to 0.50 mL of 0.020 *N* hydrochloric acid (0.05%).

Sulfate ⟨221⟩—A 0.33-g portion shows no more sulfate than corresponds to 0.10 mL of 0.020 *N* sulfuric acid (0.03%).

Arsenic ⟨211⟩: 1.5 ppm.

Iron ⟨241⟩: 0.003%.

Heavy metals, *Method I* ⟨231⟩: 0.0015%.

Assay—Transfer about 160 mg of Phenylalanine, accurately weighed, to a 125-mL flask, dissolve in a mixture of 3 mL of formic acid and 50 mL of glacial acetic acid, and titrate with 0.1 *N* perchloric acid VS, determining the end-point potentiometrically. Perform a blank determination, and make any necessary correction. Each mL of 0.1 *N* perchloric acid is equivalent to 16.52 mg of $C_9H_{11}NO_2$.

Phenylbutazone

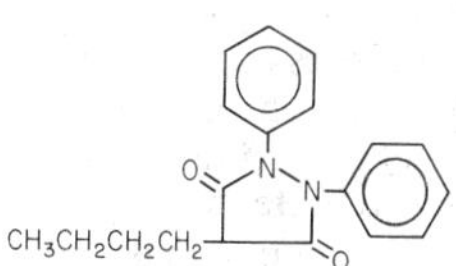

$C_{19}H_{20}N_2O_2$ 308.38
3,5-Pyrazolidinedione, 4-butyl-1,2-diphenyl-.
4-Butyl-1,2-diphenyl-3,5-pyrazolidinedione [*50-33-9*].

» Phenylbutazone contains not less than 98.0 percent and not more than 100.5 percent of $C_{19}H_{20}N_2O_2$, calculated on the dried basis.

Packaging and storage—Preserve in tight containers.

Reference standard—*USP Phenylbutazone Reference Standard*—Dry in vacuum at a pressure of 30 ± 10 mm of mercury at 80° for 4 hours before using.

Identification—

A: The infrared absorption spectrum of a potassium bromide dispersion of it exhibits maxima only at the same wavelengths as that of a similar preparation of USP Phenylbutazone RS.

B: The ultraviolet absorption spectrum of a 1 in 100,000 solution of it in sodium hydroxide solution (1 in 2500) exhibits maxima and minima at the same wavelengths as that of a similar solution of USP Phenylbutazone RS, concomitantly measured, and the respective absorptivities, calculated on the dried basis, at the wavelength of maximum absorbance at about 264 nm do not differ by more than 2.0%.

Melting range ⟨741⟩: between 104° and 107°.

Loss on drying ⟨731⟩—Dry it in vacuum at a pressure of 30 ± 10 mm of mercury at 80° for 4 hours: it loses not more than 0.5% of its weight.

Residue on ignition ⟨281⟩: not more than 0.1%, 2.0 g being used for the test.

Chloride ⟨221⟩—Boil 2.0 g with 60 mL of water for 5 minutes, cool, and filter. To a 30-mL portion of the filtrate add 1 mL of 2 *N* nitric acid and 1 mL of silver nitrate TS: the filtrate shows no more chloride than corresponds to 0.10 mL of 0.020 *N* hydrochloric acid (0.007%).

Sulfate ⟨221⟩—To a 30-mL portion of the filtrate obtained in the test for *Chloride* add 2 mL of barium chloride TS: the mixture shows no more sulfate than corresponds to 0.10 mL of 0.020 *N* sulfuric acid (0.01%).

Heavy metals, *Method II* ⟨231⟩: 0.001%.

Assay—

Acetate buffer—Transfer 2.72 g of sodium acetate (trihydrate) to a 1000-mL beaker, and dissolve in about 700 mL of water. Adjust with glacial acetic acid to a pH of 4.1. Filter through a 0.5-μm filter, dilute with filtered water to 1000 mL, and mix.

Mobile phase—Mix 440 mL of acetonitrile with 560 mL of *Acetate buffer*, and degas. Make adjustments if necessary (see *System Suitability* under *Chromatography* ⟨621⟩).

Internal standard solution—Dissolve 300 mg of desoxycorticosterone acetate in 200 mL of acetonitrile, and mix.

Standard preparation—Dissolve an accurately weighed quantity of USP Phenylbutazone RS in acetonitrile, with the aid of sonication, and dilute quantitatively with acetonitrile to obtain a solution having a concentration of about 1.4 mg per mL. Pipet 10 mL of this solution into a 50-mL volumetric flask, add 10.0 mL of *Internal standard solution*, dilute with acetonitrile to volume, and mix. [*Note—Use this solution within 8 hours of its preparation.*]

Assay preparation—Transfer about 140 mg of Phenylbutazone, accurately weighed, to a 100-mL volumetric flask, add 75 mL of acetonitrile, and sonicate to dissolve. Dilute with acetonitrile to volume, and mix. Pipet 10 mL of this solution into a 50-mL volumetric flask, add 10.0 mL of *Internal standard solution*, dilute with acetonitrile to volume, and mix. [*NOTE—Use this solution within 8 hours of its preparation.*]

Chromatographic system (see *Chromatography* ⟨621⟩)—The liquid chromatograph is equipped with a 254-nm detector and a 4.6-mm × 25-cm column that contains packing L7, preceded by a pre-column that contains packing L2. The flow rate is about 2.4 mL per minute. Chromatograph the *Standard preparation*, and record the peak responses as directed under *Procedure*. The resolution, *R*, of phenylbutazone and the internal standard is not less than 3.5, and the relative standard deviation of the ratio of their peak responses in replicate injections is not more than 2.0%.

Procedure—Separately inject equal volumes (about 25 µL) of the *Standard preparation* and the *Assay preparation* into the chromatograph, record the chromatograms, and measure the responses for the major peaks. The relative retention times are about 1.0 for the internal standard and 0.7 for phenylbutazone. Calculate the quantity, in mg, of $C_{19}H_{20}N_2O_2$ in the portion of Phenylbutazone taken by the formula:

$$500C(R_U/R_S),$$

in which C is the concentration, in mg per mL, of USP Phenylbutazone RS in the *Standard preparation*, and R_U and R_S are the ratios of the peak response of the phenylbutazone to that of the internal standard for the *Assay preparation* and the *Standard preparation*, respectively.

Phenylbutazone Capsules

» Phenylbutazone Capsules contain not less than 90.0 percent and not more than 110.0 percent of the labeled amount of $C_{19}H_{20}N_2O_2$.

Packaging and storage—Preserve in tight containers.

Reference standard—*USP Phenylbutazone Reference Standard*—Dry in vacuum at a pressure of 30 ± 10 mm of mercury at 80° for 4 hours before using.

Identification—Transfer to a 250-mL conical flask a portion of Capsule contents, equivalent to about 500 mg of phenylbutazone, and 100 mL of solvent hexane, and heat the mixture under reflux for 15 minutes. Filter the hot mixture, and allow the filtrate to cool. Separate the crystals thus formed by filtration, and dry in vacuum at 80° for 30 minutes: the phenylbutazone so obtained responds to *Identification test A* under *Phenylbutazone*, and melts between 101° and 107° when determined by the method for *Class Ia* (see *Melting Range or Temperature* ⟨741⟩).

Dissolution ⟨711⟩—

Medium: pH 7.5 phosphate buffer; 900 mL.

Apparatus 1: 100 rpm.

Time: 30 minutes.

Procedure—Determine the amount in solution on filtered portions of the solution under test suitably diluted with *Dissolution Medium*, if necessary, in 1-cm cells at the wavelength of maximum absorbance at about 264 nm, with a suitable spectrophotometer, using *Dissolution Medium* as the blank, in comparison with a solution having a known concentration of USP Phenylbutazone RS in the same medium.

Tolerances—Not less than 60% (*Q*) of the labeled amount of $C_{19}H_{20}N_2O_2$ is dissolved in 30 minutes.

Uniformity of dosage units ⟨905⟩: meet the requirements.

Procedure for content uniformity—Proceed as directed in the test for *Uniformity of dosage units* under *Phenylbutazone Tablets*, except to use the contents of 1 Capsule and to read "Capsule" instead of "Tablet" throughout.

Assay—

Acetate buffer, Mobile phase, Internal standard solution, Standard preparation, and *Chromatographic system*—Proceed as directed in the *Assay* under *Phenylbutazone*.

Assay preparation—Remove, as completely as possible, the contents of not less than 20 Phenylbutazone Capsules. Mix the combined contents, weigh accurately a portion of the contents, equivalent to about 140 mg of phenylbutazone, and transfer to a suitable flask. Pipet 100 mL of acetonitrile into the flask, and sonicate until insoluble material is dispersed into fine particles. Shake by mechanical means for 20 minutes, centrifuge a portion of this solution, and pipet 10 mL into a 50-mL volumetric flask. Add 10.0 mL of *Internal standard solution*, dilute with acetonitrile to volume, and mix. Filter a portion through a 0.5-µm filter, discarding the first few mL of the filtrate. [*Note—Use this solution within 8 hours of its preparation.*]

Procedure—Proceed as directed for *Procedure* in the *Assay* under *Phenylbutazone*. Calculate the quantity, in mg, of $C_{19}H_{20}N_2O_2$ in the portion of Capsules taken by the formula:

$$500C(R_U/R_S),$$

in which C is the concentration, in mg per mL, of USP Phenylbutazone RS in the *Standard preparation*, and R_U and R_S are the ratios of the peak response of the phenylbutazone to that of the internal standard for the *Assay preparation* and the *Standard preparation*, respectively.

Phenylbutazone Tablets

» Phenylbutazone Tablets contain not less than 93.0 percent and not more than 107.0 percent of the labeled amount of $C_{19}H_{20}N_2O_2$.

Packaging and storage—Preserve in tight containers.

Reference standard—*USP Phenylbutazone Reference Standard*—Dry in vacuum at a pressure of 30 ± 10 mm of mercury at 80° for 4 hours before using.

Identification—Transfer to a 250-mL conical flask a portion of powdered Tablets, equivalent to about 500 mg of phenylbutazone, add 100 mL of solvent hexane, and heat the mixture under reflux for 15 minutes. Filter the hot mixture, and allow the filtrate to cool. Separate the crystals thus formed by filtration, and dry in vacuum at 80° for 30 minutes: the phenylbutazone so obtained responds to *Identification test A* under *Phenylbutazone*.

Dissolution ⟨711⟩—

Medium: simulated intestinal fluid TS (without the enzyme); 900 mL.

Apparatus 1: 100 rpm.

Time: 30 minutes.

Procedure—Determine the amount in solution on filtered portions of the *Dissolution Medium*, suitably diluted with simulated intestinal fluid TS (without the enzyme), if necessary, in 1-cm cells at the wavelength of maximum absorbance at about 264 nm, with a suitable spectrophotometer, using simulated intestinal fluid TS (without the enzyme) as the blank, in comparison with a solution of known concentration of USP Phenylbutazone RS in the same medium.

Tolerances—Not less than 60% (*Q*) is dissolved in 30 minutes.

Uniformity of dosage units ⟨905⟩: meet the requirements.

Procedure for content uniformity—Transfer 1 Tablet to a 100-mL volumetric flask, add 60 mL of methanol, and shake by mechanical means for about 20 minutes or until the tablet is completely disintegrated. Dilute with methanol to volume, and mix. Filter a portion of mixture, discarding the first 10 mL of the filtrate. Dilute an accurately measured portion of the filtrate with sodium hydroxide solution (1 in 2500) to obtain a solution containing about 10 µg per mL. Prepare a solution of USP Phenylbutazone RS in methanol having a known concentration of about 1 mg per mL. Quantitatively dilute a portion of this solution with sodium hydroxide solution (1 in 2500) to obtain a Standard solution having a final known concentration of about 10 µg per mL. Concomitantly determine the absorbances of the solution from the Tablet and the Standard solution at the wavelength of maximum absorbance at about 264 nm, with a suitable spectrophotometer, using sodium hydroxide solution (1 in 2500) as the blank. Calculate the quantity, in mg, of $C_{19}H_{20}N_2O_2$ in the Tablet by the formula:

$$(TC/D)(A_U/A_S),$$

in which T is the labeled quantity, in mg, of phenylbutazone in the Tablet, C is the concentration, in µg per mL, of USP Phenylbutazone RS in the Standard solution, D is the concentration, in µg per mL, of phenylbutazone in the solution from the Tablet based on the labeled quantity per Tablet and the extent of dilution, and A_U and A_S are the absorbances of the solution from the Tablet and the Standard solution, respectively.

Assay—

Acetate buffer, Mobile phase, Internal standard solution, Standard preparation, and *Chromatographic system*—Proceed as directed in the *Assay* under *Phenylbutazone.*

Assay preparation—Weigh and finely powder not less than 20 Phenylbutazone Tablets. Weigh accurately a portion of the powder, equivalent to about 500 mg of phenylbutazone, and transfer to a 250-mL volumetric flask. Pipet 50 mL of water into the flask, and shake by mechanical means for 15 minutes. Add about 120 mL of acetonitrile, and sonicate until insoluble material is dispersed into fine particles. Shake by mechanical means for 20 minutes, dilute with acetonitrile to volume, and mix. Centrifuge a portion of this solution. Pipet 7 mL of the solution into a 50-mL volumetric flask, add 10.0 mL of *Internal standard solution,* dilute with acetonitrile to volume, and mix. Filter a portion through a 0.5-µm filter, discarding the first few mL of the filtrate. [*Note—Use this solution within 8 hours of its preparation.*]

Procedure—Proceed as directed for *Procedure* in the *Assay* under *Phenylbutazone.* Calculate the quantity, in mg, of $C_{19}H_{20}N_2O_2$ in the portion of Tablets taken by the formula:

$$1786C(R_U/R_S),$$

in which C is the concentration, in mg per mL, of USP Phenylbutazone RS in the *Standard preparation,* and R_U and R_S are the ratios of the peak response of the phenylbutazone to that of the internal standard for the *Assay preparation* and the *Standard preparation,* respectively.

Phenylephrine Bitartrate Inhalation Aerosol, Isoproterenol Hydrochloride and—*see* Isoproterenol Hydrochloride and Phenylephrine Bitartrate Inhalation Aerosol

Phenylephrine Hydrochloride

OH | C(—CH₂NHCH₃)(H) on 3-hydroxyphenyl · HCl

$C_9H_{13}NO_2 \cdot HCl$ 203.67

Benzenemethanol, 3-hydroxy-α-[(methylamino)methyl]-, hydrochloride (*R*)-.

(−)-*m*-Hydroxy-α-[(methylamino)methyl]benzyl alcohol hydrochloride [*61-76-7*].

» Phenylephrine Hydrochloride contains not less than 97.5 percent and not more than 102.5 percent of $C_9H_{13}NO_2 \cdot HCl$, calculated on the dried basis.

Packaging and storage—Preserve in tight, light-resistant containers.

Reference standard—*USP Phenylephrine Hydrochloride Reference Standard*—Dry at 105° for 2 hours before using.

Identification—

A: The infrared absorption spectrum of a potassium bromide dispersion of it exhibits maxima only at the same wavelengths as that of a similar preparation of USP Phenylephrine Hydrochloride RS.

B: A solution (1 in 100) responds to the tests for *Chloride* ⟨191⟩.

Melting range ⟨741⟩: between 140° and 145°.

Specific rotation ⟨781⟩: between −42° and −47.5°, calculated on the dried basis, determined in a solution containing 500 mg in each 10 mL.

Loss on drying ⟨731⟩—Dry it at 105° for 2 hours: it loses not more than 1.0% of its weight.

Residue on ignition ⟨281⟩: not more than 0.2%.

Sulfate ⟨221⟩—A solution of 50 mg in 25 mL of water shows no more turbidity than corresponds to 0.10 mL of 0.020 *N* sulfuric acid (0.20%).

Ketones—Dissolve 200 mg in 1 mL of water, add 2 drops of sodium nitroferricyanide TS, then add 1 mL of 1 *N* sodium hydroxide, followed by 0.6 mL of glacial acetic acid: the color of the final solution is not deeper than that obtained in a control solution prepared with 1 mL of dilute acetone (1 in 2000).

Chromatographic purity—

Standard preparations—Dissolve an accurately weighed quantity of USP Phenylephrine Hydrochloride RS in methanol to obtain a solution having a known concentration of 1 mg per mL. Dilute quantitatively with methanol to obtain *Standard preparations* having the following compositions:

Standard Preparation	Dilution	Concentration (µg RS per mL)	Percentage (%, for comparison with test specimen)
A	(1 in 2)	500	1.0
B	(1 in 4)	250	0.5
C	(1 in 10)	100	0.2
D	(1 in 20)	50	0.1

Test preparation—Dissolve an accurately weighed quantity of Phenylephrine Hydrochloride in methanol to obtain a solution containing 50 mg per mL.

Procedure—On a suitable thin-layer chromatographic plate (see *Chromatography* ⟨621⟩), coated with a 0.25-mm layer of chromatographic silica gel mixture, apply separately 5 µL of the *Test preparation* and 5 µL of each *Standard preparation.* Position the plate in a chromatographic chamber and develop the chromatograms in a solvent system consisting of a mixture of *n*-butyl alcohol, water, and formic acid (7:2:1) until the solvent front has moved about three-fourths of the length of the plate. Remove the plate from the developing chamber, mark the solvent front, and allow the solvent to evaporate in warm, circulating air. Examine the plate under short-wavelength ultraviolet light. Then spray the plate with a saturated solution of *p*-nitrobenzenediazonium tetrafluoroborate followed by sodium carbonate solution (1 in 10). Compare the intensities of any secondary spots observed in the chromatogram of the *Test preparation* with those of the principal spots in the chromatograms of the *Standard preparations:* the sum of the intensities of secondary spots obtained from the *Test preparation* corresponds to not more than 1.0% of related compounds, with no single impurity corresponding to more than 0.5%.

Chloride content—Dissolve about 300 mg, accurately weighed, in 5 mL of water. Add 5 mL of glacial acetic acid and 50 mL of methanol, then add eosin Y TS, and titrate with 0.1 *N* silver nitrate VS. Each mL of 0.1 *N* silver nitrate is equivalent to 3.545 mg of Cl. Not less than 17.0% and not more than 17.7% of Cl is found, calculated on the dried basis.

Assay—Dissolve about 100 mg of Phenylephrine Hydrochloride, accurately weighed, in 20 mL of water contained in an iodine flask, add 50.0 mL of 0.1 *N* bromine VS, then add 5 mL of hydrochloric acid, and immediately insert the stopper. Shake the flask, and allow to stand for 15 minutes. Introduce quickly 10 mL of potassium iodide solution (1 in 10), allow to stand for 5 minutes, shake thoroughly, remove the stopper, and rinse it and the neck of the flask with a small quantity of water into the flask. Titrate the liberated iodine with 0.1 *N* sodium thiosulfate VS, adding 3 mL of starch TS as the end-point is approached. Perform a blank determination (see *Residual Titrations* under *Titrimetry* ⟨541⟩). Each mL of 0.1 *N* bromine is equivalent to 3.395 mg of $C_9H_{13}NO_2.HCl$.

Phenylephrine Hydrochloride Injection

» Phenylephrine Hydrochloride Injection is a sterile solution of Phenylephrine Hydrochloride in Water for Injection. It contains not less than 90.0 percent and not more than 115.0 percent of the labeled amount of $C_9H_{13}NO_2.HCl$.

Packaging and storage—Preserve in single-dose or in multiple-dose containers, preferably of Type I glass, protected from light.

Reference standard—*USP Phenylephrine Hydrochloride Reference Standard*—Dry at 105° for 2 hours before using.

Identification—Concentrate or dilute, if necessary, a suitable volume of Injection to a concentration of about 10 mg per mL. Apply 2 µL of this solution and of a Standard solution of USP Phenylephrine Hydrochloride RS, containing about 10 mg per mL, at points about 2.5 cm from the bottom edge of a suitable thin-layer chromatographic plate (see *Chromatography* ⟨621⟩), coated with a 0.25-mm layer of chromatographic silica gel. Dry the spots in a current of warm air, and develop the chromatogram in a suitable chromatographic chamber with a mixture of methanol, water, and ammonium hydroxide (72:25:3) until the solvent front has moved about 12 cm. Dry the plate in warm air, and spray it with alcoholic potassium hydroxide TS. Dry at 60° for 15 minutes, and spray the plate with *p*-nitroaniline TS: the reddish orange spot obtained from the test solution corresponds in color, size, and intensity to that obtained from the Standard solution.

pH ⟨791⟩: between 3.0 and 6.5.

Other requirements—It meets the requirements under *Injections* ⟨1⟩.

Assay—

Adsorbant—Boil a mixture of 150 g of chromatographic siliceous earth and 1000 mL of 3 *N* hydrochloric acid for 10 minutes. Cool, filter by suction, wash thoroughly with water until the last washing is neutral to methyl orange TS, dry at 105° overnight, and ignite at 500° for 2 hours.

Buffer solution—Dissolve 10.89 g of monobasic potassium phosphate in 50 mL of water, add 3.48 g of dibasic potassium phosphate, mix to dissolve, dilute with water to 100 mL, and adjust to a pH of 5.80 ± 0.05.

Standard preparation—Dissolve an accurately weighed quantity of USP Phenylephrine Hydrochloride RS in dilute sulfuric acid (1 in 350), and dilute quantitatively and stepwise with the same solvent to obtain a solution having a known concentration of about 60 µg per mL.

Assay preparation—Dilute an accurately measured volume of Phenylephrine Hydrochloride Injection, equivalent to about 50 mg of phenylephrine hydrochloride, with *Buffer solution* to 50.0 mL, and mix.

Procedure—Using a flexible spatula, mix together in a 100-mL beaker 4 g of *Adsorbant* and 3.0 mL of *Assay preparation* until a fluffy mixture is obtained. Transfer the mixture to a 25- × 200-mm chromatographic tube (see *Chromatography* ⟨621⟩) containing a pledget of fine glass wool packed in the bottom and a trap layer consisting of 2 g of *Adsorbant* mixed with 1 mL of *Buffer solution*. Compress the mixture by moderate tamping. Use the spatula to scrub the beaker with 1 g of *Adsorbant*, transfer to the column, and pack as before. Use a pledget of glass wool to wipe the beaker, the packing rod, and the spatula, then place it on top of the tube, and press it down so as to sweep the wall of the tube. Pass 50 mL of water-saturated ether through the column, and discard. Place a small separator containing about 20 mL of dilute sulfuric acid (1 in 350) under the column, and pass 50 mL of a 1 in 50 solution of bis(2-ethylhexyl)phosphoric acid in water-saturated ether through the column, then follow this with 25 mL of water-saturated ether. Rinse the tip of the chromatographic tube with ether, combine with the effluent, and shake. Drain the acid layer into a 50-mL volumetric flask, and repeat the extraction with another 20-mL portion of dilute sulfuric acid (1 in 350). Combine the acid extracts, and add dilute sulfuric acid (1 in 350) to volume. Concomitantly determine the absorbances of this solution and the *Standard preparation* in 1-cm cells at the wavelength of maximum absorbance at about 271 nm and at the wavelengths of minimum absorbance at about 292 nm and about 238 nm, with a suitable spectrophotometer, using dilute sulfuric acid (1 in 350) as the blank. Draw a baseline between the two minima, and determine the corrected absorbances for the *Standard preparation* and the treated *Assay preparation* by measuring the difference between the absorbance at the maximum and the base-line at the same wavelength. Calculate the quantity, in mg, of $C_9H_{13}NO_2.HCl$ in each mL of the Injection taken by the formula:

$$0.833(C/V)(A_U/A_S),$$

in which C is the concentration, in µg per mL, of USP Phenylephrine Hydrochloride RS in the *Standard preparation*, V is the volume, in mL, of Injection taken, and A_U and A_S are the corrected absorbances of the solution from the *Assay preparation* and the *Standard preparation*, respectively.

Phenylephrine Hydrochloride Nasal Jelly

» Phenylephrine Hydrochloride Nasal Jelly contains not less than 90.0 percent and not more than 110.0 percent of the labeled amount of $C_9H_{13}NO_2.HCl$.

Packaging and storage—Preserve in tight containers.

Reference standard—*USP Phenylephrine Hydrochloride Reference Standard*—Dry at 105° for 2 hours before using.

Identification—Dissolve a suitable quantity in water to obtain a solution having a concentration of about 60 µg per mL, and centrifuge, if necessary: the ultraviolet absorption spectrum of the solution so obtained exhibits maxima and minima at the same wavelengths as that of a similar solution of USP Phenylephrine Hydrochloride RS, concomitantly measured.

Minimum fill ⟨755⟩: meets the requirements.

Assay—

Mobile phase—Prepare a suitable degassed and filtered mixture of methanol and water (4:1).

Standard preparation—Dissolve an accurately weighed quantity of USP Phenylephrine Hydrochloride RS in sodium chloride solution (1 in 5) to obtain a solution having a known concentration of about 1 mg per mL.

Assay preparation—Transfer an accurately weighed quantity of Nasal Jelly, equivalent to about 10 mg of phenylephrine hydrochloride, to a 10-mL volumetric flask, dilute with sodium chloride solution (1 in 5) to volume, and sonicate for 2 minutes. Filter through a 0.8-µm membrane, and use the clear filtrate.

Chromatographic system (see *Chromatography* ⟨621⟩)—The liquid chromatograph is equipped with a 280-nm detector and a 4-mm × 30-cm column that contains 5- to 10-µm packing L1. The flow rate is about 2 mL per minute. Chromatograph the *Standard preparation*, and record the peak response as directed under *Procedure:* the relative standard deviation for replicate injections is not more than 2.0%.

Procedure—Separately inject equal volumes (about 10 µL) of the *Standard preparation* and the *Assay preparation* into the chromatograph, record the chromatograms, and measure the responses for the major peaks. Calculate the quantity, in mg, of $C_9H_{13}NO_2.HCl$ in the portion of Nasal Jelly taken by the formula:

$$10C(r_U/r_S),$$

in which C is the concentration, in mg per mL, of USP Phenylephrine Hydrochloride RS in the *Standard preparation*, and r_U and r_S are the peak responses obtained from the *Assay preparation* and the *Standard preparation*, respectively.

Phenylephrine Hydrochloride Nasal Solution

» Phenylephrine Hydrochloride Nasal Solution contains not less than 90.0 percent and not more than 115.0 percent of the labeled amount of $C_9H_{13}NO_2$.-HCl.

Packaging and storage—Preserve in tight, light-resistant containers.

Reference standard—*USP Phenylephrine Hydrochloride Reference Standard*—Dry at 105° for 2 hours before using.

Identification—It responds to the *Identification test* under *Phenylephrine Hydrochloride Injection.*

Assay—Using an accurately measured volume of Phenylephrine Hydrochloride Nasal Solution as the *Assay preparation,* proceed as directed in the *Assay* under *Phenylephrine Hydrochloride Injection.*

Phenylephrine Hydrochloride Ophthalmic Solution

» Phenylephrine Hydrochloride Ophthalmic Solution is a sterile, aqueous solution of Phenylephrine Hydrochloride. It contains not less than 90.0 percent and not more than 115.0 percent of the labeled amount of $C_9H_{13}NO_2.HCl$. It may contain a suitable antimicrobial agent and buffer and may contain suitable antioxidants.

Packaging and storage—Preserve in tight, light-resistant containers of not more than 15-mL size.

Reference standard—*USP Phenylephrine Hydrochloride Reference Standard*—Dry at 105° for 2 hours before using.

Identification—It responds to the *Identification test* under *Phenylephrine Hydrochloride Injection.*

Sterility—It meets the requirements under *Sterility Tests* ⟨71⟩.

pH ⟨791⟩: between 4.0 and 7.5 for buffered Ophthalmic Solution; between 3.0 and 4.5 for unbuffered Ophthalmic Solution.

Assay—Using an accurately measured volume of Phenylephrine Hydrochloride Ophthalmic Solution or suitable dilution thereof as the *Assay preparation,* proceed as directed in the *Assay* under *Phenylephrine Hydrochloride Injection.*

Phenylephrine Hydrochloride Otic Solution, Antipyrine, Benzocaine, and—*see* Antipyrine, Benzocaine, and Phenylephrine Hydrochloride Otic Solution

Phenylephrine Hydrochlorides Injection, Procaine and—*see* Procaine and Phenylephrine Hydrochlorides Injection

Phenylethyl Alcohol

C_6H_5—CH_2CH_2OH

$C_8H_{10}O$ 122.17
Benzeneethanol.
Phenethyl alcohol [*60-12-8*].

Packaging and storage—Preserve in tight, light-resistant containers, and store in a cool, dry place.

Identification—Transfer 1 mL to a dry test tube, add 500 µL of phenyl isocyanate (*Caution—Phenyl isocyanate is a strong lacrimator*), and heat on a steam bath for 5 minutes. Cool, using ice if necessary, and induce crystallization by scratching the walls of the tube with a glass rod. After crystals have formed, add about 10 mL of solvent hexane, heat to boiling for a few minutes, and filter the solution into a warm, dry test tube. Collect the crystals that form on a filter, and wash them with cool solvent hexane: the crystals of phenethyl carbanilate so obtained melt between 78° and 80° (see *Melting Range or Temperature* ⟨741⟩).

Specific gravity ⟨841⟩: between 1.017 and 1.020.

Refractive index ⟨831⟩: between 1.531 and 1.534 at 20°.

Residue on ignition ⟨281⟩—Evaporate 10 mL in a suitable crucible, and ignite to constant weight: the limit is 0.005%.

Chlorinated compounds—Wind a 1.5- × 5-cm strip of 20-mesh copper gauze around the end of a copper wire. Heat the gauze in the nonluminous flame of a Bunsen burner until it glows without coloring the flame green. Permit the gauze to cool, and heat several times until a good coat of oxide has formed. Apply with a medicine dropper 2 drops of Phenylethyl Alcohol to the cooled gauze, ignite, and permit it to burn freely in the air. Again cool the gauze, add 2 more drops of Phenylethyl Alcohol, and burn as before. Continue this process until a total of 6 drops has been added and ignited, and then hold the gauze in the outer edge of the Bunsen flame, adjusted to a height of about 4 cm: no transient green color or other color is imparted to the flame.

Aldehyde—Shake 5 mL with 5 mL of 1 *N* sodium hydroxide, and allow to stand for 1 hour: no yellow color appears in the water layer.

Phenylmercuric Acetate—*see* Phenylmercuric Acetate NF

Phenylmercuric Nitrate—*see* Phenylmercuric Nitrate NF

Phenylpropanolamine Hydrochloride

C_6H_5—CH(OH)—CH(NH_2)—CH_3 · HCl

$C_9H_{13}NO.HCl$ 187.67
Benzenemethanol, α-(1-aminoethyl)-, hydrochloride, (R*,S*)-, (±).
(±)-Norephedrine hydrochloride [*154-41-6*].

» Phenylpropanolamine Hydrochloride contains not less than 98.0 percent and not more than 101.0 percent of $C_9H_{13}NO.HCl$, calculated on the dried basis.

Packaging and storage—Preserve in tight, light-resistant containers.

Reference standards—*USP Phenylpropanolamine Hydrochloride Reference Standard*—Dry at 105° for 2 hours before using. *USP α-Aminopropiophenone Hydrochloride Reference Standard*—Keep container tightly closed and protected from light. Dry at 105° for 2 hours before using. *USP Dextroamphetamine Sulfate Reference Standard*—Dry at 105° for 2 hours before using.

Identification—

A: The infrared absorption spectrum of a potassium bromide dispersion of it, previously dried, exhibits maxima only at the same wavelengths as that of a similar preparation of USP Phenylpropanolamine Hydrochloride RS.

B: The ultraviolet absorption spectrum of a solution (1 in 2000) exhibits maxima and minima at the same wavelengths as that of a similar solution of USP Phenylpropanolamine Hydrochloride RS, concomitantly measured, and the respective absorptivities, calculated on the dried basis, at the wavelength of maximum absorbance at about 256 nm do not differ by more than 3.0%.

C: Dissolve 1 g in 10 mL of water, add 10 mL of saturated sodium carbonate solution, and mix. Separate the precipitate by vacuum filtration, using a medium-porosity, sintered-glass filter, and wash with three 5-mL portions of ice-cold water. Dry the crystals at 80° for 1 hour: the phenylpropanolamine so obtained melts between 101° and 104° (see *Melting Range or Temperature* ⟨741⟩).

Melting range, *Class I* ⟨741⟩: between 191° and 196°.

pH ⟨791⟩: between 4.2 and 5.5, in a solution (3 in 100).

Loss on drying ⟨731⟩—Dry it at 105° for 2 hours: it loses not more than 0.5% of its weight.

Residue on ignition ⟨281⟩: not more than 0.1%.

α-Aminopropiophenone hydrochloride—Transfer 2.5 g to a 25-mL volumetric flask, add dilute hydrochloric acid (1 in 120) to volume, and mix. Concomitantly determine the absorbances of this solution and a Standard solution of USP α-Aminopropiophenone Hydrochloride RS in the same medium having a known concentration of 100 µg per mL in 1-cm cells at the wavelength of maximum absorbance at about 285 nm, with a suitable spectrophotometer, using dilute hydrochloric acid (1 in 120) as the blank: the absorbance of the test solution is not greater than that of the Standard solution (0.10%).

Amphetamine hydrochloride—

Mobile phase—Mix 20 mL of tetramethylammonium hydroxide solution (1 in 10) and 5 mL of phosphoric acid, dilute with water to 1000 mL, and mix. To 956 mL of this solution add 40 mL of methanol and 4 mL of tetrahydrofuran, and mix, making necessary adjustments (see *System Suitability* under *Chromatography* ⟨621⟩). Filter the resulting solution through a 0.5-µm or finer porosity membrane filter, and degas.

Standard preparation—Dissolve accurately weighed quantities of USP Phenylpropanolamine Hydrochloride RS and USP Dextroamphetamine Sulfate RS in water to obtain a solution having known concentrations of about 100 mg of USP Phenylpropanolamine Hydrochloride RS per mL and 1 µg of USP Dextroamphetamine Sulfate RS per mL.

Resolution solution—Dissolve suitable quantities of USP Phenylpropanolamine Hydrochloride RS and USP Dextroamphetamine Sulfate RS in water to obtain a solution containing about 5 µg of each per mL.

Test preparation—Transfer about 500 mg of Phenylpropanolamine Hydrochloride, accurately weighed, to a 5-mL volumetric flask, dilute with water to volume, and mix.

Chromatographic system (see *Chromatography* ⟨621⟩)—The liquid chromatograph is equipped with a 215-nm detector and a 3.9-mm × 15-cm column that contains spherical 5-µm packing L1. The flow rate is about 1 mL per minute. Chromatograph the *Resolution solution* and the *Standard preparation* as directed under *Procedure:* the resolution, *R*, between the phenylpropanolamine and amphetamine peaks is not less than 5.0, and the relative standard deviation for replicate injections of the *Standard preparation* is not more than 3.0%. The relative retention times are 1.0 for phenylpropanolamine and between 2.0 and 3.0 for amphetamine.

Procedure—Separately inject equal volumes (about 5 µL) of the *Standard preparation* and the *Test preparation* into the chromatograph, record the chromatograms, and measure the responses for the major peaks. Calculate the percentage of amphetamine hydrochloride in the portion of Phenylpropanolamine Hydrochloride taken by the formula:

$$(343.34/368.49)(0.5C/W)(r_U/r_S),$$

in which 343.34 is twice the molecular weight of amphetamine hydrochloride, 368.49 is the molecular weight of amphetamine sulfate, *C* is the concentration, in µg per mL, of USP Dextroamphetamine Sulfate RS in the *Standard preparation, W* is the weight, in mg, of Phenylpropanolamine Hydrochloride taken to prepare the *Test preparation*, and r_U and r_S are the amphetamine peak responses obtained from the *Test preparation* and the *Standard preparation*, respectively. The limit is 0.001%.

Heavy metals, *Method I* ⟨231⟩—Dissolve 1 g in 5 mL of water, add 1 mL of 1 *N* acetic acid, and dilute with water to 25 mL: the limit is 0.002%.

Assay—Dissolve about 500 mg of Phenylpropanolamine Hydrochloride, accurately weighed, in 50 mL of glacial acetic acid. Add 10 mL of mercuric acetate TS and 2 drops of crystal violet TS, and titrate with 0.1 *N* perchloric acid VS to a green endpoint. Perform a blank determination, and make any necessary correction. Each mL of 0.1 *N* perchloric acid is equivalent to 18.77 mg of $C_9H_{13}NO.HCl$.

Phenylpropanolamine Hydrochloride Extended-release Capsules

» Phenylpropanolamine Hydrochloride Extended-release Capsules contain not less than 90.0 percent and not more than 110.0 percent of the labeled amount of $C_9H_{13}NO.HCl$.

Packaging and storage—Preserve in tight, light-resistant containers.

Reference standard—*USP Phenylpropanolamine Hydrochloride Reference Standard*—Dry at 105° for 2 hours before using.

Identification—The retention time of the phenylpropanolamine peak in the chromatogram of the *Assay preparation* corresponds to that of the *Standard preparation*, both relative to the internal standard as obtained in the *Assay*.

Drug release ⟨724⟩—

Medium: water; 1000 mL.

Apparatus 1: 100 rpm.

Times: 0.125*D* hour, 0.250*D* hour, 0.500*D* hour.

Determine the amount of $C_9H_{13}NO.HCl$ dissolved, employing the following method.

Solvent A—Dissolve 1.9 g of sodium 1-hexanesulfonate in 700 mL of water, add 50 mL of 1 *M* monobasic sodium phosphate and 20 mL of 0.25 *N* triethylammonium phosphate (prepared by mixing 500 mL of a solution containing 25.3 g of triethylamine and 500 mL of a solution containing 9.6 g of phosphoric acid), and mix. Dilute with water to 1 liter, and mix.

Mobile phase—Prepare a filtered and degassed mixture of *Solvent A* and methanol (100:82). Make adjustments if necessary (see *Chromatography* ⟨621⟩).

Chromatographic system (see *Chromatography* ⟨621⟩)—The liquid chromatograph is equipped with a 210-nm detector and a 4.6-mm × 25-cm column that contains packing L1. The flow rate is about 1.5 mL per minute. Chromatograph replicate injections of a *Standard solution*, and record the peak responses as directed under *Procedure:* the tailing factor for the analyte peak is not more than 1.5, and the relative standard deviation for replicate injections is not more than 1%.

Procedure—Inject an accurately measured volume (about 50 µL) of a filtered portion of the solution under test into the chromatograph, record the chromatogram, and measure the response for the major peak. Calculate the quantity of $C_9H_{13}NO.HCl$ dissolved by comparison with a Standard solution having a known concentration of USP Phenylpropanolamine Hydrochloride RS in the same medium and similarly chromatographed.

Tolerances—The percentages of the labeled amount of $C_9H_{13}NO.HCl$ dissolved at the times specified conform to *Acceptance Table 1*.

Time (hours)	Amount Dissolved
0.125*D*	between 15% and 45%
0.250*D*	between 40% and 70%
0.500*D*	not less than 70%

Uniformity of dosage units ⟨905⟩: meet the requirements.

Assay—

Mobile phase—[NOTE—Prepare the *Mobile phase* one day prior to use.] Prepare a filtered and degassed mixture of water, methanol, 10% tetramethylammonium hydroxide, and phosphoric acid (700:300:14:3.5). Make adjustments if necessary (see *Chromatography* ⟨621⟩).

Internal standard solution—Prepare a solution of Theophylline in methanol having a final concentration of about 0.1 mg per mL.

Standard preparation—Prepare a solution of USP Phenylpropanolamine Hydrochloride RS in *Internal standard solution* having an accurately known concentration of 3 mg per mL.

Assay preparation—Transfer the accurately weighed contents of a counted number of Phenylpropanolamine Hydrochloride Extended-release Capsules, equivalent to about 750 mg of phenylpropanolamine hydrochloride, to a container. Add 250.0 mL of *Internal standard solution*, mix, sonicate for 30 minutes, allow to stand overnight, and filter.

Chromatographic system (see *Chromatography* ⟨621⟩)—The liquid chromatograph is equipped with a 254-nm detector and a 4-mm × 30-cm column that contains packing L11. The flow rate is about 1.5 mL per minute. Chromatograph replicate injections of the *Standard preparation*, and record the peak responses as directed under *Procedure:* the relative standard deviation is not more than 2.0%, and the resolution, *R*, between phenylpropanolamine and theophylline is not less than 5.0.

Procedure—Separately inject equal volumes (about 5 µL) of the *Standard preparation* and the *Assay preparation* into the chromatograph, record the chromatograms, and measure the responses for the major peaks. The relative retention times are about 0.6 for phenylpropanolamine and 1.0 for theophylline. Calculate the quantity, in mg, of $C_9H_{13}NO.HCl$ in each of the Capsules taken by the formula:

$$(250C/N)(R_U/R_S),$$

in which *C* is the concentration, in mg per mL, of USP Phenylpropanolamine Hydrochloride RS in the *Standard preparation*, *N* is the number of Capsules taken, and R_U and R_S are the peak response ratios obtained from the *Assay* and the *Standard preparation*, respectively.

Phenylpropanolamine Hydrochloride Extended-release Tablets

» Phenylpropanolamine Hydrochloride Extended-release Tablets contain not less than 90.0 percent and not more than 110.0 percent of the labeled amount of $C_9H_{13}NO.HCl$.

Packaging and storage—Preserve in tight, light-resistant containers.

Labeling—The labeling states the in-vitro *Drug release* test conditions of times and tolerances, as directed under *Drug release.*

Reference standard—*USP Phenylpropanolamine Hydrochloride Reference Standard*—Dry at 105° for 2 hours before using.

Identification—The retention time of the phenylpropanolamine peak in the chromatogram of the *Assay preparation* corresponds to that of the *Standard preparation*, both relative to the internal standard, as obtained in the *Assay.*

Drug release ⟨724⟩—

Medium: water; 1000 mL.

Apparatus 1: 100 rpm.

Times and *Tolerances:* as specified in the *Labeling*; use *Acceptance Table 1.*

Determine the amount of $C_9H_{13}NO.HCl$ dissolved, employing the following method.

Solvent A, Mobile phase, Chromatographic system, and *Procedure*—Proceed as directed in the test for *Drug release* under *Phenylpropanolamine Hydrochloride Extended-release Capsules.*

Uniformity of dosage units ⟨905⟩: meet the requirements.

Assay—

Mobile phase, Internal standard solution, Standard preparation, and *Chromatographic system*—Proceed as directed in the *Assay* under *Phenylpropanolamine Hydrochloride Extended-release Capsules.*

Assay preparation—Transfer an accurately weighed counted number of Phenylpropanolamine Hydrochloride Extended-release Tablets, equivalent to about 750 mg of phenylpropanolamine hydrochloride, to a container. Add 250.0 mL of *Internal standard solution*, mix, sonicate for 30 minutes, allow to stand overnight, and filter.

Procedure—Proceed as directed for *Procedure* in the *Assay* under *Phenylpropanolamine Hydrochloride Extended-release Capsules.* Calculate the quantity, in mg, of $C_9H_{13}NO.HCl$ in each of the Tablets taken by the formula:

$$(250C/N)(R_U/R_S).$$

in which *C* is the concentration, in mg per mL, of USP Phenylpropanolamine Hydrochloride RS in the *Standard preparation*, *N* is the number of Tablets taken, and R_U and R_S are the peak response ratios obtained from the *Assay preparation* and the *Standard preparation*, respectively.

Phenytoin

$C_{15}H_{12}N_2O_2$ 252.27

2,4-Imidazolidinedione, 5,5-diphenyl-.

5,5-Diphenylhydantoin [*57-41-0*].

» Phenytoin contains not less than 98.5 percent and not more than 100.5 percent of $C_{15}H_{12}N_2O_2$, calculated on the dried basis.

Packaging and storage—Preserve in tight containers.

Reference standard—*USP Phenytoin Reference Standard*—Dry at 105° for 4 hours before using.

Clarity and color of solution—Dissolve 1 g in a mixture of 5 mL of 1 *N* sodium hydroxide and 20 mL of water: the solution is clear and not darker than pale yellow.

Identification—The infrared absorption spectrum of a potassium bromide dispersion of it exhibits maxima only at the same wavelengths as that of a similar preparation of USP Phenytoin RS.

Loss on drying ⟨731⟩—Dry it at 105° for 4 hours: it loses not more than 1.0% of its weight.

Heavy metals, *Method II* ⟨231⟩: 0.002%.

Benzophenone—

Standard preparation—Dissolve an accurately weighed portion of benzophenone in *n*-heptane, and dilute quantitatively and stepwise with *n*-heptane to obtain a solution having a known concentration of about 10 µg per mL.

Test preparation—Weigh accurately about 5 g of Phenytoin into a 250-mL separator containing about 90 mL of hot (60° to 70°) water. Add 1 mL of sodium hydroxide solution (1 in 2), and shake until the substance is dissolved. Cool to room temperature, and extract with two 25-mL portions of *n*-heptane, combining the extracts in a separator, and discarding the aqueous phase. Wash the combined extracts with 30 mL of 0.1 *N* sodium hydroxide, and discard the aqueous layer. Transfer the *n*-heptane layer to a 100-mL volumetric flask, dilute with *n*-heptane to volume, and mix. Pipet 10 mL of the resulting solution into a 50-mL volumetric flask, dilute with *n*-heptane to volume, and mix.

Procedure—Concomitantly determine the absorbances of the *Standard preparation* and the *Test preparation* at the wavelength of maximum absorbance at about 248 nm, with a suitable spectrophotometer, using *n*-heptane as the blank. Calculate the percentage of benzophenone by the formula:

$$50(A_U/A_S)(C/W),$$

in which A_U and A_S are the absorbances of the *Test preparation* and the *Standard preparation*, respectively, *C* is the concentration, in µg per mL, of benzophenone in the *Standard preparation*, and *W* is the weight, in mg, of Phenytoin taken: not more than 0.1% is found.

Standard solution, 1.087 is the ratio of the molecular weight of phenytoin sodium to that of phenytoin, and A_U and A_S are the absorbances of the solution from the Extended Capsule contents and the Standard solution, respectively.

Assay—Transfer, as completely as possible, the contents of not less than 20 Extended Phenytoin Sodium Capsules to a beaker. Place the emptied capsules in another beaker, add sufficient alcohol to cover them completely, and allow to stand for 30 minutes with frequent stirring. Filter into the beaker containing the contents of the capsules, and wash the capsules and the filter with alcohol. Evaporate the liquid on a steam bath to near dryness, and dissolve the residue in a mixture of 45 mL of water and 5 mL of 1 *N* sodium hydroxide. Transfer the solution to a 200-mL volumetric flask, dilute with water to volume, mix, and if not clear, filter through a dry filter into a dry flask, rejecting the first 10 mL of the filtrate. Transfer to a separator an accurately measured aliquot of the solution, equivalent to about 300 mg of phenytoin sodium, dilute with water to about 50 mL, and add 10 mL of 3 *N* hydrochloric acid. Extract with three successive portions, measuring 100 mL, 60 mL, and 30 mL, respectively, of a 1 in 2 mixture of ether and chloroform. Evaporate the combined extracts, and dry at 105° for 4 hours. The weight of the residue of phenytoin so obtained, multiplied by 1.087, represents the corresponding weight of $C_{15}H_{11}N_2NaO_2$ in the aliquot taken.

Prompt Phenytoin Sodium Capsules

» Prompt Phenytoin Sodium Capsules contain not less than 93.0 percent and not more than 107.0 percent of the labeled amount of $C_{15}H_{11}N_2NaO_2$.

Packaging and storage—Preserve in tight containers.

Labeling—Label Prompt Phenytoin Sodium Capsules with the statement, "Not for once-a-day dosing," printed immediately under the official name, in a bold and contrasting color and/or enclosed within a box.

Reference standards—*USP Phenytoin Reference Standard*—Dry at 105° for 4 hours before using. *USP Phenytoin Sodium Reference Standard*—Dry at 105° for 4 hours before using.

Identification—

A: The contents of Prompt Capsules respond to *Identification test A* under *Phenytoin Sodium.*

B: The contents of Prompt Capsules respond to the flame test for *Sodium* ⟨191⟩.

Dissolution ⟨711⟩—

Medium: water; 900 mL.

Apparatus 1: 50 rpm.

Time: 30 minutes.

Procedure—Determine the amount of $C_{15}H_{11}N_2NaO_2$ dissolved by measuring the ultraviolet absorbance at 258 nm of filtered portions of the solution under test, suitably diluted with *Dissolution Medium*, if necessary, in comparison with a Standard solution having a known concentration of USP Phenytoin Sodium RS in the same medium.

Tolerances—Not less than 85% (*Q*) of the labeled amount of $C_{15}H_{11}N_2NaO_2$ is dissolved in 30 minutes.

Uniformity of dosage units ⟨905⟩: meet the requirements.

Procedure for content uniformity—Transfer the contents of 1 Prompt Capsule as completely as possible to a small flask, and add a quantity of alcohol, accurately measured, to obtain a solution having a concentration of about 55 mg of phenytoin sodium per 100 mL. Insert the stopper in the flask, and swirl the mixture intermittently for a few minutes. Centrifuge a portion of the mixture, in excess of 25 mL, at about 2000 rpm until the solution becomes clear. Pipet 25 mL of the supernatant liquid into a 50-mL volumetric flask, add 1.0 mL of dilute hydrochloric acid (1 in 100), then add alcohol to volume, and mix. Dissolve an accurately weighed quantity of USP Phenytoin RS in a 1 in 50 mixture of dilute hydrochloric acid (1 in 100) in alcohol, and dilute quantitatively and stepwise with the same solvent to obtain a Standard solution having a known concentration of about 250 µg per mL. Concomitantly determine the absorbances of both solutions in 1-cm cells at the wavelength of maximum absorbance at about 258 nm, with a suitable spectrophotometer, using a 1 in 50 mixture of dilute hydrochloric acid (1 in 100) in alcohol as the blank. Calculate the quantity, in mg, of $C_{15}H_{11}N_2NaO_2$ in the Prompt Capsule by the formula:

$$(T/D)(C)1.087(A_U/A_S),$$

in which T is the labeled quantity, in mg, of phenytoin sodium in the Prompt Capsule, D is the concentration, in µg per mL, of phenytoin sodium in the test solution on the basis of the labeled quantity per Prompt Capsule and the extent of the dilution, C is the concentration, in µg per mL, of USP Phenytoin RS in the Standard solution, 1.087 is the ratio of the molecular weight of phenytoin sodium to that of phenytoin, and A_U and A_S are the absorbances of the solution from the Prompt Capsule contents and the Standard solution, respectively.

Assay—Transfer, as completely as possible, the contents of not less than 20 Prompt Phenytoin Sodium Capsules to a beaker. Place the emptied capsules in another beaker, add sufficient alcohol to cover them completely, and allow to stand for 30 minutes with frequent stirring. Filter into the beaker containing the contents of the capsules, and wash the capsules and the filter with alcohol. Evaporate the liquid on a steam bath to near dryness, and dissolve the residue in a mixture of 45 mL of water and 5 mL of 1 *N* sodium hydroxide. Transfer the solution to a 200-mL volumetric flask, dilute with water to volume, mix, and if not clear, filter through a dry filter into a dry flask, rejecting the first 10 mL of filtrate. Transfer to a separator an accurately measured aliquot of the solution, equivalent to about 300 mg of phenytoin sodium, dilute with water to about 50 mL, and add 10 mL of 3 *N* hydrochloric acid. Extract with three successive portions, measuring 100 mL, 60 mL, and 30 mL, respectively, of a 1 in 2 mixture of ether and chloroform. Evaporate the combined extracts, and dry at 105° for 4 hours. The weight of the residue of phenytoin so obtained, multiplied by 1.087, represents the corresponding weight of $C_{15}H_{11}N_2NaO_2$ in the aliquot taken.

Phenytoin Sodium Injection

» Phenytoin Sodium Injection is a sterile solution of Phenytoin Sodium with Propylene Glycol and Alcohol in Water for Injection. It contains not less than 90.0 percent and not more than 110.0 percent of the labeled amount of $C_{15}H_{11}N_2NaO_2$.

Note—Do not use the Injection if it is hazy or contains a precipitate.

Packaging and storage—Preserve in single-dose or in multiple-dose containers, preferably of Type I glass, at controlled room temperature.

Reference standard—*USP Phenytoin Reference Standard*—Dry at 105° for 4 hours before using.

Identification—

A: Transfer a volume of Phenytoin Sodium Injection, equivalent to about 250 mg of phenytoin sodium, to a separator containing 25 mL of water. Extract, in the order listed, with 50-mL, 30-mL, and 30-mL portions of ethyl acetate. Wash each extract with two 20-mL portions of sodium acetate solution (1 in 100). Evaporate the combined ethyl acetate extracts, and dry the residue of phenytoin at 105° to constant weight: the infrared absorption spectrum of a potassium bromide dispersion of the residue so obtained exhibits maxima only at the same wavelengths as that of a potassium bromide dispersion of USP Phenytoin RS.

B: It responds to the flame test for *Sodium* ⟨191⟩.

pH ⟨791⟩: between 10.0 and 12.3.

Alcohol and propylene glycol content—

Internal standard solution—Pipet 8 mL of methanol and 20 mL of ethylene glycol into a 100-mL volumetric flask, dilute with water to volume, and mix.

Alcohol solution—Pipet 6 mL of dehydrated alcohol into a 100-mL volumetric flask, dilute with water to volume, and mix.

Propylene glycol solution—Pipet 20 mL of propylene glycol into a 100-mL volumetric flask, dilute with water to volume, and mix.

Standard preparation—Pipet 10 mL each of *Internal standard solution, Alcohol solution,* and *Propylene glycol solution* into a 100-mL volumetric flask, dilute with water to volume, and mix.

Test preparation—Pipet 5 mL of Phenytoin Sodium Injection and 10 mL of *Internal standard solution* into a 100-mL volumetric flask, dilute with water to volume, and mix.

Chromatographic system (see *Chromatography* ⟨621⟩)—The gas chromatograph is equipped with a flame-ionization detector and a 1.8-m × 2.0-mm (ID) glass column packed with 50- to 80-mesh silanized packing S3. The column is maintained at a temperature of 140° for 3 minutes, programmed at a rate of 6° per minute to a temperature of 190°, and maintained at 190° for 6 minutes, helium being used as the carrier gas at a flow rate of about 40 mL per minute. The injection port and detector are maintained at a temperature of 200°. Chromatograph five replicate injections of the *Standard preparation,* and record the chromatograms: the resolution factor between methanol and alcohol is not less than 2.0, the resolution factor between ethylene glycol and propylene glycol is not less than 3.0, and the relative standard deviation is not more than 2.0%.

Procedure—Separately inject equal volumes (about 2 µL) of the *Standard preparation* and the *Test preparation* into the chromatograph, record the chromatograms, and measure the responses of the major peaks. The order of elution is methanol, alcohol, ethylene glycol, and propylene glycol in order of increasing retention time: the relative retentions are about 1.0 for methanol and 2.2 for alcohol with respect to methanol and alcohol peaks, and about 1.0 for ethylene glycol and 1.4 for propylene glycol with respect to the ethylene glycol and propylene glycol peaks. Calculate the relative response ratio for the alcohol peak with respect to the methanol peak and for the propylene glycol peak with respect to the ethylene glycol peak. Calculate the alcohol content, in percentage, by the formula:

$$12(R_U / R_S),$$

in which R_U and R_S are the relative response ratios for the methanol and alcohol peaks obtained from the *Test preparation* and the *Standard preparation,* respectively. Calculate the propylene glycol content, in percentage, by the formula:

$$40(R'_U / R'_S),$$

in which R'_U and R'_S are the relative response ratios for the ethylene glycol and propylene glycol peaks obtained from the *Test preparation* and the *Standard preparation,* respectively. The alcohol content is not less than 9.0% and not more than 11.0%, and the propylene glycol content is not less than 37.0% and not more than 43.0%.

Particulate matter ⟨788⟩: meets the requirements under *Small-volume Injections.*

Other requirements—It meets the requirements under *Injections* ⟨1⟩.

Assay—

Mobile phase—Prepare a suitable degassed and filtered mixture of methanol and water (55:45).

Standard preparation—Dissolve an accurately weighed portion of USP Phenytoin RS in *Mobile phase* to obtain a solution having a known concentration of about 230 µg per mL.

Assay preparation—Transfer an accurately measured volume of Phenytoin Sodium Injection, equivalent to 250 mg of phenytoin sodium, to a volumetric flask and dilute quantitatively and stepwise with *Mobile phase* to obtain a solution having a concentration of about 250 µg of phenytoin sodium per mL.

Chromatographic system (see *Chromatography* ⟨621⟩)—The liquid chromatograph is equipped with a 254-nm detector and a 3.9-mm × 25-cm column that contains packing L1. The flow rate is about 1.5 mL per minute. Chromatograph replicate injections of the *Standard preparation,* and record the peak responses as directed under *Procedure:* the relative standard deviation is not more than 2.0%, and the tailing factor is not more than 2.0.

Procedure—Separately inject equal volumes (about 20 µL) of the *Standard preparation* and the *Assay preparation* into the chromatograph, record the chromatograms, and measure the response for the major peak. Calculate the quantity, in mg, of $C_{15}H_{11}N_2NaO_2$ in each mL of the Injection taken by the formula:

$$(274.25/252.27)(C/V)(r_U/r_S),$$

in which 274.25 and 252.27 are the molecular weights of phenytoin sodium and phenytoin, respectively, C is the concentration, in µg per mL, of USP Phenytoin RS in the *Standard preparation,* V is the volume, in mL, of Injection taken, and r_U and r_S are the peak responses obtained from the *Assay preparation* and the *Standard preparation,* respectively.

Chromic Phosphate P 32 Suspension

» Chromic Phosphate P 32 Suspension is a sterile, aqueous suspension of radioactive chromic phosphate P 32 in a 30 percent Dextrose solution suitable for intraperitoneal, intrapleural, or interstitial administration. It contains not less than 90.0 percent and not more than 110.0 percent of the labeled amount of ^{32}P as chromic phosphate expressed in megabecquerels (millicuries) per mL at the time indicated in the labeling. It may contain a preservative or a stabilizer. Other chemical forms of radioactivity do not exceed 5.0 percent of the total radioactivity.

Packaging and storage—Preserve in single-dose or in multiple-dose containers.

Labeling—Label it to include the following, in addition to the information specified for *Labeling* under *Injections* ⟨1⟩: the time and date of calibration; the amount of ^{32}P as labeled chromic phosphate expressed as total megabecquerels (millicuries) and concentration as megabecquerels (millicuries) per mL at the time of calibration; the expiration date; and the statements, "Caution—Radioactive Material," and "For intracavitary use only." The labeling indicates that in making dosage calculations, correction is to be made for radioactive decay, and also indicates that the radioactive half-life of ^{32}P is 14.3 days.

Reference standard—*USP Endotoxin Reference Standard.*

Radionuclide identification—

A: The beta radiation of the Suspension, measured according to the procedure set forth under *Radioactivity* ⟨821⟩, shows a mass absorption coefficient within ±5% of the value found for a specimen of a known standard of the same radionuclide when determined under identical counting conditions and geometry.

B: Its beta-ray spectrum is identical to that of a specimen of ^{32}P of known purity showing no distinct photopeaks and no energies greater than 1.710 MeV.

Bacterial endotoxins—It meets the requirements of the *Bacterial Endotoxins Test* ⟨85⟩, the limit of endotoxin content being not more than 175/V USP Endotoxin Unit per mL of the Injection, when compared with the USP Endotoxin RS, in which V is the maximum recommended total dose, in mL, at the expiration date or time.

pH ⟨791⟩: between 3.0 and 5.0.

Radiochemical purity—Place a measured volume of Suspension, to provide a count rate of about 20,000 counts per minute, about 2.5 cm from one end of a 25-mm × 300-mm strip of chromatographic paper (see *Chromatography* ⟨621⟩), and allow to dry. Develop the chromatogram by ascending chromatography, using water as the solvent, and air-dry: the radioactivity in the chromic phosphate is not less than 95.0% of the total radioactivity when measured at the origin.

Other requirements—It meets the requirements under *Injections* ⟨1⟩, except that the Suspension may be distributed or dispensed prior to the completion of the test for *Sterility,* the latter test being started on the day of final manufacture, and except that it is not subject to the recommendations on *Volume in Container.*

Assay for dextrose—

Periodic acid reagent solution—Dissolve 8.5 g of sodium periodate in 80 mL of 1 *N* sulfuric acid, dilute with water to 100 mL, and mix.

Assay preparation—Decant the supernatant liquid from Sterile Chromic Phosphate P 32 Suspension into a disposable centrifuge tube, and centrifuge. Pipet 1.0 mL of the clear supernatant liquid into a 25-mL volumetric flask, dilute with water to volume, and mix.

Procedure—Pipet 50 mL of *Periodic acid reagent solution* into a 250-mL conical flask, add 3.0 mL of the *Assay preparation*, swirl, cover the flask, and allow to stand at room temperature for 2 hours. Add, in the order named and with rapid stirring, 50 mL of a saturated solution of sodium bicarbonate, 50.0 mL of 0.1 *N* potassium arsenite VS, 4 mL of potassium iodide solution (1 in 5), and 20 g of sodium bicarbonate. Stir the solution at room temperature for 15 minutes. Titrate with 0.1 *N* iodine VS, using 3 mL of starch TS as the indicator. Perform a blank determination, and make any necessary correction. Each mL of 0.1 *N* iodine is equivalent to 1.802 mg of dextrose ($C_6H_{12}O_6$). Not less than 27.0% and not more than 33.0% is found.

Assay for radioactivity—Using a suitable counting assembly (see *Assay, Beta-emitting radionuclides* under *Radioactivity* ⟨821⟩), determine the radioactivity, in MBq (mCi) per mL, of Sterile Chromic Phosphate P 32 Suspension by use of a calibrated system as directed under *Radioactivity* ⟨821⟩.

Sodium Phosphate P 32 Solution

Phosphoric-^{32}P acid, disodium salt.
Dibasic sodium phosphate-^{32}P [*7635-46-3*].

» Sodium Phosphate P 32 Solution is a solution suitable for either oral or intravenous administration, containing radioactive phosphorus (^{32}P) processed in the form of Dibasic Sodium Phosphate from the neutron bombardment of elemental sulfur. Nonradioactive Dibasic Sodium Phosphate may be added during the processing.

Sodium Phosphate P 32 Solution contains not less than 90.0 percent and not more than 110.0 percent of the labeled amount of ^{32}P as phosphate expressed in megabecquerels (microcuries or millicuries) per mL at the time indicated in the labeling. Other chemical forms of radioactivity are absent.

Packaging and storage—Preserve in single-dose or in multiple-dose containers that previously have been treated to prevent adsorption.

Labeling—Label it to include the following: the time and date of calibration; the amount of ^{32}P as phosphate expressed in total megabecquerels (microcuries or millicuries) and in megabecquerels (microcuries or in millicuries) per mL at the time of calibration; the name and quantity of any added preservative or stabilizer; a statement of the intended use, whether oral or intravenous; a statement of whether the contents are intended for diagnostic or therapeutic use; the expiration date; and the statements, "Caution—Radioactive Material," and "Not for intracavitary use." The labeling indicates that in making dosage calculations, correction is to be made for radioactive decay, and also indicates that the radioactive half-life of ^{32}P is 14.3 days.

Reference standard—*USP Endotoxin Reference Standard.*

Radionuclide identification—

A: The beta radiation of the Solution, measured according to the procedure set forth under *Radioactivity* ⟨821⟩, shows a mass absorption coefficient within ±5% of the value found for a specimen of a known standard of the same radionuclide when determined under identical counting conditions and geometry.

B: Its beta-ray and/or Bremsstrahlung spectrum is identical to that of a specimen of ^{32}P of known purity showing no distinct photopeaks and no energies greater than 1.710 MeV.

Bacterial endotoxins—It meets the requirements of the *Bacterial Endotoxins Test* ⟨85⟩, the limit of endotoxin content being not more than 175/*V* USP Endotoxin Unit per mL of the Injection, when compared with the USP Endotoxin RS, in which *V* is the maximum recommended total dose, in mL, at the expiration date or time.

pH ⟨791⟩: between 5.0 and 6.0.

Radiochemical purity—Place a measured volume, appropriately diluted with phosphoric acid solution (1 in 20) such that it provides a count rate of about 20,000 counts per minute, about 45 mm from the end of a 25- × 300-mm strip of chromatographic paper (see *Chromatography* ⟨621⟩), and allow to dry. Develop the chromatogram by descending chromatography, using a mixture of tertiary butyl alcohol, water, and formic acid (40:20:5). Allow to dry, and determine the position of the phosphoric acid by spraying the paper with a solution prepared by dissolving 5 g of ammonium molybdate in 100 mL of water and pouring, with constant stirring, into a mixture of 12 mL of nitric acid and 24 mL of water. Determine the position of the radioactivity distribution by scanning with a collimated radiation detector. The radioactivity appears in one band only, corresponding in R_f value to the phosphoric acid.

Other requirements—Solution intended for intravenous use meets the requirements under *Injections* ⟨1⟩, except that the Solution may be distributed or dispensed prior to completion of the test for *Sterility*, the latter test being started on the day of final manufacture, and except that it is not subject to the recommendation on *Volume in Container*.

Assay for radioactivity—Using a suitable counting assembly (see *Assay, Beta-emitting radionuclides* under *Radioactivity* ⟨821⟩), determine the radioactivity, in MBq (mCi) per mL, of Sodium Phosphate P 32 Solution by use of a calibrated system as directed under *Radioactivity* ⟨821⟩).

Phosphoric Acid—*see* Phosphoric Acid NF

Phosphoric Acid, Diluted—*see* Phosphoric Acid, Diluted NF

Phosphoric Acid Gel, Sodium Fluoride and—*see* Sodium Fluoride and Phosphoric Acid Gel

Physostigmine

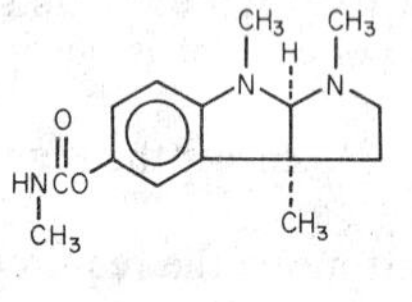

$C_{15}H_{21}N_3O_2$ 275.35

Pyrrolo[2,3-*b*]indol-5-ol, 1,2,3,3a,8,8a-hexahydro-1,3a,8-trimethyl-, methylcarbamate (ester), (3a*S*-*cis*).
Physostigmine.
1,2,3,3a*β*,8,8a*β*-Hexahydro-1,3a,8-trimethylpyrrolo[2,3-*b*]indol-5-yl methylcarbamate [*57-47-6*].

» Physostigmine is an alkaloid usually obtained from the dried ripe seed of *Physostigma venenosum* Balfour (Fam. Leguminosae). It contains not less than 97.0 percent and not more than 102.0 percent of $C_{15}H_{21}N_3O_2$, calculated on the dried basis.

Packaging and storage—Preserve in tight, light-resistant containers.

Reference standard—*USP Physostigmine Salicylate Reference Standard*—Use without drying.

Identification—It meets the requirements of the test for *Identification—Organic Nitrogenous Bases* ⟨181⟩, USP Physostigmine Salicylate RS being used, and 1 g of sodium bicarbonate being used in place of the 2 mL of 1 *N* sodium hydroxide specified.

Specific rotation ⟨781⟩: between −236° and −246°, measured at 365 nm, calculated on the dried basis, determined in a solution in methanol containing 100 mg in each 10 mL.

Loss on drying ⟨731⟩—Dry it over silica gel for 24 hours: it loses not more than 1.0% of its weight.

Residue on ignition ⟨281⟩: negligible, from 100 mg.

Readily carbonizable substances ⟨271⟩—Dissolve 100 mg in 5 mL of sulfuric acid TS: at the end of 5 minutes the solution has no more color than Matching Fluid I.

Assay—Dissolve about 175 mg of Physostigmine, accurately weighed, in 25 mL of chloroform. Add 25 mL of glacial acetic acid, and titrate with 0.02 *N* perchloric acid in dioxane VS, determining the end-point potentiometrically. Perform a blank determination, and make any necessary correction. Each mL of 0.02 *N* perchloric acid is equivalent to 5.507 mg of $C_{15}H_{21}N_3O_2$.

Physostigmine Salicylate

$C_{15}H_{21}N_3O_2 . C_7H_6O_3$ 413.47

Pyrrolo[2,3-*b*]indol-5-ol, 1,2,3,3a,8,8a-hexahydro-1,3a,8-trimethyl-, methylcarbamate (ester), (3a*S*-*cis*)-, mono-(2-hydroxybenzoate).

Physostigmine monosalicylate [*57-64-7*].

» Physostigmine Salicylate contains not less than 97.0 percent and not more than 102.0 percent of $C_{15}H_{21}N_3O_2 . C_7H_6O_3$, calculated on the dried basis.

Packaging and storage—Preserve in tight, light-resistant containers.

Reference standard—*USP Physostigmine Salicylate Reference Standard*—Use without drying.

Identification—

A: It responds to the *Identification test* under *Physostigmine.*

B: It responds to the tests for *Salicylate* ⟨191⟩.

Specific rotation ⟨781⟩: between −91° and −94°, calculated on the dried basis, determined in a solution containing 100 mg in each 10 mL.

Loss on drying ⟨731⟩—Dry it over silica gel for 24 hours: it loses not more than 1.0% of its weight.

Residue on ignition ⟨281⟩: negligible, from 100 mg.

Sulfate—Precipitate the salicylic acid from 10 mL of a cold, saturated solution of Physostigmine Salicylate with a slight excess of 3 *N* hydrochloric acid, filter, and to the filtrate add 5 drops of barium chloride TS: no turbidity is produced immediately.

Readily carbonizable substances ⟨271⟩—Dissolve 100 mg in 5 mL of sulfuric acid TS: at the end of 5 minutes the solution has no more color than Matching Fluid I.

Assay—Dissolve about 250 mg of Physostigmine Salicylate, accurately weighed, in 25 mL of chloroform. Add 25 mL of glacial acetic acid, and titrate with 0.02 *N* perchloric acid in dioxane VS, determining the end-point potentiometrically. Perform a blank determination, and make any necessary correction. Each mL of 0.02 *N* perchloric acid is equivalent to 8.270 mg of $C_{15}H_{21}N_3O_2 . C_7H_6O_3$.

Physostigmine Salicylate Injection

» Physostigmine Salicylate Injection is a sterile solution of Physostigmine Salicylate in Water for Injection. It contains not less than 90.0 percent and not more than 110.0 percent of the labeled amount of $C_{15}H_{21}N_3O_2 . C_7H_6O_3$. It may contain an antimicrobial agent and an antioxidant.

Note—Do not use the Injection if it is more than slightly discolored.

Packaging and storage—Preserve in single-dose containers, preferably of Type I glass, protected from light.

Reference standard—*USP Physostigmine Salicylate Reference Standard*—Use without drying.

Identification—

A: It responds to the *Identification test* under *Physostigmine.*

B: It responds to the tests for *Salicylate* ⟨191⟩.

Pyrogen—It meets the requirements of the *Pyrogen Test* ⟨151⟩.

pH ⟨791⟩: between 3.5 and 5.0.

Other requirements—It meets the requirements under *Injections* ⟨1⟩.

Assay—

0.05 M Ammonium acetate—Dissolve 3.85 g of ammonium acetate in 1 liter of water, and adjust, if necessary, with glacial acetic acid or ammonium hydroxide to a pH of 6 ± 0.1.

Mobile phase—Prepare a filtered and degassed mixture of equal volumes of acetonitrile and *0.05 M Ammonium acetate.* Make adjustments if necessary (see *System Suitability* under *Chromatography* ⟨621⟩).

Benzyl alcohol–benzaldehyde solution—Prepare a mixture of 100 µL of benzyl alcohol and 1 µL of benzaldehyde in each 400 mL of acetonitrile.

Standard preparation—Dissolve an accurately weighed quantity of USP Physostigmine Salicylate RS in *Benzyl alcohol–benzaldehyde solution*, and dilute quantitatively, and stepwise if necessary, with *Benzyl alcohol–benzaldehyde solution*, to obtain a solution having a known concentration of about 30 µg per mL.

Assay preparation—Transfer an accurately measured volume of Physostigmine Salicylate Injection, equivalent to about 3 mg of physostigmine salicylate, to a 100-mL volumetric flask, dilute with acetonitrile to volume, and mix.

Chromatographic system (see *Chromatography* ⟨621⟩)—The liquid chromatograph is equipped with a 254-nm detector and a 3.9-mm × 30-cm column that contains packing L1. The flow rate is about 2 mL per minute. Separately chromatograph 10-µL portions of the *Benzyl alcohol–benzaldehyde solution* and the *Standard preparation*, and record the peak responses as directed under *Procedure* [NOTE—If the components of the *Benzyl alcohol–benzaldehyde solution* co-elute, the *Standard preparation* will exhibit only 2 peaks instead of 3.]: in a suitable system, benzyl alcohol and benzaldehyde elute before physostigmine, the column efficiency determined from the analyte peak is not less than 1200 theoretical plates, the resolution, *R*, between the physostigmine peak and the adjacent peak (benzyl alcohol or benzaldehyde or the combination of these) is not less than 2.0, and the relative standard deviation for replicate injections is not more than 2.0%.

Procedure—Separately inject equal volumes (about 10 µL) of the *Standard preparation* and the *Assay preparation* into the chromatograph, record the chromatograms, and measure the responses for the major peaks. Calculate the quantity, in mg, of $C_{15}H_{21}N_3O_2 . C_7H_6O_3$ in each mL of the Injection taken by the formula:

$$0.1(C/V)(r_U/r_S),$$

in which C is the concentration, in µg per mL, of USP Physostigmine Salicylate RS in the *Standard preparation*, V is the volume, in mL, of Injection taken, and r_U and r_S are the peak responses obtained from the *Assay preparation* and the *Standard preparation*, respectively.

Physostigmine Salicylate Ophthalmic Solution

» Physostigmine Salicylate Ophthalmic Solution is a sterile, aqueous solution of Physostigmine Salicylate. It contains not less than 90.0 percent and not more than 110.0 percent of the labeled amount of $C_{15}H_{21}N_3O_2 . C_7H_6O_3$. It may contain suitable antimicrobial agents, buffers, and stabilizers, and suitable additives to increase its viscosity.

Packaging and storage—Preserve in tight, light-resistant containers.

Reference standard—*USP Physostigmine Salicylate Reference Standard*—Use without drying.

Identification—It responds to the *Identification tests* under *Physostigmine Salicylate.*

Sterility—It meets the requirements under *Sterility Tests* ⟨71⟩.

pH ⟨791⟩: between 2.0 and 4.0.

Assay—

0.05 M Ammonium acetate and *Mobile phase*—Prepare as directed in the *Assay* under *Physostigmine Salicylate Injection.*

Standard preparation—Dissolve an accurately weighed quantity of USP Physostigmine Salicylate RS in acetonitrile, and dilute quantitatively, and stepwise if necessary, with acetonitrile, to obtain a solution having a known concentration of about 30 µg per mL.

Assay preparation—Transfer an accurately measured volume of Physostigmine Salicylate Ophthalmic Solution, equivalent to about 3 mg of physostigmine salicylate, to a 100-mL volumetric flask, dilute with acetonitrile to volume, and mix.

Chromatographic system (see *Chromatography* ⟨621⟩)—The liquid chromatograph is equipped with a 254-nm detector and a 3.9-mm × 30-cm column that contains packing L1. The flow rate is about 2 mL per minute. Chromatograph the *Standard preparation* and record the peak responses as directed under *Procedure:* the column efficiency determined from the analyte peak is not less than 1200 theoretical plates, and the relative standard deviation for replicate injections is not more than 2.0%.

Procedure—Proceed as directed for *Procedure* in the *Assay* under *Physostigmine Salicylate Injection.* Calculate the quantity, in mg, of $C_{15}H_{21}N_3O_2 . C_7H_6O_3$ in each mL of the Ophthalmic Solution taken by the formula:

$$0.1(C/V)(r_U/r_S),$$

in which the terms are as defined therein.

Physostigmine Sulfate

$(C_{15}H_{21}N_3O_2)_2 . H_2SO_4$ 648.77

Pyrrolo[2,3-*b*]indol-5-ol, 1,2,3,3a,8,8a-hexahydro-1,3a,8-trimethyl-, methylcarbamate (ester), (3a*S*-*cis*)-, sulfate (2:1).

Physostigmine sulfate (2:1) [*64-47-1*].

» Physostigmine Sulfate contains not less than 97.0 percent and not more than 102.0 percent of $(C_{15}H_{21}N_3O_2)_2 . H_2SO_4$, calculated on the dried basis.

Packaging and storage—Preserve in tight, light-resistant containers.

Reference standard—*USP Physostigmine Salicylate Reference Standard*—Use without drying.

Identification—

A: It responds to the *Identification test* under *Physostigmine.*

B: A solution (1 in 100) responds to the tests for *Sulfate* ⟨191⟩.

Specific rotation ⟨781⟩: between −116° and −120°, calculated on the dried basis, determined in a solution containing 100 mg in each 10 mL.

Loss on drying ⟨731⟩—Dry it at 105° to constant weight: it loses not more than 1.0% of its weight.

Residue on ignition ⟨281⟩: negligible, from 100 mg.

Readily carbonizable substances—It meets the requirements of the test for *Readily carbonizable substances* under *Physostigmine.*

Assay—Dissolve about 200 mg of Physostigmine Sulfate, accurately weighed, in 25 mL of water. Render the solution alkaline by the addition of about 1 g of sodium bicarbonate, and extract with one 25-mL and five 10-mL portions of chloroform, each time shaking vigorously for 1 minute. Filter each extract through glass wool. Add 15 mL of glacial acetic acid and 10 mL of acetic acid anhydride to the combined chloroform extracts, and titrate with 0.02 *N* perchloric acid VS, determining the end-point potentiometrically. Perform a blank determination, and make any necessary correction. Each mL of 0.02 *N* perchloric acid is equivalent to 6.488 mg of $(C_{15}H_{21}N_3O_2)_2 . H_2SO_4$.

Physostigmine Sulfate Ophthalmic Ointment

» Physostigmine Sulfate Ophthalmic Ointment contains not less than 90.0 percent and not more than 110.0 percent of the labeled amount of $(C_{15}H_{21}N_3O_2)_2 . H_2SO_4$. It is sterile.

Packaging and storage—Preserve in collapsible ophthalmic ointment tubes.

Reference standard—*USP Physostigmine Sulfate Reference Standard*—Dry at 105° to constant weight before using.

Identification—

A: Place about 20 g of Ophthalmic Ointment in a beaker, add about 25 mL of water, and heat gently on a steam bath, with continuous stirring, until the ointment base has melted. Cool to congeal the ointment base, and decant the aqueous solution through a filter into a separator. Draw off a 2-mL portion, and reserve for *Identification test B:* the solution in the separator meets the requirements of the test for *Identification—Organic Nitrogenous Bases* ⟨181⟩, USP Physostigmine Sulfate RS being used, and 1 g of sodium bicarbonate being used in place of the 2 mL of 1 *N* sodium hydroxide specified.

B: A 2-mL portion of the aqueous solution obtained in *Identification test A* responds to the tests for *Sulfate* ⟨191⟩.

Sterility—It meets the requirements for *Ophthalmic Ointments* under *Sterility Tests* ⟨71⟩.

Metal particles—It meets the requirements of the test for *Metal Particles in Ophthalmic Ointments* ⟨751⟩.

Assay—

0.05 M Ammonium acetate and *Mobile phase*—Prepare as directed in the *Assay* under *Physostigmine Salicylate Injection.*

Standard preparation—Dissolve an accurately weighed quantity of USP Physostigmine Sulfate RS in acetonitrile, and dilute quantitatively, and stepwise if necessary, with acetonitrile, to obtain a solution having a known concentration of about 30 µg per mL.

Assay preparation—Transfer an accurately weighed quantity of Physostigmine Sulfate Ophthalmic Ointment, equivalent to about 3 mg of physostigmine sulfate, to a 60-mL separator. Add 20 mL of spectrophotometric grade *n*-hexane, and extract with four 20-mL portions of acetonitrile. Collect the acetonitrile extracts in a 100-mL volumetric flask, dilute with acetonitrile to volume, and mix.

Chromatographic system (see *Chromatography* ⟨621⟩)—Proceed as directed for *Chromatographic system* in the *Assay* under *Physostigmine Salicylate Ophthalmic Solution.*

Procedure—Proceed as directed for *Procedure* in the *Assay* under *Physostigmine Salicylate Injection.* Calculate the quantity, in mg, of $(C_{15}H_{21}N_3O_2)_2 . H_2SO_4$ in the portion of the Ophthalmic Ointment taken by the formula:

$$0.1C(r_U/r_S),$$

in which *C* is the concentration, in µg per mL, of USP Physostigmine Sulfate RS in the *Standard preparation.*

Phytonadione

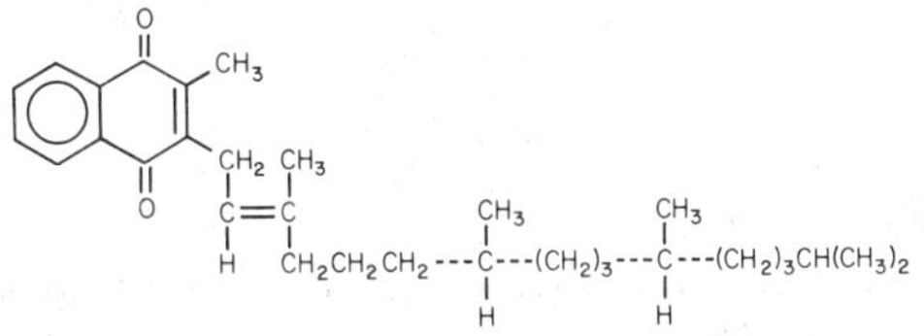

E component

$C_{31}H_{46}O_2$ 450.70
1,4-Naphthalenedione, 2-methyl-3-(3,7,11,15-tetramethyl-2-hexadecenyl)-, [*R*-[*R**,*R**-(*E*)]]-.
Phylloquinone [*84-80-0*].

» Phytonadione is a mixture of *E* and *Z* isomers containing not less than 97.0 percent and not more than 103.0 percent of $C_{31}H_{46}O_2$. It contains not more than 21.0 percent of the *Z* isomer.

Packaging and storage—Preserve in tight, light-resistant containers.

Reference standard—*USP Phytonadione Reference Standard*—Do not dry before using.

Identification—

A: The infrared absorption spectrum of a thin film of it formed between two sodium chloride plates exhibits maxima only at the same wavelengths as that of a similar preparation of USP Phytonadione RS.

B: The ultraviolet absorption spectrum of a 1 in 100,000 solution in *n*-hexane exhibits maxima and minima at the same wavelengths as that of a similar solution of USP Phytonadione RS, concomitantly measured, and the respective absorptivities at the wavelength of maximum absorbance at about 248 nm do not differ by more than 3.0%.

Refractive index ⟨831⟩: between 1.523 and 1.526.

Reaction—A 1 in 20 solution of it in dehydrated alcohol is neutral to litmus.

Menadione—Mix about 20 mg with 0.5 mL of a mixture of equal volumes of 6 *N* ammonium hydroxide and alcohol, then add 1 drop of ethyl cyanoacetate, and shake gently: no purple or blue color is produced.

***Z* isomer content**—[NOTE—Protect solutions containing Phytonadione from exposure to light.]

Mobile phase, Internal standard solution, Assay preparation, Chromatographic system, and *Procedure*—Proceed as directed in the *Assay*, except to calculate the percentage of *Z* isomer by the formula:

$$100r_Z/(r_Z + r_E),$$

in which r_Z is the peak area of the (*Z*)-phytonadione isomer peak and r_E is the peak area of the (*E*)-phytonadione isomer peak obtained from the *Assay preparation.*

Assay—[NOTE—Protect solutions containing Phytonadione from exposure to light.]

Mobile phase—Prepare a filtered and degassed solution of *n*-hexane and *n*-amyl alcohol (2000:1.5).

Internal standard solution—Dissolve cholesteryl benzoate in *Mobile phase* to obtain a solution having a concentration of 2.5 mg per mL.

Standard preparation—Transfer about 60 mg of USP Phytonadione RS, accurately weighed, to a 50-mL volumetric flask, add 20 mL of *Mobile phase*, mix, dilute with *Mobile phase* to volume, and again mix. Pipet 4 mL of the resulting solution into a 50-mL volumetric flask, dilute with *Mobile phase* to volume, and mix. Pipet 10 mL of this solution and 7 mL of *Internal standard solution* into a 25-mL volumetric flask, dilute with *Mobile phase* to volume, and mix.

Assay preparation—Prepare as directed under *Standard preparation*, using Phytonadione instead of the Reference Standard.

Chromatographic system (see *Chromatography* ⟨621⟩)—The liquid chromatograph is equipped with a 254-nm detector and a 4.6-mm × 25-cm column that contains packing L3. The flow rate is about 1 mL per minute. Chromatograph replicate injections of the *Standard preparation*, and record the peak responses as directed under *Procedure:* the relative standard deviation is not more than 2.0%, and the resolution, *R*, between (*Z*)-phytonadione and (*E*)-phytonadione is not less than 1.5.

Procedure—Separately inject equal volumes (about 50 µL) of the *Standard preparation* and the *Assay preparation* into the chromatograph, record the chromatograms, and measure the responses for the major peaks. The relative retention times are about 0.7 for the internal standard, 0.9 for (*Z*)-phytonadione, and 1.0 for (*E*)-phytonadione. Calculate the quantity, in mg, of $C_{31}H_{46}O_2$ in the portion of Phytonadione taken by the formula:

$$1.56C(R_U/R_S),$$

in which *C* is the concentration, in µg per mL, of USP Phytonadione RS in the *Standard preparation*, and R_U and R_S are the relative peak response ratios for the *Assay preparation* and the *Standard preparation*, respectively. Calculate R_U and R_S by the formula:

(response for the (*Z*)-phytonadione peak +
response for the (*E*)-phytonadione peak)/
response for the internal standard peak.

Phytonadione Injection

» Phytonadione Injection is a sterile, aqueous dispersion of Phytonadione. It contains not less than 90.0 percent and not more than 110.0 percent of the labeled amount of $C_{31}H_{46}O_2$. It contains suitable solubilizing and/or dispersing agents.

Packaging and storage—Preserve in single-dose or in multiple-dose containers, preferably of Type I glass, protected from light.

Reference standard—*USP Phytonadione Reference Standard*—Do not dry before using.

Identification—The retention time of the major peak in the chromatogram of the *Assay preparation* corresponds to that of the *Standard preparation*, as obtained in the *Assay.*

Pyrogen—It meets the requirements of the *Pyrogen Test* ⟨151⟩, the test dose being 2 mL per kg.

pH ⟨791⟩: between 3.5 and 7.0.

Other requirements—It meets the requirements under *Injections* ⟨1⟩.

Assay—[NOTE—Use low-actinic glassware throughout this assay, and otherwise protect the solutions from exposure to light.]

Mobile phase—Prepare a suitable degassed mixture of dehydrated alcohol and water (95:5).

Standard preparation—Dissolve an accurately weighed quantity of USP Phytonadione RS in *Mobile phase* to obtain a solution having a known concentration of about 1 mg per mL. Pipet 1 mL of this solution into a 10-mL volumetric flask, dilute with *Mobile phase* to volume, and mix to obtain a *Standard preparation* having a known concentration of about 0.1 mg per mL.

Assay preparation for Injection containing 10 mg or more of phytonadione per mL—Pipet a volume of Phytonadione Injection, equivalent to 10 mg of phytonadione, into a 10-mL volumetric flask, dilute with *Mobile phase* to volume, and mix. Pipet 1 mL of this solution into a 10-mL volumetric flask, dilute with *Mobile phase* to volume, and mix.

Assay preparation for Injection containing less than 10 mg of phytonadione per mL—Pipet a volume of Phytonadione Injection, equivalent to 1 mg of phytonadione, into a 10-mL volumetric flask, dilute with *Mobile phase* to volume, and mix.

Chromatographic system (see *Chromatography* ⟨621⟩)—The liquid chromatograph is equipped with a 254-nm detector and a 4-mm × 25-cm column that contains packing L1. The flow rate is about 0.7 mL per minute. Chromatograph five replicate injections of the *Standard preparation*, and record the peak responses as directed under *Procedure:* the relative standard deviation is not more than 1.5%.

Procedure—Separately inject equal volumes (about 10 µL) of the *Standard preparation* and the *Assay preparation* into the chromatograph, record the chromatograms, and measure the peak response for the major peak. Calculate the quantity, in mg, of $C_{31}H_{46}O_2$ in each mL of the Injection taken by the formula:

$$D(C/V)(r_U/r_S),$$

in which D is 100 if the Injection contains 10 mg or more of phytonadione per mL, or 10 if the Injection contains less than 10 mg of phytonadione per mL, C is the concentration, in mg per mL, of USP Phytonadione RS in the *Standard preparation*, V is the volume, in mL, of Injection taken, and r_U and r_S are the peak responses of phytonadione obtained from the *Assay preparation* and the *Standard preparation*, respectively.

Phytonadione Tablets

» Phytonadione Tablets contain not less than 90.0 percent and not more than 110.0 percent of the labeled amount of $C_{31}H_{46}O_2$.

Packaging and storage—Preserve in well-closed, light-resistant containers.

Reference standard—*USP Phytonadione Reference Standard*—Do not dry before using.

Identification—

A: Transfer a portion of finely powdered Tablets, equivalent to about 10 mg of phytonadione, to a 1000-mL volumetric flask, add 750 mL of dehydrated alcohol, and shake vigorously. Dilute with dehydrated alcohol to volume, mix, and filter: the ultraviolet absorption spectrum of the filtrate exhibits maxima and minima at the same wavelengths as that of a 1 in 100,000 solution of USP Phytonadione RS in dehydrated alcohol concomitantly measured.

B: The retention time of the major peak in the chromatogram of the *Assay preparation* corresponds to that of the *Standard preparation*, as obtained in the *Assay*.

Disintegration ⟨701⟩: 30 minutes.

Uniformity of dosage units ⟨905⟩: meet the requirements.

Assay—[NOTE—Use low-actinic glassware throughout the *Assay*, and otherwise protect the solutions from light.]

Mobile phase—Prepare a suitable filtered and degassed mixture of dehydrated alcohol and water (95:5).

Standard preparation—Prepare a solution of USP Phytonadione RS in dehydrated alcohol having a known concentration of about 0.10 mg per mL.

Assay preparation—Weigh and finely powder not less than 20 Phytonadione Tablets. Transfer a portion of the powdered Tablets, equivalent to about 5 mg of phytonadione, to a 50-mL volumetric flask, add 20 mL of dehydrated alcohol, and shake by mechanical means for 15 minutes. Dilute with dehydrated alcohol to volume, mix, and filter.

Chromatographic system (see *Chromatography* ⟨621⟩)—The liquid chromatograph is equipped with a 254-nm detector and a 4-mm × 30-cm column that contains packing L1. The flow rate is about 1.5 mL per minute. Chromatograph three replicate injections of the *Standard preparation*, and record the peak responses as directed under *Procedure:* the column efficiency determined from the analyte peak is not less than 915 theoretical plates, the relative standard deviation is not more than 2.0%, and the tailing factor is not more than 2.0.

Procedure—Separately inject equal volumes (about 10 µL) of the *Standard preparation* and the *Assay preparation* into the chromatograph, record the chromatograms, and measure the response for the major peak. Calculate the quantity, in mg, of $C_{31}H_{46}O_2$, in the portion of Tablets taken by the formula:

$$50C(r_U/r_S),$$

in which C is the concentration, in mg per mL, of USP Phytonadione RS in the *Standard preparation*, and r_U and r_S are the peak responses obtained from the *Assay preparation* and the *Standard preparation*, respectively.

Pilocarpine

$C_{11}H_{16}N_2O_2$ 208.26

2(3*H*)-Furanone, 3-ethyldihydro-4-[(1-methyl-1*H*-imidazol-5-yl)-methyl]-, (3*S*-*cis*)-.

Pilocarpine [*92-13-7*].

» Pilocarpine contains not less than 98.5 percent and not more than 100.5 percent of $C_{11}H_{16}N_2O_2$, calculated on the anhydrous basis.

Packaging and storage—Preserve in tight, light-resistant containers, in a cold place.

Reference standards—*USP Pilocarpine Reference Standard*—Do not dry before using. *USP Pilocarpine Nitrate Reference Standard*—Dry at 105° for 2 hours before using.

Identification—

A: The infrared absorption spectrum of a thin film of it exhibits maxima only at the same wavelengths as that of a similar preparation of USP Pilocarpine RS.

B: The ultraviolet absorption spectrum of a 1 in 50,000 solution in water exhibits maxima and minima at the same wavelengths as that of a similar solution of USP Pilocarpine RS, concomitantly measured.

Specific rotation ⟨781⟩: not less than +102° and not more than +107°, calculated on the anhydrous basis, determined in a solution containing 200 mg in each 10 mL of pH 6.0 phosphate buffer.

Refractive index ⟨831⟩: between 1.5170 and 1.5210 at 25°, determined in a liquid specimen. If crystals are present, first warm to about 40°.

Water, *Method I* ⟨921⟩: not more than 0.5%.

Sulfate—

Standard sulfate solution—Dissolve 148 mg of anhydrous sodium sulfate in water, and dilute with water to 100 mL. Dilute 10.0 mL of this solution with water to 1000 mL. This solution contains 10 µg of sulfate per mL.

Procedure—To about 1 g of Pilocarpine in a test tube add 1 mL of 6 *N* hydrochloric acid and 4 mL of water, and mix. For the control, transfer 4.0 mL of *Standard sulfate solution* to a test tube, add 1 mL of 6 *N* hydrochloric acid, and mix. Adjust both solutions to a pH between 2 and 3 with pH indicator paper by the dropwise addition of 3 *N* hydrochloric acid or 6 *N* ammonium hydroxide, if necessary. Add water to maintain the same volume in the control and test specimen tubes. To each tube add 1 mL of barium chloride TS, and mix: any turbidity produced in the specimen tube after 10 minutes' standing is not greater than that produced in the control (0.004%).

Nitrate—

Standard preparation—Prepare a solution of USP Pilocarpine Nitrate RS to contain 43 µg per mL. This solution contains the equivalent of 10 µg of nitrate ion per mL.

Test preparation—Prepare a solution of Pilocarpine to contain 200 mg per mL.

Procedure—Transfer 0.5-mL portions of the *Test preparation* and of the *Standard preparation*, respectively, to separate test tubes, and to each tube add 1 drop of a 1 in 100 solution of sulfanilic acid in 5 *N* acetic acid and 1 drop of a 3 in 1000 solution of *N*-(1-naphthyl)ethylenediamine dihydrochloride in 5 *N* acetic acid. Adjust the *Standard preparation* and the *Test preparation* to a pH between 2 and 3 with pH indicator paper by the dropwise addition of 3 *N* hydrochloric acid or 1 *N* ammonium hydroxide, if necessary. To each solution add a few granules of acid-washed, nitrate-free zinc. Heat the test tubes in a water bath at a temperature of about 32°. Allow 5 minutes for the development of a pink color: any pink color observed in the *Test preparation* is not greater than that observed in the *Standard preparation* (0.005%).

Readily carbonizable substances ⟨271⟩—Dissolve 100 mg in 5 mL of sulfuric acid TS: the solution has no more color than *Matching Fluid A*.

Chromatographic impurities—

Test preparation—Dissolve 100 mg, accurately weighed, of Pilocarpine in 10.0 mL of methanol (*Test preparation A*). Dilute 1.0 mL of *Test preparation A* with methanol to 10.0 mL (*Test preparation B*).

Spray reagent—

SOLUTION A—Dissolve 17 g of bismuth subnitrate and 200 g of tartaric acid in 800 mL of water.

SOLUTION B—Dissolve 160 g of potassium iodide in 400 mL of water (*Solution B*).

Mix 12.5 mL each of *Solution A* and *Solution B* with 50 g of tartaric acid and 250 mL of water. The resulting solution is stable for 1 week. *Solutions A* and *B* may be used for 30 days; store in a refrigerator.

Procedure—Using a 20- × 5-cm thin-layer chromatographic plate coated with a 0.25-mm layer of silica gel, apply 10 µL of methanol 2 cm from the lower edge of the plate. Similarly apply 10 µL of *Test preparation A* and 2 µL of *Test preparation B*, equivalent to 100 µg and 2 µg of pilocarpine, respectively. Develop the plate in a suitable pre-equilibrated chamber with a solvent system consisting of 9 volumes of methanol and 1 volume of dilute pH 6.0 phosphate buffer (1 in 100), until the solvent front has moved 15 cm above the point of application of the spots. Remove the plate, and allow to air-dry. Spray the dried plate with the *Spray reagent*, and examine the chromatogram: any secondary spots, corresponding to impurities, obtained from the 100-µg application are not greater in size or intensity than the principal spot obtained from the 2-µg application (2.0%), blank effects excluded.

Other alkaloids—Dissolve 200 mg in 20 mL of water and divide the solution into 2 portions. To one portion add a few drops of 6 *N* ammonium hydroxide, and to the other add a few drops of potassium dichromate TS: no turbidity is produced in either solution.

Assay—Dissolve about 200 mg of Pilocarpine, accurately weighed, in 25 mL of glacial acetic acid. Titrate with 0.1 *N* perchloric acid VS, determining the end-point potentiometrically. Perform a blank determination, and make any necessary correction. Each mL of 0.1 *N* perchloric acid is equivalent to 20.82 mg of $C_{11}H_{16}N_2O_2$.

Pilocarpine Ocular System

» Pilocarpine Ocular System contains not less than 85.0 percent and not more than 115.0 percent of the labeled amount of $C_{11}H_{16}N_2O_2$. It is sterile.

Packaging and storage—Preserve in single-dose containers, in a cold place.

Reference standards—*USP Pilocarpine Reference Standard*—Use as is. *USP Pilocarpine Hydrochloride Reference Standard*—Dry at 105° for 2 hours before using.

Identification—Cut around the inside margin of the Ocular System, then discard the ring encircling the Ocular System, extract the remaining portion with 0.5 mL of methanol in a small capped vial, shaking vigorously for 1 to 2 minutes. Evaporate the methanol extract on a sodium chloride plate forming a thin film: the infrared absorption spectrum of the film exhibits maxima only at the same wavelengths as that of a similar preparation of USP Pilocarpine RS.

Sterility—It meets the requirements under *Sterility Tests* ⟨71⟩.

Uniformity of dosage units ⟨905⟩: meets the requirements.

Chromatographic impurities—

Test preparations—Cut a number of Ocular Systems, containing about 20 mg of pilocarpine, in half, transfer all portions to a small, stoppered flask with 2.0 mL of methanol, and shake well to dissolve the pilocarpine (*Test preparation A*). Dilute 1.0 mL of *Test preparation A* with methanol to 10.0 mL (*Test preparation B*).

Spray reagent—Prepare as directed for *Spray reagent* in the test for *Chromatographic impurities* under *Pilocarpine*.

Procedure—Proceed as directed for *Procedure* in the test for *Chromatographic impurities* under *Pilocarpine*, except to apply 20 µL of methanol, 20 µL of *Test preparation A*, and 4 µL of *Test preparation B*, equivalent to 200 µg and 4 µg of pilocarpine: any secondary spots, corresponding to impurities, obtained from the 200-µg application are not greater in size or intensity than the principal spot obtained from the 4-µg application (2.0%), blank effects excluded.

Drug release pattern—Place each of 10 Ocular Systems in suitable porous holders made of an inert material, and suspend each from a nickel wire. To the upper end of the wire attach a tag identifying the specimen. Put each assembly into a test tube containing 27.0 mL of saline TS so that the system lies at the bottom of the tube and the identifying tag extends from the open top of the tube. Put the tubes into a horizontally reciprocating shaker in which the temperature is maintained at 37 ± 0.5°. Agitate the tubes with a horizontal amplitude of about 4 cm and a frequency of about 35 cycles per minute. At 7, 24, 48, 72, 96, and 168 hours, remove the assemblies from their tubes, and each time replace them in similar tubes containing 27.0 mL of fresh saline TS. Determine the amount of pilocarpine in solution in each tube, after adjusting the volume to 27.0 mL to make up for any evaporative losses, by measuring the ultraviolet absorbance in 1-cm cells at the wavelength of maximum absorbance at about 215 nm, with a suitable spectrophotometer, against saline TS as the blank. Concomitantly measure the absorbance of a Standard solution of USP Pilocarpine Hydrochloride RS having a known concentration of about 20 µg in each mL of saline TS. Calculate the quantity, in µg, of $C_{11}H_{16}N_2O_2$ in each solution by the formula:

$$(208.26/244.72)(A_U/A_S)27C,$$

in which 208.26 and 244.72 are the molecular weights of pilocarpine and pilocarpine hydrochloride, respectively, A_U and A_S are the absorbances of the test solution and the Standard solution, respectively, and C is the concentration, in µg per mL, of USP Pilocarpine Hydrochloride RS in the Standard solution. Calculate the amount of pilocarpine released in 168 hours by adding the pilocarpine content of each set of tubes collected over 168 hours. The average release pattern for all 10 Ocular Systems over 168 hours is not less than 80.0% and not more than 120.0% of the labeled release pattern.

Assay—Transfer a sufficient number of Pilocarpine Ocular Systems, to provide about 50 mg of pilocarpine, to a 250-mL volumetric flask. Add 10 mL of water to the flask, and heat at 90° to 100° for 1 hour or until the systems break apart visibly. Cool the solution to room temperature, dilute with water to volume, and mix. Filter a portion of the solution, dilute 5.0 mL of the filtrate with water to 50.0 mL, and mix. Concomitantly determine the absorbances of this solution and of a Standard solution of USP Pilocarpine Hydrochloride RS having a known concentration of about 20 µg per mL in 1-cm cells at the wavelength of maximum absorbance at about 215 nm, with a suitable spectrophotometer, using water as the blank. Calculate the quantity, in mg, of $C_{11}H_{16}N_2O_2$ in each Ocular System taken by the formula:

$$(208.26/244.72)(2.5C/N)(A_U/A_S),$$

in which 208.26 and 244.72 are the molecular weights of pilocarpine and pilocarpine hydrochloride, respectively, C is the concentration, in µg per mL, of USP Pilocarpine Hydrochloride RS in the Standard solution, N is the number of Ocular Systems taken for assay, and A_U and A_S are the absorbances of the solution from the assay and the Standard solution, respectively.

Pilocarpine Hydrochloride

$C_{11}H_{16}N_2O_2 \cdot HCl$ 244.72

2(3*H*)-Furanone, 3-ethyldihydro-4-[(1-methyl-1*H*-imidazol-5-yl)-methyl]-, monohydrochloride, (3*S*-*cis*)-.

Pilocarpine monohydrochloride [*54-71-7*].

» Pilocarpine Hydrochloride contains not less than 98.5 percent and not more than 101.0 percent of $C_{11}H_{16}N_2O_2.HCl$, calculated on the dried basis.

Packaging and storage—Preserve in tight, light-resistant containers.

Reference standard—*USP Pilocarpine Hydrochloride Reference Standard*—Dry at 105° for 2 hours before using.

Identification—

A: The infrared absorption spectrum of a mineral oil dispersion of it, previously dried, exhibits maxima only at the same wavelengths as that of a similar preparation of USP Pilocarpine Hydrochloride RS.

B: A solution (1 in 20) responds to the tests for *Chloride* ⟨191⟩.

Melting range ⟨741⟩: between 199° and 205°, but the range between beginning and end of melting does not exceed 3°.

Specific rotation ⟨781⟩: between +88.5° and +91.5°, calculated on the dried basis, determined in a solution containing 200 mg in each 10 mL.

Loss on drying ⟨731⟩—Dry it at 105° for 2 hours: it loses not more than 3.0% of its weight.

Readily carbonizable substances ⟨271⟩—Dissolve 250 mg in 5 mL of sulfuric acid TS: the solution has no more color than *Matching Fluid B*.

Other alkaloids—Dissolve 200 mg in 20 mL of water, and divide the solution into two portions. To one portion add a few drops of 6 *N* ammonium hydroxide, and to the other add a few drops of potassium dichromate TS: no turbidity is produced in either solution.

Assay—Dissolve about 500 mg of Pilocarpine Hydrochloride, accurately weighed, in a mixture of 20 mL of glacial acetic acid and 10 mL of mercuric acetate TS, warming slightly to effect solution. Cool the solution to room temperature, add 2 drops of crystal violet TS, and titrate with 0.1 *N* perchloric acid VS. Perform a blank determination, and make any necessary correction. Each mL of 0.1 *N* perchloric acid is equivalent to 24.47 mg of $C_{11}H_{16}N_2O_2.HCl$.

Pilocarpine Hydrochloride Ophthalmic Solution

» Pilocarpine Hydrochloride Ophthalmic Solution is a sterile, buffered, aqueous solution of Pilocarpine Hydrochloride. It contains not less than 90.0 percent and not more than 110.0 percent of the labeled amount of $C_{11}H_{16}N_2O_2.HCl$. It may contain suitable antimicrobial agents and stabilizers, and suitable additives to increase its viscosity.

Packaging and storage—Preserve in tight containers.

Reference standards—*USP Pilocarpine Hydrochloride Reference Standard*—Dry at 105° for 2 hours before using. *USP Pilocarpine Nitrate Reference Standard*—Dry at 105° for 2 hours before using.

Identification—The retention time of the major peak in the chromatogram of the *Assay preparation* corresponds to that in the chromatogram of the *Standard preparation* obtained as directed in the *Assay*.

Sterility—It meets the requirements under *Sterility Tests* ⟨71⟩.

pH ⟨791⟩: between 3.5 and 5.5.

Assay—

Mobile phase—Mix 300 mL of a 1 in 50 solution of ammonium hydroxide in isopropyl alcohol and 700 mL of *n*-hexane. Filter through a 0.5-µm filter before using.

Standard preparation—Using an accurately weighed quantity of USP Pilocarpine Hydrochloride RS, prepare a solution having a known concentration of about 1.6 mg per mL.

Assay preparation—Transfer an accurately measured volume of Pilocarpine Hydrochloride Ophthalmic Solution, equivalent to about 80 mg of pilocarpine hydrochloride, to a 50-mL volumetric flask. Dilute with methanol to volume, and mix.

Chromatographic system (see *Chromatography* ⟨621⟩)—The liquid chromatograph is equipped with a 220-nm detector and a 4.6-mm × 25-cm column that contains packing L3. The flow rate is about 2 mL per minute. Chromatograph three replicate injections of the *Standard preparation*, and record the peak responses as directed under *Procedure:* the relative standard deviation is not more than 2.0%.

Procedure—Separately inject equal volumes (about 10 µL) of the *Standard preparation* and the *Assay preparation* into the chromatograph by means of a suitable microsyringe or sampling valve, record the chromatograms, and measure the responses for the major peaks. The retention time is about 16 minutes for pilocarpine hydrochloride. Calculate the quantity, in mg, of $C_{11}H_{16}N_2O_2.HCl$ in each mL of the Ophthalmic Solution taken by the formula:

$$50(C/V)(r_U/r_S),$$

in which *C* is the concentration, in mg per mL, of USP Pilocarpine Hydrochloride RS in the *Standard preparation*, *V* is the volume, in mL, of Ophthalmic Solution taken, and r_U and r_S are the peak responses obtained from the *Assay preparation* and the *Standard preparation*, respectively.

Pilocarpine Nitrate

$C_{11}H_{16}N_2O_2.HNO_3$ 271.27

2(3*H*)-Furanone, 3-ethyldihydro-4-[(1-methyl-1*H*-imidazol-5-yl)-methyl]-, (3*S*-*cis*)-, mononitrate.

Pilocarpine mononitrate [*148-72-1*].

» Pilocarpine Nitrate contains not less than 98.5 percent and not more than 101.0 percent of $C_{11}H_{16}N_2O_2.HNO_3$, calculated on the dried basis.

Packaging and storage—Preserve in tight, light-resistant containers.

Reference standard—*USP Pilocarpine Nitrate Reference Standard*—Dry at 105° for 2 hours before using.

Identification—

A: The infrared absorption spectrum of a potassium bromide dispersion of it, previously dried, exhibits maxima only at the same wavelengths as that of a similar preparation of USP Pilocarpine Nitrate RS, similarly treated.

B: Mix a solution (1 in 10) with an equal volume of ferrous sulfate TS, and superimpose the mixture upon 5 mL of sulfuric acid contained in a test tube: the zone of contact becomes brown.

Melting range ⟨741⟩: between 171° and 176°, with decomposition, but the range between beginning and end of melting does not exceed 3°.

Specific rotation ⟨781⟩: between +79.5° and +82.5°, calculated on the dried basis, determined in a solution containing 200 mg in each 10 mL.

Loss on drying ⟨731⟩—Dry it at 105° for 2 hours: it loses not more than 2.0% of its weight.

Chloride—To 5 mL of a solution (1 in 50), acidified with nitric acid, add a few drops of silver nitrate TS: no opalescence is produced immediately.

Readily carbonizable substances ⟨271⟩—Dissolve 100 mg in 5 mL of sulfuric acid TS: the solution has no more color than *Matching Fluid A*.

Other alkaloids—Dissolve 200 mg in 20 mL of water, and divide the solution into two portions. To one portion add a few drops of 6 *N* ammonium hydroxide and to the other add a few drops of potassium dichromate TS: no turbidity is produced in either solution.

Assay—Dissolve about 600 mg of Pilocarpine Nitrate, accurately weighed, in 30 mL of glacial acetic acid, warming slightly to effect solution. Cool to room temperature, and titrate with 0.1

N perchloric acid VS, determining the end-point potentiometrically. Perform a blank determination, and make any necessary correction. Each mL of 0.1 N perchloric acid is equivalent to 27.13 mg of $C_{11}H_{16}N_2O_2 . HNO_3$.

Pilocarpine Nitrate Ophthalmic Solution

» Pilocarpine Nitrate Ophthalmic Solution is a sterile, buffered, aqueous solution of Pilocarpine Nitrate. It contains not less than 90.0 percent and not more than 110.0 percent of the labeled amount of $C_{11}H_{16}N_2O_2 . HNO_3$. It may contain suitable antimicrobial agents and stabilizers, and suitable additives to increase its viscosity.

Packaging and storage—Preserve in tight, light-resistant containers.

Reference standard—*USP Pilocarpine Nitrate Reference Standard*—Dry at 105° for 2 hours before using.

Identification—

A: The retention time of the major peak in the chromatogram of the *Assay preparation* corresponds to that in the chromatogram of the *Standard preparation* obtained as directed in the *Assay*.

B: It responds to *Identification test B* under *Pilocarpine Nitrate*.

Sterility—It meets the requirements under *Sterility Tests* ⟨71⟩.

pH ⟨791⟩: between 4.0 and 5.5.

Assay—Proceed with Pilocarpine Nitrate Ophthalmic Solution as directed in the *Assay* under *Pilocarpine Hydrochloride Ophthalmic Solution*, except to read pilocarpine nitrate instead of pilocarpine hydrochloride throughout and to calculate the quantity, in mg, of $C_{11}H_{16}N_2O_2 . HNO_3$ in each mL of the Ophthalmic Solution taken by the formula given therein.

Pimozide

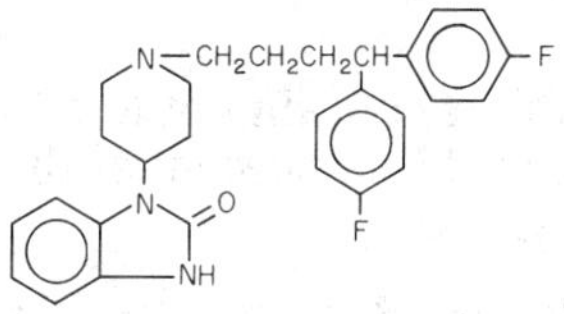

$C_{28}H_{29}F_2N_3O$ 461.55

2*H*-Benzimidazol-2-one, 1-[1-[4,4-bis(4-fluorophenyl)butyl]-4-piperidinyl]-1,3-dihydro-.

1-[1-[4,4-Bis(*p*-fluorophenyl)butyl]-4-piperidyl]-2-benzimidazolinone [*2062-78-4*].

» Pimozide contains not less than 98.0 percent and not more than 102.0 percent of $C_{28}H_{29}F_2N_3O$, calculated on the dried basis.

Packaging and storage—Preserve in tight, light-resistant containers.

Reference standard—*USP Pimozide Reference Standard*—Dry in vacuum at 80° for 4 hours before using.

Identification—

A: The infrared absorption spectrum of a potassium bromide dispersion of it, previously dried, exhibits maxima only at the same wavelengths as that of a similar preparation of USP Pimozide RS.

B: The ultraviolet absorption spectrum of a 1 in 30,000 solution in a mixture of 0.1 N hydrochloric acid and methanol (1 in 10) exhibits maxima and minima at the same wavelengths as that of a similar solution of USP Pimozide RS, concomitantly measured.

Melting range, *Class I* ⟨741⟩: between 214° and 218°.

Loss on drying ⟨731⟩—Dry it in vacuum at 80° for 4 hours: it loses not more than 0.5% of its weight.

Residue on ignition ⟨281⟩: not more than 0.2%, a 2-g portion and a platinum crucible being used for the test.

Heavy metals, *Method II* ⟨231⟩: 0.002%.

Ordinary impurities ⟨466⟩—

Test solution: chloroform.

Standard solution: chloroform.

Eluant: a mixture of cyclohexane and acetone (1:1).

Visualization: 1, then 17.

Limit—The total of any ordinary impurities observed does not exceed 1.0%.

Assay—Dissolve about 320 mg of Pimozide, accurately weighed, in 40 mL of glacial acetic acid. Titrate with 0.1 N perchloric acid VS, determining the end-point potentiometrically. Perform a blank determination, and make any necessary correction. Each mL of 0.1 N perchloric acid is equivalent to 46.16 mg of $C_{28}H_{29}F_2N_3O$.

Pimozide Tablets

» Pimozide Tablets contain not less than 90.0 percent and not more than 110.0 percent of the labeled amount of $C_{28}H_{29}F_2N_3O$.

Packaging and storage—Preserve in tight, light-resistant containers.

Reference standard—*USP Pimozide Reference Standard*—Dry in vacuum at 80° for 4 hours before using.

Identification—The retention time of the major peak in the chromatogram of the *Assay preparation* corresponds to that of the *Standard preparation*, both relative to the internal standard, as obtained in the *Assay*.

Uniformity of dosage units ⟨905⟩: meet the requirements.

Procedure for content uniformity—Place one tablet in a 50-mL flask, add 5.0 mL of 0.1 N hydrochloric acid, and shake by mechanical means for 30 minutes. Add 20.0 mL of methanol, and shake by mechanical means for 20 minutes. Dilute, if necessary, quantitatively with methanol to obtain a solution having a concentration of about 40 µg of pimozide per mL, mix, and centrifuge. Concomitantly determine the absorbance of the supernatant solution and of a solution of USP Pimozide RS in the same medium having a known concentration of about 40 µg per mL in 1-cm cells at the wavelength of maximum absorbance at about 277 nm, with a suitable spectrophotometer, using a mixture of 0.1 N hydrochloric acid and methanol (1 in 10) as the blank. Calculate the quantity, in mg, of $C_{28}H_{29}F_2N_3O$ in the Tablet by the formula:

$$(TC/D)(A_U/A_S),$$

in which T is the labeled quantity, in mg, of pimozide in the tablet, C is the concentration, in µg per mL, of USP Pimozide RS in the Standard solution, D is the concentration, in µg per mL, of pimozide in the solution from the Tablet based upon the labeled quantity per tablet and the extent of dilution, and A_U and A_S are the absorbances of the solution from the Tablet and the Standard solution, respectively.

Assay—

Ammonium acetate solution—Dissolve 500 mg of ammonium acetate in 100 mL of water, and mix.

Mobile phase—Prepare a filtered and degassed mixture of acetonitrile and *Ammonium acetate solution* (65:35), making adjustments if necessary (see *System Suitability* under *Chromatography* ⟨621⟩).

Internal standard solution—Dissolve 3,4-dimethylbenzophenone in a mixture of methanol and tetrahydrofuran (1:1) to obtain a solution having a concentration of about 1 mg per mL.

Standard preparation—Transfer about 25 mg of USP Pimozide RS, accurately weighed, to a 50-mL volumetric flask, add 10 mL of *Internal standard solution*, dilute with a mixture of methanol and tetrahydrofuran (1:1) to volume, and mix.

Assay preparation—Weigh and finely powder not less than 20 Pimozide Tablets. Transfer an accurately weighed portion of the powder, equivalent to about 25 mg of pimozide, to a 50-mL volumetric flask. Add 10 mL of *Internal standard solution* and 20 mL of a mixture of methanol and tetrahydrofuran (1:1), and shake by mechanical means for 30 minutes. Dilute with a mixture of methanol and tetrahydrofuran (1:1) to volume, and centrifuge. Use the clear supernatant liquid as the *Assay preparation.*

Chromatographic system (see *Chromatography* ⟨621⟩)—The liquid chromatograph is equipped with a 280-nm detector and a 4.6-mm × 25-cm column that contains packing L1. The flow rate is about 2 mL per minute. Chromatograph replicate injections of the *Standard preparation*, and record the peak responses as directed under *Procedure:* the relative standard deviation is not more than 2.0%, and the resolution, *R*, between the analyte and the internal standard peaks is not less than 1.3.

Procedure—Separately inject equal volumes (about 10 µL) of the *Standard preparation* and the *Assay preparation* into the chromatograph, record the chromatograms, and measure the responses for the major peaks. The relative retention times are about 0.7 for pimozide and 1.0 for the internal standard. Calculate the quantity, in mg, of $C_{28}H_{29}F_2N_3O$ in the portion of Tablets taken by the formula:

$$50C(R_U/R_S),$$

in which *C* is the concentration, in mg per mL, of USP Pimozide RS in the *Standard preparation*, and R_U and R_S are the ratios of the pimozide peak response to the internal standard peak response obtained from the *Assay preparation* and the *Standard preparation*, respectively.

Pindolol

$C_{14}H_{20}N_2O_2$ 248.32

2-Propanol,1-(1-*H*-indol-4-yloxy)-3-[(1-methylethyl)amino]-.

1-(Indol-4-yloxy)-3-(isopropylamino)-2-propanol [*13523-86-9*].

» Pindolol contains not less than 98.5 percent and not more than 101.0 percent of $C_{14}H_{20}N_2O_2$, calculated on the dried basis.

Packaging and storage—Preserve in well-closed containers, protected from light.

Reference standard—*USP Pindolol Reference Standard*—Dry at 105° for 4 hours before using.

Identification—

A: The infrared absorption spectrum of a potassium bromide dispersion of it exhibits maxima only at the same wavelengths as that of a similar preparation of USP Pindolol RS.

B: The ultraviolet absorption spectrum of a 1 in 50,000 solution in 0.01 *N* methanolic hydrochloric acid exhibits maxima and minima at the same wavelengths as that of a similar solution of USP Pindolol RS, concomitantly measured.

C: Examine the chromatograms obtained in the test for *Chromatographic purity:* the principal spot obtained from the *Test solution* is similar in R_f value, color, and intensity to that obtained from the *Stock standard solution.*

Melting range ⟨741⟩: between 169° and 173°, but the range between beginning and end of melting does not exceed 3°.

Loss on drying ⟨731⟩—Dry it at 105° for 4 hours: it loses not more than 0.5% of its weight.

Residue on ignition ⟨281⟩: not more than 0.1%.

Heavy metals, *Method II* ⟨231⟩: 0.002%.

Chromatographic purity—[NOTE—Protect solutions and chromatographic plate (after application of solutions) from light.]

p-Dimethylaminobenzaldehyde spray—Dissolve 1 g of *p*-dimethylaminobenzaldehyde in 50 mL of hydrochloric acid, add 50 mL of alcohol, and mix.

Solvent mixture—Prepare a solution of methanol and glacial acetic acid (99:1).

Stock standard solution—Dissolve an accurately weighed quantity of USP Pindolol RS in *Solvent mixture* to obtain a solution containing 10.0 mg per mL.

Standard solution 1—Dilute an accurately measured volume of *Stock standard solution* quantitatively and stepwise with *Solvent mixture* to obtain a solution containing 0.05 mg per mL.

Standard solution 2—Dilute 5.0 mL of *Standard solution 1* with *Solvent mixture* to 10.0 mL, and mix.

Test solution—[NOTE—Prepare this solution immediately before use, and apply last.] Dissolve 50 mg of Pindolol in 5.0 mL of *Solvent mixture*, and mix.

Procedure—Prepare a lined chromatographic chamber (see *Chromatography* ⟨621⟩) with a developing solvent consisting of a mixture of methylene chloride, methanol, and formic acid (75:23.5:1.5), and equilibrate for 15 minutes. Separately apply 5-µL portions of the *Stock standard solution*, each of the *Standard solutions*, and the *Test solution* to a thin-layer chromatographic plate coated with a 0.25-mm layer of chromatographic silica gel. Place the plate in the chromatographic chamber, and allow the solvent front to move about two-thirds of the length of the plate. Remove the plate from the chamber, immediately spray with the *p-Dimethylaminobenzaldehyde spray*, heat the plate at 50° for 20 minutes, and examine the chromatogram: no individual secondary spot observed in the chromatogram of the *Test solution* is greater in size or intensity than the principal spot observed in the chromatogram of *Standard solution 1*, corresponding to 0.5%, and the total of any such spots observed does not exceed 2.0%.

Assay—Dissolve about 200 mg of Pindolol, accurately weighed, in 80 mL of methanol, and titrate with 0.1 *N* hydrochloric acid VS, determining the end-point potentiometrically. Perform a blank determination (see *Titrimetry* ⟨541⟩), and make any necessary correction. Each mL of 0.1 *N* hydrochloric acid is equivalent to 24.83 mg of $C_{14}H_{20}N_2O_2$.

Pindolol Tablets

» Pindolol Tablets contain not less than 90.0 percent and not more than 110.0 percent of the labeled amount of $C_{14}H_{20}N_2O_2$.

Packaging and storage—Preserve in well-closed containers, protected from light.

Reference standard—*USP Pindolol Reference Standard*—Dry at 105° for 4 hours before using.

Identification—

A: Transfer 1 Tablet to a 10-mL volumetric flask, add 5 drops of water, and agitate until the Tablet disintegrates. Add chloroform to volume, and place in a sonic bath for 5 minutes. Mix, and filter through a funnel containing a small pledget of glass wool and anhydrous sodium sulfate. Pipet a volume of the solution, equivalent to 2 mg of pindolol, into a container with 300 mg of dried potassium bromide. Agitate to mix the contents, evaporate the chloroform with the aid of a current of air, grind the residue, and prepare a potassium bromide dispersion (*Test preparation*). Transfer 10 mg of USP Pindolol RS to a 10-mL volumetric flask, add chloroform to volume, and place in a sonic bath for 5 minutes. Mix, and pipet 2.0 mL of this solution into a container with 300 mg of dried potassium bromide. Agitate the container to mix the contents, evaporate the chloroform with the aid of a current of air, grind the residue, and prepare a potassium bromide dispersion (*Standard preparation*): the infrared absorption spectrum of the *Test preparation* exhibits maxima at the same wavelengths as that of the *Standard preparation.*

B: Examine the chromatograms obtained in the test for *Chromatographic purity:* the principal spot obtained from the *Test solution* is similar in R_f value, color, and intensity to that obtained from the *Stock standard solution.*

C: The retention time exhibited by pindolol in the chromatogram of the *Assay preparation* corresponds to that of pindolol in the chromatogram of the *Standard preparation* as obtained in the *Assay*.

Dissolution ⟨711⟩—

Medium: 0.1 *N* hydrochloric acid; 500 mL.

Apparatus 2: 50 rpm.

Time: 15 minutes.

Procedure—Determine the amount of $C_{14}H_{20}N_2O_2$ dissolved from ultraviolet absorbances at the wavelength of maximum absorbance at about 263 nm of filtered portions of the solution under test, suitably diluted with *Dissolution Medium*, if necessary, in comparison with a Standard solution having a known concentration of USP Pindolol RS in the same medium.

Tolerances—Not less than 80% (*Q*) of the labeled amount of $C_{14}H_{20}N_2O_2$ is dissolved in 15 minutes.

Uniformity of dosage units ⟨905⟩: meet the requirements.

Procedure for content uniformity—

Solvent mixture—Prepare a solution of methanol and chloroform (7:3).

Standard preparation—Dissolve an accurately weighed quantity of USP Pindolol RS in the *Solvent mixture*, and dilute quantitatively and stepwise with the *Solvent mixture* to obtain a solution having a known concentration of about 20 µg per mL.

Test preparation—Transfer 1 Tablet to a 200-mL volumetric flask. Add 5 mL of water, and sonicate until the Tablet has disintegrated. Add 100 mL of *Solvent mixture*, and shake for 45 minutes. Dilute with *Solvent mixture* to volume, and mix. Filter the solution, discarding the first portion of the filtrate. Dilute a portion of the filtrate quantitatively and stepwise with *Solvent mixture* to obtain a solution having a concentration of about 20 µg per mL.

Procedure—Concomitantly determine the absorbances of the *Standard preparation* and the *Test preparation* at the wavelength of maximum absorbance at about 266 nm, with a suitable spectrophotometer, using *Solvent mixture* as the blank. Calculate the quantity, in mg, of $C_{14}H_{20}N_2O_2$ in the Tablet taken by the formula:

$$(TC/D)(A_U/A_S),$$

in which *T* is the labeled quantity, in mg, of pindolol in the Tablet, *C* is the concentration, in µg per mL, of USP Pindolol RS in the *Standard preparation, D* is the concentration, in µg per mL, of pindolol in the *Test preparation*, based upon the labeled quantity per Tablet and the extent of dilution, and A_U and A_S are the absorbances of the solutions from the *Test preparation* and the *Standard preparation*, respectively.

Chromatographic purity—[NOTE—Protect solutions and chromatographic plate (after application of solutions) from light.]

p-Dimethylaminobenzaldehyde spray and *Solvent mixture*—Prepare as directed for *Chromatographic purity* under *Pindolol*.

Stock standard solution—Dissolve an accurately weighed quantity of USP Pindolol RS in *Solvent mixture* to obtain a solution containing 1.0 mg per mL.

Standard solution 1—Dilute an accurately measured volume of *Stock standard solution* quantitatively and stepwise with *Solvent mixture* to obtain a solution containing 0.005 mg per mL.

Standard solution 2—Dilute 5.0 mL of *Standard solution 1* with *Solvent mixture* to 10.0 mL, and mix.

Test solution—[NOTE—Prepare this solution immediately before use, and apply last.] Transfer a number of Tablets, equivalent to 100 mg of pindolol, to a 125-mL conical flask, add 50.0 mL of methanol, insert the stopper in the flask, and shake by mechanical means until the Tablets have disintegrated completely. Centrifuge a portion of the resultant suspension. Mix 10.0 mL of the clear supernatant liquid and 10.0 mL of *Solvent mixture*.

Procedure—Proceed as directed for *Procedure* in the test for *Chromatographic purity* under *Pindolol*, except to apply 40-µL portions of the *Stock standard solution, Standard solutions 1* and *2*, and the *Test solution*, and to heat the plate at 60° for 15 minutes: no individual secondary spot observed in the chromatogram of the *Test solution* is greater in size or intensity than the principal spot observed in the chromatogram of *Standard solution 1*, corresponding to 0.5%, and the total of any such spots observed does not exceed 3.0%.

Assay—

Ammonium carbonate solution—Dissolve 300 mg of ammonium carbonate in 50 mL of water, and mix.

Mobile phase—Prepare a filtered and degassed mixture of acetonitrile, methanol, and *Ammonium carbonate solution* (475:475:50), making adjustments if necessary (see *System Suitability* under *Chromatography* ⟨621⟩).

Internal standard solution—Dissolve 200 mg of nortriptyline hydrochloride in 1000 mL of methanol, and mix.

Standard preparation—Dissolve an accurately weighed quantity of USP Pindolol RS in *Internal standard solution*, and dilute quantitatively and stepwise with *Internal standard solution* to obtain a solution having a known concentration of about 0.2 mg per mL.

Assay preparation—Transfer not less than 20 Pindolol Tablets to a glass-stoppered, 500-mL conical flask, add 10 mL of water, and agitate to disintegrate the Tablets. Add 200 mL of *Internal standard solution*, mix, sonicate for 15 minutes, and filter into a 250-mL volumetric flask. Rinse the conical flask with two 20-mL portions of *Internal standard solution*, passing the rinsings through the filter and into the volumetric flask, dilute with *Internal standard solution* to volume, and mix. Transfer an accurately measured volume of this specimen solution, equivalent to about 20 mg of Pindolol, to a 100-mL volumetric flask, dilute with *Internal standard solution* to volume, and mix.

Chromatographic system (see *Chromatography* ⟨621⟩)—The liquid chromatograph is equipped with a 254-nm detector and a 4.6-mm × 25-cm column that contains packing L16. The flow rate is about 3 mL per minute. Chromatograph replicate injections of the *Standard preparation*, and record the peak responses as directed under *Procedure:* the relative standard deviation is not more than 2.0%, and the resolution, *R*, between pindolol and the internal standard is not less than 1.5.

Procedure—Separately inject equal volumes (about 10 µL) of the *Standard preparation* and the *Assay preparation* into the chromatograph, record the chromatograms, and measure the responses for the major peaks. The relative retention times are about 0.6 for pindolol and 1.0 for the internal standard. Calculate the quantity, in mg, of $C_{14}H_{20}N_2O_2$ in the volume of the specimen solution taken by the formula:

$$100C(R_U/R_S),$$

in which *C* is the concentration, in mg per mL, of USP Pindolol RS in the *Standard preparation*, and R_U and R_S are the relative response ratios obtained from the *Assay preparation* and the *Standard preparation*, respectively.

Piperacetazine

$C_{24}H_{30}N_2O_2S$ 410.57

Ethanone, 1-[10-[3-[4-(2-hydroxyethyl)-1-piperidinyl]propyl]-10*H*-phenothiazin-2-yl]-.

10-[3-[4-(2-Hydroxyethyl)piperidino]propyl]phenothiazin-2-yl methyl ketone [*3819-00-9*].

» Piperacetazine contains not less than 98.0 percent and not more than 101.5 percent of $C_{24}H_{30}N_2O_2S$, calculated on the dried basis.

Packaging and storage—Preserve in tight, light-resistant containers.

Reference standard—*USP Piperacetazine Reference Standard*—Dry in vacuum at 60° to constant weight before using.

NOTE—Throughout the following procedures, protect test or assay specimens, the Reference Standard, and solutions containing them, by conducting the procedures without delay, under subdued light, or using low-actinic glassware.

Identification—

A: The infrared absorption spectrum, determined in a 0.4-mm cell, of a 1 in 50 solution of Piperacetazine in chloroform, the specimen having been previously dried, exhibits maxima only at the same wavelengths as that of a similar solution of USP Piperacetazine RS.

B: The ultraviolet absorption spectrum of a 1 in 100,000 solution in 0.1 *N* hydrochloric acid exhibits maxima and minima at the same wavelengths as that of a similar solution of USP Piperacetazine RS, concomitantly measured, and the respective absorptivities, calculated on the dried basis, at the wavelength of maximum absorbance at about 240 nm do not differ by more than 3.0%.

Melting range, *Class I* ⟨741⟩: between 102° and 106°.

Loss on drying ⟨731⟩—Dry it in vacuum at 60° to constant weight: it loses not more than 1.0% of its weight.

Residue on ignition ⟨281⟩: not more than 0.1%.

Heavy metals, *Method II* ⟨231⟩: 0.002%.

Ordinary impurities ⟨466⟩—

Test solution: methanol.
Standard solution: methanol.
Eluant: methanol.
Visualization: 1.

Assay—Dissolve about 400 mg of Piperacetazine, accurately weighed, in about 50 mL of chloroform, add about 25 mL of glacial acetic acid and 2 drops of crystal violet TS, and titrate with 0.1 *N* perchloric acid VS to a green end-point. Perform a blank determination, and make any necessary correction. Each mL of 0.1 *N* perchloric acid is equivalent to 41.06 mg of $C_{24}H_{30}N_2O_2S$.

Piperacetazine Tablets

» Piperacetazine Tablets contain not less than 93.0 percent and not more than 107.0 percent of the labeled amount of $C_{24}H_{30}N_2O_2S$.

Packaging and storage—Preserve in well-closed, light-resistant containers.

Reference standard—*USP Piperacetazine Reference Standard*—Dry in vacuum at 60° to constant weight before using.

NOTE—Throughout the following procedures, protect test or assay specimens, the Reference Standard, and solutions containing them, by conducting the procedures without delay, under subdued light, or using low-actinic glassware.

Identification—Tablets meet the requirements under *Identification—Organic Nitrogenous Bases* ⟨181⟩.

Dissolution ⟨711⟩—

Medium: 0.1 *N* hydrochloric acid; 900 mL.
Apparatus 1: 100 rpm.
Time: 45 minutes.
Procedure—Determine the amount of $C_{24}H_{30}N_2O_2S$ dissolved from ultraviolet absorbances at the wavelength of maximum absorbance at about 240 nm of filtered portions of the solution under test, suitably diluted with *Dissolution Medium*, if necessary, in comparison with a Standard solution having a known concentration of USP Piperacetazine RS in the same medium.
Tolerances—Not less than 75% (*Q*) of the labeled amount of $C_{24}H_{30}N_2O_2S$ is dissolved in 45 minutes.

Uniformity of dosage units ⟨905⟩: meet the requirements.

Assay—Weigh and finely powder not less than 20 Piperacetazine Tablets. Transfer an accurately weighed portion of the powder, equivalent to about 8 mg of piperacetazine, to a separator containing 50 mL of water, and add 400 µL of sodium hydroxide solution (1 in 25). Extract with four 20-mL portions of chloroform, and filter the chloroform extracts through filter paper that previously has been moistened with chloroform into a 100-mL volumetric flask. Dilute with chloroform passed through the same filter to volume, and mix. Transfer 10.0 mL of this solution to a second 100-mL volumetric flask, dilute with chloroform to volume, and mix. Concomitantly determine the absorbances of this solution and of a Standard solution of USP Piperacetazine RS in the same medium, having a known concentration of about 8 µg per mL, in 1-cm cells at the wavelength of maximum absorbance at about 283 nm, with a suitable spectrophotometer, using chloroform as the blank. Calculate the quantity, in mg, of $C_{24}H_{30}N_2O_2S$ in the portion of Tablets taken by the formula:

$$C(A_U/A_S),$$

in which *C* is the concentration, in µg per mL, of USP Piperacetazine RS in the Standard solution, and A_U and A_S are the absorbances of the assay solution and the Standard solution, respectively.

Sterile Piperacillin Sodium

$C_{23}H_{26}N_5NaO_7S$ 539.54

4-Thia-1-azabicyclo[3.2.0]heptane-2-carboxylic acid, 6-[[[[(4-ethyl-2,3-dioxo-1-piperazinyl)carbonyl]amino]phenylacetyl]amino]-3,3-dimethyl-7-oxo-, monosodium salt, [2*S*-[2α,5α,6β(*S**)]].

Sodium (2*S*,5*R*,6*R*)-6-[(*R*)-2-(4-ethyl-2,3-dioxo-1-piperazinecarboxamido)-2-phenylacetamido]-3,3-dimethyl-7-oxo-4-thia-1-azabicyclo[3.2.0]heptane-2-carboxylate [*59703-84-3*].

» Sterile Piperacillin Sodium is piperacillin sodium suitable for parenteral use. It has a potency equivalent to not less than 863 µg and not more than 1007 µg of piperacillin ($C_{23}H_{27}N_5O_7S$) per mg, calculated on the anhydrous basis. In addition, where packaged for dispensing, it contains the equivalent of not less than 90.0 percent and not more than 120.0 percent of the labeled amount of piperacillin.

Packaging and storage—Preserve in *Containers for Sterile Solids* as described under *Injections* ⟨1⟩.

Reference standards—*USP Piperacillin Reference Standard*—Do not dry before using. *USP Ampicillin Reference Standard*—Do not dry before using.

Constituted solution—At the time of use, the constituted solution prepared from Sterile Piperacillin Sodium meets the requirements for *Constituted Solutions* under *Injections* ⟨1⟩.

Identification—The chromatogram obtained from the *Assay preparation* in the *Assay* exhibits a major peak for piperacillin the retention time of which corresponds to that exhibited by the *Standard preparation*, and the chromatogram compares qualitatively to that obtained from the *Standard preparation*.

Pyrogen—It meets the requirements of the *Pyrogen Test* ⟨151⟩, the test dose being 1.0 mL per kg of a solution prepared by diluting Sterile Piperacillin Sodium with pyrogen-free saline TS to a concentration of 150 mg of piperacillin per mL.

Sterility—It meets the requirements under *Sterility Tests* ⟨71⟩, when tested as directed in the section, *Test Procedures Using Membrane Filtration*.

pH ⟨791⟩: between 5.5 and 7.5, in a solution containing 400 mg per mL.

Water, *Method I* ⟨921⟩: not more than 1.0%, the method of *Test preparation* described for hygroscopic substances being used.

Particulate matter ⟨788⟩: meets the requirements under *Small-volume Injections*.

Other requirements—It meets the requirements for *Uniformity of Dosage Units* ⟨905⟩ and *Labeling* under *Injections* ⟨1⟩.

Assay—

Mobile phase—Mix 450 mL of methanol, 100 mL of 0.2 *M* monobasic sodium phosphate, and 3 mL of tetrabutylammonium hydroxide solution (1 in 10). The ratio of components may be varied to meet system suitability requirements. Dilute with water to 1000 mL, adjust with phosphoric acid to a pH of 5.5, and degas.

Standard preparation—Transfer about 20 mg of USP Piperacillin RS, accurately weighed, to a 50-mL volumetric flask. Add about 30 mL of *Mobile phase*, and shake to dissolve. Dilute with *Mobile phase* to volume, and mix.

Ampicillin-piperacillin solution—Dissolve 20 mg of USP Ampicillin RS in 70 mL of *Mobile phase*, add 10 mL of *Standard preparation*, and mix.

Assay preparation 1—Using about 20 mg of Sterile Piperacillin Sodium, accurately weighed, proceed as directed under *Standard preparation.*

Assay preparation 2—Constitute Sterile Piperacillin Sodium in a volume of water, accurately measured, corresponding to the volume of solvent specified in the labeling. Withdraw all of the withdrawable contents, using a suitable hypodermic needle and syringe, and dilute quantitatively with *Mobile phase* to obtain a solution containing about 0.4 mg of piperacillin per mL.

Chromatographic system (see *Chromatography* ⟨621⟩)—The liquid chromatograph is equipped with a 254-nm detector and a 4.6-mm × 25-cm column that contains packing L1. The flow rate is about 1 mL per minute. Chromatograph three replicate injections of the *Standard preparation*, and record the peak responses as directed under *Procedure:* the relative standard deviation is not more than 2.0%, and the resolution (R) between piperacillin and ampicillin is not less than 15.

Procedure—Separately inject equal volumes (about 10 µL) of the *Standard preparation*, the *Ampicillin-piperacillin solution*, and the *Assay preparations* into the chromatograph by means of a suitable microsyringe or sampling valve, record the chromatograms, and measure the responses for the major peaks. The retention times are about 3.5 minutes for ampicillin and 12 minutes for piperacillin. Calculate the potency, in µg of piperacillin ($C_{23}H_{27}N_5O_7S$) per mg, of the Sterile Piperacillin Sodium taken by the formula:

$$(50CP/W)(r_U/r_S),$$

in which C is the concentration, in mg per mL, of USP Piperacillin RS in the *Standard preparation*, P is the potency, in µg per mg, of USP Piperacillin RS, W is the weight, in mg, of Sterile Piperacillin Sodium taken to prepare *Assay preparation 1*, and r_U and r_S are the peak responses obtained from the *Assay preparations* and the *Standard preparation*, respectively. Calculate the quantity, in mg, of piperacillin in the container taken by the formula:

$$(L/D)(CP/1000)(r_U/r_S),$$

in which L is the labeled quantity, in mg, of piperacillin in the container, and D is the concentration, in mg per mL, of piperacillin in *Assay preparation 2*, on the basis of the labeled quantity in the container and the extent of dilution.

Piperazine

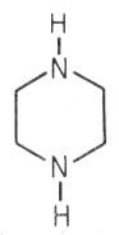

$C_4H_{10}N_2$ 86.14

Piperazine.

Piperazine [*110-85-0*].

» Piperazine contains not less than 98.0 percent and not more than 101.0 percent of $C_4H_{10}N_2$, calculated on the anhydrous basis.

Packaging and storage—Preserve in tight containers, protected from light.

Color of solution—Dissolve 10.0 g in water, and dilute with water to 50.0 mL: the solution has no more color than a standard solution prepared by adding 2.0 mL of ferric chloride CS to water and diluting with water to 50.0 mL, when compared in matched color-comparison tubes.

Identification—It responds to *Identification test A* under *Piperazine Citrate.*

Melting range ⟨741⟩: between 109° and 113°.

Water, *Method I* ⟨921⟩: not more than 2.0%.

Primary amines and ammonia—Dissolve 200 mg in 10 mL of water, add 1 mL of acetone and 0.5 mL of freshly prepared sodium nitroferricyanide solution (1 in 10), mix, and allow to stand for 10 minutes, accurately timed. Determine the absorbance of this solution at 520 nm and at 600 nm, using a reagent blank as the reference solution. The ratio A_{600}/A_{520} is not more than 0.5 (equivalent to about 0.7% of primary amines and ammonia).

Assay—Weigh accurately about 150 mg of Piperazine, and dissolve in 75 mL of glacial acetic acid. Titrate potentiometrically with 0.1 *N* perchloric acid VS, using a silver-glass electrode system. As the end-point is approached, warm the solution to 60° to 70°, then complete the titration. Perform a blank determination, and make any necessary correction. Each mL of 0.1 *N* perchloric acid is equivalent to 4.307 mg of $C_4H_{10}N_2$.

Piperazine Citrate

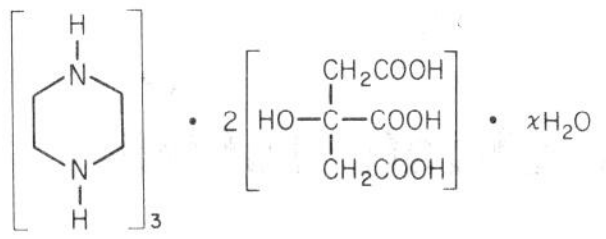

$(C_4H_{10}N_2)_3 \cdot 2C_6H_8O_7 \cdot xH_2O$ (anhydrous) 642.66

Piperazine, 2-hydroxy-1,2,3-propanetricarboxylate (3:2), hydrate.

Piperazine citrate (3:2) hydrate [*41372-10-5*].

Anhydrous 642.66 [*144-29-6*].

» Piperazine Citrate contains not less than 98.0 percent and not more than 100.5 percent of $(C_4H_{10}N_2)_3 \cdot 2C_6H_8O_7$, calculated on the anhydrous basis.

Packaging and storage—Preserve in well-closed containers.

Identification—

A: Dissolve about 200 mg in 5 mL of 3 *N* hydrochloric acid, and add, with stirring, 1 mL of sodium nitrite solution (1 in 2). Chill in an ice bath for 15 minutes, stirring if necessary, to induce crystallization, filter the precipitate on a sintered-glass funnel, wash with 10 mL of cold water, and dry at 105°: the *N,N'*-dinitrosopiperazine so obtained melts between 156° and 160°.

B: It responds to the tests for *Citrate* ⟨191⟩.

Water, *Method I* ⟨921⟩: not more than 12.0%.

Primary amines and ammonia—Dissolve 500 mg in 10 mL of water. Add 1 mL of 2.5 *N* sodium hydroxide, 1 mL of acetone, and 1 mL of sodium nitroferricyanide TS. Mix, and allow to stand for 10 minutes, accurately timed. Determine the absorbance of this solution at 520 nm and at 600 nm, using a blank consisting of the same quantities of the same reagents, but substituting water for the sodium hydroxide solution. The ratio A_{600}/A_{520} is not more than 0.50 (equivalent to about 0.7% of primary amines and ammonia).

Assay—Dissolve about 200 mg of Piperazine Citrate, accurately weighed, in 100 mL of glacial acetic acid TS, warming slightly if necessary to effect solution. Add crystal violet TS, and titrate with 0.1 *N* perchloric acid VS. Perform a blank determination, and make any necessary correction. Each mL of 0.1 *N* perchloric acid is equivalent to 10.71 mg of $(C_4H_{10}N_2)_3 \cdot 2C_6H_8O_7$.

Dilute with 0.01 *N* methanolic hydrochloric acid to volume, and mix. Centrifuge a portion of this mixture to obtain a clear solution. Transfer 10.0 mL of the solution so obtained to a 100-mL volumetric flask, dilute with 0.01 *N* methanolic hydrochloric acid to volume, and mix.

Procedure—Proceed as directed for *Procedure* in the *Assay* under *Piroxicam*. Calculate the quantity, in mg, of $C_{15}H_{13}N_3O_4S$ in the portion of the contents of Capsules taken by the formula:

$$1000C(r_U/r_S),$$

in which C is the concentration, in mg per mL, of USP Piroxicam RS in the *Standard preparation*, and r_U and r_S are the peak responses obtained from the *Assay preparation* and the *Standard preparation*, respectively.

Posterior Pituitary Injection

» Posterior Pituitary Injection is a sterile solution, in a suitable diluent, of material containing the polypeptide hormones having the property of causing the contraction of uterine, vascular, and other smooth muscle, which is prepared from the posterior lobe of the pituitary body of healthy, domestic animals used for food by man. Each mL of Posterior Pituitary Injection possesses oxytocic and pressor activities of not less than 85.0 percent and not more than 120.0 percent of those stated on the label in USP Posterior Pituitary Units.

Packaging and storage—Preserve in single-dose or in multiple-dose containers, preferably of Type I glass. Do not freeze.

Reference standard—*USP Posterior Pituitary Reference Standard*—Do not dry before using. Store at a temperature of 0° or below. Each mg represents 2.4 USP Posterior Pituitary Units of oxytocic activity and 2.1 USP Posterior Pituitary Units of vasopressor activity.

pH ⟨791⟩: between 2.5 and 4.5.

Other requirements—It meets the requirements under *Injections* ⟨1⟩.

Assay—

Oxytocic activity—Proceed with Posterior Pituitary Injection as directed in the *Assay* under *Oxytocin Injection*. Compute the potency of the Injection, in USP Units of oxytocic activity per mL, and the approximate log confidence interval *L* (see *Confidence Intervals for Individual Assays* ⟨111⟩). If *L* exceeds 0.20, repeat the assay or increase the number of sets of responses until the approximate confidence interval of the combined results is 0.20 or less.

Pressor activity—Proceed with Posterior Pituitary Injection as directed in the *Assay* under *Vasopressin Injection*. Compute the potency of the Injection, in USP Units of vasopressin activity per mL, and the log confidence interval *L* (see *Confidence Intervals for Individual Assays* ⟨111⟩). If *L* exceeds 0.15, repeat the assay or increase the number of observations until the confidence interval of the combined results is 0.15 or less.

Plague Vaccine

» Plague Vaccine conforms to the regulations of the federal Food and Drug Administration concerning biologics (see *Biologics* ⟨1041⟩). It is a sterile suspension of plague bacilli (*Yersinia pestis*) of the 195/P strain grown on E medium, harvested and killed by the addition of formaldehyde. Its potency is determined with the specific mouse protection test on the basis of the U. S. Reference Plague Vaccine. It meets the requirements of the specific mouse test for inactivation.

Packaging and storage—Preserve at a temperature between 2° and 8°.

Expiration date—The expiration date is not later than 18 months after date of issue from manufacturer's cold storage (5°, 1 year).

Labeling—Label it to state that it is to be well shaken before use and that it is not to be frozen.

Potency test—Use not less than 5 groups of 7-week-old mice, consisting of 20 animals in each group, for immunization with U. S. Reference Plague Vaccine and equivalent groups for immunization with Vaccine under test and subsequent challenge to determine the immunizing dose protective to 50% of the animals (ED_{50}). Prepare serial 3-fold dilutions of each vaccine, starting with 1:5 or 1:15 as determined for suitability by preliminary tests. Inoculate each mouse intraperitoneally with 0.2 mL of the dose to which it was assigned at the commencement of the test and again 7 days later. Seven days following the last injection challenge each mouse subcutaneously with a freshly prepared suspension of virulent *Yersinia pestis* (strain 195/P) in a dose of 0.2 mL containing 1500 ± 500 colony forming units (cfu) determined by plate count. Concurrently use similar groups of mice to determine the dose of challenge suspension lethal to 50% of the unvaccinated animals (LD_{50}), selecting suitable dilutions in preliminary tests. Observe all the animals for 14 days for death due to plague. Calculate the ED_{50} for the U. S. Reference Plague Vaccine and for the Vaccine under test by a suitable statistical procedure such as normal analysis, and the LD_{50} of the challenge suspension. The potency of the Vaccine under test, relative to that of the U. S. Reference Plague Vaccine as shown by the ratio of the respective ED_{50} is not less than 0.5, provided that the LD_{50} of the challenge suspension is not more than 12 cfu.

Plantago Seed

» Plantago Seed is the cleaned, dried, ripe seed of *Plantago psyllium* Linné, or of *Plantago indica* Linné (*Plantago arenaria* Waldstein et Kitaibel), known in commerce as Spanish or French Psyllium Seed; or of *Plantago ovata* Forskal, known in commerce as Blond Psyllium or Indian Plantago Seed (Fam. Plantaginaceae).

Packaging and storage—Preserve in well-closed containers, secure against insect attack (see *Vegetable and Animal Drugs—Preservation*, in the *General Notices*).

Botanic characteristics—

Unground Plantago psyllium Seed—Ovate to ovate-elongate, concavo-convex; mostly from 1.3 to 2.7 mm in length, rarely up to 3 mm, and from 600 µm to 1.1 mm in width. It is light brown to moderate brown, darker along the margin, and very glossy; the convex dorsal surface exhibiting a lighter colored longitudinal area extending nearly the length of the seed and representing the embryo lying beneath the seed coat, and showing a sometimes indistinct transverse groove nearer the broader end. The concave ventral surface has a deep cavity, in the center of the base of which is an oval, yellowish white hilum.

Unground Plantago indica Seed—Ovate-oblong to elliptical, concavo-convex; from 1.6 to 3 mm in length and from 1 to 1.5 mm in width. Externally it is dark reddish brown to moderate yellowish brown, occasionally somewhat glossy, often dull, rough, and reticulate; the convex dorsal surface having a longitudinal lighter colored area extending lengthwise along the center and beneath the seed coat, and a median transverse groove, dent, or fissure. The ventral surface has a deep concavity, the edges somewhat flattened and frequently forming a sharp indented angle with the base of the cavity, the latter showing a light colored oval hilum.

Unground Plantago ovata Seed—Broadly elliptical to ovate, boat-shaped, from 2 to 3.5 mm in length and from 1 to 1.5 mm in width. It is pale brown to moderate brown with a dull surface, the convex surface having a small, elongated, glossy brown spot. The concave surface has a deep cavity, in the center of the base of which occurs a hilum covered with a thin membrane.

Odor and taste—All varieties of Plantago Seed are nearly odorless and have a bland mucilaginous taste.

Histology—Plantago Seed is reniform in median transverse sections. Its seed coat has a colorless epidermis of mucilaginous cells whose radial and outer walls break down to form layers of mucilage when brought into contact with water, and a reddish brown to yellow pigment layer in the seeds of *Plantago indica* and *Plantago psyllium*, a broad endosperm with thick-walled outer palisade cells, and irregular inner endosperm cells; and a straight embryo extending lengthwise through the center. The endosperm and embryo cells contain fixed oil and aleurone grains, the latter being rounded, oval, pyriform, or irregularly shaped, from 2 to 8 µm in diameter.

Water absorption—Place 1 g of Plantago Seed in a 25-mL graduated cylinder, add water to the 20-mL mark, and shake the cylinder at intervals during 24 hours. Allow the seeds to settle for 12 hours, and note the total volume occupied by the swollen seeds: the seeds of *Plantago psyllium* occupy a volume of not less than 14 mL, those of *Plantago ovata* not less than 10 mL, and those of *Plantago indica* not less than 8 mL.

Total ash ⟨561⟩: not more than 4.0% of total ash.

Acid-insoluble ash ⟨561⟩: not more than 1.0% of acid-insoluble ash.

Foreign organic matter ⟨561⟩: not more than 0.50%.

Plasma Protein Fraction

» Plasma Protein Fraction conforms to the regulations of the federal Food and Drug Administration concerning biologics (640.90 to 640.96) (see *Biologics* ⟨1041⟩). It is a sterile preparation of serum albumin and globulin obtained by fractionating material (source blood, plasma, or serum) from healthy human donors, the source material being tested for the absence of hepatitis B surface antigen. It is made by a process that yields a product having protein components of approved composition and sedimentation coefficient content. Not less than 83 percent of its total protein is albumin and not more than 17 percent of its total protein consists of alpha and beta globulins. Not more than 1 percent of its total protein has the electrophoretic properties of gamma globulin. It is a solution containing, in each 100 mL, 5 g of protein, and it contains not less than 94 percent and not more than 106 percent of the labeled amount. It contains no added antimicrobial agent, but it contains sodium acetyltryptophanate with or without sodium caprylate as a stabilizing agent. It has a sodium content of not less than 130 mEq per liter and not more than 160 mEq per liter and a potassium content of not more than 2 mEq per liter. It has a pH between 6.7 and 7.3, measured in a solution diluted to contain 1 percent of protein with 0.15 *M* sodium chloride. It meets the requirements of the test for heat stability.

Packaging and storage—Preserve at the temperature indicated on the label.

Expiration date—The minimum expiration date is not later than 5 years after issue from manufacturer's cold storage (5°, 1 year) if labeling recommends storage between 2° and 10°; not later than 3 years after issue from manufacturer's cold storage (5°, 1 year) if labeling recommends storage at temperatures not higher than 30°.

Labeling—Label it to state that it is not to be used if it is turbid and that it is to be used within 4 hours after the container is entered. Label it also to state the osmotic equivalent in terms of plasma and the sodium content.

Plaster, Salicylic Acid—*see* Salicylic Acid Plaster

Platelet Concentrate

» Platelet Concentrate conforms to the regulations of the federal Food and Drug Administration concerning biologics (640.20 to 640.27) (see *Biologics* ⟨1041⟩). It contains the platelets taken from plasma obtained by whole blood collection, by plasmapheresis, or by plateletpheresis, from a single suitable human donor of whole blood; or from a plasmapheresis donor; or from a plateletpheresis donor who meets the criteria described in the product license application (in which case the collection procedure is as described therein), except where a licensed physician has determined that the recipient is to be transfused with the platelets from a specific donor (in which case the plateletpheresis procedure is performed under the supervision of a licensed physician who is aware of the health status of the donor and has certified that the donor's health permits such procedure). In all cases, the collection of source material is made by a single, uninterrupted venipuncture with minimal damage to and manipulation of the donor's tissue. Concentrate consists of such platelets suspended in a specified volume of the original plasma, the separation of plasma and resuspension of the platelets being done in a closed system, within 4 hours of collection of the whole blood or plasma. The separation of platelets is by a procedure shown to yield an unclumped product without visible hemolysis, with a content of not less than 5.5×10^{10} platelets per unit in not less than 75 percent of the units tested, and the volume of original plasma used for resuspension of the separated platelets is such that the product has a pH of not less than 6 during the storage period when kept at the selected storage temperature, the selected storage temperature and corresponding volume of resuspension plasma being either 30 mL to 50 mL of plasma for storage at 20° to 24°, or 20 mL to 30 mL of plasma for storage at 1° to 6°. It meets the aforementioned requirements for platelet count, pH, and actual plasma volume, when tested 72 hours after preparation.

Packaging and storage—Preserve in hermetic containers of colorless, transparent, sterile, pyrogen-free Type I or Type II glass, or of a suitable plastic material (see *Transfusion and Infusion Assemblies* ⟨161⟩). Preserve at the temperature relevant to the volume of resuspension plasma, either between 20° and 24° or between 1° and 6°, the latter except during shipment, when the temperature may be between 1° and 10°.

Expiration time—The expiration time is not more than 72 hours from the time of collection of the source material.

Labeling—In addition to the labeling requirements of Whole Blood applicable to this product, label it to state the volume of original plasma present, the kind and volume of anticoagulant solution present in the original plasma, the blood group designation of the source blood, and the hour of expiration on the stated expiration date. Where labeled for storage at 20° to 24°, label it also to state that a continuous gentle agitation shall be maintained, or where labeled for storage at 1° to 6°, to state that such agitation is optional. Label it also with the type and result of a serologic test for syphilis, or to indicate that it was nonreactive in such test; with the type and result of a test for hepatitis B surface antigen, or to indicate that it was nonreactive in such test; with a warning that it is to be used as soon as possible but not more than 4 hours after entering the container; to state that a filter is to be used in the administration equipment; and to state that the instruction circular provided is to be consulted for directions for use.

Pledgets, Erythromycin—*see* Erythromycin Pledgets

Plicamycin

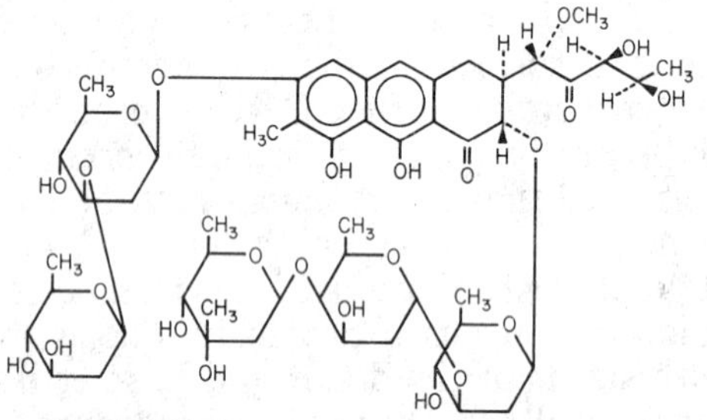

$C_{52}H_{76}O_{24}$ 1085.16

Plicamycin.

Plicamycin.

[2*S*-[2α,3β(1*R**,3*R**,4*S**)]]-6-[[2,6-Dideoxy-3-*O*-(2,6-dideoxy-β-D-*arabino*-hexopyranosyl)-β-D-*arabino*-hexopyranosyl]oxy]-2-[(*O*-2,6-dideoxy-3-*C*-methyl-β-D-*ribo*-hexopyranosyl-(1→4)-*O*-2,6-dideoxy-α-D-*lyxo*-hexopyranosyl-(1→3)-2,6-dideoxy-β-D-*arabino*-hexopyranosyl)oxy]-3-(3,4-dihydroxy-1-methyl-2-oxopentyl)-3,4-dihydro-8,9-dihydroxy-7-methyl-1(2*H*)-anthracenone [*18378-89-7*].

» Plicamycin has a potency of not less than 900 μg of $C_{52}H_{76}O_{24}$ per mg, calculated on the dried basis.

Packaging and storage—Preserve in tight, light-resistant containers, at a temperature between 2° and 8°.

Reference standard—*USP Plicamycin Reference Standard*—Dry in vacuum at a pressure not exceeding 5 mm of mercury at 25° for 4 hours before using.

Identification—

A: The infrared absorption spectrum of a potassium bromide dispersion of it, previously dried, exhibits maxima only at the same wavelengths as that of a similar preparation of USP Plicamycin RS.

B: The chromatogram obtained from the *Assay preparation* in the *Assay* exhibits a major peak for plicamycin the retention time of which corresponds to that exhibited by the *Standard preparation*, and the chromatogram compares qualitatively to that obtained from the *Standard preparation.*

Crystallinity ⟨695⟩: meets the requirements.

pH ⟨791⟩: between 4.5 and 5.5, in a solution containing 0.5 mg per mL.

Loss on drying ⟨731⟩—Dry about 100 mg, accurately weighed, in vacuum at a pressure not exceeding 5 mm of mercury at 25° for 4 hours: it loses not more than 8.0% of its weight.

Assay—[NOTE—Prepare solutions of plicamycin in low-actinic glassware.]

Mobile phase—Prepare a suitable filtered and degassed mixture of 650 mL of 0.01 *M* phosphoric acid and 350 mL of acetonitrile.

Standard preparations—Dissolve an accurately weighed quantity of USP Plicamycin RS in *Mobile phase* to obtain a solution having a concentration of 500 μg of plicamycin per mL. Dilute this solution with *Mobile phase* to obtain solutions containing 50, 100, and 150 μg of plicamycin per mL.

Assay preparation—Transfer about 5 mg of Plicamycin, accurately weighed, to a 50-mL volumetric flask. Dissolve in *Mobile phase*, dilute with *Mobile phase* to volume, and mix.

Chromatographic system (see *Chromatography* ⟨621⟩)—The liquid chromatograph is equipped with a 278-nm detector and a 4.6-mm × 25-cm column that contains packing L1. The flow rate is about 1.3 mL per minute. Chromatograph replicate injections of the *Standard preparation*, and record the peak responses as directed under *Procedure:* the relative standard deviation is not more than 2.0%.

Procedure—Separately inject equal volumes (about 10 μL) of the *Standard preparations* and the *Assay preparation* into the chromatograph by means of a suitable microsyringe or sampling valve, record the chromatograms, and measure the responses for the major peaks. The retention time is about 13 minutes for plicamycin. Plot the peak responses of the *Standard preparations* versus concentration, in μg per mL, of plicamycin, and draw the straight line best fitting the three plotted points. From the graph so obtained, determine the concentration, in μg per mL, of plicamycin in the *Assay preparation.* Calculate the potency, in μg of $C_{52}H_{76}O_{24}$ per mg, taken by the formula:

$$(50C/W),$$

in which C is the concentration, in μg per mL, of plicamycin in the *Assay preparation*, and W is the weight, in mg, of Plicamycin taken.

Plicamycin for Injection

» Plicamycin for Injection is a sterile, dry mixture of Plicamycin and Mannitol. It may contain a suitable buffer. It contains not less than 90.0 percent and not more than 110.0 percent of the labeled amount of $C_{52}H_{76}O_{24}$.

Packaging and storage—Preserve in light-resistant *Containers for Sterile Solids* as described under *Injections* ⟨1⟩, at a temperature between 2° and 8°.

Labeling—Label it with the mandatory instruction to consult the professional information for dosage and warnings, and with the warning that it is intended for hospital use only, under the direct supervision of a physician.

Reference standard—*USP Plicamycin Reference Standard*—Dry in vacuum at a pressure not exceeding 5 mm of mercury at 25° for 4 hours before using.

Constituted solution—At the time of use, the constituted solution prepared from Plicamycin for Injection meets the requirements for *Constituted Solutions* under *Injections* ⟨1⟩.

Identification—Transfer a suitable quantity to a centrifuge tube, add methanol to obtain a solution having a concentration of about 0.5 mg of plicamycin per mL, mix, and centrifuge to obtain a clear solution. On a suitable thin-layer chromatographic plate (see *Chromatography* ⟨621⟩), coated with a 0.25-mm layer of chromatographic silica gel mixture, apply 100 μL of this solution and 100 μL of a Standard solution of USP Plicamycin RS, similarly prepared. Allow the spots to dry, and develop the chromatograms in a solvent system consisting of a mixture chloroform and methanol (1:1) for about 60 minutes. Remove the plate from the developing chamber, mark the solvent front, and allow the solvent to evaporate. Spray the plate with a (1:1) mixture of ferric chloride solution (1 in 100) and potassium ferricyanide solution (1 in 100). Observe the blue spots under a long-wavelength ultraviolet light: the R_f value of the principal spot obtained from the test solution corresponds to that obtained from the Standard solution (R_f about 0.7). Spots of trace components at R_f values of about 0.4 and 0.5 are not more intense than similar spots obtained from the Standard solution.

Depressor substances—When diluted with Sterile Water for Injection to a concentration of 0.05 mg of plicamycin per mL, it meets the requirements of the *Depressor Substances Test* ⟨101⟩, the test dose being 1.0 mL per kg.

Pyrogen—When diluted with Sterile Water for Injection to a concentration of 50 µg of plicamycin per mL, it meets the requirements of the *Pyrogen Test* ⟨151⟩, the test dose being 1.0 mL per kg.

Sterility—It meets the requirements under *Sterility Tests* ⟨71⟩, when tested as directed in the section, *Test Procedures Using Membrane Filtration.*

pH ⟨791⟩: between 5.0 and 7.5, in the solution constituted as directed in the labeling.

Water, *Method I* ⟨921⟩: not more than 2.0%.

Assay—[NOTE—Prepare solutions of plicamycin in low-actinic glassware.]

Mobile phase, Standard preparation, and *Chromatographic system*—Proceed as directed in the *Assay* under *Plicamycin.*

Assay preparation—Dilute the contents of 1 container of Plicamycin for Injection quantitatively with *Mobile phase* to obtain a solution containing about 100 µg of plicamycin per mL.

Procedure—Proceed as directed for *Procedure* in the *Assay* under *Plicamycin.* Calculate the quantity, in mg, of $C_{52}H_{76}O_{24}$ in the container by the formula:

$$C(L/D),$$

in which C is the concentration, in µg per mL, of plicamycin in the *Assay preparation,* L is the labeled quantity, in mg, of plicamycin in the container, and D is the concentration, in µg per mL, of plicamycin in the *Assay preparation* on the basis of the labeled quantity in the container and the extent of dilution.

Podophyllum

» Podophyllum consists of the dried rhizomes and roots of *Podophyllum peltatum* Linné (Fam. Berberidaceae). It yields not less than 5.0 percent of podophyllum resin.

Botanic characteristics—

Podophyllum—Consists of nearly cylindrical rhizomes, jointed, compressed or flattened somewhat on upper and lower surfaces, and sometimes branched. It occurs as pieces of rhizome up to 20 cm in length, with internodes from 2 mm to 9 mm in diameter, some of the nodes being somewhat thickened. The rhizome is dusky red to light yellowish brown, longitudinally wrinkled or nearly smooth, with irregular, somewhat V-shaped scars of scale leaves; some of the nodes are annulate, the upper portion having large, circular, depressed stem-scars and buds or stem-bases. On the lower portion there are numerous root-scars or roots, the latter from 2 cm to 7 cm in length and about 2 mm in thickness. The fracture is short and weak, the fractured surface being yellowish orange to pale yellow or grayish white.

Histology—The rhizome shows an outer portion consisting of a brown epidermis, often necrosed, and 1 to 3 layers of brown to olive-brown suberized cells; a cortex about 20 cells in width, consisting chiefly of nearly isodiametric cells, the cells containing single or compound starch grains and resin masses and, in scattered cells of the nodes, rosette aggregates of calcium oxalate; a circle of from 16 to 34 open collateral vascular bundles, separated by rather wide medullary rays, each bundle containing a few lignified vessels, a more or less distinct cambium, and a rather large phloem. The pith is large, the cells being more or less rounded and containing starch grains and reddish brown resin masses. The roots show an epidermal layer of brownish suberized cells and a single row of hypodermal cells; a broad cortex of thin-walled nearly isodiametric cells; a distinct endodermis of tangentially elongated cells having uniformly thickened walls; and a 4- to 7-rayed vascular bundle.

Powdered Podophyllum—Pale brown to weak yellow. Has a slight odor and a disagreeably bitter and acrid taste. It contains numerous starch grains, simple or 2- to 6-compound, the individual grains being spheroidal, plano- to angular-convex, or polygonal, up to 20 µm in diameter; occasional rosette aggregates of calcium oxalate, up to 80 µm in diameter; vessels with simple pits or reticulate thickenings; fragments of starch- and resin-bearing parenchyma and reddish brown to yellow cork cells.

Indian podophyllum—*Podophyllum peltatum* is differentiated from *Podophyllum hexandrum* Royle (Indian podophyllum) by the reaction described in the test for *Distinction from resin of Indian podophyllum* under *Podophyllum Resin.*

Acid-insoluble ash ⟨561⟩: not more than 2.0%.

Foreign organic matter ⟨561⟩: not more than 2.0%.

Assay—Place 10 g of Podophyllum, in fine powder, in a 125-mL conical flask, add 35 mL of alcohol, and reflux on a steam bath for 3 hours. Transfer the mixture to a small percolator, and percolate slowly with warm alcohol until the percolate measures 95 mL. Cool, add sufficient alcohol to the percolate to make it measure 100.0 mL, and mix. Transfer 10.0 mL of this percolate to a separator, and add 10 mL of chloroform and 10 mL of dilute hydrochloric acid (7 in 500). Shake the mixture, allow it to separate, draw off the alcohol-chloroform layer into a second separator, and wash the acid layer with three 15-mL portions of a mixture of chloroform and alcohol (2:1), adding the washings to the second separator. Add 10 mL of dilute hydrochloric acid (7 in 500) to the combined extract and washings, again shake the mixture, allow it to separate, and draw off the alcohol-chloroform layer into a tared vessel. Wash the acid layer three times with 15-mL portions of the alcohol-chloroform mixture, adding the washings to the tared vessel. Evaporate the combined extracts on a steam bath to approximately 1 mL, add 5 mL of dehydrated alcohol, again evaporate to dryness, and dry the residue at 80° for 4 hours: the weight of this residue is the weight of resin in 1 g of the Podophyllum taken.

Podophyllum Resin

» Podophyllum Resin is the powdered mixture of resins extracted from Podophyllum by percolation with Alcohol and subsequent precipitation from the concentrated percolate upon addition to acidified water. It contains not less than 40.0 percent and not more than 50.0 percent of hexane-insoluble matter.

Caution—Podophyllum Resin is highly irritating to the eye and to mucous membranes in general.

Packaging and storage—Preserve in tight, light-resistant containers.

Identification—

A: It is soluble in 1 *N* potassium hydroxide or in 1 *N* sodium hydroxide, forming a yellow liquid, which gradually becomes darker on standing and from which the resin is precipitated by hydrochloric acid.

B: A hot solution of it deposits most of its solids on cooling, and if the cooled liquid is filtered, the filtrate turns brown upon the addition of a few drops of ferric chloride TS.

Residue on ignition ⟨281⟩: not more than 1.5%.

Distinction from resin of Indian podophyllum—Add 400 mg to 3 mL of 60 percent alcohol, then add 0.5 mL of 1 *N* potassium hydroxide, shake the mixture gently, and allow to stand for 2 hours: it does not gelatinize.

Hexane-insoluble matter—Transfer about 1 g of Podophyllum Resin, accurately weighed, to a glass-stoppered, 100-mL conical flask, add 30.0 mL of chloroform, insert the stopper tightly, and shake for 30 minutes, using a mechanical wrist-action shaker, or equivalent. Filter with suction through a medium- or fine-porosity, sintered-glass filter, into a small filter flask. Wash the conical flask and the filter with two 5-mL portions of chloroform, adding the washings to the filtrate. Transfer the filtrate with the aid of chloroform to a 50-mL volumetric flask, add chloroform to volume, and mix. Pipet 20 mL of the resulting solution into a 250-mL conical flask containing 160 mL of solvent hexane. Gently swirl, allow to stand for 10 minutes, and transfer the resulting precipitate to a tared, fine-porosity, sintered-glass filter, wash the flask and the precipitate with two 20-mL portions of solvent hexane, dry the precipitate at 70° for 1 hour, and weigh the *Hexane-*

insoluble matter so obtained. Multiply by 2.5 to find the amount present in the quantity of Podophyllum Resin taken.

Podophyllum Resin Topical Solution

» Podophyllum Resin Topical Solution is a solution in Alcohol consisting of Podophyllum Resin and an alcoholic extract of Benzoin. It contains, in each 100 mL, not less than 10 g and not more than 13 g of hexane-insoluble matter.

Caution—Podophyllum Resin Topical Solution is highly irritating to the eye and to mucous membranes in general.

Packaging and storage—Preserve in tight, light-resistant containers.

Identification—

A: A 1 in 5 solution in chloroform is levorotatory.

B: The precipitate obtained as directed in the test for *Hexane-insoluble matter* responds to *Identification test A* under *Podophyllum Resin.*

Alcohol content ⟨611⟩: between 69.0% and 72.0% of C_2H_5OH.

Hexane-insoluble matter—Using a 10.0-mL quantity of the Topical Solution, proceed as directed for *Hexane-insoluble matter* under *Podophyllum Resin.* Multiply the weight of hexane-insoluble matter found by 2.5 to find the amount present in the 10.0 mL taken.

Polacrilin Potassium—*see* Polacrilin Potassium NF

Poliovirus Vaccine Inactivated

» Poliovirus Vaccine Inactivated conforms to the regulations of the federal Food and Drug Administration concerning biologics (630.1 to 630.6) (see *Biologics* ⟨1041⟩). It is a sterile aqueous suspension of inactivated poliomyelitis virus of Types 1, 2, and 3. The virus strains are grown separately in primary cell cultures of monkey kidney tissue, and from a virus suspension with a virus titer of not less than $10^{6.5}$ $TCID_{50}$ measured in comparison with the U. S. Reference Poliovirus of the corresponding type, are inactivated so as to reduce the virus titer by a factor of 10^{-8}, and after inactivation are combined in suitable proportions.

No extraneous protein, capable of producing allergenic effects upon injection into human subjects, is added to the final virus production medium. If animal serum is used at any stage, its calculated concentration in the final medium does not exceed 1 part per million. Suitable antimicrobial agents may be used during the production. The single strain harvests or virus pools prior to inactivation meet the requirements of the specific mouse, guinea pig, and rabbit tests for absence of B virus and *Mycobacterium tuberculosis*, and the tissue culture test for absence of SV-40 virus. The single strain or trivalent virus pools after inactivation meet the requirements of the specific tissue culture and monkey tests for absence of active poliovirus, the mouse test for absence of lymphocytic choriomeningitis virus, and the tissue culture safety test for absence of SV-40 virus or other active viruses. The Vaccine meets the requirements of the specific monkey potency test by virus neutralizing antibody production, based on the U. S. Reference Poliovirus Antiserum, such that the ratio of the geometric mean titer of the group of monkey serums representing the vaccine to the mean titer value of the reference serum is not less than 1.29 for Type 1, 1.13 for Type 2, and 0.72 for Type 3.

Packaging and storage—Preserve at a temperature between 2° and 8°.

Expiration date—The expiration date is not later than 1 year after date of issue from manufacturer's cold storage (5°, 1 year).

Labeling—Label it to state that it is to be well shaken before use. Label it also to state that it was prepared in monkey tissue cultures.

Poliovirus Vaccine Live Oral

» Poliovirus Vaccine Live Oral conforms to the regulations of the federal Food and Drug Administration concerning biologics (630.10 to 630.18) (see *Biologics* ⟨1041⟩). It is a preparation of a combination of the three types of live, attenuated polioviruses derived from strains of virus tested for neurovirulence in monkeys in comparison with the U. S. Reference Attenuated Poliovirus, Type 1, for such tests, and for immunogenicity, free from all demonstrable viable microbial agents except unavoidable bacteriophage, and found suitable for human immunization. The strains are grown, for purposes of vaccine production, separately in primary cell cultures of monkey renal tissue. Monovalent lots produced with each strain meet the requirements of the specific safety tests for absence of adventitious and other infective agents including polioviruses of other types, SV_{40}, SV_5 and foamy viruses, *Mycobacterium tuberculosis*, pox virus, lymphocytic choriomeningitis virus, Echo viruses, Coxsackie viruses, and B virus. These tests are made in rabbits, adult and suckling mice, and guinea pigs, as well as in monkey kidney, human tissue, and rabbit kidney cell cultures. The lots also meet the requirements of the specific monkey neurovirulence test in comparison with the Reference Attenuated Poliovirus, and the requirements of the specific in-vitro marker tests. The Vaccine meets the requirements of the specific tissue culture tests for live virus titer, in a single immunizing dose, of not less than $10^{5.4}$ to $10^{6.4}$ for Type 1, $10^{4.5}$ to $10^{5.5}$ for Type 2, and $10^{5.2}$ to $10^{6.2}$ for Type 3, using the U. S. Reference Poliovirus, Live, Attenuated of the corresponding type for correlation of such titers. The Vaccine is filtered to prevent possible inclusion of bacteria in the final product.

Packaging and storage—Preserve in single-dose or in multiple-dose containers at a temperature that will maintain ice continuously in a solid state. Preserve thawed Vaccine at a temperature between 2° and 8°.

Expiration date—The expiration date is not later than 1 year after date of issue from manufacturer's cold storage (−10°, 1 year).

Labeling—Label the Vaccine to state that it may be thawed and refrozen not more than 10 times, provided that the thawed material is kept refrigerated and the total cumulative duration of the thaw is not more than 24 hours. Label the Vaccine to state the type of tissue in which it was prepared and to state that it is not for injection.

Poloxamer—*see* Poloxamer NF

Polycarbophil, Calcium—*see* Calcium Polycarbophil

Polyethylene Excipient—*see* Polyethylene Excipient NF

Polyethylene Glycol—*see* Polyethylene Glycol NF

Polyethylene Glycol Ointment—*see* Polyethylene Glycol Ointment NF

Polyethylene Oxide—*see* Polyethylene Oxide NF

Polymyxin B Sulfate

Polymyxin B, sulfate.
Polymyxin B sulfate [*1405-20-5*].

» Polymyxin B Sulfate is the sulfate salt of a kind of polymyxin, a substance produced by the growth of *Bacillus polymyxa* (Prazmowski) Migula (Fam. Bacillaceae), or a mixture of two or more such salts. It has a potency of not less than 6000 Polymyxin B Units per mg, calculated on the dried basis.

Packaging and storage—Preserve in tight, light-resistant containers.

Labeling—Where packaged for prescription compounding, the label states the number of Polymyxin B Units in the container and per milligram, that it is not intended for manufacturing use, that it is not sterile, and that its potency cannot be assured for longer than 60 days after opening.

Reference standard—*USP Polymyxin B Sulfate Reference Standard*—Dry in vacuum at a pressure not exceeding 5 mm of mercury at 60° for 3 hours before using.

Identification—

A: Dissolve 2 mg in 5 mL of water, add 0.5 mL of ninhydrin solution (1 in 1000) and 2 drops of pyridine, then boil for 1 minute, and cool: a blue color is produced.

B: Dissolve 2 mg in 5 mL of water, add 5 mL of 2.5 *N* sodium hydroxide, mix, and add 5 drops of cupric sulfate solution (1 in 100), shaking after the addition of each drop: a reddish violet color is produced.

pH ⟨791⟩: between 5.0 and 7.5, in a solution containing 5 mg per mL.

Loss on drying ⟨731⟩—Dry about 100 mg, accurately weighed, in a capillary-stoppered bottle in vacuum at 60° for 3 hours: it loses not more than 7.0% of its weight.

Other requirements—If for prescription compounding, it meets the requirements for *Residue on ignition* under *Sterile Polymyxin B Sulfate.*

Assay—Proceed with Polymyxin B Sulfate as directed under *Antibiotics—Microbial Assays* ⟨81⟩.

Sterile Polymyxin B Sulfate

» Sterile Polymyxin B Sulfate is Polymyxin B Sulfate suitable for parenteral use. It has a potency of not less than 6000 Polymyxin B Units per mg, calculated on the dried basis. In addition, where packaged for dispensing, it contains not less than 90.0 percent and not more than 120.0 percent of the labeled amount of polymyxin B.

Packaging and storage—Preserve in *Containers for Sterile Solids* as described under *Injections* ⟨1⟩, protected from light.

Labeling—Label it to indicate that where it is administered intramuscularly and/or intrathecally, it is to be given only to patients hospitalized so as to provide constant supervision by a physician.

Reference standard—*USP Polymyxin B Sulfate Reference Standard*—Dry in vacuum at a pressure not exceeding 5 mm of mercury at 60° for 3 hours before using.

Constituted solution—At the time of use, the constituted solution prepared from Sterile Polymyxin B Sulfate meets the requirements for *Constituted Solutions* under *Injections* ⟨1⟩.

Pyrogen—It meets the requirements of the *Pyrogen Test* ⟨151⟩, the test dose being 1.0 mL per kg of a solution in pyrogen-free saline TS containing 20,000 Polymyxin B Units per mL.

Sterility—It meets the requirements under *Sterility Tests* ⟨71⟩, when tested as directed in the section, *Test Procedures Using Membrane Filtration.*

Particulate matter ⟨788⟩: meets the requirements under *Small-volume Injections.*

Residue on ignition ⟨281⟩: not more than 5.0%, the charred residue being moistened with 2 mL of nitric acid and 5 drops of sulfuric acid.

Heavy metals, *Method II* ⟨231⟩: not more than 0.01%.

Other requirements—It responds to the *Identification tests* and meets the requirements for *pH* and *Loss on drying* under *Polymyxin B Sulfate.* Where packaged for dispensing, it meets also the requirements for *Uniformity of Dosage Units* ⟨905⟩ and for *Labeling* under *Injections* ⟨1⟩. Where intended for use in preparing sterile ophthalmic dosage forms, it is exempt from the requirements for *Pyrogen, Particulate matter,* and *Heavy metals.*

Assay—

Assay preparation 1—Dissolve a suitable quantity of Sterile Polymyxin B Sulfate, accurately weighed, in water, to obtain a solution containing about 10,000 Polymyxin B Units per mL, and dilute quantitatively with *Buffer No. 6* (see *Media and Diluents* under *Antibiotics—Microbial Assays* ⟨81⟩) to obtain a solution containing a convenient number of Polymyxin B Units per mL.

Assay preparation 2 (where it is packaged for dispensing and is represented as being in a single-dose container)—Constitute Sterile Polymyxin B Sulfate in a volume of water, accurately measured, corresponding to the volume of solvent specified in the labeling. Withdraw all of the withdrawable contents, using a suitable hypodermic needle and syringe, and dilute quantitatively with *Buffer No. 6* to obtain a solution containing a convenient number of Polymyxin B Units per mL.

Assay preparation 3 (where the label states the quantity of polymyxin B in a given volume of constituted solution)—Constitute 1 container of Sterile Polymyxin B Sulfate in a volume of water, accurately measured, corresponding to the volume of solvent specified in the labeling. Dilute an accurately measured volume of the constituted solution quantitatively with *Buffer No. 6* to obtain a solution containing a convenient number of Polymyxin B Units per mL.

Procedure—Proceed as directed under *Antibiotics—Microbial Assays* ⟨81⟩, using an accurately measured volume of *Assay preparation* diluted quantitatively with *Buffer No. 6* to yield a *Test Dilution* having a concentration assumed to be equal to the median dose level of the Standard.

Polymyxin B Sulfate Topical Aerosol, Bacitracin and—*see* Bacitracin and Polymyxin B Sulfate Topical Aerosol

Polymyxin B Sulfates and Bacitracin Ointment, Neomycin and—*see* Neomycin and Polymyxin B Sulfates and Bacitracin Ointment

Polymyxin B Sulfates, Bacitracin, and Hydrocortisone Acetate Ointment, Neomycin and—*see* Neomycin and Polymyxin B Sulfates, Bacitracin, and Hydrocortisone Acetate Ointment

Polymyxin B Sulfates, Bacitracin, and Hydrocortisone Acetate Ophthalmic Ointment, Neomycin and—*see* Neomycin and Polymyxin B Sulfates, Bacitracin, and Hydrocortisone Acetate Ophthalmic Ointment

Polymyxin B Sulfates and Bacitracin Ophthalmic Ointment, Neomycin and—*see* Neomycin and Polymyxin B Sulfates and Bacitracin Ophthalmic Ointment

Polymyxin B Sulfate Ointment, Bacitracin Zinc and—*see* Bacitracin Zinc and Polymyxin B Sulfate Ointment

Polymyxin B Sulfates, Bacitracin Zinc, and Hydrocortisone Ointment, Neomycin and—*see* Neomycin and Polymyxin B Sulfates, Bacitracin Zinc, and Hydrocortisone Ointment

Polymyxin B Sulfates, Bacitracin Zinc, and Hydrocortisone Ophthalmic Ointment, Neomycin and—*see* Neomycin and Polymyxin B Sulfates, Bacitracin Zinc, and Hydrocortisone Ophthalmic Ointment

Polymyxin B Sulfates, Bacitracin Zinc, and Hydrocortisone Acetate Ophthalmic Ointment, Neomycin and—*see* Neomycin and Polymyxin B Sulfates, Bacitracin Zinc, and Hydrocortisone Acetate Ophthalmic Ointment

Polymyxin B Sulfates, Bacitracin Zinc, and Lidocaine Ointment, Neomycin and—*see* Neomycin and Polymyxin B Sulfates, Bacitracin Zinc, and Lidocaine Ointment

Polymyxin B Sulfate and Bacitracin Zinc Topical Aerosol

» Polymyxin B Sulfate and Bacitracin Zinc Topical Aerosol contains the equivalent of not less than 90.0 percent and not more than 130.0 percent of the labeled amounts of polymyxin B and bacitracin.

Packaging and storage—Preserve in pressurized containers, and avoid exposure to excessive heat.

Reference standards—*USP Polymyxin B Sulfate Reference Standard*—Dry in vacuum at a pressure not exceeding 5 mm of mercury at 60° for 3 hours before using. *USP Bacitracin Zinc Reference Standard*—Dry in vacuum at a pressure not exceeding 5 mm of mercury at 60° for 3 hours before using.

Microbial limits—Collect aseptically in a suitable container half the contents expelled from 5 containers, dissolve in 500 mL of *Fluid A* containing 0.25 g of sodium thioglycollate and adjusted with sodium hydroxide to a pH of 6.6 ± 0.6, filter through a membrane filter as directed for *Test Procedures Using Membrane Filtration* under *Sterility Tests* ⟨71⟩, except to place the filter on the surface of Soybean-Casein Digest Agar Medium in a petri dish, incubate for 7 days at 30° to 35°, and count the number of colonies on the filter. Similarly prepare a second specimen, except to incubate at 20° to 25°. Not more than 20 colonies are observed from the two specimens. It meets also the requirements of the tests for absence of *Staphylococcus aureus* and *Pseudomonas aeruginosa* under *Microbial Limit Tests* ⟨61⟩.

Water, *Method I* ⟨921⟩—Store 1 container of Topical Aerosol in a freezer for not less than 2 hours, open the container, and transfer 10.0 mL of the freshly mixed specimen to a titration vessel containing 20 mL of a mixture of carbon tetrachloride, chloroform, and methanol (2:2:1) instead of methanol. In titrating the specimen, determine the end-point at a temperature of 10° or higher: not more than 0.5% of water is found.

Other requirements—It meets the requirements for *Leak Testing* and *Pressure Testing* under *Aerosols* ⟨601⟩.

Assay for polymyxin B—Proceed as directed for polymyxin B under *Antibiotics—Microbial Assays* ⟨81⟩, expelling the entire contents of 1 container of Polymyxin B Sulfate and Bacitracin Zinc Topical Aerosol, according to the directions in the labeling, into a 2000-mL conical flask held in a horizontal position. Add 500.0 mL of 0.01 *N* hydrochloric acid, and shake to dissolve. Immediately dilute an accurately measured volume of this acidic solution quantitatively and stepwise with *Buffer No. 6* to obtain a *Test Dilution* having a concentration of polymyxin B assumed to be equal to the median dose level of the Standard.

Assay for bacitracin—Proceed as directed for bacitracin under *Antibiotics—Microbial Assays* ⟨81⟩, using an accurately measured volume of the acidic solution obtained in the *Assay for polymyxin B* immediately diluted quantitatively and stepwise with *Buffer No. 1* to yield a *Test Dilution* having a bacitracin concentration assumed to be equal to the median dose level of the Standard. [NOTE—Add additional hydrochloric acid to each test dilution of the Standard to obtain the same concentration of hydrochloric acid as in the *Test Dilution.*]

Polymyxin B Sulfates and Bacitracin Zinc Ointment, Neomycin and—*see* Neomycin and Polymyxin B Sulfates and Bacitracin Zinc Ointment

Polymyxin B Sulfate Ophthalmic Ointment, Bacitracin Zinc and—*see* Bacitracin Zinc and Polymyxin B Sulfate Ophthalmic Ointment

Polymyxin B Sulfate and Bacitracin Zinc Topical Powder

» Polymyxin B Sulfate and Bacitracin Zinc Topical Powder contains not less than 90.0 percent and not more than 130.0 percent of the labeled amounts of polymyxin B and bacitracin.

Packaging and storage—Preserve in well-closed containers.

Reference standards—*USP Polymyxin B Sulfate Reference Standard*—Dry in vacuum at a pressure not exceeding 5 mm of mercury at 60° for 3 hours before using. *USP Bacitracin Zinc Reference Standard*—Dry in vacuum at a pressure not exceeding 5 mm of mercury at 60° for 3 hours before using.

Microbial limits—Collect aseptically in a suitable container about 1 g from not less than 5 containers, dissolve in 500 mL of *Fluid A*, filter through a membrane filter as directed for *Test Procedures Using Membrane Filtration* under *Sterility Tests* ⟨81⟩, except to place the filter on the surface of Soybean-Casein Digest Agar Medium in a petri dish, incubate for 7 days at 30° to 35°, and count the number of colonies on the filter. Similarly prepare a second specimen, except to incubate at 20° to 25°. Not more than 20 colonies are observed from the two specimens. It meets also the requirements of the tests for absence of *Staphylococcus aureus* and *Pseudomonas aeruginosa* under *Microbial Limit Tests* ⟨61⟩.

Water, *Method I* ⟨921⟩: not more than 7.0%.

Assay for polymyxin B—Proceed as directed for polymyxin B under *Antibiotics—Microbial Assays* ⟨81⟩, using an accurately weighed portion of Polymyxin B Sulfate and Bacitracin Zinc Topical Powder, equivalent to about 5000 USP Polymyxin B Units, shaken with 20 mL of water in a suitable volumetric flask. Dilute with *Buffer No. 6* to volume, and mix. Dilute an accurately measured volume of the solution so obtained quantitatively with *Buffer No. 6* to obtain a *Test Dilution* having a concentration of polymyxin B assumed to be equal to the median dose level of the Standard.

Assay for bacitracin—Proceed as directed for bacitracin under *Antibiotics—Microbial Assays* ⟨81⟩, using an accurately weighed portion of Polymyxin B Sulfate and Bacitracin Zinc Topical Powder, equivalent to about 800 USP Bacitracin Units, added to a 100-mL volumetric flask, dilute with 0.01 *N* hydrochloric acid to volume, and mix. Dilute this solution quantitatively with *Buffer No. 1* to obtain a *Test Dilution* having a concentration assumed to be equal to the median dose level of the Standard. In preparing each test dilution of the Standard, add additional hydrochloric acid to each to obtain the same concentration of hydrochloric acid as in the *Test Dilution.*

Polymyxin B Sulfates and Bacitracin Zinc Ophthalmic Ointment, Neomycin and—*see* Neomycin and Polymyxin B Sulfates and Bacitracin Zinc Ophthalmic Ointment

Polymyxin B Sulfate Ophthalmic Ointment, Chloramphenicol and—*see* Chloramphenicol and Polymyxin B Sulfate Ophthalmic Ointment

Polymyxin B Sulfates and Dexamethasone Ophthalmic Ointment, Neomycin and—*see* Neomycin and Polymyxin B Sulfates and Dexamethasone Ophthalmic Ointment

Polymyxin B Sulfates and Dexamethasone Ophthalmic Suspension, Neomycin and—*see* Neomycin and Polymyxin B Sulfates and Dexamethasone Ophthalmic Suspension

Polymyxin B Sulfates, Gramicidin, and Hydrocortisone Acetate Cream, Neomycin and—*see* Neomycin and Polymyxin B Sulfates, Gramicidin, and Hydrocortisone Acetate Cream

Polymyxin B Sulfates and Gramicidin Cream, Neomycin and—*see* Neomycin and Polymyxin B Sulfates and Gramicidin Cream

Polymyxin B Sulfates and Gramicidin Ophthalmic Solution, Neomycin and—*see* Neomycin and Polymyxin B Sulfates and Gramicidin Ophthalmic Solution

Polymyxin B Sulfate and Hydrocortisone Otic Solution

» Polymyxin B Sulfate and Hydrocortisone Otic Solution is a sterile solution containing not less than 90.0 percent and not more than 130.0 percent of the labeled amount of polymyxin B, and not less than 90.0 percent and not more than 110.0 percent of the labeled amount of hydrocortisone ($C_{21}H_{30}O_5$). It may contain one or more suitable buffers and preservatives.

NOTE—Where Polymyxin B Sulfate and Hydrocortisone Otic Solution is prescribed without reference to the quantity of polymyxin B or hydrocortisone contained therein, a product containing 10,000 Polymyxin B Units and 5 mg of hydrocortisone per mL shall be dispensed.

Packaging and storage—Preserve in tight, light-resistant containers.

Reference standards—*USP Polymyxin B Sulfate Reference Standard*—Dry in vacuum at a pressure not exceeding 5 mm of mercury at 60° for 3 hours before using. *USP Hydrocortisone Reference Standard*—Dry at 105° for 3 hours before using.

Sterility—It meets the requirements under *Sterility Tests* ⟨71⟩.

pH ⟨791⟩: between 3.0 and 5.0.

Assay for polymyxin—Proceed with Polymyxin B Sulfate and Hydrocortisone Otic Solution as directed under *Antibiotics—Microbial Assays* ⟨81⟩, using an accurately measured volume of Polymyxin B Sulfate and Hydrocortisone Otic Solution diluted quantitatively with *Buffer No. 6* to yield a *Test Dilution* having a concentration assumed to be equal to the median dose level of the Standard.

Assay for hydrocortisone—

Mobile phase—Prepare a suitable solution of about 500 volumes of methanol, 500 volumes of water, and 1 volume of glacial acetic acid, such that the retention time of hydrocortisone is between 6 and 10 minutes.

Standard preparation—Dissolve a suitable quantity of USP Hydrocortisone RS, accurately weighed, in a mixture of methanol and water (1 : 1) to obtain a solution having a known concentration of about 0.15 mg per mL.

Assay preparation—Transfer an accurately measured volume of Polymyxin B Sulfate and Hydrocortisone Otic Solution, equivalent to about 15 mg of hydrocortisone, to a 100-mL volumetric flask, dilute with a mixture of methanol and water (1 : 1) to volume, and mix.

Chromatographic system (see *Chromatography* ⟨621⟩)—The chromatograph is equipped with a 254-nm detector and a 4-mm × 30-cm column that contains packing L1. The flow rate is about 2 mL per minute. Chromatograph five replicate injections of the *Standard preparation,* and record the peak responses as directed under *Procedure:* the relative standard deviation is not more than 2.0%.

Procedure—Separately inject equal volumes (about 10 µL) of the *Standard preparation* and the *Assay preparation* into the chromatograph by means of a suitable microsyringe or sampling valve, adjusting the specimen size and other operating parameters such that the peak obtained from the *Standard preparation* is about 0.6 full-scale. Record the chromatograms, and measure the responses for the major peaks. Calculate the quantity, in mg, of $C_{21}H_{30}O_5$ in each mL of the Otic Solution taken by the formula:

$$(100C/V)(H_U/H_S),$$

in which C is the concentration, in mg per mL, of USP Hydrocortisone RS in the *Standard preparation,* V is the volume, in mL, of the portion of Otic Solution taken, and H_U and H_S are the peak responses obtained from the *Assay preparation* and the *Standard preparation,* respectively.

Polymyxin B Sulfates and Hydrocortisone Ophthalmic Suspension, Neomycin and—*see* Neomycin and Polymyxin B Sulfates and Hydrocortisone Ophthalmic Suspension

Polymyxin B Sulfates and Hydrocortisone Acetate Cream, Neomycin and—*see* Neomycin and Polymyxin B Sulfates and Hydrocortisone Acetate Cream

Polymyxin B Sulfate, and Hydrocortisone Acetate Ophthalmic Ointment, Chloramphenicol,—*see* Chloramphenicol, Polymyxin B Sulfate, and Hydrocortisone Acetate Ophthalmic Ointment

Polymyxin B Sulfates, and Hydrocortisone Acetate Topical Suspension, Penicillin G Procaine, Neomycin and—*see* Penicillin G Procaine, Neomycin and Polymyxin B Sulfates, and Hydrocortisone Acetate Topical Suspension

Polymyxin B Sulfates and Hydrocortisone Acetate Ophthalmic Suspension, Neomycin and—*see* Neomycin and Polymyxin B Sulfates and Hydrocortisone Acetate Ophthalmic Suspension

Polymyxin B Sulfates Cream, Neomycin and—*see* Neomycin and Polymyxin B Sulfates Cream

Polymyxin B Sulfates Ophthalmic Ointment, Neomycin and—*see* Neomycin and Polymyxin B Sulfates Ophthalmic Ointment

Polymyxin B Sulfates Ophthalmic Solution, Neomycin and—*see* Neomycin and Polymyxin B Sulfates Ophthalmic Solution

Polymyxin B Sulfates Solution for Irrigation, Neomycin and—*see* Neomycin and Polymyxin B Sulfates Solution for Irrigation

Polymyxin B Sulfate Ointment, Oxytetracycline Hydrochloride and—*see* Oxytetracycline Hydrochloride and Polymyxin B Sulfate Ointment

Polymyxin B Sulfate Ophthalmic Ointment, Oxytetracycline Hydrochloride and—*see* Oxytetracycline Hydrochloride and Polymyxin B Sulfate Ophthalmic Ointment

Polymyxin B Sulfate Topical Powder, Oxytetracycline Hydrochloride and—*see* Oxytetracycline Hydrochloride and Polymyxin B Sulfate Topical Powder

Polymyxin B Sulfate Vaginal Tablets, Oxytetracycline Hydrochloride and—*see* Oxytetracycline Hydrochloride and Polymyxin B Sulfate Vaginal Tablets

Polymyxin B Sulfates and Prednisolone Acetate Ophthalmic Suspension, Neomycin and—*see* Neomycin and Polymyxin B Sulfates and Prednisolone Acetate Ophthalmic Suspension

Polyoxyethylene 50 Stearate—*see* Polyoxyl 50 Stearate NF

Polyoxyl 10 Oleyl Ether—*see* Polyoxyl 10 Oleyl Ether NF

Polyoxyl 20 Cetostearyl Ether—*see* Polyoxyl 20 Cetostearyl Ether NF

Polyoxyl 35 Castor Oil—*see* Polyoxyl 35 Castor Oil

Polyoxyl 40 Hydrogenated Castor Oil—*see* Polyoxyl 40 Hydrogenated Castor Oil NF

Polyoxyl 40 Stearate—*see* Polyoxyl 40 Stearate NF

Polyoxyl 50 Stearate—*see* Polyoxyl 50 Stearate NF

Polysorbate 20—*see* Polysorbate 20 NF

Polysorbate 40—*see* Polysorbate 40 NF

Polysorbate 60—*see* Polysorbate 60 NF

Polysorbate 80—*see* Polysorbate 80 NF

Polythiazide

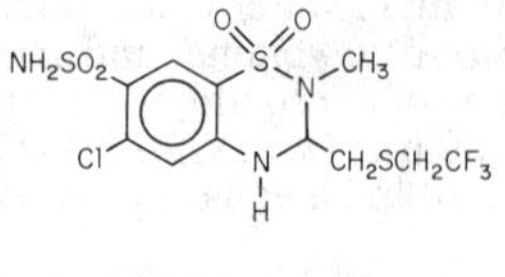

NH_2SO_2 … S(=O)(=O) … CH_3 … Cl … N … H … $CH_2SCH_2CF_3$

$C_{11}H_{13}ClF_3N_3O_4S_3$ 439.87

2*H*-1,2,4-Benzothiadiazine-7-sulfonamide, 6-chloro-3,4-dihydro-2-methyl-3-[[(2,2,2-trifluoroethyl)thio]methyl]-, 1,1-dioxide.

6-Chloro-3,4-dihydro-2-methyl-3-[[(2,2,2-trifluoroethyl)thio]-methyl]-2*H*-1,2,4-benzothiadiazine-7-sulfonamide 1,1-dioxide [*346-18-9*].

» Polythiazide, dried in vacuum at 60° for 2 hours, contains not less than 97.0 percent and not more than 101.0 percent of $C_{11}H_{13}ClF_3N_3O_4S_3$.

Packaging and storage—Preserve in tight, light-resistant containers.

Reference standards—*USP Polythiazide Reference Standard*—Dry in vacuum at 60° for 2 hours before using. *USP 4-Amino-6-chloro-1,3-benzenedisulfonamide Reference Standard*—Keep container tightly closed and protected from light. Dry over silica gel for 4 hours before using.

Identification—

A: The infrared absorption spectrum of a mineral oil dispersion of it, previously dried, exhibits maxima only at the same wavelengths as that of a similar preparation of USP Polythiazide RS.

B: The ultraviolet absorption spectrum of a 1 in 50,000 solution in methanol exhibits maxima and minima at the same wavelengths as that of a similar solution of USP Polythiazide RS, concomitantly measured.

Melting range, *Class I* ⟨741⟩: between 207° and 217°, with decomposition.

Loss on drying ⟨731⟩—Dry it in vacuum at 60° for 2 hours: it loses not more than 1.0% of its weight.

Residue on ignition ⟨281⟩: not more than 0.2%.

Heavy metals, *Method II* ⟨231⟩: 0.0025%.

Selenium ⟨291⟩: 0.003%.

Diazotizable substances—

Standard preparation—Transfer 10.0 mg of USP 4-Amino-6-chloro-1,3-benzenedisulfonamide RS, accurately weighed, to a 50-mL volumetric flask, dissolve in methanol, dilute with methanol to volume, and mix. Transfer 25.0 mL of this solution to a 100-mL volumetric flask, dilute with methanol to volume, and mix.

Test preparation—Transfer 500 mg to a 100-mL volumetric flask, dissolve in methanol, dilute with methanol to volume, and mix.

Procedure—Transfer 2.0 mL each of the *Standard preparation*, the *Test preparation*, and methanol to provide the blank, to separate 20-mL test tubes. Place the tubes in a water bath maintained at 70°, evaporate the contents under a stream of nitrogen to dryness, and dissolve each residue in 1.0 mL of acetonitrile. To each tube add 2.0 mL of dilute hydrochloric acid (1 in 120) and 1.0 mL of sodium nitrite solution (1 in 1000), and allow to stand for 15 minutes with occasional shaking. Add 2.0 mL of ammonium sulfamate solution (1 in 50), allow to stand for 5 minutes with frequent shaking, then add to each tube 1.0 mL of freshly prepared *N*-1-naphthylethylenediamine dihydrochloride solution (1 in 1000) and 10.0 mL of sodium acetate TS, and mix. Transfer each solution to a 50-mL volumetric flask with the aid of water, dilute with water to volume, and mix. Concomitantly determine the absorbances of the solutions in 1-cm cells at the wavelength of maximum absorbance at about 525 nm, with a suitable spectrophotometer, using the blank to set the instrument: the absorbance of the solution from the *Test preparation* does not exceed that from the *Standard preparation*, indicating not more than 1.0% of diazotizable substances.

Assay—Transfer about 100 mg of Polythiazide, previously dried and accurately weighed, to a 250-mL volumetric flask, add methanol to volume, and mix. Transfer 5.0 mL of this solution to a 100-mL volumetric flask, dilute with methanol to volume, and mix. Concomitantly determine the absorbances of this solution and of a solution of USP Polythiazide RS, in the same medium having a known concentration of about 20 µg per mL, in 1-cm cells at the wavelength of maximum absorbance at about 268 nm, with a suitable spectrophotometer, using methanol as the blank. Calculate the quantity, in mg, of $C_{11}H_{13}ClF_3N_3O_4S_3$ in the Polythiazide taken by the formula:

$$5C(A_U/A_S),$$

in which C is the concentration, in µg per mL, of USP Polythiazide RS in the Standard solution, and A_U and A_S are the absorbances of the solution from Polythiazide and the Standard solution, respectively.

Polythiazide Tablets

» Polythiazide Tablets contain not less than 90.0 percent and not more than 110.0 percent of the labeled amount of $C_{11}H_{13}ClF_3N_3O_4S_3$.

Packaging and storage—Preserve in tight, light-resistant containers.

Reference standard—*USP Polythiazide Reference Standard*—Dry in vacuum at 60° for 2 hours before using.

Identification—Place a quantity of finely powdered Tablets, equivalent to about 12 mg of polythiazide, in a glass-stoppered, 40-mL centrifuge tube. Add 10 mL of methanol, shake by mechanical means for 30 minutes, and centrifuge for 5 minutes. On a suitable thin-layer chromatographic plate (see *Chromatography* ⟨621⟩), coated with a 0.50-mm layer of chromatographic silica gel mixture, apply 100 µL of this solution, 100 µL of a solution of USP Polythiazide RS in methanol containing 1.2 mg per mL, and 100 µL of an equal mixture of the test solution and the Standard solution. Allow the spots to dry, and develop the chromatogram in a solvent system consisting of a mixture of chloroform and methanol (9:1) until the solvent front has moved three-fourths of the length of the plate. Remove the plate from the developing chamber, mark the solvent front, and locate the spots on the plate by viewing under short-wavelength ultraviolet light. Spray the plate with a 2 in 1000 solution of diphenylcarbazone in methanol, wait for 2 minutes, overspray with concentrated sulfuric acid, and heat for 5 to 10 minutes at 100°: the principal spots obtained from the test solution, the Standard solution, and the mixed test specimen–Standard solution have the same R_f value, and the mixed test specimen–Standard solution chromatogram exhibits a single principal spot.

Dissolution ⟨711⟩—

Medium: dilute hydrochloric acid (1 in 100); 900 mL.

Apparatus 2: 50 rpm.

Time: 90 minutes.

Standard preparation—Dissolve an accurately weighed quantity of USP Polythiazide RS in methanol to obtain a solution having a known concentration of 0.4 mg per mL. Dilute a suitable aliquot of the solution so obtained, accurately measured, with *Dissolution Medium* to obtain a *Standard preparation* having a known concentration of about 1 µg per mL for each mg of the labeled amount of polythiazide ($C_{11}H_{13}ClF_3N_3O_4S_3$).

Procedure—Determine the amount of $C_{11}H_{13}ClF_3N_3O_4S_3$ dissolved in a 25.0-mL portion of the solution under test filtered through qualitative filter paper. To each of three 25-mL volumetric flasks transfer, respectively, 15.0 mL of the *Standard preparation*, 15.0 mL of *Dissolution Medium*, and 15.0 mL of the solution under test. To each flask add 3.0 mL of dilute hydrochloric acid (1 in 2), and heat in a water bath maintained at a temperature of 65°, with periodic swirling, for 20 minutes. Cool, and add to each flask 1 mL of freshly prepared sodium nitrite solution (1 in 1000). Shake, allow to stand for 3 minutes, then add 1 mL of ammonium sulfamate solution (1 in 200), shake, and allow to stand for 3 minutes. To each flask add 1.0 mL of freshly prepared *N*-1-naphthylethylenediamine dihydrochloride solution (1 in 1000), dilute with water to volume, and mix. Immediately and concomitantly determine the absorbances of the solutions in 1-cm cells at the wavelength of maximum absorbance at about 510 nm, with a suitable spectrophotometer, using the appropriate blank to set the instrument. Calculate the quantity of $C_{11}H_{13}ClF_3N_3O_4S_3$ dissolved as percentage of the labeled amount by the formula:

$$(90C/L)(A_U/A_S),$$

in which A_U and A_S are the absorbances of the solution under test and the Standard preparation, respectively, C is the concentration, in µg per mL, of USP Polythiazide RS in the *Standard preparation*, L is the labeled amount of $C_{11}H_{13}ClF_3N_3O_4S_3$, in mg, and A_U and A_S are the absorbances of the solutions obtained from the solution under test and the *Standard preparation*, respectively.

Tolerances—Not less than 50% (Q) of the labeled amount of $C_{11}H_{13}ClF_3N_3O_4S_3$ is dissolved in 90 minutes.

Uniformity of dosage units ⟨905⟩: meet the requirements.

Procedure for content uniformity—

Standard preparation—Dissolve an accurately weighed quantity of USP Polythiazide RS in acetone, and dilute quantitatively and stepwise with acetone to obtain a solution having a known concentration of about 20 µg per mL.

Test preparation—Place 1 Tablet in a volumetric flask of such size that 50 mL is available for each mg of polythiazide in the Tablet, add an amount of water equal to about 10% of the volume of the flask, and shake by mechanical means for 15 minutes. Add acetone to about 50% of the volume of the flask, shake by mechanical means for 20 minutes, dilute with acetone to volume, and mix. Filter through glass fiber filter paper, and discard the first 10 mL of the filtrate. The subsequent filtrate contains about 20 µg of polythiazide per mL.

Procedure—To each of four 25-mL volumetric flasks transfer, respectively, 5.0 mL of the *Standard preparation*, 5.0 mL of acetone to provide the standard preparation blank, 5.0 mL of the *Test preparation*, and a mixture of 4.5 mL of acetone and 0.5 mL of water to provide the test preparation blank. To each flask add 2.5 mL of dilute hydrochloric acid (1 in 2), and heat in a boiling water bath for 3 minutes. Cool, add to each flask 15 mL of water, then add 1 mL of freshly prepared sodium nitrite solution (1 in 1000). Shake, allow to stand for 3 minutes, then add 1 mL of ammonium sulfamate solution (1 in 200), shake, and allow to stand again for 3 minutes. To each flask add 1.0 mL of freshly prepared *N*-1-naphthylethylenediamine dihydrochloride solution (1 in 1000), dilute with water to volume, and mix. Filter the solutions through separate glass fiber filters, and discard the first 10 mL of each filtrate. Immediately and concomitantly determine the absorbances of the solutions in 1-cm cells at the wavelength of maximum absorbance at about 510 nm, with a suitable spectrophotometer, using the appropriate blank to set the instrument. Calculate the quantity, in mg, of $C_{11}H_{13}ClF_3N_3O_4S_3$ in the Tablet by the formula:

$$(TC/D)(A_U/A_S),$$

in which T is the labeled quantity, in mg, of polythiazide in the Tablet, C is the concentration, in µg per mL, of USP Polythiazide RS in the *Standard preparation*, D is the concentration, in µg per mL, of polythiazide in the *Test preparation*, based upon the labeled quantity per Tablet and the extent of dilution, and A_U and A_S are the absorbances of the solutions from the *Test preparation* and the *Standard preparation*, respectively.

Assay—

Standard preparation—Dissolve an accurately weighed quantity of USP Polythiazide RS in methanol, and dilute quantitatively with methanol to obtain a solution having a known concentration of about 0.625 mg per mL.

Assay preparation—Weigh and finely powder not less than 20 Tablets. Transfer an accurately weighed portion of the powder, equivalent to about 12.5 mg of polythiazide, to a glass-stoppered, 40-mL centrifuge tube, add 20.0 mL of methanol, insert the stopper tightly, and shake by mechanical means for 30 minutes. Centrifuge at 1500 rpm for 15 minutes, and use the supernatant liquid as directed in the *Procedure*.

Procedure—Apply a 400-µL portion of the *Standard preparation* to a suitable thin-layer chromatographic plate (see *Chromatography* ⟨621⟩), coated with a 0.50-mm layer of chromatographic silica gel mixture, applying the portion as a continuous band 6 cm in length, 2 cm from the lower edge of the plate, and starting 2 cm from the side of the plate. Prepare a duplicate plate for the *Standard preparation.* Similarly, apply a 400-µL portion of the *Assay preparation* to one plate, then prepare a duplicate plate for the *Assay preparation.* Allow the spots to dry, and develop the chromatograms in a solvent system consisting of a mixture of chloroform and methanol (9:1) until the solvent front has moved about three-fourths of the length of the plate. Remove the plates from the developing chamber, and dry for 45 minutes in the air at room temperature. Observe the plates under short-wavelength ultraviolet light, and outline the absorbent bands occurring at about R_f 0.45 to 0.55. Quantitatively remove the silica gel from these areas from each plate, and transfer to separate, glass-stoppered, 50-mL conical flasks. For the blanks, transfer an equal amount of polythiazide-free silica gel at the same R_f from each plate to individual, glass-stoppered, 50-mL conical flasks. Add 25.0 mL of methanol to each flask, insert the stoppers, shake by mechanical means for 30 minutes, and filter. Concomitantly determine the absorbance of each filtrate in 1-cm cells at the wavelength of maximum absorbance at about 268 nm, with a suitable spectrophotometer, using the respective blank to set the instrument. Calculate the quantity, in mg, of $C_{11}H_{13}ClF_3N_3O_4S_3$ in the portion of Tablets taken by the formula:

$$20C(A_U/A_S),$$

in which C is the concentration, in mg per mL, of USP Polythiazide RS in the *Standard preparation,* and A_U and A_S are the averages of the absorbances of the solutions from the *Assay preparation* and the *Standard preparation,* respectively.

Polyvalent, Antivenin (Crotalidae)—*see* Antivenin (Crotalidae) Polyvalent

Polyvinyl Acetate Phthalate—*see* Polyvinyl Acetate Phthalate NF

Polyvinyl Alcohol

$$\left[CH_2-\underset{OH}{CH} \right]_n$$

$(C_2H_4O)_n$
Ethenol, homopolymer.
Vinyl alcohol polymer [*9002-89-5*].

» Polyvinyl Alcohol is a water-soluble synthetic resin, represented by the formula:

$$(C_2H_4O)_n,$$

in which the average value of n lies between 500 and 5000. It is prepared by 85 percent to 89 percent hydrolysis of polyvinyl acetate. The apparent viscosity, in centipoises, at 20°, of a solution containing 4 g of Polyvinyl Alcohol in each 100 g is not less than 85.0 percent and not more than 115.0 percent of that stated on the label.

Packaging and storage—Preserve in well-closed containers.

Viscosity—After determining the *Loss on drying,* weigh a quantity of undried Polyvinyl Alcohol, equivalent to 6.00 g on the dried basis. Over a period of seconds, transfer the test specimen with continuous slow stirring to about 140 mL of water contained in a suitable tared flask. When the specimen is well-wetted, increase the rate of stirring, avoiding mixing in excess air. Heat the mixture to 90°, and maintain the temperature at 90° for about 5 minutes. Discontinue heating, and continue stirring for 1 hour. Add water to make the mixture weigh 150 g. Resume stirring to obtain a homogenous solution. Filter the solution through a tared 100-mesh screen into a 250-mL conical flask, cool to about 15°, mix, and proceed as directed under *Viscosity* ⟨911⟩.

pH ⟨791⟩: between 5.0 and 8.0, in a solution (1 in 25).

Loss on drying ⟨731⟩—Dry it at 110° to constant weight: it loses not more than 5.0% of its weight.

Residue on ignition ⟨281⟩: not more than 2.0%.

Water-insoluble substances—Wash the tared 100-mesh screen used in the test for *Viscosity* with two 25-mL portions of water, and dry at 110° for 1 hour: not more than 6.4 mg of water-insoluble substances is found (0.1%).

Degree of hydrolysis—

Procedure—Transfer about 1 g of Polyvinyl Alcohol, previously dried at 110° to constant weight and accurately weighed, to a wide-mouth, 250-mL conical flask fitted by means of a suitable glass joint to a reflux condenser. Add 35 mL of dilute methanol (3 in 5), and mix gently to assure complete wetting of the solid. Add 3 drops of phenolphthalein TS, and add 0.2 N hydrochloric acid or 0.2 N sodium hydroxide, if necessary, to neutralize. Add 25.0 mL of 0.2 N sodium hydroxide VS, and reflux gently on a hot plate for 1 hour. Wash the condenser with 10 mL of water, collecting the washings in the flask, cool, and titrate with 0.2 N hydrochloric acid VS. Concomitantly perform a blank determination in the same manner, using the same quantity of 0.2 N sodium hydroxide VS.

Calculation of saponification value—Calculate the saponification value by the formula:

$$[(B - A)N56.11]/W,$$

in which B and A are the volumes, in mL, of 0.2 N hydrochloric acid VS consumed in the titration of the blank and the test preparation, respectively, N is the exact normality of the hydrochloric acid solution, W is the weight, in g, of the portion of Polyvinyl Alcohol taken, and 56.11 is the molecular weight of potassium hydroxide.

Calculation of degree of hydrolysis—Calculate the degree of hydrolysis, expressed as percentage of hydrolysis of polyvinyl acetate, by the formula:

$$100 - [7.84S/(100 - 0.075S)],$$

in which S is the saponification value of the Polyvinyl Alcohol taken: between 85% and 89% is found.

Sulfurated Potash

Thiosulfuric acid, dipotassium salt, mixt. with potassium sulfide (K_2S_x).
Dipotassium thiosulfate mixture with potassium sulfide (K_2S_x) [*39365-88-3*].

» Sulfurated Potash is a mixture composed chiefly of potassium polysulfides and potassium thiosulfate. It contains not less than 12.8 percent of sulfur (S) in combination as sulfide.

Packaging and storage—Preserve in tight containers. Containers from which it is to be taken for immediate use in compounding prescriptions contain not more than 120 g.

Identification—

A: To a 1 in 10 solution add an excess of 6 N acetic acid: hydrogen sulfide is evolved, and sulfur is precipitated.

B: Filter the mixture from *Identification test A,* and add to the filtrate an excess of sodium bitartrate TS: an abundant, white, crystalline precipitate is formed within 15 minutes.

Assay for sulfides—Transfer 10 to 15 pieces of Sulfurated Potash to a mortar, and reduce to a fine powder. Transfer about 1 g of the powder, accurately weighed, to a 250-mL beaker, and dissolve in 50 mL of water. Filter, if necessary, and wash or dilute with water to 75 mL. Add, with constant stirring, 50 mL of cupric sulfate solution (1 in 20), and allow the mixture to stand, with occasional stirring, for 10 minutes. Filter through a retentive filter, and wash the precipitate with 200 mL of 0.25 N hydrochloric acid, taking care to avoid breaking up the cake. (If the filtrate is not blue in color, discard the assay specimen, and start over, using a larger volume of cupric sulfate solution.) Ignite the precipitate in a tared dish at 1000° for 1 hour, cool in a desiccator, and weigh: the weight of the cupric oxide so obtained, multiplied by 0.4030, represents the weight of S in the specimen under assay.

Potassium Acetate

CH_3COOK

$C_2H_3KO_2$ 98.14
Acetic acid, potassium salt.
Potassium acetate [*127-08-2*].

» Potassium Acetate contains not less than 99.0 percent and not more than 100.5 percent of $C_2H_3KO_2$, calculated on the dried basis.

Packaging and storage—Preserve in tight containers.

Identification—A solution (1 in 10) responds to the tests for *Potassium* ⟨191⟩ and for *Acetate* ⟨191⟩.

pH ⟨791⟩: between 7.5 and 8.5, in a solution (1 in 20).

Loss on drying ⟨731⟩—Dry it at 150° for 2 hours: it loses not more than 1.0% of its weight.

Arsenic, *Method I* ⟨211⟩: 8 ppm.

Heavy metals ⟨231⟩—Dissolve 1 g in 10 mL of water, add 3.5 mL of 1 *N* acetic acid, and dilute with water to 25 mL: the limit is 0.002%.

Sodium—

Standard solution—Dissolve an accurately weighed portion of sodium acetate, previously dried at 120° to constant weight, quantitatively in water to obtain a solution containing 0.107 mg per mL. This solution contains 0.03 mg of sodium per mL.

Standard preparation—Transfer 1.0 g of Potassium Acetate to a 100-mL volumetric flask, dissolve in water, add 10.0 mL of *Standard solution*, dilute with water to volume, and mix.

Test preparation—Transfer 1.0 g of Potassium Acetate to a 100-mL volumetric flask, dissolve in water, dilute with water to volume, and mix.

Procedure—Using the *Standard preparation* set a suitable flame photometer to 100% transmittance at 589 nm. Determine the emission intensity of the *Test preparation* at this wavelength. Adjust the monochromator to 580 nm, and determine the emission intensity of the *Test preparation:* the difference between the emission intensities observed for the *Test preparation* at 589 nm and 580 nm is not greater than the difference in emission intensities at 589 nm observed for the *Test preparation* and the *Standard preparation* (0.03%).

Assay—Dissolve about 200 mg of Potassium Acetate, previously dried and accurately weighed, in 25 mL of glacial acetic acid, add 2 drops of crystal violet TS, and titrate with 0.1 *N* perchloric acid VS to a green end-point. Perform a blank determination, and make any necessary correction. Each mL of 0.1 *N* perchloric acid is equivalent to 9.814 mg of $C_2H_3KO_2$.

Potassium Acetate Injection

» Potassium Acetate Injection is a sterile solution of Potassium Acetate in Water for Injection. It contains not less than 95.0 percent and not more than 105.0 percent of the labeled amount of $C_2H_3KO_2$.

Packaging and storage—Preserve in single-dose or in multiple-dose containers, preferably of Type I or Type II glass.

Labeling—The label states the potassium acetate content in terms of weight and of milliequivalents in a given volume. Label the Injection to indicate that it is to be diluted to appropriate strength with water or other suitable fluid prior to administration. The label states also the total osmolar concentration in mOsmol per liter. Where the contents are less than 100 mL, or where the label states that the Injection is not for direct injection but is to be diluted before use, the label alternatively may state the total osmolar concentration in mOsmol per mL.

Identification—It responds to the tests for *Potassium* ⟨191⟩ and for *Acetate* ⟨191⟩.

Pyrogen—When diluted with Sodium Chloride Injection to contain 0.5% of potassium acetate, it meets the requirements of the *Pyrogen Test* ⟨151⟩.

pH ⟨791⟩: between 5.5 and 8.0, when diluted with water to 1.0% of potassium acetate.

Particulate matter ⟨788⟩: meets the requirements under *Small-volume Injections.*

Other requirements—It meets the requirements under *Injections* ⟨1⟩.

Assay—

Potassium stock solution—Dissolve 190.7 mg of potassium chloride, previously dried at 105° for 2 hours, in water. Transfer to a 1000-mL volumetric flask, dilute with water to volume, and mix. Transfer 100.0 mL of this solution to a 1000-mL volumetric flask, dilute with water to volume, and mix. This solution contains 10 µg of potassium (equivalent to 19.07 µg of potassium chloride) per mL.

Standard preparations—To separate 100-mL volumetric flasks transfer 10.0, 15.0, and 20.0 mL, respectively, of *Potassium stock solution.* To each flask add 2.0 mL of sodium chloride solution (1 in 5) and 1.0 mL of hydrochloric acid, dilute with water to volume, and mix. The *Standard preparations* contain, respectively, 1.0, 1.5, and 2.0 µg of potassium per mL.

Assay preparation—Transfer an accurately measured volume of Potassium Acetate Injection, equivalent to about 2 g of potassium acetate, to a 500-mL volumetric flask, dilute with water to volume, and mix. Transfer 5.0 mL of the solution to a 250-mL volumetric flask, dilute with water to volume, and mix. Transfer 5.0 mL of the resulting solution to a 100-mL volumetric flask, add 2.0 mL of sodium chloride solution (1 in 5) and 1.0 mL of hydrochloric acid, dilute with water to volume, and mix.

Procedure—Concomitantly determine the absorbances of the *Standard preparations* and the *Assay preparation* at the potassium emission line of 766.5 nm, with a suitable atomic absorption spectrophotometer (see *Spectrophotometry and Light-scattering* ⟨851⟩) equipped with a potassium hollow-cathode lamp and an air-acetylene flame, using water as the blank. Plot the absorbance of the *Standard preparation* versus concentration, in µg per mL, of potassium, and draw the straight line best fitting the three plotted points. From the graph so obtained, determine the concentration, in µg per mL, of potassium in the *Assay preparation.* Calculate the quantity, in mg, of $C_2H_3KO_2$ in the portion of Injection taken by the formula:

$$500C(2.510),$$

in which C is the concentration, in µg per mL, of potassium in the *Assay preparation*, and 2.510 is the ratio of the molecular weight of potassium acetate to the atomic weight of potassium.

Potassium Alum—*see* Alum, Potassium
Potassium Benzoate—*see* Potassium Benzoate NF

Potassium Bicarbonate

$KHCO_3$ 100.12
Carbonic acid, monopotassium salt.
Monopotassium carbonate [*298-14-6*].

» Potassium Bicarbonate contains not less than 99.5 percent and not more than 101.5 percent of $KHCO_3$, calculated on the dried basis.

Packaging and storage—Preserve in well-closed containers.

Identification—A solution (1 in 10) responds to the tests for *Potassium* ⟨191⟩ and for *Bicarbonate* ⟨191⟩.

Loss on drying ⟨731⟩—Dry it over silica gel for 4 hours: it loses not more than 0.3% of its weight.

Normal carbonate—Grind 3.0 g of Potassium Bicarbonate with 25 mL of alcohol and 5 mL of water in a porcelain mortar. Add 3 drops of phenolphthalein TS, and titrate slowly with barium chloride solution, prepared by dissolving 12.216 g of barium chloride in 300 mL of water and diluting with alcohol to obtain 1000 mL of solution, until the suspension becomes colorless. Continue the grinding for 2 minutes, and if the color turns pink, continue the titration with the barium chloride solution to a colorless endpoint. Repeat the grinding for 2 minutes and the addition of the barium chloride solution, if necessary, until the suspension is colorless after 2 minutes of grinding. Each mL of the barium chloride solution is equivalent to 6.911 mg of K_2CO_3: not more than 2.5% is found.

Heavy metals, *Method I* ⟨231⟩—To 2 g add 5 mL of water and 8 mL of 3 *N* hydrochloric acid, heat to boiling, and maintain that temperature for 1 minute. Add 1 drop of phenolphthalein TS and sufficient 6 *N* ammonium hydroxide, dropwise, to give the solution a faint pink color. Cool, add 2 mL of 1 *N* acetic acid, and then dilute with water to 25 mL: the limit is 0.001%.

Assay—Dissolve about 4 g of Potassium Bicarbonate, accurately weighed, in 100 mL of water, add methyl red TS, and titrate with 1 *N* hydrochloric acid VS. Add the acid slowly, with constant stirring, until the solution becomes faintly pink. Heat the solution to boiling, cool, and continue the titration until the pink color no longer fades after boiling. Each mL of 1 *N* hydrochloric acid is equivalent to 100.1 mg of $KHCO_3$.

Potassium Bicarbonate Effervescent Tablets for Oral Solution

» Potassium Bicarbonate Effervescent Tablets for Oral Solution contain not less than 90.0 percent and not more than 110.0 percent of the labeled amount of K.

Packaging and storage—Preserve in tight containers, protected from excessive heat.

Labeling—The label states the potassium content in terms of weight and in terms of milliequivalents. Where Tablets are packaged in individual pouches, the label instructs the user not to open until the time of use.

Identification—One Tablet dissolves in 100 mL of water with effervescence. The collected gas responds to the test for *Bicarbonate* ⟨191⟩, and the resulting solution responds to the test for *Potassium* ⟨191⟩.

Uniformity of dosage units ⟨905⟩: meet the requirements.

Assay—

Potassium stock solution and *Standard preparations*—Prepare as directed in the *Assay* under *Potassium Chloride Elixir.*

Assay preparation—Transfer 10 Potassium Bicarbonate Effervescent Tablets for Oral Solution to a 2000-mL volumetric flask, dissolve in 200 mL of water, swirl until effervescence ceases, dilute with water to volume, and mix. Filter, and quantitatively dilute an accurately measured volume of the filtrate with water to obtain a solution containing 30 µg of potassium per mL. Transfer 5.0 mL of the resulting solution to a 100-mL volumetric flask, add 2.0 mL of sodium chloride solution (1 in 5) and 1.0 mL of hydrochloric acid, dilute with water to volume, and mix.

Procedure—Proceed as directed for *Procedure* in the *Assay* under *Potassium Chloride Elixir.* Calculate the quantity, in mg, of K in each Tablet taken by the formula:

$$L(C/D),$$

in which L is the labeled quantity, in mg, of potassium in each Tablet, C is the concentration, in µg per mL, of potassium in the *Assay preparation*, and D is the concentration, in µg per mL, of potassium in the *Assay preparation* on the basis of the labeled quantity in each Tablet and the extent of dilution.

Potassium Bicarbonate and Potassium Chloride for Effervescent Oral Solution

» Potassium Bicarbonate and Potassium Chloride for Effervescent Oral Solution contains not less than 90.0 percent and not more than 110.0 percent of the labeled amounts of K and Cl.

Packaging and storage—Preserve in tight containers, protected from excessive heat.

Labeling—The label states the potassium and chloride contents in terms of weight and in terms of milliequivalents. Where packaged in individual pouches, the label instructs the user not to open until the time of use.

Identification—A 3-g portion dissolves in 100 mL of water with effervescence. The collected gas so obtained responds to the test for *Bicarbonate* ⟨191⟩, and the resulting solution responds to the tests for *Potassium* ⟨191⟩ and for *Chloride* ⟨191⟩.

Minimum fill ⟨755⟩—

FOR SOLID PACKAGED IN MULTIPLE-UNIT CONTAINERS: meets the requirements.

Uniformity of dosage units ⟨905⟩—

FOR SOLID PACKAGED IN SINGLE-UNIT CONTAINERS: meets the requirements.

Assay for potassium—

Potassium stock solution and *Standard preparations*—Prepare as directed in the *Assay* under *Potassium Chloride Elixir.*

Assay preparation—Weigh and mix the contents of not less than 20 containers of Potassium Bicarbonate and Potassium Chloride for Effervescent Oral Solution. Transfer an accurately weighed portion of the powder, equivalent to about 6 g of potassium, to a 1000-mL volumetric flask, dissolve in about 200 mL of water, dilute with water to volume, and mix. Transfer 5.0 mL of this solution to a second 1000-mL volumetric flask, dilute with water to volume, and mix. Transfer 5.0 mL of the resulting solution to a 100-mL volumetric flask, add 2.0 mL of sodium chloride solution (1 in 5) and 1.0 mL of hydrochloric acid, dilute with water to volume, and mix.

Procedure—Proceed as directed for *Procedure* in the *Assay* under *Potassium Chloride Elixir.* Calculate the quantity, in mg, of K in the portion of Potassium Bicarbonate and Potassium Chloride for Effervescent Oral Solution taken by the formula:

$$400C,$$

in which C is the concentration, in µg per mL, of potassium in the *Assay preparation.*

Assay for chloride—Weigh and mix the contents of not less than 20 containers of Potassium Bicarbonate and Potassium Chloride for Effervescent Oral Solution. Transfer a portion of the powder, equivalent to about 900 mg of chloride, to a 2000-mL volumetric flask. Add about 200 mL of water, swirl until effervescence ceases, dilute with water to volume, and mix. Transfer 25.0 mL of this solution to a 250-mL conical flask, add 50.0 mL of 0.1 *N* silver nitrate VS and 15 mL of nitric acid, and boil, with constant swirling, until the solution is colorless. Cool to room temperature, add water to make about 150 mL, then add 5 mL of ferric ammonium sulfate TS, and titrate the excess silver nitrate with 0.1 *N* ammonium thiocyanate VS to a permanent faint brown endpoint. Each mL of 0.1 *N* silver nitrate is equivalent to 3.545 mg of Cl.

Potassium Bicarbonate and Potassium Chloride Effervescent Tablets for Oral Solution

» Potassium Bicarbonate and Potassium Chloride Effervescent Tablets for Oral Solution contain not less than 90.0 percent and not more than 110.0 percent of the labeled amounts of K and Cl.

Packaging and storage—Preserve in tight containers, protected from excessive heat.

Labeling—The label states the potassium and chloride contents in terms of weight and in terms of milliequivalents. Where Tablets are packaged in individual pouches, the label instructs the user not to open until the time of use.

Identification—One Tablet dissolves in 100 mL of water with effervescence. The collected gas responds to the test for *Bicarbonate* ⟨191⟩, and the resulting solution responds to the tests for *Potassium* ⟨191⟩ and for *Chloride* ⟨191⟩.

Uniformity of dosage units ⟨905⟩: meet the requirements.

Assay for potassium—

Potassium stock solution and *Standard preparations*—Prepare as directed in the *Assay* under *Potassium Chloride Elixir*.

Assay preparation—Transfer 10 Potassium Bicarbonate and Potassium Chloride Effervescent Tablets for Oral Solution to a 2000-mL volumetric flask, dissolve in 200 mL of water, swirl until effervescence ceases, dilute with water to volume, and mix. Filter, and quantitatively dilute an accurately measured volume of the filtrate with water to obtain a solution containing 30 µg of potassium per mL. Transfer 5.0 mL of the resulting solution to a 100-mL volumetric flask, add 2.0 mL of sodium chloride solution (1 in 5) and 1.0 mL of hydrochloric acid, dilute with water to volume, and mix.

Procedure—Proceed as directed for *Procedure* in the *Assay* under *Potassium Chloride Elixir*. Calculate the quantity, in mg, of K in each Tablet taken by the formula:

$$L(C/D),$$

in which L is the labeled quantity, in mg, of potassium in each Tablet, C is the concentration, in µg per mL, of potassium in the *Assay preparation*, and D is the concentration, in µg per mL, of potassium in the *Assay preparation* on the basis of the labeled quantity in each Tablet and the extent of dilution.

Assay for chloride—Transfer a number of Potassium Bicarbonate and Potassium Chloride Effervescent Tablets for Oral Solution, equivalent to about 900 mg of chloride, to a 2000-mL volumetric flask. Add about 200 mL of water, swirl until effervescence ceases, dilute with water to volume, and mix. Transfer 25.0 mL of this solution to a 250-mL conical flask, add 50.0 mL of 0.1 *N* silver nitrate VS and 15 mL of nitric acid, and boil, with constant swirling, until the supernatant liquid is colorless. Cool to room temperature, add sufficient water to make a volume of about 150 mL, add 5 mL of ferric ammonium sulfate TS, and titrate the excess silver nitrate with 0.1 *N* ammonium thiocyanate VS to a permanent faint brown end-point. Each mL of 0.1 *N* silver nitrate is equivalent to 3.545 mg of Cl. Calculate the quantity, in mg, of chloride (Cl) in each Tablet by dividing the total amount of chloride in the Tablets taken by the number of Tablets taken.

Potassium Bicarbonate, and Potassium Citrate Effervescent Tablets for Oral Solution, Potassium Chloride,—*see* Potassium Chloride, Potassium Bicarbonate, and Potassium Citrate Effervescent Tablets for Oral Solution

Potassium and Sodium Bicarbonates and Citric Acid Effervescent Tablets for Oral Solution

» Potassium and Sodium Bicarbonates and Citric Acid Effervescent Tablets for Oral Solution contain not less than 90.0 percent and not more than 110.0 percent of the labeled amounts of potassium bicarbonate ($KHCO_3$), sodium bicarbonate ($NaHCO_3$), and anhydrous citric acid ($C_6H_8O_7$).

Packaging and storage—Preserve in tight containers.

Labeling—Label it to state the sodium content. The label states also that Tablets are to be dissolved in water before being taken.

Identification—One Tablet dissolves in 100 mL of water with effervescence. The collected gas responds to the test for *Bicarbonate* ⟨191⟩, and the resulting solution responds to the tests for *Potassium* ⟨191⟩ and for *Sodium* ⟨191⟩. The resulting solution responds also to the test for *Citrate* ⟨191⟩, 3 to 5 drops of it and 20 mL of the mixture of pyridine and acetic anhydride being used.

Acid-neutralizing capacity ⟨301⟩—The acid consumed by the minimum single dose recommended in the labeling is not less than 5 mEq.

Assay for potassium bicarbonate and sodium bicarbonate—

Potassium chloride stock solution—Prepare a solution of potassium chloride, previously dried at 125° for 30 minutes and accurately weighed, in water to obtain a solution having a known concentration of about 7.5 mg per mL.

Sodium chloride stock solution—Prepare a solution of sodium chloride, previously dried at 125° for 30 minutes and accurately weighed, in water to obtain a solution having a known concentration of about 7 mg per mL.

Lithium diluent solution—Transfer 1.04 g of lithium nitrate to a 1000-mL volumetric flask, add a suitable nonionic surfactant, then add water to volume, and mix.

Standard preparation—Transfer 5.0 mL of *Potassium chloride stock solution* and 5.0 mL of *Sodium chloride stock solution* to a 50-mL volumetric flask, dilute with water to volume, and mix. Each mL of this intermediate solution contains about 0.75 mg of potassium chloride and 0.7 mg of sodium chloride. Transfer 5.0 mL of this solution to a 100-mL volumetric flask, dilute with *Lithium diluent solution* to volume, and mix.

Assay preparation 1—Weigh and finely powder not less than 20 Potassium and Sodium Bicarbonates and Citric Acid Effervescent Tablets for Oral Solution. [NOTE—Tablets and powder are hygroscopic. After removal from the container, grind the Tablets promptly in an atmosphere of low relative humidity, and weigh the powder promptly.] Transfer an accurately weighed portion of the powder, equivalent to about 3000 mg of potassium bicarbonate, to a 1000-mL volumetric flask, dissolve in 500 mL of water, swirl until effervescence ceases, dilute with water to volume, and mix. Dilute an accurately measured volume of this stock solution quantitatively with water to obtain a test solution containing about 1 mg of potassium bicarbonate per mL, on the basis of the labeled quantity. Transfer 5.0 mL of this solution to a 100-mL volumetric flask, dilute with *Lithium diluent solution* to volume, and mix.

Assay preparation 2—Dilute an accurately measured volume of the stock solution used to prepare *Assay preparation 1* quantitatively with water to obtain a test solution containing about 1 mg of sodium bicarbonate per mL, on the basis of the labeled quantity. Transfer 5.0 mL of this solution to a 100-mL volumetric flask, dilute with *Lithium diluent solution* to volume, and mix.

Procedure—Using a suitable flame photometer, adjusted to read zero with *Lithium diluent solution*, concomitantly determine the potassium flame emission readings for the *Standard preparation* and *Assay preparation 1* at the wavelength of maximum emission at about 766 nm. Calculate the quantity, in mg, of potassium bicarbonate ($KHCO_3$) in each Tablet taken by the formula:

$$(100.12/74.55)(LC/D)(R_{U,K}/R_{S,K}),$$

in which 100.12 and 74.55 are the molecular weights of potassium bicarbonate and potassium chloride, respectively, L is the labeled quantity, in mg, of potassium bicarbonate in each Tablet, C is the concentration, in mg per mL, of potassium chloride in the intermediate solution used to prepare the *Standard preparation*, D is the concentration, in mg per mL, of potassium bicarbonate in the test solution used to prepare *Assay preparation 1*, on the basis of the labeled quantity of potassium bicarbonate in each Tablet and the extent of dilution, and $R_{U,K}$ and $R_{S,K}$ are the potassium emission readings obtained from *Assay preparation 1* and the *Standard preparation*, respectively. Similarly determine the sodium flame emission readings for the *Standard preparation* and *Assay preparation 2* at the wavelength of maximum emission at about 589 nm. Calculate the quantity, in mg, of sodium bicarbonate ($NaHCO_3$) in each Tablet taken by the formula:

$$(84.01/58.44)(LC/D)(R_{U,Na}/R_{S,Na}),$$

in which 84.01 and 58.44 are the molecular weights of sodium bicarbonate and sodium chloride, respectively, L is the labeled quantity, in mg, of sodium bicarbonate in each Tablet, C is the concentration, in mg per mL, of sodium chloride in the intermediate solution used to prepare the *Standard preparation*, D is the concentration, in mg per mL, of sodium bicarbonate in the test solution used to prepare *Assay preparation 2*, on the basis of the number of Tablets taken, the labeled quantity of sodium bicarbonate in each Tablet, and the extent of dilution, and $R_{U,Na}$ and $R_{S,Na}$ are the sodium emission readings obtained from *Assay preparation 2* and the *Standard preparation*, respectively.

Assay for anhydrous citric acid—

Cation-exchange column—Mix 10 g of styrenedivinylbenzene cation-exchange resin with 50 mL of water in a suitable beaker. Allow the resin to settle, and decant the supernatant liquid until a slurry of resin remains. Pour the slurry into a 14-mm × 30-cm glass chromatographic tube (having a sealed-in, coarse-porosity porous glass disk and fitted with a stopcock), and allow to settle as a homogeneous bed. Wash the resin bed with about 100 mL of water, closing the stopcock when the water level is about 2 mm above the resin bed.

Procedure—Transfer an accurately measured volume of the stock solution used to prepare *Assay preparation 1* in the *Assay for potassium bicarbonate and sodium bicarbonate*, equivalent to about 40 mg of anhydrous citric acid, carefully onto the top of the resin bed in the *Cation-exchange column*. Place a 250-mL conical flask below the column, open the stopcock, and allow to flow until the solution has entered the resin bed. Elute the column with 60 mL of water at a flow rate of about 5 mL per minute, collecting about 65 mL of the eluate in a suitable flask. Boil the eluate for 1 minute, cool, add 5 drops of phenolphthalein TS, swirl the flask, and titrate with 0.02 N sodium hydroxide VS to a pink end-point. Each mL of 0.02 N sodium hydroxide is equivalent to 1.281 mg of $C_6H_8O_7$.

Potassium Chloride

KCl 74.55

Potassium chloride.

Potassium chloride [*7447-40-7*].

» Potassium Chloride contains not less than 99.0 percent and not more than 100.5 percent of KCl, calculated on the dried basis.

Packaging and storage—Preserve in well-closed containers.

Identification—A solution (1 in 20) responds to the tests for *Potassium* ⟨191⟩ and for *Chloride* ⟨191⟩.

Acidity or alkalinity—To a solution of 5.0 g in 50 mL of carbon dioxide-free water add 3 drops of phenolphthalein TS: no pink color is produced. Then add 0.30 mL of 0.020 N sodium hydroxide: a pink color is produced.

Loss on drying ⟨731⟩—Dry it at 105° for 2 hours: it loses not more than 1.0% of its weight.

Iodide or bromide—Dissolve 2 g in 6 mL of water, add 1 mL of chloroform, and then add, dropwise, with constant agitation, 5 mL of a mixture of equal parts of chlorine TS and water: the chloroform is free from even a transient violet or a permanent orange color.

Arsenic, *Method I* ⟨211⟩: 3 ppm.

Calcium and magnesium—To 20 mL of a solution (1 in 100) add 2 mL each of 6 N ammonium hydroxide, ammonium oxalate TS, and dibasic sodium phosphate TS: no turbidity is produced within 5 minutes.

Heavy metals ⟨231⟩—Dissolve 2.0 g in 25 mL of water: the limit is 0.001%.

Sodium—A solution (1 in 20), tested on a platinum wire, does not impart a pronounced yellow color to a nonluminous flame.

Assay—Dissolve about 250 mg of Potassium Chloride, accurately weighed, in 150 mL of water. Add 1 mL of nitric acid, and immediately titrate with 0.1 N silver nitrate VS, determining the end-point potentiometrically, using silver-calomel electrodes and a salt bridge containing 4 percent agar in a saturated potassium nitrate solution. Perform a blank determination, and make any necessary correction. Each mL of 0.1 N silver nitrate is equivalent to 7.455 mg of KCl.

Potassium Chloride Extended-release Capsules

» Potassium Chloride Extended-release Capsules contain not less than 90.0 percent and not more than 110.0 percent of the labeled amount of KCl.

Packaging and storage—Preserve in tight containers at a temperature not exceeding 30°.

Identification—A portion of the filtrate obtained as directed under *Assay preparation* in the *Assay* responds to the tests for *Potassium* ⟨191⟩ and for *Chloride* ⟨191⟩.

Dissolution ⟨711⟩—

Medium: water; 900 mL.

Apparatus 1: 100 rpm.

Time: 2 hours.

Potassium stock solution and *Standard preparations*—Prepare as directed in the *Assay* under *Potassium Chloride Elixir*.

Procedure—Filter the solution under test, and dilute quantitatively with *Dissolution Medium* to obtain a test solution containing about 60 µg of potassium chloride per mL. Add 5.0 mL of the test solution to a 100-mL volumetric flask, add 2.0 mL of sodium chloride solution (1 in 5) and 1.0 mL of hydrochloric acid, dilute with water to volume, mix, and proceed as directed for *Procedure* in the *Assay* under *Potassium Chloride Elixir*. Calculate the quantity, in mg, of KCl dissolved by the formula:

$$(900F)(1.907C),$$

in which F is the extent of dilution of the solution under test, and the other terms are as defined therein.

Tolerances—Not more than 35% (Q) of the labeled amount of KCl is dissolved in 2 hours. The requirements are met if the quantities dissolved from the Extended-release Capsules tested

Acceptance Table

Stage	Number Tested	Acceptance Criteria
S_1	6	Each unit is within the range $Q \pm 30\%$.
S_2	6	Average of 12 units ($S_1 + S_2$) is within the range between $Q - 30\%$ and $Q + 35\%$, and no unit is outside the range $Q \pm 40\%$.
S_3	12	Average of 24 units ($S_1 + S_2 + S_3$) is within the range between $Q - 30\%$ and $Q + 35\%$, and not more than 2 units are outside the range $Q \pm 40\%$.

conform to the accompanying acceptance table instead of the table shown under *Dissolution* ⟨711⟩.

Uniformity of dosage units ⟨905⟩: meet the requirements.

Assay—

Potassium stock solution and *Standard preparations*—Prepare as directed in the *Assay* under *Potassium Chloride Elixir.*

Assay preparation—Place not less than 20 Potassium Chloride Extended-release Capsules in a suitable container with 400 mL of water, heat to boiling, and boil for 20 minutes. Allow to cool, transfer the solution to a 1000-mL volumetric flask, dilute with water to volume, and mix. Filter, discarding the first 20 mL of the filtrate. Transfer an accurately measured volume of the subsequent filtrate, equivalent to about 60 mg of potassium chloride, to a 1000-mL volumetric flask, dilute with water to volume, and mix. (Retain a portion of the filtrate for use in the *Identification test.*) Transfer 5.0 mL of the resulting solution to a 100-mL volumetric flask, add 2.0 mL of sodium chloride solution (1 in 5) and 1.0 mL of hydrochloric acid, dilute with water to volume, and mix.

Procedure—Proceed as directed for *Procedure* in the *Assay* under *Potassium Chloride Elixir.* Calculate the quantity, in mg, of KCl in each Capsule taken by the formula:

$$(TC/D)(1.907),$$

in which T is the labeled quantity, in mg, of potassium chloride in each Capsule, D is the concentration, in µg per mL, of potassium chloride in the *Assay preparation*, based on the labeled quantity per Capsule and the extent of dilution, and the other terms are as defined therein.

Potassium Chloride Injection

» Potassium Chloride Injection is a sterile solution of Potassium Chloride in Water for Injection. It contains not less than 95.0 percent and not more than 105.0 percent of the labeled amount of KCl.

Packaging and storage—Preserve in single-dose or in multiple-dose containers, preferably of Type I or Type II glass.

Labeling—The label states the potassium chloride content in terms of weight and of milliequivalents in a given volume. Label the Injection to indicate that it is to be diluted to appropriate strength with water or other suitable fluid prior to administration. The label states also the total osmolar concentration in mOsmol per liter. Where the contents are less than 100 mL, or where the label states that the Injection is not for direct injection but is to be diluted before use, the label alternatively may state the total osmolar concentration in mOsmol per mL.

Identification—It responds to the tests for *Potassium* ⟨191⟩ and for *Chloride* ⟨191⟩.

Pyrogen—When diluted with Sodium Chloride Injection to contain 0.3% of potassium chloride, it meets the requirements of the *Pyrogen Test* ⟨151⟩.

pH ⟨791⟩: between 4.0 and 8.0.

Particulate matter ⟨788⟩: meets the requirements under *Small-volume Injections.*

Other requirements—It meets the requirements under *Injections* ⟨1⟩.

Assay—

Potassium stock solution and *Standard preparations*—Prepare as directed in the *Assay* under *Potassium Chloride Elixir.*

Assay preparation—Transfer an accurately measured volume of Potassium Chloride Injection, equivalent to about 600 mg of potassium chloride, to a 500-mL volumetric flask, dilute with water to volume, and mix. Proceed as directed for *Assay preparation* in the *Assay* under *Potassium Chloride Elixir*, beginning with "Transfer 5.0 mL of the solution to a 100-mL volumetric flask."

Procedure—Proceed as directed for *Procedure* in the *Assay* under *Potassium Chloride Elixir.* Calculate the quantity, in mg, of KCl in the portion of Injection taken by the formula:

$$200C(1.907).$$

Potassium Chloride Injection, Dextrose in—*see* Potassium Chloride in Dextrose Injection

Potassium Chloride Oral Solution

» Potassium Chloride Oral Solution contains not less than 95.0 percent and not more than 105.0 percent of the labeled amount of KCl. It may contain alcohol.

Packaging and storage—Preserve in tight containers.

Identification—Carefully evaporate about 5 mL to dryness, and ignite the residue at dull-red heat to remove all organic matter. Cool, dissolve the residue in 10 mL of water, and filter: the filtrate responds to the tests for *Potassium* ⟨191⟩ and for *Chloride* ⟨191⟩.

Alcohol content (if present) ⟨611⟩: not less than 90.0% and not more than 115.0% of the labeled amount, the labeled amount being not more than 7.5% of C_2H_5OH, determined by the gas-liquid chromatographic procedure, acetone being used as the internal standard.

Assay—

Potassium stock solution—Dissolve 190.7 mg of potassium chloride, previously dried at 105° for 2 hours, in water. Transfer to a 1000-mL volumetric flask, dilute with water to volume, and mix. Transfer 100.0 mL of this solution to a 1000-mL volumetric flask, dilute with water to volume, and mix. This solution contains 10 µg of potassium (equivalent to 19.07 µg of potassium chloride) per mL.

Standard preparations—To separate 100-mL volumetric flasks transfer 10.0, 15.0, and 20.0 mL, respectively, of *Potassium stock solution.* To each flask add 2.0 mL of sodium chloride solution (1 in 5) and 1.0 mL of hydrochloric acid, dilute with water to volume, and mix. The *Standard preparations* contain, respectively, 1.0, 1.5, and 2.0 µg of potassium per mL.

Assay preparation—Transfer an accurately measured volume of Potassium Chloride Oral Solution, equivalent to about 600 mg of potassium chloride, to a 500-mL volumetric flask, dilute with water to volume, and mix. Transfer 5.0 mL of the solution to a 100-mL volumetric flask, dilute with water to volume, and mix. Transfer 5.0 mL of the resulting solution to a 100-mL volumetric flask, add 2.0 mL of sodium chloride solution (1 in 5) and 1.0 mL of hydrochloric acid, dilute with water to volume, and mix.

Procedure—Concomitantly determine the absorbances of the *Standard preparations* and the *Assay preparation* at the potassium emission line of 766.5 nm, with a suitable atomic absorption spectrophotometer (see *Spectrophotometry and Light-scattering* ⟨851⟩) equipped with a potassium hollow-cathode lamp and an air-acetylene flame, using water as the blank. Plot the absorbance of the *Standard preparation* versus concentration, in µg per mL, of potassium, and draw the straight line best fitting the three plotted points. From the graph so obtained, determine the concentration, in µg per mL, of potassium in the *Assay preparation.* Calculate the quantity, in mg, of KCl in the portion of Oral Solution taken by the formula:

$$200C(1.907),$$

in which C is the concentration, in µg per mL, of potassium in the *Assay preparation*, and 1.907 is the ratio of the molecular weight of potassium chloride to the atomic weight of potassium.

Potassium Chloride for Oral Solution

» Potassium Chloride for Oral Solution is a dry mixture of Potassium Chloride and one or more suitable colors, diluents, and flavors. It contains not less than 90.0 percent and not more than 110.0 percent of the labeled amount of KCl.

Packaging and storage—Preserve in tight containers.

Labeling—The label states the Potassium Chloride (KCl) content in terms of weight and in terms of milliequivalents.

Identification—Ignite about 200 mg at a temperature not above 600°, in order to remove all organic matter, cool, dissolve the residue in 10 mL of water, and filter: the filtrate responds to the tests for *Potassium* ⟨191⟩ and for *Chloride* ⟨191⟩.

Minimum fill ⟨755⟩—

FOR SOLID PACKAGED IN MULTIPLE-UNIT CONTAINERS: meets the requirements.

Uniformity of dosage units ⟨905⟩—

FOR SOLID PACKAGED IN SINGLE-UNIT CONTAINERS: meets the requirements.

Assay—

Potassium stock solution and *Standard preparations*—Prepare as directed in the *Assay* under *Potassium Chloride Elixir.*

Assay preparation 1 (where it is packaged in unit-dose containers)—Weigh and mix the contents of not less than 20 containers of Potassium Chloride for Oral Solution. Transfer an accurately weighed portion of the powder, equivalent to about 1.5 g of potassium chloride, to a 500-mL volumetric flask, dissolve in water, dilute with water to volume, and mix. Transfer 5.0 mL of the solution to a 250-mL volumetric flask, dilute with water to volume, and mix. Transfer 5.0 mL of the resulting solution to a 100-mL volumetric flask, add 2.0 mL of sodium chloride solution (1 in 5) and 1.0 mL of hydrochloric acid, dilute with water to volume, and mix.

Assay preparation 2 (where it is packaged in multiple-unit containers)—Transfer an accurately weighed portion of Potassium Chloride for Oral Solution, equivalent to about 1.5 g of potassium chloride, to a 500-mL volumetric flask, dissolve in water, dilute with water to volume, and mix. Proceed as directed for *Assay preparation 1*, beginning with "Transfer 5.0 mL of the solution."

Procedure—Proceed as directed for *Procedure* in the *Assay* under *Potassium Chloride Elixir.* Calculate the quantity of KCl, in mg, in the portion of Potassium Chloride for Oral Solution taken by the formula:

$$500C(1.907),$$

in which C is as defined therein.

Potassium Chloride for Effervescent Oral Solution, Potassium Bicarbonate and—*see* Potassium Bicarbonate and Potassium Chloride for Effervescent Oral Solution

Potassium Chloride Effervescent Tablets for Oral Solution, Potassium Bicarbonate and—*see* Potassium Bicarbonate and Potassium Chloride Effervescent Tablets for Oral Solution

Potassium Chloride Extended-release Tablets

» Potassium Chloride Extended-release Tablets contain not less than 90.0 percent and not more than 110.0 percent of the labeled amount of KCl.

Packaging and storage—Preserve in tight containers at a temperature not exceeding 30°.

Identification—A portion of the filtrate obtained as directed under *Assay preparation* in the *Assay* responds to the tests for *Potassium* ⟨191⟩ and for *Chloride* ⟨191⟩.

Dissolution ⟨711⟩—

Medium: water; 900 mL.

Apparatus 2: 50 rpm.

Time: 2 hours.

Potassium stock solution and *Standard preparations*—Prepare as directed in the *Assay* under *Potassium Chloride Elixir.*

Procedure—Filter the solution under test, and dilute quantitatively with *Dissolution Medium* to obtain a test solution containing about 60 µg of potassium chloride per mL. Place 5.0 mL of the test solution in a 100-mL volumetric flask, add 2.0 mL of sodium chloride solution (1 in 5) and 1.0 mL of hydrochloric acid, dilute with water to volume, mix, and proceed as directed for *Procedure* in the *Assay* under *Potassium Chloride Elixir.* Calculate the quantity, in mg, of KCl dissolved by the formula:

$$(900F)(1.907C),$$

in which F is the extent of dilution of the solution under test, and the other terms are as defined therein.

Tolerances—Not more than 35% (Q) of the labeled amount of KCl is dissolved in 2 hours. The requirements are met if the quantities dissolved from the Extended-release Tablets tested conform to the accompanying acceptance table instead of the table shown under *Dissolution* ⟨711⟩.

Uniformity of dosage units ⟨905⟩: meet the requirements.

Assay—

Potassium stock solution and *Standard preparations*—Prepare as directed in the *Assay* under *Potassium Chloride Elixir.*

Assay preparation—Place not less than 20 Potassium Chloride Extended-release Tablets in a suitable container with 400 mL of water, heat to boiling, and boil for 20 minutes. Allow to cool, transfer the solution to a 1000-mL volumetric flask, dilute with water to volume, and mix. Filter, discarding the first 20 mL of the filtrate. Transfer an accurately measured volume of the subsequent filtrate, equivalent to about 60 mg of potassium chloride, to a 1000-mL volumetric flask, dilute with water to volume, and mix. (Retain a portion of the filtrate for use in the *Identification test.*) Transfer 5.0 mL of the resulting solution to a 100-mL volumetric flask, add 2.0 mL of sodium chloride solution (1 in 5) and 1.0 mL of hydrochloric acid, dilute with water to volume, and mix.

Procedure—Proceed as directed for *Procedure* in the *Assay* under *Potassium Chloride Elixir.* Calculate the quantity, in mg, of KCl in each Tablet taken by the formula:

$$(TC/D)(1.907),$$

in which T is the labeled quantity, in mg, of potassium chloride in each Tablet, D is the concentration, in µg per mL, of potassium chloride in the *Assay preparation*, based on the labeled quantity per Tablet and the extent of dilution, and the other terms are as defined therein.

Acceptance Table

Stage	Number Tested	Acceptance Criteria
S_1	6	Each unit is within the range $Q \pm 30\%$.
S_2	6	Average of 12 units ($S_1 + S_2$) is within the range between $Q - 30\%$ and $Q + 35\%$, and no unit is outside the range $Q \pm 40\%$.
S_3	12	Average of 24 units ($S_1 + S_2 + S_3$) is within the range between $Q - 30\%$ and $Q + 35\%$, and not more than 2 units outside the range $Q \pm 40\%$.

Potassium Chloride in Dextrose Injection

» Potassium Chloride in Dextrose Injection is a sterile solution of Potassium Chloride and Dextrose in Water for Injection. It contains not less than 95.0 percent and not more than 110.0 percent of the labeled amount of KCl and not less than 95.0 percent and not more than 105.0 percent of the labeled amount of $C_6H_{12}O_6.H_2O$. It contains no antimicrobial agents.

Packaging and storage—Perserve in single-dose containers, preferably of Type I or Type II glass.

Labeling—The label states the total osmolar concentration in mOsmol per liter. Where the contents are less than 100 mL, or where the label states that the Injection is not for direct injection but is to be diluted before use, the label alternatively may state the total osmolar concentration in mOsmol per mL. The content of potassium, in milliequivalents, is prominently displayed on the label.

Identification—It responds to the *Identification test* under *Dextrose*, to the flame test for *Potassium* ⟨191⟩, and to the tests for *Chloride* ⟨191⟩.

Pyrogen—It meets the requirements of the test for *Pyrogen* under *Dextrose Injection.*

pH ⟨791⟩: between 3.5 and 6.5, determined on a portion diluted with water, if necessary, to a concentration of not more than 5% of dextrose.

5-Hydroxymethylfurfural and related substances—Dilute an accurately measured volume of Injection, equivalent to 1.0 g of $C_6H_{12}O_6.H_2O$, with water to 500.0 mL. Determine the absorbance of this solution in a 1-cm cell at 284 nm, with a suitable spectrophotometer, using water as the blank: the absorbance is not more than 0.25.

Other requirements—It meets the requirements of the test for *Heavy metals* under *Dextrose Injection*, and meets the requirements under *Injections* ⟨1⟩.

Assay for dextrose—Transfer an accurately measured volume of Potassium Chloride in Dextrose Injection, containing between 2 g and 5 g of dextrose, to a 100-mL volumetric flask. Add 0.2 mL of 6 *N* ammonium hydroxide, dilute with water to volume, and mix. Determine the angular rotation in a suitable polarimeter tube at 25° (see *Optical Rotation* ⟨781⟩). The observed rotation, in degrees, multiplied by 1.0425*A*, in which *A* is the ratio 200 divided by the length, in mm, of the polarimeter tube employed, represents the weight, in g, of $C_6H_{12}O_6.H_2O$ in the volume of Injection taken.

Assay for potassium chloride—Transfer an accurately measured volume of Potassium Chloride in Dextrose Injection, equivalent to between 75 mg and 150 mg of potassium chloride, to a porcelain casserole, and add 140 mL of water and 1 mL of dichlorofluorescein TS. Mix, and titrate with 0.1 *N* silver nitrate VS until the silver chloride flocculates and the mixture acquires a faint pink color. Each mL of 0.1 *N* silver nitrate is equivalent to 7.455 mg of KCl.

Potassium Chloride in Dextrose and Sodium Chloride Injection

» Potassium Chloride in Dextrose and Sodium Chloride Injection is a sterile solution of Potassium Chloride, Dextrose, and Sodium Chloride in Water for Injection. It contains not less than 95.0 percent and not more than 110.0 percent of the labeled amount of potassium (K), and not less than 95.0 percent and not more than 105.0 percent of the labeled amounts of dextrose ($C_6H_{12}O_6.H_2O$), sodium (Na), and chloride (Cl). It contains no antimicrobial agents.

Packaging and storage—Preserve in single-dose containers, preferably of Type I or Type II glass, or of a suitable plastic.

Labeling—The label states the potassium, sodium, and chloride contents in terms of milliequivalents in a given volume. The label states also the total osmolar concentration in mOsmol per liter. Where the contents are less than 100 mL, the label alternatively may state the total osmolar concentration in mOsmol per mL.

Identification—

A: It responds to the flame test for *Sodium* ⟨191⟩.

B: To 2 mL of Injection add 5 mL of sodium cobaltinitrite TS: a yellow precipitate is formed immediately (*presence of potassium*).

C: It responds to the tests for *Chloride* ⟨191⟩.

D: It responds to the *Identification test* under *Dextrose.*

Pyrogen—It meets the requirements of the *Pyrogen Test* ⟨151⟩.

pH ⟨791⟩: between 3.5 and 6.5.

Heavy metals ⟨231⟩—Transfer to a suitable vessel a volume, in mL, of Injection, calculated to two significant figures by the formula:

$$0.2/[(G_KL_K) + (G_DL_D) + (G_SL_S)],$$

in which G_K, G_D, and G_S are the labeled amounts, in g, of potassium chloride, dextrose, and sodium chloride, respectively, in each 100 mL of Injection, and L_K, L_D, and L_S are the limits, in percentage, for *Heavy metals* specified under *Potassium Chloride, Dextrose*, and *Sodium Chloride*, respectively. Adjust the volume by evaporation or addition of water to 25 mL, as necessary: it passes the test.

5-Hydroxymethylfurfural and related substances—Dilute an accurately measured volume of Injection, equivalent to 1.0 g of $C_6H_{12}O_6.H_2O$, with water to 500.0 mL. Determine the absorbance of this solution in a 1-cm cell at 284 nm, with a suitable spectrophotometer, using water as the blank: the absorbance is not more than 0.25.

Other requirements—It meets the requirements under *Injections* ⟨1⟩.

Assay for potassium and sodium—

Internal standard solution, Potassium stock solution, Sodium stock solution, Stock standard preparation, and *Standard preparation*—Prepare as directed in the *Assay for potassium and sodium* under *Potassium Chloride in Sodium Chloride Injection.*

Assay preparation—Transfer 5.0 mL of Potassium Chloride in Dextrose and Sodium Chloride Injection to a 500-mL volumetric flask, dilute with *Internal standard solution* to volume, and mix.

Procedure—Proceed as directed for *Procedure* in the Assay for *Potassium and sodium* under *Potassium Chloride in Sodium Chloride Injection.*

Assay for chloride—Transfer an accurately measured volume of Potassium Chloride in Dextrose and Sodium Chloride Injection, equivalent to about 55 mg of chloride, to a conical flask, and add 140 mL of water and 1 mL of dichlorofluorescein TS. Mix, and titrate with 0.1 *N* silver nitrate VS until the silver chloride flocculates and the mixture acquires a faint pink color when viewed against a white background. Each mL of 0.1 *N* silver nitrate is equivalent to 3.545 mg of Cl. Each mg of chloride is equivalent to 0.0282 mEq of Cl.

Assay for dextrose—Transfer an accurately measured volume of Potassium Chloride in Dextrose and Sodium Chloride Injection, containing between 2 g and 5 g of dextrose, to a 100-mL volumetric flask. Add 0.2 mL of 6 *N* ammonium hydroxide, dilute with water to volume, and mix. Determine the angular rotation in a suitable polarimeter tube at 25° (see *Optical Rotation* ⟨781⟩). The observed rotation, in degrees, multiplied by 1.0425*A*, in which *A* is the ratio 200 divided by the length, in mm, of the polarimeter tube employed, represents the weight, in g, of $C_6H_{12}O_6.H_2O$ in the volume of Injection taken.

Potassium Chloride, Potassium Bicarbonate, and Potassium Citrate Effervescent Tablets for Oral Solution

» Potassium Chloride, Potassium Bicarbonate, and Potassium Citrate Effervescent Tablets for Oral Solution contain not less than 90.0 percent and not more than 110.0 percent of the labeled amounts of K and Cl.

Packaging and storage—Preserve in tight containers, protected from excessive heat.

Labeling—The label states the potassium and chloride contents in terms of weight and in terms of milliequivalents. Where Tablets are packaged in individual pouches, the label instructs the user not to open until the time of use.

Identification—One Tablet dissolves in 100 mL of water with effervescence. The collected gas responds to the test for *Bicarbonate* ⟨191⟩, and the resulting solution responds to the tests for *Potassium* ⟨191⟩, for *Chloride* ⟨191⟩, and for *Citrate* ⟨191⟩.

Uniformity of dosage units ⟨905⟩: meet the requirements for *Weight Variation.*

Assay for potassium—

Potassium stock solution and *Standard preparations*—Prepare as directed in the *Assay* under *Potassium Chloride Elixir.*

Assay preparation—Transfer 10 Potassium Chloride, Potassium Bicarbonate, and Potassium Citrate Effervescent Tablets for Oral Solution to a 2000-mL volumetric flask, dissolve in 200 mL of water, swirl until effervescence ceases, dilute with water to volume, and mix. Filter, and quantitatively dilute an accurately measured volume of the filtrate with water to obtain a solution containing 30 µg of potassium per mL. Transfer 5.0 mL of the resulting solution to a 100-mL volumetric flask, add 2.0 mL of sodium chloride solution (1 in 5) and 1.0 mL of hydrochloric acid, dilute with water to volume, and mix.

Procedure—Proceed as directed for *Procedure* in the *Assay* under *Potassium Chloride Elixir.* Calculate the quantity, in mg, of potassium (K) in each Tablet taken by the formula:

$$L(C/D),$$

in which L is the labeled quantity, in mg, of potassium in each Tablet, C is the concentration, in µg per mL, of potassium in the *Assay preparation*, and D is the concentration, in µg per mL, of potassium in the *Assay preparation* on the basis of the labeled quantity in each Tablet and the extent of dilution.

Assay for chloride—Transfer a number of Potassium Chloride, Potassium Bicarbonate, and Potassium Citrate Effervescent Tablets for Oral Solution, equivalent to about 900 mg of chloride, to a 2000-mL volumetric flask. Add about 200 mL of water, swirl until effervescence ceases, dilute with water to volume, and mix. Transfer 25.0 mL of this solution to a 250-mL conical flask, add 50.0 mL of 0.1 N silver nitrate VS and 15 mL of nitric acid, and boil, with constant swirling, until the supernatant liquid is colorless. Cool to room temperature, add sufficient water to make a volume of about 150 mL, add 5 mL of ferric ammonium sulfate TS, and titrate the excess silver nitrate with 0.1 N ammonium thiocyanate VS to a permanent faint brown end-point. Each mL of 0.1 N silver nitrate is equivalent to 3.545 mg of Cl. Calculate the quantity, in mg, of chloride (Cl) in each Tablet by dividing the total amount of chloride in the Tablets taken by the number of Tablets taken.

Potassium Chloride Oral Solution, Potassium Gluconate and—*see* Potassium Gluconate and Potassium Chloride Oral Solution

Potassium Chloride for Oral Solution, Potassium Gluconate and—*see* Potassium Gluconate and Potassium Chloride for Oral Solution

Potassium Chloride in Sodium Chloride Injection

» Potassium Chloride in Sodium Chloride Injection is a sterile solution of Potassium Chloride and Sodium Chloride in Water for Injection. It contains not less than 95.0 percent and not more than 110.0 percent of the labeled amount of potassium (K), and not less than 95.0 percent and not more than 105.0 percent of the labeled amounts of sodium (Na) and chloride (Cl). It contains no antimicrobial agents.

Packaging and storage—Preserve in single-dose containers, preferably of Type I or Type II glass, or of a suitable plastic.

Labeling—The label states the potassium, sodium, and chloride contents in terms of milliequivalents in a given volume. The label states also the total osmolar concentration in mOsmol per liter. Where the contents are less than 100 mL, the label alternatively may state the total osmolar concentration in mOsmol per mL.

Identification—

A: It responds to the flame test for *Sodium* ⟨191⟩.

B: To 2 mL of Injection add 5 mL of sodium cobaltinitrite TS: a yellow precipitate is formed immediately (*presence of potassium*).

C: It responds to the tests for *Chloride* ⟨191⟩.

Pyrogen—It meets the requirements of the *Pyrogen Test* ⟨151⟩.

pH ⟨791⟩: between 3.5 and 6.5.

Heavy metals ⟨231⟩—Transfer to a suitable vessel a volume, in mL, of Injection, calculated to two significant figures by the formula:

$$0.2/[(G_K L_K) + (G_S L_S)],$$

in which G_K and G_S are the labeled amounts, in g, of potassium chloride and sodium chloride, respectively, in each 100 mL of Injection, and L_K and L_S are the limits, in percentage, for *Heavy metals* specified under *Potassium Chloride* and *Sodium Chloride*, respectively. Adjust the volume by evaporation or addition of water to 25 mL, as necessary: it passes the test.

Other requirements—It meets the requirements under *Injections* ⟨1⟩.

Assay for potassium and sodium—

Internal standard solution—Transfer 1.04 g of lithium nitrate to a 1000-mL volumetric flask, add a suitable nonionic surfactant, then add water to volume, and mix.

Potassium stock solution—Transfer 18.64 g of potassium chloride, previously dried at 105° for 2 hours and accurately weighed, to a 250-mL volumetric flask, add water to volume, and mix. Each mL of this stock solution contains 39.10 mg (1 mEq) of potassium.

Sodium stock solution—Transfer 14.61 g of sodium chloride, previously dried at 105° for 2 hours and accurately weighed, to a 250-mL volumetric flask, add water to volume, and mix.

Stock standard preparation—Transfer 0.1J mL of *Potassium stock solution* and 0.1J' mL of *Sodium stock solution* to a 100-mL volumetric flask, J and J' being the labeled amounts, in mEq per liter, of potassium and sodium, respectively, in the Injection. Dilute with water to volume, and mix. Each mL of this solution contains 0.0391J mg of potassium (K) and 0.02299J' mg of sodium (Na).

Standard preparation—Transfer 5.0 mL of *Stock standard preparation* to a 500-mL volumetric flask, dilute with *Internal standard solution* to volume, and mix.

Assay preparation—Transfer 5.0 mL of Potassium Chloride in Sodium Chloride Injection to a 500-mL volumetric flask, dilute with *Internal standard solution* to volume, and mix.

Procedure—Using a suitable flame photometer, adjusted to read zero with *Internal standard solution*, concomitantly determine the flame emission readings for the *Standard preparation* and the *Assay preparation* at the wavelengths of maximum emission for potassium, sodium, and lithium (766 nm, 589 nm, and 671 nm, respectively). Calculate the quantity, in mg, of K in each mL of the Injection taken by the formula:

$$C(R_{U,766}/R_{U,671})(R_{S,671}/R_{S,766}),$$

in which C is the concentration, in mg per mL, of potassium in the *Stock standard preparation*, $R_{U,766}$ and $R_{U,671}$ are the emission readings at the wavelengths identified by the subscript numbers obtained from the *Assay preparation*, and $R_{S,671}$ and $R_{S,766}$ are the emission readings at the wavelengths identified by the subscript numbers obtained from the *Standard preparation.* Each mg of potassium is equivalent to 0.02558 mEq of potassium. Calculate the quantity, in mg, of Na in each mL of the Injection taken by the formula:

$$C(R_{U,589}/R_{U,671})/(R_{S,671}/R_{S,589}),$$

in which C is the concentration, in mg per mL, of sodium in the *Stock standard preparation*, $R_{U,589}$ and $R_{U,671}$ are the emission readings at the wavelengths identified by the subscript numbers obtained for the *Assay preparation*, and $R_{S,589}$ and $R_{S,671}$ are the emission readings at the wavelengths identified by the subscript numbers obtained from the *Standard preparation.* Each mg of sodium is equivalent to 0.04350 mEq of sodium.

Assay for chloride—Transfer an accurately measured volume of Potassium Chloride in Sodium Chloride Injection, equivalent to about 55 mg of chloride, to a conical flask, and add 140 mL of water and 1 mL of dichlorofluorescein TS. Mix, and titrate with 0.1 *N* silver nitrate VS until the silver chloride flocculates and the mixture acquires a faint pink color when viewed against a white background. Each mL of 0.1 *N* silver nitrate is equivalent to 3.545 mg of Cl. Each mg of chloride is equivalent to 0.0282 mEq of Cl.

Potassium Citrate

```
     CH2COOK
     |
HO—C—COOK  · H2O
     |
     CH2COOK
```

$C_6H_5K_3O_7.H_2O$ 324.41

1,2,3-Propanetricarboxylic acid, 2-hydroxy-, tripotassium salt, monohydrate.

Tripotassium citrate monohydrate [*6100-05-6*].

Anhydrous 306.40 [*866-84-2*].

» Potassium Citrate contains not less than 99.0 percent and not more than 100.5 percent of $C_6H_5K_3O_7$, calculated on the dried basis.

Packaging and storage—Preserve in tight containers.

Identification—A solution (1 in 10) responds to the tests for *Potassium* ⟨191⟩, and for *Citrate* ⟨191⟩.

Alkalinity—A solution of 1.0 g in 20 mL of water is alkaline to litmus, but after the addition of 0.20 mL of 0.10 *N* sulfuric acid no pink color is produced by the addition of 1 drop of phenolphthalein TS.

Loss on drying ⟨731⟩—Dry it at 180° for 4 hours: it loses between 3.0% and 6.0% of its weight.

Tartrate—To a solution of 1 g in 1.5 mL of water in a test tube add 1 mL of 6 *N* acetic acid, and scratch the walls of the test tube with a glass rod: no crystalline precipitate is formed.

Heavy metals, *Method I* ⟨231⟩—Dissolve 2 g in 25 mL of water, and proceed as directed for *Test Preparation*, except to use glacial acetic acid to adjust the pH: the limit is 0.001%.

Assay—Dissolve about 200 mg of Potassium Citrate, accurately weighed, in 25 mL of glacial acetic acid. Add 2 drops of crystal violet TS, and titrate with 0.1 *N* perchloric acid VS to a green end-point. Perform a blank determination, and make any necessary correction. Each mL of 0.1 *N* perchloric acid is equivalent to 10.21 mg of $C_6H_5K_3O_7$.

Potassium Citrate, and Ammonium Chloride Oral Solution, Potassium Gluconate,—*see* Potassium Gluconate, Potassium Citrate, and Ammonium Chloride Oral Solution

Potassium Citrate and Citric Acid Oral Solution

» Potassium Citrate and Citric Acid Oral Solution is a solution of Potassium Citrate and Citric Acid in a suitable aqueous medium. It contains, in each 100 mL, not less than 7.55 g and not more than 8.35 g of potassium (K), and not less than 12.18 g and not more than 13.46 g of citrate ($C_6H_5O_7$), equivalent to not less than 20.9 g and not more than 23.1 g of potassium citrate monohydrate ($C_6H_5K_3O_7.H_2O$); and not less than 6.34 g and not more than 7.02 g of citric acid monohydrate ($C_6H_8O_7.H_2O$).

NOTE—The potassium ion content of Potassium Citrate and Citric Acid Oral Solution is approximately 2 mEq per mL.

Packaging and storage—Preserve in tight containers.

Identification—

A: To 2 mL of a dilution of Oral Solution (1 in 40) add 5 mL of sodium cobaltinitrite TS: a yellow precipitate is formed immediately (*presence of potassium*).

B: To a mixture of 1 mL of Oral Solution with 1 mL of hydrochloric acid add 10 mL of cobalt-uranyl acetate TS, and stir with a glass rod: no precipitate or turbidity forms after 15 minutes, and the solution remains clear (*absence of sodium*).

C: It responds to the test for *Citrate* ⟨191⟩, 3 to 5 drops of Oral Solution and 20 mL of the mixture of pyridine and acetic anhydride being used.

pH ⟨791⟩: between 4.9 and 5.4.

Assay for potassium—

Potassium stock solution, Sodium stock solution, Lithium diluent solution, and *Standard preparation*—Prepare as directed in the *Assay for sodium and potassium* under *Tricitrates Oral Solution.*

Assay preparation—Transfer an accurately measured volume of Potassium Citrate and Citric Acid Oral Solution, equivalent to about 2 g of potassium citrate monohydrate, to a 200-mL volumetric flask, dilute with water to volume, and mix. Transfer 50 µL of this solution to a 10-mL volumetric flask, dilute with *Lithium diluent solution* to volume, and mix.

Procedure—Using a suitable flame photometer, adjusted to read zero with *Lithium diluent solution,* concomitantly determine the potassium flame emission readings for the *Standard preparation* and the *Assay preparation* at the wavelength of maximum emission at about 766 nm. Calculate the quantity, in g, of K in the portion of Oral Solution taken by the formula:

$$(18.64/12.5)(39.10/74.55)(R_{U,K}/R_{S,K}),$$

in which 18.64 is the weight, in g, of potassium chloride in the *Potassium stock solution,* 39.10 is the atomic weight of potassium, 74.55 is the molecular weight of potassium chloride, and $R_{U,K}$ and $R_{S,K}$ are the potassium emission readings obtained for the *Assay preparation* and the *Standard preparation,* respectively.

Assay for citrate—

Cation-exchange column—Mix 10 g of styrene-divinylbenzene cation-exchange resin with 50 mL of water in a suitable beaker. Allow the resin to settle, and decant the supernatant liquid until a slurry of resin remains. Pour the slurry into a 15-mm × 30-cm glass chromatographic tube (having a sealed-in, coarse-porosity fritted disk and fitted with a stopcock), and allow to settle as a homogeneous bed. Wash the resin bed with about 100 mL of water, closing the stopcock when the water level is about 2 mm above the resin bed.

Procedure—Pipet 15 mL of Potassium Citrate and Citric Acid Oral Solution into a 250-mL volumetric flask, dilute with water to volume, and mix. Pipet 5 mL of this solution carefully onto the top of the resin bed in the *Cation-exchange column.* Place a 250-mL conical flask below the column, open the stopcock, and allow to flow until the solution has entered the resin bed. Elute the column with 60 mL of water at a flow rate of about 5 mL per minute, collecting about 65 mL of the eluate. Add 5 drops of phenolphthalein TS to the eluate, swirl the flask, and titrate with 0.02 *N* sodium hydroxide VS. Record the buret reading, and calculate the volume (*B*) of 0.02 *N* sodium hydroxide consumed. Each mL of the difference between the volume (*B*) and the volume (*A*) of 0.02 *N* sodium hydroxide consumed in the *Assay for citric acid* is equivalent to 1.261 mg of $C_6H_5O_7$.

Assay for citric acid—Transfer 15 mL of Potassium Citrate and Citric Acid Oral Solution, accurately measured, to a 250-mL volumetric flask, dilute with water to volume, and mix. Pipet 5 mL of this solution into a suitable flask, add 25 mL of water and 5 drops of phenolphthalein TS, and titrate with 0.02 *N* sodium hydroxide VS to a pink end-point. Record the buret reading, and calculate the volume (*A*) of 0.02 *N* sodium hydroxide consumed. Each mL of 0.02 *N* sodium hydroxide is equivalent to 1.401 mg of $C_6H_8O_7 \cdot H_2O$.

Potassium Citrate Effervescent Tablets for Oral Solution, Potassium Chloride, Potassium Bicarbonate, and—*see* Potassium Chloride, Potassium Bicarbonate, and Potassium Citrate Effervescent Tablets for Oral Solution

Potassium Citrate Oral Solution, Potassium Gluconate and—*see* Potassium Gluconate and Potassium Citrate Oral Solution

Potassium Gluconate

$$HOCH_2-\underset{OH}{\overset{H}{C}}-\underset{OH}{\overset{H}{C}}-\underset{H}{\overset{OH}{C}}-\underset{OH}{\overset{H}{C}}-COOK$$

$C_6H_{11}KO_7$ (anhydrous) 234.25
D-Gluconic acid, monopotassium salt.
Monopotassium D-gluconate [*299-27-4*].
Monohydrate 252.26 [*35398-15-3*].

» Potassium Gluconate is anhydrous or contains one molecule of water of hydration. It contains not less than 97.0 percent and not more than 103.0 percent of $C_6H_{11}KO_7$, calculated on the dried basis.

Packaging and storage—Preserve in tight containers.

Labeling—Label it to indicate whether it is anhydrous or the monohydrate.

Reference standard—*USP Potassium Gluconate Reference Standard*—Dry in vacuum at 105° for 4 hours before using.

Identification—

A: The infrared absorption spectrum of a mineral oil dispersion of it, previously dried, exhibits maxima only at the same wavelengths as that of a similar preparation of USP Potassium Gluconate RS.

B: It responds to the flame test for *Potassium* ⟨191⟩.

C: It responds to *Identification test B* under *Calcium Gluconate.*

Loss on drying ⟨731⟩—Dry it in vacuum at 105° for 4 hours: the anhydrous form loses not more than 3.0% of its weight, and the monohydrate loses between 6.0% and 7.5% of its weight.

Heavy metals, *Method I* ⟨231⟩—Dissolve 1 g in 10 mL of water, add 6 mL of 3 *N* hydrochloric acid, and dilute with water to 25 mL: the limit is 0.002%.

Reducing substances—Transfer 1.0 g to a 250-mL conical flask, dissolve in 10 mL of water, and add 25 mL of alkaline cupric citrate TS. Cover the flask, boil gently for 5 minutes, accurately timed, and cool rapidly to room temperature. Add 25 mL of 0.6 *N* acetic acid, 10.0 mL of 0.1 *N* iodine VS, and 10 mL of 3 *N* hydrochloric acid, and titrate with 0.1 *N* sodium thiosulfate VS, adding 3 mL of starch TS as the end-point is approached. Perform a blank determination, omitting the specimen, and note the difference in volumes required. Each mL of the difference in volume of 0.1 *N* sodium thiosulfate consumed is equivalent to 2.7 mg of reducing substances (as dextrose): the limit is 1.0%.

Assay—

Potassium stock solution—Dissolve 190.7 mg of potassium chloride, previously dried at 105° for 2 hours, in water. Transfer to a 1000-mL volumetric flask, dilute with water to volume, and mix. Transfer 100.0 mL of this solution to a 1000-mL volumetric flask, dilute with water to volume, and mix. This solution contains 10 µg of potassium (equivalent to 19.07 µg of potassium chloride) per mL.

Standard preparations—To separate 100-mL volumetric flasks transfer 10.0, 15.0, and 20.0 mL, respectively, of *Potassium stock solution.* To each flask add 2.0 mL of sodium chloride solution (1 in 5) and 1.0 mL of hydrochloric acid, dilute with water to volume, and mix. The *Standard preparations* contain, respectively, 1.0, 1.5, and 2.0 µg of potassium per mL.

Assay preparation—Transfer about 180 mg of Potassium Gluconate, accurately weighed, to a 1000-mL volumetric flask, add water to volume, and mix. Filter a portion of the solution. Transfer 5.0 mL of the filtrate to a 100-mL volumetric flask, add 2.0 mL of sodium chloride solution (1 in 5) and 1.0 mL of hydrochloric acid, dilute with water to volume, and mix.

Procedure—Concomitantly determine the absorbances of the *Standard preparations* and the *Assay preparation* at the potassium emission line of 766.5 nm, with a suitable atomic absorption spectrophotometer (see *Spectrophotometry and Light-scattering* ⟨851⟩) equipped with a potassium hollow-cathode lamp and an air-acetylene flame, using water as the blank. Plot the absorbance of the *Standard preparation* versus concentration, in µg per mL, of potassium, and draw the straight line best fitting the three plotted points. From the graph so obtained, determine the concentration, in µg per mL, of potassium in the *Assay preparation.* Calculate the weight, in mg, of $C_6H_{11}KO_7$ in the Potassium Gluconate taken by the formula:

$$20C(234.25/39.10),$$

in which *C* is the concentration, in µg per mL, of potassium in the *Assay preparation,* 234.25 is the molecular weight of potassium gluconate, and 39.10 is the atomic weight of potassium.

Potassium Gluconate Elixir

» Potassium Gluconate Elixir contains not less than 95.0 percent and not more than 105.0 percent of the labeled amount of $C_6H_{11}KO_7$.

Packaging and storage—Preserve in tight, light-resistant containers.

Identification—

A: It responds to the flame test for *Potassium* ⟨191⟩.

B: Evaporate 5 mL on a steam bath to dryness: a mineral oil dispersion of the residue exhibits an infrared absorption maximum in the spectral region between 6.2 and 6.25 µm (*carboxylic acid salt*).

Alcohol content, *Method II* ⟨611⟩: between 4.5% and 5.5% of C_2H_5OH.

Assay—

Potassium stock solution and *Standard preparations*—Prepare as directed in the *Assay* under *Potassium Gluconate.*

Assay preparation—Transfer an accurately measured volume of Potassium Gluconate Elixir, equivalent to about 1.8 g of potassium gluconate, to a 1000-mL volumetric flask, dilute with water to volume, and mix. Transfer 10.0 mL of the solution to a 100-mL volumetric flask, dilute with water to volume, and mix. Transfer 5.0 mL of the resulting solution to a 100-mL volumetric flask, add 2.0 mL of sodium chloride solution (1 in 5) and 1.0 mL of hydrochloric acid, dilute with water to volume, and mix.

Procedure—Proceed as directed for *Procedure* in the *Assay* under *Potassium Gluconate*. Calculate the quantity, in mg, of $C_6H_{11}KO_7$ in each mL of the Elixir taken by the formula:

$$(200C/V)(234.25/39.10),$$

in which V is the volume of Elixir taken, 234.25 is the molecular weight of potassium gluconate, and 39.10 is the atomic weight of potassium.

Potassium Gluconate Tablets

» Potassium Gluconate Tablets contain not less than 95.0 percent and not more than 105.0 percent of the labeled amount of $C_6H_{11}KO_7$.

Packaging and storage—Preserve in tight containers.

Reference standard—*USP Potassium Gluconate Reference Standard*—Dry in vacuum at 105° for 4 hours before using.

Identification—

A: The infrared absorption spectrum of a mineral oil dispersion prepared from finely powdered Tablets exhibits maxima only at the same wavelengths as that of a similar preparation of USP Potassium Gluconate RS.

B: Triturate a portion of powdered Tablets with a few mL of water, and filter: the filtrate responds to the flame test for *Potassium* ⟨191⟩.

Dissolution ⟨711⟩—

Medium: water; 900 mL.

Apparatus 2: 100 rpm.

Time: 45 minutes.

Procedure—Determine the amount of $C_6H_{11}KO_7$ dissolved, employing the procedure set forth in the *Assay*, making any necessary modifications.

Tolerances—Not less than 75% (Q) of the labeled amount of $C_6H_{11}KO_7$ is dissolved in 45 minutes.

Uniformity of dosage units ⟨905⟩: meet the requirements.

Assay—

Potassium stock solution and *Standard preparations*—Prepare as directed in the *Assay* under *Potassium Gluconate*.

Assay preparation—Weigh and finely powder not less than 20 Potassium Gluconate Tablets. Transfer an accurately weighed portion of the powder, equivalent to about 1.8 g of potassium gluconate, to a 1000-mL volumetric flask, add water to volume, and mix to dissolve. Filter a portion of the solution, transfer 10.0 mL of the filtrate to a 100-mL volumetric flask, dilute with water to volume, and mix. Transfer 5.0 mL of the resulting solution to a 100-mL volumetric flask, add 2.0 mL of sodium chloride solution (1 in 5) and 1.0 mL of hydrochloric acid, dilute with water to volume, and mix.

Procedure—Proceed as directed for *Procedure* in the *Assay* under *Potassium Gluconate*. Calculate the quantity, in mg, of $C_6H_{11}KO_7$ in the portion of Tablets taken by the formula:

$$200C(234.25/39.10),$$

in which 234.25 is the molecular weight of potassium gluconate and 39.10 is the atomic weight of potassium.

Potassium Gluconate and Potassium Chloride Oral Solution

» Potassium Gluconate and Potassium Chloride Oral Solution is a solution of Potassium Gluconate and Potassium Chloride in a suitable aqueous medium. It contains not less than 90.0 percent and not more than 110.0 percent of the labeled amounts of potassium (K) and chloride (Cl).

Packaging and storage—Preserve in tight containers.

Labeling—Label it to state the potassium and chloride contents in terms of milliequivalents of each in a given volume of Oral Solution.

Identification—

A: To 2 mL of a dilution of Oral Solution (1 in 40) add 5 mL of sodium cobaltinitrite TS: a yellow precipitate is formed immediately (*presence of potassium*).

B: Evaporate 5 mL on a steam bath to dryness: a mineral oil dispersion of the residue so obtained exhibits an infrared absorption maximum in the spectral region between 6.2 and 6.25 µm (*carboxylic acid salt*).

C: It responds to the tests for *Chloride* ⟨191⟩.

Assay for potassium—

Potassium stock solution—Dissolve in water 0.9535 g of potassium chloride, previously dried at 105° for 2 hours. Transfer to a 500-mL volumetric flask, dilute with water to volume, and mix. This solution contains 1000 µg of potassium per mL.

Standard preparations—To separate 200-mL volumetric flasks transfer 19.0 mL and 25.0 mL, respectively, of *Potassium stock solution*, dilute with water to volume, and mix. The *Standard preparations* contain 95.0 µg and 125.0 µg of potassium per mL, respectively.

Assay preparation—Transfer an accurately measured volume of Potassium Gluconate and Potassium Chloride Oral Solution, equivalent to about 782 mg (20 mEq) of potassium, to a 100-mL volumetric flask, dilute with water to volume, and mix. Transfer 7.0 mL of the resulting solution to a 500-mL volumetric flask, dilute with water to volume, and mix.

Procedure—Concomitantly determine the absorbances of the *Standard preparations* and the *Assay preparation* at the resonance line of 766.5 nm, with a suitable atomic absorption spectrophotometer (see *Spectrophotometry and Light-scattering* ⟨851⟩) equipped with a potassium hollow-cathode lamp and an air-acetylene flame, using water as the blank. Plot the absorbances of the *Standard preparations* versus concentration, in µg per mL, of potassium. From the graph so obtained, determine the concentration, C, in µg per mL, of potassium in the *Assay preparation*. Calculate the quantity, in mg, of potassium in each mL of the Oral Solution taken by the formula:

$$(50/7)(C/V),$$

in which V is the volume, in mL, of Oral Solution taken. Each mg of potassium is equivalent to 0.02558 mEq.

Assay for chloride—

Ionic strength adjusting solution—Use 5 M sodium nitrate.

Procedure—Transfer an accurately measured volume of Potassium Gluconate and Potassium Chloride Oral Solution, equivalent to about 100 mg (2.8 mEq) of chloride, to a suitable beaker. Add 2.0 mL of *Ionic strength adjusting solution* and water to make about 100 mL, and titrate with 0.1 N silver nitrate VS, determining the end-point potentiometrically, using a silver–sulfide specific ion-selective electrode and a double-junction reference electrode containing potassium nitrate solution (1 in 10). Perform a blank determination, and make any necessary correction. Each mL of 0.1 N silver nitrate is equivalent to 3.545 mg of chloride (Cl). Each mg of chloride is equivalent to 0.0282 mEq of Cl.

Potassium Gluconate and Potassium Chloride for Oral Solution

» Potassium Gluconate and Potassium Chloride for Oral Solution is a dry mixture of Potassium Gluconate and Potassium Chloride and one or more suitable colors, diluents, and flavors. It contains not less than 90.0 percent and not more than 110.0 percent of the labeled amounts of potassium (K) and chloride (Cl).

Packaging and storage—Preserve in tight containers.

Labeling—Label it to state the potassium and chloride contents in terms of milliequivalents. Where packaged in unit-dose pouches, the label instructs the user not to open until the time of use.

Identification—

A: Ignite about 200 mg at a temperature not above 600°, in order to remove all organic matter, cool, dissolve the residue in 10 mL of water, and filter: the filtrate responds to the tests for *Potassium* ⟨191⟩ and for *Chloride* ⟨191⟩.

B: A mineral oil dispersion of it exhibits an infrared absorption maximum in the spectral region between 6.2 and 6.25 μm (*carboxylic acid salt*).

Minimum fill ⟨755⟩: meets the requirements.

Assay for potassium—

Potassium stock solution—Dissolve in water 0.9535 g of potassium chloride, previously dried at 105° for 2 hours. Transfer to a 500-mL volumetric flask, dilute with water to volume, and mix. This solution contains 1000 μg of potassium per mL.

Standard preparations—To separate 200-mL volumetric flasks transfer 19.0 mL and 25.0 mL, respectively, of the *Potassium stock solution*, dilute with water to volume, and mix. The *Standard preparations* contain 95.0 μg and 125.0 μg of potassium per mL, respectively.

Assay preparation 1 (where it is packaged in unit-dose containers)—Weigh and mix the contents of not less than 20 containers of Potassium Gluconate and Potassium Chloride for Oral Solution. Transfer an accurately weighed portion of the powder, equivalent to about 782 mg (20 mEq) of potassium, to a 100-mL volumetric flask, dilute with water to volume, and mix. Transfer 7.0 mL of this stock solution to a 500-mL volumetric flask, dilute with water to volume, and mix.

Assay preparation 2 (where it is packaged in multiple-unit containers)—Transfer an accurately weighed portion of Potassium Gluconate and Potassium Chloride for Oral Solution, equivalent to about 780 mg (20 mEq) of potassium, to a 100-mL volumetric flask, dissolve in water, dilute with water to volume, and mix. Transfer 7.0 mL of this stock solution to a 500-mL volumetric flask, dilute with water to volume, and mix.

Procedure—Concomitantly determine the absorbances of the *Standard preparations* and the *Assay preparation* at the resonance line of 766.5 nm, with a suitable atomic absorption spectrophotometer (see *Spectrophotometry and Light-scattering* ⟨851⟩) equipped with a potassium hollow-cathode lamp and an air-acetylene flame, using water as the blank. Plot the absorbances of the *Standard preparations* versus concentration, in μg per mL, of potassium. From the graph so obtained, determine the concentration, C, in μg per mL, of potassium in the *Assay preparation*. Calculate the quantity, in mg, of potassium in the portion of Potassium Gluconate and Potassium Chloride for Oral Solution taken by the formula:

$$50C/7.$$

Each mg of potassium is equivalent to 0.02558 mEq.

Assay for chloride—

Ionic strength adjusting solution—Use 5 *M* sodium nitrate.

Assay preparation 1 (where it is packaged in unit-dose containers)—Weigh and mix the contents of not less than 20 containers of Potassium Gluconate and Potassium Chloride for Oral Solution. Transfer an accurately weighed portion of the powder, equivalent to about 100 mg (2.8 mEq) of chloride, to a suitable beaker.

Assay preparation 2 (where it is packaged in multiple-unit containers)—Transfer an accurately weighed portion of Potassium Gluconate and Potassium Chloride for Oral Solution, equivalent to about 100 mg (2.8 mEq) of chloride, to a suitable beaker.

Procedure—Add 2.0 mL of *Ionic strength adjusting solution* to *Assay preparation 1* or *Assay preparation 2*, add water to make about 100 mL, and titrate with 0.1 *N* silver nitrate VS, determining the end-point potentiometrically, using a silver–sulfide specific ion-selective electrode and a double-junction reference electrode containing potassium nitrate solution (1 in 10). Perform a blank determination, and make any necessary correction. Each mL of 0.1 *N* silver nitrate is equivalent to 3.545 mg of chloride (Cl). Each mg of chloride is equivalent to 0.0282 mEq of Cl.

Potassium Gluconate and Potassium Citrate Oral Solution

» Potassium Gluconate and Potassium Citrate Oral Solution is a solution of Potassium Gluconate and Potassium Citrate in a suitable aqueous medium. It contains not less than 90.0 percent and not more than 110.0 percent of the labeled amount of potassium (K).

Packaging and storage—Preserve in tight containers.

Labeling—Label it to state the potassium content in terms of milliequivalents in a given volume of Oral Solution.

Identification—

A: To 2 mL of a dilution of Oral Solution (1 in 40) add 5 mL of sodium cobaltinitrite TS: a yellow precipitate is formed immediately (*presence of potassium*).

B: It responds to the test for *Citrate* ⟨191⟩, 3 to 5 drops of Oral Solution and 20 mL of the mixture of pyridine and acetic anhydride being used.

Assay for potassium—

Potassium stock solution—Dissolve in water 0.9535 g of potassium chloride, previously dried at 105° for 2 hours. Transfer to a 500-mL volumetric flask, dilute with water to volume, and mix. This solution contains 1000 μg of potassium per mL.

Standard preparations—To separate 200-mL volumetric flasks transfer 19.0 mL and 25.0 mL, respectively, of *Potassium stock solution*, dilute with water to volume, and mix. The *Standard preparations* contain 95.0 μg and 125.0 μg of potassium per mL, respectively.

Assay preparation—Transfer an accurately measured volume of Potassium Gluconate and Potassium Citrate Oral Solution, equivalent to about 782 mg (20 mEq) of potassium, to a 100-mL volumetric flask, dilute with water to volume, and mix. Transfer 7.0 mL of this solution to a 500-mL volumetric flask, dilute with water to volume, and mix.

Procedure—Concomitantly determine the absorbances of the *Standard preparations* and the *Assay preparation* at the resonance line of 766.5 nm, with a suitable atomic absorption spectrophotometer (see *Spectrophotometry and Light-scattering* ⟨851⟩) equipped with a potassium hollow-cathode lamp and an air-acetylene flame, using water as the blank. Plot the absorbances of the *Standard preparations* versus concentration, in μg per mL, of potassium. From the graph so obtained, determine the concentration, C, in μg per mL, of potassium in the *Assay preparation*. Calculate the quantity, in mg, of potassium in each mL of the Oral Solution taken by the formula:

$$(50/7)(C/V),$$

in which V is the volume, in mL, of Oral Solution taken. Each mg of potassium is equivalent to 0.02558 mEq.

Potassium Gluconate, Potassium Citrate, and Ammonium Chloride Oral Solution

» Potassium Gluconate, Potassium Citrate, and Ammonium Chloride Oral Solution is a solution of Potassium Gluconate, Potassium Citrate, and Ammonium Chloride in a suitable aqueous medium. It contains not less than 90.0 percent and not more than 110.0 percent of the labeled amounts of potassium (K) and chloride (Cl).

Packaging and storage—Preserve in tight containers.

Labeling—Label it to state the potassium and chloride contents in terms of milliequivalents of each in a given volume of Oral Solution.

Identification—

A: To 2 mL of a dilution of Oral Solution (1 in 40) add 5 mL of sodium cobaltinitrite TS: a yellow precipitate is formed immediately (*presence of potassium*).

B: It responds to the test for *Citrate* ⟨191⟩, 3 to 5 drops of Oral Solution and 20 mL of the mixture of pyridine and acetic anhydride being used.

C: It responds to the tests for *Ammonium* ⟨191⟩ and for *Chloride* ⟨191⟩.

Assay for potassium—

Potassium stock solution—Dissolve in water 0.9535 g of potassium chloride, previously dried at 105° for 2 hours. Transfer to a 500-mL volumetric flask, dilute with water to volume, and mix. This solution contains 1000 µg of potassium per mL.

Standard preparations—To separate 200-mL volumetric flasks transfer 19.0 mL and 25.0 mL, respectively, of the *Potassium stock solution*, dilute with water to volume, and mix. The *Standard preparations* contain 95.0 µg and 125.0 µg of potassium per mL, respectively.

Assay preparation—Transfer an accurately measured volume of Potassium Gluconate, Potassium Citrate, and Ammonium Chloride Oral Solution, equivalent to about 782 mg (20 mEq) of potassium, to a 100-mL volumetric flask, dilute with water to volume, and mix. Transfer 7.0 mL of this solution to a 500-mL volumetric flask, dilute with water to volume, and mix.

Procedure—Concomitantly determine the absorbances of the *Standard preparations* and the *Assay preparation* at the resonance line of 766.5 nm, with a suitable atomic absorption spectrophotometer (see *Spectrophotometry and Light-scattering* ⟨851⟩) equipped with a potassium hollow-cathode lamp and an air-acetylene flame, using water as the blank. Plot the absorbances of the *Standard preparations* versus concentration, in µg per mL, of potassium. From the graph so obtained, determine the concentration, C, in µg per mL, of potassium in the *Assay preparation.* Calculate the quantity, in mg, of potassium in each mL of the Oral Solution taken by the formula:

$$(50/7)(C/V),$$

in which V is the volume, in mL, of Oral Solution taken. Each mg of potassium is equivalent to 0.02558 mEq.

Assay for chloride—

Ionic strength adjusting solution—Use 5 *M* sodium nitrate.

Procedure—Transfer an accurately measured volume of Potassium Gluconate, Potassium Citrate, and Ammonium Chloride Oral Solution, equivalent to about 100 mg (2.8 mEq) of chloride, to a suitable beaker. Add 2.0 mL of *Ionic strength adjusting solution* and water to make about 100 mL, and titrate with 0.1 *N* silver nitrate VS, determining the end-point potentiometrically, using a silver–sulfide specific ion-selective electrode and a double-junction reference electrode containing potassium nitrate solution (1 in 10). Perform a blank determination, and make any necessary correction. Each mL of 0.1 *N* silver nitrate is equivalent to 3.545 mg of chloride (Cl). Each mg of chloride is equivalent to 0.0282 mEq of Cl.

Potassium Guaiacolsulfonate

$C_7H_7KO_5S.\frac{1}{2}H_2O$ 251.29

Benzenesulfonic acid, hydroxymethoxy-, monopotassium salt, hemihydrate.

Potassium hydroxymethoxybenzenesulfonate hemihydrate [*78247-49-1*].

Anhydrous 242.29

» Potassium Guaiacolsulfonate contains not less than 98.0 percent and not more than 102.0 percent of $C_7H_7KO_5S$, calculated on the anhydrous basis.

Packaging and storage—Preserve in well-closed, light-resistant containers.

Reference standard—*USP Potassium Guaiacolsulfonate Reference Standard*—Determine the *Water* content by *Method I* before use for quantitative analyses.

Identification—

A: The infrared absorption spectrum of a mineral oil dispersion of it, previously dried at 105° for 18 hours, exhibits maxima only at the same wavelengths as that of a similar preparation of USP Potassium Guaiacolsulfonate RS, dried at 105° for 18 hours (see *USP Reference Standards* ⟨11⟩), in the spectral region between 7 µm and 13 µm.

B: The ultraviolet absorption spectrum of a 1 in 20,000 solution, prepared as directed in the *Assay*, exhibits maxima and minima at the same wavelengths as that of a similar solution of USP Potassium Guaiacolsulfonate RS.

C: A solution (1 in 10) responds to the tests for *Potassium* ⟨191⟩.

Water, *Method I* ⟨921⟩: between 3.0% and 6.0%.

Selenium ⟨291⟩: 0.003%.

Sulfate—To 10 mL of a solution (1 in 20) add 5 drops of barium chloride TS, and acidify with hydrochloric acid: no turbidity is produced in 1 minute.

Heavy metals ⟨231⟩—Dissolve 1.0 g in 1 mL of 1 *N* acetic acid, and dilute with water to 25 mL. The limit is 0.002%.

Assay—Transfer about 250 mg of Potassium Guaiacolsulfonate, accurately weighed, to a 500-mL volumetric flask, dissolve in 400 mL of water, dilute with water to volume, and mix. Dilute 10.0 mL of this solution with pH 7.0 phosphate buffer to 100.0 mL, and mix. Concomitantly determine the absorbances of this solution and a Standard solution of USP Potassium Guaiacolsulfonate RS in the same medium, having a known concentration of about 50 µg per mL, in 1-cm cells at the wavelength of maximum absorbance at about 279 nm, with a suitable spectrophotometer, using a 1 in 10 mixture of water and pH 7.0 phosphate buffer as the blank. Calculate the quantity, in mg, of $C_7H_7KO_5S$ in the Potassium Guaiacolsulfonate taken by the formula:

$$5C(A_U/A_S),$$

in which C is the concentration, in µg per mL, calculated on the anhydrous basis, of USP Potassium Guaiacolsulfonate RS in the Standard solution, and A_U and A_S are the absorbances of the preparation under assay and the Standard solution, respectively.

Potassium Hydroxide—*see* Potassium Hydroxide NF

Potassium Iodide

KI 166.00

Potassium iodide.

Potassium iodide [*7681-11-0*].

» Potassium Iodide contains not less than 99.0 percent and not more than 101.5 percent of KI, calculated on the dried basis.

Packaging and storage—Preserve in well-closed containers.

Identification—A solution of it responds to the tests for *Potassium* ⟨191⟩ and for *Iodide* ⟨191⟩.

Alkalinity—Dissolve 1.0 g in 10 mL of water, and add 0.1 mL of 0.1 *N* sulfuric acid and 1 drop of phenolphthalein TS: no color is produced.

Loss on drying ⟨731⟩—Dry it at 105° for 4 hours: it loses not more than 1.0% of its weight.

Iodate—Dissolve 1.1 g in sufficient ammonia- and carbon dioxide–free water to yield 10 mL of solution, and transfer to a color-comparison tube. Add 1 mL of starch TS and 0.25 mL of 1.0 *N* sulfuric acid, mix, and compare the color with that of a control containing, in a similar volume, 100 mg of Potassium Iodide, 1 mL of standard iodate solution [prepare by diluting 1 mL of potassium iodate solution (1 in 2500) with water to 100 mL], 1 mL of starch TS, and 0.25 mL of 1.0 *N* sulfuric acid: any color produced in the solution of the test specimen does not exceed that in the control (4 ppm).

Nitrate, nitrite, and ammonia—To a solution of 1 g in 5 mL of water contained in a test tube of about 40-mL capacity add 5 mL of 1 *N* sodium hydroxide and about 200 mg of aluminum wire. Insert a pledget of purified cotton in the upper portion of the test tube, and place a piece of moistened red litmus paper over the mouth of the tube. Heat the test tube and its contents in a steam bath for 15 minutes: no blue coloration of the paper is discernible.

Thiosulfate and barium—Dissolve 0.5 g in 10 mL of ammonia- and carbon dioxide–free water, and add 2 drops of 2 *N* sulfuric acid: no turbidity develops within 1 minute.

Heavy metals ⟨231⟩—Dissolve 2.0 g in 25 mL of water: the limit is 0.001%.

Assay—Dissolve about 500 mg of Potassium Iodide, accurately weighed, in about 10 mL of water, and add 35 mL of hydrochloric acid and 5 mL of chloroform. Titrate with 0.05 *M* potassium iodate VS until the purple color of iodine disappears from the chloroform. Add the last portions of the iodate solution dropwise, agitating vigorously and continuously. After the chloroform has been decolorized, allow the mixture to stand for 5 minutes. If the chloroform develops a purple color, titrate further with the iodate solution. Each mL of 0.05 *M* potassium iodate is equivalent to 16.60 mg of KI.

Potassium Iodide Oral Solution

» Potassium Iodide Oral Solution contains, not less than 94.0 percent and not more than 106.0 percent of the labeled amount of KI.

NOTE—If Potassium Iodide Oral Solution is not to be used within a short time, add 0.5 mg of sodium thiosulfate for each g of KI. Crystals of potassium iodide may form in Potassium Iodide Oral Solution under normal conditions of storage, especially if refrigerated.

Packaging and storage—Preserve in tight, light-resistant containers.

Identification—It responds to the tests for *Potassium* ⟨191⟩ and for *Iodide* ⟨191⟩.

Assay—Dilute an accurately measured volume of Potassium Iodide Oral Solution, quantitatively with water to obtain a solution containing about 50 mg of potassium iodide per mL. To 10.0 mL of this solution, in a 150-mL beaker, add about 40 mL of water, 25 mL of alcohol, and 1.0 mL of 1 *N* nitric acid. Titrate with 0.1 *N* silver nitrate VS, determining the end-point potentiometrically, using silver-calomel electrodes and a salt bridge containing 4 percent agar in a saturated potassium nitrate solution. Perform a blank determination, and make any necessary correction. Each mL of 0.1 *N* silver nitrate is equivalent to 16.60 mg of KI.

Potassium Iodide Tablets

» Potassium Iodide Tablets contain not less than 94.0 percent and not more than 106.0 percent of the labeled amount of KI for Tablets of 300 mg or more, and not less than 92.5 percent and not more than 107.5 percent for Tablets of less than 300 mg.

Packaging and storage—Preserve in tight containers.

Identification—A filtered solution of powdered Tablets responds to the tests for *Potassium* ⟨191⟩ and for *Iodide* ⟨191⟩.

Disintegration ⟨701⟩—

FOR ENTERIC-COATED TABLETS—Proceed as directed for *Enteric-coated Tablets:* the Tablets do not disintegrate after 1 hour of agitation in simulated gastric fluid TS, but they disintegrate within 90 minutes in simulated intestinal fluid TS.

Dissolution ⟨711⟩—

FOR UNCOATED TABLETS—

Medium: water; 900 mL.

Apparatus 2: 50 rpm.

Time: 15 minutes.

Procedure—Determine the amount of KI dissolved from ultraviolet absorbances at the wavelength of maximum absorbance at about 227 nm of filtered portions of the solutions under test, suitably diluted with *Dissolution Medium*, if necessary, in comparison with a Standard solution having a known concentration of potassium iodide in the same medium.

Tolerances—Not less than 75% (*Q*) of the labeled amount of KI is dissolved in 15 minutes.

Uniformity of dosage units ⟨905⟩: meet the requirements.

Assay—Weigh and finely powder not less than 20 Potassium Iodide Tablets. Transfer a portion of the powder, equivalent to about 1.2 g of potassium iodide, to a 250-mL volumetric flask, add 100 mL of water, shake for 20 minutes, dilute with water to volume, and mix. Filter through paper, discarding the first 20 mL of the filtrate. Transfer 100.0 mL of the filtrate, 25 mL of alcohol, and 1.0 mL of 1 *N* nitric acid to a 200-mL beaker. Titrate with 0.1 *N* silver nitrate VS, determining the end-point potentiometrically, using silver-calomel electrodes and a salt bridge containing 4 percent agar in a saturated potassium nitrate solution. Perform a blank determination, and make any necessary correction. Each mL of 0.1 *N* silver nitrate is equivalent to 16.60 mg of KI.

Potassium Metabisulfite—*see* Potassium Metabisulfite NF

Potassium Metaphosphate—*see* Potassium Metaphosphate NF

Potassium Permanganate

$KMnO_4$ 158.03

Permanganic acid ($HMnO_4$), potassium salt.

Potassium permanganate ($KMnO_4$) [*7722-64-7*].

» Potassium Permanganate contains not less than 99.0 percent and not more than 100.5 percent of $KMnO_4$, calculated on the dried basis.

Caution—Observe great care in handling Potassium Permanganate, as dangerous explosions may occur if it is brought into contact with organic or other readily oxidizable substances, either in solution or in the dry state.

Packaging and storage—Preserve in well-closed containers.

Identification—A solution of it is deep violet-red when concentrated and pink when highly diluted, and responds to the tests for *Permanganate* ⟨191⟩.

Loss on drying ⟨731⟩—Dry it over silica gel for 18 hours: it loses not more than 0.5% of its weight.

Insoluble substances—Dissolve 2.0 g in 150 mL of water that previously has been warmed to steam-bath temperature, and filter immediately through a tared, medium-porosity filtering crucible. Wash the filter with three 50-mL portions of the warm water, and dry the filtering crucible and the residue at 105° for 3 hours: not more than 4 mg of residue is obtained (0.2%).

Assay—Transfer about 1000 mg of Potassium Permanganate, accurately weighed, and for each mg of Potassium Permanganate taken, 2.13 mg of sodium oxalate, previously dried at 110° to constant weight and accurately weighed, to a 500-mL conical flask. Add 150 mL of water and 20 mL of 7 *N* sulfuric acid, heat to about 80°, and titrate the excess oxalic acid with 0.03 *N* potassium permanganate VS. Calculate the quantity, in mg, of $KMnO_4$ in the portion of Potassium Permanganate taken by the formula:

$$0.4718 W_S - 0.9482 V,$$

in which 0.4718 is the $KMnO_4$ equivalent, in mg, of each mg of sodium oxalate, W_S is the weight, in mg, of sodium oxalate taken, 0.9482 is the quantity of $KMnO_4$, in mg, in each mL of 0.03 *N* potassium permanganate, and *V* is the volume, in mL, of 0.03 *N* potassium permanganate consumed.

Dibasic Potassium Phosphate

K_2HPO_4 174.18
Phosphoric acid, dipotassium salt.
Dipotassium hydrogen phosphate [*7758-11-4*].

» Dibasic Potassium Phosphate contains not less than 98.0 percent and not more than 100.5 percent of K_2HPO_4, calculated on the dried basis.

Packaging and storage—Preserve in well-closed containers.

Identification—A solution (1 in 20) responds to the tests for *Potassium* ⟨191⟩ and for *Phosphate* ⟨191⟩.

pH ⟨791⟩: between 8.5 and 9.6, in a solution (1 in 20).

Loss on drying ⟨731⟩—Dry it at 105° to constant weight: it loses not more than 1.0% of its weight.

Insoluble substances—Dissolve 10 g in 100 mL of hot water, filter through a tared filtering crucible, wash the insoluble residue with hot water, and dry at 105° for 2 hours: the weight of the residue so obtained does not exceed 20 mg (0.2%).

Carbonate—To 1 g add 3 mL of water and 2 mL of 3 *N* hydrochloric acid: not more than a few bubbles are evolved.

Chloride ⟨221⟩—A 1.0-g portion shows no more chloride than corresponds to 0.40 mL of 0.020 *N* hydrochloric acid (0.03%).

Sulfate ⟨221⟩—A 0.20-g portion shows no more sulfate than corresponds to 0.20 mL of 0.020 *N* sulfuric acid (0.1%).

Fluoride—Proceed as directed in the test for *Fluoride* under *Dibasic Calcium Phosphate.* The limit is 0.001%.

Arsenic, *Method I* ⟨211⟩: 3 ppm.

Iron—Dissolve 0.33 g in 10 mL of water, add 6 mL of hydroxylamine hydrochloride solution (1 in 10) and 4 mL of orthophenanthroline solution prepared by dissolving 1 g of orthophenanthroline in 1000 mL of water containing 1 mL of 3 *N* hydrochloric acid, and dilute with water to 25 mL: any red color produced within 1 hour is not darker than that of a control prepared from 1 mL of *Standard Iron Solution* (see *Iron* ⟨241⟩): the limit is 0.003%.

Sodium ⟨191⟩—A solution (1 in 10) tested on a platinum wire imparts no pronounced yellow color to a nonluminous flame.

Heavy metals, *Method I* ⟨231⟩: 0.001%.

Monobasic or tribasic salt—Dissolve 3 g in 30 mL of water, cool to 20°, and add 3 drops of thymol blue TS: a blue color is produced, which is changed to yellow (with a greenish tinge) by the addition of not more than 0.4 mL of 1 *N* hydrochloric acid.

Assay—Transfer about 6.5 g of Dibasic Potassium Phosphate, accurately weighed, to a 250-mL beaker, add 50 mL of water and 50.0 mL of 1 *N* hydrochloric acid VS, stir until dissolved, and proceed as directed in the *Assay* under *Dibasic Sodium Phosphate* beginning with "Titrate the excess acid." Where *A* is equal to or less than *B*, each mL of 1 *N* hydrochloric acid is equivalent to 174.2 mg of K_2HPO_4. Where *A* is greater than *B*, each mL of the volume $2B - A$ of 1 *N* sodium hydroxide is equivalent to 174.2 mg of K_2HPO_4.

Potassium Phosphate, Monobasic—*see* Potassium Phosphate, Monobasic NF

Potassium Phosphates Injection

» Potassium Phosphates Injection is a sterile solution of Monobasic Potassium Phosphate and Dibasic Potassium Phosphate in Water for Injection. It contains not less than 95.0 percent and not more than 105.0 percent of the labeled amounts of monobasic potassium phosphate (KH_2PO_4) and dibasic potassium phosphate (K_2HPO_4). It contains no bacteriostat or other preservative.

Packaging and storage—Preserve in single-dose or in multiple-dose containers, preferably of Type I glass.

Labeling—The label states the potassium content in terms of milliequivalents in a given volume, and states also the elemental phosphorus content in terms of millimoles in a given volume. Label the Injection to indicate that it is to be diluted to appropriate strength with water or other suitable fluid prior to administration, and that once opened any unused portion is to be discarded. The label states also the total osmolar concentration in mOsmol per liter. Where the contents are less than 100 mL, or where the label states that the Injection is not for direct injection but is to be diluted before use, the label alternatively may state the total osmolar concentration in mOsmol per mL.

Identification—It responds to the tests for *Potassium* ⟨191⟩ and for *Phosphate* ⟨191⟩.

Pyrogen—After being diluted with Sodium Chloride Injection to a concentration of 4.5 mg of phosphate ion per mL, it meets the requirements of the *Pyrogen Test* ⟨151⟩, the test dose being injected over a period of about 2 minutes.

Particulate matter ⟨788⟩: meets the requirements under *Small-volume Injections.*

Other requirements—It meets the requirements under *Injections* ⟨1⟩.

Assay for monobasic potassium phosphate—Transfer an accurately measured volume of Potassium Phosphates Injection, equivalent to about 300 mg of monobasic potassium phosphate, to a 100-mL beaker, and dilute with water to about 50 mL. Place the electrodes of a suitable pH meter in the solution, and titrate with 0.1 *N* sodium hydroxide VS to the inflection point to a pH of about 9.1. Each mL of 0.1 *N* sodium hydroxide is equivalent to 13.61 mg of KH_2PO_4.

Assay for dibasic potassium phosphate—Transfer an accurately measured volume of Potassium Phosphates Injection, equivalent to about 300 mg of dibasic potassium phosphate, to a 100-mL beaker, and dilute with water to about 50 mL. Place the electrodes of a suitable pH meter in the solution, and titrate with 0.1 *N* hydrochloric acid VS to the inflection point to a pH of about 4.2. Each mL of 0.1 *N* hydrochloric acid is equivalent to 17.42 mg of K_2HPO_4.

Potassium Sodium Tartrate

COOK
H—C—OH
HO—C—H · $4H_2O$
COONa

$C_4H_4KNaO_6 \cdot 4H_2O$ 282.22
Butanedioic acid, 2,3-dihydroxy-, [*R*-(*R**,*R**)]-, monopotassium monosodium salt, tetrahydrate.
Monopotassium monosodium tartrate tetrahydrate [*6100-16-9; 6381-59-5*].
Anhydrous 210.16 [*304-59-6*].

» Potassium Sodium Tartrate contains not less than 99.0 percent and not more than 102.0 percent of $C_4H_4KNaO_6$, calculated on the anhydrous basis.

Packaging and storage—Preserve in tight containers.

Identification—

A: Ignite it: it emits the odor of burning sugar and leaves a residue that is alkaline to litmus and that effervesces with acids.

B: To 10 mL of a solution (1 in 20) add 10 mL of 6 *N* acetic acid: a white, crystalline precipitate is formed within 15 minutes.

C: A solution (1 in 10) responds to the tests for *Tartrate* ⟨191⟩.

Alkalinity—A solution of 1.0 g in 20 mL of water is alkaline to litmus, but after the addition of 0.20 mL of 0.10 *N* sulfuric acid no pink color is produced by the addition of 1 drop of phenolphthalein TS.

Water, *Method I* ⟨921⟩: between 21.0% and 27.0%.

Ammonia—Heat a 5-mL portion of a solution (1 in 10) with 5 mL of 1 *N* sodium hydroxide: the odor of ammonia is not noticeable.

Heavy metals, *Method II* ⟨231⟩: 0.001%.

Assay—Weigh accurately about 2 g of Potassium Sodium Tartrate in a tared porcelain crucible, and ignite, gently at first, until the salt is thoroughly carbonized, protecting the carbonized salt from the flame at all times. Cool the crucible, place it in a glass beaker, and break up the carbonized mass with a glass rod. Without removing the glass rod or the crucible, add 50 mL of water and 50.0 mL of 0.5 *N* sulfuric acid VS, cover the beaker, and boil the solution for 30 minutes. Filter, and wash with hot water until the last washing is neutral to litmus. Cool the combined filtrate and washings, add methyl red–methylene blue TS, and titrate the excess acid with 0.5 *N* sodium hydroxide VS. Perform a blank determination (see *Residual Titrations* under *Titrimetry* ⟨541⟩). Each mL of 0.5 *N* sulfuric acid is equivalent to 52.54 mg of $C_4H_4KNaO_6$.

Potassium Sorbate—*see* Potassium Sorbate NF

Povidone

$(C_6H_9NO)_n$

2-Pyrrolidinone, 1-ethenyl-, homopolymer.

1-Vinyl-2-pyrrolidinone polymer [*9003-39-8*].

» Povidone is a synthetic polymer consisting essentially of linear 1-vinyl-2-pyrrolidinone groups, the degree of polymerization of which results in polymers of various molecular weights. It is characterized by its viscosity in aqueous solution, relative to that of water, expressed as a K-value, ranging from 10 to 95. The K-value of Povidone having a nominal K-value of 15 or less is not less than 85.0 percent and not more than 115.0 percent of the nominal K-value, and the K-value of Povidone having a nominal K-value or nominal K-value range with an average of more than 15 is not less than 90.0 percent and not more than 108.0 percent of the nominal K-value or average of the nominal K-value range.

Packaging and storage—Preserve in tight containers.

Labeling—Label it to state, as part of the official title, the K-value or K-value range of the Povidone.

Identification—

A: To 10 mL of a solution (1 in 50) add 20 mL of 1 *N* hydrochloric acid and 5 mL of potassium dichromate TS: an orange-yellow precipitate is formed.

B: Dissolve 75 mg of cobalt nitrate and 300 mg of ammonium thiocyanate in 2 mL of water. To this solution add 5 mL of a solution of Povidone (1 in 50), and render the resulting solution acid by the addition of 3 *N* hydrochloric acid: a pale blue precipitate is formed.

C: To 5 mL of a solution (1 in 200) add a few drops of iodine TS: a deep red color is produced.

pH ⟨791⟩: between 3.0 and 7.0, in a solution (1 in 20).

Water, *Method I* ⟨921⟩: not more than 5.0%.

Residue on ignition ⟨281⟩: not more than 0.1%.

Lead ⟨251⟩—Dissolve 1.0 g in 25 mL of water: the limit is 10 ppm.

Aldehydes—Transfer 20.0 g of Povidone to a round-bottom flask containing 180 mL of 9 *N* sulfuric acid, attach a suitable water-cooled condenser, and reflux for 45 minutes. Re-assemble the apparatus, and distil until about 100 mL of distillate has been collected, receiving the distillate in a flask, placed in an ice bath, containing 20 mL of 1 *N* hydroxylamine hydrochloride previously adjusted to a pH of 3.1. Titrate the distillate solution with 0.10 *N* sodium hydroxide to a pH of 3.1. Perform a blank determination, and make any necessary correction. Each mL of 0.10 *N* sodium hydroxide is equivalent to 4.405 mg of aldehyde, calculated as acetaldehyde: not more than 0.20% is found.

Hydrazine—Transfer 2.5 g to a 50-mL centrifuge tube, add 25 mL of water, and mix to dissolve. Add 500 µL of a 1 in 20 solution of salicylaldehyde in methanol, swirl, and heat in a water bath at 60° for 15 minutes. Allow to cool, add 2.0 mL of toluene, insert a stopper in the tube, shake vigorously for 2 minutes, and centrifuge. On a suitable thin-layer chromatographic plate (see *Chromatography* ⟨621⟩), coated with a 0.25-mm layer of dimethylsilanized chromatographic silica gel mixture, apply 10 µL of the clear upper toluene layer in the centrifuge tube and 10 µL of a Standard solution of salicylaldazine in toluene containing 9.38 µg per mL. Allow the spots to dry, and develop the chromatogram in a solvent system consisting of a mixture of methanol and water (2:1) until the solvent front has moved about three-fourths of the length of the plate. Remove the plate from the developing chamber, mark the solvent front, and allow the solvent to evaporate. Locate the spots on the plate by examination under ultraviolet light at a wavelength of 365 nm: salicylaldazine appears as a fluorescent spot having an R_f value of about 0.3, and the fluorescence of any salicylaldazine spot from the test specimen is not more intense than that produced by the spot obtained from the Standard solution (1 ppm of hydrazine).

Vinylpyrrolidinone—Dissolve 10.0 g of Povidone in 80 mL of water, add 1.0 g of sodium acetate, and titrate with 0.10 *N* iodine until the color of iodine no longer fades. Add an additional 3.0 mL of 0.10 *N* iodine, allow to stand for 10 minutes, and titrate the excess iodine with 0.10 *N* sodium thiosulfate, adding 3 mL of starch TS as the end-point is approached. Perform a blank determination (see *Residual Titrations* under *Titrimetry* ⟨541⟩), using the same total volume of 0.10 *N* iodine, accurately measured, as was used for titrating the specimen: not more than 3.6 mL of 0.10 *N* iodine is consumed, corresponding to not more than 0.2% of vinylpyrrolidinone.

K-value—Weigh accurately a quantity of undried Povidone equivalent on the anhydrous basis to the amount specified in the following table:

Nominal K-value	g
≤18	5.00
>18	1.00

Dissolve it in about 50 mL of water in a 100-mL volumetric flask, dilute with water to volume, and mix. Allow to stand for 1 hour. Determine the viscosity, using a capillary-tube viscosimeter (see *Viscosity* ⟨911⟩) of this solution at 25 ± 0.2°. Calculate the K-value of Povidone by the formula:

$$[\sqrt{300c \log z + (c + 1.5c \log z)^2} + 1.5c \log z - c]/(0.15c + 0.003c^2),$$

in which c is the weight, in g, on the anhydrous basis, of the specimen tested in each 100.0 mL of solution, and z is the viscosity of the test solution relative to that of water.

Nitrogen content—Proceed as directed under *Nitrogen Determination, Method II* ⟨461⟩, using about 0.1 g of Povidone, accurately weighed. In the procedure, repeat the addition of hydrogen peroxide (usually three to six times) until a clear, light-green solution is obtained, then heat for a further 4 hours: the nitrogen content, on the anhydrous basis, is not less than 11.5% and not more than 12.8%.

Povidone-Iodine

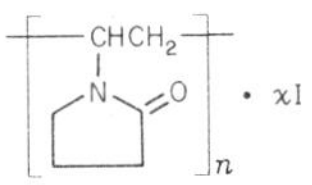

$(C_6H_9NO)_n.xI$
2-Pyrrolidinone, 1-ethenyl-, homopolymer, compd. with iodine.
1-Vinyl-2-pyrrolidinone polymer, compound with iodine *[25655-41-8]*.

» Povidone-Iodine is a complex of Iodine with Povidone. It contains not less than 9.0 percent and not more than 12.0 percent of available iodine (I), calculated on the dried basis.

Packaging and storage—Preserve in tight containers.

Identification—

A: Add 1 drop of a solution (1 in 10) to a mixture of 1 mL of starch TS and 9 mL of water: a deep blue color is produced.

B: Spread 1 mL of a solution (1 in 10) over an area of about 20 cm × 20 cm on a glass plate, and allow to air-dry at room temperature in an atmosphere of low humidity overnight: a brown, dry, non-smearing film is formed, and it dissolves readily in water.

Loss on drying ⟨731⟩—Dry 5.0 g of it at 105° until the difference between two successive weighings at 1-hour intervals is not greater than 5.0 mg: it loses not more than 8.0% of its weight.

Residue on ignition ⟨281⟩: negligible, from 2 g.

Iodide ion—

Determination of total iodine—Dissolve about 500 mg of Povidone-Iodine, accurately weighed, in 100 mL of water in a 250-mL conical flask. Add sodium bisulfite TS until the color of iodine has disappeared. Add 25.0 mL of 0.1 *N* silver nitrate VS and 10 mL of nitric acid, and mix. Titrate the excess silver nitrate with 0.1 *N* ammonium thiocyanate VS, using ferric ammonium sulfate TS as the indicator. Perform a blank determination (see *Residual Titrations* under *Titrimetry* ⟨541⟩). Each mL of 0.1 *N* silver nitrate is equivalent to 12.69 mg of I. From the percentage of total iodine, calculated on the dried basis, subtract the percentage of available iodine (see *Assay for available iodine*), to obtain the percentage of iodide ion. Not more than 6.6%, calculated on the dried basis, is found.

Heavy metals, *Method II* ⟨231⟩: 0.002%.

Nitrogen content ⟨461⟩—Not less than 9.5% and not more than 11.5% of N is found, calculated on the dried basis.

Assay for available iodine—Place about 5 g of Povidone-Iodine, accurately weighed, in a 400-mL beaker, and add 200 mL of water. Cover the beaker, and stir by mechanical means at room temperature for not more than 1 hour to dissolve as completely as possible. Titrate immediately with 0.1 *N* sodium thiosulfate VS, adding 3 mL of starch TS as the end-point is approached. Perform a blank determination, and make any necessary correction. Each mL of 0.1 *N* sodium thiosulfate is equivalent to 12.69 mg of I.

Povidone-Iodine Topical Aerosol Solution

» Povidone-Iodine Topical Aerosol Solution is a solution of Povidone-Iodine under nitrogen in a pressurized container. It contains not less than 85.0 percent and not more than 120.0 percent of the labeled amount of iodine (I).

Packaging and storage—Preserve in pressurized containers, and avoid exposure to excessive heat.

Identification—Spray Topical Aerosol Solution into a beaker or flask until about 50 mL has been collected, and allow to stand for 5 minutes to allow the entrapped propellant to escape. (Retain portions of the solution so obtained for the *pH* and *Assay* procedures.) The solution responds to the following tests.

A: Add 1 mL of a dilution containing about 0.05% of iodine to a mixture of 1 mL of starch TS and 9 mL of water: a deep blue color is produced.

B: Transfer 10 mL to a 50-mL conical flask, avoiding contact with the neck of the flask. Cover the mouth of the flask with a small disk of filter paper, and wet it with 1 drop of starch TS: no blue color appears within 60 seconds.

pH ⟨791⟩—The pH of the solution prepared for the *Identification tests* is not more than 6.0.

Other requirements—Povidone-Iodine Topical Aerosol Solution meets the requirements for *Leak Testing* and *Pressure Testing* under *Aerosols* ⟨601⟩.

Assay—Transfer an accurately measured volume of the solution of Povidone-Iodine Topical Aerosol Solution prepared for the *Identification tests*, equivalent to about 50 mg of iodine, to a 100-mL beaker, and dilute with water to a total volume of not less than 30 mL. Titrate immediately with 0.02 *N* sodium thiosulfate VS, determining the end-point potentiometrically, using a platinum-calomel electrode system. Perform a blank determination, and make any necessary correction. Each mL of 0.02 *N* sodium thiosulfate is equivalent to 2.538 mg of I.

Povidone-Iodine Ointment

» Povidone-Iodine Ointment is an emulsion, solution, or suspension of Povidone-Iodine in a suitable water-soluble ointment base. It contains not less than 85.0 percent and not more than 120.0 percent of the labeled amount of iodine (I).

Packaging and storage—Preserve in tight containers.

Identification—

A: Add 1 mL of an alcohol dilution of it containing about 0.05% of iodine to a mixture of 1 mL of starch TS and 9 mL of water: a deep blue color is produced.

B: Place 10 g in a 50-mL beaker, avoiding contact with the walls of the beaker. Cover the mouth of the beaker with a disk of filter paper, and wet it with 1 drop of starch TS: no blue color appears within 60 seconds.

Minimum fill ⟨755⟩: meets the requirements.

pH ⟨791⟩: between 1.5 and 6.5, determined in a solution (1 in 20).

Assay—Transfer an accurately weighed quantity of Povidone-Iodine Ointment, equivalent to about 50 mg of iodine, to a 100-mL beaker, add water to make a total volume of not less than 30 mL, and stir until the ointment is dissolved. Titrate immediately with 0.02 *N* sodium thiosulfate VS, determining the end-point potentiometrically, using a platinum-calomel electrode system. Perform a blank determination, and make any necessary correction. Each mL of 0.02 *N* sodium thiosulfate is equivalent to 2.538 mg of I.

Povidone-Iodine Cleansing Solution

» Povidone-Iodine Cleansing Solution is a solution of Povidone-Iodine with one or more suitable surface-active agents. It contains not less than 85.0 percent and not more than 120.0 percent of the labeled amount of iodine (I). It may contain a small amount of alcohol.

Packaging and storage—Preserve in tight containers.

Identification—

A: It responds to *Identification tests A* and *B* under *Povidone-Iodine Topical Aerosol Solution.*

B: To 2 mL of it in a glass-stoppered test tube add 1 mL of peanut oil and 4 mL of water, and shake vigorously for 10 seconds. Allow to stand for 3 minutes: a stable emulsion is formed.

pH ⟨791⟩: between 1.5 and 6.5.

Alcohol content (*if present*) ⟨611⟩: between 90.0% and 110.0% of the labeled amount of C_2H_5OH.

Assay—Transfer to a 100-mL beaker an accurately measured volume of Povidone-Iodine Cleansing Solution, equivalent to about 50 mg of iodine, and add water to make a total volume of not less than 30 mL. Titrate immediately with 0.02 *N* sodium thiosulfate VS, determining the end-point potentiometrically, using a platinum-calomel electrode system. Perform a blank determination, and make any necessary correction. Each mL of 0.02 *N* sodium thiosulfate is equivalent to 2.538 mg of I.

Povidone-Iodine Topical Solution

» Povidone-Iodine Topical Solution is a solution of Povidone-Iodine. It contains not less than 85.0 percent and not more than 120.0 percent of the labeled amount of iodine (I). It may contain a small amount of alcohol.

Packaging and storage—Preserve in tight containers.

Identification—It responds to *Identification tests A* and *B* under *Povidone-Iodine Topical Aerosol Solution.*

pH ⟨791⟩: between 1.5 and 6.5.

Alcohol content (*if present*) ⟨611⟩: between 90.0% and 110.0% of the labeled amount of C_2H_5OH.

Assay—Transfer to a 100-mL beaker an accurately measured volume of Povidone-Iodine Topical Solution, equivalent to about 50 mg of iodine, and add water to make a total volume of not less than 30 mL. Titrate immediately with 0.02 *N* sodium thiosulfate VS, determining the end-point potentiometrically, using a platinum-calomel electrode system. Perform a blank determination, and make any necessary correction. Each mL of 0.02 *N* sodium thiosulfate is equivalent to 2.538 mg of I.

Powders—*see complete list in index*

Powdered Cellulose—*see* Cellulose, Powdered NF

Powdered Digitalis—*see* Digitalis, Powdered

Powdered Ipecac—*see* Ipecac, Powdered

Powdered Opium—*see* Opium, Powdered

Powdered Rauwolfia Serpentina—*see* Rauwolfia Serpentina, Powdered

Pralidoxime Chloride

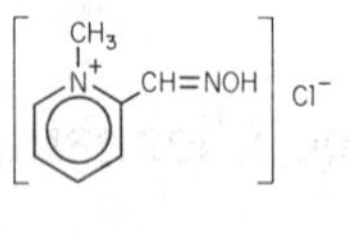

$C_7H_9ClN_2O$ 172.61

Pyridinium, 2-[(hydroxyimino)methyl]-1-methyl-, chloride.

2-Formyl-1-methylpyridinium chloride oxime [*51-15-0*].

» Pralidoxime Chloride contains not less than 97.0 percent and not more than 102.0 percent of $C_7H_9ClN_2O$, calculated on the dried basis.

Packaging and storage—Preserve in well-closed containers.

Reference standard—*USP Pralidoxime Chloride Reference Standard*—Dry at 105° for 3 hours before using.

Identification—

A: The infrared absorption spectrum of a mineral oil dispersion of it, previously dried, exhibits maxima only at the same wavelengths as that of a similar preparation of USP Pralidoxime Chloride RS.

B: A solution (1 in 10) responds to the tests for *Chloride* ⟨191⟩.

C: The retention time of the major peak in the chromatogram of the *Assay preparation* corresponds to that of the *Standard preparation*, as obtained in the *Assay*.

Melting range ⟨741⟩: between 215° and 225°, with decomposition.

Loss on drying ⟨731⟩—Dry it at 105° for 3 hours: it loses not more than 2.0% of its weight.

Residue on ignition ⟨281⟩: not more than 0.5%.

Heavy metals, *Method I* ⟨231⟩: 0.002%.

Chloride content—Dissolve about 300 mg, accurately weighed, in 150 mL of water, add 20 mL of glacial acetic acid and 10 drops of (*p-tert*-octylphenoxy)nonaethoxyethanol, and titrate with 0.1 *N* silver nitrate VS, determining the end-point potentiometrically. Perform a blank determination, and make any necessary correction. Each mL of 0.1 *N* silver nitrate is equivalent to 3.545 mg of Cl. Not less than 20.2% and not more than 20.8%, calculated on the dried basis, is found.

Assay—

Dilute phosphoric acid solution—Transfer 10 mL of phosphoric acid to a 100-mL volumetric flask containing 50 mL of water, and mix. Dilute with water to volume, and mix.

Tetraethylammonium chloride solution—Transfer about 170 mg of tetraethylammonium chloride to a 1-liter volumetric flask, add 3.4 mL of *Dilute phosphoric acid solution*, and add water to dissolve the mixture. Dilute with water to volume, and mix.

Mobile phase—Prepare a filtered and degassed mixture of acetonitrile and *Tetraethylammonium chloride solution* (52:48). Make adjustments if necessary (see *System Suitability* under *Chromatography* ⟨621⟩).

Standard preparation—Dissolve a suitable quantity of USP Pralidoxime Chloride RS, accurately weighed, in water to obtain a *Standard solution* having a known concentration of about 1.25 mg per mL. (Reserve a portion of the *Standard solution* for the *System suitability preparation.*) Pipet 2.0 mL of this solution into a 100-mL volumetric flask, dilute with *Mobile phase* to volume, and mix.

Assay preparation—Transfer about 62.5 mg of Pralidoxime Chloride, accurately weighed, to a 50-mL volumetric flask, dissolve in water, dilute with water to volume, mix, and filter. Pipet 2.0 mL of this solution into a 100-mL volumetric flask, dilute with *Mobile phase* to volume, and mix.

System suitability preparation—Prepare a solution of pyridine-2-aldoxime in water having a concentration of 0.65 mg per mL. Transfer 2.0 mL of this solution to a 100-mL volumetric flask, add 2.0 mL of the *Standard solution*, dilute with *Mobile phase* to volume, and mix.

Chromatographic system (see *Chromatography* ⟨621⟩)—The liquid chromatograph is equipped with a 270-nm detector and a 3- to 5-mm × 25-cm column containing 5-μm packing L1. The flow rate is about 1.2 mL per minute. Chromatograph the *Standard preparation* and the *System suitability preparation* by injecting about 15 μL of these preparations, and record the peak responses as directed under *Procedure:* the column efficiency determined from the analyte peak is not less than 4000 theoretical plates, the tailing factor for the analyte peak is not more than 2.5, the resolution, *R*, between the pyridine-2-aldoxime and pralidoxime chloride peaks is not less than 4.0, and the relative standard deviation for replicate injections is not more than 2.0%.

Procedure—Separately inject equal volumes (about 15 μL) of the *Standard preparation* and the *Assay preparation* into the chromatograph, record the chromatograms, and measure the responses for the major peaks. The relative retention times are about 0.6 for pyridine-2-aldoxime and 1.0 for pralidoxime chloride. Calculate the quantity, in mg, of $C_7H_9ClN_2O$ in the portion of Pralidoxime Chloride taken by the formula:

$$2.5C(r_U/r_S),$$

in which C is the concentration, in µg per mL, of USP Pralidoxime Chloride RS in the *Standard preparation*, and r_U and r_S are the peak responses obtained from the *Assay preparation* and the *Standard preparation*, respectively.

Sterile Pralidoxime Chloride

» Sterile Pralidoxime Chloride is Pralidoxime Chloride suitable for parenteral use. It contains not less than 90.0 percent and not more than 110.0 percent of the labeled amount of $C_7H_9ClN_2O$. It may contain a small amount of sodium hydroxide added for adjustment of the pH.

Packaging and storage—Preserve in *Containers for Sterile Solids* as described under *Injections* ⟨1⟩.

Reference standard—*USP Pralidoxime Chloride Reference Standard*—Dry at 105° for 3 hours before using.

Completeness of solution ⟨641⟩—The contents of 1 container dissolve in 10 mL of water to yield a clear solution.

Constituted solution—At the time of use, the constituted solution prepared from Sterile Pralidoxime Chloride meets the requirements for *Constituted Solutions* under *Injections* ⟨1⟩.

Pyrogen—It meets the requirements of the *Pyrogen Test* ⟨151⟩, the test dose being 25 mg per kg, of a solution containing 25 mg per mL.

pH ⟨791⟩: between 3.5 and 4.5, in a solution (1 in 20).

Other requirements—It conforms to the definition, responds to the *Identification tests*, and meets the requirements for *Loss on drying* and *Heavy metals* under *Pralidoxime Chloride*. It meets also the requirements for *Sterility Tests* ⟨71⟩, *Uniformity of Dosage Units* ⟨905⟩, and *Labeling* under *Injections* ⟨1⟩.

Assay—

Dilute phosphoric acid solution, Tetraethylammonium chloride solution, Mobile phase, Standard preparation, System suitability preparation, and *Chromatographic system*—Proceed as directed in the *Assay* under *Pralidoxime Chloride*.

Assay preparation—Select an accurately counted number of containers of Sterile Pralidoxime Chloride, the combined contents of which are equivalent to about 10 g of Pralidoxime Chloride. Dissolve the contents of each container in water and combine all of the solutions in a 1000-mL volumetric flask. Rinse each container with water, and add the rinsings to the volumetric flask. Dilute with water to volume, and mix. Transfer 25.0 mL of the resulting solution to a 200-mL volumetric flask, dilute with water to volume, and mix. Transfer 2.0 mL of this solution to a 100-mL volumetric flask, dilute with *Mobile phase* to volume, and mix.

Procedure—Proceed as directed for *Procedure* in the *Assay* under *Pralidoxime Chloride*. Calculate the quantity, in mg, of $C_7H_9ClN_2O$ in each container of Sterile Pralidoxime Chloride taken by the formula:

$$400(C/N)(r_U/r_S),$$

in which N is the number of containers selected for the *Assay preparation*, and the other terms are as defined therein.

Pralidoxime Chloride Tablets

» Pralidoxime Chloride Tablets contain not less than 95.0 percent and not more than 105.0 percent of the labeled amount of $C_7H_9ClN_2O$.

Packaging and storage—Preserve in well-closed containers.

Reference standard—*USP Pralidoxime Chloride Reference Standard*—Dry at 105° for 3 hours before using.

Identification—The retention time of the major peak in the chromatogram of the *Assay preparation* corresponds to that of the *Standard preparation*, as obtained in the *Assay*.

Dissolution ⟨711⟩—

Medium: water; 900 mL.

Apparatus 1: 100 rpm.

Time: 60 minutes.

Procedure—Determine the amount of $C_7H_9ClN_2O$ dissolved from ultraviolet absorbances at the wavelength of maximum absorbance at about 293 nm of filtered portions of the solution under test, suitably diluted with *Dissolution Medium*, if necessary, in comparison with a Standard solution having a known concentration of USP Pralidoxime Chloride RS in the same medium.

Tolerances—Not less than 55% (Q) of the labeled amount of $C_7H_9ClN_2O$ is dissolved in 60 minutes.

Uniformity of dosage units ⟨905⟩: meet the requirements.

Assay—

Dilute phosphoric acid solution, Tetraethylammonium chloride solution, Mobile phase, Standard preparation, System suitability preparation, and *Chromatographic system*—Proceed as directed in the *Assay* under *Pralidoxime Chloride*.

Assay preparation—Weigh and finely powder not less than 20 Pralidoxime Chloride Tablets. Transfer an accurately weighed portion of the powder, equivalent to about 250 mg of pralidoxime chloride, to a 200-mL volumetric flask. Add 150 mL of water and mechanically swirl for 30 minutes. Dilute with water to volume, and mix. Centrifuge a portion of this extract and pipet 2.0 mL of the clear supernatant liquid into a 100-mL volumetric flask. Dilute with *Mobile phase* to volume, and mix.

Procedure—Proceed as directed for *Procedure* in the *Assay* under *Pralidoxime Chloride*. Calculate the quantity, in mg, of $C_7H_9ClN_2O$ in the portion of Tablets taken by the formula:

$$10C(r_U/r_S),$$

in which the terms are as defined therein.

Pramoxine Hydrochloride

$CH_3CH_2CH_2CH_2O$–(C₆H₄)–$OCH_2CH_2CH_2$–N(morpholine)O · HCl

$C_{17}H_{27}NO_3 . HCl$ 329.87

Morpholine, 4-[3-(4-butoxyphenoxy)propyl]-, hydrochloride.

4-[3-(*p*-Butoxyphenoxy)propyl]morpholine hydrochloride [*637-58-1*].

» Pramoxine Hydrochloride, contains not less than 98.0 percent and not more than 100.5 percent of $C_{17}H_{27}NO_3 . HCl$, calculated on the dried basis.

Packaging and storage—Preserve in tight containers.

Reference standard—*USP Pramoxine Hydrochloride Reference Standard*—Dry at 105° for 1 hour before using.

Identification—

A: The infrared absorption spectrum of a potassium bromide dispersion of it, previously dried, exhibits maxima only at the same wavelengths as that of a similar preparation of USP Pramoxine Hydrochloride RS.

B: The ultraviolet absorption spectrum of a 1 in 100,000 solution in 0.1 *N* hydrochloric acid exhibits maxima and minima at the same wavelengths as that of a similar solution of USP Pramoxine Hydrochloride RS, concomitantly measured, and the respective absorptivities, calculated on the dried basis, at the wavelength of maximum absorbance at about 224 nm do not differ by more than 3.0%.

C: It responds to the tests for *Chloride* ⟨191⟩.

Melting range ⟨741⟩: between 170° and 174°.

Loss on drying ⟨731⟩—Dry it at 105° for 1 hour: it loses not more than 1.0% of its weight.

Residue on ignition ⟨281⟩: not more than 0.1%.

Assay—

pH 7.5 phosphate buffer—Dissolve 8.71 g of dibasic potassium phosphate in about 800 mL of water, adjust with dilute phosphoric acid (1 in 10) to a pH of 7.5 ± 0.1, add water to make 1000 mL, and mix.

Mobile phase—Prepare a filtered and degassed mixture of acetonitrile and *pH 7.5 phosphate buffer* (55:45). Make adjustments if necessary (see *System Suitability* under *Chromatography* ⟨621⟩).

Standard preparation—Dissolve an accurately weighed quantity of USP Pramoxine Hydrochloride RS in *Mobile phase* to obtain a solution having a known concentration of about 0.5 mg per mL.

Assay preparation—Transfer about 50 mg of Pramoxine Hydrochloride, accurately weighed, to a 100-mL volumetric flask, dissolve in *Mobile phase*, dilute with *Mobile phase* to volume, and mix.

Chromatographic system (see *Chromatography* ⟨621⟩)—The liquid chromatograph is equipped with a 224-nm detector and a 4.6-mm × 30-cm column that contains packing L1. The flow rate is about 2 mL per minute. Chromatograph the *Standard preparation*, and record the peak responses as directed under *Procedure:* the column efficiency determined from the analyte peak is not less than 3000 theoretical plates, the tailing factor for the analyte peak is not more than 1.5, and the relative standard deviation for replicate injections is not more than 1.5%.

Procedure—Separately inject equal volumes (about 20 µL) of the *Standard preparation* and the *Assay preparation* into the chromatograph, record the chromatograms, and measure the responses for the major peaks. Calculate the quantity, in mg, of $C_{17}H_{27}NO_3.HCl$ in the portion of Pramoxine Hydrochloride taken by the formula:

$$100C(r_U/r_S),$$

in which C is the concentration, in mg per mL, of USP Pramoxine Hydrochloride RS in the *Standard preparation*, and r_U and r_S are the peak responses obtained from the *Assay preparation* and the *Standard preparation*, respectively.

Pramoxine Hydrochloride Cream

» Pramoxine Hydrochloride Cream contains not less than 90.0 percent and not more than 110.0 percent of the labeled amount of $C_{17}H_{27}NO_3.HCl$ in a suitable water-miscible base.

Packaging and storage—Preserve in tight containers.

Reference standard—*USP Pramoxine Hydrochloride Reference Standard*—Dry at 105° for 1 hour before using.

Identification—

A: Dissolve a quantity of Cream, equivalent to about 50 mg of pramoxine hydrochloride, in a mixture of 25 mL of methanol and 75 mL of ether, and extract with three 25-mL portions of a mixture of equal volumes of 3 *N* hydrochloric acid and water. Discard the methanol-ether solution, render the combined extracts alkaline with 25 mL of 5 *N* sodium hydroxide, and extract the pramoxine with 50 mL of chloroform. Evaporate the clear chloroform extract with the aid of a current of air to dryness: the ultraviolet absorption spectrum of a 1 in 100,000 solution of the residue so obtained, in 0.1 *N* hydrochloric acid, exhibits maxima and minima at the same wavelengths as that of a similar solution of the residue similarly obtained from USP Pramoxine Hydrochloride RS, concomitantly measured.

B: To a 5-mg portion of the pramoxine obtained in *Identification test A* add 1 drop of nitric acid. To the yellow solution cautiously add 5 drops of ammonium hydroxide: a red-brown precipitate is formed.

Microbial limits—It meets the requirements of the tests for absence of *Staphylococcus aureus* and *Pseudomonas aeruginosa* under *Microbial Limit Tests* ⟨61⟩.

Minimum fill ⟨755⟩: meets the requirements.

Assay—

pH 7.5 phosphate buffer—Dissolve 3.5 g of dibasic potassium phosphate in 100 mL of water, and adjust the solution by the addition of phosphoric acid solution (1:1) to a pH of 7.5 ± 0.1.

Mobile phase—Prepare a suitable degassed and filtered mixture of acetonitrile, water, and *pH 7.5 phosphate buffer* (22:17:1).

Internal standard solution—Prepare a solution of dibutyl phthalate in methanol having a final concentration of about 4 µL per mL.

Standard preparation—Prepare a solution of USP Pramoxine Hydrochloride RS in methanol having a known concentration of about 2 mg per mL. Pipet 10 mL of this solution and 5 mL of *Internal standard solution* into a 100-mL volumetric flask, dilute with methanol to volume, mix, and filter.

Assay preparation—Transfer an accurately weighed portion of Pramoxine Hydrochloride Cream, equivalent to about 18 mg of pramoxine hydrochloride, to a glass-stoppered, 250-mL conical flask. Add 15.0 mL of isopropyl alcohol and 40.0 mL of methanol, heat on a steam bath, with swirling, to dissolve the Cream, add 40.0 mL of methanol and 5.0 mL of *Internal standard solution*, and mix. Cool the flask to a temperature of 10° or less to precipitate the waxes, and filter the solution.

Chromatographic system (see *Chromatography* ⟨621⟩)—The liquid chromatograph is equipped with a 224-nm detector, a 4.6-mm × 3-cm guard column that contains packing L1, and a 4-mm × 30-cm analytical column that contains packing L1. The flow rate is about 2 mL per minute. Chromatograph three replicate injections of the *Standard preparation*, and record the peak responses as directed under *Procedure:* the relative standard deviation is not more than 2.0%, and the resolution factor between pramoxine hydrochloride and dibutyl phthalate is not less than 2.4.

Procedure—Separately inject equal volumes (about 20 µL) of the *Standard preparation* and the *Assay preparation* into the chromatograph, record the chromatograms, and measure the responses for the major peaks. The relative retention times are about 0.8 for pramoxine hydrochloride and 1.0 for dibutyl phthalate. Calculate the quantity, in mg, of pramoxine hydrochloride in the portion of Cream taken by the formula:

$$100C(R_U/R_S),$$

in which C is the concentration, in mg per mL, of USP Pramoxine Hydrochloride RS in the *Standard preparation*, and R_U and R_S are the peak response ratios of pramoxine hydrochloride and internal standard obtained from the *Assay preparation* and the *Standard preparation*, respectively.

Pramoxine Hydrochloride Jelly

» Pramoxine Hydrochloride Jelly contains not less than 94.0 percent and not more than 106.0 percent of the labeled amount of $C_{17}H_{27}NO_3.HCl$.

Packaging and storage—Preserve in tight containers, preferably in collapsible tubes.

Reference standard—*USP Pramoxine Hydrochloride Reference Standard*—Dry at 105° for 1 hour before using.

Identification—Place a quantity of Jelly, equivalent to about 5 mg of pramoxine hydrochloride, in a glass-stoppered conical flask, add 25 mL of chloroform, and shake for 15 minutes. Filter into a small porcelain evaporating dish, and evaporate in a current of air on a steam bath. Add 1 drop of nitric acid to the residue, and to the resulting yellow solution cautiously add 5 drops of ammonium hydroxide: a red-brown precipitate is formed.

Microbial limits—It meets the requirements of the tests for absence of *Staphylococcus aureus* and *Pseudomonas aeruginosa* under *Microbial Limit Tests* ⟨61⟩.

Assay—

Standard preparation—Dissolve a suitable quantity of USP Pramoxine Hydrochloride RS, accurately weighed, in 0.5 *N* sulfuric acid, and dilute quantitatively with the same solvent to obtain a solution having a known concentration of about 150 µg per mL.

Assay preparation—[NOTE—If emulsions form, 2 to 5 mL of alcohol may be added to separate the phases.] Transfer an accurately weighed quantity of Pramoxine Hydrochloride Jelly, equivalent to about 15 mg of pramoxine hydrochloride, to a small beaker, and dissolve the Jelly in 0.1 *N* sulfuric acid, using four 5-mL portions, warming each portion on a steam bath, and transferring to a 125-mL separator. Shake the separator vigorously after each transfer to complete dissolution of the Jelly. To the cooled solution in the separator add 20 mL of ether, shake carefully, and proceed as directed for *Assay Preparation* under *Salts of Organic Nitrogenous Bases* ⟨501⟩, beginning with "filter the acid phase into a second 125-mL separator," except to combine the final 0.5 *N* sulfuric acid extracts in a 100-mL volumetric flask, dilute with the acid to volume, and mix.

Procedure—Proceed as directed under *Salts of Organic Nitrogenous Bases* ⟨501⟩, diluting 20.0 mL each of the *Standard preparation* and the *Assay preparation* with 0.5 *N* sulfuric acid to 50.0 mL, and determining the absorbances at the wavelength of maximum absorbance at about 286 nm. Calculate the quantity, in mg, of $C_{17}H_{27}NO_3 . HCl$ in the portion of Jelly taken by the formula:

$$0.1C(A_U / A_S),$$

in which *C* is the concentration, in µg per mL, of USP Pramoxine Hydrochloride RS in the *Standard preparation*.

Prazepam

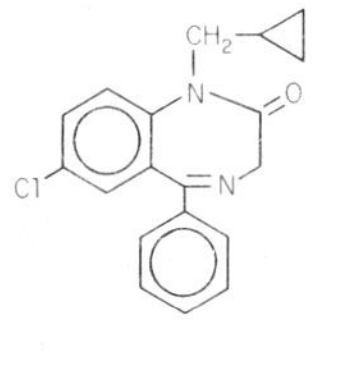

$C_{19}H_{17}ClN_2O$ 324.81

2*H*-1,4-Benzodiazepin-2-one, 7-chloro-1-(cyclopropylmethyl)-1,3-dihydro-5-phenyl-.

7-Chloro-1-(cyclopropylmethyl)-1,3-dihydro-5-phenyl-2*H*-1,4-benzodiazepin-2-one [*2955-38-6*].

» Prazepam contains not less than 98.5 percent and not more than 101.0 percent of $C_{19}H_{17}ClN_2O$, calculated on the dried basis.

Packaging and storage—Preserve in tight, light-resistant containers.

Reference standards—*USP Prazepam Reference Standard*— Dry at 105° for 4 hours before using. *USP 7-Chloro-1,3-dihydro-5-phenyl-2H-1,4-benzodiazepin-2-one Reference Standard*—Do not dry before using. *USP 2-Cyclopropylmethylamino-5-chlorobenzophenone Reference Standard*—Do not dry before using.

Identification—

A: The infrared absorption spectrum of a potassium bromide dispersion of it, previously dried at 105° for 4 hours, exhibits maxima only at the same wavelengths as that of a similar preparation of USP Prazepam RS.

B: A 1 in 165,000 solution in alcohol exhibits maxima and minima at the same wavelengths as that of a similar solution of USP Prazepam RS, concomitantly measured, and the respective absorptivities, calculated on the dried basis, at the wavelength of maximum absorbance at about 230 nm do not differ by more than 3.0%.

Melting range ⟨741⟩: between 143° and 148°, but the range between beginning and end of melting does not exceed 3°.

Loss on drying ⟨731⟩—Dry it at 105° to constant weight: it loses not more than 0.5% of its weight.

Residue on ignition ⟨281⟩: not more than 0.1%.

Heavy metals, *Method II* ⟨231⟩: 0.002%.

Related compounds—

Standard preparations—Dissolve USP Prazepam RS in acetone to obtain a solution having a concentration of 10 mg per mL (*Solution A*). Pipet 1 mL of *Solution A* into a 100-mL volumetric flask, dilute with acetone to volume, and mix (*Solution B*). Dissolve USP 7-Chloro-1,3-dihydro-5-phenyl-2*H*-1,4-benzodiazepin-2-one RS in acetone to obtain a solution having a concentration of 0.30 mg per mL (*Solution C*). Dissolve USP 2-Cyclopropylmethylamino-5-chlorobenzophenone RS in acetone to obtain a solution having a concentration of 0.10 mg per mL (*Solution D*).

Test preparation—Dissolve Prazepam in acetone to obtain a solution having a concentration of 10 mg per mL.

Procedure—On a suitable thin-layer chromatographic plate (see *Chromatography* ⟨621⟩), coated with a 0.25-mm layer of chromatographic silica gel mixture and previously washed with methanol, apply 50-µL portions of *Solution A* and the *Test preparation*, and similarly apply 5-µL portions of *Solutions B*, *C*, and *D*. Allow the spots to dry, and develop the chromatogram in a solvent system consisting of a mixture of ethyl acetate and *n*-heptane (1:1) until the solvent front has moved about three-fourths of the length of the plate. Remove the plate from the developing chamber, mark the solvent front, and dry the plate. Examine the plate under short-wavelength ultraviolet light: the chromatograms from the *Test preparation* and *Solution A* show principal spots at the same R_f value. Estimate the levels of any additional spots observed in the chromatogram of the *Test preparation* by comparison with the spots in the chromatograms of *Solutions B*, *C*, and *D*: any spots at corresponding R_f values produced by the *Test preparation* are not greater in size or intensity than the principal spots produced by *Solutions C* and *D*, corresponding to not more than 0.3% of 7-chloro-1,3-dihydro-5-phenyl-2*H*-1,4-benzodiazepin-2-one and not more than 0.1% of 2-cyclopropylmethylamino-5-chlorobenzophenone. Not more than 3 additional prazepam-related spots are produced by the *Test preparation*, of which no individual spot exhibits an intensity greater than that of the principal spot produced by *Solution B*, corresponding to not more than 0.1%, and the total of any observed spots does not exceed 0.2%.

Assay—Dissolve about 800 mg of Prazepam, accurately weighed, in 200 mL of glacial acetic acid in a 400-mL beaker. Titrate with 0.1 *N* perchloric acid VS, using crystal-violet TS as the indicator. Perform a blank determination, and make any necessary correction. Each mL of 0.1 *N* perchloric acid is equivalent to 32.48 mg of $C_{19}H_{17}ClN_2O$.

Prazepam Capsules

» Prazepam Capsules contain not less than 90.0 percent and not more than 110.0 percent of the labeled amount of $C_{19}H_{17}ClN_2O$.

Packaging and storage—Preserve in tight, light-resistant containers.

Reference standard—*USP Prazepam Reference Standard*—Dry at 105° for 4 hours before using.

Identification—

A: Shake a portion of the Capsule contents, equivalent to about 20 mg of prazepam, with 10 mL of acetone, and filter. Apply 5 µL of this solution and 5 µL of a Standard solution of USP Prazepam RS in acetone, containing 2 mg per mL, to a thin-layer chromatographic plate (see *Chromatography* ⟨621⟩) coated with a 0.25-mm layer of chromatographic silica gel mixture. Develop the chromatogram in a chamber equilibrated with a mixture of ethyl acetate and *n*-heptane (1:1) until the solvent front has moved about three-fourths of the length of the plate. Remove the plate from the developing chamber, mark the solvent front, air-dry completely, and view under short-wavelength ultraviolet light: the R_f value of the principal spot obtained from the test solution corresponds to that obtained from the Standard solution.

B: The retention time of the major peak in the chromatogram of the *Assay preparation* corresponds to that of the *Standard preparation*, as obtained in the *Assay*.

Dissolution ⟨711⟩—

Medium: 0.1 *N* hydrochloric acid; 900 mL.

Apparatus 1: 50 rpm.

Time: 60 minutes.

Procedure—Determine the amount of $C_{19}H_{17}ClN_2O$ dissolved from ultraviolet absorbances at the wavelength of maximum absorbance at about 240 nm of filtered portions of the solution under test, suitably diluted with *Dissolution Medium*, if necessary, in comparison with a Standard solution having a known concentration of USP Prazepam RS in the same medium and freshly prepared for each test.

Tolerances—Not less than 80% (*Q*) of the labeled amount of $C_{19}H_{17}ClN_2O$ is dissolved in 60 minutes.

Uniformity of dosage units ⟨905⟩: meet the requirements.

Procedure for content uniformity—Transfer the contents of 1 Capsule, as completely as possible, to a 200-mL volumetric flask. Add about 150 mL of a solution of alcohol and water (6:4), insert the stopper, and shake for 5 minutes. Dilute with the alcohol and water solution to volume, mix, and filter. Dilute the filtrate quantitatively and stepwise, if necessary, with the same alcohol and water solution to obtain a solution containing about 5 µg of prazepam per mL. Prepare a solution of USP Prazepam RS in the same medium to obtain a Standard solution having a known concentration of about 5 µg per mL. Concomitantly determine the absorbances of the solutions in 1-cm cells at the wavelength of maximum absorbance at about 228 nm, with a suitable spectrophotometer, using the alcohol and water solution as the blank. Calculate the quantity, in mg, of $C_{19}H_{17}ClN_2O$ in the Capsule by the formula:

$$(T/D)C(A_U/A_S),$$

in which T is the labeled quantity, in mg, of prazepam in the Capsule, D is the concentration, in µg per mL, of prazepam in the Test solution of the basis of the labeled quantity per Capsule, C is the concentration, in µg per mL, of USP Prazepam RS in the Standard solution, and A_U and A_S are the absorbances of the test solution and the Standard solution, respectively.

Assay—

Mobile phase—Prepare a suitable degassed solution of methanol, water, and glacial acetic acid (85:15:0.5).

Standard preparation—Transfer 25 mg of USP Prazepam RS, accurately weighed, to a 100-mL volumetric flask, dissolve in methanol, dilute with methanol to volume, and mix. Pipet 20 mL of this solution into a 100-mL volumetric flask, add 15 mL of water and 0.5 mL of glacial acetic acid, dilute with methanol to volume, and mix.

Assay preparation—Remove as completely as possible the contents of 20 Capsules, mix the combined contents, and transfer an accurately weighed portion of the powder, equivalent to about 5 mg of prazepam, to a 100-mL volumetric flask. Add 15 mL of water, and mix. Add 0.5 mL of glacial acetic acid, mix, add 60 mL of methanol, mix, and place in a sonic bath for 10 minutes. Dilute with methanol to volume, mix, and filter.

Chromatographic system (see *Chromatography* ⟨621⟩)—The liquid chromatograph is equipped with a 254-nm detector and a 4.6-mm × 30-cm column that contains packing L1. The flow rate is about 1 mL per minute. Chromatograph five replicate injections of the *Standard preparation*, and record the peak responses as directed under *Procedure:* the relative standard deviation is not more than 2%.

Procedure—Separately inject equal volumes (about 10 µL) of the *Standard preparation* and the *Assay preparation* into the chromatograph, record the chromatograms, and measure the responses for the major peak, with a retention time of about 5 minutes. Calculate the quantity, in mg, of $C_{19}H_{17}ClN_2O$ in the portion of Capsules taken by the formula:

$$0.1C(r_U/r_S),$$

in which C is the concentration, in µg per mL, of USP Prazepam RS in the *Standard preparation*, and r_U and r_S are the ratios of the peak responses of prazepam obtained from the *Assay preparation* and the *Standard preparation*, respectively.

Prazepam Tablets

» Prazepam Tablets contain not less than 90.0 percent and not more than 110.0 percent of the labeled amount of $C_{19}H_{17}ClN_2O$.

Packaging and storage—Preserve in tight, light-resistant containers.

Reference standard—*USP Prazepam Reference Standard*—Dry at 105° for 4 hours before using.

Identification—

A: Shake a portion of finely powdered Tablets, equivalent to about 20 mg of prazepam, with 10 mL of chloroform, and filter. Apply 5 µL each of this solution and of a Standard solution of USP Prazepam RS in chloroform, containing 2 mg per mL, on a thin-layer chromatographic plate coated with a 0.25-mm layer of chromatographic silica gel mixture. Develop the chromatogram in a chamber equilibrated with a mixture of ethyl acetate and *n*-heptane (1:1) until the solvent front has moved about three-fourths of the length of the plate. Remove the plate from the developing chamber, mark the solvent front, air-dry completely, and view under short-wavelength ultraviolet light: the R_f value of the principal spot obtained from the test solution corresponds to that obtained from the Standard solution.

B: The retention time of the major peak in the chromatogram of the *Assay preparation* corresponds to that of the *Standard preparation*, as obtained in the *Assay*.

Dissolution ⟨711⟩—

Medium: 0.1 *N* hydrochloric acid; 900 mL.

Apparatus 1: 50 rpm.

Time: 60 minutes.

Procedure—Determine the amount of $C_{19}H_{17}ClN_2O$ dissolved from ultraviolet absorbances at the wavelength of maximum absorbance at about 240 nm of filtered portions of the solution under test, suitably diluted with *Dissolution Medium*, if necessary, in comparison with a known concentration of USP Prazepam RS in the same medium, freshly prepared for each test.

Tolerances—Not less than 80% (*Q*) of the labeled amount of $C_{19}H_{17}ClN_2O$ is dissolved in 60 minutes.

Uniformity of dosage units ⟨905⟩: meet the requirements.

Procedure for content uniformity—Place 1 Tablet in a 200-mL volumetric flask, add 2 mL of water, mix to wet, and allow to stand with intermittent swirling until the tablet disintegrates. Add about 150 mL of a solution of alcohol and water (6:4), insert the stopper, and shake for 5 minutes. Dilute with the alcohol and water solution to volume, mix, and filter. Dilute the filtrate quantitatively and stepwise, if necessary, with the same alcohol and water solution to obtain a solution containing about 5 µg of prazepam per mL. Prepare a solution of USP Prazepam RS in the same medium to obtain a Standard solution having a known concentration of about 5 µg per mL. Concomitantly determine the absorbances of both solutions in 1-cm cells at the wavelength of maximum absorbance at about 228 nm, with a suitable spectrophotometer, using the alcohol and water solution as the blank. Calculate the quantity, in mg, of $C_{19}H_{17}ClN_2O$ in the Tablet by the formula:

$$(T/D)C(A_U/A_S),$$

in which T is the labeled quantity, in mg, of prazepam in the Tablet, D is the concentration, in µg per mL, of prazepam in the Test solution on the basis of the labeled quantity per Tablet, C is the concentration, in µg per mL, of USP Prazepam RS in the Standard solution, and A_U and A_S are the absorbances of the test solution and the Standard solution, respectively.

Assay—

Mobile phase and *Standard preparation*—Prepare as directed in the *Assay* under *Prazepam Capsules*.

Assay preparation—Weigh and finely powder not less than 20 Tablets. Transfer an accurately weighed portion of the powder, equivalent to about 5 mg of prazepam, to a 100-mL volumetric flask, add 15 mL of water, and mix. Add 0.5 mL of glacial acetic acid, mix, add 60 mL of methanol, mix, and place in a sonic bath for 10 minutes. Cool, dilute with methanol to volume, mix, and filter.

Chromatographic system and *Procedure*—Proceed as directed in the *Assay* under *Prazepam Capsules*. Calculate the quantity, in mg, of $C_{19}H_{17}ClN_2O$ in the portion of Tablets taken as directed therein.

Praziquantel

$C_{19}H_{24}N_2O_2$ 312.41

4*H*-Pyrazino[2,1-*a*]isoquinolin-4-one, 2-(cyclohexylcarbonyl)-1,2,3,6,7,11b-hexahydro-.

2-(Cyclohexylcarbonyl)-1,2,3,6,7,11b-hexahydro-4*H*-pyrazino-[2,1-*a*]isoquinolin-4-one [*55268-74-1*].

» Praziquantel contains not less than 98.5 percent and not more than 101.0 percent of $C_{19}H_{24}N_2O_2$, calculated on the dried basis.

Packaging and storage—Preserve in well-closed, light-resistant containers.

Reference standards—*USP Praziquantel Reference Standard*—Dry in vacuum at a pressure not exceeding 5 mm of mercury at 50° over phosphorus pentoxide for 2 hours before using. *USP Praziquantel Related Compound A Reference Standard*—Dry in vacuum at a pressure not exceeding 5 mm of mercury at 50° over phosphorus pentoxide for 2 hours before using. *USP Praziquantel Related Compound B Reference Standard*—Dry in vacuum at a pressure not exceeding 5 mm of mercury at 50° over phosphorus pentoxide for 2 hours before using. *USP Praziquantel Related Compound C Reference Standard*—Dry in vacuum at a pressure not exceeding 5 mm of mercury at 50° over phosphorus pentoxide for 2 hours before using.

Identification—The infrared absorption spectrum of a potassium bromide dispersion of it exhibits maxima only at the same wavelengths as that of a similar preparation of USP Praziquantel RS.

Melting range ⟨741⟩: between 136° and 140°.

Loss on drying ⟨731⟩—Dry it in vacuum at a pressure not exceeding 5 mm of mercury at 50° over phosphorus pentoxide for 2 hours: it loses not more than 0.5% of its weight.

Residue on ignition ⟨281⟩: not more than 0.1%.

Heavy metals, *Method II* ⟨231⟩: 0.002%.

Related compounds—

Mobile phase and *Chromatographic system*—Prepare as directed in the *Assay*.

Standard preparation—Dissolve accurately weighed quantities of USP Praziquantel Related Compound A RS, USP Praziquantel Related Compound B RS, and USP Praziquantel Related Compound C RS in *Mobile phase* to obtain a single solution having known concentrations of about 0.04 mg of *each* per mL.

Test preparation—Transfer about 200 mg of Praziquantel, accurately weighed, to a 10-mL volumetric flask, dissolve in *Mobile phase*, dilute with *Mobile phase* to volume, and mix.

Procedure—Separately inject equal volumes (about 10 µL) of the *Standard preparation* and the *Test preparation* into the chromatograph, record the chromatograms, and measure the responses for the peaks. The relative retention times are about 0.8 for praziquantel related compound A, 1.0 for praziquantel, 1.8 for praziquantel related compound B, and 2.1 for praziquantel related compound C. Calculate, in turn, the percentages of 2-benzoyl-1,2,3,6,7,11b-hexahydro-4*H*-pyrazino[2,1-*a*]isoquinolin-4-one (praziquantel related compound A), 2-(cyclohexylcarbonyl)-2,3,6,7-tetrahydro-4*H*-pyrazino[2,1-*a*]isoquinolin-4-one (praziquantel related compound B), and 2-(*N*-formylhexahydrohippuroyl)-1,2,3,4-tetrahydroisoquinolin-1-one (praziquantel related compound C) in the portion of Praziquantel taken by the formula:

$$1000C/W(r_U/r_S),$$

in which C is the concentration, in mg per mL, of the respective USP Reference Standard taken to prepare the *Standard preparation*, W is the weight, in mg, of Praziquantel taken to prepare the *Test preparation*, and r_U and r_S are the peak responses at corresponding retention times, obtained from the *Test preparation* and the *Standard preparation*, respectively. Not more than 0.2% of each is found.

Assay—

Mobile phase—Prepare a suitable degassed mixture of acetonitrile and water (60:40). Make adjustments if necessary (see *System Suitability* under *Chromatography* ⟨621⟩).

Standard preparation—Dissolve an accurately weighed quantity of USP Praziquantel RS in *Mobile phase*, and dilute quantitatively, and stepwise if necessary, with *Mobile phase* to obtain a solution having a known concentration of about 0.4 mg per mL.

Assay preparation—Transfer about 40 mg of Praziquantel, accurately weighed, to a 20-mL volumetric flask, dissolve in *Mobile phase*, dilute with *Mobile phase* to volume, and mix. Transfer 10.0 mL of this solution to a 50-mL volumetric flask, dilute with *Mobile phase* to volume, and mix.

Chromatographic system (see *Chromatography* ⟨621⟩)—The liquid chromatograph is equipped with a 210-nm detector and a 4-mm × 25-cm column that contains 10-µm packing L1. The flow rate is about 1.5 mL per minute. Chromatograph the *Standard preparation*, and record the peak responses as directed under *Procedure:* the tailing factor for the praziquantel peak is not more than 1.5, and the relative standard deviation for replicate injections is not more than 1.0%.

Procedure—Separately inject equal volumes (about 10 µL) of the *Standard preparation* and the *Assay preparation* into the chromatograph, record the chromatograms, and measure the responses for the major peaks. Calculate the quantity, in mg, of $C_{19}H_{24}N_2O_2$ in the portion of Praziquantel taken by the formula:

$$100C(r_U/r_S),$$

in which C is the concentration, in µg per mL, of USP Praziquantel RS in the *Standard preparation*, and r_U and r_S are the peak responses obtained from the *Assay preparation* and the *Standard preparation*, respectively.

Praziquantel Tablets

» Praziquantel Tablets contain not less than 90.0 percent and not more than 110.0 percent of the labeled amount of $C_{19}H_{24}N_2O_2$.

Packaging and storage—Preserve in tight containers.

Reference standard—*USP Praziquantel Reference Standard*—Dry in vacuum at a pressure not exceeding 5 mm of mercury at 50° over phosphorus pentoxide for 2 hours before using.

Identification—Transfer a quantity of powdered Tablets, equivalent to about 30 mg of Praziquantel, to a centrifuge tube, add 5 mL of methanol, agitate for 5 minutes, and centrifuge. Use the clear supernatant liquid as the test solution. Apply separately, as 1-cm wide bands, 10 µL each of the test solution and a Standard solution of USP Praziquantel RS in methanol containing 6 mg per mL to a thin-layer chromatographic plate (see *Chromatography* ⟨621⟩), coated with a 0.25-mm layer of chromatographic silica gel mixture. Develop the chromatogram in an unsaturated chamber, using ethyl acetate as the developing solvent, until the solvent front has moved about 8 cm. Remove the plate from the chamber, air-dry, and examine under short-wavelength ultraviolet light: the R_f value of the principal band in the chromatogram of the test solution corresponds to that obtained for the Standard solution.

Dissolution ⟨711⟩—

Medium: 0.1 *N* hydrochloric acid containing 2.0 mg of sodium lauryl sulfate per mL; 900 mL.

Apparatus 2: 50 rpm.

Time: 30 minutes.

Standard preparation—Dissolve an accurately weighed quantity of USP Praziquantel RS in methanol to obtain a solution having a known concentration of about $L/90$ mg per mL, L being the labeled quantity, in mg, of praziquantel in each Tablet. Transfer 5.0 mL of this solution to a 50-mL volumetric flask, dilute with *Dissolution Medium* to volume, and mix.

Procedure—Determine the amount of $C_{19}H_{24}N_2O_2$ dissolved from ultraviolet absorbances at the wavelength of maximum absorbance at about 263 nm of filtered portions of the solution under test in comparison with the *Standard preparation.*

Tolerances—Not less than 75% (Q) of the labeled amount of $C_{19}H_{24}N_2O_2$ is dissolved in 30 minutes.

Uniformity of dosage units ⟨905⟩: meet the requirements.

Assay—

Mobile phase and *Chromatographic system*—Proceed as directed in the *Assay* under *Praziquantel.*

Standard preparation—Dissolve an accurately weighed quantity of USP Praziquantel RS in *Mobile phase*, and dilute quantitatively, and stepwise if necessary, with *Mobile phase* to obtain a solution having a known concentration of about 0.18 mg per mL.

Assay preparation—Weigh and finely powder not less than 20 Praziquantel Tablets. Transfer an accurately weighed portion of the powder, equivalent to about 150 mg of praziquantel, to a 100-mL volumetric flask, add 70 mL of *Mobile phase*, sonicate for 5 minutes, dilute with *Mobile phase* to volume, mix, and filter. Transfer 3.0 mL of the filtrate to a 25-mL volumetric flask, dilute with *Mobile phase* to volume, and mix.

Procedure—Proceed as directed for *Procedure* in the *Assay* under *Praziquantel.* Calculate the quantity, in mg, of $C_{19}H_{24}N_2O_2$ in the portion of Tablets taken by the formula:

$$2500(C/3)(r_U/r_S),$$

in which C is the concentration, in mg per mL, of USP Praziquantel RS in the *Standard preparation*, and r_U and r_S are the peak responses obtained from the *Assay preparation* and the *Standard preparation*, respectively.

Prazosin Hydrochloride

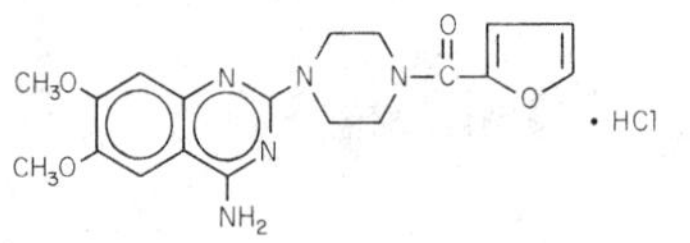

$C_{19}H_{21}N_5O_4 \cdot HCl$ 419.87

Piperazine, 1-(4-amino-6,7-dimethoxy-2-quinazolinyl)-4-(2-furanylcarbonyl)-, monohydrochloride.

1-(4-Amino-6,7-dimethoxy-2-quinazolinyl)-4-(2-furoyl)piperazine monohydrochloride [*19237-84-4*].

» Prazosin Hydrochloride contains not less than 97.0 percent and not more than 103.0 percent of $C_{19}H_{21}N_5O_4 \cdot HCl$, calculated on the anhydrous basis.

Caution—Care should be taken to prevent inhaling particles of Prazosin Hydrochloride and to prevent its contacting any part of the body.

Packaging and storage—Preserve in tight, light-resistant containers.

Labeling—Label it to indicate whether it is anhydrous or is the polyhydrate.

Reference standards—*USP Prazosin Hydrochloride Reference Standard*—Do not dry; determine the water content titrimetrically at the time of use. *USP 4-Amino-2-chloro-6,7-dimethoxyquinazoline Reference Standard*—Do not dry before using. *USP 1-(2-Furoyl)piperazine Reference Standard*—Do not dry before using. *USP 1,4-Bis(4-amino-6,7-dimethoxy-2-quinazolinyl)piperazine Reference Standard*—Do not dry before using. *USP 4-Amino-6,7-dimethoxy-2-(1-piperazinyl)quinazoline Dihydrochloride Trihydrate Reference Standard*—Do not dry before using. *USP 1,4-Bis(2-furoyl)piperazine Reference Standard*—Do not dry before using.

Identification—

A: Dissolve about 20 mg of it in 20 mL of methanol, with the aid of gentle heat, and evaporate to dryness. Dry the residue in vacuum at 130° for 3 hours: the infrared absorption spectrum of a potassium bromide dispersion of the residue so obtained exhibits maxima only at the same wavelengths as that of a similar preparation of USP Prazosin Hydrochloride RS.

B: The ultraviolet absorption spectrum of a 1 in 150,000 solution in methanolic 0.01 N hydrochloric acid exhibits maxima and minima at the same wavelengths as that of a similar solution of USP Prazosin Hydrochloride RS, concomitantly measured, and the respective absorptivities, calculated on the anhydrous basis, at the wavelengths of maximum absorbance at about 329 nm and 246 nm do not differ by more than 4.0%.

C: Prepare a test solution of it in a mixture of chloroform, methanol, and diethylamine (10:10:1) containing 5 mg per mL, and proceed as directed under *Thin-layer Chromatographic Identification Test* ⟨201⟩, using a solvent system consisting of a mixture of ethyl acetate and diethylamine (19:1).

D: It responds to the tests for *Chloride* ⟨191⟩.

Water, *Method I* ⟨921⟩: not more than 2.0% for the anhydrous form; between 8.0% and 15.0% for the polyhydrate form.

Residue on ignition ⟨281⟩: not more than 0.4%, determined on a 1-g portion, accurately weighed.

Heavy metals, *Method II* ⟨231⟩: 0.005%.

Iron—

Standard preparation—Dissolve 100 mg of iron wire, accurately weighed, in 10 mL of hydrochloric acid with boiling. Cool, transfer to a 1000-mL volumetric flask, dilute with water to volume, and mix. Dilute quantitatively and stepwise with 0.2 N nitric acid to obtain a solution containing 4.0 µg of iron per mL.

Test preparation—Dissolve the residue obtained in the test for *Residue on ignition* in 20 mL of 2 N nitric acid. Slowly evaporate this solution to approximately 5 mL, transfer to a 25-mL volumetric flask, using 0.2 N nitric acid as a wash solvent, dilute with 0.2 N nitric acid to volume, and mix.

Procedure—Concomitantly determine the absorbances of the *Standard preparation* and the *Test preparation* at the wavelength of maximum absorbance at about 248 nm, with a suitable atomic absorption spectrophotometer (see *Apparatus* under *Spectrophotometry and Light-scattering* ⟨851⟩) equipped with an iron hollow-cathode lamp and an air-acetylene flame, using water as the blank: the absorbance of the *Test preparation* is not more than that of the *Standard preparation* (0.010%).

Nickel—

Standard preparation—Dissolve 100 mg of nickel, accurately weighed, in 10 mL of nitric acid with the aid of boiling. Cool, transfer to a 1000-mL volumetric flask, dilute with water to volume, and mix. Dilute quantitatively and stepwise with 0.2 N nitric acid to obtain a solution containing 4.0 µg of nickel per mL.

Test preparation—Use the *Test preparation* prepared as directed in the test for *Iron.*

Procedure—Concomitantly determine the absorbances of the *Standard preparation* and the *Test preparation* at the wavelength of maximum absorbance at about 232 nm, with a suitable atomic absorption spectrophotometer (see *Apparatus* under *Spectrophotometry and Light-scattering* ⟨851⟩) equipped with a nickel hollow-cathode lamp and an air-acetylene flame, using water as the blank: the absorbance of the *Test preparation* is not more than that of the *Standard preparation* (0.010%).

4-Amino-2-chloro-6,7-dimethoxyquinazoline—Dissolve 60 mg, accurately weighed, in 5 mL of a solvent consisting of a mixture of chloroform, methanol, and diethylamine (10:10:1). On a suitable thin-layer chromatographic plate (see *Chromatography* ⟨621⟩), coated with a 0.25-mm layer of chromatographic silica gel, apply 100 µL of this solution as a 7.5-cm streak. Allow the streak to dry, and apply 100 µL of a Standard solution of USP 4-Amino-2-chloro-6,7-dimethoxyquinazoline RS in the same solvent containing 60 µg per mL as a 7.5-cm streak that overlaps about 2.5 cm of the streak from the solution under test. Allow the streak to dry, and develop the chromatogram in a solvent system consisting of a mixture of toluene, ethyl acetate, and glacial acetic acid (10:10:1) until the solvent front has moved about three-fourths of the length of the plate. Remove the plate from the developing chamber, mark the solvent front, allow the solvent to evaporate, and examine the plate under short-wavelength ultraviolet light: any band from the solution under test, at an R_f value corresponding to the main band from the Standard solution, is not greater in size or intensity than the main band obtained from the Standard solution (0.5% 4-Amino-2-chloro-6,7-dimethylquinazoline).

1,4-Bis(2-furoyl)piperazine—Proceed as directed under *4-Amino-2-chloro-6,7-dimethoxyquinazoline*, substituting 1,4-Bis(2-furoyl)piperazine for 4-Amino-2-chloro-6,7-dimethoxyquinazoline.

4-Amino-6,7-dimethoxy-2-(1-piperazinyl)quinazoline—Dissolve 60 mg, accurately weighed, in 5 mL of a solvent consisting of a mixture of chloroform, methanol, and diethylamine (10:10:1). On a suitable thin-layer chromatographic plate (see *Chromatography* ⟨621⟩), coated with a 0.25-mm layer of chromatographic silica gel, apply 100 µL of this solution as a 7.5-cm streak. Allow the streak to dry, and apply 100 µL of a Standard solution of USP 4-Amino-6,7-dimethoxy-2-(1-piperazinyl)quinazoline Dihydrochloride Trihydrate RS in the same solvent containing 86 µg per mL as a 7.5-cm streak that overlaps about 2.5 cm of the streak from the solution under test. Allow the streak to dry, and develop the chromatogram in a solvent system consisting of a mixture of ethyl acetate, methanol, and diethylamine (40:20:3) until the solvent front has moved about three-fourths of the length of the plate. Remove the plate from the developing chamber, mark the solvent front, allow the solvent to evaporate, and examine the plate under short-wavelength ultraviolet light: any band from the solution under test, at an R_f value corresponding to the main band from the Standard solution, is not greater in size or intensity than the main band obtained from the Standard solution [0.5% 4-amino-6,7-dimethoxy-2-(1-piperazinyl)quinazoline].

1-(2-Furoyl)piperazine—Proceed as directed under *4-Amino-6,7-dimethoxy-2-(1-piperazinyl)quinazoline*, using a Standard solution of USP 1-(2-Furoyl)piperazine RS in the same solvent containing 60 µg per mL. Not more than 0.5% of 1-(2-furoyl)-piperazine is found.

1,4-Bis(4-amino-6,7-dimethoxy-2-quinazolinyl)piperazine—Dissolve 60 mg, accurately weighed, in 5 mL of a solvent consisting of a mixture of chloroform, methanol, and diethylamine (10:10:1). On a suitable thin-layer chromatographic plate (see *Chromatography* ⟨621⟩), coated with a 0.25-mm layer of chromatographic silica gel, apply 100 µL of this solution as a 7.5-cm streak. Allow the streak to dry, and apply 100 µL of a Standard solution of USP 1,4-Bis(4-amino-6,7-dimethoxy-2-quinazolinyl)piperazine RS in the same solvent containing 60 µg per mL as a 7.5-cm streak that overlaps about 2.5 cm of the streak from the solution under test. Allow the streak to dry, and develop the chromatogram in a solvent system consisting of a mixture of ethyl acetate and diethylamine (19:1) until the solvent front has moved about three-fourths of the length of the plate. Remove the plate from the developing chamber, mark the solvent front, allow the solvent to evaporate, and examine the plate under short-wavelength ultraviolet light: any band from the solution under test, at an R_f value corresponding to the main band from the Standard solution, is not greater in size or intensity than the main band obtained from the Standard solution [0.5% 1,4-bis(4-amino-6,7-dimethoxy-2-quinazolinyl)piperazine].

Assay—

Mobile phase—Mix 700 mL of methanol, 300 mL of water, and 10 mL of glacial acetic acid. Add diethylamine in sufficient quantity (about 0.2 mL) such that the retention time of prazosin hydrochloride is between 6 and 10 minutes. Degas the solution.

Standard preparation—Transfer about 100 mg of USP Prazosin Hydrochloride RS, accurately weighed, to a 100-mL volumetric flask, dissolve in methanol, dilute with methanol to volume, and mix. Dilute this solution quantitatively with a mixture of methanol and water (7:3) to obtain a solution having a known concentration of about 30 µg per mL.

Assay preparation—Transfer about 100 mg of Prazosin Hydrochloride, accurately weighed, to a 100-mL volumetric flask, dissolve in methanol, dilute with methanol to volume, and mix. Pipet 3 mL of this solution into a 100-mL volumetric flask, dilute with a mixture of methanol and water (7:3) to volume, and mix.

Chromatographic system (see *Chromatography* ⟨621⟩)—The liquid chromatograph is equipped with a 254-nm detector and a 2-mm × 25-cm column that contains packing L3. The flow rate is adjusted to obtain a retention time of between 6 and 10 minutes for prazosin hydrochloride. Chromatograph five replicate injections of the *Standard preparation*, and record the peak responses as directed under *Procedure:* the relative standard deviation is not more than 2.0%.

Procedure—Introduce equal volumes (about 5 µL) of the *Standard preparation* and the *Assay preparation* into the chromatograph, using a suitable microsyringe or sampling valve, and measure the peak responses at identical retention times. Calculate the quantity, in mg, of $C_{19}H_{21}N_5O_4 \cdot HCl$ in the Prazosin Hydrochloride taken by the formula:

$$(C/0.3)(r_U/r_S),$$

in which C is the concentration, in µg per mL, of USP Prazosin Hydrochloride RS, calculated on the anhydrous basis, in the *Standard preparation*, and r_U and r_S are the peak responses obtained from the *Assay preparation* and the *Standard preparation*, respectively.

Prazosin Hydrochloride Capsules

» Prazosin Hydrochloride Capsules contain an amount of $C_{19}H_{21}N_5O_4 \cdot HCl$ equivalent to not less than 90.0 percent and not more than 110.0 percent of the labeled amount of prazosin ($C_{19}H_{21}N_5O_4$).

Caution—Care should be taken to prevent inhaling particles of Prazosin Hydrochloride and to prevent its contacting any part of the body.

Packaging and storage—Preserve in well-closed, light-resistant containers.

Reference standard—*USP Prazosin Hydrochloride Reference Standard*—Do not dry; determine the water content titrimetrically at the time of use.

Identification—To a portion of the contents of Capsules, equivalent to about 10 mg of prazosin, add 20 mL of a mixture of chloroform and methanol (1:1), shake by mechanical means for 10 minutes, and centrifuge. On a suitable thin-layer chromatographic plate (see *Chromatography* ⟨621⟩), coated with a 0.25-mm layer of chromatographic silica gel mixture, apply as separate 7.5-cm streaks 100 µL of this solution and 100 µL of a Standard solution of USP Prazosin Hydrochloride RS in a mixture of chloroform and methanol (1:1) containing about 0.5 mg per mL. Allow the streaks to dry, and develop the chromatogram in a solvent system consisting of a mixture of ethyl acetate and diethylamine (19:1) until the solvent front has moved about three-fourths of the length of the plate. Remove the plate from the developing chamber, mark the solvent front, allow the solvent to evaporate, and view under short-wavelength ultraviolet light: prazosin appears as a dark blue band on a yellow-green fluorescent background, and the R_f value of the prazosin band obtained from the solution from the Capsule contents corresponds to that obtained from the Standard solution. Spray the plate evenly with a 1 in 50 solution of hydrochloric acid in potassium iodoplatinate TS: the R_f value of the principal band obtained from the solution from the Capsule contents corresponds to that obtained from the Standard solution.

Dissolution ⟨711⟩—

Medium: simulated gastric fluid TS, without pepsin; 800 mL.

Apparatus—Proceed with Capsules as directed for *Uncoated Tablets* under *Disintegration* ⟨701⟩, beginning with "Place 1 Tablet in each of the six tubes of the basket," with these exceptions: (a) the disks are not used; (b) the apparatus is adjusted so that the bottom of the basket-rack assembly descends to 1.0 ± 0.1 cm from the inside bottom surface of the vessel on the downward stroke; (c) the 10-mesh, stainless-steel cloth in the basket-rack assembly is replaced with 40-mesh, stainless-steel cloth; and (d) 40-mesh, stainless-steel cloth is fitted to the top of the basket-rack assembly if necessary to prevent any dosage unit from floating out of the tubes of the assembly.

Time: 30 minutes.

Procedure—Filter a 20-mL portion of the solution under test through a 0.5-µm filter, and pipet an aliquot of the filtrate, containing about 0.1 mg of prazosin, into a 250-mL separator. Add a volume of *Dissolution Medium* so that the combined volume is 25 mL, add 5 mL of 1 *N* sodium hydroxide, and extract with 50.0 mL of methylene chloride. Transfer 20.0 mL of the lower organic phase to a second 250-mL separator containing 50 mL of hexanes and 10.0 mL of *Dissolution Medium*, shake the mixture for 1 minute, and allow the phases to separate. Determine the amount of prazosin ($C_{19}H_{21}N_5O_4$) in the lower, aqueous phase from ultraviolet absorbances at the wavelength of maximum absorbance at about 246 nm of the extract of the solution under test in comparison with a Standard solution having a known concentration of USP Prazosin Hydrochloride RS in the same medium. [NOTE—Use an amount of 0.01 *N* methanolic hydrochloric acid, not to exceed 1% of the final total volume of the Standard solution, to dissolve the USP Prazosin Hydrochloride RS initially prior to dilution with *Dissolution Medium*.] Calculate the amount of $C_{19}H_{21}N_5O_4$ dissolved per Capsule.

Tolerances—Not less than 50% of the labeled amount of $C_{19}H_{21}N_5O_4$ is dissolved in 30 minutes.

Uniformity of dosage units ⟨905⟩: meet the requirements.

Procedure for content uniformity—Transfer the contents of a Capsule to a 100-mL volumetric flask, add 50 mL of 0.01 *N* methanolic hydrochloric acid containing 30% water, shake by mechanical means for 15 minutes, adjust with the same solvent to volume, and mix. Filter through a membrane filter having a porosity of 1.2 µm and dilute, if necessary, a portion of the filtrate with the same solvent to a concentration of about 10 µg of prazosin per mL. Prepare a Standard solution of USP Prazosin Hydrochloride RS in the same solvent having a known concentration of about 11 µg per mL. Concomitantly determine the absorbances of the solution from the Capsule contents and the Standard solution in 1-cm cells at the wavelength of maximum absorbance at about 330 nm, with a suitable spectrophotometer, using the solvent as the blank. Calculate the quantity, in mg, of prazosin ($C_{19}H_{21}N_5O_4$) in the Capsule taken by the formula:

$$(383.40/419.87)(0.001DC)(A_U/A_S),$$

in which 383.40 and 419.87 are the molecular weights of prazosin and prazosin hydrochloride, respectively, *D* is the dilution factor for the Capsule contents, *C* is the concentration, in µg per mL, of USP Prazosin Hydrochloride RS, calculated on the anhydrous basis, in the Standard solution, and A_U and A_S are the absorbances of the solution from the Capsule contents and the Standard solution, respectively.

Assay—

Mobile phase, Chromatographic system, and *Procedure*—Proceed as directed in the *Assay* under *Prazosin Hydrochloride*.

Acid-methanol solution—To 300 mL of water in a 1000-mL volumetric flask, add 0.85 mL of hydrochloric acid, dilute with methanol to volume, and mix. Transfer 300 mL of this solution to a 500-mL volumetric flask, dilute with methanol to volume, and mix.

Standard preparation—Prepare a solution of USP Prazosin Hydrochloride RS in *Acid-methanol solution* having a known concentration of about 0.2 mg per mL. Pipet 5 mL of this solution into a 100-mL volumetric flask, add 45.0 mL of *Acid-methanol solution*, dilute with methanol to volume, and mix.

Assay preparation—Remove, as completely as possible, the contents of not less than 20 Prazosin Hydrochloride Capsules, and weigh. Transfer a quantity of the contents, accurately weighed, equivalent to about 1 mg of prazosin hydrochloride, to a glass-stoppered flask containing 50.0 mL of *Acid-methanol solution*, and shake by mechanical means for 30 minutes. Place the flask in an ultrasonic bath for 30 minutes, cool to room temperature, and filter the contents through a membrane filter (5-µm or finer porosity). Transfer 25.0 mL of the filtrate to a 50-mL volumetric flask, dilute with methanol to volume, and mix.

Procedure—Proceed as directed for *Procedure* in the *Assay* under *Prazosin Hydrochloride*. Calculate the quantity, in mg, of prazosin ($C_{19}H_{21}N_5O_4$) in the portion of the contents of Capsules taken by the formula:

$$(383.40/419.87)(0.1C)(r_U/r_S),$$

in which 383.40 and 419.87 are the molecular weights of prazosin and prazosin hydrochloride, respectively, *C* is the concentration, in µg per mL, of USP Prazosin Hydrochloride RS, calculated on the anhydrous basis, in the *Standard preparation*, and r_U and r_S are the peak responses obtained from the *Assay preparation* and the *Standard preparation*, respectively.

Precipitated Calcium Carbonate—*see* Calcium Carbonate, Precipitated

Precipitated Sulfur—*see* Sulfur, Precipitated

Prednisolone

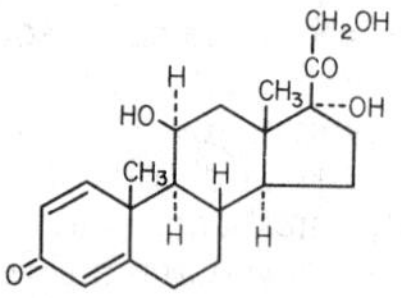

$C_{21}H_{28}O_5$ (anhydrous) 360.45
Pregna-1,4-diene-3,20-dione, 11,17,21-trihydroxy-, (11β)-.
11β,17,21-Trihydroxypregna-1,4-diene-3,20-dione (anhydrous) [*50-24-8*].
Sesquihydrate 387.48 [*52438-85-4*].

» Prednisolone is anhydrous or contains one and one-half molecules of water of hydration. It contains not less than 97.0 percent and not more than 102.0 percent of $C_{21}H_{28}O_5$, calculated on the dried basis.

Packaging and storage—Preserve in well-closed containers.

Labeling—Label it to indicate whether it is anhydrous or hydrous.

Reference standard—*USP Prednisolone Reference Standard*—Dry in vacuum at 105° for 3 hours before using.

Identification—

A: The infrared absorption spectrum of a potassium bromide dispersion of it, previously dried at 105° for 3 hours, exhibits maxima only at the same wavelengths as that of a similar preparation of USP Prednisolone RS. If a difference appears, dissolve portions of both the test specimen and the Reference Standard in ethyl acetate, evaporate the solutions to dryness, and repeat the test on the residues.

B: The ultraviolet absorption spectrum of a 1 in 100,000 solution in methanol exhibits maxima and minima at the same wavelengths as that of a similar solution of USP Prednisolone RS, concomitantly measured, and the respective absorptivities, calculated on the dried basis, at the wavelength of maximum absorbance at about 242 nm, do not differ by more than 2.5%.

Specific rotation ⟨781⟩: between +97° and +103°, calculated on the dried basis, determined in a solution in dioxane containing 100 mg in each 10 mL.

Loss on drying ⟨731⟩—Dry it in vacuum at 105° for 3 hours: anhydrous Prednisolone loses not more than 1.0%, and hydrous Prednisolone not more than 7.0%, of its weight.

Residue on ignition ⟨281⟩: negligible, from 100 mg.

Selenium ⟨291⟩: 0.003%, a 200-mg test specimen being used.

Ordinary impurities ⟨466⟩—

Test solution: a mixture of alcohol and water (1:1).

Standard solution: a mixture of alcohol and water (1:1).

Eluant: a mixture of toluene and isopropyl alcohol (70:30), in a nonequilibrated chamber.

Visualization: 1.

Assay—

Mobile phase—Prepare a solution containing a mixture of butyl chloride, water-saturated butyl chloride, tetrahydrofuran, methanol, and glacial acetic acid (95:95:14:7:6).

Internal standard solution—Prepare a solution of betamethasone in tetrahydrofuran containing 5 mg per mL. Dilute this solution with water-saturated chloroform, and mix to obtain a solution having a final concentration of 0.5 mg per mL.

Standard preparation—Dissolve about 10 mg of USP Prednisolone RS, accurately weighed, in 20.0 mL of *Internal standard solution.* Dilute with water-saturated chloroform to 100.0 mL, and mix.

Assay preparation—Dissolve about 10 mg of Prednisolone, accurately weighed, in 20.0 mL of *Internal standard solution*, dilute with water-saturated chloroform to 100.0 mL, and mix.

Chromatographic system (see *Chromatography* ⟨621⟩)—The liquid chromatograph is equipped with a 254-nm detector and a 4-mm × 30-cm column that contains packing L3. The flow rate is about 1 mL per minute. Chromatograph four replicate injections of the *Standard preparation*, and record the peak responses as directed under *Procedure:* the relative standard deviation is not more than 2.0%, and the resolution factor between prednisolone and betamethasone is not less than 3.5.

Procedure—Separately inject equal volumes (about 10 µL) of the *Standard preparation* and the *Assay preparation* into the chromatograph by means of a suitable microsyringe or sampling valve, record the chromatograms, and measure the responses for the major peaks. The retention times are about 17 minutes and 23 minutes for betamethasone and prednisolone, respectively. Calculate the quantity, in mg, of $C_{21}H_{28}O_5$ in the portion of Prednisolone taken by the formula:

$$0.1C(R_U/R_S),$$

in which *C* is the concentration, in µg per mL, of USP Prednisolone RS in the *Standard preparation*, and R_U and R_S are the peak response ratios of the prednisolone peak and the internal standard peak obtained from the *Assay preparation* and the *Standard preparation*, respectively.

Prednisolone Cream

» Prednisolone Cream contains not less than 90.0 percent and not more than 110.0 percent of the labeled amount of $C_{21}H_{28}O_5$, in a suitable cream base.

Packaging and storage—Preserve in collapsible tubes or in tight containers.

Reference standard—*USP Prednisolone Reference Standard*—Dry in vacuum at 105° for 3 hours before using.

Identification—Evaporate in a test tube to dryness about 5 mL of the dilute solution obtained in the *Assay.* Add 2 to 3 mL of warm perchloric acid, mix, and warm on a steam bath for about 2 minutes: a blood-red color develops. Add 5 mL of water: the color is not discharged.

Minimum fill ⟨755⟩: meets the requirements.

Assay—

Sulfuric acid reagent—Prepare a solution of sulfuric acid, dehydrated alcohol, and water solution (4:3:3).

Modified phenylhydrazine–sulfuric acid TS—Dissolve 65 mg of phenylhydrazine hydrochloride in 100 mL of *Sulfuric acid reagent.*

Standard preparation—Dissolve in dehydrated alcohol a suitable quantity, accurately weighed, of USP Prednisolone RS, and dilute quantitatively and stepwise with dehydrated alcohol to obtain a solution having a known concentration of about 0.1 mg per mL.

Assay preparation—To an accurately weighed portion of Prednisolone Cream, equivalent to about 20 mg of prednisolone, add 25 mL of dehydrated alcohol. Warm on a steam bath to disperse the assay specimen, then chill to congeal it. Filter through paper, previously washed with dehydrated alcohol, into a 200-mL volumetric flask. Repeat twice more, starting with "add 25 mL of dehydrated alcohol." Dilute with dehydrated alcohol to volume. (Retain about 5 mL of this dilution for the *Identification test.*)

Procedure—Pipet 2 mL of *Assay preparation* into each of two 50-mL conical flasks (identify as *Assay preparation* and *Assay preparation blank*). Pipet 2 mL of *Standard preparation* into each of two 50-mL conical flasks (identify as *Standard preparation* and *Standard preparation blank*). Pipet 2 mL of dehydrated alcohol into a 50-mL conical flask (identify as *Reagent blank*). Add 20.0 mL of *Sulfuric acid reagent* to the *Assay preparation blank* and to the *Standard preparation blank.* Add 20.0 mL of *Modified phenylhydrazine–sulfuric acid TS* to the *Assay preparation*, to the *Standard preparation*, and to the *Reagent blank*, respectively. Maintain the flasks in a water bath at 60° for about 45 minutes, then chill in a water-ice bath. Filter each through a fine-porosity, sintered-glass funnel, and identify the filtrate flasks correspondingly. Concomitantly determine the absorbances of the solutions at the wavelength of maximum absorbance at about 410 nm, with a suitable spectrophotometer, against dehydrated alcohol. Calculate the quantity, in mg, of $C_{21}H_{28}O_5$ in the portion of Cream taken by the formula:

$$200C[(A_U - A_{UB} - A_R)/(A_S - A_{SB} - A_R)],$$

in which *C* is the concentration, in mg per mL, of USP Prednisolone RS in the *Standard preparation*, A_U and A_S are the absorbances of the solutions from the *Assay preparation* and the *Standard preparation*, respectively, A_{UB} and A_{SB} are the absorbances of the solutions from the *Assay preparation blank* and the *Standard preparation blank*, respectively, and A_R is the absorbance of the *Reagent blank.*

Prednisolone Ophthalmic Ointment, Chloramphenicol and—*see* Chloramphenicol and Prednisolone Ophthalmic Ointment

Prednisolone Suspension, Sterile Penicillin G Procaine, Dihydrostreptomycin Sulfate, and—*see* Sterile Penicillin G Procaine, Dihydrostreptomycin Sulfate, and Prednisolone Suspension

Prednisolone Syrup

» Prednisolone Syrup contains not less than 90.0 percent and not more than 110.0 percent of the labeled amount of $C_{21}H_{28}O_5$. Prednisolone Syrup may contain alcohol.

Packaging and storage—Preserve in tight, light-resistant containers.

Reference standard—*USP Prednisolone Reference Standard*—Dry at 105° for 3 hours before using.

Identification—Transfer a volume of Prednisolone Syrup, equivalent to about 50 mg of prednisolone to a separator and extract with 25 mL of chloroform. Separate the layers, filter, and evaporate the chloroform layer on a steam bath to dryness. Wash the residue with two 10-mL portions of hot solvent hexane, decanting the supernatant liquid each time and discarding it. Digest the residue with 25 mL of dehydrated alcohol, warming slightly, for 15 minutes. Filter the warm solution, and evaporate the filtrate to a volume of 2 to 3 mL. Add solvent hexane until the mixture just becomes turbid, chill it to effect crystallization, collect the crystals, and dry them at 60° for 1 hour: the crystals of prednisolone so obtained respond to *Identification test A* under *Prednisolone.*

pH ⟨791⟩: between 3.0 and 4.5.

Alcohol content (if present) ⟨611⟩, *Method II:* not less than 90.0% and not more than 115.0% of the labeled amount, the labeled amount being between 2.0% and 5.0% of C_2H_5OH.

Assay—

Mobile phase—Prepare a solution containing a mixture of butyl chloride, water-saturated butyl chloride, tetrahydrofuran, methanol, and glacial acetic acid (95:95:14:7:6).

Internal standard solution—Prepare a solution of betamethasone in tetrahydrofuran containing 5 mg per mL. Dilute this solution with water-saturated chloroform, and mix to obtain a solution having a final concentration of 0.5 mg per mL.

Standard preparation—Dissolve about 10 mg of USP Prednisolone RS, accurately weighed, in 20.0 mL of *Internal standard solution*. Dilute with water-saturated chloroform to 100.0 mL, and mix.

Assay preparation—Using a "To contain" pipet, transfer a volume of Prednisolone Syrup, equivalent to about 10 mg of prednisolone with several washings with water into a separator. Extract with 20.0 mL of *Internal standard solution*, followed by three 20-mL portions of water-saturated chloroform, filtering each extract through water-saturated, chloroform-washed cotton into a 100-mL volumetric flask. Dilute with water-saturated chloroform to volume, and mix.

Chromatographic system (see *Chromatography* ⟨621⟩)—The liquid chromatograph is equipped with a 254-nm detector and a 4.6-mm × 25-cm column that contains packing L3. The flow rate is about 1 mL per minute. Chromatograph the *Standard preparation*, and record the peak responses as directed under *Procedure:* the resolution, *R*, between prednisolone and betamethasone is not less than 3.5 and the relative standard deviation for replicate injections is not more than 2.0%.

Procedure—Separately inject equal volumes (about 10 µL) of the *Standard preparation* and the *Assay preparation* into the chromatograph. Record the chromatograms, and measure the responses for the major peaks. The relative retention times are about 1.0 and 1.4 for betamethasone and prednisolone, respectively. Calculate the quantity, in mg, of $C_{21}H_{28}O_5$ in the portion of Prednisolone Syrup taken by the formula:

$$0.1C(R_U/R_S),$$

in which *C* is the concentration, in µg per mL, of USP Prednisolone RS in the *Standard preparation*, and R_U and R_S are the peak response ratios of the prednisolone peak and the internal standard peak obtained from the *Assay preparation* and the *Standard preparation*, respectively.

Prednisolone Tablets

» Prednisolone Tablets contain not less than 90.0 percent and not more than 110.0 percent of the labeled amount of $C_{21}H_{28}O_5$.

Packaging and storage—Preserve in well-closed containers.

Reference standard—*USP Prednisolone Reference Standard*—Dry in vacuum at 105° for 3 hours before using.

Identification—Pulverize a number of Tablets, equivalent to about 50 mg of prednisolone, and digest with 25 mL of chloroform for 15 minutes. Filter the mixture, and evaporate the filtrate on a steam bath to dryness. Wash the residue with two 10-mL portions of hot solvent hexane, decanting the supernatant liquid each time and discarding it. Digest the residue with 25 mL of dehydrated alcohol, warming slightly, for 15 minutes. Filter the warm solution, and evaporate the filtrate to a volume of 2 to 3 mL. Add solvent hexane until the mixture just becomes turbid, chill it to effect crystallization, collect the crystals, and dry them at 60° for 1 hour: the crystals of prednisolone so obtained respond to *Identification test A* under *Prednisolone*.

Dissolution ⟨711⟩—

Medium: water; 900 mL.

Apparatus 2: 50 rpm.

Time: 30 minutes.

Procedure—Determine the amount in solution on filtered portions of the *Dissolution Medium*, suitably diluted at the wavelength of maximum absorbance at about 246 nm, with a suitable spectrophotometer in comparison with a Standard solution having a known concentration of USP Prednisolone RS. An amount of alcohol not to exceed 5% of the total volume of the Standard solution may be used to bring the prednisolone standard into solution prior to dilution with water.

Tolerances—Not less than 70% (*Q*) of the labeled amount of $C_{21}H_{28}O_5$ is dissolved in 30 minutes.

Uniformity of dosage units ⟨905⟩: meet the requirements.

Procedure for content uniformity—

Mobile phase, Internal standard solution, Standard preparation, and *Chromatographic system*—Prepare as directed in the *Assay* under *Prednisolone*.

Test preparation—Place 1 Tablet in a suitable container. Place 0.5 mL of water directly on the tablet, and allow to stand until disintegrated (about 30 minutes). Gently agitate the container to ensure that the tablet is completely disintegrated. Add 2.0 mL of *Internal standard solution* for each mg of labeled tablet strength, and sonicate for about 10 minutes. Dilute with a quantity of water-saturated chloroform approximately four times the volume of added *Internal standard solution*. Add a few glass beads, close the container, and shake vigorously for about 30 minutes. Centrifuge, or allow to stand until a clear solution is obtained. Analyze the clear solution as directed under *Procedure*.

Procedure—Proceed as directed for *Procedure* in the *Assay* under *Prednisolone*. Calculate the quantity, in mg, of $C_{21}H_{28}O_5$ in the Tablet by the formula:

$$(FW_S)(R_U/R_S),$$

in which *F* is the ratio of the volume of the *Internal standard solution*, in mL, in the *Test preparation* to the volume, in mL, of the *Internal standard solution* in the *Standard preparation*, W_S is the weight, in mg, of USP Prednisolone RS taken for the *Standard preparation*, and the other terms are as defined therein.

Assay—

Mobile phase, Internal standard solution, and *Chromatographic system*—Prepare as directed in the *Assay* under *Prednisolone*.

Assay preparation—Weigh and finely powder not less than 10 Prednisolone Tablets. Transfer an accurately weighed portion of the powder, equivalent to about 10 mg of prednisolone, to a 100-mL volumetric flask. Add 20.0 mL of *Internal standard solution*, and sonicate for 10 minutes. Dilute with water-saturated chloroform to volume, and shake for 30 minutes. Centrifuge this mixture, and use the clear supernatant solution.

Procedure—Proceed as directed for *Procedure* in the *Assay* under *Prednisolone*. Calculate the quantity, in mg, of $C_{21}H_{28}O_5$ in the portion of Tablets taken by the formula:

$$0.1C(R_U/R_S),$$

in which the terms are as defined therein.

Prednisolone Acetate

$C_{23}H_{30}O_6$ 402.49

Pregna-1,4-diene-3,20-dione, 21-(acetyloxy)-11,17-dihydroxy-, (11β)-.

11β,17,21-Trihydroxypregna-1,4-diene-3,20-dione 21-acetate [*52-21-1*].

» Prednisolone Acetate contains not less than 97.0 percent and not more than 102.0 percent of $C_{23}H_{30}O_6$, calculated on the dried basis.

Packaging and storage—Preserve in well-closed containers.

Reference standard—*USP Prednisolone Acetate Reference Standard*—Dry at 105° for 3 hours before using.

Identification—

A: The infrared absorption spectrum of a potassium bromide dispersion of it, previously dried, exhibits maxima only at the same wavelengths as that of a similar preparation of USP Prednisolone Acetate RS.

B: The ultraviolet absorption spectrum of a 1 in 100,000 solution in methanol exhibits maxima and minima only at the same wavelengths as that of a similar solution of USP Prednisolone Acetate RS, concomitantly measured, and the respective absorptivities, calculated on the dried basis, at the wavelength of maximum absorbance at about 242 nm, do not differ by more than 2.5%.

C: To about 50 mg contained in a test tube add 2 mL of alcohol and 2 mL of dilute sulfuric acid (1 in 3.5), and boil gently for about 1 minute: the odor of ethyl acetate is perceptible.

Specific rotation ⟨781⟩: between +112° and +119°, calculated on the dried basis, determined in a solution in dioxane containing 100 mg in each 10 mL.

Loss on drying ⟨731⟩—Dry it at 105° for 3 hours: it loses not more than 1.0% of its weight.

Assay—

Mobile phase—Prepare a solution containing a mixture of *n*-butyl chloride, water-saturated *n*-butyl chloride, tetrahydrofuran, methanol, and glacial acetic acid (95:95:14:7:6).

Internal standard solution—Prepare a solution of betamethasone in tetrahydrofuran containing 10 mg per mL. Dilute this solution with water-saturated chloroform, and mix to obtain a solution having a final concentration of about 1 mg per mL.

Standard preparation—Dissolve about 5 mg of USP Prednisolone Acetate RS, accurately weighed, in 10.0 mL of *Internal standard solution.* Use sonication, if necessary, dilute with water-saturated chloroform to 200.0 mL, and mix to obtain a solution having a known concentration of about 25 µg per mL.

Assay preparation—Dissolve about 5 mg of Prednisolone Acetate, accurately weighed, in 10.0 mL of *Internal standard solution.* Use sonication, if necessary, dilute with water-saturated chloroform to 200.0 mL, and mix.

Chromatographic system (see *Chromatography* ⟨621⟩)—The liquid chromatograph is equipped with a 254-nm detector and a 4-mm × 30-cm column that contains packing L3. The flow rate is about 1 mL per minute. Chromatograph the *Standard preparation*, and record the peak responses as directed under *Procedure:* the resolution, *R*, between the analyte and internal standard peaks is not less than 3.0, and the relative standard deviation for replicate injections is not more than 2.0%.

Procedure—Separately inject equal volumes (about 10 µL) of the *Standard preparation* and the *Assay preparation* into the chromatograph, record the chromatograms, and measure the responses for the major peaks. The relative retention times are about 1.6 for betamethasone and 1.0 for prednisolone acetate. Calculate the quantity, in mg, of $C_{23}H_{30}O_6$ in the portion of Prednisolone Acetate taken by the formula:

$$0.2C(R_U / R_S),$$

in which *C* is the concentration, in µg per mL, of USP Prednisolone Acetate RS in the *Standard preparation*, and R_U and R_S are the peak response ratios obtained from the *Assay preparation* and the *Standard preparation*, respectively.

Prednisolone Acetate Ointment, Neomycin Sulfate and—*see* Neomycin Sulfate and Prednisolone Acetate Ointment

Prednisolone Acetate Ophthalmic Ointment, Neomycin Sulfate and—*see* Neomycin Sulfate and Prednisolone Acetate Ophthalmic Ointment

Prednisolone Acetate Ophthalmic Ointment, Neomycin Sulfate, Sulfacetamide Sodium, and—*see* Neomycin Sulfate, Sulfacetamide Sodium, and Prednisolone Acetate Ophthalmic Ointment

Prednisolone Acetate Ophthalmic Ointment, Sulfacetamide Sodium and—*see* Sulfacetamide Sodium and Prednisolone Acetate Ophthalmic Ointment

Prednisolone Acetate Ophthalmic Suspension

» Prednisolone Acetate Ophthalmic Suspension is a sterile, aqueous suspension of prednisolone acetate containing a suitable antimicrobial preservative. It may contain suitable buffers, stabilizers, and suspending and viscosity agents. It contains not less than 90.0 percent and not more than 115.0 percent of the labeled amount of $C_{23}H_3O_6$.

Packaging and storage—Preserve in tight containers.

Reference standard—*USP Prednisolone Acetate Reference Standard*—Dry at 105° for 3 hours before using.

Identification—Transfer a volume of Suspension, equivalent to about 7.5 mg of Prednisolone Acetate, to a test tube, add 5 mL of chloroform, and shake. Centrifuge, and apply 20 µL of the chloroform layer and 20 µL of a Standard solution of USP Prednisolone Acetate RS in chloroform containing 1.5 mg per mL on a thin-layer chromatographic plate (see *Chromatography* ⟨621⟩) coated with a 0.25-mm layer of chromatographic silica gel mixture. Develop the chromatogram in a mixture of chloroform and acetone (4:1) until the solvent front has moved about three-fourths the length of the plate. Mark the solvent front, and locate the spots on the plate by examining under ultraviolet light: the R_f value of the principal spot obtained from the solution under test corresponds to that obtained from the Standard solution.

Sterility—It meets the requirements under *Sterility Tests* ⟨71⟩.

pH ⟨791⟩: between 5.0 and 6.0.

Assay—

Mobile phase—Prepare a suitably filtered and degassed solution of acetonitrile and water (2:3). Make adjustments if necessary (see *System Suitability* under *Chromatography* ⟨621⟩).

Standard preparation—Dissolve an accurately weighed quantity of USP Prednisolone Acetate RS in a mixture of acetonitrile and water (1:1) to obtain a solution having a known concentration of about 0.1 mg per mL.

System suitability preparation—Prepare a solution of prednisolone in a mixture of acetonitrile and methanol (1:1) having a concentration of about 0.1 mg per mL. Mix equal volumes of this solution and the *Standard preparation.*

Assay preparation—Transfer an accurately measured volume of Prednisolone Acetate Ophthalmic Suspension, equivalent to about 5 mg of Prednisolone Acetate, to a 50-mL volumetric flask. Dilute with a mixture of acetonitrile and water (1:1) to volume, and mix.

Chromatographic system (see *Chromatography* ⟨621⟩)—The liquid chromatograph is equipped with a 254-nm detector and a column that contains packing L1. The flow rate is about 2 mL per minute. Chromatograph the *Standard preparation* and the *System suitability preparation*, and record the peak responses as directed under *Procedure:* the column efficiency determined from the analyte peak is not less than 7000 theoretical plates, the tailing factor for the analyte peak is not more than 2.0, and the resolution, *R*, between prednisolone and Prednisolone Acetate is not less than 2.0. The relative retention times are 0.5 for prednisolone and 1.0 for prednisolone acetate.

Procedure—Separately inject equal volumes (about 10 µL) of the *Standard preparation* and the *Assay preparation* into the chromatograph, record the chromatograms, and measure the responses for the major peaks. Calculate the quantity, in mg, of Prednisolone Acetate in each mL of the Ophthalmic Suspension taken by the formula:

$$50(C/V)(r_U / r_S),$$

in which *C* is the concentration, in mg per mL, of USP Prednisolone Acetate RS in the *Standard preparation*, *V* is the volume, in mL, of Ophthalmic Suspension taken, and r_U and r_S are the peak responses obtained from the *Assay preparation* and the *Standard preparation*, respectively.

Prednisolone Acetate Ophthalmic Suspension, Neomycin Sulfate and—*see* Neomycin Sulfate and Prednisolone Acetate Ophthalmic Suspension

Prednisolone Acetate Ophthalmic Suspension, Neomycin and Polymyxin B Sulfates and—*see* Neomycin and Polymyxin B Sulfates and Prednisolone Acetate Ophthalmic Suspension

Sterile Prednisolone Acetate Suspension

» Sterile Prednisolone Acetate Suspension is a sterile suspension of Prednisolone Acetate in a suitable aqueous medium. It contains not less than 90.0 percent and not more than 110.0 percent of the labeled amount of $C_{23}H_{30}O_6$.

Packaging and storage—Preserve in single-dose or in multiple-dose containers, preferably of Type I glass.

Reference standard—*USP Prednisolone Acetate Reference Standard*—Dry at 105° for 3 hours before using.

Identification—Allow a volume of Suspension, equivalent to about 50 mg of prednisolone acetate, to settle. Decant and discard the supernatant liquid. Dissolve the residue in 6 mL of alcohol. Evaporate the solution, with the aid of a current of air, to half its volume, when crystallization occurs. Chill, if necessary, to aid crystallization. Filter the crystals, and allow to dry with the aid of a current of air: the crystals so obtained respond to *Identification test A* under *Prednisolone Acetate*.

pH ⟨791⟩: between 5.0 and 7.5.

Other requirements—It meets the requirements under *Injections* ⟨1⟩.

Assay—

Mobile phase—Prepare a filtered and degassed mixture of water and acetonitrile (60:40). Make adjustments if necessary (see *System Suitability* under *Chromatography* ⟨621⟩).

Methanol-acetonitrile solution—Prepare a solution by mixing equal volumes of methanol and acetonitrile.

Standard preparation—Dissolve an accurately weighed quantity of USP Prednisolone Acetate RS in *Methanol-acetonitrile solution*, and dilute quantitatively, and stepwise if necessary, with *Methanol-acetonitrile solution* to obtain a solution having a known concentration of about 0.1 mg per mL.

Assay preparation—Transfer an accurately measured volume of Sterile Prednisolone Acetate Suspension, equivalent to about 50 mg of prednisolone acetate, to a 50-mL volumetric flask, add *Methanol-acetonitrile solution* to volume, and mix. Pipet 5 mL of this solution into a second 50-mL volumetric flask, dilute with *Methanol-acetonitrile solution* to volume, and mix.

Chromatographic system (see *Chromatography* ⟨621⟩)—The liquid chromatograph is equipped with a 254-nm detector and a 4.6-mm × 25-cm column that contains packing L1. The flow rate is about 1 mL per minute. Chromatograph the *Standard preparation*, and record the peak response as directed under *Procedure:* the capacity factor, k', is not less than 3.0, and the relative standard deviation for replicate injections is not more than 3.0%.

Procedure—Separately inject equal volumes (about 10 µL) of the *Standard preparation* and the *Assay preparation* into the chromatograph, record the chromatograms, and measure the responses for the major peaks. Calculate the quantity, in mg, of $C_{23}H_{30}O_6$ in each mL of Sterile Prednisolone Acetate Suspension taken by the formula:

$$500(C/V)(r_U/r_S),$$

in which C is the concentration, in mg per mL, of USP Prednisolone Acetate RS in the *Standard preparation*, V is the volume, in mL, of Sterile Suspension taken, and r_U and r_S are the peak responses for prednisolone acetate obtained from the *Assay preparation* and the *Standard preparation*, respectively.

Prednisolone Hemisuccinate

$C_{25}H_{32}O_8$ 460.52

Pregna-1,4-diene-3,20-dione, 21-(3-carboxy-1-oxopropoxy)-11,17-dihydroxy-, (11β)-.

11β,17,21-Trihydroxypregna-1,4-diene-3,20-dione 21-(hydrogen succinate) [*2920-86-7*].

» Prednisolone Hemisuccinate contains not less than 98.0 percent and not more than 102.0 percent of $C_{25}H_{32}O_8$, calculated on the dried basis.

Packaging and storage—Preserve in tight containers.

Reference standard—*USP Prednisolone Hemisuccinate Reference Standard*—Dry in vacuum at 65° for 3 hours before using.

Identification—

A: The infrared absorption spectrum of a potassium bromide dispersion of it, previously dried, exhibits maxima only at the same wavelengths as that of a similar preparation of USP Prednisolone Hemisuccinate RS.

B: The ultraviolet absorption spectrum of a 1 in 50,000 solution in methanol exhibits maxima and minima at the same wavelengths as that of a similar preparation of USP Prednisolone Hemisuccinate RS, concomitantly measured, and the respective absorptivities, calculated on the dried basis, at the wavelength of maximum absorbance at about 243 nm do not differ by more than 3.0%.

Specific rotation ⟨781⟩: between +99° and +104°, calculated on the dried basis, determined in a solution in dioxane containing 67 mg in each 10 mL.

Loss on drying ⟨731⟩—Dry it in vacuum at 65° for 3 hours: it loses not more than 0.5% of its weight.

Residue on ignition ⟨281⟩: negligible, from 100 mg.

Assay—

Standard preparation—Dissolve an accurately weighed quantity of USP Prednisolone Hemisuccinate RS in a mixture of equal volumes of alcohol and chloroform to obtain a solution having a known concentration of about 8 mg per mL.

Assay preparation—Weigh accurately about 80 mg of Prednisolone Hemisuccinate, previously dried, dissolve in a mixture of equal volumes of alcohol and chloroform to make 10.0 mL, and mix.

Procedure—Use a 20- × 20-cm chromatographic plate coated with a 0.25-mm layer of chromatographic silica gel mixture, which has been activated by heating at 105° for 1 hour. Using a solvent system consisting of a mixture of 5 parts of butyl alcohol, 4 parts of acetic acid, and 1 part of water, proceed as directed for *Procedure* under *Single-steroid Assay* ⟨511⟩, ending with "Centrifuge the tubes for 5 minutes," except to apply 50 µL in place of 200 µL of the solutions to the plate. Concomitantly determine the absorbances of the solutions in 1-cm cells at the wavelength of maximum absorbance at about 243 nm, with a suitable spectrophotometer, against the blank. Calculate the quantity, in mg, of $C_{25}H_{32}O_8$ in the portion of Prednisolone Hemisuccinate taken by the formula:

$$10C(A_U/A_S),$$

in which C is the concentration, in mg per mL, of USP Prednisolone Hemisuccinate RS in the *Standard preparation*, and A_U and A_S are the absorbances of the solutions from the *Assay preparation* and the *Standard preparation*, respectively.

Prednisolone Sodium Phosphate

$C_{21}H_{27}Na_2O_8P$ 484.39

Pregna-1,4-diene-3,20-dione, 11,17-dihydroxy-21-(phosphonooxy)-, disodium salt, (11β)-.

11β,17,21-Trihydroxypregna-1,4-diene-3,20-dione 21-(disodium phosphate) [*125-02-0*].

» Prednisolone Sodium Phosphate contains not less than 96.0 percent and not more than 102.0 percent of $C_{21}H_{27}Na_2O_8P$, calculated on the dried basis.

Packaging and storage—Preserve in tight containers.

Reference standard—*USP Prednisolone Reference Standard*—Dry in vacuum at 105° for 3 hours before using.

Identification—

A: Place 5 mL of the *Assay preparation*, obtained as directed in the *Assay*, in a glass-stoppered, 100-mL volumetric flask, mix with 5 mL of *Alkaline phosphatase solution*, prepared as directed in the *Assay*, and add 50 mL of methylene chloride. Insert the stopper, and allow to stand, with occasional gentle inversion (about once every 15 minutes), for 2 hours. Filter the methylene chloride layer through a dry paper, and evaporate 25 mL of the filtrate to dryness: the residue so obtained responds to *Identification test A* under *Prednisolone.*

B: The residue from the ignition of about 20 mg of it responds to the tests for *Sodium* ⟨191⟩ and for *Phosphate* ⟨191⟩.

Specific rotation ⟨781⟩—Transfer about 5 g, accurately weighed, to a 50-mL volumetric flask. Dissolve in carbon dioxide–free water, and dilute with carbon dioxide–free water to volume. Pipet 5 mL of this solution into a 50-mL volumetric flask, dilute with pH 7.0 phosphate buffer (see *Buffer Solutions* in the section, *Reagents, Indicators, and Solutions*) to volume, and mix. The specific rotation, as determined in this solution, is between +95° and +102°, calculated on the dried basis.

pH ⟨791⟩: between 7.5 and 10.5, in a solution (1 in 100).

Loss on drying ⟨731⟩—Dry it in vacuum at 105° for 5 hours: it loses not more than 5.0% of its weight.

Phosphate ions—

Standard phosphate solution—Dissolve 143.3 mg of dried monobasic potassium phosphate, KH_2PO_4, in water to make 1000.0 mL. This solution contains the equivalent of 0.10 mg of phosphate (PO_4) in each mL.

Phosphate reagent A—Dissolve 5 g of ammonium molybdate in 1 *N* sulfuric acid to make 100 mL.

Phosphate reagent B—Dissolve 350 mg of *p*-methylaminophenol sulfate in 50 mL of water, add 20 g of sodium bisulfite, mix to dissolve, and dilute with water to 100 mL.

Procedure—Dissolve about 50 mg of Prednisolone Sodium Phosphate, accurately weighed, in a mixture of 10 mL of water and 5 mL of 2 *N* sulfuric acid contained in a 25-mL volumetric flask, by warming if necessary. Add 1 mL each of *Phosphate reagent A* and *Phosphate reagent B*, dilute with water to 25 mL, mix, and allow to stand at room temperature for 30 minutes. Similarly and concomitantly, prepare a standard solution, using 5.0 mL of *Standard phosphate solution* instead of the 50 mg of the substance under test. Concomitantly determine the absorbances of both solutions in 1-cm cells at 730 nm, with a suitable spectrophotometer, using water as the blank. The absorbance of the test solution is not more than that of the standard solution. The limit is 1.0% of phosphate (PO_4).

Free prednisolone—Dissolve 50.0 mg of Prednisolone Sodium Phosphate, accurately weighed, in water to make 25.0 mL. Pipet 5 mL of the solution into a glass-stoppered, 50-mL tube, add 25.0 mL of methylene chloride, insert the stopper, mix by gentle shaking, and allow to stand until the methylene chloride layer is clear (about 20 minutes). Determine the absorbance of the methylene chloride solution in a 1-cm cell at 241 nm, with a suitable spectrophotometer, using methylene chloride as the blank. Calculate the quantity, in mg, of free prednisolone in the portion of Prednisolone Sodium Phosphate weighed by comparison with the absorbance of the untreated methylene chloride solution of USP Prednisolone RS obtained as directed in the *Assay:* not more than 0.5 mg is found (1.0%).

Selenium ⟨291⟩: 0.003%, a 200-mg test specimen being used.

Assay—

pH 9 buffer with magnesium—Mix 3.1 g of boric acid and 500 mL of water in a 1-liter volumetric flask, add 21 mL of 1 *N* sodium hydroxide and 10 mL of 0.1 *M* magnesium chloride, dilute with water to volume, and mix.

Alkaline phosphatase solution—Transfer 250 mg of alkaline phosphate enzyme to a 25-mL volumetric flask, dissolve by adding *pH 9 buffer with magnesium* to volume, and mix. Prepare this solution fresh daily.

Standard preparation—Dissolve a suitable, accurately weighed quantity of USP Prednisolone RS in methylene chloride, and dilute quantitatively and stepwise with methylene chloride to obtain a solution having a known concentration of about 16 µg per mL. Pipet 100 mL of the solution into a glass-stoppered, 100-mL cylinder, and add 1.0 mL of *Alkaline phosphatase solution* and 1.0 mL of water. Allow to stand, with occasional gentle inversion, for 2 hours.

Assay preparation—Dissolve about 100 mg of Prednisolone Sodium Phosphate, accurately weighed, in water that has been saturated with methylene chloride, to make 50.0 mL, and mix. Pipet 10 mL of this solution into a 125-mL separator, and shake with two 25-mL portions of water-washed methylene chloride, discarding the methylene chloride layers.

Procedure—Pipet 1 mL of the *Assay preparation* into a glass-stoppered, 100-mL cylinder, add 1.0 mL of *Alkaline phosphatase solution* and about 50 mL of methylene chloride, insert the stopper, and allow to stand, with occasional gentle inversion (about once every 15 minutes), for 2 hours. Add methylene chloride to volume, mix, and allow to stand until the methylene chloride layer is clear (about 20 minutes). Concomitantly and without delay, determine the absorbances of the methylene chloride solution obtained from the *Assay preparation* and the *Standard preparation* at 241 nm, with a suitable spectrophotometer, using methylene chloride as the blank. Calculate the quantity, in mg, of $C_{21}H_{27}Na_2O_8P$ in the portion of Prednisolone Sodium Phosphate taken by the formula:

$$1.344[5C(A_U/A_S)],$$

in which 1.344 is the ratio of the molecular weight of prednisolone sodium phosphate to that of prednisolone, *C* is the concentration, in µg per mL, of USP Prednisolone RS in the *Standard preparation*, and A_U and A_S are the absorbances of the solution from the *Assay preparation* and the *Standard preparation*, respectively.

Prednisolone Sodium Phosphate Injection

» Prednisolone Sodium Phosphate Injection is a sterile solution of Prednisolone Sodium Phosphate in Water for Injection. It contains not less than 90.0 percent and not more than 110.0 percent of the labeled amount of prednisolone phosphate ($C_{21}H_{29}O_8P$), present as the disodium salt.

Packaging and storage—Preserve in single-dose or in multiple-dose containers, preferably of Type I glass, protected from light.

Reference standard—*USP Prednisolone Reference Standard*—Dry in vacuum at 105° for 3 hours before using.

Identification—

A: Dissolve 65 mg of phenylhydrazine hydrochloride in 100 mL of dilute sulfuric acid (3 in 5), add 5 mL of isopropyl alcohol, and mix. Heat 5 mL of this solution with 1 mL of *Assay preparation* (obtained as directed in the *Assay*) at 70° for 2 hours: a yellow color develops.

B: It responds to *Identification test A* under *Prednisolone Sodium Phosphate.*

pH ⟨791⟩: between 7.0 and 8.0.

Particulate matter ⟨788⟩: meets the requirements under *Small-volume Injections.*

Other requirements—It meets the requirements under *Injections* ⟨1⟩.

Assay—

pH 9 buffer with magnesium—Prepare as directed in the *Assay* under *Prednisolone Sodium Phosphate.*

Alkaline phosphatase solution—Prepare as directed in the *Assay* under *Prednisolone Sodium Phosphate.*

Standard preparation—Prepare as directed in the *Assay* under *Prednisolone Sodium Phosphate.*

Assay preparation—Pipet a volume of Prednisolone Sodium Phosphate Injection, equivalent to about 100 mg of prednisolone phosphate, into a separator containing 20 mL of water. Wash the solution with two 10-mL portions of methylene chloride, and discard the washings. Transfer the aqueous layer to a 50-mL volumetric flask, dilute with water to volume, and mix.

Procedure—Proceed as directed for *Procedure* in the *Assay* under *Prednisolone Sodium Phosphate.* Calculate the quantity, in mg, of $C_{21}H_{29}O_8P$ in each mL of the Injection taken by the formula:

$$6.11(C/V)(A_U/A_S),$$

in which C is the concentration, in µg per mL, of USP Prednisolone RS in the *Standard preparation*, V is the volume, in mL, of Injection taken, and A_U and A_S are the absorbances of the solution from the *Assay preparation* and the *Standard preparation*, respectively.

Prednisolone Sodium Phosphate Ophthalmic Ointment, Neomycin Sulfate and—*see* Neomycin Sulfate and Prednisolone Sodium Phosphate Ophthalmic Ointment

Prednisolone Sodium Phosphate Ophthalmic Solution

» Prednisolone Sodium Phosphate Ophthalmic Solution is a sterile solution of Prednisolone Sodium Phosphate in a buffered, aqueous medium. It contains not less than 90.0 percent and not more than 115.0 percent of the labeled amount of prednisolone phosphate ($C_{21}H_{29}O_8P$), present as the disodium salt.

Packaging and storage—Preserve in tight, light-resistant containers.

Reference standard—*USP Prednisolone Reference Standard*—Dry in vacuum at 105° for 3 hours before using.

Identification—It responds to *Identification test A* under *Prednisolone Sodium Phosphate* and to *Identification test A* under *Prednisolone Sodium Phosphate Injection.*

Sterility—It meets the requirements under *Sterility Tests* ⟨71⟩.

pH ⟨791⟩: between 6.2 and 8.2.

Assay—Proceed with Prednisolone Sodium Phosphate Ophthalmic Solution as directed in the *Assay* under *Prednisolone Sodium Phosphate Injection.*

Prednisolone Sodium Succinate for Injection

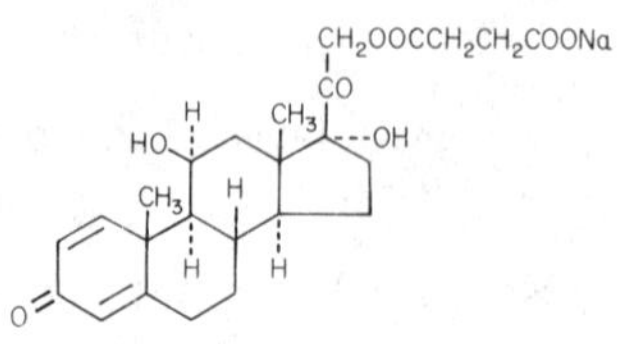

$C_{25}H_{31}NaO_8$ 482.50

Pregna-1,4-diene-3,20-dione, 21-(3-carboxy-1-oxopropoxy)-11,17-dihydroxy-, monosodium salt, (11β)-.

11β,17,21-Trihydroxypregna-1,4-diene-3,20-dione 21-(sodium succinate) [*1715-33-9*].

» Prednisolone Sodium Succinate for Injection is sterile prednisolone sodium succinate prepared from Prednisolone Succinate with the aid of Sodium Hydroxide or Sodium Carbonate. It contains the equivalent of not less than 90.0 percent and not more than 110.0 percent of the labeled amount of prednisolone ($C_{21}H_{28}O_5$). It contains suitable buffers.

Packaging and storage—Preserve in *Containers for Sterile Solids* as described under *Injections* ⟨1⟩.

Reference standard—*USP Prednisolone Succinate Reference Standard*—Dry in vacuum at 65° for 3 hours before using.

Constituted solution—At the time of use, the constituted solution prepared from Prednisolone Sodium Succinate for Injection meets the requirements for *Constituted Solutions* under *Injections* ⟨1⟩.

Identification—Place about 50 mg in a separator, add 20 mL of water and 2 mL of 3 *N* hydrochloric acid, and extract with 25 mL of chloroform. Filter the extract into a suitable beaker, evaporate on a steam bath to dryness, and dry the residue at 60° for 1 hour: the residue so obtained responds to *Identification test A* under *Prednisolone Succinate.*

pH ⟨791⟩: between 6.7 and 8.0, determined in the solution constituted as directed in the labeling.

Loss on drying ⟨731⟩—Dry it at 105° for 3 hours: it loses not more than 2.0% of its weight.

Particulate matter ⟨788⟩: meets the requirements under *Small-volume Injections.*

Other requirements—It meets the requirements for *Sterility Tests* ⟨71⟩, *Uniformity of Dosage Units* ⟨905⟩, and *Labeling* under *Injections* ⟨1⟩.

Assay—

Standard preparation—Dissolve about 64 mg of USP Prednisolone Succinate RS, accurately weighed, in about 100 mL of alcohol, add 5.0 mL of water, dilute with alcohol to 200.0 mL, and mix. Dilute 4.0 mL of this solution with alcohol to 100.0 mL, and mix. Pipet 20 mL of the resulting solution into a glass-stoppered, 50-mL conical flask.

Assay preparation—Transfer an accurately weighed portion of Prednisolone Sodium Succinate for Injection, equivalent to about 50 mg of prednisolone, to a 200-mL volumetric flask, dissolve in 5.0 mL of water, add alcohol to volume, and mix. Dilute 4.0 mL of this mixture with alcohol to 100.0 mL, and mix. Pipet 20 mL of the resulting solution into a glass-stoppered, 50-mL conical flask.

Procedure—Proceed as directed under *Assay for Steroids* ⟨351⟩. Calculate the quantity, in mg, of $C_{21}H_{28}O_5$ in the portion of Prednisolone Sodium Succinate for Injection taken by the formula:

$$5C(0.7827)(A_U/A_S),$$

in which C is the concentration, in µg per mL, of USP Prednisolone Succinate RS in the *Standard preparation*, 0.7827 is the ratio of the molecular weight of prednisolone to that of prednisolone succinate, and A_U and A_S are the absorbances of the solutions from the *Assay preparation* and the *Standard preparation*, respectively.

Prednisolone Succinate—*see* Prednisolone Hemisuccinate

Prednisolone Tebutate

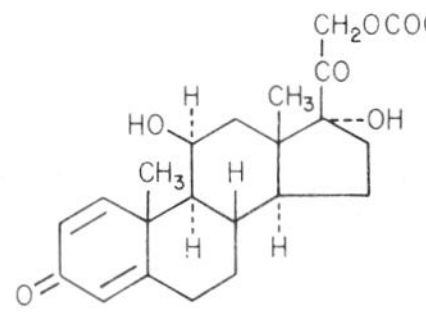

$C_{27}H_{38}O_6$ (monohydrate) 476.61

Pregna-1,4-diene-3,20-dione, 11,17-dihydroxy-21-[(3,3-dimethyl-1-oxobutyl)oxy]-, (11β)-.

11β,17,21-Trihydroxypregna-1,4-diene-3,20-dione 21-(3,3-dimethylbutyrate) [*7681-14-3*].

» Prednisolone Tebutate contains not less than 97.0 percent and not more than 103.0 percent of $C_{27}H_{38}O_6$, calculated on the dried basis.

Packaging and storage—Preserve in tight containers sealed under sterile nitrogen, in a cold place.

Reference standard—*USP Prednisolone Tebutate Reference Standard*—Dry in vacuum at a pressure of not more than 5 mm of mercury at 105° for 4 hours before using.

Identification—

A: Dissolve a portion of it in acetone, and evaporate to dryness: the infrared absorption spectrum of a mineral oil dispersion of the residue so obtained, previously dried at a pressure not exceeding 5 mm of mercury at 105° for 4 hours, exhibits maxima only at the same wavelengths as that of a similar preparation of USP Prednisolone Tebutate RS.

B: The ultraviolet absorption spectrum of a 1 in 50,000 solution in methanol exhibits maxima and minima only at the same wavelengths as that of a similar preparation of USP Prednisolone Tebutate RS, concomitantly measured, and the respective absorptivities, calculated on the dried basis, at the wavelength of maximum absorbance at about 242 nm do not differ by more than 3.0%.

Specific rotation ⟨781⟩: between +100° and +115°, calculated on the dried basis, determined in a solution in chloroform containing 100 mg in each 10 mL.

Loss on drying ⟨731⟩—Dry it at a pressure not exceeding 5 mm of mercury at 105° for 4 hours: it loses not more than 5.0% of its weight.

Residue on ignition ⟨281⟩: not more than 0.1%.

Selenium ⟨291⟩: 0.003%, a 200-mg specimen being used.

Assay—

Mobile solvent—Prepare a mixture of isooctane, tetrahydrofuran, and alcohol (89:10:1).

Standard preparation—Dissolve a suitable quantity of USP Prednisolone Tebutate RS, accurately weighed, in a mixture of tetrahydrofuran and isooctane (1:1) to obtain a solution having a known concentration of about 1 mg per mL.

Assay preparation—Weigh accurately about 50 mg of Prednisolone Tebutate, transfer to a 50-mL volumetric flask, dilute with tetrahydrofuran and isooctane (1:1) to volume, and mix.

Procedure—Inject separately 25-μL volumes of the *Standard preparation* and the *Assay preparation* into a suitable high-pressure liquid chromatograph equipped with a constant flow pump and a 30-cm × 3.9-mm column that contains packing L3, is operated at 25°, and is equipped with an ultraviolet detector capable of monitoring absorption at 254 nm. The flow rate is about 1.0 mL per minute. The retention time for Prednisolone Tebutate is about 30 minutes, but the chromatogram is run for about 45 minutes. Five replicate injections of the *Standard preparation* show a relative standard deviation of not more than 1.5%. Calculate the quantity, in mg, of $C_{27}H_{38}O_6$ in the portion of Prednisolone Tebutate taken by the formula:

$$50C(R_U/R_S),$$

in which C is the concentration, in mg per mL, of USP Prednisolone Tebutate RS in the *Standard preparation*, and R_U and R_S are the peak areas obtained from the *Assay preparation* and the *Standard preparation*, respectively.

Sterile Prednisolone Tebutate Suspension

» Sterile Prednisolone Tebutate Suspension is a sterile suspension of Prednisolone Tebutate in a suitable aqueous medium. It contains not less than 90.0 percent and not more than 110.0 percent of the labeled amount of $C_{27}H_{38}O_6$.

Packaging and storage—Preserve in single-dose or in multiple-dose containers, preferably of Type I glass.

Reference standard—*USP Prednisolone Tebutate Reference Standard*—Dry in vacuum at a pressure of not more than 5 mm of mercury at 105° for 4 hours before using.

Identification—Dilute a portion of the Sterile Suspension with methanol to obtain a solution having a concentration of about 20 μg per mL: the ultraviolet absorption spectrum of this solution exhibits maxima and minima only at the same wavelengths as that of a similar preparation of USP Prednisolone Tebutate RS, concomitantly measured.

pH ⟨791⟩: between 6.0 and 8.0.

Other requirements—It meets the requirements under *Injections* ⟨1⟩.

Assay—

Mobile solvent—Prepare a mixture of isooctane, tetrahydrofuran, and ethanol (89:10:8).

Standard preparation—Dissolve an accurately weighed quantity of USP Prednisolone Tebutate RS in a mixture of tetrahydrofuran and isooctane (1:1) to obtain a solution having a known concentration of about 1 mg per mL.

Assay preparation—Transfer to a separator an accurately measured volume, freshly mixed, of Sterile Prednisolone Tebutate Suspension, equivalent to about 100 mg of prednisolone tebutate, and dilute with about 10 mL of water. Extract with three 25-mL portions of methylene chloride, filtering each portion through methylene chloride–washed cotton into a 100-mL volumetric flask. Add methylene chloride to volume, and mix. Pipet 10 mL of this solution into a 50-mL centrifuge tube, evaporate the methylene chloride on a steam bath just to dryness, cool, and dissolve the residue in 10.0 mL of tetrahydrofuran and isooctane (1:1). Filter through a 1-μ membrane filter.

Procedure—Introduce equal volumes, about 10 μL, of the *Assay preparation* and the *Standard preparation* into a high-pressure liquid chromatograph (see *Chromatography* ⟨621⟩), operated at room temperature, by means of a suitable microsyringe or sampling valve, adjusting the sample size and other operating parameters such that the peak obtained with the *Standard preparation* is about 0.6 full-scale. Typically, the apparatus is fitted with a 3.9-mm × 30-cm column containing packing L3 and is equipped with an ultraviolet detector capable of monitoring absorption at 254 nm and a suitable recorder. In a suitable chromatogram, the coefficient of variation for five replicate injections of a single specimen is not more than 3.0%. Measure the height of the peaks, at identical retention times, obtained with the *Assay preparation* and the *Standard preparation*. Calculate the quantity, in mg, of $C_{27}H_{38}O_6$, in the volume of Suspension taken by the formula:

$$100CH_U/H_S,$$

in which C is the concentration, in mg per mL, of USP Prednisolone Tebutate RS in the *Standard preparation*, and H_U and H_S are the peak heights obtained from the *Assay preparation* and the *Standard preparation*, respectively.

Prednisone

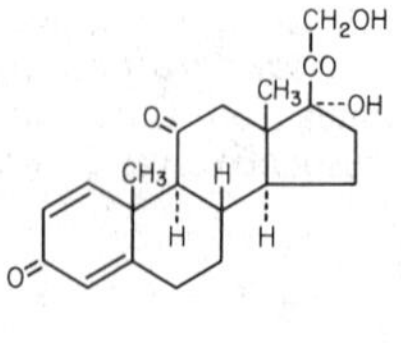

$C_{21}H_{26}O_5$ 358.43

Pregna-1,4-diene-3,11,20-trione, 17,21-dihydroxy-.

17,21-Dihydroxypregna-1,4-diene-3,11,20-trione [*53-03-2*].

» Prednisone contains not less than 97.0 percent and not more than 102.0 percent of $C_{21}H_{26}O_5$, calculated on the dried basis.

Packaging and storage—Preserve in well-closed containers.

Reference standard—*USP Prednisone Reference Standard*—Dry at 105° for 3 hours before using.

Identification—

A: The infrared absorption spectrum of a potassium bromide dispersion of it, previously dried at 105° for 30 minutes, exhibits maxima only at the same wavelengths as that of a similar preparation of USP Prednisone RS. If a difference appears, dissolve portions of both the test specimen and the Reference Standard in methanol, evaporate the solutions to dryness, and repeat the test on the residues.

B: Dissolve about 6 mg in 2 mL of sulfuric acid, and allow to stand for 5 minutes: an orange color is produced. Pour the solution into 10 mL of water: the color changes first to yellow and then, gradually, to bluish green.

Specific rotation ⟨781⟩: between +167° and +175°, calculated on the dried basis, determined in a solution in dioxane containing 50 mg in each 10 mL.

Loss on drying ⟨731⟩—Dry it at 105° for 3 hours: it loses not more than 1.0% of its weight.

Residue on ignition ⟨281⟩: negligible, from 100 mg.

Ordinary impurities ⟨466⟩—

Test solution: methanol.

Standard solution: methanol.

Eluant: a mixture of toluene, acetone, methyl ethyl ketone, and 50% formic acid (55:20:20:5), in a nonequilibrated chamber.

Visualization: 5.

Assay—

Mobile phase—Prepare a suitable filtered mixture of water, peroxide-free tetrahydrofuran, and methanol (688:250:62) such that at a flow rate of 1.0 mL per minute, the retention times of prednisone and acetanilide are about 8 and 6 minutes, respectively.

Internal standard solution—Prepare a solution of acetanilide in dilute methanol (1 in 2) having a concentration of about 110 µg per mL.

Standard preparation—Using an accurately weighed quantity of USP Prednisone RS, prepare a solution in dilute methanol (1 in 2) having a known concentration of about 0.2 mg per mL. Transfer 5.0 mL of this solution and 5.0 mL of the *Internal standard solution* to a 50-mL volumetric flask. Add dilute methanol (1 in 2) to volume, and mix to obtain a *Standard preparation* having a known concentration of about 20 µg of USP Prednisone RS per mL. Prepare this solution fresh.

Assay preparation—Using about 50 mg of Prednisone, accurately weighed, proceed as directed for *Standard preparation*, beginning with "prepare a solution in dilute methanol (1 in 2)."

Chromatographic system (see *Chromatography* ⟨621⟩)—The liquid chromatograph is equipped with a 254-nm detector and a 4-mm × 25-cm column that contains packing L1. Chromatograph five replicate injections of the *Standard preparation*, and record the peak responses as directed under *Procedure:* the relative standard deviation is not more than 2.0%, and the resolution factor between prednisone and the internal standard is not less than 3. Adjust the operating parameters so that the peak obtained from the *Standard preparation* is about one-half full-scale.

Procedure—Separately inject equal volumes (about 10 µL) of the *Standard preparation* and the *Assay preparation* into the chromatograph by means of a suitable microsyringe or sampling valve, record the chromatograms, and measure the responses at equivalent retention times. Calculate the quantity, in mg, of $C_{21}H_{26}O_5$ in the portion of Prednisone taken by the formula:

$$2.5C(R_U/R_S),$$

in which C is the concentration, in µg per mL, of USP Prednisone RS in the *Standard preparation*, and R_U and R_S are the peak response ratios of the prednisone peak to the internal standard peak obtained from the *Assay preparation* and the *Standard preparation*, respectively.

Prednisone Oral Solution

» Prednisone Oral Solution contains not less than 90.0 percent and not more than 110.0 percent of the labeled amount of $C_{21}H_{26}O_5$.

Packaging and storage—Preserve in tight containers.

Reference standard—*USP Prednisone Reference Standard*—Dry at 105° for 3 hours before using.

Identification—Shake 50 mL of Oral Solution with 25 mL of chloroform for 5 minutes. Filter the chloroform extract through a pledget of cotton and a layer of anhydrous sodium sulfate, and evaporate on a warm water bath with the aid of a current of air to about 3 mL. Continue the evaporation at room temperature to dryness. Wash the residue with two 10-mL portions of hot solvent hexane, decanting the solvent and discarding it each time. Digest the residue with 25 mL of warm dehydrated alcohol for 15 minutes. Filter the mixture, and evaporate the filtrate to about 3 mL. Add solvent hexane until the mixture becomes slightly cloudy, and chill in a freezer to promote the formation of crystals. Collect the crystals, and dry at 60° for 1 hour: the crystals respond to *Identification test A* under *Prednisone.*

pH ⟨791⟩: between 2.6 and 4.0.

Alcohol content, *Method II* ⟨611⟩: between 4.0% and 6.0% of C_2H_5OH.

Assay—

Mobile phase—Dissolve 1.36 g of monobasic potassium phosphate in 600 mL of water, add 400 mL of methanol, filter through a 0.2-µm membrane filter, and degas. Make adjustments if necessary (see *System Suitability* under *Chromatography* ⟨621⟩).

Standard preparation—Using an accurately weighed quantity of USP Prednisone RS, prepare a solution in alcohol containing 1 mg per mL. Dilute 4 volumes of this solution quantitatively with 96 volumes of water to obtain a *Standard preparation* having a known concentration of about 40 µg per mL.

Assay preparation—Transfer an accurately measured volume of Prednisone Oral Solution, equivalent to about 10 mg of prednisone, to a 250-mL volumetric flask. Dilute with water to volume, and mix.

Chromatographic system (see *Chromatography* ⟨621⟩)—The liquid chromatograph is equipped with a 254-nm detector and a 3.9-mm × 30-cm column that contains packing L1. The flow rate is about 1.5 mL per minute. Chromatograph the *Standard preparation*, and record the peak responses as directed under *Procedure:* the tailing factor for the analyte peak is not more than 2.0, and the relative standard deviation for replicate injections is not more than 2.0%.

Procedure—Separately inject equal volumes (about 10 µL) of the *Standard preparation* and the *Assay preparation* into the chromatograph by means of a sampling valve, record the chromatograms, and measure the responses for the major peaks. Calculate the quantity, in mg, of $C_{21}H_{26}O_5$ in each mL of the Oral Solution taken by the formula:

$$(0.25C/V)(r_U/r_S),$$

in which C is the concentration, in µg per mL, of USP Prednisone RS in the *Standard preparation*, V is the volume, in mL, of Oral Solution taken, and r_U and r_S are the peak responses obtained from the *Assay preparation* and the *Standard preparation*, respectively.

Prednisone Syrup

» Prednisone Syrup contains not less than 90.0 percent and not more than 110.0 percent of the labeled amount of $C_{21}H_{26}O_5$.

Packaging and storage—Preserve in tight containers.

Reference standard—*USP Prednisone Reference Standard*—Dry at 105° for 3 hours before using.

Identification—Transfer about 50 mL of Prednisone Syrup to a separator, and extract with 25 mL of chloroform. Allow the phases to separate, and filter the mixture. Evaporate the chloroform extract on a steam bath to a volume of 2 to 3 mL, then evaporate with the aid only of a current of air to dryness. Wash the residue with two 10-mL portions of hot solvent hexane, decanting the supernatant liquid each time and discarding it. Digest the residue with 25 mL of dehydrated alcohol, warming slightly for 15 minutes. Filter the warm solution, and evaporate the filtrate to a volume of 2 mL to 3 mL. Add solvent hexane until the mixture just becomes turbid, chill it to effect crystallization, collect the crystals, and dry them at 60° for 1 hour: the crystals respond to *Identification test A* under *Prednisone*.

Specific gravity ⟨841⟩: between 1.220 and 1.280 at 25°.

pH ⟨791⟩: between 3.0 and 4.5.

Alcohol content, *Method II* ⟨611⟩: between 2.0% and 5.0% of C_2H_5OH.

Assay—

Mobile phase—Dissolve 1.3609 g of monobasic potassium phosphate in 600 mL of water, add 400 mL of methanol, filter through a 0.2-µm membrane filter, and degas.

Standard preparation—Using an accurately weighed quantity of USP Prednisone RS, prepare a solution in alcohol containing 1 mg per mL. Dilute 1 volume of this solution quantitatively with 49 volumes of water to obtain a *Standard preparation* having a known concentration of about 20 µg per mL.

Assay preparation—Transfer an accurately measured volume of Prednisone Syrup, equivalent to about 5 mg of prednisone, to a 250-mL volumetric flask. Dilute with water to volume, and mix.

Chromatographic system (see *Chromatography* ⟨621⟩)—The liquid chromatograph is equipped with a 254-nm detector and a 4.6-mm × 25-cm column that contains packing L1. The flow rate is about 1.5 mL per minute. Chromatograph the *Standard preparation*, and record the peak responses as directed under *Procedure:* the relative standard deviation for replicate injections is not more than 2.0%.

Procedure—Separately inject equal volumes (about 20 µL) of the *Standard preparation* and the *Assay preparation* into the chromatograph by means of a sampling valve, record the chromatograms, and measure the responses for the major peaks. Calculate the quantity, in mg, of $C_{21}H_{26}O_5$, in each mL of the Syrup taken by the formula:

$$(0.25C/V)(r_U/r_S),$$

in which C is the concentration, in µg per mL, of USP Prednisone RS in the *Standard preparation*, V is the volume, in mL, of Syrup taken, and r_U and r_S are the peak responses obtained from the *Assay preparation* and the *Standard preparation*, respectively.

Prednisone Tablets

» Prednisone Tablets contain not less than 90.0 percent and not more than 110.0 percent of the labeled amount of $C_{21}H_{26}O_5$.

Packaging and storage—Preserve in well-closed containers.

Reference standard—*USP Prednisone Reference Standard*—Dry at 105° for 3 hours before using.

Identification—Into a 50-mL beaker containing a portion of pulverized Tablets, equivalent to about 10 mg of prednisone, add 10 mL of water, and mix to form a slurry. Transfer the slurry to a 3-cm × 13-cm column packed with diatomaceous earth, and allow to be absorbed for a period of 10 minutes. Elute the column with 60 mL of water-washed ether, and evaporate the eluate on a steam bath to dryness. Dry the residue at 105° for 30 minutes: the crystals respond to *Identification tests A* and *B* under *Prednisone*.

Dissolution ⟨711⟩—

Medium: water; use 500 mL of the *Dissolution Medium* for Tablets labeled to contain 10 mg of prednisone or less, and 900 mL for Tablets labeled to contain more than 10 mg of prednisone.

Apparatus 2: 50 rpm.

Time: 30 minutes.

Procedure—Determine the amount of $C_{21}H_{26}O_5$ dissolved from ultraviolet absorbances at the wavelength of maximum absorbance at about 242 nm of filtered portions of the solution under test, suitably diluted with *Dissolution Medium*, if necessary, in comparison with a Standard solution having a known concentration of USP Prednisone RS in the same medium. An amount of alcohol not to exceed 5% of the total volume of the Standard solution may be used to bring the prednisone standard into solution prior to dilution with water.

Tolerances—Not less than 80% (Q) of the labeled amount of $C_{21}H_{26}O_5$ is dissolved in 30 minutes.

Uniformity of dosage units ⟨905⟩: meet the requirements.

Procedure for content uniformity—

Mobile phase, Internal standard solution, Standard preparation, and *Chromatographic system*—Proceed as directed in the *Assay* under *Prednisone*.

Test preparation—Place 1 Tablet in a volumetric flask of such size that when the contents are diluted to volume the resulting solution has a concentration of about 0.2 mg of prednisone per mL. Add 5 mL of water, swirl, sonicate for 1 minute, add a volume of methanol equal to one-half the volume of the volumetric flask, and sonicate again for 1 minute. Dilute with water to volume, and mix. Transfer 5.0 mL of this solution and 5.0 mL of the *Internal standard solution* to a 50-mL volumetric flask, add dilute methanol (1 in 2) to volume, and mix. Filter through a 5-µm filter, discarding the first 20 mL of the filtrate.

Procedure—Proceed as directed for *Procedure* in the *Assay* under *Prednisone*, except to calculate the quantity, in mg, of $C_{21}H_{26}O_5$ in the Tablet by the formula:

$$DC(R_U/R_S),$$

in which D is the dilution factor for the *Test preparation* and the other terms are as defined therein.

Assay—

Mobile phase, Internal standard solution, Standard preparation, and *Chromatographic system*—Proceed as directed in the *Assay* under *Prednisone*.

Assay preparation—Weigh and finely powder not less than 20 Prednisone Tablets. Transfer an accurately weighed portion of the powder, equivalent to about 20 mg of prednisone, to a 100-mL volumetric flask. Add 5 mL of water, sonicate for 1 minute, add 50 mL of methanol, and sonicate again for 1 minute. Dilute with water to volume, and mix. Transfer 5.0 mL of this solution and 5.0 mL of the *Internal standard solution* to a 50-mL volumetric flask, add dilute methanol (1 in 2) to volume, and mix. Filter through a 5-µm filter, discarding the first 20 mL of the filtrate.

Procedure—Proceed as directed for *Procedure* in the *Assay* under *Prednisone*, except to calculate the quantity, in mg, of $C_{21}H_{26}O_5$ in the portion of Tablets taken by the formula:

$$C(R_U/R_S),$$

in which the terms are as defined therein.

Pregelatinized Starch—*see* Starch, Pregelatinized NF

Preparation, Vitamin E—*see* Vitamin E Preparation

Prilocaine Hydrochloride

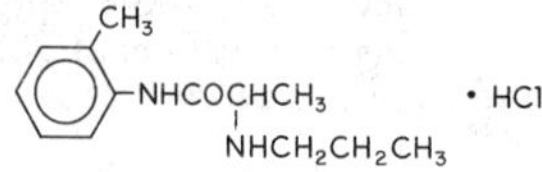

$C_{13}H_{20}N_2O.HCl$ 256.77
Propanamide, *N*-(2-methylphenyl)-2-(propylamino)-, monohydrochloride.
2-(Propylamino)-*o*-propionotoluidide monohydrochloride [*1786-81-8*].

» Prilocaine Hydrochloride contains not less than 99.0 percent and not more than 101.0 percent of $C_{13}H_{20}N_2O.HCl$, calculated on the dried basis.

Packaging and storage—Preserve in well-closed containers.

Reference standard—*USP Prilocaine Hydrochloride Reference Standard*—Dry at 105° for 4 hours before using.

Identification—

A: The infrared absorption spectrum of a potassium bromide dispersion of it, previously dried, exhibits maxima only at the same wavelengths as that of a similar preparation of USP Prilocaine Hydrochloride RS.

B: Dissolve about 300 mg in 5 mL of water, add 4 mL of 6 *N* ammonium hydroxide, and extract with 50 mL of chloroform. Filter the extract, and evaporate the filtrate on a steam bath with the aid of a current of air. Dissolve about 100 mg of the prilocaine so obtained in 1 mL of alcohol, add 10 drops of cobaltous chloride TS, and shake for 2 minutes: a bright green color develops, and a precipitate is formed.

C: Dissolve about 100 mg in 3 mL of water, render the solution alkaline with 6 *N* ammonium hydroxide, and filter: the filtrate responds to the tests for *Chloride* ⟨191⟩.

Melting range, *Class I* ⟨741⟩: between 166° and 169°.

Loss on drying ⟨731⟩—Dry it at 105° for 4 hours: it loses not more than 0.3% of its weight.

Residue on ignition ⟨281⟩: not more than 0.1%.

Heavy metals, *Method I* ⟨231⟩: 0.002%.

Assay—Dissolve about 500 mg of Prilocaine Hydrochloride, accurately weighed, in 50 mL of glacial acetic acid, add 10 mL of mercuric acetate TS and 2 drops of crystal violet TS, and titrate with 0.1 *N* perchloric acid VS to a blue-green end-point. Perform a blank determination, and make any necessary correction. Each mL of 0.1 *N* perchloric acid is equivalent to 25.68 mg of $C_{13}H_{20}N_2O.HCl$.

Prilocaine Hydrochloride Injection

» Prilocaine Hydrochloride Injection is a sterile solution of Prilocaine Hydrochloride in Water for Injection. It contains not less than 95.0 percent and not more than 105.0 percent of the labeled amount of $C_{13}H_{20}N_2O.HCl$.

Packaging and storage—Preserve in single-dose or in multiple-dose containers, preferably of Type I glass.

Reference standard—*USP Prilocaine Hydrochloride Reference Standard*—Dry at 105° for 4 hours before using.

Identification—

A: It meets the requirements under *Identification—Organic Nitrogenous Bases* ⟨181⟩.

B: It responds to *Identification test B* under *Prilocaine Hydrochloride.*

pH ⟨791⟩: between 6.0 and 7.0.

Other requirements—It meets the requirements under *Injections* ⟨1⟩.

Assay—

Mobile phase—Mix 50 mL of glacial acetic acid and 930 mL of water, and adjust with 1 *N* sodium hydroxide to a pH of 3.40. Mix about 4 volumes of this solution with 1 volume of acetonitrile, such that the retention time of prilocaine is about 4 to 6 minutes. Filter through a membrane filter (1-µm or finer porosity), and degas. Make adjustments if necessary (see *System Suitability* under *Chromatography* ⟨621⟩).

Standard preparation—Dissolve an accurately weighed quantity of USP Prilocaine Hydrochloride RS quantitatively in *Mobile phase* to obtain a solution having a known concentration of about 4 mg per mL.

Assay preparation—Transfer an accurately measured volume of Prilocaine Hydrochloride Injection, equivalent to about 200 mg of prilocaine hydrochloride, to a 50-mL volumetric flask, dilute with *Mobile phase* to volume, and mix.

Resolution preparation—Prepare a solution of procainamide hydrochloride in *Mobile phase* containing about 900 µg per mL. Mix 2 mL of this solution and 20 mL of *Standard preparation.*

Chromatographic system (see *Chromatography* ⟨621⟩)—The liquid chromatograph is equipped with a 254-nm detector and a 3.9-mm × 30-cm column that contains packing L1, and is operated at a temperature between 20° and 25° maintained at ±1.0° of the selected temperature. The flow rate is about 1.5 mL per minute. Chromatograph the *Standard preparation*, and record the peak responses as directed under *Procedure:* the relative standard deviation for replicate injections is not more than 1.5%. Chromatograph about 10 µL of the *Resolution preparation*, and record the peak responses as directed under *Procedure:* the resolution, *R*, between the prilocaine and procainamide peaks is not less than 2.0.

Procedure—[NOTE—Use peak areas where peak responses are indicated.] Separately inject equal volumes (about 10 µL) of the *Standard preparation* and the *Assay preparation* into the liquid chromatograph, record the chromatograms, and measure the responses for the major peaks. Calculate the quantity, in mg, of $C_{13}H_{20}N_2O.HCl$ in each mL of the Injection taken by the formula:

$$(50)(C/V)(r_U/r_S),$$

in which C is the concentration, in mg per mL, of USP Prilocaine Hydrochloride RS in the *Standard preparation*, V is the volume, in mL, of Injection taken, and r_U and r_S are the peak responses obtained from the *Assay preparation* and the *Standard preparation*, respectively.

Prilocaine and Epinephrine Injection

» Prilocaine and Epinephrine Injection is a sterile solution prepared from Prilocaine Hydrochloride and Epinephrine with the aid of Hydrochloric Acid in Water for Injection, or a sterile solution of Prilocaine Hydrochloride and Epinephrine Bitartrate in Water for Injection. The content of epinephrine does not exceed 0.002 percent (1 in 50,000). Prilocaine and Epinephrine Injection contains the equivalent of not less than 95.0 percent and not more than 105.0 percent of the labeled amount of prilocaine hydrochloride ($C_{13}H_{20}N_2O.HCl$) and the equivalent of not less than 90.0 percent and not more than 115.0 percent of the labeled amount of epinephrine ($C_9H_{13}NO_3$).

Packaging and storage—Preserve in single-dose or in multiple-dose, light-resistant containers, preferably of Type I glass.

Reference standards—*USP Prilocaine Hydrochloride Reference Standard*—Dry at 105° for 4 hours before using. *USP Epinephrine Bitartrate Reference Standard*—Dry in vacuum over silica gel for 3 hours before using.

Identification—

A: It responds to *Identification test B* under *Prilocaine Hydrochloride.*

B: The chromatogram of the *Assay preparation* obtained as directed in the *Assay for prilocaine hydrochloride* exhibits a major peak for prilocaine, the retention time of which corresponds to that exhibited in the chromatogram of the *Standard preparation* obtained as directed in the *Assay for prilocaine hydrochloride.*

C: The chromatogram of the *Assay preparation* obtained as directed in the *Assay for epinephrine* exhibits a major peak for epinephrine, the retention time of which corresponds to that exhibited in the chromatogram of the *Standard preparation* obtained as directed in the *Assay for epinephrine.*

pH ⟨791⟩: between 3.3 and 5.5.

Other requirements—It meets the requirements under *Injections* ⟨1⟩.

Assay for prilocaine hydrochloride—

Mobile phase, Standard preparation, Resolution preparation, and *Chromatographic system*—Proceed as directed in the *Assay* under *Prilocaine Hydrochloride Injection.*

Assay preparation—Transfer an accurately measured volume of Prilocaine and Epinephrine Injection, equivalent to about 200 mg of prilocaine hydrochloride, to a 50-mL volumetric flask, dilute with *Mobile phase* to volume, and mix.

Procedure—[NOTE—Use peak areas where peak responses are indicated.] Separately inject equal volumes (about 10 µL) of the *Assay preparation* and the *Standard preparation* into the chromatograph, record the chromatograms, and measure the responses for the major peaks. Calculate the quantity, in mg, of prilocaine hydrochloride ($C_{13}H_{20}N_2O.HCl$) in each mL of the Injection taken by the formula:

$$(50)(C/V)(r_U/r_S),$$

in which C is the concentration, in mg per mL, of USP Prilocaine Hydrochloride RS in the *Standard preparation,* V is the volume, in mL, of Injection taken, and r_U and r_S are the prilocaine peak responses obtained from the *Assay preparation* and the *Standard preparation,* respectively.

Assay for epinephrine—

Mobile phase—Mix 50 mL of glacial acetic acid and 930 mL of water, and adjust with 1 *N* sodium hydroxide to a pH of 3.40. Dissolve 1.1 g of sodium 1-heptanesulfonate in this solution, add 1.0 mL of 0.1 *M* disodium ethylenediaminetetraacetate, and mix. Mix about 9 volumes of this solution with 1 volume of methanol. Filter through a membrane filter (1-µm or finer porosity), and degas. Make adjustments if necessary (see *System Suitability* under *Chromatography* ⟨621⟩).

Standard preparation—Dissolve an accurately weighed quantity of USP Epinephrine Bitartrate RS in *Mobile phase* to obtain a solution having a known concentration of about 9 µg of epinephrine bitartrate per mL. Pipet 10 mL of this solution into a 50-mL volumetric flask, dilute with *Mobile phase* to volume, and mix to obtain a *Standard preparation* having a known concentration of about 1.8 µg of epinephrine bitartrate per mL.

Assay preparation—Transfer an accurately measured volume of Prilocaine and Epinephrine Injection, equivalent to about 50 µg of epinephrine, to a 50-mL volumetric flask, dilute with *Mobile phase* to volume, and mix.

Chromatographic system (see *Chromatography* ⟨621⟩)—The liquid chromatograph is fitted with a 30-cm × 3.9-mm stainless steel column packed with packing L1, and is equipped with an electrochemical detector held at a potential of +650 mV, a controller capable of regulating the background current, and a suitable recorder, and it is operated at a temperature between 20° and 25° maintained at ±1.0° of the selected temperature. The flow rate is about 1 mL per minute. Chromatograph the *Standard preparation* as directed under *Procedure:* the relative standard deviation of the peak responses of successive injections of the *Standard preparation* is not more than 1.5%.

Procedure—Separately inject equal volumes (about 20 µL) of the *Assay preparation* and the *Standard preparation* into the chromatograph, record the chromatograms, and measure the responses for the major peaks. Calculate the quantity, in µg, of epinephrine ($C_9H_{13}NO_3$) in each mL of the Injection taken by the formula:

$$(183.21/333.29)(50)(C/V)(r_U/r_S),$$

in which 183.21 and 333.29 are the molecular weights of epinephrine and epinephrine bitartrate, respectively, C is the concentration, in µg per mL, of USP Epinephrine Bitartrate RS in the *Standard preparation,* V is the volume, in mL, of Injection taken, and r_U and r_S are the peak responses obtained from the *Assay preparation* and the *Standard preparation,* respectively.

Primaquine Phosphate

CH(CH₃)—CH₂CH₂CH₂—NH₂ / NH / CH₃O · 2H₃PO₄

$C_{15}H_{21}N_3O.2H_3PO_4$ 455.34

1,4-Pentanediamine, N^4-(6-methoxy-8-quinolinyl)-, phosphate (1:2).

8-[(4-Amino-1-methylbutyl)amino]-6-methoxyquinoline phosphate (1:2) [*63-45-6*].

» Primaquine Phosphate contains not less than 98.0 percent and not more than 102.0 percent of $C_{15}H_{21}N_3O.2H_3PO_4$, calculated on the dried basis.

Packaging and storage—Preserve in well-closed, light-resistant containers.

Reference standard—*USP Primaquine Phosphate Reference Standard*—Dry at 105° for 2 hours before using.

Identification—

A: The infrared absorption spectrum of a potassium bromide dispersion of it, previously dried, exhibits maxima only at the same wavelengths as that of a similar preparation of USP Primaquine Phosphate RS.

B: The residue obtained by ignition responds to the test for pyrophosphate as described under *Phosphate* ⟨191⟩.

Loss on drying ⟨731⟩—Dry it at 105° for 2 hours: it loses not more than 1.0% of its weight.

Assay—Dissolve about 700 mg of Primaquine Phosphate, accurately weighed, in about 75 mL of water in a beaker, add 10 mL of hydrochloric acid, and proceed as directed under *Nitrite Titration* ⟨451⟩, beginning with "cool to about 15°." Each mL of 0.1 *M* sodium nitrite is equivalent to 45.53 mg of $C_{15}H_{21}N_3O.2H_3PO_4$.

Primaquine Phosphate Tablets

» Primaquine Phosphate Tablets contain not less than 93.0 percent and not more than 107.0 percent of the labeled amount of $C_{15}H_{21}N_3O.2H_3PO_4$.

Packaging and storage—Preserve in well-closed, light-resistant containers.

Reference standard—*USP Primaquine Phosphate Reference Standard*—Dry at 105° for 2 hours before using.

Identification—Digest a quantity of finely powdered Tablets, equivalent to about 25 mg of primaquine phosphate, with 10 mL of water for 15 minutes, and filter.

A: Dilute 0.1 mL of the filtrate with 1 mL of water, and add 1 drop of gold chloride TS: a violet-blue color is produced at once.

B: To the remainder of the filtrate add 5 mL of trinitrophenol TS: a yellow precipitate is formed. Wash the precipitate with cold water, and dry at 105° for 2 hours: the picrate melts between 208° and 215°. [*Caution—Picrates may explode.*]

Dissolution ⟨711⟩—

Medium: simulated gastric fluid TS; 900 mL.

Apparatus 2: 100 rpm.

Time: 60 minutes.

Procedure—

1-Pentanesulfonate sodium solution—Add about 961 mg of sodium 1-pentanesulfonate and 1 mL of glacial acetic acid to 400 mL of water, and mix.

Mobile phase—Prepare a filtered and degassed mixture of methanol and *1-Pentanesulfonate sodium solution* (60:40). Make adjustments if necessary (see *System Suitability* under *Chromatography* ⟨621⟩).

Chromatographic system (see *Chromatography* ⟨621⟩)—The liquid chromatograph is equipped with a 254-nm detector and a 3.9-mm × 30-cm column that contains packing L1. The flow rate is about 2 mL per minute. Chromatograph replicate injections of the Standard solution, and record the peak responses as directed under *Procedure:* the relative standard deviation is not more than 3.0%.

Procedure—Separately inject into the chromatograph equal volumes (about 20 µL) of the solution under test and a Standard solution having a known concentration of USP Primaquine Phosphate RS in the same medium, and record the chromatograms. Measure the responses for the major peaks, and calculate the amount of $C_{15}H_{21}N_3O.2H_3PO_4$ dissolved.

Tolerances—Not less than 75% (*Q*) of the labeled amount of $C_{15}H_{21}N_3O.2H_3PO_4$ is dissolved in 60 minutes.

Uniformity of dosage units ⟨905⟩: meet the requirements.

Procedure for content uniformity—Transfer 1 Tablet, previously crushed or finely powdered, to a beaker, add 5 mL of hydrochloric acid and about 25 g of crushed ice, then add water to bring the total volume to about 50 mL. Proceed as directed under *Nitrite Titration* ⟨451⟩, beginning with "slowly titrate," and using as the titrant 0.01 *M* sodium nitrite VS, freshly prepared from 0.1 *M* sodium nitrite. Concomitantly perform a blank titration, and make any necessary correction. Each mL of 0.01 *M* sodium nitrite is equivalent to 4.553 mg of $C_{15}H_{21}N_3O.2H_3PO_4$.

Assay—Weigh and finely powder not less than 30 Primaquine Phosphate Tablets. Weigh accurately a portion of the powder, equivalent to about 700 mg of primaquine phosphate, and transfer to a beaker. Add 50 mL of water and sufficient hydrochloric acid to provide about 5 mL in excess, and proceed as directed under *Nitrite Titration* ⟨451⟩, beginning with "cool to about 15°." Each mL of 0.1 *M* sodium nitrite is equivalent to 45.53 mg of $C_{15}H_{21}N_3O.2H_3PO_4$.

Primidone

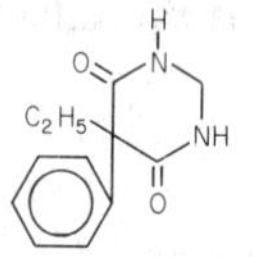

$C_{12}H_{14}N_2O_2$ 218.25

4,6(1*H*,5*H*)-Pyrimidinedione, 5-ethyldihydro-5-phenyl-.
5-Ethyldihydro-5-phenyl-4,6(1*H*,5*H*)-pyrimidinedione [*125-33-7*].

» Primidone contains not less than 98.0 percent and not more than 102.0 percent of $C_{12}H_{14}N_2O_2$, calculated on the dried basis.

Packaging and storage—Preserve in well-closed containers.

Reference standard—*USP Primidone Reference Standard*—Dry at 105° for 2 hours before using.

Identification—

A: The infrared absorption spectrum of a potassium bromide dispersion of it, previously dried, exhibits absorption maxima only at the same wavelengths as that of a similar preparation of USP Primidone RS. If a difference appears, dissolve portions of both the test specimen and the Reference Standard in alcohol, evaporate the solutions to dryness, and repeat the test on the residues.

B: The ultraviolet absorption spectrum of the solution employed for measurement of absorbance in the *Assay* exhibits maxima and minima at the same wavelengths as that of a similar solution of USP Primidone RS, concomitantly measured, and the respective absorptivities at the wavelength of maximum absorbance at about 257 nm do not differ by more than 3.0%.

C: Fuse 0.20 g with 0.20 g of anhydrous sodium carbonate: ammonia is evolved.

Melting range ⟨741⟩: between 279° and 284°.

Loss on drying ⟨731⟩—Dry it at 105° for 2 hours: it loses not more than 0.5% of its weight.

Residue on ignition ⟨281⟩: not more than 0.2%.

Ordinary impurities ⟨466⟩—

Test solution—Prepare a solution in methanol having a known concentration of 2 mg per mL.

Standard solutions—Prepare solutions in methanol having known concentrations of 2 µg per mL, 10 µg per mL, 20 µg per mL, and 40 µg per mL.

Eluant: a mixture of butyl alcohol, glacial acetic acid, and water (5:3:2).

Visualization—Expose the plate to chlorine gas for about 15 minutes, air-dry until the chlorine has dissipated (about 15 minutes), and follow with visualization technique 20.

Assay—Transfer about 40 mg of Primidone, accurately weighed, to a 100-mL volumetric flask. Add 70 mL of alcohol, and boil gently to dissolve. Cool, and add alcohol to volume. Determine the absorbances in a 2-cm cell, with a suitable spectrophotometer, using alcohol as the blank, at the minima that occur at about 254 nm and at about 261 nm, and the maximum at about 257 nm. Concomitantly determine the absorbances of a Standard solution in alcohol of USP Primidone RS, similarly prepared to have a known concentration of about 400 µg per mL. Calculate the quantity, in mg, of $C_{12}H_{14}N_2O_2$ in the Primidone taken by the formula:

$$0.1C(2A_{257} - A_{254} - A_{261})_U/(2A_{257} - A_{254} - A_{261})_S,$$

in which *C* is the concentration, in µg per mL, of USP Primidone RS in the Standard solution, and the parenthetic expressions are the differences in the absorbances of the two solutions at the wavelengths indicated by the subscripts for the solution of Primidone ($_U$) and the Standard solution ($_S$), respectively.

Primidone Oral Suspension

» Primidone Oral Suspension is a suspension of Primidone in a suitable aqueous vehicle. It contains, in each 100 mL, not less than 4.5 g and not more than 5.5 g of $C_{12}H_{14}N_2O_2$.

Packaging and storage—Preserve in tight, light-resistant containers.

Reference standard—*USP Primidone Reference Standard*—Dry at 105° for 2 hours before using.

Identification—The retention time of the major peak in the chromatogram of the *Assay preparation* corresponds to that of the *Standard preparation*, both relative to the internal standard, obtained as directed in the *Assay*.

pH ⟨791⟩: between 5.5 and 8.5.

Assay—

Internal standard solution—Dissolve a suitable quantity of androsterone in alcohol to obtain a solution having a final concentration of about 10 mg per mL.

Standard preparation—Transfer about 100 mg of USP Primidone RS, accurately weighed, to a 100-mL volumetric flask, add 65 mL of alcohol, and boil for 1 hour. Allow to cool to ambient temperature, add 10.0 mL of *Internal standard solution*, dilute with alcohol to volume, mix, and filter.

Assay preparation—Transfer an accurately weighed quantity of well-mixed Primidone Oral Suspension, equivalent to about 50 mg of primidone, to a 50-mL volumetric flask, add 35 mL of alcohol, and boil for 1 hour. Allow to cool to ambient temperature, add 5.0 mL of *Internal standard solution*, dilute with alcohol to volume, mix, and filter.

Chromatographic system (see *Chromatography* ⟨621⟩)—The gas chromatograph is equipped with a flame-ionization detector and a 4.0-mm × 120-cm column packed with 10 percent liquid phase G3 on support S1AB. Helium is used as the carrier gas at a flow rate of about 40 mL per minute. The detector and injector temperatures are maintained at about 310° and the column temperature is maintained at about 260°. Chromatograph three replicate injections of the *Standard preparation*, and record the peak responses as directed under *Procedure:* the relative standard deviation is not more than 2.0%, and the resolution factor between primidone and androsterone is not less than 1.5.

Procedure—Separately inject equal volumes (about 3 µL) of the *Standard preparation* and the *Assay preparation* into the chromatograph, record the chromatograms, and measure the responses for the major peaks. The relative retention times are about 0.8 for primidone and 1.0 for androsterone. Calculate the quantity, in mg, of $C_{12}H_{14}N_2O_2$ in each mL of the Oral Suspension taken by the formula:

$$0.5D(W_S/W_U)(R_U/R_S),$$

in which D is the density, in g per mL, of the Oral Suspension, W_S is the weight, in mg, of the Standard used, W_U is the weight, in mg, of Oral Suspension taken, and R_U and R_S are the relative response factors obtained from the *Assay preparation* and the *Standard preparation*, respectively.

Primidone Tablets

» Primidone Tablets contain not less than 95.0 percent and not more than 105.0 percent of the labeled amount of $C_{12}H_{14}N_2O_2$.

Packaging and storage—Preserve in well-closed containers.

Reference standard—*USP Primidone Reference Standard*—Dry at 105° for 2 hours before using.

Identification—The retention time of the major peak in the chromatogram of the *Assay preparation* corresponds to that of the *Standard preparation*, both relative to the internal standard, obtained as directed in the *Assay*.

Dissolution ⟨711⟩—

Medium: water; 900 mL.

Apparatus 2: 50 rpm.

Time: 60 minutes.

Procedure—Determine the amount of $C_{12}H_{14}N_2O_2$ dissolved from ultraviolet absorbances at the wavelength of maximum absorbance at about 257 nm of filtered portions of the solution under test, suitably diluted with *Dissolution Medium*, if necessary, in comparison with a Standard solution having a known concentration of USP Primidone RS in the same medium.

Tolerances—Not less than 75% (Q) of the labeled amount of $C_{12}H_{14}N_2O_2$ is dissolved in 60 minutes.

Uniformity of dosage units ⟨905⟩: meet the requirements.

Assay—

Internal standard solution—Dissolve a suitable quantity of androsterone in alcohol to obtain a solution having a final concentration of about 10 mg per mL.

Standard preparation—Transfer about 100 mg of USP Primidone RS, accurately weighed, to a 100-mL volumetric flask, add 65 mL of alcohol, and boil for one hour. Allow to cool to ambient temperature, add 10.0 mL of *Internal standard solution*, dilute with alcohol to volume, mix, and filter.

Assay preparation—Weigh and finely powder not less than 20 Primidone Tablets. Transfer an accurately weighed portion of the powder, equivalent to about 50 mg of primidone, to a 50-mL volumetric flask, add 35 mL of alcohol, and boil for one hour. Allow to cool to ambient temperature, add 5.0 mL of *Internal standard solution*, dilute with alcohol to volume, mix, and filter.

Chromatographic system (see *Chromatography* ⟨621⟩)—The gas chromatograph is equipped with a flame-ionization detector and a 4.0-mm × 120-cm column packed with 10 percent liquid phase G3 on support S1AB. Helium is used as the carrier gas at a flow rate of about 40 mL per minute. The detector and injector temperatures are maintained at about 310° and the column temperature is maintained at about 260°. Chromatograph three replicate injections of the *Standard preparation*, and record the peak responses as directed under *Procedure:* the relative standard deviation is not more than 2.0%, and the resolution factor between primidone and androsterone is not less than 1.5.

Procedure—Separately inject equal volumes (about 3 µL) of the *Standard preparation* and the *Assay preparation* into the chromatograph, record the chromatograms, and measure the responses for the major peaks. The relative retention times are about 0.8 for primidone and 1.0 for androsterone. Calculate the quantity, in mg, of $C_{12}H_{14}N_2O_2$ in the portion of Tablets taken by the formula:

$$50C(R_U/R_S),$$

in which C is the concentration, in mg per mL, of USP Primidone RS in the *Standard preparation*, and R_U and R_S are the relative response factors obtained from the *Assay preparation* and the *Standard preparation*, respectively.

Probenecid

$(CH_3CH_2CH_2)_2NSO_2-C_6H_4-COOH$

$C_{13}H_{19}NO_4S$ 285.36

Benzoic acid, 4-[(dipropylamino)sulfonyl]-.

p-(Dipropylsulfamoyl)benzoic acid [*57-66-9*].

» Probenecid contains not less than 98.0 percent and not more than 101.0 percent of $C_{13}H_{19}NO_4S$, calculated on the dried basis.

Packaging and storage—Preserve in well-closed containers.

Reference standard—*USP Probenecid Reference Standard*—Dry at 105° for 4 hours before using.

Identification—

A: The infrared absorption spectrum of a potassium bromide dispersion of it, previously dried, exhibits maxima only at the same wavelengths as that of a similar preparation of USP Probenecid RS.

B: The ultraviolet absorption spectrum of a 1 in 50,000 solution in alcohol exhibits maxima and minima at the same wavelengths as that of a similar solution of USP Probenecid RS, concomitantly measured, and the respective absorptivities, calculated on the dried basis, at the wavelength of maximum absorbance at about 248 nm do not differ by more than 3.0%.

Melting range ⟨741⟩: between 198° and 200°.

Acidity—To 2.0 g add 100 mL of water, heat on a steam bath for 30 minutes, cool, filter, and dilute with water to 100.0 mL. To 25.0 mL of this solution add 1 drop of phenolphthalein TS, and titrate with 0.10 N sodium hydroxide: not more than 0.50 mL is required to produce a pink color.

Loss on drying ⟨731⟩—Dry it at 105° for 4 hours: it loses not more than 0.5% of its weight.

Residue on ignition ⟨281⟩: not more than 0.1%.

Selenium ⟨291⟩: not more than 0.003%, a 100-mg test specimen, mixed with 100 mg of magnesium oxide, being used.

Heavy metals, *Method II* ⟨231⟩: not more than 0.002%.

Assay—

Sodium phosphate solution—Prepare in glacial acetic acid solution (1 in 100) a 0.05 M solution of monobasic sodium phosphate, and adjust with phosphoric acid to a pH of 3.0.

Mobile phase—Prepare a degassed and filtered mixture (50:50) of *Sodium phosphate solution* and a 1 in 100 solution of glacial acetic acid in acetonitrile. Make adjustments if necessary (see *System Suitability* under *Chromatography* ⟨621⟩).

Standard preparation—Dissolve an accurately weighed quantity of USP Probenecid RS in *Mobile phase* to obtain a solution having a known concentration of about 0.50 mg per mL.

Assay preparation—Transfer about 50 mg of Probenecid, accurately weighed, to a 100-mL volumetric flask, dissolve in *Mobile phase*, dilute with *Mobile phase* to volume, and mix.

Chromatographic system (see *Chromatography* ⟨621⟩)—The liquid chromatograph is equipped with a 254-nm detector and a 3.9-mm × 30-cm column that contains packing L11. The flow rate is about 1 mL per minute. Chromatograph the *Standard preparation*, and record the peak responses as directed under *Procedure:* the tailing factor is not more than 2.3, the number of theoretical plates is not less than 3900, and the relative standard deviation for replicate injections is not more than 1.5%.

Procedure—Separately inject equal volumes (about 20 µL) of the *Standard preparation* and the *Assay preparation* into the chromatograph, record the chromatograms, and measure the responses for the major peaks. Calculate the quantity, in mg, of $C_{13}H_{19}NO_4S$ in the portion of Probenecid taken by the formula:

$$100C(r_U/r_S),$$

in which C is the concentration, in mg per mL, of USP Probenecid RS in the *Standard preparation*, and r_U and r_S are the peak responses obtained from the *Assay preparation* and the *Standard preparation*, respectively.

Probenecid Capsules, Ampicillin and—*see* Ampicillin and Probenecid Capsules

Probenecid for Oral Suspension, Ampicillin and—*see* Ampicillin and Probenecid for Oral Suspension

Probenecid Tablets

» Probenecid Tablets contain not less than 93.0 percent and not more than 107.0 percent of the labeled amount of $C_{13}H_{19}NO_4S$.

Packaging and storage—Preserve in well-closed containers.

Reference standard—*USP Probenecid Reference Standard*—Dry at 105° for 4 hours before using.

Identification—

A: The ultraviolet absorption spectrum of the solution of Tablets employed for determination of absorbance in the *Assay* exhibits maxima and minima at the same wavelengths as that of a similar solution of USP Probenecid RS, concomitantly measured.

B: Finely powder a quantity of Tablets, equivalent to about 500 mg of probenecid, triturate the powder with alcohol, and filter. Evaporate the filtrate to about 20 mL, cool, acidify with hydrochloric acid until acid to litmus, remove the crystals by filtration, and recrystallize from diluted alcohol: the probenecid so obtained melts between 196° and 200°, as determined by the method for *Class Ia* under *Melting Range or Temperature* ⟨741⟩, and responds to *Identification test A* under *Probenecid.*

Dissolution ⟨711⟩—

Medium: simulated intestinal fluid TS, prepared without pancreatin, pH 7.5 ± 0.1; 900 mL.

Apparatus 2: 50 rpm.

Time: 30 minutes.

Procedure—Determine the amount of $C_{13}H_{19}NO_4S$ dissolved from ultraviolet absorbances at the wavelength of maximum absorbance at about 244 nm of filtered portions of the solution under test, suitably diluted with 0.1 *N* sodium hydroxide, if necessary, in comparison with a Standard solution having a known concentration of USP Probenecid RS.

Tolerances—Not less than 80% (*Q*) of the labeled amount of $C_{13}H_{19}NO_4S$ is dissolved in 30 minutes.

Uniformity of dosage units ⟨905⟩: meet the requirements.

Assay—Weigh and finely powder not less than 20 Probenecid Tablets. Weigh accurately a portion of the powder, equivalent to about 100 mg of probenecid, and transfer to a 250-mL volumetric flask. Add chloroform to volume, and mix. Filter a portion of the chloroform solution, discarding the first 20 to 25 mL of the filtrate, and pipet 5 mL of the filtrate into a 125-mL separator containing 10 mL of chloroform. Extract the chloroform layer with four 15-mL portions of sodium carbonate solution (1 in 100). Render the combined extracts distinctly acid with 5 *N* hydrochloric acid, and extract with four 20-mL portions of chloroform, filtering each extract through a small pledget of cotton into a 100-mL volumetric flask. Wash the cotton filter with 10 mL of chloroform, add chloroform to volume, and mix. Dissolve an accurately weighed quantity of USP Probenecid RS in chloroform, and dilute quantitatively and stepwise with chloroform to obtain a Standard solution having a known concentration of about 20 µg per mL. Concomitantly determine the absorbances of both solutions in 1-cm cells at the wavelength of maximum absorbance at about 257 nm, with a suitable spectrophotometer, using chloroform as the blank. Calculate the quantity, in mg, of $C_{13}H_{19}NO_4S$ in the portion of Tablets taken by the formula:

$$5C(A_U/A_S),$$

in which C is the concentration, in µg per mL, of USP Probenecid RS in the Standard solution, and A_U and A_S are the absorbances of the solution from the Tablets and the Standard solution, respectively.

Probenecid and Colchicine Tablets

» Probenecid and Colchicine Tablets contain not less than 90.0 percent and not more than 115.0 percent of the labeled amount of colchicine ($C_{22}H_{25}NO_6$) and not less than 90.0 percent and not more than 110.0 percent of the labeled amount of probenecid ($C_{13}H_{19}NO_4S$).

Packaging and storage—Preserve in well-closed, light-resistant containers.

Reference standards—*USP Colchicine Reference Standard*—Dry at 105° for 3 hours before using. *USP Probenecid Reference Standard*—Dry at 105° for 4 hours before using.

Identification—

Probenecid standard solution—Prepare a solution of USP Probenecid RS in chloroform having a concentration of about 1 mg per mL.

Colchicine standard solution—Prepare a solution of USP Colchicine RS in chloroform having a concentration of about 1 mg per mL.

Probenecid test solution—Using a portion of finely powdered Tablets, prepare a filtered solution in chloroform having a concentration of about 1 mg of probenecid per mL.

Colchicine test solution—Transfer a quantity of finely powdered Tablets, equivalent to about 0.5 mg of colchicine, to a container, add 15 mL of water, mix, and filter, collecting the filtrate. Extract the filtrate with 25 mL of chloroform, and evaporate the chloroform extract to a volume of about 1 mL.

Procedure (see *Chromatography* ⟨621⟩)—On a thin-layer chromatographic plate, coated with a 0.25-mm layer of chromatographic silica gel mixture, separately apply 5-µL portions of the *Probenecid test solution* and the *Probenecid standard solution*, a 7-µL portion of the *Colchicine test solution*, and a 3.5-µL portion of the *Colchicine standard solution.* Allow the spots to dry, and develop the chromatogram in a solvent system consisting of a mixture of methanol and ammonium hydroxide (100:1.5) until the solvent front has moved about three-fourths of the length of the plate. Remove the plate from the developing chamber, allow the solvent to evaporate, and view the plate under short-wavelength ultraviolet light: the R_f value of the principal spot in the chromatogram obtained from the *Probenecid test solution* corresponds to that obtained from the *Probenecid standard solution.* The R_f value of the principal spot in the chromatogram obtained from the *Colchicine test solution* corresponds to that obtained from the *Colchicine standard solution.*

Dissolution ⟨711⟩—

Medium: 0.05 *M* pH 7.5 phosphate buffer (see *Buffer Solutions* in the section, *Reagents, Indicators, and Solutions*); 900 mL.

Apparatus 2: 50 rpm.

Time: 30 minutes.

Procedure for probenecid—Determine the amount of $C_{13}H_{19}NO_4S$ dissolved from ultraviolet absorbances at the wavelength of maximum absorbance at about 244 nm of filtered portions of the solution under test, suitably diluted with 0.1 *N* sodium hydroxide, if necessary, in comparison with a Standard solution having a known concentration of USP Probenecid RS.

Procedure for colchicine—Extract a filtered 200-mL portion of the solution under test with two 25-mL portions of chloroform, collecting the chloroform extracts in a suitable flask. Evaporate the combined extracts to a small volume, and transfer to a 10-mL volumetric flask. Rinse the flask with small portions of chloroform, and add the rinsings to the 10-mL volumetric flask. Dilute with chloroform to volume, and mix. Determine the amount of $C_{22}H_{25}NO_6$ dissolved from absorbances, at the wavelength of maximum absorbance at about 350 nm, of this solution, using chloroform as the blank, in comparison with a Standard solution in chloroform having a known concentration of USP Colchicine RS.

Tolerances—Not less than 80% (*Q*) of the labeled amounts of $C_{22}H_{25}NO_6$ and $C_{13}H_{19}NO_4S$ are dissolved in 30 minutes.

Uniformity of dosage units ⟨905⟩: meet the requirements.

Assay for probenecid—

Standard preparation—Dissolve an accurately weighed quantity of USP Probenecid RS on 0.1 *N* sodium hydroxide, and dilute quantitatively and stepwise with 0.1 *N* sodium hydroxide to obtain a solution having a known concentration of about 10 µg per mL.

Assay preparation—Weigh and finely powder not less than 20 Probenecid and Colchicine Tablets. Weigh accurately a portion of the powder, equivalent to about 250 mg of probenecid, and transfer to a 250-mL volumetric flask. Add 0.1 *N* sodium hydroxide to volume, and mix. Filter a portion of the solution, discarding the first 20 mL of the filtrate, pipet 2 mL of the filtrate into a 200-mL volumetric flask, dilute with 0.1 *N* sodium hydroxide to volume, and mix.

Procedure—Concomitantly determine the absorbances of the *Assay preparation* and the *Standard preparation* at the wavelength of maximum absorbance at about 244 nm, with a suitable spectrophotometer, using 0.1 *N* sodium hydroxide as the blank. Calculate the quantity, in mg, of $C_{13}H_{19}NO_4S$ in the portion of Tablets taken by the formula:

$$25C(A_U/A_S),$$

in which *C* is the concentration, in µg per mL, of USP Probenecid RS in the *Standard preparation*, and A_U and A_S are the absorbances of the *Assay preparation* and the *Standard preparation*, respectively.

Assay for colchicine—[NOTE—Conduct this procedure without delay, under subdued light, using low-actinic glassware.]

Alcoholic sodium carbonate solution—Dissolve 5.0 g of anhydrous sodium carbonate in 900 mL of water, add 100 mL of isopropyl alcohol, and mix.

Standard preparation—Dissolve an accurately weighed quantity of USP Colchicine RS in *Alcoholic sodium carbonate solution*, and dilute quantitatively and stepwise with *Alcoholic sodium carbonate solution* to obtain a solution having a known concentration of about 10 µg per mL.

Assay preparation—Weigh and finely powder not less than 20 Probenecid and Colchicine Tablets. Weigh accurately a portion of the powder, equivalent to about 1 mg of colchicine, and transfer to a 100-mL volumetric flask. Add 75 mL of *Alcoholic sodium carbonate solution*, shake for 30 minutes, dilute with *Alcoholic sodium carbonate solution* to volume, mix, and filter, discarding the first 20 mL of the filtrate.

Procedure—Concomitantly determine the absorbances of the *Assay preparation* and the *Standard preparation* at the wavelength of maximum absorbance at about 350 nm, with a suitable spectrophotometer, using *Alcoholic sodium carbonate solution* as the blank. Calculate the quantity, in mg, of $C_{22}H_{25}NO_6$ in the portion of Tablets taken by the formula:

$$0.1C(A_U/A_S),$$

in which *C* is the concentration, in µg per mL, of USP Colchicine RS in the *Standard preparation*, and A_U and A_S are the absorbances of the *Assay preparation* and the *Standard preparation*, respectively.

Probucol

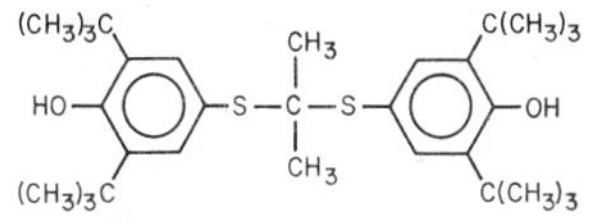

$C_{31}H_{48}O_2S_2$ 516.84

Phenol, 4,4′-[(1-methylethylidene)bis(thio)]bis[2,6-bis(1,1-dimethylethyl)-.

Acetone bis(3,5-di-*tert*-butyl-4-hydroxyphenyl) mercaptole [*23288-49-5*].

» Probucol contains not less than 98.0 percent and not more than 102.0 percent of $C_{31}H_{48}O_2S_2$, calculated on the dried basis.

Packaging and storage—Preserve in well-closed, light-resistant containers.

Reference standards—*USP Probucol Reference Standard. USP Probucol Related Compound A Reference Standard. USP Probucol Related Compound B Reference Standard. USP Probucol Related Compound C Reference Standard.*

Identification—Place about 4 drops of a 1 in 10 solution of it in carbon tetrachloride on a potassium bromide plate in such a way that a uniform film results on the surface of the plate as the solvent evaporates. Record the infrared absorption spectrum from 2.5 µm to 25 µm, using a second potassium bromide plate as the reference: the spectrum exhibits maxima only at the same wavelengths as that of a similar preparation of USP Probucol RS.

Melting range, *Class I* ⟨741⟩: between 124° and 127°, a dried specimen being used.

Loss on drying ⟨731⟩—Dry it in vacuum at 80° for 1 hour: it loses not more than 1.0% of its weight.

Residue on ignition ⟨281⟩: 0.1%.

Heavy metals, *Method II* ⟨231⟩: 0.002%.

Related compounds (see *Chromatography* ⟨621⟩)—

Mobile phase—Prepare a solution of dehydrated alcohol and *n*-hexane (1:4000). Make adjustments, if necessary (see *System Suitability* under *Chromatography* ⟨621⟩).

Reference solution A—Dissolve an accurately weighed quantity of USP Probucol Related Compound A RS (2,2′,6,6′-tetra-*tert*-butyldiphenoquinone) in *n*-hexane, and dilute quantitatively and stepwise with *n*-hexane to obtain a solution having a known concentration of about 10 µg per mL.

Reference solution B—Dissolve an accurately weighed quantity of USP Probucol Related Compound B RS [4,4′-(dithio)bis(2,6-di-*tert*-butylphenol)] in *n*-hexane, and dilute quantitatively with *n*-hexane to obtain a solution having a known concentration of about 1 mg per mL.

Reference solution C—Dissolve an accurately weighed quantity of USP Probucol Related Compound C RS (4-[(3,5-di-*tert*-butyl-2-hydroxyphenylthio)isopropylidenethio]-2,6-di-*tert*-butylphenol) in *n*-hexane, and dilute quantitatively with *n*-hexane to obtain a solution having a known concentration of about 1 mg per mL.

Standard preparation—Pipet 10 mL of *Reference solution C*, 1 mL of *Reference solution B*, and 4 mL of *Reference solution A* into a 200-mL volumetric flask, dilute with *n*-hexane to volume, and mix.

Test preparation—Transfer about 1 g of Probucol, accurately weighed, to a 25-mL volumetric flask, dissolve in *n*-hexane, dilute with *n*-hexane to volume, and mix.

System suitability preparation—Pipet 1 mL of *Reference solution B* and 1 mL of the *Test preparation* into a 200-mL volumetric flask, dilute with *n*-hexane to volume, and mix.

Chromatographic system—The liquid chromatograph is equipped with 254-nm and 420-nm detectors, connected in series, and a 4.6-mm × 25-cm column that contains packing L3. The flow rate is about 1 mL per minute. Chromatograph the *System suitability preparation*, and record the peaks, detected at 254-nm, for related compound B and probucol. Related compound B elutes first, the resolution, *R*, of the peaks is not less than 2.5, and the relative standard deviation for replicate injections, calculated for the probucol peak, is not more than 2%.

Procedure—Separately inject equal volumes (about 20 µL) of the *Standard preparation* and the *Test preparation* into the chromatograph, record the chromatograms, and measure the peak area responses. The order of elution is compound C, compound B, compound A, and finally probucol. Compound A is detected at 420 nm, and the others are detected at 254 nm. Calculate the percentage of each related compound in the Probucol taken by the formula:

$$2500(C/w)(r_U/r_S),$$

in which C is the concentration, in mg per mL, of the respective USP Probucol Related Compound RS in the *Standard preparation*, w is the weight, in mg, of Probucol taken, and r_U and r_S are the respective peak area responses for the *Test preparation* and the *Standard preparation*, respectively: not more than 0.0005% of compound A, not more than 0.02% of compound B, and not more than 0.5% of compound C are found.

Assay—

Mobile phase—Prepare a degassed and filtered solution of acetonitrile and water (85:15). Make adjustments, if necessary (see *System Suitability* under *Chromatography* ⟨621⟩).

System suitability preparation—To about 56 mg of Probucol add 10 mL of *n*-propyl alcohol, and dissolve the Probucol. Add 1.0 mL of 70 percent *tert*-butyl hydroperoxide, and mix. Cover loosely, and heat on a steam bath at about 90° for 30 minutes. Allow to cool to room temperature, dilute with a mixture of *n*-propyl alcohol and water (17:14) to 200 mL, and mix. Dilute 25 mL of this solution with *Mobile phase* to 100 mL.

Standard preparation—Dissolve an accurately weighed quantity of USP Probucol RS in *Mobile phase*, and dilute quantitatively and stepwise with *Mobile phase* to obtain a solution having a known concentration of about 63 µg per mL.

Assay preparation—Transfer about 63 mg of Probucol, accurately weighed, to a 50-mL volumetric flask. Dissolve in *Mobile phase*, dilute with *Mobile phase* to volume, and mix. Pipet 5 mL of this solution into a 100-mL volumetric flask, dilute with *Mobile phase* to volume, and mix.

Chromatographic system (see *Chromatography* ⟨621⟩)—The liquid chromatograph is equipped with a 242-nm detector and a 4.6-mm × 25-cm column that contains packing L7. The flow rate is about 2.0 mL per minute. Chromatograph the *System suitability preparation*, and record the peaks for the degradation product and probucol at relative retention times of approximately 0.8 and 1.0, respectively. The resolution, R, of the peaks is not less than 2.0. Chromatograph replicate injections of *Standard preparation:* the relative standard deviation is not more than 1.0%.

Procedure—Separately inject equal volumes (about 50 µL) of the *Standard preparation* and the *Assay preparation* into the chromatograph, record the chromatograms, and measure the area responses for the major peaks. Calculate the quantity, in mg, of $C_{31}H_{48}O_2S_2$ in the portion of Probucol taken by the formula:

$$C(r_U/r_S),$$

in which C is the concentration, in µg per mL, of USP Probucol RS in the *Standard preparation*, and r_U and r_S are the peak area responses for the *Assay preparation* and the *Standard preparation*, respectively.

Procainamide Hydrochloride

$NH_2-C_6H_4-CONHCH_2CH_2N(C_2H_5)_2 \cdot HCl$

$C_{13}H_{21}N_3O.HCl$ 271.79

Benzamide, 4-amino-*N*-[2-(diethylamino)ethyl]-, monohydrochloride.

p-Amino-*N*-[2-(diethylamino)ethyl]benzamide monohydrochloride [*614-39-1*].

» Procainamide Hydrochloride contains not less than 98.0 percent and not more than 102.0 percent of $C_{13}H_{21}N_3O.HCl$, calculated on the dried basis.

Packaging and storage—Preserve in tight containers.

Reference standard—*USP Procainamide Hydrochloride Reference Standard*—Dry at 105° for 4 hours before using.

Identification—

A: The infrared absorption spectrum of a potassium bromide dispersion of it, previously dried, exhibits maxima only at the same wavelengths as that of a similar preparation of USP Procainamide Hydrochloride RS.

B: A test solution of it, containing about 0.5 mg per mL in methanol, responds to the *Thin-layer Chromatographic Identification Test* ⟨201⟩, the chromatogram being developed with a solvent system consisting of a mixture of ethyl acetate, methanol, and ammonium hydroxide (22:2:1).

Melting range ⟨741⟩: between 165° and 169°.

Loss on drying ⟨731⟩—Dry it at 105° for 4 hours: it loses not more than 0.3% of its weight.

Residue on ignition ⟨281⟩: not more than 0.1%.

Heavy metals, *Method II* ⟨231⟩: 0.002%.

Assay—

Mobile phase—Prepare a suitable mixture of water, methanol, and triethylamine (140:60:1), adjust with phosphoric acid to a pH of 7.5 ± 0.1, filter, and degas. Make adjustments if necessary (see *System Suitability* under *Chromatography* ⟨621⟩).

Standard preparation—Dissolve an accurately weighed quantity of USP Procainamide Hydrochloride RS quantitatively in *Mobile phase* to obtain a solution having a known concentration of about 0.5 mg per mL. Dilute an accurately measured volume of this stock solution, quantitatively with *Mobile phase* to obtain a *Standard preparation* having a known concentration of about 0.05 mg per mL.

Resolution solution—Dissolve a quantity of *p*-aminobenzoic acid in *Mobile phase* to obtain a solution containing about 0.1 mg per mL. Pipet 10 mL of this solution and 10 mL of the stock solution used to prepare the *Standard preparation* to a 100-mL volumetric flask, dilute with *Mobile phase* to volume, and mix.

Assay preparation—Transfer about 50 mg of Procainamide Hydrochloride, accurately weighed, to a 100-mL volumetric flask, dissolve in *Mobile phase*, dilute with *Mobile phase* to volume, and mix. Transfer 10.0 mL of this solution to a second 100-mL volumetric flask, dilute with *Mobile phase* to volume, and mix.

Chromatographic system (see *Chromatography* ⟨621⟩)—The liquid chromatograph is equipped with a 280-nm detector and a 3.9-mm × 30-cm column that contains 10-µm packing L1. The flow rate is about 1 mL per minute. Chromatograph the *Resolution solution*, and record the peak responses as directed under *Procedure:* the resolution, R, between the *p*-aminobenzoic acid and procainamide peaks is not less than 2.0. The relative retention times are about 0.5 for procainamide hydrochloride and 1.0 for procaine hydrochloride. Chromatograph the *Standard preparation*, and record the peak responses as directed under *Procedure:* the relative standard deviation for replicate injections is not more than 2.0%.

Procedure—Separately inject equal volumes (about 20 µL) of the *Standard preparation* and the *Assay preparation* into the chromatograph, record the chromatograms, and measure the responses for the major peaks. Calculate the quantity, in mg, of $C_{13}H_{21}N_3O.HCl$ in the portion of Procainamide Hydrochloride taken by the formula:

$$1000C(r_U/r_S),$$

in which C is the concentration, in mg per mL, of USP Procainamide Hydrochloride RS in the *Standard preparation*, and r_U and r_S are the peak responses obtained from the *Assay preparation* and the *Standard preparation*, respectively.

Procainamide Hydrochloride Capsules

» Procainamide Hydrochloride Capsules contain not less than 95.0 percent and not more than 105.0 percent of the labeled amount of $C_{13}H_{21}N_3O.HCl$.

Packaging and storage—Preserve in tight containers.

Reference standard—*USP Procainamide Hydrochloride Reference Standard*—Dry at 105° for 4 hours before using.

Identification—Capsules respond to the *Thin-layer Chromatographic Identification Test* ⟨201⟩, 5 µL of the clear supernatant solution used to prepare the *Assay preparation* in the *Assay* and 5 µL of the stock solution used to prepare the *Standard preparation* in the *Assay* being applied to the plate, and a solvent system consisting of a mixture of ethyl acetate, methanol, and ammonium hydroxide (22:2:1) being used to develop the chromatogram.

Dissolution ⟨711⟩—

Medium: 0.1 *N* hydrochloric acid; 900 mL.

Apparatus 2: 50 rpm.

Time: 90 minutes.

Procedure—Determine the amount of $C_{13}H_{21}N_3O.HCl$ dissolved from ultraviolet absorbances at the wavelength of maximum absorbance at about 275 nm of filtered portions of the solution under test, suitably diluted with an amount of 0.1 *N* sodium hydroxide that is not less than twice the volume of the portion of test solution taken, in comparison with a Standard solution having a known concentration of USP Procainamide Hydrochloride RS in the same medium.

Tolerances—Not less than 75% (*Q*) of the labeled amount of $C_{13}H_{21}N_3O.HCl$ is dissolved in 90 minutes.

Uniformity of dosage units ⟨905⟩: meet the requirements.

Assay—

Mobile phase, Standard preparation, Resolution solution, and *Chromatographic system*—Prepare as directed in the *Assay* under *Procainamide Hydrochloride.*

Assay preparation—Accurately weigh and mix the contents of 20 Procainamide Hydrochloride Capsules. Transfer an accurately weighed portion of the mixture, equivalent to about 500 mg of procainamide hydrochloride, to a 500-mL volumetric flask. Add about 350 mL of methanol to the flask, and sonicate for 10 minutes in a 40° water bath. Allow the flask to cool to room temperature, dilute with methanol to volume, and mix. Centrifuge a portion of the suspension, transfer 5.0 mL of the clear supernatant solution obtained to a 100-mL volumetric flask, dilute with *Mobile phase* to volume, and mix. [NOTE—Reserve the remainder of the clear supernatant solution for the *Identification test.*]

Procedure—Proceed as directed for *Procedure* in the *Assay* under *Procainamide Hydrochloride.* Calculate the quantity, in mg, of $C_{13}H_{21}N_3O.HCl$ in the portion of Capsule contents taken by the formula:

$$10,000C(r_U / r_S),$$

in which *C* is the concentration, in mg per mL, of USP Procainamide Hydrochloride RS in the *Standard preparation,* and r_U and r_S are the peak responses obtained from the *Assay preparation* and the *Standard preparation,* respectively.

Procainamide Hydrochloride Injection

» Procainamide Hydrochloride Injection is a sterile solution of Procainamide Hydrochloride in Water for Injection. It contains not less than 95.0 percent and not more than 105.0 percent of the labeled amount of $C_{13}H_{21}N_3O.HCl$.

Packaging and storage—Preserve in single-dose or in multiple-dose containers, preferably of Type I glass.

Labeling—Label it to indicate that the Injection is not to be used if it is darker than slightly yellow, or is discolored in any other way.

Reference standard—*USP Procainamide Hydrochloride Reference Standard*—Dry at 105° for 4 hours before using.

Identification—It responds to the *Thin-layer Chromatographic Identification Test* ⟨201⟩, 5 µL of the clear supernatant solution used to prepare the *Assay preparation* in the *Assay* and 5 µL of the stock solution used to prepare the *Standard preparation* in the *Assay* being applied to the plate, and a solvent system consisting of a mixture of ethyl acetate, methanol, and ammonium hydroxide (22:2:1) being used to develop the chromatogram.

Pyrogen—It meets the requirements of the *Pyrogen Test* ⟨151⟩, the test dose being 0.5 mL of the Injection per kg.

pH ⟨791⟩: between 4.0 and 6.0.

Particulate matter ⟨788⟩: meets the requirements under *Small-volume Injections.*

Other requirements—It meets the requirements under *Injections* ⟨1⟩.

Assay—

Mobile phase, Standard preparation, Resolution solution, and *Chromatographic system*—Prepare as directed in the *Assay* under *Procainamide Hydrochloride.*

Assay preparation—Transfer an accurately measured volume of Procainamide Hydrochloride Injection, equivalent to about 500 mg of procainamide hydrochloride, to a 500-mL volumetric flask, dilute with methanol to volume, and mix. Transfer 5.0 mL of this stock solution to a 100-mL volumetric flask, reserving the remainder of the stock solution for the *Identification test.* Dilute with *Mobile phase* to volume, and mix.

Procedure—Proceed as directed for *Procedure* in the *Assay* under *Procainamide Hydrochloride.* Calculate the quantity, in mg, of $C_{13}H_{21}N_3O.HCl$ in each mL of the Injection taken by the formula:

$$10,000(C / V)(r_U / r_S),$$

in which *C* is the concentration, in mg per mL, of USP Procainamide Hydrochloride RS in the *Standard preparation, V* is the volume, in mL, of Injection taken, and r_U and r_S are the peak responses obtained from the *Assay preparation* and the *Standard preparation,* respectively.

Procainamide Hydrochloride Tablets

» Procainamide Hydrochloride Tablets contain not less than 95.0 percent and not more than 105.0 percent of the labeled amount of $C_{13}H_{21}N_3O.HCl$.

Packaging and storage—Preserve in tight containers.

Reference standard—*USP Procainamide Hydrochloride Reference Standard*—Dry at 105° for 4 hours before using.

Identification—Tablets respond to the *Thin-layer Chromatographic Identification Test* ⟨201⟩, 5 µL of the clear supernatant solution used to prepare the *Assay preparation* in the *Assay* and 5 µL of the stock solution used to prepare the *Standard preparation* in the *Assay* being applied to the plate, and a solvent system consisting of a mixture of ethyl acetate, methanol, and ammonium hydroxide (22:2:1) being used to develop the chromatogram.

Dissolution ⟨711⟩—

Medium: 0.1 *N* hydrochloric acid; 900 mL.

Apparatus 1: 100 rpm.

Time: 75 minutes.

Procedure—Determine the amount of $C_{13}H_{21}N_3O.HCl$ dissolved from ultraviolet absorbances at the wavelength of maximum absorbance at about 275 nm of filtered portions of the solution under test, suitably diluted with an amount of 0.1 *N* sodium hydroxide that is not less than twice the volume of the portion of test solution taken, in comparison with a Standard solution having a known concentration of USP Procainamide Hydrochloride RS in the same medium.

Tolerances—Not less than 80% (*Q*) of the labeled amount of $C_{13}H_{21}N_3O.HCl$ is dissolved in 75 minutes.

Uniformity of dosage units ⟨905⟩: meet the requirements.

Free *p*-aminobenzoic acid—Weigh a portion of finely ground Tablets, equivalent to 1000 mg of procainamide hydrochloride, and transfer to a 25-mL volumetric flask. Add methanol to volume, mix, and allow the particles to settle. Use the clear portion as the *Test solution.* Prepare a Standard solution of *p*-aminobenzoic acid in methanol containing 0.24 mg per mL. On a suitable thin-layer chromatographic plate (see *Chromatography* ⟨621⟩), coated with a 0.25-mm layer of chromatographic silica gel mixture, apply

range, *Acidity*, *Loss on drying*, and *Assay* under *Procaine Hydrochloride*. It meets also the requirements for *Sterility Tests* ⟨71⟩, *Uniformity of Dosage Units* ⟨905⟩, and *Labeling* under *Injections* ⟨1⟩.

Procaine Hydrochloride and Epinephrine Injection

» Procaine Hydrochloride and Epinephrine Injection is a sterile solution of Procaine Hydrochloride and epinephrine hydrochloride in Water for Injection. The content of epinephrine does not exceed 0.002 percent (1 in 50,000). It contains not less than 95.0 percent and not more than 105.0 percent of the labeled amount of procaine hydrochloride ($C_{13}H_{20}N_2O_2.HCl$), and not less than 90.0 percent and not more than 115.0 percent of the labeled amount of epinephrine ($C_9H_{13}NO_3$).

Packaging and storage—Preserve in single-dose or in multiple-dose, light-resistant containers, preferably of Type I or Type II glass.

Reference standards—*USP Procaine Hydrochloride Reference Standard*—Dry over silica gel for 18 hours before using. *USP Epinephrine Bitartrate Reference Standard*—Dry in vacuum over silica gel for 3 hours before using.

Identification—Evaporate a portion of Injection, equivalent to about 20 mg of procaine hydrochloride, on a steam bath just to dryness, and dry over silica gel for 18 hours, protected from light: the residue responds to *Identification tests A* and *B* under *Procaine Hydrochloride*.

pH ⟨791⟩: between 3.0 and 5.5.

Content of epinephrine—

Alkaline ascorbate reagent—Mix 100 mL of alcohol, 80 mL of sodium hydroxide solution (1 in 5), and 8 mL of ascorbic acid solution (1 in 50). Prepare fresh on the day of use.

Standard preparation—Place about 18 mg of USP Epinephrine Bitartrate RS, accurately weighed, in a 100-mL volumetric flask, add sodium bisulfite solution (1 in 1000) to volume, and mix. Dilute this solution quantitatively with water to obtain a solution having a known concentration of about 10 µg of epinephrine per mL.

Procedure—Pipet a volume of Procaine Hydrochloride and Epinephrine Injection, equivalent to about 10 µg of epinephrine, into a 50-mL beaker. Into another 50-mL beaker pipet 1 mL of *Standard preparation*. Treat the contents of each beaker as follows. Add 10 mL of dilute hydrochloric acid (1 in 120), and heat gently to reduce the volume of solution to about 5 mL. Allow to cool to room temperature, then add 5 mL of sodium acetate solution (1 in 5), followed by 0.5 mL of potassium ferricyanide solution (1 in 400), and mix. At 2 minutes, accurately timed, after the last addition, add 20 mL of *Alkaline ascorbate reagent*, transfer the contents to a corresponding 50-mL volumetric flask with the aid of water, add water to volume, and mix. At 15 to 20 minutes after the addition of the *Alkaline ascorbate reagent*, determine the fluorescences of each solution and of a reagent blank, with a suitable fluorometer, with an excitation wavelength setting of 420 nm and a fluorescence wavelength setting of 520 nm. Calculate the quantity, in µg, of $C_9H_{13}NO_3$ in each mL of Injection taken by the formula:

$$(C/V)[(I_U - B)/(I_S - B)],$$

in which C is the concentration of epinephrine, in µg per mL, from the USP Epinephrine Bitartrate RS in the *Standard preparation*, V is the volume, in mL, of Injection taken, and I_U, I_S, and B are the fluorescence readings of the solutions from the Injection, the *Standard preparation*, and the reagent blank, respectively.

Other requirements—It meets the requirements under *Injections* ⟨1⟩.

Assay—

Standard preparation—Transfer to a 125-mL separator about 50 mg, accurately weighed, of USP Procaine Hydrochloride RS, and dilute with water to 20 mL.

Assay preparation—Transfer to a 125-mL separator an accurately measured volume of Procaine Hydrochloride and Epinephrine Injection, equivalent to about 50 mg of procaine hydrochloride, and dilute with water to 20 mL.

Procedure—To the *Standard preparation* and also to the *Assay preparation* add 5 mL of 6 *N* ammonium hydroxide, then treat each as follows: Extract with five 25-mL portions of chloroform, and filter the combined extracts through about 1 g of anhydrous sodium sulfate supported on a pledget of glass wool. Receive the filtrate in a 200-mL volumetric flask, add chloroform to volume, and mix. Transfer 3.0 mL of this solution to a 100-mL volumetric flask, add chloroform to volume, and mix. Concomitantly determine the absorbances of both solutions at the wavelength of maximum absorbance at about 280 nm, with a suitable spectrophotometer, using chloroform as the blank. Calculate the quantity, in mg, of $C_{13}H_{20}N_2O_2.HCl$ in each mL of the Injection taken by the formula:

$$(W/V)(A_U/A_S),$$

in which W is the weight, in mg, of USP Procaine Hydrochloride RS used, V is the volume, in mL, of Injection taken, and A_U and A_S are the absorbances of the solutions from the *Assay preparation* and the *Standard preparation*, respectively.

Procaine Hydrochlorides and Levonordefrin Injection, Propoxycaine and—*see* Propoxycaine and Procaine Hydrochlorides and Levonordefrin Injection

Procaine Hydrochlorides and Norepinephrine Bitartrate Injection, Propoxycaine and—*see* Propoxycaine and Procaine Hydrochlorides and Norepinephrine Bitartrate Injection

Procaine and Phenylephrine Hydrochlorides Injection

» Procaine and Phenylephrine Hydrochlorides Injection is a sterile solution of Procaine Hydrochloride and Phenylephrine Hydrochloride in Water for Injection. It contains not less than 95.0 percent and not more than 105.0 percent of the labeled amount of procaine hydrochloride ($C_{13}H_{20}N_2O_2.HCl$), and not less than 90.0 percent and not more than 110.0 percent of the labeled amount of phenylephrine hydrochloride ($C_9H_{13}NO_2.HCl$).

Packaging and storage—Preserve in single-dose or in multiple-dose containers, preferably of Type I glass.

Reference standards—*USP Procaine Hydrochloride Reference Standard*—Dry over silica gel for 18 hours before using. *USP Phenylephrine Hydrochloride Reference Standard*—Dry at 105° for 2 hours before using.

Identification—

A: Mix 10 mL of Injection with 10 mL of 3 *N* hydrochloric acid, cool in an ice bath to 0°, add 5 mL of a solution of sodium nitrite (1 in 5), stirring gently, add 2 mL of a solution of 0.1 g of 2-naphthol in 5 mL of 1 *N* sodium hydroxide, and observe immediately: a bright orange-red precipitate is formed (*presence of a primary aminophenyl group*).

B: It responds to the tests for *Chloride* ⟨191⟩.

pH ⟨791⟩: between 3.0 and 5.5.

Other requirements—It meets the requirements under *Injections* ⟨1⟩.

Assay for procaine hydrochloride—Transfer an accurately measured volume of Procaine and Phenylephrine Hydrochlorides Injection, equivalent to about 100 mg of procaine hydrochloride, to a 125-mL separator containing 25 mL of chloroform and 10 mL of water. Add 5 mL of sodium carbonate TS, shake well, allow the layers to separate, and transfer the chloroform layer to a second separator. Extract the aqueous solution with two 10-mL portions of chloroform, adding the extracts to the second separator. Extract the combined chloroform extracts with 35 mL of dilute hydrochloric acid (1 in 6). Transfer the chloroform to another separator, and transfer the aqueous layer to a 100-mL volumetric flask. Extract the chloroform with two 30-mL portions of dilute hydrochloric acid (1 in 6), add to the volumetric flask, add dilute hydrochloric acid (1 in 6) to volume, and mix. Transfer 15.0 mL of this solution to a 100-mL volumetric flask, add dilute hydrochloric acid (1 in 6) to volume, and mix. Concomitantly determine the absorbances of this solution and of a Standard solution of USP Procaine Hydrochloride RS, in the same medium having a known concentration of about 150 µg per mL, in 1-cm cells at the wavelength of maximum absorbance at about 272 nm, using water as the blank. Calculate the quantity, in mg, of $C_{13}H_{20}N_2O_2.HCl$ in each mL of the Injection taken by the formula:

$$(0.667C/V)(A_U/A_S),$$

in which C is the concentration, in µg per mL, of USP Procaine Hydrochloride RS in the Standard solution, V is the volume, in mL, of Injection taken, and A_U and A_S are the absorbances of the solution from the Injection and the Standard solution, respectively.

Assay for phenylephrine hydrochloride—

Standard preparation—Transfer about 40 mg of USP Phenylephrine Hydrochloride RS, accurately weighed, to a 100-mL volumetric flask, add water to volume, and mix.

Assay preparation—Use Procaine and Phenylephrine Hydrochlorides Injection undiluted.

Procedure—Transfer 1.0 mL each of the *Standard preparation* and the *Assay preparation* to separate 100-mL volumetric flasks, and transfer 1.0 mL of water to a third flask to provide a blank. To each flask add 5.0 mL of potassium ferricyanide solution (1 in 50) followed by 5.0 mL of 0.025 M 4-aminoantipyrine, and mix. Dilute with 0.01 M sodium bicarbonate to volume, mix, and allow to stand for 15 minutes, accurately timed. Concomitantly determine the absorbances of the solutions in 1-cm cells at the wavelength of maximum absorbance at about 500 nm, with a suitable spectrophotometer, against the reagent blank. Calculate the quantity, in mg, of $C_9H_{13}NO_2.HCl$ in each mL of the Injection taken by the formula:

$$C(A_U/A_S),$$

in which C is the concentration, in mg per mL, of USP Phenylephrine Hydrochloride RS in the *Standard preparation*, and A_U and A_S are the absorbances of the solutions from the *Assay preparation* and the *Standard preparation*, respectively.

Procaine and Tetracaine Hydrochlorides and Levonordefrin Injection

» Procaine and Tetracaine Hydrochlorides and Levonordefrin Injection is a sterile solution of Procaine Hydrochloride, Tetracaine Hydrochloride, and Levonordefrin in Water for Injection. It contains not less than 95.0 percent and not more than 105.0 percent of the labeled amount of procaine hydrochloride ($C_{13}H_{20}N_2O_2.HCl$), not less than 95.0 percent and not more than 105.0 percent of the labeled amount of tetracaine hydrochloride ($C_{15}H_{24}N_2O_2.HCl$), and not less than 90.0 percent and not more than 110.0 percent of the labeled amount of levonordefrin ($C_9H_{13}NO_3$).

Packaging and storage—Preserve in single-dose or in multiple-dose containers, preferably of Type I glass.

Reference standard—*USP Epinephrine Bitartrate Reference Standard*—Dry in vacuum over silica gel for 18 hours before using.

Identification—

A: Mix 10 mL of Injection with 10 mL of 3 N hydrochloric acid, cool to 0° in an ice bath, add 5 mL of sodium nitrite solution (1 in 5), stirring gently, add 2 mL of a solution of 0.1 g of 2-naphthol in 5 mL of 1 N sodium hydroxide, and observe immediately: a bright orange-red precipitate is formed (*presence of a primary aminophenyl group*).

B: It responds to the tests for *Chloride* ⟨191⟩.

C: To about 25 mL of Injection add 1 drop of ferric chloride TS: the solution immediately turns a light green and changes to a light blue within 1 minute. Add 2 drops of 3 N hydrochloric acid: the color reverts to light green (*presence of levonordefrin*).

pH ⟨791⟩: between 3.5 and 5.0.

Other requirements—It meets the requirements under *Injections* ⟨1⟩.

Assay for procaine hydrochloride—Transfer 25.0 mL of Procaine and Tetracaine Hydrochlorides and Levonordefrin Injection to a 250-mL beaker, add 10 mL of dilute hydrochloric acid (1 in 2), 15 mL of water, and 1.0 mL of formaldehyde solution, and allow to stand for 3 minutes. Titrate with 0.1 M sodium nitrite VS, using starch iodide paper as an external indicator. Perform a blank determination, and make any necessary correction. Calculate the quantity, in mg, of $C_{13}H_{20}N_2O_2.HCl$ in each mL of the Injection taken by the formula:

$$(272.77/25)(V_2M_2 - 0.5V_1M_1),$$

in which V_2 is the volume, in mL, and M_2 is the exact molarity of the 0.1 M sodium nitrite solution used in the titration, V_1 and M_1 are as defined under *Assay for tetracaine hydrochloride*, and 272.77 is the molecular weight of procaine hydrochloride.

Assay for tetracaine hydrochloride—Transfer 50.0 mL of Procaine and Tetracaine Hydrochlorides and Levonordefrin Injection to a 100-mL beaker, add 5 g of sodium acetate and 5 mL of isopropyl alcohol, and stir until the sodium acetate is dissolved. Add 1.5 mL of potassium thiocyanate solution (75 in 100), and stir while cooling in an ice bath. Keep in the ice bath for 1 hour, stirring about four times during the period. Do not stir excessively. Prepare a filtering crucible with double suitable filter paper, connect to a suction flask, and moisten the paper with a few drops of the clear supernatant liquid. Remove the excess liquid by suction. Filter the mixture, and dry by suction. Transfer the entire contents of the crucible to the original beaker, and rinse the crucible with 10 mL of dilute hydrochloric acid (1 in 2), followed by 30 mL of water. Transfer the rinsings to the beaker. Cool in ice, and titrate with 0.02 M sodium nitrite VS, using starch iodide paper as an external indicator. Perform a blank determination, using 10 mL of dilute hydrochloric acid (1 in 2) and 30 mL of water, and make any necessary correction. Calculate the quantity, in mg, of $C_{15}H_{24}N_2O_2.HCl$ in each mL of the Injection taken by the formula:

$$(300.83/50)(V_1M_1),$$

in which V_1 is the volume, in mL, and M_1 is the exact molarity of the 0.02 M sodium nitrite used in the titration, and 300.83 is the molecular weight of tetracaine hydrochloride.

Assay for levonordefrin—

Ferro-citrate Solution, *Buffer Solution*, and *Standard Preparation*—Prepare as directed under *Epinephrine Assay* ⟨391⟩.

Assay preparation—Proceed with Procaine and Tetracaine Hydrochlorides and Levonordefrin Injection as directed for *Assay Preparation* under *Epinephrine Assay* ⟨391⟩, except to read "levonordefrin" where "epinephrine" is specified.

Procedure—Proceed as directed for *Procedure* under *Epinephrine Assay* ⟨391⟩. Calculate the quantity, in mg, of levonordefrin ($C_9H_{13}NO_3$) in each mL of the Injection taken by the formula:

$$(183.21/333.29)(0.05C/V)(A_U/A_S),$$

in which 183.21 and 333.29 are the molecular weights of levonordefrin and epinephrine bitartrate, respectively, C is the concentration, in μg per mL, of USP Epinephrine Bitartrate RS in the *Standard Preparation*, and *V* is the volume, in mL, of Injection taken.

Prochlorperazine

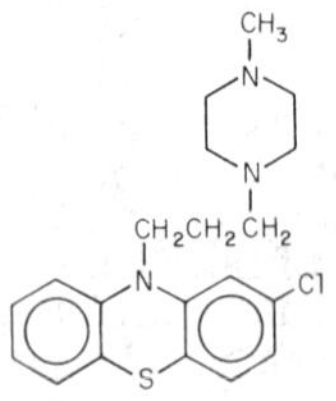

$C_{20}H_{24}ClN_3S$ 373.94
10*H*-Phenothiazine, 2-chloro-10-[3-(4-methyl-1-piperazinyl)-propyl]-.
2-Chloro-10-[3-(4-methyl-1-piperazinyl)propyl]phenothiazine [*58-38-8*].

» Prochlorperazine contains not less than 98.0 percent and not more than 101.0 percent of $C_{20}H_{24}ClN_3S$.

Packaging and storage—Preserve in tight, light-resistant containers.

Reference standard—*USP Prochlorperazine Maleate Reference Standard*—Dry in vacuum at 60° for 2 hours before using.

NOTE—Throughout the following procedures, protect test or assay specimens, the Reference Standard, and solutions containing them, by conducting the procedures without delay, under subdued light, or using low-actinic glassware.

Identification—It meets the requirements under *Identification—Organic Nitrogenous Bases* ⟨181⟩, USP Prochlorperazine Maleate RS being used as the standard for comparison.

Residue on ignition ⟨281⟩: not more than 0.1%.

Ordinary impurities ⟨466⟩—
Test solution: methanol.
Standard solution: methanol.
Eluant: a mixture of methanol and ammonium hydroxide (100:1).
Visualization: 1.

Assay—Transfer about 400 mg of Prochlorperazine, accurately weighed, to a 125-mL conical flask, add 30 mL of glacial acetic acid, and warm on a steam bath to dissolve. Add 2 drops of crystal violet TS, and titrate with 0.1 *N* perchloric acid VS to a blue end-point. Perform a blank determination, and make any necessary correction. Each mL of 0.1 *N* perchloric acid is equivalent to 18.70 mg of $C_{20}H_{24}ClN_3S$.

Prochlorperazine Suppositories

» Prochlorperazine Suppositories contain not less than 90.0 percent and not more than 110.0 percent of the labeled amount of $C_{20}H_{24}ClN_3S$.

Packaging and storage—Preserve in tight containers at a temperature below 37°. Do not expose the unwrapped Suppositories to sunlight.

Reference standard—*USP Prochlorperazine Maleate Reference Standard*—Dry in vacuum at 60° for 2 hours before using.

NOTE—Throughout the following procedures, protect test or assay specimens, the Reference Standard, and solutions containing them, by conducting the procedures without delay, under subdued light, or using low-actinic glassware.

Identification—

A: Place a quantity of Suppositories, equivalent to about 5 mg of prochlorperazine, in a test tube, add 4 mL of dilute hydrochloric acid (1 in 2), warm on a steam bath to melt the solid, and swirl to mix: a pink color develops in the aqueous layer.

B: To the solution from *Identification test A* add 10 mL of bromine TS, and mix: essentially no color change occurs (distinction from *chlorpromazine*, which immediately produces a green color).

Assay—

Standard preparation—Transfer about 40 mg of USP Prochlorperazine Maleate RS, accurately weighed, to a 250-mL separator containing 75 mL of ether. Add 0.5 mL of 6 *N* ammonium hydroxide, and proceed as directed for *Assay preparation*, beginning with "Extract with four 65-mL portions of dilute hydrochloric acid (1 in 100)."

Assay preparation—Weigh, mash, and then mix not less than 15 Prochlorperazine Suppositories. Transfer an accurately weighed quantity of the mass, equivalent to about 25 mg of prochlorperazine, to a 100-mL beaker. Dissolve in 50 mL of ether, and transfer to a 250-mL separator with the aid of three 25-mL portions of ether. Extract with four 65-mL portions of dilute hydrochloric acid (1 in 100), collecting the aqueous extracts in a 500-mL volumetric flask. Aerate the combined extracts for 15 to 20 minutes to remove dissolved ether. Add dilute hydrochloric acid (1 in 100) to volume, and mix. Filter a portion of the solution through filter paper, discarding the first 25 mL of the filtrate. To 25.0 mL of the subsequent filtrate add dilute hydrochloric acid (1 in 100) to make 200.0 mL, and mix.

Procedure—Concomitantly determine the absorbances of the *Standard preparation* and the *Assay preparation* in 1-cm cells at 254 nm and at 278 nm, with a suitable spectrophotometer, using dilute hydrochloric acid (1 in 100) as the blank. Calculate the quantity, in mg, of $C_{20}H_{24}ClN_3S$ in the portion of Suppositories taken by the formula:

$$0.617W(A_{254} - A_{278})_U/(A_{254} - A_{278})_S,$$

in which 0.617 is the ratio of the molecular weight of prochlorperazine to that of prochlorperazine maleate, *W* is the weight, in mg, of USP Prochlorperazine Maleate RS in the *Standard preparation*, and the parenthetic expressions are the differences in the absorbances of the two solutions at the wavelengths indicated by the subscripts, for the *Assay preparation* ($_U$) and the *Standard preparation* ($_S$), respectively.

Prochlorperazine Edisylate

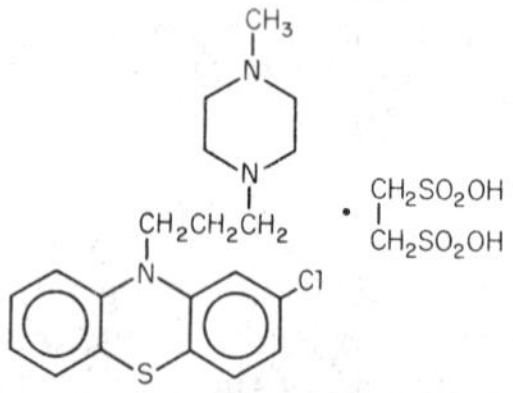

$C_{20}H_{24}ClN_3S.C_2H_6O_6S_2$ 564.13
10*H*-Phenothiazine, 2-chloro-10-[3-(4-methyl-1-piperazinyl)-propyl]-, 1,2-ethanedisulfonate (1:1).
2-Chloro-10-[3-(4-methyl-1-piperazinyl)propyl]phenothiazine 1,2-ethanedisulfonate (1:1) [*1257-78-9*].

» Prochlorperazine Edisylate contains not less than 98.0 percent and not more than 101.5 percent of $C_{20}H_{24}ClN_3S.C_2H_6O_6S_2$, calculated on the dried basis.

Packaging and storage—Preserve in tight, light-resistant containers.

Reference standard—*USP Prochlorperazine Maleate Reference Standard*—Dry in vacuum at 60° for 2 hours before using.

NOTE—Throughout the following procedures, protect test or assay specimens, the Reference Standard, and solutions containing them, by conducting the procedures without delay, under subdued light, or using low-actinic glassware.

Identification—

A: It meets the requirements under *Identification—Organic Nitrogenous Bases* ⟨181⟩, USP Prochlorperazine Maleate RS being used as the standard for comparison.

B: Fuse about 100 mg with a few pellets of sodium hydroxide: the cooled melt responds to the test for *Sulfite* ⟨191⟩.

Loss on drying ⟨731⟩—Dry it in vacuum at 100° for 3 hours: it loses not more than 0.5% of its weight.

Residue on ignition ⟨281⟩: not more than 0.1%.

Selenium ⟨291⟩: 0.003%, a 100-mg test specimen, mixed with 100 mg of magnesium oxide, being used.

Ordinary impurities ⟨466⟩—

Test solution: a mixture of methanol and 1 *N* sodium hydroxide (9:1).

Standard solution: a mixture of methanol and 1 *N* sodium hydroxide (9:1).

Eluant: a mixture of methanol and ammonium hydroxide (100:1).

Visualization: 1.

Assay—Transfer about 750 mg of Prochlorperazine Edisylate, accurately weighed, to a separator containing 40 mL of water, and shake to effect solution. Render the solution alkaline with ammonium hydroxide, and extract with three 25-mL portions of ether. Wash the combined ether extracts once with about 25 mL of water, discard the washing, and evaporate the ether solution on a steam bath to dryness. Dissolve the residue in 60 mL of glacial acetic acid, add crystal violet TS, and titrate with 0.1 *N* perchloric acid VS. Perform a blank determination, and make any necessary correction. Each mL of 0.1 *N* perchloric acid is equivalent to 28.21 mg of $C_{20}H_{24}ClN_3S.C_2H_6O_6S_2$.

Prochlorperazine Edisylate Injection

» Prochlorperazine Edisylate Injection is a sterile solution of Prochlorperazine Edisylate in Water for Injection. It contains, in each mL, an amount of prochlorperazine edisylate equivalent to not less than 4.75 mg and not more than 5.25 mg of prochlorperazine ($C_{20}H_{24}ClN_3S$).

Packaging and storage—Preserve in single-dose or in multiple-dose containers, preferably of Type I glass, protected from light.

Reference standard—*USP Prochlorperazine Maleate Reference Standard*—Dry in vacuum at 60° for 2 hours before using.

NOTE—Throughout the following procedures, protect test or assay specimens, the Reference Standard, and solutions containing them, by conducting the procedures without delay, under subdued light, or using low-actinic glassware.

Identification—It meets the requirements under *Identification—Organic Nitrogenous Bases* ⟨181⟩, USP Prochlorperazine Maleate RS being used as the standard for comparison.

pH ⟨791⟩: between 4.2 and 6.2.

Other requirements—It meets the requirements under *Injections* ⟨1⟩.

Assay—

pH 2.0 sodium phosphate solution—Dissolve 2.68 g of dibasic sodium phosphate in 950 mL of water, and adjust with dilute phosphoric acid (1 in 2) or 1 *N* sodium hydroxide to a pH of 2.0.

Mobile phase—Prepare a filtered and degassed mixture of acetonitrile and *pH 2.0 sodium phosphate solution* (65:35). Make adjustments if necessary (see *System Suitability* under *Chromatography* ⟨621⟩).

Internal standard solution—Transfer about 250 mg of chlorpromazine hydrochloride to a 250-mL volumetric flask, add *Mobile phase* to volume, and mix.

Standard preparation—Transfer about 40 mg of USP Prochlorperazine Maleate RS, accurately weighed, to a 25-mL volumetric flask, add *Mobile phase* to volume, and mix. Transfer 2.0 mL of this solution to a 25-mL volumetric flask, add 2.0 mL of *Internal standard solution*, dilute with *Mobile phase* to volume, and mix to obtain a solution having a known concentration of about 0.13 mg of USP Prochlorperazine Maleate RS per mL.

Assay preparation—Transfer an accurately measured volume of Prochlorperazine Edisylate Injection, equivalent to about 25 mg of prochlorperazine, to a 25-mL volumetric flask, add *Mobile phase* to volume, and mix. Transfer 2.0 mL of this solution to a 25-mL volumetric flask, add 2.0 mL of *Internal standard solution*, dilute with *Mobile phase* to volume, and mix.

Chromatographic system (see *Chromatography* ⟨621⟩)—The liquid chromatograph is equipped with a 254-nm detector and a 4.6-mm × 25-cm column that contains packing L10. The flow rate is about 1.5 mL per minute. Chromatograph the *Standard preparation*, and record the peak responses as directed under *Procedure:* the tailing factor for the analyte peak is not more than 1.5, the resolution, *R*, between the analyte and internal standard peaks is not less than 2.0, and the relative standard deviation for replicate injections is not more than 2.0%.

Procedure—Separately inject equal volumes (about 10 µL) of the *Standard preparation* and the *Assay preparation* into the chromatograph, record the chromatograms, and measure the responses for the major peaks. The relative retention times are about 1.0 for chlorpromazine and 1.5 for prochlorperazine. Calculate the quantity, in mg, of prochlorperazine ($C_{20}H_{24}ClN_3S$) in each mL of the Injection taken by the formula:

$$(373.94/606.09)(312.5C/V)(R_U/R_S),$$

in which 373.94 and 606.09 are the molecular weights of prochlorperazine and prochlorperazine maleate, respectively, C is the concentration, in mg per mL, of USP Prochlorperazine Maleate RS in the *Standard preparation*, V is the volume, in mL, of Injection taken, and R_U and R_S are the peak response ratios obtained from the *Assay preparation* and the *Standard preparation*, respectively.

Prochlorperazine Edisylate Oral Solution

» Prochlorperazine Edisylate Oral Solution contains not less than 92.0 percent and not more than 108.0 percent of the labeled amount of prochlorperazine ($C_{20}H_{24}ClN_3S$).

Packaging and storage—Preserve in tight, light-resistant containers.

Labeling—Label it to indicate that it is to be diluted to appropriate strength with water or other suitable fluid prior to administration.

Reference standard—*USP Prochlorperazine Maleate Reference Standard*—Dry in vacuum at 60° for 2 hours before using.

NOTE—Throughout the following procedures, protect test or assay specimens, the Reference Standard, and solutions containing them, by conducting the procedures without delay, under subdued light, or using low-actinic glassware.

Identification—A 1 in 10 dilution of it with water responds to *Identification tests A* and *B* under *Prochlorperazine Edisylate Syrup*.

Assay—

Standard preparation—Prepare as directed in the *Assay* under *Prochlorperazine Edisylate Syrup*.

Assay preparation—Dilute a volume of Prochlorperazine Edisylate Oral Solution, equivalent to about 50 mg of prochlorperazine, with dilute hydrochloric acid (1 in 100) to 250 mL, and mix. Transfer 10.0 mL of the solution to a separator (the stopcock of which is lubricated with a mixture of starch and glycerin), and proceed as directed for *Assay preparation* under *Prochlorperazine Edisylate Syrup*, except to dilute the acid extracts with dilute hydrochloric acid (1 in 100) to 250.0 mL.

Procedure—Proceed as directed in the *Assay* under *Prochlorperazine Edisylate Syrup*. Calculate the quantity, in mg, of prochlorperazine ($C_{20}H_{24}ClN_3S$) in each mL of the Oral Solution taken by the formula:

$$6.25(0.617C/V)(A_{254} - A_{278})_U/(A_{254} - A_{278})_S,$$

in which V is the volume, in mL, of Oral Solution taken.

Prochlorperazine Edisylate Syrup

» Prochlorperazine Edisylate Syrup contains, in each 100 mL, an amount of prochlorperazine edisylate equivalent to not less than 92.0 mg and not more than 108.0 mg of prochlorperazine ($C_{20}H_{24}ClN_3S$).

Packaging and storage—Preserve in tight, light-resistant containers.

Reference standard—*USP Prochlorperazine Maleate Reference Standard*—Dry in vacuum at 60° for 2 hours before using.

NOTE—Throughout the following procedures, protect test or assay specimens, the Reference Standard, and solutions containing them, by conducting the procedures without delay, under subdued light, or using low-actinic glassware.

Identification—

A: To 2 mL of Syrup add 3 mL of water and 3 or 4 drops of ferric chloride TS: a stable red color is produced.

B: To 1 mL of Syrup add 10 mL of bromine TS, previously warmed to room temperature: essentially no color change occurs (distinction from *chlorpromazine hydrochloride*, which immediately produces a green color).

Assay—

Standard preparation—Dissolve a suitable quantity of USP Prochlorperazine Maleate RS, accurately weighed, in dilute hydrochloric acid (1 in 100) to obtain a solution having a known concentration of about 325 µg per mL. Transfer 10 mL of this solution to a 250-mL separator, add about 20 mL of water, render alkaline with 6 *N* ammonium hydroxide, and extract with three 25-mL portions of ether. Extract the combined ether extracts with four 25-mL portions of dilute hydrochloric acid (1 in 100), collecting the aqueous solutions in a 250-mL volumetric flask. Aerate the combined solutions to remove dissolved ether, add dilute hydrochloric acid (1 in 100) to volume, and mix.

Assay preparation—Dilute a volume of Prochlorperazine Edisylate Syrup, equivalent to about 5 mg of prochlorperazine, with dilute hydrochloric acid (1 in 100) to 200.0 mL, and mix. Transfer 60.0 mL of the solution to a separator, add 40 mL of methanol and 1 mL of ammonium hydroxide, and extract with one 40-mL portion and then with three 20-mL portions of ether. Extract the combined ether extracts with four 20-mL portions of dilute hydrochloric acid (1 in 100), collecting the aqueous solutions in a 200-mL volumetric flask. Aerate the combined solutions to remove dissolved ether, add dilute hydrochloric acid (1 in 100) to volume, and mix.

Procedure—Concomitantly determine the absorbances of the *Standard preparation* and the *Assay preparation* in 1-cm cells, at 278 nm and at the wavelength of maximum absorbance at about 254 nm, with a suitable spectrophotometer, using dilute hydrochloric acid (1 in 100) as the blank. Calculate the quantity, in mg, of $C_{20}H_{24}ClN_3S$ in each 100 mL of the Syrup taken by the formula:

$$66.7(0.617C/V)(A_{254} - A_{278})_U/(A_{254} - A_{278})_S,$$

in which 0.617 is the ratio of the molecular weight of prochlorperazine to that of prochlorperazine maleate, C is the concentration, in µg per mL, of USP Prochlorperazine Maleate RS in the *Standard preparation*, V is the volume, in mL, of Syrup taken, and the parenthetic expressions are the differences in the absorbances of the two solutions at the wavelengths indicated by the subscripts for the *Assay preparation* ($_U$) and the *Standard preparation* ($_S$).

Prochlorperazine Maleate

$C_{20}H_{24}ClN_3S.2C_4H_4O_4$ 606.09

10*H*-Phenothiazine, 2-chloro-10-[3-(4-methyl-1-piperazinyl)-propyl]-, (*Z*)-2-butenedioate (1:2).

2-Chloro-10-[3-(4-methyl-1-piperazinyl)propyl]phenothiazine maleate (1:2) [*84-02-6*].

» Prochlorperazine Maleate contains not less than 98.0 percent and not more than 101.5 percent of $C_{20}H_{24}ClN_3S.2C_4H_4O_4$, calculated on the dried basis.

Packaging and storage—Preserve in tight, light-resistant containers.

Reference standard—*USP Prochlorperazine Maleate Reference Standard*—Dry in vacuum at 60° for 2 hours before using.

NOTE—Throughout the following procedures, protect test or assay specimens, the Reference Standard, and solutions containing them, by conducting the procedures without delay, under subdued light, or using low-actinic glassware.

Identification—The infrared absorption spectrum of a potassium bromide dispersion of it exhibits maxima only at the same wavelengths as that of a similar preparation of USP Prochlorperazine Maleate RS.

Loss on drying ⟨731⟩—Dry it in vacuum at 60° for 2 hours: it loses not more than 0.5% of its weight.

Residue on ignition ⟨281⟩: not more than 0.1%.

Ordinary impurities ⟨466⟩—

Test solution: a mixture of methanol and 1 *N* sodium hydroxide (9:1).

Standard solution: a mixture of methanol and 1 *N* sodium hydroxide (9:1).

Eluant: a mixture of methanol and ammonium hydroxide (100:1).

Visualization: 1.

Assay—Transfer to a beaker about 400 mg of Prochlorperazine Maleate, accurately weighed, and dissolve in 30 mL of chloroform, warming on a steam bath to effect solution. Add 100 mL of glacial acetic acid, cool to room temperature, and titrate with 0.05 *N* perchloric acid VS, determining the end-point potentiometrically. Perform a blank determination, and make any necessary correction. Each mL of 0.05 *N* perchloric acid is equivalent to 15.15 mg of $C_{20}H_{24}ClN_3S.2C_4H_4O_4$.

Prochlorperazine Maleate Tablets

» Prochlorperazine Maleate Tablets contain an amount of prochlorperazine maleate equivalent to not less than 95.0 percent and not more than 105.0 percent of the labeled amount of prochlorperazine ($C_{20}H_{24}ClN_3S$).

Packaging and storage—Preserve in well-closed containers, protected from light.

Reference standard—*USP Prochlorperazine Maleate Reference Standard*—Dry in vacuum at 60° for 2 hours before using.

NOTE—Throughout the following procedures, protect test or assay specimens, the Reference Standard, and solutions containing them, by conducting the procedures without delay, under subdued light, or using low-actinic glassware.

Identification—The Tablets, previously washed with several portions of water to remove the coating if necessary, meet the requirements under *Identification—Organic Nitrogenous Bases* ⟨181⟩.

Disintegration ⟨701⟩: 30 minutes.

Uniformity of dosage units ⟨905⟩: meet the requirements.

Assay—

Ion-pairing solution—Transfer 4.33 g of sodium 1-octanesulfonate, accurately weighed, to a 1-liter volumetric flask. Dissolve in 500 mL of water, add 4.0 mL of glacial acetic acid, dilute with water to volume, and mix.

Mobile phase—Prepare a suitable filtered and degassed mixture of *Ion-pairing solution*, acetonitrile, and methanol (45:40:15).

Internal standard preparation—Prepare a solution of trifluoperazine hydrochloride in a mixture of 0.2 *N* hydrochloric acid and methanol (1:1) having a concentration of about 140 µg per mL.

Standard preparation—Dissolve an accurately weighed quantity of USP Prochlorperazine Maleate RS in *Internal standard preparation* to obtain a solution having a known concentration of about 0.2 mg per mL.

Assay preparation—Weigh and finely powder not less than 20 Prochlorperazine Maleate Tablets. Transfer a portion of the powder, equivalent to about 20 mg of prochlorperazine maleate, to a 100-mL volumetric flask. Add 90 mL of *Internal standard preparation*, and sonicate for 45 minutes. Dilute with *Internal standard preparation* to volume, mix, and filter, discarding the first 20 mL of the filtrate.

Chromatographic system (see *Chromatography* ⟨621⟩)—The liquid chromatograph is equipped with a 254-nm detector and a 4-mm × 30-cm column that contains packing L1. The flow rate is about 2 mL per minute. Chromatograph the *Standard preparation*, and record the peak responses as directed under *Procedure:* the tailing factor for the analyte peak is not more than 2.0, the resolution *R*, between the analyte and internal standard peaks is not less than 4.0, and the relative standard deviation for replicate injections is not more than 2.0%.

Procedure—Separately inject equal volumes (about 10 µL) of the *Standard preparation* and the *Assay preparation* into the chromatograph, record the chromatograms, and measure the responses for the major peaks. The relative retention times are about 1.0 for prochlorperazine and 1.3 for the internal standard. Calculate the quantity, in mg, of $C_{20}H_{24}ClN_3S$ (prochlorperazine) in the portion of Tablets taken by the formula:

$$(373.94/606.09)(100C)(R_U/R_S),$$

in which 373.94 and 606.09 are the molecular weights of prochlorperazine and prochlorperazine maleate, respectively, C is the concentration, in mg per mL, of USP Prochlorperazine Maleate RS in the *Standard preparation*, and R_U and R_S are the peak response ratios of the prochlorperazine peak to the internal standard peak obtained from the *Assay preparation* and the *Standard preparation*, respectively.

Procyclidine Hydrochloride

$C_{19}H_{29}NO.HCl$ 323.91

1-Pyrrolidinepropanol, α-cyclohexyl-α-phenyl-, hydrochloride.

α-Cyclohexyl-α-phenyl-1-pyrrolidinepropanol hydrochloride [*1508-76-5*].

» Procyclidine Hydrochloride contains not less than 99.0 percent and not more than 101.0 percent of $C_{19}H_{29}NO.HCl$, calculated on the dried basis.

Packaging and storage—Preserve in tight, light-resistant containers, and store in a dry place.

Reference standard—*USP Procyclidine Hydrochloride Reference Standard*—Dry in vacuum at 105° for 4 hours before using.

Identification—

A: The infrared absorption spectrum of a potassium bromide dispersion of it, previously dried, exhibits maxima only at the same wavelengths as that of a similar preparation of USP Procyclidine Hydrochloride RS.

B: Dissolve about 250 mg in 10 mL of water in a separator, render alkaline with 6 *N* ammonium hydroxide, and extract with three 10-mL portions of ether. Filter the ether extracts slowly through a layer of about 2 g of anhydrous granular sodium sulfate supported on glass wool, evaporate the ether with a current of warm air, and scratch the surface of the container to induce crystallization of the residue: the procyclidine so obtained melts between 83° and 87°, the procedure for *Class I* being used (see *Melting Range or Temperature* ⟨741⟩).

C: A solution (1 in 100) responds to the tests for *Chloride* ⟨191⟩.

pH ⟨791⟩: between 5.0 and 6.5, in a solution (1 in 100).

Loss on drying ⟨731⟩—Dry it in vacuum at 105° for 4 hours: it loses not more than 0.5% of its weight.

Residue on ignition ⟨281⟩: not more than 0.1%.

Related compounds—Dissolve approximately 200 mg of Procyclidine Hydrochloride in 20 mL of water, and render the solution alkaline by adding 1.5 mL of 6 *N* ammonium hydroxide. Extract with three 15-mL portions of chloroform, wash the combined extracts with 20 mL of water, discard the water washing, and filter the chloroform solution through a layer of 3 to 4 g of anhydrous granular sodium sulfate supported on glass wool. Reduce the volume to 5 mL by evaporating with the aid of gentle heat and a current of air. Inject 2 µL of this solution into a suitable gas chromatograph (see *Chromatography* ⟨621⟩) equipped with a flame-ionization detector, and record the chromatogram to 2.5 relative to the retention time of the principal (procyclidine) peak. Under typical conditions, the instrument contains a 1-m × 2-mm glass column packed with 10 percent polyethylene glycol 20,000 and 2 percent potassium hydroxide on packing S1A. The column is maintained at a temperature of about 180°, the injection port is maintained at 210°, the detector block is maintained at about 220°, and dry helium is used as the carrier gas at a flow rate of about 60 mL per minute. From the total area under the curve, excluding the solvent peak, calculate the percentage of total impurities by area normalization: not more than 4.0% is found.

Assay—Dissolve about 700 mg of Procyclidine Hydrochloride, accurately weighed, in 75 mL of glacial acetic acid in a 250-mL beaker, warming, if necessary, to effect solution. Cool, add 10 mL of mercuric acetate TS, and titrate with 0.1 *N* perchloric acid VS, determining the end-point potentiometrically. Perform a blank determination, and make any necessary correction. Each mL of 0.1 *N* perchloric acid is equivalent to 32.39 mg of $C_{19}H_{29}NO.HCl$.

Procyclidine Hydrochloride Tablets

» Procyclidine Hydrochloride Tablets contain not less than 93.0 percent and not more than 107.0 percent of the labeled amount of $C_{19}H_{29}NO.HCl$.

Packaging and storage—Preserve in tight containers, and store in a dry place.

Reference standard—*USP Procyclidine Hydrochloride Reference Standard*—Dry in vacuum at 105° for 4 hours before using.

Identification—

A: Triturate a portion of finely powdered Tablets, equivalent to about 10 mg of procyclidine hydrochloride, with 20 mL of chloroform, filter, evaporate the filtrate on a steam bath to dryness, and dry the residue at 105° for 1 hour: the infrared absorption spectrum of a potassium bromide dispersion of the procyclidine hydrochloride so obtained exhibits maxima only at the same wavelengths as that of a similar preparation of USP Procyclidine Hydrochloride RS.

B: A portion of finely powdered Tablets, equivalent to about 50 mg of procyclidine hydrochloride, responds to *Identification test B* under *Procyclidine Hydrochloride*.

Dissolution ⟨711⟩—

Medium: water; 900 mL.

Apparatus 2: 50 rpm.

Time: 45 minutes.

Procedure—Determine the amount of $C_{19}H_{29}NO.HCl$ dissolved, employing the procedure set forth in the *Assay*, making any necessary modifications.

Tolerances—Not less than 75% (*Q*) of the labeled amount of $C_{19}H_{29}NO.HCl$ is dissolved in 45 minutes.

Uniformity of dosage units ⟨905⟩: meet the requirements.

Related compounds—Using a portion of powdered Tablets, equivalent to 200 mg of procyclidine hydrochloride, proceed as directed in the test for *Related compounds* under *Procyclidine Hydrochloride*.

Assay—

Bromocresol purple solution—Dissolve 250 mg of bromocresol purple in dilute glacial acetic acid (1 in 50) to make 1000 mL.

Standard preparation—Transfer about 25 mg of USP Procyclidine Hydrochloride RS, accurately weighed, to a 100-mL volumetric flask, add water to volume, and mix. Transfer 10.0 mL of this solution to a second 100-mL volumetric flask, dilute with *Bromocresol purple solution* to volume, and mix. The concentration of the *Standard preparation* is about 25 µg per mL.

Assay preparation—Weigh and finely powder not less than 20 Procyclidine Hydrochloride Tablets. Transfer an accurately weighed portion of the powder, equivalent to about 2.5 mg of procyclidine hydrochloride, to a 100-mL volumetric flask, add 10.0 mL of water, and mix. Dilute with *Bromocresol purple solution* to volume, mix, and allow the undissolved particles to settle. Use the supernatant liquid as directed in the *Procedure*.

Procedure—Transfer 5.0 mL each of the *Standard preparation* and the *Assay preparation* to individual 60-mL separators. Transfer 0.5 mL of water and 4.5 mL of *Bromocresol purple solution* to a third separator to provide the blank. Extract each solution with 20.0 mL of chloroform, and filter each extract, discarding the first 5 mL of the filtrate. Concomitantly determine the absorbance of each subsequent filtrate in a 1-cm cell at the wavelength of maximum absorbance at about 405 nm, with a suitable spectrophotometer, against the blank. Calculate the quantity, in mg, of $C_{19}H_{29}NO.HCl$ in the portion of Tablets taken by the formula:

$$0.1C(A_U/A_S),$$

in which C is the concentration, in µg per mL, of USP Procyclidine Hydrochloride RS in the *Standard preparation*, and A_U and A_S are the absorbances of the solutions from the *Assay preparation* and the *Standard preparation*, respectively.

Progesterone

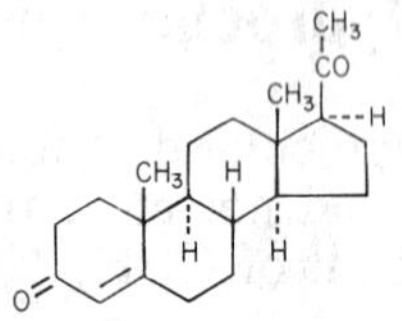

$C_{21}H_{30}O_2$ 314.47
Pregn-4-ene-3,20-dione.
Progesterone [*57-83-0*].

» Progesterone contains not less than 97.0 percent and not more than 103.0 percent of $C_{21}H_{30}O_2$, calculated on the dried basis.

Packaging and storage—Preserve in tight, light-resistant containers.

Reference standard—*USP Progesterone Reference Standard*—Dry in vacuum over silica gel for 4 hours before using.

Identification—

A: The infrared absorption spectrum of a potassium bromide dispersion of it, previously dried, exhibits maxima only at the same wavelengths as that of a similar preparation of USP Progesterone RS.

B: The ultraviolet absorption spectrum of a 1 in 100,000 solution in methanol exhibits maxima and minima at the same wavelengths as that of a similar preparation of USP Progesterone RS, concomitantly measured.

Melting range ⟨741⟩: between 126° and 131°. It may exist also in a polymorphic modification, melting at about 121°.

Specific rotation ⟨781⟩: between +175° and +183°, calculated on the dried basis, determined in a solution in dioxane containing 200 mg in each 10 mL.

Loss on drying ⟨731⟩—Dry it in vacuum over silica gel for 4 hours: it loses not more than 0.5% of its weight.

Assay—

Mobile phase—Prepare a filtered and degassed mixture of water and isopropyl alcohol (72:28). Make adjustments if necessary (see *System Suitability* under *Chromatography* ⟨621⟩).

Internal standard solution—Transfer about 66 mg of methyltestosterone to a 10-mL volumetric flask, add dilute alcohol (85 in 100) to volume, and mix.

Standard preparation—Dissolve an accurately weighed quantity of USP Progesterone RS in dilute alcohol (85 in 100) to obtain a solution having a known concentration of about 2.5 mg per mL. Transfer 4.0 mL of this solution to a 10-mL volumetric flask, add 1.0 mL of *Internal standard solution*, then add dilute alcohol (85 in 100) to volume, and mix to obtain a solution having a known concentration of about 1 mg of USP Progesterone RS per mL.

Assay preparation—Transfer about 10 mg of progesterone, accurately weighed, to a 10-mL volumetric flask, add 1.0 mL of *Internal standard solution*, then add dilute alcohol (85 in 100) to volume, and mix.

Chromatographic system (see *Chromatography* ⟨621⟩)—The liquid chromatograph is equipped with a 254-nm detector and a 4-mm × 30-cm column that contains 10-µm packing L1. The flow rate is about 1.5 mL per minute. Chromatograph the *Standard preparation*, and record the peak responses as directed under *Procedure:* the resolution, *R*, between the analyte and internal standard peaks is not less than 3.5, and the relative standard deviation for replicate injections is not more than 1.5%.

Procedure—Separately inject equal volumes (5 µL) of the *Standard preparation* and the *Assay preparation* into the chromatograph, record the chromatograms, and measure the responses for the major peaks. The relative retention times are about 2.0 for progesterone and 1.0 for methyltestosterone. Calculate the quantity, in mg, of $C_{21}H_{30}O_2$ in the portion of progesterone taken by the formula:

$$10C(R_U/R_S),$$

in which C is the concentration, in mg per mL, of USP Progesterone RS in the *Standard preparation*, and R_U and R_S are the peak response ratios obtained from the *Assay preparation* and the *Standard preparation*, respectively.

Progesterone Injection

» Progesterone Injection is a sterile solution of Progesterone in a suitable solvent. It contains not less than 90.0 percent and not more than 110.0 percent of the labeled amount of $C_{21}H_{30}O_2$.

Packaging and storage—Preserve in single-dose or in multiple-dose containers, preferably of Type I or Type III glass.

Reference standards—*USP Progesterone Reference Standard*—Dry in vacuum over silica gel for 4 hours before using. *USP Methyltestosterone Reference Standard*—Dry at 105° for 4 hours before using.

Identification—Insert a pledget of fine glass wool into the base of a chromatographic tube of about 200 × 25 mm. Mix 8.0 mL of nitromethane with 7.0 g of purified siliceous earth in a 150-mL beaker until uniform, and transfer to the chromatographic tube, packing lightly with a suitable tamping rod. Pack a pledget of glass wool on the top of the column. Dilute 1 mL of the Injection with *n*-heptane to obtain a solution having a concentration of about 1 mg of progesterone per mL. Transfer 4.0 mL of this solution to the prepared column. Pass 300 mL of *n*-heptane through the column, discarding the first 120 mL of the eluate. Collect the subsequent eluate in a 250-mL beaker. Evaporate the solution under a stream of nitrogen on a steam bath to about 50 mL, transfer to a 100-mL beaker, and evaporate to dryness. Remove the last traces of *n*-heptane by adding 1 mL of methanol and again drying. Dry the specimen over silica gel for 4 hours: the infrared absorption spectrum of a potassium bromide dispersion of the residue so obtained exhibits maxima only at the same wavelengths as that of a similar preparation of USP Progesterone RS.

Other requirements—It meets the requirements under *Injections* ⟨1⟩.

Assay—

Internal standard solution—Dissolve USP Methyltestosterone RS in dilute alcohol (85 in 100) to obtain a solution containing about 6.6 mg per mL.

Mobile solvent—Prepare a suitable degassed solution of isopropyl alcohol in water (28 in 100) such that the retention times of the internal standard and progesterone peaks are between 3 and 6 minutes and 7 and 11 minutes, respectively.

Standard preparation—Dissolve an accurately weighed quantity of USP Progesterone RS in dilute alcohol (85 in 100) to obtain a solution having a known concentration of about 2.5 mg per mL.

Assay preparation—Transfer an accurately measured volume of Progesterone Injection, equivalent to about 25 mg of progesterone, to a polytetrafluoroethylene-lined, screw-capped, 15-mL centrifuge tube. Extract with 5 mL of dilute alcohol (85 in 100), shaking the mixture gently for 10 minutes. Centrifuge for 15 minutes or until the alcohol layer is clear, and transfer the alcoholic layer to a polytetrafluoroethylene-lined, screw-capped, 25-mL test tube, rinsing the transfer device with about 1 mL of dilute alcohol (85 in 100). Repeat the process with two 5-mL portions of the same dilute alcohol, omitting the rinse. To the combined alcoholic extracts and rinsing add 2.0 mL of *Internal standard solution*, and mix.

Standard curve—To four separate polytetrafluoroethylene-lined, screw-capped centrifuge tubes, pipet 3-, 4-, 5-, and 6-mL portions, respectively, of the *Standard preparation*. Add dilute alcohol (85 in 100) to each tube to make about 8 mL of solution. Transfer 1.0 mL of *Internal standard solution* to each tube, and mix. Inject 5.0 µL from each solution, successively, into a suitable high-pressure liquid chromatograph (see *Chromatography* ⟨621⟩) of the general type equipped with a detector for monitoring ultraviolet light absorption at about 254 nm and equipped with a suitable recorder, and capable of providing column pressures of 1500 psi. Pack a stainless steel column, 1 m in length and 2.1 mm in internal diameter, with packing L2. Column and mobile phase are maintained at room temperature. Measure the peak areas, A_P and A_M, of the progesterone and methyltestosterone peaks in each chromatogram, and calculate the ratio, R_S, by the formula:

$$A_P/A_M.$$

Plot the standard curve of the values of R_S against the amount, in mg, of USP Progesterone RS contained in each portion of *Standard preparation* taken.

System suitability—Six replicate injections of a single standard solution show a relative standard deviation for the ratio, R_S, of not more than 1.5%, and the resolution factor, R (see *Chromatography* ⟨621⟩), is not less than 3.5 between the peaks from progesterone and methyltestosterone.

Procedure—Inject 5.0 µL of the *Assay preparation* into the chromatograph, obtain a chromatogram, and calculate the ratio, R_U, of the peak area of progesterone to that of methyltestosterone as directed under *Standard curve*. Read from the *Standard curve* the quantity, in mg, of progesterone, and multiply this value by 2 to obtain the quantity, in mg, of $C_{21}H_{30}O_2$ in the volume of Injection taken.

Progesterone Intrauterine Contraceptive System

» Progesterone Intrauterine Contraceptive System contains not less than 90.0 percent and not more than 110.0 percent of the labeled amount of $C_{21}H_{30}O_2$. It is sterile.

Packaging and storage—Preserve in sealed, single-unit containers.

Reference standard—*USP Progesterone Reference Standard*—Dry in vacuum over silica gel for 4 hours before using.

Identification—Cut off and discard the sealed ends of the drug-containing cores of 2 Systems, and force the contents of the tubes into a small centrifuge tube. Add 3 mL of methanol, insert the stopper in the tube, mix, centrifuge, and transfer the clear, supernatant liquid to a small beaker. Evaporate the methanol to dryness, wash the residue with two 4-mL portions of cyclohexane, and discard the washings. Dry the residue in vacuum at 50° to constant weight: the infrared absorption spectrum of a mineral oil dispersion of the dried residue so obtained exhibits maxima only at the same wavelengths as that of a similar preparation of USP Progesterone RS.

Sterility—It meets the requirements under *Sterility Tests* ⟨71⟩.

Uniformity of dosage units ⟨905⟩: meets the requirements.

Chromatographic impurities—

Test preparation—Remove the drug-containing core from 1 System, as directed in the *Assay*, transferring it to a small flask with 25 mL of methanol. Shake vigorously for several minutes, and allow the insoluble portion to settle. The resulting supernatant solution is the *Test preparation*.

Procedure—Divide a 20-cm × 20-cm thin-layer chromatographic plate, coated with a 0.25-mm layer of chromatographic silica gel mixture, into sections 2 cm apart. In successive sections of the plate, on a line 2 cm from the lower edge of the plate and parallel to it, apply 1 µL, 2 µL, 3 µL, and 100 µL of the *Test preparation*. Develop the plate in a suitable pre-equilibrated chromatographic chamber with a solvent system consisting of a mixture of chloroform and ethyl acetate (2:1) until the solvent front has moved 10 cm above the point of application of the spots. Remove the plate, and allow to air-dry. Observe the dried plate under short-wavelength ultraviolet light (254 nm). If spots other than the principal spot are observed in the lane of the 100-µL specimen, estimate the concentration of each by comparison with the 1-µL (1%), 2-µL (2%), and 3-µL (3%) spots. The requirement is met if the sum of impurities in the 100-µL specimen does not exceed 3%.

Drug-release pattern—Remove the attached sutures from 10 Systems, and secure each system to a corrosion-resistant wire of sufficient length such that the systems are completely immersed during the shaking operation but do not touch the bottoms of the flasks. Suspend each system by the attached wire from the arm of a mechanical shaker designed to travel 2.5 cm in each direction in a vertically reciprocating cycle, at a speed of 2.5 cycles per second, so that each system is immersed in a separate 250-mL volumetric flask containing 230 mL of water, pre-equilibrated to 60 ± 0.1°. Immerse the volumetric flasks in an insulated constant-temperature water bath, maintained at 60 ± 0.1° and having a suitable means of maintaining the water level, so that the water level of the bath is above the water level in the flasks. Employ a rack or other suitable means of support for the flasks in the water bath.

Operate the shaker under the conditions described above for 23.5 hours, then remove the flasks and the systems from the bath. Remove the systems from the flasks, and immerse each system in a different flask containing 230 mL of water, pre-equilibrated to 60 ± 0.1°, and immerse these flasks in the water bath. Repeat this shaking operation daily for 12 days, using different flasks each day.

Determine the quantity of progesterone in the solutions from each of the twelve days of testing as follows. Immediately add 15 mL of methanol to each solution, allow to cool to room temperature, dilute with water to volume, and mix. Concomitantly determine the ultraviolet absorbances of each test solution and of a solution of USP Progesterone RS in the same medium, having a known concentration of about 7 µg per mL, in 1-cm cells at the wavelength of maximum absorbance at about 248 nm, with a suitable spectrophotometer, against a blank of water and methanol (47:3). Calculate the progesterone release rate, in mg per day, in the solutions by the formula:

$$(A_U/A_S)(24/23.5)0.25C,$$

in which A_U and A_S are the absorbances of the test solution and the Standard solution, respectively, and C is the concentration, in µg per mL, of USP Progesterone RS in the Standard solution. Plot the drug-release pattern curve, on suitable graph paper, with *Day of Test* as the abscissa and *Progesterone Release Rate*, in mg per day, as the ordinate. Plot the data point that corresponds to day 1 as 0.5 day on the abscissa, the point for day 2 at 1.5 days, and so on. Draw the best fit line (using the method of linear regression analysis) through the twelve data points. Determine the average release rate of all systems for test days 6, 9, and 12, respectively, from the value of the ordinate, in mg per day, corresponding to the best fit line at abscissa values of 5.5, 8.5, and 11.5 days, respectively. On day 6, the average release rate of all 10 Systems is 1.35 ± 0.2 mg of progesterone per day; on day 9, the average release rate is 1.25 ± 0.2 mg of progesterone per day; and on day 12, the average release rate is 1.20 ± 0.2 mg of progesterone per day.

Assay—Cut off the lower sealed end of the drug-containing core of a number of Progesterone Intrauterine Contraceptive Systems, sufficient to provide about 400 mg of progesterone, forcing the viscous liquid core into a 1000-mL volumetric flask. Cut the core sections in half lengthwise, using a sharp blade, taking precautions not to contaminate either the core material or the outside of the membranes. Transfer all of these sections of the systems to the flask containing the core material. Add about 500 mL of methanol to the flask, shake vigorously for 5 to 10 minutes, dilute with methanol to volume, and centrifuge a portion of the solution. Dilute 10.0 mL of the clear, supernatant solution with methanol to 250 mL, and mix. Concomitantly determine the absorbances of this solution and of a Standard solution of USP Progesterone RS, previously dried and accurately weighed, in methanol having a known concentration of about 16 µg of progesterone per mL, in 1-cm cells at the wavelength of maximum absorbance at about 241 nm, with a suitable spectrophotometer, using methanol as the blank. Calculate the quantity, in mg, of $C_{21}H_{30}O_2$ in each System taken by the formula:

$$25(C/N)(A_U/A_S),$$

in which C is the concentration, in µg per mL, of USP Progesterone RS in the Standard solution, N is the number of Systems taken, and A_U and A_S are the absorbances of the test solution and the Standard solution, respectively.

Sterile Progesterone Suspension

» Sterile Progesterone Suspension is a sterile suspension of Progesterone in Water for Injection. It contains not less than 93.0 percent and not more than 107.0 percent of the labeled amount of $C_{21}H_{30}O_2$.

Packaging and storage—Preserve in single-dose or in multiple-dose containers, preferably of Type I glass.

Reference standards—*USP Progesterone Reference Standard*—Dry in vacuum over silica gel for 4 hours before using. *USP Methyltestosterone Reference Standard*—Dry at 105° for 4 hours before using.

Identification—Filter a volume of well-shaken Suspension, equivalent to not less than 100 mg of progesterone, through a medium-porosity, sintered-glass crucible, filtering again through the same crucible if the fluid is not clear. Wash with several 5-mL portions of water until 2 mL of the last washing, evaporated on a steam bath, leaves no weighable residue: the washed solid, dried at 105° to constant weight, melts between 126° and 131°, and responds to *Identification test A* under *Progesterone*.

pH ⟨791⟩: between 4.0 and 7.5.

Other requirements—It meets the requirements under *Injections* ⟨1⟩.

Assay—

Internal standard solution, Mobile solvent, Standard preparation, and *Standard curve*—Prepare as directed in the *Assay* under *Progesterone Injection*.

Assay preparation—Transfer an accurately measured volume of Sterile Progesterone Suspension, equivalent to about 25 mg of progesterone, to a polytetrafluorethylene-lined, screw-capped, 25-mL test tube. Add 16 mL of dilute alcohol (85 in 100), and shake until clear. Add 2.0 mL of *Internal standard solution*, and mix.

System suitability and *Procedure*—Proceed as directed in the *Assay* under *Progesterone Injection*.

Proline

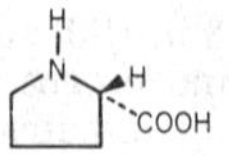

$C_5H_9NO_2$ 115.13

L-Proline.

L-Proline [*147-85-3*].

» Proline contains not less than 98.5 percent and not more than 101.5 percent of $C_5H_9NO_2$, as L-proline, calculated on the dried basis.

Packaging and storage—Preserve in well-closed containers.

Reference standard—*USP L-Proline Reference Standard*—Dry at 105° for 3 hours before using.

Identification—The infrared absorption spectrum of a potassium bromide dispersion of it, previously dried, exhibits maxima only at the same wavelengths as that of a similar preparation of USP L-Proline RS.

Specific rotation ⟨781⟩: between −83.7° and −85.7°, calculated on the dried basis, determined in a solution containing 400 mg in each 10 mL.

Loss on drying ⟨731⟩—Dry it at 105° for 3 hours: it loses not more than 0.4% of its weight.

Residue on ignition ⟨281⟩: not more than 0.4%.

Chloride ⟨221⟩—A 0.73-g portion shows no more chloride than corresponds to 0.50 mL of 0.020 *N* hydrochloric acid (0.05%).

Sulfate ⟨221⟩—A 0.33-g portion shows no more sulfate than corresponds to 0.10 mL of 0.020 *N* sulfuric acid (0.03%).

Arsenic ⟨211⟩: 1.5 ppm.

Iron ⟨241⟩: 0.003%.

Heavy metals, *Method I* ⟨231⟩: 0.0015%.

Assay—Transfer about 100 mg of Proline, accurately weighed, to a 125-mL flask, dissolve in a mixture of 3 mL of formic acid and 50 mL of glacial acetic acid, and titrate with 0.1 *N* perchloric acid VS, determining the end-point potentiometrically. Perform a blank determination, and make any necessary correction. Each mL of 0.1 *N* perchloric acid is equivalent to 11.51 mg of $C_5H_9NO_2$.

Promazine Hydrochloride

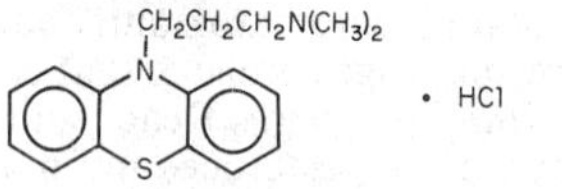

$C_{17}H_{20}N_2S.HCl$ 320.88

10*H*-Phenothiazine-10-propanamine, *N,N*-dimethyl-, monohydrochloride.

10-[3-(Dimethylamino)propyl]phenothiazine monohydrochloride [*53-60-1*].

» Promazine Hydrochloride, dried at 105° for 2 hours, contains not less than 98.0 percent and not more than 102.0 percent of $C_{17}H_{20}N_2S.HCl$.

Packaging and storage—Preserve in tight, light-resistant containers.

Reference standard—*USP Promazine Hydrochloride Reference Standard*—Dry at 105° for 2 hours before using.

NOTE—Throughout the following procedures, protect test or assay specimens, the Reference Standard, and solutions containing them, by conducting the procedures without delay, under subdued light, or using low-actinic glassware.

Completeness and clarity of solution—A solution of it (1 in 10) and a 1 in 10 solution of it in chloroform are practically clear and show not more than a light yellow color.

Identification—

A: The infrared absorption spectrum of a potassium bromide dispersion of it, previously dried, exhibits maxima only at the same wavelengths as that of a similar preparation of USP Promazine Hydrochloride RS.

B: Determine the absorbance of the assay solution employed for measurement of absorbance in the *Assay* at 301 nm, using 0.1 *N* hydrochloric acid as the blank. Similarly determine the absorbance of a 1 in 10 dilution of this solution, prepared with the same acid, at the wavelength of maximum absorbance at about 252 nm: the ratio $10(A_{252}/A_{301})$ is between 7.1 and 7.9.

C: It responds to the tests for *Chloride* ⟨191⟩.

Melting range, *Class I* ⟨741⟩: between 172° and 182°, but the range between beginning and end of melting does not exceed 3°.

pH ⟨791⟩: between 4.2 and 5.2, in a solution (1 in 20).

Loss on drying ⟨731⟩—Dry it at 105° for 2 hours: it loses not more than 0.5% of its weight.

Residue on ignition ⟨281⟩: not more than 0.1%.

Selenium ⟨291⟩—The absorbance of the solution from the *Test Solution*, prepared with 100 mg of Promazine Hydrochloride and 200 mg of magnesium oxide, is not greater than one-half that from the *Standard Solution* (0.003%).

Heavy metals, *Method II* ⟨231⟩: 0.005%.

Chromatographic purity—[NOTE—Perform this test under conditions of subdued light and with no unnecessary delays between the preparation of the solutions and the development of the chromatographic plate.]

Developing solvent—Mix 95 volumes of toluene with 15 volumes of ethanol and 1 volume of ammonium hydroxide.

Standard preparations—Dissolve an accurately weighed quantity of USP Promazine Hydrochloride RS in methanol to obtain *Standard preparation A* having a known concentration of 0.4 mg per mL. Dilute quantitatively with methanol to obtain *Standard preparations* having the following compositions:

Standard preparation	Dilution	Concentration (μg RS per mL)	Percentage (%, for comparison with test specimens)
A	(undiluted)	400	2.0
B	(1 in 2)	200	1.0
C	(3 in 10)	120	0.6
D	(1 in 10)	40	0.2

Test preparation—Dissolve an accurately weighed quantity of Promazine Hydrochloride in methanol to obtain a solution containing 20 mg per mL.

Procedure—On a suitable thin-layer chromatographic plate (see *Chromatography* ⟨621⟩), coated with a 0.25-mm layer of chromatographic silica gel mixture, apply separately 10 μL of the *Test preparation* and 10 μL of each *Standard preparation*, and allow to dry. Position the plate in a chromatographic chamber and develop the chromatograms in the *Developing solvent* until the solvent front has moved about three-fourths of the length of the plate. Remove the plate from the developing chamber, mark the solvent front, and allow the solvent to evaporate by air-drying for 15 minutes. Examine the plate under short-wavelength ultraviolet light. Compare the intensities of any secondary spots observed in the chromatogram of the *Test preparation* with those of the principal spots in the chromatograms of the *Standard preparations:* the sum of the intensities of secondary spots obtained from the *Test preparation* corresponds to not more than 2.0% of related compounds, with no single impurity corresponding to more than 1.0%.

Assay—[NOTE—Use low-actinic glassware.] Transfer about 50 mg of Promazine Hydrochloride, previously dried and accurately weighed, to a 1000-mL volumetric flask, add 0.1 *N* hydrochloric acid to volume, and mix. Without delay, concomitantly determine the absorbances of this solution and of a Standard solution of USP Promazine Hydrochloride RS in the same medium having a known concentration of about 50 μg per mL in 1-cm cells at the wavelength of maximum absorbance at about 301 nm, with a suitable spectrophotometer, using 0.1 *N* hydrochloric acid as the blank. Calculate the quantity, in mg, of $C_{17}H_{20}N_2S.HCl$ in the portion of Promazine Hydrochloride taken by the formula:

$$C(A_U/A_S),$$

in which *C* is the concentration, in μg per mL, of USP Promazine Hydrochloride RS in the Standard solution, and A_U and A_S are the absorbances of the solution of Promazine Hydrochloride and the Standard solution, respectively.

Promazine Hydrochloride Injection

» Promazine Hydrochloride Injection is a sterile solution of Promazine Hydrochloride in Water for Injection. It contains not less than 95.0 percent and not more than 110.0 percent of the labeled amount of $C_{17}H_{20}N_2S.HCl$.

Packaging and storage—Preserve in single-dose or in multiple-dose containers, preferably of Type I glass, protected from light.

Reference standard—*USP Promazine Hydrochloride Reference Standard*—Dry at 105° for 2 hours before using.

NOTE—Throughout the following procedures, protect test or assay specimens, the Reference Standard, and solutions containing them, by conducting the procedures without delay, under subdued light, or using low-actinic glassware.

Identification—

A: It meets the requirements under *Identification—Organic Nitrogenous Bases* ⟨181⟩.

B: It responds to *Identification test B* under *Promazine Hydrochloride.*

pH ⟨791⟩: between 4.0 and 5.5.

Other requirements—It meets the requirements under *Injections* ⟨1⟩.

Assay—[NOTE—Use low-actinic glassware.] Transfer a volume of Promazine Hydrochloride Injection, equivalent to about 50 mg of promazine hydrochloride, to a 100-mL volumetric flask, dilute with 0.1 *N* hydrochloric acid to volume, and mix. Transfer 10.0 mL of the solution to a 250-mL separator, add 20 mL of water, render alkaline with ammonium hydroxide, and extract with four 25-mL portions of ether. Extract the combined ether extracts with five 15-mL portions of 0.1 *N* hydrochloric acid, collecting the aqueous extracts in a 100-mL volumetric flask. Aerate to remove residual ether, dilute with 0.1 *N* hydrochloric acid to volume, and mix. Without delay, concomitantly determine the absorbances of this solution and of a Standard solution of USP Promazine Hydrochloride RS in the same medium having a known concentration of about 50 μg per mL in 1-cm cells at the wavelength of maximum absorbance at about 301 nm, with a suitable spectrophotometer, using 0.1 *N* hydrochloric acid as the blank. Calculate the quantity, in mg, of $C_{17}H_{20}N_2S.HCl$ in each mL of the Injection taken by the formula:

$$(C/V)(A_U/A_S),$$

in which *C* is the concentration, in μg per mL, of USP Promazine Hydrochloride RS in the Standard solution, *V* is the volume, in mL, of Injection taken, and A_U and A_S are the absorbances of the solution from the Injection and the Standard solution, respectively.

Promazine Hydrochloride Oral Solution

» Promazine Hydrochloride Oral Solution contains not less than 95.0 percent and not more than 110.0 percent of the labeled amount of $C_{17}H_{20}N_2S.HCl$.

Packaging and storage—Preserve in tight, light-resistant containers.

Reference standard—*USP Promazine Hydrochloride Reference Standard*—Dry at 105° for 2 hours before using.

NOTE—Throughout the following procedures, protect test or assay specimens, the Reference Standard, and solutions containing them, by conducting the procedures without delay, under subdued light, or using low-actinic glassware.

Identification—

A: Dilute a volume of Oral Solution, equivalent to about 50 mg of promazine hydrochloride, with 0.01 *N* hydrochloric acid to 25 mL, and proceed as directed under *Identification—Organic Nitrogenous Bases* ⟨181⟩, beginning with "Transfer the liquid to a separator": the Oral Solution meets the requirements of the test.

B: It responds to *Identification test B* under *Promazine Hydrochloride.*

pH ⟨791⟩: between 5.0 and 5.5.

Assay—[NOTE—Use low-actinic glassware.] Transfer an accurately measured volume of Promazine Hydrochloride Oral Solution, or a quantitative dilution of it in water, equivalent to about 10 mg of promazine hydrochloride, to a 250-mL separator. Add water to adjust the volume to about 45 mL, add 3 mL of sodium hydroxide solution (1 in 10), mix, and extract the promazine with five 25-mL portions of ether. Wash the combined ether extracts with 25 mL of water, and discard the aqueous washings. Extract the combined ether extract with one 50-mL and four 25-mL portions of 0.1 *N* hydrochloric acid. Filter the acid extracts through a pledget of cotton washed with 0.1 *N* hydrochloric acid into a 250-mL volumetric flask, dilute with the same acid to volume, and mix. Without delay, concomitantly determine the absorbances of this solution and of a Standard solution of USP Promazine Hydrochloride RS in the same medium having a known concentration of about 40 µg per mL in 1-cm cells at the wavelength of maximum absorbance at about 301 nm, with a suitable spectrophotometer, using 0.1 *N* hydrochloric acid as the blank. Calculate the quantity, in mg, of $C_{17}H_{20}N_2S.HCl$ in each mL of the Oral Solution taken by the formula:

$$(0.25C/V)(A_U/A_S),$$

in which C is the concentration, in µg per mL, of USP Promazine Hydrochloride RS in the Standard solution, V is the volume, in mL, of Oral Solution taken, and A_U and A_S are the absorbances from the assay solution and the Standard solution, respectively.

Promazine Hydrochloride Syrup

» Promazine Hydrochloride Syrup contains not less than 95.0 percent and not more than 110.0 percent of the labeled amount of $C_{17}H_{20}N_2S.HCl$.

Packaging and storage—Preserve in tight, light-resistant containers.

Reference standard—*USP Promazine Hydrochloride Reference Standard*—Dry at 105° for 2 hours before using.

NOTE—Throughout the following procedures, protect test or assay specimens, the Reference Standard, and solutions containing them, by conducting the procedures without delay, under subdued light, or using low-actinic glassware.

Identification—It responds to *Identification test B* under *Promazine Hydrochloride.*

Assay—[NOTE—Use low-actinic glassware.] Transfer a volume of Promazine Hydrochloride Syrup, equivalent to about 10 mg of promazine hydrochloride, to a 250-mL separator. Proceed as directed in the *Assay* under *Promazine Hydrochloride Oral Solution*, beginning with "Add water to adjust the volume to about 45 mL." Calculate the quantity, in mg, of $C_{17}H_{20}N_2S.HCl$ in each mL of the Syrup taken by the formula:

$$(0.25C/V)(A_U/A_S),$$

in which C is the concentration, in µg per mL, of USP Promazine Hydrochloride RS in the Standard solution, V is the volume, in mL, of Syrup taken, and A_U and A_S are the absorbances of the solution from the Syrup and the Standard solution, respectively.

Promazine Hydrochloride Tablets

» Promazine Hydrochloride Tablets contain not less than 95.0 percent and not more than 110.0 percent of the labeled amount of $C_{17}H_{20}N_2S.HCl$.

Packaging and storage—Preserve in tight, light-resistant containers.

Reference standard—*USP Promazine Hydrochloride Reference Standard*—Dry at 105° for 2 hours before using.

NOTE—Throughout the following procedures, protect test or assay specimens, the Reference Standard, and solutions containing them, by conducting the procedures without delay, under subdued light, or using low-actinic glassware.

Identification—

A: Shake a portion of powdered Tablets, equivalent to about 50 mg of promazine hydrochloride, with 25 mL of 0.01 *N* hydrochloric acid for 5 minutes, and filter: the solution meets the requirements under *Identification—Organic Nitrogenous Bases* ⟨181⟩.

B: It responds to *Identification test B* under *Promazine Hydrochloride.*

Disintegration ⟨701⟩: 30 minutes.

Uniformity of dosage units ⟨905⟩: meet the requirements.

Assay—[NOTE—Use low-actinic glassware.] Weigh and finely powder not less than 20 Promazine Hydrochloride Tablets. Transfer an accurately weighed portion of the powder, equivalent to about 50 mg of promazine hydrochloride, to a 100-mL volumetric flask. Add 50 mL of 0.1 *N* hydrochloric acid, and shake by mechanical means for about 1 hour. Dilute with 0.1 *N* hydrochloric acid to volume, mix, and centrifuge a portion of the mixture. Transfer 10.0 mL of the clear, supernatant liquid to a 250-mL separator, and proceed as directed in the *Assay* under *Promazine Hydrochloride Injection*, beginning with "add 20 mL of water." Calculate the quantity, in mg, of $C_{17}H_{20}N_2S.HCl$ in the portion of Tablets taken by the formula:

$$C(A_U/A_S),$$

in which C is the concentration, in µg per mL, of USP Promazine Hydrochloride RS in the Standard solution, and A_U and A_S are the absorbances of the solution from the Tablets and the Standard solution, respectively.

Promethazine Hydrochloride

$CH_2CH(CH_3)N(CH_3)_2$
N
S
· HCl

$C_{17}H_{20}N_2S.HCl$ 320.88

10*H*-Phenothiazine-10-ethanamine, *N*,*N*,α-trimethyl-, monohydrochloride.

10-[2-(Dimethylamino)propyl]phenothiazine monohydrochloride [*58-33-3*].

» Promethazine Hydrochloride contains not less than 97.0 percent and not more than 101.5 percent of $C_{17}H_{20}N_2S.HCl$, calculated on the dried basis.

Packaging and storage—Preserve in tight, light-resistant containers.

Reference standard—*USP Promethazine Hydrochloride Reference Standard*—Dry at 105° for 4 hours before using.

Completeness and clarity of solution—Separately prepare a 1 in 10 solution of it in water and 1 in 10 solution of it in chloroform: each solution is practically clear and show not more than a light yellow color.

NOTE—Throughout the following procedures, protect test or assay specimens, the Reference Standard, and solutions containing them, by conducting the procedures without delay, under subdued light, or using low-actinic glassware.

Identification—

A: The infrared absorption spectrum of a potassium bromide dispersion of it, previously dried, exhibits maxima only at the same wavelengths as that of a similar preparation of USP Promethazine Hydrochloride RS.

B: It responds to the tests for *Chloride* ⟨191⟩.

pH ⟨791⟩: between 4.0 and 5.0, in a solution (1 in 20).

Loss on drying ⟨731⟩—Dry it at 105° for 4 hours: it loses not more than 0.5% of its weight.

Residue on ignition ⟨281⟩: not more than 0.1%.

Related compounds—

Standard preparation and *Standard dilutions*—Dissolve an accurately weighed quantity of USP Promethazine Hydrochloride RS in methylene chloride to obtain a solution containing 10.0 mg per mL (*Standard preparation*). Prepare a series of quantitative dilutions of the *Standard preparation* in methylene chloride to contain 0.2 mg, 0.1 mg, 0.05 mg, and 0.025 mg per mL (*Standard dilutions*) corresponding to 2.0%, 1.0%, 0.5%, and 0.25% of impurities, respectively.

Test solution—Dissolve 100 mg, accurately weighed, of Promethazine Hydrochloride in 10.0 mL of methylene chloride.

Procedure—Using a 20- × 20-cm thin-layer chromatographic plate (see *Chromatography* ⟨621⟩) coated with a 0.25-mm layer of silica gel, apply 10-µL portions of the *Test preparation*, the *Standard preparation*, and each of the *Standard dilutions* 2.5 cm from the lower edge of the plate. Develop the plate in an unsaturated tank containing a mixture of ethyl acetate, acetone, alcohol, and ammonium hydroxide (90:45:2:1). After the solvent has moved not less than 10 cm, air-dry the plate, and view under short-wavelength ultraviolet light: the R_f value of the principal spot obtained from the *Test preparation* corresponds to that from the *Standard preparation*. Estimate the concentration of any other spots observed in the lane for the *Test preparation* by comparison with the *Standard dilutions:* the sum of the impurities is not greater than 2.0% and no single impurity is greater than 1.0%.

Assay—Dissolve about 700 mg of Promethazine Hydrochloride, accurately weighed, in a mixture of 75 mL of glacial acetic acid and 10 mL of mercuric acetate TS. Add 1 drop of crystal violet TS, and titrate with 0.1 *N* perchloric acid VS to a blue end-point. Perform a blank determination, and make any necessary correction. Each mL of 0.1 *N* perchloric acid is equivalent to 32.09 mg of $C_{17}H_{20}N_2S.HCl$.

Promethazine Hydrochloride Injection

» Promethazine Hydrochloride Injection is a sterile solution of Promethazine Hydrochloride in Water for Injection. It contains not less than 95.0 percent and not more than 110.0 percent of the labeled amount of $C_{17}H_{20}N_2S.HCl$.

Packaging and storage—Preserve in single-dose or in multiple-dose containers, preferably of Type I glass, protected from light.

Reference standard—*USP Promethazine Hydrochloride Reference Standard*—Dry at 105° for 4 hours before using.

NOTE—Throughout the following procedures, protect test or assay specimens, the Reference Standard, and solutions containing them, by conducting the procedures without delay, under subdued light, or using low-actinic glassware.

Identification—Add a volume of Injection, equivalent to about 50 mg of promethazine hydrochloride, to 20 mL of dilute hydrochloric acid (1 in 1000) contained in a separator. Wash the solution with a 20-mL portion of methylene chloride, discarding the washing. Add 2 mL of 1 *N* sodium hydroxide and 20 mL of methylene chloride, and shake for 2 minutes. Evaporate the methylene chloride extract on a steam bath with the aid of a stream of nitrogen to dryness. Dissolve the residue in 4 mL of carbon disulfide, filter through paper, if necessary, and determine the infrared absorption spectrum as directed under *Identification—Organic Nitrogenous Bases* ⟨181⟩, obtaining the spectrum of USP Promethazine Hydrochloride RS as directed: the Injection meets the requirements of the test.

pH ⟨791⟩: between 4.0 and 5.5.

Other requirements—It meets the requirements under *Injections* ⟨1⟩.

Assay—

Mobile phase—Dissolve 1 g of sodium 1-pentanesulfonate in 500 mL of water, add 500 mL of acetonitrile and 5 mL of glacial acetic acid, filter, and degas. Make adjustments if necessary (see *System Suitability* under *Chromatography* ⟨621⟩).

Standard preparation—Dissolve an accurately weighed quantity of USP Promethazine Hydrochloride RS in *Mobile phase*, and dilute quantitatively, and stepwise if necessary, with the same solvent to obtain a solution having a known concentration of about 0.1 mg per mL.

Assay preparation—Transfer an accurately measured volume of Promethazine Hydrochloride Injection, equivalent to about 50 mg of promethazine hydrochloride, to a 50-mL volumetric flask, dilute with *Mobile phase* to volume, and mix. Transfer 10.0 mL of the resulting solution to a 100-mL volumetric flask, dilute with *Mobile phase* to volume, and mix.

System suitability preparation—Dissolve a suitable quantity of phenothiazine in *Standard preparation* to obtain a solution containing about 10 µg of phenothiazine per mL.

Chromatographic system (see *Chromatography* ⟨621⟩)—The liquid chromatograph is equipped with a 254-nm detector and a 4.6-mm × 30-cm column that contains packing L11. The flow rate is about 1.5 mL per minute. Chromatograph the *System suitability preparation*, and record the peak responses as directed under *Procedure:* the resolution, *R*, between the promethazine and phenothiazine peaks is not less than 3.0, and the relative standard deviation for replicate injections is not more than 2.0%.

Procedure—Separately inject equal volumes (about 30 µL) of the *Standard preparation* and the *Assay preparation* into the chromatograph, record the chromatograms, and measure the responses for the major peaks. The relative retention times are 1.0 for promethazine and about 1.6 for phenothiazine. Calculate the quantity, in mg, of $C_{17}H_{20}N_2S.HCl$ in each mL of the Injection taken by the formula:

$$500(C/V)(R_U/R_S),$$

in which C is the concentration, in mg per mL, of USP Promethazine Hydrochloride RS in the *Standard preparation*, V is the volume, in mL, of Injection taken, and R_U and R_S are the peak response ratios obtained from the *Assay preparation* and the *Standard preparation*, respectively.

Promethazine Hydrochloride Suppositories

» Promethazine Hydrochloride Suppositories contain not less than 95.0 percent and not more than 110.0 percent of the labeled amount of $C_{17}H_{20}N_2S.HCl$.

Packaging and storage—Preserve in tight, light-resistant containers, and store in a cold place.

Reference standard—*USP Promethazine Hydrochloride Reference Standard*—Dry at 105° for 4 hours before using.

NOTE—Throughout the following procedures, protect test or assay specimens, the Reference Standard, and solutions containing them, by conducting the procedures without delay, under subdued light, or using low-actinic glassware.

Identification—Transfer a number of Suppositories, equivalent to about 50 mg of promethazine hydrochloride, to a 250-mL separator. Add 75 mL of solvent hexane and 25 mL of 0.01 *N* hydrochloric acid; shake to dissolve the solids. Using the aqueous phase, filtered through paper if necessary, proceed as directed under *Identification—Organic Nitrogenous Bases* ⟨181⟩, beginning with "Transfer the liquid to a separator."

Assay—

Palladium chloride solution—Add 5 mL of hydrochloric acid to a beaker containing 500 mg of palladium chloride. Warm on a steam bath to obtain a complete solution. Slowly add 200 mL of hot water. (If necessary, continue warming and add additional hydrochloric acid to maintain complete solution.) Transfer the solution to a 500-mL volumetric flask, and dilute with water to volume. To 50 mL of this solution add 250 mL of 2 *M* sodium acetate, and adjust with hydrochloric acid to a pH of 4.0. Transfer the solution to a 500-mL volumetric flask, dilute with water to volume, and mix.

Standard preparation—Using an accurately weighed quantity of USP Promethazine Hydrochloride RS, prepare a solution in 0.05 *N* hydrochloric acid containing about 0.1 mg in each mL. Protect the solution from light.

Assay preparation—Weigh 10 Suppositories and calculate the average. Carefully melt them, avoiding the use of excessive heat, and mix. Add 30 mL of hexanes to an accurately weighed portion, equivalent to about 50 mg of promethazine hydrochloride. Warm gently to dissolve, and transfer to a low-actinic separator. Rinse the transfer container with several small portions of the hexanes and 0.05 *N* hydrochloric acid, and add the rinsings to the separator. Extract with five 20-mL portions of 0.05 *N* hydrochloric acid shaking gently to avoid emulsions. Drain through glass wool prewashed with 0.05 *N* hydrochloric acid. Collect in a 500-mL low-actinic volumetric flask. Rinse the glass wool with additional 0.05 *N* hydrochloric acid, dilute the combined filtrate with 0.05 *N* hydrochloric acid to volume, and mix.

Procedure—Pipet 2.0 mL of the *Standard preparation*, *Assay preparation*, and 0.05 *N* hydrochloric acid into separate test tubes. Add 3.0 mL of *Palladium chloride solution* to each, and mix. Concomitantly determine the absorbances of the *Standard preparation* and the *Assay preparation* in 1-cm cells at the wavelength of maximum absorbance at about 450 nm, with a suitable spectrophotometer, using the hydrochloric acid-palladium chloride solution as the reagent blank in the reference cell. Calculate the quantity, in mg, of $C_{17}H_{20}N_2S.HCl$ in the portion of Suppositories taken by the formula:

$$500C(A_U/A_S),$$

in which C is the concentration, in mg per mL, of USP Promethazine Hydrochloride RS in the *Standard preparation*, and A_U and A_S are the absorbances of the solutions from the *Assay preparation* and the *Standard preparation*, respectively.

Promethazine Hydrochloride Syrup

» Promethazine Hydrochloride Syrup contains not less than 90.0 percent and not more than 110.0 percent of the labeled amount of $C_{17}H_{20}N_2S.HCl$.

Packaging and storage—Preserve in tight, light-resistant containers.

Reference standard—*USP Promethazine Hydrochloride Reference Standard*—Dry at 105° for 4 hours before using.

NOTE—Throughout the following procedures, protect test or assay specimens, the Reference Standard, and solutions containing them, by conducting the procedures without delay, under subdued light, or using low-actinic glassware.

Identification—Treat 25 mL of Syrup as directed in the *Assay*, ending with "using a current of air only." Dissolve the residue in 2.5 mL of carbon disulfide, filter through paper if necessary, and determine the infrared absorption spectrum as directed under *Identification—Organic Nitrogenous Bases* ⟨181⟩, obtaining the spectrum of USP Promethazine Hydrochloride RS as directed: the Syrup meets the requirements of the test.

Assay—[NOTE—Use low-actinic glassware in this assay.] Transfer an accurately measured volume of Promethazine Hydrochloride Syrup, equivalent to about 25 mg of promethazine hydrochloride, to a 250-mL separator. Add 10 mL of ammonium hydroxide, and extract the promethazine base with six 40-mL portions of chloroform. Wash the combined chloroform extracts with 25 mL of dilute hydrochloric acid (1 in 9). Wash the acid solution with 25 mL of chloroform, and add the washings to the main chloroform extract. Evaporate the chloroform extract on a steam bath, with the aid of a current of air, to a volume of 5 to 10 mL, and finally evaporate, using only a current of air, to dryness. Dissolve the residue, with slight warming, in dilute sulfuric acid (1 in 100), and transfer to a 500-mL volumetric flask with the aid of additional acid. Cool, add dilute sulfuric acid (1 in 100) to volume, mix, and filter, rejecting the first half of the filtrate. Dissolve an accurately weighed quantity of USP Promethazine Hydrochloride RS in dilute sulfuric acid (1 in 100), and dilute quantitatively and stepwise with the dilute acid to obtain a Standard solution having a known concentration of about 50 µg per mL. Concomitantly determine the absorbances of both solutions in 1-cm cells at the wavelength of maximum absorbance at about 298 nm, with a suitable spectrophotometer, using dilute sulfuric acid (1 in 100) as the blank. Calculate the quantity, in mg, of $C_{17}H_{20}N_2S.HCl$ in each mL of the Syrup taken by the formula:

$$500(C/V)(A_U/A_S),$$

in which C is the concentration, in mg per mL, of USP Promethazine Hydrochloride RS in the Standard solution, V is the volume, in mL, of Syrup taken, and A_U and A_S are the absorbances of the solution from the Syrup and the Standard solution, respectively.

Promethazine Hydrochloride Tablets

» Promethazine Hydrochloride Tablets contain not less than 95.0 percent and not more than 110.0 percent of the labeled amount of $C_{17}H_{20}N_2S.HCl$.

Packaging and storage—Preserve in tight, light-resistant containers.

Reference standard—*USP Promethazine Hydrochloride Reference Standard*—Dry at 105° for 4 hours before using.

NOTE—Throughout the following procedures, protect test or assay specimens, the Reference Standard, and solutions containing them, by conducting the procedures without delay, under subdued light, or using low-actinic glassware.

Identification—Shake a quantity of powdered Tablets, equivalent to 50 mg of promethazine hydrochloride, with 30 mL of chloroform, and filter into a beaker. Evaporate the chloroform, dissolve the residue in 40 mL of dilute hydrochloric acid (1 in 1000), and transfer the liquid to a separator. In a second separator, dissolve 50 mg of USP Promethazine Hydrochloride RS in 40 mL of dilute hydrochloric acid (1 in 1000). Treat each solution as follows. Add 2 mL of 1 *N* sodium hydroxide and 15 mL of carbon disulfide, and shake for 2 minutes. Centrifuge if necessary to clarify the lower phase, and filter through a dry filter, collecting the filtrate in a small flask provided with a glass stopper. Reduce the volume of the carbon disulfide extracts to 4 to 5 mL, and proceed as directed under *Identification—Organic Nitrogenous Bases* ⟨181⟩, beginning with "Determine the absorption spectra."

Dissolution ⟨711⟩—

Medium: 0.1 *N* hydrochloric acid; 900 mL.

Apparatus 1: 100 rpm.

Time: 45 minutes.

Procedure—Determine the amount of $C_{17}H_{20}N_2S.HCl$ dissolved from ultraviolet absorbances at the wavelength of maximum absorbance at about 249 nm of filtered portions of the solution under test, suitably diluted with water, in comparison with a Standard solution having a known concentration of USP Promethazine Hydrochloride RS in the same medium.

Tolerances—Not less than 75% (*Q*) of the labeled amount of $C_{17}H_{20}N_2S.HCl$ is dissolved in 45 minutes.

Uniformity of dosage units ⟨905⟩: meet the requirements.

Procedure for content uniformity—Transfer 1 finely powdered Tablet to a 100-mL volumetric flask, add 50 mL of citric acid solution (1 in 100), and shake by mechanical means for 15 minutes. Dilute with citric acid solution (1 in 100) to volume, and centrifuge about 50 mL of the mixture. Dilute an accurately measured portion of the clear solution, equivalent to 5 mg of promethazine hydrochloride, quantitatively with citric acid solution (1 in 100) to 100 mL. Concomitantly determine the absorbance of this solution and of a Standard solution of USP Promethazine Hydrochloride RS in the same medium, having a known concentration of about 50 µg per mL, in 1-cm cells at the wavelength of maximum absorbance at about 298 nm, with a suitable spectrophotometer, using citric acid solution (1 in 100) as the blank. Calculate the quantity, in mg, of $C_{17}H_{20}N_2S.HCl$ in the Tablet by the formula:

$$(TC/D)(A_U/A_S),$$

in which T is the labeled quantity, in mg, of promethazine hydrochloride in the Tablet, D is the concentration, in µg per mL, of promethazine hydrochloride in the test solution based on the labeled quantity per Tablet and the extent of dilution, C is the concentration, in µg per mL, of USP Promethazine Hydrochloride RS in the Standard solution, and A_U and A_S are the absorbances of the solution from the Tablet and the Standard solution, respectively.

Assay—

Buffered palladium chloride solution—Transfer 500 mg of palladium chloride to a 250-mL beaker, add 5 mL of hydrochloric acid, and warm on a steam bath. Add 200 mL of hot water in small quantities while stirring until solution is complete. Cool, dilute with water to 500 mL, and mix. Transfer 25 mL of this solution to a 500-mL volumetric flask. Add 50 mL of 1 *N* sodium acetate and 48 mL of 1 *N* hydrochloric acid, dilute with water to volume, and mix.

Standard preparation—Transfer about 31 mg of USP Promethazine Hydrochloride RS, accurately weighed, to a low-actinic 250-mL volumetric flask. Dissolve in 0.1 *N* hydrochloric acid, dilute with 0.1 *N* hydrochloric acid to volume, and mix.

Assay preparation—Weigh and finely powder not less than 20 Promethazine Hydrochloride Tablets. Transfer an accurately weighed portion of the powder, equivalent to about 6.25 mg of promethazine hydrochloride, to a low-actinic 125-mL separator. Add 20 mL of saturated potassium chloride solution, 10 mL of 1 *N* sodium hydroxide, and 10 mL of methanol, and extract the promethazine with three 20-mL portions of *n*-heptane. Filter the heptane extracts through anhydrous sodium sulfate and collect them in a low-actinic 125-mL separator. Extract the promethazine from the *n*-heptane solution with three 15-mL portions of 0.1 *N* hydrochloric acid, collect the acid extracts in a low-actinic 50-mL volumetric flask, dilute with 0.1 *N* hydrochloric acid to volume, and mix.

Procedure—Into separate test tubes, pipet 2-mL portions of the *Standard preparation*, the *Assay preparation*, and 0.1 *N* hydrochloric acid to provide a blank. Add 3.0 mL of *Buffered palladium chloride solution* to each tube, and mix. Concomitantly determine the absorbances of the solutions at the wavelength of maximum absorbance at about 470 nm, using a suitable spectrophotometer, and using the blank in the reference cell. Calculate the quantity, in mg, of $C_{17}H_{20}N_2S.HCl$ in the portion of Tablets taken by the formula:

$$50C(A_U/A_S),$$

in which C is the concentration, in mg per mL, of USP Promethazine Hydrochloride RS in the *Standard preparation*, and A_U and A_S are the absorbances of the solutions from the *Assay preparation* and the *Standard preparation*, respectively.

Prompt Insulin Zinc Suspension—*see* Insulin Zinc Suspension, Prompt

Prompt Phenytoin Sodium Capsules—*see* Phenytoin Sodium Capsules, Prompt

Propane—*see* Propane NF

Propantheline Bromide

$C_{23}H_{30}BrNO_3$ 448.40

2-Propanaminium, *N*-methyl-*N*-(1-methylethyl)-*N*-[2-[(9*H*-xanthen-9-ylcarbonyl)oxy]ethyl]-, bromide.

(2-Hydroxyethyl)diisopropylmethylammonium bromide xanthene-9-carboxylate [*50-34-0*].

» Propantheline Bromide contains not less than 98.0 percent and not more than 102.0 percent of $C_{23}H_{30}BrNO_3$, calculated on the dried basis.

Packaging and storage—Preserve in well-closed containers.

Reference standards—*USP Xanthanoic Acid Reference Standard*—Dry at 105° for 4 hours before using. Keep container tightly closed. *USP Xanthone Reference Standard*—Dry at 105° for 4 hours before using. Keep container tightly closed. *USP 9-Hydroxypropantheline Bromide Reference Standard*—Keep container tightly closed and protected from light at controlled room temperature. Store in a desiccator. If the container is opened frequently, seal the container under an inert gas (nitrogen or argon) atmosphere. Do not dry before use. *USP Propantheline Bromide Reference Standard*—Dry at 105° for 4 hours before using.

Identification—

A: Prepare 3 mL of a solution in chloroform having a concentration of about 6 mg per mL, and reserve a 1-mL portion for *Identification test B*. In a well-ventilated hood, apply 2 mL of this solution dropwise to a salt plate while continuously evaporating the solvent with the aid of an infrared heat lamp and a current of dry air. Heat the residue at 105° for 15 minutes: the infrared absorption spectrum of the residue on the single salt plate exhibits maxima only at the same wavelengths as that of a similar preparation of USP Propantheline Bromide RS, treated in the same manner.

B: On a suitable thin-layer chromatographic plate (see *Chromatography* ⟨621⟩), coated with a 0.25-mm layer of chromatographic silica gel, apply 5 µL of the chloroform solution retained from *Identification test A* and 5 µL of a Standard solution of USP Propantheline Bromide RS in chloroform containing 6 mg per mL. Develop the chromatogram in a solvent system consisting of a mixture of 1 *N* hydrochloric acid and acetone (1:1) until the solvent front has moved about three-fourths of the length of the plate. Remove the plate from the developing chamber, mark the solvent front, and dry at 105° for 5 minutes. Spray the plate with potassium–bismuth iodide TS, and heat at 105° for 5 minutes: the R_f value of the principal spot obtained from the test solution corresponds to that obtained from the Standard solution.

C: To 5 mL of a solution (1 in 100) add 2 mL of 2 *N* nitric acid: this solution responds to the tests for *Bromide* ⟨191⟩, except that in the test that liberates bromine, the chloroform layer may be yellow.

Loss on drying ⟨731⟩—Dry it at 105° for 4 hours: it loses not more than 0.5% of its weight.

Residue on ignition ⟨281⟩: not more than 0.1%.

Related compounds—

pH 3.5 buffer solution—Dissolve 17.3 g of dodecyl sodium sulfate in 1000 mL of water containing 10 mL of phosphoric acid in a 2000-mL volumetric flask. Carefully adjust, while stirring, with 0.5 *M* sodium hydroxide to a pH of 3.5 ± 0.05, dilute with water to volume, and mix.

Mobile phase—Prepare a filtered and degassed mixture of acetonitrile and *pH 3.5 buffer solution* (55:45). Make adjustments if necessary (see *System Suitability* under *Chromatography* ⟨621⟩).

Standard preparation—Dissolve accurately weighed quantities of USP 9-Hydroxypropantheline Bromide RS, USP Xanthanoic Acid RS, and USP Xanthone RS in *Mobile phase*, and dilute quantitatively and stepwise if necessary with *Mobile phase* to obtain a solution having a known concentration of about 6.0 µg of 9-hydroxypropantheline bromide per mL, and about 1.5 µg each of xanthanoic acid and xanthone per mL.

Test preparation—Transfer about 60 mg of Propantheline Bromide, accurately weighed, to a 200-mL volumetric flask, dissolve in *Mobile phase*, dilute with *Mobile phase* to volume, and mix.

Chromatographic system (see *Chromatography* ⟨621⟩)—The liquid chromatograph is equipped with a 254-nm detector and a 4.6-mm × 25-cm column that contains packing L7. The flow rate is about 2.0 mL per minute. Chromatograph the *Standard preparation*, and record peak responses as directed under *Procedure:* the resolution, *R*, between the least resolved peaks is not less than 1.2 and the relative standard deviation for replicate injections of the *Standard preparation* is not more than 6.0% for each component.

Procedure—Separately inject equal volumes (about 50 µL) of the *Standard preparation* and the *Test preparation* into the chromatograph, record the chromatograms for a total time of not less than 1.5 times the retention time of the propantheline bromide peak, and measure the response for each peak, except the peaks at or before the void volume. The relative retention times are 0.4, 0.5, 0.6, and 1.0 for xanthanoic acid, 9-hydroxypropantheline bromide, xanthone, and propantheline bromide, respectively. Calculate the percentage of xanthanoic acid, xanthone, and 9-hydroxypropantheline bromide greater than or equal to 0.1% in the portion of Propantheline Bromide taken by the formula:

$$20C/W(r_U/r_S),$$

in which *C* is the concentration, in µg, of xanthanoic acid, xanthone, or 9-hydroxypropantheline bromide per mL of the *Standard preparation*, *W* is the weight, in mg, of Propantheline Bromide taken, and r_U and r_S are the related compound peak responses obtained from the *Test preparation* and the *Standard preparation*, respectively: not more than 2.0% of 9-hydroxypropantheline bromide and 0.5% each of xanthone and xanthanoic acid is found. Calculate the percentage of all unknown impurities greater than or equal to 0.1% by the formula:

$$100r_i/r_t,$$

in which r_i is the response of the unknown impurity peak and r_t is the sum of the responses of all the measured peaks observed in the chromatogram: the sum total of all known and unknown impurities is not more than 3.0%.

Bromide content—Weigh accurately about 500 mg, and dissolve in 40 mL of water. Add 10 mL of glacial acetic acid and 40 mL of methanol, then add eosin Y TS, and titrate with 0.1 *N* silver nitrate VS. Each mL of 0.1 *N* silver nitrate is equivalent to 7.990 mg of Br. Not less than 17.5% and not more than 18.2% of Br, calculated on the dried basis, is found.

Assay—Dissolve about 600 mg of Propantheline Bromide, accurately weighed, in a mixture of 20 mL of glacial acetic acid and 15 mL of mercuric acetate TS, warming slightly if necessary to effect solution. Cool to room temperature, and titrate with 0.1 *N* perchloric acid VS, determining the end-point potentiometrically. Perform a blank determination, and make any necessary correction. Each mL of 0.1 *N* perchloric acid is equivalent to 44.84 mg of $C_{23}H_{30}BrNO_3$.

Sterile Propantheline Bromide

» Sterile Propantheline Bromide is Propantheline Bromide suitable for parenteral use.

Packaging and storage—Preserve in *Containers for Sterile Solids* as described under *Injections* ⟨1⟩.

Reference standards—*USP Xanthanoic Acid Reference Standard*—Dry at 105° for 4 hours before using. Keep tightly closed. *USP Xanthone Reference Standard*—Dry at 105° for 4 hours before using. Keep tightly closed. *USP Propantheline Bromide Reference Standard*—Dry at 105° for 4 hours before using.

Completeness of solution ⟨641⟩—The contents of 1 container dissolve in 1 mL of water to yield a clear solution.

Constituted solution—At the time of use, the constituted solution prepared from Sterile Propantheline Bromide meets the requirements for *Constituted Solutions* under *Injections* ⟨1⟩.

Other requirements—It conforms to the definition, responds to the *Identification tests*, and meets the requirements for *Melting range, Loss on drying, Residue on ignition, Xanthanoic acid and xanthone, Bromide content*, and *Assay* under *Propantheline Bromide.* It meets also the requirements for *Sterility Tests* ⟨71⟩, *Uniformity of Dosage Units* ⟨905⟩, and *Labeling* under *Injections* ⟨1⟩.

Propantheline Bromide Tablets

» Propantheline Bromide Tablets contain not less than 90.0 percent and not more than 110.0 percent of the labeled amount of $C_{23}H_{30}BrNO_3$.

Packaging and storage—Preserve in well-closed containers.

Reference standards—*USP Xanthanoic Acid Reference Standard*—Dry at 105° for 4 hours before using. Keep container tightly closed. *USP Xanthone Reference Standard*—Dry at 105° for 4 hours before using. Keep container tightly closed. *USP 9-Hydroxypropantheline Bromide Reference Standard*—Keep container tightly closed and protected from light at controlled room temperature. Store in a desiccator. If the container is opened frequently, seal the container under an inert gas (nitrogen or argon) atmosphere. Do not dry before use. *USP Propantheline Bromide Reference Standard*—Dry at 105° for 4 hours before using.

Identification—

A: Finely powder a number of Tablets, equivalent to about 90 mg of propantheline bromide, and triturate the powder with 10 mL of chloroform. Filter, and wash the filter with 10 mL of chloroform, collecting the filtrate and washing in a separator. Add 10 mL of water, shake, and discard the chloroform layer. Wash the aqueous layer with two 10-mL portions of ether, and discard the ether washings. Filter the aqueous solution, and evaporate on a steam bath with the aid of a current of dry air to dryness. Dissolve the residue in 5 mL of chloroform, mix, and proceed as directed in *Identification test A* under *Propantheline Bromide*, beginning with "In a well-ventilated hood:" the specified result is observed.

B: The chromatogram of the *Assay preparation* obtained as directed in the *Assay* exhibits a major peak for propantheline bromide, the retention time of which corresponds to that exhibited in the chromatogram of the *Standard preparation.*

Dissolution ⟨711⟩—

Medium: pH 4.5 (±0.05) Acetate buffer prepared by mixing 1.64 g of anhydrous sodium acetate and 1.25 mL of glacial acetic acid with 500 mL of water, and diluting with water to obtain 1000 mL of solution having a pH of 4.50 ± 0.05; 500 mL.

Apparatus 2: 50 rpm.

Time: 45 minutes.

Determine the amount of propantheline bromide dissolved using the following method.

pH 3.5 buffer solution, Mobile phase, and *Chromatographic system*—Prepare as directed under *Assay.*

Procedure—Inject a volume (about 50 µL) of a filtered portion of the solution under test into the chromatograph, record the chromatogram, and measure the response for the major peak. Calculate the quantity of $C_{23}H_{30}BrNO_3$ dissolved in comparison with a Standard solution having a known concentration of USP Propantheline Bromide RS in the same medium and similarly chromatographed.

Tolerances—Not less than 75% (*Q*) of the labeled amount of $C_{23}H_{30}BrNO_3$ is dissolved in 45 minutes.

Uniformity of dosage units ⟨905⟩: meet the requirements.

Related compounds—

pH 3.5 buffer solution and *Mobile phase*—Prepare as directed for *Related compounds* under *Propantheline Bromide.*

Standard preparation—Dissolve accurately weighed quantities of USP 9-Hydroxypropantheline Bromide RS, USP Xanthanoic Acid RS, and USP Xanthone RS in *Mobile phase*, and dilute quantitatively and stepwise if necessary with *Mobile phase* to obtain a solution having known concentrations of about 12.0 µg of 9-hydroxypropantheline bromide per mL, and about 3.0 µg each of xanthanoic acid and xanthone per mL.

Test preparation—Use the *Assay preparation* prepared as directed under *Assay*.

Chromatographic system (see *Chromatography* ⟨621⟩)—The liquid chromatograph is equipped with a 254-nm detector and a 4.6-mm × 25-cm column that contains packing L7. The flow rate is about 2.0 mL per minute. Chromatograph the *Standard preparation*, and record peak responses as directed under *Procedure:* the resolution, *R*, between the least resolved peaks is not less than 1.2 and the relative standard deviation for replicate injections of the *Standard preparation* is not more than 6.0% for each component or, if the *Assay* is performed concomitantly, the relative standard deviation for the propantheline bromide peak in the replicate injections of the *Standard preparation* is not more than 2.0%.

Procedure—Separately inject equal volumes (about 50 µL) of the *Standard preparation* and the *Test preparation* into the chromatograph, record the chromatograms, and measure the responses for the major peaks. Calculate the percentage of xanthanoic acid, xanthone, and 9-hydroxypropantheline bromide greater than or equal to 0.1% in the portion of Tablets taken by the formula:

$$100C/C_X(r_U/r_S),$$

in which *C* is the concentration, in µg, of xanthanoic acid, xanthone, or 9-hydroxypropantheline bromide per mL of the *Standard preparation*, C_X is the theoretical concentration, in µg per mL, of Propantheline Bromide in the *Test preparation*, and r_U and r_S are the related compound peak responses obtained from the *Test preparation* and the *Standard preparation*, respectively: not more than 4.0% of 9-hydroxypropantheline bromide and 1.0% each of xanthone and xanthanoic acid are found.

Assay—

pH 3.5 buffer solution and *Mobile phase*—Prepare as directed for *Related compounds* under *Propantheline Bromide*.

Standard preparation—Dissolve an accurately weighed quantity of USP Propantheline Bromide RS in *Mobile phase* to obtain a solution having a known concentration of about 0.3 mg per mL.

Assay preparation—Weigh and finely powder not less than 20 Propantheline Bromide Tablets. Transfer an accurately weighed portion of the powder, equivalent to 15 mg of propantheline bromide, to a 50-mL volumetric flask, dissolve in *Mobile phase*, dilute with *Mobile phase* to volume, mix, and filter.

Chromatographic system (see *Chromatography* ⟨621⟩)—The liquid chromatograph is equipped with a 254-nm detector and a 4.6-mm × 25-cm column that contains packing L7. The flow rate is about 2.0 mL per minute. Chromatograph the *Standard preparation*, and record peak responses as directed under *Procedure:* the relative standard deviation for replicate injections is not more than 2.0%.

Procedure—Separately inject equal volumes (about 50 µL) of the *Standard preparation* and the *Assay preparation* into the chromatograph, record the chromatograms, and measure the peak responses. Calculate the quantity, in mg, of $C_{23}H_{30}BrNO_3$ in the portion of Tablets taken by the formula:

$$50C(r_U/r_S),$$

in which *C* is the concentration, in mg per mL, of USP Propantheline Bromide RS in the *Standard preparation*, and r_U and r_S are the peak responses due to Propantheline Bromide obtained from the *Assay preparation* and the *Standard preparation*, respectively.

Proparacaine Hydrochloride

$CH_3CH_2CH_2O$–(benzene ring, NH_2)–$COOCH_2CH_2N(C_2H_5)_2$ · HCl

$C_{16}H_{26}N_2O_3 . HCl$ 330.85

Benzoic acid, 3-amino-4-propoxy-, 2-(diethylamino)ethyl ester, monohydrochloride.

2-(Diethylamino)ethyl 3-amino-4-propoxybenzoate monohydrochloride   [*5875-06-9*].

» Proparacaine Hydrochloride contains not less than 97.0 percent and not more than 103.0 percent of $C_{16}H_{26}N_2O_3 . HCl$, calculated on the dried basis.

Packaging and storage—Preserve in well-closed containers.

Reference standard—*USP Proparacaine Hydrochloride Reference Standard*—Dry at 105° for 3 hours before using.

Identification—

A: It meets the requirements under *Identification—Organic Nitrogenous Bases* ⟨181⟩.

B: Dissolve 50 mg, accurately weighed, in water to make 250.0 mL, and mix. Pipet 10 mL of this solution into a 100-mL volumetric flask, add 2 mL of 10 percent, pH 6.0 phosphate buffer (see *Buffer Solutions* in the section, *Reagents, Indicators, and Solutions*), add water to volume, and mix: the ultraviolet absorption spectrum of the solution exhibits maxima and minima at the same wavelengths as that of a similar solution of USP Proparacaine Hydrochloride RS, concomitantly measured, and the respective absorptivities, calculated on the dried basis, at the wavelength of maximum absorbance at about 310 nm do not differ by more than 3.0%.

C: A solution (1 in 50) responds to the tests for *Chloride* ⟨191⟩.

Melting range ⟨741⟩: between 178° and 185°, but the range between beginning and end of melting does not exceed 2°.

Loss on drying ⟨731⟩—Dry it at 105° for 3 hours: it loses not more than 0.5% of its weight.

Residue on ignition ⟨281⟩: not more than 0.15%.

Assay—Place 250 mg of Proparacaine Hydrochloride, accurately weighed, in a 250-mL conical flask, add 80 mL of a 1 in 20 solution of acetic anhydride in glacial acetic acid, and heat on a steam bath for 10 minutes. Cool to room temperature, add 10 mL of mercuric acetate TS and 1 or 2 drops of crystal violet TS, and titrate with 0.1 *N* perchloric acid VS to a blue-green endpoint. Perform a blank determination, and make any necessary correction. Each mL of 0.1 *N* perchloric acid is equivalent to 33.09 mg of $C_{16}H_{26}N_2O_3 . HCl$.

Proparacaine Hydrochloride Ophthalmic Solution

» Proparacaine Hydrochloride Ophthalmic Solution is a sterile, aqueous solution of Proparacaine Hydrochloride. It contains not less than 95.0 percent and not more than 110.0 percent of the labeled amount of $C_{16}H_{26}N_2O_3 . HCl$.

Packaging and storage—Preserve in tight, light-resistant containers.

Labeling—Label it to indicate that it is to be stored in a refrigerator after the container is opened.

Reference standard—*USP Proparacaine Hydrochloride Reference Standard*—Dry at 105° for 3 hours before using.

Identification—To 1 mL of Ophthalmic Solution in a test tube add 5 mL of dilute hydrochloric acid (1 in 100), mix, and cool in an ice bath for 2 minutes. Add 2 drops of sodium nitrite solution (1 in 10), stir, and again cool for 2 minutes. Add 1 mL of a solution prepared by dissolving 200 mg of 2-naphthol in 10 mL of 1 *N* sodium hydroxide: a scarlet-red precipitate is formed. Add 5 mL of acetone: the precipitate does not dissolve.

Sterility—It meets the requirements under *Sterility Tests* ⟨71⟩.

pH ⟨791⟩: between 3.5 and 6.0.

Assay—

Standard preparation—Dissolve about 20 mg of USP Proparacaine Hydrochloride RS, accurately weighed, in water to make 100 mL, and mix.

Assay preparation—Transfer an accurately measured volume of Proparacaine Hydrochloride Ophthalmic Solution, equivalent to about 10 mg of proparacaine hydrochloride, to a separator, add water to make about 50 mL, and render alkaline by the addition of 5 mL of sodium carbonate TS. Immediately extract with two 50-mL portions of ether, collecting the extracts in a separator. Wash the ether extracts with 20 mL of water, discarding the washing, then extract the ether solution successively with two 20-mL portions and one 5-mL portion of dilute hydrochloric acid (1 in 250), collecting the extracts in a 50-mL volumetric flask, add water to volume, and mix.

Procedure—Pipet 10 mL each of the *Standard preparation* and the *Assay preparation* into separate 100-mL volumetric flasks. To the flask containing the *Standard preparation* add 5 mL of dilute hydrochloric acid (1 in 250). To each flask add 10 mL of 10 percent, pH 6.0 phosphate buffer (see *Antibiotics—Microbial Assays* ⟨81⟩), then add water to volume, and mix. Concomitantly determine the absorbances of both solutions at the wavelength of maximum absorbance at about 310 nm, with a suitable spectrophotometer, using water as the blank. Calculate the quantity, in mg, of $C_{16}H_{26}N_2O_3 . HCl$ in the volume of Ophthalmic Solution taken by the formula:

$$50C(A_U/A_S),$$

in which C is the concentration, in mg per mL, of USP Proparacaine Hydrochloride RS in the *Standard preparation*, and A_U and A_S are the absorbances of the solutions from the *Assay preparation* and the *Standard preparation*, respectively.

Propiomazine Hydrochloride

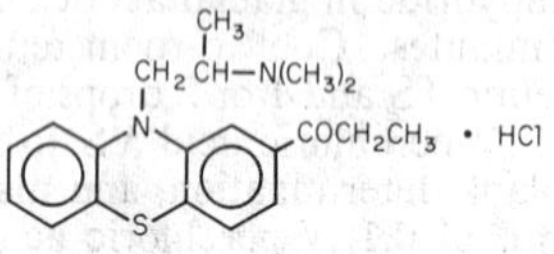

$C_{20}H_{24}N_2OS . HCl$ 376.94

1-Propanone, 1-[10-[2-(dimethylamino)propyl]-10*H*-phenothiazin-2-yl]-, monohydrochloride.

1-[10-[2-(Dimethylamino)propyl]phenothiazin-2-yl]-1-propanone monohydrochloride [*1240-15-9*].

» Propiomazine Hydrochloride contains not less than 98.0 percent and not more than 102.0 percent of $C_{20}H_{24}N_2OS . HCl$, calculated on the anhydrous basis.

Packaging and storage—Preserve in tight, light-resistant containers.

Reference standard—*USP Propiomazine Hydrochloride Reference Standard*—Dry over silica gel for 4 hours before using.

NOTE—Throughout the following procedures, protect test or assay specimens, the Reference Standard, and solutions containing them, by conducting the procedures without delay, under subdued light, or using low-actinic glassware.

Identification—

A: The infrared absorption spectrum of a potassium bromide dispersion of it, previously dried over silica gel for 4 hours, exhibits maxima only at the same wavelengths as that of a similar preparation of USP Propiomazine Hydrochloride RS.

B: The ultraviolet absorption spectrum of a solution (1 in 100,000) exhibits maxima and minima at the same wavelengths as that of a similar solution of USP Propiomazine Hydrochloride RS, concomitantly measured, and the respective absorptivities, calculated on the anhydrous basis, at the wavelength of maximum absorbance at about 240 nm do not differ by more than 3.0%.

pH ⟨791⟩: between 4.6 and 5.6, in a solution (1 in 50).

Water, *Method I* ⟨921⟩: not more than 0.5%.

Residue on ignition ⟨281⟩: not more than 0.1%.

Assay—

Buffered palladium chloride solution—Warm 500 mg of palladous chloride with 5 mL of hydrochloric acid on a steam bath, add 200 mL of hot water in small increments while stirring, and continue heating until solution is complete. Dilute with water to 500 mL, and mix. Transfer 25.0 mL of this solution to a 500-mL volumetric flask, add 50 mL of 1 *M* sodium acetate and 48 mL of 1 *N* hydrochloric acid, dilute with water to volume, and mix.

Standard preparation—Dissolve a suitable quantity of USP Propiomazine Hydrochloride RS, accurately weighed, in water, and prepare, by quantitative and stepwise dilution, if necessary, a solution in water containing about 120 µg per mL.

Assay preparation—Transfer about 150 mg of Propiomazine Hydrochloride, accurately weighed, to a 50-mL volumetric flask, add water to volume, and mix. Transfer 10.0 mL of this solution to a 250-mL volumetric flask, dilute with water to volume, and mix.

Procedure—Transfer 3.0 mL each of the *Standard preparation* and the *Assay preparation* to separate tubes, and transfer 3.0 mL of water to a third tube to provide the blank. To each tube add 4.0 mL of *Buffered palladium chloride solution*, and mix. Concomitantly determine the absorbances of the solutions in 1-cm cells at the wavelength of maximum absorbance at about 465 nm, with a suitable spectrophotometer, using the blank to set the instrument. Calculate the quantity, in mg, of $C_{20}H_{24}N_2OS . HCl$ in the Propiomazine Hydrochloride taken by the formula:

$$1.25C(A_U/A_S),$$

in which C is the concentration, in µg per mL, of USP Propiomazine Hydrochloride RS in the *Standard preparation*, and A_U and A_S are the absorbances of the solutions from the *Assay preparation* and the *Standard preparation*, respectively.

Propiomazine Hydrochloride Injection

» Propiomazine Hydrochloride Injection is a sterile solution of Propiomazine Hydrochloride in Water for Injection. It contains not less than 95.0 percent and not more than 110.0 percent of the labeled amount of $C_{20}H_{24}N_2OS . HCl$.

Note—Do not use the Injection if it is cloudy or contains a precipitate.

Packaging and storage—Preserve in single-dose containers, preferably of Type I glass, at controlled room temperature, protected from light.

Reference standard—*USP Propiomazine Hydrochloride Reference Standard*—Dry over silica gel for 4 hours before using.

NOTE—Throughout the following procedures, protect test or assay specimens, the Reference Standard, and solutions containing them, by conducting the procedures without delay, under subdued light, or using low-actinic glassware.

Identification—Dilute a volume of Injection, equivalent to about 10 mg of propiomazine hydrochloride, with water to 1000 mL, and mix. Determine the absorbance of this solution at 241 and 273 nm, with a suitable spectrophotometer, using water as the blank: the ratio $10(A_{273}/A_{241})$ is between 6.5 and 6.8.

pH ⟨791⟩: between 4.7 and 5.3.

Other requirements—It meets the requirements under *Injections* ⟨1⟩.

Assay—

Buffered palladium chloride solution—Prepare as directed in the *Assay* under *Propiomazine Hydrochloride.*

Standard preparation—Prepare as directed in the *Assay* under *Propiomazine Hydrochloride*, except to use 0.02 *N* hydrochloric acid instead of water.

Assay preparation—Transfer an accurately measured volume of Propiomazine Hydrochloride Injection, equivalent to about 60 mg of propiomazine hydrochloride, to a 250-mL separator. Add

80 mL of water and 20 mL of sodium hydroxide solution (1 in 250), mix by swirling, and extract with two 50-mL portions of ether, collecting the ether extracts in a second 250-mL separator. Wash the combined ether extracts by swirling gently with 20 mL of sodium hydroxide solution (1 in 250), and discard the aqueous washing. Extract the ether solution with two 50-mL portions of 0.1 *N* hydrochloric acid, collecting the acid extracts in a 500-mL volumetric flask, dilute with water to volume, and mix.

Procedure—Proceed as directed for *Procedure* in the *Assay* under *Propiomazine Hydrochloride.* Calculate the quantity, in mg, of $C_{20}H_{24}N_2OS.HCl$ in each mL of the Injection taken by the formula:

$$(0.5C/V)(A_U/A_S),$$

in which *C* is the concentration, in µg per mL, of USP Propiomazine Hydrochloride RS in the *Standard preparation, V* is the volume, in mL, of Injection taken, and A_U and A_S are the absorbances of the solutions from the *Assay preparation* and the *Standard preparation*, respectively.

Propionic Acid—*see* Propionic Acid NF

Propoxycaine Hydrochloride

NH_2 — $COOCH_2CH_2N(C_2H_5)_2$ · HCl

$OCH_2CH_2CH_3$

$C_{16}H_{26}N_2O_3.HCl$ 330.85

Benzoic acid, 4-amino-2-propoxy-, 2-(diethylamino)ethyl ester, monohydrochloride.

2-(Diethylamino)ethyl 4-amino-2-propoxybenzoate monohydrochloride [*550-83-4*].

» Propoxycaine Hydrochloride, dried at 105° for 3 hours, contains not less than 98.0 percent and not more than 102.0 percent of $C_{16}H_{26}N_2O_3.HCl$.

Packaging and storage—Preserve in well-closed, light-resistant containers.

Reference standard—*USP Propoxycaine Hydrochloride Reference Standard*—Dry at 105° for 3 hours before using.

Identification—

A: The infrared absorption spectrum of a potassium bromide dispersion of it, previously dried, exhibits maxima only at the same wavelengths as that of a similar preparation of USP Propoxycaine Hydrochloride RS.

B: The ultraviolet absorption spectrum of a solution (1 in 100,000) exhibits maxima and minima at the same wavelengths as that of a similar solution of USP Propoxycaine Hydrochloride RS, concomitantly measured, and the respective absorptivities, calculated on the dried basis, at the wavelength of maximum absorbance at about 303 nm do not differ by more than 3.0%.

C: Dissolve about 100 mg in 10 mL of water, heat almost to boiling, and add, with stirring, 1 mL of a saturated solution of trinitrophenol in 20 percent alcohol. Cool slowly, collect the precipitate on a filter, wash with a few small portions of water, and dry in a vacuum desiccator over phosphorus pentoxide for 18 hours: the picrate so obtained melts between 130° and 138°, with decomposition (see *Melting Range or Temperature* ⟨741⟩). [*Caution—Picrates may explode.*]

D: A solution (1 in 100) responds to the tests for *Chloride* ⟨191⟩.

Melting range, *Class I* ⟨741⟩: between 146° and 151°.

Loss on drying ⟨731⟩—Dry it at 105° for 3 hours: it loses not more than 0.5% of its weight.

Residue on ignition ⟨281⟩: not more than 0.2%.

Chromatographic purity—

Solvent—Prepare a mixture of chloroform and methanol (9:1).

Standard preparations—Dissolve a quantity of USP Propoxycaine Hydrochloride RS in *Solvent*, and dilute quantitatively, and stepwise if necessary, with *Solvent* to obtain a solution containing 0.5 mg per mL. Dilute quantitatively with *Solvent* to obtain Standard preparations having the following compositions:

Standard preparation	Dilution	Concentration (mg RS per mL)	Percentage (%, for comparison with test specimen)
A	1 in 2	0.25	0.5
B	1 in 5	0.10	0.2
C	1 in 10	0.05	0.1

Test preparation—Dissolve an accurately weighed quantity of Propoxycaine Hydrochloride in *Solvent* to obtain a solution containing 50 mg per mL.

Procedure—On a suitable thin-layer chromatographic plate (see *Chromatography* ⟨621⟩), coated with a 0.25-mm layer of chromatographic silica gel mixture, apply separately 5 µL of the *Test preparation* and 5 µL of each *Standard preparation.* Position the plate in a chromatographic chamber, and develop the chromatograms in a solvent system consisting of a mixture of chloroform, methanol, and isopropylamine (96:2:2) until the solvent front has moved about three-fourths of the length of the plate. Remove the plate from the developing chamber, mark the solvent front, and allow the solvent to evaporate, with the aid of warm circulating air. Examine the plate under short-wavelength ultraviolet light, and again after exposing the plate to iodine vapors for a few minutes and spraying it with 7 *N* sulfuric acid. In each instance, compare the intensities of any secondary spots observed in the chromatogram of the *Test preparation* with those of the principal spots in the chromatograms of the *Standard preparations:* no secondary spot is more intense than the principal spot obtained from *Standard preparation A* (0.5%), and the sum of the intensities of all secondary spots obtained from the *Test preparation* does not exceed 1.0%.

Assay—Dissolve about 500 mg of Propoxycaine Hydrochloride, previously dried and accurately weighed, in 200 mL of ice-cold water containing 1 g of potassium bromide and 2.5 mL of hydrochloric acid, and titrate with 0.1 *M* sodium nitrite VS, using starch iodide paper as an external indicator. Perform a blank determination, and make any necessary correction. Each mL of 0.1 *M* sodium nitrite is equivalent to 33.09 mg of $C_{16}H_{26}N_2O_3.HCl$.

Propoxycaine and Procaine Hydrochlorides and Levonordefrin Injection

» Propoxycaine and Procaine Hydrochlorides and Levonordefrin Injection is a sterile solution of Propoxycaine Hydrochloride, Procaine Hydrochloride, and Levonordefrin in Water for Injection. It contains not less than 95.0 percent and not more than 105.0 percent of the labeled amount of propoxycaine hydrochloride ($C_{16}H_{26}N_2O_3.HCl$), not less than 95.0 percent and not more than 105.0 percent of the labeled amount of procaine hydrochloride ($C_{13}H_{20}N_2O_2.HCl$), and not less than 90.0 percent and not more than 110.0 percent of the labeled amount of levonordefrin ($C_9H_{13}NO_3$).

Packaging and storage—Preserve in single-dose containers, preferably of Type I glass.

Reference standards—*USP Propoxycaine Hydrochloride Reference Standard*—Dry at 105° for 3 hours before using. *USP Procaine Hydrochloride Reference Standard*—Dry over silica gel for 18 hours before using. *USP Epinephrine Bitartrate Reference*

Standard—Dry in vacuum over silica gel for 18 hours before using.

Identification—

A: Cool a mixture of 10 mL of Injection and 10 mL of 3 *N* hydrochloric acid in an ice bath to 0°, add 5 mL of sodium nitrite solution (1 in 5), stirring gently, and 2 mL of a solution of 0.1 g of 2-naphthol in 5 mL of 1 *N* sodium hydroxide, and observe immediately: a bright orange-red precipitate is formed (*presence of a primary aminophenyl group*).

B: It responds to the tests for *Chloride* ⟨191⟩.

C: To about 25 mL of Injection add 1 drop of ferric chloride TS: the solution immediately turns a light green and changes to a light blue within 1 minute. Add 2 drops of 3 *N* hydrochloric acid: the color reverts to light green (*presence of levonordefrin*).

pH ⟨791⟩: between 3.5 and 5.0.

Other requirements—It meets the requirements under *Injections* ⟨1⟩.

Assay for propoxycaine and procaine hydrochlorides—

Standard propoxycaine hydrochloride preparation—Prepare, by quantitative and stepwise dilution, a solution in dilute hydrochloric acid (1 in 6) containing about 50 µg of USP Propoxycaine Hydrochloride RS, accurately weighed, in each mL.

Standard procaine hydrochloride preparation—Prepare, by quantitative and stepwise dilution, a solution in dilute hydrochloric acid (1 in 6) containing about 200 µg of USP Procaine Hydrochloride RS, accurately weighed, in each mL.

Assay preparation—Transfer an accurately measured volume of Propoxycaine and Procaine Hydrochlorides and Levonordefrin Injection, equivalent to about 100 mg of procaine hydrochloride, to a 125-mL separator containing 25 mL of chloroform and 10 mL of water. Add 5 mL of sodium carbonate TS, shake vigorously, allow the layers to separate, and transfer the chloroform layer to a second separator. Extract the aqueous layer with two 10-mL portions of chloroform, adding the extracts to the second separator. Extract the combined chloroform extracts with 35 mL of dilute hydrochloric acid (1 in 6), transfer the chloroform to a third separator, and transfer the acid extract to a 100-mL volumetric flask. Extract the chloroform with two 10-mL portions of dilute hydrochloric acid (1 in 6), add the acid extracts to the volumetric flask, add dilute hydrochloric acid (1 in 6) to volume, and mix. Transfer 15.0 mL of this solution to a second 100-mL volumetric flask, add dilute hydrochloric acid (1 in 6) to volume, and mix.

Procedure—Concomitantly determine the absorbances of the *Standard preparations* and the *Assay preparation* in 1-cm cells at 296 and 272 nm, with a suitable spectrophotometer, using water as the blank. Calculate the quantity, in mg, of propoxycaine hydrochloride ($C_{16}H_{26}N_2O_3 . HCl$) in each mL of the Injection taken by the formula:

$$(0.667C/V)(A_{296}A_4 - A_{272}A_3)/(A_1A_4 - A_2A_3),$$

in which C is the concentration, in µg per mL, of USP Propoxycaine Hydrochloride RS in the *Standard propoxycaine hydrochloride preparation*.

Calculate the quantity, in mg, of procaine hydrochloride ($C_{13}H_{20}N_2O_2 . HCl$) in each mL of the Injection taken by the formula:

$$(0.667C/V)(A_{272}A_1 - A_{296}A_2)/(A_1A_4 - A_2A_3),$$

in which C is the concentration, in µg per mL, of USP Procaine Hydrochloride RS in the *Standard procaine hydrochloride preparation*, V is the volume, in mL, of Injection taken, A_1 and A_2 are the absorbances of the *Standard propoxycaine hydrochloride preparation* at 296 and 272 nm, respectively, A_3 and A_4 are the absorbances of the *Standard procaine hydrochloride preparation* at 296 and 272 nm, respectively, and A_{296} and A_{272} are the absorbances of the *Assay preparation* at 296 and 272 nm, respectively.

Assay for levonordefrin—

Ferro-citrate Solution, *Buffer Solution*, and *Standard Preparation*—Prepare as directed under *Epinephrine Assay* ⟨391⟩.

Assay preparation—Proceed with Propoxycaine and Procaine Hydrochlorides and Levonordefrin Injection as directed for *Assay Preparation* under *Epinephrine Assay* ⟨391⟩, except to read "levonordefrin" where "epinephrine" is specified.

Procedure—Proceed as directed for *Procedure* under *Epinephrine Assay* ⟨391⟩. Calculate the quantity, in mg, of levonordefrin ($C_9H_{13}NO_3$) in each mL of the Injection taken by the formula:

$$(183.21/333.29)(0.05C/V)(A_U/A_S),$$

in which 183.21 and 333.29 are the molecular weights of levonordefrin and epinephrine bitartrate, respectively, C is the concentration, in µg per mL, of USP Epinephrine Bitartrate RS in the *Standard Preparation*, and V is the volume, in mL, of Injection taken.

Propoxycaine and Procaine Hydrochlorides and Norepinephrine Bitartrate Injection

» Propoxycaine and Procaine Hydrochlorides and Norepinephrine Bitartrate Injection is a sterile solution of Propoxycaine Hydrochloride, Procaine Hydrochloride, and Norepinephrine Bitartrate in Water for Injection. It contains not less than 95.0 percent and not more than 105.0 percent of the labeled amount of propoxycaine hydrochloride ($C_{16}H_{26}N_2O_3 . HCl$), not less than 95.0 percent and not more than 105.0 percent of the labeled amount of procaine hydrochloride ($C_{13}H_{20}N_2O_2 . HCl$), and an amount of norepinephrine bitartrate ($C_8H_{11}NO_3 . C_4H_6O_6 . H_2O$) equivalent to not less than 90.0 percent and not more than 110.0 percent of the labeled amount of norepinephrine ($C_8H_{11}NO_3$).

Packaging and storage—Preserve in single-dose or in multiple-dose containers, preferably of Type I glass.

Reference standards—*USP Propoxycaine Hydrochloride Reference Standard*—Dry at 105° for 3 hours before using. *USP Procaine Hydrochloride Reference Standard*—Dry over silica gel for 18 hours before using. *USP Norepinephrine Bitartrate Reference Standard*—Do not dry; determine the water content titrimetrically at the time of use.

Identification—

A: Cool a mixture of 10 mL of Injection and 10 mL of 3 *N* hydrochloric acid in an ice bath to 0°, add 5 mL of sodium nitrite solution (1 in 5), stirring gently, and 2 mL of a solution of 0.1 g of 2-naphthol in 5 mL of 1 *N* sodium hydroxide, and observe immediately: a bright orange-red precipitate is formed (*presence of a primary aminophenyl group*).

B: It responds to the tests for *Chloride* ⟨191⟩.

C: To about 5 mL of Injection add 1 drop of ferric chloride TS: a green color develops (*presence of norepinephrine bitartrate*).

pH ⟨791⟩: between 3.5 and 5.0.

Other requirements—It meets the requirements under *Injections* ⟨1⟩.

Assay for propoxycaine and procaine hydrochlorides—

Standard propoxycaine hydrochloride preparation—Prepare, by quantitative and stepwise dilution, a solution in 2 *N* hydrochloric acid containing about 50 µg of USP Propoxycaine Hydrochloride RS, accurately weighed, in each mL.

Standard procaine hydrochloride preparation—Prepare, by quantitative and stepwise dilution, a solution in 2 *N* hydrochloric acid containing about 200 µg of USP Procaine Hydrochloride RS, accurately weighed, in each mL.

Assay preparation—Transfer an accurately measured volume of Propoxycaine and Procaine Hydrochlorides and Norepinephrine Bitartrate Injection, equivalent to about 100 mg of procaine hydrochloride, to a 125-mL separator containing 25 mL of chloroform and 10 mL of water. Add 5 mL of sodium carbonate TS, shake vigorously, allow the layers to separate, and transfer the chloroform layer to a second separator. Extract the aqueous layer with two 10-mL portions of chloroform, adding the extracts to the second separator. Extract the combined chloroform extracts with 35 mL of 2 *N* hydrochloric acid, transfer the chloroform to a third separator, and transfer the acid extract to a 100-mL volumetric flask. Extract the chloroform with two 10-mL portions of 2 *N* hydrochloric acid, add the acid extracts to the volumetric flask, dilute with 2 *N* hydrochloric acid to volume, and mix. Transfer 15.0 mL of this solution to a second 100-mL volumetric flask, dilute with 2 *N* hydrochloric acid to volume, and mix.

Procedure—Concomitantly determine the absorbances of the *Standard preparations* and the *Assay preparation*, in 1-cm cells, at 296 nm and 272 nm, with a suitable spectrophotometer, using water as the blank.

Calculate the quantity, in mg, of propoxycaine hydrochloride ($C_{16}H_{26}N_2O_3 . HCl$) in each mL of the Injection taken by the formula:

$$(0.667C/V)(A_{296}A_4 - A_{272}A_3)/(A_1A_4 - A_2A_3),$$

in which C is the concentration, in µg per mL, of USP Propoxycaine Hydrochloride RS in the *Standard propoxycaine hydrochloride preparation*, V is the volume, in mL, of Injection taken, A_1 and A_2 are the absorbances of the *Standard propoxycaine hydrochloride preparation* at 296 nm and 272 nm, respectively, A_3 and A_4 are the absorbances of the *Standard procaine hydrochloride preparation* at 296 nm and 272 nm, respectively, and A_{296} and A_{272} are the absorbances of the *Assay preparation* at 296 nm and 272 nm, respectively.

Calculate the quantity, in mg, of procaine hydrochloride ($C_{13}H_{20}N_2O_2 . HCl$) in each mL of the Injection taken by the formula:

$$(0.667C/V)(A_{272}A_1 - A_{296}A_2)/(A_1A_4 - A_2A_3),$$

in which C is the concentration, in µg per mL, of USP Procaine Hydrochloride RS in the *Standard procaine hydrochloride preparation*, and V is the volume, in mL, of Injection taken.

Assay for norepinephrine—

Ferro-citrate Solution, *Buffer Solution*, and *Standard Preparation*—Prepare as directed under *Epinephrine Assay* ⟨391⟩, except to read "USP Norepinephrine Bitartrate RS" for "USP Epinephrine Bitartrate RS" throughout.

Assay preparation—Prepare as directed for *Assay Preparation* under *Epinephrine Assay* ⟨391⟩, except to read "norepinephrine" where "epinephrine" is specified.

Procedure—Proceed as directed for *Procedure* under *Epinephrine Assay* ⟨391⟩. Calculate the quantity, in mg, of norepinephrine ($C_8H_{11}NO_3$) in each mL of the Injection taken by the formula:

$$(169.18/319.27)(0.05C/V)(A_U/A_S),$$

in which 169.18 and 319.27 are the molecular weights of norepinephrine and norepinephrine bitartrate, respectively, C is the concentration, in µg per mL, calculated on the anhydrous basis, of USP Norepinephrine Bitartrate RS in the *Standard Preparation*, and V is the volume, in mL, of Injection taken.

Propoxyphene Hydrochloride

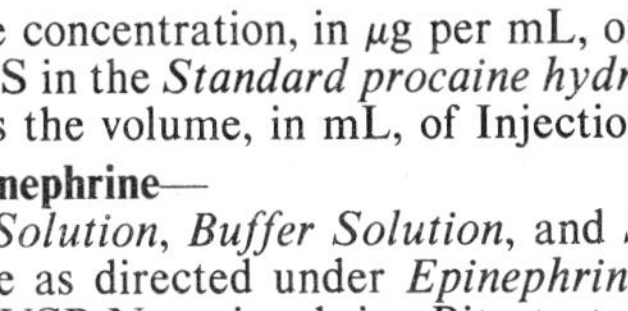

$C_{22}H_{29}NO_2 . HCl$ 375.94

Benzeneethanol, α-[2-(dimethylamino)-1-methylethyl]-α-phenyl-, propanoate (ester), hydrochloride, [*S*-(*R**,*S**)]-.

(2*S*,3*R*)-(+)-4-(Dimethylamino)-3-methyl-1,2-diphenyl-2-butanol propionate (ester) hydrochloride   [*1639-60-7*].

» Propoxyphene Hydrochloride contains not less than 98.0 percent and not more than 101.0 percent of $C_{22}H_{29}NO_2 . HCl$, calculated on the dried basis.

Packaging and storage—Preserve in tight containers.

Reference standards—*USP Propoxyphene Hydrochloride Reference Standard*—Dry at 105° for 3 hours before using. *USP α-d-2-Acetoxy-4-dimethylamino-1,2-diphenyl-3-methylbutane Reference Standard*—Keep container tightly closed. Do not dry before using. *USP α-d-4-Dimethylamino-1,2-diphenyl-3-methyl-2-butanol Hydrochloride Reference Standard*—Keep container tightly closed. Do not dry before using.

Identification—

A: The infrared absorption spectrum of a 1 in 20 solution in chloroform in 0.1-mm cells exhibits maxima only at the same wavelengths as that of a similarly prepared solution of USP Propoxyphene Hydrochloride RS.

B: It responds to the tests for *Chloride* ⟨191⟩.

Melting range ⟨741⟩: between 163.5° and 168.5°, but the range between beginning and end of melting does not exceed 3°.

Specific rotation ⟨781⟩: between +52° and +57°, calculated on the dried basis, determined in a freshly prepared solution containing 100 mg in each 10 mL.

Loss on drying ⟨731⟩—Dry it at 105° for 3 hours: it loses not more than 1.0% of its weight.

Related compounds—

Internal standard solution—Dissolve a quantity of *n*-tricosane in chloroform to obtain a solution containing 1 mg per mL.

Solution A—Weigh accurately 12.5 mg of the acetoxy analog standard, USP α-*d*-2-Acetoxy-4-dimethylamino-1,2-diphenyl-3-methylbutane RS, and 12.5 mg of the carbinol hydrochloride standard, USP α-*d*-4-Dimethylamino-1,2-diphenyl-3-methyl-2-butanol Hydrochloride RS, and transfer to the same 50-mL volumetric flask. Add chloroform to volume, and mix.

Standard preparation—Transfer about 200 mg of USP Propoxyphene Hydrochloride RS, accurately weighed, to a 125-mL separator, and add about 25 mL of water and 5.0 mL of sodium hydroxide solution (1 in 10). Add 3.0 mL of *Solution A* and 30 mL of chloroform to the contents of the separator, shake for 1 minute, and allow the phases to separate. Drain the chloroform layer, through a small portion of anhydrous sodium sulfate previously washed with chloroform, into a 100-mL volumetric flask containing 2.0 mL of *Internal standard solution*. Extract the aqueous phase in the separator with two 25-mL portions of chloroform, draining each through the anhydrous sodium sulfate into the volumetric flask. Rinse the sodium sulfate with chloroform, dilute with chloroform to volume, and mix. Evaporate 10.0 mL of this chloroform solution under a stream of nitrogen to about 2 mL.

Test preparation—Prepare as directed for *Standard preparation*, except to use 200 mg of Propoxyphene Hydrochloride and to omit the addition of the 3.0 mL of *Solution A*.

Procedure—Inject separately a suitable volume of the *Standard preparation* and an identical volume of the *Test preparation* into a suitable gas chromatograph equipped with a flame-ionization detector and in which the 0.6-m × 3-mm column is packed with 3 percent phase G2 on packing S1AB. [NOTE—Frequent resilylation of the column packing may be necessary to prevent on-column decomposition of propoxyphene.] The temperatures of the injection port and columns are maintained at 160°, and the detector temperature is maintained at 190°. [NOTE—Do not allow the temperature to exceed 200°.] The carrier gas is helium. In a suitable chromatogram, the retention times for the carbinol and acetoxy compounds are about 0.3 and 0.4, respectively, relative to the internal standard, the resolution factor, R, between the carbinol and the acetoxy peaks is not less than 1.5, five replicate injections of the *Standard preparation* show a coefficient of variation of not more than 6.0% in the ratios of the peak heights of the acetoxy analog to that of *n*-tricosane and of the carbinol analog to that of *n*-tricosane, and there is no evidence of spurious peaks resulting from the decomposition of propoxyphene in the system.

Calculate the peak heights of the carbinol, acetoxy, and *n*-tricosane peaks in each chromatogram. The order of elution is carbinol, acetoxy, propoxyphene, and *n*-tricosane.

Calculate the percentage of carbinol hydrochloride in the Propoxyphene Hydrochloride taken by the formula:

$$300(C/W)(R_U/R_S),$$

in which C is the concentration, in mg per mL, of the carbinol hydrochloride standard in *Solution A*, W is the weight, in mg, of the specimen taken, and R_U and R_S are the ratios of the peak heights of carbinol hydrochloride to those of *n*-tricosane of the *Test preparation* and the *Standard preparation*, respectively: the limit is 0.5%.

Calculate the percentage of acetoxy hydrochloride in the Propoxyphene Hydrochloride taken by the formula:

$$300(C/W)1.112(R_U/R_S),$$

in which C is the concentration, in mg per mL, of the acetoxy analog standard in *Solution A*, W is the weight, in mg, of the specimen taken, 1.112 is the ratio of the molecular weight of the acetoxy hydrochloride to that of the acetoxy free base, and R_U and R_S are the ratios of the peak heights of the acetoxy free base to that of *n*-tricosane of the *Test preparation* and the *Standard preparation*, respectively: the limit is 0.6%.

Assay—Dissolve about 600 mg of Propoxyphene Hydrochloride, accurately weighed, in 40 mL of glacial acetic acid, and add 10 mL of mercuric acetate TS. Add crystal violet TS, and titrate with 0.1 *N* perchloric acid VS. Perform a blank determination, and make any necessary correction. Each mL of 0.1 *N* perchloric acid is equivalent to 37.59 mg of $C_{22}H_{29}NO_2 . HCl$.

Propoxyphene Hydrochloride Capsules

» Propoxyphene Hydrochloride Capsules contain not less than 92.5 percent and not more than 107.5 percent of the labeled amount of $C_{22}H_{29}NO_2 . HCl$.

Packaging and storage—Preserve in tight containers.

Reference standards—*USP Propoxyphene Hydrochloride Reference Standard*—Dry at 105° for 3 hours before using. *USP α-d-2-Acetoxy-4-dimethylamino-1,2-diphenyl-3-methylbutane Reference Standard*—Keep container tightly closed. Do not dry before using. *USP α-d-4-Dimethylamino-1,2-diphenyl-3-methyl-2-butanol Hydrochloride Reference Standard*—Keep container tightly closed. Do not dry before using.

Identification—Remove, as completely as possible, the contents of not less than 20 Propoxyphene Hydrochloride Capsules. Transfer a quantity of the contents, equivalent to about 130 mg of propoxyphene hydrochloride, to a 125-mL separator containing 20 mL of water. Add 5 mL of sodium carbonate solution (1 in 10), and swirl the mixture for 3 minutes. Add 10 mL of chloroform, insert the stopper, and shake the mixture for 3 minutes. Allow the mixture to stand until most of the emulsion has broken, and filter the chloroform extract through a layer of anhydrous sodium sulfate into a 50-mL volumetric flask. Extract the alkaline water mixture with three more 10-mL portions of chloroform, filtering each extract through the anhydrous sodium sulfate, finally wash the filter with sufficient chloroform to make up to volume, and mix. Use the chloroform solution so obtained for the following tests *A* and *B*. (Reserve the alkaline water mixture for test *C*.)

A: The infrared absorption spectrum of the chloroform solution of the contents of Capsules, concentrated if necessary by evaporating a portion on a steam bath with the aid of a current of air to about one-fifth of its volume, exhibits maxima only at the same wavelengths as that of a similar preparation of USP Propoxyphene Hydrochloride RS.

B: Transfer 40 mL of the chloroform solution to a suitable beaker, and evaporate on a steam bath with the aid of a current of air to about 5 mL. Remove the beaker from the steam bath, and continue evaporation with the aid of a current of air until the chloroform is completely evaporated. Add 5 mL of 0.1 *N* hydrochloric acid to the beaker, and dissolve the residue. Transfer the solution to a 10-mL volumetric flask with the aid of additional 0.1 *N* hydrochloric acid, dilute with 0.1 *N* hydrochloric acid to volume, and mix: the specific rotation of this solution is not less than +52°.

C: Filter a portion of the extracted alkaline water mixture, render 5 mL of the clear filtrate acid to litmus with nitric acid, and add 10 drops of silver nitrate TS: a white precipitate is formed, and it is soluble in an excess of 6 *N* ammonium hydroxide.

Dissolution ⟨711⟩—

Medium: pH 4.5 acetate buffer, prepared as directed in the test for *Dissolution* under *Propoxyphene Hydrochloride, Aspirin, and Caffeine Capsules;* 500 mL.

Apparatus 1: 100 rpm.

Time: 60 minutes.

Procedure—Proceed as directed for *Determination of dissolved propoxyphene hydrochloride* in the test for *Dissolution* under *Propoxyphene Hydrochloride, Aspirin, and Caffeine Capsules.*

Tolerances—Not less than 85% (*Q*) of the labeled amount of $C_{22}H_{29}NO_2 . HCl$ is dissolved in 60 minutes.

Uniformity of dosage units ⟨905⟩: meet the requirements.

Procedure for content uniformity—Transfer, with the aid first of 2 mL of alcohol and then of about 50 mL of cold water, the contents of each Capsule to a 100-mL volumetric flask. Bring the flask and contents to room temperature, shake for 15 minutes, and add water to volume. Filter, discarding the first portion of the filtrate, and dilute, if necessary, a portion of the filtrate to a concentration of about 300 µg of propoxyphene hydrochloride per mL. Prepare a Standard solution of USP Propoxyphene Hydrochloride RS in water having a known concentration of about 300 µg per mL. Concomitantly determine the absorbances of the solution from the Capsule contents and the Standard solution in 1-cm cells at the wavelength of maximum absorbance at about 257 nm, with a suitable spectrophotometer, using water as the blank. Calculate the quantity, in mg, of $C_{22}H_{29}NO_2 . HCl$ in the Capsule by the formula:

$$BC(A_U/A_S),$$

in which B is either 0.1 or 0.2, respectively (depending upon whether the filtrate was used undiluted or was diluted with 1 part of water), C is the concentration, in µg per mL, of USP Propoxyphene Hydrochloride RS in the Standard solution, and A_U and A_S are the absorbances of the solution from the Capsule contents and the Standard solution, respectively.

Assay—

Internal standard solution—Dissolve *n*-tricosane in chloroform to obtain a solution containing about 0.6 mg per mL.

Standard preparation—Transfer 16 mg of USP Propoxyphene Hydrochloride RS, accurately weighed, to a 25-mL volumetric flask. Dissolve in 5 mL of acetone. Dilute with water to volume, and mix.

Assay preparation—Remove as completely as possible the contents of 20 Propoxyphene Hydrochloride Capsules, and transfer an accurately weighed portion of the powder, equivalent to 65 mg of propoxyphene hydrochloride, to a 100-mL volumetric flask containing 20 mL of acetone. Sonicate for about 1 minute. Dilute the milky solution with water to volume, and mix. Filter, discarding the first 20 mL of the filtrate.

Procedure—Proceed as directed for *Procedure* in the *Assay for propoxyphene hydrochloride and caffeine* under *Propoxyphene Hydrochloride, Aspirin, and Caffeine Capsules.* Calculate the quantity, in mg, of $C_{22}H_{29}NO_2 . HCl$ in the portion of Capsules taken by the formula:

$$100C(R_U/R_S),$$

in which C is the concentration, in mg per mL, of USP Propoxyphene Hydrochloride RS in the *Standard preparation*, and R_U and R_S are the ratios of the peak areas of the propoxyphene hydrochloride peak to the internal standard peak from the *Assay preparation* and the *Standard preparation*, respectively.

Propoxyphene Hydrochloride and Acetaminophen Tablets

» Propoxyphene Hydrochloride and Acetaminophen Tablets contain not less than 90.0 percent and not more than 110.0 percent of the labeled amounts of propoxyphene hydrochloride ($C_{22}H_{29}NO_2 . HCl$) and acetaminophen ($C_8H_9NO_2$).

Packaging and storage—Preserve in tight containers.

Reference standards—*USP Propoxyphene Hydrochloride Reference Standard*—Dry at 105° for 3 hours before using. *USP Acetaminophen Reference Standard*—Dry over silica gel for 18 hours before using.

Identification—Transfer 1 finely ground Tablet to a test tube. If the Tablets are coated, first immerse the Tablet in acetone for 1½ minutes, remove the shell, and grind. Add 5 mL of methanol, shake for 5 minutes, and centrifuge. Use the clear supernatant liquid as the *Test solution.* Prepare a *Standard solution* in methanol containing, in each mL, 130 mg of USP Acetaminophen RS and 13 mg of USP Propoxyphene Hydrochloride RS. Apply 5 µL of the *Test solution* on a line parallel to and about 2 cm from the bottom edge of a 20- × 5-cm thin-layer chromatographic plate (see *Chromatography* ⟨621⟩) coated with chromatographic silica gel mixture, and apply 5 µL of the *Standard solution* separately on the starting line. Place the plate in a developing chamber containing a mixture of butyl acetate, chloroform, and formic acid (60:40:20), and develop the chromatogram until the solvent front has moved about 15 cm above the line of application. Remove the plate, allow to dry in a hood, and view under short-wavelength ultraviolet light: the R_f value of the principal spot from the *Test solution* corresponds to that from the *Standard solution.* Spray the plate with iodoplatinate TS: the R_f value of the orange-brown spot from the *Test solution* corresponds to that from the *Standard solution.*

Dissolution ⟨711⟩—

Medium: pH 4.5 acetate buffer; 700 mL.

Apparatus 2: 50 rpm.

Time: 30 minutes.

pH 4.5 acetate buffer—Prepare as directed in the test for *Dissolution* under *Propoxyphene Hydrochloride, Aspirin, and Caffeine Capsules.*

pH 5.3 phosphate buffer—Dissolve 19.0 g of monobasic sodium phosphate and 1.0 g of anhydrous dibasic sodium phosphate in water to make 250.0 mL. Adjust by the addition of small amounts of either salt to a pH of 5.3.

Dye solution—Dissolve 200 mg of bromocresol purple in 250.0 mL of water. Filter before use.

Standard solution—Transfer about 23 mg of USP Propoxyphene Hydrochloride RS, accurately weighed, to a 250-mL volumetric flask, dilute with *pH 4.5 acetate buffer* to volume, and mix.

Procedure—Determine the amount of propoxyphene hydrochloride ($C_{22}H_{29}NO_2 . HCl$) dissolved in 30 minutes for 6 individual Tablets. Pipet 5 mL each of the filtered *Dissolution Medium* and the *Standard solution* into separate 125-mL separators, and treat each as follows: Add 10.0 mL of *pH 5.3 phosphate buffer*, and mix. Add 5.0 mL of *Dye solution*, and shake. Extract with two 20-mL portions of chloroform, filter, and collect the chloroform extracts in 50-mL volumetric flasks. Dilute with chloroform to volume, and mix. Determine the absorbances of both solutions at the wavelength of maximum absorbance at about 400 nm, with a suitable spectrophotometer, using chloroform as the blank. Calculate the quantity, in µg, of $C_{22}H_{29}NO_2 . HCl$ in the 5-mL portion of solution taken by the formula:

$$5C(A_U / A_S),$$

in which C is the concentration, in µg per mL, of USP Propoxyphene Hydrochloride RS in the *Standard solution*, and A_U and A_S are the absorbances of the solutions from the *Dissolution Medium* and the *Standard solution*, respectively.

Tolerances—Not less than 75% (Q) of the labeled amount of $C_{22}H_{29}NO_2 . HCl$ is dissolved in 30 minutes.

Uniformity of dosage units ⟨905⟩: meet the requirements for *Content Uniformity* with respect to propoxyphene hydrochloride.

Assay for propoxyphene hydrochloride—

Internal standard solution—Dissolve *n*-tricosane in chloroform to obtain a solution containing about 0.6 mg per mL.

Standard preparation—Transfer 16 mg of USP Propoxyphene Hydrochloride RS, accurately weighed, to a 25-mL volumetric flask. Dilute with acetone to volume, and mix.

Assay preparation—Weigh and finely powder not less than 20 Propoxyphene Hydrochloride and Acetaminophen Tablets. If the Tablets are coated, first immerse them in acetone for 1.5 minutes, remove the shells, and air-dry. Transfer a portion of the powder, equivalent to about 65 mg of propoxyphene hydrochloride, to a 100-mL volumetric flask. Add 50 mL of acetone, and shake by mechanical means for 15 minutes. Dilute with acetone to volume, and mix. Centrifuge, and use the clear supernatant solution.

Procedure—Proceed as directed for *Procedure* in the *Assay for propoxyphene hydrochloride and caffeine* under *Propoxyphene Hydrochloride, Aspirin, and Caffeine Capsules.* Calculate the quantity, in mg, of $C_{22}H_{29}NO_2 . HCl$ in the portion of Tablets taken by the formula:

$$100C(R_U / R_S),$$

in which C is the concentration, in mg per mL, of USP Propoxyphene Hydrochloride RS in the *Standard preparation*, and R_U and R_S are the ratios of the peak areas of the propoxyphene hydrochloride peak to the internal standard peak from the *Assay preparation* and the *Standard preparation*, respectively.

Assay for acetaminophen—

Standard preparation—Transfer about 13 mg of USP Acetaminophen RS, accurately weighed, to a 10-mL volumetric flask. Add methanol to volume, and mix. Transfer 2.0 mL of this solution to a 100-mL volumetric flask, dilute with water to volume, and mix.

Chromatographic column—Mix a portion of 100- to 200-mesh sulfonic acid cation-exchange resin with dilute hydrochloric acid (1 in 4) in the ratio of 1 g of resin to 5 mL of acid. After 15 minutes, decant the acid, and wash the resin with water until free from acid. Pack a pledget of glass wool in the base of a 1- × 20-cm chromatographic tube fitted with a stopcock. Transfer an amount of acid-treated resin to it sufficient to make a settled column of 6 cm to 7 cm in height. Stir to remove air bubbles, place a second pledget of glass wool on top of the column, then drain and discard the excess water without allowing air to enter the column.

Assay preparation—Weigh and finely powder not less than 20 Propoxyphene Hydrochloride and Acetaminophen Tablets. If the Tablets are coated, first immerse them in acetone for 1.5 minutes, remove the shells, and air-dry. Transfer a portion of the powder, accurately weighed, equivalent to about 650 mg of acetaminophen, to a 50-mL volumetric flask. Add 10 mL of water, and heat gently on a steam bath for 10 minutes, with occasional swirling. Cool the mixture to room temperature, dilute with methanol to volume, and mix. Centrifuge, and use the clear supernatant solution. Transfer 5.0 mL of this solution to a 50-mL volumetric flask, dilute with methanol to volume, and mix. Transfer 2.0 mL of the second solution to a 100-mL beaker, add 20 mL of water, and transfer this solution to the *Chromatographic column* with the aid of two 5-mL portions of water. Allow it to pass through the column slowly, and collect the eluate in a 100-mL volumetric flask. Without allowing air to enter the column, rinse with three 20-mL portions of water, and collect these eluates in the 100-mL volumetric flask. Dilute with water to volume, and mix.

Procedure—Transfer 5.0-mL portions of *Standard preparation* and *Assay preparation* to separate 25-mL volumetric flasks. Dilute with methanol to volume, and mix. Concomitantly measure the absorbances of both solutions in 1-cm cells at the wavelength of maximum absorbance at about 249 nm, with a suitable spectrophotometer, using a 1 in 5 solution of water in methanol as the blank. Calculate the quantity, in mg, of $C_8H_9NO_2$ in the portion of Tablets taken by the formula:

$$50C(A_U / A_S),$$

in which C is the concentration, in mg per mL, of USP Acetaminophen RS in the *Standard preparation*, and A_U and A_S are the absorbances of the solutions from the *Assay preparation* and the *Standard preparation*, respectively.

Propoxyphene Hydrochloride, Aspirin, and Caffeine Capsules

» Propoxyphene Hydrochloride, Aspirin, and Caffeine Capsules contain not less than 90.0 percent and not more than 110.0 percent of the labeled amounts of propoxyphene hydrochloride ($C_{22}H_{29}NO_2.HCl$), aspirin ($C_9H_8O_4$), and caffeine ($C_8H_{10}N_4O_2$).

NOTE—Where Propoxyphene Hydrochloride, Aspirin, and Caffeine Capsules are prescribed, the quantity of propoxyphene hydrochloride is to be specified. Where the Capsules are prescribed without reference to the quantity of aspirin or caffeine contained therein, a product containing 389 mg of aspirin and 32.4 mg of caffeine shall be dispensed.

Packaging and storage—Preserve in tight containers at controlled room temperature.

Reference standards—*USP Propoxyphene Hydrochloride Reference Standard*—Dry at 105° for 3 hours before using. *USP Aspirin Reference Standard*—Dry over silica gel for 5 hours before using. *USP Caffeine Reference Standard*—Dry at 80° for 4 hours before using. *USP Salicylic Acid Reference Standard*—Dry over silica gel for 3 hours before using. *USP α-d-2-Acetoxy-4-dimethylamino-1,2-diphenyl-3-methylbutane Reference Standard*—Keep container tightly closed. Do not dry before using. *USP α-d-4-Dimethylamino-1,2-diphenyl-3-methyl-2-butanol Hydrochloride Reference Standard*—Keep container tightly closed. Do not dry before using.

Identification—

A: Place an amount of the finely powdered contents of Capsules, equivalent to about 65 mg of propoxyphene hydrochloride, in a test tube, add 5 mL of methanol, shake for 5 minutes, and centrifuge. The clear supernatant liquid is the *Test solution.* Dissolve weighed amounts of USP Propoxyphene Hydrochloride RS, USP Aspirin RS, and USP Caffeine RS corresponding, proportionately, to the amounts of propoxyphene hydrochloride, aspirin, and caffeine in the Capsules to obtain a *Standard solution* having a known concentration of about 13 mg of propoxyphene hydrochloride per mL. On a suitable thin-layer chromatographic plate (see *Chromatography* ⟨621⟩), coated with a 0.25-mm layer of chromatographic silica gel mixture, apply 10 µL each of the *Test solution* and the *Standard solution.* Develop the chromatograms in a solvent system consisting of a mixture of chloroform, butyl acetate, and formic acid (30:20:10) until the solvent front has moved about three-fourths of the length of the plate. Remove the plate from the developing chamber, mark the solvent front, allow it to dry in a fume hood, and examine the plate under short-wavelength ultraviolet light: the chromatogram of the *Test solution* exhibits 3 principal spots that correspond in R_f value and intensity with those obtained from the *Standard solution.*

B: The *Assay preparation* prepared as directed in the *Assay for propoxyphene hydrochloride and caffeine* is dextrorotatory.

Dissolution ⟨711⟩—

Medium: pH 4.5 acetate buffer; 500 mL.

Apparatus 1: 100 rpm.

Time: 60 minutes.

pH 4.5 acetate buffer—Dissolve 2.99 g of sodium acetate trihydrate in 200 mL of water, add 1.66 mL of glacial acetic acid, dilute with water to 1000 mL, and mix.

Determination of dissolved aspirin—Transfer 5.0 mL of a filtered portion of the solution under test to a 25-mL volumetric flask. Concurrently pipet 5 mL of a standard solution, prepared by dissolving 35 mg of accurately weighed USP Aspirin RS in 50.0 mL of *pH 4.5 acetate buffer,* into a second 25-mL volumetric flask. Proceed as directed for *Procedure* in the *Assay for aspirin,* beginning with "Into each flask pipet 5 mL of *Sodium hydroxide reagent.*" Concomitantly determine the absorbances of both solutions against a similarly treated blank of 5.0 mL of *pH 4.5 acetate buffer.* Determine the amount of aspirin ($C_9H_8O_4$) in solution by comparison with the Standard solution.

Determination of dissolved propoxyphene hydrochloride—

INTERNAL STANDARD SOLUTION—Dissolve *n*-tricosane in chloroform to obtain a solution containing about 0.06 mg per mL.

STANDARD PREPARATION—Dissolve an accurately weighed quantity of USP Propoxyphene Hydrochloride RS in *pH 4.5 acetate buffer* to obtain a solution having a known concentration (between 0.06 and 0.13 mg per mL) that corresponds to the concentration of propoxyphene hydrochloride that is estimated for the solution under test, based on the labeled content of the Capsules and the extent of dissolution.

PROCEDURE—Transfer equal, accurately measured, volumes (between 5.0 and 10.0 mL) of filtered portions of the solution under test and the *Standard preparation* into separate, screw-capped centrifuge tubes. To each tube add 5.0 mL of sodium carbonate solution (1 in 5), and shake. Extract with one 5.0-mL portion of *Internal standard solution,* and one 5.0-mL portion of chloroform, and shake each extract for 5 minutes. Pass the organic layers through phase-separating paper. Evaporate to about 1 mL, and inject a 10-µL portion into a suitable gas chromatograph equipped with a flame-ionization detector. Proceed as directed for *Procedure* in the *Assay for propoxyphene hydrochloride and caffeine,* beginning with "The column is typically 60 cm × 4 mm." Determine the amount of propoxyphene hydrochloride ($C_{22}H_{29}NO_2.HCl$) in solution by comparison with the *Standard preparation.*

Tolerances—Not less than 75% (*Q*) of the labeled amount of $C_9H_8O_4$ is dissolved in 60 minutes, and not less than 85% (*Q*) of the labeled amount of $C_{22}H_{29}NO_2.HCl$ is dissolved in 60 minutes.

Uniformity of dosage units ⟨905⟩: meet the requirements for *Content Uniformity* with respect to propoxyphene hydrochloride and to caffeine.

Free salicylic acid—

Ferric chloride–urea reagent—Dissolve by swirling, without the aid of heat, 60 g of urea in a mixture of 8 mL of ferric chloride solution (6 in 10) and 42 mL of 0.05 *N* hydrochloric acid. Adjust this solution, if necessary, with 6 *N* hydrochloric acid to a pH of 3.2.

Standard preparation—Transfer 15.0 mg of USP Salicylic Acid RS, accurately weighed, to a 100-mL volumetric flask, add chloroform to volume, and mix. Transfer 10.0 mL of this solution to a 100-mL volumetric flask containing 20 mL of methanol, 4 drops of hydrochloric acid, and 20 mL of a 1 in 10 solution of glacial acetic acid in ether, dilute with chloroform to volume, and mix.

Test preparation—Pack a pledget of glass wool in the base of a 25-mm × 200-mm chromatographic tube. In a beaker, prepare a mixture of 6 g of chromatographic siliceous earth, 2 mL of freshly prepared *Ferric chloride–urea reagent,* and 40 mL of chloroform. Transfer the mixture to the chromatographic tube. Rinse the beaker with 15 mL of chloroform, transfer to the column, and pack tightly. Place a small amount of glass wool at the top of the column. Weigh accurately a quantity of the finely powdered contents of Capsules, equivalent to about 50 mg of aspirin, mix with 10 mL of chloroform by stirring for 3 minutes, and transfer to the chromatographic column with the aid of 10 mL of chloroform. Pass 40 mL of chloroform through the column, rinse the tip of the chromatographic tube with chloroform, and discard the eluate. Prepare as a receiver a 100-mL volumetric flask containing 20 mL of methanol and 4 drops of hydrochloric acid, and elute any salicylic acid from the column with 20 mL of a 1 in 10 solution of glacial acetic acid in ether that recently has been saturated with water, followed by 30 mL of chloroform. Dilute the eluate with chloroform to volume, and mix.

Procedure—Concomitantly determine the absorbances of the *Standard preparation* and the *Test preparation* in 1-cm cells at the wavelength of maximum absorbance at about 306 nm, with a suitable spectrophotometer, using as the blank a solvent mixture of the same composition as that used for the *Standard preparation:* the absorbance of the *Test preparation* does not exceed that of the *Standard preparation* (3.0%, calculated on the basis of the labeled content of aspirin).

Assay for propoxyphene hydrochloride and caffeine—

Internal standard solution—Dissolve *n*-tricosane in chloroform to obtain a solution having a concentration of about 0.6 mg per mL.

Standard preparation—Transfer to a 50-mL volumetric flask accurately weighed quantities of about 32 mg of USP Propoxyphene Hydrochloride RS, about 32*J* mg of USP Aspirin RS, and about 32*J′* mg of USP Caffeine RS, where *J* is the ratio of the labeled amount, in mg, of aspirin to the labeled amount, in mg, of propoxyphene hydrochloride per Tablet, and *J′* is the ratio of the labeled amount, in mg, of caffeine to the labeled amount, in mg, of propoxyphene hydrochloride per Tablet. Add 10 mL of acetone, and swirl to dissolve the reference standards completely. Dilute with water to volume, and mix.

Assay preparation—Remove as completely as possible the contents of 20 Propoxyphene Hydrochloride, Aspirin, and Caffeine Capsules, and transfer an accurately weighed portion of the powder, equivalent to 65 mg of propoxyphene hydrochloride, to a 100-mL volumetric flask containing 20 mL of acetone. If the Capsules contain a pellet (propoxyphene hydrochloride) as well as a powder, finely grind the pellets, then mix with the powder before proceeding. Sonicate for about 1 minute. Dilute the milky solution with water to volume, and mix. Filter, discarding the first 20 mL of the filtrate.

Procedure for propoxyphene hydrochloride and caffeine—Transfer 5.0-mL aliquots of the *Assay preparation* and the *Standard preparation* to separate 60-mL separators. To each add 5.0 mL of sodium carbonate solution (1 in 5) and 5.0 mL of *Internal standard solution.* Shake vigorously for 5 minutes, and allow the layers to separate. Drain the chloroform layer through phase-separating paper, suitably supported in a funnel, into a screw-capped test tube. Extract with one 5-mL portion of chloroform, and drain the chloroform layer through phase-separating paper. Evaporate the combined chloroform extracts, using a stream of dry nitrogen, to a final volume of about 2 mL. Inject separately a suitable volume, equivalent to about 6.4 µg of propoxyphene, of the chloroform extracts from the *Assay preparation* and the *Standard preparation* into a suitable gas chromatograph equipped with a flame-ionization detector. The column is typically 120 cm × 3 mm and is packed with 3 percent methyl phenyl silicone, liquid phase on 80- to 100-mesh chromatographic siliceous earth. The temperature of the injection port is 200°, the column temperature is 175°, and the carrier gas, nitrogen, has a flow rate of about 60 mL per minute. Relative retention times are about 0.65 for caffeine, 1.0 for the internal standard, and 1.7 for propoxyphene. In a suitable chromatogram, the resolution factor is not less than 1.0 between any two peaks, the relative standard deviation for five replicate injections of the *Standard preparation* is not more than 2.0, and the tailing factor for caffeine is not greater than 1.5. Calculate the quantities, in mg, of propoxyphene hydrochloride ($C_{22}H_{29}NO_2 . HCl$) and caffeine ($C_8H_{10}N_4O_2$), respectively, in the portion taken for the *Assay preparation* by the same formula:

$$100C(R_U/R_S),$$

in which C is the concentration, in mg per mL, of the appropriate USP Reference Standard in the *Standard preparation,* and R_U and R_S are the ratios of the peak areas of the corresponding analyte to those of the internal standard obtained from the *Assay preparation* and the *Standard preparation,* respectively.

Assay for aspirin—

Sodium hydroxide reagent—Dissolve 1 g of polyoxyethylene (23) lauryl ether in about 100 mL of hot water contained in a 1000-mL volumetric flask. Dilute with water to about 600 mL, and dissolve 10 g of sodium hydroxide in this solution. Dilute with water to volume, and mix.

Ferric nitrate reagent—Mix 70 mL of nitric acid with about 600 mL of water contained in a 1000-mL volumetric flask. Dissolve 40 g of ferric nitrate [$Fe(NO_3)_3 . 9H_2O$] in this solution, dilute with water to volume, and mix.

Standard preparation and *Assay preparation*—Prepare as directed in the *Assay for propoxyphene hydrochloride and caffeine* to obtain solutions having concentrations of about 4 mg of aspirin per mL.

Procedure—Into separate 25-mL volumetric flasks pipet 2 mL each of the *Standard preparation* and the *Assay preparation,* and 2 mL of dilute acetone (1 in 5) to provide the blank. Into each flask pipet 5 mL of *Sodium hydroxide reagent,* mix by gentle swirling, and allow to stand at room temperature for 8 minutes. Dilute with *Ferric nitrate reagent* to volume, and mix. Concomitantly determine the absorbances of both solutions against the blank in 1-cm cells at the wavelength of maximum absorbance at about 530 nm, taking care to allow the solutions to reach an equilibrium temperature in the cell compartment. The color intensity is temperature-dependent. Calculate the quantity, in mg, of aspirin ($C_9H_8O_4$) in the portion taken for the *Assay preparation* by the formula:

$$100C(A_U/A_S),$$

in which C is the concentration, in mg per mL, of USP Aspirin RS in the *Standard preparation,* and A_U and A_S are the absorbances of the solutions from the *Assay preparation* and the *Standard preparation,* respectively.

Propoxyphene Napsylate

$C_{22}H_{29}NO_2 . C_{10}H_8O_3S . H_2O$ 565.72

Benzeneethanol, α-[2-(dimethylamino)-1-methylethyl]-α-phenyl-, propanoate (ester), [*S*-(*R**,*S**)]-, compd. with 2-naphthalenesulfonic acid (1:1), monohydrate.

(α*S*,1*R*)-α-[2-(Dimethylamino)-1-methylethyl]-α-phenylphenethyl propionate compound with 2-naphthalenesulfonic acid (1:1) monohydrate [*26570-10-5*].

Anhydrous 547.71 [*23239-43-2; 17140-78-2*].

» Propoxyphene Napsylate contains not less than 97.0 percent and not more than 103.0 percent of $C_{22}H_{29}NO_2 . C_{10}H_8O_3S$, calculated on the anhydrous basis.

Packaging and storage—Preserve in tight containers.

Reference standards—*USP Propoxyphene Napsylate Reference Standard*—This is the monohydrate form. Do not dry; determine the water content titrimetrically at time of use for quantitative analyses. *USP α-d-2-Acetoxy-4-dimethylamino-1,2-diphenyl-3-methylbutane Reference Standard*—Keep container tightly closed. Do not dry before using. *USP α-d-4-Dimethylamino-1,2-diphenyl-3-methyl-2-butanol Hydrochloride Reference Standard*—Keep container tightly closed. Do not dry before using.

Identification—

A: The infrared absorption spectrum of a potassium bromide dispersion of it, previously dried at 105° for 3 hours, exhibits maxima only at the same wavelengths as that of a similar preparation of USP Propoxyphene Napsylate RS.

B: The ultraviolet absorption spectrum of a 1 in 25,000 solution in methanol exhibits maxima and minima at the same wavelengths as that of a similar solution of USP Propoxyphene Napsylate RS, concomitantly measured, and the respective absorptivities, calculated on the anhydrous basis, at the wavelength of maximum absorbance at about 275 nm do not differ by more than 3.0%.

Melting range, *Class I* ⟨741⟩: between 158° and 165°, but the range between beginning and end of melting does not exceed 4°, determined after drying at 105° for 3 hours.

Specific rotation ⟨781⟩: between +35° and +43°, determined in a solution in chloroform containing 100 mg in each 10 mL.

Water, *Method I* ⟨921⟩: between 2.5% and 5.0%.

Residue on ignition ⟨281⟩: not more than 0.5%.

Heavy metals, *Method II* ⟨231⟩: 0.003%.

Related compounds—

Internal standard solution and *Solution A*—Prepare as directed in the test for *Related compounds* under *Propoxyphene Hydrochloride.*

Standard preparation—Transfer about 150 mg of USP Propoxyphene Napsylate RS, accurately weighed, to a 125-mL separator, add 15 mL of water, and mix. Add 5.0 mL of sodium hydroxide solution (1 in 10), 2.0 mL of *Solution A,* and 15 mL of chloroform to the contents of the separator, shake for 1 minute, and allow the phases to separate. Drain the chloroform extract through a layer of about 2 g of anhydrous granular sodium sulfate, supported on glass wool and previously washed with chloroform, into a 50-mL volumetric flask containing 1.0 mL of *Internal standard solution.* Extract the aqueous phase in the separator with two additional 15-mL portions of chloroform, draining each chloroform extract through the same sodium sulfate filter into the same volumetric flask. Rinse the sodium sulfate filter with 1 to 2 mL of chloroform, collect the rinsing with the combined chloroform extracts, dilute with chloroform to volume, and mix.

Evaporate 10.0 mL of this chloroform solution under a stream of nitrogen to about 2 mL.

Test preparation—Prepare as directed for *Standard preparation*, except to use 150 mg of Propoxyphene Napsylate and to omit the addition of the 2.0 mL of *Solution A*.

Chromatographic system (see *Chromatography* ⟨621⟩)—Under typical conditions, the instrument is equipped with a flame-ionization detector and contains a 0.6-m × 3-mm glass column packed with 3 percent phase G2 on packing S1AB. [NOTE—Frequent resilylation of the column packing may be necessary to prevent on-column decomposition of propoxyphene.] The temperatures of the column and the injector port are maintained at about 160°, and the temperature of the detector block is maintained at about 190°. [NOTE—Do not allow the temperature to exceed 200°.] Dry helium is used as the carrier gas at a flow rate of about 75 mL per minute.

System suitability—Chromatograph five injections of the *Standard preparation*, and record peak heights as directed under *Procedure*. The development time for one chromatogram is about 10 minutes, and retention times for the carbinol precursor and acetoxy analog are, respectively, about 0.3 and 0.4 relative to the internal standard: the analytical system is suitable for conducting the test if the resolution factor, *R*, between the carbinol and the acetoxy peaks is not less than 1.5, the coefficients of variation for the ratios R_A and R_C do not exceed 6.0%, and there is no evidence of spurious peaks resulting from the decomposition of propoxyphene in the system.

Procedure—Inject separately a suitable volume of the *Standard preparation* and an identical volume of the *Test preparation* into a suitable gas chromatograph, and record the chromatogram. Calculate the peak heights of the carbinol precursor, the acetoxy analog, and the *n*-tricosane peaks in each chromatogram.

Calculate the percentage of carbinol napsylate precursor in the portion of Propoxyphene Napsylate taken by the formula:

$$(509.66/319.87)(200C_C/W)(r_C/R_C),$$

in which 509.66 and 319.87 are the molecular weights of carbinol napsylate monohydrate precursor and anhydrous carbinol hydrochloride precursor, respectively, C_C is the concentration, in mg per mL, of USP *α-d*-4-Dimethylamino-1,2-diphenyl-3-methyl-2-butanol Hydrochloride RS in *Solution A*, *W* is the weight, in mg, of Propoxyphene Napsylate taken, and r_C and R_C are the ratios of the peak heights of the carbinol precursor to those of *n*-tricosane in the *Test preparation* and the *Standard preparation*, respectively: the limit is 0.5%.

Calculate the percentage of acetoxy analog napsylate in the Propoxyphene Napsylate taken by the formula:

$$(551.70/325.45)(200C_A/W)(r_A/R_A),$$

in which 551.70 and 325.45 are the molecular weights of acetoxy analog napsylate monohydrate and anhydrous acetoxy analog, respectively, C_A is the concentration, in mg per mL, of USP *α-d*-2-Acetoxy-4-dimethylamino-1,2-diphenyl-3-methylbutane RS in *Solution A*, *W* is the weight, in mg, of Propoxyphene Napsylate taken, and r_A and R_A are the ratios of the peak heights of the acetoxy analog to those of *n*-tricosane in the *Test preparation* and the *Standard preparation*, respectively: the limit is 0.6%.

Assay—

pH 12.5 borate buffer—Dissolve 6.18 g of boric acid and 7.5 g of potassium chloride in water to make 500 mL. Dissolve 17.5 g of sodium hydroxide in water to make 500 mL. Mix the two solutions, dilute with water to 2000 mL, and mix.

Procedure—Transfer about 375 mg of Propoxyphene Napsylate, accurately weighed, to a 250-mL separator, and dissolve in 40 mL of chloroform. Add 25 mL of *pH 12.5 borate buffer*, shake for 2 minutes, and allow the phases to separate. Filter the chloroform extract through a layer of about 3 g of anhydrous granular sodium sulfate supported on a pledget of cotton into a 250-mL beaker. Extract the aqueous phase in the separator with three additional 40-mL portions of chloroform, pass the extracts through the same sodium sulfate filter, and combine the chloroform extracts in the same beaker. Evaporate the chloroform from the solution on a steam bath with the aid of a current of air nearly to dryness, and continue the evaporation, with gentle heat and the aid of a current of air, to remove all traces of chloroform. Add a magnetic stirring bar and 40 mL of glacial acetic acid, and stir with a magnetic stirrer until the residue is dissolved. Add crystal violet TS, and titrate with 0.1 *N* perchloric acid VS. Perform a blank determination, and make any necessary correction. Each mL of 0.1 *N* perchloric acid is equivalent to 54.77 mg of $C_{22}H_{29}NO_2 . C_{10}H_8O_3S$.

Propoxyphene Napsylate Oral Suspension

» Propoxyphene Napsylate Oral Suspension contains not less than 90.0 percent and not more than 110.0 percent of the labeled amount of $C_{22}H_{29}NO_2 . C_{10}H_8O_3S . H_2O$.

Packaging and storage—Preserve in tight containers, protected from light. Avoid freezing.

Reference standards—*USP Propoxyphene Napsylate Reference Standard*—This is the monohydrate form. Do not dry; determine the water content titrimetrically at time of use for quantitative analyses. *USP α-d-2-Acetoxy-4-dimethylamino-1,2-diphenyl-3-methylbutane Reference Standard*—Keep container tightly closed. Do not dry before using. *USP α-d-4-Dimethylamino-1,2-diphenyl-3-methyl-2-butanol Hydrochloride Reference Standard*—Keep container tightly closed. Do not dry before using.

Identification—Transfer a volume of Oral Suspension, equivalent to about 100 mg of propoxyphene napsylate, to a small flask, mix with 10 mL of chloroform, and filter: the chloroform solution is dextrorotatory (see *Optical Rotation* ⟨781⟩).

Alcohol content, *Method II* ⟨611⟩: between 0.5% and 1.5% of C_2H_5OH.

Assay—

Internal standard solution—Dissolve a quantity of *n*-tricosane in ethyl acetate to obtain a solution containing about 1 mg per mL.

Standard preparation—Transfer about 50 mg of USP Propoxyphene Napsylate RS, accurately weighed, to a 100-mL volumetric flask, add 10 mL of acetone, mix, and sonicate for about 15 minutes. Dilute with water to volume, and mix. Transfer 10.0 mL of this solution to a 125-mL separator, and add 5 mL of sodium carbonate solution (1 in 5). Add 5.0 mL of *Internal standard solution* to the contents of the separator, shake for 5 minutes, and allow the phases to separate. Discard the lower, aqueous phase, and transfer the remaining liquid to a small centrifuge tube. Centrifuge, transfer the upper layer to a small flask containing about 1 g of anhydrous granular sodium sulfate, and mix.

Assay preparation—Transfer an accurately weighed quantity of well-mixed Propoxyphene Napsylate Oral Suspension, equivalent to about 50 mg of propoxyphene napsylate, to a 100-mL volumetric flask, add 10 mL of acetone, swirl gently (do not shake) to mix, and proceed as directed under *Standard preparation*, beginning with "sonicate for about 15 minutes."

Chromatographic system (see *Gas Chromatography* ⟨621⟩)—Proceed as directed for *Chromatographic system* in the test for *Related compounds* under *Propoxyphene Napsylate*.

System suitability—Chromatograph five injections of the *Standard preparation*, and record peak heights as directed under *Procedure*. The retention time for propoxyphene is about 0.5 relative to the internal standard: the analytical system is suitable for conducting the test if the coefficient of variation for the ratio R_S does not exceed 2.0%, the resolution factor between P_1 and P_2 is not less than 3.0, and the tailing factor (the sum of the distances from peak center to the leading edge and to the tailing edge divided by twice the distance from peak center to the leading edge), measured at 5.0% of peak heights, P_1 and P_2, does not exceed 1.5 for each component.

Procedure—Inject a portion of the *Standard preparation* into a suitable gas chromatograph, and record the chromatogram. Measure the heights of the propoxyphene peak and the *n*-tricosane peak, and record the values as P_1 and P_2, respectively. Calculate the standard propoxyphene ratio, R_S, by the equation:

$$R_S = P_1/P_2.$$

Inject a portion of the *Assay preparation* into the chromatograph, and record the chromatogram. Measure the heights of the propoxyphene peak and the *n*-tricosane peak, and record the values as p_1 and p_2, respectively. Calculate the propoxyphene ratio, R_U, of the assay specimen by the equation:

$$R_U = p_1/p_2.$$

Calculate the quantity, in mg, of $C_{22}H_{29}NO_2 . C_{10}H_8O_3S . H_2O$ in each mL of the Oral Suspension taken by the formula:

$$(565.72/547.71)(W_S/W_U)(D)(R_U/R_S),$$

in which 565.72 and 547.71 are the molecular weights of propoxyphene napsylate monohydrate and anhydrous propoxyphene napsylate, respectively, W_S is the weight, in mg, of (anhydrous) USP Propoxyphene Napsylate RS taken, as determined from the weight of USP Propoxyphene Napsylate RS corrected for moisture content by a titrimetric water determination, W_U is the weight, in g, of Oral Suspension taken, D is the density, in g per mL, of Oral Suspension, and R_U and R_S are the peak response ratios obtained from the *Assay preparation* and the *Standard preparation*, respectively.

Propoxyphene Napsylate Tablets

» Propoxyphene Napsylate Tablets contain not less than 90.0 percent and not more than 110.0 percent of the labeled amount of $C_{22}H_{29}NO_2 . C_{10}H_8O_3S . H_2O$.

Packaging and storage—Preserve in tight containers.

Reference standards—*USP Propoxyphene Napsylate Reference Standard*—This is the monohydrate form. Do not dry; determine the water content titrimetrically at time of use for quantitative analyses. *USP α-d-2-Acetoxy-4-dimethylamino-1,2-diphenyl-3-methylbutane Reference Standard*—Keep container tightly closed. Do not dry before using. *USP α-d-4-Dimethylamino-1,2-diphenyl-3-methyl-2-butanol Hydrochloride Reference Standard*—Keep container tightly closed. Do not dry before using.

Identification—Transfer a portion of finely powdered Tablets, equivalent to about 100 mg of propoxyphene napsylate, to a small flask, and mix with 10 mL of chloroform. Add 10 mL of *pH 12.5 borate buffer* (prepared as directed in the *Assay* under *Propoxyphene Napsylate*), shake for 3 minutes, allow to stand until most of the emulsion has broken, and filter the chloroform solution: the chloroform solution is dextrorotatory (see *Optical Rotation* ⟨781⟩).

Dissolution ⟨711⟩—

Medium: pH 4.5 acetate buffer, prepared as directed in the *Dissolution* test under *Propoxyphene Hydrochloride, Aspirin, and Caffeine Capsules;* 500 mL.

Apparatus 1: 100 rpm.

Time: 60 minutes.

Standard preparation—Transfer 20 mg of USP Propoxyphene Napsylate RS, accurately weighed, to a 100-mL volumetric flask. Dissolve in *pH 4.5 acetate buffer*, dilute with the same solvent to volume, and mix.

Procedure—Proceed as directed for *Determination of dissolved propoxyphene hydrochloride* in the *Dissolution* test under *Propoxyphene Hydrochloride, Aspirin, and Caffeine Capsules.*

Tolerances—Not less than 75% (Q) of the labeled amount of $C_{22}H_{29}NO_2 . C_{10}H_8O_3S . H_2O$ is dissolved in 60 minutes.

Uniformity of dosage units ⟨905⟩: meet the requirements.

Assay—

Internal standard solution—Dissolve a quantity of *n*-tricosane in chloroform to obtain a solution containing about 1 mg per mL.

Standard preparation—Transfer about 100 mg of USP Propoxyphene Napsylate RS, accurately weighed, to a 50-mL volumetric flask. Add 40 mL of methanol, mix, sonicate for about 15 minutes, dilute with methanol to volume, and mix. Transfer 5.0 mL of this solution to a 125-mL separator, and add 5 mL of sodium carbonate solution (1 in 20). Add 5.0 mL of *Internal standard solution* to the contents of the separator, shake for 5 minutes, and allow the phases to separate. Carefully drain the chloroform layer, discarding the first mL, collect the next 2 mL of chloroform extract in a glass-stoppered small conical flask containing about 100 mg of anhydrous granular sodium sulfate, and insert the stopper.

Assay preparation—Weigh and finely powder not less than 20 Propoxyphene Napsylate Tablets. Transfer an accurately weighed portion of the powder, equivalent to about 500 mg of propoxyphene napsylate, to a 250-mL volumetric flask. Add 200 mL of methanol, mix, and proceed as directed for *Standard preparation*, beginning with "sonicate for about 15 minutes."

Chromatographic system, System suitability, and *Procedure*—Proceed as directed in the *Assay* under *Propoxyphene Napsylate Oral Suspension.* Calculate the quantity, in mg, of $C_{22}H_{29}NO_2 . C_{10}H_8O_3S . H_2O$ in the portion of Tablets taken by the formula:

$$(565.72/547.71)(5W_S)(R_U/R_S),$$

in which 565.72 and 547.71 are the molecular weights of propoxyphene napsylate monohydrate and anhydrous propoxyphene napsylate, respectively, W_S is the weight, in mg, of (anhydrous) USP Propoxyphene Napsylate RS taken, as determined from the weight of USP Propoxyphene Napsylate RS corrected for moisture content by a titrimetric water determination, and R_U and R_S are the peak response ratios obtained from the *Assay preparation* and the *Standard preparation*, respectively.

Propoxyphene Napsylate and Acetaminophen Tablets

» Propoxyphene Napsylate and Acetaminophen Tablets contain not less than 90.0 percent and not more than 110.0 percent of the labeled amounts of propoxyphene napsylate ($C_{22}H_{29}NO_2 . C_{10}H_8O_3S . H_2O$) and acetaminophen ($C_8H_9NO_2$).

Packaging and storage—Preserve in tight containers, at controlled room temperature.

Reference standards—*USP Propoxyphene Napsylate Reference Standard*—This is the monohydrate form. Do not dry; determine the water content titrimetrically at time of use for quantitative analyses. *USP Acetaminophen Reference Standard*—Dry over silica gel for 18 hours before using. *USP α-d-2-Acetoxy-4-dimethylamino-1,2-diphenyl-3-methylbutane Reference Standard*—Keep container tightly closed. Do not dry before using. *USP α-d-4-Dimethylamino-1,2-diphenyl-3-methyl-2-butanol Hydrochloride Reference Standard*—Keep container tightly closed. Do not dry before using.

Identification—

A: Transfer the finely ground contents of 1 Tablet to a test tube, add 5 mL of methanol, shake for 5 minutes, and centrifuge. Use the clear supernatant liquid as the *Test solution.* Prepare a *Standard solution* in methanol containing, in each mL, about 20 mg of USP Propoxyphene Napsylate RS and 130 mg of USP Acetaminophen RS. Apply 10 µL of the *Test solution* on a line parallel to and about 2 cm from the bottom edge of a 20- × 5-cm thin-layer chromatographic plate (see *Chromatography* ⟨621⟩) coated with chromatographic silica gel mixture, and apply 10 µL of the *Standard solution* separately on the starting line. Place the plate in a developing chamber containing a mixture of chloroform, butyl acetate, and formic acid (60:40:20), and develop the chromatogram until the solvent front has moved about 15 cm above the line of application. Remove the plate, allow to dry in a hood, and view under short-wavelength ultraviolet light: the solution under test exhibits two principal spots having intensities and R_f values identical to those of the two principal spots obtained from the *Standard solution.*

B: The *Assay preparation* prepared as directed in the *Assay* is dextrorotatory.

Dissolution ⟨711⟩—

Medium: pH 4.5 acetate buffer, prepared as directed in the test for *Dissolution* under *Propoxyphene Hydrochloride, Aspirin, and Caffeine Capsules;* 500 mL.

Apparatus 1: 100 rpm.

Time: 60 minutes.

Determination of dissolved propoxyphene napsylate—

Standard preparation—Transfer 20 mg of USP Propoxyphene Napsylate RS, accurately weighed, to a 100-mL volumetric flask. Dissolve in pH 4.5 acetate buffer, dilute with the same solvent to volume, and mix.

Procedure—Proceed as directed for *Determination of dissolved propoxyphene hydrochloride* in the test for *Dissolution* under *Propoxyphene Hydrochloride, Aspirin, and Caffeine Capsules.*

Determination of dissolved acetaminophen procedure—Pipet a filtered portion of the solution under test to a suitable flask, and dilute with *Dissolution Medium* to an expected concentration of 6.5 µg of acetaminophen per mL. Determine the amount of $C_8H_9NO_2$ dissolved from ultraviolet absorbance at the wavelength of maximum absorbance at about 244 nm, using *Dissolution Medium* as the blank, in comparison with a Standard solution having a known concentration of USP Acetaminophen RS and USP Propoxyphene Napsylate RS in the same medium.

Tolerances—Not less than 75% (*Q*) of the labeled amount of $C_{22}H_{29}NO_2 . C_{10}H_8O_3S . H_2O$ and not less than 75% (*Q*) of the labeled amount of $C_8H_9NO_2$ are dissolved in 60 minutes.

Uniformity of dosage units ⟨905⟩: meet the requirements for *Content Uniformity* with respect to propoxyphene napsylate.

Assay for propoxyphene napsylate—

Internal standard solution—Dissolve *n*-tricosane in chloroform to obtain a solution containing about 0.6 mg per mL.

Standard preparation—Transfer 50 mg of USP Propoxyphene Napsylate RS and 325 mg of USP Acetaminophen RS, all accurately weighed, to a 50-mL volumetric flask. Add 40 mL of methanol, and swirl to dissolve the Reference Standards completely. Dilute with methanol to volume, and mix.

Assay preparation—Weigh and finely powder not less than 20 Propoxyphene Napsylate and Acetaminophen Tablets. Transfer an accurately weighed portion of the powder, equivalent to about 100 mg of propoxyphene napsylate, to a 100-mL volumetric flask. Add 50 mL of methanol, and swirl to dissolve the powder. Dilute with methanol to volume, mix, and filter, discarding the first 20 mL of the filtrate.

Procedure—Transfer 5.0-mL aliquots of the *Assay preparation* and the *Standard preparation* to separate 60-mL separators. To each add 5.0 mL of sodium carbonate solution (1 in 5) and 5.0 mL of *Internal standard solution.* Shake vigorously for 5 minutes, and allow the layers to separate. Drain the chloroform layer through phase-separating paper, suitably supported in a funnel, into a screw-capped test tube. Extract with one 5-mL portion of chloroform, and drain the chloroform layer through phase-separating paper. Evaporate the combined chloroform extracts, using a stream of dry nitrogen, to a final volume of about 2 mL. Chromatograph separately volumes of the chloroform extracts from the *Assay preparation* and the *Standard preparation* that are equivalent to about 6.4 µg of propoxyphene. Use a gas chromatograph equipped with a flame-ionization detector and a 120-cm × 3-mm column packed with 3 percent phase G3 on support S1A. The temperature of the injection port is 200°, the column temperature is 175°, and the carrier gas, nitrogen, flows at about 60 mL per minute. Relative retention times are 1.0 for the internal standard, and 1.7 for propoxyphene. In a suitable chromatogram, the resolution factor is not less than 1.0 and the relative standard deviation for five replicate injections of the *Standard preparation* is not more than 2.0. Calculate the quantity, in mg, of $C_{22}H_{29}NO_2 . C_{10}H_8O_3S . H_2O$ in the portion of Tablets taken by the formula:

$$(565.72/547.71)(100C)(R_U/R_S),$$

in which 565.72 and 547.71 are the molecular weights of propoxyphene napsylate monohydrate and anhydrous propoxyphene napsylate, respectively, *C* is the concentration, in mg per mL, of anhydrous propoxyphene napsylate in the *Standard preparation*, as determined from the concentration of USP Propoxyphene Napsylate RS corrected for moisture content by a titrimetric water determination, and R_U and R_S are the peak response ratios obtained from the *Assay preparation* and the *Standard preparation*, respectively.

Assay for acetaminophen—

Internal standard solution—Dissolve 1.0 mL of *n*-tetradecane in chloroform contained in a 100-mL volumetric flask, dilute with chloroform to volume, and mix.

Standard preparation and *Assay preparation*—Proceed as directed in the *Assay for propoxyphene napsylate.*

Procedure—Transfer 1.0-mL aliquots of the *Assay preparation* and the *Standard preparation* to separate flasks, and evaporate on a steam bath with the aid of a current of air to dryness. To the cooled flasks add 1.0 mL of the *Internal standard solution*, and swirl to dissolve the residue. Add 0.5 mL of bis(trimethylsilyl)acetamide, and immediately insert the stopper in the flask. Mix, and heat at 80° for 60 minutes. Inject suitable volumes of each into a suitable gas chromatograph equipped with a flame-ionization detector. The column is typically 180 cm × 3 mm and is packed with 4 percent phase G2 on support S1A. The temperature of the injection port is 195°, the column temperature is 140°, and the carrier gas is helium. Relative retention times are 1.0 for the internal standard and 3.2 for acetaminophen. In a suitable chromatogram, the resolution factor is not less than 1.0 and the relative standard deviation for five replicate injections of the *Standard preparation* is not more than 2.0. Calculate the quantity, in mg, of $C_8H_9NO_2$ in the portion taken by the formula:

$$100C(R_U/R_S),$$

in which *C* is the concentration, in mg per mL, of USP Acetaminophen RS in the *Standard preparation*, and R_U and R_S are the peak response ratios obtained from the *Assay preparation* and the *Standard preparation*, respectively.

Propoxyphene Napsylate and Aspirin Tablets

» Propoxyphene Napsylate and Aspirin Tablets contain not less than 90.0 percent and not more than 110.0 percent of the labeled amounts of propoxyphene napsylate ($C_{22}H_{29}NO_2 . C_{10}H_8O_3S . H_2O$) and aspirin ($C_9H_8O_4$).

Packaging and storage—Preserve in tight containers, at controlled room temperature.

Reference standards—*USP Propoxyphene Napsylate Reference Standard*—This is the monohydrate form. Do not dry; determine the water content titrimetrically at time of use for quantitative analyses. *USP Aspirin Reference Standard*—Dry over silica gel for 5 hours before using. *USP Salicylic Acid Reference Standard*—Dry over silica gel for 3 hours before using. *USP α-d-2-Acetoxy-4-dimethylamino-1,2-diphenyl-3-methylbutane Reference Standard*—Keep container tightly closed. Do not dry before using. *USP α-d-4-Dimethylamino-1,2-diphenyl-3-methyl-2-butanol Hydrochloride Reference Standard*—Keep container tightly closed. Do not dry before using.

Identification—

A: Transfer the finely ground contents of 1 Tablet to a test tube, add 5 mL of methanol, shake for 5 minutes, and centrifuge. Use the clear supernatant liquid as the *Test solution.* Prepare a *Standard solution* in methanol containing, in each mL, about 65 mg of USP Aspirin RS and 20 mg of USP Propoxyphene Napsylate RS. Apply 10 µL of the *Test solution* on a line parallel to and about 2 cm from the bottom edge of a 20- × 5-cm thin-layer chromatographic plate (see *Chromatography* ⟨621⟩) coated with chromatographic silica gel mixture, and apply 10 µL of the *Standard solution* separately on the starting line. Place the plate in a developing chamber containing a mixture of chloroform, butyl acetate, and formic acid (60:40:20), and develop the chromatogram until the solvent front has moved about 15 cm above the line of application. Remove the plate, allow to dry in a hood, and view under short-wavelength ultraviolet light: the solution under test exhibits two principal spots having intensities and R_f values identical to those of the two principal spots obtained from the *Standard solution.*

B: The *Assay preparation* prepared as directed in the *Assay* is dextrorotatory.

Dissolution ⟨711⟩—

Medium: pH 4.5 acetate buffer, prepared as directed in the test for *Dissolution* under *Propoxyphene Hydrochloride, Aspirin, and Caffeine Capsules;* 500 mL.

Apparatus 1: 100 rpm.

Time: 60 minutes.

Determination of dissolved propoxyphene napsylate—

Internal standard solution and *pH 9.0 buffer*—Prepare as directed in the test for *Dissolution* under *Propoxyphene Hydrochloride, Aspirin, and Caffeine Capsules.*

Standard preparation and *Procedure*—Proceed as directed in the test for *Dissolution* under *Propoxyphene Napsylate Tablets.*

Determination of dissolved aspirin—Proceed as directed for *Determination of dissolved aspirin* in the test for *Dissolution* under *Propoxyphene Hydrochloride, Aspirin, and Caffeine Capsules.*

Tolerances—Not less than 75% (*Q*) of the labeled amount of $C_{22}H_{29}NO_2 . C_{10}H_8O_3S . H_2O$ and not less than 75% (*Q*) of the labeled amount of $C_9H_8O_4$ are dissolved in 60 minutes.

Uniformity of dosage units ⟨905⟩: meet the requirements for *Content Uniformity* with respect to propoxyphene napsylate.

Free salicylic acid—

Ferric chloride–urea reagent—Dissolve by swirling, without the aid of heat, 60 g of urea in a mixture of 8 mL of ferric chloride solution (6 in 10) and 42 mL of 0.05 *N* hydrochloric acid. Adjust this solution by the addition of 6 *N* hydrochloric acid to a pH of 3.2, if necessary.

Standard preparation—Transfer 15.0 mg of USP Salicylic Acid RS, accurately weighed, to a 100-mL volumetric flask, add chloroform to volume, and mix. Transfer 10.0 mL of this solution to a 100-mL volumetric flask containing 20 mL of methanol, 4 drops of hydrochloric acid, and 20 mL of a 1 in 10 solution of glacial acetic acid in ether, dilute with chloroform to volume, and mix.

Test preparation—Pack a pledget of glass wool in the base of a 25- × 200-mm chromatographic tube. In a beaker, prepare a mixture of 6 g of chromatographic siliceous earth, 2 mL of freshly prepared *Ferric chloride–urea reagent*, and 40 mL of chloroform. Transfer the mixture to the chromatographic tube. Rinse the beaker with 15 mL of chloroform, transfer to the column, and pack tightly. Place a small amount of glass wool at the top of the column. Weigh accurately a quantity of the finely powdered contents of Tablets, equivalent to about 50 mg of aspirin, mix with 10 mL of chloroform by stirring for 3 minutes, and transfer with the aid of 10 mL of chloroform to the chromatographic column. Pass 40 mL of chloroform through the column, rinse the tip of the chromatographic tube with chloroform, and discard the eluate. Prepare as a receiver a 100-mL volumetric flask containing 20 mL of methanol and 4 drops of hydrochloric acid, and elute any salicylic acid from the column by passing 20 mL of a 1 in 10 solution of glacial acetic acid in ether that recently has been saturated with water, followed by 30 mL of chloroform. Dilute the eluate with chloroform to volume, and mix.

Procedure—Concomitantly determine the absorbances of the *Standard preparation* and the *Test preparation* in 1-cm cells at the wavelength of maximum absorbance at about 306 nm, with a suitable spectrophotometer, using as the blank a solvent mixture of the same composition as that used for the *Standard preparation:* the absorbance of the *Test preparation* does not exceed that of the *Standard preparation* (3.0%, calculated on the basis of the labeled content of aspirin).

Assay for propoxyphene napsylate—

Internal standard solution—Dissolve *n*-tricosane in chloroform to obtain a solution containing about 0.6 mg per mL.

Standard preparation—Transfer 50 mg of USP Propoxyphene Napsylate RS and 163 mg of USP Aspirin RS, all accurately weighed, to a 50-mL volumetric flask. Add 10 mL of acetone, and swirl to dissolve the Reference Standards completely. Dilute with water to volume, and mix.

Assay preparation—Weigh and finely powder not less than 20 Propoxyphene Napsylate and Aspirin Tablets. Transfer an accurately weighed portion of the powder, equivalent to about 100 mg of propoxyphene napsylate, to a 100-mL volumetric flask containing 20 mL of acetone, and sonicate for about 1 minute. Dilute the milky solution with water to volume, mix, and filter, discarding the first 20 mL of the filtrate.

Procedure—Transfer 5.0-mL aliquots of the *Assay preparation* and the *Standard preparation* to separate 60-mL separators. To each add 5.0 mL of sodium carbonate solution (1 in 5) and 5.0 mL of *Internal standard solution.* Shake vigorously for 5 minutes, and allow the layers to separate. Drain the chloroform layer through phase-separating paper, suitably supported in a funnel, into a screw-capped test tube. Extract with one 5-mL portion of chloroform, and drain the chloroform layer through phase-separating paper. Evaporate the combined chloroform extracts, using a stream of dry nitrogen, to a final volume of about 2 mL. Inject separately a suitable volume, equivalent to about 6.4 µg of propoxyphene, of the chloroform extracts from the *Assay preparation* and the *Standard preparation* into a suitable gas chromatograph equipped with a flame-ionization detector. The column is typically 120 cm × 3 mm and is packed with 3 percent phase G3 on support S1A. The temperature of the injection port is 200°, the column temperature is 175°, and the carrier gas is nitrogen flowing at the rate of about 60 mL per minute. Relative retention times are 1.0 for the internal standard, and 1.7 for propoxyphene. In a suitable chromatogram, the resolution factor is not less than 1.0 and the relative standard deviation for five replicate injections of the *Standard preparation* is not more than 2.0. Calculate the quantity, in mg, of $C_{22}H_{29}NO_2 . C_{10}H_8O_3S . H_2O$ in the portion of Tablets taken by the formula:

$$(565.72/547.71)(100C)(R_U/R_S),$$

in which 565.72 and 547.71 are the molecular weights of propoxyphene napsylate monohydrate and anhydrous propoxyphene napsylate, respectively, *C* is the concentration, in mg per mL, of anhydrous propoxyphene napsylate in the *Standard preparation*, as determined from the concentration of USP Propoxyphene Napsylate RS corrected for moisture content by a titrimetric water determination, and R_U and R_S are the peak response ratios obtained from the *Assay preparation* and the *Standard preparation*, respectively.

Assay for aspirin—

Sodium hydroxide reagent—Dissolve 1 g of polyoxyethylene (23) lauryl ether in about 100 mL of hot water contained in a 1000-mL volumetric flask. Dilute with water to about 600 mL, and dissolve 10 g of sodium hydroxide in this solution. Dilute with water to volume, and mix.

Ferric nitrate reagent—Mix 70 mL of nitric acid with about 600 mL of water contained in a 1000-mL volumetric flask. Dissolve 40 g of ferric nitrate $[Fe(NO_3)_3 . 9H_2O]$ in this solution, dilute with water to volume, and mix.

Standard preparation and *Assay preparation*—Prepare as directed in the *Assay for propoxyphene napsylate.*

Procedure—Into separate 25-mL volumetric flasks pipet 2 mL each of the *Standard preparation* and the *Assay preparation*, and 2 mL of dilute acetone (2 in 10) to provide the blank. Into each flask pipet 5 mL of *Sodium hydroxide reagent*, mix by gentle swirling, and allow to stand at room temperature for 8 minutes. Dilute with *Ferric nitrate reagent* to volume, and mix. Concomitantly determine the absorbances of both solutions against the blank in 1-cm cells at the wavelength of maximum absorbance at about 530 nm, taking care to allow the solutions to reach an equilibrium temperature in the cell compartment. The color intensity is temperature-dependent. Calculate the quantity, in mg, of $C_9H_8O_4$ in the portion taken for the *Assay preparation* by the formula:

$$100C(A_U/A_S),$$

in which *C* is the concentration, in mg per mL, of USP Aspirin RS in the *Standard preparation*, and A_U and A_S are the absorbances of the solutions from the *Assay preparation* and *Standard preparation*, respectively.

Propranolol Hydrochloride

OH
OCH₂CHCH₂NHCH(CH₃)₂ · HCl

$C_{16}H_{21}NO_2 . HCl$ 295.81

2-Propanol, 1-[(1-methylethyl)amino]-3-(1-naphthalenyloxy)-, hydrochloride.
1-(Isopropylamino)-3-(1-naphthyloxy)-2-propanol hydrochloride [*318-98-9*].

» Propranolol Hydrochloride contains not less than 98.0 percent and not more than 101.5 percent of $C_{16}H_{21}NO_2.HCl$, calculated on the dried basis.

Packaging and storage—Preserve in well-closed containers.

Reference standard—*USP Propranolol Hydrochloride Reference Standard*—Dry at 105° for 4 hours before using.

Identification—

A: The infrared absorption spectrum of a mineral oil dispersion of it, previously dried, exhibits maxima only at the same wavelengths as that of a similar preparation of USP Propranolol Hydrochloride RS.

B: The chromatogram of the *Assay preparation* obtained as directed in the *Assay* exhibits a major peak for propranolol the retention time of which corresponds to that exhibited in the chromatogram of the *Standard preparation* obtained as directed in the *Assay*.

C: It responds to the tests for *Chloride* ⟨191⟩.

Melting range, *Class Ia* ⟨741⟩: between 162° and 165°.

Specific rotation ⟨781⟩: between −1.0° and +1.0°, calculated on the dried basis, in a solution containing 1.0 g in each 25 mL.

Loss on drying ⟨731⟩—Dry it at 105° for 4 hours: it loses not more than 0.5% of its weight.

Residue on ignition ⟨281⟩: not more than 0.1%.

Assay—

Mobile phase—Dissolve 0.5 g of dodecyl sodium sulfate in 18 mL of 0.15 *M* phosphoric acid, add 90 mL of acetonitrile and 90 mL of methanol, dilute with water to make 250 mL, mix, and filter through a filter of 0.5-µm or finer porosity. Make adjustments if necessary (see *System Suitability* under *Chromatography* ⟨621⟩).

Standard preparation—Dissolve an accurately weighed quantity of USP Propranolol Hydrochloride RS quantitatively in methanol to obtain a stock solution having a known concentration of about 1 mg per mL. Transfer 5.0 mL of this solution to a 25-mL volumetric flask, dilute with methanol to volume, mix, and filter through a filter of 0.7-µm or finer porosity. This solution contains about 0.2 mg of USP Propranolol Hydrochloride RS per mL.

Resolution solution—Prepare a solution of procainamide hydrochloride in methanol containing about 0.25 mg per mL. Transfer 5 mL of this solution and 5 mL of the stock solution used to prepare the *Standard preparation* to a 25-mL volumetric flask, dilute with methanol to volume, and mix.

Assay preparation—Transfer about 50 mg of Propranolol Hydrochloride, accurately weighed, to a 50-mL volumetric flask, add 45 mL of methanol, shake, and sonicate for 5 minutes. Dilute with methanol to volume, mix, and filter through a 0.7-µm or finer porosity filter. Transfer 5.0 mL of this solution to a 25-mL volumetric flask, dilute with methanol to volume, and mix.

Chromatographic system (see *Chromatography* ⟨621⟩)—The liquid chromatograph is equipped with a 290-nm detector and a 4.6-mm × 25-cm column that contains 5-µm packing L7. The flow rate is about 1.5 mL per minute. Chromatograph the *Resolution solution* and record the peak responses as directed under *Procedure:* the relative retention times are about 0.6 for procainamide and 1.0 for propranolol, and the resolution, *R*, between the procainamide and the propranolol peaks is not less than 2.0. Chromatograph the *Standard preparation*, and record the peak responses as directed under *Procedure:* the tailing factor for the propranolol peak is not more than 3.0, and the relative standard deviation for replicate injections is not more than 2.0%.

Procedure—Separately inject equal volumes (about 20 µL) of the *Standard preparation* and the *Assay preparation* into the chromatograph, record the chromatograms, and measure the responses for the major peaks. Calculate the quantity, in mg, of $C_{16}H_{21}NO_2.HCl$ in the portion of Propranolol Hydrochloride taken by the formula:

$$250C(r_U/r_S),$$

in which *C* is the concentration, in mg per mL, of USP Propranolol Hydrochloride RS in the *Standard preparation*, and r_U and r_S are the propranolol peak responses obtained from the *Assay preparation* and the *Standard preparation*, respectively.

Propranolol Hydrochloride Extended-release Capsules

» Propranolol Hydrochloride Extended-release Capsules contain not less than 90.0 percent and not more than 110.0 percent of the labeled amount of $C_{16}H_{21}.HCl$.

Packaging and storage—Preserve in well-closed containers.

Reference standard—*USP Propranolol Hydrochloride Reference Standard*—Dry at 105° for 4 hours before using.

Identification—Transfer the contents of a number of Capsules, equivalent to about 160 mg of propranolol hydrochloride, to a glass mortar. Add about 5 mL of water, and triturate the mixture with a glass pestle. Transfer the suspension to a centrifuge tube with the aid of about 10 mL of water. Add 1 mL of 1 *N* sodium hydroxide, and mix. Add 15 mL of ether, and shake by mechanical means for about 5 minutes. Centrifuge the mixture, and transfer as much of the ether layer as possible to a second centrifuge tube. Add 0.1 mL of hydrochloric acid to the ether extract, and shake. Centrifuge, and discard the ether layer. Add 15 mL of ether to the precipitate, and shake by mechanical means for about 5 minutes. Centrifuge, and discard the ether layer. Dry the precipitate in vacuum at about 45° for 30 minutes. Transfer a small amount of the dried precipitate to a mortar, and grind to a fine powder: the infrared absorption spectrum of a mineral oil dispersion of the powder so obtained exhibits maxima only at the same wavelengths as that of a similar preparation of USP Propranolol Hydrochloride RS.

Drug release ⟨724⟩—

pH 1.5 buffer solution—Dissolve 2.0 g of sodium chloride in water, add 7.0 mL of hydrochloric acid, dilute with water to 1 liter, and mix.

pH 6.8 buffer solution—Dissolve 21.72 g of dibasic sodium phosphate and 4.94 g of citric acid in water, dilute with water to 1 liter, and mix.

Media—Proceed as directed under *Method B* for *Delayed-release (Enteric-coated) Articles*, using 900 mL of *pH 1.5 buffer solution* during the *Acid stage*, run for 1.5 hours (0.0625*D* hours), and use the *Acceptance Criteria* given under *Tolerances*. For the *Buffer Stage*, use 900 mL of *pH 6.8 buffer solution*, run for the time specified, and use the *Acceptance Criteria* given under *Tolerances*.

Apparatus 1: 100 rpm.

Times: 0.0625*D* hours, 0.167*D* hours, 0.333*D* hours, 0.583*D* hours, 1.00*D* hours.

Procedure—Using filtered portions of the solution under test, diluted if necessary, determine the amount of $C_{16}H_{21}NO_2.HCl$ dissolved, using ultraviolet absorbances at the wavelength of maximum absorbance at about 320 nm, with respect to a baseline drawn from 355 nm through 340 nm, by comparison with a Standard solution in water having a known concentration of USP Propranolol RS.

Tolerances—The percentages of the labeled amount of $C_{16}H_{21}NO_2.HCl$ dissolved at the times specified conform to *Acceptance Table 1*.

Time (hours)	Amount Dissolved (%)
0.0625*D*	Not more than 30%
0.167*D*	Between 35% and 60%
0.333*D*	Between 55% and 80%
0.583*D*	Between 70% and 95%
1.00*D*	Between 81% and 110%

Uniformity of dosage units ⟨905⟩: meet the requirements.

Procedure for content uniformity—Transfer the contents of 1 Capsule to a 100-mL volumetric flask, add about 70 mL of methanol, swirl occasionally for 30 minutes, and sonicate for about 1 minute, and then swirl occasionally for an additional 30 minutes. Dilute with methanol to volume, mix, and centrifuge a portion of the solution. Dilute an accurately measured volume of the clear solution quantitatively with methanol to obtain a solution containing about 40 µg of propranolol hydrochloride per mL. Concomitantly determine the absorbances of this solution and a Standard solution of USP Propranolol Hydrochloride RS in methanol having a known concentration of about 40 µg per mL, in 1-cm cells at the wavelength of maximum absorbance at about 290 nm, with a suitable spectrophotometer, using methanol as the blank. Calculate the quantity, in mg, of $C_{16}H_{21}NO_2 \cdot HCl$ in the Capsule taken by the formula:

$$(LC/D)(A_U/A_S),$$

in which L is the labeled quantity, in mg, of propranolol hydrochloride in the Capsule, C is the concentration, in µg per mL, of USP Propranolol Hydrochloride RS in the Standard solution, D is the concentration, in µg per mL, of the solution from the Capsule, based on the labeled quantity per Capsule and the extent of dilution, and A_U and A_S are the absorbances of the solution from the Capsule and the Standard solution, respectively.

Assay—

Phosphate buffer—Dissolve 13.6 g of monobasic potassium phosphate in 2 liters of water, and mix. Filter the solution through a 0.5-µm or finer porosity filter before use.

Mobile phase—Prepare a suitable degassed mixture of *Phosphate buffer* and acetonitrile (650:350). Make adjustments if necessary (see *System Suitability* under *Chromatography* ⟨621⟩).

Diluting solvent—Mix 650 mL of water with 350 mL of acetonitrile.

Standard preparation—Dissolve an accurately weighed quantity of USP Propranolol Hydrochloride RS in methanol, and dilute quantitatively and stepwise with methanol to obtain a solution having a known concentration of about 200 µg per mL. Transfer 5.0 mL of this solution to a 50.0-mL volumetric flask, add *Diluting solvent* to volume, and mix.

Assay preparation—Transfer the contents of 10 Propranolol Hydrochloride Extended-release Capsules, accurately counted, to a 500-mL volumetric flask. Add 300 mL of methanol, and swirl by mechanical means for 1 hour. Dilute with methanol to volume, mix, and centrifuge a portion of the solution. Dilute an accurately measured volume of the clear solution quantitatively with *Diluting solvent* to obtain a solution having a concentration of about 20 µg of propranolol hydrochloride per mL.

Chromatographic system (see *Chromatography* ⟨621⟩)—The liquid chromatograph is equipped with a 220-nm detector and a 4-mm × 15-cm column that contains 5-µm packing L1. The flow rate is about 2 mL per minute. Chromatograph the *Standard preparation*, and record the peak responses as directed under *Procedure:* the column efficiency determined from the analyte peak is not less than 1000 theoretical plates, the tailing factor for the analyte peak is not more than 3, and the relative standard deviation for replicate injections is not more than 2%.

Procedure—Separately inject equal volumes (about 20 µL) of the *Standard preparation* and the *Assay preparation* into the chromatograph, record the chromatograms, and measure the responses for the major peaks. The retention time of propranolol is about 5 to 9 minutes. Calculate the quantity, in mg, of $C_{16}H_{21}NO_2 \cdot HCl$ in each Capsule taken by the formula:

$$(LC/D)(r_U/r_S),$$

in which L is the labeled quantity, in mg, of propranolol hydrochloride in each Capsule, C is the concentration, in µg per mL, of USP Propranolol Hydrochloride RS in the *Standard preparation*, D is the concentration, in µg of propranolol hydrochloride per mL, of the *Assay preparation*, based on the labeled quantity per Capsule, the number of Capsules taken, and the extent of dilution, and r_U and r_S are the peak responses obtained from the *Assay preparation* and the *Standard preparation*, respectively.

Propranolol Hydrochloride Injection

» Propranolol Hydrochloride Injection is a sterile solution of Propranolol Hydrochloride in Water for Injection. It contains not less than 90.0 percent and not more than 110.0 percent of the labeled amount of $C_{16}H_{21}NO_2 \cdot HCl$.

Packaging and storage—Preserve in single-dose, light-resistant containers, preferably of Type I glass.

Reference standard—*USP Propranolol Hydrochloride Reference Standard*—Dry at 105° for 4 hours before using.

Identification—The chromatogram of the *Assay preparation* obtained as directed in the *Assay* exhibits a major peak for propranolol the retention time of which corresponds to that exhibited in the chromatogram of the *Standard preparation* obtained as directed in the *Assay*.

pH ⟨791⟩: between 2.8 and 4.0.

Other requirements—It meets the requirements under *Injections* ⟨1⟩.

Assay—

Mobile phase, *Standard preparation*, and *Resolution solution*—Prepare as directed in the *Assay* under *Propranolol Hydrochloride*.

Assay preparation—Transfer an accurately measured volume of Propranolol Hydrochloride Injection, equivalent to about 5 mg of propranolol hydrochloride, to a 25-mL volumetric flask, dilute with methanol to volume, and mix.

Procedure—Proceed as directed for *Procedure* in the *Assay* under *Propranolol Hydrochloride*. Calculate the quantity, in mg, of $C_{16}H_{21}NO_2 \cdot HCl$ in each mL of the Injection taken by the formula:

$$250(C/V)(r_U/r_S),$$

in which C is the concentration, in mg per mL, of USP Propranolol Hydrochloride in the *Standard preparation*, V is the volume, in mL, of Injection taken, and r_U and r_S are the propranolol peak responses obtained from the *Assay preparation* and the *Standard preparation*, respectively.

Propranolol Hydrochloride Tablets

» Propranolol Hydrochloride Tablets contain not less than 90.0 percent and not more than 110.0 percent of the labeled amount of $C_{16}H_{21}NO_2 \cdot HCl$.

Packaging and storage—Preserve in well-closed, light-resistant containers.

Reference standard—*USP Propranolol Hydrochloride Reference Standard*—Dry at 105° for 4 hours before using.

Identification—The chromatogram of the *Assay preparation* obtained as directed in the *Assay* exhibits a major peak for propranolol the retention time of which corresponds to that exhibited in the chromatogram of the *Standard preparation* obtained as directed in the *Assay*.

Dissolution ⟨711⟩—

Medium: dilute hydrochloric acid (1 in 100); 1000 mL.

Apparatus 1: 100 rpm.

Time: 30 minutes.

Procedure—Determine the amount of $C_{16}H_{21}NO_2 \cdot HCl$ dissolved from ultraviolet absorbances at the wavelength of maximum absorbance at about 289 nm of filtered portions of the solution under test, suitably diluted with *Dissolution Medium*, if necessary, in comparison with a Standard solution having a known concentration of USP Propranolol Hydrochloride RS in the same medium.

Tolerances—Not less than 75% (Q) of the labeled amount of $C_{16}H_{21}NO_2 \cdot HCl$ is dissolved in 30 minutes.

Uniformity of dosage units ⟨905⟩: meet the requirements.

Procedure for content uniformity—Transfer 1 Tablet to a 100-mL volumetric flask, add 5 mL of dilute hydrochloric acid (1 in 100), and let stand, swirling occasionally, until it is disintegrated. Add about 70 mL of methanol, and sonicate for about 1 minute. Dilute with methanol to volume, mix, and centrifuge a portion of the solution. Dilute an aliquot of the clear solution quantitatively with methanol to provide a solution containing about 40 µg of propranolol hydrochloride per mL. Concomitantly determine the absorbances of this solution and of a solution of USP Propranolol Hydrochloride RS in methanol, at a known concentration of about 40 µg per mL, in 1-cm cells at the wavelength of maximum absorbance at about 290 nm, with a suitable spectrophotometer, using methanol as the blank. Calculate the quantity, in mg, of $C_{16}H_{21}NO_2 \cdot HCl$ in the Tablet by the formula:

$$(T/D)C(A_U/A_S),$$

in which T is the labeled quantity, in mg, of propranolol hydrochloride in the Tablet, D is the concentration, in µg per mL, of the solution from the Tablet, based on the labeled quantity per Tablet and the extent of dilution, C is the concentration, in µg per mL, of USP Propranolol Hydrochloride RS in the Standard solution, and A_U and A_S are the absorbances of the solution from the Tablet and the Standard solution, respectively.

Assay—

Mobile phase, Standard preparation, and *Resolution solution*—Prepare as directed in the *Assay* under *Propranolol Hydrochloride.*

Assay preparation—Weigh and finely powder not less than 20 Propranolol Hydrochloride Tablets. Transfer an accurately weighed portion of the powder, equivalent to about 50 mg of propranolol hydrochloride, to a 50-mL volumetric flask, add 40 mL of methanol, shake, and sonicate for 5 minutes. Dilute with methanol to volume, mix, and filter through a 0.7-µm or finer porosity filter. Transfer 5.0 mL of this solution to a 25-mL volumetric flask, dilute with methanol to volume, and mix.

Procedure—Proceed as directed for *Procedure* in the *Assay* under *Propranolol Hydrochloride.* Calculate the quantity, in mg, of $C_{16}H_{21}NO_2 \cdot HCl$ in the portion of Tablets taken by the formula:

$$250C(r_U/r_S),$$

in which C is the concentration, in mg per mL, of USP Propranolol Hydrochloride in the *Standard preparation,* and r_U and r_S are the propranolol peak responses obtained from the *Assay preparation* and the *Standard preparation,* respectively.

Propranolol Hydrochloride and Hydrochlorothiazide Extended-release Capsules

» Propranolol Hydrochloride and Hydrochlorothiazide Extended-release Capsules contain not less than 90.0 percent and not more than 110.0 percent of the labeled amounts of propranolol hydrochloride ($C_{16}H_{21}NO_2 \cdot HCl$) and hydrochlorothiazide ($C_7H_8ClN_3O_4S_2$).

Packaging and storage—Preserve in well-closed containers.

Reference standards—*USP Propranolol Hydrochloride Reference Standard*—Dry at 105° for 4 hours before using. *USP Hydrochlorothiazide Reference Standard*—Dry at 105° for 1 hour before using. *USP 4-Amino-6-chloro-1,3-benzenedisulfonamide Reference Standard*—Keep container tightly closed and protected from light. Dry over silica gel for 4 hours before using.

Identification—

A: Transfer the contents of a number of Capsules, equivalent to about 100 mg of hydrochlorothiazide, to a 20-mesh sieve. Break up any large lumps with the aid of a spatula, and collect the powder that passes through the sieve. [NOTE—Retain the material on the screen for *Identification test B.*] Transfer the powder that passed through the sieve to a screw-capped, 35-mL centrifuge tube, add 5 mL of solvent hexane, and shake for 5 minutes. Centrifuge, and discard the solvent. To the residue in the centrifuge tube add 10 mL of 1 *N* sodium hydroxide, shake, and filter, collecting the filtrate in a separator. Wash the filter with 5 mL of water, and collect the washing in the separator. Add 50 mL of ether to the separator, shake for 2 minutes, and allow the phases to separate. Drain the aqueous layer into a beaker, adjust with 6 *N* hydrochloric acid to a pH of about 2, induce crystallization by scratching the inner surface of the beaker with a glass rod, and allow to stand until crystallization is complete. Collect the crystals on a filter, and dry at 105° for 30 minutes. Grind the crystals to a fine powder: the infrared absorption spectrum of a mineral oil dispersion of the powder so obtained exhibits maxima only at the same wavelengths as that of a similar preparation of USP Hydrochlorothiazide RS.

B: Wash the material retained on the screen from *Identification test A* with a small amount of water, discarding the washings. Transfer the particles remaining on the screen to a glass mortar, add about 5 mL of water, and triturate the mixture with a glass pestle. Transfer the suspension, with the aid of about 10 mL of water, to a 35-mL screw-capped centrifuge tube, add about 1 mL of 1 *N* sodium hydroxide, and mix. Add about 15 mL of ether, and shake by mechanical means for 5 minutes. Centrifuge the mixture, and transfer as much of the ether layer as possible to a second centrifuge tube. Add 0.1 mL of hydrochloric acid to the ether extract, and shake. Centrifuge, and discard the ether. Add about 15 mL of ether to the residue, and shake by mechanical means for 5 minutes. Centrifuge, and discard the ether layer. Dry the residue in vacuum at 45° for 30 minutes. Grind the crystals to a fine powder: the infrared absorption spectrum of a mineral oil dispersion of the powder so obtained exhibits maxima only at the same wavelengths as that of a similar preparation of USP Propranolol Hydrochloride RS.

C: The chromatogram of the *Assay preparation* obtained as directed under *Assay and limit of 4-amino-6-chloro-1,3-benzenedisulfonamide* exhibits major peaks for propranolol hydrochloride and hydrochlorothiazide, the retention times of which correspond to those exhibited in the chromatogram of the *Standard preparation* obtained as directed under *Assay and limit of 4-amino-6-chloro-1,3-benzenedisulfonamide.*

Drug release ⟨724⟩—

pH 1.5 buffer solution, pH 6.8 buffer solution, Media, and *Apparatus*—Proceed as directed in the test for *Drug release* under *Propranolol Hydrochloride Extended-release Capsules.*

Analytical method—Determine the amounts of hydrochlorothiazide ($C_7H_8ClN_3O_4S_2$) and propranolol hydrochloride ($C_{16}H_{21}NO_2 \cdot HCl$) dissolved, using the following method.

Stock propranolol hydrochloride standard solution—Prepare a solution of USP Propranolol Hydrochloride RS in dilute hydrochloric acid (1 in 100) having a known concentration of about 0.4 mg per mL.

Stock hydrochlorothiazide standard solution—Dissolve an accurately weighed quantity of USP Hydrochlorothiazide RS in 0.25 *N* sodium hydroxide to obtain a solution having a concentration of about 25 mg per mL. Dilute this solution quantitatively with water to obtain a solution having a known concentration of about 0.5 mg per mL.

Standard solutions—Prepare, by combining aliquots of the two *Stock standard solutions,* and diluting with dilute hydrochloric acid (1 in 100), solutions bracketing the expected concentration of the samples at the various time points.

Times: 30 minutes, 0.0625*D* hour, 0.167*D* hour, 0.333*D* hour, 0.583*D* hour, 1.00*D* hour.

Procedure—Use an automatic analyzer consisting of a liquid sampler, a proportioning pump, two ultraviolet spectrophotometers, and a manifold consisting of the components illustrated in the diagram under *Automated Methods of Analysis* ⟨16⟩. Start the sampler and conduct determinations at a rate of 30 per hour, using a ratio of about 1:1 for the sample to wash time. Calculate the amounts of $C_7H_8ClN_3O_4S_2$ and $C_{16}H_{21}NO_2 \cdot HCl$ dissolved by comparison with the *Standard solutions.*

Tolerances (*Hydrochlorothiazide*)—Use the *Acceptance Table* under *Dissolution* ⟨711⟩. Not less than 60% (*Q*) of the labeled amount of $C_7H_8ClN_3O_4S_2$ is dissolved in 30 minutes.

Tolerances (*Propranolol hydrochloride*)—The percentages of the labeled amount of $C_{16}H_{21}NO_2 . HCl$ dissolved at the times specified conform to *Acceptance Table 1* under *Drug Release* ⟨724⟩.

Time (hours)	Amount Dissolved (%)
0.0625*D*	Not more than 30%
0.167*D*	Between 35% and 60%
0.333*D*	Between 55% and 80%
0.583*D*	Between 70% and 95%
1.00*D*	Between 83% and 108%

Uniformity of dosage units ⟨905⟩: meet the requirements for *Content Uniformity* with respect to propranolol hydrochloride and to hydrochlorothiazide.

Procedure for content uniformity—

Apparatus—Use an automatic analyzer consisting of (1) a 20-channel peristaltic pump; (2) an ultraviolet spectrophotometer equipped with a 10-mm flow cell and a 293-nm detector; (3) an ultraviolet spectrophotometer equipped with a 10-mm flow cell and a 273-nm detector; (4) recording devices for each of the two aforementioned detectors; and (5) a manifold consisting of components illustrated in the pertinent diagram in the chapter, *Automated Methods of Analysis* ⟨16⟩.

Standard hydrochlorothiazide stock solution—Dissolve an accurately weighed quantity of USP Hydrochlorothiazide RS in methanol to obtain a solution having a known concentration of about 5 mg per mL.

Standard propranolol hydrochloride stock solution—Dissolve an accurately weighed quantity of USP Propranolol Hydrochloride RS in methanol to obtain a solution having a known concentration of about 5*J* mg per mL, *J* being the ratio of the labeled amount, in mg, of propranolol hydrochloride to the labeled amount, in mg, of hydrochlorothiazide per Capsule.

Standard preparation—Transfer 10.0 mL of the *Standard hydrochlorothiazide stock solution*, 10.0 mL of the *Standard propranolol hydrochloride stock solution*, and 10.0 mL of methanol to a 100-mL volumetric flask, dilute with 0.12 *N* hydrochloric acid to volume, and mix.

Test preparations—Transfer the contents of an appropriate number of individual Capsules to separate 100-mL volumetric flasks, rinsing each empty Capsule shell with 2 mL of methanol, and adding the rinsings to the respective volumetric flasks. Add 28 mL of methanol to each flask, and mix by mechanical means for 30 minutes. Dilute the contents of each flask with 0.12 *N* hydrochloric acid to volume, and mix.

Procedure—With the sampler in the standby position, pump all reagents through the system until a stable baseline is achieved. Activate the sampler, and allow one cycle to pass without introducing the *Standard preparation* or the *Test preparations*, then introduce a 5-mL portion of the *Standard preparation* into the sampler for the next two cycles and for every sixth cycle thereafter. Disregard the first value for the *Standard preparation*. Add the *Test preparations* to the sampler at the rate of 30 per hour, using a ratio of about 1:1 for the sample to wash time, to follow the second 5-mL portion of the *Standard preparation*. Record the absorbance values, and calculate each peak value by the difference between peak height and baseline. Calculate the quantity, in mg, of hydrochlorothiazide ($C_7H_8ClN_3O_4S_2$) per Capsule by the formula:

$$100C(A_U/A_S),$$

in which *C* is the concentration, in mg per mL, of USP Hydrochlorothiazide RS in the *Standard preparation*, A_U is the absorbance at 273 nm of the individual *Test preparation*, and A_S is the averaged absorbance at 273 nm of the *Standard preparations*. Make any necessary correction of the results obtained as follows.

(1) Calculate the correction, F', by the formula:

$$F' = A/P',$$

in which *A* is the weight of active ingredient equivalent to 1 average dosage unit obtained by the *Assay* procedure, and P' is the weight of active ingredient equivalent to 1 average dosage unit calculated as the mean of the dosage units tested by the *Content Uniformity* procedure.

(2) If F' is between 0.970 and 1.030, no correction is required.

(3) If F' is not between 0.970 and 1.030, calculate the weight of active ingredient in each dosage unit by multiplying each of the weights found using the special procedure by F'.

Calculate the quantity, in mg, of propranolol hydrochloride ($C_{16}H_{21}NO_2 . HCl$) per Capsule by the same formula, in which *C* is the concentration, in mg per mL, of USP Propranolol Hydrochloride RS in the *Standard preparation*, A_U is the absorbance at 293 nm of the individual *Test preparation*, and A_S is the averaged absorbance at 293 nm of the *Standard preparations*. Make any necessary correction of the results obtained as directed above.

Assay and limit of 4-amino-6-chloro-1,3-benzenedisulfonamide—

Tetrabutylammonium hydroxide solution—Use a suitable aqueous or methanolic solution having a known concentration of tetrabutylammonium hydroxide.

Buffer—Dissolve 31.25 g of monobasic potassium phosphate in 500 mL of water in a 1000-mL volumetric flask. Add 18.75 mL of phosphoric acid, mix, and add a volume of *Tetrabutylammonium hydroxide solution* equivalent to about 13 g of tetrabutylammonium hydroxide. Dilute with water to volume, and mix. Dilute 100 mL of this solution with water to obtain 1000 mL of solution, adjusting, if necessary, with 10 *N* potassium hydroxide to a pH of 2.4 ± 0.1.

Mobile phase—Prepare a suitable mixture of *Buffer* and methanol (850:150). Make adjustments if necessary (see *System Suitability* under *Chromatography* ⟨621⟩).

Standard hydrochlorothiazide stock solution—Transfer about 25 mg of USP Hydrochlorothiazide RS, accurately weighed, to a 100-mL volumetric flask, add 15 mL of methanol, and sonicate for 5 minutes, adding ice to the bath, if necessary, to maintain the temperature at not more than 20°. Dilute with *Buffer* to volume, and mix.

Standard propranolol hydrochloride stock solution—Dissolve an accurately weighed quantity of USP Propranolol Hydrochloride RS in *Mobile phase* to obtain a solution having a known concentration of about 0.25*J* mg per mL, *J* being the ratio of the labeled quantity, in mg, of propranolol hydrochloride to the labeled quantity, in mg, of hydrochlorothiazide per Capsule.

Standard preparation—Transfer 5.0 mL of *Standard hydrochlorothiazide stock solution* and 5.0 mL of *Standard propranolol hydrochloride stock solution* to a 25-mL volumetric flask, dilute with *Mobile phase* to volume, and mix. This solution contains about 50 µg of hydrochlorothiazide and 50*J* µg of propranolol hydrochloride per mL.

4-Amino-6-chloro-1,3-benzenedisulfonamide standard solution—Transfer about 25 mg of USP 4-Amino-6-chloro-1,3-benzenedisulfonamide RS, accurately weighed, to a 200-mL volumetric flask, add 30 mL of methanol, and swirl to dissolve. Dilute with *Buffer* to volume, and mix. Dilute an accurately measured volume of this solution quantitatively and stepwise, if necessary, with *Mobile phase* to obtain a solution having a known concentration of about 0.5 µg per mL.

Assay preparation—Carefully open an accurately counted number of Propranolol Hydrochloride and Hydrochlorothiazide Extended-release Capsules, equivalent to about 500 mg of hydrochlorothiazide, and transfer the contents and the Capsule shells to a 500-mL volumetric flask. Add 5.0 mL of water to the flask, and allow to stand for 5 minutes. Dilute with methanol to volume, mix, and sonicate for 10 minutes, adding ice to the bath, if necessary, to maintain the temperature at not more than 20°. Remove the flask from the bath, and shake it occasionally for 1 hour. Centrifuge a portion of the contents of the flask, if necessary, to obtain a clear solution. Transfer 5.0 mL of the clear solution to a 100-mL volumetric flask, add 10.0 mL of methanol, dilute with *Buffer* to volume, and mix.

Chromatographic system (see *Chromatography* ⟨621⟩)—The liquid chromatograph is equipped with a 220-nm detector and a 4-mm × 15-cm column that contains packing L1. The flow rate is about 1.5 mL per minute. Chromatograph the *Standard preparation*, and record the peak responses as directed under *Procedure:* the column efficiency determined from the propranolol peak is not less than 2500 theoretical plates when calculated by the formula:

$$5.545(t/W_{h/2})^2,$$

the tailing factor for the hydrochlorothiazide and propranolol peaks is not more than 1.5, and the relative standard deviation for replicate injections is not more than 2.0%. Chromatograph the *4-Amino-6-chloro-1,3-benzenedisulfonamide standard solution*, and record the peak responses as directed under *Procedure:* the relative standard deviation for replicate injections is not more than 5%.

Procedure—[NOTE—Use peak areas where peak responses are indicated.] Separately inject equal volumes (about 50 µL) of the *Standard preparation*, the *4-Amino-6-chloro-1,3-benzenedisulfonamide standard solution*, and the *Assay preparation* into the chromatograph, record the chromatograms, and measure the responses for the major peaks. The retention time for propranolol is between 12 and 25 minutes, and the relative retention times are about 0.25 for 4-amino-6-chloro-1,3-benzenedisulfonamide, 0.4 for hydrochlorothiazide, and 1.0 for propranolol. Calculate the quantities, in mg, of hydrochlorothiazide ($C_7H_8ClN_3O_4S_2$) and propranolol hydrochloride ($C_{16}H_{21}NO_2.HCl$) in each Capsule taken by the same formula:

$$10(C/N)(r_U/r_S),$$

in which C is the concentration, in µg per mL, of USP Propranolol Hydrochloride RS or USP Hydrochlorothiazide RS in the *Standard preparation*, N is the number of Capsules taken to prepare the *Assay preparation*, and r_U and r_S are the peak responses of the corresponding analyte obtained from the *Assay preparation* and the *Standard preparation*, respectively. Calculate the percentage of 4-amino-6-chloro-1,3-benzenedisulfonamide in the Capsules taken by the formula:

$$1000(C/NL)(r_U/r_S),$$

in which C is the concentration, in µg per mL, of USP 4-Amino-6-chloro-1,3-benzenedisulfonamide RS in the *4-Amino-6-chloro-1,3-benzenedisulfonamide standard solution*, N is the number of Capsules taken to prepare the *Assay preparation*, L is the labeled amount, in mg, of hydrochlorothiazide in each Capsule taken, and r_U and r_S are the peak responses of 4-amino-6-chloro-1,3-benzenedisulfonamide obtained from the *Assay preparation* and the *4-Amino-6-chloro-1,3-benzenedisulfonamide standard solution*, respectively: not more than 1.0% is present.

Propranolol Hydrochloride and Hydrochlorothiazide Tablets

» Propranolol Hydrochloride and Hydrochlorothiazide Tablets contain not less than 90.0 percent and not more than 110.0 percent of the labeled amounts of propranolol hydrochloride ($C_{16}H_{21}NO_2.HCl$) and hydrochlorothiazide ($C_7H_8ClN_3O_4S_2$).

Packaging and storage—Preserve in well-closed containers.

Reference standards—*USP Propranolol Hydrochloride Reference Standard*—Dry at 105° for 4 hours before using. *USP Hydrochlorothiazide Reference Standard*—Dry at 105° for 1 hour before using. *USP 4-Amino-6-chloro-1,3-benzenedisulfonamide Reference Standard*—Keep container tightly closed and protected from light. Dry over silica gel for 4 hours before using.

Identification—

A: Transfer a quantity of finely powdered Tablets, equivalent to about 100 mg of propranolol hydrochloride, to a 50-mL centrifuge tube, add 15 mL of water and 1 mL of 1 N sodium hydroxide, and mix. Add 20 mL of ether, cap the tube, and shake by mechanical means for 5 minutes. Centrifuge the mixture, and transfer as much of the ether layer as possible to a second centrifuge tube. Add 0.05 mL of hydrochloric acid to the ether extract, and shake. Centrifuge, and discard the ether layer. Add 20 mL of ether to the residue in the tube, and shake by mechanical means for 5 minutes. Centrifuge, and discard the ether layer. Dry the residue in the tube in vacuum at about 50° for 30 minutes. Transfer a small amount of the dried residue to a mortar, and grind to fine powder: the infrared absorption spectrum of a mineral oil dispersion of the powder exhibits maxima only at the same wavelengths as that of a similar preparation of USP Propranolol Hydrochloride RS.

B: Transfer a quantity of finely powdered Tablets, equivalent to about 100 mg of hydrochlorothiazide, to a 35-mL centrifuge tube, add 30 mL of acetone, mix, and allow to stand for 30 minutes, with occasional shaking. Centrifuge, and decant the acetone extract into a beaker, discarding the residue in the centrifuge tube. Evaporate the acetone extract on a steam bath to dryness, add 10 mL of 0.1 N sodium hydroxide, and mix, using a spatula to dislodge any residue from the beaker. Transfer the suspension to a 125-mL separator, and wash the beaker with about 5 mL of water, adding the washing to the separator. Add 50 mL of ether to the separator, shake for 2 minutes, and allow the phases to separate. Draw off the clear lower layer, filtering it into a beaker. Add 1 N hydrochloric acid dropwise with stirring until a pH of about 2 is reached. [NOTE—Precipitation occurs during the addition of the acid.] When precipitation is complete, decant the supernatant liquid, and wash the precipitate with 5 mL of water. Dry the precipitate at 105° for 30 minutes: the infrared absorption spectrum of a mineral oil dispersion of the dried precipitate exhibits maxima only at the same wavelengths as a similar preparation of USP Hydrochlorothiazide RS.

Uniformity of dosage units ⟨905⟩: meet the requirements for *Content Uniformity* with respect to propranolol hydrochloride and hydrochlorothiazide.

Procedure for content uniformity—Transfer a Tablet to a suitable container, and add 500.0 mL of 0.1 N hydrochloric acid. Shake until the Tablet has disintegrated, sonicate for 30 seconds, shake by mechanical means for 30 minutes, and then repeat the sonication and shaking. Centrifuge a portion of the solution, and transfer 6.0 mL of the clear supernatant liquid and 15.0 mL of water to a suitable capped bottle. Add 1.0 mL of 5 N sodium hydroxide and 25.0 mL of n-heptane, cap the bottle, shake by mechanical means for 5 minutes, and allow the layers to separate, centrifuging, if necessary, to obtain clear upper (n-heptane) and lower (aqueous) layers (test solutions). Prepare a similar Standard solution by mixing 6.0 mL of 0.1 N hydrochloric acid, 3.0 mL of water, 6.0 mL of an aqueous solution having a known concentration of USP Propranolol Hydrochloride RS, and 6.0 mL of a 0.04 N sodium hydroxide solution having a known concentration of USP Hydrochlorothiazide RS, and proceeding as directed for the Test solutions, beginning with "Add 1.0 mL of 5 N sodium hydroxide." Prepare similar blank n-heptane and aqueous extracts by adding 6.0 mL of 0.1 N hydrochloric acid to 15.0 mL of water, and proceeding as directed for the Test solutions, beginning with "Add 1.0 mL of 5 N sodium hydroxide." Concomitantly determine the absorbances of the n-heptane Test solution and the n-heptane Standard solution at the wavelength of maximum absorbance at about 293 nm, using the n-heptane blank extract to set the instrument. Calculate the quantity, in mg, of propranolol hydrochloride ($C_{16}H_{21}NO_2.HCl$) in the Tablet by the formula:

$$(12.5/6)(C)(A_U/A_S),$$

in which C is the concentration, in µg per mL, of USP Propranolol Hydrochloride RS in the n-heptane Standard solution, and A_U and A_S are the absorbances at 293 nm of the n-heptane Test solution and the n-heptane Standard solution, respectively. Concomitantly determine the absorbances of the aqueous Test solution and the aqueous Standard solution at the wavelength of maximum absorbance at about 273 nm, using the aqueous blank extract to set the instrument. Calculate the quantity, in mg, of hydrochlorothiazide ($C_7H_8ClN_3O_4S_2$) in the Tablet by the formula:

$$(11/6)(C)(A_U/A_S),$$

in which C is the concentration, in µg per mL, of USP Hydrochlorothiazide RS in the aqueous Standard solution, and A_U and A_S are the absorbances at 273 nm of the aqueous Test solution and the aqueous Standard solution, respectively.

Assay and limit of 4-amino-6-chloro-1,3-benzenedisulfonamide—

Tetrabutylammonium hydroxide solution—Use a suitable aqueous or methanolic solution having a known concentration of tetrabutylammonium hydroxide.

Buffer—Dissolve 6.8 g of monobasic potassium phosphate in 1000 mL of water in a 2000-mL volumetric flask. Add 3.4 mL of phosphoric acid and a volume of *Tetrabutylammonium hydroxide solution* equivalent to about 2.6 g of tetrabutylammonium hydroxide, dilute with water to volume, and mix. Adjust, if necessary, with phosphoric acid or 10 *N* potassium hydroxide to a pH of 2.5 ± 0.1, and filter through a 0.5-µm or finer porosity filter.

Mobile phase—Prepare a suitable mixture of *Buffer* and methanol (850:150). Make adjustments if necessary (see *System Suitability* under *Chromatography* ⟨621⟩) so that the retention time of propranolol is between 12 and 25 minutes.

Standard hydrochlorothiazide stock solution—Transfer about 25 mg of USP Hydrochlorothiazide RS, accurately weighed, to a 100-mL volumetric flask, add 15 mL of methanol, and mix to dissolve. Dilute with *Buffer* to volume, and mix.

Standard propranolol hydrochloride stock solution—Dissolve an accurately weighed quantity of USP Propranolol Hydrochloride RS in *Mobile phase* to obtain a solution having a known concentration of about 0.25*J* mg per mL, *J* being the ratio of the labeled quantity, in mg, of propranolol hydrochloride to the labeled quantity, in mg, of hydrochlorothiazide per Tablet.

Standard preparation—Transfer 5.0 mL of *Standard hydrochlorothiazide stock solution* and 5.0 mL of *Standard propranolol hydrochloride stock solution* to a 25-mL volumetric flask, dilute with *Mobile phase* to volume, and mix. This solution contains about 50 µg of hydrochlorothiazide and 50*J* µg of propranolol hydrochloride per mL.

Standard 4-amino-6-chloro-1,3-benzenedisulfonamide solution—Dissolve an accurately weighed quantity of USP 4-Amino-6-chloro-1,3-benzenedisulfonamide RS in methanol to obtain a solution having a known concentration of about 0.5 mg per mL. Dilute an accurately measured volume of this solution quantitatively and stepwise, if necessary, with *Mobile phase* to obtain a solution having a known concentration of about 0.5 µg per mL.

Assay preparation—Weigh and finely powder not less than 20 Propranolol Hydrochloride and Hydrochlorothiazide Tablets. Transfer an accurately weighed portion of the powder, equivalent to about 25 mg of hydrochlorothiazide, to a 500-mL volumetric flask. Add 5 mL of water, mix, and allow to stand for 5 minutes, with occasional swirling. Add 75 mL of methanol, mix, and sonicate for 10 minutes, with occasional swirling, adding ice to the bath, if necessary, to maintain the temperature at not more than 20°. Add about 350 mL of *Buffer* to the flask, and sonicate for 10 minutes, with occasional swirling, maintaining the temperature of the bath at not more than 20°. Dilute with *Buffer* to volume, and mix. Centrifuge a portion of this solution, if necessary, to obtain a clear solution (*Assay preparation*).

Chromatographic system (see *Chromatography* ⟨621⟩)—The liquid chromatograph is equipped with a 270-nm detector and a 4-mm × 15-cm column that contains 5-µm packing L1. The flow rate is about 1.5 mL per minute. Chromatograph the *Standard preparation*, and record the peak responses as directed under *Procedure:* the column efficiency determined from the propranolol peak is not less than 2500 theoretical plates, the tailing factor for the propranolol and hydrochlorothiazide peaks is not more than 1.5, and the relative standard deviation for replicate injections is not more than 2.0%. Chromatograph the *Standard 4-amino-6-chloro-1,3-benzenedisulfonamide solution*, and record the peak responses as directed under *Procedure:* the relative standard deviation for replicate injections is not more than 5.0%.

Procedure—[NOTE—Use peak areas where peak responses are indicated.] Separately inject equal volumes (about 50 µL) of the *Standard preparation*, the *Standard 4-amino-6-chloro-1,3-benzenedisulfonamide solution*, and the *Assay preparation* into the chromatograph, record the chromatograms, and measure the responses for the major peaks. The relative retention times are about 0.25 for 4-amino-6-chloro-1,3-benzenedisulfonamide, 0.4 for hydrochlorothiazide, and 1.0 for propranolol. Calculate the quantities, in mg, of propranolol hydrochloride ($C_{16}H_{21}NO_2 . HCl$) and hydrochlorothiazide ($C_7H_8ClN_3O_4S_2$) in the portion of Tablets taken by the same formula:

$$0.5C(r_U / r_S),$$

in which *C* is the concentration, in µg per mL, of the appropriate Reference Standard in the *Standard preparation*, and r_U and r_S are the peak responses of the corresponding analyte obtained from the *Assay preparation* and the *Standard preparation*, respectively. Calculate the percentage of 4-amino-6-chloro-1,3-benzenedisulfonamide in the portion of Tablets taken by the formula:

$$50(C/L)(r_U / r_S),$$

in which *C* is the concentration, in µg per mL, of USP 4-Amino-6-chloro-1,3-benzenedisulfonamide RS in the *Standard 4-amino-6-chloro-1,3-benzenedisulfonamide solution*, *L* is the amount, in mg, of hydrochlorothiazide in the portion of Tablets taken, based on the labeled amount, and r_U and r_S are the peak responses of 4-amino-6-chloro-1,3-benzenedisulfonamide obtained from the *Assay preparation* and the *Standard 4-amino-6-chloro-1,3-benzenedisulfonamide solution*, respectively: not more than 1.0% is found.

Propyl Gallate—*see* Propyl Gallate NF

Propylene Carbonate—*see* Propylene Carbonate NF

Propylene Glycol

$CH_3CH(OH)CH_2OH$

$C_3H_8O_2$ 76.09

1,2-Propanediol.

1,2-Propanediol [*57-55-6*].

» Propylene Glycol contains not less than 99.5 percent of $C_3H_8O_2$.

Packaging and storage—Preserve in tight containers.

Reference standard—*USP Propylene Glycol Reference Standard*—Do not dry before using.

Identification—The infrared absorption spectrum of a thin film of it exhibits maxima only at the same wavelengths as that of a similar preparation of USP Propylene Glycol RS.

Specific gravity ⟨841⟩: between 1.035 and 1.037.

Acidity—Add 1 mL of phenolphthalein TS to 50 mL of water, then add 0.1 *N* sodium hydroxide until the solution remains pink for 30 seconds. Then add 10 mL of Propylene Glycol, accurately measured, and titrate with 0.10 *N* sodium hydroxide until the original pink color returns and remains for 30 seconds: not more than 0.20 mL of 0.10 *N* sodium hydroxide is required.

Water, *Method I* ⟨921⟩: not more than 0.2%.

Residue on ignition—Heat 50 g in a tared 100-mL shallow dish until it ignites, and allow it to burn without further application of heat in a place free from drafts. Cool, moisten the residue with 0.5 mL of sulfuric acid, and ignite to constant weight: the weight of the residue does not exceed 3.5 mg.

Chloride ⟨221⟩—A 1-mL portion shows no more chloride than corresponds to 0.10 mL of 0.020 *N* hydrochloric acid (0.007%).

Sulfate ⟨221⟩—A 5.0-mL portion shows no more sulfate than corresponds to 0.30 mL of 0.020 *N* sulfuric acid (0.006%).

Arsenic, *Method I* ⟨211⟩: 3 ppm.

Heavy metals ⟨231⟩—Mix 4.0 mL with water to make 25 mL: the limit is 5 ppm.

Assay—

Chromatographic system—The gas chromatograph is equipped with a thermal conductivity detector, and contains a 1-m × 4-mm column packed with 5 percent G16 on support S5. The injection port temperature is 240°, the detector temperature is 250°, and the column temperature is programmed at a rate of 5° per minute from 120° to 200°, and helium is used as the carrier gas. The approximate retention time for propylene glycol is 5.7 minutes, and the approximate retention times for the 3 isomers of dipropylene glycol, when present, are 8.2, 9.0, and 10.2 minutes, respectively.

Procedure—Inject a suitable volume, typically about 10 µL, of Propylene Glycol into a suitable gas chromatograph, and record the chromatogram. Calculate the percentage of $C_3H_8O_2$ in the Propylene Glycol by dividing the area under the propylene glycol peak by the sum of the areas under all of the peaks, excluding those due to air and water, and multiplying by 100.

Propylene Glycol Alginate—*see* Propylene Glycol Alginate NF

Propylene Glycol Diacetate—*see* Propylene Glycol Diacetate NF

Propylene Glycol Monostearate—*see* Propylene Glycol Monostearate NF

Propylhexedrine

[Structure: cyclohexane ring–$CH_2CH(CH_3)NHCH_3$]

$C_{10}H_{21}N$ 155.28

Cyclohexaneethanamine, *N*,α-dimethyl-, (±)-.

(±)-*N*,α-Dimethylcyclohexaneethylamine [*101-40-6*].

» Propylhexedrine contains not less than 98.0 percent and not more than 101.0 percent of $C_{10}H_{21}N$.

Packaging and storage—Preserve in tight containers.

Identification—

A: To 3 mL of water contained in a small flask add about 0.1 mL of it and 0.5 mL of 1 *N* hydrochloric acid, and agitate the mixture until clear. Add 20 mL of trinitrophenol TS, insert the stopper in the flask, shake vigorously for a few minutes, and allow to stand for 2 hours. Filter, wash the precipitate with about 20 mL of cold water, and dry in vacuum at 60° for 4 hours: the picrate so obtained melts between 108° and 110° (see *Melting Range or Temperature* ⟨741⟩). (*Caution—Picrates may explode.*)

B: A solution, prepared as directed in *Identification test A*, yields a brown precipitate with iodine TS and a white precipitate with mercuric–potassium iodide TS.

Specific gravity ⟨841⟩: between 0.848 and 0.852.

Assay—Tare a glass-stoppered conical flask containing about 15 mL of water, add quickly about 0.5 mL of Propylhexedrine, and again weigh. Add to the contents of the flask 30 mL of neutralized alcohol, then add methyl red TS, and titrate with 0.1 *N* sulfuric acid VS. Perform a blank determination, and make any necessary correction. Each mL of 0.1 *N* sulfuric acid is equivalent to 15.53 mg of $C_{10}H_{21}N$.

Propylhexedrine Inhalant

» Propylhexedrine Inhalant consists of cylindrical rolls of suitable fibrous material impregnated with Propylhexedrine, usually aromatized, and contained in a suitable inhaler. The inhaler contains not less than 90.0 percent and not more than 125.0 percent of the labeled amount of $C_{10}H_{21}N$.

Packaging and storage—Preserve in tight containers (inhalers), and avoid exposure to excessive heat.

Identification—Place the contents of 1 inhaler in a glass-stoppered flask, add 50 mL of methanol, and allow to stand for 1 hour with frequent agitation. Filter, pressing out the roll on the filter. Add to the filtrate 1 *N* hydrochloric acid until it is slightly acid to moistened litmus paper, then add 30 mL of water, and evaporate to about 20 mL. Cool, transfer to a small separator, and shake with 10 mL of ether. Withdraw the water layer, warm it on a steam bath to expel any ether, and dilute to about 25 mL. From 10 mL of the solution, precipitate the propylhexedrine with trinitrophenol TS as directed in *Identification test A* under *Propylhexedrine:* the propylhexedrine picrate so obtained melts between 108° and 110° (see *Melting Range or Temperature* ⟨741⟩). (*Caution—Picrates may explode.*)

Assay—Place the contents of 2 inhalers of Propylhexedrine Inhalant in the thimble of a continuous-extraction apparatus, and quickly assemble the apparatus. Rinse each of the emptied inhalers with about 20 mL of methanol, pouring the rinsings through the condenser into the extraction flask. Add through the condenser 20 mL to 30 mL of methanol, and extract for 15 to 20 cycles. Cool the extract, transfer it completely with the aid of small portions of methanol to a 100-mL volumetric flask, dilute with methanol to volume, and mix. To 50.0 mL of the solution add 25.0 mL of 0.1 *N* sulfuric acid VS, and evaporate to about 40 mL. Cool, add methyl red TS, and titrate the excess acid with 0.1 *N* sodium hydroxide VS. Each mL of 0.1 *N* sulfuric acid is equivalent to 15.53 mg of $C_{10}H_{21}N$.

Propyliodone

[Structure: 3,5-diiodo-4-oxo-pyridine ring–N–$CH_2COOCH_2CH_2CH_3$]

$C_{10}H_{11}I_2NO_3$ 447.01

1(4*H*)-Pyridineacetic acid, 3,5-diiodo-4-oxo-, propyl ester.

Propyl 3,5-diiodo-4-oxo-1(4*H*)pyridineacetate [*587-61-1*].

» Propyliodone contains not less than 99.0 percent and not more than 101.0 percent of $C_{10}H_{11}I_2NO_3$, calculated on the dried basis.

Packaging and storage—Preserve in tight, light-resistant containers.

Identification—

A: Heat 100 mg with a few drops of sulfuric acid: violet vapors are evolved.

B: Reflux 1 g with 10 mL of 1 *N* sodium hydroxide for 30 minutes, add 10 mL of water, and acidify to litmus paper with hydrochloric acid: the precipitate of 3,5-diiodo-4-oxo-1(4*H*)-pyridineacetic acid, after being washed with water and dried at 105°, melts at about 245°.

Melting range ⟨741⟩: between 187° and 190°.

Acidity—Dissolve 1.0 g in 40 mL of hot *n*-propyl alcohol previously neutralized to phenolphthalein TS, cool, and allow to stand in an ice bath for 15 minutes with frequent shaking. Filter, wash the residue with neutralized *n*-propyl alcohol, combine the filtrate and washings, add phenolphthalein TS, and titrate with 0.050 *N* sodium hydroxide to a pink color that persists for 15 seconds: not more than 0.15 mL of 0.050 *N* sodium hydroxide is required for neutralization.

Loss on drying ⟨731⟩—Dry it at 105° to constant weight: it loses not more than 0.5% of its weight.

Residue on ignition ⟨281⟩: not more than 0.1%.

Iodine and iodide—Shake 2.4 g with 30 mL of water for 15 minutes, filter, and to 10 mL of filtrate add 1 mL of 2 *N* nitric acid, 1 mL of sodium nitrite solution (1 in 500), and 2 mL of chloroform. Shake, and centrifuge: any purple color in the chloroform layer is not darker than that obtained with a mixture of 6 mL of water and 4 mL of potassium iodide solution (2.6 in 100,000) treated in the same manner (0.01% of I).

Heavy metals, *Method II* ⟨231⟩: 0.002%.

Assay—Using about 15 mg of Propyliodone, accurately weighed, proceed as directed in the *Assay* under *Iodoquinol.* Each mL of 0.02 *N* sodium thiosulfate is equivalent to 0.7450 mg of $C_{10}H_{11}I_2NO_3$.

Sterile Propyliodone Oil Suspension

» Sterile Propyliodone Oil Suspension is a sterile suspension of Propyliodone in Peanut Oil. It contains not less than 57.0 percent and not more than 63.0 percent of $C_{10}H_{11}I_2NO_3$.

Packaging and storage—Preserve in single-dose, light-resistant containers.

Identification—Mix 1 g of Suspension with 20 mL of solvent hexane, filter the diluted suspension through a fine-porosity, sintered-glass crucible, and wash the residue free from peanut oil with the solvent hexane: the residue responds to the *Identification tests* under *Propyliodone*, and, after being dried at 105° to constant weight, melts between 187° and 190°.

Weight per mL—Transfer 60 to 70 mL of well-shaken Suspension to a 250-mL beaker, place in a vacuum desiccator, and cautiously apply vacuum. When vigorous frothing has ceased, apply a pressure of about 10 mm of mercury for 15 minutes. Remove the specimen, mix gently with a spatula without stirring in any air bubbles, adjust its temperature to about 20°, and fill a clean, dry, tared 50-mL pycnometer with it. Adjust the temperature of the filled pycnometer to 25°, remove any excess of the specimen, weigh, and calculate the net weight. The weight per mL is between 1.236 g and 1.276 g.

Iodine and iodide—Disperse 3.3 mL in 125 mL of alcohol-free chloroform, add 25 mL of sodium hydroxide solution (1 in 2500), and shake. Separate the aqueous layer, shake it with 125 mL of alcohol-free chloroform, and discard the chloroform. Proceed with 10 mL of the aqueous layer as directed in the test for *Iodine and iodide* under *Propyliodone*, beginning with "add 1 mL of 2 *N* nitric acid."

Other requirements—It meets the requirements under *Injections* ⟨1⟩.

Assay—Open 1 container of Sterile Propyliodone Oil Suspension, stir the contents with a glass rod until mixed, replace the closure, and shake. Quickly transfer about 30 mg of the Suspension to a tared combustion capsule, and weigh accurately. Using this as the assay specimen, proceed as directed in the *Assay* under *Iodoquinol*. Each mL of 0.02 *N* sodium thiosulfate is equivalent to 0.7450 mg of $C_{10}H_{11}I_2NO_3$. From the weight of Suspension taken and the observed *Weight per mL*, calculate the weight of propyliodone in each mL of the Suspension.

Propylparaben—*see* Propylparaben NF

Propylparaben Sodium—*see* Propylparaben Sodium NF

Propylthiouracil

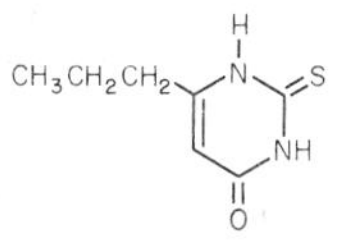

$C_7H_{10}N_2OS$ 170.23

4(1*H*)-Pyrimidinone, 2,3-dihydro-6-propyl-2-thioxo-.

6-Propyl-2-thiouracil [*51-52-5*].

» Propylthiouracil contains not less than 98.0 percent and not more than 100.5 percent of $C_7H_{10}N_2OS$, calculated on the dried basis.

Packaging and storage—Preserve in well-closed, light-resistant containers.

Reference standard—*USP Propylthiouracil Reference Standard*—Dry at 105° for 2 hours before using.

Identification—The infrared absorption spectrum of a potassium bromide dispersion of it, previously dried, exhibits maxima only at the same wavelengths as that of a similar preparation of USP Propylthiouracil RS.

Melting range ⟨741⟩: between 218° and 221°.

Loss on drying ⟨731⟩—Dry it at 105° for 2 hours: it loses not more than 0.5% of its weight.

Residue on ignition ⟨281⟩: not more than 0.1%.

Selenium ⟨291⟩: 0.003%, a 200-mg specimen being used.

Heavy metals, *Method II* ⟨231⟩: 0.002%.

Ordinary impurities ⟨466⟩—

Test solution: methanol.

Standard solution: methanol.

Application volume: 10 µL.

Eluant: a mixture of toluene, ethyl acetate, and formic acid (50:45:5), in a nonequilibrated chamber.

Visualization: 1.

Assay—Weigh accurately about 300 mg of Propylthiouracil, transfer to a 500-mL conical flask, and add 30 mL of water. Add from a buret about 30 mL of 0.1 *N* sodium hydroxide VS, heat to boiling, and agitate until solution is complete. Wash down any particles on the wall of the flask with a few mL of water, then add about 50 mL of 0.1 *N* silver nitrate while mixing, and boil gently for 7 minutes. Cool to room temperature, and continue to titrate with 0.1 *N* sodium hydroxide VS, determining the endpoint potentiometrically, using a glass-calomel electrode system. Each mL of 0.1 *N* sodium hydroxide is equivalent to 8.512 mg of $C_7H_{10}N_2OS$.

Propylthiouracil Tablets

» Propylthiouracil Tablets contain not less than 93.0 percent and not more than 107.0 percent of the labeled amount of $C_7H_{10}N_2OS$.

Packaging and storage—Preserve in well-closed containers.

Reference standard—*USP Propylthiouracil Reference Standard*—Dry at 105° for 2 hours before using.

Identification—Boil a quantity of finely powdered Tablets, equivalent to about 100 mg of propylthiouracil, with 10 mL of alcohol under a reflux condenser for 20 minutes. Filter while hot, and evaporate the filtrate on a steam bath to dryness: a portion of the residue responds to the *Identification tests* under *Propylthiouracil.*

Dissolution ⟨711⟩—

Medium: water; 900 mL.

Apparatus 1: 100 rpm.

Time: 30 minutes.

Procedure—Determine the amount of $C_7H_{10}N_2OS$ dissolved from ultraviolet absorbances at the wavelength of maximum absorbance at about 274 nm of filtered portions of the solution under test, suitably diluted with *Dissolution Medium*, in comparison with a Standard solution having a known concentration of USP Propylthiouracil RS in the same medium.

Tolerances—Not less than 85% (*Q*) of the labeled amount of $C_7H_{10}N_2OS$ is dissolved in 30 minutes.

Uniformity of dosage units ⟨905⟩: meet the requirements.

Assay—Weigh and finely powder not less than 20 Propylthiouracil Tablets. Transfer an accurately weighed portion of the powder, equivalent to about 50 mg of propylthiouracil, to a 100-mL volumetric flask, add 50 mL of methanol, and shake by mechanical means for 30 minutes. Dilute with methanol to volume, and filter, discarding the first 20 mL of the filtrate. Transfer 4.0 mL of the subsequent filtrate to a 500-mL volumetric flask, add 5 mL of 0.1 *N* hydrochloric acid, and dilute with water to volume. Concomitantly determine the absorbances of this solution and of a Standard solution of USP Propylthiouracil RS in the same medium having a known concentration of about 4 µg per mL, in 1-cm cells at the wavelength of maximum absorbance at about 274 nm, with a suitable spectrophotometer, using water as the blank. Calculate the quantity, in mg, of $C_7H_{10}N_2OS$ in the portion of Tablets taken by the formula:

$$12.5C(A_U/A_S),$$

in which C is the concentration, in µg per mL, of USP Propylthiouracil RS in the Standard solution, and A_U and A_S are the absorbances of the solution from the Tablets and the Standard solution, respectively.

Protamine Sulfate

» Protamine Sulfate is a purified mixture of simple protein principles obtained from the sperm or testes of suitable species of fish, which has the property of neutralizing heparin. Each mg of Protamine Sulfate, calculated on the dried basis, neutralizes not less than 100 USP Heparin Units.

Packaging and storage—Preserve in tight containers, in a refrigerator.

Reference standard—*USP Heparin Sodium Reference Standard*—Store in a cool place and do not freeze.

Loss on drying ⟨731⟩—Dry it at 105° for 3 hours: it loses not more than 5% of its weight.

Sulfate—Dissolve about 150 mg, accurately weighed, in 75 mL of water, add 5 mL of 3 *N* hydrochloric acid, heat to boiling, and while maintaining at the boiling point, slowly add 10 mL of barium chloride TS. Cover the vessel, and allow the mixture to stand on a steam bath for 1 hour. Filter, wash the precipitate with several portions of hot water, dry, and ignite to constant weight. The weight of the barium sulfate, multiplied by 0.4117, represents the weight of sulfate in the portion of Protamine Sulfate taken. Not less than 16% and not more than 22%, calculated on the dried basis, is found.

Nitrogen content—Determine the nitrogen content as directed under *Method II* (see *Nitrogen Determination* ⟨461⟩). Not less than 22.5% and not more than 25.5% of N, calculated on the dried basis, is found.

Assay—

Assay preparation—Dissolve a suitable quantity of Protamine Sulfate, accurately weighed, in Water for Injection to obtain a solution having a concentration of 1 mg per mL, calculated on the dried basis.

Plasma—Prepare as directed for *Preparation of plasma* in the *Assay* under *Heparin Sodium*.

Heparin preparation—On the day of the assay, prepare a solution of USP Heparin Sodium RS in saline TS to give a final concentration of 115 USP Heparin Units per mL.

Calcium-thromboplastin solution—Dissolve in calcium chloride solution (1 in 50) a quantity of thromboplastin that is sufficient, as determined by preliminary trial if necessary, to produce clotting in about 35 seconds in a mixture consisting of equal volumes of plasma and a mixture of 4 volumes of saline TS and 1 volume of the prepared calcium-thromboplastin solution.

Procedure—Into each of ten meticulously cleansed, 13- × 100-mm test tubes pipet 2.5 mL of *Plasma*. Place the tubes in a water bath at 37 ± 0.2°, and to each of nine of them add 0.5 mL of *Assay preparation*. Into the tenth tube, to provide the control, pipet 2 mL of saline TS and 0.5 mL of *Calcium-thromboplastin solution*, noting the time, to the nearest second, of adding the latter. While mixing with a wire loop, note the time of the first appearance of fibrin fibers, and record it to the nearest second. The elapsed time is the normal clotting time of the plasma. Pipet into the nine remaining tubes the following volumes, in mL, of *Heparin preparation:* 0.43, 0.45, 0.47, 0.49, 0.50, 0.51, 0.53, 0.55, and 0.57, respectively. To each tube add saline TS to make 4.5 mL. Taking the tubes in random order, add 0.5 mL of *Calcium-thromboplastin solution*, and note the clotting time in each tube in the same manner as for the control tube.

Calculation—Calculate the number of USP Heparin Units neutralized per mg, by the formula:

$$N_S/W_U,$$

in which N_S is the number of USP Heparin Units, and W_U is the number of mg of Protamine Sulfate in the last tube prior to the first one in which the clotting time is not less than 2 seconds longer than that in the control tube.

Protamine Sulfate Injection

» Protamine Sulfate Injection is a sterile, isotonic solution of Protamine Sulfate. It contains not less than 90.0 percent and not more than 120.0 percent of the labeled amount of protamine sulfate.

Packaging and storage—Preserve in single-dose containers, preferably of Type I glass. Store in a refrigerator.

Labeling—Label it to indicate the approximate neutralization capacity in USP Heparin Units.

Reference standard—*USP Heparin Sodium Reference Standard*—Store in a cool place and do not freeze.

Identification—It responds to the tests for *Sulfate* ⟨191⟩.

Pyrogen—It meets the requirements of the *Pyrogen Test* ⟨151⟩, the test dose per kg being 0.5 mL of Injection containing 10 mg per mL.

Other requirements—It meets the requirements under *Injections* ⟨1⟩.

Assay—Using as the *Assay preparation* a solution prepared by diluting with Water for Injection an accurately measured volume of Protamine Sulfate Injection to give an estimated final concentration of protamine sulfate of 1 mg per mL, proceed as directed in the *Assay* under *Protamine Sulfate*. The potency, in mg of protamine sulfate in each mL of the Injection, is given by the formula:

$$v/V,$$

in which v and V are the volumes, in mL, respectively, of the *Heparin preparation* and of the Injection present in the last tube prior to the first one in which the clotting time is not less than 2 seconds longer than that in the control tube.

Protamine Sulfate for Injection

» Protamine Sulfate for Injection is a sterile mixture of Protamine Sulfate with one or more suitable, dry diluents. It contains not less than 90.0 percent and not more than 120.0 percent of the labeled amount of protamine sulfate.

Packaging and storage—Preserve in *Containers for Sterile Solids* as described under *Injections* ⟨1⟩. Preserve the accompanying solvent in single-dose or in multiple-dose containers, preferably of Type I glass.

Labeling—Label it to indicate the approximate neutralization capacity in USP Heparin Units.

Reference standard—*USP Heparin Sodium Reference Standard*—Store in a cool place and do not freeze.

Constituted solution—At the time of use, the constituted solution prepared from *Protamine Sulfate for Injection* meets the requirements for *Constituted Solutions* under *Injections* ⟨1⟩.

Pyrogen—It meets the requirements of the *Pyrogen Test* ⟨151⟩, the test dose per kg being 1.0 mL of a solution prepared to contain 5 mg of protamine sulfate per mL in pyrogen-free saline TS.

pH and clarity of solution—Dissolve it in the solvent recommended in the labeling: the pH of the solution is between 6.5 and 7.5, and the solution is clear.

Other requirements—Both it and the accompanying solvent meet the requirements for *Sterility Tests* ⟨71⟩, and *Labeling* under

Injections ⟨1⟩. It meets the requirements for *Uniformity of Dosage Units* ⟨905⟩.

Assay—Using as the *Assay preparation* a solution prepared by dissolving the contents of 1 container of Protamine Sulfate for Injection in Water for Injection to give a final concentration of about 1 mg of protamine sulfate per mL, proceed as directed in the *Assay* under *Protamine Sulfate*. Calculate the potency, in mg, of protamine sulfate in each mL of the *Assay preparation* by the formula:

$$v/V,$$

in which v and V are the volumes, in mL, respectively, of the *Heparin preparation* and the *Assay preparation* present in the last tube prior to the first one in which the clotting time is not less than 2 seconds longer than in the control tube.

Protamine Zinc Insulin Suspension—*see* Insulin Suspension, Protamine Zinc

Protein Hydrolysate Injection

» Protein Hydrolysate Injection is a sterile solution of amino acids and short-chain peptides which represent the approximate nutritive equivalent of the casein, lactalbumin, plasma, fibrin, or other suitable protein from which it is derived by acid, enzymatic, or other method of hydrolysis. It may be modified by partial removal and restoration or addition of one or more amino acids. It may contain alcohol, dextrose, or other carbohydrate suitable for intravenous infusion. Not less than 50.0 percent of the total nitrogen present is in the form of α-amino nitrogen.

Packaging and storage—Preserve in single-dose containers, preferably of Type I or Type II glass, and avoid excessive heat.

Labeling—The label of the immediate container bears in a subtitle the name of the protein from which the hydrolysate has been derived and the word "modified" if one or more of the "essential" amino acids has been partially removed, restored, or added. The label bears a statement of the pH range; the name and percentage of any added other nutritive ingredient; the method of hydrolysis; the nature of the modification, if any, in amino acid content after hydrolysis; the percentage of each essential amino acid or its equivalent; the approximate protein equivalent, in g per liter; the approximate number of calories per liter; the percentage of the total nitrogen in the form of α-amino nitrogen; and the quantity of the sodium and of the potassium ions present in each 100 mL of the Injection. Injection that contains not more than 30 mg of sodium per 100 mL may be labeled "Protein Hydrolysate Injection, Low Sodium," or by a similar title the approximate equivalent thereof.

The label states the total osmolar concentration in mOsmol per liter. Where the contents are less than 100 mL, or where the label states that the Injection is not for direct injection but is to be diluted before use, the label alternatively may state the total osmolar concentration in mOsmol per mL.

Non-antigenicity—

Sensitizing solution—Select a suitable quantity of the protein identical in nature and quality with that from which the hydrolysate was manufactured, and subject it to the same hydrolytic process used in manufacturing the hydrolysate but reduce the time of hydrolysis to one-third. For purposes of preservation add, if desired, 0.5% of chlorobutanol, and package the sensitizing solution in 100-mL multiple-dose containers. Store in a cold place.

Preparation of animals—Select healthy guinea pigs each weighing between 420 and 480 g. Inject each animal intraperitoneally with 6 mL of the sensitizing solution on the second, fourth, and sixth days of each of two successive weeks. Use the sensitized animals not less than 30 days and not more than 37 days after the last sensitizing dose. Re-sensitize any animals not used during the 7-day period by injecting intraperitoneally a booster dose of 6 mL of the sensitizing solution, and use re-sensitized animals not less than 9 days and not more than 16 days after the injection of the booster dose.

Procedure—Inject, intravenously, 3 mL of Protein Hydrolysate Injection, at the rate of 2 mL per minute, using a 5-mL syringe fitted with a 27-gauge needle, into each of five guinea pigs prepared as described above. During the injection and during the 15 minutes following, observe the animals for any of the following symptoms: (1) licking the nose or rubbing the nose with forefeet; (2) ruffling of the fur; (3) labored breathing; (4) sneezing or coughing (three or more times); (5) retching. The requirements of the test are met if none of the injected animals show more than two of the listed symptoms and none show rales, convulsions, or prostration, or die. If none of the listed symptoms are observed, test the sensitivity of the animals by injecting into one of them 3 mL of the sensitizing solution intravenously at the rate of 2 mL per minute: the animal shows positive signs of anaphylaxis such as rales, convulsions, prostration, and/or death in addition to one or more of the lesser reaction symptoms.

Pyrogen—It meets the requirements of the *Pyrogen Test* ⟨151⟩.

Biological adequacy—It meets the requirements under *Protein—Biological Adequacy Test* ⟨141⟩.

pH ⟨791⟩: between 4.0 and 7.0, determined potentiometrically, but the variation from the pH range stated on the label is not greater than ±0.5 pH unit.

Nitrogen content—Using 0.1 mL of Injection, determine the nitrogen content as directed under *Method II* (see *Nitrogen Determination* ⟨461⟩).

α-Amino nitrogen—Dilute 5.0 mL of a 5 percent Injection, or an appropriate volume of any other concentration, to 25 mL. Adjust to a pH of 7, potentiometrically, by the addition of 0.1 *N* sodium hydroxide or 0.1 *N* hydrochloric acid. Add 10 mL of formaldehyde TS, previously adjusted to a pH of 9.0, potentiometrically, then while stirring the solution, preferably with a mechanical stirrer, and with a suitable glass electrode in the system, add 0.1 *N* sodium hydroxide VS slowly toward the end, until a pH of 9.0 is reached. Continue stirring for 2 minutes, check the pH, and adjust if necessary. Record the volume of 0.1 *N* sodium hydroxide VS added in the titration. Each mL of 0.10 *N* sodium hydroxide corresponds to 1.4 mg of α-amino nitrogen.

Potassium content—

Standard solutions—Prepare four standard solutions (numbered 1, 2, 3, and 4) each containing a suitable wetting agent and 25.0 mEq of sodium (1.46 g of sodium chloride) per liter, and to the solutions add, respectively, 0-, 2.0-, 3.0-, and 4.0-mg supplements of potassium, in the form of the chloride, per liter. If necessary, because of changes in the sensitivity of the instrument mentioned below, vary the levels of concentration of the potassium, keeping the ratios between solutions approximately as given.

Standard graph—Set a suitable flame photometer to a wavelength of 766 nm. Adjust the instrument to zero transmittance with Solution 1. Then adjust the instrument to 100% transmittance with Solution 4. Read the percent transmittance of Solutions 2 and 3. Plot the observed transmittance of Solutions 2, 3, and 4 as the ordinate and the concentration as the abscissa on arithmetic coordinate paper.

Procedure—Pipet a portion of the Injection containing approximately 300 µg of potassium, or a quantity corresponding to the concentration of the *Standard solutions*, into a 100-mL volumetric flask. Add a small amount of wetting agent, and make to volume with a sodium solution of such strength that the final sodium concentration is 25.0 mEq per liter. Adjust the instrument to zero transmittance with Solution 1 and to 100% transmittance with Solution 4. Read the percent transmittance of the test solution and, by reference to the standard potassium graph, calculate the potassium content of the Injection in mg of potassium per mL.

Sodium content—Proceed as directed under *Potassium content*, with the following modifications: (1) prepare the *Standard solutions* to contain 6.00 mEq of potassium (447 mg of potassium chloride), and substitute sodium for the stated quantities of potassium; (2) prepare the *Standard graph* with the flame photometer set at 589 nm instead of 766 nm; and (3) under *Procedure* read "sodium" for "potassium" throughout, but in the second sentence read "a potassium solution of such strength that the final potassium concentration is 6.00 mEq per liter" in place of "a sodium solution of such strength that the final sodium concentration is 25.0 mEq per liter."

Other requirements—It meets the requirements under *Injections* ⟨1⟩.

Protriptyline Hydrochloride

· HCl

H CH₂CH₂CH₂NHCH₃

$C_{19}H_{21}N.HCl$ 299.84

5*H*-Dibenzo[*a,d*]cycloheptene-5-propanamine, *N*-methyl-, hydrochloride.

N-Methyl-5*H*-dibenzo[*a,d*]cycloheptene-5-propylamine hydrochloride [*1225-55-4*].

» Protriptyline Hydrochloride contains not less than 99.0 percent and not more than 101.0 percent of $C_{19}H_{21}N.HCl$, calculated on the dried basis.

Packaging and storage—Preserve in well-closed containers.

Reference standard—*USP Protriptyline Hydrochloride Reference Standard*—Dry at a pressure below 5 mm of mercury at 60° to constant weight before using.

Identification—

A: The infrared absorption spectrum of a mineral oil dispersion of it exhibits maxima only at the same wavelengths as that of a similar preparation of USP Protriptyline Hydrochloride RS.

B: The ultraviolet absorption spectrum of a 1 in 100,000 solution in 0.1 *N* methanolic hydrochloric acid exhibits maxima and minima at the same wavelengths as that of a similar solution of USP Protriptyline Hydrochloride RS, concomitantly measured, and the respective absorptivities, calculated on the dried basis, at the wavelength of maximum absorbance at about 292 nm do not differ by more than 3.0%.

C: It responds to the test for *Chloride* ⟨191⟩, when tested as specified for alkaloidal hydrochlorides.

D: Its X-ray diffraction pattern (see *X-ray Diffraction* ⟨941⟩) conforms to that of USP Protriptyline Hydrochloride RS.

pH ⟨791⟩: between 5.0 and 6.5, in a solution (1 in 100).

Loss on drying ⟨731⟩—Dry it at a pressure below 5 mm of mercury at 60° to constant weight: it loses not more than 0.3% of its weight.

Residue on ignition ⟨281⟩: not more than 0.1%.

Heavy metals, *Method II* ⟨231⟩: 0.001%.

Assay—Transfer about 700 mg of Protriptyline Hydrochloride, accurately weighed, to a 125-mL conical flask, and dissolve in 30 mL of glacial acetic acid. Add crystal violet TS and 10 mL of mercuric acetate TS, and titrate with 0.1 *N* perchloric acid VS to a green end-point. Perform a blank determination, and make any necessary correction. Each mL of 0.1 *N* perchloric acid is equivalent to 29.98 mg of $C_{19}H_{21}N.HCl$.

Protriptyline Hydrochloride Tablets

» Protriptyline Hydrochloride Tablets contain not less than 90.0 percent and not more than 110.0 percent of the labeled amount of $C_{19}H_{21}N.HCl$.

Packaging and storage—Preserve in tight containers.

Reference standard—*USP Protriptyline Hydrochloride Reference Standard*—Dry at a pressure below 5 mm of mercury at 60° to constant weight before using.

Identification—

A: Place a portion of finely powdered Tablets, equivalent to about 10 mg of protriptyline hydrochloride, in a 100-mL volumetric flask, add about 50 mL of methanol, shake, dilute with methanol to volume, and mix. Filter, dilute 10 mL of the filtrate with methanol to 100 mL, and mix: the ultraviolet absorption spectrum of this solution exhibits maxima and minima at the same wavelengths as a 1 in 100,000 solution of USP Protriptyline Hydrochloride RS in methanol, concomitantly measured.

B: A filtered solution of finely powdered Tablets, equivalent to protriptyline hydrochloride solution (1 in 20), responds to the tests for *Chloride* ⟨191⟩, when tested as specified for alkaloidal hydrochlorides.

Dissolution ⟨711⟩—

Medium: water; 900 mL.

Apparatus 1: 100 rpm.

Time: 45 minutes.

Procedure—Determine the amount of $C_{19}H_{21}N.HCl$ dissolved from ultraviolet absorbances at the wavelength of maximum absorbance at about 290 nm of filtered portions of the solution under test, suitably diluted with *Dissolution Medium,* if necessary, in comparison with a Standard solution having a known concentration of USP Protriptyline Hydrochloride RS in the same medium.

Tolerances—Not less than 75% (*Q*) of the labeled amount of $C_{19}H_{21}N.HCl$ is dissolved in 45 minutes.

Uniformity of dosage units ⟨905⟩: meet the requirements.

Procedure for content uniformity—Transfer 1 finely powdered Tablet to a 100-mL volumetric flask, add about 50 mL of 0.1 *N* hydrochloric acid, and mix. Dilute with 0.1 *N* hydrochloric acid, to volume, mix, and filter. Dilute a portion of the subsequent filtrate quantitatively with 0.1 *N* hydrochloric acid to provide a solution containing approximately 10 µg of protriptyline hydrochloride per mL. Concomitantly determine the absorbances of this solution and a Standard solution of USP Protriptyline Hydrochloride RS in the same medium at a concentration of about 10 µg per mL, in 1-cm cells at the maximum at about 290 nm, with a suitable spectrophotometer, using 0.1 *N* hydrochloric acid as the blank. Calculate the quantity, in mg, of $C_{19}H_{21}N.HCl$ in the Tablet by the formula:

$$(TC/D)(A_U/A_S),$$

in which *T* is the labeled quantity, in mg, of protriptyline hydrochloride in the Tablet, *C* is the concentration, in µg per mL, of USP Protriptyline Hydrochloride RS in the Standard solution, *D* is the concentration, in µg per mL, of protriptyline hydrochloride in the solution from the Tablet, based on the labeled quantity per Tablet and the extent of dilution, and A_U and A_S are the absorbances of the solution from the Tablet and the Standard solution, respectively.

Assay—Proceed with Protriptyline Hydrochloride Tablets as directed under *Salts of Organic Nitrogenous Bases* ⟨501⟩, diluting the *Standard Preparation* and the *Assay Preparation* with an equal volume of 0.5 *N* sulfuric acid and determining the absorbance at the maximum at about 292 nm. Calculate the quantity, in mg, of $C_{19}H_{21}N.HCl$ in the portion of Tablets taken by the formula:

$$0.05C(A_U/A_S),$$

in which *C* is the concentration, in µg per mL, of USP Protriptyline Hydrochloride RS in the *Standard Preparation.*

Pseudoephedrine Hydrochloride

H NHCH₃ — C—C—CH₃ · HCl — OH H

$C_{10}H_{15}NO.HCl$ 201.70

Benzenemethanol, α-[1-(methylamino)ethyl]-, [*S*-(*R**,*R**)]-, hydrochloride.

(+)-Pseudoephedrine hydrochloride [*345-78-8*].

» Pseudoephedrine Hydrochloride contains not less than 98.0 percent and not more than 100.5 percent of $C_{10}H_{15}NO.HCl$, calculated on the dried basis.

Packaging and storage—Preserve in tight, light-resistant containers.

Reference standard—*USP Pseudoephedrine Hydrochloride Reference Standard*—Dry at 105° for 3 hours before using.

Identification—

A: The infrared absorption spectrum of a potassium bromide dispersion of it, previously dried, exhibits maxima only at the same wavelengths as that of a similar preparation of USP Pseudoephedrine Hydrochloride RS.

B: A solution responds to the tests for *Chloride* ⟨191⟩.

Melting range, *Class I* ⟨741⟩: between 182° and 186°, but the range between beginning and end of melting does not exceed 2°.

Specific rotation ⟨781⟩: between +61.0° and +62.5°, calculated on the dried basis, determined in a solution containing 500 mg in each 10 mL.

pH ⟨791⟩: between 4.6 and 6.0, in a solution (1 in 20).

Loss on drying ⟨731⟩—Dry it at 105° for 3 hours: it loses not more than 0.5% of its weight.

Residue on ignition ⟨281⟩: not more than 0.1%.

Assay—Dissolve about 400 mg of Pseudoephedrine Hydrochloride, accurately weighed, in a mixture of 50 mL of glacial acetic acid and 10 mL of mercuric acetate TS, add 1 drop of crystal violet TS, and titrate with 0.1 *N* perchloric acid VS to a blue-green end-point. Perform a blank determination, and make any necessary correction. Each mL of 0.1 *N* perchloric acid is equivalent to 20.17 mg of $C_{10}H_{15}NO.HCl$.

Pseudoephedrine Hydrochloride Syrup

» Pseudoephedrine Hydrochloride Syrup contains not less than 90.0 percent and not more than 110.0 percent of the labeled amount of $C_{10}H_{15}NO.HCl$.

Packaging and storage—Preserve in tight, light-resistant containers.

Reference standard—*USP Pseudoephedrine Hydrochloride Reference Standard*—Dry at 105° for 3 hours before using.

Identification—Extract a volume of Syrup, equivalent to about 120 mg of pseudoephedrine hydrochloride, with two 30-mL portions of ether, discard the extracts, and add 4 mL of 1 *N* sodium hydroxide. Extract with 30 mL of chloroform, and evaporate the chloroform on a steam bath, avoiding overheating: the pseudoephedrine so obtained melts at about 118°, the procedure for *Class I* being used (see *Melting Range or Temperature* ⟨741⟩), and when 50 mg is dissolved in 10 mL of 0.1 *N* hydrochloric acid, the resulting solution is dextrorotatory.

Reaction—It is acid to litmus.

Assay—

Mobile phase—Prepare a suitable degassed and filtered mixture of alcohol and ammonium acetate solution (1 in 250) (17:3).

Standard preparation—Dissolve an accurately weighed quantity of USP Pseudoephedrine Hydrochloride RS in 0.01 *N* hydrochloric acid to obtain a solution having a known concentration of about 1.2 mg per mL.

Assay preparation—Transfer an accurately measured volume of Pseudoephedrine Hydrochloride Syrup, equivalent to about 60 mg of pseudoephedrine hydrochloride, to a 50-mL volumetric flask, dilute with 0.01 *N* hydrochloric acid to volume, mix, and filter.

Chromatographic system (see *Chromatography* ⟨621⟩)—The liquid chromatograph is equipped with a 254-nm detector and a 4.2-mm × 25-cm column that contains packing L3. The flow rate is about 1.5 mL per minute. Chromatograph five replicate injections of the *Standard preparation*, and record the peak responses as directed under *Procedure:* the relative standard deviation is not more than 2.0%, and the tailing factor is not more than 1.5.

Procedure—Separately inject equal volumes (about 10 µL) of the *Standard preparation* and the *Assay preparation* into the chromatograph, record the chromatograms, and measure the responses for the major peaks. Calculate the quantity, in mg, of $C_{10}H_{15}NO.HCl$ in each mL of the Syrup taken by the formula:

$$50(C/V)(r_U/r_S),$$

in which C is the concentration, in mg per mL, of USP Pseudoephedrine Hydrochloride RS in the *Standard preparation*, V is the volume, in mL, of Syrup taken, and r_U and r_S are the peak responses obtained from the *Assay preparation* and the *Standard preparation*, respectively.

Pseudoephedrine Hydrochlorides Syrup, Triprolidine and—*see* Triprolidine and Pseudoephedrine Hydrochlorides Syrup

Pseudoephedrine Hydrochloride Tablets

» Pseudoephedrine Hydrochloride Tablets contain not less than 93.0 percent and not more than 107.0 percent of the labeled amount of $C_{10}H_{15}NO.HCl$.

Packaging and storage—Preserve in tight containers.

Reference standard—*USP Pseudoephedrine Hydrochloride Reference Standard*—Dry at 105° for 3 hours before using.

Identification—Triturate a number of Tablets, equivalent to about 180 mg of pseudoephedrine hydrochloride, with chloroform, filter through paper backed by a pledget of cotton, and evaporate the chloroform on a steam bath, avoiding overheating. Recrystallize the residue from a small amount of dehydrated alcohol: the pseudoephedrine hydrochloride so obtained responds to *Identification test A* under *Pseudoephedrine Hydrochloride.*

Dissolution ⟨711⟩—

Medium: water; 900 mL.

Apparatus 2: 50 rpm.

Time: 45 minutes.

Chromatographic system (see *Chromatography* ⟨621⟩)—The liquid chromatograph is equipped with a 214-nm detector and a 4.2-mm × 25-cm column that contains packing L3. The mobile phase is a suitable degassed and filtered mixture of alcohol and 0.40 percent ammonium acetate (17:3), and the flow rate is about 1.5 mL per minute. Chromatograph five replicate injections of the *Standard preparation*, and record the peak responses as directed under *Procedure:* the relative standard deviation is not more than 2.0%, and the tailing factor is not more than 1.5.

Procedure—Separately inject equal volumes (about 10 µL) of the *Standard preparation* and filtered portions of the solution under test into the chromatograph, record the chromatograms, and measure the responses for the major peaks. Calculate the quantity of $C_{10}H_{15}NO.HCl$ dissolved in comparison with a Standard solution having a known concentration of USP Pseudoephedrine Hydrochloride RS in the same medium and similarly chromatographed.

Tolerances—Not less than 75% (*Q*) of the labeled amount of $C_{10}H_{15}NO.HCl$ is dissolved in 45 minutes.

Uniformity of dosage units ⟨905⟩: meet the requirements.

Assay—

Mobile phase and *Standard preparation*—Prepare as directed in the *Assay* under *Pseudoephedrine Hydrochloride Syrup.*

Assay preparation—Weigh and finely powder not less than 20 Pseudoephedrine Hydrochloride Tablets. Transfer an accurately weighed portion of the powder, equivalent to about 120 mg of pseudoephedrine hydrochloride, to a 100-mL volumetric flask, add about 10 mL of 0.01 *N* hydrochloric acid, and sonicate for 10 minutes. Cool to room temperature. Dilute with 0.01 *N* hydrochloric acid to volume, mix, and filter.

Chromatographic system and *Procedure*—Proceed as directed in the *Assay* under *Pseudoephedrine Hydrochloride Syrup*, except to calculate the quantity, in mg, of $C_{10}H_{15}NO.HCl$ in the portion of Tablets taken by the formula:

$$100C(r_U/r_S),$$

in which C is the concentration, in mg per mL, of USP Pseudoephedrine Hydrochloride RS in the *Standard preparation*, and r_U and r_S are the peak responses obtained from the *Assay preparation* and the *Standard preparation*, respectively.

Pseudoephedrine Hydrochlorides Tablets, Triprolidine and—*see* Triprolidine and Pseudoephedrine Hydrochlorides Tablets

Pseudoephedrine Sulfate

$(C_{10}H_{15}NO)_2.H_2SO_4$ 428.54

Benzenemethanol, α-[1-(methylamino)ethyl]-, [*S*-(*R**,*R**)]-, sulfate (2:1) (salt).

(+)-Pseudoephedrine sulfate (2:1) (salt) [*7460-12-0*].

» Pseudoephedrine Sulfate contains not less than 98.0 percent and not more than 100.5 percent of $(C_{10}H_{15}NO)_2.H_2SO_4$, calculated on the dried basis.

Packaging and storage—Preserve in tight, light-resistant containers.

Reference standard—*USP Pseudoephedrine Sulfate Reference Standard*—Dry at 105° for 2 hours before using.

Identification—

A: The infrared absorption spectrum of a potassium bromide dispersion of it, previously dried, exhibits maxima only at the same wavelengths as that of a similar preparation of USP Pseudoephedrine Sulfate RS.

B: The ultraviolet absorption spectrum of a solution (1 in 2000) exhibits maxima and minima at the same wavelengths as that of a similar solution of USP Pseudoephedrine Sulfate RS, concomitantly measured, and the respective absorptivities, calculated on the dried basis, at the wavelength of maximum absorbance at about 257 nm do not differ by more than 3.0%.

C: A solution of it responds to the test for *Sulfate* ⟨191⟩.

Melting range, *Class I* ⟨741⟩: between 174° and 179°, but the range between beginning and end of melting does not exceed 2°.

Specific rotation ⟨781⟩: between +56.0° and +59.0°, calculated on the dried basis, determined in a solution having a known concentration of 500 mg in each 10 mL.

pH ⟨791⟩: between 5.0 and 6.5 in a solution (1 in 20).

Loss on drying ⟨731⟩—Dry it at 105° for 2 hours: it loses not more than 2.0% of its weight.

Residue on ignition ⟨281⟩: not more than 0.1%.

Heavy metals—Treat a solution of 1.0 g in 20 mL of dilute alcohol (1 in 2) with 5 mL of sodium hydroxide solution (1 in 20) and 5 drops of sodium sulfide TS: the color developed is not darker than that of a blank determination performed simultaneously, containing 10 ppm of standard lead solution (see *Heavy Metals* ⟨231⟩).

Chloride ⟨221⟩—A 200-mg portion shows no more chloride than corresponds to 0.4 mL of 0.02 *N* hydrochloric acid (0.14%).

Assay—Dissolve about 150 mg of Pseudoephedrine Sulfate, accurately weighed, in 50 mL of glacial acetic acid. Titrate with 0.1 *N* perchloric acid VS, determining the end-point potentiometrically. Perform a blank determination, and make any necessary correction. Each mL of 0.1 *N* perchloric acid is equivalent to 42.85 mg of $(C_{10}H_{15}NO)_2.H_2SO_4$.

Psyllium Husk

» Psyllium Husk is the cleaned, dried seed coat (epidermis) separated by winnowing and thrashing from the seeds of *Plantago ovata* Forskal, known in commerce as Blond Psyllium or Indian Psyllium or Ispaghula, or from *Plantago psyllium* Linné or from *Plantago indica* Linné (*Plantago arenaria* Waldstein et Kitaibel) known in commerce as Spanish or French Psyllium (Fam. Plantaginaceae), in whole or in powdered form.

Packaging and storage—Preserve in well-closed containers, secured against insect attack (see under *Vegetable and Animal Drugs* in the *General Notices*).

Botanic characteristics—

Histology—*Husk*—The epidermis is composed of large cells having transparent walls filled with mucilage, and the cells swell rapidly in aqueous mounts and appear polygonal to slightly rounded in a surface view, when viewed from above (from below they appear elongated to rectangular). The swelling takes place mainly in the radial direction. The mucilage of the epidermal cells stains red with ruthenium red and lead acetate TS. The very occasional starch granules that are present in some of the epidermal cells, and that may be found embedded in the mucilage, are small and simple or compounded with four or more components.

Powdered Psyllium Seed Husk—Pale to medium buff-colored powder, having a slight pinkish tinge, a weak characteristic odor, and a very mucilaginous taste. Occasional single and 2- to 4-compound starch granules, the individual grains being spheroidal plano to angular convex from 2 µm to 10 µm in diameter, are found embedded in the mucilage. Entire or broken epidermal cells are filled with mucilage. In surface view, the epidermal cells appear polygonal to slightly rounded. Mucilage stains red with ruthenium red and lead acetate TS. Some of the elongated and rectangular cells representing the lower part of epidermis and also radially swollen epidermal cells can be found.

Identification—

A: *Mounted in cresol*—Cells, viewed microscopically, are composed of polygonal prismatic cells having 4 to 6 straight or slightly wavy walls.

B: *Mounted in alcohol and irrigated with water*—Viewed microscopically, the mucilage in the outer part of the epidermal cells swells rapidly and goes into solution.

Microbial limits ⟨61⟩—The total combined molds and yeasts count does not exceed 1000 per g, and it meets the requirements of the test for absence of *Salmonella* species and *Escherichia coli.*

Total ash ⟨561⟩: not more than 4.0%.

Acid-insoluble ash ⟨561⟩: not more than 1.0%.

Water, *Method II* ⟨921⟩: not more than 12.0%.

Light extraneous matter—[NOTE—Perform this test in a well-ventilated hood.] Transfer 15.0 g to a 250-mL separator. Add about 90 mL of a liquid mixture of carbon tetrachloride and ethylene dichloride (about 2:1), having a specific gravity of 1.45. Shake for 30 seconds, and allow to settle for 30 seconds. Repeat the shaking and settling twice more. Drain all the material and liquid except the floating layer. Add 25 mL of the liquid mixture, stir carefully, allow to settle, and drain as before. Repeat the washing of the floating layer twice more, but use only 10 mL of the liquid mixture each time. Transfer the washed floating layer to a tared beaker, heat on a steam bath until the odor of the liquid no longer persists, dry at 40° for 3 hours, allow to cool in a desiccator, and weigh: the limit is 15%.

Heavy extraneous matter—[NOTE—Perform this test in a well-ventilated hood.] Transfer 10.0 g to a 250-mL separator. Add about 80 mL of carbon tetrachloride, and shake for 1 minute. Allow to stand for 5 minutes. Drain into a tared 1000-mL beaker the non-mucilaginous material that sinks to the bottom, taking care not to drain any of the floating material. Heat in a hot air oven, at a temperature not exceeding 90°, until the odor of the liquid no longer persists, allow to cool in a desiccator, and weigh: the limit is 1.1%.

Insect infestation—Transfer 25 g to a 250-mL beaker, add sufficient solvent hexane to saturate, add an additional 75 to 100 mL of solvent hexane, and allow to stand for 10 minutes, stirring occasionally with a stirring rod. Wet a sheet of filter paper with alcohol, and filter the mixture with the aid of vacuum. Discard the filtrate. Transfer the residue to the original beaker with the aid of alcohol. Add alcohol to bring the volume to 150 mL above the level of the transferred residue. Boil for 10 minutes. Filter through alcohol-wetted paper as above. Prepare a trap flask, consisting of a 2000-mL graduated, narrow-mouth conical flask into which is inserted a rubber disk supported on a stiff metal rod about 4 mm in diameter and longer than the height of the flask, the rod being threaded at the lower end and furnished with nuts and washers to hold the disk in place, and the disk being of the proper shape and size to prevent liquid in the body of the flask from spilling when it is pressed up against the neck from the inside. Transfer the residue to the trap flask, completing the transfer with the aid of hot water. Add sufficient hot water to bring the volume to 1000 mL. Add 20 mL of hydrochloric acid. Raise the rod, and support it so that the rubber disk is held above the liquid level. Rinse the rubber disk with hot water. Spray the inside of the neck of the flask with an antifoam spray. Boil for 30 minutes, and cool to room temperature. Add 40 mL of solvent hexane, and agitate for 1 minute by tilting the flask and moving the rod vertically with wrist action. Allow to stand for 5 minutes. Add water to bring the level of liquid to the neck of the flask, and allow to stand for 20 minutes. Simultaneously rotate the disk to free it from settled material, and raise it as far as possible into the neck of the flask. Prepare a sheet of ruled filter paper, with lines approximately 5 mm apart for filtration by moistening it with water and placing it on a vacuum funnel. Transfer the material trapped in the neck of the flask to the filter with the aid of water. If necessary, wash the paper with alcohol to remove traces of hexane. Place the paper on a 100-mm petri dish that has been wetted with a solution containing equal volumes of glycerin and alcohol. Add 35 mL of solvent hexane to the flask, and gently stir with the trapping rod. Add water to bring the liquid level into the neck of the flask. Allow to stand for 15 minutes. Using the same technique as before, transfer the trapped material onto a separate paper. Examine the papers at 30× magnification: in the case of powdered Psyllium Husk, not more than 400 insect fragments, including mites and psocids, can be seen; in the case of whole Psyllium Husk, not more than 100 insect fragments, including mites and psocids, can be seen.

Swell volume—Transfer 250 mL of simulated intestinal fluid TS without enzymes to a glass-stoppered, 500-mL graduated cylinder. Gradually, with shaking, add 3.5 g of the Husk until a uniform, smooth suspension is obtained. Dilute with the same fluid to 500 mL. Shake the cylinder for about 1 minute every 30 minutes for 8 hours. Allow the gel to settle for 16 hours (total time 24 hours). Determine the volume of the gel: it is not less than 40 mL per g for powdered Psyllium Husk, and not less than 35 mL per g for whole Psyllium Husk.

Psyllium Hydrophilic Mucilloid for Oral Suspension

» Psyllium Hydrophilic Mucilloid for Oral Suspension is a dry mixture of Psyllium Husk with suitable additives.

Packaging and storage—Preserve in tight containers.

Identification—Microscopically, it shows the presence of fragmented Psyllium Husk, as described for *Histology—Husk* in the section, *Botanic characteristics*, under *Psyllium Husk*.

Microbial limits ⟨61⟩—It meets the requirements of the tests for absence of *Salmonella* species and of *Escherichia coli*.

Swell volume—Transfer 250 mL of simulated intestinal fluid TS without enzymes to a glass-stoppered, 500-mL graduated cylinder. Gradually, with shaking, add an amount of Psyllium Hydrophilic Mucilloid for Oral Suspension, equivalent to 3.5 g of psyllium husk, and shake until a uniform, smooth suspension is obtained. Dilute with the same fluid to 500 mL. Shake the cylinder for about 1 minute every 30 minutes for 8 hours. Allow the gel to settle for 16 hours (total time 24 hours). Determine the volume of the gel: it is not less than 110 mL.

Pumice

» Pumice is a substance of volcanic origin, consisting chiefly of complex silicates of aluminum, potassium, and sodium.

Packaging and storage—Preserve in well-closed containers.

Labeling—Label powdered Pumice to indicate, in descriptive terms, the fineness of the powder.

Powdered Pumice meets the following requirements:

"Pumice Flour" or "Superfine Pumice": not less than 97.0% of pumice flour or superfine pumice passes through a No. 200 standard mesh sieve.

"Fine Pumice": not less than 95.0% of fine pumice passes through a No. 150 standard mesh sieve and not more than 75.0% passes through a No. 200 standard mesh sieve.

"Coarse Pumice": not less than 95.0% of coarse pumice passes through a No. 60 standard mesh sieve and not more than 5.0% passes through a No. 200 standard mesh sieve.

Water-soluble substances—Boil 10 g with 50 mL of water for 30 minutes, adding water from time to time to maintain approximately the original volume, and then filter: the filtrate is neutral to litmus, and one-half of this filtrate, when evaporated and dried at 105° for 1 hour, yields not more than 10 mg of residue (0.20%).

Acid-soluble substances—Boil 1 g of Pumice with 25 mL of 3 *N* hydrochloric acid for 30 minutes, adding water from time to time to maintain approximately the original volume, then filter the liquid. Add 5 drops of sulfuric acid to the filtrate, evaporate to dryness, ignite, and weigh the residue: not more than 60 mg of residue is obtained (6.0%).

Iron—Acidify the remaining half of the filtrate from the test for *Water-soluble substances* with hydrochloric acid, and add a few drops of potassium ferrocyanide TS: no blue color is produced.

Purified Bentonite—*see* Bentonite, Purified NF

Purified Cotton—*see* Cotton, Purified

Purified Rayon—*see* Rayon, Purified

Purified Siliceous Earth—*see* Siliceous Earth, Purified NF

Purified Stearic Acid—*see* Stearic Acid, Purified NF

Purified Water—*see* Water, Purified

Pyrantel Pamoate

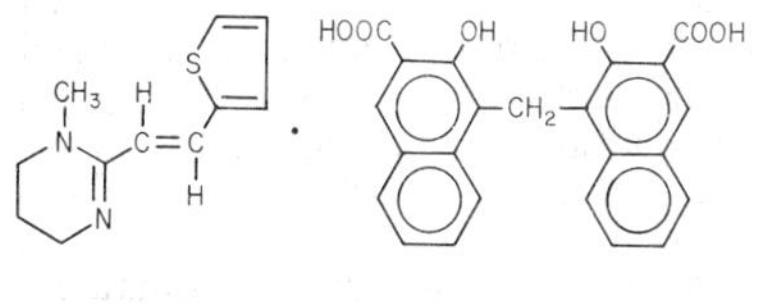

$C_{11}H_{14}N_2S.C_{23}H_{16}O_6$ 594.68

Pyrimidine, 1,4,5,6-tetrahydro-1-methyl-2-[2-(2-thienyl)-ethenyl]-, (*E*)-, compd. with 4,4′-methylenebis[3-hydroxy-2-naphthalenecarboxylic acid] (1:1).
(*E*)-1,4,5,6-Tetrahydro-1-methyl-2-[2-(2-thienyl)vinyl]pyrimidine 4,4′-methylenebis[3-hydroxy-2-naphthoate] (1:1) [*22204-24-6*].

» Pyrantel Pamoate contains not less than 97.0 percent and not more than 103.0 percent of $C_{34}H_{30}N_2O_6S$, calculated on the dried basis.

Packaging and storage—Preserve in well-closed, light-resistant containers.

Reference standards—*USP Pamoic Acid Reference Standard*—Dry in vacuum at 100° for 3 hours before using. *USP Pyrantel Pamoate Reference Standard*—Dry in vacuum at 60° for 3 hours before using.

Identification—

A: The infrared absorption spectrum of a potassium bromide dispersion of it, previously dried, exhibits maxima only at the same wavelengths as that of a similar preparation of USP Pyrantel Pamoate RS.

B: The ultraviolet absorption spectrum of a 1 in 62,500 solution in methanol exhibits maxima and minima at the same wavelengths as that of a similar preparation of USP Pyrantel Pamoate RS, concomitantly measured.

C: The chromatogram of the *Assay preparation* obtained as directed in the *Assay* exhibits major peaks due to pyrantel base and pamoic acid, the retention times of which correspond to those exhibited in the chromatogram of the *Standard preparation* obtained as directed in the *Assay*.

Loss on drying ⟨731⟩—Dry it in vacuum at 60° for 3 hours: it loses not more than 2.0% of its weight.

Residue on ignition ⟨281⟩: not more than 0.5%, from 1.33 g.

Heavy metals, *Method II* ⟨231⟩: 0.005%.

Iron—To the residue obtained in the test for *Residue on ignition* add 3 mL of hydrochloric acid and 2 mL of nitric acid, and evaporate on a steam bath to dryness. Dissolve the residue in 2 mL of hydrochloric acid with the aid of gentle heat. Add 18 mL of hydrochloric acid, dilute with water to 50 mL, and mix. Dilute 5 mL of this solution with water to 47 mL: the limit is 0.0075%.

Related compounds—

A: Impregnate 18- × 56-cm filter paper (Whatman No. 1 or equivalent) with a freshly prepared solution obtained by mixing 7 volumes of acetone and 3 volumes of glycine–sodium chloride–hydrochloric acid buffer solution (prepared by mixing 3 volumes of a solution that is 0.3 *M* with respect to both glycine and sodium chloride with 7 volumes of 0.3 *M* hydrochloric acid). Press the impregnated paper uniformly between white, non-fluorescent blotters to remove the excess solvent. Prepare solutions of Pyrantel Pamoate and USP Pyrantel Pamoate RS in chloroform, methanol, and ammonium hydroxide (50:50:5) having concentrations of 20 mg and 0.2 mg of pyrantel per mL. Apply 20-µL portions of the solutions to the prepared sheet. Place the sheet immediately in a suitable chromatographic chamber, and develop by descending chromatography (see *Chromatography* ⟨621⟩) for 16 to 20 hours, using as the solvent system a mixture of ethyl acetate, butanol, and water (10:1:1). Remove the sheet from the chamber, air-dry for 10 minutes, transfer to an air-circulating oven, and dry at 60° for 30 minutes. Examine the dried chromatogram on a 254-nm ultraviolet scanner screen: the R_f value of the principal spot from the solution under test corresponds to that obtained from the USP Pyrantel Pamoate RS solution, and no spot in the chromatogram of the more concentrated test solution, other than the principal spot, is larger or more intense than the principal spot from the diluted test solution.

B: Prepare solutions of Pyrantel Pamoate and USP Pyrantel Pamoate RS in chloroform, methanol, and ammonium hydroxide (50:50:5) at concentrations of 20 and 0.2 mg per mL. Apply 100-µL portions to a thin-layer chromatographic plate (see *Chromatography* ⟨621⟩) coated with chromatographic silica gel mixture. Develop the chromatogram in a suitable chromatographic chamber, and develop by ascending chromatography, using a solvent mixture of ethyl acetate, methanol, and diethylamine (200:50:15) until the solvent front is about 2 cm from the top edge of the plate. Air-dry the plate for 10 minutes, and examine under ultraviolet light: the R_f value of the principal spot obtained from the solution under test corresponds to that obtained from the USP Pyrantel Pamoate RS solution, and no spot in the chromatogram of the more concentrated test solution, other than the principal spot, is larger or more intense than the principal spot from the diluted test solution.

Pamoic acid content—

Mobile phase and *Chromatographic system*—Prepare as directed under *Assay*.

Standard preparation—Dissolve an accurately weighed quantity of *USP Pamoic Acid RS* in *Mobile phase* to obtain a solution having a known concentration of about 0.52 mg per mL. Transfer 1.0 mL of this solution to a 10-mL volumetric flask, dilute with *Mobile phase* to volume, and mix.

Test preparation—Use the *Assay preparation*.

Procedure—Inject only *Standard preparation* (about 20 µL) into the chromatograph, record the chromatograms, and record the peak responses as directed under *Assay*. Calculate the quantity, in mg, of $C_{23}H_{16}O_3$ in the portion of pyrantel pamoate taken by the formula:

$$1000C(r_U/r_S),$$

in which C is the concentration, in mg per mL, of USP Pamoic Acid RS in the *Standard preparation* and r_U is the peak response due to pamoic acid in the chromatogram of the *Test preparation* obtained in the *Assay*, and r_S is the peak response due to pamoic acid in the *Standard preparation*: the content of pamoic acid is between 63.4% and 67.3% calculated on the dried basis.

Assay—[NOTE—Use low-actinic glassware in preparing solutions of pyrantel pamoate, and otherwise protect the solutions from unnecessary exposure to bright light. Complete the *Assay* without prolonged interruption.]

Mobile phase—Prepare a mixture of acetonitrile, acetic acid, water, and diethylamine (94:2.5:2.5:1), filter, and degas. Make adjustments if necessary (see *System Suitability* under *Chromatography* ⟨621⟩). [NOTE—Increasing the amount of acetonitrile in *Mobile phase* increases retention times. Increasing the amount of acetic acid, water, and diethylamine decreases retention times. Should the *Mobile phase* need to be adjusted, maintain the ratios among acetic acid, water, and diethylamine (1:1:0.4).]

Standard preparation—Prepare a solution in *Mobile phase* having an accurately known concentration of about 80 µg of USP Pyrantel Pamoate RS per mL.

Assay preparation—Transfer about 80 mg of pyrantel pamoate, accurately weighed, to a 100-mL volumetric flask, dissolve in *Mobile phase*, dilute with *Mobile phase* to volume, and mix. Dilute 1.0 mL of this solution with *Mobile phase* to 10.0 mL, and mix.

Chromatographic system (see *Chromatography* ⟨621⟩)—The liquid chromatograph is equipped with a 288-nm detector and 4.6-mm × 25-cm column that contains packing L3. The flow rate is about 1.0 mL per minute. Chromatograph the *Standard preparation*, and record the peak responses as directed under *Procedure:* the resolution, R, between pyrantel base and pamoic acid is not less than 10.0; the number of theoretical plates, n, for the pyrantel base peak is not less than 8000; the tailing factor, T, for the pyrantel base peak is not greater than 1.1, and the relative standard deviation for replicate injections is not more than 1.0%.

Procedure—Separately inject equal volumes (about 20 µL) of the *Standard preparation* and the *Assay preparation* into the chromatograph, record the chromatograms obtained for a period of not less than 2.5 times the retention times of pyrantel base, and measure the responses for the major peaks. The relative retention times for pamoic acid and pyrantel base are about 0.6 and 1.0, respectively. Calculate the quantity, in mg, of $C_{34}H_{30}N_2O_6S$ in the portion of Pyrantel Pamoate taken by the formula:

$$1000C(r_U/r_S),$$

in which C is the concentration, in mg, of USP Pyrantel Pamoate RS in the *Standard preparation*, and r_U and r_S are the peak responses for pyrantel base obtained from the *Assay preparation* and the *Standard preparation*, respectively.

Pyrantel Pamoate Oral Suspension

» Pyrantel Pamoate Oral Suspension is a suspension of Pyrantel Pamoate in a suitable aqueous vehicle. It contains not less than 90.0 percent and not more than 110.0 percent of the labeled amount of pyrantel ($C_{11}H_{14}N_2S$).

Packaging and storage—Preserve in tight, light-resistant containers.

Reference standard—*USP Pyrantel Pamoate Reference Standard*—Dry in vacuum at 60° for 3 hours before using.

Identification—[See *Note* in the *Assay*.]

A: Dilute a suitable volume of Oral Suspension with 0.05 *N* methanolic ammonium hydroxide to obtain a solution having a concentration of about 8 mg of pyrantel pamoate per mL. Similarly prepare a Standard solution of USP Pyrantel Pamoate RS. Shake both solutions by mechanical means, and centrifuge to obtain clear solutions. Apply a 100-µL portion of each solution to a 20- × 20-cm thin-layer chromatographic plate (see *Chromatography* ⟨621⟩) coated with a 0.50-mm layer of silica gel mixture. Develop the plate in a suitable chromatographic chamber containing the upper phase obtained by shaking together methyl isobutyl ketone, formic acid, and water (2:1:1). Develop the plate until the solvent front is about 2 cm from the top edge of the plate. Remove the plate, allow the solvent to evaporate, and examine the plate under ultraviolet light at about 365 nm: the R_f value of the principal spot from the test solution corresponds to that obtained from the Standard solution.

B: Dilute a suitable volume of Oral Suspension with 0.05 *N* methanolic ammonium hydroxide to obtain a solution having a concentration of about 16 mg of pyrantel pamoate per mL. Similarly prepare a Standard solution of USP Pyrantel Pamoate RS. Shake both solutions by mechanical means, and centrifuge to obtain clear solutions. Apply a 20-µL portion of each solution to an 18- × 24-cm sheet of chromatographic paper (Whatman No. 1 or equivalent) that previously has been prepared as follows: Impregnate the paper with a freshly prepared solution obtained by mixing 7 volumes of acetone and 3 volumes of glycine–sodium chloride–hydrochloric acid buffer solution (prepared by mixing 3 volumes of a solution that is 0.3 *M* with respect to both glycine and sodium chloride with 7 volumes of 0.3 *M* hydrochloric acid). Press the impregnated paper uniformly between white, nonfluorescent blotters to remove the excess solvent. Place the spotted chromatographic paper in a suitable chromatographic chamber, and develop by descending chromatography (see *Chromatography* ⟨621⟩), using as the solvent system the upper phase obtained by mixing ethyl acetate, butanol, and water (10:1:1). After developing for 20 hours, remove the paper from the chamber, air-dry for 10 minutes, transfer to an air-circulating oven, and dry at 60° for 30 minutes. The R_f value of the principal spot from the solution under test corresponds to that obtained from the Standard solution.

C: The chromatogram of the *Assay preparation* obtained as directed in the *Assay* exhibits major peaks due to pyrantel base and pamoic acid, the retention times of which correspond to those exhibited in the chromatogram of the *Standard preparation* obtained as directed in the *Assay*.

pH ⟨791⟩: between 4.5 and 6.0.

Assay—[NOTE—Use low-actinic glassware in preparing solutions of pyrantel pamoate, and otherwise protect the solutions from unnecessary exposure to bright light. Complete the assay without prolonged interruption.]

Mobile phase, *Standard preparation*, and *Chromatographic system*—Proceed as directed for the *Assay* under *Pyrantel Pamoate*.

Assay preparation—Transfer by means of a pipet an accurately measured volume of Pyrantel Pamoate Oral Suspension equivalent to about 200 mg of pyrantel pamoate into a 100-mL volumetric flask, disperse, and dilute with water to volume. While stirring the dispersion with a magnetic stirrer, transfer 1.0 mL of the aliquot to a 25-mL volumetric flask, dissolve in *Mobile phase*, dilute with *Mobile phase* to volume, mix, and filter.

Procedure—Separately inject equal volumes (about 20 µL) of the *Standard preparation* and the *Assay preparation* into the chromatograph, record the chromatograms obtained for a period of not less than 2.5 times the retention times of pyrantel base, and measure the responses for the major peaks. The relative retention times for pamoic acid and pyrantel base are about 0.6 and 1.0, respectively. Calculate the quantity, in mg, of pyrantel ($C_{11}H_{14}N_2S$) in each mL of Pyrantel Pamoate Oral Suspension taken by the formula:

$$2500(0.347)C/V(r_U/r_S),$$

in which 0.347 is the ratio of the molecular weight of pyrantel to that of pyrantel pamoate, *C* is the concentration, in mg per mL, of USP Pyrantel Pamoate RS in the *Standard preparation*, *V* is the volume, in mL, of Oral Suspension taken, and r_U and r_S are the peak responses for pyrantel base obtained from the *Assay preparation* and the *Standard preparation*, respectively.

Pyrazinamide

$C_5H_5N_3O$ 123.11

Pyrazinecarboxamide.

Pyrazinecarboxamide [*98-96-4*].

» Pyrazinamide contains not less than 99.0 percent and not more than 101.0 percent of $C_5H_5N_3O$, calculated on the anhydrous basis.

Packaging and storage—Preserve in well-closed containers.

Reference standard—*USP Pyrazinamide Reference Standard*—Dry over silica gel for 18 hours before using.

Identification—

A: The infrared absorption spectrum of a mineral oil dispersion of it exhibits maxima only at the same wavelengths as that of a similar preparation of USP Pyrazinamide RS.

B: The ultraviolet absorption spectrum of a 1 in 100,000 solution exhibits maxima and minima at the same wavelengths as that of a similar solution of USP Pyrazinamide RS, concomitantly measured, and the respective absorptivities, calculated on the dried basis, at the wavelength of maximum absorbance at about 268 nm do not differ by more than 3.0%.

C: Boil 20 mg with 5 mL of 5 *N* sodium hydroxide: the odor of ammonia is perceptible.

Melting range ⟨741⟩: between 188° and 191°.

Water, *Method I* ⟨921⟩: not more than 0.5%.

Residue on ignition ⟨281⟩: not more than 0.1%.

Heavy metals, *Method II* ⟨231⟩: 0.001%.

Assay—Place about 300 mg of Pyrazinamide, accurately weighed, in a 500-mL Kjeldahl flask, dissolve in 100 mL of water, and add 75 mL of 5 *N* sodium hydroxide. Connect the flask by means of a distillation trap to a well-cooled condenser, the delivery tube of which dips into 20 mL of boric acid solution (1 in 25) contained in a suitable receiver. Boil gently for 20 minutes, avoiding insofar as possible distilling any of the liquid, then boil vigorously to complete the distillation of the ammonia. Cool the liquid in the receiver if necessary, add methyl purple TS, and titrate with 0.1 *N* hydrochloric acid VS. Perform a blank determination, and make any necessary correction. Each mL of 0.1 *N* hydrochloric acid is equivalent to 12.31 mg of $C_5H_5N_3O$.

Pyrazinamide Tablets

» Pyrazinamide Tablets contain not less than 93.0 percent and not more than 107.0 percent of the labeled amount of $C_5H_5N_3O$.

Packaging and storage—Preserve in well-closed containers.

Reference standard—*USP Pyrazinamide Reference Standard*—Dry over silica gel for 18 hours before using.

Identification—

A: To a quantity of powdered Tablets, equivalent to about 1 g of pyrazinamide, add about 75 mL of isopropyl alcohol, heat on a steam bath, and filter while hot. Allow to cool, filter the crystals that form, and dry at 105° for 1 hour: the infrared absorption spectrum of a mineral oil dispersion of the dried crystals so obtained exhibits maxima only at the same wavelengths as that of a similar preparation of USP Pyrazinamide RS. If a difference appears, dissolve portions of both the dried crystals and the Reference Standard in acetone, evaporate the solutions to dryness, and repeat the test on the residues.

B: The dried crystals obtained in *Identification test A* respond to *Identification test B* under *Pyrazinamide.*

C: To 20 mg of the dried crystals obtained in *Identification test A* add 5 mL of 5 *N* sodium hydroxide, and heat gently over an open flame: the odor of ammonia is perceptible.

Dissolution ⟨711⟩—

Medium: water; 900 mL.

Apparatus 2: 50 rpm.

Time: 45 minutes.

Procedure—Determine the amount of $C_5H_5N_3O$ dissolved from ultraviolet absorbances at the wavelength of maximum absorbance at about 268 nm of filtered portions of the solution under test, suitably diluted with *Dissolution Medium*, if necessary, in comparison with a Standard solution having a known concentration of USP Pyrazinamide RS in the same medium.

Tolerances—Not less than 75% (*Q*) of the labeled amount of $C_5H_5N_3O$ is dissolved in 45 minutes.

Uniformity of dosage units ⟨905⟩: meet the requirements.

Assay—Weigh and finely powder not less than 20 Pyrazinamide Tablets. Weigh accurately a portion of the powder, equivalent to about 100 mg of pyrazinamide, and transfer with the aid of 200 mL of water to a 500-mL volumetric flask. Allow to stand for about 10 minutes, with occasional swirling, add water to volume, and mix. Filter a portion through a dry filter into a dry flask, discarding the first portion of the filtrate. Pipet 5 mL of the subsequent filtrate into a 100-mL volumetric flask, add water to volume, and mix. Dissolve an accurately weighed quantity of USP Pyrazinamide RS in water, and dilute quantitatively and stepwise with water to obtain a Standard solution having a known concentration of about 10 µg per mL. Concomitantly determine the absorbances of both solutions in 1-cm cells at the wavelength of maximum absorbance at about 268 nm, with a suitable spectrophotometer, using water as the blank. Calculate the quantity, in mg, of $C_5H_5N_3O$ in the portion of Tablets taken by the formula:

$$10C(A_U/A_S),$$

in which *C* is the concentration, in µg per mL, of USP Pyrazinamide RS in the Standard solution, and A_U and A_S are the absorbances of the solution from Pyrazinamide Tablets and the Standard solution, respectively.

Pyridostigmine Bromide

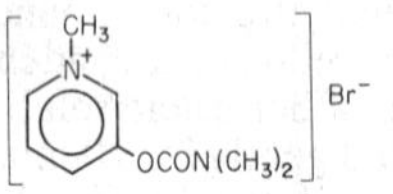

$C_9H_{13}BrN_2O_2$ 261.12

Pyridinium, 3-[[(dimethylamino)carbonyl]oxy]-1-methyl-, bromide.

3-Hydroxy-1-methylpyridinium bromide dimethylcarbamate [*101-26-8*].

» Pyridostigmine Bromide contains not less than 98.5 percent and not more than 100.5 percent of $C_9H_{13}BrN_2O_2$, calculated on the dried basis.

Packaging and storage—Preserve in tight containers.

Reference standard—*USP Pyridostigmine Bromide Reference Standard*—Dry in a suitable vacuum drying tube, using phosphorus pentoxide as the desiccant, at 100° for 4 hours before using.

Identification—

A: The infrared absorption spectrum of a potassium bromide dispersion of it, previously dried, exhibits maxima only at the same wavelengths as that of a similar preparation of USP Pyridostigmine Bromide RS.

B: The ultraviolet absorption spectrum of a 1 in 30,000 solution in 0.1 *N* hydrochloric acid exhibits maxima and minima at the same wavelengths as that of a similar solution of USP Pyridostigmine Bromide RS, concomitantly measured, and the respective absorptivities, calculated on the dried basis, at the wavelength of maximum absorbance at about 269 nm do not differ by more than 3.0%.

C: To about 100 mg in a test tube add 0.6 mL of 1 *N* sodium hydroxide: an orange color develops. When the mixture is heated, the color changes to yellow, and a strip of moistened red litmus paper held over the top of the test tube turns blue.

D: A solution (1 in 50) responds to tests for *Bromide* ⟨191⟩.

Melting range ⟨741⟩: between 154° and 157°, the test specimen having been previously dried.

Loss on drying ⟨731⟩—Dry it in a suitable vacuum drying tube, using phosphorus pentoxide as the desiccant, at 100° for 4 hours: it loses not more than 2.0% of its weight.

Residue on ignition ⟨281⟩: not more than 0.1%.

Assay—Dissolve about 850 mg of Pyridostigmine Bromide, accurately weighed, in 80 mL of glacial acetic acid. Add 25 mL of mercuric acetate TS and 2 drops of quinaldine red TS, and titrate with 0.1 *N* perchloric acid in dioxane VS to a colorless end-point. Perform a blank determination, and make any necessary correction. Each mL of 0.1 *N* perchloric acid is equivalent to 26.11 mg of $C_9H_{13}BrN_2O_2$.

Pyridostigmine Bromide Injection

» Pyridostigmine Bromide Injection is a sterile solution of Pyridostigmine Bromide in a suitable medium. It contains not less than 90.0 percent and not more than 110.0 percent of the labeled amount of $C_9H_{13}BrN_2O_2$.

Packaging and storage—Preserve in single-dose containers, preferably of Type I glass, protected from light.

Reference standard—*USP Pyridostigmine Bromide Reference Standard*—Dry in a suitable vacuum drying tube, using phosphorus pentoxide as the desiccant, at 100° for 4 hours before using.

Identification—

A: The solution prepared for measurement of absorbance in the *Assay* exhibits maxima and minima at the same wavelengths as that of a similar solution of USP Pyridostigmine Bromide RS, concomitantly measured.

B: To 2 mL of Injection add 1 mL of 2 *N* nitric acid: the solution so obtained responds to the tests for *Bromide* ⟨191⟩.

Pyrogen—It meets the requirements of the *Pyrogen Test* ⟨151⟩, the test dose being 0.25 mg per kg.

pH ⟨791⟩: between 4.5 and 5.5.

Other requirements—It meets the requirements under *Injections* ⟨1⟩.

Assay—Transfer to a suitable separator an accurately measured volume of Pyridostigmine Bromide Injection, equivalent to about

20 mg of pyridostigmine bromide. Add 10 mL of 1 *N* hydrochloric acid, and mix. Extract with four 20-mL portions and one 15-mL portion of ethyl ether, and discard the extracts. Transfer the aqueous layer to a 100-mL volumetric flask, using about 25 mL of water to aid the transfer. Place the flask on a steam bath and warm, with the aid of a stream of nitrogen, to evaporate any residual ether, then cool the flask to room temperature, dilute with water to volume, and mix. Dilute 20.0 mL of the resulting solution with 0.1 *N* hydrochloric acid to 100.0 mL, and mix. Dissolve an accurately weighed quantity of USP Pyridostigmine Bromide RS in 0.1 *N* hydrochloric acid to obtain a Standard solution having a known concentration of about 40 µg per mL. Concomitantly determine the absorbances of both solutions in 1-cm cells at the wavelength of maximum absorbance at about 269 nm, with a suitable spectrophotometer, using 0.1 *N* hydrochloric acid as the blank. Calculate the quantity, in mg, of $C_9H_{13}BrN_2O_2$ in each mL of the Injection taken by the formula:

$$(0.5C/V)(A_U/A_S),$$

in which C is the concentration, in µg per mL, of USP Pyridostigmine Bromide RS in the Standard solution, V is the volume, in mL, of Injection taken, and A_U and A_S are the absorbances of the solution from the Injection and the Standard solution, respectively.

Pyridostigmine Bromide Syrup

» Pyridostigmine Bromide Syrup contains, in each 100 mL, not less than 1.08 g and not more than 1.32 g of $C_9H_{13}BrN_2O_2$.

Packaging and storage—Preserve in tight, light-resistant containers.

Reference standard—*USP Pyridostigmine Bromide Reference Standard*—Dry in a suitable vacuum drying tube, using phosphorus pentoxide as the desiccant, at 100° for 4 hours before using.

Identification—To 5 mL of Syrup in a separator add 100 mL of 2.5 *N* hydrochloric acid, and mix. Extract with five 20-mL portions of chloroform, place 2 mL of the aqueous solution in a 50-mL volumetric flask, and add water to volume: the ultraviolet absorption spectrum of this solution exhibits maxima and minima at the same wavelengths as that of a similar solution of USP Pyridostigmine Bromide RS, concomitantly measured.

Assay—

Phosphate solution—Dissolve 38 g of monobasic sodium phosphate and 2 g of anhydrous dibasic sodium phosphate in water to make 1000 mL. Adjust the pH, if necessary, by slight variation of the ratio of the two ingredients, to 5.3 ± 0.1.

Bromocresol green solution—Dissolve 250 mg of bromocresol green sodium salt in 250 mL of *Phosphate solution.*

Standard preparation—Dissolve a suitable quantity of USP Pyridostigmine Bromide RS, accurately weighed, in water, to obtain a solution having a known concentration of about 1.2 mg per mL.

Assay preparation—Transfer an accurately measured volume of Pyridostigmine Bromide Syrup, equivalent to about 120 mg of pyridostigmine bromide, to a 100-mL volumetric flask, add water to volume, and mix.

Procedure—Transfer 10 mL each, accurately measured, of the *Assay preparation* and the *Standard preparation* to separate 125-mL separators. Into each separator pipet 10 mL of water, 20 mL of *Phosphate solution*, and 5 mL of *Bromocresol green solution.* Extract each solution with six 15-mL portions of chloroform, collecting the chloroform extracts in respective 100-mL volumetric flasks, then add chloroform to volume in each flask, and mix. Concomitantly determine the absorbances of both solutions in 1-cm cells at the wavelength of maximum absorbance at about 415 nm, with a suitable spectrophotometer, using chloroform as the blank. Calculate the quantity, in mg, of $C_9H_{13}BrN_2O_2$ in the portion of Syrup taken by the formula:

$$100C(A_U/A_S),$$

in which C is the concentration, in mg per mL, of USP Pyridostigmine Bromide RS in the *Standard preparation*, and A_U and A_S are the absorbances of the solutions from the *Assay preparation* and the *Standard preparation*, respectively.

Pyridostigmine Bromide Tablets

» Pyridostigmine Bromide Tablets contain not less than 95.0 percent and not more than 105.0 percent of the labeled amount of $C_9H_{13}BrN_2O_2$.

Packaging and storage—Preserve in tight containers.

Reference standard—*USP Pyridostigmine Bromide Reference Standard*—Dry in a suitable vacuum drying tube, using phosphorus pentoxide as the desiccant, at 100° for 4 hours before using.

Identification—

A: The ultraviolet absorption spectrum of the solution employed for measurement of absorbance in the *Assay* exhibits maxima and minima at the same wavelengths as that of a similar solution of USP Pyridostigmine Bromide RS, concomitantly measured.

B: Shake a quantity of finely powdered Tablets, equivalent to about 100 mg of pyridostigmine bromide, with 20 mL of water for 5 minutes, and filter the mixture: the filtrate responds to the tests for *Bromide* ⟨191⟩.

Dissolution ⟨711⟩—

Medium: water; 900 mL.

Apparatus 2: 50 rpm.

Time: 45 minutes.

Procedure—Determine the amount of $C_9H_{13}BrN_2O_2$ dissolved from ultraviolet absorbances at the wavelength of maximum absorbance at about 270 nm of filtered portions of the solution under test, suitably diluted with water, in comparison with a Standard solution having a known concentration of USP Pyridostigmine Bromide RS in the same medium.

Tolerances—Not less than 75% (*Q*) of the labeled amount of $C_9H_{13}BrN_2O_2$ is dissolved in 45 minutes.

Uniformity of dosage units ⟨905⟩: meet the requirements.

Assay—Weigh and finely powder not less than 20 Pyridostigmine Bromide Tablets. Weigh accurately, without delay, a portion of the powder, equivalent to about 300 mg of pyridostigmine bromide, and place in a glass-stoppered, 50-mL centrifuge tube. Add about 25 mL of dehydrated alcohol, shake for about 5 minutes, centrifuge, and pour the supernatant liquid into a 100-mL volumetric flask. Extract further with three 20-mL portions of dehydrated alcohol, then add dehydrated alcohol to volume, and mix. Dilute 10 mL of the resulting solution with 0.1 *N* hydrochloric acid in a 100-mL volumetric flask to volume, mix, and filter through filter paper, discarding the first 10 mL to 20 mL of the filtrate. Dilute 10 mL of the clear filtrate with 0.1 *N* hydrochloric acid in a 100-mL volumetric flask to volume, and mix. Dissolve about 30 mg of USP Pyridostigmine Bromide RS, accurately weighed, in dehydrated alcohol, mix, and dilute quantitatively and stepwise with 0.1 *N* hydrochloric acid to obtain a Standard solution having a known concentration of about 30 µg per mL. Concomitantly determine the absorbances of both solutions at the wavelength of maximum absorbance at about 269 nm, with a suitable spectrophotometer, using 0.1 *N* hydrochloric acid as the blank. Calculate the quantity, in mg, of $C_9H_{13}BrN_2O_2$ in the portion of Tablets taken by the formula:

$$10C(A_U/A_S),$$

in which C is the concentration, in µg per mL, of USP Pyridostigmine Bromide RS in the Standard solution, and A_U and A_S are the absorbances of the solution from Pyridostigmine Bromide Tablets and the Standard solution, respectively.

Pyridoxine Hydrochloride

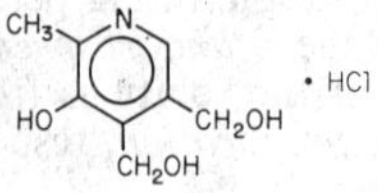

$C_8H_{11}NO_3 . HCl$ 205.64

3,4-Pyridinedimethanol, 5-hydroxy-6-methyl-, hydrochloride.

Pyridoxol hydrochloride [*58-56-0*].

» Pyridoxine Hydrochloride contains not less than 98.0 percent and not more than 102.0 percent of $C_8H_{11}NO_3 . HCl$, calculated on the dried basis.

Packaging and storage—Preserve in tight, light-resistant containers.

Reference standard—*USP Pyridoxine Hydrochloride Reference Standard*—Dry in vacuum over silica gel for 4 hours before using.

Identification—

A: The infrared absorption spectrum of a mineral oil dispersion of it exhibits maxima only at the same wavelengths as that of a similar preparation of USP Pyridoxine Hydrochloride RS.

B: It responds to the tests for *Chloride* ⟨191⟩.

Loss on drying ⟨731⟩—Dry it in vacuum over silica gel for 4 hours: it loses not more than 0.5% of its weight.

Residue on ignition ⟨281⟩: not more than 0.1%.

Heavy metals, *Method II* ⟨231⟩: 0.003%.

Chloride content—Dissolve about 500 mg, accurately weighed, in 50 mL of methanol in a glass-stoppered flask. Add 5 mL of glacial acetic acid and 2 to 3 drops of eosin Y TS, and titrate with 0.1 *N* silver nitrate VS. Each mL of 0.1 *N* silver nitrate is equivalent to 3.545 mg of Cl. Not less than 16.9% and not more than 17.6% of Cl, calculated on the dried basis, is found.

Assay—

Mobile phase—Mix 20 mL of glacial acetic acid, 1.2 g of sodium 1-hexanesulfonate, and about 1400 mL of water in a 2000-mL volumetric flask. Adjust with glacial acetic acid or 1 *N* sodium hydroxide to a pH of 3.0. Add 470 mL of methanol, dilute with water to volume, mix, and filter through a 0.5-μm filter. Make adjustments if necessary (see *System Suitability* under *Chromatography* ⟨621⟩).

Internal standard solution—Dissolve *p*-hydroxybenzoic acid in *Mobile phase* to obtain a solution having a concentration of 5 mg per mL.

Standard preparation—Dissolve about 50 mg of USP Pyridoxine Hydrochloride RS, accurately weighed, in *Mobile phase* in a 100-mL volumetric flask, dilute with *Mobile phase* to volume, and mix. Transfer 10.0 mL of the resulting solution to a 100-mL volumetric flask, add 1.0 mL of *Internal standard solution*, dilute with *Mobile phase* to volume, and mix to obtain a solution having a known concentration of about 0.05 mg per mL.

Assay preparation—Dissolve about 50 mg of Pyridoxine Hydrochloride, accurately weighed, in *Mobile phase* in a 100-mL volumetric flask, dilute with *Mobile phase* to volume, and mix. Transfer 10.0 mL of the resulting solution to a 100-mL volumetric flask, add 1.0 mL of *Internal standard solution*, dilute with *Mobile phase* to volume, and mix.

Chromatographic system (see *Chromatography* ⟨621⟩)—The liquid chromatograph is equipped with a 280-nm detector and a 4.6-mm × 25-cm column that contains packing L1. The flow rate is about 1.5 mL per minute. Chromatograph the *Standard preparation*, and record the peak responses as directed under *Procedure:* the resolution, *R*, of the pyridoxine and *p*-hydroxybenzoic acid peaks is not less than 2.5, and the relative standard deviation for replicate injections is not more than 3.0%.

Procedure—Separately inject equal volumes (about 20 μL) of the *Standard preparation* and the *Assay preparation* into the chromatograph, record the chromatograms, and measure the responses for the major peaks. The relative retention times are about 0.7 for pyridoxine and 1.0 for *p*-hydroxybenzoic acid. Calculate the quantity, in mg, of $C_8H_{11}NO_3 . HCl$ in the portion of Pyridoxine Hydrochloride taken by the formula:

$$1000C(R_U / R_S),$$

in which *C* is the concentration, in mg per mL, of USP Pyridoxine Hydrochloride RS in the *Standard preparation*, and R_U and R_S are the ratios of the peak responses of pyridoxine to internal standard obtained from the *Assay preparation* and the *Standard preparation*, respectively.

Pyridoxine Hydrochloride Injection

» Pyridoxine Hydrochloride Injection is a sterile solution of Pyridoxine Hydrochloride in Water for Injection. It contains not less than 95.0 percent and not more than 115.0 percent of the labeled amount of $C_8H_{11}NO_3 . HCl$.

Packaging and storage—Preserve in single-dose or in multiple-dose containers, preferably of Type I glass, protected from light.

Reference standard—*USP Pyridoxine Hydrochloride Reference Standard*—Dry in vacuum over silica gel for 4 hours before using.

Identification—Evaporate a volume of Injection, equivalent to about 50 mg of pyridoxine hydrochloride, on a steam bath to dryness. Add 5 mL of dehydrated alcohol, and again evaporate to dryness. Dry the residue at 105° for 3 hours: the residue so obtained responds to *Identification tests A*, and *B* under *Pyridoxine Hydrochloride*.

pH ⟨791⟩: between 2.0 and 3.8.

Other requirements—It meets the requirements under *Injections* ⟨1⟩.

Assay—Dilute an accurately measured volume of Pyridoxine Hydrochloride Injection, equivalent to about 100 mg of pyridoxine hydrochloride, quantitatively and stepwise with water to a concentration of about 10 μg of pyridoxine hydrochloride per mL. Using this as the *Assay preparation*, proceed as directed in the *Assay for pyridoxine hydrochloride* under *Decavitamin Capsules*. Calculate the quantity, in mg, of $C_8H_{11}NO_3 . HCl$ in each mL of the Injection taken by the formula:

$$10(C/V)(A_U - A_U')/(A_S - A_S'),$$

in which *C* is the concentration, in μg per mL, of USP Pyridoxine Hydrochloride RS in the *Standard preparation*, and *V* is the volume, in mL, of Injection taken, and A_S and A_U are as defined therein.

Pyridoxine Hydrochloride Tablets

» Pyridoxine Hydrochloride Tablets contain not less than 95.0 percent and not more than 115.0 percent of the labeled amount of $C_8H_{11}NO_3 . HCl$.

Packaging and storage—Preserve in well-closed containers, protected from light.

Reference standard—*USP Pyridoxine Hydrochloride Reference Standard*—Dry in vacuum over silica gel for 4 hours before using.

Identification—To a quantity of powdered Tablets, equivalent to about 100 mg of pyridoxine hydrochloride, add about 5 mL of water. Shake the mixture well, filter into a test tube, and add 2 or 3 drops of ferric chloride TS: an orange to deep red color is produced.

Disintegration ⟨701⟩: 30 minutes.

Uniformity of dosage units ⟨905⟩: meet the requirements.

Procedure for content uniformity—Transfer 1 Tablet, previously finely powdered, to a 500-mL volumetric flask containing about 300 mL of water, shake for about 30 minutes, and dilute with water to volume. Filter a portion of the mixture, discarding the first 25 mL of the filtrate. Dilute a suitable aliquot of the subsequent filtrate quantitatively and stepwise with dilute hydrochloric acid (1 in 100) so that the concentration of pyridoxine hydrochloride is about 10 μg per mL. Dissolve an accurately weighed quantity of USP Pyridoxine Hydrochloride RS in dilute hydrochloric acid (1 in 100), and dilute quantitatively and stepwise with the same solvent to obtain a Standard solution having a known concentration of about 10 μg per mL. Concomitantly determine the absorbances of both solutions in 1-cm cells at the wavelength of maximum absorbance at about 290 nm, with a suitable spectrophotometer, using dilute hydrochloric acid (1 in 100) as the blank. Calculate the quantity, in mg, of $C_8H_{11}NO_3.HCl$ in the Tablet by the formula:

$$(T/D)C(A_U/A_S),$$

in which T is the labeled quantity, in mg, of pyridoxine hydrochloride in the Tablet, D is the dilution factor, C is the concentration, in μg per mL, of USP Pyridoxine Hydrochloride RS in the Standard solution, and A_U and A_S are the absorbances of the solution from the Tablet and the Standard solution, respectively.

Assay—Weigh and finely powder not less than 20 Pyridoxine Hydrochloride Tablets. Weigh accurately a portion of the powder, equivalent to about 10 mg of pyridoxine hydrochloride, and transfer, with the aid of water, to a conical flask. Add 5 mL of hydrochloric acid, then dilute with water to about 250 mL, and heat on a steam bath until disintegration is complete. Cool, transfer to a 1000-mL volumetric flask, dilute with water to volume, mix, and centrifuge a portion of the mixture. Using the clear supernatant liquid as the *Assay preparation*, proceed as directed in the *Assay for pyridoxine hydrochloride* under *Decavitamin Capsules*, calculating the quantity, in mg, of $C_8H_{11}NO_3.HCl$ in the portion of Tablets taken by the formula:

$$C(A_U - A_U')/(A_S - A_S'),$$

in which C is the concentration, in μg per mL, of USP Pyridoxine Hydrochloride RS in the *Standard preparation.*

Pyrilamine Maleate

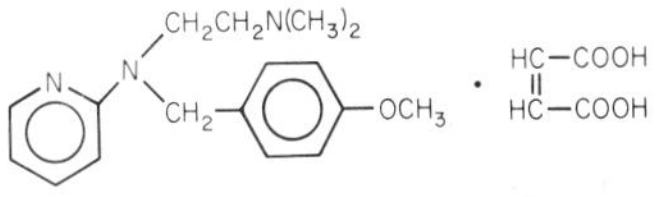

$C_{17}H_{23}N_3O.C_4H_4O_4$ 401.46

1,2-Ethanediamine, *N*-[(4-methoxyphenyl)methyl]-*N'*,*N'*-dimethyl-*N*-2-pyridinyl-, (*Z*)-2-butenedioate (1:1).

2-[[2-(Dimethylamino)ethyl](*p*-methoxybenzyl)amino]pyridine maleate (1:1) [*59-33-6*].

» Pyrilamine Maleate, dried in vacuum over phosphorus pentoxide for 5 hours, contains not less than 98.0 percent and not more than 100.5 percent of $C_{17}H_{23}N_3O.C_4H_4O_4$.

Packaging and storage—Preserve in tight, light-resistant containers.

Reference standard—*USP Pyrilamine Maleate Reference Standard*—Dry in vacuum over phosphorus pentoxide for 5 hours before using.

Identification—

A: The infrared absorption spectrum of a potassium bromide dispersion of it, previously dried, exhibits maxima only at the same wavelengths as that of a similar preparation of USP Pyrilamine Maleate RS.

B: The ultraviolet absorption spectrum of a 1 in 100,000 solution in 0.5 *N* sulfuric acid exhibits maxima and minima at the same wavelengths as that of a similar solution of USP Pyrilamine Maleate RS, concomitantly measured, and the respective absorptivities, calculated on the dried basis, at the wavelengths of maximum absorbance at about 236 and 312 nm do not differ by more than 3.0%.

Melting range, *Class I* ⟨741⟩: between 99° and 103°.

Loss on drying ⟨731⟩—Dry it in vacuum over phosphorus pentoxide for 5 hours: it loses not more than 0.5% of its weight.

Residue on ignition ⟨281⟩: not more than 0.1%.

Assay—Dissolve about 400 mg of Pyrilamine Maleate, previously dried and accurately weighed, in 50 mL of glacial acetic acid. Add 1 drop of crystal violet TS, and titrate with 0.1 *N* perchloric acid VS to a blue-green end-point. Perform a blank determination, and make any necessary correction. Each mL of 0.1 *N* perchloric acid is equivalent to 20.07 mg of $C_{17}H_{23}N_3O.C_4H_4O_4$.

Pyrilamine Maleate Tablets

» Pyrilamine Maleate Tablets contain not less than 93.0 percent and not more than 107.0 percent of the labeled amount of $C_{17}H_{23}N_3O.C_4H_4O_4$.

Packaging and storage—Preserve in well-closed containers.

Reference standard—*USP Pyrilamine Maleate Reference Standard*—Dry in vacuum over phosphorus pentoxide for 5 hours before using.

Identification—Tablets meet the requirements under *Identification—Organic Nitrogenous Bases* ⟨181⟩.

Dissolution ⟨711⟩—

Medium: water; 900 mL.

Apparatus 2: 50 rpm.

Time: 45 minutes.

Procedure—Determine the amount of $C_{17}H_{23}N_3O.C_4H_4O_4$ dissolved, employing the procedure set forth in the *Assay*, making any necessary modifications.

Tolerances—Not less than 75% (*Q*) of the labeled amount of $C_{17}H_{23}N_3O.C_4H_4O_4$ is dissolved in 45 minutes.

Uniformity of dosage units ⟨905⟩: meet the requirements.

Assay—Proceed with Pyrilamine Maleate Tablets as directed under *Salts of Organic Nitrogenous Bases* ⟨501⟩, determining the absorbance at the wavelength of maximum absorbance at about 312 nm. Calculate the quantity, in mg, of $C_{17}H_{23}N_3O.C_4H_4O_4$ in the portion of Tablets taken by the formula:

$$0.05C(A_U/A_S),$$

in which C is the concentration, in μg per mL, calculated on the dried basis, of USP Pyrilamine Maleate RS in the *Standard Preparation.*

Pyrimethamine

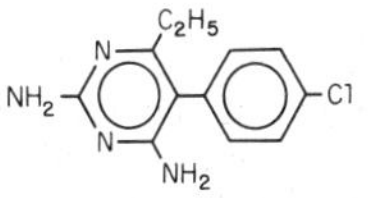

$C_{12}H_{13}ClN_4$ 248.71

2,4-Pyrimidinediamine, 5-(4-chlorophenyl)-6-ethyl-.

2,4-Diamino-5-(*p*-chlorophenyl)-6-ethylpyrimidine [*58-14-0*].

» Pyrimethamine contains not less than 99.0 percent and not more than 101.0 percent of $C_{12}H_{13}ClN_4$, calculated on the dried basis.

Packaging and storage—Preserve in tight, light-resistant containers.

Reference standard—*USP Pyrimethamine Reference Standard*—Dry at 105° for 4 hours before using.

Identification—

A: The infrared absorption spectrum of a potassium bromide dispersion of it exhibits maxima only at the same wavelengths as that of a similar preparation of USP Pyrimethamine RS.

B: Mix about 100 mg with 500 mg of anhydrous sodium carbonate, and ignite the mixture. Cool, add 5 mL of hot water, heat for 5 minutes on a steam bath, filter, and neutralize the filtrate with nitric acid: the solution responds to the test for *Chloride* ⟨191⟩.

Melting range ⟨741⟩: between 238° and 242°.

Loss on drying ⟨731⟩—Dry it at 105° for 4 hours: it loses not more than 0.5% of its weight.

Residue on ignition ⟨281⟩: not more than 0.1%.

Ordinary impurities ⟨466⟩—

Test solution: a mixture of methanol and chloroform (1:1).

Standard solution: a mixture of methanol and chloroform (1:1).

Eluant: a mixture of *n*-propyl alcohol, glacial acetic acid, and water (8:1:1).

Visualization: 2.

Assay—Dissolve about 200 mg of Pyrimethamine, accurately weighed, in 25 mL of glacial acetic acid, warming slightly to effect solution. Cool the solution to room temperature, add 4 drops of quinaldine red TS, and titrate with 0.1 *N* perchloric acid VS. Perform a blank determination, and make any necessary correction. Each mL of 0.1 *N* perchloric acid is equivalent to 24.87 mg of $C_{12}H_{13}ClN_4$.

Pyrimethamine Tablets

» Pyrimethamine Tablets contain not less than 93.0 percent and not more than 107.0 percent of the labeled amount of $C_{12}H_{13}ClN_4$.

Packaging and storage—Preserve in tight, light-resistant containers.

Reference standard—*USP Pyrimethamine Reference Standard*—Dry at 105° for 4 hours before using.

Identification—

A: The ultraviolet absorption spectrum of the *Assay Preparation* exhibits maxima at the same wavelengths as that of a similar solution of USP Pyrimethamine RS.

B: To a quantity of powdered Tablets, equivalent to about 250 mg of pyrimethamine, add 25 mL of acetone, boil for 2 minutes, and filter through a sintered-glass crucible. Repeat this treatment three times with 25-mL portions of acetone. Evaporate the combined filtrates carefully on a steam bath with the aid of a current of air to dryness: the residue responds to *Identification test A* under *Pyrimethamine*, and melts between 237° and 242° (see *Melting Range or Temperature* ⟨741⟩).

Dissolution ⟨711⟩—

Medium: 0.1 *N* hydrochloric acid; 900 mL.

Apparatus 2: 50 rpm.

Time: 45 minutes.

Procedure—Determine the amount of $C_{12}H_{13}ClN_4$ dissolved from ultraviolet absorbances at the wavelength of maximum absorbance at about 273 nm of filtered portions of the solution under test, suitably diluted with *Dissolution Medium*, if necessary, in comparison with a Standard solution having a known concentration of USP Pyrimethamine RS in the same medium.

Tolerances—Not less than 75% (*Q*) of the labeled amount of $C_{12}H_{13}ClN_4$ is dissolved in 45 minutes.

Uniformity of dosage units ⟨905⟩: meet the requirements.

Procedure for content uniformity—Transfer 1 Tablet to a 100-mL volumetric flask, add 25 mL of 0.1 *N* hydrochloric acid, warm the flask on a steam bath for 5 minutes, cool, dilute with 0.1 *N* hydrochloric acid to volume, mix, and filter, rejecting the first few mL of the filtrate. Pipet a portion of the clear filtrate, equivalent to about 2.5 mg of pyrimethamine, into a 250-mL volumetric flask, dilute with 0.1 *N* hydrochloric acid to volume, and mix. Dissolve an accurately weighed quantity of USP Pyrimethamine RS in 0.1 *N* hydrochloric acid, and dilute quantitatively and stepwise with 0.1 *N* hydrochloric acid to obtain a Standard solution having a known concentration of about 10 µg per mL. Concomitantly determine the absorbances of both solutions in 1-cm cells at the wavelength of maximum absorbance at about 273 nm, with a suitable spectrophotometer, using 0.1 *N* hydrochloric acid as the blank. Calculate the quantity, in mg, of $C_{12}H_{13}ClN_4$ in the Tablet by the formula:

$$(T/D)C(A_U/A_S),$$

in which T is the labeled quantity, in mg, of pyrimethamine in the Tablet, D is the concentration, in µg per mL, of pyrimethamine in the solution from the Tablet, on the basis of the labeled quantity per Tablet and the extent of dilution, C is the concentration, in µg per mL, of USP Pyrimethamine RS in the Standard solution, and A_U and A_S are the absorbances of the solution from the Tablet and the Standard solution, respectively.

Assay—Proceed with Pyrimethamine Tablets as directed under *Salts of Organic Nitrogenous Bases* ⟨501⟩, diluting 5.0 mL each of the *Standard Preparation* and *Assay Preparation*, respectively, with 0.5 *N* sulfuric acid to 200.0 mL and determining the absorbances of both solutions in 1-cm cells at the wavelength of maximum absorbance at about 273 nm. Calculate the quantity, in mg, of $C_{12}H_{13}ClN_4$ in the portion of Tablets taken by the formula:

$$50C(A_U/A_S),$$

in which C is the concentration, in mg per mL, of USP Pyrimethamine RS in the *Standard Preparation.*

Pyrimethamine Tablets, Sulfadoxine and—*see* Sulfadoxine and Pyrimethamine Tablets

Pyrophosphate Injection, Technetium Tc 99m—*see under "technetium"*

(Pyro- and trimeta-) Phosphates Injection, Technetium Tc 99m—*see under "technetium"*

Pyrvinium Pamoate

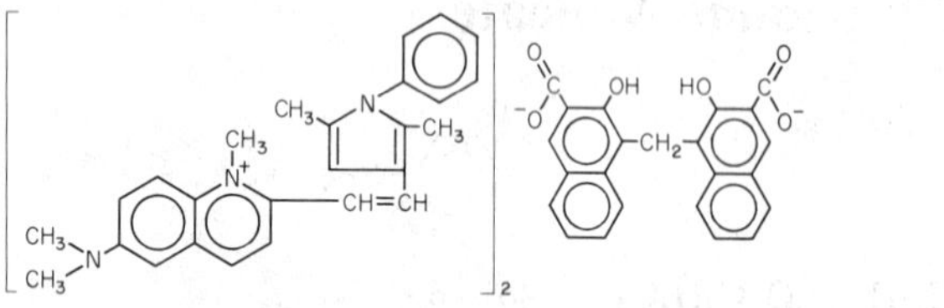

$C_{75}H_{70}N_6O_6$ 1151.41

Quinolinium, 6-(dimethylamino)-2-[2-(2,5-dimethyl-1-phenyl-1*H*-pyrrol-3-yl)ethenyl]-1-methyl-, salt with 4,4′-methylenebis[3-hydroxy-2-naphthalenecarboxylic acid] (2:1).

6-(Dimethylamino)-2-[2-(2,5-dimethyl-1-phenylpyrrol-3-yl)vinyl]-1-methylquinolinium 4,4′-methylenebis[3-hydroxy-2-naphthoate] (2:1) *[3546-41-6]*.

» Pyrvinium Pamoate contains not less than 96.0 percent and not more than 104.0 percent of $C_{75}H_{70}N_6O_6$, calculated on the anhydrous basis.

Packaging and storage—Preserve in tight, light-resistant containers.

Reference standard—*USP Pyrvinium Pamoate Reference Standard*—Do not dry; determine the *Water* content by *Method I* ⟨921⟩ before using.

Identification—

A: The infrared absorption spectrum of a potassium bromide dispersion of it exhibits maxima only at the same wavelengths as a similar preparation of USP Pyrvinium Pamoate RS.

B: A solution of it in a 1 in 200 solution of glacial acetic acid in methanol, prepared as directed in the *Assay*, exhibits absorbance maxima at about 358 nm and at about 505 nm, and the ratio A_{505}/A_{358} is between 1.93 and 2.07.

Water, *Method I* ⟨921⟩: not more than 6.0%, a 200-mg specimen being used for the test, and a mixture of 10 mL of methanol and 10 mL of chloroform being used as the solvent.

Residue on ignition ⟨281⟩: not more than 0.5%.

Assay—[NOTE—Use low-actinic flasks in preparing the solutions, and otherwise protect the solutions from unnecessary exposure to bright light. Complete the assay without prolonged interruption.] Dissolve about 250 mg of Pyrvinium Pamoate, accurately weighed, in 125 mL of glacial acetic acid in a 250-mL volumetric flask, dilute with methanol to volume, and mix. Transfer 5 mL of this solution to a 500-mL volumetric flask, dilute with methanol to volume, and mix. Similarly dissolve an accurately weighed quantity of USP Pyrvinium Pamoate RS in glacial acetic acid, using 1 mL for each 2 mg taken, and dilute quantitatively and stepwise with methanol to obtain a Standard solution having a known concentration of about 10 µg per mL. Concomitantly determine the absorbances of both solutions in 1-cm cells at the wavelength of maximum absorbance at about 505 nm, with a suitable spectrophotometer, using methanol as the blank. Calculate the quantity, in mg, of $C_{75}H_{70}N_6O_6$ in the portion of Pyrvinium Pamoate taken by the formula:

$$25C(A_U/A_S),$$

in which C is the concentration, in µg per mL, of USP Pyrvinium Pamoate RS in the Standard solution, calculated on the anhydrous basis, and A_U and A_S are the absorbances of the solution of Pyrvinium Pamoate and the Standard solution, respectively.

Pyrvinium Pamoate Oral Suspension

» Pyrvinium Pamoate Oral Suspension contains, in each 100 mL, an amount of pyrvinium pamoate equivalent to not less than 0.90 g and not more than 1.10 g of pyrvinium ($C_{26}H_{28}N_3^+$).

Packaging and storage—Preserve in tight, light-resistant containers.

Reference standard—*USP Pyrvinium Pamoate Reference Standard*—Do not dry; determine the *Water* content by *Method I* ⟨921⟩ before using.

Identification—The absorption spectrum, between 300 nm and 600 nm, of the solution employed for measurement of absorbance in the *Assay* exhibits maxima and minima at the same wavelengths as that of a similar solution of USP Pyrvinium Pamoate RS, concomitantly measured.

pH ⟨791⟩: between 6.0 and 8.0, determined potentiometrically.

Assay—[NOTE—Use low-actinic flasks in preparing the solutions, and otherwise protect the solutions from unnecessary exposure to bright light. Complete the assay without prolonged interruption.] Using a pipet calibrated "to contain," transfer 5 mL of Pyrvinium Pamoate Oral Suspension, freshly mixed and free from air bubbles, to a 250-mL volumetric flask. Complete the transfer by rinsing the pipet with 10 mL of methanol. Add 100 mL of glacial acetic acid, and mix to dissolve the pyrvinium pamoate. Dilute this solution with methanol to volume, and mix. Transfer 3 mL of the resulting solution to a 100-mL volumetric flask, dilute with methanol to volume, and mix. Dissolve an accurately weighed quantity of USP Pyrvinium Pamoate RS in glacial acetic acid, using 4 mL for each 3 mg taken, and dilute quantitatively and stepwise with methanol to obtain a Standard solution having a known concentration of about 9 µg per mL. Concomitantly determine the absorbances of both solutions in 1-cm cells at the wavelength of maximum absorbance at about 505 nm, with a suitable spectrophotometer, using methanol as the blank. Calculate the quantity, in g per 100 mL, of pyrvinium ($C_{26}H_{28}N_3^+$) in the Oral Suspension taken by the formula:

$$0.1667C(0.6644A_U/A_S),$$

in which C is the concentration, in µg per mL, of USP Pyrvinium Pamoate RS in the Standard solution, calculated on the anhydrous basis, 0.6644 is the ratio of the molecular weight of pyrvinium to one-half the molecular weight of pyrvinium pamoate, and A_U and A_S are the absorbances of the solution prepared from the Oral Suspension and the Standard solution, respectively.

Pyrvinium Pamoate Tablets

» Pyrvinium Pamoate Tablets contain an amount of pyrvinium pamoate equivalent to not less than 92.0 percent and not more than 108.0 percent of the labeled amount of pyrvinium ($C_{26}H_{28}N_3^+$).

Packaging and storage—Preserve in tight, light-resistant containers.

Reference standard—*USP Pyrvinium Pamoate Reference Standard*—Do not dry; determine the *Water* content by *Method I* ⟨921⟩, before using.

Identification—Tablets respond to the *Identification test* under *Pyrvinium Pamoate Oral Suspension.*

Disintegration ⟨701⟩: 30 minutes.

Uniformity of dosage units ⟨905⟩: meet the requirements.

Assay—[NOTE—Use low-actinic flasks in preparing the solutions, and otherwise protect the solutions from unnecessary exposure to bright light. Complete the assay without prolonged interruption.] Place a number of Pyrvinium Pamoate Tablets, equivalent to 500 mg of pyrvinium, in a 500-mL volumetric flask, and add 25 mL of water and 25 mL of acetone. Completely disintegrate the tablets by heating on the steam bath for 10 minutes with frequent mixing, allowing part of the acetone to boil off slowly. To the hot mixture add 250 mL of glacial acetic acid, and mix occasionally for 5 minutes without further heating, to dissolve the pyrvinium pamoate. Dilute with methanol to volume at room temperature, and mix. Centrifuge a portion of the mixture until a clear solution is obtained. Transfer 3 mL of the clear, supernatant solution to a 500-mL volumetric flask, dilute with methanol to volume, and mix. Dissolve an accurately weighed quantity of USP Pyrvinium Pamoate RS in glacial acetic acid, using 1 mL for each 3 mg taken, and dilute quantitatively and stepwise with methanol to obtain a Standard solution having a known concentration of about 9 µg per mL. Concomitantly determine the absorbances of both solutions in 1-cm cells at the wavelength of maximum absorbance at about 505 nm, with a suitable spectrophotometer, using methanol as the blank. Calculate the quantity, in mg, of pyrvinium ($C_{26}H_{28}N_3^+$) in the portion of Tablets taken by the formula:

$$83.3C(0.6644A_U/A_S),$$

in which C is the concentration, in µg per mL, of USP Pyrvinium Pamoate RS in the Standard solution, calculated on the anhydrous basis, 0.6644 is the ratio of the molecular weight of pyrvinium to one-half the molecular weight of pyrvinium pamoate, and A_U and A_S are the absorbances of the solution from the Tablets and the Standard solution, respectively.

Quinacrine Hydrochloride

NHCH(CH₃)CH₂CH₂CH₂N(C₂H₅)₂ · OCH₃ · Cl · N · 2HCl · 2H₂O

$C_{23}H_{30}ClN_3O.2HCl.2H_2O$ 508.91

1,4-Pentanediamine, N^4-(6-chloro-2-methoxy-9-acridinyl)-N^1,N^1-diethyl-, dihydrochloride, dihydrate.

6-Chloro-9-[[4-(diethylamino)-1-methylbutyl]amino]-2-methoxyacridine dihydrochloride dihydrate [*6151-30-0*].

Anhydrous 472.88 [*69-05-6*].

» Quinacrine Hydrochloride contains not less than 99.0 percent and not more than 101.0 percent of $C_{23}H_{30}ClN_3O.2HCl$, calculated on the anhydrous basis.

Packaging and storage—Preserve in tight, light-resistant containers.

Reference standard—*USP Quinacrine Hydrochloride Reference Standard*—Dry at 105° for 4 hours before using.

Identification—

A: The infrared absorption spectrum of a potassium bromide dispersion of it exhibits maxima only at the same wavelengths as that of a similar preparation of USP Quinacrine Hydrochloride RS.

B: The absorption spectra in the ultraviolet and visible ranges, measured in a 1 in 125,000 solution and a 1 in 25,000 solution, respectively, the solvent being dilute hydrochloric acid (1 in 1000), exhibit maxima and minima only at the same wavelengths as the corresponding spectra of similarly prepared solutions of USP Quinacrine Hydrochloride RS. The ratio A_{425}/A_{445} is between 1.02 and 1.08.

C: To 5 mL of a solution (1 in 40) add a slight excess of 6 *N* ammonium hydroxide: a yellow to orange, oily precipitate is formed, and it is soluble in ether. Filter: the filtrate, after acidification with nitric acid, responds to the tests for *Chloride* ⟨191⟩.

Water, *Method III* ⟨921⟩—Dry it at 105° for 4 hours: it loses between 6.0% and 8.0% of its weight.

Residue on ignition ⟨281⟩: not more than 0.1%.

Ordinary impurities ⟨466⟩—

Test solution: methanol.

Standard solution: methanol.

Eluant: a mixture of alcohol, glacial acetic acid, and water (5:3:2).

Visualization: 1.

Assay—Transfer about 500 mg of Quinacrine Hydrochloride, accurately weighed, to a beaker, add 60 mL of glacial acetic acid, and 10 mL of mercuric acetate TS, and titrate with 0.1 *N* perchloric acid VS, determining the end-point potentiometrically. Perform a blank determination, and make any necessary correction. Each mL of 0.1 *N* perchloric acid is equivalent to 23.64 mg of $C_{23}H_{30}ClN_3O.2HCl$.

Quinacrine Hydrochloride Tablets

» Quinacrine Hydrochloride Tablets contain not less than 93.0 percent and not more than 107.0 percent of the labeled amount of $C_{23}H_{30}ClN_3O.2HCl.2H_2O$.

Packaging and storage—Preserve in tight containers.

Reference standard—*USP Quinacrine Hydrochloride Reference Standard*—Dry at 105° for 4 hours before using.

Identification—

A: Portions of the solution of Tablets prepared for the *Assay* respond to *Identification test B* under *Quinacrine Hydrochloride.*

B: Apply 5 µL of the initial, undiluted solution of Tablets prepared for the *Assay* on a suitable thin-layer chromatographic plate, coated with a 0.25-mm layer of chromatographic silica gel. In a similar starting position on the same plate, apply 5 µL of a solution of USP Quinacrine Hydrochloride RS (1 in 500). Dry the spots in a current of warm air, and develop with a mixture of *n*-hexane, ether, and isopropylamine (50:48:2). Remove the plate, dry, and examine: only one spot of significant size appears in the test specimen channel, and it has the same yellow color, size, and R_f value (about 0.2) as the spot in the Reference Standard channel.

Dissolution ⟨711⟩—

Medium: water; 900 mL.

Apparatus 2: 50 rpm.

Time: 45 minutes.

Procedure—Determine the amount of $C_{23}H_{30}ClN_3O.2HCl.2H_2O$ dissolved from ultraviolet absorbances at the wavelength of maximum absorbance at about 425 nm of filtered portions of the solution under test, suitably diluted with water, in comparison with a Standard solution having a known concentration of USP Quinacrine Hydrochloride RS in the same medium.

Tolerances—Not less than 75% (*Q*) of the labeled amount of $C_{23}H_{30}ClN_3O.2HCl.2H_2O$ is dissolved in 45 minutes.

Uniformity of dosage units ⟨905⟩: meet the requirements.

Assay—Weigh and finely powder not less than 20 Quinacrine Hydrochloride Tablets. Weigh accurately a portion of the powder, equivalent to about 200 mg of quinacrine hydrochloride, and transfer to a 100-mL volumetric flask. Add about 70 mL of water, shake thoroughly, dilute with water to volume, mix, and filter, discarding the first 10 mL of the filtrate. Pipet 10 mL of the subsequent filtrate into a 500-mL volumetric flask, add 5 mL of dilute hydrochloric acid (1 in 10), add water to volume, and mix. Dissolve an accurately weighed quantity of USP Quinacrine Hydrochloride RS, in dilute hydrochloric acid (1 in 1000), and dilute quantitatively and stepwise with the same solvent to obtain a Standard solution having a known concentration of about 40 µg per mL. Concomitantly determine the absorbances of both solutions in 1-cm cells at the wavelength of maximum absorbance at about 425 nm, with a suitable spectrophotometer, using dilute hydrochloric acid (1 in 1000) as the blank. Calculate the quantity, in mg, of $C_{23}H_{30}ClN_3O.2HCl.2H_2O$ in the portion of Tablets taken by the formula:

$$5(1.076)C(A_U/A_S),$$

in which the factor 1.076 is the ratio of the respective molecular weights of the dihydrate and anhydrous forms of quinacrine dihydrochloride, *C* is the concentration, in µg per mL, of the dried USP Quinacrine Hydrochloride RS in the Standard solution, and A_U and A_S are the absorbances of the solution from the Tablets and the Standard solution, respectively.

Quinestrol

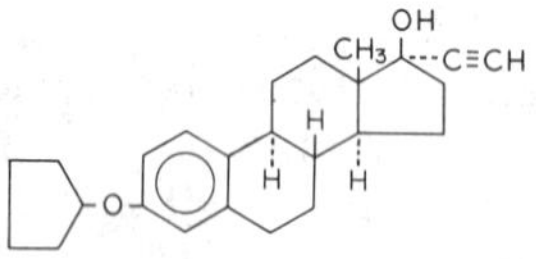

$C_{25}H_{32}O_2$ 364.53

19-Norpregna-1,3,5(10)-trien-20-yn-17-ol, 3-(cyclopentyloxy)-, (17α)-.

3-(Cyclopentyloxy)-19-nor-17α-pregna-1,3,5(10)-trien-20-yn-17-ol [*152-43-2*].

» Quinestrol contains not less than 98.0 percent and not more than 102.0 percent of $C_{25}H_{32}O_2$, calculated on the dried basis.

Packaging and storage—Preserve in well-closed containers.

Reference standard—*USP Quinestrol Reference Standard*—Dry in vacuum over silica gel for 4 hours before using.

Identification—

A: The infrared absorption spectrum of a potassium bromide dispersion of it exhibits maxima only at the same wavelengths as that of a similar preparation of USP Quinestrol RS.

B: The ultraviolet absorption spectrum of a 1 in 10,000 solution in alcohol, measured between 220 nm and 300 nm, exhibits maxima and minima at the same wavelengths as that of a similar solution of USP Quinestrol RS, concomitantly measured.

Melting range ⟨741⟩: between 107° and 110.5°.

Specific rotation ⟨781⟩: between +3° and +5°, calculated on the dried basis, determined in a 1-dm polarimeter tube of a solution in dry, peroxide-free dioxane containing 50 mg per mL.

Loss on drying ⟨731⟩—Dry about 500 mg in vacuum over silica gel for 4 hours: it loses not more than 0.5% of its weight.

Ordinary impurities ⟨466⟩—

Test solution: chloroform.

Standard solution: chloroform.

Eluant: a mixture of cyclohexane and ethyl acetate (2:1).

Visualization: 5; followed by visualization under long-wavelength ultraviolet light.

Assay—[NOTE—In this procedure, use scrupulously dry glassware and separators fitted with solvent-resistant stopcocks, and use isooctane that gives no color when shaken with an equal volume of sulfuric acid.]

Chromogenic reagent—Cautiously add sulfuric acid to 30 mL of chilled anhydrous methanol in a 100-mL volumetric flask, in small increments and with mixing. Adjust to room temperature, dilute with sulfuric acid to volume, and mix. Prepare this solution fresh.

Standard preparation—Dissolve about 20 mg of USP Quinestrol RS, accurately weighed, in anhydrous methanol contained in a 50-mL volumetric flask, add anhydrous methanol to volume, and mix. Pipet 5 mL of this solution into a 100-mL volumetric flask, add isooctane to volume, and mix. Pipet 5 mL of the resulting solution into a 50-mL volumetric flask, add isooctane to volume, and mix.

Assay preparation—Using about 20 mg of Quinestrol, accurately weighed, proceed as directed for *Standard preparation.*

Procedure—Pipet 10 mL of the *Standard preparation* and 10 mL of the *Assay preparation,* respectively, into separators each containing 40 mL of isooctane, and treat each solution as follows: Add 5.0 mL of *Chromogenic reagent,* shake the mixture for 2 minutes, allow the phases to separate, and discard 5 drops of the fluorescent, pink lower phase through the stopcock. Deliver as much as possible of each of the lower phases into separate test tubes, and pipet 4 mL of the colored solution from the *Standard preparation* and the *Assay preparation,* respectively, into glass-stoppered centrifuge tubes. To each tube add 0.4 mL, accurately measured, of anhydrous methanol, insert the stopper, and shake vigorously for a few seconds. Centrifuge, if necessary, to dispel air bubbles. Concomitantly determine the absorbances of both solutions in 1-cm cells at the wavelength of maximum absorbance at about 538 nm, with a suitable spectrophotometer, against a reagent blank prepared by adding 5.0 mL of *Chromogenic reagent* to 50 mL of isooctane and proceeding as directed for *Procedure,* beginning with "allow the phases to separate." Calculate the quantity, in mg, of $C_{25}H_{32}O_2$ in the portion of Quinestrol taken by the formula:

$$10C(A_U / A_S),$$

in which C is the concentration, in µg per mL, of USP Quinestrol RS in the *Standard preparation,* and A_U and A_S are the absorbances of the solutions from the *Assay preparation* and the *Standard preparation,* respectively.

Quinestrol Tablets

» Quinestrol Tablets contain not less than 90.0 percent and not more than 110.0 percent of the labeled amount of $C_{25}H_{32}O_2$.

Packaging and storage—Preserve in well-closed containers.

Reference standard—*USP Quinestrol Reference Standard*—Dry in vacuum over silica gel for 4 hours before using.

Identification—Add 10 mL of chloroform to a quantity of finely powdered Tablet, equivalent to about 100 µg of quinestrol, in a glass-stoppered, 15-mL centrifuge tube. Insert the stopper in the tube, and shake with the aid of a vibrating mixing device for 2 minutes. Centrifuge the mixture at 3000 rpm for 5 minutes, and decant the clear supernatant solution into a second tube. Concentrate the extract to a volume of about 1 mL by evaporation in a warm water bath with the aid of a stream of nitrogen. Apply 10 µL of this solution and 10 µL of a solution of USP Quinestrol RS in chloroform, containing about 100 µg per mL, to a point about 2.5 cm above one edge of a thin-layer chromatographic plate (see *Chromatography* ⟨621⟩) coated with a 0.25-mm layer of chromatographic silica gel and previously activated by heating at 105° for 1 hour. Develop the plate in a suitable chamber containing a mixture of 2 volumes of cyclohexane and 1 volume of ethyl acetate until the solvent has moved about 15 cm. Remove the plate, and dry in an oven at 105° for 1 minute. Spray the plate lightly with sulfuric acid–methanol (prepared as directed for *Chromogenic reagent* in the test for *Content uniformity*): the spot in the chromatogram from the solution under test corresponds in R_f value and pink color to that obtained from the Standard solution.

Disintegration ⟨701⟩: 30 minutes.

Uniformity of dosage units ⟨905⟩: meet the requirements.

Procedure for content uniformity—

Chromogenic reagent—Add, in small increments and with frequent agitation, sulfuric acid to volume in a 100-mL volumetric flask containing 30 mL of chilled anhydrous methanol. Prepare this solution fresh.

Standard preparation—Using a suitable quantity of USP Quinestrol RS, accurately weighed, prepare a solution in methanol having a known concentration of about 200 µg per mL.

Test preparation—Transfer 1 Tablet to a 250-mL separator containing 20 mL of water and 1 mL of methanol, and shake vigorously to disintegrate and dissolve the tablet.

Procedure—Pipet 1 mL of the *Standard preparation* into a separator containing 20 mL of water. Add 10 drops of hydrochloric acid and 100 mL of isooctane to the separators containing the *Test preparation* and the *Standard preparation.* Shake vigorously for 2 minutes. When the layers have separated completely, draw off and discard the lower, aqueous phases. Shake each isooctane layer with 10 mL of 1 *N* sodium hydroxide for 1 minute, and discard the lower, aqueous phases. Filter the isooctane solutions through solvent-wet cotton into separate, dry 200-mL volumetric flasks. Use 40 mL to 50 mL of fresh isooctane to effect the transfers, dilute with isooctane to volume, and mix. Separately pipet 20 mL of the isooctane extract from the *Standard preparation* and a portion of the isooctane extract from the *Test preparation,* equivalent to about 20 µg of quinestrol, into dry 125-mL separators. Add isooctane, if necessary, to make the volumes in the separators equal to 40.0 mL. Then add 5.0 mL of *Chromogenic reagent* to each separator, and shake vigorously for 2 minutes. When the layers have separated completely, drain the lower, pink phases into separate, dry centrifuge tubes. Transfer 4.0 mL of these solutions into separate, dry, glass-stoppered centrifuge tubes, add 0.4 mL of anhydrous methanol to each, mix, and centrifuge. Concomitantly determine the absorbances of the solutions from the *Test preparation* and the *Standard preparation,* against water, in 1-cm cells, with a suitable spectrophotometer, at the wavelength of maximum absorbance at about 540 nm. Calculate the quantity, in µg, of $C_{25}H_{32}O_2$ in the Tablet by the formula:

$$(TC / D)(A_U / A_S),$$

in which T is the labeled quantity, in µg, of quinestrol in the Tablet, C is the concentration, in µg per mL, of USP Quinestrol RS in the *Standard preparation, D* is the concentration, in µg per mL, of quinestrol in the solution from the Tablet, based upon the labeled quantity per Tablet and the extent of dilution, and A_U and A_S are the absorbances of the solutions from the *Test preparation* and the *Standard preparation,* respectively.

Assay—

Mobile phase—Prepare a suitable, filtered and degassed solution of spectrophotometric acetonitrile and water (4:1) such that the retention time for quinestrol is approximately 8.5 minutes.

Standard preparation—Accurately weigh about 10 mg of USP Quinestrol RS, transfer to a 50-mL volumetric flask, dissolve in acetonitrile, dilute with acetonitrile to volume, and mix. Dilute 5.0 mL of this solution with *Mobile phase* to volume in a 50-mL volumetric flask to obtain a solution having a known concentration of about 20 µg per mL.

Assay preparation—Weigh and pulverize not less than 20 Tablets. Weigh a portion of the powder, equivalent to about 100 µg of quinestrol, transfer to a 5-mL volumetric flask, suspend in 1 mL of water, and heat at the steam-bath temperature for about 5 minutes. Dilute with acetonitrile to volume, and mix on a vibrating device for about 2 minutes.

Chromatographic system (see *Chromatography* ⟨621⟩)—The liquid chromatograph is equipped with a 281-nm detector and a 4.6-mm × 25-cm column that contains packing L1. The flow rate is about 2 mL per minute. Chromatograph six replicate injections of the *Standard preparation,* and record the peak responses as directed under *Procedure:* the relative standard deviation is not more than 1.0%.

Procedure—Separately inject equal volumes (about 50 µL) of the *Standard preparation* and the *Assay preparation* into the chromatograph by means of a suitable microsyringe or sampling valve, record the chromatograms for 15 minutes, and measure the responses for the major peaks. Calculate the quantity, in µg, of $C_{25}H_{32}O_2$ in the portion of Tablets taken by the formula:

$$5C(r_U / r_S),$$

in which C is the concentration, in µg per mL, of USP Quinestrol RS in the *Standard preparation*, and r_U and r_S are the peak responses in the chromatograms from the *Assay preparation* and the *Standard preparation*, respectively.

Quinethazone

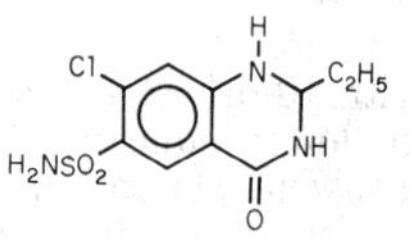

$C_{10}H_{12}ClN_3O_3S$ 289.74

6-Quinazolinesulfonamide, 7-chloro-2-ethyl-1,2,3,4-tetrahydro-4-oxo-.

7-Chloro-2-ethyl-1,2,3,4-tetrahydro-4-oxo-6-quinazolinesulfonamide [*73-49-4*].

» Quinethazone contains not less than 98.0 percent and not more than 102.0 percent of $C_{10}H_{12}ClN_3O_3S$, calculated on the dried basis.

Packaging and storage—Preserve in well-closed containers.

Reference standard—*USP Quinethazone Reference Standard*—Dry in vacuum at 60° for 3 hours before using.

Identification—

A: The infrared absorption spectrum of a potassium bromide dispersion of it, previously dried, exhibits maxima only at the same wavelengths as that of a similar preparation of USP Quinethazone RS.

B: The ultraviolet absorption spectrum of a 1 in 100,000 solution in alcohol exhibits maxima and minima at the same wavelengths as that of a similar solution of USP Quinethazone RS, concomitantly measured, and the respective absorptivities, calculated on the dried basis, at the wavelength of maximum absorbance at about 345 nm do not differ by more than 3.0%.

Loss on drying ⟨731⟩—Dry it in vacuum at 60° for 3 hours: it loses not more than 1.0% of its weight.

Residue on ignition ⟨281⟩: not more than 0.1%.

Selenium ⟨291⟩: 0.003%.

Heavy metals, *Method II* ⟨231⟩: 0.002%. [NOTE—The treatment with nitric and sulfuric acids may be repeated, if necessary, to remove unburned carbon.]

Assay—

0.1 N Tributylethylammonium hydroxide—Prepare an approximately 0.1 *N* solution of tributylethylammonium hydroxide, using as the solvent a mixture of 10 volumes of benzene and 1 volume of methanol, and standardize the solution against about 400 mg of primary standard benzoic acid, accurately weighed, that has been dissolved in 80 mL of dimethylformamide, using 3 drops of a 1 in 100 solution of thymol blue in dimethylformamide, and titrating to a blue end-point. Perform a blank determination, and make any necessary correction. Each mL of 0.1 *N* tributylethylammonium hydroxide is equivalent to 12.21 mg of benzoic acid.

Procedure—Transfer about 500 mg of Quinethazone, accurately weighed, to a 250-mL conical flask, and dissolve in 100 mL of pyridine. Titrate with *0.1 N Tributylethylammonium hydroxide*, using a suitable titrimeter equipped with a glass electrode and a methanolic calomel electrode of which the outer jacket contains a saturated solution of potassium chloride in methanol, and determining the end-point potentiometrically. Perform a blank determination, and make any necessary correction. Each mL of *0.1 N Tributylethylammonium hydroxide* is equivalent to 28.97 mg of $C_{10}H_{12}ClN_3O_3S$.

Quinethazone Tablets

» Quinethazone Tablets contain not less than 92.5 percent and not more than 107.5 percent of the labeled amount of $C_{10}H_{12}ClN_3O_3S$.

Packaging and storage—Preserve in tight containers.

Reference standard—*USP Quinethazone Reference Standard*—Dry in vacuum at 60° for 3 hours before using.

Identification—Place a quantity of finely powdered Tablets, equivalent to about 150 mg of quinethazone, in a glass-stoppered centrifuge tube, add 10.0 mL of pyridine, insert the stopper, shake by mechanical means for 15 minutes, and centrifuge. Place 1 mL of the supernatant liquid on a watch glass, and evaporate on a steam bath to dryness: the quinethazone so obtained responds to *Identification test A* under *Quinethazone*.

Disintegration ⟨701⟩: 30 minutes.

Uniformity of dosage units ⟨905⟩: meet the requirements.

Procedure for content uniformity—Place 1 Tablet in a glass mortar, moisten it with several drops of alcohol, and triturate to obtain a smooth paste. Transfer the mixture to a 200-mL volumetric flask with the aid of alcohol, and add alcohol to obtain a volume of about 120 mL. Place on a steam bath, and boil with occasional swirling until solution of quinethazone is complete (about 10 minutes). Cool, dilute with alcohol to volume, and mix. Centrifuge a portion until clear, transfer an accurately measured volume of the supernatant liquid, equivalent to about 2.5 mg of quinethazone, to a 50-mL volumetric flask, dilute with alcohol to volume, and mix. Concomitantly determine the absorbances of this solution and a solution of USP Quinethazone RS, in the same medium having a known concentration of about 50 µg per mL, in 1-cm cells, at the wavelength of maximum absorbance at about 345 nm, with a suitable spectrophotometer, using alcohol as the blank. Calculate the quantity, in mg, of $C_{10}H_{12}ClN_3O_3S$ in the Tablet by the formula:

$$(TC/D)(A_U / A_S),$$

in which T is the labeled quantity, in mg, of quinethazone in the Tablet, C is the concentration, in µg per mL, of USP Quinethazone RS in the Standard solution, D is the concentration, in µg per mL, of the test solution, based upon the labeled quantity per Tablet and the extent of dilution, and A_U and A_S are the absorbances of the solution from the Tablet and the Standard solution, respectively.

Assay—Weigh and finely powder not less than 20 Quinethazone Tablets. Transfer an accurately weighed portion of the powder, equivalent to about 150 mg of quinethazone, to a glass-stoppered centrifuge tube, add 10.0 mL of pyridine, insert the stopper, shake by mechanical means for 15 minutes, centrifuge, and use the clear, supernatant liquid. Concomitantly determine the absorbances of this solution and a solution of USP Quinethazone RS in the same medium having a known concentration of about 15 mg per mL, in 0.1-mm cells at the wavelength of maximum absorbance at about 7.45 µm, with a suitable infrared spectrophotometer, using pyridine as the blank. Calculate the quantity, in mg, of $C_{10}H_{12}ClN_3O_3S$ in the portion of Tablets taken by the formula:

$$10C(A_U / A_S),$$

in which C is the concentration, in mg per mL, of USP Quinethazone RS in the Standard solution, and A_U and A_S are the absorbances of the solution from the Tablets and the Standard solution, respectively.

Quinidine Gluconate

$C_{20}H_{24}N_2O_2 . C_6H_{12}O_7$ 520.58
Cinchonan-9-ol, 6′-methoxy-, (9*S*)-, mono-D-gluconate (salt).
Quinidine mono-D-gluconate (salt) [*7054-25-3*].

» Quinidine Gluconate is the gluconate of an alkaloid that may be obtained from various species of *Cinchona* and their hybrids, or from *Remijia pedunculata* Flückiger (Fam. Rubiaceae), or prepared from quinine. Quinidine Gluconate contains not less than 99.0 percent and not more than 100.5 percent of total alkaloid salt, calculated as $C_{20}H_{24}N_2O_2 . C_6H_{12}O_7$, on the dried basis.

Packaging and storage—Preserve in well-closed, light-resistant containers.

Reference standards—*USP Quinidine Gluconate Reference Standard*—Dry at 105° for 1 hour before using. *USP Quininone Reference Standard*—Do not dry. Keep container tightly closed and protected from light.

Identification—

A: A 1 in 2000 solution in dilute sulfuric acid (1 in 350) exhibits a vivid blue fluorescence. On the addition of a few drops of hydrochloric acid, the fluorescence disappears.

B: In the test for *Chromatographic purity*, the R_f value of the principal spot obtained from the *Test preparation* corresponds to that from the *Standard preparation.*

C: A solution (1 in 50) is dextrorotatory.

D: Dissolve 700 mg in 5 mL of water, with the aid of heat, and add 1 mL of glacial acetic acid and 200 mg of phenylhydrazine hydrochloride. Heat in a water bath for 15 minutes, cool, and scratch the inner surface of the tube with a glass rod: orange crystals are formed.

Loss on drying ⟨731⟩—Dry it at 105° for 1 hour: it loses not more than 0.5% of its weight.

Residue on ignition ⟨281⟩: not more than 0.15%.

Heavy metals, *Method II* ⟨231⟩: 0.001%.

Chromatographic purity—

Standard preparation—Prepare a solution of USP Quinidine Gluconate RS in diluted alcohol to contain 6 mg per mL.

Diluted standard preparation—Dilute a portion of the *Standard preparation* with diluted alcohol to a concentration of 0.06 mg per mL.

Related substances preparation—Prepare a solution in diluted alcohol containing in each mL 0.04 mg each of USP Quininone RS (corresponding to 0.06 mg of the gluconate), and 0.08 mg of cinchonine (corresponding to 0.12 mg of the gluconate).

Test preparation—Prepare a solution of Quinidine Gluconate in diluted alcohol to contain 6 mg per mL.

Procedure—On a suitable thin-layer chromatographic plate (see *Chromatography* ⟨621⟩), coated with a 0.25-mm layer of chromatographic silica gel, apply 10-µL portions of the *Test preparation*, the *Standard preparation*, the *Diluted standard preparation*, and the *Related substances preparation.* Allow to dry, and develop the chromatogram in a solvent system consisting of a mixture of chloroform, acetone, and diethylamine (5:4:1), the solvent chamber being used without previous equilibration. When the solvent front has moved about 15 cm, remove the plate from the chamber, mark the solvent front, and allow the solvent to evaporate. Spray the chromatogram with glacial acetic acid. Locate the spots on the plate by examination under long-wavelength ultraviolet light. Any spot produced by the *Test preparation* at the R_f value of a spot produced by the *Related substances preparation* is not greater in size or intensity than that corresponding spot. Apart from these spots and from the spots appearing at the R_f value of Quinidine Gluconate and dihydroquinidine gluconate (the two spots most evident from the *Standard preparation*), any additional fluorescent spot is not greater in size or intensity than the principal spot of the *Diluted standard preparation.* Spray the plate with potassium iodoplatinate TS. Any spot produced by the *Test preparation* is not greater in size or intensity than a corresponding spot from the *Related substances preparation.*

Dihydroquinidine gluconate—

Methanesulfonic acid solution—Add 35.0 mL of methanesulfonic acid to 20.0 mL of glacial acetic acid, dilute with water to 500 mL, and mix.

Diethylamine solution—Dissolve 10.0 mL of diethylamine in water to obtain 100 mL of solution.

Mobile phase—Prepare a suitable filtered and degassed mixture of water, acetonitrile, *Methanesulfonic acid solution*, and *Diethylamine solution* (860:100:20:20). Adjust with *Diethylamine solution* to a pH of 2.6 if found to be lower.

System suitability preparation—Transfer about 10 mg each of quinidine gluconate and dihydroquinidine sulfate to a 50-mL volumetric flask. Dissolve in about 5 mL of methanol, dilute with *Mobile phase* to volume, and mix.

System suitability test—Chromatograph injections of the *System suitability preparation* as directed under *Procedure.* Typical relative retention times for quinidine, and dihydroquinidine are 1 and 1.5, respectively. The resolution, *R*, between the quinidine and dihydroquinidine peaks is not less than 1.2. The relative standard deviation for the peak response of quinidine is not more than 2.0%.

Test preparation—Transfer about 26 mg of Quinidine Gluconate to a 100-mL volumetric flask, dissolve in *Mobile phase*, dilute with *Mobile phase* to volume, and mix.

Procedure (see *Chromatography* ⟨621⟩)—Inject about 50 µL of the *Test preparation* into a chromatograph equipped with a 235-nm detector and a 3.9-mm × 30-cm column that contains packing L1. The response of the dihydroquinidine peak is not greater than 0.25 that of the quinidine peak (20.0%).

Assay—Dissolve about 150 mg of Quinidine Gluconate, accurately weighed, in 10 mL of glacial acetic acid, heating gently if necessary. Cool the solution, add 20 mL of acetic anhydride and 4 drops of *p*-naphtholbenzein TS, and titrate with 0.1 *N* perchloric acid VS from a 10-mL microburet to a green endpoint. Perform a blank determination, and make any necessary correction. Each mL of 0.1 *N* perchloric acid is equivalent to 26.03 mg of total alkaloid salt, calculated as $C_{20}H_{24}N_2O_2 . C_6H_{12}O_7$.

Quinidine Gluconate Injection

» Quinidine Gluconate Injection is a sterile solution of Quinidine Gluconate in Water for Injection. It contains, in each mL, amounts of quinidine gluconate and dihydroquinidine gluconate totaling not less than 76 mg and not more than 84 mg of quinidine gluconate, calculated as $C_{20}H_{24}N_2O_2 . C_6H_{12}O_7$.

Packaging and storage—Preserve in single-dose or in multiple-dose containers, preferably of Type I glass.

Reference standards—*USP Quinidine Gluconate Reference Standard*—Dry at 105° for 1 hour before using. *USP Quininone Reference Standard*—Do not dry. Keep container tightly closed and protected from light.

Identification—

A: A 1 in 150 solution of Injection in dilute sulfuric acid (1 in 350) exhibits a vivid blue fluorescence. On the addition of a few drops of hydrochloric acid, the fluorescence disappears.

B: A solution of Injection (1 in 4) is dextrorotatory.

C: In the test for *Chromatographic purity*, the R_f value of the principal spot obtained from the *Test preparation* corresponds to that from the *Standard preparation.*

Chromatographic purity—Mix an accurately measured volume of Injection, equivalent to 80 mg of quinidine gluconate, with 25 mL of water, add 2 drops of 2 *N* sulfuric acid, and extract with

50 mL of ether, discarding the ether extract. To the aqueous solution add 2 mL of 1 *N* sodium hydroxide, extract with 50 mL of ether, wash the extract with 25 mL of water, and discard the aqueous solutions. Evaporate the ether extract just to dryness, and dissolve the residue in 10 mL of alcohol. Using this as the test solution, proceed as directed in the test for *Chromatographic purity* under *Quinidine Gluconate*, beginning with "On a suitable thin-layer chromatographic plate."

Other requirements—It meets the requirements under *Injections* ⟨1⟩.

Assay—

Methanesulfonic acid solution, Diethylamine solution, Mobile phase, System suitability preparation, and *System suitability test*—Proceed as directed in the test for *Dihydroquinidine* gluconate under *Quinidine Gluconate.*

Standard preparation—Transfer about 26 mg of USP Quinidine Gluconate RS, accurately weighed, to a 100-mL volumetric flask, dissolve in *Mobile phase*, dilute with *Mobile phase* to volume, and mix.

Assay preparation—Transfer an accurately measured volume of Quinidine Gluconate Injection, equivalent to about 400 mg of quinidine gluconate, to a 50-mL volumetric flask, add methanol to volume, and mix. Transfer 3.0 mL of this solution to a 100-mL volumetric flask, dilute with *Mobile phase* to volume, and mix.

Procedure (see *Chromatography* ⟨621⟩)—Inject equal volumes (about 50 µL) of the *Standard preparation* and the *Assay preparation* into a chromatograph equipped with a 235-nm detector and a 3.9-mm × 30-cm column that contains packing L1. Calculate the quantity, in mg, of the *Sum of quinidine gluconate and dihydroquinidine gluconate* in each mL of the Injection taken by the formula:

$$(5000/3)(C/V)(r_{b,U} + r_{d,U})/(r_{b,S} + r_{d,S}),$$

in which *C* is the concentration, in mg per mL, of USP Quinidine Gluconate RS in the *Standard preparation*, *V* is the volume, in mL, of Injection taken, $r_{b,U}$ and $r_{b,S}$ are the peak responses of quinidine obtained from the *Assay preparation* and the *Standard preparation*, respectively, and $r_{d,U}$ and $r_{d,S}$ are the peak responses of dihydroquinidine obtained from the *Assay preparation* and the *Standard preparation*, respectively.

Quinidine Sulfate

$(C_{20}H_{24}N_2O_2)_2 . H_2SO_4 . 2H_2O$ 782.95

Cinchonan-9-ol, 6′-methoxy-, (9*S*)-, sulfate (2:1) (salt), dihydrate.

Quinidine sulfate (2:1) (salt) dihydrate [*6591-63-5*].

Anhydrous 746.92 [*50-54-4*].

» Quinidine Sulfate is the sulfate of an alkaloid obtained from various species of *Cinchona* and their hybrids and from *Remijia pedunculata* Flückiger (Fam. Rubiaceae), or prepared from quinine. It contains not less than 99.0 percent and not more than 101.0 percent of total alkaloid salt, calculated as $(C_{20}H_{24}N_2O_2)_2 . H_2SO_4$, on the anhydrous basis.

Packaging and storage—Preserve in well-closed, light-resistant containers.

Reference standards—*USP Quinidine Sulfate Reference Standard*—Do not dry. *USP Quininone Reference Standard*—Do not dry. Keep container tightly closed and protected from light.

Identification—

A: A 1 in 2000 solution in dilute sulfuric acid (1 in 350) exhibits a vivid blue fluorescence. On the addition of a few drops of hydrochloric acid, the fluorescence disappears.

B: In the test for *Chromatographic purity*, the R_f value of the principal spot obtained from the *Test preparation* corresponds to that from the *Standard preparation.*

C: A 1 in 50 solution made with the aid of a few drops of hydrochloric acid responds to the tests for *Sulfate* ⟨191⟩.

Specific rotation ⟨781⟩: between +275° and +288°, calculated on the anhydrous basis, determined in a solution in dilute hydrochloric acid (1 in 100) containing 200 mg in each 10 mL.

Water, *Method I* ⟨921⟩: between 4.0% and 5.5%.

Residue on ignition ⟨281⟩: not more than 0.1%.

Heavy metals, *Method II* ⟨231⟩: 0.001%.

Chloroform-alcohol-insoluble substances—Warm 2 g with 15 mL of a mixture of chloroform and dehydrated alcohol (2:1) at about 50° for 10 minutes. Filter through a tared, sintered-glass filter, using gentle suction. Wash the filter with five 10-mL portions of the chloroform-alcohol mixture, dry at 105° for 1 hour, and weigh: the weight of the residue does not exceed 2 mg (0.1%).

Chromatographic purity—

Standard preparation—Prepare a solution of USP Quinidine Sulfate RS in diluted alcohol to contain 6 mg per mL.

Diluted standard preparation—Dilute a portion of the *Standard preparation* with diluted alcohol to a concentration of 0.06 mg per mL.

Related substances preparation—Prepare a solution in diluted alcohol containing in each mL 0.05 mg each of USP Quininone RS (corresponding to 0.06 mg of the sulfate), and 0.10 mg of cinchonine (corresponding to 0.12 mg of the sulfate).

Test preparation—Prepare a solution of Quinidine Sulfate in diluted alcohol to contain 6 mg per mL.

Procedure—On a suitable thin-layer chromatographic plate (see *Chromatography* ⟨621⟩), coated with a 0.25-mm layer of chromatographic silica gel, apply 10-µL portions of the *Test preparation*, the *Standard preparation*, the *Diluted standard preparation*, and the *Related substances preparation.* Allow to dry, and develop the chromatogram in a solvent system consisting of a mixture of chloroform, acetone, and diethylamine (5:4:1), the solvent chamber being used without previous equilibration. When the solvent front has moved about 15 cm, remove the plate from the chamber, mark the solvent front, and allow the solvent to evaporate. Spray the chromatogram with glacial acetic acid. Locate the spots on the plate by examination under long-wavelength ultraviolet light. Any spot produced by the *Test preparation* at the R_f value of a spot produced by the *Related substances preparation* is not greater in size or intensity than that corresponding spot. Apart from these spots and from the spots appearing at the R_f value of Quinidine Sulfate and dihydroquinidine sulfate (the two spots most evident from the *Standard preparation*), any additional fluorescent spot is not greater in size or intensity than the principal spot of the *Diluted standard preparation.* Spray the plate with potassium iodoplatinate TS. Any spot produced by the *Test preparation* is not greater in size or intensity than a corresponding spot from the *Related substances preparation.*

Dihydroquinidine sulfate—

Methanesulfonic acid solution—Add 35.0 mL of methanesulfonic acid to 20.0 mL of glacial acetic acid, dilute with water to 500 mL, and mix.

Diethylamine solution—Dissolve 10.0 mL of diethylamine in water to obtain 100 mL of solution.

Mobile phase—Prepare a suitable filtered and degassed mixture of water, acetonitrile, *Methanesulfonic acid solution*, and *Diethylamine solution* (860:100:20:20). Adjust with *Diethylamine solution* to a pH of 2.6 if found to be lower.

System suitability preparation—Transfer about 10 mg each of quinidine sulfate and dihydroquinidine sulfate to a 50-mL volumetric flask. Dissolve in about 5 mL of methanol, dilute with *Mobile phase* to volume, and mix.

System suitability test—Chromatograph injections of the *System suitability preparation* as directed under *Procedure*: typical relative retention times for quinidine and dihydroquinidine are 1 and 1.5, respectively. The resolution between the quinidine and dihydroquinidine peaks is not less than 1.2. The relative standard deviation for the peak response of quinidine is not more than 2.0%.

Test preparation—Transfer about 20 mg of Quinidine Sulfate to a 100-mL volumetric flask, dissolve in *Mobile phase*, dilute with *Mobile phase* to volume, and mix.

Procedure (see *Chromatography* ⟨621⟩)—Inject about 50 µL of the *Test preparation* into a chromatograph equipped with a 235-nm detector and a 3.9-mm × 30-cm column that contains packing L1. The response of the dihydroquinidine peak is not greater than 0.25 that of the quinidine peak (20.0%).

Assay—Dissolve about 200 mg of Quinidine Sulfate, accurately weighed, in 20 mL of acetic anhydride, and proceed as directed in the *Assay* under *Quinidine Gluconate*, beginning with "[add] 4 drops of *p*-naphtholbenzein TS." Each mL of 0.1 *N* perchloric acid is equivalent to 24.90 mg of total alkaloid salt, calculated as $(C_{20}H_{24}N_2O_2)_2 \cdot H_2SO_4$.

Quinidine Sulfate Capsules

» Quinidine Sulfate Capsules contain amounts of quinidine sulfate and dihydroquinidine sulfate totaling not less than 90.0 percent and not more than 110.0 percent of the labeled amount of quinidine sulfate, calculated as $(C_{20}H_{24}N_2O_2)_2 \cdot H_2SO_4 \cdot 2H_2O$.

Packaging and storage—Preserve in tight, light-resistant containers.

Reference standards—*USP Quinidine Sulfate Reference Standard*—Do not dry. *USP Quininone Reference Standard*—Do not dry. Keep container tightly closed and protected from light.

Identification—

A: Shake a quantity of the contents of Capsules, equivalent to about 100 mg of quinidine sulfate, with 10 mL of dilute sulfuric acid (1 in 350), and filter: an appropriate dilution of the filtrate exhibits a vivid blue fluorescence. On the addition of a few drops of hydrochloric acid the fluorescence disappears.

B: In the test for *Chromatographic purity*, the R_f value of the principal spot obtained from the *Test preparation* corresponds to that from the *Standard preparation*.

C: Shake a quantity of the contents of Capsules, equivalent to about 100 mg of quinidine sulfate, with 10 mL of dilute hydrochloric acid (1 in 100), and filter: the filtrate responds to the tests for *Sulfate* ⟨191⟩.

Dissolution ⟨711⟩—

Medium: 0.1 *N* hydrochloric acid; 900 mL.

Apparatus 1: 100 rpm.

Time: 30 minutes.

Procedure—Measure the amount in solution in filtered portions of the *Dissolution Medium*, suitably diluted, in 1-cm cells at the wavelength of maximum absorbance at about 248 nm, with a suitable spectrophotometer, in comparison with the Standard solution of known concentration of USP Quinidine Sulfate RS.

Tolerances—Not less than 85% (*Q*) of the labeled amount of $(C_{20}H_{24}N_2O_2)_2 \cdot H_2SO_4 \cdot 2H_2O$ is dissolved in 30 minutes.

Uniformity of dosage units ⟨905⟩: meet the requirements.

Procedure for content uniformity—Transfer the contents of 1 Capsule to a 250-mL volumetric flask, add about 175 mL of dilute hydrochloric acid (1 in 100), and shake by mechanical means for 30 minutes. Add dilute hydrochloric acid (1 in 100) to volume, and mix. Filter a portion of the mixture, discarding the first 20 mL of the filtrate. Concomitantly determine the absorbances of this solution, quantitatively diluted, if necessary, and a Standard solution of USP Quinidine Sulfate RS in dilute hydrochloric acid (1 in 100) having a known concentration of about 400 µg per mL, in 1-cm cells, at the wavelength of maximum absorbance at about 345 nm, with a suitable spectrophotometer, using water as the blank. Calculate the quantity, in mg, of active ingredients, calculated as quinidine sulfate $[(C_{20}H_{24}N_2O_2)_2 \cdot H_2SO_4 \cdot 2H_2O]$, in the Capsule by the formula:

$$(TC/D)(A_U/A_S),$$

in which T is the labeled quantity, in mg, of quinidine sulfate in the Capsule, D is the concentration, in µg per mL, of quinidine sulfate in the solution from the Capsule, based on the labeled quantity per Capsule and the extent of dilution, C is the concentration, in µg per mL, of USP Quinidine Sulfate RS in the Standard solution, and A_U and A_S are the absorbances of the solution from the Capsule and the Standard solution, respectively.

Chromatographic purity—Shake a quantity of the contents of Capsules, equivalent to about 150 mg of quinidine sulfate, with 25 mL of diluted alcohol for 10 minutes, and filter. Using this as the test solution, proceed as directed in the test for *Chromatographic purity* under *Quinidine Sulfate*.

Assay—

Methanesulfonic acid solution, Diethylamine solution, Mobile phase, System suitability preparation, and *System suitability test*—Proceed as directed for *Dihydroquinidine sulfate* under *Quinidine Sulfate*.

Standard preparation—Transfer about 20 mg of USP Quinidine Sulfate RS, accurately weighed, to a 100-mL volumetric flask, dissolve in *Mobile phase*, dilute with *Mobile phase* to volume, and mix.

Assay preparation—Transfer the contents of not less than 20 Quinidine Sulfate Capsules to a container, and mix. Transfer an accurately weighed portion of the powder, equivalent to about 160 mg of quinidine sulfate, to a 100-mL volumetric flask, add 80 mL of methanol, and shake the flask by mechanical means for 30 minutes. Dilute with methanol to volume, mix, and filter, discarding the first 10 mL of the filtrate. Transfer 3.0 mL of the filtrate to a 25-mL volumetric flask, dilute with *Mobile phase* to volume, and mix.

Procedure (see *Chromatography* ⟨621⟩)—Inject equal volumes (about 50 µL) of the *Standard* and *Assay preparations* into a chromatograph equipped with a 235-nm detector and a 3.9-mm × 30-cm column that contains packing L1. Calculate the quantity, in mg, of the sum of quinidine sulfate and dihydroquinidine sulfate in the portion of Capsules taken by the formula:

$$(2500/3)C(r_{b,U} + r_{d,U})/(r_{b,S} + r_{d,S}),$$

in which C is the concentration, in mg per mL, of USP Quinidine Sulfate RS in the *Standard preparation*, $r_{b,U}$ and $r_{b,S}$ are the peak responses of quinidine obtained from the *Assay preparation* and the *Standard preparation*, respectively, and $r_{d,U}$ and $r_{d,S}$ are the peak responses of dihydroquinidine obtained from the *Assay preparation* and the *Standard preparation*, respectively.

Quinidine Sulfate Tablets

» Quinidine Sulfate Tablets contain amounts of quinidine sulfate and dihydroquinidine sulfate totaling not less than 90.0 percent and not more than 110.0 percent of the labeled amount of quinidine sulfate, calculated as $(C_{20}H_{24}N_2O_2)_2 \cdot H_2SO_4 \cdot 2H_2O$.

Packaging and storage—Preserve in well-closed, light-resistant containers.

Reference standards—*USP Quinidine Sulfate Reference Standard*—Do not dry. *USP Quininone Reference Standard*—Do not dry. Keep container tightly closed and protected from light.

Identification—The Tablets respond to the *Identification tests* under *Quinidine Sulfate Capsules*, powdered Tablets being used in place of the contents of capsules.

Dissolution ⟨711⟩—

Medium: 0.1 *N* hydrochloric acid; 900 mL.

Apparatus 1: 100 rpm.

Time: 30 minutes.

Procedure—Measure the amount in solution in filtered portions of the *Dissolution Medium*, suitably diluted, in 1-cm cells at the wavelength of maximum absorbance at about 248 nm, with a suitable spectrophotometer, in comparison with a Standard solution of known concentration of USP Quinidine Sulfate RS.

Tolerances—Not less than 85% (*Q*) of the labeled amount of $(C_{20}H_{24}N_2O_2)_2 \cdot H_2SO_4 \cdot 2H_2O$ is dissolved in 30 minutes.

Uniformity of dosage units ⟨905⟩: meet the requirements.

Procedure for content uniformity—Proceed as directed for *Procedure for content uniformity* in the test for *Uniformity of dosage units* under *Quinidine Sulfate Capsules*, using 1 powdered Tablet instead of the contents of 1 Capsule.

Chromatographic purity—Shake a quantity of powdered Tablets, equivalent to about 150 mg of quinidine sulfate, with 25 mL of diluted alcohol for 10 minutes, and filter. Using the filtrate as the test solution, proceed as directed in the test for *Chromatographic purity* under *Quinidine Sulfate*.

Assay—Proceed as directed in the *Assay* under *Quinidine Sulfate Capsules*, using powdered Quinidine Sulfate Tablets.

Quinidine Sulfate Extended-release Tablets

» Quinidine Sulfate Extended-release Tablets contain amounts of quinidine sulfate and dihydroquinidine sulfate totaling not less than 90.0 percent and not more than 110.0 percent of the labeled amount of quinidine sulfate, calculated as $(C_{20}H_{24}N_2O_2)_2 \cdot H_2SO_4 \cdot 2H_2O$.

Packaging and storage—Preserve in well-closed, light-resistant containers.

Reference standards—*USP Quinidine Sulfate Reference Standard*—Do not dry. *USP Quininone Reference Standard*—Do not dry. Keep container tightly closed and protected from light.

Identification—

A: Shake a quantity of powdered Tablets, equivalent to about 50 mg of quinidine sulfate, with 100 mL of dilute sulfuric acid (1 in 350), and filter: the filtrate so obtained exhibits a vivid blue fluorescence. On the addition of a few drops of hydrochloric acid, the fluorescence disappears.

B: In the test for *Chromatographic purity*, the R_f value of the principal spot obtained from the *Test preparation* corresponds to that from the *Standard preparation*.

C: Shake a quantity of the powdered Tablets, equivalent to about 100 mg of quinidine sulfate, with 10 mL of dilute hydrochloric acid (1 in 100), and filter: the filtrate so obtained responds to the tests for *Sulfate* ⟨191⟩.

Drug release ⟨724⟩—

Medium: 0.1 *N* hydrochloric acid; 900 mL.

Apparatus 1: 100 rpm.

Time: 0.125*D* hours, 0.500*D* hours, 1.50*D* hours.

Procedure—Using filtered portions of the solution under test, diluted with 0.1 *N* hydrochloric acid if necessary, determine the amount of quinidine sulfate dissolved from ultraviolet absorbances at the wavelength of maximum absorbance at about 248 nm by comparison with a Standard solution having a known concentration of USP Quinidine Sulfate RS in the same medium.

Tolerances—The percentages of the labeled amount of quinidine sulfate dissolved at the times specified conform to *Acceptance Table 1*.

Time (hours)	Amount Dissolved
0.125*D*	between 20% and 50%
0.500*D*	between 43% and 73%
1.50*D*	not less than 70%

Uniformity of dosage units ⟨905⟩: meet the requirements.

Procedure for content uniformity—Proceed as directed for *Procedure for content uniformity* in the test for *Uniformity of dosage units* under *Quinidine Sulfate Capsules*, using 1 powdered Tablet instead of the contents of 1 Capsule.

Chromatographic purity—Shake a quantity of powdered Tablets, equivalent to about 150 mg of quinidine sulfate, with 25 mL of diluted alcohol for 10 minutes, and filter. Using the filtrate as the test solution, proceed as directed in the test for *Chromatographic purity* under *Quinidine Sulfate*.

Assay—Proceed as directed in the *Assay* under *Quinidine Sulfate Capsules*, using powdered Quinidine Sulfate Extended-release Tablets.

Quinine Sulfate

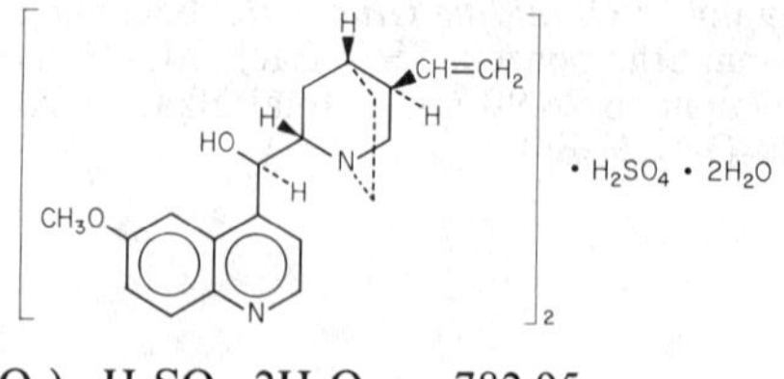

$(C_{20}H_{24}N_2O_2)_2 \cdot H_2SO_4 \cdot 2H_2O$ 782.95

Cinchonan-9-ol, 6′-methoxy-, (8α,9*R*)-, sulfate (2:1) (salt), dihydrate.

Quinine sulfate (2:1) (salt) dihydrate [*6119-70-6*].

Anhydrous 746.92 [*804-63-7*].

» Quinine Sulfate is the sulfate of an alkaloid obtained from the bark of species of *Cinchona*. It contains not less than 99.0 percent and not more than 101.0 percent of total alkaloid salt, calculated as $(C_{20}H_{24}N_2O_2)_2 \cdot H_2SO_4$, on the dried basis.

Packaging and storage—Preserve in well-closed, light-resistant containers.

Reference standards—*USP Quinine Sulfate Reference Standard*—Do not dry. *USP Quininone Reference Standard*—Do not dry. Keep container tightly closed and protected from light.

Identification—

A: A 1 in 2000 solution in dilute sulfuric acid (1 in 350) exhibits a vivid blue fluorescence. On the addition of a few drops of hydrochloric acid, the fluorescence disappears.

B: In the test for *Chromatographic purity*, the R_f value of the principal spot obtained from the *Test preparation* corresponds to that from the *Standard preparation*.

C: A solution (1 in 50) made with the aid of a few drops of hydrochloric acid responds to the tests for *Sulfate* ⟨191⟩.

Specific rotation ⟨781⟩: between −235° and −244°, calculated on the anhydrous basis, determined in a solution in 0.1 *N* hydrochloric acid containing 200 mg in each 10 mL.

Water, *Method I* ⟨921⟩: between 4.0% and 5.5%.

Residue on ignition ⟨281⟩: not more than 0.1%.

Heavy metals, *Method II* ⟨231⟩: 0.001%.

Chloroform-alcohol–insoluble substances—Warm 2 g with 15 mL of a mixture of chloroform and dehydrated alcohol (2:1) at about 50° for 10 minutes. Filter through a tared, sintered-glass filter, using gentle suction. Wash the filter with five 10-mL portions of the chloroform-alcohol mixture, dry at 105° for 1 hour, and weigh: the weight of the residue does not exceed 2 mg (0.1%).

Chromatographic purity—

Standard preparation—Prepare a solution of USP Quinine Sulfate RS in diluted alcohol to contain 6 mg per mL.

Diluted standard preparation—Dilute a portion of the *Standard preparation* with diluted alcohol to a concentration of 0.06 mg per mL.

Related substances preparation—Prepare a solution in diluted alcohol containing in each mL 0.05 mg each of USP Quininone RS (corresponding to 0.06 mg of the sulfate), and 0.10 mg of cinchonidine (corresponding to 0.12 mg of the sulfate).

Test preparation—Prepare a solution of Quinine Sulfate in diluted alcohol to contain 6 mg per mL.

Procedure—On a suitable thin-layer chromatographic plate (see *Chromatography* ⟨621⟩), coated with a 0.25-mm layer of chromatographic silica gel, apply 10-µL portions of the *Test preparation*, the *Standard preparation*, the *Diluted standard preparation*, and the *Related substances preparation*. Allow to dry, and develop the chromatogram in a solvent system consisting of a mixture of chloroform, acetone, and diethylamine (5:4:1), the solvent chamber being used without previous equilibration. When the solvent front has moved about 15 cm, remove the plate from the chamber, mark the solvent front, and allow the solvent to evaporate. Spray the chromatogram with glacial acetic acid. Locate the spots on the plate by examination under long-wavelength ultraviolet light. Any spot produced by the *Test preparation* at the R_f value of a spot produced by the *Related substances preparation* is not greater in size or intensity than that corresponding spot. Apart from these spots and from the spot appearing at the R_f value of Quinine Sulfate, any additional fluorescent spot is not greater in size or intensity than the spot of the *Diluted standard preparation.* Spray the plate with potassium iodoplatinate TS. Any spot produced by the *Test preparation* is not greater in size or intensity than a corresponding spot from the *Related substances preparation.*

Dihydroquinine Sulfate—

Methanesulfonic acid solution—Add 35.0 mL of methanesulfonic acid to 20.0 mL of glacial acetic acid, dilute with water to 500 mL, and mix.

Diethylamine solution—Dissolve 10.0 mL of diethylamine in water to obtain 100 mL of solution.

Mobile phase—Prepare a suitable filtered and degassed mixture of water, acetonitrile, *Methanesulfonic acid solution*, and *Diethylamine solution* (860:100:20:20). Adjust with *Diethylamine solution* to a pH of 2.6 if found to be lower.

System suitability preparation—Transfer about 10 mg each of quinine sulfate and dihydroquinine sulfate to a 50-mL volumetric flask. Dissolve in about 5 mL of methanol, dilute with *Mobile phase* to volume, and mix.

System suitability test—Chromatograph injections of the *System suitability preparation* as directed under *Procedure*: the relative retention times for quinine and dihydroquinine are about 1 and 1.5, respectively. The resolution between the quinine and dihydroquinine peaks is not less than 1.2. The relative standard deviation for the peak response of quinine is not more than 2.0%.

Test preparation—Transfer about 20 mg of Quinine Sulfate to a 100-mL volumetric flask, dissolve in *Mobile phase*, dilute with *Mobile phase* to volume, and mix.

Procedure (see *Chromatography* ⟨621⟩)—Inject about 50 µL of the *Test preparation* into a chromatograph equipped with a 235-nm detector and a 3.9-mm × 30-cm column that contains packing L1. The response of the dihydroquinine peak is not greater than one-ninth that of the quinine peak (10.0%).

Assay—Dissolve about 200 mg of Quinine Sulfate, accurately weighed, in 20 mL of acetic anhydride, add 4 drops of *p*-naphtholbenzein TS, and titrate with 0.1 *N* perchloric acid VS from a 10-mL microburet to a green end-point. Perform a blank determination, and make any necessary correction. Each mL of 0.1 *N* perchloric acid is equivalent to 24.90 mg of total alkaloid salt, calculated as $(C_{20}H_{24}N_2O_2)_2 \cdot H_2SO_4$.

Quinine Sulfate Capsules

» Quinine Sulfate Capsules contain amounts of quinine sulfate and dihydroquinine sulfate totaling not less than 90.0 percent and not more than 110.0 percent of the labeled amount of quinine sulfate, calculated as $(C_{20}H_{24}N_2O_2)_2 \cdot H_2SO_4 \cdot 2H_2O$.

Packaging and storage—Preserve in tight containers.

Reference standards—*USP Quinine Sulfate Reference Standard*—Do not dry. *USP Quininone Reference Standard*—Do not dry. Keep container tightly closed and protected from light.

Identification—

A: Shake well a quantity of the contents of Capsules, equivalent to about 100 mg of quinine sulfate, with 100 mL of dilute sulfuric acid (1 in 350), and filter. An appropriate dilution of the filtrate exhibits a vivid blue fluorescence. On the addition of a few drops of hydrochloric acid the fluorescence disappears.

B: In the test for *Chromatographic purity*, the R_f value of the principal spot obtained from the *Test preparation* corresponds to that from the *Standard preparation.*

C: Shake a quantity of the contents of Capsules, equivalent to about 20 mg of quinine sulfate, with 10 mL of dilute hydrochloric acid (1 in 100), and filter: the filtrate responds to the tests for *Sulfate* ⟨191⟩.

D: The retention time of the major peak in the chromatogram of the *Assay preparation* corresponds to that in the chromatogram of the *Standard preparation*, obtained as directed in the *Assay.*

Dissolution ⟨711⟩—

Medium: 0.1 *N* hydrochloric acid; 900 mL.

Apparatus 1: 100 rpm.

Time: 45 minutes.

Procedure—Determine the amount of $(C_{20}H_{24}N_2O_2)_2 \cdot H_2SO_4 \cdot 2H_2O$ dissolved from ultraviolet absorbances at the wavelength of maximum absorbance at about 248 nm of filtered portions of the solution under test, suitably diluted with 0.1 *N* hydrochloric acid, in comparison with a Standard solution having a known concentration of USP Quinine Sulfate RS in the same medium.

Tolerances—Not less than 75% (*Q*) of the labeled amount of $(C_{20}H_{24}N_2O_2)_2 \cdot H_2SO_4 \cdot 2H_2O$ is dissolved in 45 minutes.

Uniformity of dosage units ⟨905⟩: meet the requirements.

Procedure for content uniformity—Transfer the contents of 1 Capsule to a 250-mL volumetric flask, add about 175 mL of dilute hydrochloric acid (1 in 100), and shake by mechanical means for 30 minutes. Add dilute hydrochloric acid (1 in 100) to volume, and mix. Filter a portion of the mixture, discarding the first 20 mL of the filtrate. Concomitantly determine the absorbances of this solution, quantitatively diluted, if necessary, and a Standard solution of USP Quininc Sulfate RS in dilute hydrochloric acid (1 in 100) having a known concentration of about 40 µg per mL, in 1-cm cells, at the wavelength of maximum absorbance at about 345 nm, with a suitable spectrophotometer, using water as the blank. Calculate the quantity, in mg, of active ingredients, calculated as quinine sulfate $[(C_{20}H_{24}N_2O_2)_2 \cdot H_2SO_4 \cdot 2H_2O]$, in the Capsule by the formula:

$$(TC/D)(A_U/A_S),$$

in which *T* is the labeled quantity, in mg, of quinine sulfate in the Capsule, *D* is the concentration, in µg per mL, of quinine sulfate in the solution from the Capsule, based on the labeled quantity per Capsule and the extent of dilution, *C* is the concentration, in µg per mL, of USP Quinine Sulfate in the Standard solution, and A_U and A_S are the absorbances of the solution from the Capsule and the Standard solution, respectively.

Chromatographic purity—Shake a quantity of the contents of Capsules, equivalent to about 150 mg of quinine sulfate, with 25 mL of diluted alcohol for 10 minutes, and filter. Using this as the test solution, proceed as directed in the test for *Chromatographic purity* under *Quinine Sulfate.*

Assay—[NOTE—Where peak responses are indicated, use peak areas.]

Methanesulfonic acid solution, Diethylamine solution, Mobile phase, System suitability preparation, and System suitability test—Proceed as directed in the test for *Dihydroquinine sulfate* under *Quinine Sulfate.*

Standard preparation—Transfer about 20 mg of USP Quinine Sulfate RS, accurately weighed, to a 100-mL volumetric flask, dissolve in *Mobile phase*, dilute with *Mobile phase* to volume, and mix.

Assay preparation—Transfer the contents of not less than 20 Quinine Sulfate Capsules to a container, and mix. Transfer an accurately weighed portion of the powder, equivalent to about 160 mg of quinine sulfate, to a 100-mL volumetric flask, add 80 mL of methanol, and shake the flask by mechanical means for 30 minutes. Dilute with methanol to volume, and filter, discarding the first 10 mL of the filtrate. Transfer 3.0 mL of the filtrate to a 25-mL volumetric flask, dilute with *Mobile phase* to volume, and mix.

Procedure (see *Chromatography* ⟨621⟩)—Inject equal volumes (about 50 µL) of the *Standard preparation* and the *Assay preparation* into a chromatograph equipped with a 235-nm detector and a 3.9-mm × 30-cm column that contains packing L1. Calculate the quantity, in mg, of the sum of quinine sulfate and dihydroquinine sulfate in the portion of Capsules taken by the formula:

$$(2500/3)C(r_{b,U} + r_{d,U})/(r_{b,S} + r_{d,S}),$$

in which *C* is the concentration, in mg per mL, of USP Quinine Sulfate RS in the *Standard preparation*, $r_{b,U}$ and $r_{b,S}$ are the peak responses of quinine obtained from the *Assay preparation* and the *Standard preparation*, respectively, and $r_{d,U}$ and $r_{d,S}$ are the peak responses of dihydroquinine obtained from the *Assay preparation* and the *Standard preparation*, respectively.

Quinine Sulfate Tablets

» Quinine Sulfate Tablets contain amounts of quinine sulfate and dihydroquinine sulfate totaling not less than 90.0 percent and not more than 110.0 percent of the labeled amount of quinine sulfate, calculated as $(C_{20}H_{24}N_2O_2)_2 . H_2SO_4 . 2H_2O$.

Packaging and storage—Preserve in well-closed containers.

Reference standards—*USP Quinine Sulfate Reference Standard*—Do not dry. *USP Quininone Reference Standard*—Do not dry. Keep container tightly closed and protected from light.

Identification—

A: Shake well a quantity of powdered Tablets, equivalent to about 100 mg of quinine sulfate, with 100 mL of dilute sulfuric acid (1 in 350), and filter. An appropriate dilution of the filtrate exhibits a vivid blue fluorescence. On the addition of a few drops of hydrochloric acid the fluorescence disappears.

B: In the test for *Chromatographic purity*, the R_f value of the principal spot obtained from the *Test preparation* corresponds to that from the *Standard preparation.*

C: Shake a quantity of powdered Tablets, equivalent to about 20 mg of quinine sulfate, with 10 mL of dilute hydrochloric acid (1 in 100), and filter: the filtrate responds to the tests for *Sulfate* ⟨191⟩.

D: The retention time of the major peak in the chromatogram of the *Assay preparation* corresponds to that in the chromatogram of the *Standard preparation*, obtained as directed in the *Assay.*

Dissolution ⟨711⟩—

Medium: 0.1 *N* hydrochloric acid; 900 mL.

Apparatus 1: 100 rpm.

Time: 45 minutes.

Procedure—Determine the amount of $(C_{20}H_{24}N_2O_2)_2 . H_2SO_4 . 2H_2O$ dissolved from ultraviolet absorbances at the wavelength of maximum absorbance at about 248 nm of filtered portions of the solution under test, suitably diluted with 0.1 *N* hydrochloric acid, in comparison with a Standard solution having a known concentration of USP Quinine Sulfate RS in the same medium.

Tolerances—Not less than 75% (*Q*) of the labeled amount of $(C_{20}H_{24}N_2O_2)_2 . H_2SO_4 . 2H_2O$ is dissolved in 45 minutes.

Uniformity of dosage units ⟨905⟩: meet the requirements.

Procedure for content uniformity—Proceed as directed for *Procedure for content uniformity* under *Uniformity of dosage units* under *Quinine Sulfate Capsules*, using 1 powdered Tablet instead of the contents of 1 Capsule.

Chromatographic purity—Shake a quantity of powdered Tablets, equivalent to about 150 mg of quinine sulfate, with 25 mL of diluted alcohol for 10 minutes, and filter. Using this as the test solution, proceed as directed in the test for *Chromatographic purity* under *Quinine Sulfate.*

Assay—[NOTE—Where peak responses are indicated, use peak areas.]

Methanesulfonic acid solution, Diethylamine solution, Mobile phase, System suitability preparation, and *System suitability test*—Proceed as directed in the test for *Dihydroquinine sulfate* under *Quinine Sulfate.*

Standard preparation—Transfer about 20 mg of USP Quinine Sulfate RS, accurately weighed, to a 100-mL volumetric flask, dissolve in *Mobile phase*, dilute with *Mobile phase* to volume, and mix.

Assay preparation—Weigh and finely powder not less than 20 Quinine Sulfate Tablets. Transfer an accurately weighed portion of the powder, equivalent to about 160 mg of quinine sulfate, to a 100-mL volumetric flask, add 80 mL of methanol, and shake the flask by mechanical means for 30 minutes. Dilute with methanol to volume, and filter, discarding the first 10 mL of the filtrate. Transfer 3.0 mL of the filtrate to a 25-mL volumetric flask, dilute with *Mobile phase* to volume, and mix.

Procedure (see *Chromatography* ⟨621⟩)—Inject equal volumes (about 50 µL) of the *Standard preparation* and the *Assay preparation* into a chromatograph equipped with a 235-nm detector and a 3.9-mm × 30-cm column that contains packing L1. Calculate the quantity, in mg, of the sum of quinine sulfate and dihydroquinine sulfate in the portion of Tablets taken by the formula:

$$(2500/3)C(r_{b,U} + r_{d,U})/(r_{b,S} + r_{d,S}),$$

in which C is the concentration, in mg per mL, of USP Quinine Sulfate RS in the *Standard preparation*, $r_{b,U}$ and $r_{b,S}$ are the peak responses of quinine obtained from the *Assay preparation* and the *Standard preparation*, respectively, and $r_{d,U}$ and $r_{d,S}$ are the peak responses of dihydroquinine obtained from the *Assay preparation* and the *Standard preparation*, respectively.

Rabies Immune Globulin

» Rabies Immune Globulin conforms to the regulations of the federal Food and Drug Administration concerning biologics (see *Biologics* ⟨1041⟩). It is a sterile, nonpyrogenic, slightly opalescent solution consisting of globulins derived from blood plasma or serum that has been tested for the absence of hepatitis B surface antigen, derived from selected adult human donors who have been immunized with rabies vaccine and have developed high titers of rabies antibody. It has a potency such that when labeled as 150 International Units (IU) per mL, it has a geometric mean lower limit (95% confidence) potency value of not less than 110 IU per mL, and proportionate lower limit potency values for other labeled potencies, based on the U.S. Standard Rabies Immune Globulin and using the CVS Virus challenge, by neutralization test in mice or tissue culture. It contains not less than 10 g and not more than 18 g of protein per 100 mL, of which not less than 80 percent is monomeric immunoglobulin G, having a sedimentation coefficient in the range of 6.0 to 7.5S, with no fragments having a sedimentation coefficient less than 6S and no aggregates having a sedimentation coefficient greater than 12S. It contains 0.3 *M* glycine as a stabilizing agent, and it contains a suitable preservative. It has a pH between 6.4 and 7.2, measured in a solution diluted to contain 1 percent of protein with 0.15 *M* sodium chloride. It meets the requirements of the test for heat stability.

Packaging and storage—Preserve at a temperature between 2° and 8°.

Expiration date—The expiration date is not later than 1 year after date of issue from manufacturer's cold storage (5°, 1 year).

Labeling—Label it to state that it is not for intravenous injection.

Rabies Vaccine

» Rabies Vaccine conforms to the regulations of the federal Food and Drug Administration concerning biologics (see *Biologics* ⟨1041⟩). It is a sterile preparation, in dried or liquid form, of inactivated rabies virus harvested from inoculated diploid cell cultures. The cell cultures are shown to consist of diploid cells by tests of karyology, to be non-tumorigenic by tests in hamsters treated with anti-lymphocytic serum (ALS) and to be free from extraneous agents by tests in animals or cell-culture systems. The harvested virus meets the requirements for identity by serological tests, for absence of infectivity by tests in mice or cell-culture systems, and for absence of extraneous agents by tests in animals or cell-culture systems. The Vaccine meets the requirements for absence of live virus by tests using a suitable virus amplification system involving inoculation and incubation of sensitive cell cultures for not less than 14 days followed by inoculation of the cell-culture fluid thereafter into not less than 20 adult mice. It has a potency of rabies antigen equivalent to not less than 2.5 International Units for Rabies Vaccine, per dose, determined with the specific mouse protection test using the U.S. Standard Rabies Vaccine. It meets the requirements for general safety (see *Safety Tests—General* under *Biological Reactivity Tests, In-vivo* ⟨88⟩).

Packaging and storage—Preserve at a temperature between 2° and 8°.

Expiration date—The expiration date is not later than 2 years after date of issue from manufacturer's cold storage (5°, 1 year).

Labeling—Label it to state that it contains rabies antigen equivalent to not less than 2.5 IU per dose and that it is intended for intramuscular injection only.

Racemic Calcium Pantothenate—*see* Calcium Pantothenate, Racemic

Racepinephrine

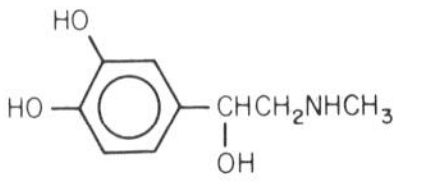

HO
HO—⟨ring⟩—CHCH$_2$NHCH$_3$
OH

$C_9H_{13}NO_3$ 183.21

1,2-Benzenediol, 4-[1-hydroxy-2-(methylamino)ethyl]-, (±)-.

(±)-3,4-Dihydroxy-α-[(methylamino)methyl]benzyl alcohol [*329-65-7*].

» Racepinephrine is a racemic mixture of the enantiomorphs of epinephrine. It contains not less than 97.0 percent and not more than 102.0 percent of $C_9H_{13}NO_3$, calculated on the dried basis.

Packaging and storage—Preserve in tight, light-resistant containers.

Reference standards—*USP Epinephrine Bitartrate Reference Standard*—Dry in vacuum over silica gel for 3 hours before using. *USP Norepinephrine Bitartrate Reference Standard*—Do not dry before using.

Identification—To 5 mL of pH 4.0 acid phthalate buffer (see *Buffer Solutions* in the section, *Reagents, Indicators, and Solutions*) add 0.5 mL of a solution of Racepinephrine (1 in 1000) and 1.0 mL of 0.1 *N* iodine. Mix, and allow to stand for 5 minutes. Add 2 mL of sodium thiosulfate solution (1 in 40): a deep red color is produced.

Loss on drying ⟨731⟩—Dry it in vacuum over silica gel for 18 hours: it loses not more than 2.0% of its weight.

Residue on ignition ⟨281⟩: not more than 0.5%.

Specific rotation ⟨781⟩: between −1° and +1°, determined on a 1 in 100 solution in dilute hydrochloric acid (1 in 20).

Adrenalone—Its absorptivity (see *Spectrophotometry and Light-scattering* ⟨851⟩) at 310 nm, determined in a solution in dilute hydrochloric acid (1 in 200) containing 2 mg per mL, is not more than 0.2.

Limit of norepinephrine—

Epinephrine standard solution—Dilute with methanol an accurately measured volume of a solution of USP Epinephrine Bitartrate RS in formic acid containing about 364 mg per mL to obtain a solution having a concentration of about 20 mg per mL.

Norepinephrine standard solution—Dilute with methanol an accurately measured volume of a solution of USP Norepinephrine Bitartrate RS in formic acid containing 16 mg per mL to obtain a solution having a known concentration of 1.6 mg per mL.

Test solution—Dissolve 200 mg of Racepinephrine in 1.0 mL of formic acid, dilute with methanol to 10.0 mL, and mix.

Procedure—On a suitable thin-layer chromatographic plate (see *Chromatography* ⟨621⟩), coated with a 0.25-mm layer of chromatographic silica gel mixture, apply 5-µL portions of *Epinephrine standard solution, Norepinephrine standard solution,* and *Test solution.* Allow to dry, and develop the chromatogram in an unsaturated tank using a solvent system consisting of a mixture of *n*-butanol, water, and formic acid (7:2:1) until the solvent front has moved about three-fourths of the length of the plate. Remove the plate from the developing chamber, mark the solvent front, and allow the solvent to evaporate in warm circulating air. Spray with Folin-Ciocalteu Phenol TS, followed by sodium carbonate solution (1 in 10): the R_f value of the principal spot obtained from the *Test solution* corresponds to that obtained from the *Epinephrine standard solution.* Any spot obtained from the *Test solution* is not larger nor more intense than the spot with the same R_f value obtained from the *Norepinephrine standard solution*, corresponding to not more than 4.0% of norepinephrine.

Assay—

Ferro-citrate solution, Buffer solution, and *Standard preparation*—Prepare as directed under *Epinephrine Assay* ⟨391⟩.

Assay preparation—Transfer about 10 mg of Racepinephrine, accurately weighed, to a 1-liter volumetric flask. Dilute with sodium bisulfite solution (1 in 500) to volume, and mix.

Procedure—Proceed as directed for *Procedure* under *Epinephrine Assay* ⟨391⟩. Calculate the quantity, in mg, of $C_9H_{13}NO_3$ in the portion of Racepinephrine taken by the formula:

$$(183.21/333.29)C(A_U/A_S),$$

in which 183.21 and 333.29 are the molecular weights of racepinephrine and epinephrine bitartrate, respectively, C is the concentration, in µg per mL, of USP Epinephrine Bitartrate RS in the *Standard preparation*, and A_U and A_S are the absorbances of the solutions from the *Assay preparation* and the *Standard preparation*, respectively.

Racepinephrine Inhalation Solution

» Racepinephrine Inhalation Solution is a solution of Racepinephrine in Purified Water prepared with the aid of Hydrochloric Acid or of Racepinephrine Hydrochloride in Purified Water. It contains not less than 90.0 percent and not more than 110.0 percent of the labeled amount of racepinephrine ($C_9H_{13}NO_3$).

Note—Do not use the Inhalation Solution if it is brown or contains a precipitate.

Packaging and storage—Preserve in tight, light-resistant containers. Do not freeze.

Reference standard—*USP Epinephrine Bitartrate Reference Standard*—Dry in vacuum over silica gel for 3 hours before using.

Identification—To 5 mL of pH 4.0 acid phthalate buffer (see *Buffer Solutions* in the section, *Reagents, Indicators, and Solutions*) add 0.5 mL of Racepinephrine Inhalation Solution and 1.0 mL of 0.1 *N* iodine. Mix, and allow to stand for 5 minutes. Add 2 mL of sodium thiosulfate solution (1 in 40): a deep red color is produced.

pH ⟨791⟩: between 2.0 and 3.5.

Assay—

Mobile phase—Prepare a filtered and degassed mixture of 0.05 *M* monobasic sodium phosphate and methanol (85:15). Dissolve a suitable quantity of sodium 1-octanesulfonate in this mixture to obtain a solution that is 0.005 *M* with respect to sodium 1-octanesulfonate. Make adjustments if necessary (see *System Suitability* under *Chromatography* ⟨621⟩).

Standard preparation—Dissolve an accurately weighed quantity of USP Epinephrine Bitartrate RS in a mixture of 0.05 *M* monobasic sodium phosphate and methanol (85:15) and dilute quantitatively, and stepwise if necessary, with the same solvent mixture to obtain a solution having a known concentration of about 0.1 mg per mL.

Assay preparation—Transfer an accurately measured volume of Racepinephrine Inhalation Solution, equivalent to about 11 mg of racepinephrine, to a 200-mL volumetric flask, dilute with a mixture of 0.05 *M* monobasic sodium phosphate and methanol (85:15) to volume, and mix.

System suitability preparation—Dissolve a suitable quantity of dopamine hydrochloride in a mixture of 0.05 *M* monobasic sodium phosphate and methanol (85:15) to obtain a solution containing about 0.1 mg per mL. Transfer 20.0 mL of this solution to a 50-mL volumetric flask. Add 5.0 mL of the *Standard preparation*, dilute with a mixture of 0.05 *M* monobasic sodium phosphate and methanol (85:15) to volume, and mix.

Chromatographic system (see *Chromatography* ⟨621⟩)—The liquid chromatograph is equipped with a 280-nm detector and a 4.6-mm × 25-cm column that contains 5-µm packing L1. The flow rate is about 1 mL per minute. Chromatograph the *Standard preparation*, and the *System suitability preparation*, and record the peak responses as directed under *Procedure:* the tailing factor for the analyte peak is not more than 1.2, the resolution, *R*, between the dopamine and analyte peak is not less than 5.0, and the relative standard deviation for replicate injections is not more than 2.0%.

Procedure—Separately inject equal volumes (about 10 µL) of the *Standard preparation* and the *Assay preparation* into the chromatograph, record the chromatograms, and measure the responses for the major peaks. Calculate the quantity, in mg, of $C_9H_{13}NO_3$ in each mL of the Inhalation Solution taken by the formula:

$$(183.21/333.29)200(C/V)(r_U/r_S),$$

in which 183.21 and 333.29 are the molecular weights of racepinephrine and epinephrine bitartrate, respectively, *C* is the concentration, in mg per mL, of USP Epinephrine Bitartrate RS in the *Standard preparation*, *V* is the volume, in mL, of Inhalation Solution taken, and r_U and r_S are the peak responses obtained from the *Assay preparation* and the *Standard preparation*, respectively.

Racepinephrine Hydrochloride

$C_9H_{13}NO_3.HCl$ 219.67

» Racepinephrine Hydrochloride is a racemic mixture of the hydrochlorides of the enantiomorphs of epinephrine. It contains not less than 97.0 percent and not more than 102.0 percent of $C_9H_{13}NO_3.HCl$, calculated on the anhydrous basis.

Packaging and storage—Preserve in tight, light-resistant containers.

Reference standards—*USP Epinephrine Bitartrate Reference Standard*—Dry in vacuum over silica gel for 3 hours before using. *USP Norepinephrine Bitartrate Reference Standard*—Do not dry before using.

Identification—

A: The ultraviolet absorption spectrum of a 1 in 20,000 solution exhibits maxima and minima at the same wavelengths as that of a similar solution of USP Epinephrine Bitartrate RS, concomitantly measured.

B: A solution (1 in 100) responds to the tests for *Chloride* ⟨191⟩.

Melting range ⟨741⟩: between 155° and 160°.

Water, *Method I* ⟨921⟩: not more than 0.5%.

Specific rotation ⟨781⟩: between −1° and +1°, determined on a 1 in 100 solution in dilute hydrochloric acid (1 in 100).

Residue on ignition ⟨281⟩: not more than 0.5%.

Other requirements—It meets the requirements for *Adrenalone* and *Limit of norepinephrine* under *Racepinephrine.*

Assay—

Ferro-citrate solution, Buffer solution, and *Standard preparation*—Prepare as directed under *Epinephrine Assay* ⟨391⟩.

Assay preparation—Transfer about 10 mg of Racepinephrine Hydrochloride, accurately weighed, to a 1-liter volumetric flask. Dilute with sodium bisulfite solution (1 in 500) to volume, and mix.

Procedure—Proceed as directed for *Procedure* under *Epinephrine Assay* ⟨391⟩. Calculate the quantity, in mg, of $C_9H_{13}NO_3.HCl$ in the portion of racepinephrine hydrochloride taken by the formula:

$$(219.67/333.29)C(A_U/A_S),$$

in which 219.67 and 333.29 are the molecular weights of racepinephrine hydrochloride and epinephrine bitartrate, respectively, *C* is the concentration, in µg per mL, of USP Epinephrine Bitartrate RS in the *Standard preparation*, and A_U and A_S are the absorbances of the solutions from the *Assay preparation* and the *Standard preparation*, respectively.

Ranitidine Hydrochloride

$(CH_3)_2NCH_2$–(furan)–$CH_2SCH_2CH_2NHC(=CHNO_2)NHCH_3$ · HCl

$C_{13}H_{22}N_4O_3S.HCl$ 350.87

1,1-Ethenediamine, *N*-[2-[[[5-[(dimethylamino)methyl]-2-furanyl]methyl]thio]ethyl]-*N'*-methyl-2-nitro-,monohydrochloride.

N-[2-[[[5-[(Dimethylamino)methyl]-2-furanyl]methyl]thio]-ethyl]-*N'*-methyl-2-nitro-1,1-ethenediamine, hydrochloride. [*66357-59-3*].

» Ranitidine Hydrochloride contains not less than 97.5 percent and not more than 102.0 percent of $C_{13}H_{22}N_4O_3S.HCl$, calculated on the dried basis.

Packaging and storage—Preserve in tight, light-resistant containers.

Reference standards—*USP Ranitidine Hydrochloride Reference Standard*—Dry at 60° in vacuum for 3 hours before using. *USP Ranitidine Related Compound A Reference Standard*—Keep container tightly closed and protected from light. Do not dry before using. *USP Ranitidine Related Compound B Reference Standard*—Keep container tightly closed and protected from light. Do not dry before using. *USP Ranitidine Related Compound C Reference Standard*—Keep container tightly closed and protected from light. Do not dry before using.

Identification—

A: The infrared absorption spectrum of a mineral oil dispersion of it, previously dried, exhibits maxima only at the same wavelengths as that of a similar preparation of USP Ranitidine Hydrochloride RS.

B: The ultraviolet absorption spectrum of a solution (1 in 100,000) exhibits maxima and minima at the same wavelengths as that of a similar solution of USP Ranitidine Hydrochloride RS, concomitantly measured, and the respective absorptivities, calculated on the dried basis, at the wavelength of maximum absorbance at about 229 nm and 315 nm do not differ by more than 3.0%.

C: It responds to the tests for *Chloride* ⟨191⟩.

pH ⟨791⟩: between 4.5 and 6.0, in a solution (1 in 100).

Loss on drying ⟨731⟩—Dry it in vacuum at 60° for 3 hours: it loses not more than 0.75% of its weight.

Residue on ignition ⟨281⟩: not more than 0.1%.

Chromatographic purity—

Test preparation—Prepare a solution in methanol containing 22.3 mg of Ranitidine Hydrochloride per mL.

Standard preparations—Dissolve USP Ranitidine Hydrochloride RS in methanol to obtain a solution having a known concentration of 0.22 mg per mL. Dilute portions of this *Standard preparation* quantitatively with methanol to obtain solutions having concentrations of 110 µg per mL (*Diluted standard preparation A*), and 66 µg per mL (*Diluted standard preparation B*), and 11 µg per mL (*Diluted Standard preparation C*), respectively.

Resolution preparation—Dissolve USP Ranitidine Related Compound A RS, 5-[[(2-aminoethyl)thio]methyl]-*N*,*N*-dimethyl-2-furanmethanamine, hemifumarate salt, in methanol to obtain a solution having a known concentration of 1.27 mg per mL.

Identification preparation—Dissolve USP Ranitidine Related Compound B RS, *N*,*N'*-bis[2-[[[5-[(dimethylamino)methyl]-2-furanyl]methyl]thio]ethyl]-2-nitro-1,1-ethenediamine, in methanol to obtain a solution having a known concentration of 1 mg per mL.

Procedure—Apply separately 10 µL of the *Test preparation*, the *Standard preparation, Diluted standard preparation A, Diluted standard preparation B, Diluted standard preparation C*, and the *Identification preparation* to a suitable thin-layer chromatographic plate (see *Chromatography* ⟨621⟩), coated with a 0.25-mm layer of chromatographic silica gel mixture. In addition, apply separately 10 µL of the *Test preparation* to the same plate, and on top of this application, apply 10 µL of the *Resolution preparation*. Allow the spots to dry, and develop the chromatograms in a solvent system consisting of a mixture of ethyl acetate, isopropyl alcohol, ammonium hydroxide, and water (25:15:5:1) until the solvent front has moved not less than 15 cm from the origin. Remove the plate from the developing chamber, mark the solvent front, and allow to air-dry. Expose the plate to iodine vapor in a closed chamber until the chromatogram is fully revealed. Examine the plate, and compare the intensities of any secondary spots observed in the chromatogram of the *Test preparation* with those of the principal spots in the chromatograms of the *Standard preparation*; *Diluted standard preparations A, B*, and *C*; and the *Identification preparation:* the system suitability requirements are met if there is complete resolution between the primary spots in the chromatogram of the combined *Test preparation* and *Resolution preparation*, and if a spot is observed in the chromatogram of *Diluted standard preparation C*. If a spot is observed in the chromatogram of the *Test preparation*, at the R_f value corresponding to that of the principal spot produced by the *Identification preparation*, it is not greater in size or intensity than the principal spot produced by *Diluted standard preparation A*, corresponding to not more than 0.5% of related compound B, and no other spot in the chromatogram of the *Test preparation* exceeds the size or intensity due to the principal spot produced by *Diluted standard preparation B* (0.3%). The sum of the intensities of all secondary spots obtained from the *Test preparation* corresponds to not more than 1.0%.

Assay—[NOTE—Where peak responses are indicated, use peak areas.]

Mobile phase—Prepare a filtered and degassed mixture of methanol and 0.1 *M* aqueous ammonium acetate (85:15). Make adjustments if necessary (see *System Suitability* under *Chromatography* ⟨621⟩).

Standard preparation—Dissolve an accurately weighed quantity of USP Ranitidine Hydrochloride RS in *Mobile phase* to obtain a solution having a known concentration of about 0.112 mg (equivalent to 0.100 mg of ranitidine base) per mL.

System suitability solution—Dissolve accurately weighed quantities of USP Ranitidine Hydrochloride RS and USP Ranitidine Related Compound C RS in *Mobile phase* to obtain a solution having known concentrations of about 0.112 mg per mL and 0.002 mg per mL, respectively.

Assay preparation—Transfer about 112 mg of Ranitidine Hydrochloride, accurately weighed, to a 100-mL volumetric flask. Dissolve in *Mobile phase*, dilute with *Mobile phase* to volume, and mix. Transfer 1.0 mL of this solution to a 10-mL volumetric flask, dilute with *Mobile phase* to volume, and mix.

Chromatographic system (see *Chromatography* ⟨621⟩)—The liquid chromatograph is equipped with a 322-nm detector and a 4.6-mm × 20- to 30-cm column that contains packing L1. The flow rate is about 2 mL per minute. Chromatograph the *System suitability solution*, and record the peak responses as directed under *Procedure:* the resolution, *R*, between the ranitidine hydrochloride and *N*-[2-[[[5-[(dimethylamino)methyl]-2-furanyl]methyl]sulfinyl]ethyl]-*N'*-methyl-2-nitro-1,1-ethenediamine (ranitidine related compound C) peaks is not less than 1.5. Chromatograph the *Standard preparation*, and record the peak responses as directed under *Procedure:* the tailing factor for the ranitidine hydrochloride peak is not greater than 2.0, the column efficiency determined from the ranitidine hydrochloride peak is not less than 700 theoretical plates, and the relative standard deviation for replicate injections is not more than 2%.

Procedure—Separately inject equal volumes (about 10 µL) of the *Standard preparation* and the *Assay preparation* into the chromatograph, record the chromatograms, and measure the responses for the major peaks. Calculate the quantity, in mg, of $C_{13}H_{22}N_4O_3S.HCl$ in the portion of Ranitidine Hydrochloride taken by the formula:

$$1000C(r_U / r_S),$$

in which *C* is the concentration, in mg per mL, of USP Ranitidine Hydrochloride RS in the *Standard preparation*, and r_U and r_S are the peak responses obtained from the *Assay preparation* and the *Standard preparation*, respectively.

Ranitidine Injection

» Ranitidine Injection is a sterile solution of Ranitidine Hydrochloride in Water for Injection. It contains the equivalent of not less than 90.0 percent and not more than 110.0 percent of the labeled amount of ranitidine ($C_{13}H_{22}N_4O_3S$).

Packaging and storage—Preserve in single-dose or in multiple-dose containers of Type I glass, protected from light. Store below 30°. Do not freeze.

Labeling—Label Injection to state both the content of the active moiety and the content of the salt used in formulating the article.

Reference standards—*USP Ranitidine Hydrochloride Reference Standard*—Dry in vacuum at 60° for 3 hours before using. *USP Ranitidine Related Compound A Reference Standard*—Keep container tightly closed and protected from light. Do not dry before using. *USP Ranitidine Related Compound C Reference Standard*—Keep container tightly closed and protected from light. Do not dry before using.

Identification—

A: The R_f value of the principal spot observed in the chromatogram of the *Test preparation* obtained as directed in the *Chromatographic purity* test corresponds to that obtained from the *Standard preparation*.

B: The retention time of the major peak in the chromatogram of the *Assay preparation* corresponds to that in the chromatogram of the *Standard preparation*, as obtained in the *Assay*.

Bacterial endotoxins—When tested as directed under *Bacterial Endotoxins Test* ⟨85⟩, it contains not more than 7.00 USP Endotoxin Units per mg of ranitidine.

pH ⟨791⟩: between 6.7 and 7.3.

Particulate matter ⟨788⟩: meets the requirements under *Small-volume Injections*.

Chromatographic purity—

Test preparation—Dilute Ranitidine Injection quantitatively with water, if necessary, to obtain a solution containing 25 mg of ranitidine per mL. [NOTE—Use Injection of lower concentration without dilution as directed under *Procedure*.]

Standard preparation—Dissolve USP Ranitidine Hydrochloride RS in water to obtain a solution having a known concentration of 560 µg per mL. Dilute portions of this *Standard preparation* quantitatively with water to obtain solutions having concentrations of 280 µg per mL (*Diluted standard preparation A*), 140 µg per mL (*Diluted standard preparation B*), 84 µg per mL (*Diluted standard preparation C*), 28 µg per mL (*Diluted *standard preparation D*), and 14 µg per mL (*Diluted standard preparation E*), respectively.

Resolution preparation—Dissolve USP Ranitidine Related Compound A RS, 5-[[(2-aminoethyl)thio]methyl]-*N*,*N*-dimethyl-2-furanmethanamine, hemifumarate salt, in methanol to obtain a solution having a known concentration of 1.27 mg per mL.

Procedure—Apply separately 10 µL of the *Standard preparation, Diluted standard preparations A, B, C, D* and *E*, and the required volume of the *Test preparation,* equivalent to 250 µg of ranitidine, to a suitable thin-layer chromatographic plate (see *Chromatography* ⟨621⟩), coated with a 0.25-mm layer of chromatographic silica gel mixture. In addition, apply separately a further loading of the same volume of the *Test preparation* to the same plate, and on top of this application, apply 10 µL of the *Resolution preparation.* Perform the chromatography as described in the *Chromatographic purity* test under *Ranitidine Hydrochloride.* Examine the plate and compare the intensities of any secondary spots observed in the chromatogram of the *Test preparation* with those of the principal spots in the chromatograms of the *Standard preparation* and *Diluted standard preparations* (*A, B, C, D* and *E*): the system suitability requirements are met when there is complete resolution between the primary spots of the *Test preparation* and the *Resolution preparation* and if a spot is observed in the chromatogram of *Diluted standard preparation E.* The major secondary spot is not greater in size or intensity than the principal spot produced by the *Standard preparation* (2.0%) and no other secondary spot is greater in size or intensity than the principal spot produced by *Diluted standard preparation A* (1.0%). The sum of the intensities of all secondary spots obtained from the *Test preparation* corresponds to not more than 5.0%.

Other requirements—It meets the requirements under *Injections* ⟨1⟩.

Assay—[NOTE—Where peak responses are indicated, use peak areas.]

Mobile phase, Standard preparation, System suitability solution, and *Chromatographic system*—Prepare as directed in the *Assay* under *Ranitidine Hydrochloride.*

Assay preparation—Dilute an accurately measured volume of Ranitidine Injection, quantitatively and stepwise if necessary, with *Mobile phase* to obtain a solution having a concentration of 0.1 mg of ranitidine per mL.

Procedure—Separately inject equal volumes (about 10 µL) of the *Standard preparation* and the *Assay preparation* into the chromatograph, record the chromatograms, and measure the responses for the major peaks. Calculate the quantity, in mg, of $C_{13}H_{22}N_4O_3S$ in the portion of Injection taken by the formula:

$$(314.40/350.86)(L/D)(C)(r_U/r_S),$$

in which 314.40 and 350.86 are the molecular weights of ranitidine and ranitidine hydrochloride, respectively, *L* is the labeled quantity of ranitidine in the Injection taken, *D* is the concentration, in mg per mL, of ranitidine in the *Assay preparation* on the basis of the labeled quantity and the extent of dilution, *C* is the concentration, in mg per mL, of USP Ranitidine Hydrochloride RS in the *Standard preparation,* and r_U and r_S are the peak responses obtained from the *Assay preparation* and the *Standard preparation,* respectively.

Ranitidine Tablets

» Ranitidine Tablets contain an amount of ranitidine hydrochloride ($C_{13}H_{22}N_4O_3S.HCl$) equivalent to not less than 90.0 percent and not more than 110.0 percent of the labeled amount of ranitidine ($C_{13}H_{22}N_4O_3S$).

Packaging and storage—Preserve in tight, light-resistant containers.

Labeling—Label Tablets to state both the content of the active moiety and the content of the salt used in formulating the article.

Reference standards—*USP Ranitidine Hydrochloride Reference Standard*—Dry in vacuum at 60° for 3 hours before using. *USP Ranitidine Related Compound A Reference Standard*—Keep container tightly closed and protected from light. Do not dry before using. *USP Ranitidine Related Compound C Reference Standard*—Keep container tightly closed and protected from light. Do not dry before using.

Identification—

A: The R_f value of the principal spot observed in the chromatogram of the *Test preparation* obtained as directed in the *Chromatographic purity* test corresponds to that obtained from the *Standard preparation.*

B: The retention time of the major peak in the chromatogram of the *Assay preparation* corresponds to that of the major peak in the chromatogram of the *Standard preparation,* as obtained in the *Assay.*

C: Shake a quantity of crushed Tablets, equivalent to about 100 mg of ranitidine, with 2 mL of water, and filter: the filtrate responds to the tests for *Chloride* ⟨191⟩.

Dissolution ⟨711⟩—

Medium: water; 900 mL.

Apparatus 2: 50 rpm.

Time: 45 minutes.

Procedure—Determine the amount of $C_{13}H_{22}N_4O_3S$ dissolved from ultraviolet absorbances at the wavelength of maximum absorbance at about 314 nm using filtered portions of the solution under test, suitably diluted with water, if necessary, in comparison with a Standard solution having a known concentration of USP Ranitidine Hydrochloride RS in the same medium.

Tolerances—Not less than 80% (*Q*) of the labeled amount of $C_{13}H_{22}N_4O_3S$ is dissolved in 45 minutes.

Uniformity of dosage units ⟨905⟩: meet the requirements.

Chromatographic purity—

Test preparation—Prepare a filtered solution in methanol containing 20 mg of ranitidine per mL (equivalent to 22.4 mg of ranitidine hydrochloride per mL) by shaking an appropriate number of Ranitidine Tablets in a suitable volume of methanol until the tablets have disintegrated completely.

Standard preparations—Dissolve USP Ranitidine Hydrochloride RS in methanol to obtain a solution having a known concentration of 0.22 mg per mL. Dilute portions of this *Standard preparation* quantitatively with methanol to obtain solutions having concentrations of 110 µg per mL (*Diluted standard preparation A*), 66 µg per mL (*Diluted standard preparation B*), 22 µg per mL (*Diluted standard preparation C*), and 11 µg per mL (*Diluted standard preparation D*), respectively.

Resolution preparation—Dissolve USP Ranitidine Related Compound A RS, 5-[[(2-aminoethyl)thio]methyl]-*N*,*N*-dimethyl-2-furanmethanamine, hemifumarate salt, in methanol to obtain a solution having a known concentration of 1.27 mg per mL.

Procedure—Apply separately 10 µL of the *Test preparation,* the *Standard preparation,* and *Diluted standard preparations A, B, C,* and *D* to a suitable thin-layer chromatographic plate (see *Chromatography* ⟨621⟩), coated with a 0.25-mm layer of chromatographic silica gel mixture. In addition, apply separately 10 µL of the *Test preparation* to the same plate, and on top of this application, apply 10 µL of the *Resolution preparation.* Allow the spots to dry, and develop the chromatograms in a solvent system consisting of a mixture of ethyl acetate, isopropyl alcohol, ammonium hydroxide, and water (25:15:5:1) until the solvent front has moved not less than 15 cm from the origin. Remove the plate from the developing chamber, mark the solvent front, and air-dry. Expose the plate to iodine vapor in a closed chamber until the chromatogram is fully revealed. Examine the plate, and compare the intensities of any secondary spots observed in the chromatogram of the *Test preparation* with those of the principal spots in the chromatograms of the *Standard preparation* and *Diluted standard preparations A, B, C,* and *D:* the system suitability requirements are met if there is complete resolution between the primary spots in the chromatogram of the combined *Test preparation* and the *Resolution preparation,* and if a spot is observed in the chromatogram of *Diluted standard preparation D.* No single secondary spot exhibits an intensity greater than that of *Diluted standard preparation A* (0.5%), and no other secondary spot exhibits an intensity greater than that of *Diluted standard preparation B* (0.3%). The sum of the intensities of all secondary spots obtained from the *Test preparation* correspond to not more than 1.2%.

Assay—[NOTE—Where peak responses are indicated, use peak areas.]

Mobile phase, Standard preparation, System suitability solution, and *Chromatographic system*—Prepare as directed in the *Assay* under *Ranitidine Hydrochloride.*

Assay preparation—Transfer 10 Ranitidine Tablets to a minimum of 250 mL of *Mobile phase*, accurately measured. Shake the mixture until the Tablets have disintegrated completely, and filter. Dilute the filtrate quantitatively, and stepwise if necessary, with *Mobile phase* to obtain a solution having a concentration of ranitidine similar to that of the *Standard preparation.*

Procedure—Separately inject equal volumes (about 10 µL) of the *Standard preparation* and the *Assay preparation* into the chromatograph, record the chromatograms, and measure the responses for the major peaks. Calculate the quantity, in mg, of $C_{13}H_{22}N_4O_3S$ in the portion of Tablets taken by the formula:

$$(314.40/350.86)(L/D)(C)(r_U/r_S),$$

in which 314.40 and 350.86 are the molecular weights of ranitidine and ranitidine hydrochloride, respectively, L is the labeled amount, in mg, of ranitidine in each tablet, D is the concentration, in mg per mL, of ranitidine in the *Assay preparation*, based on the labeled quantity per Tablet and the extent of dilution, C is the concentration, in mg per mL, of USP Ranitidine Hydrochloride RS in the *Standard preparation*, and r_U and r_S are the peak responses obtained from the *Assay preparation* and the *Standard preparation*, respectively.

Rauwolfia Serpentina

» Rauwolfia Serpentina is the dried root of *Rauwolfia serpentina* (Linné) Bentham ex Kurz (Fam. Apocynaceae), sometimes having fragments of rhizome and aerial stem bases attached. It contains not less than 0.15 percent of reserpine-rescinnamine group alkaloids, calculated as reserpine.

Packaging and storage—Preserve in well-closed containers, and store at controlled room temperature, in a dry place, secure against insect attack (see *Preservation* under *Vegetable and Animal Drugs*, in the *General Notices*).

Reference standards—*USP Rauwolfia Serpentina Reference Standard*—Do not dry before using. *USP Reserpine Reference Standard*—Keep container tightly closed and protected from light. Dry at 60° for 3 hours before using.

Botanic characteristics—

Unground Rauwolfia Serpentina root—This occurs as segments usually from 5 cm to 15 cm in length (pieces sometimes shorter) and from 3 mm to 20 mm in diameter. The pieces are subcylindrical to tapering, rather tortuous or curved, rarely branched, but bearing occasional twisted rootlets, which are larger, more abundant, and more rigid and woody on the thicker parts of the roots. Externally, light brown to grayish yellow to grayish brown, dull, rough or slightly wrinkled longitudinally yet peculiarly smooth to the touch, occasionally showing small circular rootlet scars in the larger pieces, with some exfoliation of the bark in small areas to reveal the paler wood beneath. When scraped, the bark separates readily from the wood. Fracture short, but irregular, the longer pieces readily breaking with a snap, slightly fibrous marginally. The freshly fractured surfaces show a rather thin layer of grayish yellow bark, with the pale yellowish white wood constituting about 80% of the radius. The smoothed transverse surface of larger pieces shows a finely radiate stele with three or more clearly marked growth rings; a small knob-like protuberance is frequently noticeable at the center. The wood is hard and of relatively low density. The odor is indistinct, earthy, reminiscent of stored white potatoes, and the taste is bitter.

Histology—A transverse section of Rauwolfia Serpentina root shows externally two to eight alternating strata of cork cells, the strata with larger cells alternating with strata made up of markedly smaller cells (*distinction from R. canescens*). Each stratum composed of smaller cells includes from three to five tangentially arranged cell layers, while each stratum made up of larger cells includes from one to six tangential layers. In a cross-sectional view, the largest central cells of the larger cell group measure 40 µm to 90 µm radially and up to 75 µm tangentially (although usually smaller), while the cells of the smaller cell groups measure about 5 µm to 20 µm radially and up to 75 µm tangentially. The walls are thin and suberized. The secondary cortex consists of several rows of tangentially elongated to isodiametric parenchyma cells, most being densely filled with starch grains; others (the short latex cells) occur singly or in short series and contain brown resin masses. The secondary phloem is relatively narrow and is made up of phloem parenchyma (bearing starch grains and less commonly tabular to angular calcium oxalate crystals up to 20 µm in length; also, occasionally, with some brown resin masses in outer cells and phloem rays) interlaid with scattered sieve tissue and traversed by phloem rays two to four cells in width. Sclerenchyma cells (stone cells and fibers) are absent in root (*distinction from other species of Rauwolfia*). Cambium is indistinct, narrow, dark, and wavering. The secondary xylem represents the large bulk of the root and shows one or more prominent annual rings with a denser core of wood about 500 µm across at the center. The xylem is composed of many wood wedges separated by xylem rays, and on closer examination reveals vessels in interrupted radial rows, much xylem parenchyma, many large-celled xylem rays, few wood fibers, and tracheids, all lignified-walled. The xylem fibers occur in both tangential and radial rows. The xylem rays are 1 to 12, occasionally up to 16, cells in width.

Rauwolfia Serpentina rhizome—Histology—This is similar to that of root except for the presence of a prominent cortex, pericycle fibers, bicollateral vascular bundles, and a small central pith. The pericycle fibers occur singly or in groups of two to five, have thick, nonlignified walls, tapering, often lobed ends, with subterminal enlargements having thin walls and broad lumina. Vessel elements up to 485 µm are found. The xylem rays are one to four cells in width, with lignified and pitted walls. Internal phloem strands occur embedded in the outer region of the pith. The xylem fibers are somewhat less wavy than those of the root. The pith consists of starch parenchyma cells, among which are scattered short latex cells with yellowish contents stained brown with iodine TS.

Ground Rauwolfia Serpentina root—This is brownish to reddish gray in color. Present are very numerous starch grains (mostly simple, two- to three-compound, occasionally four-compound); simple grains spheroid, ovate, muller-shaped, plano- to angular-convex, or irregular; hilum simple, Y-shaped, stellate, or irregularly cleft; unaltered grains 6 µm to 34 µm (average 20 µm) in diameter, mostly in the lower range (maximum sizes larger than in *R. canescens* and *R. micrantha*); altered grains up to about 50 µm in diameter; large unaltered grains show polarization cross clearly; calcium oxalate prisms and cluster crystals scattered, about 10 µm to 15 µm in size; brown resin masses and yellowish granular secretion masses occur occasionally; isolated cork cells elongated, up to 90 µm in length; phelloderm and phloem parenchyma cells similar in appearance; vessels subcylindrical, up to 360 µm in length and from about 20 µm up to 57 µm in diameter (narrower than in *R. canescens*) (the wall markings generally consist of simple pits, with bordered pits adjacent to xylem ray cells), the vessel end walls oblique to transverse, generally with openings in the end walls, some vessels showing tyloses; tracheids pitted, with moderately thick, tapering, beaded walls, with relatively broad lumina, polygonal in cross-section; xylem parenchyma cells with moderately thick walls with simple circular pits, cells polygonal in cross-section, bearing considerable starch; phloem and xylem-ray cells with pitted walls bearing much starch, sometimes with brown resin masses, xylem fibers with thick heavily lignified walls showing small transverse and oblique linear pits and pointed simple to bifurcate ends, measuring from 200 µm to 750 µm in length (shorter than in *R. micrantha* and *R. canescens*). No phloem fibers or sclereids are present in root (colorless nonlignified pericycle or primary phloem fibers, single or in small groups, may be present from rhizome or stem tissues).

Loss on drying ⟨731⟩—Dry it at 100° to constant weight: it loses not more than 12.0% of its weight.

Microbial limit—Rauwolfia Serpentina (as the ground root) meets the requirements of the test for absence of *Salmonella* species under *Microbial Limit Tests* ⟨61⟩.

Acid-insoluble ash ⟨561⟩: not more than 2.0%.

Stems and other foreign organic matter ⟨561⟩—It contains not more than 2.0% of stems and not more than 3.0% of other foreign organic matter.

Chemical identification—[NOTE—In this procedure, use formamide treated as directed in the specifications for Formamide (see under *Reagents* in the section, *Reagents, Indicators, and Solutions*) if it has an ammoniacal odor.]

Immobile solvent—Dilute 30 mL of formamide with acetone to 100 mL.

Mobile solvent A—Mix 90 mL of isooctane, 60 mL of carbon tetrachloride, 4 mL of piperidine, and 2 mL of tertiary butyl alcohol.

Mobile solvent B—Mix 75 mL of chloroform, 75 mL of isooctane, and 2 mL of tertiary butyl alcohol.

Spray solution—Dissolve 25 g of trichloroacetic acid in 100 mL of methanol.

Standard solution—Warm a 1-g portion of USP Rauwolfia Serpentina RS with 5 mL of alcohol at 55° to 65° for 30 minutes, with occasional mixing. Cool, and filter.

Test preparation—Reduce 10 g of Rauwolfia Serpentina root to a fine powder. Treat a 1-g portion as in the preparation of the *Standard solution*.

Procedure A—Line the sides of a chromatographic chamber, suitable for ascending chromatography (see *Chromatography* ⟨621⟩), with blotting paper. Transfer *Mobile solvent A* to the bottom of the container, and cover the chamber. Immerse a 20- × 20-cm sheet of filter paper (Whatman No. 1 or equivalent) in the *Immobile solvent*, and blot between paper toweling. Allow the acetone solvent to evaporate completely. On a line 2.5 cm from the bottom of the filter paper, apply about 1-µL portions of the *Test preparation* and of the *Standard solution*. Allow to dry. Apply a 2-µL portion of the *Immobile solvent* to each spot, allow to dry, and suspend the paper so that it dips into the *Mobile solvent*. Cover the chamber, and after about 1 hour, when the *Mobile solvent* has risen approximately seven-eighths of the height of the paper, remove the chromatogram, and dry at 90° in a current of air. Spray the paper lightly and evenly with the *Spray solution*, and dry at 90° for 10 minutes.

Procedure B—Use the apparatus described in *Procedure A*, but containing a glass trough with about 2 mL of ammonium hydroxide to saturate the atmosphere of the tank with NH_3. Transfer *Mobile solvent B* to the bottom of the tank outside the trough. Complete the test as described in *Procedure A*, omitting the trichloroacetic acid spray. Examine both chromatograms under ultraviolet light, and note the fluorescent spots. In both chromatograms the *Test preparation* yields spots corresponding in position and color to those of the *Standard solution*.

Assay—

Apparatus—A medium-sized continuous-extraction apparatus provided with a 250-mL flask and a 35- × 80-mm thimble is convenient, although a smaller apparatus may be used.

Solvents—Alcohol, chloroform, and 1,1,1-trichloroethane—Redistil the trichloroethane in an all-glass apparatus, and collect the fraction boiling between 73° and 76°.

Standard solution—Dissolve 20.0 mg of USP Reserpine RS in 25 mL of hot alcohol, cool, dilute with alcohol to 50.0 mL, and mix. When stored in a tightly-stoppered, light-resistant bottle in the dark, this solution is chromogenically stable for several weeks. Dilute 5.0 mL with alcohol to 100.0 mL, and mix before using.

Procedure—Extract about 2.5 g of finely powdered Rauwolfia Serpentina, accurately weighed, in a continuous-extraction apparatus for 4 hours. Use about 100 mL of vigorously boiling alcohol as solvent, and a few boiling chips to prevent bumping. Protect the flask and thimble and all solutions of Rauwolfia Serpentina alkaloids from direct or strong light. Wash the extract into a 100-mL volumetric flask with alcohol, cool, dilute with alcohol to volume, and mix. Transfer 20.0 mL to a separator containing 200 mL of 0.5 *N* sulfuric acid, mix, and extract with three 25-mL portions of methyl chloroform. Lubricate stopcocks only with lubricants insoluble in trichloroethane or chloroform (polytetrafluoroethylene stopcocks are satisfactory). Drain the lower phase as completely as possible. Wash each of the methyl chloroform extracts in a second separator containing 50 mL of 0.5 *N* sulfuric acid, and discard the trichloroethane extracts. Extract the weakly basic alkaloids from the first acid solution with 25-, 15-, 15-, 10-, 10-, and 10-mL portions of chloroform. Wash each chloroform extract with the acid in the second separator, then with two 10-mL portions of sodium bicarbonate solution (1 in 50) in two additional separators. Filter the chloroform extracts through chloroform-washed cotton into a 100-mL volumetric flask containing 10 mL of alcohol. Dilute with alcohol to volume, and mix. Transfer duplicate 10.0-mL aliquots to glass-stoppered, 25-mL conical flasks, and mix with 4 mL of alcohol. Evaporate with gentle heating almost to dryness, place in a vacuum desiccator, and evaporate to dryness. Dissolve the residues by agitating with 5.0 mL of alcohol. Transfer duplicate 5.0-mL aliquots of the *Standard solution* to flasks. Add 2.0 mL of 0.5 *N* sulfuric acid to one of the test specimen flasks and to one of the standard flasks (the blanks). Add to the other flasks 1.0 mL of 0.5 *N* sulfuric acid and 1.0 mL of sodium nitrite solution (3 in 1000). Mix the contents of each flask, and warm in a water bath at 50° to 60° for 20 minutes. Cool, add to each flask 500 µL of sulfamic acid solution (1 in 20), and mix. After stabilization of the solution colors, determine their absorbances in 1-cm cells at 390 nm, relative to a blank consisting of a mixture of alcohol and water (2:1). The quantity, in mg, of reserpine-rescinnamine group alkaloids in the specimen taken is given by the formula:

$$5(A - A_0)/(S - S_0),$$

in which A and A_0 are the absorbances of the nitrite-treated specimen and specimen blank, respectively, and S and S_0 are the corresponding absorbances for the solutions from the respective *Standard solution* aliquots.

Powdered Rauwolfia Serpentina

» Powdered Rauwolfia Serpentina is Rauwolfia Serpentina reduced to a fine or a very fine powder, and adjusted, if necessary, to conform to the requirements for reserpine-rescinnamine group alkaloids by admixture with lactose or starch or with a powdered rauwolfia serpentina containing a higher or lower content of these alkaloids. It contains not less than 0.15 percent and not more than 0.20 percent of reserpine-rescinnamine group alkaloids, calculated as reserpine.

Packaging and storage—Preserve in well-closed containers, and store at controlled room temperature, in a dry place, secure against insect attack (see *Preservation* under *Vegetable and Animal Drugs*, and the *General Notices*).

Reference standards—*USP Rauwolfia Serpentina Reference Standard*—Do not dry before using. *USP Reserpine Reference Standard*—Keep container tightly closed and protected from light. Dry at 60° for 3 hours before using.

Identification—It conforms to the requirements for *Ground Rauwolfia Serpentina root* under *Botanic characteristics* and meets the requirements of the test for *Chemical identification* under *Rauwolfia Serpentina*.

Microbial limit—Powdered Rauwolfia Serpentina meets the requirements of the test for absence of *Salmonella* species under *Microbial Limit Tests* ⟨61⟩.

Acid-insoluble ash ⟨561⟩: not more than 2.0%.

Assay—Proceed with Powdered Rauwolfia Serpentina as directed in the *Assay* under *Rauwolfia Serpentina*.

Rauwolfia Serpentina Tablets

» Rauwolfia Serpentina Tablets contain an amount of reserpine-rescinnamine group alkaloids, calculated as reserpine, equivalent to not less than 0.15 percent and not more than 0.20 percent of the labeled amount of powdered rauwolfia serpentina.

Packaging and storage—Preserve in tight, light-resistant containers.

Reference standards—*USP Rauwolfia Serpentina Reference Standard*—Do not dry before using. *USP Reserpine Reference Standard*—Keep container tightly closed and protected from light. Dry at 60° for 3 hours before using.

Identification—The powdered rauwolfia serpentina in the Tablets conforms to the requirements for *Ground Rauwolfia Serpentina root* under *Botanic characteristics* and meets the requirements of the test for *Chemical identification* under *Rauwolfia Serpentina.*

Microbial limit—Rauwolfia Serpentina Tablets meet the requirements of the test for absence of *Salmonella* species under *Microbial Limit Tests* ⟨61⟩.

Disintegration ⟨701⟩: 30 minutes.

Uniformity of dosage units ⟨905⟩: meet the requirements.

Assay—Weigh and finely powder not less than 20 Rauwolfia Serpentina Tablets. Weigh accurately a portion of this powder, equivalent to 2.5 g of powdered rauwolfia serpentina, and proceed as directed for *Procedure* in the *Assay* under *Rauwolfia Serpentina.*

Purified Rayon

» Purified Rayon is a fibrous form of bleached, regenerated cellulose. It may contain not more than 1.25 percent of titanium dioxide.

Alkalinity or acidity—Immerse about 10 g in 100 mL of recently boiled and cooled water, and decant 25-mL portions of the water, with the aid of a glass rod, into each of two dishes. To one portion add 3 drops of phenolphthalein TS, and to the other portion add 1 drop of methyl orange TS: neither portion appears pink when viewed against a white background without special illumination.

Residue on ignition ⟨281⟩: not more than 1.50%, from 5.0 g.

Acid-insoluble ash—Boil the residue obtained in the test for *Residue on ignition* with 25 mL of 3 *N* hydrochloric acid for 5 minutes, collect the insoluble matter on a tared filtering crucible, wash with hot water, ignite, and weigh: the residue weighs not more than 63.0 mg (1.25%).

Water-soluble substances—Proceed as directed in the test for *Water-soluble substances* under *Purified Cotton:* the residue weighs not more than 100 mg (1.0%).

Fiber length and absorbency—Remove it from its wrappings, and condition it for not less than 4 hours in a standard atmosphere of 65 ± 2% relative humidity at 21 ± 1.1° (70 ± 2° F), before determining the *Fiber length* and *Absorbency*.

Fiber length—Determine the fiber length of Purified Rayon as directed under *Cotton—Fiber Length* ⟨691⟩: not less than 70.0%, by weight, of the fibers are 19 mm or greater in length, and not more than 5.0%, by weight, of the fibers are 6.3 mm or less in length.

Absorbency—Proceed as directed under *Cotton—Absorbency Test* ⟨691⟩: submersion is complete in 5 seconds, and the rayon retains not less than 24 times its weight of water.

Other requirements—It meets the requirements for *Dyes* and *Other foreign matter* under *Purified Cotton.*

Red Blood Cells—*see* Blood Groupings

Oral Rehydration Salts

» Oral Rehydration Salts is a dry mixture of Sodium Chloride, Potassium Chloride, Sodium Bicarbonate, and Dextrose (anhydrous). Alternatively, it may contain Sodium Citrate (anhydrous or dihydrate) instead of Sodium Bicarbonate. It may contain Dextrose (monohydrate) instead of Dextrose (anhydrous), provided that the Sodium Bicarbonate or Sodium Citrate is packaged in a separate, accompanying container. It contains the equivalent of not less than 90.0 percent and not more than 110.0 percent of the amounts of sodium (Na^+), potassium (K^+), chloride (Cl^-), and bicarbonate (HCO_3^-) or citrate ($C_6H_5O_7^{-3}$), calculated from the labeled amounts of Sodium Chloride, Potassium Chloride, and Sodium Bicarbonate [or Sodium Citrate (anhydrous or dihydrate)]. It contains not less than 90.0 percent and not more than 110.0 percent of the labeled amounts of anhydrous dextrose ($C_6H_{12}O_6$), or dextrose monohydrate ($C_6H_{12}O_6 \cdot H_2O$). It may contain suitable flavors.

Packaging and storage—Preserve in tight containers, and avoid exposure to temperatures in excess of 30°. The Sodium Bicarbonate or Sodium Citrate component may be omitted from the mixture and packaged in a separate, accompanying container.

Labeling—The label indicates prominently whether Sodium Bicarbonate or Sodium Citrate is a component by the placement of the word "Bicarbonate" or "Citrate," as appropriate, in juxtaposition to the official title. The label states the name and quantity, in g, of each component in each unit-dose container, or in a stated quantity, in g, of Salts in a multiple-unit container. The label states the net weight in each container, and provides directions for constitution. Where packaged in individual unit-dose pouches, the label instructs the user not to open until the time of use. The label states also that any solution that remains unused 24 hours after constitution is to be discarded.

Identification—

A: It responds to the flame tests for *Sodium* ⟨191⟩ and for *Potassium* ⟨191⟩.

B: It responds to the tests for *Chloride* ⟨191⟩.

C: Where it contains Sodium Bicarbonate, it dissolves with effervescence, and the collected gas so obtained responds to the test for *Bicarbonate* ⟨191⟩.

D: Where it contains Sodium Citrate, it responds to the tests for *Citrate* ⟨191⟩, 3 to 5 drops of the solution constituted as directed in the labeling and 20 mL of the mixture of pyridine and acetic anhydride being used.

E: Where it contains Dextrose, add a few drops of the solution constituted as directed in the labeling to 5 mL of hot alkaline cupric tartrate TS: a copious red precipitate of cuprous oxide is formed (*presence of dextrose*).

F: When heated it melts, swells, and burns, yielding the odor of burnt sugar.

Loss on drying ⟨731⟩—Dry it at 50° to constant weight: it loses not more than 1.0% of its weight.

Minimum fill ⟨755⟩—Proceed as directed, except to change the requirements following "The average net weight of the contents of the 10 containers is not less than the labeled amount, and the net weight of the contents of any single container is not less than" to read "95% and not more than 105% of the labeled amount. If the contents of not more than 1 container are less than 95% but not less than 90% of the labeled amount or more than 105% but not more than 110% of the labeled amount, determine the net weight of the contents of 20 additional containers. The average net weight of the contents of 30 containers is not less than the labeled amount, and the net weight of the contents of not more than 1 of the 30 containers is less than 95% but not less than 90% of the labeled amount or more than 105% but not more than 110% of the labeled amount."

NOTE—In performing the *Assay for sodium and potassium*, the *Assay for chloride*, the *Assay for bicarbonate*, and the *Assay for citrate*, calculate from the labeled amounts of sodium chloride, potassium chloride, and sodium bicarbonate or sodium citrate the total equivalent amounts of sodium (Na^+), potassium (K^+), chloride (Cl^-), and bicarbonate (HCO_3^-), [or citrate ($C_6H_5O_7^{-3}$)] contained therein. (See accompanying table.)

pH ⟨791⟩: between 7.0 and 8.8, in the solution constituted as directed in the labeling.

Assay for dextrose—Transfer the contents of 1 or more unit-dose containers of Oral Rehydration Salts, or an accurately weighed portion of the contents of 1 multiple-unit container, equivalent to about 20 g of dextrose, to a 100-mL volumetric flask, dilute with water to volume, and mix. Transfer 50.0 mL of this stock solution to a 100-mL volumetric flask. Add 0.2 mL of 6 *N* ammonium hydroxide, dilute with water to volume, and mix. [NOTE—Reserve the remaining stock solution for the *Assay for sodium and potassium*, the *Assay for chloride*, the *Assay for bicarbonate*, and the *Assay for citrate*.] Determine the angular rotation in a suitable polarimeter tube at 25° (see *Optical Rotation* ⟨781⟩). Calculate the quantity, in g, of anhydrous dextrose ($C_6H_{12}O_6$) in the unit-dose container or containers taken or in the portion of powder taken from the multiple-unit container, by the formula:

$$(200/52.7)(a/l),$$

in which 52.7 is the specific rotation of anhydrous dextrose, a is the corrected observed rotation, in degrees, and l is the length, in dm, of the polarimeter tube. Where the Oral Rehydration Salts is labeled to contain Dextrose Monohydrate, calculate the quantity of dextrose monohydrate ($C_6H_{12}O_6 . H_2O$) by the same formula, substituting the figure 47.9, the specific rotation of dextrose monohydrate, in place of 52.7.

Assay for sodium and potassium—

Sodium stock solution—Transfer 14.61 g of sodium chloride, previously dried at 105° for 2 hours and accurately weighed, to a 250-mL volumetric flask, add water to volume, and mix.

Potassium stock solution—Transfer 18.64 g of potassium chloride, previously dried at 105° for 2 hours and accurately weighed, to a 250-mL volumetric flask, add water to volume, and mix.

Lithium diluent solution—Transfer 1.04 g of lithium nitrate to a 1000-mL volumetric flask, add a suitable nonionic surfactant, then add water to volume, and mix.

Standard preparation—Transfer 5.0 mL of *Sodium stock solution* and 5.0 mL of *Potassium stock solution* to a 500-mL volumetric flask, dilute with water to volume, and mix. Transfer 5.0 mL of the resulting solution to a 100-mL volumetric flask, dilute with *Lithium diluent solution* to volume, and mix. Each mL of this solution contains 0.01150 mg of sodium (Na^+) and 0.01955 mg of potassium (K^+).

Assay preparation 1—Dilute an accurately measured volume of the stock solution remaining from the *Assay for dextrose*, quantitatively and stepwise, if necessary, with water to obtain a solution containing about 0.23 mg of sodium (Na^+) per mL. Transfer 5.0 mL of the resulting solution to a 100-mL volumetric flask, dilute with *Lithium diluent solution* to volume, and mix.

Assay preparation 2—Dilute an accurately measured volume of the stock solution remaining from the *Assay for dextrose* quantitatively and stepwise, if necessary, with water to obtain a solution containing about 0.39 mg of potassium per mL. Transfer 5.0 mL of the resulting solution to a 100-mL volumetric flask, dilute with *Lithium diluent solution* to volume, and mix.

Procedure—Using a suitable flame photometer, adjusted to read zero with *Lithium diluent solution*, concomitantly determine the sodium flame emission readings for the *Standard preparation* and *Assay preparation 1* at the wavelength of maximum emission at about 589 nm. Calculate the quantity, in mg, of Na^+ in the unit-dose container or containers taken or in the portion of powder taken from the multiple-unit container by the formula:

$$0.23(L_{Na}/D_{Na})(R_{U,Na}/R_{S,Na}),$$

in which L_{Na} is the quantity, in mg, of sodium (Na^+) in the unit-dose container or containers taken or in the portion of powder taken from the multiple-unit container, calculated from the labeled quantities of sodium chloride and sodium bicarbonate (or sodium citrate), D_{Na} is the concentration, in mg per mL, of sodium in *Assay preparation 1*, based on the volume taken of the stock solution remaining from the *Assay for dextrose* and the extent of dilution, and $R_{U,Na}$ and $R_{S,Na}$ are the sodium emission readings obtained from *Assay preparation 1* and the *Standard preparation*, respectively. Similarly determine the potassium flame emission readings from the *Standard preparation* and *Assay preparation 2* at the wavelength of maximum emission at about 766 nm. Calculate the quantity, in mg, of K^+ in the unit-dose container or containers taken or in the portion of powder taken from the multiple-unit container by the formula:

$$0.391(L_K/D_K)(R_{U,K}/R_{S,K}),$$

in which L_K is the quantity, in mg, of potassium in the unit-dose container or containers taken or in the portion of powder taken from the multiple-unit container, calculated from the labeled quantity of potassium chloride, D_K is the concentration, in mg per mL, of potassium in *Assay preparation 2*, based on the volume taken of the stock solution remaining from the *Assay for dextrose* and the extent of dilution, and $R_{U,K}$ and $R_{S,K}$ are the potassium emission readings obtained from *Assay preparation 2* and the *Standard preparation*, respectively.

Assay for chloride—Transfer an accurately measured volume of the stock solution remaining from the *Assay for dextrose*, equivalent to about 55 mg of chloride (Cl^-), to a porcelain casserole, and add 140 mL of water and 1 mL of dichlorofluorescein TS. Mix, and titrate with 0.1 *N* silver nitrate VS until the silver chloride flocculates and the mixture acquires a faint pink color. Calculate the quantity, in mg, of Cl^- in the unit-dose container or containers taken or in the portion of powder taken from the multiple-unit container by the formula:

$$354.5T/v,$$

in which T is the volume, in mL, of 0.1 *N* silver nitrate consumed, and v is the volume, in mL, of stock solution taken.

Assay for bicarbonate (*if present*)—Transfer an accurately measured volume of the stock solution remaining from the *Assay for dextrose*, equivalent to about 100 mg of bicarbonate (HCO_3^-), to a suitable beaker, add 25 mL of water and 3 drops of methyl orange TS, and titrate with 0.1 *N* hydrochloric acid VS. Calculate the quantity, in mg, of HCO_3^- in the unit-dose container or containers taken or in the portion of powder taken from the multiple-unit container, by the formula:

$$610.2T/v,$$

in which T is the volume, in mL, of 0.1 *N* hydrochloric acid consumed, and v is the volume, in mL, of the stock solution taken.

Assay for citrate (*if present*)—

Mobile phase—Dissolve 20 g of ammonium sulfate in a mixture of water and acetonitrile (980:20). Make adjustments if necessary (see *System Suitability* under *Chromatography* ⟨621⟩).

Standard preparation—Dissolve an accurately weighed quantity of sodium citrate, previously dried at 180° for 18 hours, in water to obtain a solution having a known concentration of about 2.5 mg of anhydrous sodium citrate per mL.

	mg equivalent of each g of component				
Component	Na^+	K^+	Cl^-	HCO_3^-	$C_6H_5O_7^{-3}$
Sodium Chloride	393.4		606.6		
Potassium Chloride		524.4	475.6		
Sodium Bicarbonate	273.6			726.4	
Anhydrous Sodium Citrate	267.2				732.8
Sodium Citrate Dihydrate	234.5				643.0

Assay preparation—Transfer an accurately measured volume of the stock solution remaining from the *Assay for dextrose*, equivalent to about 180 mg of citrate ($C_6H_5O_7^{-3}$), to a 100-mL volumetric flask, dilute with water to volume, and mix.

Chromatographic system (see *Chromatography* ⟨621⟩)—The liquid chromatograph is equipped with a 220-nm detector and a 4.8-mm × 20-cm column that contains packing L8. The flow rate is about 2 mL per minute. Chromatograph the *Standard preparation*, and record the peak response as directed under *Procedure:* the retention time for the citrate peak is about 3 minutes, the column efficiency is not less than 1000 theoretical plates, the tailing factor is not more than 2.0, and the relative standard deviation for replicate injections is not more than 2.0%. [NOTE—The column may be equilibrated before use by making a series of injections of the *Standard preparation* over a period of several hours. If the tailing factor is greater than 2, the equilibration may be facilitated by adding 1 g of sodium citrate to each 1000 mL of the *Mobile phase* and pumping this solution through the column at about 0.5 mL per minute for several hours. The column must then be washed with *Mobile phase* for a few minutes before use.]

Procedure—Separately inject equal volumes (about 20 µL) of the *Standard preparation* and the *Assay preparation* into the chromatograph, record the chromatograms, and measure the responses for the major peaks. Calculate the quantity, in mg, of $C_6H_5O_7^{-3}$ in the unit-dose container or containers taken or in the portion of powder taken from the multiple-unit container by the following formula:

$$(189.12/258.07)(10{,}000C/v)(r_U/r_S),$$

in which 189.12 and 258.07 are the molecular weights of citrate ($C_6H_5O_7^{-3}$) and anhydrous sodium citrate, respectively, C is the concentration, in mg per mL, of anhydrous sodium citrate in the *Standard preparation*, v is the volume, in mL, of the stock solution taken to prepare the *Standard preparation*, and r_U and r_S are the citrate peak responses obtained from the *Assay preparation* and the *Standard preparation*, respectively.

Repository Corticotropin Injection—*see* Corticotropin Injection, Repository

Reserpine

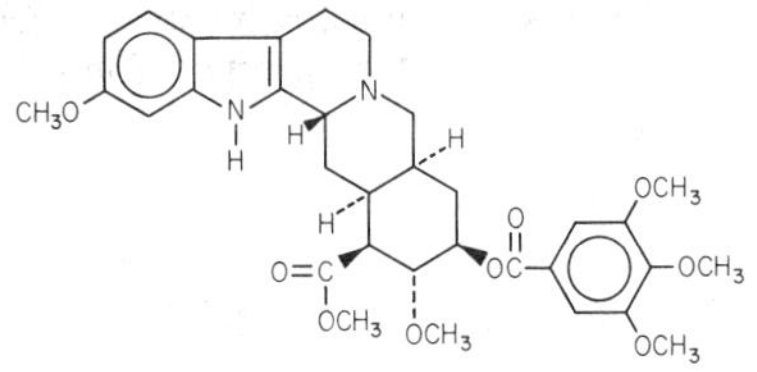

$C_{33}H_{40}N_2O_9$ 608.69

Yohimban-16-carboxylic acid, 11,17-dimethoxy-18-[(3,4,5-trimethoxybenzoyl)oxy]-, methyl ester, (3β,16β,17α,18β,20α)-.

Methyl 18β-hydroxy-11,17α-dimethoxy-3β,20α-yohimban-16β-carboxylate 3,4,5-trimethoxybenzoate (ester) [*50-55-5*].

» Reserpine contains not less than 97.0 percent and not more than 101.0 percent of $C_{33}H_{40}N_2O_9$, calculated on the dried basis.

Packaging and storage—Preserve in tight, light-resistant containers.

Reference standard—*USP Reserpine Reference Standard*—Dry at 60° for 3 hours before using.

Identification—

A: The infrared absorption spectrum of a potassium bromide dispersion of it, previously dried, exhibits maxima only at the same wavelengths as that of a similar preparation of USP Reserpine RS.

B: [NOTE—Conduct this test promptly, with a minimum exposure to light.] Dissolve 25.0 mg of it, previously dried, in 0.25 mL of chloroform, mix with about 30 mL of methanol previously warmed to 50°, transfer the mixture with the aid of warm methanol to a 250-mL volumetric flask, cool the solution to room temperature, dilute with methanol to volume, and mix. Pipet 10 mL of this solution into a 50-mL volumetric flask, add 36 mL of chloroform, dilute with methanol to volume, and mix: the ultraviolet absorption spectrum of a 1 in 50,000 solution so obtained exhibits the same maxima in the range of 255 nm to 350 nm as that of a similar solution of USP Reserpine RS, concomitantly measured, and the respective absorptivities, determined with reference to a mixture of 36 volumes of chloroform and 14 volumes of methanol as the blank, at the wavelength of maximum absorbance at about 268 nm, do not differ by more than 3.0%.

C: To about 0.5 mL of glacial acetic acid add 1 drop of *Solution 2*, obtained as directed under *Assay preparation* in the *Assay*, mix, and add 1 mL of a 1 in 50 solution of vanillin in hydrochloric acid: a pink color is produced, and it becomes deep violet-red within a few minutes or as a result of warming the solution for 10 to 20 seconds.

Loss on drying ⟨731⟩—Dry it at 60° for 3 hours: it loses not more than 0.5% of its weight.

Residue on ignition ⟨281⟩: not more than 0.1%.

Assay—

Standard preparation—Dissolve 25.0 mg of USP Reserpine RS, accurately weighed, in 0.25 mL of chloroform, mix with about 30 mL of methanol previously warmed to 50°, transfer the mixture to a 250-mL volumetric flask with the aid of warm methanol, cool the solution to room temperature, dilute with methanol to volume, and mix (*Solution 1*). Protect the solution from light. Just prior to use in the *Assay*, pipet 10 mL of *Solution 1* into a 50-mL volumetric flask, add 36 mL of chloroform, and dilute with methanol to volume (*Standard preparation*).

Assay preparation—Dissolve 25.0 mg of Reserpine, accurately weighed, in chloroform, and transfer the solution to a 25-mL volumetric flask, diluting with chloroform to volume (*Solution 2*). Dilute quantitatively 5 mL of *Solution 2* with chloroform to 50 mL (*Assay preparation*). Protect the solutions from light.

Procedure—[NOTE—Conduct the entire procedure quickly, without exposure to direct sunlight. In this procedure, perform the directed extractions in a suitable separator or in a suitably stoppered, 50-mL centrifuge tube with separation being effected by centrifuging, the portion to be retained being withdrawn into a hypodermic syringe fitted with a square-tipped, 14-gauge, 15-cm needle.]

Pipet 10 mL of the *Assay preparation* into the extraction vessel, add 10 mL of citric acid solution (1 in 50), and shake thoroughly for 2 minutes. Separate and withdraw the chloroform. Wash the citric acid solution with two 10-mL portions of chloroform, adding the washings to the main chloroform solution. To the combined chloroform solutions add 10 mL of sodium bicarbonate solution (1 in 100), shake for 2 minutes, and separate. Withdraw the chloroform, filtering it through a pledget of cotton, into a 50-mL volumetric flask containing 14.0 mL of methanol. Extract the aqueous bicarbonate layer in the extraction vessel with two 2-mL portions of chloroform, passing each portion successively through the filter into the volumetric flask. Add chloroform to volume, and mix.

Pipet duplicate 5-mL aliquots of the chloroform-methanol solution and of the *Standard preparation* into separate, 10-mL volumetric flasks. Add 2.0 mL of a 1 in 10 solution of hydrochloric acid in methanol to each flask. To one flask of each pair of duplicates (representing the *Standard preparation* and the extracted *Assay preparation*) add 1.0 mL of a 3 in 1000 solution of sodium nitrite in dilute methanol (1 in 2). To the second flask of each pair (constituting the blanks) add 1 mL of dilute methanol (1 in 2). Mix, and allow to stand for 30 minutes. Add 0.5 mL of ammonium sulfamate solution (1 in 20) to each flask, add methanol to volume, mix, and allow to stand for 10 minutes. Determine the absorbance of each solution in a 1-cm cell at the wavelength of maximum absorbance at about 390 nm, with a suitable spectrophotometer, using a mixture of methanol, chloroform, and water (5.4:3.6:1) as the blank.

Calculate the quantity, in mg, of $C_{33}H_{40}N_2O_9$ in the portion of Reserpine taken by the formula:

$$25(A - A°)_U/(A - A°)_S,$$

in which the parenthetic expressions are the differences in absorbances of the nitrite-treated and blank solutions, respectively, from the *Assay preparation* ($_U$) and the *Standard preparation* ($_S$).

Reserpine Elixir

» Reserpine Elixir contains not less than 90.0 percent and not more than 110.0 percent of the labeled amount of $C_{33}H_{40}N_2O_9$.

Packaging and storage—Preserve in tight, light-resistant containers.

Reference standard—*USP Reserpine Reference Standard*—Dry at 60° for 3 hours before using.

Identification—Evaporate about 2 mL of the chloroform-methanol solution obtained from the *Assay preparation* as directed under *Procedure* in the *Assay*, in a test tube to dryness, add to the residue 0.5 mL of glacial acetic acid, swirl for 1 to 2 minutes, and add 1 mL of a 1 in 50 solution of vanillin in hydrochloric acid: a pink color is produced, and it becomes deep violet-red within a few minutes or as a result of warming the solution for 10 to 20 seconds.

Alcohol content ⟨611⟩: between 11.0% and 13.0% of C_2H_5OH.

Assay—

Standard preparation—Prepare as directed in the *Assay* under *Reserpine.*

Procedure—Transfer an accurately measured volume of Reserpine Elixir, equivalent to about 1 mg of reserpine, into a separator or a suitably stoppered, 50-mL centrifuge tube, add 5 mL of citric acid solution (1 in 50) and 10 mL of chloroform, and shake for 2 minutes. Proceed as directed in the second paragraph of the *Procedure* in the *Assay* under *Reserpine*, beginning with "Separate and withdraw the chloroform." Calculate the quantity, in mg, of $C_{33}H_{40}N_2O_9$ in each mL of the Elixir taken by the formula:

$$(A - A°)_U/V(A - A°)_S,$$

in which V is the volume, in mL, of Elixir taken, and the parenthetic expressions are the differences in absorbances of the nitrite-treated and blank solutions, respectively, from the *Assay preparation* ($_U$) and the *Standard preparation* ($_S$).

Reserpine Injection

» Reserpine Injection is a sterile solution of Reserpine in Water for Injection, prepared with the aid of a suitable acid. It contains not less than 90.0 percent and not more than 110.0 percent of the labeled amount of $C_{33}H_{40}N_2O_9$. It contains suitable antioxidants.

Packaging and storage—Preserve in single-dose (or, if stabilizers are present, in multiple-dose), light-resistant containers, preferably of Type I glass.

Reference standard—*USP Reserpine Reference Standard*—Dry at 60° for 3 hours before using.

Identification—It responds to the *Identification test* under *Reserpine Elixir.*

pH ⟨791⟩: between 3.0 and 4.0.

Other alkaloids—[NOTE—Conduct this test promptly after preparation of the test and standard solutions.] Pipet 10 mL each of the citric acid solution of the Injection, and of *Solution 1* used in preparing the *Standard preparation*, respectively, obtained as directed in the *Assay*, into two separators. To the Injection solution add 100 mL of saturated sodium bicarbonate solution, and to *Solution 1* add 10 mL of water, 10 drops of saturated sodium bicarbonate solution, and 90 mL of water, and extract both of the resulting solutions with 50 mL of ether. Transfer the aqueous phase to another separator, extract with a second 50-mL portion of ether, and discard the aqueous layers. Wash the ether layers in succession with two 25-mL portions of water, and discard the washings. Extract the combined ether layers with three 15-mL portions of 2 *N* sulfuric acid, collect the extracts in a 50-mL volumetric flask, add 2 *N* sulfuric acid to volume, and mix. The absorption spectrum of the solution from the Injection, in the range of 255 nm to 350 nm, measured in a 1-cm cell, 2 *N* sulfuric acid being used as the blank, exhibits maxima and minima only at the same wavelengths as that of the solution from the *Standard preparation*, concomitantly measured. Calculate the quantity, in mg, of total alkaloids in each mL of Injection taken by the formula:

$$10(I/SV),$$

in which I is the absorbance of the solution from the Injection at the wavelength of maximum absorbance at about 268 nm, S is that of the solution from the *Standard preparation*, and V is the volume, in mL, of Injection taken. The content of total alkaloids is not more than 114.0% of the labeled amount of $C_{33}H_{40}N_2O_9$, and does not differ by more than 10.0% from the amount of $C_{33}H_{40}N_2O_9$ determined in the *Assay.*

Other requirements—It meets the requirements under *Injections* ⟨1⟩.

Assay—

Standard preparation—Prepare as directed in the *Assay* under *Reserpine.*

Assay preparation—Transfer to a 100-mL volumetric flask an accurately measured volume of Reserpine Injection, equivalent to about 10 mg of reserpine, and dilute with citric acid solution (1 in 50) to volume.

Procedure—Pipet 10 mL of the *Assay preparation* into a separator or a suitably stoppered, 50-mL centrifuge tube, add 10 mL of chloroform, and shake for 2 minutes. Then proceed as directed for *Procedure* in the *Assay* under *Reserpine*, beginning with "Separate and withdraw the chloroform." Calculate the quantity, in mg, of $C_{33}H_{40}N_2O_9$ in each mL of the Injection taken by the formula:

$$10(A - A°)_U/V(A - A°)_S,$$

in which V is the volume, in mL, of Injection taken, and the parenthetic expressions are the differences in absorbances of the nitrite-treated and blank solutions, respectively, from the *Assay preparation* ($_U$) and the *Standard preparation* ($_S$).

Reserpine Tablets

» Reserpine Tablets contain not less than 90.0 percent and not more than 110.0 percent of the labeled amount of $C_{33}H_{40}N_2O_9$.

Packaging and storage—Preserve in tight, light-resistant containers.

Reference standard—*USP Reserpine Reference Standard*—Dry at 60° for 3 hours before using.

Identification—Evaporate about 2 mL of the chloroform-methanol solution, obtained from the *Assay preparation* as directed under *Procedure* in the *Assay*, in a test tube to dryness, add to the residue 0.5 mL of the glacial acetic acid, swirl for 1 to 2 minutes, and add 1 mL of a 1 in 50 solution of vanillin in hydrochloric acid: a pink color is produced, and it becomes deep violet-red within a few minutes or as a result of warming the solution for 10 to 20 seconds.

Dissolution ⟨711⟩—[NOTE—Do not substitute membrane filters for filter paper where the filtration of reserpine-containing solutions is indicated. Reserpine has been shown to be adsorbed onto membranes.]

Medium: 0.1 *N* acetic acid; 900 mL.

Apparatus 1: 100 rpm.

Time: 45 minutes.

ANALYTICAL METHOD—Determine the amount of $C_{33}H_{40}N_2O_9$ dissolved using the following method.

Apparatus—Use an automatic analyzer consisting of a liquid sampler, a proportioning pump, a fluorometer with excitation set at 375 nm and emission measured at 495 nm, and a manifold consisting of the components illustrated in the diagram under *Automated Methods of Analysis* ⟨16⟩.

Diluting solvent—Dilute 19 mL of phosphoric acid with methanol to 1 liter, and mix.

Vanadium pentoxide reagent—Prepare a saturated solution of vanadium pentoxide in phosphoric acid by shaking by mechanical means vanadium pentoxide in phosphoric acid for 2 hours. Filter the resultant solution through a medium-porosity, sintered-glass funnel.

Stock standard solution—Prepare a solution in *Diluting solvent* of USP Reserpine RS having a known concentration of about 0.222 mg per mL.

Standard solution I (for Tablets labeled to contain 0.25 mg or more)—Dilute 10.0 mL of *Stock standard solution* with 0.1 *N* acetic acid to 100 mL (*Solution A*). Dilute 25.0 mL of *Solution A* quantitatively and stepwise with 0.1 *N* acetic acid to obtain a final solution having a known concentration of about 277.5 ng per mL.

Standard solution II (for Tablets labeled to contain 0.1 mg)—Dilute 10.0 mL of *Solution A* with 0.1 *N* acetic acid to 100 mL. Dilute 5.0 mL of the resulting solution with 0.1 *N* acetic acid to 100 mL.

Test solution—Use a filtered portion of the solution under test.

Procedure—With the *Apparatus* set as described under *Automated Methods of Analysis* ⟨16⟩, start the sampler, and conduct determinations at a rate of 10 per hour, using a ratio of about 1:4 for the sample and wash time. Calculate the amount of reserpine dissolved.

Tolerances—Not less than 75% (*Q*) of the labeled amount of $C_{33}H_{40}N_2O_9$ is dissolved in 45 minutes.

Uniformity of dosage units ⟨905⟩: meet the requirements.

Procedure for content uniformity—

Phosphoric acid–methanol solution—Dilute 20 mL of phosphoric acid with 200 mL of methanol, then dilute with water to 1000 mL, and mix.

Vanadium pentoxide reagent—Prepare a saturated solution of vanadium pentoxide in phosphoric acid by shaking by mechanical means for 1 hour, 100 mg of vanadium pentoxide with 100 mL of phosphoric acid. Allow undissolved solids to settle overnight, and filter the supernatant solution through a medium-porosity, sintered-glass funnel.

Standard solution—Using an accurately weighed quantity of USP Reserpine RS, prepare a solution in *Phosphoric acid–methanol solution* having a known concentration of about 0.002 mg per mL.

Test preparation—Place 1 Tablet in a suitable container, and add an accurately measured volume of *Phosphoric acid–methanol solution* so that the concentration of the final solution is about 0.002 mg of reserpine per mL. Shake by mechanical means until the tablet is completely disintegrated. Filter before using.

Procedure—With the *Apparatus* set as described for *Automated Dissolution Test for Reserpine Tablets* under *Automated Methods of Analysis* ⟨16⟩, conduct determinations of one Standard per five Tablets. Calculate the quantity, in mg, of $C_{33}H_{40}N_2O_9$ in the Tablet taken by the formula:

$$CD(I_U / I_S),$$

in which *C* is the concentration, in mg per mL, of USP Reserpine RS in the *Standard solution*, *D* is the dilution factor for the *Test preparation*, and I_U and I_S are the fluorescence intensities obtained from the *Test preparation* and the *Standard solution*, respectively.

Other alkaloids—Prepare a blank chloroform-methanol solution as directed in the *Assay*, replacing the *Assay preparation* with 1 mL of dimethylsulfoxide and 2 g of *Adsorbant*. Determine the ultraviolet absorption spectrum of the chloroform-methanol solution obtained from the *Assay preparation*, between 255 nm and 350 nm, using the blank in the reference cell. In a similar manner, determine the ultraviolet absorption spectrum of the *Standard preparation* obtained in the *Assay*, between 255 nm and 350 nm, using a solution of 3.6 volumes of chloroform and 1.4 volumes of methanol as the blank. The two spectra are similar, and the ratio A_{268}/A_{295} for the solution representing the Tablets does not differ by more than 4.0% from the corresponding ratio for the solution representing the USP Reserpine RS. Calculate the quantity, in mg, of $C_{33}H_{40}N_2O_9$ in the portion of Tablets taken by the formula:

$$T/S,$$

in which *T* and *S* are the absorbances of the solutions representing the Tablets and the USP Reserpine RS, respectively, at the absorption maximum at about 268 nm. The result so obtained does not differ from the result obtained in the *Assay* by more than 6.0%.

Reserpine and Chlorothiazide Tablets

» Reserpine and Chlorothiazide Tablets contain not less than 90.0 percent and not more than 110.0 percent of the labeled amount of reserpine ($C_{33}H_{40}N_2O_9$) and not less than 93.0 percent and not more than 107.0 percent of the labeled amount of chlorothiazide ($C_7H_6ClN_3O_4S_2$).

Packaging and storage—Preserve in tight, light-resistant containers.

Reference standards—*USP Reserpine Reference Standard*—Dry at 60° for 3 hours before using. *USP Chlorothiazide Reference Standard*—Dry at 105° for 1 hour before using.

Identification—

A: Transfer a quantity of powdered Tablets, equivalent to about 1 mg of reserpine, to a stoppered, 50-mL centrifuge tube. Add 20 mL of citric acid solution (1 in 50), and shake until the powder is suspended. Extract the mixture with two 20-mL portions of chloroform, centrifuge, and withdraw the chloroform, filtering each extract through a pledget of cotton into a 50-mL volumetric flask. Dilute with chloroform to volume, and mix: the ultraviolet absorption spectrum of the solution so obtained exhibits maxima and minima at the same wavelengths as that of a similar solution of USP Reserpine RS, concomitantly measured (*presence of reserpine*).

B: Transfer a quantity of powdered Tablets, equivalent to about 50 mg of chlorothiazide, to a test tube containing 10 mL of acetone, agitate for 5 minutes, and centrifuge. Use the clear supernatant liquid as the Test solution. Separately apply 10 µL each of the Test solution and a Standard solution of USP Chlorothiazide RS in acetone containing 5 mg per mL to a thin-layer chromatographic plate (see *Chromatography* ⟨621⟩), coated with a 0.25-mm layer of chromatographic silica gel mixture and previously washed with methanol. Develop the chromatogram in a solvent system consisting of a mixture of ethyl acetate and isopropyl alcohol (17:3) until the solvent front has moved about three-fourths of the length of the plate. Remove the plate from the chamber, air-dry, and examine under short-wavelength ultraviolet light: the R_f value of the principal spot in the chromatogram of the Test solution corresponds to that obtained from the Standard solution (*presence of chlorothiazide*).

Dissolution ⟨711⟩—

Medium: mixture of pH 8.0 phosphate buffer (see *Buffer Solutions* in the section, *Reagents, Indicators, and Solutions*) and *n*-propyl alcohol (3:2); 900 mL.

Apparatus 2: 75 rpm.

Time: 60 minutes.

Determination of dissolved reserpine—

STANDARD PREPARATION—Dissolve about 34 mg of USP Reserpine RS, accurately weighed, in a 50-mL volumetric flask containing 5 mL of chloroform, dilute with *n*-propyl alcohol to volume (*Solution 1*), and mix. Pipet 1 mL of *Solution 1* into a 50-mL volumetric flask, dilute with *n*-propyl alcohol to volume (*Solution 2*), and mix. Pipet 1 mL of *Solution 2* into a 100-mL volumetric flask, dilute with *Dissolution Medium* to volume, and mix to obtain the *Standard preparation*.

PROCEDURE—Filter a portion of the solution under test through paper, and transfer 5.0 mL of the clear filtrate to a 25-mL volumetric flask. Pipet 5 mL of the *Standard preparation* into a separate 25-mL volumetric flask. Treat each flask as follows: Add 5 drops of hydrochloric acid and 5 mL of a mixture of water, alcohol, and sulfuric acid (29:20:1), and mix. Add by pipet 5 mL of sodium nitrite solution (3 in 1000), dilute with alcohol to volume, mix, and allow to stand for 30 minutes. Concomitantly determine the fluorescences of the solution under test and the *Standard preparation* in a suitable spectrophotometer arranged to deliver activation radiation at 405 nm and to measure the resultant fluorescence at the emission wavelength of about 500 nm. Calculate the amount of reserpine ($C_{33}H_{40}N_2O_9$) dissolved.

Determination of dissolved chlorothiazide—

STANDARD PREPARATION—Transfer about 27 mg of USP Chlorothiazide RS, accurately weighed, to a 50-mL volumetric flask containing 5 mL of methanol, swirl to dissolve, dilute with *Dissolution Medium* to volume, and mix. Pipet 2 mL of this solution into a 100-mL volumetric flask, dilute with *Dissolution Medium* to volume, and mix.

PROCEDURE—Determine the amount of chlorothiazide ($C_7H_6ClN_3O_4S_2$) dissolved from ultraviolet absorbances at the wavelength of maximum absorbance at about 292 nm of filtered portions of the solution under test, suitably diluted with *Dissolution Medium*, in comparison with the *Standard preparation*.

Tolerances—Not less than 75% (*Q*) of the labeled amount of $C_{33}H_{40}N_2O_9$ is dissolved in 60 minutes, and not less than 75% (*Q*) of the labeled amount of $C_7H_6ClN_3O_4S_2$ is dissolved in 60 minutes.

Uniformity of dosage units ⟨905⟩: meet the requirements with respect to chlorothiazide and to reserpine.

Procedure for content uniformity for reserpine—

Standard preparation—Prepare as directed for *Standard preparation* in the *Assay for reserpine*.

Test preparation—Weigh 1 Tablet, grind to a fine powder, and transfer to a stoppered, 50-mL centrifuge tube. Add 25.0 mL of chloroform and methanol solution (1:1), shake by mechanical means for 15 minutes, and centrifuge. Pipet 4 mL of the clear supernatant liquid into a 100-mL volumetric flask, dilute with the chloroform and methanol solution to volume, and mix.

Procedure—Proceed as directed for *Procedure* in the *Assay for reserpine*. Calculate the quantity, in mg, of $C_{33}H_{40}N_2O_9$, in the Tablet by the formula:

$$(W_t/W_U)(TC/D)(I_U/I_S),$$

in which W_t is the weight, in mg, of the Tablet, W_U is the weight, in mg, of the portion of Tablet taken, T is the labeled quantity, in mg, of reserpine in the Tablet, C is the concentration, in µg per mL, of USP Reserpine RS in the *Standard preparation*, D is the concentration, in µg per mL, of reserpine in the *Test preparation*, based upon the labeled quantity per Tablet and the extent of dilution, and I_U and I_S are the fluorescence intensities of the solutions from the *Test preparation* and the *Standard preparation*, respectively.

Procedure for content uniformity for chlorothiazide—

Standard preparation—Prepare as directed for *Standard preparation* under *Assay for chlorothiazide*.

Test preparation—Transfer 1 Tablet to a 500-mL volumetric flask, add 300 mL of 0.1 *N* sodium hydroxide, and sonicate, swirling the flask intermittently, until the Tablet is dissolved. Dilute with 0.1 *N* sodium hydroxide to volume, mix, and filter, discarding the first 15 mL of the filtrate. Dilute a portion of the clear filtrate quantitatively with 0.1 *N* sodium hydroxide to obtain a solution having a concentration of about 10 µg of chlorothiazide per mL.

Procedure—Proceed as directed for *Procedure* in the *Assay for chlorothiazide*. Calculate the quantity, in mg, of $C_7H_6ClN_3O_4S_2$, in the Tablet by the formula:

$$(TC/D)(A_U/A_S),$$

in which T is the labeled quantity, in mg, of chlorothiazide in the Tablet, C is the concentration, in µg per mL, of USP Chlorothiazide RS in the *Standard preparation*, D is the concentration, in µg per mL, of chlorothiazide in the *Test preparation*, based upon the labeled quantity per Tablet and the extent of dilution, and A_U and A_S are the absorbances of the solutions from the *Test preparation* and the *Standard preparation*, respectively.

Assay for reserpine—

Standard preparation—Dissolve about 25 mg of USP Reserpine RS, accurately weighed, in 1 mL of chloroform contained in a 50-mL volumetric flask, dilute with methanol to volume, and mix. Dilute a portion of this solution quantitatively and stepwise with chloroform and methanol solution (1:1) to obtain a solution having a known concentration of about 0.2 µg of reserpine per mL.

Assay preparation—Weigh and finely powder not less than 20 Reserpine and Chlorothiazide Tablets. Transfer an accurately weighed portion of the powder, equivalent to about 1 mg of reserpine, to a stoppered, 50-mL centrifuge tube, add 25.0 mL of chloroform and methanol solution (1:1), shake by mechanical means for 15 minutes, and centrifuge. Dilute a portion of the clear supernatant liquid quantitatively and stepwise with chloroform and methanol solution (1:1) to obtain a solution having a concentration of about 0.2 µg of reserpine per mL.

Procedure—Separately transfer 5.0 mL of the *Assay preparation*, 5.0 mL of the *Standard preparation*, and 5.0 mL of chloroform and methanol solution (1:1) to provide the reagent blank, respectively, to three 25-mL volumetric flasks. To each flask add 0.5 mL of hydrochloric acid, 1.0 mL of a 3 in 1000 solution of sodium nitrite in dilute methanol (1 in 2), mix, and allow to stand for 30 minutes. Add 1 mL of ammonium sulfamate solution (1 in 20) to each flask, add chloroform and methanol solution (1:1) to volume, mix, and allow to stand for 10 minutes. Concomitantly determine the fluorescence intensities of the solutions in a suitable spectrophotometer arranged to deliver activation radiation at 405 nm and to measure the resultant fluorescence at the emission wavelength of about 500 nm. Calculate the quantity, in mg, of $C_{33}H_{40}N_2O_9$ in the portion of Tablets taken by the formula:

$$5C(I_U/I_S),$$

in which C is the concentration, in µg per mL, of USP Reserpine RS in the *Standard preparation*, and I_U and I_S are the fluorescence intensities of the solutions from the *Assay preparation* and the *Standard preparation*, respectively.

Assay for chlorothiazide—

Standard preparation—Dissolve an accurately weighed quantity of USP Chlorothiazide RS in 0.1 *N* sodium hydroxide, and dilute quantitatively and stepwise with 0.1 *N* sodium hydroxide to obtain a solution having a known concentration of about 10 µg per mL. Use a freshly prepared solution.

Assay preparation—Weigh and finely powder not less than 20 Reserpine and Chlorothiazide Tablets. Transfer an accurately weighed portion of the powder, equivalent to about 250 mg of chlorothiazide, to a 500-mL volumetric flask, add about 300 mL of 0.1 *N* sodium hydroxide, and shake by mechanical means for 15 minutes. Dilute with 0.1 *N* sodium hydroxide to volume, and mix. Filter through paper, discarding the first 15 mL of the filtrate. Pipet 2 mL of the clear filtrate into a 100-mL volumetric flask, dilute with 0.1 *N* sodium hydroxide to volume, and mix.

Procedure—Concomitantly determine the absorbances of the *Standard preparation* and the *Assay preparation* in 1-cm cells at the wavelength of maximum absorbance at about 292 nm, with a suitable spectrophotometer, using 0.1 *N* sodium hydroxide as the blank. Calculate the quantity, in mg, of $C_7H_6ClN_3O_4S_2$, in the portion of Tablets taken by the formula:

$$25C(A_U/A_S),$$

in which C is the concentration, in µg per mL, of USP Chlorothiazide RS in the *Standard preparation*, and A_U and A_S are the absorbances of the *Assay preparation* and the *Standard preparation*, respectively.

Reserpine, Hydralazine Hydrochloride, and Hydrochlorothiazide Tablets

» Reserpine, Hydralazine Hydrochloride, and Hydrochlorothiazide Tablets contain not less than 90.0 percent and not more than 110.0 percent of the labeled amount of reserpine ($C_{33}H_{40}N_2O_9$), and not less than 93.0 percent and not more than 107.0 percent of the labeled amounts of hydralazine hydrochloride ($C_8H_8N_4 . HCl$) and hydrochlorothiazide ($C_7H_8ClN_3O_4S_2$).

NOTE—Where Reserpine, Hydralazine Hydrochloride, and Hydrochlorothiazide Tablets are prescribed, without reference to the quantity of reserpine, hydralazine hydrochloride, or hydrochlorothiazide contained therein, a product containing 0.1 mg of reserpine, 25 mg of hydralazine hydrochloride, and 15 mg of hydrochlorothiazide shall be dispensed.

Packaging and storage—Preserve in tight, light-resistant containers.

Reference standards—*USP Reserpine Reference Standard*—Dry at 60° for 3 hours before using. *USP Hydralazine Hydrochloride Reference Standard*—Dry in vacuum over phosphorus pentoxide for 8 hours before using. *USP Hydrochlorothiazide Reference Standard*—Dry at 105° for 1 hour before using. *USP 4-Amino-6-chloro-1,3-benzenedisulfonamide Reference Standard*—Dry at 105° for 1 hour before using.

Identification—

A: Transfer a quantity of finely powdered Tablets, equivalent to about 100 mg of hydralazine hydrochloride, to a glass-stoppered flask. Add 40 mL of dilute hydrochloric acid (1 in 12), shake by mechanical means for 5 minutes, and filter, discarding the first few mL of the filtrate. Transfer 20 mL of the filtrate to a separator. Extract with a 10-mL portion of methylene chloride, and discard the methylene chloride. To the aqueous phase add 2 mL of sodium nitrite solution (14 in 1000) and 10 mL of methylene chloride, and shake by mechanical means for 5 minutes. Allow the layers to separate, and drain the methylene chloride through sodium sulfate that has been pre-washed with methylene chloride, into a 50-mL beaker. Evaporate over gentle heat with the aid of nitrogen to dryness: the infrared absorption spectrum of a potassium bromide dispersion of the residue so obtained exhibits maxima only at the same wavelengths as that of a similar preparation of USP Hydralazine Hydrochloride RS, similarly treated (*presence of hydralazine hydrochloride*).

B: Transfer a quantity of finely powdered Tablets, equivalent to about 1 mg of reserpine, to a 50-mL centrifuge tube. Add 20 mL of cyclohexane, shake by mechanical means for 15 minutes, centrifuge, and discard the cyclohexane. Repeat the extraction with two additional 20-mL portions of cyclohexane, shaking by mechanical means for 2 minutes each time. To the residue add 10 mL of chloroform, shake for 2 minutes, and filter through a medium-porosity, sintered-glass funnel into another 50-mL centrifuge tube. Extract the chloroform with 10 mL of 1.0 *N* hydrochloric acid, discarding the aqueous acid layer. Extract the chloroform with 10 mL of 0.5 *N* sodium hydroxide. Centrifuge for 5 minutes, and withdraw the chloroform with a syringe, passing the chloroform through cotton into a 50-mL volumetric flask containing 40 mL of methanol. Dilute with chloroform to volume, and mix: the ultraviolet absorption spectrum of the solution so obtained exhibits maxima and minima at the same wavelengths as that of a similar solution of USP Reserpine RS, concomitantly measured (*presence of reserpine*).

C: The ultraviolet absorption spectrum of the solution from the *Assay preparation*, obtained as directed under *Procedure* in the *Assay for hydrochlorothiazide*, exhibits maxima and minima at the same wavelengths as that of the solution from the *Standard preparation*, prepared as directed in the *Assay for hydrochlorothiazide*, similarly measured (*presence of hydrochlorothiazide*).

Disintegration ⟨701⟩: 30 minutes.

Uniformity of dosage units ⟨905⟩: meet the requirements for *Content Uniformity* with respect to reserpine, to hydralazine hydrochloride, and to hydrochlorothiazide.

Procedure for content uniformity—

Apparatus—Use an automatic analyzer consisting of (1) a solid sampler with 100-mL dissolution capability; (2) a 20-channel peristaltic pump; (3) a continuous filtering device; (4) a colorimeter equipped with a 2-mm flow cell and analysis capability at 530 nm; (5) an ultraviolet spectrophotometer equipped with a 10-mm flow cell and analysis capability at 271 nm; (6) a spectrofluorometer equipped with a 2-mm flow cell and analysis capability of 365 nm activation energy and 495 nm fluorescence; (7) recording devices for each of the three aforementioned detectors; and (8) a manifold consisting of components illustrated in the pertinent diagram in the chapter, *Automated Methods of Analysis* ⟨16⟩. Prepare the ion-exchange column listed in the manifold as follows. Wash sulfonic acid cation-exchange resin (40- to 60-mesh) with isopropyl alcohol until the alcohol shows no appreciable ultraviolet absorption at 271 nm, then add sufficient 1 *N* hydrochloric acid to cover the resin. Drain off the acid, and wash the resin with an equivalent volume of 1 *N* sodium hydroxide and then again with 1 *N* hydrochloric acid. Finally wash the resin with water until the effluent is neutral to litmus. Draw the resin into a 10-mm length of "solvaflex" tubing (1 mm ID) by vacuum, and plug each end with glass wool.

Saturated vanadium pentoxide (V_2O_5)—Stir for about 6 hours or shake by mechanical means for 2 hours about 100 mg of vanadium pentoxide powder with 100 mL of phosphoric acid. Filter through a medium-porosity, sintered-glass funnel. This solution is stable for 1 month.

0.3% Hydrogen peroxide (H_2O_2)—Dilute 1 mL of 30 percent hydrogen peroxide with water to 100 mL. Prepare fresh daily.

Blue tetrazolium reagent (B.T.)—Dissolve 760 mg of blue tetrazolium in 3.8 liters of a mixture of dehydrated alcohol and methanol (19:1).

Tetramethylammonium hydroxide (T.M.A.H.)—Dilute 38 mL of tetramethylammonium hydroxide solution (1 in 10) with dehydrated alcohol and methanol (19:1) to 3.8 liters.

Solvent, wash, and diluent—Prepare by mixing equal volumes of methanol and water. Add 1 mL of phosphoric acid to each 3.8 liters.

Standard preparation—Accurately weigh about 42.0 mg of USP Reserpine RS into a 200-mL volumetric flask, dissolve in methanol, and dilute with methanol to volume. Pipet 10 mL of this solution into a 100-mL volumetric flask containing about 315.0 mg of USP Hydrochlorothiazide RS and about 525.0 mg of USP Hydralazine Hydrochloride RS, accurately weighed. Dissolve in *Solvent*, and dilute with *Solvent* to volume. (A 5-mL aliquot represents 1 standard Tablet.)

Procedure—With the sampler in the standby position, pump all reagents through the system until a stable baseline is achieved. Activate the sampler, and allow one cycle to pass without introducing the Tablets or the *Standard preparation*, then pipet a 5-mL aliquot of the *Standard preparation* into the hopper at the solvent addition portion for the next two cycles and for every sixth cycle thereafter. Disregard the first value for the *Standard preparation*. Add the Tablets to the sampler at the rate of 20 per hour to follow the second 5-mL aliquot of the *Standard preparation*. Record the absorbance and fluorescence values, and calculate each peak value by the difference between peak height and baseline. Calculate the quantity, in mg per Tablet, by the formula:

$$(0.005/1.05)C(A_U/A_S),$$

in which C is the concentration, in µg per mL, of the appropriate Reference Standard in the *Standard preparation*, A_U is the absorbance of the Tablet, and A_S is the averaged absorbance of the *Standard preparations*.

Diazotizable substances—

Standard preparation—Weigh accurately 25 mg of USP 4-Amino-6-chloro-1,3-benzenedisulfonamide RS, dissolve in 5 mL of methanol contained in a 50-mL volumetric flask, dilute with water to volume, and mix. Pipet 4 mL of this solution into a 100-mL volumetric flask, dilute with water to volume, and mix.

Test preparation—Transfer a portion of the powdered Tablets prepared for the *Assay for hydrochlorothiazide*, accurately weighed and equivalent to about 100 mg of hydrochlorothiazide, to a 100-mL volumetric flask, and add a mixture of 20 mL of methanol and 20 mL of water. Shake continuously for 5 to 10 minutes, dilute with water to volume, mix, and filter. Use the filtrate as the *Test preparation*.

Procedure—Pipet 5 mL each of the *Standard preparation* and the *Test preparation* into separate, 50-mL volumetric flasks. Pipet 5 mL of water into a third 50-mL volumetric flask to provide the blank. To each flask add 1 mL of freshly prepared sodium nitrite solution (1 in 100) and 5 mL of dilute hydrochloric acid (1 in 12), and allow to stand for 5 minutes. Add 2 mL of ammonium sulfamate solution (1 in 50), allow to stand for 5 minutes with frequent swirling, then add 2 mL of freshly prepared disodium chromotropate solution (1 in 100) and 10 mL of sodium acetate TS. Dilute with water to volume, and mix. Concomitantly determine the absorbances of the solutions from the *Standard preparation* and the *Test preparation* in 1-cm cells at the wavelength of maximum absorbance at about 500 nm, with a suitable spectrophotometer, against the blank. The absorbance of the solution from the *Test preparation* does not exceed that of the solution from the *Standard preparation*, corresponding to not more than 2.0% of diazotizable substances.

Assay for reserpine—

Standard preparation—Dissolve about 25 mg of USP Reserpine RS, accurately weighed, in chloroform in a 50-mL volumetric flask, and dilute with chloroform to volume (*Solution 1*). Pipet 10 mL of *Solution 1* into a 50-mL volumetric flask, and dilute with chloroform to volume. Protect the solution from light.

Assay preparation—Weigh and finely powder not less than 20 Reserpine, Hydralazine Hydrochloride, and Hydrochlorothiazide Tablets. Transfer an accurately weighed portion of the powder, equivalent to about 1 mg of reserpine, to a separator or a stoppered, 50-mL centrifuge tube containing 10 mL of citric acid solution (1 in 50). Shake vigorously until the powder is completely suspended. Add 10 mL of chloroform, and proceed as directed under *Procedure*, beginning with "shake thoroughly for 2 minutes."

Procedure—[NOTE—Conduct the entire procedure quickly, without exposure to direct sunlight. Perform the extractions in a suitable separator or in a suitably stoppered, 50-mL centrifuge tube, and separate by centrifuging, withdrawing the portion to be retained into a hypodermic syringe fitted with a square-tipped, 14-gauge, 15-cm needle.] Pipet 10 mL of the *Standard preparation* into the extraction vessel, add 10 mL of citric acid solution (1 in 50), and shake thoroughly for 2 minutes. Separate and withdraw the chloroform. Wash the citric acid solution with two 10-mL portions of chloroform, adding the washings to the main chloroform solution. To the combined chloroform solutions add 10 mL of citric acid solution (1 in 50), shake for 2 minutes, and separate and withdraw the chloroform. To the combined chloroform solutions add 10 mL of sodium bicarbonate solution (1 in 100), shake for 2 minutes, and separate. Withdraw the chloroform, filtering it through a pledget of cotton, into a 50-mL volumetric flask containing 14.0 mL of methanol. Extract the aqueous bicarbonate layer in the extraction vessel with two 2-mL portions of chloroform, passing each portion successively through the filter into the volumetric flask. Add chloroform to volume, and mix. Pipet duplicate 5-mL aliquots of the chloroform-methanol solutions into separate, 10-mL volumetric flasks. Add 2.0 mL of a 1 in 10 solution of hydrochloric acid in methanol to each flask. To one flask of each pair of duplicates (representing the extracted *Standard preparation* and the extracted *Assay preparation*) add 1.0 mL of a 3 in 1000 solution of sodium nitrite in dilute methanol (1 in 2). To the second flask of each pair (constituting the blanks) add 1 mL of dilute methanol (1 in 2). Mix, and allow to stand for 30 minutes. Add 0.5 mL of ammonium sulfamate solution (1 in 20) to each flask, add methanol to volume, mix, and allow to stand for 10 minutes. Determine the absorbance of each solution in a 1-cm cell at the wavelength of maximum absorbance at about 390 nm, with a suitable spectrophotometer, relative to the absorbance of a mixture of methanol, chloroform, and water (5.4:3.6:1).

Calculate the quantity, in mg, of $C_{33}H_{40}N_2O_9$ in the portion of Tablets taken by the formula:

$$0.01C(A - A°)_U/(A - A°)_S,$$

in which C is the concentration, in µg per mL, of USP Reserpine RS in the *Standard preparation*, and the parenthetic expressions are the differences in absorbances of the nitrite-treated and blank solutions, respectively, from the *Assay preparation* ($_U$) and the *Standard preparation* ($_S$).

Assay for hydralazine hydrochloride—

Sodium acetate solution—Dissolve 27.2 g of sodium acetate trihydrate in 50 mL of water, bring to room temperature, and dilute with water to 100 mL.

Ferric ammonium sulfate solution—Dissolve 1.8 g of ferric ammonium sulfate in 4 mL of dilute hydrochloric acid (1 in 12), and dilute with water to 100 mL. Filter, and use the clear filtrate. Prepare fresh.

1,10-Phenanthroline solution—Shake 300 mg of 1,10-phenanthroline with 100 mL of water for 1 hour. Filter, and use the clear filtrate. Prepare fresh.

Standard preparation—Dissolve about 50 mg of USP Hydralazine Hydrochloride RS, accurately weighed, in water in a 100-mL volumetric flask, and dilute with water to volume. Pipet 10 mL of this solution into a 100-mL volumetric flask, dilute with water to volume, and mix.

Assay preparation—Weigh and finely powder not less than 20 Reserpine, Hydralazine Hydrochloride, and Hydrochlorothiazide Tablets. Transfer to a 200-mL volumetric flask an accurately weighed portion of the powder, equivalent to about 100 mg of hydralazine hydrochloride, add 100 mL of water, and shake by mechanical means for 30 minutes. Dilute with water to volume, mix, and filter through paper, discarding the first 15 mL of the filtrate. Pipet 10 mL of the clear filtrate into a 100-mL volumetric flask, and dilute with water to volume.

Procedure—Pipet 10 mL each of the *Assay preparation*, the *Standard preparation*, and water to provide the blank, into separate, 200-mL volumetric flasks. To each flask add 5 mL of acetic acid solution (12 in 100), 5 mL of *Sodium acetate solution*, 2 mL of *1,10-Phenanthroline solution*, and 1 mL of *Ferric ammonium sulfate solution*, mix, and allow to stand in the dark for 30 minutes. Dilute with water to volume, and mix. Concomitantly determine the absorbances of both solutions in 1-cm cells at the wavelength of maximum absorbance at about 510 nm, with a suitable spectrophotometer, against the blank. Calculate the quantity, in mg, of $C_8H_8N_4 . HCl$ in the portion of Tablets taken by the formula:

$$2C(A_U/A_S),$$

in which C is the concentration, in µg per mL, of USP Hydralazine Hydrochloride RS in the *Standard preparation*, and A_U and A_S are the absorbances of the solutions from the *Assay preparation* and the *Standard preparation*, respectively.

Assay for hydrochlorothiazide—

Methanolic sodium hydroxide solution—Dissolve 40 g of sodium hydroxide in 125 mL of water, cool, and dilute with methanol to 500 mL. Filter before use.

Column preparation—Weigh about 10 g of sulfonic acid cation exchange resin into a 250-mL beaker. Add 75 mL of methanol, and stir the mixture with a magnetic stirrer for 30 minutes. Place a glass wool plug at the lower end of a glass chromatographic column (15 mm ID × 45 cm long), equipped with a stopcock to regulate the eluant flow. Transfer and pack the slurry in portions into the prepared column to a height of 10 cm. If air is trapped, remove by tapping, stirring, or back-flushing the column. Place a glass wool plug on top of the resin bed after packing. Wash the resin with consecutive 100-mL portions of a 1 in 5 solution of hydrochloric acid in methanol, and of methanol, then of *Methanolic sodium hydroxide solution* and of methanol, respectively, using a flow rate of approximately 4 mL per minute. Wash the sodium form of the resin with 150 mL of a 1 in 5 solution of hydrochloric acid in methanol, followed by not less than 100 mL of methanol. The ultraviolet spectrum of the last few mL of the methanol washing conforms to that of the methanol.

Standard preparation—Mix about 40 mg of USP Hydrochlorothiazide RS, accurately weighed, with 4 mL of water in a 200-mL volumetric flask. Add 150 mL of methanol, warm gently on a steam bath for a few minutes, and shake by mechanical means for 20 minutes. Dilute with methanol to volume, and mix.

Assay preparation—Weigh and finely powder not less than 20 Reserpine, Hydralazine Hydrochloride, and Hydrochlorothiazide Tablets. Transfer an accurately weighed portion of the powder, equivalent to about 20 mg of hydrochlorothiazide, to a 100-mL volumetric flask. Add 2 mL of water, mix, and add 75 mL of methanol. Warm gently on a steam bath for a few minutes, and shake by mechanical means for 20 minutes. Dilute with methanol to volume, and mix. Centrifuge a portion of the mixture, and use the clear solution as the *Assay preparation.*

Procedure—Keeping the column stopcocks closed, pipet 4 mL of the clear *Assay preparation* on the top of the resin bed of 1 column, and pipet 4 mL of the *Standard preparation* on the top of the second column. Place a 100-mL volumetric flask beneath each of the columns, and collect the eluate. Adjust the flow rate to approximately 2 mL per minute, and allow the solution to flow until it just disappears into the upper glass wool plug. Close the stopcock, carefully add 25 mL of methanol, and further elute each column with the same flow rate as before. Repeat with two additional 25-mL portions of methanol. Dilute the contents of each flask with methanol to volume. Concomitantly determine the absorbances of both solutions at the wavelength of maximum absorbance at about 271 nm, with a suitable spectrophotometer, using methanol as the blank. Calculate the quantity, in mg, of $C_7H_8ClN_3O_4S_2$ in the portion of Tablets taken by the formula:

$$0.1C(A_U/A_S),$$

in which C is the concentration, in µg per mL, of USP Hydrochlorothiazide RS in the *Standard preparation*, and A_U and A_S are the absorbances of the solution from the *Assay preparation* and the *Standard preparation*, respectively.

Reserpine and Hydrochlorothiazide Tablets

» Reserpine and Hydrochlorothiazide Tablets contain not less than 90.0 percent and not more than 110.0 percent of the labeled amount of reserpine ($C_{33}H_{40}N_2O_9$), and not less than 93.0 percent and not more than 107.0 percent of the labeled amount of hydrochlorothiazide ($C_7H_8ClN_3O_4S_2$).

Packaging and storage—Preserve in tight, light-resistant containers.

Reference standards—*USP Reserpine Reference Standard*—Dry at 60° for 3 hours before using. *USP Hydrochlorothiazide Reference Standard*—Dry at 105° for 1 hour before using. *USP 4-Amino-6-chloro-1,3-benzenedisulfonamide Reference Standard*—Dry at 105° for 1 hour before using.

Identification—

A: Transfer a quantity of powdered Tablets, equivalent to 1 mg of reserpine, to a stoppered, 50-mL centrifuge tube. Add 20 mL of citric acid solution (1 in 50), and shake until the powder is suspended. Extract the mixture with two 20-mL portions of chloroform, centrifuge, and withdraw the chloroform, filtering each extract through a pledget of cotton into a 50-mL volumetric flask. Dilute with chloroform to volume, and mix: the ultraviolet absorption spectrum of the solution exhibits maxima and minima at the same wavelengths as that of a similar solution of USP Reserpine RS, concomitantly measured (*presence of reserpine*).

B: Transfer a quantity of powdered Tablets, equivalent to about 50 mg of hydrochlorothiazide, to a test tube containing 10 mL of acetone, agitate for 5 minutes, and centrifuge. Use the clear supernatant liquid as the Test solution. Separately apply 10 µL each of the Test solution and a Standard solution of USP Hydrochlorothiazide RS in acetone containing 5 mg per mL to a thin-layer chromatographic plate (see *Chromatography* ⟨621⟩) coated with a 0.25-mm layer of chromatographic silica gel mixture and previously washed with methanol. Develop the chromatogram in a solvent system consisting of a mixture of ethyl acetate and isopropyl alcohol (17:3) until the solvent front has moved about three-fourths of the length of the plate. Remove the plate from the chamber, air-dry, and examine under short-wavelength ultraviolet light: the R_f value of the principal spot in the chromatogram of the Test solution corresponds to that obtained from the Standard solution (*presence of hydrochlorothiazide*).

Disintegration ⟨701⟩: 30 minutes.

Dissolution ⟨711⟩—

Medium: a mixture of 0.1 *N* hydrochloric acid and *n*-propyl alcohol (3:2); 900 mL.

Apparatus 2: 50 rpm.

Times: 45 minutes, 60 minutes.

Determination of dissolved reserpine—

STANDARD PREPARATION—Dissolve about 34 mg of USP Reserpine RS, accurately weighed, in a 50-mL volumetric flask containing 5 mL of chloroform, dilute with *n*-propyl alcohol to volume (*Solution 1*), and mix. Pipet 1 mL of *Solution 1* into a 50-mL volumetric flask, dilute with *n*-propyl alcohol to volume (*Solution 2*), and mix. Pipet 1 mL of *Solution 2* into a 100-mL volumetric flask, dilute with *Dissolution Medium* to volume, and mix to obtain the *Standard preparation.*

PROCEDURE—Filter a portion of the solution under test through paper, and transfer 5.0 mL of the clear filtrate to a 25-mL volumetric flask. Pipet 5 mL of the *Standard preparation* into a separate 25-mL volumetric flask. Treat each flask as follows. Add 5 drops of hydrochloric acid and 5 mL of a mixture of water, alcohol, and sulfuric acid (29:20:1), and mix. Add by pipet 5 mL of sodium nitrite solution (3 in 1000), dilute with alcohol to volume, mix, and allow to stand for 30 minutes. Concomitantly determine the fluorescences of the solution under test and the *Standard preparation* in a suitable spectrophotometer arranged to deliver activation radiation at 405 nm and to measure the resultant fluorescence at the emission wavelength of about 500 nm. Calculate the amount of reserpine ($C_{33}H_{40}N_2O_9$) dissolved.

Determination of dissolved hydrochlorothiazide—

STANDARD PREPARATION—Dissolve about 27 mg of USP Hydrochlorothiazide RS, accurately weighed, in a 50-mL volumetric flask containing 5 mL of methanol, dilute with *Dissolution Medium* to volume, and mix. Pipet 2 mL of this solution into a 100-mL volumetric flask, dilute with *Dissolution Medium* to volume, and mix.

PROCEDURE—Determine the amount of hydrochlorothiazide ($C_7H_8ClN_3O_4S_2$) dissolved from ultraviolet absorbances, at the wavelength of maximum absorbance at about 271 nm, of filtered portions of the solution under test, suitably diluted with *Dissolution Medium*, in comparison with the *Standard preparation.*

Tolerances—Not less than 80% (Q) of the labeled amount of $C_{33}H_{40}N_2O_9$ is dissolved in 45 minutes and not less than 80% (Q) of the labeled amount of $C_7H_8ClN_3O_4S_2$ is dissolved in 60 minutes.

Uniformity of dosage units ⟨905⟩: meet the requirements with respect to reserpine and to hydrochlorothiazide.

Procedure for content uniformity for reserpine—

Standard preparation—Prepare as directed for *Standard preparation* under *Assay for reserpine.*

Test solution—Weigh 1 Tablet, grind to a fine powder, and transfer to a stoppered, 50-mL centrifuge tube. Add 25.0 mL of a mixture of chloroform and methanol solution (1:1), shake by mechanical means for 15 minutes, and centrifuge. Pipet 4 mL of the clear supernatant liquid into a 100-mL volumetric flask, dilute with the chloroform and methanol solution to volume, and mix.

Procedure—Proceed as directed for *Procedure* under *Assay for reserpine.* Calculate the quantity, in mg, of $C_{33}H_{40}N_2O_9$, in the Tablet by the formula:

$$(W_t/W_U)(TC/D)(I_U/I_S),$$

in which W_t is the weight, in mg, of the Tablet, W_U is the weight, in mg, in the portion of Tablet taken, T is the labeled quantity, in mg, of reserpine in the Tablet, C is the concentration, in µg per mL, of USP Reserpine RS in the *Standard preparation*, D is the concentration, in µg per mL, of reserpine in the *Test solution*, based upon the labeled quantity per Tablet and the extent of dilution, and I_U and I_S are the fluorescence intensities of the solutions from the *Test solution* and the *Standard preparation*, respectively.

Procedure for content uniformity for hydrochlorothiazide—

Standard preparation—Prepare as directed for *Standard preparation* under *Assay for hydrochlorothiazide.*

Test solution—Transfer 1 Tablet to a 250-mL volumetric flask, add 150 mL of 0.1 *N* sodium hydroxide, and sonicate, swirling the flask intermittently, until the tablet is dissolved. Dilute with

0.1 *N* sodium hydroxide to volume, mix, and filter, discarding the first 15 mL of the filtrate. Dilute a portion of the clear filtrate quantitatively and stepwise with 0.1 *N* sodium hydroxide to obtain a solution having a concentration of about 10 µg of hydrochlorothiazide per mL.

Procedure—Proceed as directed for *Procedure* under *Assay for hydrochlorothiazide.* Calculate the quantity, in mg, of $C_7H_8ClN_3O_4S_2$, in the Tablet by the formula:

$$(TC/D)(A_U/A_S),$$

in which *T* is the labeled quantity, in mg, of hydrochlorothiazide in the Tablet, *C* is the concentration, in µg per mL, of USP Hydrochlorothiazide RS in the *Standard preparation, D* is the concentration, in µg per mL, of hydrochlorothiazide in the *Test solution*, based upon the labeled quantity per Tablet and the extent of dilution, and A_U and A_S are the absorbances of the solutions from the *Test solution* and the *Standard preparation*, respectively.

Diazotizable substances—

Standard preparation—Weigh accurately 25 mg of USP 4-Amino-6-chloro-1,3-benzenedisulfonamide RS, dissolve in 5 mL of methanol contained in a 100-mL volumetric flask, dilute with water to volume, and mix. Pipet 4 mL of this solution into a 100-mL volumetric flask, dilute with water to volume, and mix.

Test preparation—Transfer a portion of finely powdered Tablets, accurately weighed and equivalent to about 100 mg of hydrochlorothiazide, to a 100-mL volumetric flask, and add 20 mL of methanol and 20 mL of water. Shake continuously for 5 to 10 minutes, dilute with water to volume, mix, and filter. Use the filtrate as the *Test preparation.*

Procedure—Pipet 5 mL each of the *Standard preparation* and the *Test preparation* into separate, 50-mL volumetric flasks. Pipet 5 mL of water into a third 50-mL volumetric flask to provide the blank. To each flask add 1 mL of freshly prepared sodium nitrite solution (1 in 100) and 5 mL of dilute hydrochloric acid (1 in 12), and allow to stand for 5 minutes. Add 2 mL of ammonium sulfamate solution (1 in 50), shake vigorously, allow to stand for 5 minutes, then add 2 mL of freshly prepared disodium chromotropate solution (1 in 100) and 10 mL of sodium acetate TS. Dilute with water to volume, and mix. Concomitantly determine the absorbances of the solutions from the *Standard preparation* and the *Test preparation* in 1-cm cells at the wavelength of maximum absorbance at about 500 nm, with a suitable spectrophotometer, against the blank. The absorbance of the solution from the *Test preparation* does not exceed that of the solution from the *Standard preparation*, corresponding to not more than 1.0% of diazotizable substances.

Assay for reserpine—

Standard preparation—Dissolve about 25 mg of USP Reserpine RS, accurately weighed, in 1 mL of chloroform contained in a 50-mL volumetric flask, dilute with methanol to volume, and mix. Dilute a portion of this solution quantitatively and stepwise with chloroform-methanol solution (1:1) to obtain a solution having a known concentration of about 0.2 µg of reserpine per mL.

Assay preparation—Weigh and finely powder not less than 20 Reserpine and Hydrochlorothiazide Tablets. Transfer to a stoppered, 50-mL centrifuge tube an accurately weighed portion of the powder, equivalent to about 1 mg of reserpine, add 25.0 mL of chloroform-methanol solution (1:1), shake by mechanical means for 15 minutes, and centrifuge. Dilute a portion of the clear supernatant liquid quantitatively and stepwise with chloroform-methanol solution (1:1) to obtain a solution having a concentration of about 0.2 µg of reserpine per mL.

Procedure—Separately transfer 5.0 mL of the *Assay preparation*, 5.0 mL of the *Standard preparation*, and 5.0 mL of chloroform-methanol solution (1:1) to provide the reagent blank, respectively, to three 25-mL volumetric flasks. To each flask add 0.5 mL of hydrochloric acid, 1.0 mL of a 3 in 1000 solution of sodium nitrite in dilute methanol (1 in 2), mix, and allow to stand for 30 minutes. Add 1 mL of ammonium sulfamate solution (1 in 20) to each flask, add chloroform-methanol solution (1:1) to volume, mix, and allow to stand for 10 minutes. Concomitantly determine the fluorescence intensities of the solutions in a suitable spectrophotometer arranged to deliver activation radiation at 405 nm and to measure the resultant fluorescence at the emission wavelength of about 500 nm. Calculate the quantity, in mg, of $C_{33}H_{40}N_2O_9$ in the portion of Tablets taken by the formula:

$$5C(I_U/I_S),$$

in which *C* is the concentration, in µg per mL, of USP Reserpine RS in the *Standard preparation*, and I_U and I_S are the fluorescence intensities of the solutions from the *Assay preparation* and the *Standard preparation*, respectively.

Assay for hydrochlorothiazide—

Standard preparation—Dissolve an accurately weighed quantity of USP Hydrochlorothiazide RS in 0.1 *N* sodium hydroxide, and dilute quantitatively and stepwise with 0.1 *N* sodium hydroxide to obtain a solution having a known concentration of about 10 µg of hydrochlorothiazide per mL. Use a freshly prepared solution.

Assay preparation—Weigh and finely powder not less than 20 Reserpine and Hydrochlorothiazide Tablets. Weigh accurately a portion of the powder, equivalent to about 50 mg of hydrochlorothiazide, and transfer to a 500-mL volumetric flask. Add 200 mL of 0.1 *N* sodium hydroxide, shake by mechanical means for 15 minutes, dilute with the same solvent to volume, and mix. Filter a portion of the solution through paper, discarding the first 15 mL of the filtrate, and transfer 10.0 mL of the clear filtrate to a 100-mL volumetric flask. Dilute with 0.1 *N* sodium hydroxide to volume, and mix.

Procedure—Concomitantly determine the absorbances of the *Standard preparation* and the *Assay preparation* in 1-cm cells at the wavelength of maximum absorbance, at about 274 nm, with a suitable spectrophotometer, using 0.1 *N* sodium hydroxide as the blank. Calculate the quantity, in mg, of $C_7H_8ClN_3O_4S_2$ in the portion of Tablets taken by the formula:

$$5C(A_U/A_S),$$

in which *C* is the concentration, in µg per mL, of USP Hydrochlorothiazide RS in the *Standard preparation*, and A_U and A_S are the absorbances of the solutions from the *Assay preparation* and the *Standard preparation*, respectively.

Resin, Cholestyramine—*see* Cholestyramine Resin
Resin, Podophyllum—*see* Podophyllum Resin

Resorcinol

$C_6H_6O_2$ 110.11
1,3-Benzenediol.
Resorcinol [*108-46-3*].

» Resorcinol contains not less than 99.0 percent and not more than 100.5 percent of $C_6H_6O_2$, calculated on the dried basis.

Packaging and storage—Preserve in well-closed, light-resistant containers.

Reference standard—*USP Resorcinol Reference Standard*—Dry over silica gel for 4 hours before using.

Identification—

A: The infrared absorption spectrum of a potassium bromide dispersion of it exhibits maxima only at the same wavelengths as that of a similar preparation of USP Resorcinol RS. If a difference appears, dissolve portions of both the test specimen and the Reference Standard in dehydrated alcohol, evaporate the solutions to dryness, and repeat the test on the residues.

B: Dissolve 100 mg in 2 mL of 1 *N* sodium hydroxide, add 1 drop of chloroform, and heat the mixture: an intense crimson color is produced. Then add a slight excess of hydrochloric acid: the color changes to pale yellow.

C: To 10 mL of a solution (1 in 100) add 1 drop of ferric chloride TS: a blue-violet color is produced, and it fades slowly.

Melting range ⟨741⟩: between 109° and 111°.

Loss on drying ⟨731⟩—Dry it over silica gel for 4 hours: it loses not more than 1.0% of its weight.

Residue on ignition ⟨281⟩: not more than 0.05%.

Phenol—Heat gently a solution (1 in 20): the odor of phenol is not perceptible.

Catechol—To 10 mL of a solution (1 in 20) previously mixed with 2 drops of 1 *N* acetic acid add 0.5 mL of lead acetate TS: no turbidity is produced.

Ordinary impurities ⟨466⟩—

Test solution: methanol.

Standard solution: methanol.

Eluant: a mixture of hexanes and ethyl acetate (70:30).

Visualization: 17 and then 1.

Application volume: 10 µL.

Limits: not more than 1.0%.

Assay—Dissolve about 1.5 g of Resorcinol, accurately weighed, in water to make 500.0 mL. Transfer to an iodine flask 25.0 mL of the resulting solution, add 50.0 mL of 0.1 *N* bromine VS, dilute with 50 mL of water, add 5 mL of hydrochloric acid, and at once insert the stopper in the flask. Shake for 1 minute, allow to stand for 2 minutes, and add 5 mL of potassium iodide TS while slightly loosening the stopper. Shake thoroughly, allow to stand for 5 minutes, remove the stopper, and rinse it and the neck of the flask with 20 mL of water into the flask. Titrate the liberated iodine with 0.1 *N* sodium thiosulfate VS, adding starch TS as the end-point is approached. Perform a blank determination. From the volume of 0.1 *N* sodium thiosulfate VS used, calculate the volume, in mL, of 0.1 *N* bromine VS consumed by the resorcinol. Each mL of 0.1 *N* bromine is equivalent to 1.835 mg of $C_6H_6O_2$.

Compound Resorcinol Ointment

» Prepare Compound Resorcinol Ointment as follows:

Resorcinol	60 g
Zinc Oxide	60 g
Bismuth Subnitrate	60 g
Juniper Tar	20 g
Yellow Wax	100 g
Petrolatum	290 g
Anhydrous Lanolin	280 g
Glycerin	130 g
To make	1000 g

Melt the Yellow Wax and the Anhydrous Lanolin in a dish on a steam bath. Triturate the Zinc Oxide and the Bismuth Subnitrate with the Petrolatum until smooth, and add it to the melted mixture. Dissolve the Resorcinol in the Glycerin, incorporate the solution with the warm mixture just prepared, then add the Juniper Tar, and stir the Ointment until it congeals.

Packaging and storage—Preserve in tight containers, and avoid prolonged exposure to temperatures exceeding 30°.

Resorcinol and Sulfur Lotion

» Resorcinol and Sulfur Lotion is Resorcinol and Sulfur in a suitable hydroalcoholic vehicle. It contains not less than 90.0 percent and not more than 110.0 percent of the labeled amount of resorcinol ($C_6H_6O_2$) and not less than 95.0 percent and not more than 110.0 percent of the labeled amount of sulfur (S).

Packaging and storage—Preserve in tight containers.

Reference standard—*USP Resorcinol Reference Standard*—Dry over silica gel for 4 hours before using.

Identification—

A: Transfer a quantity of Lotion, equivalent to about 20 mg of resorcinol, to a 15-mL centrifuge tube, add 5 mL of 5 *N* sodium hydroxide, mix, and centrifuge the mixture for 5 minutes. Decant the supernatant liquid into a test tube, and retain the residue for *Identification test B.* Add 0.5 mL of chloroform, mix, and heat on a steam bath: an intense crimson color is produced. Add a slight excess of hydrochloric acid: the color changes to pale yellow (presence of *resorcinol*).

B: Place a small portion of the residue from the centrifuge tube in *Identification test A* on the tip of a spatula, and burn it: sulfur dioxide, which turns moistened starch-iodate paper blue, is formed (presence of *sulfur*).

Alcohol content ⟨611⟩—Determine by the gas-liquid chromatographic method, acetone being used as the internal standard: it contains between 90.0% and 110.0% of the labeled amount of C_2H_5OH.

Assay for resorcinol—

Mobile phase—Prepare a suitable degassed solution of water, acetonitrile, and methanol (about 55:7:6) such that the retention times of resorcinol and caffeine are about 3 minutes and 4 minutes, respectively.

Internal standard solution—Dissolve about 140 mg of caffeine in 2 mL of chloroform, add methanol to make 100 mL, and mix.

Standard preparation—Transfer 50 mg of USP Resorcinol RS, accurately weighed, to a 25-mL volumetric flask, dilute with methanol to volume, and mix. Transfer 10.0 mL of this solution and 5.0 mL of *Internal standard solution* to a 100-mL volumetric flask, dilute with methanol to volume, and mix.

Assay preparation—Transfer an accurately weighed portion of Resorcinol and Sulfur Lotion, equivalent to about 20 mg of resorcinol, to a 150-mL beaker. Add 40 mL of methanol and 5.0 mL of *Internal standard solution*, and heat on a steam bath for 5 minutes. Cool the mixture to room temperature, and decant the liquid into a 100-mL volumetric flask. Wash the residue in the beaker by adding 20 mL of methanol to the beaker. Heat on a steam bath for 5 minutes, cool the mixture to room temperature, and decant the liquid into the volumetric flask. Repeat the washing, heating, cooling, and decanting. Dilute the contents of the volumetric flask with methanol to volume, and mix.

Procedure—Introduce equal volumes (about 10 µL) of the *Assay preparation* and the *Standard preparation* into a high-pressure liquid chromatograph (see *Chromatography* ⟨621⟩), operated at room temperature, by means of a suitable microsyringe or sampling valve, adjusting the specimen size and other operating parameters such that the peak obtained from the *Standard preparation* is about 0.6 full scale. Typically, the apparatus is fitted with a 30-cm × 4-mm column packed with packing L1 and is equipped with an ultraviolet detector capable of monitoring absorption at 280 nm, and a suitable recorder. In a suitable chromatogram the coefficient of variation for five replicate injections of the *Standard preparation* is not more than 3.0%. Measure the peak responses at equivalent retention times, obtained from the *Assay preparation* and the *Standard preparation*, and calculate the quantity, in mg, of $C_6H_6O_2$ in the portion of Lotion taken by the formula:

$$(100C)(R_U/R_S),$$

in which C is the concentration, in mg per mL, of USP Resorcinol RS in the *Standard preparation*, and R_U and R_S are the ratios of the responses of the resorcinol and caffeine peaks obtained from the *Assay preparation* and the *Standard preparation*, respectively.

Assay for sulfur—Transfer an accurately weighed portion of Resorcinol and Sulfur Lotion, equivalent to about 85 mg of sulfur, to a suitable flask, add 40 mL of sodium sulfite solution (1 in 20), a few drops of antifoam, and a few boiling chips, and boil under a reflux condenser for 1 hour. Cool to room temperature, add 10 mL of formaldehyde solution and 6 mL of 6 *N* acetic acid, and dilute with water to 150 mL. Add 3 mL of starch TS, and titrate with 0.1 *N* iodine VS until a permanent blue color is produced. Each mL of 0.1 *N* iodine is equivalent to 3.206 mg of S.

Resorcinol Monoacetate

$C_8H_8O_3$ 152.15
1,3-Benzenediol, monoacetate.
Resorcinol monoacetate [*102-29-4*].

Packaging and storage—Preserve in tight, light-resistant containers.

Identification—

A: Fuse 3 drops with about 300 mg of phthalic anhydride and about 50 mg of zinc chloride: a small portion of the fused mixture when dissolve in 10 mL of 1 *N* sodium hydroxide produces an intense yellow-green fluorescence typical of fluorescein.

B: Dissolve 0.5 mL in 3 mL of alcohol, add 3 drops of sulfuric acid, and boil: ethyl acetate, recognizable by its odor, is evolved.

Specific gravity ⟨841⟩: between 1.203 and 1.207.

Acidity—A solution of 10 mL of Resorcinol Monoacetate in 20 mL of benzene, when shaken with 100 mL of water, requires not more than 0.50 mL of 0.10 *N* sodium hydroxide for neutralization, methyl orange TS being used as the indicator.

Loss on drying ⟨731⟩—Dry it on a steam bath for 1 hour: it loses not more than 2.5% of its weight.

Residue on ignition ⟨281⟩: not more than 0.1%.

Rh$_0$ (D) Immune Globulin—*see* Globulin, Rh$_0$ (D) Immune

Riboflavin

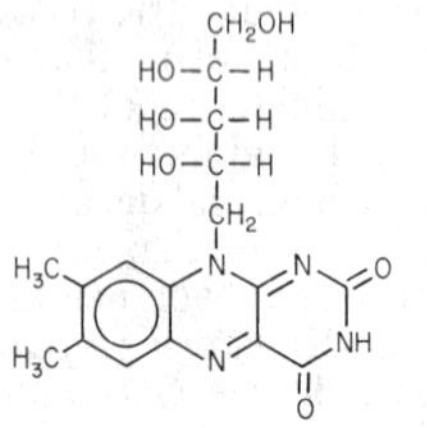

$C_{17}H_{20}N_4O_6$ 376.37
Riboflavine.
Riboflavine [*83-88-5*].

» Riboflavin contains not less than 98.0 percent and not more than 102.0 percent of $C_{17}H_{20}N_4O_6$, calculated on the dried basis.

Packaging and storage—Preserve in tight, light-resistant containers.

Reference standard—*USP Riboflavin Reference Standard*—Dry at 105° for 2 hours before using.

Identification—A solution of 1 mg in 100 mL of water is pale greenish yellow by transmitted light, and has an intense yellowish green fluorescence which disappears upon the addition of mineral acids or alkalies.

Specific rotation ⟨781⟩: between +56.5° and +59.5°, calculated on the dried basis, determined in a solution in hydrochloric acid containing 50 mg in each 10 mL.

Loss on drying ⟨731⟩—Dry about 500 mg at 105° for 2 hours: it loses not more than 1.5% of its weight.

Residue on ignition ⟨281⟩: not more than 0.3%.

Lumiflavin—Prepare alcohol-free chloroform just prior to use, as follows. Shake 20 mL of chloroform gently but thoroughly with 20 mL of water for 3 minutes, draw off the chloroform layer, and wash twice more with 20-mL portions of water. Finally filter the chloroform through a dry filter paper, shake it for 5 minutes with 5 g of powdered anhydrous sodium sulfate, allow the mixture to stand for 2 hours, and decant or filter the clear chloroform. Shake 25 mg of Riboflavin with 10 mL of the alcohol-free chloroform for 5 minutes, and filter: the absorbance of the filtrate, determined in 1-cm cells at a wavelength of 440 nm, with a suitable spectrophotometer, alcohol-free chloroform being used as the blank, does not exceed 0.025.

Assay—[NOTE—Conduct the entire procedure without exposure to direct sunlight.] Place about 50 mg of Riboflavin, accurately weighed, in a 1000-mL volumetric flask containing about 50 mL of water. Add 5 mL of 6 *N* acetic acid and sufficient water to make about 800 mL. Heat on a steam bath, protected from light, with frequent agitation until dissolved. Cool to about 25°, dilute with water to volume, and mix. Dilute this solution quantitatively and stepwise with water to bring it within the operating sensitivity of the fluorometer used. In the same manner prepare a standard solution to contain, in each mL, a quantity, accurately weighed, of USP Riboflavin RS, equivalent to that of the Riboflavin solution prepared as directed above, and measure the intensity of its fluorescence in a fluorometer at about 530 nm. (An excitation wavelength of about 444 nm is preferable.) Immediately after the reading, add to the solution about 10 mg of sodium hydrosulfite, stirring with a glass rod until dissolved, and at once measure the fluorescence again. The difference between the two readings represents the intensity of the fluorescence due to the Standard. Similarly, measure the intensity of the fluorescence of the final solution of the Riboflavin being assayed at about 530 nm, before and after the addition of sodium hydrosulfite. Calculate the quantity, in µg per mL, of $C_{17}H_{20}N_4O_6$ in the final solution of Riboflavin by the formula:

$$C(I_U / I_S),$$

in which *C* is the concentration, in µg per mL, of USP Riboflavin RS in the final solution of the Standard, and I_U and I_S are the corrected fluorescence values observed for the solutions of the Riboflavin and Standard, respectively.

Riboflavin Injection

» Riboflavin Injection is a sterile solution of Riboflavin in Water for Injection. It contains not less than 95.0 percent and not more than 120.0 percent of the labeled amount of $C_{17}H_{20}N_4O_6$. It may contain niacinamide or other suitable solubilizers.

Packaging and storage—Preserve in light-resistant, in single-dose or in multiple-dose containers, preferably of Type I glass.

Reference standard—*USP Riboflavin Reference Standard*—Dry at 105° for 2 hours before using.

Identification—It responds to the *Identification test* under *Riboflavin.*

pH ⟨791⟩: between 4.5 and 7.0.

Other requirements—It meets the requirements under *Injections* ⟨1⟩.

Assay—Dilute an accurately measured volume of not less than 1 mL of Riboflavin Injection to make a solution containing approximately 0.1 µg of riboflavin per mL. Using this as the *Assay Preparation*, proceed as directed under *Riboflavin Assay* ⟨481⟩. Calculate the quantity, in mg, of $C_{17}H_{20}N_4O_6$ in each mL of the Injection by the formula:

$$C(A/B),$$

in which C is the concentration, in mg per mL, of $C_{17}H_{20}N_4O_6$ obtained for the *Assay Preparation*, A is the test specimen dilution volume, in mL, and B is the volume, in mL, of Injection taken.

Riboflavin Tablets

» Riboflavin Tablets contain not less than 95.0 percent and not more than 115.0 percent of the labeled amount of $C_{17}H_{20}N_4O_6$.

Packaging and storage—Preserve in tight, light-resistant containers.

Reference standard—*USP Riboflavin Reference Standard*—Dry at 105° for 2 hours before using.

Disintegration ⟨701⟩: 30 minutes.

Uniformity of dosage units ⟨905⟩: meet the requirements.

Assay—Weigh and finely powder not less than 20 Riboflavin Tablets. Weigh accurately a portion of the powder, equivalent to about 20 mg of riboflavin, transfer to a 250-mL flask, and add 150 mL of 0.1 *N* hydrochloric acid. Shake vigorously, and wash down the sides of the flask with sufficient 0.1 *N* hydrochloric acid to ensure that the pH remains below 1.5 during the subsequent period of heating. Heat the mixture on a steam bath, with frequent agitation, until the riboflavin has dissolved, or in an autoclave at 121° for 30 minutes. Cool, adjust the mixture, with vigorous agitation, to a pH of 5 to 6 with 1 *N* sodium hydroxide, transfer to a 1000-mL volumetric flask, dilute with water to volume, and mix. If the solution is not clear, filter through paper known not to adsorb riboflavin. Dilute an aliquot of the clear solution quantitatively and stepwise with water to make a final measured volume that contains approximately 0.1 µg of riboflavin per mL. Using this as the *Assay Preparation*, proceed as directed under *Riboflavin Assay* ⟨481⟩. Calculate the quantity, in mg, of $C_{17}H_{20}N_4O_6$ in the portion of powdered Tablets taken by the formula:

$$200{,}000C/W,$$

in which C is the concentration, in mg per mL, of $C_{17}H_{20}N_4O_6$ obtained for the *Assay Preparation*, and W is the weight, in mg, of powdered Tablets. Calculate the quantity, in mg, of $C_{17}H_{20}N_4O_6$ in each Tablet.

Riboflavin 5′-Phosphate Sodium

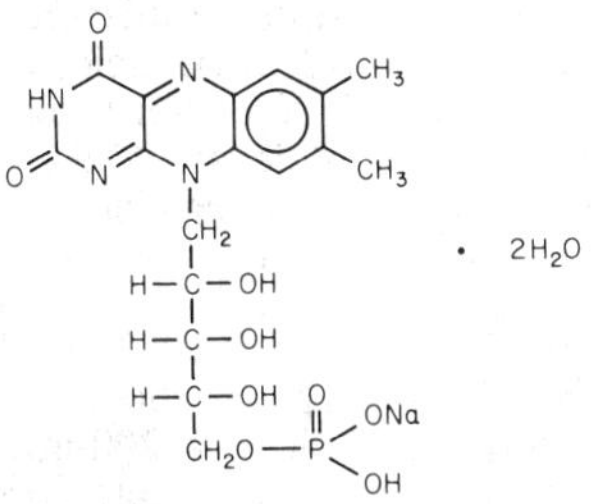

$C_{17}H_{20}N_4NaO_9P.2H_2O$ 514.36

Riboflavin 5′-(dihydrogen phosphate), monosodium salt, dihydrate.

Riboflavine 5′-(sodium hydrogen phosphate), dihydrate.

Anhydrous 478.33 [*130-40-5*].

» Riboflavin 5′-Phosphate Sodium contains not less than the equivalent of 73.0 percent and not more than the equivalent of 79.0 percent of riboflavin ($C_{17}H_{20}N_4O_6$), calculated on the dried basis.

Packaging and storage—Preserve in tight, light-resistant containers.

Reference standards—*USP Riboflavin Reference Standard*—Dry at 105° for 2 hours before using. *USP Phosphated Riboflavin Reference Standard.*

Identification—

A: A solution of 1 mg in 100 mL of water is pale greenish yellow by transmitted light, and has an intense yellowish green fluorescence which disappears upon the addition of mineral acids or alkalies.

B: To 0.5 g add 10 mL of nitric acid, evaporate the mixture on a water bath to dryness, ignite the residue until the carbon is removed, dissolve the residue in 5 mL of water, and filter: the filtrate so obtained responds to the tests for *Sodium* ⟨191⟩ and for *Phosphate* ⟨191⟩.

Specific rotation ⟨781⟩: between +37.0° and +42.0°, calculated on the dried basis, determined within 15 minutes in a solution in 5 *N* hydrochloric acid containing 150 mg in each 10 mL.

pH ⟨791⟩: between 5.0 and 6.5, in a solution (1 in 100).

Loss on drying ⟨731⟩—Dry it in vacuum over phosphorus pentoxide at 100° for 5 hours: it loses not more than 7.5% of its weight.

Residue on ignition ⟨281⟩: not more than 25.0%.

Free phosphate—

Acid molybdate solution—Dilute 25 mL of ammonium molybdate solution (7 in 100) with water to 200 mL. To this dilution add slowly 25 mL of 7.5 *N* sulfuric acid, and mix.

Ferrous sulfate solution—Just prior to use, prepare a 1 in 10 solution of ferrous sulfate in 0.15 *N* sulfuric acid.

Standard preparation—Prepare a solution in water containing 44.0 µg of monobasic potassium phosphate in each mL.

Test preparation—Transfer 300 mg of Riboflavin 5′-Phosphate Sodium to a 100-mL volumetric flask, dissolve in water, dilute with water to volume, and mix.

Procedure—Transfer 10.0 mL each of the *Standard preparation* and the *Test preparation* to separate 50-mL conical flasks, add 10.0 mL of *Acid molybdate solution* and 5.0 mL of *Ferrous sulfate solution* to each flask, and mix. Concomitantly determine the absorbances of the solutions, in 1-cm cells, at the wavelength of maximum absorbance at about 700 nm, with a suitable spectrophotometer, using as the blank a mixture of 10.0 mL of water, 10.0 mL of *Acid molybdate solution*, and 5.0 mL of *Ferrous sulfate solution:* the absorbance of the solution from the *Test preparation* is not greater than that from the *Standard preparation* (1% as PO_4).

Free riboflavin and riboflavin diphosphates—[NOTE—Conduct this test so that all solutions are protected from actinic light at all stages, preferably by using low-actinic glassware.]

Mobile phase—Mix 850 mL of 0.054 *M* monobasic potassium phosphate with 150 mL of methanol, filter, and degas the solution. Make adjustments if necessary (see *System Suitability* under *Chromatography* ⟨621⟩).

Standard preparation—Transfer 60 mg of USP Riboflavin RS, accurately weighed, to a 250-mL volumetric flask, dissolve carefully in 1 mL of hydrochloric acid, dilute with water to volume, and mix. Pipet a 4-mL aliquot into a 100-mL volumetric flask, dilute with *Mobile phase* to volume, and mix.

Test preparation—Transfer 100.0 mg of Riboflavin 5′-Phosphate Sodium to a 100-mL volumetric flask, dissolve in 50 mL of water, dilute with *Mobile phase* to volume, and mix. Pipet 8 mL of this solution into a 50-mL volumetric flask, dilute with *Mobile phase* to volume, and mix.

System suitability preparation—Dissolve USP Phosphated Riboflavin RS in water to obtain a solution containing 2 mg per mL. Add an equal volume of *Mobile phase*, and mix. Dilute 8 mL of this solution with *Mobile phase* to 50 mL, and mix.

Chromatographic system (see *Chromatography* ⟨621⟩)—The liquid chromatograph is equipped with a fluorometric detector set at 440 nm excitation wavelength and provided with a 470-nm emission filter or set at about 530 nm for a fluorescence detector that uses a monochromator for emission wavelength selection, and a 3.9-mm × 30-cm column that contains packing L1. The flow rate is about 2.0 mL per minute. Chromatograph the *System suitability preparation*, and record the peak responses. The retention time for riboflavin 5′-monophosphate is about 20 to 25 minutes, and the approximate relative retention times for the components are:

Riboflavin 3′4′-diphosphate:	0.23
Riboflavin 3′5′-diphosphate:	0.39
Riboflavin 4′5′-diphosphate:	0.58
Riboflavin 3′-monophosphate:	0.70
Riboflavin 4′-monophosphate:	0.87
Riboflavin 5′-monophosphate:	1.00
Riboflavin:	1.63

The resolution, *R*, between the peaks for riboflavin 4′-monophosphate and riboflavin 5′-monophosphate is not less than 1.0, and the relative standard deviation of the response for riboflavin 5′-monophosphate in replicate injections is not more than 1.5%.

Procedure—Separately inject equal volumes (about 100 µL) of the *Standard preparation*, the *Test preparation*, and the *System suitability preparation* into the chromatograph. Measure the peak responses obtained from the *Standard preparation* and the *Test preparation*, identifying the peaks to be measured in the chromatogram of the *Test preparation* by comparison of retention times with those of the peaks in the chromatogram of the *System suitability preparation.* Calculate the percentage of free riboflavin by the formula:

$$625C(r_F/r_S),$$

and calculate the percentage of riboflavin in the form of riboflavin diphosphates by the formula:

$$625C(r_D/r_S),$$

in which C is the concentration, in mg per mL, of USP Riboflavin RS in the *Standard preparation*, r_F is the riboflavin peak response, if any, obtained from the *Test preparation*, r_D is the sum of the responses for any of the 3 riboflavin diphosphate peaks obtained from the *Test preparation*, and r_S is the riboflavin peak response obtained from the *Standard preparation.* Not more than 6.0% of free riboflavin and not more than 6.0% of riboflavin diphosphates, as riboflavin, calculated on the dried basis, are found.

Lumiflavin—Prepare alcohol-free chloroform just prior to use, as follows: Shake 20 mL of chloroform gently but thoroughly with 20 mL of water for 3 minutes, draw off the chloroform layer, and wash twice more with 20-mL portions of water. Finally filter the chloroform through a dry filter paper, shake it for 5 minutes with 5 g of powdered anhydrous sodium sulfate, allow the mixture to stand for 2 hours, and decant or filter the clear chloroform. Shake 35 mg of Riboflavin 5′-Phosphate Sodium with 10 mL of the alcohol-free chloroform for 5 minutes, and filter: the absorbance of the filtrate so obtained, determined in 1-cm cells at a wavelength of 440 nm, with a suitable spectrophotometer, alcohol-free chloroform being used as the blank, does not exceed 0.025.

Assay—[NOTE—Conduct the assay so that all solutions are protected from actinic light at all stages, preferably by using low-actinic glassware.]

Standard preparation—Transfer about 35 mg of USP Riboflavin RS, accurately weighed, to a 250-mL conical flask, add 20 mL of pyridine and 75 mL of water, and dissolve the riboflavin by frequent shaking. Transfer the solution to a 1000-mL volumetric flask, dilute with water to volume, and mix. Transfer 10.0 mL of this solution to a second 1000-mL volumetric flask, add sufficient 0.1 *N* sulfuric acid (about 4 mL) so that the final pH of the solution is between 5.9 and 6.1, dilute with water to volume, and mix. The *Standard preparation* so obtained contains about 0.35 µg of riboflavin per mL.

Assay preparation—Transfer about 50 mg of Riboflavin 5′-Phosphate Sodium, accurately weighed, to a 250-mL conical flask, add 20 mL of pyridine and 75 mL of water, and dissolve by frequent shaking. Transfer the solution to a 1000-mL volumetric flask, dilute with water to volume, and mix. Transfer 10.0 mL of this solution to a second 1000-mL volumetric flask, add sufficient 0.1 *N* sulfuric acid (about 4 mL) so that the final pH of the solution is between 5.9 and 6.1, dilute with water to volume, and mix.

Procedure—With a suitable fluorometer, determine the maximum fluorescence intensities, I_S and I_U, of the *Standard preparation* and the *Assay preparation*, respectively, at about 530 nm, using an excitation wavelength of about 440 nm. Calculate the quantity, in mg, of $C_{17}H_{20}N_4O_6$ in the portion of Riboflavin 5′-Phosphate Sodium taken by the formula:

$$100C(I_U/I_S),$$

in which C is the concentration, in µg per mL, of USP Riboflavin RS in the *Standard preparation.*

Rifampin

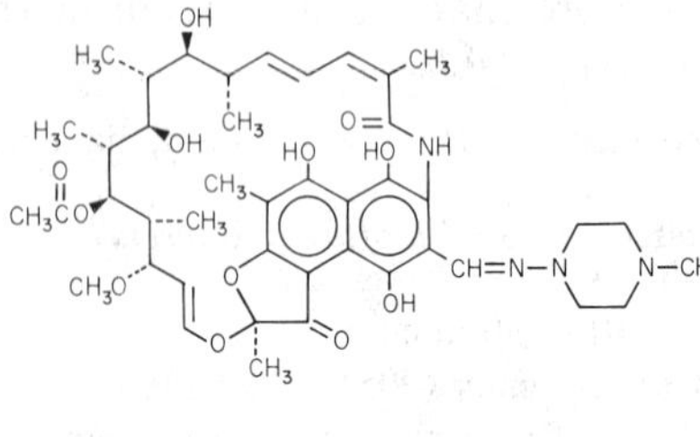

$C_{43}H_{58}N_4O_{12}$ 822.95

Rifamycin, 3-[[(4-methyl-1-piperazinyl)imino]methyl]-.

5,6,9,17,19,21-Hexahydroxy-23-methoxy-2,4,12,16,18,20,22-heptamethyl-8-[*N*-(4-methyl-1-piperazinyl)formimidoyl]-2,7-(epoxypentadeca[1,11,13]trienimino)naphtho[2,1-*b*]furan-1,11-(2*H*)-dione 21-acetate [*13292-46-1*].

» Rifampin contains not less than 95.0 percent and not more than 103.0 percent of $C_{43}H_{58}N_4O_{12}$ per mg, calculated on the dried basis.

Packaging and storage—Preserve in tight, light-resistant containers, protected from excessive heat.

Reference standards—*USP Rifampin Reference Standard*—Do not dry; determine the *Loss on drying* of a separate portion at the time of use. *USP Rifampin Quinone Reference Standard*—Do not dry before using.

Identification—The infrared absorption spectrum of a mineral oil dispersion of it exhibits maxima only at the same wavelengths as that of a similar preparation of USP Rifampin RS.

Crystallinity ⟨695⟩: meets the requirements.

pH ⟨791⟩: between 4.5 and 6.5, in a suspension (1 in 100).

Loss on drying ⟨731⟩—Dry about 100 mg in a capillary-stoppered bottle in vacuum at 60° for 4 hours: it loses not more than 2.0% of its weight.

Assay and chromatographic purity—

Phosphate buffer—Dissolve 136.1 g of monobasic potassium phosphate in about 500 mL of water, add 6.3 mL of phosphoric acid, dilute with water to 1000 mL, and mix.

Mobile phase—Prepare a suitable mixture of water, acetonitrile, *Phosphate buffer*, 1.0 *M* citric acid, and 0.5 *M* sodium perchlorate (510:350:100:20:20), filter through a suitable filter of 0.7-µm or finer porosity, and degas. Make adjustments if necessary (see *System Suitability* under *Chromatography* ⟨621⟩).

Solvent mixture—Prepare a mixture of water, acetonitrile, 1.0 *M* dibasic potassium phosphate, 1.0 *M* monobasic potassium phosphate, and 1.0 *M* citric acid (640:250:77:23:10).

Standard preparation—Transfer about 40 mg of USP Rifampin RS, accurately weighed, to a 200-mL volumetric flask. Dissolve in acetonitrile, dilute with acetonitrile to volume, and mix. Sonicate for about 30 seconds, if necessary to ensure dissolution. [NOTE—Use this solution within 5 hours.] Transfer 10.0 mL of this solution to a 100-mL volumetric flask, dilute with *Solvent mixture* to volume, and mix. [NOTE—Prepare this final dilution immediately prior to injection into the chromatograph.]

Assay preparation—Using Rifampin, proceed as directed for *Standard preparation.*

Stock test preparation—Transfer about 200 mg of Rifampin to a 100-mL volumetric flask, dissolve in acetonitrile, dilute with acetonitrile to volume, and mix. Sonicate for about 30 seconds, if necessary to ensure dissolution. [NOTE—Use this solution within 2 hours.]

Test preparation—Transfer 5.0 mL of *Stock test preparation* to a 50-mL volumetric flask, dilute with *Solvent mixture* to volume, and mix. [NOTE—Prepare this solution immediately prior to injection into the chromatograph.]

Dilute test preparation—Transfer 10.0 mL of *Stock test preparation* to a 100-mL volumetric flask, dilute with acetonitrile to volume, and mix. Transfer 5.0 mL of the resulting solution to a 50-mL volumetric flask, dilute with acetonitrile to volume, and mix. Transfer 5.0 mL of this solution to another 50-mL volumetric flask, dilute with *Solvent mixture* to volume, and mix. [NOTE—Prepare this final dilution immediately prior to injection into the chromatograph.]

Resolution solution—Dissolve suitable quantities of USP Rifampin RS and USP Rifampin Quinone RS in acetonitrile to obtain a solution containing about 0.1 mg of each per mL. Transfer 1.0 mL of this solution to a 10-mL volumetric flask, dilute with *Solvent mixture* to volume, and mix.

Chromatographic system (see *Chromatography* ⟨621⟩)—The liquid chromatograph is equipped with a 254-nm detector and a 4.6-mm × 10-cm column that contains 5-µm packing L7. The flow rate is about 1.5 mL per minute. Chromatograph the *Resolution solution*, and record the peak responses as directed under *Procedure:* the resolution, *R*, between the rifampin quinone and rifampin peaks is not less than 4.0. Chromatograph the *Standard preparation*, and record the peak responses as directed under *Procedure:* the column efficiency determined from the rifampin peak is not less than 1000 theoretical plates, and the relative standard deviation for replicate injections is not more than 1.0%.

Procedure—Separately inject equal volumes (about 50 µL) of the *Standard preparation*, the *Assay preparation*, the *Test preparation*, and the *Dilute test preparation* into the chromatograph, record the chromatograms, and measure the responses for all of the peaks. The relative retention times are about 0.6 for rifampin quinone and 1.0 for rifampin. Calculate the quantity, in mg, of rifampin ($C_{43}H_{58}N_4O_{12}$) in the portion of Rifampin taken to prepare the *Assay preparation* by the formula:

$$2000C(r_U / r_S),$$

in which *C* is the concentration, in mg per mL, calculated on the dried basis, of USP Rifampin RS in the *Standard preparation*, and r_U and r_S are the areas of the rifampin peaks obtained from the *Assay preparation* and the *Standard preparation*, respectively. Calculate the percentage of each related substance by the formula:

$$r_{Ti}/(r_D + 0.01\Sigma r_{Ti}),$$

in which r_{Ti} is the area of the peak of the individual related substance in the chromatogram obtained from the *Test preparation*, r_D is the area of the rifampin peak in the chromatogram obtained from the *Dilute test preparation*, and Σr_{Ti} is the sum of the areas of all of the peaks of the related substances obtained in the chromatogram of the *Test preparation:* not more than 1.5% of rifampin quinone is found, not more than 1.0% of any other individual related substance is found, and a total of not more than 3.5% of all individual related substances, other than rifampin quinone, having retention times of up to 3 in relation to the retention time of rifampin is found.

Rifampin Capsules

» Rifampin Capsules contain not less than 90.0 percent and not more than 110.0 percent of the labeled amount of $C_{43}H_{58}N_4O_{12}$.

Packaging and storage—Preserve in tight, light-resistant containers, protected from excessive heat.

Reference standards—*USP Rifampin Reference Standard*—Do not dry before using. *USP Rifampin Quinone Reference Standard*—Do not dry before using.

Dissolution ⟨711⟩—

Medium: 0.1 *N* hydrochloric acid; 900 mL.

Apparatus 1: 50 rpm.

Time: 45 minutes.

Procedure—Determine the amount of $C_{43}H_{58}N_4O_{12}$ dissolved from absorbances at the wavelength of maximum absorbance at about 475 nm of filtered portions of the solution under test, suitably diluted, if necessary, with *Dissolution Medium*, in comparison with a Standard solution having a known concentration of USP Rifampin RS, calculated on the dried basis, in the same medium, prepared concomitantly and held in the water bath for the *Time* specified.

Tolerances—Not less than 75% (*Q*) of the labeled amount of $C_{43}H_{58}N_4O_{12}$ is dissolved in 45 minutes.

Uniformity of dosage units ⟨905⟩: meet the requirements.

Procedure for content uniformity—

Phosphate buffer, Mobile phase, Solvent mixture, Diluent, Standard preparation, Resolution solution, and *Chromatographic system*—Proceed as directed in the *Assay*.

Test preparation—Transfer the contents of 1 Capsule to a suitable volumetric flask so that when diluted to volume as directed below, each mL of solution contains about 1.5 mg of rifampin. Rinse the Capsule shell with a small quantity of *Solvent mixture*, and add the washing to the volumetric flask. Add *Solvent mixture* until the flask is about four-fifths full. Proceed as directed for *Assay preparation* in the *Assay*, beginning with "Sonicate for about 5 minutes."

Procedure—Proceed as directed for *Procedure* in the *Assay*. Calculate the quantity, in mg, of $C_{43}H_{58}N_4O_{12}$ in the Capsule content by the formula:

$$(LC/D)(r_U/r_S),$$

in which *L* is the labeled quantity, in mg, of rifampin in the Capsule, *C* is the concentration, in mg per mL, of USP Rifampin RS, calculated on the dried basis, in the *Standard preparation*, *D* is the concentration, in mg per mL, of rifampin in the *Test preparation*, based on the labeled quantity per Capsule and the extent of dilution, and r_U and r_S are the rifampin peak responses obtained from the *Test preparation* and the *Standard preparation*, respectively.

Loss on drying ⟨731⟩—Dry about 100 mg of Capsule contents in a capillary-stoppered bottle in vacuum at 60° for 3 hours: it loses not more than 3.0% of its weight.

Assay—

Phosphate buffer—Dissolve 136.1 g of monobasic potassium phosphate in about 500 mL of water, add 6.3 mL of phosphoric acid, dilute with water to 1000 mL, and mix (pH 3.1 ± 0.1).

Mobile phase—Prepare a suitable mixture of water, acetonitrile, *Phosphate buffer*, 1.0 *M* citric acid, and 0.5 *M* sodium perchlorate (510:350:100:20:20), filter through a suitable filter of 0.7-µm or finer porosity, and degas. Make adjustments if necessary (see *System Suitability* under *Chromatography* ⟨621⟩).

Solvent mixture—Prepare a mixture of acetonitrile and methanol (1:1).

Diluent—Prepare a suitable mixture of water, acetonitrile, 1.0 *M* dibasic sodium phosphate, 1.0 *M* monobasic potassium phosphate, and 1.0 *M* citric acid (640:250:77:23:10).

Standard preparation—Dissolve an accurately weighed quantity of USP Rifampin RS in *Solvent mixture* to obtain a solution having a known concentration of about 1.5 mg per mL, sonicating for about 30 seconds, if necessary, to ensure dissolution. Transfer 10.0 mL of this solution to a 50-mL volumetric flask, dilute with acetonitrile to volume, and mix. [NOTE—Use this working solution within 5 hours.] Transfer 5.0 mL of the working solution to a 50-mL volumetric flask, dilute with *Diluent* to volume, and mix. Each mL of this solution contains about 0.03 mg of USP Rifampin RS. [NOTE—Inject this *Standard preparation* into the chromatograph within 30 to 60 seconds after preparation.]

Assay preparation—Remove, as completely as possible, the contents of not less than 20 Rifampin Capsules, and weigh accurately. Mix the Capsule contents, and transfer an accurately weighed portion of the powder, equivalent to about 300 mg of rifampin, to a 200-mL volumetric flask, and add about 180 mL of *Solvent mixture*. Sonicate for about 5 minutes, allow to equilibrate to room temperature, dilute with *Solvent mixture* to volume, and mix. Transfer 10.0 mL of the resulting solution to a 50-mL volumetric flask, dilute with acetonitrile to volume, and mix. [NOTE—Use this solution within 5 hours.] Transfer 5.0 mL of this solution to a 50-mL volumetric flask, dilute with *Diluent* to volume, and mix. [NOTE—Inject this *Assay preparation* into the chromatograph within 30 to 60 seconds after preparation.]

Resolution solution—Dissolve USP Rifampin Quinone RS in *Solvent mixture* to obtain a solution containing about 0.1 mg per mL. Transfer 1.5 mL of this solution and 5.0 mL of the working solution used to prepare the *Standard preparation* to a 50-mL volumetric flask, dilute with *Diluent* to volume, and mix.

Chromatographic system (see *Chromatography* ⟨621⟩)—The liquid chromatograph is equipped with a 254-nm detector and a 4.6-mm × 10-cm column that contains 5-μm packing L7. The flow rate is about 1.5 mL per minute. Chromatograph the *Resolution solution*, and record the peak responses as directed under *Procedure:* the relative retention times are about 0.6 for rifampin quinone and 1.0 for rifampin, and the resolution, *R*, between the rifampin quinone and rifampin peaks is not less than 4.0. Chromatograph the *Standard preparation*, and record the peak responses as directed under *Procedure:* the relative standard deviation for replicate injections is not more than 1.0%.

Procedure—[NOTE—Use peak areas where peak responses are indicated.] Separately inject equal volumes (about 50 μL) of the *Standard preparation* and the *Assay preparation* into the chromatograph, record the chromatograms, and measure the responses for the major peaks. Calculate the quantity, in mg, of $C_{43}H_{58}N_4O_{12}$ in the portion of Capsules taken by the formula:

$$10{,}000C(r_U/r_S),$$

in which *C* is the concentration, in mg per mL, of USP Rifampin RS, calculated on the dried basis, in the *Standard preparation*, and r_U and r_S are the rifampin peak responses obtained from the *Assay preparation* and the *Standard preparation*, respectively.

Rifampin and Isoniazid Capsules

» Rifampin and Isoniazid Capsules contain not less than 90.0 percent and not more than 130.0 percent of the labeled amount of rifampin ($C_{43}H_{58}N_4O_{12}$) and not less than 90.0 percent and not more than 110.0 percent of the labeled amount of isoniazid ($C_6H_7N_3O$).

NOTE—Where Rifampin and Isoniazid Capsules are prescribed without reference to the quantity of rifampin or isoniazid contained therein, a product containing 300 mg of rifampin and 150 mg of isoniazid shall be dispensed.

Packaging and storage—Preserve in tight, light-resistant containers, and avoid exposure to excessive heat.

Reference standard—*USP Rifampin Reference Standard*—Do not dry before using.

Loss on drying ⟨731⟩—Dry about 100 mg of Capsule contents in a capillary-stoppered bottle in vacuum at 60° for 3 hours: it loses not more than 3.0% of its weight.

Assay for rifampin—Proceed with Rifampin and Isoniazid Capsules as directed in the *Assay* under *Rifampin Capsules*.

Assay for isoniazid—Weigh the contents of not less than 10 Rifampin and Isoniazid Capsules, mix, and transfer an accurately weighed portion of the powder, equivalent to about 100 mg of isoniazid, to a 125-mL separator. Add 20 mL of 0.1 *N* hydrochloric acid, and shake. Extract the acidic solution with six 25-mL portions of chloroform, retaining any interfacial emulsion with the aqueous phase, and discarding the chloroform extracts. Transfer the acidic aqueous phase to a 100-mL volumetric flask, dilute with 0.1 *N* hydrochloric acid to volume, and mix. Pipet 25 mL of this solution into a titration vessel, add 10 mL of hydrochloric acid, and adjust the volume with water to about 50 mL. Titrate with 0.1 *N* bromine VS, determining the end-point potentiometrically. Each mL of 0.1 *N* bromine is equivalent to 3.429 mg of $C_6H_7N_3O$.

Ringer's Injection

» Ringer's Injection is a sterile solution of Sodium Chloride, Potassium Chloride, and Calcium Chloride in Water for Injection. It contains, in each 100 mL, not less than 323.0 mg and not more than 354.0 mg of sodium (Na, equivalent to not less than 820.0 mg and not more than 900.0 mg of NaCl); not less than 14.9 mg and not more than 16.5 mg of potassium (K, equivalent to not less than 28.5 mg and not more than 31.5 mg of KCl); not less than 8.20 mg and not more than 9.80 mg of calcium (Ca, equivalent to not less than 30.0 mg and not more than 36.0 mg of $CaCl_2.2H_2O$); and not less than 523.0 mg and not more than 580.0 mg of chloride (Cl, as NaCl, KCl, and $CaCl_2.2H_2O$). Ringer's Injection contains no antimicrobial agents.

NOTE—The calcium, chloride, potassium, and sodium ion contents of Ringer's Injection are approximately 4.5, 156, 4, and 147.5 milliequivalents per liter, respectively.

Sodium Chloride	8.6 g
Potassium Chloride	0.3 g
Calcium Chloride	0.33 g
Water for Injection, a sufficient quantity, to make	1000 mL

Dissolve the three salts in the Water for Injection, filter until clear, place in suitable containers, and sterilize.

Packaging and storage—Preserve in single-dose containers, preferably of Type I or Type II glass.

Labeling—The label states the total osmolar concentration in mOsmol per liter. Where the contents are less than 100 mL, or where the label states that the Injection is not for direct injection but is to be diluted before use, the label alternatively may state the total osmolar concentration in mOsmol per mL.

Identification—It responds to the tests for *Sodium* ⟨191⟩ and for *Chloride* ⟨191⟩, and when concentrated to one-half its original volume, to the test for *Calcium* ⟨191⟩ and to the flame test for *Potassium* ⟨191⟩.

Pyrogen—It meets the requirements of the *Pyrogen Test* ⟨151⟩.

pH ⟨791⟩: between 5.0 and 7.5.

Heavy metals ⟨231⟩—Evaporate 67 mL to a volume of about 20 mL, add 2 mL of 1 *N* acetic acid, and dilute with water to 25 mL: the limit is 0.3 ppm.

Other requirements—It meets the requirements under *Injections* ⟨1⟩.

Assay for calcium—Pipet 50 mL of Ringer's Injection into a 250-mL beaker, add 15 mL of 1 *N* sodium hydroxide and 300 mg of hydroxy naphthol blue trituration, and titrate immediately with 0.005 *M* disodium ethylenediaminetetraacetate VS to a deep blue end-point. Each mL of 0.005 *M* disodium ethylenediaminetetraacetate is equivalent to 200.4 μg of Ca^{++}.

Assay for potassium—

Standard stock solution—Dissolve 190.7 mg of potassium chloride, previously dried at 105° for 2 hours, in 50 mL of water, transfer to a 1000-mL volumetric flask, dilute with water to volume, and mix. Each mL of this solution contains 100 μg of potassium.

Standard preparations—Dissolve 1.093 g of sodium chloride in 100.0 mL of water, and transfer 10 mL of this solution to each of five 100-mL volumetric flasks containing 10.0 mL of a solution of a suitable nonionic wetting agent (1 in 500). Dilute the contents of one of the flasks with water to volume to provide a blank. To the remaining flasks add, respectively, 5.0, 10.0, 15.0, and 20.0 mL of *Standard stock solution*, dilute with water to volume, and mix.

Assay preparation—Pipet 10 mL of Ringer's Injection into a 100-mL volumetric flask, add 10.0 mL of a solution of a suitable wetting agent (1 in 500), dilute with water to volume, and mix.

Standard graph—Set a suitable flame photometer for maximum transmittance at a wavelength of about 766 nm. Adjust the instrument to zero transmittance with the blank. Adjust the instrument to 100% transmittance with the most concentrated of the *Standard preparations*. Read the percentage transmittance of the other *Standard preparations*, and plot transmittances versus concentration of potassium.

Procedure—Adjust the instrument as directed under *Standard graph*, read the percentage transmittance of the *Assay preparation*, and calculate the potassium content, in mg per 100 mL, of Ringer's Injection.

Assay for sodium—

Standard stock solution—Dissolve 254.2 mg of sodium chloride, previously dried at 105° for 2 hours, in 50 mL of water, transfer to a 1000-mL volumetric flask, dilute with water to volume, and mix. Each mL of this solution contains 100 µg of sodium.

Standard preparations—Transfer to each of five 100-mL volumetric flasks 10 mL of a solution of a suitable nonionic wetting agent (1 in 500). Dilute the contents of one of the flasks with water to volume to provide a blank. To the remaining flasks add, respectively, 5.0, 10.0, 15.0, and 20.0 mL of *Standard stock solution*, dilute with water to volume, and mix.

Assay preparation—Pipet 5 mL of Ringer's Injection into a 1000-mL volumetric flask containing 100 mL of a solution of a suitable wetting agent (1 in 500), dilute with water to volume, and mix.

Procedure—Proceed as directed for *Standard graph* and for *Procedure* in the *Assay for potassium*, setting the flame photometer for maximum transmittance at a wavelength of about 589 nm, instead of about 766 nm. Calculate the sodium content, in mg per 100 mL, of Ringer's Injection.

Assay for chloride—Pipet 10 mL of Ringer's Injection into a porcelain casserole, and add 140 mL of water and 1 mL of dichlorofluorescein TS. Mix, and titrate with 0.1 *N* silver nitrate VS until the silver chloride flocculates and the mixture turns a faint pink color. Each mL of 0.1 *N* silver nitrate is equivalent to 3.545 mg of Cl.

Lactated Ringer's Injection

» Lactated Ringer's Injection is a sterile solution of Calcium Chloride, Potassium Chloride, Sodium Chloride, and Sodium Lactate in Water for Injection. It contains, in each 100 mL, not less than 285.0 mg and not more than 315.0 mg of sodium (as NaCl and $C_3H_5NaO_3$), not less than 14.1 mg and not more than 17.3 mg of potassium (K, equivalent to not less than 27.0 mg and not more than 33.0 mg of KCl), not less than 4.90 mg and not more than 6.00 mg of calcium (Ca, equivalent to not less than 18.0 mg and not more than 22.0 mg of $CaCl_2.2H_2O$), not less than 368.0 mg and not more than 408.0 mg of chloride (Cl, as NaCl, KCl, and $CaCl_2.2H_2O$), and not less than 231.0 mg and not more than 261.0 mg of lactate ($C_3H_5O_3$, equivalent to not less than 290.0 mg and not more than 330.0 mg of $C_3H_5NaO_3$). Lactated Ringer's Injection contains no antimicrobial agents.

NOTE—The calcium, potassium, and sodium contents of Lactated Ringer's Injection are approximately 2.7, 4, and 130 milliequivalents per liter, respectively.

Packaging and storage—Preserve in single-dose containers, preferably of Type I or Type II glass.

Labeling—Solutions of higher concentrations of the specified constituents of Lactated Ringer's Injection, in which the ratios of the components are consistent with the official limits, may be labeled "For the preparation of Lactated Ringer's Injection USP," provided that the label indicates also that the solution is a concentrate, and provided that directions are given for dilution to the official strength.

The label states the total osmolar concentration in mOsmol per liter. Where the contents are less than 100 mL, or where the label states that the Injection is not for direct injection but is to be diluted before use, the label alternatively may state the total osmolar concentration in mOsmol per mL. The label includes also the warning: "Not for use in the treatment of lactic acidosis."

Identification—It responds to the tests for *Sodium* ⟨191⟩, for *Chloride* ⟨191⟩, and for *Lactate* ⟨191⟩, and when concentrated to one-half its original volume, to the test for *Calcium* ⟨191⟩ and to the flame test for *Potassium* ⟨191⟩.

Pyrogen—It meets the requirements of the *Pyrogen Test* ⟨151⟩.

pH ⟨791⟩: between 6.0 and 7.5.

Heavy metals ⟨231⟩—Evaporate 67 mL to a volume of 20 mL, add 2 mL of 1 *N* acetic acid, then dilute with water to 25 mL: the limit is 0.3 ppm.

Other requirements—It meets the requirements under *Injections* ⟨1⟩.

Assay for calcium—Proceed with Lactated Ringer's Injection as directed in the *Assay for calcium* under *Ringer's Injection*.

Assay for potassium—Proceed with Lactated Ringer's Injection as directed in the *Assay for potassium* under *Ringer's Injection*.

Assay for sodium—Proceed with Lactated Ringer's Injection as directed in the *Assay for sodium* under *Ringer's Injection*.

Assay for chloride—Proceed with Lactated Ringer's Injection as directed in the *Assay for chloride* under *Ringer's Injection*.

Assay for lactate—Evaporate 50.0 mL of Lactated Ringer's Injection in a suitable crucible or dish, and ignite gently until thoroughly charred. Break up the charred mass well with a glass rod, add 25 mL of water and 25.0 mL of 0.1 *N* sulfuric acid VS, and heat on a steam bath for 30 minutes, breaking up any lumps with a glass rod during the heating. Filter, wash well with hot water until the last washing is neutral to litmus paper, then cool the combined filtrate and washings, add methyl orange TS, and titrate the excess acid with 0.1 *N* sodium hydroxide VS. Each mL of 0.1 *N* sulfuric acid is equivalent to 8.907 mg of $C_3H_5O_3$.

Ringer's Irrigation

» Ringer's Irrigation is Ringer's Injection that has been suitably packaged, and it contains no antimicrobial agents.

Packaging and storage—Preserve in single-dose containers, preferably of Type I or Type II glass. The container may be designed to empty rapidly and may contain a volume of more than 1 liter.

Labeling—The designation "not for injection" appears prominently on the label.

Sterility—It meets the requirements under *Sterility Tests* ⟨71⟩.

Other requirements—It responds to the *Identification tests*, and meets the requirements for *pH*, *Pyrogen*, *Heavy metals*, *Assay for calcium*, *Assay for potassium*, *Assay for sodium*, and *Assay for chloride* under *Ringer's Injection*.

Ritodrine Hydrochloride

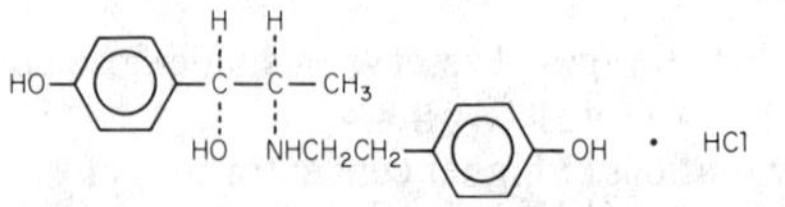

$C_{17}H_{21}NO_3.HCl$ 323.82

Benzenemethanol, 4-hydroxy-α-[1-[[2-(4-hydroxyphenyl)ethyl]amino]ethyl]-, hydrochloride, (R*, S*)-.

erythro-p-Hydroxy-α-[1-[(*p*-hydroxyphenethyl)amino]ethyl]benzyl alcohol hydrochloride [*23239-51-2*].

» Ritodrine Hydrochloride contains not less than 97.0 percent and not more than 103.0 percent of $C_{17}H_{21}NO_3.HCl$, calculated on the dried basis.

Packaging and storage—Preserve in tight containers.

Reference standard—*USP Ritodrine Hydrochloride Reference Standard*—Dry at 105° for 1 hour before using.

Identification—

A: The infrared absorption spectrum of a potassium bromide dispersion of it exhibits maxima only at the same wavelengths as that of a similar preparation of USP Ritodrine Hydrochloride RS.

B: The retention time of the ritodrine hydrochloride in the *Assay preparation* obtained in the *Assay* corresponds to that of the *Standard preparation* obtained in the *Assay*.

C: A solution (1 in 100) responds to the tests for *Chloride* ⟨191⟩.

pH ⟨791⟩: between 4.5 and 6.0, in a solution (1 in 50).

Loss on drying ⟨731⟩: Dry it at 105° for 2 hours: it loses not more than 1.0% of its weight.

Residue on ignition ⟨281⟩: not more than 0.2%.

Heavy metals, *Method II* ⟨231⟩: not more than 0.002%.

Related substances—

Mobile phase—Prepare as directed in the *Assay*.

Test preparation—Prepare a solution containing about 1 mg of Ritodrine Hydrochloride in each mL of *Mobile phase*.

Procedure—Chromatograph the *Test preparation* as directed in the *Assay*. Typical retention times are 5 minutes for tyramine, 12 minutes for ritodrine, 14 minutes for the *threo* diastereomer of ritodrine, and 21 minutes for the amino ketone, 1-(4-hydroxyphenyl)-2-[[2-(4-hydroxyphenyl)ethyl]amino]-1-propanone hydrochloride. Measure the peak responses, and designate them as R_a, R_b, R_c, and R_d, respectively. R_a/R_b is not more than 0.004, R_c/R_b is not more than 0.015, and R_d/R_b is not more than 0.027.

Assay—

Ammonium acetate buffer solution—Dissolve 1.81 g of ammonium acetate in a mixture of 120 mL of glacial acetic acid and about 500 mL of water. Add water to make 1000 mL, and mix.

0.25 M Sodium 1-heptanesulfonate solution—Dissolve 13.8 g of sodium 1-heptanesulfonate in 25 mL of glacial acetic acid and enough water to make 250 mL, and mix.

Mobile phase—Prepare a suitable degassed and filtered mixture of water, methanol, *Ammonium acetate buffer solution*, and *0.25 M Sodium 1-heptanesulfonate solution* (63:25:10:2).

Standard preparation—Dissolve an accurately weighed quantity of USP Ritodrine Hydrochloride RS in *Mobile phase* to obtain a solution having a known concentration of about 0.2 mg per mL.

Assay preparation—Transfer about 200 mg of Ritodrine Hydrochloride, accurately weighed, to a 100-mL volumetric flask, dissolve in *Mobile phase*, dilute with *Mobile phase* to volume, and mix. Transfer 10.0 mL of this solution to a 100-mL volumetric flask, dilute with *Mobile phase* to volume, and mix.

System suitability preparation—Dissolve about 20 mg of Ritodrine Hydrochloride in about 50 mL of *Mobile phase*. Add 5.6 mL of sulfuric acid, dilute with *Mobile phase* to 100 mL, and mix. Heat a portion of this solution for about 2 hours at about 85°, and then cool to room temperature. Cautiously mix 10.0 mL of the cooled solution with 8.0 mL of sodium hydroxide solution (1 in 10), and allow to cool. This solution contains ritodrine and its *threo* diastereomer.

Chromatographic system—The chromatograph is equipped with a 30-cm × 4.6-mm stainless steel column that contains packing L7 and an ultraviolet detector that monitors absorption at 254 nm. Chromatograph about 50 µL of the *System suitability preparation:* the resolution between ritodrine and its *threo* diastereomer is not less than 1.0. [NOTE—Chromatograms obtained as directed for this test exhibit relative retention times of 1.0 for ritodrine and approximately 1.2 for the *threo* diastereomer.]

Procedure—Separately inject equal volumes (20 to 50 µL) of the *Standard preparation* and the *Assay preparation* into the chromatograph (see *Chromatography* ⟨621⟩) by means of a suitable sampling valve. Record the chromatograms and measure the peak responses. Calculate the quantity, in mg, of $C_{17}H_{21}NO_3.HCl$ in the portion of Ritodrine Hydrochloride taken by the formula:

$$1000C(r_U/r_S),$$

in which C is the concentration, in mg per mL, of USP Ritodrine Hydrochloride RS in the *Standard preparation*, and r_U and r_S are the peak responses for Ritodrine Hydrochloride obtained from the *Assay preparation* and the *Standard preparation*, respectively.

Ritodrine Hydrochloride Injection

» Ritodrine Hydrochloride Injection is a sterile solution of Ritodrine Hydrochloride in Water for Injection. It contains not less than 90.0 percent and not more than 110.0 percent of the labeled amount of $C_{17}H_{21}NO_3.HCl$.

Packaging and storage—Preserve in single-dose containers, preferably of Type I glass. Store at room temperature, preferably below 30°.

Reference standard—*USP Ritodrine Hydrochloride Reference Standard*—Dry at 105° for 1 hour before using.

Identification—The retention time of the ritodrine hydrochloride in the *Assay preparation* obtained in the *Assay* corresponds to that of the *Standard preparation* obtained in the *Assay*.

Pyrogen—It meets the requirements of the *Pyrogen Test* ⟨151⟩, the test dose being 10 mg of ritodrine hydrochloride per kg.

pH ⟨791⟩: between 4.8 and 5.5.

Other requirements—It meets the requirements under *Injections* ⟨1⟩.

Assay—

Ammonium acetate buffer solution, 0.25 M Sodium 1-heptanesulfonate solution, Mobile phase, Standard preparation, Chromatographic system, and *System suitability preparation*—Proceed as directed in the *Assay* under *Ritodrine Hydrochloride*.

Assay preparation—Dilute an accurately measured volume of Ritodrine Hydrochloride Injection, equivalent to about 20 mg of ritodrine hydrochloride, with *Mobile phase* to 100.0 mL, and mix.

Procedure—Proceed as directed in the *Assay* under *Ritodrine Hydrochloride*. Calculate the quantity, in mg, of $C_{17}H_{21}NO_3.HCl$ in each mL of the Injection taken by the formula:

$$(100C/V)(r_U/r_S),$$

in which V is the volume, in mL, of Injection taken.

Ritodrine Hydrochloride Tablets

» Ritodrine Hydrochloride Tablets contain not less than 90.0 percent and not more than 110.0 percent of the labeled amount of $C_{17}H_{21}NO_3.HCl$.

Packaging and storage—Preserve in tight containers. Store at room temperature, preferably below 30°.

Reference standard—*USP Ritodrine Hydrochloride Reference Standard*—Dry at 105° for 1 hour before using.

Identification—The retention time of the Tablets in the *Assay preparation* obtained in the *Assay* corresponds to that of the *Standard preparation* obtained in the *Assay*.

Dissolution ⟨711⟩—

Medium: 0.1 *N* hydrochloric acid; 900 mL.

Apparatus 2: 50 rpm.

Time: 30 minutes.

Procedure—Determine the amount of $C_{17}H_{21}NO_3.HCl$ dissolved from ultraviolet absorbances at the wavelength of maximum absorbance at about 224 nm of filtered portions of the solution under test, suitably diluted with *Dissolution Medium*, in comparison with a Standard solution having a known concentration of USP Ritodrine Hydrochloride RS in the same medium.

Tolerances—Not less than 80% (*Q*) of the labeled amount of $C_{17}H_{21}NO_3.HCl$ is dissolved in 30 minutes.

Uniformity of dosage units ⟨905⟩: meet the requirements.

Procedure for content uniformity—Transfer 1 Tablet to a 200-mL volumetric flask, add about 5 mL of water, and stir until disintegration is complete. Add about 150 mL of methanolic sulfuric acid solution (0.005 *M* sulfuric acid in methanol), continue stirring for an additional 15 minutes, dilute with methanolic sulfuric acid solution to volume, and mix. Filter a portion of the mixture, discarding the first 20 mL of thc filtrate. Dissolve an accurately weighed quantity of USP Ritodrine Hydrochloride RS in methanolic sulfuric acid solution, and dilute quantitatively and stepwise with the same solvent to obtain a Standard solution having a known concentration of about 50 µg per mL. Concomitantly determine the absorbances of both solutions in 1-cm cells at the wavelength of maximum absorbance at about 276 nm, with a suitable spectrophotometer, using methanolic sulfuric acid solution as the blank. Calculate the quantity, in mg, of $C_{17}H_{21}NO_3.HCl$ in the Tablet by the formula:

$$(TC/D)(A_U/A_S),$$

in which *T* is the labeled quantity, in mg, of ritodrine hydrochloride in the Tablet, *C* is the concentration, in µg per mL, of USP Ritodrine Hydrochloride RS in the Standard solution, *D* is the concentration, in µg per mL, of ritodrine hydrochloride in the solution from the Tablet based upon the labeled quantity per Tablet and the extent of dilution, and A_U and A_S are the absorbances of the solution from the Tablet and the Standard solution, respectively.

Assay—

Ammonium acetate buffer solution, 0.25 M Sodium 1-heptanesulfonate solution, Mobile phase, Standard preparation, Chromatographic system, and *System suitability preparation*—Proceed as directed in the *Assay* under *Ritodrine Hydrochloride*.

Assay preparation—Weigh and finely powder not less than 20 Ritodrine Hydrochloride Tablets. Transfer an accurately weighed portion of the powder, equivalent to about 20 mg of ritodrine hydrochloride, to a 100-mL volumetric flask. Add about 70 mL of *Mobile phase*, stir for about 15 minutes, dilute with *Mobile phase* to volume, mix, and filter a portion to remove any particulate matter.

Procedure—Proceed as directed in the *Assay* under *Ritodrine Hydrochloride*. Calculate the quantity, in mg, of $C_{17}H_{21}NO_3.HCl$ in the portion of Tablets taken by the formula:

$$100C(r_U/r_S),$$

in which the terms are as defined therein.

Rolitetracycline for Injection

» Rolitetracycline for Injection is a sterile dry mixture of Sterile Rolitetracycline and one or more suitable buffers, and if intended for intramuscular use, one or more suitable anesthetics. It contains not less than 90.0 percent and not more than 115.0 percent of the labeled amount of $C_{27}H_{33}N_3O_8$.

Packaging and storage—Preserve in *Containers for Sterile Solids* as described under *Injections* ⟨1⟩, protected from light.

Reference standard—*USP Rolitetracycline Reference Standard*—Dry in vacuum at a pressure not exceeding 5 mm of mercury at 60° for 3 hours before using.

Constituted solution—At the time of use, the constituted solution prepared from Rolitetracycline for Injection meets the requirements for *Constituted Solutions* under *Injections* ⟨1⟩.

Depressor substances—When dissolved in sterile saline TS and the solution is diluted with the same solvent to a concentration of 5.0 mg of rolitetracycline ($C_{27}H_{33}N_3O_8$) per mL, it meets the requirements of the *Depressor Substances Test* ⟨101⟩, the test dose being 0.6 mL per kg. Rolitetracycline for Injection intended for intramuscular use only is exempt from this requirement.

Pyrogen—When dissolved in Sterile Water for Injection and the solution is diluted with the same solvent to a concentration of 5.0 mg of rolitetracycline ($C_{27}H_{33}N_3O_8$) per mL, it meets the requirements of the *Pyrogen Test* ⟨151⟩, the test dose being 1.0 mL per kg.

Sterility—It meets the requirements under *Sterility Tests* ⟨71⟩, when tested as directed in the section, *Test Procedures Using Membrane Filtration*, the specimen being dissolved in *Fluid D*.

pH ⟨791⟩: between 3.0 and 4.5, in the solution constituted as directed in the labeling.

Loss on drying ⟨731⟩—Dry about 100 mg, accurately weighed, in a capillary-stoppered bottle in vacuum at a pressure not exceeding 5 mm of mercury at 110° for 3 hours: it loses not more than 5.0% of its weight.

Assay—Proceed as directed under *Antibiotics—Microbial Assays* ⟨81⟩, using Rolitetracycline for Injection constituted as directed in the labeling. Remove an accurately measured portion, and dilute quantitatively with water so that the stock solution so obtained contains a convenient concentration. Dilute an accurately measured volume of this stock solution quantitatively and stepwise with water to obtain a *Test Dilution* having a concentration assumed to be equal to the median dose level of the Standard.

Sterile Rolitetracycline

OH O OH O OH CONHCH2N OH H H HO CH3 H N(CH3)2

$C_{27}H_{33}N_3O_8$ 527.57

2-Naphthacenecarboxamide, 4-(dimethylamino)-1,4,4a,5,5a,6,11,12a-octahydro-3,6,10,12,12a-pentahydroxy-6-methyl-1,11-dioxo-*N*-(1-pyrrolidinylmethyl)-, [4*S*-(4α,4aα,5aα,6β,12aα)]-.

4-(Dimethylamino)-1,4,4a,5,5a,6,11,12a-octahydro-3,6,10,12,12a-pentahydroxy-6-methyl-1,11-dioxo-*N*-(1-pyrrolidinylmethyl)-2-naphthacenecarboxamide [*751-97-3*].

» Sterile Rolitetracycline is rolitetracycline suitable for parenteral use. It has a potency of not less than 900 µg of $C_{27}H_{33}N_3O_8$ per mg, calculated on the anhydrous basis.

Packaging and storage—Preserve in *Containers for Sterile Solids* as described under *Injections* ⟨1⟩, protected from light.

Reference standard—*USP Rolitetracycline Reference Standard*—Dry in vacuum at a pressure not exceeding 5 mm of mercury at 60° for 3 hours before using.

Identification—

A: The ultraviolet absorption spectrum of a 1 in 60,000 solution in 0.25 *N* sodium hydroxide exhibits maxima and minima at the same wavelengths as that of a similar solution of USP

Rolitetracycline RS, concomitantly measured, and the absorptivity, calculated on the anhydrous basis, at the wavelength of maximum absorbance at about 380 nm is between 95.6% and 104.4% of that of the USP Rolitetracycline RS, the potency of the Reference Standard being taken into account.

B: To 0.1 g of Sterile Rolitetracycline in a test tube add 5 mL of 1 *N* sodium hydroxide, and heat gently to boiling for about 15 seconds: the musty, amine-like odor of pyrrolidine is detectable, and on cooling to room temperature a deep burgundy-red color is produced.

Crystallinity ⟨695⟩: meets the requirements.

Depressor substances—When dissolved in sterile saline TS and the solution is diluted with sterile saline TS to a concentration of 5.0 mg of rolitetracycline ($C_{27}H_{33}N_3O_8$) per mL, it meets the requirements of the *Depressor Substances Test* ⟨101⟩, the test dose being 0.6 mL per kg.

Pyrogen—When dissolved in Sterile Water for Injection and the solution is diluted with the same solvent to a concentration of 5.0 mg of rolitetracycline ($C_{27}H_{33}N_3O_8$) per mL, it meets the requirements of the *Pyrogen Test* ⟨151⟩, the test dose being 1.0 mL per kg.

Sterility—It meets the requirements under *Sterility Tests* ⟨71⟩, when tested as directed in the section, *Test Procedures Using Membrane Filtration*, 6 g of specimen aseptically dissolved in 200 mL of *Fluid D* being used.

pH ⟨791⟩: between 7 and 9, in a solution containing 10 mg per mL (the solution is substantially clear).

Water, *Method I* ⟨921⟩: not more than 3.0%.

Assay—Proceed with Sterile Rolitetracycline as directed under *Antibiotics—Microbial Assays* ⟨81⟩.

Rose Bengal Sodium I 131 Injection—*see under "iodine"*

Rose Oil—*see* Rose Oil NF

Rose Water Ointment

» Prepare Rose Water Ointment as follows:

Cetyl Esters Wax	125 g
White Wax	120 g
Almond Oil	560 g
Sodium Borate	5 g
Stronger Rose Water	25 mL
Purified Water	165 mL
Rose Oil	200 μL
To make about	1000 g

Reduce the cetyl esters wax and the white wax to small pieces, melt them on a steam bath, add the almond oil, and continue heating until the temperature of the mixture reaches 70°. Dissolve the sodium borate in the purified water and the stronger rose water, warmed to 70°, and gradually add the warm aqueous phase to the melted oil phase, stirring rapidly and continuously until it has cooled to about 45°. Then incorporate the rose oil.

NOTE—Rose Water Ointment is free from rancidity. If the Ointment has been chilled, warm it slightly before attempting to incorporate other ingredients (see *Ointments and Suppositories* in the section, *Added Substances*, under *Ingredients and Processes* in the *General Notices*).

Packaging and storage—Preserve in tight, light-resistant containers.

Rose Water, Stronger—*see* Rose Water, Stronger NF

Rubbing Alcohol—*see* Alcohol, Rubbing

Rubbing Alcohol, Isopropyl—*see* Isopropyl Rubbing Alcohol

Rubella Virus Vaccine Live

» Rubella Virus Vaccine Live conforms to the regulations of the federal Food and Drug Administration concerning biologics (630.60 to 630.67) (see *Biologics* ⟨1041⟩). It is a bacterially sterile preparation of live virus derived from a strain of rubella virus that has been tested for neurovirulence in monkeys, and for immunogenicity, that is free from all demonstrable viable microbial agents except unavoidable bacteriophage, and that has been found suitable for human immunization. The strain is grown, for purposes of vaccine production, on primary cell cultures of duck embryo tissue, derived from pathogen-free flocks, or on primary cell cultures of a designated strain of human tissue, provided that the same cell culture system is used as that in which the strain was tested. The strain meets the requirements of the specific safety tests in adult and suckling mice; and the requirements of the tests in monkey kidney, chicken embryo, and human tissue cell cultures and embryonated eggs. In the case of virus grown in duck embryo cell cultures, the strain meets the requirements of the test by inoculation of embryonated duck eggs, and of the tests for absence of *Mycobacterium tuberculosis* and of avian leucosis. In the case of virus grown in rabbit kidney cell cultures, the strain meets the requirements of the tests by inoculation of rabbits and guinea pigs, and of the tests for absence of *Mycobacterium tuberculosis* and of known adventitious agents of rabbits. In the case of virus grown in human tissue cell cultures, the strain meets the requirements of the specific safety tests and tests for absence of *Mycobacterium tuberculosis* or other adventitious agents tests by inoculation of rabbits and guinea pigs and the requirements for karyology and of the tests for absence of adventitious and other infective agents, including hemadsorption viruses and *Mycoplasma*, in human diploid cell cultures. The strain cultures are treated to remove all intact tissue cells. The Vaccine meets the requirements of the specific tissue culture test for live virus titer, in a single immunizing dose, of not less than the equivalent of 1000 $TCID_{50}$ (quantity of virus estimated to infect 50% of inoculated cultures × 1000) when tested in parallel with the U.S. Reference Rubella Virus, Live.

Packaging and storage—Preserve in single-dose containers, or in light-resistant, multiple-dose containers, at a temperature between 2° and 8°. Multiple-dose containers for 50 doses are adapted for use only in jet injectors, and those for 10 doses for use by jet or syringe injection.

Expiration date—The expiration date is 1 to 2 years, depending on the manufacturer's data, after date of issue from manufacturer's cold storage (−20°, 1 year).

Labeling—Label the Vaccine in multiple-dose containers to indicate that the contents are intended solely for use by jet injector or for use by either jet or syringe injection, whichever is applicable. Label the Vaccine in single-dose containers, if such containers are not light-resistant, to state that it should be protected from sunlight. Label it also to state that constituted Vaccine should be discarded if not used within 8 hours.

Rubella Virus Vaccine Live, Measles and—*see* Measles and Rubella Virus Vaccine Live

Rubella Virus Vaccine Live, Measles, Mumps, and—*see* Measles, Mumps, and Rubella Virus Vaccine Live

Rubella and Mumps Virus Vaccine Live

» Rubella and Mumps Virus Vaccine Live conforms to the regulations of the federal Food and Drug Administration concerning biologics (see *Biologics* ⟨1041⟩). It is a bacterially sterile preparation of a combination of live rubella virus and live mumps virus such that each component is prepared in conformity with and meets the requirements for Rubella Virus Vaccine Live, and for Mumps Virus Vaccine Live, whichever is applicable. It may contain suitable antimicrobial agents.

Packaging and storage—Preserve in single-dose containers, or in light-resistant, multiple-dose containers, at a temperature between 2° and 8°. Multiple-dose containers for 50 doses are adapted for use only in jet injectors, and those for 10 doses for use by jet or syringe injection.

Expiration date—The expiration date is 1 to 2 years, depending on the manufacturer's data, after date of issue from manufacturer's cold storage (−20°, 1 year).

Labeling—Label the Vaccine in multiple-dose containers to indicate that the contents are intended solely for use by jet injector or for use by either jet or syringe injection, whichever is applicable. Label the Vaccine in single-dose containers, if such containers are not light-resistant, to state that it should be protected from sunlight. Label it also to state that constituted Vaccine should be discarded if not used within 8 hours.

Saccharin—*see* Saccharin NF

Saccharin Calcium

$$\left[C_7H_4NO_3S \right]_2 Ca \cdot 3\tfrac{1}{2}H_2O$$

$C_{14}H_8CaN_2O_6S_2 \cdot 3½H_2O$ 467.48

1,2-Benzisothiazol-3(2*H*)-one, 1,1-dioxide, calcium salt, hydrate (2:7).

1,2-Benzisothiazolin-3-one 1,1-dioxide calcium salt hydrate (2:7) [*6381-91-5*].

Anhydrous 404.43 [*6485-34-3*].

» Saccharin Calcium contains not less than 98.0 percent and not more than 101.0 percent of $C_{14}H_8CaN_2O_6S_2$, calculated on the anhydrous basis.

Packaging and storage—Preserve in well-closed containers.

Labeling—Where the quantity of saccharin calcium is indicated in the labeling of any preparation containing Saccharin Calcium, this shall be understood to be in terms of saccharin ($C_7H_5NO_3S$).

Reference standards—*USP o-Toluenesulfonamide Reference Standard*—Do not dry before using. Keep container tightly closed. *USP p-Toluenesulfonamide Reference Standard*—Do not dry before using. Keep container tightly closed.

Identification—

A: Dissolve about 100 mg in 5 mL of sodium hydroxide solution (1 in 20), evaporate to dryness, and gently fuse the residue over a small flame until it no longer evolves ammonia. Allow the residue to cool, dissolve in 20 mL of water, neutralize with 3 *N* hydrochloric acid, and filter: the addition of a drop of ferric chloride TS to the filtrate produces a violet color.

B: Mix 20 mg with 40 mg of resorcinol, add 10 drops of sulfuric acid, and heat the mixture in a suitable liquid bath at 200° for 3 minutes. Allow it to cool, and add 10 mL of water and an excess of 1 *N* sodium hydroxide: a fluorescent green liquid results.

C: A solution (1 in 10) responds to the tests for *Calcium* ⟨191⟩.

D: To 10 mL of a solution (1 in 10) add 1 mL of hydrochloric acid: a crystalline precipitate of saccharin is formed. Wash the precipitate well with cold water, and dry at 105° for 2 hours: it melts between 226° and 230°, the procedure for *Class I* being used (see *Melting Range or Temperature* ⟨741⟩).

Water, *Method I* ⟨921⟩: not more than 15.0%.

Benzoate and salicylate—To 10 mL of a solution (1 in 20), previously acidified with 5 drops of 6 *N* acetic acid, add 3 drops of ferric chloride TS: no precipitate or violet color appears.

Arsenic, *Method II* ⟨211⟩: 3 ppm.

Selenium ⟨291⟩: 0.003%.

Toluenesulfonamides—

Internal standard solution, Standard stock solution, and *Standard preparations*—Prepare as directed for *Internal standard solution, Standard stock solution,* and *Standard preparations* in the test for *Toluenesulfonamides* under *Saccharin* (see NF monograph).

Test preparation—Prepare as directed under *Column Partition Chromatography* (see *Chromatography* ⟨621⟩), employing a chromatographic tube fitted with a porous glass disk in its base, a plastic stopcock on the delivery tube, and a reservoir on the top. Add a mixture consisting of 10 g of *Solid Support* and a solution of 2.0 g, accurately weighed, of Saccharin Calcium in 8.0 mL of sodium carbonate solution (1 in 20), and proceed as directed under *Test preparation* in the test for *Toluenesulfonamides* under *Saccharin* (see NF monograph), beginning with "Pack the contents."

Chromatographic system and *Procedure*—Proceed as directed for *Chromatographic system* and *Procedure* in the test for *Toluenesulfonamides* under *Saccharin* (see NF monograph).

Heavy metals, *Method I* ⟨231⟩—Dissolve 4 g in 46 mL of water, add 4 mL of dilute hydrochloric acid (1 in 12), mix, and rub the inner wall of the vessel with a glass rod until crystallization begins. Allow the solution to stand for 1 hour, then filter through a dry filter, discarding the first 10 mL of the filtrate, and use 25 mL of the subsequent filtrate for the *Test Preparation:* the limit is 0.001%.

Readily carbonizable substances ⟨271⟩—Dissolve 200 mg in 5 mL of sulfuric acid TS, and maintain at a temperature of 48° to 50° for 10 minutes: the solution has no more color than *Matching Fluid A*.

Assay—Weigh accurately about 500 mg of Saccharin Calcium, and transfer completely to a separator with the aid of 10 mL of water. Add 2 mL of 3 *N* hydrochloric acid, and extract the precipitated saccharin first with 30 mL, then with five 20-mL portions, of a solvent composed of chloroform and alcohol (9:1). Evaporate the combined extracts on a steam bath to dryness, with the aid of a current of air, then dissolve the residue in 40 mL of alcohol, add 40 mL of water, mix, add phenolphthalein TS, and titrate with 0.1 *N* sodium hydroxide VS. Perform a blank determination on a mixture of 40 mL of alcohol and 40 mL of water, and make any necessary correction. Each mL of 0.1 *N* sodium hydroxide is equivalent to 20.22 mg of $C_{14}H_8CaN_2O_6S_2$.

Saccharin Sodium

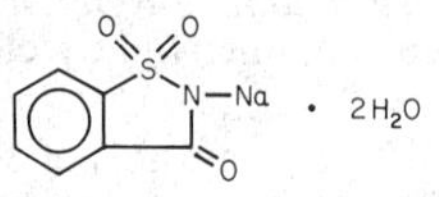

$C_7H_4NNaO_3S.2H_2O$ 241.19
1,2-Benzisothiazol-3(2*H*)-one, 1,1-dioxide, sodium salt, dihydrate.
1,2-Benzisothiazolin-3-one 1,1-dioxide sodium salt dihydrate [*6155-57-3*].
Anhydrous 205.16 [*128-44-9*].

» Saccharin Sodium contains not less than 98.0 percent and not more than 101.0 percent of $C_7H_4NNaO_3S$, calculated on the anhydrous basis.

Packaging and storage—Preserve in well-closed containers.

Labeling—Where the quantity of saccharin sodium is indicated in the labeling of any preparation containing Saccharin Sodium, this shall be understood to be in terms of saccharin ($C_7H_5NO_3S$).

Reference standards—*USP o-Toluenesulfonamide Reference Standard*—Do not dry before using. Keep container tightly closed. *USP p-Toluenesulfonamide Reference Standard*—Do not dry before using. Keep container tightly closed.

Identification—

A: The residue obtained by igniting it responds to the tests for *Sodium* ⟨191⟩.

B: To 10 mL of a solution (1 in 10) add 1 mL of hydrochloric acid: a crystalline precipitate of saccharin is formed. Wash the precipitate well with cold water until the last washing is free from chloride, and dry at 105° for 2 hours: it melts between 226° and 230°, the procedure for *Class I* being used (see *Melting Range or Temperature* ⟨741⟩).

Alkalinity—A solution (1 in 10) is neutral or alkaline to litmus, but no red color is produced with phenolphthalein TS.

Toluenesulfonamides—

Internal standard solution, Standard stock solution, and *Standard preparations*—Prepare as directed for *Internal standard solution, Standard stock solution,* and *Standard preparations* in the test for *Toluenesulfonamides* under *Saccharin* (see NF monograph).

Test preparation—Prepare as directed under *Column Partition Chromatography* (see *Chromatography* ⟨621⟩), employing a chromatographic tube fitted with a porous glass disk in its base, a plastic stopcock on the delivery tube, and a reservoir on the top. Add a mixture consisting of 10 g of *Solid Support* and a solution of 2.0 g, accurately weighed, of Saccharin Sodium in 8.0 mL of sodium carbonate solution (1 in 20), and proceed as directed under *Test preparation* in the test for *Toluenesulfonamides* under *Saccharin* (see NF monograph), beginning with "Pack the contents."

Chromatographic system and *Procedure*—Proceed as directed for *Chromatographic system* and *Procedure* in the test for *Toluenesulfonamides* under *Saccharin* (see NF monograph).

Heavy metals, *Method I* ⟨231⟩—Dissolve 4 g in 46 mL of water, add 4 mL of 1 *N* hydrochloric acid, mix, and rub the inner wall of the vessel with a glass rod until crystallization begins. Allow the solution to stand for 1 hour, and then filter through a dry filter, discarding the first 10 mL of the filtrate: the limit, determined on 25 mL of the subsequent filtrate, is 0.001%.

Other requirements—It responds to *Identification tests A* and *B*, and meets the requirements of the tests for *Water, Benzoate and salicylate, Arsenic, Selenium,* and *Readily carbonizable substances* under *Saccharin Calcium.*

Assay—Proceed with Saccharin Sodium as directed in the *Assay* under *Saccharin Calcium.* Each mL of 0.1 *N* sodium hydroxide is equivalent to 20.52 mg of $C_7H_4NNaO_3S$.

Saccharin Sodium Oral Solution

» Saccharin Sodium Oral Solution contains the equivalent of not less than 95.0 percent and not more than 105.0 percent of the labeled amount of saccharin ($C_7H_5NO_3S$).

Packaging and storage—Preserve in tight containers.

Reference standard—*USP Saccharin Reference Standard*—Keep container tightly closed. Dry at 105° for 2 hours before using.

Identification—

A: Transfer a volume of Oral Solution, equivalent to about 100 mg of saccharin, to a small dish, evaporate to dryness, and gently fuse the residue over a small flame until it no longer evolves ammonia. Allow the residue to cool, dissolve in 20 mL of water, neutralize the solution with 3 *N* hydrochloric acid, and filter: the addition of 1 drop of ferric chloride TS to the filtrate produces a violet color.

B: It responds to the tests for *Sodium* ⟨191⟩.

pH ⟨791⟩: between 3.0 and 5.0.

Assay—

Mobile phase—Mix 5 mL of 10 percent tetramethylammonium hydroxide solution and 400 mL of water in a 500-mL volumetric flask, adjust with phosphoric acid to a pH of 4.0, and mix. Add 50 mL of methanol to the solution, dilute with water to volume, mix, and degas the solution.

Standard preparation—Add to an accurately weighed quantity of USP Saccharin RS an amount of 0.1 *N* sodium hydroxide sufficient to dissolve the solid, and dilute quantitatively with water to obtain a solution having a known concentration of about 1.2 mg of saccharin per mL.

Assay preparation—Transfer an accurately measured volume of Saccharin Sodium Oral Solution, equivalent to about 120 mg of saccharin, to a 100-mL volumetric flask, dilute with water to volume, and mix.

Chromatographic system (see *Chromatography* ⟨621⟩)—The liquid chromatograph is equipped with a 257-nm detector and a 4.6-mm × 25-cm column that contains packing L1. The flow rate is about 2 mL per minute. Chromatograph three replicate injections of the *Standard preparation,* and record the peak responses as directed under *Procedure:* the relative standard deviation is not more than 2.0%.

Procedure—Separately inject equal volumes (about 10 µL) of the *Standard preparation* and the *Assay preparation* into the chromatograph using a suitable microsyringe or sampling valve, record the chromatograms, and measure the responses for the major peaks. Calculate the quantity, in mg, of saccharin ($C_7H_5NO_3S$) in each mL of the Oral Solution taken by the formula:

$$(100C/V)(r_U/r_S),$$

in which C is the concentration, in mg per mL, of USP Saccharin RS in the *Standard preparation,* V is the volume, in mL, of Oral Solution taken, and r_U and r_S are the peak responses obtained from the *Assay preparation* and the *Standard preparation,* respectively.

Saccharin Sodium Tablets

» Saccharin Sodium Tablets contain the equivalent of not less than 95.0 percent and not more than 110.0 percent of the labeled amount of saccharin ($C_7H_5NO_3S$).

Packaging and storage—Preserve in well-closed containers.

Reference standard—*USP Saccharin Reference Standard*—Keep container tightly closed. Dry at 105° for 2 hours before using.

Completeness of solution—Place 5 Tablets in a 250-mL beaker containing 150 mL of water at 25°. Stir for 5 minutes: all of the Tablets dissolve to give a clear or practically clear solution.

Identification—Dissolve a quantity of Tablets, equivalent to about 1 g of saccharin, in 10 mL of water, filter if necessary, and to the solution add 5 mL of 3 *N* hydrochloric acid: a white precipitate of saccharin is formed. Collect the precipitate on a filter, wash with small portions of cold water until the last washing is practically free from chloride, and dry it at 105° for 2 hours: the saccharin so obtained melts between 226° and 230°, the procedure for *Class I* being used (*see Melting Range or Temperature* ⟨741⟩), and responds to *Identification tests A* and *B* under *Saccharin Calcium*.

Uniformity of dosage units ⟨905⟩: meet the requirements.

Ammonium salts—Dissolve a quantity of powdered Tablets, equivalent to about 300 mg of saccharin, in 5 mL of water, and warm with 3 mL of 1 *N* sodium hydroxide: the odor of ammonia is not perceptible.

Assay—Weigh and finely powder not less than 20 Saccharin Sodium Tablets. Transfer an accurately weighed portion of the powder, equivalent to about 100 mg of saccharin, to a 100-mL volumetric flask, add about 50 mL of water, agitate until the tablet material is dissolved or evenly dispersed, dilute with water to volume, and mix. Transfer 10.0 mL to a separator, add 2 mL of 3 *N* hydrochloric acid, and extract with five 20-mL portions of a solvent composed of chloroform and alcohol (9:1). Collect the combined extracts in a beaker, and evaporate on a steam bath to dryness with the aid of a current of air. Dissolve the residue in 10 mL of sodium hydroxide solution (1 in 250), transfer the solution with the aid of water to a 200-mL volumetric flask, dilute with water to volume, and mix. Concomitantly determine the absorbances of this solution and a Standard solution of USP Saccharin RS in the same medium having a known concentration of about 50 µg per mL, in 1-cm cells at the wavelength of maximum absorbance at about 269 nm, with a suitable spectrophotometer, using water as the blank. Calculate the quantity, in mg, of saccharin ($C_7H_5NO_3S$) in the portion of Tablets taken by the formula:

$$(2C)(A_U/A_S),$$

in which *C* is the concentration, in µg per mL, of USP Saccharin RS in the Standard solution, and A_U and A_S are the absorbances of the solution from the Tablets and the Standard solution, respectively.

Safflower Oil

» Safflower Oil is the refined fixed oil obtained from the seed of *Carthamus tinctorius* Linné (Fam. Compositae).

Packaging and storage—Preserve in tight, light-resistant containers.

Fatty acid composition—Place about 1 g of Safflower Oil in a small conical flask fitted with a reflux attachment. Add 10 mL of methanol and 0.5 mL of 1 *N* methanolic potassium hydroxide solution prepared by dissolving 34 g of potassium hydroxide in sufficient methanol to produce 500 mL, allow to settle for 24 hours, and decant the clear solution. Reflux the mixture for 10 minutes, cool, transfer to a separator with the aid of 15 mL of *n*-heptane, shake with 10 mL of saturated sodium chloride solution, and allow to separate. Transfer the lower layer to another separator, and shake it with 10 mL of *n*-heptane. Wash the combined organic layers with 10 mL of water, dry over anhydrous sodium sulfate, and filter. Introduce a suitable portion of the filtrate into a gas chromatograph equipped with a flame-ionization detector and a column, preferably glass, 1.5 mm in length and 4 mm in internal diameter packed with 10 percent liquid phase G4 on support S1A, maintained at a temperature of 175°. The carrier gas is nitrogen. Measure the 4 main peak areas of the methyl esters of the fatty acids. The order of elution is palmitate, stearate, oleate, and linoleate, and their relative areas, expressed as percentages of the total area of the 4 main peaks, are in the ranges 2 to 10, 1 to 10, 7 to 42, and 72 to 84, respectively.

Free fatty acids ⟨401⟩—The free fatty acids in 10 g require for neutralization not more than 2.5 mL of 0.020 *N* sodium hydroxide.

Iodine value, *Method II* ⟨401⟩: between 135 and 150.

Heavy metals, *Method II* ⟨231⟩: 0.001%.

Unsaponifiable matter ⟨401⟩: not more than 1.5%.

Peroxide—

Mixed solvent—Mix 60 mL of glacial acetic acid with 40 mL of chloroform.

Potassium iodide solution—Prepare a saturated solution of potassium iodide in freshly boiled and cooled water, and store it protected from light. Discard it if it gives a color on addition of *Mixed solvent* and starch TS.

Procedure—Transfer about 10 g of Safflower Oil, accurately weighed, to a conical flask, add 30 mL of *Mixed solvent*, swirl to dissolve, add 0.5 mL of *Potassium iodide solution*, swirl the flask for 1 minute, accurately timed, add 30 mL of water, and titrate with 0.01 *N* sodium thiosulfate VS, with vigorous agitation, to a light yellow color. Add 0.5 mL of starch TS, and continue the titration until the blue color has disappeared. Perform a blank test, and make any necessary correction. Calculate the peroxide content, in mEq per kg, by the formula:

$$1000VN/W,$$

in which *V* is the volume, in mL, of sodium thiosulfate required, *N* is its normality, and *W* is the weight, in g, of Safflower Oil taken. The limit is 10.0.

Salicylamide

CONH2
OH

$C_7H_7NO_2$ 137.14
Benzamide, 2-hydroxy-.
2-Hydroxybenzamide [*65-45-2*].

» Salicylamide contains not less than 98.0 percent and not more than 102.0 percent of $C_7H_7NO_2$, calculated on the anhydrous basis.

Packaging and storage—Preserve in well-closed containers.

Reference standard—*USP Salicylamide Reference Standard*—Dry over silica gel for 18 hours before using.

Identification—

A: The infrared absorption spectrum of a potassium bromide dispersion of it, previously dried, exhibits maxima only at the same wavelengths as that of a similar preparation of USP Salicylamide RS.

B: The ultraviolet absorption spectrum of a 1 in 62,500 solution of it in methanol exhibits maxima and minima at the same wavelengths as that of USP Salicylamide RS, concomitantly measured, and the respective absorptivities, calculated on the anhydrous basis, at the wavelength of maximum absorbance at about 302 nm, do not differ by more than 3%.

C: Dissolve about 100 mg in 5 mL of alcohol, and add a few drops of ferric chloride TS: a violet color develops.

Melting range ⟨741⟩: between 139° and 142°.

Water, *Method I* ⟨921⟩: not more than 0.5%.

Residue on ignition ⟨281⟩: not more than 0.1%.

Heavy metals, *Method II* ⟨231⟩: 0.001%.

Chromatographic purity—

Standard preparations—Dissolve USP Salicylamide RS quantitatively in methanol, and mix to obtain a solution having a concentration of 1.0 mg per mL. Dilute quantitatively with methanol to obtain *Standard preparations,* designated by letter, having the following compositions:

Standard preparation	Dilution	Concentration (μg RS per mL)	Percentage (%, for comparison with test specimen)
A	(1 in 5)	200	1.0
B	(3 in 20)	150	0.75
C	(1 in 10)	100	0.5
D	(1 in 20)	50	0.25

Test preparation—Dissolve 200 mg of Salicylamide in 10.0 mL of methanol, and mix.

Developing solvent system—Prepare a mixture of normal butyl acetate, chloroform, and formic acid (6:4:2).

Procedure—On a suitable thin-layer chromatographic plate (see *Chromatography* ⟨621⟩), coated with a 0.25-mm layer of chromatographic silica gel mixture, apply separately 10 μL of the *Test preparation* and 10 μL of each of the *Standard preparations,* and dry the spots with the aid of a current of air. Place the plate in a suitable chromatographic chamber, and develop the chromatograms with the *Developing solvent system* until the solvent front has moved about three-fourths of the length of the plate. Remove the plate from the developing chamber, mark the solvent front, allow the plate to dry, and locate the spots under short-wavelength ultraviolet light. Compare the intensities of any secondary spots observed in the chromatogram of the *Test preparation* with those of the principal spots in the chromatograms of the *Standard preparations:* the total of the intensities of all secondary spots obtained from the *Test preparation* does not exceed that of the principal spot obtained from *Standard preparation B* (1%).

Assay—Transfer about 500 mg of Salicylamide, accurately weighed, to a 100-mL beaker equipped with a mechanical stirrer and a suitable cover with a single hole for the buret tip. Add 30 mL of freshly neutralized dimethylformamide containing a few drops of thymol blue TS. Titrate with 0.1 *N* sodium methoxide VS in toluene to the same blue end-point obtained in the standardization of the sodium methoxide solution. Perform a blank determination, and make any necessary correction. Each mL of 0.1 *N* sodium methoxide is equivalent to 13.71 mg of $C_7H_7NO_2$.

Salicylic Acid

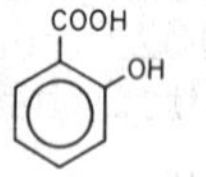

$C_7H_6O_3$ 138.12
Benzoic acid, 2-hydroxy-.
Salicylic acid [*69-72-7*].

» Salicylic Acid contains not less than 99.5 percent and not more than 101.0 percent of $C_7H_6O_3$, calculated on the dried basis.

Packaging and storage—Preserve in well-closed containers.

Identification—It responds to the tests for *Salicylate* ⟨191⟩.

Melting range ⟨741⟩: between 158° and 161°.

Loss on drying ⟨731⟩—Dry it over silica gel for 3 hours: it loses not more than 0.5% of its weight.

Residue on ignition ⟨281⟩: not more than 0.05%.

Chloride ⟨221⟩—Heat 1.5 g with 75 mL of water until the acid is dissolved, cool, add water to restore the original volume, and filter: a 25-mL portion of the filtrate shows no more chloride than corresponds to 0.10 mL of 0.020 *N* hydrochloric acid (0.014%).

Sulfate—Dissolve 12.0 g in 37 mL of acetone, and add 3 mL of water. Titrate potentiometrically with 0.02 *M* lead perchlorate, prepared by dissolving 9.20 g of lead perchlorate in water to make 1000 mL of solution, using a pH meter capable of a minimum reproducibility of ±0.1 mV (see *pH* ⟨791⟩) equipped with an electrode system consisting of a lead-specific electrode and a silver–silver chloride reference glass-sleeved electrode containing a 1 in 44 solution of tetraethylammonium perchlorate in glacial acetic acid: not more than 1.25 mL of 0.02 *M* lead perchlorate is consumed (0.02%).

Heavy metals—Dissolve 1 g in 25 mL of acetone, and add 2 mL of water and 10 mL of hydrogen sulfide TS: any color produced is not darker than that of a control made with 25 mL of acetone, 2 mL of standard lead solution (see *Heavy Metals* ⟨231⟩), and 10 mL of hydrogen sulfide TS (0.002%).

Readily carbonizable substances ⟨271⟩—Dissolve 500 mg in 5 mL of sulfuric acid TS: the solution has no more color than *Matching Fluid C.*

Chromatographic purity—

Solvent—Prepare a mixture of chloroform and methanol (9:1).

Standard preparations—Prepare a solution of USP Salicylic Acid RS in *Solvent* to obtain a solution having a concentration of 2.5 mg per mL. Dilute quantitatively with *Solvent* to obtain Standard preparations having the following compositions:

Standard preparation	Dilution	Concentration (mg RS per mL)	Percentage (%, for comparison with test specimen)
A	(3 in 20)	0.375	0.75
B	(1 in 10)	0.25	0.5
C	(1 in 50)	0.05	0.1

Test preparation—Dissolve an accurately weighed quantity of Salicylic Acid in *Solvent* to obtain a solution containing 50 mg per mL.

Procedure—On a suitable thin-layer chromatographic plate (see *Chromatography* ⟨621⟩), coated with a 0.25-mm layer of chromatographic silica gel mixture, apply separately as 1-cm streaks, 20 μL of the *Test preparation* and 20 μL of each *Standard preparation.* Position the plate in a chromatographic chamber, and develop the chromatograms in a solvent system consisting of equal volumes of butyl alcohol, previously saturated with ammonium hydroxide, and acetone until the solvent front has moved about three-fourths of the length of the plate. Remove the plate from the developing chamber, mark the solvent front, and allow the solvent to evaporate, with the aid of warm circulating air. View the plate under short- and long-wavelength ultraviolet light, then spray the plate with ferric chloride solution (1 in 60), and heat at 60° for 3 minutes. At each visualization step, compare the intensities of any secondary spots observed in the chromatogram of the *Test preparation* with those of the principal spots in the chromatograms of the *Standard preparations:* no secondary spot is more intense than the principal spot obtained from *Standard preparation A* (0.75%), and the sum of the intensities of all secondary spots obtained from the *Test preparation* does not exceed 2.0%.

Assay—Dissolve about 500 mg of Salicylic Acid, accurately weighed, in 25 mL of diluted alcohol that previously has been neutralized with 0.1 *N* sodium hydroxide, add phenolphthalein TS, and titrate with 0.1 *N* sodium hydroxide VS. Each mL of 0.1 *N* sodium hydroxide is equivalent to 13.81 mg of $C_7H_6O_3$.

Salicylic Acid Collodion

» Salicylic Acid Collodion contains not less than 9.5 percent and not more than 11.5 percent of $C_7H_6O_3$.

Salicylic Acid	100 g
Flexible Collodion, a sufficient quantity, to make	1000 mL

Dissolve the Salicylic Acid in about 750 mL of Flexible Collodion, add sufficient of the latter to make the product measure 1000 mL, and mix.

Packaging and storage—Preserve in tight containers, at controlled room temperature, remote from fire.

Assay—Transfer to a 400-mL beaker 5 mL of Salicylic Acid Collodion, accurately measured, rinse the measuring device with three 10-mL portions of a mixture of ether and alcohol (3:1) that previously has been neutralized to bromothymol blue TS, and add the rinsings to the beaker. Add, with stirring, 5 mL of sodium lauryl sulfate solution (1 in 10) that previously has been neutralized with 0.1 *N* hydrochloric acid to the distinct yellow color of bromothymol blue TS, and finally add 20 mL of water. Mix, add 5 drops of bromothymol blue TS, and titrate the mixture with 0.1 *N* sodium hydroxide VS. Each mL of 0.1 *N* sodium hydroxide is equivalent to 13.81 mg of $C_7H_6O_3$.

Salicylic Acid Topical Foam

» Salicylic Acid Topical Foam contains not less than 90.0 percent and not more than 110.0 percent of the labeled amount of $C_7H_6O_3$.

Packaging and storage—Preserve in tight containers.

Reference standard—*USP Salicylic Acid Reference Standard*—Dry over silica gel for 3 hours before using.

Identification—The retention time of the salicylic acid peak of the *Assay preparation* observed in the chromatogram obtained in the *Assay* corresponds to the retention time of the salicylic acid peak of the *Standard preparation*.

pH ⟨791⟩: between 5.0 and 6.0.

Assay—

Mobile phase—To 225 mg of tetramethylammonium hydroxide pentahydrate add 700 mL of water, 150 mL of methanol, 150 mL of acetonitrile, and 1.0 mL of glacial acetic acid, mix, filter, and degas.

Internal standard solution—Dissolve benzoic acid in methanol to obtain a solution having a concentration of about 8 mg per mL.

Standard preparation—Transfer about 20 mg of USP Salicylic Acid RS, accurately weighed, to a 100-mL volumetric flask, add 10.0 mL of *Internal standard solution*, dilute with *Mobile phase* to volume, and mix.

Assay preparation—Transfer an accurately weighed portion of Salicylic Acid Topical Foam, equivalent to about 20 mg of salicylic acid, to a 100-mL volumetric flask, add 10.0 mL of *Internal standard solution*, dilute with *Mobile phase* to volume, and mix. Cool in an ice bath to below room temperature and filter, discarding the first few mL of the filtrate.

Chromatographic system (see *Chromatography* ⟨621⟩)—The liquid chromatograph is equipped with a 280-nm detector and a 4-mm × 30-cm column that contains packing L1. The flow rate is about 2 mL per minute. Chromatograph four replicate injections of the *Standard preparation*, and record the peak responses as directed under *Procedure:* the relative standard deviation is not more than 3.0%, the resolution factor between salicylic acid and benzoic acid is not less than 3.0, and the tailing factors for the salicylic acid and benzoic acid peaks are not more than 2.0.

Procedure—Separately inject equal volumes (about 5 µL) of the *Standard preparation* and the *Assay preparation* into the chromatograph, record the chromatograms, and measure the responses for the major peaks. The retention times are about 2.5 minutes for salicylic acid and 5.5 minutes for benzoic acid. Calculate the quantity, in mg, of $C_7H_6O_3$ in the portion of Topical Foam taken by the formula:

$$(100C)(R_U/R_S),$$

in which C is the concentration, in mg per mL, of USP Salicylic Acid RS in the *Standard preparation*, and R_U and R_S are the ratios of the peak responses for salicylic acid to the peak responses for benzoic acid obtained from the *Assay preparation* and the *Standard preparation*, respectively.

Salicylic Acid Gel

» Salicylic Acid Gel is Salicylic Acid in a suitable viscous hydrophilic vehicle. It contains not less than 90.0 percent and not more than 110.0 percent of the labeled amount of $C_7H_6O_3$. It may contain alcohol.

Packaging and storage—Preserve in collapsible tubes or in tight containers, preferably at controlled room temperature.

Identification—Filter 5 mL of the solution obtained by titration in the *Assay*. Add 1 mL of ferric chloride TS to the filtrate: a violet color is produced. Add 1 mL of 6 *N* acetic acid: the violet color does not change. Add 1 mL of 6 *N* hydrochloric acid: the violet color is discharged. A small amount of white precipitate may appear.

Water, *Method I* ⟨921⟩: not more than 15.0%.

Alcohol content (if present) ⟨611⟩: from 90.0% to 110.0% of the labeled amount of C_2H_5OH.

Assay—To 25 mL of diluted alcohol add 1 drop of phenolphthalein TS and sufficient 0.1 *N* sodium hydroxide to produce a faint pink color. Add 5.0 g of Salicylic Acid Gel, accurately weighed, and stir. Titrate the dispersion with 0.1 *N* sodium hydroxide VS until a pink color is produced. [NOTE—Reserve this solution for the *Identification test*.] Each mL of 0.1 *N* sodium hydroxide is equivalent to 13.81 mg of $C_7H_6O_3$.

Salicylic Acid Paste, Zinc Oxide and—*see* Zinc Oxide and Salicylic Acid Paste

Salicylic Acid Plaster

» Salicylic Acid Plaster is a uniform mixture of Salicylic Acid in a suitable base, spread on paper, cotton cloth, or other suitable backing material. The plaster mass contains not less than 90.0 percent and not more than 110.0 percent of the labeled amount of $C_7H_6O_3$.

Packaging and storage—Preserve in well-closed containers, preferably at controlled room temperature.

Assay—Weigh accurately an amount of Salicylic Acid Plaster, corresponding to about 500 mg of salicylic acid, cut the portion into small strips, place them in a small flask, add 50 mL of chloroform, and shake the mixture until the plaster mass is disintegrated. Decant the chloroform extract into a 250-mL beaker, and wash the plaster backing with two 25-mL portions of chloroform, receiving the washings in the same beaker. Then wash the backing with 50 mL of alcohol to which has been added 1 mL of 6 *N* ammonium hydroxide, and add the washing to the chloroform extract. Again wash the backing with 40 mL of alcohol, and add the washing to the chloroform extract. Dry the backing, weigh, and subtract the weight from the weight of Plaster taken for the assay to obtain the weight of plaster mass. Stir the chloroform extract until any coagulum has separated into a compact mass, and filter the extract through purified cotton into a separator. Knead the coagulum, if any, with a glass rod to expel the solvent, and rinse the coagulum and the beaker with 10 mL of alcohol. Pour the rinsing through the cotton, then press the cotton with a glass rod to expel the solvent. Extract the filtrate with three 10-mL portions of 1 *N* sodium hydroxide, drawing off each portion into a 500-mL volumetric flask, and finally wash with two 25-mL portions of water, receiving the washings in the same flask. Dilute with water to volume, and pipet a 25-mL aliquot into a 500-mL iodine flask. Add 30.0 mL of 0.1 *N* bromine VS, then add 5 mL of hydrochloric acid, and immediately insert the stopper. Shake the flask repeatedly during 30 minutes, allow it to stand for 15 minutes, add quickly 5 mL of potassium iodide solution (1 in 5), taking precautions against the escape of bromine vapor, and at once insert the stopper in the flask. Shake thoroughly, remove the stopper, and rinse it and the neck of the flask with a small quantity of water, so that the washing flows into the flask. Add 1 mL of chloroform, shake the mixture, and titrate the liberated iodine with 0.1 *N* sodium thiosulfate VS, adding 3 mL of starch TS as the end-point is approached. Perform a blank determination (see *Residual Titrations* ⟨541⟩). Each mL of 0.1 *N* bromine is equivalent to 2.302 mg of $C_7H_6O_3$.

Salicylic Acids Ointment, Benzoic and—*see* Benzoic and Salicylic Acids Ointment

Schick Test Control

» Schick Test Control conforms to the regulations of the federal Food and Drug Administration concerning biologics (see *Biologics* ⟨1041⟩). It is Diphtheria Toxin for Schick Test that has been inactivated by heat for use as control for the Schick Test. It meets the requirements of the specific guinea pig test for detoxification by injection of not less than 2.0 mL into each of at least four guinea pigs. The animals are observed daily for 30 days and during this period show no evidence of diphtheria toxin poisoning (extensive necrosis, paralysis, or specific lethality).

Packaging and storage—Preserve at a temperature between 2° and 8°.

Expiration date—The expiration date is not later than 1 year after date of issue from manufacturer's cold storage (5°, 1 year).

Scopolamine Hydrobromide

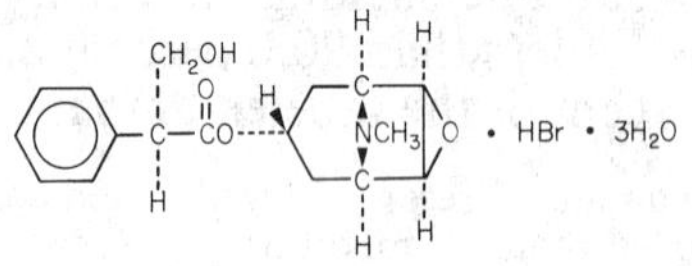

$C_{17}H_{21}NO_4 . HBr . 3H_2O$ 438.31

Benzeneacetic acid, α-(hydroxymethyl)-, 9-methyl-3-oxa-9-azatricyclo[3.3.1.0^{2,4}]non-7-yl ester, hydrobromide, trihydrate, [7(*S*)-(1α,2β,4β,5α,7β)]-.

6β,7β-Epoxy-1αH,5αH-tropan-3α-ol (−)-tropate (ester) hydrobromide trihydrate [*6533-68-2*].

Anhydrous 384.27 [*114-49-8*].

» Scopolamine Hydrobromide contains not less than 98.5 percent and not more than 102.0 percent of $C_{17}H_{21}NO_4 . HBr$, calculated on the anhydrous basis.

Caution—Handle Scopolamine Hydrobromide with exceptional care, since it is highly potent.

Packaging and storage—Preserve in tight, light-resistant containers.

Reference standard—*USP Scopolamine Hydrobromide Reference Standard*—Dry at 105° for 3 hours before using.

Identification—

A: Dissolve 3 mg in 1 mL of alcohol, and evaporate the solution on a steam bath to dryness. Dissolve the residue in 0.5 mL of chloroform, add 200 mg of potassium bromide and 15 mL of ether, and stir frequently during 5 minutes. Decant the solvent, dry the residue on a steam bath until the odor of the solvent no longer is perceptible, and compress the residue to a disk: the infrared absorption spectrum of the resulting potassium bromide dispersion, previously dried at 105° for 3 hours, exhibits maxima only at the same wavelengths as that of a similar preparation of USP Scopolamine Hydrobromide RS, treated in the same manner.

B: To 1 mL of a solution (1 in 20) add a few drops of chlorine TS, and shake the mixture with 1 mL of chloroform: the latter assumes a brownish color.

Melting range ⟨741⟩: between 195° and 199°, determined after drying under vacuum for 24 hours and then at 105° for 2 hours.

Specific rotation ⟨781⟩: between −24° and −26°, calculated on the anhydrous basis, determined in a solution containing the equivalent of 500 mg of anhydrous Scopolamine Hydrobromide in 10 mL.

pH ⟨791⟩: between 4.0 and 5.5, in a solution (1 in 20).

Water, *Method III* ⟨921⟩—Dry it at 105° for 3 hours: it loses not more than 13.0% of its weight.

Residue on ignition ⟨281⟩: negligible, from 100 mg.

Apoatropine—To 15 mL of a solution (1 in 100) add 0.05 mL of 0.1 *N* potassium permanganate VS: the solution is not completely decolorized within 5 minutes.

Other foreign alkaloids—To 1 mL of a solution (1 in 20) add a few drops of 6 *N* ammonium hydroxide: no turbidity is produced. Add 1 *N* potassium hydroxide to another 1-mL portion of the solution: only a transient whitish turbidity is produced.

Assay—Dissolve about 750 mg of Scopolamine Hydrobromide, accurately weighed, in a mixture of 30 mL of glacial acetic acid and 10 mL of mercuric acetate TS, warming slightly to effect solution. Cool the solution to room temperature, add 2 drops of crystal violet TS, and titrate with 0.1 *N* perchloric acid VS. Perform a blank determination, and make any necessary correction. Each mL of 0.1 *N* perchloric acid is equivalent to 38.43 mg of $C_{17}H_{21}NO_4 . HBr$.

Scopolamine Hydrobromide Injection

» Scopolamine Hydrobromide Injection is a sterile solution of Scopolamine Hydrobromide in Water for Injection. It contains not less than 90.0 percent and not more than 110.0 percent of the labeled amount of $C_{17}H_{21}NO_4 . HBr . 3H_2O$.

Packaging and storage—Preserve in light-resistant, single-dose or multiple-dose containers, preferably of Type I glass.

Reference standard—*USP Scopolamine Hydrobromide Reference Standard*—Dry at 105° for 3 hours before using.

Identification—

A: Transfer a volume of Injection, equivalent to about 3 mg of scopolamine hydrobromide, to a 50-mL separator, dilute with water, if necessary, to 10 mL, add 0.2 mL of ammonium hydroxide, and extract with 25 mL of chloroform. Add 50 mL of ether to the chloroform solution, and pass the mixture through a 25- × 250-mm chromatographic tube fitted with a pledget of glass wool at the base and packed with 2 g of purified siliceous earth that previously has been mixed with 1 mL of 0.2 *N* phosphoric acid saturated with sodium bromide. Discard the eluate, and pass 25 mL of water-saturated ether through the column and discard. Elute with 100 mL of water-saturated chloroform, collect the eluate in a suitable receiver, and evaporate just to dryness. Dissolve the residue in 1 mL of alcohol, and proceed as directed in *Identification test A* under *Scopolamine Hydrobromide*, beginning with "and evaporate the solution on a steam bath to dryness."

B: Add to the Injection silver nitrate TS: a yellowish white precipitate, insoluble in nitric acid but slightly soluble in 6 *N* ammonium hydroxide, is formed.

pH ⟨791⟩: between 3.5 and 6.5.

Other requirements—It meets the requirements under *Injections* ⟨1⟩.

Assay—

pH 9.0 buffer—Dissolve 34.8 g of dibasic potassium phosphate in 900 mL of water, and adjust to a pH of 9.0, determined electrometrically, by the addition of 3 *N* hydrochloric acid or 1 *N* sodium hydroxide, as necessary, with mixing.

Internal standard solution—Dissolve about 25 mg of homatropine hydrobromide, in water, contained in a 50-mL volumetric flask, add water to volume, and mix. Prepare fresh daily.

Standard solution—Dissolve about 10 mg of USP Scopolamine Hydrobromide RS, accurately weighed, in water, contained in a 100-mL volumetric flask, add water to volume, and mix. Prepare fresh daily.

Assay solution—Transfer an accurately measured volume of Scopolamine Hydrobromide Injection, equivalent to about 10 mg of scopolamine hydrobromide, to a 100-mL volumetric flask. Dilute with water to volume, and mix.

Assay preparation and *Standard preparation*—Pipet 10 mL of *Assay solution* and 10 mL of *Standard solution*, respectively, into two different separators, and treat them identically as follows: Add 2.0 mL of *Internal standard solution* and 5.0 mL of *pH 9.0 buffer*, and carefully adjust the solution in the separator with 1 *N* sodium hydroxide to a pH of 9.0, avoiding any excess. Immediately extract with two 10-mL portions of methylene chloride, filter the methylene chloride extracts through 1 g of anhydrous sodium sulfate supported by a small cotton plug in a funnel into a 50-mL beaker, and evaporate under nitrogen to approximately 2.0 mL.

Chromatographic system—Under typical conditions the instrument contains a 1.8-m × 2-mm glass column packed with 3 percent liquid phase G3 on support S1AB, conditioned as directed under *Gas Chromatography* (see *Chromatography* ⟨621⟩). Maintain the column at 225°, and use nitrogen as the carrier gas at a flow rate of 25 mL per minute.

System suitability—Chromatograph six to ten injections of the *Standard preparation*, and record peak areas as directed under *Procedure*. The analytical system is suitable for conducting this assay if the relative standard deviation for the ratio, R_A, does not exceed 2.0%; the resolution factor between A_H and A_A is not less than 5; and the tailing factor does not exceed 2.0.

Procedure—Inject 1-µL portions of the *Assay preparation* and the *Standard preparation* successively into the gas chromatograph. Measure the areas under the peaks for scopolamine hydrobromide and homatropine hydrobromide in each chromatogram. Calculate the ratio, A_U, of the area of the scopolamine hydrobromide peak to the area of the internal standard peak in the chromatogram from the *Assay preparation*, and similarly calculate the ratio, A_S, in the chromatogram from the *Standard preparation*. Calculate the quantity, in mg, of $C_{17}H_{21}NO_4 . HBr . 3H_2O$ in the volume of Injection taken by the formula:

$$1.141 W(A_U/A_S),$$

in which W is the weight, in mg, of USP Scopolamine Hydrobromide RS in the *Standard solution*, and 1.141 is the ratio of the molecular weight of scopolamine hydrobromide trihydrate to that of anhydrous scopolamine hydrobromide.

Scopolamine Hydrobromide Ophthalmic Ointment

» Scopolamine Hydrobromide Ophthalmic Ointment is Scopolamine Hydrobromide in a suitable ophthalmic ointment base. It contains not less than 90.0 percent and not more than 110.0 percent of the labeled amount of $(C_{17}H_{21}NO_4) . HBr . 3H_2O$. It is sterile.

Packaging and storage—Preserve in collapsible ophthalmic ointment tubes.

Reference standard—*USP Scopolamine Hydrobromide Reference Standard*—Dry at 105° for 3 hours before using. Keep container tightly closed and protected from light. [*Caution—Avoid contact.*]

Identification—

A: Transfer a portion of Ophthalmic Ointment, equivalent to about 50 mg of scopolamine hydrobromide, to a suitable separator, and dissolve in 25 mL of ether. Add 25 mL of 0.01 *N* hydrochloric acid, shake vigorously, allow the layers to separate, and discard the organic phase. Proceed as directed under *Identification—Organic Nitrogenous Bases* ⟨181⟩, beginning with "In a second separator dissolve 50 mg."

B: Transfer about 5 g of Ophthalmic Ointment to a separator, dissolve in 50 mL of ether, and extract with 20 mL of water: the extracted solution so obtained responds to the tests for *Bromide* ⟨191⟩.

Sterility—It meets the requirements under *Sterility Tests* ⟨71⟩.

Metal particles—It meets the requirements of the test for *Metal Particles in Ophthalmic Ointments* ⟨751⟩.

Assay—Proceed with Scopolamine Hydrobromide Ophthalmic Ointment as directed in the *Assay* under *Scopolamine Hydrobromide Injection*, but to prepare the *Assay solution*, weigh accurately a portion of Ophthalmic Ointment equivalent to about 10 mg of scopolamine hydrobromide into a separator containing 50 mL of ether, shake to dissolve, extract with three 25-mL portions of 0.2 *N* sulfuric acid, collect the acid extracts in a 100-mL volumetric flask, dilute with 0.2 *N* sulfuric acid to volume, and mix. Calculate the quantity, in mg, of $C_{17}H_{21}NO_4 . HBr . 3H_2O$ in the portion of Ophthalmic Ointment taken by the formula given therein.

Scopolamine Hydrobromide Ophthalmic Solution

» Scopolamine Hydrobromide Ophthalmic Solution is a sterile, buffered, aqueous solution of Scopolamine Hydrobromide. It contains not less than 90.0 percent and not more than 110.0 percent of the labeled amount of $C_{17}H_{21}NO_4 . HBr . 3H_2O$. It may contain suitable antimicrobial agents and stabilizers, and may contain suitable additives for the purpose of increasing its viscosity.

Packaging and storage—Preserve in tight containers.

Reference standard—*USP Scopolamine Hydrobromide Reference Standard*—Dry at 105° for 3 hours before using.

Identification—

A: A volume of Ophthalmic Solution, equivalent to about 3 mg of scopolamine hydrobromide, responds to *Identification test A* under *Scopolamine Hydrobromide Injection.*

B: Add to the Ophthalmic Solution silver nitrate TS: a yellowish white precipitate, insoluble in nitric acid but slightly soluble in 6 *N* ammonium hydroxide, is formed.

Sterility—It meets the requirements for *Sterility Tests* ⟨71⟩.

pH ⟨791⟩: between 4.0 and 6.0.

Assay—Transfer an accurately measured volume of Scopolamine Hydrobromide Ophthalmic Solution, equivalent to about 10 mg of scopolamine hydrobromide, to a 100-mL volumetric flask, dilute with water to volume, and mix. Using this as the *Assay solution*, proceed as directed in the *Assay* under *Scopolamine Hydrobromide Injection*. Calculate the quantity, in mg, of $C_{17}H_{21}NO_4 . HBr . 3H_2O$ in the volume of Ophthalmic Solution taken by the formula given therein.

Scopolamine Hydrobromide Tablets

» Scopolamine Hydrobromide Tablets contain not less than 90.0 percent and not more than 110.0 percent of the labeled amount of $C_{17}H_{21}NO_4 . HBr . 3H_2O$.

Packaging and storage—Preserve in tight, light-resistant containers.

Reference standard—*USP Scopolamine Hydrobromide Reference Standard*—Dry at 105° for 3 hours before using.

Identification—Place an amount of powdered Tablets, equivalent to about 3 mg of scopolamine hydrobromide, in a 50-mL separator, add 10 mL of water, and shake for 2 minutes. Proceed as directed in *Identification test A* under *Scopolamine Hydrobromide Injection*, beginning with "add 0.2 mL of ammonium hydroxide."

Disintegration ⟨701⟩: 15 minutes, the use of disks being omitted.

Uniformity of dosage units ⟨905⟩: meet the requirements.

Assay—Weigh and finely powder not less than 20 Scopolamine Hydrobromide Tablets. Transfer an accurately weighed portion of the powder, equivalent to about 1.0 mg of scopolamine hydrobromide, to a separator containing 5 mL of *pH 9.0 buffer*, and add, by pipet, 2.0 mL of *Internal standard solution* (prepared as directed in the *Assay* under *Scopolamine Hydrobromide Injection*). Adjust with 1 *N* sodium hydroxide to a pH of 9.0, extract with two 10-mL portions of methylene chloride, filter the methylene chloride extracts through 1 g of anhydrous sodium sulfate supported by a small cotton plug in a funnel into a 50-mL beaker, and evaporate under nitrogen to approximately 2.0 mL. Using this as the *Assay preparation*, proceed as directed in the *Assay* under *Scopolamine Hydrobromide Injection.* Calculate the quantity, in mg, of $C_{17}H_{21}NO_4 . HBr . 3H_2O$ in the portion of Tablets taken by the formula:

$$1.141(W/10)(A_U/A_S),$$

in which W is the weight, in mg, of USP Scopolamine Hydrobromide RS in the *Standard solution*, and 1.141 is the ratio of the molecular weight of scopolamine hydrobromide trihydrate to that of anhydrous scopolamine hydrobromide and A_U and A_S are as defined therein.

Secobarbital

$C_{12}H_{18}N_2O_3$ 238.29

2,4,6(1*H*,3*H*,5*H*)-Pyrimidinetrione, 5-(1-methylbutyl)-5-(2-propenyl)-.

5-Allyl-5-(1-methylbutyl)barbituric acid [*76-73-3*].

» Secobarbital contains not less than 97.5 percent and not more than 100.5 percent of $C_{12}H_{18}N_2O_3$, calculated on the dried basis.

Packaging and storage—Preserve in tight containers.

Reference standard—*USP Secobarbital Reference Standard*—Dry over silica gel for 18 hours before using.

Identification—The infrared absorption spectrum of a mineral oil dispersion of it exhibits maxima only at the same wavelengths as that of a similar preparation of USP Secobarbital RS.

Loss on drying ⟨731⟩—Dry it over silica gel for 18 hours: it loses not more than 1.0% of its weight.

Residue on ignition ⟨281⟩: not more than 0.1%.

Isomer content—Dissolve about 300 ± 5 mg in 5 mL of sodium hydroxide solution (1 in 100), add a solution of 300 ± 5 mg of *p*-nitrobenzyl bromide in 10 mL of alcohol, reflux for 30 minutes, cool, collect the precipitate on a small filter, and wash well with water: the precipitate, recrystallized from 25 mL of alcohol and dried at 105° for 30 minutes, melts between 156° and 161°.

Assay—Add about 450 mg of Secobarbital, accurately weighed, to 60 mL of dimethylformamide in a 125-mL conical flask. Add 4 drops of thymol blue TS, and titrate with 0.1 *N* sodium methoxide VS, using a magnetic stirrer and taking precautions against the absorption of atmospheric carbon dioxide. Perform a blank determination, and make any necessary correction. Each mL of 0.1 *N* sodium methoxide is equivalent to 23.83 mg of $C_{12}H_{18}N_2O_3$.

Secobarbital Elixir

» Secobarbital Elixir contains, in each 100 mL, not less than 417 mg and not more than 461 mg of $C_{12}H_{18}N_2O_3$, in a suitable, flavored vehicle.

Packaging and storage—Preserve in tight containers.

Reference standard—*USP Secobarbital Reference Standard*—Dry over silica gel for 18 hours before using.

Identification—Place 10 mL of Elixir in a separator containing 20 mL of water, add 5 mL of 1 *N* sodium hydroxide, and extract with two 10-mL portions of chloroform, discarding the chloroform extracts. Add 5 mL of 3 *N* hydrochloric acid, and extract with two 25-mL portions of chloroform, filtering the extracts through paper into a beaker. Remove the chloroform by evaporation on a steam bath, and dry the residue at 105° for 2 hours: the residue so obtained responds to the *Identification test* under *Secobarbital.*

Alcohol content ⟨611⟩: between 10.0% and 14.0% of C_2H_5OH.

Assay—

Internal standard—Butabarbital.

Internal standard solution—Dissolve an accurately weighed quantity of Butabarbital in chloroform, and dilute quantitatively with chloroform to obtain a solution having a known concentration of about 0.7 mg per mL.

Standard preparation—Dissolve accurately weighed quantities of USP Secobarbital RS and Butabarbital in chloroform, and dilute quantitatively with chloroform to obtain a solution that contains, in each mL, known amounts of about 1.2 mg of USP Secobarbital RS and about 0.9 mg of Butabarbital.

Assay preparation—Transfer an accurately measured volume of Secobarbital Elixir, equivalent to about 22 mg of secobarbital, to a separator, add 1 mL of dilute hydrochloric acid (1 in 5), and extract with four 10-mL portions of chloroform. Filter the extracts through about 15 g of anhydrous sodium sulfate that is supported on a funnel by a small pledget of glass wool. Collect the combined filtrate in a 50-mL volumetric flask, wash the sodium sulfate with 5 mL of chloroform, dilute with chloroform to volume, and mix. Combine 4.0 mL of this solution with 2.0 mL of *Internal standard solution* in a suitable container, and reduce the volume to about 1.5 mL by evaporation, with the aid of a stream of dry nitrogen, at room temperature.

Chromatographic system and *System suitability*—Proceed as directed for *Chromatographic System* and *System Suitability* under *Barbiturate Assay* ⟨361⟩, the resolution, *R*, between secobarbital and butabarbital being not less than 3.0. [NOTE—Relative retention times are, approximately, 0.6 for butabarbital and 1.0 for secobarbital.]

Procedure—Proceed as directed for *Procedure* under *Barbiturate Assay* ⟨361⟩. Calculate the quantity, in mg, of $C_{12}H_{18}N_2O_3$ in each mL of the Elixir taken by the the formula:

$$25(R_U)(Q_S)(C_i)/V(R_S),$$

in which V is the volume, in mL, of Elixir taken, and the other terms are as defined therein.

Secobarbital Sodium

$C_{12}H_{17}N_2NaO_3$ 260.27

2,4,6(1*H*,3*H*,5*H*)-Pyrimidinetrione, 5-(1-methylbutyl)-5-(2-propenyl)-, monosodium salt.

Sodium 5-allyl-5-(1-methylbutyl)barbiturate [*309-43-3*].

» Secobarbital Sodium contains not less than 98.5 percent and not more than 100.5 percent of $C_{12}H_{17}N_2NaO_3$, calculated on the dried basis.

Packaging and storage—Preserve in tight containers.

Reference standard—*USP Secobarbital Reference Standard*—Dry over silica gel for 18 hours before using.

Completeness of solution—Mix 1.0 g with 10 mL of carbon dioxide–free water: after 1 minute, the solution is clear and free from undissolved solid.

Identification—

A: The infrared absorption spectrum of a chloroform solution of the residue of secobarbital obtained as directed in the *Assay* exhibits maxima only at the same wavelengths as that of a similar preparation of USP Secobarbital RS.

B: Ignite about 500 mg: the residue effervesces with acids, and responds to the tests for *Sodium* ⟨191⟩.

pH ⟨791⟩: between 9.7 and 10.5, in the solution prepared in the test for *Completeness of solution.*

Loss on drying ⟨731⟩—Dry it at 80° for 5 hours: it loses not more than 3.0% of its weight.

Heavy metals, *Method II* ⟨231⟩: 0.003%.

Isomer content—Dissolve about 300 ± 5 mg in 5 mL of sodium hydroxide solution (1 in 100), add a solution of 300 ± 5 mg of *p*-nitrobenzyl bromide in 10 mL of alcohol, reflux for 30 minutes, cool, collect the precipitate on a small filter, and wash well with water: the precipitate, recrystallized from 25 mL of alcohol and dried at 105° for 30 minutes, melts between 156° and 161°.

Assay—Dissolve about 500 mg of Secobarbital Sodium, accurately weighed, in 15 mL of water in a separator. To the solution add 2 mL of hydrochloric acid, shake, and extract the liberated secobarbital with eight 25-mL portions of chloroform. Test for completeness of extraction by extracting with an additional 10-mL portion of chloroform and evaporating the solvent: not more than 0.5 mg of residue remains. Filter the extracts into a tared beaker, and finally rinse the separator and the filter with several small portions of chloroform. Evaporate the combined filtrate and washings on a steam bath with the aid of a current of air just to dryness. Dissolve the residue in 2 mL of dehydrated alcohol, and evaporate to dryness. Repeat the dissolution and evaporation with 2 mL of dehydrated alcohol, and dry the residue at 100° for 2 hours. The weight of the residue, multiplied by 1.092, represents the weight of $C_{12}H_{17}N_2NaO_3$.

Secobarbital Sodium Capsules

» Secobarbital Sodium Capsules contain not less than 92.5 percent and not more than 107.5 percent of the labeled amount of $C_{12}H_{17}N_2NaO_3$.

Packaging and storage—Preserve in tight containers.

Reference standard—*USP Secobarbital Reference Standard*—Dry over silica gel for 18 hours before using.

Identification—

A: Dissolve a portion of the contents of Capsules, equivalent to about 50 mg of secobarbital sodium, in 10 mL of water, and filter into a separator. Add 2 mL of 3 *N* hydrochloric acid, extract the liberated secobarbital with 20 mL of chloroform, and evaporate the extract to dryness: the infrared absorption spectrum of a 1 in 20 solution of the residue in chloroform exhibits maxima only at the same wavelengths as that of a similar preparation of USP Secobarbital RS.

B: Ignite a portion of the contents of Capsules, equivalent to about 500 mg of secobarbital sodium: the residue so obtained responds to the tests for *Sodium* ⟨191⟩.

Dissolution ⟨711⟩—

Medium: water; 500 mL.

Apparatus 1: 100 rpm.

Time: 60 minutes.

Procedure—Determine the amount of $C_{12}H_{17}N_2NaO_3$ dissolved from ultraviolet absorbances at the wavelength of maximum absorbance at about 243 nm of filtered portions of the solution under test, mixed with sufficient sodium hydroxide to provide a concentration of 0.1 *N* sodium hydroxide, and suitably diluted with *Dissolution Medium,* if necessary, in comparison with a Standard solution having a known concentration of USP Secobarbital RS in 0.1 *N* sodium hydroxide.

Tolerances—Not less than 75% (*Q*) of the labeled amount of $C_{12}H_{17}N_2NaO_3$ is dissolved in 60 minutes.

Uniformity of dosage units ⟨905⟩: meet the requirements.

Procedure for content uniformity—Transfer the contents of 1 Capsule to a 250-mL volumetric flask, with the aid of about 5 mL of alcohol. Add 10 mL of freshly prepared dilute ammonium hydroxide (1 in 200), and without delay dilute with the same solution to volume. Mix, filter if necessary, and discard the first 20 mL of filtrate. Dilute a portion of the clear solution with dilute ammonium hydroxide (1 in 200) to obtain a solution having a concentration of about 10 µg of secobarbital sodium per mL. Dissolve a suitable quantity of USP Secobarbital RS in dilute ammonium hydroxide (1 in 200) to obtain a Standard solution having a known concentration of about 10 µg per mL. Concomitantly determine the absorbances of both solutions in 1-cm cells at the wavelength of maximum absorbance at about 240 nm, with a suitable spectrophotometer, using dilute ammonium hydroxide (1 in 200) as the blank. Calculate the quantity, in mg, of $C_{12}H_{17}N_2NaO_3$ in the Capsule by the formula:

$$(260.27/238.29)(T/C_U)C_S(A_U/A_S),$$

in which 260.27 and 238.29 are the molecular weights of secobarbital sodium and secobarbital, respectively, T is the labeled quantity, in mg, of secobarbital sodium in the Capsule, C_U is the concentration, in µg per mL, of secobarbital sodium in the solution from the Capsule contents, on the basis of the labeled quantity per Capsule and the extent of dilution, C_S is the concentration, in µg per mL, of USP Secobarbital RS in the Standard solution, and A_U and A_S are the absorbances of the solution from the Capsule contents and the Standard solution, respectively.

Assay—

Internal standard—Butabarbital.

Internal standard solution—Dissolve an accurately weighed quantity of Butabarbital in chloroform, and dilute quantitatively with chloroform to obtain a solution having a known concentration of about 0.8 mg per mL.

Standard preparation—Dissolve accurately weighed quantities of USP Secobarbital RS and Butabarbital in chloroform, and dilute quantitatively with chloroform to obtain a solution that contains, in each mL, known amounts of about 0.9 mg of USP Secobarbital RS and about 0.8 mg of Butabarbital.

Assay preparation—Weigh not less than 20 Secobarbital Sodium Capsules, and transfer the contents as completely as possible to a suitable container. Remove any residual powder from the empty capsules with the aid of a current of air, and weigh the capsule shells, determining the weight of the contents by difference. Mix the contents of the Capsules, and transfer an accurately weighed portion of the powder, equivalent to about 100 mg of secobarbital sodium, to a separator. Add 15 mL of water, 1 mL of hydrochloric acid, and 100 mL of chloroform, and shake for 3 minutes. Filter a portion of the chloroform layer through about 15 g of anhydrous sodium sulfate that is supported on a funnel by a small pledget of glass wool, combine 2.0 mL of this solution with 2.0 mL internal standard solution in a suitable container, and reduce the volume to about 2 mL by evaporation, with the aid of a stream of dry nitrogen, at room temperature.

Chromatographic system and *System suitability*—Proceed as directed for *Chromatographic System* and *System Suitability* under *Barbiturate Assay* ⟨361⟩, the resolution, *R*, between secobarbital and butabarbital being not less than 3.0. [NOTE—Relative retention times are, approximately, 0.6 for butabarbital and 1.0 for secobarbital.]

Procedure—Proceed as directed for *Procedure* under *Barbiturate Assay* ⟨361⟩. Calculate the quantity, in mg, of $C_{12}H_{17}N_2NaO_3$ in the portion of Capsules taken by the formula:

$$(260.27/238.29)(100)(R_U)(Q_S)(C_i)/(R_S),$$

in which 260.27 and 238.29 are the molecular weights of secobarbital sodium and secobarbital, respectively, and the other terms are as defined therein.

Secobarbital Sodium Injection

» Secobarbital Sodium Injection is a sterile solution of Secobarbital Sodium in a suitable solvent. It contains not less than 90.0 percent and not more than 110.0 percent of the labeled amount of $C_{12}H_{17}N_2NaO_3$.

Packaging and storage—Preserve in single-dose or in multiple-dose containers, preferably of Type I glass, protected from light, in a refrigerator.

Labeling—The label indicates that the Injection is not to be used if it contains a precipitate.

Reference standard—*USP Secobarbital Reference Standard*—Dry over silica gel for 18 hours before using.

Identification—

A: Transfer a volume of Injection, equivalent to about 100 mg of secobarbital sodium, to a separator containing 15 mL of water. Render the mixture distinctly acid to litmus with hydrochloric acid, extract the liberated secobarbital with 25 mL of ether, collect the ether extract in a separator, and wash with 10 mL of water. Discard the water solution. Filter the ether extract into a beaker, and evaporate on a steam bath with the aid of a current of air just to dryness. Dissolve the residue in 3 mL of alcohol, and evaporate to dryness. Repeat the dissolution and evaporation with 3 mL of alcohol, and dry the residue at 100° for 2 hours. The infrared absorption spectrum of a solution prepared by dissolving the residue of secobarbital so obtained in chloroform to a concentration of about 50 mg per mL, 0.1-mm sodium chloride cells being used and chloroform being used as the blank, exhibits maxima only at the same wavelengths as that of a similar preparation of USP Secobarbital RS.

B: It responds to the flame test for *Sodium* ⟨191⟩.

pH ⟨791⟩: between 9.0 and 10.5.

Other requirements—It meets the requirements under *Injections* ⟨1⟩.

Assay—

Buffer solution—Dissolve 6.19 g of boric acid and 14.91 g of potassium chloride in water, dilute with water to 200 mL, and mix. After 24 hours, filter if necessary to obtain a clear solution.

Standard preparation 1—Dissolve a suitable quantity of USP Secobarbital RS, accurately weighed, in 0.5 *N* sodium hydroxide to obtain a solution having a known concentration of about 23 µg per mL.

Standard preparation 2—Mix 5.0 mL of *Standard preparation 1* with 5.0 mL of *Buffer solution.*

Assay preparation 1—Transfer an accurately measured volume of Secobarbital Sodium Injection, equivalent to about 50 mg of secobarbital sodium, to a 100-mL volumetric flask, dilute with 0.5 *N* sodium hydroxide to volume, and mix. Pipet 5 mL of this solution into a 100-mL volumetric flask, add 0.5 *N* sodium hydroxide to volume, and mix.

Assay preparation 2—Mix 5.0 mL of *Assay preparation 1* with 5.0 mL of *Buffer solution.*

Procedure—Concomitantly determine the absorbances of *Assay preparation 1* and *Standard preparation 1* in 1-cm cells at 260 nm, with a suitable spectrophotometer, using 0.5 *N* sodium hydroxide as the blank. Similarly determine the absorbances of *Assay preparation 2* and *Standard preparation 2*, using as the blank a mixture of equal volumes of 0.5 *N* sodium hydroxide and *Buffer solution.* Calculate the quantity, in mg, of $C_{12}H_{17}N_2NaO_3$ in the volume of the Injection taken by the formula:

$$1.092(2C)(A_U - 2a_U)/(A_S - 2a_S),$$

in which 1.092 is the ratio of the molecular weight of sodium secobarbital to that of secobarbital, C is the concentration, in µg per mL, of USP Secobarbital RS in *Standard preparation 1*, A_U and A_S are the absorbances of *Assay preparation 1* and *Standard preparation 1*, respectively, and a_U and a_S are the absorbances of *Assay preparation 2* and *Standard preparation 2*, respectively.

Sterile Secobarbital Sodium

» Sterile Secobarbital Sodium is Secobarbital Sodium suitable for parenteral use. It contains not less than 90.0 percent and not more than 110.0 percent of the labeled amount of $C_{12}H_{17}N_2NaO_3$.

Packaging and storage—Preserve in *Containers for Sterile Solids* as described under *Injections* ⟨1⟩.

Reference standard—*USP Secobarbital Reference Standard*—Dry over silica gel for 18 hours before using.

Constituted solution—At the time of use, the constituted solution prepared from Sterile Secobarbital Sodium meets the requirements for *Constituted Solutions* under *Injections* ⟨1⟩.

Other requirements—It conforms to the definition, responds to the *Identification tests*, and meets the requirements for *pH, Completeness of solution, Loss on drying, Heavy metals*, and *Assay* under *Secobarbital Sodium.* It meets also the requirements for *Sterility Tests* ⟨71⟩, *Uniformity of Dosage Units* ⟨905⟩, and *Labeling* under *Injections* ⟨1⟩.

Secobarbital Sodium and Amobarbital Sodium Capsules

» Secobarbital Sodium and Amobarbital Sodium Capsules contain not less than 90.0 percent and not more than 110.0 percent of the labeled amounts of secobarbital sodium ($C_{12}H_{17}N_2NaO_3$) and amobarbital sodium ($C_{11}H_{17}N_2NaO_3$).

Packaging and storage—Preserve in well-closed containers.

Reference standards—*USP Amobarbital Reference Standard*—Dry at 105° for 4 hours before using. *USP Secobarbital Reference Standard*—Dry over silica gel for 18 hours before using.

Identification—

A: Suspend the contents of 1 Capsule in 10 mL of water, and filter: the filtrate responds to the flame test for *Sodium* ⟨191⟩.

B: The retention times of the major peaks in the chromatogram of the *Assay preparation* correspond to those of the *Standard preparation* obtained in the *Assay.*

Dissolution ⟨711⟩—

Medium: water; 500 mL.

Apparatus 1: 100 rpm.

Time: 60 minutes.

Procedure—Determine the total amount of $C_{12}H_{17}N_2NaO_3$ and $C_{11}H_{17}N_2NaO_3$ dissolved from ultraviolet absorbances at the wavelength of maximum absorbance at about 239 nm of filtered portions of the solution under test, suitably diluted with 0.1 *N* sodium hydroxide, in comparison with a Standard solution having known concentrations of about 7.5 µg each per mL, of USP Secobarbital RS and USP Amobarbital RS in the same medium. An amount of alcohol not to exceed 1% of the total volume of the Standard solution may be used to dissolve the Reference Standards prior to dilution with water and 0.1 *N* sodium hydroxide.

Tolerances—Not less than 60% (*Q*) of the labeled total amount of $C_{12}H_{17}N_2NaO_3$ and $C_{11}H_{17}N_2NaO_3$ is dissolved in 60 minutes.

Uniformity of dosage units ⟨905⟩: meet the requirements for *Content Uniformity* with respect to secobarbital sodium and to amobarbital.

Assay—

Internal standard solution—Dissolve aprobarbital in chloroform to obtain a solution having a concentration of about 0.75 mg per mL.

Standard preparation—Transfer about 92 mg of USP Secobarbital RS, and about 91 mg of USP Amobarbital RS, both accurately weighed, to a 100-mL volumetric flask, and dissolve in 50 mL of chloroform. Dilute with chloroform to volume, and mix.

Assay preparation—Remove, as completely as possible, the contents of not less than 20 Capsules. Transfer an accurately weighed portion of the powder, equivalent to about 100 mg of secobarbital sodium, to a separator, add 20 mL of water, 1 mL of hydrochloric acid, and 100.0 mL of chloroform, and shake for 3 minutes. Remove the chloroform layer, and use as directed in the *Procedure.*

Chromatographic system (see *Chromatography* ⟨621⟩)—The gas chromatograph is equipped with a flame-ionization detector and contains a 0.6-m × 3.5-mm glass column packed with 3 percent liquid phase G10 on 100- to 120-mesh support S1AB. The column is maintained at about 175°, the injection port at about 235°, the detector block at about 245°, and dry helium is used as the carrier gas at a flow rate of about 55 mL per minute. Chromatograph five replicate injections of the *Standard preparation,* and record the peak responses as directed under *Procedure:* the relative standard deviation is not more than 2%; the resolution factor between amobarbital and the internal standard is not less than 1.5; the resolution factor between amobarbital and secobarbital is not less than 2.5; and the tailing factor does not exceed 1.5 for any of the three peaks.

Procedure—Mix 5.0 mL of the *Standard preparation* with 5.0 mL of the *Internal standard solution.* Mix 5.0 mL of the *Assay preparation* with 5.0 mL of the *Internal standard solution.* Separately inject equal volumes (about 3 μL) of the resulting solutions into the chromatograph, and record the chromatograms. Measure the responses for the major peaks. The relative retention times with respect to the internal standard are about 1.3 for amobarbital and 1.8 for secobarbital. Calculate the quantity, in mg, of secobarbital sodium ($C_{12}H_{17}N_2NaO_3$) in the portion of Capsules taken by the formula:

$$(260.27/238.29)W(R_U/R_S),$$

in which 260.27 and 238.29 are the molecular weights of secobarbital sodium and secobarbital, respectively, W is the weight, in mg, of USP Secobarbital RS taken for the *Standard preparation,* and R_U and R_S are the ratios of the peak response of secobarbital to that of the internal standard in the *Assay preparation* and the *Standard preparation,* respectively. Similarly calculate the quantity, in mg, of amobarbital sodium ($C_{11}H_{17}N_2NaO_3$) in the portion of Capsules taken by the formula:

$$(248.26/226.27)W'(R'_U/R'_S),$$

in which 248.26 and 226.27 are the molecular weights of amobarbital sodium and amobarbital, respectively, W' is the weight, in mg, of USP Amobarbital RS taken for the *Standard preparation,* and R'_U and R'_S are the ratios of the peak response of amobarbital to that of the internal standard obtained from *Assay preparation* and the *Standard preparation,* respectively.

Selenious Acid

H_2SeO_3 128.97
Selenium dioxide, monohydrated.
Selenious acid [*7783-00-8*].

» Selenious Acid contains not less than 93.0 percent and not more than 101.0 percent of H_2SeO_3.

Packaging and storage—Preserve in tight containers.

Identification—

A: Dissolve about 50 mg in 5 mL of water, add about 100 mg of sodium bicarbonate, and mix: gas bubbles develop.

B: Dissolve about 50 mg in 5 mL of 0.1 *N* hydrochloric acid, and add about 50 mg of stannous chloride: a curdy tan-orange precipitate is formed.

Residue on ignition ⟨281⟩: not more than 1.0 mg, from 10.0 g (0.01%).

Insoluble matter—Dissolve 1 g in 5 mL of water: it dissolves completely and the solution is clear.

Selenate and sulfate—Dissolve about 0.5 g in 10 mL of water, add 0.1 mL of hydrochloric acid and 1 mL of barium chloride TS, and mix: no turbidity or precipitate is formed in 10 minutes.

Assay—Transfer about 100 mg of Selenious Acid, accurately weighed, to a suitable glass-stoppered flask, and dissolve in 50 mL of water. Add 10 mL of potassium iodide solution (3 in 10) and 5 mL of hydrochloric acid, mix, insert the stopper in the flask, and allow to stand for 10 minutes. Add 50 mL of water and 3 mL of starch TS, and titrate with 0.1 *N* sodium thiosulfate VS until the solution is colorless, then titrate with 0.1 *N* iodine VS to a blue end-point. Subtract the volume of 0.1 *N* iodine from the volume of 0.1 *N* sodium thiosulfate to obtain the volume of 0.1 *N* sodium thiosulfate equivalent to selenious acid. Each mL of 0.1 *N* sodium thiosulfate is equivalent to 3.225 mg of H_2SeO_3.

Selenious Acid Injection

» Selenious Acid Injection is a sterile solution of Selenious Acid in Water for Injection. It contains not less than 95.0 percent and not more than 105.0 percent of the labeled amount of selenium (Se).

Packaging and storage—Preserve in single-dose or in multiple-dose containers, preferably of Type I or Type II glass.

Labeling—Label the Injection to indicate that it is to be diluted to the appropriate strength with Sterile Water for Injection or other suitable fluid prior to administration.

Identification—The *Assay preparation,* prepared as directed in the *Assay,* exhibits an absorption maximum at about 196 nm when tested as directed for *Procedure* in the *Assay.*

Pyrogen—When diluted with Sodium Chloride Injection to contain 4 μg of selenium per mL, it meets the requirements of the *Pyrogen Test* ⟨151⟩.

pH ⟨791⟩: between 1.8 and 2.4.

Particulate matter ⟨788⟩: meets the requirements under *Small-volume Injections.*

Other requirements—It meets the requirements under *Injections* ⟨1⟩.

Assay—

Selenium stock solution—Dissolve about 1 g of metallic selenium, accurately weighed, in a minimum volume of nitric acid. [NOTE—Selenium is toxic; handle it with care.] Evaporate to dryness, add 2 mL of water, and evaporate to dryness. Repeat the addition of water and the evaporation to dryness three times. Dissolve the residue in 3 *N* hydrochloric acid, transfer to a 1000-mL volumetric flask, dilute with 3 *N* hydrochloric acid to volume, and mix. This solution contains about 1000 μg of selenium per mL.

Standard preparations—Transfer 3.0, 4.0, and 5.0 mL, respectively, of *Selenium stock solution* to separate 100-mL volumetric flasks, dilute the contents of each flask with water to volume, and mix. These *Standard preparations* contain, respectively, about 30, 40, and 50 μg of selenium per mL.

Assay preparation—Using water as the diluent, quantitatively dilute an accurately measured volume of Selenious Acid Injection to obtain a solution containing about 40 μg of selenium per mL.

Procedure—Concomitantly determine the absorbances of the *Standard preparations* and the *Assay preparation* at the selenium emission line of 196 nm, with a suitable atomic absorption spectrophotometer (see *Spectrophotometry and Light-scattering* ⟨851⟩) equipped with a selenium electrodeless discharge lamp and an air-acetylene flame, using water as the blank. Plot the absorbances of the *Standard preparations* versus concentration, in μg per mL, of selenium, and draw the straight line best fitting the three plotted points. From the graph so obtained, determine the concentration, C, in μg per mL, of selenium in the *Assay preparation.* Calculate the quantity, in mg, of selenium in each mL of the Injection taken by the formula:

$$LC/D,$$

in which L is the labeled quantity, in mg per mL, of selenium in the Injection taken, and D is the concentration, in μg of selenium per mL, of the *Assay preparation* on the basis of the labeled quantity in the Injection and the extent of dilution.

Selenomethionine Se 75 Injection

$CH_3SeCH_2CH_2-C(H)(NH_2)-COOH$

$C_5H_{11}NO_2{}^{75}Se$
Butanoic acid, 2-amino-4-(methylseleno-^{75}*Se*)-, (*S*)-.
(*S*)-2-Amino-4-(methylselenyl-^{75}*Se*) butyric acid
[*1187-56-0*].

» Selenomethionine Se 75 Injection is a sterile, aqueous solution of radioactive L-selenomethionine which is the analog of the essential amino acid, methionine, in which the sulfur atom is replaced by a selenium atom.

Selenomethionine Se 75 Injection contains not less than 90.0 percent and not more than 110.0 percent of the labeled amount of ^{75}Se expressed in megabecquerels (microcuries or millicuries) per mL at the time indicated in the labeling. Its specific activity is not less than 37.0 MBq (1.0 mCi) per mg of selenium at the time of manufacture. It may contain not more than 3 mg per mL of L-methionine as carrier and may contain suitable anti-oxidants and preservatives.

Packaging and storage—Preserve in single-dose or in multiple-dose containers, at a temperature between 2° and 8°, unless otherwise specified by the manufacturer.

Labeling—Label it to include the following, in addition to the information specified for *Labeling* under *Injections* ⟨1⟩: the date of calibration; the amount of ^{75}Se as selenomethionine expressed as total megabecquerels (microcuries or millicuries), and concentration as megabecquerels (microcuries or millicuries) per mL at the time of calibration; the expiration date; and the statement, "Caution—Radioactive Material." The labeling indicates that in making dosage calculations, correction is to be made for radioactive decay, and also indicates that the radioactive half-life of ^{75}Se is 120 days.

Reference standard—*USP Endotoxin Reference Standard.*

Radionuclide identification (see *Radioactivity* ⟨821⟩)—Its gamma-ray spectrum is identical to that of a specimen of ^{75}Se of known purity that exhibits a major photopeak having an energy of 0.265 MeV.

Bacterial endotoxins—It meets the requirements of the *Bacterial Endotoxins Test* ⟨85⟩, the limit of endotoxin content being not more than 175/*V* USP Endotoxin Unit per mL of the Injection, when compared with the USP Endotoxin RS, in which *V* is the maximum recommended total dose, in mL, at the expiration date or time.

pH ⟨791⟩: between 3.5 and 8.0.

Radiochemical purity—Mix a measured volume of Injection, appropriately diluted, such that it provides a count rate of about 20,000 counts per minute, with an equal volume of a solution of nonradioactive selenomethionine (2 mg per mL). Apply the mixture about 25 mm from one end of a 25- × 300-mm strip of chromatographic paper (see *Chromatography* ⟨621⟩), and allow to dry. Develop the chromatogram by descending chromatography, using the upper layer obtained by shaking together a mixture of ethyl acetate, water, and glacial acetic acid (50:30:15) until the solvent front has descended about three-quarters of the length of the paper. Use the developing solvent to equilibrate the apparatus prior to the start of development. Dry the chromatogram in air, then spray it with a solution prepared by dissolving 0.4 g of triketohydrindene hydrate in water-saturated *n*-butanol, to locate the selenomethionine spot. Determine the radioactivity distribution by scanning with a suitable collimated radiation detector. The radioactivity under the selenomethionine band is not less than 90.0% of the radioactivity and the R_f value of the main radioactive band is within ±5% of the standard selenomethionine band, which is indicated by a purple coloration.

Other requirements—It meets the requirements under *Injections* ⟨1⟩, except that it is not subject to the recommendation on *Volume in Container*.

Assay for radioactivity—Using a suitable counting assembly (see *Selection of a Counting Assembly* under *Radioactivity* ⟨821⟩), determine the radioactivity, in MBq (μCi) per mL, of Selenomethionine Se 75 Injection by use of a calibrated system as directed under *Radioactivity* ⟨821⟩.

Selenium Sulfide

SeS_2 143.08
Selenium sulfide (SeS_2).
Selenium sulfide (SeS_2) [*7488-56-4*].

» Selenium Sulfide contains not less than 52.0 percent and not more than 55.5 percent of selenium (Se).

Packaging and storage—Preserve in well-closed containers.

Identification—

A: Filter 20 mL of the solution of Selenium Sulfide prepared as directed in the *Assay*, and to 10 mL of the filtrate add 5 mL of water and 5 g of urea. Heat to boiling, cool, and add 2 mL of potassium iodide solution (1 in 10): a yellowish orange to orange color is produced, and it darkens rapidly (*presence of selenium*).

B: Allow the solution obtained in *Identification test A* to stand for 10 minutes, filter, and to the filtrate add 10 mL of barium chloride TS: the solution becomes turbid (*presence of sulfur*).

Residue on ignition ⟨281⟩: not more than 0.2%.

Soluble selenium compounds—

Test solution—Mix 10.0 g of Selenium Sulfide with 100.0 mL of water in a 250-mL flask, allow to stand for 1 hour, with frequent agitation, and filter. To 10.0 mL of the filtrate add 2 mL of 2.5 *M* formic acid, dilute with water to 50 mL, mix, and adjust, if necessary, to a pH of 2.5 ± 0.5. Add 2 mL of freshly prepared 3,3′-diaminobenzidine hydrochloride solution (1 in 200), mix, allow to stand for 45 minutes, and adjust with 6 *N* ammonium hydroxide to a pH of 6.5 ± 0.5. Transfer to a separator, add 10.0 mL of toluene, shake vigorously for 1 minute, allow the layers to separate, and discard the aqueous phase.

Standard solution—Using 10.0 mL of a solution of selenious acid containing 0.5 μg of selenium per mL, prepare a solution as directed under *Test solution*, beginning with "add 2 mL of 2.5 *M* formic acid."

Procedure—Concomitantly determine the absorbances of the toluene layers of the *Test solution* and the *Standard solution* in 1-cm cells at 420 nm, with a suitable spectrophotometer, using a blank consisting of the same quantities of the same reagents treated in the same manner as the *Test solution:* the absorbance of the *Test solution* is not greater than that of the *Standard solution* (5 ppm).

Assay—Place about 100 mg of Selenium Sulfide, accurately weighed, in a suitable container, add 25 mL of fuming nitric acid, and digest over gentle heat until no further solution occurs. Cool, transfer the solution to a 250-mL volumetric flask containing 100 mL of water, cool again, dilute with water to volume, and mix. Pipet 50 mL of the solution into a suitable flask, add 25 mL of water and 10 g of urea, and heat to boiling. Cool, add 3 mL of starch TS, then add 10 mL of potassium iodide solution (1 in 10), and immediately titrate with 0.05 *N* sodium thiosulfate VS. Perform a blank determination, and make any necessary correction. Each mL of 0.05 *N* sodium thiosulfate is equivalent to 987.0 μg of Se.

Selenium Sulfide Lotion

» Selenium Sulfide Lotion is an aqueous, stabilized suspension of Selenium Sulfide. It contains not less than 90.0 percent and not more than 110.0 percent of the labeled amount of SeS_2. It contains suitable buffering and dispersing agents.

NOTE—Where labeled for use as a shampoo, it contains a detergent. Where labeled for other uses, it may contain a detergent.

Packaging and storage—Preserve in tight containers.

Identification—Digest about 2 g with 5 mL of nitric acid over gentle heat for 1 hour, dilute with water to about 50 mL, and filter: the solution responds to *Identification test A* under *Selenium Sulfide*, when tested as directed, beginning with "to 10 mL of the filtrate add 5 mL of water."

pH ⟨791⟩: between 2.0 and 6.0.

Assay—Place a portion of well-mixed Lotion, equivalent to about 100 mg of selenium sulfide and accurately weighed, in a suitable flask. Cautiously digest with 25 mL of fuming nitric acid over gentle heat for 2 hours, and proceed as directed in the *Assay* under *Selenium Sulfide*, beginning with "Cool, transfer the solution to a 250-mL volumetric flask." Each mL of 0.05 *N* sodium thiosulfate is equivalent to 1.789 mg of SeS_2. Where the Lotion is labeled in terms of percentage (w/v) or of the amount of SeS_2 in a given volume of Lotion, determine the density of the Lotion as follows: Using a tared, 100-mL volumetric flask, weigh 100 mL of Selenium Sulfide Lotion that previously has been shaken to ensure homogeneity, allowed to stand until the entrapped air rises, and finally inverted carefully just prior to transfer to the volumetric flask. From the observed weight of 100 mL of the Lotion, calculate the quantity of SeS_2 in each 100 mL.

Senna

» Senna consists of the dried leaflet of *Cassia acutifolia* Delile, known in commerce as Alexandria Senna, or of *Cassia angustifolia* Vahl, known in commerce as Tinnevelly Senna (Fam. Leguminosae).

Packaging and storage—Preserve against attack by insects and rodents (see *Vegetable and Animal Drugs—Preservation* in the *General Notices*).

Botanic characteristics—

Unground Alexandria Senna—Inequilaterally lanceolate or lance-ovate leaflets, frequently broken; from 1.5 cm to 3.5 cm in length and from 5 mm to 10 mm in width, unequal at the base, with very short, stout petiolules. The leaflets are acutely cuspidate, entire, brittle, and subcoriaceous, with short and somewhat appressed hairs, few on the upper surface, more numerous on the lower surface, where they occur spreading on the midrib, especially on its lower part. The color is weak yellow to light grayish green to pale olive. The odor is characteristic, and the taste is mucilaginous and slightly bitter.

Unground Tinnevelly Senna—Usually unbroken leaflets, from 2 cm to 5 cm in length and from 6 mm to 15 mm in width; acute at the apex; and slightly hairy. The color of the leaves is weak yellow to pale olive.

Histology—Senna shows polygonal epidermal cells with straight walls and frequently containing mucilage; numerous, broadly elliptical stomata mostly from 20 µm to 35 µm in length, usually bordered by two neighbor-cells with their long axes parallel to that of the stoma, and rarely, though more frequently in Alexandria Senna, a third epidermal cell at the end of the stoma. The hairs are nonglandular, one-celled, conical, often curved, with thick papillose walls, from 100 µm to 350 µm in length. Palisade cells in a single layer underlie both surfaces except in the midrib region where they occur only beneath the upper epidermis. A meristele occurs in the midrib composed of several radially arranged fibrovascular bundles, the latter separated by narrow vascular rays and supported above and below by arcs of lignified pericyclic fibers. Calcium oxalate occurs in rosette aggregates in the spongy parenchyma and in six- to eight-sided prisms in the crystal fibers, which lie on the outer surface of each group of pericyclic fibers.

Powdered Senna—Dusky greenish yellow to light olive-brown, displaying fragments of veins bearing lignified vessels, tracheids, and crystal fibers, isolated hairs, masses of palisade and spongy parenchyma, fragments of epidermis with stomata, free calcium oxalate rosette aggregates, and prisms from 10 µm to 20 µm in length. In powdered Alexandria Senna, the hairs are more numerous than in powdered Tinnevelly Senna.

Identification—Mix 500 mg with 10 mL of a 1 in 10 solution of potassium hydroxide in alcohol, boil for about 2 minutes, dilute with 10 mL of water, and filter. Acidify the filtrate with hydrochloric acid, shake it with ether, remove the ether layer, and shake it with 5 mL of 6 *N* ammonium hydroxide: the latter is colored orange or bluish red.

Senna stems, pods, or other foreign organic matter ⟨561⟩—The amount of senna stems does not exceed 8.0%, and the amount of senna pods or other foreign organic matter does not exceed 2.0%.

Acid-insoluble ash ⟨561⟩: not more than 3.0%.

Senna Fluidextract

» Prepare Senna Fluidextract as follows.

Mix 1000 g of Senna, in coarse powder, with a sufficient quantity (600 mL to 800 mL) of menstruum consisting of a mixture of 1 volume of alcohol and 2 volumes of water to make it evenly and distinctly damp. After 15 minutes, pack the mixture firmly into a suitable percolator, and cover the drug with additional menstruum. Macerate for 24 hours, then percolate at a moderate rate, adding fresh menstruum, until the drug is practically exhausted of its active principles. Reserve the first 800 mL of percolate, and use it to dissolve the residue from the additional percolate that has been concentrated to a soft extract at a temperature not to exceed 60°. Add water and alcohol to make the product measure 1000 mL, and mix.

Packaging and storage—Preserve in tight, light-resistant containers, and avoid exposure to direct sunlight and to excessive heat.

Alcohol content, *Method I* ⟨611⟩: between 23.0% and 27.0% of C_2H_5OH.

Senna Syrup

» Prepare Senna Syrup as follows:

Senna Fluidextract	250 mL
Suitable essential oil(s)	
Sucrose	635 g
Purified Water, a sufficient quantity, to make	1000 mL

Mix the oil(s) with the Senna Fluidextract, and gradually add 330 mL of Purified Water. Allow the mixture to stand for 24 hours in a cool place, with occasional agitation, then filter, and pass enough Purified Water through the filter to obtain 580 mL of filtrate. Dissolve the Sucrose in this liquid, and add sufficient Purified Water to make the product measure 1000 mL. Mix, and strain.

Packaging and storage—Preserve in tight containers, at a temperature not exceeding 25°.

Alcohol content, *Method I* ⟨611⟩: between 5.0% and 7.0% of C_2H_5OH.

Sennosides

» Sennosides is a partially purified natural complex of anthraquinone glucosides found in senna, isolated from *Cassia angustifolia* or *C. acutifolia* as calcium salts. It contains not less than 90.0 percent and not more than 110.0 percent of sennosides, calculated on the dried basis, or, if the sennosides is in higher concentration, not less than 90.0 percent and not more than 110.0 percent of the concentration indicated on the label.

Packaging and storage—Preserve in well-closed containers.

Reference standard—*USP Sennosides Reference Standard*—Dry in vacuum over phosphorus pentoxide to constant weight before using.

Identification—Place equal volumes of ethyl acetate, *n*-propyl alcohol, and water in a separator, shake, and discard the upper layer. Add a sufficient quantity of Sennosides to obtain a solution having a concentration of 1 mg per mL. Prepare similarly a Standard solution of USP Sennosides RS. Apply 20-µL portions of each solution, as 1-cm streaks, on a line about 2.5 cm from the bottom edge of a thin-layer chromatographic plate (see *Chromatography* ⟨621⟩) coated with a 0.25-mm layer of chromatographic silica gel mixture. Develop the chromatogram in a solvent system consisting of a mixture of ethyl acetate, *n*-propyl alcohol, and water (4:4:3) until the solvent front has moved about 15 cm. Remove the plate from the developing chamber, air-dry, and examine under long-wavelength ultraviolet light. Expose the plate to ammonium hydroxide vapor until color develops (about 5 minutes). Cover the plate with a piece of glass, and heat at 120° for 5 minutes: the 2 most prominent spots from the test solution corresponds in color and mobility to those from the Standard solution.

pH ⟨791⟩: between 6.3 and 7.3, in a solution (1 in 10).

Loss on drying ⟨731⟩—Dry it in vacuum over phosphorus pentoxide to constant weight: it loses not more than 5.0% of its weight.

Residue on ignition ⟨281⟩: between 5.0% and 8.0%, the use of sulfuric acid being omitted.

Heavy metals, *Method II* ⟨231⟩: 0.006%.

Assay—

pH 7.0 phosphate buffer—Dissolve 4.54 g of monobasic potassium phosphate in water to make 500 mL of solution. Dissolve 4.73 g of anhydrous dibasic sodium phosphate in water to make 500 mL of solution. Mix 38.9 mL of the monobasic potassium phosphate solution with 61.1 mL of the dibasic sodium phosphate solution. Adjust dropwise, if necessary, with the dibasic sodium phosphate solution to a pH of 7.0.

Borate solution—Dissolve 75.80 g of sodium borate in water, dilute with water to 2000 mL, and mix.

Sodium dithionite solution—Prepare a 1.5 in 100 solution of sodium dithionite in water.

Standard preparation—Dissolve about 25 mg of USP Sennosides RS, accurately weighed, in *pH 7.0 phosphate buffer* in a 25-mL volumetric flask with the aid of an ultrasonic bath, dilute with *pH 7.0 phosphate buffer* to volume, and mix.

Assay preparation—Dissolve about 25 mg of Sennosides, accurately weighed, in *pH 7.0 phosphate buffer* in a 25-mL volumetric flask, with the aid of an ultrasonic bath, dilute with *pH 7.0 phosphate buffer* to volume, and mix.

Procedure—Pipet 1-mL portions of the *Standard preparation* and the *Assay preparation* into separate 100-mL volumetric flasks, dilute with *Borate solution* to volume, and mix. Transfer 5.0-mL portions of each of the resulting solutions to separate low-actinic glass 50-mL volumetric flasks, and add 15 mL of *Borate solution* and 15.0 mL of *Sodium dithionite solution.* Pass nitrogen through the solutions, seal the flasks with nitrogen-filled balloons, and heat in a water bath for 30 minutes. Cool the flasks for 15 minutes in a water bath thermostatically controlled at 20°. Dilute the solutions with *Borate solution* to volume, and mix. Determine without delay the fluorescence intensities of the resulting solutions in a fluorometer at an excitation wavelength of 392 nm and an emission wavelength of 505 nm, the time elapsed between the addition of the *Sodium dithionite solution* and the measurement being the same for the two solutions. Calculate the quantity, in mg, of sennosides in the Sennosides taken by the formula:

$$25C(I_U/I_S),$$

in which C is the concentration, in mg per mL, of USP Sennosides RS in the *Standard preparation,* and I_U and I_S are the fluorescence values observed for the solutions from the *Assay preparation* and the *Standard preparation,* respectively.

Sennosides Tablets

» Sennosides Tablets contain not less than 90.0 percent and not more than 110.0 percent of the labeled amount of Sennosides.

Packaging and storage—Preserve in well-closed containers.

Reference standard—*USP Sennosides Reference Standard*—Dry in vacuum over phosphorus pentoxide to constant weight before using.

Identification—A portion of finely powdered Tablets, equivalent to about 20 mg of sennosides, responds to the *Identification test* under *Sennosides.*

Dissolution ⟨711⟩—

Medium: water; 900 mL.

Apparatus 1: 100 rpm.

Time: 120 minutes.

Procedure—Determine the amount of sennosides dissolved, employing the procedure set forth in the *Assay,* making any necessary volumetric adjustments.

Tolerances—Not less than 75% (*Q*) of the labeled amount of sennosides is dissolved in 120 minutes.

Uniformity of dosage units ⟨905⟩: meet the requirements.

Assay—

pH 7.0 phosphate buffer, Borate solution, Sodium dithionite solution, and *Standard preparation*—Prepare as directed in the *Assay* under *Sennosides.*

Assay preparation—Weigh and finely powder not less than 20 Sennosides Tablets. Transfer an accurately weighed portion of the powder, equivalent to about 25 mg of sennosides, to a 25-mL volumetric flask. Add 20 mL of *pH 7.0 phosphate buffer,* sonicate to dissolve the sennosides, add *pH 7.0 phosphate buffer* to volume, and mix. Centrifuge the resulting suspension for 15 minutes at 3500 rpm. The supernatant solution is the *Assay preparation.*

Procedure—Proceed as directed for *Procedure* in the *Assay* under *Sennosides.* Calculate the quantity, in mg, of sennosides in the portion of Tablets taken by the formula:

$$25C(I_U/I_S),$$

in which C is the concentration, in mg per mL, of USP Sennosides RS in the *Standard preparation,* and I_U and I_S are the fluorescence values observed for the solutions from the *Assay preparation* and the *Standard preparation,* respectively.

Serine

$$HOCH_2-\underset{NH_2}{\overset{H}{C}}-COOH$$

$C_3H_7NO_3$ 105.09
L-Serine.
L-Serine [*56-45-1*].

» Serine contains not less than 98.5 percent and not more than 101.5 percent of $C_3H_7NO_3$, as L-serine, calculated on the dried basis.

Packaging and storage—Preserve in well-closed containers.

Reference standard—*USP L-Serine Reference Standard*—Dry at 105° for 3 hours before using.

Identification—The infrared absorption spectrum of a potassium bromide dispersion of it, previously dried, exhibits maxima only at the same wavelengths as that of a similar preparation of USP L-Serine RS.

Specific rotation ⟨781⟩: between +13.6° and +15.6°, calculated on the dried basis, determined in a solution in 2 *N* hydrochloric acid containing 1.00 g in each 10 mL.

Loss on drying ⟨731⟩—Dry it at 105° for 3 hours: it loses not more than 0.2% of its weight.

Residue on ignition ⟨281⟩: not more than 0.1%.

Chloride ⟨221⟩—A 0.73-g portion shows no more chloride than corresponds to 0.50 mL of 0.020 *N* hydrochloric acid (0.05%).

Sulfate ⟨221⟩—A 0.33-g portion shows no more sulfate than corresponds to 0.10 mL of 0.020 *N* sulfuric acid (0.03%).

Arsenic ⟨211⟩: 1.5 ppm.

Iron ⟨241⟩: 0.003%.

Heavy metals, *Method I* ⟨231⟩: 0.0015%.

Assay—Transfer about 100 mg of Serine, accurately weighed, to a 125-mL flask, dissolve in a mixture of 3 mL of formic acid and 50 mL of glacial acetic acid, and titrate with 0.1 *N* perchloric acid VS, determining the end-point potentiometrically. Perform a blank determination, and make any necessary correction. Each mL of 0.1 *N* perchloric acid is equivalent to 10.51 mg of $C_3H_7NO_3$.

Serums—*see complete list in index*

Sesame Oil—*see* Sesame Oil NF

Shellac—*see* Shellac NF

Siliceous Earth, Purified—*see* Siliceous Earth, Purified NF

Silicon Dioxide—*see* Silicon Dioxide NF

Silicon Dioxide, Colloidal—*see* Silicon Dioxide, Colloidal NF

Silver Nitrate

$AgNO_3$ 169.87
Nitric acid silver(1+) salt.
Silver(1+) nitrate [*7761-88-8*].

» Silver Nitrate, powdered and then dried in the dark over silica gel for 4 hours, contains not less than 99.8 percent and not more than 100.5 percent of $AgNO_3$.

Packaging and storage—Preserve in tight, light-resistant containers.

Clarity and color of solution—A solution of 2 g in 20 mL of water is clear and colorless.

Identification—

A: A solution (1 in 50) responds to the tests for *Silver* ⟨191⟩.

B: Mix a solution (1 in 10) in a test tube with 1 drop of diphenylamine TS, and then carefully superimpose it upon sulfuric acid: a deep blue color appears at the zone of contact.

Copper—To 5 mL of a solution (1 in 10) add 6 *N* ammonium hydroxide, dropwise, until a precipitate first formed is just dissolved: no blue color is produced.

Assay—Powder about 1 g of Silver Nitrate, and dry in the dark over silica gel for 4 hours. Weigh accurately about 700 mg of this dried salt, dissolve in 50 mL of water, add 2 mL of nitric acid and 2 mL of ferric ammonium sulfate TS, and titrate with 0.1 *N* ammonium thiocyanate VS. Each mL of 0.1 *N* ammonium thiocyanate is equivalent to 16.99 mg of $AgNO_3$.

Silver Nitrate Ophthalmic Solution

» Silver Nitrate Ophthalmic Solution is a solution of Silver Nitrate in a water medium. It contains not less than 0.95 percent and not more than 1.05 percent of $AgNO_3$. The solution may be buffered by the addition of Sodium Acetate.

Packaging and storage—Preserve it protected from light, in inert, collapsible capsules or in other suitable single-dose containers.

Clarity and color of solution—It is clear and colorless.

Identification—It responds to the tests for *Silver* ⟨191⟩ and for *Nitrate* ⟨191⟩.

Sterility—It meets the requirements under *Sterility Tests* ⟨71⟩.

pH ⟨791⟩: between 4.5 and 6.0.

Assay—Place 5 mL of Silver Nitrate Ophthalmic Solution, accurately measured, in a conical flask, dilute with 20 mL of water, add 1 mL of nitric acid and 1 mL of ferric ammonium sulfate TS, and titrate with 0.02 *N* ammonium thiocyanate VS. Each mL of 0.02 *N* ammonium thiocyanate is equivalent to 3.397 mg of $AgNO_3$.

Toughened Silver Nitrate

» Toughened Silver Nitrate contains not less than 94.5 percent of $AgNO_3$, the remainder consisting of silver chloride (AgCl).

Packaging and storage—Preserve in tight, light-resistant containers.

Identification—

A: A solution (1 in 50) responds to the tests for *Silver* ⟨191⟩.

B: Mix a solution (1 in 10) in a test tube with 1 drop of diphenylamine TS, then carefully superimpose it upon sulfuric acid: a deep blue color appears at the zone of contact.

Copper—A solution (1 in 10) shows no trace of blue coloration when treated with an excess of 6 *N* ammonium hydroxide.

Assay—Add about 700 mg of Toughened Silver Nitrate, accurately weighed, to 50 mL of water, and when the silver nitrate has dissolved, filter the solution. Thoroughly wash the filter and sediment with water, add 2 mL of nitric acid and 2 mL of ferric ammonium sulfate TS to the combined filtrate and washings, and titrate with 0.1 *N* ammonium thiocyanate VS. Each mL of 0.1 *N* ammonium thiocyanate is equivalent to 16.99 mg of $AgNO_3$.

Simethicone

$$(CH_3)_3Si\left[OSi(CH_3)_2\right]_n CH_3 \quad + \quad SiO_2$$

Simethicone.
α-(Trimethylsilyl)-ω-methylpoly[oxy(dimethylsilylene)], mixture with silicon dioxide [*8050-81-5*].

» Simethicone is a mixture of fully methylated linear siloxane polymers containing repeating units of the formula $[-(CH_3)_2SiO-]_n$, stabilized with trimethylsiloxy end-blocking units of the formula $[(CH_3)_3SiO-]$, and silicon dioxide. It contains not less than 90.5 percent and not more than 99.0 percent of polydimethylsiloxane ($[-(CH_3)_2SiO-]_n$), and not less than 4.0 percent and not more than 7.0 percent of silicon dioxide.

Packaging and storage—Preserve in tight containers.

Reference standard—*USP Polydimethylsiloxane Reference Standard*—Keep container tightly closed. Do not dry before using.

Identification—The infrared absorption spectrum, determined in a 0.5-mm cell, of the solution of Simethicone prepared as directed in the *Assay* exhibits maxima only at the same wavelengths as that of the Standard solution prepared as directed in the *Assay*.

Loss on heating—Heat about 15 g, accurately weighed, in an open, tared vessel having a diameter of 5.5 ± 0.5 cm and a wall height of 2.5 ± 1.0 cm at 200° in a circulating air oven for 4 hours, and allow to come to room temperature in a desiccator before weighing: it loses not more than 18.0% of its weight.

Heavy metals, *Method II* ⟨231⟩: 0.001%.

Defoaming activity—

Foaming solution—Dissolve 1 g of octoxynol 9 in 100 mL of water.

Test preparation—Transfer 200 mg of Simethicone to a 60-mL bottle, add 50 mL of tertiary butyl alcohol, cap the bottle, and shake vigorously. [NOTE—Warm slightly, if necessary, to effect solution.]

Procedure—[NOTE—For each test, employ a clean, unused, 250-mL glass jar.] Add, dropwise, 500 µL of *Test preparation* to a clean, unused, cylindrical 250-mL glass jar, fitted with a 50-mm cap, containing 100 mL of *Foaming solution*. Cap the jar, and clamp it in an upright position on a wrist-action shaker. Employing a radius of 13.3 ± 0.4 cm (measured from center of shaft to center of bottle), shake for 10 seconds through an arc of 10 degrees at a frequency of 300 ± 30 strokes per minute. Record the time required for the foam to collapse. The time, in seconds, for foam collapse is determined at the instant the first portion of foam-free liquid surface appears, measured from the end of the shaking period. The defoaming activity time does not exceed 15 seconds.

Silicon dioxide content—Transfer about 1.25 g, accurately weighed, to a platinum combustion boat, the tare weight of which includes the weight of a small borosilicate glass melting-point capillary tube, the fused end of which serves as a mixing aid. Add 2 drops of sulfuric acid, and mix. Place the combustion boat containing the mixture and the capillary tube in a combustion tube that is capable of being purged with compressed gas while being heated in a furnace. Purge the system with nitrogen flowing at a rate sufficient to expel volatilized material (e.g., 125 to 150 mL per minute), and heat the mixture slowly and cautiously to avoid spattering, at a rate such that the temperature rises to 800 ± 25° in 60 minutes. Heat isothermally at 800 ± 25° for 60 minutes, maintaining a steady flow of nitrogen, then replace the nitrogen gas with compressed air, and continue heating for 30 minutes. Remove the boat, cool it in a desiccator, and determine the weight of the residue of silicon dioxide.

Assay—Transfer about 50 mg of Simethicone, accurately weighed, to a suitable round, narrow-mouth, screw-capped, 120-mL bottle, add 25.0 mL of carbon tetrachloride, and swirl to disperse. Add 50 mL of dilute hydrochloric acid (2 in 5), close the bottle securely with a cap having an inert liner, and shake for 5 minutes, accurately timed, on a reciprocating shaker at a suitable rate (e.g., about 200 oscillations per minute and a stroke of 38 ± 2 mm). Transfer the mixture to a 125-mL separator, and remove about 5 mL of the lower, organic layer to a 15-mL screw-capped test tube containing 0.5 g of anhydrous sodium sulfate. Close the tube with a screw-cap having an inert liner, agitate vigorously, and centrifuge the mixture until a clear supernatant liquid (*Assay preparation*) is obtained. Prepare a *Standard preparation* by similarly treating a 25.0-mL portion of a solution of USP Polydimethylsiloxane RS in carbon tetrachloride having a known concentration of about 2 mg per mL. Prepare a blank by mixing 10 mL of carbon tetrachloride with 0.5 g of anhydrous sodium sulfate and centrifuging to obtain a clear supernatant liquid. Concomitantly determine the absorbances of the solutions in 0.5-mm cells at the wavelength of maximum absorbance at about 7.9 µm, with a suitable infrared spectrophotometer, using the blank to set the instrument. Calculate the quantity, in mg, of $[-(CH_3)_2SiO-]_n$ in the Simethicone taken by the formula:

$$25C(A_U/A_S),$$

in which C is the concentration, in mg per mL, of USP Polydimethylsiloxane RS in the *Standard preparation*, and A_U and A_S are the absorbances of the *Assay preparation* and the *Standard preparation*, respectively.

Simethicone Emulsion

» Simethicone Emulsion is a water-dispersible form of Simethicone composed of Simethicone, suitable emulsifiers, preservatives, and water. It may contain suitable viscosity-increasing agents. It contains an amount of polydimethylsiloxane ($[-(CH_3)_2SiO-]_n$) that is not less than 85.0 percent and not more than 110.0 percent of the labeled amount of simethicone.

Packaging and storage—Preserve in tight containers.

Reference standard—*USP Polydimethylsiloxane Reference Standard*—Keep container tightly closed. Do not dry before using.

Identification—The infrared absorption spectrum, determined in a 0.5-mm cell, of the solution of Simethicone Emulsion prepared as directed in the *Assay*, exhibits maxima only at the same wavelengths as that of the Standard solution prepared as directed in the *Assay*.

Microbial limits ⟨61⟩—Its total aerobic microbial count does not exceed 100 per g.

Heavy metals, *Method II* ⟨231⟩: 0.001%.

Defoaming activity—

Foaming solution—Dissolve 1 g of octoxynol 9 in 100 mL of water.

Test preparation—Transfer an accurately weighed quantity of Emulsion, equivalent to 300 mg of simethicone, to a 60-mL bottle, dilute with water to 30 g, cap the bottle, and shake vigorously.

Procedure—Proceed as directed for *Procedure* in the test for *Defoaming activity* under *Simethicone*. The defoaming activity time does not exceed 15 seconds.

Assay—Transfer an accurately weighed quantity of Simethicone Emulsion, equivalent to about 50 mg of simethicone, to a suitable round, narrow-mouth, screw-capped, 120-mL bottle, add 25.0 mL of carbon tetrachloride, and swirl to disperse. Proceed as directed in the *Assay* under *Simethicone*, beginning with "Add 50 mL of dilute hydrochloric acid." Calculate the quantity, in mg, of $[-(CH_3)_2SiO-]_n$ in each g of the Emulsion taken by the formula:

$$(25C/S)(A_U/A_S),$$

in which C is the concentration, in mg per mL, of USP Polydimethylsiloxane RS in the *Standard preparation*, S is the weight, in g, of Emulsion taken, and A_U and A_S are the absorbances of the *Assay preparation* and the *Standard preparation*, respectively.

Simethicone Oral Suspension

» Simethicone Oral Suspension is a suspension of Simethicone in Water. It contains an amount of polydimethylsiloxane ($[-(CH_3)_2SiO-]_n$) that is not less than 85.0 percent and not more than 115.0 percent of the labeled amount of simethicone.

Packaging and storage—Preserve in tight, light-resistant containers.

Reference standard—*USP Polydimethylsiloxane Reference Standard*—Keep container tightly closed. Do not dry before using.

Identification—It responds to the *Identification test* under *Simethicone*.

pH ⟨791⟩: between 4.4 and 4.6.

Defoaming activity—

Foaming solution—Dissolve 500 µg of FD&C Blue No. 1 and 1 g of octoxynol 9 in 100 mL of water.

Procedure—[NOTE—For each test, employ a clean, unused, 250-mL glass jar.] Transfer a volume of Oral Suspension, equivalent to 20 mg of simethicone, to a clean, unused, cylindrical 250-mL glass jar, fitted with a 50-mm cap, containing 100 mL of *Foaming solution* that has been warmed to 37°. Proceed as directed for *Procedure* in the test for *Defoaming activity* under *Simethicone*, beginning with "Cap the jar." The defoaming activity time does not exceed 45 seconds.

Assay—Transfer an accurately measured volume of Simethicone Oral Suspension, equivalent to about 50 mg of simethicone, to a suitable round, narrow-mouth, screw-capped, 120-mL bottle, and proceed as directed in the *Assay* under *Simethicone*, beginning with "add 25.0 mL of carbon tetrachloride." Calculate the quantity, in mg, of $[-(CH_3)_2SiO-]_n$ in each mL of the Oral Suspension taken by the formula:

$$(25C/V)(A_U/A_S),$$

in which C is the concentration, in mg per mL, of USP Polydimethylsiloxane RS in the Standard solution, V is the volume, in mL, of Oral Suspension taken, and A_U and A_S are the absorbances of the solution from the Oral Suspension and the Standard solution, respectively.

Simethicone Oral Suspension, Alumina, Magnesia, and—*see* Alumina, Magnesia, and Simethicone Oral Suspension

Simethicone Oral Suspension, Magaldrate and—*see* Magaldrate and Simethicone Oral Suspension

Simethicone Tablets

» Simethicone Tablets contain an amount of polydimethylsiloxane ($[-(CH_3)_2SiO-]_n$) that is not less than 85.0 percent and not more than 115.0 percent of the labeled amount of simethicone.

Packaging and storage—Preserve in well-closed containers.

Reference standard—*USP Polydimethylsiloxane Reference Standard*—Keep container tightly closed. Do not dry before using.

Identification—Tablets respond to the *Identification test* under *Simethicone*.

Disintegration ⟨701⟩: 30 minutes.

Uniformity of dosage units ⟨905⟩: meet the requirements.

Defoaming activity—It meets the requirements of the test for *Defoaming activity* under *Simethicone Oral Suspension*, a quantity of finely powdered Tablets, equivalent to 20 mg of simethicone, being used.

Assay—Weigh and finely powder not less than 20 Simethicone Tablets. Transfer an accurately weighed portion of the powder, equivalent to about 50 mg of simethicone, to a suitable round, narrow-mouth, screw-capped, 120-mL bottle, and proceed as directed in the *Assay* under *Simethicone*, beginning with "add 25.0 mL of carbon tetrachloride." Calculate the quantity, in mg, of $[-(CH_3)_2SiO-]_n$ in the portion of Tablets taken by the formula:

$$25C(A_U/A_S),$$

in which C is the concentration, in mg per mL, of USP Polydimethylsiloxane RS in the Standard solution, and A_U and A_S are the absorbances of the solution from the Tablets and the Standard solution, respectively.

Simethicone Tablets, Alumina, Magnesia, and—*see* Alumina, Magnesia, and Simethicone Tablets

Simethicone Tablets, Magaldrate and—*see* Magaldrate and Simethicone Tablets

Sisomicin Sulfate

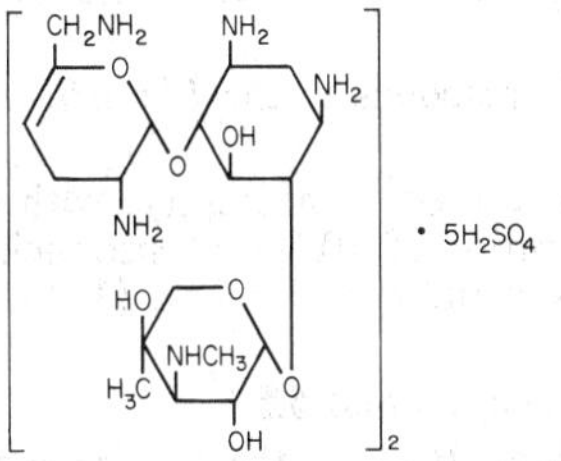

$(C_{19}H_{37}N_5O_7)_2 \cdot 5H_2SO_4$ 1385.43

D-Streptamine, (2*S*-*cis*)-4-*O*-[3-amino-6-(aminomethyl)-3,4-dihydro-2*H*-pyran-2-yl]-2-deoxy-6-*O*-[3-deoxy-4-*C*-methyl-3-(methylamino)-β-L-arabinopyranosyl]-, sulfate (2:5)(salt).

O-3-Deoxy-4-*C*-methyl-3-(methylamino)-β-L-arabinopyranosyl-(1 → 4)-*O*-[2,6-diamino-2,3,4,6-tetradeoxy-α-D-*glycero*-hex-4-enopyranosyl-(1 → 6)]-2-deoxy-L-streptamine sulfate (2:5)(salt) [*53179-09-2*].

» Sisomicin Sulfate has a potency equivalent to not less than 580 µg of sisomicin ($C_{19}H_{37}N_5O_7$) per mg, calculated on the dried basis.

Packaging and storage—Preserve in tight containers.

Reference standard—*USP Sisomicin Sulfate Reference Standard*—Do not dry before using; calculate its weight on the dried basis.

Identification—Prepare a solution containing 10 mg of sisomicin per mL. On a suitable thin-layer chromatographic plate (see *Chromatography* ⟨621⟩), coated with a 0.25-mm layer of chromatographic silica gel mixture, apply 5 µL of this solution, 5 µL of a solution of USP Sisomicin Sulfate RS containing 10 mg of sisomicin per mL, and 5 µL of a mixture of the two solutions (1:1). Allow the spots to dry, place the plate in a developing chamber fitted for continuous-flow thin-layer chromatography, develop the chromatogram in a solvent system consisting of a mixture of methanol, ammonium hydroxide, and chloroform (60:30:25) for 3 hours. Remove the plate from the developing chamber, allow the solvent to evaporate, and heat the plate at 110° for 15 minutes. Spray the plate with a 1 in 100 solution of ninhydrin in butanol to which 1 mL of pyridine has been added: sisomicin appears as a brown spot, and the spots obtained from the test solution and from the mixture of Test solution and Standard solution, respectively, correspond in distance from the origin to that of the spot from the Standard solution.

Specific rotation ⟨781⟩: between +100° and +110°, calculated on the dried basis, in a solution containing 10 mg per mL.

pH ⟨791⟩: between 3.5 and 5.5, in a solution containing 40 mg of sisomicin per mL.

Loss on drying ⟨731⟩—Dry about 100 mg in vacuum at a pressure not exceeding 5 mm of mercury at 110° for 3 hours: it loses not more than 15.0% of its weight.

Residue on ignition ⟨281⟩: not more than 1.0%, the charred residue being moistened with 2 mL of nitric acid and 5 drops of sulfuric acid.

Assay—Proceed as directed under *Antibiotics—Microbial Assays* ⟨81⟩.

Sisomicin Sulfate Injection

» Sisomicin Sulfate Injection is a sterile solution of Sisomicin Sulfate in Water for Injection. It contains the equivalent of not less than 90.0 percent and not more than 120.0 percent of the labeled amount of sisomicin ($C_{19}H_{37}N_5O_7$). It may contain one or more suitable buffers, chelating agents, and preservatives.

Packaging and storage—Preserve in single-dose or in multiple-dose containers, preferably of Type I glass.

Reference standard—*USP Sisomicin Sulfate Reference Standard*—Do not dry before using; calculate its weight on the dried basis.

Identification—It responds to the *Identification test* under *Sisomicin Sulfate.*

Pyrogen—When diluted, if necessary, with pyrogen-free saline TS to a concentration of 10 mg of sisomicin per mL, it meets the requirements of the *Pyrogen Test* ⟨151⟩, the test dose being 1 mL per kg.

pH ⟨791⟩: between 2.5 and 5.5.

Other requirements—It meets the requirements under *Injections* ⟨1⟩.

Assay—Proceed as directed under *Antibiotics—Microbial Assays* ⟨81⟩, using an accurately measured volume of Sisomicin Sulfate Injection diluted quantitatively and stepwise with *Buffer No. 3* to yield a *Test Dilution* having a concentration assumed to be equal to the median dose level of the Standard (0.1 µg of sisomicin per mL).

Smallpox Vaccine

» Smallpox Vaccine conforms to the regulations of the federal Food and Drug Administration concerning biologics (630.70 to 630.76) (see *Biologics* ⟨1041⟩). It is a suspension or solid containing the living virus of vaccinia of a strain of approved origin and manipulation, that has been grown in the skin of a vaccinated bovine calf. It meets the requirements of the specific potency test using embryonated chicken eggs in comparison with the U.S. Reference Smallpox Vaccine in the case of Vaccine intended for multiple-puncture administration or with such Reference Vaccine diluted (1:30) in the case of Vaccine intended for jet injection, and the requirements for the tests for absence of specific microorganisms. It may contain a suitable preservative.

Packaging and storage—Preserve and dispense in the containers in which it was placed by the manufacturer. Keep liquid Vaccine during storage and in shipment at a temperature below 0°. Keep dried Vaccine at a temperature between 2° and 8°.

Expiration date—The expiration date for liquid Vaccine is not later than 3 months after date of issue from manufacturer's cold storage (−10°, 9 months as glycerinated or equivalent preparation). The expiration date for dried Vaccine is not later than 18 months after date of issue from manufacturer's cold storage (5°, 6 months).

Labeling—Label it to state that it contains not more than 200 microorganisms per mL in the case of Vaccine intended for multiple-puncture administration, or that it contains not more than 1 microorganism per 100 doses in the case of Vaccine intended for jet injection, unless it meets the requirements for sterility. In the case of Vaccine intended for jet injection, so state on the label. In the case of dried Vaccine, label it to state that after constitution it is to be well shaken before use. Label it also to state that it was prepared in the bovine calf.

Soap, Green—*see* Green Soap
Soap, Hexachlorophene Liquid—*see* Hexachlorophene Liquid Soap
Soda Lime—*see* Soda Lime NF

Sodium Acetate

$CH_3COONa.3H_2O$

$C_2H_3NaO_2.3H_2O$ 136.08
Acetic acid, sodium salt, trihydrate.
Sodium acetate trihydrate [*6131-90-4*].
Anhydrous 82.03 [*127-09-3*].

» Sodium Acetate contains three molecules of water of hydration, or is anhydrous. It contains not less than 99.0 percent and not more than 101.0 percent of $C_2H_3NaO_2$, calculated on the dried basis.

Packaging and storage—Preserve in tight containers.

Labeling—Label it to indicate whether it is the trihydrate or is anhydrous.

Identification—A solution responds to the test for *Sodium* ⟨191⟩ and for *Acetate* ⟨191⟩.

pH ⟨791⟩: between 7.5 and 9.2, in a 1 in 20 solution in carbon dioxide-free water containing the equivalent of 30 mg of anhydrous sodium acetate per mL.

Loss on drying ⟨731⟩—Dry at 120° to constant weight: the hydrous form loses between 38.0% and 41.0% of its weight, and the anhydrous form loses not more than 1.0% of its weight.

Insoluble matter—Dissolve the equivalent of 20 g of anhydrous sodium acetate in 150 mL of water, heat to boiling, and digest in a covered beaker on a steam bath for 1 hour. Filter through a tared filtering crucible, wash thoroughly, and dry at 105°: the weight of the residue does not exceed 10 mg (0.05%).

Chloride ⟨221⟩—A portion equivalent to 1.0 g of anhydrous sodium acetate shows no more chloride than corresponds to 0.50 mL of 0.020 *N* hydrochloric acid (0.035%).

Sulfate ⟨221⟩—A portion equivalent to 10 g of anhydrous sodium acetate shows no more sulfate than corresponds to 0.50 mL of 0.020 *N* sulfuric acid (0.005%).

Arsenic, *Method I* ⟨211⟩—Dissolve a portion equivalent to 1.0 g of anhydrous sodium acetate in 35 mL of water: the limit is 3 ppm.

Calcium and magnesium—To 20 mL of a solution containing the equivalent of 10 mg of anhydrous sodium acetate per mL add 2 mL each of 6 *N* ammonium hydroxide, ammonium oxalate TS, and dibasic sodium phosphate TS: no turbidity is produced within 5 minutes.

Potassium—Dissolve the equivalent of 3 g of anhydrous sodium acetate in 5 mL of water, and add 0.2 mL of sodium bitartrate TS: no turbidity is produced within 5 minutes.

Heavy metals, *Method I* ⟨231⟩: 0.001%, calculated on the dried basis, glacial acetic acid being used instead of diluted acetic acid for adjustment of the pH.

Assay—Weigh accurately the equivalent of about 200 mg of anhydrous sodium acetate, and dissolve in 25 mL of glacial acetic acid, warming gently if necessary to effect complete solution. Add 2 drops of *p*-naphtholbenzein TS, and titrate with 0.1 *N* perchloric acid VS. Perform a blank determination, and make any necessary correction. Each mL of 0.1 *N* perchloric acid is equivalent to 8.203 mg of $C_2H_3NaO_2$.

Sodium Acetate Injection

» Sodium Acetate Injection is a sterile solution of Sodium Acetate in Water for Injection. It contains not less than 95.0 percent and not more than 105.0 percent of the labeled amount of CH_3COONa.

Packaging and storage—Preserve in single-dose containers, preferably of Type I glass.

Labeling—The label states the sodium acetate content in terms of weight and of milliequivalents in a given volume. Label the Injection to indicate that it is to be diluted to appropriate strength with water or other suitable fluid prior to administration. The label states also the total osmolar concentration in mOsmol per liter. Where the contents are less than 100 mL, or where the label states that the Injection is not for direct injection but is to be diluted before use, the label alternatively may state the total osmolar concentration in mOsmol per mL.

Identification—It responds to the tests for *Sodium* ⟨191⟩ and for *Acetate* ⟨191⟩.

Pyrogen—When diluted with Sodium Chloride Injection to contain 0.2% of sodium acetate, it meets the requirements of the *Pyrogen Test* ⟨151⟩.

pH ⟨791⟩: between 6.0 and 7.0.

Particulate matter ⟨788⟩: meets the requirements under *Small-volume Injections.*

Other requirements—It meets the requirements under *Injections* ⟨1⟩.

Assay—

Standard stock solution—Dissolve 570.0 mg of sodium chloride, previously dried at 105° for 2 hours, in 100 mL of water, transfer to a 1000-mL volumetric flask, dilute with water to volume, and mix. Each mL of this solution contains 224 µg of sodium, equivalent to 800 µg of anhydrous sodium acetate.

Standard preparations—Transfer to each of four 100-mL volumetric flasks 10 mL of a nonionic wetting agent (1 in 500). Dilute the contents of one of the flasks with water to volume to provide a blank. To the remaining flasks add, respectively, 5.0, 10.0, and 15.0 mL of *Standard stock solution*, dilute with water to volume, and mix.

Assay preparation—Transfer an accurately measured volume of Sodium Acetate Injection, equivalent to about 800 mg of anhydrous sodium acetate, to a 1000-mL volumetric flask, dilute with water to volume, and mix. Pipet 10 mL of this solution into a 100-mL volumetric flask containing 10 mL of a nonionic wetting agent (1 in 500), dilute with water to volume, and mix.

Standard graph—Set a flame photometer for maximum transmittance at a wavelength of about 589 nm. Adjust the instrument to zero transmittance with the blank, and to 100% transmittance with the most concentrated of the *Standard preparations.* Read the transmittances of the other *Standard preparations*, and plot transmittances versus equivalent concentration of sodium acetate.

Procedure—Adjust the instrument as directed under *Standard graph*, read the transmittance of the *Assay preparation*, and calculate the sodium acetate content, in mg per mL, of Sodium Acetate Injection.

Sodium Acetate Solution

» Sodium Acetate Solution is an aqueous solution of Sodium Acetate. It contains not less than 97.0 percent and not more than 103.0 percent (w/w) of the labeled amount of $C_2H_3NaO_2$.

Packaging and storage—Preserve in tight containers.

Identification—It responds to the tests for *Sodium* ⟨191⟩ and for *Acetate* ⟨191⟩.

pH ⟨791⟩: between 7.5 and 9.2, when diluted with carbon dioxide–free water to contain 5 percent of solids.

Insoluble matter—Dilute a quantity of Solution, equivalent to 20 g of anhydrous sodium acetate, with water to 150 mL, heat to boiling, and digest in a covered beaker on a steam bath for 1 hour. Filter through a tared filtering crucible, wash thoroughly, and dry at 105°: the weight of the residue does not exceed 1 mg (0.005%).

Chloride ⟨221⟩—A quantity of Solution, equivalent to 1.0 g of anhydrous sodium acetate, shows no more chloride than corresponds to 0.50 mL of 0.020 *N* hydrochloric acid (0.035%).

Sulfate ⟨221⟩—A quantity of Solution, equivalent to 10 g of anhydrous sodium acetate, shows no more sulfate than corresponds to 0.50 mL of 0.020 *N* sulfuric acid (0.005%).

Arsenic, *Method I* ⟨211⟩—Dilute a quantity of Solution, equivalent to 1.0 g of anhydrous sodium acetate, with water to 35 mL: the limit is 3 ppm.

Calcium and magnesium—Dilute a quantity of Solution, equivalent to 1.0 g of anhydrous sodium acetate, to 100 mL with water. To 20 mL of this solution add 2 mL each of 6 *N* ammonium hydroxide, ammonium oxalate TS, and sodium phosphate TS: no turbidity is produced within 5 minutes.

Potassium—To a quantity of Solution, equivalent to 3.0 g of anhydrous sodium acetate, add 0.2 mL of sodium bitartrate TS: no turbidity is produced within 5 minutes.

Heavy metals, *Method I* ⟨231⟩—Dilute a quantity of Solution, equivalent to 2.0 g of anhydrous sodium acetate, with water to 25 mL, and use glacial acetic acid instead of 1 *N* acetic acid for adjustment of the pH: the limit is 0.001%.

Assay—Weigh accurately about 1 g of Sodium Acetate Solution into a 250-mL conical flask, cautiously add (in a fume hood) 2.6 mL of acetic anhydride, mix, and allow to stand for 5 minutes. Add 25 mL of glacial acetic acid and 2 drops of *p*-naphtholbenzein TS, and titrate with 0.1 *N* perchloric acid VS. Perform a blank determination, using 0.5 mL of water, and make any necessary correction. Each mL of 0.1 *N* perchloric acid is equivalent to 8.203 mg of $C_2H_3NaO_2$.

Sodium Alginate—*see* Sodium Alginate NF

Sodium Ascorbate

CH_2OH / H—C—OH / NaO OH

$C_6H_7NaO_6$ 198.11

L-Ascorbic acid, monosodium salt.

Monosodium L-ascorbate [*134-03-2*].

» Sodium Ascorbate contains not less than 99.0 percent and not more than 101.0 percent of $C_6H_7NaO_6$, calculated on the dried basis.

Packaging and storage—Preserve in tight, light-resistant containers.

Reference standard—*USP Sodium Ascorbate Reference Standard.*

Identification—

A: The infrared absorption spectrum of a mineral oil dispersion of it exhibits maxima only at the same wavelengths as that of a similar preparation of USP Sodium Ascorbate RS.

B: Add 1 mL of 0.1 *N* hydrochloric acid to 4 mL of a solution (1 in 50): the solution reduces alkaline cupric tartrate TS slowly at room temperature but more readily upon heating.

C: A solution (1 in 50) responds to the tests for *Sodium* ⟨191⟩.

Specific rotation ⟨781⟩: between +103° and +108°, calculated on the dried basis, determined in a solution in carbon dioxide–free water (1 in 10), the optical rotation being measured immediately following the preparation of the solution.

pH ⟨791⟩: between 7.0 and 8.0, in a solution (1 in 10).

Loss on drying ⟨731⟩—Dry it in a suitable vacuum drying tube, phosphorus pentoxide being used as the desiccant, at 60° for 4 hours: it loses not more than 0.25% of its weight.

Heavy metals, *Method II* ⟨231⟩: 0.002%.

Assay—Dissolve about 400 mg of Sodium Ascorbate, accurately weighed, in a mixture of 100 mL of carbon dioxide–free water and 25 mL of 2 *N* sulfuric acid. Titrate immediately with 0.1 *N* iodine VS, adding 3 mL of starch TS as the end-point is approached. Each mL of 0.1 *N* iodine is equivalent to 9.905 mg of $C_6H_7NaO_6$.

Sodium Benzoate—*see* Sodium Benzoate NF

Sodium Benzoate Injection, Caffeine and—*see* Caffeine and Sodium Benzoate Injection

Sodium Bicarbonate

$NaHCO_3$ 84.01

Carbonic acid monosodium salt.

Monosodium carbonate [*144-55-8*].

» Sodium Bicarbonate contains not less than 99.0 percent and not more than 100.5 percent of $NaHCO_3$, calculated on the dried basis.

Packaging and storage—Preserve in well-closed containers.

Labeling—Where Sodium Bicarbonate is intended for use in hemodialysis, it is so labeled.

Identification—A solution of it responds to the tests for *Sodium* ⟨191⟩, and for *Bicarbonate* ⟨191⟩.

Loss on drying ⟨731⟩—Dry about 4 g, accurately weighed, over silica gel for 4 hours: it loses not more than 0.25% of its weight.

Insoluble substances—Dissolve 1 g in 20 mL of water: the resulting solution is complete and clear.

Carbonate ⟨218⟩ (where it is labeled as intended for use in hemodialysis)—

Apparatus—The apparatus (see illustration) consists of a 50-mL flask with a side arm connected to a source of carbon dioxide humidified by bubbling through a saturated solution of sodium bicarbonate and equipped with a top-mounted stopper fitted with zan exit tube connected via a T-tube to a system vent and a leveling buret and reservoir.

Reagents—

Saturated sodium bicarbonate solution—Mix about 20 g of sodium bicarbonate and 100 mL of water, and allow any undissolved crystals to settle. Use the clear supernatant solution.

Displacement solution—Dissolve 100 g of sodium chloride in 350 mL of water, add about 1 g of sodium bicarbonate and 1 mL of methyl orange TS. After the sodium bicarbonate has dissolved, add 6 *N* sulfuric acid until the solution turns pink. Use this solution to fill the reservoir of the apparatus.

Procedure—Add 25 mL of *Saturated sodium bicarbonate solution* to the 50-mL flask, and flush the system by allowing humidified carbon dioxide to enter through the side arm. Close the carbon dioxide inlet and the system vent, and stir the *Saturated sodium bicarbonate solution* until no further carbon dioxide absorption is noted from successive buret readings. Maintain atmospheric pressure in the apparatus by adjusting the *Displacement solution* to the same level in both the reservoir and the buret, noting the buret reading. Open the system vent, and reintroduce humidified carbon dioxide through the side arm of the flask. Close the carbon dioxide inlet and the system vent, and stir the *Saturated sodium bicarbonate solution* vigorously until no further carbon dioxide absorption is noted. Repeat the carbon dioxide absorption procedure starting with "Open the system vent" until no more than a 0.2-mL change in buret reading is noted. Discontinue stirring, reintroduce humidified carbon dioxide through the side arm of the flask, remove the top-mounted stopper from the flask briefly, and promptly add about 10 g of Sodium Bicarbonate, accurately weighed, to the flask. Replace the stopper, continue the flow of humidified carbon dioxide for about 30 seconds, and then close the carbon dioxide inlet and the system vent. Stir the solution in the flask vigorously until carbon dioxide absorption ceases, noting the volume absorbed from the buret reading. Restore atmospheric pressure in the apparatus by leveling the *Displacement solution* in the reservoir and the buret, and discontinue stirring. Open the system vent, and flush humidified carbon dioxide through the system. Close the carbon dioxide inlet and the system vent, and stir the solution in the flask vigorously until carbon dioxide absorption ceases. Determine the total volume, *V*, in mL, of carbon dioxide absorbed after the addition of the specimen to the flask, and calculate the percentage of carbonate in the portion of specimen tested by the formula:

$$273V(6001P)/[22400(273 + T)(760W)],$$

in which *P* is the ambient atmospheric pressure, in mm of mercury, *T* is the ambient temperature, and *W* is the quantity, in g, of specimen taken. [NOTE—Maintain a constant temperature during the measurement of the volume of carbon dioxide absorbed.] The limit of carbonate is not more than 0.23%.

Normal carbonate—Add 2.0 mL of 0.10 *N* hydrochloric acid and 2 drops of phenolphthalein TS to 1.0 g of Sodium Bicarbonate, previously dissolved without agitation in 20 mL of water at a temperature not exceeding 5°: the solution does not assume more than a faint pink color immediately.

Chloride—Proceed as directed in the test for *Chloride* ⟨221⟩: a 0.35-g portion shows no more chloride than corresponds to 1.48 mL of 0.0010 *N* hydrochloric acid (0.015%).

Sulfate—Proceed as directed in the test for *Sulfate* ⟨221⟩: a 1.0-g portion shows no more sulfate than corresponds to 0.15 mL of 0.020 *N* sulfuric acid (0.015%).

Ammonia—Heat about 1 g in a test tube: no odor of ammonia is evolved.

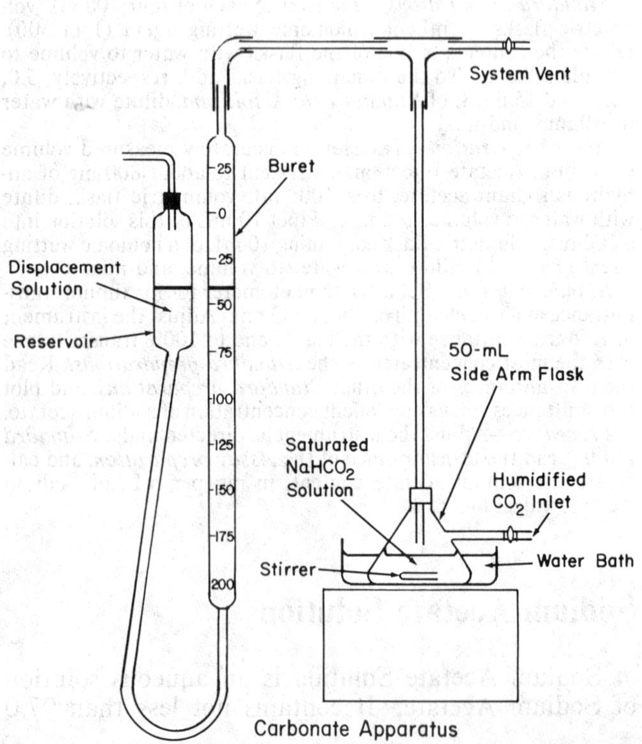

Carbonate Apparatus

Aluminum (where it is labeled as intended for use in hemodialysis)—[NOTE—The *Standard preparations* and the *Test preparation* may be modified, if necessary, to obtain solutions, of suitable concentrations, adaptable to the linear or working range of the instrument.]

Nitric acid diluent—Dilute 40 mL of nitric acid to 1000 mL with water.

Standard preparations—Transfer 2.000 g of aluminum metal to a 1000-mL volumetric flask, add 50 mL of 6 *N* hydrochloric acid, swirl to assure contact of the aluminum and the acid, and allow the reaction to proceed until all of the aluminum has dissolved. Dilute with water to volume, and mix. Transfer 5.0 mL of this solution to a 1000-mL volumetric flask, dilute with water to volume, and mix. Transfer 10.0 mL of this solution to a 100-mL volumetric flask, dilute with *Nitric acid diluent* to volume, and mix. Transfer 1.0-, 2.0-, and 4.0-mL portions of this solution to separate 100-mL volumetric flasks, dilute with *Nitric acid diluent* to volume, and mix. These solutions contain 0.01, 0.02, and 0.04 µg of Al per mL, respectively.

Test preparation—Transfer 1.0 g of Sodium Bicarbonate to a 100-mL plastic volumetric flask, and carefully add 4 mL of nitric acid. Sonicate for 30 minutes, dilute with water to volume, and mix.

Procedure—Determine the absorbances of the *Standard preparations* and the *Test preparation* at the aluminum emission line at 309.3 nm with a suitable atomic absorption spectrophotometer (see *Spectrophotometry and Light-scattering* ⟨851⟩) equipped with an aluminum hollow-cathode lamp and a flameless electrically heated furnace, using *Nitric acid diluent* as the blank. Plot the absorbances of the *Aluminum standard preparations* versus the contents of Al, in µg per mL, drawing the straight line best fitting the three points. From the graph so obtained determine the quantity, in µg, of Al in each mL of the *Test preparation.* Calculate the ppm of Al in the specimen taken by multiplying this value by 100: the limt is 2 ppm.

Arsenic, *Method I* ⟨211⟩—Prepare the *Test Preparation* by dissolving 1.5 g in 20 mL of 7 *N* sulfuric acid, and adding 35 mL of water: the resulting solution meets the requirements of the test, the addition of 20 mL of 7 *N* sulfuric acid specified under *Procedure* being omitted. The limit is 2 ppm.

Calcium and magnesium (where it is labeled as intended for use in hemodialysis)—[NOTE—The *Standard preparations* and the *Test preparation* may be modified, if necessary, to obtain solutions, of suitable concentrations, adaptable to the linear or working range of the instrument.]

Potassium choride solution—Dissolve 10 g of potassium chloride in 1000 mL of 0.36 *N* hydrochloric acid.

Calcium standard preparations—Transfer 249.7 mg of calcium carbonate, previously dried at 300° for 3 hours and cooled in a desiccator for 2 hours, to a 100-mL volumetric flask. Dissolve in 6 mL of 6 *N* hydrochloric acid, add 1 g of potassium chloride, dilute with water to volume, and mix. Transfer 10.0 mL of this solution to a second 100-mL volumetric flask, dilute with *Potassium chloride solution* to volume and mix. This solution contains 100 µg of Ca per mL. Transfer 2.0-, 3.0-, 4.0-, and 5.0-mL portions of this solution to separate 100-mL volumetric flasks (each containing 6 mL of 6 *N* hydrochloric acid), dilute with *Potassium chloride solution* to volume, and mix. These *Calcium standard preparations* contain 2.0, 3.0, 4.0, and 5.0 µg of Ca per mL, respectively.

Magnesium standard preparations—Place 1.000 g of magnesium in a 250-mL beaker containing 20 mL of water and carefully add 20 mL of hydrochloric acid, warming if necessary to complete the reaction. Transfer this solution to a 1000-mL volumetric flask containing 10 g of potassium chloride, dilute with water to volume, and mix. Transfer 10.0 mL of this solution to a 100-mL volumetric flask containing [illegible] g of potassium chloride, dilute with water to volume, and mix. Transfer 10.0 mL of this solution to a second 100-mL volumetric flask, dilute with *Potassium chloride solution* to volume, and mix. This solution contains 10.0 µg of Mg per mL. Transfer 2.0-, 3.0-, 4.0-, and 5.0-mL portions of this solution to separate 100-mL volumetric flasks (each containing 6 mL of 6 *N* hydrochloric acid), dilute with *Potassium chloride solution* to volume and mix. These *Magnesium standard preparations* contain 0.2, 0.3, 0.4, and 0.5 µg of Mg per mL, respectively.

Test preparation—Transfer 3.0 g of Sodium Bicarbonate to a 100-mL volumetric flask, add 6 mL of 6 *N* hydrochloric acid and 1 g of potassium chloride, dilute with water to volume, and mix.

Procedure for calcium—Concomitantly determine the absorbances of the *Calcium standard preparations* and the *Test preparation* at the calcium emission line at 422.7 nm with a suitable atomic absorption spectrophotometer (see *Spectrophotometry and Light-scattering* ⟨851⟩) equipped with a calcium hollow-cathode lamp and a nitrous oxide–acetylene flame, using *Potassium chloride solution* as the blank. Plot the absorbances of the *Calcium standard preparations* versus their contents of calcium, in µg per mL, by drawing a straight line best fitting the four plotted points. From the graph so obtained determine the quantity, in µg, of Ca in each mL of the *Test preparation.* Calculate the percentage of Ca in the specimen taken by dividing this value by 300: the limit is 0.01%.

Procedure for magnesium—Concomitantly determine the absorbances of the *Magnesium standard preparations* and the *Test preparation* at the magnesium emission line at 285.2 nm with a suitable atomic absorption spectrophotometer (see *Spectrophotometry and Light-scattering* ⟨851⟩) equipped with a magnesium hollow-cathode lamp and a reducing air-acetylene flame, using *Potassium chloride solution* as the blank. Plot the absorbances of the *Magnesium standard preparations* versus their contents of magnesium, in µg per mL, by drawing a straight line best fitting the four plotted points. From the graph so obtained determine the quantity, in µg, of Mg in each mL of the *Test preparation.* Calculate the percentage of Mg in the specimen taken by dividing this value by 300: the limit is 0.004%.

Copper (where it is labeled as intended for use in hemodialysis)—[NOTE—The *Standard preparation* and the *Test preparation* may be modified, if necessary, to obtain solutions, of suitable concentrations, adaptable to the linear or working range of the instrument.]

Nitric acid diluent—Dilute 40 mL of nitric acid to 1000 mL with water.

Standard preparation—Transfer 1.000 g of copper to a 1000-mL volumetric flask, dissolve in 20 mL of nitric acid, dilute with 0.2 *N* nitric acid to volume, and mix. Transfer 10.0 mL of this solution to a second 1000-mL volumetric flask, dilute with 0.2 *N* nitric acid to volume, and mix. This solution contains 10.0 µg of copper per mL. Store in a polyethylene bottle.

Test preparation—Transfer 5.0 g of Sodium Bicarbonate to a 100-mL plastic volumetric flask, and carefully add 4 mL of nitric acid. Sonicate for 30 minutes, dilute with water to volume, and mix.

Procedure—To 10.0 mL of the *Test preparation* add 20 µL of *Standard preparation*, and mix. This *Spiked test preparation* contains 0.02 µg of added Cu per mL. Concomitantly determine the absorbances of the *Test preparation* and the *Spiked test preparation* at the copper emission line at 324.7 nm with a suitable atomic absorption spectrophotometer (see *Spectrophotometry and Light-scattering* ⟨851⟩), equipped with a copper hollow-cathode lamp and a flameless electrically heated furnace, using *Nitric acid diluent* as the blank. Plot the absorbances of the *Test preparation* and the *Spiked test preparation* versus their contents of added Cu, in µg per mL, draw a line connecting the two points, and extrapolate the line until it intercepts the concentration axis. From the intercept determine the quantity, in µg, of Cu in each mL of the *Test preparation.* Calculate the ppm of Cu in the specimen tested by multiplying this value by 20: the limit is 1 ppm.

Iron ⟨241⟩ (where it is labeled as intended for use in hemodialysis)—Place 2.0 g of Sodium Bicarbonate in a beaker, and neutralize with hydrochloric acid, noting the volume of acid consumed. Transfer this solution to a 25-mL volumetric flask with the aid of water (*Test Preparation*). Prepare the *Standard Preparation* by transferring 1.0 mL of *Standard iron solution* to a 25-mL volumetric flask and adding the same volume of hydrochloric acid as used to prepare the *Test Preparation.* Prepare a *Blank* by adding the same volume of hydrochloric acid to a third 25-mL volumetric flask. To each of the flasks containing the *Standard Preparation*, the *Test Preparation*, and the *Blank* add 50 mg of ammonium peroxydisulfate crystals and 2 mL of *Ammonium Thiocyanate Solution*, dilute with water to volume, and mix. Concomitantly determine the absorbances of the solutions from the *Standard Preparation* and the *Test Preparation* at the wavelength of maximum absorbance at about 480 nm with a suitable spectrophotometer, using the solution from the *Blank* to set the instrument to zero: the absorbance of the solution from the *Test Preparation* is not greater than that of the solution from the *Standard Preparation* (5 ppm).

Heavy metals ⟨231⟩—Mix 4.0 g with 5 mL of water and 19 mL of 3 *N* hydrochloric acid, heat to boiling, and maintain that temperature for 1 minute. Add 1 drop of phenolphthalein TS, then add sufficient 6 *N* ammonium hydroxide, dropwise, to give the solution a faint pink color. Cool, and dilute with water to 25 mL: the limit is 5 ppm.

Organics (where it is labeled as intended for use in hemodialysis)—

Silver sulfate solution—Dissolve 22 g of silver sulfate in 2000 mL of sulfuric acid.

Indicator solution—Dissolve 1.485 g of 1,10-phenanthrolene and 695 mg of ferrous sulfate in water to make 100 mL of solution.

Standard preparation—Transfer 850.3 mg of potassium biphthalate, previously crushed lightly and dried at 120° for 2 hours, to a 1000-mL volumetric flask, dilute with water to volume, and mix. Transfer 6.0 mL of this solution to a 100-mL volumetric flask, dilute with water to volume, and mix. This solution contains the equivalent of 0.06 mg of organics equivalents per mL. Transfer 40.0 mL of this solution to a 500-mL reflux flask.

Test preparation—Transfer about 20 g of Sodium Bicarbonate, accurately weighed, to a 500-mL reflux flask. Add 20 mL of water, and swirl. Cautiously add 20 mL of sulfuric acid, and swirl. [*Caution—Perform this operation under a hood.*]

Blank—Add 40 mL of water to a 500-mL reflux flask.

Procedure—Concomitantly treat the *Standard preparation*, the *Test preparation*, and the *Blank* as follows: Add 1 g of mercuric sulfate and about 5 glass beads. Cool the flask in an ice bath, and add 5 mL of *Silver sulfate solution*. While gently swirling the flask in the ice bath, add 25.0 mL of 0.025 *N* potassium dichromate VS and, slowly, 70 mL of *Silver sulfate solution*. Fit a cold water condenser on the reflux flask, and reflux for 2 hours. Allow the contents of the flask to cool for 10 minutes, and wash the condenser with 50 mL of water, collecting the washings in the flask. Add water to the flask to obtain a volume of about 350 mL. Add 3 drops of *Indicator solution*, and titrate, at room temperature, with 0.07 *N* ferrous sulfate VS until the solution changes from greenish-blue to reddish-brown. Calculate the amount, in mg, of organics equivalent in the *Standard preparation* by the formula:

$$8N(V_B - V_S),$$

in which *N* is the normality of the ferrous ammonium sulfate VS, and V_B and V_S are the volumes, in mL, of 0.07 *N* ferrous ammonium sulfate VS consumed by the *Blank* and the *Standard preparation*, respectively. In a suitable system, between 2.328 and 2.424 mg is found. Calculate the amount, in mg, of organics equivalent in the portion of Sodium Bicarbonate taken by the formula:

$$8N(V_B - V_U),$$

in which V_U is the volume, in mL, of 0.07 *N* ferrous ammonium sulfate VS consumed by the *Test preparation:* the limit is 0.01%.

Assay—Weigh accurately about 3 g of Sodium Bicarbonate, mix with 100 mL of water, add methyl red TS, and titrate with 1 *N* hydrochloric acid VS. Add the acid slowly, with constant stirring, until the solution becomes faintly pink. Heat the solution to boiling, cool, and continue the titration until the faint pink color no longer fades after boiling. Each mL of 1 *N* hydrochloric acid is equivalent to 84.01 mg of $NaHCO_3$.

Sodium Bicarbonate Injection

» Sodium Bicarbonate Injection is a sterile solution of Sodium Bicarbonate in Water for Injection, the pH of which may be adjusted by means of added Carbon Dioxide. It contains not less than 95.0 percent and not more than 105.0 percent of the labeled amount of $NaHCO_3$.

Note—Do not use the Injection if it contains a precipitate.

Packaging and storage—Preserve in single-dose containers, of Type I glass.

Labeling—The label states the total osmolar concentration in mOsmol per liter. Where the contents are less than 100 mL, or where the label states that the Injection is not for direct injection but is to be diluted before use, the label alternatively may state the total osmolar concentration in mOsmol per mL.

Identification—It responds to the tests for *Sodium* ⟨191⟩ and for *Bicarbonate* ⟨191⟩.

Pyrogen—It meets the requirements of the *Pyrogen Test* ⟨151⟩, the test dose being 5 mL per kg. [NOTE—Dilute, if necessary, the Injection to a concentration of 5% of sodium bicarbonate.]

pH ⟨791⟩: between 7.0 and 8.5.

Particulate matter ⟨788⟩: meets the requirements under *Small-volume Injections*.

Other requirements—It meets the requirements under *Injections* ⟨1⟩.

Assay—Measure accurately a volume of Sodium Bicarbonate Injection, equivalent to about 3 g of sodium bicarbonate, add methyl red TS, and titrate with 1 *N* hydrochloric acid VS. Add the acid slowly, with constant stirring, until the solution becomes faintly pink. Heat the solution to boiling, cool, and continue the titration until the faint pink color no longer fades after boiling. Each mL of 1 *N* hydrochloric acid is equivalent to 84.01 mg of $NaHCO_3$.

Sodium Bicarbonate Oral Powder

» Sodium Bicarbonate Oral Powder contains Sodium Bicarbonate and suitable added substances. It contains not less than 98.5 percent and not more than 100.5 percent of $NaHCO_3$, calculated on the dried basis.

Packaging and storage—Preserve in well-closed containers.

Labeling—Label Oral Powder to indicate that it is for oral use only.

Other requirements—It meets the requirements for *Identification* and *Loss on drying*, under *Sodium Bicarbonate*.

Assay—Proceed with Sodium Bicarbonate Oral Powder as directed in the *Assay* under *Sodium Bicarbonate*.

Sodium Bicarbonate for Oral Suspension, Magnesium Carbonate and—*see* Magnesium Carbonate and Sodium Bicarbonate for Oral Suspension

Sodium Bicarboate Tablets

» Sodium Bicarbonate Tablets contain not less than 95.0 percent and not more than 105.0 percent of the labeled amount of $NaHCO_3$.

Packaging and storage—Preserve in well-closed containers.

Identification—A solution of Tablets responds to the tests for *Sodium* ⟨191⟩ and for *Bicarbonate* ⟨191⟩.

Disintegration ⟨701⟩: 30 minutes, simulated gastric fluid TS being substituted for water in the test.

Uniformity of dosage units ⟨905⟩: meet the requirements.

Assay—Weigh and finely powder not less than 20 Sodium Bicarbonate Tablets. Weigh accurately a portion of the powder, equivalent to about 2 g of sodium bicarbonate, dissolve in 100 mL of water, add methyl red TS, and titrate with 1 *N* hydrochloric acid VS. Add the acid slowly, with constant stirring, until the solution becomes faintly pink. Heat the solution to boiling, cool, and continue the titration until the pink color no longer fades after boiling. Each mL of 1 *N* hydrochloric acid is equivalent to 84.01 mg of $NaHCO_3$.

Sodium Bicarbonates and Citric Acid Effervescent Tablets for Oral Solution, Potassium and—*see* Potassium and Sodium Bicarbonates and Citric Acid Effervescent Tablets for Oral Solution

Sodium Biphosphate—*see* Sodium Phosphate, Monobasic

Sodium Biphosphate Tablets, Methenamine and—*see* Methenamine and Monobasic Sodium Phosphate Tablets

Sodium Borate—*see* Sodium Borate NF

Sodium Carbonate—*see* Sodium Carbonate NF

Sodium Carbonate Irrigation, Citric Acid, Magnesium Oxide, and—*see* Citric Acid, Magnesium Oxide, and Sodium Carbonate Irrigation

Sodium Chloride

NaCl 58.44
Sodium chloride.
Sodium chloride [*7647-14-5*].

» Sodium Chloride contains not less than 99.0 percent and not more than 101.0 percent of NaCl, calculated on the dried basis. It contains no added substance.

Packaging and storage—Preserve in well-closed containers.

Identification—A solution (1 in 20) responds to the tests for *Sodium* ⟨191⟩ and for *Chloride* ⟨191⟩.

Acidity or alkalinity—Dissolve 50.0 g in 200 mL of carbon dioxide–free water, and add 10 drops of bromothymol blue pH indicator. If the solution is yellow, it requires not more than 1.0 mL of 0.020 *N* sodium hydroxide to produce a blue color. If the solution is blue or green, it requires not more than 3.12 mL of 0.020 *N* hydrochloric acid to produce a yellow color.

Loss on drying ⟨731⟩—Dry it at 105° for 2 hours: it loses not more than 0.5% of its weight.

Arsenic, *Method I* ⟨211⟩: 3 ppm.

Barium—Dissolve 4.0 g in 20 mL of water, filter if necessary, and divide the solution into two portions. To one portion add 2 mL of 2 *N* sulfuric acid, and to the other add 2 mL of water: the solutions are equally clear after standing for 2 hours.

Iodide or bromide—Digest 2.0 g of finely powdered Sodium Chloride with 25 mL of warm alcohol for 3 hours, cool the mixture, and remove the undissolved salt by filtration. Evaporate the filtrate to dryness, dissolve the residue in 5 mL of water, add 1 mL of chloroform, and cautiously introduce, dropwise, with constant agitation, 5 drops of dilute chlorine TS (1 in 3): the chloroform does not acquire a violet, yellow, or orange color.

Calcium and magnesium—Dissolve 20 g in 200 mL of water, and add 0.1 mL of hydrochloric acid, 5 mL of ammonia–ammonium chloride buffer TS, and 5 drops of eriochrome black TS. Titrate with 0.005 *M* disodium ethylenediaminetetraacetate VS to a pure blue end-point. Each mL of 0.005 *M* disodium ethylenediaminetetraacetate is equivalent to 0.2004 mg of Ca. Not more than 0.005% of calcium and magnesium (as Ca) is found.

Iron ⟨241⟩—Dissolve 5.0 g in 45 mL of water and 2 mL of hydrochloric acid: the limit is 2 ppm.

Sulfate ⟨221⟩—A 1.0-g portion shows no more sulfate than corresponds to 0.15 mL of 0.020 *N* sulfuric acid (0.015%).

Sodium ferrocyanide—Dissolve 25 g in 80 mL of water in a glass-stoppered, 100-mL graduated cylinder or flask. Add 2 mL of ferrous sulfate TS and 1 mL of 2 *N* sulfuric acid, dilute with water to 100 mL, and mix. For a control, place 80 mL of water in a glass-stoppered, 100-mL graduated cylinder or flask, add 2 mL of ferrous sulfate TS and 1 mL of 2 *N* sulfuric acid, dilute with water to 100 mL, and mix. Transfer 50-mL portions of the respective solutions to matched color-comparison tubes: the test solution shows no more blue color than the control, indicating the absence of sodium ferrocyanide.

Heavy metals, *Method I* ⟨231⟩: 5 ppm.

Assay—Transfer about 250 mg of Sodium Chloride, accurately weighed, to a porcelain casserole, and add 140 mL of water and 1 mL of dichlorofluorescein TS. Mix, and titrate with 0.1 *N* silver nitrate VS until the silver chloride flocculates and the mixture acquires a faint pink color. Each mL of 0.1 *N* silver nitrate is equivalent to 5.844 mg of NaCl.

Sodium Chloride Inhalation—*see* Sodium Chloride Inhalation Solution

Sodium Chloride Injection

» Sodium Chloride Injection is a sterile solution of Sodium Chloride in Water for Injection. It contains no antimicrobial agents. It contains not less than 95.0 percent and not more than 105.0 percent of the labeled amount of NaCl.

Packaging and storage—Preserve in single-dose containers, preferably of Type I or Type II glass.

Labeling—The label states the total osmolar concentration in mOsmol per liter. Where the contents are less than 100 mL, or where the label states that the Injection is not for direct injection but is to be diluted before use, the label alternatively may state the total osmolar concentration in mOsmol per mL.

Identification—It responds to the tests for *Sodium* ⟨191⟩ and for *Chloride* ⟨191⟩.

Pyrogen—It meets the requirements of the *Pyrogen Test* ⟨151⟩. [NOTE—Dilute, with Water for Injection, Injections containing more than 0.9% of sodium chloride to give a concentration of 0.9% of sodium chloride.]

pH ⟨791⟩: between 4.5 and 7.0.

Particulate matter ⟨788⟩: meets the requirements under *Small-volume Injections.*

Iron ⟨241⟩—Dilute 5.0 mL of Injection with water to 45 mL, and add 2 mL of hydrochloric acid: the limit is 2 ppm.

Heavy metals, *Method I* ⟨231⟩—Place a volume of Injection, equivalent to 1.0 g of sodium chloride, in a suitable vessel, if necessary evaporate to a volume of about 20 mL, add 2 mL of 1 *N* acetic acid, then dilute with water to 25 mL. Proceed as directed, except to use 1 mL of *Standard Lead Solution* (10 µg of Pb) in the *Standard Preparation* and in the *Monitor Preparation:* the limit is 0.001%, based on the amount of sodium chloride.

Other requirements—It meets the requirements under *Injections* ⟨1⟩.

Assay—Pipet a volume of Sodium Chloride Injection, equivalent to about 90 mg of sodium chloride, into a porcelain casserole, and add 140 mL of water and 1 mL of dichlorofluorescein TS. Mix, and titrate with 0.1 *N* silver nitrate VS until the silver chloride flocculates and the mixture acquires a faint pink color. Each mL of 0.1 *N* silver nitrate is equivalent to 5.844 mg of NaCl.

Bacteriostatic Sodium Chloride Injection

» Bacteriostatic Sodium Chloride Injection is a sterile, isotonic solution of Sodium Chloride in Water for Injection, and it contains one or more suitable antimicrobial agents. It contains not less than 0.85 percent and not more than 0.95 percent of NaCl.

Note—Use Bacteriostatic Sodium Chloride Injection with due regard for the compatibility of the antimicrobial agent or agents it contains with the particular medicinal substance that is to be dissolved or diluted.

Packaging and storage—Preserve in single-dose or in multiple-dose containers, of not larger than 30-mL size, preferably of Type I or Type II glass.

Labeling—Label it to indicate the name(s) and proportion(s) of the added antimicrobial agent(s). Label it also to include the statement, "NOT FOR USE IN NEWBORNS," in boldface capital letters, on the label immediately under the official name, printed in a contrasting color, preferably red. Alternatively, the statement may be placed prominently elsewhere on the label if the statement is enclosed within a box.

Reference standards—*USP Methylparaben Reference Standard*—Dry over silica gel for 5 hours before using. *USP Propylparaben Reference Standard*—Dry over silica gel for 5 hours before using.

Antimicrobial agent(s)—It meets the requirements under *Antimicrobial Preservatives—Effectiveness* ⟨51⟩, and meets the labeled claim for content of the antimicrobial agent(s) as determined by the method set forth under *Antimicrobial Agents—Content* ⟨341⟩, except to use the following procedure when methylparaben and propylparaben are used as the antimicrobial agents.

Mobile phase—Prepare a mixture of filtered methanol and water (about 70:30), such that capacity factors (k') of about 0.52 and 1.05 are established for methylparaben and propylparaben, respectively, with a minimum separation factor (α) of about 2.0.

Standard preparation—Accurately weigh 30 mg of USP Propylparaben RS into a 25-mL volumetric flask, dilute with methanol to volume (*Solution 1*), and mix. Accurately weigh 30 mg of USP Methylparaben RS into a 25-mL volumetric flask, pipet 2.5 mL of *Solution 1* into the flask, and dilute with methanol to volume (*Solution 2*). Pipet 5 mL of *Solution 2* into a 50-mL volumetric flask, add by pipet 30 mL of methanol, dilute with water to volume, and mix.

Test preparation—Pipet 1 mL of Injection into a 10-mL volumetric flask, add by pipet 7 mL of methanol, dilute with water to volume, and mix.

Procedure—Introduce equal volumes (about 12 µL) of the *Test preparation* and the *Standard preparation* into a high-performance liquid chromatograph (see *Chromatography* ⟨621⟩), by means of a suitable microsyringe or sampling valve, adjusting the specimen size and other operating parameters such that the peak obtained with the *Standard preparation* is about 0.7 full scale. Typically, the apparatus is fitted with a 4-mm × 30-cm column containing packing L1, and is equipped with an ultraviolet detector capable of monitoring absorbance at 254 nm and a suitable recorder. Measure the height of the peaks, at identical retention times, obtained with the *Test preparation* and the *Standard preparation*, and calculate the concentration of methylparaben or propylparaben, in mg per mL, by the formula:

$$C(H_U/H_S),$$

in which C is the concentration, in mg per mL, of USP Methylparaben RS or USP Propylparaben RS in the *Standard preparation*, and H_U and H_S are the peak heights obtained from the *Test preparation* and the *Standard preparation*, respectively.

Pyrogen—It meets the requirements of the *Pyrogen Test* ⟨151⟩, the test dose being 5 mL per kg, injected very slowly.

Particulate matter ⟨788⟩: meets the requirements under *Small-volume Injections.*

Other requirements—It responds to the *Identification test* and meets the requirements for *pH*, *Iron*, *Heavy metals*, and *Assay* under *Sodium Chloride Injection.* It meets also the requirements under *Injections* ⟨1⟩.

Sodium Chloride Injection, Dextrose and—*see* Dextrose and Sodium Chloride Injection

Sodium Chloride Injection, Fructose and—*see* Fructose and Sodium Chloride Injection

Sodium Chloride Injection, Inulin and—*see* Inulin and Sodium Chloride Injection

Sodium Chloride Injection, Mannitol and—*see* Mannitol and Sodium Chloride Injection

Sodium Chloride Injection, Potassium Chloride in—*see* Potassium Chloride in Sodium Chloride Injection

Sodium Chloride Injection, Potassium Chloride in Dextrose and—*see* Potassium Chloride in Dextrose and Sodium Chloride Injection

Sodium Chloride Irrigation

» Sodium Chloride Irrigation is Sodium Chloride Injection that has been suitably packaged, and it contains no antimicrobial agents. It contains not less than 95.0 percent and not more than 105.0 percent of the labeled amount of NaCl.

Packaging and storage—Preserve in single-dose containers, preferably of Type I or Type II glass. The container may be designed to empty rapidly and may contain a volume of more than 1 liter.

Labeling—The designation "not for injection" appears prominently on the label.

Identification—It responds to the tests for *Sodium* ⟨191⟩ and for *Chloride* ⟨191⟩.

Sterility—It meets the requirements under *Sterility Tests* ⟨71⟩.

Other requirements—It meets the requirements for *Pyrogen, pH, Iron, Heavy metals*, and *Assay* under *Sodium Chloride Injection.*

Sodium Chloride Ophthalmic Ointment

» Sodium Chloride Ophthalmic Ointment is Sodium Chloride in a suitable ophthalmic ointment base. It contains not less than 90.0 percent and not more than 110.0 percent of the labeled amount of NaCl. It is sterile.

Packaging and storage—Preserve in collapsible ophthalmic ointment tubes.

Identification—Transfer a quantity of Ophthalmic Ointment, equivalent to about 200 mg of sodium chloride, to a separator containing about 25 mL of ether, and extract with 5 mL of water: the aqueous extract so obtained responds to the tests for *Sodium* ⟨191⟩, and for *Chloride* ⟨191⟩.

Sterility—It meets the requirements for *Ophthalmic Ointments* under *Sterility Tests* ⟨71⟩.

Minimum fill ⟨755⟩: meets the requirements.

Particulate matter—It meets the requirements of the test for *Metal Particles in Ophthalmic Ointments* ⟨751⟩.

Assay—Transfer an accurately weighed quantity of Sodium Chloride Ophthalmic Ointment, equivalent to about 100 mg of sodium chloride, to a separator containing about 50 mL of ether, and extract with four 20-mL portions of water. Combine the aqueous extracts in a porcelain casserole, and add 140 mL of water and 1 mL of dichlorofluorescein TS. Mix, and titrate with

0.1 *N* silver nitrate VS until the silver chloride flocculates and the mixture acquires a faint pink color. Each mL of 0.1 *N* silver nitrate is equivalent to 5.844 mg of NaCl.

Sodium Chloride Inhalation Solution

» Sodium Chloride Inhalation Solution is a sterile solution of Sodium Chloride in water purified by distillation or by reverse osmosis and rendered sterile. It contains not less than 90.0 percent and not more than 110.0 percent of the labeled amount of NaCl. It contains no antimicrobial agents or other added substances.

Packaging and storage—Preserve in single-dose containers.

Identification—It responds to the test for *Sodium* ⟨191⟩ and for *Chloride* ⟨191⟩.

Sterility—It meets the requirements under *Sterility Tests* ⟨71⟩.

pH ⟨791⟩: between 4.5 and 7.0.

Assay—Pipet a volume of Sodium Chloride Inhalation Solution, equivalent to about 90 mg of sodium chloride, into a porcelain casserole, and add 140 mL of water and 1 mL of dichlorofluorescein TS. Mix, and titrate with 0.1 *N* silver nitrate VS until the silver chloride flocculates and the mixture acquires a faint pink color. Each mL of 0.1 *N* silver nitrate is equivalent to 5.844 mg of NaCl.

Sodium Chloride Ophthalmic Solution

» Sodium Chloride Ophthalmic Solution is a sterile solution of Sodium Chloride. It contains not less than 90.0 percent and not more than 110.0 percent of the labeled amount of sodium chloride. It may contain suitable antimicrobial and stabilizing agents. It contains a buffer.

Packaging and storage—Preserve in tight containers.

Identification—Heat a portion of Ophthalmic Solution to boiling, and filter while hot. After cooling, the filtrate responds to the tests for *Sodium* ⟨191⟩ and for *Chloride* ⟨191⟩.

Sterility—It meets the requirements under *Sterility Tests* ⟨71⟩.

pH ⟨791⟩: between 6.0 and 8.0.

Assay—Transfer an accurately measured volume of Sodium Chloride Ophthalmic Solution, equivalent to about 90 mg of sodium chloride, to a porcelain casserole, and add 140 mL of water and 1 mL of dichlorofluorescein TS. Mix, and titrate with 0.1 *N* silver nitrate VS until the silver chloride flocculates and the mixture acquires a faint pink color. Each mL of 0.1 *N* silver nitrate is equivalent to 5.844 mg of NaCl.

Sodium Chloride Tablets

» Sodium Chloride Tablets contain not less than 95.0 percent and not more than 105.0 percent of the labeled amount of NaCl.

Packaging and storage—Preserve in well-closed containers.

Identification—A filtered extract of Tablets responds to the tests for *Sodium* ⟨191⟩ and for *Chloride* ⟨191⟩.

Disintegration ⟨701⟩: 30 minutes.

Uniformity of dosage units ⟨905⟩: meet the requirements.

Iodide or bromide—Digest 2.0 g of powdered Tablets with 25 mL of warm alcohol for 3 hours, cool, and filter. Evaporate the filtrate to dryness, dissolve the residue in 5 mL of water, filter if necessary, and add 1 mL of chloroform. Cautiously introduce, dropwise, with constant agitation, 5 drops of dilute chlorine TS (1 in 3): the chloroform does not acquire a violet, yellow, or orange color.

Barium—Digest 4.0 g of powdered Tablets with 20 mL of water, filter, and divide the solution into two equal portions. To one portion add 2 mL of 2 *N* sulfuric acid and to the other add 2 mL of water: the solutions are equally clear after standing for 2 hours.

Calcium and magnesium—Digest 1 g of powdered Tablets with 50 mL of water, and filter. Add 4 mL of 6 *N* ammonium hydroxide to the filtrate, and divide the mixture into two equal portions. Treat one portion with 1 mL of ammonium oxalate TS and the other portion with 1 mL of dibasic sodium phosphate TS: neither mixture becomes turbid within 5 minutes.

Assay—Dissolve a counted number of not less than 20 Sodium Chloride Tablets in about 100 mL of water, filter into a 500-mL volumetric flask, and wash the original container and the filter with 100 mL of water in divided portions, adding the washings to the original filtrate. Dilute with water to volume. Pipet a volume of the solution, equivalent to about 250 mg of sodium chloride, into a glass-stoppered flask, and dilute with water to about 50 mL. Add 50.0 mL of 0.1 *N* silver nitrate VS, 3 mL of nitric acid, 5 mL of nitrobenzene, and 2 mL of ferric ammonium sulfate TS. Shake well, and titrate the excess silver nitrate with 0.1 *N* ammonium thiocyanate VS. Each mL of 0.1 *N* silver nitrate is equivalent to 5.844 mg of NaCl.

Sodium Chloride Tablets for Solution

» Sodium Chloride Tablets for Solution are composed of Sodium Chloride in compressed form, containing no added substance. Sodium Chloride Tablets for Solution contain not less than 95.0 percent and not more than 105.0 percent of the labeled amount of NaCl.

Other requirements—The Tablets respond to the *Identification test* and meet the requirements for *Packaging and storage*, *Iodide or bromide*, *Barium*, *Calcium and magnesium*, *Disintegration*, *Uniformity of dosage units*, and *Assay* under *Sodium Chloride Tablets*.

Sodium Chloride and Dextrose Tablets

» Sodium Chloride and Dextrose Tablets contain not less than 92.5 percent and not more than 107.5 percent of the labeled amount of sodium chloride (NaCl) and of dextrose ($C_6H_{12}O_6 \cdot H_2O$).

Packaging and storage—Preserve in well-closed containers.

Identification—

A: A filtered solution of Tablets responds to the tests for *Sodium* ⟨191⟩ and for *Chloride* ⟨191⟩.

B: Add a few drops of the filtered solution of Tablets to 5 mL of hot alkaline cupric tartrate TS: a copious red precipitate of cuprous oxide is formed.

Disintegration ⟨701⟩: 30 minutes.

Uniformity of dosage units ⟨905⟩: meet the requirements.

Assay for sodium chloride—Transfer 20.0 mL of the solution prepared for the *Assay for dextrose* to a 100-mL volumetric flask, dilute with water to volume, mix, and proceed as directed in the *Assay* under *Sodium Chloride Tablets*, beginning with "Pipet a volume of the solution."

Assay for dextrose—Dissolve not less than 10 Sodium Chloride and Dextrose Tablets, containing from 2 to 5 g of dextrose, in about 75 mL of water in a 100-mL volumetric flask, add several drops of 6 *N* ammonium hydroxide, dilute with water to volume, and mix. After 30 minutes, filter through a dry filter, and determine the angular rotation in a 200-mm tube at 25°, retaining the excess of the solution for the *Assay for sodium chloride.* The observed rotation in degrees, multiplied by 1.0425, represents the weight, in g, of $C_6H_{12}O_6 . H_2O$ in the specimen taken.

Sodium Chromate Cr 51 Injection—*see* Chromate Cr 51 Injection, Sodium

Sodium Citrate

$CH_2(COONa)C(OH)(COONa)CH_2COONa$

$C_6H_5Na_3O_7$ (anhydrous) 258.07
1,2,3-Propanetricarboxylic acid, 2-hydroxy-, trisodium salt.
Trisodium citrate (anhydrous) [*68-04-2*].
Trisodium citrate dihydrate 294.10 [*6132-04-3*].

» Sodium Citrate is anhydrous or contains two molecules of water of hydration. It contains not less than 99.0 percent and not more than 100.5 percent of $C_6H_5Na_3O_7$, calculated on the anhydrous basis.

Packaging and storage—Preserve in tight containers.

Labeling—Label it to indicate whether it is anhydrous or hydrous.

Identification—

A: A solution (1 in 20) responds to the tests for *Sodium* ⟨191⟩ and for *Citrate* ⟨191⟩.

B: Upon ignition, it yields an alkaline residue which effervesces when treated with 3 *N* hydrochloric acid.

Alkalinity—A solution of 1.0 g in 20 mL of water is alkaline to litmus paper, but after the addition of 0.20 mL of 0.10 *N* sulfuric acid no pink color is produced by 1 drop of phenolphthalein TS.

Water, *Method III* ⟨921⟩—Dry it at 180° for 18 hours: the anhydrous form loses not more than 1.0%, and the hydrous form between 10.0% and 13.0%, of its weight.

Tartrate—To a solution of 1 g in 2 mL of water add 1 mL of potassium acetate TS and 1 mL of 6 *N* acetic acid. Rub the wall of the tube with a glass rod: no crystalline precipitate is formed.

Heavy metals ⟨231⟩—Dissolve 2.0 g in 25 mL of water, and proceed as directed for *Test Preparation,* except to use glacial acetic acid to adjust the pH: the limit is 0.001%.

Assay—Transfer about 350 mg of Sodium Citrate, previously dried at 180° for 18 hours and accurately weighed, to a 250-mL beaker. Add 100 mL of glacial acetic acid, stir until completely dissolved, and titrate with 0.1 *N* perchloric acid VS, determining the end-point potentiometrically. Perform a blank determination, and make any necessary correction. Each mL of 0.1 *N* perchloric acid is equivalent to 8.602 mg of $C_6H_5Na_3O_7$.

Sodium Citrate and Citric Acid Oral Solution

» Sodium Citrate and Citric Acid Oral Solution is a solution of Sodium Citrate and Citric Acid in a suitable aqueous medium. It contains, in each 100 mL, not less than 2.23 g and not more than 2.46 g of sodium (Na), and not less than 6.11 g and not more than 6.75 g of citrate ($C_6H_5O_7$), equivalent to not less than 9.5 g and not more than 10.5 g of sodium citrate dihydrate ($C_6H_5Na_3O_7 . 2H_2O$); and not less than 6.34 g and not more than 7.02 g of citric acid monohydrate ($C_6H_8O_7 . H_2O$).

Packaging and storage—Preserve in tight containers.

Identification—

A: It responds to the flame test for *Sodium* ⟨191⟩.

B: To a mixture of 1 mL of Oral Solution with 1 mL of hydrochloric acid add 10 mL of cobalt uranylacetate TS, and stir well with a glass rod: a pale yellow, fine crystalline precipitate is formed within several minutes (*presence of sodium*).

C: To 2 mL of a dilution of Oral Solution (1 in 20) add 5 mL of sodium cobaltinitrite TS: a yellow precipitate is not formed immediately (*absence of potassium*).

D: It responds to the tests for *Citrate* ⟨191⟩, 3 to 5 drops of Oral Solution and 20 mL of the mixture of pyridine and acetic anhydride being used.

pH ⟨791⟩: between 4.0 and 4.4.

Assay for sodium—

Potassium stock solution, Sodium stock solution, Lithium diluent solution, and *Standard preparation*—Prepare as directed in the *Assay for sodium and potassium* under *Tricitrates Oral Solution.*

Assay preparation—Transfer an accurately measured volume of Sodium Citrate and Citric Acid Oral Solution, equivalent to about 1 g of sodium citrate dihydrate, to a 100-mL volumetric flask, dilute with water to volume, and mix. Transfer 50 µL of this solution to a 10-mL volumetric flask, dilute with *Lithium diluent solution* to volume, and mix.

Procedure—Using a suitable flame photometer, adjusted to read zero with *Lithium diluent solution,* concomitantly determine the sodium flame emission readings for the *Standard preparation* and the *Assay preparation* at the wavelength of maximum emission at about 589 nm. Calculate the quantity, in g, of Na in each mL of Oral Solution taken by the formula:

$$(14.61/25V)(22.99/58.44)(R_{U,Na}/R_{S,Na}),$$

in which 14.61 is the weight, in g, of sodium chloride in the *Sodium stock solution, V* is the volume, in mL, of Oral Solution taken, 22.99 is the atomic weight of sodium, 58.44 is the molecular weight of sodium chloride, and $R_{U,Na}$ and $R_{S,Na}$ are the sodium emission readings obtained for the *Assay preparation* and the *Standard preparation,* respectively.

Assay for sodium citrate—

Cation-exchange column—Mix 10 g of styrene-divinylbenzene cation-exchange resin with 50 mL of water in a suitable beaker. Allow the resin to settle, and decant the supernatant liquid until a slurry of resin remains. Pour the slurry into a 15-mm × 30-cm glass chromatographic tube (having a sealed-in, coarse-porosity fritted disk and fitted with a stopcock), and allow to settle as a homogeneous bed. Wash the resin bed with about 100 mL of water, closing the stopcock when the water level is about 2 mm above the resin bed.

Procedure—Transfer an accurately measured volume of Sodium Citrate and Citric Acid Oral Solution, equivalent to about 1 g of sodium citrate dihydrate, to a 100-mL volumetric flask, dilute with water to volume, and mix. Pipet 5 mL of this solution carefully onto the top of the resin bed in the *Cation-exchange column.* Place a 250-mL conical flask below the column, open the stopcock, and allow to flow until the solution has entered the resin bed. Elute the column with 60 mL of water at a flow rate of about 5 mL per minute, collecting about 65 mL of the eluate. Add 5 drops of phenolphthalein TS to the eluate, swirl the flask, and titrate with 0.02 *N* sodium hydroxide VS. Record the buret reading, and calculate the volume (*B*) of 0.02 *N* sodium hydroxide consumed. Calculate the quantity, in mg, of $C_6H_5Na_3O_7 . 2H_2O$ in each mL of the Oral Solution taken by the formula:

$$[1.961B(20/V)] - [(294.10/210.14)C],$$

in which 1.961 is the equivalent, in mg, of $C_6H_5Na_3O_7 . 2H_2O$, of each mL of 0.02 *N* sodium hydroxide, *V* is the volume, in mL, of Oral Solution taken, 294.10 and 210.14 are the molecular weights of sodium citrate dihydrate and citric acid monohydrate, respectively, and *C* is the concentration, in mg per mL, of citric acid monohydrate in the Oral Suspension, as obtained in the *Assay for citric acid.*

Assay for citric acid—Transfer an accurately measured volume of Sodium Citrate and Citric Acid Oral Solution, equivalent to about 0.67 g of citric acid monohydrate, to a 100-mL volumetric flask, dilute with water to volume, and mix. Pipet 5 mL of this solution into a suitable flask, add 25 mL of water and 5 drops of phenolphthalein TS, and titrate with 0.02 *N* sodium hydroxide VS to a pink end-point. Record the buret reading, and calculate the volume (*A*) of 0.02 *N* sodium hydroxide consumed. Calculate the quantity, in mg, of $C_6H_8O_7 . H_2O$ in each mL of the Oral Solution taken by the formula:

$$1.401A(20/V),$$

in which 1.401 is the equivalent, in mg, of $C_6H_8O_7 . H_2O$, of each mL of 0.02 *N* sodium hydroxide, and *V* is the volume, in mL, of Oral Solution taken.

Sodium Citrate Solution, Anticoagulant—*see* Anticoagulant Sodium Citrate Solution

Sodium Dehydroacetate—*see* Sodium Dehydroacetate NF

Sodium Fluoride

NaF 41.99
Sodium fluoride.
Sodium fluoride [*7681-49-4*].

» Sodium Fluoride contains not less than 98.0 percent and not more than 102.0 percent of NaF, calculated on the dried basis.

Packaging and storage—Preserve in well-closed containers.

Identification—

A: Place 1 g in a platinum crucible *in a well-ventilated hood*, add 15 mL of sulfuric acid, and cover the crucible with a piece of clear, polished glass. Heat the crucible on a steam bath for 1 hour, remove the glass cover, rinse it in water, and wipe dry: the surface of the glass is etched.

B: A solution (1 in 25) responds to the tests for *Sodium* ⟨191⟩.

Acidity or alkalinity—Dissolve 2.0 g in 40 mL of water in a platinum dish, add 10 mL of a saturated solution of potassium nitrate, cool the solution to 0°, and add 3 drops of phenolphthalein TS. If no color appears, a pink color persisting for 15 seconds is produced by not more than 2.0 mL of 0.10 *N* sodium hydroxide. If the solution is colored pink by the addition of phenolphthalein TS, it is rendered colorless by not more than 0.50 mL of 0.10 *N* sulfuric acid. Save the neutralized solution for the test for *Fluosilicate*.

Loss on drying ⟨731⟩—Dry it at 150° for 4 hours: it loses not more than 1.0% of its weight.

Fluosilicate—After the solution from the test for *Acidity or alkalinity* has been neutralized, heat to boiling, and titrate while hot with 0.10 *N* sodium hydroxide until a permanent pink color is obtained: not more than 1.5 mL of 0.10 *N* sodium hydroxide is required.

Chloride—Dissolve 300 mg in 20 mL of water, and add 200 mg of boric acid, 1 mL of nitric acid, and 1 mL of 0.1 *N* silver nitrate: any turbidity produced is not greater than that of a blank to which has been added 1.0 mL of 0.0010 *N* hydrochloric acid (0.012%).

Heavy metals ⟨231⟩—To 1 g, in a platinum dish or crucible, add 1 mL of water and 3 mL of sulfuric acid, and heat under a hood at as low a temperature as practicable until all of the sulfuric acid has been expelled. Dissolve the residue in 20 mL of water, neutralize the solution to phenolphthalein TS with ammonium hydroxide, add 1 mL of glacial acetic acid, dilute with water to 45 mL, filter, and use 30 mL of the filtrate for the test: the limit is 0.003%.

Assay—Prepare a fresh solution of ferric chloride by dissolving 10 g of ferric chloride in water, diluting the solution with water to 1000 mL, and mixing. Standardize the solution as follows: Pipet 40 mL of the solution into a glass-stoppered, 250-mL conical flask, add 5 g of potassium iodide and 5 mL of hydrochloric acid, and allow to stand stoppered for 15 minutes. Titrate the liberated iodine with 0.1 *N* sodium thiosulfate VS, adding 3 mL of starch TS as the end-point is approached. Calculate the normality of the solution from the volume of 0.1 *N* sodium thiosulfate VS consumed. Dissolve about 350 mg of Sodium Fluoride, accurately weighed, in 25 mL of water in a glass-stoppered, 125-mL conical flask, warm to about 70°, neutralize to phenolphthalein TS, with 0.1 *N* alkali or acid solution, then cool to room temperature, and add 20 g of sodium chloride and 5 mL of potassium thiocyanate solution (1 in 5). Titrate with the ferric chloride solution until the solution becomes faintly yellow, and then add 15 mL of alcohol and 15 mL of ether. Insert the stopper in the flask, and shake the mixture: provided the end-point has not been passed, the alcohol-ether layer does not become red. Cautiously continue the titration, shaking the mixture after each small addition of ferric chloride solution, until the alcohol-ether mixture acquires a slight, permanent, red color. Calculate the quantity, in mg, of NaF in the portion of Sodium Fluoride taken by the formula:

$$252FV,$$

in which *F* is the normality of the ferric chloride solution and *V* is the volume, in mL, of it consumed.

Sodium Fluoride Oral Solution

» Sodium Fluoride Oral Solution contains not less than 90.0 percent and not more than 110.0 percent of the labeled amount of NaF.

Packaging and storage—Preserve in tight containers, plastic containers being used for Oral Solution having a pH below 7.5.

Labeling—Label Oral Solution in terms of the content of sodium fluoride (NaF) and in terms of the content of fluoride ion.

Reference standard—*USP Sodium Fluoride Reference Standard*—Dry at 150° for 4 hours before using.

Identification—

A: Transfer 0.1 mL of Oral Solution to a small test tube, and add 0.1 mL of a freshly prepared mixture (1:1) of sodium alizarinsulfonate solution (1 in 1000) and zirconyl nitrate solution (1 in 1000) in 7 *N* hydrochloric acid: a yellow color is produced.

B: It responds to the tests for *Sodium* ⟨191⟩.

Assay—[NOTE—Store all solutions, except *Buffer solution*, in plastic containers.]

Buffer solution—Dissolve 57 mL of glacial acetic acid, 58 g of sodium chloride, and 4 g of (1,2-cyclohexylenedinitrilo)-tetraacetic acid in 500 mL of water. Adjust with 5 *N* sodium hydroxide to a pH of 5.25 ± 0.25, dilute with water to 1000 mL, and mix.

Standard preparations—Dissolve an accurately weighed quantity of USP Sodium Fluoride RS quantitatively in water to obtain a solution containing 420 µg per mL. Each mL of this solution (*Standard preparation A*) contains 190 µg of fluoride ion (10^{-2} *M*). Transfer 25.0 mL of *Standard preparation A* to a 250-mL volumetric flask, dilute with water to volume, and mix. This solution (*Standard preparation B*) contains 19 µg of fluoride ion per mL (10^{-3} *M*). Transfer 25.0 mL of *Standard preparation B* to a 250-mL volumetric flask, dilute with water to volume, and mix. This solution (*Standard preparation C*) contains 1.9 µg of fluoride ion per mL (10^{-4} *M*).

Assay preparation—Transfer an accurately measured volume of Sodium Fluoride Oral Solution, equivalent to about 10 mg of fluoride, to a 500-mL volumetric flask, dilute with water to volume, and mix.

Procedure—Pipet 20 mL of each *Standard preparation* and of the *Assay preparation* into separate plastic beakers each containing a plastic-coated stirring bar. Pipet 20 mL of *Buffer solution* into each beaker. Concomitantly measure the potentials (see *pH* ⟨791⟩), in mV, of the solutions from the *Standard preparations* and of the solution from the *Assay preparation*, with a pH meter capable of a minimum reproducibility of ±0.2 mV and equipped with a fluoride-specific ion-indicating electrode and a calomel reference electrode. [NOTE—When taking measurements, immerse the electrodes in the solution, stir on a magnetic stirrer having an insulated top until equilibrium is attained (1 to

2 minutes), and record the potential. Rinse and dry the electrodes between measurements, taking care to avoid damaging the crystal of the specific-ion electrode.] Plot the logarithms of the fluoride-ion concentrations, in µg per mL, of the *Standard preparations* versus potential, in mV. From the measured potential of the *Assay preparation* and the standard response line, determine the concentration, *C*, in µg per mL, of fluoride ion in the *Assay preparation.* Calculate the quantity, in mg, of fluoride ion in each mL of the Oral Solution taken by the formula:

$$0.5(C/V),$$

in which *C* is the determined concentration of fluoride, in µg per mL, in the *Assay preparation*, and *V* is the volume, in mL, of Oral Solution taken. Multiply the quantity of fluoride ion by 2.21 to obtain the quantity of NaF.

Sodium Fluoride Tablets

» Sodium Fluoride Tablets contain not less than 90.0 percent and not more than 110.0 percent of the labeled amount of NaF.

Packaging and storage—Preserve in tight containers.

Labeling—Label Tablets in terms of the content of sodium fluoride (NaF) and in terms of the content of fluoride ion. Tablets that are to be chewed may be labeled as Sodium Fluoride Chewable Tablets.

Reference standard—*USP Sodium Fluoride Reference Standard*—Dry at 150° for 4 hours before using.

Identification—

A: Disperse 20 finely powdered Tablets in 25 mL of water, shake, and filter: a portion of the filtrate responds to the tests for *Sodium* ⟨191⟩.

B: Evaporate a 10-mL portion of the filtrate obtained in *Identification test A* to dryness. To the residue add a mixture of 0.1 mL of freshly prepared sodium alizarinsulfonate solution (1 in 1000) and 0.1 mL of a 1 in 1000 solution of zirconyl nitrate in 7 *N* hydrochloric acid: a yellow color is produced.

Disintegration ⟨701⟩: 15 minutes.

Uniformity of dosage units ⟨905⟩: meet the requirements.

Assay—[NOTE—Store all solutions, except *Buffer solution*, in plastic containers.]

Buffer solution and *Standard preparations*—Prepare as directed in the *Assay* under *Sodium Fluoride Oral Solution.*

Assay preparation—Weigh and finely powder not less than 20 Sodium Fluoride Tablets. Transfer an accurately weighed portion of the powder, equivalent to about 10 mg of fluoride, to a plastic 500-mL conical flask containing 400 mL of water. Heat on a steam bath for 25 minutes with occasional shaking, cool to room temperature, transfer to a 500-mL volumetric flask, dilute with water to volume, and mix.

Procedure—Proceed as directed for *Procedure* in the *Assay* under *Sodium Fluoride Oral Solution.* Calculate the quantity, in mg, of fluoride ion in the portion of Tablets taken by the formula:

$$0.5C,$$

in which *C* is the determined concentration, in µg per mL, of fluoride ion in the *Assay preparation.* Multiply the quantity of fluoride ion by 2.21 to obtain the quantity of NaF.

Sodium Fluoride and Phosphoric Acid Gel

» Sodium Fluoride and Phosphoric Acid Gel contains not less than 90.0 percent and not more than 110.0 percent of the labeled amount of fluoride ion, in an aqueous medium containing a suitable viscosity-inducing agent.

Packaging and storage—Preserve in tight, plastic containers.

Labeling—Label Gel in terms of the content of sodium fluoride (NaF) and in terms of the content of fluoride ion.

Reference standard—*USP Sodium Fluoride Reference Standard*—Dry at 150° for 4 hours before using.

Identification—

A: Place a quantity of Gel, equivalent to about 500 mg of fluoride ion, in a platinum crucible in a well-ventilated hood, and add 15 mL of sulfuric acid. Cover the crucible with a piece of clear, polished glass, and heat on a steam bath for 1 hour. Remove the glass cover, rinse it in water, and dry: the glass surface exposed to vapors from the crucible is etched.

B: It responds to the tests for *Phosphate* ⟨191⟩.

Viscosity ⟨911⟩—Place a quantity of Gel in a suitable plastic container, insert the stopper securely, and allow to stand until the gel is free from air bubbles. Place it in a water bath maintained at a temperature of 25 ± 0.5° until it adjusts to the temperature of the water bath (30 minutes or longer). Do not stir the gel while it is in the bath. Remove the sample from the bath, stir the gel gently for 5 seconds, and without delay, using a rotational viscosimeter, determine the viscosity using the appropriate spindle to obtain a scale reading between 10% and 90% of full scale at a speed of 60 rpm or of 30 rpm. Calculate the viscosity, in centipoises, by multiplying the scale reading by the constant for the spindle and speed used: the viscosity is between 7000 and 20,000 centipoises.

pH ⟨791⟩—Place about 40 mL in a plastic beaker, add about 250 mg of quinhydrone, and stir for 1 minute, leaving some of the quinhydrone undissolved. Determine the pH using a hydrofluoric acid frit-junction calomel reference electrode and a gold, hydrofluoric acid–resistant, metallic electrode: the pH is between 3.0 and 4.0.

Assay—[NOTE—Store all solutions, except *Buffer solution*, in plastic containers.]

Buffer solution and *Standard preparations*—Prepare as directed in the *Assay* under *Sodium Fluoride Oral Solution.*

Assay preparation—Transfer a quantity of Sodium Fluoride and Phosphoric Acid Gel, equivalent to about 20 mg of fluoride ion, accurately weighed, to a 1000-mL volumetric flask, add water to dissolve, dilute with water to volume, and mix.

Procedure—Proceed as directed for *Procedure* in the *Assay* under *Sodium Fluoride Oral Solution.* The quantity, in mg, of fluoride ion in the portion of Gel taken is equivalent to *C*, the determined concentration of fluoride ion, in µg per mL, in the *Assay preparation.*

Sodium Fluoride and Phosphoric Acid Topical Solution

» Sodium Fluoride and Phosphoric Acid Topical Solution contains not less than 90.0 percent and not more than 110.0 percent of the labeled amount of fluoride ion.

Packaging and storage—Preserve in tight, plastic containers.

Labeling—Label Topical Solution in terms of the content of sodium fluoride (NaF) and in terms of the content of fluoride ion.

Reference standard—*USP Sodium Fluoride Reference Standard*—Keep the container tightly closed. Dry at 150° for 4 hours before using.

Other requirements—It responds to the *Identification tests* and meets the requirements of the test for *pH* under *Sodium Fluoride and Phosphoric Acid Gel.*

Assay—[NOTE—Store all solutions, except *Buffer solution*, in plastic containers.]

Buffer solution, and *Standard preparations*—Prepare as directed in the *Assay* under *Sodium Fluoride Oral Solution.*

Assay preparation—Transfer an accurately measured volume of Sodium Fluoride and Phosphoric Acid Topical Solution, equivalent to about 20 mg of fluoride ion, to a 1000-mL volumetric flask, add water to dissolve, dilute with water to volume, and mix.

Procedure—Proceed as directed for *Procedure* in the *Assay* under *Sodium Fluoride Oral Solution.* Calculate the quantity, in mg, of fluoride ion in each mL of the Topical Solution taken by the formula:

$$C/V,$$

in which C is the determined concentration of fluoride ion, in µg per mL, in the *Assay preparation*, and V is the volume, in mL, of Topical Solution taken.

Sodium Formaldehyde Sulfoxylate—*see* Sodium Formaldehyde Sulfoxylate NF

Sodium Gluceptate Injection, Technetium Tc 99m—*see* Technetium Tc 99m Gluceptate Injection

Sodium Gluconate

```
         H  H  OH H
         |  |  |  |
HOCH2—C—C—C—C—COONa
         |  |  |  |
         OH OH H  OH
```

$C_6H_{11}NaO_7$ 218.14
D-Gluconic acid, monosodium salt.
Monosodium D-gluconate [*527-07-1*].

» Sodium Gluconate contains not less than 98.0 percent and not more than 102.0 percent of $C_6H_{11}NaO_7$.

Packaging and storage—Preserve in well-closed containers.

Reference standard—*USP Potassium Gluconate Reference Standard*—Dry in vacuum at 105° for 4 hours before using.

Identification—

A: A solution (1 in 20) responds to the tests for *Sodium* ⟨191⟩.

B: It responds to *Identification test B* under *Calcium Gluconate.*

Chloride ⟨221⟩—A 1.0-g portion shows no more chloride than corresponds to 1 mL of 0.020 *N* hydrochloric acid (0.07%).

Sulfate ⟨221⟩—A 2.0-g portion dissolved in boiling water shows no more sulfate than corresponds to 1 mL of 0.020 *N* sulfuric acid (0.05%).

Arsenic, *Method I* ⟨211⟩—Dissolve 1.0 g in 35 mL of water: the limit is 3 ppm.

Lead ⟨251⟩—Dissolve 1.0 g in 25 mL of water: the limit is 0.001%.

Heavy metals, *Method I* ⟨231⟩—Dissolve 1.0 g in 10 mL of water, add 6 mL of 3 *N* hydrochloric acid, and dilute with water to 25 mL: the limit is 0.002%.

Reducing substances—Transfer 1.0 g to a 250-mL conical flask, dissolve in 10 mL of water, and add 25 mL of alkaline cupric citrate TS. Cover the flask, boil gently for 5 minutes, accurately timed, and cool rapidly to room temperature. Add 25 mL of 0.6 *N* acetic acid, 10.0 mL of 0.1 *N* iodine VS, and 10 mL of 3 *N* hydrochloric acid, and titrate with 0.1 *N* sodium thiosulfate VS, adding 3 mL of starch TS as the end-point is approached. Perform a blank determination, omitting the specimen, and note the difference in volumes required. Each mL of the difference in volume of 0.1 *N* sodium thiosulfate consumed is equivalent to 2.7 mg of reducing substances (as dextrose): the limit is 0.5%.

Assay—Transfer about 150 mg of Sodium Gluconate, accurately weighed, to a suitable conical flask, and dissolve in 75 mL of glacial acetic acid, warming if necessary to effect complete solution. Cool, add quinaldine red TS, and titrate with 0.1 *N* perchloric acid VS to a colorless end-point. Each mL of 0.1 *N* perchloric acid is equivalent to 21.81 mg of $C_6H_{11}NaO_7$.

Sodium Hydroxide—*see* Sodium Hydroxide NF

Sodium Hypochlorite Solution

NaClO 74.44
Hypochlorous acid, sodium salt.
Sodium hypochlorite [*7681-52-9*].

» Sodium Hypochlorite Solution contains not less than 4.0 percent and not more than 6.0 percent, by weight, of NaClO.

Caution—This Solution is not suitable for application to wounds.

Packaging and storage—Preserve in tight, light-resistant containers, at a temperature not exceeding 25°.

Identification—

A: The Solution at first colors red litmus blue and then bleaches it.

B: The addition of 3 *N* hydrochloric acid to the Solution causes an evolution of chlorine.

C: The solution obtained in *Identification test B* responds to the flame test for *Sodium* ⟨191⟩.

Assay—Weigh accurately, in a glass-stoppered flask, about 3 mL of Sodium Hypochlorite Solution, and dilute it with 50 mL of water. Add 2 g of potassium iodide and 10 mL of 6 *N* acetic acid, and titrate the liberated iodine with 0.1 *N* sodium thiosulfate VS, adding 3 mL of starch TS as the end-point is approached. Perform a blank determination, and make any necessary correction. Each mL of 0.1 *N* sodium thiosulfate is equivalent to 3.722 mg of NaClO.

Sodium Iodide

NaI 149.89
Sodium iodide.
Sodium iodide [*7681-82-5*].

» Sodium Iodide contains not less than 99.0 percent and not more than 101.5 percent of NaI, calculated on the anhydrous basis.

Packaging and storage—Preserve in tight containers.

Identification—A solution (1 in 20) responds to the tests for *Sodium* ⟨191⟩ and for *Iodide* ⟨191⟩.

Alkalinity—Dissolve 1.0 g in 10 mL of water, and add 0.15 mL of 0.10 *N* sulfuric acid: no red color is produced by the addition of 1 drop of phenolphthalein TS.

Water, *Method I* ⟨921⟩: not more than 2.0%.

Iodate—It meets the requirements of the test for *Iodate* under *Potassium Iodide*, 100 mg of Sodium Iodide being used in the control.

Nitrate, nitrite, and ammonia—Add 5 mL of 1 *N* sodium hydroxide and about 200 mg of aluminum wire to a solution of 1.0 g of Sodium Iodide in 5 mL of water, contained in a test tube of about 40-mL capacity. Insert a pledget of purified cotton in the upper portion of the test tube, and place a piece of moistened red litmus paper over the mouth of the tube. Heat in a steam bath for 15 minutes: no blue coloration of the paper is discernible.

Thiosulfate and barium—It meets the requirements of the test for *Thiosulfate and barium* under *Potassium Iodide.*

Potassium—A solution of 1.0 g in 2 mL of water yields no precipitate with 1.0 mL of sodium bitartrate TS.

Heavy metals ⟨231⟩—Dissolve 2.0 g in 25 mL of water: the limit is 0.001%.

Assay—Weigh accurately about 500 mg of Sodium Iodide, and dissolve in about 10 mL of water. Add 35 mL of hydrochloric acid and 5 mL of chloroform, and titrate with 0.05 *M* potassium iodate VS until the purple color of iodine disappears from the chloroform. Add the last portions of the iodate solution, dropwise, agitating the mixture vigorously. After the chloroform has been decolorized, allow the mixture to stand for 5 minutes. If the chloroform develops a purple color, titrate further with the iodate solution. Each mL of 0.05 *M* potassium iodate is equivalent to 14.99 mg of NaI.

Sodium Iodide I 123, I 125, I 131 Capsules and Solutions—*see under "iodine"*

Sodium Lactate Injection

CH3CH(OH)COONa

$C_3H_5NaO_3$ 112.06
Propanoic acid, 2-hydroxy-, monosodium salt.
Sodium lactate [*72-17-3*].

» Sodium Lactate Injection is sterile Sodium Lactate Solution in Water for Injection, or a sterile solution of Lactic Acid in Water for Injection prepared with the aid of Sodium Hydroxide. It contains not less than 95.0 percent and not more than 110.0 percent of the labeled amount of $C_3H_5NaO_3$.

Packaging and storage—Preserve in single-dose containers, preferably of Type I or Type II glass.

Labeling—The label states the total osmolar concentration in mOsmol per liter. Where the contents are less than 100 mL, or where the label states that the Injection is not for direct injection but is to be diluted before use, the label alternatively may state the total osmolar concentration in mOsmol per mL. The label includes also the warning: "Not for use in the treatment of lactic acidosis."

Identification—

A: Overlay 2 mL of Injection on 5 mL of a 1 in 100 solution of catechol in sulfuric acid: a deep red color is produced at the zone of contact.

B: To 2 mL of Injection add 5 mL of 2 *N* sulfuric acid and 2 mL of potassium permanganate TS, and heat: the odor of acetaldehyde is evolved.

Pyrogen—Diluted, if necessary, with water for injection to approximately 0.16 *M* (20 mg per mL), it meets the requirements of the *Pyrogen Test* ⟨151⟩.

pH ⟨791⟩: between 6.0 and 7.3, the Injection being diluted with water, if necessary, to approximately 0.16 *M* (20 mg per mL).

Particulate matter ⟨788⟩: meets the requirements under *Small-volume Injections.*

Heavy metals ⟨231⟩—Evaporate a volume of Injection, equivalent to 2.0 g of sodium lactate, to 5 mL, and dilute with 1 *N* acetic acid to 25 mL: the limit is 0.001%.

Other requirements—It meets the requirements under *Injections* ⟨1⟩.

Assay—Pipet into a small beaker a volume of Sodium Lactate Injection, equivalent to about 300 mg of sodium lactate, and evaporate to dryness. Add to the residue 60 mL of a 1 in 5 mixture of acetic anhydride in glacial acetic acid, and stir until the residue is completely dissolved. Titrate with 0.1 *N* perchloric acid VS, determining the end-point potentiometrically. Perform a blank determination, and make any necessary correction. Each mL of 0.1 *N* perchloric acid is equivalent to 11.21 mg of $C_3H_5NaO_3$.

Sodium Lactate Solution

» Sodium Lactate Solution is an aqueous solution containing not less than 50.0 percent, by weight, of monosodium lactate. It contains not less than 98.0 percent and not more than 102.0 percent of the labeled amount of $C_3H_5NaO_3$.

Packaging and storage—Preserve in tight containers.

Labeling—Label it to indicate its content of sodium lactate.

Identification—It responds to the tests for *Sodium* ⟨191⟩ and for *Lactate* ⟨191⟩.

pH ⟨791⟩: between 5.0 and 9.0.

Chloride—⟨221⟩—A portion, equivalent to 1 g of sodium lactate, shows no more chloride than corresponds to 0.7 mL of 0.020 *N* hydrochloric acid (0.05%).

Citrate, oxalate, phosphate, or tartrate—Dilute 5 mL with recently boiled and cooled water to 50 mL. To 4 mL of this solution add 6 *N* ammonium hydroxide or 3 *N* hydrochloric acid, if necessary, to bring the pH to between 7.3 and 7.7. Add 1 mL of calcium chloride TS, and heat in a boiling water bath for 5 minutes: the solution remains clear.

Sulfate—To 10 mL of a solution (1 in 100) add 2 drops of hydrochloric acid and 1 mL of barium chloride TS: no turbidity is produced.

Heavy metals, *Method I* ⟨231⟩—Dilute a quantity of Solution, equivalent to 2.0 g of sodium lactate, with 1 *N* acetic acid to 25 mL: the limit is 0.001%.

Sugars—To 10 mL of hot alkaline cupric tartrate TS add 5 drops of Solution: no red precipitate is formed.

Methanol and methyl esters—

Potassium permanganate and phosphoric acid solution—Dissolve 3 g of potassium permanganate in a mixture of 15 mL of phosphoric acid and 70 mL of water. Dilute with water to 100 mL.

Oxalic acid and sulfuric acid solution—Cautiously add 50 mL of sulfuric acid to 50 mL of water, mix, cool, add 5 g of oxalic acid, and mix to dissolve.

Standard preparation—Prepare a solution containing 10.0 mg of methanol in 100 mL of dilute alcohol (1 in 10).

Test preparation—Place 40.0 g in a glass-stoppered, round-bottom flask, add 10 mL of water, and add cautiously 30 mL of 5 *N* potassium hydroxide. Connect a condenser to the flask, and steam-distil, collecting the distillate in a suitable 100-mL graduated vessel containing 10 mL of alcohol. Continue the distillation until the volume in the receiver reaches approximately 95 mL, and dilute the distillate with water to 100.0 mL.

Procedure—Transfer 10.0 mL each of the *Standard preparation* and the *Test preparation* to 25-mL volumetric flasks, to each add 5.0 mL of *Potassium permanganate and phosphoric acid solution*, and mix. After 15 minutes, to each add 2.0 mL of *Oxalic acid and sulfuric acid solution*, stir with a glass rod until the solution is colorless, add 5.0 mL of fuchsin–sulfurous acid TS, and dilute with water to volume. After 2 hours, concomitantly determine the absorbances of both solutions in 1-cm cells at the wavelength of maximum absorbance at about 575 nm, with a suitable spectrophotometer, using water as the blank: the absorbance of the solution from the *Test preparation* is not greater than that from the *Standard preparation* (0.025%).

Assay—Weigh accurately into a suitable flask a volume of Sodium Lactate Solution, equivalent to about 300 mg of sodium lactate, add 60 mL of a 1 in 5 mixture of acetic anhydride in glacial acetic acid, mix, and allow to stand for 20 minutes. Titrate with 0.1 *N* perchloric acid VS, determining the end-point potentiometrically. Perform a blank determination, and make any necessary correction. Each mL of 0.1 *N* perchloric acid is equivalent to 11.21 mg of $C_3H_5NaO_3$.

Sodium Lauryl Sulfate—*see* Sodium Lauryl Sulfate NF

Sodium Metabisulfite—*see* Sodium Metabisulfite NF

Sodium Monofluorophosphate

$FPO(ONa)_2$

Na_2PFO_3 143.95
Phosphorofluoridic acid, disodium salt.
Disodium phosphorofluoridate [*10163-15-2*].

» Sodium Monofluorophosphate contains not less than 91.7 percent and not more than 100.5 percent of Na_2PFO_3, calculated on the dried basis.

Packaging and storage—Preserve in well-closed containers.

Reference standard—*USP Sodium Fluoride Reference Standard*—Dry at 150° for 4 hours before using.

Identification—

A: Place about 1 g in a platinum crucible in a well-ventilated hood, and add 15 mL of sulfuric acid. Cover the crucible with a piece of clear, polished glass, and heat on a steam bath for 1 hour. Remove the glass cover, rinse it in water, and dry: the glass surface exposed to vapors from the crucible is etched.

B: A solution (1 in 10) yields with silver nitrate TS a white precipitate, which is soluble in diluted nitric acid and in dilute ammonium hydroxide (1 in 2).

C: A solution responds to the tests for *Sodium* ⟨191⟩.

pH ⟨791⟩: between 6.5 and 8.0, in a solution (1 in 50).

Loss on drying ⟨731⟩—Dry it at 105° to constant weight: it loses not more than 0.2% of its weight.

Arsenic, *Method I* ⟨211⟩: 3 ppm.

Limit of fluoride ion—[NOTE—Use plasticware throughout this test.]

Buffer solution—To 55 g of sodium chloride in a 1000-mL volumetric flask add 500 mg of sodium citrate, 255 g of sodium acetate, and 300 mL of water. Shake to dissolve, and cautiously add 115 mL of glacial acetic acid with mixing. Cool to room temperature, add 300 mL of isopropyl alcohol, dilute with water to volume, and mix: the pH of this solution is between 5.0 and 5.5.

Standard stock solution—Transfer 221 mg of USP Sodium Fluoride RS to a 200-mL volumetric flask, add water to dissolve, dilute with water to volume, and mix. Each mL of this solution contains 500 µg of fluoride ion. Store in a tightly closed, plastic container.

Standard preparations—To four 100-mL volumetric flasks transfer, respectively, 2.0-, 4.0-, 10.0-, and 20.0-mL portions of the *Standard stock solution*, dilute each with *Buffer solution* to volume, and mix to obtain solutions having fluoride ion concentrations of 10, 20, 50, and 100 µg per mL, respectively.

Test preparation—Transfer about 1.8 g, accurately weighed, to a 100-mL volumetric flask, add water to dissolve, dilute with water to volume, and mix. Transfer 20.0 mL of this solution to a second 100-mL volumetric flask, dilute with *Buffer solution* to volume, and mix.

Procedure—Concomitantly measure the potential (see *pH* ⟨791⟩), in mV, of the *Standard preparations* and of the *Test preparation* with a pH meter capable of a minimum reproducibility of ±0.2 millivolt, equipped with a glass-sleeved calomel-fluoride specific-ion electrode system. [NOTE—When taking measurements, immerse the electrodes in the solution which has been transferred to a 150-mL plastic beaker containing a plastic-coated stirring bar. Allow to stir on a magnetic stirrer having an insulated top until equilibrium is attained (1 to 2 minutes), and record the potential. Rinse and dry the electrodes between measurements, taking care to avoid damaging the crystal of the specific-ion electrode.] Plot the logarithm of the fluoride-ion concentrations, in µg per mL, of the *Standard preparation* versus potential, in mV. From the measured potential of the *Test preparation* and the standard curve, determine the concentration, in µg per mL, of fluoride ion in the *Test preparation:* not more than 1.2% is found.

Heavy metals, *Method I* ⟨231⟩—Dissolve 400 mg in 25 mL of water: the limit is 0.005%.

Assay—

Monochloroacetate buffer—Dissolve 189 g of monochloroacetic acid and 55 g of sodium hydroxide in about 1500 mL of water. Cool, dilute with water to 2000 mL, and mix.

0.025 M Thorium nitrate—Dissolve 13.8 g of thorium nitrate in about 800 mL of water, and filter the solution into a 1000-mL volumetric flask. Dilute with water to volume, and mix. Standardize this solution as follows. Transfer 20.0 mL of *Standard stock solution*, prepared as directed in the test for *Limit of fluoride ion*, to a 150-mL beaker containing 50 mL of water. Add 3 drops of sodium alizarinsulfonate TS, mix, and adjust the acidity by the careful addition, successively, of sodium hydroxide solution (1 in 50) and dilute hydrochloric acid (1 in 160), until the pink color has just been discharged. Add 1 mL of *Monochloroacetate buffer*, and titrate with *0.025 M Thorium nitrate* to a permanent pink color. Calculate the molarity of the thorium nitrate titrant by the equation:

$$M = (S)/[(4)(41.99V)],$$

in which S is the weight, in mg, of USP Sodium Fluoride RS in the portion of *Standard stock solution* taken, 41.99 is the molecular weight of sodium fluoride, and V is the volume, in mL, of titrant consumed.

Procedure—Transfer about 1.8 g of Sodium Monofluorophosphate, accurately weighed, to a 100-mL volumetric flask. Add about 50 mL of water, mix to effect solution, dilute with water to volume, and mix. Transfer 20.0 mL of this solution to the reaction flask of a suitable fluoride steam distilling apparatus containing 10 lime-glass beads measuring 5 mm in diameter and 70 mL of dilute sulfuric acid (1 in 2). Distil, collecting the distillate in a 250-mL volumetric flask. Change receivers when the flask is filled to volume with distillate, and continue distilling into a 400-mL beaker until an additional 150 mL to 200 mL of distillate tailing has been collected. Mix the solution in the 250-mL volumetric flask, and transfer 50.0 mL to a 150-mL beaker. To the 150-mL beaker add 50 mL of water and 3 drops of sodium alizarinsulfonate TS, and mix. Adjust the acidity of this solution by the careful addition, successively, of sodium hydroxide solution (1 in 50) and dilute hydrochloric acid (1 in 160), until the pink color has just been discharged. Add 1 mL of *Monochloroacetate buffer*, titrate with *0.025 M Thorium nitrate* to a permanent pink color, and record the volume, in mL, of titrant consumed as V_a. Add 3 drops of sodium alizarinsulfonate TS to the distillation tail in the 400-mL beaker, proceed as directed previously with the adjustment of the acidity, the addition of *Monochloroacetate buffer*, and the titration with *0.025 M Thorium nitrate*, and record the volume, in mL, of titrant consumed as V_b. Calculate the percentage of Na_2PFO_3 in the Sodium Monofluorophosphate taken by the formula:

$$[(2)(143.95)(5V_a + V_b)(M)/(W)] - (143.95/18.9984)(F),$$

in which M is the molarity of *0.025 M Thorium nitrate*, W is the weight, in g, of the Sodium Monofluorophosphate taken, F is the percentage of fluoride ions determined as directed in the test for *Limit of fluoride ion*, 143.95 is the molecular weight of sodium monofluorophosphate, and 18.9984 is the atomic weight of fluorine.

Sodium Nitrite

$NaNO_2$ 69.00
Nitrous acid, sodium salt.
Sodium nitrite [*7632-00-0*].

» Sodium Nitrite contains not less than 97.0 percent and not more than 101.0 percent of $NaNO_2$, calculated on the dried basis.

Packaging and storage—Preserve in tight containers.

Identification—A solution of it responds to the tests for *Sodium* ⟨191⟩ and for *Nitrite* ⟨191⟩.

Loss on drying ⟨731⟩—Dry it over silica gel for 4 hours: it loses not more than 0.25% of its weight.

Heavy metals ⟨231⟩—Dissolve 1 g in 6 mL of 3 *N* hydrochloric acid, and evaporate on a steam bath to dryness. Reduce the residue to a coarse powder, and continue heating on the steam bath until the odor of hydrochloric acid no longer is perceptible. Dissolve the residue in 23 mL of water, and add 2 mL of 1 *N* acetic acid: the limit is 0.002%.

Assay—Dissolve about 1 g of Sodium Nitrite, accurately weighed, in water to make 100.0 mL. Pipet 10 mL of this solution into a mixture of 50.0 mL of 0.1 *N* potassium permanganate VS, 100 mL of water, and 5 mL of sulfuric acid. When adding the Sodium Nitrite solution, immerse the tip of the pipet beneath the surface of the permanganate mixture. Warm the liquid to 40°, allow it to stand for 5 minutes, and add 25.0 mL of 0.1 *N* oxalic acid VS. Heat the mixture to about 80°, and titrate with 0.1 *N* potassium permanganate VS. Each mL of 0.1 *N* potassium permanganate is equivalent to 3.450 mg of $NaNO_2$.

Sodium Nitrite Injection

» Sodium Nitrite Injection is a sterile solution of Sodium Nitrite in Water for Injection. It contains not less than 95.0 percent and not more than 105.0 percent of the labeled amount of $NaNO_2$.

Packaging and storage—Preserve in single-dose containers, of Type I glass.

Identification—It responds to the *Identification tests* under *Sodium Nitrite.*

Pyrogen—Dilute, if necessary, with *Water for Injection* to a concentration of 15 mg per mL. It meets the requirements of the *Pyrogen Test* ⟨151⟩, the test dose being 1 mL per kg.

pH ⟨791⟩: between 7.0 and 9.0.

Other requirements—It meets the requirements under *Injections* ⟨1⟩.

Assay—Pipet an accurately measured volume of Sodium Nitrite Injection, containing about 150 mg of sodium nitrite into a mixture of 50.0 mL of 0.1 *N* potassium permanganate VS, 100 mL of water, and 5 mL of sulfuric acid, immersing the tip of the pipet beneath the surface of the mixture during the addition. Warm the liquid to 40°, allow it to stand for 5 minutes, and add 25.0 mL of 0.1 *N* oxalic acid VS. Heat the mixture to about 80°, and titrate with 0.1 *N* potassium permanganate VS. Each mL of 0.1 *N* potassium permanganate is equivalent to 3.450 mg of $NaNO_2$.

Sodium Nitroprusside

$Na_2[Fe(CN)_5NO].2H_2O$ 297.95

Ferrate(2-), pentakis(cyano-*C*)nitrosyl-, disodium, dihydrate, (*OC*-6-22)-.

Disodium pentacyanonitrosylferrate(2-) dihydrate [*13755-38-9*].

Sodium nitroferricyanide dihydrate.

Anhydrous 261.92 [*14402-89-2*].

» Sodium Nitroprusside contains not less than 99.0 percent of $Na_2[Fe(CN)_5NO].2H_2O$.

Packaging and storage—Preserve in tight, light-resistant containers.

Reference standard—*USP Sodium Nitroprusside Reference Standard*—Do not dry before using.

Identification—

A: [NOTE—Use low-actinic glassware in this test.] The absorption spectrum in the range from 350 nm to 700 nm of a 1 in 135 solution exhibits maxima and minima at the same wavelengths as that of a similar solution of USP Sodium Nitroprusside RS, concomitantly measured.

B: Dissolve 5 mg in 2 mL of water, and add 2 drops of acetone and 0.5 mL of 2 *N* sodium hydroxide: an orange color is produced. Add 2 mL of acetic acid: the color changes to purple.

C: A solution (1 in 4) responds to the tests for *Sodium* ⟨191⟩.

Water, *Method I* ⟨921⟩: between 9.0% and 15.0%.

Insoluble substances—Dissolve 10.0 g in 50 mL of water, heat the solution on a steam bath for 30 minutes, filter, wash the residue with water, and dry at 105° to constant weight: the weight of the residue is not greater than 1 mg (0.01%).

Chloride—

Standard chloride solution—Dissolve 42.4 mg of potassium chloride in water to make 100.0 mL of solution. Each mL of this solution contains 0.2 mg of chloride.

Procedure—Transfer 1.0 g of Sodium Nitroprusside to a 250-mL conical flask, transfer 1.0 mL of *Standard chloride solution* to a similar flask, and add to each flask 85 mL of water. To the flask containing the substance under test, add 15 mL of cupric sulfate solution (83 in 1000), mix, and allow any undissolved particles to settle. Carefully add cupric sulfate solution (83 in 1000) to the flask containing the diluted *Standard chloride solution,* with mixing, so that its color matches that of the test solution in the first flask. Filter the contents of each flask, and discard the first 25 mL of filtrate. To 10 mL of the subsequent filtrate from each flask add 2 mL of nitric acid, and mix. Add 1 mL of 1 *N* silver nitrate to each, and again mix: the test solution so treated becomes no more turbid than the treated *Standard chloride solution* (0.02%).

Ferricyanide—Dissolve 500 mg in 20 mL of ammonium acetate TS, previously adjusted with 1 *N* acetic acid to a pH of 4.62. Divide this solution into halves, and transfer each half to a separate 50-mL volumetric flask, identified as *A* and *B*, respectively. To flask *B* add 1.0 mL of a freshly prepared solution of potassium ferricyanide containing 78 µg per mL. To both flasks add 5 mL of ferrous ammonium sulfate solution (1 in 1000), dilute with water to volume, and mix. Allow the flasks to stand for 1 hour, and concomitantly determine the absorbance of the solutions at the wavelength of maximum absorbance at about 720 nm, using as a blank a solution prepared by dissolving 250 mg of the specimen in 10 mL of the pH 4.62 ammonium acetate TS and diluting with water to 50 mL. The absorbance of the solution in flask *A* is not greater than the absorbance of the solution in flask *B* minus the absorbance of the solution in flask *A* (0.02% of ferricyanide).

Ferrocyanide—Dissolve 2.0 g in 40 mL of water, divide the solution into halves, and transfer each half to a separate 50-mL volumetric flask, identified as *A* and *B*, respectively. To flask *B* add 2 mL of a freshly prepared solution of potassium ferrocyanide containing 200 µg per mL. To both flasks add 0.2 mL of ferric chloride TS, dilute with water to volume, and mix. Allow to stand for 20 minutes, accurately timed, and concomitantly measure the absorbance of the solutions at the wavelength of maximum absorbance at about 695 nm, using as a blank a solution prepared by dissolving 1.0 g of the specimen in water to make 50 mL: the absorbance of the solution in flask *A* is not greater than the absorbance of the solution in flask *B* minus the absorbance of the solution in flask *A* (0.02% of ferrocyanide).

Sulfate—

Standard sulfate solution—Dissolve 15 mg of anhydrous sodium sulfate in water to make 100.0 mL of solution. Each mL of this solution contains 0.1 mg of sulfate.

Procedure—Dissolve 5.0 g of Sodium Nitroprusside in water to make 250.0 mL of solution, and filter the solution into a flat-bottom, 250-mL graduated flask. Transfer 5.0 mL of *Standard sulfate solution* to a similar flask, and dilute to the same volume as the test solution. To each flask add 10 drops of glacial acetic acid and 5 mL of 1 *N* barium chloride, and allow to stand for 10 minutes. Place both flasks over a fluorescent light source, and observe: the turbidity in the treated test solution is not more intense than that of the treated *Standard sulfate solution* (0.01%).

Assay—Dissolve about 500 mg of Sodium Nitroprusside, accurately weighed, in 130 mL of chloride-free water. Titrate with 0.1 *N* silver nitrate VS, determining the end-point potentiometrically, using a silver–silver chloride electrode system. Each mL of 0.1 *N* silver nitrate is equivalent to 14.90 mg of $Na_2[Fe(CN)_5NO].2H_2O$.

Sterile Sodium Nitroprusside

» Sterile Sodium Nitroprusside is Sodium Nitroprusside suitable for parenteral use. It contains not less than 90.0 percent and not more than 110.0 percent of the labeled amount of $Na_2[Fe(CN)_5NO].2H_2O$.

Packaging and storage—Preserve protected from light in *Containers for Sterile Solids* as described under *Injections* ⟨1⟩.

Reference standard—*USP Sodium Nitroprusside Reference Standard*—Do not dry before using.

Constituted solution—At the time of use, the constituted solution prepared from Sterile Sodium Nitroprusside meets the requirements for *Constituted Solutions* under *Injections* ⟨1⟩.

Identification—To 50 mg contained in a small test tube add 10 mL of ascorbic acid solution (1 in 50), and mix. Add 1 mL of dilute hydrochloric acid (1 in 10), mix, and add dropwise, while mixing, 1 to 2 mL of 1 *N* sodium hydroxide: a transient blue color is produced.

Pyrogen—It meets the requirements of the *Pyrogen Test* ⟨151⟩, the test dose being 1 mg per kg.

Water, *Method I* ⟨921⟩: not more than 15.0%.

Other requirements—It responds to the *Identification test* under *Sodium Nitroprusside.* It meets also the requirements for *Sterility Tests* ⟨71⟩, *Uniformity of Dosage Units* ⟨905⟩, and *Labeling* under *Injections* ⟨1⟩.

Assay—

pH 7.1 buffer—Dissolve 1.36 g of monobasic potassium phosphate and 5.2 mL of a 1 in 4 solution of tetrabutylammonium hydroxide in methanol in water to make 1000 mL, and adjust with phosphoric acid to a pH of 7.1.

Mobile phase—Prepare a suitable filtered mixture of *pH 7.1 buffer* and acetonitrile (about 70:30).

NOTE—Use low-actinic glassware throughout the following sections.

Standard preparation—Dissolve an accurately weighed quantity of USP Sodium Nitroprusside RS, in *Mobile phase* to obtain a solution having a known concentration of about 0.05 mg per mL.

Assay preparation 1 (where the label states only the total contents of the container)—Transfer the contents of 1 container of Sterile Sodium Nitroprusside to a 100-mL volumetric flask with the aid of *Mobile phase*, dilute with *Mobile phase* to volume, and mix. Dilute an accurately measured volume of this solution quantitatively with *Mobile phase* to obtain a solution containing about 0.05 mg of $Na_2[Fe(CN)_5NO].2H_2O$ per mL.

Assay preparation 2 (where the label states the quantity of $Na_2[Fe(CN_5)NO].2H_2O$ in a given volume of constituted solution)—Constitute Sterile Sodium Nitroprusside as directed in the labeling. Dilute an accurately measured volume of the constituted solution thus obtained quantitatively and stepwise with *Mobile phase* to obtain a solution containing about 0.05 mg of $Na_2[Fe(CN)_5NO].2H_2O$ per mL.

Chromatographic system (see *Chromatography* ⟨621⟩)—The liquid chromatograph is equipped with a 210-nm detector and a 3.9-mm × 30-cm column that contains 10-µm packing L11. The flow rate is about 2 mL per minute. Chromatograph the *Standard preparation*, and record the peak responses as directed under *Procedure:* the tailing factor for the analyte peak is not more than 2.0, and the relative standard deviation for replicate injections is not more than 1.5%.

Procedure—Separately inject equal volumes (about 25 µL) of the *Standard preparation* and the *Assay preparation* into the chromatograph, record the chromatograms, and measure the responses for the major peaks. Calculate the quantity, in mg, of $Na_2[Fe(CN)_5NO].2H_2O$ in the container or in the portion of constituted solution taken by the formula:

$$L(C/D)(r_U/r_S),$$

in which L is the labeled quantity, in mg of $Na_2[Fe(CN)_5NO].2H_2O$, in the container, or in the volume of constituted solution taken, C is the concentration, in mg per mL, of USP Sodium Nitroprusside RS in the *Standard preparation, D* is the concentration, in mg of $Na_2[Fe(CN)_5NO].2H_2O$ per mL, of *Assay preparation 1*, or of *Assay preparation 2*, on the basis of the labeled quantity in the container, or in the volume of constituted solution taken, respectively, and the extent of dilution, and r_U and r_S are the peak responses obtained from the *Assay preparation* and the *Standard preparation*, respectively.

Sodium Pertechnetate Tc 99m Injection—*see under "technetium"*

Sodium Phosphate Cream, Neomycin Sulfate and Dexamethasone—*see* Neomycin Sulfate and Dexamethasone Sodium Phosphate Cream

Dibasic Sodium Phosphate

$Na_2HPO_4.7H_2O$ 268.07
Phosphoric acid, disodium salt, heptahydrate.
Disodium hydrogen phosphate heptahydrate [*7782-85-6*].
Phosphoric acid, disodium salt, hydrate.
Disodium hydrogen phosphate hydrate [*10140-65-5*].
Anhydrous 141.96 [*7558-79-4*].

» Dibasic Sodium Phosphate is dried or contains seven molecules of water of hydration. It contains not less than 98.0 percent and not more than 100.5 percent of Na_2HPO_4, calculated on the dried basis.

Packaging and storage—Preserve in tight containers.

Labeling—Label it to indicate whether it is dried or is the heptahydrate.

Identification—A solution (the equivalent of 1 part of Na_2HPO_4 in 30) responds to the tests for *Sodium* ⟨191⟩ and for *Phosphate* ⟨191⟩.

Loss on drying ⟨731⟩—Dry it at 105° to constant weight: the dried form loses not more than 5.0% of its weight, and the heptahydrate loses between 43.0% and 50.0% of its weight.

Insoluble substances—Dissolve the equivalent of 5.0 g of Na_2HPO_4 in 100 mL of hot water, filter through a tared filtering crucible, wash the insoluble residue with hot water, and dry at 105° for 2 hours: the weight of the residue so obtained does not exceed 20 mg (0.4%).

Chloride ⟨221⟩—A portion equivalent to 0.5 g of Na_2HPO_4 shows no more chloride than corresponds to 0.42 mL of 0.020 *N* hydrochloric acid (0.06%).

Sulfate ⟨221⟩—A portion equivalent to 0.1 g of Na_2HPO_4 shows no more sulfate than corresponds to 0.2 mL of 0.020 *N* sulfuric acid (0.2%).

Arsenic, *Method I* ⟨211⟩—Prepare a *Test Preparation* by dissolving a portion equivalent to 187.5 mg of Na_2HPO_4 in 35 mL of water: the limit is 16 ppm.

Heavy metals ⟨231⟩—Dissolve a portion equivalent to 1.0 g of Na_2HPO_4 in 1 mL of 1 *N* hydrochloric acid and enough water to make 25 mL of solution, and proceed as directed for *Test Preparation*, except to use glacial acetic acid to adjust the pH: the limit is 0.002%.

Assay—Transfer a portion of Dibasic Sodium Phosphate, accurately weighed, equivalent to about 3.2 g of Na_2HPO_4, to a 250-mL beaker, add 50.0 mL of 1 *N* hydrochloric acid VS and 50 mL of water, and stir until dissolved. Titrate the excess acid potentiometrically with 1 *N* sodium hydroxide VS to the inflection point at about pH 4. Record the buret reading, and calculate the volume (A) of 1 *N* hydrochloric acid consumed by the assay solution. Continue the titration with 1 *N* sodium hydroxide VS to the inflection point at about pH 8.8, record the buret reading, and calculate the volume (B) of 1 *N* sodium hydroxide required in the titration between the two inflection points (pH 4 to pH 8.8). Where A is equal to or less than B, each mL of the volume A of 1 *N* hydrochloric acid is equivalent to 142.0 mg of Na_2HPO_4.

Where A is greater than B, each mL of the volume $2B - A$ of 1 N sodium hydroxide is equivalent to 142.0 mg of Na_2HPO_4.

Monobasic Sodium Phosphate

$NaH_2PO_4.xH_2O$ (anhydrous) 119.98
Phosphoric acid, monosodium salt, monohydrate.
Monosodium phosphate monohydrate 137.99 [*10049-21-5*].
Phosphoric acid, monosodium salt, dihydrate.
Monosodium phosphate dihydrate 156.01 [*10028-24-7*].
Anhydrous [*7558-80-7*].

» Monobasic Sodium Phosphate contains one or two molecules of water of hydration, or is anhydrous. It contains not less than 98.0 percent and not more than 103.0 percent of NaH_2PO_4, calculated on the anhydrous basis.

Packaging and storage—Preserve in well-closed containers.

Labeling—Label it to indicate whether it is anhydrous or is the monohydrate or the dihydrate.

Identification—A solution (1 in 20) responds to the tests for *Sodium* ⟨191⟩ and for *Phosphate* ⟨191⟩.

pH ⟨791⟩: between 4.1 and 4.5, in a solution containing the equivalent of 1.0 g of $NaH_2PO_4.H_2O$ in 20 mL of water.

Water, *Method I* ⟨921⟩: less than 2.0% (anhydrous form); between 10.0% and 15.0% (monohydrate); between 18.0% and 26.5% (dihydrate).

Insoluble substances—Dissolve a portion equivalent to 10.0 g of $NaH_2PO_4.H_2O$ in 100 mL of hot water, filter through a tared filtering crucible, wash the insoluble residue with hot water, and dry at 105° for 2 hours: the weight of the residue so obtained does not exceed 20 mg (0.2%).

Chloride ⟨221⟩—A portion equivalent to 1.0 g of $NaH_2PO_4.H_2O$ shows no more chloride than corresponds to 0.20 mL of 0.020 N hydrochloric acid (0.014%).

Sulfate ⟨221⟩—A portion equivalent to 0.20 g of $NaH_2PO_4.H_2O$ shows no more sulfate than corresponds to 0.30 mL of 0.020 N sulfuric acid (0.15%).

Aluminum, calcium, and related elements—A solution containing the equivalent of 1.0 g of $NaH_2PO_4.H_2O$ in 10 mL of water does not become turbid when rendered slightly alkaline to litmus paper with 6 N ammonium hydroxide.

Arsenic, *Method I* ⟨211⟩—Dissolve a portion equivalent to 0.375 g of $NaH_2PO_4.H_2O$ in 35 mL of water: the limit is 8 ppm.

Heavy metals ⟨231⟩—Dissolve a portion equivalent to 1.0 g of $NaH_2PO_4.H_2O$ in 20 mL of water, and add 1 mL of 3 N hydrochloric acid and water to make 25 mL: the limit is 0.002%.

Assay—Dissolve about 2.5 g of Monobasic Sodium Phosphate, accurately weighed, in 10 mL of cold water, add 20 mL of a cold, saturated solution of sodium chloride, then add phenolphthalein TS, and titrate with 1 N sodium hydroxide VS, keeping the temperature of the solution between 10° and 15° during the entire titration. Perform a blank determination, and make any necessary correction. Each mL of 1 N sodium hydroxide is equivalent to 120.0 mg of NaH_2PO_4.

Sodium Phosphate P 32 Solution—*see* Phosphate P 32 Solution, Sodium

Sodium Phosphates Enema

» Sodium Phosphates Enema is a solution of Dibasic Sodium Phosphate and Monobasic Sodium Phosphate, or Dibasic Sodium Phosphate and Phosphoric Acid, in Purified Water. It contains, in each 100 mL, not less than 5.7 g and not more than 6.3 g of dibasic sodium phosphate ($Na_2HPO_4.7H_2O$), and not less than 15.2 g and not more than 16.8 g of monobasic sodium phosphate ($NaH_2PO_4.H_2O$).

Packaging and storage—Preserve in well-closed, single-unit containers.

Identification—It responds to the tests for *Sodium* ⟨191⟩ and for *Phosphate* ⟨191⟩.

Specific gravity ⟨841⟩: between 1.121 and 1.128.

pH ⟨791⟩: between 5.0 and 5.8.

Chloride ⟨221⟩—A 5.0-mL portion shows no more chloride than corresponds to 0.60 mL of 0.020 N hydrochloric acid (0.008%).

Arsenic, *Method I* ⟨211⟩—Prepare the *Test Preparation* by diluting 5.0 mL with water to 50 mL, then diluting 15 mL of the solution with water to 35 mL. The limit is 2 ppm.

Heavy metals ⟨231⟩—Use a 2-mL specimen for the test: the limit is 0.001%.

Assay—Pipet 5 mL of Sodium Phosphates Enema into a 250-mL beaker, and add 15.0 mL of 0.5 N sodium hydroxide VS and 95 mL of water. Place the electrodes of a suitable pH meter in the solution, and titrate the excess base with 0.5 N hydrochloric acid VS to the first inflection point, to a pH of about 9.7. Record the volume (A), in mL, of 0.5 N hydrochloric acid consumed. Continue the titration to the second inflection point, to a pH of about 4.7, and record the total volume (B), in mL, of 0.5 N hydrochloric acid required in the titration. Each mL of the volume $(15.0 - A)$ of 0.5 N hydrochloric acid is equivalent to 69.0 mg of monobasic sodium phosphate ($NaH_2PO_4.H_2O$). Each mL of the volume $(B - 15.0)$ of 0.5 N hydrochloric acid is equivalent to 134.0 mg of dibasic sodium phosphate ($Na_2HPO_4.7H_2O$).

Sodium Phosphates Injection

» Sodium Phosphates Injection is a sterile solution of Monobasic Sodium Phosphate and Dibasic Sodium Phosphate in Water for Injection. It contains not less than 95.0 percent and not more than 105.0 percent of the labeled amounts of monobasic sodium phosphate ($NaH_2PO_4.H_2O$) and dibasic sodium phosphate ($Na_2HPO_4.7H_2O$). It contains no bacteriostat or other preservative.

Packaging and storage—Preserve in single-dose or in multiple-dose containers, preferably of Type I glass.

Labeling—The label states the sodium content in terms of milliequivalents in a given volume, and states also the phosphorus content in terms of millimoles in a given volume. Label the Injection to indicate that it is to be diluted to appropriate strength with water or other suitable fluid prior to administration, and that once opened any unused portion is to be discarded. The label states also the total osmolar concentration in mOsmol per liter. Where the contents are less than 100 mL, or where the label states that the Injection is not for direct injection but is to be diluted before use, the label alternatively may state the total osmolar concentration in mOsmol per mL.

Identification—It responds to the tests for *Sodium* ⟨191⟩ and for *Phosphate* ⟨191⟩.

Pyrogen—After being diluted with Sodium Chloride Injection to a concentration of 40 mg of phosphate ion per mL, it meets the requirements of the *Pyrogen Test* ⟨151⟩, the test dose being 1 mL per kg.

Particulate matter ⟨788⟩: meets the requirements under *Small-volume Injections*.

Other requirements—It meets the requirements under *Injections* ⟨1⟩.

Assay for monobasic sodium phosphate—Transfer an accurately measured volume of Sodium Phosphates Injection, equivalent to about 300 mg of anhydrous monobasic sodium phosphate, to a 100-mL beaker, and dilute with water to about 50 mL. Place the electrodes of a suitable pH meter in the solution, and titrate with 0.1 *N* sodium hydroxide VS to the inflection point to a pH of about 8.8. Each mL of 0.1 *N* sodium hydroxide is equivalent to 13.80 mg of $NaH_2PO_4 . H_2O$.

Assay for dibasic sodium phosphate—Transfer an accurately measured volume of Sodium Phosphates Injection, equivalent to about 300 mg of anhydrous dibasic sodium phosphate, to a 100-mL beaker, and dilute with water to about 50 mL. Place the electrodes of a suitable pH meter in the solution, and titrate with 0.1 *N* hydrochloric acid VS to the inflection point to a pH of about 4.0. Each mL of 0.1 *N* hydrochloric acid is equivalent to 26.81 mg of $Na_2HPO_4 . 7H_2O$.

Sodium Phosphates Oral Solution

» Sodium Phosphates Oral Solution is a solution of Dibasic Sodium Phosphate and Monobasic Sodium Phosphate, or Dibasic Sodium Phosphate and Phosphoric Acid, in Purified Water. It contains, in each 100 mL, not less than 17.1 g and not more than 18.9 g of dibasic sodium phosphate ($Na_2HPO_4 . 7H_2O$), and not less than 45.6 g and not more than 50.4 g of monobasic sodium phosphate ($NaH_2PO_4 . H_2O$).

Packaging and storage—Preserve in tight containers.

Identification—It responds to the tests for *Sodium* ⟨191⟩ and for *Phosphate* ⟨191⟩.

pH ⟨791⟩: between 4.4 and 5.2.

Chloride ⟨221⟩—A 5.0-mL portion shows no more chloride than corresponds to 0.75 mL of 0.020 *N* hydrochloric acid (0.01%).

Arsenic, *Method I* ⟨211⟩—It meets the requirements of the test for *Arsenic* under *Sodium Phosphates Enema*.

Heavy metals ⟨231⟩—Use a 2-mL specimen for the test: the limit is 0.001%.

Assay—Pipet 2 mL of Sodium Phosphates Oral Solution into a 250-mL beaker, and proceed as directed in the *Assay* under *Sodium Phosphates Enema*, beginning with "and add 15.0 mL of 0.5 *N* sodium hydroxide VS."

Sodium Polystyrene Sulfonate

Benzene, diethenyl-, polymer with ethenylbenzene, sulfonated, sodium salt.
Divinylbenzene copolymer with styrene, sulfonated, sodium salt.

» Sodium Polystyrene Sulfonate is a cation-exchange resin prepared in the sodium form. Each g exchanges not less than 110 mg and not more than 135 mg of potassium, calculated on the anhydrous basis.

Packaging and storage—Preserve in well-closed containers.

Labeling—Sodium Polystyrene Sulfonate that is intended for preparing suspensions for oral or rectal administration may be labeled Sodium Polystyrene Sulfonate for Suspension.

Water, *Method I* ⟨921⟩: not more than 10.0%.

Ammonia—Place 1 g in a 50-mL beaker, add 5 mL of 1 *N* sodium hydroxide, cover the beaker with a watch glass having a moistened strip of red litmus paper on the underside, and allow to stand for 15 minutes: the litmus paper shows no blue color.

Sodium content—

Sodium solution—Dissolve in water an accurately weighed quantity of sodium chloride to make a solution containing 5.00 mg of sodium per mL.

Standard graph—Into four 1-liter flasks pipet, respectively, 0, 1, 2, and 3 mL of *Sodium solution*. To each flask add 0.1 mL of nitric acid, 0.1 mL of sulfuric acid, and 10 mL of low-sodium, low-potassium, nonionic surfactant solution (1 in 50), dilute with water to volume, and mix. Adjust the scale of a suitable flame spectrophotometer to a reading of 100 at a wavelength of 588 nm with the solution containing 15 mg of sodium per liter. Determine the instrument readings on the other three solutions, and plot the observed readings, on ruled coordinate paper, as the ordinate, and the concentration of sodium, in mg per liter, as the abscissa. The line intersects the ordinate at, or below, a scale reading of 25 ("blank reading").

Procedure—Ash the equivalent of 1 g of Sodium Polystyrene Sulfonate, accurately weighed, with a slight excess of sulfuric acid. Add 1 mL of nitric acid and a few mL of water to the residue. Warm to dissolve, and transfer with water to a 1-liter volumetric flask, dilute with water to volume, and mix. Pipet 10 mL of this solution into a 100-mL volumetric flask, add 1 mL of low-sodium, low-potassium, non-ionic surfactant solution (1 in 50), dilute with water to volume, and mix. Determine the instrument reading concomitantly with the readings obtained for plotting the *Standard graph*, and determine the sodium concentration, in mg per liter, by interpolation from the *Standard graph*. Calculate the percentage of sodium by the formula:

$$A/W,$$

in which A is the weight, in mg, of sodium found per liter and W is the weight, in g, of Sodium Polystyrene Sulfonate taken. The sodium content is not less than 9.4% and not more than 11.0%, calculated on the anhydrous basis.

Potassium exchange capacity—

Potassium solution—Dissolve an accurately weighed quantity of potassium chloride in water to make a solution containing 5.00 mg of potassium per mL.

Sodium solution—Dissolve an accurately weighed quantity of sodium chloride in water to make a solution containing 4.00 mg of sodium per mL.

Standard graph—Identify five 1-liter volumetric flasks by the numbers *1*, *2*, *3*, *4*, and *5*. In that order pipet into the flasks 4, 3, 2, 1, and 0 mL, respectively, of *Sodium solution*, and in the same order 0, 1, 2, 3, and 4 mL, respectively, of *Potassium solution*. To each flask add 10 mL of low-sodium, low-potassium, nonionic surfactant solution (1 in 50), dilute with water to volume, and mix. Adjust the scale of a suitable flame spectrophotometer to 100 with solution from flask *5* at 766 nm. Determine the instrument readings with solutions from flasks *4*, *3*, *2*, and *1*. On ruled coordinate paper, plot the observed instrument readings as the ordinate, and the concentrations, in mg per liter, of potassium as the abscissa.

Procedure—Pipet 100 mL of *Potassium solution* into a glass-stoppered flask containing about 1.6 g of Sodium Polystyrene Sulfonate, accurately weighed, shake by mechanical means for 15 minutes, filter, and discard the first 20 mL of the filtrate. Pipet 5 mL of the filtrate into a 1-liter volumetric flask, add 10 mL of low-sodium, low-potassium, nonionic surfactant solution (1 in 50), dilute with water to volume, and mix. Observe the flame spectrophotometer readings of the exchanged solution concomitantly with those obtained for plotting the *Standard graph*, and determine the potassium concentration, in mg per liter, by interpolation from the *Standard graph*. Calculate the quantity, in mg per g, of potassium adsorbed on the resin by the formula:

$$(X - 20Y)/W,$$

in which X is the weight, in mg, of potassium in 100 mL of *Potassium solution* before exchange, Y is the weight, in mg, of potassium per liter as interpolated from the *Standard graph*, and W is the weight, in g, of Sodium Polystyrene Sulfonate taken, expressed on the anhydrous basis.

Sodium Polystyrene Sulfonate Suspension

» Sodium Polystyrene Sulfonate Suspension is a suspension of Sodium Polystyrene Sulfonate in an aqueous vehicle containing a suitable quantity of sorbitol. It contains not less than 90.0 percent and not more than 110.0 percent of the labeled amount of sorbitol ($C_6H_{14}O_6$), and exchanges not less than 110 mg and not more than 135 mg of potassium for each g of the labeled amount of sodium polystyrene sulfonate.

Packaging and storage—Preserve in well-closed containers, protected from freezing and from excessive heat.

Labeling—Label it to state the quantity of sorbitol in a given volume of Suspension.

Microbial limits ⟨61⟩—Its total aerobic microbial count does not exceed 100 per mL, its total combined molds and yeasts count does not exceed 100 per mL, and it meets the requirements of the test for absence of *Pseudomonas aeruginosa*.

Sodium content—

Sodium solution and *Standard graph*—Prepare as directed in the test for *Sodium content* under *Sodium Polystyrene Sulfonate*.

Procedure—Transfer an accurately measured quantity of Sodium Polystyrene Sulfonate Suspension, freshly mixed and free from air bubbles, equivalent to about 1 g of sodium polystyrene sulfonate, to a suitable crucible, heat at 80° until dry, and ash the residue with a slight excess of sulfuric acid. Proceed as directed for *Procedure* in the test for *Sodium content* under *Sodium Polystyrene Sulfonate*, beginning with "Add 1 mL of nitric acid." Calculate the percentage of sodium by the formula:

$$A/L,$$

in which A is the quantity, in mg, of sodium found per liter, and L is the quantity, in g, of sodium polystyrene sulfonate in the portion of Suspension taken, based on the labeled amount: the sodium content is not less than 9.4% and not more than 11.0%.

Potassium exchange capacity—

Potassium solution, *Sodium solution*, and *Standard graph*—Prepare as directed in the test for *Potassium exchange capacity* under *Sodium Polystyrene Sulfonate*.

Procedure—Transfer an accurately measured quantity of Sodium Polystyrene Sulfonate Suspension, freshly mixed and free from air bubbles, equivalent to about 1.6 g of sodium polystyrene sulfonate, to a suitable glass-stoppered flask, add 100.0 mL of *Potassium solution*, shake by mechanical means for 15 minutes, filter, and discard the first 20 mL of the filtrate. Proceed as directed for *Procedure* in the test for *Potassium exchange capacity* under *Sodium Polystyrene Sulfonate*, beginning with "Pipet 5 mL of the filtrate." Calculate the quantity, in mg, of potassium adsorbed on each g of the sodium polystyrene sulfonate taken by the formula:

$$(X - 20Y)/L,$$

in which X is the quantity, in mg, of potassium in 100 mL of *Potassium solution* before exchange, Y is the quantity, in mg, of potassium per liter as interpolated from the *Standard graph*, and L is the labeled quantity, in g, of sodium polystyrene sulfonate in the portion of Suspension taken.

Assay for sorbitol—

Mobile phase, *Resolution solution*, *Standard preparation*, and *Chromatographic system*—Proceed as directed in the *Assay* under *Sorbitol* (see NF monograph).

Assay preparation—Dilute an accurately measured volume of Sodium Polystyrene Sulfonate Suspension, freshly mixed and free from air bubbles, quantitatively with water to obtain a solution containing about 4.8 mg of sorbitol per mL. Filter, and use the filtrate as the *Assay preparation*.

Procedure—Separately inject equal volumes (about 20 µL) of the *Assay preparation* and the *Standard preparation* into the chromatograph, record the chromatograms, and measure the responses for the major peaks. Proceed as directed for *Procedure* in the *Assay* under *Sorbitol* (see NF monograph). Calculate the quantity, in g, of $C_6H_{14}O_6$, in each mL of the Suspension taken by the formula:

$$(L/D)(C)(r_U/r_S),$$

in which L is the labeled quantity, in g, of sorbitol in each mL of Suspension, D is the quantity, in mg, of sorbitol in each mL of the *Assay preparation* based on the labeled quantity and the extent of dilution, C is the concentration, in mg per mL, of USP Sorbitol RS in the *Standard preparation*, and r_U and r_S are the peak responses obtained from the *Assay preparation* and the *Standard preparation*, respectively.

Sodium Propionate—*see* Sodium Propionate NF

Sodium Salicylate

COONa
OH

$C_7H_5NaO_3$ 160.10

Benzoic acid, 2-hydroxy-, monosodium salt.

Monosodium salicylate [*54-21-7*].

» Sodium Salicylate contains not less than 99.5 percent and not more than 100.5 percent of $C_7H_5NaO_3$, calculated on the anhydrous basis.

Packaging and storage—Preserve in well-closed, light-resistant containers.

Identification—A solution (1 in 20) responds to the tests for *Sodium* ⟨191⟩ and for *Salicylate* ⟨191⟩.

Water, *Method I* ⟨921⟩: not more than 0.5%.

Sulfite or thiosulfate—Add 1 mL of hydrochloric acid to a solution of 1.0 g in 20 mL of water, and filter the liquid: not more than 0.15 mL of 0.10 *N* iodine is required to produce a yellow color in the filtrate.

Heavy metals, *Method I* ⟨231⟩—Dissolve 2 g in 46 mL of water. Add, with constant stirring, 4 mL of 3 *N* hydrochloric acid, filter, and use 25 mL of the filtrate: the limit is 0.002%.

Assay—Transfer about 700 mg of Sodium Salicylate, accurately weighed, to a 250-mL beaker. Add 100 mL of glacial acetic acid, stir until the sample is completely dissolved, add crystal violet TS, and titrate with 0.1 *N* perchloric acid VS. Perform a blank determination, and make any necessary correction. Each mL of 0.1 *N* perchloric acid is equivalent to 16.01 mg of $C_7H_5NaO_3$.

Sodium Salicylate Tablets

» Sodium Salicylate Tablets contain not less than 95.0 percent and not more than 105.0 percent of the labeled amount of $C_7H_5NaO_3$.

Packaging and storage—Preserve in well-closed containers.

Identification—

A: Digest a quantity of powdered Tablets, equivalent to about 1 g of sodium salicylate, with 20 mL of water, and filter: the filtrate responds to the flame test for *Sodium* ⟨191⟩ and to the tests for *Salicylate* ⟨191⟩.

B: To 10 mL of the filtrate obtained in *Identification test A* add a slight excess of 3 *N* hydrochloric acid, collect the precipitate on a filter, wash it with small portions of cold water until the last washing is free from chloride, and dry at about 105° for 1 hour: the salicylic acid so obtained melts between 158° and 161° (see *Melting Range or Temperature* ⟨741⟩).

Dissolution ⟨711⟩—

Medium: water; 900 mL.

Apparatus 1: 100 rpm.

Time: 45 minutes.

Procedure—Determine the amount of $C_7H_5NaO_3$ dissolved from ultraviolet absorbances at the wavelength of maximum absorbance at about 266 nm, using filtered portions of the solution under test, diluted with water, if necessary, in comparison with a Standard solution having a known concentration of Sodium Salicylate in the same medium.

Tolerances—Not less than 75% (Q) of the labeled amount of $C_7H_5NaO_3$ is dissolved in 45 minutes.

Uniformity of dosage units ⟨905⟩: meet the requirements.

Assay—Place not less than 20 Sodium Salicylate Tablets in a 200-mL volumetric flask, add 100 mL of water, and allow to stand, with frequent agitation, until the tablets disintegrate completely. Dilute with water to volume, and mix. Filter through a dry filter into a dry flask, discarding the first 10 mL of the filtrate. Transfer an accurately measured volume of the subsequent filtrate, equivalent to about 500 mg of sodium salicylate, to a separator, and dilute with water, if necessary, to make about 25 mL. Add 75 mL of ether and 10 drops of bromophenol blue TS, and titrate with 0.1 *N* hydrochloric acid VS, mixing the water and ether layers by vigorous shaking until a permanent, pale green color is produced in the water layer. Draw off the water layer into a small flask, wash the ether layer once with 5 mL of water, and add this to the water layer. Add 20 mL of ether to the combined water solutions, and mix. Continue the titration with vigorous shaking until a permanent, pale green color is produced in the water layer. Each mL of 0.1 *N* hydrochloric acid is equivalent to 16.01 mg of $C_7H_5NaO_3$.

Sodium Starch Glycolate—*see* Sodium Starch Glycolate NF

Sodium Stearate—*see* Sodium Stearate NF

Sodium Stearyl Fumarate—*see* Sodium Stearyl Fumarate NF

Sodium Sulfate

$Na_2SO_4 . 10H_2O$ 322.19
Sulfuric acid disodium salt, decahydrate.
Disodium sulfate decahydrate [*7727-73-3*].
Anhydrous 142.04 [*7757-82-6*].

» Sodium Sulfate contains ten molecules of water of hydration, or is anhydrous. It contains not less than 99.0 percent of Na_2SO_4, calculated on the dried basis.

Packaging and storage—Preserve in tight containers, preferably at a temperature not exceeding 30°.

Labeling—Label it to indicate whether it is the decahydrate or is anhydrous.

Identification—A solution (1 in 20) responds to the tests for *Sodium* ⟨191⟩ and for *Sulfate* ⟨191⟩.

Acidity or alkalinity—To 10 mL of a solution, containing the equivalent of 1.0 g of $Na_2SO_4.10H_2O$ in 20 mL of water, add 1 drop of bromothymol blue TS: not more than 0.50 mL of either 0.010 *N* hydrochloric acid or 0.010 *N* sodium hydroxide is required to change the color of the solution.

Loss on drying ⟨731⟩—Dry at 105° for 4 hours: the decahydrate loses between 51.0% and 57.0% of its weight, and the anhydrous form loses not more than 0.5% of its weight.

Chloride ⟨221⟩—A portion equivalent to 1.0 g of $Na_2SO_4.10H_2O$ shows no more chloride than corresponds to 0.30 mL of 0.020 *N* hydrochloric acid (0.02%).

Arsenic, *Method I* ⟨211⟩—Dissolve a portion equivalent to 0.3 g of $Na_2SO_4 . 10H_2O$ in 35 mL of water: the limit is 0.001%.

Heavy metals ⟨231⟩—Dissolve a portion containing the equivalent of 2.0 g of $Na_2SO_4 . 10H_2O$ in 10 mL of water, add 2 mL of 0.1 *N* hydrochloric acid, then add water to make 25 mL: the limit is 0.001%.

Assay—Weigh accurately a portion of Sodium Sulfate, equivalent to about 400 mg of anhydrous sodium sulfate, dissolve in 200 mL of water, and add 1 mL of hydrochloric acid. Heat to boiling, and gradually add, in small portions and while constantly stirring, an excess of hot barium chloride TS (about 8 mL). Heat the mixture on a steam bath for 1 hour, collect the precipitate of barium sulfate on a tared filtering crucible, wash until free from chloride, dry, ignite, and weigh. The weight of the barium sulfate so obtained, multiplied by 0.6086, represents its equivalent of Na_2SO_4.

Sodium Sulfate Injection

» Sodium Sulfate Injection is a sterile, concentrated solution of Sodium Sulfate in Water for Injection, which upon dilution is suitable for parenteral use. It contains not less than 95.0 percent and not more than 105.0 percent of the labeled amount of $Na_2SO_4 . 10H_2O$.

Packaging and storage—Preserve in single-dose containers, preferably of Type I glass.

Labeling—Label it to indicate that it is to be diluted before injection to render it isotonic (3.89% of $Na_2SO_4 . 10H_2O$).

Identification—It responds to the tests for *Sodium* ⟨191⟩ and for *Sulfate* ⟨191⟩.

Pyrogen—When diluted with water for injection to contain 3.89% of $Na_2SO_4 . 10H_2O$, it meets the requirements of the *Pyrogen Test* ⟨151⟩.

pH ⟨791⟩: between 5.0 and 6.5.

Other requirements—It meets the requirements under *Injections* ⟨1⟩.

Assay—Transfer an accurately measured volume of Sodium Sulfate Injection, equivalent to about 400 mg of sodium sulfate ($Na_2SO_4 . 10H_2O$), to a suitable vessel. Dilute if necessary, to 200 mL, and proceed as directed in the *Assay* under *Sodium Sulfate*, beginning with "add 1 mL of hydrochloric acid." The weight of the barium sulfate so obtained, multiplied by 1.3804, represents its equivalent of $Na_2SO_4 . 10H_2O$.

Sodium Thiosulfate

$Na_2S_2O_3 . 5H_2O$ 248.17
Thiosulfuric acid, disodium salt, pentahydrate.
Disodium thiosulfate pentahydrate [*10102-17-7*].
Anhydrous 158.10 [*7772-98-7*].

» Sodium Thiosulfate contains not less than 99.0 percent and not more than 100.5 percent of $Na_2S_2O_3$, calculated on the anhydrous basis.

Packaging and storage—Preserve in tight containers.

Identification—

A: To a solution (1 in 10) add a few drops of iodine TS: the color is discharged.

B: A solution (1 in 10) responds to the tests for *Sodium* ⟨191⟩ and for *Thiosulfate* ⟨191⟩.

Water ⟨921⟩—Dry about 1.0 g, accurately weighed, in vacuum at 40° to 45° for 16 hours: it loses between 32.0% and 37.0% of its weight.

Arsenic, *Method I* ⟨211⟩—Prepare the *Test Preparation* as follows. Mix 1.0 g with 10 mL of water in the arsine generator flask. Add 15 mL of nitric acid and 5 mL of perchloric acid, mix, and heat cautiously to the production of strong fumes of perchloric acid. Cool, wash down the sides of the flask with water, and again heat to strong fumes. Again cool, wash down the sides of the flask, and heat to fumes. Cool, dilute with water to 52 mL, and add 3 mL of hydrochloric acid: the resulting solution meets the requirements of the test, the addition of 20 mL of 7 *N* sulfuric acid specified under *Procedure* being omitted. The limit is 3 ppm.

Calcium—Dissolve 1 g in 20 mL of water, and add a few mL of ammonium oxalate TS: no turbidity is produced.

Heavy metals ⟨231⟩—Dissolve 1 g in 10 mL of water, slowly add 5 mL of 3 *N* hydrochloric acid, evaporate on a steam bath nearly to dryness, and heat the residue at 150° for 1 hour. Add 15 mL of water to the residue, boil gently for 2 minutes, and filter. Heat the filtrate to boiling, and add sufficient bromine TS to produce a clear solution and provide a slight excess of bromine. Boil the solution to expel the excess bromine, cool to room temperature, add 1 drop of phenolphthalein TS, and neutralize with 1 *N* sodium hydroxide. Dilute with water to 25 mL: the limit is 0.002%.

Assay—Dissolve about 800 mg of Sodium Thiosulfate, accurately weighed, in 30 mL of water, and titrate with 0.1 *N* iodine VS, adding 3 mL of starch TS as the end-point is approached. Each mL of 0.1 *N* iodine is equivalent to 15.81 mg of $Na_2S_2O_3$.

Sodium Thiosulfate Injection

» Sodium Thiosulfate Injection is a sterile solution of Sodium Thiosulfate in freshly boiled Water for Injection. It contains not less than 95.0 percent and not more than 105.0 percent of the labeled amount of $Na_2S_2O_3 . 5H_2O$.

Packaging and storage—Preserve in single-dose containers, of Type I glass.

Identification—It responds to the *Identification tests* under *Sodium Thiosulfate.*

pH ⟨791⟩: between 6.0 and 9.5.

Other requirements—It meets the requirements under *Injections* ⟨1⟩.

Assay—Transfer to a suitable container an accurately measured volume of Sodium Thiosulfate Injection, containing about 1 g of sodium thiosulfate, and adjust to a pH of about 7 by the addition of 3 *N* hydrochloric acid. Dilute with water to about 20 mL, and titrate with 0.1 *N* iodine VS, adding 3 mL of starch TS as the end-point is approached. Each mL of 0.1 *N* iodine is equivalent to 24.82 mg of $Na_2S_2O_3 . 5H_2O$.

Solutions—*see complete list in index*
Sorbic Acid—*see* Sorbic Acid NF
Sorbitan Monolaurate—*see* Sorbitan Monolaurate NF
Sorbitan Monooleate—*see* Sorbitan Monooleate NF
Sorbitan Monopalmitate—*see* Sorbitan Monopalmitate NF
Sorbitan Monostearate—*see* Sorbitan Monostearate NF
Sorbitol—*see* Sorbitol NF

Sorbitol Solution

» Sorbitol Solution is a water solution containing, in each 100.0 g, not less than 64.0 g of D-sorbitol ($C_6H_{14}O_6$).

Packaging and storage—Preserve in tight containers.

Reference standard—*USP Sorbitol Reference Standard*—Use without drying.

Identification—

A: To 3 mL of a 1 in 75 dilution of it in a 15-cm test tube add 3 mL of freshly prepared catechol solution (1 in 10), mix, add 6 mL of sulfuric acid, again mix, then gently heat the tube in a flame for about 30 seconds: a faint pink color appears.

B: To 6 mL of it add 7 mL of methanol, 1 mL of benzaldehyde, and 1 mL of hydrochloric acid, and shake by mechanical means until crystals appear. Filter with the aid of suction, dissolve the crystals in 20 mL of boiling water containing 1 g of sodium bicarbonate, filter while hot, cool the filtrate, filter with suction, wash with 5 mL of a mixture of equal parts of methanol and water, and air-dry: the sorbitol monobenzylidene derivative so obtained melts between 174° and 179°.

Specific gravity ⟨841⟩: not less than 1.285.

Refractive index ⟨831⟩: between 1.455 and 1.465 at 20°.

Water, *Method I* ⟨921⟩: between 28.5% and 31.5%.

Residue on ignition ⟨281⟩: not more than 0.1%.

Chloride ⟨221⟩—A 1.5-g portion shows no more chloride than corresponds to 0.10 mL of 0.020 *N* hydrochloric acid (0.005%).

Sulfate ⟨221⟩—A 1.0-g portion shows no more sulfate than corresponds to 0.10 mL of 0.020 *N* sulfuric acid (0.010%).

Arsenic, *Method II* ⟨211⟩: 2.5 ppm.

Heavy metals, *Method II* ⟨231⟩: 0.001%.

Reducing sugars—A 10.0-g portion, accurately weighed, meets the requirements of the test for *Reducing sugars* under *Sorbitol.*

Assay—

Mobile phase, Resolution solution, Standard preparation, and *Chromatographic system*—Proceed as directed in the *Assay* under *Sorbitol.*

Assay preparation—Transfer an accurately weighed portion of Sorbitol Solution, equivalent to about 0.24 g of sorbitol, to a 50-mL volumetric flask, dilute with water to volume, and mix.

Procedure—Proceed as directed for *Procedure* in the *Assay* under *Sorbitol* (see NF monograph). Calculate the quantity, in mg, of $C_6H_{14}O_6$ in the portion of Solution taken by the formula:

$$50C(r_U / r_S),$$

in which C is the concentration, in mg per mL, of USP Sorbitol RS in the *Standard preparation,* and r_U and r_S are the peak responses obtained from the *Assay preparation* and the *Standard preparation,* respectively.

Soybean Oil

» Soybean Oil is the refined fixed oil obtained from the seeds of the soya plant *Glycine soja* (Fam. Leguminosae).

Packaging and storage—Preserve in tight, light-resistant containers, and avoid exposure to excessive heat.

Specific gravity ⟨841⟩: between 0.916 and 0.922.

Refractive index ⟨831⟩: between 1.465 and 1.475.

Heavy metals, *Method II* ⟨231⟩: 0.001%.

Free fatty acids ⟨401⟩—The free fatty acids in 10 g require for neutralization not more than 2.5 mL of 0.020 *N* sodium hydroxide.

Fatty acid composition—Place about 1 g of Soybean Oil in a small conical flask fitted with a reflux attachment. Add 10 mL of methanol and 0.5 mL of 1 *N* methanolic potassium hydroxide solution prepared by dissolving 34 g of potassium hydroxide in sufficient methanol to produce 500 mL, allowing to settle for 24 hours, and decanting the clear solution. Reflux the mixture for 10 minutes, cool, transfer to a separator with the aid of 15 mL of *n*-heptane, shake with 10 mL of saturated sodium chloride solution, and allow to separate. Transfer the lower layer to another separator, and shake it with 10 mL of *n*-heptane. Wash the combined organic layers with 10 mL of water, dry over anhydrous sodium sulfate, and filter. Introduce a suitable portion of the filtrate into a gas chromatograph equipped with a flame-ionization detector and a column, preferably glass, 1.5 mm in length and 4 mm in internal diameter packed with 10 percent liquid phase G4 on support S1A, maintained at a temperature of 175°. The carrier gas is nitrogen. Measure the 5 main peak areas of the methyl esters of the fatty acids. The order of elution is palmitate, stearate, oleate, linoleate, and linolenate, and their relative areas, expressed as percentages of the total area of the 5 main peaks, are in the ranges 7 to 14, 1 to 6, 19 to 30, 44 to 62, and 4 to 11, respectively.

Iodine value, *Method II* ⟨401⟩: between 120 and 141.

Saponification value ⟨401⟩: between 180 and 200.

Unsaponifiable matter ⟨401⟩: not more than 1.0%.

Cottonseed oil—Mix 10 mL in a 250-mm × 25-mm test tube with 10 mL of a mixture of equal volumes of amyl alcohol and a 1 in 100 solution of sulfur in carbon disulfide. Warm the mixture carefully. [*Caution—Carbon disulfide vapors may be ignited with a hot bath or hot steam pipe.*] When the carbon disulfide has been expelled, immerse the tube to one-third of its length in a boiling, saturated solution of sodium chloride: no red color develops in the mixture within 15 minutes.

Peroxide—

Mixed solvent—Mix 60 mL of glacial acetic acid with 40 mL of chloroform.

Potassium iodide solution—Prepare a saturated solution of potassium iodide in freshly boiled and cooled water, and store it protected from light. Discard it if it gives a color on addition of *Mixed solvent* and starch TS.

Procedure—Transfer about 10 g of Soybean Oil, accurately weighed, to a conical flask, add 30 mL of *Mixed solvent*, swirl to dissolve, add 0.5 mL of *Potassium iodide solution*, swirl the flask for 1 minute, accurately timed, add 30 mL of water, and titrate with 0.01 *N* sodium thiosulfate VS, with vigorous agitation, to a light yellow color. Add 0.5 mL of starch TS, and continue the titration until the blue color has disappeared. Perform a blank test, and make any necessary correction. Calculate the peroxide content, in mEq per kg, by the formula:

$$1000VN/W,$$

in which V is the volume, in mL, of sodium thiosulfate required and N is its normality, and W is the weight, in g, of Soybean Oil taken. The limit is 10.0.

Sterile Spectinomycin Hydrochloride

· 2HCl · 5H$_2$O

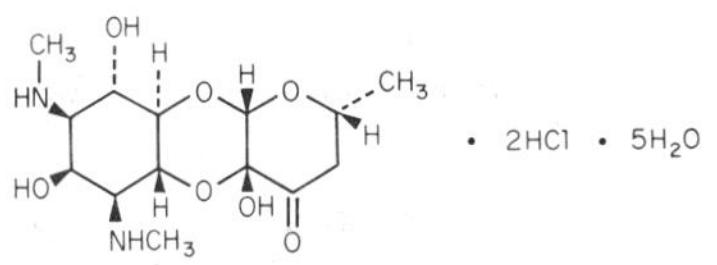

$C_{14}H_{24}N_2O_7 . 2HCl . 5H_2O$ 495.35

4*H*-Pyrano[2,3-*b*][1,4]benzodioxin-4-one, decahydro-4a,7,9-trihydroxy-2-methyl-6,8-bis(methylamino)-, dihydrochloride, pentahydrate, [2*R*-(2α,4aβ,5aβ,6β,7β,8β,9α,9aα,10aβ)]-.

(2*R*,4a*R*,5a*R*,6*S*,7*S*,8*R*,9*S*,9a*R*,10a*S*)-Decahydro-4a,7,9-trihydroxy-2-methyl-6,8-bis(methylamino)-4*H*-pyrano[2,3-*b*]-[1,4]benzodioxin-4-one dihydrochloride pentahydrate [*22189-32-8*].

Anhydrous 405.27 [*21736-83-4*].

» Sterile Spectinomycin Hydrochloride has a potency equivalent to not less than 603 µg of spectinomycin ($C_{14}H_{24}N_2O_7$) per mg.

Packaging and storage—Preserve in *Containers for Sterile Solids* as described under *Injections* ⟨1⟩.

Reference standard—*USP Spectinomycin Hydrochloride Reference Standard*—Do not dry before using.

Identification—The infrared absorption spectrum of a mineral oil dispersion of it exhibits maxima only at the same wavelengths as that of a similar preparation of USP Spectinomycin Hydrochloride RS.

Crystallinity ⟨695⟩: meets the requirements.

Depressor substances—It meets the requirements of the *Depressor Substances Test* ⟨101⟩, the test dose being 1.0 mg per kg of a solution in sterile saline TS containing 15.0 mg of spectinomycin per mL.

Pyrogen—It meets the requirements of the *Pyrogen Test* ⟨151⟩, the test dose being 1.0 mL per kg of a solution in pyrogen-free saline TS containing 50 mg of spectinomycin per mL.

Sterility—It meets the requirements under *Sterility Tests* ⟨71⟩, when tested as directed in the section, *Test Procedures Using Membrane Filtration*.

pH ⟨791⟩: between 3.8 and 5.6, in a solution containing 10 mg per mL (where packaged for dispensing, between 4.0 and 7.0, in the solution constituted as directed in the labeling).

Water, *Method I* ⟨921⟩: between 16.0% and 20.0%.

Residue on ignition ⟨281⟩: not more than 1.0%, the charred residue being moistened with 2 mL of nitric acid and 5 drops of sulfuric acid.

Assay—

Internal standard solution—Dissolve triphenylantimony in dimethylformamide to obtain a solution containing about 2 mg per mL.

Standard preparation—Transfer about 30 mg of USP Spectinomycin Hydrochloride RS, accurately weighed, to a glass-stoppered, 25-mL conical flask. Add 10.0 mL of *Internal standard solution* and 1.0 mL of hexamethyldisilazane, and shake intermittently for 1 hour.

Assay preparation—Using Sterile Spectinomycin Hydrochloride, proceed as directed under *Standard preparation*.

Chromatographic system (see *Chromatography* ⟨621⟩)—The gas chromatograph is equipped with a flame-ionization detector and contains a 60-cm × 3-mm glass column packed with 5 percent phase G27 on 80- to 100-mesh support S1AB. The column and detector are maintained at about 190° and 220°, respectively, and the injection port at about 215°, and dry helium is used as the carrier gas at a flow rate of about 45 mL per minute.

Procedure—Separately inject equal volumes (about 1 µL) of the *Standard preparation* and the *Assay preparation* into the chromatograph, record the chromatograms, and measure the responses for the major peaks. Calculate the ratio, R_U, of the response of the spectinomycin peak to the response of the internal standard peak in the chromatogram from the *Assay preparation*, and similarly calculate the ratio, R_S, in the chromatogram from the *Standard preparation*. In a suitable chromatogram, the resolution, R, between the peaks is not less than 2.0, and the relative standard deviation of the peak response ratios, R_S, from replicate injections of the *Standard preparation* is not more than 3.5%. Calculate the quantity, in µg, of $C_{14}H_{24}N_2O_7$ in the portion of Sterile Spectinomycin Hydrochloride taken to prepare the *Assay preparation* by the formula:

$$P(W_S)(R_U/R_S),$$

in which P is the potency of USP Spectinomycin Hydrochloride RS, in µg of spectinomycin per mg, and W_S is the weight, in mg, of USP Spectinomycin Hydrochloride RS taken from the *Standard preparation*.

Sterile Spectinomycin Hydrochloride for Suspension

» Sterile Spectinomycin Hydrochloride for Suspension is Sterile Spectinomycin Hydrochloride packaged for dispensing. It contains the equivalent of not less than 90.0 percent and not more than 120.0 percent of the labeled amount of spectinomycin ($C_{14}H_{24}N_2O_7$).

Packaging and storage—Preserve in *Containers for Sterile Solids* as described under *Injections* ⟨1⟩.

Reference standard—*USP Spectinomycin Hydrochloride Reference Standard*—Do not dry before using.

Other requirements—It conforms to the definition, responds to the *Identification test*, and meets the requirements for *Depressor substances, Crystallinity, Pyrogen, Sterility, pH, Water, and Residue on ignition* under *Sterile Spectinomycin Hydrochloride*. It meets also the requirements for *Uniformity of Dosage Units* ⟨905⟩ and *Labeling* under *Injections* ⟨1⟩.

Assay—

Internal standard solution, Standard preparation, and *Chromatographic system*—Prepare as directed in the *Assay* under *Sterile Spectinomycin Hydrochloride.*

Assay preparation 1—Suspend the contents of 1 container of Sterile Spectinomycin Hydrochloride for Suspension in water, and dilute quantitatively with water to obtain a stock solution containing about 20 mg of spectinomycin per mL. Transfer 1.0 mL of this solution to a glass-stoppered, 25-mL conical flask, and freeze-dry. Add 10.0 mL of *Internal standard solution* and 1.0 mL of hexamethyldisilazane, and shake intermittently for 1 hour.

Assay preparation 2 (where the label states the quantity of spectinomycin in a given volume of constituted suspension)—Constitute 1 container of Sterile Spectinomycin Hydrochloride for Suspension in a volume of water, accurately measured, corresponding to the volume of diluent specified in the labeling. Dilute an accurately measured portion of the constituted suspension quantitatively with water to obtain a stock solution containing about 20 mg of spectinomycin per mL. Transfer 1.0 mL of this solution to a glass-stoppered, 25-mL conical flask, and freeze-dry. Add 10.0 mL of *Internal standard solution* and 1.0 mL of hexamethyldisilazane, and shake intermittently for 1 hour.

Procedure—Proceed as directed in the *Assay* under *Sterile Spectinomycin Hydrochloride.* Calculate the quantity, in g, of $C_{14}H_{24}N_2O_7$ in the container of Sterile Spectinomycin Hydrochloride for Suspension taken to prepare *Assay preparation 1* by the formula:

$$(L_1/D_1)(P/1000)(W_S)(R_U/R_S),$$

in which L_1 is the labeled quantity, in g, of $C_{14}H_{24}N_2O_7$ in the container, and D_1 is the concentration, in mg per mL, of spectinomycin in the stock solution used to prepare *Assay preparation 1*, on the basis of the labeled quantity in the container and the extent of dilution, and the other terms are as defined therein. Calculate the quantity, in mg, of $C_{14}H_{24}N_2O_7$ in each mL of constituted Suspension taken to prepare *Assay preparation 2* by the formula:

$$(L_2/D_2)(P/1000)(W_S)(R_U/R_S),$$

in which L_2 is the labeled quantity, in mg, of $C_{14}H_{24}N_2O_7$ in each mL of constituted suspension of Sterile Spectinomycin Hydrochloride for Suspension, and D_2 is the concentration, in mg per mL, of spectinomycin in the stock solution used to prepare *Assay preparation 2*, on the basis of the labeled quantity in each mL of constituted suspension and the extent of dilution.

Spirit, Aromatic Ammonia—*see* Ammonia Spirit, Aromatic

Spirit, Camphor—*see* Camphor Spirit

Spirit, Peppermint—*see* Peppermint Spirit

Spironolactone

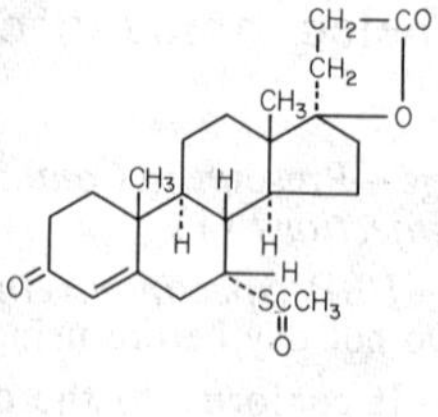

$C_{24}H_{32}O_4S$ 416.57

Pregn-4-ene-21-carboxylic acid, 7-(acetylthio)-17-hydroxy-3-oxo-, γ-lactone, (7α,17α)-.

17-Hydroxy-7α-mercapto-3-oxo-17α-pregn-4-ene-21-carboxylic acid γ-lactone acetate [*52-01-7*].

» Spironolactone contains not less than 97.0 percent and not more than 103.0 percent of $C_{24}H_{32}O_4S$, calculated on the dried basis.

Packaging and storage—Preserve in well-closed containers.

Reference standard—*USP Spironolactone Reference Standard*—Dry at 105° for 2 hours before using.

Identification—

A: The infrared absorption spectrum of a 1 in 20 solution in chloroform, the test specimen having been previously dried, exhibits maxima only at the same wavelengths as that of a similar preparation of USP Spironolactone RS.

B: The ultraviolet absorption spectrum of a 1 in 100,000 solution in methanol exhibits maxima at the same wavelengths as that of a similar solution of USP Spironolactone RS, concomitantly measured, and the respective absorptivities, calculated on the dried basis, at the wavelength of maximum absorbance at about 238 nm do not differ by more than 3.0%.

C: Add 100 mg to a mixture of 10 mL of water and 2 mL of 1 *N* sodium hydroxide, boil the mixture for 3 minutes, cool, and add 1 mL of glacial acetic acid and 1 mL of lead acetate TS: a brown to black precipitate of lead sulfide is formed.

Melting range ⟨741⟩: between 198° and 207°, with decomposition. Occasionally it may show preliminary melting at about 135°, followed by resolidification.

Specific rotation ⟨781⟩: between −33° and −37°, calculated on the dried basis, determined in a solution in chloroform containing 10 mg per mL.

Loss on drying ⟨731⟩—Dry it at 105° for 2 hours: it loses not more than 0.5% of its weight.

Mercapto compounds—Shake 2.0 g with 30 mL of water, filter, then to 15 mL of the filtrate add 3 mL of starch TS, and titrate with 0.010 *N* iodine. Perform a blank determination, and make any necessary correction. Not more than 0.10 mL of 0.010 *N* iodine is consumed.

Ordinary impurities ⟨466⟩—

Test solution: chloroform.

Standard solution: chloroform.

Eluant: butyl acetate.

Visualization: 5.

Assay—

Mobile phase—Prepare a mixture of acetonitrile and aqueous 0.02 *M* dibasic ammonium phosphate (55:45). Degas the mixture under vacuum with continuous stirring for about 30 minutes before use.

Standard preparation—Dissolve an accurately weighed quantity of USP Spironolactone RS in a mixture of acetonitrile and water (9:1), and dilute quantitatively with the same mixture to obtain a solution having a known concentration of about 250 µg of USP Spironolactone RS per mL.

Assay preparation—Transfer about 25 mg of Spironolactone, accurately weighed, to a 100-mL volumetric flask, add a mixture of acetonitrile and water (9:1) to volume, and mix.

Chromatographic system (see *Chromatography* ⟨621⟩)—The liquid chromatograph is equipped with a 254-nm detector and a 4-mm × 30-cm column that contains packing L1. The flow rate is about 1 mL per minute. Chromatograph six replicate injections of the *Standard preparation*, and record the peak responses as directed under *Procedure:* the relative standard deviation is not more than 1.5%.

Procedure—Separately inject equal volumes (about 20 µL) of the *Standard preparation* and the *Assay preparation* into the chromatograph by means of a suitable microsyringe or sampling valve, record the chromatograms, and measure the responses for the major peaks. The retention time for spironolactone is about 5 minutes. Calculate the quantity, in mg, of $C_{24}H_{32}O_4S$ in the portion of Spironolactone taken by the formula:

$$0.1C(r_U/r_S),$$

in which C is the concentration, in µg per mL, of USP Spironolactone RS in the *Standard preparation*, and r_U and r_S are the peak responses obtained for spironolactone from the *Assay preparation* and the *Standard preparation*, respectively.

Spironolactone Tablets

» Spironolactone Tablets contain not less than 95.0 percent and not more than 105.0 percent of the labeled amount of $C_{24}H_{32}O_4S$.

Packaging and storage—Preserve in tight, light-resistant containers.

Reference standard—*USP Spironolactone Reference Standard*—Dry at 105° for 2 hours before using.

Identification—Mix a quantity of finely powdered Tablets, equivalent to about 100 mg of spironolactone, with 25 mL of methanol, and filter. On a suitable thin-layer chromatographic plate (see *Chromatography* ⟨621⟩), coated with a 0.25-mm layer of chromatographic silica gel mixture, apply 10 µL of this solution and 10 µL of a solution of USP Spironolactone RS in methanol containing 4 mg per mL. Develop the chromatogram in a solvent system consisting of a mixture of chloroform, ethyl acetate, and methanol (2:2:1) until the solvent front has moved about three-fourths of the length of the plate. Remove the plate from the developing chamber, mark the solvent front, and allow the solvent to evaporate. Locate the spots on the plate by viewing under short-wavelength ultraviolet light: the R_f value of the principal spot obtained from the solution under test corresponds to that obtained from the Standard solution.

Dissolution ⟨711⟩—

Medium: 0.1 *N* hydrochloric acid containing 0.1% of sodium lauryl sulfate; 1000 mL.

Apparatus 2: 75 rpm.

Time: 60 minutes.

Procedure—Determine the amount of $C_{24}H_{32}O_4S$ dissolved using ultraviolet absorbances at the wavelength of maximum absorbance at about 242 nm obtained on filtered portions of the solution under test, diluted with *Dissolution Medium*, if necessary, in comparison with a Standard solution having a known concentration of USP Spironolactone RS in the same medium. [NOTE—A volume of alcohol not exceeding 1% of the final volume of the solution may be used to prepare the Standard solution.]

Tolerances—Not less than 75% (*Q*) of the labeled amount of $C_{24}H_{32}O_4S$ is dissolved in 60 minutes.

Uniformity of dosage units ⟨905⟩: meet the requirements.

Assay—

Mobile phase, Standard preparation, and *Chromatographic system*—Prepare as directed in the *Assay* under *Spironolactone.*

Assay preparation—Weigh and finely powder not less than 20 Spironolactone Tablets. Transfer an accurately weighed portion of the powder, equivalent to about 25 mg of spironolactone, to a 100-mL volumetric flask, add 10.0 mL of water, and swirl gently for about 10 minutes. Add 70 mL of acetonitrile, sonicate for 30 minutes with occasional shaking to dissolve the soluble substances in the mixture, dilute with acetonitrile to volume, and mix. Centrifuge a portion of the mixture at about 3000 rpm for 10 minutes, and use the supernatant liquid.

Procedure—Proceed as directed for *Procedure* in the *Assay* under *Spironolactone.* Calculate the quantity, in mg, of $C_{24}H_{32}O_4S$ in the portion of Tablets taken by the formula:

$$0.1C(r_U/r_S),$$

in which *C* is the concentration, in µg per mL, of USP Spironolactone RS in the *Standard preparation*, and r_U and r_S are the peak responses obtained for spironolactone from the *Assay preparation* and the *Standard preparation*, respectively.

Spironolactone and Hydrochlorothiazide Tablets

» Spironolactone and Hydrochlorothiazide Tablets contain not less than 90.0 percent and not more than 110.0 percent of the labeled amounts of spironolactone ($C_{24}H_{32}O_4S$) and hydrochlorothiazide ($C_7H_8ClN_3O_4S_2$).

Packaging and storage—Preserve in tight, light-resistant containers.

Reference standards—*USP Spironolactone Reference Standard*—Dry at 105° for 2 hours before using. *USP Hydrochlorothiazide Reference Standard*—Dry at 105° for 1 hour before using.

Identification—The retention times of the major peaks in the chromatogram of the *Assay preparation* correspond to those of the *Standard preparation*, obtained as directed in the *Assay.*

Dissolution ⟨711⟩—

Medium: 0.1 *N* hydrochloric acid containing 0.1% sodium lauryl sulfate; 900 mL.

Apparatus 1: 100 rpm.

Time: 60 minutes.

Determine the amounts of Spironolactone and Hydrochlorothiazide dissolved using the following method.

Standard solution—Prepare a solution of USP Spironolactone RS and USP Hydrochlorothiazide RS in a mixture of methanol and water (1:1) having accurately known concentrations of about 0.0125 mg of each per mL.

Test solution—Transfer a 5.0-mL portion of the solution under test to a 10-mL volumetric flask, dilute with methanol to volume, and mix.

Mobile phase, Chromatographic system, and *Procedure*—Proceed as directed in the *Assay.*

Tolerances—Not less than 75% (*Q*) of each of the labeled amounts of $C_{24}H_{32}O_4S$, and $C_7H_8ClN_3O_4S_2$ are dissolved in 60 minutes.

Uniformity of dosage units ⟨905⟩: meet the requirements for *Content uniformity* with respect to spironolactone and to hydrochlorothiazide.

Assay—

Mobile phase—Prepare a filtered and degassed mixture of methanol and water (7:3). Make adjustments if necessary (see *System Suitability* under *Chromatography* ⟨621⟩).

Standard preparation—Dissolve accurately weighed quantities of USP Spironolactone RS and USP Hydrochlorothiazide RS in methanol to obtain a solution having known concentrations of about 50 µg of each per mL.

Assay preparation—Weigh and finely powder not less than 20 Spironolactone and Hydrochlorothiazide Tablets. Transfer an accurately weighed portion of the powder, equivalent to about 25 mg of spironolactone, to a 100-mL volumetric flask, add about 70 mL of methanol, shake by mechanical means for 30 minutes, dilute with methanol to volume, mix, and centrifuge. Transfer 20.0 mL of the resultant clear liquid to a 100-mL volumetric flask, dilute with methanol to volume, and mix.

Chromatographic system (see *Chromatography* ⟨621⟩)—The liquid chromatograph is equipped with a 254-nm detector and a 4-mm × 30-cm column that contains packing L1. The flow rate is about 1 mL per minute. Chromatograph the *Standard preparation*, and record the peak responses as directed under *Procedure:* the resolution, *R*, between hydrochlorothiazide and spironolactone is not less than 2.0, and the relative standard deviation for replicate injections is not more than 2.0%.

Procedure—Separately inject equal volumes (about 10 µL) of the *Standard preparation* and the *Assay preparation* into the chromatograph, record the chromatograms, and measure the responses for the major peaks. The relative retention times are about 0.5 for hydrochlorothiazide and 1.0 for spironolactone. Calculate the quantity, in mg, of spironolactone ($C_{24}H_{32}O_4S$) in the portion of Tablets taken by the formula:

$$0.5C(r_U/r_S),$$

in which *C* is the concentration, in µg per mL, of USP Spironolactone RS in the *Standard preparation*, and r_U and r_S are the responses of the spironolactone peak obtained from the *Assay preparation* and the *Standard preparation*, respectively. Calculate the quantity, in mg, of hydrochlorothiazide ($C_7H_8ClN_3O_4S_2$) by the same formula, changing the terms to refer to hydrochlorothiazide.

Sponge, Absorbable Gelatin—*see* Gelatin Sponge, Absorbable

Squalane—*see* Squalane NF

Stannous Fluoride

SnF_2 156.69
Tin fluoride (SnF_2).
Tin fluoride (SnF_2) [*7783-47-3*].

» Stannous Fluoride contains not less than 71.2 percent of stannous tin (Sn^{++}), and not less than 22.3 percent and not more than 25.5 percent of fluoride (F), calculated on the dried basis.

Packaging and storage—Preserve in well-closed containers.

Reference standard—*USP Sodium Fluoride Reference Standard*—Dry at 150° for 4 hours before using.

Identification—

A: To 5 mL of a solution (1 in 100) in a test tube add 2 mL of calcium chloride TS: a fine, white precipitate of calcium fluoride is formed.

B: Mix on a spot plate 2 drops of a solution (1 in 100) with 2 drops of silver nitrate TS: a brown-black precipitate is formed.

C: Add 1 drop of a solution (1 in 100) to 2 drops of mercuric chloride TS: a white, silky precipitate is formed. On further addition of the solution (1 in 100), a brown-black precipitate is formed.

pH ⟨791⟩: between 2.8 and 3.5, in a freshly prepared 0.4% solution.

Loss on drying ⟨731⟩—Dry it at 105° for 4 hours: it loses not more than 0.5% of its weight.

Water-insoluble substances—Transfer about 10 g, accurately weighed, to a 400-mL plastic beaker, add 200 mL of water, and stir with a plastic rod for 3 minutes, or until no more solid dissolves. Filter through a tared porcelain filtering crucible tightly packed with asbestos fiber, and wash thoroughly, first with ammonium fluoride solution (1 in 100), then with water. [NOTE—Prepare and use the filtering crucible in a well-ventilated hood.] Dry the residue at 105° for 4 hours, cool, and weigh: the weight of the residue does not exceed 0.2%.

Antimony—

Rhodamine B solution—Dissolve 20 mg of rhodamine B in 200 mL of 0.5 *N* hydrochloric acid.

Standard preparation—Transfer 55.0 mg of antimony potassium tartrate, accurately weighed, to a 200-mL volumetric flask, dissolve in water, dilute with water to volume, and mix. Transfer 5.0 mL of this solution to a 500-mL volumetric flask, add 6 *N* hydrochloric acid to volume, and mix.

Test preparation—Transfer 1.0 g of Stannous Fluoride, accurately weighed, to a 50-mL volumetric flask, add 6 *N* hydrochloric acid to volume, and mix.

Procedure—Pipet 5 mL each of the *Standard preparation* and the *Test preparation* into separate 125-mL separators, add 15 mL of hydrochloric acid and 1 g of ceric sulfate, and allow to stand for 5 minutes, with occasional shaking. Add 500 mg of hydroxylamine hydrochloride, and shake for 1 minute. Pipet 15 mL of isopropyl ether into the mixture, shake for 30 seconds, add 7 mL of water, and mix. Cool in a water bath at room temperature for 10 minutes, shake for 30 seconds, allow the layers to separate, and discard the aqueous phase. Add 20 mL of *Rhodamine B solution*, shake for 30 seconds, and discard the aqueous layer. Decant the ether layer from the top of the separator, and centrifuge, if necessary, to obtain a clear solution. Concomitantly determine the absorbances of the ether solutions from the *Test preparation* and the *Standard preparation* at the wavelength of maximum absorbance at about 550 nm, with a suitable spectrophotometer, using water as the blank: the absorbance of the *Test preparation* does not exceed that of the *Standard preparation* (0.005%).

Assay for stannous ion—

0.1 N Potassium iodide–iodate—In a 1000-mL volumetric flask, dissolve 3.567 g of potassium iodate, previously dried at 110° to constant weight, in 200 mL of oxygen-free water containing 1 g of sodium hydroxide and 10 g of potassium iodide, dilute with oxygen-free water to volume, and mix. Standardize this solution by titrating a solution prepared from an accurately weighed quantity of reagent tin (Sn) and hydrochloric acid. Each mL of *0.1 N Potassium iodide–iodate* is equivalent to 5.935 mg of Sn.

Procedure—Transfer about 250 mg of Stannous Fluoride, accurately weighed, to a 500-mL conical flask, and add 300 mL of hot, recently boiled 3 *N* hydrochloric acid. While passing a stream of an oxygen-free inert gas over the surface of the liquid, swirl the flask to dissolve the Stannous Fluoride, and cool to room temperature. Add 5 mL of potassium iodide TS, and titrate in an inert atmosphere with *0.1 N Potassium iodide–iodate*, adding 3 mL of starch TS as the end-point is approached. Each mL of *0.1 N Potassium iodide–iodate* is equivalent to 5.935 mg of Sn^{++}.

Assay for fluoride—

Solution A—Dissolve 3.16 g of 4,5-dihydroxy-3-(*p*-sulfophenylazo)-2,7-naphthalenedisulfonic acid trisodium salt in 550 mL of water.

Solution B—Transfer to a 500-mL volumetric flask 113 mg of zirconium oxychloride, and dissolve in 50 mL of water. Add 350 mL of hydrochloric acid, dilute with water to volume, and mix.

Solution C—Dilute 50 mL of *Solution A* with 500 mL of water and 35 mL of hydrochloric acid.

Solution D—Mix equal volumes of *Solution A* and *Solution B*, and store in an amber bottle.

Solution E—Dilute 10.0 mL of *Solution D* with water in a 100-mL volumetric flask to volume, and mix.

Standard preparation—Dissolve a suitable quantity of USP Sodium Fluoride RS, accurately weighed, in water, and dilute quantitatively and stepwise with water to obtain a Standard solution having a known concentration of about 10 µg of fluoride per mL.

Assay preparation—Transfer to a 250-mL volumetric flask about 100 mg of Stannous Fluoride, accurately weighed. Add 50 mL of water, mix vigorously for 5 minutes, dilute with water to volume, and mix. Pipet 5 mL of this solution into a 100-mL volumetric flask, dilute with water to volume, and mix.

Procedure—Pipet 1.0, 2.0, 3.0, and 4.0 mL, respectively, of the *Standard preparation* into separate 100-mL volumetric flasks, and pipet 5.0 mL of the *Assay preparation* into another 100-mL volumetric flask. To each flask add 10.0 mL of *Solution D*, dilute with water to volume, and mix. Concomitantly determine the absorbances of *Solution E*, the solutions from the *Standard preparation*, and the solution from the *Assay preparation* at the wavelength of maximum absorbance at about 590 nm, with a suitable spectrophotometer, using *Solution C* as the blank. Subtract the absorbances obtained with the fluoride-containing solutions from the absorbance obtained with *Solution E*. Prepare a standard curve by plotting the amount of fluoride added, in µg, against the absorbance differences. From the standard curve, determine the concentration, in µg per 5 mL, of F in the *Assay preparation*. This value is equal to the quantity, in mg, of F in the portion of Stannous Fluoride taken.

Stannous Fluoride Gel

» Stannous Fluoride Gel contains not less than 95.0 percent and not more than 115.0 percent of the labeled amount of SnF_2 in a glycerin medium containing a suitable viscosity-inducing agent.

NOTE—In the preparation of this Gel, use Glycerin that has a low water content, that is, Glycerin having a specific gravity of not less than 1.2607, corresponding to a concentration of 99.5%.

Packaging and storage—Preserve in well-closed containers.

Reference standard—*USP Sodium Fluoride Reference Standard*—Dry at 150° for 4 hours before using.

Identification—It responds to the *Identification tests* under *Stannous Fluoride*, a solution of it in water containing about 1 mg of stannous fluoride per mL being used instead of a 1 in 100 solution.

Viscosity ⟨911⟩—Place a quantity of Gel in a suitable plastic container, insert the stopper securely, and allow to stand until the gel is free from air bubbles. Place it in a water bath maintained at a temperature of 25 ± 0.5° until it adjusts to the temperature of the water bath (4 hours or longer). Do not stir the gel while it is in the bath. Remove the specimen from the bath, stir the gel gently for 2 minutes, and without delay, using a rotational viscosimeter, determine the viscosity using a spindle having a cylinder 1.27 cm in diameter and 0.16 cm high attached to a shaft 0.32 cm in diameter, the distance from the top of the cylinder to the lower tip of the shaft being 2.54 cm and the immersion depth being 5.00 cm (No. 3 spindle). Operate the viscosimeter at 12 rpm, and record the scale reading at 1-minute intervals for 4 minutes. Calculate the viscosity, in centipoises, by multiplying the scale reading by 100: the viscosity is between 600 and 170,000 centipoises.

pH ⟨791⟩: between 2.8 and 4.0, in a freshly prepared mixture with water (1:1).

Stannous ion content—

0.1 N Potassium iodide–iodate—Prepare as directed in the *Assay for stannous ion* under *Stannous Fluoride*.

Procedure—Transfer an accurately weighed quantity of Stannous Fluoride Gel, equivalent to about 80 mg of stannous fluoride, to a capped plastic vessel equipped for titration in an inert atmosphere. Add a plastic coated stirring bar, 20 mL of recently boiled 3 *N* hydrochloric acid, and 5 mL of potassium iodide TS. Close the vessel, purge the system with an oxygen-free inert gas, and titrate immediately with *0.1 N Potassium iodide–iodate* adding 2 mL of starch TS as the end-point is approached. Each mL of *0.1 N Potassium iodide–iodate* is equivalent to 5.935 mg of Sn^{++}: an amount of stannous ion equivalent to not less than 68.2% of the labeled amount of stannous fluoride is found.

Assay—[NOTE—Store all solutions, except *Buffer solution*, in plastic containers.]

Buffer solution and *Standard preparations*—Prepare as directed in the *Assay* under *Sodium Fluoride Oral Solution*.

Assay preparation—Transfer an accurately weighed quantity of Stannous Fluoride Gel, equivalent to about 8 mg of stannous fluoride, to a 100-mL volumetric flask, dilute with water to volume, and mix.

Procedure—Proceed as directed for *Procedure* in the *Assay* under *Sodium Fluoride Oral Solution*. Calculate the percentage of stannous fluoride (SnF_2) in the Stannous Fluoride Gel taken by the formula:

$$(156.69/38.0)(C/100W),$$

in which 156.69 is the molecular weight of stannous fluoride, 38.0 is twice the atomic weight of fluorine, and *W* is the weight, in g, of the Stannous Fluoride Gel taken.

Stanozolol

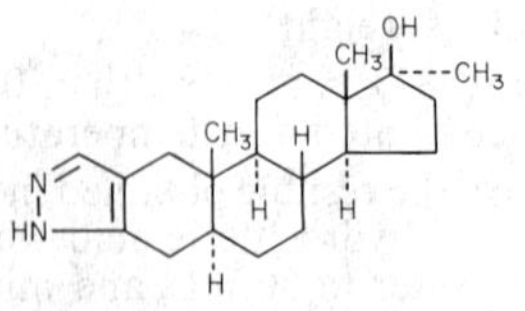

$C_{21}H_{32}N_2O$ 328.50

2′*H*-Androst-2-eno[3,2-*c*]pyrazol-17-ol, 17-methyl-, (5α,17β)-.
17-Methyl-2′*H*-5α-androst-2-eno[3,2-*c*]pyrazol-17β-ol [*10418-03-8*].

» Stanozolol contains not less than 98.0 percent and not more than 100.5 percent of $C_{21}H_{32}N_2O$, calculated on the dried basis.

Packaging and storage—Preserve in tight, light-resistant containers.

Reference standard—*USP Stanozolol Reference Standard*—Dry at a pressure not above 5 mm of mercury at 100° to constant weight before using.

Identification—

A: The infrared absorption spectrum of a potassium bromide dispersion of it, previously dried, exhibits maxima only at the same wavelengths as that of a similar preparation of USP Stanozolol RS.

B: The ultraviolet absorption spectrum of a 1 in 20,000 solution in alcohol exhibits maxima and minima at the same wavelengths as that of a similar solution of USP Stanozolol RS, concomitantly measured, and the respective absorptivities, calculated on the dried basis, at the wavelength of maximum absorbance at about 224 nm, do not differ by more than 3.0%.

Specific rotation ⟨781⟩: between +34° and +40°, calculated on the dried basis, determined in a solution in chloroform containing 100 mg in each 10 mL.

Loss on drying ⟨731⟩—Dry it at a pressure not above 5 mm of mercury at 100° to constant weight: it loses not more than 1.0% of its weight.

Chromatographic impurities—

Standard dilutions—Dissolve a suitable quantity of USP Stanozolol RS in a mixture of chloroform and methanol (9:1) to obtain a solution having a known concentration of about 20 mg per mL. Dilute this solution with the same medium to obtain *Standard dilutions* having known concentrations of 50, 100, 200, and 400 µg per mL, respectively.

Procedure—Score a 20- × 20-cm thin-layer chromatographic plate coated with a 0.25-mm layer of chromatographic silica gel mixture (binder-free) into channels 10 mm wide. Apply 10-µL portions, in two 5-µL increments, of a test solution prepared by dissolving Stanozolol in a mixture of chloroform and methanol (9:1) to obtain a solution containing about 20 mg per mL, and of each of the four *Standard dilutions* in the center of the channels at points about 2.5 cm from one edge of the plate. Develop the plate in a suitable chamber, lined with filter paper and previously equilibrated with 200 mL of a mixture of chloroform and methanol (188:12), for 15 minutes, taking care to ensure that the filter paper has been wetted completely with the solvent mixture. Allow the plate to develop until the solvent front has moved about 15 cm above the line of application. Remove the plate, and allow the solvent to evaporate completely. Spray it with 20 percent sulfuric acid, and heat in an oven at 100° for 15 minutes. Examine the plate under long-wavelength ultraviolet light: the channel for the test solution exhibits its principal spot at the same R_f value as the spots for the *Standard dilutions*. Estimate the concentration of any spots in the channel for the test solution, other than the principal spot, by comparison with the spots from the *Standard dilutions*. The spots from the 50-, 100-, 200-, and 400-µg-per-mL dilutions correspond to 0.25%, 0.5%, 1.0%, and 2.0% of chromatographic impurities, respectively, and the sum of the chromatographic impurities in the test solution is not greater than 2.0%.

Assay—Dissolve about 700 mg of Stanozolol, accurately weighed, in 50 mL of glacial acetic acid, add 1 drop of crystal violet TS, and titrate with 0.1 *N* perchloric acid VS to a green end-point. Perform a blank determination, and make any necessary correction. Each mL of 0.1 *N* perchloric acid is equivalent to 32.85 mg of $C_{21}H_{32}N_2O$.

Stanozolol Tablets

» Stanozolol Tablets contain not less than 90.0 percent and not more than 110.0 percent of the labeled amount of $C_{21}H_{32}N_2O$.

Packaging and storage—Preserve in tight, light-resistant containers.

Reference standard—*USP Stanozolol Reference Standard*—Dry at a pressure not above 5 mm of mercury at 100° to constant weight before using.

Identification—Boil an amount of powdered Tablets, equivalent to about 2 mg of stanozolol, with 5 mL of benzene, filter, and evaporate on a steam bath to dryness. Add 3 mL of *p*-dimethylaminobenzaldehyde TS to the residue: a yellow color develops, which exhibits a green fluorescence under long-wavelength ultraviolet light.

Dissolution ⟨711⟩—

Medium: 0.1 *N* hydrochloric acid; 500 mL.

Apparatus 2: 50 rpm.

Time: 45 minutes.

Bromocresol purple solution—Mix 1.0 g of bromocresol purple with 1000 mL of dilute glacial acetic acid (1 in 50), and filter if necessary to obtain a clear solution.

Standard preparations—[NOTE—Prepare *Standard preparations* on the day of use.] Transfer about 50 mg of USP Stanozolol RS, accurately weighed, to a 50-mL volumetric flask, add 15.0 mL of methanol, and mix to dissolve. Add 5.0 mL of 1.0 *N* hydrochloric acid, dilute with water to volume, and mix. Transfer 5.0 mL of the resulting solution to a 200-mL volumetric flask, dilute with 0.1 *N* hydrochloric acid to volume, and mix. Separately pipet 2-mL, 4-mL, and 6-mL portions of the solution into three 60-mL separators, add accurately measured volumes of 0.1 *N* hydrochloric acid to adjust the volumes in each to 25.0 mL, and pipet 25 mL of 0.1 *N* hydrochloric acid into a fourth 60-mL separator.

Procedure—Pipet 25 mL of a filtered portion of the solution under test into a 60-mL separator. To this separator and to each of the four separators containing *Standard preparations* add 1.0 mL of *Bromocresol purple solution* and 10.0 mL of chloroform. Insert the stopper in each, shake gently for 1 minute, allow the phases to separate, and swirl if necessary to break up emulsions. Transfer the lower chloroform layers to separate glass-stoppered, 50-mL centrifuge tubes, insert the stoppers, and centrifuge for 5 minutes to clarify the solutions. Concomitantly determine the absorbances of the solutions in 1-cm cells at the wavelength of maximum absorbance at about 420 nm, with a suitable spectrophotometer, using chloroform as the blank. Construct a standard plot of absorbances versus the concentrations of the *Standard preparations*. Determine the amount of $C_{21}H_{32}N_2O$ dissolved from the ultraviolet absorbance of the solution obtained from the solution under test in comparison with the standard plot.

Tolerances—Not less than 75% (*Q*) of the labeled amount of $C_{21}H_{32}N_2O$ is dissolved in 45 minutes.

Uniformity of dosage units ⟨905⟩—[NOTE—Maintain the acid concentration at a uniform level in the solutions being compared spectrophotometrically; the same acidic alcohol solution is to be used throughout this procedure. Also, take precautions throughout this procedure to minimize evaporation.] Transfer 1 Tablet to a 25-mL volumetric flask, add 0.5 mL of water, and shake to disintegrate. Add about 20 mL of alcohol, heat on a steam bath, with occasional swirling, for 10 to 15 minutes, then cool, dilute with alcohol to volume, and mix. Filter through medium-porosity filter paper, taking precautions to minimize evaporation, discard the first 5 mL of the filtrate, and proceed as directed for *Assay preparations* in the *Assay*, beginning with "Transfer 5.0 mL of the filtrate."

Assay—[NOTE—Maintain the acid concentration at a uniform level in the solutions being compared spectrophotometrically; the same acidic alcohol solution is to be used throughout this procedure.]

Standard preparations—Dissolve a suitable quantity of USP Stanozolol RS, accurately weighed, in alcohol, and dilute quantitatively and stepwise with alcohol, if necessary, to obtain a stock solution having a known concentration of about 80 µg per mL. Transfer 5.0 mL of this stock solution to a 10-mL volumetric flask, dilute with alcohol to volume, and mix to prepare the *Neutral standard preparation.* Transfer another 5.0-mL portion of the stock solution to a second 10-mL volumetric flask, dilute with acidic alcohol (1.5 mL of hydrochloric acid in 100 mL of alcohol) to volume, and mix to prepare the *Acidic standard preparation.* The concentration of USP Stanozolol RS in the *Standard preparations* is about 40 µg per mL.

Assay preparations—Weigh and finely powder not less than 20 Stanozolol Tablets. Transfer an accurately weighed portion of the powder, equivalent to about 4 mg of stanozolol, to a 50-mL volumetric flask, add about 25 mL of alcohol, and heat on a steam bath, with frequent swirling, for 15 minutes. Cool, dilute with alcohol to volume, mix, filter through medium-porosity filter paper, taking precautions to minimize evaporation, and discard the first 10 mL of the filtrate. Transfer 5.0 mL of the filtrate to a 10-mL volumetric flask, dilute with alcohol to volume, and mix to prepare the *Neutral assay preparation.* Transfer another 5.0-mL portion of the filtrate to a second 10-mL volumetric flask, dilute with acidic alcohol (1.5 mL of hydrochloric acid in 100 mL of alcohol) to volume, and mix to prepare the *Acidic assay preparation.*

Procedure—Concomitantly determine the absorbances of the acidic alcohol solution, the *Acidic standard preparation*, and the *Acidic assay preparation* in 1-cm cells at the wavelength of maximum absorbance at about 235 nm, with a suitable spectrophotometer, using alcohol, the *Neutral standard preparation*, and the *Neutral assay preparation*, respectively, as the blanks. Calculate the quantity, in mg, of $C_{21}H_{32}N_2O$ in the portion of Tablets taken by the formula:

$$0.1C(A_U - A_O)/(A_S - A_O),$$

in which *C* is the concentration, in µg per mL, of USP Stanozolol RS in the *Standard preparations*, and A_U, A_S, and A_O are the absorbances of the *Acidic assay preparation*, the *Acidic standard preparation*, and the acidic alcohol solution, respectively.

Starch—*see* Starch NF

Starch, Pregelatinized—*see* Starch, Pregelatinized NF

Topical Starch

» Topical Starch consists of the granules separated from the mature grain of corn [*Zea mays* Linné (Fam. Gramineae)].

Packaging and storage—Preserve in well-closed containers.

Botanic characteristics—Polygonal, rounded or spheroidal granules up to about 35 µm in diameter and usually having a circular or several-rayed central cleft.

Identification—

A: Prepare a smooth mixture of 1 g of it with 2 mL of cold water, stir it into 15 mL of boiling water, boil gently for 2 minutes, and cool: a translucent, whitish jelly is produced.

B: A water slurry of it is colored reddish violet to deep blue by iodine TS.

Microbial limits ⟨61⟩—The total aerobic microbial count does not exceed 500 per g and the total combined molds and yeasts count does not exceed 50 per g.

pH ⟨791⟩—Prepare a slurry by weighing 20.0 g ± 100 mg of Topical Starch, transferring to a suitable nonmetallic container, and adding 100 mL of water. Agitate continuously at a moderate rate for 5 minutes, then stop agitation, and immediately determine the pH to the nearest 0.1 unit: the pH, determined potentiometrically, is between 4.5 and 7.0.

Loss on drying ⟨731⟩—Dry it at 120° for 4 hours: it loses not more than 14.0% of its weight.

Residue on ignition ⟨281⟩: not more than 0.5%, determined on a 2.0-g test specimen ignited at a temperature of 575 ± 25°.

Iron ⟨241⟩—Dissolve the residue obtained in the test for *Residue on ignition* in 4 mL of hydrochloric acid with the aid of gentle heating, dilute with water to 50 mL, and mix. Dilute 25 mL of the resulting solution with water to 47 mL: the limit is 0.001%.

Oxidizing substances—Transfer 4.0 g to a glass-stoppered, 125-mL conical flask, and add 50.0 mL of water. Insert the stopper, and swirl for 5 minutes. Decant into a glass-stoppered, 50-mL centrifuge tube, and spin to clarify. Transfer 30.0 mL of clear supernatant liquid to a glass-stoppered, 125-mL conical flask. Add 1 mL of glacial acetic acid and 0.5 g to 1.0 g of potassium iodide. Insert the stopper, swirl, and allow to stand for 25 to 30 minutes in the dark. Add 1 mL of starch TS, and titrate with 0.002 *N* sodium thiosulfate VS to the disappearance of the starch-

iodine color. Each mL of 0.002 *N* sodium thiosulfate is equivalent to 34 μg of oxidant, calculated as hydrogen peroxide. Not more than 12.6 mL of 0.002 *N* sodium thiosulfate is required (0.018%).

Sulfur dioxide—Mix 20 g with 200 mL of water to obtain a smooth suspension, and filter. To 100 mL of the clear filtrate add 3 mL of starch TS, and titrate with 0.010 *N* iodine to the first permanent blue color: not more than 2.7 mL is consumed (0.008%).

Steam Sterilization, Paper Strip, Biological Indicator for—*see* Biological Indicator for Steam Sterilization, Paper Strip

Stearic Acid—*see* Stearic Acid NF

Stearic Acid, Purified—*see* Stearic Acid, Purified NF

Stearyl Alcohol—*see* Stearyl Alcohol NF

Storax

» Storax is a balsam obtained from the trunk of *Liquidambar orientalis* Miller, known in commerce as Levant Storax, or of *Liquidambar styraciflua* Linné, known in commerce as American Storax (Fam. Hamamelidaceae).

Packaging and storage—Preserve in well-closed containers.

Loss on drying ⟨731⟩—Dry about 2 g, accurately weighed, at 105° for 2 hours: it loses not more than 20.0% of its weight.

Alcohol-insoluble substances—Accurately weigh about 10 g of mixed Storax in a beaker, heat at 105° for 30 minutes, take up the residue in 100 mL of hot alcohol, filter through counterbalanced filters or a tared filter crucible, and wash the residue with small portions of hot alcohol until the last washing is colorless or practically so: the weight of the residue so obtained, after drying at 105° for 1 hour, does not exceed 5.0% of the weight of Storax taken.

Alcohol-soluble substances—Evaporate the combined alcohol filtrate and washings obtained in the test for *Alcohol-insoluble substances* at a temperature not exceeding 60°, and dry the residue at 105° for 1 hour: the weight of the yellow to brown residue of purified Storax so obtained is not less than 70.0% of the weight of the Storax taken.

Acid value, Saponification value, Cinnamic acid—The purified Storax obtained in the test for *Alcohol-soluble substances* meets the requirements of the following tests.

Acid value ⟨401⟩—Dissolve about 1 g of the purified Storax, accurately weighed, in 50 mL of neutralized alcohol, add 0.5 mL of phenolphthalein TS, and titrate with 0.5 *N* sodium hydroxide VS: the acid value is between 50 and 85 for Levant Storax and between 36 and 85 for American Storax.

Saponification value ⟨401⟩—Place about 2 g of the purified Storax, accurately weighed, in a 250-mL flask, mix it with 50 mL of solvent hexane, add 25.0 mL of 0.5 *N* alcoholic potassium hydroxide VS, and allow the mixture to stand for 24 hours with frequent agitation. Then add 0.5 mL of phenolphthalein TS, and titrate the excess alkali with 0.5 *N* hydrochloric acid VS: the saponification value thus determined is between 160 and 200.

Cinnamic acid—Add about 2 g of the purified Storax, accurately weighed, to 25 mL of 0.5 *N* alcoholic potassium hydroxide, and boil the mixture for 1 hour under a reflux condenser. Add 0.5 mL of phenolphthalein TS, neutralize with 0.5 *N* sulfuric acid, and evaporate the alcohol on a steam bath. Dissolve the residue in 50 mL of water, and shake the solution with 20 mL of ether. Shake the separated ether with 5 mL of water, adding the washing to the water solution, and reject the ether extract. Add to the water solution 10 mL of diluted sulfuric acid, and shake with four 20-mL portions of ether. Wash the combined ether extracts with 5 mL of water, rejecting the water washing, transfer to a flask, and distil off the ether. Add to the residue 100 mL of water, and boil the mixture vigorously for 15 minutes under a reflux condenser. Filter while hot, and allow the filtrate to cool to about 25°: white crystals of cinnamic acid separate. Collect and dry the cinnamic acid by vacuum filtration. Repeat the extraction of the residue twice by boiling each time under a reflux condenser, as previously described, with the filtrate from the preceding crystallization, and collect the additional cinnamic acid in the same crucible. Finally wash the cinnamic acid with two 10-mL portions of ice-cold water, dry at 80°, and weigh. The weight of the cinnamic acid so obtained is not less than 25.0% of the weight of purified Storax taken. A portion of the acid recrystallized from hot water melts between 134° and 135°.

To about 50 mg of the cinnamic acid obtained as directed above add 5 mL of 2 *N* sulfuric acid, heat, and add potassium permanganate TS: the odor of benzaldehyde is perceptible.

Streptomycin Sulfate Injection

» Streptomycin Sulfate Injection contains the equivalent of not less than 90.0 percent and not more than 115.0 percent of the labeled amount of streptomycin ($C_{21}H_{39}N_7O_{12}$). It may contain one or more suitable buffers, preservatives, and stabilizers.

Packaging and storage—Preserve in single-dose or in multiple-dose containers, preferably of Type I glass.

Reference standard—*USP Streptomycin Sulfate Reference Standard*—Dry in vacuum at a pressure not exceeding 5 mm of mercury at 60° for 3 hours before using.

pH ⟨791⟩: between 5.0 and 8.0.

Other requirements—It responds to the *Identification test* and meets the requirements for *Depressor substances*, *Pyrogen*, and *Sterility* under *Sterile Streptomycin Sulfate*. It meets also the requirements under *Injections* ⟨1⟩.

Assay—

Assay preparation 1 (where it is represented as being in a single-dose container)—Withdraw all of the withdrawable contents of Streptomycin Sulfate Injection, using a suitable hypodermic needle and syringe, and dilute quantitatively with water to obtain a solution containing a convenient quantity of streptomycin per mL.

Assay preparation 2 (where the label states the quantity of streptomycin in a given volume of solution)—Dilute an accurately measured volume of Streptomycin Sulfate Injection quantitatively with water to obtain a solution containing a convenient quantity of streptomycin per mL.

Procedure—Proceed as directed under *Antibiotics—Microbial Assays* ⟨81⟩, using an accurately measured volume of *Assay preparation* diluted quantitatively with water to yield a *Test Dilution* having a concentration assumed to be equal to the median dose level of the Standard.

Sterile Streptomycin Sulfate

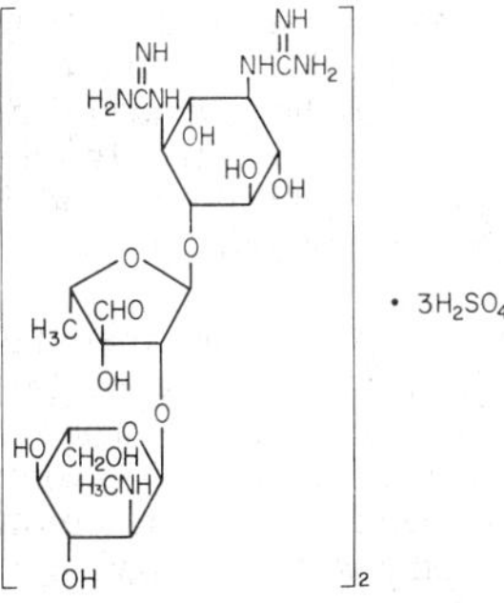

$(C_{21}H_{39}N_7O_{12})_2 \cdot 3H_2SO_4$ 1457.38

D-Streptamine, *O*-2-deoxy-2-(methylamino)-α-L-glucopyranosyl-(1→2)-*O*-5-deoxy-3-*C*-formyl-α-L-lyxofuranosyl-(1→4)-*N,N'*-bis(aminoiminomethyl)-, sulfate (2:3) (salt).
Streptomycin sulfate (2:3) (salt) [*3810-74-0*].

» Sterile Streptomycin Sulfate has a potency equivalent to not less than 650 μg and not more than 850 μg of streptomycin ($C_{21}H_{39}N_7O_{12}$) per mg. In addition, where packaged for dispensing, it contains the equivalent of not less than 90.0 percent and not more than 115.0 percent of the labeled amount of streptomycin ($C_{21}H_{39}N_7O_{12}$).

Packaging and storage—Preserve in *Containers for Sterile Solids* as described under *Injections* ⟨1⟩.

Reference standard—*USP Streptomycin Sulfate Reference Standard*—Dry in vacuum at a pressure not exceeding 5 mm of mercury at 60° for 3 hours before using.

Constituted solution—At the time of use, the constituted solution prepared from Sterile Streptomycin Sulfate meets the requirements for *Constituted Solutions* under *Injections* ⟨1⟩.

Identification—

Iron reagent—Dissolve 5 g of ferric chloride in 50 mL of 0.1 *N* hydrochloric acid. Transfer 2.5 mL of this stock solution to a 100-mL volumetric flask, dilute with 0.01 *N* hydrochloric acid to volume, and mix. Prepare this solution at the time of use.

Procedure—Dissolve the specimen in water, and dilute with water to obtain a solution containing about 1 mg of streptomycin per mL. To 5 mL of this solution add 2.0 mL of 1 *N* sodium hydroxide, and heat in a water bath for 10 minutes. Cool in ice water for 3 minutes, then add 2.0 mL of 1.2 *N* hydrochloric acid, and mix. Add 5 mL of *Iron reagent*, and mix: a violet color is produced.

Depressor substances—It meets the requirements of the *Depressor Substances Test* ⟨101⟩, the test dose being 1.0 mg per kg of a solution in sterile saline TS, containing 3.0 mg of streptomycin per mL.

Pyrogen—It meets the requirements of the *Pyrogen Test* ⟨151⟩, the test dose being 1.0 mL per kg of a solution in Sterile Water for Injection containing 10 mg of streptomycin per mL.

Sterility—It meets the requirements under *Sterility Tests* ⟨71⟩, when tested as directed in the section, *Test Procedures Using Membrane Filtration*.

pH ⟨791⟩: between 4.5 and 7.0, in a solution containing 200 mg of streptomycin per mL.

Loss on drying ⟨731⟩—Dry about 100 mg, accurately weighed, in a capillary-stoppered bottle in vacuum at a pressure not exceeding 5 mm of mercury at 60° for 3 hours: it loses not more than 5.0% of its weight.

Other requirements—It meets the requirements for *Uniformity of Dosage Units* ⟨905⟩ and *Labeling* under *Injections* ⟨1⟩.

Assay—

Assay preparation 1—Dissolve a suitable quantity of Sterile Streptomycin Sulfate, accurately weighed, in water, and dilute quantitatively with water to obtain a solution containing a convenient quantity of streptomycin per mL.

Assay preparation 2 (where it is packaged for dispensing and is represented as being in a single-dose container)—Constitute Sterile Streptomycin Sulfate in a volume of water, accurately measured, corresponding to the volume of solvent specified in the labeling. Withdraw all of the withdrawable contents, using a suitable hypodermic needle and syringe, and dilute quantitatively with water to obtain a solution containing a convenient quantity of streptomycin per mL.

Assay preparation 3 (where the label states the quantity of streptomycin in a given volume of constituted solution)—Constitute Sterile Streptomycin Sulfate in a volume of water, accurately measured, corresponding to the volume of solvent specified in the labeling. Dilute an accurately measured volume of the constituted solution quantitatively with water to obtain a solution containing a convenient quantity of streptomycin per mL.

Procedure—Proceed as directed under *Antibiotics—Microbial Assays* ⟨81⟩, using an accurately measured volume of *Assay preparation* diluted quantitatively with water to yield a *Test Dilution* having a concentration assumed to be equal to the median dose level of the Standard.

Strips, Fluorescein Sodium Ophthalmic—*see* Fluorescein Sodium Ophthalmic Strips
Strong Ammonia Solution—*see* Ammonia Solution, Strong NF
Strong Iodine Solution—*see* Iodine Solution, Strong
Strong Iodine Tincture—*see* Iodine Tincture, Strong
Stronger Rose Water—*see* Rose Water, Stronger NF
Sublingual Tablets, Isosorbide Dinitrate—*see* Isosorbide Dinitrate Sublingual Tablets

Succinylcholine Chloride

$$\left[\begin{array}{l} COOCH_2CH_2N^+(CH_3)_3 \\ (CH_2)_2 \\ COOCH_2CH_2N^+(CH_3)_3 \end{array}\right] 2Cl^-$$

$C_{14}H_{30}Cl_2N_2O_4$ (anhydrous) 361.31
Ethanaminium, 2,2'-[(1,4-dioxo-1,4-butanediyl)bis(oxy)]-bis-[*N,N,N*-trimethyl-], dichloride.
Choline chloride succinate (2:1) [*71-27-2*].
Dihydrate 397.34 [*6101-15-1*].

» Succinylcholine Chloride usually contains approximately two molecules of water of hydration. It contains not less than 96.0 percent and not more than 102.0 percent of $C_{14}H_{30}Cl_2N_2O_4$, calculated on the anhydrous basis.

Packaging and storage—Preserve in tight containers.

Labeling—Label it in terms of its anhydrous equivalent.

Reference standards—*USP Succinylcholine Chloride Reference Standard*—Do not dry; determine the *Water* content by *Method I* before using for quantitative analyses. *USP Succinylmonocholine Chloride Reference Standard*—Dry in vacuum at 55° for 16 hours.

Identification—

A: The infrared absorption spectrum of a potassium bromide dispersion of it exhibits maxima only at the same wavelengths as that of a similar preparation of USP Succinylcholine Chloride RS.

B: The retention time of the major peak in the chromatogram of the *Assay preparation* is the same as that of the *Standard preparation* obtained in the *Assay*.

C: Dissolve a portion in water to obtain a solution containing 1 mg per mL. Applying 1-μL portions on a plate coated with a 0.25-mm layer of chromatographic silica gel (see *Chromatography* ⟨621⟩), and using a solvent system consisting of a mixture of acetone and 1 *N* hydrochloric acid (1:1), proceed as directed under *Thin-layer Chromatographic Identification Test* ⟨201⟩. Use the following procedure to locate the spots: heat the plate at 105° for 5 minutes, cool, and spray with potassium bismuth iodide TS, then heat again at 105° for 5 minutes.

Water, *Method I* ⟨921⟩: not more than 10.0%.

Residue on ignition ⟨281⟩: not more than 0.2%.

Ammonium salts—To about 200 mg add 5 mL of sodium carbonate TS, and bring to a boil: no odor of ammonia is evolved.

Chromatographic purity—

Standard solution—Transfer 18.75 mg each of choline chloride and USP Succinylmonocholine Chloride RS, accurately weighed, to a 50-mL volumetric flask, dissolve in 40 mL of methanol, dilute with methanol to volume, and mix.

Diluted Standard solution—Transfer 4.0 mL of *Standard solution* to a 10-mL volumetric flask, dilute with methanol to volume, and mix.

Test solution—Prepare, immediately prior to use, a solution of Succinylcholine Chloride in methanol having a known concentration of about 50 mg per mL.

Procedure—Separately apply 2 µL of the *Test solution*, the *Standard solution*, and the *Diluted Standard solution* to a suitable high-performance 10- × 10-cm thin-layer chromatographic plate (see *Chromatography* ⟨621⟩), coated with a 0.10-mm layer of chromatographic cellulose. Allow the spots to dry, and immediately place the plate, its coated surface toward the nearer wall, in the dry trough of a twin-trough chromatographic chamber whose other trough contains a solvent system consisting of the upper layer of a mixture of butyl alcohol, water, and 96% formic acid (65:35:15) that has been shaken and allowed to stand for 24 hours until the phases have separated. Equilibrate the chromatographic chamber for 30 minutes, and tilt the chamber to introduce the developing solvent into the trough containing the plate. Develop the chromatogram until the solvent front has moved about three-fourths of the length of the plate, remove the plate from the developing chamber, quickly and thoroughly evaporate the solvent with the aid of a current of air, and dry at 105° for 15 minutes. [NOTE—During the drying, support the plate in such a manner that only the upper and lower edges of the plate, outside the chromatographic zone, are in direct contact with any heated surface.] Spray the plate with potassium iodoplatinate TS, dry at 105° for about 2 minutes, and allow to cool to room temperature: any spots from the *Test solution* are not greater in size or intensity than the spots, occurring at the respective R_f values (approximately 0.4 for succinylmonocholine chloride, and 0.3 for choline chloride), produced by 2 µL of the *Diluted Standard solution*, corresponding to 0.75% of each compound. Estimate the size and intensities of any other spots detected by comparison with the spots produced by succinylmonocholine chloride in the *Standard solution*, and in the *Diluted Standard solution*. The total of any such spots detected is not more than 1.5%.

Chloride content—Dissolve about 400 mg, accurately weighed, in 5 mL of water. Add 5 mL of glacial acetic acid, 50 mL of methanol, and 1 drop of eosin Y TS, and titrate with 0.1 *N* silver nitrate VS. Each mL of 0.1 *N* silver nitrate is equivalent to 3.545 mg of Cl. Not less than 19.3% and not more than 19.8% of Cl, calculated on the anhydrous basis, is found.

Assay—[NOTE—Since the *Mobile phase* employed in this procedure has a fairly high concentration of chloride ion and a low pH, it may be advisable to rinse the entire system with water following the use of this *Mobile phase*.]

Mobile phase—Prepare a 1 in 10 solution of 1 *N* aqueous tetramethylammonium chloride in methanol. Filter this solution through a 0.45-µm membrane filter, and adjust with hydrochloric acid to a pH of about 3.0.

Standard preparation—Transfer about 88 mg of USP Succinylcholine Chloride RS, accurately weighed, to a 10-mL volumetric flask, add 4.0 mL of water, and dilute with *Mobile phase* to volume while mixing. Prepare the *Standard preparation* concurrently with the *Assay preparation*.

Assay preparation—Transfer about 88 mg of Succinylcholine Chloride, accurately weighed, to a 10-mL volumetric flask, add 4.0 mL of water, and dilute with *Mobile phase* to volume while mixing.

Chromatographic system (see *Chromatography* ⟨621⟩)—The liquid chromatograph is equipped with a 214-nm detector and a 4-mm × 25-cm column that contains packing L3. The flow rate is about 0.75 mL per minute. Chromatograph five replicate injections of the *Standard preparation*, and record the peak responses as directed under *Procedure:* the relative standard deviation is not more than 1.5%, and the tailing factor is not greater than 2.5.

Procedure—Separately inject equal volumes (about 10 µL) of the *Standard preparation* and the *Assay preparation* into the chromatograph by means of a suitable microsyringe or sampling valve, record the chromatograms, and measure the responses for the major peaks. Calculate the quantity, in mg, of $C_{14}H_{30}Cl_2N_2O_4$ in the Succinylcholine Chloride taken by the formula:

$$10C(r_U/r_S),$$

in which *C* is the concentration, in mg per mL, of anhydrous succinylcholine chloride in the *Standard preparation*, as determined from the concentration of USP Succinylcholine Chloride RS corrected for moisture content by a titrimetric water determination, r_U is the peak response of the *Assay preparation*, and r_S is the average peak response of the *Standard preparation*.

Succinylcholine Chloride Injection

» Succinylcholine Chloride Injection is a sterile solution of Succinylcholine Chloride in a suitable aqueous vehicle. It contains not less than 90.0 percent and not more than 110.0 percent of the labeled amount of anhydrous succinylcholine chloride ($C_{14}H_{30}Cl_2N_2O_4$).

Packaging and storage—Preserve in single-dose or in multiple-dose containers, preferably of Type I or Type II glass, in a refrigerator.

Labeling—Label it to indicate, as its expiration date, the month and year not more than 2 years from the month during which the Injection was last assayed and released by the manufacturer.

Reference standard—*USP Succinylcholine Chloride Reference Standard*—Do not dry; determine the *Water* content by *Method I* before using for quantitative analyses.

Identification—It responds to *Identification tests B* and *C* under *Succinylcholine Chloride*.

pH ⟨791⟩: between 3.0 and 4.5.

Other requirements—It meets the requirements under *Injections* ⟨1⟩.

Assay—[NOTE—Since the *Mobile phase* employed in this procedure has a fairly high concentration of chloride ion and a low pH, it may be advisable to rinse the entire system with water following the use of this *Mobile phase*.]

Mobile phase and *Chromatographic system*—Prepare as directed in the *Assay* under *Succinylcholine Chloride*.

Standard preparation—Transfer about 88 mg of USP Succinylcholine Chloride RS, accurately weighed, to a 10-mL volumetric flask, add a volume of water to correspond to the solvent composition of the *Assay preparation*, and dilute with *Mobile phase* to volume while mixing. Prepare the *Standard preparation* concurrently with the *Assay preparation*.

Assay preparation—Transfer a volume of Succinylcholine Chloride Injection, equivalent to about 80 mg of anhydrous succinylcholine chloride, to a 10-mL volumetric flask, and dilute with *Mobile phase* to volume while mixing.

Procedure—Proceed as directed for *Procedure* in the *Assay* under *Succinylcholine Chloride*. Calculate the quantity, in mg, of anhydrous succinylcholine chloride ($C_{14}H_{30}Cl_2N_2O_4$) in each mL of the Injection taken by the formula:

$$(10C/V)(r_U/r_S),$$

in which *V* is the volume, in mL, of Injection taken.

Sterile Succinylcholine Chloride

» Sterile Succinylcholine Chloride is Succinylcholine Chloride suitable for parenteral use.

Packaging and storage—Preserve in *Containers for Sterile Solids* as described under *Injections* ⟨1⟩.

Reference standards—*USP Succinylcholine Chloride Reference Standard*—Do not dry; determine the *Water* content by *Method I* before using for quantitative analyses. *USP Succinylmonocholine Chloride Reference Standard*—Dry in vacuum at 55° for 16 hours.

Completeness of solution ⟨641⟩—A 500-mg portion dissolves in 10 mL of carbon dioxide–free water to yield a clear and colorless solution.

Constituted solution—At the time of use, the constituted solution prepared from Sterile Succinylcholine Chloride meets the requirements for *Constituted Solutions* under *Injections* ⟨1⟩.

Chromatographic purity—

Standard solution—Prepare as directed in the test for *Chromatographic purity* under *Succinylcholine Chloride*, using 20 mg of choline chloride and of USP Succinylmonocholine Chloride RS.

Test solution and *Procedure*—Proceed as directed for *Test solution* and *Procedure* in the test for *Chromatographic purity* under *Succinylcholine Chloride:* any spots from the *Test solution* are not greater in size or intensity than the spots, occurring at the respective R_f values (approximately 0.6 for succinylmonocholine chloride and 0.5 for choline chloride), produced by 5 µL of the *Standard solution*, corresponding to 1.0% of each compound. Estimate the size and intensities of any other spots detected by comparison with the spot produced by succinylmonocholine chloride in the *Standard solution.* The total of any such spots detected is not more than 2.0%.

Other requirements—It conforms to the definition, responds to the *Identification tests*, and meets the requirements for *Water, Residue on ignition, Ammonium salts, Chloride content,* and *Assay* under *Succinylcholine Chloride.* It meets also the requirements for *Sterility Tests* ⟨71⟩, *Uniformity of Dosage Units* ⟨905⟩, and *Labeling* under *Injections* ⟨1⟩.

Sucrose—*see* Sucrose NF

Sucrose Octaacetate—*see* Sucrose Octaacetate NF

Sugar, Compressible—*see* Sugar, Compressible NF

Sugar, Confectioner's—*see* Sugar, Confectioner's NF

Invert Sugar Injection

» Invert Sugar Injection is a sterile solution of a mixture of equal amounts of Dextrose and Fructose in Water for Injection, or an equivalent sterile solution produced by the hydrolysis of Sucrose, in Water for Injection. It contains not less than 95.0 percent and not more than 105.0 percent of the labeled amount of $C_6H_{12}O_6$. It contains no antimicrobial agents.

NOTE—Invert Sugar Injection that is produced by mixing Dextrose and Fructose is exempt from the requirement of the test for *Completeness of inversion.*

Packaging and storage—Preserve in single-dose containers, preferably of Type I or Type II glass, or of a suitable plastic material.

Labeling—The label states the total osmolar concentration in mOsmol per liter.

Identification—Add a few drops of Injection to 5 mL of hot alkaline cupric tartrate TS: a copious red precipitate of cupric oxide is formed.

Pyrogen—It meets the requirements of the *Pyrogen Test* ⟨151⟩. [NOTE—Dilute, with Water for Injection, Injection containing more than 10% of invert sugar to give a concentration of 10% of invert sugar.]

pH ⟨791⟩: between 3.0 and 6.5.

Chloride ⟨221⟩—A 2.0-mL portion shows no more chloride than corresponds to 0.34 mL of 0.020 *N* hydrochloric acid (0.012%).

Heavy metals ⟨231⟩—Transfer a volume of Injection, equivalent to 4.0 g of invert sugar, to a suitable vessel, and adjust the volume to 25 mL by evaporation: the limit is 0.0005*C*%, in which *C* is the labeled amount, in g, of invert sugar per mL of Injection.

5-Hydroxymethylfurfural and related substances—Dilute an accurately measured volume of Injection, equivalent to 1.0 g of invert sugar, with water to 500.0 mL. Determine the absorbance of this solution in a 1-cm cell at 284 nm, with a suitable spectrophotometer, using water as the blank: the absorbance is not more than 0.25.

Completeness of inversion—

Mobile phase—Use filtered, degassed water.

Standard preparation—Prepare a solution in water containing known concentrations of about 0.25 mg of sucrose and about 12.5 mg of dextrose per mL.

Test preparation—Transfer a volume of Injection, equivalent to about 2.5 g of invert sugar, to a 100-mL volumetric flask, dilute with water to volume, and mix.

Chromatographic system (see *Chromatography* ⟨621⟩)—The liquid chromatograph is equipped with a refractive index detector and a 7.8-mm × 30-cm column that contains 9-µm packing L19, maintained at a constant temperature of about 40°. Chromatograph the *Standard preparation*, and record the peak responses as directed under *Procedure:* the sucrose elutes first, and the peak is baseline separated from the dextrose peak. The relative standard deviation for replicate injections is not more than 2.0%.

Procedure—Separately inject equal volumes (about 20 µL) of the *Standard preparation* and the *Test preparation* into the chromatograph, record the chromatograms, and measure the responses for the sucrose peaks. Calculate the quantity, in mg, of sucrose, in the volume of Injection taken by the formula:

$$100C(r_U/r_S),$$

in which *C* is the concentration, in mg per mL, of sucrose in the *Standard preparation*, and r_U and r_S are the peak responses for sucrose obtained from the *Test preparation* and the *Standard preparation*, respectively: not more than 1.5% of the quantity of invert sugar in the volume of Injection taken, based on the value stated on the label, is found.

Other requirements—It meets the requirements under *Injections* ⟨1⟩.

Assay—Pipet 50 mL of alkaline cupric tartrate TS into a 400-mL beaker, add 48 mL of water, mix, and pipet into the mixture 2 mL of Invert Sugar Injection that has been diluted quantitatively with water, if necessary, to a 5.0% concentration. Cover the beaker with a watch glass, heat the solution, regulating the heat so that boiling begins in 4 minutes, and continue boiling for 2.0 minutes. Filter the hot solution at once through a tared porcelain filtering crucible, wash the precipitate with water maintained at 60°, then with 10 mL of alcohol. Dry at 105° to constant weight. Perform a blank determination, and make any necessary correction. The corrected weight of the precipitate so obtained is not less than 204.0 mg and not more than 224.4 mg, corresponding to between 95.0 and 105.0 mg of $C_6H_{12}O_6$.

Sugar Spheres—*see* Sugar Spheres NF

Sterile Sulbactam Sodium

$C_8H_{10}NNaO_5S$ 255.22

4-Thia-1-azabicyclo[3.2.0]heptane-2-carboxylic acid, 3,3-dimethyl-7-oxo-, 4,4-dioxide, sodium salt, (2*S-cis*)-.

Sodium (2*S*,5*R*)-3,3-dimethyl-7-oxo-4-thia-1-azabicyclo[3.2.0]-heptane-2-carboxylate 4,4-dioxide [*69388-84-7*].

» Sterile Sulbactam Sodium is sulbactam sodium suitable for parenteral use. It contains not less than 886 µg and not more than 941 µg of sulbactam ($C_8H_{11}NO_5S$), calculated on the anhydrous basis.

Packaging and storage—Preserve in tight containers.

Reference standards—*USP Sulbactam Reference Standard*—Do not dry before using. *USP Penicillanic Acid Sodium Salt Reference Standard*—Do not dry before using.

Identification—The retention time of the major peak in the chromatogram of the *Assay preparation* corresponds to that of the *Standard preparation*, obtained as directed in the *Assay*.

Crystallinity ⟨695⟩: meets the requirements.

Pyrogen—It meets the requirements of the *Pyrogen Test* ⟨151⟩, the test dose being 1.0 mL per kg of a solution prepared by diluting Sterile Sulbactam Sodium with Sterile Water for Injection to a concentration of 20 mg of sulbactam per mL.

Sterility—It meets the requirements under *Sterility Tests* ⟨71⟩, when tested as directed in the section, *Test Procedures Using Membrane Filtration*.

Water, *Method I* ⟨921⟩: not more than 1.0%.

Assay—

0.005 M Tetrabutylammonium hydroxide—Dilute 6.6 mL of a 40% solution of tetrabutylammonium hydroxide with water to obtain 1800 mL of solution. Adjust with 1 *M* phosphoric acid to a pH of 5.0 ± 0.1, dilute with water to 2000 mL, and mix.

Mobile phase—Prepare a suitable filtered and degassed mixture of *0.005 M Tetrabutylammonium hydroxide* and acetonitrile (1650:350). Make adjustments if necessary (see *System Suitability* under *Chromatography* ⟨621⟩).

Standard preparation—Dissolve an accurately weighed quantity of USP Sulbactam RS quantitatively in *Mobile phase* to obtain a solution having a known concentration of about 1 mg per mL.

Resolution solution—Prepare a solution of USP Penicillanic Acid Sodium Salt RS in *Standard preparation* containing about 1 mg per mL.

Assay preparation—Transfer about 110 mg of Sterile Sulbactam Sodium to a 100-mL volumetric flask, dilute with *Mobile phase* to volume, and mix.

Chromatographic system (see *Chromatography* ⟨621⟩)—The liquid chromatograph is equipped with a 230-nm detector and a 4-mm × 30-cm column containing packing L1. The flow rate is about 2 mL per minute. Chromatograph the *Resolution solution*, and record the responses as directed under *Procedure:* the relative retention times are about 0.8 for sulbactam and 1.0 for penicillanic acid, and the resolution, *R*, between the sulbactam and penicillanic acid peaks is not less than 3.8. Chromatograph the *Standard preparation*, and record the responses as directed under *Procedure:* the column efficiency determined from the sulbactam peak is not less than 3500 theoretical plates, the tailing factor is not more than 1.5, and the relative standard deviation for replicate injections is not more than 2.0%.

Procedure—[NOTE—Use peak areas where peak responses are specified.] Separately inject equal volumes (about 10 µL) of the *Standard preparation* and the *Assay preparation* into the chromatograph, record the chromatograms, and measure the responses for the major peaks. Calculate the quantity, in µg, of sulbactam ($C_8H_{11}NO_5S$) in each mg of the Sterile Sulbactam Sodium taken by the formula:

$$100(CP/W)(r_U/r_S),$$

in which *C* is the concentration, in mg per mL, of USP Sulbactam Sodium RS in the *Standard preparation*, *P* is the assigned sulbactam content, in µg per mg, of the USP Sulbactam RS, *W* is the quantity, in mg, of Sterile Sulbactam Sodium taken, and r_U and r_S are the sulbactam peak responses obtained from the *Assay preparation* and the *Standard preparation*, respectively.

Triple Sulfa Vaginal Cream

» Triple Sulfa Vaginal Cream contains not less than 90.0 percent and not more than 110.0 percent of the labeled amounts of sulfathiazole ($C_9H_9N_3O_2S_2$), sulfacetamide ($C_8H_{10}N_2O_3S$), and sulfabenzamide ($C_{13}H_{12}N_2O_3S$).

Packaging and storage—Preserve in well-closed, light-resistant containers, or in collapsible tubes.

Reference standards—*USP Sulfathiazole Reference Standard*—Dry at 105° for 2 hours before using. *USP Sulfacetamide Reference Standard*—Dry at 105° for 2 hours before using. *USP Sulfabenzamide Reference Standard*—Dry at 105° for 2 hours before using.

Identification—The retention times of the major peaks in the chromatogram of the *Assay preparation* correspond to those of the *Standard preparation*, as obtained in the Assay.

Minimum fill ⟨755⟩: meets the requirements.

pH ⟨791⟩: between 3.0 and 4.0.

Assay—

Mobile phase—Prepare a suitably degassed solution of water, acetonitrile, and 1 *M* tetrabutylammonium hydroxide (78:22:10). Adjust, dropwise with dilute phosphoric acid (1 in 10) to a pH of 7.7 ± 0.2, and mix for 5 minutes. If necessary, adjust to a pH of 7.7 ± 0.2, using dilute phosphoric acid (1 in 50) or 1 *M* tetrabutylammonium hydroxide.

Internal standard solution—Dissolve sulfapyridine in acetone to obtain a solution having a concentration of about 10 mg per mL.

Standard preparation—Weigh accurately about 29 mg of USP Sulfacetamide RS, 34 mg of USP Sulfathiazole RS, and 37 mg of USP Sulfabenzamide RS, and transfer to a 50-mL volumetric flask. Add 2.0 mL of *Internal standard solution* and 30 mL of acetone, and shake for 10 minutes. If necessary, sonicate to effect solution. Dilute with acetone to volume, and mix. Pipet 5 mL of this solution into a 50-mL volumetric flask, evaporate on a steam bath with the aid of a gentle stream of nitrogen to dryness, dissolve the residue in *Mobile phase*, dilute with *Mobile phase* to volume, and mix.

Assay preparation—Using a plastic syringe equipped with a suitable cannula, transfer an accurately weighed quantity of Triple Sulfa Vaginal Cream, equivalent to about 144 mg of sulfacetamide, 184 mg of sulfabenzamide, and 173 mg of sulfathiazole, to a 250-mL volumetric flask. Add 10.0 mL of *Internal standard solution* and 100 mL of acetone, and warm the flask on a steam bath while swirling the contents to dissolve the cream. Cool to room temperature, dilute with acetone to volume, and mix. Filter the solution through filter paper, discarding the first 10 mL of the filtrate. Pipet 5 mL of the filtrate so obtained into a 50-mL volumetric flask, and evaporate on a steam bath with the aid of a gentle stream of nitrogen to dryness. Dissolve the residue in *Mobile phase*, dilute with *Mobile phase* to volume, and mix. Cool the solution in an ice bath for 10 minutes, filter the cold solution through filter paper, discarding the first 10 mL to 15 mL of the filtrate, and collect 5 mL for analysis.

Chromatographic system (see *Chromatography* ⟨621⟩)—The liquid chromatograph is equipped with a 280-nm detector and a 3.9-mm × 30-cm column that contains packing L1. The flow rate is about 1.0 mL per minute. Chromatograph five replicate injections of the *Standard preparation*, and record the peak responses as directed under *Procedure:* the relative standard deviation is not more than 3.0%, and the resolution factor is not less than 2.0.

Procedure—Separately inject equal volumes (about 10 µL) of the *Standard preparation* and the *Assay preparation* into the chromatograph, record the chromatograms, and measure the responses for the major peaks. Chromatograms exhibit relative retention times of about 0.8 for sulfacetamide, 1.0 for sulfapyridine, 1.8 for sulfathiazole, and 2.5 for sulfabenzamide. Calculate the quantity, in mg, of $C_8H_{10}N_2O_3S$, $C_9H_9N_3O_2S_2$, and $C_{13}H_{12}N_2O_3S$ in the portion of the Vaginal Cream taken by the formula:

$$2.5C(R_U/R_S),$$

in which *C* is the concentration, in µg per mL, of the appropriate USP Reference Standard in the *Standard preparation*, and R_U and R_S are the ratios of the peak responses of the corresponding sulfonamides to those of the internal standard obtained from the *Assay preparation* and the *Standard preparation*, respectively.

Triple Sulfa Vaginal Tablets

» Triple Sulfa Vaginal Tablets contain not less than 90.0 percent and not more than 110.0 percent of the labeled amounts of sulfathiazole ($C_9H_9N_3O_2S_2$), sulfacetamide ($C_8H_{10}N_2O_3S$), and sulfabenzamide ($C_{13}H_{12}N_2O_3S$).

Packaging and storage—Preserve in well-closed, light-resistant containers.

Reference standards—*USP Sulfacetamide Reference Standard*—Dry at 105° for 2 hours before using. *USP Sulfabenzamide Reference Standard*—Dry at 105° for 2 hours before using. *USP Sulfathiazole Reference Standard*—Dry at 105° for 2 hours before using.

Identification—The retention times of the major peaks in the chromatogram of the *Assay preparation* correspond to those of the *Standard preparation*, as obtained in the *Assay*.

Disintegration ⟨701⟩: 30 minutes.

Uniformity of dosage units ⟨905⟩: meet the requirements for *Weight Variation* with respect to sulfathiazole, to sulfacetamide, and to sulfabenzamide.

Assay—

Mobile phase, Internal standard solution, and *Standard preparation*—Prepare as directed in the *Assay* under Triple Sulfa Vaginal Cream.

Assay preparation—Weigh and finely powder not less than 10 Triple Sulfa Vaginal Tablets. Transfer an accurately weighed portion of the powder, equivalent to about 144 mg of sulfacetamide, 184 mg of sulfabenzamide, and 173 mg of sulfathiazole, to a 250-mL volumetric flask. Add 10.0 mL of water, and shake for 10 minutes. Add 10.0 mL of *Internal standard solution* and 100 mL of acetone, and shake for 30 minutes at low speed on a mechanical shaker. Dilute with acetone to volume, mix, and allow to stand for 30 minutes. Pipet 5 mL of the clear supernatant solution into a 50-mL volumetric flask, and evaporate on a steam bath with the aid of a gentle stream of nitrogen to dryness. Dissolve the residue in *Mobile phase*, dilute with *Mobile phase* to volume, and mix.

Chromatographic system and *Procedure*—Proceed with the Vaginal Tablets as directed in the *Assay* under *Triple Sulfa Vaginal Cream.*

Sulfabenzamide

$H_2N-C_6H_4-SO_2NHC(=O)-C_6H_5$

$C_{13}H_{12}N_2O_3S$ 276.31
Benzamide, *N*-[(4-aminophenyl)sulfonyl]-.
N-Sulfanilylbenzamide [*127-71-9*].

» Sulfabenzamide contains not less than 99.0 percent and not more than 100.5 percent of $C_{13}H_{12}N_2O_3S$, calculated on the dried basis.

Packaging and storage—Preserve in well-closed, light-resistant containers.

Reference standard—*USP Sulfabenzamide Reference Standard*—Dry at 105° for 2 hours before using.

Clarity and color of solution—Dissolve 2.0 g in 15 mL of 1 *N* sodium hydroxide, with warming: a colorless to pale yellow solution having not more than a slight turbidity is produced.

Identification—

A: The infrared absorption spectrum of a potassium bromide dispersion of it, previously dried, exhibits maxima only at the same wavelengths as that of a similar preparation of USP Sulfabenzamide RS.

B: To about 100 mg, suspended in 2 mL of water, add 100 mg of sodium bicarbonate: it dissolves with effervescence (distinction from *sulfanilamide, sulfapyridine, sulfathiazole, sulfadiazine*, and *sulfaguanidine*).

Melting range, *Class I* ⟨741⟩: between 180° and 184°.

Loss on drying ⟨731⟩—Dry it at 105° for 2 hours: it loses not more than 0.5% of its weight.

Selenium ⟨291⟩: 0.001%, a 300-mg test specimen and 3 mL of *Stock Solution* being used.

Heavy metals, *Method II* ⟨231⟩: 0.002%.

Ordinary impurities ⟨466⟩—

Test solution: methanol.

Standard solution: methanol.

Eluant: a mixture of chloroform, methanol, and glacial acetic acid (90:5:5).

Visualization: 1.

Assay—Transfer about 800 mg of Sulfabenzamide, accurately weighed, to a 125-mL conical flask, and dissolve in 25 mL of dimethylformamide. Add 3 drops of thymol blue TS (prepared with methanol), and titrate with 0.1 *N* sodium methoxide VS to a blue end-point. Perform a blank determination, and make any necessary correction. Each mL of 0.1 *N* sodium methoxide is equivalent to 27.63 mg of $C_{13}H_{12}N_2O_3S$.

Sulfacetamide

$NH_2-C_6H_4-SO_2NHCOCH_3$

$C_8H_{10}N_2O_3S$ 214.24
Acetamide, *N*-[(4-aminophenyl)sulfonyl]-.
N-Sulfanilylacetamide [*144-80-9*].

» Sulfacetamide contains not less than 99.0 percent and not more than 100.5 percent of $C_8H_{10}N_2O_3S$, calculated on the dried basis.

Packaging and storage—Preserve in well-closed, light-resistant containers.

Reference standard—*USP Sulfacetamide Reference Standard*—Dry at 105° for 2 hours before using.

Clarity and color of solution—Dissolve about 200 mg in 5 mL of 1 *N* sodium hydroxide: a yellow to faintly yellow solution having not more than a trace of turbidity is produced.

Identification—

A: The infrared absorption spectrum of a potassium bromide dispersion of it, previously dried, exhibits maxima only at the same wavelengths as that of a similar preparation of USP Sulfacetamide RS.

B: Place about 500 mg in a test tube, heat gently until it boils, and cool: an oily liquid, which has the characteristic odor of acetamide, condenses on the walls of the test tube (distinction from the sublimates of *sulfadiazine, sulfamerazine, sulfamethazine*, and *sulfapyrazine*, which are solids at room temperature).

Melting range, *Class I* ⟨741⟩: between 181° and 184°.

Reaction—A solution (1 in 150) is acid to litmus.

Loss on drying ⟨731⟩—Dry it at 105° for 2 hours: it loses not more than 0.5% of its weight.

Residue on ignition ⟨281⟩: not more than 0.1%.

Selenium ⟨291⟩: 0.003%, a 200-mg test specimen being used.

Sulfate ⟨221⟩—Digest 1 g with 50 mL of water at about 70° for 5 minutes. Cool immediately to room temperature, and filter. A 25-mL portion of the filtrate so obtained shows no more sulfate than corresponds to 0.2 mL of 0.02 *N* sulfuric acid (0.04%).

Heavy metals, *Method II* ⟨231⟩: 0.002%.

Assay—Proceed with Sulfacetamide as directed under *Nitrite Titration* ⟨451⟩. Each mL of 0.1 *M* sodium nitrite is equivalent to 21.42 mg of $C_8H_{10}N_2O_3S$.

Sulfacetamide Sodium

$C_8H_9N_2NaO_3S.H_2O$ 254.24
Acetamide, *N*-[(4-aminophenyl)sulfonyl]-, monosodium salt, monohydrate.
N-Sulfanilylacetamide monosodium salt monohydrate [*6209-17-2*].
Anhydrous 236.22 [*127-56-0*].

» Sulfacetamide Sodium contains not less than 99.0 percent and not more than 100.5 percent of $C_8H_9N_2NaO_3S$, calculated on the anhydrous basis.

Packaging and storage—Preserve in tight, light-resistant containers.

Identification—

A: Dissolve about 1 g in 25 mL of water, adjust with 6 *N* acetic acid to a pH of between 4 and 5, and filter. Wash the precipitate with water, and dry at 105° for 2 hours: the sulfacetamide so obtained melts between 180° and 184°.

B: Place about 500 mg of the sulfacetamide obtained in *Identification test A* in a test tube, and heat gently until it boils: an oily liquid, which has the characteristic odor of acetamide, condenses on the walls of the test tube (distinction from the sublimates of *sulfadiazine*, *sulfamerazine*, and *sulfamethazine*, which are solids at room temperature).

C: The filtrate obtained in *Identification test A* responds to the tests for *Sodium* ⟨191⟩.

D: Dissolve about 100 mg in 5 mL of water, and add 5 drops of cupric sulfate TS: a light bluish green precipitate is formed, and it remains unchanged on standing.

E: Dissolve about 500 mg in 10 mL of dilute hydrochloric acid (1 in 10). To about one-half of the solution add 2 mL of trinitrophenol TS: a very heavy flocculent or almost gelatinous precipitate is formed. To the remainder of the solution add 3 drops of formaldehyde TS: a white precipitate is formed, and it changes to orange on standing (distinction from *sulfamethoxypyridazine*).

pH ⟨791⟩: between 8.0 and 9.5, in a solution (1 in 20).

Water, *Method I* ⟨921⟩: not more than 8.1%.

Selenium ⟨291⟩: 0.003%, a 200-mg test specimen being used.

Heavy metals—Dissolve 1.0 g in 25 mL of water, and add 5 drops of freshly prepared sodium sulfide TS: any color produced is not darker than that of a control made with 25 mL of water, 2.0 mL of *Standard Lead Solution* (see *Heavy Metals* ⟨231⟩), and 5 drops of sodium sulfide TS (0.002%).

Assay—Proceed with Sulfacetamide Sodium as directed under *Nitrite Titration* ⟨451⟩. Each mL of 0.1 *M* sodium nitrite is equivalent to 23.62 mg of $C_8H_9N_2NaO_3S$.

Sulfacetamide Sodium, and Prednisolone Acetate Ophthalmic Ointment, Neomycin Sulfate,—*see* Neomycin Sulfate, Sulfacetamide Sodium, and Prednisolone Acetate Ophthalmic Ointment

Sulfacetamide Sodium Ophthalmic Ointment

» Sulfacetamide Sodium Ophthalmic Ointment contains not less than 95.0 percent and not more than 105.0 percent of the labeled amount of $C_8H_9N_2NaO_3S.H_2O$. It is sterile.

Packaging and storage—Preserve in collapsible ophthalmic ointment tubes.

Identification—Dissolve a quantity of Ophthalmic Ointment, equivalent to about 1 g of sulfacetamide sodium, in 100 mL of ether in a separator, and extract the mixture with 25 mL of water. Wash the extract with 25 mL of ether, and warm the water extract on a steam bath to remove the last traces of ether. Adjust with 6 *N* acetic acid to a pH of between 4 and 5, and filter. Wash the precipitate with water, and dry at 105° for 2 hours: the sulfacetamide so obtained melts between 180° and 184°, and responds to *Identification tests B, D,* and *E* under *Sulfacetamide Sodium.*

Sterility—It meets the requirements for *Ophthalmic Ointments* under *Sterility Tests* ⟨71⟩.

Metal particles—It meets the requirements of the test for *Metal Particles in Ophthalmic Ointments* ⟨751⟩.

Assay—Weigh accurately an amount of Sulfacetamide Sodium Ophthalmic Ointment, equivalent to about 500 mg of sulfacetamide sodium, and transfer to a 125-mL separator. Dissolve the ointment in 50 mL of ether, and extract the mixture with six 25-mL portions of water. Warm the combined extracts on a steam bath to remove the last traces of ether, add 20 mL of hydrochloric acid, and proceed as directed under *Nitrite Titration* ⟨451⟩, beginning with "cool to 15°." Each mL of 0.1 *M* sodium nitrite is equivalent to 25.42 mg of $C_8H_9N_2NaO_3S.H_2O$.

Sulfacetamide Sodium Ophthalmic Solution

» Sulfacetamide Sodium Ophthalmic Solution is a sterile solution containing not less than 95.0 percent and not more than 110.0 percent of the labeled amount of $C_8H_9N_2NaO_3S.H_2O$. It may contain suitable buffers, stabilizers, and antimicrobial agents.

Packaging and storage—Preserve in tight, light-resistant containers, in a cool place.

Reference standard—*USP Sulfacetamide Sodium Reference Standard*—Determine the *Water* content by *Method I* ⟨921⟩ before using.

Identification—Transfer to a beaker a volume of Solution, equivalent to about 1 g of sulfacetamide sodium, and dilute with water to 25 mL. Adjust with 6 *N* acetic acid to a pH of between 4 and 5, and filter. Wash the precipitate with water, and dry at 105° for 2 hours: the sulfacetamide so obtained melts between 180° and 184°, and responds to *Identification tests B* and *E* under *Sulfacetamide Sodium.*

Sterility—It meets the requirements under *Sterility Tests* ⟨71⟩.

Assay—

Standard preparation—Dissolve a suitable quantity of USP Sulfacetamide Sodium RS, accurately weighed, in water to obtain a solution having a known concentration of about 2 mg (as the monohydrate) per mL.

Assay preparation—Transfer an accurately measured volume of Sulfacetamide Sodium Ophthalmic Solution, equivalent to about 500 mg of sulfacetamide sodium, to a 250-mL volumetric flask, dilute with water to volume, and mix.

Procedure—Coat a 20- × 20-cm chromatographic plate with a 0.25-mm layer of chromatographic silica gel mixture, dry at room temperature for 15 minutes, heat at 105° for 1 hour, and cool in a desiccator. Divide the area on the plate into three equal sections, the left and right sections to be used for the *Assay preparation* and the *Standard preparation*, respectively, and the center section for the blank. Apply 30 µL each of the *Assay preparation* and of the *Standard preparation* as streaks 2.5 cm from the bottom of the appropriate section of the plate, and dry the streaks with a current of air. Using butyl alcohol previously saturated with a mixture of water and ammonium hydroxide (8:1), develop the chromatogram in a suitable chromatographic chamber, previously equilibrated and lined with absorbent paper, until the solvent front has moved about 15 cm above the initial streaks. Remove the plate, evaporate the solvent, and locate the principal band occupied by the *Standard preparation* by viewing under short-wavelength ultraviolet light. Mark this band, as well as corresponding bands in the *Assay preparation* and blank sections of the plate. Quantitatively remove the silica gel from each band separately, and transfer it to a glass-stoppered, 50-mL centrifuge tube. Add 10.0 mL of dilute hydrochloric acid (1 in 80) to each tube, shake gently for 30 minutes, and centrifuge. Pipet 5 mL of each of the clear supernatant solutions into separate 10-mL volumetric flasks. Into each flask pipet 1 mL of freshly prepared sodium nitrite solution (1 in 1000), swirl the flask intermittently during 3 minutes, add by pipet 1 mL of ammonium sulfamate solution (1 in 200), shake gently for 2 minutes, then add by pipet 1 mL of a freshly prepared solution of *N*-1-naphthylethylenediamine dihydrochloride (1 in 1000). Allow to stand, with occasional shaking, for 15 minutes, add water to volume, and mix. Concomitantly determine the absorbances of the solutions from the *Assay preparation* and the *Standard preparation* at about 545 nm, with a suitable spectrophotometer, using the blank to set the instrument. Calculate the quantity, in g, of $C_8H_9N_2NaO_3S \cdot H_2O$ in each 100 mL of the Ophthalmic Solution taken by the formula:

$$25(C/V)(A_U/A_S),$$

in which C is the concentration, in mg per mL, of USP Sulfacetamide Sodium RS in the *Standard preparation*, V is the volume, in mL, of the Ophthalmic Solution taken, and A_U and A_S are the absorbances of the solutions from the *Assay preparation* and the *Standard preparation*, respectively.

Sulfacetamide Sodium and Prednisolone Acetate Ophthalmic Ointment

» Sulfacetamide Sodium and Prednisolone Acetate Ophthalmic Ointment is a sterile ointment containing not less than 90.0 percent and not more than 110.0 percent of the labeled amounts of sulfacetamide sodium ($C_8H_9N_2NaO_3S \cdot H_2O$) and prednisolone acetate ($C_{23}H_{30}O_6$).

Packaging and storage—Preserve in collapsible ophthalmic ointment tubes that are tamper-proof so that sterility is assured at time of first use.

Reference standards—*USP Sulfacetamide Sodium Reference Standard*—Determine the *Water* content by *Method I* ⟨921⟩ before using. *USP Prednisolone Acetate Reference Standard*—Dry at 105° for 3 hours before using.

Identification—Transfer the contents of 1 tube of Ophthalmic Ointment to a 100-mL beaker, add about 25 mL of alcohol, and stir by mechanical means for about 15 minutes. Filter, and use the clear solution as the *Test preparation.* Prepare Standard solutions in alcohol containing 0.7 mg per mL of USP Prednisolone Acetate RS and 14 mg per mL of USP Sulfacetamide Sodium RS. On a suitable thin-layer chromatographic plate (see *Chromatography* ⟨621⟩), coated with a 0.25-mm layer of chromatographic silica gel mixture, apply separately 10 µL of the *Test preparation* and 10 µL of each Standard solution. Position the plate in a chromatographic chamber, and develop the chromatograms in a solvent system consisting of a mixture of chloroform, heptane, alcohol, and water (50:50:50:2) until the solvent front has moved about three-fourths of the length of the plate. Remove the plate from the developing chamber, mark the solvent front, and allow the solvent to evaporate. Examine the plate under ultraviolet light: the *Test preparation* exhibits two spots whose R_f values and intensities correspond to the respective spots from the Standard solutions.

Minimum fill ⟨755⟩: meets the requirements.

Sterility—It meets the requirements under *Sterility Tests* ⟨71⟩.

Metal particles—It meets the requirements of the test for *Metal Particles in Ophthalmic Ointments* ⟨751⟩.

Assay for sulfacetamide sodium—

Mobile phase—Prepare a filtered and degassed mixture of water, methanol, and glacial acetic acid (890:100:10). Make adjustments if necessary (see *System Suitability* under *Chromatography* ⟨621⟩).

Standard preparation—Transfer about 50 mg of USP Sulfacetamide Sodium RS, accurately weighed, to a 40-mL centrifuge tube. Add 10.0 mL of dilute methanol (1 in 5), insert the stopper in the tube, and vortex for about 3 minutes to dissolve the standard. Add 7.5 mL of heptane, insert the stopper in the tube, and vortex for another 3 minutes. Centrifuge to effect separation of the phases. Withdraw and discard the upper, heptane layer. Transfer 3.0 mL of the bottom layer to a 500-mL volumetric flask, add dilute methanol (1 in 5) to volume, and mix.

Assay preparation—Transfer an accurately weighed quantity of Sulfacetamide Sodium and Prednisolone Acetate Ophthalmic Ointment, equivalent to about 100 mg of sulfacetamide sodium, to a 40-mL centrifuge tube. Add 15.0 mL of heptane, insert the stopper in the tube, and vortex for about 3 minutes to dissolve the Ointment. Add 20.0 mL of dilute methanol (1 in 5), insert the stopper in the tube, and vortex for 3 minutes. Centrifuge to effect separation of the phases. Withdraw and discard the upper, heptane layer. Transfer 3.0 mL of the bottom layer into a 500-mL volumetric flask, add dilute methanol (1 in 5) to volume, and mix.

System suitability preparation—Dissolve 3 mg of sulfanilamide in 100 mL of the *Standard preparation,* and mix.

Chromatographic system (see *Chromatography* ⟨621⟩)—The liquid chromatograph is equipped with a 254-nm detector and a 4.6-mm × 25-cm column that contains packing L1. The flow rate is about 1.5 mL per minute. Chromatograph the *Standard preparation* and the *System suitability preparation,* and record the peak responses as directed under *Procedure:* the column efficiency determined for the analyte peak is not less than 1500 theoretical plates, the resolution, R, between the sulfacetamide and sulfanilamide peaks is not less than 3, and the relative standard deviation for replicate injections is not more than 2.0%.

Procedure—Separately inject equal volumes (about 90 µL) of the *Standard preparation* and the *Assay preparation* into the chromatograph, record the chromatograms, and measure the responses for the major peaks. Calculate the quantity, in mg, of $C_8H_9N_2NaO_3S \cdot H_2O$ in the portion of Ophthalmic Ointment taken by the formula:

$$3.33C(r_U/r_S),$$

in which C is the concentration, in µg per mL, of USP Sulfacetamide Sodium RS in the *Standard preparation,* and r_U and r_S are the peak responses obtained from the *Assay preparation* and the *Standard preparation,* respectively.

Assay for prednisolone acetate—

Mobile phase—Prepare a filtered and degassed mixture of water and acetonitrile (60:40). Make adjustments if necessary (see *System Suitability* under *Chromatography* ⟨621⟩).

Internal standard solution—Transfer about 70 mg of norethindrone to a 100-mL volumetric flask, add dilute methanol (9 in 10) to volume, and mix.

Standard preparation—Transfer about 20 mg of USP Prednisolone Acetate RS, accurately weighed, to a 25-mL volumetric flask, add dilute methanol (9 in 10) to volume, and mix. Transfer 5.0 mL of this solution to a 100-mL volumetric flask, add 5.0 mL of *Internal standard solution,* add dilute methanol (9 in 10) to volume, and mix.

Assay preparation—Transfer an accurately weighed quantity of Sulfacetamide Sodium and Prednisolone Acetate Ophthalmic Ointment, equivalent to about 4 mg of prednisolone acetate, to a 50-mL centrifuge tube. Add 10.0 mL of heptane, and vortex for about 2 minutes to dissolve the Ointment. Add 5.0 mL of *Internal standard solution* and 20.0 mL of dilute methanol (9 in 10), and vortex for 2 minutes. Centrifuge to effect separation of the phases. Withdraw and discard the upper, heptane layer. Transfer the lower layer to a 100-mL volumetric flask. Add dilute methanol (9 in 10) to volume, and mix to obtain the *Assay preparation.*

Chromatographic system (see *Chromatography* ⟨621⟩)—The liquid chromatograph is equipped with a 254-nm detector and a 3.9-mm × 30-cm column that contains packing L1. The flow rate is about 1.5 mL per minute. Chromatograph the *Standard preparation*, and record the peak responses as directed under *Procedure:* the column efficiency determined from the analyte peak is not less than 3000, the tailing factor for the analyte peak is not more than 2.5, the resolution, *R*, between the analyte and internal standard peaks is not less than 4.5, and the relative standard deviation for replicate injections is not more than 1.5%.

Procedure—Separately inject equal volumes (about 40 µL) of the *Standard preparation* and the *Assay preparation* into the chromatograph, record the chromatograms, and measure the responses for the major peaks. The relative retention times are about 1.0 for prednisolone acetate and 1.5 for norethindrone. Calculate the quantity, in mg, of $C_{23}H_{30}O_6$ in the portion of the Ophthalmic Ointment taken by the formula:

$$100C(R_U/R_S),$$

in which *C* is the concentration, in mg per mL, of USP Prednisolone Acetate RS in the *Standard preparation* taken, and R_U and R_S are the peak response ratios obtained from the *Assay preparation* and the *Standard preparation*, respectively.

Sulfacetamide Sodium and Prednisolone Acetate Ophthalmic Suspension

» Sulfacetamide Sodium and Prednisolone Acetate Ophthalmic Suspension is a sterile, aqueous suspension containing not less than 90.0 percent and not more than 110.0 percent of the labeled amounts of sulfacetamide sodium ($C_8H_9N_2NaO_3S.H_2O$) and prednisolone acetate ($C_{23}H_{30}O_6$). It may contain suitable preservatives, buffers, stabilizers, and suspending agents.

Packaging and storage—Preserve in tight containers. The containers or individual cartons are sealed and tamper-proof so that sterility is assured at time of first use.

Reference standards—*USP Sulfacetamide Sodium Reference Standard*—Determine the *Water* content by *Method I* ⟨921⟩ before using. *USP Prednisolone Acetate Reference Standard*—Dry at 105° for 3 hours before using.

Identification—

A: Filter about 25 mL of the well-mixed Ophthalmic Suspension through a fine, sintered-glass filter, saving the filtrate. Wash the crystals in the funnel with a small amount of water. Dry the crystals at 105° for 3 hours: the infrared absorption spectrum of a potassium bromide dispersion of the crystals exhibits maxima only at the same wavelengths as that of a similar preparation of USP Prednisolone Acetate RS.

B: To the filtrate saved from *Identification test A*, add 6 *N* acetic acid dropwise until the pH is between 4 and 5. Allow crystals of sulfacetamide to develop. Filter the crystals, wash with a small amount of water, and dry at 105° for 2 hours: the infrared absorption spectrum of a potassium bromide dispersion of the crystals so obtained exhibits maxima only at the same wavelengths as a preparation of USP Sulfacetamide Sodium RS, similarly treated.

Sterility—It meets the requirements under *Sterility Tests* ⟨71⟩.

pH ⟨791⟩: between 6.0 and 7.4.

Assay for sulfacetamide sodium—

Mobile phase—Prepare a filtered and degassed mixture of water, methanol, and glacial acetic acid (890:100:10). Make adjustments if necessary (see *System Suitability* under *Chromatography* ⟨621⟩).

Standard preparation—Dissolve an accurately weighed quantity of USP Sulfacetamide Sodium RS in a mixture of water and methanol (4:1), and dilute quantitatively, and stepwise if necessary, with the same solvent mixture to obtain a solution having a known concentration of about 30 µg per mL.

Assay preparation—Transfer an accurately measured volume of Sulfacetamide Sodium and Prednisolone Acetate Ophthalmic Suspension, freshly mixed and free from air bubbles, equivalent to about 100 mg of sulfacetamide sodium, to a 100-mL volumetric flask, dilute with a mixture of water and methanol (4:1) to volume, and mix. Dilute 3.0 mL of this solution with the same solvent mixture to 100.0 mL, and mix.

System suitability preparation—Dissolve about 9 mg of sulfanilamide in 100 mL of the *Standard preparation*, and mix.

Chromatographic system (see *Chromatography* ⟨621⟩)—The liquid chromatograph is equipped with a 254-nm detector and a 4.6-mm × 25-cm column that contains packing L1. The flow rate is about 1.5 mL per minute. Chromatograph the *Standard preparation* and the *System suitability preparation*, and record the peak responses as directed under *Procedure:* the column efficiency determined for the analyte peak is not less than 1500 theoretical plates, the resolution, *R*, between the sulfacetamide and sulfanilamide peaks is not less than 3, and the relative standard deviation for replicate injections is not more than 2.0%.

Procedure—Separately inject equal volumes (about 90 µL) of the *Standard preparation* and the *Assay preparation* into the chromatograph, record the chromatograms, and measure the responses for the major peaks. Calculate the quantity, in mg, of $C_8H_9N_2NaO_3S.H_2O$ in each mL of the Ophthalmic Suspension taken by the formula:

$$3.33(C/V)(r_U/r_S),$$

in which *C* is the concentration, in µg per mL, of USP Sulfacetamide Sodium RS in the *Standard preparation*, *V* is the volume, in mL, of Ophthalmic Suspension taken, and r_U and r_S are the peak responses obtained from the *Assay preparation* and the *Standard preparation*, respectively.

Assay for prednisolone acetate—

Mobile phase—Prepare a filtered and degassed mixture of water and acetonitrile (60:40). Make adjustments if necessary (see *System Suitability* under *Chromatography* ⟨621⟩).

Standard preparation—Dissolve an accurately weighed quantity of USP Prednisolone Acetate RS in methanol to obtain a solution containing about 2 mg per mL. Transfer 2.0 mL of this solution to a 100-mL volumetric flask, and dilute with a solvent mixture containing 70 volumes of methanol and 30 volumes of a monobasic potassium phosphate solution prepared by dissolving 2.72 g in 300 mL of water and 700 mL of methanol. The *Standard preparation* has a known concentration of about 0.04 mg per mL.

Assay preparation—Using a "To contain" pipet, transfer an accurately measured volume of Sulfacetamide Sodium and Prednisolone Acetate Ophthalmic Suspension, freshly mixed and free from air bubbles, equivalent to about 10 mg of prednisolone acetate, to a 250-mL volumetric flask. Rinse the pipet with the solvent mixture described under *Standard preparation*, collecting the rinsings in the flask, dilute with the same solvent mixture to volume, and mix.

Chromatographic system (see *Chromatography* ⟨621⟩)—The liquid chromatograph is equipped with a 254-nm detector and a 4.0-mm × 30-cm column that contains packing L1. The flow rate is about 1.5 mL per minute. Chromatograph the *Standard preparation*, and record the peak responses as directed under *Procedure:* the column efficiency determined from the analyte peak is not less than 3000 theoretical plates, and the relative standard deviation for replicate injections is not more than 2.0%.

Procedure—Separately inject equal volumes (about 30 µL) of the *Standard preparation* and the *Assay preparation* into the chromatograph, record the chromatograms, and measure the responses for the major peaks. Calculate the quantity, in mg, of $C_{23}H_{30}O_6$ in each mL of the Ophthalmic Suspension taken by the formula:

$$250(C/V)(r_U/r_S),$$

in which *C* is the concentration, in mg per mL, of USP Prednisolone Acetate RS in the *Standard preparation*, *V* is the volume, in mL, of Ophthalmic Suspension taken, and r_U and r_S are the peak responses obtained from the *Assay preparation* and the *Standard preparation*, respectively.

Sulfadiazine

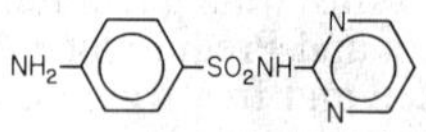

$C_{10}H_{10}N_4O_2S$ 250.27
Benzenesulfonamide, 4-amino-*N*-2-pyrimidinyl-.
N^1-2-Pyrimidinylsulfanilamide [*68-35-9*].

» Sulfadiazine contains not less than 98.0 percent and not more than 102.0 percent of $C_{10}H_{10}N_4O_2S$, calculated on the dried basis.

Packaging and storage—Preserve in well-closed, light-resistant containers.

Reference standard—*USP Sulfadiazine Reference Standard*—Dry at 105° for 2 hours before using.

Clarity and color of solution—Dissolve 1 g in a mixture of 20 mL of water and 5 mL of 1 *N* sodium hydroxide. The solution is clear and not more deeply colored than pale yellow.

Identification—

A: The infrared absorption spectrum of a potassium bromide dispersion of it, previously dried at 105° for 2 hours, exhibits maxima only at the same wavelengths as that of a similar preparation of USP Sulfadiazine RS.

B: Carefully melt about 50 mg in a small test tube: a reddish brown color develops. The fumes evolved during the decomposition do not discolor moistened lead acetate test paper (distinction from *sulfathiazole*).

C: Gently heat about 1 g in a small test tube until a sublimate is formed. Collect a few mg of the sublimate with a glass rod, and mix in a test tube with 1 mL of a 1 in 20 solution of resorcinol in alcohol. Add 1 mL of sulfuric acid, and mix by shaking: a deep red color appears at once. Cautiously dilute the mixture with 25 mL of ice-cold water, and add an excess of 6 *N* ammonium hydroxide: a blue or reddish blue color is produced.

Acidity—Digest 2.00 g with 100 mL of water at about 70° for 5 minutes. Cool at once to room temperature, and filter. To 25.0 mL of the filtrate add 2 drops of phenolphthalein TS, and titrate with 0.10 *N* sodium hydroxide: not more than 0.20 mL is required to produce a pink color.

Loss on drying ⟨731⟩—Dry it at 105° for 2 hours: it loses not more than 0.5% of its weight.

Residue on ignition ⟨281⟩: not more than 0.1%.

Selenium ⟨291⟩: 0.003%, a 200-mg test specimen being used.

Heavy metals, *Method II* ⟨231⟩: 0.002%.

Assay—

Mobile phase—Prepare a suitable degassed solution consisting of a mixture of water, acetonitrile, and glacial acetic acid (87: 12:1).

Standard preparation—Dissolve an accurately weighed quantity of USP Sulfadiazine RS in 0.025 *N* sodium hydroxide to obtain a solution having a known concentration of about 1 mg per mL.

Assay preparation—Transfer about 100 mg of Sulfadiazine, accurately weighed, to a 100-mL volumetric flask, add 0.025 *N* sodium hydroxide to volume, and mix.

Chromatographic system (see *Chromatography* ⟨621⟩)—The liquid chromatograph is equipped with a 254-nm detector and a 4-mm × 30-cm column that contains packing L1. The flow rate is about 2 mL per minute. Chromatograph five replicate injections of the *Standard preparation*, and record the peak responses as directed under *Procedure:* the relative standard deviation is not more than 2.0%, and the tailing factor for sulfadiazine is not more than 1.5.

Procedure—Separately inject equal volumes (about 10 µL) of the *Standard preparation* and the *Assay preparation* into the chromatograph, record the chromatograms, and measure the responses for the major peaks. Calculate the quantity, in mg, of $C_{10}H_{10}N_4O_2S$ in the portion of Sulfadiazine taken by the formula:

$$100C(r_U/r_S),$$

in which *C* is the concentration, in mg per mL, of USP Sulfadiazine RS in the *Standard preparation*, and r_U and r_S are the peak responses for sulfadiazine obtained from the *Assay preparation* and the *Standard preparation*, respectively.

Sulfadiazine Tablets

» Sulfadiazine Tablets contain not less than 95.0 percent and not more than 105.0 percent of the labeled amount of $C_{10}H_{10}N_4O_2S$.

Packaging and storage—Preserve in well-closed, light-resistant containers.

Reference standard—*USP Sulfadiazine Reference Standard*—Dry at 105° for 2 hours before using.

Identification—Triturate a quantity of finely powdered Tablets, equivalent to about 500 mg of sulfadiazine, with 5 mL of chloroform, and transfer to a small filter. Wash with another 5-mL portion of chloroform, and discard the filtrate. Triturate the residue with 10 mL of 6 *N* ammonium hydroxide for 5 minutes, add 10 mL of water, and filter. Warm the filtrate until most of the ammonia is expelled, cool, and add 6 *N* acetic acid until the reaction is distinctly acid: a precipitate of sulfadiazine is formed. Collect the precipitate on a filter, wash it well with cold water, and dry at 105° for 1 hour: the sulfadiazine so obtained melts between 250° and 254°, as determined by the method for *Class Ia* under *Melting Range or Temperature* ⟨741⟩ and responds to *Identification test A* under *Sulfadiazine.*

Dissolution ⟨711⟩—

Medium: 0.1 *N* hydrochloric acid; 900 mL.

Apparatus 2: 100 rpm.

Time: 60 minutes.

Procedure—Determine the amount of $C_{10}H_{10}N_4O_2S$ dissolved from ultraviolet absorbances at the wavelength of maximum absorbance at about 254 nm of filtered portions of the solution under test, suitably diluted with 0.01 *N* sodium hydroxide, in comparison with a Standard solution having a known concentration of USP Sulfadiazine RS in the same medium.

Tolerances—Not less than 70% (*Q*) of the labeled amount of $C_{10}H_{10}N_4O_2S$ is dissolved in 60 minutes.

Uniformity of dosage units ⟨905⟩: meet the requirements.

Assay—

Mobile phase and *Standard preparation*—Prepare as directed in the *Assay* under *Sulfadiazine.*

Assay preparation—Weigh and finely powder not less than 20 Sulfadiazine Tablets. Transfer an accurately weighed portion of the powder, equivalent to about 100 mg of sulfadiazine, to a 100-mL volumetric flask, add 75 mL of 0.025 *N* sodium hydroxide, shake for 30 minutes, dilute with 0.025 *N* sodium hydroxide to volume, mix, and centrifuge.

Chromatographic system and *Procedure*—Proceed as directed for *Chromatographic system* and for *Procedure* in the *Assay* under *Sulfadiazine.* Calculate the quantity, in mg, of $C_{10}H_{10}N_4O_2S$ in the portion of Tablets taken by the formula:

$$100C(r_U/r_S),$$

in which the terms are as defined therein.

Sulfadiazine Sodium

$C_{10}H_9N_4NaO_2S$ 272.26
Benzenesulfonamide, 4-amino-*N*-2-pyrimidinyl-, monosodium salt.
N^1-2-Pyrimidinylsulfanilamide monosodium salt [*547-32-0*].

» Sulfadiazine Sodium contains not less than 99.0 percent and not more than 100.5 percent of $C_{10}H_9N_4NaO_2S$, calculated on the dried basis.

Packaging and storage—Preserve in tight, light-resistant containers.

Reference standard—*USP Sulfadiazine Reference Standard*—Dry at 105° for 2 hours before using.

Identification—

A: Dissolve about 1.5 g in 25 mL of water, and add 3 mL of 6 *N* acetic acid: a white precipitate of sulfadiazine is formed. Collect the precipitate on a filter, wash it well with cold water, and dry at 105° for 1 hour: the sulfadiazine so obtained melts between 250° and 254° as determined by the method for *Class Ia* under *Melting Range or Temperature* ⟨741⟩, and responds to *Identification test A* under *Sulfadiazine.*

B: Ignite about 500 mg: the residue responds to the tests for *Sodium* ⟨191⟩.

Loss on drying ⟨731⟩—Dry it at 105° for 2 hours: it loses not more than 0.5% of its weight.

Selenium ⟨291⟩: 0.003%, a 200-mg test specimen being used.

Heavy metals—Dissolve 1.0 g in 25 mL of water, and add 5 drops of sodium sulfide TS: any dark color produced is not darker than that in a blank to which 2 mL of *Standard Lead Solution* (see *Heavy Metals* ⟨231⟩) has been added (0.002%).

Assay—Proceed with Sulfadiazine Sodium as directed under *Nitrite Titration* ⟨451⟩. Each mL of 0.1 *M* sodium nitrite is equivalent to 27.23 mg of $C_{10}H_9N_4NaO_2S$.

Sulfadiazine Sodium Injection

» Sulfadiazine Sodium Injection is a sterile solution of Sulfadiazine Sodium in Water for Injection. It contains, in each mL, not less than 237.5 mg and not more than 262.5 mg of $C_{10}H_9N_4NaO_2S$.

Packaging and storage—Preserve in single-dose, light-resistant containers, of Type I glass.

Reference standard—*USP Sulfadiazine Reference Standard*—Dry at 105° for 2 hours before using.

Identification—It responds to *Identification test A* under *Sulfadiazine Sodium.*

pH ⟨791⟩: between 8.5 and 10.5.

Particulate matter ⟨788⟩: meets the requirements under *Small-volume Injections.*

Other requirements—It meets the requirements under *Injections* ⟨1⟩.

Assay—Pipet a volume of Sulfadiazine Sodium Injection, equivalent to about 500 mg of sulfadiazine sodium, into a beaker or a casserole, and proceed as directed under *Nitrite Titration* ⟨451⟩, beginning with "Add 20 mL of hydrochloric acid." Each mL of 0.1 *M* sodium nitrite is equivalent to 27.23 mg of $C_{10}H_9N_4NaO_2S$.

Sulfadoxine

CH_3O OCH_3
NH_2–C_6H_4–SO_2NH–(pyrimidine, N, N)

$C_{12}H_{14}N_4O_4S$ 310.33

Benzenesulfonamide, 4-amino-*N*-(5,6-dimethoxy-4-pyrimidinyl)-.

N^1-(5,6-Dimethoxy-4-pyrimidinyl)sulfanilamide [*2447-57-6*].

» Sulfadoxine contains not less than 99.0 percent and not more than 101.0 percent of $C_{12}H_{14}N_4O_4S$, calculated on the dried basis.

Packaging and storage—Preserve in well-closed, light-resistant containers.

Reference standard—*USP Sulfadoxine Reference Standard*—Dry at 105° for 4 hours.

Identification—

A: The infrared absorption spectrum of a potassium bromide dispersion of it, previously dried, exhibits maxima at the same wavelengths as that of a similar preparation of USP Sulfadoxine RS.

B: The ultraviolet absorption spectrum of a 1 in 167,000 solution in 0.1 *N* sodium hydroxide exhibits maxima and minima at the same wavelengths as that of a similar preparation of USP Sulfadoxine RS.

Melting range ⟨741⟩: between 197° and 200°.

Loss on drying ⟨731⟩—Dry it at 105° for 4 hours: it loses not more than 0.5% of its weight.

Residue on ignition ⟨281⟩: not more than 0.1%.

Heavy metals, *Method II* ⟨231⟩: 0.002%.

Chromatographic purity—Prepare a solution in a mixture of alcohol and ammonium hydroxide (9:1) having a concentration of about 20 mg per mL. Prepare a Standard solution of USP Sulfadoxine RS having a concentration of about 0.10 mg per mL in a mixture of alcohol and ammonium hydroxide (9:1). Separately apply 10 µL each of the test solution and the Standard solution to a suitable chromatographic plate (see *Chromatography* ⟨621⟩) coated with a 0.25-mm layer of chromatographic silica gel. Develop the chromatogram in a solvent system consisting of a mixture of chloroform, methanol, and dimethylformamide (20:2:1) until the solvent front has moved about three-fourths of the length of the plate. Remove the plate from the chamber, and air-dry. Spray the dried plate with a 1 in 10 solution of sulfuric acid in alcohol, and expose to nitrous fumes generated by adding 7 *M* sulfuric acid dropwise to a solution containing 10% sodium nitrite and 3% potassium iodide. Dry the plate in a current of warm air for 15 minutes, and spray with a 1 in 200 solution of *N*-(1-naphthyl)ethane-1,2-diammonium dichloride in alcohol. No spot in the test solution, other than the principal spot, is greater in size and intensity than the spot obtained from the Standard solution.

Assay—Proceed with Sulfadoxine as directed under *Nitrite Titration* ⟨451⟩. Each mL of 0.1 *M* sodium nitrite is equivalent to 31.03 mg of $C_{12}H_{14}N_4O_4S$.

Sulfadoxine and Pyrimethamine Tablets

» Sulfadoxine and Pyrimethamine Tablets contain not less than 90.0 percent and not more than 110.0 percent of the labeled amount of sulfadoxine ($C_{12}H_{14}N_4O_4S$) and not less than 90.0 percent and not more than 110.0 percent of the labeled amount of pyrimethamine ($C_{12}H_{13}ClN_4$).

Packaging and storage—Preserve in well-closed, light-resistant containers.

Reference standards—*USP Sulfadoxine Reference Standard*—Dry at 105° for 4 hours before using. *USP Pyrimethamine Reference Standard*—Dry at 105° for 4 hours before using.

Identification—

A: The retention times of the major peaks in the chromatogram of the *Assay preparation* correspond to those of the *Standard preparations* of sulfadoxine and pyrimethamine, relative to the internal standard, as obtained in the *Assay*.

B: Vigorously shake 700 mg of finely ground Tablet powder with 50 mL of a 1 in 50 solution of ammonium hydroxide in methanol for 3 minutes, and filter. Separately apply 10 µL each of the test solution, a Standard solution of USP Sulfadoxine RS similarly prepared, containing 10 mg per mL, and a Standard solution of USP Pyrimethamine RS similarly prepared, containing 0.5 mg per mL, to a suitable thin-layer chromatographic plate (see *Chromatography* ⟨621⟩) coated with a 0.25-mm layer of chromatographic silica gel mixture. Dry the spots in a current of warm air, and develop the plate in a solvent system consisting of a mixture of heptane, chloroform, a 1 in 20 solution of methanol in alcohol, and glacial acetic acid (4:4:4:1). Allow the solvent front to move about two-thirds of the length of the plate, remove the plate, dry, and examine under short-wavelength ultraviolet light: the R_f values of the principal spots from the solution under test correspond to the R_f values of the principal spots from the corresponding Standard solutions.

Dissolution ⟨711⟩—

Medium: pH 6.8 phosphate buffer, prepared as directed under *Buffer Solutions* in the section, *Reagents, Indicators, and Solutions;* 1000 mL.

Apparatus 2: 75 rpm.

Time: 30 minutes.

Procedure—Determine the amounts of $C_{12}H_{14}N_4O_4S$ and $C_{12}H_{13}ClN_4$ dissolved, employing the procedure set forth in the *Assay*, making any necessary modifications.

Tolerances—Not less than 60% (*Q*) of the labeled amount of each of $C_{12}H_{14}N_4O_4S$ and $C_{12}H_{13}ClN_4$ is dissolved in 30 minutes.

Uniformity of dosage units ⟨905⟩: meet the requirements for *Content Uniformity* with respect to sulfadoxine and to pyrimethamine.

Assay—

Mobile phase—Prepare a suitable degassed and filtered mixture of dilute glacial acetic acid (1 in 100) and acetonitrile (4:1).

Internal standard solution—Prepare a solution of phenacetin in acetonitrile having a concentration of 1 mg per mL.

Standard stock solution—Transfer about 500 mg, accurately weighed, of USP Sulfadoxine RS and 25 mg, accurately weighed, of USP Pyrimethamine RS to a 100-mL volumetric flask, dissolve in 35 mL of acetonitrile, dilute with *Mobile phase* to volume, and mix.

Standard preparation I—Pipet 25 mL of *Standard stock solution* and 2 mL of *Internal standard solution* into a 50-mL volumetric flask, dilute with *Mobile phase* to volume, and mix.

Standard preparation II—Pipet 2 mL of *Standard stock solution* and 10 mL of *Internal standard solution* into a 250-mL volumetric flask, dilute with *Mobile phase* to volume, and mix.

Assay preparations—Weigh and finely powder not less than 20 Sulfadoxine and Pyrimethamine Tablets. Transfer an accurately weighed portion of the finely ground powder, equivalent to about 500 mg of sulfadoxine and 25 mg of pyrimethamine, to a 100-mL volumetric flask, add 35 mL of acetonitrile, shake for 30 minutes, dilute with *Mobile phase* to volume, mix, and filter. Pipet 25 mL of the filtrate and 2 mL of *Internal standard solution* into a 50-mL volumetric flask, dilute with *Mobile phase* to volume, and mix (*Assay preparation I*). Pipet 2 mL of the filtrate and 10 mL of the *Internal standard solution* into a 250-mL volumetric flask, dilute with *Mobile phase* to volume, and mix (*Assay preparation II*).

Chromatographic system (see *Chromatography* ⟨621⟩)—The liquid chromatograph is equipped with a 254-nm detector and a 3.9-mm × 30-cm column that contains packing L1. The flow rate is about 2.0 mL per minute. Chromatograph five replicate injections of the *Standard preparation*, and record the peak responses as directed under *Procedure:* the relative standard deviation is not more than 2.5%, and the resolution factor between sulfadoxine and phenacetin is not less than 1.0, and between pyrimethamine and phenacetin is not less than 1.0.

Procedure—Separately inject equal volumes (about 10 µL) of the *Standard preparations* and the *Assay preparations* into the chromatograph, record the chromatograms, and measure the responses for the major peaks. The relative retention times are about 0.7 for sulfadoxine and 1.0 for phenacetin and 1.3 for pyrimethamine. Calculate the quantity, in mg, of sulfadoxine in the portion of Tablets taken by the formula:

$$12.5C(R_U/R_S),$$

in which *C* is the concentration, in µg per mL, of USP Sulfadoxine RS in *Standard preparation II*, and R_U and R_S are the relative peak response ratios obtained from *Assay preparation II* and *Standard preparation II*, respectively. Calculate the quantity, in mg, of pyrimethamine in the portion of Tablets taken by the formula:

$$0.2C'(R'_U/R'_S),$$

in which *C'* is the concentration, in µg per mL, of USP Pyrimethamine RS in *Standard preparation I*, and R'_U and R'_S are the relative peak response ratios obtained from *Assay preparation I* and *Standard preparation I*, respectively.

Sulfamerazine

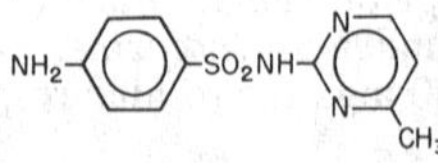

$C_{11}H_{12}N_4O_2S$ 264.30

Benzenesulfonamide, 4-amino-*N*-(4-methyl-2-pyrimidinyl)-.

N^1-(4-Methyl-2-pyrimidinyl)sulfanilamide [*127-79-7*].

» Sulfamerazine contains not less than 99.0 percent and not more than 100.5 percent of $C_{11}H_{12}N_4O_2S$, calculated on the dried basis.

Packaging and storage—Preserve in well-closed, light-resistant containers.

Reference standard—*USP Sulfamerazine Reference Standard*—Dry at 105° for 2 hours before using.

Clarity and color of solution—Dissolve 1.0 g in a mixture of 20 mL of water and 5 mL of 1 *N* sodium hydroxide: the solution is clear and not more deeply colored than pale yellow.

Identification—

A: The infrared absorption spectrum of a potassium bromide dispersion of it exhibits maxima only at the same wavelengths as that of a similar preparation of USP Sulfamerazine RS.

B: To about 20 mg suspended in 5 mL of water add, dropwise, 1 *N* sodium hydroxide until dissolved, then add 3 drops of cupric sulfate TS: an olive-green precipitate is formed, and it becomes dark gray on standing.

Melting range ⟨741⟩: between 234° and 239°.

Acidity—Digest 2.0 g with 100 mL of water at about 70° for 5 minutes, cool at once to about 20°, and filter. To 25 mL of the filtrate add 2 drops of phenolphthalein TS, and titrate with 0.10 *N* sodium hydroxide: not more than 0.50 mL is required to produce a pink color.

Loss on drying ⟨731⟩—Dry it at 105° for 2 hours: it loses not more than 0.5% of its weight.

Residue on ignition ⟨281⟩: not more than 0.1%.

Selenium ⟨291⟩: 0.003%, a 200-mg test specimen being used.

Heavy metals, *Method II* ⟨231⟩: 0.002%.

Assay—Proceed with Sulfamerazine as directed under *Nitrite Titration* ⟨451⟩. Each mL of 0.1 *M* sodium nitrite is equivalent to 26.43 mg of $C_{11}H_{12}N_4O_2S$.

Sulfamerazine Tablets

» Sulfamerazine Tablets contain not less than 95.0 percent and not more than 105.0 percent of the labeled amount of $C_{11}H_{12}N_4O_2S$.

Packaging and storage—Preserve in well-closed containers.

Reference standard—*USP Sulfamerazine Reference Standard*—Dry at 105° for 2 hours before using.

Identification—Triturate a quantity of finely powdered Tablets, equivalent to about 500 mg of sulfamerazine, with two 5-mL portions of chloroform, and discard the chloroform. Triturate the residue with 10 mL of 1 *N* sodium hydroxide for 5 minutes, then add 10 mL of water, and filter. To the filtrate add 6 *N* acetic acid with stirring until the mixture is distinctly acid: a white precipitate is formed. Collect the precipitate on a filter, wash it well with cold water, dry at 105° for 1 hour, and use the sulfamerazine so obtained for the following tests.

A: The infrared absorption spectrum of a potassium bromide dispersion of the sulfamerazine exhibits maxima at the same wavelengths as that of a similar preparation of USP Sulfamerazine RS.

B: To about 20 mg of the sulfamerazine add 5 mL of water, shake, and add 1 *N* sodium hydroxide dropwise until dissolved, then add 2 or 3 drops of cupric sulfate TS: an olive-green precipitate forms which becomes dark gray on standing.

C: The sulfamerazine melts between 233° and 239°, the procedure for *Class I* being used (see *Melting Range or Temperature* ⟨741⟩).

Dissolution ⟨711⟩—

Medium: water; 900 mL.

Apparatus 1: 100 rpm.

Time: 45 minutes.

Procedure—Determine the amount of $C_{11}H_{12}N_4O_2S$ dissolved from ultraviolet absorbances at the wavelength of maximum absorbance at about 243 nm of filtered portions of the solution under test, suitably diluted with *Dissolution Medium*, if necessary, in comparison with a Standard solution having a known concentration of USP Sulfamerazine RS in the same medium.

Tolerances—Not less than 75% (*Q*) of the labeled amount of $C_{11}H_{12}N_4O_2S$ is dissolved in 45 minutes.

Uniformity of dosage units ⟨905⟩: meet the requirements.

Assay—Weigh and finely powder not less than 20 Sulfamerazine Tablets. Transfer an accurately weighed portion of the powder, equivalent to about 500 mg of sulfamerazine, to a beaker or casserole. Add 50 mL of water and 20 mL of hydrochloric acid, stir until dissolved, cool to 15°, add about 25 g of crushed ice, and slowly titrate with 0.1 *M* sodium nitrite VS, stirring vigorously, until a blue color is produced immediately when the titrated solution is spotted on starch iodide paper. When the titration is complete, the end-point is reproducible after the mixture has been allowed to stand for 1 minute. Each mL of 0.1 *M* sodium nitrite is equivalent to 26.43 mg of $C_{11}H_{12}N_4O_2S$.

Sulfamethazine

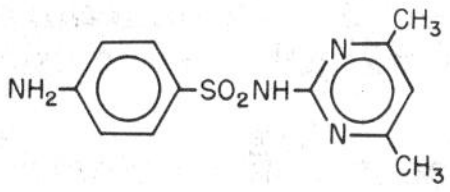

$C_{12}H_{14}N_4O_2S$ 278.33

Benzenesulfonamide, 4-amino-*N*-(4,6-dimethyl-2-pyrimidinyl)-.

N^1-(4,6-Dimethyl-2-pyrimidinyl)sulfanilamide [*57-68-1*].

» Sulfamethazine contains not less than 99.0 percent and not more than 100.5 percent of $C_{12}H_{14}N_4O_2S$, calculated on the dried basis.

Packaging and storage—Preserve in well-closed, light-resistant containers.

Reference standard—*USP Sulfamethazine Reference Standard*—Dry at 105° for 2 hours before using.

Clarity and color of solution—Dissolve 1.0 g in a mixture of 20 mL of water and 5 mL of 1 *N* sodium hydroxide: the solution is clear and not more deeply colored than pale yellow.

Identification—

A: The infrared absorption spectrum of a potassium bromide dispersion of it exhibits maxima only at the same wavelengths as that of a similar preparation of USP Sulfamethazine RS.

B: To 0.10 g add 10 mL of water and just sufficient sodium hydroxide solution (1 in 250) to give a faint pink spot on phenolphthalein paper. Add 5 drops of cupric sulfate TS: a yellow-green precipitate is formed, and it becomes brown on standing.

Melting range ⟨741⟩: between 197° and 200°.

Acidity—Digest 3.0 g with 150 mL of carbon dioxide-free water at 70° for about 5 minutes, stirring occasionally to maintain suspension of the test specimen. Cool the mixture rapidly in an ice bath to 20 ± 0.5°, with mechanical stirring. Filter immediately, with vacuum, omitting washing of the solid but drying it thoroughly by suction. Add 2 drops of thymolphthalein TS to 25.0 mL of the filtrate, and titrate with 0.1 *N* sodium hydroxide VS. To a second 25.0-mL portion of the filtrate add 10 mL of hydrochloric acid. Cool the mixture to 15°, and titrate with 0.1 *M* sodium nitrite VS as directed under *Nitrite Titration* ⟨451⟩: the volume of 0.1 *N* sodium hydroxide consumed does not exceed the volume of 0.1 *M* sodium nitrite consumed by more than 0.5 mL.

Loss on drying ⟨731⟩—Dry about 1 g, accurately weighed, at 105° for 2 hours: it loses not more than 0.5% of its weight.

Residue on ignition ⟨281⟩: not more than 0.1%.

Selenium ⟨291⟩: 0.003%, a 200-mg test specimen being used.

Heavy metals, *Method II* ⟨231⟩: 0.002%.

Ordinary impurities ⟨466⟩—

Test solution: acetone.

Standard solution: acetone.

Eluant: a mixture of ethyl acetate, methanol, and ammonium hydroxide (17:6:5).

Visualization: 11.

Assay—Proceed with Sulfamethazine as directed under *Nitrite Titration* ⟨451⟩. Each mL of 0.1 *M* sodium nitrite is equivalent to 27.83 mg of $C_{12}H_{14}N_4O_2S$.

Sulfamethazine Bisulfates Soluble Powder, Chlortetracycline and—*see* Chlortetracycline and Sulfamethazine Bisulfates Soluble Powder

Sulfamethizole

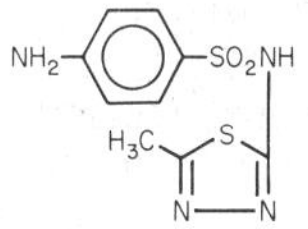

$C_9H_{10}N_4O_2S_2$ 270.32

Benzenesulfonamide, 4-amino-*N*-(5-methyl-1,3,4-thiadiazol2-yl)-.

N^1-(5-Methyl-1,3,4-thiadiazol-2-yl)sulfanilamide [*144-82-1*].

» Sulfamethizole contains not less than 98.0 percent and not more than 101.0 percent of $C_9H_{10}N_4O_2S_2$, calculated on the dried basis.

Packaging and storage—Preserve in well-closed, light-resistant containers.

Reference standard—*USP Sulfamethizole Reference Standard*—Dry at 105° for 2 hours before using.

Clarity and color of solution—Dissolve 1.0 g in 20 mL of water and 5 mL of 1 *N* sodium hydroxide: the solution is clear and not more than pale yellow.

Identification—

A: The infrared absorption spectrum of a mineral oil dispersion of it, previously dried, exhibits maxima only at the same wavelengths as that of a similar preparation of USP Sulfamethizole RS.

B: To about 0.1 g add 5 mL of 3 *N* hydrochloric acid, and boil gently for about 5 minutes. Cool in an ice bath, then add 4 mL of a sodium nitrite solution (1 in 100), add water to make 10 mL, and place the mixture in an ice bath for 10 minutes. To 5 mL of the cooled mixture add a solution of 50 mg of 2-naphthol in 2 mL of sodium hydroxide solution (1 in 10): an orange-red precipitate is formed, and it darkens on standing.

C: To about 20 mg suspended in 5 mL of water add, dropwise, 1 *N* sodium hydroxide until dissolved, then add 2 or 3 drops of cupric sulfate TS: a light green precipitate is formed, and it does not change on standing.

D: The retention time of the major peak in the chromatogram of the *Assay preparation* corresponds to that of the *Standard preparation* as obtained in the *Assay*.

Melting range ⟨741⟩: between 208° and 212°.

Acidity—Digest 2.0 g with 100 mL of water at about 70° for 5 minutes, cool immediately to about 20°, and filter. To 25.0 mL of the filtrate add 2 drops of phenolphthalein TS, and titrate with

0.10 *N* sodium hydroxide: not more than 0.50 mL is required for neutralization. Save the remainder of the filtrate for the tests for *Chloride* and for *Sulfate.*

Loss on drying ⟨731⟩—Dry it at 105° for 2 hours: it loses not more than 0.5% of its weight.

Residue on ignition ⟨281⟩: not more than 0.1%.

Chloride ⟨221⟩—A 25.0-mL portion of the filtrate prepared in the test for *Acidity* shows no more chloride than corresponds to 0.10 mL of 0.020 *N* hydrochloric acid (0.014%).

Sulfate ⟨221⟩—A 25.0-mL portion of the filtrate prepared in the test for *Acidity* shows no more sulfate than corresponds to 0.20 mL of 0.020 *N* sulfuric acid (0.04%).

Selenium ⟨291⟩: 0.003%.

Heavy metals, *Method II* ⟨231⟩: 0.002%.

Ordinary impurities ⟨466⟩—

Test solution: methanol.
Standard solution: methanol.
Eluant: acetone.
Visualization: 1.

Assay—

Mobile phase—Prepare a filtered and degassed mixture of water, methanol, and glacial acetic acid (69:30:1). Make adjustments if necessary (see *System Suitability* under *Chromatography* ⟨621⟩).

Standard preparation—Dissolve an accurately weighed quantity of USP Sulfamethizole RS in methanol to obtain a solution having a known concentration of about 0.4 mg per mL. Quantitatively dilute a volume of this solution with *Mobile phase* to obtain the *Standard preparation* having a known concentration of about 8 µg per mL.

Assay preparation—Transfer about 20 mg of sulfamethizole, accurately weighed, to a 50-mL volumetric flask, dilute with methanol to volume, and mix. Quantitatively dilute a volume of this solution with *Mobile phase* to obtain the *Assay preparation* having a concentration of about 8 µg per mL.

Chromatographic system (see *Chromatography* ⟨621⟩)—The liquid chromatograph is equipped with a 254-nm detector and a 3.9-mm × 30-cm column that contains 10-µm packing L1. The flow rate is about 1.0 mL per minute. Chromatograph the *Standard preparation*, and record the peak responses as directed under *Procedure:* the column efficiency determined from the analyte peak is not less than 2000 theoretical plates, the tailing factor for the analyte peak is not more than 2.0, and the relative standard deviation for replicate injections is not more than 2.0%.

Procedure—Separately inject equal volumes (about 50 µL) of the *Standard preparation* and the *Assay preparation* into the chromatograph, record the chromatograms, and measure the responses for the major peaks. Calculate the quantity, in mg, of $C_9H_{10}N_4O_2S_2$ in the portion of Sulfamethizole taken by the formula:

$$2.5C(r_U/r_S),$$

in which C is the concentration, in µg per mL, of USP Sulfamethizole RS in the *Standard preparation*, and r_U and r_S are the peak responses obtained from the *Assay preparation* and the *Standard preparation*, respectively.

Sulfamethizole Capsules, Oxytetracycline and Phenazopyridine Hydrochlorides and—*see* Oxytetracycline and Phenazopyridine Hydrochlorides and Sulfamethizole Capsules

Sulfamethizole Oral Suspension

» Sulfamethizole Oral Suspension contains not less than 90.0 percent and not more than 110.0 percent of the labeled amount of $C_9H_{10}N_4O_2S_2$, in a buffered aqueous suspension.

Packaging and storage—Preserve in tight, light-resistant containers.

Reference standard—*USP Sulfamethizole Reference Standard*—Dry at 105° for 2 hours before using.

Identification—Place a quantity of Oral Suspension, equivalent to about 500 mg of sulfamethizole, in a 50-mL centrifuge tube, add about 30 mL of water, mix, and centrifuge the mixture. Decant and discard the supernatant liquid. Resuspend the residue in 15 mL of water, mix, and centrifuge the mixture. Decant and discard the clear, supernatant liquid. Repeat the washing procedure an additional two times. Dissolve the residue in a mixture of 10 mL of 6 *N* ammonium hydroxide and 10 mL of water, and filter. Warm the filtrate until most of the ammonia is expelled, cool, and add 6 *N* acetic acid until the mixture is distinctly acid: a precipitate of sulfamethizole is formed. Collect the precipitate on a filter, wash it well with cold water, and dry at 105° for 2 hours: the sulfamethizole so obtained responds to *Identification test A* under *Sulfamethizole.*

Assay—Transfer an accurately measured volume of Sulfamethizole Oral Suspension, equivalent to about 500 mg of sulfamethizole, to a beaker. Add 50 mL of water and 20 mL of hydrochloric acid, stir until dissolved, cool to 15°, and slowly titrate with 0.1 *M* sodium nitrite VS, determining the end-point potentiometrically, using suitable electrodes. Each mL of 0.1 *M* sodium nitrite is equivalent to 27.03 mg of $C_9H_{10}N_4O_2S_2$.

Sulfamethizole Tablets

» Sulfamethizole Tablets contain not less than 95.0 percent and not more than 105.0 percent of the labeled amount of $C_9H_{10}N_4O_2S_2$.

Packaging and storage—Preserve in well-closed containers.

Reference standard—*USP Sulfamethizole Reference Standard*—Dry at 105° for 2 hours before using.

Identification—

A: Triturate a quantity of finely powdered Tablets, equivalent to about 500 mg of sulfamethizole, with 5 mL of chloroform, and transfer to a small filter. Wash with another 5-mL portion of chloroform, and discard the filtrate. Triturate the residue with 10 mL of 6 *N* ammonium hydroxide for 5 minutes, add 10 mL of water, filter, and proceed as directed in the *Identification test* under *Sulfamethizole Oral Suspension*, beginning with "Warm the filtrate."

B: The retention time of the major peak in the chromatogram of the *Assay preparation* corresponds to that of the *Standard preparation* as obtained in the *Assay.*

Dissolution ⟨711⟩—

Medium: 0.1 *N* hydrochloric acid; 900 mL.
Apparatus 2: 50 rpm.
Time: 30 minutes.

Procedure—Determine the amount of $C_9H_{10}N_4O_2S_2$ dissolved from ultraviolet absorbances at the wavelength of maximum absorbance at about 267 nm of filtered portions of the solution under test, suitably diluted with 0.1 *N* hydrochloric acid, if necessary, in comparison with a Standard solution having a known concentration of USP Sulfamethizole RS in the same medium.

Tolerances—Not less than 75% (*Q*) of the labeled amount of $C_9H_{10}N_4O_2S_2$ is dissolved in 30 minutes.

Uniformity of dosage units ⟨905⟩: meet the requirements.

Assay—

Mobile phase, Standard preparation, and *Chromatographic system*—Prepare as directed in the *Assay* under *Sulfamethizole.*

Assay preparation—Weigh and finely powder not less than 20 Sulfamethizole Tablets. Transfer an accurately weighed portion of the powder, equivalent to about 40 mg of sulfamethizole, to a 250-mL screw-capped bottle. Transfer 100.0 mL of methanol to the bottle, cap the bottle, shake by mechanical means for 30 minutes, and filter. Quantitatively dilute a portion of the filtered solution with *Mobile phase* to obtain the *Assay preparation* having a concentration of about 8 µg per mL.

Procedure—Separately inject equal volumes (about 50 µL) of the *Standard preparation* and the *Assay preparation* into the chromatograph, record the chromatograms, and measure the responses for the major peaks. Calculate the quantity, in mg, of $C_9H_{10}N_4O_2S_2$ in the portion of Sulfamethizole Tablets taken by the formula:

$$5C(r_U/r_S),$$

in which C is the concentration, in µg per mL, of USP Sulfamethizole RS in the *Standard preparation*, and r_U and r_S are the peak responses obtained from the *Assay preparation* and the *Standard preparation*, respectively.

Sulfamethoxazole

$C_{10}H_{11}N_3O_3S$ 253.28
Benzenesulfonamide, 4-amino-*N*-(5-methyl-3-isoxazolyl)-.
N^1-(5-Methyl-3-isoxazolyl)sulfanilamide [*723-46-6*].

» Sulfamethoxazole contains not less than 98.5 percent and not more than 101.0 percent of $C_{10}H_{11}N_3O_3S$, calculated on the dried basis.

Packaging and storage—Preserve in well-closed, light-resistant containers.

Reference standards—*USP Sulfamethoxazole Reference Standard*—Dry at 105° for 4 hours before using. *USP Sulfanilamide Reference Standard*—Keep container tightly closed and protected from light. Dry at 105° for 2 hours before using.

Identification—

A: The infrared absorption spectrum of a potassium bromide dispersion of it, previously dried, exhibits maxima only at the same wavelengths as that of a similar preparation of USP Sulfamethoxazole RS.

B: The ultraviolet absorption spectrum of a 1 in 100,000 solution in sodium hydroxide solution (1 in 250) exhibits maxima and minima at the same wavelengths as that of a similar solution of USP Sulfamethoxazole RS, concomitantly measured, and the respective absorptivities, calculated on the dried basis, at the wavelength of maximum absorbance at about 257 nm do not differ by more than 2.0%.

C: Dissolve about 100 mg in 2 mL of hydrochloric acid, and add 3 mL of sodium nitrite solution (1 in 100) and 1 mL of sodium hydroxide solution (1 in 10) containing 10 mg of 2-naphthol: a red-orange precipitate is formed.

Melting range, *Class I* ⟨741⟩: between 168° and 172°.

Loss on drying ⟨731⟩—Dry it at 105° for 4 hours: it loses not more than 0.5% of its weight.

Residue on ignition ⟨281⟩: not more than 0.1%.

Selenium ⟨291⟩: 0.003%, a 200-mg test specimen being used.

Sulfanilamide and sulfanilic acid—

Standard preparation—Dissolve 100 mg of USP Sulfamethoxazole RS in 0.10 mL of ammonium hydroxide, dilute with methanol to 10.0 mL, and mix.

Reference preparation—Dissolve 20 mg of USP Sulfanilamide RS and 20 mg of sulfanilic acid in 10 mL of ammonium hydroxide, and dilute with methanol to 100 mL. Transfer 2.0 mL of the solution to a 50-mL volumetric flask, add 10 mL of ammonium hydroxide, dilute with methanol to volume, and mix.

Test preparation—Dissolve 100 mg in 0.10 mL of ammonium hydroxide, dilute with methanol to 10.0 mL, and mix.

Procedure—On a suitable thin-layer chromatographic plate (see *Chromatography* ⟨621⟩), coated with a 0.25-mm layer of chromatographic silica gel, apply 10 µL of the *Standard preparation*, 25 µL of the *Reference preparation*, and 10 µL of the *Test preparation*. Allow the spots to dry, and develop the chromatogram in a solvent system consisting of a mixture of alcohol, *n*-heptane, chloroform, and glacial acetic acid (25:25:25:7) until the solvent front has moved about three-fourths of the length of the plate. Remove the plate from the chamber, and allow it to air-dry. Spray the plate with a solution prepared by dissolving 0.10 g of *p*-dimethylaminobenzaldehyde in 1 mL of hydrochloric acid and diluting with alcohol to 100 mL. Sulfamethoxazole produces a spot at about R_f 0.7, sulfanilamide at about R_f 0.5, and sulfanilic acid at about R_f 0.1. Any spots produced by sulfanilamide or sulfanilic acid from the *Test preparation* do not exceed in size or intensity similar spots, occurring at the respective R_f values, produced by sulfanilamide or sulfanilic acid from the *Reference preparation* (0.2%).

Assay—Dissolve about 500 mg of Sulfamethoxazole, accurately weighed, in a mixture of 20 mL of glacial acetic acid and 40 mL of water, and add 15 mL of hydrochloric acid. Cool to 15°, and titrate immediately with 0.1 *M* sodium nitrite VS, determining the end-point potentiometrically using a calomel-platinum electrode system. Each mL of 0.1 *M* sodium nitrite is equivalent to 25.33 mg of $C_{10}H_{11}N_3O_3S$.

Sulfamethoxazole Oral Suspension

» Sulfamethoxazole Oral Suspension contains not less than 95.0 percent and not more than 110.0 percent of the labeled amount of $C_{10}H_{11}N_3O_3S$.

Packaging and storage—Preserve in tight, light-resistant containers.

Reference standard—*USP Sulfamethoxazole Reference Standard*—Dry at 105° for 4 hours before using.

Identification—

A: Place a quantity of Oral Suspension, equivalent to about 100 mg of sulfamethoxazole, in a 50-mL centrifuge tube, add 5 mL of ammonium hydroxide, and shake gently. Add 25 mL of methanol, shake thoroughly for 3 minutes, centrifuge, decant the supernatant liquid into a 50-mL volumetric flask, dilute with methanol to volume, and mix. On a suitable thin-layer chromatographic plate (see *Chromatography* ⟨621⟩), coated with a 0.25-mm layer of chromatographic silica gel, apply 50 µL of this solution and 50 µL of a solution prepared by dissolving 100 mg of USP Sulfamethoxazole RS in 5 mL of ammonium hydroxide and diluting with methanol to 50.0 mL. Allow the spots to dry, and develop the chromatogram in a solvent system consisting of a mixture of alcohol, *n*-heptane, chloroform, and glacial acetic acid (25:25:25:7) until the solvent front has moved about three-fourths of the length of the plate. Remove the plate from the developing chamber, mark the solvent front, and allow the solvent to evaporate. Locate the spots on the plate by lightly spraying with a solution prepared by dissolving 0.10 g of *p*-dimethylaminobenzaldehyde in 1 mL of hydrochloric acid and diluting to 100 mL with alcohol: the R_f value of the principal spot obtained from the test solution corresponds to that obtained from the Standard solution.

B: Transfer a quantity of Oral Suspension, equivalent to about 500 mg of sulfamethoxazole, to a 50-mL centrifuge tube, add about 25 mL of water, mix, and centrifuge. Decant and discard the supernatant liquid, resuspend the residue in 25 mL of water, mix, and centrifuge again. Decant and discard the clear, supernatant liquid. Repeat the washing procedure an additional two times. Dissolve the residue in 10 mL of hydrochloric acid, and add 15 mL of sodium nitrite solution (1 in 100) and 5 mL of sodium hydroxide solution (1 in 10) containing 10 mg of 2-naphthol: a red-orange precipitate is produced.

Assay—Mix an accurately measured volume of Sulfamethoxazole Oral Suspension, equivalent to about 1 g of sulfamethoxazole, with 20 mL of glacial acetic acid and 40 mL of water, and add 15 mL of hydrochloric acid. Cool to 15°, and titrate immediately with 0.1 *M* sodium nitrite VS, determining the end-point potentiometrically using a calomel-platinum electrode system. Each mL of 0.1 *M* sodium nitrite is equivalent to 25.33 mg of $C_{10}H_{11}N_3O_3S$.

Sulfamethoxazole Tablets

» Sulfamethoxazole Tablets contain not less than 95.0 percent and not more than 105.0 percent of the labeled amount of $C_{10}H_{11}N_3O_3S$.

Packaging and storage—Preserve in well-closed, light-resistant containers.

Reference standard—*USP Sulfamethoxazole Reference Standard*—Dry at 105° for 4 hours before using.

Identification—

A: Place a quantity of finely powdered Tablets, equivalent to about 100 mg of sulfamethoxazole, in a 50-mL centrifuge tube, and proceed as directed for *Identification test A* under *Sulfamethoxazole Oral Suspension*, beginning with "add 5 mL of ammonium hydroxide."

B: To a portion of finely powdered Tablets, equivalent to about 100 mg of sulfamethoxazole, add 2 mL of hydrochloric acid, 3 mL of sodium nitrite solution (1 in 100), and 1 mL of sodium hydroxide solution (1 in 10) containing 10 mg of 2-naphthol: a red-orange precipitate is formed.

Dissolution ⟨711⟩—

Medium: dilute hydrochloric acid (7 in 100); 900 mL.

Apparatus 1: 100 rpm.

Time: 20 minutes.

Procedure—Determine the amount of $C_{10}H_{11}N_3O_3S$ dissolved from ultraviolet absorbances at the wavelength of maximum absorbance at about 265 nm of filtered portions of the solution under test, suitably diluted with *Dissolution Medium*, if necessary, in comparison with a Standard solution having a known concentration of USP Sulfamethoxazole RS in the same medium.

Tolerances—Not less than 50% (*Q*) of the labeled amount of $C_{10}H_{11}N_3O_3S$ is dissolved in 20 minutes.

Uniformity of dosage units ⟨905⟩: meet the requirements.

Assay—Weigh and finely powder not less than 20 Sulfamethoxazole Tablets. Weigh accurately a portion of the powder, equivalent to about 500 mg of sulfamethoxazole, dissolve in a mixture of 20 mL of glacial acetic acid and 40 mL of water, and add 15 mL of hydrochloric acid. Cool to 15°, and titrate immediately with 0.1 *M* sodium nitrite VS, determining the end-point potentiometrically using a calomel-platinum electrode system. Each mL of 0.1 *M* sodium nitrite is equivalent to 25.33 mg of $C_{10}H_{11}N_3O_3S$.

Sulfamethoxazole and Trimethoprim Concentrate for Injection

» Sulfamethoxazole and Trimethoprim Concentrate for Injection is a sterile solution of Sulfamethoxazole and Trimethoprim in Water for Injection which, when diluted with Dextrose Injection, is suitable for intravenous infusion. It contains not less than 90.0 percent and not more than 110.0 percent of the labeled amounts of sulfamethoxazole ($C_{10}H_{11}N_3O_3S$) and trimethoprim ($C_{14}H_{18}N_4O_3$).

Packaging and storage—Preserve in single-dose, light-resistant containers, preferably of Type I glass. Concentrate for Injection may be packaged in 50-mL multiple-dose containers.

Labeling—Label it to indicate that it is to be diluted with 5 percent Dextrose Injection prior to administration.

Reference standards—*USP Trimethoprim Reference Standard*—Dry in vacuum at 105° for 4 hours before using. *USP Sulfamethoxazole Reference Standard*—Dry at 105° for 4 hours before using. *USP Sulfanilamide Reference Standard*—Dry at 105° for 4 hours before using. *USP Sulfanilic Acid Reference Standard*—Dry at 105° for 4 hours before using.

Identification—In tests *A* and *B* under *Related compounds*, the respective *Test preparations* exhibit principal spots whose R_f values correspond to those spots produced by the *Standard preparations* of USP Trimethoprim RS (R_f about 0.5) and USP Sulfamethoxazole RS (R_f about 0.7).

Pyrogen—It meets the requirements of the *Pyrogen Test* ⟨151⟩, the test dose being 0.5 mL per kg.

pH ⟨791⟩: between 9.5 and 10.5.

Particulate matter ⟨788⟩: meets the requirements under *Small-volume Injections*.

Related compounds—

TEST A (FOR TRIMETHOPRIM DEGRADATION PRODUCT)—

Hydrochloric acid solution—Dilute 2 mL of 3 *N* hydrochloric acid with water to 100 mL.

Test preparation—Transfer an accurately measured volume of Concentrate for Injection, equivalent to about 48 mg of trimethoprim and 240 mg of sulfamethoxazole, to a glass-stoppered, 50-mL centrifuge tube. Add 15 mL of *Hydrochloric acid solution*, and mix. Add 15 mL of chloroform, shake for 30 seconds, and centrifuge at high speed for 3 minutes. Transfer the supernatant layer to a 125-mL separator. Extract the chloroform layer in the centrifuge tube with 15 mL of *Hydrochloric acid solution*, centrifuge at high speed, and add the extract to the separator. Add 2 mL of sodium hydroxide solution (1 in 10) to the solution in the separator, and extract with three 20-mL portions of chloroform, collecting the organic layer in a 125-mL conical flask. Evaporate the chloroform under a stream of nitrogen to dryness. Dissolve the residue in 1 mL of a mixture of chloroform and methanol (1:1).

Standard preparation A—Using an accurately weighed quantity of USP Trimethoprim RS, prepare a solution in chloroform and methanol (1:1) having a known concentration of 48 mg per mL.

Standard preparation B—Dilute an accurately measured volume of *Standard preparation A* with a mixture of chloroform and methanol (1:1) to obtain a solution having a known concentration of 240 µg per mL.

Procedure—Apply 10 µL each of the *Test preparation*, *Standard preparation A*, and *Standard preparation B* to separate points on a thin-layer chromatographic plate (see *Chromatography* ⟨621⟩) coated with a 0.25-mm layer of chromatographic silica gel. Develop the chromatogram in a solvent system consisting of a mixture of chloroform, methanol, and ammonium hydroxide (97:7.5:1) until the solvent front has moved at least 12 cm. Remove the plate from the developing chamber, and air-dry. Locate the bands by viewing under short-wavelength ultraviolet light and by spraying with a freshly prepared mixture of ferric chloride solution (1 in 10) and potassium ferricyanide solution (1 in 20) (1:1). Trimethoprim produces a spot at about R_f 0.5, and the trimethoprim degradation product produces a spot at about R_f 0.6 to 0.7. Any spot from the *Test preparation* at about R_f 0.6 to 0.7 is not greater in size and intensity than the spot produced by *Standard preparation B* at about R_f 0.5, corresponding to not more than 0.5%. [NOTE—There may be spots due to concentrate excipients at about R_f 0.1.]

TEST B (FOR SULFANILAMIDE AND SULFANILIC ACID)—

Alcohol-methanol mixture—Mix dehydrated alcohol and methanol (95:5).

Ammonium hydroxide solution—Dilute 1 mL of ammonium hydroxide with *Alcohol-methanol mixture* to 100 mL.

Modified Ehrlich's reagent—Dissolve 100 mg of *p*-dimethylaminobenzaldehyde in 1 mL of hydrochloric acid, dilute with alcohol to 100 mL, and mix.

Test preparation—Transfer an accurately measured volume of Concentrate for Injection, equivalent to about 32 mg of trimethoprim and 160 mg of sulfamethoxazole, to a 25-mL graduated cylinder, dilute with *Ammonium hydroxide solution* to 16 mL, and mix.

Standard preparation A—Weigh 50 mg of USP Sulfamethoxazole RS into a 5-mL volumetric flask, dissolve in *Ammonium hydroxide solution*, dilute with the same solvent to volume, and mix.

Standard preparation B—Weigh 25 mg of USP Sulfanilamide RS into a 250-mL volumetric flask, dissolve in *Ammonium hydroxide solution*, dilute with the same solvent to volume, and mix. Pipet 5 mL of this solution into a 10-mL volumetric flask, dilute with *Ammonium hydroxide solution* to volume, and mix.

Standard preparation C—Weigh 25 mg of USP Sulfanilic Acid RS into a 250-mL volumetric flask, dissolve in *Ammonium hy-*

droxide solution, and mix. Pipet 3 mL of this solution into a 10-mL volumetric flask, dilute with *Ammonium hydroxide solution* to volume, and mix.

Procedure—Apply 10 µL each of the *Test preparation* and *Standard preparations A, B*, and *C* to separate points on a thin-layer chromatographic plate (see *Chromatography* ⟨621⟩) coated with a 0.25-mm layer of chromatographic silica gel. Develop the chromatogram in a solvent system consisting of a mixture of *Alcohol-methanol mixture*, heptane, chloroform and glacial acetic acid (30:30:30:10) until the solvent front has moved not less than 12 cm. Remove the plate from the developing chamber, air-dry, spray with *Modified Ehrlich's reagent*, and allow the plate to stand for 15 minutes: sulfamethoxazole produces a spot at about R_f 0.7. Any spots from the *Test preparation* at about R_f 0.5 or 0.1 are not greater in size or intensity than spots produced by *Standard preparations B* and *C*, respectively, corresponding to not more than 0.5% of sulfanilamide and 0.3% of sulfanilic acid.

Other requirements—It meets the requirements under *Injections* ⟨1⟩.

Assay—

Mobile phase, Standard preparation, and *Chromatographic system*—Proceed as directed in the *Assay* under *Sulfamethoxazole and Trimethoprim Oral Suspension.*

Assay preparation—Transfer an accurately measured volume of Sulfamethoxazole and Trimethoprim Concentrate for Injection, equivalent to about 80 mg of sulfamethoxazole, to a 50-mL volumetric flask. Add methanol to volume, and mix. Transfer 5.0 mL of this solution to a 50-mL volumetric flask, dilute with *Mobile phase* to volume, mix, and filter.

Procedure—Separately inject equal volumes (about 20 µL) of the *Standard preparation* and the *Assay preparation* into the chromatograph, record the chromatograms, and measure the responses for the major peaks. Calculate the quantities, in mg, of trimethoprim ($C_{14}H_{18}N_4O_3$) and sulfamethoxazole ($C_{10}H_{11}N_3O_3S$) in each mL of the Concentrate for Injection taken by the formula:

$$(500C/V)(r_U/r_S),$$

in which C is the concentration, in mg per mL, of the appropriate USP Reference Standard in the *Standard preparation*, and r_U and r_S are the responses of the corresponding analyte obtained from the *Assay preparation* and the *Standard preparation*, respectively.

Sulfamethoxazole and Trimethoprim Oral Suspension

» Sulfamethoxazole and Trimethoprim Oral Suspension contains not less than 90.0 percent and not more than 110.0 percent of the labeled amounts of sulfamethoxazole ($C_{10}H_{11}N_3O_3S$) and trimethoprim ($C_{14}H_{18}N_4O_3$).

Packaging and storage—Preserve in tight, light-resistant containers.

Reference standards—*USP Trimethoprim Reference Standard*—Dry in vacuum at 105° for 4 hours before using. *USP Sulfamethoxazole Reference Standard*—Dry at 105° for 4 hours before using. *USP Sulfamethoxazole N_4-glucoside Reference Standard*—Do not dry; use as is. *USP Sulfanilamide Reference Standard*—Dry at 105° for 4 hours before using. *USP Sulfanilic Acid Reference Standard*—Dry at 105° for 4 hours before using.

Identification—In tests *A* and *B* under *Related compounds*, the respective *Test preparations* exhibit spots whose R_f values correspond to those spots produced by the *Standard preparations* of USP Trimethoprim RS (R_f about 0.4) and USP Sulfamethoxazole RS (R_f about 0.7).

pH ⟨791⟩: between 5.0 and 6.5.

Alcohol content, *Method II* ⟨611⟩: not more than 0.5% of C_2H_5OH.

Related compounds—

TEST A (FOR TRIMETHOPRIM DEGRADATION PRODUCT)—

Iodine-modified Dragendorff reagent—Dissolve 850 mg of bismuth nitrate in a mixture of 10 mL of glacial acetic acid and 40 mL of water. Dissolve 20 g of potassium iodide in 50 mL of water. Mix the two solutions, dilute with dilute sulfuric acid (1 in 10) to 500 mL, and mix. Add 7 g of iodine, and mix.

Test preparation—Using a syringe, transfer an accurately measured volume of Oral Suspension, equivalent to 40 mg of trimethoprim, to a low-actinic, glass-stoppered centrifuge tube containing 40 mL of a mixture of 0.1 *N* sodium hydroxide and methanol (1:1). Shake by mechanical means for 10 minutes, and then centrifuge at high speed for 10 minutes. Decant the supernatant solution into a 125-mL separator. Extract the residue in the centrifuge tube with an additional 40 mL of the solvent mixture, and add the extract to the separator. Extract the solution in the separator with three 30-mL portions of chloroform. Pass the extracts through anhydrous sodium sulfate into a 125-mL conical flask. Evaporate the chloroform on a steam bath under a stream of nitrogen to about 2 mL and then at room temperature to dryness. Dissolve the residue in 2.0 mL of a mixture of chloroform and methanol (8:2).

Standard preparation A—Using an accurately weighed quantity of USP Trimethoprim RS, prepare a solution in a mixture of chloroform and methanol (8:2) having a known concentration of 20 mg per mL.

Standard preparation B—Dilute an accurately measured volume of *Standard preparation A* with a mixture of chloroform and methanol (8:2) to obtain a solution having a known concentration of 100 µg per mL.

Procedure—Apply 20 µL each of the *Test preparation, Standard preparation A*, and *Standard preparation B* to separate points on a thin-layer chromatographic plate (see *Chromatography* ⟨621⟩) coated with a 0.25-mm layer of chromatographic silica gel. Place the plate in an unsaturated chromatographic chamber, and develop the chromatogram in a solvent system consisting of a mixture of chloroform, methanol, and ammonium hydroxide (97:7.5:1) until the solvent front has moved at least 12 cm. Remove the plate from the developing chamber, air-dry, and spray with *Iodine-modified Dragendorff reagent:* trimethoprim produces a spot at about R_f 0.4, and the trimethoprim degradation product produces a spot at about R_f 0.6 to 0.7. Any spot from the *Test preparation* at about R_f 0.6 to 0.7 is not greater in size and intensity than the spot produced by *Standard preparation B* at about R_f 0.4 corresponding to not more than 0.5%. [NOTE—There may be spots due to suspension excipients at about R_f 0.05 and 0.1.]

TEST B (FOR SULFANILAMIDE, SULFANILIC ACID, AND SULFAMETHOXAZOLE N_4-GLUCOSIDE)—

Alcohol-methanol mixture—Mix dehydrated alcohol and methanol (95:5).

Modified Ehrlich's reagent—Dissolve 100 mg of *p*-dimethylaminobenzaldehyde in 1 mL of hydrochloric acid, and dilute with alcohol to 100 mL.

Test preparation—Using a syringe, transfer an accurately measured volume of Oral Suspension, equivalent to 200 mg of sulfamethoxazole, to a 100-mL volumetric flask containing 10 mL of ammonium hydroxide, and add 50 mL of methanol. Shake for 3 minutes, and dilute with methanol to volume. Centrifuge a portion of the solution for 3 minutes.

Standard preparation A—Weigh 20 mg of USP Sulfamethoxazole RS into a 10-mL volumetric flask, dissolve in 1 mL of ammonium hydroxide, dilute with methanol to volume, and mix.

Standard preparation B—Weigh 10 mg of USP Sulfanilamide RS into a 50-mL volumetric flask, dissolve in 5 mL of ammonium hydroxide, and dilute with methanol to volume. Pipet 5 mL of this solution into a 100-mL volumetric flask, add 10 mL of ammonium hydroxide, and dilute with methanol to volume.

Standard preparation C—Weigh 10 mg of USP Sulfanilic Acid RS into a 50-mL volumetric flask, dissolve in 5 mL of ammonium hydroxide, and dilute with methanol to volume. Pipet 3 mL of this solution into a 100-mL volumetric flask, add 10 mL of ammonium hydroxide, and dilute with methanol to volume.

Standard preparation D—Weigh 3.0 mg of USP Sulfamethoxazole N_4-glucoside RS into a 50-mL volumetric flask, dissolve in 5 mL of ammonium hydroxide, and dilute with methanol to volume.

Procedure—Apply 50 µL each of the *Test preparation* and *Standard preparations A, B, C,* and *D* to separate points on a thin-layer chromatographic plate (see *Chromatography* ⟨621⟩) coated with 0.25-mm layer of chromatographic silica gel. Place the plate in an unsaturated chromatographic chamber, and develop the chromatogram in a solvent system consisting of a mixture of *Alcohol-methanol mixture*, heptane, chloroform, and glacial acetic acid (25:25:25:7) until the solvent front has moved at least 12 cm. Remove the plate from the developing chamber, air-dry, spray with *Modified Ehrlich's reagent*, and allow the plate to stand for 15 minutes: sulfamethoxazole produces a spot at about R_f 0.7. Any spots from the *Test preparation* at about R_f 0.5, 0.1, and 0.3 are not greater in size and intensity than spots produced by *Standard preparations B, C,* and *D*, respectively, corresponding to not more than 0.5% of sulfanilamide, 0.3% of sulfanilic acid, and 3.0% of sulfamethoxazole N_4-glucoside.

Assay—

Mobile phase—Mix 1400 mL of water, 400 mL of acetonitrile, and 2.0 mL of triethylamine in a 2000-mL volumetric flask. Allow to equilibrate to room temperature, and adjust with 0.2 *N* sodium hydroxide or dilute glacial acetic acid (1 in 100) to a pH of 5.9 ± 0.1. Dilute with water to volume, and filter through a 0.45-µm membrane, making adjustments if necessary (see *System Suitability* under *Chromatography* ⟨621⟩).

Standard preparation—Dissolve accurately weighed quantities of USP Trimethoprim RS and USP Sulfamethoxazole RS in methanol, and dilute quantitatively with methanol to obtain a solution containing, in each mL, about 0.32 mg and 0.32*J* mg, respectively, *J* being the ratio of the labeled amount, in mg, of sulfamethoxazole to the labeled amount, in mg, of trimethoprim in the dosage form. Transfer 5.0 mL of this solution to a 50-mL volumetric flask, dilute with *Mobile phase* to volume, and mix to obtain a *Standard preparation* having known concentrations of about 0.032 mg of USP Trimethoprim RS per mL and 0.032*J* mg of USP Sulfamethoxazole RS per mL.

Assay preparation—Transfer an accurately measured volume of Sulfamethoxazole and Trimethoprim Oral Suspension, equivalent to about 80 mg of Sulfamethoxazole, to a 50-mL volumetric flask with the aid of about 30 mL of methanol. Sonicate the mixture for about 10 minutes with occasional shaking. Allow to equilibrate to room temperature, dilute with methanol to volume, mix, and centrifuge. Transfer 5.0 mL of the supernatant liquid to a second 50-mL volumetric flask, dilute with *Mobile phase* to volume, mix, and filter.

Chromatographic system (see *Chromatography* ⟨621⟩)—The liquid chromatograph is equipped with a 254-nm detector and a 3.9-mm × 30-cm column that contains packing L1. The flow rate is about 2 mL per minute. Chromatograph the *Standard preparation*, and record the peak responses as directed under *Procedure:* the resolution, *R*, between sulfamethoxazole and trimethoprim is not less than 5.0 and the relative standard deviation for replicate injections is not more than 2.0%. Relative retention times are 1.0 for trimethoprim and 1.8 for sulfamethoxazole.

Procedure—Separately inject equal volumes (about 20 µL) of the *Standard preparation* and the *Assay preparation* into the chromatograph, record the chromatograms, and measure the responses for the major peaks. Calculate the quantities, in mg, of trimethoprim ($C_{14}H_{18}N_4O_3$) and sulfamethoxazole ($C_{10}H_{11}N_3O_3S$) in each mL of the Oral Suspension taken by the formula:

$$(500C/V)(r_U/r_S),$$

in which *C* is the concentration, in mg per mL, of the appropriate USP Reference Standard in the *Standard preparation*, *V* is the volume, in mL, of Oral Suspension taken, and r_U and r_S are the responses of the corresponding analyte obtained from the *Assay preparation* and the *Standard preparation*, respectively.

Sulfamethoxazole and Trimethoprim Tablets

» Sulfamethoxazole and Trimethoprim Tablets contain not less than 93.0 percent and not more than 107.0 percent of the labeled amounts of sulfamethoxazole ($C_{10}H_{11}N_3O_3S$) and trimethoprim ($C_{14}H_{18}N_4O_3$).

Packaging and storage—Preserve in well-closed, light-resistant containers.

Reference standards—*USP Trimethoprim Reference Standard*—Dry in vacuum at 105° for 4 hours before using. *USP Sulfamethoxazole Reference Standard*—Dry at 105° for 4 hours before using.

Identification—Transfer an amount of finely ground Tablets, equivalent to 4 mg of trimethoprim, to a 10-mL volumetric flask, add 8 mL of methanol, warm for several minutes on a steam bath with frequent shaking, cool, dilute with methanol to volume, mix, and centrifuge briefly. Apply 5 µL each of this solution, a Standard solution of USP Trimethoprim RS in methanol containing 0.4 mg per mL, and a Standard solution of USP Sulfamethoxazole RS in methanol containing 2 mg per mL to separate points about 3 cm from one end of a thin-layer chromatographic plate coated with chromatographic silica gel mixture. Dry the spots in a current of warm air, and develop the plate with a mixture of chloroform, isopropyl alcohol, and diethylamine (6:5:1) in a chamber lined with filter paper. Remove the plate, dry, and examine under short-wavelength ultraviolet light: the trimethoprim and sulfamethoxazole spots from the solution under test have the same R_f values as the spots from the corresponding Standard solutions, about 0.5 and about 0.3, respectively.

Dissolution ⟨711⟩—

Medium: 0.1 *N* hydrochloric acid; 900 mL.

Apparatus 2: 75 rpm.

Time: 60 minutes.

Procedure—Determine the amounts of sulfamethoxazole ($C_{10}H_{11}N_3O_3S$) and trimethoprim ($C_{14}H_{18}N_4O_3$) dissolved, employing the procedure set forth in the *Assay*, making any necessary volumetric adjustments (see *Chromatography* ⟨621⟩). Calculate the percentage of each active component dissolved by comparison of the peak responses obtained from a filtered aliquot of the solution under test with the peak responses from the corresponding component obtained from the *Standard preparation*.

Tolerances—Not less than 70% (*Q*) of the labeled amounts of $C_{10}H_{11}N_3O_3S$ and $C_{14}H_{18}N_4O_3$ are dissolved in 60 minutes.

Uniformity of dosage units ⟨905⟩: meet the requirements.

Assay—

Mobile phase, *Standard preparation*, and *Chromatographic system*—Prepare as directed in the *Assay* under *Sulfamethoxazole and Trimethoprim Oral Suspension*.

Assay preparation—Weigh and finely powder not less than 20 Sulfamethoxazole and Trimethoprim Tablets. Transfer an accurately weighed portion of the powder, equivalent to about 160 mg of sulfamethoxazole, to a 100-mL volumetric flask. Add about 50 mL of methanol and sonicate, with intermittent shaking, for 5 minutes. Allow to equilibrate to room temperature, dilute with methanol to volume, mix, and filter. Transfer 5.0 mL of the clear filtrate to a 50-mL volumetric flask, dilute with *Mobile phase* to volume, and mix.

Procedure—Separately inject equal volumes (about 20 µL) of the *Standard preparation* and the *Assay preparation* into the chromatograph, record the chromatograms, and measure the responses for the major peaks. Calculate the quantities, in mg, of trimethoprim ($C_{14}H_{18}N_4O_3$) and sulfamethoxazole ($C_{10}H_{11}N_3O_3S$) in the portion of Tablets taken by the formula:

$$1000C(r_U/r_S),$$

in which *C* is the concentration, in mg per mL, of the appropriate USP Reference Standard in the *Standard preparation*, and r_U and r_S are the responses of the corresponding analyte obtained from the *Assay preparation* and the *Standard preparation*, respectively.

Sulfapyridine

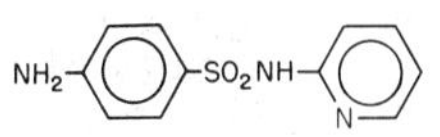

$C_{11}H_{11}N_3O_2S$ 249.29
Benzenesulfonamide, 4-amino-*N*-2-pyridinyl-.
N^1-2-Pyridylsulfanilamide [*144-83-2*].

» Sulfapyridine contains not less than 99.0 percent and not more than 100.5 percent of $C_{11}H_{11}N_3O_2S$, calculated on the dried basis.

Packaging and storage—Preserve in well-closed, light-resistant containers.

Reference standard—*USP Sulfapyridine Reference Standard*—Dry at 105° for 2 hours before using.

Clarity and color of solution—A solution of 1.0 g in a mixture of 20 mL of water and 5 mL of 1 *N* sodium hydroxide is clear and not more deeply colored than pale yellow.

Identification—

A: The infrared absorption spectrum of a potassium bromide dispersion of it exhibits maxima only at the same wavelengths as that of a similar preparation of USP Sulfapyridine RS.

B: Add 5 mL of 3 *N* hydrochloric acid to about 0.1 g of Sulfapyridine, and boil gently for about 5 minutes. Cool in an ice bath, add 4 mL of sodium nitrite solution (1 in 100), dilute with water to 10 mL, and place the mixture in the ice bath for 10 minutes. To 5 mL of the cooled mixture add a solution of 50 mg of 2-naphthol in 2 mL of sodium hydroxide solution (1 in 10): an orange-red precipitate is formed, and it darkens on standing.

Melting range ⟨741⟩: between 190° and 193°.

Acidity—Digest 2.0 g with 100 mL of water at about 70° for 5 minutes, cool at once to about 20°, and filter. To 25.0 mL of the filtrate add 2 drops of phenolphthalein TS, and titrate with 0.10 *N* sodium hydroxide: not more than 0.20 mL is required for neutralization.

Loss on drying ⟨731⟩—Dry it at 105° for 2 hours: it loses not more than 0.5% of its weight.

Residue on ignition ⟨281⟩: not more than 0.1%.

Selenium ⟨291⟩: 0.003%, a 200-mg test specimen being used.

Heavy metals, *Method II* ⟨231⟩: 0.002%.

Assay—Proceed with Sulfapyridine as directed under *Nitrite Titration* ⟨451⟩. Each mL of 0.1 *N* sodium nitrite is equivalent to 24.93 mg of $C_{11}H_{11}N_3O_2S$.

Sulfapyridine Tablets

» Sulfapyridine Tablets contain not less than 95.0 percent and not more than 105.0 percent of the labeled amount of $C_{11}H_{11}N_3O_2S$.

Packaging and storage—Preserve in well-closed, light-resistant containers.

Reference standard—*USP Sulfapyridine Reference Standard*—Dry at 105° for 2 hours before using.

Identification—Triturate a quantity of finely powdered Tablets, equivalent to about 500 mg of sulfapyridine, with 5 mL of chloroform, and transfer to a small filter. Wash with another 5-mL portion of chloroform, and discard the filtrate. Triturate the residue with 10 mL of 6 *N* ammonium hydroxide for 5 minutes, add 10 mL of water, and filter. Warm the filtrate until most of the ammonia is expelled, cool, and add 6 *N* acetic acid until the reaction is distinctly acid: a precipitate of sulfapyridine is formed. Collect the precipitate on a filter, wash it well with cold water, and dry at 105° for 1 hour: the sulfapyridine so obtained melts between 189° and 192° as determined by the method for *Class Ia* under *Melting Range or Temperature* ⟨741⟩, and responds to the *Identification tests* under *Sulfapyridine*.

Dissolution ⟨711⟩—

Medium: 0.1 *N* hydrochloric acid; 900 mL.

Apparatus 2: 50 rpm.

Time: 60 minutes.

Procedure—Determine the amount of $C_{11}H_{11}N_3O_2S$ dissolved from ultraviolet absorbances at the wavelength of maximum absorbance at about 254 nm of filtered portions of the solution under test, suitably diluted with 0.01 *N* sodium hydroxide, in comparison with a Standard solution having a known concentration of USP Sulfapyridine RS in the same medium.

Tolerances—Not less than 70% (*Q*) of the labeled amount of $C_{11}H_{11}N_3O_2S$ is dissolved in 60 minutes.

Uniformity of dosage units ⟨905⟩: meet the requirements.

Assay—Weigh and finely powder not less than 20 Sulfapyridine Tablets. Weigh accurately a portion of the powder, equivalent to about 500 mg of sulfapyridine, and proceed as directed under *Nitrite Titration* ⟨451⟩, beginning with "and transfer to a suitable open vessel." Each mL of 0.1 *M* sodium nitrite is equivalent to 24.93 mg of $C_{11}H_{11}N_3O_2S$.

Sulfasalazine

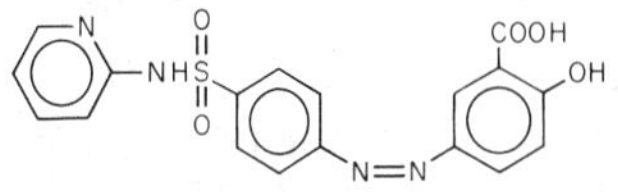

$C_{18}H_{14}N_4O_5S$ 398.39
Benzoic acid, 2-hydroxy-5-[[4-[(2-pyridinylamino)sulfonyl]-phenyl]azo]-.
5-[[*p*-(2-Pyridylsulfamoyl)phenyl]azo]salicylic acid [*599-79-1*].

» Sulfasalazine contains not less than 97.0 percent and not more than 101.5 percent of $C_{18}H_{14}N_4O_5S$, calculated on the dried basis.

Packaging and storage—Preserve in tight, light-resistant containers.

Reference standard—*USP Sulfasalazine Reference Standard*—Dry at 105° for 2 hours before using.

Identification—

A: The infrared absorption spectrum of a potassium bromide dispersion of it, previously dried, exhibits maxima only at the same wavelengths as that of a similar preparation of USP Sulfasalazine RS.

B: The visible absorption spectrum of the solution from the *Assay preparation*, prepared as directed in the *Assay*, exhibits maxima and minima at the same wavelengths as that of the solution from the *Standard preparation*, prepared as directed in the *Assay*, concomitantly measured.

Loss on drying ⟨731⟩—Dry it at 105° for 2 hours: it loses not more than 1.0% of its weight.

Residue on ignition ⟨281⟩: not more than 0.5%.

Chloride ⟨221⟩—Digest 2.0 g of Sulfasalazine with 100 mL of water at 70° for 5 minutes. Cool immediately to room temperature, and filter. Transfer a 25-mL portion of the filtrate to a 50-mL beaker (retain the remainder of this filtrate for the *Sulfate test*), add 1 mL of nitric acid, mix, and allow to stand for 5 minutes. Filter through a fine texture, retentive filter paper (Whatman No. 42, or equivalent): the filtrate shows no more chloride than corresponds to 0.10 mL of 0.020 *N* hydrochloric acid (0.014%).

Sulfate ⟨221⟩—Transfer a 25-mL portion of the filtrate from the *Chloride test* to a 50-mL beaker. Add 1 mL of 3 *N* hydrochloric acid, mix, and allow to stand for 5 minutes. Filter through a fine texture, retentive filter paper (Whatman No. 42, or equivalent): the filtrate shows no more sulfate than corresponds to 0.20 mL of 0.020 *N* sulfuric acid (0.04%).

Heavy metals, *Method II* ⟨231⟩: 0.002%.

Chromatographic purity—Prepare a solution of Sulfasalazine in a mixture of alcohol and 2 *M* ammonium hydroxide (4:1) having a concentration of 10 mg per mL. Similarly prepare a Standard solution of USP Sulfasalazine RS in the same medium having a concentration of 10 mg per mL. Dilute aliquots of the Standard solution quantitatively and stepwise with the same medium to obtain solutions having concentrations of 200 µg per mL, 150 µg per mL, 100 µg per mL, and 20 µg per mL, corresponding to 2.0%, 1.5%, 1.0%, and 0.2%, respectively (diluted Standard solutions A, B, C, and D). On a suitable thin-layer chromatographic plate (see *Chromatography* ⟨621⟩), coated with a 0.25-mm layer of chromatographic silica gel mixture, separately apply 10-µL each of the test solution, the Standard solution, and the diluted Standard solutions. Allow the spots to dry, and develop the chromatogram in an unequilibrated chamber with a solvent system consisting of a mixture of chloroform, acetone, and formic acid (60:30:5) until the solvent front has moved about three-fourths of the length of the plate. Remove the plate from the chamber, mark the solvent front, dry with the aid of a current of hot air, and examine the plate under short-wavelength ultraviolet light: the R_f value of the principal spot obtained from the test solution corresponds to that obtained from the Standard solution, and no spots, other than the principal spot, in the chromatogram of the test solution are larger or more intense than the principal spot obtained from diluted Standard solution A (2%), and the sum of the intensities of any secondary spots detected does not exceed 4%.

Assay—Transfer about 150 mg of Sulfasalazine, accurately weighed, to a 100-mL volumetric flask. Dissolve in 0.1 *N* sodium hydroxide, dilute with 0.1 *N* sodium hydroxide to volume, and mix. Transfer 5.0 mL of this solution to a 1000-mL volumetric flask containing about 750 mL of water, mix, add 20.0 mL of 0.1 *N* acetic acid, dilute with water to volume, and mix. Concomitantly determine the absorbances of this solution and of a Standard solution of USP Sulfasalazine RS in the same medium having a known concentration of about 7.5 µg per mL, at the wavelength of maximum absorbance at about 359 nm, using water as the blank. Calculate the quantity, in mg, of $C_{18}H_{14}N_4O_5S$ in the Sulfasalazine taken by the formula:

$$20C(A_U/A_S),$$

in which *C* is the concentration, in µg per mL, of USP Sulfasalazine RS in the *Standard solution,* and A_U and A_S are the absorbances obtained from the assay solution and the Standard solution, respectively.

Sulfasalazine Tablets

» Sulfasalazine Tablets contain not less than 95.0 percent and not more than 105.0 percent of the labeled amount of $C_{18}H_{14}N_4O_5S$.

Packaging and storage—Preserve in well-closed containers.

Reference standard—*USP Sulfasalazine Reference Standard*—Dry at 105° for 2 hours before using.

Identification—The visible absorption spectrum of the solution from the *Assay preparation*, prepared as directed in the *Assay*, exhibits maxima and minima at the same wavelengths as that of the *Standard preparation*, prepared as directed in the *Assay* and concomitantly measured.

Disintegration ⟨701⟩: 15 minutes, for Tablets that are enteric-coated, determined as directed under *Enteric-coated Tablets.*

Dissolution ⟨711⟩—

Medium: pH 7.5 phosphate buffer (see under *Buffer Solutions* in the section, *Reagents, Indicators, and Solutions*); 900 mL.

Apparatus 1: 100 rpm.

Time: 60 minutes.

Procedure—Determine the amount of $C_{18}H_{14}N_4O_5S$ dissolved from ultraviolet absorbances at the wavelength of maximum absorbance at about 358 nm of filtered portions of the solution under test, suitably diluted with *Dissolution Medium,* if necessary, in comparison with a Standard solution having a known concentration of USP Sulfasalazine RS in the same medium.

Tolerances—Not less than 85% (*Q*) of the labeled amount of $C_{18}H_{14}N_4O_5S$ is dissolved in 60 minutes.

Uniformity of dosage units ⟨905⟩: meet the requirements.

Assay—Weigh and finely powder not less than 20 Sulfasalazine Tablets. Transfer an accurately weighed portion of the powder, equivalent to about 150 mg of sulfasalazine, to a 100-mL volumetric flask, add 50 mL of 0.1 *N* sodium hydroxide, and mix. Dilute with 0.1 *N* sodium hydroxide to volume, mix, and filter, discarding the first 20 mL of the filtrate. Transfer 5.0 mL of the filtrate to a 1000-mL volumetric flask containing about 750 mL of water, mix, add 20.0 mL of 0.1 *N* acetic acid, mix, dilute with water to volume, and mix. Concomitantly determine the absorbances of this solution and of a Standard solution of USP Sulfasalazine RS in the same medium having a known concentration of about 7.5 µg per mL, at the wavelength of maximum absorbance at about 359 nm, using water as the blank. Calculate the quantity, in mg, of $C_{18}H_{14}N_4O_5S$ in the portion of Tablets taken by the formula:

$$20C(A_U/A_S),$$

in which *C* is the concentration, in µg per mL, of USP Sulfasalazine RS in the Standard solution, and A_U and A_S are the absorbances of the solution from the Tablets and the Standard solution, respectively.

Sulfathiazole

$C_9H_9N_3O_2S_2$ 255.31
Benzenesulfonamide, 4-amino-*N*-2-thiazolyl-.
N^1-2-Thiazolylsulfanilamide [*72-14-0*].

» Sulfathiazole contains not less than 99.0 percent and not more than 100.5 percent of $C_9H_9N_3O_2S_2$, calculated on the dried basis.

Packaging and storage—Preserve in well-closed, light-resistant containers.

Reference standard—*USP Sulfathiazole Reference Standard*—Dry at 105° for 2 hours before using.

Identification—

A: The infrared absorption spectrum of a potassium bromide dispersion of it, previously dried, exhibits maxima only at the same wavelengths as that of a similar preparation of USP Sulfathiazole RS.

B: To 100 mg add 5 mL of 3 *N* hydrochloric acid, and boil gently for about 5 minutes. Cool in an ice bath, add 4 mL of sodium nitrite solution (1 in 100), dilute with water to 10 mL, and place the mixture in an ice bath for 10 minutes. To 5 mL of the cooled mixture add a solution of 50 mg of 2-naphthol in 2 mL of sodium hydroxide solution (1 in 10): an orange-red precipitate is formed, and it darkens on standing.

Melting range, *Class I* ⟨741⟩: between 200° and 204°.

Acidity—Digest 2 g with 100 mL of water at about 70° for 5 minutes, cool at once to about 20°, and filter. Titrate 25 mL of the filtrate with 0.10 *N* sodium hydroxide, using 2 drops of phenolphthalein TS as the indicator: not more than 0.5 mL is required to produce a pink color. (Retain the remainder of the filtrate for the tests for *Chloride* ⟨221⟩ and for *Sulfate* ⟨221⟩.)

Loss on drying ⟨731⟩—Dry it at 105° for 2 hours: it loses not more than 0.5% of its weight.

Residue on ignition ⟨281⟩: not more than 0.1%.

Chloride ⟨221⟩—A 25-mL portion of the filtrate prepared as directed in the test for *Acidity* shows no more chloride than corresponds to 0.10 mL of 0.020 *N* hydrochloric acid (0.014%).

Sulfate ⟨221⟩—A 25-mL portion of the filtrate prepared as directed in the test for *Acidity* shows no more sulfate than corresponds to 0.2 mL of 0.02 *N* sulfuric acid (0.04%).

Heavy metals, *Method II* ⟨231⟩: 0.002%.

Ordinary impurities ⟨466⟩—
Test solution: methanol.
Standard solution: methanol.
Eluant: a mixture of 1-propanol and 1 *N* ammonium hydroxide (22:3).
Visualization: 1.

Assay—Proceed with Sulfathiazole as directed under *Nitrite Titration* ⟨451⟩. Each mL of 0.1 *M* sodium nitrite is equivalent to 25.53 mg of $C_9H_9N_3O_2S_2$.

Sulfinpyrazone

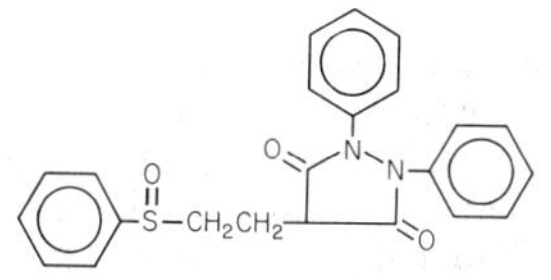

$C_{23}H_{20}N_2O_3S$ 404.48

3,5-Pyrazolidinedione, 1,2-diphenyl-4-[2-(phenylsulfinyl)ethyl]-.
1,2-Diphenyl-4-[2-(phenylsulfinyl)ethyl]-3,5-pyrazolidinedione [57-96-5].

» Sulfinpyrazone contains not less than 98.5 percent and not more than 101.5 percent of $C_{23}H_{20}N_2O_3S$, calculated on the dried basis.

Packaging and storage—Preserve in well-closed containers.

Reference standard—*USP Sulfinpyrazone Reference Standard*—Dry at 105° for 2 hours before using.

Solubility in acetone—A 250-mg portion dissolves in 5.0 mL of acetone to yield a clear, practically colorless solution.

Solubility in 0.50 *N* sodium hydroxide—A 0.50-g portion dissolves in 10.0 mL of 0.50 *N* sodium hydroxide to yield a clear, practically colorless solution.

Identification—The infrared absorption spectrum of a potassium bromide dispersion of it exhibits maxima only at the same wavelengths as that of a similar preparation of USP Sulfinpyrazone RS.

Melting range ⟨741⟩: between 130.5° and 134.5°.

Loss on drying ⟨731⟩—Dry it at 105° for 2 hours: it loses not more than 0.5% of its weight.

Residue on ignition ⟨281⟩: not more than 0.1%.

Heavy metals, *Method II* ⟨231⟩: 0.001%.

Chromatographic purity—
Standard solution—Prepare a solution in acetone having a known concentration of 20 mg of USP Sulfinpyrazone RS per mL.
Standard dilution 1—Transfer 1.0 mL of *Standard solution* to a 100-mL volumetric flask, dilute with acetone to volume, and mix (equivalent to 1.0% of impurity).
Standard dilution 2—Transfer 2.0 mL of *Standard dilution 1* to a 10-mL volumetric flask, dilute with acetone to volume, and mix (equivalent to 0.2% of impurity).
Test solution—Prepare a solution in acetone having a concentration of 20 mg per mL of Sulfinpyrazone.
Procedure—Place a volume of a solvent system consisting of a mixture of chloroform and glacial acetic acid (4:1) sufficient to develop a plate in a suitable lined chromatographic chamber (see *Chromatography* ⟨621⟩), and allow the chamber to equilibrate. To a thin-layer chromatographic plate coated with a 0.25-mm layer of chromatographic silica gel mixture that has been purged with nitrogen, apply, under nitrogen, 5-μL portions of the *Test solution, Standard solution,* and each of the *Standard dilutions.* Develop the chromatogram until the solvent front has moved about three-fourths of the length of the plate, remove the plate, and dry under a current of air. Expose the plate to high-intensity ultraviolet light for 8 minutes, and examine the chromatogram under long-wavelength ultraviolet light: the R_f value and intensity of the principal spot obtained from the *Test solution* correspond to those obtained from the *Standard solution.* Determine the intensity of any spot other than the principal spot detected in the *Test preparation* by comparison with the *Standard dilutions.* The sum of the intensities of any spots detected is not more than 2.0%.

Assay—Dissolve about 600 mg of Sulfinpyrazone, accurately weighed, in 50 mL of neutralized alcohol, with slight heating. Add phenolphthalein TS, and titrate with 0.1 *N* sodium hydroxide VS. Each mL of 0.1 *N* sodium hydroxide is equivalent to 40.45 mg of $C_{23}H_{20}N_2O_3S$.

Sulfinpyrazone Capsules

» Sulfinpyrazone Capsules contain not less than 93.0 percent and not more than 107.0 percent of the labeled amount of $C_{23}H_{20}N_2O_3S$.

Packaging and storage—Preserve in well-closed containers.

Reference standard—*USP Sulfinpyrazone Reference Standard*—Dry at 105° for 2 hours before using.

Identification—To a quantity of the contents of Capsules, equivalent to about 400 mg of sulfinpyrazone, add 20 mL of a 1 in 20 solution of methanol in dehydrated alcohol, boil, and filter. Add water to the filtrate until the solution becomes turbid, allow to stand until crystals form (1 to 2 hours), remove the crystals by filtration, and dry at 105° for 1 hour: the sulfinpyrazone so obtained melts between 128° and 134°, and responds to the *Identification test* under *Sulfinpyrazone.*

Dissolution ⟨711⟩—
Medium: pH 7.5 phosphate buffer (see under *Buffer Solutions* in the section, *Reagents, Indicators, and Solutions*); 900 mL.
Apparatus 1: 100 rpm.
Time: 45 minutes.
Procedure—Determine the amount of $C_{23}H_{20}N_2O_3S$ dissolved from ultraviolet absorbances at the wavelength of maximum absorbance at about 259 nm of filtered portions of the solution under test, suitably diluted with *Dissolution Medium,* in comparison with a Standard solution having a known concentration of USP Sulfinpyrazone RS in the same medium.
Tolerances—Not less than 75% (*Q*) of the labeled amount of $C_{23}H_{20}N_2O_3S$ is dissolved in 45 minutes.

Uniformity of dosage units ⟨905⟩: meet the requirements.

Assay—
Acetonitrile and tetrahydrofuran mixture—Prepare a mixture of acetonitrile and tetrahydrofuran (4:1).
Mobile phase—Prepare a degassed and filtered solution of dilute phosphoric acid (3 in 1000) and *Acetonitrile and tetrahydrofuran mixture* (65:35).
Internal standard solution—Prepare a solution of Benzoic Acid in acetonitrile having a concentration of about 2 mg per mL.
Standard preparation—Transfer about 100 mg, accurately weighed, of USP Sulfinpyrazone RS to a 200-mL volumetric flask, add 100 mL of acetonitrile, mix, add 10.0 mL of *Internal standard solution,* dilute with acetonitrile to volume, and mix. Filter, discarding the first 5 mL of the filtrate.
Assay preparation—Remove as completely as possible, and weigh, the contents of not less than 20 Sulfinpyrazone Capsules, and mix. Transfer an accurately weighed portion of the powder, equivalent to about 100 mg of sulfinpyrazone, to a 200-mL volumetric flask, add 100 mL of acetonitrile, mix, add 10.0 mL of *Internal standard solution,* dilute with acetonitrile to volume, and mix. Filter, discarding the first 5 mL of the filtrate.
Chromatographic system (see *Chromatography* ⟨621⟩)—The liquid chromatograph is equipped with a 235-nm detector and a 4.6-mm × 10-cm column that contains packing L1. The flow rate is about 3 mL per minute. Chromatograph replicate injections of the *Standard preparation,* and record the peak responses as directed under *Procedure*: the relative standard deviation is not more than 2.0%, and the resolution *R* between benzoic acid and sulfinpyrazone is not less than 13.
Procedure—Separately inject equal volumes (about 10 μL) of the *Standard preparation* and the *Assay preparation* into the chromatograph, record the chromatograms, and measure the responses for the major peaks. The relative retention times are about 0.2 for benzoic acid and 1.0 for sulfinpyrazone. Calculate

the quantity, in mg, of $C_{23}H_{20}N_2O_3S$ in the portion of Capsule contents taken by the formula:

$$200C(R_U/R_S),$$

in which C is the concentration, in mg per mL, of USP Sulfinpyrazone RS in the *Standard preparation*, and R_U and R_S are the relative peak response ratios obtained from the *Assay preparation* and the *Standard preparation*, respectively.

Sulfinpyrazone Tablets

» Sulfinpyrazone Tablets contain not less than 93.0 percent and not more than 107.0 percent of the labeled amount of $C_{23}H_{20}N_2O_3S$.

Packaging and storage—Preserve in well-closed containers.

Reference standard—*USP Sulfinpyrazone Reference Standard*—Dry at 105° for 2 hours before using.

Identification—Finely powder a quantity of Tablets, equivalent to about 400 mg of sulfinpyrazone, boil with 20 mL of alcohol, and filter. Add water to the filtrate until the solution becomes turbid, allow to stand until crystals form (1 to 2 hours), remove the crystals by filtration, and dry at 105° for 1 hour: the sulfinpyrazone so obtained melts between 128° and 134°, and responds to the *Identification test* under *Sulfinpyrazone.*

Dissolution ⟨711⟩—

Medium: pH 7.5 phosphate buffer (see under *Buffer Solutions* in the section, *Reagents, Indicators, and Solutions*); 900 mL.

Apparatus 1: 100 rpm.

Time: 45 minutes.

Procedure—Determine the amount of $C_{23}H_{20}N_2O_3S$ dissolved from ultraviolet absorbances at the wavelength of maximum absorbance at about 259 nm of filtered portions of the solution under test, suitably diluted with *Dissolution Medium*, in comparison with a Standard solution having a known concentration of USP Sulfinpyrazone RS in the same medium.

Tolerances—Not less than 75% (Q) of the labeled amount of $C_{23}H_{20}N_2O_3S$ is dissolved in 45 minutes.

Uniformity of dosage units ⟨905⟩: meet the requirements.

Assay—

Acetonitrile and tetrahydrofuran mixture, Mobile phase, Internal standard solution, Standard preparation, Chromatographic system, and *Procedure*—Proceed as directed in the *Assay* for *Sulfinpyrazone Capsules.*

Assay preparation—Weigh and finely powder not less than 20 Sulfinpyrazone Tablets. Transfer an accurately weighed portion of the powder, equivalent to about 100 mg of sulfinpyrazone, to a 200-mL volumetric flask, add 100 mL of acetonitrile, mix, add 10.0 mL of *Internal standard solution*, dilute with acetonitrile to volume, and mix. Filter, discarding the first 5 mL of the filtrate.

Sulfisoxazole

$C_{11}H_{13}N_3O_3S$ 267.30

Benzenesulfonamide, 4-amino-*N*-(3,4-dimethyl-5-isoxazolyl)-.

N^1-(3,4-Dimethyl-5-isoxazolyl)sulfanilamide [*127-69-5*].

» Sulfisoxazole contains not less than 99.0 percent and not more than 101.0 percent of $C_{11}H_{13}N_3O_3S$, calculated on the dried basis.

Packaging and storage—Preserve in tight, light-resistant containers.

Reference standard—*USP Sulfisoxazole Reference Standard*—Dry at 105° for 2 hours before using.

Identification—

A: The infrared absorption spectrum of a potassium bromide dispersion of it, previously dried, exhibits maxima only at the same wavelengths as that of a similar preparation of USP Sulfisoxazole RS.

B: The ultraviolet absorption spectrum of a 1 in 100,000 solution in pH 7.5 phosphate buffer (see *Buffer Solutions* in the section, *Reagents, Indicators, and Solutions*), prepared by dissolving about 100 mg of Sulfisoxazole, accurately weighed, in 10 mL of sodium hydroxide solution (1 in 250) and adding sufficient of the buffer solution to make 100 mL, then diluting 10 mL of the resulting solution with the buffer solution to 1000 mL, exhibits maxima and minima at the same wavelengths as that of a similar solution of USP Sulfisoxazole RS, concomitantly measured.

C: Dissolve about 10 mg in 2 mL of 3 N hydrochloric acid with the aid of heat, cool for 5 minutes in an ice bath, add 3 drops of sodium nitrite solution (1 in 100), and dilute with water to 4 mL: a yellow solution is produced. Add 1 mL of sodium hydroxide solution (1 in 10) containing 10 mg of 2-naphthol: an orange-red precipitate is formed.

Melting range ⟨741⟩: between 194° and 199°.

Loss on drying ⟨731⟩—Dry it at 105° for 2 hours: it loses not more than 0.5% of its weight.

Residue on ignition ⟨281⟩: not more than 0.1%.

Selenium ⟨291⟩: 0.003%, a 200-mg test specimen being used.

Heavy metals, *Method II* ⟨231⟩: 0.002%.

Ordinary impurities ⟨466⟩—

Test solution: ethyl acetate.

Standard solution: ethyl acetate.

Eluant: a mixture of acetone, cyclohexane, and glacial acetic acid (5:4:1).

Visualization: 1.

Assay—Place about 800 mg of Sulfisoxazole, accurately weighed, in a 250-mL conical flask, add 50 mL of dimethylformamide, shake thoroughly to dissolve the solid, add 5 drops of a 1 in 100 solution of thymol blue in dimethylformamide, and titrate with 0.1 N lithium methoxide VS to a blue end-point, taking precautions against absorption of atmospheric carbon dioxide. Perform a blank determination, and make any necessary correction. Each mL of 0.1 N lithium methoxide is equivalent to 26.73 mg of $C_{11}H_{13}N_3O_3S$.

Sulfisoxazole Tablets

» Sulfisoxazole Tablets contain not less than 95.0 percent and not more than 105.0 percent of the labeled amount of $C_{11}H_{13}N_3O_3S$.

Packaging and storage—Preserve in well-closed, light-resistant containers.

Reference standard—*USP Sulfisoxazole Reference Standard*—Dry at 105° for 2 hours before using.

Identification—Extract a quantity of powdered Tablets, equivalent to about 1 g of sulfisoxazole, with 50 mL of alcohol by boiling on a steam bath for 3 minutes, then immediately filter into a beaker. Allow to stand until a quantity of fine, needle-like crystals form. Cool, filter off the crystals, recrystallize from a small volume of alcohol, and dry at 105°: the crystals respond to *Identification test A* under *Sulfisoxazole.*

Dissolution ⟨711⟩—

Medium: dilute hydrochloric acid (1 in 12.5); 900 mL.

Apparatus 1: 100 rpm.

Time: 30 minutes.

Procedure—[NOTE—Because of the pH-dependent nature of the ultraviolet absorption spectrum, prepare the standard and specimen solutions in the same strength of acid at approximately equal concentrations.] Determine the amount of $C_{11}H_{13}N_3O_3S$ dissolved from ultraviolet absorbances at the wavelength of maximum absorbance at about 267 nm of filtered portions of the solution under test, suitably diluted with water, if necessary, in comparison with a Standard solution having a known concentration of USP Sulfisoxazole RS in the same medium.

Tolerances—Not less than 70% (*Q*) of the labeled amount of $C_{11}H_{13}N_3O_3S$ is dissolved in 30 minutes.

Uniformity of dosage units ⟨905⟩: meet the requirements.

Assay—Weigh and finely powder not less than 20 Sulfisoxazole Tablets. Weigh accurately a portion of the powder, equivalent to about 800 mg of sulfisoxazole, transfer to a 250-mL conical flask, and proceed as directed in the *Assay* under *Sulfisoxazole*, beginning with "add 50 mL of dimethylformamide."

Sulfisoxazole Acetyl

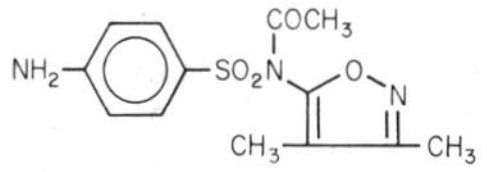

$C_{13}H_{15}N_3O_4S$ 309.34

Acetamide, *N*-[(4-aminophenyl)sulfonyl]-*N*-(3,4-dimethyl-5-isoxazolyl)-.

N-(3,4-Dimethyl-5-isoxazolyl)-*N*-sulfanilylacetamide [*80-74-0*].

» Sulfisoxazole Acetyl contains not less than 98.0 percent and not more than 100.5 percent of $C_{13}H_{15}N_3O_4S$, calculated on the dried basis.

Packaging and storage—Preserve in tight, light-resistant containers.

Reference standard—*USP Sulfisoxazole Acetyl Reference Standard*—Dry at 105° for 3 hours before using.

Identification—

A: The infrared absorption spectrum of a potassium bromide dispersion of it, previously dried, exhibits maxima only at the same wavelengths as that of a similar preparation of USP Sulfisoxazole Acetyl RS.

B: The ultraviolet absorption spectrum of a 1 in 100,000 solution in alcohol exhibits maxima and minima at the same wavelengths as that of a similar preparation of USP Sulfisoxazole Acetyl RS, concomitantly measured, and the respective absorptivities, calculated on the dried basis, at the wavelength of maximum absorbance at about 290 nm do not differ by more than 3.0%.

Melting range ⟨741⟩: between 192° and 195°.

Loss on drying ⟨731⟩—Dry it at 105° for 3 hours: it loses not more than 0.5% of its weight.

Ordinary impurities ⟨466⟩—

Test solution: methanol.

Standard solution: methanol.

Eluant: a mixture of toluene and acetone (1:1).

Visualization: 1.

Other requirements—It meets the requirements for *Residue on ignition*, *Selenium*, and *Heavy metals* under *Sulfisoxazole*.

Assay—Transfer about 1 g of Sulfisoxazole Acetyl, accurately weighed, to a 250-mL beaker. Add 15 mL of glacial acetic acid, swirl to dissolve, then add 25 mL of hydrochloric acid and 80 mL of water. Proceed as directed under *Nitrite Titration* ⟨451⟩, beginning with "Cool to about 15°." Each mL of 0.1 *M* sodium nitrite is equivalent to 30.94 mg of $C_{13}H_{15}N_3O_4S$.

Sulfisoxazole Acetyl Oral Suspension

» Sulfisoxazole Acetyl Oral Suspension contains an amount of Sulfisoxazole Acetyl equivalent to not less than 93.0 percent and not more than 107.0 percent of the labeled amount of sulfisoxazole ($C_{11}H_{13}N_3O_3S$).

Packaging and storage—Preserve in tight, light-resistant containers.

Reference standard—*USP Sulfisoxazole Acetyl Reference Standard*—Dry at 105° for 3 hours before using.

Identification—Centrifuge a portion of it, wash the separated solid by centrifugation with several portions of water, and recrystallize from hot alcohol: the crystals so obtained respond to *Identification tests A* and *B* under *Sulfisoxazole Acetyl*.

pH ⟨791⟩: between 5.0 and 5.5.

Assay—Transfer an accurately measured volume of Sulfisoxazole Acetyl Oral Suspension, previously mixed, equivalent to about 1 g of sulfisoxazole, to a 250-mL beaker. Add 40 mL of hydrochloric acid and 25 mL of glacial acetic acid, and swirl to dissolve. Cautiously add 100 mL of water, and proceed as directed under *Nitrite Titration* ⟨451⟩, beginning with "Cool to about 15°." Each mL of 0.1 *M* sodium nitrite is equivalent to 26.73 mg of sulfisoxazole ($C_{11}H_{13}N_3O_3S$).

Sulfisoxazole Acetyl for Oral Suspension, Erythromycin Ethylsuccinate and—*see* Erythromycin Ethylsuccinate and Sulfisoxazole Acetyl for Oral Suspension

Sulfisoxazole Diolamine

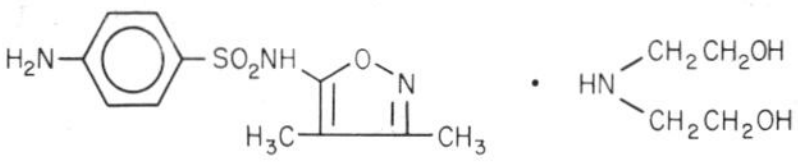

$C_{11}H_{13}N_3O_3S.C_4H_{11}NO_2$ 372.44

Benzenesulfonamide, 4-amino-*N*-(3,4-dimethyl-5-isoxazolyl)-, compd. with 2,2′-iminobis[ethanol] (1:1).

N^1-(3,4-Dimethyl-5-isoxazolyl)sulfanilamide compound with 2,2′-iminodiethanol (1:1) [*4299-60-9*].

» Sulfisoxazole Diolamine contains not less than 99.0 percent and not more than 101.0 percent of $C_{11}H_{13}N_3O_3S.C_4H_{11}NO_2$, calculated on the dried basis.

Packaging and storage—Preserve in tight, light-resistant containers.

Reference standard—*USP Sulfisoxazole Diolamine Reference Standard*—Dry in a suitable vacuum drying tube, using phosphorus pentoxide as the desiccant, at 80° for 4 hours before using.

Identification—

A: The ultraviolet absorption spectrum of a 1 in 100,000 solution in pH 7.5 buffer [prepared by mixing 50.0 mL of monobasic potassium phosphate solution (27.2 in 1000.0) and 40.5 mL of sodium hydroxide solution (1 in 125) with water to make 200 mL of solution, and adjusting to a pH of 7.5 ± 0.1, if necessary] exhibits maxima and minima at the same wavelengths as that of a similar solution of USP Sulfisoxazole Diolamine RS, concomitantly measured.

B: Dissolve about 15 mg in 2 mL of water. Add 2 mL of 3 *N* hydrochloric acid, 5 drops of sodium nitrite solution (1 in 100), and 1 mL of 2-naphthol TS: an orange-red precipitate is produced.

Melting range, *Class I* ⟨741⟩: between 119° and 124°.

Loss on drying ⟨731⟩—Dry it in a suitable vacuum drying tube, using phosphorus pentoxide as the desiccant, at 80° for 4 hours: it loses not more than 0.2% of its weight.

Residue on ignition ⟨281⟩: not more than 0.1%.

Heavy metals, *Method II* ⟨231⟩: 0.002%.

Assay—Dissolve about 1.2 g of Sulfisoxazole Diolamine, accurately weighed, in about 50 mL of dimethylformamide, add 2 drops of a solution of thymol blue in dimethylformamide (1 in 100), and titrate with 0.1 *N* lithium methoxide VS to a light blue end-point, taking precautions against the absorption of atmospheric carbon dioxide. Perform a blank determination, and make any necessary correction. Each mL of 0.1 *N* lithium methoxide is equivalent to 37.25 mg of $C_{11}H_{13}N_3O_3S.C_4H_{11}NO_2$.

Sulfisoxazole Diolamine Injection

» Sulfisoxazole Diolamine Injection is a sterile solution of Sulfisoxazole Diolamine in Water for Injection. It contains sulfisoxazole diolamine ($C_{11}H_{13}N_3O_3S.C_4H_{11}NO_2$) equivalent to not less than 90.0 percent and not more than 110.0 percent of the labeled amount of sulfisoxazole ($C_{11}H_{13}N_3O_3S$).

Packaging and storage—Preserve in single-dose or in multiple-dose containers, preferably of Type I glass, protected from light.

Reference standard—*USP Sulfisoxazole Diolamine Reference Standard*—Dry in a suitable vacuum drying tube, using phosphorus pentoxide as the desiccant, at 80° for 4 hours before using.

Identification—A solution of 1 mL in 2 mL of water responds to *Identification test B* under *Sulfisoxazole Diolamine.*

Pyrogen—It meets the requirements of the *Pyrogen Test* ⟨151⟩, the test dose being 1 mL per kg, of a solution containing the equivalent of 80 mg of sulfisoxazole in each mL.

pH ⟨791⟩: between 7.0 and 8.5.

Particulate matter ⟨788⟩: meets the requirements under *Small-volume Injections.*

Other requirements—It meets the requirements under *Injections* ⟨1⟩.

Assay—Transfer an accurately measured volume of Sulfisoxazole Diolamine Injection, equivalent to about 800 mg of sulfisoxazole, to a 250-mL conical flask, and evaporate on a steam bath with the aid of a current of air to dryness. Dissolve the residue in 50 mL of dimethylformamide, add 2 drops of a 1 in 100 solution of thymol blue in dimethylformamide and titrate with 0.1 *N* lithium methoxide VS to a light blue end-point, taking precautions against the absorption of atmospheric carbon dioxide. Perform a blank determination, and make any necessary correction. Each mL of 0.1 *N* lithium methoxide is equivalent to 26.73 mg of $C_{11}H_{13}N_3O_3S$.

Sulfisoxazole Diolamine Ophthalmic Ointment

» Sulfisoxazole Diolamine Ophthalmic Ointment is a sterile ointment containing sulfisoxazole diolamine ($C_{11}H_{13}N_3O_3S.C_4H_{11}NO_2$) equivalent to not less than 90.0 percent and not more than 110.0 percent of the labeled amount of sulfisoxazole ($C_{11}H_{13}N_3O_3S$).

Packaging and storage—Preserve in collapsible ophthalmic ointment tubes.

Reference standard—*USP Sulfisoxazole Diolamine Reference Standard*—Dry in a suitable vacuum drying tube, using phosphorus pentoxide as the desiccant, at 80° for 4 hours before using.

Identification—To 1 g add 2 mL of 3 *N* hydrochloric acid, heat to boiling, and cool. Add 2 mL of water, 5 drops of sodium nitrite solution (1 in 100), and 1 mL of 2-naphthol TS: an orange-red precipitate is formed.

Sterility—It meets the requirements under *Sterility Tests* ⟨71⟩.

Minimum fill ⟨755⟩: meets the requirements.

Leakage—It meets the requirements for *Leakage* under *Ophthalmic Ointments* ⟨771⟩.

Metal particles—It meets the requirements under *Metal Particles in Ophthalmic Ointments* ⟨751⟩.

Assay—

Standard preparation—Transfer about 70 mg of USP Sulfisoxazole Diolamine RS, accurately weighed, to a 100-mL volumetric flask, add 5 mL of sodium hydroxide solution (1 in 250), and mix, warming if necessary, to dissolve. Cool, add 20 mL of dilute sulfuric acid (1 in 9), cool, dilute with water to volume, and mix. Transfer 10.0 mL of this solution to a 1000-mL volumetric flask, dilute with water to volume, and mix. The concentration of USP Sulfisoxazole Diolamine RS in the *Standard preparation* is about 7 µg per mL.

Assay preparation—Transfer an accurately weighed portion of Sulfisoxazole Diolamine Ophthalmic Ointment, equivalent to about 100 mg of sulfisoxazole, to a 250-mL separator, add 100 mL of *n*-heptane, and shake thoroughly to disperse the ointment base. Extract with 60 mL of sodium hydroxide solution (1 in 10), shaking well, and transfer the aqueous extract to a 1000-mL volumetric flask. Extract the heptane layer with three 50-mL portions of water, and combine these extracts with the main extract in the volumetric flask. Dilute with water to volume, and mix. Transfer 10.0 mL of this solution to a 200-mL volumetric flask, add 20 mL of dilute sulfuric acid (1 in 10), cool, dilute with water to volume, and mix.

Procedure—Transfer 10.0 mL each of the *Standard preparation*, of the *Assay preparation*, and of water to serve as the blank to separate 25-mL volumetric flasks. To each flask add 1 mL of dilute sulfuric acid (1 in 9), 4 mL of water, and 1 mL of sodium nitrite solution (1 in 1000). Insert the stoppers, shake, and allow to stand for 5 minutes. Add 1.0 mL of freshly prepared ammonium sulfamate solution (1 in 200), mix, and allow to stand for 2 minutes. Add 1.0 mL of *N*-1-naphthylethylenediamine dihydrochloride solution (1 in 1000), insert the stoppers, shake, dilute with water to volume, mix, and allow to stand in the dark for 15 minutes. Concomitantly determine the absorbances of the solutions in 1-cm cells at the wavelength of maximum absorbance at about 540 nm, with a suitable spectrophotometer, using the blank to set the instrument. Calculate the quantity, in mg, of $C_{11}H_{13}N_3O_3S$ in the portion of Ophthalmic Ointment taken by the formula:

$$(267.30/372.44)(20C)(A_U/A_S),$$

in which 267.30 and 372.44 are the molecular weights of sulfisoxazole and sulfisoxazole diolamine, respectively, C is the concentration, in µg per mL, of USP Sulfisoxazole Diolamine RS in the *Standard preparation*, and A_U and A_S are the absorbances of the solutions from the *Assay preparation* and the *Standard preparation*, respectively.

Sulfisoxazole Diolamine Ophthalmic Solution

» Sulfisoxazole Diolamine Ophthalmic Solution is a sterile solution containing sulfisoxazole diolamine ($C_{11}H_{13}N_3O_3S.C_4H_{11}NO_2$) equivalent to not less than 90.0 percent and not more than 115.0 percent of the labeled amount of sulfisoxazole ($C_{11}H_{13}N_3O_3S$).

Packaging and storage—Preserve in tight, light-resistant containers.

Reference standard—*USP Sulfisoxazole Diolamine Reference Standard*—Dry in a suitable vacuum drying tube, using phosphorus pentoxide as the desiccant, at 80° for 4 hours before using.

Identification—A solution of 1 mL in 2 mL of water responds to *Identification test B* under *Sulfisoxazole Diolamine.*

Sterility—It meets the requirements under *Sterility Tests* ⟨71⟩.

pH ⟨791⟩: between 7.2 and 8.2.

Assay—Transfer an accurately measured volume of Sulfisoxazole Diolamine Ophthalmic Solution, equivalent to about 800 mg of sulfisoxazole, to a 250-mL conical flask, and evaporate on a steam bath with the aid of a current of air to dryness. Dissolve the residue in 50 mL of dimethylformamide, add 2 drops of a 1 in 100 solution of thymol blue in dimethylformamide, and titrate with 0.1 *N* lithium methoxide VS to a light blue end-point, taking precautions against the absorption of atmospheric carbon dioxide. Perform a blank determination, and make any necessary correction. Each mL of 0.1 *N* lithium methoxide is equivalent to 26.73 mg of $C_{11}H_{13}N_3O_3S$.

Sulfobromophthalein Sodium

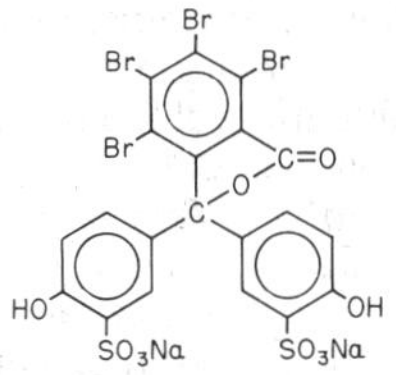

$C_{20}H_8Br_4Na_2O_{10}S_2$ 837.99

Benzenesulfonic acid, 3,3′-(4,5,6,7-tetrabromo-3-oxo-1(3*H*)-isobenzofuranylidene)bis[6-hydroxy-, disodium salt.

4,5,6,7-Tetrabromo-3′,3″-disulfophenolphthalein disodium salt [*71-67-0*].

» Sulfobromophthalein Sodium contains not less than 95.0 percent and not more than 105.0 percent of $C_{20}H_8Br_4Na_2O_{10}S_2$, calculated on the dried basis.

Packaging and storage—Preserve in tight containers.

Reference standard—*USP Sulfobromophthalein Sodium Reference Standard*—Dry at 105° for 3 hours before using.

Color and completeness of solution—A solution of 0.20 g in 10 mL of water is complete and colorless or practically colorless.

Identification—

A: Dissolve 10 mg in 5 mL of sodium carbonate solution (1 in 100): an intense bluish purple color is produced, and it disappears upon the addition of acid.

B: Mix about 100 mg with 0.5 g of sodium carbonate, and ignite until thoroughly charred. Cool, add 5 mL of hot water, heat for 5 minutes on a steam bath, and filter: the solution responds to the tests for *Bromide* ⟨191⟩.

C: It responds to the flame test for *Sodium* ⟨191⟩.

Loss on drying ⟨731⟩—Dry about 500 mg, accurately weighed, at 105° for 3 hours: it loses not more than 5.0% of its weight.

Halide ion—To 5 mL of a solution (1 in 100) add 1 mL of 2 *N* nitric acid and 1 mL of silver nitrate TS: not more than a slight opalescence is produced.

Sulfate—To 10 mL of a solution (1 in 500) add 5 drops of 3 *N* hydrochloric acid, heat to boiling, and add 1 mL of barium chloride TS: the solution remains clear while hot (on cooling, crystals of a barium salt of sulfobromophthalein may form).

Calcium—Ignite 2.0 g in a platinum crucible over a flame until completely carbonized. Complete the ignition in a furnace, cool the residue, and leach it with three 5-mL portions of 0.1 *N* hydrochloric acid, filtering the portions into a 100-mL beaker. Add 2.5 mL of a pH 4 buffer solution prepared by dissolving 10 g of sodium acetate in 6 *N* acetic acid to make 100 mL. Warm the buffered solution, add 20 mL of hot, filtered sodium iodohydroxyquinolinesulfonate TS, heat at about 100° for 15 minutes, and allow to stand overnight. Collect the precipitate on a tared, medium-porosity glass filter crucible, wash with several portions of the reagent, followed by small portions of alcohol and acetone, and dry in vacuum to constant weight: the residue weighs not more than 20.0 mg (0.05% as Ca).

Bromine content—Place about 50 mg, accurately weighed, in halide-free filter paper measuring about 4 cm square, and fold the paper to enclose it. Proceed as directed under *Oxygen Flask Combustion* ⟨471⟩, using a 1-liter flask and using a mixture of 10 mL of 0.1 *N* sodium hydroxide, 5 mL of hydrogen peroxide TS, and 5 mL of water as the absorbing liquid. When the combustion is complete, place a few mL of water in the cup, loosen the stopper, then rinse the stopper, the sample holder, and the sides of the flask with about 20 mL of water. Boil the solution for 5 minutes, cool, acidify with 2 *N* nitric acid, and add 20.0 mL of 0.05 *N* silver nitrate VS. Shake, add 2 mL of ferric ammonium sulfate TS, and titrate the excess silver nitrate with 0.05 *N* ammonium thiocyanate VS. Each mL of 0.05 *N* silver nitrate is equivalent to 3.995 mg of bromine (Br). Not less than 36.0% and not more than 39.0%, calculated on the dried basis, of Br is found.

Sulfur content—Place about 50 mg, accurately weighed, in halide-free filter paper measuring about 4 cm square, and fold the paper to enclose it. Proceed as directed under *Oxygen Flask Combustion* ⟨471⟩, using a 1-liter flask and using a mixture of 25 mL of water and 5 mL of hydrogen peroxide TS as the absorbing liquid. When the combustion is complete, place a few mL of water in the cup, loosen the stopper, then rinse the stopper, the sample holder, and the sides of the flask with about 20 mL of water. Add 2 mL of hydrochloric acid, dilute with water to 250 mL, heat to boiling, and slowly add 10 mL of barium chloride TS. Heat the mixture on a steam bath for 1 hour, collect the precipitate of barium sulfate on a filter, wash it until free from chloride, dry, ignite, and weigh. Each g of residue is equivalent to 137.4 mg of sulfur (S). Not less than 7.4% and not more than 8.2%, calculated on the dried basis, of S is found.

Assay—Transfer about 100 mg of Sulfobromophthalein Sodium, accurately weighed, to a 500-mL volumetric flask, dilute with water to volume, and mix. Pipet 5 mL of this solution into a 200-mL volumetric flask, dilute with sodium carbonate solution (1 in 100) to volume, and mix. Dissolve an accurately weighed quantity of USP Sulfobromophthalein Sodium RS in sodium carbonate solution (1 in 100), and dilute quantitatively and stepwise with sodium carbonate solution (1 in 100) to obtain a Standard solution having a known concentration of about 5 µg per mL. Concomitantly determine the absorbances of both solutions in 1-cm cells at the wavelength of maximum absorbance at about 580 nm, with a suitable spectrophotometer, using sodium carbonate solution (1 in 100) as the blank. Calculate the quantity, in mg, of $C_{20}H_8Br_4Na_2O_{10}S_2$ in the Sulfobromophthalein Sodium taken by the formula:

$$20C(A_U/A_S),$$

in which C is the concentration, in µg per mL, of USP Sulfobromophthalein Sodium RS in the Standard solution, and A_U and A_S are the absorbances of the solution of Sulfobromophthalein Sodium and the Standard solution, respectively.

Sulfobromophthalein Sodium Injection

» Sulfobromophthalein Sodium Injection is a colorless, or almost colorless, sterile solution of Sulfobromophthalein Sodium in Water for Injection. It contains, in each mL, not less than 47 mg and not more than 53 mg of $C_{20}H_8Br_4Na_2O_{10}S_2$.

Packaging and storage—Preserve in single-dose containers, preferably of Type I glass.

Reference standard—*USP Sulfobromophthalein Sodium Reference Standard*—Dry at 105° for 3 hours before using.

Identification—

A: To a few mL of Injection add 1 *N* sodium hydroxide: an intense bluish purple color is produced, and it disappears upon the addition of acid.

B: To a volume of Injection containing about 50 mg of sulfobromophthalein sodium, add 250 mg of sodium carbonate, evaporate to dryness, and ignite: the residue responds to *Identification test B* under *Sulfobromophthalein Sodium.*

pH ⟨791⟩: between 5.0 and 6.5, a suitable agar–potassium nitrate salt bridge being used.

Pyrogen—It meets the requirements of the *Pyrogen Test* ⟨151⟩, the test dose being 5 mL per kg, of a solution in pyrogen-free saline TS containing 0.1 mL of the Injection per mL.

Other requirements—It meets the requirements under *Injections* ⟨1⟩.

Assay—Transfer an accurately measured volume of Sulfobromophthalein Sodium Injection, equivalent to 100 mg of sulfobromophthalein sodium, to a 500-mL volumetric flask, dilute with water to volume, and mix. Proceed as directed in the *Assay* under *Sulfobromophthalein Sodium*, beginning with "Pipet 5 mL of this solution." Calculate the quantity, in mg, of $C_{20}H_8Br_4Na_2O_{10}S_2$ in each mL of the Injection taken by the formula:

$$20(C/V)(A_U/A_S),$$

in which C is the concentration, in µg per mL, of USP Sulfobromophthalein Sodium RS in the Standard solution, V is the volume, in mL, of Injection taken, and A_U and A_S are the absorbances of the solution from the Injection and the Standard solution, respectively.

Sulfoxone Sodium

NaO_2SCH_2NH–C₆H₄–SO_2–C₆H₄–$NHCH_2SO_2Na$

$C_{14}H_{14}N_2Na_2O_6S_3$ 448.43

Methanesulfinic acid, [sulfonylbis(1,4-phenyleneimino)]bis-, disodium salt.

Disodium [sulfonylbis(*p*-phenyleneimino)]dimethanesulfinate [*144-75-2*].

» Sulfoxone Sodium is a mixture of disodium [sulfonylbis(*p*-phenyleneimino)]dimethanesulfinate ($C_{14}H_{14}N_2Na_2O_6S_3$) and suitable buffers and inert ingredients. It contains not less than 73.0 percent and not more than 81.0 percent of $C_{14}H_{14}N_2Na_2O_6S_3$, calculated on the dried basis.

Packaging and storage—Preserve in tight, light-resistant containers, protected by a nitrogen overlay, in a freezer.

Reference standard—*USP Dapsone Reference Standard*—Dry at 105° for 3 hours before using.

Identification—

A: Transfer about 350 mg to a 100-mL volumetric flask, dissolve in water, dilute with water to volume, and mix. Transfer 10.0 mL of this solution to a 100-mL volumetric flask, dilute with water to volume, and mix. Transfer 4.0 mL of the resulting solution to a suitable flask, add 0.5 mL of 3 *N* hydrochloric acid, and hydrolyze in a boiling water bath for 30 minutes. Transfer the contents of the flask to a separator with the aid of water, add 2 mL of sodium hydroxide solution (1 in 10), and extract with three 25-mL portions of ethylene dichloride. Filter the extracts through a pledget of glass wool into a 100-mL volumetric flask, and dilute with ethylene dichloride to volume: the ultraviolet absorption spectrum of the solution so obtained exhibits maxima and minima at the same wavelengths as that of a solution of USP Dapsone RS in ethylene dichloride having a known concentration of 5 µg per mL, concomitantly measured.

B: To 10 mL of a solution (1 in 100) add 1 mL of iodine TS and 2 mL of chloroform, shake vigorously, and allow the layers to separate: no color appears in either layer.

C: Ignite 200 mg: the residue responds to the tests for *Sodium* ⟨191⟩.

Safety—Prepare a solution of Sulfoxone Sodium (1 in 10). Select 15 mice, each weighing between 20 and 25 g, and divide them into three groups of five each. Administer orally to each mouse of the first group 20 µL of the solution per g of body weight. In the same manner administer to each mouse of the second group and to each mouse of the third group 25 and 30 µL, respectively, per g of body weight: all of the mice of the first group, at least four of the second group, and at least two of the third group survive for 5 days.

pH ⟨791⟩: between 10.5 and 11.5 in a solution (1 in 10).

Loss on drying ⟨731⟩—Dry it in vacuum at 60° for 18 hours: it loses not more than 5.0% of its weight.

Selenium ⟨291⟩: 0.003%.

Heavy metals, *Method II* ⟨231⟩: 0.002%.

Suitability test—

p-Toluenesulfonic acid solution—Dissolve 20 g of *p*-toluenesulfonic acid in water to make 100 mL, and mix.

Sodium nitrite solution—Dissolve 100 mg of sodium nitrite in water to make 100 mL, and mix. Prepare this solution fresh daily.

Ammonium sulfamate solution—Dissolve 500 mg of ammonium sulfamate in water to make 100 mL, and mix.

Dimethyl-α-naphthylamine solution—Pipet 1 mL of dimethyl-α-naphthylamine into a 250-mL volumetric flask, dissolve in alcohol, dilute with alcohol to volume, and mix.

Test preparation—Weigh accurately 1.00 g of Sulfoxone Sodium, transfer to a 2000-mL volumetric flask, dissolve in water, dilute with water to volume, and mix.

Procedure—Pipet 4 mL of *Test preparation* into each of two 100-mL volumetric flasks. To one of the flasks (*A*) add 1.0 mL of *p-Toluenesulfonic acid solution* and 0.5 mL of 3 *N* hydrochloric acid, and place the flask in a boiling water bath for 30 minutes. Cool, dilute with water to volume, and mix. To the second flask (*B*) add water to volume, and mix. Pipet 1 mL of the solution in flask *A* into a small conical flask (A_2), and pipet 1 mL of the solution in flask *B* into a second small conical flask (B_2). Treat each flask as follows: Add 1.0 mL of *p-Toluenesulfonic acid solution*, 8.0 mL of water, and 1.0 mL of *Sodium nitrite solution*. After 3 minutes, accurately timed, add 0.5 mL of *Ammonium sulfamate solution*, and mix. Allow to stand for 2 minutes, accurately timed, add 5.0 mL of *Dimethyl-α-naphthylamine solution*, and mix. Allow to stand for 15 minutes, and determine the absorbances of the solutions, at 540 nm, using 2-cm cells, with a suitable spectrophotometer, and using water to set the instrument: the ratio of the absorbances, B_2/A_2, is not less than 0.85.

Assay—

p-Toluenesulfonic acid solution, Sodium nitrite solution, and *Ammonium sulfamate solution*—Prepare as directed in the *Suitability test*.

Dapsone stock solution—Dissolve an accurately weighed quantity of USP Dapsone RS in acetone, and dilute quantitatively with acetone to obtain a solution having a known concentration of about 1 mg per mL. Store in a refrigerator, and prepare fresh weekly.

Dapsone standard solution—Pipet 5 mL of *Dapsone stock solution* into a 500-mL volumetric flask, dilute with water to volume, and mix. Each mL of this solution contains a known quantity of about 10 µg of dapsone, equivalent to 18.06 µg of $C_{14}H_{14}N_2Na_2O_6S_3$.

N-1-Naphthylethylenediamine dihydrochloride solution—Dissolve 500 mg of *N*-1-naphthylethylenediamine dihydrochloride in water to make 100 mL, and mix.

Assay preparation—Transfer about 1 g, accurately weighed, of Sulfoxone Sodium to a 2000-mL volumetric flask, dissolve in water, dilute with water to volume, and mix. Pipet 4 mL of the solution into a 100-mL volumetric flask, add 1.0 mL of *p-Toluenesulfonic acid solution* and 0.5 mL of 3 *N* hydrochloric acid, and heat the flask in a boiling water bath for 30 minutes. Cool, dilute with water to volume, and mix.

Procedure—Pipet 2 mL each of *Dapsone standard solution*, the *Assay preparation*, and water to provide the blank, into separate 25-mL volumetric flasks. To each flask add 2.0 mL of 3 *N* hydrochloric acid, 10.0 mL of water, and 2.0 mL of *Sodium nitrite solution*, and mix. After 3 minutes, accurately timed, add 1.0 mL of *Ammonium sulfamate solution*, and mix. Allow to stand for 2 minutes, accurately timed, dilute with *N-1-Naphthylethylenediamine dihydrochloride solution* to volume, insert the stoppers in the flasks, and mix. Allow to stand for 10 minutes, and concomitantly determine the absorbances of the solutions in 2-cm cells, at the wavelength of maximum absorbance about at 535 nm, with a suitable spectrophotometer, using the blank to set the instrument. Calculate the quantity, in mg, of $C_{14}H_{14}N_2Na_2O_6S_3$ in the Sulfoxone Sodium taken by the formula:

$$(448.43/248.30)(50C)(A_U/A_S),$$

in which 448.43 and 248.30 are the molecular weights of $C_{14}H_{14}N_2Na_2O_6S_3$ and dapsone ($C_{12}H_{12}N_2O_2S$), respectively, C is the concentration, in µg per mL, of USP Dapsone RS in the *Dapsone standard solution*, and A_U and A_S are the absorbances of the solutions from the *Assay preparation* and *Dapsone standard solution*, respectively.

Sulfoxone Sodium Tablets

» Sulfoxone Sodium Tablets contain an amount of $C_{14}H_{14}N_2Na_2O_6S_3$ equivalent to not less than 95.0 percent and not more than 105.0 percent of the labeled amount of sulfoxone sodium. Sulfoxone Sodium Tablets are enteric coated.

Packaging and storage—Preserve in tight, light-resistant containers.

Reference standard—*USP Dapsone Reference Standard*—Dry at 105° for 3 hours before using.

Identification—Transfer 2.0 mL of the test solution prepared as directed in the *Assay* to a suitable flask, add 2.0 mL of water, and proceed as directed in *Identification test A* under *Sulfoxone Sodium*, beginning with "add 0.5 mL of 3 *N* hydrochloric acid": the solution so obtained gives the specified result.

Disintegration ⟨701⟩—The Tablets do not disintegrate after 1 hour of agitation in simulated gastric fluid TS, but then disintegrate within 2 hours in simulated intestinal fluid TS, the procedure for *Enteric-coated Tablets* being used.

Uniformity of dosage units ⟨905⟩: meet the requirements.

Suitability test—

p-Toluenesulfonic acid solution, *Sodium nitrite solution*, *Ammonium sulfamate solution*, and *Dimethyl-α-naphthylamine solution*—Prepare as directed in the *Suitability test* under *Sulfoxone Sodium*.

Test preparation—Use the test solution prepared as directed in the *Assay*.

Procedure—Proceed as directed for *Procedure* in the *Suitability test* under *Sulfoxone Sodium*. The ratio, B_2/A_2, of the absorbances is not less than 0.75.

Assay—

p-Toluenesulfonic acid solution, *Sodium nitrite solution*, and *Ammonium sulfamate solution*—Prepare as directed in the *Suitability test* under *Sulfoxone Sodium*.

N-1-Naphthylethylenediamine dihydrochloride solution, *Dapsone stock solution*, and *Dapsone standard solution*—Prepare as directed in the *Assay* under *Sulfoxone Sodium*.

Assay preparation—Blend a counted number of Tablets, equivalent to about 1.16 g of sulfoxone sodium, with about 250 mL of water for about 5 minutes in a suitable high-speed blender. Transfer the mixture to a 2000-mL volumetric flask, using small portions of methanol (up to 20 mL) if necessary to control foaming, dilute with water to volume, and mix (test solution). Proceed as directed for *Assay preparation* in the *Assay* under *Sulfoxone Sodium*, beginning with "Pipet 4 mL of the solution."

Procedure—Proceed as directed for *Procedure* in the *Assay* under *Sulfoxone Sodium*. Calculate the quantity, in mg, of sulfoxone sodium in the number of Tablets taken by the formula:

$$(448.43/248.30)(50C)(A_U/A_S)(1.3),$$

in which 1.3 is the factor converting $C_{14}H_{14}N_2Na_2O_6S_3$ to sulfoxone sodium.

Sulfur Colloid Injection, Technetium Tc 99m—*see under "technetium"*

Sulfur Dioxide—*see* Sulfur Dioxide NF

Sulfur Lotion, Resorcinol and—*see* Resorcinol and Sulfur Lotion

Precipitated Sulfur

S 32.06
Sulfur.
Sulfur [*7704-34-9*].

» Precipitated Sulfur contains not less than 99.5 percent and not more than 100.5 percent of S, calculated on the anhydrous basis.

Packaging and storage—Preserve in well-closed containers.

Identification—It burns in the air, forming sulfur dioxide, which can be recognized by its characteristic odor.

Reaction—Agitate 2.0 g with 10 mL of water, and filter: the filtrate is neutral to litmus.

Water, *Method I* ⟨921⟩: not more than 0.5%.

Residue on ignition ⟨281⟩: not more than 0.3%.

Other forms of sulfur—Shake 1.0 g with 5 mL of carbon disulfide: it dissolves quickly, with the exception of a small amount of insoluble matter that is usually present.

Assay—Using about 60 mg of Precipitated Sulfur, accurately weighed, proceed as directed under *Oxygen Flask Combustion* ⟨471⟩, using a 1000-mL flask and using a mixture of 10 mL of water and 5.0 mL of hydrogen peroxide TS as the absorbing liquid. When the combustion is complete, fill the lip of the flask with water, loosen the stopper, then rinse the stopper, the sample holder, and the sides of the flask with water, and remove the stopper assembly. Heat the contents of the flask to boiling, and boil for about 2 minutes. Cool to room temperature, add phenolphthalein TS, and titrate with 0.1 *N* sodium hydroxide VS. Perform a blank determination, and make any necessary correction. Each mL of 0.1 *N* sodium hydroxide is equivalent to 1.603 mg of S.

Sulfur Ointment

» Sulfur Ointment contains not less than 9.5 percent and not more than 10.5 percent of S.

Precipitated Sulfur	100 g
Mineral Oil	100 g
White Ointment	800 g
To make	1000 g

Levigate the sulfur with the Mineral Oil to a smooth paste, and then incorporate with the White Ointment.

Packaging and storage—Preserve in well-closed containers, and avoid prolonged exposure to excessive heat.

Assay—Transfer about 500 mg of Sulfur Ointment, accurately weighed, to a suitable conical flask, add 5 mL of nitric acid and 3 mL of bromine, warm slightly, and allow to stand overnight. Heat gently on a steam bath until the excess bromine has been dissipated. Add 30 mL of water and then 30 mL of ether, and swirl to dissolve most of the ointment base. Transfer the mixture to a separator, with the aid of a 20-mL and a 10-mL portion of ether, followed by two 10-mL portions of water. Shake the mixture, and draw off the water layer into a suitable beaker or flask. Extract the ether layer with two 40-mL portions of water each containing 1 mL of hydrochloric acid. Heat the combined water extracts on a steam bath to remove all traces of ether. Dilute the solution with water to about 200 mL, heat to boiling, and add slowly, with constant stirring, about 20 mL of hot barium chloride TS. Heat on a steam bath for 1 hour, then collect the precipitate on a filter, wash it well with hot water, dry, and ignite to constant weight. The weight of the barium sulfate so obtained, multiplied by 0.1374, represents the weight of S.

Sublimed Sulfur

S 32.06
Sulfur.
Sulfur [*7704-34-9*].

» Sublimed Sulfur, dried over phosphorus pentoxide for 4 hours, contains not less than 99.5 percent and not more than 100.5 percent of S.

Packaging and storage—Preserve in well-closed containers.

Solubility in carbon disulfide—One g dissolves slowly and usually incompletely in about 2 mL of carbon disulfide.

Identification—It burns in the air to sulfur dioxide, which can be recognized by its characteristic odor.

Residue on ignition ⟨281⟩: not more than 0.5%.

Arsenic, *Method I* ⟨211⟩—Prepare the *Test Preparation* as follows. Digest 750 mg of Sublimed Sulfur with 20 mL of 6 *N* ammonium hydroxide for 3 hours, filter, and evaporate the clear filtrate on a steam bath to dryness. Add 15 mL of 2 *N* sulfuric acid and 1 mL of 30 percent hydrogen peroxide solution, evaporate to strong fumes of sulfur trioxide, cool, add cautiously 10 mL of water, and again evaporate to strong fumes, repeating, if necessary, to remove any trace of hydrogen peroxide. Cool, and dilute cautiously with water to 35 mL. The limit is 4 ppm.

Assay—Proceed as directed under *Oxygen Flask Combustion* ⟨471⟩, using a 1000-mL flask and using about 60 mg of Sublimed Sulfur, previously dried over phosphorus pentoxide for 4 hours and accurately weighed, as the sample and a mixture of 10 mL of water and 5.0 mL of hydrogen peroxide TS as the absorbing liquid. When the combustion is complete, fill the lip of the flask with water, loosen the stopper, then rinse the stopper, sample holder, and sides of the flask with water, and remove the stopper assembly. Heat the contents of the flask to boiling, and boil for about 2 minutes. Cool to room temperature, add phenolphthalein TS, and titrate with 0.1 *N* sodium hydroxide VS. Perform a blank determination, and make any necessary correction. Each mL of 0.1 *N* sodium hydroxide is equivalent to 1.603 mg of S.

Sulfurated Potash—*see* Potash, Sulfurated
Sulfuric Acid—*see* Sulfuric Acid NF

Sulindac

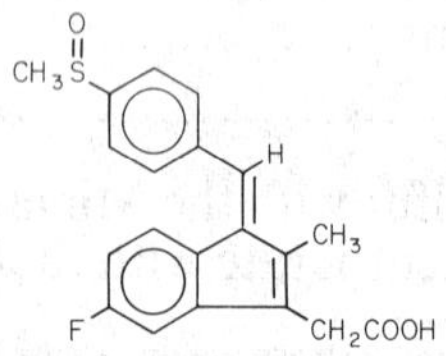

$C_{20}H_{17}FO_3S$ 356.41
1*H*-Indene-3-acetic acid, 5-fluoro-2-methyl-1-[[4-(methylsulfinyl)phenyl]methylene]-, (*Z*)-.
cis-5-Fluoro-2-methyl-1-[(*p*-methylsulfinyl)benzylidene]indene-3-acetic acid [*38194-50-2*].

» Sulindac contains not less than 99.0 percent and not more than 101.0 percent of $C_{20}H_{17}FO_3S$, calculated on the dried basis.

Packaging and storage—Preserve in well-closed containers.

Reference standard—*USP Sulindac Reference Standard*—Dry in vacuum at 100° for 2 hours before using.

Identification—

A: The infrared absorption spectrum of a mineral oil dispersion of it, previously dried, exhibits maxima only at the same wavelengths as that of a similar preparation of USP Sulindac RS.

B: The ultraviolet absorption spectrum of a 1 in 65,000 solution in 0.1 *N* hydrochloric acid in methanol exhibits maxima and minima at the same wavelengths as that of a similar solution of USP Sulindac RS, concomitantly measured, and the respective absorptivities, calculated on the dried basis, at the wavelength of maximum absorbance at about 284 nm do not differ by more than 3.0%.

Loss on drying ⟨731⟩—Dry it in vacuum at 100° for 2 hours: it loses not more than 0.5% of its weight.

Residue on ignition ⟨281⟩: not more than 0.1%.

Heavy metals, *Method II* ⟨231⟩: 0.001%.

Chromatographic purity ⟨621⟩—

Standard preparation—Prepare a solution of USP Sulindac RS in a mixture of chloroform and methanol (1:1) having a concentration of 25 mg per mL (*Solution A*). Prepare a second solution by diluting 1.0 volume of *Solution A* with the mixture of chloroform and methanol (1:1) to obtain 250 volumes of solution (*Solution B*).

Test preparation—Prepare a solution of the sample in a mixture of chloroform and methanol (1:1) having a concentration of 25 mg per mL.

System suitability—From the chromatograms obtained as directed under *Procedure*, estimate the intensity of the origin spot, if any, in the chromatogram of *Solution A*. The system is satisfactory if any spot observed at the origin is less intense than that obtained from the principal spot in the chromatogram of 2 µL of *Solution B*.

Procedure—On a suitable thin-layer chromatographic plate (see *Chromatography* ⟨621⟩), coated with a 0.25-mm layer of chromatographic silica gel mixture, apply 4-µL portions of *Solution A* and the *Test preparation* and 2-, 4-, 6-, 8-, and 10-µL portions of *Solution B*. Allow the spots to dry, and develop the chromatogram in a solvent system consisting of a mixture of chloroform, ethyl acetate, and glacial acetic acid (16:5:1) until the solvent front has moved about three-fourths of the length of the plate. Remove the plate from the developing chamber, mark the solvent front, allow the solvent to evaporate, and examine the plate under short-wavelength ultraviolet light: the chromatograms show principal spots at about the same R_f value. Estimate the levels of any additional spots observed in the chromatogram of the *Test preparation* by comparison with the spots in the series of chromatograms of *Solution B:* the sum of the levels is not greater than that of the principal spot obtained from the 10-µL portion of *Solution B* (1%).

Assay—Dissolve about 700 mg of Sulindac, accurately weighed, in about 80 mL of methanol, and titrate with 0.1 *N* sodium hydroxide VS, determining the end-point potentiometrically, using a glass-calomel electrode system (see *Titrimetry* ⟨541⟩). During the titration, and just prior to reaching the end-point, wash down the walls of the titration vessel with small amounts of methanol. Each mL of 0.1 *N* sodium hydroxide is equivalent to 35.64 mg of $C_{20}H_{17}FO_3S$.

Sulindac Tablets

» Sulindac Tablets contain not less than 90.0 percent and not more than 110.0 percent of the labeled amount of $C_{20}H_{17}FO_3S$.

Packaging and storage—Preserve in well-closed containers.

Reference standard—*USP Sulindac Reference Standard*—Dry in vacuum at 100° for 2 hours before using.

Identification—Compare the chromatograms obtained in the *Assay:* the *Assay preparation* exhibits a major peak for sulindac the retention time of which corresponds to that exhibited by the *Standard preparation.*

Dissolution ⟨711⟩—

Medium: 0.1 *M* pH 7.2 phosphate buffer prepared as directed under *Solutions* in the section, *Reagents, Indicators, and Solutions,* except to use twice the stated quantities of the monobasic potassium phosphate solution and of the sodium hydroxide solution; 900 mL.

Apparatus 2: 50 rpm.
Time: 45 minutes.
Procedure—Filter 20 mL of the solution under test, and transfer 10.0 mL of the filtrate to a 100-mL volumetric flask. Dilute with *Dissolution Medium* to volume, and mix. Determine the absorbances of this solution and of a Standard solution prepared from USP Sulindac RS in the same medium, having a known concentration of about 20 µg per mL, in 1-cm cells at the wavelength of maximum absorbance at about 326 nm, using *Dissolution Medium* as the blank. Calculate the amount of $C_{20}H_{17}FO_3S$ dissolved by the formula:

$$10C(A_U/A_S),$$

in which C is the concentration, in mg per mL, of sulindac in the Standard solution, and A_U and A_S are the absorbances of the solutions obtained from the specimen under test and the Reference Standard, respectively.
Tolerances—Not less than 80% (Q) of the labeled amount of $C_{20}H_{17}FO_3S$ is dissolved in 45 minutes.

Uniformity of dosage units ⟨905⟩: meet the requirements.
Procedure for content uniformity—Transfer 1 finely powdered Tablet to a 100-mL volumetric flask. Add 0.1 *N* sodium hydroxide with gentle agitation, dilute with 0.1 *N* sodium hydroxide to volume, and mix. Filter, discarding the first 20 mL of the filtrate. Transfer an accurately measured portion of the filtrate, equivalent to about 4 mg of sulindac, to a 200-mL volumetric flask, dilute with 0.1 *N* sodium hydroxide to volume, and mix. Promptly and concomitantly determine the absorbances of this solution and a solution of USP Sulindac RS in the same medium, having a known concentration of about 20 µg per mL, in 1-cm cells at the wavelength of maximum absorbance at about 327 nm, with a suitable spectrophotometer, using 0.1 *N* sodium hydroxide as the blank. Calculate the quantity, in mg, of $C_{20}H_{17}FO_3S$ in the Tablet by the formula:

$$(TC/D)(A_U/A_S),$$

in which T is the labeled quantity, in mg, of sulindac in the Tablet, C is the concentration, in µg per mL, of USP Sulindac RS in the Standard solution, D is the concentration, in µg per mL, of the solution from the Tablet, based upon the labeled quantity per Tablet and the extent of dilution, and A_U and A_S are the absorbances of the solution from the Tablet and the Standard solution, respectively.

Related compounds—
Mobile phase—Prepare as directed in the *Assay.*
Standard preparation—Dilute the *Standard preparation* prepared as directed in the *Assay* with *Mobile phase* to obtain a solution having a known concentration of about 15 µg per mL.
Test preparation—Prepare as directed for *Assay preparation* in the *Assay.*
Procedure—Proceed as directed for *Procedure* in the *Assay.* Measure the responses of the sulindac peak of the *Standard preparation* and of all peaks other than that of sulindac in the *Test preparation.* Calculate the amount, in mg, of related compounds in the portion of Tablets taken by the formula:

$$0.1C(r_U/r_S),$$

in which C is the concentration, in µg per mL, of USP Sulindac RS in the *Standard preparation,* and r_U and r_S are the peak responses of the *Test preparation* and the *Standard preparation,* respectively: the limit is 3.0%, calculated on the basis of the *Assay* of sulindac in the portion of the Tablets taken.

Assay—
Mobile phase—Prepare a suitable solution consisting of a mixture of chloroform, ethyl acetate, and acetic acid (approximately 38:5:1) such that the retention time of sulindac is approximately 11 minutes. Degas the solution.
Standard preparation—Dissolve an accurately weighed quantity of USP Sulindac RS in *Mobile phase* to obtain a solution having a known concentration of about 0.5 mg per mL.
Assay preparation—Weigh and finely powder not less than 20 Sulindac Tablets. Transfer an accurately weighed portion of the powder, equivalent to about 50 mg of sulindac, to a 100-mL volumetric flask. Add about 60 mL of *Mobile phase,* and shake by mechanical means for about 15 minutes. Dilute with *Mobile phase* to volume, mix, and centrifuge a portion of the mixture to obtain a clear supernatant liquid.
Chromatographic system (see *Chromatography* ⟨621⟩)—The liquid chromatograph is equipped with a 332-nm detector and a 3.9-mm × 30-cm column that contains packing L3. The flow rate is about 2 mL per minute. Chromatograph three replicate injections of the *Standard preparation,* and record the peak responses as directed under *Procedure:* the relative standard deviation is not more than 1.0% and the tailing factor is not more than 1.5.
Procedure—Using a suitable microsyringe or sampling valve, separately inject equal volumes (about 10 µL) of the *Standard preparation* and the *Assay preparation* into the chromatograph, and record the chromatograms. Measure the peak responses at an approximate retention time of 11 minutes for sulindac. Calculate the quantity, in mg, of $C_{20}H_{17}FO_3S$ in the portion of Tablets taken by the formula:

$$100C(r_U/r_S),$$

in which C is the concentration, in µg per mL, of USP Sulindac RS in the *Standard preparation,* and r_U and r_S are the peak responses obtained from the *Assay preparation* and the *Standard preparation,* respectively.

Suppositories—*see complete list in index*
Suspensions—*see complete list in index*

Sutilains

» Sutilains is a substance, containing proteolytic enzymes, derived from the bacterium *Bacillus subtilis.* When assayed as directed herein, it contains not less than 2,500,000 USP Casein Units of proteolytic activity per g, calculated on the dried basis.

NOTE—One USP Casein Unit of proteolytic activity is contained in the amount of sutilains which, when incubated with 35 mg of denatured casein at 37°, produces in 1 minute a hydrolysate whose absorbance at 275 nm is equal to that of a tyrosine solution containing 1.5 µg of USP Tyrosine Reference Standard per mL.

Packaging and storage—Preserve in tight containers, and store in a refrigerator. Allow to reach room temperature before opening container.

Reference standard—*USP Tyrosine Reference Standard*—Keep container tightly closed. Dry at 105° for 3 hours before using.

Solubility test—An amount of Sutilains, equivalent to 300,000 USP Casein Units of proteolytic activity, is soluble in 10.0 mL of water.

pH ⟨791⟩: between 6.1 and 7.1, in a solution (1 in 100).

Loss on drying ⟨731⟩—Dry it in vacuum over phosphorus pentoxide for 24 hours: it loses not more than 5.0% of its weight.

Nitrogen, ***Method II*** ⟨461⟩: between 11.0% and 13.5%.

Assay—
pH 7.0 buffer—Dissolve 12.1 g of tris(hydroxymethyl)aminomethane in 700 mL of water, and adjust with 1 *N* hydrochloric acid to a pH of 7.0 ± 0.1. Add 10.0 mL of 0.01 *M* zinc acetate, readjust, if necessary, to a pH of 7.0 ± 0.1, and dilute with water to 1000 mL.
Precipitant solution—Dissolve 18 g of trichloroacetic acid and 19 g of sodium acetate in 20 mL of glacial acetic acid and sufficient water to make 1000 mL of solution.
Substrate solution—Dissolve 6.05 g of tris(hydroxymethyl)-aminomethane in 500 mL of water, and adjust to a pH of 8.8 ± 0.1 with about 8 mL of 1 *N* hydrochloric acid. Suspend 7.0 g of casein (Hammarsten grade containing about 13.9% of nitrogen and 7.5% of moisture) in the solution. Heat, with occasional

stirring, in a boiling water bath for 30 minutes. Cool, adjust to a pH of 7.0 ± 0.1 by the slow addition, with constant stirring, of 0.2 *N* hydrochloric acid, and add 10.0 mL of 0.01 *M* zinc acetate. Dilute with water to about 950 mL, readjust, if necessary, to a pH of 7.0 ± 0.1, and dilute with water to 1000 mL. This solution is usable for 5 days if stored at a temperature of 5° to 8°.

Standard preparations—Transfer 50.0 mg of USP Tyrosine RS to a 500-mL volumetric flask, dissolve in 30 mL of 0.1 *N* hydrochloric acid, and dilute with water to volume. This solution contains 100 µg of tyrosine per mL. By quantitative dilution of this solution with water, prepare solutions containing 25, 50, and 75 µg of tyrosine per mL. Determine the absorbances of the four solutions, in 1-cm cells, at the wavelength of maximum absorbance at about 275 nm, with a suitable spectrophotometer, and prepare a plot of absorbance *versus* tyrosine concentration, in µg per mL. The absorbance (A_S) of a solution containing 1.5 µg of tyrosine per mL, determined from the standard curve, is about 0.0115.

Assay preparations—Transfer 50.0 mg of Sutilains to a 100-mL volumetric flask, dissolve in *pH 7.0 buffer*, and dilute with *pH 7.0 buffer to volume.* Transfer 2.0 mL of this solution to a 200-mL volumetric flask, dilute with *pH 7.0 buffer* to volume, and mix: this solution is *Assay preparation I.* Prepare a quantitative dilution of *Assay preparation I* with an equal volume of *pH 7.0 buffer:* this dilution is *Assay preparation II.*

Procedure—Transfer 5.0 mL of *Substrate solution* into each of five 18- × 150-mm test tubes, labeled, respectively, *I*, *I*, *II*, *II*, and *B* to serve as a substrate blank. In two additional tubes place, respectively, 5 mL of *Assay preparation I* and 5 mL of *Assay preparation II.* Place all tubes in a 37 ± 0.5° bath, and allow to stand for 5 minutes to reach the bath temperature. At time zero, transfer 1.0 mL of *pH 7.0 buffer* to tube *B*, and at accurately timed intervals (30 to 60 seconds) thereafter, transfer 1.0-mL portions of 37°-equilibrated *Assay preparation I* to each of the tubes labeled *I* and 1.0-mL portions of 37°-equilibrated *Assay preparation II* to each of the tubes labeled *II*, mixing each tube in the bath immediately following each addition. After 30 minutes, accurately timed, transfer 5.0 mL of *Precipitant solution* to tube *B*, shake vigorously, and return to the bath, and, at the same accurately timed intervals (30 to 60 seconds) observed at the start of the procedure, add successively to the sets of tubes labeled *I* and *II* 5.0 mL of *Precipitant solution*, shaking each vigorously following the addition, and immediately returning each to the bath. After 30 minutes, accurately timed and, successively, at the same accurately timed intervals, shake each tube vigorously, and filter each through a separate 11-cm filter (Whatman No. 42 paper, or equivalent). Pass each filtrate through the same filter a second time, and determine the absorbance of each filtrate from the sets of tubes labeled *I* and *II*, in 1-cm cells, at 275 nm, with a suitable spectrophotometer, using the filtrate from tube *B* to set the instrument. Average the absorbances (A_U) obtained for solutions obtained from individual tubes labeled *I*, and similarly average those obtained for solutions obtained from individual tubes labeled *II.* Calculate the number of USP Casein Units of proteolytic activity per g of Sutilains by the formula:

$$(1{,}000{,}000)(11/30W)(A_U/A_S),$$

in which 11 is the volume, in mL, of the total contents of each tube, and 30 is the incubation time, in minutes, for each tube, and *W* is the weight, in µg, of Sutilains in the tube for which the corresponding absorbance value, A_U, is recorded.

Sutilains Ointment

» Sutilains Ointment contains not less than 85.0 percent and not more than 125.0 percent of the labeled potency of sutilains in a suitable ointment base.

NOTE—One USP Casein Unit of proteolytic activity is contained in the amount of sutilains which, when incubated with 35 mg of denatured casein at 37°, produces in 1 minute a hydrolysate whose absorbance at 275 nm is equal to that of a tyrosine solution containing 1.5 µg of USP Tyrosine Reference Standard per mL.

Packaging and storage—Preserve in collapsible tubes or in tight containers, and store in a refrigerator.

Reference standard—*USP Tyrosine Reference Standard*—Keep container tightly closed. Dry at 105° for 3 hours before using.

Sterility—It meets the requirements under *Sterility Tests* ⟨71⟩, the test specimens being prepared as follows. Expel about one-third (about 5 g) of the contents of each of 20 tubes into 20 mL of filter-sterilized solvent hexane. Allow to stand for 5 to 10 minutes with occasional swirling to disperse the ointment base. Transfer aseptically a 10-mL portion of the suspension to 800 mL of Fluid Thioglycollate Medium and a second 10-mL portion to 800 mL of Soybean Casein Digest Medium, contained in 1000-mL bottles. Swirl the media units to disperse the solvent. Incubate the test specimens as directed under *Selection of Test Specimens and Incubation* ⟨71⟩. Swirl all sterility test units twice daily to assure continued distribution of the solvent and ointment in the medium.

Assay—

pH 7.0 buffer, *Precipitant solution*, *Substrate solution*, and *Standard preparations*—Prepare as directed in the *Assay* under *Sutilains.*

Assay preparation—Transfer 1.00 g of Sutilains Ointment, accurately weighed, to a beaker, add 5 mL of solvent hexane, and stir for several minutes with a glass rod: a granular precipitate remains undissolved. Add 20 mL of a solution of nonoxynol 9 in *pH 7.0 buffer* (1 in 100), stir to form an emulsion, dilute with the nonoxynol 9 in *pH 7.0 buffer* solution (1 in 100) to 100.0 mL, and mix. Remove a 1.0-mL portion from the center of the fluid before separation of the phases occurs, transfer to a 50-mL volumetric flask, and dilute with cold *pH 7.0 buffer* to volume: this solution is *Assay preparation I.* Transfer 15.0 mL of *Assay preparation I* to a 25-mL volumetric flask, and dilute with cold *pH 7.0 buffer* to volume: this solution is *Assay preparation II.*

Procedure—Proceed as directed for *Procedure* in the *Assay* under *Sutilains*, except to include an enzyme blank correction as follows: Transfer 5.0 mL of *Substrate solution* into an 18- × 150-mm test tube, and allow to stand in the 37 ± 0.5° bath for 30 minutes. Add 5.0 mL of *Precipitant solution* and 1.0 mL of *Assay preparation I*, and shake vigorously. Shake again to disperse the precipitate, and filter (11-cm Whatman No. 42 paper, or equivalent). Pour the filtrate through the same filter a second time, and determine the absorbance of the filtrate, in a 1-cm cell, at 275 nm, concomitantly and in the same manner as for the other solutions. Subtract this absorbance from the average of the absorbances A_U, and use the resulting value in place of A_U in the formula for the calculation of the proteolytic activity of Sutilains Ointment.

Absorbable Surgical Suture

» Absorbable Surgical Suture is a sterile, flexible strand prepared from collagen derived from healthy mammals, or from a synthetic polymer. Suture prepared from synthetic polymer may be in either monofilament or multifilament form. It is capable of being absorbed by living mammalian tissue, but may be treated to modify its resistance to absorption. Its diameter and tensile strength correspond to the size designation indicated on the label, within the limits prescribed herein. It may be modified with respect to body or texture. It may be impregnated or treated with a suitable coating, softening, or antimicrobial agent. It may be colored by a color additive approved by the federal Food and Drug Administration. The collagen suture is designated as either *Plain Suture* or *Chromic Suture.* Both types consist of processed

strands of collagen, but *Chromic Suture* is processed by physical or chemical means so as to provide greater resistance to absorption in living mammalian tissue.

Packaging and storage—Preserve dry or in fluid, in containers (packets) so designed that sterility is maintained until the container is opened. A number of such containers may be placed in a box.

Labeling—The label of each individual container (packet) of Suture indicates the size, length, type of Suture, kind of needle (if a needle is included), number of sutures (if multiple), lot number, and name of the manufacturer or distributor. If removable needles are used, the labeling so indicates. Suture size is designated by the metric size (gauge number) and the corresponding USP size. The label of the box indicates also the address of the manufacturer, packer, or distributor, and the composition of any packaging fluids used.

Note—If the Suture is packaged with a fluid, make the required measurements for the first four of the following tests within 2 minutes after removing it from the fluid.

Length—Determine the length of Suture without stretching: the length of each strand is not less than 95.0% of the length stated on the label.

Diameter—Determine the diameter of 10 strands of Suture as directed under *Sutures—Diameter* ⟨861⟩.

Collagen suture—The average diameter, and not fewer than 20, of the 30 measurements on the 10-strand sample are within the limits on average diameter prescribed in Table 1 for the respective size. None of the individual measurements is less than the midpoint of the range for the next smaller size or greater than the midpoint of the range for the next larger size.

Synthetic suture—The average diameter of the strands being measured is within the tolerances prescribed in Table 2 for the respective size. None of the observed measurements is less than the midpoint of the range for the next smaller size or greater than the midpoint of the range for the next larger size.

Tensile strength—Determine the tensile strength on not less than 10 strands of Suture as directed under *Tensile Strength—Surgical Sutures* ⟨881⟩.

Collagen suture—The tensile strength, determined as the minimum strength for each individual strand tested, and calculated as the average strength from any one lot, is as set forth in Table 1. If not more than one strand fails to meet the limit on individual strands, repeat the test with not less than 20 additional strands: the requirements of the test are met if none of the additional strands falls below the limit on individual strands, and if the average strength of all the strands tested does not fall below the stated limit in Table 1.

Table 1. Collagen Suture.

USP Size	Metric Size (Gauge No.)	Limits on Average Diameter (mm)		Knot-pull Tensile Strength (kgf)	
		Min.	Max.	Limit on Average Min.	Limit on Individual Strand Min.
9–0	0.4	0.040	0.049		
8–0	0.5	0.050	0.069	0.045	0.025
7–0	0.7	0.070	0.099	0.07	0.055
6–0	1	0.10	0.149	0.18	0.10
5–0	1.5	0.15	0.199	0.38	0.20
4–0	2	0.20	0.249	0.77	0.40
3–0	3	0.30	0.339	1.25	0.68
2–0	3.5	0.35	0.399	2.00	1.04
0	4	0.40	0.499	2.77	1.45
1	5	0.50	0.599	3.80	1.95
2	6	0.60	0.699	4.51	2.40
3	7	0.70	0.799	5.90	2.99
4	8	0.80	0.899	7.00	3.49

Synthetic suture—The minimum tensile strength of each size of synthetic suture, calculated as the average strength from any one lot, is as set forth in Table 2.

Needle attachment—Suture on which eyeless needles are swaged meets the requirements under *Sutures—Needle Attachment* ⟨871⟩.

Sterility—It meets the requirements under *Sterility Tests* ⟨71⟩.

Extractable color (if Suture is dyed)—Prepare the *Matching Solution* that corresponds to the extractable color of the Suture by combining the Colorimetric Solutions in the proportions indicated in Table 3, and adding water, if necessary, to make 10.0 parts. [See under *Solutions* in the section, *Reagents, Indicators, and Solutions*, for composition of the Colorimetric Solutions (CS).]

Weigh a quantity of Suture, equivalent to not less than 250 mg, and place in a conical flask containing 1.0 mL of water for each 10 mg of the sample. Close the flask, and allow it to stand at 37 ± 0.5° for 24 hours. Cool, decant the water from the Suture, and compare it with the *Matching Solution:* any color present is not more intense than that of the appropriate *Matching Solution.*

Soluble chromium compounds—Pipet 5 mL of the fluid prepared as directed in the test for *Extractable color* into a small test tube. Into a similar tube pipet 5 mL of a standard solution of potassium dichromate having a concentration of 2.83 µg per mL. To both tubes add 2 mL of a 1 in 100 solution of diphenylcarbazide in

Table 2. Synthetic Suture.

USP Size	Metric Size (Gauge No.)	Limits on Average Diameter (mm)		Knot-pull Tensile Strength (kgf) (except where otherwise specified)* Limit on Average Min.
		Min.	Max.	
12–0	0.01	0.001	0.009	
11–0	0.1	0.010	0.019	
10–0	0.2	0.020	0.029	0.025*
9–0	0.3	0.030	0.039	0.050*
8–0	0.4	0.040	0.049	0.07
7–0	0.5	0.050	0.069	0.14
6–0	0.7	0.070	0.099	0.25
5–0	1	0.10	0.149	0.68
4–0	1.5	0.15	0.199	0.95
3–0	2	0.20	0.249	1.77
2–0	3	0.30	0.339	2.68
0	3.5	0.35	0.399	3.90
1	4	0.40	0.499	5.08
2	5	0.50	0.599	6.35
3 and 4	6	0.60	0.699	7.29
5	7	0.70	0.799	

* The tensile strength of the specified USP size is measured by straight pull.

Table 3. Matching Solutions.

Color of Suture (Extractable Color)	Parts of Each CS per 10 Parts of Total Volume		
	Cobaltous Chloride CS	Ferric Chloride CS	Cupric Sulfate CS
Yellow-brown	0.2	1.2	—
Pink-red	1.0	—	—
Green-blue	—	—	2.0
Violet	1.6	—	8.4

alcohol and 2 mL of 2 *N* sulfuric acid: any color that develops in the test solution is not more intense than that in the standard solution (0.0001% of Cr).

Nonabsorbable Surgical Suture

» Nonabsorbable Surgical Suture is a flexible strand of material that is suitably resistant to the action of living mammalian tissue. It may be in either monofilament or multifilament form. If it is a multifilament strand, the individual filaments may be combined by spinning, twisting, braiding, or any combination thereof. It may be either sterile or nonsterile. Its diameter and tensile strength correspond to the size designation indicated on the label, within the limits prescribed herein. It may be modified with respect to body or texture, or to reduce capillarity, and may be suitably bleached. It may be impregnated or treated with a suitable coating, softening, or antimicrobial agent. It may be colored by a color additive approved by the federal Food and Drug Administration.

Nonabsorbable Surgical Suture is classed and typed as follows. *Class I* Suture is composed of silk or synthetic fibers of monofilament, twisted, or braided construction where the coating, if any, does not significantly affect thickness (e.g., braided silk, polyester, or nylon; monofilament nylon or polypropylene). *Class II* Suture is composed of cotton or linen fibers or coated natural or synthetic fibers where the coating significantly affects thickness but does not contribute significantly to strength (e.g., virgin silk sutures). *Class III* Suture is composed of monofilament or multifilament metal wire.

Packaging and storage—Preserve nonsterilized Suture in well-closed containers. Preserve sterile Suture dry or in fluid, in containers (packets) so designed that sterility is maintained until the container is opened. A number of such containers may be placed in a box.

Labeling—The label of each individual container (packet) of Suture indicates the material from which the Suture is made, the size, construction, and length of the Suture, whether it is sterile or non-sterile, kind of needle (if a needle is included), number of sutures (if multiple), lot number, and name of the manufacturer or distributor. If removable needles are used, the labeling so indicates. Suture size is designated by the metric size (gauge number) and the corresponding USP size. The label of the box indicates also the address of the manufacturer, packer, or distributor, and the composition of any packaging fluids used.

NOTE—If the Suture is packaged with a fluid, make the required measurements for the first four of the following tests within 2 minutes after removing it from the fluid.

Length—Determine the length of Suture while the strand is laid out smooth, without tension, on a plane surface: the length of the strand is not less than 95.0% of the length stated on the label.

Diameter—Determine the diameter of 10 strands of Suture as directed under *Sutures—Diameter* ⟨861⟩. The average diameter of the strands being measured is within the tolerances prescribed in the accompanying table for the size stated on the label. In the case of braided or twisted Suture, none of the observed diameters is less than the midpoint of the range for the next smaller size or is greater than the midpoint of the range for the next larger size.

Tensile strength—Determine the tensile strength on not less than 10 strands of Suture as directed under *Tensile Strength—Surgical Sutures* ⟨881⟩. Average all observations obtained: the average tensile strength is not less than that set forth in the accompanying table for the class and the size stated on the label.

USP Size	Metric Size (Gauge No.)	Limits on Average Diameter (mm)		Limits on Average Knot-pull (except where otherwise specified)† Tensile Strength (Kgf)*		
		Min.	Max.	Class I Min.	Class II Min.	Class III Min.
12–0	0.01	0.001	0.009	0.001†	—	0.002†
11–0	0.1	0.010	0.019	0.006†	0.005†	0.02†
10–0	0.2	0.020	0.029	0.019†	0.014†	0.06†
9–0	0.3	0.030	0.039	0.043†	0.029†	0.07†
8–0	0.4	0.040	0.049	0.06	0.04	0.11
7–0	0.5	0.050	0.069	0.11	0.06	0.16
6–0	0.7	0.070	0.099	0.20	0.11	0.27
5–0	1	0.10	0.149	0.40	0.23	0.54
4–0	1.5	0.15	0.199	0.60	0.46	0.82
3–0	2	0.20	0.249	0.96	0.66	1.36
2–0	3	0.30	0.339	1.44	1.02	1.80
0	3.5	0.35	0.399	2.16	1.45	3.40†
1	4	0.40	0.499	2.72	1.81	4.76†
2	5	0.50	0.599	3.52	2.54	5.90†
3 and 4	6	0.60	0.699	4.88	3.68	9.11†
5	7	0.70	0.799	6.16	—	11.4†
6	8	0.80	0.899	7.28	—	13.6†
7	9	0.90	0.999	9.04	—	15.9†
8	10	1.00	1.099	—	—	18.2†
9	11	1.100	1.199	—	—	20.5†
10	12	1.200	1.299	—	—	22.8†

* The limits on knot-pull tensile strength apply to Nonabsorbable Surgical Suture that has been sterilized. For non-sterile Sutures of *Class I* and *Class II*, the limits are 25% higher.

† The tensile strength of sizes smaller than USP size 8–0 (metric size 0.4) is measured by straight pull. The tensile strength of sizes larger than USP size 2–0 (metric size 3) of monofilament *Class III* (metallic) Nonabsorbable Surgical Suture is measured by straight pull.

Silver wire meets the tensile strength values of *Class I* Sutures but is tested in the same manner as *Class III* Sutures.

Needle attachment—Suture on which eyeless needles are swaged meets the requirements under *Sutures—Needle Attachment* ⟨871⟩.

Sterility—Suture that is claimed to be sterile meets the requirements under *Sterility Tests* ⟨71⟩.

Extractable color (if Suture is dyed)—Proceed as directed in the test for *Extractable color* under *Absorbable Surgical Suture*, but instead of allowing to stand at 37 ± 0.5° for 24 hours, cover the flask with a short-stemmed funnel, heat the contents of the flask at the boiling point for 15 minutes, cool, and restore the volume by the addition of water, if necessary, to replace that lost by evaporation.

Syrup—*see* Syrup NF

Syrups—*see complete list in index*

Tablets—*see complete list in index*

Talbutal

$C_{12}H_{16}N_2O_3$ 224.26

2,4,6(1*H*,3*H*,5*H*)-Pyrimidinetrione, 5-(1-methylpropyl)-5-(2-propenyl)-.
5-Allyl-5-*sec*-butylbarbituric acid [*115-44-6*].

» Talbutal contains not less than 98.0 percent and not more than 102.0 percent of $C_{11}H_{16}N_2O_3$, calculated on the dried basis.

Packaging and storage—Preserve in tight containers.

Reference standard—*USP Talbutal Reference Standard*—Dry in vacuum at 60° for 4 hours before using.

Identification—

A: The infrared absorption spectrum of a potassium bromide dispersion of it, previously dried, exhibits maxima only at the same wavelengths as that of a similar preparation of USP Talbutal RS.

B: The ultraviolet absorption spectrum of a 1 in 67,000 solution in pH 9.6 alkaline borate buffer (see under *Solutions* in the section, *Reagents, Indicators, and Solutions*) exhibits maxima and minima at the same wavelengths as that of a similar solution of USP Talbutal RS, concomitantly measured, and the respective absorptivities, calculated on the dried basis, at the wavelength of maximum absorbance at about 241 nm do not differ by more than 3.0%.

Loss on drying ⟨731⟩—Dry it in vacuum at 60° for 4 hours: it loses not more than 1.0% of its weight.

Residue on ignition ⟨281⟩: not more than 0.2%.

Heavy metals, *Method II* ⟨231⟩: 0.002%.

Assay—Transfer about 500 mg of Talbutal, accurately weighed, to a 125-mL conical flask, and dissolve in 25 mL of dimethylformamide. Add 5 drops of a freshly prepared 1 in 1000 solution of azo violet in dimethylformamide, and titrate with 0.1 *N* lithium methoxide VS to a blue-violet end-point, taking precautions against the absorption of atmospheric carbon dioxide. Perform a blank determination, and make any necessary correction. Each mL of 0.1 *N* lithium methoxide is equivalent to 22.43 mg of $C_{11}H_{16}N_2O_3$.

Talbutal Tablets

» Talbutal Tablets contain not less than 90.0 percent and not more than 110.0 percent of the labeled amount of $C_{11}H_{16}N_2O_3$.

Packaging and storage—Preserve in tight containers.

Reference standard—*USP Talbutal Reference Standard*—Dry in vacuum at 60° for 4 hours before using.

Identification—Shake a quantity of finely powdered Tablets, equivalent to about 200 mg of talbutal, with 10 mL of pentane for 5 minutes, and filter through a medium-porosity, sintered-glass filter. Discard the filtrate, and shake the residue with 10 mL of chloroform for 15 minutes. Filter through the same filter, evaporate the filtrate with the aid of gentle heat to dryness, and use the residue of talbutal so obtained for the following tests.

A: A portion of the residue responds to *Identification test A* under *Talbutal.*

B: To the remainder of the residue add 1 mL of glacial acetic acid and 10 mL of water, mix, then add bromine TS dropwise: the bromine color is discharged on shaking.

Dissolution ⟨711⟩—

Medium: water; 900 mL.

Apparatus 2: 50 rpm.

Time: 45 minutes.

Procedure—Determine the amount of $C_{11}H_{16}N_2O_3$ dissolved from ultraviolet absorbances at the wavelength of maximum absorbance at about 241 nm of filtered portions of the solution under test, suitably diluted with pH 9.6 alkaline borate buffer (see under *Buffer Solutions* in the section, *Reagents, Indicators, and Solutions*), in comparison with a Standard solution having a known concentration of USP Talbutal RS in the same medium.

Tolerances—Not less than 75% (*Q*) of the labeled amount of $C_{11}H_{16}N_2O_3$ is dissolved in 45 minutes.

Uniformity of dosage units ⟨905⟩: meet the requirements.

Assay—

Standard preparation—Dissolve an accurately weighed quantity of USP Talbutal RS in 5 mL of alcohol contained in a 100-mL volumetric flask, dilute with pH 9.6 alkaline borate buffer (see under *Solutions* in the section, *Reagents, Indicators, and Solutions*) to volume, mix, and dilute quantitatively and stepwise with the same alcohol-buffer mixture to obtain a solution having a known concentration of about 10 µg per mL.

Assay preparation—Weigh and finely powder not less than 20 Talbutal Tablets. Transfer an accurately weighed portion of the powder, equivalent to about 50 mg of talbutal, to a separator with the aid of 15 mL of water, and add 5 mL of 3 *N* hydrochloric acid. Extract with four 25-mL portions of chloroform, filter each portion through chloroform-washed cotton into a 250-mL volumetric flask, dilute with chloroform to volume, and mix. Transfer 5.0 mL of this solution to a beaker, and evaporate just to dryness. Transfer the residue to a 100-mL volumetric flask with the aid of, first, 5 mL of alcohol, and then pH 9.6 alkaline borate buffer. Dilute with the buffer to volume, and mix.

Procedure—Concomitantly determine the absorbances of the *Standard preparation* and the *Assay preparation* in 1-cm cells at the wavelength of maximum absorbance at about 241 nm, with a suitable spectrophotometer, using a 1 in 20 solution of alcohol in pH 9.6 alkaline borate buffer as the blank. Calculate the quantity, in mg, of $C_{11}H_{16}N_2O_3$ in the portion of Tablets taken by the formula:

$$5C(A_U/A_S),$$

in which *C* is the concentration, in µg per mL, of USP Talbutal RS in the *Standard preparation*, and A_U and A_S are the absorbances of the *Assay preparation* and the *Standard preparation*, respectively.

Talc

» Talc is a native, hydrous magnesium silicate, sometimes containing a small proportion of aluminum silicate.

Packaging and storage—Preserve in well-closed containers.

Identification—Mix about 200 mg of anhydrous sodium carbonate with 2 g of anhydrous potassium carbonate, and melt in a platinum crucible. To the melt add 100 mg of the substance under test, and continue heating until fusion is complete. Cool, and transfer the fused mixture to a dish or beaker with the aid of about 50 mL of hot water. Add hydrochloric acid to the liquid until effervescence ceases, then add 10 mL more of the acid, and evaporate the mixture on a steam bath to dryness. Cool, add 20 mL of water, boil, and filter the mixture: an insoluble residue of silica remains. Dissolve in the filtrate about 2 g of ammonium chloride, and add 5 mL of 6 *N* ammonium hydroxide. Filter, if necessary, and add dibasic sodium phosphate TS to the filtrate: a white, crystalline precipitate of magnesium ammonium phosphate separates.

Microbial limit—The total bacterial count does not exceed 500 per g.

Loss on ignition ⟨733⟩—Weigh accurately about 1 g, and ignite at 1000° to constant weight: it loses not more than 6.5% of its weight.

Acid-soluble substances—Digest 1.00 g with 20 mL of 3 *N* hydrochloric acid at 50° for 15 minutes, add water to restore the original volume, mix, and filter. To 10 mL of the filtrate add 1 mL of 2 *N* sulfuric acid, evaporate to dryness, and ignite to constant weight: the weight of the residue does not exceed 10 mg (2.0%).

Reaction and soluble substances—Boil 10 g with 50 mL of water for 30 minutes, adding water from time to time to maintain approximately the original volume, and filter: the filtrate is neutral to litmus paper. Evaporate one-half of the filtrate to dryness, and dry at 105° for 1 hour: the weight of the residue does not exceed 5 mg (0.1%).

Water-soluble iron—Slightly acidify with hydrochloric acid the remaining half of the filtrate obtained in the test for *Reaction* *and soluble substances*, and add 1 mL of potassium ferrocyanide TS: the liquid does not acquire a blue color.

Arsenic, Heavy metals, and Lead—

Test solution—Transfer 10.0 g to a 250-mL flask, and add 50 mL of 0.5 *N* hydrochloric acid. Attach a reflux condenser to the flask, heat on a steam bath for 30 minutes, cool, transfer the mixture to a beaker, and allow the undissolved material to settle. Decant the supernatant liquid through thick, strong, medium-speed filter paper into a 100-mL volumetric flask, retaining as much as possible of the insoluble material in the beaker. Wash the slurry and beaker with three 10-mL portions of hot water, decanting each washing through the filter into the flask. Finally, wash the filter paper with 15 mL of hot water, cool the filtrate to room temperature, dilute with water to volume, and mix. Use this *Test solution* for the following tests.

Arsenic, Method I ⟨211⟩—Use 10 mL of the *Test solution* in preparing the *Test Preparation*. The limit is 3 ppm.

Heavy metals ⟨231⟩—Use 5 mL of the *Test solution* in preparing the *Test Preparation*. The limit is 0.004%.

Lead ⟨251⟩—A 5-mL portion of the *Test solution* contains not more than 5 µg of lead (0.001%).

Tamoxifen Citrate

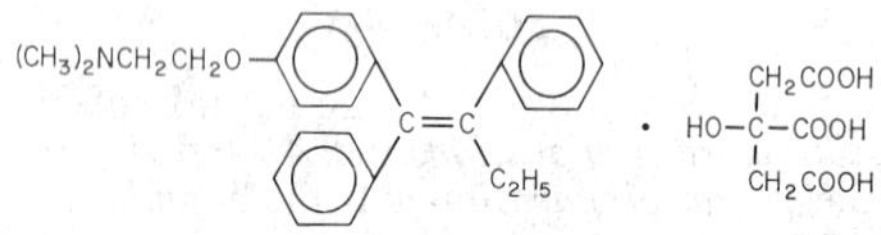

$C_{26}H_{29}NO.C_6H_8O_7$ 563.65

Ethanamine, 2-[4-(1,2-diphenyl-1-butenyl)phenoxy]-*N*,*N*-dimethyl-, (*Z*)-, 2-hydroxy-1,2,3-propanetricarboxylate (1:1).

(*Z*)-2-[*p*-(1,2-Diphenyl-1-butenyl)phenoxy]-*N*,*N*-dimethylethylamine citrate (1:1) [*54965-24-1*].

» Tamoxifen Citrate contains not less than 99.0 percent and not more than 101.0 percent of $C_{26}H_{29}NO.C_6H_8O_7$, calculated on the dried basis.

Packaging and storage—Preserve in well-closed, light-resistant containers.

Reference standard—*USP Tamoxifen Citrate Reference Standard*—Dry at 105° for 4 hours before using.

Identification—

A: The infrared absorption spectrum of a potassium bromide dispersion of it exhibits maxima only at the same wavelengths as that of a similar preparation of USP Tamoxifen Citrate RS, exhibiting a single band in the 1700 to 1740 cm^{-1} region of the spectrum.

B: The ultraviolet absorption spectrum of a 1 in 50,000 solution in methanol exhibits maxima and minima at the same wavelengths as that of a similar solution of USP Tamoxifen Citrate RS, concomitantly measured.

Melting range ⟨741⟩: melts at about 142°, with decomposition.

Loss on drying ⟨731⟩—Dry it at 105° for 4 hours: it loses not more than 0.5% of its weight.

Residue on ignition ⟨281⟩: not more than 0.2%.

***E*-isomer**—

Mobile phase—Prepare a methanol solution containing, in each liter, 320 mL of water, 2 mL of glacial acetic acid, and 1.08 g of sodium 1-octanesulfonate.

Standard preparation—Dissolve a suitable quantity, accurately weighed, of USP Tamoxifen Citrate RS in *Mobile phase* to obtain a solution having a known concentration of about 600 µg per mL.

Test preparation—Using about 30 mg of Tamoxifen Citrate, accurately weighed, proceed as directed under *Standard preparation.*

Chromatographic system (see *Chromatography* ⟨621⟩)—The liquid chromatograph is equipped with a 254-nm detector and a 4-mm × 30-cm column that contains packing L11. The flow rate is about 0.7 mL per minute. Chromatograph five replicate injections of the *Standard preparation*, and record the responses of the major peak: the relative standard deviation is not more than 3.0% and the relative retention time of the minor *E*-isomer peak to that of the *Z*-isomer peak is not greater than 0.93.

Procedure—Separately introduce equal volumes (about 20 µL) of the *Test preparation* and the *Standard preparation* into the liquid chromatograph by means of a suitable sampling valve. Measure the minor peak responses for the *E*-isomer obtained from the *Standard preparation* and the *Assay preparation*. Calculate the quantity, in mg, of *E*-isomer ($C_{26}H_{29}NO.C_6H_8O_7$) in the portion of Tamoxifen Citrate taken by the formula:

$$0.05C(r_U/r_S),$$

in which *C* is the concentration, in µg per mL, of the *E*-isomer as the citrate, based on its declared content in USP Tamoxifen Citrate RS in the *Standard preparation*, and the r_U and r_S are the minor peak responses obtained from the *Assay preparation* and the *Standard preparation*, respectively. The *E*-isomer content is not more than 1.0% of tamoxifen citrate ($C_{26}H_{29}NO.C_6H_8O_7$).

Related impurities—

Test preparation A—Disperse about 3 g in 100 mL of water in a separator. Over a 10-minute period add 50 mL of 0.5 *N* sodium hydroxide, with mixing. Extract with two 50-mL portions of ether, and combine the extracts. Wash with 20 mL of water, remove the water layer, and dry the ether layer over anhydrous sodium sulfate. Evaporate the ether layer under nitrogen, and dry in vacuum at room temperature for 2 hours. Accurately weigh 1.5 g of the residue into a 10-mL volumetric flask, add 5.0 mL of a mixture of 5 volumes of acetic anhydride and 95 volumes of pyridine, and heat at 60° for 10 to 15 minutes. Cool, dilute with the same solvent mixture to volume, and mix.

Test preparation B—Using the same acetic anhydride–pyridine mixture, prepare a 1:200 dilution of *Test preparation A*.

Chromatographic system (see *Chromatography* ⟨621⟩)—Typically, the gas chromatograph is equipped with a flame-ionization detector, and contains a 1-m × 4-mm glass column packed with 5 percent liquid phase G17 on 100- to 120-mesh support S1AB conditioned at 300° for 24 hours. The column and injection port are maintained at about 260° and the detector at about 300°. Dry helium is used as the carrier gas at a flow rate of about 60 mL per minute. In a suitable chromatogram, five replicate injections of *Test preparation B* show a relative standard deviation of not more than 3.0%.

Procedure—Inject equal portions (about 2 µL), accurately measured, of *Test preparation A* and *Test preparation B* into the chromatograph, and record the chromatograms from 0.1 to 5.0 relative to the retention time of the major peak. Measure the individual areas of the peaks other than those produced by the solvent and the tamoxifen on the chromatograms obtained from *Test preparation A*, and calculate their sum. No single peak area is greater than total area of the tamoxifen peak on the chromatogram obtained from *Test preparation B* (0.5%), and the sum of the peak areas is not greater than twice the total area of the tamoxifen peak on the chromatogram obtained from *Test preparation B* (1.0%).

Iron ⟨241⟩—Accurately weigh 1.0 g, and transfer to a suitable crucible. Add sufficient sulfuric acid to wet the substance, and carefully ignite at a low temperature until thoroughly charred. (The crucible may be loosely covered with a suitable lid during the charring.) Add to the carbonized mass 2 mL of nitric acid and 5 drops of sulfuric acid, and heat cautiously until white fumes no longer are evolved. Ignite, preferably in a muffle furnace, at 500° to 600°, until the carbon is completely burned off. Cool, add 10 mL of warm 0.1 *N* hydrochloric acid, and digest for about 5 minutes. Transfer the contents of the crucible with the aid of small portions of water to a 50-mL volumetric flask, dilute with water to volume, and mix. Pipet 10 mL from the volumetric flask into a color-comparison tube, dilute with water to 45 mL, add 2 mL of hydrochloric acid, and mix. The limit is 0.005%.

Arsenic, *Method II* ⟨211⟩—Use 10 mL of dilute sulfuric acid (1 in 2) instead of 5 mL of sulfuric acid. The limit is 2 ppm.

Heavy metals, *Method II* ⟨231⟩: 0.001%.

Assay—Weigh accurately about 1 g of Tamoxifen Citrate, and dissolve in 150 mL of glacial acetic acid. Titrate the solution with 0.1 *N* perchloric acid VS, determining the end-point potentiometrically, using a glass indicator electrode and a silver–silver chloride reference electrode. Each mL of 0.1 *N* perchloric acid is equivalent to 56.36 mg of $C_{26}H_{29}NO.C_6H_8O_7$.

Tamoxifen Citrate Tablets

» Tamoxifen Citrate Tablets contain not less than 90.0 percent and not more than 110.0 percent of the labeled amount of tamoxifen ($C_{26}H_{29}NO$).

Packaging and storage—Preserve in well-closed, light-resistant containers.

Reference standard—*USP Tamoxifen Citrate Reference Standard*—Dry at 105° for 4 hours before using.

Identification—

A: The ultraviolet absorption spectrum of the *Test preparation*, obtained as directed in the test for *Content uniformity*, exhibits maxima and minima at the same wavelengths as that of the *Standard preparation*, concomitantly measured.

B: To 1 Tablet contained in a 15-mL tube add 4 mL of pyridine and 2 mL of acetic anhydride: an immediate yellow color is produced on shaking. Then heat gently on a steam bath: a rose-pink to a deep red color develops, indicating the presence of citrate ion.

Dissolution ⟨711⟩—

Medium: 0.02 *N* hydrochloric acid; 1000 mL.

Apparatus 1: 100 rpm.

Time: 30 minutes.

Procedure—Determine the amount of tamoxifen ($C_{26}H_{29}NO$) dissolved from ultraviolet absorbances at the wavelength of maximum absorbance at about 275 nm of filtered portions of the solution under test, suitably diluted with *Dissolution Medium*, if necessary, in comparison with a Standard solution having a known concentration of USP Tamoxifen Citrate RS in the same medium.

Tolerances—Not less than 75% (*Q*) of the labeled amount of $C_{26}H_{29}NO$ is dissolved in 30 minutes.

Uniformity of dosage units ⟨905⟩: meet the requirements.

Procedure for content uniformity—

Standard preparation—Dissolve an accurately weighed quantity of USP Tamoxifen Citrate RS in methanol to obtain a solution having a known concentration of about 15 µg per mL.

Test preparation—Place 1 Tablet in a 100-mL volumetric flask, and crush with a stirring rod. Add about 75 mL of methanol, and shake for about 5 minutes. Dilute with methanol to volume, mix, and filter the solution through paper. Pipet 10 mL of the filtrate into a 100-mL volumetric flask, dilute with methanol to volume, and mix.

Procedure—Determine the absorbances of the *Test preparation* and the *Standard preparation* in 1-cm cells at the wavelength of maximum absorbance at about 275 nm, with a suitable spectrophotometer, using methanol as the blank. Calculate the quantity, in mg, of tamoxifen ($C_{26}H_{29}NO$) in the Tablets by the formula:

$$(371.52/563.65)(TC/D)(A_U/A_S),$$

in which 371.52 is the molecular weight of tamoxifen and 563.65 is the molecular weight of tamoxifen citrate, *T* is the labeled quantity, in mg, of tamoxifen in the Tablet, *C* is the concentration, in µg per mL, of USP Tamoxifen Citrate RS in the *Standard preparation*, *D* is the concentration, in µg per mL, of tamoxifen in the solution from the Tablet, based upon the labeled quantity per Tablet and the extent of dilution, and A_U and A_S are the absorbances of the *Test preparation* and the *Standard preparation*, respectively.

Assay—

Mobile solvent—Prepare a methanol solution containing, in each liter, 320 mL of water, 2 mL of glacial acetic acid, and 1.08 g of sodium 1-octanesulfonate.

Standard preparation—Dissolve a suitable quantity, accurately weighed, of USP Tamoxifen Citrate RS in methanol to obtain a solution containing about 200 µg per mL.

Assay preparation—Weigh accurately and finely powder not less than 20 Tamoxifen Citrate Tablets. Weigh accurately a portion of the powder, equivalent to about 20 mg of tamoxifen, and transfer to a stoppered, 50-mL centrifuge tube. Pipet 30 mL of methanol into the tube, and shake by mechanical means for not less than 15 minutes. Centrifuge at about 1000 rpm, pipet 5 mL of the clear supernatant liquid into a 25-mL volumetric flask, dilute with methanol to volume, and mix.

Chromatographic system (see *Chromatography* ⟨621⟩)—The liquid chromatograph is equipped with a 254-nm detector and a 4-mm × 30-cm column that contains packing L11. The flow rate is about 1.5 mL per minute. Chromatograph five replicate injections of the *Standard preparation*, and record the peak areas as directed under *Procedure:* the relative standard deviation is not more than 3.0%.

Procedure—Separately introduce equal volumes (about 25 µL) of the *Assay preparation* and the *Standard preparation* into the liquid chromatograph by means of a suitable sampling valve. Measure the peak areas obtained with the *Standard preparation* and the *Assay preparation*. Calculate the quantity, in mg, of tamoxifen ($C_{26}H_{29}NO$) in the portion of Tablets taken by the formula:

$$0.15C(371.52/563.65)(A_U/A_S),$$

in which 371.52 is the molecular weight of tamoxifen and 563.65 is the molecular weight of tamoxifen citrate, *C* is the concentration, in µg per mL, of USP Tamoxifen Citrate RS in the *Standard preparation*, and A_U and A_S are the peak areas obtained with the *Assay preparation* and the *Standard preparation*, respectively.

Tannic Acid

Tannin.
Tannic acid; Tannin [*1401-55-4*].

» Tannic Acid is a tannin usually obtained from nutgalls, the excrescences produced on the young twigs of *Quercus infectoria* Oliver, and allied species of *Quercus* Linné (Fam. Fagaceae), from the seed pods of Tara (*Caesalpinia spinosa*), or from the nutgalls or leaves of sumac (any of a genus *Rhus*).

Packaging and storage—Preserve in tight, light-resistant containers.

Identification—

A: To 2 mL of a solution (1 in 10) add 1 drop of ferric chloride TS: a bluish black color or precipitate results.

B: To a solution (1 in 10) add an equal volume of gelatin solution (1 in 100): a precipitate is formed.

Loss on drying ⟨731⟩—Dry it at 105° for 2 hours: it loses not more than 12.0% of its weight.

Residue on ignition ⟨281⟩: not more than 1.0%.

Arsenic, *Method II* ⟨211⟩: 3 ppm.

Heavy metals, *Method II* ⟨231⟩: 0.004%.

Gum or dextrin—Dissolve 2 g in 10 mL of hot water: the solution is not more than slightly turbid. Cool, filter, and divide the filtrate into two equal portions. To one portion add 10 mL of alcohol: no turbidity is produced.

Resinous substances—To a portion of the filtrate obtained in the test for *Gum or dextrin* add 10 mL of water: no turbidity is produced.

Adhesive Tape

» Adhesive Tape consists of fabric and/or film evenly coated on one side with a pressure-sensitive, adhesive mixture. Its length is not less than 98.0 percent of that declared on the label, and its average width is not less than 95.0 percent of the declared width. If

Adhesive Tape has been rendered sterile, it is protected from contamination by appropriate packaging.

Packaging and storage—Preserve in well-closed containers, and prevent exposure to excessive heat and to sunlight. Adhesive Tape that has been rendered sterile is so packaged that the sterility of the contents of the package is maintained until the package is opened for use.

Labeling—The package label of Adhesive Tape that has been rendered sterile indicates that the contents may not be sterile if the package bears evidence of damage or previously has been opened. The package label indicates the length and width of the Tape, and the name of the manufacturer, packer, or distributor.

Dimensions—Measure its length: it is not less than 98.0% of the labeled length. Measure its width at 5 locations evenly spaced along the center line of the Tape: the average of 5 measurements is not more than 1.6 mm less than the labeled width of the Tape.

Tensile strength—Determine the tensile strength of Adhesive Tape, after previously unrolling and conditioning it for not less than 4 hours in a standard atmosphere of 65 ± 2% relative humidity, at 21 ± 1.1° (70 ± 2°F), with a pendulum-type testing machine, as described under *Tensile Strength* ⟨881⟩. The Tape made from fabric has a tensile strength, determined warpwise, of not less than 20.41 kg (45 pounds) per 2.54 cm of width. The Tape made from film has a tensile strength of not less than 3 kg per 2.54 cm of width.

Adhesive strength—Determine the adhesive strength of Adhesive Tape that is made from fabric by cutting a strip of the Tape 2.54 cm wide and approximately 15 cm long, and applying 12.90 sq cm, 2.54 cm by 5.08 cm, of one end of the strip to a clean plastic or glass surface by means of a rubber roller under a pressure of 850 g, passing the roller twice over the Tape at a rate of 30 cm per minute. Adjust the temperature of the plastic or glass surface and the Tape to 37°, and conduct the test immediately thereafter as directed under *Tensile Strength* ⟨881⟩, using a pendulum-type testing machine, the pull being exerted parallel with the warp and the plastic or glass surface: the average of not less than 10 tests is not less than 18 kg.

Sterility—Adhesive Tape that has been rendered sterile meets the requirements under *Sterility Tests* ⟨71⟩.

Tape, Flurandrenolide—*see* Flurandrenolide Tape

Tar, Coal—*see* Coal Tar

Tar, Juniper—*see* Juniper Tar

Tartaric Acid—*see* Tartaric Acid NF

Technetium Tc 99m Albumin Injection

» Technetium Tc 99m Albumin Injection is a sterile, aqueous solution, suitable for intravenous administration, of Albumin Human that is labeled with ^{99m}Tc. It contains not less than 90.0 percent and not more than 110.0 percent of the labeled amount of ^{99m}Tc as albumin expressed in megabecquerels (microcuries or millicuries) per mL at the time indicated in the labeling. It may contain antimicrobial agents, buffers, reducing agents, and stabilizers. Other chemical forms of radioactivity do not exceed 10.0 percent of the total radioactivity. Its production and distribution are subject to federal regulations (see *Biologics* ⟨1041⟩ and *Radioactivity* ⟨821⟩).

Packaging and storage—Preserve in single-dose or in multiple-dose containers, at a temperature between 2° and 8°.

Labeling—Label it to include the following in addition to the information specified for *Labeling* under *Injections* ⟨1⟩: the time and date of calibration; the amount of ^{99m}Tc as albumin expressed as total megabecquerels (microcuries or millicuries) and concentration as megabecquerels (microcuries or millicuries) per mL at the time of calibration; the expiration date; and the statement, "Caution—Radioactive Material." The labeling indicates that in making dosage calculations, correction is to be made for radioactive decay, and also indicates that the radioactive half-life of ^{99m}Tc is 6.0 hours.

Reference standard—*USP Endotoxin Reference Standard.*

Bacterial endotoxins—It meets the requirements of the *Bacterial Endotoxins Test* ⟨85⟩, the limit of endotoxin content being not more than $175/V$ USP Endotoxin Unit per mL of the Injection, when compared with the USP Endotoxin RS, in which V is the maximum recommended total dose, in mL, at the expiration date or time.

pH ⟨791⟩: between 2.5 and 5.0.

Radiochemical purity—Not more than 10.0% of unbound Tc 99m (free pertechnetate) and reduced Tc 99m is present, determined as follows.

System A—Place a measured volume of Injection, such that it provides a count rate of about 100,000 counts per minute, about 3 cm from the end of a 2- × 17-cm thin-layer chromatographic strip impregnated with silica gel (see *Chromatography* ⟨621⟩), and allow to dry. Develop the chromatogram over a suitable period (approximately 30 minutes) by ascending chromatography, using acetone, and air-dry. Determine the radioactivity distribution by scanning the chromatograph with a suitable collimated radiation detector. Reduced Tc 99m and Tc 99m labeled albumin are located at the origin (R_f 0 to 0.1). Not more than 5.0% of the total radioactivity is found as unbound Tc 99m (free pertechnetate) at the solvent front (R_f 0.9 to 1.0).

System B—Prepare a 4- × 30-cm strip of chromatographic paper by immersing in 1% human serum albumin for 30 minutes and air-drying for 2 to 4 hours. Place a measured volume of Injection (about 3 µL) on the strip, mark the origin, and allow to dry. Develop the chromatogram by descending chromatography (see *Chromatography* ⟨621⟩), using nitrogen-purged saline TS until the solvent front has moved approximately 15 cm from the origin, mark the solvent front, and air-dry. Determine the radioactivity distribution by scanning the chromatogram with a suitable collimated radiation detector. Unbound Tc 99m and Tc 99m labeled albumin are at the solvent front (R_f 0.7 to 1.0). Not more than 5.0% of the total radioactivity is found as reduced Tc 99m at the origin (R_f 0 to 0.1).

Biological distribution—Inject intravenously between 0.075 MBq and 185 MBq (2 µCi and 5 mCi) of the Injection in a volume not exceeding 0.2 mL into the caudal vein of each of two 20- to 25-g mice. Approximately 30 minutes after the injection, sacrifice the animals, and drain the blood into a suitable container. Dissect the animals, and place the liver, stomach, and a 1-g specimen of blood, accurately weighed, in separate, suitable counting containers, and discard the tail cut above the injection site. Determine the radioactivity, in counts per minute, in each container with an appropriate detector, using the same counting geometry. Determine the percentage of radioactivity in the liver and stomach by the formula:

$$100(Ai/A),$$

in which Ai is the net radioactivity, in counts per minute, in the organ, and A is the total radioactivity, in counts per minute, in the liver, stomach, blood, and carcass. Determine the percentage of radioactivity in the blood by the formula:

$$[100(B/W_S)0.078(W_R)]/A,$$

in which B is the net radioactivity, in counts per minute, in the specimen of blood, W_S is the weight, in g, of the blood specimen, W_R is the weight, in g, of the mouse, and 0.078 is the assumption that the total blood weight of the mouse is 7.8% of the total body weight. Not more than 15.0% of the radioactivity is found in the liver, and not more than 1.0% of the radioactivity is found in the stomach. Not less than 30.0% of the radioactivity is found in the blood.

Other requirements—It meets the requirements of the tests for *Radionuclide identification* and *Radionuclidic purity* under *Sodium Pertechnetate Tc 99m Injection.* It meets also the requirements under *Injections* ⟨1⟩, except that it may be distributed or dispensed prior to completion of the test for *Sterility*, the latter test being started on the day of final manufacture, and except that it is not subject to the recommendation on *Volume in Container.*

Assay for radioactivity—Using a suitable counting assembly (see *Selection of a Counting Assembly* under *Radioactivity* ⟨821⟩), determine the radioactivity, in MBq (μCi) per mL, of Technetium Tc 99m Albumin Injection by use of a calibrated system as directed under *Radioactivity* ⟨821⟩.

Technetium Tc 99m Albumin Aggregated Injection

Albumins, blood serum, metastable technetium-99 labeled.

» Technetium Tc 99m Albumin Aggregated Injection is a sterile, aqueous suspension of Albumin Human that has been denatured to produce aggregates of controlled particle size that are labeled with ^{99m}Tc. It is suitable for intravenous administration. It contains not less than 90.0 percent and not more than 110.0 percent of the labeled amount of ^{99m}Tc as aggregated albumin expressed in megabecquerels (microcuries or millicuries) per mL at the time indicated in the labeling. It may contain antimicrobial, reducing, chelating, and stabilizing agents, buffers, and nonaggregated albumin human. Other chemical forms of radioactivity do not exceed 10 percent of the total radioactivity. Its production and distribution are subject to federal regulations (see *Biologics* ⟨1041⟩ and *Radioactivity* ⟨821⟩).

Packaging and storage—Preserve in single-dose or in multiple-dose containers, at a temperature between 2° and 8°.

Labeling—Label it to include the following, in addition to the information specified for *Labeling* under *Injections* ⟨1⟩: the time and date of calibration; the amount of ^{99m}Tc as aggregated albumin expressed as total megabecquerels (millicuries or microcuries) and concentration as megabecquerels (microcuries or millicuries) per mL at the time of calibration; the expiration date; and the statement, "Caution—Radioactive Material." The labeling indicates that in making dosage calculations, correction is to be made for radioactive decay, and also indicates that the radioactive half-life of ^{99m}Tc is 6.0 hours. In addition, the labeling states that it is not to be used if clumping of the albumin is observed and directs that the container be agitated before the contents are withdrawn into a syringe.

Reference standard—*USP Endotoxin Reference Standard.*

Particle size—Shake the Injection well, and determine the dimension of not less than 100 particles of a representative test specimen, using a suitable counting chamber, such as a hemacytometer grid, by optical microscopy. Not less than 90.0% of the observed aggregated particles have a diameter between 10 μm and 90 μm, and none of the observed particles have a diameter greater than 150 μm.

Bacterial endotoxins—It meets the requirements of the *Bacterial Endotoxins Test* ⟨85⟩, the limit of endotoxin content being not more than 175/*V* USP Endotoxin Unit per mL of the Injection, when compared with the USP Endotoxin RS, in which *V* is the maximum recommended total dose, in mL, at the expiration date or time.

pH ⟨791⟩: between 3.8 and 8.0.

Radiochemical purity—Place a measured volume of Technetium Tc 99m Albumin Aggregated Injection, appropriately diluted, such that it provides a count rate of about 20,000 counts per minute, about 25 mm from one end of a 25- × 300-mm strip of chromatographic paper (see *Chromatography* ⟨621⟩), and allow to dry. Develop the chromatogram over a period of about 3 to 4 hours by ascending chromatography, using dilute methanol (7 in 10), and air-dry. Determine the radioactivity distribution by scanning the chromatogram with a suitable collimated radiation detector. Not less than 90% of the total radioactivity is found as aggregated albumin (at the point of application).

Place a measured volume of Technetium Tc 99m Albumin Aggregated Injection in a centrifuge tube, and determine the net radioactivity in a suitable counting assembly. Centrifuge at approximately 2000 rpm for 5 to 10 minutes. Separate the supernatant liquid by aspiration, and determine its net radioactivity in a suitable counting assembly. Determine the percentage of radioactivity in the supernatant liquid taken by the formula:

$$100(A/B),$$

in which A is the net radioactivity, in counts per minute, in the supernatant aliquot, and B is the total radioactivity, in counts per minute, in the tube prior to centrifugation. Not more than 10.0% of the radioactivity is found in the supernatant liquid, which contains soluble and dispersed radiochemical impurities, following centrifugation.

Protein concentration—

Test preparation—Transfer 2.0 mL of Technetium Tc 99m Albumin Aggregated Injection to a suitable centrifuge tube, and centrifuge at about 2000 rpm for 5 to 10 minutes. Decant the supernatant liquid, and add 2.0 mL of Sodium Chloride Injection to wash the centrifuged aggregate. Centrifuge again at 2000 rpm for 5 to 10 minutes, decant the supernatant liquid, and add 2.0 mL of Sodium Chloride Injection.

Standard preparation—To a second test tube add 2.0 mL of a solution containing 2.0 mg of Albumin Human per mL in 0.9 percent sodium chloride solution.

Procedure—To a third test tube add 2.0 mL of Sodium Chloride Injection to provide a blank. To each of the three tubes containing the *Test preparation, Standard preparation,* and blank, add 4.0 mL of biuret reagent TS, mix, and allow to stand for 30 minutes, accurately timed, for maximum color development. Additional mixing or slight heating may be required to dissolve the aggregated albumin completely, but the *Test preparation, Standard preparation,* and blank are to be treated identically. Determine the absorbances of the solutions from the *Test preparation* and the *Standard preparation* in 1-cm cells at the wavelength of maximum absorbance at about 540 nm, with a suitable spectrophotometer, against the blank. Calculate the quantity, in mg, of aggregated albumin in each mL of the Injection by the formula:

$$2(A_U/A_S),$$

in which A_U and A_S are the absorbances of the solutions from the *Test preparation* and the *Standard preparation*, respectively. The protein concentration is not more than 1 mg, as aggregated albumin, per 37 MBq (1 mCi) of Tc 99m at the time of administration.

Biological distribution—Inject intravenously between 0.075 MBq and 0.75 MBq (2 μCi and 20 μCi) of Technetium Tc 99m Albumin Aggregated Injection, in a volume not exceeding 0.2 mL, into the caudal vein of each of three 20- to 25-g mice. [NOTE—Other animal species, such as Sprague-Dawley rats (weighing 100 g to 175 g), may be used.] Five to 10 minutes after the injection, sacrifice the animals, and carefully remove the liver and lungs of each by dissection. Place each organ and the remaining carcass in separate, suitable counting containers, and determine the radioactivity, in counts per minute, in each container with an appropriate detector, using the same counting geometry. Determine the percentage of radioactivity in the liver and the lungs by the formula:

$$100(A/B),$$

in which A is the net radioactivity, in counts per minute, in the organ, and B is the total radioactivity, in counts per minute, in the lungs, liver, and carcass. Not less than 80.0% of the radioactivity is found in the lungs, and not more than 5.0% of the radioactivity is found in the liver, in not less than two of the animals.

Other requirements—It meets the requirements of the tests for *Radionuclide identification* and *Radionuclidic purity* under *Sodium Pertechnetate Tc 99m Injection.* It meets the requirements under *Injections* ⟨1⟩, except that it may be distributed or dispensed prior to completion of the test for *Sterility*, the latter test being started on the day of final manufacture, and except that it is not subject to the recommendation on *Volume in Container.*

Assay for radioactivity—Using a suitable counting assembly (see *Selection of a Counting Assembly* under *Radioactivity* ⟨821⟩), determine the radioactivity, in MBq (μCi) per mL, of Technetium Tc 99m Albumin Aggregated Injection by use of a calibrated system as directed under *Radioactivity* ⟨821⟩.

Technetium Tc 99m Disofenin Injection

» Technetium Tc 99m Disofenin Injection is a sterile, aqueous solution, suitable for intravenous administration, of disofenin that is labeled with ^{99m}Tc. It contains not less than 90.0 percent and not more than 110.0 percent of the labeled amount of ^{99m}Tc as a disofenin complex, expressed in megabecquerels (microcuries or millicuries) per mL at the time indicated on the labeling. It contains a suitable reducing agent.

Packaging and storage—Preserve in single-dose or in multiple-dose containers sealed under a suitable inert atmosphere.

Labeling—Label it to include the following, in addition to the information specified for *Labeling* under *Injections* ⟨1⟩: the time and date of preparation; the amount of ^{99m}Tc as expressed as total megabecquerels (microcuries or millicuries) and concentration as megabecquerels (microcuries or millicuries) per mL at the time of preparation; the expiration date and time; and a statement, "Caution—Radioactive Material." The labeling indicates that in making dosage calculations, correction is to be made for radioactive decay, and also indicates that the radioactive half-life of ^{99m}Tc is 6.0 hours.

pH ⟨791⟩: between 5.8 and 6.6.

Radiochemical purity—The determination of radiochemical purity for this Injection requires the use of two separate chromatography systems.

System A—Apply a measured volume of Injection, appropriately diluted, such that it provides a count rate of about 20,000 counts per minute, about 13 mm from one end of a 10- × 51-mm thin-layer chromatographic strip impregnated with silica gel (see *Chromatography* ⟨621⟩), and immediately develop the chromatogram over a suitable period by ascending chromatography, using methanol as the solvent. Allow the chromatogram to dry. Cut the chromatogram in two along a line about 13 mm from the top. Place the top and bottom strips into separate counting containers. Determine the radioactivity of each, using a suitable ionization chamber. Hydrolyzed Tc 99m and technetium-tin colloid are located at the origin of the bottom strip (R_f 0 to 0.1).

System B—Pretreat a 1.3- × 51-mm strip of chromatographic paper by soaking it for 1 minute in a 0.3 *M* carbonate-bicarbonate buffer, pH 9.0, prepared by dissolving 0.3 g of anhydrous sodium carbonate and 1.9 g of anhydrous sodium bicarbonate in 100 mL of water. Remove the paper strip, blot lightly with absorbent paper, and dry in an oven for about 45 minutes at 85°. Proceed as directed for *System A* using the pretreated chromatographic paper and a developing solvent of methyl ethyl ketone. Free pertechnetate is located at the solvent front (top strip). The sum of the percentage of radioactivity at the origin in *System A* and the percentage of radioactivity at the solvent front in *System B* is not greater than 10.0%.

Biological distribution—Inject between 75 MBq and 111 MBq (2 mCi and 3 mCi) of Technetium Tc 99m Disofenin Injection, in a volume not exceeding 0.1 mL, into the caudal vein of each of three 25- to 40-g male albino mice. One hour after the injection, anesthetize and then decapitate the animals. Drain the blood of each separately into pre-weighed counting containers. Dissect the animals, and place the kidneys, liver, stomach (exclusive of duodenum), and gall bladder, and intestines, of each mouse into separate counting containers. Using a suitable counting assembly, determine the radioactivity of each container. Correct all radioactivity measurements for decay. Determine the percentage of radioactivity in the kidneys, liver, stomach, and gall bladder and intestines by the formula:

$$100(A/B),$$

in which A is the radioactivity, in counts per minute, in the organ, and B is the total radioactivity, in counts per minute, injected. Determine the percentage of radioactivity in the blood by the formula:

$$[100(C/W_S)0.07(W_R)]/B,$$

in which C is the radioactivity, in counts per minute, in the specimen of blood, W_S is the weight, in g, of the blood specimen, 0.07 is the assumption that the total blood weight of the mouse is 7% of the total body weight, and W_R is the weight, in g, of the mouse. Not less than 70% of the injected radioactivity is present in the gall bladder and intestines, not more than 10.0% of the injected radioactivity is present in the liver, not more than 10.0% of the injected radioactivity is present in the kidneys, not more than 3.0% of the injected radioactivity is present in the stomach, and not more than 3.0% of the injected radioactivity is present in the blood, in not less than two mice.

Other requirements—It meets the requirements of the tests for *Radionuclide identification, Radionuclidic purity*, and *Bacterial endotoxins* under *Sodium Pertechnetate Tc 99m Injection.* It also meets the requirements under *Injections* ⟨1⟩, except that it may be distributed or dispensed prior to completion of the test for *Sterility*, the latter test being started on the date of preparation, and except that it is not subject to the recommendation on *Volume in Container.*

Assay for radioactivity—Using a suitable counting assembly (see *Selection of a Counting Assembly* under *Radioactivity* ⟨821⟩), determine the radioactivity, in MBq (μCi or mCi) per mL, of Technetium Tc 99m Disofenin Injection by the use of a calibrated system as directed under *Radioactivity* ⟨821⟩.

Technetium Tc 99m Etidronate Injection

» Technetium Tc 99m Etidronate Injection is a sterile, clear, colorless solution, suitable for intravenous administration, of radioactive technetium (^{99m}Tc) in the form of a chelate of etidronate sodium. It contains not less than 90.0 percent and not more than 110.0 percent of the labeled amount of ^{99m}Tc as chelate expressed in megabecquerels (microcuries or millicuries) per mL at the time indicated in the labeling. It may contain buffers, preservatives, reducing agents, and stabilizers. Other chemical forms of radioactivity do not exceed 10.0 percent of the total radioactivity.

Packaging and storage—Preserve in single-dose or in multiple-dose containers.

Labeling—Label it to include the following, in addition to the information specified for *Labeling* under *Injections* ⟨1⟩: the time and date of calibration; the amount of ^{99m}Tc as labeled etidronate expressed as total megabecquerels (microcuries or millicuries) and concentration as megabecquerels (microcuries or millicuries) per mL at the time of calibration; the expiration date and time; and the statement, "Caution—Radioactive Material." The labeling indicates that in making dosage calculations, correction is to be made for radioactive decay, and also indicates that the radioactive half-life of ^{99m}Tc is 6.0 hours.

Reference standard—*USP Endotoxin Reference Standard.*

pH ⟨791⟩: between 2.5 and 7.0.

Other requirements—It meets the requirements for *Bacterial endotoxins*, *Radiochemical purity*, *Biological distribution*, *Other requirements*, and *Assay for radioactivity* under *Technetium Tc 99m Pyrophosphate Injection.*

Technetium Tc 99m Ferpentetate Injection

Iron, ascorbic acid and *N,N*-bis[2-[bis(carboxymethyl)amino]-ethyl]glycine complex, metastable technetium-99 labeled.
Iron, ascorbic acid and *N,N*-bis[2-[bis(carboxymethyl)amino]-ethyl]glycine complex, metastable technetium-99 labeled.

» Technetium Tc 99m Ferpentetate Injection is a sterile, aqueous solution of iron ascorbate pentetic acid that is complexed with ^{99m}Tc. It is suitable for intravenous administration, and may contain buffers. It contains not less than 90.0 percent and not more than 110.0 percent of the labeled amount of ^{99m}Tc as the ferpentetate expressed in megabecquerels (microcuries or millicuries) per mL at the time indicated in the labeling. Other chemical forms of radioactivity do not exceed 10.0 percent of the total radioactivity.

Packaging and storage—Preserve in single-dose or in multiple-dose containers, at a temperature between 2° and 8°. Protect from light.

Labeling—Label it to include the following, in addition to the information specified for *Labeling* under *Injections* ⟨1⟩: the time and date of calibration; the amount of ^{99m}Tc as labeled ferpentetate expressed as total megabecquerels (millicuries or microcuries) and concentration as megabecquerels (microcuries or millicuries) per mL at the time of calibration; the expiration date; and the statement, "Caution—Radioactive Material." The labeling indicates that in making dosage calculations, correction is to be made for radioactive decay, and also indicates that the radioactive half-life of ^{99m}Tc is 6.0 hours.

Reference standard—*USP Endotoxin Reference Standard.*

Bacterial endotoxins—It meets the requirements of the *Bacterial Endotoxins Test* ⟨85⟩, the limit of endotoxin content being not more than 175/*V* USP Endotoxin Unit per mL of the Injection, when compared with the USP Endotoxin RS, in which *V* is the maximum recommended total dose, in mL, at the expiration date or time.

pH ⟨791⟩: between 4.0 and 5.5.

Radiochemical purity—Proceed with Technetium Tc 99m Ferpentetate Injection as directed in the test for *Radiochemical purity* under *Technetium Tc 99m Sulfur Colloid Injection.* Not less than 90.0% of the total radioactivity is found as the labeled ferpentetate (at the point of application).

Biological distribution—Inject intravenously between 0.075 MBq and 0.75 MBq (2 µCi and 20 µCi) of Technetium Tc 99m Ferpentetate Injection, in a volume not exceeding 0.2 mL, into the caudal vein of each of three 20- to 25-g mice. Approximately 10 to 15 minutes after injection, place each animal in a suitable container and count the radioactivity of each with an appropriate radiation detector, using the same counting geometry, in order to determine the total activity injected. Approximately 24 hours after injection, repeat the radioactivity counting of each animal, as above. After correcting for decay, determine the percentage of radioactivity retained in each animal 24 hours after injection by the formula:

$$100(A/B),$$

in which *A* is the net radioactivity, in counts per minute (corrected for decay), in the animal at 24 hours, and *B* is the total radioactivity, in net counts per minute, at the time of injection. The percentage of injected radioactivity retained 24 hours after administration is not greater than 5.0% of the injected activity in any of the animals.

Other requirements—It meets the requirements for the tests for *Radionuclide identification* and *Radionuclidic purity* under *Sodium Pertechnetate Tc 99m Injection.* It meets the requirements under *Injections* ⟨1⟩, except that it may be distributed or dispensed prior to completion of the test for *Sterility*, the latter test being started on the day of manufacture, and except that it is not subject to the recommendation on *Volume in Container.*

Assay for radioactivity—Using a suitable counting assembly (see *Selection of a Counting Assembly* under *Radioactivity* ⟨821⟩), determine the radioactivity, in MBq (µCi) per mL, of Technetium Tc 99m Ferpentetate Injection by use of a calibrated system as directed under *Radioactivity* ⟨821⟩.

Technetium Tc 99m Gluceptate Injection

D-*glycero*-D-*gulo*-Heptonic acid, technetium-^{99m}*Tc* complex.
Technetium-^{99m}*Tc* D-*glycero*-D-*gulo*-heptonate complex.

» Technetium Tc 99m Gluceptate Injection is a sterile, aqueous solution, suitable for intravenous administration, of sodium gluceptate and stannous chloride that is labeled with ^{99m}Tc. It contains not less than 90.0 percent and not more than 110.0 percent of the labeled amount of ^{99m}Tc as stannous gluceptate complex expressed in megabecquerels (microcuries or millicuries) per mL at the time indicated in the labeling. It may contain antimicrobial agents and buffers. Other chemical forms of radioactivity do not exceed 10.0 percent of the total radioactivity.

Packaging and storage—Preserve in single-dose or in multiple-dose containers, at a temperature between 2° and 8°.

Labeling—Label it to include the following, in addition to the information specified for *Labeling* under *Injections* ⟨1⟩: the time and date of calibration; the amount of ^{99m}Tc as labeled stannous gluceptate expressed as total megabecquerels (microcuries or millicuries) and concentration as megabecquerels (microcuries or millicuries) per mL at the time of calibration; the expiration date and time; and the statement, "Caution—Radioactive Material." The labeling indicates that in making dosage calculations, correction is to be made for radioactive decay, and also indicates that the radioactive half-life of ^{99m}Tc is 6.0 hours.

Reference standard—*USP Endotoxin Reference Standard.*

Bacterial endotoxins—It meets the requirements of the *Bacterial Endotoxins Test* ⟨85⟩, the limit of endotoxin content being not more than 175/*V* USP Endotoxin Unit per mL of the Injection, when compared with the USP Endotoxin RS, in which *V* is the maximum recommended total dose, in mL, at the expiration date or time.

pH ⟨791⟩: between 4.0 and 8.0.

Radiochemical purity—Place a measured volume of Injection, appropriately diluted, such that it provides a count rate of about 20,000 counts per minute, about 25 mm from one end of a 25- × 300-mm strip of chromatographic paper (see *Chromatography* ⟨621⟩), and allow to air-dry. With no delay, develop the chromatogram over a suitable period of time by ascending chromatography, using acetone that has been purged with oxygen-free nitrogen for not less than 10 minutes. Allow the chromatogram to dry, and determine the radioactivity distribution by scanning with a suitable collimated radiation detector. Not less than 90.0% of the total radioactivity is found as stannous gluceptate (at the point of application).

Biological distribution—Constitute 1 vial of Technetium Tc 99m Gluceptate Injection with 5 mL to 10 mL of Sodium Pertechnetate Tc 99m Injection. Inject intravenously 0.25 mL of the resulting undiluted Injection into the caudal vein of each of three 150- to 250-g rats. Discard any injections in which extravasation occurs. Approximately 1 hour after the injection, anesthetize the animals and exsanguinate, collecting the blood into a suitable container. Dissect the animals, and place the kidney, liver, gastrointestinal tract, and a specimen of blood, accurately weighed, in separate, suitable counting containers, and determine the radioactivity, in counts per minute, in each container with an appropriate detector using the same counting geometry. Determine the percentage of radioactivity in the kidney, liver, and gastrointestinal tract by the formula:

$$100Ai/A,$$

in which Ai is the net radioactivity in the organ, and A is the total radioactivity, in MBq (μCi), injected, both corrected to injection time. Determine the percentage of radioactivity in the blood by the formula:

$$[100(B/W_S)0.07(W_R)]/A,$$

in which B is the net radioactivity in the specimen of blood, and A is the total radioactivity, injected, both corrected to injection time, W_S is the weight, in g, of the blood specimen, W_R is the weight, in g, of the rat, and 0.07 is the assumption that the total blood weight of the rat is 7% of the total body weight. Not less than 15.0% of the radioactivity is found in the kidney, not more than 5.0% of the radioactivity is found in the blood, not more than 15.0% of the radioactivity is found in the entire gastrointestinal tract and not more than 5.0% of the radioactivity is found in the liver, in not fewer than 2 of the rats.

Other requirements—It meets the requirements for *Radionuclide identification* and *Radionuclidic purity* under *Sodium Pertechnetate Tc 99m Injection.* It meets also the requirements under *Injections* ⟨1⟩, except that it may be distributed or dispensed prior to completion of the test for *Sterility*, the latter test being started on the date of manufacture, and except that it is not subject to the recommendation on *Volume in Container*.

Assay for radioactivity—Using a suitable counting assembly (see *Selection of a Counting Assembly* under *Radioactivity* ⟨821⟩), determine the radioactivity, in MBq (μCi) per mL, of Technetium Tc 99m Gluceptate Injection by use of a calibrated system as directed under *Radioactivity* ⟨821⟩.

Technetium Tc 99m Medronate Injection

» Technetium Tc 99m Medronate Injection is a sterile, aqueous solution, suitable for intravenous administration, of sodium medronate and stannous chloride or stannous fluoride that is labeled with radioactive Tc 99m. It contains not less than 90.0 percent and not more than 110.0 percent of the labeled amount of Tc 99m as stannous medronate complex expressed in megabecquerels (microcuries or millicuries) per mL at the date and time indicated in the labeling. It may contain antimicrobial agents, antioxidants, and buffers. Other chemical forms of radioactivity do not exceed 10.0 percent of the total radioactivity.

Packaging and storage—Preserve in single-dose or in multiple-dose containers at a temperature specified in the labeling.

Reference standard—*USP Endotoxin Reference Standard.*

Bacterial endotoxins—It meets the requirements under *Bacterial Endotoxins Test* ⟨85⟩, the limit of endotoxin content being not more than $175/V$ USP Endotoxin Unit per mL of the Injection, when compared with the USP Endotoxin RS, in which V is the maximum recommended total dose, in mL, at the expiration date or time.

pH ⟨791⟩: between 4.0 and 7.8.

Radiochemical purity—Not more than 10.0% of unbound Tc 99m (free pertechnetate), hydrolyzed Tc 99m, and technetium-tin colloid is present, determined as follows.

System A: Under an atmosphere of nitrogen, place a measured volume of Injection, such that it provides a count rate of about 20,000 counts per minute, about 25 mm from one end of a paper chromatographic strip (see *Chromatography* ⟨621⟩). Immediately develop the chromatogram over a suitable period by ascending chromatography, using sodium chloride solution (0.9 in 100), and dry it under nitrogen. Determine the radioactivity distribution by scanning the chromatogram with a suitable collimated radiation detector. Hydrolyzed Tc 99m and technetium-tin colloid are located at the origin (R_f 0 to 0.1).

System B: Proceed as directed for *System A*, except to develop the chromatogram in dilute methanol (85 in 100). Free pertechnetate is located at an R_f of 0.6 to 0.8. The sum of the percentage of radioactivity at the origin in *System A* plus the percentage of radioactivity at an R_f of 0.6 to 0.8 in *System B* is not greater than 10.0%.

Other requirements—It meets the requirements of the tests for *Radionuclide identification* and *Radionuclidic purity* under *Sodium Pertechnetate Tc 99m Injection*, and meets the requirements for *Labeling, Biological distribution*, and *Assay for radioactivity* under *Technetium Tc 99m Pyrophosphate Injection.* It meets also the requirements under *Injections* ⟨1⟩, except that it may be distributed or dispensed prior to completion of the test for *Sterility*, the latter test being started on the day of manufacture, and except that it is not subject to the recommendation on *Volume in Container.*

Technetium Tc 99m Oxidronate Injection

» Technetium Tc 99m Oxidronate Injection is a sterile, clear, colorless solution, suitable for intravenous administration, of radioactive technetium (^{99m}Tc) in the form of a chelate of oxidronate sodium. It contains not less than 90.0 percent and not more than 110.0 percent of the labeled amount of ^{99m}Tc as chelate expressed in megabecquerels (microcuries or millicuries) per mL at the time indicated in the labeling. It may contain buffers, preservatives, reducing agents, and stabilizers. Other chemical forms of radioactivity do not exceed 10.0 percent of the total radioactivity.

Packaging and storage—Preserve in single-dose or in multiple-dose containers.

Labeling—Label it to include the following, in addition to the information specified for *Labeling* under *Injections* ⟨1⟩: the time and date of calibration; the amount of ^{99m}Tc as labeled oxidronate expressed as total megabecquerels (microcuries or millicuries) and concentration as megabecquerels (microcuries or millicuries) per mL at the time of calibration; the expiration date and time; and the statement, "Caution—Radioactive Material." The labeling indicates that in making dosage calculations, correction is to be made for radioactive decay, and also indicates that the radioactive half-life of ^{99m}Tc is 6.0 hours.

Reference standard—*USP Endotoxin Reference Standard.*

pH ⟨791⟩: between 2.5 and 7.0.

Other requirements—It meets the requirements for *Bacterial endotoxins, Radiochemical purity, Biological distribution, Other requirements*, and *Assay for radioactivity* under *Technetium Tc 99m Pyrophosphate Injection.*

Technetium Tc 99m Pentetate Injection

$$\left[{}^{99m}Tc^{4+} \quad \begin{matrix} {}^{-}OOCCH_2 & & CH_2COO^{-} & CH_2COO^{-} \\ | & & | & | \\ & NCH_2CH_2NCH_2CH_2N & & \\ | & & & | \\ {}^{-}OOCCH_2 & & & CH_2COO^{-} \end{matrix} \right]^{-} Na^{+}$$

$C_{14}H_{18}N_3NaO_{10}{}^{99m}Tc$

Technetate(1-)^{99m}Tc, [*N,N*-bis[2-[bis(carboxymethyl)amino]-ethyl]glycinato(5-)]-, sodium.
Sodium [*N,N*-bis[2-[bis(carboxymethyl)amino]ethyl]-glycinato(5-)]-technetate(1-)-^{99m}Tc [*65454-61-7*].

» Technetium Tc 99m Pentetate Injection is a sterile solution of pentetic acid that is complexed with ^{99m}Tc in Sodium Chloride Injection. It is suitable for intravenous administration, and may contain buffers. It contains not less than 90.0 percent and not more than 110.0 percent of the labeled amount of ^{99m}Tc as the pentetic acid complex expressed in megabecquerels (microcuries or millicuries) per mL at the time indicated in the labeling. Other chemical forms of radioactivity do not exceed 10.0 percent of the total radioactivity.

Packaging and storage—Preserve in single-dose or in multiple-dose containers, at a temperature between 2° and 8°.

Labeling—Label it to include the following, in addition to the information specified for *Labeling* under *Injections* ⟨1⟩: the time and date of calibration; the amount of ^{99m}Tc as labeled pentetic acid complex expressed as total megabecquerels (millicuries or microcuries) and concentration as megabecquerels (microcuries or millicuries) per mL at the time of calibration; the expiration date; and the statement, "Caution—Radioactive Material." The labeling indicates that in making dosage calculations, correction is to be made for radioactive decay and also indicates that the radioactive half-life of ^{99m}Tc is 6.0 hours.

Reference standard—*USP Endotoxin Reference Standard.*

pH ⟨791⟩: between 3.8 and 7.5.

Radiochemical purity—

Phosphate buffer—Dissolve 2.6 g of monobasic sodium phosphate and 3.0 g of anhydrous dibasic sodium phosphate in water to make 1000 mL. Adjust, if necessary, by the addition of 0.2 *M* phosphoric acid or 0.2 *M* sodium hydroxide, to a pH of 6.8 ± 0.05.

Procedure (see *Electrophoresis* ⟨726⟩)—Soak a 2.5- × 17.0-cm cellulose polyacetate strip in 100 mL of *Phosphate buffer* for 10 to 60 minutes. Remove the strip with forceps, taking care to handle the outer edges only. Place the strip between two absorbent pads, and blot to remove excess solution. Attach the strip to the support bridge of an electrophoresis chamber containing *Phosphate buffer* in each side of the chamber. Ensure that each end of the strip is in contact with *Phosphate buffer*. Mark the application line near the cathode end of the strip. Apply as a spot 1 µL of a 1 in 1000 solution of amaranth in *Phosphate buffer*. Adjacent to this, apply as a narrow streak 5 µL of Technetium Tc 99m Pentetate Injection having an activity of about 1 mCi per mL, previously diluted with saline TS, if necessary. Attach the chamber cover, and perform the electrophoresis at 300 volts until the amaranth (red dye) has moved about four-fifths of the length of the strip. Remove the strip from the chamber, and blot the ends on paper towels. Using a suitable scanner and counting assembly, determine the radioactivity distribution. Labeled technetium pentetate migrates a distance that is 0.9 relative to that of amaranth. Hydrolyzed technetium Tc 99m is located at the origin and pertechnetate migrates a distance that is 0.7 relative to that of amaranth: the sum of the percentages of radioactivity of hydrolyzed technetium Tc 99m and pertechnetate is not greater than 10.0% of the total radioactivity on the strip.

Other requirements—It meets the requirements of the tests for *Radionuclide identification* and *Radionuclidic purity* under *Sodium Pertechnetate Tc 99m Injection.* It meets the requirements under *Injections* ⟨1⟩, except that it may be distributed or dispensed prior to completion of the test for *Sterility*, the latter test being started on the day of manufacture, and except that it is not subject to the recommendation on *Volume in Container.* It meets also the requirements for *Bacterial endotoxins*, *Biological distribution*, and *Assay for radioactivity* under *Technetium Tc 99m Ferpentetate Injection.*

Sodium Pertechnetate Tc 99m Injection

Pertechnetic acid ($H^{99m}TcO_4$), sodium salt.
Sodium pertechnetate ($Na^{99m}TcO_4$) [*23288-60-0*].

» Sodium Pertechnetate Tc 99m Injection is a sterile solution, suitable for intravenous or oral administration, containing radioactive technetium (^{99m}Tc) in the form of sodium pertechnetate and sufficient Sodium Chloride to make the solution isotonic. Technetium 99m is a radioactive nuclide formed by the radioactive decay of molybdenum 99. Molybdenum 99 is a radioactive isotope of molybdenum and may be formed by the neutron bombardment of molybdenum 98 or as a product of uranium fission.

Sodium Pertechnetate Tc 99m Injection contains not less than 90.0 percent and not more than 110.0 percent of the labeled amount of ^{99m}Tc at the date and hour stated on the label. Other chemical forms of ^{99m}Tc do not exceed 5 percent of the total radioactivity.

Packaging and storage—Preserve in single-dose or in multiple-dose containers.

Labeling—If intended for intravenous use, label it with the information specified for *Labeling* under *Injections* ⟨1⟩. Label it also to include the following: the time and date of calibration; the amount of ^{99m}Tc as sodium pertechnetate expressed as total megabecquerels (millicuries) and as megabecquerels (millicuries) per mL on the date and at the time of calibration; a statement of the intended use, whether oral or intravenous; the expiration date; and the statement, "Caution—Radioactive Material." If the Injection has been prepared from molybdenum 99 produced from uranium fission, the label so states. The labeling indicates that in making dosage calculations, correction is to be made for radioactive decay, and also indicates that the radioactive half-life of ^{99m}Tc is 6.0 hours.

Reference standard—*USP Endotoxin Reference Standard.*

Radionuclide identification (see *Radioactivity* ⟨821⟩)—Its gamma-ray spectrum is identical to that of a specimen of ^{99m}Tc that exhibits a major photopeak having an energy of 0.140 MeV.

Bacterial endotoxins—It meets the requirements of the *Bacterial Endotoxins Test* ⟨85⟩, the limit of endotoxin content being not more than 175/*V* USP Endotoxin Unit per mL of the Injection, when compared with the USP Endotoxin RS, in which *V* is the maximum recommended total dose, in mL, at the expiration date or time.

pH ⟨791⟩: between 4.5 and 7.5.

Radiochemical purity—Place a volume of Injection, appropriately diluted, such that it provides a count rate of about 20,000 counts per minute, about 25 mm from one end of a 25- × 300-mm strip of chromatographic paper (see *Chromatography* ⟨621⟩). Develop the chromatogram over a suitable period of time by ascending chromatography, using a mixture of acetone and 2 *N* hydrochloric acid (80:20). Allow the chromatogram to air-dry. Determine the radioactivity distribution by scanning the chromatogram with a suitable collimated radiation detector. The radioactivity of the pertechnetate band is not less than 95% of the total radioactivity in the test specimen. The R_f value for the pertechnetate band (approximately 0.9) falls within ±10.0% of the value found for a known sodium pertechnetate Tc 99m specimen when determined under identical conditions.

Radionuclidic purity—Using a suitable counting assembly (see *Selection of a Counting Assembly* under *Radioactivity* ⟨821⟩), determine the radioactivity of each radionuclidic impurity, in kBq per MBq (µCi per mCi) of technetium 99m, in the Injection by use of a calibrated system as directed under *Radioactivity* ⟨821⟩.

For Injection prepared from technetium 99m derived from parent molybdenum 99 formed as a result of neutron bombardment of stable molybdenum—

MOLYBDENUM 99—The presence of molybdenum 99 in the Injection is shown by its characteristic gamma-ray spectrum. The most prominent photopeaks of this radioactive nuclide have energies of 0.181, 0.740, and 0.780 MeV. Molybdenum 99 decays with a radioactive half-life of 66.0 hours. The amount of molybdenum 99 is not greater than 0.15 kBq per MBq (0.15 μCi per mCi) of technetium 99m per administered dose in the Injection, at the time of administration.

OTHER GAMMA-EMITTING RADIONUCLIDIC IMPURITIES—The total amount of other gamma-emitting radionuclidic impurities does not exceed 0.5 kBq per MBq (0.5 μCi per mCi) of technetium 99m, and does not exceed 92 kBq (2.5 μCi) per administered dose of the Injection, at the time of administration.

For Injection prepared from technetium 99m derived from parent molybdenum 99 formed as a result of uranium fission—Gamma- and beta-emitting impurities—

MOLYBDENUM 99—The Injection meets the requirements set forth for the Injection prepared by neutron irradiation of stable molybdenum (see foregoing).

IODINE 131—The most prominent photopeak of this radioactive nuclide has an energy of 0.364 MeV. Iodine 131 decays with a radioactive half-life of 8.08 days. The concentration of iodine 131 is not more than 0.05 kBq per MBq (0.05 μCi per mCi) of technetium 99m, at the time of administration.

RUTHENIUM 103—The most prominent photopeak of this radioactive nuclide has an energy of 0.497 MeV. Ruthenium 103 decays with a radioactive half-life of 39.5 days. The concentration of ruthenium 103 is not more than 0.05 kBq per MBq (0.05 μCi per mCi) of technetium 99m, at the time of administration.

STRONTIUM 89—Determine the presence of strontium 89 in the Injection by a counting system appropriate for the detection of particulate radiations. Strontium 89 decays by a beta emission with a maximum energy of 1.463 MeV, and a radioactive half-life of 52.7 days. Strontium 89 may be present in a concentration of not more than 0.0006 kBq per MBq (0.0006 μCi per mCi) of technetium 99m, at the time of administration.

STRONTIUM 90—Determine the presence of strontium 90 in the Injection by a counting system appropriate for the detection of particulate radiations. Strontium 90 decays by a beta emission with a maximum energy of 0.546 MeV, and a radioactive half-life of 27.7 years. Strontium 90 may be present in a concentration of not more than 0.00006 kBq per MBq (0.00006 μCi per mCi) of technetium 99m, at the time of administration.

ALL OTHER RADIONUCLIDIC IMPURITIES—Not more than 0.01% of all other beta and gamma emitters is present at the time of administration. Not more than 0.001 Bq of gross alpha impurity per 1 MBq (or 0.001 nCi of gross alpha impurity per 1 mCi) of technetium 99m is present at the time of administration.

Chemical purity—

Aluminum (To be determined if separation is accomplished by an alumina column in the preparation of the Injection)—

ALUMINUM STANDARD SOLUTION—Dissolve 35.17 mg, accurately weighed, of aluminum potassium sulfate dodecahydrate in water to make 1000.0 mL. Each mL of this solution contains 2 μg of Al.

PROCEDURE—Pipet 10 mL of *Aluminum standard solution* into each of two 50-mL volumetric flasks. To each flask add 3 drops of methyl orange TS and 2 drops of 6 N ammonium hydroxide, then add 0.5 N hydrochloric acid, dropwise, until the solution turns red. To one flask add 25 mL of sodium thioglycolate TS, and to the other flask add 1 mL of disodium ethylenediaminetetraacetate TS. To each flask add 5 mL of eriochrome cyanine TS and 5 mL of acetate buffer TS, and add water to volume. Immediately determine the absorbance of the solution containing sodium thioglycolate TS at the wavelength of maximum absorbance at about 535 nm, with a suitable spectrophotometer, using the solution containing the disodium ethylenediaminetetraacetate TS as a blank. Repeat the procedure using two 1.0-mL aliquots of Sodium Pertechnetate Tc 99m Injection. Calculate the quantity, in μg per mL, of aluminum in the Injection taken by the formula:

$$20(T_U/T_S),$$

in which T_U and T_S are the absorbances of the solution from the Injection and the solution containing the aluminum standard, respectively. The concentration of aluminum ion in the Injection is not greater than 10 μg per mL.

Methyl ethyl ketone (To be determined if separation is accomplished by liquid-liquid extraction in the preparation of the Injection)—Place 1.0 mL of the Injection in a suitable container, and dilute with water to 20.0 mL. Add 2.0 mL of 1 N sodium hydroxide, mix, then add 2.0 mL of 0.1 N iodine, dropwise, and again mix. At the same time, prepare a standard by placing 1.0 mL of a solution of methyl ethyl ketone (1 in 1000) in a similar container and diluting with water to 20.0 mL. Add 2.0 mL of 1 N sodium hydroxide, mix, then add 2.0 mL of 0.1 N iodine, dropwise, and again mix. After 2 minutes, the turbidity of the test specimen does not exceed that of the standard (0.1%).

Other requirements—It meets the requirements under *Injections* ⟨1⟩, except that the Injection may be distributed or dispensed prior to the completion of the test for *Sterility*, the latter test being started on the day of manufacture, and except that it is not subject to the recommendation on *Volume in Container*.

Assay for radioactivity—Using a suitable counting assembly (see *Selection of a Counting Assembly* under *Radioactivity* ⟨821⟩), determine the radioactivity, in MBq (mCi) per mL, in Sodium Pertechnetate Tc 99m Injection by use of a calibrated system as directed under *Radioactivity* ⟨821⟩.

Technetium Tc 99m Pyrophosphate Injection

» Technetium Tc 99m Pyrophosphate Injection is a sterile, aqueous solution, suitable for intravenous administration, of pyrophosphate that is labeled with ^{99m}Tc. It contains not less than 90.0 percent and not more than 110.0 percent of the labeled amount of ^{99m}Tc as pyrophosphate expressed in megabecquerels (microcuries or millicuries) per mL at the time indicated in the labeling. It may contain antimicrobial agents, buffers, reducing agents, and stabilizers. Other chemical forms of radioactivity do not exceed 10.0 percent of the total radioactivity.

Packaging and storage—Preserve in single-dose or in multiple-dose containers, at a temperature between 2° and 8°.

Labeling—Label it to include the following, in addition to the information specified for *Labeling* under *Injections* ⟨1⟩: the time and date of calibration; the amount of ^{99m}Tc as labeled tetrasodium pyrophosphate expressed as total megabecquerels (microcuries or millicuries) and concentration as megabecquerels (microcuries or millicuries) per mL at the time of calibration; the expiration date and time; and the statement, "Caution—Radioactive Material." The labeling indicates that in making dosage calculations, correction is to be made for radioactive decay, and also indicates that the radioactive half-life of ^{99m}Tc is 6.0 hours.

Reference standard—*USP Endotoxin Reference Standard.*

Bacterial endotoxins—It meets the requirements of the *Bacterial Endotoxins Test* ⟨85⟩, the limit of endotoxin content being not more than 175/V USP Endotoxin Unit per mL of the Injection, when compared with the USP Endotoxin RS, in which V is the maximum recommended total dose, in mL, at the expiration date or time.

pH ⟨791⟩: between 4.0 and 7.5.

Radiochemical purity—The determination of radiochemical purity for this product requires the use of two separate chromatography systems.

System A—Under an atmosphere of nitrogen, place a measured volume of Injection, appropriately diluted, such that it provides a count rate of about 20,000 counts per minute, about 20 mm from one end of a thin-layer chromatographic strip impregnated with silica gel (see *Chromatography* ⟨621⟩), and allow to dry. Develop the chromatogram over a suitable period by ascending chromatography, using saline TS, and dry it under nitrogen. Determine the radioactivity distribution by scanning the chromatogram with a suitable collimated radiation detector. Hydrolyzed Tc 99m and technetium–tin colloid are located at the origin (R_f 0 to 0.1).

System B—Proceed as directed for *System A*, except to develop the chromatogram in a mixture of methanol and acetone (1:1). Free pertechnetate is located at the solvent front. The sum of the percentage of radioactivity at the origin in *System A* plus the percentage of radioactivity at the solvent front in *System B* is not greater than 10.0%.

Biological distribution—Inject intravenously between 0.075 MBq and 75 MBq (2 μCi and 2 mCi) of Technetium Tc 99m Pyrophosphate Injection, in a volume not exceeding 0.2 mL, into the caudal or external jugular vein of each of three 175- to 250-g rats. Approximately 1 hour after the injection, sacrifice the animals, and carefully remove the liver, both kidneys, and one femur of each by dissection, freeing the femur from soft tissue. Remove the tail 20 to 30 mm above the injection site, and discard. Place each organ, both kidneys, and the remaining carcass in separate, suitable counting containers, and determine the radioactivity, in counts per minute, in each container with an appropriate detector, using the same counting geometry. Determine the percentage of radioactivity in the liver, kidneys, and femur by the formula:

$$100(A/B),$$

in which A is the net radioactivity, in counts per minute, in the organ, and B is the total radioactivity, in counts per minute, in the liver, kidneys, femur, and carcass. Not more than 5.0% of the total radioactivity is found in the liver or in the kidneys, and not less than 1.0% of the total radioactivity is found in the femur, in not fewer than 2 of the rats.

Other requirements—It meets the requirements of the tests for *Radionuclide identification* and *Radionuclidic purity* under *Sodium Pertechnetate Tc 99m Injection*. It meets also the requirements under *Injections* ⟨1⟩, except that it may be distributed or dispensed prior to completion of the test for *Sterility*, the latter test being started on the day of final manufacture, and except that it is not subject to the recommendation on *Volume in Container*.

Assay for radioactivity—Using a suitable counting assembly (see *Selection of a Counting Assembly* under *Radioactivity* ⟨821⟩), determine the radioactivity, in MBq (μCi) per mL, of Technetium Tc 99m Pyrophosphate Injection by use of a calibrated system as directed under *Radioactivity* ⟨821⟩.

Technetium Tc 99m (Pyro- and trimeta-) Phosphates Injection

» Technetium Tc 99m (Pyro- and trimeta-) Phosphates Injection is a sterile, aqueous solution, suitable for intravenous administration, composed of sodium pyrophosphate, sodium trimetaphosphate, and stannous chloride labeled with radioactive Tc 99m. It contains not less than 90.0 percent and not more than 110.0 percent of the labeled amount of ^{99m}Tc as phosphate expressed in megabecquerels (microcuries or millicuries) per mL at the time indicated in the labeling. Other chemical forms of radioactivity do not exceed 10.0 percent of the total radioactivity.

Reference standard—*USP Endotoxin Reference Standard.*

pH ⟨791⟩: between 4.0 and 7.0.

Radiochemical purity—Place a measured volume of Technetium Tc 99m (Pyro- and trimeta-) Phosphates Injection, appropriately diluted, such that it provides a count rate of about 20,000 counts per minute, about 10 mm from one end of a 10- × 50-mm strip of chromatographic paper (see *Chromatography* ⟨621⟩), and immediately develop the chromatogram over a period of 3 to 5 minutes by ascending chromatography, using saline TS, and air-dry. Determine the radioactivity distribution by scanning with a suitable collimated radiation detector. Not more than 10.0% of the total radioactivity is found at the point of application. Similarly prepare a second strip, and thoroughly dry the spot with nitrogen. Develop the chromatogram over a period of 35 to 45 seconds by ascending chromatography, using freshly distilled methyl ethyl ketone, and allow to dry. Determine the radioactivity distribution as previously directed. Not less than 90.0% of the total radioactivity is found at the point of application.

Other requirements—It meets the requirements of the tests for *Radionuclide identification* and *Radionuclidic purity* under *Sodium Pertechnetate Tc 99m Injection*, and the requirements of the tests for *Packaging and storage*, *Labeling*, *Bacterial endotoxins*, *Biological distribution*, and *Assay for radioactivity* under *Technetium Tc 99m Pyrophosphate Injection*. It meets the requirements under *Injections* ⟨1⟩, except that it may be distributed or dispensed prior to completion of the test for *Sterility*, the latter test being started on the day of manufacture, and except that it is not subject to the recommendation on *Volume in Container*.

Technetium Tc 99m Succimer Injection

meso-2,3-Dimercaptosuccinic acid, ^{99m}Tc complex.

» Technetium Tc 99m Succimer Injection is a sterile, clear, colorless, aqueous solution of succimer complexed with ^{99m}Tc. It is suitable for intravenous administration. It contains not less than 85.0 percent of the labeled amount of ^{99m}Tc as the succimer complex expressed in megabecquerels (microcuries or millicuries) per mL at the time indicated in the labeling. It may contain reducing agents. Other chemical forms of radioactivity do not exceed 15.0 percent of the total radioactivity.

Packaging and storage—Preserve in single-dose containers, at a temperature between 15° and 30°. Do not freeze or store above 30°. Protect from light.

Labeling—Label it to include the following, in addition to the information specified for *Labeling* under *Injections* ⟨1⟩: the time and date of calibration; the amount of ^{99m}Tc as labeled succimer expressed as total megabecquerels (microcuries or millicuries) and concentration as megabecquerels (microcuries or millicuries) per mL at the time of calibration; the expiration date and time; and the statement, "Caution—Radioactive Material." The labeling indicates that in making dosage calculations, correction is to be made for radioactive decay, and also indicates that the radioactive half-life of ^{99m}Tc is 6.0 hours. In addition, the labeling states that it is not to be used if discoloration or particulate matter is observed. [NOTE—A beyond-use time of 30 minutes shall be stated on the label upon constitution with Sodium Pertechnetate Tc 99m Injection.]

Reference standard—*USP Endotoxin Reference Standard.*

Bacterial endotoxins—It meets the requirements of the *Bacterial Endotoxins Test* ⟨85⟩, the limit of endotoxin content being not more than $175/V$ USP Endotoxin Unit per mL of the Injection, when compared with the USP Endotoxin RS, in which V is the maximum recommended total dose, in mL, at the expiration date or time.

pH ⟨791⟩: between 2.0 and 3.0.

Radiochemical purity—Activate a 65- × 95-mm silicic acid thin-layer chromatographic plate by heating at 100–110° for 30 minutes. Cool over silica gel and immediately apply 1 μL of Technetium Tc 99m Succimer Injection, appropriately diluted, if necessary, to a radioactive concentration of 18.5 to 370 MBq (0.5 to 10 mCi) per mL, about 17 mm from one end of the chromatographic plate, and allow to dry. Develop the chromatogram over a period of about 30 to 45 minutes by ascending chromatography, using *n*-butanol saturated with 0.3 *N* hydrochloric acid, and air-dry. Determine the radioactive distribution by scanning the chromatogram with a suitable collimated radiation detector. Not less than 85% of the total radioactivity is found as succimer at an R_f between 0.45 and 0.70. Hydrolyzed ^{99m}Tc is located at the origin (R_f 0 to 0.15) and the unbound ^{99m}Tc is located at the solvent front (R_f 1.0).

Biological distribution—Inject intravenously between 3.7 MBq and 92.5 MBq (100 μCi and 2500 μCi) of Technetium Tc 99m Succimer Injection, in a volume of 0.2 to 0.25 mL, into the caudal vein of each of three 125-g to 225-g anesthetized Sprague-Dawley female rats. Clamp the opening of the urethra with a hemostat. Sacrifice the animals 1 hour after the injection, and carefully remove the kidneys, bladder, *and* liver and spleen of each as *three* separate organs by dissections. Place each organ and the remaining carcass (excluding the tail) in separate, suitable counting containers, and determine the radioactivity, in counts per minute, in each container with an appropriate detector, using the same counting geometry. Determine the percentage of administered radioactive dose in each organ: not less than 40% of the administered radioactive dose is found in the kidneys and a ratio of not less than 6:1 of the administered dose is found in the ratio kidneys/(liver and spleen), in not fewer than two of the animals.

Other requirements—It meets the requirements of the tests for *Radionuclide identification* and *Radionuclidic purity* under *Sodium Pertechnetate Tc 99m Injection.* It meets also the requirements under *Injections* ⟨1⟩, except that it may be distributed or dispensed prior to completion of the test for *Sterility*, the latter test being started on the day of final manufacture, and except that it is not subject to the recommendation on *Volume in Container.*

Assay for radioactivity—Using a suitable counting assembly (see *Selection of a Counting Assembly* under *Radioactivity* ⟨821⟩), determine the radioactivity, in MBq (μCi) per mL, of Technetium Tc 99m Succimer Injection by use of a calibrated system as directed under *Radioactivity* ⟨821⟩.

Technetium Tc 99m Sulfur Colloid Injection

Sulfur, colloidal, metastable technetium-99 labeled [*7704-34-9*].

» Technetium Tc 99m Sulfur Colloid Injection is a sterile, colloidal dispersion of sulfur labeled with radioactive ^{99m}Tc, suitable for intravenous administration.

Technetium Tc 99m Sulfur Colloid Injection contains not less than 90.0 percent and not more than 110.0 percent of the labeled concentration of ^{99m}Tc as sulfur colloid expressed in megabecquerels (microcuries or millicuries) per mL at the time indicated in the labeling. It may contain chelating agents, buffers, and stabilizing agents. Other chemical forms of radioactivity do not exceed 8 percent of the total radioactivity.

Note—Agitate the container before withdrawing the Injection into a syringe.

Packaging and storage—Store in single-dose or in multiple-dose containers.

Labeling—Label it to include the following, in addition to the information specified for *Labeling* under *Injections* ⟨1⟩: the time and date of calibration; the amount of ^{99m}Tc as sulfur colloid expressed as total megabecquerels (microcuries or millicuries) and concentration as megabecquerels (microcuries or millicuries) per mL at the time of calibration; the expiration date; and the statement, "Caution—Radioactive Material." The labeling indicates that in making dosage calculations, correction is to be made for radioactive decay, and also indicates that the radioactive half-life of ^{99m}Tc is 6.0 hours; in addition, the labeling states that it is not to be used if flocculent material is visible and directs that the container be agitated before the Injection is withdrawn into a syringe.

Reference standard—*USP Endotoxin Reference Standard.*

Radionuclide identification (see *Radioactivity* ⟨821⟩)—It meets the requirements of the test for *Radionuclide identification* under *Sodium Pertechnetate Tc 99m Injection.*

Bacterial endotoxins—It meets the requirements of the *Bacterial Endotoxins Test* ⟨85⟩, the limit of endotoxin content being not more than 175/*V* USP Endotoxin Unit per mL of the Injection, when compared with the USP Endotoxin RS, in which *V* is the maximum recommended total dose, in mL, at the expiration date or time.

pH ⟨791⟩: between 4.5 and 7.5.

Radionuclidic purity—It meets the requirements of the test for *Radionuclidic purity* under *Sodium Pertechnetate Tc 99m Injection.*

Radiochemical purity—Place a measured volume of Injection, appropriately diluted, such that it provides a count rate of about 20,000 counts per minute, about 25 mm from one end of a 25- × 300-mm strip of chromatographic paper (see *Chromatography* ⟨621⟩), and allow to dry. Develop the chromatogram over a suitable period by ascending chromatography, using dilute methanol (8.5 in 10), and air-dry. Determine the radioactivity distribution by scanning the chromatogram with a suitable collimated radiation detector. Not less than 92% of the total radioactivity is found as sulfur colloid (at the point of application).

Biological distribution—Inject intravenously between 0.075 MBq and 0.75 MBq (2 μCi and 20 μCi) of the Injection in a volume not exceeding 0.2 mL into the caudal vein of each of three 20-g to 25-g mice. Ten to 30 minutes after the injection, sacrifice the animals, and carefully remove the liver and lungs of each by dissection. Place each organ and remaining carcass in separate, suitable counting containers, and determine the radioactivity, in counts per minute, in each container in an appropriate scintillation well counter, using the same counting geometry. Determine the percentage of radioactivity in the liver and the lungs by the formula:

$$100(A/B),$$

in which *A* is the net radioactivity, in counts per minute, in the organ, and *B* is the total radioactivity, in counts per minute, in the lungs, liver, and carcass. Not less than 80% of the radioactivity is found in the liver and not more than 5% of the radioactivity is found in the lungs, in two of the mice.

Other requirements—It meets the requirements under *Injections* ⟨1⟩, except that the Injection may be distributed or dispensed prior to completion of the test for *Sterility*, the latter test being started on the day of final manufacture, and except that it is not subject to the recommendation on *Volume in Container.*

Assay for radioactivity—Using a suitable counting assembly (see *Selection of a Counting Assembly* under *Radioactivity* ⟨821⟩), determine the radioactivity, in MBq (μCi) per mL, of Technetium Tc 99m Sulfur Colloid Injection by use of a calibrated system as directed under *Radioactivity* ⟨821⟩.

Terbutaline Sulfate

$$\left[(HO)_2C_6H_3-CH(OH)CH_2NHC(CH_3)_3 \right]_2 \cdot H_2SO_4$$

$(C_{12}H_{19}NO_3)_2 . H_2SO_4$ 548.65

1,3-Benzenediol, 5-[2-[(1,1-dimethylethyl)amino]-1-hydroxyethyl]-, sulfate (2:1) (salt).

α-[(*tert*-Butylamino)methyl]-3,5-dihydroxybenzyl alcohol sulfate (2:1) (salt) [*23031-32-5*].

» Terbutaline Sulfate contains not less than 98.0 percent and not more than 101.0 percent of $(C_{12}H_{19}NO_3)_2 . H_2SO_4$, calculated on the dried basis.

Packaging and storage—Preserve in well-closed, light-resistant containers, at controlled room temperature.

Reference standard—*USP Terbutaline Sulfate Reference Standard*—Dry at 105° for 3 hours before using.

Identification—

A: The infrared absorption spectrum of a potassium bromide dispersion of it, previously dried, exhibits maxima only at the same wavelengths as that of a similar preparation of USP Terbutaline Sulfate RS.

B: The ultraviolet absorption spectrum of a 1 in 10,000 solution in 0.1 *N* hydrochloric acid exhibits maxima and minima at the same wavelengths as that of a similar preparation of USP Terbutaline Sulfate RS, concomitantly measured.

Acidity—Dissolve 0.20 g in 10 mL of carbon dioxide–free water, and titrate with 0.020 *N* sodium hydroxide from a microburet to a pH of about 6, determining the end-point potentiometrically, using a calomel-glass electrode system: not more than 0.50 mL of 0.020 *N* sodium hydroxide is required (0.3% as acetic acid).

Loss on drying ⟨731⟩—Dry it at 105° for 3 hours: it loses not more than 0.5% of its weight.

Residue on ignition ⟨281⟩: not more than 0.2%.

Heavy metals, *Method II* ⟨231⟩: 0.0025%.

3,5-Dihydroxy-ω-*tert*-butylaminoacetophenone sulfate—Its absorbance in 1-cm cells at 330 nm, determined in 0.01 *N* hydrochloric acid containing 20 mg per mL, is not more than 0.47.

Assay—Dissolve, by warming, about 0.4 g of Terbutaline Sulfate, accurately weighed, in 60 mL of glacial acetic acid. Cool to room temperature, add 60 mL of acetonitrile, and titrate with 0.1 *N* perchloric acid VS, determining the end-point potentiometrically, using a glass electrode and a calomel electrode containing lithium chloride in glacial acetic acid (see *Titrimetry* ⟨541⟩). Perform a blank determination, and make any necessary correction. Each mL of 0.1 *N* perchloric acid is equivalent to 54.87 mg of $(C_{12}H_{19}NO_3)_2 . H_2SO_4$.

4-Aminoantipyrine solution—On the day of use, prepare a solution of 4-aminoantipyrine in water having a concentration of 20 mg per mL.

Potassium ferricyanide solution—On the day of use, prepare a solution of potassium ferricyanide in water having a concentration of 80 mg per mL.

Standard preparation—Dissolve an accurately weighed quantity of USP Terbutaline Sulfate RS in water, and dilute quantitatively and stepwise, if necessary, with water to obtain a solution having a known concentration of about 100 µg per mL.

Assay preparation—Transfer an accurately measured volume of Terbutaline Sulfate Injection, equivalent to about 5 mg of terbutaline sulfate, to a 50-mL volumetric flask, dilute with water to volume, and mix.

Procedure—Transfer 5.0 mL each of the *Standard preparation*, the *Assay preparation*, and water to provide a blank, to separate 50-mL volumetric flasks. To each flask add 35 mL of *pH 9.5 buffer* and 1.0 mL of *4-Aminoantipyrine solution*, and mix. Add 1.0 mL of *Potassium ferricyanide solution* while vigorously swirling the flask, dilute with *pH 9.5 buffer* to volume, and mix. Seventy-five seconds, accurately timed, following the addition of the *Potassium ferricyanide solution*, determine the absorbances of the solutions in 1-cm cells relative to the blank at the wavelength of maximum absorbance at about 550 nm, with a suitable spectrophotometer. Calculate the quantity, in mg, of $(C_{12}H_{19}NO_3)_2 . H_2SO_4$ in each mL of the Injection taken by the formula:

$$0.05C/V(A_U/A_S),$$

in which C is the concentration, in µg per mL, of USP Terbutaline Sulfate RS in the *Standard preparation*, V is the volume, in mL, of Injection taken, and A_U and A_S are the absorbances of the solutions from the *Assay preparation* and the *Standard preparation*, respectively.

Terbutaline Sulfate Injection

» Terbutaline Sulfate Injection is a sterile solution of Terbutaline Sulfate in Water for Injection. It contains not less than 90.0 percent and not more than 110.0 percent of the labeled amount of $(C_{12}H_{19}NO_3)_2 . H_2SO_4$.

Note—Do not use the Injection if it is discolored.

Packaging and storage—Preserve in single-dose containers, preferably of Type I glass, protected from light, at controlled room temperature.

Reference standard—*USP Terbutaline Sulfate Reference Standard*—Dry at 105° for 3 hours before using.

Identification—On a suitable thin-layer chromatographic plate (see *Chromatography* ⟨621⟩), coated with a 0.25-mm layer of chromatographic silica gel, apply 2 µL of Injection and 2 µL of a solution of USP Terbutaline Sulfate RS in sodium chloride solution (0.9 in 100) containing 1 mg per mL. Develop the chromatogram in a solvent system consisting of a mixture of isopropyl alcohol, cyclohexane, and formic acid (13:5:1) until the solvent front has moved about three-fourths of the length of the plate. Remove the plate from the developing chamber, mark the solvent front, and dry with a current of air. Spray the plate with a 1 in 50 solution of 4-aminoantipyrine in methanol, allow to air-dry, and spray with a 2 in 25 solution of potassium ferricyanide in a solvent prepared by mixing ammonium hydroxide with water (4:1): the R_f value of the principal spot obtained from the Injection corresponds to that obtained from the Standard solution.

pH ⟨791⟩: between 3.0 and 5.0.

Other requirements—It meets the requirements under Injections ⟨1⟩.

Assay—

pH 9.5 buffer—Dissolve 36.3 g of tris(hydroxymethyl)aminomethane in water to make 1000 mL of solution. Adjust the solution with 1 *N* hydrochloric acid to a pH of 9.5 ± 0.1.

Terbutaline Sulfate Tablets

» Terbutaline Sulfate Tablets contain not less than 90.0 percent and not more than 110.0 percent of the labeled amount of $(C_{12}H_{19}NO_3)_2 . H_2SO_4$.

Packaging and storage—Preserve in tight containers, at controlled room temperature.

Reference standard—*USP Terbutaline Sulfate Reference Standard*—Dry at 105° for 3 hours before using.

Identification—Place a quantity of finely powdered Tablets, equivalent to 10 mg of terbutaline sulfate, in a centrifuge tube, add 1 mL of a mixture of alcohol and water (1:1), shake for 10 minutes, and centrifuge. On a suitable thin-layer chromatographic plate (see *Chromatography* ⟨621⟩), coated with a 0.25-mm layer of chromatographic silica gel, apply 2 µL of this solution and 2 µL of a Standard solution of USP Terbutaline Sulfate RS in a mixture of alcohol and water (1:1) containing 10 mg per mL, and allow the spots to dry. Proceed as directed in the *Identification test* under *Terbutaline Sulfate Injection*, beginning with "Develop the chromatogram": the R_f value of the principal spot obtained from the test solution corresponds to that obtained from the Standard solution.

Dissolution ⟨711⟩—

Medium: water; 900 mL.

Apparatus 1: 100 rpm.

Time: 45 minutes.

Procedure—Determine the amount of $(C_{12}H_{19}NO_3)_2 . H_2SO_4$ dissolved, employing the procedure set forth in the *Assay*, making any necessary modifications.

Tolerances—Not less than 75% (*Q*) of the labeled amount of $(C_{12}H_{19}NO_3)_2 . H_2SO_4$ is dissolved in 45 minutes.

Uniformity of dosage units ⟨905⟩: meet the requirements.

Assay—

pH 9.5 buffer, 4-Aminoantipyrine solution, and *Potassium ferricyanide solution*—Prepare as directed in the *Assay* under *Terbutaline Sulfate Injection*.

Standard preparation—Dissolve an accurately weighed quantity of USP Terbutaline Sulfate RS in 0.01 *N* hydrochloric acid, and dilute quantitatively and stepwise, if necessary, with the same solvent to obtain a solution having a known concentration of about 100 µg per mL.

Assay preparation—Weigh and finely powder not less than 20 Terbutaline Sulfate Tablets. Transfer an accurately weighed portion of the powder, equivalent to about 5 mg of terbutaline sulfate, to a 50-mL volumetric flask, add 30 mL of 0.01 *N* hydrochloric acid, insert the stopper, and shake by mechanical means for 10 minutes. Add 0.01 *N* hydrochloric acid to volume, mix, and filter.

Procedure—Transfer 5.0 mL each of the *Standard preparation*, the *Assay preparation*, and 0.01 *N* hydrochloric acid to provide a blank, to separate 50-mL volumetric flasks. Proceed as directed for *Procedure* in the *Assay* under *Terbutaline Sulfate Injection*, beginning with "To each flask add 35 mL of *pH 9.5 buffer*." Calculate the quantity, in mg, of $(C_{12}H_{19}NO_3)_2 \cdot H_2SO_4$ in the portion of Tablets taken by the formula:

$$0.05C(A_U/A_S),$$

in which *C* is the concentration, in µg per mL, of USP Terbutaline Sulfate RS in the *Standard preparation*, and A_U and A_S are the absorbances of the solutions from the *Assay preparation* and the *Standard preparation*, respectively.

Terpin Hydrate

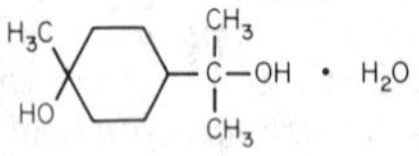

$C_{10}H_{20}O_2 \cdot H_2O$ 190.28

Cyclohexanemethanol, 4-hydroxy-α,α,4-trimethyl-, monohydrate.

p-Menthane-1,8-diol monohydrate [*2451-01-6*].

Anhydrous 172.27 [*80-53-5*].

» Terpin Hydrate contains not less than 98.0 percent and not more than 100.5 percent of $C_{10}H_{20}O_2$, calculated on the anhydrous basis.

Packaging and storage—Preserve in tight containers.

Reference standard—*USP Terpin Hydrate Reference Standard*—Do not dry; determine the water content at the time of use.

Identification—

A: The infrared absorption spectrum of a potassium bromide dispersion of it, previously dried over a suitable desiccant for 4 hours, exhibits maxima only at the same wavelengths as that of a similar preparation of USP Terpin Hydrate RS.

B: Add a few drops of sulfuric acid to a hot solution: the liquid becomes turbid and develops a strongly aromatic odor.

Water, *Method I* ⟨921⟩: between 9.0% and 10.0%.

Residue on ignition ⟨281⟩: not more than 0.1%.

Residual turpentine—It has no odor of turpentine oil.

Assay—

Internal standard solution—Dissolve a quantity of biphenyl in chloroform to obtain a solution containing about 20 mg per mL.

Standard preparation—Transfer about 170 mg of USP Terpin Hydrate RS, accurately weighed, to a 100-mL volumetric flask, dissolve in 5 mL of alcohol, add 5.00 mL of *Internal standard solution*, dilute with chloroform to volume, and mix.

Assay preparation—Transfer about 170 mg of Terpin Hydrate, accurately weighed, to a 100-mL volumetric flask, and proceed as directed under *Standard preparation*, beginning with "dissolve in 5 mL of alcohol."

Chromatographic system—Under typical conditions, the gas chromatograph is equipped with a flame-ionization detector, and it contains a 1.2-m × 3.5-mm (ID) glass column packed with 6 percent G1 on support S1A; the injection port and the detector are maintained at 260°, and the column at 120°; and nitrogen is used as the carrier gas, at a flow rate necessary to yield approximate retention times of 7 minutes for terpin and 11 minutes for biphenyl.

System suitability test—The chromatogram obtained for the *Standard preparation* as directed in the *Procedure* exhibits a resolution factor, *R* (see *Chromatography* ⟨621⟩), of not less than 2.0 between the peaks for terpin and for biphenyl.

Procedure—Inject about 1 µL of the *Standard preparation* into a suitable gas chromatograph, and record the chromatogram. Similarly, inject about 1 µL of the *Assay preparation*, and record the chromatogram. Calculate the quantity, in mg, of $C_{10}H_{20}O_2$ in the portion of Terpin Hydrate taken by the formula:

$$W_S(R_U/R_S),$$

in which W_S is the weight, in mg, of USP Terpin Hydrate RS, calculated on the anhydrous basis, and R_U and R_S are the area-ratios of terpin to internal standard obtained from the chromatograms for the *Assay preparation* and the *Standard preparation*, respectively.

Terpin Hydrate Elixir

» Terpin Hydrate Elixir contains, in each 100 mL, not less than 1.53 g and not more than 1.87 g of $C_{10}H_{20}O_2 \cdot H_2O$.

Packaging and storage—Preserve in tight containers.

Reference standard—*USP Terpin Hydrate Reference Standard*—Do not dry; determine the water content at the time of use.

Alcohol content, *Method II* ⟨611⟩: between 39.0% and 44.0% of C_2H_5OH.

Assay—

Internal standard solution, Standard preparation, Chromatographic system, and *System suitability*—Proceed as directed in the *Assay* under *Terpin Hydrate*.

Assay preparation—Pipet 10 mL of Terpin Hydrate Elixir into a separator, add 20 mL of water and 10 mL of 5 *N* sodium hydroxide, and extract with three 25-mL portions of chloroform, filtering each, successively, through cotton. Rinse the cotton with chloroform. To the combined rinse and extracts add 5.00 mL of *Internal standard solution*, and mix.

Procedure—Inject about 1 µL of the *Standard preparation* into a suitable gas chromatograph, and record the chromatogram. Similarly, inject about 1 µL of the *Assay preparation*, and record the chromatogram. Calculate the quantity, in mg, of $C_{10}H_{20}O_2 \cdot H_2O$ in each mL of the Elixir taken by the formula:

$$0.1(190.28/172.27)W_S(R_U/R_S),$$

in which 190.28 and 172.27 are the molecular weights of terpin hydrate ($C_{10}H_{20}O_2 \cdot H_2O$) and anhydrous terpin ($C_{10}H_{20}O_2$), respectively, W_S is the weight, in mg, of USP Terpin Hydrate RS, calculated on the anhydrous basis, and R_U and R_S are the area-ratios of terpin to biphenyl obtained from the chromatograms for the *Assay preparation* and the *Standard preparation*, respectively.

Terpin Hydrate and Codeine Elixir

» Terpin Hydrate and Codeine Elixir contains, in each 100 mL, not less than 1.53 g and not more than 1.87 g of $C_{10}H_{20}O_2 \cdot H_2O$ (terpin hydrate), and not less than 180 mg and not more than 220 mg of $C_{18}H_{21}NO_3 \cdot H_2O$ (codeine).

Packaging and storage—Preserve in tight containers.

Reference standards—*USP Terpin Hydrate Reference Standard*—Do not dry; determine the water content at the time of use. *USP Codeine Phosphate Reference Standard*—This is the hemihydrate form of codeine phosphate. Dry at 105° for 18 hours before using, to obtain anhydrous codeine phosphate.

Identification—Transfer 1 mL of Elixir to a 125-mL separator containing 10 mL of dilute hydrochloric acid (1 in 120), extract with two 20-mL portions of chloroform, and discard the chloroform extracts. Add 3 mL of sodium hydroxide solution (1 in 10), extract with 20 mL of chloroform, filter the chloroform extract through anhydrous granular sodium sulfate into a porcelain dish, and evaporate on a steam bath to dryness. To the residue add 1 mL of sulfuric acid containing 2 drops of formaldehyde TS: an intense violet-blue color is produced (*presence of codeine*).

Alcohol content, *Method II* ⟨611⟩: between 39.0% and 44.0% of C_2H_5OH.

Assay for terpin hydrate—

Internal standard solution—Prepare a chloroform solution containing 20 mg of biphenyl and 2.6 mg of *N*-phenylcarbazole in each mL.

Standard preparation—Transfer about 26 mg of USP Codeine Phosphate RS and about 170 mg of USP Terpin Hydrate RS, both accurately weighed, to a separator, add 5 mL of alcohol, shake to dissolve the terpin hydrate, add 25 mL of water to dissolve the codeine phosphate, add 10 mL of 5 *N* sodium hydroxide, and extract with three 25-mL portions of chloroform, filtering each, successively, through cotton. Rinse the cotton with chloroform. To the combined rinse and extracts add 5.00 mL of *Internal standard solution*, and mix.

Assay preparation—Pipet 10 mL of Terpin Hydrate and Codeine Elixir into a separator, add 20 mL of water and 10 mL of 5 *N* sodium hydroxide, and extract with three 25-mL portions of chloroform, filtering each, successively, through cotton. Rinse the cotton with chloroform. To the combined rinse and extracts add 5.00 mL of *Internal standard solution*, and mix.

Chromatographic system and *System suitability*—Proceed as directed in the *Assay* under *Terpin Hydrate*. [NOTE—Heat the column to 230° to remove the *N*-phenylcarbazole and codeine from prior injections.]

Procedure—Inject about 1 µL of the *Standard preparation* into a suitable gas chromatograph, and record the chromatogram. Similarly, inject about 1 µL of the *Assay preparation*, and record the chromatogram. Calculate the quantity, in mg, of $C_{10}H_{20}O_2 . H_2O$ in each mL of the Elixir taken by the formula:

$$0.1(190.28/172.27)W_S(R_U/R_S),$$

in which 190.28 and 172.27 are the molecular weights of terpin hydrate ($C_{10}H_{20}O_2 . H_2O$) and anhydrous terpin ($C_{10}H_{20}O_2$), respectively, W_S is the weight, in mg, of USP Terpin Hydrate RS, calculated on the anhydrous basis, and R_U and R_S are the area-ratios of terpin to biphenyl obtained from the chromatograms for the *Assay preparation* and the *Standard preparation*, respectively.

Assay for codeine—

Internal standard solution—Prepare as directed under *Assay for terpin hydrate*.

Standard preparation—Evaporate the remaining *Standard preparation* for Terpin Hydrate and Codeine Elixir from the *Assay for terpin hydrate* nearly to dryness, and dissolve the residue in about 20 mL of chloroform.

Assay preparation—Evaporate the remaining *Assay preparation* for Terpin Hydrate and Codeine Elixir from the *Assay for terpin hydrate* nearly to dryness, and dissolve the residue in about 20 mL of chloroform.

Chromatographic system and *System suitability*—Proceed as directed in the *Assay* under *Terpin Hydrate*, except to maintain the temperature of the column at 230° instead of 120°.

Procedure—Proceed as directed under *Assay for terpin hydrate*, except to maintain the temperature of the column at 230° instead of 120°. The retention times for *N*-phenylcarbazole and codeine are about 7 minutes and 10 minutes, respectively. Calculate the quantity, in mg, of $C_{18}H_{21}NO_3 . H_2O$ in each mL of the Elixir taken by the formula:

$$0.1(317.38/397.36)W_S(R_U/R_S),$$

in which 317.38 and 397.36 are the molecular weights of codeine ($C_{18}H_{21}NO_3 . H_2O$) and codeine phosphate ($C_{18}H_{21}NO_3 . H_3PO_4$), respectively, W_S is the weight, in mg, of USP Codeine Phosphate RS, and R_U and R_S are the area-ratios of codeine to *N*-phenylcarbazole obtained from the chromatograms for the *Assay preparation* and the *Standard preparation*, respectively.

Terpin Hydrate and Dextromethorphan Hydrobromide Elixir

» Terpin Hydrate and Dextromethorphan Hydrobromide Elixir contains, in each 100 mL, not less than 1.53 g and not more than 1.87 g of terpin hydrate ($C_{10}H_{20}O_2 . H_2O$), and not less than 180 mg and not more than 220 mg of dextromethorphan hydrobromide ($C_{18}H_{25}NO . HBr . H_2O$).

Packaging and storage—Preserve in tight containers.

Reference standards—*USP Terpin Hydrate Reference Standard*—Do not dry; determine the water content at the time of use. *USP Dextromethorphan Hydrobromide Reference Standard*—Do not dry; determine the *Water* content by *Method I* before using.

Identification—Transfer 1 mL of Elixir to a 125-mL separator containing 20 mL of a saturated solution of sodium sulfate, extract with a 20-mL portion of chloroform, transfer the chloroform extract to a conical flask, and evaporate on a steam bath to dryness. Dissolve the residue in 1 mL of 3 *N* hydrochloric acid, add 20 mL of water, mix, transfer the solution to a separator, add 4 mL of ammonia TS, and extract with a 20-mL portion of solvent hexane. Wash the hexane layer with two 20-mL portions of water, filter the washed hexane solution through anhydrous granular sodium sulfate, and evaporate on a steam bath to dryness. To the residue add 1 mL of sulfuric acid containing 2 drops of formaldehyde TS: a green color is produced (*presence of dextromethorphan hydrobromide*).

Alcohol content, *Method II* ⟨611⟩: between 39.0% and 44.0% of C_2H_5OH, determined by the *Gas-liquid Chromatographic Method*.

Assay for terpin hydrate—

Internal standard solution—Dissolve about 1 g of dodecyl alcohol, accurately weighed, in dehydrated alcohol to make 100.0 mL.

Standard preparation—Transfer about 34 mg of USP Terpin Hydrate RS, accurately weighed, to a 10-mL volumetric flask, and dissolve in about 6 mL of dehydrated alcohol. Add 2.0 mL of *Internal standard solution*, dilute with dehydrated alcohol to volume, and mix.

Assay preparation—Transfer an accurately measured volume of Terpin Hydrate and Dextromethorphan Hydrobromide Elixir, equivalent to about 170 mg of terpin hydrate, to a separator, add 25 mL of a saturated solution of sodium acetate, and shake for 1 minute. Allow to stand for 30 minutes with occasional shaking, and extract with four 25-mL portions of chloroform, filtering each extract through chloroform-moistened cotton into a beaker. Evaporate the combined extracts to a volume of about 12 mL by warming on a steam bath with the aid of a current of air, then complete the evaporation to dryness without heating. Dissolve the residue in 15 mL of dehydrated alcohol, and transfer the solution to a 50-mL volumetric flask containing 10.0 mL of *Internal standard solution*. Wash the beaker with sufficient dehydrated alcohol to bring to volume, and mix.

Calibration—Inject an appropriate volume of the *Standard preparation* into a suitable gas chromatograph equipped with a flame-ionization detector. Under typical conditions, the instrument contains a 1.8-m × 3-mm stainless steel column packed with 10 percent G16 on support S1A. The column is maintained at 170°, the injection port and the detector are maintained at 170°, and dry helium is used as the carrier gas at a flow rate of 40 mL per minute. In the detector, hydrogen is introduced at a rate of 44 mL per minute, and air at a rate of 360 mL per minute. Measure the heights of the dodecyl alcohol and terpin hydrate peaks, and record the values as P_1 and P_2, respectively. Calculate the relative response factor, F, of equal weights of dodecyl alcohol and USP Terpin Hydrate RS by the formula:

$$(P_1/P_2)(W_2/W_1),$$

in which W_2 represents the weight percentage of USP Terpin Hydrate RS in the *Standard preparation*, calculated on the anhydrous basis, and W_1 represents the weight percentage of dodecyl alcohol in the *Standard preparation*.

Procedure—Inject a 0.5-µL portion of the *Assay preparation* into a suitable chromatograph as described under *Calibration*. Measure the heights of the dodecyl alcohol and terpin hydrate peaks, and record the values as p_1 and p_2, respectively. Calculate the quantity, in mg, of $C_{10}H_{20}O_2 . H_2O$ in each mL of the Elixir taken by the formula:

$$(190.28/172.27)(10CF/V)(p_2/p_1),$$

in which 190.28 and 172.27 are the molecular weights of terpin hydrate ($C_{10}H_{20}O_2 . H_2O$) and anhydrous terpin ($C_{10}H_{20}O_2$), respectively, C is the concentration, in mg per mL, of the *Internal standard solution*, F is the response factor as described under *Calibration*, and V is the volume, in mL, of Elixir taken.

Assay for dextromethorphan hydrobromide—

Standard preparation—Dissolve an accurately weighed quantity of USP Dextromethorphan Hydrobromide RS in chloroform, and dilute quantitatively and stepwise with chloroform to obtain a solution having a known concentration of about 100 µg of anhydrous dextromethorphan hydrobromide per mL.

Assay preparation—Transfer a 10.0-mL portion of Terpin Hydrate and Dextromethorphan Hydrobromide Elixir to a 125-mL separator, add 5 mL of 1 *N* hydrochloric acid, and extract the solution with four 25-mL portions of solvent hexane. Wash each hexane extract with 5 mL of water, adding the washings to the aqueous solution. Discard the washed hexane extracts. To the aqueous solution, add 10 mL of 1 *N* sodium hydroxide, and allow to stand for 5 minutes. Extract with six 30-mL portions of chloroform, collecting the extracts in a 200-mL volumetric flask, dilute with chloroform to volume, and mix.

Procedure—Transfer 5.0 mL of the *Standard preparation*, 5.0 mL of the *Assay preparation*, and 5.0 mL of chloroform to provide the blank, into three individual separators. To each separator add 25 mL of pH 5.3 buffer solution (prepared by dissolving 38 g of monobasic sodium phosphate and 3.8 g of dibasic sodium phosphate in sufficient water to make 1000 mL) and 5 mL of sodium bromocresol green solution (prepared by dissolving 100 mg of bromocresol green in 1.4 mL of 0.1 *N* sodium hydroxide and diluting with pH 5.3 buffer solution to 100 mL). Extract the aqueous solutions in the separators with three 25-mL portions of chloroform, combining the extracts in respective 100-mL volumetric flasks. Dilute with chloroform to volume, and mix. Determine the absorbances of the chloroform solutions in 1-cm cells at the wavelength of maximum absorbance at about 420 nm, with a suitable spectrophotometer, using the blank to set the instrument. Calculate the quantity, in mg, of $C_{18}H_{25}NO . HBr . H_2O$ in each mL of the Elixir taken by the formula:

$$(370.33/352.31)(0.2C/V)(A_U/A_S),$$

in which 370.33 and 352.31 are the molecular weights of dextromethorphan hydrobromide monohydrate and anhydrous dextromethorphan hydrobromide, respectively, C is the concentration, in µg per mL, of USP Dextromethorphan Hydrobromide RS on the anhydrous basis in the *Standard preparation*, V is the volume, in mL, of Elixir taken, and A_U and A_S are the absorbances of the solutions from the *Assay preparation* and the *Standard preparation*, respectively.

Test Strip, Glucose Enzymatic—*see* Glucose Enzymatic Test Strip

—*see* Glucose Enzymatic Test Strip

Testolactone

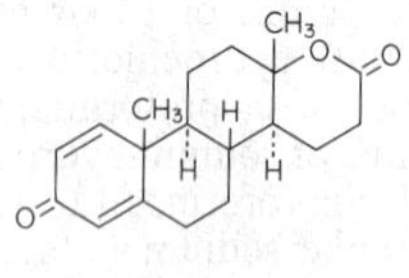

$C_{19}H_{24}O_3$ 300.40

D-Homo-17a-oxaandrosta-1,4-diene-3,17-dione.

13-Hydroxy-3-oxo-13,17-secoandrosta-1,4-dien-17-oic acid δ-lactone [*968-93-4*].

» Testolactone contains not less than 95.0 percent and not more than 105.0 percent of $C_{19}H_{24}O_3$, calculated on the dried basis.

Packaging and storage—Preserve in tight containers.

Reference standard—*USP Testolactone Reference Standard*—Dry in vacuum at 100° for 3 hours before using.

Identification—

A: The infrared absorption spectrum of a potassium bromide dispersion of it, previously dried, exhibits maxima only at the same wavelengths as that of a similar preparation of USP Testolactone RS.

B: The ultraviolet absorption spectrum of a 1 in 100,000 solution in methanol exhibits maxima and minima at the same wavelengths as that of a similar solution of USP Testolactone RS, concomitantly measured.

Specific rotation 〈781〉: between −44° and −52°, calculated on the dried basis, determined in a solution in chloroform containing 125 mg of Testolactone in each 10 mL.

Loss on drying 〈731〉—Dry it in vacuum at 100° for 3 hours: it loses not more than 1.0% of its weight.

Residue on ignition 〈281〉: not more than 0.1%.

Heavy metals, *Method II* 〈231〉: 0.003%.

Chromatographic impurities—

Standard preparation—Transfer 10 mg of USP Testolactone RS, accurately weighed, to a 50-mL volumetric flask, dissolve in acetone, dilute with acetone to volume, and mix.

Test preparation—Transfer 250 mg of Testolactone, accurately weighed, to a 50-mL volumetric flask, dissolve in acetone, dilute with acetone to volume, and mix.

Procedure—Coat a 20- × 20-cm thin-layer chromatographic plate (see *Chromatography* 〈621〉) with a 0.25-mm layer of chromatographic silica gel mixture, dry for 15 minutes at room temperature, heat at 105° for 1 hour, and cool in a desiccator. Divide the area of the plate into three approximately equal sections, the left and right sections to be used for the *Test preparation* and the *Standard preparation*, respectively, and the center section for the blank. Apply 10 µL of the *Standard preparation* and 20 µL of the *Test preparation* 2.5 cm from the bottom of the designated sections of the plate, and dry the spots with a current of air. Develop the chromatogram in a solvent system consisting of a mixture of butyl acetate and acetone (4:1) until the solvent front has moved to within about 1 cm of the top of the plate. Remove the plate from the developing chamber, mark the solvent front, and allow the solvent to evaporate. Locate the spots on the plate by viewing under short-wavelength ultraviolet light: the R_f value of the principal spot obtained from the *Test preparation* corresponds to that obtained from the *Standard preparation*, not more than two impurities are found in the chromatogram of the *Test preparation*, and the size and color of the spot representing any impurity obtained from the *Test preparation* are not greater or more intense than those of the principal spot obtained from the *Standard preparation*.

Ordinary impurities 〈466〉—

Test solution: methanol.

Standard solution: methanol.

Eluant: a mixture of butyl acetate and acetone (4:1).

Visualization: 6.

Assay—

Isoniazid reagent—Dissolve 1.0 g of isoniazid in about 500 mL of methanol, add 1.25 mL of hydrochloric acid, dilute with methanol to 1000 mL, and mix.

Standard preparation—Dissolve a suitable quantity of USP Testolactone RS, accurately weighed, in chloroform, and prepare, by quantitative and stepwise dilution, if necessary, a solution in chloroform having a known concentration of about 30 µg per mL.

Assay preparation—Transfer about 60 mg of Testolactone, accurately weighed, to a 100-mL volumetric flask, dissolve in chloroform, dilute with chloroform to volume, and mix. Transfer 5.0 mL of this solution to a second 100-mL volumetric flask, dilute with chloroform to volume, and mix.

Procedure—Transfer 5.0 mL each of the *Standard preparation*, the *Assay preparation*, and chloroform to provide the blank, to separate 25-mL volumetric flasks, add 10.0 mL of *Isoniazid reagent* to each flask, and mix. Place the flasks in a water bath maintained at a temperature of 55 ± 2°, and allow them to stand for 70 minutes. Cool, dilute each solution with chloroform to volume, and mix. Determine the absorbances of the solutions in 1-cm cells at the wavelength of maximum absorbance at about 415 nm, with a suitable spectrophotometer, against the blank. Calculate the quantity, in mg, of $C_{19}H_{24}O_3$ in the Testolactone taken by the formula:

$$2C(A_U / A_S),$$

in which C is the concentration, in µg per mL, of USP Testolactone RS in the *Standard preparation*, and A_U and A_S are the absorbances of the solutions from the *Assay preparation* and the *Standard preparation*, respectively.

Sterile Testolactone Suspension

» Sterile Testolactone Suspension is a sterile suspension of Testolactone in a suitable aqueous medium. It contains not less than 90.0 percent and not more than 120.0 percent of the labeled amount of $C_{19}H_{24}O_3$.

Packaging and storage—Preserve in single-dose or in multiple-dose containers, preferably of Type I glass.

Reference standard—*USP Testolactone Reference Standard*—Dry in vacuum at 100° for 3 hours before using.

Identification—

Hydroxylamine reagent—Mix 5 mL of a 3.5 in 100 solution of hydroxylamine hydrochloride in methanol with 15 mL of a 5.6 in 100 solution of potassium hydroxide in methanol, and use this reagent within 2 hours after mixing. [NOTE—Store the 3.5 in 100 methanol–hydroxylamine hydrochloride solution and the 5.6 in 100 methanol–potassium hydroxide solution in a refrigerator, and discard both after 1 month.]

Iron reagent—Dissolve 1.5 g of ferric chloride in 30 mL of water, add 3 mL of perchloric acid and 15 mL of nitric acid, and heat the solution until dense, white fumes are produced. Cool, add 40 mL of water and 10 mL of nitric acid, and dilute with perchloric acid to 100 mL. [NOTE—This concentrate may be stored and used for several months.] Transfer 4.0 mL of this concentrate to a 100-mL volumetric flask, add 36 mL of alcohol, dilute with water to volume, and mix.

Procedure—Place 1 drop of well-mixed Suspension in a small test tube, add 0.1 mL of *Hydroxylamine reagent*, heat in a water bath at about 60° for 1 minute, then add 1 mL of *Iron reagent:* a rose-violet color is produced.

Uniformity of dosage units ⟨905⟩: meets the requirements.

pH ⟨791⟩: between 5.0 and 7.5.

Other requirements—It meets the requirements under *Injections* ⟨1⟩.

Assay—

Isoniazid reagent and *Standard preparation*—Prepare as directed in the *Assay* under *Testolactone*.

Assay preparation—Transfer an accurately measured volume of well-mixed Sterile Testolactone Suspension, equivalent to about 300 mg of testolactone, to a 100-mL volumetric flask, add 85 mL of alcohol, insert the stopper, and shake by mechanical means for 30 minutes. Add alcohol to volume, mix, and filter through a medium-porosity, sintered-glass filter with the aid of gentle positive pressure. Transfer 10.0 mL of the clear filtrate to a 100-mL volumetric flask, add chloroform to volume, mix, and filter through a fine-porosity, sintered-glass filter. Transfer 10.0 mL of this clear filtrate to another 100-mL volumetric flask, dilute with chloroform to volume, and mix.

Procedure—Proceed as directed for *Procedure* in the *Assay* under *Testolactone*. Calculate the quantity, in mg, of $C_{19}H_{24}O_3$ in each mL of Suspension taken by the formula:

$$(10C/V)(A_U / A_S),$$

in which V is the volume, in mL, of Suspension taken, and C, A_U, and A_S are as defined therein.

Testolactone Tablets

» Testolactone Tablets contain not less than 90.0 percent and not more than 110.0 percent of the labeled amount of $C_{19}H_{24}O_3$.

Packaging and storage—Preserve in tight containers.

Reference standard—*USP Testolactone Reference Standard*—Dry in vacuum at 100° for 3 hours before using.

Identification—

A: *Hydroxylamine reagent* and *Iron reagent*—Prepare as directed in the *Identification test* under *Sterile Testolactone Suspension*.

Procedure—Place a portion of finely powdered Tablets, equivalent to about 250 mg of testolactone, in a 100-mL volumetric flask, add 85 mL of chloroform, and shake by mechanical means for 30 minutes. Add chloroform to volume, mix, and centrifuge the solution. Transfer 2 mL of the supernatant solution to a small test tube, and evaporate with the aid of a gentle current of air to dryness. Add 0.1 mL of *Hydroxylamine reagent*, heat in a water bath at about 60° for 1 minute, then add 1 mL of *Iron reagent:* a rose-violet color is produced.

B: Place a portion of finely powdered Tablets, equivalent to about 100 mg of testolactone, in a glass-stoppered, 35-mL centrifuge tube. Add 20 mL of chloroform, insert the stopper, and shake by mechanical means for 10 minutes. Centrifuge the mixture, and filter the supernatant liquid into a 50-mL beaker. Evaporate on a steam bath to dryness, and dry the residue in vacuum at 100° for 3 hours: the infrared absorption spectrum of a potassium bromide dispersion of the testolactone so obtained exhibits maxima only at the same wavelengths as that of a similar preparation of USP Testolactone RS.

Disintegration ⟨701⟩: 30 minutes.

Uniformity of dosage units ⟨905⟩: meet the requirements.

Assay—

Isoniazid reagent and *Standard preparation*—Prepare as directed in the *Assay* under *Testolactone*.

Assay preparation—Weigh and finely powder not less than 10 Testolactone Tablets. Transfer an accurately weighed portion of the powder, equivalent to about 50 mg of testolactone, to a 100-mL volumetric flask, add 85 mL of chloroform, and shake by mechanical means for 30 minutes. Add chloroform to volume, mix, and centrifuge a portion of about 10 mL of the mixture. Transfer 3.0 mL of the clear supernatant liquid to a 50-mL volumetric flask, dilute with chloroform to volume, and mix.

Procedure—Proceed as directed for *Procedure* in the *Assay* under *Testolactone*. Calculate the quantity, in mg, of $C_{19}H_{24}O_3$ in the portion of Tablets taken by the formula:

$$1.667C(A_U / A_S).$$

Testosterone

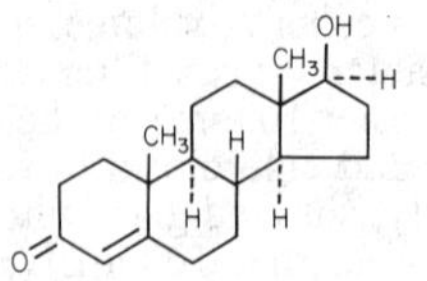

$C_{19}H_{28}O_2$ 288.43
Androst-4-en-3-one, 17-hydroxy-, (17β)-.
17β-Hydroxyandrost-4-en-3-one [*58-22-0*].

» Testosterone contains not less than 97.0 percent and not more than 103.0 percent of $C_{19}H_{28}O_2$, calculated on the dried basis.

Packaging and storage—Preserve in well-closed containers.

Reference standard—*USP Testosterone Reference Standard*—Dry in vacuum over phosphorus pentoxide for 4 hours before using.

Identification—

A: The infrared absorption spectrum of a potassium bromide dispersion of it, previously dried, exhibits maxima only at the same wavelengths as that of a similar preparation of USP Testosterone RS.

B: The ultraviolet absorption spectrum of a 1 in 100,000 solution in methanol exhibits maxima and minima at the same wavelengths as that of a similar solution of USP Testosterone RS, concomitantly measured.

Melting range ⟨741⟩: between 153° and 157°.

Specific rotation ⟨781⟩: between +101° and +105°, calculated on the dried basis, determined in a solution in dioxane containing 100 mg of Testosterone in each 10 mL.

Loss on drying ⟨731⟩—Dry it in vacuum over phosphorus pentoxide for 4 hours: it loses not more than 1.0% of its weight.

Assay—

Standard preparation—Prepare as directed under *Single-steroid Assay* ⟨511⟩, using USP Testosterone RS.

Assay preparation—Weigh accurately about 20 mg of Testosterone, previously dried, dissolve in a sufficient quantity of a mixture of equal volumes of alcohol and chloroform to make 10.0 mL, and mix.

Procedure—Proceed as directed for *Procedure* under *Single-steroid Assay* ⟨511⟩, using a solvent system consisting of a mixture of benzene and ethyl acetate (1:1), through the fourth sentence of the second paragraph under *Procedure.* Then centrifuge the tubes for 5 minutes, and determine the absorbances of the supernatant solutions in 1-cm cells at the wavelength of maximum absorbance at about 241 nm, with a suitable spectrophotometer, against the blank. Calculate the quantity, in mg, of $C_{19}H_{28}O_2$ in the portion of Testosterone taken by the formula:

$$10C(A_U/A_S),$$

in which C is the concentration, in mg per mL, of USP Testosterone RS in the *Standard preparation,* and A_U and A_S are the absorbances of the solutions from the *Assay preparation* and the *Standard preparation*, respectively.

Testosterone Pellets

» Testosterone Pellets are sterile pellets composed of Testosterone in compressed form, without the presence of any binder, diluent, or excipient. They contain not less than 97.0 percent and not more than 103.0 percent of $C_{19}H_{28}O_2$.

Packaging and storage—Preserve in tight containers holding one pellet each and suitable for maintaining sterile contents.

Reference standard—*USP Testosterone Reference Standard*—Dry in vacuum over phosphorus pentoxide for 4 hours before using.

Identification, Melting range, and Specific rotation—Pellets respond to *Identification tests A* and *B* and meet the requirements of the tests for *Melting range* and *Specific rotation* under *Testosterone.*

Solubility in chloroform—A solution of 50 mg of powdered Pellets in 2 mL of chloroform is clear and practically free from insoluble residue.

Sterility—Pellets meet the requirements under *Sterility Tests* ⟨71⟩.

Weight variation—Weigh 5 Pellets singly, and calculate the average weight. The average weight is between 95.0% and 105.0% of the labeled weight of testosterone, and each Pellet weighs between 90.0% and 110.0% of the labeled weight of testosterone.

Assay—Weigh and finely powder not less than 10 Testosterone Pellets. Transfer about 50 mg of the powder, accurately weighed, to a 100-mL volumetric flask, add methanol to volume, and mix. Transfer 5.0 mL of the solution to a 250-mL volumetric flask, dilute with methanol to volume, and mix. Proceed as directed in the *Assay* under *Sterile Testosterone Suspension*, beginning with "Concomitantly determine the absorbances." Calculate the quantity, in mg, of $C_{19}H_{28}O_2$ in the Testosterone Pellets taken by the formula:

$$5C(A_U/A_S),$$

in which C is the concentration, in µg per mL, of USP Testosterone RS in the Standard solution, and A_U and A_S are the absorbances of the solution from the Pellets and the Standard solution, respectively.

Sterile Testosterone Suspension

» Sterile Testosterone Suspension is a sterile suspension of Testosterone in an aqueous medium. It contains not less than 90.0 percent and not more than 110.0 percent of the labeled amount of $C_{19}H_{28}O_2$.

Packaging and storage—Preserve in single-dose or in multiple-dose containers, preferably of Type I glass.

Reference standard—*USP Testosterone Reference Standard*—Dry in vacuum over phosphorus pentoxide for 4 hours before using.

Identification—Testosterone obtained by filtration and washing, as directed in the *Assay*, and dried at 105° to constant weight, meets the requirements for *Identification tests A* and *B* under *Testosterone.*

Uniformity of dosage units ⟨905⟩: meets the requirements.

pH ⟨791⟩: between 4.0 and 7.5.

Other requirements—It meets the requirements under *Injections* ⟨1⟩.

Assay—Transfer an accurately measured volume of previously well-mixed Sterile Testosterone Suspension, equivalent to about 100 mg of testosterone, to a fine-porosity, sintered-glass filtering crucible, previously dried at 105° for 1 hour, and filter with suction. If the filtrate is not clear, again pass it through the same filter into a second receiver. Wash the residue in the filter with several 5-mL portions of water until 2 mL of the last washing, when evaporated on a steam bath, leaves a negligible residue. [NOTE—If the Suspension is passed through the filter twice, rinse the first receiver with the portions of water before passing them through the filter.] Dry the crucible and the collected testosterone at 105° for 1 hour. Completely dissolve the testosterone with five 25-mL portions of methanol, passing each portion through the crucible under gentle suction, and transfer the combined methanol solution to a 200-mL volumetric flask. Rinse the crucible and receiver with two 25-mL portions of methanol, add the rinsings to the main solution, dilute with methanol to volume, and mix. Transfer 5.0 mL of this solution to a 250-mL volumetric flask, dilute with methanol to volume, and mix. Concomitantly determine the absorbances of this solution and a Standard solution of USP Testosterone RS, in the same medium having a known concentration of about 10 µg per mL in 1-cm cells at the wavelength of maximum absorbance at about 241 nm, with a suitable spectrophotometer, using methanol as the blank. Calculate the quantity, in mg, of $C_{19}H_{28}O_2$ in each mL of Suspension taken by the formula:

$$(10C/V)(A_U/A_S),$$

in which C is the concentration, in µg per mL, of USP Testosterone RS in the Standard solution, V is the volume, in mL, of Suspension taken, and A_U and A_S are the absorbances of the solution from the Suspension and the Standard solution, respectively.

Testosterone Cypionate

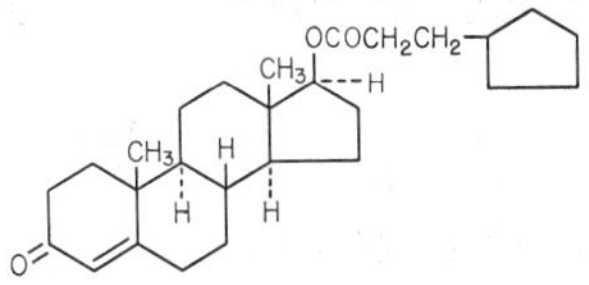

$C_{27}H_{40}O_3$ 412.61

Androst-4-en-3-one, 17-(3-cyclopentyl-1-oxopropoxy)-, (17β)-.

Testosterone cyclopentanepropionate [58-20-8].

» Testosterone Cypionate contains not less than 97.0 percent and not more than 103.0 percent of $C_{27}H_{40}O_3$, calculated on the dried basis.

Packaging and storage—Preserve in well-closed, light-resistant containers.

Reference standards—*USP Testosterone Cypionate Reference Standard*—Dry in vacuum over silica gel for 4 hours before using. *USP Cholesteryl Caprylate Reference Standard*—Dry in vacuum over silica gel for 4 hours before using. Keep container tightly closed and protected from light.

Identification—The infrared absorption spectrum of a potassium bromide dispersion of it, the test specimen having been previously dried, exhibits maxima only at the same wavelengths as that of a similar preparation of USP Testosterone Cypionate RS.

Melting range ⟨741⟩: between 98° and 104°.

Specific rotation ⟨781⟩: between +85° and +92°, calculated on the dried basis, determined in a solution in chloroform containing 200 mg in each 10 mL.

Loss on drying ⟨731⟩—Dry it in vacuum over silica gel for 4 hours: it loses not more than 0.5% of its weight.

Residue on ignition ⟨281⟩: not more than 0.2%.

Free cyclopentanepropionic acid—Dissolve 500 mg in 10 mL of alcohol that previously has been neutralized to a faint blue color following the addition of 2 or 3 drops of bromothymol blue TS, and promptly titrate with 0.01 *N* sodium hydroxide VS: not more than 0.70 mL of 0.01 *N* sodium hydroxide is required (0.20% of cyclopentanepropionic acid).

Assay—

Internal standard solution—Dissolve 80 mg of USP Cholesteryl Caprylate RS in a mixture of methanol and chloroform (4:1) in a 100-mL volumetric flask, then add the same solvent mixture to volume.

Standard preparation—Weigh accurately about 10 mg of USP Testosterone Cypionate RS into a suitable vial, add by pipet 10 mL of *Internal standard solution*, and mix.

Assay preparation—Prepare as directed for *Standard preparation*, using an accurately weighed portion of about 10 mg of Testosterone Cypionate instead of the Reference Standard.

Procedure—Inject 1 µL of the *Assay preparation* and the *Standard preparation*, successively, into a suitable gas chromatograph fitted with a flame-ionization detector. Under typical conditions, the instrument contains a 1.2-m × 3-mm glass column packed with 1 percent (w/w) phase G6 on packing S1AB. The column is maintained at 260° and the helium carrier gas flow at 50 mL per minute. In a suitable chromatogram, the resolution factor, *R* (see *Chromatography* ⟨621⟩), is not less than 3 between the internal standard and testosterone cypionate peaks, and five replicate injections of a single *Standard preparation* show a coefficient of variation of not more than 2% in the peak area ratio of testosterone cypionate to internal standard. Measure the areas under the peaks for testosterone cypionate and cholesteryl caprylate in each chromatogram. Calculate the ratio, R_U, of the area of the testosterone cypionate peak to the area of the internal standard peak in the chromatogram from the *Assay preparation*, and similarly calculate the ratio, R_S, in the chromatogram from the *Standard preparation*. Calculate the quantity, in mg, of $C_{27}H_{40}O_3$ in the portion of Testosterone Cypionate taken by the formula:

$$W(R_U/R_S),$$

in which W is the weight, in mg, of USP Testosterone Cypionate RS in the *Standard preparation* and the other terms are as defined therein.

Testosterone Cypionate Injection

» Testosterone Cypionate Injection is a sterile solution of Testosterone Cypionate in a suitable vegetable oil. It contains not less than 90.0 percent and not more than 110.0 percent of the labeled amount of $C_{27}H_{40}O_3$. It may contain a suitable solubilizing agent.

Packaging and storage—Preserve in single-dose or in multiple-dose containers, preferably of Type I glass, protected from light.

Reference standards—*USP Testosterone Cypionate Reference Standard*—Dry in vacuum over silica gel for 4 hours before using. *USP Cholesteryl Caprylate Reference Standard*—Dry in vacuum over silica gel for 4 hours before using. Keep container tightly closed and protected from light.

Identification—Dilute a suitable volume of Injection with chloroform to obtain a solution having a concentration of about 400 µg of testosterone cypionate per mL. Prepare a 20- × 20-cm thin-layer chromatographic plate (see *Chromatography* ⟨621⟩), coated with a 0.25-mm layer of chromatographic siliceous earth, by placing it in a developing chamber containing and equilibrated with a mixture of chloroform and corn oil (90:10), and allowing the solvent front to move about three-fourths of the length of the plate. Remove the plate, and allow the chloroform to evaporate. Apply 10 µL each of the solution under test and of a solution of USP Testosterone Cypionate RS in chloroform containing about 400 µg per mL on the plate, on a line about 2.5 cm from the bottom edge and about 1.5 cm apart. Place the plate in a developing chamber that contains and has been equilibrated with a mixture of methanol and water (90:10) previously saturated with corn oil. Develop the chromatogram until the solvent front has moved to about 10 cm above the line of application. Remove the plate, and heat in an oven at 105° for a few minutes. Spray the plate with a mixture of alcohol and sulfuric acid (3:1), and heat in an oven at 105° for 1 to 2 minutes. Observe the plate under long-wavelength ultraviolet light: the R_f value of the principal spot obtained from the solution under test corresponds to that obtained from the Standard solution.

Other requirements—It meets the requirements under *Injections* ⟨1⟩.

Assay—

Internal standard solution and *Standard preparation*—Prepare as directed in the *Assay* under *Testosterone Cypionate*.

Assay preparation—Transfer 1 mL of Testosterone Cypionate Injection, accurately measured, into a glass-stoppered, 50-mL centrifuge tube. Add 30 mL of a mixture of methanol and water (9:1), insert the stopper, and shake for 15 minutes. Centrifuge, remove the dilute methanol layer without disturbing the oil, and transfer it to a 200-mL volumetric flask. Repeat the extraction with three additional 30-mL portions of the dilute methanol, collecting the combined extracts in the volumetric flask. Dilute the combined extracts with the dilute methanol to volume, mix, and chill the contents of the flask to −8°. Remove the flask from the freezer, and immediately filter a portion of the contents. Allow the filtrate to reach room temperature, transfer a portion of it, equivalent to about 3 mg of testosterone cypionate, to a suitable vial, and evaporate to dryness. Add by pipet 3 mL of *Internal standard solution*, and shake vigorously to dissolve the residue.

Procedure—Proceed as directed for *Procedure* in the *Assay* under *Testosterone Cypionate.* Calculate the quantity, in mg, of $C_{27}H_{40}O_3$ in the portion of Injection taken by the formula:

$$600(C/V)(R_U/R_S),$$

in which C is the concentration, in mg per mL, of USP Testosterone Cypionate RS in the *Standard preparation*, and V is the volume, in mL, of filtrate used in the *Assay preparation.*

Testosterone Enanthate

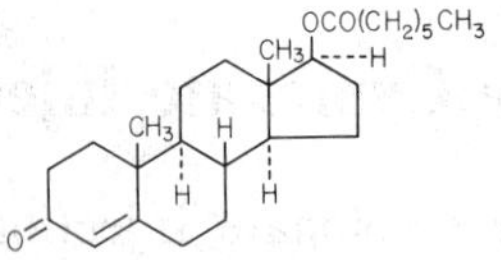

$C_{26}H_{40}O_3$ 400.60
Androst-4-en-3-one, 17-[(1-oxoheptyl)oxy]-, (17β)-.
Testosterone heptanoate [*315-37-7*].

» Testosterone Enanthate contains not less than 97.0 percent and not more than 103.0 percent of $C_{26}H_{40}O_3$.

Packaging and storage—Preserve in well-closed containers, in a cool place.

Reference standard—*USP Testosterone Enanthate Reference Standard*—Do not dry; use as is.

Identification—

A: The infrared absorption spectrum of a potassium bromide dispersion of it exhibits maxima only at the same wavelengths as that of a similar preparation of USP Testosterone Enanthate RS.

B: The ultraviolet absorption spectrum of a 1 in 100,000 solution in alcohol exhibits maxima and minima at the same wavelengths as that of a similar solution of USP Testosterone Enanthate RS, concomitantly measured, and the respective absorptivities, calculated on the anhydrous basis, at the wavelength of maximum absorbance at about 240 nm do not differ by more than 3.0%.

C: Reflux 25 mg with 2 mL of a 1 in 100 solution of potassium hydroxide in methanol for 1 hour. Cool the mixture, add 10 mL of water, filter, and wash the precipitate with water until the last washing is neutral to litmus. Dry the precipitate in vacuum at 60° for 3 hours: the testosterone so obtained melts between 151° and 157°.

Melting range ⟨741⟩: between 34° and 39°, the initial temperature of the bath not exceeding 20°.

Specific rotation ⟨781⟩: between +77° and +82°, calculated on the anhydrous basis, determined in a solution in dioxane containing 200 mg in each 10 mL.

Water, *Method I* ⟨921⟩: not more than 0.05%.

Free heptanoic acid—Dissolve 500 mg in 10 mL of alcohol that previously has been neutralized to a faint blue color following the addition of 2 or 3 drops of bromothymol blue TS, and promptly titrate with 0.01 *N* sodium hydroxide VS: not more than 0.6 mL of 0.01 *N* sodium hydroxide is required (0.16% of heptanoic acid).

Assay—Dissolve about 40 mg of Testosterone Enanthate, accurately weighed, in chloroform to make 100 mL, and mix. Pipet 10 mL of this solution into a 100-mL volumetric flask, add chloroform to volume, and mix. Dissolve a suitable quantity of USP Testosterone Enanthate RS, accurately weighed, in chloroform, and dilute quantitatively and stepwise with chloroform to obtain a Standard solution having a known concentration of about 40 µg per mL. Pipet 5 mL each of the solution of Testosterone Enanthate and the Standard solution into separate, glass-stoppered, 50-mL conical flasks, and place 5.0 mL of chloroform in a similar flask to provide a blank. Treat each flask as follows. Add 10.0 mL of a solution of 375 mg of isoniazid and 0.47 mL of hydrochloric acid in 500 mL of methanol, mix, and allow to stand for 45 minutes. Concomitantly determine the absorbances of the solutions at the wavelength of maximum absorbance at about 380 nm, with a suitable spectrophotometer, using the blank to set the instrument. Calculate the quantity, in mg, of $C_{26}H_{40}O_3$ in the Testosterone Enanthate taken by the formula:

$$C(A_U/A_S),$$

in which C is the concentration, in µg per mL, of USP Testosterone Enanthate RS in the Standard solution, and A_U and A_S are the absorbances of the solution of Testosterone Enanthate and the Standard solution, respectively.

Testosterone Enanthate Injection

» Testosterone Enanthate Injection is a sterile solution of Testosterone Enanthate in a suitable vegetable oil. It contains not less than 90.0 percent and not more than 110.0 percent of the labeled amount of $C_{26}H_{40}O_3$.

Packaging and storage—Preserve in single-dose or in multiple-dose containers, preferably of Type I glass.

Reference standard—*USP Testosterone Enanthate Reference Standard*—Do not dry; use as is.

Identification—Dilute a suitable volume of Injection with chloroform to obtain a solution having a concentration of about 400 µg of testosterone enanthate per mL. Proceed as directed in the *Identification test* under *Testosterone Cypionate Injection,* beginning with "Prepare a 20- × 20-cm thin-layer chromatographic plate," but using USP Testosterone Enanthate RS. The R_f value of the principal spot obtained from the solution under test corresponds to that obtained from the Reference Standard solution.

Other requirements—It meets the requirements under *Injections* ⟨1⟩.

Assay—

Chromatographic solvent—Equilibrate, by shaking in a separator, 95 mL of alcohol, 5 mL of water, and 50 mL of chromatographic *n*-heptane. Allow the layers to separate.

Isoniazid reagent—Dissolve 375 mg of isoniazid and 0.47 mL of hydrochloric acid in 500 mL of methanol.

Standard preparation—Dissolve a suitable quantity of USP Testosterone Enanthate RS, accurately weighed, in methanol, and dilute quantitatively and stepwise with methanol to obtain a solution having a known concentration of about 40 µg per mL.

Assay preparation—Transfer to a 10-mL volumetric flask an accurately measured volume of Testosterone Enanthate Injection, equivalent to about 100 mg of testosterone enanthate, add chromatographic *n*-heptane to volume, and mix. Pipet 5 mL of this solution into a 100-mL volumetric flask, add chromatographic *n*-heptane to volume, and mix.

Procedure—Mix in a beaker 3 g of silanized chromatographic siliceous earth and 3 mL of the upper layer of the *Chromatographic solvent.* Pack the mixture into a 250- × 25-mm chromatographic tube that contains a small pledget of glass wool above the stem constriction. Mix in a beaker 3 g of silanized chromatographic siliceous earth and 2.0 mL of *Assay preparation*, transfer the mixture to the tube, and pack. Dry-wash the beaker with 1 g of silanized chromatographic siliceous earth, and transfer to the tube. Place a small pad of glass wool above the column packing. Pass 35 mL of the lower layer of the *Chromatographic solvent* through the column, and collect the eluate in a 50-mL volumetric flask. Add alcohol to volume, and mix. Pipet 10 mL of the resulting solution into a glass-stoppered, 50-mL conical flask, and evaporate on a water bath to dryness. Pipet 5 mL of methanol into the flask, and swirl to dissolve the residue. Pipet 5 mL of *Standard preparation* into a similar flask. To each flask add 10.0 mL of *Isoniazid reagent*, mix, and allow to stand for about 45 minutes. Concomitantly determine the absorbances of both solutions at the wavelength of maximum absorbance at about 380 nm, with a suitable spectrophotometer, using as a blank 5 mL of methanol that has been treated similarly with *Isoniazid reagent*. Calculate the quantity, in mg, of $C_{26}H_{40}O_3$ in each mL of the Injection taken by the formula:

$$2.5(C/V)(A_U/A_S),$$

in which C is the concentration, in µg per mL, of USP Testosterone Enanthate RS in the *Standard preparation*, V is the volume, in mL, of Injection taken, and A_U and A_S are the absorbances of the solutions from the *Assay preparation* and the *Standard preparation*, respectively.

Testosterone Propionate

$C_{22}H_{32}O_3$ 344.49
Androst-4-en-3-one, 17-(1-oxopropoxy)-, (17β)-.
Testosterone propionate [*57-85-2*].

» Testosterone Propionate contains not less than 97.0 percent and not more than 103.0 percent of $C_{22}H_{32}O_3$, calculated on the dried basis.

Packaging and storage—Preserve in well-closed, light-resistant containers.

Reference standard—*USP Testosterone Propionate Reference Standard*—Dry in vacuum over silica gel for 4 hours before using.

Identification—

A: The infrared absorption spectrum of a potassium bromide dispersion of it exhibits maxima only at the same wavelengths as that of a similar preparation of USP Testosterone Propionate RS.

B: The ultraviolet absorption spectrum of a 1 in 100,000 solution in alcohol exhibits maxima and minima at the same wavelengths as that of a similar solution of USP Testosterone Propionate RS, concomitantly measured, and the respective absorptivities, calculated on the dried basis, at the wavelength of maximum absorbance at about 241 nm do not differ by more than 3.0%.

C: It responds to *Identification test C* under *Testosterone Enanthate*.

Melting range ⟨741⟩: between 118° and 123°.

Specific rotation ⟨781⟩: between +83° and +90°, previously dried, determined in a solution in dioxane containing 200 mg in each 10 mL.

Loss on drying ⟨731⟩—Dry it in vacuum over silica gel for 4 hours: it loses not more than 0.5% of its weight.

Assay—Proceed with Testosterone Propionate as directed in the *Assay* under *Testosterone Enanthate*, except to use USP Testosterone Propionate RS and otherwise substitute Testosterone Propionate throughout. Calculate the quantity, in mg, of $C_{22}H_{32}O_3$ in the Testosterone Propionate taken by the formula given therein.

Testosterone Propionate Injection

» Testosterone Propionate Injection is a sterile solution of Testosterone Propionate in a suitable vegetable oil. It contains not less than 88.0 percent and not more than 112.0 percent of the labeled amount of $C_{22}H_{32}O_3$.

Packaging and storage—Preserve in single-dose or in multiple-dose containers, preferably of Type I glass.

Reference standard—*USP Testosterone Propionate Reference Standard*—Dry in vacuum over silica gel for 4 hours before using.

Identification—Dilute a suitable volume of Injection with chloroform to obtain a solution having a concentration of about 400 µg of testosterone propionate per mL. Proceed as directed in the *Identification test* under *Testosterone Cypionate Injection*, beginning with "Prepare a 20- × 20-cm thin-layer chromatographic plate," except to use USP Testosterone Propionate RS. The R_f value of the principal spot obtained from the solution under test corresponds to that obtained from the Standard solution.

Other requirements—It meets the requirements under *Injections* ⟨1⟩.

Assay—

Chromatographic solvent and *Isoniazid reagent*—Prepare as directed in the *Assay* under *Testosterone Enanthate Injection*.

Standard preparation—Prepare as directed in the *Assay* under *Testosterone Enanthate Injection*, using USP Testosterone Propionate RS.

Assay preparation—Transfer to a 10-mL volumetric flask an accurately measured volume of Testosterone Propionate Injection, equivalent to about 100 mg of testosterone propionate, add chromatographic *n*-heptane to volume, and mix. Pipet 5 mL of this solution into a 100-mL volumetric flask, add chromatographic *n*-heptane to volume, and mix.

Procedure—Proceed as directed for *Procedure* in the *Assay* under *Testosterone Enanthate Injection*. Calculate the quantity, in mg, of $C_{22}H_{32}O_3$ in each mL of Injection taken by the formula:

$$2.5(C/V)(A_U/A_S),$$

in which C is the concentration, in µg per mL, of USP Testosterone Propionate RS in the *Standard preparation*, V is the volume, in mL, of Injection taken, and A_U and A_S are the absorbances of the solutions from the *Assay preparation* and the *Standard preparation*, respectively.

Tetanus Antitoxin

» Tetanus Antitoxin conforms to the regulations of the federal Food and Drug Administration concerning biologics (see *Biologics* ⟨1041⟩). It is a sterile, non-pyrogenic solution of the refined and concentrated proteins, chiefly globulins, containing antitoxic antibodies obtained from the blood serum or plasma of healthy horses that have been immunized against tetanus toxin or toxoid. It has a potency of not less than 400 antitoxin units per mL based on the U.S. Standard Tetanus Antitoxin and the U.S. Control Tetanus Test Toxin, tested in guinea pigs.

Packaging and storage—Preserve at a temperature between 2° and 8°.

Expiration date—The expiration date for Antitoxin containing a 20% excess of potency is not later than 5 years after date of issue from manufacturer's cold storage (5°, 1 year; or 0°, 2 years).

Labeling—Label it to state that it was prepared from horse serum or plasma.

Tetanus Immune Globulin

» Tetanus Immune Globulin conforms to the regulations of the federal Food and Drug Administration concerning biologics (see *Biologics* ⟨1041⟩). It is a sterile, non-pyrogenic solution of globulins derived from the blood plasma of adult human donors who have been immunized with tetanus toxoid. It has a potency of not less than 50 antitoxin units per mL based on the U.S. Standard Tetanus Antitoxin and the U.S. Control Tetanus Test Toxin, tested in guinea pigs. It contains not less than 10 g and not more than 18 g of protein per 100 mL, of which not less than 90 percent is gamma globulin. It contains 0.3 *M* glycine as a stabilizing agent, and it contains a suitable preservative.

Packaging and storage—Preserve at a temperature between 2° and 8°.

Expiration date—The expiration date for Tetanus Immune Globulin containing a 10% excess of potency is not later than 3 years after date of issue from manufacturer's cold storage (5°, 1 year).

Labeling—Label it to state that it is not for intravenous injection.

Tetanus Toxoid

» Tetanus Toxoid conforms to the regulations of the federal Food and Drug Administration concerning biologics (see *Biologics* ⟨1041⟩). It is a sterile solution of the formaldehyde-treated products of growth of the tetanus bacillus (*Clostridium tetani*). It meets the requirements of the specific guinea pig potency test of antitoxin production based on the U.S. Standard Tetanus Antitoxin and the U.S. Control Tetanus Test Toxin. It meets the requirements of the specific guinea pig detoxification test. It contains not more than 0.02 percent of residual free formaldehyde. It contains a preservative other than a phenoloid compound.

Packaging and storage—Preserve at a temperature between 2° and 8°.

Expiration date—The expiration date is not later than 2 years after date of issue from manufacturer's cold storage (5°, 1 year).

Labeling—Label it to state that it is not to be frozen.

Tetanus Toxoid Adsorbed

» Tetanus Toxoid Adsorbed conforms to the regulations of the federal Food and Drug Administration concerning biologics (see *Biologics* ⟨1041⟩). It is a sterile preparation of plain tetanus toxoid that meets all of the requirements for that product with the exception of those for potency, and that has been precipitated or adsorbed by alum, aluminum hydroxide, or aluminum phosphate adjuvants. It meets the requirements of the specific mouse or guinea pig potency test of antitoxin production based on the U.S. Standard Tetanus Antitoxin and the U.S. Control Test Tetanus Toxin. It meets the requirements of the specific guinea pig detoxification test.

Packaging and storage—Preserve at a temperature between 2° and 8°.

Expiration date—The expiration date is not later than 2 years after date of issue from manufacturer's cold storage (5°, 1 year).

Labeling—Label it to state that it is to be well shaken before use and that it is not to be frozen.

Aluminum content—It contains not more than 0.85 mg per single injection, determined by analysis, or not more than 1.14 mg calculated on the basis of the amount of aluminum compound added.

Tetanus Toxoids, Diphtheria and—*see* Diphtheria and Tetanus Toxoids

Tetanus and Diphtheria Toxoids Adsorbed for Adult Use

» Tetanus and Diphtheria Toxoids Adsorbed for Adult Use conforms to the regulations of the federal Food and Drug Administration concerning biologics (see *Biologics* ⟨1041⟩). It is a sterile suspension prepared by mixing suitable quantities of adsorbed diphtheria toxoid and adsorbed tetanus toxoid using the same precipitating or adsorbing agent for both toxoids. The antigenicity or potency and the proportions of the toxoids are such as to provide, in each dose prescribed in the labeling, an immunizing dose of Tetanus Toxoid Adsorbed as defined for that product, and one-tenth of the immunizing dose of Diphtheria Toxoid Adsorbed as defined for that product for children, such that in the specific guinea pig antigenicity test it meets the requirement of production of not less than 0.5 unit of diphtheria antitoxin per mL and each immunizing dose has an antigen content of not more than 2 Lf (flocculating units) value as measured with the U.S. Reference Diphtheria Antitoxin for Flocculation Test. Each component meets the other requirements for those products. It contains not more than 0.02 percent of residual free formaldehyde.

Packaging and storage—Preserve at a temperature between 2° and 8°.

Expiration date—The expiration date is not later than 2 years after date of issue from manufacturer's cold storage (5°, 1 year).

Labeling—Label it to state that it is to be well shaken before use and that it is not to be frozen.

Tetanus Toxoids Adsorbed, Diphtheria and—*see* Diphtheria and Tetanus Toxoids Adsorbed

Tetanus Toxoids and Pertussis Vaccine Adsorbed, Diphtheria and—*see* Diphtheria and Tetanus Toxoids and Pertussis Vaccine Adsorbed

Tetanus Toxoids and Pertussis Vaccine, Diphtheria and—*see* Diphtheria and Tetanus Toxoids and Pertussis Vaccine

Tetracaine

$CH_3(CH_2)_3NH-C_6H_4-COOCH_2CH_2N(CH_3)_2$

$C_{15}H_{24}N_2O_2$ 264.37

Benzoic acid, 4-(butylamino)-, 2-(dimethylamino)ethyl ester.

2-(Dimethylamino)ethyl *p*-(butylamino)benzoate [*94-24-6*].

» Tetracaine contains not less than 98.0 percent and not more than 101.0 percent of $C_{15}H_{24}N_2O_2$, calculated on the dried basis.

Packaging and storage—Preserve in tight, light-resistant containers.

Reference standard—*USP Tetracaine Hydrochloride Reference Standard*—Dry in vacuum over phosphorus pentoxide for 18 hours before using.

Identification—

A: Dissolve 100 mg in 10 mL of dilute hydrochloric acid (1 in 120), and add 1 mL of potassium thiocyanate solution (1 in 4): a crystalline precipitate is formed. Recrystallize the precipitate from water, and dry at 80° for 2 hours: it melts between 130° and 132° (see *Melting Range or Temperature* ⟨741⟩).

B: Dissolve about 90 mg, accurately weighed, in 10 mL of dilute hydrochloric acid (1 in 120) in a 500-mL volumetric flask, dilute with water to volume, and mix. Transfer 5.0 mL of this solution to a 100-mL volumetric flask, add 2 mL of *Buffer No. 6*, 10 percent, pH 6.0 (see *Phosphate Buffers* ⟨81⟩), dilute with water to volume, and mix: the ultraviolet absorption spectrum of the solution so obtained exhibits maxima and minima at the same wavelengths as that of a 1 in 100,000 solution of USP Tetracaine Hydrochloride RS in a mixture of water and *Buffer No. 6* (50:1), 10 percent, pH 6.0 (see *Phosphate Buffers* ⟨81⟩), and the respective molar absorptivities, calculated on the dried basis, at the wavelength of maximum absorbance at about 310 nm do not differ by more than 2.0%. [NOTE—The molecular weight of tetracaine hydrochloride ($C_{15}H_{24}N_2O_2 \cdot HCl$) is 300.83.]

Melting range, *Class I* ⟨741⟩: between 41° and 46°.

Loss on drying ⟨731⟩—Dry it in vacuum over phosphorus pentoxide for 18 hours: it loses not more than 0.5% of its weight.

Residue on ignition ⟨281⟩: not more than 0.1%.

Chromatographic purity—Dissolve an accurately weighed quantity of Tetracaine in chloroform to obtain a test solution containing 50 mg per mL. Prepare a Standard solution of 4-(butylamino) benzoic acid in methanol containing 0.2 mg per mL. Apply separate 5-µL portions of the test solution and the Standard solution on a suitable thin-layer chromatographic plate (see *Chromatography* ⟨621⟩), coated with a 0.25-mm layer of chromatographic silica gel mixture. Develop the plate in a suitable chromatographic chamber containing a solvent system consisting of a mixture of chloroform, methanol, and isopropylamine (98:7:2) until the solvent front has moved about three-fourths of the length of the plate. Remove the plate from the chamber, and dry in a current of warm air. Examine the plate under short-wavelength ultraviolet light: any spot obtained from the test solution, other than the principal spot, is not more intense than the principal spot obtained from the Standard solution (0.4%), and the sum of the intensities of any such spots is not greater than 0.8%.

Assay—Transfer about 500 mg of Tetracaine, accurately weighed, to a suitable vessel. Add 5 mL of hydrochloric acid and 50 mL of water, cool to 15°, add about 25 g of crushed ice, and slowly titrate with 0.1 *M* sodium nitrite VS, stirring vigorously, until a glass rod dipped into the titrated solution produces an immediate blue ring when touched to starch iodide paper. When the titration is complete, the end-point is reproducible after the mixture has been allowed to stand for 1 minute. Perform a blank determination, and make any necessary correction. Each mL of 0.1 *M* sodium nitrite is equivalent to 26.44 mg of $C_{15}H_{24}N_2O_2$.

Tetracaine Ointment

» Tetracaine Ointment contains not less than 90.0 percent and not more than 110.0 percent of the labeled amount of $C_{15}H_{24}N_2O_2$ in a suitable ointment base.

Packaging and storage—Preserve in collapsible ointment tubes.

Reference standard—*USP Tetracaine Hydrochloride Reference Standard*—Dry in vacuum over phosphorus pentoxide for 18 hours before using.

Identification—

A: The solution employed for measurement of absorbance in the *Assay* exhibits a maximum at 310 ± 2 nm.

B: Dissolve 5 g in 50 mL of ether, extract the ether solution with 5 mL of 3 *N* hydrochloric acid, and filter the acid extract. Add 2 mL of potassium thiocyanate solution (1 in 2) to the filtrate: a crystalline precipitate is formed, and when recrystallized from water and dried at 80° for 2 hours, it melts between 130° and 132° (see *Melting Range or Temperature* ⟨741⟩).

Microbial limits—It meets the requirements of the tests for absence of *Staphylococcus aureus* and *Pseudomonas aeruginosa* under *Microbial Limit Tests* ⟨61⟩.

Minimum fill ⟨755⟩: meets the requirements.

Assay—

Standard preparation—Transfer about 20 mg of USP Tetracaine Hydrochloride RS, accurately weighed, to a 100-mL volumetric flask, dissolve in water, add water to volume, and mix. Transfer 5.0 mL of this solution to a second 100-mL volumetric flask, add 5 mL of dilute hydrochloric acid (1 in 240) and 10 mL of *Buffer No. 6*, 10 percent, pH 6.0 (see *Phosphate Buffers* ⟨81⟩), dilute with water to volume, and mix. The concentration of USP Tetracaine Hydrochloride RS in the *Standard preparation* is about 10 µg per mL.

Assay preparation—Transfer an accurately weighed portion of Tetracaine Ointment, equivalent to about 9 mg of tetracaine, to a separator, and dissolve in 15 mL of ether. Extract with one 20-mL portion and two 10-mL portions of dilute hydrochloric acid (1 in 240), collecting the acid extracts in a second separator. Render the aqueous solution alkaline by the addition of 5 mL of sodium carbonate TS, and extract immediately with two 50-mL portions of ether, collecting the ether extracts in another separator. Wash the ether solution with 20 mL of water, discard the washing, and extract the ether solution with two 20-mL portions and one 5-mL portion of dilute hydrochloric acid (1 in 240), collecting the acid extracts in a 50-mL volumetric flask. Dilute with water to volume, and mix. Transfer 5.0 mL of this solution to a 100-mL volumetric flask, add 10 mL of *Buffer No. 6*, 10 percent, pH 6.0 (see *Phosphate Buffers* ⟨81⟩), dilute with water to volume, and mix.

Procedure—Concomitantly determine the absorbances of the *Assay preparation* and the *Standard preparation* in 1-cm cells at the wavelength of maximum absorbance at about 310 nm, with a suitable spectrophotometer, using water as the blank. Calculate the quantity, in mg, of $C_{15}H_{24}N_2O_2$ in the portion of Ointment taken by the formula:

$$(264.37/300.83)(C)(A_U/A_S),$$

in which 264.37 and 300.83 are the molecular weights of tetracaine and tetracaine hydrochloride, respectively, C is the concentration, in µg per mL, of USP Tetracaine Hydrochloride RS in the *Standard preparation*, and A_U and A_S are the absorbances of the *Assay preparation* and the *Standard preparation*, respectively.

Tetracaine Ophthalmic Ointment

» Tetracaine Ophthalmic Ointment is a sterile ointment containing not less than 0.45 percent and not more than 0.55 percent of $C_{15}H_{24}N_2O_2$ in White Petrolatum.

Packaging and storage—Preserve in collapsible ophthalmic ointment tubes.

Reference standard—*USP Tetracaine Hydrochloride Reference Standard*—Dry in vacuum over phosphorus pentoxide for 18 hours before using.

Identification—

A: The solution employed for measurement of absorbance in the *Assay* exhibits a maximum at 310 ± 2 nm.

B: Dissolve 5 g in 50 mL of ether, extract the ether solution with 5 mL of 3 *N* hydrochloric acid, and filter the extract. To the extract add 2 mL of potassium thiocyanate solution (1 in 2): a crystalline precipitate is formed, and when recrystallized from water and dried at 80° for 2 hours, it melts between 130° and 132° (see *Melting Range or Temperature* ⟨741⟩).

Sterility—It meets the requirements under *Sterility Tests* ⟨71⟩.

Minimum fill ⟨755⟩: meets the requirements.

Metal particles—It meets the requirements of the test for *Metal Particles in Ophthalmic Ointments* ⟨751⟩.

Assay—

Standard preparation—Prepare as directed in the *Assay* under *Tetracaine Ointment.*

Assay preparation—Using an accurately weighed portion of Tetracaine Ophthalmic Ointment, prepare as directed in the *Assay* under *Tetracaine Ointment.*

Procedure—Proceed as directed for *Procedure* in the *Assay* under *Tetracaine Ointment.*

Tetracaine and Menthol Ointment

» Tetracaine and Menthol Ointment contains not less than 90.0 percent and not more than 110.0 percent of the labeled amounts of tetracaine ($C_{15}H_{24}N_2O_2$) and menthol ($C_{10}H_{20}O$) in a suitable ointment base.

Packaging and storage—Preserve in collapsible ointment tubes.

Reference standards—*USP Tetracaine Hydrochloride Reference Standard*—Dry in vacuum over phosphorus pentoxide for 18 hours before using. *USP Menthol Reference Standard.*

Identification—

A: The solution employed for measurement of absorbance in the *Assay for tetracaine* exhibits a maximum at 310 ± 2 nm (*presence of tetracaine*).

B: Dissolve 5 g in 50 mL of ether, extract the ether solution with 5 mL of 3 *N* hydrochloric acid, and filter the acid extract. Add 2 mL of potassium thiocyanate solution (1 in 2) to the filtrate: a crystalline precipitate is formed, and when recrystallized from water and dried at 80° for 2 hours, it melts between 130° and 132° (see *Melting Range or Temperature* ⟨741⟩) (*presence of tetracaine*).

C: When chromatographed as directed in the *Assay for menthol*, the *Assay preparation* exhibits a major peak for menthol, the retention time of which corresponds to that exhibited by menthol in the *Standard preparation.*

Minimum fill ⟨755⟩: meets the requirements.

Assay for tetracaine—

Standard preparation—Prepare as directed in the *Assay* under *Tetracaine Ointment.*

Assay preparation—Using Tetracaine and Menthol Ointment, proceed as directed in the *Assay* under *Tetracaine Ointment.*

Procedure—Proceed as directed for *Procedure* in the *Assay* under *Tetracaine Ointment.*

Assay for menthol—

Internal standard solution—Dissolve decanol in *n*-hexane to obtain a solution having a concentration of about 1 mg per mL.

Standard preparation—Dissolve an accurately weighed quantity of USP Menthol RS in *n*-hexane to obtain a solution having a known concentration of about 1 mg per mL. Transfer 5.0 mL of this solution and 5.0 mL of *Internal standard solution* to a 50-mL volumetric flask, dilute with ether to volume, and mix. Combine 2.0 mL of this solution and 2.0 mL of ether in a suitable container, and mix. This *Standard preparation* has a known concentration of about 0.05 mg per mL.

Assay preparation—Transfer an accurately weighed quantity of Tetracaine and Menthol Ointment, equivalent to about 5 mg of menthol, to a 50-mL volumetric flask, add 5.0 mL of *Internal standard solution*, dilute with *n*-hexane to volume, mix, and sonicate. Using a suitable syringe attached firmly to a 25-mm × 12.5-mm chromatographic cartridge containing packing L4, force 2.0 mL of the solution through the cartridge at a rate of 1 mL per 12 seconds. Wash the cartridge at the same rate with two 5-mL portions of *n*-hexane, and discard the washings. Force two 2.0-mL portions of ether through the cartridge, combine the ether eluates in a suitable container, and mix.

Chromatographic system (see *Chromatography* ⟨621⟩)—The gas chromatograph is equipped with a flame-ionization detector and contains a 1.8-m × 2-mm column packed with 10 percent phase G16 on support S1AB. The column is maintained isothermally at about 170°, the injection port is maintained at about 260°, and the detector block is maintained at about 240°. Dry helium is used as the carrier gas at a flow rate of about 50 mL per minute.

System suitability—Chromatograph three injections of the *Standard preparation*, and record the peak responses as directed under *Procedure*: the retention time of menthol is about 0.7 relative to decanol. In a suitable chromatogram, the resolution, *R*, between the 2 peaks is not less than 2.5, and the relative standard deviation of the ratio of the peak response obtained with menthol to that obtained with decanol is not more than 2%.

Procedure—Separately inject equal volumes (about 2 µL) of the *Assay preparation* and the *Standard preparation* into the gas chromatograph, and measure the peak responses for menthol and decanol in each chromatogram. Calculate the quantity, in mg, of $C_{10}H_{20}O$ in the portion of Ointment taken by the formula:

$$100C(R_U/R_S),$$

in which *C* is the concentration, in mg per mL, of USP Menthol RS in the *Standard preparation*, and R_U and R_S are the peak response ratios of menthol to decanol obtained from the *Assay preparation* and the *Standard preparation*, respectively.

Tetracaine Hydrochloride

$C_{15}H_{24}N_2O_2 . HCl$ 300.83

Benzoic acid, 4-(butylamino)-, 2-(dimethylamino)ethyl ester, monohydrochloride.

2-(Dimethylamino)ethyl *p*-(butylamino)benzoate monohydrochloride [*136-47-0*].

» Tetracaine Hydrochloride contains not less than 98.5 percent and not more than 101.0 percent of $C_{15}H_{24}N_2O_2 . HCl$, calculated on the anhydrous basis.

Packaging and storage—Preserve in tight, light-resistant containers.

Reference standard—*USP Tetracaine Hydrochloride Reference Standard*—Dry in vacuum over phosphorus pentoxide for 18 hours before using.

Identification—

A: Dissolve about 50 mg, accurately weighed, in water to make 250.0 mL. Pipet 5 mL of this solution into a 100-mL volumetric flask, add 2 mL of *Buffer No. 6*, 10 percent, pH 6.0 (see *Phosphate Buffers* ⟨81⟩), then dilute with water to volume, and mix: the ultraviolet absorption spectrum of the solution so obtained exhibits maxima and minima at the same wavelengths as that of a similar solution of USP Tetracaine Hydrochloride RS, concomitantly measured, and the respective absorptivities, calculated on the anhydrous basis, at the wavelength of maximum absorbance at about 310 nm do not differ by more than 2.0%.

B: Dissolve 100 mg in 10 mL of water, and add 1 mL of potassium thiocyanate solution (1 in 4): a crystalline precipitate is formed. Recrystallize the precipitate from water, and dry at 80° for 2 hours: it melts between 130° and 132°.

C: A solution of 100 mg in 5 mL of water responds to the tests for *Chloride* ⟨191⟩.

Water, *Method I* ⟨921⟩: not more than 2.0%.

Residue on ignition ⟨281⟩: not more than 0.1%.

Chromatographic purity—Dissolve an accurately weighed quantity in water to obtain a test solution containing 50 mg per mL, and proceed as directed in the test for *Chromatographic purity* under *Tetracaine*, beginning with "Prepare a Standard solution."

Assay—Transfer about 500 mg of Tetracaine Hydrochloride, accurately weighed, to a suitable vessel, add 5 mL of hydrochloric acid and 50 mL of water, and proceed as directed under *Nitrite Titration* ⟨451⟩, beginning with "cool to 15°." Each mL of 0.1 *M* sodium nitrite is equivalent to 30.08 mg of $C_{15}H_{24}N_2O_2 . HCl$.

Tetracaine Hydrochloride Cream

» Tetracaine Hydrochloride Cream contains tetracaine hydrochloride ($C_{15}H_{24}N_2O_2.HCl$) equivalent to not less than 90.0 percent and not more than 110.0 percent of the labeled amount of tetracaine ($C_{15}H_{24}N_2O_2$) in a suitable water-miscible base.

Packaging and storage—Preserve in collapsible, lined metal tubes.

Reference standard—*USP Tetracaine Hydrochloride Reference Standard*—Dry in vacuum over phosphorus pentoxide for 18 hours before using.

Identification—The ultraviolet absorption spectrum of the *Assay preparation*, prepared as directed in the *Assay*, exhibits maxima and minima at the same wavelengths as that of the *Standard preparation*, prepared as directed in the *Assay*.

Microbial limits—It meets the requirements of the tests for absence of *Staphylococcus aureus* and *Pseudomonas aeruginosa* under *Microbial Limit Tests* ⟨61⟩.

Minimum fill ⟨755⟩: meets the requirements.

pH ⟨791⟩: between 3.2 and 3.8.

Assay—

pH 6 acetate buffer—Dissolve 250 g of sodium acetate in about 500 mL of water in a 1000-mL volumetric flask, add 5.0 mL of glacial acetic acid, dilute with water to volume, and mix.

Standard preparation—Transfer about 25 mg of USP Tetracaine Hydrochloride RS, accurately weighed, to a 100-mL volumetric flask, dissolve in isopropyl alcohol, add isopropyl alcohol to volume, and mix. Transfer 2.0 mL of this solution to another 100-mL volumetric flask, add 2.0 mL of *pH 6 acetate buffer*, dilute with isopropyl alcohol to volume, and mix. The concentration of USP Tetracaine Hydrochloride RS in the *Standard preparation* is about 5 µg per mL.

Assay preparation—Transfer an accurately weighed portion of Tetracaine Hydrochloride Cream, equivalent to about 4.5 mg of tetracaine, to a 50-mL beaker, add 25 mL of isopropyl alcohol, and warm on a steam bath to dissolve the specimen completely. Transfer the solution with the aid of isopropyl alcohol to a 100-mL volumetric flask, dilute with isopropyl alcohol to volume, and mix. Transfer 10.0 mL of this solution to another 100-mL volumetric flask, add 2.0 mL of *pH 6 acetate buffer*, dilute with isopropyl alcohol to volume, and mix.

Procedure—Concomitantly determine the absorbances of the *Assay preparation* and the *Standard preparation* in 1-cm cells at the wavelength of maximum absorbance at about 310 nm, with a suitable spectrophotometer, using a 1 in 50 solution of *pH 6 acetate buffer* in isopropyl alcohol as the blank. Calculate the quantity, in mg, of $C_{15}H_{24}N_2O_2$ in the portion of Cream taken by the formula:

$$(264.37/300.83)(C)(A_U/A_S),$$

in which 264.37 and 300.83 are the molecular weights of tetracaine and tetracaine hydrochloride, respectively, C is the concentration, in µg per mL, of USP Tetracaine Hydrochloride RS in the *Standard preparation*, and A_U and A_S are the absorbances of the *Assay preparation* and the *Standard preparation*, respectively.

Tetracaine Hydrochloride Injection

» Tetracaine Hydrochloride Injection is a sterile solution of Tetracaine Hydrochloride in Water for Injection. It contains not less than 95.0 percent and not more than 105.0 percent of the labeled amount of $C_{15}H_{24}N_2O_2.HCl$.

Packaging and storage—Preserve in single-dose or in multiple-dose containers, preferably of Type I glass, under refrigeration and protected from light. It may be packaged in 100-mL multiple-dose containers. Injection supplied as a component of spinal anesthesia trays may be stored at room temperature for 12 months.

Labeling—Label it to indicate that the Injection is not to be used if it contains crystals, or if it is cloudy or discolored.

Reference standard—*USP Tetracaine Hydrochloride Reference Standard*—Dry in vacuum over phosphorus pentoxide for 18 hours before using.

Identification—It responds to *Identification tests B* and *C* under *Tetracaine Hydrochloride.*

pH ⟨791⟩: between 3.2 and 6.0.

Particulate matter ⟨788⟩: meets the requirements under *Small-volume Injections.*

Other requirements—It meets the requirements under *Injections* ⟨1⟩.

Assay—

Standard preparation—Dissolve about 20 mg of USP Tetracaine Hydrochloride RS, accurately weighed, in water to make 100.0 mL, and mix. Pipet 5 mL of this solution into a 100-mL volumetric flask, add 5 mL of dilute hydrochloric acid (1 in 200) and 10 mL of *Buffer No. 6*, 10 percent, pH 6.0 (see *Phosphate Buffers* ⟨81⟩), dilute with water to volume, and mix.

Assay preparation—Transfer an accurately measured volume of Tetracaine Hydrochloride Injection, equivalent to about 10 mg of tetracaine hydrochloride, to a separator, dilute with water to about 50 mL, and render alkaline by the addition of 5 mL of sodium carbonate TS. Extract immediately with two 50-mL portions of ether, collecting the extracts in a separator. Wash the ether extracts with 20 mL of water, discarding the wash solution, and extract the ether solution with two 20-mL portions and one 5-mL portion of dilute hydrochloric acid (1 in 200), collecting the extracts in a 50-mL volumetric flask. Dilute with water to volume, and mix. Transfer a 5.0-mL aliquot to a 100-mL volumetric flask, add 10 mL of *Buffer No. 6*, 10 percent, pH 6.0 (see *Phosphate Buffers* ⟨81⟩), dilute with water to volume, and mix.

Procedure—Concomitantly determine the absorbances of the *Assay preparation* and the *Standard preparation* at the wavelength of maximum absorbance at about 310 nm, with a suitable spectrophotometer, using water as the blank. Calculate the quantity, in mg, of $C_{15}H_{24}N_2O_2.HCl$ in the volume of Injection taken by the formula:

$$C(A_U/A_S),$$

in which C is the concentration, in µg per mL, of USP Tetracaine Hydrochloride RS in the *Standard preparation*, and A_U and A_S are the absorbances of the *Assay preparation* and the *Standard preparation*, respectively.

Tetracaine Hydrochloride Ophthalmic Solution

» Tetracaine Hydrochloride Ophthalmic Solution is a sterile, aqueous solution of Tetracaine Hydrochloride. It contains not less than 90.0 percent and not more than 110.0 percent of the labeled amount of $C_{15}H_{24}N_2O_2.HCl$. It may contain suitable antimicrobial and thickening agents.

Packaging and storage—Preserve in tight, light-resistant containers.

Labeling—Label it to indicate that the Ophthalmic Solution is not to be used if it contains crystals, or if it is cloudy or discolored.

Reference standard—*USP Tetracaine Hydrochloride Reference Standard*—Dry in vacuum over phosphorus pentoxide for 18 hours before using.

Identification—Add 5 mL of Ophthalmic Solution to 5 mL of water in a test tube, then add 1 mL of potassium thiocyanate solution (1 in 4): a crystalline precipitate is formed. Recrystallize the precipitate from water, and dry at 80° for 2 hours: the crystals so obtained melt between 130° and 132°.

Sterility—It meets the requirements under *Sterility Tests* ⟨71⟩.

pH ⟨791⟩: between 3.7 and 6.0.

Assay—

Standard preparation—Transfer about 25 mg of USP Tetracaine Hydrochloride RS, accurately weighed, to a 25-mL volumetric flask, dilute with water to volume, and mix. Pipet 5 mL of this solution into a 100-mL volumetric flask, dilute with water to volume, and mix. Pipet 10 mL of the resulting solution into a 50-mL volumetric flask, add 10 mL of dilute hydrochloric acid (1 in 200) and 5 mL of *Buffer No. 6*, 10 percent, pH 6.0 (see *Phosphate Buffers* ⟨81⟩), dilute with water to volume, and mix.

Assay preparation—Transfer an accurately measured volume of Tetracaine Hydrochloride Ophthalmic Solution, equivalent to about 10 mg of tetracaine hydrochloride, to a 100-mL volumetric flask, dilute with water to volume, and mix. Pipet 25 mL of this mixture into a 125-mL separator, add 50 mL of benzene, then add 2 mL of sodium carbonate TS, insert the stopper, and shake for 1 minute. Allow the phases to separate, discard the lower, aqueous phase, swirl the separator briefly, allow to stand, and draw off any additional aqueous phase (a slight amount of emulsified solution may be present). Extract the benzene phase for 1 minute, first with 25 mL and then with 20 mL of dilute hydrochloric acid (1 in 200), collecting the aqueous extracts in a 50-mL volumetric flask. After collecting the last extract, swirl the separator, and transfer any additional aqueous phase to the volumetric flask. Add dilute hydrochloric acid (1 in 200) to volume, and mix. Pipet 10 mL of this solution into a 50-mL volumetric flask, add 5 mL of *Buffer No. 6*, 10 percent, pH 6.0 (see *Phosphate Buffers* ⟨81⟩), dilute with water to volume, and mix.

Procedure—Concomitantly determine the absorbances of the *Assay preparation* and the *Standard preparation* at the wavelength of maximum absorbance at about 310 nm, with a suitable spectrophotometer, using water as the blank. Calculate the quantity, in mg, of $C_{15}H_{24}N_2O_2 . HCl$ in each mL of Ophthalmic Solution taken by the formula:

$$(C/V)(A_U/A_S),$$

in which C is the concentration, in µg per mL, of USP Tetracaine Hydrochloride RS in the *Standard preparation*, V is the volume, in mL, of the Ophthalmic Solution taken, and A_U and A_S are the absorbances of the *Assay preparation* and the *Standard preparation*, respectively.

Tetracaine Hydrochloride Topical Solution

» Tetracaine Hydrochloride Topical Solution is an aqueous solution of Tetracaine Hydrochloride. It contains not less than 95.0 percent and not more than 105.0 percent of the labeled amount of $C_{15}H_{24}N_2O_2 . HCl$. It contains a suitable antimicrobial agent.

Packaging and storage—Preserve in tight, light-resistant containers.

Labeling—Label it to indicate that the Topical Solution is not to be used if it contains crystals, or if it is cloudy or discolored.

Reference standard—*USP Tetracaine Hydrochloride Reference Standard*—Dry in vacuum over phosphorus pentoxide for 18 hours before using.

Identification—

A: The ultraviolet absorption spectrum of the solution employed for measurement of absorbance in the *Assay* exhibits maxima and minima at the same wavelengths as that of a similar solution of USP Tetracaine Hydrochloride RS, concomitantly measured.

B: It responds to the tests for *Chloride* ⟨191⟩.

pH ⟨791⟩: between 4.5 and 6.0.

Assay—

Standard preparation—Prepare as directed in the *Assay* under *Tetracaine Hydrochloride Injection.*

Assay preparation—Using an accurately measured volume of Tetracaine Hydrochloride Topical Solution, prepare as directed in the *Assay* under *Tetracaine Hydrochloride Injection.*

Procedure—Proceed as directed for *Procedure* in the *Assay* under *Tetracaine Hydrochloride Injection.* Calculate the quantity, in mg, of $C_{15}H_{24}N_2O_2 . HCl$ in the volume of Topical Solution taken by the formula:

$$C(A_U/A_S),$$

in which C is the concentration, in µg per mL, of USP Tetracaine Hydrochloride RS in the *Standard preparation*, and A_U and A_S are the absorbances of the *Assay preparation* and the *Standard preparation*, respectively.

Sterile Tetracaine Hydrochloride

» Sterile Tetracaine Hydrochloride is Tetracaine Hydrochloride suitable for parenteral use. It contains not less than 98.0 percent and not more than 101.0 percent of $C_{15}H_{24}N_2O_2 . HCl$, calculated on the anhydrous basis.

Packaging and storage—Preserve in *Containers for Sterile Solids* as described under *Injections* ⟨1⟩, preferably of Type I glass.

Reference standard—*USP Tetracaine Hydrochloride Reference Standard*—Dry in vacuum over phosphorus pentoxide for 18 hours before using.

Completeness of solution ⟨641⟩—A 10-mg portion dissolves in 1 mL of water in not more than 2 seconds to yield a colorless solution free from undissolved solid.

Constituted solution—At the time of use, the constituted solution prepared from Sterile Tetracaine Hydrochloride meets the requirements for *Constituted Solutions* under *Injections* ⟨1⟩.

Identification—

A: The ultraviolet absorption spectrum of the solution employed for measurement of absorbance in the *Assay* exhibits maxima and minima at the same wavelengths as that of a similar solution of USP Tetracaine Hydrochloride RS, concomitantly measured.

B: It responds to *Identification test B* under *Tetracaine Hydrochloride.*

Container content variation—It meets the requirements for *Uniformity of Dosage Units* ⟨905⟩, the contents of each of the selected containers being determined as follows. Transfer the contents of the container, with the aid of water, to a 200-mL volumetric flask, add water to volume, and mix. Pipet a portion of this solution, equivalent to about 1 mg of tetracaine hydrochloride, to a 100-mL volumetric flask, add 5 mL of dilute hydrochloric acid (1 in 200) and 10 mL of *Buffer No. 6*, 10 percent, pH 6.0 (see *Phosphate Buffers* ⟨81⟩), add water to volume, and mix. Using this as the "*Assay preparation*," proceed as directed for *Procedure* in the *Assay*. Calculate the quantity, in mg, of $C_{15}H_{24}N_2O_2 . HCl$ in each container by the formula:

$$20(C/V)(A_U/A_S),$$

in which V is the volume, in mL, of the portion used in the "*Assay preparation*," and C, A_U, and A_S are as defined in the *Assay*.

pH ⟨791⟩: between 5.0 and 6.0, in a solution (1 in 100).

Water, *Method I* ⟨921⟩: not more than 2.0%.

Residue on ignition—Weigh accurately about 500 mg, transfer to a beaker, and dissolve in 10 mL of methanol. Filter through paper previously washed with methanol, collecting the filtrate in an ignited and tared crucible and washing the beaker and the filter paper with 25 mL to 30 mL of methanol. Evaporate with the aid of heat and a current of air to dryness, and proceed as directed under *Residue on Ignition* ⟨281⟩, beginning with "Ignite, gently at first." Not more than 0.1% of residue is found.

Chromatographic purity—Dissolve an accurately weighed quantity of Sterile Tetracaine Hydrochloride in water to obtain a test solution containing 50 mg per mL, and proceed as directed in the test for *Chromatographic purity* under *Tetracaine*, beginning with "Prepare a Standard solution."

Other requirements—It meets the requirements for *Sterility Tests* ⟨71⟩ and *Labeling* under *Injections* ⟨1⟩.

Assay—

Standard preparation—Prepare as directed in the *Assay* under *Tetracaine Hydrochloride Injection.*

Assay preparation—Transfer to a tared 20-mL beaker the contents of a sufficient number of containers of Sterile Tetracaine Hydrochloride to yield about 100 mg of tetracaine hydrochloride. Weigh immediately, and transfer with the aid of water to a 500-mL volumetric flask. Add water to volume, and mix. Transfer 5.0 mL to a 100-mL volumetric flask, add 5 mL of dilute hydrochloric acid (1 in 200) and 10 mL of *Buffer No. 6*, 10 percent, pH 6.0 (see *Phosphate Buffers* ⟨81⟩), then add water to volume, and mix.

Procedure—Concomitantly determine the absorbances of the *Assay preparation* and the *Standard preparation* at the wavelength of maximum absorbance at about 310 nm, with a suitable spectrophotometer, using water as the blank. Calculate the quantity, in mg, of $C_{15}H_{24}N_2O_2 . HCl$ in the portion of Sterile Tetracaine Hydrochloride taken by the formula:

$$10C(A_U/A_S),$$

in which C is the concentration, in µg per mL, of USP Tetracaine Hydrochloride RS in the *Standard preparation*, and A_U and A_S are the absorbances of the *Assay preparation* and the *Standard preparation*, respectively.

Tetracaine Hydrochloride in Dextrose Injection

» Tetracaine Hydrochloride in Dextrose Injection is a sterile solution of Tetracaine Hydrochloride and Dextrose in Water for Injection. It contains not less than 95.0 percent and not more than 105.0 percent of the labeled amounts of tetracaine hydrochloride ($C_{15}H_{24}N_2O_2 . HCl$) and dextrose ($C_6H_{12}O_6$).

Packaging and storage—Preserve in single-dose or in multiple-dose containers, preferably of Type I glass, under refrigeration and protected from light. It may be packaged in 100-mL multiple-dose containers. Injection supplied as a component of spinal anesthesia trays may be stored at room temperature for 12 months.

Labeling—Label it to indicate that the Injection is not to be used if it contains crystals, or if it is cloudy or discolored.

Reference standard—*USP Tetracaine Hydrochloride Reference Standard*—Dry in vacuum over phosphorus pentoxide for 18 hours before using.

Identification—

A: The ultraviolet absorption spectrum of the *Assay preparation*, prepared as directed in the *Assay*, exhibits maxima and minima at the same wavelengths as that of the *Standard preparation*, prepared as directed in the *Assay*.

B: It responds to *Identification test C* under *Tetracaine Hydrochloride.*

C: It responds to the *Identification test* under *Dextrose.*

pH ⟨791⟩: between 3.5 and 6.0.

Particulate matter ⟨788⟩: meets the requirements under *Small-volume Injections.*

Other requirements—It meets the requirements under *Injections* ⟨1⟩.

Assay—

Standard preparation—Dissolve about 20 mg of USP Tetracaine Hydrochloride RS, accurately weighed, in water to make 100.0 mL, and mix. Pipet 5 mL of this solution into a 100-mL volumetric flask, add 5 mL of dilute hydrochloric acid (1 in 200) and 10 mL of *Buffer No. 6*, 10 percent, pH 6.0 (see *Phosphate Buffers* ⟨81⟩), dilute with water to volume, and mix.

Assay preparation—Transfer an accurately measured volume of Tetracaine Hydrochloride in Dextrose Injection, equivalent to about 10 mg of tetracaine hydrochloride, to a separator, dilute with water to about 50 mL, and render alkaline by the addition of 5 mL of sodium carbonate TS. Extract immediately with two 50-mL portions of ether, collecting the extracts in a separator. Wash the ether extracts with 20 mL of water, discarding the wash solution, and extract the ether solution with two 20-mL portions and one 5-mL portion of dilute hydrochloric acid (1 in 200), collecting the extracts in a 50-mL volumetric flask. Dilute with water to volume, and mix. Transfer a 5.0-mL aliquot to a 100-mL volumetric flask, add 10 mL of *Buffer No. 6*, 10 percent, pH 6.0 (see *Phosphate Buffers* ⟨81⟩), dilute with water to volume, and mix.

Procedure—Concomitantly determine the absorbances of the *Assay preparation* and the *Standard preparation* at the wavelength of maximum absorbance at about 310 nm, with a suitable spectrophotometer, using water as the blank. Calculate the quantity, in mg, of tetracaine hydrochloride ($C_{15}H_{24}N_2O_2 . HCl$) in the volume of Injection taken by the formula:

$$C(A_U/A_S),$$

in which C is the concentration, in µg per mL, of USP Tetracaine Hydrochloride RS in the *Standard preparation*, and A_U and A_S are the absorbances of the *Assay preparation* and the *Standard preparation*, respectively.

Assay for dextrose—Determine the angular rotation of Tetracaine Hydrochloride in Dextrose Injection in a suitable polarimeter tube (see *Optical Rotation* ⟨781⟩). The observed rotation, in degrees, multiplied by $9.452A$, in which A is the ratio of 200 divided by the length, in mm, of the polarimeter tube employed, represents the weight, in mg, of dextrose ($C_6H_{12}O_6$) in each mL of the Injection.

Tetracaine Hydrochlorides and Levonordefrin Injection, Procaine and—*see* Procaine and Tetracaine Hydrochlorides and Levonordefrin Injection

Tetracycline

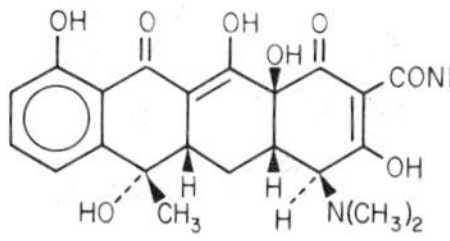

$C_{22}H_{24}N_2O_8$ 444.44

2-Naphthacenecarboxamide, 4-(dimethylamino)-1,4,4a,5,5a,6,-11,12a-octahydro-3,6,10,12,12a-pentahydroxy-6-methyl-1,11-dioxo-, [4*S*-(4α,4aα,5aα,6β,12aα)]-.

4-(Dimethylamino)-1,4,4a,5,5a,6,11,12a-octahydro-3,6,10,12,-12a-pentahydroxy-6-methyl-1,11-dioxo-2-naphthacene-carboxamide [*60-54-8*].

Trihydrate 498.49 [*6416-04-2*].

» Tetracycline has a potency equivalent to not less than 975 µg of tetracycline hydrochloride ($C_{22}H_{24}N_2O_8 . HCl$) per mg, calculated on the anhydrous basis.

Packaging and storage—Preserve in tight, light-resistant containers.

Labeling—Label it to indicate that it is to be used in the manufacture of nonparenteral drugs only.

Reference standards—*USP Tetracycline Hydrochloride Reference Standard*—Do not dry before using. *USP 4-Epianhydrotetracycline Hydrochloride Reference Standard*—Dry in vacuum at 60° for 3 hours before using.

Identification—

A: The ultraviolet absorption spectrum of a 1 in 50,000 solution in 0.25 *N* sodium hydroxide exhibits maxima and minima 6 minutes after its preparation at the same wavelengths as that of a similar preparation of USP Tetracycline Hydrochloride RS, concomitantly measured, and the absorptivity, calculated on the anhydrous basis, at the wavelength of maximum absorbance at about 380 nm is between 104.45% and 111.95% of that of the

USP Tetracycline Hydrochloride RS, the potency of the Reference Standard being taken into account.

B: The chromatogram of the *Assay preparation* obtained as directed in the *Assay* exhibits a major peak for tetracycline, the retention time of which corresponds to that exhibited in the chromatogram of the *Standard preparation* obtained as directed in the *Assay*.

C: To 0.5 mg add 2 mL of sulfuric acid: a purplish red color is produced.

Crystallinity ⟨695⟩: meets the requirements.

pH ⟨791⟩: between 3.0 and 7.0, in an aqueous suspension containing 10 mg per mL.

Water, *Method I* ⟨921⟩: not more than 13.0%.

4-Epianhydrotetracycline—Using the *Diluting solvent, Chromatographic system,* and *Procedure* set forth in the *Assay*, chromatograph a Standard solution prepared by dissolving an accurately weighed quantity of USP 4-Epianhydrotetracycline Hydrochloride RS in *Diluting solvent* to obtain a solution having a known concentration of about 10 µg per mL. Using the chromatogram so obtained and the chromatogram of the *Assay preparation* obtained as directed in the *Assay*, calculate the percentage of 4-epianhydrotetracycline in the Tetracycline taken by the formula:

$$10(C_E/W)(r_U/r_S),$$

in which C_E is the concentration, in µg per mL, of USP 4-Epianhydrotetracycline Hydrochloride RS in the Standard solution, W is the weight, in mg, of Tetracycline taken to prepare the *Assay preparation*, and r_U and r_S are the 4-epianhydrotetracycline peak responses obtained from the *Assay preparation* and the Standard solution, respectively: not more than 2.0% is found.

Assay—

Diluting solvent, Mobile phase, Standard preparation, Resolution solution, and *Chromatographic system*—Prepare as directed in the *Assay* under *Tetracycline Hydrochloride*.

Assay preparation—Transfer about 45 mg of Tetracycline, accurately weighed, to a 100-mL volumetric flask, dissolve in *Diluting solvent*, dilute with the same solvent to volume, and mix.

Procedure—Proceed as directed in the *Assay* under *Tetracycline Hydrochloride*. Calculate the quantity, in µg, of tetracycline hydrochloride ($C_{22}H_{24}N_2O_8 . HCl$) equivalent in each mg of Tetracycline taken by the formula:

$$100(CP/W)(r_U/r_S),$$

in which W is the weight, in mg, of Tetracycline taken to prepare the *Assay preparation*, and the other terms are as defined therein.

Tetracycline Boluses

» Tetracycline Boluses contain the equivalent of not less than 90.0 percent and not more than 120.0 percent of the labeled amount of tetracycline hydrochloride ($C_{22}H_{24}N_2O_8 . HCl$).

Packaging and storage—Preserve in tight containers.

Labeling—Label Boluses to indicate that they are intended for veterinary use only.

Reference standard—*USP Tetracycline Hydrochloride Reference Standard*—Do not dry before using.

Identification—Shake a suitable quantity of finely powdered Boluses with methanol to obtain a solution containing the equivalent of 1 mg of tetracycline hydrochloride per mL, and filter. Using the filtrate as the *Test Solution*, proceed as directed under *Identification—Tetracyclines* ⟨193⟩.

Uniformity of dosage units ⟨905⟩: meet the requirements for *Weight Variation*.

Loss on drying ⟨731⟩—Dry about 100 mg of finely powdered Boluses, accurately weighed, in a capillary-stoppered bottle in vacuum at a pressure not exceeding 5 mm of mercury at 60° for 3 hours: it loses not more than 3.0% of its weight, or where the Boluses have a diameter of greater than 15 mm, it loses not more than 6.0% of its weight.

Assay—Transfer not less than 2 Tetracycline Boluses to a high-speed blender jar containing an accurately measured volume of 0.1 *N* hydrochloric acid, so that the solution so obtained contains not less than 150 µg of tetracycline hydrochloride per mL, and blend for about 3 to 5 minutes. Proceed as directed for tetracycline under *Antibiotics—Microbial Assays* ⟨81⟩, using an accurately measured volume of this solution diluted quantitatively and stepwise with water to obtain a *Test Dilution* having a concentration assumed to be equal to the median dose level of the Standard.

Tetracycline Oral Suspension

» Tetracycline Oral Suspension is Tetracycline with or without one or more suitable buffers, preservatives, stabilizers, and suspending agents. It contains the equivalent of not less than 90.0 percent and not more than 125.0 percent of the labeled amount of tetracycline hydrochloride ($C_{22}H_{24}N_2O_8 . HCl$).

Packaging and storage—Preserve in tight, light-resistant containers.

Reference standards—*USP Tetracycline Hydrochloride Reference Standard*—Do not dry before using. *USP 4-Epianhydrotetracycline Hydrochloride Reference Standard*—Dry in vacuum at 60° for 3 hours before using.

Identification—The chromatogram of the *Assay preparation* obtained as directed in the *Assay* exhibits a major peak for tetracycline, the retention time of which corresponds to that exhibited in the chromatogram of the *Standard preparation* obtained as directed in the *Assay*.

Uniformity of dosage units ⟨905⟩—

FOR SUSPENSION PACKAGED IN SINGLE-UNIT CONTAINERS: meets the requirements.

pH ⟨791⟩: between 3.5 and 6.0.

4-Epianhydrotetracycline—Using the *Diluting solvent, Chromatographic system,* and *Procedure* set forth in the *Assay*, chromatograph a Standard solution prepared by dissolving an accurately weighed quantity of USP 4-Epianhydrotetracycline Hydrochloride RS in *Diluting solvent* to obtain a solution having a known concentration of about 10 µg per mL. Using the chromatogram so obtained and the chromatogram of the *Assay preparation* obtained as directed in the *Assay*, calculate the percentage of 4-epianhydrotetracycline in the Oral Suspension taken by the formula:

$$(25C_E/T)(r_U/r_S),$$

in which C_E is the concentration, in µg per mL, of USP 4-Epianhydrotetracycline Hydrochloride RS in the Standard solution, T is the quantity, in mg, of tetracycline hydrochloride equivalent in the portion of Oral Suspension taken, based on the labeled quantity, and r_U and r_S are the 4-epianhydrotetracycline peak responses obtained from the *Assay preparation* and the Standard solution, respectively: not more than 5.0% is found.

Assay—

Diluting solvent, Mobile phase, Standard preparation, Resolution solution, and *Chromatographic system*—Prepare as directed in the *Assay* under *Tetracycline Hydrochloride*.

Assay preparation—Transfer an accurately measured volume of Tetracycline Oral Suspension, equivalent to about 125 mg of tetracycline hydrochloride, to a 250-mL volumetric flask, add 200 mL of *Diluting solvent*, and shake. Add *Diluting solvent* to volume, mix, and filter.

Procedure—Proceed as directed in the *Assay* under *Tetracycline Hydrochloride*. Calculate the quantity, in mg per mL, of $C_{22}H_{24}N_2O_8 . HCl$ equivalent in the Oral Suspension taken by the formula:

$$(CP/4V)(r_U/r_S),$$

in which V is the volume, in mL, of Oral Suspension taken, and the other terms are as defined therein.

Tetracycline Hydrochloride

$C_{22}H_{24}N_2O_8 . HCl$ 480.90
2-Naphthacenecarboxamide, 4-(dimethylamino)-1,4,4a,5,5a,6,-11,12a-octahydro-3,6,10,12,12a-pentahydroxy-6-methyl-1,-11-dioxo-, monohydrochloride, [4*S*-(4α,4aα,5aα,6β,12aα)]-.
4-(Dimethylamino)-1,4,4a,5,5a,6,11,12a-octahydro-3,6,10,12,-12a-pentahydroxy-6-methyl-1,11-dioxo-2-naphthacenecarboxamide monohydrochloride [*64-75-5*].

» Tetracycline Hydrochloride has a potency of not less than 900 µg of $C_{22}H_{24}N_2O_8 . HCl$ per mg.

Packaging and storage—Preserve in tight, light-resistant containers.

Reference standards—*USP Tetracycline Hydrochloride Reference Standard*—Do not dry before using. *USP 4-Epianhydrotetracycline Hydrochloride Reference Standard*—Dry in vacuum at 60° for 3 hours before using.

Identification—

A: The infrared absorption spectrum of a potassium bromide dispersion of it exhibits maxima only at the same wavelengths as that of a similar preparation of USP Tetracycline Hydrochloride RS.

B: The ultraviolet absorption spectrum of a 1 in 50,000 solution in 0.25 *N* sodium hydroxide exhibits maxima and minima 6 minutes after its preparation at the same wavelengths as that of a similar preparation of USP Tetracycline Hydrochloride RS, concomitantly measured, and the absorptivity, calculated on the dried basis, at the wavelength of maximum absorbance at about 380 nm is between 96.0% and 104.0% of that of the USP Tetracycline Hydrochloride RS, the potency of the Reference Standard being taken into account.

C: The chromatogram of the *Assay preparation* obtained as directed in the *Assay* exhibits a major peak for tetracycline, the retention time of which corresponds to that exhibited in the chromatogram of the *Standard preparation* obtained as directed in the *Assay*.

Crystallinity ⟨695⟩: meets the requirements.

pH ⟨791⟩: between 1.8 and 2.8, in a solution containing 10 mg per mL.

Loss on drying ⟨731⟩—Dry about 100 mg, accurately weighed, in a capillary-stoppered bottle in vacuum at a pressure not exceeding 5 mm of mercury at 60° for 3 hours: it loses not more than 2.0% of its weight.

4-Epianhydrotetracycline—Using the *Diluting solvent*, *Chromatographic system*, and *Procedure* set forth in the *Assay*, chromatograph a Standard solution prepared by dissolving an accurately weighed quantity of USP 4-Epianhydrotetracycline Hydrochloride RS in *Diluting solvent* to obtain a solution having a known concentration of about 10 µg per mL. Using the chromatogram so obtained and the chromatogram of the *Assay preparation* obtained as directed in the *Assay*, calculate the percentage of 4-epianhydrotetracycline hydrochloride in the Tetracycline Hydrochloride taken by the formula:

$$10(C_E/W)(r_U/r_S),$$

in which C_E is the concentration, in µg per mL, of USP 4-Epianhydrotetracycline Hydrochloride RS in the Standard solution, W is the weight, in mg, of Tetracycline Hydrochloride taken to prepare the *Assay preparation*, and r_U and r_S are the 4-epianhydrotetracycline peak responses obtained from the *Assay preparation* and the Standard solution, respectively: not more than 2.0% is found.

Assay—

Diluting solvent—Mix 680 mL of 0.1 *M* ammonium oxalate and 270 mL of dimethylformamide.

Mobile phase—Mix 680 mL of 0.1 *M* ammonium oxalate, 270 mL of dimethylformamide, and 50 mL of 0.2 *M* dibasic ammonium phosphate. Adjust, if necessary, with 3 *N* ammonium hydroxide or 3 *N* phosphoric acid to a pH of 7.6 to 7.7. Make any other necessary adjustments (see *System Suitability* under *Chromatography* ⟨621⟩). Filter through a membrane filter of 0.5-µm or finer porosity.

Standard preparation—Dissolve an accurately weighed quantity of USP Tetracycline Hydrochloride RS in *Diluting solvent*, and dilute quantitatively with *Diluting solvent* to obtain a solution having a known concentration of about 0.5 mg per mL.

Assay preparation—Transfer about 50 mg of Tetracycline Hydrochloride, accurately weighed, to a 100-mL volumetric flask, dissolve in *Diluting solvent*, dilute with the same solvent to volume, and mix.

Resolution solution—Prepare a solution in *Diluting solvent* containing about 100 µg of tetracycline hydrochloride and 25 µg of USP 4-Epianhydrotetracycline Hydrochloride RS per mL.

Chromatographic system (see *Chromatography* ⟨621⟩)—The liquid chromatograph is equipped with a 280-nm detector, a 4.6-mm × 3-cm guard column that contains 10-µm packing L7, and a 4.6-mm × 25-cm analytical column that contains 5- to 10-µm packing L7. The flow rate is about 2 mL per minute. Chromatograph the *Resolution solution*, and record the peak responses as directed under *Procedure:* the relative retention times are about 0.9 for 4-epianhydrotetracycline and 1.0 for tetracycline, and the resolution, *R*, between the 4-epianhydrotetracycline and tetracycline peaks is not less than 1.2. Chromatograph the *Standard preparation*, and record the peak responses as directed under *Procedure:* the relative standard deviation for replicate injections is not more than 2.0%.

Procedure—Separately inject equal volumes (about 20 µL) of the *Standard preparation* and the *Assay preparation* into the chromatograph, record the chromatograms, and measure the responses for the major peaks. Calculate the quantity, in µg, of tetracycline hydrochloride ($C_{22}H_{24}N_2O_8 . HCl$) in each mg of the Tetracycline Hydrochloride taken by the formula:

$$100(CP/W)(r_U/r_S),$$

in which C is the concentration, in mg per mL, of USP Tetracycline Hydrochloride RS in the *Standard preparation*, P is the potency, in µg per mg, of USP Tetracycline Hydrochloride RS, W is the weight, in mg, of Tetracycline Hydrochloride taken to prepare the *Assay preparation*, and r_U and r_S are the peak responses obtained from the *Assay preparation* and the *Standard preparation*, respectively.

Tetracycline Hydrochloride Capsules

» Tetracycline Hydrochloride Capsules contain not less than 90.0 percent and not more than 125.0 percent of the labeled amount of $C_{22}H_{24}N_2O_8 . HCl$.

Packaging and storage—Preserve in tight, light-resistant containers.

Reference standards—*USP Tetracycline Hydrochloride Reference Standard*—Do not dry before using. *USP 4-Epianhydrotetracycline Hydrochloride Reference Standard*—Dry in vacuum at 60° for 3 hours before using.

Identification—The chromatogram of the *Assay preparation* obtained as directed in the *Assay* exhibits a major peak for tetracycline, the retention time of which corresponds to that exhibited in the chromatogram of the *Standard preparation* obtained as directed in the *Assay*.

Dissolution ⟨711⟩—

Medium: water; 900 mL.

Apparatus 2: 75 rpm. Maintain a distance of 45 ± 5 mm between the blade and the inside bottom of the vessel.

Time: 60 minutes.

Procedure—Determine the amount of $C_{22}H_{24}N_2O_8 . HCl$ dissolved from ultraviolet absorbances at the wavelength of maximum absorbance at about 276 nm of filtered portions of the solution under test, suitably diluted with *Dissolution Medium*, if necessary, in comparison with a Standard solution having a known concentration of USP Tetracycline Hydrochloride RS in the same medium.

Tolerances—Not less than 70% (*Q*) of the labeled amount of $C_{22}H_{24}N_2O_8 \cdot HCl$ is dissolved in 60 minutes.

Uniformity of dosage units ⟨905⟩: meet the requirements.

Loss on drying ⟨731⟩—Dry about 100 mg of Capsule contents, accurately weighed, in a capillary-stoppered bottle in vacuum at a pressure not exceeding 5 mm of mercury at 60° for 3 hours: it loses not more than 4.0% of its weight.

4-Epianhydrotetracycline—Using the *Diluting solvent*, *Chromatographic system*, and *Procedure* set forth in the *Assay*, chromatograph a Standard solution prepared by dissolving an accurately weighed quantity of USP 4-Epianhydrotetracycline Hydrochloride RS in *Diluting solvent* to obtain a solution having a known concentration of about 10 µg per mL. Using the chromatogram so obtained and the chromatogram of the *Assay preparation* obtained as directed in the *Assay*, calculate the percentage of 4-epianhydrotetracycline hydrochloride in the Capsules taken by the formula:

$$(10C_E/T)(r_U/r_S),$$

in which C_E is the concentration, in µg per mL, of USP 4-Epianhydrotetracycline Hydrochloride RS in the Standard solution, T is the quantity, in mg, of tetracycline hydrochloride in the portion of Capsules taken to prepare the *Assay preparation*, based on the labeled quantity, and r_U and r_S are the 4-epianhydrotetracycline peak responses obtained from the *Assay preparation* and the Standard solution, respectively: not more than 3.0% is found.

Assay—

Diluting solvent, *Mobile phase*, *Standard preparation*, *Resolution solution*, and *Chromatographic system*—Prepare as directed in the *Assay* under *Tetracycline Hydrochloride*.

Assay preparation—Weigh accurately not less than 20 Tetracycline Hydrochloride Capsules. Empty the Capsule contents into a mortar. Clean and accurately weigh the Capsule shells, and calculate the net weight of the Capsule contents. Mix the powder in the mortar, using a pestle, and transfer an accurately weighed quantity of the freshly mixed powder, equivalent to about 50 mg of tetracycline hydrochloride, to a 100-mL volumetric flask. Add about 50 mL of *Diluting solvent*, mix, and sonicate for about 5 minutes. Allow to cool, dilute with *Diluting solvent* to volume, mix, and filter.

Procedure—Proceed as directed in the *Assay* under *Tetracycline Hydrochloride*. Calculate the quantity, in mg, of $C_{22}H_{24}N_2O_8 \cdot HCl$ in the portion of Capsules taken by the formula:

$$(CP/10)(r_U/r_S),$$

in which the terms are as defined therein.

Tetracycline Hydrochloride for Injection

» Tetracycline Hydrochloride for Injection is a sterile, dry mixture of Sterile Tetracycline Hydrochloride, one form of which contains Magnesium Chloride or magnesium ascorbate and one or more suitable buffers, and may contain one or more suitable preservatives, solubilizers, stabilizers, and anesthetic agents, and the other form of which contains one or more suitable stabilizing agents. It contains not less than 90.0 percent and not more than 115.0 percent of the labeled amount of $C_{22}H_{24}N_2O_8 \cdot HCl$.

Packaging and storage—Preserve in *Containers for Sterile Solids* as described under *Injections* ⟨1⟩, protected from light.

Labeling—Label Tetracycline Hydrochloride for Injection that contains an anesthetic agent to indicate that it is intended for intramuscular administration only.

Reference standards—*USP Tetracycline Hydrochloride Reference Standard*—Do not dry before using. *USP 4-Epianhydrotetracycline Hydrochloride Reference Standard*—Dry in vacuum at 60° for 3 hours before using.

Constituted solution—At the time of use, the constituted solution prepared from Tetracycline Hydrochloride for Injection meets the requirements for *Constituted Solutions* under *Injections* ⟨1⟩.

Identification—The chromatogram of the *Assay preparation* obtained as directed in the *Assay* exhibits a major peak for tetracycline, the retention time of which corresponds to that exhibited in the chromatogram of the *Standard preparation* obtained as directed in the *Assay*.

Pyrogen—It meets the requirements of the *Pyrogen Test* ⟨151⟩, the test dose being 1.0 mL per kg of a solution prepared to contain 5.0 mg of tetracycline hydrochloride per mL of Sterile Water for Injection.

Sterility—It meets the requirements under *Sterility Tests* ⟨71⟩, when tested as directed in the section, *Test Procedures Using Membrane Filtration*, *Fluid D* being used instead of *Fluid A*.

pH ⟨791⟩: between 2.0 and 3.0, in a solution containing 10 mg per mL.

Loss on drying ⟨731⟩—Dry about 100 mg, accurately weighed, in a capillary-stoppered bottle in vacuum at a pressure not exceeding 5 mm of mercury at 60° for 3 hours: it loses not more than 5.0% of its weight.

Particulate matter ⟨788⟩: meets the requirements under *Small-volume Injections*.

4-Epianhydrotetracycline—Using the *Diluting solvent*, *Chromatographic system*, and *Procedure* set forth in the *Assay*, chromatograph a Standard solution prepared by dissolving an accurately weighed quantity of USP 4-Epianhydrotetracycline Hydrochloride RS in *Diluting solvent* to obtain a solution having a known concentration of about 15 µg per mL. Using the chromatogram so obtained and the chromatogram of the *Assay preparation* obtained as directed in the *Assay*, calculate the percentage of 4-epianhydrotetracycline hydrochloride in the Tetracycline Hydrochloride for Injection taken by the formula:

$$(10C_E/T)(r_U/r_S),$$

in which C_E is the concentration, in µg per mL, of USP 4-Epianhydrotetracycline Hydrochloride RS in the Standard solution, T is the quantity, in mg, of tetracycline hydrochloride in the portion of Tetracycline Hydrochloride for Injection taken to prepare the *Assay preparation*, based on the labeled quantity, and r_U and r_S are the 4-epianhydrotetracycline peak responses obtained from the *Assay preparation* and the Standard solution, respectively: not more than 3.0% is found.

Other requirements—It meets the requirements for *Uniformity of Dosage Units* ⟨905⟩ and for *Labeling* under *Injections* ⟨1⟩. Where it is labeled for intravenous use it meets the requirements for *Depressor substances* under *Sterile Tetracycline Hydrochloride*.

Assay—

Diluting solvent, *Mobile phase*, *Standard preparation*, *Resolution solution*, and *Chromatographic system*—Prepare as directed in the *Assay* under *Tetracycline Hydrochloride*.

Assay preparation 1 (where it is represented as being in a single-dose container)—Constitute Tetracycline Hydrochloride for Injection as directed in the labeling. Withdraw all of the withdrawable contents, using a suitable hypodermic needle and syringe, and dilute quantitatively and stepwise with *Diluting solvent* to obtain a solution containing about 0.5 mg of tetracycline hydrochloride per mL.

Assay preparation 2 (where the label states the quantity of tetracycline hydrochloride in a given volume of constituted solution)—Constitute Tetracycline Hydrochloride for Injection as directed in the labeling. Dilute an accurately measured volume of the constituted solution quantitatively with *Diluting solvent* to obtain a solution containing about 0.5 mg of tetracycline hydrochloride per mL.

Procedure—Proceed as directed in the *Assay* under *Tetracycline Hydrochloride*. Calculate the quantity, in mg, of $C_{22}H_{24}N_2O_8 \cdot HCl$ withdrawn from the container, or in the portion of constituted solution taken by the formula:

$$(L/D)(CP/1000)(r_U/r_S),$$

in which L is the labeled quantity, in mg, of $C_{22}H_{24}N_2O_8 . HCl$ in the container, or in the volume of constituted solution taken, D is the concentration, in mg of tetracycline hydrochloride per mL, of *Assay preparation 1* or *Assay preparation 2*, based on the labeled quantity in the container or in the portion of constituted solution taken, respectively, and the extent of dilution, and the other terms are as defined therein.

Tetracycline Hydrochloride Ointment

» Tetracycline Hydrochloride Ointment contains not less than 90.0 percent and not more than 125.0 percent of the labeled amount of $C_{22}H_{24}N_2O_8 . HCl$.

Packaging and storage—Preserve in well-closed containers, preferably at controlled room temperature.

Reference standards—*USP Tetracycline Hydrochloride Reference Standard*—Do not dry before using. *USP 4-Epianhydrotetracycline Hydrochloride Reference Standard*—Dry in vacuum at 60° for 3 hours before using.

Identification—The chromatogram of the *Assay preparation* obtained as directed in the *Assay* exhibits a major peak for tetracycline, the retention time of which corresponds to that exhibited in the chromatogram of the *Standard preparation* obtained as directed in the *Assay*.

Minimum fill ⟨755⟩: meets the requirements.

Water, *Method I* ⟨921⟩: not more than 1.0%, 20 mL of a mixture of carbon tetrachloride, chloroform, and methanol (2:2:1) being used in place of methanol in the titration vessel.

Assay—

Diluting solvent, Mobile phase, Standard preparation, Resolution solution, and *Chromatographic system*—Prepare as directed in the *Assay* under *Tetracycline Hydrochloride*.

Assay preparation—Transfer an accurately weighed portion of Tetracycline Hydrochloride Ointment, equivalent to about 50 mg of tetracycline hydrochloride, with the aid of 30 mL of cyclohexane to a 125-mL separator. Add 30 mL of *Diluting solvent*, insert the stopper, and shake. Allow to separate, and collect the lower layer in a 100-mL volumetric flask. Repeat the extraction with two additional 30-mL portions of *Diluting solvent*, combining the extracts in the 100-mL volumetric flask. Add *Diluting solvent* to volume, mix, and filter.

Procedure—Proceed as directed in the *Assay* under *Tetracycline Hydrochloride*. Calculate the quantity, in mg, of $C_{22}H_{24}N_2O_8 . HCl$ in the portion of Ointment taken by the formula:

$$(CP/10)(r_U/r_S),$$

in which the terms are as defined therein.

Tetracycline Hydrochloride Ophthalmic Ointment

» Tetracycline Hydrochloride Ophthalmic Ointment contains not less than 90.0 percent and not more than 125.0 percent of the labeled amount of $C_{22}H_{24}N_2O_8 . HCl$.

Packaging and storage—Preserve in collapsible ophthalmic ointment tubes.

Reference standards—*USP Tetracycline Hydrochloride Reference Standard*—Do not dry before using. *USP 4-Epianhydrotetracycline Hydrochloride Reference Standard*—Dry in vacuum at 60° for 3 hours before using.

Sterility—It meets the requirements for *Ophthalmic Ointments* under *Sterility Tests* ⟨71⟩.

Minimum fill ⟨755⟩: meets the requirements.

Water, *Method I* ⟨921⟩: not more than 0.5%, 20 mL of a mixture of carbon tetrachloride, chloroform, and methanol (2:2:1) being used in place of methanol in the titration vessel.

Metal particles—It meets the requirements of the test for *Metal Particles in Ophthalmic Ointments* ⟨751⟩.

Assay—Proceed with Tetracycline Hydrochloride Ophthalmic Ointment as directed in the *Assay* under *Tetracycline Hydrochloride Ointment*.

Tetracycline Hydrochloride Soluble Powder

» Tetracycline Hydrochloride Soluble Powder contains not less than 90.0 percent and not more than 125.0 percent of the labeled amount of $C_{22}H_{24}N_2O_8 . HCl$.

Packaging and storage—Preserve in tight containers.

Labeling—Label it to indicate that it is intended for veterinary use only.

Reference standard—*USP Tetracycline Hydrochloride Reference Standard*—Do not dry before using.

Identification—Shake a suitable quantity of Soluble Powder with methanol to obtain a solution containing the equivalent of 1 mg of tetracycline hydrochloride per mL, and filter. Using the filtrate so obtained as the *Test Solution*, proceed as directed under *Identification—Tetracyclines* ⟨193⟩.

Loss on drying ⟨731⟩—Dry about 100 mg, accurately weighed, in a capillary-stoppered bottle in vacuum at a pressure not exceeding 5 mm of mercury at 60° for 3 hours: it loses not more than 2.0% of its weight.

Assay—Transfer an accurately weighed quantity of Tetracycline Hydrochloride Soluble Powder to a high-speed blender jar containing an accurately measured volume of 0.1 N hydrochloric acid, so that the solution so obtained contains not less than 150 µg of tetracycline hydrochloride per mL, and blend for about 3 to 5 minutes. Proceed as directed for tetracycline under *Antibiotics—Microbial Assays* ⟨81⟩, diluting an accurately measured volume of this solution quantitatively and stepwise with water to yield a *Test Dilution* having a concentration assumed to be equal to the median dose level of the Standard.

Tetracycline Hydrochloride for Topical Solution

» Tetracycline Hydrochloride for Topical Solution is a dry mixture of Tetracycline Hydrochloride and Epitetracycline Hydrochloride with Sodium Metabisulfite packaged in conjunction with a suitable aqueous vehicle. It contains not less than 90.0 percent and not more than 130.0 percent of the labeled amount of tetracycline hydrochloride, when constituted as directed.

Packaging and storage—Preserve in tight, light-resistant containers.

Reference standard—*USP Tetracycline Hydrochloride Reference Standard*—Do not dry before using.

Identification—Dissolve a suitable quantity in methanol to obtain a solution containing 1 mg of tetracycline hydrochloride per mL, and filter, if necessary, to obtain a clear solution. Using the clear solution as the *Test solution*, proceed as directed under *Identification—Tetracyclines* ⟨193⟩: the specified result is obtained.

pH ⟨791⟩: between 1.9 and 3.5, in the solution constituted as directed in the labeling.

Loss on drying ⟨731⟩—Dry the contents of 1 container, accurately weighed, in a capillary-stoppered bottle in vacuum at a pressure not exceeding 5 mm of mercury at 60° for 3 hours: it loses not more than 5.0% of its weight.

Epitetracycline hydrochloride content and Assay for tetracycline hydrochloride—

Edetate disodium solution—Dissolve 74.4 g of edetate disodium in about 1800 mL of water, adjust with ammonium hydroxide to a pH of 7.0, dilute with water to 2000 mL, and mix.

Stationary phase—Mix 95 mL of *Edetate disodium solution* and 5 mL of a mixture of glycerine and polyethylene glycol 400 (4:1).

Alkaline methanol solution—On the day of use, prepare a mixture of methanol and ammonium hydroxide (19:1).

Column support—Suspend 300 g of chromatographic siliceous earth in 2000 mL of 6 *N* hydrochloric acid in a suitable vessel, and mix for about 15 minutes. Filter, and wash the siliceous earth with water until the last washing is neutral to moistened litmus paper. Suspend the washed siliceous earth in 2000 mL of a mixture of ethyl acetate and methanol (1:1), and mix for about 15 minutes. Filter, and dry the siliceous earth in vacuum at 60° for about 16 hours. Weigh, add 0.5 mL of *Stationary phase* for each g of dried siliceous earth, and shake until the column packing is uniformly moist. Store in a tight container.

Chromatographic column—Prepare as directed under *Column Partition Chromatography* ⟨621⟩, using a 10- × 300-mm chromatographic tube equipped with a solvent reservoir on the top and a stopcock on the bottom and packed with 8 ± 0.1 g of *Column support*.

Standard preparation—Transfer about 22 mg of USP Tetracycline Hydrochloride RS, accurately weighed, to a 25-mL volumetric flask, add 1 mL of methanol, and swirl to dissolve. Dilute with *Stationary phase* to volume, and mix. Transfer 2.0 mL of this solution to a 10-mL volumetric flask, dilute with *Stationary phase* to volume, and mix. Pipet 2.0 mL of the resulting solution into the *Chromatographic column*, and allow it to permeate the *Column support*. Add 20 mL of benzene to the solvent reservoir, and collect the eluate at the rate of about 1 mL per minute, using a 50-mL graduated cylinder as a receiver. When the benzene level reaches the top of *Column support*, add 60 mL of chloroform to the solvent reservoir, and continue collecting the eluate until 30 mL has been collected. Discard this eluate, and continue collecting the eluate in a low-actinic, 50-mL volumetric flask. When the chloroform level reaches the top of the *Column support*, add 10 mL of a mixture of butyl alcohol and chloroform (1:1) to the solvent reservoir, and replace the low-actinic, 50-mL volumetric flask with a 10-mL graduated cylinder. Collect 8 mL of the eluate, close the column stopcock, and transfer the eluate to the low-actinic, 50-mL volumetric flask. Rinse the graduated cylinder with 2 mL of chloroform, and add the rinsing to the volumetric flask. The eluate in the 50-mL volumetric flask is the *Standard preparation*.

Assay preparation—Constitute Tetracycline Hydrochloride for Topical Solution as directed in the labeling. Transfer an accurately measured volume of the constituted Topical Solution, equivalent to about 4.4 mg of tetracycline hydrochloride, to a 25-mL volumetric flask, dilute with *Stationary phase* to volume, and mix. Pipet 2.0 mL of this solution into the *Chromatographic column*, and allow it to penetrate the *Column support*. Add 20 mL of benzene to the solvent reservoir, and collect the eluate at the rate of about 1 mL per minute, using a 50-mL graduated cylinder as a receiver. When the benzene level reaches the top of the *Column support*, add 60 mL of chloroform to the solvent reservoir, and continue collecting the eluate until 30 mL has been collected. Discard this eluate, and continue collecting the eluate in a low-actinic, 50-mL volumetric flask. When the chloroform level reaches the top of the *Column support*, add 50 mL of a mixture of butyl alcohol and chloroform (1:1) to the solvent reservoir, and replace the low-actinic, 50-mL volumetric flask with a 10-mL graduated cylinder. Collect 8 mL of the eluate, close the column stopcock, and transfer the eluate to the low-actinic, 50-mL volumetric flask. Rinse the graduated cylinder with 2 mL of chloroform, and add the rinsing to the volumetric flask. The eluate in the 50-mL volumetric flask is the *Assay preparation*.

Test solution for epitetracycline hydrochloride content—Open the column stopcock of the *Chromatographic column* remaining from the *Assay preparation*, and collect the eluate in a low-actinic, 50-mL volumetric flask until the column runs dry.

Procedure—Add 2.0 mL of *Alkaline methanol solution* to the *Standard preparation*, to the *Assay preparation*, and to the *Test solution for epitetracycline hydrochloride content*, dilute each with chloroform to volume, and mix. Concomitantly, within 10 minutes of preparation, determine the absorbances of these solutions at the wavelength of maximum absorbance at about 366 nm, with a suitable spectrophotometer, using chloroform as the blank. Calculate the quantity, in mg, of tetracycline hydrochloride in each mL of constituted Topical Solution taken by the formula:

$$0.0002(WP/V)(A_U/A_S),$$

in which W is the weight, in mg, of USP Tetracycline Hydrochloride RS taken, P is the potency, in µg per mg, of the USP Tetracycline Hydrochloride RS, V is the volume, in mL, of constituted Topical Solution taken, and A_U and A_S are the absorbances of the solutions from the *Assay preparation* and the *Standard preparation*, respectively. Calculate the quantity, in mg, of epitetracycline hydrochloride ($C_{22}H_{24}N_2O_8 \cdot HCl$) in each mL of constituted Topical Solution taken by the same formula, in which A_U is the absorbance of the solution from the *Test solution for epitetracycline hydrochloride content*. The quantity of epitetracycline hydrochloride found in each mL of constituted Topical Solution is between 115.0% and 140.0% of the quantity of tetracycline hydrochloride found.

Sterile Tetracycline Hydrochloride

» Sterile Tetracycline Hydrochloride is Tetracycline Hydrochloride suitable for parenteral use. It has a potency of not less than 900 µg of $C_{22}H_{24}N_2O_8 \cdot HCl$ per mg and, where packaged for dispensing, it contains the equivalent of not less than 90.0 percent and not more than 115.0 percent of the labeled amount of tetracycline hydrochloride ($C_{22}H_{24}N_2O_8 \cdot HCl$).

Packaging and storage—Preserve in *Containers for Sterile Solids* as described under *Injections* ⟨1⟩, protected from light.

Reference standards—*USP Tetracycline Hydrochloride Reference Standard*—Do not dry before using. *USP 4-Epianhydrotetracycline Hydrochloride Reference Standard*—Dry in vacuum at 60° for 3 hours before using.

Constituted solution—At the time of use, the constituted solution prepared from Sterile Tetracycline Hydrochloride meets the requirements for *Constituted Solutions* under *Injections* ⟨1⟩.

Depressor substances—It meets the requirements of the *Depressor Substances Test* ⟨101⟩, the test dose being 0.6 mL per kg of a solution prepared to contain 5.0 mg of tetracycline hydrochloride per mL in sterile saline TS.

Pyrogen—It meets the requirements of the *Pyrogen Test* ⟨151⟩, the test dose being 1.0 mL per kg of a solution prepared to contain 5.0 mg of tetracycline hydrochloride per mL of Sterile Water for Injection.

Sterility—It meets the requirements under *Sterility Tests* ⟨71⟩, when tested as directed in the section, *Test Procedures Using Membrane Filtration, Fluid D* being used instead of *Fluid A*.

Particulate matter ⟨788⟩: meets the requirements under *Small-volume Injections*.

Other requirements—It responds to the *Identification tests* and meets the requirements for *Crystallinity, pH, Loss on drying*, and *4-Epianhydrotetracycline* under *Tetracycline Hydrochloride* and, where packaged for dispensing, it meets the requirements for *Uniformity of Dosage Units* ⟨905⟩ and *Labeling* under *Injections* ⟨1⟩.

Assay—

Diluting solvent, Mobile phase, Standard preparation, Resolution solution, and *Chromatographic system*—Prepare as directed in the *Assay* under *Tetracycline Hydrochloride*.

Assay preparation 1—Dissolve an accurately weighed quantity of Sterile Tetracycline Hydrochloride in an accurately measured volume of *Diluting solvent* to obtain a solution having a concentration of about 0.5 mg of tetracycline hydrochloride per mL.

Assay preparation 2 (where it is packaged for dispensing)—Constitute Sterile Tetracycline Hydrochloride as directed in the labeling. Withdraw all of the withdrawable contents, using a suitable hypodermic needle and syringe, and dilute quantitatively and stepwise with *Diluting solvent* to obtain a solution having a concentration of about 0.5 mg of tetracycline hydrochloride per mL.

Assay preparation 3 (where the label states the quantity of tetracycline hydrochloride in a given volume of constituted solution)—Constitute Sterile Tetracycline Hydrochloride as directed in the labeling. Dilute an accurately measured volume of the constituted solution quantitatively with *Diluting solvent* to obtain a solution having a concentration of about 0.5 mg of tetracycline hydrochloride per mL.

Procedure—Proceed in the *Assay* under *Tetracycline Hydrochloride.* Calculate the quantity, in µg, of $C_{22}H_{24}N_2O_8 . HCl$ in each mg of Sterile Tetracycline Hydrochloride taken by the formula:

$$(CP/M)(r_U/r_S),$$

in which M is the concentration, in mg per mL, of *Assay preparation 1*, based on the weight of Sterile Tetracycline Hydrochloride taken and the extent of dilution, and the other terms are as defined therein. Calculate the quantity, in mg, of $C_{22}H_{24}N_2O_8 . HCl$ withdrawn from the container, or in the portion of constituted solution taken by the formula:

$$(L/D)(CP/1000)(r_U/r_S),$$

in which L is the labeled quantity, in mg, of $C_{22}H_{24}N_2O_8 . HCl$ in the container, or in the volume of constituted solution taken, and D is the concentration, in mg of tetracycline hydrochloride per mL, of *Assay preparation 2* or *Assay preparation 3*, based on the labeled quantity in the container or in the portion of constituted solution taken, respectively, and the extent of dilution, and the other terms are as defined therein.

Tetracycline Hydrochloride Ophthalmic Suspension

» Tetracycline Hydrochloride Ophthalmic Suspension is a sterile suspension of Sterile Tetracycline Hydrochloride in a suitable oil. It contains not less than 90.0 percent and not more than 125.0 percent of the labeled amount of $C_{22}H_{24}N_2O_8 . HCl$.

Packaging and storage—Preserve in tight, light-resistant containers of glass or plastic, containing not more than 15 mL. The containers or individual cartons are sealed and tamper-proof so that sterility is assured at time of first use.

Reference standards—*USP Tetracycline Hydrochloride Reference Standard*—Do not dry before using. *USP 4-Epianhydrotetracycline Hydrochloride Reference Standard*—Dry in vacuum at 60° for 3 hours before using.

Identification—The chromatogram of the *Assay preparation* obtained as directed in the *Assay* exhibits a major peak for tetracycline, the retention time of which corresponds to that exhibited in the chromatogram of the *Standard preparation* obtained as directed in the *Assay*.

Sterility—It meets the requirements for *Ophthalmic Ointments* under *Sterility Tests* ⟨71⟩.

Water, *Method I* ⟨921⟩: not more than 0.5%, 20 mL of a mixture of carbon tetrachloride, chloroform, and methanol (2:2:1) being used in place of methanol in the titration vessel.

Assay—

Diluting solvent, Mobile phase, Standard preparation, Resolution solution, and *Chromatographic system*—Prepare as directed in the *Assay* under *Tetracycline Hydrochloride.*

Assay preparation—Transfer an accurately measured volume of Tetracycline Hydrochloride Ophthalmic Suspension, equivalent to about 50 mg of tetracycline hydrochloride, with the aid of 30 mL of cyclohexane to a 125-mL separator, add 30 mL of *Diluting solvent*, insert the stopper, and shake. Allow to separate, and collect the lower layer in a 100-mL volumetric flask. Repeat the extraction with two additional 30-mL portions of *Diluting solvent*, combining the extracts in the 100-mL volumetric flask. Add *Diluting solvent* to volume, mix, and filter.

Procedure—Proceed as directed for *Procedure* in the *Assay* under *Tetracycline Hydrochloride.* Calculate the quantity, in mg per mL, of $C_{22}H_{24}N_2O_8 . HCl$ equivalent in the Ophthalmic Suspension taken by the formula:

$$(CP/10V)(r_U/r_S),$$

in which V is the volume, in mL, of Ophthalmic Suspension taken, and the other terms are as defined therein.

Tetracycline Hydrochloride Tablets

» Tetracycline Hydrochloride Tablets contain not less than 90.0 percent and not more than 125.0 percent of the labeled amount of $C_{22}H_{24}N_2O_8 . HCl$.

Packaging and storage—Preserve in tight, light-resistant containers.

Reference standards—*USP Tetracycline Hydrochloride Reference Standard*—Do not dry before using. *USP 4-Epianhydrotetracycline Hydrochloride Reference Standard*—Dry in vacuum at 60° for 3 hours before using.

Identification—The chromatogram of the *Assay preparation* obtained as directed in the *Assay* exhibits a major peak for tetracycline, the retention time of which corresponds to that exhibited in the chromatogram of the *Standard preparation* obtained as directed in the *Assay*.

Dissolution ⟨711⟩—

Medium: water; 900 mL.

Apparatus 2: 75 rpm.

Time: 60 minutes.

Procedure—Determine the amount of $C_{22}H_{24}N_2O_8 . HCl$ dissolved from ultraviolet absorbances at the wavelength of maximum absorbance at about 276 nm of filtered portions of the solution under test, suitably diluted with *Dissolution Medium*, if necessary, in comparison with a Standard solution having a known concentration of USP Tetracycline Hydrochloride RS in the same medium.

Tolerances—Not less than 70% (Q) of the labeled amount of $C_{22}H_{24}N_2O_8 . HCl$ is dissolved in 60 minutes.

Uniformity of dosage units ⟨905⟩: meet the requirements.

Loss on drying ⟨731⟩—Dry about 100 mg, accurately weighed, in vacuum at a pressure not exceeding 5 mm of mercury at 60° for 3 hours: it loses not more than 3.0% of its weight.

4-Epianhydrotetracycline—Using the *Diluting solvent, Chromatographic system*, and *Procedure* set forth in the *Assay*, chromatograph a Standard solution prepared by dissolving an accurately weighed quantity of USP 4-Epianhydrotetracycline Hydrochloride RS in *Diluting solvent* to obtain a solution having a known concentration of about 15 µg per mL. Using the chromatogram so obtained and the chromatogram of the *Assay preparation* obtained as directed in the *Assay*, calculate the percentage of 4-epianhydrotetracycline hydrochloride in the Tablets taken by the formula:

$$(10C_E/T)(r_U/r_S),$$

in which C_E is the concentration, in µg per mL, of USP 4-Epianhydrotetracycline Hydrochloride RS in the Standard solution, T is the quantity, in mg, of tetracycline hydrochloride in the portion of Tablets taken to prepare the *Assay preparation*, based on the labeled quantity, and r_U and r_S are the 4-epianhydrotetracycline peak responses obtained from the *Assay preparation* and the Standard solution, respectively: not more than 3.0% is found.

Assay—

Diluting solvent, Mobile phase, Standard preparation, Resolution solution, and *Chromatographic system*—Prepare as directed in the *Assay* under *Tetracycline Hydrochloride.*

Assay preparation—Weigh and finely powder not less than 20 Tetracycline Hydrochloride Tablets. Transfer an accurately weighed portion of the powder, equivalent to about 50 mg of tetracycline hydrochloride, to a 100-mL volumetric flask, add 50 mL of *Diluting solvent,* mix, and sonicate for 5 minutes. Allow to cool, add *Diluting solvent* to volume, mix, and filter.

Procedure—Proceed as directed for *Procedure* in the *Assay* under *Tetracycline Hydrochloride.* Calculate the quantity, in mg, of $C_{22}H_{24}N_2O_8.HCl$ in the portion of Tablets taken by the formula:

$$(CP/10)(r_U/r_S),$$

in which the terms are as defined therein.

Tetracycline Hydrochloride and Novobiocin Sodium Tablets

» Tetracycline Hydrochloride and Novobiocin Sodium Tablets contain the equivalent of not less than 90.0 percent and not more than 125.0 percent of the labeled amounts of tetracycline hydrochloride ($C_{22}H_{24}N_2O_8.HCl$) and novobiocin ($C_{31}H_{36}N_2O_{11}$).

Packaging and storage—Preserve in tight containers.

Labeling—Label Tablets to indicate that they are intended for veterinary use only.

Reference standards—*USP Tetracycline Hydrochloride Reference Standard*—Do not dry before using. *USP Novobiocin Reference Standard*—Dry in vacuum at a pressure not exceeding 5 mm of mercury at 100° for 4 hours before using.

Identification—Shake a suitable quantity of finely powdered Tablets with methanol to obtain a solution containing 1 mg of tetracycline hydrochloride per mL, and filter. Using the filtrate as the *Test Solution,* proceed as directed under *Identification—Tetracyclines* ⟨193⟩.

Disintegration ⟨701⟩: 60 minutes, simulated gastric fluid TS being substituted for water in the test.

Uniformity of dosage units ⟨905⟩: meet the requirements for *Weight Variation* with respect to tetracycline hydrochloride and to novobiocin sodium.

Loss on drying ⟨731⟩—Dry about 100 mg, accurately weighed, of finely powdered Tablets in a capillary-stoppered bottle in vacuum at a pressure not exceeding 5 mm of mercury at 60° for 3 hours: it loses not more than 6.0% of its weight.

4-Epianhydrotetracycline ⟨226⟩—To an accurately weighed quantity of finely powdered Tablets, equivalent to about 250 mg of tetracycline hydrochloride, add 10 mL of 0.1 *N* hydrochloric acid, and adjust with 6 *N* ammonium hydroxide to a pH of 7.8. Transfer this solution with the aid of *EDTA buffer* to a 50-mL volumetric flask, dilute with *EDTA buffer* to volume, and mix. Use this solution, without delay, as the *Test Solution:* not more than 2.0% is found.

Assay for tetracycline hydrochloride—Proceed as directed for tetracycline under *Antibiotics—Microbial Assays* ⟨81⟩, except to use *Escherichia coli* ATCC 10536 as the test organism instead of *Staphylococcus aureus* ATCC 29737 and an inoculum composition of about 0.2 mL of stock suspension in each 100 mL of Medium 3. Transfer not less than 5 Tetracycline Hydrochloride and Novobiocin Sodium Tablets to a high-speed blender jar containing an accurately measured volume of 0.1 *N* hydrochloric acid, so that, after blending for about 3 to 5 minutes, the solution so obtained contains not less than 150 µg of tetracycline hydrochloride per mL. Dilute an accurately measured volume of this solution quantitatively and stepwise with water to obtain a *Test Dilution* having a concentration of tetracycline hydrochloride assumed to be equal to the median dose level of the Standard.

Assay for novobiocin—Proceed as directed for novobiocin under *Antibiotics—Microbial Assays* ⟨81⟩, blending not less than 5 Tetracycline Hydrochloride and Novobiocin Sodium Tablets for 3 to 5 minutes in a high-speed glass blender jar containing 1.0 mL of polysorbate 80 and a sufficient accurately measured volume of *Buffer No. 3* to provide a stock solution of convenient concentration. Dilute an accurately measured volume of this stock solution quantitatively and stepwise with *Buffer No. 6* to obtain a *Test Dilution* having a concentration of novobiocin assumed to be equal to the median dose level of the Standard.

Tetracycline Hydrochloride, Novobiocin Sodium, and Prednisolone Tablets

» Tetracycline Hydrochloride, Novobiocin Sodium, and Prednisolone Tablets contain not less than 90.0 percent and not more than 125.0 percent of the labeled amounts of tetracycline hydrochloride ($C_{22}H_{24}N_2O_8.HCl$) and novobiocin ($C_{31}H_{36}N_2O_{11}$), and not less than 90.0 percent and not more than 110.0 percent of the labeled amount of prednisolone ($C_{21}H_{28}O_5$).

Packaging and storage—Preserve in tight containers.

Labeling—Label Tablets to indicate that they are intended for veterinary use only.

Reference standards—*USP Tetracycline Hydrochloride Reference Standard*—Do not dry before using. *USP Novobiocin Reference Standard*—Dry in vacuum at a pressure not exceeding 5 mm of mercury at 100° for 4 hours before using. *USP Prednisolone Reference Standard*—Dry at 105° for 3 hours before using.

Disintegration ⟨701⟩: 60 minutes, simulated gastric fluid TS being substituted for water in the test.

Uniformity of dosage units ⟨905⟩: meet the requirements for *Weight Variation* with respect to tetracycline hydrochloride and to novobiocin sodium and for *Content Uniformity* with respect to prednisolone.

Procedure for content uniformity for prednisolone—

Standard preparation—Prepare as directed for *Standard Preparation* under *Assay for Steroids* ⟨351⟩, using USP Prednisolone RS.

Test preparation—Place 1 Tablet in a separator with 15 mL of water, and swirl to disintegrate the Tablet completely. Extract with three 25-mL portions and finally with one 20-mL portion of chloroform, filtering each portion through chloroform-washed cotton into a 100-mL volumetric flask. Add chloroform to volume, and mix. Pipet an accurately measured volume of this solution, equivalent to 200 µg of prednisolone, into a suitable glass-stoppered flask or tube, evaporate the chloroform on a steam bath just to dryness, cool, and dissolve the residue in 20.0 mL of alcohol. Use this where *Assay Preparation* is specified in the *Procedure.*

Procedure—Proceed as directed for *Procedure* under *Assay for Steroids* ⟨351⟩.

4-Epianhydrotetracycline ⟨226⟩—To an accurately weighed quantity of finely powdered Tablets, equivalent to about 250 mg of tetracycline hydrochloride, add 10 mL of 0.1 *N* hydrochloric acid, and adjust with 6 *N* ammonium hydroxide to a pH of 7.8. Transfer this solution with the aid of *EDTA buffer* to a 50-mL volumetric flask, dilute with *EDTA buffer* to volume, and mix. Use this solution, without delay, as the *Test Solution:* not more than 2.0% is found.

Other requirements—Tablets respond to the *Identification test* and meet the requirements of the test for *Loss on drying* under *Tetracycline Hydrochloride and Novobiocin Sodium Tablets.*

Assay for tetracycline hydrochloride and Assay for novobiocin—Using Tetracycline Hydrochloride, Novobiocin Sodium, and Prednisolone Tablets, proceed as directed in the *Assay for tetracycline hydrochloride* and the *Assay for novobiocin* under *Tetracycline Hydrochloride and Novobiocin Sodium Tablets.*

Assay for prednisolone—

Standard preparation—Prepare as directed for *Standard Preparation* under *Single-steroid Assay* ⟨511⟩, using USP Prednisolone RS.

Assay preparation—Weigh and finely powder not less than 20 Tetracycline Hydrochloride, Novobiocin Sodium, and Prednisolone Tablets. Weigh accurately a portion of the powder, equivalent to about 20 mg of prednisolone, and transfer, with the aid of 15 mL of water, to a separator. Extract with three 25-mL portions and finally with one 20-mL portion of chloroform, filtering each portion through chloroform-washed cotton into a 100-mL volumetric flask. Add chloroform to volume, and mix. Pipet 20 mL of this solution into a suitable glass-stoppered flask or tube, evaporate the chloroform on a steam bath just to dryness, cool, and dissolve the residue in 2.0 mL, accurately measured, of a mixture of equal volumes of chloroform and alcohol.

Procedure—Proceed as directed for *Procedure* under *Single-steroid Assay* ⟨511⟩, using *Solvent A* to develop the chromatogram. Calculate the quantity, in mg, of $C_{21}H_{28}O_5$ in the portion of Tablets taken by the formula:

$$0.01C(A_U/A_S).$$

Tetracycline Hydrochloride and Nystatin Capsules

» Tetracycline Hydrochloride and Nystatin Capsules contain not less than 90.0 percent and not more than 125.0 percent of the labeled amount of tetracycline hydrochloride ($C_{22}H_{24}N_2O_8.HCl$), and not less than 90.0 percent and not more than 135.0 percent of the labeled amount of USP Nystatin Units.

Packaging and storage—Preserve in tight, light-resistant containers.

Reference standards—*USP Tetracycline Hydrochloride Reference Standard*—Do not dry before using. *USP Nystatin Reference Standard*—Dry in vacuum at a pressure not exceeding 5 mm of mercury at 40° for 2 hours before using.

Identification—Shake a suitable quantity of Capsule contents with methanol to obtain a solution containing about 1 mg of tetracycline hydrochloride per mL, and filter. Using the filtrate as the *Test Solution*, proceed as directed under *Identification—Tetracyclines* ⟨193⟩.

Dissolution ⟨711⟩—

Medium: water; 900 mL.

Apparatus 2: 75 rpm.

Time: 60 minutes.

Procedure—Determine the amount of tetracycline hydrochloride ($C_{22}H_{24}N_2O_8.HCl$) dissolved from ultraviolet absorbances at the wavelength of maximum absorbance at about 276 nm of filtered portions of the solution under test, suitably diluted with *Dissolution Medium*, if necessary, in comparison with a Standard solution having a known concentration of USP Tetracycline Hydrochloride RS in the same medium.

Tolerances—Not less than 70% (Q) of the labeled amount of $C_{22}H_{24}N_2O_8.HCl$ is dissolved in 60 minutes.

Loss on drying ⟨731⟩—Dry about 100 mg of Capsule contents, accurately weighed, in a capillary-stoppered bottle in vacuum at a pressure not exceeding 5 mm of mercury at 60° for 3 hours: it loses not more than 4.0% of its weight.

4-Epianhydrotetracycline ⟨226⟩—To a quantity of Capsule contents, equivalent to about 250 mg of tetracycline hydrochloride, add 10 mL of 0.1 *N* hydrochloric acid, and adjust with 6 *N* ammonium hydroxide to a pH of 7.8. Transfer this solution with the aid of *EDTA buffer* to a 50-mL volumetric flask, dilute with *EDTA buffer* to volume, and mix. Use this solution, without delay, as the *Test Solution:* not more than 3.0% is found.

Assay for tetracycline hydrochloride—Proceed with Tetracycline Hydrochloride and Nystatin Capsules as directed in the *Assay* under *Tetracycline Hydrochloride Capsules*.

Assay for nystatin—Proceed as directed for Nystatin under *Antibiotics—Microbial Assays* ⟨81⟩, blending not less than 5 Tetracycline Hydrochloride and Nystatin Capsules for 3 to 5 minutes in a high-speed blender with a sufficient, accurately measured, volume of dimethylformamide to obtain a solution of convenient concentration. Dilute an accurately measured portion of this solution quantitatively with dimethylformamide to obtain a stock solution containing about 400 USP Nystatin Units per mL. Dilute this stock solution quantitatively with *Buffer No. 6* to obtain a *Test Dilution* having a concentration of nystatin assumed to be equal to the median dose level of the Standard.

Tetracycline Phosphate Complex

2-Naphthacenecarboxamide, 4-(dimethylamino)-1,4,4a,5,5a,6,-11,12a-octahydro-3,6,10,12,12a-pentahydroxy-6-methyl-1,-11-dioxo, [4*S*-(4α,4aα,5aα,6β,12aα)]-, phosphate complex.

4-(Dimethylamino)-1,4,4a,5,5a,6,11,12a-octahydro-3,6,10,12,-12a-pentahydroxy-6-methyl-1,11-dioxo-2-naphthacenecarboxamide phosphate complex [*1336-20-5*].

» Tetracycline Phosphate Complex has a potency equivalent to not less than 750 µg of tetracycline hydrochloride ($C_{22}H_{24}N_2O_8.HCl$) per mg, calculated on the anhydrous basis.

Packaging and storage—Preserve in tight, light-resistant containers.

Labeling—Label it to indicate that it is to be used in the manufacture of nonparenteral drugs only.

Reference standards—*USP Tetracycline Hydrochloride Reference Standard*—Do not dry before using. *USP 4-Epianhydrotetracycline Hydrochloride Reference Standard*—Dry in vacuum at 60° for 3 hours before using.

Identification—

A: Dissolve about 40 mg, accurately weighed, in 2 mL of 0.1 *N* hydrochloric acid, transfer to a 250-mL volumetric flask, dilute with water to volume, and mix. Transfer 10.0 mL of this solution to a 100-mL volumetric flask, add about 75 mL of water and 5.0 mL of 5 *N* sodium hydroxide, dilute with water to volume, and mix: the ultraviolet absorption spectrum of the solution so obtained exhibits maxima and minima at the same wavelengths as that of a similar preparation of USP Tetracycline Hydrochloride RS, concomitantly measured, and the absorptivity, determined at 6.0 minutes after the addition of the 5 *N* sodium hydroxide and calculated on the anhydrous basis, at the wavelength of maximum absorbance at about 380 nm is between 77.1% and 86.9% of that of the USP Tetracycline Hydrochloride RS, the potency of the Reference Standard being taken into account.

B: The chromatogram of the *Assay preparation* obtained as directed in the *Assay* exhibits a major peak for tetracycline, the retention time of which corresponds to that exhibited in the chromatogram of the *Standard preparation* obtained as directed in the *Assay*.

C: Shake 100 mg with 10 mL of water, and filter. To 1 mL of the filtrate in a glass-stoppered cylinder add 10 mL of water, 2 mL of ammonium molybdate TS, 1 mL of stannous chloride TS, and 10 mL of a mixture of equal volumes of benzene and isobutyl alcohol. Shake vigorously for 1 minute, and allow the layers to separate: a blue color is produced in the upper layer (presence of phosphate).

Crystallinity ⟨695⟩: meets the requirements.

pH ⟨791⟩: between 2.0 and 4.0, in an aqueous suspension containing 10 mg per mL.

Water, *Method I* ⟨921⟩: not more than 9.0%.

Chloride ⟨221⟩—Shake 100 mg with 10.0 mL of water, and filter. To 1.0 mL of the filtrate add 1 drop of nitric acid and 1 drop of silver nitrate TS: any turbidity produced is not greater than that obtained from 1.0 mL of 0.00057 *N* hydrochloric acid, similarly treated (0.2%).

Tetracycline—Transfer about 1 g, accurately weighed, to a 50-mL conical flask, add 10.0 mL of dioxane, and shake for about 2 minutes. Allow to settle, decant the supernatant liquid into a

50-mL centrifuge tube, and centrifuge. Transfer 5.0 mL of the clear liquid to a 50-mL beaker, add 2 drops of methyl red TS, and titrate with 0.01 *N* perchloric acid VS in dioxane to an orange end-point. Perform a blank determination, and make any necessary correction. Each mL of 0.01 *N* perchloric acid is equivalent to 4.444 mg of tetracycline ($C_{22}H_{24}N_2O_8$): not more than 1.0% of $C_{22}H_{24}N_2O_8$ is found.

4-Epianhydrotetracycline—Using the *Diluting solvent*, *Chromatographic system*, and *Procedure* set forth in the *Assay*, chromatograph a Standard solution prepared by dissolving an accurately weighed quantity of USP 4-Epianhydrotetracycline Hydrochloride RS in *Diluting solvent* to obtain a solution having a known concentration of about 10 µg per mL. Using the chromatogram so obtained and the chromatogram of the *Assay preparation* obtained as directed in the *Assay*, calculate the percentage of 4-epianhydrotetracycline hydrochloride equivalent in the Tetracycline Phosphate Complex taken by the formula:

$$10(C_E/W)(r_U/r_S),$$

in which C_E is the concentration, in µg per mL, of USP 4-Epianhydrotetracycline Hydrochloride RS, calculated on the anhydrous basis, in the Standard solution, W is the weight, in mg, of Tetracycline Phosphate Complex taken to prepare the *Assay preparation*, and r_U and r_S are the 4-epianhydrotetracycline peak responses obtained from the *Assay preparation* and the Standard solution, respectively: not more than 2.0% is found.

Assay—

Diluting solvent, *Mobile phase*, *Standard preparation*, *Resolution solution*, and *Chromatographic system*—Prepare as directed in the *Assay* under *Tetracycline Hydrochloride*.

Assay preparation—Transfer about 60 mg of Tetracycline Phosphate Complex, accurately weighed, to a 100-mL volumetric flask, dissolve in *Diluting solvent*, dilute with the same solvent to volume, and mix.

Procedure—Proceed as directed in the *Assay* under *Tetracycline Hydrochloride*. Calculate the quantity, in µg, of tetracycline hydrochloride ($C_{22}H_{24}N_2O_8 . HCl$) equivalent in each mg of Tetracycline Phosphate Complex taken by the formula:

$$100(CP/W)(r_U/r_S),$$

in which W is the weight, in mg, of Tetracycline Phosphate Complex taken to prepare the *Assay preparation*, and the other terms are as defined therein.

Tetracycline Phosphate Complex Capsules

» Tetracycline Phosphate Complex Capsules contain the equivalent of not less than 90.0 percent and not more than 125.0 percent of the labeled amount of tetracycline hydrochloride ($C_{22}H_{24}N_2O_8 . HCl$).

Packaging and storage—Preserve in tight, light-resistant containers.

Reference standards—*USP Tetracycline Hydrochloride Reference Standard*—Do not dry before using. *USP 4-Epianhydrotetracycline Hydrochloride Reference Standard*—Dry in vacuum at 60° for 3 hours before using.

Identification—The chromatogram of the *Assay preparation* obtained as directed in the *Assay* exhibits a major peak for tetracycline, the retention time of which corresponds to that exhibited in the chromatogram of the *Standard preparation* obtained as directed in the *Assay*.

Dissolution ⟨711⟩—

Medium: 0.1 *N* hydrochloric acid; 900 mL.

Apparatus 1: 100 rpm.

Time: 30 minutes.

Procedure—Determine the amount of $C_{22}H_{24}N_2O_8 . HCl$ dissolved from ultraviolet absorbances at the wavelength of maximum absorbance at about 353 nm of filtered portions of the solution under test, suitably diluted with *Dissolution Medium*, in comparison with a Standard solution having a known concentration of USP Tetracycline Hydrochloride RS in the same medium.

Tolerances—Not less than 75% (Q) of the labeled amount of $C_{22}H_{24}N_2O_8 . HCl$ is dissolved in 30 minutes.

Uniformity of dosage units ⟨905⟩: meet the requirements.

Loss on drying ⟨731⟩—Dry about 100 mg of Capsule contents, accurately weighed, in a capillary-stoppered bottle in vacuum at a pressure not exceeding 5 mm of mercury at 60° for 3 hours: it loses not more than 9.0% of its weight.

4-Epianhydrotetracycline—Using the *Diluting solvent*, *Chromatographic system*, and *Procedure* set forth in the *Assay*, chromatograph a Standard solution prepared by dissolving an accurately weighed quantity of USP 4-Epianhydrotetracycline Hydrochloride RS in *Diluting solvent* to obtain a solution having a known concentration of about 15 µg per mL. Using the chromatogram so obtained and the chromatogram of the *Assay preparation* obtained as directed in the *Assay*, calculate the percentage of 4-epianhydrotetracycline hydrochloride equivalent in the Capsules taken by the formula:

$$(10C_E/T)(r_U/r_S),$$

in which C_E is the concentration, in µg per mL, of USP 4-Epianhydrotetracycline Hydrochloride RS in the Standard solution, T is the quantity, in mg, of tetracycline hydrochloride equivalent in the portion of Capsules taken to prepare the *Assay preparation*, based on the labeled quantity, and r_U and r_S are the 4-epianhydrotetracycline peak responses obtained from the *Assay preparation* and the Standard solution, respectively: not more than 3.0% is found.

Assay—

Diluting solvent, *Mobile phase*, *Standard preparation*, *Resolution solution*, and *Chromatographic system*—Prepare as directed in the *Assay* under *Tetracycline Hydrochloride*.

Assay preparation—Weigh accurately not less than 20 Tetracycline Phosphate Complex Capsules. Empty the Capsule contents into a mortar. Clean and accurately weigh the Capsule shells, and calculate the net weight of the Capsule contents. Mix the powder in the mortar, using a pestle, and transfer an accurately weighed quantity of the freshly mixed powder, equivalent to about 50 mg of tetracycline hydrochloride, to a 100-mL volumetric flask. Add about 50 mL of *Diluting solvent*, mix, and sonicate for about 5 minutes. Allow to cool, dilute with *Diluting solvent* to volume, mix, and filter.

Procedure—Proceed as directed in the *Assay* under *Tetracycline Hydrochloride*. Calculate the quantity, in mg, of $C_{22}H_{24}N_2O_8 . HCl$ equivalent in the portion of Capsules taken by the formula:

$$(CP/10)(r_U/r_S),$$

in which the terms are as defined therein.

Tetracycline Phosphate Complex for Injection

» Tetracycline Phosphate Complex for Injection is a sterile, dry mixture of Sterile Tetracycline Phosphate Complex and Magnesium Chloride or magnesium ascorbate, and one or more suitable buffers, and may contain one or more suitable anesthetics, preservatives, solubilizers, and stabilizers. It contains the equivalent of not less than 90.0 percent and not more than 115.0 percent of the labeled amount of tetracycline hydrochloride ($C_{22}H_{24}N_2O_8 . HCl$).

Packaging and storage—Preserve in *Containers for Sterile Solids* as described under *Injections* ⟨1⟩, protected from light.

Reference standards—*USP Tetracycline Hydrochloride Reference Standard*—Do not dry before using. *USP 4-Epianhydro-*

tetracycline Hydrochloride Reference Standard—Dry in vacuum at 60° for 3 hours before using.

Constituted solution—At the time of use, the constituted solution prepared from Tetracycline Phosphate Complex for Injection meets the requirements for *Constituted Solutions* under *Injections* ⟨1⟩.

Identification—The chromatogram of the *Assay preparation* obtained as directed in the *Assay* exhibits a major peak for tetracycline, the retention time of which corresponds to that exhibited in the chromatogram of the *Standard preparation* obtained as directed in the *Assay*.

Pyrogen—It meets the requirements of the *Pyrogen Test* ⟨151⟩, the test dose being 1.0 mL per kg of a solution prepared to contain the equivalent of 5.0 mg of tetracycline hydrochloride per mL in Sterile Water for Injection.

Sterility—It meets the requirements under *Sterility Tests* ⟨71⟩, when tested as directed in the section, *Test Procedures Using Membrane Filtration*, the equivalent of 50 mg of tetracycline hydrochloride being tested from each container.

pH ⟨791⟩: between 2.0 and 3.0, in a solution containing 10 mg per mL.

Loss on drying ⟨731⟩—Dry about 100 mg, accurately weighed, in a capillary-stoppered bottle in vacuum at a pressure not exceeding 5 mm of mercury at 60° for 3 hours: it loses not more than 5.0% of its weight.

Particulate matter ⟨788⟩: meets the requirements under *Small-volume Injections*.

4-Epianhydrotetracycline—Using the *Diluting solvent*, *Chromatographic system*, and *Procedure* set forth in the *Assay*, chromatograph a Standard solution prepared by dissolving an accurately weighed quantity of USP 4-Epianhydrotetracycline Hydrochloride RS in *Diluting solvent* to obtain a solution having a known concentration of about 15 µg per mL. Using the chromatogram so obtained and the chromatogram of the *Assay preparation* obtained as directed in the *Assay*, calculate the percentage of 4-epianhydrotetracycline hydrochloride equivalent in the Tetracycline Phosphate Complex for Injection taken by the formula:

$$(10C_E/T)(r_U/r_S),$$

in which C_E is the concentration, in µg per mL, of USP 4-Epianhydrotetracycline Hydrochloride RS in the Standard solution, T is the quantity, in mg, of tetracycline hydrochloride equivalent in the portion of constituted Tetracycline Phosphate Complex for Injection taken to prepare the *Assay preparation*, based on the labeled quantity, and r_U and r_S are the 4-epianhydrotetracycline peak responses obtained from the *Assay preparation* and the Standard solution, respectively: not more than 3.0% is found.

Other requirements—It meets the requirements for *Uniformity of Dosage Units* ⟨905⟩ and for *Labeling* under *Injections* ⟨1⟩.

Assay—

Diluting solvent, *Mobile phase*, *Standard preparation*, *Resolution solution*, and *Chromatographic system*—Prepare as directed in the *Assay* under *Tetracycline Hydrochloride*.

Assay preparation 1 (where it is represented as being in a single-dose container)—Constitute Tetracycline Phosphate Complex for Injection as directed in the labeling. Withdraw all of the withdrawable contents, using a suitable hypodermic needle and syringe, and dilute quantitatively and stepwise with *Diluting solvent* to obtain a solution containing the equivalent of about 0.5 mg of tetracycline hydrochloride per mL.

Assay preparation 2 (where the label states the quantity of tetracycline hydrochloride equivalent in a given volume of constituted solution)—Constitute Tetracycline Phosphate Complex for Injection as directed in the labeling. Dilute an accurately measured volume of the constituted solution quantitatively with *Diluting solvent* to obtain a solution containing the equivalent of about 0.5 mg of tetracycline hydrochloride per mL.

Procedure—Proceed as directed for *Procedure* in the *Assay* under *Tetracycline Hydrochloride*. Calculate the quantity, in mg, of $C_{22}H_{24}N_2O_8.HCl$ equivalent withdrawn from the container, or in the portion of constituted solution taken by the formula:

$$(L/D)(CP/1000)(r_U/r_S),$$

in which L is the labeled quantity, in mg, of $C_{22}H_{24}N_2O_8.HCl$ equivalent in the container, or in the volume of constituted solution taken, D is the concentration, in mg of tetracycline hydrochloride equivalent per mL, of *Assay preparation 1* or *Assay preparation 2*, based on the labeled quantity in the container or in the portion of constituted solution taken, respectively, and the extent of dilution, and the other terms are as defined therein.

Sterile Tetracycline Phosphate Complex

» Sterile Tetracycline Phosphate Complex is Tetracycline Phosphate Complex suitable for parenteral use. It has a potency equivalent to not less than 750 µg of tetracycline hydrochloride ($C_{22}H_{24}N_2O_8.HCl$) per mg, calculated on the anhydrous basis.

Packaging and storage—Preserve in *Containers for Sterile Solids* as described under *Injections* ⟨1⟩, protected from light.

Reference standards—*USP Tetracycline Hydrochloride Reference Standard*—Do not dry before using. *USP 4-Epianhydrotetracycline Hydrochloride Reference Standard*—Dry in vacuum at 60° for 3 hours before using.

Depressor substances—Dissolve 40 mg in 2.0 mL of 0.1 *N* hydrochloric acid, and dilute with sterile water to obtain a solution containing the equivalent of 5.0 mg of tetracycline hydrochloride per mL: it meets the requirements of the *Depressor Substances Test* ⟨101⟩, the test dose being 0.6 mL per kg.

Sterility—It meets the requirements under *Sterility Tests* ⟨71⟩, when tested as directed in the section, *Test Procedures Using Membrane Filtration*, 1 g of specimen aseptically dissolved in 200 mL of Fluid A being used.

Other requirements—It responds to the *Identification tests* and meets the requirements for *pH*, *Water*, *Chloride*, *Crystallinity*, *Tetracycline*, and *4-Epianhydrotetracycline* under *Tetracycline Phosphate Complex*.

Assay—Proceed with Sterile Tetracycline Phosphate Complex as directed for the *Assay* under *Tetracycline Phosphate Complex*.

Tetracycline Phosphate Complex and Novobiocin Sodium Capsules

» Tetracycline Phosphate Complex and Novobiocin Sodium Capsules contain the equivalent of not less than 90.0 percent and not more than 120.0 percent of the labeled amounts of tetracycline hydrochloride ($C_{22}H_{24}N_2O_8.HCl$) and novobiocin ($C_{31}H_{36}N_2O_{11}$).

Packaging and storage—Preserve in tight containers.

Labeling—Label Capsules to indicate that they are intended for veterinary use only.

Reference standards—*USP Tetracycline Hydrochloride Reference Standard*—Do not dry before using. *USP Novobiocin Reference Standard*—Dry in vacuum at a pressure not exceeding 5 mm of mercury at 100° for 4 hours before using.

Identification—Shake a suitable quantity of Capsule contents with methanol to obtain a solution containing the equivalent of 1 mg of tetracycline hydrochloride per mL, and filter. Using the filtrate as the *Test Solution*, proceed as directed under *Identification—Tetracyclines* ⟨193⟩.

Uniformity of dosage units ⟨905⟩: meet the requirements for *Weight Variation* with respect to tetracycline hydrochloride equivalent and to novobiocin sodium.

Loss on drying ⟨731⟩—Dry about 100 mg, accurately weighed, of the powder obtained from 4 Capsules, in vacuum at a pressure not exceeding 5 mm of mercury at 60° for 3 hours: it loses not more than 9.0% of its weight.

4-Epianhydrotetracycline ⟨226⟩—To a quantity of Capsule contents, equivalent to about 250 mg of tetracycline hydrochloride, add 10 mL of 0.1 *N* hydrochloric acid, and adjust with 6 *N* ammonium hydroxide to a pH of 7.8. Transfer this solution with the aid of *EDTA buffer* to a 50-mL volumetric flask, dilute with *EDTA buffer* to volume, and mix. Use this solution, without delay, as the *Test Solution:* not more than 3.0% is found.

Assay for tetracycline hydrochloride and Assay for novobiocin—Using Tetracycline Phosphate Complex and Novobiocin Sodium Capsules, proceed as directed in the *Assay for tetracycline hydrochloride* and in the *Assay for novobiocin* under *Tetracycline Hydrochloride and Novobiocin Sodium Tablets.*

Tetrahydrozoline Hydrochloride

· HCl

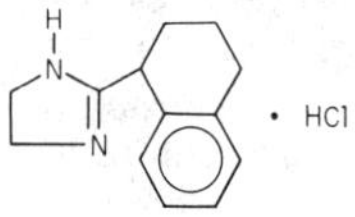

$C_{13}H_{16}N_2 \cdot HCl$ 236.74

1*H*-Imidazole, 4,5-dihydro-2-(1,2,3,4-tetrahydro-1-naphthalenyl)-, monohydrochloride.

2-(1,2,3,4-Tetrahydro-1-naphthyl)-2-imidazoline monohydrochloride [*522-48-5*].

» Tetrahydrozoline Hydrochloride contains not less than 98.0 percent and not more than 100.5 percent of $C_{13}H_{16}N_2 \cdot HCl$, calculated on the dried basis.

Packaging and storage—Preserve in tight containers.

Reference standard—*USP Tetrahydrozoline Hydrochloride Reference Standard*—Dry at 105° for 2 hours before using.

Identification—

A: The infrared absorption spectrum of a potassium bromide dispersion of it, previously dried at 105° for 2 hours, exhibits maxima only at the same wavelengths as that of a similar preparation of USP Tetrahydrozoline Hydrochloride RS.

B: The ultraviolet absorption spectrum of a 1 in 4000 solution exhibits maxima and minima at the same wavelengths as that of a similar preparation of USP Tetrahydrozoline Hydrochloride RS, concomitantly measured, and the respective absorptivities, calculated on the dried basis, at the wavelengths of maximum absorbance at about 264 nm and 271 nm do not differ by more than 4.0%.

C: A solution (1 in 200) responds to the tests for *Chloride* ⟨191⟩.

Loss on drying ⟨731⟩—Dry it at 105° for 2 hours: it loses not more than 1.0% of its weight.

Residue on ignition ⟨281⟩: not more than 0.1%.

Heavy metals ⟨231⟩—Dissolve 0.40 g in 23 mL of water, and add 2 mL of 1 *N* acetic acid: the limit is 0.005%.

Ordinary impurities ⟨466⟩—

Test solution: methanol.

Standard solution: methanol.

Eluant: a mixture of methanol, glacial acetic acid, and water (8:1:1).

Visualization: 2, followed by 1.

Assay—Transfer about 400 mg of Tetrahydrozoline Hydrochloride, accurately weighed, to a 250-mL beaker, and dissolve in 60 mL of glacial acetic acid, heating if necessary. Add 5 mL of acetic anhydride, 5 mL of mercuric acetate TS, and 3 drops of quinaldine red TS, and titrate with 0.1 *N* perchloric acid VS. Perform a blank determination, and make any necessary correction. Each mL of 0.1 *N* perchloric acid is equivalent to 23.67 mg of $C_{13}H_{16}N_2 \cdot HCl$.

Tetrahydrozoline Hydrochloride Nasal Solution

» Tetrahydrozoline Hydrochloride Nasal Solution is a solution of Tetrahydrozoline Hydrochloride in water adjusted to a suitable tonicity. It contains not less than 90.0 percent and not more than 110.0 percent of the labeled amount of $C_{13}H_{16}N_2 \cdot HCl$.

Packaging and storage—Preserve in tight containers.

Reference standard—*USP Tetrahydrozoline Hydrochloride Reference Standard*—Dry at 105° for 2 hours before using.

Identification—The ultraviolet absorption spectrum of the Nasal Solution, diluted with water, if necessary, to a concentration of about 1 in 4000, exhibits maxima and minima at the same wavelengths as that of a similar solution of USP Tetrahydrozoline Hydrochloride RS, concomitantly measured.

Microbial limits—It meets the requirements of the tests for absence of *Staphylococcus aureus* and *Pseudomonas aeruginosa* under *Microbial Limit Tests* ⟨61⟩.

pH ⟨791⟩: between 5.3 and 6.5.

Assay—

Oxidized nitroprusside reagent—Dissolve 1.0 g of sodium nitroferricyanide in water to make 10.0 mL (*Solution A*). Dissolve 1.0 g of potassium ferricyanide in water to make 10.0 mL (*Solution B*). Transfer 1.0 mL each of *Solution A* and *Solution B* to a 100-mL volumetric flask, add 1 mL of sodium hydroxide solution (1 in 10), and allow to stand until the solution changes to a light yellow color (about 20 to 30 minutes). Dilute with water to volume, and mix. Store in a refrigerator or keep in an ice bath, and use within 4 hours.

Standard preparation—Dissolve a suitable quantity of USP Tetrahydrozoline Hydrochloride RS, accurately weighed, in water, and dilute quantitatively with water to obtain a solution having a known concentration of about 100 µg per mL.

Assay preparation—Transfer an accurately measured volume of Tetrahydrozoline Hydrochloride Nasal Solution, equivalent to about 10 mg of tetrahydrozoline hydrochloride, to a 100-mL volumetric flask, dilute with water to volume, and mix.

Procedure—Transfer 5.0 mL each of the *Standard preparation* and the *Assay preparation* to separate glass-stoppered test tubes. Pipet 5 mL of water into a third tube to provide a blank. To each tube add 4.0 mL of *Oxidized nitroprusside reagent*, mix, and allow to stand at 30° for 15 minutes. Concomitantly determine the absorbances of the solutions in 1-cm cells at the wavelength of maximum absorbance at about 570 nm, with a suitable spectrophotometer, using the blank to set the instrument. Calculate the quantity, in mg, of $C_{13}H_{16}N_2 \cdot HCl$ in each mL of the Nasal Solution taken by the formula:

$$0.1(C/V)(A_U/A_S),$$

in which *C* is the concentration, in µg per mL, of USP Tetrahydrozoline Hydrochloride RS in the *Standard preparation*, *V* is the volume, in mL, of Nasal Solution taken, and A_U and A_S are the absorbances of the solutions from the *Assay preparation* and the *Standard preparation*, respectively.

Tetrahydrozoline Hydrochloride Ophthalmic Solution

» Tetrahydrozoline Hydrochloride Ophthalmic Solution is a sterile, isotonic solution of Tetrahydrozoline Hydrochloride in water. It contains not less than 90.0 percent and not more than 110.0 percent of the labeled amount of $C_{13}H_{16}N_2 \cdot HCl$.

Packaging and storage—Preserve in tight containers.

Reference standard—*USP Tetrahydrozoline Hydrochloride Reference Standard*—Dry at 105° for 2 hours before using.

Identification—The ultraviolet absorption spectrum of the Ophthalmic Solution, diluted with dilute hydrochloric acid (1 in 100) to a concentration of about 1 in 4000, exhibits maxima and minima at the same wavelengths as that of a similar solution of USP Tetrahydrozoline Hydrochloride RS, concomitantly measured.

Sterility—It meets the requirements under *Sterility Tests* ⟨71⟩.

pH ⟨791⟩: between 5.8 and 6.5.

Assay—

Standard preparation—Dissolve a suitable quantity of USP Tetrahydrozoline Hydrochloride RS, accurately weighed, in water, and dilute quantitatively with water to obtain a solution having a known concentration of about 500 µg per mL.

Procedure—Transfer 2.0 mL of *Standard preparation* to a 50-mL volumetric flask. Transfer an accurately measured volume of Tetrahydrozoline Hydrochloride Ophthalmic Solution, equivalent to about 1 mg of tetrahydrozoline hydrochloride, to a second 50-mL flask, and transfer 2 mL of water to a third 50-mL volumetric flask to provide a blank. To each flask add 5.0 mL of bromophenol blue sodium salt solution (1 in 1000), dilute each flask with potassium biphthalate solution (1 in 100) to volume, and mix. Allow to stand for 20 minutes, and filter each mixture through a suitable filter paper (Whatman No. 42 or the equivalent) that does not absorb the dye, discarding the first 15 mL of the filtrate. Transfer 20.0 mL of the subsequent filtrate to separate 125-mL separators, and extract each solution with four 20-mL portions of chloroform, filtering each extract through a pledget of glass wool into a 100-mL volumetric flask. Dilute the combined extracts from each solution with chloroform to volume, and mix. Concomitantly determine the absorbances of the solutions in 1-cm cells at the wavelength of maximum absorbance at about 415 nm, with a suitable spectrophotometer, using the blank to set the instrument. Calculate the quantity, in µg, of $C_{13}H_{16}N_2 \cdot HCl$ in each mL of the Ophthalmic Solution taken by the formula:

$$2(C/V)(A_U/A_S),$$

in which C is the concentration, in µg per mL, of USP Tetrahydrozoline Hydrochloride RS in the *Standard preparation*, V is the volume, in mL, of Solution taken, and A_U and A_S are the absorbances of the solutions from the Ophthalmic Solution and the *Standard preparation*, respectively.

Thallous Chloride Tl 201 Injection

» Thallous Chloride Tl 201 Injection is a sterile, isotonic, aqueous solution of radioactive thallium (^{201}Tl) in the form of thallous chloride suitable for intravenous administration. It contains not less than 90.0 percent and not more than 110.0 percent of the labeled amount of ^{201}Tl as chloride, expressed in megabecquerels (microcuries or millicuries) per mL, at the time indicated in the labeling. Other chemical forms of radioactivity do not exceed 5.0 percent of the total radioactivity. It may contain a preservative or stabilizer.

Packaging and storage—Preserve in single-dose or in multiple-dose containers.

Labeling—Label it to include the following, in addition to the information specified for *Labeling* under *Injections* ⟨1⟩: the time and date of calibration; the amount of ^{201}Tl as labeled thallous chloride expressed as total megabecquerels (microcuries or millicuries) and concentration as megabecquerels (microcuries or millicuries) per mL at the time of calibration; the expiration date and time; and the statement, "Caution—Radioactive Material." The labeling indicates that in making dosage calculations, correction is to be made for radioactive decay, and also indicates that the radioactive half-life of ^{201}Tl is 73.1 hours.

Reference standard—*USP Endotoxin Reference Standard.*

Radionuclide identification (see *Radioactivity* ⟨821⟩)—Its gamma-ray spectrum is identical to that of a specimen of ^{201}Tl of known purity that exhibits a major photopeak at an energy of 167 KeV and a minor photopeak of 135 KeV.

Bacterial endotoxins—It meets the requirements of the *Bacterial Endotoxins Test* ⟨85⟩, the limit of endotoxin content being not more than $175/V$ USP Endotoxin Unit per mL of the Injection, when compared with the USP Endotoxin RS, in which V is the maximum recommended total dose, in mL, at the expiration date or time.

pH ⟨791⟩: between 4.5 and 7.5.

Radiochemical purity—Soak a 2.5- × 15.0-cm cellulose polyacetate strip in 0.05 *M* disodium ethylenediaminetetraacetate for 45 to 60 minutes. Remove the strip with forceps, taking care to handle the outer edges only. Place the strip between two absorbent pads, and blot to remove excess solution. Apply not less than 5 µL of a previously mixed solution consisting of equal volumes of Thallous Chloride Tl 201 Injection and 0.05 *M* disodium ethylenediaminetetraacetate to the center of the blotted strip, and mark the point of application. Attach the strip to the support bridge of an electrophoresis chamber containing equal portions of 0.05 *M* disodium ethylenediaminetetraacetate in each side of the chamber. Ensure that each end of the strip is in contact with the 0.05 *M* disodium ethylenediaminetetraacetate. Attach the chamber cover, and perform the electrophoresis at 250 volts for 30 minutes. Remove the strip from the chamber, and allow to air-dry without blotting. Using a suitable scanner and counting assembly, determine the radioactivity. Not less than 95.0% of the radioactivity on the strip migrates toward the cathode as a single peak.

Radionuclidic purity—Using a suitable counting assembly (see *Selection of a Counting Assembly* under *Radioactivity* ⟨821⟩), determine the radioactivity of each radionuclidic impurity in the Injection by use of a calibrated system. Not less than 95.0% of the total radioactivity is present as thallium 201. In addition, not more than 2.0% of thallium 200 (half-life is 26.1 hours), not more than 0.3% of lead 203 (half-life is 52.02 hours), and not more than 2.7% of thallium 202 (half-life is 12.23 days) are present.

Thallium—

Standard thallium solution—Transfer 235 mg of thallous chloride, accurately weighed, to a 1000-mL volumetric flask, dilute with water to volume, and mix. Transfer 1.0 mL of the resulting solution to a 100-mL volumetric flask, dilute with saline TS containing 0.9% benzyl alcohol to volume, and mix. This standard solution contains 2 µg of thallium per mL.

Procedure—Transfer 1.0-mL portions of the *Standard thallium solution* and the Injection to separate screw-cap test tubes. To each tube add 2 drops of a solution, prepared by carefully mixing 18 mL of nitric acid and 82 mL of hydrochloric acid, and mix. Then add to each tube 1.0 mL of sulfosalicylic acid solution (1 in 10), and mix. Add 2 drops of 12 *N* hydrochloric acid to each tube, and mix. To each tube add 4 drops of rhodamine B solution (50 mg of rhodamine B diluted with hydrochloric acid to 100.0 mL), and mix. Add 1.0 mL of diisopropyl ether. Screw the caps on tightly, shake the tubes by hand for 1 minute, accurately timed, releasing any pressure build-up by loosening the caps slightly. Recap the tubes and allow the phases to separate. Transfer 0.5 mL of the diisopropyl ether layer from each tube to clean tubes. Visually compare the ether layers: the color of the ether layer from the Injection is not darker than that from the *Standard thallium solution.*

Iron—Into separate cavities of a spot plate, place 0.1 mL of the Injection and 0.1 mL of *Standard iron solution* (see *Iron* ⟨241⟩) diluted with water to a concentration of 5 µg per mL. Add to each cavity 0.1 mL of hydroxylamine hydrochloride solution (1 in 10), 1 mL of sodium acetate solution (1 in 4), and 0.1 mL of 0.5% dipyridyl solution (0.5 g of 2,2′-dipyridyl dissolved in 100 mL of water containing 0.15 mL of hydrochloric acid), and mix. After 5 minutes, the color of the specimen of Thallous Chloride Tl 201 Injection is not darker than that of the *Standard iron solution.*

Copper—

Standard copper solution—Dissolve 0.982 of $CuSO_4 \cdot 5H_2O$ in 1000 mL of 0.1 *N* hydrochloric acid. Transfer 2.0 mL of this solution to a 100-mL volumetric flask, dilute with 0.1 *N* hydrochloric acid to volume, and mix to obtain a Standard solution containing 5 µg of copper per mL.

Procedure—Into separate cavities of a spot plate, place 0.2 mL of the Injection and 0.2 mL of *Standard copper solution*. Add to each cavity 0.2 mL of water and 0.1 mL of iron thiocyanate solution (1.5 g ferric chloride and 2 g potassium thiocyanate dissolved in water and diluted with water to 100.0 mL). Mix, then add 0.1 mL of sodium thiosulfate solution (1 in 100), and again mix. The time required for the specimen of Thallous Chloride Tl 201 Injection to decolorize is equal to or longer than that observed for the *Standard copper solution*.

Other requirements—It meets the requirements under *Injections* ⟨1⟩, except that the Injection may be distributed or dispensed prior to the completion of the test for *Sterility*, the latter test being started on the day of final manufacture, and except that it is not subject to the recommendations on *Volume in Container*.

Assay for radioactivity—Using a suitable counting assembly (see *Selection of a Counting Assembly* under *Radioactivity* ⟨821⟩), determine the radioactivity in MBq (μCi or mCi) per mL of Thallous Chloride Tl 201 Injection by use of a calibrated system as directed under *Radioactivity* ⟨821⟩.

Theophylline

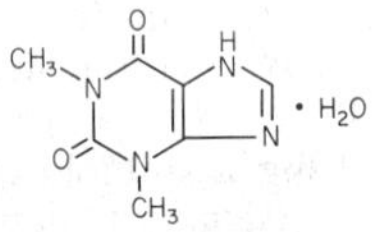

$C_7H_8N_4O_2.H_2O$ 198.18
1*H*-Purine-2,6-dione, 3,7-dihydro-1,3-dimethyl-, monohydrate.
Theophylline monohydrate [*5967-84-0*].
Anhydrous 180.17 [*58-55-9*].

» Theophylline contains one molecule of water of hydration or is anhydrous. It contains not less than 97.0 percent and not more than 102.0 percent of $C_7H_8N_4O_2$, calculated on the dried basis.

Packaging and storage—Preserve in well-closed containers.

Labeling—Label it to indicate whether it is hydrous or anhydrous.

Reference standard—*USP Theophylline Reference Standard*—Dry at 105° for 4 hours before using.

Identification—

A: The infrared absorption spectrum of a potassium bromide dispersion of it, previously dried, exhibits maxima only at the same wavelengths as that of a similar preparation of USP Theophylline RS.

B: The retention time of the major peak in the chromatogram of the *Assay preparation* corresponds to that in the chromatogram of the *Standard preparation*, both relative to the internal standard, as obtained in the *Assay*.

Melting range ⟨741⟩: between 270° and 274°, but the range between beginning and end of melting does not exceed 3°.

Acidity—Dissolve 500 mg in 75 mL of water, and add 1 drop of methyl red TS: not more than 1.0 mL of 0.020 *N* sodium hydroxide is required to change the red color to yellow.

Loss on drying ⟨731⟩—Dry it at 105° for 4 hours: the hydrous form loses between 7.5% and 9.5% of its weight, and the anhydrous form loses not more than 0.5% of its weight.

Residue on ignition ⟨281⟩: not more than 0.15%.

Assay—

Buffer solution—Transfer 2.72 g of sodium acetate trihydrate to a 2000-mL volumetric flask, add about 200 mL of water, and shake until dissolution is complete. Add 10.0 mL of glacial acetic acid, dilute with water to volume, and mix.

Mobile phase—Transfer 70.0 mL of acetonitrile to a 1000-mL volumetric flask, dilute with *Buffer solution* to volume, and mix. Degas, and filter before using. Make adjustments if necessary (see *System Suitability* under *Chromatography* ⟨621⟩).

Internal standard solution—Transfer about 50 mg of theobromine, accurately weighed, to a 100-mL volumetric flask, dissolve in 10.0 mL of 6 *N* ammonium hydroxide, dilute with *Mobile phase* to volume, and mix.

Standard preparation—Dissolve an accurately weighed quantity of USP Theophylline RS in *Mobile phase*, and dilute quantitatively, and stepwise if necessary, with *Mobile phase* to obtain a solution having a known concentration of about 1 mg per mL. Transfer 10.0 mL of this solution to a 100-mL volumetric flask, add 20.0 mL of *Internal standard solution*, dilute with *Mobile phase* to volume, and mix to obtain a solution having a known concentration of about 0.1 mg of USP Theophylline RS per mL.

Assay preparation—Transfer about 100 mg of Theophylline, accurately weighed, to a 100-mL volumetric flask, add about 50 mL of *Mobile phase*, shake by mechanical means until solution is complete, dilute with *Mobile phase* to volume, and mix. Transfer 10.0 mL of this solution to a second 100-mL volumetric flask, add 20.0 mL of *Internal standard solution*, dilute with *Mobile phase* to volume, and mix.

Chromatographic system (see *Chromatography* ⟨621⟩)—The liquid chromatograph is equipped with a 280-nm detector and a 4-mm × 30-cm column that contains packing L1. The flow rate is about 1.0 mL per minute. The relative standard deviation of R_S for replicate injections of *Standard preparation* is not more than 1.5%, the resolution, *R*, between theophylline and theobromine is not less than 1.5, and the tailing factor, *T*, is not greater than *R* and under no circumstances higher than 2.5.

Procedure—Separately inject equal volumes (between 10 μL and 25 μL) of the *Standard preparation* and the *Assay preparation* into the chromatograph, and measure the peak responses for the major peaks. The retention time of theophylline relative to that of theobromine is about 1.6. Calculate the quantity, in mg, of $C_7H_8N_4O_2$ in the portion of Theophylline taken by the formula:

$$1000C(R_U/R_S),$$

in which *C* is the concentration, in mg per mL, of USP Theophylline RS in the *Standard preparation*, and R_U and R_S are the response ratios of the theophylline peak to the internal standard peak obtained from the *Assay preparation* and the *Standard preparation*, respectively.

Theophylline Capsules

» Theophylline Capsules contain not less than 90.0 percent and not more than 110.0 percent of the labeled amount of anhydrous theophylline ($C_7H_8N_4O_2$).

Packaging and storage—Preserve in well-closed containers.

Reference standard—*USP Theophylline Reference Standard*—Dry at 105° for 4 hours before using.

Identification—

A: The contents of the Capsules respond to *Identification tests A* and *B* under *Theophylline Tablets*.

B: The retention time of the major peak in the chromatogram of the *Assay preparation* corresponds to that in the chromatogram of the *Standard preparation* as obtained in the *Assay*.

Dissolution ⟨711⟩—

Medium: water; 900 mL.
Apparatus 2: 50 rpm.
Time: 60 minutes.

Procedure—Determine the amount of $C_7H_8N_4O_2$ dissolved from ultraviolet absorbances at the wavelength of maximum absorbance at about 268 nm of filtered portions of the solution under test, suitably diluted with 0.1 *N* hydrochloric acid, if necessary, in comparison with a Standard solution having a known concentration of USP Theophylline RS in the same medium.

Tolerances—Not less than 80% (*Q*) of the labeled amount of $C_7H_8N_4O_2$ is dissolved in 60 minutes.

Uniformity of dosage units ⟨905⟩: meet the requirements.

Assay—

Mobile phase—Prepare a solution containing a mixture of water, methanol, and glacial acetic acid (64:35:1).

Standard preparation—Dissolve an accurately weighed quantity of USP Theophylline RS in methanol to obtain a solution having a known concentration of about 400 µg per mL.

Assay preparation for hard Capsules—Remove, as completely as possible, the contents of not less than 20 Theophylline Capsules, weigh, and mix. Transfer an accurately weighed portion of the powder, equivalent to about 100 mg of anhydrous theophylline, to a 250-mL volumetric flask, add about 150 mL of methanol, and shake to dissolve. Dilute with methanol to volume, mix, and filter, using a membrane filter.

Assay preparation for soft Capsules—Cut open 20 Theophylline Capsules, and place them in a 200-mL volumetric flask. Add 50 mL of 6 *N* ammonium hydroxide, shake to dissolve the contents, add water to volume, mix, and filter, discarding the first 20 mL of the filtrate. Transfer an accurately measured portion of the filtrate, equivalent to about 100 mg of anhydrous theophylline, to a 250-mL volumetric flask, add methanol to volume, mix, and filter through a membrane filter.

Chromatographic system (see *Chromatography* ⟨621⟩)—The liquid chromatograph is equipped with a 254-nm detector and a 4-mm × 30-cm column that contains packing L1. The flow rate is about 2 mL per minute. Chromatograph three replicate injections of the *Standard preparation*, and record the peak responses as directed under *Procedure:* the relative standard deviation is not more than 2%.

Procedure—Separately inject equal volumes (about 20 µL) of the *Standard preparation* and the *Assay preparation* into the chromatograph, record the chromatograms, and measure the responses. Calculate the quantity, in mg, of anhydrous theophylline in the portion of Capsule contents taken by the formula:

$$0.25C(r_U / r_S),$$

in which C is the concentration, in µg per mL, of USP Theophylline RS in the *Standard preparation*, and r_U and r_S are the peak responses obtained from the *Assay preparation* and the *Standard preparation*, respectively.

Theophylline Extended-release Capsules

» Theophylline Extended-release Capsules contain not less than 90.0 percent and not more than 110.0 percent of the labeled amount of anhydrous theophylline ($C_7H_8N_4O_2$).

Packaging and storage—Preserve in well-closed containers.

Labeling—The labeling indicates steady-state blood level profiles and confidence intervals and states the in-vitro *Drug release* test conditions of medium, apparatus and speed, times and tolerances, as directed under *Drug release.*

Reference standard—*USP Theophylline Reference Standard*—Dry at 105° for 4 hours before using.

Identification—

A: Transfer a quantity of Capsule contents, equivalent to about 100 mg of anhydrous theophylline, to a suitable conical flask. Add 150 mL of methanol, and sonicate until insoluble material is dispersed into fine particles. Shake by mechanical means for 15 minutes, and filter into a 250-mL volumetric flask. Dilute with water to volume, and mix. Pipet 5 mL of this solution into a 200-mL volumetric flask, dilute with 0.1 *N* hydrochloric acid to volume, and mix: the ultraviolet absorption spectrum of the solution so obtained exhibits maxima and minima at the same wavelengths as that of a similar solution of USP Theophylline RS, concomitantly measured.

B: The retention time of the major peak in the chromatogram of the *Assay preparation* corresponds to that of the *Standard preparation*, as obtained in the *Assay*.

Drug release ⟨724⟩—

Medium, Times, and *Tolerances:* as specified in the *Labeling;* use *Acceptance Table 1.*

Apparatus 1 or *2:* as specified in the *Labeling* operated at the speed specified in the *Labeling.*

Procedure—Determine the amount of $C_7H_8N_4O_2$ dissolved from ultraviolet absorbances at the wavelength of maximum absorbance at about 271 nm of filtered portions of the solution under test, diluted with *Dissolution Medium*, if necessary, in comparison with a Standard solution having a known concentration of USP Theophylline RS in the same medium.

Uniformity of dosage units ⟨905⟩: meet the requirements.

Assay—

Phosphate buffer—Dissolve 1.36 g of monobasic potassium phosphate in 1000 mL of water, and adjust with 1 *N* potassium hydroxide to a pH of 6.50 ± 0.05.

Mobile phase—Prepare a suitable mixture of *Phosphate buffer* and methanol (7:3), filter, and degas. Make adjustments if necessary (see *System Suitability* under *Chromatography* ⟨621⟩).

Standard preparation—Dissolve an accurately weighed quantity of USP Theophylline RS in *Phosphate buffer* to obtain a solution having a known concentration of about 1 mg per mL (*Stock standard solution*). Dilute an accurately measured volume of this solution quantitatively with *Mobile phase* to obtain a solution having a known concentration of about 0.1 mg per mL (*Standard preparation*).

Assay preparation—Remove, as completely as possible, the contents of not less than 20 Theophylline Extended-release Capsules, weigh accurately, and grind the contents to a fine powder. Transfer an accurately weighed portion of the powder, equivalent to about 100 mg of anhydrous theophylline, to a 100-mL volumetric flask. Add 2 mL of methanol, and swirl until the powder is wetted. Add 80 mL of *Phosphate buffer*, sonicate with frequent swirling until the powder is completely dispersed, and heat on a steam bath for 15 minutes. Cool to room temperature, dilute with *Phosphate buffer* to volume, and mix. Filter a portion of this solution, discarding the first 10 mL of the filtrate. Transfer 10.0 mL of the filtrate to a 100-mL volumetric flask, dilute with *Mobile phase* to volume, and mix.

Resolution solution—Prepare a solution of anhydrous caffeine in a mixture of methanol and water (9:1) containing about 2 mg per mL. Pipet 6 mL of this solution and 10 mL of the *Stock standard solution* into a 100-mL volumetric flask, dilute with *Mobile phase* to volume, and mix.

Chromatographic system (see *Chromatography* ⟨621⟩)—The liquid chromatograph is equipped with a 254-nm detector and a 4-mm × 30-cm column that contains packing L1. The flow rate is about 1 mL per minute. Chromatograph the *Resolution solution*, and record the peak responses as directed under *Procedure:* the relative retention times are about 0.7 for theophylline and 1.0 for caffeine, and the resolution, R, between the theophylline and caffeine peaks is not less than 2.5. Chromatograph the *Standard preparation*, and record the peak responses as directed under *Procedure:* the relative standard deviation for replicate injections is not more than 2.0%.

Procedure—Separately inject equal volumes (about 20 µL) of the *Standard preparation* and the *Assay preparation* into the chromatograph, record the chromatograms, and measure the responses for the major peaks. Calculate the quantity, in mg, of $C_7H_8N_4O_2$ in the portion of Capsules taken by the formula:

$$1000C(r_U / r_S),$$

in which C is the concentration, in mg per mL, of USP Theophylline RS in the *Standard preparation*, and r_U and r_S are the peak responses obtained from the *Assay preparation* and the *Standard preparation*, respectively.

Theophylline Tablets

» Theophylline Tablets contain not less than 94.0 percent and not more than 106.0 percent of the labeled amount of anhydrous theophylline ($C_7H_8N_4O_2$).

Packaging and storage—Preserve in well-closed containers.

Identification—

A: Triturate a quantity of finely powdered Tablets, equivalent to about 500 mg of theophylline, with 10-mL and 5-mL portions of solvent hexane, and discard the solvent hexane. Triturate the residue with two 10-mL portions of a mixture of equal volumes of 6 *N* ammonium hydroxide and water, and filter each time.

Evaporate the combined filtrates to about 5 mL, neutralize, if necessary, with 6 *N* acetic acid, using litmus, and then cool to about 15°, with stirring. Collect the precipitate on a filter, wash it with cold water, and dry at 105° for 2 hours: the theophylline so obtained melts between 270° and 274°, the procedure for *Class I* being used (see *Melting Range or Temperature* ⟨741⟩). Retain the remaining portion of the theophylline for use in *Identification test B*.

B: The infrared absorption spectrum of a potassium bromide dispersion of the residue obtained in *Identification test A* exhibits maxima only at the same wavelengths as that of a potassium bromide dispersion of USP Theophylline RS.

C: The retention time of the major peak in the chromatogram of the *Assay preparation* corresponds to that in the chromatogram of the *Standard preparation*, both relative to the internal standard, as obtained in the *Assay*.

Dissolution ⟨711⟩—

Medium: water; 900 mL.

Apparatus 2: 50 rpm.

Time: 45 minutes.

Procedure—Determine the amount of $C_7H_8N_4O_2$ dissolved from ultraviolet absorbances at the wavelength of maximum absorbance at about 272 nm of filtered portions of the solution under test, suitably diluted with water, if necessary, in comparison with a Standard solution having a known concentration of USP Theophylline RS in the same medium.

Tolerances—Not less than 80% (*Q*) of the labeled amount of $C_7H_8N_4O_2$ is dissolved in 45 minutes.

Uniformity of dosage units ⟨905⟩: meet the requirements.

Assay—

Mobile phase, *Internal standard solution*, and *Standard preparation*—Prepare as directed in the *Assay* under *Theophylline*.

Assay preparation—Place 10 Theophylline Tablets in a 500-mL volumetric flask, add 50 mL of water, and when the Tablets have disintegrated add 50 mL of 6 *N* ammonium hydroxide. Shake until no more dissolves, dilute with water to volume, mix, and filter through a dry filter with the aid of suction, if necessary, into a dry flask, discarding the first 20 mL of the filtrate. Transfer an accurately measured aliquot portion of this concentrate, equivalent to about 10 mg of theophylline, to a 100-mL volumetric flask. Add 20.0 mL of *Internal standard solution*, dilute with *Mobile phase* to volume, and mix.

Chromatographic system—Proceed as directed in the *Assay* under *Theophylline*.

Procedure—Proceed as directed for *Procedure* in the *Assay* under *Theophylline*. Calculate the quantity, in mg, of $C_7H_8N_4O_2$ per Tablet taken by the formula:

$$5000(C/V)(R_U/R_S),$$

in which *C* is the concentration, in mg per mL, of USP Theophylline RS in the *Standard preparation*, *V* is the volume, in mL, of concentrate taken for the *Assay preparation*, and R_U and R_S are the response ratios of the theophylline peak to the internal standard peak obtained from the *Assay preparation* and the *Standard preparation*, respectively.

Theophylline in Dextrose Injection

» Theophylline in Dextrose Injection is a sterile solution of Theophylline and Dextrose in Water for Injection. It contains not less than 93.0 percent and not more than 107.0 percent of the labeled amount of anhydrous theophylline ($C_7H_8N_4O_2$) and not less than 95.0 percent and not more than 105.0 percent of the labeled amount of dextrose ($C_6H_{12}O_6 . H_2O$).

Packaging and storage—Preserve in single-dose containers, preferably of Type I or Type II glass, or of a suitable plastic material.

Reference standards—*USP Theophylline Reference Standard*—Dry at 105° for 4 hours before using. *USP Endotoxin Reference Standard*.

Identification—

A: The ultraviolet absorption spectrum of the Injection, diluted with 0.5 *N* sodium hydroxide to a concentration of about 8 µg of anhydrous theophylline per mL, exhibits maxima and minima at the same wavelengths as that of a similar solution of USP Theophylline RS, concomitantly measured.

B: Add a few drops of the Injection to 5 mL of hot alkaline cupric tartrate TS: a red to orange precipitate of cuprous oxide is formed.

Bacterial endotoxin—When tested as directed under *Bacterial Endotoxins Test* ⟨85⟩, it contains not more than 1.0 USP Endotoxin Unit per mg of anhydrous theophylline.

pH ⟨791⟩: between 3.5 and 6.5, determined on a portion diluted with water, if necessary, to a concentration of not more than 5% of dextrose.

5-Hydroxymethylfurfural and related substances—

Anion-exchange column—Mix 8 g of 20- to 50-mesh styrene-divinylbenzene anion-exchange resin in the hydroxide form with 25 mL of water in a suitable beaker. Allow the resin to settle, and decant the supernatant liquid until a slurry of resin remains. Pour the slurry into a 1.1-cm × 66-cm glass chromatographic tube (having a sealed-in, coarse-porosity fritted disk or a glass wool plug and fitted with a stopcock), and allow to settle as a homogeneous bed having a bed volume of about 15 mL. Wash the resin bed at a flow rate of about 3 mL per minute with 100 mL of an ammonium carbonate solution containing 5.7 g of ammonium carbonate per 100 mL, followed by washing with water until the eluate has a neutral pH. Close the stopcock when the water level is about 3 mm above the resin bed.

Procedure—Dilute an accurately measured volume of Injection, equivalent to 1.0 g of $C_6H_{12}O_6 . H_2O$, with water to 250.0 mL. Pass this solution through the resin bed in the *Anion-exchange column* at a flow rate of about 3.5 mL per minute, discarding the first 50 mL of the eluate. Determine the absorbance of the eluate in a 1-cm cell at 284 nm, with a suitable spectrophotometer, using water as the blank: the absorbance is not more than 0.25.

Other requirements—It meets the requirements under *Injections* ⟨1⟩.

Assay for theophylline—

Buffer solution, *Mobile phase*, and *Internal standard solution*—Prepare as directed in the *Assay* under *Theophylline*.

Standard preparation—Dissolve an accurately weighed quantity of USP Theophylline RS in *Mobile phase*, and dilute quantitatively, and stepwise if necessary, with *Mobile phase* to obtain a solution having a known concentration of about 1 mg per mL. Transfer 8.0 mL of this solution to a 50-mL volumetric flask, add 10.0 mL of *Internal standard solution*, dilute with *Mobile phase* to volume, and mix to obtain a solution having a known concentration of about 0.16 mg of USP Theophylline RS per mL.

Assay preparation—Transfer an accurately measured volume of Theophylline in Dextrose Injection, equivalent to about 16 mg of theophylline, to a 100-mL volumetric flask, add 20.0 mL of *Internal standard solution*, dilute with *Mobile phase* to volume, and mix.

Chromatographic system—Proceed as directed in the *Assay* under *Theophylline*.

Procedure—Proceed as directed for *Procedure* in the *Assay* under *Theophylline*. Calculate the quantity, in mg, of $C_7H_8N_4O_2$ in each mL of the Injection taken by the formula:

$$(100C/V)(R_U/R_S),$$

in which *V* is the volume, in mL, of Injection taken, and the other terms are as defined therein.

Assay for dextrose—Transfer an accurately measured volume of Theophylline in Dextrose Injection, containing 2 to 5 g of dextrose, to a 100-mL volumetric flask. Add 0.2 mL of 6 *N* ammonium hydroxide, dilute with water to volume, and mix. Determine the angular rotation in a suitable polarimeter tube at 25° (see *Optical Rotation* ⟨781⟩). The observed rotation, in degrees multiplied by 1.0425*A*, in which *A* is the ratio 200 divided by the length, in mm, of the polarimeter tube employed, represents the weight, in g, of $C_6H_{12}O_6 . H_2O$ in the volume of Injection taken.

Theophylline, Ephedrine Hydrochloride, and Phenobarbital Tablets

» Theophylline, Ephedrine Hydrochloride, and Phenobarbital Tablets contain not less than 90.0 percent and not more than 110.0 percent of the labeled amounts of anhydrous theophylline ($C_7H_8N_4O_2$), ephedrine hydrochloride ($C_{10}H_{15}NO.HCl$), and phenobarbital ($C_{12}H_{12}N_2O_3$).

Packaging and storage—Preserve in tight containers.

Reference standards—*USP Theophylline Reference Standard*—Dry at 105° for 4 hours before using. *USP Ephedrine Sulfate Reference Standard*—Dry at 105° for 3 hours before using. *USP Phenobarbital Reference Standard*—Dry at 105° for 2 hours before using.

Identification—Place a quantity of finely powdered Tablets, equivalent to about 24 mg of ephedrine hydrochloride, in a 15-mL centrifuge tube, add 4.0 mL of a mixture of chloroform and methanol (4:1), mix by sonication for 10 minutes, and filter to obtain the test solution. Prepare separate Standard solutions containing known concentrations of about 36 mg of USP Theophylline RS per mL, about 9.6 mg of USP Ephedrine Sulfate RS per mL, and about 5 mg of USP Phenobarbital RS per mL, respectively, in the mixture of chloroform and methanol (4:1). Apply 2 µL of the test solution and of each of the three Standard solutions at equidistant points about 2.5 cm above one edge of a suitable thin-layer chromatographic plate (see *Chromatography* ⟨621⟩), coated with a 0.25-mm layer of chromatographic silica gel mixture. Allow the spots to dry, and develop the chromatogram in a suitable chamber previously equilibrated with a solvent system consisting of a mixture of chloroform, acetone, methanol, and ammonium hydroxide (50:10:10:1) until the solvent front has moved about three-fourths of the length of the plate. Remove the plate from the developing chamber, mark the solvent front, air-dry, and locate the spots on the plate by viewing under short-wavelength ultraviolet light: the R_f values of the spots obtained from the test solution correspond to those obtained from the Standard solutions.

Dissolution ⟨711⟩—

Medium: water; 900 mL.

Apparatus 1: 100 rpm.

Time: 30 minutes.

Determination of dissolved theophylline, ephedrine hydrochloride, and phenobarbital—

MOBILE PHASE—Dissolve accurately weighed quantities of monobasic potassium phosphate and sodium 1-hexanesulfonate in water, and dilute quantitatively with water to obtain a solution having concentrations of 0.953 mg per mL (0.007 *M* monobasic potassium phosphate) and 0.564 mg per mL (0.003 *M* sodium 1-hexanesulfonate), respectively. Adjust, if necessary, with 0.3 *M* phosphoric acid or 0.2 *M* monobasic potassium phosphate to a pH of 3.0 ± 0.05, to obtain a *Phosphate buffer*. The *Mobile phase* is a mixture of *Phosphate buffer* and methanol (75:25).

STANDARD PREPARATION—Dissolve accurately weighed quantities of USP Theophylline RS, USP Ephedrine Sulfate RS, and USP Phenobarbital RS in water, and dilute quantitatively and stepwise with water to obtain a solution having known concentrations of about 145 µg of anhydrous theophylline per mL, 28 µg of ephedrine sulfate per mL, and 9 µg of phenobarbital per mL.

CHROMATOGRAPHIC SYSTEM (see *Chromatography* ⟨621⟩)—The liquid chromatograph is equipped with a 215- nm detector and a 3.9-mm × 30-cm column that contains packing L1. The flow rate is about 3.0 mL per minute. Chromatograph the *Standard preparation*, and record the peak responses as directed under *Procedure:* the resolution, *R*, between the theophylline and phenobarbital peaks is not less than 4.0, the tailing factors for the ephedrine and phenobarbital peaks are not more than 3.0 and 2.0, respectively, and the relative standard deviation for replicate injections is not more than 2.0%.

PROCEDURE—Inject an accurately measured volume (about 75 µL) of a filtered portion of the solution under test into the chromatograph, record the chromatogram, and measure the responses for the major peaks. The elution order is theophylline, ephedrine, and phenobarbital (last). Calculate the quantity, in mg, of $C_7H_8N_4O_2$ dissolved by the formula:

$$(0.9C)(r_U / r_S),$$

in which C is the concentration, in µg per mL, of USP Theophylline RS in the *Standard preparation*, and r_U and r_S are the peak responses for theophylline obtained from the solution under test and the similarly chromatographed *Standard preparation*, respectively. Calculate the quantity, in mg, of $C_{10}H_{15}NO.HCl$ dissolved by the formula:

$$(201.70/214.27)(0.9C')(r'_U / r'_S),$$

in which 201.70 is the molecular weight of ephedrine hydrochloride, 214.27 is one-half the molecular weight of ephedrine sulfate, C' is the concentration, in µg per mL, of USP Ephedrine Sulfate RS in the *Standard preparation*, and r'_U and r'_S are the peak responses for ephedrine obtained from the solution under test and the similarly chromatographed *Standard preparation*, respectively. Calculate the quantity, in mg, of $C_{12}H_{12}N_2O_3$ dissolved by the formula:

$$(0.9C'')(r''_U / r''_S),$$

in which C'' is the concentration, in µg per mL, of USP Phenobarbital RS in the *Standard preparation*, and r''_U and r''_S are the peak responses for phenobarbital obtained from the solution under test and the similarly chromatographed *Standard preparation*, respectively.

Tolerances—Not less than 75% (*Q*) of the labeled amounts of $C_7H_8N_4O_2$, $C_{10}H_{15}NO.HCl$, and $C_{12}H_{12}N_2O_3$ are dissolved in 30 minutes.

Uniformity of dosage units ⟨905⟩: meet the requirements.

Assay—

Mobile phase—Mix 240 mL of acetonitrile with 760 mL of 0.01 *M*, pH 7.8 phosphate buffer (see *Buffer Solutions* under *Solutions* in the section, *Reagents, Indicators, and Solutions*).

Internal standard solution—Dissolve 25 mg of butabarbital sodium in 50 mL of dibasic potassium phosphate solution (17 in 1000). Mix 40 mL of this solution with 10 mL of sodium metaperiodate solution (1 in 100).

Standard preparation—Transfer about 120 mg of USP Theophylline RS, about 8 mg of USP Phenobarbital RS, and about 25 mg of USP Ephedrine Sulfate RS, each accurately weighed, to a 200-mL volumetric flask. Add 10 mL of methanol and 100 mL of chloroform, mix to dissolve, add chloroform to volume, and mix. Pipet 15 mL of this solution into a 50-mL volumetric flask, add chloroform to volume, and mix. The concentrations of anhydrous theophylline, phenobarbital, and ephedrine sulfate in the *Standard preparation* are about 180 µg per mL, 12 µg per mL, and 37.5 µg per mL, respectively.

Assay preparation—Weigh and finely powder not less than 20 Theophylline, Ephedrine Hydrochloride, and Phenobarbital Tablets. Transfer an accurately weighed portion of the powder, equivalent to about 120 mg of anhydrous theophylline, to a 200-mL volumetric flask, and shake by mechanical means for 20 minutes with a mixture of 10 mL of methanol and 100 mL of chloroform. Add chloroform to volume, mix, and filter. Pipet 15 mL of the clear filtrate into a 50-mL volumetric flask, dilute with chloroform to volume, and mix.

Standard solution and *Assay solution*—Pipet 10-mL portions of the *Standard preparation* and the *Assay preparation* into separate glass-stoppered, 25-mL conical flasks, and treat each as follows: Evaporate in a warm-water bath with the aid of a current of air to dryness. Add by pipet 4 mL of *Internal standard solution*, insert the stopper in the flask, mix to dissolve the residue, and allow to stand at room temperature for 30 minutes. Pipet 1 mL of propylene glycol solution (1 in 100) into the flask, insert the stopper, and mix.

Chromatographic system (see *Chromatography* ⟨621⟩)—The liquid chromatograph is equipped with a 241-nm detector and a 30-cm × 4-mm stainless steel column that contains packing L1. Set a flow rate of 1.0 mL per minute for the *Mobile phase*, and allow the system to equilibrate until a stable baseline is obtained on the recorder. Chromatograph a 10-µL portion of the *Standard solution*, and record the peak responses as directed under *Pro-*

cedure: the peaks are completely resolved, and the resolution factors between each two neighboring peaks are not less than 1.5. Five replicate injections of the *Standard preparation* show a relative standard deviation of not more than 2.0%.

Procedure—Separately inject equal volumes (about 10 μL) of the *Standard solution* and the *Assay solution* into the chromatograph, record the chromatograms, and measure the responses for the major peaks. In the order of increasing elution times, the five peaks correspond to the reagent (iodate), theophylline, phenobarbital, butabarbital, and benzaldehyde (from ephedrine). Designate the peak response ratio of each component to the internal standard in the *Standard solution* as R_S, and that of each component to the internal standard in the *Assay solution* as R_U.

Calculation for theophylline—Calculate the quantity, in mg, of anhydrous theophylline ($C_7H_8N_4O_2$) in the portion of Tablets taken by the formula:

$$0.667C(R_U/R_S),$$

in which C is the concentration, in μg per mL, of USP Theophylline RS in the *Standard preparation.*

Calculation for ephedrine hydrochloride—Calculate the quantity, in mg, of ephedrine hydrochloride ($C_{10}H_{15}NO.HCl$) in the portion of Tablets taken by the formula:

$$(201.70/214.27)(0.667C)(R_U/R_S),$$

in which 201.70 is the molecular weight of ephedrine hydrochloride, 214.27 is one-half the molecular weight of ephedrine sulfate, and C is the concentration, in μg per mL, of USP Ephedrine Sulfate RS in the *Standard preparation.*

Calculation for phenobarbital—Calculate the quantity, in mg, of phenobarbital ($C_{12}H_{12}N_2O_3$) in the portion of Tablets taken by the formula:

$$0.667C(R_U/R_S),$$

in which C is the concentration, in mg per mL, of USP Phenobarbital RS in the *Standard preparation.*

Theophylline and Guaifenesin Capsules

» Theophylline and Guaifenesin Capsules contain not less than 90.0 percent and not more than 110.0 percent of the labeled amounts of anhydrous theophylline ($C_7H_8N_4O_2$) and guaifenesin ($C_{10}H_{14}O_4$).

Packaging and storage—Preserve in tight containers.

Reference standards—*USP Theophylline Reference Standard*—Dry at 105° for 4 hours before using. *USP Guaifenesin Reference Standard*—Dry in vacuum at a pressure not below 10 mm of mercury at 60° to constant weight before using.

Identification—Transfer a quantity of the contents of the Capsules, equivalent to about 150 mg of theophylline, to a separator, and add 15 mL of water. To a second separator transfer 15 mL of an aqueous Standard solution containing USP Theophylline RS and USP Guaifenesin RS corresponding, proportionately, to the amounts of theophylline and guaifenesin in the Capsules and having a known concentration of about 10 mg of theophylline per mL. Treat the contents of each separator as follows. Add 25 mL of chloroform, and shake vigorously for 0.5 minute. Allow the layers to separate, filter the lower chloroform layer through glass wool, and evaporate the filtrate to dryness. Dissolve the residue in 10 mL of chloroform. On a thin-layer chromatographic cellulose sheet with fluorescent indicator (see *Chromatography* ⟨621⟩), apply separately 10 μL of each solution so obtained, allow the spots to dry, and develop the chromatogram in a solvent system consisting of a mixture of methanol and water (95:5) until the solvent front has moved about 10 cm above the starting line. Remove the sheet from the developing chamber, mark the solvent front, and allow the solvent to evaporate. Expose the sheet to short-wavelength ultraviolet light: the R_f values of the spots obtained from the test preparation correspond to those obtained from the Standard solution.

Uniformity of dosage units ⟨905⟩: meet the requirements.

Assay—

pH 6.5 buffer solution—Dissolve 1.36 g of monobasic potassium phosphate in water to make 1000 mL. Carefully adjust with 2.5 *N* sodium hydroxide to a pH of 6.5, and filter.

Mobile phase—Prepare a degassed solution of *pH 6.5 buffer solution* and methanol (70:30).

Internal standard solution—Dissolve about 400 mg of caffeine in 1000 mL of a solution of methanol and water (90:10), and mix.

Standard preparation—Dissolve an accurately weighed quantity of USP Theophylline RS in *pH 6.5 buffer solution*, and dilute quantitatively with *pH 6.5 buffer solution* to obtain a solution (*Solution T*) having a known concentration of about 900J μg per mL, in which J is the ratio of the labeled amount of theophylline to that of guaifenesin. Transfer about 90 mg of USP Guaifenesin RS, accurately weighed, to a 200-mL volumetric flask, add about 150 mL of *pH 6.5 buffer solution*, shake to dissolve, dilute with *pH 6.5 buffer solution* to volume, and mix. Pipet 10 mL of this solution, 10 mL of *Internal standard solution*, and 5 mL of *Solution T* into a 50-mL volumetric flask, dilute with *Mobile phase* to volume, and mix to obtain a *Standard preparation* having known concentrations of about 90 μg of guaifenesin and about 90J μg of theophylline per mL.

Assay preparation—Transfer a number of Theophylline and Guaifenesin Capsules, equivalent to about 900 mg of guaifenesin, to a 200-mL volumetric flask, add about 160 mL of *pH 6.5 buffer solution*, heat to dissolve completely, cool to room temperature, dilute with *pH 6.5 buffer solution* to volume, and mix. Dilute 10.0 mL of this solution with *pH 6.5 buffer solution* to 100.0 mL. Pipet 10 mL of the diluted solution and 10 mL of *Internal standard solution* into a 50-mL volumetric flask, dilute with *Mobile phase* to volume, and mix.

Chromatographic system—The liquid chromatograph is equipped with a 280-nm detector and a 3.9-mm × 30-cm column that contains packing L1. The flow rate is about 1.0 mL per minute. Chromatograph six replicate injections of the *Standard preparation*, and record the peak responses as directed under *Procedure:* the relative standard deviation of the ratio of peak responses (peak response of ingredient/peak response of internal standard) is not more than 2.0% for theophylline and not more than 2.5% for guaifenesin. The resolution, R, between theophylline and caffeine is not less than 3.0.

Procedure—Separately inject equal volumes (about 25 μL) of the *Standard preparation* and the *Assay preparation* into the chromatograph, record the chromatograms, and measure the responses for the major peaks. The relative retention times are about 0.7 for theophylline, 1.0 for caffeine, and 1.5 for guaifenesin. Calculate the quantities, in mg, of anhydrous theophylline ($C_7H_8N_4O_2$) and guaifenesin ($C_{10}H_{14}O_4$) in the portion of Capsules taken by the formula:

$$10C(R_U/R_S),$$

in which C is the concentration, in μg per mL, of the appropriate USP Reference Standard in the *Standard preparation*, and R_U and R_S are the ratios of the peak responses of the corresponding analyte to those of caffeine in the *Assay preparation* and the *Standard preparation*, respectively.

Theophylline and Guaifenesin Oral Solution

» Theophylline and Guaifenesin Oral Solution contains not less than 90.0 percent and not more than 110.0 percent of the labeled amount of anhydrous theophylline ($C_7H_8N_4O_2$) and not less than 86.7 percent and not more than 113.3 percent of the labeled amount of guaifenesin ($C_{10}H_{14}O_4$).

Packaging and storage—Preserve in tight containers.

Reference standards—*USP Theophylline Reference Standard*—Dry at 105° for 4 hours before using. *USP Guaifenesin Reference*

Standard—Dry in vacuum at a pressure not below 10 mm of mercury at 60° to constant weight before using.

Identification—Transfer a volume of Oral Solution, equivalent to about 150 mg of theophylline, to a separator, add 15 mL of water, and proceed as directed in the *Identification test* under *Theophylline and Guaifenesin Capsules*, beginning with "To a second separator."

Assay—

pH 6.5 buffer solution, and *Mobile phase*—Prepare as directed in the *Assay* under *Theophylline and Guaifenesin Capsules*.

Caffeine solution—Dissolve about 400 mg of caffeine in 1000 mL of a mixture of methanol and water (90:10), and mix.

Standard preparation—Dissolve an accurately weighed quantity of USP Theophylline RS in *pH 6.5 buffer solution*, and dilute quantitatively with *pH 6.5 buffer solution* to obtain a solution (*Solution T*) having a known concentration of about 900J µg per mL, in which J is the ratio of the labeled amount of theophylline to that of guaifenesin. Transfer about 90 mg of USP Guaifenesin RS, accurately weighed, to a 200-mL volumetric flask, add about 150 mL of *pH 6.5 buffer solution*, shake to dissolve, dilute with *pH 6.5 buffer solution* to volume, and mix. Pipet 10 mL of this solution and 5 mL of *Solution T* into a 50-mL volumetric flask, dilute with *Mobile phase* to volume, and mix to obtain a *Standard preparation* having known concentrations of about 90 µg of guaifenesin and about 90J µg of theophylline per mL.

Assay preparation—Transfer an accurately measured volume of Theophylline and Guaifenesin Oral Solution, equivalent to about 90 mg of guaifenesin, to a 200-mL volumetric flask, dilute with *pH 6.5 buffer solution* to volume, and mix. Transfer 10.0 mL of this solution to a 50-mL volumetric flask, dilute with *Mobile phase* to volume, and mix.

Chromatographic system (see *Chromatography* ⟨621⟩)—The liquid chromatograph is equipped with a 280-nm detector and a 3.9-mm × 30-cm column that contains packing L1. The flow rate is about 1.0 mL per minute. Chromatograph a mixture of 4 mL of *Standard preparation* and 1 mL of *Caffeine solution*, and record the peak responses as directed under *Procedure:* the resolution, R, between the theophylline and caffeine peaks is not less than 3.0, and the relative standard deviation for replicate injections is not more than 2.0 for theophylline and not more than 2.5% for guaifenesin.

Procedure—Proceed as directed in the *Assay* under *Theophylline and Guaifenesin Capsules*. Calculate the quantities, in mg, of anhydrous theophylline ($C_7H_8N_4O_2$) and guaifenesin ($C_{10}H_{14}O_4$) per mL of the Oral Solution taken by the formula:

$$(C/V)(r_U/r_S),$$

in which C is the concentration, in µg per mL, of the appropriate USP Reference Standard in the *Standard preparation*, V is the volume, in mL, of Oral Solution taken, and r_U and r_S are the peak responses of the corresponding analyte in the *Assay preparation* and the *Standard preparation*, respectively.

Theophylline Sodium Glycinate

Glycine, mixt. with 3,7-dihydro-1,3-dimethyl-1*H*-purine-2,6-dione, monosodium salt.

Theophylline sodium mixture with glycine [*8000-10-0*].

» Theophylline Sodium Glycinate is an equilibrium mixture containing Theophylline Sodium ($C_7H_7N_4NaO_2$) and Glycine ($C_2H_5NO_2$) in approximately equimolecular proportions buffered with an additional mole of Glycine. Dried at 105° for 4 hours, it contains theophylline sodium glycinate equivalent to not less than 44.5 percent and not more than 47.3 percent of anhydrous theophylline ($C_7H_8N_4O_2$).

Packaging and storage—Preserve in tight containers.

Reference standard—*USP Glycine Reference Standard*—Dry at 105° for 2 hours before using.

Identification—

A: Dissolve about 1 g in 20 mL of warm water, and neutralize the solution with 6 *N* acetic acid: a white, crystalline precipitate of theophylline is formed. Filter, wash the precipitate with small portions of cold water, and dry it at 105° for 1 hour: the theophylline so obtained melts between 270° and 274°, the procedure for *Class I* being used (see *Melting Range or Temperature* ⟨741⟩). Retain the remaining portion of the theophylline for use in *Identification test B*.

B: The infrared absorption spectrum of a potassium bromide dispersion of the residue obtained in *Identification test A* exhibits maxima only at the same wavelengths as that of a potassium bromide dispersion of USP Theophylline RS.

C: Ignite it: the residue colors a nonluminous flame intensely yellow and effervesces with acids.

pH ⟨791⟩: between 8.5 and 9.5, in a saturated solution.

Loss on drying ⟨731⟩—Dry it at 105° for 4 hours: it loses not more than 2.0% of its weight.

Glycine content—

Mobile phase—Prepare a solution containing 470 mg of sodium acetate trihydrate and 1 mL of glacial acetic acid in 2 liters of water. Adjust with 10 *N* sodium hydroxide to a pH of 4.3. Mix 3 volumes of the resulting solution with 7 volumes of acetonitrile. Make adjustments if necessary (see *System Suitability* under *Chromatography* ⟨621⟩).

Standard preparation—Dissolve an accurately weighed quantity of USP Glycine RS in *Mobile phase*, and dilute quantitatively with *Mobile phase* to obtain a solution having a known concentration of about 1.3 mg per mL.

Test preparation—Transfer about 153 mg of Theophylline Sodium Glycinate, accurately weighed, to a 50-mL volumetric flask, dissolve in *Mobile phase*, dilute with *Mobile phase* to volume, and mix.

Chromatographic system (see *Chromatography* ⟨621⟩)—The liquid chromatograph is equipped with a 200-nm detector and a 3.9-mm × 30-cm column that contains packing L8. The flow rate is about 1.5 mL per minute. Chromatograph the *Standard preparation*, and record the peak responses as directed under *Procedure:* the column efficiency determined from the analyte peak is not less than 2000 theoretical plates, the tailing factor for the glycine peak is not more than 2.0, and the relative standard deviation for replicate injections is not more than 2.0%.

Procedure—Separately inject equal volumes (about 20 µL) of the *Standard preparation* and the *Test preparation* into the chromatograph, record the chromatograms, and measure the responses for the glycine peaks. Calculate the quantity, in mg, of $C_2H_5NO_2$ in the portion of Theophylline Sodium Glycinate taken by the formula:

$$50C(r_U/r_S),$$

in which C is the concentration, in mg per mL, of USP Glycine RS in the *Standard preparation*, and r_U and r_S are the peak responses obtained from the *Test preparation* and the *Standard preparation*, respectively: not less than 42.0 percent and not more than 48.0 percent, on the dried basis, is found.

Assay—Transfer about 1 g of Theophylline Sodium Glycinate, previously dried and accurately weighed, to a 500-mL conical flask. Add 100 mL of water and 16 mL of 6 *N* ammonium hydroxide, and warm on a steam bath until solution is complete. Then add 50.0 mL of 0.1 *N* silver nitrate VS, mix, and continue to warm on a steam bath for 15 minutes. While the solution is still warm, filter off the precipitate, and wash the precipitate and the filter with four 20-mL portions of warm water, adding the washings to the filtrate. Acidify the combined washings and filtrate with nitric acid, and add an excess of 7 mL of the acid. Cool, add 4 mL of ferric ammonium sulfate TS, and titrate the excess silver nitrate with 0.1 *N* ammonium thiocyanate VS. Each mL of 0.1 *N* silver nitrate is equivalent to 18.02 mg of $C_7H_8N_4O_2$.

Theophylline Sodium Glycinate Elixir

» Theophylline Sodium Glycinate Elixir contains an amount of theophylline sodium glycinate equivalent to not less than 93.0 percent and not more than 107.0 percent of the labeled amount of anhydrous theophylline ($C_7H_8N_4O_2$).

Packaging and storage—Preserve in tight containers.

Labeling—Label Elixir to state both the content of theophylline sodium glycinate and the content of anhydrous theophylline.

Identification—Mix a volume of Elixir, equivalent to about 500 mg of theophylline, with 10 mL of 6 *N* ammonium hydroxide, and evaporate on a steam bath to a volume of about 20 mL. Neutralize with 6 *N* acetic acid to litmus, and cool, with stirring, to about 15°. Collect the precipitate on a filter, wash with cold water, and dry at 105° for 4 hours: the theophylline so obtained melts between 270° and 274°, the procedure for *Class I* being used (see *Melting Range or Temperature* ⟨741⟩), and responds to *Identification test B* under *Theophylline Sodium Glycinate*.

pH ⟨791⟩: between 8.7 and 9.1.

Alcohol content, *Method I* ⟨611⟩: between 17.0% and 23.0% of C_2H_5OH.

Assay—Transfer an accurately measured volume of Theophylline Sodium Glycinate Elixir, equivalent to about 220 mg of theophylline, to a 250-mL beaker, add 10 mL of water and 15 mL of 6 *N* ammonium hydroxide, and mix. Add 25 mL of 0.1 *N* silver nitrate VS, mix, and heat on a steam bath until the precipitate begins to settle. Cool in a cold water bath, and filter through a filtering crucible, passing the filtrate back through the filter, if necessary, until the filtrate is clear. Wash the beaker and precipitate with three 10-mL portions of water, and discard the filtrate and washings. Mix 30 mL of water and 15 mL of nitric acid in the washed beaker, and use this mixture to dissolve the precipitate on the filter. Wash the beaker and the filter with three 10-mL portions of water. Add 3 mL of ferric ammonium sulfate TS to the combined filtrate and washings, and titrate with 0.1 *N* potassium thiocyanate VS. Each mL of 0.1 *N* potassium thiocyanate is equivalent to 18.02 mg of $C_7H_8N_4O_2$.

Theophylline Sodium Glycinate Tablets

» Theophylline Sodium Glycinate Tablets contain an amount of theophylline sodium glycinate equivalent to not less than 93.0 percent and not more than 107.0 percent of the labeled amount of anhydrous theophylline ($C_7H_8N_4O_2$).

Packaging and storage—Preserve in well-closed containers.

Labeling—Label Tablets to state both the content of theophylline sodium glycinate and the content of anhydrous theophylline.

Identification—Triturate a quantity of finely powdered Tablets, equivalent to about 500 mg of theophylline, with 10-mL and 15-mL portions of solvent hexane, and discard the solvent hexane. Triturate the residue with two 10-mL portions of a mixture of equal volumes of 6 *N* ammonium hydroxide and water, and filter each time. Evaporate the combined filtrates to about 5 mL, neutralize, if necessary, with 6 *N* acetic acid, using litmus, and cool to about 15°, with stirring. Collect the precipitate on a filter, wash it with cold water, and dry at 105° for 1 hour: the theophylline so obtained melts between 270° and 274°, the procedure for *Class I* being used (see *Melting Range or Temperature* ⟨741⟩), and responds to *Identification test B* under *Theophylline Sodium Glycinate*.

Dissolution ⟨711⟩—

Medium: water; 900 mL.
Apparatus 1: 100 rpm.
Time: 45 minutes.

Procedure—Determine the amount of anhydrous theophylline ($C_7H_8N_4O_2$) dissolved from ultraviolet absorbances at the wavelength of maximum absorbance at about 272 nm of filtered portions of the solution under test, suitably diluted with *Dissolution Medium*, if necessary, in comparison with a Standard solution having a known concentration of USP Theophylline RS in the same medium.

Tolerances—Not less than 75% (*Q*) of the labeled amount of anhydrous $C_7H_8N_4O_2$ is dissolved in 45 minutes.

Uniformity of dosage units ⟨905⟩: meet the requirements.

Assay—Place 20 Theophylline Sodium Glycinate Tablets in a 200-mL volumetric flask, add 50 mL of water, and when the tablets have disintegrated add 50 mL of 6 *N* ammonium hydroxide. Shake until no more dissolves, then dilute with water to volume, mix, and filter through a dry filter into a dry flask, discarding the first 20 mL of the filtrate. Transfer an accurately measured aliquot of the subsequent filtrate, equivalent to about 250 mg of theophylline, to a 250-mL conical flask, add 20.0 mL of 0.1 *N* silver nitrate VS, and heat on a steam bath for 15 minutes. Filter through a filter crucible under reduced pressure, and wash the precipitate with three 10-mL portions of water. Acidify the combined filtrate and washings with nitric acid, and add an excess of 3 mL of the acid. Cool, add 2 mL of ferric ammonium sulfate TS, and titrate the excess silver nitrate with 0.1 *N* ammonium thiocyanate VS. Each mL of 0.1 *N* silver nitrate is equivalent to 18.02 mg of $C_7H_8N_4O_2$.

Thiabendazole

$C_{10}H_7N_3S$ 201.25
1*H*-Benzimidazole, 2-(4-thiazolyl)-.
2-(4-Thiazolyl)benzimidazole [*148-79-8*].

» Thiabendazole contains not less than 98.0 percent and not more than 101.0 percent of $C_{10}H_7N_3S$, calculated on the anhydrous basis.

NOTE—Thiabendazole labeled solely for veterinary use is exempt from the requirements of the tests for *Residue on ignition*, *Selenium*, *Heavy metals*, and *Chromatographic purity*.

Packaging and storage—Preserve in well-closed containers.

Reference standard—*USP Thiabendazole Reference Standard*—Dry in vacuum at 100° for 2 hours before using.

Identification—

A: The infrared absorption spectrum of a potassium bromide dispersion of it exhibits maxima only at the same wavelengths as that of a similar preparation of USP Thiabendazole RS.

B: The ultraviolet absorption spectrum of a 1 in 200,000 solution in 0.1 *N* hydrochloric acid exhibits maxima and minima at the same wavelengths as that of a similar solution of USP Thiabendazole RS, concomitantly measured.

C: Dissolve about 5 mg in 5 mL of 0.1 *N* hydrochloric acid, add 3 mg of *p*-phenylenediamine dihydrochloride, and shake to dissolve. Add about 0.1 g of zinc dust, mix, and allow to stand for 2 minutes. Add 5 mL of a solution prepared by dissolving 20 g of ferric ammonium sulfate in 75 mL of water, adding 10 mL of 1 *N* sulfuric acid, and diluting with water to 100 mL: a blue or blue-violet color develops.

D: The R_f value of the principal spot in the chromatogram of the *Identification test preparation* corresponds to that of *Standard preparation A*, as obtained in the test for *Chromatographic purity*.

Melting range ⟨741⟩: between 296° and 303°.

Water, *Method I* ⟨921⟩: not more than 0.5%.

Residue on ignition ⟨281⟩: not more than 0.1%.

Selenium ⟨291⟩: 0.003%, a 200-mg test specimen being used.

Heavy metals, *Method II* ⟨231⟩: 0.001%.

Chromatographic purity—

Standard preparations—Dissolve USP Thiabendazole RS in glacial acetic acid, and mix to obtain a solution having a known concentration of 1.0 mg per mL. Dilute quantitatively with glacial acetic acid to obtain *Standard preparations A, B,* and *C* having the following compositions:

Standard preparation	Dilution	Concentration (μg RS per mL)	Percentage (%, for comparison with test specimen)
A	(1 in 4)	250	0.5
B	(3 in 20)	150	0.3
C	(1 in 20)	50	0.1

Test preparation—Dissolve an accurately weighed quantity of Thiabendazole in glacial acetic acid to obtain a solution containing 50 mg per mL.

Identification preparation—Dilute a portion of the *Test preparation* quantitatively with glacial acetic acid to obtain a solution containing 0.25 mg per mL.

Procedure—On a suitable thin-layer chromatographic plate (see *Chromatography* ⟨621⟩), coated with a 0.25-mm layer of chromatographic silica gel mixture, apply separately 10 μL of the *Test preparation,* 10 μL of the *Identification preparation,* and 10 μL of each *Standard preparation.* Position the plate in a chromatographic chamber, and develop the chromatograms in a solvent system consisting of a mixture of toluene, glacial acetic acid, acetone, and water (60:20:8:2) until the solvent front has moved about three-fourths of the length of the plate. Remove the plate from the developing chamber, mark the solvent front, and allow the solvent to evaporate. Examine the plate under short-wavelength ultraviolet light, and compare the intensities of any secondary spots observed in the chromatogram of the *Test preparation* with those of the principal spots in the chromatograms of the *Standard preparations.* No secondary spot from the chromatogram of the *Test preparation* is larger or more intense than the principal spot obtained from *Standard preparation A* (0.5%), and the sum of the intensities of all secondary spots obtained from the *Test preparation* corresponds to not more than 1.0%.

Assay—Dissolve about 160 mg of Thiabendazole, accurately weighed, in 10 mL of glacial acetic acid. Add 50 mL of acetic anhydride, 1 mL of mercuric acetate TS, and 2 drops of crystal violet TS, and titrate with 0.1 *N* perchloric acid VS (the color change at the end-point is from blue to blue-green). Perform a blank determination, and make any necessary correction. Each mL of 0.1 *N* perchloric acid is equivalent to 20.13 mg of $C_{10}H_7N_3S$.

Thiabendazole Oral Suspension

» Thiabendazole Oral Suspension contains not less than 90.0 percent and not more than 110.0 percent of the labeled amount of $C_{10}H_7N_3S$.

Packaging and storage—Preserve in tight containers.

Reference standard—*USP Thiabendazole Reference Standard*—Dry in vacuum at 100° for 2 hours before using.

Identification—

A: Mix a portion of Oral Suspension, equivalent to about 0.5 g of thiabendazole, with about 20 mL of water, and filter. Wash the residue with 20 mL of water, discard the washing, dissolve the residue in 30 mL of 0.1 *N* hydrochloric acid, and filter. Collect the filtrate in a separator, render it alkaline with 1 *N* sodium hydroxide, and extract with 10 mL of carbon disulfide. Filter the carbon disulfide layer through a dry filter, collecting the filtrate in an evaporating dish. Evaporate the solvent with the aid of gentle heat and a stream of nitrogen. [*Caution—Do not overheat the residue.*] The residue so obtained responds to *Identification test A* under *Thiabendazole.*

B: The retention time of the major peak in the chromatogram of the *Assay preparation* corresponds to that of the *Standard preparation,* as obtained in the *Assay.*

pH ⟨791⟩: between 3.4 and 4.2.

Assay—

pH 3.1 phosphate buffer—Dissolve 13.8 g of monobasic sodium phosphate in water to obtain 2000 mL of solution. Adjust this solution with phosphoric acid to a pH of 3.1 ± 0.05.

Mobile phase—Prepare a filtered and degassed mixture of *pH 3.1 phosphate buffer* and methanol (65:35). Make adjustments if necessary (see *System Suitability* under *Chromatography* ⟨621⟩).

Standard preparation—Dissolve an accurately weighed quantity of USP Thiabendazole RS in 0.1 *N* hydrochloric acid, and dilute quantitatively, and stepwise if necessary, with 0.1 *N* hydrochloric acid to obtain a solution containing about 2 mg per mL. Transfer 5.0 mL of this solution to a 50-mL volumetric flask, dilute with water to volume, and mix to obtain a *Standard preparation* having a known concentration of about 0.2 mg of USP Thiabendazole RS per mL.

Assay preparation—Transfer an accurately measured volume of Thiabendazole Oral Suspension, equivalent to about 0.5 g of thiabendazole, to a 250-mL volumetric flask, dilute with 0.1 *N* hydrochloric acid to volume, and mix. Transfer 5.0 mL of this solution to a 50-mL volumetric flask, dilute with water to volume, and mix.

Chromatographic system (see *Chromatography* ⟨621⟩)—The liquid chromatograph is equipped with a 254-nm detector and a 4-mm × 30-cm column that contains packing L1. The flow rate is about 2 mL per minute. Chromatograph the *Standard preparation,* and record the peak responses as directed under *Procedure:* the column efficiency determined from the analyte peak is not less than 960 theoretical plates, the tailing factor for the analyte peak is not more than 2.0, and the relative standard deviation for replicate injections is not more than 2%.

Procedure—Separately inject equal volumes (about 20 μL) of the *Standard preparation* and the *Assay preparation* into the chromatograph, record the chromatograms, and measure the responses for the major peaks. Calculate the quantity, in g, of $C_{10}H_7N_3S$ in each mL of the Oral Suspension taken by the formula:

$$2.5(C/V)(r_U/r_S),$$

in which *C* is the concentration, in mg per mL, of USP Thiabendazole RS in the *Standard preparation, V* is the volume, in mL, of Oral Suspension taken, and r_U and r_S are the peak responses obtained from the *Assay preparation* and the *Standard preparation,* respectively.

Thiabendazole Tablets

» Thiabendazole Tablets contain not less than 90.0 percent and not more than 110.0 percent of the labeled amount of $C_{10}H_7N_3S$.

Packaging and storage—Preserve in tight containers.

Labeling—Label the Tablets to indicate that they are to be chewed before swallowing.

Reference standard—*USP Thiabendazole Reference Standard*—Dry in vacuum at 100° for 2 hours before using.

Identification—

A: Triturate a quantity of powdered Tablets, equivalent to about 0.5 g of thiabendazole, with about 20 mL of water, and filter. Wash the residue with 20 mL of water, discard the washing, dissolve the residue in 30 mL of 0.1 *N* hydrochloric acid, and filter. Collect the filtrate in a separator, render it alkaline with 1 *N* sodium hydroxide, and extract with 10 mL of carbon disulfide. Filter the carbon disulfide layer through a dry filter, collecting the filtrate in an evaporating dish. Evaporate the solvent with the aid of gentle heat and a stream of nitrogen. [*Caution—Do not overheat the residue.*] The residue so obtained responds to *Identification test A* under *Thiabendazole.*

B: The retention time of the major peak in the chromatogram of the *Assay preparation* corresponds to that of the *Standard preparation,* as obtained in the *Assay.*

Uniformity of dosage units ⟨905⟩: meet the requirements.

Procedure for content uniformity—

Standard preparation—Dissolve an accurately weighed quantity of USP Thiabendazole RS in 0.1 *N* hydrochloric acid, and dilute quantitatively and stepwise with 0.1 *N* hydrochloric acid to obtain a solution having a known concentration of about 5 µg per mL.

Test preparation—Transfer 1 Tablet to a 1000-mL volumetric flask, add about 75 mL of 0.1 *N* hydrochloric acid, and heat on a steam bath for about 1 hour. Cool to room temperature, dilute with 0.1 *N* hydrochloric acid to volume, mix, and filter a portion of the solution, discarding the first 20 mL of the filtrate. Pipet 5 mL of the filtrate into a 500-mL volumetric flask, dilute with 0.1 *N* hydrochloric acid to volume, and mix.

Procedure—Concomitantly determine the absorbances of the *Standard preparation* and the *Test preparation* at the wavelength of maximum absorbance at about 302 nm, with a suitable spectrophotometer, using 0.1 *N* hydrochloric acid as the blank. Calculate the quantity, in mg, of $C_{10}H_7N_3S$ in the Tablet by the formula:

$$(TC/D)(A_U/A_S),$$

in which T is the labeled quantity, in mg, of thiabendazole in the Tablet, C is the concentration, in µg per mL, of USP Thiabendazole RS in the *Standard preparation, D* is the concentration, in µg per mL, of thiabendazole in the *Test preparation*, based upon the labeled quantity per Tablet and the extent of dilution, and A_U and A_S are the absorbances of the *Test preparation* and the *Standard preparation*, respectively.

Assay—

Standard preparation and *Chromatographic system*—Prepare as directed in the *Assay* under *Thiabendazole Oral Suspension.*

pH 3.5 phosphate buffer—Dissolve 13.8 g of monobasic sodium phosphate in water to obtain 2000 mL of solution. Adjust this solution with phosphoric acid to a pH of 3.5 ± 0.05.

Mobile phase—Prepare a filtered and degassed mixture of *pH 3.5 phosphate buffer* and methanol (54:46). Make adjustments if necessary (see *System Suitability* under *Chromatography* ⟨621⟩).

Assay preparation—Weigh and finely powder not less than 20 Thiabendazole Tablets. Transfer an accurately weighed portion of the powder, equivalent to about 200 mg of thiabendazole, to a 1000-mL volumetric flask, add 100 mL of 0.1 *N* hydrochloric acid, mix, and warm the solution for a minimum of 30 minutes. Allow to cool to room temperature, dilute with water to volume, mix, and filter, discarding the first 20 mL of the filtrate.

Procedure—Proceed as directed for *Procedure* in the *Assay* under *Thiabendazole Oral Suspension.* Calculate the quantity, in mg, of $C_{10}H_7N_3S$ in the portion of Tablets taken by the formula:

$$1000C(r_U/r_S),$$

in which C is the concentration, in mg per mL, of USP Thiabendazole RS in the *Standard preparation*, and r_U and r_S are the peak responses obtained from the *Assay preparation* and the *Standard preparation*, respectively.

Thiamine Hydrochloride

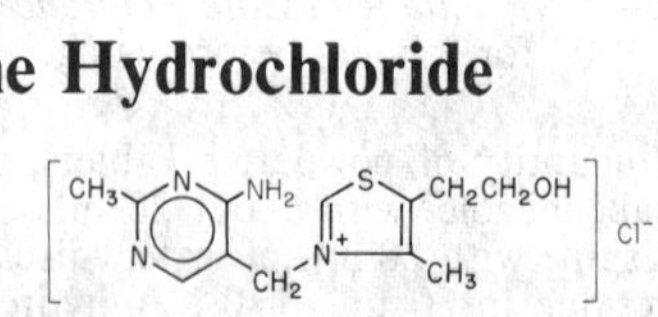

$C_{12}H_{17}ClN_4OS.HCl$ 337.27

Thiazolium, 3-[(4-amino-2-methyl-5-pyrimidinyl)methyl]-5-(2-hydroxyethyl)-4-methyl-, chloride, monohydrochloride.

Thiamine monohydrochloride [*67-03-8*].

» Thiamine Hydrochloride contains not less than 98.0 percent and not more than 102.0 percent of $C_{12}H_{17}ClN_4OS.HCl$, calculated on the anhydrous basis.

Packaging and storage—Preserve in tight, light-resistant containers.

Reference standard—*USP Thiamine Hydrochloride Reference Standard*—Do not dry; determine the water content titrimetrically at the time of use.

Identification—

A: The infrared absorption spectrum of a potassium bromide dispersion of it, previously dried at 105° for 2 hours, exhibits maxima only at the same wavelengths as that of a similar preparation of USP Thiamine Hydrochloride RS. If a difference appears, dissolve portions of both the test specimen and the Reference Standard in water, evaporate the solutions to dryness, and repeat the test on the residues.

B: A solution (1 in 50) responds to the tests for *Chloride* ⟨191⟩.

pH ⟨791⟩: between 2.7 and 3.4, in a solution (1 in 100).

Water, *Method I* ⟨921⟩: not more than 5.0%.

Residue on ignition ⟨281⟩: not more than 0.2%.

Absorbance of solution—Dissolve 1.0 g in water to make 10 mL. The absorbance of this solution, after filtration through a fine-porosity, sintered-glass funnel, determined in 1-cm cells at a wavelength of 400 nm, with a suitable spectrophotometer, water being used as the blank, does not exceed 0.025.

Nitrate—To 2 mL of a solution (1 in 50) add 2 mL of sulfuric acid, cool, and superimpose 2 mL of ferrous sulfate TS: no brown ring is produced at the junction of the two layers.

Assay—

Standard preparation—Prepare as directed for *Standard Preparation* under *Thiamine Assay* ⟨531⟩.

Assay preparation—Dissolve about 25 mg of Thiamine Hydrochloride, accurately weighed, in 0.2 *N* hydrochloric acid to make 500.0 mL, and mix. Dilute 5 mL of this solution quantitatively and stepwise with 0.2 *N* hydrochloric acid to obtain a solution containing about 0.2 µg per mL. Using this solution as the *Assay Preparation*, proceed as directed for *Procedure* under *Thiamine Assay* ⟨531⟩. Calculate the quantity, in mg, of $C_{12}H_{17}ClN_4OS.HCl$ in the portion of Thiamine Hydrochloride taken by the formula:

$$125C(A - b)/(S - d),$$

in which C is the concentration, in µg per mL, of USP Thiamine Hydrochloride RS in the *Standard preparation*, and A, b, S, and d are as defined under *Calculation* in the *Thiamine Assay* ⟨531⟩.

Thiamine Hydrochloride Elixir

» Thiamine Hydrochloride Elixir contains not less than 95.0 percent and not more than 135.0 percent of the labeled amount of $C_{12}H_{17}ClN_4OS.HCl$.

Packaging and storage—Preserve in tight, light-resistant containers.

Reference standard—*USP Thiamine Hydrochloride Reference Standard*—Do not dry; determine the water content titrimetrically at the time of use.

Identification—It responds to *Identification test B* under *Thiamine Hydrochloride Injection.*

Alcohol content, *Method II* ⟨611⟩: between 8.5% and 20.0% of C_2H_5OH, acetone being used as the internal standard.

Assay—

Standard preparation—Prepare as directed for *Standard Preparation* under *Thiamine Assay* ⟨531⟩.

Assay preparation—Dilute an accurately measured volume of not less than 1 mL of Thiamine Hydrochloride Elixir quantitatively and stepwise with 0.2 *N* hydrochloric acid to obtain a solution containing about 0.2 µg of thiamine hydrochloride per mL.

Procedure—Proceed as directed for *Procedure* under *Thiamine Assay* ⟨531⟩.

Thiamine Hydrochloride Injection

» Thiamine Hydrochloride Injection is a sterile solution of Thiamine Hydrochloride in Water for Injection. It contains not less than 90.0 percent and not more than 110.0 percent of the labeled amount of $C_{12}H_{17}ClN_4OS.HCl$.

Packaging and storage—Preserve in single-dose or in multiple-dose containers, preferably of Type I glass, protected from light.

Reference standard—*USP Thiamine Hydrochloride Reference Standard*—Do not dry; determine the water content titrimetrically at the time of use.

Identification—

A: It yields a white precipitate with mercuric chloride TS, and a red-brown precipitate with iodine TS. It also yields a precipitate with mercuric-potassium iodide TS, and with trinitrophenol TS.

B: Dilute a portion of Injection with water to a concentration of about 10 mg of thiamine hydrochloride per mL. To 0.5 mL of this solution add 5 mL of 0.5 *N* sodium hydroxide, then add 0.5 mL of potassium ferricyanide TS and 5 mL of isobutyl alcohol, shake the mixture vigorously for 2 minutes, and allow the liquid layers to separate: when illuminated from above by a vertical beam of ultraviolet light and viewed at a right angle to this beam, the air-liquid meniscus shows a vivid blue fluorescence, which disappears when the mixture is slightly acidified, but reappears when it is again made alkaline.

C: It responds to the tests for *Chloride* ⟨191⟩.

pH ⟨791⟩: between 2.5 and 4.5.

Other requirements—It meets the requirements under *Injections* ⟨1⟩.

Assay—

Standard preparation—Prepare as directed for *Standard Preparation* under *Thiamine Assay* ⟨531⟩.

Assay preparation—Dilute an accurately measured volume of not less than 1 mL of Thiamine Hydrochloride Injection quantitatively and stepwise with 0.2 *N* hydrochloric acid to obtain a solution containing about 0.2 µg of thiamine hydrochloride per mL. Using this solution as the *Assay Preparation*, proceed as directed for *Procedure* under *Thiamine Assay* ⟨531⟩.

Thiamine Hydrochloride Tablets

» Thiamine Hydrochloride Tablets contain not less than 90.0 percent and not more than 110.0 percent of the labeled amount of $C_{12}H_{17}ClN_4OS.HCl$.

Packaging and storage—Preserve in tight, light-resistant containers.

Reference standard—*USP Thiamine Hydrochloride Reference Standard*—Do not dry; determine the water content titrimetrically at the time of use.

Identification—

A: Triturate a quantity of powdered Tablets, equivalent to about 10 mg of thiamine hydrochloride, with 10 mL of 0.5 *N* sodium hydroxide, and filter. Using a 5-mL portion of the filtrate, proceed as directed in *Identification test B* under *Thiamine Hydrochloride Injection*, beginning with "then add 0.5 mL of potassium ferricyanide TS": the specified reaction is observed.

B: Triturate a quantity of powdered Tablets, equivalent to about 10 mg of thiamine hydrochloride, with 10 mL of water, and filter: treated separately, 2-mL portions of the filtrate yield a red-brown precipitate with iodine TS and a white precipitate with mercuric chloride TS, and respond to the tests for *Chloride* ⟨191⟩.

C: To the remainder of the filtrate prepared for the preceding test add 1 mL of lead acetate TS and 1 mL of 2.5 *N* sodium hydroxide: a yellow color is produced. Heat the mixture for several minutes on a steam bath: the color changes to brown, and, on standing, a precipitate of lead sulfide separates.

Disintegration ⟨701⟩: 30 minutes.

Uniformity of dosage units ⟨905⟩: meet the requirements.

Assay—

Standard preparation—Prepare as directed for *Standard Preparation* under *Thiamine Assay* ⟨531⟩.

Assay preparation—Place not less than 20 Thiamine Hydrochloride Tablets in a flask of suitable size, half fill the flask with 0.2 *N* hydrochloric acid, and heat on a steam bath, with frequent agitation, until the tablets have dissolved or have disintegrated so that a uniform dispersion is obtained. Cool, transfer the contents of the flask to a volumetric flask, and dilute with 0.2 *N* hydrochloric acid to volume. If the mixture is not clear, either centrifuge it or filter it through paper known not to adsorb thiamine. Dilute a portion of the clear solution quantitatively and stepwise with 0.2 *N* hydrochloric acid to obtain a solution containing about 0.2 µg of thiamine hydrochloride per mL. Using this as the *Assay Preparation*, proceed as directed for *Procedure* under *Thiamine Assay* ⟨531⟩.

Thiamine Mononitrate

CH3 N NH2 S CH2CH2OH N N+ CH2 CH3 NO_3^-

$C_{12}H_{17}N_5O_4S$ 327.36

Thiazolium, 3-[(4-amino-2-methyl-5-pyrimidinyl)methyl]-5-(2-hydroxyethyl)-4-methyl-, nitrate (salt).

Thiamine nitrate (salt) [*532-43-4*].

» Thiamine Mononitrate contains not less than 98.0 percent and not more than 102.0 percent of $C_{12}H_{17}N_5O_4S$, calculated on the dried basis.

Packaging and storage—Preserve in tight, light-resistant containers.

Reference standard—*USP Thiamine Hydrochloride Reference Standard*—Do not dry; determine the water content titrimetrically at the time of use.

Identification—

A: To 2 mL of a solution (1 in 50) add 2 mL of sulfuric acid, cool, and superimpose 2 mL of ferrous sulfate TS: a brown ring is produced at the junction of the two liquids.

B: Dissolve about 5 mg in a mixture of 1 mL of lead acetate TS and 1 mL of 2.5 *N* sodium hydroxide: a yellow color is produced. Heat the mixture for several minutes on a steam bath: the color changes to brown, and, on standing, a precipitate of lead sulfide separates.

C: A solution of it responds to *Identification test A* under *Thiamine Hydrochloride Injection.*

D: Dissolve about 5 mg in 5 mL of 0.5 *N* sodium hydroxide, and proceed as directed in *Identification test B* under *Thiamine Hydrochloride Injection*, beginning with "then add 0.5 mL of potassium ferricyanide TS": the specified reaction is observed.

pH ⟨791⟩: between 6.0 and 7.5, in a solution (1 in 50).

Loss on drying ⟨731⟩—Dry about 500 mg, accurately weighed, at 105° for 2 hours: it loses not more than 1.0% of its weight.

Residue on ignition ⟨281⟩: not more than 0.2%.

Chloride ⟨221⟩—A 500-mg portion shows no more chloride than corresponds to 0.40 mL of 0.020 *N* hydrochloric acid (0.06%).

Assay—

Standard preparation—Prepare as directed for *Standard Preparation* under *Thiamine Assay* ⟨531⟩.

Assay preparation—Using Thiamine Mononitrate instead of thiamine hydrochloride, prepare the *Assay Preparation* as directed in the *Assay* under *Thiamine Hydrochloride*, and proceed as directed for *Procedure* under *Thiamine Assay* ⟨531⟩. Calculate the quantity, in mg, of $C_{12}H_{17}N_5O_4S$ in the portion of Thiamine Mononitrate taken by the formula:

$$125C[0.9706(A - b)/(S - d)],$$

in which *C* is the concentration, in µg per mL, of USP Thiamine

Hydrochloride RS in the *Standard preparation*, 0.9706 is the ratio of the molecular weight of thiamine mononitrate to that of thiamine hydrochloride, and *A*, *b*, *S*, and *d* are as defined under *Calculation* in the *Thiamine Assay* ⟨531⟩.

Thiamine Mononitrate Elixir

» Thiamine Mononitrate Elixir contains not less than 95.0 percent and not more than 115.0 percent of the labeled amount of $C_{12}H_{17}N_5O_4S$.

Packaging and storage—Preserve in tight, light-resistant containers.

Reference standard—*USP Thiamine Hydrochloride Reference Standard*—Do not dry; determine the water content titrimetrically at the time of use.

Identification—

A: It responds to *Identification test B* under *Thiamine Hydrochloride Injection.*

B: To 5 mL of Elixir add 2 mL of sulfuric acid, cool, and superimpose 2 mL of ferrous sulfate TS: a brown ring is produced at the junction of the two liquids.

Alcohol content, *Method II* ⟨611⟩: between 7.5% and 10.5% of C_2H_5OH, acetone being used as the internal standard.

Assay—

Standard preparation—Prepare as directed for *Standard Preparation* under *Thiamine Assay* ⟨531⟩.

Assay preparation—Dilute an accurately measured volume of not less than 1 mL of Thiamine Mononitrate Elixir quantitatively and stepwise with 0.2 *N* hydrochloric acid to obtain a solution containing about 0.2 µg of thiamine mononitrate per mL.

Procedure—Proceed as directed for *Procedure* under *Thiamine Assay* ⟨531⟩.

Thiamylal Sodium for Injection

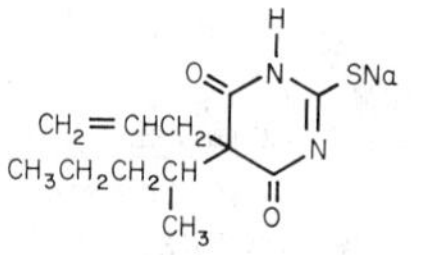

$C_{12}H_{17}N_2NaO_2S$ 276.33

4,6-(1*H*,5*H*)-Pyrimidinedione, dihydro-5-(1-methylbutyl)-5-(2-propenyl)-2-thioxo-, monosodium salt.

Sodium 5-allyl-5-(1-methylbutyl)-2-thiobarbiturate [*337-47-3*].

» Thiamylal Sodium for Injection is a sterile mixture of thiamylal sodium with anhydrous Sodium Carbonate as a buffer. It contains not less than 93.0 percent and not more than 107.0 percent of the labeled amount of $C_{12}H_{17}N_2NaO_2S$, calculated on the dried basis.

Packaging and storage—Preserve in *Containers for Sterile Solids* as described under *Injections* ⟨1⟩.

Labeling—It meets the requirements for *Labeling* under *Injections* ⟨1⟩.

Completeness of solution—Mix 0.50 g with 10 mL of carbon dioxide-free water: after 1 minute, the solution is clear and free from undissolved solid.

Constituted solution—At the time of use, the constituted solution prepared from Thiamylal Sodium for Injection meets the requirements for *Constituted Solutions* under *Injections* ⟨1⟩.

Identification—

A: Dissolve about 500 mg in 10 mL of water, and add an excess of 3 *N* hydrochloric acid: a white precipitate of thiamylal is formed.

B: Ignite about 500 mg: the residue so obtained responds to the tests for *Carbonate* ⟨191⟩ and for *Sodium* ⟨191⟩.

C: The residue of thiamylal obtained in the *Assay* melts between 133° and 139°, when tested by the procedure for *Class I* (see *Melting Range or Temperature* ⟨741⟩).

Sterility—It meets the requirements under *Sterility Tests* ⟨71⟩.

Uniformity of dosage units ⟨905⟩: meets the requirements.

pH ⟨791⟩: between 10.7 and 11.5, in the solution prepared as directed in the test for *Completeness of solution.*

Loss on drying ⟨731⟩—Dry it at 105° for 1 hour: it loses not more than 2.0% of its weight.

Selenium ⟨291⟩: 0.003%.

Heavy metals, *Method II* ⟨231⟩: 0.003%.

Assay—Dissolve the contents of 10 containers of Thiamylal Sodium for Injection in water, and dilute quantitatively with water to obtain a solution containing about 50 mg of thiamylal sodium in each mL. Transfer 10.0 mL of the solution to a separator, and add 40 mL of water and 5 mL of 3 *N* hydrochloric acid. Completely extract the liberated acid barbiturate with five 25-mL portions of chloroform. Test for completeness of extraction by extracting with an additional 10-mL portion of chloroform and evaporating the solvent: not more than 500 µg of residue remains. Wash the combined chloroform extracts with 10 mL of water acidified with 1 drop of hydrochloric acid, then extract the water with 10 mL of chloroform, adding the latter to the main chloroform solution. Filter the chloroform solution through a pledget of cotton or other suitable filter, previously washed with chloroform, into a tared beaker, and finally wash the separator and the filter with three 5-mL portions of chloroform. Evaporate the combined chloroform solution and washings on a steam bath to dryness with the aid of a current of air, dry the residue at 105° for 30 minutes, cool, and weigh. The weight of the dried thiamylal, multiplied by 1.086, represents the weight of $C_{12}H_{17}N_2NaO_2S$ in the 10-mL portion of the specimen taken.

Thiethylperazine Maleate

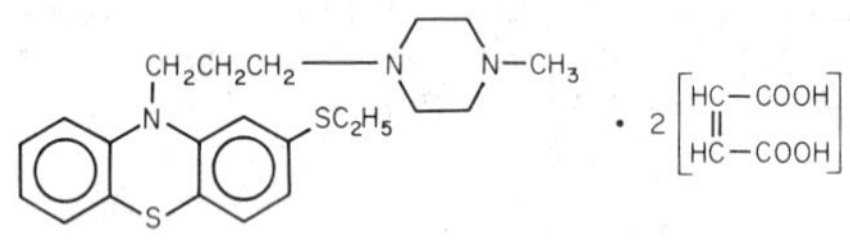

$C_{22}H_{29}N_3S_2 \cdot 2C_4H_4O_4$ 631.76

10*H*-Phenothiazine, 2-(ethylthio)-10-[3-(4-methyl-1-piperazinyl)propyl]-, (*Z*)-2-butenedioate (1:2).

2-(Ethylthio)-10-[3-(4-methyl-1-piperazinyl)propyl]phenothiazine maleate (1:2) [*1179-69-7*].

» Thiethylperazine Maleate contains not less than 98.0 percent and not more than 101.5 percent of $C_{22}H_{29}N_3S_2 \cdot 2C_4H_4O_4$, calculated on the dried basis.

Packaging and storage—Preserve in tight, light-resistant containers.

Reference standards—*USP Thiethylperazine Maleate Reference Standard*—Dry at 105° for 4 hours before using. *USP Maleic Acid Reference Standard*—Use as is.

NOTE—Throughout the following procedures, protect test or assay specimens, the Reference Standard, and solutions containing them, by conducting the procedures without delay, under subdued light, or using low-actinic glassware.

Identification—The infrared absorption spectrum of a potassium bromide dispersion of it, previously dried, exhibits maxima only at the same wavelengths as that of a similar preparation of USP Thiethylperazine Maleate RS.

pH ⟨791⟩—Dissolve 100 mg in 100 mL of water, warming, if necessary, to effect solution: the pH of this solution is between 2.8 and 3.8.

Loss on drying ⟨731⟩—Dry it at 105° for 4 hours: it loses not more than 0.5% of its weight.

Residue on ignition ⟨281⟩: not more than 0.1%.

Selenium ⟨291⟩—The absorbance of the solution from the *Test Solution*, prepared with 100 mg of Thiethylperazine Maleate and 100 mg of magnesium oxide, is not greater than one-half that from the *Standard Solution* (0.003%).

Chromatographic purity—[NOTE—Conduct this test promptly without exposure to daylight and with minimum exposure to artificial light.]

Developing solvent—Mix 85 volumes of methylene chloride, 15 volumes of isopropyl alcohol, and 1.5 volumes of ammonium hydroxide.

Standard stock solution—Transfer 20.0 mg of USP Thiethylperazine Maleate RS to a 25-mL volumetric flask, add methanol and 5 drops of ammonium hydroxide, and dilute with methanol to volume to obtain a solution having a known concentration of 0.8 mg per mL.

Standard preparations—Dilute accurately measured volumes of *Standard stock solution* quantitatively with methanol (designated below as parts by volume of *Standard stock solution* in total parts by volume of the finished *Standard preparation*) to obtain *Standard preparations*, designated below by letter, having the following compositions:

Dilution	Concentration (μg RS per mL)	Percentage (%, for comparison with test specimen)
A (1 in 8)	100	0.5
B (1 in 16)	50	0.25
C (1 in 32)	25	0.125

Test preparation—Transfer 200 mg of Thiethylperazine Maleate, accurately weighed, to a 10-mL volumetric flask, and add 5 mL of methanol to dissolve. Add 2 drops of ammonium hydroxide, add methanol to volume, and mix. Use this solution immediately.

Procedure—On a suitable thin-layer chromatographic plate (see *Chromatography* ⟨621⟩), coated with a 0.25-mm layer of chromatographic silica gel mixture, apply separately 5 μL of the *Test preparation* and 5 μL of each *Standard preparation*. Dry the applied spots in a current of cold air. Position the plate in a chromatographic chamber, and develop the chromatograms in the *Developing solvent* until the solvent front has moved about three-fourths of the length of the plate. Remove the plate from the developing chamber, mark the solvent front, and allow the solvent to evaporate. Examine under long-wavelength ultraviolet light for any fluorescent impurities. Compare the intensities of any secondary spots observed in the chromatogram of the *Test preparation* with those of the principal spots in the chromatograms of the *Standard preparations:* the sum of the intensities of all secondary spots obtained from the *Test preparation* corresponds to not more than 2.0% of related compounds, with no individual impurity being greater than 0.5%.

Assay—Dissolve about 250 mg of Thiethylperazine Maleate, accurately weighed, in 30 mL of glacial acetic acid, and warm on a steam bath to effect solution. Add 1 drop of crystal violet TS, and titrate with 0.1 *N* perchloric acid VS to a blue-green endpoint. Perform a blank determination, and make any necessary correction. Each mL of 0.1 *N* perchloric acid is equivalent to 31.59 mg of $C_{22}H_{29}N_3S_2.2C_4H_4O_4$.

Thiethylperazine Maleate Suppositories

» Thiethylperazine Maleate Suppositories contain not less than 90.0 percent and not more than 110.0 percent of the labeled amount of $C_{22}H_{29}N_3S_2.2C_4H_4O_4$.

Packaging and storage—Preserve in tight containers at temperatures below 25°. Do not expose unwrapped Suppositories to sunlight.

Reference standard—*USP Thiethylperazine Maleate Reference Standard*—Dry at 105° for 4 hours before using.

NOTE—Throughout the following procedures, protect test or assay specimens, the Reference Standard, and solutions containing them, by conducting the procedures without delay, under subdued light, or using low-actinic glassware.

Identification—

Standard preparation—Prepare a solution containing about 2 mg per mL of Thiethylperazine Maleate RS in chloroform and methanol (1:1).

Test preparation—Transfer 100 mL of the chloroform solution obtained as directed in the *Assay* to a conical flask, and evaporate under vacuum to dryness. Dissolve the residue in 2.0 mL of chloroform and methanol (1:1).

Procedure—Proceed as directed for *Procedure* in the *Identification test* under *Thiethylperazine Malate Injection*, except to apply 20 μL each of the *Test preparation* and the *Standard preparation*. The specified result is obtained.

Assay—[NOTE—Conduct this procedure with minimum exposure to light.] Weigh not less than 10 Thiethylperazine Maleate Suppositories, break them up into small pieces, and mix. Transfer an accurately weighed portion of the mass, equivalent to about 10 mg of thiethylperazine maleate, to a 250-mL separator, add 50 mL of tartaric acid solution (1 in 50) and 50 mL of ether, and shake until the mass is dissolved. Transfer the acid layer to a second 250-mL separator, wash with two 50-mL portions of ether, and combine these washings with the first ether layer. Extract the combined ether solution with 10 mL of tartaric acid solution (1 in 50), combine this with the first acid extract, and discard the ether layer. Add 6 *N* ammonium hydroxide to render the mixture alkaline to litmus paper, add 1 mL in excess, and extract with five 45-mL portions of chloroform, passing each extract through anhydrous sodium sulfate into a 250-mL volumetric flask. Dilute with chloroform to volume, and mix. Evaporate 20.0 mL of this solution to dryness, dissolve the residue in 0.1 *N* hydrochloric acid to make 100.0 mL, and filter if necessary. Concomitantly determine the absorbances of this solution and of a Standard solution of USP Thiethylperazine Maleate RS in the same medium having a known concentration of about 8 μg per mL in 1-cm cells at the wavelength of maximum absorbance at about 263 nm, with a suitable spectrophotometer, using 0.1 *N* hydrochloric acid as the blank. Calculate the quantity, in mg, of $C_{22}H_{29}N_3S_2.2C_4H_4O_4$ in the portion of Suppositories taken by the formula:

$$1.25C(A_U/A_S),$$

in which *C* is the concentration, in μg per mL, of USP Thiethylperazine Maleate RS in the Standard solution, and A_U and A_S are the absorbances of the solution from the Suppositories and the Standard solution, respectively.

Thiethylperazine Maleate Tablets

» Thiethylperazine Maleate Tablets contain not less than 90.0 percent and not more than 110.0 percent of the labeled amount of $C_{22}H_{29}N_3S_2.2C_4H_4O_4$.

Packaging and storage—Preserve in tight, light-resistant containers.

Reference standard—*USP Thiethylperazine Maleate Reference Standard*—Dry at 105° for 4 hours before using.

NOTE—Throughout the following procedures, protect test or assay specimens, the Reference Standard, and solutions containing them, by conducting the procedures without delay, under subdued light, or using low-actinic glassware.

Identification—

Standard preparation—Prepare a solution containing about 2 mg per mL of USP Thiethylperazine Maleate RS in a mixture of chloroform and methanol (1:1).

Test preparation—Transfer 50 mL of the chloroform solution obtained in the *Assay* to a conical flask, and evaporate under vacuum to dryness. Dissolve the residue in 2.0 mL of a mixture of chloroform and methanol (1:1).

Procedure—Proceed as directed for *Procedure* in the *Identification test* under *Thiethylperazine Malate Injection*, except to apply 20 μL each of the *Test preparation* and the *Standard preparation*. The specified result is obtained.

Dissolution ⟨711⟩—

Medium: 0.1 *N* hydrochloric acid; 1000 mL.

Apparatus 1: 120 rpm.

Time: 30 minutes.

Procedure—Determine the amount of $C_{22}H_{29}N_3S_2 . 2C_4H_4O_4$ dissolved from ultraviolet absorbances at the wavelength of maximum absorbance at about 263 nm of filtered portions of the solution under test, suitably diluted with *Dissolution Medium*, if necessary, in comparison with a Standard solution having a known concentration of USP Thiethylperazine Maleate RS in the same medium.

Tolerances—Not less than 75% (*Q*) of the labeled amount of $C_{22}H_{29}N_3S_2 . 2C_4H_4O_4$ is dissolved in 30 minutes.

Uniformity of dosage units ⟨905⟩: meet the requirements.

Procedure for content uniformity—Transfer 1 finely powdered Tablet to a 200-mL volumetric flask, and add about 10 mL of water. Shake for 10 minutes, add 125 mL of methanol, and continue shaking for not less than 15 minutes. Dilute with methanol to volume, mix, and filter. Dilute a portion of this solution quantitatively with methanol to obtain a solution containing approximately 10 µg of thiethylperazine maleate per mL. Concomitantly determine the absorbances of the resulting solution and of a Standard solution of USP Thiethylperazine Maleate RS in the same medium having a known concentration of about 10 µg per mL, in 1-cm cells at the wavelength of maximum absorbance at about 264 nm, with a suitable spectrophotometer, using methanol as the blank. Calculate the quantity, in mg, of $C_{22}H_{29}N_3S_2 . 2C_4H_4O_4$ in the Tablet taken by the formula:

$$(TC/D)(A_U/A_S),$$

in which *T* is the labeled quantity, in mg, of thiethylperazine maleate in the Tablet, *C* is the concentration, in µg per mL, of USP Thiethylperazine Maleate RS in the Standard solution, *D* is the concentration, in µg per mL, of the solution from the Tablet, based on the labeled quantity per Tablet and the extent of dilution, and A_U and A_S are the absorbances of the solution from the Tablet and the Standard solution, respectively.

Assay—

Mobile phase—Prepare a mixture of acetonitrile, water, and triethylamine (400:100:1), by mixing the acetonitrile and water, filtering, and degassing under vacuum. Add the triethylamine, and mix. Make adjustments if necessary (see *System Suitability* under *Chromatography* ⟨621⟩).

Solvent mixture—Use acetonitrile and water (9:1).

Standard preparation—Dissolve an accurately weighed quantity of USP Thiethylperazine Maleate RS in *Solvent mixture*, and dilute quantitatively, and stepwise if necessary, with *Solvent mixture* to obtain a solution having a known concentration of about 0.25 mg per mL.

Assay preparation—Weigh and finely powder not less than 20 Thiethylperazine Maleate Tablets. Transfer an accurately weighed portion of the powder, equivalent to about 50 mg of thiethylperazine maleate, to a 200-mL volumetric flask. Add 150 mL of *Solvent mixture*, and shake to disperse the powder. Sonicate the mixture for 10 minutes, then shake by mechanical means for 1 hour. Dilute with *Solvent mixture* to volume, and mix. Filter through a 0.45-µm disk, discarding the first 20 mL of the filtrate.

Chromatographic system (see *Chromatography* ⟨621⟩)—The liquid chromatograph is equipped with a 263-nm detector and a 3.9-mm × 30-cm column that contains packing L1. The flow rate is about 3 mL per minute. Chromatograph the *Standard preparation*, and record the peak heights as directed under *Procedure:* the column efficiency determined from the analyte peak is not less than 630 theoretical plates, baseline resolution is achieved for all peaks, and the relative standard deviation for replicate injections is not more than 2.0%.

Procedure—Separately inject equal volumes (about 20 µL) of the *Standard preparation* and the *Assay preparation* into the chromatograph, record the chromatograms, and measure the heights of the major peaks. Calculate the quantity, in mg, of $C_{22}H_{29}N_3S_2 . 2C_4H_4O_4$ in the portion of Tablets taken by the formula:

$$200C(r_U/r_S),$$

in which *C* is the concentration, in mg per mL, of USP Thiethylperazine Maleate RS in the *Standard preparation*, and r_U and r_S are the peak heights obtained from the *Assay preparation* and the *Standard preparation*, respectively.

Thimerosal

$C_9H_9HgNaO_2S$ 404.81

Mercury, ethyl(2-mercaptobenzoato-*S*)-, sodium salt.

Ethyl (sodium *o*-mercaptobenzoato)mercury [*54-64-8*].

» Thimerosal contains not less than 97.0 percent and not more than 101.0 percent of $C_9H_9HgNaO_2S$, calculated on the dried basis.

Packaging and storage—Preserve in tight, light-resistant containers.

Reference standard—*USP Thimerosal Reference Standard*—Dry in vacuum over phosphorus pentoxide to constant weight before using.

Identification—

A: The infrared absorption spectrum of a potassium bromide dispersion of it, previously dried, exhibits maxima only at the same wavelengths as that of a similar preparation of USP Thimerosal RS.

B: To a solution (1 in 100) add a few drops of silver nitrate TS: a white precipitate is formed.

Loss on drying ⟨731⟩—Dry it in vacuum over phosphorus pentoxide to constant weight: it loses not more than 0.5% of its weight.

Ether-soluble substances—Shake 500 mg, accurately weighed, with 20 mL of anhydrous ethyl ether for 10 minutes. Filter, evaporate the ether in a tared container, dry the residue in vacuum over phosphorus pentoxide, and weigh: the weight of residue does not exceed 4 mg (0.8%).

Mercury ions—

Iodide reagent—[NOTE—Prepare fresh daily; keep the stopper in the flask and protect from light.] Dissolve 33.20 g of potassium iodide in 75 mL of water in a 100-mL volumetric flask, dilute with water to volume, and mix.

Standard preparation—Transfer about 190 mg of mercuric chloride, accurately weighed, to a 200-mL volumetric flask, and dissolve in 100 mL of water. Dilute with water to volume, and mix. Transfer 5.0 mL of this solution to a 50-mL volumetric flask, dilute with water to volume, and mix. The concentration of mercuric chloride in the *Standard preparation* is about 95 µg per mL.

Test solution—Transfer about 500 mg of Thimerosal, accurately weighed, to a 100-mL volumetric flask, add water to volume, and mix to dissolve.

Test preparation A—Transfer 10.0 mL of the *Test solution* to a 50-mL volumetric flask, dilute with water to volume, and mix.

Test preparation B—Transfer 10.0 mL of the *Test solution* to a 50-mL volumetric flask, add 5.0 mL of the *Standard preparation*, dilute with water to volume, and mix.

Procedure—[NOTE—Protect all solutions from light prior to determining their absorbances.] Label five 10-mL volumetric flasks *C, D, E, F,* and *R*. Transfer 5.0 mL of *Test preparation A* to flasks *C* and *D*, transfer 5.0 mL of *Test preparation B* to flasks *E* and *F*, and transfer 5.0 mL of water to flask *R*. Dilute flasks *C* and *E* with water to volume, and mix. Dilute flasks *D*, *F*, and *R* with the *Iodide reagent* to volume, and mix. Concomitantly determine the absorbances of the solutions in 1-cm cells at the wavelength of maximum absorbance for the tetraiodomercurate ion at about 323 nm (located from the similarly determined ultraviolet spectrum of a solution prepared by mixing 1.0 mL of *Standard preparation* with 5.0 mL of *Iodide reagent* and diluting with water to 10.0 mL), with a suitable spectrophotometer, using water as the blank. Record the absorbances of the solutions in flasks *C, D, E, F,* and *R* as A_C, A_D, A_E, A_F, and

A_R, respectively. Calculate the percentage of mercury ions by the formula:

$$(200.59/271.50)(5C/W)(A_U/A_S),$$

in which 200.59 is the atomic weight of mercury, 271.50 is the molecular weight of mercuric chloride, A_U is the absorbance of the test specimen obtained by the equation:

$$A_U = (A_D - A_R - A_C),$$

A_S is the absorbance of the Standard obtained by the equation:

$$A_S = (A_F - A_R - A_E - A_U),$$

C is the concentration, in µg per mL, of mercuric chloride in the *Standard preparation*, and W is the weight, in mg, of Thimerosal taken: the limit is 0.70%.

Readily carbonizable substances ⟨271⟩—Dissolve about 200 mg in 5 mL of sulfuric acid TS: the solution has no more color than Matching Fluid J.

Assay—[NOTE—The *Standard preparation* and *Assay preparation* may be diluted quantitatively with water, if necessary, to obtain solutions, of suitable concentrations, adaptable to the linear or working range of the instrument.]

Standard preparation—Transfer about 100 mg of USP Thimerosal RS, accurately weighed, to a 100-mL volumetric flask, add 25 mL of water, and mix to dissolve. Dilute with water to volume, and mix.

Assay preparation—Transfer about 100 mg of Thimerosal, accurately weighed, to a 100-mL volumetric flask, add 25 mL of water, and mix to dissolve. Dilute with water to volume, and mix.

Procedure—Concomitantly determine the absorbances of the *Standard preparation* and the *Assay preparation*, at the mercury resonance line of 254 nm, with a suitable atomic absorption spectrophotometer (see *Spectrophotometry and Light-scattering* ⟨851⟩), equipped with a mercury hollow-cathode lamp and an air-acetylene flame, using water as the blank. Calculate the quantity, in mg, of $C_9H_9HgNaO_2S$ in the Thimerosal taken by the formula:

$$100C(A_U/A_S),$$

in which C is the concentration, in mg per mL, of USP Thimerosal RS in the *Standard preparation*, and A_U and A_S are the absorbances of the *Assay preparation* and the *Standard preparation*, respectively.

Thimerosal Topical Aerosol

» Thimerosal Topical Aerosol is an alcoholic solution of Thimerosal mixed with suitable propellants in a pressurized container. It contains not less than 85.0 percent and not more than 115.0 percent of the labeled amount of $C_9H_9HgNaO_2S$.

NOTE—Thimerosal Topical Aerosol is sensitive to some metals.

Packaging and storage—Preserve in tight, light-resistant, pressurized containers, and avoid exposure to excessive heat.

Reference standard—*USP Thimerosal Reference Standard*—Dry in vacuum over phosphorus pentoxide to constant weight before using.

Identification—To about 10 mL of solution sprayed into a suitable container from Thimerosal Topical Aerosol add 10 mL of water, and heat on a steam bath until the odor of alcohol is no longer perceptible. Cool, and pass hydrogen sulfide through the solution: no black discoloration or black precipitate is formed. To 15 mL of solution sprayed into a suitable container from Thimerosal Topical Aerosol add 15 mL of water, heat on a steam bath until the odor of alcohol is no longer perceptible, and add 2 drops of bromine. Mix with 1.5 mL of 3 *N* hydrochloric acid, filter, evaporate the excess bromine with a current of air, and pass hydrogen sulfide through the filtrate: a black precipitate is formed.

Alcohol content—Weigh, chill, and open 1 Topical Aerosol container, and remove the propellant as directed for *Assay preparation* in the *Assay*, continuing until the bulk of the propellant has evaporated. Determine the alcohol content of the specimen thus prepared by the *Gas-liquid Chromatographic Method* (see *Water Determination* ⟨611⟩), using methyl ethyl ketone as the internal standard in place of acetone: between 18.7% and 25.3% (w/w) of C_2H_5OH is found.

Other requirements—It meets the requirements for *Leak Testing* and *Pressure Testing* under *Aerosols* ⟨601⟩.

Assay—[NOTE—The *Standard preparations* and *Assay preparation* may be diluted quantitatively with water, if necessary, to yield solutions, of suitable concentration, adaptable to the linear or working range of the instrument.]

Stannous chloride solution—Dissolve 50 g of stannous chloride in 100 mL of hydrochloric acid on a steam bath, cool, dilute with water to 500 mL, and mix. Use within 3 months.

Standard solutions—Prepare aqueous solutions of USP Thimerosal RS of known concentrations of about 1.8, 2.0, and 2.2 µg per mL.

Standard preparations—Pipet 20 mL of each *Standard solution* into separate 100-mL volumetric flasks, and treat each flask as follows. Add 5 mL of sulfuric acid, cool, add 3 mL of nitric acid, and mix. Add potassium permanganate crystals, while mixing, until the purple color persists for not less than 15 minutes. Add about 200 mg of potassium persulfate, mix, and heat on a steam bath for 2 hours. Cool, dilute with water to volume, and mix.

Assay solution—Weigh accurately a filled Thimerosal Topical Aerosol container, and record the weight. Place the container in a dry ice–alcohol bath, and cool for 60 minutes. Remove the container from the bath, and carefully remove the spray cap with wire cutters, taking precautions to save all pieces of the spray head and cap. With the aid of three 5-mL portions of water, transfer the contents of the container to a beaker previously chilled in the bath. Dry the rinsed empty container and all of its parts in an oven at 105° for 2 hours, cool, and weigh. Calculate the weight of the container contents. Add a few boiling chips to the beaker, and carefully stir to help evaporate the propellant. After the bulk of the propellant has evaporated, place the beaker on a steam bath, evaporate the volatile solvents, and cool. Transfer the residual liquid with the aid of 35.0 mL of alcohol to a 50-mL volumetric flask, dilute with water to volume, and mix. Dilute quantitatively with water a volume, v mL, of this solution to V mL of *Assay solution* containing about 2 µg of thimerosal per mL.

Assay preparation—Pipet 20 mL of *Assay solution* into a 100-mL volumetric flask, and proceed as directed for *Standard preparations*, beginning with "Add 5 mL of sulfuric acid."

Blank preparation—Pipet 20 mL of water into a 100-mL volumetric flask, and proceed as directed for *Standard preparation*, beginning with "Add 5 mL of sulfuric acid."

Procedure—Proceed with each of the *Standard preparations*, the *Assay preparation*, and the *Blank preparation* as follows: Pipet 3 mL into the scrubbing chamber of a suitable system designed for determination of mercury by flameless atomic absorption, using a mercury hollow-cathode lamp, dilute with water to about 150 mL, and add hydroxylamine hydrochloride solution (1 in 10) just to reduce the excess permanganate. Add 5 mL of *Stannous chloride solution*, and immediately attach the scrubbing chamber to the system. Concomitantly determine the absorbance of the vapor from each solution at an integration time of 15 seconds. Use the absorbance of the *Blank preparation* to correct the absorbances of the *Standard preparations* and the *Assay preparation.* Plot the corrected absorbances of the standards versus the respective concentrations of the *Standard solutions*, in µg per mL, and from the curve so obtained determine the concentration, C, in µg per mL, of the *Assay solution.* Calculate the quantity, in mg, of $C_9H_9HgNaO_2S$ in the weight of the container contents by the formula:

$$50C(V/v),$$

in which the terms are as defined therein.

Thimerosal Topical Solution

» Thimerosal Topical Solution contains, in each 100 mL, not less than 95 mg and not more than 105 mg of $C_9H_9HgNaO_2S$.

NOTE—Thimerosal Topical Solution is sensitive to some metals.

Packaging and storage—Preserve in tight, light-resistant containers, and avoid exposure to excessive heat.

Reference standard—*USP Thimerosal Reference Standard*—Dry in vacuum over phosphorus pentoxide to constant weight before using.

Identification—

A: Pass hydrogen sulfide through 50 mL of it: no black discoloration or black precipitate is formed. To 50 mL of Topical Solution add 3 or 4 drops of bromine, mix, and warm on a steam bath to expel the excess bromine. Add 5 mL of 3 *N* hydrochloric acid, filter, and pass hydrogen sulfide through the filtrate: a black precipitate is formed.

B: To 1 mL of Topical Solution add 9 mL of water, mix, and add 1 mL of cupric sulfate TS: a green color is produced immediately and is followed by the gradual precipitation of flocculent, greenish brown particles.

pH ⟨791⟩: between 9.6 and 10.2.

Assay—[NOTE—The *Standard preparations* and *Assay preparation* may be diluted quantitatively with water, if necessary, to yield solutions, of suitable concentration, adaptable to the linear or working range of the instrument.]

Stannous chloride solution—Dissolve 50 g of stannous chloride in 100 mL of hydrochloric acid on a steam bath, cool, dilute with water to 500 mL, and mix. Use within 3 months.

Standard solutions—Prepare aqueous solutions of USP Thimerosal RS of known concentrations of about 1.8, 2.0, and 2.2 µg per mL.

Standard preparations—Pipet 20 mL of each *Standard solution* into separate 100-mL volumetric flasks, and treat each flask as follows. Add 5 mL of sulfuric acid, cool, add 3 mL of nitric acid, and mix. Add potassium permanganate crystals, while mixing, until the purple color persists for not less than 15 minutes. Add about 200 mg of potassium persulfate, mix, and heat on a steam bath for 2 hours. Cool, dilute with water to volume, and mix.

Assay solution—Pipet 2 mL of Thimerosal Topical Solution into a 1000-mL volumetric flask, dilute with water to volume, and mix.

Assay preparation—Pipet 20 mL of *Assay solution* into a 100-mL volumetric flask, and proceed as directed for *Standard preparations*, beginning with "Add 5 mL of sulfuric acid."

Blank preparation—Pipet 20 mL of water into a 100-mL volumetric flask, and proceed as directed for *Standard preparation*, beginning with "Add 5 mL of sulfuric acid."

Procedure—Proceed with each of the *Standard preparations*, the *Assay preparation*, and the *Blank preparation* as follows: Pipet 3 mL into the scrubbing chamber of a suitable system designed for determination of mercury by flameless atomic absorption, using a mercury hollow-cathode lamp, dilute with water to about 150 mL, and add hydroxylamine hydrochloride solution (1 in 10) just to reduce the excess permanganate. Add 5 mL of *Stannous chloride solution*, and immediately attach the scrubbing chamber to the system. Concomitantly determine the absorbance of each solution at an integration time of 15 seconds. Use the absorbance of the *Blank preparation* to correct the absorbances of the *Standard preparations* and the *Assay preparation*. Plot the corrected absorbances of the standards versus the respective concentrations of the *Standard solutions*, in µg per mL, and from the curve so obtained determine the concentration, *C*, in µg per mL, of the *Assay solution*. Calculate the quantity, in mg, of $C_9H_9HgNaO_2S$ in each 100 mL of Topical Solution by the formula:

$$50C,$$

in which the terms are as defined therein.

Thimerosal Tincture

» Thimerosal Tincture contains, in each 100 mL, not less than 90 mg and not more than 110 mg of $C_9H_9HgNaO_2S$.

NOTE—Thimerosal Tincture is sensitive to some metals.

Packaging and storage—Preserve in tight, light-resistant containers, and avoid exposure to excessive heat.

Reference standard—*USP Thimerosal Reference Standard*—Dry in vacuum over phosphorus pentoxide to constant weight before using.

Identification—Heat 25 mL on a steam bath until the odors of alcohol and acetone are no longer perceptible. Cool and pass hydrogen sulfide through the solution: no black discoloration or black precipitate is formed. Evaporate 50 mL of Tincture on a steam bath to a volume of approximately 20 mL, cool, and add 3 or 4 drops of bromine. Add 5 mL of 3 *N* hydrochloric acid, filter, and pass hydrogen sulfide through the filtrate: a black precipitate is formed.

Alcohol content, *Method II* ⟨611⟩: between 45.0% and 55.0% of C_2H_5OH.

Assay—[NOTE—The *Standard preparations* and *Assay preparation* may be diluted quantitatively with water, if necessary, to yield solutions, of suitable concentration, adaptable to the linear or working range of the instrument.]

Stannous chloride solution—Dissolve 50 g of stannous chloride in 100 mL of hydrochloric acid on a steam bath, cool, dilute with water to 500 mL, and mix. Use within 3 months.

Standard solutions—Prepare aqueous solutions of USP Thimerosal RS of known concentrations of about 1.8, 2.0, and 2.2 µg per mL.

Standard preparations—Pipet 20 mL of each *Standard solution* into separate 100-mL volumetric flasks, and treat each flask as follows. Add 5 mL of sulfuric acid, cool, add 3 mL of nitric acid, and mix. Add potassium permanganate crystals, while mixing, until the purple color persists for not less than 15 minutes. Add about 200 mg of potassium persulfate, mix, and heat on a steam bath for 2 hours. Cool, dilute with water to volume, and mix.

Assay solution—Pipet 2 mL of Thimerosal Tincture into a 1000-mL volumetric flask, dilute with water to volume, and mix.

Assay preparation—Pipet 20 mL of *Assay solution* into a 100-mL volumetric flask, and proceed as directed for *Standard preparations*, beginning with "Add 5 mL of sulfuric acid."

Blank preparation—Pipet 20 mL of water into a 100-mL volumetric flask, and proceed as directed for *Standard preparation*, beginning with "Add 5 mL of sulfuric acid."

Procedure—Proceed with each of the *Standard preparations*, the *Assay preparation*, and the *Blank preparation* as follows: Pipet 3 mL into the scrubbing chamber of a suitable system designed for determination of mercury by flameless atomic absorption, using a mercury hollow-cathode lamp, dilute with water to about 150 mL, and add hydroxylamine hydrochloride solution (1 in 10) just to reduce the excess permanganate. Add 5 mL of *Stannous chloride solution*, and immediately attach the scrubbing chamber to the system. Concomitantly determine the absorbance of the vapor from each solution at an integration time of 15 seconds. Use the absorbance of the *Blank preparation* to correct the absorbances of the *Standard preparations* and the *Assay preparation*. Plot the corrected absorbances of the standards versus the respective concentrations of the *Standard solutions*, in µg per mL, and from the curve so obtained determine the concentration, *C*, in µg per mL, of the *Assay solution*. Calculate the quantity, in mg, of $C_9H_9HgNaO_2S$ in each 100 mL of Tincture by the formula:

$$50C,$$

in which the terms are as defined therein.

Thioguanine

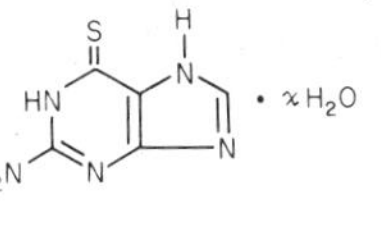

$C_5H_5N_5S.xH_2O$ (anhydrous) 167.19
6*H*-Purine-6-thione, 2-amino-1,7-dihydro-.
2-Aminopurine-6(1*H*)-thione [*154-42-7*].
Hemihydrate 176.20 [*5580-03-0*].

» Thioguanine is anhydrous or contains one-half molecule of water of hydration. It contains not less than 97.0 percent and not more than 100.5 percent of $C_5H_5N_5S$, calculated on the dried basis.

Packaging and storage—Preserve in tight containers.

Labeling—Label it to indicate its state of hydration.

Reference standard—*USP Thioguanine Reference Standard*—Dry in vacuum at 105° for 5 hours before using.

Identification—

A: The infrared absorption spectrum of a potassium bromide dispersion of it, previously dried in vacuum at 105° for 5 hours, exhibits maxima only at the same wavelengths as that of a similar preparation of USP Thioguanine RS.

B: The ultraviolet absorption spectrum of a 1 in 200,000 solution of it, prepared as directed in the *Assay*, exhibits maxima and minima at the same wavelengths as that of a similar solution of USP Thioguanine RS, concomitantly measured.

Loss on drying ⟨731⟩—Dry it in vacuum at 105° for 5 hours: it loses not more than 6.0% of its weight.

Selenium ⟨291⟩: 0.003%, 200 mg being used for the test.

Phosphorous-containing substances—

Ammonium molybdate solution—Dissolve 8.3 g of ammonium molybdate in 40 mL of water, add 33 mL of dilute sulfuric acid (2 in 7), dilute with water to 100.0 mL, and mix. This solution is stable for about two weeks.

Procedure—Transfer 50.0 mg, accurately weighed, to a large test tube, add 1 mL of dilute sulfuric acid (2 in 7), and heat in a boiling water bath for 5 minutes. Cautiously add nitric acid, dropwise, continue heating until the mixture becomes colorless, and then heat for 1 minute longer. Cool, dilute with water to about 10 mL, and transfer the solution to a 25-mL volumetric flask with the aid of a few mL of water. To the flask add 0.75 mL of *Ammonium molybdate solution* and 1.0 mL of aminonaphtholsulfonic acid TS, dilute with water to volume, and mix. Determine the absorbance of this solution in a 1-cm cell, with a suitable spectrophotometer, at a wavelength of about 620 nm, using a reagent blank to set the instrument: the absorbance is not greater than that produced by 1.5 mL of a similar solution of monobasic potassium phosphate in water having a known concentration of 10 µg of phosphate (PO_4) in each mL, concomitantly measured (0.03% as phosphate).

Free sulfur—Dissolve 50 mg in 5 mL of 1 *N* sodium hydroxide: the solution is clear.

Nitrogen content—Determine the nitrogen content as directed under *Nitrogen Determination, Method II* ⟨461⟩, using about 100 mg, accurately weighed. Each mL of 0.1 *N* sulfuric acid is equivalent to 1.401 mg of N. Not less than 40.6% and not more than 43.1%, calculated on the dried basis, is found.

Assay—Transfer about 100 mg of Thioguanine, previously dried and accurately weighed, to a 100-mL volumetric flask, dissolve in a mixture of 15 mL of water and 1.5 mL of 1 *N* sodium hydroxide, dilute with water to volume, and mix. Transfer 10.0 mL of this solution to a second 100-mL volumetric flask, add dilute hydrochloric acid (1 in 100) to volume, and mix. Finally, transfer 5.0 mL of the last solution to a third 100-mL volumetric flask, then add dilute hydrochloric acid (1 in 100) to volume, and mix. Concomitantly determine the absorbances of this solution and a solution of USP Thioguanine RS, in the same medium, having a known concentration of about 5 µg per mL, in 1-cm cells at the wavelength of maximum absorbance at about 348 nm, with a suitable spectrophotometer, using dilute hydrochloric acid (1 in 100) as the blank. Calculate the quantity, in mg, of $C_5H_5N_5S$ in the portion of Thioguanine taken by the formula:

$$20C(A_U/A_S),$$

in which C is the concentration, in µg per mL, of USP Thioguanine RS in the Standard solution, and A_U and A_S are the absorbances of the solution from Thioguanine and the Standard solution, respectively.

Thioguanine Tablets

» Thioguanine Tablets contain not less than 93.0 percent and not more than 107.0 percent of the labeled amount of $C_5H_5N_5S$.

Packaging and storage—Preserve in tight containers.

Reference standard—*USP Thioguanine Reference Standard*—Dry in vacuum at 105° for 5 hours before using.

Identification—The ultraviolet absorption spectrum of the *Acidic assay preparation* employed for measurement of absorbance in the *Assay* exhibits maxima and minima at the same wavelengths as that of a similar solution of USP Thioguanine RS, concomitantly measured.

Dissolution ⟨711⟩—

Medium: water; 900 mL.

Apparatus 2: 50 rpm.

Time: 45 minutes.

Standard preparation—Dissolve an accurately weighed quantity of USP Thioguanine RS in 1 *N* sodium hydroxide to obtain a solution having a known concentration of about 4.5 µg per mL.

Procedure—Determine the amount of $C_5H_5N_5S$ dissolved from ultraviolet absorbances at the wavelength of maximum absorbance at about 348 nm of filtered portions of the solution under test, suitably diluted with 0.1 *N* hydrochloric acid, in comparison with the *Standard preparation.*

Tolerances—Not less than 75% (*Q*) of the labeled amount of $C_5H_5N_5S$ is dissolved in 45 minutes.

Uniformity of dosage units ⟨905⟩: meet the requirements.

Assay—

Standard preparations—Dissolve an accurately weighed quantity of USP Thioguanine RS in sodium hydroxide solution (1 in 250), and dilute quantitatively and stepwise with the sodium hydroxide solution to obtain a *Standard preparation* having a known concentration of 80 µg per mL. Transfer 5.0 mL of this solution to a 100-mL volumetric flask, add dilute hydrochloric acid (1 in 10) to volume, and mix, to obtain the *Acidic standard preparation.* Transfer another 5.0-mL portion of the solution to a second 100-mL volumetric flask, add 10.0 mL of 1 *N* sodium hydroxide, dilute with water to volume, and mix, to obtain the *Basic standard preparation.*

Assay preparations—Weigh and finely powder not less than 20 Thioguanine Tablets. Transfer an accurately weighed portion of the powder, equivalent to about 40 mg of thioguanine, to a 500-mL volumetric flask, add 50 mL of 1 *N* sodium hydroxide, and allow to stand for 10 minutes, with frequent swirling. Dilute with water to volume, mix, and filter a portion of the solution through a pledget of glass wool. Transfer 5.0 mL of the filtrate to a 100-mL volumetric flask, add dilute hydrochloric acid (1 in 10) to volume, and mix, to obtain the *Acidic assay preparation.* Transfer another 5.0-mL portion of the filtrate to a second 100-mL volumetric flask, add 10.0 mL of 1 *N* sodium hydroxide, dilute with water to volume, and mix, to obtain the *Basic assay preparation.*

Procedure—Concomitantly determine the absorbances of the *Acidic assay preparation* and the *Acidic standard preparation* in 1-cm cells at the wavelength of maximum absorbance at about 348 nm, with a suitable spectrophotometer, using the *Basic assay preparation* and the *Basic standard preparation*, respectively, as the blanks. Calculate the quantity, in mg, of $C_5H_5N_5S$ in the portion of Tablets taken by the formula:

$$10C(A_U/A_S),$$

in which C is the concentration, in µg per mL, of USP Thioguanine RS in the *Acidic standard preparation*, and A_U and A_S are the absorbances of the *Acidic assay preparation* and the *Acidic standard preparation*, respectively.

Thiopental Sodium

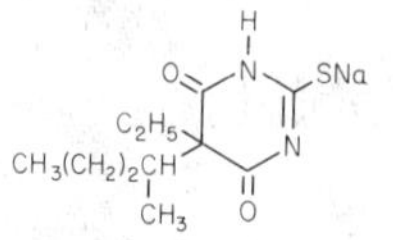

$C_{11}H_{17}N_2NaO_2S$ 264.32
4,6(1*H*,5*H*)-Pyrimidinedione, 5-ethyldihydro-5-(1-methylbutyl)-2-thioxo-, monosodium salt.
Sodium 5-ethyl-5-(1-methylbutyl)-2-thiobarbiturate [*71-73-8*].

» Thiopental Sodium contains not less than 97.0 percent and not more than 102.0 percent of $C_{11}H_{17}N_2NaO_2S$, calculated on the dried basis.

Packaging and storage—Preserve in tight containers.

Reference standard—*USP Thiopental Reference Standard*—Dry at 105° for 2 hours before using.

Identification—

A: Dissolve about 500 mg in 10 mL of water in a separator, add 10 mL of 3 *N* hydrochloric acid, and extract the liberated thiopental with two 25-mL portions of chloroform. Evaporate the combined chloroform extracts to dryness. Add 10 mL of ether, evaporate again, and dry at 105° for 2 hours: the infrared absorption spectrum of a potassium bromide dispersion of the residue so obtained exhibits maxima only at the same wavelengths as that of a similar preparation of USP Thiopental RS.

B: Ignite about 500 mg: the residue responds to the tests for *Sodium* ⟨191⟩.

C: Dissolve about 200 mg in 5 mL of 1 *N* sodium hydroxide, and add 2 mL of lead acetate TS: a white precipitate is formed, and it gradually darkens when the mixture is boiled. Acidify the darkened mixture with hydrochloric acid: hydrogen sulfide is evolved, and it is recognizable by its odor and by its darkening of moistened lead acetate test paper held in the vapor.

Loss on drying ⟨731⟩—Dry it at 80° for 4 hours: it loses not more than 2.0% of its weight.

Heavy metals, *Method II* ⟨231⟩: 0.002%.

Ordinary impurities ⟨466⟩—

Test solution: 10 mg of Thiopental Sodium per mL of methanol.

Standard solution: 9.2 mg of USP Thiopental RS per mL of methanol.

Application volume: 40 µL.

Eluant: a mixture of toluene and methanol (85:15).

Visualization: 1.

Assay—Transfer about 100 mg of Thiopental Sodium, accurately weighed, to a 200-mL volumetric flask, add sodium hydroxide solution (1 in 250) to volume, and mix. Pipet 5 mL of the solution into a 500-mL volumetric flask, add sodium hydroxide solution (1 in 250) to volume, and mix. Dissolve an accurately weighed quantity of USP Thiopental RS in sodium hydroxide solution (1 in 250), and dilute quantitatively and stepwise with sodium hydroxide solution (1 in 250) to obtain a Standard solution having a known concentration of about 5 µg per mL. Concomitantly determine the absorbances of both solutions in 1-cm cells at the wavelength of maximum absorbance at about 304 nm, with a suitable spectrophotometer, using sodium hydroxide solution (1 in 250) as the blank. Calculate the quantity, in mg, of $C_{11}H_{17}N_2NaO_2S$ in the Thiopental Sodium taken by the formula:

$$20C(1.091A_U/A_S),$$

in which C is the concentration, in µg per mL, of USP Thiopental RS in the Standard solution, 1.091 is the ratio of the molecular weight of thiopental sodium to that of thiopental, and A_U and A_S are the absorbances of the solution of Thiopental Sodium and the Standard solution, respectively.

Thiopental Sodium for Injection

» Thiopental Sodium for Injection is a sterile mixture of Thiopental Sodium and anhydrous Sodium Carbonate as a buffer. It contains not less than 93.0 percent and not more than 107.0 percent of the labeled amount of $C_{11}H_{17}N_2NaO_2S$.

Packaging and storage—Preserve in *Containers for Sterile Solids* as described under *Injections* ⟨1⟩, preferably of Type III glass.

Reference standard—*USP Thiopental Reference Standard*—Dry at 105° for 2 hours before using.

Completeness of solution ⟨641⟩—Mix 800 mg with 10 mL of carbon dioxide–free water: after 1 minute, the solution is clear and free from undissolved solid.

Constituted solution—At the time of use, the constituted solution prepared from Thiopental Sodium for Injection meets the requirements for *Constituted Solutions* under *Injections* ⟨1⟩.

pH ⟨791⟩: between 10.2 and 11.2, in the solution prepared in the test for *Completeness of solution.*

Other requirements—It responds to the *Identification tests* and meets the requirements of the test for *Heavy metals* under *Thiopental Sodium.* It meets also the requirements for *Sterility Tests* ⟨71⟩, *Uniformity of Dosage Units* ⟨905⟩, and for *Labeling* under *Injections* ⟨1⟩.

Assay—Dissolve the contents of 10 containers of Thiopental Sodium for Injection in sufficient water, diluting to an accurately measured volume, to obtain a solution containing about 50 mg of thiopental sodium per mL. Dilute this solution quantitatively and stepwise with sodium hydroxide solution (1 in 250) to obtain a solution having a concentration of about 5 µg of thiopental sodium per mL. Proceed as directed in the *Assay* under *Thiopental Sodium*, beginning with "Dissolve an accurately weighed quantity of USP Thiopental RS." Calculate the average quantity, in mg, of $C_{11}H_{17}N_2NaO_2S$ in each container of Thiopental Sodium for Injection taken by the formula:

$$VC(1.091A_U/A_S),$$

in which V is the volume, in mL, of the solution prepared to contain about 50 mg of thiopental sodium per mL, C is the concentration, in µg per mL, of USP Thiopental RS in the Standard solution, 1.091 is the ratio of the molecular weight of thiopental sodium to that of thiopental, and A_U and A_S are the absorbances of the solution of Thiopental Sodium for Injection and the Standard solution, respectively.

Thioridazine

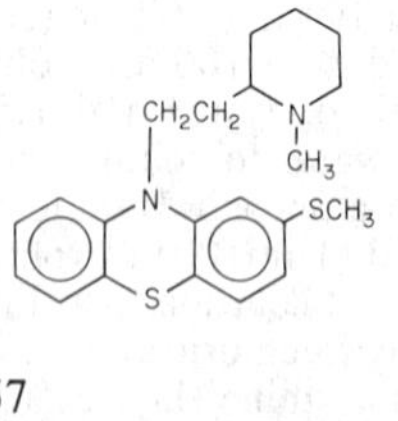

$C_{21}H_{26}N_2S_2$ 370.57
10*H*-Phenothiazine, 10-[2-(1-methyl-2-piperidinyl)ethyl]-2-(methylthio)-.
10-[2-(1-Methyl-2-piperidyl)ethyl]-2-(methylthio)phenothiazine [*50-52-2*].

» Thioridazine contains not less than 99.0 percent and not more than 101.0 percent of $C_{21}H_{26}N_2S_2$, calculated on the dried basis.

Packaging and storage—Preserve in well-closed, light-resistant containers.

Reference standard—*USP Thioridazine Reference Standard*—Dry in vacuum at 50° for 4 hours before using.

NOTE—Throughout the following procedures, protect test, or assay specimens, the Reference Standard, and solutions containing them, by conducting the procedures without delay, under subdued light, or using low-actinic glassware.

Identification—The infrared absorption spectrum of a potassium bromide dispersion of it exhibits maxima only at the same wavelengths as that of a similar preparation of USP Thioridazine RS.

Loss on drying ⟨731⟩—Dry it in vacuum at 50° for 4 hours: it loses not more than 0.5% of its weight.

Residue on ignition ⟨281⟩: not more than 0.1%.

Chromatographic purity—[NOTE—Conduct this procedure without delay, under subdued light.] Transfer 100 mg of Thioridazine to a 10-mL volumetric flask, add a mixture of methanol and ammonium hydroxide (49:1) to volume, and mix to obtain the *Test solution.* Using an accurately weighed quantity of USP Thioridazine RS, prepare two solutions in the same solvent system containing 50 µg per mL (*Solution A*, equivalent to 0.5%) and 20 µg per mL (*Solution B*, equivalent to 0.2%). On a thin-layer chromatographic plate (see *Chromatography* ⟨621⟩), coated with a 0.25-mm layer of chromatographic silica gel mixture, apply 5-µL portions of the *Test solution* and each of the two Standard solutions. Immediately develop the chromatogram in a solvent system consisting of a mixture of chloroform, isopropyl alcohol, and ammonium hydroxide (74:25:1) until the solvent front has moved about three-fourths of the length of the plate. Remove the plate from the developing chamber, mark the solvent front, allow the solvent to evaporate, and examine the plate under short-wavelength ultraviolet light: the chromatograms show principal spots at about the same R_f value; no secondary spot, if present in the chromatogram from the *Test solution*, is more intense than the principal spot obtained from *Solution A* (0.5%); and the sum of the intensities of all secondary spots, if present in the chromatogram from the *Test solution*, is not greater than 0.5%.

Assay—Dissolve about 300 mg of Thioridazine, accurately weighed, in 60 mL of glacial acetic acid, and titrate with 0.1 *N* perchloric acid VS, determining the end-point potentiometrically. Perform a blank determination, and make any necessary correction. Each mL of 0.1 *N* perchloric acid is equivalent to 37.06 mg of $C_{21}H_{26}N_2S_2$.

Thioridazine Oral Suspension

» Thioridazine Oral Suspension contains not less than 90.0 percent and not more than 110.0 percent of the labeled amount of $C_{21}H_{26}N_2S_2$.

Packaging and storage—Preserve in tight, light-resistant containers, at a temperature not exceeding 30°.

Reference standard—*USP Thioridazine Reference Standard*—Dry in vacuum at 50° for 4 hours before using.

NOTE—Throughout the following procedures, protect test or assay specimens, the Reference Standard, and solutions containing them, by conducting the procedures without delay, under subdued light, or using low-actinic glassware.

Identification—[NOTE—Conduct this test without exposure to daylight, and with a minimum of exposure to artificial light.] Transfer 40 mL of the combined chloroform extracts obtained for the *Assay* to a 50-mL beaker, and reduce the volume to 1 mL by evaporating the chloroform with the aid of a stream of nitrogen. On a thin-layer chromatographic plate (see *Chromatography* ⟨621⟩), coated with a 0.25-mm layer of chromatographic silica gel mixture, apply 25 µL of this test solution and 25 µL of a Standard solution of USP Thioridazine RS in chloroform containing 4 mg per mL. Allow the spots to dry, and develop the chromatogram in a solvent system consisting of a mixture of toluene, acetone, solvent hexane, and diethylamine (15:15:15:1) until the solvent front has moved about 10 cm from the origin. Remove the plate from the developing chamber, mark the solvent front, and locate the spots on the plate by viewing under short-wavelength and long-wavelength ultraviolet light: the R_f value of the principal spot obtained from the test solution corresponds to that obtained from the Standard solution.

Specific gravity ⟨841⟩: between 1.180 and 1.310.

pH ⟨791⟩: between 8.0 and 10.0.

Assay—Transfer an accurately measured volume of Thioridazine Oral Suspension, freshly mixed but free from air bubbles, equivalent to about 100 mg of thioridazine, to a 125-mL separator, using a pipet calibrated to contain the required volume. Rinse the pipet with 10 to 15 mL of water, adding the rinsing to the separator. Render the mixture alkaline by adding several drops of ammonium hydroxide, and mix. Extract with six 30-mL portions of chloroform, and filter the extracts through anhydrous sodium sulfate, collecting the combined filtrates in a 200-mL volumetric flask. Dilute quantitatively and stepwise with chloroform to obtain a solution having a concentration of about 100 µg of thioridazine per mL. [NOTE—Reserve a 40-mL portion of this solution for the *Identification* test.] Transfer 5.0 mL of this solution to a 100-mL volumetric flask, dilute with chloroform to volume, and mix. Concomitantly determine the absorbances of this Assay solution and a Standard solution of USP Thioridazine RS in chloroform having a known concentration of about 5 µg per mL, in 1-cm cells at the wavelength of maximum absorbance at about 266 nm, with a suitable spectrophotometer, using chloroform as the blank. Calculate the quantity, in mg, of $C_{21}H_{26}N_2S_2$ in each mL of the Oral Suspension taken by the formula:

$$20(C/V)(A_U/A_S),$$

in which C is the concentration, in µg per mL, of USP Thioridazine RS in the Standard solution, V is the volume, in mL, of Oral Suspension taken, and A_U and A_S are the absorbances of the Assay solution and the Standard solution, respectively.

Thioridazine Hydrochloride

$C_{21}H_{26}N_2S_2 \cdot HCl$ 407.03

10*H*-Phenothiazine, 10-[2-(1-methyl-2-piperidinyl)ethyl]-2-(methylthio)-, monohydrochloride.

10-[2-(1-Methyl-2-piperidyl)ethyl]-2-(methylthio)phenothiazine monohydrochloride [*130-61-0*].

» Thioridazine Hydrochloride contains not less than 99.0 percent and not more than 101.0 percent of $C_{21}H_{26}N_2S_2 \cdot HCl$, calculated on the dried basis.

Packaging and storage—Preserve in tight, light-resistant containers.

Reference standard—*USP Thioridazine Hydrochloride Reference Standard*—Dry at 105° for 4 hours before using.

NOTE—Throughout the following procedures, protect test or assay specimens, the Reference Standard, and solutions containing them, by conducting the procedures without delay, under subdued light, or using low-actinic glassware.

Identification—

A: The infrared absorption spectrum of a potassium bromide dispersion of it exhibits maxima only at the same wavelengths as that of a similar preparation of USP Thioridazine Hydrochloride RS.

B: A solution (1 in 10) responds to the tests for *Chloride* ⟨191⟩.

Melting range ⟨741⟩: between 159° and 165°, but the range between beginning and end of melting does not exceed 3°.

pH ⟨791⟩: between 4.2 and 5.2, in a solution (1 in 100).

Loss on drying ⟨731⟩—Dry it at 105° for 4 hours: it loses not more than 0.4% of its weight.

Residue on ignition ⟨281⟩: not more than 0.1%.

Selenium ⟨291⟩: 0.003%, a 200-mg specimen being used.

Chromatographic purity—[NOTE—Conduct this procedure without delay, under subdued light.] Prepare a *Test solution* in methanol and ammonium hydroxide (49:1) containing 10 mg of Thioridazine Hydrochloride per mL. Using accurately weighed quantities of USP Thioridazine Hydrochloride RS, prepare solutions in methanol and ammonium hydroxide (49:1) containing 50 µg per mL (*Solution A*, equivalent to 0.5%) and 20 µg per mL (*Solution B*, equivalent to 0.2%). On a suitable thin-layer chromatographic plate (see *Chromatography* ⟨621⟩), coated with a 0.25-mm layer of chromatographic silica gel mixture, apply 5-µL portions of the *Test solution* and each of the two Standard solutions. Immediately develop the chromatogram in a suitable chamber containing a mixture of chloroform, isopropyl alcohol, and ammonium hydroxide (74:25:1) until the solvent front has moved about three-fourths of the length of the plate. Remove the plate from the chamber, air-dry, and view under short-wavelength ultraviolet light: no spot from the *Test solution*, other than the principal spot, is greater in size or intensity than the spot obtained from *Solution A* (0.5%), and the sum of the impurities is not greater than 0.5%.

Assay—Dissolve about 350 mg of Thioridazine Hydrochloride, accurately weighed, in 80 mL of a solution of equal parts of glacial acetic acid and acetic anhydride, and titrate with 0.1 *N* perchloric acid VS, determining the end-point potentiometrically. Perform a blank determination, and make any necessary correction. Each mL of 0.1 *N* perchloric acid is equivalent to 40.70 mg of $C_{21}H_{26}N_2S_2 \cdot HCl$.

Thioridazine Hydrochloride Oral Solution

» Thioridazine Hydrochloride Oral Solution contains not less than 90.0 percent and not more than 110.0 percent of the labeled amount of $C_{21}H_{26}N_2S_2 \cdot HCl$.

Packaging and storage—Preserve in tight, light-resistant containers, at controlled room temperature.

Labeling—Label it to indicate that it is to be diluted to appropriate strength with water or other suitable fluid prior to administration.

Reference standard—*USP Thioridazine Hydrochloride Reference Standard*—Dry at 105° for 4 hours before using.

NOTE—Throughout the following procedures, protect test or assay specimens, the Reference Standard, and solutions containing them, by conducting the procedures without delay, under subdued light, or using low-actinic glassware.

Identification—A volume of Oral Solution containing 50 mg of thioridazine hydrochloride, diluted with water to 25 mL, meets the requirements under *Identification—Organic Nitrogenous Bases* ⟨181⟩, 2 mL of sodium bicarbonate solution (1 in 12) being used in place of the 2 mL of 1 *N* sodium hydroxide specified in the test.

Alcohol content ⟨611⟩: not more than 4.75% of C_2H_5OH.

Assay—[NOTE—Conduct this procedure with a minimum of exposure to light.]

Ammoniacal chloroform—Shake 125 mL of chloroform with 5 mL of ammonium hydroxide in a separator, slowly filter the bottom layer through filter paper containing anhydrous sodium sulfate, and discard the top layer.

Standard preparation—Dissolve a suitable quantity of USP Thioridazine Hydrochloride RS, accurately weighed, in *Ammoniacal chloroform* to obtain a solution having a known concentration of about 6 µg per mL.

Assay preparation—Pipet a portion of Thioridazine Hydrochloride Oral Solution, equivalent to about 120 mg of thioridazine hydrochloride, into a separator containing 15 mL of water. Render alkaline with ammonium hydroxide, and extract with three 25-mL portions of *Ammoniacal chloroform*. Filter the extracts through a pledget of glass wool into a 200-mL volumetric flask. Rinse the filter, add chloroform to volume, and mix. Dilute 2.0 mL of this solution with *Ammoniacal chloroform* to 200.0 mL, and mix.

Procedure—Concomitantly determine the absorbances of the *Standard preparation* and the *Assay preparation* in 1-cm cells at the wavelength of maximum absorbance at about 265 nm, with a suitable spectrophotometer, using *Ammoniacal chloroform* as the blank. Calculate the quantity, in mg, of $C_{21}H_{26}N_2S_2 \cdot HCl$ in each mL of the Oral Solution taken by the formula:

$$20(C/V)(A_U/A_S),$$

in which *C* is the concentration, in µg per mL, of USP Thioridazine Hydrochloride RS in the *Standard preparation*, *V* is the volume, in mL, of Oral Solution taken, and A_U and A_S are the absorbances of the *Assay preparation* and the *Standard preparation*, respectively.

Thioridazine Hydrochloride Tablets

» Thioridazine Hydrochloride Tablets contain not less than 90.0 percent and not more than 110.0 percent of the labeled amount of $C_{21}H_{26}N_2S_2 \cdot HCl$.

Packaging and storage—Preserve in tight, light-resistant containers.

Reference standard—*USP Thioridazine Hydrochloride Reference Standard*—Dry at 105° for 4 hours before using.

NOTE—Throughout the following procedures, protect test or assay specimens, the Reference Standard, and solutions containing them, by conducting the procedures without delay, under subdued light, or using low-actinic glassware.

Identification—Tablets meet the requirements under *Identification—Organic Nitrogenous Bases* ⟨181⟩, 2 mL of sodium bicarbonate solution (1 in 12) being used in place of the 2 mL of 1 *N* sodium hydroxide specified in the test.

Dissolution ⟨711⟩—

Medium: 0.1 *N* hydrochloric acid; 1000 mL.

Apparatus 2: 75 rpm.

Time: 60 minutes.

Procedure—Determine the amount of $C_{21}H_{26}N_2S_2 \cdot HCl$ dissolved from ultraviolet absorbances at the wavelength of maximum absorbance at about 262 nm of filtered portions of the solution under test, suitably diluted with *Dissolution Medium*, if necessary, in comparison with a Standard solution having a known concentration of USP Thioridazine Hydrochloride RS in the same medium.

Tolerances—Not less than 75% (*Q*) of the labeled amount of $C_{21}H_{26}N_2S_2 \cdot HCl$ is dissolved in 60 minutes.

Uniformity of dosage units ⟨905⟩: meet the requirements.

Assay—

Mobile phase—Prepare a filtered and degassed mixture of acetonitrile, water, and triethylamine (850:150:1). Make adjustments if necessary (see *System Suitability* under *Chromatography* ⟨621⟩).

Standard preparation—Dissolve an accurately weighed quantity of USP Thioridazine Hydrochloride RS in methanol with the aid of sonication, and dilute quantitatively and stepwise, if necessary, with methanol to obtain a solution having a known concentration of about 125 µg per mL.

Assay preparation—Weigh and finely powder not less than 20 Thioridazine Hydrochloride Tablets. Transfer an accurately weighed portion of the powder, equivalent to about 100 mg of thioridazine hydrochloride, to a 100-mL volumetric flask. Add about 80 mL of methanol, and shake by mechanical means for 30 minutes. Dilute with methanol to volume, and sonicate for 45 minutes with intermittent shaking. Allow the undissolved solids to settle, and filter, discarding the first 20 mL of the filtrate. Transfer 25.0 mL of the clear filtrate to a 200-mL volumetric flask, dilute with methanol to volume, and mix. Filter through a 0.45-µm disk before injecting into the chromatograph.

System suitability preparation—Dissolve 100 mg of mesoridazine besylate in 100 mL of methanol. Mix 1.0 mL of this solution with 9.0 mL of the *Standard preparation.*

Chromatographic system (see *Chromatography* ⟨621⟩)—The liquid chromatograph is equipped with a 265-nm detector and a 4.6-mm × 25-cm column that contains packing L1. The flow rate is about 2.5 mL per minute. Chromatograph the *Standard preparation* and the *System suitability preparation*, and record the peak responses as directed under *Procedure:* the resolution, *R*, between the mesoridazine and thioridazine peaks is not less than 1.0, and the relative standard deviation for replicate injections of the *Standard preparation* is not more than 2.0%.

Procedure—Separately inject equal volumes (about 10 µL) of the *Standard preparation* and the *Assay preparation* into the chromatograph, record the chromatograms, and measure the responses for the major peaks. Calculate the quantity, in mg, of $C_{21}H_{26}N_2S_2 \cdot HCl$ in the portion of Tablets taken by the formula:

$$0.8C(r_U/r_S),$$

in which *C* is the concentration, in µg per mL, of USP Thioridazine Hydrochloride RS in the *Standard preparation*, and r_U and r_S are the peak responses obtained from the *Assay preparation* and the *Standard preparation*, respectively.

Thiotepa

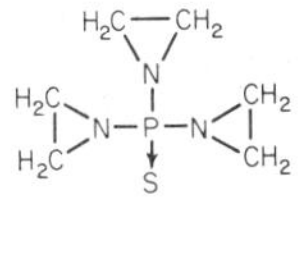

$C_6H_{12}N_3PS$ 189.21

Aziridine, 1,1′,1″-phosphinothioylidynetris-.

Tris(1-aziridinyl)phosphine sulfide [*52-24-4*].

» Thiotepa contains not less than 97.0 percent and not more than 102.0 percent of $C_6H_{12}N_3PS$, calculated on the anhydrous basis.

Caution—Great care should be taken to prevent inhaling particles of Thiotepa or exposing the skin to it.

Packaging and storage—Preserve in tight, light-resistant containers, and store in a refrigerator.

Reference standard—*USP Thiotepa Reference Standard*—Dry over silica gel for 24 hours before using.

Identification—The infrared absorption spectrum, determined in a 0.1-mm cell, of a 3 in 400 solution in carbon disulfide of Thiotepa, previously dried over silica gel for 24 hours, exhibits maxima only at the same wavelengths as that of a similar solution of USP Thiotepa RS.

Melting range ⟨741⟩: between 52° and 57°.

Water, *Method I* ⟨921⟩: not more than 2.0%.

Assay—Transfer about 200 mg of Thiotepa, accurately weighed, to a 250-mL iodine flask with the aid of 50 mL of sodium thiosulfate solution (1 in 5), and add 1 drop of methyl orange TS. Titrate the solution immediately with 0.1 *N* hydrochloric acid VS until a faint red color appears and the end-point persists for not less than 10 seconds. Insert the stopper in the flask, and allow to stand for 30 minutes. Add 4 drops of phenolphthalein TS, and titrate with 0.1 *N* sodium hydroxide VS. Repeat the titration (blank determination) on 50 mL of sodium thiosulfate solution (1 in 5). Determine the number of mL of 0.1 *N* hydrochloric acid consumed by the assay specimen by subtracting the number of mL of 0.1 *N* sodium hydroxide used from the number of mL of 0.1 *N* hydrochloric acid used, and correct it by subtracting the corresponding difference in consumption of 0.1 *N* hydrochloric acid and 0.1 *N* sodium hydroxide for the blank. Each mL of 0.1 *N* hydrochloric acid is equivalent to 6.307 mg of $C_6H_{12}N_3PS$.

Thiotepa for Injection

» Thiotepa for Injection is a sterile mixture of 1 part of Thiotepa, 5.33 parts of Sodium Chloride, and 3.33 parts of Sodium Bicarbonate. It contains not less than 95.0 percent and not more than 110.0 percent of the labeled amount of $C_6H_{12}N_3PS$.

Packaging and storage—Preserve in *Containers for Sterile Solids* as described under *Injections* ⟨1⟩, and store in a refrigerator, protected from light.

Reference standard—*USP Thiotepa Reference Standard*—Dry over silica gel for 24 hours before using.

Completeness of solution ⟨641⟩—The contents of 1 container dissolve in 4 mL of water to yield a clear solution.

Identification—The infrared absorption spectrum of the solution employed for measurements of absorbance in the *Assay* exhibits maxima only at the same wavelengths as that of the Standard solution, prepared as directed in the *Assay.*

pH ⟨791⟩: between 7.0 and 8.2, in a solution, reconstituted in Sterile Water for Injection, containing 10 mg of thiotepa per mL.

Loss on drying ⟨731⟩—Dry the contents of 1 container, accurately weighed, over silica gel for 24 hours: it loses not more than 0.5% of its weight.

Bacterial endotoxins—When tested as directed under *Bacterial Endotoxins Test* ⟨85⟩, the USP Endotoxin RS being used, it contains not more than 6.25 USP Endotoxin Units for each mg of Thiotepa.

Other requirements—It meets the requirements for *Sterility Tests* ⟨71⟩, *Uniformity of Dosage Units* ⟨905⟩, and *Labeling* under *Injections* ⟨1⟩.

Assay—Remove, as completely as possible, the contents of not less than 20 containers of Thiotepa for Injection, weigh, and mix. Transfer an accurately weighed portion of the powder, equivalent to about 75 mg of thiotepa, to a suitable container, extract with three 5-mL portions of carbon disulfide, and filter the carbon disulfide extract with the aid of vacuum. Concentrate the combined filtrates under vacuum to approximately 5 mL. Transfer the carbon disulfide solution to a 10-mL volumetric flask with the aid of a few mL of carbon disulfide, and dilute with carbon disulfide to volume. Concomitantly determine the absorbances of this solution and a Standard solution of USP Thiotepa RS in the same medium having a known concentration of about 7.5 mg per mL, in 0.1-mm cells, at the wavelength of maximum absorbance at about 10.75 µm, with a suitable infrared spectrophotometer, using carbon disulfide as the blank. Calculate the quantity, in mg, of $C_6H_{12}N_3PS$ in the portion of Thiotepa for Injection taken by the formula:

$$10C(A_U/A_S),$$

in which *C* is the concentration, in mg per mL, of USP Thiotepa RS in the Standard solution, and A_U and A_S are the absorbances of the Assay solution and the Standard solution, respectively.

Thiothixene

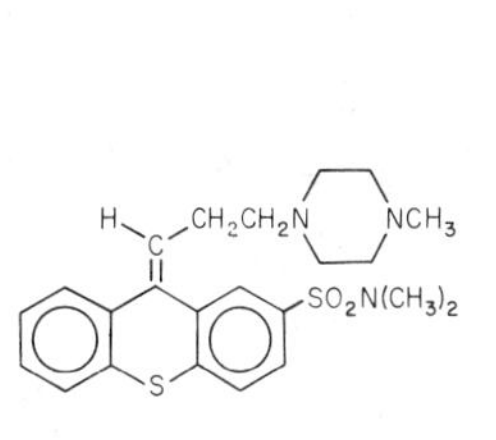

$C_{23}H_{29}N_3O_2S_2$ 443.62

9*H*-Thioxanthene-2-sulfonamide, *N*,*N*-dimethyl-9-[3-(4-methyl-1-piperazinyl)propylidene]-, (*Z*)-.

N,*N*-Dimethyl-9-[3-(4-methyl-1-piperazinyl)propylidene]thioxanthene-2-sulfonamide [*5591-45-7; 3313-26-6*].

» Thiothixene contains not less than 98.0 percent and not more than 104.0 percent of $C_{23}H_{29}N_3O_2S_2$, calculated on the dried basis.

Packaging and storage—Preserve in tight, light-resistant containers.

Reference standards—*USP Thiothixene Reference Standard*—Dry in vacuum at 100° for 3 hours before using. *USP (E)-Thiothixene Reference Standard*—Keep container tightly closed and protected from light. Dry in vacuum at 100° for 3 hours before using.

Identification—

A: The infrared absorption spectrum of a 1 in 20 solution of previously dried Thiothixene in chloroform, determined in a 0.1-mm cell, exhibits maxima only at the same wavelengths as that of a similar solution of USP Thiothixene RS.

B: The ultraviolet absorption spectrum of a 1 in 100,000 solution in methanol exhibits maxima and minima at the same wavelengths as that of a similar solution of USP Thiothixene RS, concomitantly measured, and the respective absorptivities, calculated on the dried basis, at the wavelengths of maximum absorbance at about 230 nm and 307 nm do not differ by more than 4.0%.

Melting range, *Class I* ⟨741⟩: between 147° and 153.5°.

Loss on drying ⟨731⟩—Dry it in vacuum at 100° for 3 hours: it loses not more than 2.0% of its weight.

Residue on ignition ⟨281⟩: not more than 0.2%.

Selenium ⟨291⟩: 0.003%.

Heavy metals, *Method II* ⟨231⟩: 0.0025%.

Limit of (*E*)-thiothixene—[NOTE—Prepare all solutions in low-actinic glassware.]

Mobile phase—Transfer 6.9 g of sodium dihydrogen phosphate monohydrate to a 1-liter volumetric flask, dissolve in deionized water, dilute with deionized water to volume, and mix. Filter through a suitable membrane filter. Mix 4 volumes of this solution with 6 volumes of methanol. The concentration of methanol may be adjusted to meet the system suitability requirements.

Standard preparations—

A—Using accurately weighed quantities of USP (*E*)-Thiothixene RS and USP Thiothixene RS, prepare a solution in methanol containing, in each mL, 0.4 mg and 1.2 mg, respectively.

B—Transfer 5.0 mL of *Standard preparation A* to a 100-mL volumetric flask, dilute with methanol to volume, and mix.

C—Transfer about 200 mg of thiothixene, accurately weighed, to a 100-mL volumetric flask. Transfer 5.0 mL of *Standard preparation A* to the same flask, dissolve in methanol, dilute with methanol to volume, and mix.

Test preparation—Transfer about 200 mg of Thiothixene, accurately weighed, to a 100-mL volumetric flask. Dissolve in methanol, dilute with methanol to volume, and mix.

Procedure—Concomitantly introduce equal volumes (typically 20 µL) of *Standard preparation C* and *Test preparation* into a high-pressure liquid chromatograph operated at room temperature and equipped with a suitable microsyringe or sampling valve, a column containing packing L9 (typically 25 cm × 4.6 mm), an ultraviolet detector capable of monitoring absorption at 254 nm, and a suitable recorder. The *Mobile phase* is maintained at a flow rate of about 1 to 1.5 mL per minute. In a suitable chromatographic system, three replicate injections of *Standard preparation B* show a resolution factor of not less than 2.2 between the thiothixene and (*E*)-thiothixene peaks, their retention times being 13 and 15 minutes, and between 16 and 18 minutes, respectively. Calculate the quantity, in mg, of (*E*)-thiothixene in the portion of Thiothixene taken by the formula:

$$5CH_U/(H_C - H_U),$$

in which C is the concentration of USP (*E*)-Thiothixene RS, in mg per mL, in *Standard preparation A*, and H_C and H_U are the peak responses of the (*E*)-thiothixene peaks corrected for the tailing of the main peak, obtained from *Standard preparation C* and the *Test preparation*, respectively: the limit of (*E*)-thiothixene is 1.0%.

Assay—[NOTE—Perform the dilution operations in low-actinic glassware.]

Mobile phase—Mix 0.5 mL of ethanolamine with 3780 mL of methanol, mix 1400 mL of this solution with 200 mL of water, filter, and degas. Make adjustments if necessary (see *System Suitability* under *Chromatography* ⟨621⟩).

Standard preparation—Using an accurately weighed quantity of USP Thiothixene RS, prepare a solution in methanol having a known concentration of about 0.02 mg per mL.

Assay preparation—Transfer about 100 mg of Thiothixene, accurately weighed, to a 100-mL volumetric flask, dissolve in methanol, dilute with methanol to volume, and mix. Pipet 2 mL of the resulting solution into a 100-mL volumetric flask, dilute with methanol to volume, and mix.

Chromatographic system (see *Chromatography* ⟨621⟩)—The liquid chromatograph is equipped with a 254-nm detector and a 3.9-mm × 30-cm column that contains packing L3. The flow rate is about 0.5 mL per minute. Chromatograph the *Standard preparation*, and record the peak responses as directed under *Procedure:* the column efficiency determined from the analyte peak is not less than 2000 theoretical plates, and the relative standard deviation for replicate injections is not more than 1.5%.

Procedure—Separately inject equal volumes (about 20 µL) of the *Standard preparation* and the *Assay preparation* into the chromatograph, record the chromatograms, and measure the responses for the major peaks. Calculate the quantity, in mg, of $C_{23}H_{29}N_3O_2S_2$ in the portion of Thiothixene taken by the formula:

$$5000C(r_U / r_S),$$

in which C is the concentration, in mg per mL, of USP Thiothixene RS in the *Standard preparation*, and r_U and r_S are the peak responses obtained from the *Assay preparation* and the *Standard preparation*, respectively.

Thiothixene Capsules

» Thiothixene Capsules contain not less than 90.0 percent and not more than 110.0 percent of the labeled amount of $C_{23}H_{29}N_3O_2S_2$.

Packaging and storage—Preserve in well-closed, light-resistant containers.

Reference standard—*USP Thiothixene Reference Standard*—Dry in vacuum at 100° for 3 hours before using.

Identification—Dissolve a portion of the contents of Capsules in a solvent consisting of equal volumes of chloroform and methanol to obtain a solution containing 1 mg of thiothixene per mL. Shake by mechanical means for 10 minutes, clarify a portion of the mixture by centrifugation, filter, if necessary, and use the clear supernatant solution or filtrate for the test. On a suitable thin-layer chromatographic plate (see *Chromatography* ⟨621⟩), coated with a 0.25-mm layer of chromatographic silica gel mixture, apply 10 µL of this test solution, 10 µL of a Standard solution containing 1 mg of USP Thiothixene RS per mL in the same medium, and 10 µL of a mixture of equal volumes of the test solution and the Standard solution. Allow the spots to dry, and develop the chromatogram in a solvent system consisting of a mixture of ethyl acetate, methanol, and diethylamine (65:35:5) until the solvent front has moved about three-fourths of the length of the plate. Remove the plate from the developing chamber, mark the solvent front, and locate the spots on the plate by viewing under short- and long-wavelength ultraviolet light. Spray the plate lightly with acidified iodoplatinate spray reagent (prepared by mixing 1 volume of hydrochloric acid with 50 volumes of potassium iodoplatinate TS): the R_f value of the principal spot obtained from the test solution corresponds to that obtained from the Standard solution and the mixed test-Standard solution.

Dissolution ⟨711⟩—

Medium—Dissolve 2.0 g of sodium chloride and 7 mL of hydrochloric acid in water to make 1000 mL, and mix; 1000 mL.

Apparatus 1: 150 rpm.

Time: 15 minutes.

Buffer solution—On the day of use, prepare a mixture of 55 volumes of dibasic potassium phosphate solution (87 in 1000), 20 volumes of citric acid monohydrate solution (21 in 200), and 40 volumes of sodium hydroxide solution (1 in 25).

Methyl orange solution—Transfer 15.5 g of boric acid and 2.0 g of methyl orange to a glass-stoppered, 1000-mL flask. Add 500 mL of water, insert the stopper, and shake by mechanical means for not less than 3 hours. Filter through retentive filter paper, and wash the filtrate with two 100-mL portions of chloroform, discarding the chloroform washings. Store the *Methyl orange solution* over 50 mL of chloroform in a glass-stoppered bottle.

Procedure—Prepare a *Test preparation* by passing a 60-mL portion of the dissolution solution through a suitable filter, discarding the first 5 mL of the filtrate, and diluting the subsequent filtrate quantitatively, if necessary, to obtain a concentration of about 1 µg of thiothixene per mL. Prepare a *Standard preparation* of USP Thiothixene RS in *Dissolution Medium* having a known concentration of about 1 µg per mL. Transfer 40.0 mL each of the *Test preparation*, the *Standard preparation*, and *Dissolution Medium* to provide the blank, to individual separators, each containing 8.0 mL of *Buffer solution*, 10.0 mL of *Methyl orange solution*, and 50.0 mL of chloroform. Shake for 3 minutes, allow the layers to separate, transfer 40.0 mL of the chloroform layer, clarified by centrifugation, to a 60-mL separator containing 8.0 mL of dilute hydrochloric acid (1 in 120), shake for 1 minute, and allow the layers to separate. Concomitantly determine the absorbances of the aqueous layers in 1-cm cells, at the wavelength of maximum absorbance at about 508 nm, with a suitable spectrophotometer, using the blank to set the instrument. Calculate the quantity, in µg per mL, of $C_{23}H_{29}N_3O_2S_2$ in the *Test preparation* by the formula:

$$C(A_U/A_S),$$

in which *C* is the concentration, in µg per mL, of USP Thiothixene RS in the *Standard preparation*, and A_U and A_S are the absorbances of the solutions from the *Test preparation* and the *Standard preparation*, respectively, and, from the known extent of dilution, determine the amount of it in the dissolution solution.

Tolerances—Not less than 75% (*Q*) of the labeled amount of $C_{23}H_{29}N_3O_2S_2$ is dissolved in 15 minutes.

Uniformity of dosage units ⟨905⟩: meet the requirements.

Assay—[NOTE—Perform the dilution operations in low-actinic glassware.]

Mobile phase, *Standard preparation*, and *Chromatographic system*—Prepare as directed in the *Assay* under *Thiothixene*.

Assay preparation—Transfer, as completely as possible, the contents of not less than 20 Thiothixene Capsules to a tared beaker, and weigh. Mix, and transfer an accurately weighed portion of the powder, equivalent to about 10 mg of thiothixene, to a 500-mL volumetric flask. Add about 400 mL of methanol, shake by mechanical means for 10 minutes, place in an ultrasonic bath for 5 minutes, dilute with methanol to volume, and filter the suspension through a 5-µm polytetrafluoroethylene membrane filter.

Procedure—Proceed as directed for *Procedure* in the *Assay* under *Thiothixene*. Calculate the quantity, in mg, of $C_{23}H_{29}N_3O_2S_2$ in the portion of Capsules taken by the formula:

$$500C(r_U/r_S),$$

in which *C* is the concentration, in mg per mL, of USP Thiothixene RS in the *Standard preparation*, and r_U and r_S are the peak responses obtained from the *Assay preparation* and the *Standard preparation*, respectively.

Thiothixene Hydrochloride

$C_{23}H_{29}N_3O_2S_2 . 2HCl . 2H_2O$ 552.57

9*H*-Thioxanthene-2-sulfonamide, *N,N*-dimethyl-9-[3-(4-methyl-1-piperazinyl)propylidene]-, dihydrochloride, dihydrate, (*Z*)-.

N,N-Dimethyl-9-[3-(4-methyl-1-piperazinyl)propylidene]thioxanthene-2-sulfonamide dihydrochloride dihydrate [*22189-31-7; 49746-09-0*].

Anhydrous 516.54 [*58513-59-0; 49746-04-5*].

» Thiothixene Hydrochloride contains two molecules of water of hydration or is anhydrous. It contains not less than 97.0 percent and not more than 102.5 percent of $C_{23}H_{29}N_3O_2S_2 . 2HCl$, calculated on the anhydrous basis.

Packaging and storage—Preserve in tight, light-resistant containers.

Reference standards—*USP Thiothixene Reference Standard*—Dry in vacuum at 100° for 3 hours before using. *USP (E)-Thiothixene Reference Standard*—Keep container tightly closed and protected from light. Dry in vacuum at 100° for 3 hours before using.

Identification—

A: It meets the requirements under *Identification—Organic Nitrogenous Bases* ⟨181⟩, USP Thiothixene RS being used as the standard for comparison, chloroform being used instead of carbon disulfide, and a reagent blank being used in the matched reference cell.

B: The ultraviolet absorption spectrum of a 1 in 100,000 solution in dilute methanolic hydrochloric acid (1 in 1200) exhibits maxima and minima at the same wavelengths as that of a similar solution of USP Thiothixene RS, concomitantly measured, and the respective molar absorptivities, calculated on the anhydrous basis, at the wavelength of maximum absorbance at about 307 nm do not differ by more than 3.0%. [NOTE—The molecular weight of thiothixene ($C_{23}H_{29}N_3O_2S_2$) is 443.62.]

C: A solution (1 in 100) responds to the tests for *Chloride* ⟨191⟩.

Water, *Method I* ⟨921⟩: between 6.2% and 7.5% for the dihydrate; not more than 1.0% for the anhydrous form.

Residue on ignition ⟨281⟩: not more than 0.2%.

Heavy metals, *Method II* ⟨231⟩: 0.0025%.

Selenium ⟨291⟩: 0.003%.

Limit of (*E*)-thiothixene—[NOTE—Prepare all solutions in low-actinic glassware.]

Mobile phase—Prepare as directed in the test for *Limit of (E)-thiothixene* under *Thiothixene*.

Standard preparations and *Test preparation*—Prepare as directed in the test for *Limit of (E)-thiothixene* under *Thiothixene*, except to use 250 mg of Thiothixene Hydrochloride instead of 200 mg of Thiothixene.

Procedure—Proceed as directed for *Procedure* in the test for *Limit of (E)-thiothixene* under *Thiothixene*. Calculate the quantity, in mg, of (*E*)-thiothixene hydrochloride in the portion of Thiothixene Hydrochloride taken by the formula:

$$5CH_U(516.54/443.62)/(H_C - H_U),$$

in which 516.54 and 443.62 are the molecular weights of anhydrous thiothixene hydrochloride and thiothixene, respectively, and the other terms are as previously defined. The limit of (*E*)-thiothixene, as the hydrochloride, is 1.0%.

Assay—[NOTE—Perform the dilution operations in low-actinic glassware.]

Mobile phase, *Standard preparation*, and *Chromatographic system*—Prepare as directed in the *Assay* under *Thiothixene*.

Assay preparation—Transfer about 116 mg of Thiothixene Hydrochloride, accurately weighed, to a 100-mL volumetric flask, dissolve in methanol, dilute with methanol to volume, and mix. Pipet 2 mL of this solution into a 100-mL volumetric flask, dilute with methanol to volume, and mix.

Procedure—Proceed as directed for *Procedure* in the *Assay* under *Thiothixene*. Calculate the quantity, in mg, of $C_{23}H_{29}N_3O_2S_2 . 2HCl$ in the portion of Thiothixene Hydrochloride taken by the formula:

$$(516.54/443.62)(5000C)(r_U/r_S),$$

in which 516.54 and 443.62 are the molecular weights of anhydrous thiothixene hydrochloride and thiothixene, respectively, *C* is the concentration, in mg per mL, of USP Thiothixene RS in the *Standard preparation*, and r_U and r_S are the peak responses obtained from the *Assay preparation* and the *Standard preparation*, respectively.

Thiothixene Hydrochloride Injection

» Thiothixene Hydrochloride Injection is a sterile solution of Thiothixene Hydrochloride in Water for Injection. It contains an amount of thiothixene hydrochloride equivalent to not less than 90.0 percent and not more than 110.0 percent of the labeled amount of thiothixene ($C_{23}H_{29}N_3O_2S_2$).

Packaging and storage—Preserve in single-dose containers, preferably of Type I glass, protected from light.

Reference standard—*USP Thiothixene Reference Standard*—Dry in vacuum at 100° for 3 hours before using.

Identification—Transfer a volume of Injection, equivalent to about 20 mg of thiothixene hydrochloride, to a separator containing 20.0 mL of chloroform. Render the aqueous layer just basic with ammonium hydroxide, shake for 1 minute, and allow the layers to separate. Pass a portion of the chloroform layer through filter paper, previously washed with chloroform, and use the clear filtrate for the test. Proceed as directed in the *Identification* test under *Thiothixene Capsules*, beginning with "On a suitable thin-layer chromatographic plate."

pH ⟨791⟩: between 2.5 and 3.5.

Other requirements—It meets the requirements under *Injections* ⟨1⟩.

Assay—[NOTE—Perform the dilution operations in low-actinic glassware.]

Mobile phase, *Standard preparation*, and *Chromatographic system*—Prepare as directed in the *Assay* under *Thiothixene*.

Assay preparation—Transfer an accurately measured volume of Thiothixene Hydrochloride Injection, equivalent to about 2 mg of thiothixene, to a 100-mL volumetric flask, dilute with methanol to volume, and mix.

Procedure—Proceed as directed for *Procedure* in the *Assay* under *Thiothixene*. Calculate the quantity, in mg, of $C_{23}H_{29}N_3O_2S_2$ in each mL of the Injection taken by the formula:

$$(100C/V)(r_U/r_S),$$

in which C is the concentration, in mg per mL, of USP Thiothixene RS in the *Standard preparation*, V is the volume, in mL, of Injection taken, and r_U and r_S are the peak responses obtained from the *Assay preparation* and the *Standard preparation*, respectively.

Thiothixene Hydrochloride for Injection

» Thiothixene Hydrochloride for Injection is a sterile, dry mixture of Thiothixene Hydrochloride and Mannitol. It contains not less than 90.0 percent and not more than 110.0 percent of the labeled amount of $C_{23}H_{29}N_3O_2S_2$.

Packaging and storage—Preserve in light-resistant *Containers for Sterile Solids* as described under *Injections* ⟨1⟩.

Reference standard—*USP Thiothixene Reference Standard*—Dry in vacuum at 100° for 3 hours before using.

Identification—Constitute a vial of Thiothixene Hydrochloride for Injection with 2.2 mL of water. Further dilute one volume of the constituted solution with 4 volumes of methanol to obtain a test solution containing 1 mg of thiothixene per mL. On a suitable thin-layer chromatographic plate (see *Chromatography* ⟨621⟩), coated with a 0.25-mm layer of chromatographic silica gel mixture, apply 10 µL of this test solution and 10 µL of a Standard solution containing 1 mg of USP Thiothixene RS per mL in the same medium. Allow the spots to dry, and develop the chromatogram in a solvent system consisting of a mixture of isopropyl alcohol and diethylamine (25:1) until the solvent front has moved about three-fourths of the length of the plate. Remove the plate from the developing chamber, mark the solvent front, air-dry for 30 minutes, and then oven-dry at 110° for 30 minutes. Locate the spots on the plate by viewing under short-wavelength ultraviolet light. Spray the plate with alkaline permanganate spray reagent (prepared by dissolving 0.5 g of potassium permanganate and 2.0 g of sodium carbonate in 100 mL of water): the R_f value of the principal spot obtained from the test solution, under both detection conditions, corresponds to that obtained from the Standard solution.

pH ⟨791⟩: between 2.3 and 3.7, in the solution constituted as directed in the labeling.

Water, *Method I* ⟨921⟩: not more than 4.0%.

Other requirements—It meets the requirements under *Injections* ⟨1⟩.

Assay—[NOTE—Perform the dilution operations in low-actinic glassware.]

Mobile phase, *Standard preparation*, and *Chromatographic system*—Prepare as directed in the *Assay* under *Thiothixene*.

Assay preparation—Constitute a vial of Thiothixene Hydrochloride for Injection with an accurately measured volume (V_L mL) of water as directed in the labeling. Transfer an accurately measured volume (V_F mL) of the constituted solution, equivalent to about 5 mg of thiothixene, to a 250-mL volumetric flask, dilute with methanol to volume, and mix.

Procedure—Proceed as directed for *Procedure* in the *Assay* under *Thiothixene*. Calculate the quantity, in mg, of $C_{23}H_{29}N_3O_2S_2$ in each mL of the constituted solution taken by the formula:

$$(250CV_L/V_F)(r_U/r_S),$$

in which C is the concentration, in mg per mL, of USP Thiothixene RS in the *Standard preparation*, and r_U and r_S are the peak responses obtained from the *Assay preparation* and the *Standard preparation*, respectively.

Thiothixene Hydrochloride Oral Solution

» Thiothixene Hydrochloride Oral Solution contains an amount of thiothixene hydrochloride equivalent to not less than 90.0 percent and not more than 110.0 percent of the labeled amount of thiothixene ($C_{23}H_{29}N_3O_2S_2$).

Packaging and storage—Preserve in tight, light-resistant containers.

Reference standard—*USP Thiothixene Reference Standard*—Dry in vacuum at 100° for 3 hours before using.

Identification—Transfer a portion of Oral Solution, equivalent to about 20 mg of thiothixene hydrochloride, to a separator containing 20.0 mL of chloroform. Render the aqueous layer just basic with ammonium hydroxide, shake for 1 minute, and allow the layers to separate. Pass a portion of the chloroform layer through filter paper, previously washed with chloroform, and use the clear filtrate for the test. Proceed as directed in the *Identification* test under *Thiothixene Capsules*, beginning with "On a suitable thin-layer chromatographic plate."

pH ⟨791⟩: between 2.0 and 3.0.

Alcohol content, *Method I* ⟨611⟩: between 6.3% and 7.7% of C_2H_5OH.

Assay—[NOTE—Perform the dilution operations in low-actinic glassware.]

Mobile phase, *Standard preparation*, and *Chromatographic system*—Prepare as directed in the *Assay* under *Thiothixene*.

Assay preparation—Transfer an accurately measured volume of Thiothixene Hydrochloride Oral Solution, equivalent to about 25 mg of thiothixene, to a 25-mL volumetric flask, dissolve in methanol, dilute with methanol to volume, and mix. Pipet 2 mL of this solution into a 100-mL volumetric flask, dilute with methanol to volume, and mix.

Procedure—Proceed as directed for *Procedure* in the *Assay* under *Thiothixene*. Calculate the quantity, in mg, of $C_{23}H_{29}N_3O_2S_2$ in each mL of the Oral Solution taken by the formula:

$$(1250C/V)(r_U/r_S),$$

in which C is the concentration, in mg per mL, of USP Thiothixene RS in the *Standard preparation*, V is the volume, in mL, of Oral Solution taken, and r_U and r_S are the peak responses obtained from the *Assay preparation* and the *Standard preparation*, respectively.

Thonzonium Bromide

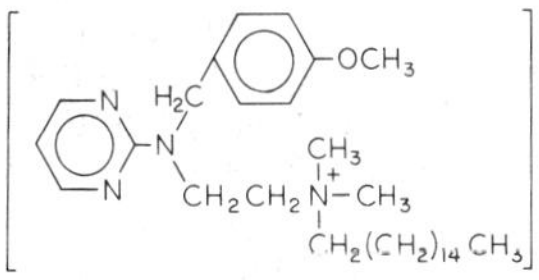

$C_{32}H_{55}BrN_4O$ 591.72
1-Hexadecanaminium, *N*-[2-[[(4-methoxyphenyl)methyl]-2-pyrimidinylamino]ethyl]-*N*,*N*-dimethyl-, bromide.
Hexadecyl[2-[(*p*-methoxybenzyl)-2-pyrimidinylamino]ethyl]-dimethylammonium bromide [*553-08-2*].

» Thonzonium Bromide contains not less than 97.0 percent and not more than 103.0 percent of $C_{32}H_{55}BrN_4O$, calculated on the anhydrous basis.

Packaging and storage—Preserve in tight containers.

Reference standard—*USP Thonzonium Bromide Reference Standard*—Do not dry before using.

Identification—

A: The infrared absorption spectrum of a potassium bromide dispersion of it exhibits maxima only at the same wavelengths as that of a similar preparation of USP Thonzonium Bromide RS.

B: Transfer 20 to 30 mg to the well of a porcelain spot plate, wet the powder with 5 drops of freshly prepared chlorine TS, and immediately add 5 drops of sodium fluorescein solution (3 in 1000): a pinkish orange or red color is developed at once. View the spot plate under long-wavelength ultraviolet light: the spot exhibits a bright orange fluorescence.

Melting range ⟨741⟩: between 93° and 97°.

Water, *Method I* ⟨921⟩: not more than 0.5%.

Residue on ignition ⟨281⟩: not more than 0.2%.

Ordinary impurities ⟨466⟩—

Test solution: chloroform.

Standard solution: chloroform.

Eluant: a mixture of 1-butanol, water, and glacial acetic acid (12:5:3).

Visualization: 1.

Assay—Dissolve about 750 mg of Thonzonium Bromide, accurately weighed, in 80 mL of glacial acetic acid, and swirl until the solution is clear. Add mercuric acetate TS until the precipitate first formed dissolves. Add an additional 10 mL of mercuric acetate TS and 2 drops of crystal violet TS, and titrate with 0.1 *N* perchloric acid VS to an emerald-green end-point. Each mL of 0.1 *N* perchloric acid is equivalent to 29.59 mg of $C_{32}H_{55}BrN_4O$.

Threonine

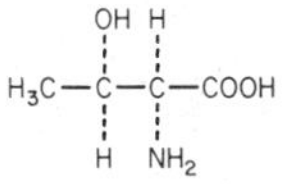

$C_4H_9NO_3$ 119.12
L-Threonine.
L-Threonine [*72-19-5*].

» Threonine contains not less than 98.5 percent and not more than 101.5 percent of $C_4H_9NO_3$, as L-threonine, calculated on the dried basis.

Packaging and storage—Preserve in well-closed containers.

Reference standard—*USP L-Threonine Reference Standard*—Dry at 105° for 3 hours before using.

Identification—The infrared absorption spectrum of a potassium bromide dispersion of it, previously dried, exhibits maxima only at the same wavelengths as that of a similar preparation of USP L-Threonine RS.

Specific rotation ⟨781⟩: between −26.7° and −29.1°, calculated on the dried basis, determined in a solution containing 600 mg in each 10 mL.

pH ⟨791⟩: between 5.0 and 6.5 in a solution (1 in 20).

Loss on drying ⟨731⟩—Dry it at 105° for 3 hours: it loses not more than 0.2% of its weight.

Residue on ignition ⟨281⟩: not more than 0.4%.

Chloride ⟨221⟩—A 0.73-g portion shows no more chloride than corresponds to 0.50 mL of 0.020 *N* hydrochloric acid (0.05%).

Sulfate ⟨221⟩—A 0.33-g portion shows no more sulfate than corresponds to 0.10 mL of 0.020 *N* sulfuric acid (0.03%).

Arsenic ⟨211⟩: 1.5 ppm.

Iron ⟨241⟩: 0.003%.

Heavy metals, *Method I* ⟨231⟩: 0.0015%.

Assay—Transfer about 110 mg of Threonine, accurately weighed, to a 125-mL flask, dissolve in a mixture of 3 mL of formic acid and 50 mL of glacial acetic acid, and titrate with 0.1 *N* perchloric acid VS, determining the end-point potentiometrically. Perform a blank determination, and make any necessary correction. Each mL of 0.1 *N* perchloric acid is equivalent to 11.91 mg of $C_4H_9NO_3$.

Thrombin

» Thrombin conforms to the regulations of the federal Food and Drug Administration concerning biologics (see *Biologics* ⟨1041⟩). It is a sterile, freeze-dried powder derived from bovine plasma containing the protein substance prepared from prothrombin through interaction with added thromboplastin in the presence of calcium. It is capable, without the addition of other substances, of causing the clotting of whole blood, plasma, or a solution of fibrinogen. Its potency is determined in U.S. Units in terms of the U.S. Standard Thrombin in a test comparing clotting times of fibrinogen solution.

Packaging and storage—Preserve at a temperature between 2° and 8°. Dispense it in the unopened container in which it was placed by the manufacturer.

Expiration date—The expiration date is not more than 3 years after date of manufacture.

Labeling—Label it to indicate that solutions of Thrombin are to be used within a few hours after preparation, and are not to be injected into or otherwise allowed to enter large blood vessels.

Thymol—*see* Thymol NF

Thyroglobulin

» Thyroglobulin is an extract obtained by the fractionation of thyroid glands from the hog, *Sus scrofa* Linné var. *domesticus* Gray (Fam. Suidae). On hydrolysis it yields not less than 90.0 percent and not more than 110.0 percent of the labeled amounts of levothyroxine ($C_{15}H_{11}I_4NO_4$) and liothyronine ($C_{15}H_{12}I_3NO_4$). It is free from iodine in inorganic or any form of combination other than that peculiar to thyroglobulin. It may contain a suitable diluent such as Lactose, Sodium Chloride, Starch, Sucrose, or Dextrose.

Packaging and storage—Preserve in tight containers.

Reference standards—*USP Liothyronine Reference Standard*—Keep container tightly closed and protected from light. Use without drying; correct for moisture, determined by drying a separate portion in vacuum at 60° for 3 hours. *USP Levothyroxine Reference Standard*—Keep container tightly closed and protected from light. Use without drying; correct for moisture, determined by drying a separate portion in vacuum at 60° for 4 hours.

Identification—Mix about 500 mg with 5 mL of diluted alcohol, add 2 drops of hydrochloric acid and 2 drops of sodium nitrite solution (1 in 10), and mix. Boil gently on a steam bath for 2 minutes, cool, and render alkaline with 1 mL of 6 *N* ammonium hydroxide. Shake for 30 seconds: the precipitate develops a uniform salmon color.

Microbial limits—It meets the requirements of the test for absence of *Salmonella* species and *Escherichia coli* under *Microbial Limit Tests* ⟨61⟩.

Loss on drying ⟨731⟩—Dry it at 105° for 3 hours: it loses not more than 5.0% of its weight.

Residue on ignition ⟨281⟩: not more than 4.0%.

Inorganic iodides—To 1 g in a dry test tube add 10 mL of a saturated solution of zinc sulfate. Shake for about 5 minutes, and filter through a sintered-glass filter. To 5 mL of the filtrate add 0.5 mL of starch TS, 4 drops of sodium nitrite solution (1 in 10), and 4 drops of 2 *N* sulfuric acid, shaking after each addition: no blue color is produced.

Assay—

Mobile phase—Prepare a degassed and filtered mixture of water and acetonitrile (7:3) that contains 5 mL of phosphoric acid in each 1000 mL of solution. Make adjustments if necessary (see *System Suitability* under *Chromatography* ⟨621⟩).

Reducing buffer solution—Freshly prepare a solution in 0.11 *M* sodium chloride that is 0.04 *M* with respect to tris(hydroxymethyl)aminomethane and 0.05 *M* with respect to methimazole. Adjust, if necessary, with 6 *N* hydrochloric acid or 0.1 *N* sodium hydroxide to a pH of 8.4 ± 0.05.

Proteolytic enzyme—Freshly prepare a solution containing 15 mg of bacterial protease* in each 5 mL of *Reducing buffer solution*.

Enzyme deactivating solution—Prepare a 1 in 100 mixture of phosphoric acid in acetonitrile.

Standard preparation—[NOTE—Protect solutions from light.] Transfer accurately weighed quantities of about 9 mg of USP Liothyronine RS and about 27 mg of USP Levothyroxine RS to a 100-mL volumetric flask, add 50 mL of a mixture of water, acetonitrile, and ammonium hydroxide (500:500:1) and swirl to dissolve. Dilute with a mixture of water and acetonitrile (1:1) to volume, and mix (stock solution). On the day of use, pipet 5 mL of the freshly prepared stock solution into a 250-mL volumetric flask, dilute with *Reducing buffer solution* to volume, and mix to obtain a solution having known concentrations of about 1.8 µg of liothyronine per mL and about 5.4 µg of levothyroxine per mL. Pipet 5 mL of this solution into a screw-capped 16-mm × 125-mm culture tube. Pipet 2 mL of *Enzyme deactivating solution* into the tube, place the cap on the tube, and shake the mixture vigorously.

Assay preparation—Transfer an accurately weighed portion of finely powdered Thyroglobulin, equivalent to about 27 µg of levothyroxine, to a screw-capped 16-mm × 125-mm culture tube that previously has been flushed with nitrogen. Taking precautions to avoid unnecessary exposure to air, pipet 5 mL of *Proteolytic enzyme* into the tube. Allow nitrogen to flow gently over the mixture for 5 minutes. Place the cap on the tube, mix to disperse the contents, and place in a covered water bath maintained at a temperature of 37 ± 1° for 28 hours. Protect the contents of the tubes from light. Examine occasionally, and mix as necessary to ensure dispersion. At the end of the incubation period, pipet 2 mL of *Enzyme deactivating solution* into the tube, place the cap on the tube, mix vigorously, and centrifuge at about 2000 rpm for 5 minutes. Filter the supernatant liquid through a 0.45-µm porosity filter, discarding the first 1 mL of the filtrate.

Chromatographic system (see *Chromatography* ⟨621⟩)—The liquid chromatograph is equipped with a 230-nm detector and a 25- to 30-cm column that contains packing L10. The flow rate is about 1 mL per minute. Chromatograph the *Standard preparation*, and record the peak responses as directed under *Procedure:* the tailing factors for the liothyronine and levothyroxine peaks are not more than 1.8, and the relative standard deviation for replicate injections is not more than 2.0%.

Procedure—Separately inject equal volumes (about 50 µL) of the *Assay preparation*, and the *Standard preparation*, record the chromatograms, and measure the responses for the major peaks. Calculate the quantity, in µg, of liothyronine ($C_{15}H_{12}I_3NO_4$) and levothyroxine ($C_{15}H_{11}I_4NO_4$) in the portion of Thyroglobulin taken by the formula:

$$7C(r_U/r_S),$$

in which C is the concentration, in µg per mL, of the corresponding USP Reference Standard in the *Standard preparation*, and r_U and r_S are the peak responses for the corresponding analytes obtained from the *Assay preparation* and the *Standard preparation*, respectively.

* A suitable grade is available as "Pronase" (Catalog number 53702) from Calbiochem-Behring, P. O. Box 12087, San Diego, CA 92112.

Thyroglobulin Tablets

» Thyroglobulin Tablets contain not less than 85.0 percent and not more than 115.0 percent of the labeled amount of levothyroxine and not less than 90.0 percent and not more than 110.0 percent of the labeled amount of liothyronine, the labeled amounts being 36 µg of levothyroxine and 12 µg of liothyronine for each 65 mg (1 grain) of the labeled content of thyroglobulin.

Packaging and storage—Preserve in tight containers.

Reference standards—*USP Liothyronine Reference Standard*—Keep container tightly closed and protected from light. Use without drying; correct for moisture, determined by drying a separate portion in vacuum at 60° for 3 hours. *USP Levothyroxine Reference Standard*—Keep container tightly closed and protected from light. Use without drying; correct for moisture, determined by drying a separate portion in vacuum at 60° for 4 hours.

Identification—Mix about 1 g of powdered Tablets with 10 mL of diluted alcohol in a 50-mL centrifuge tube. Add 4 drops of hydrochloric acid and 4 drops of sodium nitrite solution (1 in 10), mixing after each addition. Boil gently on a steam bath for 2 minutes, cool in an ice bath, and centrifuge at 2000 rpm for 2 minutes. Discard the supernatant liquid, and render the precipitate alkaline with 1 mL of 6 *N* ammonium hydroxide. Shake for 30 seconds: the precipitate develops a uniform salmon color.

Microbial limits—Tablets meet the requirements of the tests for absence of *Salmonella* species and *Escherichia coli* under *Microbial Limit Tests* ⟨61⟩.

Disintegration ⟨701⟩: 15 minutes.

Uniformity of dosage units ⟨905⟩: meet the requirements.

Procedure for content uniformity—

Standard preparation—Transfer about 80 mg of potassium iodide, accurately weighed, to a 100-mL volumetric flask, add water to volume, and mix. Dilute quantitatively and stepwise with 0.06 *N* sodium hydroxide to provide a solution containing about 1.6 µg of potassium iodide per mL.

Test preparation—Follow the procedure described under *Oxygen Flask Combustion* ⟨471⟩, using a single Tablet as the test specimen and 15 mL of sodium hydroxide solution (1 in 250) as the absorbing liquid. Transfer the contents of the flask to a 25-mL volumetric flask with the aid of several small volumes of water, and dilute with water to volume. Dilute quantitatively and stepwise, if necessary, with 0.06 *N* sodium hydroxide, to provide a solution containing about 1.2 µg of iodine per mL.

Procedure—Transfer 25.0 mL each of the *Standard preparation*, the *Test preparation*, and 0.06 *N* sodium hydroxide to separate 50-mL volumetric flasks. Acidify each solution with 1 mL of 5 *M* phosphoric acid, mix, and add 1.0 mL of bromine TS. Mix, allow the solutions to stand for 5 minutes, add to each 1.0 mL of a solution of salicylic acid in alcohol (1 in 100), and mix. Filter, if necessary, add 1.0 mL of potassium iodide solution (6 in 10,000) to each flask, and mix. Allow the solutions to stand in the dark for 30 minutes, dilute with pH 7.0 phosphate buffer (see *Buffer Solutions* under *Reagents*, *Indicators*, and *Solutions*) to volume, and mix. Transfer 1.0 mL of each solution to separate 10-mL volumetric flasks, add to each 1.0 mL of a solution of sodium fluorescein in pH 7.0 phosphate buffer (1 in 200,000), dilute with pH 7.0 phosphate buffer to volume, and mix. Concomitantly determine the fluorescence of the solutions in 1-cm fluorometer cells with a suitable fluorometer, setting the excitation wavelength at the wavelength of maximum fluorescence at about 330 nm and the fluorescence monochromator at the wavelength of maximum fluorescence at about 505 nm. Calculate the amount of iodine, in µg per Tablet, by the formula:

$$(126.90/166.00)(TC/D)(F_U - F_B)/(F_S - F_B),$$

in which 126.90 is the atomic weight of iodine, 166.00 is the molecular weight of potassium iodide, T is the labeled quantity, in µg, of organically bound iodine in the Tablet, C is the concentration, in µg per mL, of the potassium iodide standard solution, D is the concentration, in µg per mL, of iodine in the test solution, based upon the labeled quantity per Tablet and the extent of dilution, F_U is the fluorescence of the test solution, F_S is the fluorescence of the standard solution, and F_B is the fluorescence of the blank solution. Proceed as directed under *Content Uniformity*, using the results obtained by this procedure for determining the total iodine content of individual Tablets, but omit the composite *Assay* determination and application of the correction factor (F).

Assay—Proceed with Thyroglobulin Tablets as directed in the *Assay* under *Thyroglobulin*, except to use an accurately weighed portion of powdered Tablets, equivalent to about 27 µg of levothyroxine, for the *Assay preparation* and equivalent quantities of the USP Reference Standards for the *Standard preparation.*

Thyroid

» Thyroid is the cleaned, dried, and powdered thyroid gland previously deprived of connective tissue and fat. It is obtained from domesticated animals that are used for food by man.

On hydrolysis it yields not less than 90.0 percent and not more than 110.0 percent each of the labeled amounts of levothyroxine ($C_{15}H_{11}I_4NO_4$) and liothyronine ($C_{15}H_{12}I_3NO_4$), calculated on the dried basis. It is free from iodine in inorganic or any form of combination other than that peculiar to the thyroid gland. It may contain a suitable diluent such as Lactose, Sodium Chloride, Starch, Sucrose or Dextrose.

Packaging and storage—Preserve in tight containers.

Reference standards—*USP Liothyronine Reference Standard*—Keep container tightly closed and protected from light. Use without drying; correct for moisture, determined by drying a separate portion in vacuum at 60° for 3 hours. *USP Levothyroxine Reference Standard*—Keep container tightly closed and protected from light. Use without drying; correct for moisture, determined by drying a separate portion in vacuum at 60° for 4 hours.

Identification—When suitably mounted and examined under a microscope, it shows numerous smooth to striated hyaline fragments of colloid of angular to irregular shape, which are colorless to pale yellow in water mounts, brown in Mallory's stain, and pink in eosin solution, some of these fragments containing granules, minute vacuoles, crystalloidal bodies, and cells; numerous irregular fragments of follicular epithelium staining brown with Mallory's stain, the individual cells more or less polygonal to rounded-angular or irregularly cuboidal, often with prominent nuclei staining dark blue, their cytoplasm purplish with Delafield's hematoxylin TS; slender, glistening segments of capillaries of closely undulate outline; numerous slender segments of neuraxons; numerous aggregates of particles of intercellular substance and slender, mostly straight, connective tissue fibers staining blue to greenish blue with a mixture of Mallory's stain and phosphotungstic acid TS, the bundles of fibers often appearing reddish in Mallory's stain; few glistening fragments of blood vessels with serrated or crenated ends as viewed in water mounts.

Microbial limits—It meets the requirements of the tests for absence of *Salmonella* species and *Escherichia coli* under *Microbial Limit Tests* ⟨61⟩.

Loss on drying ⟨731⟩—Dry it in vacuum at 60° for 4 hours: it loses not more than 6.0% of its weight.

Inorganic iodides—Transfer 1 g, accurately weighed, or proportionately less, if the iodine content is greater than 0.20 percent, to a dry test tube. To a second tube add 0.5 mL of a solution prepared by dissolving 105 mg of potassium iodide in water to make 1000.0 mL. To each tube add 10 mL of a saturated solution of zinc sulfate. Shake for about 5 minutes, and filter through a sintered-glass filter. To 5 mL of the filtrates add 0.5 mL of starch TS, 4 drops of sodium nitrite solution (1 in 10), and 4 drops of 2 *N* sulfuric acid, shaking after each addition: any blue color produced in the solution prepared from Thyroid does not exceed that produced in the solution prepared from the potassium iodide. (If the treated potassium iodide solution does not produce a distinct blue color, repeat the test using fresh starch TS.) The limit is 0.004%.

Assay—

Mobile phase—Prepare a degassed and filtered mixture of water, acetonitrile, and phosphoric acid (650:350:5). Make adjustments if necessary (see *System Suitability* under *Chromatography* ⟨621⟩).

Reducing buffer solution—Freshly prepare a solution in 0.11 *M* sodium chloride that is 0.04 *M* with respect to tris(hydroxymethyl)aminomethane and 0.05 *M* with respect to methimazole. Adjust, if necessary, with 6 *N* hydrochloric acid or 0.1 *N* sodium hydroxide to a pH of 8.4 ± 0.05.

Proteolytic enzyme—Freshly prepare a solution containing 15 mg of bacterial protease* in each 5 mL of *Reducing buffer solution.*

Enzyme deactivating solution—Prepare a 1 in 100 mixture of phosphoric acid in acetonitrile.

Standard preparation—[NOTE—Protect solutions from light.] Transfer accurately weighed quantities of about 9 mg of USP Liothyronine RS and about 38 mg of USP Levothyroxine RS to a 100-mL volumetric flask, add 50 mL of a mixture of water, acetonitrile, and ammonium hydroxide (500:500:1), and swirl to dissolve. Dilute with a mixture of water and acetonitrile (1:1) to volume, and mix (stock solution). On the day of use, pipet 5 mL of the freshly prepared stock solution into a 250-mL volumetric flask, dilute with *Reducing buffer solution* to volume, and mix to obtain a solution having known concentrations of about 1.8 µg of liothyronine per mL and about 7.6 µg of levothyroxine per mL. Pipet 5 mL of this solution into a screw-capped 16-mm × 125-mm culture tube. Pipet 2 mL of *Enzyme deactivating solution* into the tube, place the cap on the tube, and shake the mixture vigorously.

Assay preparation—Transfer an accurately weighed portion of finely powdered Thyroid, equivalent to about 38 µg of levothyroxine, to a screw-capped 16-mm × 125-mm culture tube that previously has been flushed with nitrogen. Taking precautions to avoid unnecessary exposure to air, pipet 5 mL of *Proteolytic*

* A suitable grade is available as "Pronase" (Catalog number 53702) from Calbiochem-Behring, P. O. Box 12087, San Diego, CA 92112.

enzyme into the tube. Allow nitrogen to flow gently over the mixture for 5 minutes. Place the cap on the tube, mix to disperse the contents, and place in a covered water bath maintained at a temperature of 37 ± 1° for 28 hours. Protect the contents of the tubes from light. Examine occasionally, and mix as necessary to ensure dispersion. At the end of the incubation period, pipet 2 mL of *Enzyme deactivating solution* into the tube, place the cap on the tube, mix vigorously, and centrifuge at about 2000 rpm for 5 minutes. Filter the supernatant liquid through a 0.45-µm porosity filter, discarding the first 1 mL of the filtrate.

Chromatographic system (see *Chromatography* ⟨621⟩)—The liquid chromatograph is equipped with a 230-nm detector and a 4.6-mm × 25-cm column that contains packing L1. The flow rate is about 1.5 mL per minute. Chromatograph the *Standard preparation*, and record the peak responses as directed under *Procedure:* the tailing factors for the liothyronine and levothyroxine peaks are not more than 1.8, and the relative standard deviation for replicate injections is not more than 2.0%.

Procedure—Separately inject equal volumes (about 200 µL) of the *Assay preparation*, and the *Standard preparation*, record the chromatograms, and measure the responses for the major peaks. Calculate the quantity, in µg, of liothyronine ($C_{15}H_{12}I_3NO_4$) and levothyroxine ($C_{15}H_{11}I_4NO_4$) in the portion of Thyroid taken by the formula:

$$7C(r_U / r_S),$$

in which C is the concentration, in µg per mL, of the corresponding USP Reference Standard in the *Standard preparation*, and r_U and r_S are the peak responses for the corresponding analytes obtained from the *Assay preparation* and the *Standard preparation*, respectively.

Thyroid Tablets

» Thyroid Tablets contain not less than 85.0 percent and not more than 115.0 percent of the labeled amount of levothyroxine and not less than 90.0 percent and not more than 110.0 percent of the labeled amount of liothyronine, the labeled amounts being 38 µg of levothyroxine and 9 µg of liothyronine for each 65 mg (1 grain) of the labeled content of thyroid.

Packaging and storage—Preserve in tight containers.

Reference standards—*USP Liothyronine Reference Standard*—Keep container tightly closed and protected from light. Use without drying; correct for moisture, determined by drying a separate portion in vacuum at 60° for 3 hours. *USP Levothyroxine Reference Standard*—Keep container tightly closed and protected from light. Use without drying; correct for moisture, determined by drying a separate portion in vacuum at 60° for 4 hours.

Microbial limit—Tablets meet the requirements of the tests for absence of *Salmonella* species and *Escherichia coli* under *Microbial Limit Tests* ⟨61⟩.

Disintegration ⟨701⟩: 15 minutes.

Uniformity of dosage units ⟨905⟩: meet the requirements.

Procedure for content uniformity—

pH 3.9 buffer—Dissolve 15.0 g of sodium acetate in 100 mL of 6 *N* acetic acid.

Standard preparation—Dilute 2.0 mL of 0.05 *M* potassium iodate with water to 200.0 mL. Dilute 2.0 mL of this solution with *pH 3.9 buffer* to 100.0 mL to obtain a solution containing 1.269 µg of iodine per mL. To 10.0 mL of the resulting solution add 2.0 mL of potassium iodide TS, and mix.

Test preparation—*For Tablets labeled to contain less than 130 mg of thyroid*, wrap 1 Tablet in a halide-free filter paper fuse-strip, place the charge in a platinum holder, and proceed as directed under *Oxygen Flask Combustion* ⟨471⟩, using a 1000-mL flask containing 15 mL of 10 *N* acetic acid. *For Tablets labeled to contain 130 mg or more of thyroid*, prepare the combustion charge using either 1 Tablet or an accurately weighed fraction of a Tablet, equivalent to a multiple of 32.5 mg of thyroid, and proceed as directed under *Oxygen Flask Combustion* ⟨471⟩, using a 1000-mL flask containing 15 mL of 10 *N* acetic acid.

Flush the flask thoroughly with oxygen, remove the oxygen inlet, quickly inject 1 mL to 2 mL of bromine–sodium acetate TS into the flask with a suitable dispensing device, and ignite the specimen under test. When combustion is completed, swirl the contents for about 5 minutes, allowing the solution to wet all internal surfaces repeatedly. Chill the flask in cold water, place it under a hood, and carefully release the stopper. Rinse the stopper and the walls of the flask with 10 mL of water, add a few silicon carbide boiling chips, and gently boil the mixture until the vapors are negative to a moistened potassium iodide–starch test paper, evaporating to a volume of not less than 10 mL. *For Tablets labeled to contain not more than 16.2 mg of thyroid*, transfer the mixture in the combustion flask to a 125-mL conical flask with the aid of two 5-mL portions of 6 *N* acetic acid and three 5-mL portions of water. Add a few silicon carbide boiling chips, and boil the mixture gently, evaporating to a volume of about 10 mL. Transfer the cooled mixture to a 25-mL volumetric flask with the aid of three 4-mL portions of *pH 3.9 buffer*, dilute with *pH 3.9 buffer* to volume, and mix. Filter a portion of the mixture, add 10.0 mL of the clear filtrate to 2.0 mL of potassium iodide TS, and mix. *For Tablets labeled to contain more than 16.2 mg of thyroid*, cool the mixture in the combustion flask, use several 10-mL portions of *pH 3.9 buffer* to transfer the mixture to a suitable volumetric flask to obtain a solution containing about 65 mg of thyroid per 100 mL, dilute with *pH 3.9 buffer* to volume, and mix. Filter a portion of the mixture, add 10.0 mL of the clear filtrate to 2.0 mL of potassium iodide TS, and mix.

Blank preparation—Mix 10.0 mL of *pH 3.9 buffer* with 2.0 mL of potassium iodide TS.

Procedure—Concomitantly determine the absorbances of the *Test preparation* and the *Standard preparation* in 1-cm cells at the wavelength of maximum absorbance at about 352 nm, and at the base-line absorbance at about 450 nm, with a suitable spectrophotometer, using the *Blank preparation* as the blank. Calculate the quantity, in µg, of iodine in the combustion charge by the formula:

$$1.269V(A_{352} - A_{450})_U / (A_{352} - A_{450})_S,$$

in which V is the capacity, in mL, of the volumetric flask, and $(A_{352} - A_{450})_U$ and $(A_{352} - A_{450})_S$ are the corrected absorbances of the *Test preparation* and the *Standard preparation*, respectively. Proceed as directed under *Content Uniformity*, using the results obtained by this procedure for determining the total iodine content of individual Tablets, but omit the composite *Assay* determination and application of the correction factor (F).

Inorganic iodides—Place a quantity of finely powdered Tablets, equivalent to 1.00 g of thyroid, in a dry test tube and proceed as directed in the test for *Inorganic iodides* under *Thyroid*, beginning with "To a second tube add 0.5 mL." The limit of inorganic iodide is 0.004%.

Assay—Proceed with Thyroid Tablets as directed in the *Assay* under *Thyroid*, except to use the following *Assay preparation*.

Assay preparation—Weigh and finely powder not less than 20 Thyroid Tablets. Using an accurately weighed portion of the powder, proceed as directed for *Assay preparation* in the *Assay* under *Thyroid*, beginning with "equivalent to about 38 µg of levothyroxine."

Sterile Ticarcillin Disodium

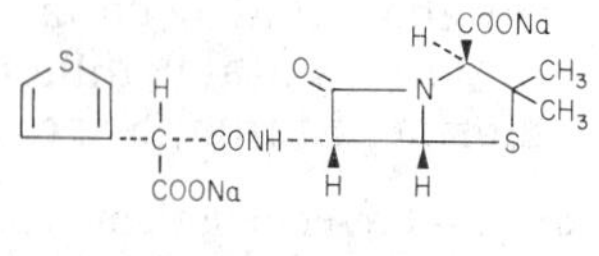

$C_{15}H_{14}N_2Na_2O_6S_2$ 428.38

4-Thia-1-azabicyclo[3.2.0]heptane-2-carboxylic acid, 6-[(carboxy-3-thienylacetyl)amino]-3,3-dimethyl-7-oxo-, disodium salt, [2*S*-[2α,5α,6β(*S**)]]-.

N-(2-Carboxy-3,3-dimethyl-7-oxo-4-thia-1-azabicyclo[3.2.0]hept-6-yl)-3-thiophenemalonamic acid disodium salt [*4697-14-7*].

» Sterile Ticarcillin Disodium is ticarcillin disodium suitable for parenteral use. It has a potency equivalent to not less than 800 µg of ticarcillin ($C_{15}H_{16}N_2O_6S_2$) per mg, calculated on the anhydrous basis. In addition, where packaged for dispensing, it contains the equivalent of not less than 90.0 percent and not more than 115.0 percent of the labeled amount of ticarcillin.

Packaging and storage—Preserve in *Containers for Sterile Solids* as described under *Injections* ⟨1⟩.

Reference standard—*USP Ticarcillin Monosodium Monohydrate Reference Standard*—Do not dry before using.

Constituted solution—At the time of use, the constituted solution prepared from Sterile Ticarcillin Disodium meets the requirements for *Constituted Solutions* under *Injections* ⟨1⟩.

Identification—The ultraviolet absorption spectrum, between 200 nm and 300 nm, of the test solution prepared as directed in the test for *Ticarcillin content* exhibits maxima and minima at the same wavelengths as that of the Standard solution prepared as directed in the test for *Ticarcillin content,* concomitantly measured.

Pyrogen—It meets the requirements of the *Pyrogen Test* ⟨151⟩, the test dose being 1.0 mL per kg of a solution in Sterile Water for Injection containing 100 mg of ticarcillin per mL.

Sterility—It meets the requirements under *Sterility Tests* ⟨71⟩, when tested as directed in the section, *Test Procedures Using Membrane Filtration.*

pH ⟨791⟩: between 6.0 and 8.0, in a solution containing 10 mg of ticarcillin per mL (or in the solution constituted as directed in the labeling).

Water, *Method I* ⟨921⟩: not more than 6.0%.

Particulate matter ⟨788⟩: meets the requirements under *Small-volume Injections.*

Ticarcillin content—Transfer about 40 mg, accurately weighed, to a 100-mL volumetric flask, dissolve in water, dilute with water to volume, and mix. Transfer 5.0 mL of this solution to another 100-mL volumetric flask, dilute with 0.1 *N* methanolic hydrochloric acid (0.8 mL of hydrochloric acid diluted with methanol to 100 mL) to volume, and mix. Concomitantly determine the absorbances of this test solution with a similarly prepared Standard solution of USP Ticarcillin Monosodium Monohydrate RS, at the wavelength of maximum absorbance at about 230 nm, using a reagent blank. Calculate the percentage of ticarcillin ($C_{15}H_{16}N_2O_6S_2$) by the formula:

$$P(W_S/W_U)(A_U/A_S),$$

in which P is the percentage content of ticarcillin in the USP Ticarcillin Monosodium Monohydrate RS, W_S and W_U are the amounts of USP Ticarcillin Monosodium Monohydrate RS and Sterile Ticarcillin Disodium taken, respectively, and A_U and A_S are the absorbances of the test solution and the Standard solution, respectively: between 80.0% and 94.0%, calculated on the anhydrous basis, is found.

Assay—

Assay preparation 1—Dissolve a suitable quantity of Sterile Ticarcillin Disodium, accurately weighed, in *Buffer No. 1,* and dilute quantitatively with *Buffer No. 1* to obtain a solution having a convenient concentration of ticarcillin.

Assay preparation 2 (where it is packaged for dispensing and is represented as being in a single-dose container)—Constitute Sterile Ticarcillin Disodium as directed in the labeling. Withdraw all of the withdrawable contents, and dilute quantitatively with *Buffer No. 1* to obtain a solution having a convenient concentration of ticarcillin.

Assay preparation 3 (where the label states the quantity of ticarcillin in a given volume of constituted solution)—Constitute Sterile Ticarcillin Disodium as directed in the labeling. Dilute an accurately measured volume of the constituted solution quantitatively with *Buffer No. 1* to obtain a solution having a convenient concentration of ticarcillin.

Procedure—Proceed as directed under *Antibiotics—Microbial Assays* ⟨81⟩, using an accurately measured volume of *Assay preparation* diluted quantitatively with *Buffer No. 1* to yield a *Test Dilution* having a concentration assumed to be equal to the median dose level of the Standard.

Sterile Ticarcillin Disodium and Clavulanate Potassium

» Sterile Ticarcillin Disodium and Clavulanate Potassium is a sterile, dry mixture of Sterile Ticarcillin Disodium and Sterile Clavulanate Potassium. It contains the equivalent of not less than 90.0 percent and not more than 115.0 percent of the labeled amount of ticarcillin ($C_{15}H_{16}N_2O_6S_2$) and the equivalent of not less than 85.0 percent and not more than 120.0 percent of the labeled amount of clavulanic acid ($C_8H_9NO_5$), the labeled amounts representing proportions of ticarcillin to clavulanic acid of 15:1 or 30:1. Where the proportion is 15:1, it contains not less than 733 µg of ticarcillin ($C_{15}H_{16}N_2O_6S_2$) per mg, calculated on the anhydrous basis. Where the proportion is 30:1, it contains not less than 755 µg of ticarcillin ($C_{15}H_{16}N_2O_6S_2$) per mg, calculated on the anhydrous basis.

Packaging and storage—Preserve in *Containers for Sterile Solids* as described under *Injections* ⟨1⟩.

Reference standards—*USP Ticarcillin Monosodium Monohydrate Reference Standard*—Do not dry before using. *USP Clavulanate Lithium Reference Standard*—Keep container tightly closed and protected from light. Do not dry before using.

Constituted solution—At the time of use, the constituted solution prepared from Sterile Ticarcillin Disodium and Clavulanate Potassium meets the requirements for *Constituted Solutions* under *Injections* ⟨1⟩.

Identification—The retention times of the major peaks in the chromatogram of the *Assay preparation* correspond to those of the *Standard preparation,* as obtained in the *Assay.*

Pyrogen—It meets the requirements of the *Pyrogen Test* ⟨151⟩, the test dose being 1.0 mL per kg of a solution in Sterile Water for Injection containing 100 mg of it per mL.

Sterility—It meets the requirements under *Sterility Tests* ⟨71⟩, when tested as directed in the section, *Test Procedures Using Membrane Filtration.*

pH ⟨791⟩: between 5.5 and 7.5, in a solution (1 in 10).

Water, *Method I* ⟨921⟩: not more than 4.2%.

Particulate matter ⟨788⟩: meets the requirements under *Small-volume Injections.*

Assay—

pH 4.3 sodium phosphate buffer—Dissolve 13.8 g of monobasic sodium phosphate in 900 mL of water, adjust with phosphoric acid or 10 *N* sodium hydroxide to a pH of 4.3 ± 0.1, dilute with water to make 1000 mL, and mix.

Mobile phase—Prepare a suitable mixture of *pH 4.3 sodium phosphate buffer* and acetonitrile (95:5), and filter through a membrane filter of 0.5-µm or finer porosity. Make adjustments if necessary (see *System Suitability* under *Chromatography* ⟨621⟩).

pH 6.4 sodium phosphate buffer—Dissolve 6.9 g of monobasic sodium phosphate in 900 mL of water, adjust with 10 *N* sodium hydroxide to a pH of 6.4 ± 0.1, dilute with water to make 1000 mL, and mix.

Clavulanate lithium stock standard solution—Dissolve an accurately weighed quantity of USP Clavulanate Lithium RS in *pH 6.4 sodium phosphate buffer* to obtain a solution having a known concentration of about 0.6 mg per mL.

Standard preparation—Transfer about 100 mg of USP Ticarcillin Monosodium Monohydrate RS, accurately weighed, to a 100-mL volumetric flask, add $150/J$ mL of *Clavulanate lithium stock standard solution,* accurately measured, J being the ratio of the labeled amount, in mg, of ticarcillin to the labeled amount, in mg, of clavulanic acid in the Sterile Ticarcillin Disodium and Clavulanate Potassium, dilute with *pH 6.4 sodium phosphate buffer* to volume, and mix.

Assay preparation 1—Accurately weigh the contents of 1 container of Sterile Ticarcillin Disodium and Clavulanate Potassium, dissolve in *pH 6.4 sodium phosphate buffer*, and dilute quantitatively with the same solvent to obtain a solution containing the equivalent of about 0.9 mg of ticarcillin ($C_{15}H_{16}N_2O_6S_2$) per mL.

Assay preparation 2—Dissolve the contents of 1 container of Sterile Ticarcillin Disodium and Clavulanate Potassium in a volume of water, accurately measured, corresponding to the volume of solvent specified in the labeling. Using a suitable hypodermic needle and syringe, remove all of the withdrawable contents from the container, and dilute quantitatively and stepwise with *pH 6.4 sodium phosphate buffer* to obtain a solution having a concentration of about 0.9 mg of ticarcillin ($C_{15}H_{16}N_2O_6S_2$) per mL.

Chromatographic system (see *Chromatography* ⟨621⟩)—The liquid chromatograph is equipped with a 220-nm detector and a 4-mm × 30-cm column that contains 3- to 10-μm packing L1. The flow rate is about 2 mL per minute. Chromatograph the *Standard preparation*, and record the peak responses as directed under *Procedure:* the column efficiency determined from the analyte peaks is not less than 1000 theoretical plates, the tailing factors for the analyte peaks are not more than 2.0, the resolution, *R*, between the ticarcillin and clavulanic acid peaks is not less than 5.0, and the relative standard deviation for replicate injections is not more than 2.0%.

Procedure—Separately inject equal volumes (about 20 μL) of the *Standard preparation* and the *Assay preparations* into the chromatograph, record the chromatograms, and measure the responses for the major peaks. The relative retention times are about 0.2 for clavulanic acid and 1.0 for ticarcillin. Calculate the quantity, in μg of ticarcillin per mg, in the Sterile Ticarcillin Disodium and Clavulanate Potassium taken by the formula:

$$(CP/W)(r_U/r_S),$$

in which *C* is the concentration, in mg per mL, of USP Ticarcillin Monosodium Monohydrate RS in the *Standard preparation*, *P* is the designated potency, in μg of ticarcillin per mg, of the USP Ticarcillin Monosodium Monohydrate RS, *W* is the quantity, in mg, of the Sterile Ticarcillin Disodium and Clavulanate Potassium taken in each mL of *Assay preparation 1*, and r_U and r_S are the ticarcillin peak responses obtained from *Assay preparation 1* and the *Standard preparation*, respectively.

Calculate the quantity, in mg, of ticarcillin ($C_{15}H_{16}N_2O_6S_2$) in the container of Sterile Ticarcillin Disodium and Clavulanate Potassium taken by the formula:

$$(L/D)(CP/1000)(r_U/r_S),$$

in which *L* is the labeled quantity, in mg, of ticarcillin in the container, *D* is the concentration, in mg per mL, of ticarcillin in *Assay preparation 2* on the basis of the labeled quantity of ticarcillin in the container and the extent of dilution, *C* is the concentration, in mg per mL, of USP Ticarcillin Monosodium Monohydrate RS in the *Standard preparation*, *P* is the designated potency, in μg of ticarcillin per mg, of USP Ticarcillin Monosodium Monohydrate RS, and r_U and r_S are the ticarcillin peak responses obtained from *Assay preparation 2* and the *Standard preparation*, respectively.

Calculate the quantity, in mg, of clavulanic acid ($C_8H_9NO_5$) in the container of Sterile Ticarcillin Disodium and Clavulanate Potassium taken by the formula:

$$(L/D)(CP/1000)(r_U/r_S),$$

in which *L* is the labeled quantity, in mg, of clavulanic acid in the container, *D* is the concentration, in mg per mL, of clavulanic acid in *Assay preparation 2* on the basis of the labeled quantity of clavulanic acid in the container and the extent of dilution, *C* is the concentration, in mg per mL, of USP Clavulanate Lithium RS in the *Standard preparation*, *P* is the designated potency, in μg of clavulanic acid per mg, of the USP Clavulanate Lithium RS, and r_U and r_S are the clavulanic acid peak responses obtained from *Assay preparation 2* and the *Standard preparation*, respectively.

Timolol Maleate

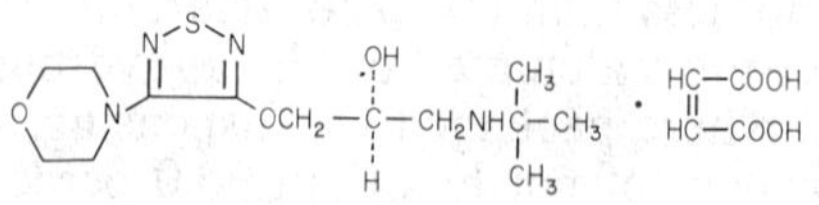

$C_{13}H_{24}N_4O_3S.C_4H_4O_4$ 432.49

2-Propanol, 1-[(1,1-dimethylethyl)amino]-3-[[4-(4-morpholinyl)-1,2,5-thiadiazol-3-yl]oxy]-, (*S*)-, (*Z*)-2-butenedioate (1:1) (salt).

(−)-1-(*tert*-Butylamino)-3-[(4-morpholino-1,2,5-thiadiazol-3-yl)oxy]-2-propanol maleate (1:1) (salt) [*26921-17-5*].

» Timolol Maleate contains not less than 98.0 percent and not more than 101.0 percent of $C_{13}H_{24}N_4O_3S.C_4H_4O_4$, calculated on the dried basis.

Packaging and storage—Preserve in well-closed containers.

Reference standard—*USP Timolol Maleate Reference Standard*—Dry in vacuum at 100° to constant weight.

Identification—

A: The infrared absorption spectrum of a mineral oil dispersion of it, previously dried, exhibits maxima only at the same wavelengths as that of a similar preparation of USP Timolol Maleate RS.

B: The ultraviolet absorption spectrum of a 1 in 40,000 solution in dilute hydrochloric acid (1 in 100) exhibits maxima and minima at the same wavelengths as that of a similar solution of USP Timolol Maleate RS, concomitantly measured, and the respective absorptivities, calculated on the dried basis, at the wavelength of maximum absorbance at about 294 nm do not differ by more than 3.0%.

Specific rotation ⟨781⟩: between −11.7° and −12.5°, calculated on the dried basis, determined with a mercury source at 405 nm in a solution in 1.0 *N* hydrochloric acid containing 500 mg in each 10 mL.

Loss on drying ⟨731⟩—Dry it in vacuum at 100° to constant weight: it loses not more than 0.5% of its weight.

Residue on ignition ⟨281⟩: not more than 0.1%.

Heavy metals, *Method II* ⟨231⟩: 0.002%.

Chromatographic purity—Dissolve 500 mg in methanol to obtain 10.0 mL of test solution. Dissolve an accurately weighed quantity of USP Timolol Maleate RS in methanol, and dilute quantitatively and stepwise with methanol to obtain Standard solutions having the following compositions:

Standard solution	Concentration (μg RS per mL)	Percentage (%, for comparison with test specimen)
A	200	0.4
B	100	0.2
C	50	0.1

On a suitable thin-layer chromatographic plate (see *Chromatography* ⟨621⟩), coated with a 0.25-mm layer of chromatographic silica gel mixture, separately apply 10-μL portions of the solutions. Allow the spots to dry, and develop the chromatogram in a solvent system consisting of a mixture of chloroform, methanol, and ammonium hydroxide (80:20:1) until the solvent front has moved about three-fourths of the length of the plate. Remove the plate from the developing chamber, mark the solvent front, and allow the solvent to evaporate. Expose the plate to iodine vapors for 2 hours, and locate the spots on the plate by examination under short-wavelength ultraviolet light. Compare the intensities of any secondary spots observed in the chromatogram of the test solution, excluding the origin spot due to the maleate anion, with those of the principal spots in the chromatograms of the Standard solutions: no secondary spot is more intense than the principal spot obtained from Standard solution A (0.4%), and the sum of the intensities of all secondary spots, excluding any having intensities less than the principal spot obtained from Standard solution C, does not exceed 1.0%.

Assay—Dissolve about 800 mg of Timolol Maleate, accurately weighed, in about 90 mL of glacial acetic acid, and titrate with 0.1 *N* perchloric acid VS, determining the end-point potentiometrically, using a platinum ring electrode and a sleeve-type calomel electrode containing 0.1 *N* lithium perchlorate in acetic anhydride (see *Titrimetry* ⟨541⟩). Perform a blank determination, and make any necessary correction. Each mL of 0.1 *N* perchloric acid is equivalent to 43.25 mg of $C_{13}H_{24}N_4O_3S.C_4H_4O_4$.

Timolol Maleate Ophthalmic Solution

» Timolol Maleate Ophthalmic Solution is a sterile, aqueous solution of Timolol Maleate. It contains an amount of $C_{13}H_{24}N_4O_3S.C_4H_4O_4$ equivalent to not less than 90.0 percent and not more than 110.0 percent of the labeled amount of timolol ($C_{13}H_{24}N_4O_3S$).

Packaging and storage—Preserve in tight, light-resistant containers.

Reference standard—*USP Timolol Maleate Reference Standard*—Dry in vacuum at 100° to constant weight.

Identification—Dilute a suitable quantity of Ophthalmic Solution with water to obtain a solution having a concentration of about 20 µg of timolol per mL: the ultraviolet absorption spectrum of the solution so obtained exhibits maxima and minima at the same wavelengths as that of a similar preparation of USP Timolol Maleate RS, concomitantly measured.

Sterility—It meets the requirements under *Sterility Tests* ⟨71⟩.

pH ⟨791⟩: between 6.5 and 7.5.

Assay—

pH 9.7 buffer—Dissolve 8.4 g of sodium bicarbonate and 10.6 g of sodium carbonate in water to make 500 mL.

Standard preparation—Dissolve an accurately weighed quantity of USP Timolol Maleate RS in water, and dilute quantitatively and stepwise with water to obtain a *Standard preparation* having a known concentration of about 0.7 mg of timolol maleate per mL.

Assay preparation—Transfer an accurately measured volume of Timolol Maleate Ophthalmic Solution, equivalent to about 25 mg of timolol, to a 50-mL volumetric flask, dilute with water to volume, and mix to obtain a solution having a concentration of about 0.5 mg of timolol per mL.

Procedure—Transfer 5.0 mL each of the *Standard preparation* and the *Assay preparation* to individual 125-mL separators. To each separator (*Set I*) add 15 mL of *pH 9.7 buffer* and 20 mL of toluene, shake for 1 minute, and allow the layers to separate. Transfer the aqueous phase of each separator (*Set I*) to a second set of separators (*Set II*) containing 20 mL of toluene each, shake for 1 minute, and discard the aqueous phases. Add 10 mL of *pH 9.7 buffer* to each of the toluene solutions in the separators from *Set I*, shake for 1 minute, and transfer the aqueous phases to the corresponding separators in *Set II*. Shake for 1 minute, and discard the aqueous phases. Combine the toluene layers by adding the toluene layers from the separators in *Set II* to the separators in *Set I*. Wash each of the separators in *Set II* with 2 mL of toluene, adding the washings to the corresponding separators in *Set I*. Extract the combined toluene solutions with four 20-mL portions of 0.1 *N* sulfuric acid. Collect the sulfuric acid extracts in separate 100-mL volumetric flasks, dilute each with 0.1 *N* sulfuric acid to volume, mix, and centrifuge. Concomitantly determine the absorbances of the *Standard preparation* and the *Assay preparation* in 1-cm cells at the wavelength of maximum absorbance at about 294 nm, with a suitable spectrophotometer, using 0.1 *N* sulfuric acid as the blank to set the instrument. Calculate the quantity, in mg, of timolol ($C_{13}H_{24}N_4O_3S$) in each mL of the Ophthalmic Solution taken by the formula:

$$316.42/432.49(50C/V)(A_U/A_S),$$

in which 316.42 and 432.49 are the molecular weights of timolol and timolol maleate, respectively, *C* is the concentration, in mg per mL, of USP Timolol Maleate RS in the *Standard preparation*, *V* is the volume, in mL, of Ophthalmic Solution taken, and A_U and A_S are the absorbances of the solutions from the *Assay preparation* and the *Standard preparation*, respectively.

Timolol Maleate Tablets

» Timolol Maleate Tablets contain not less than 90.0 percent and not more than 110.0 percent of the labeled amount of $C_{13}H_{24}N_4O_3S.C_4H_4O_4$.

Packaging and storage—Preserve in well-closed, light-resistant containers.

Reference standard—*USP Timolol Maleate Reference Standard*—Dry in vacuum at 100° to constant weight before using.

Identification—Transfer a portion of powdered Tablets, equivalent to about 30 mg of timolol maleate, to a 50-mL volumetric flask, add about 2 mL of 0.1 *N* hydrochloric acid, and shake gently. Add about 30 mL of methanol, agitate for 20 minutes, add methanol to volume, mix, and centrifuge. Similarly prepare a Standard solution containing 0.6 mg of USP Timolol Maleate RS per mL. Separately apply 10 µL of the test solution and 10 µL of the Standard solution to a thin-layer chromatographic plate (see *Chromatography* ⟨621⟩) coated with a 0.25-mm layer of chromatographic silica gel mixture. Develop the chromatogram using a solvent system consisting of a mixture of chloroform, methanol, and ammonium hydroxide (80:20:1) until the solvent front has moved about three-fourths of the length of the plate. Air-dry, and examine under short-wavelength ultraviolet light: the R_f values of the principal spots obtained from the test solution correspond to those obtained from the Standard solution.

Dissolution ⟨711⟩—

Medium: 0.1 *N* hydrochloric acid; 500 mL.

Apparatus 1: 100 rpm.

Time: 20 minutes.

Procedure—Determine the amount of timolol maleate in solution in filtered portions of the solution under test, in comparison with a Standard solution having a known concentration of USP Timolol Maleate RS in the same medium, employing the procedure set forth in the *Assay*, making any necessary modifications.

Tolerances—Not less than 80% (*Q*) of the labeled amount of timolol maleate ($C_{13}H_{24}N_4O_3S.C_4H_4O_4$) is dissolved in 20 minutes.

Uniformity of dosage units ⟨905⟩: meet the requirements.

Assay—

pH 2.8 phosphate buffer—Transfer 22.08 g of monobasic sodium phosphate to a 2-liter volumetric flask, dilute with water to volume, adjust with phosphoric acid to a pH of 2.8 ± 0.05, and filter.

Mobile phase—Prepare a suitable degassed and filtered mixture of *pH 2.8 phosphate buffer* and methanol (3:2).

Standard preparation—Transfer about 50 mg of USP Timolol Maleate RS, accurately weighed, to a 500-mL volumetric flask. Add 50 mL of 0.05 *M* monobasic sodium phosphate. Sonicate until the standard is dissolved, add 100 mL of acetonitrile, shake, dilute with water to volume, and mix.

Assay preparation—Weigh and finely powder not less than 20 Timolol Maleate Tablets. Transfer an accurately weighed portion of the powder, equivalent to about 10 mg of timolol maleate, to a 100-mL volumetric flask, add 10 mL of 0.05 *M* monobasic sodium phosphate, sonicate for 5 minutes, and add 20 mL of acetonitrile. Sonicate for 5 minutes, add 20 mL of water, shake for 10 minutes, dilute with water to volume, and mix.

Chromatographic system (see *Chromatography* ⟨621⟩)—The liquid chromatograph is equipped with a 295-nm detector and a 3.9-mm × 30-cm column that contains packing L1. The flow rate is about 1.8 mL per minute. Chromatograph five replicate injections of the *Standard preparation*, and record the peak responses as directed under *Procedure:* the relative standard deviation is not more than 2.0%, and the tailing factor for the main peak is not greater than 2.0.

Procedure—Separately inject equal volumes (about 15 µL) of the *Standard preparation* and the *Assay preparation* into the chromatograph by means of a suitable microsyringe or sampling valve, record the chromatograms, and measure the responses for the major peaks. Calculate the quantity, in mg, of $C_{13}H_{24}N_4O_3S.C_4H_4O_4$ in the portion of Tablets taken by the formula:

$$100C(r_U/r_S),$$

in which C is the concentration, in mg per mL, of USP Timolol Maleate RS in the *Standard preparation*, and r_U and r_S are the peak responses obtained for timolol maleate from the *Assay preparation* and the *Standard preparation*, respectively.

Timolol Maleate and Hydrochlorothiazide Tablets

» Timolol Maleate and Hydrochlorothiazide Tablets contain not less than 90.0 percent and not more than 110.0 percent of the labeled amounts of timolol maleate ($C_{17}H_{28}N_4O_7S$) and hydrochlorothiazide ($C_7H_8ClN_3O_4S_2$).

Packaging and storage—Preserve in well-closed, light-resistant containers.

Reference standards—*USP Timolol Maleate Reference Standard*—Dry in vacuum at 100° to constant weight. *USP Hydrochlorothiazide Reference Standard*—Dry at 105° for 1 hour.

Identification—Transfer a portion of powdered Tablets, equivalent to about 20 mg of timolol maleate, to a suitable centrifuge tube containing about 5 mL of methanol. Agitate for 20 minutes, and centrifuge. Separately dissolve suitable quantities of USP Timolol Maleate RS and USP Hydrochlorothiazide RS in methanol to obtain Standard solutions each having a concentration of 10 mg per mL. Separately apply 3 µL of the test solution and of each Standard solution to a thin-layer chromatographic plate (see *Chromatography* ⟨621⟩) coated with a 0.25-mm layer of chromatographic silica gel mixture. Develop the chromatogram using a solvent system consisting of a mixture of chloroform, methanol, and ammonium hydroxide (80:20:1) until the solvent front has moved about three-fourths of the length of the plate. Air-dry, and examine under short-wavelength ultraviolet light: the R_f values of the principal spots obtained from the Standard solutions correspond to those obtained from the test solution.

Dissolution ⟨711⟩—

Medium: 0.1 *N* hydrochloric acid; 900 mL.

Apparatus 2: 50 rpm.

Time: 20 minutes.

Procedure—Determine the amount of timolol maleate ($C_{13}H_{24}N_4O_3S.C_4H_4O_4$) dissolved, employing the following procedure. Prepare a Standard solution of USP Timolol Maleate RS in 0.1 *N* hydrochloric acid having a known concentration of about 11 µg per mL. Filter a portion of the solution under test, and transfer 10.0 mL of the clear filtrate to a suitable separator. Transfer 10.0 mL each of the Standard solution and 0.1 *N* hydrochloric acid, to provide the blank, to individual separators, and treat each of three separators as follows. Add 20.0 mL of ethyl acetate, mix for 1 minute, allow the phases to separate, and filter the aqueous layer into a suitable vessel, retaining the ethyl acetate layer from the solution under test for the hydrochlorothiazide determination. Determine the amount of $C_{13}H_{24}N_4O_3S.C_4H_4O_4$ dissolved from ultraviolet absorbances of the aqueous layer from the solution under test at the wavelength of maximum absorbance at about 293 nm in comparison with the aqueous layer from the Standard solution.

Determine the amount of hydrochlorothiazide ($C_7H_8ClN_3O_4S_2$) dissolved, employing the following procedure. Filter the ethyl acetate layer obtained previously from the solution under test through filter paper. Determine the amount of $C_7H_8ClN_3O_4S_2$ dissolved from ultraviolet absorbances at the wavelength of maximum absorbance at about 270 nm of the ethyl acetate layer from the solution under test in comparison with a Standard solution in ethyl acetate having a known concentration of USP Hydrochlorothiazide RS.

Tolerances—Not less than 80% (*Q*) of each of the labeled amounts of $C_{13}H_{24}N_4O_3S.C_4H_4O_4$ and $C_7H_8ClN_3O_4S_2$, respectively, is dissolved in 20 minutes.

Uniformity of dosage units ⟨905⟩: meet the requirements for *Content Uniformity* with respect to timolol maleate and to hydrochlorothiazide.

Assay—

pH 3.0 phosphate buffer—Dissolve 13.6 g of monobasic potassium phosphate in 100 mL of water, adjust with phosphoric acid to a pH of 3.0 ± 0.05, and filter.

Mobile phase—Prepare a suitable filtered and degassed mixture of water, acetonitrile, methanol, and *pH 3.0 phosphate buffer* (38:8:2:2), making adjustments if necessary (see *System Suitability* under *Chromatography* ⟨621⟩).

Standard preparation—Transfer about 50 mg of USP Timolol Maleate RS, accurately weighed, to a 500-mL volumetric flask. Add 50*J* mg of USP Hydrochlorothiazide RS, accurately weighed, *J* being the ratio of the labeled amount, in mg, of hydrochlorothiazide to the labeled amount, in mg, of timolol maleate per Tablet. Add 50 mL of 0.05 *M* monobasic sodium phosphate, and 125 mL of acetonitrile, sonicate for 4 minutes, dilute with water to volume, and mix. Pipet 5 mL into a 25-mL volumetric flask, dilute with acetonitrile solution (1 in 10) to volume, and mix.

Assay preparation—Weigh and finely powder not less than 20 Timolol Maleate and Hydrochlorothiazide Tablets. Transfer an accurately weighed portion of the powder, equivalent to about 20 mg of timolol maleate, to a 1-liter volumetric flask, add about 100 mL of 0.05 *M* monobasic sodium phosphate, 125 mL of acetonitrile, and 100 mL of water, and mix by mechanical means. Allow to stand for 16 hours, dilute with water to volume, mix, and filter.

Chromatographic system (see *Chromatography* ⟨621⟩)—The liquid chromatograph is equipped with a 295-nm detector and a 4-mm × 30-cm column that contains packing L1. The flow rate is about 1.5 mL per minute. Chromatograph replicate injections of the *Standard preparation*, and record the peak responses as directed under *Procedure:* the relative standard deviation is not more than 1.5% and the resolution, *R*, between hydrochlorothiazide and timolol maleate is not less than 4.0.

Procedure—Separately inject equal volumes (about 50 µL) of the *Standard preparation* and the *Assay preparation* into the chromatograph by means of a suitable microsyringe or sampling valve, record the chromatograms, and measure the responses for the major peaks. The relative retention times are about 0.6 for hydrochlorothiazide and 1.0 for timolol maleate. Calculate the quantity, in mg, of hydrochlorothiazide ($C_7H_8ClN_3O_4S_2$) in the portion of Tablets taken by the formula:

$$1000C(r_U/r_S),$$

in which C is the concentration, in mg per mL, of USP Hydrochlorothiazide RS in the *Standard preparation*, and r_U and r_S are the responses of the hydrochlorothiazide peak obtained from the *Assay preparation* and the *Standard preparation*, respectively. Calculate the quantity, in mg, of timolol maleate ($C_{17}H_{28}N_4O_7S$) by the same formula, changing the terms to refer to timolol maleate.

Tinctures—*see complete list in index*

Tioconazole

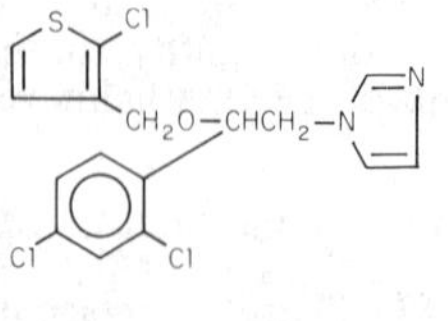

$C_{16}H_{13}Cl_3N_2OS$ 387.71

1*H*-Imidazole, 1-[2-[(2-chloro-3-thienyl)methoxy]-2-(2,4-dichlorophenyl)ethyl]-.
1-[2,4-Dichloro-β-[(2-chloro-3-thenyl)-oxy]phenethyl]imidazole [*65899-73-2*].

» Tioconazole contains not less than 97.0 percent and not more than 103.0 percent of $C_{16}H_{13}Cl_3N_2OS$.

Packaging and storage—Preserve in tight containers.

Reference standards—*USP Tioconazole Reference Standard*—Do not dry before using. *USP Tioconazole Related Compound A Reference Standard*—Do not dry before using. *USP Tioconazole Related Compound B Reference Standard—Do not dry before using. USP Tioconazole Related Compound C Reference Standard*—Do not dry before using.

Identification—

A: The infrared absorption spectrum of a mineral oil dispersion of it exhibits maxima only at the same wavelengths as that of a similar preparation of USP Tioconazole RS.

B: *Visualizing solution*—Dissolve 0.85 g of bismuth subnitrate in 10 mL of glacial acetic acid, dilute with water to 50 mL, and mix. Mix 10 mL of this solution, 50 mL of potassium iodide solution (2 in 25), and 20 mL of glacial acetic acid, dilute with water to 100 mL, and mix.

Procedure—Prepare a test solution by dissolving 50 mg of Tioconazole in 1 mL of methanol. On a thin-layer chromatographic plate (see *Chromatography* ⟨621⟩), coated with a 0.25-mm layer of chromatographic silica gel mixture, separately apply 10 μL of the test solution and 10 μL of a Standard solution of USP Tioconazole RS, similarly prepared. Allow the spots to dry, and develop the chromatogram using a solvent system consisting of a mixture of chloroform, methanol, and glacial acetic acid (40:5:1) until the solvent front has moved about three-fourths of the length of the plate. Remove the plate from the developing chamber, mark the solvent front, and locate the spots on the plate by viewing under short- and long-wavelength ultraviolet light after drying the plate at 80° for 5 minutes. Spray the plate with *Visualizing solution*, air-dry for 2 minutes, and overspray with sodium nitrite solution (1 in 20). Air-dry the plate for 5 minutes, and examine it for brown spots on a pale yellow background: the R_f value of the principal spot from the test solution corresponds to that obtained from the Standard solution.

C: The chromatogram of the *Assay preparation* obtained as directed in the *Assay* exhibits a major peak for tioconazole, the retention time of which corresponds to that exhibited in the chromatogram of the *Standard preparation* obtained as directed in the *Assay*.

Water, *Method I* ⟨921⟩: not more than 0.5%.

Residue on ignition ⟨281⟩: not more than 0.2%.

Chloride ⟨221⟩—A 0.7-g portion dissolved in methanol shows no more chloride than corresponds to 0.50 mL of 0.020 *N* hydrochloric acid (0.05%).

Heavy metals, *Method II* ⟨231⟩: 0.005%.

Related compounds—

Mobile phase and *Chromatographic system*—Prepare as directed in the *Assay*.

Standard preparation—Transfer about 1 mg, accurately weighed, each of USP Tioconazole Related Compound A RS, USP Tioconazole Related Compound B RS, and USP Tioconazole Related Compound C RS to a 25-mL flask. Add 15.0 mL of methanol, and shake until the contents are completely dissolved.

Test preparation—Transfer about 100 mg of Tioconazole, accurately weighed, to a 25-mL flask, add 15.0 mL of methanol, and shake until the substance is completely dissolved.

Procedure—Separately inject equal volumes (about 20 μL) of the *Standard preparation* and the *Test preparation* into the chromatograph, record the chromatograms, and measure the responses for the peaks. Calculate, in turn, the percentages of 1-[2,4-dichloro-β-[(3-thenyl)-oxy]phenethyl]imidazole hydrochloride (tioconazole related compound A), 1-[2,4-dichloro-β-[(2,5-dichloro-3-thenyl)-oxy]phenethyl]imidazole hydrochloride (tioconazole related compound B), and 1-[2,4-dichloro-β-[(5-bromo-2-chloro-3-thenyl)-oxy]phenethyl]imidazole hydrochloride (tioconazole related compound C) in the portion of Tioconazole taken by the same formula:

$$100(W_I/W_U)(r_U/r_S),$$

in which W_I is the weight, in mg, of the respective USP Reference Standard taken to prepare the *Standard preparation*, W_U is the weight, in mg, of Tioconazole taken to prepare the *Test preparation*, and r_U and r_S are the peak responses at corresponding retention times, obtained from the *Test preparation* and the *Standard preparation*, respectively. The limit of each related compound is 1.0%.

Assay—

Mobile phase—[NOTE—Prepare the *Mobile phase* fresh daily.] Mix 440 mL of acetonitrile, 400 mL of methanol, and 280 mL of water. Degas the solution. Add 2.0 mL of ammonium hydroxide, and mix. Make adjustments if necessary (see *System Suitability* under *Chromatography* ⟨621⟩).

Standard preparation—Dissolve an accurately weighed quantity of USP Tioconazole RS in methanol, and dilute quantitatively and stepwise, if necessary, with methanol to obtain a solution having a known concentration of about 200 μg per mL.

Assay preparation—Transfer about 100 mg of Tioconazole, accurately weighed, to a 100-mL volumetric flask, dissolve in methanol, dilute with methanol to volume, and mix. Transfer 10.0 mL of the resulting solution to a 50-mL volumetric flask, dilute with methanol to volume, and mix.

Chromatographic system (see *Chromatography* ⟨621⟩)—The liquid chromatograph is equipped with a 219-nm detector, a 4-mm × 10-cm pre-column that contains packing L4, installed between the pump and the injector, and a 5-mm × 25-cm analytical column that contains packing L1. [NOTE—Replace the pre-column daily.] The flow rate is adjusted to obtain a retention time of between 12 and 17 minutes for tioconazole. Chromatograph the *Standard preparation*, and record the peak responses as directed under *Procedure*. The column efficiency determined from the analyte peak is not less than 1000 theoretical plates, the tailing factor for the analyte peak is not more than 2.0, and the relative standard deviation for replicate injections is not more than 2.0%.

Procedure—Separately inject equal volumes (about 20 μL) of the *Standard preparation* and the *Assay preparation* into the chromatograph, record the chromatograms, and measure the responses for the major peaks. Calculate the quantity, in mg, of $C_{16}H_{13}Cl_3N_2OS$ in the Tioconazole taken by the formula:

$$(0.5C)(r_U/r_S),$$

in which C is the concentration, in μg per mL, of USP Tioconazole RS, calculated on the anhydrous basis, in the *Standard preparation*, and r_U and r_S are the peak responses obtained from the *Assay preparation* and the *Standard preparation*, respectively.

Tioconazole Cream

» Tioconazole Cream contains not less than 90.0 percent and not more than 110.0 percent of the labeled amount of $C_{16}H_{13}Cl_3N_2OS$ in a suitable cream base.

Packaging and storage—Preserve in tight containers.

Reference standard—*USP Tioconazole Reference Standard*—Do not dry before using.

Identification—Mix a quantity of Cream, equivalent to about 10 mg of tioconazole, and 4 mL of isopropyl alcohol, and filter. Use the filtrate as the test solution. On a suitable thin-layer chromatographic plate (see *Chromatography* ⟨621⟩), coated with a 0.25-mm layer of chromatographic silica gel mixture, apply separately 40 μL of the test solution and 40 μL of a Standard solution of USP Tioconazole RS in isopropyl alcohol containing about 2.5 mg per mL, and allow the spots to air-dry. Position the plate in a chromatographic chamber, and develop the chromatograms in a solvent system consisting of a mixture of chloroform, methanol, and ammonium hydroxide (40:5:1), until the solvent front has moved about three-fourths of the length of the plate. Remove the plate from the developing chamber, mark the solvent front, and dry at 80° for 5 minutes. Observe the plate under short-wavelength ultraviolet light: the R_f value of the principal spot obtained from the test solution corresponds to that obtained from the Standard solution.

Microbial limits—It meets the requirements of the tests for absence of *Staphylococcus aureus* and *Pseudomonas aeruginosa* under *Microbial Limit Tests* ⟨61⟩.

Minimum fill ⟨755⟩: meets the requirements.

pH ⟨791⟩: between 3.0 and 6.0, in a 1:1 aqueous suspension of the Cream.

Assay—

Mobile phase—Prepare a filtered and degassed mixture of methanol and 0.005 *M* tetrabutylammonium phosphate (4:1). Make adjustments if necessary (see *System Suitability* under *Chromatography* ⟨621⟩).

Standard preparation—Dissolve an accurately weighed quantity of USP Tioconazole RS in *Mobile phase*, and dilute quantitatively and stepwise, if necessary, with *Mobile phase* to obtain a solution having a known concentration of about 0.2 mg per mL.

Assay preparation—Transfer an accurately weighed quantity of Tioconazole Cream, equivalent to about 10 mg of tioconazole, to a 100-mL conical flask, add 50.0 mL of *Mobile phase*, shake until the cream is completely dispersed, and filter, discarding the first 10 mL of the filtrate.

Chromatographic system (see *Chromatography* ⟨621⟩)—The liquid chromatograph is equipped with a 219-nm detector, a 4.6-mm × 25-mm pre-column that contains packing L4 installed between the pump and the injector, and a 4.6-mm × 25-cm analytical column that contains packing L1. The flow rate is about 1 mL per minute. Chromatograph the *Standard preparation*, and record the peak responses as directed under *Procedure:* the column efficiency determined from the analyte peak is not less than 1000 theoretical plates, the tailing factor for the analyte peak is not more than 2.0, and the relative standard deviation for replicate injections is not more than 2.0%.

Procedure—Separately inject equal volumes (about 20 µL) of the *Standard preparation* and the *Assay preparation* into the chromatograph, record the chromatograms, and measure the responses for the major peaks. Calculate the quantity, in mg, of $C_{16}H_{13}Cl_3N_2OS$ in the portion of Cream taken by the formula:

$$50C(r_U/r_S),$$

in which C is the concentration, in mg per mL, of USP Tioconazole RS in the *Standard preparation*, and r_U and r_S are the peak responses obtained from the *Assay preparation* and the *Standard preparation*, respectively.

Titanium Dioxide

TiO_2 79.88
Titanium oxide (TiO_2).
Titanium oxide (TiO_2) [*13463-67-7*].

» Titanium Dioxide contains not less than 99.0 percent and not more than 100.5 percent of TiO_2, calculated on the dried basis. In addition, it contains not more than 0.001 percent of lead, not more than 2 ppm of antimony, and not more than 1 ppm of mercury, to comply with the U.S. Food and Drug Administration regulation on trace impurities limits (21 CFR 73.575).

Packaging and storage—Preserve in well-closed containers.

Identification—To 500 mg add 5 mL of sulfuric acid, and heat gently until fumes of sulfur trioxide appear. Cool the suspension, and cautiously dilute with water to 100 mL. Filter, and to 5 mL of the clear filtrate add a few drops of hydrogen peroxide TS: an orange-red color develops immediately.

Loss on drying ⟨731⟩—Dry it at 105° for 3 hours: it loses not more than 0.5% of its weight.

Loss on ignition ⟨733⟩—Ignite 2 g, previously dried and accurately weighed, at 800 ± 25° to constant weight: it loses not more than 0.5% of its weight.

Water-soluble substances—Suspend 4.0 g in 50 mL of water, mix, and allow to stand overnight. Transfer to a 200-mL volumetric flask, add 2 mL of ammonium chloride TS, and mix. If the Titanium Dioxide does not settle, add another 2-mL portion of ammonium chloride TS. Allow the suspension to settle, dilute with water to volume, mix, and filter through a double thickness of fine-porosity filter paper, discarding the first 10 mL of the filtrate. Collect 100 mL of the clear filtrate, transfer to a tared platinum dish, evaporate on a hot plate to dryness, and ignite at a dull red heat to constant weight: the residue weighs not more than 5 mg (0.25%).

Acid-soluble substances—Suspend 5.0 g in 100 mL of 0.5 *N* hydrochloric acid, and heat on a steam bath for 30 minutes, with occasional stirring. Filter through an appropriate filter medium until clear. Wash with three 10-mL portions of 0.5 *N* hydrochloric acid. Evaporate the combined filtrate and washings to dryness, and ignite at a dull red heat to constant weight: the residue weighs not more than 25 mg (0.5%).

Arsenic, *Method I* ⟨211⟩—Prepare the *Test Preparation* as follows. Add 3.0 g to a 250-mL conical flask fitted with a thermometer and a vapor outlet. Add 50 mL of water, 500 mg of hydrazine sulfate, 500 mg of potassium bromide, 20 g of sodium chloride, and 25 mL of sulfuric acid. Arrange to collect the evolved vapors in 52 mL of water contained in the arsine generator flask, then heat the test specimen to 90°, and maintain the temperature at 90° to 100° for 15 minutes. Add 3 mL of hydrochloric acid to the solution in the generator flask: the resulting solution meets the requirements of the test, the addition of 20 mL of 7 *N* sulfuric acid specified under *Procedure* being omitted. The limit is 1 ppm.

Assay—Accurately weigh about 300 mg of Titanium Dioxide, transfer to a 250-mL beaker, and add 20 mL of sulfuric acid and 7 g to 8 g of ammonium sulfate. Mix, heat on a hot plate until fumes of sulfur trioxide appear, and continue heating over a strong flame until solution is complete or it is apparent that the undissolved residue is siliceous matter. Cool, cautiously dilute with 100 mL of water, stir, heat carefully to boiling while stirring, and allow the insoluble matter to settle. Filter, transfer the entire residue to the filter, and wash thoroughly with cold 2 *N* sulfuric acid. Dilute the filtrate with water to 200 mL, and cautiously add about 10 mL of ammonium hydroxide.

Prepare a zinc amalgam column in a 25-cm Jones reductor tube, placing a pledget of glass wool in the bottom of the tube, and filling the constricted portion of the tube with zinc amalgam prepared as follows. Add 20- to 30-mesh zinc to mercuric chloride solution (1 in 50), using about 100 mL of the solution for each 100 g of zinc, and after about 10 minutes, decant the solution from the zinc, then wash the zinc by decantation. Wash the zinc amalgam column with 100-mL portions of 2 *N* sulfuric acid until 100 mL of the washing does not decolorize 1 drop of 0.1 *N* potassium permanganate.

Place 50 mL of ferric ammonium sulfate TS in a 1000-mL suction flask, and add 0.1 *N* potassium permanganate until a faint pink color persists for 5 minutes. Attach the Jones reductor tube to the neck of the flask, and pass 50 mL of 2 *N* sulfuric acid through the reductor at a rate of about 30 mL per minute. Pass the prepared titanium solution through the reductor at the same rate, and follow with 100 mL each of 2 *N* sulfuric acid and of water. During these operations, keep the reductor filled with solution or water above the upper level of the amalgam. Taking precautions against the admission of atmospheric oxygen, gradually release the suction, wash down the outlet tube of the reductor and the sides of the receiver, and titrate immediately with 0.1 *N* potassium permanganate VS. Perform a blank determination, substituting 200 mL of 2 *N* sulfuric acid for the assay solution, and make any necessary correction. Each mL of 0.1 *N* potassium permanganate is equivalent to 7.988 mg of TiO_2.

Tobramycin

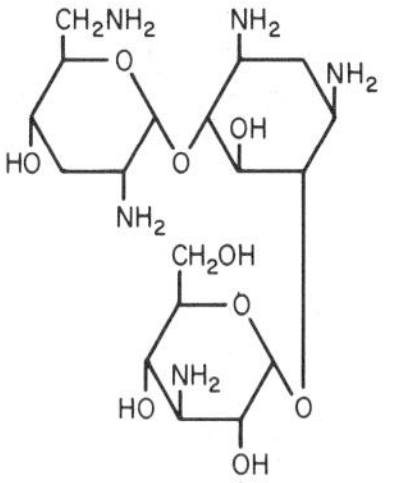

$C_{18}H_{37}N_5O_9$ 467.52

D-Streptamine, *O*-3-amino-3-deoxy-α-D-glucopyranosyl-(1→ 6)-*O*-[2,6-diamino-2,3,6-trideoxy-α-D-*ribo*-hexopyranosyl-(1→4)]-2-deoxy-.

O-3-Amino-3-deoxy-α-D-glucopyranosyl-(1→4)-*O*-[2,6-diamino-2,3,6-trideoxy-α-D-*ribo*-hexopyranosyl-(1→6)]-2-deoxy-L-streptamine [*32986-56-4*].

» Tobramycin has a potency of not less than 900 µg of $C_{18}H_{37}N_5O_9$ per mg, calculated on the anhydrous basis.

Packaging and storage—Preserve in tight containers.

Reference standard—*USP Tobramycin Reference Standard*—Do not dry before using.

Identification—Prepare a solution of it in water containing 6 mg per mL. On a suitable thin-layer chromatographic plate (see *Chromatography* ⟨621⟩), coated with a 0.25-mm layer of chromatographic silica gel mixture, apply 3 µL of this test solution, 3 µL of a Standard solution of USP Tobramycin RS containing 6 mg per mL, and 3 µL of a mixture of equal volumes of the two solutions. Place the plate in a suitable chromatographic chamber, and develop the chromatogram by continuous flow with a mixture of methanol, ammonium hydroxide, and chloroform (60:30:25) for 5.5 hours. Remove the plate from the chamber, allow the solvent to evaporate, and heat the plate at 110° for 15 minutes. Immediately locate the spots on the plate by spraying with a 1 in 100 solution of ninhydrin in a mixture of butyl alcohol and pyridine (100:1): tobramycin appears as a pink spot, and the spots obtained from the test solution and from the mixture of test solution and Standard solution, respectively, correspond in distance from the origin to that obtained from the Standard solution.

pH ⟨791⟩: between 9 and 11, in a solution (1 in 10).

Water, *Method I* ⟨921⟩: not more than 8.0%.

Residue on ignition ⟨281⟩: not more than 1.0%, the charred residue being moistened with 2 mL of nitric acid and 5 drops of sulfuric acid.

Heavy metals, *Method II* ⟨231⟩: 0.003%.

Assay—Proceed with Tobramycin as directed under *Antibiotics—Microbial Assays* ⟨81⟩.

Tobramycin Ophthalmic Ointment

» Tobramycin Ophthalmic Ointment contains the equivalent of not less than 90.0 percent and not more than 120.0 percent of the labeled amount of $C_{18}H_{37}N_5O_9$.

Packaging and storage—Preserve in collapsible ophthalmic ointment tubes.

Reference standard—*USP Tobramycin Reference Standard*—Do not dry before using.

Identification—Vigorously shake by mechanical means a quantity of Ophthalmic Ointment, equivalent to about 3 mg of tobramycin, with 2 mL of chloroform. Add 1 mL of water, shake vigorously by mechanical means for 1 minute, and centrifuge for 15 minutes: the clear upper, aqueous layer so obtained meets the requirements of the *Identification* test under *Tobramycin.*

Sterility—It meets the requirements under *Sterility Tests* ⟨71⟩, when tested as directed in the section, *Test Procedures Using Membrane Filtration.*

Minimum fill ⟨755⟩: meets the requirements.

Water, *Method I* ⟨921⟩: not more than 1.0%, 20 mL of a mixture of carbon tetrachloride, chloroform, and methanol (2:2:1) being used in place of methanol in the titration vessel.

Metal particles—It meets the requirements of the test for *Metal Particles in Ophthalmic Ointments* ⟨751⟩.

Assay—Proceed as directed under *Antibiotics—Microbial Assays* ⟨81⟩, using an accurately weighed quantity of Tobramycin Ophthalmic Ointment, equivalent to about 3 mg of tobramycin, shaken with about 50 mL of ether in a separator, and extracted with four 20-mL portions of water. Combine the aqueous extracts, and dilute quantitatively and stepwise with water to obtain a *Test Dilution* having a concentration assumed to be equal to the median dose level of the Standard.

Tobramycin Ophthalmic Solution

» Tobramycin Ophthalmic Solution contains the equivalent of not less than 90.0 percent and not more than 120.0 percent of the labeled amount of tobramycin. It may contain one or more suitable buffers, dispersants, preservatives, and tonicity agents.

Packaging and storage—Preserve in tight containers, and avoid exposure to excessive heat.

Reference standard—*USP Tobramycin Reference Standard*—Do not dry before using.

Identification—Allow any suspended matter in the Solution to settle, decant the clear liquid into a test tube, and heat on a steam bath for 5 minutes. Using a suction device, remove any coagulated material that may float to the top of the hot solution, heat the remaining solution on a steam bath, and filter while hot: the filtrate so obtained responds to the *Identification* test under *Tobramycin.*

Sterility—It meets the requirements under *Sterility Tests* ⟨71⟩ when tested as directed in the section, *Test Procedures Using Membrane Filtration.*

pH ⟨791⟩: between 7.0 and 8.0.

Assay—Proceed with Tobramycin Ophthalmic Solution as directed under *Antibiotics—Microbial Assays* ⟨81⟩, using an accurately measured volume of Ophthalmic Solution diluted quantitatively with water to yield a *Test Dilution* having a concentration assumed to be equal to the median dose level of the Standard.

Tobramycin Sulfate Injection

» Tobramycin Sulfate Injection is a sterile solution of Tobramycin in Water for Injection prepared with the aid of Sulfuric Acid. It contains not less than 90.0 percent and not more than 120.0 percent of the labeled amount of tobramycin ($C_{18}H_{37}N_5O_9$). It contains one or more suitable antioxidants, chelating agents, and preservatives.

Packaging and storage—Preserve in single-dose or in multiple-dose containers, preferably of Type I glass.

Reference standard—*USP Tobramycin Reference Standard*—Do not dry before using.

Identification—Dilute the Injection with water to obtain a solution containing 6 mg of tobramycin per mL, and proceed as directed in the *Identification test* under *Tobramycin*, beginning with "On a suitable thin-layer chromatographic plate."

Pyrogen—When diluted, if necessary, with pyrogen-free saline TS to a concentration of 10.0 mg of tobramycin per mL, it meets the requirements of the *Pyrogen Test* ⟨151⟩, the test dose being 1 mL per kg.

Sterility—It meets the requirements under *Sterility Tests* ⟨71⟩ when tested as directed in the section, *Test Procedures Using Membrane Filtration.*

pH ⟨791⟩: between 3.0 and 6.5.

Particulate matter ⟨788⟩: meets the requirements under *Small-volume Injections.*

Other requirements—It meets the requirements under *Injections* ⟨1⟩.

Assay—Proceed with Tobramycin Sulfate Injection as directed under *Antibiotics—Microbial Assays* ⟨81⟩, using an accurately measured volume of Injection diluted quantitatively with water to yield a *Test Dilution* having a concentration assumed to be equal to the median dose level of the Standard.

Sterile Tobramycin Sulfate

$(C_{18}H_{37}N_5O_9)_2 . 5H_2SO_4$ 1425.39

D-Streptamine, *O*-3-amino-3-deoxy-α-D-glucopyranosyl-(1→6)-*O*-[2,6-diamino-2,3,6-trideoxy-α-D-*ribo*-hexopyranosyl-(1→4)]-2-deoxy-, sulfate (2:5) (salt).

O-3-Amino-3-deoxy-α-D-glucopyranosyl-(1→4)-*O*-[2,6-diamino-2,3,6-trideoxy-α-D-*ribo*-hexopyranosyl-(1→6)]-2-deoxy-L-streptamine, sulfate (2:5) (salt).

Tobramycin sulfate (2:5) (salt) [*79645-27-5*].

» Sterile Tobramycin Sulfate is tobramycin sulfate suitable for parenteral use. It has a potency of not less than 634 μg and not more than 739 μg of tobramycin ($C_{18}H_{37}N_5O_9$) per mg. In addition, where packaged for dispensing, it contains not less than 90.0 percent and not more than 115.0 percent of the labeled amount of tobramycin.

Packaging and storage—Preserve in *Containers for Sterile Solids* as described under *Injections* ⟨1⟩.

Reference standard—*USP Tobramycin Reference Standard*—Do not dry before using.

Constituted solution—At the time of use, the constituted solution prepared from Sterile Tobramycin Sulfate meets the requirements for *Constituted Solutions* under *Injections* ⟨1⟩.

Pyrogen—It meets the requirements of the *Pyrogen Test* ⟨151⟩, the test dose being 1.0 mL per kg of a solution prepared to contain 10.0 mg of tobramycin ($C_{18}H_{37}N_5O_9$) per mL in pyrogen-free saline TS.

Sterility—It meets the requirements under *Sterility Tests* ⟨71⟩ when tested as directed in the section, *Test Procedures Using Membrane Filtration*, 6 g being used if it is not packaged for dispensing.

pH ⟨791⟩: between 6.0 and 8.0, in a solution containing 40 mg per mL (or, where packaged for dispensing, in the solution constituted as directed in the labeling).

Water, *Method I* ⟨921⟩: not more than 2.0%.

Particulate matter ⟨788⟩: meets the requirements under *Small-volume Injections.*

Other requirements—It responds to the *Identification test* and meets the requirements for *Residue on ignition* and *Heavy metals* under *Tobramycin.* It meets also the requirements for *Uniformity of Dosage Units* ⟨905⟩ (where packaged for dispensing), and *Labeling* under *Injections* ⟨1⟩.

Assay—Using an accurately weighed quantity of Sterile Tobramycin Sulfate, proceed as directed under *Antibiotics—Microbial Assays* ⟨81⟩. Where it is packaged for dispensing, use the contents of a container, and if the label states the quantity of tobramycin in a given volume of constituted solution, use the solution obtained when a container is constituted in a volume of water, accurately measured, corresponding to the volume of solvent specified in the labeling.

Tocainide Hydrochloride

$CH_3CHCONH$ (NH_2) — 2,6-dimethylphenyl (CH_3, CH_3) · HCl

$C_{11}H_{16}N_2O.HCl$ 228.72

Propanamide, 2-amino-*N*-(2,6-dimethylphenyl)-, hydrochloride.

Amino-2′,6′-propionoxylidide hydrochloride.

» Tocainide Hydrochloride contains not less than 98.0 percent and not more than 101.0 percent of $C_{11}H_{16}N_2O.HCl$, calculated on the dried basis.

Packaging and storage—Preserve in well-closed containers.

Reference standard—*USP Tocainide Hydrochloride Reference Standard*—Dry at 105° for 2 hours before using.

Identification—

A: The infrared absorption spectrum of a potassium bromide dispersion of it exhibits maxima only at the same wavelengths as that of a similar preparation of USP Tocainide Hydrochloride RS.

B: It responds to the tests for *Chloride* ⟨191⟩.

Loss on drying ⟨731⟩—Dry it at 105° for 2 hours: it loses not more than 0.5% of its weight.

Residue on ignition ⟨281⟩: not more than 0.1%.

Heavy metals, *Method II* ⟨231⟩: 0.002%.

Chromatographic purity—Prepare a test solution of it in methanol containing 100 mg per mL. Prepare solutions of USP Tocainide Hydrochloride RS in methanol having concentrations of 100, 1.0, 0.5, 0.25, and 0.1 mg per mL, respectively (*Standard solutions A, B, C, D,* and *E*). Apply 20-μL volumes of all six solutions on a suitable thin-layer chromatographic plate (see *Chromatography* ⟨621⟩), coated with a 0.25-mm layer of chromatographic silica gel mixture and previously washed with methanol. Allow the spots to dry, and develop the chromatogram in a freshly prepared solvent system consisting of a mixture of toluene and alcohol (4:1) in a paper-lined equilibrated tank in an atmosphere of ammonia vapors, until the solvent front has moved about three-fourths of the length of the plate. Remove the plate from the developing chamber, mark the solvent front, allow to air-dry, and examine the plate under short-wavelength ultraviolet light. Expose the plate to iodine vapors, and observe again under white light: the chromatograms show principal spots at about the same R_f value. Estimate the concentration of any spot observed in the chromatogram of the test solution, other than the principal spot and that observed at the origin (which may appear because of the presence of ammonium chloride), by comparison with the principal spots in the chromatograms of *Standard solutions B, C, D,* and *E:* the intensity of any secondary spot is not greater than that of the principal spot obtained from *Standard solution C* (0.5%), and the sum of all secondary spots is not greater than the intensity of the principal spot obtained from *Standard solution B* (1.0%).

Assay—Dissolve about 180 mg of Tocainide Hydrochloride, accurately weighed, in about 40 mL of glacial acetic acid and 15 mL of a 6 in 100 solution of mercuric acetate in glacial acetic acid, and titrate with 0.1 *N* perchloric acid VS, determining the end-point potentiometrically, using a platinum ring electrode and a sleeve-type calomel electrode containing 0.1 *N* lithium perchlorate in acetic anhydride (see *Titrimetry* ⟨541⟩). Perform a blank determination, and make any necessary correction. Each mL of 0.1 *N* perchloric acid is equivalent to 22.87 mg of $C_{11}H_{16}N_2O.HCl$.

Tocainide Hydrochloride Tablets

» Tocainide Hydrochloride Tablets contain not less than 95.0 percent and not more than 105.0 percent of the labeled amount of $C_{11}H_{16}N_2O . HCl$.

Packaging and storage—Preserve in well-closed containers.

Reference standard—*USP Tocainide Hydrochloride Reference Standard*—Dry at 105° for 2 hours before using.

Identification—

A: Transfer a quantity of finely powdered Tablets, equivalent to about 150 mg of tocainide hydrochloride, to a 100-mL volumetric flask, add 75 mL of water, shake for 15 minutes, dilute with water to volume, and mix. Filter a portion of this solution, and dilute 10 mL of the filtrate with water to 50 mL: the ultraviolet absorption spectrum of the solution so obtained exhibits a maximum at the same wavelength as that of a similar solution of USP Tocainide Hydrochloride RS, concomitantly measured.

B: Transfer about 100 mg of finely powdered Tablets to a suitable separator, and add 10 mL of water and 2 mL of 2 *M* sodium carbonate. Extract with 20 mL of methylene chloride. Add 0.3 mL of filtered methylene chloride extract to 300 mg of potassium bromide, and grind in an agate mortar. Evaporate to dryness under a current of air: the infrared absorption spectrum of the potassium bromide dispersion so obtained exhibits maxima only at the same wavelengths as that of a similar preparation of USP Tocainide Hydrochloride RS.

Dissolution ⟨711⟩—

Medium: water; 750 mL.

Apparatus 2: 50 rpm.

Time: 30 minutes.

Procedure—Determine the amount of $C_{11}H_{16}N_2O.HCl$ dissolved from ultraviolet absorbances at the wavelength of maximum absorbance at about 263 nm of filtered portions of the solution under test, suitably diluted with *Dissolution Medium*, if necessary, in comparison with a Standard solution having a known concentration of USP Tocainide Hydrochloride RS in the same medium.

Tolerances—Not less than 80% (*Q*) of the labeled amount of $C_{11}H_{16}N_2O.HCl$ is dissolved in 30 minutes.

Uniformity of dosage units ⟨905⟩: meet the requirements.

Procedure for content uniformity—Transfer 1 Tablet to a 100-mL volumetric flask, add 50 mL of water, place in an ultrasonic bath for 20 minutes, dilute with water to volume, and mix. Filter, discarding the first few mL of the filtrate. Transfer an accurately measured volume of the filtrate, equivalent to about 30 mg of tocainide hydrochloride, to a 100-mL volumetric flask, dilute with water to volume, and mix. Dissolve an accurately weighed quantity of USP Tocainide Hydrochloride RS in water, and dilute quantitatively and stepwise with water to obtain a Standard solution having a known concentration of about 300 µg per mL. Concomitantly determine the absorbances of both solutions at the wavelength of maximum absorbance at about 263 nm, with a suitable spectrophotometer, using water as the blank. Calculate the quantity, in mg, of $C_{11}H_{16}N_2O.HCl$ in the Tablet by the formula:

$$(TC/D)(A_U/A_S),$$

in which T is the labeled quantity, in mg, of tocainide hydrochloride in the Tablet, C is the concentration, in µg per mL, of USP Tocainide Hydrochloride RS in the Standard solution, D is the concentration, in µg per mL, of tocainide hydrochloride in the solution from the Tablet on the basis of the labeled quantity per Tablet and the extent of dilution, and A_U and A_S are the absorbances of the solution from the Tablet and the Standard solution, respectively.

Assay—

Mobile phase—Dissolve 2.16 g of sodium 1-octanesulfonate in 500 mL of 0.67 *N* acetic acid, add 500 mL of methanol, and mix. Degas, and filter the solution. Make adjustments if necessary (see *System Suitability* under *Chromatography* ⟨621⟩).

Standard preparation—Dissolve an accurately weighed quantity of USP Tocainide Hydrochloride RS quantitatively in water to obtain a solution having a known concentration of about 0.5 mg per mL.

Assay preparation—Weigh and finely powder not less than 20 Tocainide Hydrochloride Tablets. Transfer an accurately weighed portion of the powder, equivalent to about 100 mg of tocainide hydrochloride, to a 200-mL volumetric flask, add 100 mL of water, and place in an ultrasonic bath for 20 minutes. Dilute with water to volume, and mix. Filter the solution through a membrane filter, and use the filtrate as the *Assay preparation*.

Chromatographic system (see *Chromatography* ⟨621⟩)—The liquid chromatograph is equipped with a 254-nm detector and a 3.9-mm × 30-cm column that contains packing L1. The flow rate is about 2 mL per minute. Chromatograph the *Standard preparation*, and record the peak responses as directed under *Procedure:* the capacity factor, k', is greater than 1.6, the column efficiency determined from the analyte peak is not less than 1500 theoretical plates, the tailing factor for the analyte peak is not more than 2, and the relative standard deviation for replicate injections is not more than 2.0%.

Procedure—Separately inject equal volumes (about 40 µL) of the *Standard preparation* and the *Assay preparation* into the chromatograph, record the chromatograms, and measure the responses for the major peaks. Calculate the quantity, in mg, of $C_{11}H_{16}N_2O.HCl$ in the portion of Tablets taken by the formula:

$$200C(r_U/r_S),$$

in which C is the concentration, in mg per mL, of USP Tocainide Hydrochloride RS in the *Standard preparation*, and r_U and r_S are the peak responses obtained from the *Assay preparation* and the *Standard preparation*, respectively.

Tocopherols Excipient—*see* Tocopherols Excipient NF

Tolazamide

$CH_3-C_6H_4-SO_2NHCONH-N(CH_2)_6$

$C_{14}H_{21}N_3O_3S$ 311.40

Benzenesulfonamide, *N*-[[(hexahydro-1*H*-azepin-1-yl)amino]-carbonyl]-4-methyl-.

1-(Hexahydro-1*H*-azepin-1-yl)-3-(*p*-tolylsulfonyl)urea [*1156-19-0*].

» Tolazamide contains not less than 97.0 percent and not more than 103.0 percent of $C_{14}H_{21}N_3O_3S$, calculated on the dried basis.

Packaging and storage—Preserve in well-closed containers.

Reference standard—*USP Tolazamide Reference Standard*—Dry at a pressure not exceeding 5 mm of mercury at 60° for 3 hours before using.

Identification—

A: The infrared absorption spectrum of a potassium bromide dispersion of it exhibits maxima only at the same wavelengths as that of a similar preparation of USP Tolazamide RS.

B: The ultraviolet absorption spectrum of a 1 in 5000 solution in alcohol exhibits maxima and minima at the same wavelengths as that of a similar solution of USP Tolazamide RS, concomitantly measured, and the respective absorptivities, calculated on the dried basis, at the wavelength of maximum absorbance at about 262 nm do not differ by more than 2.5%.

Loss on drying ⟨731⟩—Dry it at a pressure not exceeding 5 mm of mercury at 60° for 3 hours: it loses not more than 0.5% of its weight.

Residue on ignition ⟨281⟩: not more than 0.2%.

Selenium ⟨291⟩: 0.003%, a 200-mg specimen being used.

Heavy metals, *Method II* ⟨231⟩: 0.002%.

***N*-Aminohexamethyleneimine**—

Trisodium pentacyanoaminoferroate solution—Mix 1.0 g of sodium nitroferricyanide and 3.2 mL of ammonium hydroxide in a glass-stoppered flask, insert the stopper in the flask, and refrigerate the mixture overnight. Pour the solution into 10 mL of dehydrated alcohol, and collect the yellow precipitate that is formed on coarse filter paper in a Buchner-type funnel by filtration under reduced pressure. Wash the residue on the filter with anhydrous ether, and store the dry solid in a desiccator. Dissolve a portion of the dry solid in water to obtain a solution containing 1.0 mg per mL, store in a refrigerator, and use within 7 days.

Buffer solution—Dissolve 0.96 g of anhydrous citric acid and 2.92 g of dibasic sodium phosphate in 200 mL of water. Adjust by adding phosphoric acid or 1 *N* sodium hydroxide, if necessary, to a pH of 5.4 ± 0.1.

Standard solution—Transfer, with the aid of a syringe, 100 mg of *N*-aminohexamethyleneimine to a 200-mL volumetric flask, dilute with acetone to volume, and mix. Dilute the resulting solution quantitatively with acetone to obtain a solution containing 12.5 µg per mL. Pipet 2 mL of this solution into a 25-mL glass-stoppered flask, add 8.0 mL of *Buffer solution*, shake the mixture, allow to stand for 15 minutes, and filter. Collect the filtrate in a suitable glass-stoppered tube, and use the filtrate as the *Standard solution.*

Test solution—Transfer 0.50 g of Tolazamide to a glass-stoppered, 25-mL flask, add 2.0 mL of acetone, insert the stopper in the flask, and shake the mixture vigorously for 15 minutes. Add 8.0 mL of *Buffer solution*, shake the mixture, allow to stand for 15 minutes, and filter. Collect the filtrate in a suitable glass-stoppered tube, and use the filtrate as the *Test solution.*

Procedure—Add 1.0 mL of *Trisodium pentacyanoaminoferroate solution* to the *Standard solution* and to the *Test solution*, and mix both solutions: the intensity of any pink color that may develop in the *Test solution* within 30 minutes does not exceed that produced in the *Standard solution* within 30 minutes (0.005%).

Assay—

Internal standard preparation—Dissolve a suitable quantity of Tolbutamide in alcohol-free chloroform to obtain a solution containing about 1.5 mg per mL.

Mobile phase—Prepare a solution containing hexane, water-saturated hexane, tetrahydrofuran, alcohol, and glacial acetic acid (475:475:20:15:9).

Standard preparation—Dissolve an accurately weighed quantity of USP Tolazamide RS in *Internal standard preparation* to obtain a solution having a known concentration of about 3 mg per mL.

Assay preparation—Transfer about 30 mg of Tolazamide, accurately weighed, to a 10-mL volumetric flask, add *Internal standard preparation* to volume, and mix.

Procedure—Using a suitable microsyringe or sampling valve, inject suitable volumes of the *Assay preparation* and the *Standard preparation* into a suitable high-pressure liquid chromatograph of the general type (see *Chromatography* ⟨621⟩) capable of providing column pressure up to about 1000 psi operated at room temperature and equipped with an ultraviolet detector capable of monitoring absorbance at 254 nm, a suitable recorder, and a 4-mm × 30-cm stainless steel column that contains 10-µm packing L3. In a suitable chromatogram, the resolution factor, *R*, is not less than 2.0 between peaks for tolazamide and the internal standard. Four replicate injections of the *Standard preparation* show a relative standard deviation for the peak response ratio of not more than 2.0%. Calculate the quantity, in mg, of $C_{14}H_{21}N_3O_3S$ in the portion of Tolazamide taken by the formula:

$$10C(R_U/R_S),$$

in which *C* is the concentration, in mg per mL, of USP Tolazamide RS in the *Standard preparation*, and R_U and R_S are the peak response ratios of the tolazamide and the tolbutamide peaks obtained from the *Assay preparation* and the *Standard preparation*, respectively.

Tolazamide Tablets

» Tolazamide Tablets contain not less than 95.0 percent and not more than 105.0 percent of the labeled amount of $C_{14}H_{21}N_3O_3S$.

Packaging and storage—Preserve in tight containers.

Reference standard—*USP Tolazamide Reference Standard*—Dry at a pressure not exceeding 5 mm of mercury at 60° for 3 hours before using.

Identification—Triturate a quantity of Tablets, equivalent to about 250 mg of tolazamide, with 50 mL of chloroform, and filter. Evaporate the filtrate to dryness, and dry in vacuum at 60° for 3 hours: the residue so obtained responds to *Identification test A* under *Tolazamide.*

Dissolution ⟨711⟩—

Medium: 0.05 *M* Tris(hydroxymethyl)aminomethane, pH 7.6, adjusted, if necessary, with hydrochloric acid to a pH of 7.6; 900 mL.

Apparatus 2: 75 rpm.

Time: 30 minutes.

Procedure—Determine the amount of $C_{14}H_{21}N_3O_3S$ dissolved from ultraviolet absorbances at the wavelength of maximum absorbance at about 224 nm of filtered portions of the solution under test, suitably diluted with *Dissolution Medium*, in comparison with a Standard solution having a known concentration of USP Tolazamide RS in the same medium. [NOTE—Sonicate the Standard solution until the Reference Standard is dissolved.]

Tolerances—Not less than 70% (*Q*) of the labeled amount of $C_{14}H_{21}N_3O_3S$ is dissolved in 30 minutes.

Uniformity of dosage units ⟨905⟩: meet the requirements.

Assay—

Internal standard preparation, Mobile phase, and *Standard preparation*—Prepare as directed in the *Assay* under *Tolazamide.*

Assay preparation—Weigh and finely powder not less than 10 Tolazamide Tablets. Weigh accurately a portion of the powder, equivalent to about 300 mg of tolazamide, and transfer to a suitable container. Add 100.0 mL of *Internal standard solution* and about 20 glass beads. Securely close the container, and shake vigorously for approximately 30 minutes. Centrifuge, and use the clear liquid as the *Assay preparation.*

Procedure—Proceed as directed for *Procedure* in the *Assay* under *Tolazamide.* Calculate the quantity, in mg, of $C_{14}H_{21}N_3O_3S$ in the portion of Tablets taken by the formula:

$$100C(R_U/R_S),$$

in which the terms are as defined therein.

Tolazoline Hydrochloride

$C_{10}H_{12}N_2.HCl$ 196.68

1*H*-Imidazole, 4,5-dihydro-2-(phenylmethyl)-, monohydrochloride.

2-Benzyl-2-imidazoline monohydrochloride [*59-97-2*].

» Tolazoline Hydrochloride contains not less than 98.0 percent and not more than 101.0 percent of $C_{10}H_{12}N_2.HCl$, calculated on the dried basis.

Packaging and storage—Preserve in well-closed containers.

Reference standard—*USP Tolazoline Hydrochloride Reference Standard*—Dry in vacuum over silica gel for 4 hours before using.

Identification—

A: The infrared absorption spectrum of a mineral oil dispersion of it exhibits maxima only at the same wavelengths as that of a similar preparation of USP Tolazoline Hydrochloride RS.

B: The R_f value of the principal spot in the chromatogram of the *Identification preparation* corresponds to that of *Standard preparation A*, as obtained in the test for *Chromatographic purity*.

Melting range ⟨741⟩: between 172.0° and 176.0°.

Loss on drying ⟨731⟩—Dry it in vacuum over silica gel for 4 hours: it loses not more than 0.2% of its weight.

Residue on ignition ⟨281⟩: not more than 0.1%.

Heavy metals, *Method II* ⟨231⟩: 0.001%.

Chromatographic purity—

Standard preparations—Dissolve USP Tolazoline Hydrochloride RS in methanol, and mix to obtain *Standard preparation A* having a known concentration of 100 µg per mL. Dilute quantitatively with methanol to obtain *Standard preparations*, designated below by letter, having the following compositions:

Dilution	Concentration (µg RS per mL)	Percentage (%, for comparison with test specimen)
A (undiluted)	100	0.5
B (4 in 5)	80	0.4
C (3 in 5)	60	0.3
D (2 in 5)	40	0.2
E (1 in 5)	20	0.1

Test preparation—Dissolve an accurately weighed quantity of Tolazoline Hydrochloride in methanol to obtain a solution containing 20 mg per mL.

Identification preparation—Dilute a portion of the *Test preparation* quantitatively with methanol to obtain a solution containing 100 µg per mL.

Detection reagent—Prepare (1) a solution of 0.5 g of potassium iodide in 50 mL of water, and (2) a solution of 1.5 g of soluble starch in 50 mL of boiling water. Just prior to use, mix 10 mL of each solution with 3 mL of alcohol.

Procedure—On a suitable thin-layer chromatographic plate (see *Chromatography* ⟨621⟩), coated with a 0.25-mm layer of chromatographic silica gel, apply separately 5 µL of the *Test preparation*, 5 µL of the *Identification preparation*, and 5 µL of each *Standard preparation*. Position the plate in a chromatographic chamber, and develop the chromatograms in a solvent system consisting of a mixture of methanol and ammonium hydroxide (95:5) until the solvent front has moved about three-fourths of the length of the plate. Remove the plate from the developing chamber, mark the solvent front, and allow the plate to dry under a current of warm air for at least 30 minutes. Expose the plate to chlorine gas for not more than 5 minutes, and air-dry until the chlorine has dissipated (about 15 minutes). Spray the plate with *Detection reagent*, and immediately compare the intensities of any secondary spots observed in the chromatogram of the *Test preparation* with those of the principal spots in the chromatograms of the *Standard preparations:* the sum of the intensities of all secondary spots obtained from the *Test preparation* corresponds to not more than 1.0%.

Assay—Dissolve about 300 mg of Tolazoline Hydrochloride, accurately weighed, in 100 mL of glacial acetic acid, add 25 mL of mercuric acetate TS, and titrate with 0.1 *N* perchloric acid VS, determining the end-point potentiometrically (see *Titrimetry* ⟨541⟩), using a calomel-glass electrode system. Perform a blank determination, and make any necessary correction. Each mL of 0.1 *N* perchloric acid is equivalent to 19.67 mg of $C_{10}H_{12}N_2 \cdot HCl$.

Tolazoline Hydrochloride Injection

» Tolazoline Hydrochloride Injection is a sterile solution of Tolazoline Hydrochloride in Water for Injection. It contains not less than 95.0 percent and not more than 105.0 percent of the labeled amount of $C_{10}H_{12}N_2 \cdot HCl$.

Packaging and storage—Preserve in single-dose or in multiple-dose containers, preferably of Type I glass.

Reference standard—*USP Tolazoline Hydrochloride Reference Standard*—Dry in vacuum over silica gel for 4 hours before using.

Identification—

A: Steam-distil a volume of Injection, equivalent to about 250 mg of tolazoline hydrochloride, for 5 to 10 minutes, and discard the distillate. Transfer the remaining solution to a separator, add about 2 mL of 1 *N* sodium hydroxide, and extract with 20 mL of ether. Filter the ether extract through cotton into a beaker, evaporate to dryness, and dry the residue in vacuum over silica gel for 4 hours: the infrared absorption spectrum of a potassium bromide dispersion of the tolazoline so obtained exhibits maxima only at the same wavelengths as that of a similar preparation of USP Tolazoline Hydrochloride RS.

B: It responds to *Identification test B* under *Tolazoline Hydrochloride*.

C: To 1 mL of Injection add 1 mL of ammonium reineckate TS: a pink precipitate is formed.

pH ⟨791⟩: between 3.0 and 4.0.

Other requirements—It meets the requirements under *Injections* ⟨1⟩.

Assay—

Standard preparation—Transfer about 30 mg of USP Tolazoline Hydrochloride RS, accurately weighed, to a 100-mL volumetric flask, dissolve in methanol, dilute with methanol to volume, and mix.

Assay preparation—Transfer an accurately measured volume of Tolazoline Hydrochloride Injection, equivalent to about 150 mg of tolazoline hydrochloride, to a 100-mL volumetric flask, dilute with methanol to volume, and mix. Transfer 20.0 mL of this solution to a second 100-mL volumetric flask, dilute with methanol to volume, and mix.

Procedure—Transfer 3.0 mL each of the *Standard preparation*, the *Assay preparation*, and methanol to provide the blank, to separate 25-mL volumetric flasks. To each flask add 1 mL of 0.5 *N* sodium hydroxide and 1 mL of dilute sodium nitroferricyanide TS (1 in 2), mix, and allow to stand for 10 minutes. Add 3 mL of sodium bicarbonate solution (1 in 12) to each flask, dilute with water to volume, mix, and allow to stand for 10 minutes. Concomitantly determine the absorbances of the solutions in 1-cm cells at the wavelength of maximum absorbance at about 565 nm, with a suitable spectrophotometer, against the blank. Calculate the quantity, in mg, of $C_{10}H_{12}N_2 \cdot HCl$ in each mL of the Injection taken by the formula:

$$(0.5C/V)(A_U/A_S),$$

in which *C* is the concentration, in µg per mL, of USP Tolazoline Hydrochloride RS in the *Standard preparation*, *V* is the volume, in mL, of Injection taken, and A_U and A_S are the absorbances of the solutions from the *Assay preparation* and the *Standard preparation*, respectively.

Tolazoline Hydrochloride Tablets

» Tolazoline Hydrochloride Tablets contain not less than 95.0 percent and not more than 105.0 percent of the labeled amount of $C_{10}H_{12}N_2 \cdot HCl$.

Packaging and storage—Preserve in well-closed containers.

Reference standard—*USP Tolazoline Hydrochloride Reference Standard*—Dry in vacuum over silica gel for 4 hours before using.

Identification—Place a quantity of finely powdered Tablets, equivalent to about 200 mg of tolazoline hydrochloride, in a glass-stoppered centrifuge tube containing about 25 mL of water, shake thoroughly, and centrifuge. Decant the clear supernatant liquid into a separator, and proceed as directed for *Identification test A* under *Tolazoline Hydrochloride Injection*, beginning with "add about 2 mL of 1 *N* sodium hydroxide."

Dissolution ⟨711⟩—

Medium: water; 900 mL.

Apparatus 2: 50 rpm.

Time: 45 minutes.

Procedure—Determine the amount of $C_{10}H_{12}N_2 \cdot HCl$ dissolved, employing the procedure set forth in the *Assay*, making any necessary modifications.

Tolerances—Not less than 75% (*Q*) of the labeled amount of $C_{10}H_{12}N_2 \cdot HCl$ is dissolved in 45 minutes.

Uniformity of dosage units ⟨905⟩: meet the requirements.

Assay—

Standard preparation—Prepare as directed in the *Assay* under *Tolazoline Hydrochloride Injection.*

Assay preparation—Weigh and finely powder not less than 20 Tolazoline Hydrochloride Tablets. Transfer an accurately weighed portion of the powder, equivalent to about 300 mg of tolazoline hydrochloride, to a 100-mL volumetric flask, add methanol to volume, mix, and centrifuge the mixture. Transfer 10 mL of the clear supernatant solution to a 100-mL volumetric flask, dilute with methanol to volume, and mix.

Procedure—Proceed as directed for *Procedure* in the *Assay* under *Tolazoline Hydrochloride Injection.* Calculate the quantity, in mg, of $C_{10}H_{12}N_2.HCl$ in the portion of Tablets taken by the formula:

$$C(A_U/A_S).$$

Tolbutamide

$CH_3-C_6H_4-SO_2-NHCONH(CH_2)_3CH_3$

$C_{12}H_{18}N_2O_3S$ 270.35

Benzenesulfonamide, *N*-[(butylamino)carbonyl]-4-methyl-.

1-Butyl-3-(*p*-tolylsulfonyl)urea [*64-77-7*].

» Tolbutamide contains not less than 97.0 percent and not more than 103.0 percent of $C_{12}H_{18}N_2O_3S$, calculated on the dried basis.

Packaging and storage—Preserve in well-closed containers.

Reference standard—*USP Tolbutamide Reference Standard*—Dry at 105° for 3 hours before using.

Identification—The infrared absorption spectrum of a mineral oil dispersion of it exhibits maxima only at the same wavelengths as that of a similar preparation of USP Tolbutamide RS.

Melting range ⟨741⟩: between 126° and 130°.

Loss on drying ⟨731⟩—Dry it at 105° for 3 hours: it loses not more than 0.5% of its weight.

Selenium ⟨291⟩: 0.003%, a 100-mg specimen, mixed with 100 mg of magnesium oxide, being used.

Heavy metals, *Method II* ⟨231⟩: 0.002%.

Non-sulfonyl urea—Dissolve 500 mg in 10 mL of 0.5 *N* ammonium hydroxide: not more than a faint opalescence occurs.

Assay—

Mobile phase—Prepare a filtered and degassed mixture of hexane, water-saturated–hexane, tetrahydrofuran, alcohol, and glacial acetic acid (475:475:20:15:9). Make adjustments if necessary (see *System Suitability* under *Chromatography* ⟨621⟩).

Internal standard solution—Dissolve a suitable quantity of tolazamide in alcohol-free chloroform to obtain a solution containing about 3 mg per mL.

Standard preparation—Dissolve an accurately weighed quantity of USP Tolbutamide RS in *Internal standard solution* to obtain a solution having a known concentration of about 1.5 mg per mL.

Assay preparation—Transfer about 15 mg of Tolbutamide, accurately weighed, to a 10-mL volumetric flask. Dissolve in *Internal standard solution*, dilute with *Internal standard solution* to volume, and mix.

Chromatographic system (see *Chromatography* ⟨621⟩)—The liquid chromatograph is equipped with a 254-nm detector and a 4.0-mm × 30-cm column that contains packing L3. The flow rate is about 1.5 mL per minute. Chromatograph the *Standard preparation*, and record the peak responses as directed under *Procedure:* the relative standard deviation for replicate injections is not more than 2.0%, and the resolution, *R*, between tolbutamide and tolazamide is not less than 2.0.

Procedure—Separately inject equal volumes (about 10 µL) of the *Standard preparation* and the *Assay preparation* into the chromatograph, record the chromatograms, and measure the responses for the major peaks. The relative retention times are about 0.6 for tolbutamide and 1.0 for tolazamide. Calculate the quantity, in mg, of $C_{12}H_{18}N_2O_3S$ in the portion of Tolbutamide taken by the formula:

$$10C(R_U/R_S),$$

in which *C* is the concentration, in mg per mL, of USP Tolbutamide RS in the *Standard preparation*, and R_U and R_S are the peak response ratios obtained from the *Assay preparation* and the *Standard preparation*, respectively.

Tolbutamide Tablets

» Tolbutamide Tablets contain not less than 90.0 percent and not more than 110.0 percent of the labeled amount of $C_{12}H_{18}N_2O_3S$.

Packaging and storage—Preserve in well-closed containers.

Reference standard—*USP Tolbutamide Reference Standard*—Dry at 105° for 3 hours before using.

Identification—Triturate a quantity of finely powdered Tablets, equivalent to about 500 mg of tolbutamide, with 50 mL of chloroform, and filter. Evaporate the clear filtrate on a steam bath to dryness: the residue so obtained responds to *Identification test A* under *Tolbutamide.*

Dissolution ⟨711⟩—

Medium: pH 7.4 phosphate buffer (see *Buffer Solutions* in the section, *Reagents, Indicators, and Solutions*); 900 mL.

Apparatus 2: 75 rpm.

Time: 30 minutes.

Procedure—Measure the amount in solution in filtered portions of the *Dissolution Medium*, suitably diluted with water, if necessary, at the wavelength of maximum absorbance at about 226 nm, with a suitable spectrophotometer, in comparison with a solution having a known concentration of USP Tolbutamide RS. An amount of alcohol not to exceed 1% of the total volume of the Standard solution may be used to bring the Reference Standard into solution prior to dilution with *Dissolution Medium.*

Tolerances—Not less than 70% (*Q*) of the labeled amount of $C_{12}H_{18}N_2O_3S$ is dissolved in 30 minutes.

Uniformity of dosage units ⟨905⟩: meet the requirements.

Assay—

Mobile phase, Internal standard solution, Standard preparation, and *Chromatographic system*—Prepare as directed in the *Assay* under *Tolbutamide.*

Assay preparation—Weigh and finely powder not less than 10 Tolbutamide Tablets. Transfer an accurately weighed portion of the powder, equivalent to about 150 mg of tolbutamide, to a suitable container. Add 100.0 mL of *Internal standard solution* and about 20 glass beads. Securely close the container, and shake vigorously by mechanical means for approximately 30 minutes. Centrifuge and use the clear supernatant liquid.

Procedure—Proceed as directed for *Procedure* in the *Assay* under *Tolbutamide.* Calculate the quantity, in mg, of $C_{12}H_{18}N_2O_3S$ in the portion of Tablets taken by the formula:

$$100C(R_U/R_S),$$

in which the terms are as defined therein.

Sterile Tolbutamide Sodium

» Sterile Tolbutamide Sodium is prepared from Tolbutamide with the aid of Sodium Hydroxide. It is suitable for parenteral use. It contains an amount of tolbutamide sodium equivalent to not less than 95.0 percent and not more than 105.0 percent of the labeled amount of tolbutamide ($C_{12}H_{18}N_2O_3S$).

Packaging and storage—Preserve in *Containers for Sterile Solids* as described under *Injections* ⟨1⟩.

Reference standard—*USP Tolbutamide Reference Standard*—Dry at 105° for 3 hours before using.

Constituted solution—At the time of use, the constituted solution prepared from Sterile Tolbutamide Sodium, as directed in the labeling, meets the requirements for *Constituted Solutions* under *Injections* ⟨1⟩.

Identification—Place about 200 mg in a suitable container, dissolve in about 20 mL of water, add 2 mL of 2 *N* sulfuric acid, and extract with 10 mL of chloroform. Filter, and evaporate an aliquot of the chloroform layer. Dry the residue at 105° for 3 hours: the infrared absorption spectrum of a mineral oil dispersion of the residue so obtained exhibits maxima only at the same wavelengths as that of a similar preparation of USP Tolbutamide RS.

Pyrogen—It meets the requirements of the *Pyrogen Test* ⟨151⟩, the test dose being 100 mg of tolbutamide sodium per kg.

pH ⟨791⟩: between 8.0 and 9.8, in a solution containing 50 mg per mL.

Loss on drying ⟨731⟩—Dry it at 105° for 4 hours: it loses not more than 1.0% of its weight.

Other requirements—It meets the requirements for *Sterility Tests* ⟨71⟩, *Uniformity of Dosage Units* ⟨905⟩, and *Labeling* under *Injections* ⟨1⟩.

Assay—

Mobile phase and *Chromatographic system*—Prepare as directed in the *Assay* under *Tolbutamide*.

Internal standard solution—Prepare a solution of tolazamide in alcohol-free chloroform containing about 15 mg per mL.

Diluting solution—Prepare an alcohol-free chloroform solution containing 3% (v/v) of glacial acetic acid.

Standard preparation—Dissolve an accurately weighed quantity of USP Tolbutamide RS in *Internal standard solution* to obtain a known concentration of about 7.5 mg per mL. Add *Diluting solution* to obtain a *Standard preparation* having a final known concentration of about 1.5 mg of tolbutamide per mL.

Assay preparation—Add about 15 mL of water to 1 container of Sterile Tolbutamide Sodium, and shake vigorously to dissolve the contents. Transfer the contents, using adequate rinsing with water, to a 50-mL volumetric flask. Dilute with water to volume, and mix. Transfer an accurately measured portion of this solution, equivalent to about 75 mg of tolbutamide, to a 50 mL volumetric flask, add 10.0 mL of *Internal standard solution*, and dilute with *Diluting solution* to volume. Shake vigorously for about 15 minutes, and centrifuge or allow to stand for about 15 minutes. Use the lower, clear layer, as the *Assay preparation*.

Procedure—Separately inject equal volumes (about 20 µL) of the *Standard preparation* and the *Assay preparation* into the chromatograph, record the chromatograms, and measure the responses for the major peaks. The relative retention times are about 0.6 for tolbutamide and 1.0 for tolazamide. Calculate the quantity, in mg, of $C_{12}H_{18}N_2O_3S$ in the portion of solution taken for the *Assay preparation* by the formula:

$$50C(R_U/R_S),$$

in which *C* is the concentration, in mg per mL, of USP Tolbutamide RS in the *Standard preparation*, and R_U and R_S are the peak response ratios of the tolbutamide and internal standard peaks obtained from the *Assay preparation* and the *Standard preparation*, respectively.

Tolmetin Sodium

$C_{15}H_{14}NNaO_3 . 2H_2O$ 315.30

1*H*-Pyrrole-2-acetic acid, 1-methyl-5-(4-methylbenzoyl)-, sodium salt, dihydrate.

Sodium 1-methyl-5-*p*-toluoylpyrrole-2-acetate dihydrate [*64490-92-2*].

Anhydrous 279.27 [*35711-34-3*].

» Tolmetin Sodium contains not less than 98.0 percent and not more than 102.0 percent of $C_{15}H_{14}NNaO_3$, calculated on the dried basis.

Packaging and storage—Preserve in well-closed containers.

Reference standard—*USP Tolmetin Sodium Reference Standard*—Dry in vacuum at 60° for 4 hours before using.

Identification—

A: The infrared absorption spectrum of a potassium bromide dispersion of it, previously dried, exhibits maxima only at the same wavelengths as that of a similar preparation of USP Tolmetin Sodium RS.

B: The ultraviolet absorption spectrum of a 1 in 100,000 solution in pH 7 phosphate buffer (see *Buffer solutions* in the section, *Reagents, Indicators, and Solutions*) exhibits maxima and minima at the same wavelengths as that of a similar solution of USP Tolmetin Sodium RS, concomitantly measured.

C: A solution (1 in 20) responds to the tests for *Sodium* ⟨191⟩.

Loss on drying ⟨731⟩—Dry it in vacuum at 60° for 4 hours: it loses between 10.4% and 12.4% of its weight.

Heavy metals, *Method II* ⟨231⟩: 0.002%.

Chromatographic purity—Dissolve 125 mg in 10 mL of methanol to obtain the *Test solution*. Dissolve USP Tolmetin Sodium RS in methanol to obtain a *Standard solution* having a concentration of 12.5 mg per mL. Dilute a portion of this *Standard solution* quantitatively with methanol to obtain a *Diluted standard solution* having a concentration of 62.5 µg per mL. Apply separate 20-µL portions of the three solutions on the starting line of a suitable thin-layer chromatographic plate (see *Chromatography* ⟨621⟩), coated with 0.25-mm layer of chromatographic silica gel mixture. Develop the chromatogram in a solvent system consisting of a mixture of chloroform and glacial acetic acid (95:5) until the solvent front has moved about three-fourths of the length of the plate. Remove the plate from the chamber, mark the solvent front, air-dry, and view under short-wavelength ultraviolet light: the R_f value of the principal spot from the *Test solution* corresponds to that from the *Standard solution*. Any other spot obtained from the *Test solution* does not exceed in size or intensity the principal spot obtained from the *Diluted standard solution* (0.5%), and the sum of the total impurities based on a comparison of the intensities of all such other spots with the *Diluted standard solution* does not exceed 2.0%.

Assay—Dissolve, by warming, about 300 mg of Tolmetin Sodium, accurately weighed, in 150 mL of glacial acetic acid. Cool to room temperature, and titrate with 0.1 *N* perchloric acid VS, determining the end-point electrometrically. Perform a blank determination, and make any necessary correction. Each mL of 0.1 *N* perchloric acid is equivalent to 27.93 mg of $C_{15}H_{14}NNaO_3$.

Tolmetin Sodium Capsules

» Tolmetin Sodium Capsules contain an amount of tolmetin sodium equivalent to not less than 93.0 percent and not more than 107.0 percent of the labeled amount of tolmetin ($C_{15}H_{15}NO_3$).

Packaging and storage—Preserve in tight containers.

Reference standard—*USP Tolmetin Sodium Reference Standard*—Dry in vacuum at 60° for 4 hours before using.

Identification—

A: The contents of Capsules respond to *Identification test A* under *Tolmetin Sodium*.

B: The chromatogram of the *Assay preparation* obtained as directed in the *Assay* exhibits a major peak for tolmetin the retention time of which corresponds to that exhibited in the chromatogram of the *Standard preparation* obtained as directed in the *Assay*.

Dissolution—

Medium—Dissolve 2.0 g of sodium chloride in 7.0 mL of hydrochloric acid, and add water to make 1000 mL (*Solution A*). Dissolve 6.8 g of monobasic potassium phosphate in 250 mL of water, mix, and add 190 mL of 0.2 *N* sodium hydroxide and 400 mL of water. Adjust the solution with 0.2 *N* sodium hydroxide to a pH of 7.4 to 7.6. Dilute with water to 1000 mL (*Solution B*). Add 336 mL of *Solution A* to 664 mL of *Solution B*, mix, and adjust the solution with small amounts of either solution to a pH of 4.5; 900 mL.

Apparatus 2: 50 rpm.

Time: 30 minutes.

Procedure—Determine the amount of $C_{15}H_{15}NO_3$ dissolved from ultraviolet absorbances at the wavelength of maximum absorbance at about 322 nm of filtered portions of the solution under test, suitably diluted with 0.1 *N* sodium hydroxide, in comparison with a Standard solution having a known concentration of USP Tolmetin Sodium RS in the same medium.

Tolerances—Not less than 85% (*Q*) of the labeled amount of $C_{15}H_{15}NO_3$ is dissolved in 30 minutes.

Uniformity of dosage units ⟨905⟩: meet the requirements.

Assay—

pH 2.7 buffer solution—Dissolve 1.7 g of tetrabutylammonium phosphate in 1000 mL of water, and adjust by the addition of phosphoric acid to a pH of 2.7 ± 0.1.

Mobile phase—Prepare a filtered and degassed mixture of 64 parts of *pH 2.7 buffer solution* and 36 parts of acetonitrile, making adjustments if necessary (see *System Suitability* under *Chromatography* ⟨621⟩).

Solvent mixture—Mix 400 mL of acetonitrile with 600 mL of 0.01 *N* sodium hydroxide.

Standard preparation—Dissolve an accurately weighed quantity of USP Tolmetin Sodium RS in *Solvent mixture* to obtain a solution having a known concentration of about 0.65 mg of anhydrous tolmetin sodium per mL.

Assay preparation—Remove, as completely as possible, the contents of not less than 20 Tolmetin Sodium Capsules, weigh, and mix. Transfer an accurately weighed portion of the powder, equivalent to about 60 mg of tolmetin, to a 100-mL volumetric flask. Add about 75 mL of *Solvent mixture*, and shake by mechanical means for 30 minutes. Dilute with *Solvent mixture* to volume, mix, and filter through a filter having a porosity of 0.45 µm or less.

Resolution solution—Dissolve suitable quantities of *p*-toluic acid and USP Tolmetin Sodium RS in *Solvent mixture* to obtain a solution containing 10 µg and 500 µg, respectively, in each mL.

Chromatographic system (see *Chromatography* ⟨621⟩)—The liquid chromatograph is equipped with a 254-nm detector and a 4.6-mm × 15-cm column that contains 5-µm packing L7, and is operated at a temperature of 40 ± 1.0°. The flow rate is about 3 mL per minute. Chromatograph the *Standard preparation* and the *Resolution solution*, and record the peak responses as directed under *Procedure:* the resolution, *R*, between the *p*-toluic acid and tolmetin sodium peaks is not less than 1.2, and the relative standard deviation for replicate injections of the *Standard preparation* is not more than 3.0%.

Procedure—Separately inject equal volumes (about 20 µL) of the *Standard preparation* and the *Assay preparation* into the chromatograph, record the chromatograms, and measure the responses for the major peaks. Calculate the quantity, in mg, of $C_{15}H_{15}NO_3$ in the portion of Capsule contents taken by the formula:

$$100C(257.29/279.27)(r_U/r_S),$$

in which *C* is the concentration, in mg per mL, of USP Tolmetin Sodium RS in the *Standard preparation*, 257.29 and 279.27 are the molecular weights of tolmetin and anhydrous tolmetin sodium, respectively, and r_U and r_S are the peak responses obtained from the *Assay preparation* and the *Standard preparation*, respectively.

Tolmetin Sodium Tablets

» Tolmetin Sodium Tablets contain an amount of tolmetin sodium equivalent to not less than 90.0 percent and not more than 110.0 percent of the labeled amount of tolmetin ($C_{15}H_{15}NO_3$).

Packaging and storage—Preserve in well-closed containers.

Reference standard—*USP Tolmetin Sodium Reference Standard*—Dry in vacuum at 60° for 4 hours before using.

Identification—

A: The infrared absorption spectrum of a portion of powdered Tablets, the multiple internal reflectance technique being used, exhibits maxima at the same wavelengths as that of USP Tolmetin Sodium RS.

B: The chromatogram of the *Assay preparation* obtained as directed in the *Assay* exhibits a major peak for tolmetin the retention time of which corresponds to that exhibited in the chromatogram of the *Standard preparation* obtained as directed in the *Assay*.

Dissolution ⟨711⟩—

Medium, Apparatus, Time, and *Procedure*—Proceed as directed in the test for *Dissolution* under *Tolmetin Sodium Capsules*.

Tolerances—Not less than 75% (*Q*) of the labeled amount of $C_{15}H_{15}NO_3$ is dissolved in 30 minutes.

Uniformity of dosage units ⟨905⟩: meet the requirements.

Assay—

pH 2.7 buffer solution, Mobile phase, Solvent mixture, Standard preparation, Resolution solution, and *Chromatographic system*—Prepare as directed in the *Assay* under *Tolmetin Sodium Capsules*.

Assay preparation—Weigh and finely powder not less than 20 Tolmetin Sodium Tablets. Transfer an accurately weighed portion of the powder, equivalent to about 60 mg of tolmetin, to a 100-mL volumetric flask. Add about 75 mL of *Solvent mixture*, and shake by mechanical means for 30 minutes. Dilute with *Solvent mixture* to volume, mix, and filter through a filter having a porosity of 0.45 µm or less.

Procedure—Separately inject equal volumes (about 20 µL) of the *Standard preparation* and the *Assay preparation* into the chromatograph, record the chromatograms, and measure the responses for the major peaks. Calculate the quantity, in mg, of $C_{15}H_{15}NO_3$ in the portion of Tablets taken by the formula:

$$100C(257.29/279.27)(r_U/r_S),$$

in which *C* is the concentration, in mg per mL, of USP Tolmetin Sodium RS in the *Standard preparation*, 257.29 and 279.27 are the molecular weights of tolmetin and anhydrous tolmetin sodium, respectively, and r_U and r_S are the peak responses obtained from the *Assay preparation* and the *Standard preparation*, respectively.

Tolnaftate

$C_{19}H_{17}NOS$ 307.41

Carbamothioic acid, methyl(3-methylphenyl)-, *O*-2-naphthalenyl ester.

O-2-Naphthyl *m,N*-dimethylthiocarbanilate [*2398-96-1*].

» Tolnaftate contains not less than 98.0 percent and not more than 102.0 percent of $C_{19}H_{17}NOS$, calculated on the dried basis.

Packaging and storage—Preserve in tight containers.

Reference standard—*USP Tolnaftate Reference Standard*—Dry in vacuum at 65° for 3 hours.

Identification—

A: The infrared absorption spectrum of a potassium bromide dispersion of it exhibits maxima only at the same wavelengths as that of a similar preparation of USP Tolnaftate RS.

B: The ultraviolet absorption spectrum of the solution employed for measurement of absorbance in the *Assay* exhibits maxima and minima at the same wavelengths as that of a similar solution of USP Tolnaftate RS, concomitantly measured.

C: Prepare a test solution by dissolving 10 mg in 10 mL of alcohol. On a thin-layer chromatographic plate (see *Chromatography* ⟨621⟩), coated with a 0.25-mm layer of chromatographic silica gel, apply 10 µL of this test solution and 10 µL of a Standard solution of USP Tolnaftate RS in alcohol having a concentration of 1.0 mg per mL. Allow the spots to dry, and develop the chromatogram, using toluene as the solvent system, until the solvent front has moved about three-fourths of the length of the plate. Remove the plate from the developing chamber, allow the solvent to evaporate, and view under short-wavelength ultraviolet light: the R_f value of the principal spot obtained from the test solution corresponds to that obtained from the Standard solution.

Melting range ⟨741⟩: between 110° and 113°.

Loss on drying ⟨731⟩: Dry it in vacuum at 65° for 3 hours: it loses not more than 0.5% of its weight.

Residue on ignition ⟨281⟩: not more than 0.1%.

Heavy metals, *Method II* ⟨231⟩: 0.002%.

Assay—Dissolve about 50 mg of Tolnaftate, accurately weighed, in methanol, and dilute the solution quantitatively and stepwise with methanol to obtain a concentration of about 10 µg per mL. Dissolve an accurately weighed quantity of USP Tolnaftate RS in methanol, and dilute quantitatively and stepwise with methanol to obtain a Standard solution having a known concentration of about 10 µg per mL. Concomitantly determine the absorbances of both solutions in 1-cm cells at the wavelength of maximum absorbance at about 258 nm, with a suitable spectrophotometer, using methanol as the blank. Calculate the quantity, in mg, of $C_{19}H_{17}NOS$ in the portion of Tolnaftate taken by the formula:

$$5C(A_U/A_S),$$

in which C is the concentration, in µg per mL, of USP Tolnaftate RS in the Standard solution, and A_U and A_S are the absorbances of the solution of Tolnaftate and the Standard solution, respectively.

Tolnaftate Topical Aerosol Powder

» Tolnaftate Topical Aerosol Powder is a suspension of powder in suitable propellants in a pressurized container. The powder contains not less than 90.0 percent and not more than 110.0 percent of the labeled amount of $C_{19}H_{17}NOS$.

Packaging and storage—Preserve in tight, pressurized containers, and avoid exposure to excessive heat.

Reference standard—*USP Tolnaftate Reference Standard*—Dry in vacuum at 65° for 3 hours before using.

Identification—It responds to the *Identification test* under *Tolnaftate Powder.*

Other requirements—It meets the requirements for *Leak Testing* and *Pressure Testing* under *Aerosols* ⟨601⟩.

Assay—

Mobile phase, Internal standard solution, Standard preparation, and *Chromatographic system*—Proceed as directed in the *Assay* under *Tolnaftate Powder.*

Assay preparation—Remove the actuator button, and replace it with an actuator button that has a small-diameter, stiff polyethylene tube about 15 cm in length fitted tightly into the orifice. Deliver the entire contents of the Topical Aerosol Powder into a conical flask, and heat the flask gently to expel the liquid phase. Cool, mix, and transfer an accurately weighed portion of the powder, equivalent to about 5 mg of tolnaftate, to a screw-capped, 50-mL centrifuge tube. Proceed as directed for *Assay preparation* in the *Assay* under *Tolnaftate Powder,* beginning with "Add 25.0 mL of methanol."

Procedure—Proceed as directed for *Procedure* in the *Assay* under *Tolnaftate Powder.*

Tolnaftate Cream

» Tolnaftate Cream contains not less than 90.0 percent and not more than 110.0 percent of the labeled amount of $C_{19}H_{17}NOS$.

Packaging and storage—Preserve in tight containers.

Reference standard—*USP Tolnaftate Reference Standard*—Dry in vacuum at 65° for 3 hours.

Identification—Evaporate 10 mL of the next-to-final chloroform solution prepared in the *Assay* on a steam bath just to dryness, and dissolve the residue in 1 mL of alcohol. Using this as the test solution, proceed as directed in *Identification test C* under *Tolnaftate:* the specified result is observed.

Minimum fill ⟨755⟩: meets the requirements.

Assay—Transfer a portion of Tolnaftate Cream, equivalent to about 10 mg of tolnaftate and accurately weighed, to a 250-mL separator containing about 75 mL of chloroform. Wash the chloroform solution successively with two 25-mL portions of 0.1 *N* sodium hydroxide, two 25-mL portions of 0.1 *N* hydrochloric acid, and 25 mL of water. Filter the chloroform layer through a chloroform-washed cotton pledget into a 100-mL volumetric flask. Add chloroform to volume, and mix. [NOTE—Reserve a 10-mL portion of this solution for the *Identification test.*] Dilute 5.0 mL of the solution with chloroform to 50.0 mL, and mix. Dissolve an accurately weighed quantity of USP Tolnaftate RS in chloroform, and dilute quantitatively and stepwise with chloroform to obtain a Standard solution having a known concentration of about 10 µg per mL. Concomitantly determine the absorbances of both solutions in 1-cm cells at the wavelength of maximum absorbance at about 258 nm, with a suitable spectrophotometer, using chloroform as the blank. Calculate the quantity, in mg, of $C_{19}H_{17}NOS$ in the portion of Cream taken by the formula:

$$C(A_U/A_S),$$

in which C is the concentration, in µg per mL, of USP Tolnaftate RS in the Standard solution, and A_U and A_S are the absorbances of the solution from the Cream and the Standard solution, respectively.

Tolnaftate Gel

» Tolnaftate Gel contains not less than 90.0 percent and not more than 110.0 percent of the labeled amount of $C_{19}H_{17}NOS$.

Packaging and storage—Preserve in tight containers.

Reference standard—*USP Tolnaftate Reference Standard*—Dry in vacuum at 65° for 3 hours.

Identification—It responds to the *Identification test* under *Tolnaftate Cream.*

Minimum fill ⟨755⟩: meets the requirements.

Assay—Proceed with Tolnaftate Gel as directed in the *Assay* under *Tolnaftate Cream,* except to omit the filtration of the chloroform layer through a chloroform-washed cotton pledget.

Tolnaftate Powder

» Tolnaftate Powder contains not less than 90.0 percent and not more than 110.0 percent of the labeled amount of $C_{19}H_{17}NOS$.

Packaging and storage—Preserve in tight containers.

Reference standard—*USP Tolnaftate Reference Standard*—Dry in vacuum at 65° for 3 hours before using.

Identification—Evaporate the 5-mL portion of the methanol solution, reserved from the *Assay preparation,* on a steam bath just to dryness, and dissolve the residue in 1 mL of alcohol. Using this as the test solution, proceed as directed in *Identification test C* under *Tolnaftate:* the specified result is observed.

Minimum fill ⟨755⟩: meets the requirements.

Assay—

Mobile phase—Prepare a filtered and degassed mixture of acetonitrile and water (2:1). Make adjustments if necessary (see *System Suitability* under *Chromatography* ⟨621⟩).

Internal standard solution—Dissolve progesterone in methanol to obtain a solution containing about 1 mg per mL.

Standard preparation—Dissolve an accurately weighed quantity of USP Tolnaftate RS in methanol to obtain a solution having a known concentration of about 0.22 mg per mL. Transfer 20.0 mL of this solution to a 50-mL volumetric flask, add 5.0 mL of *Internal standard solution*, dilute with methanol to volume, and mix to obtain a solution having a known concentration of about 0.088 mg of USP Tolnaftate RS per mL.

Assay preparation—Transfer an accurately weighed quantity of Tolnaftate Powder, equivalent to about 5 mg of tolnaftate, to a screw-capped, 50-mL centrifuge tube. Add 25.0 mL of methanol, place the cap on the tube, rotate on a rotating device for 10 minutes, and centrifuge at about 2000 rpm for 5 minutes. Pass the supernatant liquid through a 0.45-μm porosity filter, and transfer 20.0 mL of the filtrate to a 50-mL volumetric flask, retaining the remaining portion of the filtrate (about 5 mL) for the *Identification test.* Add 5.0 mL of *Internal standard solution* to the volumetric flask, dilute with methanol to volume, and mix.

Chromatographic system (see *Chromatography* ⟨621⟩)—The liquid chromatograph is equipped with a 254-nm detector and a 4-mm × 30-cm column that contains 10-μm packing L1. The flow rate is about 1 mL per minute. Chromatograph the *Standard preparation*, and record the peak responses as directed under *Procedure:* the resolution, *R*, between the analyte and internal standard peaks is not less than 3.0, and the relative standard deviation for replicate injections is not more than 3.0%.

Procedure—Separately inject equal volumes (about 10 μL) of the *Standard preparation* and the *Assay preparation* into the chromatograph, record the chromatograms, and measure the responses for the major peaks. The relative retention times are about 0.7 for progesterone and 1.0 for tolnaftate. Calculate the quantity, in mg, of $C_{19}H_{17}NOS$ in the portion of Powder taken by the formula:

$$62.5C(R_U/R_S),$$

in which *C* is the concentration, in mg per mL, of USP Tolnaftate RS in the *Standard preparation*, and R_U and R_S are the peak response ratios obtained from the *Assay preparation* and the *Standard preparation*, respectively.

Tolnaftate Topical Solution

» Tolnaftate Topical Solution contains not less than 90.0 percent and not more than 115.0 percent of the labeled amount of $C_{19}H_{17}NOS$.

Packaging and storage—Preserve in tight containers.

Reference standard—*USP Tolnaftate Reference Standard*—Dry in vacuum at 65° for 3 hours.

Identification—Evaporate 25 mL of the next-to-final chloroform solution prepared in the *Assay* on a steam bath just to dryness, and dissolve the residue in 1 mL of alcohol. Using this as the test solution, proceed as directed in *Identification test C* under *Tolnaftate:* the specified result is observed.

Assay—Pipet into a separator a volume of Tolnaftate Topical Solution, equivalent to about 10 mg of tolnaftate, add 50 mL of chloroform, and extract with 50 mL of 0.1 *N* sodium hydroxide. Filter the chloroform layer through a chloroform-washed cotton pledget into a 250-mL volumetric flask, and extract the aqueous layer with two 45-mL portions of chloroform, filtering each portion into the flask. Add chloroform to volume, and mix. [NOTE—Reserve a 25-mL portion of this solution for the *Identification test.*] Dilute 25.0 mL of the solution with chloroform to 100.0 mL, and mix. Dissolve an accurately weighed quantity of USP Tolnaftate RS in chloroform, and dilute quantitatively and stepwise with chloroform to obtain a Standard solution having a known concentration of about 10 μg per mL. Concomitantly determine the absorbances of both solutions in 1-cm cells at the wavelength of maximum absorbance at about 258 nm, with a suitable spectrophotometer, using chloroform as the blank. Calculate the quantity, in mg, of $C_{19}H_{17}NOS$ in each mL of Topical Solution taken by the formula:

$$(C/V)(A_U/A_S),$$

in which *C* is the concentration, in μg per mL, of USP Tolnaftate RS in the Standard solution, *V* is the volume, in mL, of Topical Solution taken, and A_U and A_S are the absorbances of the solution from Tolnaftate Topical Solution and the Standard solution, respectively.

Tolu Balsam

» Tolu Balsam is a balsam obtained from *Myroxylon balsamum* (Linné) Harms (Fam. Leguminosae).

Packaging and storage—Preserve in tight containers, and avoid exposure to excessive heat.

Rosin, rosin oil, and copaiba—Place 1 g, powdered or crushed if necessary, in a small mortar, and add 10 mL of petroleum benzin. Triturate well for 1 to 2 minutes, filter into a test tube, and to the filtrate so obtained add 10 mL of freshly prepared cupric acetate solution (1 in 200). Shake, and allow the phases to separate: the petroleum benzin layer does not show a green color.

Acid value ⟨401⟩—Dissolve about 1 g, accurately weighed, in 50 mL of neutralized alcohol, add phenolphthalein TS, and titrate with 0.5 *N* alcoholic potassium hydroxide VS: the acid value is between 112 and 168.

Saponification value ⟨401⟩—Add sufficient 0.5 *N* alcoholic potassium hydroxide VS to the neutralized liquid obtained in the test for *Acid value* to make the total volume of the alkali solution added 20.0 mL, heat the liquid on a steam bath for 30 minutes under a reflux condenser, and cool. Add about 200 mL of water, or more if necessary, and titrate the excess potassium hydroxide with 0.5 *N* hydrochloric acid VS. Perform a blank determination (see *Residual Titrations* under *Titrimetry* ⟨541⟩). The total volume of 0.5 *N* alcoholic potassium hydroxide consumed, including that required to neutralize the free acid in the determination of *Acid value*, is equivalent to a saponification value of between 154 and 220.

Tolu Balsam Syrup—*see* Tolu Balsam Syrup NF

Tolu Balsam Tincture—*see* Tolu Balsam Tincture NF

Topical Light Mineral Oil—*see* Mineral Oil, Topical Light

Toughened Silver Nitrate—*see* Silver Nitrate, Toughened

Toxin for Schick Test, Diphtheria—*see* Diphtheria Toxin for Schick Test

Toxoids—*see complete list in index*

Tragacanth—*see* Tragacanth NF

Tretinoin

$C_{20}H_{28}O_2$ 300.44

Retinoic acid.
all *trans*-Retinoic acid [*302-79-4*].

» Tretinoin contains not less than 97.0 percent and not more than 103.0 percent of $C_{20}H_{28}O_2$, calculated on the dried basis.

Packaging and storage—Preserve in tight containers, preferably under an atmosphere of an inert gas, protected from light.

Reference standards—*USP Isotretinoin Reference Standard*—Store ampuls at a temperature below 0°, allow to reach room temperature before opening, and use the contents promptly after opening ampuls. *USP Tretinoin Reference Standard*—Store ampuls at a temperature below 0°, allow to reach room temperature before opening, and use the contents promptly after opening ampuls.

NOTE—Avoid exposure to strong light, and use low-actinic glassware in the performance of the following procedures.

Identification—

A: The infrared absorption spectrum of a mineral oil dispersion of it exhibits maxima only at the same wavelengths as that of a similar preparation of USP Tretinoin RS.

B: The ultraviolet absorption spectrum of a 1 in 250,000 solution in acidified isopropyl alcohol, prepared by diluting 1 mL of 0.01 *N* hydrochloric acid with isopropyl alcohol to 1000 mL, exhibits maxima and minima at the same wavelengths as that of a similar solution of USP Tretinoin RS, concomitantly measured, and the respective absorptivities, calculated on the dried basis, at the wavelength of maximum absorbance at about 352 nm do not differ by more than 3.0%.

Loss on drying ⟨731⟩—Dry it at 105° for 3 hours: it loses not more than 0.5% of its weight.

Residue on ignition ⟨281⟩: not more than 0.1%.

Limit of isotretinoin—

Mobile phase—Prepare a suitable filtered and degassed mixture of isooctane, isopropyl alcohol, and glacial acetic acid (99.65:0.25:0.1), making adjustments if necessary (see *System Suitability* under *Chromatography* ⟨621⟩).

System suitability solution—Dissolve a quantity of USP Tretinoin RS in a minimum amount of methylene chloride, add a suitable amount of isooctane to obtain a solution having a tretinoin concentration of about 250 µg per mL, and mix.

Standard solution—Dissolve an accurately weighed quantity of USP Isotretinoin RS in a minimum quantity of methylene chloride, and add isooctane to obtain a solution having a known concentration of about 250 µg per mL.

System suitability preparation—Pipet 5 mL of *Standard solution* into a 100-mL volumetric flask, add *System suitability solution* to volume, and mix.

Standard preparation—Pipet 5 mL of *Standard solution* into a 100-mL volumetric flask, add isooctane to volume, and mix.

Test preparation—Transfer about 25 mg of Tretinoin, accurately weighed, to a 100-mL volumetric flask, dissolve in a minimum quantity of methylene chloride, add isooctane to volume, and mix.

Chromatographic system (see *Chromatography* ⟨621⟩)—The liquid chromatograph is equipped with a 352-nm detector and a 4.0-mm × 25-cm column containing packing L3. The flow rate is about 1 mL per minute. Chromatograph about 20 µL of *System suitability preparation*, and record the peak responses. The relative retention times for isotretinoin and tretinoin are about 0.84 and 1.00, respectively. The relative standard deviation of the isotretinoin peak response in replicate injections is not more than 2.0%, and the resolution, *R*, of isotretinoin and tretinoin is not less than 2.0.

Procedure—Separately inject equal volumes (about 20 µL) of the *Standard preparation* and the *Test preparation* into the chromatograph, record the chromatograms, and measure the responses for the major peaks. Calculate the percentage of isotretinoin by the formula:

$$10(C/W)(r_U/r_S),$$

in which *C* is the concentration, in µg per mL, of USP Isotretinoin RS in the *Standard preparation*, *W* is the weight, in mg, of Tretinoin taken, and r_U and r_S are the peak responses of the isotretinoin peaks obtained from the *Test preparation* and the *Standard preparation*, respectively. The content of isotretinoin is not more than 5.0%.

Heavy metals, *Method II* ⟨231⟩: 0.002%.

Assay—Dissolve about 240 mg of Tretinoin, accurately weighed, in 50 mL of dimethylformamide, add 3 drops of a 1 in 100 solution of thymol blue in dimethylformamide, and titrate with 0.1 *N* sodium methoxide VS to a greenish end-point. Perform a blank determination, and make any necessary correction. Each mL of 0.1 *N* sodium methoxide is equivalent to 30.04 mg of $C_{20}H_{28}O_2$.

Tretinoin Cream

» Tretinoin Cream contains not less than 90.0 percent and not more than 130.0 percent of the labeled amount of $C_{20}H_{28}O_2$.

Packaging and storage—Preserve in collapsible tubes or in tight, light-resistant containers.

Reference standard—*USP Tretinoin Reference Standard*—Store ampuls at a temperature below 0°, allow to reach room temperature before opening, and use the contents promptly after opening ampuls.

Identification—The absorption spectrum, obtained between wavelengths of 300 nm and 450 nm, of the solution employed for measurement of absorbance in the *Assay* exhibits maxima and minima at the same wavelengths as that of a similar solution of USP Tretinoin RS, concomitantly measured.

Minimum fill ⟨755⟩: meets the requirements.

Assay—[NOTE—Avoid exposure to strong light, and use low-actinic glassware in the performance of the following procedure.] Transfer to a 100-mL volumetric flask an accurately weighed quantity of Tretinoin Cream, equivalent to about 375 µg of tretinoin, and dissolve in about 70 mL of a chloroform–isopropyl alcohol solvent (prepared by mixing equal volumes of chloroform and isopropyl alcohol). Dilute with the chloroform–isopropyl alcohol solvent to volume, mix, and centrifuge. Dissolve an accurately weighed quantity of USP Tretinoin RS in the chloroform–isopropyl alcohol solvent, and dilute quantitatively and stepwise with the same solvent to obtain a Standard solution having a known concentration of about 3.75 µg per mL. Concomitantly determine the absorbances of both solutions in 1-cm cells at the wavelength of maximum absorbance at about 358 nm, with a suitable spectrophotometer, using the chloroform–isopropyl alcohol solvent as the blank. Calculate the quantity, in µg, of $C_{20}H_{28}O_2$ in the portion of Cream taken by the formula:

$$100C(A_U/A_S),$$

in which *C* is the concentration, in µg per mL, of USP Tretinoin RS in the Standard solution, and A_U and A_S are the absorbances of the solution from the Cream and the Standard solution, respectively.

Tretinoin Gel

» Tretinoin Gel contains not less than 90.0 percent and not more than 130.0 percent of the labeled amount of $C_{20}H_{28}O_2$.

Packaging and storage—Preserve in tight containers, protected from light.

Reference standard—*USP Tretinoin Reference Standard*—Store ampuls at a temperature below 0°, allow to reach room temperature before opening, and use the contents promptly after opening ampuls.

Identification—The absorption spectrum, obtained between wavelengths of 300 nm and 450 nm, of the solution employed for measurement of absorbance in the *Assay* exhibits maxima and minima at the same wavelengths as that of a similar solution of USP Tretinoin RS, concomitantly measured.

Minimum fill ⟨755⟩: meets the requirements.

Assay—[NOTE—Avoid exposure to strong light, and use low-actinic glassware in the performance of the following procedure.] Transfer to a 100-mL volumetric flask an accurately weighed quantity of Tretinoin Gel, equivalent to about 375 µg of tretinoin, and dissolve in about 70 mL of chloroform, dilute with chloroform to volume, and mix. Dissolve an accurately weighed quantity of USP Tretinoin RS in chloroform, and dilute quantitatively and stepwise with chloroform to obtain a Standard solution having a known concentration of about 3.75 µg per mL. Concomitantly determine the absorbances of both solutions in 1-cm cells at the wavelength of maximum absorbance at about 365 nm, with a suitable spectrophotometer, using chloroform as the blank. Calculate the quantity, in µg, of $C_{20}H_{28}O_2$ in the portion of Gel taken by the formula:

$$100C(A_U/A_S),$$

in which C is the concentration, in µg per mL, of USP Tretinoin RS in the Standard solution, and A_U and A_S are the absorbances of the solution from the Gel and the Standard solution, respectively.

Tretinoin Topical Solution

» Tretinoin Topical Solution is a solution of Tretinoin in a suitable nonaqueous, hydrophilic solvent. It contains not less than 90.0 percent and not more than 135.0 percent of the labeled amount of $C_{20}H_{28}O_2$ (w/w).

Packaging and storage—Preserve in tight, light-resistant containers.

Reference standard—*USP Tretinoin Reference Standard*—Store ampuls at a temperature below 0°, allow to reach room temperature before opening, and use the contents promptly after opening ampuls.

Identification—

A: Mix 5 mL of Topical Solution with 5 mL of water: a voluminous yellow precipitate is formed.

B: The absorption spectrum, obtained between wavelengths of 300 nm and 450 nm, of the solution employed for measurement of absorbance in the *Assay* exhibits maxima and minima at the same wavelengths as that of a similar solution of USP Tretinoin RS, concomitantly measured.

Alcohol content ⟨611⟩: between 50.0% and 60.0% of C_2H_5OH.

Assay—[NOTE—Avoid exposure to strong light, and use low-actinic glassware in the performance of the following procedure.] Transfer to a 100-mL volumetric flask an accurately weighed quantity of Tretinoin Topical Solution, equivalent to about 375 µg of tretinoin, dilute with acidified isopropyl alcohol (prepared by mixing 1 mL of 0.01 *N* hydrochloric acid with isopropyl alcohol to make 1000 mL of solution) to volume, and mix. Dissolve an accurately weighed quantity of USP Tretinoin RS in the acidified isopropyl alcohol, and dilute quantitatively and stepwise with the same solvent to obtain a Standard solution having a known concentration of about 3.75 µg per mL. Concomitantly determine the absorbances of both solutions in 1-cm cells at the wavelength of maximum absorbance at about 352 nm, with a suitable spectrophotometer, using the acidified isopropyl alcohol as the blank. Calculate the quantity, in µg, of $C_{20}H_{28}O_2$ in the portion of Topical Solution taken by the formula:

$$100C(A_U/A_S),$$

in which C is the concentration, in µg per mL, of USP Tretinoin RS in the Standard solution, and A_U and A_S are the absorbances of the solution from the Topical Solution and the Standard solution, respectively.

Triacetin

$CH_2(OCOCH_3)CH(OCOCH_3)CH_2OCOCH_3$

$C_9H_{14}O_6$ 218.21

1,2,3-Propanetriol triacetate.

Triacetin.

Glyceryl triacetate [*102-76-1*].

» Triacetin contains not less than 97.0 percent and not more than 100.5 percent of $C_9H_{14}O_6$, calculated on the anhydrous basis.

Packaging and storage—Preserve in tight containers.

Reference standard—*USP Triacetin Reference Standard*—Do not dry before using.

Identification—

A: The infrared absorption spectrum of a 1 in 100 solution of it, previously dried over a suitable desiccant, in carbon disulfide, exhibits maxima only at the same wavelengths as that of a similar solution of USP Triacetin RS.

B: Heat a few drops in a test tube with about 500 mg of potassium bisulfate: pungent vapors of acrolein are evolved.

C: The solution prepared as directed in the *Assay* responds to the tests for *Acetate* ⟨191⟩.

Specific gravity ⟨841⟩: not less than 1.152 and not more than 1.158.

Refractive index ⟨831⟩: not less than 1.429 and not more than 1.430.

Acidity—Dilute 25 g of Triacetin, accurately weighed, with 50 mL of neutralized alcohol, add 5 drops of phenolphthalein TS, and titrate with 0.020 sodium hydroxide: not more than 1.0 mL of 0.020 *N* sodium hydroxide is required for neutralization.

Water, *Method I* ⟨921⟩—Determine the water content by the titrimetric method: not more than 0.2% is found.

Assay—Transfer about 1 g of Triacetin, accurately weighed, to a 250-mL boiling flask, add 50.0 mL of 0.5 *N* alcoholic potassium hydroxide VS, connect the flask to a water-jacketed condenser, and reflux on a steam bath for 45 minutes, swirling frequently. Cool, add 5 drops of phenolphthalein TS, and titrate the excess alkali with 0.5 *N* hydrochloric acid VS. Perform a blank determination (see *Residual Titrations* ⟨541⟩). Each mL of 0.5 *N* alcoholic potassium hydroxide is equivalent to 36.37 mg of $C_9H_{14}O_6$.

Triamcinolone

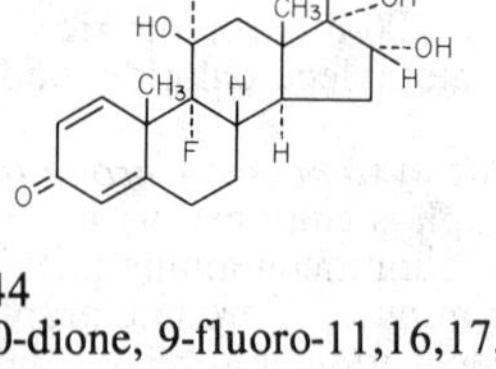

$C_{21}H_{27}FO_6$ 394.44

Pregna-1,4-diene-3,20-dione, 9-fluoro-11,16,17,21-tetrahydroxy-, (11β,16α).

9-Fluoro-11β,16α,17,21-tetrahydroxypregna-1,4-diene-3,20-dione [*124-94-7*].

» Triamcinolone contains not less than 97.0 percent and not more than 102.0 percent of $C_{21}H_{27}FO_6$, calculated on the dried basis.

Packaging and storage—Preserve in well-closed containers.

Reference standard—*USP Triamcinolone Reference Standard*—Dry in vacuum at 60° for 4 hours before using.

Identification—

A: The infrared absorption spectrum of a potassium bromide dispersion of it, previously dried, exhibits maxima only at the same wavelengths as that of a similar preparation of USP Triamcinolone RS.

B: The ultraviolet absorption spectrum of a 1 in 50,000 solution of it in methanol exhibits maxima and minima at the same wavelengths as that of a similar preparation of USP Triamcinolone RS, concomitantly measured, and the respective absorptivities, calculated on the dried basis, at the wavelength of maximum absorbance at about 238 nm, do not differ by more than 3.0%.

Specific rotation ⟨781⟩: between +65° and +72°, calculated on the dried basis, determined in a solution in dimethylformamide containing 20 mg in each 10 mL.

Loss on drying ⟨731⟩—Dry it in vacuum at 60° for 4 hours: it loses not more than 2.0% of its weight.

Residue on ignition ⟨281⟩: 0.5%.

Heavy metals, *Method II* ⟨231⟩: 0.0025%.

Assay—

Mobile phase—Prepare a suitable degassed solution containing about 60 volumes of methanol and 40 volumes of water such that the retention times for triamcinolone and hydrocortisone are about 5 and 10 minutes, respectively.

Internal standard solution—Dissolve hydrocortisone in *Mobile phase* to obtain a solution having a concentration of about 0.3 mg per mL.

Standard preparation—Transfer about 10 mg of USP Triamcinolone RS, accurately weighed, to a 50-mL volumetric flask, dissolve in *Internal standard solution*, dilute with the same solvent to volume, and mix.

Assay preparation—Using about 10 mg of Triamcinolone, accurately weighed, prepare as directed under *Standard preparation*.

Chromatographic system (see *Chromatography* ⟨621⟩)—The liquid chromatograph is equipped with a 254-nm detector and a 3.9-mm × 30-cm column that contains packing L1. The flow rate is about 1.5 mL per minute. Chromatograph five replicate injections of the *Standard preparation*, and record the peak responses as directed under *Procedure:* the relative standard deviation is not more than 2.0%, and the resolution factor between triamcinolone and hydrocortisone is not less than 3.0.

Procedure—Separately inject equal volumes (about 10 µL) of the *Standard preparation* and the *Assay preparation* into the chromatograph by means of a suitable microsyringe or sampling valve, record the chromatograms, and measure the responses for the major peaks. Calculate the quantity, in mg, of $C_{21}H_{27}FO_6$ in the portion of Triamcinolone taken by the formula:

$$50C(R_U/R_S),$$

in which *C* is the concentration, in mg per mL, of USP Triamcinolone RS in the *Standard preparation*, and R_U and R_S are the peak response ratios of triamcinolone to hydrocortisone obtained from the *Assay preparation* and the *Standard preparation*, respectively.

Triamcinolone Tablets

» Triamcinolone Tablets contain not less than 90.0 percent and not more than 110.0 percent of the labeled amount of $C_{21}H_{27}FO_6$.

Packaging and storage—Preserve in well-closed containers.

Reference standard—*USP Triamcinolone Reference Standard*—Dry in vacuum at 60° for 4 hours before using.

Identification—Powder a number of Tablets, equivalent to about 25 mg of triamcinolone, and digest with 25 mL of acetone for 15 minutes. Filter through a fine-porosity, sintered-glass filtering funnel into about 100 mL of solvent hexane, swirl the liquid, and allow to stand for 30 minutes. Collect the crystals that form, wash the crystals with three 10-mL portions of water followed by 2 mL of acetone, and dry at 60° for 1 hour: the dried crystals so obtained respond to *Identification test A* under *Triamcinolone*.

Dissolution ⟨711⟩—

Medium: 0.1 *N* hydrochloric acid; 900 mL.

Apparatus 1: 100 rpm.

Time: 45 minutes.

Procedure—Determine the amount of $C_{21}H_{27}FO_6$ dissolved from ultraviolet absorbances at the wavelength of maximum absorbance at about 238 nm of filtered portions of the solution under test, suitably diluted with *Dissolution Medium*, in comparison with a Standard solution having a known concentration of USP Triamcinolone RS in the same medium.

Tolerances—Not less than 75% (*Q*) of the labeled amount of $C_{21}H_{27}FO_6$ is dissolved in 45 minutes.

Uniformity of dosage units ⟨905⟩: meet the requirements.

Assay—

Mobile phase, Internal standard solution, Standard preparation, and *Chromatographic system*—Prepare as directed in the *Assay* under *Triamcinolone*.

Assay preparation—Weigh and finely powder not less than 20 Triamcinolone Tablets. Transfer an accurately weighed portion of the powder, equivalent to about 10 mg of triamcinolone, to a suitable container. Add 50.0 mL of *Internal standard solution*, and shake vigorously by mechanical means for 10 minutes. Centrifuge for 10 minutes or until a clear supernatant solution is obtained.

Procedure—Proceed as directed for *Procedure* in the *Assay* under *Triamcinolone*. The relative retention times are about 1.0 for triamcinolone and 1.9 for hydrocortisone. Calculate the quantity, in mg, of $C_{21}H_{27}FO_6$ in the portion of Tablets taken by the formula:

$$50C(R_U/R_S),$$

in which the terms are as defined therein.

Triamcinolone Acetonide

$C_{24}H_{31}FO_6$ 434.50

Pregna-1,4-diene-3,20-dione, 9-fluoro-11,21-dihydroxy-16,17-[(1-methylethylidene)bis(oxy)]-, (11β,16α)-.

9-Fluoro-11β,16α,17,21-tetrahydroxypregna-1,4-diene-3,20-dione cyclic 16,17-acetal with acetone [*76-25-5*].

» Triamcinolone Acetonide contains not less than 97.0 percent and not more than 102.0 percent of $C_{24}H_{31}FO_6$, calculated on the dried basis.

Packaging and storage—Preserve in well-closed containers.

Reference standards—*USP Triamcinolone Acetonide Reference Standard*—Do not dry; use as is. *USP Fluoxymesterone Reference Standard*—Dry at 105° for 3 hours before using.

Identification—

A: The infrared absorption spectrum of a potassium bromide dispersion of it, recrystallized from methanol, exhibits maxima only at the same wavelengths as that of a similar preparation of USP Triamcinolone Acetonide RS.

B: The ultraviolet absorption spectrum of a 1 in 50,000 solution in methanol exhibits maxima and minima at the same wavelengths as that of a similar solution of USP Triamcinolone Acetonide RS, concomitantly measured.

Specific rotation ⟨781⟩: between +118° and +130°, calculated on the dried basis, determined in a solution in dimethylformamide containing 50 mg in each 10 mL.

Loss on drying ⟨731⟩—Dry it in vacuum at 60° for 4 hours: it loses not more than 1.5% of its weight.

Heavy metals—Carefully ignite 1.0 g in a muffle furnace at about 550° until thoroughly charred. Cool, add to the contents of the crucible 5 drops of sulfuric acid and 2 mL of nitric acid, cautiously heat until reaction has ceased, then ignite in a muffle furnace at 500° to 600° until the carbon is entirely burned off. Cool, add 2 mL of hydrochloric acid, and slowly evaporate on a steam bath to dryness. Moisten the residue with 1 drop of hydrochloric acid and 5 mL of hot water and digest for 2 minutes. Add 1 drop of phenolphthalein TS, then add 6 *N* ammonium hydroxide dropwise until the reaction is alkaline. Render the solution acid with 1 *N* acetic acid, then add 1 mL of excess, transfer to a beaker, and add water to make 10 mL. Pipet 2.5 mL (equivalent to 25 µg of lead) of *Standard Lead Solution* (see *Lead* ⟨231⟩) into a second beaker, add 3 mL of water and 1 drop of phenolphthalein TS, render just alkaline with 6 *N* ammonium hydroxide, then render acid with 1 *N* acetic acid, and add 1 mL in excess. Dilute with water to 10 mL. To each beaker add 5 mL of freshly prepared hydrogen sulfide TS, mix, and allow to stand for 5 minutes. Filter each solution through a separate, acid-resistant, white, plain membrane filter of 0.22-µm pore size and 25 mm in diameter, collecting the precipitates on the filter disks: the color of the precipitate from the solution under test is not darker than that from the control. The heavy metals limit is 0.0025%.

Assay—

Mobile phase—Prepare a solution of acetonitrile in water containing approximately 30 percent (v/v) of acetonitrile.

Internal standard solution—Dissolve USP Fluoxymesterone RS in methanol to obtain a solution having a concentration of about 50 µg per mL.

Standard preparation—Dissolve an accurately weighed quantity of USP Triamcinolone Acetonide RS in *Internal standard solution* to obtain a solution having a known concentration of about 75 µg per mL. Mix an accurately measured volume of the resulting solution with an equal volume of *Mobile phase* to obtain a *Standard preparation* containing about 37.5 µg of USP Triamcinolone Acetonide RS per mL.

Assay preparation—Using about 37 mg of Triamcinolone Acetonide, accurately weighed, proceed as directed for *Standard preparation.*

Procedure—Introduce equal volumes (between 15 µL and 25 µL) of the *Assay preparation* and the *Standard preparation* into a high-pressure liquid chromatograph (see *Chromatography* ⟨621⟩) operated at room temperature, by means of a suitable microsyringe or sampling valve. Adjust the operating parameters with *Mobile phase* on the column, such that the separation of triamcinolone acetonide and internal standard is optimized, with a retention time of about 14.5 minutes for triamcinolone acetonide. Typically, the apparatus is fitted with a 30-cm × 4-mm column containing packing L1, and is equipped with an ultraviolet detector capable of monitoring absorbance at 254 nm, and a suitable recorder. In a suitable chromatogram, the coefficient of variation for five replicate injections of a single specimen is not more than 3.0%, and the resolution factor, *R* (see *Chromatography* ⟨621⟩), between the peaks for triamcinolone acetonide and fluoxymesterone is not less than 2.0. Measure the heights of the internal standard and triamcinolone acetonide peaks, at the same retention times obtained from *Assay preparation* and the *Standard preparation.* Calculate the quantity, in mg, of $C_{24}H_{31}FO_6$ in the portion of Triamcinolone Acetonide taken by the formula:

$$1000C(R_U/R_S),$$

in which *C* is the concentration, in mg per mL, of USP Triamcinolone Acetonide RS in the *Standard preparation*, and R_U and R_S are the ratios of the peak heights of triamcinolone acetonide to the internal standard obtained from the *Assay preparation* and the *Standard preparation*, respectively.

Triamcinolone Acetonide Topical Aerosol

» Triamcinolone Acetonide Topical Aerosol is a solution of Triamcinolone Acetonide in a suitable propellant in a pressurized container. It contains not less than 90.0 percent and not more than 115.0 percent of the labeled amount of $C_{24}H_{31}FO_6$.

Packaging and storage—Preserve in pressurized containers, and avoid exposure to excessive heat.

Reference standards—*USP Triamcinolone Acetonide Reference Standard*—Do not dry; use as is. *USP Fluoxymesterone Reference Standard*—Dry at 105° for 3 hours before using.

Identification—Apply 20 µL of a solution prepared as directed for *Assay preparation* in the *Assay* but without the addition of the *Internal standard solution*, and 20 µL of a solution of USP Triamcinolone Acetonide RS in methanol containing 30 µg per mL, on a line parallel to and about 1.5 cm from the bottom edge of a thin-layer chromatographic plate (see *Chromatography* ⟨621⟩) coated with a 0.25-mm layer of chromatographic silica gel. Proceed as directed in the *Identification test* under *Triamcinolone Acetonide Cream*, beginning with "Place the plate in a developing chamber." The specified result is obtained.

Microbial limits—It meets the requirements of the tests for absence of *Staphylococcus aureus* and *Pseudomonas aeruginosa* under *Microbial Limit Tests* ⟨61⟩.

Other requirements—It meets the requirements for *Leak Testing* and *Pressure Testing* under *Aerosols* ⟨601⟩.

Assay—

Mobile phase—Approximately 30% acetonitrile in water.

Internal standard solution—Dissolve USP Fluoxymesterone RS in methanol to obtain a solution having a concentration of about 25 µg per mL.

Standard preparation—Dissolve an accurately weighed quantity of USP Triamcinolone Acetonide RS in methanol to obtain a solution having a concentration of about 100 µg per mL. Transfer 15.0 mL of this solution to a 50-mL volumetric flask, add 25.0 mL of *Internal standard solution*, dilute with methanol to volume, and mix. This solution has a known concentration of about 30 µg per mL.

Assay preparation—Fit the valve of a previously weighed Triamcinolone Acetonide Aerosol container with a suitable tube assembly so that the contents can be sprayed directly into the bulb portion of a 100-mL volumetric flask containing 50.0 mL of *Internal standard solution* and 20 mL of methanol. Spray a portion of the contents, equivalent to about 3 mg of triamcinolone acetonide, into the flask, determining the exact amount sprayed by difference. Place in a sonic bath for about 5 minutes to expel the propellant. Dilute with methanol to volume, and mix. [NOTE—The propellant is extremely flammable. When evaporating, observe proper precautions and work under an explosion-proof hood.]

Procedure—Introduce equal volumes (between 15 µL and 25 µL) of the *Assay preparation* and the *Standard preparation* into a chromatograph (see *Chromatography* ⟨621⟩) operated at room temperature and fitted with a 3.9-mm × 30-cm column, packed with packing L1, and equipped with a 254-nm detector. Adjust the operating parameters and the *Mobile phase* composition such that the separation of triamcinolone acetonide and internal standard is optimized, with a retention time of about 14 minutes for triamcinolone acetonide. In a suitable system, the relative standard deviation for five replicate injections of the *Standard preparation* is not more than 3.0%. Measure the responses of the internal standard and triamcinolone acetonide peaks at the same retention times obtained from the *Assay preparation* and the *Standard preparation.* Calculate the quantity, in µg, of $C_{24}H_{31}FO_6$ in the portion of Aerosol taken by the formula:

$$100C(R_U/R_S),$$

in which *C* is the concentration, in µg per mL, of USP Triamcinolone Acetonide RS in the *Standard preparation*, and R_U and R_S are the ratios of the peak responses of triamcinolone acetonide to internal standard obtained from the *Assay preparation* and the *Standard preparation*, respectively.

Triamcinolone Acetonide Cream

» Triamcinolone Acetonide Cream is Triamcinolone Acetonide in a suitable cream base. It contains not less than 90.0 percent and not more than 115.0 percent of the labeled amount of $C_{24}H_{31}FO_6$.

Packaging and storage—Preserve in tight containers.

Reference standards—*USP Triamcinolone Acetonide Reference Standard*—Do not dry; use as is. *USP Fluoxymesterone Reference Standard*—Dry at 105° for 3 hours before using.

Identification—Place a 2-g quantity of Cream in a conical flask, add 50 mL of chloroform and 15 g of anhydrous sodium sulfate, and swirl to dissolve the specimen. Filter the solution and clarify the filtrate, if necessary, by the further addition of anhydrous sodium sulfate and a second filtration. Evaporate the filtrate to near dryness, and dissolve the residue in chloroform to obtain a solution containing about 100 µg per mL. Apply 10 µL of this solution, and 10 µL of a solution of USP Triamcinolone Acetonide RS in chloroform containing 100 µg per mL, on a line parallel to and about 1.5 cm from the bottom edge of a thin-layer chromatographic plate (see *Chromatography* ⟨621⟩) coated with a 0.25-mm layer of chromatographic silica gel. Place the plate in a developing chamber containing and equilibrated with a mixture of chloroform, benzene, and methanol (100:40:20). Develop the chromatogram until the solvent front has moved about 12 cm above the line of application. Remove the plate, allow the solvent to evaporate, and spray with a mixture of equal volumes of sodium hydroxide solution (1 in 5) and a 1 in 500 solution of blue tetrazolium in methanol: the intensity of the blue color and the R_f of the spot obtained with the solution under test are similar to those of the spot obtained with the Standard solution.

Microbial limits—It meets the requirements of the tests for absence of *Staphylococcus aureus* and *Pseudomonas aeruginosa* under *Microbial Limit Tests* ⟨61⟩.

Assay—

Mobile phase—Prepare a solution of acetonitrile in water containing approximately 30 percent (v/v) of acetonitrile.

Internal standard solution—Dissolve USP Fluoxymesterone RS in isopropyl alcohol to obtain a solution having a concentration of about 50 µg per mL.

Standard preparation—Dissolve an accurately weighed quantity of USP Triamcinolone Acetonide RS in *Internal standard solution* to obtain a solution having a known concentration of about 75 µg per mL. Mix an accurately measured volume of the resulting solution with an equal volume of *Mobile phase* to obtain a *Standard preparation* containing about 37.5 µg of USP Triamcinolone Acetonide RS per mL.

Assay preparation—Transfer an accurately weighed quantity of Triamcinolone Acetonide Cream, equivalent to about 1.5 mg of triamcinolone acetonide, to a screw-cap tube. Add 20.0 mL of *Internal standard solution*, and cap securely. Heat for 5 minutes at 60°, then swirl vigorously for not less than 30 seconds. Repeat the heating and swirling sequence three times. Cool in a methanol-ice bath for 15 to 20 minutes, then centrifuge for 15 minutes at −5°. Dilute an accurately measured volume of the supernatant solution with an equal volume of *Mobile phase*. Cool in a methanol-ice bath for 10 to 15 minutes, with occasional agitation. Filter first through a pledget of glass wool or a prefilter disk and then through a 0.45-µm porosity membrane to obtain a clear solution.

Procedure—Introduce equal volumes (between 15 and 25 µL) of the *Assay preparation* and the *Standard preparation* into a high-pressure liquid chromatograph (see *Chromatography* ⟨621⟩), operated at room temperature, by means of a suitable microsyringe or sampling valve. Adjust the operating parameters with *Mobile phase* on the column, such that the separation of triamcinolone acetonide and internal standard is optimized, with a retention time of about 14.5 minutes for triamcinolone acetonide. Typically, the apparatus is fitted with a 30-cm × 4-mm column containing packing L1, and is equipped with an ultraviolet detector capable of monitoring absorbance at 254 nm, and a suitable recorder. In a suitable chromatogram, the coefficient of variation for five replicate injections of a single specimen is not more than 3.0%, and the resolution factor, *R* (see *Chromatography* ⟨621⟩), between the peaks for triamcinolone acetonide and fluoxymesterone is not less than 2.0. Measure the heights of the internal standard and triamcinolone acetonide peaks, at the same retention times obtained from the *Assay preparation* and the *Standard preparation*. Calculate the quantity, in mg, of $C_{24}H_{31}FO_6$ in the portion of Cream taken by the formula:

$$40C(R_U/R_S),$$

in which C is the concentration, in mg per mL, of USP Triamcinolone Acetonide RS in the *Standard preparation*, and R_U and R_S are the ratios of the peak heights of triamcinolone acetonide to the internal standard obtained from the *Assay preparation* and the *Standard preparation*, respectively.

Triamcinolone Acetonide Cream, Neomycin Sulfate and—*see* Neomycin Sulfate and Triamcinolone Acetonide Cream

Triamcinolone Acetonide Cream, Nystatin and—*see* Nystatin and Triamcinolone Acetonide Cream

Triamcinolone Acetonide Cream, Nystatin, Neomycin Sulfate, Gramicidin, and—*see* Nystatin, Neomycin Sulfate, Gramicidin, and Triamcinolone Acetonide Cream

Triamcinolone Acetonide Lotion

» Triamcinolone Acetonide Lotion is Triamcinolone Acetonide in a suitable lotion base. It contains not less than 90.0 percent and not more than 110.0 percent of the labeled amount of $C_{24}H_{31}FO_6$.

Packaging and storage—Preserve in tight containers.

Reference standards—*USP Triamcinolone Acetonide Reference Standard*—Do not dry; use as is. *USP Fluoxymesterone Reference Standard*—Dry at 105° for 3 hours before using.

Identification—It responds to the *Identification test* under *Triamcinolone Acetonide Cream.*

Microbial limits—It meets the requirements of the tests for absence of *Staphylococcus aureus* and *Pseudomonas aeruginosa* under *Microbial Limit Tests* ⟨61⟩.

Assay—Proceed with Triamcinolone Acetonide Lotion as directed in the *Assay* under *Triamcinolone Acetonide Cream*, except to read "Lotion" in place of "Cream" throughout.

Triamcinolone Acetonide Ointment

» Triamcinolone Acetonide Ointment is Triamcinolone Acetonide in a suitable ointment base. It contains not less than 90.0 percent and not more than 115.0 percent of the labeled amount of $C_{24}H_{31}FO_6$.

Packaging and storage—Preserve in well-closed containers.

Reference standard—*USP Triamcinolone Acetonide Reference Standard*—Do not dry; use as is.

Identification—Place 2 g of Ointment in a conical flask, add 5.0 mL of chloroform, and shake for 10 minutes. Add 15 mL of alcohol, and shake for an additional 10 minutes. Filter the solution into a centrifuge tube, and evaporate the filtrate to dryness. Dissolve the residue in alcohol to obtain a solution containing about 250 µg per mL. Apply 10 µL of this solution, and 10 µL of a solution of USP Triamcinolone Acetonide RS in alcohol containing 250 µg per mL, on a line parallel to and about 1.5 cm from the bottom edge of a thin-layer chromatographic plate (see *Chromatography* ⟨621⟩) coated with a 0.25-mm layer of chromatographic silica gel. Place the plate in a developing chamber containing and equilibrated with a mixture of chloroform, benzene, and methanol (100:40:20). Develop the chromatogram until the solvent front has moved about 12 cm above the line of application. Remove the plate, allow the solvent to evaporate, and spray with a mixture of equal volumes of sodium hydroxide solution (1 in 5) and a 1 in 500 solution of blue tetrazolium in methanol: the intensity of the blue color and the R_f of the spot obtained from the solution under test are similar to those of the spot obtained from the Standard solution.

Microbial limits—It meets the requirements of the tests for absence of *Staphylococcus aureus* and *Pseudomonas aeruginosa* under *Microbial Limit Tests* ⟨61⟩.

Assay—Proceed with Triamcinolone Acetonide Ointment as directed in the *Assay* under *Triamcinolone Acetonide Cream*, except to read "Ointment" in place of "Cream" throughout.

Triamcinolone Acetonide Ointment, Nystatin and—*see* Nystatin and Triamcinolone Acetonide Ointment

Triamcinolone Acetonide Ointment, Nystatin, Neomycin Sulfate, Gramicidin, and—*see* Nystatin, Neomycin Sulfate, Gramicidin, and Triamcinolone Acetonide Ointment

Triamcinolone Acetonide Ophthalmic Ointment, Neomycin Sulfate and—*see* Neomycin Sulfate and Triamcinolone Acetonide Ophthalmic Ointment

Triamcinolone Acetonide Dental Paste

» Triamcinolone Acetonide Dental Paste is Triamcinolone Acetonide in a suitable emollient paste. It contains not less than 90.0 percent and not more than 115.0 percent of the labeled amount of $C_{24}H_{31}FO_6$.

Packaging and storage—Preserve in tight containers.

Reference standard—*USP Triamcinolone Acetonide Reference Standard*—Do not dry; use as is.

Identification—It responds to the *Identification test* under *Triamcinolone Acetonide Cream.*

Microbial limits—It meets the requirements of the tests for absence of *Staphylococcus aureus* and *Pseudomonas aeruginosa* under *Microbial Limit Tests* ⟨61⟩.

Assay—

Mobile phase—Prepare as directed in the *Assay* under *Triamcinolone Acetonide Cream.*

Internal standard solution and *Standard preparation*—Prepare as directed in the *Assay* under *Triamcinolone Acetonide Cream.*

Assay preparation—Transfer an accurately weighed quantity of Triamcinolone Acetonide Dental Paste, equivalent to about 1.5 mg of triamcinolone acetonide, to a screw-cap tube. Add 20.0 mL of *Internal standard solution*, cap, and place in a sonic bath for 15 to 20 minutes. Place in a water bath at 70° for 5 minutes, then swirl for 1 minute. Repeat the heating and swirling sequence once. Cool in an ice-methanol bath for 15 minutes, then centrifuge for 10 minutes at −5°. Mix an accurately measured volume of the supernatant solution with an equal volume of *Mobile phase.* Place in an ice-methanol bath for 15 minutes, then centrifuge for 15 minutes. Draw off and discard the upper phase. Filter the lower phase to obtain a clear solution.

Procedure—Proceed as directed for *Procedure* in the *Assay* under *Triamcinolone Acetonide Cream.*

Sterile Triamcinolone Acetonide Suspension

» Sterile Triamcinolone Acetonide Suspension is a sterile suspension of Triamcinolone Acetonide in a suitable aqueous medium. It contains not less than 90.0 percent and not more than 115.0 percent of the labeled amount of $C_{24}H_{31}FO_6$.

Packaging and storage—Preserve in single-dose or in multiple-dose containers, preferably of Type I glass, protected from light.

Reference standard—*USP Triamcinolone Acetonide Reference Standard*—Do not dry; use as is.

Identification—Extract a volume of Suspension, equivalent to about 50 mg of triamcinolone acetonide, with two 10-mL portions of peroxide-free ether, and discard the ether extracts. Filter with the aid of suction, wash with small portions of water, and dry the precipitate at 105° for 1 hour: the triamcinolone acetonide so obtained responds to the *Identification tests* under *Triamcinolone Acetonide.*

pH ⟨791⟩: between 5.0 and 7.5.

Other requirements—It meets the requirements under *Injections* ⟨1⟩.

Assay—

Mobile phase—Approximately 30% acetonitrile in water.

Internal standard solution—Dissolve Fluoxymesterone RS in methanol to obtain a solution having a concentration of about 84 µg per mL.

Standard preparation—Dissolve an accurately weighed quantity of USP Triamcinolone Acetonide RS in methanol to obtain a solution having a known concentration of about 200 µg per mL. Pipet 20 mL of this solution into a 50-mL volumetric flask, dilute with *Internal standard solution* to volume, and mix. The *Standard preparation* has a known concentration of about 80 µg of USP Triamcinolone Acetonide RS per mL.

Assay preparation—Dissolve an accurately measured volume of freshly mixed Sterile Triamcinolone Acetonide Suspension in methanol, and dilute quantitatively with methanol to obtain a solution having an expected concentration of about 200 µg of triamcinolone acetonide per mL. Pipet 20 mL of this solution into a 50-mL volumetric flask, dilute with *Internal standard solution* to volume, and mix.

Procedure—Proceed as directed for *Procedure* in the *Assay* under *Triamcinolone Acetonide Cream*, except to use peak responses in the calculation. Calculate the quantity, in mg, of $C_{24}H_{31}FO_6$ in each mL of the Suspension taken by the formula:

$$(CD/V)(R_U/R_S),$$

in which C is the concentration, in mg per mL, of USP Triamcinolone Acetonide RS in the *Standard preparation*, D is the dilution factor used in the *Assay preparation*, V is the volume, in mL, of Suspension taken, and R_U and R_S are the ratios of the peak responses of triamcinolone acetonide to the internal standard, obtained from the *Assay preparation* and the *Standard preparation*, respectively.

Triamcinolone Diacetate

$C_{25}H_{31}FO_8$ 478.51

Pregna-1,4-diene-3,20-dione, 16,21-bis(acetyloxy)-9-fluoro-11,17-dihydroxy-, (11β,16α)-.

9-Fluoro-11β,16α,17,21-tetrahydroxypregna-1,4-diene-3,20-dione 16,21-diacetate [*67-78-7*].

» Triamcinolone Diacetate contains not less than 97.0 percent and not more than 103.0 percent of $C_{25}H_{31}FO_8$, calculated on the dried basis.

Packaging and storage—Preserve in well-closed containers.

Reference standard—*USP Triamcinolone Diacetate Reference Standard*—Dry in vacuum at 60° for 4 hours before using.

Identification—

A: The infrared absorption spectrum of a potassium bromide dispersion of it, previously dried, exhibits maxima only at the same wavelengths as that of a similar preparation of USP Triamcinolone Diacetate RS.

B: The ultraviolet absorption spectrum of a 1 in 50,000 solution in dehydrated alcohol exhibits maxima and minima at the same wavelengths as that of a similar solution of USP Triamcinolone Diacetate RS, concomitantly measured, and the respective absorptivities, calculated on the dried basis, at the wavelength of maximum absorbance at about 238 nm do not differ by more than 3.0%.

Specific rotation ⟨781⟩: between +39° and +45°, calculated on the dried basis, determined in a solution in dimethylformamide containing 50 mg in each 10 mL.

Loss on drying ⟨731⟩—Dry it in vacuum at 60° for 4 hours: it loses not more than 6.0% of its weight.

Residue on ignition ⟨281⟩: not more than 0.5%.

Heavy metals, *Method II* ⟨231⟩: 0.0025%.

Assay—

Mobile phase—Prepare a mixture of acetonitrile, water, and glacial acetic acid (50:50:0.5), and degas the solution for 10 minutes in an ultrasonic bath before use. Make adjustments if necessary (see *System Suitability* under *Chromatography* ⟨621⟩).

Internal standard solution—Dissolve a quantity of testosterone in acetonitrile to obtain a solution having a concentration of about 1 mg per mL.

Standard preparation—Transfer about 5 mg of USP Triamcinolone Diacetate RS, accurately weighed, to a 10-mL volumetric flask. Add acetonitrile to volume, and sonicate for 5 minutes. Transfer 2.0 mL of this solution to a 10-mL volumetric flask containing 1.0 mL of *Internal standard solution*. Add 3 mL of water, dilute with *Mobile phase* to volume, and mix to obtain a *Standard preparation* having a known concentration of about 0.1 mg per mL.

Assay preparation—Prepare a solution with about 5 mg of Triamcinolone Diacetate, previously dried and accurately weighed, as directed under *Standard preparation*.

Chromatographic system (see *Chromatography* ⟨621⟩)—The liquid chromatograph is equipped with a 254-nm detector and a 4-mm × 25-cm column that contains packing L1. The flow rate is about 1 mL per minute. Chromatograph the *Standard preparation*, and record the peak responses as directed under *Procedure:* the resolution, *R*, between the analyte and internal standard peaks is not less than 1.5, and the relative standard deviation for replicate injections is not more than 2.0%.

Procedure—Separately inject equal volumes (about 20 µL) of the *Standard preparation* and the *Assay preparation* into the chromatograph, record the chromatograms, and measure the responses for the major peaks. The relative retention times are about 1.0 for triamcinolone diacetate and 1.4 for testosterone. Calculate the quantity, in mg, of $C_{25}H_{31}FO_8$ in the portion of Triamcinolone Diacetate taken by the formula:

$$50C(R_U/R_S),$$

in which *C* is the concentration, in mg, of USP Triamcinolone Diacetate RS in the *Standard preparation*, and R_U and R_S are the peak response ratios obtained from the *Assay preparation* and the *Standard preparation*, respectively.

Sterile Triamcinolone Diacetate Suspension

» Sterile Triamcinolone Diacetate Suspension is a sterile suspension of Triamcinolone Diacetate in a suitable aqueous medium. It contains not less than 90.0 percent and not more than 115.0 percent of the labeled amount of $C_{25}H_{31}FO_8$.

Packaging and storage—Preserve in single-dose or in multiple-dose containers, preferably of Type I glass.

Reference standard—*USP Triamcinolone Diacetate Reference Standard*—Dry in vacuum at 60° for 4 hours before using.

Identification—Filter a volume of Suspension, equivalent to about 50 mg of triamcinolone diacetate, through a medium-porosity, sintered-glass funnel, wash well with water, and dry the crystals at 60° for 1 hour in vacuum. Dissolve 2 mg of the dried crystals in 1 mL of methanol in a small mortar. Evaporate with the aid of gentle heat and a stream of nitrogen to dryness: a potassium bromide dispersion of the crystals so obtained responds to *Identification test A* under *Triamcinolone Diacetate*.

Uniformity of dosage units ⟨905⟩: meets the requirements.

pH ⟨791⟩: between 4.5 and 7.5.

Other requirements—It meets the requirements under *Injections* ⟨1⟩.

Assay—

Mobile phase, Internal standard solution, Standard preparation, and *Chromatographic system*—Proceed as directed in the *Assay* under *Triamcinolone Diacetate*.

Barium hydroxide solution—Add 2.5 g of barium hydroxide to 50 mL of a saturated solution of sodium chloride in a glass-stoppered, 125-mL flask. Insert the stopper, and shake the solution for 1 minute. Prepare this solution fresh daily.

Assay preparation—Weigh 1 intact vial of Sterile Triamcinolone Diacetate Suspension, and record the weight. Transfer a portion of the vial contents to an appropriate size tared volumetric flask to make to volume, and determine the density. Transfer, with the aid of five 5-mL portions of water-washed chloroform, the contents of the volumetric flask and the remaining suspension in the vial to a 125-mL separator for assay. Dry and weigh the empty suspension vial. Subtract the weight of the empty vial from the weight of intact vial, and determine the original volume of the suspension in the vial from the calculated density of the suspension. Add 20 mL of *Barium hydroxide solution* to the separator. Shake vigorously for 30 seconds, and allow the chloroform–barium hydroxide layer to separate. Drain the chloroform layer into a clean 125-mL separator containing 20 mL of dilute phosphoric acid solution (1 in 20). Shake the second separator to remove traces of the barium hydroxide solution, and drain the chloroform layer through a pledget of glass wool into a 100-mL volumetric flask. Complete the extraction of the triamcinolone diacetate from the barium hydroxide solution, using three additional 25-mL portions of water-washed chloroform. Wash each extract with phosphoric acid solution, combine the extract in a 100-mL volumetric flask, dilute with chloroform to volume, and mix. Transfer an aliquot of the chloroform solution, equivalent to about 1 mg of triamcinolone diacetate, to a glass-stoppered, 50-mL conical flask and add 1.0 mL of *Internal standard solution*. Evaporate the organic solvent, using a gentle stream of nitrogen. Dissolve the residue in 10.0 mL of *Mobile phase*, using sonication.

Procedure—Proceed as directed for *Procedure* in the *Assay* under *Triamcinolone Diacetate*. Calculate the quantity, in mg, of $C_{25}H_{31}FO_8$ in the portion of Suspension taken by the formula:

$$10C(R_U/R_S),$$

in which *C* is the concentration, in mg per mL, of USP Triamcinolone Diacetate RS in the *Standard preparation*, and R_U and R_S are the peak response ratios obtained from the *Assay preparation* and the *Standard preparation*, respectively.

Triamcinolone Diacetate Syrup

» Triamcinolone Diacetate Syrup contains not less than 90.0 percent and not more than 110.0 percent of the labeled amount of $C_{25}H_{31}FO_8$. It contains a suitable preservative.

Packaging and storage—Preserve in tight, light-resistant containers.

Reference standard—*USP Triamcinolone Diacetate Reference Standard*—Dry in vacuum at 60° for 4 hours before using.

Identification—Transfer a quantity of Syrup, equivalent to about 10 mg of triamcinolone diacetate, to a separator, and extract with three 10-mL portions of chloroform. Evaporate the combined chloroform extracts on a steam bath to dryness, and dissolve the residue in 5.0 mL of chloroform. On a suitable thin-layer chromatographic plate (see *Chromatography* ⟨621⟩), coated with a 0.25-mm layer of chromatographic silica gel, apply 10 µL each of this solution and a solution of USP Triamcinolone Diacetate RS in chloroform containing 2 mg per mL. Allow the spots to dry, and develop the chromatogram in a solvent system consisting of a mixture of ethyl acetate and chloroform (9:1) until the solvent front has moved about three-fourths of the length of the plate. Remove the plate from the developing chamber, mark the solvent front, and allow the solvent to evaporate. Locate the spots on the plate by lightly spraying with dilute sulfuric acid (1 in 2) and heating on a hot plate or under a lamp until spots appear: the R_f value of the principal spot obtained from the test solution corresponds to that obtained from the Standard solution.

Assay—

Mobile phase, Internal standard solution, Standard preparation, and *Chromatographic system*—Proceed as directed in the *Assay* under *Triamcinolone Diacetate.*

Barium hydroxide solution—Add 2.5 g of barium hydroxide to 50 mL of a saturated solution of sodium chloride in a glass-stoppered, 125-mL flask. Insert the stopper, and shake the solution for 1 minute. Prepare this solution fresh daily.

Assay preparation—Using a 5-mL volumetric flask, transfer 5.0 mL of Triamcinolone Diacetate Syrup to a 125-mL separator. Wash the flask with three 5-mL portions of water, and combine the washes in the separator. Wash the flask with 5 mL of a water-washed chloroform-hexane mixture (9:1), and transfer to the separator. Add an additional 15 mL of the chloroform-hexane mixture to the separator, and shake vigorously for 30 seconds. Allow the phases to separate, and transfer the chloroform-hexane layer (bottom) to a second 125-mL separator containing 25 mL of *Barium hydroxide solution.* Shake the second separator vigorously for 30 seconds, allow the phases to separate, and transfer the bottom layer to a third separator containing 25 mL of dilute phosphoric acid solution (1 in 20). Shake the third separator vigorously, allow the phases to separate, and drain the organic layer into a 100-mL volumetric flask. Repeat the extraction procedure using four additional 20-mL portions of water-washed chloroform-hexane mixture. Combine the extracts in the 100-mL volumetric flask, dilute with the same chloroform-hexane mixture to volume, if necessary, and mix. Transfer an aliquot of the chloroform-hexane solution, equivalent to about 1 mg of triamcinolone diacetate, to a 250-mL beaker and add 1.0 mL of *Internal standard solution.* Evaporate the solvent, using low heat and a gentle stream of nitrogen. Dissolve the residue in 10.0 mL of *Mobile phase,* using sonication.

Procedure—Proceed as directed for *Procedure* in the *Assay* under *Triamcinolone Diacetate.* Calculate the quantity, in mg, of $C_{25}H_{31}FO_8$ in the portion of Syrup taken by the formula:

$$10C(R_U/R_S),$$

in which C is the concentration, in mg per mL, of USP Triamcinolone Diacetate RS in the *Standard preparation,* and R_U and R_S are the peak response ratios obtained from the *Assay preparation* and the *Standard preparation,* respectively.

Triamcinolone Hexacetonide

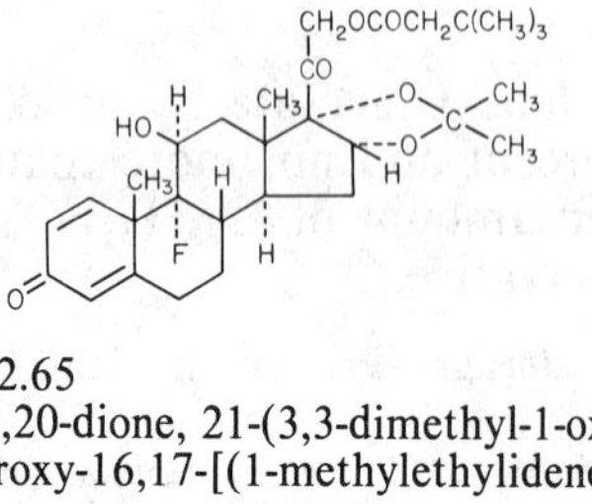

$C_{30}H_{41}FO_7$ 532.65

Pregna-1,4-diene-3,20-dione, 21-(3,3-dimethyl-1-oxobutoxy)-9-fluoro-11-hydroxy-16,17-[(1-methylethylidene)bis(oxy)]-, (11β,16α)-.

9-Fluoro-11β,16α,17,21-tetrahydroxypregna-1,4-diene-3,20-dione cyclic 16,17-acetal with acetone 21-(3,3-dimethylbutyrate) [*5611-51-8*].

» Triamcinolone Hexacetonide contains not less than 97.0 percent and not more than 102.0 percent of $C_{30}H_{41}FO_7$, calculated on the dried basis.

Packaging and storage—Preserve in well-closed containers.

Reference standard—*USP Triamcinolone Hexacetonide Reference Standard*—Dry in vacuum at 60° for 4 hours before using.

Identification—The infrared absorption spectrum of a potassium bromide dispersion of it exhibits maxima only at the same wavelengths as that of a similar preparation of USP Triamcinolone Hexacetonide RS.

Specific rotation ⟨781⟩: between +85° and +95°, calculated on the dried basis, determined in a solution in chloroform containing 100 mg in each 10 mL.

Loss on drying ⟨731⟩—Dry it in vacuum at 60° for 4 hours: it loses not more than 2.0% of its weight.

Heavy metals, *Method II* ⟨231⟩: 0.002%.

Assay—

Mobile solvent—Prepare a suitable solution of isopropyl alcohol in dichloromethane [up to about 5% (v/v)], such that the retention time of triamcinolone hexacetonide is about 2.5 to 3.0 minutes when the chromatograph is operated at ambient temperature.

Standard preparation—Transfer about 10 mg of USP Triamcinolone Hexacetonide RS, accurately weighed, to a 100-mL volumetric flask, add 50 mL of *Mobile solvent,* shake, if necessary, on a mechanical shaker until solution is complete, dilute with *Mobile solvent* to volume, and mix.

Assay preparation—Weigh accurately about 10 mg of Triamcinolone Hexacetonide, and prepare a solution as directed for *Standard preparation.*

Procedure—Introduce equal volumes (between 4 µL and 20 µL) of the *Assay preparation* and the *Standard preparation,* respectively, into a high-pressure liquid chromatograph (see *Chromatography* ⟨621⟩) by means of a suitable microsyringe or sampling valve, adjusting the specimen size and other operating parameters such that the peak obtained with the *Standard preparation* is about 0.5 full scale. Typically, the apparatus is fitted with a 2-mm × 60- to 100-cm column containing packing L4 and equipped with an ultraviolet detector capable of monitoring absorption at 254 nm and a suitable recorder. Measure the height of the peaks, at identical retention times, obtained with the *Assay preparation* and the *Standard preparation.* Calculate the quantity, in mg, of $C_{30}H_{41}FO_7$ in the portion of Triamcinolone Hexacetonide taken by the formula:

$$100C(H_U/H_S),$$

in which C is the concentration, in mg per mL, of USP Triamcinolone Hexacetonide RS in the *Standard preparation,* and H_U and H_S are the peak heights obtained with the *Assay preparation* and the *Standard preparation,* respectively.

Sterile Triamcinolone Hexacetonide Suspension

» Sterile Triamcinolone Hexacetonide Suspension is a sterile suspension of Triamcinolone Hexacetonide in a suitable aqueous medium. It contains not less than 90.0 percent and not more than 115.0 percent of the labeled amount of $C_{30}H_{41}FO_7$.

Packaging and storage—Preserve in single-dose or in multiple-dose containers, preferably of Type I glass.

Reference standard—*USP Triamcinolone Hexacetonide Reference Standard*—Dry in vacuum at 60° for 4 hours before using.

Identification—Place a volume of Suspension, equivalent to about 25 mg of triamcinolone hexacetonide, and 2 mL of water in a membrane filter having a pore size of 0.20 µm. Apply vacuum to the filter, wash the residue with two 5-mL portions of water, and air-dry the filter and the precipitate. Place the dried filter and precipitate in a small beaker with 5 mL of alcohol, and dissolve the precipitate. Decant the alcohol solution into a small beaker, and evaporate, with the aid of low heat and a stream of air, to dryness: the triamcinolone hexacetonide so obtained responds to the *Identification test* under *Triamcinolone Hexacetonide.*

pH ⟨791⟩: between 4.0 and 8.0.

Other requirements—It meets the requirements under *Injections* ⟨1⟩.

Assay—

Mobile solvent and *Standard preparation*—Prepare as directed in the *Assay* under *Triamcinolone Hexacetonide.*

Assay preparation—Transfer to a separator an accurately measured volume of Sterile Triamcinolone Hexacetonide Suspension, equivalent to about 20 mg of triamcinolone hexacetonide, and add 20 mL of water. Extract with four 35-mL portions of chloroform, filtering each portion through chloroform-soaked anhydrous sodium sulfate into a 200-mL volumetric flask. Add chloroform to volume, and mix. Evaporate a 10.0-mL portion of this solution at a temperature below 50°, with the aid of a stream of nitrogen, to dryness. Dissolve the residue in 10.0 mL of *Mobile solvent.*

Procedure—Proceed as directed for *Procedure* in the *Assay* under *Triamcinolone Hexacetonide.* Calculate the quantity, in mg, of $C_{30}H_{41}FO_7$ in each mL of the Suspension taken by the formula:

$$200(C/V)(H_U/H_S),$$

in which C is the concentration, in mg per mL, of USP Triamcinolone Hexacetonide RS in the *Standard preparation,* V is the volume, in mL, of Suspension taken, and H_U and H_S are the peak heights obtained with the *Assay preparation* and the *Standard preparation,* respectively.

Triamterene

$C_{12}H_{11}N_7$ 253.27

2,4,7-Pteridinetriamine, 6-phenyl-.

2,4,7-Triamino-6-phenylpteridine [*396-01-0*].

» Triamterene contains not less than 98.0 percent and not more than 102.0 percent of $C_{12}H_{11}N_7$, calculated on the dried basis.

Packaging and storage—Preserve in tight, light-resistant containers.

Reference standard—*USP Triamterene Reference Standard*—Dry in vacuum at 105° for 2 hours before using.

Identification—

A: The infrared absorption spectrum of a mineral oil dispersion of it, previously dried, exhibits maxima only at the same wavelengths as that of a similar preparation of USP Triamterene RS.

B: A solution in formic acid solution (1 in 1000) shows an intense, bluish fluorescence.

Loss on drying ⟨731⟩—Dry it in vacuum at 105° for 2 hours: it loses not more than 1.0% of its weight.

Ordinary impurities ⟨466⟩—

Test solution: 2-methoxyethanol.

Standard solution: 2-methoxyethanol.

Eluant: a mixture of ethyl acetate, glacial acetic acid, and water (8:1:1).

Visualization: 1.

Limit for 2,4,6-triamino-5-nitrosopyrimidine—

Mobile phase—Prepare a filtered and degassed mixture of 0.01 *N* potassium dihydrogen phosphate (adjusted to a pH of 3.0) and methanol (80:20). Make adjustments if necessary (see *System Suitability* under *Chromatography* ⟨621⟩).

Internal standard solution—Prepare a solution of furosemide in methanol having a known concentration of 0.25 mg per mL.

Standard preparation—[NOTE—Heating to 50° and sonication may be used to dissolve the triamino-nitrosopyrimidine.] Dissolve an accurately weighed quantity of 2,4,6-triamino-5-nitrosopyrimidine in methanol, and dilute quantitatively if necessary with methanol to obtain a Standard solution having a known concentration of about 50 µg per mL. Mix 10.0 mL of this solution with 2.0 mL of *Internal standard solution.*

Test preparation—Transfer about 5 g of Triamterene, accurately weighed, to a 250-mL conical flask. Add 100.0 mL of methanol, and stir for 30 minutes with heating to 50°. Cool, and filter to obtain a solution having a concentration of 50.0 mg per mL. Mix 10.0 mL of this solution with 2.0 mL of *Internal standard solution.*

Chromatographic system—The liquid chromatograph is equipped with a 330-nm detector and a 30-cm × 3.9-mm column that contains 10-µm packing L10. The flow rate is about 1.5 mL per minute. Chromatograph the *Standard preparation,* and record the peak responses as directed under *Procedure:* the tailing factor is not more than 1.1 and the relative standard deviation for replicate injections is not more than 2.0%.

Procedure—Separately inject equal volumes (about 20 µL) of the *Standard preparation* and the *Test preparation* into the chromatograph, record the chromatograms, and measure the responses for the major peaks. The relative retention times are about 0.27 for 2,4,6-triamino-5-nitrosopyrimidine, 0.64 for triamterene and 1.0 for furosemide. Calculate the percentage of 2,4,6-triamino-5-nitrosopyrimidine in the portion of Triamterene taken by the formula:

$$10C/W(R_U/R_S),$$

in which the C is the concentration, in µg per mL, of 2,4,6-triamino-5-nitrosopyrimidine in the Standard solution, W is the weight, in mg of triamterene taken, and R_U and R_S are the ratios of the responses of the 2,4,6-triamino-5-nitrosopyrimidine peak to that of the internal standard peak obtained from the *Test preparation* and the *Standard preparation,* respectively: not more than 0.1% of 2,4,6-triamino-5-nitrosopyrimidine is found.

Assay—Transfer about 0.5 g of Triamterene, accurately weighed, to a 400-mL beaker, and dissolve in 250 mL of a solvent previously prepared by mixing, in the order named and with cooling prior to use, 1 volume of formic acid, 1 volume of acetic anhydride, and 2 volumes of glacial acetic acid. Titrate with 0.1 *N* perchloric acid VS, determining the end-point potentiometrically. Perform a blank determination, and make any necessary correction. Each mL of 0.1 *N* perchloric acid is equivalent to 25.33 mg of $C_{12}H_{11}N_7$.

Triamterene Capsules

» Triamterene Capsules contain not less than 93.0 percent and not more than 107.0 percent of the labeled amount of $C_{12}H_{11}N_7$.

Packaging and storage—Preserve in tight, light-resistant containers.

Reference standard—*USP Triamterene Reference Standard*—Dry in vacuum at 105° for 2 hours before using.

Identification—

A: The ultraviolet absorption spectrum of the solution of Capsules employed for measurement of absorbance in the *Assay* exhibits maxima and minima at the same wavelengths as that of the solution of USP Triamterene RS used in the *Assay.*

B: The solution of Capsules prepared for the *Assay* shows an intense, bluish fluorescence.

Uniformity of dosage units ⟨905⟩: meet the requirements.

Assay—Remove, as completely as possible, the contents of not less than 20 Triamterene Capsules, weigh, and mix. Transfer an accurately weighed portion of the powder, equivalent to about 0.1 g of triamterene, to a 250-mL volumetric flask. Add 100 mL of methoxyethanol, shake until dissolved, dilute with water nearly to volume, mix, and cool to room temperature. Add water to volume, and mix. Transfer 5.0 mL of this solution to a 200-mL volumetric flask, add 5 mL of formic acid, add water to volume, and mix. Using an accurately weighed quantity of USP Triamterene RS and a solvent of the same composition as that for the solution from the Capsules, prepare a Standard solution having a known concentration of about 10 µg per mL. Concomitantly determine the absorbances of both solutions in 1-cm cells from 400 nm to 300 nm, with a suitable spectrophotometer, using water as the blank. Draw a base-line between the absorbance values on each curve at 390 nm and at 320 nm, and determine the corrected absorbances for the Standard solution and the solution from the Capsule contents at the wavelength of maximum absorbance at about 357.5 nm, by subtracting the baseline absorbance at 357.5 nm from the maximum absorbance at 357.5 nm. Calculate the quantity, in mg, of $C_{12}H_{11}N_7$ in the portion of Capsules taken by the formula:

$$10C(A_U/A_S),$$

in which C is the concentration, in µg per mL, of USP Triamterene RS in the Standard solution, and A_U and A_S are the absorbances of the solution from the Capsules and the Standard solution, respectively.

Triamterene and Hydrochlorothiazide Extended Capsules

» Triamterene and Hydrochlorothiazide Extended Capsules contain not less than 90.0 percent and not more than 110.0 percent of the labeled amounts of triamterene ($C_{12}H_{11}N_7$) and hydrochlorothiazide ($C_7H_8ClN_3O_4S_2$).

Packaging and storage—Preserve in tight, light-resistant containers.

Labeling—Label Extended Capsules to state that Triamterene and Hydrochlorothiazide Extended Capsules may not be interchangeable with either Triamterene and Hydrochlorothiazide Prompt Capsules or Triamterene and Hydrochlorothiazide Tablets without retitration of the patient. The labeling indicates steady-state blood level profiles and confidence intervals and states the dissolution *Tolerances* (% dissolved) at the time points stated under *Dissolution*, using the test method described therein.

Reference standards—*USP Triamterene Reference Standard*—Dry in vacuum at 105° for 2 hours before using. *USP Hydrochlorothiazide Reference Standard*—Dry at 105° for 1 hour before using.

Identification—The retention times of the major peaks in the chromatogram of the *Assay preparation* correspond to those of the *Standard preparation*, both relative to the internal standard, as obtained in the *Assay*.

Dissolution ⟨711⟩—

Medium: 0.1 *N* hydrochloric acid; 900 mL.

Apparatus 1: 100 rpm.

Times: 2 hours; 6 hours.

Procedure—Determine the amounts of triamterene ($C_{12}H_{11}N_7$) and hydrochlorothiazide ($C_7H_8ClN_3O_4S_2$) dissolved from ultraviolet absorbances at the wavelength of maximum absorbance at about 357 nm for triamterene and 271 nm for hydrochlorothiazide (corrected for interference from triamterene on the basis of the absorbances of triamterene at 271 nm and 357 nm) by comparison with a Standard solution having known concentrations of USP Triamterene RS and USP Hydrochlorothiazide RS in the same medium.

Tolerances—Meet the *Tolerances* specified in the labeling; use the Acceptance Criteria in *Acceptance Table I*, under *Drug Release* ⟨724⟩.

Uniformity of dosage units ⟨905⟩: meet the requirements for *Content uniformity* with respect to triamterene and to hydrochlorothiazide.

Assay—

Mobile phase—Prepare a filtered and degassed mixture of water, acetonitrile, methanol, and pH 5.0, 0.2 *M* sodium acetate buffer (78:15:5:2). Make adjustments if necessary (see *System Suitability* under *Chromatography* ⟨621⟩).

Internal standard solution—Transfer about 375 mg of *m*-hydroxyacetophenone to a 250-mL volumetric flask, dissolve in 125 mL of acetonitrile, add water to volume, and mix.

Standard preparation—Transfer about 20 mg of USP Hydrochlorothiazide RS, accurately weighed, to a 100-mL volumetric flask. Add 20*J* mg of USP Triamterene RS, accurately weighed, *J* being the ratio of the labeled amount, in mg, of triamterene to the labeled amount, in mg, of hydrochlorothiazide per Capsule. Add 25 mL of acetonitrile and 4 mL of glacial acetic acid, and mix. Dilute with water to volume, and mix. Transfer 25.0 mL of this solution to a 100-mL volumetric flask, add 10.0 mL of *Internal standard solution*, dilute with water to volume, and mix.

Assay preparation—Remove, as completely as possible, the contents of not less than 20 Triamterene and Hydrochlorothiazide Extended Capsules, and weigh accurately. Mix the combined contents, and transfer an accurately weighed portion of the powder, equivalent to about 20 mg of hydrochlorothiazide, to a 100-mL volumetric flask. Add 25 mL of acetonitrile and 4 mL of glacial acetic acid, and mix. Dilute with water to volume, and mix. Filter a portion, discarding the first 15 mL of the filtrate. Transfer 25.0 mL of the filtrate to a 100-mL volumetric flask, add 10.0 mL of *Internal standard solution*, dilute with water to volume, and mix.

Chromatographic system (see *Chromatography* ⟨621⟩)—The liquid chromatograph is equipped with a 280-nm detector and a 4-mm × 30-cm column that contains 10-µm packing L1. The flow rate is about 2 mL per minute. Chromatograph the *Standard preparation*, and record the peak responses as directed under *Procedure:* the tailing factor for the triamterene peak is not more than 2.5, the resolution, *R*, between hydrochlorothiazide and *m*-hydroxyacetophenone is not less than 2.0, the resolution, *R*, between *m*-hydroxyacetophenone and triamterene is not less than 1.5, and the relative standard deviation for replicate injections is not more than 2.0%.

Procedure—Separately inject equal volumes (about 25 µL) of the *Standard preparation* and the *Assay preparation* into the chromatograph, record the chromatograms, and measure the responses for the major peaks. The relative retention times are about 0.4 for hydrochlorothiazide, 0.8 for *m*-hydroxyacetophenone, and 1.0 for triamterene. Calculate the quantity, in mg, of triamterene ($C_{12}H_{11}N_7$) in the portion of Capsules taken by the formula:

$$400C(R_U/R_S),$$

in which C is the concentration, in mg per mL, of USP Triamterene RS in the *Standard preparation*, and R_U and R_S are the peak response ratios obtained from the *Assay preparation* and the *Standard preparation*, respectively. Calculate the quantity, in mg, of hydrochlorothiazide ($C_7H_8ClN_3O_4S_2$) by the same formula, changing the terms to refer to hydrochlorothiazide.

Triamterene and Hydrochlorothiazide Tablets

» Triamterene and Hydrochlorothiazide Tablets contain not less than 90.0 percent and not more than 110.0 percent of the labeled amounts of triamterene ($C_{12}H_{11}N_7$) and hydrochlorothiazide ($C_7H_8ClN_3O_4S_2$).

Packaging and storage—Preserve in tight, light-resistant containers.

Labeling—Label Tablets to state that Triamterene and Hydrochlorothiazide Tablets may not be interchangeable with either

Triamterene and Hydrochlorothiazide Prompt Capsules or Triamterene and Hydrochlorothiazide Extended Capsules without retitration of the patient.

Reference standards—*USP Triamterene Reference Standard*—Dry in vacuum at 105° for 2 hours before using. *USP Hydrochlorothiazide Reference Standard*—Dry at 105° for 1 hour before using.

Identification—The retention times of the major peaks in the chromatogram of the *Assay preparation* correspond to those of the *Standard preparation*, both relative to the internal standard, as obtained in the *Assay*.

Dissolution ⟨711⟩—

Medium: 0.1 *N* hydrochloric acid; 900 mL.

Apparatus 2: 75 rpm.

Time: 30 minutes.

Mobile phase and *Chromatographic system*—Prepare as directed in the *Assay*.

Internal standard solution—Transfer about 30 mg of dipyridamole to a 250-mL volumetric flask. Add 50 mL of acetonitrile and 25 mL of water, and mix to dissolve the dipyridamole. Dilute with pH 2.8, 0.01 *M* sodium acetate buffer to volume, and mix.

Standard preparation—Transfer about 50 mg of USP Hydrochlorothiazide RS and about 50*J* mg of USP Triamterene RS, *J* being the ratio as defined in the *Assay*, accurately weighed, to a 900-mL volumetric flask. Add 50 mL of acetonitrile and 50 mL of pH 2.8, 0.01 *M* sodium acetate buffer, mix to dissolve the standards, and dilute with 0.1 *N* hydrochloric acid to volume. Pipet 5 mL of this solution and 5 mL of *Internal standard solution* into a container, and mix.

Procedure—Separately inject equal volumes (about 10 µL) of the *Standard solution* and a filtered portion of the solution under test mixed with an equal volume of *Internal standard solution* into the chromatograph, record the chromatograms, and measure the responses for the major peaks. The relative retention times are about 0.5 for hydrochlorothiazide, 0.8 for triamterene, and 1.0 for dipyridamole. Calculate the amounts of triamterene ($C_{12}H_{11}N_7$) and hydrochlorothiazide ($C_7H_8ClN_3O_4S_2$) dissolved by comparison of the corresponding peak responses obtained from the solution under test and the *Standard solution*.

Tolerances—Not less than 80% (*Q*) each of the labeled amounts of $C_{12}H_{11}N_7$ and $C_7H_8ClN_3O_4S_2$ are dissolved in 30 minutes.

Uniformity of dosage units ⟨905⟩: meet the requirements for *Content uniformity* with respect to triamterene and to hydrochlorothiazide.

Assay—

Mobile phase—Prepare a filtered and degassed mixture of pH 4.5, 0.01 *M* sodium acetate buffer and acetonitrile (3:2). Make adjustments if necessary (see *System Suitability* under *Chromatography* ⟨621⟩).

Internal standard solution—Transfer about 100 mg of dipyridamole to a 100-mL volumetric flask, add 50 mL of acetonitrile and 25 mL of *Mobile phase*, sonicate for 5 minutes, dilute with *Mobile phase* to volume, and mix.

Standard preparation—Transfer about 50 mg of USP Hydrochlorothiazide RS, accurately weighed, to a 100-mL volumetric flask. Add 50*J* mg of USP Triamterene RS, accurately weighed, *J* being the ratio of the labeled amount, in mg, of triamterene to the labeled amount, in mg, of hydrochlorothiazide per Tablet. Add 50 mL of acetonitrile, 5 mL of glacial acetic acid, and 40 mL of *Mobile phase*, sonicate for 10 minutes after each addition, dilute with *Mobile phase* to volume, and mix. Pipet 5 mL of this solution into a 50-mL volumetric flask, add 5.0 mL of *Internal standard solution*, dilute with *Mobile phase* to volume, and mix.

Assay preparation—Weigh and finely powder not less than 20 Triamterene and Hydrochlorothiazide Tablets. Transfer an accurately weighed portion of the powder, equivalent to about 50 mg of hydrochlorothiazide, to a 100-mL volumetric flask, add 50 mL of acetonitrile, 5 mL of glacial acetic acid, and 40 mL of *Mobile phase*, sonicating for 10 minutes after each addition, dilute with *Mobile phase* to volume, mix, and filter, discarding the first 15 mL of the filtrate. Pipet 5 mL of the filtrate into a 50-mL volumetric flask, add 5.0 mL of *Internal standard solution*, dilute with *Mobile phase* to volume and mix.

Chromatographic system (see *Chromatography* ⟨621⟩)—The liquid chromatograph is equipped with a 272-nm detector and a 3.9-mm × 15-cm column that contains packing L10. The flow rate is about 1 mL per minute. Chromatograph the *Standard preparation*, and record the peak responses as directed under *Procedure:* the resolution, *R*, between hydrochlorothiazide and triamterene is not less than 2.0, the resolution, *R*, between triamterene and dipyridamole is not less than 1.5, and the relative standard deviation for replicate injections is not more than 2.0%.

Procedure—Separately inject equal volumes (about 10 µL) of the *Standard preparation* and the *Assay preparation* into the chromatograph, record the chromatograms, and measure the responses for the major peaks. The relative retention times are about 0.5 for hydrochlorothiazide, 0.8 for triamterene, and 1.0 for dipyridamole. Calculate the quantity, in mg, of triamterene ($C_{12}H_{11}N_7$) in the portion of Tablets taken by the formula:

$$1000C(R_U / R_S),$$

in which *C* is the concentration, in mg per mL, of USP Triamterene RS in the *Standard preparation*, and R_U and R_S are the peak response ratios obtained from the *Assay preparation* and the *Standard preparation*, respectively. Calculate the quantity, in mg, of hydrochlorothiazide ($C_7H_8ClN_3O_4S_2$) by the same formula, changing the terms to refer to hydrochlorothiazide.

Triazolam

$C_{17}H_{12}Cl_2N_4$ 343.21

4*H*-[1,2,4]Triazolo[4,3-*a*][1,4]benzodiazepine, 8-chloro-6-(2-chlorophenyl)-1-methyl-.

8-Chloro-6-(*o*-chlorophenyl)-1-methyl-4*H*-*s*-triazolo[4,3-*a*][1,4]-benzodiazepine [*28911-01-5*].

» Triazolam contains not less than 97.0 percent and not more than 103.0 percent of $C_{17}H_{12}Cl_2N_4$, calculated on the dried basis.

Caution—Exercise care to prevent inhaling particles of triazolam and to prevent its contacting any part of the body.

Packaging and storage—Preserve in well-closed containers.

Reference standard—*USP Triazolam Reference Standard*—Do not dry before using.

Identification—

A: The infrared absorption spectrum of a mineral oil dispersion of it exhibits maxima only at the same wavelengths as that of a similar preparation of USP Triazolam RS.

B: The ultraviolet spectrum of a 1 in 250,000 solution in alcohol exhibits maxima and minima at the same wavelengths as that of a similar solution of USP Triazolam RS, concomitantly measured, and the respective absorptivities, calculated on the dried basis, at the wavelength of maximum absorbance at about 220 nm do not differ by more than 3%.

Loss on drying ⟨731⟩—Dry it at 60° and at a pressure not exceeding 5 mm of mercury for 16 hours: it loses not more than 0.5% of its weight.

Residue on ignition ⟨281⟩: not more than 0.5%.

Heavy metals, *Method II* ⟨231⟩: 0.002%.

Chromatographic purity—

Test solution—Prepare a solution of Triazolam in chloroform containing about 2 mg per mL.

Chromatographic system (see *Chromatography* ⟨621⟩)—The gas chromatograph is equipped with a flame-ionization detector, and contains a 3-mm × 120-cm glass column packed with 3 percent phase G6 on support S1AB. The column and injection port are maintained at a temperature of about 240°. The detector is maintained at a temperature of about 20° to 50° above column temperature. The carrier gas is helium.

Procedure—[NOTE—Allow about three times the elution time of the major component before making another injection.] Chromatograph about 4 µL of the *Test solution*. Calculate the total percentage of impurities by the formula:

$$100S/(S + A),$$

in which S is the sum of the areas of each of the minor component peaks detected, and A is the area of the major component peak. The total amount of impurities present is not more than 1.5%.

Assay—

Internal standard solution—Prepare a solution of alprazolam in acetonitrile having a concentration of about 0.25 mg per mL.

Standard preparation—Transfer about 3.2 mg of USP Triazolam RS, accurately weighed, to a 100-mL volumetric flask. Add 10.0 mL of *Internal standard solution*, dilute with acetonitrile to volume, and mix.

Assay preparation—Prepare as directed under *Standard preparation*, using a 3.2-mg portion of Triazolam.

Mobile phase—Prepare a solution consisting of a mixture of acetonitrile, chloroform, butyl alcohol, water, and glacial acetic acid (850:80:50:20:0.5). Make adjustments if necessary (see *System Suitability* under *Chromatography* ⟨621⟩).

Chromatographic system (see *Chromatography* ⟨621⟩)—The liquid chromatograph is equipped with a detector capable of monitoring ultraviolet absorbance at 254 nm and a 4- to 4.6-mm × 25- to 30-cm stainless steel column containing packing L3. The flow rate is about 2 mL per minute. The resolution factor is not less than 2.0 between the internal standard and triazolam. The relative standard deviation of the peak response ratios of a minimum of four replicate injections of the *Standard preparation* is less than 2.0%. The relative retention times are about 1.0 for triazolam and 1.4 for the internal standard.

Procedure—Separately inject equal volumes (about 20 µL) of the *Assay preparation* and the *Standard preparation* into the liquid chromatograph. Calculate the quantity, in mg, of $C_{17}H_{12}Cl_2N_4$ in the portion of Triazolam taken by the formula:

$$100C(R_U/R_S),$$

in which C is the concentration, in mg per mL, of USP Triazolam RS in the *Standard preparation*, and R_U and R_S are the peak response ratios from the *Assay preparation* and the *Standard preparation*, respectively.

Triazolam Tablets

» Triazolam Tablets contain not less than 90.0 percent and not more than 110.0 percent of the labeled amount of $C_{17}H_{12}Cl_2N_4$.

Packaging and storage—Preserve in tight, light-resistant containers.

Reference standard—*USP Triazolam Reference Standard*—Do not dry before using.

Identification—The retention time of the major peak in the chromatogram of the *Assay preparation* obtained as directed in the *Assay* corresponds to that of the *Standard preparation*, relative to the internal standard.

Dissolution ⟨711⟩—

Medium: water; 500 mL.

Apparatus 2: 50 rpm.

Time: 30 minutes.

Stock standard solution—Weigh accurately about 5 mg of USP Triazolam RS, dissolve in methanol, dilute with methanol to 200.0 mL, and mix.

Working standard solution—For each 0.125 mg of the labeled amount of triazolam per Tablet, add 2.0 mL of *Stock standard solution* to a 200-mL volumetric flask. Dilute with water to volume, and mix.

Procedure—After 30 minutes, withdraw a portion of the solution under test, and filter immediately. Inject equal volumes (about 200 µL) of this solution and the *Working standard solution* into a liquid chromatograph (see *Chromatography* ⟨621⟩) equipped with a detector capable of monitoring ultraviolet absorbance at 222 nm and a 4.6-mm × 10-cm stainless steel column containing packing L7. The mobile phase is a mixture of water and acetonitrile (60:40). The flow rate is about 1 mL per minute. The relative standard deviation of the peak response for the *Working standard solution* is not more than 3.0%, and the number of theoretical plates is not less than 500. Calculate the percentage of $C_{17}H_{12}Cl_2N_4$ dissolved by the formula:

$$50{,}000(r_U/r_S)(C/L),$$

in which r_U and r_S are the peak responses of the solution under test and the *Working standard solution*, C is the concentration, in mg per mL, of USP Triazolam RS in the *Working standard solution*, and L is the labeled amount, in mg, of triazolam in the Tablet.

Tolerances—Not less than 70% (Q) of the labeled amount of $C_{17}H_{12}Cl_2N_4$ is dissolved in 30 minutes.

Uniformity of dosage units ⟨905⟩: meet the requirements.

Procedure for content uniformity—

Mobile phase and *Chromatographic system*—Proceed as directed in the *Assay* under *Triazolam*.

Internal standard solution—Prepare a solution of alprazolam in acetonitrile having a concentration of about 0.025 mg per mL.

Standard preparation—Dissolve about 3.2 mg of USP Triazolam RS, accurately weighed, in 100.0 mL of *Internal standard solution*, and mix.

Test preparation—Transfer 1 Tablet to a container, add about 0.4 mL of water directly onto the Tablet, allow to stand for about 2 minutes, and swirl the container to disperse the Tablet. For each 0.25 mg of the labeled amount of triazolam in the Tablet, add 10.0 mL of *Internal standard solution* to the container. Shake, and centrifuge if necessary.

Procedure—Proceed as directed for *Procedure* in the *Assay* under *Triazolam*. Calculate the quantity, in mg, of $C_{17}H_{12}Cl_2N_4$ in the Tablet taken by the formula:

$$CV(R_U/R_S),$$

in which V is the volume, in mL, of *Internal standard solution* in the *Test preparation*.

Water, *Method I* ⟨921⟩: not more than 8%.

Assay—

Mobile phase, Internal standard solution, Standard preparation, and *Chromatographic system*—Proceed as directed in the *Assay* under *Triazolam*.

Assay preparation—Transfer 6 Tablets to a container, add about 2 mL of water, and swirl to disperse the Tablets. For each 0.25 mg of the labeled amount of triazolam per Tablet, add 5.0 mL of *Internal standard solution*, and dilute quantitatively with acetonitrile to obtain a solution having a concentration of about 0.03 mg of triazolam per mL. Shake for about 10 minutes, and centrifuge if necessary.

Procedure—Proceed as directed for *Procedure* in the *Assay* under *Triazolam*. Calculate the quantity, in mg, of $C_{17}H_{12}Cl_2N_4$ in the Tablets taken by the formula:

$$CV(R_U/R_S),$$

in which V is the final volume, in mL, of the *Assay preparation*, and the other terms are as defined in the *Assay* under *Triazolam*.

Tribasic Calcium Phosphate—*see* Calcium Phosphate, Tribasic NF

Trichlormethiazide

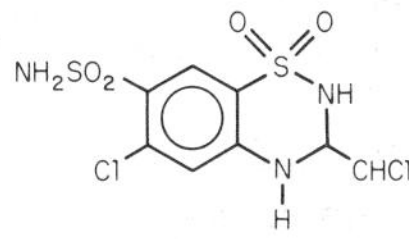

$C_8H_8Cl_3N_3O_4S_2$ 380.65

2*H*-1,2,4-Benzothiadiazine-7-sulfonamide, 6-chloro-3-(dichloromethyl)-3,4-dihydro-, 1,1-dioxide.

6-Chloro-3-(dichloromethyl)-3,4-dihydro-2*H*-1,2,4-benzothiadiazine-7-sulfonamide 1,1-dioxide [*133-67-5*].

» Trichlormethiazide, dried at 105° for 3 hours, contains not less than 98.0 percent and not more than 102.0 percent of $C_8H_8Cl_3N_3O_4S_2$.

Packaging and storage—Preserve in well-closed containers.

Reference standards—*USP Trichlormethiazide Reference Standard*—Dry at 105° for 3 hours before using. *USP 4-Amino-6-chloro-1,3-benzenedisulfonamide Reference Standard*—Keep container tightly closed and protected from light. Dry over silica gel for 4 hours before using.

Identification—

A: The infrared absorption spectrum of a mineral oil dispersion of it, previously dried, exhibits maxima only at the same wavelengths as that of a similar preparation of USP Trichlormethiazide RS.

B: Prepare a solution in a mixture of equal volumes of toluene and alcohol containing 1 mg per mL. On a suitable thin-layer chromatographic plate (see *Thin-layer Chromatography* under *Chromatography* ⟨621⟩), coated with a 0.25-mm layer of chromatographic silica gel mixture, apply 10 µL of this solution and 10 µL of a Standard solution of USP Trichlormethiazide RS in the same medium having a known concentration of 1 mg per mL. Allow the spots to dry, and develop the chromatogram, using ethyl acetate as the solvent, until the solvent front has moved about three-fourths of the length of the plate. Remove the plate from the developing chamber, mark the solvent front, and allow the solvent to evaporate. Locate the spots using short-wavelength ultraviolet light: the R_f value of the principal spot obtained from the test solution corresponds to that obtained from the Standard solution.

Loss on drying ⟨731⟩—Dry it at 105° for 3 hours: it loses not more than 0.5% of its weight.

Residue on ignition ⟨281⟩: not more than 0.1%.

Selenium ⟨291⟩: 0.003%.

Heavy metals, *Method II* ⟨231⟩: 0.002%.

Diazotizable substances—

Standard preparation—Prepare a solution in a mixture of equal volumes of toluene and alcohol containing 250 µg of USP 4-Amino-6-chloro-1,3-benzenedisulfonamide RS in each mL.

Test preparation—Transfer 100.0 mg of Trichlormethiazide, accurately weighed, to a 10-mL volumetric flask, dissolve in 1 mL of acetone, dilute with a mixture of equal volumes of toluene and alcohol to volume, and mix.

Procedure—Apply 5 µL each of the *Standard preparation* and of the *Test preparation* to a suitable thin-layer chromatographic plate (see *Thin-layer Chromatography*, under *Chromatography* ⟨621⟩), coated with a 0.25-mm layer of chromatographic silica gel mixture. Allow the spots to dry, and develop the chromatogram in ethyl acetate until the solvent front has moved about three-fourths of the length of the plate. Remove the plate from the developing chamber, mark the solvent front, and locate the spots on the plate by spraying first with a 1 in 20 solution of sodium nitrite in dilute hydrochloric acid (1 in 12), and then with a 1 in 1000 solution of *N*-1-naphthylethylenediamine dihydrochloride in alcohol: after 3 minutes, any spot from the *Test preparation*, occurring at the R_f value corresponding to that produced by the *Standard preparation*, is not greater in size or intensity than the spot produced by the *Standard preparation*, corresponding to not more than 2.5% of diazotizable substances.

Assay—

Mobile phase—Prepare a filtered and degassed mixture of 0.05 *M* monobasic potassium phosphate and methanol (7:3). Make adjustments if necessary (see *System Suitability* under *Chromatography* ⟨621⟩).

Internal standard solution—Transfer about 120 mg of methylparaben to a 50-mL volumetric flask, add methanol to volume, and mix.

Standard preparation—Transfer about 25 mg of USP Trichlormethiazide RS, accurately weighed, to a 50-mL volumetric flask, add 4.0 mL of *Internal standard solution*, dilute with methanol to volume, and mix to obtain a solution having a known concentration of about 0.5 mg of USP Trichlormethiazide RS per mL.

Assay preparation—Transfer about 50 mg of Trichlormethiazide, accurately weighed, to a 100-mL volumetric flask, add 8.0 mL of *Internal standard solution*, dilute with methanol to volume, and mix.

Chromatographic system (see *Chromatography* ⟨621⟩)—The liquid chromatograph is equipped with a 254-nm detector and a 3.9-mm × 30-cm column that contains packing L1. The flow rate is about 2.3 mL per minute. Chromatograph the *Standard preparation*, and record the peak responses as directed under *Procedure:* the resolution, *R*, between the analyte and internal standard peaks is not less than 2.0, and the relative standard deviation for the response ratios calculated for replicate injections is not more than 2%.

Procedure—Separately inject equal volumes (5 to 25 µL) of the *Standard preparation* and the *Assay preparation* into the chromatograph, record the chromatograms, and measure the responses for the major peaks. The relative retention times are about 0.56 for trichlormethiazide and 1.0 for methylparaben. Calculate the quantity, in mg, of $C_8H_8Cl_3N_3O_4S_2$ in the portion of Trichlormethiazide taken by the formula:

$$100C(R_U/R_S),$$

in which *C* is the concentration, in mg per mL, of USP Trichlormethiazide RS in the *Standard preparation*, and R_U and R_S are the peak response ratios obtained from the *Assay preparation* and the *Standard preparation*, respectively.

Trichlormethiazide Tablets

» Trichlormethiazide Tablets contain not less than 90.0 percent and not more than 110.0 percent of the labeled amount of $C_8H_8Cl_3N_3O_4S_2$.

Packaging and storage—Preserve in tight containers.

Reference standard—*USP Trichlormethiazide Reference Standard*—Dry at 105° for 3 hours before using.

Identification—Evaporate 25 mL of the combined ethyl acetate extracts obtained in the *Assay* on a steam bath to dryness, and dissolve the residue in 1 mL of a mixture of equal volumes of benzene and alcohol: a 10-µL portion of this solution responds to *Identification test B* under *Trichlormethiazide.*

Dissolution ⟨711⟩—

Medium: water; 900 mL.

Apparatus 2: 50 rpm.

Time: 60 minutes.

Procedure—[Note—Conduct the analysis of the specimen promptly after the specimen aliquot is withdrawn from the vessel, to minimize hydrolysis.] Determine the amount of $C_8H_8Cl_3N_3O_4S_2$ dissolved, employing the procedure set forth in the *Assay*, making any necessary volumetric adjustments.

Tolerances—Not less than 65% (*Q*) of the labeled amount of $C_8H_8Cl_3N_3O_4S_2$ is dissolved in 60 minutes.

Uniformity of dosage units ⟨905⟩: meet the requirements.

Assay—

Mobile phase and *Chromatographic system*—Prepare as directed in the *Assay* under *Trichlormethiazide*.

Acidic methanol—Add 2.8 mL of sulfuric acid to 100 mL of water, mix, and cool. Dilute with methanol to 1000 mL, and mix.

Internal standard solution—Transfer about 20 mg of methylparaben to a 200-mL volumetric flask, add *Acidic methanol* to volume, and mix.

Standard preparation—Transfer about 20 mg of USP Trichlormethiazide RS, accurately weighed, to a 50-mL volumetric flask, add *Acidic methanol* to volume, and mix. Transfer 5.0 mL of this solution to a centrifuge tube, add 5.0 mL of *Internal standard solution*, and mix to obtain a solution having a known concentration of about 0.2 mg of USP Trichlormethiazide RS per mL.

Assay preparation—Weigh and finely powder not less than 20 Trichlormethiazide Tablets. Transfer an accurately weighed portion of the powder, equivalent to 2 mg of trichlormethiazide, to a centrifuge tube. Add 5.0 mL of *Acidic methanol* and 5.0 mL of *Internal standard solution*, and mix to disperse the powder. Insert the stopper in the tube, rotate by mechanical means for 20 minutes, and centrifuge to obtain a clear supernatant liquid.

Procedure—Separately inject equal volumes (5 µL to 25 µL) of the *Standard preparation* and the *Assay preparation* into the chromatograph, record the chromatograms, and measure the responses for the major peaks. The relative retention times are about 0.58 for trichlormethiazide and 1.0 for methylparaben. Calculate the quantity, in mg, of $C_8H_8Cl_3N_3O_4S_2$ in the portion of Tablets taken by the formula:

$$10C(R_U/R_S),$$

in which C is the concentration, in mg per mL, of USP Trichlormethiazide RS in the *Standard preparation*, and R_U and R_S are the peak response ratios obtained from the *Assay preparation* and the *Standard preparation*, respectively.

Trichloromonofluoromethane—*see* Trichloromonofluoromethane NF

Tricitrates Oral Solution

» Tricitrates Oral Solution is a solution of Sodium Citrate, Potassium Citrate, and Citric Acid in a suitable aqueous medium. It contains, in each 100 mL, not less than 2.23 g and not more than 2.46 g of sodium (Na), equivalent to not less than 9.5 g and not more than 10.5 g of sodium citrate dihydrate ($C_6H_5Na_3O_7 . 2H_2O$); not less than 3.78 g and not more than 4.18 g of potassium (K), equivalent to not less than 10.45 g and not more than 11.55 g of potassium citrate monohydrate ($C_6H_5K_3O_7 . H_2O$); not less than 12.20 g and not more than 13.48 g of citrate ($C_6H_5O_7$) as sodium citrate and potassium citrate; and not less than 6.34 g and not more than 7.02 g of citric acid monohydrate ($C_6H_8O_7 . H_2O$).

NOTE—The sodium and potassium ion contents of Tricitrates Oral Solution are each approximately 1 mEq per mL.

Packaging and storage—Preserve in tight containers.

Identification—

A: It responds to the flame test for *Sodium* ⟨191⟩.

B: To a mixture of 1 mL of Oral Solution with 1 mL of hydrochloric acid add 10 mL of cobalt uranylacetate TS, and stir well with a glass rod: a pale yellow, fine crystalline precipitate is formed within several minutes (*presence of sodium*).

C: To 2 mL of a dilution of Oral Solution (1 in 20) add 5 mL of sodium cobaltinitrite TS: a yellow precipitate is formed immediately (*presence of potassium*).

D: It responds to the tests for *Citrate* ⟨191⟩, 3 to 5 drops of Oral Solution and 20 mL of the mixture of pyridine and acetic anhydride being used.

pH ⟨791⟩: between 4.9 and 5.4.

Assay for sodium and potassium—

Sodium stock solution—Transfer 14.61 g of sodium chloride, previously dried at 105° for 2 hours and accurately weighed, to a 250-mL volumetric flask, add water to volume, and mix.

Potassium stock solution—Transfer 18.64 g of potassium chloride, previously dried at 105° for 2 hours and accurately weighed, to a 250-mL volumetric flask, add water to volume, and mix.

Lithium diluent solution—Transfer 1.04 g of lithium nitrate to a 1000-mL volumetric flask, add a suitable nonionic surfactant, then add water to volume, and mix. This solution contains 15 mEq of Li per 1000 mL.

Standard preparation—Pipet 50 mL of *Sodium stock solution* and 50 mL of *Potassium stock solution* into a 500-mL volumetric flask, dilute with water to volume, and mix. Each mL of this solution contains 0.1 mEq of Na and 0.1 mEq of K. Transfer 50 µL of this solution to a 10-mL volumetric flask, dilute with *Lithium diluent solution* to volume, and mix.

Assay preparation—Transfer an accurately measured volume of Tricitrates Oral Solution, equivalent to about 2 g of combined citrates, to a 100-mL volumetric flask, dilute with water to volume, and mix. Transfer 50 µL of this solution to a 10-mL volumetric flask, dilute with *Lithium diluent solution* to volume, and mix.

Procedure—Using a suitable flame photometer, adjusted to read zero with *Lithium diluent solution*, concomitantly determine the sodium flame emission readings for the *Standard preparation* and the *Assay preparation* at the wavelength of maximum emission at about 589 nm. Similarly determine the potassium flame emission readings for the same solutions at the wavelength of maximum emission at about 766 nm. Calculate the quantity, in g, of Na in the portion of Oral Solution taken by the formula:

$$(14.61/25)(22.99/58.44)(R_{U,Na}/R_{S,Na}),$$

in which 14.61 is the weight, in g, of sodium chloride in the *Sodium stock solution*, 22.99 is the atomic weight of sodium, 58.44 is the molecular weight of sodium chloride, and $R_{U,Na}$ and $R_{S,Na}$ are the sodium emission readings obtained for the *Assay preparation* and the *Standard preparation*, respectively. Calculate the quantity, in g, of K in the portion of Oral Solution taken by the formula:

$$(18.64/25)(39.10/74.55)(R_{U,K}/R_{S,K}),$$

in which 18.64 is the weight, in g, of potassium chloride in the *Potassium stock solution*, 39.10 is the atomic weight of potassium, 74.55 is the molecular weight of potassium chloride, and $R_{U,K}$ and $R_{S,K}$ are the potassium emission readings obtained from the *Assay preparation* and the *Standard preparation*, respectively.

Assay for citrate—

Cation-exchange column—Mix 10 g of styrene-divinylbenzene cation-exchange resin with 50 mL of water in a suitable beaker. Allow the resin to settle, and decant the supernatant liquid until a slurry of resin remains. Pour the slurry into a 15-mm × 30-cm glass chromatographic tube (having a sealed-in, coarse-porosity fritted disk and fitted with a stopcock), and allow to settle as a homogeneous bed. Wash the resin bed with about 100 mL of water, closing the stopcock when the water level is about 2 mm above the resin bed.

Procedure—Pipet 15 mL of Tricitrates Oral Solution into a 250-mL volumetric flask, dilute with water to volume, and mix. Pipet 5 mL of this solution carefully onto the top of the resin bed in the *Cation-exchange column*. Place a 250-mL conical flask below the column, open the stopcock, and allow to flow until the solution has entered the resin bed. Elute the column with 60 mL of water at a flow rate of about 5 mL per minute, collecting about 65 mL of eluate. Add 5 drops of phenolphthalein TS to the eluate, swirl the flask, and titrate with 0.02 *N* sodium hydroxide VS. Record the buret reading, and calculate the volume (B) of 0.02 *N* sodium hydroxide consumed. Each mL of the difference between the volume (B) and the volume (A) of 0.02

N sodium hydroxide consumed in the *Assay for citric acid* is equivalent to 1.261 mg of $C_6H_5O_7$.

Assay for citric acid—Transfer 15 mL of Tricitrates Oral Solution, accurately measured, to a 250-mL volumetric flask, dilute with water to volume, and mix. Pipet 5 mL of this solution into a suitable flask, add 25 mL of water and 5 drops of phenolphthalein TS, and titrate with 0.02 *N* sodium hydroxide VS to a pink end-point. Record the buret reading, and calculate the volume (*A*) of 0.02 *N* sodium hydroxide consumed. Each mL of 0.02 *N* sodium hydroxide is equivalent to 1.401 mg of $C_6H_8O_7 . H_2O$.

Tridihexethyl Chloride

[structural formula: OH, C—$CH_2CH_2N^+(C_2H_5)_3$ with phenyl and cyclohexyl, Cl^-]

$C_{21}H_{36}ClNO$ 353.97

Benzenepropanaminium, γ-cyclohexyl-*N*,*N*,*N*-triethyl-γ-hydroxy-, chloride.

(3-Cyclohexyl-3-hydroxy-3-phenylpropyl)triethylammonium chloride [*4310-35-4*].

» Tridihexethyl Chloride, dried at 105° for 2 hours, contains not less than 98.0 percent and not more than 100.5 percent of $C_{21}H_{36}ClNO$.

Packaging and storage—Preserve in tight containers.

Reference standard—*USP Tridihexethyl Chloride Reference Standard*—Dry at 105° for 2 hours before using.

Identification—

A: The infrared absorption spectrum of a potassium bromide dispersion of it, previously dried, exhibits maxima only at the same wavelengths as that of a similar preparation of USP Tridihexethyl Chloride RS.

B: *pH 5.3 phosphate buffer*—Dissolve 38.0 g of monobasic sodium phosphate and 2.0 g of anhydrous dibasic sodium phosphate in water, and dilute with water to 1000 mL.

Bromocresol purple solution—Dissolve 400 mg of bromocresol purple in 30 mL of water and 6.4 mL of 0.1 *N* sodium hydroxide, and dilute with water to 500 mL.

Procedure—Dissolve 25 mg of Tridihexethyl Chloride in 10 mL of water. To 5 mL of this solution add 2 mL of *pH 5.3 phosphate buffer*, 1 mL of *Bromocresol purple solution*, and 5 mL of chloroform. Mix, and allow the layers to separate: a yellow color appears in the chloroform layer.

C: The remaining 5 mL of the solution prepared in *Identification test A* responds to the tests for *Chloride* ⟨191⟩.

D: Dissolve about 200 mg in 10 mL of water, and add, with stirring, 50 mL of sodium tetraphenylboron TS. Allow the mixture to stand for 30 minutes, collect the precipitate on a sintered-glass filter, wash it with three 5-mL portions of water, and draw air through the filter for 30 minutes. Dissolve the precipitate in 5 mL of acetone, filter the solution, if necessary, and pour the solution, with stirring, into 50 mL of water. Allow the mixture to stand for 30 minutes, collect the precipitate on a sintered-glass filter, and wash it with three 5-mL portions of water. Dry the precipitate in a vacuum desiccator over silica gel for 18 hours: the tetraphenylboron derivative so obtained melts between 163° and 167° with slight decomposition, as determined by the method for *Class Ia* under *Melting Range or Temperature* ⟨741⟩.

Melting range ⟨741⟩: between 196° and 202°.

Loss on drying ⟨731⟩—Dry it at 105° for 2 hours: it loses not more than 0.5% of its weight.

Residue on ignition ⟨281⟩: not more than 0.1%.

Sulfate ⟨221⟩—Digest 0.50 g in 25 mL of water by heating to 70°, cool to room temperature, and filter: the filtrate shows no more sulfate than corresponds to 0.20 mL of 0.020 *N* sulfuric acid (0.04%).

Heavy metals, *Method II* ⟨231⟩: 0.002%.

Ordinary impurities ⟨466⟩—

Test solution: chloroform.

Standard solution: chloroform.

Eluant: a mixture of butyl alcohol, water, and glacial acetic acid (13:5:3).

Visualization: 5, followed by viewing under long-wavelength ultraviolet light.

Chloride content—Dissolve about 500 mg, previously dried and accurately weighed, in 200 mL of water and 10 mL of nitric acid, add 25.0 mL of 0.1 *N* silver nitrate VS, and mix. Add 5 mL of nitrobenzene and 5 mL of ferric ammonium sulfate TS, and titrate the excess silver nitrate with 0.1 *N* ammonium thiocyanate VS. Each mL of 0.1 *N* silver nitrate is equivalent to 3.545 mg of Cl: the content is between 9.81% and 10.06%.

Assay—Dissolve about 600 mg of Tridihexethyl Chloride, previously dried and accurately weighed, in 100 mL of chloroform, add 5 mL of mercuric acetate TS and a few crystals of thymol blue, and titrate to a pink end-point with 0.1 *N* perchloric acid in dioxane VS. Perform a blank determination, and make any necessary correction. Each mL of 0.1 *N* perchloric acid is equivalent to 35.40 mg of $C_{21}H_{36}ClNO$.

Tridihexethyl Chloride Injection

» Tridihexethyl Chloride Injection is a sterile solution of Tridihexethyl Chloride in Water for Injection. It contains not less than 90.0 percent and not more than 110.0 percent of the labeled amount of $C_{21}H_{36}ClNO$.

Packaging and storage—Preserve in single-dose containers, preferably of Type I glass.

Reference standard—*USP Tridihexethyl Chloride Reference Standard*—Dry at 105° for 2 hours before using.

Identification—Place a volume of Injection, equivalent to about 10 mg of tridihexethyl chloride, in a glass mortar, evaporate at 100° to dryness, and dry the residue at 105° for 2 hours: a potassium bromide dispersion of the residue so obtained responds to *Identification test A* under *Tridihexethyl Chloride.*

Pyrogen—Tridihexethyl Chloride Injection, diluted with sodium chloride injection to a concentration of 1 mg of tridihexethyl chloride per mL, meets the requirements of the *Pyrogen Test* ⟨151⟩, the test dose being 3 mL per kg.

pH ⟨791⟩: between 5.0 and 7.5.

Other requirements—It meets the requirements under *Injections* ⟨1⟩.

Assay—

pH 5.3 phosphate buffer and *Bromocresol purple solution*—Prepare as directed in *Identification test B* under *Tridihexethyl Chloride.*

Standard preparation—Dissolve a suitable quantity of USP Tridihexethyl Chloride RS, accurately weighed, in water, and dilute quantitatively and stepwise with water to obtain a solution having a known concentration of about 100 µg per mL.

Assay preparation—Transfer an accurately measured volume of Tridihexethyl Chloride Injection, equivalent to about 20 mg of tridihexethyl chloride, to a 200-mL volumetric flask, dilute with water to volume, and mix.

Procedure—Transfer 5.0 mL each of the *Standard preparation* and the *Assay preparation*, respectively, and of water to provide the blank, to individual 125-mL separators. Add 5 mL of *pH 5.3 phosphate buffer*, 5 mL of *Bromocresol purple solution*, and

10 mL of chloroform to each, and shake gently. Filter the chloroform layers through separate chloroform-washed cotton filters into separate 50-mL volumetric flasks, and repeat the extraction of the aqueous layers with three additional 10-mL portions of chloroform, filtering the chloroform extracts into the respective flasks. Dilute the solution in each flask with chloroform to volume, and mix. Concomitantly determine the absorbances of the solutions in 1-cm cells at the wavelength of maximum absorbance at about 408 nm, with a suitable spectrophotometer, using the blank to set the instrument. Calculate the quantity, in mg, of $C_{21}H_{36}ClNO$ in each mL of Injection taken by the formula:

$$(0.2C/V)(A_U/A_S),$$

in which C is the concentration, in µg per mL, of USP Tridihexethyl Chloride RS in the *Standard preparation*, V is the volume, in mL, of Injection taken, and A_U and A_S are the absorbances of the solutions from the *Assay preparation* and the *Standard preparation*, respectively.

Tridihexethyl Chloride Tablets

» Tridihexethyl Chloride Tablets contain not less than 90.0 percent and not more than 110.0 percent of the labeled amount of $C_{21}H_{36}ClNO$.

Packaging and storage—Preserve in tight containers.

Reference standard—*USP Tridihexethyl Chloride Reference Standard*—Dry at 105° for 2 hours before using.

Identification—

A: Place a quantity of finely powdered Tablets, equivalent to about 25 mg of tridihexethyl chloride, in a glass-stoppered, conical flask, add 10 mL of water, shake, filter, and evaporate 5 mL of the filtrate to dryness: the residue so obtained, previously dried, responds to *Identification test B* under *Tridihexethyl Chloride*.

B: The remaining 5 mL of the filtrate obtained in *Identification test A* responds to the tests for *Chloride* ⟨191⟩.

Dissolution ⟨711⟩—

Medium: water; 900 mL.

Apparatus 1: 100 rpm.

Time: 45 minutes.

Procedure—Determine the amount of $C_{21}H_{36}ClNO$ dissolved, employing the procedure set forth in the *Assay*, making any necessary modifications.

Tolerances—Not less than 75% (Q) of the labeled amount of $C_{21}H_{36}ClNO$ is dissolved in 45 minutes.

Uniformity of dosage units ⟨905⟩: meet the requirements.

Assay—

pH 5.3 phosphate buffer and *Bromocresol purple solution*—Prepare as directed in *Identification test B* under *Tridihexethyl Chloride*.

Standard preparation—Prepare as directed in the *Assay* under *Tridihexethyl Chloride Injection*.

Assay preparation—Weigh and finely powder not less than 20 Tridihexethyl Chloride Tablets. Transfer an accurately weighed portion of the powder, equivalent to about 25 mg of tridihexethyl chloride, to a 250-mL volumetric flask, add 100 mL of water, shake, dilute with water to volume, and mix. Centrifuge a portion of this mixture.

Procedure—Proceed as directed for *Procedure* in the *Assay* under *Tridihexethyl Chloride Injection*. Calculate the quantity, in mg, of $C_{21}H_{36}ClNO$ in the portion of Tablets taken by the formula:

$$0.25C(A_U/A_S),$$

in which C is the concentration, in µg per mL, of USP Tridihexethyl Chloride RS in the *Standard preparation* and the other terms are as defined therein.

Trientine Hydrochloride

$H_2N(CH_2)_2NH(CH_2)_2NH(CH_2)_2NH_2 \cdot 2HCl$

$C_6H_{18}N_4.2HCl$ 219.16
1,2-Ethanediamine, *N,N′*-bis(2-aminoethyl)-, dihydrochloride.
Triethylenetetramine dihydrochloride [*38260-01-4*].

» Trientine Hydrochloride contains not less than 97.0 percent and not more than 103.0 percent of $C_6H_{18}N_4.2HCl$, calculated on the dried basis.

Packaging and storage—Preserve under an inert gas in tight, light-resistant containers, and store in a refrigerator.

Reference standard—*USP Trientine Hydrochloride Reference Standard*—Dry in vacuum at a pressure not exceeding 5 mm of mercury at 40° for 4 hours before using.

Identification—The infrared absorption spectrum of a mineral oil dispersion of it exhibits maxima only at the same wavelengths as that of a similar preparation of USP Trientine Hydrochloride RS.

pH ⟨791⟩: between 7.0 and 8.5, in a solution (1 in 100).

Loss on drying ⟨731⟩—Dry it in vacuum at a pressure not exceeding 5 mm of mercury at 40° for 4 hours: it loses not more than 2.0% of its weight.

Residue on ignition ⟨281⟩: not more than 0.15%.

Heavy metals, *Method II* ⟨231⟩: 0.001%.

Chromatographic purity—The sum of the intensities of all secondary spots obtained from the *Test preparation* in *Part I* and *Part II* corresponds to not more than 2.0%.

Part I—

Spray reagent—Dissolve 300 mg of ninhydrin in a mixture of 100 mL of butyl alcohol and 3 mL of glacial acetic acid.

Standard preparation A—[Note—Use low-actinic glassware.] Dissolve an accurately weighed quantity of USP Trientine Hydrochloride RS in methanol to obtain a solution containing 10 mg per mL.

Standard preparation B—[Note—Use low-actinic glassware.] Dissolve an accurately weighed quantity of diethylenetriamine in methanol to obtain a solution containing 1.0 mg per mL. Transfer 3.0 mL of this solution to a 100-mL volumetric flask, dilute with methanol to volume, and mix.

Standard preparation C—[Note—Use low-actinic glassware.] Dissolve an accurately weighed quantity of 1-(2-aminoethyl)piperazine in methanol to obtain a solution containing 1.0 mg per mL. Transfer 10.0 mL of this solution to a 100-mL volumetric flask, dilute with methanol to volume, and mix.

Standard preparation D—[Note—Use low-actinic glassware.] Transfer 5.0 mL of *Standard preparation C* to a 10-mL volumetric flask, dilute with methanol to volume, and mix.

Test preparation—[Note—Use low-actinic glassware.] Dissolve an accurately weighed quantity of Trientine Hydrochloride in methanol to obtain a solution containing 10 mg per mL.

Procedure—On a suitable unwashed, high performance thin-layer chromatographic plate (see *Chromatography* ⟨621⟩), having a 1.5-cm preadsorbent zone and coated with a 0.15-mm layer of chromatographic silica gel mixture, apply separately 3 µL of the *Test preparation*, 3 µL of *Standard preparation B*, and 3 µL of *Standard preparation C*. To a fourth spot, apply 3 µL each of *Standard preparations A, B,* and *C*. To a fifth spot, apply 3 µL each of *Standard preparations A, B,* and *D*. Allow the spots to dry, place the plate in a chromatographic chamber, and develop the chromatograms in a solvent system consisting of a mixture of isopropyl alcohol and ammonium hydroxide (3:2) until the solvent front has moved about three-fourths of the length of the plate. Remove the plate from the developing chamber, mark the solvent front, and dry the plate with the aid of a current of air. Spray the plate with *Spray reagent*, dry at 105° for 5 minutes, and observe the plate under long-wavelength ultraviolet light. Determine the locus of the diethylenetriamine and the 1-(2-aminoethyl)piperazine spots from the chromatograms of *Standard preparations B* and *C*, respectively. Determine the concentration of diethylenetriamine in the *Test preparation* by comparing the size and intensity of any secondary spot from the chromatogram of the *Test preparation* having an R_f value corresponding to the R_f value of diethylenetriamine with the diethylenetriamine spots obtained from the chromatograms of the *Standard preparation* mixtures. Determine the concentration of any other observed impurities in the *Test preparation* by comparing the size and intensity of any other secondary spots from the chromatogram of the *Test preparation* with the 1-(2-aminoethyl)piperazine spots obtained from the chromatograms of the *Standard preparation* mixtures.

Part II—

Spray reagent—Dissolve 200 mg of ninhydrin in 100 mL of alcohol.

Tris(2-aminoethyl)amine stock solution—[NOTE—Use low-actinic glassware.] Dissolve an accurately weighed quantity of tris(2-aminoethyl)amine in methanol to obtain a solution containing 1.0 mg per mL.

Standard preparation A—[NOTE—Use low-actinic glassware.] Dissolve an accurately weighed quantity of USP Trientine Hydrochloride RS in methanol to obtain a solution containing 10 mg per mL.

Standard preparation B—[NOTE—Use low-actinic glassware.] Transfer 1.0 mL of *Tris(2-aminoethyl)amine stock solution* to a 10-mL volumetric flask, dilute with methanol to volume, and mix.

Standard preparation C—[NOTE—Use low-actinic glassware.] Transfer 0.5 mL of *Tris(2-aminoethyl)amine stock solution* to a 10-mL volumetric flask, dilute with methanol to volume, and mix.

Test preparation—[NOTE—Use low-actinic glassware.] Dissolve an accurately weighed quantity of Trientine Hydrochloride in methanol to obtain a solution containing 10 mg per mL.

Procedure—On a suitable thin-layer chromatographic plate (see *Chromatography* ⟨621⟩), coated with a 0.25-mm layer of chromatographic silica gel mixture and previously washed with methanol, apply separately 3 µL of the *Test preparation*, and 3 µL of *Standard preparation A*. To a third spot, apply 3 µL each of *Standard preparations A* and *B*. To a fourth spot, apply 3 µL each of *Standard preparations A* and *C*. Allow the spots to dry, place the plate in a chromatographic chamber, and develop the chromatograms in a solvent system consisting of a mixture of ammonium hydroxide and alcohol (2:1) at a temperature of 2° to 6° until the solvent front has moved about three-fourths of the length of the plate. Remove the plate from the developing chamber, mark the solvent front, and dry the plate with the aid of a current of air. Spray the plate with *Spray reagent*, dry at 105° for 5 minutes, and observe the plate under long-wavelength ultraviolet light. Determine the concentration of tris(2-aminoethyl)amine in the *Test preparation* by comparing the size and intensity of any secondary spot from the chromatogram of the *Test preparation* having an R_f value corresponding to the R_f value of tris(2-aminoethyl)amine with the tris(2-aminoethyl)amine spots obtained from the chromatograms of the *Standard preparation* mixtures.

Assay—Dissolve about 220 mg of Trientine Hydrochloride, accurately weighed, in 10 mL of 0.1 *N* hydrochloric acid. Add 2 mL of 5 *M* sodium nitrate, 10 mL of acetic acid–ammonium acetate buffer TS, and dilute with water to about 70 mL. Titrate with 0.1 *N* cupric nitrate VS, determining the end-point potentiometrically, using an electrode system consisting of a cupric ion-selective electrode and a double junction reference electrode with an outer filling solution such as 1 *M* potassium nitrate. Perform a blank determination (see *Titrimetry* ⟨541⟩), and make any necessary correction. Each mL of 0.1 *N* cupric nitrate is equivalent to 21.92 mg of $C_6H_{18}N_4 . 2HCl$.

Trientine Hydrochloride Capsules

» Trientine Hydrochloride Capsules contain not less than 90.0 percent and not more than 110.0 percent of the labeled amount of $C_6H_{18}N_4 . 2HCl$.

Packaging and storage—Preserve in well-closed containers, and store in a refrigerator.

Reference standard—*USP Trientine Hydrochloride Reference Standard*—Dry in vacuum at a pressure not exceeding 5 mm of mercury at 40° for 4 hours before using.

Identification—Triturate an amount of the contents of Capsules, equivalent to about 1.5 mg of trientine hydrochloride, with 0.5 mL of acetone in an agate or mullite mortar. Evaporate in a gentle current of air to dryness. Repeat the acetone addition, trituration, and drying steps: the infrared absorption spectrum of a potassium bromide dispersion of the residue so obtained exhibits maxima only at the same wavelengths as that of a similar preparation of USP Trientine Hydrochloride RS.

Dissolution ⟨711⟩—

Medium: water; 500 mL.

Apparatus 2: 50 rpm.

Time: 30 minutes.

pH 8.2 buffer—Prepare as directed under *Assay*.

Copper sulfate reagent—Mix 10 mL of copper sulfate solution (5 g copper sulfate pentahydrate in 100 mL of water) with 40 mL of *pH 8.2 buffer*. [Note—The solution must be clear.]

Standard preparation—Dissolve an accurately weighed quantity of Trientine Hydrochloride RS in water to obtain a solution having a known concentration of about 0.5 mg per mL.

Procedure—Pipet an aliquot of a filtered portion of the solution under test, estimated to contain about 5 mg of trientine hydrochloride, into a 50-mL centrifuge tube. Into a similar centrifuge tube, pipet an equivalent volume of water to provide a reagent blank, and into a third centrifuge tube pipet 10 mL of *Standard preparation*. Into each tube, pipet 5 mL of *Copper sulfate reagent*. Stopper and mix immediately using a vortex mixer. Determine the absorbances of the solutions from the *Standard preparation* and the test solution at 580 and 410 nm, with a suitable spectrophotometer, against the reagent blank. Calculate the quantity in mg of trientine hydrochloride dissolved by the formula:

$$5000(C/V)[(A_U - A_{UX})/(A_S - A_{SX})],$$

in which C is the concentration, in mg per mL, of USP Trientine Hydrochloride RS in the *Standard preparation*, V is the volume, in mL, of the aliquot of test solution used, A_U and A_S are the absorbances at 580 nm of test and standard solutions, respectively, and A_{UX} and A_{SX} are the absorbances at 410 nm of test and standard solutions, respectively.

Tolerances—Not less than 80% (Q) of the labeled amount of $C_6H_{18}N_4 . 2HCl$ is dissolved in 30 minutes.

Uniformity of dosage units ⟨905⟩: meet the requirements.

Assay—

Copper reagent—Dissolve 5 g of copper sulfate pentahydrate in water to make 100 mL, and mix.

pH 8.2 buffer—Dissolve 20.74 g of anhydrous dibasic sodium phosphate, 6.72 g of anhydrous citric acid, and 0.535 g of monobasic sodium phosphate in 400 mL of water, adjust with sodium hydroxide solution (1 in 2) to a pH of 8.2 ± 0.05, dilute with water to make 500 mL, and mix.

Standard preparation—Dissolve an accurately weighed quantity of USP Trientine Hydrochloride RS in methanol to obtain a solution having a known concentration of about 2.5 mg per mL. Transfer 5.0 mL of this solution to a glass-stoppered, 50-mL conical flask.

Assay preparation—Remove, as completely as possible, the contents of not less than 20 Trientine Hydrochloride Capsules. Weigh the contents, and determine the average weight per capsule. Mix the combined contents, and transfer an accurately weighed quantity of the powder, equivalent to about 250 mg of trientine hydrochloride, to a 100-mL volumetric flask. Add about 70 mL of methanol, and shake or sonicate to dissolve. Dilute with methanol to volume, mix, and filter, discarding the first few mL of the filtrate. Transfer 5.0 mL of this solution to a glass-stoppered, 50-mL conical flask.

Procedure—To each of the flasks containing the *Standard preparation* and the *Assay preparation* and to a similar flask containing 5.0 mL of methanol to provide the blank, add 10.0 mL of *pH 8.2 buffer* and 1.0 mL of *Copper reagent*, and mix. Concomitantly determine the absorbances of the solutions at the wavelength of maximum absorbance at about 580 nm, with a suitable spectrophotometer, against the blank. Calculate the quantity, in mg, of $C_6H_{18}N_4.2HCl$ in the portion of Capsules taken by the formula:

$$100C(A_U / A_S),$$

in which C is the concentration, in mg per mL, of USP Trientine Hydrochloride RS in the *Standard preparation*, and A_U and A_S are the absorbances of the solutions from the *Assay preparation* and the *Standard preparation*, respectively.

Triethyl Citrate—*see* Triethyl Citrate NF

Trifluoperazine Hydrochloride

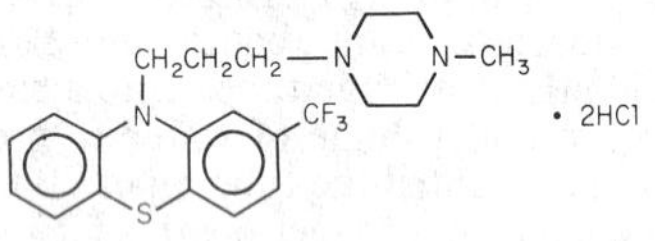

$C_{21}H_{24}F_3N_3S.2HCl$ 480.42

10*H*-Phenothiazine, 10-[3-(4-methyl-1-piperazinyl)propyl]-2-(trifluoromethyl)-, dihydrochloride.

10-[3-(4-Methyl-1-piperazinyl)propyl]-2-(trifluoromethyl)phenothiazine dihydrochloride [*440-17-5*].

» Trifluoperazine Hydrochloride, dried in vacuum at 60° for 4 hours, contains not less than 98.0 percent and not more than 101.0 percent of $C_{21}H_{24}F_3N_3S.2HCl$.

Packaging and storage—Preserve in tight, light-resistant containers.

Reference standard—*USP Trifluoperazine Hydrochloride Reference Standard*—Dry in vacuum at 60° for 4 hours before using.

NOTE—Throughout the following procedures, protect test or assay specimens, the Reference Standard, and solutions containing them, by conducting the procedures without delay, under subdued light, or using low-actinic glassware.

Identification—

A: The infrared absorption spectrum of a mineral oil dispersion of it, previously dried, exhibits maxima only at the same wavelengths as that of a similar preparation of USP Trifluoperazine Hydrochloride RS.

B: The ultraviolet absorption spectrum of a 1 in 100,000 solution in 0.1 *N* hydrochloric acid exhibits maxima and minima at the same wavelengths as that of a similar solution of USP Trifluoperazine Hydrochloride RS, concomitantly measured, and the respective absorptivities, calculated on the dried basis, at the wavelength of maximum absorbance at about 255 nm do not differ by more than 2.0%.

C: A solution (1 in 100) responds to the tests for *Chloride* ⟨191⟩.

D: Prepare a solution in methanol containing 1.2 mg per mL. On a suitable thin-layer chromatographic plate (see *Chromatography* ⟨621⟩), coated with a 0.25-mm layer of chromatographic silica gel, apply 5 µL each of this solution and a solution of USP Trifluoperazine Hydrochloride RS in methanol containing 1.2 mg per mL. Allow the spots to dry, and develop the chromatogram in a solvent system consisting of a mixture of acetone and ammonium hydroxide (200:1) until the solvent front has moved about three-fourths of the length of the plate. Remove the plate from the developing chamber, mark the solvent front, and allow the solvent to evaporate. Locate the spots on the plate by lightly spraying with a solution of iodoplatinic acid prepared by dissolving 100 mg of chloroplatinic acid in 1 mL of 1 *N* hydrochloric acid, adding 25 mL of potassium iodide solution (1 in 25), diluting with water to 100 mL, and then adding 0.5 mL of formic acid: the R_f value of the principal spot from the test solution corresponds to that from the Standard solution.

pH ⟨791⟩: between 1.7 and 2.6, in a solution (1 in 20).

Loss on drying ⟨731⟩—Dry it in vacuum at 60° for 4 hours: it loses not more than 1.5% of its weight.

Residue on ignition ⟨281⟩: not more than 0.1%.

Assay—Dissolve about 500 mg of Trifluoperazine Hydrochloride, previously dried and accurately weighed, in 50 mL of glacial acetic acid, and add crystal violet TS and 15 mL of mercuric acetate TS. Titrate with 0.1 *N* perchloric acid VS to a blue-green end-point. Perform a blank determination, and make any necessary correction. Each mL of 0.1 *N* perchloric acid is equivalent to 24.02 mg of $C_{21}H_{24}F_3N_3S.2HCl$.

Trifluoperazine Hydrochloride Injection

» Trifluoperazine Hydrochloride Injection is a sterile solution of Trifluoperazine Hydrochloride in Water for Injection. It contains an amount of trifluoperazine hydrochloride ($C_{21}H_{24}F_3N_3S.2HCl$) equivalent to not less than 90.0 percent and not more than 110.0 percent of the labeled amount of trifluoperazine ($C_{21}H_{24}F_3N_3S$).

Packaging and storage—Preserve in multiple-dose containers, preferably of Type I glass, protected from light.

Reference standard—*USP Trifluoperazine Hydrochloride Reference Standard*—Dry in vacuum at 60° for 4 hours before using.

NOTE—Throughout the following procedures, protect test or assay specimens, the Reference Standard, and solutions containing them, by conducting the procedures without delay, under subdued light, or using low-actinic glassware.

Identification—

A: The solution employed for measurement of absorbance in the *Assay* exhibits ultraviolet maxima and minima at the same wavelengths as that of a similar solution of USP Trifluoperazine Hydrochloride RS, concomitantly measured.

B: Mix 5 mL of it with 5 mL of methanol: a 5-µL portion of this solution responds to *Identification test D* under *Trifluoperazine Hydrochloride*.

pH ⟨791⟩: between 4.0 and 5.0.

Other requirements—It meets the requirements under *Injections* ⟨1⟩.

Assay—[NOTE—Use low-actinic glassware.] Transfer an accurately measured volume of Trifluoperazine Hydrochloride Injection, equivalent to about 20 mg of trifluoperazine, to a 250-mL separator. Add 10 mL of 4 *N* sulfuric acid, and extract with three 25-mL portions of carbon tetrachloride. Discard the carbon tetrachloride after each extraction. Add 10 mL of ammonium hydroxide, and extract with five 40-mL portions of cyclohexane. Extract the combined cyclohexane extracts with five 50-mL portions of 0.1 *N* hydrochloric acid, collecting the aqueous extracts in a 500-mL volumetric flask. Dilute with 0.1 *N* hydrochloric acid to volume, and mix. Transfer 25.0 mL of this solution to a 100-mL volumetric flask, dilute with 0.1 *N* hydrochloric acid to volume, and mix. Concomitantly determine the absorbances of this solution and of a Standard solution of USP Trifluoperazine

Hydrochloride RS in the same medium having a known concentration of about 12 µg per mL in 1-cm cells at 278 nm and at the maximum at about 255 nm, with a suitable spectrophotometer, using 0.1 *N* hydrochloric acid as the blank. Calculate the quantity, in mg, of trifluoperazine ($C_{21}H_{24}F_3N_3S$) in each mL of the Injection taken by the formula:

$$(407.50/480.42)(2C/V)(A_{255} - A_{278})_U/(A_{255} - A_{278})_S,$$

in which 407.50 and 480.42 are the molecular weights of trifluoperazine and trifluoperazine hydrochloride, respectively, *C* is the concentration, in µg per mL, of USP Trifluoperazine Hydrochloride RS in the Standard solution, *V* is the volume, in mL, of Injection taken, and the parenthetic expressions are the differences in the absorbances of the two solutions at the wavelengths indicated by the subscripts, for the assay solution ($_U$) and the Standard solution ($_S$), respectively.

Trifluoperazine Hydrochloride Syrup

» Trifluoperazine Hydrochloride Syrup contains an amount of trifluoperazine hydrochloride ($C_{21}H_{24}F_3N_3S.2HCl$) equivalent to not less than 93.0 percent and not more than 107.0 percent of the labeled amount of trifluoperazine ($C_{21}H_{24}F_3N_3S$).

Packaging and storage—Preserve in tight, light-resistant containers.

Reference standard—*USP Trifluoperazine Hydrochloride Reference Standard*—Dry in vacuum at 60° for 4 hours before using.

NOTE—Throughout the following procedures, protect test or assay specimens, the Reference Standard, and solutions containing them, by conducting the procedures without delay, under subdued light, or using low-actinic glassware.

Identification—

A: The ultraviolet absorption spectrum of the solution employed for measurement of absorbance in the *Assay* exhibits maxima and minima at the same wavelengths as that of a similar solution of USP Trifluoperazine Hydrochloride RS, concomitantly measured.

B: Mix 1 mL of Syrup with 5 mL of methanol: a 5-µL portion of this solution responds to *Identification test D* under *Trifluoperazine Hydrochloride.*

pH ⟨791⟩: between 2.0 and 3.2.

Assay—[NOTE—Use low-actinic glassware.] Transfer an accurately measured volume of Trifluoperazine Hydrochloride Syrup, equivalent to about 50 mg of trifluoperazine, to a 250-mL separator with the aid of about 100 mL of water. Add 10 mL of sodium hydroxide solution (1 in 10), and extract with three 50-mL portions of cyclohexane. Wash the combined cyclohexane extracts with about 20 mL of water, and discard the water washing. Extract the combined cyclohexane extracts with four 50-mL portions of 0.1 *N* hydrochloric acid, collecting the aqueous extracts in a 500-mL volumetric flask. Dilute with 0.1 *N* hydrochloric acid to volume, and mix. Transfer 10.0 mL of this solution to a 100-mL volumetric flask, dilute with 0.1 *N* hydrochloric acid to volume, and mix. Concomitantly determine the absorbances of this solution and of a Standard solution of USP Trifluoperazine Hydrochloride RS in the same medium having a known concentration of about 12 µg per mL in 1-cm cells at 278 nm and at the wavelength of maximum absorbance at about 255 nm, with a suitable spectrophotometer, using 0.1 *N* hydrochloric acid as the blank. Calculate the quantity, in mg, of trifluoperazine ($C_{21}H_{24}F_3N_3S$) in each mL of the Syrup taken by the formula:

$$(407.50/480.42)(5C/V)(A_{255} - A_{278})_U/(A_{255} - A_{278})_S,$$

in which 407.50 and 480.42 are the molecular weights of trifluoperazine and trifluoperazine hydrochloride, respectively, *C* is the concentration, in µg per mL, of USP Trifluoperazine Hydrochloride RS in the Standard solution, *V* is the volume, in mL, of Syrup taken, and the parenthetic expressions are the differences in the absorbances of the two solutions at the wavelengths indicated by the subscripts, for the assay solution ($_U$) and the Standard solution ($_S$), respectively.

Trifluoperazine Hydrochloride Tablets

» Trifluoperazine Hydrochloride Tablets contain an amount of trifluoperazine hydrochloride ($C_{21}H_{24}F_3N_3S.2HCl$) equivalent to not less than 93.0 percent and not more than 107.0 percent of the labeled amount of trifluoperazine ($C_{21}H_{24}F_3N_3S$).

Packaging and storage—Preserve in well-closed, light-resistant containers.

Reference standard—*USP Trifluoperazine Hydrochloride Reference Standard*—Dry in vacuum at 60° for 4 hours before using.

NOTE—Throughout the following procedures, protect test or assay specimens, the Reference Standard, and solutions containing them, by conducting the procedures without delay, under subdued light, or using low-actinic glassware.

Identification—

A: The ultraviolet absorption spectrum of the solution employed for measurement of absorbance in the *Assay* exhibits maxima and minima at the same wavelengths as that of a similar solution of USP Trifluoperazine Hydrochloride RS, concomitantly measured.

B: Triturate a portion of powdered Tablets, equivalent to about 10 mg of trifluoperazine, with 10 mL of methanol, and centrifuge. A 5-µL portion of this solution responds to *Identification test D* under *Trifluoperazine Hydrochloride.*

Dissolution ⟨711⟩—

Medium: 0.1 *N* hydrochloric acid; 900 mL.

Apparatus 1: 50 rpm.

Time: 30 minutes.

Procedure—Determine the amount of trifluoperazine ($C_{21}H_{24}F_3N_3S$) dissolved from ultraviolet absorbances at the wavelength of maximum absorbance at about 255 nm (determine the analytical value to be used for the absorbance at 255 nm by subtracting the absorbance at 278 nm from the observed maximum absorbance at 255 nm), using filtered portions of the solution under test, suitably diluted with 0.1 *N* hydrochloric acid, in comparison with a Standard solution having a known concentration of USP Trifluoperazine Hydrochloride RS in the same medium.

Tolerances—Not less than 75% (*Q*) of the labeled amount of $C_{21}H_{24}F_3N_3S$ is dissolved in 30 minutes.

Uniformity of dosage units ⟨905⟩: meet the requirements.

Procedure for content uniformity—Transfer 1 Tablet to a 100-mL volumetric flask, add about 50 mL of 0.1 *N* hydrochloric acid, and shake by mechanical means until the tablet is completely disintegrated. Dilute with 0.1 *N* hydrochloric acid to volume, mix, and filter, discarding the first 25 mL of the filtrate. Dilute a portion of the subsequent filtrate quantitatively and stepwise, if necessary, with 0.1 *N* hydrochloric acid to provide a solution containing approximately 10 µg of trifluoperazine per mL. Concomitantly determine the absorbances of this solution and of a Standard solution of USP Trifluoperazine Hydrochloride RS in the same medium having a known concentration of about 12 µg per mL in 1-cm cells at 278 nm and at the wavelength of maximum absorbance at about 255 nm, with a suitable spectrophotometer, using 0.1 *N* hydrochloric acid as the blank. Calculate the quantity, in mg, of trifluoperazine ($C_{21}H_{24}F_3N_3S$) in the portion of Tablets taken by the formula:

$$(407.50/480.42)(TC/D)(A_{255} - A_{278})_U/(A_{255} - A_{278})_S,$$

in which 407.50 and 480.42 are the molecular weights of trifluoperazine and trifluoperazine hydrochloride, respectively, *T* is the labeled quantity, in mg, of trifluoperazine in the Tablet, *C* is the concentration, in µg per mL, of USP Trifluoperazine Hydrochloride RS in the Standard solution, *D* is the concentration, in μg per mL, of trifluoperazine in the test solution, based upon the labeled quantity per Tablet and the extent of dilution, and the parenthetic expressions are the differences in the absorbances of the two solutions at the wavelengths indicated by the subscripts, for the test solution ($_U$) and the Standard solution ($_S$), respectively.

Assay—[NOTE—Use low-actinic glassware.] Weigh and finely powder not less than 20 Trifluoperazine Hydrochloride Tablets. Transfer an accurately weighed portion of the powder, equivalent to about 20 mg of trifluoperazine, to a 250-mL volumetric flask, add 150 mL of 0.1 *N* hydrochloric acid, and shake by mechanical means for 15 minutes. Dilute with 0.1 *N* hydrochloric acid to volume, mix, and filter a portion of the suspension, discarding the first 25 mL of the filtrate. Transfer 25.0 mL of the subsequent filtrate to a 250-mL separator, render alkaline with ammonium hydroxide, and extract with four 25-mL portions of ether. Extract the combined ether extracts with four 25-mL portions of 0.1 *N* hydrochloric acid, collecting the aqueous extracts in a 200-mL volumetric flask. Aerate to remove residual ether, dilute with 0.1 *N* hydrochloric acid to volume, and mix. Concomitantly determine the absorbances of this solution and of a Standard solution of USP Trifluoperazine Hydrochloride RS in the same medium having a known concentration of about 12 μg per mL in 1-cm cells at 278 nm and at the wavelength of maximum absorbance at about 255 nm, with a suitable spectrophotometer, using 0.1 *N* hydrochloric acid as the blank. Calculate the quantity, in mg, of trifluoperazine ($C_{21}H_{24}F_3N_3S$) in the portion of Tablets taken by the formula:

$$(407.50/480.42)(2C)(A_{255} - A_{278})_U/(A_{255} - A_{278})_S,$$

in which 407.50 and 480.42 are the molecular weights of trifluoperazine and trifluoperazine hydrochloride, respectively, *C* is the concentration, in μg per mL, of USP Trifluoperazine Hydrochloride RS in the Standard solution, and the parenthetic expressions are the differences in the absorbances of the two solutions at the wavelengths indicated by the subscripts, for the assay solution ($_U$) and the Standard solution ($_S$), respectively.

Triflupromazine

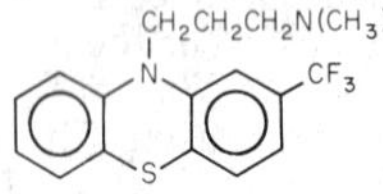

$C_{18}H_{19}F_3N_2S$ 352.42

10*H*-Phenothiazine-10-propanamine, *N,N*-dimethyl-2-(trifluoromethyl)-.

10-[3-(Dimethylamino)propyl]-2-(trifluoromethyl)phenothiazine [*146-54-3*].

» Triflupromazine contains not less than 97.0 percent and not more than 103.0 percent of $C_{18}H_{19}F_3N_2S$.

Packaging and storage—Preserve in tight, light-resistant containers.

Reference standard—*USP Triflupromazine Hydrochloride Reference Standard*—Dry at 100° for 2 hours before using.

NOTE—Throughout the following procedures, protect test or assay specimens, the Reference Standard, and solutions containing them, by conducting the procedures without delay, under subdued light, or using low-actinic glassware.

Identification—

A: It meets the requirements under *Identification—Organic Nitrogenous Bases* ⟨181⟩, USP Triflupromazine Hydrochloride RS being used, and 0.01 *N* hydrochloric acid being used in place of water to dissolve the specimen.

B: The ultraviolet absorption spectrum of a 1 in 150,000 solution in 0.5 *N* sulfuric acid exhibits maxima and minima at the same wavelengths as that of a similar solution of USP Triflupromazine Hydrochloride RS, concomitantly measured, and the respective molar absorptivities, at the wavelength of maximum absorbance at about 255 nm do not differ by more than 3.0%. [NOTE—The molecular weight of triflupromazine hydrochloride ($C_{18}H_{19}F_3N_2S.HCl$) is 388.88.]

Residue on ignition ⟨281⟩: not more than 0.2%.

Ordinary impurities ⟨466⟩—

Test solution: acetone.

Standard solution: acetone.

Eluant: a mixture of chloroform and methanol (4:1).

Visualization: 1.

Assay—Dissolve about 800 mg of Triflupromazine, accurately weighed, in 100 mL of glacial acetic acid. Add crystal violet TS, and titrate with 0.1 *N* perchloric acid VS to a blue end-point. Perform a blank determination, and make any necessary correction. Each mL of 0.1 *N* perchloric acid is equivalent to 35.24 mg of $C_{18}H_{19}F_3N_2S$.

Triflupromazine Oral Suspension

» Triflupromazine Oral Suspension contains an amount of triflupromazine ($C_{18}H_{19}F_3N_2S$) equivalent to not less than 90.0 percent and not more than 110.0 percent of the labeled amount of triflupromazine hydrochloride ($C_{18}H_{19}F_3N_2S.HCl$).

Packaging and storage—Preserve in tight, light-resistant, glass containers.

Reference standard—*USP Triflupromazine Hydrochloride Reference Standard*—Dry at 100° for 2 hours before using.

NOTE—Throughout the following procedures, protect test or assay specimens, the Reference Standard, and solutions containing them, by conducting the procedures without delay, under subdued light, or using low-actinic glassware.

Identification—

A: Transfer about 1.0 mL of Oral Suspension, accurately weighed, to a glass-stoppered, low-actinic 35-mL centrifuge tube, add 10.0 mL of methanol, and shake vigorously by mechanical means for 3 minutes. Centrifuge for 5 minutes, and use the methanol layer for the test. On a suitable thin-layer chromatographic plate (see *Chromatography* ⟨621⟩), coated with a 0.25-mm layer of chromatographic silica gel, apply 50 μL of the test solution in streaks 4 to 5 cm in length and 0.2 cm in width and 50 μL of a Standard solution containing 1 mg of USP Triflupromazine Hydrochloride RS per mL of methanol. Allow the streaks to dry, and develop the chromatogram in a solvent system consisting of a mixture of chloroform and methanol (4:1) until the solvent front has moved about four-fifths of the length of the plate. Remove the plate from the developing chamber, mark the solvent front, and allow the solvent to evaporate. Locate the streaks by viewing the plate under short-wavelength and long-wavelength ultraviolet light: the R_f value and fluorescence of the streak obtained from the test solution correspond to those obtained from the Standard solution.

B: The ultraviolet absorption spectrum of the solution employed for measurement of absorbance in the *Assay* exhibits maxima and minima at the same wavelengths as that of a similar solution of USP Triflupromazine Hydrochloride RS, concomitantly measured.

Assay—

Mixed solvent—Mix 25 mL of isoamyl alcohol with 10 mL of sodium hydroxide solution (1 in 25) in a separator, shake, and discard the aqueous washing. Add 10 mL of 0.1 *N* hydrochloric acid, shake, and discard the aqueous washing. Prepare 500 mL of a 3 in 100 solution of the washed isoamyl alcohol in *n*-heptane.

pH 5.6 acetate buffer—Dissolve 1.4 g of sodium acetate in 100 mL of water, and adjust by the addition of glacial acetic acid to a pH of 5.6 ± 0.1.

Procedure—Transfer an accurately measured volume of well-mixed Triflupromazine Oral Suspension, equivalent to about 20 mg of triflupromazine hydrochloride, to a glass-stoppered, 150-mL centrifuge bottle. In a second, similar bottle dissolve about 20 mg of USP Triflupromazine Hydrochloride RS, accurately weighed, in 1.0 mL of water, then add 1.0 mL of sodium hydroxide solution (1 in 25), and mix. To a third, similar bottle add 2 mL of water to provide the blank. Treat the two preparations and the blank as follows. Add 100.0 mL of ether, shake by mechanical means for 15 minutes, and allow the layers to separate. Transfer 50.0 mL of the ether layer to another glass-stoppered, 150-mL centrifuge bottle, and evaporate on a water bath maintained at about 35°, with the aid of a current of air, to dryness. Add 5.0 mL of sodium hydroxide solution (1 in 50), mix, add 100.0 mL of *Mixed solvent*, shake by mechanical means for 10 minutes, and centrifuge. Transfer 20.0 mL of the nonaqueous phase to a glass-stoppered centrifuge tube, add 10.0 mL of *pH 5.6 acetate buffer*, shake by mechanical means for 5 minutes, and centrifuge. Transfer 2.0 mL of the nonaqueous phase to a glass-stoppered centrifuge tube containing 25.0 mL of 0.5 *N* sulfuric acid, shake by mechanical means for 5 minutes, and centrifuge. Concomitantly determine the absorbances of the aqueous solutions in 1-cm cells at the wavelength of maximum absorbance at about 255 nm, with a suitable spectrophotometer, using the blank to set the instrument. Calculate the quantity, in mg, of triflupromazine hydrochloride ($C_{18}H_{19}F_3N_2S.HCl$) equivalent to the triflupromazine ($C_{18}H_{19}F_3N_2S$) in each mL of the Oral Suspension taken by the formula:

$$(W/V)(A_U/A_S),$$

in which W is the weight, in mg, of USP Triflupromazine Hydrochloride RS taken, V is the volume, in mL, of Oral Suspension taken, and A_U and A_S are the absorbances of the solution from the Oral Suspension and the Standard solution, respectively.

Triflupromazine Hydrochloride

$C_{18}H_{19}F_3N_2S.HCl$ 388.88

10*H*-Phenothiazine-10-propanamine, *N,N*-dimethyl-2-(trifluoromethyl)-, monohydrochloride.

10-[3-(Dimethylamino)propyl]-2-(trifluoromethyl)phenothiazine monohydrochloride [*1098-60-8*].

» Triflupromazine Hydrochloride contains not less than 97.0 percent and not more than 103.0 percent of $C_{18}H_{19}F_3N_2S.HCl$, calculated on the dried basis.

Packaging and storage—Preserve in well-closed, light-resistant glass containers.

Reference standard—*USP Triflupromazine Hydrochloride Reference Standard*—Dry at 100° for 2 hours before using.

NOTE—Throughout the following procedures, protect test or assay specimens, the Reference Standard, and solutions containing them, by conducting the procedures without delay, under subdued light, or using low-actinic glassware.

Identification—

A: The infrared absorption spectrum of a potassium bromide dispersion of it, previously dried, exhibits maxima only at the same wavelengths as that of a similar preparation of USP Triflupromazine Hydrochloride RS.

B: The ultraviolet absorption spectrum of a 1 in 100,000 solution in 0.5 *N* sulfuric acid exhibits maxima and minima at the same wavelengths as that of a similar solution of USP Triflupromazine Hydrochloride RS, concomitantly measured, and the respective absorptivities, calculated on the dried basis, at the wavelength of maximum absorbance at about 255 nm do not differ by more than 3.0%.

C: [NOTE—Conduct the following procedure with minimum exposure to light.] Prepare a solution of Triflupromazine Hydrochloride in methanol containing 10 mg per mL. On a suitable thin-layer chromatographic plate (see *Chromatography* ⟨621⟩), coated with a 0.25-mm layer of chromatographic silica gel mixture, apply 10 µL each of this solution and a Standard solution of USP Triflupromazine Hydrochloride RS in methanol containing 10 mg per mL. Allow the spots to dry, and develop the chromatogram in a solvent system consisting of a mixture of *n*-propyl alcohol, water, and ammonium hydroxide (88:11:1) until the solvent front has moved about three-fourths of the length of the plate. Remove the plate from the developing chamber, mark the solvent front, and allow the solvent to evaporate. Locate the spots on the plate by spraying lightly with dilute methanolic sulfuric acid (4 in 10) and then heating for 15 minutes: the R_f value and color (pink-orange) of the principal spot obtained from the test solution correspond to that obtained from the Standard solution.

Loss on drying ⟨731⟩—Dry it at 100° for 2 hours: it loses not more than 0.5% of its weight.

Residue on ignition ⟨281⟩: not more than 0.1%.

Ordinary impurities ⟨466⟩—

Test solution: acetone.

Standard solution: acetone.

Eluant: a mixture of chloroform and methanol (4:1).

Visualization: 1.

Assay—Dissolve about 800 mg of Triflupromazine Hydrochloride, accurately weighed, in 100 mL of glacial acetic acid. Add 10 mL of mercuric acetate TS and 1 drop of crystal violet TS, and titrate with 0.1 *N* perchloric acid VS to a blue end-point. Perform a blank determination, and make any necessary correction. Each mL of 0.1 *N* perchloric acid is equivalent to 38.89 mg of $C_{18}H_{19}F_3N_2S.HCl$.

Triflupromazine Hydrochloride Injection

» Triflupromazine Hydrochloride Injection is a sterile solution of Triflupromazine Hydrochloride in Water for Injection. It contains not less than 90.0 percent and not more than 112.0 percent of the labeled amount of $C_{18}H_{19}F_3N_2S.HCl$.

Packaging and storage—Preserve in single-dose or in multiple-dose containers, preferably of Type I glass, protected from light.

Reference standard—*USP Triflupromazine Hydrochloride Reference Standard*—Dry at 100° for 2 hours before using.

NOTE—Throughout the following procedures, protect test or assay specimens, the Reference Standard, and solutions containing them, by conducting the procedures without delay, under subdued light, or using low-actinic glassware.

Identification—

A: Place a volume of Injection, equivalent to about 100 mg of triflupromazine hydrochloride, in a test tube, add 5 mL of 8 *N* nitric acid, and mix: a peach to amber color develops, quickly turns dark brown, and then changes to a clear solution having a yellow tint.

B: A volume of Injection, equivalent to about 50 mg of triflupromazine hydrochloride, meets the requirements under *Identification—Organic Nitrogenous Bases* ⟨181⟩.

C: The ultraviolet absorption spectrum of the *Assay preparation*, prepared as directed in the *Assay*, exhibits maxima and minima at the same wavelengths as that of the *Standard preparation*, prepared as directed in the *Assay*.

pH ⟨791⟩: between 3.5 and 5.2.

Other requirements—It meets the requirements under *Injections* ⟨1⟩.

Assay—

Standard preparation—Transfer about 50 mg of USP Triflupromazine Hydrochloride RS, accurately weighed, to a 50-mL volumetric flask, dissolve in 0.5 *N* sulfuric acid, dilute with 0.5 *N* sulfuric acid to volume, and mix. Proceed as directed under *Assay preparation*, beginning with "Transfer 10.0 mL of this solution to a 100-mL volumetric flask." The concentration of USP Triflupromazine Hydrochloride RS in the *Standard preparation* is about 5 µg per mL.

Assay preparation—Transfer an accurately measured volume of Triflupromazine Hydrochloride Injection, equivalent to about 50 mg of triflupromazine hydrochloride, to a 50-mL volumetric flask, dilute with 0.5 *N* sulfuric acid to volume, and mix. Transfer 10.0 mL of this solution to a 100-mL volumetric flask, dilute with the same acid to volume, and mix. Transfer 10.0 mL of this solution to a glass-stoppered, 50-mL centrifuge tube containing 10 mL of ether previously chilled in an ice bath, insert the stopper, shake for 3 minutes, and centrifuge at 1500 rpm for 5 minutes. Transfer 5.0 mL of the aqueous layer to a 100-mL volumetric flask, dilute with 0.5 *N* sulfuric acid to volume, and mix.

Procedure—Concomitantly determine the absorbances of the *Assay preparation* and the *Standard preparation* in 1-cm cells at the wavelength of maximum absorbance at about 255 nm, with a suitable spectrophotometer, using 0.5 *N* sulfuric acid as the blank. Calculate the quantity, in mg, of $C_{18}H_{19}F_3N_2S.HCl$ in each mL of the Injection taken by the formula:

$$(10C/V)(A_U/A_S),$$

in which C is the concentration, in µg per mL, of USP Triflupromazine Hydrochloride RS in the *Standard preparation*, V is the volume, in mL, of Injection taken, and A_U and A_S are the absorbances of the *Assay preparation* and the *Standard preparation*, respectively.

Triflupromazine Hydrochloride Tablets

» Triflupromazine Hydrochloride Tablets contain not less than 90.0 percent and not more than 110.0 percent of the labeled amount of $C_{18}H_{19}F_3N_2S.HCl$.

Packaging and storage—Preserve in well-closed, light-resistant containers.

Reference standard—*USP Triflupromazine Hydrochloride Reference Standard*—Dry at 100° for 2 hours before using.

NOTE—Throughout the following procedures, protect test or assay specimens, the Reference Standard, and solutions containing them, by conducting the procedures without delay, under subdued light, or using low-actinic glassware.

Identification—

A: Triturate a portion of powdered Tablets, equivalent to about 50 mg of triflupromazine hydrochloride, with 5 mL of methanol, and centrifuge. A 10-µL portion of the supernatant liquid responds to *Identification test C* under *Triflupromazine Hydrochloride*.

B: The solution prepared from the Tablets for measurement of absorbance in the *Assay* exhibits an absorbance maximum at 255 ± 2 nm.

Dissolution ⟨711⟩—

Medium: 0.1 *N* hydrochloric acid; 900 mL.

Apparatus 1: 100 rpm.

Time: 45 minutes.

Procedure—Determine the amount of $C_{18}H_{19}F_3N_2S.HCl$ dissolved from ultraviolet absorbances at the wavelength of maximum absorbance at about 305 nm of filtered portions of the solution under test, suitably diluted with *Dissolution Medium*, if necessary, in comparison with a Standard solution having a known concentration of USP Triflupromazine Hydrochloride RS in the same medium.

Tolerances—Not less than 75% (*Q*) of the labeled amount of $C_{18}H_{19}F_3N_2S.HCl$ is dissolved in 45 minutes.

Uniformity of dosage units ⟨905⟩: meet the requirements.

Assay—Weigh and finely powder not less than 20 Triflupromazine Hydrochloride Tablets. Transfer an accurately weighed portion of the powder, equivalent to about 20 mg of triflupromazine hydrochloride, to a separator, add 10 mL of 0.1 *N* hydrochloric acid and 20 mL of water, and mix. Add 6 *N* ammonium hydroxide to render the mixture alkaline to litmus, add 1 mL in excess, and extract with five 50-mL portions of chloroform, passing each extract through anhydrous sodium sulfate into a 250-mL volumetric flask. Dilute with chloroform to volume, and mix. Evaporate 10.0 mL of this solution under reduced pressure to dryness, and dissolve the residue in 0.1 *N* hydrochloric acid to make 100.0 mL. Concomitantly determine the absorbances of this solution and of a Standard solution of USP Triflupromazine Hydrochloride RS in the same medium having a known concentration of about 8 µg per mL in 1-cm cells at the wavelength of maximum absorbance at about 255 nm, with a suitable spectrophotometer, using 0.1 *N* hydrochloric acid as the blank. Calculate the quantity, in mg, of $C_{18}H_{19}F_3N_2S.HCl$ in the portion of Tablets taken by the formula:

$$2.5C(A_U/A_S),$$

in which C is the concentration, in µg per mL, of USP Triflupromazine Hydrochloride RS in the Standard solution, and A_U and A_S are the absorbances of the solution from the Tablets and the Standard solution, respectively.

Trihexyphenidyl Hydrochloride

$C_{20}H_{31}NO.HCl$ 337.93

1-Piperidinepropanol, α-cyclohexyl-α-phenyl-, hydrochloride.

α-Cyclohexyl-α-phenyl-1-piperidinepropanol hydrochloride [*52-49-3*].

» Trihexyphenidyl Hydrochloride contains not less than 98.0 percent and not more than 102.0 percent of $C_{20}H_{31}NO.HCl$, calculated on the dried basis.

Packaging and storage—Preserve in tight containers.

Reference standard—*USP Trihexyphenidyl Hydrochloride Reference Standard*—Dry at 105° for 3 hours before using.

Identification—

A: The infrared absorption spectrum of a potassium bromide dispersion of it, previously dried, exhibits maxima only at the same wavelengths as that of a similar preparation of USP Trihexyphenidyl Hydrochloride RS.

B: It responds to the tests for *Chloride* ⟨191⟩.

C: The retention time exhibited by trihexyphenidyl hydrochloride in the chromatogram of the *Assay preparation* corresponds to that of the *Standard preparation*, both relative to the internal standard, as obtained in the *Assay*.

Loss on drying ⟨731⟩—Dry it at 105° for 3 hours: it loses not more than 0.5% of its weight.

Residue on ignition ⟨281⟩: not more than 0.1%.

Heavy metals, *Method II* ⟨231⟩: 0.002%.

Chloride content—Dissolve about 1.2 g, accurately weighed, in a mixture consisting of 50 mL of methanol, 5 mL of glacial acetic acid, and 5 mL of water. Add 3 drops of eosin Y TS, and mix. Stir, preferably with a magnetic stirrer, and titrate with 0.1 *N* silver nitrate VS until the straw-orange suspension that forms during the titration changes sharply to red. Each mL of 0.1 *N* silver nitrate is equivalent to 3.545 mg of Cl. Not less than 10.3% and not more than 10.7% of Cl, calculated on the dried basis, is found.

Chromatographic purity—

Standard preparations—Dissolve USP Trihexyphenidyl Hydrochloride RS in a mixture of chloroform and isopropylamine (98:2), and mix to obtain a solution having a known concentration of 2.5 mg per mL. Dilute quantitatively with a mixture of chloroform and isopropylamine (98:2) to obtain *Standard preparation A*, containing 500 µg of the Reference Standard per mL, and *Standard preparation B*, containing 250 µg of the Reference Standard per mL.

Test preparation—Dissolve an accurately weighed quantity of Trihexyphenidyl Hydrochloride in a mixture of chloroform and isopropylamine (98:2) to obtain a solution containing 50 mg per mL.

Spray reagent—Dissolve 0.8 g of bismuth subnitrate in a mixture of 40 mL of water and 10 mL of glacial acetic acid (*Solution A*). Dissolve 8 g of potassium iodide in 20 mL of water (*Solution B*). On the day of use, mix equal volumes of *Solution A* and *Solution B*.

Procedure—On a suitable thin-layer chromatographic plate (see *Chromatography* ⟨621⟩), coated with a 0.25-mm layer of chromatographic silica gel mixture, apply separately 10 µL of the *Test preparation* and 10 µL of each *Standard preparation*. Position the plate in a chromatographic chamber, and develop the chromatograms in a solvent system consisting of a mixture of hexane and isopropylamine (98:2) until the solvent front has moved about three-fourths of the length of the plate. Remove the plate from the developing chamber, mark the solvent front, and allow the solvent to evaporate. Examine the plate under short-wavelength ultraviolet light, and spray the plate, first with the *Spray reagent*, and then with sodium nitrite solution (4 in 100). Compare the intensities of any secondary spots observed in the chromatogram of the *Test preparation* with those of the principal spots in the chromatograms of the *Standard preparations.* No secondary spot from the chromatogram of the *Test preparation* is larger or more intense than the principal spot obtained from *Standard preparation B* (0.5%), and the sum of the intensities of all secondary spots obtained from the *Test preparation* corresponds to not more than 1.0%.

Assay—

Internal standard solution—Dissolve *n*-tricosane in chloroform to obtain a solution having a concentration of about 1 mg per mL.

Standard preparation—Dissolve an accurately weighed quantity of USP Trihexyphenidyl Hydrochloride RS in a mixture of pH 6.0, 0.2 *M* phosphate buffer and water (1:1) to obtain a solution having a known concentration of about 0.2 mg per mL. Transfer 50.0 mL of this solution to a 125-mL separator containing about 10 mL of water, and extract with five 25-mL portions of chloroform. Filter the combined chloroform extracts through a chloroform-washed cotton pledget into a 150-mL beaker, and evaporate carefully, with the aid of a current of dry air, on a steam bath to a volume of about 25 mL. Transfer with the aid of two 5-mL portions of chloroform to a 50-mL volumetric flask that contains 10.0 mL of *Internal standard solution*, add chloroform to volume, and mix. The concentration of USP Trihexyphenidyl Hydrochloride RS in the *Standard preparation* is about 0.2 mg per mL.

Assay preparation—Transfer about 40 mg of Trihexyphenidyl Hydrochloride, accurately weighed, to a 200-mL volumetric flask. Dissolve in a mixture of pH 6.0, 0.2 *M* phosphate buffer and water (1:1), dilute with the same buffer and water mixture to volume, mix, and proceed as directed for *Standard preparation*, beginning with "Transfer 50.0 mL of this solution to a 125-mL separator."

Chromatographic system—The gas chromatograph is equipped with a flame-ionization detector and contains a 1.8-m × 4-mm glass column packed with 3 percent liquid phase G1 on support S1AB, and it is cured as directed (see *Chromatography* ⟨621⟩). The column is maintained at a temperature of about 210°, and dry nitrogen flowing at a rate of about 60 mL per minute is used as the carrier gas. Chromatograph the *Standard preparation*, and record the peak responses as directed under *Procedure:* the resolution, *R*, between the trihexyphenidyl and *n*-tricosane peaks is not less than 1.8 (see *Chromatography* ⟨621⟩), and the relative standard deviation for replicate injections is not greater than 2%.

Procedure—Inject separate, suitable portions (4 µL to 5 µL) of the *Assay preparation* and the *Standard preparation*, respectively, into the gas chromatograph. Record the peak responses, and calculate the ratios of the peak responses for trihexyphenidyl to the peak responses for *n*-tricosane from the *Assay preparation* and the *Standard preparation*, respectively. Calculate the quantity, in mg, of $C_{20}H_{31}NO.HCl$ in the portion of Trihexyphenidyl Hydrochloride taken by the formula:

$$200C(R_U/R_S),$$

in which *C* is the concentration, in mg per mL, of USP Trihexyphenidyl Hydrochloride RS in the *Standard preparation*, and R_U and R_S are the ratios of the peak responses for trihexyphenidyl to the peak responses for *n*-tricosane obtained from the *Assay preparation* and the *Standard preparation*, respectively.

Trihexyphenidyl Hydrochloride Extended-release Capsules

» Trihexyphenidyl Hydrochloride Extended-release Capsules contain not less than 90.0 percent and not more than 110.0 percent of the labeled amount of $C_{20}H_{31}NO.HCl$.

Packaging and storage—Preserve in tight containers.

Reference standard—*USP Trihexyphenidyl Hydrochloride Reference Standard*—Dry at 105° for 3 hours before using.

Identification—

A: Reduce the contents of a number of Capsules, equivalent to 20 mg of trihexyphenidyl hydrochloride, to a fine powder, and triturate with 25 mL of chloroform. Filter the mixture, and evaporate the filtrate, by gently heating, to about 10 mL. Add the solution to 100 mL of *n*-hexane: a white precipitate is formed. Allow the mixture to stand for 30 minutes, and collect the precipitate on a solvent-resistant membrane filter to 1-µm pore size. Wash the crystals with a small portion of *n*-hexane, and air-dry: the infrared absorption spectrum of a potassium bromide dispersion of the crystals so obtained exhibits maxima only at the same wavelengths as that of a similar preparation of USP Trihexyphenidyl Hydrochloride RS.

B: The precipitate obtained in *Identification test A* responds to the tests for *Chloride* ⟨191⟩.

Drug release ⟨724⟩—

Medium: water; 500 mL.

Apparatus 1: 100 rpm.

Time: 0.25*D* hours, 0.50*D* hours, 1.00*D* hours.

Determine the amount of $C_{20}H_{31}NO.HCl$ dissolved, using the following method.

Mobile phase—Prepare a filtered and degassed mixture of acetonitrile, water, and triethylamine (920:80:0.2), and adjust with phosphoric acid to a pH of 4.0.

Standard preparation—Dissolve an accurately weighed quantity of USP Trihexyphenidyl Hydrochloride RS in water, and dilute quantitatively and stepwise with water to obtain a solution having a known concentration of about 5 µg per mL.

Chromatographic system (see *Chromatography* ⟨621⟩)—The liquid chromatograph is equipped with a 210-nm detector and a 4.6-mm × 8.3-cm column that contains packing L1. The flow rate is about 2 mL per minute. Chromatograph the *Standard preparation*, and record the peak responses as directed under *Procedure:* the tailing factor for the trihexyphenidyl peak is not more than 2.8, and the relative standard deviation for replicate injections is not more than 2.0%.

Procedure—Dilute the *Standard preparation* and the solution under test with acetonitrile (1:1). Separately inject equal volumes (about 20 µL) of these solutions into the chromatograph, record the chromatograms, and measure the responses for trihexyphenidyl. Calculate the percentage of the labeled amount of $C_{20}H_{31}NO.HCl$ dissolved.

Tolerances—The percentages of the labeled amount of $C_{20}H_{31}NO.HCl$ dissolved at the times specified conform to *Acceptance Table 1.*

Time (hours)	Amount Dissolved
0.25*D*	between 20% and 50%
0.50*D*	between 40% and 60%
1.00*D*	not less than 60%

Uniformity of dosage units ⟨905⟩: meet the requirements.

Assay—

Internal standard solution, *Standard preparation*, and *Chromatographic system*—Prepare as directed in the *Assay* under *Trihexyphenidyl Hydrochloride.*

Assay preparation—Weigh and finely powder the contents of not less than 20 Trihexyphenidyl Hydrochloride Extended-release Capsules. Transfer an accurately weighed portion of the powder, equivalent to about 10 mg of trihexyphenidyl hydrochloride, to a 125-mL separator, add 25 mL of dilute hydrochloric acid (1 in 1000), and shake vigorously by mechanical means for about 30 minutes. Pour the mixture into a 100-mL beaker, taking precautions to avoid loss of material, adjust dropwise with dibasic potassium phosphate solution (1 in 100) to a pH of 6 ± 0.5, and transfer, with the aid of two 5-mL portions of water, back into a 125-mL separator. Extract with five 25-mL portions of chloroform, and proceed as directed for *Standard preparation*, beginning with "Filter the combined chloroform extracts."

Procedure—Proceed as directed for *Procedure* in the *Assay* under *Trihexyphenidyl Hydrochloride.* Calculate the quantity, in mg, of $C_{20}H_{31}NO.HCl$ in the portion of Capsule contents taken by the formula:

$$50C(R_U/R_S),$$

in which C is the concentration, in mg per mL, of USP Trihexyphenidyl Hydrochloride RS in the *Standard preparation*, and R_U and R_S are the ratios of the peak responses for trihexyphenidyl to the peak responses for *n*-tricosane obtained from the *Assay preparation* and the *Standard preparation*, respectively.

Trihexyphenidyl Hydrochloride Elixir

» Trihexyphenidyl Hydrochloride Elixir contains not less than 90.0 percent and not more than 110.0 percent of $C_{20}H_{31}NO.HCl$.

Packaging and storage—Preserve in tight containers.

Reference standard—*USP Trihexyphenidyl Hydrochloride Reference Standard*—Dry at 105° for 3 hours before using.

Identification—

A: To 50 mL of Elixir add 50 mL of water and 50 mL of 1 *N* sodium hydroxide, and stir. Cool the mixture at 4° to 5° for 30 minutes: a white precipitate or cloudiness is observed. Add 100 mL of water to the cooled mixture, stir, and filter by means of vacuum through a 47-mm membrane filter of 1-μm pore size. Wash the crystals with about 100 mL of water, and allow to dry in air: the infrared absorption spectrum of a potassium bromide dispersion of the crystals so obtained exhibits maxima only at the same wavelengths as that of the crystalline base obtained from about 20 mg of USP Trihexyphenidyl Hydrochloride RS, similarly prepared and measured.

B: The retention time exhibited by trihexyphenidyl hydrochloride in the chromatogram of the *Assay preparation* corresponds to that of the *Standard preparation*, both relative to the internal standard, as obtained in the *Assay*.

pH ⟨791⟩: between 2.0 and 3.0.

Alcohol content ⟨611⟩: between 4.5% and 5.5% of C_2H_5OH.

Assay—

Internal standard solution—Dissolve *n*-tricosane in chloroform to obtain a solution having a concentration of about 2 mg per mL.

Standard preparation—Dissolve an accurately weighed quantity of USP Trihexyphenidyl Hydrochloride RS in a mixture of pH 6.0, 0.2 *M* phosphate buffer, and water (1:1) to obtain a solution having a known concentration of about 0.5 mg per mL. Transfer 20.0 mL of this solution to a 125-mL separator containing about 20 mL of water. Extract with 25 mL of chloroform, and filter the chloroform layer through a chloroform-washed cotton pledget into a 100-mL volumetric flask containing 4.0 mL of *Internal standard solution.* Repeat the extraction with three 20-mL portions of chloroform. Filter each portion through the cotton pledget into the 100-mL volumetric flask, dilute with chloroform to volume, and mix. Add about 4 g of granular, anhydrous sodium sulfate, shake, and allow to settle. The concentration of USP Trihexyphenidyl Hydrochloride RS in the *Standard preparation* is about 0.1 mg per mL.

Assay preparation—Transfer an accurately measured volume of Trihexyphenidyl Hydrochloride Elixir, equivalent to about 10 mg of trihexyphenidyl hydrochloride, to a 100-mL beaker, and add water to a volume of about 40 mL. Adjust dropwise with dibasic potassium phosphate solution (1 in 10) to a pH of 6 ± 0.5, and transfer, with the aid of two 5-mL portions of water, to a 125-mL separator. Proceed as directed for *Standard preparation*, beginning with "Extract with 25 mL of chloroform."

Chromatographic system—Prepare as directed in the *Assay* under *Trihexyphenidyl Hydrochloride*.

Procedure—Proceed as directed for *Procedure* in the *Assay* under Trihexyphenidyl Hydrochloride. Calculate the quantity, in mg, of $C_{20}H_{31}NO.HCl$ in each mL of the Elixir taken by the formula:

$$100(C/V)(R_U/R_S),$$

in which C is the concentration, in mg per mL, of USP Trihexyphenidyl Hydrochloride RS in the *Standard preparation*, V is the volume, in mL, of Elixir taken, and R_U and R_S are the ratios of the peak responses for trihexyphenidyl to the peak responses for *n*-tricosane obtained from the *Assay preparation* and the *Standard preparation*, respectively.

Trihexyphenidyl Hydrochloride Tablets

» Trihexyphenidyl Hydrochloride Tablets contain not less than 90.0 percent and not more than 110.0 percent of the labeled amount of $C_{20}H_{31}NO.HCl$.

Packaging and storage—Preserve in tight containers.

Reference standard—*USP Trihexyphenidyl Hydrochloride Reference Standard*—Dry at 105° for 3 hours before using.

Identification—

A: Reduce a number of Tablets, equivalent to 20 mg of trihexyphenidyl hydrochloride, to a fine powder, and triturate with 25 mL of chloroform. Filter the mixture, and evaporate the filtrate, by gently heating, to about 10 mL. Add the solution to 100 mL of *n*-hexane: a white precipitate is formed. Allow the mixture to stand for 30 minutes, and collect the precipitate on a solvent-resistant membrane filter of 1-μm pore size. Wash the crystals with a small portion of *n*-hexane, and allow them to air-dry: the infrared absorption spectrum of a potassium bromide dispersion of the crystals so obtained exhibits maxima only at the same wavelengths as that of a similar preparation of USP Trihexyphenidyl Hydrochloride RS.

B: The precipitate obtained in *Identification test A* responds to the tests for *Chloride* ⟨191⟩.

C: The retention time exhibited by trihexyphenidyl hydrochloride in the chromatogram of the *Assay preparation* corresponds to that of the *Standard preparation*, both relative to the internal standard, as obtained in the *Assay*.

Dissolution ⟨711⟩—

Medium: pH 4.5 acetate buffer, prepared by mixing 2.99 g of sodium acetate trihydrate and 1.66 mL of glacial acetic acid with water to obtain 1000 mL of solution having a pH of 4.50 ± 0.05; 900 mL.

Apparatus 1: 100 rpm.

Time: 45 minutes.

Determination of dissolved trihexyphenidyl hydrochloride—

Bromocresol green solution—Dissolve 250 mg of bromocresol green in a mixture of 15 mL of water and 5 mL of 0.1 *N* sodium hydroxide, dilute with *Dissolution Medium* to 500 mL, and mix. Extract 250-mL portions of this solution with two 100-mL portions of chloroform, and discard the chloroform extracts.

Procedure—Transfer an accurately measured, filtered portion of the solution under test, estimated to contain about 50 μg of trihexyphenidyl hydrochloride, to a 50-mL centrifuge tube. Transfer an equal, accurately measured volume of a Standard solution, having a known concentration of USP Trihexyphenidyl Hydrochloride RS in *Dissolution Medium*, to a second 50-mL centrifuge tube, and transfer an equal, accurately measured volume of *Dissolution Medium* to a third 50-mL centrifuge tube to provide a blank. Add 5 mL of *Bromocresol green solution* and 10.0 mL of chloroform to each tube, insert the stoppers into the tubes, and shake vigorously for not less than 20 seconds. Centrifuge the mixtures to separate the layers, and aspirate and discard the upper aqueous layers. Filter each chloroform layer through a separate phase-separating filter paper. Determine the amount of $C_{20}H_{31}NO.HCl$ dissolved from absorbances, at the wavelength of maximum absorbance at about 415 nm, of the filtrate from the solution under test in comparison with that from the Standard solution, the filtrate from the blank being used to set the instrument.

Tolerances—Not less than 75% (*Q*) of the labeled amount of $C_{20}H_{31}NO.HCl$ is dissolved in 45 minutes.

Uniformity of dosage units ⟨905⟩: meet the requirements.

Assay—

Internal standard solution, Standard preparation, and *Chromatographic system*—Prepare as directed in the *Assay* under *Trihexyphenidyl Hydrochloride.*

Assay preparation—Weigh and finely powder not less than 20 Trihexyphenidyl Hydrochloride Tablets. Transfer an accurately weighed portion of the powder, equivalent to about 10 mg of trihexyphenidyl hydrochloride, to a 125-mL separator, add 25 mL of dilute hydrochloric acid (1 in 1000), and shake vigorously by mechanical means for about 30 minutes. Pour the mixture into a 100-mL beaker, taking precautions to avoid loss of material, adjust dropwise with dibasic potassium phosphate solution (1 in 100) to a pH of 6 ± 0.5, and transfer, with the aid of two 5-mL portions of water, back into a 125-mL separator. Extract with five 25-mL portions of chloroform, and proceed as directed for *Standard preparation,* beginning with "Filter the combined chloroform extracts."

Procedure—Proceed as directed for *Procedure* in the *Assay* under *Trihexyphenidyl Hydrochloride.* Calculate the quantity, in mg, of $C_{20}H_{31}NO.HCl$ in the portion of Tablets taken by the formula:

$$50C(R_U/R_S),$$

in which C is the concentration, in mg per mL, of USP Trihexyphenidyl Hydrochloride RS in the *Standard preparation,* and R_U and R_S are the ratios of the peak responses for trihexyphenidyl to the peak responses for *n*-tricosane obtained from the *Assay preparation* and the *Standard preparation,* respectively.

Trikates Oral Solution

» Trikates Oral Solution is a solution of Potassium Acetate, Potassium Bicarbonate, and Potassium Citrate in Purified Water. It contains not less than 90.0 percent and not more than 110.0 percent of the labeled amount of potassium.

Packaging and storage—Preserve in tight, light-resistant containers.

Identification—

A: Two mL, diluted with water to 12 mL, responds to the ferric chloride test for *Acetate* ⟨191⟩.

B: It responds to the tests for *Bicarbonate* ⟨191⟩.

C: To 5 mL add 1 mL of hydrochloric acid, and heat to near boiling for 5 minutes to evolve carbon dioxide. Cool, add 1 mL of calcium chloride TS, and render just alkaline to bromothymol blue TS with 1 *N* sodium hydroxide. Boil for 3 minutes with gentle agitation: a white, crystalline precipitate appears that is insoluble in 1 *N* sodium hydroxide but dissolves in 3 *N* hydrochloric acid.

Assay—

Potassium stock solution—Dissolve, in water, 191 mg of potassium chloride, previously dried at 105° for 2 hours. Transfer to a 1000-mL volumetric flask, dilute with water to volume, and mix. Transfer 100.0 mL of this solution to a second 1000-mL volumetric flask, dilute with water to volume, and mix. This solution contains 10 µg of potassium per mL.

Standard preparations—To separate 100-mL volumetric flasks transfer 10.0 mL and 13.0 mL, respectively, of the *Potassium stock solution.* To each flask add 2.0 mL of sodium chloride solution (1 in 5) and 1.0 mL of hydrochloric acid, dilute with water to volume, and mix. The *Standard preparations* contain 1.0 µg and 1.3 µg of potassium per mL, respectively.

Assay preparation—Transfer 1.0 mL of Trikates Oral Solution to a 100-mL volumetric flask, dilute with water to volume, and mix. Transfer 1.0 mL of this solution to a second 100-mL volumetric flask, dilute with water to volume, and mix. Transfer 10.0 mL of this solution to a third 100-mL volumetric flask, add 2.0 mL of sodium chloride solution (1 in 5) and 1.0 mL of hydrochloric acid, dilute with water to volume, and mix.

Procedure—Concomitantly determine the absorbances of the *Standard preparations* and the *Assay preparation* at the resonance line of 766.5 nm, with a suitable atomic absorption spectrophotometer (see *Spectrophotometry and Light-scattering* ⟨851⟩) equipped with a potassium hollow-cathode lamp and an air-acetylene flame, using water as the blank. Plot the absorbances of the *Standard preparations* versus concentration, in µg per mL, of potassium. From the graph so obtained, determine the concentration, C, in µg per mL, of potassium in the *Assay preparation.* Calculate the quantity, in mEq, of potassium in the Oral Solution by the formula:

$$100C/39.10,$$

in which 39.10 is the atomic weight of potassium.

Trimeprazine Tartrate

$$\left[\text{CH}_3 / \text{CH}_2\text{CHCH}_2\text{N(CH}_3)_2 \text{ on N of phenothiazine} \right]_2 \cdot \text{HOOC–CH(OH)–CH(OH)–COOH}$$

$(C_{18}H_{22}N_2S)_2.C_4H_6O_6$ 746.98

10*H*-Phenothiazine-10-propanamine *N,N,β*-trimethyl-, [*R*-(*R**,*R**)]-2,3-dihydroxybutanedioate (2:1).

10-[3-(Dimethylamino)-2-methylpropyl]phenothiazine tartrate (2:1) [*4330-99-8; 41375-66-0*].

» Trimeprazine Tartrate contains not less than 98.0 percent and not more than 101.0 percent of $(C_{18}H_{22}N_2S)_2.C_4H_6O_6$, calculated on the dried basis.

Packaging and storage—Preserve in tight, light-resistant containers.

Reference standard—*USP Trimeprazine Tartrate Reference Standard*—Dry in vacuum at 60° for 4 hours before using.

NOTE—Throughout the following procedures, protect test or assay specimens, the Reference Standard, and solutions containing them, by conducting the procedures without delay, under subdued light, or using low-actinic glassware.

Identification—

A: The infrared absorption spectrum of a mineral oil dispersion of it, previously dried, exhibits maxima only at the same wavelengths as that of a similar preparation of USP Trimeprazine Tartrate RS.

B: The ultraviolet absorption spectrum of a 1 in 150,000 solution in 0.1 *N* hydrochloric acid exhibits maxima and minima at the same wavelengths as that of a similar solution of USP Trimeprazine Tartrate RS, concomitantly measured, and the respective absorptivities, calculated on the dried basis, at the wavelength of maximum absorbance at about 251 nm do not differ by more than 2.0%.

C: Prepare a solution of it in methanol containing 6 mg in each 5 mL. On a suitable thin-layer chromatographic plate (see *Chromatography* ⟨621⟩), coated with a 0.25-mm layer of chromatographic silica gel, apply 5 µL of this solution and 5 µL of a solution of USP Trimeprazine Tartrate RS in methanol containing 6 mg in each 5 mL. Proceed as directed under *Thin-layer Chromatographic Identification Test* ⟨201⟩, beginning with "Allow the spots to dry," using as the solvent system a mixture of 0.15 mL of ammonium hydroxide and 100 mL of acetone. Locate the spots on the plate by lightly spraying with iodoplatinic acid solution [prepared by dissolving 100 mg of chloroplatinic acid in 1 mL of 1 *N* hydrochloric acid, adding 25 mL of potassium iodide solution (1 in 25), diluting with water to 100 mL, and adding 0.5 mL of formic acid]: the R_f value of the principal spot obtained from the test solution corresponds to that obtained from the Standard solution.

Loss on drying ⟨731⟩—Dry it in vacuum at 60° for 4 hours: it loses not more than 0.5% of its weight.

Residue on ignition ⟨281⟩: not more than 0.1%.

Heavy metals, *Method II* ⟨231⟩: 0.002%.

Ordinary impurities ⟨466⟩—

Test solution: methanol.

Standard solution: methanol.

Eluant: a mixture of ethyl acetate saturated with ammonium hydroxide and ether (1:1).

Visualization: 1.

Assay—Dissolve about 1 g of Trimeprazine Tartrate, accurately weighed, in a mixture of 50 mL of chloroform and 50 mL of glacial acetic acid, add 3 drops of crystal violet TS, and titrate with 0.1 *N* perchloric acid VS to a blue end-point. Perform a blank determination, and make any necessary correction. Each mL of 0.1 *N* perchloric acid is equivalent to 37.35 mg of $(C_{18}H_{22}N_2S)_2 \cdot C_4H_6O_6$.

Trimeprazine Tartrate Syrup

» Trimeprazine Tartrate Syrup contains an amount of trimeprazine tartrate $[(C_{18}H_{22}N_2S)_2 \cdot C_4H_6O_6]$ equivalent to not less than 90.0 percent and not more than 110.0 percent of the labeled amount of trimeprazine $(C_{18}H_{22}N_2S)$.

Packaging and storage—Preserve in tight, light-resistant containers.

Reference standard—*USP Trimeprazine Tartrate Reference Standard*—Dry in vacuum at 60° for 4 hours before using.

NOTE—Throughout the following procedures, protect test or assay specimens, the Reference Standard, and solutions containing them, by conducting the procedures without delay, under subdued light, or using low-actinic glassware.

Identification—

A: The ultraviolet absorption spectrum of the solution employed for measurement of absorbance in the *Assay* exhibits maxima and minima at the same wavelengths as that of a similar solution of USP Trimeprazine Tartrate RS, concomitantly measured.

B: Mix 10 mL of Syrup with about 30 mL of water in a separator, render the solution alkaline with 1 *N* sodium hydroxide, and extract with two 30-mL portions of ether. Transfer the ether extracts to a beaker, evaporate the ether by warming, and dissolve the residue in 5 mL of methanol: 5 µL of this solution responds to *Identification test C* under *Trimeprazine Tartrate*.

Alcohol content ⟨611⟩: between 4.5% and 6.5% of C_2H_5OH.

Assay—Transfer an accurately measured volume of Trimeprazine Tartrate Syrup, equivalent to about 2.5 mg of trimeprazine, to a 250-mL separator, add about 100 mL of water and 10 mL of 2.5 *N* sodium hydroxide, mix, and extract with four 40-mL portions of cyclohexane. Wash the combined cyclohexane extracts with 20 mL of water, discarding the water washing. Extract the combined cyclohexane extracts with three 40-mL portions of 0.1 *N* hydrochloric acid, collecting the acid extracts in a 500-mL volumetric flask, add 0.1 *N* hydrochloric acid to volume, and mix. Concomitantly determine the absorbances of this solution and a Standard solution of USP Trimeprazine Tartrate RS in the same medium having a known concentration of about 6 µg per mL, in 1-cm cells, at 276 nm and at the wavelength of maximum absorbance at about 251 nm, with a suitable spectrophotometer, using 0.1 *N* hydrochloric acid as the blank. Calculate the quantity, in mg, of $C_{18}H_{22}N_2S$ in each mL of the Syrup taken by the formula:

$$0.5(C/V)(0.7991)(A_{251} - A_{276})_U/(A_{251} - A_{276})_S,$$

in which *C* is the concentration, in µg per mL, of USP Trimeprazine Tartrate RS in the Standard solution, *V* is the volume, in mL, of Syrup taken, 0.7991 is the factor converting trimeprazine tartrate to trimeprazine, and the parenthetic expressions are the differences in the absorbances of the two solutions at the wavelengths indicated by the subscripts, for the solution from the Syrup ($_U$) and the Standard solution ($_S$), respectively.

Trimeprazine Tartrate Tablets

» Trimeprazine Tartrate Tablets contain an amount of trimeprazine tartrate $[(C_{18}H_{22}N_2S)_2 \cdot C_4H_6O_6]$ equivalent to not less than 93.0 percent and not more than 107.0 percent of the labeled amount of trimeprazine $(C_{18}H_{22}N_2S)$.

Packaging and storage—Preserve in well-closed, light-resistant containers.

Reference standard—*USP Trimeprazine Tartrate Reference Standard*—Dry in vacuum at 60° for 4 hours before using.

NOTE—Throughout the following procedures, protect test or assay specimens, the Reference Standard, and solutions containing them, by conducting the procedures without delay, under subdued light, or using low-actinic glassware.

Identification—

A: The ultraviolet absorption spectrum of the solution employed for measurement of absorbance in the *Assay* exhibits maxima and minima at the same wavelengths as that of a similar solution of USP Trimeprazine Tartrate RS, concomitantly measured.

B: Triturate a portion of powdered Tablets, equivalent to about 10 mg of trimeprazine, with 10 mL of methanol, and centrifuge: 5 µL of this solution responds to *Identification test C* under *Trimeprazine Tartrate*.

Dissolution ⟨711⟩—

Medium: 0.1 *N* hydrochloric acid; 500 mL.

Apparatus 1: 100 rpm.

Time: 45 minutes.

Procedure—Determine the amount of trimeprazine $(C_{18}H_{22}N_2S)$ dissolved from ultraviolet absorbances at the wavelength of maximum absorbance at about 251 nm of filtered portions of the solution under test, suitably diluted with *Dissolution Medium*, if necessary, in comparison with a Standard solution having a known concentration of USP Trimeprazine Tartrate RS in the same medium.

Tolerances—Not less than 75% (*Q*) of the labeled amount of $C_{18}H_{22}N_2S$ is dissolved in 45 minutes.

Uniformity of dosage units ⟨905⟩: meet the requirements.

Procedure for content uniformity—Transfer 1 Tablet to a 100-mL volumetric flask, add about 50 mL of 0.1 *N* hydrochloric acid, and shake by mechanical means until the tablet is completely disintegrated. Add 0.1 *N* hydrochloric acid to volume, mix, and filter, discarding the first 20 mL of the filtrate. Dilute a portion of the subsequent filtrate, quantitatively and stepwise if necessary, with 0.1 *N* hydrochloric acid to obtain a solution having a known concentration of about 5 µg of trimeprazine per mL. Concomitantly determine the absorbances of this solution and a solution of USP Trimeprazine Tartrate RS in the same medium, having a known concentration of about 6 µg per mL, in 1-cm cells, at 276 nm and at the wavelength of maximum absorbance at about 251 nm, with a suitable spectrophotometer, using 0.1 *N* hydrochloric acid as the blank. Calculate the quantity, in mg, of trimeprazine in the Tablet by the formula:

$$(T/D)(0.7991C)(A_{251} - A_{276})_U/(A_{251} - A_{276})_S,$$

in which *T* is the labeled quantity, in mg, of trimeprazine in the Tablet, *D* is the concentration, in µg per mL, of trimeprazine in the test solution, based on the labeled quantity per Tablet and the extent of dilution, *C* is the concentration, in µg per mL, of USP Trimeprazine Tartrate RS in the Standard solution, 0.7991 is the factor converting trimeprazine tartrate to trimeprazine, and the parenthetic expressions are the differences in the absorbances of the two solutions at the wavelengths indicated by the subscripts, for the solution from the Tablets ($_U$) and the Standard solution ($_S$), respectively.

Assay—Weigh and finely powder not less than 20 Trimeprazine Tartrate Tablets. Transfer an accurately weighed portion of the powder, equivalent to about 25 mg of trimeprazine, to a 250-mL volumetric flask, add about 100 mL of water and 2.5 mL of hydrochloric acid, and shake by mechanical means for about 15 minutes. Dilute with water to volume, mix, and filter, discarding the first 20 mL of the filtrate. Transfer 10.0 mL of the subsequent filtrate to a 250-mL separator, add about 20 mL of water, render alkaline with ammonium hydroxide, and extract with four 25-mL portions of ether. Extract the combined ether extracts with four 25-mL portions of 0.1 *N* hydrochloric acid, collecting the aqueous acid extracts in a 200-mL volumetric flask. Aerate to remove residual ether, add 0.1 *N* hydrochloric acid to volume, and mix. Concomitantly determine the absorbances of this solution and a solution of USP Trimeprazine Tartrate RS in the same medium having a known concentration of about 6 µg per mL, in 1-cm cells, at 276 nm and at the wavelength of maximum absorbance at about 251 nm, with a suitable spectrophotometer, using 0.1 *N* hydrochloric acid as the blank. Calculate the quantity, in mg, of $C_{18}H_{22}N_2S$ in the portion of Tablets taken by the formula:

$$5C(0.7991)(A_{251} - A_{276})_U/(A_{251} - A_{276})_S,$$

in which *C* is the concentration, in µg per mL, of USP Trimeprazine Tartrate RS in the Standard solution, 0.7991 is the factor converting trimeprazine tartrate to trimeprazine, and the parenthetic expressions are the differences in the absorbances of the two solutions at the wavelengths indicated by the subscripts, for the solution from the Tablets ($_U$) and the Standard solution ($_S$), respectively.

Trimethadione

$C_6H_9NO_3$ 143.14
2,4-Oxazolidinedione, 3,5,5-trimethyl-.
3,5,5-Trimethyl-2,4-oxazolidinedione [*127-48-0*].

» Trimethadione contains not less than 98.0 percent and not more than 102.0 percent of $C_6H_9NO_3$, calculated on the dried basis.

Packaging and storage—Preserve in tight containers, preferably at controlled room temperature.

Reference standard—*USP Trimethadione Reference Standard*—Dry over silica gel for 6 hours before using.

Identification—

A: The infrared absorption spectrum of a 1 in 50 solution of it in chloroform exhibits maxima only at the same wavelengths as that of a similar solution of USP Trimethadione RS.

B: To 5 mL of a solution (1 in 50) add 2 mL of barium hydroxide TS: a precipitate is formed immediately.

Melting range ⟨741⟩: between 45° and 47°.

Loss on drying ⟨731⟩—Dry it over silica gel for 6 hours: it loses not more than 0.5% of its weight.

Residue on ignition ⟨281⟩: not more than 0.1%.

Urethane—

Standard preparation—Dissolve about 10 mg of urethane, accurately weighed, in about 10 mL of methanol, dilute with methanol to 200.0 mL, and mix to obtain a solution having a known urethane concentration of about 50 µg per mL. Transfer 5.0 µL of this solution to a suitable vessel containing about 250 mg of USP Trimethadione RS, accurately weighed. Warm the mixture to about 65° to liquefy the USP Trimethadione RS, and mix to obtain a solution having a known urethane concentration of about 1 ppm. Continue warming the solution to about 65° and cautiously bubble nitrogen through the solution for about 7 minutes until all traces of methanol have dissipated from the solution.

Test preparation—Transfer about 2 g of Trimethadione to an appropriate size vial equipped with a pressure-tight, septum-type closure, warm to about 65° until the specimen has liquefied, and mix.

Chromatographic system (see *Chromatography* ⟨621⟩)—The gas chromatograph is equipped with a flame-ionization detector and a 3-m × 2-mm glass column packed with a liquid phase consisting of 13.3 percent phase G2 plus 1.7 percent phase G16 on support S1A. The column temperature is programmed as follows. Start at 115° and maintain this temperature for 13 minutes after injection, then program to increase the temperature, at the rate of 25° per minute, to 220°, and maintain this temperature for 20 minutes. The injection port and detector are maintained at temperatures of 200° and 250°, respectively. Helium is used as the carrier gas at a flow rate of about 20 mL per minute. [NOTE—The *Standard preparation* and the *Test preparation* must be in the liquid state and mixed immediately prior to injection into the chromatograph.] Chromatograph the *Standard preparation* and the 50 µg urethane per mL methanol solution used in preparing the *Standard preparation*, to assure correct identification of the urethane peak by its retention time, as directed under *Procedure*: the relative standard deviation of the urethane peak response for replicate injections of the *Standard preparation* is not more than 5.0%. Under typical conditions, the approximate retention time of urethane is 4 minutes.

Procedure—Separately inject equal volumes (about 2 µL) of the *Standard preparation* and the *Test preparation* into the chromatograph, record the chromatograms, and measure the responses for the urethane peak. Calculate the concentration of urethane, in ppm, in the Trimethadione taken by the formula:

$$C(R_U/R_S),$$

in which *C* is the concentration, in ppm, of urethane in the *Standard preparation*, and R_U and R_S are the ratios of the urethane peak responses obtained from the *Test preparation* and the *Standard preparation*, respectively. Not more than 1 ppm is found.

Assay—

Internal standard solution—Transfer about 2.5 g of 1-decanol, accurately weighed, to a 500-mL volumetric flask, add dehydrated alcohol to volume, and mix.

Standard preparation—Dissolve an accurately weighed portion of USP Trimethadione RS in *Internal standard solution* to obtain a solution having a known concentration of about 10 mg per mL.

Assay preparation—Transfer about 500 mg of Trimethadione, accurately weighed, to a 50-mL volumetric flask, dilute with *Internal standard solution* to volume, and mix.

Procedure—Inject 2 µL or other suitable volume of *Standard preparation* into a suitable gas chromatograph equipped with a flame-ionization detector, and record the chromatogram. Under typical conditions, the instrument contains a 75-cm × 3-mm (OD) stainless steel column packed with a suitable 100- to 120-mesh column support, preferably of type S4. The injection port and detector are maintained at 220°, the column temperature is 210°, helium is used as the carrier gas at a flow rate of 45 mL per minute, and in the detector, hydrogen is introduced at a rate of 40 mL per minute, and air at a rate of 350 mL per minute. Under these conditions, the trimethadione and the 1-decanol retention times are at about 6 minutes and 25 minutes, respectively. In a suitable chromatogram, the resolution factor, *R*, is not less than 7.0, between the trimethadione and the internal standard peaks. Record the areas of the peaks of USP Trimethadione RS and the internal standard. Under the same conditions as described for *Standard preparation*, inject 2 µL or other suitable volume of *Assay preparation* into the chromatograph, and record the areas of the peaks of trimethadione and the internal standard. Calculate the quantity, in mg, of $C_6H_9NO_3$ in the Trimethadione taken by the formula:

$$50C(R_U/R_S),$$

in which *C* is the concentration, in mg per mL, of USP Trimethadione RS in the *Standard preparation*, and R_U and R_S are the ratios of the peak areas of trimethadione to those of the internal standard from the *Assay preparation* and the *Standard preparation*, respectively.

Trimethadione Capsules

» Trimethadione Capsules contain not less than 94.0 percent and not more than 106.0 percent of the labeled amount of $C_6H_9NO_3$.

Packaging and storage—Preserve in tight containers, preferably at controlled room temperature.

Reference standard—*USP Trimethadione Reference Standard*—Dry over silica gel for 6 hours before using.

Identification—Digest the contents of a number of Capsules, equivalent to about 1 g of trimethadione, with 25 mL of ether in a stoppered flask for 20 minutes, decant the ether through a filter, and if an insoluble residue remains, digest it with another 10-mL portion of ether as before, and filter into the first ether filtrate. Evaporate the ether solution, preferably with the aid of a current of air, to dryness, and dry the residue in vacuum for 2 hours: the trimethadione so obtained responds to the *Identification tests* under *Trimethadione.*

Dissolution ⟨711⟩—

Medium: water; 750 mL.

Apparatus 1: 100 rpm.

Time: 30 minutes.

Internal standard solution—Transfer about 500 mg of Paramethadione to a 50-mL volumetric flask, dissolve in methanol, and dilute with methanol to volume. Transfer 4.0 mL of this solution to a 100-mL volumetric flask, dilute with water to volume, and mix.

Standard preparation—Transfer about 250 mg of USP Trimethadione RS, accurately weighed, to a 50-mL volumetric flask, dissolve in methanol, and dilute with methanol to volume. Transfer 4.0 mL of this solution to a 50-mL volumetric flask, dilute with water to volume, and mix. Combine 4.0 mL of this solution with 2.0 mL of the *Internal standard solution*, and mix.

Test preparation—Combine 4.0 mL of the solution under test, suitably filtered, and 2.0 mL of the *Internal standard solution*, and mix.

Chromatographic system (see *Chromatography* ⟨621⟩)—The gas chromatograph is equipped with a flame-ionization detector and a 2-mm × 1.8-m glass column that is packed with 2.5 percent phase G26 on support S1A, maintained at a temperature of 110°. The injection port and detector temperatures are maintained at 140° and 200°, respectively. Chromatograph replicate injections of the *Standard preparation*, and record the peak responses as directed under *Procedure:* the relative standard deviation is not more than 3.0%, and the resolution factor between trimethadione and paramethadione is not less than 1.5.

Procedure—Separately inject equal volumes (about 2 µL) of the *Standard preparation* and the *Test preparation* into the chromatograph, record the chromatograms, and measure the responses for the major peaks. The relative retention times are about 0.8 for trimethadione and 1.0 for paramethadione. Calculate the amount of $C_6H_9NO_3$ dissolved based on the peak response ratios obtained.

Tolerances—Not less than 80% (*Q*) of the labeled amount of $C_6H_9NO_3$ is dissolved in 30 minutes.

Uniformity of dosage units ⟨905⟩: meet the requirements.

Assay—

Internal standard solution—Dilute 1.0 mL of cyclohexanol with dehydrated alcohol to make 50.0 mL of solution.

Standard preparation—Dissolve an accurately weighed quantity of USP Trimethadione RS in dehydrated alcohol to obtain a solution having a known concentration of about 15 mg per mL.

Assay preparation—Remove, as completely as possible, the contents of not less than 20 Trimethadione Capsules, weigh, and mix the contents. Transfer an accurately weighed portion of the powder, equivalent to about 750 mg of trimethadione, to a 50-mL volumetric flask, add dehydrated alcohol to volume, and mix. Filter through a dry filter, discarding the first 5 to 10 mL of the filtrate. The subsequent clear filtrate is the *Assay preparation.*

Procedure—Mix 5.0 mL of *Standard preparation* with 2.0 mL of *Internal standard solution*, inject 2 µL or other suitable volume of this mixture into a suitable gas chromatograph equipped with a thermal-conductivity detector, and record the chromatogram. Under typical conditions, the instrument contains a 1.5-m × 3-mm (OD) stainless steel column packed with a suitable 100- to 120-mesh column support, preferably of the type S6. The injection port and detector are maintained at 220°, the column temperature is 190°, and helium is used as the carrier gas at a flow rate of 30 mL per minute. Under these conditions, the cyclohexanol and trimethadione retention times are about 4 and 7 minutes, respectively. In a suitable chromatogram, the resolution factor, *R* (see *Chromatography* ⟨621⟩), is not less than 3.0 between the trimethadione and the internal standard peaks. Measure the areas of the peaks of USP Trimethadione RS and the internal standard. Mix 5.0 mL of *Assay preparation* with 2.0 mL of *Internal standard solution*, inject 2 µL or other suitable volume into the chromatograph, and record the chromatogram under the same conditions as described for the *Standard preparation.* Calculate the quantity, in mg, of $C_6H_9NO_3$ in the portion of the contents of Capsules taken by the formula:

$$50C(R_U/R_S),$$

in which *C* is the concentration, in mg per mL, of USP Trimethadione RS in the *Standard preparation*, and R_U and R_S are the ratios of the peak areas of trimethadione to those of the internal standard from the *Assay preparation* and the *Standard preparation*, respectively.

Trimethadione Oral Solution

» Trimethadione Oral Solution is an aqueous solution of trimethadione containing not less than 94.0 percent and not more than 106.0 percent of the labeled amount of $C_6H_9NO_3$.

Packaging and storage—Preserve in tight containers, preferably at controlled room temperature.

Reference standard—*USP Trimethadione Reference Standard*—Dry over silica gel for 6 hours before using.

Identification—The chloroform solution obtained in the *Assay* responds to *Identification test A* under *Trimethadione.*

pH ⟨791⟩: between 3.0 and 5.0.

Assay—

Internal standard solution and *Standard preparation*—Prepare as directed in the *Assay* under *Trimethadione Capsules.*

Assay preparation—Pipet 10 mL of Trimethadione Oral Solution into a 125-mL separator, add 10 mL of chloroform, shake for 2 minutes, allow the layers to separate, and drain the chloroform layer into a 25-mL volumetric flask. Repeat the extraction with a second 10-mL portion of chloroform, and add the chloroform layer to the same volumetric flask. Dilute with chloroform to volume, and mix.

Procedure—Proceed as directed for *Procedure* in the *Assay* under *Trimethadione Capsules.* Calculate the quantity, in mg, of $C_6H_9NO_3$ in each mL of the Oral Solution taken by the formula:

$$25(C/V)(R_U/R_S),$$

in which *C* is the concentration, in mg per mL, of USP Trimethadione RS in the *Standard preparation*, *V* is the volume, in mL, of Oral Solution taken, and R_U and R_S are the ratios of the peak areas of trimethadione to those of the internal standard of the *Assay preparation* and *Standard preparation*, respectively.

Trimethadione Tablets

» Trimethadione Tablets contain not less than 94.0 percent and not more than 106.0 percent of the labeled amount of $C_6H_9NO_3$.

Packaging and storage—Preserve in tight containers, preferably at a temperature not exceeding 25°.

Reference standard—*USP Trimethadione Reference Standard*—Dry over silica gel for 6 hours before using.

Identification—To a portion of powdered Tablets, equivalent to about 1 g of trimethadione, add 10 mL of solvent hexane, and digest with frequent agitation for 15 minutes. Decant the solvent hexane as completely as possible, and repeat the extraction in the same manner. Transfer the residue to a stoppered flask, digest it with 25 mL of ether for 20 minutes, decant the ether through a filter, and if an insoluble residue remains digest it with 10 mL of ether for 20 minutes, and filter into the first ether filtrate. Evaporate the ether solution to dryness, preferably with a current of air, and dry the residue in a vacuum desiccator for 2 hours: the trimethadione so obtained responds to the *Identification tests* under *Trimethadione*.

Disintegration ⟨701⟩: 30 minutes.

Uniformity of dosage units ⟨905⟩: meet the requirements.

Assay—

Internal standard solution and *Standard preparation*—Prepare as directed in the *Assay* under *Trimethadione Capsules*.

Assay preparation—Weigh and finely powder not less than 20 Trimethadione Tablets. Transfer an accurately weighed portion of the powder, equivalent to about 750 mg of trimethadione, to a 50-mL volumetric flask, add dehydrated alcohol to volume, and mix. Filter through a dry filter, discarding the first 5 to 10 mL of the filtrate. The subsequent clear filtrate is the *Assay preparation*.

Procedure—Proceed as directed for *Procedure* in the *Assay* under *Trimethadione Capsules*. Calculate the quantity, in mg, of $C_6H_9NO_3$ in the portion of Tablets taken by the formula:

$$50C(R_U / R_S),$$

in which C is the concentration, in mg per mL, of USP Trimethadione RS in the *Standard preparation*, and R_U and R_S are the ratios of the peak areas of trimethadione to those of the internal standard of the *Assay preparation* and *Standard preparation*, respectively.

Trimethaphan Camsylate

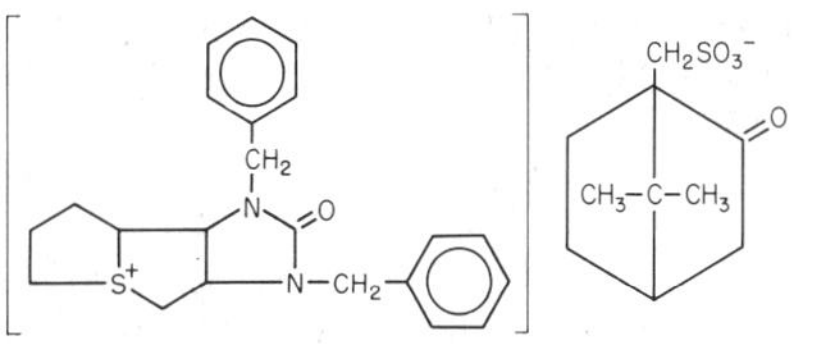

$C_{32}H_{40}N_2O_5S_2$ 596.80

Thieno[1′,2′:1,2]thieno[3,4-*d*]imidazol-5-ium, decahydro-2-oxo-1,3-bis(phenylmethyl)-, salt with (+)-7,7-dimethyl-2-oxobicyclo[2.2.1]heptane-1-methanesulfonic acid (1:1).

(+)-1,3-Dibenzyldecahydro-2-oxoimidazo[4,5-*c*]thieno[1,2-*a*]-thiolium 2-oxo-10-bornanesulfonate (1:1) [*68-91-7*].

» Trimethaphan Camsylate contains not less than 99.0 percent and not more than 101.5 percent of $C_{32}H_{40}N_2O_5S_2$, calculated on the dried basis.

Packaging and storage—Preserve in tight containers, in a cold place.

Reference standard—*USP Trimethaphan Camsylate Reference Standard*—Dry in vacuum over phosphorus pentoxide at 80° for 4 hours before using.

Identification—

A: The infrared absorption spectrum of a 1 in 20 solution in chloroform exhibits maxima only at the same wavelengths as that of a similar preparation of USP Trimethaphan Camsylate RS.

B: The ultraviolet absorption spectrum of a 1 in 2000 solution exhibits maxima and minima at the same wavelengths as that of a similar solution of USP Trimethaphan Camsylate RS, concomitantly measured, and the respective absorptivities, calculated on the dried basis, at the wavelength of maximum absorbance at about 257 nm do not differ by more than 3.0%.

Specific rotation ⟨781⟩: between +20° and +23°, determined in a solution containing 400 mg in each 10 mL.

Loss on drying ⟨731⟩—Dry it in vacuum over phosphorus pentoxide in a suitable drying tube at 80° for 4 hours: it loses not more than 0.1% of its weight.

Residue on ignition ⟨281⟩: not more than 0.1%.

Selenium ⟨291⟩: 0.003%, a 100-mg specimen, mixed with 100 mg of magnesium oxide, being used.

Assay—Dissolve about 1 g of Trimethaphan Camsylate, accurately weighed, in 100 mL of acetic anhydride, and titrate with 0.1 *N* perchloric acid VS, determining the end-point potentiometrically. Perform a blank determination, and make any necessary correction. Each mL of 0.1 *N* perchloric acid is equivalent to 59.68 mg of $C_{32}H_{40}N_2O_5S_2$.

Trimethaphan Camsylate Injection

» Trimethaphan Camsylate Injection is a sterile solution of Trimethaphan Camsylate in Water for Injection. It contains not less than 93.0 percent and not more than 107.0 percent of the labeled amount of $C_{32}H_{40}N_2O_5S_2$.

Packaging and storage—Preserve in single-dose or in multiple-dose containers, preferably of Type I glass. Store in a refrigerator, but avoid freezing.

Labeling—Label it to indicate that it is to be appropriately diluted prior to administration.

Reference standard—*USP Trimethaphan Camsylate Reference Standard*—Dry in vacuum over phosphorus pentoxide at 80° for 4 hours before using.

Identification—To a volume of Injection, equivalent to about 150 mg of trimethaphan camsylate, add 10 mL of cold, saturated trinitrophenol solution, mix, and chill until the picrate has crystallized. Collect the precipitate on a filter, wash with several 10-mL portions of ether, and dry at 80°: the picrate so obtained melts between 138° and 140°. [*Caution—Picrates may explode.*]

Pyrogen—When diluted with Sodium Chloride Injection to a concentration of 5 mg per mL, it meets the requirements of the *Pyrogen Test* ⟨151⟩, the test dose being 1 mL per kg.

pH ⟨791⟩: between 4.9 and 5.6.

Particulate matter ⟨788⟩: meets the requirements under *Small-volume Injections*.

Other requirements—It meets the requirements under *Injections* ⟨1⟩.

Assay—

Phosphate solution—Dissolve 152 g of monobasic sodium phosphate and 27 g of anhydrous dibasic sodium phosphate in water to make 1000 mL, and adjust, if necessary, to a pH of 5.3 ± 0.1.

Bromocresol green solution—Dissolve 250 mg of bromocresol green sodium salt in 250 mL of *Phosphate solution*.

Standard preparation—Dissolve a suitable quantity of USP Trimethaphan Camsylate RS, accurately weighed, in water, and dilute quantitatively and stepwise with water to obtain a solution having a known concentration of 100 µg per mL.

Assay preparation—Dilute a volume of Trimethaphan Camsylate Injection, equivalent to about 100 mg of trimethaphan camsylate and accurately measured, with water to 1000.0 mL, and mix.

Procedure—Pipet 10 mL each of the *Assay preparation* and the *Standard preparation* into respective 125-mL separators containing 10 mL of water, and treat each solution as follows. Add, by pipet, 25 mL of *Phosphate solution* and 5 mL of *Bromocresol green solution*, and extract with two 25-mL portions of chloroform, collecting the chloroform extracts in a 100-mL volumetric flask. Add chloroform to volume, and mix. Concomitantly determine the absorbances of both solutions in 1-cm cells at the wavelength of maximum absorbance at about 420 nm, with a suitable spectrophotometer, using a reagent blank. Calculate the quantity, in mg, of $C_{32}H_{40}N_2O_5S_2$ in the portion of Injection taken by the formula:

$$C(A_U / A_S),$$

in which C is the concentration, in µg per mL, of USP Trimethaphan Camsylate RS in the *Standard preparation*, and A_U and

A_S are the absorbances of the solutions from the *Assay preparation* and the *Standard preparation*, respectively.

Trimethobenzamide Hydrochloride

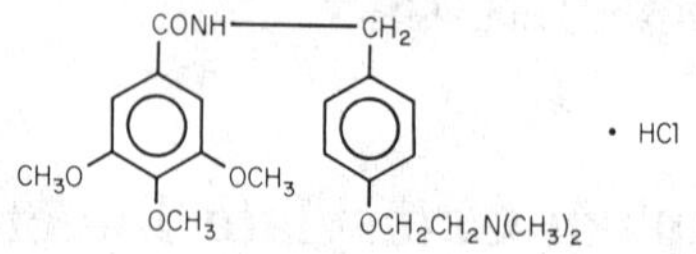

$C_{21}H_{28}N_2O_5.HCl$ 424.92

Benzamide, *N*-[[4-[2-(dimethylamino)ethoxy]phenyl]methyl]-3,4,5-trimethoxy-, monohydrochloride.

N-[*p*-[2-(Dimethylamino)ethoxy]benzyl]-3,4,5-trimethoxybenzamide monohydrochloride [*554-92-7*].

» Trimethobenzamide Hydrochloride, dried at 105° for 4 hours, contains not less than 98.5 percent and not more than 100.5 percent of $C_{21}H_{28}N_2O_5.HCl$.

Packaging and storage—Preserve in well-closed containers.

Reference standard—*USP Trimethobenzamide Hydrochloride Reference Standard*—Dry at 105° for 4 hours before using.

Identification—

A: The infrared absorption spectrum of a potassium bromide dispersion of it, previously dried, exhibits maxima only at the same wavelengths as that of a similar preparation of USP Trimethobenzamide Hydrochloride RS.

B: The ultraviolet absorption spectrum of a 1 in 50,000 solution in 0.1 *N* hydrochloric acid exhibits maxima and minima at the same wavelengths as that of a similar solution of USP Trimethobenzamide Hydrochloride RS, concomitantly measured, and the respective absorptivities, calculated on the dried basis, at the wavelengths of maximum absorbance at about 258 nm do not differ by more than 3.0%.

C: It responds to the *Thin-layer Chromatographic Identification Test* ⟨201⟩. Prepare the test solution by dissolving 10 mg of Trimethobenzamide Hydrochloride in 10.0 mL of methanol. Apply 10-µL portions of the test solution and the Standard solution to the plate, and develop in a solvent system consisting of a mixture of ethyl acetate, alcohol, and ammonium hydroxide (90:10:5).

D: It responds to the tests for *Chloride* ⟨191⟩.

Melting range, *Class I* ⟨741⟩: between 186° and 190°.

Loss on drying ⟨731⟩—Dry it at 105° for 4 hours: it loses not more than 0.5% of its weight.

Residue on ignition ⟨281⟩: not more than 0.1%.

Heavy metals, *Method I* ⟨231⟩—Dissolve 1.0 g in 20 mL of water, add 2 mL of 1 *N* acetic acid, and dilute with water to 25 mL: the limit is 0.002%.

Assay—Dissolve about 1.3 g of Trimethobenzamide Hydrochloride, previously dried and accurately weighed, in 80 mL of glacial acetic acid and 15 mL of mercuric acetate TS. Titrate with 0.1 *N* perchloric acid VS, determining the end-point potentiometrically using a calomel-glass electrode system. Perform a blank determination, and make any necessary correction. Each mL of 0.1 *N* perchloric acid is equivalent to 42.49 mg of $C_{21}H_{28}N_2O_5.HCl$.

Trimethobenzamide Hydrochloride Capsules

» Trimethobenzamide Hydrochloride Capsules contain not less than 90.0 percent and not more than 110.0 percent of the labeled amount of $C_{21}H_{28}N_2O_5.HCl$.

Packaging and storage—Preserve in well-closed containers.

Reference standard—*USP Trimethobenzamide Hydrochloride Reference Standard*—Dry at 105° for 4 hours before using.

Identification—

A: The ultraviolet absorption spectrum of the solution employed for measurement of absorbance in the *Assay* exhibits maxima and minima at the same wavelengths as that of the Standard solution.

B: Transfer a portion of the contents of Capsules, equivalent to about 20 mg of trimethobenzamide hydrochloride, to a suitable vessel, dissolve in 15 mL of 0.1 *N* hydrochloric acid, and filter. Transfer the filtrate to a separator, and add 5 mL of 1 *N* sodium hydroxide. Extract with 15 mL of chloroform, filtering the chloroform extract through anhydrous sodium sulfate into a suitable vessel, and evaporate to dryness. Allow to cool to room temperature, add a small portion of ether, and evaporate at room temperature to dryness. Dry the residue at 60° for 1 hour: the infrared absorption spectrum of a potassium bromide dispersion of the residue so obtained exhibits maxima only at the same wavelengths as that of a similar preparation of USP Trimethobenzamide Hydrochloride RS.

C: Place a portion of the contents of Capsules, equivalent to about 25 mg of trimethobenzamide hydrochloride, in a 10-mL volumetric flask, add methanol to volume, mix, and filter: the filtrate so obtained responds to the *Thin-layer Chromatographic Identification Test* ⟨201⟩, a solvent system consisting of a mixture of ethyl acetate, alcohol, and ammonium hydroxide (90:10:5) being used.

Dissolution ⟨711⟩—

Medium: water; 900 mL.

Apparatus 1: 100 rpm.

Time: 45 minutes.

Procedure—Determine the amount of $C_{21}H_{28}N_2O_5.HCl$ dissolved from ultraviolet absorbances at the wavelength of maximum absorbance at about 258 nm of filtered portions of the solution under test, suitably diluted with *Dissolution Medium*, if necessary, in comparison with a Standard solution having a known concentration of USP Trimethobenzamide Hydrochloride RS in the same medium.

Tolerances—Not less than 75% (*Q*) of the labeled amount of $C_{21}H_{28}N_2O_5.HCl$ is dissolved in 45 minutes.

Uniformity of dosage units ⟨905⟩: meet the requirements.

Assay—Transfer, as completely as possible, the contents of not less than 20 Trimethobenzamide Hydrochloride Capsules to a suitable tared container, and determine the average weight per Capsule. Mix the combined contents, and transfer an accurately weighed portion of the powder, equivalent to about 50 mg of trimethobenzamide hydrochloride, to a 100-mL volumetric flask. Add 50 mL of dilute hydrochloric acid (1 in 120), shake the mixture for several minutes, then add dilute hydrochloric acid (1 in 120) to volume, and mix. Filter through small retentive filter paper, discarding the first 20 mL of the filtrate. Transfer 4.0 mL of the subsequent filtrate to a 100-mL volumetric flask, add dilute hydrochloric acid (1 in 120) to volume, and mix. Concomitantly determine the absorbances of this solution and a Standard solution of USP Trimethobenzamide Hydrochloride RS in the same medium having a known concentration of about 20 µg per mL, in 1-cm cells at the wavelength of maximum absorbance at about 258 nm, with a suitable spectrophotometer, using dilute hydrochloric acid (1 in 120) as the blank. Calculate the quantity, in mg, of $C_{21}H_{28}N_2O_5.HCl$ in the portion of Capsules taken by the formula:

$$2.5C(A_U/A_S),$$

in which C is the concentration, in µg per mL, of USP Trimethobenzamide RS in the Standard solution, and A_U and A_S are the absorbances of the solution from the Capsules and the Standard solution, respectively.

Trimethobenzamide Hydrochloride Injection

» Trimethobenzamide Hydrochloride Injection is a sterile solution of Trimethobenzamide Hydrochloride in Water for Injection. It contains not less than 95.0 percent and not more than 105.0 percent of the labeled amount of $C_{21}H_{28}N_2O_5 \cdot HCl$.

Packaging and storage—Preserve in single-dose or in multiple-dose containers, preferably of Type I glass.

Reference standard—*USP Trimethobenzamide Hydrochloride Reference Standard*—Dry at 105° for 4 hours before using.

Identification—

A: It responds to *Identification test A* under *Trimethobenzamide Hydrochloride Capsules.*

B: Transfer a volume of Injection, equivalent to about 100 mg of trimethobenzamide hydrochloride, to a separator containing 20 mL of water. Add 2 mL of 1 *N* sodium hydroxide, and proceed as directed in *Identification test B* under *Trimethobenzamide Hydrochloride Capsules*, beginning with "Extract with 15 mL of chloroform."

C: Dilute a portion of Injection quantitatively and stepwise with methanol to obtain a solution containing 2.5 mg of trimethobenzamide hydrochloride per mL: this solution responds to the *Thin-layer Chromatographic Identification Test* ⟨201⟩, a solvent system consisting of a mixture of ethyl acetate, alcohol, and ammonium hydroxide (90:10:5) being used.

Pyrogen—When diluted with pyrogen-free saline TS to a concentration of 2 g of trimethobenzamide hydrochloride in 100 mL, it meets the requirements of the *Pyrogen Test* ⟨151⟩, the test dose being 1 mL per kg.

pH ⟨791⟩: between 4.5 and 5.5.

Other requirements—It meets the requirements under *Injections* ⟨1⟩.

Assay—Transfer to a suitable separator an accurately measured volume of Trimethobenzamide Hydrochloride Injection, equivalent to about 200 mg of trimethobenzamide hydrochloride. Add 5 mL of water and 3 mL of dilute hydrochloric acid (1 in 12), and extract with four 20-mL portions of ether, collecting the ether extracts in a second separator, and transferring the aqueous layer to a 500-mL volumetric flask. Wash the combined ether extracts with one 20-mL portion of water, transfer the aqueous layer to the 500-mL volumetric flask, dilute with water to volume, and mix. Dilute 5.0 mL of the solution with dilute hydrochloric acid (1 in 120) to 100.0 mL, and mix. Concomitantly determine the absorbances of this solution and a Standard solution of USP Trimethobenzamide Hydrochloride RS in the same medium having a known concentration of about 20 µg per mL, in 1-cm cells at the wavelength of maximum absorbance at about 258 nm, with a suitable spectrophotometer, using dilute hydrochloric acid (1 in 120) as the blank. Calculate the quantity, in mg, of $C_{21}H_{28}N_2O_5 \cdot HCl$ in each mL of the Injection taken by the formula:

$$(10C/V)(A_U/A_S),$$

in which C is the concentration, in µg per mL, of USP Trimethobenzamide RS in the Standard solution, V is the volume, in mL, of Injection taken, and A_U and A_S are the absorbances of the solution from the Injection and the Standard solution, respectively.

Trimethoprim

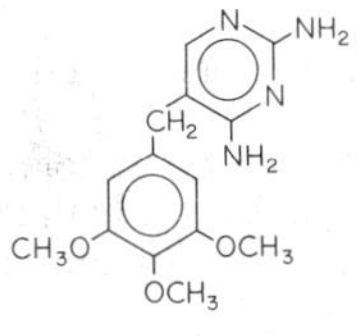

$C_{14}H_{18}N_4O_3$ 290.32

2,4-Pyrimidinediamine, 5-[(3,4,5-trimethoxyphenyl)methyl]-.

2,4-Diamino-5-(3,4,5-trimethoxybenzyl)pyrimidine [*738-70-5*].

» Trimethoprim contains not less than 98.5 percent and not more than 101.0 percent of $C_{14}H_{18}N_4O_3$, calculated on the dried basis.

Packaging and storage—Preserve in tight, light-resistant containers.

Reference standards—*USP Trimethoprim Reference Standard*—Dry in vacuum at 105° for 4 hours before using. *USP 3-Anilino-2-(3,4,5-trimethoxybenzyl)acrylonitrile Reference Standard*—Do not dry; use as is.

Identification—

A: The infrared absorption spectrum of a 1 in 100 solution in chloroform exhibits maxima only at the same wavelengths as that of a similar preparation of USP Trimethoprim RS.

B: Transfer about 100 mg of it, accurately weighed, to a 100-mL volumetric flask, and dissolve in 25 mL of alcohol. Dilute quantitatively and stepwise with sodium hydroxide solution (1 in 250) to obtain a 1 in 50,000 solution: the ultraviolet absorption spectrum of this solution exhibits maxima and minima only at the same wavelengths as that of a similar solution of USP Trimethoprim RS, concomitantly measured, and the respective absorptivities, calculated on the dried basis, at the wavelength of maximum absorbance at about 287 nm do not differ by more than 3.0%.

Melting range ⟨741⟩: between 199° and 203°.

Loss on drying ⟨731⟩—Dry it in vacuum at 105° for 4 hours: it loses not more than 0.5% of its weight.

Residue on ignition ⟨281⟩: not more than 0.1%.

Chromatographic impurities—

Test preparation—Prepare a solution of Trimethoprim in a mixture of chloroform and methanol (9:1) to contain 20 mg per mL.

Standard solution—Prepare a solution of USP 3-Anilino-2-(3,4,5-trimethoxybenzyl)acrylonitrile RS in chloroform to contain 100 µg per mL.

Procedure—Apply separately 10-µL portions of the *Test preparation* and the *Standard solution* on a thin-layer chromatographic plate coated with chromatographic silica gel mixture. Place the plate in a suitable chromatographic chamber containing a mixture of chloroform, methanol, and 6 *N* ammonium hydroxide (95:7.5:1). When the solvent front has moved four-fifths of the length of the plate, remove the plate, air-dry, and spray with a freshly prepared mixture of 1.9 g of ferric chloride in 20 mL of water, and 0.5 g of potassium ferricyanide $[K_3Fe(CN)_6]$ in 10 mL of water. The *Test preparation* exhibits a principal spot, and may exhibit a minor spot that corresponds in R_f value to, and is not greater in intensity than, the spot exhibited by the *Standard solution* (0.5%).

Assay—Transfer about 300 mg of Trimethoprim, accurately weighed, to a conical flask, add 60 mL of glacial acetic acid, and titrate with 0.1 *N* perchloric acid VS, determining the end-point potentiometrically. Perform a blank determination, and make any necessary correction. Each mL of 0.1 *N* perchloric acid is equivalent to 29.03 mg of $C_{14}H_{18}N_4O_3$.

Trimethoprim Concentrate for Injection, Sulfamethoxazole and—*see* Sulfamethoxazole and Trimethoprim Concentrate for Injection

Trimethoprim Oral Suspension, Sulfamethoxazole and—*see* Sulfamethoxazole and Trimethoprim Oral Suspension

Trimethoprim Tablets

» Trimethoprim Tablets contain not less than 90.0 percent and not more than 110.0 percent of the labeled amount of $C_{14}H_{18}N_4O_3$.

Packaging and storage—Preserve in tight, light-resistant containers.

Reference standard—*USP Trimethoprim Reference Standard*—Dry in vacuum at 105° for 4 hours before using.

Identification—Triturate a quantity of finely powdered Tablets, equivalent to about 100 mg of trimethoprim, with 2.5 mL of methanol. Add 2.5 mL of chloroform, triturate again, and centrifuge. Apply 25 µL of this test solution and 25 µL of a Standard solution of USP Trimethoprim RS in a mixture of methanol and chloroform (1:1) containing 20 mg per mL to a suitable thin-layer chromatographic plate (see *Chromatography* ⟨621⟩) coated with a 0.25-mm layer of chromatographic silica gel mixture. Allow the spots to dry, and develop the chromatogram in an unsaturated chamber with a solvent system consisting of a mixture of chloroform, methanol, and ammonium hydroxide (95:7.5:1), until the solvent front has moved approximately 15 cm from the origin. Remove the plate from the developing chamber, mark the solvent front, and allow the solvent to evaporate. Locate the spots on the plate by viewing under short-wavelength ultraviolet light: the R_f value of the principal spot obtained from the test solution corresponds to that obtained from the Standard solution.

Dissolution ⟨711⟩—

Medium: 0.01 *N* hydrochloric acid; 900 mL.

Apparatus 2: 50 rpm.

Time: 45 minutes.

Procedure—Determine the amount of $C_{14}H_{18}N_4O_3$ dissolved from ultraviolet absorbances at the wavelength of maximum absorbance at about 271 nm of filtered portions of the solution under test, suitably diluted with 0.01 *N* hydrochloric acid to a concentration of about 20 µg per mL, in comparison with a Standard solution having a known concentration of USP Trimethoprim RS in the same medium.

Tolerances—Not less than 75% (*Q*) of the labeled amount of $C_{14}H_{18}N_4O_3$ is dissolved in 45 minutes.

Uniformity of dosage units ⟨905⟩: meet the requirements.

Assay—

Mobile phase—Prepare a suitable solution of 84 volumes of dilute glacial acetic acid (1 in 100) and 16 volumes of acetonitrile such that the retention time of trimethoprim is about 7 minutes.

Standard preparation—Using an accurately weighed quantity of USP Trimethoprim RS, prepare a solution in methanol having a known concentration of about 0.2 mg per mL.

Assay preparation—Weigh and finely powder not less than 20 Trimethoprim Tablets. Transfer an accurately weighed portion of the powder, equivalent to about 100 mg of trimethoprim, to a 100-mL volumetric flask, add 50 mL of methanol, and sonicate for 5 minutes, with intermittent swirling. Dilute with methanol to volume, and mix. Centrifuge, pipet 10 mL of the supernatant solution into a 50-mL volumetric flask, dilute with methanol to volume, and mix.

Chromatographic system (see *Chromatography* ⟨621⟩)—Use a high-performance liquid chromatograph, operated at room temperature, equipped with a 25-cm × 4.2-mm column that contains packing L1. The chromatograph is equipped with an ultraviolet detector capable of monitoring absorption at 254 nm and a suitable recorder. The *Mobile phase* flow rate is about 2 mL per minute. Chromatograph six replicate injections of the *Standard preparation*, and measure the peak response as directed under *Procedure:* the relative standard deviation is not more than 2.0%.

Procedure—By means of a suitable sampling valve, inject equal volumes (about 10 µL) of the *Standard preparation* and the *Assay preparation* into the chromatograph. Calculate the quantity, in mg, of $C_{14}H_{18}N_4O_3$, in the portion of Tablets taken by the formula:

$$500C(r_U/r_S),$$

in which *C* is the concentration, in mg per mL, of USP Trimethoprim RS in the *Standard preparation*, and r_U and r_S are the peak responses obtained from the *Assay preparation* and the *Standard preparation*, respectively.

Trimethoprim Tablets, Sulfamethoxazole and—*see* Sulfamethoxazole and Trimethoprim Tablets

Trioxsalen

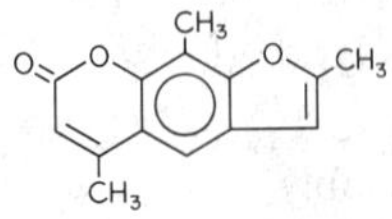

$C_{14}H_{12}O_3$ 228.25

7*H*-Furo[3,2-*g*][1]benzopyran-7-one, 2,5,9-trimethyl-.

2,5,9-Trimethyl-7*H*-furo[3,2-*g*][1]benzopyran-7-one [*3902-71-4*].

» Trioxsalen contains not less than 97.0 percent and not more than 103.0 percent of $C_{14}H_{12}O_3$, calculated on the dried basis.

Caution—Avoid exposing the skin to Trioxsalen.

Packaging and storage—Preserve in well-closed, light-resistant containers.

Reference standard—*USP Trioxsalen Reference Standard*—Dry at 105° for 6 hours before using.

Identification—

A: The infrared absorption spectrum of a mineral oil dispersion of it, previously dried, exhibits maxima only at the same wavelengths as that of a similar preparation of USP Trioxsalen RS.

B: The ultraviolet absorption spectrum of a 1 in 200,000 solution in chloroform exhibits maxima and minima at the same wavelengths as that of a similar solution of USP Trioxsalen RS, concomitantly measured.

C: The retention time of the major peak in the chromatogram of the *Assay preparation* corresponds to that of the *Standard preparation*, as obtained in the *Assay*.

Loss on drying ⟨731⟩—Dry it at 105° for 6 hours: it loses not more than 0.5% of its weight.

Residue on ignition ⟨281⟩: not more than 0.5%.

Related substances—In the chromatogram obtained from the *Assay preparation* in the *Assay*, the sum of the responses of any peaks detected, other than the major peak due to trioxsalen, is not more than 2.0% of the total of all the peak responses and the response of the peak occurring at retention time relative to trioxsalen of about 0.75 is not more than 1.5% of the total of all responses.

Assay—

Mobile phase—Prepare a filtered and degassed mixture of methanol and water (70:30). Make adjustments if necessary (see *System Suitability* under *Chromatography* ⟨621⟩).

Standard preparation—Dissolve an accurately weighed quantity of USP Trioxsalen RS in tetrahydrofuran to obtain a solution having a known concentration of about 1 mg per mL. Transfer 5.0 mL of this solution to a 100-mL volumetric flask, dilute with *Mobile phase* to volume, and mix.

Assay preparation—Transfer about 100 mg of Trioxsalen, accurately weighed, to a 100-mL volumetric flask, dissolve in tetrahydrofuran, dilute with tetrahydrofuran to volume, mix, and filter. Transfer 5.0 mL of this solution to a 100-mL volumetric flask, dilute with *Mobile phase* to volume, and mix.

Chromatographic system (see *Chromatography* ⟨621⟩)—The liquid chromatograph is equipped with a 254-nm detector and a 4.6-nm × 25-cm column that contains packing L1. The flow rate is about 1 mL per minute. Chromatograph the *Standard preparation*, and record the peak responses as directed under *Procedure:* the tailing factor for the trioxsalen peak is not more than 2.0, and the relative standard deviation for replicate injections is not more than 2.0%.

Procedure—Separately inject equal volumes (about 20 µL) of the *Standard preparation* and the *Assay preparation* into the chromatograph, record the chromatograms, and measure the responses for the major peaks. Calculate the quantity, in mg, of $C_{14}H_{12}O_3$ in the portion of Trioxsalen taken by the formula:

$$2000C(r_U/r_S),$$

in which *C* is the concentration, in mg per mL, of USP Trioxsalen RS in the *Standard preparation*, and r_U and r_S are the peak

responses obtained from the *Assay preparation* and the *Standard preparation*, respectively.

Trioxsalen Tablets

» Trioxsalen Tablets contain not less than 93.0 percent and not more than 107.0 percent of the labeled amount of $C_{14}H_{12}O_3$.

Packaging and storage—Preserve in well-closed, light-resistant containers.

Reference standard—*USP Trioxsalen Reference Standard*—Dry at 105° for 6 hours before using.

Identification—Triturate an amount of finely powdered Tablets, equivalent to about 10 mg of trioxsalen, with 100 mL of chloroform, and filter. On a suitable thin-layer chromatographic plate, coated with a 0.25-mm layer of chromatographic silica gel, apply 5 µL each of this solution and a Standard solution of USP Trioxsalen RS in chloroform having a known concentration of 100 µg per mL. Allow the spots to dry, and develop the chromatogram, using methanol as the solvent, until the solvent front has moved about three-fourths of the length of the plate. Remove the plate from the developing chamber, mark the solvent front, and allow the solvent to evaporate. Locate the spots on the plate by viewing under an ultraviolet lamp: the R_f value of the principal spot obtained from the test solution corresponds to that obtained from the Standard solution.

Dissolution ⟨711⟩—

Apparatus 2: 100 rpm.

Time: 60 minutes.

Dilute simulated intestinal fluid—Prepare a 1 in 100 solution of simulated intestinal fluid TS and water.

Procedure—Assemble the apparatus, adding 225 mL of *Dilute simulated intestinal fluid* to each vessel, and operate the apparatus for 40 minutes. At the end of the 40 minutes, immediately add 675 mL of dehydrated alcohol to each of the vessels. Continue to operate the apparatus for an additional 20 minutes. Determine the amount of $C_{14}H_{12}O_3$ dissolved from ultraviolet absorbance determined at the wavelength of maximum absorbance at about 252 nm, filtered portions of the solution under test, in comparison with a Standard solution having a known concentration of USP Trioxsalen RS in the same medium.

Tolerances—Not less than 75% (*Q*) of the labeled amount of $C_{14}H_{12}O_3$ is dissolved in 60 minutes.

Uniformity of dosage units ⟨905⟩: meet the requirements.

Assay—Weigh and finely powder not less than 20 Trioxsalen Tablets. Transfer an accurately weighed portion of the powder, equivalent to about 5 mg of trioxsalen, to a separator containing 25 mL of water. Extract with three 25-mL portions of chloroform, filtering each extract into a 100-mL volumetric flask. Wash the filter with chloroform, dilute with chloroform to volume, and mix. Transfer 10.0 mL of this solution to a second 100-mL volumetric flask, dilute with chloroform to volume, and mix. Concomitantly determine the absorbances of this solution and a solution of USP Trioxsalen RS in the same medium having a known concentration of about 5 µg per mL in 1-cm cells at the wavelength of maximum absorbance at about 252 nm, with a suitable spectrophotometer, using chloroform as the blank. Calculate the quantity, in mg, of $C_{14}H_{12}O_3$ in the portion of Tablets taken by the formula:

$$C(A_U / A_S),$$

in which *C* is the concentration, in µg per mL, of USP Trioxsalen RS in the Standard solution, and A_U and A_S are the absorbances of the solution from the Tablets and the Standard solution, respectively.

Tripelennamine Citrate

$C_{16}H_{21}N_3 . C_6H_8O_7$ 447.49

1,2-Ethanediamine, *N,N*-dimethyl-*N'*-(phenylmethyl)-*N'*-2-pyridinyl-, 2-hydroxy-1,2,3-propanetricarboxylate (1:1).

2-[Benzyl[2-(dimethylamino)ethyl]amino]pyridine citrate (1:1) [*6138-56-3*].

» Tripelennamine Citrate contains not less than 98.0 percent and not more than 100.5 percent of $C_{16}H_{21}N_3 . C_6H_8O_7$, calculated on the dried basis.

Packaging and storage—Preserve in well-closed, light-resistant containers.

Reference standard—*USP Tripelennamine Citrate Reference Standard*—Dry in vacuum over phosphorus pentoxide for 24 hours before using.

Identification—

A: It meets the requirements under *Identification—Organic Nitrogenous Bases* ⟨181⟩.

B: Dissolve about 500 mg in 5 mL of water, render the mixture alkaline with 1 *N* sodium hydroxide, and remove the liberated base by extraction with three 5-mL portions of chloroform: the aqueous liquid responds to the tests for *Citrate* ⟨191⟩.

Loss on drying ⟨731⟩—Dry it in vacuum over phosphorus pentoxide for 24 hours: it loses not more than 0.5% of its weight.

Residue on ignition ⟨281⟩: not more than 0.1%.

Ordinary impurities ⟨466⟩—

Test solution: methanol.

Standard solution: methanol.

Eluant: a mixture of methanol and ammonium hydroxide (100:1.5).

Visualization: 1.

Assay—Dissolve about 500 mg of Tripelennamine Citrate, accurately weighed, in 20 mL of glacial acetic acid. Add 2 drops of crystal violet TS, and titrate with 0.1 *N* perchloric acid VS. Perform a blank determination, and make any necessary correction. Each mL of 0.1 *N* perchloric acid is equivalent to 22.37 mg of $C_{16}H_{21}N_3 . C_6H_8O_7$.

Tripelennamine Citrate Elixir

» Tripelennamine Citrate Elixir contains, in each 100 mL, not less than 705 mg and not more than 795 mg of $C_{16}H_{21}N_3 . C_6H_8O_7$.

Packaging and storage—Preserve in tight, light-resistant containers.

Reference standard—*USP Tripelennamine Hydrochloride Reference Standard*—Dry at 105° for 3 hours before using.

Identification—The ultraviolet absorption spectrum of the solution of Elixir employed for measurement of absorbance in the *Assay* exhibits maxima and minima at the same wavelengths as that of the solution made from the *Standard Preparation.*

Alcohol content ⟨611⟩: between 11.0% and 13.0% of C_2H_5OH.

Assay—Proceed with Tripelennamine Citrate Elixir as directed under *Salts of Organic Nitrogenous Bases* ⟨501⟩, using a 5.0-mL specimen, determining the absorbance at 313 nm, and using USP Tripelennamine Hydrochloride RS. Calculate the quantity, in mg, of $C_{16}H_{21}N_3 . C_6H_8O_7$ in the portion of Elixir taken by the formula:

$$1.533(50C)(A_U / A_S),$$

in which 1.533 is the ratio of the molecular weight of tripelennamine citrate to that of tripelennamine hydrochloride, and *C* is the concentration, in mg per mL, of USP Tripelennamine Hydrochloride RS in the *Standard Preparation.*

Tripelennamine Hydrochloride

$C_{16}H_{21}N_3.HCl$ 291.82
1,2-Ethanediamine, *N,N*-dimethyl-*N'*-(phenylmethyl)-*N'*-2-pyridinyl-, monohydrochloride.
2-[Benzyl[2-(dimethylamino)ethyl]amino]pyridine monohydrochloride [*154-69-8*].

» Tripelennamine Hydrochloride contains not less than 98.0 percent and not more than 100.5 percent of $C_{16}H_{21}N_3.HCl$, calculated on the dried basis.

Packaging and storage—Preserve in well-closed, light-resistant containers.

Reference standard—*USP Tripelennamine Hydrochloride Reference Standard*—Dry at 105° for 3 hours before using.

Identification—

A: It meets the requirements under *Identification—Organic Nitrogenous Bases* ⟨181⟩.

B: It responds to the tests for *Chloride* ⟨191⟩.

Melting range ⟨741⟩: between 188° and 192°.

Loss on drying ⟨731⟩—Dry it at 105° for 3 hours: it loses not more than 1.0% of its weight.

Residue on ignition ⟨281⟩: not more than 0.1%.

Ordinary impurities ⟨466⟩—

Test solution: methanol.

Standard solution: methanol.

Eluant: a mixture of methanol and ammonium hydroxide (100:1.5).

Visualization: 1.

Assay—Dissolve about 300 mg of Tripelennamine Hydrochloride, accurately weighed, in a mixture of 10 mL of glacial acetic acid and 10 mL of mercuric acetate TS. Add 2 drops of crystal violet TS, and titrate with 0.1 *N* perchloric acid VS. Perform a blank determination, and make any necessary correction. Each mL of 0.1 *N* perchloric acid is equivalent to 14.59 mg of $C_{16}H_{21}N_3.HCl$.

Tripelennamine Hydrochloride Tablets

» Tripelennamine Hydrochloride Tablets contain not less than 95.0 percent and not more than 105.0 percent of the labeled amount of $C_{16}H_{21}N_3.HCl$.

Packaging and storage—Preserve in well-closed containers.

Reference standard—*USP Tripelennamine Hydrochloride Reference Standard*—Dry at 105° for 3 hours before using.

Identification—Tablets meet the requirements under *Identification—Organic Nitrogenous Bases* ⟨181⟩.

Dissolution ⟨711⟩—

Medium: water; 900 mL.

Apparatus 1: 100 rpm.

Time: 45 minutes.

Procedure—Determine the amount of $C_{16}H_{21}N_3.HCl$ dissolved from ultraviolet absorbances at the wavelength of maximum absorbance at about 306 nm of filtered portions of the solution under test, suitably diluted with *Dissolution Medium*, if necessary, in comparison with a Standard solution having a known concentration of USP Tripelennamine Hydrochloride RS in the same medium.

Tolerances—Not less than 75% (*Q*) of the labeled amount of $C_{16}H_{21}N_3.HCl$ is dissolved in 45 minutes.

Uniformity of dosage units ⟨905⟩: meet the requirements.

Assay—Proceed with Tripelennamine Hydrochloride Tablets as directed under *Salts of Organic Nitrogenous Bases* ⟨501⟩, determining the absorbance at 313 nm. Calculate the quantity, in mg, of $C_{16}H_{21}N_3.HCl$ in the portion of Tablets taken by the formula:

$$50C(A_U/A_S),$$

in which *C* is the concentration, in mg per mL, of USP Tripelennamine Hydrochloride RS in the *Standard Preparation.*

Triprolidine Hydrochloride

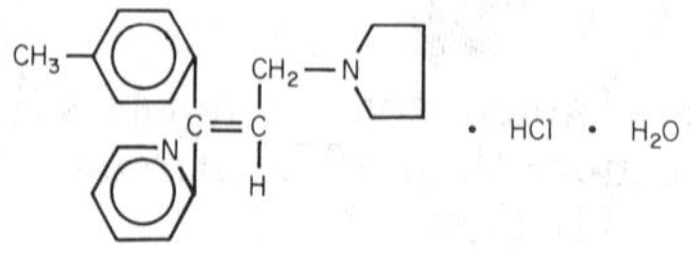

$C_{19}H_{22}N_2.HCl.H_2O$ 332.87
Pyridine, 2-[1-(4-methylphenyl)-3-(1-pyrrolidinyl)-1-propenyl]-, monohydrochloride, monohydrate, (*E*)-.
(*E*)-2-[3-(1-Pyrrolidinyl)-1-*p*-tolylpropenyl]pyridine monohydrochloride monohydrate [*6138-79-0*].
Anhydrous 314.86 [*550-70-9*].

» Triprolidine Hydrochloride contains not less than 98.0 percent and not more than 101.0 percent of $C_{19}H_{22}N_2.HCl$, calculated on the anhydrous basis.

Packaging and storage—Preserve in tight, light-resistant containers.

Reference standards—*USP Triprolidine Hydrochloride Reference Standard*—Do not dry; determine the water content at the time of use for quantitative analyses. *USP Triprolidine Hydrochloride Z-isomer Reference Standard.*

Identification—

A: The infrared absorption spectrum of a potassium bromide dispersion of it exhibits maxima only at the same wavelengths as that of a similar preparation of USP Triprolidine Hydrochloride RS.

B: The ultraviolet absorption spectrum of a 1 in 100,000 solution in 0.1 *N* hydrochloric acid exhibits maxima and minima at the same wavelengths as that of a similar solution of USP Triprolidine Hydrochloride RS, concomitantly measured, and the respective absorptivities, calculated on the anhydrous basis, at the wavelength of maximum absorbance at about 290 nm do not differ by more than 3.0%.

C: A solution of it responds to the tests for *Chloride* ⟨191⟩.

Water, *Method I* ⟨921⟩: between 4.0% and 6.0%.

Residue on ignition ⟨281⟩: not more than 0.1%.

Arsenic, *Method II* ⟨211⟩: 4 ppm.

Heavy metals, *Method II* ⟨231⟩: 0.002%.

Chromatographic purity—

Standard preparations—Dissolve USP Triprolidine Hydrochloride RS in chloroform, and mix to obtain a solution having a known concentration of 1.0 mg per mL. Dilute quantitatively with chloroform to obtain four diluted *Standard preparations* (*A*, *B*, *C*, and *D*) having the following compositions:

Dilution	Concentration (µg RS per mL)	Percentage (%, for comparison with test specimen)
A (1 in 5)	200	2.0
B (15 in 100)	150	1.5
C (1 in 10)	100	1.0
D (5 in 100)	50	0.5

Standard Z-isomer preparations—Proceed as directed for *Standard preparations*, using USP Triprolidine Hydrochloride Z-isomer RS to obtain four diluted *Standard preparations* having the same compositions as in the table shown therein.

Test preparation—Dissolve an accurately weighed quantity of Triprolidine Hydrochloride in chloroform to obtain a solution containing 10 mg per mL.

Procedure—On a suitable thin-layer chromatographic plate (see *Chromatography* ⟨621⟩), coated with a 0.25-mm layer of chromatographic silica gel mixture, apply separately 5 µL of the *Test preparation* and 5 µL of each of the eight diluted *Standard preparations.* Position the plate in a chromatographic chamber, and develop the chromatograms, protected from light, in a solvent system consisting of a mixture of chloroform and diethylamine (95:5) until the solvent front has moved about three-fourths of the length of the plate. Remove the plate from the developing chamber, mark the solvent front, and allow the solvent to evaporate. Examine the plate under long- and short-wavelength ultraviolet light. Compare the intensities of any secondary spots observed in the chromatogram of the *Test preparation* with those of the principal spots in the chromatograms of the *Standard preparations:* the intensity of the *Z*-isomer triprolidine hydrochloride spot (R_f value about 1.2 relative to the R_f value for triprolidine hydrochloride) obtained from the *Test preparation* corresponds to not more than 2.0%, and the sum of the intensities of all secondary spots obtained from the *Test preparation* corresponds to not more than 3.0%.

Assay Dissolve about 400 mg of Triprolidinc Hydrochloridc, accurately weighed, in 80 mL of glacial acetic acid, warming, if necessary, to effect solution. Add 15 mL of mercuric acetate TS, and titrate with 0.1 *N* perchloric acid VS, determining the end-point potentiometrically. Perform a blank determination, and make any necessary correction. Each mL of 0.1 *N* perchloric acid is equivalent to 15.74 mg of $C_{19}H_{22}N_2 . HCl$.

Triprolidine Hydrochloride Syrup

» Triprolidine Hydrochloride Syrup contains not less than 90.0 percent and not more than 110.0 percent of the labeled amount of $C_{19}H_{22}N_2 . HCl . H_2O$.

Packaging and storage—Preserve in tight, light-resistant containers.

Reference standard—*USP Triprolidine Hydrochloride Reference Standard*—Do not dry; determine the water content at the time of use for quantitative analyses.

Identification—

A: Transfer a volume of Triprolidine Hydrochloride Syrup, equivalent to about 12 mg of triprolidine hydrochloride, to a 125-mL separator, add 25 mL of water, then add 4 mL of sodium hydroxide solution (1 in 2), and mix. Add 10 mL of cyclohexane, shake, allow the phases to separate completely, and discard the aqueous layer. Transfer 8 mL of the cyclohexane solution to a glass-stoppered, 25-mL conical flask, evaporate on a steam bath with the aid of a current of air to dryness, and continue to heat the flask for about 1 minute after the solvent has completely evaporated. Cool, add 2 mL of cyclohexane, and mix: the infrared absorption spectrum of the cyclohexane solution so obtained exhibits maxima only at the same wavelengths as that of a similar preparation of USP Triprolidine Hydrochloride RS.

B: The retention time of the major peak in the chromatogram of the *Assay preparation* corresponds to that in the chromatogram of the *Standard preparation* as obtained in the *Assay.*

pH ⟨791⟩: between 5.6 and 6.6.

Alcohol content, *Method II* ⟨611⟩: between 3.0% and 5.0% of C_2H_5OH.

Assay—

Mobile phase—Prepare a suitable degassed and filtered mixture of alcohol and ammonium acetate solution (1 in 250) (17:3).

Standard preparation—Dissolve an accurately weighed quantity of USP Triprolidine Hydrochloride RS in 0.01 *N* hydrochloric acid, and dilute quantitatively and stepwise with 0.01 *N* hydrochloric acid to obtain a solution having a known concentration of about 0.05 mg of anhydrous USP Triprolidine Hydrochloride RS per mL.

Assay preparation—Transfer an accurately measured volume of Triprolidine Hydrochloride Syrup, equivalent to about 2.5 mg of triprolidine hydrochloride, to a 50-mL volumetric flask, dilute with 0.01 *N* hydrochloric acid to volume, and mix.

Chromatographic system (see *Chromatography* ⟨621⟩)—The liquid chromatograph is equipped with a 254-nm detector and a 4.2-mm × 25-cm column that contains packing L3. The flow rate is about 1.5 mL per minute. Chromatograph five replicate injections of the *Standard preparation*, and record the peak responses as directed under *Procedure:* the relative standard deviation is not more than 2.0%, and the tailing factor is not more than 1.5.

Procedure—Separately inject equal volumes (about 10 µL) of the *Standard preparation* and the *Assay preparation* into the chromatograph, record the chromatograms, and measure the responses for the major peaks. Calculate the quantity, in mg, of $C_{19}H_{22}N_2 . HCl . H_2O$ in the portion of Syrup taken by the formula:

$$(332.87/314.86)(50C)(r_U/r_S),$$

in which 332.87 and 314.86 are the molecular weights of triprolidine hydrochloride monohydrate and anhydrous triprolidine hydrochloride, respectively, *C* is the concentration, in mg per mL, of USP Triprolidine Hydrochloride RS in the *Standard preparation*, and r_U and r_S are the peak responses obtained from the *Assay preparation* and the *Standard preparation*, respectively.

Triprolidine Hydrochloride Tablets

» Triprolidine Hydrochloride Tablets contain not less than 90.0 percent and not more than 110.0 percent of the labeled amount of $C_{19}H_{22}N_2 . HCl . H_2O$.

Packaging and storage—Preserve in tight, light-resistant containers.

Reference standard—*USP Triprolidine Hydrochloride Reference Standard*—Do not dry; determine the water content at the time of use for quantitative analyses.

Identification—

A: Weigh and finely powder not less than 20 Triprolidine Hydrochloric Tablets. Transfer a portion of the powder, equivalent to about 20 mg of triprolidine hydrochloride, to a glass-stoppered test tube, add 20 mL of water, and shake for 3 minutes. Add 2 mL of 1 *N* sodium hydroxide, mix, then add 3 mL of cyclohexane, shake for 3 minutes, and centrifuge for 5 minutes: the infrared absorption spectrum of the clear supernatant liquid so obtained exhibits maxima only at the same wavelengths as that of a similar preparation of USP Triprolidine Hydrochloride RS.

B: The retention time of the major peak in the chromatogram of the *Assay preparation* corresponds to that in the chromatogram of the *Standard preparation* as obtained in the *Assay.*

Dissolution ⟨711⟩—

Medium: pH 4.0 ± 0.05 acetate buffer, prepared by mixing 4.9 g of glacial acetic acid and 2.45 g of sodium acetate trihydrate with water to obtain 1000 mL of solution; 500 mL.

Apparatus 1: 50 rpm.

Time: 30 minutes.

Procedure—Determine the amount of $C_{19}H_{22}N_2 . HCl . H_2O$ dissolved from ultraviolet absorbances at the wavelength of maximum absorbance at about 277 nm of filtered portions of the solution under test, in comparison with a Standard solution having a known concentration of USP Triprolidine Hydrochloride RS in the same medium.

Tolerances—Not less than 80% (*Q*) of the labeled amount of $C_{19}H_{22}N_2 . HCl . H_2O$ is dissolved in 30 minutes.

Uniformity of dosage units ⟨905⟩: meet the requirements.

Procedure for content uniformity—Transfer 1 Tablet to a 100-mL volumetric flask, add 70 mL of water, and sonicate, swirling the flask intermittently, until the tablet is dissolved. Dilute with water to volume, mix, and filter, discarding the first 50 mL of the filtrate. Dilute a portion of the filtrate quantitatively and stepwise with 0.1 *N* sulfuric acid to obtain a solution having a concentration of about 1.25 µg of triprolidine hydrochloride per mL. Concomitantly determine the fluorescence intensities of this solution and a similarly prepared Standard solution having a known concentration of about 1.25 µg of USP Tripolidone Hydrochloride RS per mL, at the excitation wavelength of 300 nm with a slit width of 2 mm, and an emission wavelength of 460 nm with a slit width of 2 mm, with a suitable spectrophotometer, using 0.1 *N* sulfuric acid as the blank. Calculate the quantity, in mg, of $C_{19}H_{22}N_2.HCl.H_2O$ in the Tablet taken by the formula:

$$(332.87/314.86)(TC/D)(I_U/I_S),$$

in which 332.87 and 314.86 are the molecular weights of the monohydrate and anhydrous forms of triprolidine hydrochloride, respectively, *T* is the labeled quantity, in mg, of triprolidine hydrochloride in the Tablet, *C* is the concentration, in µg per mL, of USP Triprolidine Hydrochloride RS in the Standard solution, *D* is the concentration, in µg per mL, of triprolidine hydrochloride in the solution from the Tablet, on the basis of the labeled quantity per Tablet and the extent of dilution, and I_U and I_S are the fluorescence intensities of the solution from the Tablet and the Standard solution, respectively.

Assay—

Mobile phase and *Standard preparation*—Prepare as directed in the *Assay* under *Triprolidine Hydrochloride Syrup*.

Assay preparation—Weigh and finely powder not less than 20 Triprolidine Hydrochloride Tablets. Transfer an accurately weighed portion of the powder, equivalent to about 5.0 mg of triprolidine hydrochloride, to a 100-mL volumetric flask. Add about 10 mL of 0.01 *N* hydrochloric acid, and sonicate for 10 minutes. Cool to room temperature. Dilute with 0.01 *N* hydrochloric acid to volume, mix, and filter.

Chromatographic system and *Procedure*—Proceed as directed in the *Assay* under *Triprolidine Hydrochloride Syrup*, except to calculate the quantity, in mg, of $C_{19}H_{22}N_2.HCl.H_2O$ in the portion of Tablets taken by the formula:

$$(332.87/314.86)(100C)(r_U/r_S),$$

in which 332.87 and 314.86 are the molecular weights of triprolidine hydrochloride monohydrate and anhydrous triprolidine hydrochloride, respectively, *C* is the concentration, in mg per mL, of USP Triprolidine Hydrochloride RS in the *Standard preparation*, and r_U and r_S are the peak responses obtained from the *Assay preparation* and the *Standard preparation*, respectively.

Triprolidine and Pseudoephedrine Hydrochlorides Syrup

» Triprolidine and Pseudoephedrine Hydrochlorides Syrup contains not less than 90.0 percent and not more than 110.0 percent of the labeled amounts of triprolidine hydrochloride ($C_{19}H_{22}N_2.HCl.H_2O$) and pseudoephedrine hydrochloride ($C_{10}H_{15}NO.HCl$).

Packaging and storage—Preserve in tight, light-resistant containers.

Reference standards—*USP Triprolidine Hydrochloride Reference Standard*—Do not dry; determine the water content titrimetrically at the time of use for quantitative analyses. *USP Pseudoephedrine Hydrochloride Reference Standard*—Dry at 105° for 3 hours before using.

Identification—

A: The retention times of the major peaks in the chromatogram of the *Assay preparation* correspond to those of the *Standard preparation*, as obtained in the *Assay*.

B: Transfer 10 mL of Syrup to a suitable glass-stoppered tube, add 10 mL of ether and 2 mL of 1 *N* sodium hydroxide, shake for 5 minutes, and allow the layers to separate. The ether layer is the test solution. Prepare a Standard solution in water of USP Pseudoephedrine Hydrochloride RS and USP Triprolidine Hydrochloride RS having known concentrations of 6 mg per mL and 250 µg per mL, respectively. On a suitable thin-layer chromatographic plate (see *Chromatography* ⟨621⟩), coated with a 0.25-mm layer of chromatographic silica gel mixture, separately apply 10-µL portions of the test solution and the Standard solution. Allow the spots to dry, and develop the chromatogram in a solvent system consisting of a mixture of butyl alcohol, glacial acetic acid, and water (8:2:2) until the solvent front has moved about three-fourths of the length of the plate. Remove the plate, mark the solvent front, allow the solvent to evaporate, and examine the plate under short-wavelength and long-wavelength ultraviolet light: the R_f values of the principal spots obtained from the test solution correspond to those obtained from the Standard solution.

Assay—

Mobile phase—Prepare a suitable degassed and filtered mixture of alcohol and ammonium acetate solution (1 in 250) (17:3).

Standard preparation—Dissolve accurately weighed quantities of USP Pseudoephedrine Hydrochloride RS and USP Triprolidine Hydrochloride RS in 0.01 *N* hydrochloric acid, and dilute quantitatively and stepwise with 0.01 *N* hydrochloric acid to obtain a solution having known concentrations of about 1.2 mg of USP Pseudoephedrine Hydrochloride RS per mL and about 0.05 mg of anhydrous USP Triprolidine Hydrochloride RS per mL, and filter.

Assay preparation—Transfer an accurately measured volume of Triprolidine and Pseudoephedrine Hydrochlorides Syrup, equivalent to about 60 mg of pseudoephedrine hydrochloride, to a 50-mL volumetric flask, dilute with 0.01 *N* hydrochloric acid to volume, and mix.

Chromatographic system (see *Chromatography* ⟨621⟩)—The liquid chromatograph is equipped with a 254-nm detector and a 4.2-mm × 25-cm column that contains packing L3. The flow rate is about 1.5 mL per minute. Chromatograph replicate injections of the *Standard preparation*, and record the peak responses as directed under *Procedure:* the relative standard deviation is not more than 2.0%, and the resolution factor between triprolidine and pseudoephedrine is not less than 2.0. The tailing factor for the triprolidine peak is not more than 1.5 and for pseudoephedrine peak is not more than 1.5.

Procedure—Separately inject equal volumes (about 10 µL) of the *Standard preparation* and the *Assay preparation* into the chromatograph, record the chromatograms, and measure the responses for the major peaks. The relative retention times are about 0.68 for pseudoephedrine hydrochloride and 1.0 for triprolidine hydrochloride. Calculate the quantity, in mg, of pseudoephedrine hydrochloride ($C_{10}H_{15}NO.HCl$) in the portion of Syrup taken by the formula:

$$50C(r_U/r_S),$$

in which *C* is the concentration, in mg per mL, of USP Pseudoephedrine Hydrochloride RS in the *Standard preparation*, and r_U and r_S are the peak responses for pseudoephedrine hydrochloride obtained from the *Assay preparation* and the *Standard preparation*, respectively. Calculate the quantity, in mg, of triprolidine hydrochloride ($C_{19}H_{22}N_2.HCl.H_2O$) in the portion of Syrup taken by the formula:

$$(332.87/314.86)(50C)(r_U/r_S),$$

in which 332.87 and 314.86 are the molecular weights of triprolidine hydrochloride monohydrate and anhydrous triprolidine hydrochloride, respectively, *C* is the concentration, in mg per mL, of USP Triprolidine Hydrochloride RS in the *Standard preparation*, and r_U and r_S are the peak responses for triprolidine hydrochloride obtained from the *Assay preparation* and the *Standard preparation*, respectively.

Triprolidine and Pseudoephedrine Hydrochlorides Tablets

» Triprolidine and Pseudoephedrine Hydrochlorides Tablets contain not less than 90.0 percent and not more than 110.0 percent of the labeled amounts of triprolidine hydrochloride ($C_{19}H_{22}N_2.HCl.H_2O$) and pseudoephedrine hydrochloride ($C_{10}H_{15}NO.HCl$).

Packaging and storage—Preserve in tight, light-resistant containers.

Reference standards—*USP Triprolidine Hydrochloride Reference Standard*—Do not dry; determine the water content titrimetrically at the time of use for quantitative analyses. *USP Pseudoephedrine Hydrochloride Reference Standard*—Dry at 105° for 3 hours before using.

Identification—

A: The retention times of the major peaks in the chromatogram of the *Assay preparation* correspond to those of the *Standard preparation*, as obtained in the *Assay*.

B: Transfer 1 Tablet to a suitable glass-stoppered tube, add 10 mL of water, shake for 5 minutes, and allow the solids to settle. Prepare a Standard solution in water of USP Pseudoephedrine Hydrochloride RS and USP Triprolidine Hydrochloride RS having known concentrations of 6 mg per mL and 250 µg per mL, respectively. On a suitable thin-layer chromatographic plate (see *Chromatography* ⟨621⟩), coated with a 0.25-mm layer of chromatographic silica gel mixture, separately apply 10-µL portions of the test solution and the Standard solution. Allow the spots to dry, and develop the chromatogram in a solvent system consisting of a mixture of butyl alcohol, glacial acetic acid, and water (8:2:2) until the solvent front has moved about three-fourths of the length of the plate. Remove the plate, mark the solvent front, allow the solvent to evaporate, and examine the plate under short-wavelength and long-wavelength ultraviolet light: the R_f values of the principal spots obtained from the test solution correspond to those obtained from the Standard solution.

Dissolution ⟨711⟩—

Medium: water, 900 mL.

Apparatus 2: 50 rpm.

Time: 45 minutes.

Procedure for pseudoephedrine hydrochloride—Determine the amount of $C_{10}H_{15}NO.HCl$ dissolved from ultraviolet absorbances at the wavelength of maximum absorbance at about 208 nm of filtered portions of the solution under test, suitably diluted with 0.1 *N* hydrochloric acid, if necessary, in comparison with a Standard solution having a known concentration of USP Pseudoephedrine Hydrochloride RS, similarly prepared.

Procedure for triprolidine hydrochloride—Determine the amount of $C_{19}H_{22}N_2.HCl.H_2O$ dissolved from fluorescence intensities at the excitation wavelength of 300 nm, and an emission wavelength of 460 nm with 2-mm slit widths of filtered portions of the test solution, suitably diluted with 0.1 *N* sulfuric acid, if necessary, in comparison with a Standard solution having a known concentration of USP Triprolidine Hydrochloride RS, similarly prepared.

Tolerances—Not less than 75% (*Q*) of the labeled amounts of $C_{10}H_{15}NO.HCl$ and $C_{19}H_{22}N_2.HCl.H_2O$ is dissolved in 45 minutes.

Uniformity of dosage units ⟨905⟩: meet the requirements for *Content Uniformity* with respect to triprolidine hydrochloride and to pseudoephedrine hydrochloride.

Assay—

Mobile phase and *Standard preparation*—Prepare as directed in the *Assay* under *Triprolidine and Pseudoephedrine Hydrochlorides Syrup.*

Assay preparation—Weigh and finely powder not less than 20 Triprolidine and Pseudoephedrine Hydrochlorides Tablets. Transfer an accurately weighed portion of the powder, equivalent to about 120 mg of pseudoephedrine hydrochloride, to a 100-mL volumetric flask. Add about 10 mL of 0.01 *N* hydrochloric acid, and sonicate for 10 minutes. Cool to room temperature. Dilute with 0.01 *N* hydrochloric acid to volume, mix, and filter.

Chromatographic system (see *Chromatography* ⟨621⟩) and *Procedure*—Proceed as directed in the *Assay* under *Triprolidine and Pseudoephedrine Hydrochlorides Syrup*, except to calculate the quantity, in mg, of pseudoephedrine hydrochloride ($C_{10}H_{15}NO.HCl$) in the portion of Tablets taken by the formula:

$$100C(r_U/r_S),$$

in which *C* is the concentration, in mg per mL, of USP Pseudoephedrine Hydrochloride RS in the *Standard preparation*, and r_U and r_S are the peak responses for pseudoephedrine hydrochloride obtained from the *Assay preparation* and the *Standard preparation*, respectively. Calculate the quantity, in mg, of triprolidine hydrochloride ($C_{19}H_{22}N_2.HCl.H_2O$) in the portion of Tablets taken by the formula:

$$(332.87/314.86)(100C)(r_U/r_S),$$

in which 332.87 and 314.86 are the molecular weights of triprolidine hydrochloride monohydrate and anhydrous triprolidine hydrochloride, respectively, *C* is the concentration, in mg per mL, of USP Triprolidine Hydrochloride RS in the *Standard preparation*, and r_U and r_S are the peak responses for triprolidine hydrochloride obtained from the *Assay preparation* and the *Standard preparation*, respectively.

Trisulfapyrimidines Oral Suspension

» Trisulfapyrimidines Oral Suspension contains, in each 100 mL, not less than 3.0 g and not more than 3.7 g of sulfadiazine ($C_{10}H_{10}N_4O_2S$), of sulfamerazine ($C_{11}H_{12}N_4O_2S$), and of sulfamethazine ($C_{12}H_{14}N_4O_2S$). It may contain either Sodium Citrate or Sodium Lactate, and may contain a suitable antimicrobial agent.

Packaging and storage—Preserve in tight containers, at a temperature above freezing.

Labeling—Its label indicates the presence and proportion of any sodium citrate or sodium lactate and any antimicrobial agent.

Reference standards—*USP Sulfadiazine Reference Standard*—Dry at 105° for 2 hours before using. *USP Sulfamerazine Reference Standard*—Dry at 105° for 2 hours before using. *USP Sulfamethazine Reference Standard*—Dry at 105° for 2 hours before using.

Identification—The retention times of the three individual sulfapyrimidines obtained in the *Assay* correspond to the retention times of the respective USP Reference Standards.

Assay—

Mobile phase—Prepare a suitable degassed solution of water, acetonitrile, and glacial acetic acid (86:13:1) such that the relative retention times of sulfadiazine, sulfamerazine, and sulfamethazine are approximately 0.6, 0.8, and 1.0, respectively. (If the retention times are excessive, the concentration of acetonitrile may be increased.)

Standard preparation—Transfer 33 mg each of USP Sulfadiazine RS, USP Sulfamerazine RS, and USP Sulfamethazine RS, accurately weighed, to a 100-mL volumetric flask, dissolve in 25 mL of 0.1 *N* sodium hydroxide, dilute with water to volume, and mix. Pipet 3 mL into a 25-mL volumetric flask, dilute with water to volume, and mix to obtain a *Standard preparation* having a known concentration of about 40 µg of each Reference Standard per mL.

Assay preparation—Determine the specific gravity of the Oral Suspension, using a tared, 50-mL volumetric flask, by weighing 50 mL of Trisulfapyrimidines Oral Suspension that previously has been shaken in the original container to ensure homogeneity, allowed to stand long enough for entrapped air to rise, and finally inverted carefully just prior to transfer to the volumetric flask. Transfer an accurately weighed quantity to Trisulfapyrimidines Oral Suspension, well-shaken and free from entrapped air, equivalent to about 100 mg of total sulfapyrimidines, to a 100-mL volumetric flask, add 25 mL of 0.1 *N* sodium hydroxide, and swirl for several minutes to dissolve the sulfapyrimidines. Dilute with water to volume, and mix. Filter the mixture, discarding the first several mL of the filtrate. Pipet 3 mL of the clear filtrate into a 25-mL volumetric flask, dilute with water to volume, and mix.

Chromatographic system (see *Chromatography* ⟨621⟩)—The liquid chromatograph is equipped with a 254-nm detector and a 3.9-mm × 30-cm column that contains packing L1. The flow rate is about 2 mL per minute. Chromatograph five replicate injections of the *Standard preparation*, and record the peak responses as directed under *Procedure:* the relative standard deviation is not more than 2.0%, and the resolution factors between sulfadiazine and sulfamerazine and between sulfamerazine and sulfamethazine are each not less than 3.0.

Procedure—Separately inject equal volumes (about 20 µL) of the *Standard preparation* and the *Assay preparation* into the

chromatograph, record the chromatograms, and measure the responses for the major peaks. The relative retention times for sulfadiazine, sulfamerazine, and sulfamethazine are approximately 0.6, 0.8, and 1.0, respectively. Calculate the quantity, in mg, of sulfadiazine in the portion of Oral Suspension taken by the formula:

$$0.833C(R_U/R_S),$$

in which C is the concentration, in µg per mL, of USP Sulfadiazine RS in the *Standard preparation*, and R_U and R_S are the peak responses obtained from the *Assay preparation* and the *Standard preparation*, respectively. Similarly measure the responses of the sulfamerazine and sulfamethazine peaks, and calculate the quantity, in mg, of each in the portion of Oral Suspension taken.

Trisulfapyrimidines Tablets

» Trisulfapyrimidines Tablets contain not less than 95.0 percent and not more than 105.0 percent of the labeled amount of each of the sulfapyrimidines, consisting of equal amounts of sulfadiazine ($C_{10}H_{10}N_4O_2S$), sulfamerazine ($C_{11}H_{12}N_4O_2S$), and sulfamethazine ($C_{12}H_{14}N_4O_2S$).

Packaging and storage—Preserve in well-closed containers.

Reference standards—*USP Sulfadiazine Reference Standard*—Dry at 105° for 2 hours before using. *USP Sulfamerazine Reference Standard*—Dry at 105° for 2 hours before using. *USP Sulfamethazine Reference Standard*—Dry at 105° for 2 hours before using.

Identification—The Tablets respond to the *Identification test* under *Trisulfapyrimidines Oral Suspension.*

Dissolution ⟨711⟩—

Medium: 0.1 *N* hydrochloric acid; 900 mL.

Apparatus 2: 50 rpm.

Time: 60 minutes.

Procedure—Determine the amount of total sulfapyrimidines dissolved from ultraviolet absorbances at the wavelength of maximum absorbance at about 254 nm of filtered portions of the solution under test, suitably diluted with 0.01 *N* sodium hydroxide, in comparison with a Standard solution having approximately equal, known, concentrations of USP Sulfadiazine RS, USP Sulfamerazine RS, and USP Sulfamethazine RS in the same medium.

Tolerances—Not less than 70% (*Q*) of the labeled amount of total sulfapyrimidines is dissolved in 60 minutes.

Uniformity of dosage units ⟨905⟩: meet the requirements.

Assay—

Mobile phase, Standard preparation, and *Chromatographic system*—Proceed as directed in the *Assay* under *Trisulfapyrimidines Oral Suspension.*

Assay preparation—Weigh and finely powder not less than 20 Trisulfapyrimidines Tablets. Transfer an accurately weighed portion of the powder, equivalent to about 250 mg of total sulfapyrimidines, to a 250-mL volumetric flask, add 50 mL of 0.1 *N* sodium hydroxide, swirl for several minutes to dissolve the sulfapyrimidines, dilute with water to volume, and mix. Filter the mixture, discarding the first several mL of the filtrate. Pipet 3 mL of the clear filtrate into a 25-mL volumetric flask, dilute with water to volume, and mix.

Procedure—Proceed as directed for *Procedure* in the *Assay* under *Trisulfapyrimidines Oral Suspension.* Calculate the quantity, in mg, of sulfadiazine in the portion of Tablets taken by the formula:

$$2.08C(R_U/R_S),$$

in which C is the concentration, in µg per mL, of USP Sulfadiazine RS in the *Standard preparation*, and R_U and R_S are the peak responses obtained from the *Assay preparation* and the *Standard preparation*, respectively. Similarly measure the responses of the sulfamerazine and sulfamethazine peaks, and calculate the quantity, in mg, of each in the portion of Tablets taken.

Trolamine—*see* Trolamine NF

Troleandomycin

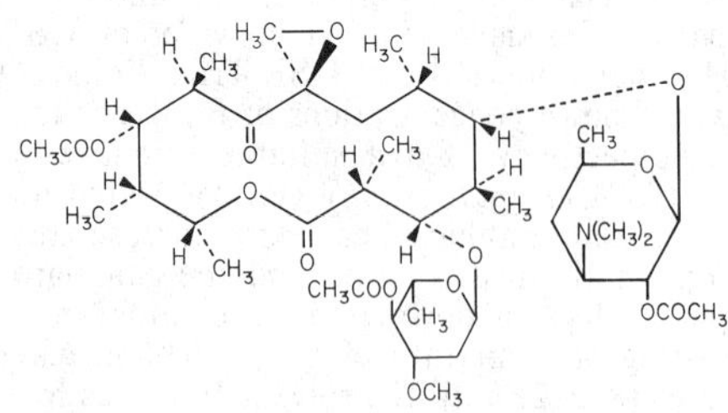

$C_{41}H_{67}NO_{15}$ 813.98
Oleandomycin, triacetate (ester).
Triacetyloleandomycin [*2751-09-9*].

» Troleandomycin contains the equivalent of not less than 750 µg of oleandomycin ($C_{35}H_{61}NO_{12}$) per mg.

Packaging and storage—Preserve in tight containers.

Reference standard—*USP Troleandomycin Reference Standard*—Dry in vacuum at a pressure not exceeding 5 mm of mercury at 60° for 3 hours before using.

Identification—

A: Dissolve about 10 mg in 5 mL of hydrochloric acid, and heat in a water bath: a greenish yellow color is produced.

B: Prepare a solution of it in methanol containing 10 mg per mL. On a suitable thin-layer chromatographic plate (see *Chromatography* ⟨621⟩), coated with a 0.25-mm layer of chromatographic cellulose, apply 5 µL of this test solution, 5 µL of a methanol solution of USP Troleandomycin RS containing 10 mg per mL (Standard solution), and 5 µL of a mixture of the two solutions (1:1). Allow the spots to dry, and develop the chromatogram in a solvent system of ammonium carbonate solution (1 in 100) until the solvent front has moved three-fourths of the length of the plate. Remove the plate from the developing chamber, mark the solvent front, and allow the plate to dry. Expose the plate to iodine vapors in a closed chamber for about 20 minutes, and locate the spots: the R_f value of the principal spot obtained from the test solution and from the mixture of the test solution and the Standard solution corresponds to that obtained from the Standard solution.

Crystallinity ⟨695⟩: meets the requirements.

pH ⟨791⟩: between 7.0 and 8.5, in a solution of alcohol and water (1:1) containing 100 mg per mL.

Loss on drying ⟨731⟩—Dry about 100 mg in vacuum at a pressure not exceeding 5 mm of mercury at 60° for 3 hours: it loses not more than 1.0% of its weight.

Residue on ignition ⟨281⟩: not more than 0.1%.

Acetyl content—Transfer about 30 mg, accurately weighed, to a three-neck, ground-glass jointed 50-mL flask fitted with a glass-stoppered funnel in the center neck, and a condenser and a gas inlet with a bubble counter in the other two necks. Add 2 mL of methanol to the flask to dissolve the Troleandomycin, and, slowly with swirling, add 1 mL of 2 *N* sodium hydroxide and a boiling chip. Allow nitrogen to flow into the flask at a rate of about 2 bubbles per second. Add about 5 mL of water to the funnel, and heat the flask. Allow to reflux for 30 minutes. Allow the assembly to cool slightly, and rinse the condenser with about 3 mL of water, collecting the rinsings in the flask. Change the condenser to the distillation position, and add water from the funnel to make a total of 5 mL added to the flask. Heat the flask, and collect about 5 mL of distillate in about 10 minutes. Discard the distillate, and allow the flask to cool slightly. Add 1 mL of 12 *N* sulfuric acid to the flask through the funnel. Heat the flask, and collect about 20 mL of distillate in about 20 minutes, adding more water from time to time through the funnel to maintain the volume in the flask at about 2 to 3 mL. As the distillation proceeds, treat the first fraction as follows. Boil gently for about 20 seconds, and add a few drops of barium chloride TS: no turbidity is produced. Add 1 drop of phenolphthalein TS, and titrate the solution with 0.015 *N* sodium hydroxide VS until a permanent pale pink color is produced. Collect a second, 10-mL, fraction, and treat it as directed for the first fraction, beginning with "Boil gently for about 20 seconds." If the second fraction consumes more than 0.1 mL of 0.015 *N* sodium hydroxide, collect a third, 10-mL, fraction, and treat as directed for the first fraction, beginning with "Boil gently for about 20 seconds." Each mL of 0.015 *N* sodium hydroxide is equivalent to 0.6458 mg of CH_3CO: between 15.3% and 16.0% is found.

Assay—Proceed with Troleandomycin as directed for troleandomycin under *Antibiotics—Microbial Assays* ⟨81⟩.

Troleandomycin Capsules

» Troleandomycin Capsules contain the equivalent of not less than 90.0 percent and not more than 120.0 percent of the labeled amount of oleandomycin ($C_{35}H_{61}NO_{12}$).

Packaging and storage—Preserve in tight containers.

Reference standard—*USP Troleandomycin Reference Standard*—Dry in vacuum at a pressure not exceeding 5 mm of mercury at 60° for 3 hours before using.

Identification—Suspend mixed Capsule contents, equivalent to about 200 mg of oleandomycin ($C_{35}H_{61}NO_{12}$), in 20 mL of chloroform, allow to settle, and filter. Using the filtrate so obtained as the test solution, proceed as directed in *Identification test B* under *Troleandomycin:* the specified result is obtained.

Loss on drying ⟨731⟩—Dry about 100 mg, accurately weighed, of Capsule contents in a capillary-stoppered bottle in vacuum at 60° for 3 hours: it loses not more than 5.0% of its weight.

Assay—Place not less than 5 Troleandomycin Capsules in a high-speed glass blender jar containing 500.0 mL of a mixture of isopropyl alcohol and water (4:1), and blend for 4 ± 1 minutes. Dilute this solution quantitatively with the same solvent to obtain a stock test solution containing the equivalent of about 1 mg of oleandomycin ($C_{35}H_{61}NO_{12}$) per mL. Proceed as directed for troleandomycin under *Antibiotics—Microbial Assays* ⟨81⟩, using an accurately measured volume of this stock test solution diluted quantitatively with water to yield a *Test Dilution* having a concentration assumed to be equal to the median dose level of the Standard.

Troleandomycin Oral Suspension

» Troleandomycin Oral Suspension contains the equivalent of not less than 90.0 percent and not more than 125.0 percent of the labeled amount of oleandomycin ($C_{35}H_{61}NO_{12}$). It contains one or more suitable buffers, colors, dispersants, and preservatives.

Packaging and storage—Preserve in tight containers, in a cool place.

Reference standard—*USP Troleandomycin Reference Standard*—Dry in vacuum at a pressure not exceeding 5 mm of mercury at 60° for 3 hours before using.

Identification—Shake a quantity of Oral Suspension, equivalent to about 200 mg of oleandomycin ($C_{35}H_{61}NO_{12}$), with 10 mL of chloroform in a separator, and allow to separate. Using the chloroform extract as the test solution, proceed as directed in *Identification test B* under *Troleandomycin:* the specified result is obtained.

Uniformity of dosage units ⟨905⟩—

FOR SUSPENSION PACKAGED IN SINGLE-UNIT CONTAINERS: meets the requirements.

pH ⟨791⟩: between 5.0 and 8.0.

Assay—Transfer an accurately measured volume of Troleandomycin Oral Suspension, previously freshly mixed and free from air bubbles, equivalent to about 100 mg of oleandomycin ($C_{35}H_{61}NO_{12}$), to a 100-mL volumetric flask, dissolve in a mixture of isopropyl alcohol and water (4:1), dilute with the same solvent to volume, and mix. Proceed as directed for troleandomycin under *Antibiotics—Microbial Assays* ⟨81⟩, using an accurately measured volume of this stock test solution diluted quantitatively with water to yield a *Test Dilution* having a concentration assumed to be equal to the median dose level of the Standard.

Tromethamine

$$\begin{array}{c} CH_2OH \\ | \\ HOCH_2CCH_2OH \\ | \\ NH_2 \end{array}$$

$C_4H_{11}NO_3$ 121.14

1,3-Propanediol, 2-amino-2-(hydroxymethyl)-.

2-Amino-2-(hydroxymethyl)-1,3-propanediol [*77-86-1*].

» Tromethamine contains not less than 99.0 percent and not more than 101.0 percent of $C_4H_{11}NO_3$, calculated on the dried basis.

Packaging and storage—Preserve in tight containers.

Reference standard—*USP Tromethamine Reference Standard*—Dry at 105° for 3 hours before using.

Identification—

A: The infrared absorption spectrum of a mineral oil dispersion of it, previously dried, exhibits maxima only at the same wavelengths as that of a similar preparation of USP Tromethamine RS.

B: To 4.5 mL of a saturated solution of salicylaldehyde add 0.5 mL of glacial acetic acid, and mix. Add 4.0 mL of a solution of Tromethamine (1 in 5), and mix: a yellow color is produced.

C: To 0.5 mL of a 4 in 10 solution of ceric ammonium nitrate in 2 *N* nitric acid add 3 mL of water and 0.5 mL of a solution of Tromethamine (1 in 5), and mix: the color changes from light yellow to orange.

Melting range ⟨741⟩: between 168° and 172°.

pH ⟨791⟩: between 10.0 and 11.5, in a solution (1 in 20).

Loss on drying ⟨731⟩—Dry it at 105° for 3 hours: it loses not more than 1.0% of its weight.

Residue on ignition ⟨281⟩: not more than 0.1%.

Heavy metals, *Method II* ⟨231⟩: 0.001%.

Assay—Dissolve about 250 mg of Tromethamine, accurately weighed, in 100 mL of water, add bromocresol purple TS, and titrate with 0.1 *N* hydrochloric acid VS to a yellow end-point. Each mL of 0.1 *N* hydrochloric acid is equivalent to 12.11 mg of $C_4H_{11}NO_3$.

Tromethamine for Injection

» Tromethamine for Injection is a sterile, lyophilized mixture of tromethamine with Potassium Chloride and Sodium Chloride. It contains not less than 93.0 percent and not more than 107.0 percent of the labeled amount of tromethamine ($C_4H_{11}NO_3$), and not less than 90.0 percent and not more than 110.0 percent of the labeled amounts of potassium chloride (KCl) and of sodium chloride (NaCl).

Packaging and storage—Preserve in *Containers for Sterile Solids* as described under *Injections* ⟨1⟩.

Reference standard—*USP Tromethamine Reference Standard*—Dry at 105° for 3 hours before using.

Constituted solution—At the time of use, the solution prepared from Tromethamine for Injection meets the requirements for *Constituted Solutions* under *Injections* ⟨1⟩.

Identification—

A: The infrared absorption spectrum of a mineral oil dispersion of it exhibits maxima only at the same wavelengths as that of a similar preparation of USP Tromethamine RS.

B: A solution, prepared as directed in the labeling, responds to the tests for *Chloride* ⟨191⟩, for *Sodium* ⟨191⟩, and for *Potassium* ⟨191⟩.

Pyrogen—When diluted with Water for Injection to a concentration of 50 mg of tromethamine per mL, it meets the requirements of the *Pyrogen Test* ⟨151⟩, the test dose being 10 mL per kg, injected slowly.

pH ⟨791⟩: between 10.0 and 11.5, in a solution prepared as directed in the labeling.

Water, *Method I* ⟨921⟩—Add 5 mL of glacial acetic acid prior to the titration: the content is not more than 1.0%.

Particulate matter ⟨788⟩: meets the requirements under *Small-volume Injections.*

Potassium chloride content—

Standard solutions—Prepare five standard solutions (*1, 2, 3, 4*, and *5*) each containing 0.60 mEq of sodium (35 mg of sodium chloride) per liter, and to the solutions add, respectively, 0-, 2-, 4-, 6-, and 8-mg supplements of potassium, in the form of the chloride, per liter. If necessary, because of changes in the sensitivity of the photometer, vary the levels of concentration of the potassium, keeping the ratios between solutions approximately as given.

Standard graph—Set a suitable flame photometer for maximum emittance at a wavelength of 766 nm to 767 nm. (The exact wavelength setting will vary slightly with the instrument.) Adjust the instrument to zero emittance with solution *1*. Then adjust the instrument to 100% emittance with solution *5*. Read the percentage emittance of solutions *2, 3*, and *4*. Plot the observed emittance of solutions *2, 3, 4*, and *5* as the ordinate and the concentration, in µg per mL, of potassium as the abscissa on arithmetic coordinate paper.

Procedure—Dissolve the entire contents of 1 container of Tromethamine for Injection in sufficient water, and dilute quantitatively and stepwise with water to obtain a solution containing about 4 µg of potassium per mL, or a quantity corresponding to the concentration of the *Standard solutions*. Adjust the instrument to zero emittance with solution *1* and to 100% emittance with solution *5*. Read the percentage emittance of the test solution. By reference to the *Standard graph*, determine the concentration, in µg per mL, of potassium in the test solution, apply the dilution factor, and calculate the quantity, in mg, of potassium in the container of Tromethamine for Injection. Each mg of potassium is equivalent to 1.907 mg of potassium chloride (KCl).

Sodium chloride content—Proceed as directed under *Potassium chloride content*, with the following modifications: (1) Prepare the *Standard solutions* to contain 0, 2, 4, 6, and 8 mg of sodium, in the form of the chloride, per 1000 mL, without added potassium; (2) prepare the *Standard graph* with the flame photometer set at 588 nm to 589 nm; and (3) under *Procedure* read "sodium" for "potassium" throughout. Each mg of sodium is equivalent to 2.542 mg of sodium chloride (NaCl).

Other requirements—It meets the requirements for *Sterility Tests* ⟨71⟩, *Uniformity of Dosage Units* ⟨905⟩, and *Labeling* under *Injections* ⟨1⟩.

Assay for tromethamine—Dissolve the entire contents of 1 container of Tromethamine for Injection in sufficient water, diluting with water to an accurately measured volume to obtain a solution containing about 36 mg of tromethamine per mL. Transfer to a beaker an accurately measured volume of the solution, equivalent to about 180 mg of tromethamine, dilute with water to about 100 mL, add bromocresol purple TS, and titrate with 0.1 *N* hydrochloric acid VS to a yellow end-point. Each mL of 0.1 *N* hydrochloric acid is equivalent to 12.11 mg of $C_4H_{11}NO_3$.

Tropicamide

$C_{17}H_{20}N_2O_2$ 284.36

Benzeneacetamide, *N*-ethyl-α-(hydroxymethyl)-*N*-(4-pyridinylmethyl)-.

N-Ethyl-2-phenyl-*N*-(4-pyridylmethyl)hydracrylamide [*1508-75-4*].

» Tropicamide contains not less than 99.0 percent and not more than 101.0 percent of $C_{17}H_{20}N_2O_2$, calculated on the dried basis.

Packaging and storage—Preserve in tight, light-resistant containers.

Reference standard—*USP Tropicamide Reference Standard*—Dry in vacuum over phosphorus pentoxide at 80° for 4 hours before using.

Identification—

A: The infrared absorption spectrum of a potassium bromide dispersion of it exhibits maxima only at the same wavelengths as that of a similar preparation of USP Tropicamide RS.

B: The ultraviolet absorption spectrum of a 1 in 40,000 solution in 3 *N* hydrochloric acid exhibits maxima and minima at the same wavelengths as that of a similar solution of USP Tropicamide RS, concomitantly measured.

Melting range, *Class I* ⟨741⟩: between 96° and 100°.

Loss on drying ⟨731⟩—Dry about 500 mg, accurately weighed, in vacuum over phosphorus pentoxide at 80° for 4 hours: it loses not more than 0.5% of its weight.

Heavy metals, *Method II* ⟨231⟩: 0.002%.

Assay—Dissolve about 750 mg of Tropicamide, accurately weighed, in 80 mL of glacial acetic acid, add 4 drops of crystal violet TS, and titrate with 0.1 *N* perchloric acid VS to a blue-green end-point. Perform a blank determination, and make any necessary correction. Each mL of 0.1 *N* perchloric acid is equivalent to 28.44 mg of $C_{17}H_{20}N_2O_2$.

Tropicamide Ophthalmic Solution

» Tropicamide Ophthalmic Solution is a sterile, aqueous solution of Tropicamide. It contains not less than 95.0 percent and not more than 105.0 percent of the labeled amount of $C_{17}H_{20}N_2O_2$. It contains a suitable antimicrobial agent, and may contain suitable substances to increase its viscosity.

Packaging and storage—Preserve in tight containers, and avoid freezing.

Reference standard—*USP Tropicamide Reference Standard*—Dry in vacuum over phosphorus pentoxide at 80° for 4 hours before using.

Identification—

A: Extract 10 mL of it with 25 mL of chloroform, filter the chloroform extract through dry, folded filter paper, and evaporate the filtrate to dryness: the residue so obtained responds to *Identification test A* under *Tropicamide.*

B: The ultraviolet absorption spectrum of the solution employed for measurement of absorbance in the *Assay* exhibits maxima and minima at the same wavelengths as that of a similar solution of USP Tropicamide RS, concomitantly measured.

Sterility—It meets the requirements under *Sterility Tests* ⟨71⟩.

pH ⟨791⟩: between 4.0 and 5.8.

Assay—Transfer an accurately measured volume of Tropicamide Ophthalmic Solution, equivalent to about 30 mg of tropicamide, to a 100-mL volumetric flask, add water to volume, and mix. Transfer 10.0 mL of this solution to a separator, add 2 mL of sodium carbonate solution (1 in 10), extract with four 20-mL portions of chloroform, and combine the extracts in a second separator. Wash the combined extracts with a 25-mL portion of pH 6.5 phosphate buffer (see *Buffer Solutions* in the section, *Reagents, Indicators, and Solutions*), and transfer to another separator. Wash the aqueous layer with 10 mL of chloroform, and add it to the extracts. Extract the chloroform solution with four 20-mL portions of dilute sulfuric acid (1 in 6), combine the acid extracts in a 100-mL volumetric flask, and add the dilute acid to volume. Dissolve an accurately weighed quantity of USP Tropicamide RS in dilute sulfuric acid (1 in 6), and dilute quantitatively and stepwise with the same solvent to obtain a Standard solution having a known concentration of about 30 µg per mL. Concomitantly determine the absorbances of both solutions in 1-cm cells at the wavelength of maximum absorbance at about 253 nm, with a suitable spectrophotometer, using dilute sulfuric acid (1 in 6) as the blank. Calculate the quantity, in mg, of $C_{17}H_{20}N_2O_2$ in each mL of the Ophthalmic Solution taken by the formula:

$$(C/V)(A_U/A_S),$$

in which C is the concentration, in µg per mL, of USP Tropicamide RS in the Standard solution, V is the volume, in mL, of Ophthalmic Solution taken, and A_U and A_S are the absorbances of the solution from the Ophthalmic Solution and the Standard solution, respectively.

Crystallized Trypsin

» Crystallized Trypsin is a proteolytic enzyme crystallized from an extract of the pancreas gland of the ox, *Bos taurus* Linné (Fam. Bovidae). When assayed as directed herein, it contains not less than 2500 USP Trypsin Units in each mg, calculated on the dried basis, and not less than 90.0 percent and not more than 110.0 percent of the labeled potency.

NOTE—Determine the suitability of the substrates and check the adjustment of the spectrophotometer by performing the *Assay* using USP Crystallized Trypsin Reference Standard.

Packaging and storage—Preserve in tight containers, and avoid exposure to excessive heat.

Reference standard—*USP Crystallized Trypsin Reference Standard*—Keep container tightly closed, and store in a refrigerator. Allow container to reach room temperature before opening, and do not dry before using.

Solubility test—An amount, equivalent to 500,000 USP Trypsin Units, is soluble in 10 mL of water and in 10 mL of saline TS.

Microbial limit—It meets the requirements of the test for absence of *Pseudomonas aeruginosa* and *Salmonella* species and *Staphylococcus aureus* under *Microbial Limit Tests* ⟨61⟩.

Loss on drying ⟨731⟩—Dry it in vacuum at 60° for 4 hours: it loses not more than 5.0% of its weight.

Residue on ignition ⟨281⟩: not more than 2.5%.

Chymotrypsin—

0.067 M Phosphate buffer, pH 7.0—Dissolve 4.54 g of monobasic potassium phosphate in water to make 500 mL of solution. Dissolve 4.73 g of anhydrous dibasic sodium phosphate in water to make 500 mL of solution. Mix 38.9 mL of the monobasic potassium phosphate solution with 61.1 mL of dibasic sodium phosphate solution. Adjust dropwise, if necessary, with dibasic sodium phosphate solution to a pH of 7.0.

Substrate solution—Dissolve 23.7 mg of *N*-acetyl-L-tyrosine ethyl ester, suitable for use in determining chymotrypsin, in about 50 mL of *0.067 M Phosphate buffer, pH 7.0* with warming. When cool, dilute with additional pH 7.0 buffer to 100 mL. (*Substrate solution* may be stored in the frozen state and used after thawing; it is important, however, to freeze immediately after preparation.)

Crystallized Trypsin solution—Dissolve a sufficient quantity of Crystallized Trypsin, accurately weighed, in 0.0010 *N* hydrochloric acid to obtain a solution containing 650 USP Trypsin Units per mL.

Procedure—Conduct the test in a suitable spectrophotometer equipped to maintain a temperature of 25 ± 0.1° in the cell compartment. Determine the temperature in the reaction cell before and after the measurement of absorbance to ensure that the temperature does not change by more than 0.5°. Pipet 200 µL of 0.0010 *N* hydrochloric acid and 3.0 mL of the *Substrate solution* into a 1-cm cell. Place this cell in the spectrophotometer, and adjust the instrument so that the absorbance reads 0.200 at 237 nm. Pipet 200 µL of *Crystallized Trypsin solution* into another 1-cm cell, add 3.0 mL of the *Substrate solution*, and place the cell in the spectrophotometer. [NOTE—This order of addition is to be followed.] At the time the *Substrate solution* is added, start a stopwatch, and read the absorbance at 30-second intervals for not less than 5 minutes. Repeat the procedure on the same dilution at least once. Absolute absorbance values are of less importance than the constancy of the rate of change of absorbance. If the rate of change does not remain constant for at least 3 minutes, repeat the run, and if necessary, use a lower concentration. The duplicate run at the same dilution should match the first run in rate of absorbance change. Determine the average absorbance change per minute, using only the values within the 3-minute portion of the curve where the rate of absorbance is constant. Plot a curve of absorbance against time. One USP Chymotrypsin Unit is the activity causing a change in absorbance of 0.0075 per minute under the conditions specified in this test. Calculate the number of USP Chymotrypsin Units per mg of Crystallized Trypsin by the formula:

$$(A_2 - A_1)/(0.0075TW),$$

in which A_2 is the absorbance straight-line initial reading, A_1 is the absorbance straight-line final reading, T is the elapsed time, in minutes, between the initial and final readings, and W is the weight, in mg, of Crystallized Trypsin in the volume of solution used in determining the absorbance. Not more than 50 USP Chymotrypsin Units per 2500 USP Trypsin Units is found, indicating the presence of not more than approximately 5% of chymotrypsin.

Assay—

0.067 M Phosphate buffer, pH 7.6—Dissolve 4.54 g of monobasic potassium phosphate in water to make 500 mL of solution. Dissolve 4.73 g of anhydrous dibasic sodium phosphate in water to make 500 mL of solution. Mix 13 mL of the monobasic potassium phosphate solution with 87 mL of the anhydrous dibasic sodium phosphate solution.

Substrate solution—Dissolve 85.7 mg of *N*-benzoyl-L-arginine ethyl ester hydrochloride, suitable for use in assaying Crystallized Trypsin (see NOTE), in water to make 100 mL. Dilute 10 mL of this solution with *0.067 M Phosphate buffer, pH 7.6* to 100 mL. Determine the absorbance of this solution, in a 1-cm cell, at 253 nm, in a suitable spectrophotometer equipped with thermospacers to maintain a temperature of 25 ± 0.1°, using water as the blank. By the addition of *0.067 M Phosphate buffer, pH 7.6*, or of the *Substrate solution* before dilution, adjust the absorbance so that it measures not less than 0.575 and not more than 0.585. Use this *Substrate solution* within 2 hours.

Crystallized Trypsin solution—Dissolve a sufficient quantity of Crystallized Trypsin, accurately weighed, in 0.0010 *N* hydrochloric acid to obtain a solution containing about 50 to 60 USP Trypsin Units per mL.

Procedure—Pipet 200 µL of 0.0010 *N* hydrochloric acid and 3.0 mL of the *Substrate solution* into a 1-cm cell. Place this cell in a spectrophotometer, and adjust the instrument so that the absorbance reads 0.050 at 253 nm. Pipet 200 µL of *Crystallized Trypsin solution*, containing 10 to 12 USP Trypsin Units, into another 1-cm cell, add 3.0 mL of *Substrate solution*, and place the cell in the spectrophotometer. At the time the *Substrate solution* is added, start a stopwatch, and read the absorbance at 30-second intervals for 5 minutes. Repeat the procedure on the same dilution at least once. Plot a curve of absorbance against time, and use only those values that form a straight line to determine the activity of the Crystallized Trypsin. If the rate of change does not remain constant for at least 3 minutes, repeat the run, and if necessary, use a lower concentration. One USP Trypsin Unit is the activity causing a change in absorbance of

0.003 per minute under the conditions specified in this *Assay*. Calculate the number of USP Trypsin Units per mg by the formula:

$$(A_1 - A_2)/(0.003TW),$$

in which A_1 is the absorbance straight-line final reading, A_2 is the absorbance straight-line initial reading, T is the elapsed time, in minutes, between the initial and final readings, and W is the weight, in mg, of Crystallized Trypsin in the volume of solution used in determining the absorbances.

Crystallized Trypsin for Inhalation Aerosol

» Crystallized Trypsin for Inhalation Aerosol is prepared by cryodesiccation. It contains not less than 90.0 percent and not more than 110.0 percent of the labeled potency of trypsin.

Packaging and storage—Preserve in single-dose containers, preferably of Type I glass, and avoid exposure to excessive heat.

Reference standard—*USP Crystallized Trypsin Reference Standard*—Keep container tightly closed, and store in a refrigerator. Allow container to reach room temperature before opening, and do not dry before using.

Identification—Prepare a substrate solution as follows. Transfer 85.7 mg of *N*-benzoyl-L-arginine ethyl ester hydrochloride, suitable for use in assaying trypsin crystallized, to a 100-mL volumetric flask. Add 20 mL of *0.067 M Phosphate buffer, pH 7.6*, prepared as directed in the *Assay* under *Crystallized Trypsin*, add 1 mL of methyl red–methylene blue TS, and dilute with water to volume. Mix 0.01 mL of this solution with 0.01 mL of a solution of Crystallized Trypsin for Inhalation Aerosol containing 250,000 USP Units in 6 mL of saline on a spot plate: a purple color is produced (*distinction from chymotrypsin, which produces no purple color within 3 minutes*).

Solubility test—Crystallized Trypsin for Inhalation Aerosol containing 500,000 USP Trypsin Units is soluble in 10 mL of water and in 10 mL of saline TS.

Assay—

0.067 M Phosphate buffer, pH 7.6 and *Substrate solution*—Prepare as directed in the *Assay* under *Crystallized Trypsin.*

Crystallized Trypsin solution—Dissolve the contents of one vial of Crystallized Trypsin for Inhalation Aerosol in 10.0 mL of 0.0010 *N* hydrochloric acid. Dilute this solution quantitatively with the same dilute acid to obtain a solution containing 50 to 60 USP Trypsin Units per mL.

Procedure—Proceed with Crystallized Trypsin for Inhalation Aerosol as directed for *Procedure* in the *Assay* under *Crystallized Trypsin.*

Tryptophan

H N CH₂—C—COOH H NH₂

$C_{11}H_{12}N_2O_2$ 204.23

L-Tryptophan.

L-Tryptophan [*73-22-3*].

» Tryptophan contains not less than 98.5 percent and not more than 101.5 percent of $C_{11}H_{12}N_2O_2$, as L-tryptophan, calculated on the dried basis.

Packaging and storage—Preserve in well-closed containers.

Reference standard—*USP L-Tryptophan Reference Standard*—Dry at 105° for 3 hours before using.

Identification—The infrared absorption spectrum of a potassium bromide dispersion of it, previously dried, exhibits maxima only at the same wavelengths as that of a similar preparation of USP L-Tryptophan RS.

Specific rotation ⟨781⟩: between −29.4° and −32.8°, calculated on the dried basis, determined in a solution containing 100 mg in each 10 mL. (Heat gently to dissolve, if necessary.)

pH ⟨791⟩: between 5.5 and 7.0, in a solution (1 in 100).

Loss on drying ⟨731⟩—Dry it at 105° for 3 hours: it loses not more than 0.3% of its weight.

Residue on ignition ⟨281⟩: not more than 0.1%.

Chloride ⟨221⟩—A 0.73-g portion shows no more chloride than corresponds to 0.50 mL of 0.020 *N* hydrochloric acid (0.05%).

Sulfate ⟨221⟩—A 0.33-g portion shows no more sulfate than corresponds to 0.10 mL of 0.020 *N* sulfuric acid (0.03%).

Arsenic ⟨211⟩: 1.5 ppm.

Iron ⟨241⟩: 0.003%.

Heavy metals, *Method II* ⟨231⟩: 0.0015%.

Assay—Transfer about 200 mg of Tryptophan, accurately weighed, to a 125-mL flask, dissolve in a mixture of 3 mL of formic acid and 50 mL of glacial acetic acid, and titrate with 0.1 *N* perchloric acid VS, determining the end-point potentiometrically. Perform a blank determination, and make any necessary correction. Each mL of 0.1 *N* perchloric acid is equivalent to 20.42 mg of $C_{11}H_{12}N_2O_2$.

Tuaminoheptane

$CH_3(CH_2)_4CHNH_2$ | CH_3

$C_7H_{17}N$ 115.22

2-Heptanamine.

1-Methylhexylamine [*123-82-0*].

» Tuaminoheptane contains not less than 99.0 percent and not more than 100.5 percent of $C_7H_{17}N$.

Packaging and storage—Preserve in tight containers, and store in a cool place.

Reference standard—*USP Tuaminoheptane Sulfate Reference Standard*—Dry at 105° to constant weight before using.

Identification—

Standard preparation—Dissolve 150 mg of USP Tuaminoheptane Sulfate RS in 5 mL of water. Render the solution alkaline to litmus with 1 *N* sodium hydroxide, and extract the solution with 2 mL of chloroform. Filter the chloroform extract through a layer of 2 g of granular anhydrous sodium sulfate supported on glass wool.

Procedure—The infrared absorption spectrum, determined in a 0.1-mm cell, of a 1 in 20 solution of Tuaminoheptane in chloroform exhibits maxima only at the same wavelengths as that of the *Standard preparation.*

Specific gravity ⟨841⟩: between 0.760 and 0.763.

Refractive index ⟨831⟩: between 1.415 and 1.417.

Nonvolatile residue—Weigh accurately about 1 g in a tared, small-diameter weighing bottle. Evaporate *in a well-ventilated hood*, on a steam bath, with the aid of a stream of nitrogen, to dryness, and dry the residue at 105° for 2 hours: the weight of the residue does not exceed 2 mg (0.2%).

Assay—Weigh accurately about 1 g of Tuaminoheptane in a tared, glass-stoppered flask containing 25 mL of neutralized alcohol. Add methyl red TS, and titrate with 0.5 *N* hydrochloric acid VS. Each mL of 0.5 *N* hydrochloric acid is equivalent to 57.61 mg of $C_7H_{17}N$.

Tuaminoheptane Inhalant

» Tuaminoheptane Inhalant consists of cylindrical rolls of suitable fibrous material impregnated with Tuaminoheptane (as the carbonate), usually aromatized, and contained in a suitable inhaler. The inhaler contains not less than 90.0 percent and not more than 125.0 percent of the labeled amount of $C_7H_{17}N$.

Packaging and storage—Preserve in tight containers (inhalers), and avoid exposure to excessive heat.

Identification—Dismantle 1 inhaler, and express a portion of the Inhalant from the fibrous roll. Inject a portion of this liquid into a suitable gas chromatograph (see *Chromatography* ⟨621⟩), equipped with a thermal conductivity detector. Under typical conditions, the instrument contains a 2-m × 4-mm stainless steel column packed with 15 percent polyethylene glycol 4000 on sodium hydroxide-treated (5 percent sodium hydroxide) chromatographic siliceous earth. The column is maintained at about 80°, the injection port and detector block are maintained about 10° above the temperature of the column, and helium is used as the carrier gas. The major peak of the specimen chromatogram compares qualitatively to that of the chromatogram obtained from Tuaminoheptane (see monograph), similarly measured.

Assay—Dismantle 1 inhaler containing Tuaminoheptane Inhalant, and immediately remove and slit the fibrous roll with a clean scalpel while holding the roll with forceps on a small watch glass. Immediately transfer the slit roll to a 250-mL beaker containing 100 mL of glacial acetic acid, rinse the scalpel, forceps, watch glass, and the inside of the inhaler shell with chloroform, and add the chloroform rinsings to the beaker. Carefully pull the fibrous material apart with two glass stirring rods, rinse the stirring rods with chloroform, and add the rinsings to the solution in the beaker. Add 2 drops of crystal violet TS, and titrate with 0.2 *N* perchloric acid VS to an emerald-green end-point. Perform a blank determination, and make any necessary correction. Each mL of 0.2 *N* perchloric acid is equivalent to 23.04 mg of $C_7H_{17}N$.

Tuberculin

» Tuberculin conforms to the regulations of the federal Food and Drug Administration concerning biologics (650.10 to 650.15) (see *Biologics* ⟨1041⟩). It is a sterile solution derived from the concentrated, soluble products of growth of the tubercle bacillus (*Mycobacterium tuberculosis* or *Mycobacterium bovis*) prepared in a special medium. It is provided either as Old Tuberculin, a culture filtrate adjusted to the standard potency based on the U.S. Standard Tuberculin, Old, by addition of glycerin and isotonic sodium chloride solution, or as Purified Protein Derivative (PPD), a further purified protein fraction standardized with the U.S. Standard Tuberculin, Purified Protein Derivative. It has a potency, tested by comparison with the corresponding U.S. Standard Tuberculin, on intradermal injection of sensitized guinea pigs, of between 80 percent and 120 percent of that stated on the label. It is free from viable *Mycobacteria* as shown by injection into guinea pigs.

Packaging and storage—Preserve at a temperature between 2° and 8°. Multiple-puncture devices may be stored at a temperature not exceeding 30°.

Expiration date—The expiration date of concentrated Old Tuberculin containing 50 percent of glycerin is not later than 5 years after date of issue from manufacturer's cold storage (5°, 1 year; or 0°, 2 years). The expiration date of diluted Old Tuberculin is not later than 1 year after date of issue from manufacturer's cold storage (5°, 1 year; or 0°, 2 years). The expiration date of concentrated PPD containing 50 percent of glycerin is not later than 2 years after date of issue from manufacturer's cold storage (5°, 1 year). The expiration date of diluted PPD is not later than 1 year after date of issue by the manufacturer. The expiration date of Old Tuberculin and PPD dried on multiple-puncture devices is not later than 2 years after date of issue from manufacturer's cold storage (30°, 1 year), provided the recommended storage is at a temperature not exceeding 30°.

Tubocurarine Chloride

$Cl^- \cdot HCl \cdot 5H_2O$

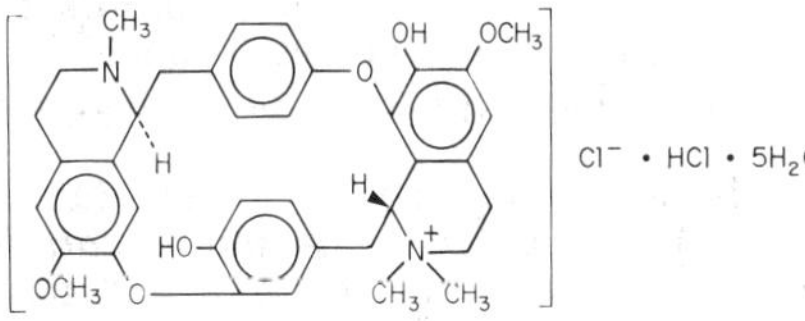

$C_{37}H_{41}ClN_2O_6.HCl.5H_2O$ 771.73

Tubocuraranium, 7′,12′-dihydroxy-6,6′-dimethoxy-2,2′,2′-trimethyl-, chloride, hydrochloride, pentahydrate.

(+)-Tubocurarine chloride hydrochloride pentahydrate [*41354-45-4*].

Anhydrous 681.65 [*57-94-3*].

» Tubocurarine Chloride contains not less than 95.0 percent and not more than 105.0 percent of $C_{37}H_{41}ClN_2O_6.HCl$, calculated on the anhydrous basis.

Packaging and storage—Preserve in tight containers.

Reference standard—*USP Tubocurarine Chloride Reference Standard*—Use as directed on the label.

Clarity of alcohol solution—A solution of 100 mg in 10 mL of alcohol is clear.

Identification—

A: The infrared absorption spectrum of a potassium bromide dispersion of it exhibits maxima only at the same wavelengths as that of a similar preparation of USP Tubocurarine Chloride RS.

B: The chromatogram of the *Assay preparation* obtained as directed in the *Assay* exhibits a major peak, the retention time of which corresponds to that exhibited in the chromatogram of the *Standard preparation.*

C: A solution (1 in 100) responds to the tests for *Chloride* ⟨191⟩.

Specific rotation ⟨781⟩: between +210° and +224°, calculated on the anhydrous basis, determined in a solution containing 100 mg in each 10 mL, which has been allowed to stand for 3 hours.

Water, *Method I* ⟨921⟩: not more than 12.0%.

Residue on ignition ⟨281⟩: not more than 0.25%.

Related substances—In the chromatogram obtained from the *Assay preparation* in the *Assay*, the sum of the responses of any peaks detected, other than the peak due to tubocurarine, is not more than 5.0% of the total of all peak responses.

Chloride content—Dissolve about 300 mg, accurately weighed, in 5 mL of water, warming slightly to effect solution. Add 5 mL of glacial acetic acid and 50 mL of methanol, and cool to room temperature. Add 1 drop of eosin Y TS, and titrate with 0.1 *N* silver nitrate VS. Each mL of 0.1 *N* silver nitrate is equivalent to 3.545 mg of Cl. Not less than 9.9% and not more than 10.7% of Cl is found, calculated on the anhydrous basis.

Assay—

Mobile phase—Mix 3 volumes of acetonitrile and 2 volumes of methanol, and cool to room temperature. To 270 mL of this solution in a 1-liter graduated cylinder add 20.0 mL of 25% tetramethylammonium hydroxide solution in methanol, and add water to make 1 liter. Adjust with phosphoric acid to a pH of 4.0, filter, and degas.

Standard preparation—Dissolve an accurately weighed quantity of USP Tubocurarine Chloride RS in *Mobile phase* to obtain a solution having a known concentration of about 0.3 mg per mL.

Assay preparation—Transfer 30 mg of Tubocurarine Chloride, accurately weighed, to a 100-mL volumetric flask. Dissolve in *Mobile phase*, dilute with *Mobile phase* to volume, and mix.

System suitability preparation—Dissolve suitable quantities of metocurine iodide, tubocurarine chloride, and phenol in *Mobile phase* to obtain a solution containing about 0.20 mg, 0.30 mg, and 0.50 mg per mL, respectively.

Chromatographic system (see *Chromatography* ⟨621⟩)—The liquid chromatograph is equipped with a 220-nm detector, and a 4-mm × 25-cm column that contains packing L1. The flow rate is about 1 mL per minute. Chromatograph the *System suitability preparation*, and record the peak responses as directed under *Procedure:* the resolution, *R*, between any two major peaks is not less than 2.0 and the tailing factor, *T*, for tubocurarine chloride is not more than 2.0. The relative standard deviation for replicate injections of the *Standard preparation* is not more than 2.0%. The relative retention times are about 0.35, 0.50, and 1.0 for tubocurarine chloride, phenol, and metocurine iodide, respectively.

Procedure—Separately inject equal volumes (about 10 µL) of the *Standard preparation* and the *Assay preparation* into the chromatograph, record the chromatograms, and measure the responses for the major peaks. Calculate the quantity, in mg, of $C_{37}H_{41}ClN_2O_6 . HCl$ in the portion of Tubocurarine Chloride taken by the formula:

$$100C(r_U / r_S),$$

in which C is the concentration, in mg per mL, of USP Tubocurarine Chloride RS in the *Standard preparation*, and r_U and r_S are the peak responses obtained from the *Assay preparation* and the *Standard preparation*, respectively.

Tubocurarine Chloride Injection

» Tubocurarine Chloride Injection is a sterile solution of Tubocurarine Chloride in Water for Injection. It contains not less than 93.0 percent and not more than 107.0 percent of the labeled amount of $C_{37}H_{41}ClN_2O_6 . HCl . 5H_2O$.

Packaging and storage—Preserve in single-dose or in multiple-dose containers.

Reference standard—*USP Tubocurarine Chloride Reference Standard*—Use as directed on label.

Identification—

A: It responds to *Identification test C* under *Tubocurarine Chloride*.

B: The chromatogram of the *Assay preparation* obtained as directed in the *Assay* exhibits a major peak, the retention time of which corresponds to that exhibited in the chromatogram of the *Standard preparation*.

Angular rotation ⟨781⟩: between +0.32° and +0.48° for each mg of tubocurarine chloride per mL claimed on the label, determined in a suitable polarimeter tube and the observed reading being multiplied by the factor 200/*L*, in which *L* is the length, in mm, of the tube.

pH ⟨791⟩: between 2.5 and 5.0.

Other requirements—It meets the requirements under *Injections* ⟨1⟩.

Assay—

Mobile phase, Standard preparation, System suitability preparation, and *Chromatographic system*—Prepare as directed in the *Assay* under *Tubocurarine Chloride*.

Assay preparation—Transfer an accurately measured volume of Tubocurarine Chloride Injection, equivalent to about 15 mg of tubocurarine chloride, to a 50-mL volumetric flask, dilute with *Mobile phase* to volume, and mix.

Procedure—Separately inject equal volumes (about 10 µL) of the *Standard preparation* and the *Assay preparation* into the chromatograph, record the chromatograms, and measure the responses for the major peaks. Calculate the quantity, in mg, of $C_{37}H_{41}ClN_2O_6 . HCl . 5H_2O$ in each mL of the Injection taken by the formula:

$$50C(r_U / r_S),$$

in which C is the concentration, in mg per mL, of USP Tubocurarine Chloride RS in the *Standard preparation*, and r_U and r_S are the peak responses obtained from the *Assay preparation* and the *Standard preparation*, respectively.

Tyloxapol

[*R* is $CH_2CH_2O(CH_2CH_2O)_mCH_2CH_2OH$; *m* is 6 to 8; *n* is not more than 5]

Phenol, 4-(1,1,3,3-tetramethylbutyl)-, polymer with formaldehyde and oxirane.

p-(1,1,3,3-Tetramethylbutyl)phenol polymer with ethylene oxide and formaldehyde [*25301-02-4*].

» Tyloxapol is a nonionic liquid polymer of the alkyl aryl polyether alcohol type.

Note—Precautions should be exercised to prevent contact of Tyloxapol with metals.

Packaging and storage—Preserve in tight containers.

Reference standard—*USP Tyloxapol Reference Standard*—Do not dry before using.

Identification—The infrared absorption spectrum of a thin film of it formed between two sodium chloride plates exhibits maxima only at the same wavelengths as that of a similar preparation of USP Tyloxapol RS.

Cloud point—Transfer 1.0 g of it, previously mixed, to a 150-mL beaker. Add 100.0 mL of water, and mix until solution is effected. Warm the solution while mixing: transient turbidity may be observed as the solution is warmed. Determine the temperature at which the mixture becomes completely turbid: the cloud point is between 92° and 97°.

pH ⟨791⟩: between 4.0 and 7.0, in a solution (1 in 20).

Residue on ignition ⟨281⟩: not more than 1.0%.

Free phenol—To 10 mL of a solution (1 in 100) add 1 mL of bromine TS, and mix: no cloudiness or precipitation is observed immediately.

Limit of anionic detergents—Mix 20 mL of a solution (1 in 100) with 30 mL of water in a 125-mL separator. In a second 125-mL separator mix 50 mL of water and 1 mL of a solution of sodium lauryl sulfate containing 150 µg per mL. To both separators add 2 drops of 3 *N* hydrochloric acid, 1 drop of methylene blue solution (1 in 25), and 25 mL of chloroform. Shake both separators gently for 2 minutes, allow to stand for 10 minutes, and transfer the chloroform layers to individual separators. Wash the chloroform extracts with separate 25-mL portions of water, transfer the chloroform solutions to matched 50-mL color-comparison tubes, and view downward over a white surface: the chloroform solution from the Tyloxapol preparation is not darker than that from the sodium lauryl sulfate preparation, corresponding to not more than 0.075% of anionic detergents (as sodium lauryl sulfate).

Absence of cationic detergents—Place 10 mL of a solution (1 in 100) in a glass-stoppered, 50-mL graduated cylinder, and make distinctly alkaline to litmus with sodium carbonate TS (about 1 mL). Add 4 mL of aqueous bromophenol blue solution (1 in 2500), mix, and add 10 mL of a 1 in 10 solvent mixture of ethylene dichloride in toluene. Shake gently, and allow the layers to separate: no blue color is observed in the organic solvent layer.

Limit of formaldehyde—

Standard preparations—Weigh 2.7 g of formaldehyde solution into a 100-mL volumetric flask, dilute with water to volume, and mix. Transfer 1.0 mL of this solution to a second 100-mL volumetric flask, dilute with water to volume, and mix. Transfer 10.0 mL of this second solution to a third 100-mL volumetric flask, dilute with water to volume, and mix. Transfer 750 µL of this solution to a 25-mL volumetric flask containing 5 mL of a solution of isopropyl alcohol (4 in 10).

Test preparation—Transfer 2.00 g of Tyloxapol to a 10-mL volumetric flask, and dissolve in a solution of isopropyl alcohol (4 in 10), then dilute with a solution of isopropyl alcohol (4 in 10) to volume, and mix. Transfer 500 µL of this solution to a 25-mL volumetric flask containing 5 mL of isopropyl alcohol solution (4 in 10).

Procedure—To the *Standard preparation*, the *Test preparation*, and a blank, prepared by placing 5 mL of isopropyl alcohol solution (4 in 10) in a 25-mL volumetric flask, add 500 µL of phenylhydrazine hydrochloride solution (7.5 in 100), mix, and allow to stand for 10 ± 1 minutes. Add 300 µL of potassium ferricyanide solution (1 in 20) to each flask, mix, and allow to stand for 5 minutes ± 30 seconds. Then add 2.0 mL of 2.5 *N* sodium hydroxide to each, mix, and allow to stand for 4 ± 1 minutes. Dilute each flask with isopropyl alcohol solution (4 in 10) to volume, mix, and after 10 ± 3 minutes determine the absorbances of the preparations, in 1-cm cells, at the wavelength of maximum absorbance at about 520 nm, with a suitable spectrophotometer, using the blank to set the instrument. The absorbance of the solution from the *Test preparation* does not exceed that of the solution from the *Standard preparation*, corresponding to not more than 0.0075% of formaldehyde.

Typhoid Vaccine

» Typhoid Vaccine conforms to the regulations of the federal Food and Drug Administration concerning biologics (620.10 to 620.15) (see *Biologics* ⟨1041⟩). It is a sterile suspension or solid containing killed typhoid bacilli (*Salmonella typhosa*) of the Ty 2 strain. It has a labeled potency of 8 units per mL. Its geometric mean potency, determined by the specific mouse potency test based on the U.S. Standard Typhoid Vaccine using the Ty 2 strain for challenge, from at least two assays is not less than 3.9 units per mL. Aqueous vaccine and any constituting fluid supplied with dried vaccine contains a preservative. Dried vaccine contains no preservative.

Packaging and storage—Preserve at a temperature between 2° and 8°.

Expiration date—The expiration date is not later than 18 months after date of issue from manufacturer's cold storage (5°, 1 year).

Labeling—Label it to state that it is to be well shaken before use and that it is not to be frozen.

Nitrogen content, *Method II* ⟨461⟩—The total nitrogen content of the Vaccine does not exceed 35.0 µg per mL for non-extracted bacteria preparations and does not exceed 23.0 µg per mL for acetone-extracted bacteria preparations.

Tyropanoate Sodium

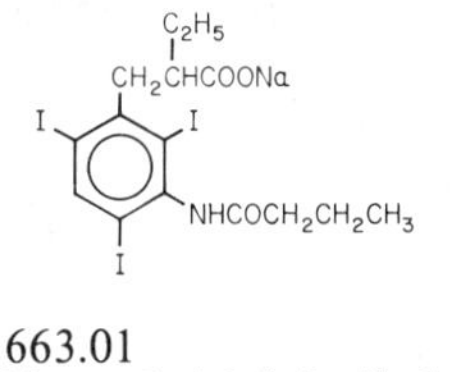

$C_{15}H_{17}I_3NNaO_3$ 663.01

Benzenepropanoic acid, α-ethyl-2,4,6-triiodo-3-[(1-oxobutyl)amino]-, monosodium salt.

Sodium 3-butyramido-α-ethyl-2,4,6-triiodohydrocinnamate [*7246-21-1*].

» Tyropanoate Sodium contains not less than 98.0 percent and not more than 102.0 percent of $C_{15}H_{17}I_3NNaO_3$, calculated on the anhydrous basis.

Packaging and storage—Preserve in tight, light-resistant containers.

Reference standard—*USP Tyropanoate Sodium Reference Standard*—Dry at 105° to constant weight before using.

Identification—

A: Dissolve about 500 mg in 10 to 15 mL of water, add about 10 mL of 3 *N* hydrochloric acid, and mix. Transfer to a 60-mL separator, and extract with three 15-mL portions of chloroform, filtering the extracts through a glass wool plug. Combine the chloroform extracts, and evaporate with the aid of a rotary evaporator to dryness. Reserve the aqueous portion for *Identification test C*. Dissolve the residue in about 5 mL of chloroform, and allow to stand for 2 hours. Filter the precipitate with the aid of vacuum, and wash with 2 portions of chloroform. Dry the precipitate in vacuum at 50° for 1 hour: the infrared absorption spectrum of a potassium bromide dispersion of the dried precipitate (tyropanoic acid) so obtained exhibits maxima only at the same wavelengths as that of a similar preparation of USP Tyropanoate Sodium RS.

B: It responds to the *Thin-layer Chromatographic Identification Test* ⟨201⟩, the test solution and the Standard solution of USP Tyropanoate Sodium RS being prepared at a concentration of 20 mg per mL in a 1 in 10 solution of methanol in chloroform, the solvent mixture being a mixture of chloroform, methanol, and formic acid (90:5:5), and short-wavelength ultraviolet light being used to locate the spots.

C: The filtrate from *Identification test A* responds to the flame test for *Sodium* ⟨191⟩.

Water, *Method I* ⟨921⟩: not more than 3.0%.

Iodine and iodide—

Standard iodide solution—Dissolve in water, and dilute with water a quantity of potassium iodide to obtain a solution containing 0.10 mg of iodide per mL.

Test preparation—Transfer 1.0 g of Tyropanoate Sodium to a stoppered, 50-mL centrifuge tube, dilute with water to 24 mL, and shake to dissolve.

Procedure—To the *Test preparation* add 5 mL of toluene and 5 mL of 2 *N* sulfuric acid, shake, and centrifuge: the toluene layer shows no red color (absence of free iodine). Add 1 mL of sodium nitrite solution (1 in 50), shake, and centrifuge: any red color is not darker than that obtained when a mixture of 1.0 mL of *Standard iodide solution* and 23 mL of water is substituted for the *Test preparation* (0.01% of iodide).

Heavy metals—

Standard preparation—Transfer 3.0 mL of *Standard Lead Solution* (30 µg of Pb; see *Heavy Metals* ⟨231⟩) to a 50-mL color-comparison tube, add 5 mL of 1 *N* sodium hydroxide, dilute with water to 40 mL, and mix.

Test preparation—Dissolve 1.0 g of Tyropanoate Sodium in 20 mL of water and 5 mL of 1 *N* sodium hydroxide, transfer the solution to a 50-mL color-comparison tube, dilute with water to 40 mL, and mix.

Procedure—To each of the tubes containing the *Standard preparation* and the *Test preparation* add 10 mL of sodium sulfide TS, mix, allow to stand for 5 minutes, and view downward over a white surface: the color of the solution from the *Test preparation* is not darker than that of the solution from the *Standard preparation* (0.003%).

Assay—Transfer about 500 mg of Tyropanoate Sodium, accurately weighed, to a 125-mL conical flask. Add 30 mL of 1.25 *N* sodium hydroxide and 500 mg of powdered zinc, connect the flask to a reflux condenser, and reflux the mixture for 1 hour. Cool the flask to room temperature, rinse the condenser with four 5-mL portions of water, disconnect the flask from the condenser, and filter the mixture. Rinse the flask and the filter thoroughly, adding the rinsings to the filtrate. Add 5 mL of glacial acetic acid and 1 mL of freshly prepared tetrabromophenolphthalein ethyl ester TS, and titrate with 0.1 *N* silver nitrate VS until the yellow precipitate just turns green. Each mL of 0.1 *N* silver nitrate is equivalent to 22.10 mg of $C_{15}H_{17}I_3NNaO_3$.

Tyropanoate Sodium Capsules

» Tyropanoate Sodium Capsules contain not less than 94.0 percent and not more than 106.0 percent of the labeled amount of $C_{15}H_{17}I_3NNaO_3$.

Packaging and storage—Preserve in tight, light-resistant containers.

Reference standard—*USP Tyropanoate Sodium Reference Standard*—Dry at 105° to constant weight before using.

Identification—

A: Shake a quantity of the contents of Capsules, equivalent to about 500 mg of tyropanoate sodium, with about 12 mL of water, and filter to obtain a clear solution. Proceed as directed in *Identification test A* under *Tyropanoate Sodium*, beginning with "Add about 10 mL of 3 *N* hydrochloric acid."

B: Shake a quantity of the contents of Capsules, equivalent to 0.5 g of tyropanoate sodium, with 25 mL of a 1 in 10 solution of methanol in chloroform, and filter. Using the filtrate as the test solution, proceed as directed in *Identification test B* under *Tyropanoate Sodium*.

C: The filtrate obtained in *Identification test A* responds to the flame test for *Sodium* ⟨191⟩.

Uniformity of dosage units ⟨905⟩: meet the requirements.

Iodine and iodide—Shake a quantity of the contents of Capsules, equivalent to 2.0 g of tyropanoate sodium, with 48 mL of water, and filter. Using 24 mL of the filtrate as the *Test preparation*, proceed as directed in the test for *Iodine and iodide* under *Tyropanoate Sodium*.

Assay—Transfer, as completely as possible, the contents of not less than 20 Tyropanoate Sodium Capsules to a suitable tared container, determine the average content weight per capsule, and mix the combined contents. Transfer an accurately weighed portion of the capsule contents, equivalent to about 500 mg of tyropanoate sodium, to a 500-mL volumetric flask. Dilute with 0.01 *N* sodium hydroxide to volume, mix, and filter, discarding the first 20 mL of the filtrate. Transfer 5.0 mL of the filtrate to a 500-mL volumetric flask, dilute with 0.01 *N* sodium hydroxide to volume, and mix. Concomitantly determine the absorbances of this solution and a Standard solution of USP Tyropanoate Sodium RS in the same medium having a known concentration of about 10 µg per mL, in 1-cm cells at the wavelength of maximum absorbance at about 237 nm, with a suitable spectrophotometer, using 0.01 *N* sodium hydroxide as the blank. Calculate the quantity, in mg, of $C_{15}H_{17}I_3NNaO_3$ in the portion of Capsule contents taken by the formula:

$$50C(A_U / A_S),$$

in which *C* is the concentration, in µg per mL, of USP Tyropanoate Sodium RS in the Standard solution, and A_U and A_S are the absorbances of the solution from the Capsules and the Standard solution, respectively.

Tyrosine

HO–C_6H_4–CH_2–CH(NH_2)–COOH

$C_9H_{11}NO_3$ 181.19

L-Tyrosine.

L-Tyrosine [*60-18-4*].

» Tyrosine contains not less than 98.5 percent and not more than 101.5 percent of $C_9H_{11}NO_3$, as L-tyrosine, calculated on the dried basis.

Packaging and storage—Preserve in well-closed containers.

Reference standard—*USP L-Tyrosine Reference Standard*—Dry at 105° for 3 hours before using.

Identification—The infrared absorption spectrum of a potassium bromide dispersion of it, previously dried, exhibits maxima only at the same wavelengths as that of a similar preparation of USP L-Tyrosine RS.

Specific rotation ⟨781⟩: between −9.8° and −11.2°, calculated on the dried basis, determined in a solution in 1 *N* hydrochloric acid containing 500 mg in each 10 mL.

Loss on drying ⟨731⟩—Dry it at 105° for 3 hours: it loses not more than 0.3% of its weight.

Residue on ignition ⟨218⟩: not more than 0.4%.

Chloride ⟨221⟩—A solution containing 0.35 g shows no more chloride than corresponds to 0.20 mL of 0.020 *N* hydrochloric acid (0.04%).

Sulfate ⟨221⟩—A solution containing 1.2 g shows no more sulfate than corresponds to 0.50 mL of 0.020 *N* sulfuric acid (0.04%).

Arsenic ⟨211⟩: 1.5 ppm.

Iron ⟨241⟩: 0.003%.

Heavy metals, *Method I* ⟨231⟩: 0.0015%.

Assay—Transfer about 180 mg of Tyrosine, accurately weighed, to a 125-mL flask, dissolve in 6 mL of formic acid, add 50 mL of glacial acetic acid, and titrate with 0.1 *N* perchloric acid VS, determining the end-point potentiometrically. Perform a blank determination, and make any necessary correction. Each mL of 0.1 *N* perchloric acid is equivalent to 18.12 mg of $C_9H_{11}NO_3$.

Tyrothricin

» Tyrothricin is an antibacterial substance produced by the growth of *Bacillus brevis* Dubos (Fam. *Bacteriaceae*). It consists principally of gramicidin and tyrocidine, the tyrocidine usually being present as the hydrochloride. It contains not less than 900 µg and not more than 1400 µg of tyrothricin per mg.

Packaging and storage—Preserve in tight containers.

Reference standard—*USP Gramicidin Reference Standard*—Dry in vacuum at a pressure not exceeding 5 mm of mercury at 60° for 3 hours before using.

Identification—Add about 5 mg to 5 mL of *p*-dimethylaminobenzaldehyde TS, and shake for 2 minutes. Add 2 drops of sodium nitrite solution (1 in 150) and 5 mL of water, and mix: a blue color is produced.

Loss on drying ⟨731⟩—Dry about 100 mg, accurately weighed, in a capillary-stoppered bottle in vacuum at a pressure not exceeding 5 mm of mercury at 60° for 3 hours: it loses not more than 5.0% of its weight.

Assay—Proceed as directed for gramicidin under *Antibiotics—Microbial Assays* ⟨81⟩, using a suitable, accurately weighed portion of Tyrothricin dissolved quantitatively in alcohol to yield a stock solution of convenient concentration. Dilute an accurately measured volume of this solution quantitatively and stepwise with alcohol to obtain a *Test Dilution* having a concentration of tyrothricin assumed to be 5.0 times the median dose level of the gramicidin Standard. Multiply the result of the gramicidin assay by 5.0 to obtain the amount of tyrothricin, in µg, in each mL of the *Test Dilution*.

Ultramicrosize Griseofulvin Tablets—*see*
Griseofulvin Tablets, Ultramicrosize

Undecylenic Acid

$CH_2{=}CHCH_2(CH_2)_6CH_2COOH$

$C_{11}H_{20}O_2$ 184.28
10-Undecenoic acid.
10-Undecenoic acid [*112-38-9*].

» Undecylenic Acid contains not less than 97.0 percent and not more than 100.5 percent of $C_{11}H_{20}O_2$.

Packaging and storage—Preserve in tight, light-resistant containers.

Identification—

A: To 1 mL add potassium permanganate TS, dropwise: the permanganate color is discharged.

B: Place 3 mL of it and 3 mL of freshly distilled aniline in a tall test tube, and heat for 10 minutes at a rate such that the ring of condensate remains just below the mouth of the tube. Cool, add 10 mL of alcohol and 10 mL of ether, and transfer to a separator. Wash the ether solution with four 20-mL portions of water, and discard the water washings. Heat on a steam bath until the odor of ether no longer is perceptible, then add a few mg of activated carbon, mix, and filter. Evaporate the filtrate nearly to dryness, and recrystallize the residue from 70 percent alcohol: the anilide so obtained melts between 66° and 67.5°.

Specific gravity ⟨841⟩: between 0.910 and 0.913.

Congealing range ⟨651⟩: not lower than 21°.

Refractive index ⟨831⟩: between 1.447 and 1.448.

Residue on ignition ⟨281⟩: not more than 0.15%.

Water-soluble acids—Shake 5 mL with 5 mL of water, and filter the water layer through a filter paper previously moistened with water. Add 1 drop of methyl orange TS, and titrate with 0.01 *N* sodium hydroxide VS: not more than 1.0 mL of 0.010 *N* sodium hydroxide is required to match the color produced by 1 drop of methyl orange TS in 5 mL of water.

Heavy metals, *Method II* ⟨231⟩: 0.001%.

Iodine value ⟨401⟩: between 131 and 138.

Assay—Dissolve about 750 mg of Undecylenic Acid, accurately weighed, in 50 mL of alcohol, add 3 drops of phenolphthalein TS, and titrate with 0.1 *N* sodium hydroxide VS to the first pink color that persists for not less than 30 seconds. Perform a blank determination, and make any necessary correction. Each mL of 0.1 *N* sodium hydroxide is equivalent to 18.43 mg of $C_{11}H_{20}O_2$.

Compound Undecylenic Acid Ointment

» Compound Undecylenic Acid Ointment contains not less than 18.0 percent and not more than 22.0 percent of zinc undecylenate ($C_{22}H_{38}O_4Zn$), and not less than 4.5 percent and not more than 5.5 percent of free undecylenic acid ($C_{11}H_{20}O_2$), in a suitable ointment base.

Packaging and storage—Preserve in tight containers, and avoid prolonged exposure to temperatures exceeding 30°.

Reference standard—*USP Undecylenic Acid Reference Standard*—Do not dry before using.

Assay for zinc undecylenate—

Standard preparations—Prepare a solution of freshly ignited zinc oxide in dilute hydrochloric acid (1 in 60) to obtain the equivalent of 1.0 mg of zinc per mL. Dilute quantitatively with water to obtain separate solutions containing the equivalent of 15 and 30 µg of zinc per mL.

Assay preparation—Transfer about 1.0 g of Compound Undecylenic Acid Ointment, accurately weighed, to a 100-mL beaker. Add 25 mL of dilute hydrochloric acid (1 in 20), swirl, and heat carefully until the mixture is liquefied. Cool, and transfer the mixture to a 250-mL separator. Complete the transfer of the waxy residue by thoroughly rinsing the beaker with 50 mL of water and two 50-mL portions of chloroform and adding the rinsings to the separator. Equilibrate the mixture, and transfer the chloroform extract to a 500-mL separator. Extract the aqueous phase with another 100-mL portion of chloroform, combine the second chloroform extract with the main extract in the 500-mL separator, and transfer the aqueous phase to a 200-mL volumetric flask. Wash the combined chloroform extracts with three 25-mL portions of water, add the aqueous washings to the 200-mL volumetric flask, dilute with water to volume, and mix to obtain a specimen stock solution. [NOTE—Retain the chloroform extract for the *Assay for undecylenic acid*.] Transfer 15.0 mL or other suitable volume (see *Procedure*) of this specimen stock solution to a 100-mL volumetric flask, dilute with water to volume, and mix.

Procedure—Aspirate each *Standard preparation* and the *Assay preparation* into the flame of a suitable atomic absorption spectrophotometer, and determine the absorbances of the solutions at 214 nm. Typically, an acetylene-air mixture is adjusted to obtain a blue flame about 7 mm in height with a suitable burner that is rotated to a position perpendicular to the light path. [NOTE—If the absorbance of the *Assay preparation* is outside the central 70% of the range between the absorbances of the *Standard preparations*, discard the *Assay preparation* and prepare another by diluting the specimen stock solution quantitatively as necessary to obtain a suitable absorbance.] Calculate the percentage of zinc undecylenate in the Ointment by the formula:

$$(431.92/65.38)(0.2/W)\ C_L + \frac{(C_H - C_L)(A_U - A_L)}{(A_H - A_L)},$$

in which 431.92 is the molecular weight of zinc undecylenate, 65.38 is the atomic weight of zinc, W is the weight, in g, of Ointment taken, A_U, A_H, and A_L are the absorbances of the *Assay preparation* and the high- and low-concentration *Standard preparations*, respectively, and C_H and C_L are the concentrations, in µg per mL, of the high- and low-concentration *Standard preparations*, respectively.

Assay for undecylenic acid—

Internal standard solution—Prepare a solution in chloroform containing 10 mg of tridecanoic acid in each mL.

Standard preparation—Dissolve an accurately weighed quantity of USP Undecylenic Acid RS in chloroform to obtain a solution having a known concentration of about 3.8 mg per mL. Transfer 5.0 mL of this solution to a 50-mL volumetric flask, add 3.0 mL of *Internal standard solution*, dilute with chloroform to volume, and mix.

Assay preparation—Pass the chloroform extract prepared from the Ointment as directed under *Assay for zinc undecylenate* through phase-separating filter paper into a 250-mL volumetric flask. Rinse the separator with three 15-mL portions of chloroform, passing the rinsings through the filter and combining them with the main chloroform solution, add chloroform to volume, and mix. Transfer 20.0 mL of this solution to a 50-mL volumetric flask, add 3.0 mL of *Internal standard solution*, dilute with chloroform to volume, and mix.

Chromatographic system—Under typical conditions, the gas chromatograph is equipped with a flame-ionization detector and contains a 1.8-m × 2-mm glass column packed with 3 percent liquid phase G1 on 100- to 200-mesh support S1A. The column is maintained at a temperature of about 165°. Dry helium is used as the carrier gas at a flow rate of about 30 mL per minute.

System suitability—Chromatograph five injections of the silylated *Standard preparation*, and record peak responses as directed under *Procedure*. The resolution factor, *R* (see *Chromatography* ⟨621⟩), is not less than 3.0.

Procedure—Transfer 1.0-mL portions of the *Standard preparation* and the *Assay preparation* to separate, stoppered test tubes. To each tube add 50 µL of bis(trimethylsilyl)trifluoroacetamide, insert the stopper, mix, and allow to stand for 30 minutes. Inject a suitable portion (2 to 5 µL) of the *Standard preparation* into a suitable gas chromatograph, and record the chromatogram so as to obtain not less than 50% of maximum recorder response. Similarly inject a suitable portion of the *Assay preparation*, and record the chromatogram. Measure the peak responses for the first (undecylenic acid) and second (tridecanoic acid) peaks of the chromatograms. [NOTE—Relative retention times are, approximately, 0.43 for undecylenic acid and 1.0 for tridecanoic acid.] Calculate the percentage of total undecylenic acid in the Ointment by the formula:

$$62.5(C/W)(R_U/R_S),$$

in which *C* is the concentration, in mg per mL, of USP Undecylenic Acid RS in the *Standard preparation*, *W* is the weight, in g, of Ointment taken, and R_U and R_S are the ratios of the peak responses of undecylenic acid to those of tridecanoic acid from the *Assay preparation* and the *Standard preparation*, respectively. The difference between the total undecylenic acid and the undecylenic acid equivalent to the determined zinc undecylenate (the weight of zinc undecylenate multiplied by 0.8533 gives the equivalent of undecylenic acid), both expressed as a percentage of the Ointment, gives the percentage of free undecylenic acid in the Ointment.

Uracil Mustard

$C_8H_{11}Cl_2N_3O_2$ 252.10
2,4(1*H*,3*H*)-Pyrimidinedione, 5-[bis(2-chloroethyl)amino]-.
5-[Bis(2-chloroethyl)amino]uracil [*66-75-1*].

» Uracil Mustard contains not less than 97.0 percent and not more than 103.0 percent of $C_8H_{11}Cl_2N_3O_2$, calculated on the dried basis.

Caution—Handle Uracil Mustard with exceptional care since it is a highly potent cytotoxic agent. Perform operations with the material in a well-ventilated hood, using a protective mask and gloves, and thoroughly clean the work area afterward. Rinse hands in water only, for several minutes, then complete the washing with soap and water.

Packaging and storage—Preserve in tight containers.

Reference standard—*USP Uracil Mustard Reference Standard*—[*Caution—Avoid contact.*] Dry in vacuum over silica gel for 18 hours before using.

Identification—

A: The infrared absorption spectrum of a mineral oil dispersion of it, previously dried, exhibits maxima only at the same wavelengths as that of a similar preparation of USP Uracil Mustard RS.

B: The ultraviolet absorption spectrum of a 1 in 40,000 solution in 0.01 *N* alcoholic sulfuric acid exhibits maxima and minima at the same wavelengths as that of a similar solution of USP Uracil Mustard RS, concomitantly measured, and the respective absorptivities, calculated on the dried basis, at the wavelength of maximum absorbance at about 256 nm do not differ by more than 3.0%.

Loss on drying ⟨731⟩—Dry it in vacuum over silica gel for 18 hours: it loses not more than 0.5% of its weight.

Residue on ignition ⟨281⟩: not more than 0.5%.

Assay—

Standard preparation—Dissolve in alcohol a suitable quantity of USP Uracil Mustard RS, accurately weighed, to obtain a solution having a known concentration of about 2 mg per mL, and mix.

Assay preparation—Dissolve about 0.1 g of Uracil Mustard, accurately weighed, in sufficient alcohol to make 50.0 mL, and mix.

Procedure—Transfer 20.0 mL each of the *Standard preparation*, the *Assay preparation*, and alcohol to provide the blank, to separate 50-mL volumetric flasks. Add 20 mL of water, 1 mL of 1 *N* acetic acid, and 6 mL of a freshly prepared 1 in 20 solution of 8-hydroxyquinoline in alcohol to each of the three flasks, and mix. Add 3 mL of sodium carbonate solution (1 in 10) to each of the three flasks, dilute with water to volume, and mix. After 150 minutes, concomitantly determine the absorbances of the solutions obtained from the *Assay preparation* and the *Standard preparation* in 1-cm cells at the wavelength of maximum absorbance at about 466 nm, with a suitable spectrophotometer, using the blank to set the instrument. Calculate the quantity, in mg, of $C_8H_{11}Cl_2N_3O_2$ in the portion of Uracil Mustard taken by the formula:

$$50C(A_U/A_S),$$

in which *C* is the concentration, in mg per mL, of USP Uracil Mustard RS in the *Standard preparation*, and A_U and A_S are the absorbances of the solutions from the *Assay preparation* and the *Standard preparation*, respectively.

Uracil Mustard Capsules

» Uracil Mustard Capsules contain not less than 90.0 percent and not more than 110.0 percent of the labeled amount of $C_8H_{11}Cl_2N_3O_2$.

Packaging and storage—Preserve in tight containers.

Reference standard—*USP Uracil Mustard Reference Standard*—[*Caution—Avoid contact.*] Dry in vacuum over silica gel for 18 hours before using.

Identification—

A: The ultraviolet absorption spectrum of the solution employed for measurement of absorbance in the *Assay* exhibits maxima and minima at the same wavelengths as that of a similar solution of USP Uracil Mustard RS, concomitantly measured.

B: Mix a portion of the contents of Capsules with alcohol to obtain a concentration of 1 mg of uracil mustard per mL. Heat on a steam bath for 15 minutes with frequent agitation, cool, and filter, if necessary, to obtain a clear solution. On a suitable thin-layer chromatographic plate (see *Chromatography* ⟨621⟩), coated with 0.25-mm layer of chromatographic silica gel mixture, apply, with the aid of a stream of nitrogen, 20 µL of this solution and 20 µL of an alcohol solution of USP Uracil Mustard RS containing 1 mg per mL. Allow the spots to dry, and develop the chromatogram in methanol until the solvent front has moved three-fourths of the length of the plate. Remove the plate from the developing chamber, mark the solvent front, and allow the solvent to evaporate. Locate the spots on the plate by examination under short-wavelength ultraviolet light: the R_f value of the principal spot obtained from the test solution corresponds to that obtained from the Standard solution.

Disintegration ⟨701⟩: 15 minutes, the use of disks being omitted.

Uniformity of dosage units ⟨905⟩: meet the requirements.

Assay—Transfer, as completely as possible, the contents of not less than 20 Uracil Mustard Capsules to a tared weighing bottle, and weigh accurately. Transfer an accurately weighed portion of the powder, equivalent to about 5 mg of uracil mustard, to a 200-mL volumetric flask with 50 mL of 0.01 *N* alcoholic sulfuric acid. Heat on a steam bath for 15 minutes with frequent agitation, cool, add the dilute alcoholic sulfuric acid to volume, and mix. Concomitantly determine the absorbances of this solution and of a Standard solution of USP Uracil Mustard RS, in the same medium having a known concentration of about 25 µg per mL, in 1-cm cells at the wavelength of maximum absorbance at about 256 nm, with a suitable spectrophotometer, using 0.01 *N* alcoholic sulfuric acid as the blank. Calculate the quantity, in mg, of $C_8H_{11}Cl_2N_3O_2$ in the portion of Capsules taken by the formula:

$$0.2C(A_U/A_S),$$

in which *C* is the concentration, in µg per mL, of USP Uracil Mustard RS in the Standard solution, and A_U and A_S are the absorbances of the solution from the Capsules and the Standard solution, respectively.

Urea

CO$(NH_2)_2$

CH_4N_2O 60.06
Urea.
Carbamide [*57-13-6*].

» Urea contains not less than 99.0 percent and not more than 100.5 percent of CH_4N_2O.

Packaging and storage—Preserve in well-closed containers.

Identification—

A: Heat about 500 mg in a test tube: it liquefies, and ammonia is evolved. Continue the heating until the liquid becomes turbid, then cool. Dissolve the fused mass in a mixture of 10 mL of water and 1 mL of sodium hydroxide solution (1 in 10), and add 1 drop of cupric sulfate TS: the solution acquires a reddish violet color.

B: Dissolve 100 mg in 1 mL of water, and add 1 mL of nitric acid: a white crystalline precipitate of urea nitrate is formed.

Melting range ⟨741⟩: between 132° and 135°.

Residue on ignition ⟨281⟩: not more than 0.1%.

Alcohol-insoluble matter—Dissolve 5.0 g in 50 mL of warm alcohol, and if any insoluble residue remains, filter the solution on a tared filter, wash the residue and the filter with 20 mL of warm alcohol, and dry at 105° for 1 hour: the weight of the residue does not exceed 2 mg (0.04%).

Chloride ⟨221⟩—A 2.0-g portion shows no more chloride than corresponds to 0.20 mL of 0.020 *N* hydrochloric acid (0.007%).

Sulfate ⟨221⟩—A 2.0-g portion shows no more sulfate than corresponds to 0.20 mL of 0.020 *N* sulfuric acid (0.010%).

Heavy metals ⟨231⟩—Dissolve 1.0 g in 20 mL of water, and add 5 mL of 0.1 *N* hydrochloric acid: the limit is 0.002%.

Assay—Transfer about 500 mg of Urea, accurately weighed, to a 200-mL volumetric flask, dissolve in water, dilute with water to volume, and mix. Pipet 2 mL of this solution into a micro-Kjeldahl digestion flask, and proceed as directed under *Nitrogen Determination, Method II* ⟨461⟩, beginning with "Add 1 g of a powdered mixture." [NOTE—In this procedure, continue heating the flask until fuming begins, then heat for 1 additional hour.] Each mL of 0.01 *N* acid is equivalent to 0.3003 mg of CH_4N_2O.

Sterile Urea

» Sterile Urea is Urea suitable for parenteral use.

Packaging and storage—Preserve in *Containers for Sterile Solids* as described under *Injections* ⟨1⟩.

Completeness of solution ⟨641⟩—A 1.0-g portion dissolves in 10 mL of carbon dioxide–free water to yield a clear solution.

Constituted solution—At the time of use, the constituted solution prepared from Sterile Urea meets the requirements for *Constituted Solutions* under *Injections* ⟨1⟩.

Pyrogen—When dissolved in 5 percent Dextrose Injection to yield a concentration of 80 mg per mL, it meets the requirements of the *Pyrogen Test* ⟨151⟩, the test dose being the equivalent of 800 mg per kg, and the solution being injected over a period of 2 minutes.

Other requirements—It responds to the *Identification tests* and meets the requirements for *Melting range, Residue on ignition, Alcohol-insoluble matter, Chloride, Sulfate, Heavy metals,* and *Assay* under *Urea.* It meets also the requirements for *Sterility Tests* ⟨71⟩, *Uniformity of Dosage Units* ⟨905⟩, and *Labeling* under *Injections* ⟨1⟩.

Vaccines—*see complete list in index*

Vaccinia Immune Globulin

» Vaccinia Immune Globulin conforms to the regulations of the federal Food and Drug Administration concerning biologics (see *Biologics* ⟨1041⟩). It is a sterile, non-pyrogenic solution of globulins derived from the blood plasma of adult human donors who have been immunized with vaccinia virus (Smallpox Vaccine). It is standardized for viral neutralizing activity in eggs or tissue culture with the U.S. Reference Vaccinia Immune Globulin and a specified vaccinia virus. It contains not less than 15 g and not more than 18 g of protein per 100 mL, not less than 90.0 percent of which is gamma globulin. It contains 0.3 *M* glycine as a stabilizing agent, and contains a suitable antimicrobial agent.

Packaging and storage—Preserve at a temperature between 2° and 8°.

Expiration date—The expiration date is not later than 3 years after date of issue.

Labeling—Label it to state that it is not intended for intravenous injection.

Valine

H
$(CH_3)_2$CH—C—COOH
NH_2

$C_5H_{11}NO_2$ 117.15
L-Valine.
L-Valine [*72-18-4*].

» Valine contains not less than 98.5 percent and not more than 101.5 percent of $C_5H_{11}NO_2$, as L-valine, calculated on the dried basis.

Packaging and storage—Preserve in well-closed containers.

Reference standard—*USP L-Valine Reference Standard*—Dry at 105° for 3 hours before using.

Identification—The infrared absorption spectrum of a potassium bromide dispersion of it, previously dried, exhibits maxima only at the same wavelengths as that of a similar preparation of USP L-Valine RS.

Specific rotation ⟨781⟩: between +26.6° and +28.4°, calculated on the dried basis, determined in a solution in 6 *N* hydrochloric acid containing 800 mg in each 10 mL.

pH ⟨791⟩: between 5.5 and 7.0, in a solution (1 in 20).

Loss on drying ⟨731⟩—Dry it at 105° for 3 hours: it loses not more than 0.3% of its weight.

Residue on ignition ⟨281⟩: not more than 0.1%.

Chloride ⟨221⟩—A 0.73-g portion shows no more chloride than corresponds to 0.50 mL of 0.020 *N* hydrochloric acid (0.05%).

Sulfate ⟨221⟩—A 0.33-g portion shows no more sulfate than corresponds to 0.10 mL of 0.020 *N* sulfuric acid (0.03%).

Arsenic ⟨211⟩: 1.5 ppm.

Iron ⟨241⟩: 0.003%

Heavy metals, *Method I* ⟨231⟩: 0.0015%.

Assay—Transfer about 110 mg of Valine, accurately weighed, to a 125-mL flask, dissolve in a mixture of 3 mL of formic acid and 50 mL of glacial acetic acid, and titrate with 0.1 *N* perchloric acid VS, determining the end-point potentiometrically. Perform a blank determination, and make any necessary correction. Each mL of 0.1 *N* perchloric acid is equivalent to 11.72 mg of $C_5H_{11}NO_2$.

Valproic Acid

$$\begin{array}{l} CH_3CH_2CH_2CHCOOH \\ \quad\quad\quad\quad\ | \\ \quad\ \ CH_3CH_2CH_2 \end{array}$$

$C_8H_{16}O_2$ 144.21
Pentanoic acid, 2-propyl-.
Propylvaleric acid [*99-66-1*].

» Valproic Acid contains not less than 98.0 percent and not more than 102.0 percent of $C_8H_{16}O_2$, calculated on the anhydrous basis.

Packaging and storage—Preserve in tight, glass containers.

Reference standard—*USP Valproic Acid Reference Standard*—Use without drying.

Identification—

A: Its infrared absorption spectrum, obtained by spreading a capillary film of it between sodium chloride plates, exhibits maxima only at the same wavelengths as that of a similar preparation of USP Valproic Acid RS.

B: Add 0.5 mL each of potassium iodide solution (1 in 50) and potassium iodate solution (1 in 25) to a test tube, and mix. Add 2 drops of Valproic Acid, and mix: a yellow color is produced.

Water, *Method I* ⟨921⟩: not more than 1.0%.

Residue on ignition ⟨281⟩: not more than 0.1%.

Heavy metals, *Method II* ⟨231⟩: 0.002%.

Chromatographic purity—

Chromatographic system (see *Chromatography* ⟨621⟩)—The gas chromatograph is equipped with a flame-ionization detector that contains a 1.8-m × 2-mm (ID) column packed with 10 percent phase G34 on 80- to 100-mesh support S1A. The column is maintained at about 140°, the injector is maintained at about 225°, and the detector is maintained at about 235°. Dry helium is used as the carrier gas at a flow rate of about 50 mL per minute. Chromatograph duplicate portions (about 1 µL) of USP Valproic Acid RS, and measure the peak responses as directed under *Procedure:* the peak responses and retention times for valproic acid in the chromatograms do not differ by more than 1% and 3%, respectively.

Procedure—Inject a volume (about 1 µL) of Valproic Acid into the chromatograph, record the chromatogram, and measure the responses for the peaks: the ratio of the peak response for valproic acid to the sum of all peak responses is not less than 0.98.

Assay—

Tetrabutylammonium hydroxide titrant—Dilute 1 volume of 1 *M* tetrabutylammonium hydroxide solution in methanol with 9 volumes of chlorobenzene, flush the solution for 10 minutes with dry, carbon dioxide–free nitrogen, and mix. Store in a reservoir protected from carbon dioxide and moisture, and discard after 60 days. Determine the molarity of the titrant on the day of use as follows: Dissolve about 125 mg of primary standard benzoic acid, accurately weighed, in 100 mL of acetone. Titrate with the *Tetrabutylammonium hydroxide titrant*, taking precautions against the absorption of atmospheric carbon dioxide, determining the end-point potentiometrically, using a glass electrode and a calomel electrode containing 1.0 *M* tetrabutylammonium chloride (aqueous) (see *Titrimetry* ⟨541⟩). Calculate the molarity of the titrant by the formula:

$$W/122.12V,$$

in which W is the weight, in mg, of benzoic acid taken, 122.12 is the molecular weight of benzoic acid, and V is the volume, in mL, of *Tetrabutylammonium hydroxide titrant* consumed.

Procedure—Dissolve about 160 mg of Valproic Acid, accurately weighed, in 100 mL of acetone. Titrate the solution with *Tetrabutylammonium hydroxide titrant*, taking precautions against the absorption of atmospheric carbon dioxide, determining the end-point potentiometrically, using a glass electrode and a calomel electrode containing 1.0 *M* tetrabutylammonium chloride (aqueous) (see *Titrimetry* ⟨541⟩). Perform a blank determination, and make any necessary correction. Each mL of 0.1 *M Tetrabutylammonium hydroxide titrant* is equivalent to 14.42 mg of $C_8H_{16}O_2$.

Valproic Acid Capsules

» Valproic Acid Capsules contain not less than 90.0 percent and not more than 110.0 percent of the labeled amount of $C_8H_{16}O_2$.

Packaging and storage—Preserve in tight containers, at controlled room temperature.

Reference standard—*USP Valproic Acid Reference Standard*—Use without drying.

Identification—

A: The retention time ratios of the valproic acid peak to the internal standard peak obtained from the solution from the *Standard preparation* and from the *Assay preparation* as directed in the *Assay* do not differ by more than 2.0%.

B: Place a portion of Capsule contents, equivalent to about 250 mg of valproic acid, in a separator. Add 20 mL of 1 *N* sodium hydroxide, shake, and allow the layers to separate. Transfer the aqueous layer to a second separator, add 4 mL of hydrochloric acid, mix, and extract with 40 mL of *n*-heptane. Filter the *n*-heptane layer through glass wool into a beaker, and evaporate the solvent completely on a steam bath with the aid of a current of air: the residue so obtained responds to *Identification test B* under *Valproic Acid.*

Disintegration ⟨701⟩: 15 minutes, determined as directed for *Soft Gelatin Capsules.*

Uniformity of dosage units ⟨905⟩: meet the requirements, chloroform being used as the solvent in the procedure for *Soft Capsules.*

Assay—

Internal standard solution—Dissolve a quantity of biphenyl in chromatographic *n*-heptane to obtain a solution having a concentration of about 5 mg per mL.

Standard preparation—Dissolve an accurately weighed quantity of USP Valproic Acid RS in chromatographic *n*-heptane to obtain a solution having a known concentration of about 2.5 mg per mL.

Assay preparation—Transfer not less than 20 Valproic Acid Capsules to a blender jar or other container, add about 150 mL of methylene chloride, cool in a solid carbon dioxide–acetone mixture until the contents have solidified. If necessary, transfer the mixture of Capsules and methylene chloride to a blender jar, and blend with a high-speed blender until all the solids are reduced to fine particles. Transfer the mixture to a 500-mL volumetric flask, add chromatographic *n*-heptane to volume, mix, and allow solids to settle. Transfer an accurately measured volume of this solution, equivalent to 250 mg of valproic acid, to a 100-mL volumetric flask, dilute with chromatographic *n*-heptane to volume, and mix.

Chromatographic system (see *Chromatography* ⟨621⟩)—The gas chromatograph is equipped with a flame-ionization detector, and contains a 1.8-m × 2-mm (ID) glass column packed with 10 percent phase G34 on 80- to 100-mesh support S1A. The column is maintained at about 150°, and the injection port and the detector block are maintained at about 250°. Dry helium is used as the carrier gas at a flow rate of about 40 mL per minute. Chromatograph 2 µL portions of the solution from the *Standard preparation*, measure the peak responses, and calculate the ratio, R_S, as directed under *Procedure.* The relative standard deviation for replicate injections is not more than 2.0%, and the resolution, R, between the 2 peaks is not less than 3.0.

Procedure—Transfer 5.0 mL of the *Standard preparation* and the *Assay preparation* to separate containers equipped with closures. To each container add 2.0 mL of *Internal standard solution*, close the containers, and mix. Inject, successively, into the gas chromatograph 2 µL each of the solution from the *Standard preparation* and from the *Assay preparation*, and measure the peak responses of the components in each chromatogram. The relative retention times are about 0.5 for valproic acid and

1.0 for biphenyl. Calculate the quantity, in mg, of $C_8H_{16}O_2$ in the portion of Capsules taken by the formula:

$$100C(R_U/R_S),$$

in which C is the concentration, in mg per mL, of USP Valproic Acid RS in the *Standard preparation*, and R_U and R_S are the peak response ratios obtained from the solutions from the *Assay preparation* and the *Standard preparation*, respectively.

Valproic Acid Syrup

» Valproic Acid Syrup contains not less than 90.0 percent and not more than 110.0 percent of the labeled amount of $C_8H_{16}O_2$. It is prepared with the aid of Sodium Hydroxide.

Packaging and storage—Preserve in tight containers.

Reference standard—*USP Valproic Acid Reference Standard*—Use without drying.

Identification—

A: The retention time ratios of the valproic acid peak to the internal standard peak obtained from the solutions from the *Standard preparation* and from the *Assay preparation* as directed in the *Assay* do not differ by more than 2.0%.

B: Place a volume of Syrup, equivalent to about 250 mg of valproic acid, in a separator. Add 40 mL of water and 2 mL of hydrochloric acid, mix, and extract with 40 mL of *n*-heptane. Filter the *n*-heptane layer through glass wool into a beaker, and evaporate the solvent completely on a steam bath with the aid of a current of air: the residue so obtained responds to *Identification test B* under *Valproic Acid*.

pH ⟨791⟩: between 7.0 and 8.0.

Assay—

Internal standard solution, *Standard preparation*, and *Chromatographic system*—Prepare as directed in the *Assay* under *Valproic Acid Capsules*.

Assay preparation—Transfer an accurately measured volume of Valproic Acid Syrup, equivalent to about 250 mg of valproic acid, to a separator. Add 40 mL of water and 2 mL of hydrochloric acid, mix, and extract gently with 80 mL of chromatographic *n*-heptane until the aqueous layer is clear (about 3 minutes). Filter the *n*-heptane layer through glass wool, collecting the filtrate in a 100-mL volumetric flask. Rinse the separator and the glass wool with small portions of chromatographic *n*-heptane, add the rinsings to the flask, dilute with chromatographic *n*-heptane to volume, and mix.

Procedure—Transfer 5.0 mL of the *Standard preparation* and the *Assay preparation* to separate containers equipped with closures. To each container add 2.0 mL of *Internal standard solution*, close the containers, and mix. Inject, successively, into the gas chromatograph 2 µL each of the solutions from the *Standard preparation* and from the *Assay preparation*, and measure the peak responses of the components in each chromatogram. The relative retention times are about 0.5 for valproic acid and 1.0 for biphenyl. Calculate the quantity, in mg, of $C_8H_{16}O_2$ in each mL of the Syrup taken by the formula:

$$100(C/V)(R_U/R_S),$$

in which C is the concentration, in mg per mL, of USP Valproic Acid RS in the *Standard preparation*, V is the volume, in mL, of Syrup taken, and R_U and R_S are the peak response ratios obtained from the solutions from the *Assay preparation* and the *Standard preparation*, respectively.

Vancomycin Hydrochloride

$C_{66}H_{75}Cl_2N_9O_{24}\cdot HCl$ 1485.73

Vancomycin, monohydrochloride.

Vancomycin monohydrochloride.

(S_a)-(3*S*,6*R*,7*R*,22*R*,23*S*,26*S*,36*R*,38a*R*)-44-[[2-*O*-(3-Amino-2,3,6-trideoxy-3-*C*-methyl-α-L-*lyxo*-hexopyranosyl)-β-D-glucopyranosyl]oxy]-3-(carbamoylmethyl)-10,19-dichloro-2,3,4,5,6,7,23,24,25,26,36,37,38,38a-tetradecahydro-7,22,28,30,32-pentahydroxy-6-[(2*R*)-4-methyl-2-(methylamino)valeramido]-2,5,24,38,39-pentaoxo-22*H*-8,11:18,21-dietheno-23,36-(iminomethano)-13,16:31,35-dimetheno-1*H*,16*H*-[1,6,9]oxadiazacyclohexadecino[4,5-*m*][10,2,16]-benzoxadiazacyclotetracosine-26-carboxylic acid, monohydrochloride.

[3*S*-[3*R**,6*S**(*S**),7*S**,22*S**,23*R**,26*R**,36*S**,38a*S**]]-3-(2-Amino-2-oxoethyl)-44-[[2-*O*-(3-amino-2,3,6-trideoxy-3-*C*-methyl-α-L-*lyxo*-hexopyranosyl)-β-D-glucopyranosyl]oxy]-10,19-dichloro-2,3,4,5,6,7,23,24,25,26,36,37,38,38a-tetradecahydro-7,22,28,30,32-pentahydroxy-6-[[4-methyl-2-(methylamino)-1-oxopentyl]amino]-2,5,24,38,39-pentaoxo-22*H*-8,11:18,21-dietheno-23,36-(iminomethano)-13,16:31,35-dimetheno-1*H*,16*H*-[1,6,9]oxadiazacyclohexadecino[4,5-*m*][10,2,16]-benzoxadiazacyclotetracosine-26-carboxylic acid, monohydrochloride [*1404-93-9*].

» Vancomycin Hydrochloride is the hydrochloride salt of a kind of vancomycin, a substance produced by the growth of *Streptomyces orientalis* (Fam. Streptomycetaceae), or a mixture of two or more such salts. It has a potency equivalent to not less than 900 µg of vancomycin per mg, calculated on the anhydrous basis.

Packaging and storage—Preserve in tight containers.

Reference standard—*USP Vancomycin Hydrochloride Reference Standard*—Do not dry. Constitute the entire contents, without weighing, for *Assay*.

Identification—The infrared absorption spectrum of a potassium bromide dispersion of an undried portion of it exhibits maxima only at the same wavelengths as that of a similar preparation of undried USP Vancomycin Hydrochloride RS.

pH ⟨791⟩: between 2.5 and 4.5, in a solution containing 50 mg per mL.

Water, *Method I* ⟨921⟩: not more than 5.0%.

Assay—Proceed with Vancomycin Hydrochloride as directed under *Antibiotics—Microbial Assays* ⟨81⟩.

Vancomycin Hydrochloride Capsules

» Vancomycin Hydrochloride Capsules contain a dispersion of Vancomycin Hydrochloride in Polyethylene Glycol. Capsules contain the equivalent of not less than 90.0 percent and not more than 115.0 percent of the labeled amount of vancomycin.

Packaging and storage—Preserve in tight containers.

Reference standard—*USP Vancomycin Hydrochloride Reference Standard*—Do not dry. Constitute the entire contents, without weighing, for *Assay*.

Identification—Place 1 or more Capsules in a high-speed glass blender jar containing a volume of water sufficient to yield a solution containing the equivalent of 1 mg of vancomycin per mL, and blend for 3 to 5 minutes. To a suitable sheet of chromatographic filter paper apply 5 μL of this solution and 5 μL of a solution of USP Vancomycin Hydrochloride RS containing the equivalent of 1 mg of vancomycin per mL. Develop by descending chromatography (see *Chromatography* ⟨621⟩) with a mixture of butyl alcohol, water, and pyridine (6:4:3) for 7 hours. Allow the paper to dry, and place it on an inoculated agar surface of sufficient area to accommodate the paper and prepared for vancomycin assay as directed under *Antibiotics—Microbial Assays* ⟨81⟩, except to use *Medium 2*. Remove the paper from the agar surface after 30 minutes, and incubate the agar medium at 37° for 18 hours: clear zones of inhibition are produced at corresponding positions on the two chromatograms.

Dissolution ⟨711⟩—

Medium: water; 900 mL.

Apparatus 1: 100 rpm.

Time: 45 minutes.

Procedure—Determine the amount of vancomycin dissolved by assaying a filtered portion of the solution under test as directed for vancomycin under *Antibiotics—Microbial Assays* ⟨81⟩.

Tolerances—Not less than 85% (*Q*) of the labeled amount of vancomycin is dissolved in 45 minutes.

Uniformity of dosage units ⟨905⟩: meet the requirements.

Water, *Method I* ⟨921⟩: not more than 8.0%.

Assay—Proceed as directed for vancomycin under *Antibiotics—Microbial Assays* ⟨81⟩, using not less than 5 Vancomycin Hydrochloride Capsules blended at high speed in a glass blender jar for 3 to 5 minutes with a sufficient, accurately measured, volume of *Buffer No. 4* to yield a stock solution having a convenient concentration of vancomycin. Dilute an accurately measured volume of this stock solution quantitatively and stepwise with *Buffer No. 4* to obtain a *Test Dilution* having a concentration assumed to be equal to the median dose level of the Standard.

Vancomycin Hydrochloride for Oral Solution

» Vancomycin Hydrochloride for Oral Solution contains the equivalent of not less than 90.0 percent and not more than 115.0 percent of the labeled amount of vancomycin.

Packaging and storage—Preserve in tight containers.

Reference standard—*USP Vancomycin Hydrochloride Reference Standard*—Do not dry. Constitute the entire contents, without weighing, for *Assay*.

pH ⟨791⟩: between 2.5 and 4.5, for the solution constituted as directed in the labeling.

Water, *Method I* ⟨921⟩: not more than 5.0%.

Assay—Proceed as directed under *Antibiotics—Microbial Assays* ⟨81⟩, dissolving the contents of 1 container of Vancomycin Hydrochloride for Oral Solution in water as directed in the labeling. Dilute a portion of this solution quantitatively with *Buffer No. 4* to obtain a *Test Dilution* having a concentration assumed to be equal to the median dose level of the Standard.

Sterile Vancomycin Hydrochloride

» Sterile Vancomycin Hydrochloride has a potency equivalent to not less than 900 μg of vancomycin per mg, calculated on the anhydrous basis. In addition, where packaged for dispensing, it contains the equivalent of not less than 90.0 percent and not more than 115.0 percent of the labeled amount of vancomycin.

Packaging and storage—Preserve in *Containers for Sterile Solids* as described under *Injections* ⟨1⟩.

Reference standard—*USP Vancomycin Hydrochloride Reference Standard*—Do not dry. Constitute the entire contents, without weighing, for *Assay*.

Constituted solution—At the time of use, the constituted solution prepared from Sterile Vancomycin Hydrochloride meets the requirements for *Constituted Solutions* under *Injections* ⟨1⟩.

Pyrogen—It meets the requirements of the *Pyrogen Test* ⟨151⟩, the test dose being 1 mL per kg of a solution in pyrogen-free saline TS containing 5 mg of vancomycin hydrochloride per mL.

Sterility—It meets the requirements under *Sterility Tests* ⟨71⟩, when tested as directed in the section, *Test Procedures Using Membrane Filtration*, except to dissolve the specimen in water, instead of in *Fluid A*.

Particulate matter ⟨788⟩: meets the requirements under *Small-volume Injections*.

Heavy metals, *Method II* ⟨231⟩: not more than 0.003%.

Other requirements—It responds to the *Identification test* and meets the requirements of the tests for *pH* and *Water*, under *Vancomycin Hydrochloride*. It meets also the requirements for *Uniformity of Dosage Units* ⟨905⟩, and *Labeling* under *Injections* ⟨1⟩.

Assay—

Assay preparation 1—Dissolve a suitable quantity of Sterile Vancomycin Hydrochloride, accurately weighed, in water, and dilute quantitatively with water to obtain a solution containing about 1 mg of vancomycin per mL.

Assay preparation 2 (where it is packaged for dispensing)—Dissolve the contents of 1 container of Sterile Vancomycin Hydrochloride in water, and dilute quantitatively with water to obtain a solution having a concentration of about 1 mg of vancomycin per mL.

Assay preparation 3 (where the label states the quantity of vancomycin in a given volume of constituted solution)—Constitute 1 container of Sterile Vancomycin Hydrochloride in a volume of water, accurately measured, corresponding to the volume of solvent specified in the labeling. Dilute an accurately measured volume of the constituted solution quantitatively with water to obtain a solution having a concentration of about 1 mg of vancomycin per mL.

Procedure—Proceed as directed under *Antibiotics—Microbial Assays* ⟨81⟩, using an accurately measured volume of *Assay preparation* diluted quantitatively with *Buffer No. 4* to yield a *Test Dilution* having a concentration assumed to be equal to the median dose level of the Standard.

Vanilla—*see* Vanilla NF

Vanilla Tincture—*see* Vanilla Tincture NF

Vanillin—*see* Vanillin NF

Varicella-Zoster Immune Globulin

» Varicella-Zoster Immune Globulin conforms to the regulations of the federal Food and Drug Administration concerning biologics (see *Biologics* ⟨1041⟩). It is a sterile 15 percent to 18 percent solution of pH 7.0 containing the globulin fraction of human plasma consisting of not less than 99 percent of immunoglobulin G with traces of immunoglobulin A and immunoglobulin M, in 0.3 *M* glycine as a stabilizer and 1:10,000 thimerosal as a preservative. It is derived from adult human plasma selected for high titers of varicella-zoster antibodies. Each unit of blood or plasma has been found non-reactive for hepatitis B surface antigen by a suitable method. The proteins of the plasma pools are fractionated by the cold ethanol precipitation method. The content of specific antibody is not less than 125 units, deliverable from a vial containing not more than 2.5 mL solution. The unit is defined as equivalent to 0.01 mL of a Varicella-Zoster Immune Globulin lot found effective in clinical trials and used as a reference for potency determinations, based on a fluorescent-antibody membrane antigen (FAMA) method for antibody titration.

Packaging and storage—Preserve at a temperature between 2° and 8°.

Expiration date—The expiration date is not later than 2 years after date of issue from manufacturer's cold storage.

Labeling—Label it to state that it is to be administered by intramuscular injection, in the recommended dose based on body weight.

Vasopressin Injection

H-Cys-Tyr-Phe-Glu(NH$_2$)-Asp(NH$_2$)-Cys-Pro-Arg*-Gly-NH$_2$ (residues numbered 1–9; Cys 1 and Cys 6 joined)

(*In pig vasopressin, Arg is Lys)

$C_{46}H_{65}N_{15}O_{12}S_2$ 1084.23
Vasopressin, 8-L-arginine- [*113-79-1*].
$C_{46}H_{65}N_{13}O_{12}S_2$ 1056.22
Vasopressin, 8-L-lysine- [*50-57-7*].

» Vasopressin Injection is a sterile solution, in a suitable diluent, of material containing the polypeptide hormone having the properties of causing the contraction of vascular and other smooth muscle, and of antidiuresis, which is prepared by synthesis or obtained from the posterior lobe of the pituitary of healthy, domestic animals used for food by man. Each mL of Vasopressin Injection possesses a pressor activity of not less than 85.0 percent and not more than 120.0 percent of that stated on the label in USP Posterior Pituitary Units.

Packaging and storage—Preserve in single-dose or in multiple-dose containers, preferably of Type I glass. Do not freeze.

Reference standard—*USP Posterior Pituitary Reference Standard*—Do not dry before using. Store at a temperature of 0° or below. Each mg represents 2.4 USP Posterior Pituitary Units of oxytocic activity and 2.1 USP Posterior Pituitary Units of vasopressor activity.

Oxytocic activity—

The animal preparation—Select a healthy, nulliparous, anestrous female guinea pig weighing between 175 g and 350 g. Sacrifice the animal by a blow on the head or by decapitation, and immediately remove the entire uterus from the body. Suspend one horn of the uterus in a chamber, containing oxygenated Locke-Ringer's TS, of an apparatus used for recording the activity of isolated smooth muscle. Maintain the solution surrounding the uterus at a temperature between 37° and 38°, with a variation of not more than 0.1°, during the course of the test. Allow the uterus to relax to its normal level of tone after suspending it. Weigh the lever as necessary, but do not change the weight while the contractions constituting the test are being obtained.

Preparation of the test dilutions—Prepare a dilution from the *Standard preparation* described in the *Assay* under *Oxytocin Injection*, using sufficient saline TS to give a concentration such that the addition of 0.5 mL of the test dilution to the bath will elicit a substantial contraction of the uterus. Similarly, dilute the Vasopressin Injection to a corresponding extent, assuming that each mL of the Injection contains an oxytocic activity of 1.0 USP Posterior Pituitary Unit corresponding to a potency of not more than 1.2 Units per mL on assay.

Conduct of the test—Determine the respective quantities of the dilution of the Standard and of the test dilution of the Injection which when administered alternately will elicit a series of four contractions of which the contractions elicited by the two equal doses of the Standard are submaximal. Then administer a third dose of the dilution of the Standard 25% larger than the two preceding doses of it. Measure the record of each of the five contractions. Conclude that the first two contractions to the dilution of the Standard are submaximal if the difference between the lower of them and the response to the increased dose of the dilution of the Standard is greater than twice the difference between the responses to the two equal doses of the dilution of the Standard. If the contractions to the dilution of the Standard are not submaximal, repeat the test. If the responses to the Standard are submaximal in the dilution used, and are not exceeded in height by the mean response to the dose of the Injection in the dilution used, it can be concluded that the Injection contains an amount of oxytocic activity of not more than 1.2 USP Posterior Pituitary Units for each 20 USP Units of pressor activity found in the *Assay*, and the Injection meets the requirement with respect to oxytocic activity.

pH ⟨791⟩: between 2.5 and 4.5.

Other requirements—It meets the requirements under *Injections* ⟨1⟩.

Assay—

Standard preparation—Prepare as directed in the *Assay* under *Oxytocin Injection.*

Assay preparation—To an accurately measured quantity of Vasopressin Injection add sufficient saline TS so that each mL of the resulting *Assay preparation* is expected to have the same potency as the *Standard preparation.*

The animal—About 18 hours prior to the assay, select a male rat weighing between 275 g and 325 g. Inject, subcutaneously, 1 mL per kg of body weight of a solution prepared by dissolving 50 mg of phenoxybenzamine hydrochloride in 0.1 mL of alcohol, acidifying with 1 drop of hydrochloric acid, and diluting with saline TS to 5 mL.

On the day of the assay, anesthetize the rat, using an anesthetic substance that favors the maintenance of a uniform blood pressure. Suitably secure the animal, and cannulate the trachea for artificial respiration. Arrange to obtain a continuous record of the blood pressure from the carotid artery. Arrange for intravenous injections through a suitable cannula approximately 1 mm in external diameter inserted in a femoral or jugular vein. Keep the animal warm during its preparation and during the assay.

Determination of the sensitivity of the animal preparation—Determine by trial the dose of the *Standard preparation* which, when injected intravenously at regular intervals of 12 to 15 minutes, will produce consistent blood pressure elevations of between 20 and 70 mm of mercury. From these observations select two doses of the *Standard preparation* that are in the ratio of approximately 2 to 3, and designate them as S_1 and S_2, respectively. After similar trial, select two doses (U_1 and U_2) of the *Assay preparation* that are in the same ratio as and that correspond in activity to the doses selected for the *Standard preparation.* If necessary, alter the ratio between the two doses during the assay, but maintain the same ratio for both preparations within a set, and retain the same assumed potency R throughout a given assay, and make sufficient complete sets so that there are not less than two sets for each ratio.

Procedure—Inject the selected doses of the *Standard preparation* and of the *Assay preparation* intravenously at regular intervals in sets of four paired injections. Wash in each injection with 0.2 mL of saline TS. Randomize the order in which the four pairs of doses are given within each set, but maintain the indicated sequence of the doses within pairs. Pair 1: S_2, U_1; Pair 2: S_1, U_2; Pair 3: U_2, S_1; Pair 4: U_1, S_2. Make a total of four or more sets of observations on two or more rats. One or more complete sets are to be made on each rat used for the assay.

Calculation—Record the increase in blood pressure following each dose. For each pair of doses in each replicate subtract the increase at the low dose from that at the high dose to obtain the response y. Tabulate the values of y for each pair. If an individual y is missing adjust to groups of equal size by suitable means (see *Replacement of Missing Values* ⟨111⟩). Where the ratio between the higher and lower doses is unchanged between sets, total the values of y for each dosage pair to obtain the total responses T_1, T_2, T_3, and T_4. Compute $T_a = -T_1 + T_2 + T_3 - T_4$ and $T_b = T_1 + T_2 + T_3 + T_4$. The logarithm of the relative potency of the *Assay preparation* is $M' = iT_a/T_b$, in which $i = \log (S_2/S_1) = \log (U_2/U_1)$. The potency in USP Units of vasopressin activity is the antilog ($\log R + M'$).

Compute the log confidence interval L of the log-potency (see *Confidence Intervals for Individual Assays* ⟨111⟩). If this exceeds 0.15, repeat the assay or increase the number of observations until the confidence interval of the combined results is 0.15 or less.

Where the ratio between doses has been changed, calculate the log potencies for the sets with the same ratios separately. The combined log potency is the weighted mean $\bar{M}$ of the individual potencies so calculated. Similarly compute the log confidence interval L of each individual log-potency and determine the combined log confidence interval L_c; this does not exceed 0.15.

Vegetable Oil, Hydrogenated—*see* Vegetable Oil, Hydrogenated NF

Verapamil Hydrochloride

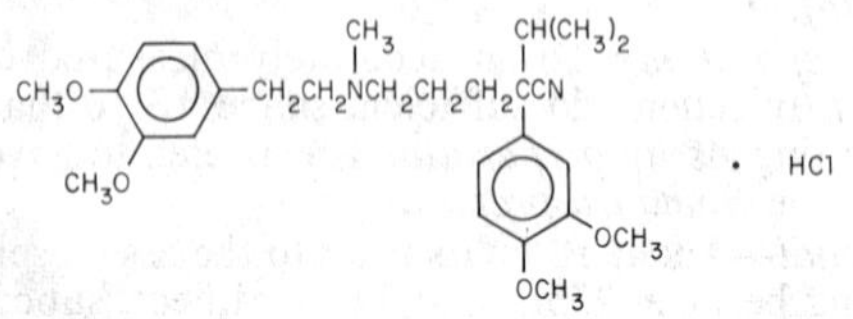

$C_{27}H_{38}N_2O_4 \cdot HCl$ 491.07

Benzeneacetonitrile, α-[3-[[2-(3,4-dimethoxyphenyl)ethyl]-methylamino]propyl]-3,4-dimethoxy-α(1-methylethyl)-, monohydrochloride.

5-[(3,4-Dimethoxyphenethyl)methylamino]-2-(3,4-dimethoxy-phenyl)-2-isopropylvaleronitrile monohydrochloride [*152-11-4*].

» Verapamil Hydrochloride contains not less than 99.0 percent and not more than 100.5 percent of $C_{27}H_{38}N_2O_4 \cdot HCl$, calculated on the dried basis.

Packaging and storage—Preserve in tight, light-resistant containers.

Reference standards—*USP Verapamil Hydrochloride Reference Standard*—Dry at 105° for 2 hours before using. *USP Verapamil Hydrochloride Related Compound B* [Benzeneacetonitrile, α-[2-[[2-(*3,4-dimethoxyphenyl*)*ethyl*]*methylamino*]*ethyl*]*-3,4-dimethoxy-α-*(*1-methylethyl*)*-monohydrochloride*] *Reference Standard*—Do not dry before using.

Identification—

A: The infrared absorption spectrum of a potassium bromide dispersion of it, previously dried, exhibits maxima only at the same wavelengths as that of a similar preparation of USP Verapamil Hydrochloride RS.

B: The chromatogram of the *Assay preparation* obtained as directed in the *Chromatographic purity* test exhibits a major peak for Verapamil Hydrochloride the retention time of which corresponds to that exhibited in the chromatogram of the *Standard preparation* obtained as directed in the test for *Chromatographic purity*.

C: It responds to the tests for *Chloride* ⟨191⟩.

Melting range ⟨741⟩: between 140° and 144°.

pH ⟨791⟩: between 4.5 and 6.5, in a solution, prepared with gentle heating, containing 50 mg per mL.

Loss on drying ⟨731⟩—Dry it at 105° for 2 hours: it loses not more than 0.5% of its weight.

Residue on ignition ⟨281⟩: not more than 0.1%.

Chromatographic purity—

Aqueous solvent mixture—Prepare a 0.01 *N* sodium acetate solution containing about 33 mL of glacial acetic acid per liter.

Mobile phase—Prepare a filtered and degassed mixture of *Aqueous solvent mixture*, acetonitrile, and 2-aminoheptane (70:30:0.5). Make adjustments if necessary (see *System Suitability* under *Chromatography* ⟨621⟩).

Standard preparations—Dissolve an accurately weighed quantity of USP Verapamil Hydrochloride RS in *Mobile phase*, and dilute quantitatively, and stepwise if necessary, with *Mobile phase* to obtain *Standard preparation A* and *Standard preparation B* having known concentrations of about 7.5 µg and 12.5 µg per mL, respectively.

Test preparation—Prepare a solution of Verapamil Hydrochloride in *Mobile phase* having a known concentration of about 2.5 mg per mL.

System suitability solution—Dissolve suitable quantities of USP Verapamil Hydrochloride RS and USP Verapamil Hydrochloride Related Compound B RS in *Mobile phase* to obtain a *System suitability solution* having known concentrations of about 2.5 mg and 2.0 mg, respectively, in each mL.

Chromatographic system (see *Chromatography* ⟨621⟩)—The liquid chromatograph is equipped with a 278-nm detector and a 4.6-mm × 12.5- to 15-cm column that contains 3-µm packing L1. The flow rate is about 0.9 mL per minute. Chromatograph the *System suitability solution*, and record the peak responses as directed under *Procedure:* the resolution, *R*, between the Verapamil Hydrochloride Related Compound B and verapamil hydrochloride peaks is not less than 1.5. The relative standard deviation for replicate injections is not more than 2.0%.

Procedure—Separately inject equal volumes (about 10 µL) of *Standard preparations A* and *B* and the *Test preparation* into the chromatograph, and allow the *Test preparation* to elute for not less than four times the retention time of Verapamil Hydrochloride. Record the chromatograms, and measure all of the peak responses. The relative retention times are about 0.88 for Verapamil Hydrochloride Related Compound B and 1.0 for verapamil hydrochloride. The sum of the peak responses, excluding that of Verapamil Hydrochloride, from the *Test preparation* is not more than that of the verapamil hydrochloride response from *Standard preparation B* (0.5%), and no single peak response is greater than that of the verapamil hydrochloride response from *Standard preparation A* (0.3%).

Assay—Dissolve about 400 mg of Verapamil Hydrochloride, accurately weighed, in 40 mL of glacial acetic acid, and add 10 mL of mercuric acetate TS and 5 mL of acetic anhydride. Titrate (see *Titrimetry* ⟨541⟩) with 0.10 *N* perchloric acid VS, determining the end-point potentiometrically. Perform a blank determination, and make any necessary correction. Each mL of 0.10 *N* perchloric acid is equivalent to 49.11 mg of $C_{27}H_{38}N_2O_4 \cdot HCl$.

Verapamil Injection

» Verapamil Injection is a sterile solution of Verapamil Hydrochloride in Water for Injection. It contains not less than 90.0 percent and not more than 110.0 percent of the labeled amount of verapamil hydrochloride ($C_{27}H_{38}N_2O_4 \cdot HCl$).

Packaging and storage—Preserve in single-dose containers, preferably of Type I glass, protected from light.

Labeling—Label Injection to state both the content of the active moiety and the content of the salt used in formulating the article.

Reference standards—*USP Verapamil Hydrochloride Reference Standard*—Dry at 105° for 2 hours before using. *USP Verapamil Hydrochloride Related Compound A* [*3,4-Dimethoxy-α-*[*3-*(*methylamino*)*-propyl*]*-α-*(*1-methylethyl*)*-benzeneacetonitrile*, *monohydrochloride*] *Reference Standard*—Do not dry before using.

Identification—

A: It meets the requirements under *Identification—Organic Nitrogenous Bases* ⟨181⟩, a volume of Injection equivalent to 100 mg of verapamil hydrochloride being used, chloroform being used in place of carbon disulfide, and a 0.1-mm cell being used in place of a 1-mm cell.

B: The chromatogram of the *Assay preparation* obtained as directed in the *Assay* exhibits a major peak for verapamil hydrochloride, the retention time of which corresponds to that exhibited in the chromatogram of the *Standard preparation* obtained as directed in the *Assay*.

C: It responds to the tests for *Chloride* ⟨191⟩.

pH ⟨791⟩: between 4.0 and 6.5.

Particulate matter ⟨788⟩: meets the requirements under *Small-volume Injections*.

Other requirements—It meets the requirements under *Injections* ⟨1⟩.

Assay for verapamil hydrochloride and limit for related compounds—

Aqueous solvent mixture, Mobile phase, System suitability solution, and *Chromatographic system*—Proceed as directed for *Chromatographic purity* under *Verapamil Hydrochloride*.

Standard preparation—Dissolve accurately weighed quantities of USP Verapamil Hydrochloride RS, USP Verapamil Hydrochloride Related Compound A RS, 3,4-Dimethoxybenzaldehyde and 3,4-Dimethoxybenzyl alcohol in *Mobile phase* to obtain a solution having known concentrations of about 2.5 mg of verapamil hydrochloride per mL, and 0.0075 mg of each related compound *A*, 3,4-dimethoxybenzaldehyde and 3,4-dimethoxybenzyl alcohol per mL.

Assay preparation—Use Verapamil Injection. Dilute quantitatively, if necessary, with *Mobile phase* to obtain a solution having a concentration not exceeding 2.5 mg of verapamil hydrochloride per mL.

Procedure—Proceed as directed for *Chromatographic purity* under *Verapamil Hydrochloride*. The retention times, relative to verapamil hydrochloride, are about 0.29 for 3,4-dimethoxybenzyl alcohol, 0.33 for *Related compound A*, and 0.53 for 3,4-dimethoxybenzaldehyde. Calculate the quantity, in mg, of the individual related compounds in each mL of the Injection taken by the formula:

$$C(L/D)(r_U/r_S),$$

in which C is the concentration, in mg per mL, of the appropriate related compound in the *Standard preparation*, L is the labeled quantity, in mg per mL, of verapamil hydrochloride in the Injection, D is the concentration, in mg per mL, of verapamil hydrochloride in the *Assay preparation*, on the basis of the labeled quantity in each mL and the extent of dilution, and r_U and r_S are the peak responses due to the appropriate related compound in the *Assay preparation* and the *Standard preparation*, respectively: not more than 0.3% of any related compound is found. Calculate the percentage of any other species, if present, by the formula:

$$100r_i/r_t,$$

in which r_i is the area of the unknown impurity peak and r_t is the sum of the areas of all the measured peaks observed in the chromatogram: the sum of all known and unknown impurities found is not greater than 1.0%. Calculate the quantity, in mg, of $C_{27}H_{38}N_2O_4.HCl$, in each mL of the Injection taken by the formula:

$$C(L/D)(r_U/r_S),$$

in which C is the concentration, in mg per mL, of USP Verapamil Hydrochloride RS in the *Standard preparation*, L is the labeled quantity, in mg per mL, of verapamil hydrochloride in the Injection, D is the concentration, in mg per mL, of verapamil hydrochloride in the *Assay preparation*, on the basis of the labeled quantity in each mL and the extent of dilution, and r_U and r_S are the peak areas due to verapamil hydrochloride in the *Assay preparation* and the *Standard preparation*, respectively.

Verapamil Tablets

» Verapamil Tablets contain not less than 90.0 percent and not more than 110.0 percent of the labeled amount of verapamil hydrochloride ($C_{27}H_{38}N_2O_4.HCl$).

Packaging and storage—Preserve in tight, light-resistant containers.

Labeling—Label Tablets to state both the content of the active moiety and the content of the salt used in formulating the article.

Reference standards—*USP Verapamil Hydrochloride Reference Standard*—Dry at 105° for 2 hours before using. *USP Verapamil Hydrochloride Related Compound A* [*3,4-Dimethoxy-α-[3-(methylamino)-propyl]-α-(1-methylethyl)-benzeneacetonitrile, monohydrochloride*] *Reference Standard*—Do not dry before using.

Identification—

A: Transfer a portion of finely powdered Tablets, equivalent to about 25 mg of verapamil hydrochloride, to a separator. Add 25 mL of water and shake by mechanical means for 30 minutes. Add 1 mL of 1 *N* sodium hydroxide and extract with 25 mL of chloroform, shaking by mechanical means for 10 minutes. Filter the chloroform extract through a filter containing anhydrous sodium sulfate. Triturate the chloroform extract with 400 mg of potassium bromide and evaporate to dryness. Dry at 105° for 2 hours: the infrared absorption spectrum of a potassium bromide dispersion of the sample so obtained exhibits maxima only at the same wavelengths as that of a similar preparation of USP Verapamil Hydrochloride RS.

B: The retention time of the major peak in the chromatogram of the *Assay preparation* corresponds to that in the chromatogram of the *Standard preparation* as obtained in the *Assay*.

Dissolution ⟨711⟩—

Medium: 0.1 *N* hydrochloric acid; 900 mL.

Apparatus 2: 50 rpm.

Time: 30 minutes.

Procedure—Determine the amount of $C_{27}H_{38}N_2O_4.HCl$ dissolved from the difference between ultraviolet absorbances at the wavelengths of maximum absorbance at about 278 nm and 300 nm using filtered portions of the solution under test, suitably diluted with 0.1 *N* hydrochloric acid if necessary, in comparison with a Standard solution having a known concentration of USP Verapamil Hydrochloride RS in the same medium.

Tolerances—Not less than 75% (*Q*) of the labeled amount of $C_{27}H_{38}N_2O_4.HCl$ is dissolved in 30 minutes.

Uniformity of dosage units ⟨905⟩: meet the requirements.

Procedure for content uniformity—Transfer one Tablet to a 100-mL volumetric flask, add 50 mL of 0.01 *N* hydrochloric acid, and heat on a steam bath for 50 minutes. Sonicate the heated solution for about 10 minutes, cool, dilute with 0.01 *N* hydrochloric acid to volume, mix, and filter. Dilute an accurately measured portion of the filtrate quantitatively with 0.01 *N* hydrochloric acid to obtain a *Test preparation* containing about 48 µg of verapamil hydrochloride per mL. Dissolve an accurately weighed quantity of USP Verapamil Hydrochloride RS in 0.01 *N* hydrochloric acid to obtain a *Standard preparation* having a known concentration of about 48 µg per mL. Concomitantly determine the absorbances of the *Test preparation* and the *Standard preparation* in 1-cm cells at the wavelength of maximum absorbance at about 278 nm and the absorbance of the *Test preparation* at 300 nm, with a suitable spectrophotometer using 0.01 *N* hydrochloric acid as the blank. Calculate the quantity, in mg, of $C_{27}H_{38}N_2O_4.HCl$ in the Tablet by the formula:

$$(TC/D)(A_U/A_S),$$

in which T is the labeled quantity, in mg, of verapamil hydrochloride in the Tablet, C is the concentration, in µg per mL, of USP Verapamil Hydrochloride RS in the *Standard preparation*, D is the concentration, in µg per mL, of verapamil hydrochloride in the *Test preparation*, on the basis of the labeled quantity per Tablet and the extent of dilution, and A_U is the difference between absorbances at 278 nm and 300 nm of the *Test preparation* and A_S is the absorbance of the *Standard preparation* at 278 nm.

Assay for verapamil hydrochloride and limit for related compounds—

Aqueous solvent mixture, Mobile phase, System suitability solution, and *Chromatographic system*—Proceed as directed for *Chromatographic purity* under *Verapamil Hydrochloride.*

Standard preparation—Dissolve accurately weighed quantities of USP Verapamil Hydrochloride RS, USP Verapamil Hydrochloride Related Compound A RS, 3,4-Dimethoxybenzaldehyde and 3,4-Dimethoxybenzyl alcohol in *Mobile phase* to obtain a solution having known concentrations of about 1.6 mg of verapamil hydrochloride per mL and 0.0048 mg of each related compound A, 3,4-dimethoxybenzaldehyde, and 3,4-dimethoxybenzyl alcohol per mL.

Assay preparation—Weigh and finely powder not less than 20 Verapamil Tablets. Transfer an accurately weighed portion of the powder, equivalent to about 40 mg of verapamil hydrochloride, to a stoppered centrifuge tube and add 25 mL of *Mobile phase*. Shake by mechanical means for 15 minutes, centrifuge, and if necessary, filter the supernatant liquid.

Procedure—Proceed as directed for *Chromatographic purity* under *Verapamil Hydrochloride.* The retention times, relative to verapamil hydrochloride, are about 0.29 for 3,4-dimethoxybenzyl alcohol, 0.33 for *Related compound A*, and 0.53 for 3,4-dimethoxybenzaldehyde. Calculate the quantity, in mg, of the individual related compounds in the portion of Tablets taken by the formula:

$$25C(r_U / r_S),$$

in which C is the concentration, in mg per mL, of *Related compound A*, 3,4-dimethoxybenzaldehyde, and 3,4-dimethoxybenzyl alcohol in the *Standard preparation*, and r_U and r_S are the related compound peak responses in the *Assay preparation* and the *Standard preparation*, respectively: not more than 0.3% of any related compound is found. Calculate the percentage of any other species, if present, by the formula:

$$100r_i / r_t,$$

in which r_i is the area of the unknown impurity peak and r_t is the sum of the areas of all the measured peaks observed in the chromatogram: the sum of all known and unknown impurities found is not greater than 1.0%. Calculate the quantity, in mg, of verapamil hydrochloride ($C_{27}H_{38}N_2O_4.HCl$) in the portion of Tablets taken by the formula:

$$25C(r_U / r_S),$$

in which C is the concentration, in mg per mL, of USP Verapamil Hydrochloride RS in the *Standard preparation*, and r_U and r_S are the verapamil hydrochloride peak areas in the *Assay preparation* and the *Standard preparation*, respectively.

Vidarabine Concentrate for Injection

» Vidarabine Concentrate for Injection contains not less than 90.0 percent and not more than 120.0 percent of the labeled amount of anhydrous vidarabine ($C_{10}H_{13}N_5O_4$) in a sterile, aqueous suspension intended for solubilization with a suitable parenteral vehicle prior to intravenous infusion. It contains suitable buffers and preservatives.

Packaging and storage—Preserve in single-dose or in multiple-dose containers, preferably of Type I glass.

Labeling—Label it to indicate that it is to be solubilized in a suitable parenteral vehicle prior to intravenous infusion.

Reference standard—*USP Vidarabine Reference Standard*—Do not dry before using.

Depressor substances—It meets the requirements of the *Depressor Substances Test* ⟨101⟩, the test dose being 1.0 mg per kg of a solution in sterile saline TS containing 1 mg of vidarabine per mL.

Pyrogen—When diluted with sterile, pyrogen-free saline TS to a concentration of 10 mg of vidarabine per mL, it meets the requirements of the *Pyrogen Test* ⟨151⟩, the test dose being 1.0 mL per kg.

Sterility—It meets the requirements under *Sterility Tests* ⟨71⟩.

pH ⟨791⟩: between 5.0 and 6.2, in the undiluted suspension.

Assay—

Mobile phase, Standard preparation, and *Chromatographic system*—Proceed as directed in the *Assay* under *Sterile Vidarabine.*

Assay preparation—Transfer an accurately measured volume of well-shaken Vidarabine Concentrate for Injection, equivalent to about 375 mg of vidarabine, to a 500-mL volumetric flask, add 50 mL of water and 5 mL of glacial acetic acid, and dissolve by heating on a steam bath for 15 minutes. Cool to room temperature, dilute with water to volume, and mix. Transfer 4.0 mL of this solution to a 25-mL volumetric flask, dilute with water to volume, and mix.

Procedure—Proceed as directed for *Procedure* in the *Assay* under *Sterile Vidarabine.* Calculate the quantity, in mg, of $C_{10}H_{13}N_5O_4$ in each mL of the Vidarabine Concentrate for Injection by the formula:

$$(0.25F/16)(r_U / r_S)(W_S / V_U),$$

in which V_U is the volume, in mL, of Vidarabine Concentrate for Injection taken, and the other terms are as defined therein.

Vidarabine Ophthalmic Ointment

» Vidarabine Ophthalmic Ointment contains not less than 90.0 percent and not more than 120.0 percent of the labeled amount of anhydrous vidarabine ($C_{10}H_{13}N_5O_4$).

Packaging and storage—Preserve in collapsible ophthalmic ointment tubes.

Reference standard—*USP Vidarabine Reference Standard*—Do not dry before using.

Sterility—It meets the requirements under *Sterility Tests* ⟨71⟩, when tested as directed in the section, *Test Procedures Using Membrane Filtration.*

Minimum fill ⟨755⟩: meets the requirements.

Metal particles—It meets the requirements of the test for *Metal Particles in Ophthalmic Ointments* ⟨751⟩.

Assay—

Mobile phase, Standard preparation, and *Chromatographic system*—Proceed as directed in the *Assay* under *Sterile Vidarabine.*

Assay preparation—Transfer an accurately weighed portion of Vidarabine Ophthalmic Ointment, equivalent to about 12 mg of vidarabine, to a 100-mL volumetric flask, add 80 mL of water, and heat on a steam bath for 15 minutes. Shake, and add 10 mL of *n*-heptane to the hot suspension. Swirl, and cool to room temperature. Remove the *n*-heptane layer, and discard it. Dilute the aqueous phase with water to volume, and mix.

Procedure—Proceed as directed for *Procedure* in the *Assay* under *Sterile Vidarabine.* Calculate the potency, in mg, of $C_{10}H_{13}N_5O_4$ per g of the Vidarabine Ophthalmic Ointment taken by the formula:

$$0.5F(r_U / r_S)(W_S / W_U),$$

in which W_U is the amount, in mg, of Ophthalmic Ointment taken, and the other terms are as defined therein.

Sterile Vidarabine

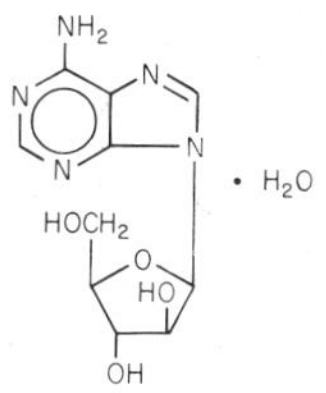

$C_{10}H_{13}N_5O_4 . H_2O$ 285.26
9*H*-Purin-6-amine, 9-β-D-arabinofuranosyl-, monohydrate.
9-β-D-Arabinofuranosyladenine monohydrate [*24356-66-9*].
Anhydrous 267.24 [*5536-17-4*].

» Sterile Vidarabine has a potency equivalent to not less than 845 μg and not more than 985 μg of $C_{10}H_{13}N_5O_4$ per mg.

Packaging and storage—Preserve in tight containers.

Reference standard—*USP Vidarabine Reference Standard*—Do not dry before using.

Identification—The infrared absorption spectrum of a potassium bromide dispersion of it exhibits maxima only at the same wavelengths as that of a similar preparation of USP Vidarabine RS.

Specific rotation ⟨781⟩: between −56.0° and −65.0°, in a solution containing 10 mg of anhydrous vidarabine per mL in dimethylformamide, determined at 365 nm.

Sterility—It meets the requirements under *Sterility Tests* ⟨71⟩, when tested as directed in the section, *Test Procedures for Direct Transfer to Test Media*, except to transfer 2 g of solid specimen to each Test Medium.

Loss on drying ⟨731⟩—Dry about 100 mg in vacuum at 100° and at a pressure not exceeding 5 mm of mercury for 4 hours: it loses between 5.0% and 7.0% of its weight.

Assay—

Mobile phase—Dissolve 2.2 g of docusate sodium in 10 mL of glacial acetic acid and 500 mL of methanol in a 1000-mL volumetric flask. Dilute with water to volume, and mix. Filter this solution through a membrane filter (1-μm or finer porosity).

Standard preparation—Dissolve about 24 mg of USP Vidarabine RS, accurately weighed, in 150 mL of water in a 200-mL volumetric flask by heating to 100° for 10 minutes. Cool, dilute with water to volume, and mix.

Assay preparation—Using Sterile Vidarabine, prepare as directed under *Standard preparation.*

Chromatographic system (see *Chromatography* ⟨621⟩)—The chromatograph has a detector set at 254 nm, and a 30-cm × 4-mm column that contains packing L1. Chromatograph three replicate injections of the *Standard preparation*, and record the peak responses as directed under *Procedure:* the relative standard deviation is not more than 3.0%.

Procedure—Introduce equal volumes (approximately 10 μL) of the *Assay preparation* and the *Standard preparation* into the instrument, operated at room temperature, by means of a suitable microsyringe or sampling valve. Adjust the operating conditions so that satisfactory chromatography and peak responses are obtained. Use a detector sensitivity setting that gives a peak height for vidarabine that is at least 50% of scale. Measure peak responses at the same retention times obtained with the *Assay preparation* and the *Standard preparation.* Calculate the potency, in μg of $C_{10}H_{13}N_5O_4$ per mg, of the Sterile Vidarabine taken, by the formula:

$$F(r_U / r_S)(W_S / W_U),$$

in which F is the potency of the USP Vidarabine RS, in μg of vidarabine per mg, r_U and r_S are the peak responses obtained with the *Assay preparation* and the *Standard preparation*, respectively, and W_U and W_S are the amounts, in mg, of USP Vidarabine RS and Sterile Vidarabine taken, respectively.

Vinblastine Sulfate

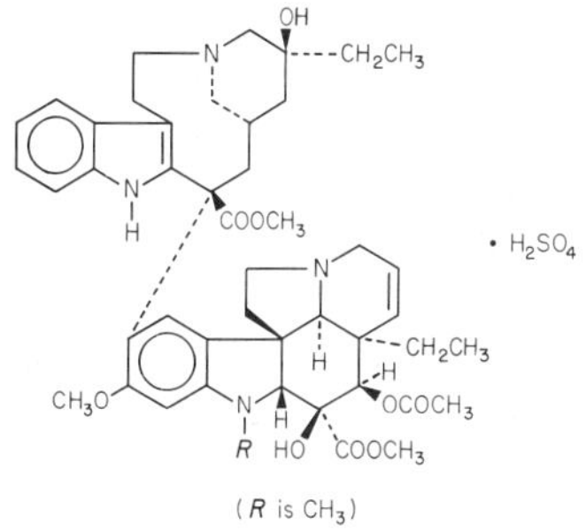

(*R* is CH_3)

$C_{46}H_{58}N_4O_9 . H_2SO_4$ 909.06
Vincaleukoblastine, sulfate (1:1) (salt).
Vincaleukoblastine sulfate (1:1) (salt) [*143-67-9*].

» Vinblastine Sulfate contains not less than 96.0 percent and not more than 102.0 percent of $C_{46}H_{58}N_4O_9 . H_2SO_4$, corrections being applied for loss in weight.

Caution—Handle Vinblastine Sulfate with great care since it is a potent cytotoxic agent.

Packaging and storage—Preserve in tight, light-resistant containers, in a freezer.

Reference standards—*USP Vinblastine Sulfate Reference Standard*—[*Caution—Avoid contact.*] After opening ampul, allow the contents to equilibrate for 30 minutes with the ambient humidity before weighing for analyses. Using thermogravimetric analysis, heat a separate equilibrated 10-mg portion at 5° per minute between ambient temperature and 200° under nitrogen flowing at 40 mL per minute (see *Thermal Analysis* ⟨891⟩). From the thermogram, determine the accumulated loss in weight between ambient temperature and a point on the plateau before decomposition is indicated (at about 160°). Keep the containers tightly closed and protected from light, and store in a cold place. [NOTE—Prepare dilutions with water at the time of use; solutions for assay that are stored in a refrigerator are to be used within 7 days.] *USP Vincristine Sulfate Reference Standard*—[NOTE—No *Loss on drying* determination is needed.]

Identification—

A: The infrared absorption spectrum of a potassium bromide dispersion of it, previously dried in vacuum at 60° for 16 hours, exhibits maxima only at the same wavelengths as that of a similar preparation of USP Vinblastine Sulfate RS.

B: A solution (1 in 10) responds to the test for *Sulfate* ⟨191⟩.

pH ⟨791⟩: between 3.5 and 5.0, in a solution prepared by dissolving 3 mg in 2 mL of water.

Loss on drying (see *Thermal Analysis* ⟨891⟩)—[NOTE—In this procedure, perform weighings rapidly with minimum exposure of the substances to air.] Determine the percentage of volatile substances by thermogravimetric analysis on an appropriately calibrated instrument, using about 10 mg of Vinblastine Sulfate, accurately weighed. Heat the specimen at the rate of 5° per minute between ambient temperature and 200° in an atmosphere of nitrogen at a flow rate of 40 mL per minute. From the thermogram, determine the accumulated loss in weight between ambient temperature and a point on the plateau before decomposition is indicated (at about 160°): it loses not more than 15.0% of its weight.

Related substances—

Mobile phase, System suitability preparation, and *Chromatographic system*—Prepare as directed in the *Assay.*

High load test preparation—Prepare as directed for *Assay preparation* in the *Assay.*

Low load test preparation—Pipet 1 mL of *High load test preparation* into a 25-mL volumetric flask, dilute with water to volume, and mix.

Procedure—Separately inject 200 μL of the *Low load test preparation* and of the *High load test preparation* into the chromatograph, and record the chromatograms. Measure the peak responses, r_i, of any related substances appearing after the solvent peak in the chromatogram of the *High load test preparation.* Calculate the total percentage of responses due to related substances by the formula:

$$100r_t/(r_t + 25r_v),$$

in which r_t is the sum of the r_i responses, and r_v is the vinblastine peak response in the chromatogram of the *Low load test preparation.* Not more than 4.0% is found. Calculate the percentage response of each related substance by the formula:

$$100r_i/(r_t + 25r_v).$$

Not more than 2.0% of response due to any individual related substance is found.

Assay—

Mobile phase—Mix 14 mL of diethylamine with 986 mL of water, and adjust with phosphoric acid to a pH of 7.5 (*Solution A*). Mix 200 mL of acetonitrile with 800 mL of methanol (*Solution B*). Mix 380 mL of *Solution A* with 620 mL of *Solution B*, filter through a 0.5-μm filter, and degas under vacuum. The ratio of *Solutions A* and *B* may be varied to meet system suitability requirements and to provide a suitable elution time for vinblastine sulfate.

Standard preparation—Dissolve an accurately weighed quantity of USP Vinblastine Sulfate RS in water to obtain a solution having a known concentration of about 0.4 mg per mL.

Assay preparation—Transfer about 4 mg of Vinblastine Sulfate, accurately weighed, to a 10-mL volumetric flask, dissolve in water, dilute with water to volume, and mix.

System suitability preparation—Dissolve an amount of USP Vincristine Sulfate RS in a portion of *Standard preparation* to obtain a solution having concentrations of about 0.4 mg of each Reference Standard per mL.

Chromatographic system (see *Chromatography* ⟨621⟩)—The liquid chromatograph is equipped with a 262-nm detector, a precolumn packed with porous silica gel installed between the pump and the injector, and a 4.6-mm × 15-cm analytical column that contains packing L1. The *Mobile phase* is maintained at a pressure and flow rate (about 2 mL per minute) capable of producing the required resolution and a suitable elution time. Chromatograph replicate injections of the *Standard preparation*, and record the peak responses as directed under *Procedure:* the relative standard deviation is not more than 2.0%. Similarly chromatograph 20 μL of the *System suitability preparation*, and record the peak responses: the resolution, *R*, between the vincristine and vinblastine is not less than 4.0 [NOTE—For a particular column, the resolution may be increased by increasing the proportion of *Solution A* in the *Mobile phase*].

Procedure—Separately inject equal volumes (about 20 μL) of the *Standard preparation* and the *Assay preparation* into the chromatograph, record the chromatograms, and measure the responses for the major peaks. Calculate the quantity, in mg, of $C_{46}H_{58}N_4O_9 \cdot H_2SO_4$ in the portion of Vinblastine Sulfate taken by the formula:

$$10C(r_U/r_S),$$

in which *C* is the concentration, in mg per mL, of USP Vinblastine Sulfate RS (corrected for loss in weight) in the *Standard preparation*, and r_U and r_S are the peak responses obtained from the *Assay preparation* and the *Standard preparation*, respectively.

Sterile Vinblastine Sulfate

» Sterile Vinblastine Sulfate is Vinblastine Sulfate suitable for parenteral use. It contains not less than 90.0 percent and not more than 110.0 percent of the labeled amount of $C_{46}H_{58}N_4O_9 \cdot H_2SO_4$.

Caution—Handle Sterile Vinblastine Sulfate with great care since it is a potent cytotoxic agent.

Packaging and storage—Preserve in *Containers for Sterile Solids* as described under *Injections* ⟨1⟩, in a refrigerator.

Reference standards—*USP Vinblastine Sulfate Reference Standard*—[*Caution—Avoid contact.*] After opening ampul, allow the contents to equilibrate for 30 minutes with the ambient humidity before weighing for analyses. Using thermogravimetric analysis, heat a separate equilibrated 10-mg portion at 5° per minute between ambient temperature and 200° under nitrogen flowing at 40 mL per minute (see *Thermal Analysis* ⟨891⟩). From the thermogram, determine the accumulated loss in weight between ambient temperature and a point on the plateau before decomposition is indicated (at about 160°). Keep the containers tightly closed and protected from light, and store in a cold place. [NOTE—Prepare dilutions with water at the time of use; solutions for assay that are stored in a refrigerator are to be used within 7 days.] *USP Vincristine Sulfate Reference Standard*—[NOTE—No *Loss on drying* determination is needed.]

Completeness of solution ⟨641⟩—A 10-mg portion dissolves in 10 mL of *Water for Injection* to yield a clear solution.

Constituted solution—At the time of use, the constituted solution prepared from Sterile Vinblastine Sulfate meets the requirements for *Constituted Solutions* under *Injections* ⟨1⟩.

Uniformity of dosage units ⟨905⟩: meets the requirements.

Procedure for content uniformity—

Buffer solution—Dissolve 13.61 g of sodium acetate in about 900 mL of water in a 1000-mL volumetric flask, adjust with glacial acetic acid to a pH of 5.0 while stirring, dilute with water to volume, and mix.

Standard preparation—Dissolve an accurately weighed quantity of USP Vinblastine Sulfate RS in *Buffer solution*, and dilute quantitatively and stepwise with *Buffer solution* to obtain a solution having a known concentration of about 40 μg per mL.

Test preparation—Dissolve the contents of 1 container of Sterile Vinblastine Sulfate in an accurately measured volume of *Buffer solution* to obtain a solution having a concentration between 40 μg per mL and 50 μg per mL.

Procedure—Concomitantly determine the absorbances of the *Test preparation* and the *Standard preparation* in 1-cm cells at the wavelength of maximum absorbance at about 269 nm versus the *Buffer solution* as the blank. Calculate the quantity, in mg, of $C_{46}H_{58}N_4O_9 \cdot H_2SO_4$ in the container by the formula:

$$0.001CV(A_U/A_S),$$

in which *C* is the concentration, in μg per mL, of USP Vinblastine Sulfate RS (corrected for loss on drying) in the *Standard preparation*, *V* is the volume, in mL, of *Buffer solution* taken for the *Test preparation*, and A_U and A_S are the absorbances of the *Test preparation* and the *Standard preparation*, respectively.

Related substances—Proceed as directed in the test for *Related substances* under *Vinblastine Sulfate*. The total of the responses due to related substances does not exceed 8.0%, and no single related substance response exceeds 2.0%.

Other requirements—It responds to the *Identification test* under *Vinblastine Sulfate*. It meets the requirements for *Sterility Tests* ⟨71⟩ and *Labeling* under *Injections* ⟨1⟩.

Assay—

Mobile phase, Standard preparation, System suitability preparation, and *Chromatographic system*—Prepare as directed in the *Assay* under *Vinblastine Sulfate*.

Assay preparation—Pipet a suitable volume of water into each of 5 containers of Sterile Vinblastine Sulfate to obtain a solution in each having a concentration of about 1 mg per mL. Insert the stopper, shake to mix, and combine the solutions from the 5 containers. Quantitatively dilute this solution with water to obtain a solution having a concentration of about 0.4 mg per mL, and mix.

Procedure—Separately inject equal volumes (about 20 μL) of the *Standard preparation* and the *Assay preparation* into the chromatograph, record the chromatograms, and measure the responses for the major peaks. Calculate the quantity, in mg, of $C_{46}H_{58}N_4O_9 \cdot H_2SO_4$ in each container of Sterile Vinblastine Sulfate taken by the formula:

$$0.2CV(r_U/r_S),$$

in which *C* is the concentration, in mg per mL, of USP Vinblastine Sulfate RS (corrected for loss in weight) in the *Standard preparation*, *V* is the volume, in mL, of the *Assay preparation*, and

r_U and r_S are the peak responses obtained from the *Assay preparation* and the *Standard preparation*, respectively.

Vincristine Sulfate

$C_{46}H_{56}N_4O_{10} \cdot H_2SO_4$ 923.04
Vincaleukoblastine, 22-oxo-, sulfate (1:1) (salt).
Leurocristine sulfate (1:1) (salt) [*2068-78-2*].

» Vincristine Sulfate contains not less than 95.0 percent and not more than 105.0 percent of $C_{46}H_{56}N_4O_{10} \cdot H_2SO_4$, corrections being applied for loss in weight.

Caution—Handle Vincristine Sulfate with great care since it is a potent cytotoxic agent.

Packaging and storage—Preserve in tight, light-resistant, containers, in a freezer.

Reference standards—*USP Vincristine Sulfate Reference Standard*—[*Caution—Avoid contact.*] After opening ampul, allow the contents to equilibrate for 30 minutes with the ambient humidity, before weighing for analysis. Using thermogravimetric analysis, heat a separate equilibrated 10-mg portion at 5° per minute between ambient temperature and 200° under nitrogen flowing at 40 mL per minute (see *Thermal Analysis* ⟨891⟩). From the thermogram, determine the accumulated loss in weight between ambient temperature and a point on the plateau before decomposition is indicated (at about 160°). Keep the containers tightly closed and protected from light, and store in a cold place. [NOTE—Prepare dilutions with water at the time of use; solutions for assay that are stored in a refrigerator are to be used within 15 days.] *USP Vinblastine Reference Standard*—[NOTE—No *Loss on drying* determination is needed.]

Identification—

A: The infrared absorption spectrum of a potassium bromide dispersion of it, previously dried in vacuum at 40° for 16 hours, exhibits maxima only at the same wavelengths as that of a similar preparation of USP Vincristine Sulfate RS.

B: A solution (1 in 10) responds to the test for *Sulfate* ⟨191⟩.

pH ⟨791⟩: between 3.5 and 4.5, in a solution (1 in 1000).

Loss on drying (see *Thermal Analysis* ⟨891⟩)—[NOTE—In this procedure, perform weighings rapidly with minimum exposure of the substances to air.] Determine the percentage of volatile substances by thermogravimetric analysis on an appropriately calibrated instrument, using about 10 mg of Vincristine Sulfate, accurately weighed. Heat the specimen at the rate of 5° per minute between ambient temperature and 200° in an atmosphere of nitrogen at a flow rate of 40 mL per minute. From the thermogram, determine the accumulated loss in weight between ambient temperature and a point on the plateau before decomposition is indicated (at about 160°): it loses not more than 12.0% of its weight.

Related substances—

Mobile phase—Prepare separate, suitable filtered and degassed components consisting of a mixture of water and diethylamine (985:15), adjusted with phosphoric acid to a pH of 7.5 (*Component A*), and methanol (*Component B*).

High load test preparation—Prepare as directed for *Assay preparation* in the *Assay*.

Low load test preparation—Pipet 1 mL of *High load test preparation* into a 25-mL volumetric flask, dilute with water to volume, and mix.

Chromatographic system (see *Chromatography* ⟨621⟩)—Use the liquid chromatograph equipped as directed in the *Assay*. The *Mobile phase* is maintained at a flow rate of about 2 mL per minute, with an initial gradient of 62% of *Component B* and 38% of *Component A* for 12 minutes, then changed to increase *Component B* at a rate of 2% per minute, so that after 15 minutes it will comprise 92% of the mixture, then changed to decrease *Component B* at a rate of 15% per minute, so that after 2 minutes it will again comprise 62% of the mixture, then maintained at this ratio for 5 minutes.

Procedure—Separately inject 200 µL of the *Low load test preparation* and of the *High load test preparation* into the chromatograph, and record the chromatograms. Measure the peak responses, r_i, of any related substances appearing after the solvent peak in the chromatogram of the *High load test preparation.* Calculate the total percentage of responses due to related substances by the formula:

$$100r_t/(r_t + 25r_v),$$

in which r_t is the sum of the r_i responses, and r_v is the vincristine peak response in the chromatogram of the *Low load test preparation.* Not more than 5.0% is found. Calculate the percentage response of each related substance by the formula:

$$100r_i/(r_t + 25r_v).$$

Not more than 2.0% of response due to any individual related substance is found.

Assay—

Mobile phase—Mix 5 mL of diethylamine with 295 mL of water, and adjust with phosphoric acid to a pH of 7.5. Add methanol to obtain 1 liter of solution, mix, filter through a 0.5-µm filter, and degas under vacuum. The methanol concentration may be varied to meet system suitability requirements and to provide a suitable elution time for vincristine sulfate.

Standard preparation—Dissolve an accurately weighed quantity of USP Vincristine Sulfate RS in water to obtain a solution having a known concentration of about 1 mg per mL.

Assay preparation—Equilibrate a portion of Vincristine Sulfate for 30 minutes with the ambient humidity, transfer about 10 mg, accurately weighed, to a 10-mL volumetric flask, dissolve in water, dilute with water to volume, and mix. Using another portion of the equilibrated specimen, determine the loss in weight as directed for *USP Vincristine Sulfate Reference Standard* under *Reference standards.*

System suitability preparation—Transfer 5 mg of USP Vincristine Sulfate RS and 5 mg of Vinblastine Sulfate RS, each accurately weighed, to a 5-mL volumetric flask, dissolve in water, dilute with water to volume, and mix.

Chromatographic system (see *Chromatography* ⟨621⟩)—The liquid chromatograph is equipped with a 297-nm detector, a pre-column packed with porous silica gel installed between the pump and the injector, a 2- to 5-cm guard column containing packing L1 installed between the injector and the analytical column, and a 4.6-mm × 25-cm analytical column that contains packing L7 (5-µm particles). The *Mobile phase* is maintained at a pressure and flow rate (about 2 mL per minute) capable of producing the required resolution and a suitable elution time. Chromatograph five replicate injections of the *Standard preparation*, and record the peak responses as directed under *Procedure:* the relative standard deviation is not more than 2.0%. Similarly chromatograph 10 µL of the *System suitability preparation*, and record the peak responses: the resolution factor between vincristine sulfate and vinblastine sulfate is not less than 4.0. [NOTE—For a particular column, the resolution may be increased by increasing the proportion of water in the *Mobile phase.*]

Procedure—Separately inject equal volumes (about 10 µL) of the *Standard preparation* and the *Assay preparation* into the chromatograph, record the chromatograms, and measure the responses for the major peaks. Calculate the quantity, in mg, of $C_{46}H_{56}N_4O_{10} \cdot H_2SO_4$ in the portion of Vincristine Sulfate taken, corrected for loss in weight in the *Assay preparation*, by the formula:

$$10C(r_U/r_S),$$

in which C is the concentration, in mg per mL, of USP Vincristine Sulfate RS, corrected for loss in weight in the *Standard preparation*, and r_U and r_S are the peak responses obtained from the *Assay preparation* and the *Standard preparation*, respectively.

Vincristine Sulfate Injection

» Vincristine Sulfate Injection is a sterile solution of Vincristine Sulfate in Water for Injection. It contains not less than 90.0 percent and not more than 110.0 percent of the labeled amount of $C_{46}H_{56}N_4O_{10} \cdot H_2SO_4$.

Caution—Handle Vincristine Sulfate Injection with great care since it is a potent cytotoxic agent.

Packaging and storage—Preserve in light-resistant, glass containers, in a refrigerator.

Reference standards—*USP Vincristine Sulfate Reference Standard*—[*Caution—Avoid contact.*] After opening ampul, allow the contents to equilibrate for 30 minutes with the ambient humidity, before weighing for analysis. Using thermogravimetric analysis, heat a separate equilibrated 10-mg portion at 5° per minute between ambient temperature and 200° under nitrogen flowing at 40 mL per minute (see *Thermal Analysis* ⟨891⟩). From the thermogram, determine the accumulated loss in weight between ambient temperature and a point on the plateau before decomposition is indicated (at about 160°). Keep the containers tightly closed and protected from light, and store in a cold place. [NOTE—Prepare dilutions with water at the time of use; solutions for assay that are stored in a refrigerator are to be used within 15 days.] *USP Vinblastine Sulfate Reference Standard*—[NOTE—No *Loss on drying* determination is needed.]

Identification—

Spray reagent—Dissolve 2.0 g of ceric ammonium sulfate in 100 mL of water with heating and stirring, and slowly add 100 mL of phosphoric acid. Filter if necessary.

Procedure—Transfer a volume of Injection, equivalent to 2 mg of vincristine sulfate, to a small centrifuge tube. For each mL of solution add 1 drop of ammonium hydroxide. Add 0.2 mL of dichloromethane. Place the cap on the tube, shake it vigorously for not less than 1 minute, and centrifuge for 1 minute. Carefully withdraw the dichloromethane layer, and transfer to a small stoppered vial. Proceed as directed for *Procedure* in the test for *Identification* under *Vincristine Sulfate for Injection*, beginning with "Also prepare a 10-mg-per-mL solution of USP Vincristine Sulfate RS."

pH ⟨791⟩: between 3.5 and 5.5.

Related substances—Proceed as directed in the test for *Related substances* under *Vincristine Sulfate*. Also inject into the chromatograph the same volume of a suitable dilution of any preservative present in the Injection, as identified in the labeling, and determine the retention time. The sum of the responses at retention times other than the retention time of vincristine and the retention times of preservatives does not exceed 6.0% of the total of all responses. The response due to *N*-desformylvincristine, eluting at 1.4 ± 0.1 of the retention time of vincristine, is not more than 3.0% of all responses, and the response due to any other related substance is not more than 2.0% of all responses.

Other requirements—It meets the requirements for *Sterility Tests* ⟨71⟩ and for *Labeling* under *Injections* ⟨1⟩.

Assay—

Mobile phase, Standard preparation, System suitability preparation, and *Chromatographic system*—Proceed as directed in the *Assay* under *Vincristine Sulfate*.

Assay preparation—Dilute, if necessary, an accurately measured volume of Vincristine Sulfate Injection quantitatively with water to obtain a solution having a concentration of about 1 mg per mL, insert the stopper, and shake to mix.

Procedure—Proceed as directed for *Procedure* in the *Assay* under *Vincristine Sulfate*. Calculate the quantity, in mg, of $C_{46}H_{56}N_4O_{10} \cdot H_2SO_4$ in each mL of the Injection taken by the formula:

$$C(L/D)(r_U/r_S),$$

in which C is the concentration, in mg per mL, of USP Vincristine Sulfate RS corrected for loss in weight in the *Standard preparation*, L is the labeled quantity, in mg per mL, of vincristine sulfate in the Vincristine Sulfate Injection, D is the concentration, in mg per mL, of vincristine sulfate in the *Assay preparation* on the basis of the labeled quantity and the extent of dilution, if any, and r_U and r_S are the peak responses obtained from the *Assay preparation* and the *Standard preparation*, respectively.

Vincristine Sulfate for Injection

» Vincristine Sulfate for Injection is a sterile mixture of Vincristine Sulfate with suitable diluents. It contains not less than 90.0 percent and not more than 110.0 percent of the labeled amount of $C_{46}H_{56}N_4O_{10} \cdot H_2SO_4$.

Caution—Handle Vincristine Sulfate for Injection with great care since it is a potent cytotoxic agent.

Packaging and storage—Preserve in *Containers for Sterile Solids* as described under *Injections* ⟨1⟩, in a refrigerator.

Reference standards—*USP Vincristine Sulfate Reference Standard*—[*Caution—Avoid contact.*] After opening ampul, allow the contents to equilibrate for 30 minutes with the ambient humidity, before weighing for analyses. Using thermogravimetric analysis, heat a separate equilibrated 10-mg portion at 5° per minute between ambient temperature and 200° under nitrogen flowing at 40 mL per minute (see *Thermal Analysis* ⟨891⟩). From the thermogram, determine the accumulated loss in weight between ambient temperature and a point on the plateau before decomposition is indicated (at about 160°). Keep the containers tightly closed and protected from light, and store in a cold place. [NOTE—Prepare dilutions with water at the time of use; solutions for assay that are stored in a refrigerator are to be used within 15 days.] *USP Vinblastine Sulfate Reference Standard*—[NOTE—No *Loss on drying* determination is needed.]

Constituted solution—At the time of use, the constituted solution prepared from Vincristine Sulfate for Injection meets the requirements for *Constituted Solutions* under *Injections* ⟨1⟩.

Identification—

Spray reagent—Dissolve 2.0 g of ceric ammonium sulfate in 100 mL of water with heating and stirring, and slowly add 100 mL of phosphoric acid. Filter if necessary.

Procedure—Dissolve a sufficient quantity in water to obtain a solution containing 25 mg per mL. Further dilute the solution to 10 mg per mL with methanol, and mix. Also prepare a 10-mg-per-mL solution of USP Vincristine Sulfate RS in a mixture of dichloromethane and methanol (3:1), and mix. Use a thin-layer chromatographic plate coated with a 0.25-mm layer of chromatographic silica gel mixture (see *Chromatography* ⟨621⟩). Develop it in a methanol prewash tank, and dry it, for maximum sensitivity, not more than 2 hours before use. Score it about 15 cm above the points of application. Apply 20 µL of each solution at points about 2.5 cm from the lower edge of the plate, and dry thoroughly (a stream of cool air may be used to help dry the spots). Prepare the developing solvent system consisting of a mixture of fresh ether, methanol, and methylamine solution (2 in 5) (95:10:5) immediately prior to development. Place the plate in the nonequilibrated developing chamber that contains a paper liner around the back and sides and developing solvent to a depth of about 2 cm. Remove the plate when the solvent moves to the scored line (about 80 minutes), and discard the solvent system. Dry the plate in a fume hood at room temperature, heat on a metal plate on a steam bath for about 15 minutes, and spray the plate while still hot with *Spray reagent*. Continue heating the plate for 15 minutes to stabilize the spots: the R_f value and the color of the principal spot obtained from the test specimen correspond to those obtained from the Reference Standard.

Uniformity of dosage units ⟨905⟩—It meets the requirements for solids.

Procedure for content uniformity—

Buffer solution—Dissolve 6.3 g of ammonium formate in about 900 mL of water in a 1000-mL volumetric flask, adjust with formic acid to a pH of 5.0 while stirring, dilute with water to volume, and mix.

Standard preparation—Dissolve an accurately weighed quantity of USP Vincristine Sulfate RS in *Buffer solution*, and dilute quantitatively and stepwise with *Buffer solution* to obtain a solution having a known concentration of about 40 µg per mL.

Test preparation—Dissolve the contents of 1 container of Vincristine Sulfate for Injection in an accurately measured volume of *Buffer solution* to obtain a solution having a concentration between 40 µg and 50 µg per mL.

Procedure—Concomitantly determine the absorbances of the *Test preparation* and the *Standard preparation* in 1-cm cells at the wavelength of maximum absorbance at about 262 nm versus the *Buffer solution* as the blank. Calculate the quantity, in mg, of $C_{46}H_{56}N_4O_{10} \cdot H_2SO_4$ in the container by the formula:

$$0.001CV(A_U/A_S),$$

in which C is the concentration, in µg per mL, of USP Vincristine Sulfate RS (corrected for loss in weight) in the *Standard preparation*, V is the volume, in mL, to which the contents of the container are diluted, and A_U and A_S are the absorbances of the *Test preparation* and the *Standard preparation*, respectively.

Related substances—Proceed as directed in the test for *Related substances* under *Vincristine Sulfate*. The total of the responses due to related substances does not exceed 5.0%, and no single related substance response exceeds 2.0%.

Other requirements—It meets the requirements for *Sterility Tests* ⟨71⟩, and for *Labeling* under *Injections* ⟨1⟩.

Assay—

Mobile phase, Standard preparation, System suitability preparation, and *Chromatographic system*—Prepare as directed in the *Assay* under *Vincristine Sulfate*.

Assay preparation—Pipet a suitable volume of water into a container of Vincristine Sulfate for Injection to obtain a solution having a concentration of about 1 mg of vincristine sulfate per mL. Insert the stopper, and shake to mix.

Procedure—Proceed as directed for *Procedure* in the *Assay* under *Vincristine Sulfate*. Calculate the quantity, in mg, of $C_{46}H_{56}N_4O_{10} \cdot H_2SO_4$ in the portion of Vincristine Sulfate for Injection taken by the formula:

$$10C(r_U/r_S),$$

in which C is the concentration, in mg per mL, of USP Vincristine Sulfate RS corrected for loss in weight in the *Standard preparation*, and r_U and r_S are the peak responses for vincristine sulfate obtained from the *Assay preparation* and the *Standard preparation*, respectively.

Vitamin A

» Vitamin A contains a suitable form of retinol ($C_{20}H_{30}O$; vitamin A alcohol) and possesses vitamin A activity equivalent to not less than 95.0 percent of that declared on the label. It may consist of retinol or esters of retinol formed from edible fatty acids, principally acetic and palmitic acids. It may be diluted with edible oils, or it may be incorporated in solid, edible carriers or excipients, and it may contain suitable antimicrobial agents, dispersants, and antioxidants.

Packaging and storage—Preserve in tight containers, preferably under an atmosphere of an inert gas, protected from light.

Labeling—Label it to indicate the form in which the vitamin is present, and to indicate the presence of any antimicrobial agent, dispersant, antioxidant, or other added substance, and to indicate the vitamin A activity in terms of the equivalent amount of retinol, in mg per g. The vitamin A activity may be stated also in USP Units, on the basis that 1 USP Vitamin A Unit equals the biological activity of 0.3 µg of the all-*trans* isomer of retinol.

Reference standard—*USP Vitamin A Reference Standard*—Discard the unused portion after opening individual capsules. Keep container tightly closed, and store in a cool, dry place, or in a refrigerator, protected from light.

Identification for vitamin A—

A: To 1 mL of a chloroform solution of it containing the equivalent of approximately 6 µg of retinol, add 10 mL of antimony trichloride TS: a transient blue color appears at once.

B: Assemble an apparatus for *Thin-layer Chromatography* (see *Chromatography* ⟨621⟩), using chromatographic silica gel as the adsorbant, and a mixture of cyclohexane and ether (4:1) as the solvent system. Prepare a Standard solution by dissolving the contents of 1 capsule of USP Vitamin A RS in chloroform to make 25.0 mL.

If the Vitamin A is in liquid form, dissolve a volume representing approximately 15,000 USP Units in chloroform to make 10 mL. If the Vitamin A is in solid form, weigh a quantity representing approximately 15,000 USP Units, place in a separator, add 75 mL of water, shake vigorously for 1 minute, extract with 10 mL of chloroform by shaking for 1 minute, and centrifuge to clarify the chloroform extract.

Apply at the starting point of the chromatogram 0.015 mL of the Standard solution and 0.01 mL of the solution of Vitamin A. Develop the chromatogram in the chromatographic chamber, lined with filter paper, dipping into the solvent mixture. When the solvent has ascended for a distance of 10 cm, remove the plate, and allow it to dry in air. Spray with phosphomolybdic acid TS: the blue-green spot formed is indicative of the presence of retinol. The approximate R_f values of the predominant spots, corresponding to the different forms of retinol, are: (*alcohol* form) 0.1; (*acetate*) 0.45; and (*palmitate*) 0.7.

Absorbance ratio—The ratio of the corrected absorbance (A_{325}) to the observed absorbance A_{325} determined as directed under *Vitamin A Assay* ⟨571⟩ is not less than 0.85.

Assay—Using a suitable quantity of Vitamin A, accurately weighed, proceed as directed under *Vitamin A Assay* ⟨571⟩.

Vitamin A Capsules

» Vitamin A Capsules contain not less than 95.0 percent and not more than 120.0 percent of the labeled amount of vitamin A.

Packaging and storage—Preserve in tight, light-resistant containers.

Labeling—Label Capsules to indicate the form in which the vitamin is present, and to indicate the vitamin A activity in terms of the equivalent amount of retinol in mg. The vitamin A activity may be stated also in USP Units per Capsule, on the basis that 1 USP Vitamin A Unit equals the biological activity of 0.3 µg of the all-*trans* isomer of retinol.

Reference standard—*USP Vitamin A Reference Standard*—Discard the unused portion after opening individual capsules. Keep container tightly closed, and store in a cool, dry place, or in a refrigerator, protected from light.

Uniformity of dosage units ⟨905⟩: meet the requirements.

Other requirements—The contents of Capsules respond to the tests for *Identification for vitamin A* and meet the requirements of the test for *Absorbance ratio* under *Vitamin A*.

Assay—Using not less than 5 Vitamin A Capsules, proceed as directed under *Vitamin A Assay* ⟨571⟩. Calculate the content of retinol ($C_{20}H_{30}O$) in mg and in USP Vitamin A Units per Capsule.

Vitamin E

» Vitamin E is a form of alpha tocopherol ($C_{29}H_{50}O_2$). It includes the following: *d*- or *dl*-alpha tocopherol ($C_{29}H_{50}O_2$); *d*- or *dl*-alpha tocopheryl acetate ($C_{31}H_{52}O_3$); *d*- or *dl*-alpha tocopheryl acid succinate ($C_{33}H_{54}O_5$). It contains not less than 96.0 percent and not more than 102.0 percent of $C_{29}H_{50}O_2$, $C_{31}H_{52}O_3$, or $C_{33}H_{54}O_5$, respectively.

Packaging and storage—Preserve in tight containers, protected from light. Protect *d*- or *dl*-alpha tocopherol with a blanket of an inert gas.

Labeling—Label Vitamin E to indicate the chemical form and to indicate whether it is the *d*- or the *dl*-form.

Reference standards—*USP Alpha Tocopherol Reference Standard*—Keep container tightly closed and protected from light. Do not dry before using. *USP Alpha Tocopheryl Acetate Reference Standard*—Keep container tightly closed and protected from light. Do not dry before using. *USP Alpha Tocopheryl Acid Succinate Reference Standard*—Keep container tightly closed and protected from light. Dry over silica gel for 18 hours before using.

Identification—

Test solution for alpha tocopheryl acetate—[NOTE—Use low-actinic glassware.] Transfer about 220 mg of *d*- or *dl*-alpha tocopheryl acetate, accurately weighed, to a round-bottom, glass-stoppered, 150-mL flask, and dissolve in 25 mL of dehydrated alcohol. Add 20 mL of dilute sulfuric acid in alcohol (1 in 7), and reflux in an all-glass apparatus for 3 hours, protected from sunlight. Cool, transfer to a 200-mL volumetric flask, add dilute sulfuric acid in alcohol (1 in 72) to volume, and mix.

Test solution for alpha tocopheryl acid succinate—[NOTE—Use low-actinic glassware.] Transfer an accurately weighed amount of the sample, equivalent to about 200 mg of alpha tocopherol, to a round-bottom, glass-stoppered, 250-mL flask, dissolve in 50 mL of dehydrated alcohol, and reflux for 1 minute. While the solution is boiling, add, through the condenser, 1 g of potassium hydroxide pellets, one at a time to avoid overheating. [*Caution—Wear safety goggles.*] Continue refluxing for 20 minutes and, without cooling, add 2 mL of hydrochloric acid dropwise through the condenser. [NOTE—This technique is essential to prevent oxidative action by air while the sample is in an alkaline medium.] Cool, and transfer the contents of the flask to a 500-mL separator, rinsing the flask with 100 mL each of water and of ether, and adding the rinsings to the separator. Shake vigorously, allow the layers to separate, and collect each of the two layers in individual separators. Extract the aqueous layer with two 50-mL portions of ether, and add these extracts to the main ether extract. Wash the combined ether extracts with four 100-mL portions of water, then evaporate the ether solution on a water bath under reduced pressure or in an atmosphere of nitrogen until about 7 or 8 mL remains. Complete the evaporation, removing the last traces of ether without the application of heat. Immediately dissolve the residue in dilute sulfuric acid in alcohol (1 in 72), transfer to a 200-mL volumetric flask, dilute with the alcoholic sulfuric acid to volume, and mix.

A: Prepare a solution in dehydrated alcohol containing 10 mg of unesterified alpha tocopherol in 10 mL, or use 10 mL of *Test solution for alpha tocopheryl acetate* or of *Test solution for alpha tocopheryl acid succinate*. Add, with swirling, 2 mL of nitric acid, and heat at about 75° for 15 minutes: a bright red or orange color develops.

B: Prepare a solution of about 100 mg, accurately weighed, of unesterified alpha tocopherol in 50 mL of ether, or in the case of esterified *d*-tocopherols, transfer an accurately measured volume of *Test solution for alpha tocopheryl acetate* or of *Test solution for alpha tocopheryl acid succinate*, equivalent to about 100 mg of the test specimen, to a separator, and add 200 mL of water. Extract first with 75 mL, then with 25 mL, of ether, and combine the ether extracts in another separator. To the ether solution of unesterified or hydrolyzed alpha tocopherol, add 20 mL of a 1 in 10 solution of potassium ferricyanide in sodium hydroxide solution (1 in 125), and shake for 3 minutes. Wash the ether solution with four 50-mL portions of water, discard the washings, and dry over anhydrous sodium sulfate. Evaporate the dried ether solution on a water bath under reduced pressure or in an atmosphere of nitrogen until about 7 or 8 mL remains, then complete the evaporation, removing the last traces of ether without the application of heat. Immediately dissolve the residue in 5.0 mL of isooctane, and determine the optical rotation. Calculate the specific rotation (see *Optical Rotation* ⟨781⟩), using as *c* the number of g of total tocopherols, determined in the *Assay*, in each 100 mL of solution employed for the test: the *d*-isomers have a specific rotation of not less than +24°. The *dl*-forms show essentially no optical rotation.

C: The retention time of the major peak in the chromatogram of the *Assay preparation* is the same as that of the *Standard preparation*, both relative to the internal standard, as obtained in the *Assay*.

Acidity—Dissolve 1.0 g of the test specimen in 25 mL of a mixture of equal volumes of alcohol and ether (which has been neutralized to phenolphthalein with 0.1 *N* sodium hydroxide), add 0.5 mL of phenolphthalein TS, and titrate with 0.10 *N* sodium hydroxide until the solution remains faintly pink after shaking for 30 seconds: alpha tocopheryl acid succinate requires between 18.0 and 19.3 mL of 0.10 *N* sodium hydroxide; the other forms of Vitamin E require not more than 1.0 mL of 0.10 *N* sodium hydroxide.

Assay for alpha tocopherol—

Internal standard solution—Dissolve an accurately weighed quantity of hexadecyl hexadecanoate in *n*-hexane to obtain a solution having a known concentration of about 1 mg per mL.

Standard preparation—[NOTE—Use low-actinic glassware.] Dissolve in *Internal standard solution* a suitable quantity of USP Alpha Tocopherol RS, accurately weighed, to obtain a solution having a known concentration of about 1 mg of the Reference Standard in each mL.

Assay preparation—[NOTE—Use low-actinic glassware.] Transfer about 50 mg of Vitamin E (*d*- or *dl*-alpha tocopherol), accurately weighed, to a 50-mL volumetric flask, dissolve in *Internal standard solution*, dilute with *Internal standard solution* to volume, and mix.

Chromatographic system (see *Chromatography* ⟨621⟩)—Under typical conditions, the instrument is equipped with a flame-ionization detector, and contains a 2-m × 4-mm borosilicate glass column packed with 2 percent to 5 percent liquid phase G2 on 80- to 100-mesh support S1AB utilizing either a glass-lined sample introduction system or on-column injection. The column is maintained isothermally at a temperature between 245° and 265°, and the injection port and detector block are maintained at about 10° higher than the column temperature; the flow rate of dry carrier gas is adjusted to obtain a hexadecyl hexadecanoate peak approximately 18 to 20 minutes after sample introduction when a 2% column is used, or 30 to 32 minutes when a 5% column is used. [NOTE—Cure and condition the column as necessary (see *Chromatography* ⟨621⟩).]

Interference check—Dissolve an accurately weighed quantity of the specimen in *n*-hexane to obtain a solution having a known concentration of about 1 mg per mL. Chromatograph an accurately measured volume of this solution to obtain a chromatogram in which the principal peak exhibits not less than 50% of maximum recorder response. Similarly chromatograph an accurately measured volume of *Internal standard solution.* If a peak observed in the chromatogram for the specimen has the same retention time as that for hexadecyl hexadecanoate, make any necessary correction for factors of dilution or attenuation, and determine the area due to the interfering component that must be subtracted from the area of the internal standard peak appearing in the chromatogram recorded for the *Assay preparation* as directed under *Procedure*.

System suitability—Chromatograph a sufficient number of injections of a mixture, in *n*-hexane, of 1 mg per mL each of USP Alpha Tocopherol RS and USP Alpha Tocopheryl Acetate RS as directed under *Procedure* to ensure that the resolution factor, *R* (see *Chromatography* ⟨621⟩), is not less than 1.0.

Calibration—Chromatograph a portion of the *Standard preparation*, and record peak areas as directed under *Procedure*. Calculate the relative response factor, *F*, for the *Standard preparation* by the formula:

$$(A_S/A_D)(C_D/C_S),$$

in which C_D and C_S are the concentrations, in mg per mL, of hexadecyl hexadecanoate and of USP Alpha Tocopherol RS, respectively, in the *Standard preparation.* Successively chromatograph a sufficient number of portions of the *Standard preparation* to ensure that the relative response factor, *F*, is constant within a range of 2.0%.

Procedure—Inject a suitable portion (2 to 5 µL) of the *Assay preparation* into a suitable gas chromatograph, and record the chromatogram so as to obtain at least 50% of maximum recorder response. Measure the areas under the first (alpha tocopherol) and second major (hexadecyl hexadecanoate) peaks, record the values as a_U and a_D, respectively. Calculate the quantity, in mg, of alpha tocopherol in the Vitamin E taken by the formula:

$$(50C_D/F)(a_U/a_D),$$

in which C_D is the concentration, in mg per mL, of hexadecyl hexadecanoate in the *Standard preparation,* and F is the relative response factor (see *Calibration*).

Assay for alpha tocopheryl acetate—Proceed as directed in the *Assay for alpha tocopherol*, substituting alpha tocopheryl acetate for alpha tocopherol and USP Alpha Tocopheryl Acetate RS for USP Alpha Tocopherol RS.

Assay for alpha tocopheryl acid succinate—Proceed as directed in the *Assay for alpha tocopherol*, substituting alpha tocopheryl acid succinate for alpha tocopherol and USP Alpha Tocopheryl Acid Succinate RS for USP Alpha Tocopherol RS.

NOTE—Chromatograms obtained as directed in the foregoing *Assays* exhibit relative retention times of approximately 0.53 for alpha tocopherol, 0.62 for alpha tocopheryl acetate, 0.54 for alpha tocopheryl acid succinate, and 1.0 for hexadecyl hexadecanoate.

Vitamin E Preparation

» Vitamin E Preparation is a combination of a single form of Vitamin E with one or more inert substances. It may be in a liquid or solid form. It contains not less than 95.0 percent and not more than 120.0 percent of the labeled amount of Vitamin E. Vitamin E Preparation labeled to contain a *dl*-form of Vitamin E may contain also a small amount of a *d*-form occurring as a minor constituent of an added substance.

Packaging and storage—Preserve in tight containers, protected from light. Protect Preparation containing *d*- or *dl*-alpha tocopherol with a blanket of an inert gas.

Labeling—Label it to indicate the chemical form of Vitamin E present, and to indicate whether the *d*- or the *dl*-form is present, excluding any different forms that may be introduced as a minor constituent of the vehicle. Designate the quantity of Vitamin E present.

Reference standards—*USP Alpha Tocopherol Reference Standard*—Keep container tightly closed and protected from light. Do not dry before using. *USP Alpha Tocopheryl Acetate Reference Standard*—Keep container tightly closed and protected from light. Do not dry before using. *USP Alpha Tocopheryl Acid Succinate Reference Standard*—Keep container tightly closed and protected from light. Dry over silica gel for 18 hours before using.

Identification—

Test solution—Proceed with the extraction and isolation of the residue obtained by hydrolysis as directed for *Test solution for alpha tocopheryl acid succinate* in the *Identification test* under *Vitamin E.* Immediately dissolve the residue in dehydrated alcohol, transfer to a 250-mL volumetric flask, dilute with dehydrated alcohol to volume, and mix.

A: To 10 mL of *Test solution* add, with swirling, 2 mL of nitric acid, and heat at about 75° for 15 minutes: a bright red or orange color develops.

B: Transfer an accurately measured volume of *Test solution*, equivalent to about 100 mg of the test specimen, to a separator, and add 200 mL of water. Proceed as directed in *Identification test B* under *Vitamin E*, beginning with "Extract first with 75 mL."

C: The retention time of the major peak in the chromatogram of the *Assay preparation* is the same as that of the *Standard preparation*, both relative to the internal standard, as obtained in the *Assay*.

Acidity—*Liquid forms of Vitamin E Preparation*—Dissolve 1.0 g in 25 mL of a mixture of equal volumes of alcohol and ether (which has been neutralized to phenolphthalein with 0.1 N sodium hydroxide), add 0.5 mL of phenolphthalein TS, and titrate with 0.10 N sodium hydroxide until the solution remains faintly pink after shaking for 30 seconds: not more than 1.0 mL of 0.10 N sodium hydroxide is required.

Assay—Proceed with Vitamin E Preparation as directed for the appropriate *Assay* under *Vitamin E*, substituting the following for the *Assay preparation.*

Assay preparation—[NOTE—Use low-actinic glassware.]

If the Preparation is in the liquid form, transfer an accurately weighed portion of Vitamin E Preparation, equivalent to about 50 mg of the specified form, to a 50-mL volumetric flask, dissolve in *Internal standard solution*, dilute with *Internal standard solution* to volume, and mix.

If the Preparation is in the solid form, transfer an accurately weighed portion of Vitamin E Preparation, equivalent to about 50 mg of Vitamin E, into a flask suitable for refluxing. Add about 5 mL of water, and heat on a water bath at 60° for 10 minutes. Add about 25 mL of alcohol, and reflux for 30 minutes. Cool, and transfer to a separator with the aid of 50 mL of water and 50 mL of ether. Shake vigorously, allow the layers to separate, and collect each in individual separators. Extract the aqueous layer with two 25-mL portions of ether, combining the extracts with the original ether layer. Wash the combined ether extracts with one 25-mL portion of water, filter the ether solution through 1 g of granular anhydrous sodium sulfate, and evaporate the ether solution on a water bath, controlled at a temperature that will not cause the ether solution to boil over, with the aid of a stream of nitrogen. Remove the container from the water bath when 5 mL remains, and complete the evaporation without the application of heat. Dissolve the residue in 50.0 mL of *Internal standard solution*, and mix.

Vitamin E Capsules

» Vitamin E Capsules contain Vitamin E or Vitamin E Preparation. They contain not less than 95.0 percent and not more than 120.0 percent of the labeled amount of vitamin E.

Packaging and storage—Preserve in tight containers, protected from light.

Labeling—Vitamin E Capsules meet the requirements for *Labeling* under *Vitamin E Preparation.*

Reference standards—*USP Alpha Tocopherol Reference Standard*—Keep container tightly closed and protected from light. Do not dry before using. *USP Alpha Tocopheryl Acetate Reference Standard*—Keep container tightly closed and protected from light. Do not dry before using. *USP Alpha Tocopheryl Acid Succinate Reference Standard*—Keep container tightly closed and protected from light. Dry over silica gel for 18 hours before using.

Identification—The contents of Capsules respond to the *Identification tests* under *Vitamin E* or under *Vitamin E Preparation.*

Uniformity of dosage units ⟨905⟩: meet the requirements.

Assay—Proceed as directed in the *Assay* for the labeled form under *Vitamin E*, substituting the following for the *Assay preparation.*

Assay preparation—Weigh accurately not less than 10 Vitamin E Capsules in a tared weighing bottle. With a sharp blade, or by other appropriate means, carefully open the capsules, without loss of shell material, and transfer the combined capsule contents to a 100-mL beaker. Remove any adhering substance from the emptied capsules by washing with several small portions of *n*-hexane. Discard the washings, and allow the empty capsules to dry in a current of dry air until the odor of *n*-hexane is no longer perceptible. Weigh the empty capsules in the original tared weighing bottle, and calculate the average net weight per capsule. Transfer an accurately weighed portion of the combined capsule contents, equivalent to the quantity of Vitamin E specified for *Assay preparation* in the *Assay* for the labeled form under *Vitamin E*, dissolve in *Internal standard solution*, dilute with *Internal standard solution* to volume, and mix.

Warfarin Sodium

$C_{19}H_{15}NaO_4$ 330.31

2*H*-1-Benzopyran-2-one, 4-hydroxy-3-(3-oxo-1-phenylbutyl)-, sodium salt.

3-(α-Acetonylbenzyl)-4-hydroxycoumarin sodium salt
[*129-06-6*].

» Warfarin Sodium is an amorphous solid or a crystalline clathrate. The clathrate form consists principally of warfarin sodium and isopropyl alcohol, in a 2:1 molecular ratio; it contains not less than 8.0 percent and not more than 8.5 percent of isopropyl alcohol. Warfarin Sodium contains not less than 97.0 percent and not more than 102.0 percent of $C_{19}H_{15}NaO_4$, calculated on the anhydrous and isopropyl alcohol–free basis.

Packaging and storage—Preserve in well-closed, light-resistant containers.

Labeling—Label it to indicate whether it is the amorphous or the crystalline form.

Reference standard—*USP Warfarin Reference Standard*—Keep container tightly closed and protected from light. Dry in vacuum over phosphorus pentoxide for 4 hours before using.

Identification—

A: The infrared absorption spectrum of a potassium bromide dispersion of the residue obtained in *Identification test B* exhibits maxima only at the same wavelengths as that of a similar preparation of USP Warfarin RS.

B: Dissolve about 100 mg in 25 mL of water, and adjust with hydrochloric acid to a pH of less than 3, using short-range pH indicator paper. Filter the mixture, wash the precipitate with four 5-mL portions of water, and dry in vacuum over phosphorus pentoxide for 4 hours: the warfarin so obtained melts between 157° and 167°, but the range between beginning and end of melting does not exceed 4°.

C: The filtrate obtained in *Identification test B* responds to the tests for *Sodium* ⟨191⟩.

pH ⟨791⟩: between 7.2 and 8.3, in a solution (1 in 100).

Water, *Method I* ⟨921⟩: not more than 4.5% for the amorphous form; not more than 0.075% for the crystalline clathrate form.

Heavy metals ⟨231⟩—Dissolve 4.0 g in 45 mL of water, add 5 mL of glacial acetic acid, stir until the precipitate agglomerates, filter, and use 25 mL of the filtrate, employing glacial acetic acid, if necessary, to make the pH adjustment: the limit is 0.001%.

Isopropyl alcohol content (crystalline clathrate form)—

Internal standard stock solution—Dilute 3 mL of dehydrated alcohol with water to 100 mL, and mix.

Internal standard solution—Pipet 5 mL of *Internal standard stock solution* into a 100-mL volumetric flask, add water to volume, and mix.

Standard preparation—Dissolve an accurately weighed quantity of isopropyl alcohol in water, and dilute quantitatively with water to obtain a solution having a known concentration of about 33 mg per mL. Pipet 10 mL of this solution and 10 mL of *Internal standard stock solution* into a 200-mL volumetric flask, dilute with water to volume, and mix.

Test preparation—Transfer about 500 mg of Warfarin Sodium, accurately weighed, to a 50-mL centrifuge tube, add 25.0 mL of *Internal standard solution*, and shake to effect complete solution.

Chromatographic system (see *Chromatography* ⟨621⟩)—The gas chromatograph is equipped with a flame-ionization detector and a 2-mm × 1.8-m column packed with support S2. The column temperature is maintained at about 115°, and the retention times of ethanol and isopropyl alcohol are about 2.2 minutes and 3.6 minutes, respectively. The resolution is not less than 2.0, and the relative standard deviation of the warfarin responses is not more than 2.0%.

Procedure—Separately inject equal volumes (about 5 µL) of the *Standard preparation* and the *Test preparation* into the chromatograph, record the chromatograms, and measure the responses for the major peaks. Calculate the weight, in mg, of isopropyl alcohol in the portion of Warfarin Sodium taken by the formula:

$$25C(R_U/R_S),$$

in which *C* is the concentration, in mg per mL, of isopropyl alcohol in the *Standard preparation*, and R_U and R_S are the peak response ratios of isopropyl alcohol to alcohol obtained from the *Test preparation* and the *Standard preparation*, respectively.

Assay—

pH 7.4 buffer—Transfer 1.36 g of monobasic potassium phosphate to a 200-mL volumetric flask, and dissolve in 50 mL of water. Add 39.1 mL of 0.2 *N* sodium hydroxide, and dilute with water to volume. Adjust with sodium hydroxide or phosphoric acid to a pH of 7.4 ± 0.1.

Mobile phase—Prepare a degassed solution containing a mixture of methanol, water, and glacial acetic acid (64:36:1). Adjust the ratio as necessary.

Internal standard solution—Dissolve propylparaben in a mixed solvent consisting of acetonitrile and glacial acetic acid (988:12), to obtain a solution having a concentration of about 0.2 mg per mL.

Standard preparation—Transfer about 94 mg of USP Warfarin RS, accurately weighed, to a 250-mL volumetric flask, and dissolve in 97.8 mL of 0.1 *N* sodium hydroxide. Add 62.5 mL of 0.2 *M* monobasic potassium phosphate, dilute with water to volume, and mix. Pipet 5 mL of this solution, 5 mL of *pH 7.4 buffer*, and 10 mL of *Internal standard solution* into a conical flask, and mix.

Assay preparation—Using about 100 mg of Warfarin Sodium, accurately weighed, prepare as directed under *Standard preparation.*

Chromatographic system (see *Chromatography* ⟨621⟩)—The liquid chromatograph is equipped with a 280-nm detector and a 4.6-mm × 25-cm column that contains packing L7. The flow rate is about 1.4 mL per minute. Chromatograph five replicate injections of the *Standard preparation*, and record the peak responses as directed under *Procedure:* the relative retention times of propylparaben and warfarin are about 0.75 and 1.0, respectively, the resolution of the two peaks is not less than 2.0, and the relative standard deviation of the warfarin responses is not more than 2.0%.

Procedure—Separately inject equal volumes (about 20 µL) of the *Standard preparation* and the *Assay preparation* into the chromatograph, record the chromatograms, and measure the responses for the major peaks. Calculate the quantity, in mg, of $C_{19}H_{15}NaO_4$ in the portion of Warfarin Sodium taken by the formula:

$$1.071C(R_U/R_S),$$

in which 1.071 is the ratio of the molecular weight of warfarin sodium to that of warfarin, *C* is the concentration, in µg per mL, of USP Warfarin RS in the *Standard preparation*, and R_U and R_S are the peak response ratios of warfarin to propylparaben obtained from the *Assay preparation* and the *Standard preparation*, respectively.

Warfarin Sodium for Injection

» Warfarin Sodium for Injection is a sterile, freeze-dried mixture of Warfarin Sodium and Sodium Chloride. It contains not less than 95.0 percent and not more than 105.0 percent of the labeled amount of $C_{19}H_{15}NaO_4$. It may contain a suitable buffer.

Packaging and storage—Preserve in light-resistant *Containers for Sterile Solids* as described under *Injections* ⟨1⟩.

Reference standard—*USP Warfarin Reference Standard*—Keep container tightly closed and protected from light. Dry in vacuum over phosphorus pentoxide for 4 hours before using.

Completeness of solution ⟨641⟩—A 1.0-g portion dissolves in 10 mL of carbon dioxide–free water to yield a clear solution.

Constituted solution—At the time of use, the constituted solution prepared from Warfarin Sodium for Injection meets the requirements for *Constituted Solutions* under *Injections* ⟨1⟩.

Water, *Method I* ⟨921⟩: not more than 4.5%.

Other requirements—It responds to *Identification tests A* and *B*, and meets the requirements for *pH* and *Heavy metals* under

Warfarin Sodium. It meets also the requirements for *Sterility Tests* ⟨71⟩, *Uniformity of Dosage Units* ⟨905⟩, and *Labeling* under *Injections* ⟨1⟩.

Assay—

pH 7.4 buffer, Mobile phase, and *Chromatographic system*—Prepare as directed in the *Assay* under *Warfarin Sodium.*

Internal standard solution—Dissolve propylparaben in a solvent consisting of mixture of acetonitrile and glacial acetic acid (988:12) to obtain a solution having a concentration of about 1.0 mg per mL.

Standard preparation—Transfer about 94 mg of USP Warfarin RS, accurately weighed, to a 100-mL volumetric flask, and dissolve in 39.1 mL of 0.1 *N* sodium hydroxide. Add 25.0 mL of 0.2 *M* monobasic potassium phosphate, dilute with water to volume, and mix. Pipet 5 mL of this solution and 5 mL of *Internal standard solution* into a 50-mL volumetric flask, dilute with *pH 7.4 buffer* to volume, and mix.

Assay preparation—Dissolve the contents of not less than 10 containers of Warfarin Sodium for Injection in a sufficient volume, accurately measured, of *pH 7.4 buffer solution* to obtain a solution containing about 1 mg of warfarin sodium per mL. Pipet 5 mL of the resulting solution and 5 mL of *Internal standard solution* into a 50-mL volumetric flask, dilute with *pH 7.4 buffer* to volume, and mix.

Procedure—Proceed as directed for *Procedure* in the *Assay* under *Warfarin Sodium.* Calculate the average quantity, in mg, of $C_{19}H_{15}NaO_4$ in each container of Warfarin Sodium for Injection taken by the formula:

$$(10)(1.071)(VC/N)(R_U/R_S),$$

in which 1.071 is the ratio of the molecular weight of warfarin sodium to that of warfarin, V is the volume, in mL, of the solution prepared from the contents of the 10 or more containers, C is the concentration, in mg per mL, of USP Warfarin RS in the *Standard preparation,* N is the number of containers taken, and R_U and R_S are the peak response ratios of warfarin to propylparaben obtained from the *Assay preparation* and the *Standard preparation,* respectively.

Warfarin Sodium Tablets

» Warfarin Sodium Tablets contain not less than 95.0 percent and not more than 105.0 percent of the labeled amount of $C_{19}H_{15}NaO_4$.

Packaging and storage—Preserve in tight, light-resistant containers.

Reference standard—*USP Warfarin Reference Standard*—Keep container tightly closed and protected from light. Dry in vacuum over phosphorus pentoxide for 4 hours before using.

Identification—

A: The retention time of the major peak obtained from the *Assay preparation* corresponds to that obtained from the *Standard preparation,* both relative to the internal standard, obtained as directed in the *Assay.*

B: Triturate a quantity of finely powdered Tablets, equivalent to about 200 mg of warfarin sodium, with 50 mL of water, centrifuge, and filter the supernatant liquid. Extract with 50 mL of ether, transfer the aqueous layer to a second separator, and discard the ether. Adjust with hydrochloric acid to a pH of less than 3, using short-range pH indicator paper, and extract with 50 mL of chloroform. Transfer the chloroform layer to another separator, extract with 50 mL of sodium hydroxide solution (1 in 250), and discard the chloroform. Transfer the aqueous layer to a beaker, and adjust with hydrochloric acid to a pH of less than 3 (using the pH indicator paper) to precipitate the warfarin. Filter, and wash the precipitate with four 5-mL portions of water. If the precipitate is not white or practically white, dissolve it in a minimum volume of sodium hydroxide solution (1 in 250), dilute with water to 50 mL, and repeat the foregoing procedure, beginning with "Extract with 50 mL of ether." Dry the precipitate in vacuum over phosphorus pentoxide for 4 hours: the infrared absorption spectrum of the warfarin so obtained exhibits maxima only at the same wavelengths as that of a similar preparation of USP Warfarin RS.

C: Dissolve a quantity of finely powdered Tablets, equivalent to about 100 mg of warfarin sodium, in 25 mL of water, and filter, if necessary: a 5-mL portion of the filtrate responds to the tests for *Sodium* ⟨191⟩.

Dissolution ⟨711⟩—

Medium: water; 900 mL.

Apparatus 2: 50 rpm.

Time: 30 minutes.

Standard preparation—[NOTE—Prepare at the time of use.]—Transfer about 25 mg of USP Warfarin RS, accurately weighed, to a 50-mL volumetric flask, dissolve in 5 mL of sodium hydroxide solution (1 in 250), dilute with *Dissolution Medium* to volume, and mix. Transfer 2.0 mL of this solution to a 100-mL volumetric flask, dilute with *Dissolution Medium* to volume, and mix. The concentration of USP Warfarin RS in the resulting solution, multiplied by 330.31/308.33, represents the equivalent concentration of warfarin sodium.

Procedure—Determine the amount of $C_{19}H_{15}NaO_4$ dissolved from ultraviolet absorbances at the wavelength of maximum absorbance at about 308 nm of filtered portions of the solution under test in comparison with the *Standard preparation.*

Tolerances—Not less than 80% (*Q*) of the labeled amount of $C_{19}H_{15}NaO_4$ is dissolved in 30 minutes.

Uniformity of dosage units ⟨905⟩: meet the requirements.

Procedure for content uniformity—

Standard preparation—Dissolve an accurately weighed quantity of USP Warfarin RS in chloroform, and dilute quantitatively and stepwise with chloroform to obtain a solution having a known concentration of about 14 µg per mL.

Test preparation—Dissolve 1 finely powdered Tablet in an accurately measured volume of sodium hydroxide solution (1 in 2500) to obtain a solution having a concentration of about 0.15 mg per mL. Filter, and pipet 10 mL of the filtrate into a 125-mL separator, add 2 mL of water, adjust with hydrochloric acid to a pH of less than 3, using short-range pH indicator paper, and mix. Extract with one 25-mL and two 15-mL portions of chloroform, filtering each through a pledget of glass wool into a 100-mL volumetric flask. Rinse the pledget, and add chloroform to volume.

Procedure—Concomitantly determine the absorbances of the *Test preparation* and the *Standard preparation* at the wavelength of maximum absorbance at about 307 nm, with a suitable spectrophotometer, using chloroform as the blank. Calculate the quantity, in mg, of $C_{19}H_{15}NaO_4$ in the Tablet by the formula:

$$1.071(TC/D)(A_U/A_S),$$

in which 1.071 is the ratio of the molecular weight of warfarin sodium to that of warfarin, T is the labeled quantity, in mg, of warfarin sodium in the Tablet, C is the concentration, in µg per mL, of USP Warfarin RS in the *Standard preparation,* D is the concentration, in µg per mL, of warfarin sodium in the *Test preparation,* based on the labeled quantity per Tablet and the extent of dilution, and A_U and A_S are the absorbances of the *Test preparation* and the *Standard preparation,* respectively.

Assay—

pH 7.4 buffer, Mobile phase, Internal standard solution, Standard preparation, and *Chromatographic system*—Prepare as directed in the *Assay* under *Warfarin Sodium.*

Assay preparation—Place 20 Warfarin Sodium Tablets in a 500-mL volumetric flask. Add 300 mL of *pH 7.4 buffer,* place in a sonic bath for 10 minutes, then shake by mechanical means for 60 minutes, add *pH 7.4 buffer* to volume, and mix. Allow the undissolved material to settle. Filter a portion of the supernatant solution through a suitable filter medium. If necessary, dilute a portion, v mL, of the clear solution quantitatively with 0.01 *N* sodium hydroxide to obtain V mL of a solution containing about 0.2 mg of warfarin per mL. Pipet 5 mL of the warfarin solution and 5 mL of *Internal standard solution* into a conical flask, and mix.

Procedure—Proceed as directed for *Procedure* in the *Assay* under *Warfarin Sodium.* Calculate the quantity, in mg, of $C_{19}H_{15}NaO_4$ in the 20 Tablets taken by the formula:

$$1.071(0.4CV/v)(R_U/R_S),$$

in which 1.071 is the ratio of the molecular weight of warfarin sodium to that of warfarin, C is the concentration, in µg per mL, of USP Warfarin RS in the *Standard preparation*, and R_U and R_S are the peak response ratios of warfarin to propylparaben obtained from the *Assay preparation* and the *Standard preparation*, respectively.

Water for Injection

» Water for Injection is water purified by distillation or by reverse osmosis. It contains no added substance.

Note—Water for Injection is intended for use as a solvent for the preparation of parenteral solutions. Where used for the preparation of parenteral solutions subject to final sterilization, use suitable means to minimize microbial growth, or first render the Water for Injection sterile and thereafter protect it from microbial contamination. For parenteral solutions that are prepared under aseptic conditions and are not sterilized by appropriate filtration or in the final container, first render the Water for Injection sterile and, thereafter, protect it from microbial contamination.

Packaging and storage—Where packaged, preserve in tight containers. Where packaged, it may be stored at a temperature below or above the range in which microbial growth occurs.

Reference standard—*USP Endotoxin Reference Standard.*

Bacterial endotoxins—When tested as directed under *Bacterial Endotoxins Test* ⟨85⟩, the USP Endotoxin RS being used, it contains not more than 0.25 USP Endotoxin Unit per mL.

Other requirements—It meets the requirements of the tests under *Purified Water*, with the exception of the test for *Bacteriological purity*.

Bacteriostatic Water for Injection

» Bacteriostatic Water for Injection is Sterile Water for Injection containing one or more suitable antimicrobial agents.

Note—Use Bacteriostatic Water for Injection with due regard for the compatibility of the antimicrobial agent or agents it contains with the particular medicinal substance that is to be dissolved or diluted.

Packaging and storage—Preserve in single-dose or in multiple-dose containers, preferably of Type I or Type II glass, of not larger than 30-mL size.

Labeling—Label it to indicate the name(s) and proportion(s) of the added antimicrobial agent(s). Label it also to include the statement, "NOT FOR USE IN NEWBORNS," in boldface capital letters, on the label immediately under the official name, printed in a contrasting color, preferably red. Alternatively, the statement may be placed prominently elsewhere on the label if the statement is enclosed within a box.

Reference standard—*USP Endotoxin Reference Standard.*

Antimicrobial agent(s)—It meets the requirements under *Antimicrobial Preservatives—Effectiveness* ⟨51⟩, and meets the labeled claim for content of the antimicrobial agent(s), as deter method set forth under *Antimicrobial Agents—Content* ⟨341⟩.

Bacterial endotoxins—When tested as directed under *Bacterial Endotoxins Test* ⟨85⟩, the USP Endotoxin RS being used, it contains not more than 0.5 USP Endotoxin Unit per mL.

Sterility—It meets the requirements under *Sterility Tests* ⟨71⟩.

pH ⟨791⟩: between 4.5 and 7.0, determined potentiometrically in a solution prepared by the addition of 0.30 mL of saturated potassium chloride solution to 100 mL of test specimen.

Particulate matter ⟨788⟩: meets the requirements under *Small-volume Injections.*

Other requirements—It meets the requirements of the tests for *Sulfate, Calcium, Carbon dioxide*, and *Heavy metals* under *Sterile Water for Injection.*

Sterile Water for Inhalation

» Sterile Water for Inhalation is water purified by distillation or by reverse osmosis and rendered sterile. It contains no antimicrobial agents, except where used in humidifiers or other similar devices and where liable to contamination over a period of time, or other added substances.

Note—Do not use Sterile Water for Inhalation for parenteral administration or for other sterile compendial dosage forms.

Packaging and storage—Preserve in single-dose containers.

Labeling—Label it to indicate that it is for inhalation therapy only and that it is not for parenteral administration.

Reference standard—*USP Endotoxin Reference Standard.*

Bacterial endotoxins—When tested as directed under *Bacterial Endotoxins Test* ⟨85⟩, the USP Endotoxin RS being used, it contains not more than 0.5 USP Endotoxin Unit per mL.

Sterility—It meets the requirements under *Sterility Tests* ⟨71⟩.

pH ⟨791⟩: between 4.5 and 7.5, in a solution containing 0.30 mL of saturated potassium chloride solution per 100 mL of test specimen.

Chloride—To 20 mL in a color-comparison tube add 5 drops of nitric acid and 1 mL of silver nitrate TS, and gently mix: any turbidity formed within 10 minutes is not greater than that produced in a similarly treated control consisting of 20 mL of *High-purity Water* (see under *Chemical Resistance—Glass Containers* ⟨661⟩), containing 10 µg of Cl, viewed downward over a dark surface with light entering the tubes from the sides (0.5 ppm).

Other requirements—It meets the requirements of the tests for *Sulfate, Calcium, Carbon dioxide*, and *Heavy metals* under *Purified Water* and of the tests for *Ammonia, Oxidizable substances*, and *Total solids* under *Sterile Water for Injection.*

Sterile Water for Injection

» Sterile Water for Injection is Water for Injection sterilized and suitably packaged. It contains no antimicrobial agent or other added substance.

Packaging and storage—Preserve in single-dose containers, preferably of Type I or Type II glass, of not larger than 1-liter size.

Labeling—Label it to indicate that no antimicrobial or other substance has been added, and that it is not suitable for intravascular injection without its first having been made approximately isotonic by the addition of a suitable solute.

Reference standard—*USP Endotoxin Reference Standard.*

Bacterial endotoxins—When tested as directed under *Bacterial Endotoxins Test* ⟨85⟩, the USP Endotoxin RS being used, it contains not more than 0.25 USP Endotoxin Unit per mL.

Sterility—It meets the requirements under *Sterility Tests* ⟨71⟩.

Particulate matter ⟨788⟩: meets the requirements under *Small-volume Injections.*

Ammonia—For Sterile Water for Injection in glass containers holding a volume up to 50 mL, dilute 50 mL with 50 mL of *High-purity Water* (see *Reagents* under *Containers* ⟨661⟩), and use this dilution as the test solution; where larger volumes are

held, use 100 mL of Sterile Water for Injection as the test solution. To 100 mL of the test solution add 2 mL of alkaline mercuric-potassium iodide TS: any yellow color produced immediately is not darker than that of a control containing 30 µg of added NH_3 in *High-purity Water* (see *Reagents* under *Containers* ⟨661⟩) (0.6 ppm for Sterile Water for Injection packaged in volumes up to 50 mL in containers; 0.3 ppm for larger volumes).

Chloride—To 20 mL in a color-comparison tube add 5 drops of nitric acid and 1 mL of silver nitrate TS, and gently mix: any turbidity formed within 10 minutes is not greater than that produced in a similarly treated control consisting of 20 mL of *High-purity Water* (see under *Reagents* in *Containers* ⟨661⟩) containing 10 µg of Cl (0.5 ppm), viewed downward over a dark surface with light entering the tubes from the sides.

Oxidizable substances—To 100 mL add 10 mL of 2 *N* sulfuric acid, and heat to boiling. For Sterile Water for Injection in glass containers holding a volume up to 50 mL, add 0.4 mL of 0.1 *N* potassium permanganate, and boil for 5 minutes; for larger volumes, add 0.2 mL of 0.1 *N* potassium permanganate, and boil for 5 minutes: the pink color does not completely disappear.

Total solids—Proceed as directed in the test for *Total solids* under *Purified Water*. The following limits apply for Sterile Water for Injection in glass containers holding up to 30 mL, 0.004%; from 30 mL up to 100 mL, 0.003%; and for larger volumes, 0.002%.

Other requirements—It meets the requirements of the tests for *pH*, *Sulfate*, *Calcium*, *Carbon dioxide*, and *Heavy metals* under *Purified Water*.

Sterile Water for Irrigation

» Sterile Water for Irrigation is Water for Injection sterilized and suitably packaged. It contains no antimicrobial agent or other added substance.

Packaging and storage—Preserve in single-dose containers, preferably of Type I or Type II glass. The container may contain a volume of more than 1 liter, and may be designed to empty rapidly.

Labeling—Label it to indicate that no antimicrobial or other substance has been added. The designations "For irrigation only" and "Not for injection" appear prominently on the label.

Reference standard—*USP Endotoxin Reference Standard.*

Other requirements—It meets the requirements of all of the tests under *Sterile Water for Injection* except the test for *Particulate matter.*

Purified Water

H_2O 18.02

» Purified Water is water obtained by distillation, ion-exchange treatment, reverse osmosis, or other suitable process. It is prepared from water complying with the regulations of the federal Environmental Protection Agency with respect to drinking water. It contains no added substance.

Note—Purified Water is intended for use as an ingredient in the preparation of compendial dosage forms. Where used for sterile dosage forms, other than for parenteral administration, process the article to meet the requirements under Sterility Tests ⟨71⟩, or first render the Purified Water sterile and thereafter protect it from microbial contamination. Do not use Purified Water in preparations intended for parenteral administration. For such purposes use Water for Injection, Bacteriostatic Water for Injection, or Sterile Water for Injection.

Packaging and storage—Where packaged, preserve in tight containers.

Labeling—Where packaged, label it to indicate the method of preparation.

pH ⟨791⟩: between 5.0 and 7.0, determined potentiometrically in a solution prepared by the addition of 0.30 mL of saturated potassium chloride solution to 100 mL of test specimen.

Chloride—To 100 mL add 5 drops of nitric acid and 1 mL of silver nitrate TS: no opalescence is produced.

Sulfate—To 100 mL add 1 mL of barium chloride TS: no turbidity is produced.

Ammonia—To 100 mL add 2 mL of alkaline mercuric-potassium iodide TS: any yellow color produced immediately is not darker than that of a control containing 30 µg of added NH_3 in *High-purity Water* (see under *Reagents* in *Containers* ⟨661⟩) [0.3 ppm].

Calcium—To 100 mL add 2 mL of ammonium oxalate TS: no turbidity is produced.

Carbon dioxide—To 25 mL add 25 mL of calcium hydroxide TS: the mixture remains clear.

Heavy metals—Adjust 40 mL of Purified Water with 1 *N* acetic acid to a pH of 3.0 to 4.0 (using short-range pH indicator paper), add 10 mL of freshly prepared hydrogen sulfide TS, and allow the liquid to stand for 10 minutes: the color of the liquid, when viewed downward over a white surface, is not darker than the color of a mixture of 50 mL of the same Purified Water with the same amount of 1 *N* acetic acid as was added to the test specimen, matched color-comparison tubes being used for the comparison.

Oxidizable substances—To 100 mL add 10 mL of 2 *N* sulfuric acid, and heat to boiling. Add 0.1 mL of 0.1 *N* potassium permanganate, and boil for 10 minutes: the pink color does not completely disappear.

Total solids—Evaporate 100 mL on a steam bath to dryness, and dry the residue at 105° for 1 hour: not more than 1 mg of residue remains (0.001%).

Bacteriological purity—It complies with the federal Environmental Protection Agency regulations for drinking water with respect to bacteriological purity (40 CFR 141.14; 141.21).

Wax, Carnauba—*see* Wax, Carnauba NF
Wax, Cetyl Esters—*see* Cetyl Esters Wax NF
Wax, Emulsifying—*see* Wax, Emulsifying NF
Wax, Microcrystalline—*see* Wax, Microcrystalline NF
Wax, White—*see* Wax, White NF
Wax, Yellow—*see* Wax, Yellow NF

White Lotion

» Prepare White Lotion as follows:

Zinc Sulfate	40 g
Sulfurated Potash	40 g
Purified Water, a sufficient quantity, to make	1000 mL

Dissolve the Zinc Sulfate and the Sulfurated Potash separately, each in 450 mL of Purified Water, and filter each solution. Add the sulfurated potash solution slowly to the zinc sulfate solution with constant stirring. Then add the required amount of purified water, and mix.

NOTE—Prepare the Lotion fresh, and shake it thoroughly before dispensing.

Packaging—Dispense in tight containers.

White Ointment—*see* Ointment, White

White Petrolatum—*see* Petrolatum, White

White Wax—*see* Wax, White NF

Whole Blood—*see* Blood Groupings

Widow Spider Species Antivenin—*see* Antivenin (Latrodectus Mactans)

Xanthan Gum—*see* Xanthan Gum NF

Xenon Xe 127

» Xenon Xe 127 is a gas suitable for inhalation in diagnostic studies. Xenon 127 is a radioactive nuclide that may be prepared from the bombardment of a cesium 133 target with high-energy protons. It contains not less than 85.0 percent and not more than 115.0 percent of the labeled amount of ^{127}Xe at the calibration date indicated on the labeling.

Packaging and storage—Preserve in single-dose vials having leak-proof stoppers, at room temperature. The vials are enclosed in appropriate lead radiation shields. The vial content may be diluted with air and is packaged at atmospheric pressure.

Labeling—Label it to include the following: the name of the preparation; the container volume, MBq (mCi) of ^{127}Xe per container; the amount of ^{127}Xe expressed as megabecquerels (millicuries) per mL; the intended route of administration; recommended storage conditions; the date of calibration; the expiration date; the name, address, and batch number of the manufacturer; the statement, "Caution—Radioactive Material"; and a radioactive symbol. The labeling contains a statement of radionuclide purity, identifies probable radionuclidic impurities, and indicates permissible quantities of each impurity. The labeling indicates that in making dosage calculations, correction is to be made for radioactive decay, and also indicates that the radioactive half-life of ^{127}Xe is 36.41 days.

Radionuclide identification (see *Radioactivity* ⟨821⟩)—Its gamma-ray spectrum is identical to that of a known specimen of xenon 127 that exhibits major photopeaks at 202.8 keV, and 172.1 keV, and 375.0 keV. Minor photopeaks from other xenon radioisotopes, namely Xe 129m (197 keV) and Xe 131m (164 keV) may also be present.

Radionuclidic purity—Using a suitable counting assembly (see *Selection of a Counting Assembly* under *Radioactivity* ⟨821⟩), determine the radioactivity of the Xe 127 in the gas by use of a calibrated system as directed under *Radioactivity* ⟨821⟩. Using the gamma-ray spectrum, determine the energy of each gamma photopeak. Identify each radionuclide present, and using the established detector efficiency and known gamma abundance, calculate the quantity of each radionuclide present in the specimen in MBq (mCi). The amount of Xe 127 present in the specimen is not less than 80%; the quantity of either Xe 131m or Xe 129m does not exceed 10%, and no other radioisotope exceeds 1%.

Assay for radioactivity—Using a suitable counting assembly (see *Selection of a Counting Assembly* under *Radioactivity* ⟨821⟩), determine the radioactivity, in MBq (mCi), of Xe 127 in each container by use of a calibrated system as directed under *Radioactivity* ⟨821⟩.

Xenon Xe 133

Xenon, isotope of mass 133.
Xenon, isotope of mass 133 [*14932-42-4*].

» Xenon Xe 133 is a gas suitable for inhalation in diagnostic studies. Xenon 133 is a radioactive nuclide that may be prepared from the fission of uranium 235. It contains not less than 85.0 percent and not more than 115.0 percent of the labeled amount of ^{133}Xe at the date and time indicated in the labeling.

Packaging and storage—Preserve in single-dose or in multiple-dose vials having leak-proof stoppers, at room temperature.

Other requirements—It meets the requirements for *Labeling*, except for the information specified for *Labeling* under *Injections;* for *Radionuclide identification;* and for *Radionuclidic purity*, and *Assay for radioactivity* under *Xenon Xe 133 Injection*, except to determine the radioactivity in MBq (mCi) per container.

Xenon Xe 133 Injection

» Xenon Xe 133 Injection is a sterile, isotonic solution of Xenon 133 in Sodium Chloride Injection suitable for intravenous administration. Xenon 133 is a radioactive nuclide prepared from the fission of uranium 235. It contains not less than 90.0 percent and not more than 110.0 percent of the labeled amount of Xenon 133 at the date and time stated on the label.

Packaging and storage—Preserve in single-dose containers that are totally filled, so that any air present occupies not more than 0.5% of the total volume of the container. Store at a temperature between 2° and 8°. If there is free space above the solution, a significant amount of the xenon 133 is present in the gaseous phase. Glass containers may darken under the effects of radiation.

Labeling—Label it to include the following, in addition to the information specified for *Labeling* under *Injections* ⟨1⟩: the time and date of calibration; the amount of xenon 133 expressed as total megabecquerels (microcuries or millicuries), and concentration as megabecquerels (microcuries or millicuries), per mL at the time of calibration; the expiration date; the name and amount of any added bacteriostatic agent; and the statement, "Caution—Radioactive Material." The labeling indicates that in making dosage calculations, correction is to be made for radioactive decay, and also indicates that the radioactive half-life of ^{133}Xe is 5.24 days.

Reference standard—*USP Endotoxin Reference Standard.*

Radionuclide identification (see *Radioactivity* ⟨821⟩)—Its gamma-ray and X-ray spectra are identical to those of a known specimen of xenon 133 that exhibits two major photopeaks having energies of 0.081 MeV and 0.031 MeV (X-ray peak).

Bacterial endotoxins—It meets the requirements of the *Bacterial Endotoxins Test* ⟨85⟩, the limit of endotoxin content being not more than 175/*V* USP Endotoxin Unit per mL of the Injection, when compared with the USP Endotoxin RS, in which *V* is the maximum recommended total dose, in mL, at the expiration date or time.

pH ⟨791⟩: between 4.5 and 8.0.

Radionuclidic purity—Using a suitable counting assembly (see *Selection of a Counting Assembly* under *Radioactivity* ⟨821⟩), determine the radioactivity of Xe 133 in the Injection by use of a calibrated system as directed under *Radioactivity* ⟨821⟩. The radioactivity exhibited at 0.081 MeV and 0.031 MeV is not less than 95.0% of the total radioactivity of the specimen.

Other requirements—It meets the requirements under *Injections* ⟨1⟩, except that the Injection may be distributed or dispensed prior to the completion of the test for *Sterility*, the latter test being started on the day of manufacture, and except that it is not subject to the recommendation on *Volume in Container*.

Assay for radioactivity—Using a suitable counting assembly (see *Selection of a Counting Assembly* under *Radioactivity* ⟨821⟩), determine the radioactivity, in MBq (mCi) per mL, of Xenon Xe 133 Injection by use of a calibrated system as directed under *Radioactivity* ⟨821⟩.

Xylometazoline Hydrochloride

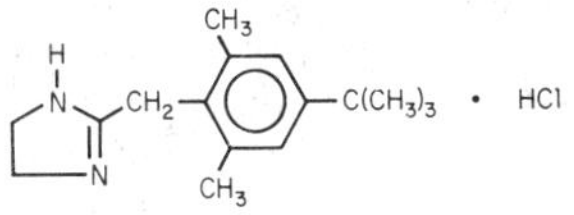

$C_{16}H_{24}N_2 \cdot HCl$ 280.84

1*H*-Imidazole, 2-[[4-(1,1-dimethylethyl)-2,6-dimethylphenyl]-methyl]-4,5-dihydro-, monohydrochloride.

2-(4-*tert*-Butyl-2,6-dimethylbenzyl)-2-imidazoline monohydrochloride [*1218-35-5*].

» Xylometazoline Hydrochloride contains not less than 99.0 percent and not more than 101.0 percent of $C_{16}H_{24}N_2 \cdot HCl$, calculated on the dried basis.

Packaging and storage—Preserve in tight, light-resistant containers.

Reference standard—*USP Xylometaxoline Hydrochloride Reference Standard*—Dry at 105° for 4 hours before using.

Identification—

A: The infrared absorption spectrum of a mineral oil dispersion of it, previously dried, exhibits maxima only at the same wavelengths as that of a similar preparation of USP Xylometazoline Hydrochloride RS.

B: The R_f value of the principal spot in the chromatogram of the *Identification preparation* corresponds to that of *Standard preparation A*, as obtained in the test for *Chromatographic purity*.

pH ⟨791⟩: between 5.0 and 6.6, in a solution (1 in 20).

Loss on drying ⟨731⟩—Dry it at 105° for 4 hours: it loses not more than 0.5% of its weight.

Residue on ignition ⟨281⟩: not more than 0.1%.

Chromatographic purity—

Standard preparations—Dissolve USP Xylometazoline Hydrochloride RS in methanol, and mix to obtain *Standard preparation A* having a known concentration of 100 µg per mL. Dilute quantitatively with methanol to obtain *Standard preparations*, designated below by letter, having the following compositions:

Dilution	Concentration (µg RS per mL)	Percentage (%, for comparison with test specimen)
A (undiluted)	100	0.5
B (4 in 5)	80	0.4
C (3 in 5)	60	0.3
D (2 in 5)	40	0.2
E (1 in 5)	20	0.1

Test preparation—Dissolve an accurately weighed quantity of Xylometazoline Hydrochloride in methanol to obtain a solution containing 20 mg per mL.

Identification preparation—Dilute a portion of the *Test preparation* quantitatively with methanol to obtain a solution containing 100 µg per mL.

Detection reagent—Prepare (1) a solution of 0.5 g of potassium iodide in 50 mL of water, and (2) a solution of 1.5 g of soluble starch in 50 mL of boiling water. Just prior to use, mix 10 mL of each solution with 3 mL of alcohol.

Procedure—On a suitable thin-layer chromatographic plate (see *Chromatography* ⟨621⟩), coated with a 0.25-mm layer of chromatographic silica gel, apply separately 5 µL of the *Test preparation*, 5 µL of the *Identification preparation*, and 5 µL of each *Standard preparation*. Position the plate in a chromatographic chamber, and develop the chromatograms in a solvent system consisting of a mixture of methanol and ammonium hydroxide (20:1) until the solvent front has moved about three-fourths of the length of the plate. Remove the plate from the developing chamber, mark the solvent front, and allow the plate to dry under a current of warm air for at least 30 minutes. Expose the plate to chlorine gas for not more than 5 minutes, and air-dry until the chlorine has dissipated (about 15 minutes). Spray the plate with *Detection reagent*, and immediately compare the intensities of any secondary spots observed in the chromatogram of the *Test preparation* with those of the principal spots in the chromatograms of the *Standard preparations:* the sum of the intensities of all secondary spots obtained from the *Test preparation* corresponds to not more than 1.0%.

Assay—Dissolve about 500 mg of Xylometazoline Hydrochloride, accurately weighed, in 70 mL of glacial acetic acid, add 10 mL of mercuric acetate TS, and titrate with 0.1 *N* perchloric acid VS, determining the end-point potentiometrically (see *Titrimetry* ⟨541⟩), using a calomel-glass electrode system. Perform a blank determination, and make any necessary correction. Each mL of 0.1 *N* perchloric acid is equivalent to 28.08 mg of $C_{16}H_{24}N_2 \cdot HCl$.

Xylometazoline Hydrochloride Nasal Solution

» Xylometazoline Hydrochloride Nasal Solution is an isotonic solution of Xylometazoline Hydrochloride in Water. It contains not less than 90.0 percent and not more than 110.0 percent of the labeled amount of $C_{16}H_{24}N_2 \cdot HCl$.

Packaging and storage—Preserve in tight, light-resistant containers.

Reference standard—*USP Xylometazoline Hydrochloride Reference Standard*—Dry at 105° for 4 hours before using.

Identification—

Standard solution—Dissolve 50 mg of USP Xylometazoline Hydrochloride RS in 50 mL of water, and proceed as directed for *Test solution*.

Test solution—Transfer 10 mL to a suitable separator, add 2 mL of sodium carbonate solution (1 in 10), and extract with 10 mL of chloroform, filtering the extract through anhydrous sodium sulfate. Evaporate the chloroform extract on a steam bath to dryness, and dissolve the residue in 1 mL of a mixture of chloroform and methanol (1:1).

Procedure—On a suitable thin-layer chromatographic plate coated with a 0.25-mm layer of chromatographic silica gel mixture (see *Chromatography* ⟨621⟩) apply, separately, 5-µL portions of the *Test solution* and the *Standard solution*. Allow the spots to dry, and develop the chromatogram in a solvent system consisting of a mixture of chloroform, methanol, and isopropylamine (92:3:3). Remove the plate from the developing chamber, mark the solvent front, and allow the solvent to evaporate. Spray the plate with *p*-nitrobenzenediazonium tetrafluoroborate solution, prepared by adding 250 mg to 5 mL of water, mixing, and filtering. Spray the plate with sodium carbonate solution (1 in 10): the R_f value of the principal spot obtained from the *Test solution* corresponds to that obtained from the *Standard solution*.

pH ⟨791⟩: between 5.0 and 7.5.

Assay—

Standard preparation—Transfer about 50 mg of USP Xylometazoline Hydrochloride RS, accurately weighed, to a 100-mL volumetric flask, add water to dissolve, dilute with water to volume, and mix. Transfer 10.0 mL of this solution to a 125-mL separator, and proceed as directed under *Assay preparation*, beginning with "add 10 mL each of water and dilute hydrochloric acid (1 in 6), respectively." The concentration of USP Xylometazoline Hydrochloride RS in the *Standard preparation* is about 100 µg per mL.

Assay preparation—Transfer an accurately measured volume of Xylometazoline Hydrochloride Nasal Solution, equivalent to about 5 mg of xylometazoline hydrochloride, to a 125-mL separator, add 10 mL each of water and dilute hydrochloric acid (1 in 6), respectively, and extract with three 10-mL portions of methylene chloride. Discard the methylene chloride extracts, add 10 mL of sodium hydroxide solution (1 in 5) to the separator, and extract with three 15-mL portions of methylene chloride. Filter the combined extracts through glass wool into a 50-mL volumetric flask, dilute with methylene chloride to volume, and mix. (Reserve a 25-mL portion of this solution for the *Identification test.*)

Procedure—Transfer 5.0 mL each of the *Standard preparation* and the *Assay preparation*, respectively, to separate 10-mL volumetric flasks, and evaporate in a water bath maintained at 40°, with the aid of a stream of nitrogen, to dryness. Dissolve the residue in each flask in 0.50 mL of dehydrated alcohol, and add 0.50 mL of dehydrated alcohol to a third 10-mL volumetric flask to provide the blank. To each flask add 0.50 mL of sodium hydroxide solution (1 in 25), swirl, to each add 5.0 mL of sodium nitroferricyanide solution (1 in 200), and mix. After 10 minutes, accurately timed, add 1.0 mL of a saturated solution of sodium bicarbonate to each flask, swirl, and allow to stand for 10 minutes. Dilute each with water to volume, mix, and allow to stand for 15 minutes. Concomitantly determine the absorbances of the solutions in 1-cm cells at the wavelength of maximum absorbance at about 565 nm, with a suitable spectrophotometer, using the blank to set the instrument. Calculate the quantity, in mg, of $C_{16}H_{24}N_2.HCl$ in each mL of the Nasal Solution taken by the formula:

$$(0.05C/V)(A_U/A_S),$$

in which C is the concentration, in µg per mL, of USP Xylometazoline Hydrochloride RS in the *Standard preparation*, V is the volume, in mL, of Nasal Solution taken, and A_U and A_S are the absorbances of the solutions from the *Assay preparation* and the *Standard preparation*, respectively.

Xylose

D-Xylopyranose

$C_5H_{10}O_5$ 150.13

D-Xylose.

D-Xylose [*58-86-6; 6763-34-4*].

» Xylose contains not less than 98.0 percent and not more than 102.0 percent of $C_5H_{10}O_5$, calculated on the dried basis.

Packaging and storage—Preserve in tight containers at controlled room temperature.

Reference standard—*USP Xylose Reference Standard*—Dry in vacuum at 60° to constant weight before using.

Color of solution—A freshly prepared solution (1 in 10) is clear and colorless.

Identification—

A: *Solvent system*—Mix 60 mL of butyl alcohol with 40 mL of pyridine and 30 mL of water.

Standard preparation—Prepare a solution of USP Xylose RS in water to obtain a solution having a concentration of 100 mg per mL.

Test preparation—Dissolve 1 g of Xylose in water, and add water to make 10 mL.

Spray reagent—Dissolve 1.66 g of phthalic acid and 0.93 g of freshly distilled aniline in 100 mL of water-saturated butyl alcohol. The solution may be stored in a brown glass bottle in a cold place, but is to be discarded if darkening becomes marked.

Chromatographic sheet—Use filter paper (Whatman No. 1 or equivalent). Draw a spotting line 6 cm from one edge of the sheet.

Procedure—Line a suitable chromatographic chamber, prepared for descending chromatography (see *Chromatography* ⟨621⟩), with blotting paper. Fill the solvent trough with *Solvent system*, and place a sufficient amount of *Solvent system* in the bottom of the chamber to permit the lining to be in contact with it. Allow the chamber to equilibrate for not less than 16 hours. On the spotting line apply 2 µL of the *Standard preparation* stepwise so that the spot is not more than 3 mm in diameter. Similarly apply 2 µL of the *Test preparation* on the spotting line and 4 cm from the *Standard preparation* spot. Expose the sheet to the atmosphere of the *Solvent system* in the closed chamber for 4 hours, then dip the edge of the sheet into the *Solvent system* in the trough, and develop until the liquid front has reached about 2.5 cm from the end of the sheet. Remove the sheet from the chamber, dry it with the aid of a gentle current of air, apply the *Spray reagent*, and dry the sheet at 105° to 110° for 5 to 10 minutes. If the spots are faint, respray and redry, and if necessary view under ultraviolet light: the R_f value of the spot from the *Test preparation* corresponds to that from the *Standard preparation.*

B: *Standard preparation*—Transfer 10 mg of USP Xylose RS to a suitable vial, and add 1 mL of pyridine, 0.2 mL of hexamethyldisilazane, and 0.1 mL of chlorotrimethylsilane. Cap the vial, shake vigorously for 30 seconds, and allow to stand for 5 minutes.

Test preparation—Using 10 mg of Xylose, proceed as directed under *Standard preparation.*

Procedure—Use a gas chromatograph equipped with a flame-ionization detector and a 1.8-m × 3-mm stainless steel column packed with 10 percent phase G2 on support S1A. Under typical conditions, nitrogen being used as the carrier gas, the column is operated at 170°, and the injector block and detector at 300°. Inject 0.5 µL each of the *Test preparation* and the *Standard preparation:* the retention times correspond.

Specific rotation ⟨781⟩: between +18.2° and +19.4°, calculated on the dried basis, determined in a solution containing 10 g of Xylose and 0.2 mL of 6 *N* ammonium hydroxide in each 100 mL.

Loss on drying ⟨731⟩—Dry 2 g to 5 g at a pressure not exceeding 50 mm of mercury at 60° to constant weight, a current of dried air being passed through the oven during the drying period to remove water vapor: it loses not more than 0.1% of its weight.

Residue on ignition ⟨281⟩: not more than 0.05%.

Arsenic, *Method I* ⟨211⟩: 1 ppm.

Iron ⟨241⟩—Dissolve 2.0 g in 45 mL of water, and add 2 mL of hydrochloric acid: the limit is 5 ppm.

Heavy metals ⟨231⟩—Dissolve 2.0 g in water to make 25 mL of solution: the limit is 0.001%.

Chromatographic impurities—The paper chromatogram of the *Test preparation* in *Identification test A* shows no foreign spot greater than any foreign spot from the *Standard preparation*, and the gas chromatogram of the *Test preparation* in *Identification test B* shows no foreign peak greater than any foreign peak from the *Standard preparation.*

Assay—

p-Bromoaniline solution—Dissolve 2 g of *p*-bromoaniline in 100 mL of thiourea-saturated glacial acetic acid. Store in an amber glass bottle, and prepare weekly.

Standard preparation—Dissolve a suitable quantity of USP Xylose RS, accurately weighed, in saturated benzoic acid solution to obtain a solution having a known concentration of about 100 µg per mL.

Assay preparation—Dissolve about 1000 mg of Xylose, accurately weighed, in saturated benzoic acid solution in a 100-mL volumetric flask, and dilute with saturated benzoic acid solution to volume. Pipet 1 mL of this solution into a second 100-mL volumetric flask, dilute with saturated benzoic acid solution to volume, and mix.

Procedure—[NOTE—In this procedure, keep strict control of time between steps.] Pipet 1-mL portions of the *Standard preparation* into each of two test tubes, and pipet 1-mL portions of the *Assay preparation* into each of two other test tubes. Into each tube pipet 5 mL of *p-Bromoaniline solution*, and mix. Loosely stopper one tube from each pair, place in a water bath at 70° for 10 minutes, remove, cool rapidly to room temperature, and mix. Set the tubes in the dark for 70 minutes. Concomitantly determine the absorbances of the treated solutions at the wavelength of maximum absorbance at 520 nm, with a suitable spectrophotometer, using the respective untreated solutions as blanks. Calculate the quantity, in mg, of $C_5H_{10}O_5$ in the portion of Xylose taken by the formula:

$$10C(A_U/A_S),$$

in which C is the concentration, in µg per mL, of USP Xylose RS in the *Standard preparation*, and A_U and A_S are the absorbances of the solutions from the *Assay preparation* and the *Standard preparation*, respectively.

Yellow Ferric Oxide—*see* Ferric Oxide, Yellow NF

Yellow Fever Vaccine

» Yellow Fever Vaccine conforms to the regulations of the federal Food and Drug Administration concerning biologics (see *Biologics* ⟨1041⟩). It is the attenuated strain that has been tested in monkeys for viscerotropism, immunogenicity, and neurotropism, of living yellow fever virus selected for high antigenic activity and safety. It is prepared by the culturing of the virus in the living embryos of chicken eggs, from which a suspension is prepared, processed with aseptic precautions, and finally dried from the frozen state. It meets the requirements of the specific mouse potency test in titer of mouse LD_{50} (quantity of virus estimated to produce fatal specific encephalitis in 50% of the mice) or the requirements for plaque-forming units in a suitable cell culture system, such as a Vero cell system for which the relationship between mouse LD_{50} and plaque-forming units has been established, in which cell monolayers in 35 mm petri dishes are inoculated for a specified time with dilutions of Vaccine, after which the dilutions are replaced with 0.5% agarose-containing medium. Following adsorption and incubation for five days an overlay is added of the 0.5% agarose medium containing 1:50,000 neutral red and the plaques are counted on the sixth day following inoculation. It is sterile, and contains no human serum and no antimicrobial agent.

Yellow Fever Vaccine is constituted, with Sodium Chloride Injection containing no antimicrobial agent, just prior to use.

Packaging and storage—Preserve in nitrogen-filled, flame-sealed ampuls or suitable stoppered vials at a temperature preferably below 0° but never above 5°, throughout the dating period. Preserve it during shipment in a suitable container adequately packed in solid carbon dioxide, or provided with other means of refrigeration, so as to insure a temperature constantly below 0°.

Expiration date—The expiration date is not later than 1 year after date of issue from manufacturer's cold storage (−20°, 1 year).

Labeling—Label it to state that it is to be well shaken before use and that the constituted vaccine is to be used entirely or discarded within 1 hour of opening the container. Label it also to state that it is the living yellow fever vaccine virus prepared from chicken embryos and that the dose is the same for persons of all ages, but that it is not recommended for infants under six months of age.

Yellow Ointment—*see* Ointment, Yellow

Yellow Phenolphthalein—*see* Phenolphthalein, Yellow

Yellow Wax—*see* Wax, Yellow NF

Ytterbium Yb 169 Pentetate Injection

$$\left[{}^{169}Yb^{3+} \quad \begin{matrix} {}^{-}OOCCH_2 & & CH_2COO^{-} & & CH_2COO^{-} \\ | & & | & & | \\ N-CH_2-CH_2- & & N-CH_2-CH_2- & & N \\ | & & & & | \\ {}^{-}OOCCH_2 & & & & CH_2COO^{-} \end{matrix} \right]^{2-} 2Na^{+}$$

$C_{14}H_{18}N_3Na_2O_{10}{}^{169}Yb$

Ytterbate (2-)-^{169}*Yb*, [*N,N*-bis[2-[bis(carboxymethyl)amino]-ethyl]glycinato(5-)]-, disodium.

Disodium [*N,N*-bis[2-[bis(carboxymethyl)amino]ethyl]glycinato(5-)]ytterbate(2-)-^{169}*Yb* [*81098-59-1*].

» Ytterbium Yb 169 Pentetate Injection is a sterile, isotonic solution of pentetic acid labeled with radioactive ^{169}Yb, suitable for intrathecal injection. It contains not less than 90.0 percent and not more than 110.0 percent of the labeled amount of ^{169}Yb as pentetic acid expressed in megabecquerels (millicuries) per mL at the time indicated in the labeling. It contains a stabilizing agent and trace quantities of calcium, sodium, and chloride ions as products of the chelating reaction. Other chemical forms of radioactivity do not exceed 2.0 percent of the total radioactivity.

Packaging and storage—Store in single-dose containers, at room temperature.

Labeling—Label it to include the following, in addition to the information specified for *Labeling* under *Injections* ⟨1⟩: the date of calibration; the amount of ^{169}Yb as pentetic acid expressed in megabecquerels (millicuries) and concentration as megabecquerels (millicuries) per mL at time of calibration; and the statement, "Caution—Radioactive Material." The labeling indicates that in making dosage calculations, correction is to be made for radioactive decay. The radioactive half-life of ^{169}Yb is 32.0 days.

Reference standard—*USP Endotoxin Reference Standard.*

Radionuclide identification (see *Radioactivity* ⟨821⟩)—Its gamma-ray spectrum is identical to a spectrum of ^{169}Yb of known purity that exhibits photoelectric peaks at 0.063, 0.110, 0.198, and 0.308 MeV.

Bacterial endotoxins—It meets the requirements of the *Bacterial Endotoxins Test* ⟨85⟩, the limit of endotoxin content being not more than $14/V$ USP Endotoxin Unit per mL of the Injection, when compared with the USP Endotoxin RS, in which V is the maximum recommended total dose, in mL, at the expiration date or time.

pH ⟨791⟩: between 5.0 and 7.0.

Radiochemical purity—

Test A—Place a measured volume (5 µL to 10 µL) of Ytterbium Yb 169 Pentetate Injection, such that it provides a count rate of about 1,000,000 counts per minute, about 20 mm from one end of a 20- × 200-mm strip of chromatographic paper (see *Chromatography* ⟨621⟩), and allow to dry. Develop the chromatogram over a suitable period by ascending chromatography, using a mixture of a solution of 0.025 *M* ammonium sulfate (pH 8.1 to 8.2 adjusted with ammonium hydroxide solution (1 in 10)) and acetone (2:1), and air-dry. Determine the radioactivity distribution by scanning the chromatogram with a suitable collimated radiation detector. Not less than 96.0% of the total radioactivity is found as labeled pentetic acid, at an R_f value of about 0.75 to 0.95. Not more than 2.0% of the total radioactivity is found as free Yb 169 at the origin of the chromatographic strip.

Test B (Biological elimination)—Inject intravenously 0.1 mL of Ytterbium Yb 169 Pentetate Injection into the caudal vein of each of three 20-g to 25-g mice. Immediately after injection, determine the amount of radioactivity in each mouse, and maintain the animals alive for 24 hours. Sacrifice the animals, determine the amount of radioactivity in each mouse, and calculate the percentage of radioactivity remaining after 24 hours by the formula:

$$100(A/B),$$

in which A is the radioactivity remaining at 24 hours after injection, and B is the total radioactivity injected. Not more than 2.0% of the injected radioactivity is retained by any test animal 24 hours following intravenous injection.

Other requirements—It meets the requirements under *Injections* ⟨1⟩, except that it is not subject to the recommendation on *Volume in Container.*

Assay for radioactivity—Using a suitable counting assembly (see *Selection of a Counting Assembly* under *Radioactivity* ⟨821⟩), determine the radioactivity, in MBq (μCi) per mL, of Ytterbium Yb 169 Pentetate Injection by use of a calibrated system as directed under *Radioactivity* ⟨821⟩.

Zein—*see* Zein NF

Zinc Acetate

$(CH_3COO-)_2Zn \cdot 2H_2O$

$C_4H_6O_4Zn \cdot 2H_2O$ 219.50
Acetic acid, zinc salt, dihydrate.
Zinc acetate dihydrate [*5970-45-6*].
Anhydrous 183.47 [*557-34-6*].

» Zinc Acetate contains not less than 98.0 percent and not more than 102.0 percent of $C_4H_6O_4Zn \cdot 2H_2O$.

Packaging and storage—Preserve in tight containers.

Identification—A solution (1 in 20) responds to the tests for *Zinc* ⟨191⟩, and for *Acetate* ⟨191⟩.

pH ⟨791⟩: between 6.0 and 8.0, in a solution (1 in 20).

Insoluble matter—A 20-g portion, dissolved in 150 mL of water containing 1 mL of glacial acetic acid, shows not more than 1.0 mg of insoluble matter (0.005%).

Arsenic, *Method I* ⟨211⟩: 3 ppm.

Lead ⟨251⟩—Dissolve 0.5 g in 1 mL of a mixture of equal parts, by volume, of nitric acid and water in a separator. Add 3 mL of *Ammonium Citrate Solution* and 0.5 mL of *Hydroxylamine Hydrochloride Solution,* and render alkaline, with ammonium hydroxide, to phenol red TS. Add 10 mL of *Potassium Cyanide Solution,* and immediately extract the solution with successive 5-mL portions of *Dithizone Extraction Solution,* draining off each extract into another separator, until the last portion of dithizone solution retains its green color. Shake the combined extracts for 30 seconds with 20 mL of dilute nitric acid (1 in 100), and discard the chloroform layer. Add to the acid solution 4.0 mL of *Ammonia-cyanide Solution* and 2 drops of *Hydroxylamine Hydrochloride Solution.* Add 10.0 mL of *Standard Dithizone Solution,* and shake the mixture for 30 seconds. Filter the chloroform layer through acid-washed filter paper into a color-comparison tube, and compare the color with that of a standard prepared as follows: to 20 mL of dilute nitric acid (1 in 100) add 0.01 mg of lead, 4 mL of *Ammonia-cyanide Solution,* and 2 drops of *Hydroxylamine Hydrochloride Solution,* and shake for 30 seconds with 10.0 mL of *Standard Dithizone Solution.* Filter through acid-washed filter paper into a color-comparison tube: the color of the sample solution does not exceed that of the control (0.002%).

Chloride ⟨221⟩—A 1.5-g portion shows no more chloride than corresponds to 0.10 mL of 0.020 N hydrochloric acid (0.005%).

Sulfate ⟨221⟩—A 1.0-g portion shows no more sulfate than corresponds to 0.10 mL of 0.020 N sulfuric acid (0.010%).

Alkalies and alkaline earths—Dissolve 2.0 g in about 150 mL of water contained in a 200-mL volumetric flask, add sufficient ammonium sulfide TS to precipitate the zinc completely, dilute with water to volume, and mix. Filter through a dry filter, rejecting the first portion of the filtrate. To 100 mL of the subsequent filtrate add 5 drops of sulfuric acid, evaporate to dryness, and ignite: the weight of the residue does not exceed 2 mg (0.2%).

Assay—Dissolve about 400 mg of Zinc Acetate, accurately weighed, in 100 mL of water. Add 5 mL of ammonia–ammonium chloride buffer TS and 0.1 mL of eriochrome black TS, and titrate with 0.05 M disodium ethylenediaminetetraacetate VS until the solution is deep blue in color. Each mL of 0.05 M disodium ethylenediaminetetraacetate is equivalent to 10.98 mg of $C_4H_6O_4Zn \cdot 2H_2O$.

Zinc, Bacitracin—*see* Bacitracin Zinc

Zinc Chloride

$ZnCl_2$ 136.29
Zinc chloride.
Zinc chloride [*7646-85-7*].

» Zinc Chloride contains not less than 97.0 percent and not more than 100.5 percent of $ZnCl_2$.

Packaging and storage—Preserve in tight containers.

Identification—A solution of it responds to the tests for *Zinc* ⟨191⟩ and for *Chloride* ⟨191⟩.

Oxychloride—Dissolve 1.0 g in 20 mL of water, add 20 mL of alcohol, and mix. To 10 mL of the mixture add 0.30 mL of 1.0 N hydrochloric acid: the solution becomes perfectly clear.

Sulfate ⟨221⟩—Dissolve 1.0 g in 30 mL of water: 20 mL of this solution shows no more sulfate than corresponds to 0.20 mL of 0.020 N sulfuric acid (0.03%).

Alkalies and alkaline earths—Dissolve 2.0 g in about 150 mL of water contained in a 200-mL volumetric flask. Add sufficient ammonium sulfide TS to precipitate the zinc completely, dilute with water to volume, and mix. Filter through a dry filter, and reject the first portion of the filtrate. To 100 mL of the subsequent filtrate add 5 drops of sulfuric acid, evaporate to dryness, and ignite: the weight of the residue does not exceed 10 mg (1.0%).

Ammonium salts—To 5 mL of a solution (1 in 10) add 1 N sodium hydroxide until the precipitate first formed is redissolved, and then warm the solution: no odor of ammonia is perceptible.

Lead ⟨251⟩—Dissolve 0.50 g in 5 mL of water, and transfer the solution to a color-comparison tube (A). Add 15 mL of *Potassium Cyanide Solution* (1 in 10), mix, and allow the mixture to become clear. In a similar, matched color-comparison tube (B) place 5 mL of water, and add 2.50 mL of *Standard Lead Solution* (see *Heavy Metals* ⟨231⟩) and 15 mL of *Potassium Cyanide Solution* (1 in 10). Add to the solution in each tube 0.1 mL of sodium sulfide TS. Mix the contents of each tube, and allow to stand for 5 minutes: viewed downward over a white surface, the solution in tube A is not darker than that in tube B (indicating not more than 0.005% of lead).

Assay—Dissolve about 12 g of Zinc Chloride, accurately weighed, in about 500 mL of water in a 1-liter volumetric flask, add 12 g of ammonium chloride, dilute with water to volume, and mix. Pipet 25 mL of the solution into a 400-mL beaker, add 100 mL of water, 10 mL of ammonia–ammonium chloride buffer TS, and 1 mL of eriochrome black TS solution (1 in 2000), and titrate with 0.05 *M* disodium ethylenediaminetetraacetate VS to a deep blue end-point. Each mL of 0.05 *M* disodium ethylenediaminetetraacetate is equivalent to 6.815 mg of $ZnCl_2$.

Zinc Chloride Injection

» Zinc Chloride Injection is a sterile solution of Zinc Chloride in Water for Injection. It contains not less than 95.0 percent and not more than 105.0 percent of the labeled amount of zinc (Zn).

Packaging and storage—Preserve in single-dose or in multiple-dose containers, preferably of Type I or Type II glass.

Labeling—Label the Injection to indicate that it is to be diluted with Water for Injection or other suitable fluid to appropriate strength prior to administration.

Identification—The *Assay preparation* prepared as directed in the *Assay* exhibits an absorption maximum at about 213.8 nm, when determined as directed in the *Assay*.

Pyrogen—When diluted with Sodium Chloride Injection to contain 20 µg of zinc per mL, it meets the requirements of the *Pyrogen Test* ⟨151⟩.

pH ⟨791⟩: between 1.5 and 2.5.

Particulate matter ⟨788⟩: meets the requirements under *Small-volume Injections*.

Other requirements—It meets the requirements under *Injections* ⟨1⟩.

Assay—[NOTE—The *Standard preparations* and the *Assay preparation* may be diluted quantitatively with 0.012 *N* hydrochloric acid, if necessary, to obtain solutions of suitable concentrations adaptable to the linear or working range of the instrument.]

Zinc stock solution—Transfer 0.50 g of zinc metal powder, accurately weighed, to a 1000-mL volumetric flask, dissolve in a minimum volume of 6 *N* hydrochloric acid, dilute with 0.12 *N* hydrochloric acid to volume, and mix. This solution contains 500 µg of zinc per mL. Store in a polyethylene bottle.

Standard preparations—Pipet 10 mL of *Zinc stock solution* into a 1000-mL volumetric flask, dilute with water to volume, and mix. Transfer 10.0, 15.0, and 20.0 mL, respectively, of this solution into separate 100-mL volumetric flasks. Dilute the contents of each flask with 0.012 *N* hydrochloric acid to volume, and mix. The *Standard preparations* contain, respectively, 0.50, 0.75, and 1.0 µg of zinc per mL.

Assay preparation—Transfer an accurately measured volume of Zinc Chloride Injection, equivalent to about 7.5 mg of zinc, to a 1000-mL volumetric flask, dilute with water to volume, and mix. Transfer 10.0 mL of this solution to a 100-mL volumetric flask, dilute with 0.012 *N* hydrochloric acid to volume, and mix.

Procedure—Concomitantly determine the absorbances of the *Standard preparations* and the *Assay preparation* at the zinc emission line at 213.8 nm, with a suitable atomic absorption spectrophotometer (see *Spectrophotometry and Light-scattering* ⟨851⟩) equipped with a zinc hollow-cathode lamp and an air-acetylene flame, using 0.012 *N* hydrochloric acid as the blank. Plot the absorbances of the *Standard preparations* versus concentration, in µg per mL, of zinc, and draw the straight line best fitting the three plotted points. From the graph so obtained, determine the concentration, in µg per mL, of zinc in the *Assay preparation*. Calculate the quantity, in mg, of zinc in each mL of the Injection taken by the formula:

$$10C/V,$$

in which *C* is the concentration, in µg per mL, of zinc in the *Assay preparation*, and *V* is the volume, in mL, of Injection taken.

Zinc Gluconate

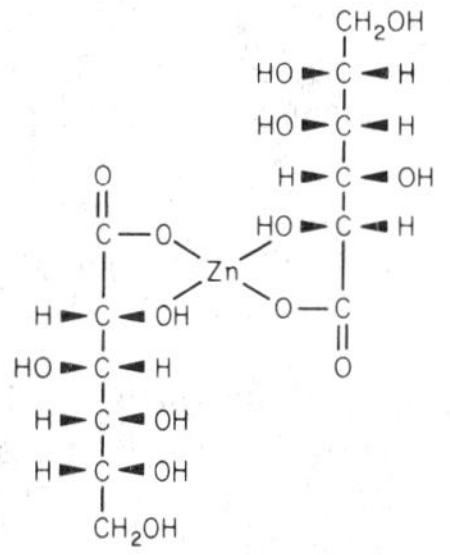

$C_{12}H_{22}O_{14}Zn$ 455.68

Bis(D-gluconato-O^1,O^2) zinc.

Zinc D-gluconate (1:2) [*4468-02-4*].

» Zinc Gluconate contains not less than 97.0 percent and not more than 102.0 percent of $C_{12}H_{22}O_{14}Zn$, calculated on the anhydrous basis.

Packaging and storage—Preserve in well-closed containers.

Reference standard—*USP Potassium Gluconate Reference Standard*—Dry in vacuum at 105° for 4 hours before using.

Identification—

A: A solution (1 in 10) responds to the tests for *Zinc* ⟨191⟩.

B: It responds to *Identification test B* under *Calcium Gluconate*.

pH ⟨791⟩: between 5.5 and 7.5, in a solution (1 in 100).

Water, *Method Ib* ⟨921⟩: not more than 11.6%.

Chloride ⟨221⟩—A 1.0-g portion shows no more chloride than corresponds to 0.70 mL of 0.020 *N* hydrochloric acid (0.05%).

Sulfate ⟨221⟩—A 2.0-g portion shows no more sulfate than corresponds to 1.0 mL of 0.020 *N* sulfuric acid (0.05%).

Arsenic, *Method I* ⟨211⟩—Dissolve 1.0 g in 35 mL of water: the limit is 3 ppm.

Reducing substances—Transfer 1.0 g to a 250-mL conical flask, dissolve in 10 mL of water, and add 25 mL of alkaline cupric citrate TS. Cover the flask, boil gently for 5 minutes, accurately timed, and cool rapidly to room temperature. Add 25 mL of 0.6 *N* acetic acid, 10.0 mL of 0.1 *N* iodine VS, and 10 mL of 3 *N* hydrochloric acid, and titrate with 0.1 *N* sodium thiosulfate VS, adding 3 mL of starch TS as the end-point is approached. Perform a blank determination, omitting the specimen, and note the difference in volumes required. Each mL of the difference in volume of 0.1 *N* sodium thiosulfate consumed is equivalent to 2.7 mg of reducing substances (as dextrose): the limit is 1.0%.

Cadmium—

Standard preparation—Transfer 137.2 mg of cadmium nitrate to a 1000-mL volumetric flask, dissolve in water, dilute with water to volume, and mix. Pipet 25 mL of the resulting solution into a 100-mL volumetric flask, add 1 mL of hydrochloric acid, dilute with water to volume, and mix. Each mL of this *Standard preparation* contains 12.5 µg of Cd.

Test preparation—Transfer 10.0 g of Zinc Gluconate to a 50-mL volumetric flask, dissolve in water, dilute with water to volume, and mix.

Procedure—To three separate 25-mL volumetric flasks add 0 mL, 2.0 mL, and 4.0 mL of *Standard preparation*, respectively. To each flask add 5.0 mL of *Test preparation*, dilute with water to volume, and mix. These test solutions contain, respectively, 0, 1.0, and 2.0 µg per mL of cadmium from the *Standard preparation*. Concomitantly determine the absorbances of the test solutions at the cadmium emission line at 228.8 nm, with a suitable atomic absorption spectrophotometer (see *Spectrophotometry and Light-scattering* ⟨851⟩), equipped with a cadmium hollow-cathode lamp and an air-acetylene flame, using water as the blank. Plot the absorbances of the test solutions versus their contents of cadmium, in µg per mL, as furnished by the *Standard preparation*, draw the straight line best fitting the three points, and extrapolate the line until it intercepts the concentration axis. From the intercept determine the amount, in µg, of cadmium in each mL of the test solution containing 0 mL of the *Standard preparation.* Calculate the quantity, in ppm, of Cd in the specimen by multiplying this value by 25: the limit is 5 ppm.

Lead—[NOTE—For the preparation of all aqueous solutions and for the rinsing of glassware before use, employ water that has been passed through a strong-acid, strong-base, mixed-bed ion-exchange resin before use. Select all reagents to have as low a content of lead as practicable, and store all reagent solutions in containers of borosilicate glass. Cleanse glassware before use by soaking in warm 8 *N* nitric acid for 30 minutes and by rinsing with deionized water.]

Ascorbic acid–sodium iodide solution—Dissolve 20 g of ascorbic acid and 38.5 g of sodium iodide in water in a 200-mL volumetric flask, dilute with water to volume, and mix.

Trioctylphosphine oxide solution—[*Caution—This solution causes irritation. Avoid contact with eyes, skin, and clothing. Take special precautions in disposing of unused portions of solutions to which this reagent is added.*] Dissolve 5.0 g of trioctylphosphine oxide in 4-methyl-2-pentanone in a 100-mL volumetric flask, dilute with the same solvent to volume, and mix.

Standard preparation and *Blank*—Transfer 5.0 mL of *Lead Nitrate Stock Solution*, prepared as directed in the test for *Heavy Metals* ⟨231⟩, to a 100-mL volumetric flask, dilute with water to volume, and mix. Transfer 2.0 mL of the resulting solution to a 50-mL volumetric flask. To this volumetric flask and to a second, empty 50-mL volumetric flask (*Blank*) add 10 mL of 9 *N* hydrochloric acid and about 10 mL of water. To each flask add 20 mL of *Ascorbic acid–sodium iodide solution* and 5.0 mL of *Trioctylphosphine oxide solution*, shake for 30 seconds, and allow to separate. Add water to bring the organic solvent layer into the neck of each flask, shake again, and allow to separate. The organic solvent layers are the *Blank* and the *Standard preparation*, and they contain 0.0 µg and 2.0 µg of lead per mL, respectively.

Test preparation—Add 1.0 g of Zinc Gluconate, 10 mL of 9 *N* hydrochloric acid, about 10 mL of water, 20 mL of *Ascorbic acid–sodium iodide solution*, and 5.0 mL of *Trioctylphosphine oxide solution* to a 50-mL volumetric flask, shake for 30 seconds, and allow to separate. Add water to bring the organic solvent layer into the neck of the flask, shake again, and allow to separate. The organic solvent layer is the *Test preparation.*

Procedure—Concomitantly determine the absorbances of the *Blank*, the *Standard preparation*, and the *Test preparation* at the lead emission line at 283.3 nm, with a suitable atomic absorption spectrophotometer (see *Spectrophotometry and Light-scattering* ⟨851⟩) equipped with a lead hollow-cathode lamp and an air-acetylene flame, using 4-methyl-2-pentanone to set the instrument to zero. In a suitable analysis, the absorbance of the *Blank* is not greater than 20% of the difference between the absorbance of the *Standard preparation* and the absorbance of the *Blank:* the absorbance of the *Test preparation* does not exceed that of the *Standard preparation* (0.001%).

Assay—Dissolve about 700 mg of Zinc Gluconate, accurately weighed, in 100 mL of water. Add 5 mL of ammonia–ammonium chloride buffer TS and 0.1 mL of eriochrome black TS, and titrate with 0.05 *M* disodium ethylenediaminetetraacetate VS until the solution is deep blue in color. Each mL of 0.05 *M* disodium ethylenediaminetetraacetate is equivalent to 22.78 mg of $C_{12}H_{22}O_{14}Zn$.

Zinc Oxide

ZnO 81.38
Zinc oxide.
Zinc oxide [*1314-13-2*].

» Zinc Oxide, freshly ignited, contains not less than 99.0 percent and not more than 100.5 percent of ZnO.

Packaging and storage—Preserve in well-closed containers.

Identification—

A: When strongly heated, it assumes a yellow color that disappears on cooling.

B: A solution of it in a slight excess of 3 *N* hydrochloric acid responds to the tests for *Zinc* ⟨191⟩.

Alkalinity—Mix 1.0 g with 10 mL of hot water, add 2 drops of phenolphthalein TS, and filter: if a red color is produced, not more than 0.30 mL of 0.10 *N* hydrochloric acid is required to discharge it.

Loss on ignition ⟨733⟩—Weigh accurately about 2 g, and ignite at 500° to constant weight: it loses not more than 1.0% of its weight.

Carbonate and color of solution—Mix 2.0 g with 10 mL of water, add 30 mL of 2 *N* sulfuric acid, and heat on a steam bath, with constant stirring: no effervescence occurs and the resulting solution is clear and colorless.

Arsenic, *Method I* ⟨211⟩: 6 ppm.

Iron and other heavy metals—Cooled 5-mL portions of the solution obtained in the test for *Carbonate and color of solution* yield white precipitates with potassium ferrocyanide TS and with sodium sulfide TS.

Lead—Add 2 g to 20 mL of water, stir well, add 5 mL of glacial acetic acid, and warm on a steam bath until solution is effected: the addition of 5 drops of potassium chromate TS produces no turbidity or precipitate.

Assay—Dissolve about 1.5 g of freshly ignited Zinc Oxide, accurately weighed, and 2.5 g of ammonium chloride in 50.0 mL of 1 *N* sulfuric acid VS with the aid of gentle heat, if necessary. When solution is complete, add methyl orange TS, and titrate the excess sulfuric acid with 1 *N* sodium hydroxide VS. Each mL of 1 *N* sulfuric acid is equivalent to 40.69 mg of ZnO.

Zinc Oxide Ointment

» Zinc Oxide Ointment contains not less than 18.5 percent and not more than 21.5 percent of ZnO.

It may be prepared as follows:

Zinc Oxide	200 g
Mineral Oil	150 g
White Ointment	650 g
To make	1000 g

Levigate the Zinc Oxide with the Mineral Oil to a smooth paste, and then incorporate the White Ointment [see *Ointments and Suppositories* under *Added Substances* (*Ingredients and Processes*) in the *General Notices*].

Packaging and storage—Preserve in well-closed containers, and avoid prolonged exposure to temperatures exceeding 30°.

Identification—The residue obtained in the *Assay* is yellow when hot and white when cool.

Minimum fill ⟨755⟩: meets the requirements.

Calcium, magnesium, and other foreign substances—Heat about 2 g gently until melted, and continue the heating, gradually raising the temperature until the mass is thoroughly charred. Ignite the mass until the residue is uniformly yellow. To the residue add 6 mL of 3 *N* hydrochloric acid: no effervescence occurs. Heat the mixture on a steam bath for 10 to 15 minutes: not more than a trace of insoluble residue remains. Filter the solution, dilute with water to 10 mL, add 6 *N* ammonium hydroxide until the precipitate first formed redissolves, then add 2 mL each of ammonium oxalate TS and dibasic sodium phosphate TS: not more than a slight turbidity is produced in 5 minutes.

Assay—Weigh accurately in a porcelain crucible about 700 mg of Zinc Oxide Ointment, heat gently until melted, and continue the heating, gradually raising the temperature until the mass is thoroughly charred. Ignite the mass until the residue is uniformly yellow, and cool. Dissolve the residue in 10 mL of 2 *N* sulfuric acid, warming if necessary to effect complete solution, transfer the solution to a beaker, and rinse the crucible with small portions of water until the combined solution and rinsings measure 50 mL. Add 15 mL of ammonia–ammonium chloride buffer TS and 1 mL of eriochrome black TS, and titrate with 0.05 *M* disodium ethylenediaminetetraacetate VS until the solution is blue in color.

Each mL of 0.05 *M* disodium ethylenediaminetetraacetate is equivalent to 4.069 mg of ZnO.

Zinc Oxide Paste

» Zinc Oxide Paste contains not less than 24.0 percent and not more than 26.0 percent of ZnO.

It may be prepared as follows:

Zinc Oxide	250 g
Starch	250 g
White Petrolatum	500 g
To make	1000 g

Mix the ingredients.

Packaging and storage—Preserve in well-closed containers, and avoid prolonged exposure to temperatures exceeding 30°.

Identification—The residue obtained in the *Assay* is yellow when hot and white when cool.

Minimum fill ⟨755⟩: meets the requirements.

Assay—Using about 600 mg of Zinc Oxide Paste, proceed as directed in the *Assay* under *Zinc Oxide Ointment.*

Zinc Oxide and Salicylic Acid Paste

» Zinc Oxide and Salicylic Acid Paste contains not less than 23.5 percent and not more than 25.5 percent of zinc oxide (ZnO), and not less than 1.9 percent and not more than 2.1 percent of salicylic acid ($C_7H_6O_3$).

It may be prepared as follows:

Salicylic Acid, in fine powder	20 g
Zinc Oxide Paste, a sufficient quantity, to make	1000 g

Thoroughly triturate the Salicylic Acid with a portion of the paste, then add the remaining paste, and triturate until a smooth mixture is obtained.

Packaging and storage—Preserve in well-closed containers.

Identification—

A: The residue obtained in the *Assay for zinc oxide* is yellow when hot and white when cool.

B: Shake 1 g of it with 10 mL of water, and filter. To the filtrate add 1 mL of ferric chloride TS: an intense reddish violet color is produced. To this solution add 1 mL of acetic acid: the color is not dispersed. To this solution add 2 mL of 2 *N* hydrochloric acid: the color is dispersed and a white crystalline precipitate is formed.

Minimum fill ⟨755⟩: meets the requirements.

Assay for zinc oxide—Weigh accurately in a tared porcelain crucible about 500 mg of Zinc Oxide and Salicylic Acid Paste, heat gently until melted, and continue the heating, gradually raising the temperature until the mass is thoroughly charred. Ignite the mass strongly until all of the carbonaceous material has been dissipated, the residue is uniformly yellow, and the weight of the cooled residue is constant. The weight of the residue represents the quantity of ZnO in the weight of the Paste taken for the assay.

Assay for salicylic acid—Transfer to a 100-mL beaker about 3 g of Zinc Oxide and Salicylic Acid Paste, accurately weighed. Add 30 mL of dehydrated benzene and 30 mL of dimethylformamide, cover the beaker to protect against the absorption of atmospheric carbon dioxide, and stir, using a magnetic stirrer, until a homogeneous dispersion is obtained, using gentle heat, if necessary. Add thymol blue TS, and titrate with 0.1 *N* sodium methoxide VS, with the buret tip inserted through an opening in the cover, to the first permanent blue color. Perform a blank determination, and make any necessary correction. Each mL of 0.1 *N* sodium methoxide is equivalent to 13.81 mg of $C_7H_6O_3$.

Zinc Stearate

Octadecanoic acid, zinc salt.
Zinc stearate [*557-05-1*].

» Zinc Stearate is a compound of zinc with a mixture of solid organic acids obtained from fats, and consists chiefly of variable proportions of zinc stearate and zinc palmitate. It contains the equivalent of not less than 12.5 percent and not more than 14.0 percent of ZnO.

Packaging and storage—Preserve in well-closed containers.

Identification—

A: Mix 25 g with 200 mL of hot water, add 60 mL of 2 *N* sulfuric acid, and boil until the fatty acids separate as a transparent layer. Cool the mixture, and remove the solidified layer of fatty acids: a portion of the water layer responds to the tests for *Zinc* ⟨191⟩.

B: Place the separated fatty acids obtained in *Identification test A* in a filter wetted with water, and wash with boiling water until free from sulfate. Collect the fatty acids in a small beaker, allow to cool, pour off the separated water, then melt the acids, filter into a dry beaker while hot, and dry at 105° for 20 minutes: the fatty acids congeal (see *Congealing Temperature* ⟨651⟩) at a temperature not below 54°.

Alkalies and alkaline earths—Mix 2.0 g with 50 mL of water, add 10 mL of hydrochloric acid, boil until the solution is clear, filter while hot, and wash the separated fatty acids with about 50 mL of hot water. Render the combined filtrate and washings alkaline with 6 *N* ammonium hydroxide, add ammonium sulfide TS to precipitate the zinc completely, dilute with water to 200 mL, mix, and filter. To 100 mL of the clear filtrate add 0.5 mL of sulfuric acid, evaporate to dryness, and ignite to constant weight: the weight of the residue does not exceed 10 mg (1.0%).

Arsenic, *Method I* ⟨211⟩—Prepare the *Test Preparation* as follows. Mix 5.0 g with 50 mL of water, cautiously add 5 mL of sulfuric acid, and boil gently until the fatty acids layer is clear and the volume is reduced to about 25 mL. Filter while hot, cool the filtrate, and dilute with water to 50 mL. Transfer a 20-mL aliquot to the arsine generator flask, and dilute with water to 35 mL. The limit is 1.5 ppm.

Lead—Ignite 0.50 g in a platinum crucible for 15 to 20 minutes in a muffle furnace at 475° to 500°. Cool, add 3 drops of nitric acid, evaporate over a low flame to dryness, and ignite again at 475° to 500° for 30 minutes. Dissolve the residue in 1 mL of 8 *N* nitric acid, and proceed as directed in the test for *Lead* under *Magnesium Stearate.* The limit is 0.001%.

Assay—Boil about 1 g of Zinc Stearate, accurately weighed, with 50 mL of 0.1 *N* sulfuric acid for at least 10 minutes, or until the fatty acids layer is clear, adding more water as necessary to maintain the original volume, cool, and filter. Wash the filter and the flask thoroughly with water until the last washing is not acid to litmus paper. Add to the combined filtrate and washings 15 mL of ammonia–ammonium chloride buffer TS and 0.2 mL of eriochrome black TS, heat the solution to about 40°, and titrate with 0.05 *M* disodium ethylenediaminetetraacetate VS until the solution is deep blue in color. Each mL of 0.05 *M* disodium ethylenediaminetetraacetate is equivalent to 4.069 mg of ZnO.

Zinc Sulfate

$ZnSO_4.xH_2O$.
Sulfuric acid, zinc salt (1:1), hydrate.
Zinc sulfate (1:1) monohydrate 179.46.
Zinc sulfate (1:1) heptahydrate 287.54 [*7446-20-0*].
Anhydrous 161.44 [*7733-02-0*].

» Zinc Sulfate contains one or seven molecules of water of hydration. The monohydrate contains not less than 89.0 percent and not more than 90.4 percent of $ZnSO_4$, corresponding to not less than 99.0 percent and not more than 100.5 percent of $ZnSO_4.H_2O$, and the heptahydrate contains not less than 55.6 percent and not more than 61.0 percent of $ZnSO_4$, corresponding to not less than 99.0 percent and not more than 108.7 percent of $ZnSO_4.7H_2O$.

Packaging and storage—Preserve in tight containers.

Labeling—The label indicates whether it is the monohydrate or the heptahydrate. Label any oral or parenteral preparations containing Zinc Sulfate to state the content of elemental zinc.

Identification—A solution of it responds to the tests for *Zinc* ⟨191⟩ and for *Sulfate* ⟨191⟩.

Acidity—A solution containing the equivalent of 28 mg of $ZnSO_4$ per mL is not colored pink by methyl orange TS.

Alkalies and alkaline earths—Dissolve the equivalent of 1.12 g of $ZnSO_4$ in about 150 mL of water contained in a 200-mL volumetric flask. Precipitate the zinc completely by means of ammonium sulfide TS, and dilute with water to volume. Mix, and filter through a dry filter, rejecting the first portion of the filtrate. To 100 mL of the subsequent filtrate add a few drops of sulfuric acid, evaporate to dryness in a tared dish, and ignite: the weight of the residue does not exceed 5 mg (0.9%).

Arsenic, *Method I* ⟨211⟩—Prepare a *Test Preparation* by dissolving a portion equivalent to 215 mg of $ZnSO_4$ in 35 mL of water: the limit is 14 ppm.

Lead ⟨251⟩—Dissolve a portion equivalent to 0.25 g of $ZnSO_4$ in 5 mL of water, and transfer the solution to a color-comparison tube (*A*). Add 10 mL of *Potassium Cyanide Solution* (1 in 10), mix, and allow the mixture to become clear. In a similar, matched color-comparison tube (*B*) place 5 mL of water, and add 0.50 mL of *Standard Lead Solution* (see *Heavy Metals* ⟨231⟩) and 10 mL of *Potassium Cyanide Solution* (1 in 10). Add to the solution in each tube 0.1 mL of sodium sulfide TS. Mix the contents of each tube, and allow to stand for 5 minutes: viewed downward over a white surface, the solution in tube *A* is no darker than that in tube *B*. The lead limit is 0.002%.

Assay—Dissolve an accurately weighed quantity of Zinc Sulfate, equivalent to about 170 mg of $ZnSO_4$, in 100 mL of water. Add 5 mL of ammonia–ammonium chloride buffer TS and 0.1 mL of eriochrome black TS, and titrate with 0.05 *M* disodium ethylenediaminetetraacetate VS until the solution is deep blue in color. Each mL of 0.05 *M* disodium ethylenediaminetetraacetate is equivalent to 8.072 mg of $ZnSO_4$.

Zinc Sulfate Injection

» Zinc Sulfate Injection is a sterile solution of Zinc Sulfate in Water for Injection. It contains not less than 95.0 percent and not more than 105.0 percent of the labeled amount of zinc (Zn).

Packaging and storage—Preserve in single-dose or in multiple-dose containers.

Labeling—Label the Injection in terms of its content of anhydrous zinc sulfate ($ZnSO_4$) and in terms of its content of elemental zinc. Label it to state that it is not intended for direct injection but is to be added to other intravenous solutions.

Identification—It responds to the tests for *Zinc* ⟨191⟩ and for *Sulfate* ⟨191⟩.

Pyrogen—When diluted with Sodium Chloride Injection to contain 20 µg of zinc per mL, it meets the requirements of the *Pyrogen Test* ⟨151⟩.

pH ⟨791⟩: between 2.0 and 4.0.

Particulate matter ⟨788⟩: meets the requirements under *Small-volume Injections.*

Other requirements—It meets the requirements under *Injections* ⟨1⟩.

Assay—[NOTE—The *Standard preparations* and the *Assay preparation* may be diluted quantitatively with water, if necessary, to yield solutions of suitable concentrations adaptable to the linear or working range of the instrument.]

Standard preparations—Transfer 3.11 g of zinc oxide, accurately weighed, to a 250-mL volumetric flask, add 80 mL of 1 *N* sulfuric acid, warm to dissolve, cool, dilute with water to volume, and mix. This solution contains 10 mg of zinc per mL. Further dilute this solution quantitatively with water to obtain three solutions containing 1.5, 2.0, and 2.5 µg of zinc per mL, respectively.

Assay preparation—Transfer an accurately measured volume of Zinc Sulfate Injection, equivalent to about 20 mg of zinc, to a 500-mL volumetric flask, dilute with water to volume, and mix. Transfer 5.0 mL of this solution to a 100-mL volumetric flask, dilute with water to volume, and mix.

Procedure—Concomitantly determine the absorbances of the *Standard preparations* and the *Assay preparation* at the zinc emission line of 213.8 nm with a suitable atomic absorption spectrophotometer (see *Spectrophotometry and Light-scattering* ⟨851⟩), equipped with a zinc hollow-cathode lamp and an air-acetylene flame, using water as the blank. Plot the absorbances of the *Standard preparations* versus concentration, in µg per mL, of zinc, and draw the straight line best fitting the three plotted points. From the graph so obtained, determine the concentration, in µg per mL, of zinc in the *Assay preparation.* Determine the concentration, in mg per mL, of Zn in the Injection taken by the formula:

$$10C/V,$$

in which C is the concentration, in µg per mL, of zinc in the *Assay preparation*, and V is the volume, in mL, of Injection taken.

Zinc Sulfate Ophthalmic Solution

» Zinc Sulfate Ophthalmic Solution is a sterile solution of Zinc Sulfate in Water rendered isotonic by the addition of suitable salts. It contains not less than 95.0 percent and not more than 105.0 percent of the labeled amount of $ZnSO_4$.

Packaging and storage—Preserve in tight containers.

Identification—It responds to the tests for *Zinc* ⟨191⟩ and for *Sulfate* ⟨191⟩.

Sterility—It meets the requirements under *Sterility Tests* ⟨71⟩.

pH ⟨791⟩: between 5.8 and 6.2; or, if it contains sodium citrate, between 7.2 and 7.8.

Assay—Pipet into a beaker a volume of Zinc Sulfate Ophthalmic Solution, equivalent to about 25 mg of zinc sulfate. Add 1 mL of glacial acetic acid, and adjust by the dropwise addition of 6 *N* ammonium hydroxide to a pH of between 5.0 and 5.5. Add 1 drop of copper ethylenediaminetetraacetate solution [prepared by mixing 1 mL of cupric sulfate solution (1 in 40) and 1 mL of 0.1 *M* disodium ethylenediaminetetraacetate] and 3 drops of a 1 in 1000 solution of 1-(2-pyridylazo)-2-naphthol in anhydrous methanol, and titrate with 0.01 *M* disodium ethylenediaminetetraacetate VS. Each mL of 0.01 *M* disodium ethylenediaminetetraacetate is equivalent to 1.614 mg of $ZnSO_4$.

Zinc Undecylenate

$[CH_2{=}CHCH_2(CH_2)_6CH_2COO{-}]_2Zn$

$C_{22}H_{38}O_4Zn$ 431.92
10-Undecenoic acid, zinc(2+) salt.
Zinc 10-undecenoate [*557-08-4*].

» Zinc Undecylenate contains not less than 98.0 percent and not more than 102.0 percent of $C_{22}H_{38}O_4Zn$, calculated on the dried basis.

Packaging and storage—Preserve in well-closed containers.

Identification—

A: Acidify about 5 g with 25 mL of 2 *N* sulfuric acid, add 20 mL of water, and extract in a separator with two 25-mL portions of ether. Evaporate the ether solution until the odor of ether no longer is perceptible. Add potassium permanganate TS dropwise to a 1-mL portion of this residue: the permanganate color is discharged.

B: A 3-mL portion of the residue of undecylenic acid obtained in *Identification test A* responds to *Identification test B* under *Undecylenic Acid*.

C: Dissolve about 100 mg in a mixture of 10 mL of water and 1 mL of ammonium hydroxide, and add a few drops of sodium sulfide TS: a white, flocculent precipitate of zinc sulfide is formed.

Loss on drying ⟨731⟩—Dry it at 105° for 2 hours: it loses not more than 1.25% of its weight.

Alkalies and alkaline earths—Boil 1.50 g with a mixture of 50 mL of water and 10 mL of hydrochloric acid, filter while hot, and wash the separated acid with about 50 mL of hot water. Render the combined filtrate and washings alkaline with 6 *N* ammonium hydroxide, add ammonium sulfide TS to precipitate the zinc completely, dilute with water to 200 mL, mix, and filter. To 100 mL of the clear filtrate add 0.5 mL of sulfuric acid, evaporate to dryness, and ignite over a low flame to constant weight: the weight of the residue does not exceed 7.5 mg (1.0%).

Assay—Boil 50.0 mL of 0.1 *N* sulfuric acid VS with about 1 g of Zinc Undecylenate, accurately weighed, for 10 minutes, or until the undecylenic acid layer is clear, adding water, as necessary, to maintain the original volume. Cool, and transfer the mixture, with the aid of water, to a 500-mL separator. Dilute with water to about 250 mL, and extract with two 100-mL portions of solvent hexane. Wash the combined extracts with water until the last washing is neutral to litmus, add the washings to the original water layer, and evaporate on a steam bath to about 100 mL. Cool, add 3 drops of methyl orange TS, and titrate the excess sulfuric acid with 0.1 *N* sodium hydroxide VS. Perform a blank determination (see *Residual Tritations* under *Titrimetry* ⟨541⟩). Each mL of 0.1 *N* sulfuric acid is equivalent to 21.60 mg of $C_{22}H_{38}O_4Zn$.

Guide to GENERAL CHAPTERS

(For complete alphabetic list of all general chapters in this Pharmacopeia, see under "General chapters" in the index.)

General Tests and Assays

General Requirements for Tests and Assays

Apparatus for Tests and Assays

Microbiological Tests

Biological Tests and Assays

Chemical Tests and Assays

IDENTIFICATION TESTS

LIMIT TESTS

OTHER TESTS AND ASSAYS

Physical Tests and Determinations

General Information

GENERAL CHAPTERS

General Tests *and* Assays

General Requirements for Tests and Assays

⟨1⟩ INJECTIONS

Every care should be exercised in the preparation of all products intended for injection, to prevent contamination with microorganisms and foreign material. Good pharmaceutical practice requires also that each final container of Injection be subjected individually to a physical inspection, whenever the nature of the container permits, and that every container whose contents show evidence of contamination with visible foreign material be rejected.

Definitions—In this Pharmacopeia, the sterile preparations for parenteral use are grouped into five distinct classes, defined as follows: (1) medicaments or solutions or emulsions thereof suitable for injection, bearing titles of the form, ____ *Injection;* (2) dry solids or liquid concentrates containing no buffers, diluents, or other added substances, and which, upon the addition of suitable solvents, yield solutions conforming in all respects to the requirements for Injections, and which are distinguished by titles of the form, *Sterile* ____; (3) preparations the same as those described under (2) except that they contain one or more buffers, diluents, or other added substances, and which are distinguished by titles of the form, ____ *for Injection;* (4) solids which are suspended in a suitable fluid medium and which are not to be injected intravenously or into the spinal canal, distinguished by titles of the form, *Sterile* ____ *Suspension;* and (5) dry solids which, upon the addition of suitable vehicles, yield preparations conforming in all respects to the requirements for Sterile Suspensions, and which are distinguished by titles of the form, *Sterile* ____ *for Suspension.*

A Pharmacy bulk package is a container of a sterile preparation for parenteral use that contains many single doses. The contents are intended for use in a pharmacy admixture program and are restricted to the preparation of admixtures for infusion or, through a sterile transfer device, for the filling of empty sterile syringes.

The closure shall be penetrated only one time after constitution with a suitable sterile transfer device or dispensing set which allows measured dispensing of the contents. The *Pharmacy bulk package* is to be used only in a suitable work area such as a laminar flow hood (or an equivalent clean air compounding area).

Designation as a *Pharmacy bulk package* is limited to preparations from classes 1, 2, or 3 as defined above. *Pharmacy bulk packages*, although containing more than one single dose, are exempt from the multiple-dose container volume limit of 30 mL and the requirement that they contain a substance or suitable mixture of substances to prevent the growth of microorganisms.

Where a container is offered as a *Pharmacy bulk package*, the label shall (a) state prominently "Pharmacy Bulk Package—Not for direct infusion," (b) contain or refer to information on proper techniques to help assure safe use of the product, and (c) bear a statement limiting the time frame in which the container may be used once it has been entered, provided it is held under the labeled storage conditions.

Where used in this Pharmacopeia, the designation *Large-volume intravenous solution* applies to a single-dose injection that is intended for intravenous use and is packaged in containers labeled as containing more than 100 mL. The designation *Small-volume Injection* applies to an Injection that is packaged in containers labeled as containing 100 mL or less.

The Pharmacopeial definitions for sterile preparations for parenteral use generally do not apply in the case of the biologics, because of their special nature and licensing requirements (see *Biologics* ⟨1041⟩).

Aqueous Vehicles—The vehicles for aqueous Injections meet the requirements of the *Pyrogen Test* ⟨151⟩ or the *Bacterial Endotoxins Test* ⟨85⟩, whichever is specified. *Water for Injection* generally is used as the vehicle, unless otherwise specified in the individual monograph. Sodium chloride may be added in amounts sufficient to render the resulting solution isotonic; and *Sodium Chloride Injection*, or *Ringer's Injection*, may be used in whole or in part instead of *Water for Injection* unless otherwise specified in the individual monograph. For conditions applying to other adjuvants, see *Added Substances*, in this chapter.

Other Vehicles—Fixed oils used as vehicles for nonaqueous injections are of vegetable origin, are odorless or nearly so, and have no odor or taste suggesting rancidity. They meet the requirements of the test for *Solid paraffin* under *Mineral Oil*, the cooling bath being maintained at 10°, have a *Saponification value* of between 185 and 200 (see *Fats and Fixed Oils* ⟨401⟩), have an *Iodine value* of between 79 and 128 (see *Fats and Fixed Oils* ⟨401⟩), and meet the requirements of the following tests:

Unsaponifiable Matter—Reflux on a steam bath 10 mL of the oil with 15 mL of sodium hydroxide solution (1 in 6) and 30 mL of alcohol, with occasional shaking until the mixture becomes clear. Transfer the solution to a shallow dish, evaporate the alcohol on a steam bath, and mix the residue with 100 mL of water: a clear solution results.

Free Fatty Acids—The free fatty acids in 10 g of oil require for neutralization not more than 2.0 mL of 0.020 *N* sodium hydroxide (see *Fats and Fixed Oils* ⟨401⟩).

Synthetic mono- or diglycerides of fatty acids may be used as vehicles, provided they are liquid and remain clear when cooled to 10° and have an *Iodine value* of not more than 140 (see *Fats and Fixed Oils* ⟨401⟩).

These and other nonaqueous vehicles may be used, provided they are safe in the volume of injection administered, and also provided they do not interfere with the therapeutic efficacy of the preparation or with its response to prescribed assays and tests.

Added Substances—Suitable substances may be added to increase stability or usefulness, unless proscribed in the individual monograph, provided they are harmless in the amounts administered and do not interfere with the therapeutic efficacy or with the responses to the specified assays and tests. No coloring agent may be added, solely for the purpose of coloring the finished preparation, to a solution intended for parenteral administration (see also *Added Substances* under *General Notices*, and *Antimicrobial Preservatives—Effectiveness* ⟨51⟩).

Observe special care in the choice and use of added substances in preparations for injection that are administered in a volume exceeding 5 mL. The following maximum limits prevail unless otherwise directed: for agents containing mercury and the cationic, surface-active compounds, 0.01%; for those of the types of chlorobutanol, cresol, and phenol, 0.5%; and for sulfur dioxide, or an equivalent amount of the sulfite, bisulfite, or metabisulfite of potassium or sodium, 0.2%.

A suitable substance or mixture of substances to prevent the growth of microorganisms must be added to preparations intended for injection that are packaged in multiple-dose containers, regardless of the method of sterilization employed, unless otherwise directed in the individual monograph, or unless the active ingredients are themselves antimicrobial. Such substances are used in concentrations that will prevent the growth of or kill microorganisms in the preparations for injection (see also *Antimicrobial Preservatives—Effectiveness* ⟨51⟩ and *Antimicrobial Agents—Content* ⟨341⟩). Sterilization processes are employed even though such substances are used (see also *Parenteral and Topical Preparations* in the section, *Added Substances*, under *General Notices*, and *Sterilization and Sterility Assurance of Compendial Articles* ⟨1211⟩). The air in the container may be evacuated or be displaced by a chemically inert gas.

Containers for Injections—Containers, including the closures, for preparations for injection do not interact physically or chemically with the preparations in any manner to alter the strength, quality, or purity beyond the official requirements under the ordinary or customary conditions of handling, shipment, storage, sale, and use. The container is made of material that permits inspection of the contents. The type of glass preferable for each parenteral preparation is usually stated in the individual monograph.

For definitions of single-dose and multiple-dose containers, see *Containers* under *General Notices*. Containers meet the requirements under *Containers* ⟨661⟩.

Containers are closed by fusion, or by application of suitable closures, in such manner as to prevent contamination or loss of contents. Closures for multiple-dose containers permit the withdrawal of the contents without removal or destruction of the closure. The closure permits penetration by a needle, and, upon withdrawal of the needle, at once recloses the container against contamination.

Containers for Sterile Solids—Containers, including the closures, for dry solids intended for parenteral use do not interact physically or chemically with the preparation in any manner to alter the strength, quality, or purity beyond the official requirements under the ordinary or customary conditions of handling, shipment, storage, sale, and use.

A container for a sterile solid permits the addition of a suitable solvent and withdrawal of portions of the resulting solution or suspension in such manner that the sterility of the product is maintained.

Where the *Assay* in a monograph provides a procedure for *Assay preparation* in which the total withdrawable contents are to be withdrawn from a single-dose container with a hypodermic needle and syringe, the contents are to be withdrawn as completely as possible into a dry hypodermic syringe of a rated capacity not exceeding three times the volume to be withdrawn and fitted with a 21-gauge needle not less than 2.5 cm (1 inch) in length, care being taken to expel any air bubbles, and discharged into a container for dilution and assay.

Volume in Container—Each container of an Injection is filled with a volume in slight excess of the labeled "size" or that volume which is to be withdrawn. The excess volumes recommended in the accompanying table are usually sufficient to permit withdrawal and administration of the labeled volumes.

DETERMINATION OF VOLUME OF INJECTION IN CONTAINERS—Select 1 or more containers if the volume is 10 mL or more, 3 or more if the volume is more than 3 mL and less than 10 mL, or 5 or more if the volume is 3 mL or less. Take up individually the contents of each container selected into a dry hypodermic syringe of a rated capacity not exceeding three times the volume to be measured, and fitted with a 21-gauge needle not less than 2.5 cm (1 inch) in length. Expel any air bubbles from the syringe and needle, and then discharge the contents of the syringe, without emptying the needle, into a standardized, dry cylinder (graduated to contain rather than to deliver the designated volumes) of such size that the volume to be measured occupies at least 40% of its rated volume. Alternatively, the contents of the syringe may be discharged into a dry, tared beaker, the volume, in mL, being calculated as the weight, in g, of Injection taken divided by its density. The contents of two or three 1-mL or 2-mL containers may be pooled for the measurement, provided that a separate, dry syringe assembly is used for each container. The content of containers holding 10 mL or more may be determined by means of opening them and emptying the contents directly into the graduated cylinder or tared beaker.

	Recommended Excess Volume	
Labeled Size	For Mobile Liquids	For Viscous Liquids
0.5 mL	0.10 mL	0.12 mL
1.0 mL	0.10 mL	0.15 mL
2.0 mL	0.15 mL	0.25 mL
5.0 mL	0.30 mL	0.50 mL
10.0 mL	0.50 mL	0.70 mL
20.0 mL	0.60 mL	0.90 mL
30.0 mL	0.80 mL	1.20 mL
50.0 mL or more	2%	3%

The volume is not less than the labeled volume in the case of containers examined individually or, in the case of 1-mL and 2-mL containers, is not less than the sum of the labeled volumes of the containers taken collectively.

For Injections in multiple-dose containers labeled to yield a specific number of doses of a stated volume, proceed as directed in the foregoing, using the same number of separate syringes as the number of doses specified. The volume is such that each syringe delivers not less than the stated dose.

For Injections containing oil, warm the containers, if necessary, and thoroughly shake them immediately before removing the contents. Cool to 25° before measuring the volume.

Particulate Matter—All large-volume Injections for single-dose infusion, and those small-volume Injections for which the monographs specify such requirements, are subject to the particulate matter limits set forth under *Particulate Matter in Injections* ⟨788⟩. An article packaged as both a large-volume and a small-volume Injection meets the requirements set forth for *Small-volume Injections* where the container is labeled as containing 100 mL or less if the individual monograph includes a test for *Particulate matter;* it meets the requirements set forth for *Large-volume Injections for Single-dose Infusion* where the container is labeled as containing more than 100 mL. Injections packaged and labeled for use as irrigating solutions are exempt from requirements for *Particulate matter*.

Sterility Tests—Preparations for injection meet the requirements under *Sterility Tests* ⟨71⟩.

Labeling—[NOTE—See definitions of "label" and "labeling" under *Preservation, Packaging, Storage, and Labeling—Labeling* in the *General Notices.*]

The label states the name of the preparation; in the case of a liquid preparation, the percentage content of drug or amount of drug in a specified volume; in the case of a dry preparation, the amount of *active* ingredient; the route of administration; a statement of storage conditions and an expiration date; the name of the manufacturer and distributor; and an identifying lot number. The lot number is capable of yielding the complete manufacturing history of the specific package, including all manufacturing, filling, sterilizing, and labeling operations.

Where the individual monograph permits varying concentrations of active ingredients in the large-volume parenteral, the concentration of each ingredient named in the official title is stated as if part of the official title, e.g., Dextrose Injection 5%, or Dextrose (5%) and Sodium Chloride (0.2%) Injection.

The labeling includes the following information, if the complete formula is not specified in the individual monograph: (1) In the case of a liquid preparation, the percentage content of each ingredient or the amount of each ingredient in a specified volume, except that ingredients added to adjust to a given pH or to make the solution isotonic may be declared by name and a statement of their effect; and (2) in the case of a dry preparation or other preparation to which a diluent is intended to be added before use, the amount of each ingredient, the composition of recommended diluent(s) [the name(s) alone, if the formula is specified in the individual monograph], the amount to be used to attain a specific concentration of active ingredient and the final volume of solution so obtained, a brief description of the physical appearance of the constituted solution, directions for proper storage of the constituted solution, and an expiration date limiting the period during which the constituted solution may be expected to have the required or labeled potency if it has been stored as directed.

Containers for Injections that are intended for use as dialysis, hemofiltration, or irrigation solutions and that contain a volume of more than 1 liter are labeled to indicate that the contents are not intended for use by intravenous infusion.

Injections intended for veterinary use are labeled to that effect.

The container is so labeled that a sufficient area of the container remains uncovered for its full length or circumference to permit inspection of the contents.

Packaging and Storage—The volume of Injection in single-dose containers provides the amount specified for parenteral administration at one time and in no case is more than sufficient to permit the withdrawal and administration of 1 liter.

Preparations intended for intraspinal, intracisternal, or peridural administration are packaged only in single-dose containers.

Unless otherwise specified in the individual monograph, no multiple-dose container contains a volume of Injection more than sufficient to permit the withdrawal of 30 mL.

Injections packaged for use as irrigation solutions or for hemofiltration or dialysis or for parenteral nutrition are exempt from the 1-liter restriction of the foregoing requirements relating to packaging. Containers for Injections packaged for use as hemofiltration or irrigation solutions may be designed to empty rapidly and may contain a volume of more than 1 liter.

Injections labeled for veterinary use are exempt from packaging and storage requirements concerning the limitation to single-dose containers and the limitation on the volume of multiple-dose containers.

CONSTITUTED SOLUTIONS

Sterile dosage forms from which constituted solutions are prepared for injection bear titles of the form, *Sterile* ____ or ____ *for Injection.* Since these dosage forms are constituted at the time of use by the health-care practitioner, tests and standards pertaining to the solution as constituted for administration are not included in the individual monographs on sterile dry solids or liquid concentrates. However, in the interest of assuring the quality of injection preparations as they are actually administered, the following nondestructive tests are provided for demonstrating the suitability of constituted solutions when they are prepared just prior to use.

Completeness and Clarity of Solution—Constitute the solution as directed in the labeling supplied by the manufacturer for the sterile dry dosage form.

A: The solid dissolves completely, leaving no visible residue as undissolved matter.

B: The constituted solution is not significantly less clear than an equal volume of the diluent or of Purified Water contained in a similar vessel and examined similarly.

Particulate Matter—Constitute the solution as directed in the labeling supplied by the manufacturer for the sterile dry dosage form: the solution is essentially free from particles of foreign matter that can be observed on visual inspection.

⟨11⟩ USP REFERENCE STANDARDS

USP Reference Standards are established and released under the authority of the USPC Board of Trustees upon recommendation of the USP Reference Standards Committee, which passes on the selection and suitability of each lot. The critical characteristics of each lot of specimen selected for the standard are usually determined independently in three or more laboratories. The USP Drug Research and Testing Laboratory (see *Preface*) and the Food and Drug Administration laboratories participate in testing almost all new Standards and replacements for existing Standards. In addition, laboratories throughout the nation, both academic and industrial, participate in the testing.

Reference Standards are specifically required in many Pharmacopeial assays and tests and are provided solely for such use; suitability for other non-official application(s) rests with the purchase. Originally introduced for the biological assays of USP X, reference standards are now required for numerous other procedures as well. This reflects the extensive use of modern chromatography and spectrophotometry, which require measurements relative to a reference standard to attain accurate and reproducible results.

USP Reference Standards are substances selected for their high purity, critical characteristics, and suitability for the intended purpose. Heterogeneous substances, of natural origin, also are designated "Reference Standards" where needed. Usually these are the counterparts of international standards.

Antibiotic reference standards distributed by the USPC have been designated by the U.S. Food and Drug Administration as identical to FDA working standards under the FDA certification procedures. USPC distributes both U.S. Reference Standards and USP Reference Standards for antibiotic substances. This difference in labeling the Standards is in effect only temporarily, and eventually all vials will bear the same title. Where a USP Reference Standard is called for, the corresponding substance labeled as a "U.S. Reference Standard" may be used, and vice versa.

Reference Standards currently labeled as "NF Reference Standards" will eventually all be designated and labeled as "USP Reference Standards" pursuant to the consolidation of USP and NF within the USPC as of January 2, 1975. Meanwhile, where a USP Reference Standard is called for, the corresponding substance labeled as an "NF Reference Standard" may be used.

Other Reference Substances

As a service, the USPC tests and distributes additional authenticated substances not currently required as USP or NF Reference Standards. These also are provided under the supervision of the USP Reference Standards Committee. These additional substances fall into three groups: (1) former USP and NF Reference Standards, not required in the current USP or NF but for which sufficient demand remains; (2) FCC Reference Standards, specified in the current edition of the Food Chemicals Codex; and (3) Authentic Substances (AS), which are highly purified samples of chemicals, including substances of abuse, that are collaboratively tested and made available as a service primarily to analytical, clinical, pharmaceutical, and research laboratories.

The distribution of controlled substances is subject to the regulations and licensing provisions of the Drug Enforcement Administration of the Department of Justice.

As an additional service, the USPC distributes several noncommercial reagents required in certain USP monographs. These reagents are specially prepared for their intended use and will be distributed by USPC only until they become commercially available.

A program to provide international biological standards and chemical reference substances is maintained by the World Health Organization, an agency of the United Nations. The WHO program is concerned with reference materials for antibiotics, biologicals, and chemotherapeutic agents. As a rule, an International Standard for a material of natural origin is discontinued once the substance responsible for its characteristic activity has been isolated, identified, and prepared in such form that it can be completely characterized by chemical and physical means. The USP Reference Standards Committee collaborates closely with the WHO in order to minimize unavoidable differences in the actual units of potency, and in some cases to share in the preparation of a reference standard. Since some USP Reference Standards are standardized in terms of the corresponding International Standards, the relevant USP Units and the International Units of potency are generally identical.

Proper Use of USP Reference Standards—To serve its intended purpose, each USP Reference Standard must be properly stored, handled, and used. Generally, Reference Standards should be stored in their original stoppered containers away from heat and protected from light. Avoid humid storage areas in particular. Where special storage conditions are necessary, directions are given on the label.

Neither Reference Standards nor Authentic Substances are intended for use as drugs or as medical devices.

Many Pharmacopeial tests and assays are based on comparison of a test specimen with a USP Reference Standard. In such cases, measurements are made on preparations of both the test specimen and the Reference Standard. Where it is directed that a Standard solution or a *Standard preparation* be prepared for a quantitative determination by stepwise dilution or otherwise, it is intended that the Reference Standard substance shall be accurately weighed (see *Weights and Balances* ⟨41⟩ and *Volumetric Apparatus* ⟨31⟩). Due account should also be taken of the relatively large errors associated with weighing small masses (see also *Dilution* under *General Notices*).

Assay and test results are determined on the basis of comparisons of the specimen under test with a USP Reference Standard that has been freed from or corrected for volatile residues or water content as instructed on the label. The same directions are given in the *Reference standard(s)* sections of the applicable monographs. Where specific label instruction differ from the text in the applicable **Reference standard** section(s), the actual labe on the current distributed item will take precedence. such a situation reflects the ability to immediately effect a necessary change, on scientific ground in advance of the written change to the **Reference standard** section(s) via periodic official revisions. Where special drying requirements for Reference Standards are found in specific sections of USP or NF monographs, those supersede the usual instructions (see *Procedures* under *Tests and Assays* in the *General Notices*). Where a USP Reference Standard is required to be dried before using, transfer an amount, sufficient after drying, to a clean and dry vessel. Do not use the original container as the drying vessel, and do not dry a specimen repeatedly at temperatures above 25°. Where the titrimetric determination of water is required at the time a Reference Standard is to be used, proceed as directed for *Method I* under *Water Determination* ⟨921⟩. Instrumental or microanalytical methods are acceptable for this purpose. When using typical amounts, about 50 mg, of the Reference Standard as the test specimen, titrate with a fourfold dilution of the *Reagent*.

Unless a Reference Standard label states a specific potency or content, the Reference Standard is taken as being 100.0% pure for compendial purposes. The suitability of a USP Reference Stadard for noncompendial application is left up to the user.

Current Lots

It is the responsibility of each analyst to ascertain that his particular supply of USP Reference Standard is current. Only sufficient quantity for immedite use should be purchased, and long-term storage should be avoided.

To ensure ready access to the latest information, the USPC publishes the Official Catalog of Reference Standards and Authentic Substances, and the lot designations, bimonthly in *Pharmacopeial Forum*.* This system offers more positive control and flexibility in responding to revisions in Reference Standard usage than would expiration dates. The Catalog in the most recent *Pharmacopeial Forum* identifies items that are official in the USP Reference Standards collection at the time of publication.

Two columns appear in the Catalog to identify the current official lots. One column identifies the official lot currently being shipped by USPC. In some cases, the previous lot may still be considered official. If so, it is identified in the second column. Ordinarily the previous lot is carried in official status for about one year after the current lot entered distribution unless, because of a change in monograph requirements or stability limitations, the previous lot is found to be no longer suitable.

* For nonsubscribers, the most recent Official Catalog is available from: U.S. Pharamcopeial Convention, Inc., Reference Standards Order Department, 12601 Twinbrook Parkway, Rockville, MD 20852. Telephone 1-301-881-0666. FAX 1-301-881-5021. Toll-free telephone 1-800-227-USPC.

Apparatus for Tests and Assays

⟨16⟩ AUTOMATED METHODS OF ANALYSIS

Where a sufficiently large number of similar units are to be subjected routinely to the same type of examination, automated methods of analysis may be far more efficient and precise than manual methods. Such automated methods have been found especially useful in testing the content uniformity of tablets and capsules and in facilitating methods requiring precisely controlled experimental conditions. Many manufacturing establishments, as well as the laboratories of regulatory agencies, have found it convenient to utilize automated methods as alternatives to Pharmacopeial methods (see *Procedures*, under *General Notices*). In addition, the detection system and calculation of results for automated methods are often computerized.

Before an automated method for testing an article is adopted as an alternative, it is advisable to ascertain that the results obtained by the automated method are equivalent in accuracy and precision to those obtained by the prescribed Pharmacopeial method, bearing in mind the further principle stated in the General Notices that "where a difference appears, or in the event of dispute, only the result obtained by the procedure given in this Pharmacopeia is conclusive."

It is necessary to monitor the performance of the automated analytical system continually, by assaying standard preparations of known composition frequently interspersed among the test preparations. Where immiscible solvents are employed in the automated apparatus for rapid extractions, they are often separated for analysis before complete extraction is attained, and the chemical reactions utilized in automated methods rarely are stoichiometric. Both the accuracy and the precision of the determinations depend upon precise adjustment of the equipment, so maintained that all standard and test preparations are exposed to identical physical and chemical manipulations for identical time intervals. Excessive variability in the response of the standard preparations indicates that the analytical system is malfunctioning and that the test results are therefore invalid. However, where automated systems are shown to operate reliably, the precision of the automated method may surpass that of the manual procedure employing the same basic chemistry.

Many of the manual methods given in this Pharmacopeia can be adapted for use in automated equipment incorporating either discrete analyzers or continuous flow systems and operating under a variety of conditions. On the other hand, an analytical scheme devised for a particular automated system may not be readily transposable for use either in a manual procedure or in other types of automated equipment.

The apparatus required for manual methods is, in general, less complicated than the apparatus of automated systems, even those systems used for the direct automated measurement of a single analyte (i.e., the substance being determined or analyzed for) in a binary mixture. However, because of their versatility, automated systems designed for the rapid determination of a specified substance often can be readily modified by the addition of suitable modules and accessories to permit the determination of one or more additional substances in a dosage form. Such extended systems have been utilized, for example, in the automated analysis of articles containing both estrogens and progestogens.

The accompanying pertinent diagrams represent examples of automated methods. Diagrams for official methods are reproduced here rather than in the individual monographs. The descriptions of the procedural details in these methods exemplify the general approach in automated analysis applicable to dosage forms. It should be noted that the diagrams, with many minutiae, are an indispensable part of the directions for conducting the analysis.

Antibiotics—Hydroxylamine Assay

The following procedure is applicable for the assay of those Pharmacopeial antibiotics, such as cephalosporins and penicillins, that possess the beta-lactam structure.

Apparatus—Automatic analyzer consisting of (1) a liquid sampler, (2) a proportioning pump, (3) suitable spectrophotometers equipped with matched flow cells and analysis capability at 480 nm, (4) a means of recording spectrophotometric readings, and/or computer for data retrieval and calculation, and (5) a manifold consisting of the components illustrated in the accompanying pertinent diagram.

Reagents—

Hydroxylamine hydrochloride solution—Dissolve 20 g of hydroxylamine hydrochloride in 5 mL of polyoxyethylene (23) lauryl ether solution (1 in 1000), and add water to make 1000 mL.

Acetate buffer—Dissolve 173 g of sodium hydroxide and 20.6 g of sodium acetate in water to make 1000 mL. Dilute 75 mL of this solution with water to 500 mL, and mix.

Ferric nitrate solution—Suspend 233 g of ferric nitrate in about 600 mL of water, add 2.8 mL of sulfuric acid, stir until the ferric nitrate is dissolved, add 1 mL of polyoxyethylene (23) lauryl ether, dilute with water to 1000 mL, and mix.

Reference Standard—Use the USP Reference Standard as directed in the individual monograph.

Standard Preparation—Unless otherwise directed in the individual monograph, dissolve an accurately weighed quantity of the Reference Standard in water, and dilute quantitatively with water to obtain a solution having a known concentration of about 1 mg per mL.

Assay Preparation—Unless otherwise directed in the individual monograph, using the specimen under test, prepare as directed under *Standard Preparation*.

Procedure—With the sample line pumping water, the other lines pumping their respective reagents, and the spectrophotometer set at 480 nm, standardize the system until a steady absorbance baseline has been established. Transfer portions of the *Standard Preparation* and the *Assay Preparation* to sampler cups, and place in the sampler. Start the sampler, and conduct determinations of the *Standard Preparation* and the *Assay Preparation* typically at the rate of 40 per hour, using a ratio of about 2:1 for sample and wash time. Calculate the potency by the formula given in the individual monograph, in which C is the concentration, in mg per mL, of USP Reference Standard in the *Standard Preparation*, P is the potency, in µg per mg, of the USP Reference Standard, and A_U and A_S are the absorbances, corrected for the absorbances of the respective blanks, of the solutions from the *Assay Preparation* and the *Standard Preparation*, respectively.

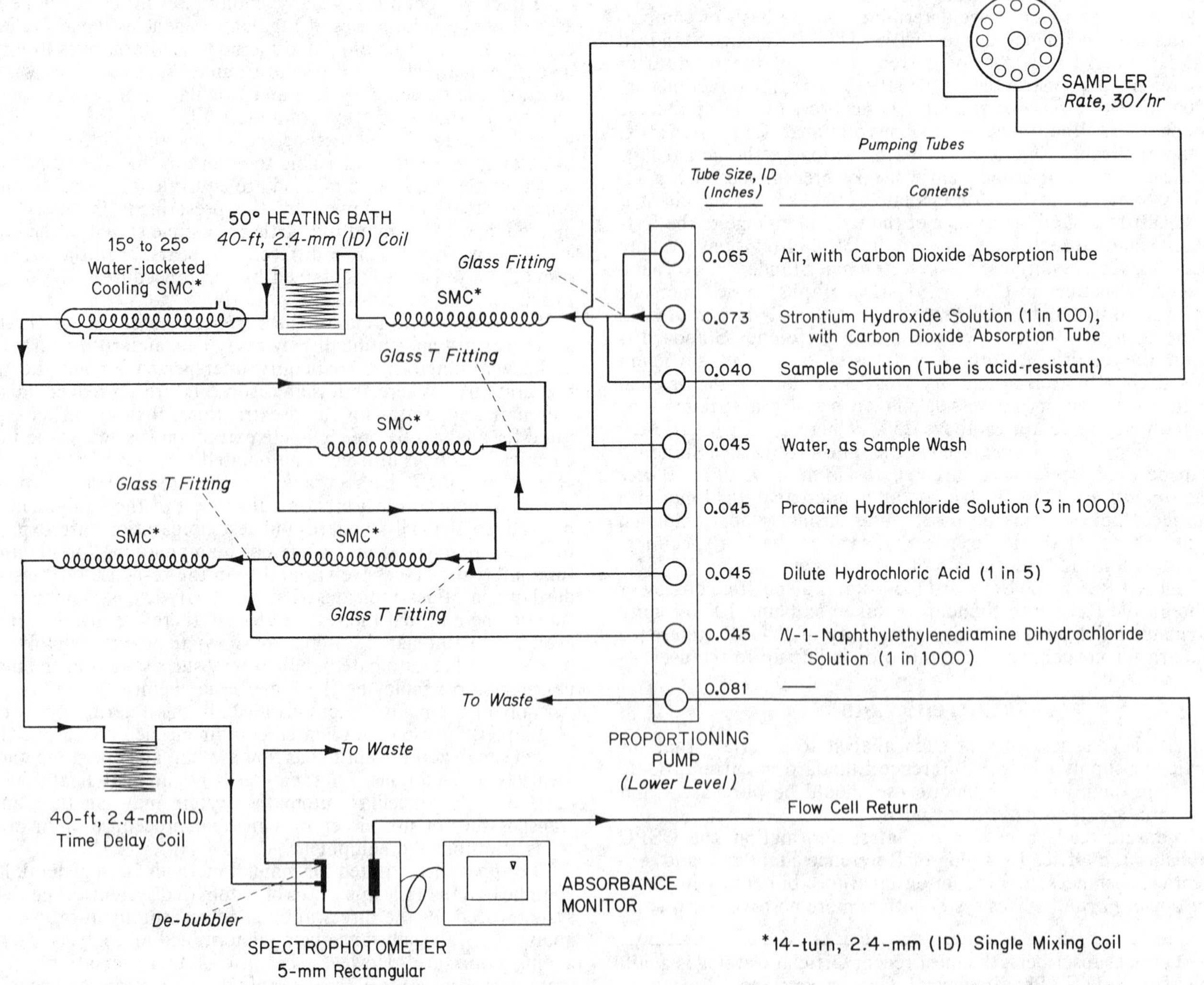

Diagram for Automated Assay for Nitroglycerin Tablets

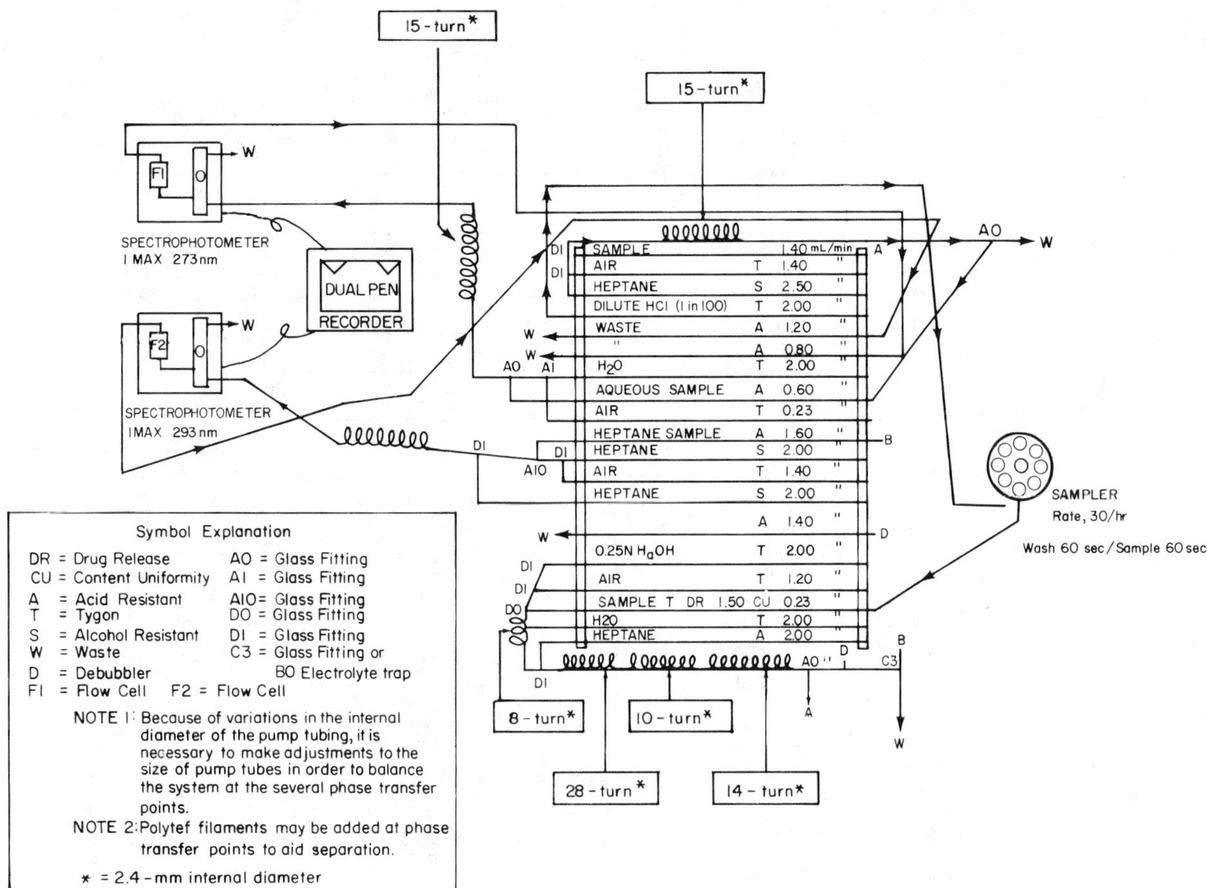

Diagram for Automated Drug Release and Content Uniformity Test for Propranolol Hydrochloride and Hydrochlorothiazide Extended-release Capsules

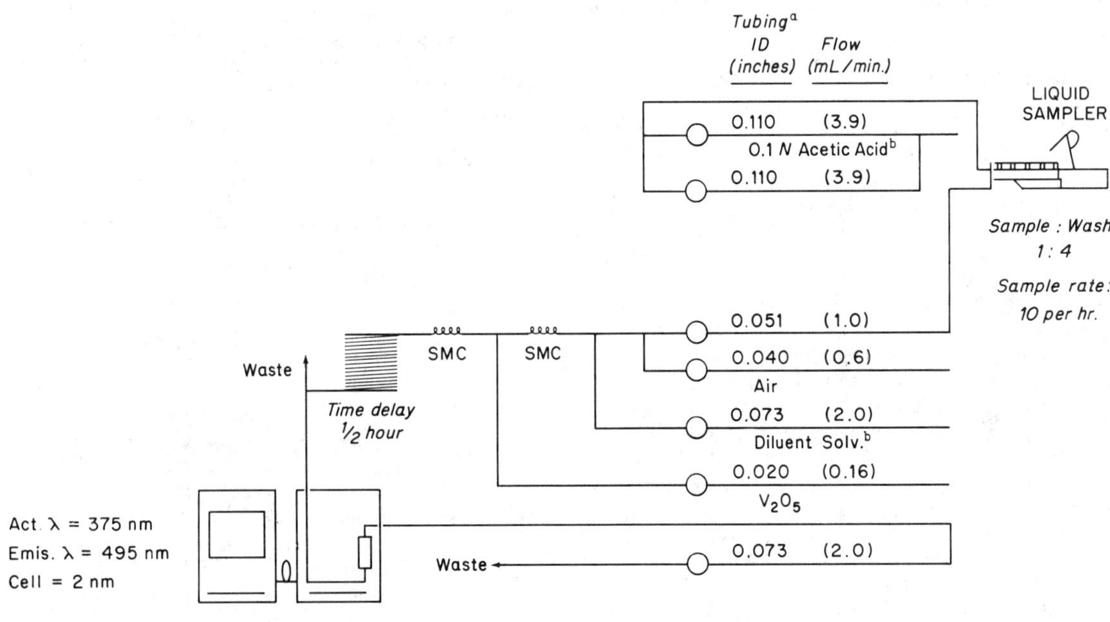

[a]All tubing is Tygon except for the 0.073-in. tubing, which is silicone.
[b]For the *Content Uniformity* test, use *Phosphoric acid–methanol solution*.

Diagram for Automated Dissolution and Content Uniformity Test for Reserpine Tablets

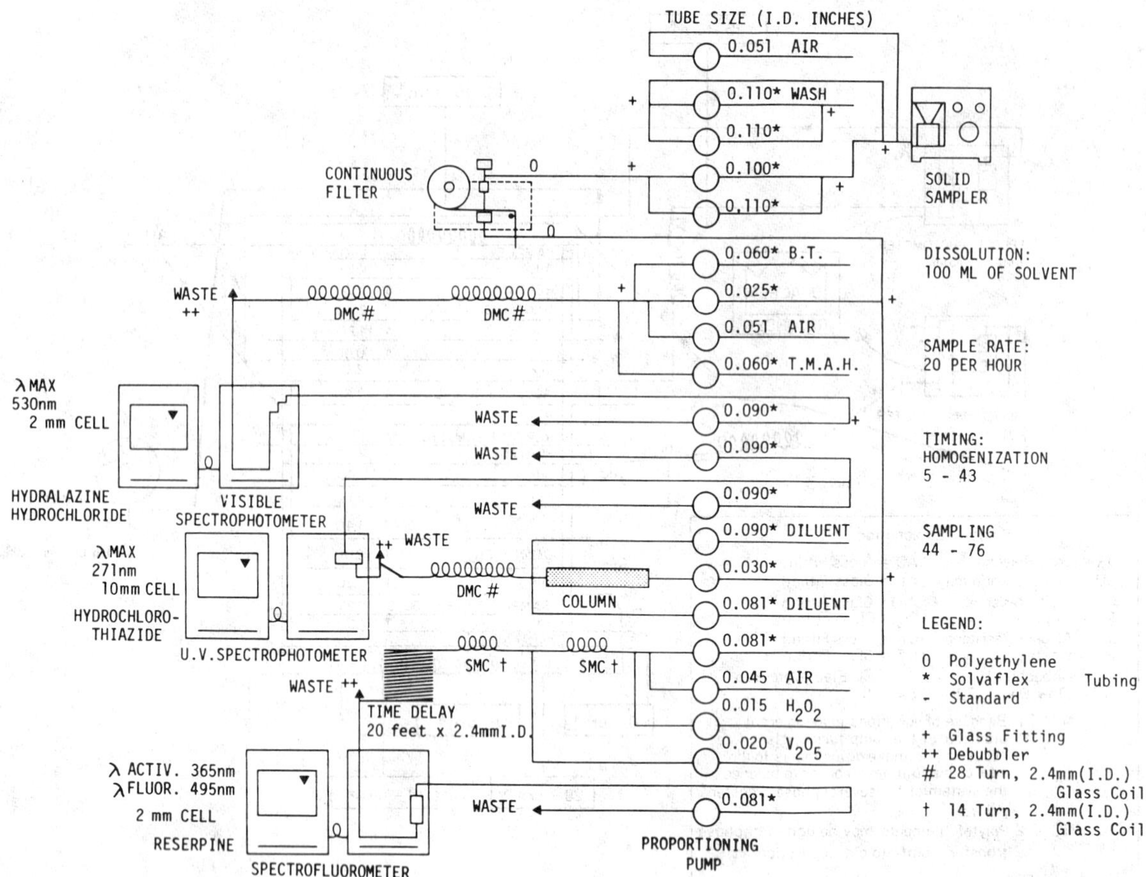

Diagram for Automated Content Uniformity Test for Reserpine, Hydralazine Hydrochloride, and Hydrochlorothiazide Tablets

Content Uniformity of Nitroglycerin Tablets

This is not to be considered as the official method. It is detailed here for further illustration of descriptions of automated methods.

Apparatus—Automatic analyzer consisting of (1) a liquid sampler, (2) a proportioning pump, (3) a heating bath, (4) a suitable spectrophotometer equipped with a 5-mm flow cell and analysis capability at 545 nm, (5) a means of recording spectrophotometric readings, and (6) a manifold consisting of the components illustrated in the accompanying pertinent diagram.

Reagents—

1 Percent strontium hydroxide solution—Dissolve 20.0 g of strontium hydroxide [$Sr(OH)_2.8H_2O$] in 1800 mL of carbon dioxide-free water, heating if necessary. Cool to room temperature, dilute with carbon dioxide-free water to 2000 mL, and mix. Allow to stand overnight, and filter. Store the clear solution in tightly closed containers, protected from carbon dioxide.

0.3 Percent procaine hydrochloride solution—Dissolve 3.0 g of procaine hydrochloride in water to make 1000 mL.

0.1 Percent N-1-naphthylethylenediamine dihydrochloride solution—Dissolve 1.0 g of *N*-1-naphthylethylenediamine dihydrochloride in water to make 1000 mL. Prepare fresh each week.

Standard Preparation—Dissolve an accurately weighed portion of 10 percent nitroglycerin–betalactose absorbate, previously standardized, in water, and dilute quantitatively and stepwise with water to obtain a solution having a known concentration of about 30 µg per mL.

Test Preparation—Dissolve 1 Nitroglycerin Tablet in water to obtain a solution having a concentration of about 30 µg of nitroglycerin per mL.

Procedure—With the sample line pumping water, the other lines pumping their respective reagents, and the spectrophotometer set at 545 nm, standardize the system by pumping until a

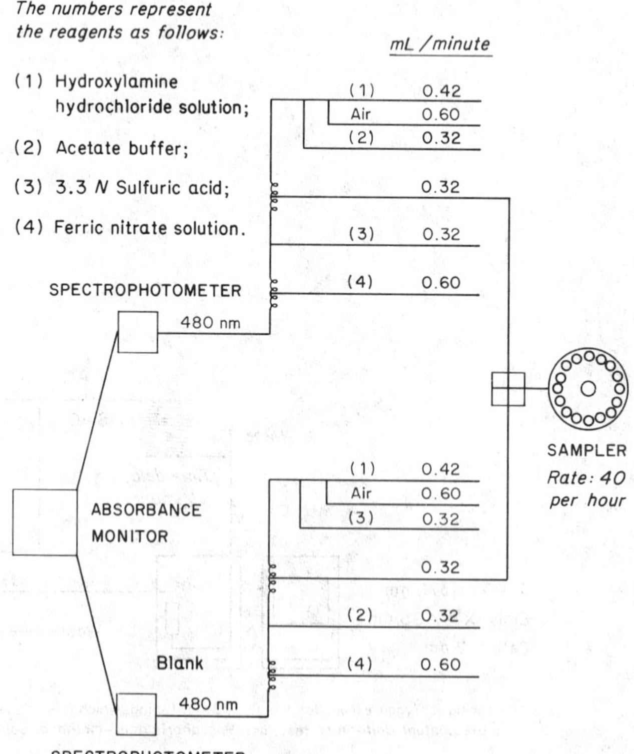

Diagram for Automated Hydroxylamine Assay for Antibiotics

steady absorbance baseline has been established. Transfer portions of the *Standard Preparation* and the *Test Preparation* to sampler cups, and place in the sampler. Start the sampler, and conduct determinations of the *Standard Preparation* and the *Test Preparation* at a rate of 30 per hour, using a ratio of 1:1 for sample and wash time. First, run 2 standards, discarding the first value, then continue the run using 1 standard after each 5 samples, recording the absorbance values. Calculate the quantity, in mg, of $C_3H_5N_3O_9$ in the Tablet by the formula:

$$(T/D)C(A_U/A_S),$$

in which T is the labeled quantity, in mg, of nitroglycerin in the Tablet, D is the concentration, in µg per mL, of nitroglycerin in the solution from the Tablet, based on the labeled quantity per Tablet and the extent of dilution, C is the concentration, in µg per mL, of nitroglycerin in the *Standard Preparation*, A_U is the absorbance of the *Test Preparation*, and A_S is the average of the absorbances of the two *Standard Preparations* that bracket the *Test Preparation*.

Diagrams

The preceding diagrams are arranged in alphabetic order by the name of the drug first mentioned, where the diagram is for a procedure for a specific article. Diagrams pertaining to general classes of articles, e.g., *Diagram for Automated Hydroxylamine Assay for Antibiotics*, appear after that alphabetic series.

⟨21⟩ THERMOMETERS

Thermometers suitable for Pharmacopeial tests conform to specifications of the American Society for Testing and Materials, ASTM Standards E 1, and are standardized in accordance with ASTM Method E 77.

The thermometers are of the mercury-in-glass type, and the column above the liquid is filled with nitrogen. Thermometers may be standardized for total immersion or for partial immersion. Insofar as practicable, each thermometer should be employed according to the condition of immersion under which it was standardized.

Standardization for total immersion involves immersion of the thermometer to the top of the mercury column, with the remainder of the stem and the upper expansion chamber exposed to ambient temperature. Standardization for partial immersion involves immersion of the thermometer to the indicated immersion line etched on the front of the thermometer, with the remainder of the stem exposed to ambient temperature. For use under other conditions of immersion, an emergent stem correction is necessary to obtain correct temperature readings.

In the selection of a thermometer, careful consideration of the conditions under which it is to be used is essential. The accompanying table lists specifications for a number of thermometers suitable for use in Pharmacopeial tests. The lower and upper limits of temperature range specified in the table are to be regarded as inclusive.

Thermometer Specifications

ASTM No. E 1	Temperature Range (°C)	Graduations (°C)	Immersion (mm)
Thermometers for General Use, Including Melting Range Determinations			
1C	−20 to 150	1	76
2C	−5 to 300	1	76
3C	−5 to 400	1	76
Thermometers for Boiling or Distilling Range or Temperature Determinations			
37C	−2 to 52	0.2	100
38C	24 to 78	0.2	100
39C	48 to 102	0.2	100
40C	72 to 126	0.2	100
41C	98 to 152	0.2	100
102C	123 to 177	0.2	100
103C	148 to 202	0.2	100
104C	173 to 227	0.2	100
105C	198 to 252	0.2	100
106C	223 to 277	0.2	100
107C	248 to 302	0.2	100
Thermometers for Congealing Range or Temperature Determinations			
89C	−20 to 10	0.1	76
90C	0 to 30	0.1	76
91C	20 to 50	0.1	76
92C	40 to 70	0.1	76
93C	60 to 90	0.1	76
94C	80 to 110	0.1	76
95C	100 to 130	0.1	76
96C	120 to 150	0.1	76

NOTE—The revised centigrade scale attributed to Celsius differs slightly from the so-called centigrade scale heretofore accepted. However, the magnitude of the difference is negligible, and the two scales are considered equally suitable for Pharmacopeial purposes.

⟨31⟩ VOLUMETRIC APPARATUS

Most of the volumetric apparatus available in the United States is calibrated at 20°, although the temperatures generally prevailing in laboratories more nearly approach 25°, which is the temperature specified generally for Pharmacopeial tests and assays. This discrepancy is inconsequential provided the room temperature is reasonably constant.

Use—To attain the degree of precision required in many Pharmacopeial assays involving volumetric measurements and directing that a quantity be "accurately measured," the apparatus must be chosen and used with care. A buret should be of such size that the titrant volume represents not less than 30% of the nominal volume. Where less than 10 mL of titrant is to be measured, a 10-mL buret or a microburet generally is required.

The design of volumetric apparatus is an important factor in assuring accuracy. For example, the length of the graduated portions of graduated cylinders should be not less than five times the inside diameter, and the tips of burets and pipets should restrict the outflow rate to not more than 500 µL per second.

Standards of Accuracy—The capacity tolerances for volumetric flasks, transfer pipets, and burets are those accepted by the National Institute of Standards and Technology (Class A),* as indicated in the accompanying tables found at the top of the next page.

The capacity tolerances for measuring (i.e., "graduated") pipets of up to and including 10-mL capacity are somewhat larger than those for the corresponding sizes of transfer pipets, namely, 10, 20, and 30 µL for the 2-, 5-, and 10-mL sizes, respectively.

Transfer and measuring pipets calibrated "to deliver" should be drained in a vertical position and then touched against the wall of the receiving vessel to drain the tips. Volume readings on burets should be estimated to the nearest 0.01 mL for 25- and 50-mL burets, and to the nearest 0.005 mL for 5- and 10-mL burets. Pipets calibrated "to contain" are called for in special cases, generally for measuring viscous fluids like syrups; however, a volumetric flask may be substituted for a "to contain" pipet. In such cases, the pipet or flask should be washed clean, after draining, and the washings added to the measured portion.

⟨41⟩ WEIGHTS AND BALANCES

Pharmacopeial tests and assays require the use of balances that vary in capacity, sensitivity, and reproducibility. The accuracy needed for a weighing dictates the type of balance and the class of weights required for that weighing. Where substances are to be "accurately weighed," the weighing is to be performed so as to limit the error to not more than 0.1%. For example, a quantity of 50 mg is to be weighed so that the error does not exceed 50

* See "Testing of Glass Volumetric Apparatus," N.B.S. Circ. 602, April 1, 1959, and NTIS COM-73-10504, National Technical Information Service.

Volumetric Flasks

Designated volume, mL	10	25	50	100	250	500	1000
Limit of error, mL	0.02	0.03	0.05	0.08	0.12	0.15	0.30
Limit of error, %	0.20	0.12	0.10	0.08	0.05	0.03	0.03

Transfer Pipets

Designated volume, mL	1	2	5	10	25	50	100
Limit of error, mL	0.006	0.006	0.01	0.02	0.03	0.05	0.08
Limit of error, %	0.60	0.30	0.20	0.20	0.12	0.10	0.08

Burets

Designated volume, mL	10 ("micro" type)	25	50
Subdivisions, mL	0.02	0.10	0.10
Limit of error, mL	0.02	0.03	0.05

µg. A balance should be chosen such that the value of three times the standard deviation of the reproducibility of the instrument, divided by the amount to be weighed, does not exceed 0.001.

A weight classification should be chosen so as to limit the error to 0.1%. This generally means that Class P weights can be used for quantities greater than 100 mg, Class S-1 for quantities greater than 50 mg, Class S for quantities greater than 20 mg, and Class M for quantities greater than 10 mg. Quantities of less than 10 mg may be weighed on balances having appropriate reproducibilities and designed to afford electrical or optical methods for accurately subdividing a 10-mg, full-scale range, after calibration with a 10-mg, Class M weight.

The tolerances shown in the accompanying table are for new or newly adjusted weights. For weights that have been in use, the tolerances are somewhat larger, as follows:

Class M: 100-, 200-, 300-, and 500-mg denominations—10.5 µg; and 20.0 µg for the group.

Class S: 100-mg and heavier denominations—Twice the values shown in the accompanying table (for individual and group).

Class S-1: Same as shown in the accompanying table.

Class P: For all weights—Twice the values shown in the accompanying table.

Weights should be calibrated periodically, preferably against an absolute standard weight.

Tolerances for New Weights in Sets

	Class M		Class S			
Denomination g	Individual µg	Group µg	Individual µg	Group µg	Class S-1 Individual µg	Class P Individual µg
100	500		250		1000	2000
50	250		120		600	1200
30	150		74	154	450	900
20	100		74	"	350	700
10	50		74	"	250	500
5	34	65	54	105	180	360
3	34	"	54	"	150	300
2	34	"	54	"	130	260
1	34	"	54	"	100	200
mg						
500	5.4	10.5	25	55	80	160
300	5.4	"	25	"	70	140
200	5.4	"	25	"	60	120
100	5.4	"	25	"	50	100
50	5.4	10.5	14	34	42	85
30	5.4	"	14	"	38	75
20	5.4	"	14	"	35	70
10	5.4	"	14	"	30	60
5	5.4	10.5	14	34	28	55
3	5.4	"	14	"	26	52
2	5.4	"	14	"	25	50
1	5.4	"	14	"	25	50

NOTE—Not more than one-third of Class S-1 weights are in error by more than one-half of the tabulated tolerances.

Microbiological Tests

⟨51⟩ ANTIMICROBIAL PRESERVATIVES—EFFECTIVENESS

Antimicrobial preservatives are substances added to dosage forms to protect them from microbial contamination. They are used primarily in multiple-dose containers to inhibit the growth of microorganisms that may be introduced inadvertently during or subsequent to the manufacturing process. Antimicrobial agents should not be used solely to reduce the viable microbial count as a substitute for good manufacturing practice. Situations may arise, however, where their use may be required to minimize proliferation of microorganisms. It should be recognized that the presence of dead microorganisms or the metabolic by-products of living microorganisms may cause adverse reactions in sensitized persons.

Any antimicrobial agent may exhibit the protective properties of a preservative. However, all useful antimicrobial agents are toxic substances. For maximum protection of the consumer, the concentration of the preservative shown to be effective in the final packaged product should be considerably below the concentrations of the preservative that may be toxic to human beings.

The following tests are provided to demonstrate, in multiple-dose parenteral, otic, nasal, and ophthalmic products made with aqueous bases or vehicles, the effectiveness of any added antimicrobial preservative(s), the presence of which is declared on the label of the product concerned. The tests and standards apply only to the product in the original, unopened container in which it was distributed by the producer.[1]

Test Organisms—Use cultures of the following microorganisms:[2] *Candida albicans* (ATCC No. 10231), *Aspergillus niger* (ATCC No. 16404), *Escherichia coli* (ATCC No. 8739), *Pseudomonas aeruginosa* (ATCC No. 9027), and *Staphylococcus aureus* (ATCC No. 6538). Other microorganisms, in addition to those listed, may be included in the test on an optional basis, especially if it appears likely that such microorganisms may represent contaminants likely to be introduced during use of the article.

Media—For the initial cultivation of the test organisms, select an agar medium that is favorable to vigorous growth of the respective stock culture, such as Soybean-Casein Digest Agar Medium (see under *Microbial Limit Tests* ⟨61⟩).

Preparation of Inoculum—Preparatory to the test, inoculate the surface of a suitable volume of solid agar medium from a recently grown stock culture of each of the specified microorganisms. Incubate the bacterial cultures at 30° to 35° for 18 to

[1] For products made with nonaqueous (anhydrous) bases or vehicles, a suitable test may be feasible only at a particular stage in manufacture.

[2] Available from American Type Culture Collection, 12301 Parklawn Drive, Rockville, MD 20852.

24 hours, the culture of *C. albicans* at 20° to 25° for 48 hours, and the culture of *A. niger* at 20° to 25° for 1 week.

To harvest the bacterial and *C. albicans* cultures, use sterile saline TS, washing the surface growth into a suitable vessel, and add sufficient additional saline TS to reduce the microbial count to about 100 million microorganisms per mL. To harvest the *A. niger* culture, use sterile saline TS containing 0.05% of polysorbate 80, and adjust the spore count to about 100 million per mL by adding more sterile saline TS.

Alternatively, the stock culture organisms may be grown in a suitable liquid medium, and the cells may be harvested by centrifugation, washed, and resuspended in sterile saline TS to give the required microbial or spore count.

Determine the number of colony-forming units per mL in each suspension. This value serves to determine the size of inoculum to use in the test. If the standardized suspensions are not used promptly, periodically monitor the suspensions by the plate-count method to determine any loss of viability.

For the plate-count monitoring of inoculated test preparations, use an agar medium corresponding to that used for the initial cultivation of the respective microorganism. Where a specific inactivator of the preservative(s) is available, add a suitable amount of it to the agar plate count medium.

Procedure—Where the product container can be entered aseptically, such as with a needle and syringe through a rubber stopper, conduct the test in five original product containers. If the product container is such that it cannot be entered aseptically, transfer 20-mL samples of the product to each of five sterile, capped bacteriological tubes of suitable size. Inoculate each tube or product container with one of the standardized microbial suspensions, using a ratio equivalent to 0.10 mL of inoculum to 20 mL of product, and mix. A suitable concentration of test microorganisms should be added so that the concentration in the test preparation immediately after inoculation is between 100,000 and 1,000,000 microorganisms per mL. Determine the number of viable microorganisms in each inoculum suspension, and calculate the initial concentration of microorganisms per mL of product under test by the plate-count method.

Incubate the inoculated containers or tubes at 20° to 25°. Examine the containers or tubes at 7, 14, 21, and 28 days subsequent to inoculation. Record any changes observed in appearance, and determine by the plate-count procedure the number of viable microorganisms present at each of these time intervals. Using the theoretical concentrations of microorganisms present at the start of the test, calculate the percentage change in the concentration of each microorganism during the test.

Interpretation—The preservative is effective in the product examined if (a) the concentrations of viable bacteria are reduced to not more than 0.1% of the initial concentrations by the fourteenth day; (b) the concentrations of viable yeasts and molds remain at or below the initial concentrations during the first 14 days; and (c) the concentration of each test microorganism remains at or below these designated levels during the remainder of the 28-day test period.

⟨61⟩ MICROBIAL LIMIT TESTS

This chapter provides tests for the estimation of the number of viable aerobic microorganisms present and for freedom from designated microbial species in pharmaceutical articles of all kinds, from raw materials to the finished forms. An automated method may be substituted for the tests presented here, provided it has been properly validated as giving equivalent or better results. In preparing for and in applying the tests, observe aseptic precautions in handling the specimens. Unless otherwise directed, where the procedure specifies simply "incubate," hold the container in air that is thermostatically controlled at a temperature between 30° and 35°, for a period of 24 to 48 hours. The term "growth" is used in a special sense herein, i.e., to designate *the presence and presumed proliferation of viable microorganisms.*

Preparatory Testing

The validity of the results of the tests set forth in this chapter rests largely upon the adequacy of a demonstration that the test specimens to which they are applied do not, of themselves, inhibit the multiplication, under the test conditions, of microorganisms that may be present. Therefore, preparatory to conducting the tests on a regular basis and as circumstances require subsequently, inoculate diluted specimens of the material to be tested with separate viable cultures of *Staphylococcus aureus, Escherichia coli, Pseudomonas aeruginosa*, and *Salmonella.* This can be done by adding 1 mL of not less than 10^{-3} dilution of a 24-hour broth culture of the microorganism to the first dilution (in pH 7.2 Phosphate Buffer, Fluid Soybean–Casein Digest Medium, or Fluid Lactose Medium) of the test material and following the test procedure. Failure of the organism(s) to grow in the relevant medium invalidates that portion of the examination and necessitates a modification of the procedure by (1) an increase in the volume of diluent, the quantity of test material remaining the same, or by (2) the incorporation of a sufficient quantity of suitable inactivating agent(s) in the diluents, or by (3) an appropriate combination of modifications (1) and (2) so as to permit growth of the inocula.

The following are examples of ingredients and their concentrations that may be added to the culture medium to neutralize inhibitory substances present in the sample: soy lecithin, 0.5%; and polysorbate 20, 4.0%. Alternatively, repeat the test as described in the preceding paragraph, using Fluid Casein Digest–Soy Lecithin–Polysorbate 20 Medium to demonstrate neutralization of preservatives or other antimicrobial agents in the test material. Where inhibitory substances are contained in the product and the latter is soluble, a suitable, validated adaptation of a procedure set forth in the *Test Procedures Using Membrane Filtration*, under *Sterility Tests* ⟨71⟩, may be used.

If in spite of the incorporation of suitable inactivating agents and a substantial increase in the volume of diluent it is still not possible to recover the viable cultures described above and where the article is not suitable for employment of membrane filtration, it can be assumed that the failure to isolate the inoculated organism is attributable to the bactericidal activity of the product. This information serves to indicate that the article is not likely to be contaminated with the given species of microorganism. Monitoring should be continued in order to establish the spectrum of inhibition and bactericidal activity of the article.

Buffer Solution and Media

Culture media may be prepared as follows, or dehydrated culture media may be used provided that, when reconstituted as directed by the manufacturer or distributor, they have similar ingredients and/or yield media comparable to those obtained from the formulas given herein.

In preparing media by the formulas set forth herein, dissolve the soluble solids in the water, using heat, if necessary, to effect complete solution, and add solutions of hydrochloric acid or sodium hydroxide in quantities sufficient to yield the desired pH in the medium when it is ready for use. Determine the pH at 25 ± 2°.

Where agar is called for in a formula, use agar that has a moisture content of not more than 15%. Where water is called for in a formula, use *Purified Water.*

pH 7.2 Phosphate Buffer

Stock Solution—Dissolve 34 g of monobasic potassium phosphate in about 500 mL of water contained in a 1000-mL volumetric flask. Adjust to pH 7.2 ± 0.1 by the addition of sodium hydroxide TS (about 175 mL), add water to volume, and mix. Dispense and sterilize. Store under refrigeration.

For use, dilute the *Stock Solution* with water in the ratio of 1 to 800, and sterilize.

MEDIA

Unless otherwise indicated, the media should be sterilized by heating in an autoclave (see *Steam Sterilization* under *Sterilization* ⟨1211⟩), the exposure time depending on the volume to be sterilized.

I. Fluid Casein Digest–Soy Lecithin–Polysorbate 20 Medium

Pancreatic Digest of Casein	20 g
Soy Lecithin	5 g
Polysorbate 20	40 mL
Water	960 mL

Dissolve the pancreatic digest of casein and soy lecithin in 960 mL of water, heating in a water bath at 48° to 50° for about 30 minutes to effect solution. Add 40 mL of polysorbate 20. Mix, and dispense as desired.

II. Soybean-Casein Digest Agar Medium

Pancreatic Digest of Casein	15.0 g
Papaic Digest of Soybean Meal	5.0 g
Sodium Chloride	5.0 g
Agar	15.0 g
Water	1000 mL

pH after sterilization: 7.3 ± 0.2.

III. Fluid Soybean-Casein Digest Medium

Prepare as directed for *Soybean-Casein Digest Medium* under *Sterility Tests* ⟨71⟩.

IV. Mannitol-Salt Agar Medium

Pancreatic Digest of Casein	5.0 g
Peptic Digest of Animal Tissue	5.0 g
Beef Extract	1.0 g
D-Mannitol	10.0 g
Sodium Chloride	75.0 g
Agar	15.0 g
Phenol Red	0.025 g
Water	1000 mL

Mix, then heat with frequent agitation, and boil for 1 minute to effect solution.

pH after sterilization: 7.4 ± 0.2.

V. Baird-Parker Agar Medium

Pancreatic Digest of Casein	10.0 g
Beef Extract	5.0 g
Yeast Extract	1.0 g
Lithium Chloride	5.0 g
Agar	20.0 g
Glycine	12.0 g
Sodium Pyruvate	10.0 g
Water	950 mL

Heat with frequent agitation, and boil for 1 minute. Sterilize, cool to between 45° and 50°, and add 10 mL of sterile potassium tellurite solution (1 in 100) and 50 mL of egg-yolk emulsion. Mix intimately but gently, and pour into plates. (Prepare the egg-yolk emulsion by disinfecting the surface of whole shell eggs, aseptically cracking the eggs, and separating out intact yolks into a sterile graduated cylinder. Add sterile saline TS to obtain a 3 to 7 ratio of egg yolk to saline. Add to a sterile blender cup, and mix at high speed for 5 seconds.)

pH after sterilization: 6.8 ± 0.2.

VI. Vogel-Johnson Agar Medium

Pancreatic Digest of Casein	10.0 g
Yeast Extract	5.0 g
Mannitol	10.0 g
Dibasic Potassium Phosphate	5.0 g
Lithium Chloride	5.0 g
Glycine	10.0 g
Agar	16.0 g
Phenol Red	25.0 mg
Water	1000 mL

Boil the solution of solids for 1 minute. Sterilize, cool to between 45° and 50°, and add 20 mL of sterile potassium tellurite solution (1 in 100).

pH after sterilization: 7.2 ± 0.2.

VII. Cetrimide Agar Medium

Pancreatic Digest of Gelatin	20.0 g
Magnesium Chloride	1.4 g
Potassium Sulfate	10.0 g
Agar	13.6 g
Cetyl Trimethylammonium Bromide (Cetrimide)	0.3 g
Glycerin	10.0 mL
Water	1000 mL

Dissolve all solid components in the water, and add the glycerin. Heat, with frequent agitation, and boil for 1 minute to effect solution.

pH after sterilization: 7.2 ± 0.2.

VIII. Pseudomonas Agar Medium for Detection of Fluorescin

Pancreatic Digest of Casein	10.0 g
Peptic Digest of Animal Tissue	10.0 g
Anhydrous Dibasic Potassium Phosphate	1.5 g
Magnesium Sulfate ($MgSO_4 \cdot 7H_2O$)	1.5 g
Glycerin	10.0 mL
Agar	15.0 g
Water	1000 mL

Dissolve the solid components in the water before adding the glycerin. Heat, with frequent agitation, and boil for 1 minute to effect solution.

pH after sterilization: 7.2 ± 0.2.

IX. Pseudomonas Agar Medium for Detection of Pyocyanin

Pancreatic Digest of Gelatin	20.0 g
Anhydrous Magnesium Chloride	1.4 g
Anhydrous Potassium Sulfate	10.0 g
Agar	15.0 g
Glycerin	10.0 mL
Water	1000 mL

Dissolve the solid components in the water before adding the glycerin. Heat, with frequent agitation, and boil for 1 minute to effect solution.

pH after sterilization: 7.2 ± 0.2.

X. Fluid Lactose Medium

Beef Extract	3.0 g
Pancreatic Digest of Gelatin	5.0 g
Lactose	5.0 g
Water	1000 mL

Cool as quickly as possible after sterilization.

pH after sterilization: 6.9 ± 0.2.

XI. Fluid Selenite-Cystine Medium

Pancreatic Digest of Casein	5.0 g
Lactose	4.0 g
Sodium Phosphate	10.0 g
Sodium Acid Selenite	4.0 g
L-Cystine	10.0 mg
Water	1000 mL

Final pH: 7.0 ± 0.2.

Mix, and heat to effect solution. Heat in flowing steam for 15 minutes. *Do not sterilize.*

XII. Fluid Tetrathionate Medium

Pancreatic Digest of Casein	2.5 g
Peptic Digest of Animal Tissue	2.5 g
Bile Salts	1.0 g
Calcium Carbonate	10.0 g
Sodium Thiosulfate	30.0 g
Water	1000 mL

Heat the solution of solids to boiling. On the day of use, add a solution prepared by dissolving 5 g of potassium iodide and 6 g of iodine in 20 mL of water. Then add 10 mL of a solution of brilliant green (1 in 1000), and mix. Do not heat the medium after adding the brilliant green solution.

XIII. Brilliant Green Agar Medium

Yeast Extract	3.0 g
Peptic Digest of Animal Tissue	5.0 g
Pancreatic Digest of Casein	5.0 g
Lactose	10.0 g
Sodium Chloride	5.0 g
Sucrose	10.0 g
Phenol Red	80 mg
Agar	20.0 g
Brilliant Green	12.5 mg
Water	1000 mL

Boil the solution of solids for 1 minute. Sterilize just prior to use, melt the medium, pour into Petri dishes, and allow to cool.

pH after sterilization: 6.9 ± 0.2.

XIV. Xylose-Lysine-Desoxycholate Agar Medium

Xylose	3.5 g
L-Lysine	5.0 g
Lactose	7.5 g
Sucrose	7.5 g
Sodium Chloride	5.0 g
Yeast Extract	3.0 g
Phenol Red	80 mg
Agar	13.5 g
Sodium Desoxycholate	2.5 g
Sodium Thiosulfate	6.8 g
Ferric Ammonium Citrate	800 mg
Water	1000 mL

Final pH: 7.4 ± 0.2.

Heat the mixture of solids and water, with swirling, just to the boiling point. *Do not overheat or sterilize.* Transfer at once to a water bath maintained at about 50°, and pour into plates as soon as the medium has cooled.

XV. Bismuth Sulfite Agar Medium

Beef Extract	5.0 g
Pancreatic Digest of Casein	5.0 g
Peptic Digest of Animal Tissue	5.0 g
Dextrose	5.0 g
Sodium Phosphate	4.0 g
Ferrous Sulfate	300 mg
Bismuth Sulfite Indicator	8.0 g
Agar	20.0 g
Brilliant Green	25 mg
Water	1000 mL

Final pH: 7.6 ± 0.2.

Heat the mixture of solids and water, with swirling, just to the boiling point. *Do not overheat or sterilize.* Transfer at once to a water bath maintained at about 50°, and pour into plates as soon as the medium has cooled.

XVI. Triple Sugar-Iron-Agar Medium

Pancreatic Digest of Casein	10.0 g
Pancreatic Digest of Animal Tissue	10.0 g
Lactose	10.0 g
Sucrose	10.0 g
Dextrose	1.0 g
Ferrous Ammonium Sulfate	200 mg
Sodium Chloride	5.0 g
Sodium Thiosulfate	200 mg
Agar	13.0 g
Phenol Red	25 mg
Water	1000 mL

pH after sterilization: 7.3 ± 0.2.

XVII. MacConkey Agar Medium

Pancreatic Digest of Gelatin	17.0 g
Pancreatic Digest of Casein	1.5 g
Peptic Digest of Animal Tissue	1.5 g
Lactose	10.0 g
Bile Salts Mixture	1.5 g
Sodium Chloride	5.0 g
Agar	13.5 g
Neutral Red	30 mg
Crystal Violet	1.0 mg
Water	1000 mL

Boil the mixture of solids and water for 1 minute to effect solution.

pH after sterilization: 7.1 ± 0.2.

XVIII. Levine Eosin–Methylene Blue Agar Medium

Pancreatic Digest of Gelatin	10.0 g
Dibasic Potassium Phosphate	2.0 g
Agar	15.0 g
Lactose	10.0 g
Eosin Y	400 mg
Methylene Blue	65 mg
Water	1000 mL

Dissolve the pancreatic digest of gelatin, the dibasic potassium phosphate, and the agar in the water, with warming, and allow to cool. Just prior to use, liquefy the gelled agar solution, add the remaining ingredients, as solutions, in the following amounts, and mix: for each 100 mL of the liquefied agar solution—5 mL of lactose solution (1 in 5), 2 mL of the eosin Y solution (1 in 50), and 2 mL of methylene blue solution (1 in 300). The finished medium may not be clear.

pH after sterilization: 7.1 ± 0.2.

XIX. Sabouraud Dextrose Agar Medium

Dextrose	40 g
Mixture of equal parts of Peptic Digest of Animal Tissue and Pancreatic Digest of Casein	10 g
Agar	15 g
Water	1000 mL

Mix, and boil to effect solution.

pH after sterilization: 5.6 ± 0.2.

XX. Potato Dextrose Agar Medium

Cook 300 g of peeled and diced potatoes in 500 mL of water prepared by distillation, filter through cheesecloth, add water prepared by distillation to make 1000 mL, and add the following:

Agar	15 g
Glucose	20 g

Dissolve by heating, and sterilize.

pH after sterilization: 5.6 ± 0.2.

For use, just prior to pouring the plates, adjust the melted and cooled to 45° medium with sterile tartaric acid solution (1 in 10) to a pH of 3.5 ± 0.1. Do not reheat the pH 3.5 medium.

Sampling

Provide separate 10-mL or 10-g specimens for each of the tests called for in the individual monograph.

Procedure

Prepare the specimen to be tested, by treatment that is appropriate to its physical characteristics and that does not alter the number and kind of microorganisms originally present, in order to obtain a solution or suspension of all or part of it in a form suitable for the test procedure(s) to be carried out.

For a solid that dissolves to an appreciable extent but not completely, reduce the substance to a moderately fine powder, suspend it in the vehicle specified, and proceed as directed under *Total Aerobic Microbial Count*, and under *Test for Staphylococcus aureus and Pseudomonas aeruginosa* and *Test for Salmonella Species and Escherichia coli.*

For a fluid specimen that consists of a true solution, or a suspension in water or a hydroalcoholic vehicle containing less than 30 percent of alcohol, and for a solid that dissolves readily and practically completely in 90 mL of *pH 7.2 Phosphate Buffer* or the media specified, proceed as directed under *Total Aerobic Microbial Count*, and under *Test for Staphylococcus aureus and Pseudomonas aeruginosa* and *Test for Salmonella Species and Escherichia coli.*

For water-immiscible fluids, ointments, creams, and waxes, prepare a suspension with the aid of a minimal quantity of a suitable, sterile emulsifying agent (such as one of the polysorbates), using a mechanical blender and warming to a temperature not exceeding 45°, if necessary, and proceed with the suspension as directed under *Total Aerobic Microbial Count*, and under *Test for Staphylococcus aureus and Pseudomonas aeruginosa* and *Test for Salmonella Species and Escherichia coli.*

For a fluid specimen in aerosol form, chill the container in an alcohol–dry ice mixture for approximately 1 hour, cut open the container, allow it to reach room temperature, permit the propellant to escape, or warm to drive off the propellant if feasible, and transfer the quantity of test material required for the procedures specified in one of the two preceding paragraphs, as appropriate. Where 10.0 g or 10.0 mL of the specimen, whichever is applicable, cannot be obtained from 10 containers in aerosol form, transfer the entire contents from 10 chilled containers to the culture medium, permit the propellant to escape, and proceed with the test on the residues. If the results of the test are inconclusive or doubtful, repeat the test with a specimen from 20 more containers.

Total Aerobic Microbial Count—For specimens that are sufficiently soluble or translucent to permit use of the *Plate Method*, use that method; otherwise, use the *Multiple-tube Method*. With either method, first dissolve or suspend 10.0 g of the specimen if it is a solid, or 10 mL, accurately measured, if the specimen is a liquid, in pH 7.2 Phosphate Buffer, Fluid Soybean-Casein Digest Medium, or Fluid Casein Digest–Soy Lecithin–Polysorbate 20 Medium to make 100 mL. For viscous specimens that cannot be pipeted at this initial 1:10 dilution, dilute the specimen until a suspension is obtained, i.e., 1:50 or 1:100, etc., that can be pipeted. Perform the test for absence of inhibitory (antimicrobial) properties as described under *Preparatory Testing* before the determination of *Total Aerobic Microbial Count*. Add the specimen to the medium not more than 1 hour after preparing the appropriate dilutions for inoculation.

PLATE METHOD—Dilute further, if necessary, the fluid so that 1 mL will be expected to yield between 30 and 300 colonies. Pipet 1 mL of the final dilution onto each of two sterile petri dishes. Promptly add to each dish 15 to 20 mL of Soybean-Casein Digest Agar Medium that previously has been melted and cooled to approximately 45°. Cover the petri dishes, mix the sample with the agar by tilting or rotating the dishes, and allow the contents to solidify at room temperature. Invert the petri dishes, and incubate for 48 to 72 hours. Following incubation, examine the plates for growth, count the number of colonies, and express the average for the two plates in terms of the number of microorganisms per g or per mL of specimen. If no microbial colonies are recovered from the dishes representing the initial 1:10 dilution of the specimen, express the results as "less than 10 microorganisms per g or per mL of specimen."

Table 1. Most Probable Total Count by Multiple-tube Method.

Observed Combinations of Numbers of Tubes Showing Growth in Each Set — No. of mg (or mL) of Specimen per Tube: 100 (100 µL)	10 (10 µL)	1 (1 µL)	Most Probable Number of Microorganisms per g or per mL
3	3	3	>1100
3	3	2	1100
3	3	1	500
3	3	0	200
3	2	3	290
3	2	2	210
3	2	1	150
3	2	0	90
3	1	3	160
3	1	2	120
3	1	1	70
3	1	0	40
3	0	3	95
3	0	2	60
3	0	1	40
3	0	0	23

MULTIPLE-TUBE METHOD—Into each of fourteen test tubes of similar size place 9.0 mL of sterile Fluid Soybean-Casein Digest Medium. Arrange twelve of the tubes in four sets of three tubes each. Put aside one set of three tubes to serve as the controls. Into each of three tubes of one set ("100") and into a fourth tube (*A*) pipet 1 mL of the solution or suspension of the specimen, and mix. From tube *A*, pipet 1 mL of its contents into the one remaining tube (*B*) not included in a set, and mix. These two tubes contain 100 mg (or 100 µL) and 10 mg (or 10 µL) of the specimen, respectively. Into each of the second set ("10") of three tubes pipet 1 mL from tube *A*, and into each tube of the third set ("1") pipet 1 mL from tube *B*. Discard the unused contents of tubes *A* and *B*. Close well, and incubate all of the tubes. Following the incubation period, examine the tubes for growth: the three control tubes remain clear and the observations in the tubes containing the specimen, when interpreted by reference to Table 1, indicate the most probable number of microorganisms per g or per mL of specimen.

Test for *Staphylococcus aureus* and *Pseudomonas aeruginosa*—To the specimen add Fluid Soybean-Casein Digest Medium to make 100 mL, mix, and incubate. Examine the medium for growth, and if growth is present, use an inoculating loop to streak a portion of the medium on the surface of Vogel-Johnson Agar Medium (or Baird-Parker Agar Medium, or Mannitol-Salt Agar Medium) and of Cetrimide Agar Medium, each plated on petri dishes. Cover and invert the dishes, and incubate. If, upon examination, none of the plates contains colonies having the characteristics listed in Tables 2 and 3 for the media used, the test specimen meets the requirements for freedom from *Staphylococcus aureus* and *Pseudomonas aeruginosa.*

COAGULASE TEST (FOR *Staphylococcus aureus*)—With the aid of an inoculating loop, transfer representative suspect colonies from the agar surfaces of the Vogel-Johnson Agar Medium (or Baird-Parker Agar Medium, or Mannitol-Salt Agar Medium) to individual tubes, each containing 0.5 mL of mammalian, preferably rabbit or horse, plasma with or without suitable additives. Incubate in a water bath at 37°, examining the tubes at 3 hours and subsequently at suitable intervals up to 24 hours. Test positive and negative controls simultaneously with the unknown specimens. If no coagulation in any degree is observed, the specimen meets the requirements of the test for absence of *Staphylococcus aureus.*

OXIDASE AND PIGMENT TESTS (FOR *Pseudomonas aeruginosa*)—With the aid of an inoculating loop, streak representative suspect colonies from the agar surface of Cetrimide Agar Medium on the agar surfaces of Pseudomonas Agar Medium for Detection of Fluorescin and Pseudomonas Agar Medium for Detection of Pyocyanin contained in petri dishes. If numerous colonies are to be transferred, divide the surface of each plate into quadrants, each of which may be inoculated from a separate colony. Cover and invert the inoculated media, and incubate at 35 ± 2° for not less than three days. Examine the streaked surfaces under ultraviolet light. Examine the plates to determine

Table 2. Morphologic Characteristics of *Staphylococcus aureus* on Selective Agar Media.

Selective Medium	Vogel-Johnson Agar Medium	Mannitol-Salt Agar Medium	Baird-Parker Agar Medium
Characteristic Colonial Morphology	Black surrounded by yellow zone	Yellow colonies with yellow zones	Black, shiny, surrounded by clear zones 2 to 5 mm
Gram Stain	Positive cocci (in clusters)	Positive cocci (in clusters)	Positive cocci (in clusters)

Table 3. Morphologic Characteristics of *Pseudomonas aeruginosa* on Selective and Diagnostic Agar Media.

Medium	Cetrimide Agar Medium	Pseudomonas Agar Medium for Detection of Fluorescin	Pseudomonas Agar Medium for Detection of Pyocyanin
Characteristic Colonial Morphology	Generally greenish	Generally colorless to yellowish	Generally greenish
Fluorescence in Ultraviolet Light	Greenish	Yellowish	Blue
Oxidase Test	Positive	Positive	Positive
Gram Stain	Negative rods	Negative rods	Negative rods

Table 4. Morphologic Characteristics of *Salmonella* Species on Selective Agar Media.

Medium	Description of Colony
Brilliant Green Agar Medium	Small, transparent, colorless or pink to white opaque (frequently surrounded by pink to red zone)
Xylose-Lysine-Desoxycholate Agar Medium	Red, with or without black centers
Bismuth Sulfite Agar Medium	Black or green

Table 5. Morphologic Characteristics of *Escherichia coli* on MacConkey Agar Medium.

Characteristic Colonial Morphology	Brick-red; may have surrounding zone of precipitated bile
Gram Stain	Negative rods (cocco-bacilli)

whether colonies having the characteristics listed in Table 3 are present.

Confirm any suspect colonial growth on one or more of the media as *Pseudomonas aeruginosa* by means of the oxidase test. Upon the colonial growth place or transfer colonies to strips or disks of filter paper that previously has been impregnated with *N,N*-dimethyl-*p*-phenylenediamine dihydrochloride: if there is no development of a pink color, changing to purple, the specimen meets the requirements of the test for the absence of *Pseudomonas aeruginosa.* The presence of *Pseudomonas aeruginosa* may be confirmed by other suitable cultural and biochemical tests, if necessary.

Test for *Salmonella* Species and *Escherichia coli*—To the specimen, contained in a suitable vessel, add a volume of Fluid Lactose Medium to make 100 mL, and incubate. Examine the medium for growth, and if growth is present, mix by gently shaking. Pipet 1-mL portions into vessels containing, respectively, 10 mL of Fluid Selenite-Cystine Medium and Fluid Tetrathionate Medium, mix, and incubate for 12 to 24 hours. (Retain the remainder of the Fluid Lactose Medium.)

TEST FOR *Salmonella* SPECIES—By means of an inoculating loop, streak portions from both the selenite-cystine and tetrathionate media on the surface of Brilliant Green Agar Medium, Xylose-Lysine-Desoxycholate Agar Medium, and Bismuth Sulfite Agar Medium contained in petri dishes. Cover and invert the dishes, and incubate. Upon examination, if none of the colonies conforms to the description given in Table 4, the specimen meets the requirements of the test for absence of the genus *Salmonella.*

If colonies of Gram-negative rods matching the description in Table 4 are found, proceed with further identification by transferring representative suspect colonies individually, by means of an inoculating wire, to a butt-slant tube of Triple Sugar-Iron-Agar Medium by first streaking the surface of the slant and then stabbing the wire well beneath the surface. Incubate. If examination discloses no evidence of tubes having alkaline (red) slants and acid (yellow) butts (with or without concomitant blackening of the butt from hydrogen sulfide production), the specimen meets the requirements of the test for the absence of the genus *Salmonella.**

TEST FOR *Escherichia coli*—By means of an inoculating loop, streak a portion from the remaining Fluid Lactose Medium on the surface of MacConkey Agar Medium. Cover and invert the dishes, and incubate. Upon examination, if none of the colonies conforms to the description given in Table 5 for this medium, the specimen meets the requirements of the test for absence of *Escherichia coli.*

If colonies matching the description in Table 5 are found, proceed with further identification by transferring the suspect colonies individually, by means of an inoculating loop, to the surface of Levine Eosin–Methylene Blue Agar Medium, plated on petri dishes. If numerous colonies are to be transferred, divide the surface of each plate into quadrants, each of which may be seeded from a separate colony. Cover and invert the plates, and incubate. Upon examination, if none of the colonies exhibits both a characteristic metallic sheen under reflected light and a blue-black appearance under transmitted light, the specimen meets the requirements of the test for the absence of *Escherichia coli.* The presence of *Escherichia coli* may be confirmed by further suitable cultural and biochemical tests.

* Additional, confirmatory evidence may be obtained by use of procedures set forth in *Official Methods of Analysis of the AOAC*, 12th ed. (1975), sections 46.013–46.026.

Total Combined Molds and Yeasts Count—Proceed as for the *Plate Method* under *Total Aerobic Microbial Count*, except for using the same amount of Sabouraud Dextrose Agar Medium or Potato Dextrose Agar Medium, instead of Soybean Casein Digest Medium, and except for incubating the inverted petri dishes for 5 to 7 days at 20° to 25°.

Retest—For the purpose of confirming a doubtful result by any of the procedures outlined in the foregoing tests following their application to a 10.0-g specimen, a retest on a 25-g specimen of the product may be conducted. Proceed as directed under *Procedure*, but make allowance for the larger specimen size.

⟨71⟩ STERILITY TESTS

The following procedures are applicable for determining whether a Pharmacopeial article purporting to be sterile complies with the requirements set forth in the individual monograph with respect to the test for *Sterility.* (For the use of sterility test procedures as part of quality control in manufacture, see *Sterilization and Sterility Assurance of Compendial Articles* ⟨1211⟩.) In view of the possibility that positive results may be due to faulty aseptic techniques or environmental contamination in testing, provisions are included under *Interpretation of Sterility Test Results* for two stages of testing.

Alternative procedures may be employed to demonstrate that an article is sterile, provided the results obtained are at least of equivalent reliability. (See *Procedures* under *Tests and Assays* in the *General Notices.*) Where a difference appears, or in the event of a dispute, when evidence of microbial contamination is obtained by the procedure given in this Pharmacopeia, the result so obtained is conclusive of failure of the article to meet the requirements of the test. Similarly, failure to demonstrate microbial contamination by the procedure given in this Pharmacopeia is evidence that the article meets the requirements of the test. For additional interpretive information, see *Sterilization and Sterility Assurance of Compendial Articles* ⟨1211⟩.

Media

Media for the tests may be prepared as described below, or dehydrated mixtures yielding similar formulations may be used provided that, when reconstituted as directed by the manufacturer or distributor, they have growth-promoting properties equal or superior to those obtained from the formulas given herein.

I. Fluid Thioglycollate Medium

L-Cystine	0.5 g
Sodium Chloride	2.5 g
Dextrose ($C_6H_{12}O_6 \cdot H_2O$)	5.5 g
Agar, granulated (moisture content not in excess of 15%)	0.75 g
Yeast Extract (water-soluble)	5.0 g
Pancreatic Digest of Casein	15.0 g
Sodium Thioglycollate	0.5 g
or Thioglycollic Acid	0.3 mL
Resazurin Sodium Solution (1 in 1000), freshly prepared	1.0 mL
Water	1000 mL

pH after sterilization: 7.1 ± 0.2.

Mix, and heat until solution is effected. Adjust the solution with 1 *N* sodium hydroxide so that, after sterilization, it will have a pH of 7.1 ± 0.2. Filter while hot through filter paper, if necessary. Place the medium in suitable vessels, which provide a ratio of surface to depth of medium such that not more than the upper half of the medium has undergone a color change indicative of oxygen uptake at the end of the incubation period. Sterilize in an autoclave. If more than the upper one-third has acquired a pink color, the medium may be restored once by heating on a steam bath or in free-flowing steam until the pink color disappears. When ready for use, not more than the upper one-tenth of the medium should have a pink color.

Use Fluid Thioglycollate Medium by incubating it under aerobic conditions.

II. Alternative Thioglycollate Medium for Devices Having Tubes with Small Lumina

L-Cystine	0.5 g
Sodium Chloride	2.5 g
Dextrose ($C_6H_{12}O_6 . H_2O$)	5.5
Yeast Extract (water-soluble)	5.0
Pancreatic Digest of Casein	15.0
Sodium Thioglycollate	0.5 g
or Thioglycollic Acid	0.3 mL
Water	1000 mL

pH after sterilization: 7.1 ± 0.2.

Heat the ingredients in a suitable container until solution is effected. Mix, and, if necessary, adjust the solution with 1 *N* sodium hydroxide so that, after sterilization, it will have a pH of 7.1 ± 0.2. Filter, if necessary, place in suitable vessels, and sterilize by steam. The medium is freshly prepared or heated in a steam bath and allowed to cool just prior to use. Do not reheat.

Use Alternative Thioglycollate Medium in a manner that will assure anaerobic conditions for the duration of the incubation period.

III. Soybean-Casein Digest Medium

Pancreatic Digest of Casein	17.0 g
Papaic Digest of Soybean Meal	3.0 g
Sodium Chloride	5.0 g
Dibasic Potassium Phosphate	2.5 g
Dextrose ($C_6H_{12}O_6 . H_2O$)	2.5 g
Water	1000 mL

pH after sterilization: 7.3 ± 0.2.

Dissolve the solids in the water, warming slightly to effect solution. Cool the solution to room temperature, and adjust with 1 *N* sodium hydroxide, if necessary, to obtain a pH of 7.3 ± 0.2 after sterilization. Filter, if necessary, and dispense into suitable vessels. Sterilize by steam.

Use Soybean-Casein Digest Medium by incubating it under aerobic conditions.

Diluting and Rinsing Fluids

FLUID A—Dissolve 1 g of peptic digest of animal tissue (see *Reagent Specifications* in the section, *Reagents, Indicators, and Solutions*) in water to make 1 liter, filter or centrifuge to clarify, adjust to a pH of 7.1 ± 0.2, dispense into containers in 100-mL quantities, and sterilize by steam. NOTE—Where *Fluid A* is to be used in performing the test for *Sterility* on a specimen of the penicillin or cephalosporin class of antibiotics, aseptically add a quantity of sterile penicillinase to the *Fluid A* to be used to rinse the membrane(s) sufficient to inactivate any residual antibiotic activity on the membrane(s) after the solution of the specimen has been filtered.

FLUID D—If the test specimen contains lecithin or oil, or for device sterile pathway tests using membrane filtration, use *Fluid A* to each liter of which has been added 1 mL of polysorbate 80, adjust to a pH of 7.1 ± 0.2, dispense into flasks, and sterilize by steam.

FLUID K—

Peptic Digest of Animal Tissue (see *Reagent Specifications* in the section, *Reagents, Indicators, and Solutions*)	5.0 g
Beef Extract	3.0 g
Polysorbate 80	10.0 g
Water	1000 mL

pH after sterilization: 6.9 ± 0.2.

Sterilize by steam.

NOTE—A sterile fluid shall not have antibacterial or antifungal properties if it is to be considered suitable for dissolving, diluting, or rinsing an article under test for sterility.

Growth Promotion Test

Confirm the sterility of each lot of medium by incubation of representative containers, at the temperature and for the length of time specified in the test.

Test each autoclaved load of each lot of medium for its growth-promoting qualities by separately inoculating duplicate test containers of each medium with 10 to 100 viable microorganisms of each of the strains listed in the accompanying table, and incubating according to the conditions specified.

The test media are satisfactory if clear evidence of growth appears in all inoculated media containers within 7 days. The tests may be conducted simultaneously with the use of the test media for sterility test purposes. The sterility test is considered invalid if the test medium shows inadequate growth response.

If freshly prepared media are not used within 2 days, store them in the dark, preferably at 2° to 25°.

Finished media, if stored in unsealed containers, may be used for not more than one month, provided that they are tested within one week of the time of use and if the color indicator requirements are met. If stored in suitable sealed containers, the media may be used for not more than one year, provided they are tested for growth promotion every three months and if the color indicator requirements are met.

Bacteriostasis and Fungistasis

Before initiating direct transfer sterility tests on an article, determine the level of bacteriostatic and fungistatic activity by the following procedures. Prepare dilute cultures of bacteria and fungi from at least the strains of microorganisms cited under *Growth Promotion Test*. Inoculate the sterility test media with 10 to 100 viable microorganisms, employing volumes of medium listed in the table of Quantities for Liquid Articles under *Selection of Test Specimens and Incubation*. Add the specified portion of article to half of a suitable number of the containers already containing the inoculum and culture medium. Incubate the containers at the appropriate temperatures and under the conditions listed in the table for not less than 7 days.

If growth of the test organisms in the article-medium mixture is visually comparable to that in the control vessels, use the amounts

Medium	Test Microorganisms*	Incubation	
		Temperature (°)	Conditions
Fluid Thioglycollate	(1) *Bacillus subtilis* (ATCC No. 6633)†	30 to 35	
	(2) *Candida albicans* (ATCC No. 10231)	30 to 35	Aerobic
	(3) *Bacteroides vulgatus* (ATCC No. 8482)‡	30 to 35	
Alternative Thioglycollate	(1) *Bacteroides vulgatus* (ATCC No. 8482)‡	30 to 35	Anaerobic
Soybean-Casein Digest	(1) *Bacillus subtilis* (ATCC No. 6633)†	20 to 25	
	(2) *Candida albicans* (ATCC No. 10231)	20 to 25	Aerobic

* Available from the American Type Culture Collection, 12301 Parklawn Drive, Rockville, MD 20852.

NOTE—Seed lot culture maintenance techniques should be employed so that the viable microorganisms used for inoculation are not more than 5 passages removed from the ATCC cultures.

† If a spore-forming organism is not desired, use *Micrococcus luteus* (ATCC No. 9341) at the incubation temperatures indicated in the table.

‡ If a spore-forming organism is desired, use *Clostridium sporogenes* (ATCC No. 11437) at the incubation temperature indicated in the table.

Quantities for Liquid Articles

Container content (mL)	Minimum volume taken from each container for each medium	Minimum Volume of Each Medium: Used for direct transfer of volume taken from each container (mL)	Minimum Volume of Each Medium: Used for membrane or half membrane representing total volume from the appropriate number of containers (mL)	No. of containers per medium
Less than 10	1 mL, or entire contents if less than 1 mL	15	100	20 (40 if each does not contain sufficient volume for both media)
10 to less than 50	5 mL	40	100	20
50 to less than 100	10 mL	80	100	20
50 to less than 100, intended for intravenous administration	Entire contents	—	100	10
100 to 500	Entire contents	—	100	10
Over 500	500 mL	—	100	10

of article and medium regularly specified in the table of Quantities for Liquid Articles under *Selection of Test Specimens and Incubation.*

If the article is bacteriostatic and/or fungistatic when tested as described above, use a suitable sterile neutralizing agent, if available. Suitability of such an agent is determined as in the test described below. If a neutralizing agent is not available, establish, as described below, suitable amounts of article and medium to be used.

Repeat the tests set forth above, using the specified amount of article and larger volumes of the medium to determine the ratio of article to medium in which growth of the test organisms is not adversely affected.

If the specified amount of article is bacteriostatic or fungistatic in 250 mL of the medium, decrease the amount of the article to find the maximum amount that does not adversely affect the growth of the test organism in 250 mL of the medium. For liquids and suspensions, if this amount is less than 1 mL, increase the quantity of medium so that the 1 mL is sufficiently diluted to prevent inhibition of growth. For solids that are not readily soluble or dispersible, if the amount is less than 50 mg, increase the quantity of medium so that the 50 mg of the article is sufficiently diluted to prevent inhibition of growth. In either case, use the amounts of the article and medium established in this ratio for sterility testing.

Where membrane filtration is used, make similar comparisons using the specified portions of the article under test and similar quantities of a suitable diluting and rinsing fluid, rinsing the membrane in each case with three 100-mL portions of the diluting and rinsing fluid. Inoculate the stated quantities of viable microorganisms into each final portion of diluting and rinsing fluid used to filter the article under test and to filter the diluting and rinsing fluid only. The growth of the test organism in each case from the membrane(s) used to filter the article under test followed by the inoculated final diluting and rinsing fluid is visually comparable to that from the membrane(s) used to filter only the inoculated diluting and rinsing fluid.

General Procedure

The test procedures include (1) direct transfer to test media and (2) membrane filtration techniques. Sterility testing of Pharmacopeial articles using membrane filtration of the test specimens, where feasible, is the method of choice. The procedure is particularly useful for liquids and soluble powders possessing bacteriostatic or fungistatic properties, so as to permit separation of possible contaminating microorganisms from such growth inhibitors. The procedure is to be validated for such use. For similar reasons, it is very useful where the article is an oil, an ointment, or a cream that can be put into solution with non-bacteriostatic or non-fungistatic diluting fluids. Its use is also entirely appropriate and preferable in the sterility testing of non-bacteriostatic or non-fungistatic liquids or soluble powders. Certain devices also may be appropriately tested for sterility of surfaces or the critical pathways by the membrane filtration technique.

Because of diversity in the nature of articles to be tested and other factors affecting the conduct of the sterility test, it is important to observe the following considerations in performing sterility tests.

OPENING CONTAINERS

Cleanse the exterior surfaces of ampuls and closures of vials and bottles with a suitable decontaminating agent, and gain access to the contents in an aseptic manner. If the vial contents are packaged under vacuum, admit sterile air by means of a suitable sterile device, such as a needle attached to a syringe containing sterilizing grade filter material.

For purified cotton, gauze, surgical dressings, sutures, and related Pharmacopeial articles, open the package or container aseptically.

SELECTION OF TEST SPECIMENS AND INCUBATION

For liquid articles, use not less than the volumes of article and medium for each unit and the number of containers per medium specified in the table of Quantities for Liquid Articles, in this section. If the contents are of sufficient quantity, they may be divided so that portions are added to each of the two specified media. If each container does not contain sufficient volume for both media, use double the number of containers. For articles other than liquids, test 20 units of the article with each medium. For such articles in which only the lumen must be sterile, flush the lumen with a suitable quantity of appropriate medium to yield a recovery of not less than 15 mL of medium.

Unless otherwise directed in the individual monograph or in a section of this chapter, incubate the test mixture for 14 days with Fluid Thioglycollate Medium (or Alternative Thioglycollate Medium, where so indicated) at 30° to 35°, and with Soybean-Casein Digest Medium at 20° to 25°.

Test Procedures for Direct Transfer to Test Media

LIQUIDS

Remove liquids from test containers with a sterile pipet or with a sterile syringe and needle. Aseptically transfer the specified volume of the material from each test container to a vessel of culture medium. Mix the liquid with the medium, but do not aerate excessively. Incubate in the specified media as directed under *General Procedure*, for not less than 14 days. Examine the media visually for growth at least as often as on the third or fourth or fifth day, on the seventh or eighth day, and on the last day of the test period.

Where the material being tested renders the medium turbid, so that the presence or absence of microbial growth cannot be determined readily by visual examination, transfer suitable portions of the medium to fresh vessels of the same medium at least once during the period from the third to the seventh day after the test is started. Continue incubation of the original and of the transfer vessels for a total of not less than 14 days from the original inoculation.

OINTMENTS AND OILS INSOLUBLE IN ISOPROPYL MYRISTATE

Select 20 representative containers, assign them to 2 groups of 10 containers, and treat each group as follows. Aseptically transfer 100 mg from each of the 10 containers to a flask containing 100 mL of a sterile, aqueous vehicle capable of dispersing the test material homogeneously throughout the fluid mixture. [NOTE—The choice of dispersing agent incorporated in the aqueous vehicle may differ according to the nature of the ointment or oil. Before initiating routine use of a given dispersing agent, test the dispersing agent to ascertain that in the concentration used it has no significant antimicrobial effects during the time interval for all transfers employing test procedures set forth under *Bacteriostasis and Fungistasis*.] Mix an aliquot of 10 mL of the fluid mixture so obtained with 80 mL of each medium, and proceed as directed under *Liquids*, beginning with "Incubate in the specified media."

SOLIDS

Take a quantity of the product in the form of a dry solid (or of a solution or a suspension of the product prepared by adding sterile diluent to the immediate container), corresponding to not less than 300 mg from each container being tested, or to the entire contents if each contains less than 300 mg of solids. Transfer it to not less than 40 mL of Fluid Thioglycollate Medium and to not less than 40 mL of Soybean-Casein Digest Medium, respectively, and mix, the number of containers and the conditions of incubation being the same as for liquids. Proceed as directed under *Liquids*, beginning with "Examine the media visually."

PURIFIED COTTON, GAUZE, SURGICAL DRESSINGS, SUTURES, AND RELATED ARTICLES

From each package of cotton, rolled gauze, or gauze bandage being tested, remove aseptically two or more portions of 100 mg to 500 mg each from the innermost part of the sample. From individually packaged single-use materials such as gauze pads, remove aseptically a single portion of 250 mg to 500 mg or the entire article in the case of small, i.e., 25- × 75-mm or smaller, adhesive absorbent bandages, or sutures.

Aseptically transfer these portions of the article to the specified number of containers of appropriate media and incubate as directed under *General Procedure*. Proceed as directed under *Liquids*, beginning with "Examine the media visually."

STERILIZED DEVICES

The following considerations apply to sterilized devices manufactured in lots, each consisting of a number of units. Special considerations apply to sterile devices manufactured in small lots or in individual units where the self-destructive nature of the Sterility Test renders the conventional Sterility Test impracticable. For these articles, appropriate and acceptable modifications to the Sterility Test must be made.

For articles of such size and shape as to permit complete immersion in not more than 1000 mL of culture medium, test the intact article, using the appropriate media, and incubate as directed under *General Procedure*. Proceed as directed under *Liquids*, beginning with "Examine the media visually."

For devices having hollow tubes, such as transfusion or infusion assemblies, or where the size of an item makes immersion impracticable and where only the fluid pathway must be sterile, flush the lumen of each of 20 units with a sufficient quantity of Fluid Thioglycollate Medium and the lumen of each of 20 units with a sufficient quantity of Soybean-Casein Digest Medium to yield a recovery of not less than 15 mL of each medium, and incubate with not less than 100 mL of each of the two media as directed under *General Procedure*. For devices in which the lumen is so small that Fluid Thioglycollate Medium will not pass through, substitute Alternative Thioglycollate Medium for Fluid Thioglycollate Medium, but incubate the medium anaerobically.

Where the entire intact article, because of its size and shape, cannot be tested for sterility by immersion in not more than 1000 mL of culture medium, expose that portion of the article most difficult to sterilize, and test that portion, or where practicable remove two or more portions each from the innermost portion of the article. Aseptically transfer these portions of the article to the specified number of vessels of appropriate media in a volume of not more than 1000 mL, and incubate as directed under *General Procedure*. Proceed as directed under *Liquids*, beginning with "Examine the media visually."

Where the presence of the test specimen in the medium interferes with the test because of bacteriostatic or fungistatic action, rinse the article thoroughly with a minimal amount of rinse fluid (see under *Diluting and Rinsing Fluids*). Recover the rinse fluid, and test as directed for *Devices* under *Test Procedures using Membrane Filtration*.

STERILE EMPTY OR PREFILLED SYRINGES

Sterility testing of prefilled syringes is performed by employing the same techniques used in testing sterile products in vials or ampuls. The direct transfer technique may be employed if the *Bacteriostasis and Fungistasis* determination has indicated no adverse activity under the test conditions. Where appropriate, the membrane filtration procedure may be employed. For prefilled syringes containing a sterile needle, flush the contained product through the lumen. For syringes packaged with a separate needle, aseptically attach the needle, and expel the product into the appropriate media. Pay special attention toward demonstrating that the outside of the attached needle (that portion which will enter the patient's tissues) is sterile. For empty sterile syringes, take up sterile medium or diluent into the barrel through the needle if attached, or if not attached, through a sterile needle attached for the purpose of the test, and express the contents into the appropriate media.

Test Procedures Using Membrane Filtration

Where the membrane filtration technique is used for liquid articles that may be tested by direct transfer to test media, test not less than the volumes and numbers specified under *Selection of Test Specimens and Incubation.*

Apparatus—A suitable membrane filter unit consists of an assembly that facilitates the aseptic handling of the test articles and that allows the processed membrane to be removed aseptically for inoculation of appropriate media or an assembly where sterile media can be added to the sealed filter and the membrane incubated in situ. A membrane generally suitable for sterility testing has a nominal porosity of 0.45 µm, a diameter of approximately 47 mm, and a flow rate of 55 to 75 mL of water per minute at a pressure of 70 cm of mercury. The entire unit may be assembled and sterilized with the membrane(s) in place prior to use in the test, or the membranes may be sterilized separately by any means that maintains the performance characteristics of the filter and assures the sterility of the filter and the assembly.

Where the article to be tested is an oil, the membrane may be sterilized separately, and after thorough drying, the unit assembled, using aseptic precautions.

LIQUIDS MISCIBLE WITH AQUEOUS VEHICLES

Aseptically transfer the volumes required for both media, as indicated in the table of Quantities for Liquid Articles under *Selection of Test Specimens and Incubation*, either directly into one or two separate membrane filter funnels or to separate sterile pooling vessel(s) prior to transfer. In the case of liquid articles in containers in which the volume of liquid is either less than 50 mL, or 50 mL to less than 100 mL, and not intended for intravenous administration, the required volumes from not less than 20 containers are thus represented by one membrane, or membrane half, transferred to each medium. If the volume of liquid in the article is 50 mL to less than 100 mL per container and is intended for intravenous administration, or is 100 mL or more up to 500 mL, aseptically transfer the entire contents of not less than 10 containers through each of two filter assemblies, or not less than 20 containers if only one filter assembly is used. If the volume of the liquid in the article is more than 500 mL, aseptically transfer not less than 500 mL from each of not less than 10 containers through each of two filter assemblies, or not less than 20 containers if only one filter assembly is used. Immediately pass each specimen through the filter with the aid of vacuum or pressure.

In some cases, where the liquid is highly viscous and not readily filterable through one or two membranes, more than two filter assemblies may be needed. In such cases, half the number of membranes used are incubated in each medium, provided that the volumes and requirements for numbers of containers per medium are complied with. If the product is bacteriostatic or fungistatic, rinse the membrane(s) with three 100-mL portions of *Fluid A*.

Aseptically remove the membrane(s) from the holder(s), cut the membrane in half (if only one is used), immerse the membrane, or one-half of the membrane, in 100 mL of Soybean-Casein Digest Medium, and incubate at 20° to 25° for not less than 7 days. Similarly, immerse the other membrane, or other half of the membrane, in 100 mL of Fluid Thioglycollate Medium, and incubate at 30° to 35° for not less than 7 days.

NOTE—Where the product under test has inherent bacteriostatic properties, use hydrophobic membrane filter disks, or after the specimen has been filtered, cut a disk comprising about one-half of the filtering area from the center of the membrane using a sterile cutting device, aseptically transferring the disk cut from the center of the membrane to Fluid Thioglycollate Medium, and aseptically transferring the remainder of the disk to Soybean-Casein Digest Medium.

LIQUIDS IMMISCIBLE WITH AQUEOUS VEHICLES (LESS THAN 100 mL PER CONTAINER)

Using the contents of not less than 20 containers (40 containers, if each one does not contain sufficient volume for both media), aseptically transfer the volumes required for both media, as indicated in the table of Quantities for Liquid Articles under *Selection of Test Specimens and Incubation*, either directly into one or two separate membrane filter funnels or to separate sterile pooling vessels prior to transfer. The required volumes from not less than 20 containers are thus represented by the membrane, or membrane half to be transferred to each medium. Immediately pass each specimen through the filter with the aid of vacuum or pressure.

If the substance is a viscous liquid or suspension and not adaptable to rapid filtration, aseptically add a sufficient quantity of diluting fluid to the pooled specimen prior to filtration to increase the flow rate.

If the product under test has inherent bacteriostatic or fungistatic properties or contains a preservative, wash the filter with from one to three 100-mL portions of *Fluid A*. If the substance under test contains lecithin or oil, substitute *Fluid D* for *Fluid A*.

Upon completion of the filtration, and rinsing, treat the membrane(s) as directed under *Liquids Miscible with Aqueous Vehicles*, beginning with "Aseptically remove the membrane(s)."

FILTERABLE SOLIDS

Take about 6 g of the product in the form of a dry solid (or a portion of a solution or suspension of the product, prepared by adding sterile diluent to the immediate container(s), corresponding to 6 g of solid), or not less than 300 mg from each container being tested, or the entire contents of each container if each contains less than 300 mg of solids, unless otherwise specified in the individual monograph, the number of containers being the same as specified for *Liquids Miscible with Aqueous Vehicles*. Transfer the specimen aseptically to a vessel containing 200 mL of *Fluid A*, and swirl to dissolve. If the specimen does not dissolve completely, use 400 mL of *Fluid A*, or divide the specimen aseptically into two portions and test each using 200 mL of *Fluid A*. Aseptically transfer the solution(s) into one or two membrane funnels, and immediately filter with the aid of vacuum or pressure. If the product under test has inherent bacteriostatic or fungistatic properties, rinse the membrane(s) with three 100-mL portions of *Fluid A*. Upon completion of the filtration and rinsing, treat the membrane(s) as directed under *Liquids Miscible with Aqueous Vehicles*, beginning with "Aseptically remove the membrane(s)."

OINTMENTS AND OILS SOLUBLE IN ISOPROPYL MYRISTATE

Dissolve not less than 100 mg from each of not less than 20 containers (40 containers, if each one does not contain sufficient volume for both media) in not less than 100 mL of isopropyl myristate with a pH of water extract not less than 6.5 (see under *Reagent Specifications* in the section, *Reagents, Indicators, and Solutions*) which previously has been rendered sterile by filtration through a 0.22-μm membrane filter. [NOTE—Warm the sterilized solvent, and if necessary the test material, to not more than 44° just prior to use.] Swirl the flask to dissolve the ointment or oil, taking care to expose a large surface of the material to the solvent. Filter the dissolved ointment promptly following dissolution. Aseptically transfer the mixture into one or two membrane filter funnels. Immediately pass each specimen through the filter with the aid of vacuum or pressure. Keep filter membrane(s) covered with liquid throughout the filtration for maximum efficiency of the filter.

Following filtration of the specimen, wash the membrane(s) with two 200-mL portions of *Fluid D*, then wash with 100 mL of *Fluid A*. Treat the test membrane(s) as directed under *Liquids Miscible with Aqueous Vehicles*, beginning with "Aseptically remove the membrane(s)," except to provide that the sterility test medium to be used contains 1 g of polysorbate 80 per liter.

If the substance under test contains petrolatum, use *Fluid K*. Moisten the membrane(s) with approximately 200 μL of the rinse medium before the filtration operation begins, and keep the membrane(s) covered with liquid throughout the filtration operation for maximum efficiency of the filter.

Following filtration of the specimen, wash the membrane(s) with three 100-mL portions of the rinse medium. Treat the test membrane(s) as directed above.

NOTE—For ointments and oils that are insoluble in isopropyl myristate, proceed as directed for *Ointments and Oils Insoluble in Isopropyl Myristate* under *Test Procedures for Direct Transfer to Test Media*.

NON-FILTERABLE SOLIDS

The sterility testing of these articles by membrane filtration is considered inadvisable unless it can be demonstrated that filter blockage does not occur. Proceed as directed for *Solids* under *Test Procedures for Direct Transfer to Test Media*.

DEVICES

Devices that are purported to contain sterile pathways may be tested for sterility by the membrane filtration technique as follows.

Aseptically pass a sufficient volume of *Fluid D* through each of not less than 20 devices so that not less than 100 mL is recovered from each device. Collect the fluids in aseptic containers, and filter the entire volume collected through membrane filter funnel(s) as directed under *Liquids Miscible with Aqueous Vehicles*, beginning with "Aseptically remove the membrane(s)."

Where the devices are large, and lot sizes are small, test an appropriate number of units as described for similar cases in the section, *Sterilized Devices*, under *Test Procedures for Direct Transfer to Test Media*.

Interpretation of Sterility Test Results

FIRST STAGE

At the prescribed intervals during and at the conclusion of the incubation period, examine the contents of all of the vessels for evidence of microbial growth, such as the development of turbidity and/or surface growth. If no growth is observed, the article tested meets the requirements of the test for sterility.

If microbial growth is found, but a review in the sterility testing facility of the monitoring, materials used, testing procedure, and negative controls indicates that inadequate or faulty aseptic technique was used in the test itself, the *First Stage* is declared invalid and may be repeated.

If microbial growth is observed but there is no evidence invalidating the *First Stage* of the test, proceed to the *Second Stage*.

SECOND STAGE

The minimum number of specimens selected is double the number tested in the *First Stage*. The minimum volumes tested from each specimen and the media and incubation periods are the same as those indicated for the *First Stage*. If no microbial growth is found, the article tested meets the requirements of the test for sterility. If growth is found, the result so obtained is conclusive that the article tested fails to meet the requirements of the test for sterility. If, however, it can be demonstrated that the *Second Stage* was invalid because of faulty or inadequate aseptic technique in the performance of the test the *Second Stage* may be repeated.

NOTE—Where sterility testing is used as part of an assessment of a production lot or batch or as one of the quality control criteria for release of such lot or batch, see *Sterilization and Sterility Assurance of Compendial Articles* ⟨1211⟩.

Biological Tests and Assays

⟨81⟩ ANTIBIOTICS—MICROBIAL ASSAYS

The activity (potency) of antibiotics may be demonstrated under suitable conditions by their inhibitory effect on microorganisms. A reduction in antimicrobial activity also will reveal subtle changes not demonstrable by chemical methods. Accordingly, microbial or biological assays remain generally the standard for resolving doubt with respect to possible loss of activity. This chapter summarizes these procedures for the antibiotics recognized in this Pharmacopeia for which microbiological assay remains the definitive method.

Two general methods are employed, the cylinder-plate or "plate" assay and the turbidimetric or "tube" assay. The first depends upon diffusion of the antibiotic from a vertical cylinder through a solidified agar layer in a petri dish or plate to an extent such that growth of the added microorganism is prevented entirely in a circular area or "zone" around the cylinder containing a solution of the antibiotic. The turbidimetric method depends upon the inhibition of growth of a microbial culture in a uniform solution of the antibiotic in a fluid medium that is favorable to its rapid growth in the absence of the antibiotic.

Apparatus

All equipment is to be thoroughly cleaned before and after each use. Glassware for holding and transferring test organisms is sterilized by dry heat, or by steam.

TEMPERATURE CONTROL

Thermostatic control is required in several stages of a microbial assay, when culturing a microorganism and preparing its inoculum, and during incubation in plate and tube assays. Maintain the temperature of assay plates at ±0.5° of the temperature selected. Closer control of the temperature (±0.1° of the selected temperature) is imperative during incubation in a tube assay, and may be achieved in either circulated air or water, the greater heat capacity of water lending it some advantage over circulating air.

SPECTROPHOTOMETER

Measuring transmittance within a fairly narrow frequency band requires a suitable spectrophotometer in which the wavelength of the light source can be varied or restricted by the use of a 580-nm filter for preparing inocula of the required density, or with a 530-nm filter for reading the absorbance in a tube assay. For the latter purpose, the instrument may be arranged to accept the tube in which incubation takes place, to accept a modified cell fitted with a drain that facilitates rapid change of content, or preferably, fixed with a flow-through cell for a continuous flow-through analysis; set the instrument at zero absorbance with clear, uninoculated broth prepared as specified for the particular antibiotic, including the same amount of test solution and formaldehyde as found in each sample.

NOTE—Either absorbance or transmittance measurement may be used for preparing inocula.

CYLINDER-PLATE ASSAY RECEPTACLES

For assay plates, use glass or plastic petri dishes (approximately 20 × 100 mm) having covers of suitable material. For assay cylinders, use stainless steel or porcelain cylinders with the following dimensions, each dimension having a tolerance of ±0.1 mm: outside diameter 8 mm; inside diameter 6 mm; and length 10 mm. Carefully clean cylinders to remove all residues. An occasional acid bath, e.g., with about 2 *N* nitric acid or with chromic acid (see *Cleaning Glass Apparatus* ⟨1051⟩) is needed.

TURBIDIMETRIC ASSAY RECEPTACLES

For assay tubes, use glass or plastic test tubes, e.g., 16 × 125 mm or 18 × 150 mm that are relatively uniform in length, diameter, and thickness and substantially free from surface blemishes and scratches. Tubes that are to be placed in the spectrophotometer are matched and are without scratches or blemishes. Cleanse thoroughly, to remove all antibiotic residues and traces of cleaning solution, and sterilize tubes that have been used previously, before subsequent use.

Media and Diluents

MEDIA

The media required for the preparation of test organism inocula are made from the ingredients listed herein. Minor modifications of the individual ingredients, or reconstituted dehydrated media, may be substituted, provided the resulting media possess equal or better growth-promoting properties and give a similar standard curve response.

Dissolve the ingredients in water to make 1 liter, and adjust the solutions with either 1 *N* sodium hydroxide or 1 *N* hydrochloric acid as required, so that after steam sterilization the pH is as specified.

Medium 1

Peptone	6.0 g
Pancreatic Digest of Casein	4.0 g
Yeast Extract	3.0 g
Beef Extract	1.5 g
Dextrose	1.0 g
Agar	15.0 g
Water	1000 mL

pH after sterilization: 6.6 ± 0.1.

Medium 2

Peptone	6.0 g
Yeast Extract	3.0 g
Beef Extract	1.5 g
Agar	15.0 g
Water	1000 mL

pH after sterilization: 6.6 ± 0.1.

Medium 3

Peptone	5.0 g
Yeast Extract	1.5 g
Beef Extract	1.5 g
Sodium Chloride	3.5 g
Dextrose	1.0 g
Dipotassium Phosphate	3.68 g
Potassium Dihydrogen Phosphate	1.32 g
Water	1000 mL

pH after sterilization: 7.0 ± 0.05.

Medium 5

Same as Medium 2, except that the final pH after sterilization is 7.9 ± 0.1.

Medium 8

Same as Medium 2, except that the final pH after sterilization is 5.9 ± 0.1.

Medium 9

Pancreatic Digest of Casein	17.0 g
Papaic Digest of Soybean	3.0 g
Sodium Chloride	5.0 g
Dipotassium Phosphate	2.5 g
Dextrose	2.5 g
Agar	20.0 g
Water	1000 mL

pH after sterilization: 7.2 ± 0.1.

Medium 10

Same as Medium 9, except to use 12.0 g of Agar instead of 20.0 g, and to add 10 mL of Polysorbate 80 after boiling the medium to dissolve the agar.

pH after sterilization: 7.2 ± 0.1.

Medium 11

Same as Medium 1, except that the final pH after sterilization is 8.3 ± 0.1.

Medium 13

Dextrose	20.0 g
Peptone	10.0 g
Water	1000 mL

pH after sterilization: 5.6 ± 0.1.

Medium 19

Peptone	9.4 g
Yeast Extract	4.7 g
Beef Extract	2.4 g
Sodium Chloride	10.0 g
Dextrose	10.0 g
Agar	23.5 g
Water	1000 mL

pH after sterilization: 6.1 ± 0.1.

Medium 32

Same as Medium 1, except for the additional ingredient 0.3 g of Manganese Sulfate.

Medium 34

Glycerol	10.0 g
Peptone	10.0 g
Beef Extract	10.0 g
Sodium Chloride	3.0 g
Water	1000 mL

pH after sterilization: 7.0 ± 0.1.

Medium 35

Same as Medium 34, except for the additional ingredient 17.0 g of Agar.

Medium 36

Pancreatic Digest of Casein	15.0 g
Papaic Digest of Soybean	5.0 g
Sodium Chloride	5.0 g
Agar	15.0 g
Water	1000 mL

pH after sterilization: 7.3 ± 0.1.

Medium 37

Pancreatic Digest of Casein	17.0 g
Soybean Peptone	3.0 g
Dextrose	2.5 g
Sodium Chloride	5.0 g
Dipotassium Phosphate	2.5 g
Water	1000 mL

pH after sterilization: 7.3 ± 0.1.

Medium 38

Peptone	15.0 g
Papaic Digest of Soybean Meal	5.0 g
Sodium Chloride	4.0 g
Sodium Sulfite	0.2 g
L-Cystine	0.7 g
Dextrose	5.5 g
Agar	15.0 g
Water	1000 mL

pH after sterilization: 7.0 ± 0.1.

PHOSPHATE BUFFERS AND OTHER SOLUTIONS

Prepare as follows or by other suitable means the potassium phosphate buffers required for the antibiotic under assay. The buffers are sterilized after preparation, and the pH specified in each case is that after sterilization.

Buffer No. 1, 1 percent, pH 6.0—Dissolve 2.0 g of dibasic potassium phosphate and 8.0 g of monobasic potassium phosphate in 1000 mL of water. Adjust the pH with 18 *N* phosphoric acid or 10 *N* potassium hydroxide to 6.0 ± 0.05.

Buffer No. 3, 0.1 *M*, pH 8.0—Dissolve 16.73 g of dibasic potassium phosphate and 0.523 g of monobasic potassium phosphate in 1000 mL of water. Adjust the pH with 18 *N* phosphoric acid or 10 *N* potassium hydroxide to 8.0 ± 0.1.

Buffer No. 4, 0.1 *M*, pH 4.5—Dissolve 13.61 g of monobasic potassium phosphate in 1000 mL of water. Adjust the pH with 18 *N* phosphoric acid or 10 *N* potassium hydroxide to 4.5 ± 0.05.

Buffer No. 6, 10 percent, pH 6.0—Dissolve 20.0 g of dibasic potassium phosphate and 80.0 g of monobasic potassium phosphate in 1000 mL of water. Adjust the pH with 18 *N* phosphoric acid or 10 *N* potassium hydroxide to 6.0 ± 0.05.

Buffer No. 10, 0.2 *M*, pH 10.5—Dissolve 35.0 g of dibasic potassium phosphate in 1000 mL of water, and add 2 mL of 10 *N* potassium hydroxide. Adjust the pH with 18 *N* phosphoric acid or 10 *N* potassium hydroxide to 10.5 ± 0.1.

Buffer No. 16, 0.1 *M*, pH 7.0—Dissolve 13.6 g of dibasic potassium phosphate and 4.0 g of monobasic potassium phosphate in 1000 mL of water. Adjust with 18 *N* phosphoric acid or 10 *N* potassium hydroxide to a pH of 7.0 ± 0.2.

Other solutions—Use the substances specified under *Reagents, Indicators, and Solutions*. For water, use *Purified Water*. For saline, use *Sodium Chloride Injection*. Dilute formaldehyde is *Formaldehyde Solution* diluted 1:3 with water.

Units and Reference Standards

The potency of antibiotics is designated in either "Units" or "µg" of activity. In each case the "Unit" or "µg" of antibiotic activity is established and defined by the designated federal master standard for that antibiotic. The corresponding USP Reference Standard is calibrated in terms of the master standard. USP Reference Standards for antibiotic substances are held and distributed by the U. S. Pharmacopeial Convention, Inc.

The concept of "µg" of activity originated from the situation where the antibiotic preparation selected as the reference standard was thought to consist entirely of a single chemical entity

Table 1. Units of Potency of Reference Standards Available in 1988.

Antibiotic	Weight of Material (in µg) containing the Unit as defined by the master standard (and corresponding number of Units per mg)		Number of USP Units per mg of the USP Reference Standard (1988)
Bacitracin Zinc	13.51	(74)	58.5 (Lot L)
Nystatin	0.2817	(3550)	6044 (Lot L)
Penicillin G Sodium	0.600	(1667)	1590 (Lot G)
Penicillin V	0.590	(1695)	1520 (Lot F)
Polymyxin B Sulfate	0.1274	(7849)	8300 (Lot I)

NOTE—Each mg of penicillin G benzathine contains 1211 USP Penicillin G Units, and each mg of penicillin G procaine contains 1009 USP Penicillin G Units.

For any of the antibiotics listed, when stocks of a batch of USP Reference Standard are depleted, the replacement is calibrated to maintain continuity of the USP Unit. The number of USP Units per mg of the USP Reference Standard may therefore differ from that shown above.

Table 2. Preparation of Stock Solutions and Test Dilutions of Reference Standards.

	Stock Solution				Test Dilution	
Antibiotic and Type of Assay [Cylinder-plate (CP) or Turbidimetric (T)]	Prior Drying	Initial Solvent (and initial concentration where specified); [Further Diluent, if different]	Final Stock Concentration per mL	Use Within	Final Diluent	Median Dose (µg of activity or Units per mL)
Amikacin (T)	No	Water	1 mg	14 days	Water	10 µg
Amphotericin B (CP)	Yes	Dimethyl sulfoxide	1 mg	Same day	*B.* 10	1.0 µg
Ampicillin (CP)	No	Water	100 µg	7 days	*B.* 3	0.1 µg
Bacitracin Zinc (CP)	Yes	0.01 *N* hydrochloric acid	100 U	Same day	*B.* 1	1.0 U
Bleomycin (CP)	Yes	*B.* 16	2 U	14 days	*B.* 16	0.04 U
Candicidin (T)	Yes	Dimethyl sulfoxide	1 mg	Same day	Water	0.06 µg
Capreomycin (T)	Yes	Water	1 mg	7 days	Water	100 µg
Carbenicillin (CP)	No	*B.* 1	1 mg	14 days	*B.* 1	20 µg
Cephalexin (CP)	No	*B.* 1	1 mg	7 days	*B.* 1	20 µg
Cephalothin (CP)	Yes	*B.* 1	1 mg	5 days	*B.* 1	1.0 µg
Cephapirin (CP)	No	*B.* 1	1 mg	3 days	*B.* 1	1.0 µg
Cephradine (CP)	No	*B.* 1	1 mg	5 days	*B.* 1	10 µg
Chloramphenicol (T)	No	Alcohol (10 mg/mL); [Water]	1 mg	30 days	Water	2.5 µg
Chlortetracycline (T)	No	0.01 *N* hydrochloric acid	1 mg	4 days	Water	0.06 µg
Clindamycin (CP)	No	Water	1 mg	30 days	*B.* 3	1.0 µg
Cloxacillin (CP)	No	*B.* 1	1 mg	7 days	*B.* 1	5.0 µg
Colistimethate Sodium (CP)	Yes	Water (10 mg/mL); [*B.* 6]	1 mg	Same day	*B.* 6	1.0 µg
Colistin (CP)	Yes	Water (10 mg/mL); [*B.* 6]	1 mg	14 days	*B.* 6	1.0 µg
Cycloserine (T)	Yes	Water	1 mg	30 days	Water	50 µg
Dactinomycin (CP)	Yes	Methanol (10 mg/mL); [*B.* 3]	1 mg	90 days	*B.* 3	1.0 µg
Demeclocycline (T)	Yes	0.1 *N* hydrochloric acid	1 mg	4 days	Water	0.1 µg
Dicloxacillin (CP)	No	*B.* 1	1 mg	7 days	*B.* 1	5.0 µg
Dihydrostreptomycin (CP)	Yes	*B.* 3	1 mg	30 days	*B.* 3	1.0 µg
Dihydrostreptomycin (T)	Yes	Water	1 mg	30 days	Water	30 µg
Doxycycline (T)	No	0.1 *N* hydrochloric acid	1 mg	5 days	Water	0.1 µg
Erythromycin (CP)	Yes	Methanol (10 mg/mL); [*B.* 3]	1 mg	14 days	*B.* 3	1.0 µg
Gentamicin (CP)	Yes	*B.* 3	1 mg	30 days	*B.* 3	0.1 µg
Gramicidin (T)	Yes	Alcohol 95%	1 mg	30 days	Alcohol 95%	0.04 µg
Kanamycin (T)	No	Water	1 mg	30 days	Water	10 µg
Lincomycin (T)	No	Water	1 mg	30 days	Water	0.5 µg
Methacycline (T)	Yes	Water	1 mg	7 days	Water	0.06 µg
Methicillin (CP)	No	*B.* 1	1 mg	4 days	*B.* 1	10 µg
Minocycline (T)	No	0.1 *N* hydrochloric acid	1 mg	2 days	Water	0.085 µg
Mitomycin (CP)	No	*B.* 1	1 mg	14 days	*B.* 1	1.0 µg
Nafcillin (CP)	No	*B.* 1	1 mg	2 days	*B.* 1	2.0 µg
Natamycin (CP)	No	Dimethyl sulfoxide	1 mg	Same day	*B.* 10	5.00 µg
Neomycin (CP)	Yes	*B.* 3	1 mg	14 days	*B.* 3	1.0 µg
Netilmicin (CP)	No	*B.* 3	1 mg	7 days	*B.* 3	0.1 µg
Novobiocin (CP)	Yes	Alcohol (10 mg/mL); [*B.* 3]	1 mg	5 days	*B.* 6	0.5 µg
Nystatin (CP)	Yes	Dimethylformamide	1,000 U	Same day	*B.* 6	20 U
Oxacillin (CP)	No	*B.* 1	1 mg	3 days	*B.* 1	5.0 µg
Oxytetracycline (T)	No	0.1 *N* hydrochloric acid	1 mg	4 days	Water	0.24 µg
Paromomycin (CP)	Yes	*B.* 3	1 mg	21 days	*B.* 3	1.0 µg
Penicillin G (CP)	No	*B.* 1	1,000 U	4 days	*B.* 1	1.0 U
Plicamycin (CP)	Yes	Water	100 µg	1 day	*B.* 1	1.0 µg
Polymyxin B (CP)	Yes	Water; [*B.* 6]	10,000 U	14 days	*B.* 6	10 U
Rifampin (CP)	No	Methanol	1 mg	1 day	*B.* 1	5.0 µg
Rolitetracycline (T)	Yes	Water	1 mg	1 day	Water	0.24 µg
Sisomicin (CP)	No	*B.* 3	1 mg	14 days	*B.* 3	0.1 µg
Spectinomycin (T)	No	Water	1 mg	30 days	Water	30 µg
Streptomycin (T)	Yes	Water	1 mg	30 days	Water	30 µg
Tetracycline (T)	No	0.1 *N* hydrochloric acid	1 mg	1 day	Water	0.24 µg
Ticarcillin (CP)	No	*B.* 1	1 mg	1 day	*B.* 1	5.0 µg
Tobramycin (T)	No	Water	1 mg	14 days	Water	2.5 µg
Troleandomycin (T)	Yes	Isopropyl alcohol–water (4:1)	1 mg	Same day	Water	25 µg
Vancomycin (CP)	No	Water	1 mg	7 days	*B.* 4	10 µg
Viomycin (T)	Yes	Water	1 mg	7 days	Water	100 µg

"*B*" denotes "buffer," and the number following refers to the potassium phosphate buffers defined in this chapter.

For Amphotericin B, Colistimethate Sodium, and Nystatin, prepare the reference standard solutions and the sample test solution simultaneously.

For Ampicillin, prepare the test dilutions of the reference standard and the sample simultaneously.

For Amphotericin B, further dilute the stock solution with dimethyl sulfoxide to give concentrations of 12.8, 16, 20, 25, and 31.2 µg per mL prior to making the test dilutions. The *Test Dilution* of the sample prepared from the *Assay Preparation* should contain the same amount of dimethyl sulfoxide as the test dilutions of the Standard.

For Bacitracin Zinc, each of the Standard test dilutions should contain the same amount of hydrochloric acid as the *Test Dilution* of the sample.

For Natamycin, further dilute the stock solution with dimethyl sulfoxide to give concentrations of 64.0, 80.0, 100, 125, and 156 µg per mL prior to making the test dilutions. Prepare the standard response line solutions simultaneously with dilutions of the specimen to be tested. Use red low-actinic glassware. The *Test Dilution* of the sample prepared from the *Assay Preparation* should contain the same amount of dimethyl sulfoxide as the test dilutions of the Standard.

For Nystatin, further dilute the stock solution with dimethylformamide to give concentrations of 256, 320, 400, 500, and 624 Units per mL prior to making the test dilutions. Prepare the standard response line solutions simultaneously with dilutions of the sample to be tested. The *Test Dilution* of the sample prepared from the *Assay Preparation* should contain the same amount of dimethylformamide as the test dilutions of the Standard. Use red low-actinic glassware.

When making the stock solution of Polymyxin B, add 2 mL of water for each 5 mg of weighed reference standard material.

For Sterile Penicillin G Procaine with Aluminum Stearate Suspension cylinder-plate assay only, use Penicillin G Reference Standard.

Where indicated, about 100 mg of the Reference Standard is dried before use in a vacuum oven at a pressure of 5 mm or less of mercury at a temperature of 60° for 3 hours, except in the case of Bleomycin and Plicamycin (dry at 25° for 4 hours), Capreomycin, Dihydrostreptomycin, and Novobiocin (dry at 100° for 4 hours), Gentamicin (dry at 110° for 3 hours), Candicidin (dry at 40° for 3 hours), and Nystatin (dry at 40° for 2 hours). The Sisomicin Reference Standard and the Netilmicin Reference Standard are not dried; the weight is calculated on the dried basis from the *Loss on drying* value of a separate portion taken concomitantly.

and was therefore assigned a potency of 1000 "µg" per mg. In several such instances, as a result of the development of manufacturing and purification methods for particular antibiotics, preparations became available that contained more than 1000 "µg" of activity per mg. It was then understood that such preparations had an activity equivalent to a given number of "µg" of the original reference standard. In most instances, however, the "µg" of activity is exactly equivalent numerically to the µg (weight) of the pure substance. Complications arise in some situations, e.g., where an antibiotic exists as the free base and in salt form, and the "µg" of activity has been defined in terms of one such form; where the antibiotic substance consists of a number of components having close chemical similarity but differing antibiotic activity; or where the potencies of a family of antibiotics are expressed in terms of a reference standard consisting of a single member which, however, might itself be heterogeneous. In such cases the "µg" of activity defined in terms of a "Master Standard" is tantamount to a "Unit." The "µg" of activity should therefore not be assumed necessarily to correspond to the µg (weight) of the antibiotic substance.

Table 3. Test Organisms for Antibiotics Assayed by the Procedure Indicated in Table 2.

Antibiotic	Test Organism	ATCC* Number
Amikacin	*Staphylococcus aureus*	29737
Amphotericin B	*Saccharomyces cerevisiae*	9763
Ampicillin	*Micrococcus luteus*	9341
Bacitracin	*Micrococcus luteus*	10240
Bleomycin	*Mycobacterium smegmatis*	607
Candicidin	*Saccharomyces cerevisiae*	9763
Capreomycin	*Klebsiella pneumoniae*	10031
Carbenicillin	*Pseudomonas aeruginosa*	25619
Cephalexin	*Staphylococcus aureus*	29737
Cephalothin	*Staphylococcus aureus*	29737
Cephapirin	*Staphylococcus aureus*	29737
Cephradine	*Staphylococcus aureus*	29737
Chloramphenicol	*Escherichia coli*	10536
Chlortetracycline	*Staphylococcus aureus*	29737
Clindamycin	*Micrococcus luteus*	9341
Cloxacillin	*Staphylococcus aureus*	29737
Colistimethate Sodium	*Bordetella bronchiseptica*	4617
Colistin	*Bordetella bronchiseptica*	4617
Cycloserine	*Staphylococcus aureus*	29737
Dactinomycin	*Bacillus subtilis*	6633
Demeclocycline	*Staphylococcus aureus*	29737
Dicloxacillin	*Staphylococcus aureus*	29737
Dihydrostreptomycin (CP)	*Bacillus subtilis*	6633
Dihydrostreptomycin (T)	*Klebsiella pneumoniae*	10031
Doxycycline	*Staphylococcus aureus*	29737
Erythromycin	*Micrococcus luteus*	9341
Gentamicin	*Staphylococcus epidermidis*	12228
Gramicidin	*Streptococcus faecium*	10541
Kanamycin	*Staphylococcus aureus*	29737
Lincomycin	*Staphylococcus aureus*	29737
Methacycline	*Staphylococcus aureus*	29737
Methicillin	*Staphylococcus aureus*	29737
Minocycline	*Staphylococcus aureus*	29737
Mitomycin	*Bacillus subtilis*	6633
Nafcillin	*Staphylococcus aureus*	29737
Natamycin	*Saccharomyces cerevisiae*	9763
Neomycin	*Staphylococcus epidermidis*	12228
Netilmicin	*Staphylococcus epidermidis*	12228
Novobiocin	*Staphylococcus epidermidis*	12228
Nystatin	*Saccharomyces cerevisiae*	2601
Oxacillin	*Staphylococcus aureus*	29737
Oxytetracycline	*Staphylococcus aureus*	29737
Paromomycin	*Staphylococcus epidermidis*	12228
Penicillin G	*Staphylococcus aureus*	29737
Plicamycin	*Staphylococcus aureus*	29737
Polymyxin B	*Bordetella bronchiseptica*	4617
Rifampin	*Bacillus subtilis*	6633
Rolitetracycline	*Staphylococcus aureus*	29737
Sisomicin	*Staphylococcus epidermidis*	12228
Spectinomycin	*Escherichia coli*	10536
Streptomycin (T)	*Klebsiella pneumoniae*	10031
Tetracycline	*Staphylococcus aureus*	29737
Ticarcillin	*Pseudomonas aeruginosa*	29336
Tobramycin	*Staphylococcus aureus*	29737
Troleandomycin	*Klebsiella pneumoniae*	10031
Vancomycin	*Bacillus subtilis*	6633
Viomycin	*Klebsiella pneumoniae*	10031

* American Type Culture Collection, 21301 Parklawn Drive, Rockville, MD 20852.

Preparation of the Standard

To prepare a stock solution, dissolve a quantity of the Reference Standard of a given antibiotic, accurately weighed, and previously dried where so indicated in Table 2, or the entire contents of a vial of Reference Standard, where appropriate, in the solvent specified in that table, and then dilute to the required concentration as indicated. Store in a refrigerator, and use within the period indicated. On the day of the assay, prepare from the stock solution 5 or more test dilutions, the successive solutions increasing stepwise in concentration, usually in the ratio of 1:1.25 for a cylinder-plate assay or smaller for a turbidimetric assay. Use the final diluent specified and a sequence such that the middle or median has the concentration designated.

Preparation of the Sample

From the information available for the preparation to be assayed (the "Unknown"), assign to it an assumed potency per unit weight or volume, and on this assumption prepare on the day of the assay a stock solution and test dilution as specified for each antibiotic but with the same final diluent as used for the Reference Standard. The assay with 5 levels of the Standard requires only one level of the Unknown at a concentration assumed equal to the median level of the Standard.

Organisms and Inoculum

TEST ORGANISMS

The test organism for each antibiotic is listed in Table 3, together with its identification number in the American Type Culture Collection. The method of assay is given for each in Table 2. Maintain a culture on slants of the medium and under the incubation conditions specified in Table 4, and transfer weekly to fresh slants. For *K. pneumoniae* use a non-capsulated culture.

PREPARATION OF INOCULUM

Preparatory to an assay, inoculate, from a recently grown slant or culture of the organism, the surface of 250 mL of the agar medium specified for that organism in Table 4 and contained in a Roux bottle except in the case of *Streptococcus faecium*, which is grown in a liquid medium. Spread the suspension evenly over the surface of the agar with the aid of sterile glass beads, and incubate at the temperature shown for approximately the indicated length of time. At the end of this period, prepare the stock suspension by collecting the surface growth in 50 mL of sterile saline, except for Bleomycin (use 50 mL of Medium 34) and for Ticarcillin (use 50 mL of Medium 37).

For the assay, dilute a portion of the stock suspension by adding a volume of sterile, purified water or sterile saline, in the dilution indicated in Table 4, and determine the transmittance of this trial dilution at 580 nm, with a spectrophotometer. Adjust the proportion in such a way that the *Inoculum* will have a transmittance of 25% against saline as the blank. For the turbidimetric assay, vary the composition of the stock suspension, if necessary, to obtain the optimum dose-response relationship.

For the cylinder-plate assay, determine by trial the proportions of stock suspension to be incorporated in the *Inoculum*, starting with the volumes indicated in Table 4, that result in satisfactory demarcation of the zones of inhibition of about 14 to 16 mm in diameter and giving a reproducible dose relationship. Prepare the inoculum by adding a portion of stock suspension to a sufficient amount of agar medium that has been melted and cooled to 45° to 50°, and swirling to attain a homogeneous suspension.

Table 4. Preparation of Inoculum.

Test Organism & (ATCC No.)	Incubation Conditions			Dilution of Stock Suspension to obtain 25% Transmittance	Suggested Inoculum Composition		Antibiotics Assayed
	Medium	Temp. (°)	Time		Medium	Amount (mL per 100 mL)	
Bacillus subtilis (6633)	32	32 to 35	5 days		5	As required	Dactinomycin, Dihydrostreptomycin, Rifampin
					8	0.5	Mitomycin
					8	As required	Vancomycin
Bordetella bronchiseptica (4617)	1	32 to 35	24 hr.	1:20	10	0.1	Colistimethate Sodium, Colistin, Polymyxin B
Escherichia coli (10536)	1	32 to 35	24 hr.	1:20	3	0.7	Chloramphenicol
						0.1	Spectinomycin
Klebsiella pneumoniae (10031)	1	36 to 37.5	24 hr.	1:25	3	0.05	Capreomycin
						0.1	Streptomycin, Troleandomycin, Viomycin, Dihydrostreptomycin
Micrococcus luteus (9341)	1	32 to 35	24 hr.	1:40	11	0.5	Ampicillin, Clindamycin,
					11	1.5	Erythromycin
Micrococcus luteus (10240)	1	32 to 35	24 hr.	1:35	1	0.3	Bacitracin
Mycobacterium smegmatis (607)	36	36 to 37.5	48 hr.	As determined	35	1.0	Bleomycin
Pseudomonas aeruginosa (25619)	1	36 to 37.5	24 hr.	1:25	10	0.5	Carbenicillin
Pseudomonas aeruginosa (29336)	36	36 to 37.5	24 hr.	1:50	38	1.5	Ticarcillin
Saccharomyces cerevisiae (9763)	19	29 to 31	48 hr.	As determined	13	0.2	Candicidin
					19	1.0	Amphotericin B
					19	0.8	Natamycin
Saccharomyces cerevisiae (2601)	19	29 to 31	48 hr.	1:30	19	1.0	Nystatin
Staphylococcus aureus (29737)	1	32 to 35	24 hr.	1:20	1	0.05	Cephalexin, Cephradine
					1	0.1	Cephalothin, Cephapirin, Cloxacillin, Dicloxacillin
					1	0.3	Nafcillin, Oxacillin, Methicillin
					1	1.0	Penicillin G
					3	0.1	Amikacin, Chlortetracycline, Demeclocycline, Doxycycline, Lincomycin, Methacycline, Oxytetracycline, Rolitetracycline, Tetracycline
					3	0.2	Kanamycin, Minocycline
					3	0.4	Cycloserine
					3	0.15	Tobramycin
					8	0.1	Plicamycin
Staphylococcus epidermidis (12228)	1	32 to 35	24 hr.	1:14	11	0.25	Netilmicin
					1	4.0	Novobiocin
					11	0.03	Gentamicin, Sisomicin
					11	0.4	Neomycin
				1:25	11	2.0	Paromomycin
Streptococcus faecium (10541)	3	36 to 37.5	16 to 18 hr.	As determined	3	1.0	Gramicidin

The incubated test organism *Pseudomonas aeruginosa* (ATCC 29336) for the assay of Ticarcillin is suspended in Medium 37, instead of saline, for the determination of light transmittance.

For *Pseudomonas aeruginosa* (ATCC 25619) in the assay of Carbenicillin, use the dilution yielding 25% light transmission, rather than the stock suspension, for preparing the inoculum suspension.

The incubated test organism *Mycobacterium smegmatis* (ATCC 607) for the assay of Bleomycin is suspended in Medium 34, instead of saline, for the determination of light transmittance.

NOTE—The dilution of the Stock Suspension to obtain 25% transmittance is for checking the quality of the suspension and is not used for preparing the inoculum suspension, except as specified above. In all other cases, incorporate the stated amount of undiluted stock suspension in the Medium indicated by number, for preparing the inoculum of suggested starting composition.

Procedure

ASSAY DESIGNS

Microbial assays gain markedly in precision by the segregation of relatively large sources of potential error and bias through suitable experimental designs. In a cylinder-plate assay, the essential comparisons are restricted to relationships between zone diameter measurements within plates, exclusive of the variation between plates in their preparation and subsequent handling. To conduct a turbidimetric assay so that the differences in observed turbidity will reflect the differences in the antibiotic concentration requires both greater uniformity in the environment created for the tubes through closer thermostatic control of the incubator and the avoidance of systematic bias by use of a random placement of replicate tubes in separate tube racks, each rack containing one complete set of treatments. The essential comparisons are then restricted to relationships between the observed turbidities within racks.

NOTE—For some purposes, the practice is to design the assay so that a set of treatments consists of not fewer than three tubes for each sample and standard concentration, and each set is placed in a single rack.

Within these restrictions, the assay design recommended is a 1-level assay with a standard curve. For this assay with a standard curve, prepare solutions of 5, 6, or more test dilutions, provided they include one corresponding to the reference concentration (S_3), of the Standard and a solution of a single median test level of the Unknown as described under *Preparation of Standard* and *Preparation of the Sample*. Consider an assay as preliminary if its computed potency with either design is less than 80 percent or more than 125 percent of that assumed in preparing the stock solution of the Unknown. In such a case, adjust its assumed potency accordingly and repeat the assay.

Microbial determinations of potency are subject to inter-assay as well as intra-assay variables, so that two or more independent assays are required for a reliable estimate of the potency of a given assay preparation or Unknown. Starting with separately prepared stock solutions and test dilutions of both the Standard and the Unknown, repeat the assay of a given Unknown on a different day. If the estimated potency of the second assay differs significantly, as indicated by the calculated standard error, from that of the first, conduct one or more additional assays. The combined result of a series of smaller, independent assays spread over a number of days is a more reliable estimate of potency than that from a single large assay with the same total number of plates or tubes.

CYLINDER-PLATE METHOD

To prepare assay plates using petri dishes, place 21 mL of Medium 2 in each of the required number of plates, and allow it to harden into a smooth base layer of uniform depth, except for Amphotericin B, Natamycin, and Nystatin, where no separate base layer is used. For Ampicillin, Clindamycin, Erythromycin, Gentamicin, Lincomycin, Neomycin B, Paromomycin, and Sisomicin, use Medium 11. For Bleomycin, use 10 mL of Medium 35. For Dihydrostreptomycin use Medium 5. For Dactinomycin, use 10 mL of Medium 5. For Plicamycin, Mitomycin, and Vancomycin, use 10 mL of Medium 8. For Carbenicillin, Colistimethate Sodium, Colistin, and Polymyxin B, use Medium 9. For Netilmicin, use 20 mL of Medium 11. Add 4.0 mL of seed layer inoculum (see *Preparation of Inoculum* and Table 4), prepared as directed for the given antibiotic, except for Bleomycin (use 6 mL), for Netilmicin (use 5 mL), and for Natamycin, Nystatin and Amphotericin B (use 8 mL), tilting the plate back and forth to spread the inoculum evenly over the surface, and allow it to harden. Drop 6 assay cylinders on the inoculated surface from a height of 12 mm, using a mechanical guide or other device to insure even spacing on a radius of 2.8 cm, and cover the plates to avoid contamination. After filling the 6 cylinders on each plate with dilutions of antibiotic containing the test levels specified below, incubate the plates at 32° to 35°, or at the temperature specified below for the individual case, for 16 to 18 hours, remove the cylinders, and measure and record the diameter of each zone of growth inhibition to the nearest 0.1 mm. Incubate the plates at 29° to 31° for Amphotericin B, Natamycin, Nystatin, and Rifampin. Incubate at 34° to 36° for Novobiocin. Incubate at 36° to 37.5° for Carbenicillin, Clindamycin, Colistimethate Sodium, Colistin, Dactinomycin, Dihydrostreptomycin, Gentamicin, Mitomycin, Neomycin, Netilmicin, Paromomycin, Polymyxin B, Sisomicin, Ticarcillin, and Vancomycin.

For the 1-level assay with a standard curve, prepare dilutions representing 5 test levels of the Standard (S_1 to S_5) and a single test level of the Unknown U_3 corresponding to S_3 of the standard curve, as defined under *Preparation of the Standard* and *Preparation of the Sample*. For deriving the standard curve, fill alternate cylinders on each of 3 plates with the median test dilution (S_3) of the Standard and each of the remaining 9 cylinders with one of the other four dilutions of the Standard. Repeat the process for the three dilutions of the Standard. For each Unknown, fill alternate cylinders on each of 3 plates with the median test dilution of the Standard (S_3), and the remaining 9 cylinders with the corresponding test dilution (U_3) of the Unknown.

TURBIDIMETRIC METHOD

On the day of the assay, prepare the necessary doses by dilution of stock solutions of the Standard and of each Unknown as defined under *Preparation of the Standard* and *Preparation of the Sample*. Add 1 mL of each dose to each of 3 prepared test tubes, and place the 3 replicate tubes in a position selected at random, in a tube rack. Include similarly in each rack 1 or 2 control tubes containing 1 mL of the test diluent (see Table 4) but no antibiotic. Upon completion of the rack of test solutions (with candicidin within 30 minutes of the time when water is added to the methyl sulfoxide stock solution), add 9.0 mL of inoculum to each tube in the rack in turn, and place the completed rack immediately in an incubator or a water bath maintained at 36° to 37.5° for 2 to 4 hours, except for Candicidin (incubate at 27° to 29° for 16 to 18 hours). After incubation add 0.5 mL of dilute formaldehyde to each tube, taking one rack at a time, and read its transmittance or absorbance in a suitable spectrophotometer fitted with a 530-nm filter.

For the 1-level assay with a standard curve, prepare dilutions representing 5 test levels of the Standard (S_1 to S_5) and a single test level (U_3) of each of up to 20 Unknowns corresponding to S_3 of the Standard. Prepare also an extra S_3 as a test of growth. Add 1 mL of each test dilution, except for Gramicidin (use 0.1 mL) to 3 tubes and 1 mL of antibiotic-free diluent to 6 tubes as controls. Distribute one complete set, including 2 tubes of controls, to a tube rack, intermingling them at random. Add 9.0 mL of inoculum, incubate, add 0.5 mL of dilute formaldehyde, and complete the assay as directed above. Determine the exact duration of incubation by observation of growth in the reference concentration (median dose) of the dilutions of the standard (S_3).

Calculation

To calculate the potency from the data obtained either by the cylinder-plate or by the turbidimetric method, proceed in each case as directed under *Potencies Interpolated from a Standard Curve* (see *Design and Analysis of Biological Assays* ⟨111⟩), using a log transformation, straight-line method with a least squares fitting procedure, and a test for linearity. Where a number of assays of the same material are made with the same standard curve, calculate the coefficient of variation of results of all of the assays of the material. Where more than one assay is made of the same material with different standard curves, average the two or more values of the potency.

⟨85⟩ BACTERIAL ENDOTOXINS TEST

This chapter provides a test for estimating the concentration of bacterial endotoxins that may be present in or on the sample of the article(s) to which the test is applied using Limulus Amebocyte Lysate (LAL) which has been obtained from aqueous extracts of the circulating amebocytes of the horseshoe crab, *Limulus polyphemus*, and which has been prepared and characterized for use as a LAL reagent for gel-clot formation.

Where the test is conducted as a limit test, the specimen is determined to be positive or negative to the test judged against the endotoxin concentration specified in the individual monograph. Where the test is conducted as an assay of the concentration of endotoxin, with calculation of confidence limits of the result obtained, the specimen is judged to comply with the requirements if the result does not exceed (a) the concentration limit specified in the individual monograph, and (b) the specified confidence limits for the assay. In either case the determination of the reaction end-point is made with dilutions from the material under test in direct comparison with parallel dilutions of a reference endotoxin, and quantities of endotoxin are expressed in defined Endotoxin Units.

Since LAL reagents have also been formulated to be used for turbidimetric (including kinetic assays) or colorimetric readings, such tests may be used if shown to comply with the requirements for alternative methods. These tests require the establishment of a standard regression curve and the endotoxin content of the test material is determined by interpolation from the curve. The procedures include incubation for a pre-selected time of reacting endotoxin and control solutions with LAL Reagent and reading of the spectrophotometric light absorbance at suitable wavelengths. In the case of the turbidimetric procedure the reading is made immediately at the end of the incubation period, or in the kinetic assays, the absorbance is measured throughout the reaction period and rate values are determined from those readings. In the colorimetric procedure the reaction is arrested at the end of the pre-selected time by the addition of an appropriate amount of acetic acid solution, prior to the readings. A possible advantage in the mathematical treatment of results, if the test be otherwise validated and the assay suitably designed, could be the application of tests of assay validity and the calculation of the confidence interval and limits of potency from the internal evidence of each assay itself (see *Design and Analysis of Biological Assays* ⟨111⟩).

Reference Standard and Control Standard Endotoxins

The reference standard endotoxin (RSE) is the USP Endotoxin Reference Standard which has a defined potency of 10,000 USP Endotoxin Units (EU) per vial. Constitute the entire contents of 1 vial of the RSE with 5 mL of LAL Reagent Water,[1] vortex for not less than 20 minutes, and use this concentrate for making appropriate serial dilutions. Preserve the concentrate in a refrigerator, for making subsequent dilutions, for not more than 14 days. Allow it to reach room temperature, if applicable, and vortex it vigorously for not less than 5 minutes before use. Vortex each dilution for not less than 1 minute before proceeding to make the next dilution. Do not use stored dilutions. A control standard endotoxin (CSE) is an endotoxin preparation other than the RSE that has been standardized against the RSE. If a CSE is a preparation not already adequately characterized, its evaluation should include characterizing parameters both for endotoxin quality and performance (such as reaction in the rabbit), and for suitability of the material to serve as a reference (such as uniformity and stability). Detailed procedures for its weighing and/or constitution and use to assure consistency in performance should also be included. Standardization of a CSE against the RSE using a LAL Reagent for the gel-clot procedure may be effected by assaying a minimum of 4 vials of the CSE or 4 corresponding aliquots, where applicable, of the bulk CSE and 1 vial of the RSE, as directed under *Test Procedure*, but using 4 replicate reaction tubes at each level of the dilution series for the RSE and 4 replicate reaction tubes similarly for each vial or aliquot of the CSE. If all of the dilutions for the 4 vials or aliquots of the CSE cannot be accommodated with the dilutions for the 1 vial of the RSE on the same rack for incubation, additional racks may be used for accommodating some of the replicate dilutions for the CSE, but all of the racks containing the dilutions of the RSE and the CSE are incubated as a block. However, in such cases, the replicate dilution series from the 1 vial of the RSE are accommodated together on a single rack and the replicate dilution series from any one of the 4 vials or aliquots of the CSE are not divided between racks. The antilog of the difference between the mean $\log_{10}$ end-point of the RSE and the mean $\log_{10}$ end-point of the CSE is the standardized potency of the CSE which then is to be converted to and expressed in Units per ng under stated drying conditions for the CSE, or in Units per container, whichever is appropriate. Standardize each new lot of CSE prior to use in the test. Calibration of a CSE in terms of the RSE must be with the specific lot of LAL Reagent and the test procedure with which it is to be used. Subsequent lots of LAL Reagent from the same source and with similar characteristics need only checking of the potency ratio. The inclusion of one or more dilution series made from the RSE when the CSE is used for testing will enable observation of whether or not the relative potency shown by the latter remains within the determined confidence limits. A large lot of a CSE may, however, be characterized by a collaborative assay of a suitable design to provide a representative relative potency and the within-laboratory and between-laboratory variance.

A suitable CSE has a potency of not less than 2 Endotoxin Units per ng and not more than 50 Endotoxin Units per ng, where in bulk form, under adopted uniform drying conditions, e.g., to a particular low moisture content and other specified conditions of use, and a potency within a corresponding range where filled in vials of a homogeneous lot.

Preparatory Testing

Use a LAL reagent of confirmed label or determined sensitivity. In addition, where there is to be a change in lot of CSE, LAL Reagent or another reagent, conduct tests of a prior satisfactory lot of CSE, LAL and/or other reagent in parallel on changeover. Treat any containers or utensils employed so as to destroy extraneous surface endotoxins that may be present, such as by heating in an oven at 250° or above for sufficient time.[2]

The validity of test results for bacterial endotoxins requires an adequate demonstration that specimens of the article, or of solutions, washings, or extracts thereof to which the test is to be applied do not of themselves inhibit or enhance the reaction or otherwise interfere with the test. Validation is accomplished by testing untreated specimens or appropriate dilutions thereof, concomitantly with and without known and demonstrable added amounts of RSE or a CSE, and comparing the results obtained. Appropriate negative controls are included. Validation must be repeated if the LAL Reagent source or the method of manufacture or formulation of the article is changed.

Test for confirmation of labeled LAL Reagent sensitivity—Confirm the labeled sensitivity of the particular LAL reagent with the RSE (or CSE) using not less than 4 replicate vials, under conditions shown to achieve an acceptable variability of the test, viz., the antilog of the geometric mean $\log_{10}$ lysate gel-clot sensitivity is within 0.5λ to 2.0λ, where λ is the labeled sensitivity in Endotoxin Units per mL. The RSE (or CSE) concentrations selected in confirming the LAL reagent label potency should bracket the stated sensitivity of the LAL reagent. Confirm the labeled sensitivity of each new lot of LAL reagent prior to use in the test.

Inhibition or Enhancement Test—Conduct assays with standard endotoxin, of untreated specimens in which there is no endogeneous endotoxin detectable, and of the same specimens to which endotoxin has been added, as directed under *Test Procedure*, but using not less than 4 replicate reaction tubes at each level of the dilution series for each untreated specimen and for each specimen to which endotoxin has been added. Record the end-points (E, in Units per mL) observed in the replicates. Take the logarithms (e) of the end-points, and compute the geometric means of the log end-points for the RSE (or CSE), for the untreated specimens and for specimens containing endotoxin by the formula antilog:

$$\Sigma e/f,$$

in which Σe is the sum of the log end-points of the dilution series used and f is the number of replicate end-points in each case. Compute the amount of endotoxin in the specimen to which endotoxin has been added. The test is valid for the article if this result is within twofold of the known added amount of endotoxin. Alternatively, if the test has been appropriately set up, the test is valid for the article if the geometric mean end-point dilution for the specimen to which endotoxin has been added is within one 2-fold dilution of the corresponding geometric mean end-point dilution of the standard endotoxin.

If the result obtained for the specimens to which endotoxin has been added is outside the specified limit, the article is unsuitable for the *Bacterial Endotoxins Test*, or, in the case of Injections or solutions for parenteral administration, it may be rendered suitable by diluting specimens appropriately.

Repeat the test for inhibition or enhancement using specimens diluted by a factor not exceeding that given by the formula:

$$x/\lambda$$

(see *Maximum Valid Dilution*, below). Use the least dilution sufficient to overcome the inhibition or enhancement of the known added endotoxin, for subsequent assays of endotoxin in test specimens.

If endogeneous endotoxin is detectable in the untreated specimens under the conditions of the test, the article is unsuitable

[1] LAL Reagent Water—Sterile Water for Injection or other water that shows no reaction with the specific LAL Reagent with which it is to be used, at the limit of sensitivity of such reagent.

[2] For a test for validity of procedure for inactivation of endotoxins, see "Dry-heat Sterilization" under *Sterilization and Sterility Assurance of Compendial Articles* ⟨1211⟩. Use a LAL Reagent having a sensitivity of not less than 0.15 Endotoxin Unit per mL.

for the *Inhibition or Enhancement Test*, or, it may be rendered suitable by removing the endotoxin present by ultra-filtration, or by appropriate dilution. Dilute the untreated specimen (as constituted, where applicable, for administration or use), to a level not exceeding the maximum valid dilution, at which no endotoxin is detectable. Repeat the test for *Inhibition or Enhancement* using the specimens at those dilutions.

Test Procedure

In preparing for and applying the test, observe precautions in handling the specimens in order to avoid gross microbial contamination. Washings or rinsings of devices must be with LAL Reagent Water in volumes appropriate to their use and, where applicable, of the surface area which comes into contact with body tissues or fluids. Use such washings or rinsings if the extracting fluid has been in contact with the relevant pathway or surface for not less than 1 hour at controlled room temperature (15° to 30°). Such extracts may be combined, where appropriate. The ultimate rinse or wash volume is such as to result in possible dilution of any contained endotoxin to a level not less than that suitable for use in the *Pyrogen Test* ⟨151⟩ under *Transfusion and Infusion Assemblies* ⟨161⟩.

For validating the test for an article, for endotoxin limit tests or assays, or for special purposes where so specified, testing of specimens is conducted quantitatively to determine response end-points for gel-clot readings. Usually graded strengths of the specimen and standard endotoxin are made by multifold dilutions. Select dilutions so that they correspond to a geometric series in which each step is greater than the next lower by a constant ratio. Do not store diluted endotoxin, because of loss of activity by adsorption. In the absence of supporting data to the contrary, negative and positive controls are incorporated in the test.

Use not less than 2 replicate reaction tubes at each level of the dilution series for each specimen under test. Whether the test is employed as a limit test or as a quantitative assay, a standard endotoxin dilution series involving not less than 2 replicate reaction tubes is conducted in parallel. A set of standard endotoxin dilution series is included for each block of tubes, which may consist of a number of racks for incubation together, provided the environmental conditions within blocks are uniform.

Preparation—Since the form and amount per container of standard endotoxin and of LAL reagent may vary, constitution and/or dilution of contents should be as directed in the labeling. The pH of the test mixture of the specimen and the LAL Reagent is in the range 6.0 to 7.5 unless specifically directed otherwise in the individual monograph. The pH may be adjusted by the addition of sterile, endotoxin-free sodium hydroxide or hydrochloric acid or suitable buffers to the specimen prior to testing.

Maximum Valid Dilution (MVD)—The Maximum Valid Dilution is appropriate to Injections or to solutions for parenteral administration in the form constituted or diluted for administration, or where applicable, to the amount of drug by weight if the volume of the dosage form for administration could be varied. Where the endotoxin limit concentration is specified in the individual monograph in terms of volume (in EU per mL), divide the limit by λ, which is the labeled sensitivity (in EU per mL) of the lysate employed in the assay, to obtain the MVD factor. Where the endotoxin limit concentration is specified in the individual monograph in terms of weight or of Units of active drug (in EU per mg or in EU per Unit), multiply the limit by the concentration (in mg per mL or in Units per mL) of the drug in the solution tested or of the drug constituted according to the label instructions, whichever is applicable, and divide the product of the multiplication by λ, to obtain the MVD factor. The MVD factor so obtained is the limit dilution factor for the preparation for the test to be valid.

Procedure—To 10- × 75-mm test tubes add aliquots of the appropriately constituted LAL reagent, and the specified volumes of specimens, endotoxin standard, negative controls, and a positive product control consisting of the article, or of solutions, washings or extracts thereof to which the RSE (or a standardized CSE) has been added at a concentration of endotoxin of 2λ for that LAL reagent (see under *Test for confirmation of labeled LAL Reagent sensitivity*). Swirl each gently to mix, and place in an incubating device such as a water bath or heating block, accurately recording the time at which the tubes are so placed. Incubate each tube, undisturbed, for 60 ± 2 minutes at 37 ± 1°, and carefully remove it for observation. A positive reaction is characterized by the formation of a firm gel that remains when inverted through 180°. Record such a result as positive (+). A negative result is characterized by the absence of such a gel or by the formation of a viscous gel that does not maintain its integrity. Record such a result as negative (−). Handle the tubes with care, and avoid subjecting them to unwanted vibrations, or false negative observations may result. The test is invalid if the positive product control or the endotoxin standard does not show the end-point concentration to be within ±1 twofold dilutions from the label claim sensitivity of the LAL Reagent or if any negative control shows a gel-clot end-point.

Calculation and Interpretation

Calculation—Calculate the concentration of endotoxin (in Units per mL or in Units per g or mg) in or on the article under test by the formula:

$$\rho S/U,$$

in which S is the antilog of the geometric mean $\log_{10}$ of the end-points, expressed in Endotoxin Units (EU) per mL for the Standard Endotoxin, U is the antilog of $\Sigma e/f$, where e is the $\log_{10}$ of the end-point dilution factors, expressed in decimal fractions, f is the number of replicate reaction tubes read at the end-point level for the specimen under test, and ρ is the correction factor for those cases where a specimen of the article cannot be taken directly into test but is processed as an extract, solution, or washing.

Where the test is conducted as an assay with sufficient replication to provide a suitable number of independent results, calculate for each replicate assay the concentration of endotoxin in or on the article under test from the antilog of the geometric mean log end-point ratios. Calculate the mean and the confidence limits from the replicate logarithmic values of all the obtained assay results by a suitable statistical method (see *Calculation of Potency from a Single Assay* ⟨111⟩).

Interpretation—The article meets the requirements of the test if the concentration of endotoxin does not exceed that specified in the individual monograph, and where so specified in the individual monograph or in this chapter, the confidence limits of the assay do not exceed those specified.

⟨87⟩ BIOLOGICAL REACTIVITY TESTS, IN-VITRO

The following tests are designed to determine the biological reactivity of mammalian cell cultures following contact with the elastomeric plastics and other polymeric materials with direct or indirect patient contact or of specific extracts prepared from the materials under test. It is essential to make available the specific surface area for extraction. When the surface area of the specimen cannot be determined, use 0.1 g of elastomer or 0.2 g of plastic or other material for every mL of extraction fluid. Also it is essential to exercise care in the preparation of the materials to prevent contamination with microorganisms and other foreign matter.

Three tests are described; i.e., the *Agar Diffusion Test*, the *Direct Contact Test*, and the *Elution Test*.* The decision as to which type of test or the number of tests to be performed to assess the potential biological response of a specific sample or extract depends upon the material, the final product, and its intended use. Other factors that may also affect the suitability of sample for a specific use are the polymeric composition; processing and cleaning procedures; contacting media; inks; adhesives; absorption, adsorption, and permeability of preservatives; and conditions of storage. Evaluation of such factors should be

* Further details are given in the following publications of the American Society for Testing and Materials, 1916 Race St., Philadelphia, PA 19103: "Standard Test Method for Agar Diffusion Cell Culture Screening for Cytotoxicity," ASTM Designation F 895-84; "Standard Practice for Direct Contact Cell Culture Evaluation of Materials for Medical Devices," ASTM Designation F 813-83.

made by appropriate additional specific tests before determining that a product made from a specific material is suitable for its intended use.

Reference Standards—*USP Negative Control Plastic Reference Standard. USP Positive Bioreaction Solid Reference Standard. USP Positive Bioreaction Extract Reference Standard.*

Cell Culture Preparation—Prepare multiple cultures of L-929 (ATCC cell line CCL 1, NCTC clone 929) mammalian fibroblast cells in serum-supplemented minimum essential medium having a seeding density of about 10^5 cells per mL. Incubate the cultures at 37 ± 1° for not less than 24 hours in a 5 ± 1% carbon dioxide atmosphere, until a monolayer, with greater than 80% confluence, is obtained. Examine the prepared cultures under a microscope to ensure uniform, near-confluent monolayers. [NOTE—The reproducibility of the *In-vitro Biological Reactivity Tests* depends upon obtaining uniform cell culture density.]

Extraction Solvents—Sodium Chloride Injection (see monograph—use Sodium Chloride Injection containing 0.9 percent of NaCl); Alternatively, serum-free mammalian cell culture media or serum-supplemented mammalian cell culture media may be used. Serum supplementation is used when extraction is done at 37° for 24 hours.

Apparatus—

Autoclave—Employ an autoclave capable of maintaining a temperature of 121 ± 2°, equipped with a thermometer, a pressure gauge, a vent cock, a rack adequate to accommodate the test containers above the water level, and a water cooling system that will allow for cooling of the test containers to about 20°, but not below 20°, immediately following the heating cycle.

Oven—Use an oven, preferably a mechanical convection model, that will maintain operating temperatures in the range of 50° to 70° within ±2°.

Incubator—Use an incubator capable of maintaining a temperature of 37 ± 1° and an atmosphere of 5 ± 1% carbon dioxide in air. [NOTE—If capped culture tubes are used, it is unnecessary to maintain a carbon dioxide atmosphere in the incubator.]

Extraction Containers—Use only containers, such as ampuls or screw-cap culture test tubes, or their equivalent, of Type I glass. If used, culture test tubes, or their equivalent, are closed with a screw cap having a suitable elastomeric liner. The exposed surface of the elastomeric liner is completely protected with an inert solid disk 50 to 75 µm in thickness. A suitable disk can be fabricated from polytetrafluoroethylene (polytef).

Preparation of Apparatus—Cleanse all glassware thoroughly with chromic acid cleansing mixture and, if necessary, with hot nitric acid followed by prolonged rinsing with Sterile Water for Injection. Make containers and devices used for extraction, transfer, or administration of test material sterile and dry by a suitable process. If ethylene oxide is used as the sterilizing agent, allow not less than 48 hours for complete degassing.

Procedure—

Preparation of Sample for Extracts—Follow the procedure in chapter ⟨88⟩ *Biological Reactivity Tests, In-Vivo.*

Preparation of Extracts—Prepare as directed for *Preparation of Extracts* in chapter ⟨88⟩, *Biological Reactivity Tests, In-vivo* using either Sodium Chloride Injection (0.9 percent NaCl) or serum-free mammalian cell culture media as *Extraction Solvents.* [NOTE—If extraction is done at 37° for 24 hours, in an incubator, use cell culture media supplemented by serum. The extraction conditions should not in any instance cause physical changes such as fusion or melting of the material pieces other than a slight adherence.]

Agar Diffusion Test

This test is designed for elastomeric closures in a variety of shapes. The agar layer acts as a cushion to protect the cells from mechanical damage while allowing the diffusion of leachable chemicals from the polymeric specimens. Extracts of materials that are to be tested are applied to a piece of filter paper.

Sample Preparation—Use extracts, prepared as directed or use portions of the test specimens having flat surfaces not less than 100 mm² in surface area.

Procedure—Prepare the monolayers in 60-mm diameter plates using 7 mL of *Cell Culture Preparation.* Aspirate the culture medium from the monolayers, and replace it with serum-supplemented culture medium containing not more than 2% of agar. Place the flat surfaces of *Sample Preparation,* USP Negative Control Plastic RS (to provide a *Negative Control*), and either USP Positive Bioreaction Extract RS or USP Positive Bioreaction Solid RS (to provide a *Positive Control*) in duplicate cultures in contact with the solidified agar surface. Incubate all cultures for not less than 24 hours at 37 ± 1°, preferably in a humidified incubator containing 5 ± 1% of carbon dioxide. Examine each culture around each *Sample, Negative Control,* and *Positive Control,* under a microscope, using cytochemical stains, if desired.

Interpretation of Results—The biological reactivity (cellular degeneration and malformation) is described and rated on a scale of 0 to 4 (see Table 1). Measure the responses obtained from the *Negative Control* and the *Positive Control.* The test system is suitable if the observed response corresponds to the labeled biological reactivity grade of the relevant Reference Standard. Measure the response obtained from the *Sample Preparation.* The *Sample* meets the requirements of the test if none of the cell culture exposed to the *Sample* shows greater than a mild reactivity (Grade 2). Repeat the test if the suitability of the system is not confirmed.

Direct Contact Test

This test is designed for materials in a variety of shapes. The procedure allows for simultaneous extraction and testing of leachable chemicals from the specimen with a serum-supplemented medium. The procedure is not appropriate for very low- or high-density materials that could cause mechanical damage to the cells.

Sample Preparation—Use portions of the test specimen having flat surfaces not less than 100 mm² in surface area.

Procedure—Prepare the monolayers in 35-mm diameter plates using 2 mL of cell suspension. Aspirate the culture medium from the cultures, and replace it with 0.8 mL of fresh culture medium. Place a single *Sample Preparation,* USP Negative Control Plastic RS (to provide a *Negative Control*), and USP Positive Bioreaction Solid RS (to provide a *Positive Control*) in each of duplicate cultures. Incubate all cultures for not less than 24 hours at 37 ± 1° in a humidified incubator preferably containing 5 ± 1%

Table 1. Reactivity Grades for Agar Diffusion Test.

Grade	Reactivity	Description of Reactivity Zone
0	None	No detectable zone around or under specimen
1	Slight	Zone limited to area under specimen
2	Mild	Zone extends less than 0.5 cm beyond specimen
3	Moderate	Zone extends 0.5 to 1.0 cm beyond specimen
4	Severe	Zone extends greater than 1.0 cm beyond specimen but does not involve entire dish

Table 2. Reactivity Grades for Direct Contact Test and for Elution Test.

Grade	Reactivity	Conditions of all Cultures
0	None	Discrete intracytoplasmic granules; no cell lysis
1	Slight	More than 20% of the cells are round, loosely attached, and without intracytoplasmic granules; occasional lysed cells are present
2	Mild	More than 50% of the cells are round and devoid of intracytoplasmic granules; extensive cell lysis and empty areas between cells
3	Moderate	Greater than 70% of the cell layers; contain rounded cells and/or are lysed
4	Severe	Nearly complete destruction of the cell layers

held, use 100 mL of Sterile Water for Injection as the test solution. To 100 mL of the test solution add 2 mL of alkaline mercuric-potassium iodide TS: any yellow color produced immediately is not darker than that of a control containing 30 µg of added NH_3 in *High-purity Water* (see *Reagents* under *Containers* ⟨661⟩) (0.6 ppm for Sterile Water for Injection packaged in volumes up to 50 mL in containers; 0.3 ppm for larger volumes).

Chloride—To 20 mL in a color-comparison tube add 5 drops of nitric acid and 1 mL of silver nitrate TS, and gently mix: any turbidity formed within 10 minutes is not greater than that produced in a similarly treated control consisting of 20 mL of *High-purity Water* (see under *Reagents* in *Containers* ⟨661⟩) containing 10 µg of Cl (0.5 ppm), viewed downward over a dark surface with light entering the tubes from the sides.

Oxidizable substances—To 100 mL add 10 mL of 2 *N* sulfuric acid, and heat to boiling. For Sterile Water for Injection in glass containers holding a volume up to 50 mL, add 0.4 mL of 0.1 *N* potassium permanganate, and boil for 5 minutes; for larger volumes, add 0.2 mL of 0.1 *N* potassium permanganate, and boil for 5 minutes: the pink color does not completely disappear.

Total solids—Proceed as directed in the test for *Total solids* under *Purified Water.* The following limits apply for Sterile Water for Injection in glass containers holding up to 30 mL, 0.004%; from 30 mL up to 100 mL, 0.003%; and for larger volumes, 0.002%.

Other requirements—It meets the requirements of the tests for *pH*, *Sulfate*, *Calcium*, *Carbon dioxide*, and *Heavy metals* under *Purified Water.*

Sterile Water for Irrigation

» Sterile Water for Irrigation is Water for Injection sterilized and suitably packaged. It contains no antimicrobial agent or other added substance.

Packaging and storage—Preserve in single-dose containers, preferably of Type I or Type II glass. The container may contain a volume of more than 1 liter, and may be designed to empty rapidly.

Labeling—Label it to indicate that no antimicrobial or other substance has been added. The designations "For irrigation only" and "Not for injection" appear prominently on the label.

Reference standard—*USP Endotoxin Reference Standard.*

Other requirements—It meets the requirements of all of the tests under *Sterile Water for Injection* except the test for *Particulate matter.*

Purified Water

H_2O 18.02

» Purified Water is water obtained by distillation, ion-exchange treatment, reverse osmosis, or other suitable process. It is prepared from water complying with the regulations of the federal Environmental Protection Agency with respect to drinking water. It contains no added substance.

Note—Purified Water is intended for use as an ingredient in the preparation of compendial dosage forms. Where used for sterile dosage forms, other than for parenteral administration, process the article to meet the requirements under Sterility Tests ⟨71⟩, or first render the Purified Water sterile and thereafter protect it from microbial contamination. Do not use Purified Water in preparations intended for parenteral administration. For such purposes use Water for Injection, Bacteriostatic Water for Injection, or Sterile Water for Injection.

Packaging and storage—Where packaged, preserve in tight containers.

Labeling—Where packaged, label it to indicate the method of preparation.

pH ⟨791⟩: between 5.0 and 7.0, determined potentiometrically in a solution prepared by the addition of 0.30 mL of saturated potassium chloride solution to 100 mL of test specimen.

Chloride—To 100 mL add 5 drops of nitric acid and 1 mL of silver nitrate TS: no opalescence is produced.

Sulfate—To 100 mL add 1 mL of barium chloride TS: no turbidity is produced.

Ammonia—To 100 mL add 2 mL of alkaline mercuric-potassium iodide TS: any yellow color produced immediately is not darker than that of a control containing 30 µg of added NH_3 in *High-purity Water* (see under *Reagents* in *Containers* ⟨661⟩) [0.3 ppm].

Calcium—To 100 mL add 2 mL of ammonium oxalate TS: no turbidity is produced.

Carbon dioxide—To 25 mL add 25 mL of calcium hydroxide TS: the mixture remains clear.

Heavy metals—Adjust 40 mL of Purified Water with 1 *N* acetic acid to a pH of 3.0 to 4.0 (using short-range pH indicator paper), add 10 mL of freshly prepared hydrogen sulfide TS, and allow the liquid to stand for 10 minutes: the color of the liquid, when viewed downward over a white surface, is not darker than the color of a mixture of 50 mL of the same Purified Water with the same amount of 1 *N* acetic acid as was added to the test specimen, matched color-comparison tubes being used for the comparison.

Oxidizable substances—To 100 mL add 10 mL of 2 *N* sulfuric acid, and heat to boiling. Add 0.1 mL of 0.1 *N* potassium permanganate, and boil for 10 minutes: the pink color does not completely disappear.

Total solids—Evaporate 100 mL on a steam bath to dryness, and dry the residue at 105° for 1 hour: not more than 1 mg of residue remains (0.001%).

Bacteriological purity—It complies with the federal Environmental Protection Agency regulations for drinking water with respect to bacteriological purity (40 CFR 141.14; 141.21).

Wax, Carnauba—*see* Wax, Carnauba NF

Wax, Cetyl Esters—*see* Cetyl Esters Wax NF

Wax, Emulsifying—*see* Wax, Emulsifying NF

Wax, Microcrystalline—*see* Wax, Microcrystalline NF

Wax, White—*see* Wax, White NF

Wax, Yellow—*see* Wax, Yellow NF

White Lotion

» Prepare White Lotion as follows:

Zinc Sulfate	40 g
Sulfurated Potash	40 g
Purified Water, a sufficient quantity, to make	1000 mL

Dissolve the Zinc Sulfate and the Sulfurated Potash separately, each in 450 mL of Purified Water, and filter each solution. Add the sulfurated potash solution slowly to the zinc sulfate solution with constant stirring. Then add the required amount of purified water, and mix.

NOTE—Prepare the Lotion fresh, and shake it thoroughly before dispensing.

Packaging—Dispense in tight containers.

White Ointment—*see* Ointment, White

White Petrolatum—*see* Petrolatum, White

White Wax—*see* Wax, White NF

Whole Blood—*see* Blood Groupings

Widow Spider Species Antivenin—*see* Antivenin (Latrodectus Mactans)

Xanthan Gum—*see* Xanthan Gum NF

Xenon Xe 127

» Xenon Xe 127 is a gas suitable for inhalation in diagnostic studies. Xenon 127 is a radioactive nuclide that may be prepared from the bombardment of a cesium 133 target with high-energy protons. It contains not less than 85.0 percent and not more than 115.0 percent of the labeled amount of ^{127}Xe at the calibration date indicated on the labeling.

Packaging and storage—Preserve in single-dose vials having leak-proof stoppers, at room temperature. The vials are enclosed in appropriate lead radiation shields. The vial content may be diluted with air and is packaged at atmospheric pressure.

Labeling—Label it to include the following: the name of the preparation; the container volume, MBq (mCi) of ^{127}Xe per container; the amount of ^{127}Xe expressed as megabecquerels (millicuries) per mL; the intended route of administration; recommended storage conditions; the date of calibration; the expiration date; the name, address, and batch number of the manufacturer; the statement, "Caution—Radioactive Material"; and a radioactive symbol. The labeling contains a statement of radionuclide purity, identifies probable radionuclidic impurities, and indicates permissible quantities of each impurity. The labeling indicates that in making dosage calculations, correction is to be made for radioactive decay, and also indicates that the radioactive half-life of ^{127}Xe is 36.41 days.

Radionuclide identification (see *Radioactivity* ⟨821⟩)—Its gamma-ray spectrum is identical to that of a known specimen of xenon 127 that exhibits major photopeaks at 202.8 keV, and 172.1 keV, and 375.0 keV. Minor photopeaks from other xenon radioisotopes, namely Xe 129m (197 keV) and Xe 131m (164 keV) may also be present.

Radionuclidic purity—Using a suitable counting assembly (see *Selection of a Counting Assembly* under *Radioactivity* ⟨821⟩), determine the radioactivity of the Xe 127 in the gas by use of a calibrated system as directed under *Radioactivity* ⟨821⟩. Using the gamma-ray spectrum, determine the energy of each gamma photopeak. Identify each radionuclide present, and using the established detector efficiency and known gamma abundance, calculate the quantity of each radionuclide present in the specimen in MBq (mCi). The amount of Xe 127 present in the specimen is not less than 80%; the quantity of either Xe 131m or Xe 129m does not exceed 10%, and no other radioisotope exceeds 1%.

Assay for radioactivity—Using a suitable counting assembly (see *Selection of a Counting Assembly* under *Radioactivity* ⟨821⟩), determine the radioactivity, in MBq (mCi), of Xe 127 in each container by use of a calibrated system as directed under *Radioactivity* ⟨821⟩.

Xenon Xe 133

Xenon, isotope of mass 133.
Xenon, isotope of mass 133 [*14932-42-4*].

» Xenon Xe 133 is a gas suitable for inhalation in diagnostic studies. Xenon 133 is a radioactive nuclide that may be prepared from the fission of uranium 235. It contains not less than 85.0 percent and not more than 115.0 percent of the labeled amount of ^{133}Xe at the date and time indicated in the labeling.

Packaging and storage—Preserve in single-dose or in multiple-dose vials having leak-proof stoppers, at room temperature.

Other requirements—It meets the requirements for *Labeling*, except for the information specified for *Labeling* under *Injections;* for *Radionuclide identification;* and for *Radionuclidic purity*, and *Assay for radioactivity* under *Xenon Xe 133 Injection*, except to determine the radioactivity in MBq (mCi) per container.

Xenon Xe 133 Injection

» Xenon Xe 133 Injection is a sterile, isotonic solution of Xenon 133 in Sodium Chloride Injection suitable for intravenous administration. Xenon 133 is a radioactive nuclide prepared from the fission of uranium 235. It contains not less than 90.0 percent and not more than 110.0 percent of the labeled amount of Xenon 133 at the date and time stated on the label.

Packaging and storage—Preserve in single-dose containers that are totally filled, so that any air present occupies not more than 0.5% of the total volume of the container. Store at a temperature between 2° and 8°. If there is free space above the solution, a significant amount of the xenon 133 is present in the gaseous phase. Glass containers may darken under the effects of radiation.

Labeling—Label it to include the following, in addition to the information specified for *Labeling* under *Injections* ⟨1⟩: the time and date of calibration; the amount of xenon 133 expressed as total megabecquerels (microcuries or millicuries), and concentration as megabecquerels (microcuries or millicuries), per mL at the time of calibration; the expiration date; the name and amount of any added bacteriostatic agent; and the statement, "Caution—Radioactive Material." The labeling indicates that in making dosage calculations, correction is to be made for radioactive decay, and also indicates that the radioactive half-life of ^{133}Xe is 5.24 days.

Reference standard—*USP Endotoxin Reference Standard.*

Radionuclide identification (see *Radioactivity* ⟨821⟩)—Its gamma-ray and X-ray spectra are identical to those of a known specimen of xenon 133 that exhibits two major photopeaks having energies of 0.081 MeV and 0.031 MeV (X-ray peak).

Bacterial endotoxins—It meets the requirements of the *Bacterial Endotoxins Test* ⟨85⟩, the limit of endotoxin content being not more than 175/*V* USP Endotoxin Unit per mL of the Injection, when compared with the USP Endotoxin RS, in which *V* is the maximum recommended total dose, in mL, at the expiration date or time.

pH ⟨791⟩: between 4.5 and 8.0.

Radionuclidic purity—Using a suitable counting assembly (see *Selection of a Counting Assembly* under *Radioactivity* ⟨821⟩), determine the radioactivity of Xe 133 in the Injection by use of a calibrated system as directed under *Radioactivity* ⟨821⟩. The radioactivity exhibited at 0.081 MeV and 0.031 MeV is not less than 95.0% of the total radioactivity of the specimen.

Other requirements—It meets the requirements under *Injections* ⟨1⟩, except that the Injection may be distributed or dispensed prior to the completion of the test for *Sterility*, the latter test

Standard Preparation—On the day of the assay, dilute a measured volume of *Standard Stock Solution of Calcium Pantothenate* with sufficient water so that it contains, in each mL, between 0.01 µg and 0.04 µg of calcium pantothenate, the exact concentration being such that the responses obtained as directed under *Procedure*, 2.0 and 4.0 mL of the *Standard Preparation* being used, are within the linear portion of the log-concentration response curve.

Assay Preparation—Proceed as directed in the individual monograph for preparing a solution expected to contain approximately the equivalent of the calcium pantothenate concentration in the *Standard Preparation.*

Basal Medium Stock Solution—

Acid-hydrolyzed Casein Solution	25 mL
Cystine-Tryptophane Solution	25 mL
Polysorbate 80 Solution	0.25 mL
Dextrose, Anhydrous	10 g
Sodium Acetate, Anhydrous	5 g
Adenine-Guanine-Uracil Solution	5 mL
Riboflavin–Thiamine Hydrochloride–Biotin Solution	5 mL
Para-aminobenzoic Acid–Niacin–Pyridoxine Hydrochloride Solution	5 mL
Salt Solution A	5 mL
Salt Solution B	5 mL

Dissolve the anhydrous dextrose and sodium acetate in the solutions previously mixed, and adjust with 1 *N* sodium hydroxide to a pH of 6.8. Finally, dilute with water to 250 mL, and mix.

Acid-hydrolyzed Casein Solution—Mix 100 g of vitamin-free casein with 500 mL of 6 *N* hydrochloric acid, and reflux the mixture for 8 to 12 hours. Remove the hydrochloric acid from the mixture by distillation under reduced pressure until a thick paste remains. Redissolve the resulting paste in water, adjust the solution with 1 *N* sodium hydroxide to a pH of 3.5 ± 0.1, and add water to make 1000 mL. Add 20 g of activated charcoal, stir for 1 hour, and filter. Repeat the treatment with activated charcoal. Store under toluene in a refrigerator at a temperature not below 10°. Filter the solution if a precipitate forms during storage.

Cystine-Tryptophane Solution—Suspend 4.0 g of L-cystine and 1.0 g of L-tryptophane (or 2.0 g of D,L-tryptophane) in 700 to 800 mL of water, heat to 70° to 80°, and add dilute hydrochloric acid (1 in 2) dropwise, with stirring, until the solids are dissolved. Cool, and add water to make 1000 mL. Store under toluene in a refrigerator at a temperature not below 10°.

Adenine-Guanine-Uracil Solution—Dissolve 200 mg each of adenine sulfate, guanine hydrochloride, and uracil, with the aid of heat, in 10 mL of 4 *N* hydrochloric acid, cool, and add water to make 200 mL. Store under toluene in a refrigerator.

Polysorbate 80 Solution—Dissolve 25 g of polysorbate 80 in alcohol to make 250 mL.

Riboflavin–Thiamine Hydrochloride–Biotin Solution—Prepare a solution containing, in each mL, 20 µg of riboflavin, 10 µg of thiamine hydrochloride, and 0.04 µg of biotin, by dissolving riboflavin, thiamine hydrochloride, and biotin in 0.02 *N* acetic acid. Store, protected from light, under toluene in a refrigerator.

Para-aminobenzoic Acid–Niacin–Pyridoxine Hydrochloride Solution—Prepare a solution in neutral 25 percent alcohol to contain 10 µg of para-aminobenzoic acid, 50 µg of niacin, and 40 µg of pyridoxine hydrochloride in each mL. Store in a refrigerator.

Salt Solution A—Dissolve 25 g of monobasic potassium phosphate and 25 g of dibasic potassium phosphate in water to make 500 mL. Add 5 drops of hydrochloric acid, and store under toluene.

Salt Solution B—Dissolve 10 g of magnesium sulfate, 0.5 g of sodium chloride, 0.5 g of ferrous sulfate, and 0.5 g of manganese sulfate in water to make 500 mL. Add 5 drops of hydrochloric acid, and store under toluene.

Stock Culture of Lactobacillus plantarum—Dissolve 2.0 g of water-soluble yeast extract in 100 mL of water, add 500 mg of anhydrous dextrose, 500 mg of anhydrous sodium acetate, and 1.5 g of agar, and heat the mixture, with stirring, on a steam bath, until the agar dissolves. Add approximately 10-mL portions of the hot solution to test tubes, suitably close or cover the tubes, sterilize at 121°, and allow the tubes to cool in an upright position. Prepare stab cultures in 3 or more of the tubes, using a pure culture of *Lactobacillus plantarum*,* incubating for 16 to 24 hours at any selected temperature between 30° and 37° but held constant to within ±0.5°, and finally store in a refrigerator. Prepare a fresh stab of the stock culture every week, and do not use for inoculum if the culture is more than 1 week old.

Culture Medium—To each of a series of test tubes containing 5.0 mL of *Basal Medium Stock Solution* add 5.0 mL of water containing 0.2 µg of calcium pantothenate. Plug the tubes with cotton, sterilize in an autoclave at 121°, and cool.

Inoculum—Make a transfer of cells from the stock culture of *Lactobacillus plantarum* to a sterile tube containing 10 mL of culture medium. Incubate this culture for 16 to 24 hours at any selected temperature between 30° and 37° but held constant to within ±0.5°. The cell suspension so obtained is the inoculum.

Procedure—To similar test tubes add, in duplicate, 1.0 and/or 1.5, 2.0, 3.0, 4.0, and 5.0 mL, respectively, of the *Standard Preparation*. To each tube and to 4 similar tubes containing no *Standard Preparation* add 5.0 mL of *Basal Medium Stock Solution* and sufficient water to make 10 mL.

To similar test tubes add, in duplicate, volumes of the *Assay Preparation* corresponding to 3 or more of the levels listed above for the *Standard Preparation*, including the levels of 2.0, 3.0, and 4.0 mL. To each tube add 5.0 mL of the *Basal Medium Stock Solution* and sufficient water to make 10 mL. Place one complete set of Standard and Assay tubes together in one tube rack and the duplicate set in a second rack or section of a rack, preferably in random order.

Cover the tubes of both series suitably to prevent contamination, and heat in an autoclave at 121° for 5 minutes. Cool, add 1 drop of inoculum to each tube, except 2 of the 4 tubes containing no *Standard Preparation* (to serve as the uninoculated blanks), and mix. Incubate the tubes at a temperature between 30° and 37°, held constant to within ±0.5° until, following 16 to 24 hours of incubation, there has been no substantial increase in turbidity in the tubes containing the highest level of standard during a 2-hour period.

Determine the transmittance of the tubes in the following manner: Mix the contents of each tube, and transfer to an optical container if necessary. Place the container in a spectrophotometer that has been set at a specific wavelength between 540 nm and 660 nm, and read the transmittance when a steady state is reached. This steady state is observed a few seconds after agitation when the galvanometer reading remains constant for 30 seconds or more. Allow approximately the same time interval for the reading on each tube.

With the transmittance set at 1.00 for the uninoculated blank, read the transmittance of the inoculated blank. With the transmittance set at 1.00 for the inoculated blank, read the transmittance for each of the remaining tubes. If there is evidence of contamination with a foreign microorganism, disregard the result of the assay.

Calculation—Prepare a standard concentration-response curve as follows: For each level of the standard, calculate the response from the sum of the duplicate values of the transmittance as the difference, $y = 2.00 - \Sigma$ (of transmittance). Plot this response on the ordinate of cross-section paper against the logarithm of the mL of *Standard Preparation* per tube on the abscissa, using for the ordinate either an arithmetic or a logarithmic scale, whichever gives the better approximation to a straight line. Draw the straight line or smooth curve that best fits the plotted points.

Calculate the response, y, adding together the two transmittances for each level of the *Assay Preparation*. Read from the standard curve the logarithm of the volume of the *Standard Preparation* corresponding to each of those values of y that fall within the range of the lowest and highest points plotted for the standard. Subtract from each logarithm so obtained the logarithm of the volume, in mL, of the *Assay Preparation* to obtain the difference, x, for each dosage level. Average the values of x for each of three or more dosage levels to obtain $\bar{x} = M'$, the log-relative potency of the *Assay Preparation*. Determine the quantity, in mg, of USP Calcium Pantothenate RS corresponding

* American Type Culture Collection No. 8014 is suitable. This strain formerly was known as *Lactobacillus arabinosus* 17–5.

to the calcium pantothenate in the portion of material taken for assay as antilog:

$$M = \text{antilog } (M' + \log R),$$

in which R is the number of mg of calcium pantothenate that was assumed to be present in each mg (or capsule or tablet) of the material taken for assay.

Replication—Repeat the entire determination at least once, using separately prepared *Assay Preparations.* If the difference between the two log-potencies M is not greater than 0.08, their mean, $\bar{M}$, is the assayed log-potency of the test material (see *The Confidence Interval and Limits of Potency* ⟨111⟩). If the two determinations differ by more than 0.08, conduct one or more additional determinations. From the mean of two or more values of M that do not differ by more than 0.15, compute the mean potency of the preparation under assay.

⟨101⟩ DEPRESSOR SUBSTANCES TEST

Reference Standard—*USP Histamine Dihydrochloride Reference Standard*—Keep container tightly closed and protected from light. Dry over silica gel for 2 hours before using.

Standard Solution of Histamine—Dissolve a suitable quantity of USP Histamine Dihydrochloride RS, accurately weighed, in water, and dilute with water to obtain a solution having a known concentration of the equivalent of 1.0 µg of histamine base per mL.

The Animal—Weigh and anesthetize a healthy, and if female, nonpregnant adult cat by intraperitoneal injection of an anesthetic substance, such as sodium phenobarbital, that is favorable to maintenance of uniform blood pressure. Immobilize the animal, and make provisions to prevent excess loss of body heat. If preferable, insert a tracheal cannula. Expose a carotid or other suitable artery, separate it from surrounding tissues, and arrange for continuous blood-pressure recording with a manometer or other apparatus of at least equivalent sensitivity. Then expose a femoral vein to facilitate intravenous injection.

Determine the sensitivity of the animal to histamine by injecting, at uniform time intervals of not less than 5 minutes, doses of the *Standard Solution of Histamine* corresponding to 0.05, 0.1, and 0.15 µg of histamine base per kg of body weight of the animal. Repeat these injections, and disregard the first set of responses. Determine the extent of variation in depressor response to the same dose by repeating the injection of 0.1 µg per kg. Use the animal for the test only if the responses to the graded doses are clearly different and the responses to several injections of the dose of 0.1 µg per kg are approximately the same and correspond to a decrease in pressure of not less than 20 mm of mercury. If in this test the doses of the *Standard Solution of Histamine* and of the solution under test are to be given through a single common cannula, each injection in the preliminary test and in the succeeding test is to be followed immediately by an injection of about 2.0 mL of Sodium Chloride Injection to flush in any residual activity.

Procedure—Dissolve the substance under test in the diluent designated so as to give the concentration specified in the individual monograph. Follow the same time schedule established during the injection of the *Standard Solution of Histamine.* Inject a series of three doses, of which two doses of 0.1 µg of histamine base per kg are alternated with an intervening dose of the solution under test in the dosage specified in the individual monograph. Measure the change in blood pressure following each of the three injections. The depressor response to the solution under test is not greater than one-half the mean depressor response to the two associated doses representing 0.1 µg of histamine base per kg. If this requirement is not met, continue the series of injections similarly until it consists of five doses, of which three doses of 0.1 µg of histamine base per kg are alternated with two doses of the solution under test in the dosage specified in the individual monograph. Measure the change in blood pressure following each of the additional injections. The depressor response to each dose of the solution under test is not greater than the mean of the respective depressor responses to the associated doses, representing 0.1 µg of histamine base per kg.

If the depressor response to either dose of the solution under test is greater than the mean of the depressor responses to the associated doses representing 0.1 µg of histamine base per kg, the test may be continued in the same animal, or in another animal similarly prepared and tested for responses to the *Standard Solution of Histamine.* If the test is continued in the same animal, after the last dose of the *Standard Solution of Histamine* of the initial series, administer four more injections, of which two are doses of the solution under test and two are doses representing 0.1 µg of histamine base per kg, alternately in sequence. If the test is continued in another animal, prepare a fresh solution of the substance under test from an independent container or containers of test substance, and inject a series of five doses comprising the *Standard Solution of Histamine* and the solution under test in accordance with the initial injection sequence. Measure the change in blood pressure following each of the additional injections. Compute the difference between each response to the dose of the solution under test and the mean of the associated doses representing 0.1 µg of histamine base per kg in the entire series, initial and additional, and calculate the average of all such differences. The requirements of the test are met if the average of the differences is such that in the specified dose the depressor response to the solution under test is not greater than the depressor response to the dose representing 0.1 µg of histamine base per kg, and if not more than one-half of the depressor responses to the solution under test are greater than the mean of the respective depressor responses to the associated doses, representing 0.1 µg of histamine base per kg.

⟨111⟩ DESIGN AND ANALYSIS OF BIOLOGICAL ASSAYS

General

The potency of several Pharmacopeial drugs must be determined by bioassay. A controlling factor in assay design and analysis is the variability of the biological test system, which may vary in its mean response from one laboratory to another, and from time to time in the same laboratory. To control this type of variation, the response to a Pharmacopeial drug is compared with that to a USP Reference Standard or other suitable standard. For convenience, each such preparation will be called the "Standard" and each preparation under assay, or Sample, the "Unknown," and these will be designated respectively by the symbols S and U. (The Sample is sometimes referred to as the "test preparation.")

After elimination of extraneous variables from the comparison of the Standard and the Unknown, an error variance is computed from the remaining variation, which, while uncontrolled, can nevertheless be measured. The error variance is required in calculating the confidence interval of the assayed potency. The confidence interval, known also as the fiducial interval, is so computed that its upper and lower limits are expected to enclose the true potency of the Unknown in 19 out of 20 assays. Many assay procedures fix the acceptable width of the confidence interval, and two or more independent assays may be needed to meet the specified limit. The confidence limits of the individual component assays usually overlap.

The aim of this chapter is to present a concise account of biometrical procedures for the USP bioassays. Its various sections are interrelated. Although the procedures are planned primarily for the assay of a single Unknown, equations for the joint assay of several Unknowns are given in context throughout the chapter and are summarized in the last section. Proof that an assayed potency meets its required confidence limits may be based also upon other recognized biometric methods that have a precision equivalent to that of the methods outlined herein.

A glossary of the terms used in the equations is provided at the end of this chapter.

Steps Preceding the Calculation of Potency

Designs for Minimizing the Error Variance—Variation in response is reduced as much as is practicable by the limitations imposed on body weight, age, previous handling, environment, and similar factors. In a number of assays, the test animals or their equivalent are then assigned at random but in equal numbers to the different doses of the Standard and Unknown. This implies an objective random process, such as throwing dice, shuffling cards, or using a table of random numbers. Assigning the same number of individuals to each treatment simplifies the subsequent calculations materially, and usually leads to the shortest confidence interval for a given number of observations.

In some assays, the potential responses can be assembled into homogeneous sets in advance of treatment. The differences between sets are later segregated, so that they do not affect adversely either the computed potency or its confidence interval. One unit within each set, picked at random, receives each treatment. Examples of randomized sets are the cleared areas on a single plate in the plate assay of an antibiotic, and four successive paired readings in the same rat in the Vasopressin Injection assay. Sets of two occur where each test animal is used twice, as in the assays of Tubocurarine Chloride Injection and Insulin Injection. In these cases, neither the average differences between individuals nor the order of treatment can bias the potency or precision. In the microbial assays for vitamin B_{12} activity and for calcium pantothenate, replicate tubes are assigned to two or more separate, complete sets, preferably with the tubes arranged at random within each set. This restricts the variation due to position or order within a set to the differences within each complete replicate.

Rejection of Outlying or Aberrant Observations—A response that is questionable because of failure to comply with the procedure during the course of an assay is rejected. Other aberrant values may be discovered only after the responses have been tabulated, but can then be traced to assay irregularities, which justify their omission. The arbitrary rejection *or* retention of an apparently aberrant response can be a serious source of bias. In general, the rejection of observations solely on the basis of their relative magnitudes is a procedure to be used sparingly. When this is unavoidable, each suspected aberrant response or outlier may be tested against one of two criteria:

1. The first criterion is based upon the variation within a single group of supposedly equivalent responses. On the average, it will reject a valid observation once in 25 or once in 50 trials, provided that relatively few, if any, responses within the group are identical. Beginning with the supposedly erratic value or outlier, designate the responses in order of magnitude from y_1 to y_N, where N is the number of observations in the group. Compute the relative gap $G_1 = (y_2 - y_1)/(y_N - y_1)$ when $N = 3$ to 7, $G_2 = (y_3 - y_1)/(y_{N-1} - y_1)$ when $N = 8$ to 13, or $G_3 = (y_3 - y_1)/(y_{N-2} - y_1)$ when $N = 14$ to 24. If G_1, G_2, or G_3 exceeds the critical value in Table 1 for the observed N, there is a statistical basis for omitting the outlier.

This criterion is applicable also in a microbial assay where each treatment is represented by a transmittance in each of two separate complete sets. Subtract each transmittance in the first set from its paired value in the second set, and record each difference with its sign, either plus or minus. Beginning with the most divergent difference, designate the N differences in order of magnitude from y_1 to y_N and compute the relative gap G_1, G_2, or G_3. If this exceeds its critical value in Table 1, one of the two transmittances giving the aberrant difference is suspect and may be identified on inspection or by comparison with its expectation (see next column). Repeat the process with the remaining differences if an outlier is suspected in a second pair.

2. The second criterion compares the ranges from a series of $k = 2$ or more groups. Different groups may receive different treatments, but all f responses within each group represent the same treatment. Compute the range from each group by subtracting the smallest response from the largest within each of the k groups. Divide the largest of the k ranges by the sum of all the ranges in the series. Refer this ratio R_* to Table 2. If k is not larger than 10, use the tabular values in the upper part of Table 2; if k is larger than 10, multiply R_* by $(k + 2)$ and interpolate, if necessary, between the tabular values in the lower

Table 1

Test for outliers. In samples from a normal population, gaps equal to or larger than the following values of G_1, G_2, and G_3 occur with a probability $P = 0.02$ where outliers can occur only at one end, or with $P = 0.04$ where they may occur at either end.

N	3	4	5	6	7						
G_1	.976	.846	.729	.644	.586						
N	8	9	10	11	12	13					
G_2	.780	.725	.678	.638	.605	.578					
N	14	15	16	17	18	19	20	21	22	23	24
G_3	.602	.579	.559	.542	.527	.514	.502	.491	.481	.472	.464

Table 2

Test for groups containing outliers. Compute the range from the f observations in each of k groups, where all groups in the series are equal in size. The observed ratio R_* of the largest range to the sum of the k ranges will equal or exceed the following critical values at a probability of $P = 0.05$.

No. of Ranges k	Critical R_* for Ranges Each from f Observations								
	2	3	4	5	6	7	8	9	10
2	0.962	0.862	0.803	0.764	0.736	0.717	0.702	0.691	0.682
3	.813	.667	.601	.563	.539	.521	.507	.498	.489
4	.681	.538	.479	.446	.425	.410	.398	.389	.382
5	.581	.451	.398	.369	.351	.338	.328	.320	.314
6	0.508	0.389	0.342	0.316	0.300	0.288	0.280	0.273	0.267
7	.451	.342	.300	.278	.263	.253	.245	.239	.234
8	.407	.305	.267	.248	.234	.225	.218	.213	.208
9	.369	.276	.241	.224	.211	.203	.197	.192	.188
10	.339	.253	.220	.204	.193	.185	.179	.174	.172

No. of Ranges k	Critical $(k + 2)R_*$ for Ranges Each from f Observations								
	2	3	4	5	6	7	8	9	10
10	4.06	3.04	2.65	2.44	2.30	2.21	2.14	2.09	2.05
12	4.06	3.03	2.63	2.42	2.29	2.20	2.13	2.07	2.04
15	4.06	3.02	2.62	2.41	2.28	2.18	2.12	2.06	2.02
20	4.13	3.03	2.62	2.41	2.28	2.18	2.11	2.05	2.01
50	4.26	3.11	2.67	2.44	2.29	2.19	2.11	2.06	2.01

part of Table 2. If R_* exceeds the tabular or interpolated value, the group with the largest range is suspect and inspection of its components will usually identify the observation, which is then assumed to be aberrant or an outlier. The process may be repeated with the remaining ranges if an outlier is suspected in a second group.

Replacement of Missing Values—As directed in the monographs and in this section, the calculation of potency and its confidence interval from the total response for each dose of each preparation requires the same number of observations in each total. When observations are lost or additional responses have been obtained with the Standard, the balance may be restored by one of the following procedures, so that the usual equations apply.

1. Reduce the number of observations in the larger groups until the number of responses is the same for each treatment. If animals have been assigned at random to each treatment group, either omit one or more responses, selected at random, from each larger group, or subtract the mean of each larger group from its initial total as often as may be necessary. The latter technique is preferred when extra animals have been assigned deliberately to the Standard. When the assay consists of randomized sets, retain only the complete sets.

2. Alternatively, an occasional smaller group may be brought up to size when the number of missing responses is not more than one in any one treatment or 10% in the entire assay. Estimate a replacement for each missing value by either method *a* or method *b*. One degree of freedom (n) is lost from the error variance s^2 for each replacement by either method, except in a microbial assay where each response is based on the sum of two or more transmittances and only one transmittance is replaced.

(*a*) If animals have been assigned to treatments at random, add the mean of the remaining responses in the incomplete group to their total. In a microbial assay, when one of two transmittances is missing for a given treatment, add the mean difference between sets, computed from all complete pairs, to the remaining transmittance to obtain the replacement.

(*b*) If the assay consists of randomized sets, replace the missing value by

$$y' = \frac{fT_r' + kT_t' - T'}{(f-1)(k-1)}, \qquad (1)$$

where f is the number of sets, k is the number of treatments or doses, and T_r', T_t', and T' are the incomplete totals for the randomized set, treatment, and assay from which an observation is missing.

If the assay consists of n' Latin squares with k rows in common, replace a missing value by

$$y' = \frac{k(n'T_c' + T_r' + T_t') - 2T'}{(k-1)(n'k-2)}, \qquad (1a)$$

where n' is the number of Latin squares with k rows in common, k is the number of treatments or doses, and T_c', T_r', T_t', and T' are respectively the incomplete totals for the column, row, treatment, and assay from which an observation is missing.

If more than one value is missing, substitute the treatment mean temporarily in all but one of the empty places, and compute y' for the other by Equation 1. Replace each of the initial substitutions in turn by Equation 1, and repeat the process in successive approximations until a stable y' is obtained for each missing observation.

Calculation of Potency from a Single Assay

Directions for calculating potency from the data of a single assay are given in the individual monographs. In those assays which specify graphical interpolation from dosage-response curves but which meet the conditions for assay validity set forth herein, potency may be computed alternatively by the appropriate method in this section.

Planning the assay involves assigning to the Unknown an assumed potency, to permit administering it in dosages equivalent to those of the Standard. The closer the agreement between this original assumption and the result of the assay, the more precise is the calculated potency. The ratio of a given dose of the Standard, in µg or in USP Units, to the corresponding dose of the Unknown, measured as specified in the monograph, is designated uniformly by R. The log-relative potency in quantities assumed initially to equal those of the Standard is designated as M'. Ideally, M' should not differ significantly from zero. The log-potency is

$$M = M' + \log R \qquad (2)$$

or

$$\text{Potency} = P_* = \text{antilog } M = (\text{antilog } M')R.$$

Assay from Direct Determinations of the Threshold Dose—Tubocurarine Chloride Injection and Metocurarine Iodide are assayed from the threshold dose that just produces a characteristic biological response. The ratio of the mean threshold dose for the Standard to that for the Unknown gives the potency directly. The threshold dose is determined twice in each animal, once with the Standard and once with the Unknown. Each dose is converted to its logarithm, the difference (x) between the two log-doses is determined for each animal, and potency is calculated from the average of these differences.

In the *Bacterial Endotoxins Test* ⟨85⟩, the geometric mean dilution end-point for the Unknown corresponding to the geometric mean dilution end-point for the Standard (multiplied by a dilution factor, where applicable) gives the concentration of endotoxin in the test material.

In these assays, the confidence interval depends upon the variability in the threshold dose.

Indirect Assays from the Relationship between the Log-dose and the Response—Generally, the threshold dose cannot be measured directly; therefore, potency is determined indirectly by comparing the responses following known doses of the Standard with the responses following one or more similar doses of the Unknown. Within a restricted dosage range, a suitable measure of the response usually can be plotted as a straight line against the log-dose, a condition that simplifies the calculation of potency and its confidence interval. Both the slope and position of the log-dose response relationship are determined in each assay by the use of two or more levels of the Standard, or, preferably, of both the Standard and the Unknown.

In the assay of Heparin Sodium, the interval between the dose at which clotting occurs and that which produces no clotting is so small that the dosage-response curve is not determined explicitly. Moving averages are used instead to interpolate the log-dose corresponding to 50% clotting for both the Standard and the Unknown, leading to the log-potency (see *Calculation* under *Heparin Sodium*). The precision of the potency is estimated from the agreement between independent assays of the same Unknown.

For a drug that is assayed biologically, the response should plot as a straight line against the log-dose over an adequate range of doses. Where a preliminary test is required or the assay depends upon interpolation from a multi-dose Standard curve, plot on coordinate paper the mean response of the *Standard* at each dosage level on the ordinate against the log-dose x on the abscissa. If the trend is basically linear over the required dosage range, the initial response unit may be used directly as y; if, instead, the trend is clearly curvilinear, a suitable transformation of each initial reading may bring linearity.

One possible transformation is to logarithms; another, in microbial tube assays, where y = (100 − % transmittance) does not plot linearly against the log-dose x, is to probits. In this case, if absorbance cannot be read directly, the % transmittance for each tube or test solution is first converted to absorbance, $A = 2 - \log(\%\ \text{transmittance})$. Each absorbance value, in turn, is converted to % reduction in bacterial growth as

$$\%\ \text{reduction} = 100(\bar{A}_c - A)/\bar{A}_c,$$

where $\bar{A}_c$ is the mean density for the control tubes (without antibiotic or with excess of vitamin) in the same set or tube rack. Percent reduction is then transformed to a probit (see Table 3) to obtain a new y for all later calculation. The probit transformation offers the advantage of extending the working range of linearity even where a portion of the dosage-response relationship is non-linear in the original units of percent transmittance, provided that the incubation period does not extend beyond the logarithmic phase of growth of the control tubes.

The LD_{50} in the *Safety* test for Iron Dextran Injection is calculated with log-doses and probits. The four doses of the Injection

Table 3

Probits (normal deviates + 5) corresponding to percentages in the margins.

	0	1	2	3	4	5	6	7	8	9
0	—	2.67	2.95	3.12	3.25	3.36	3.45	3.52	3.59	3.66
10	3.72	3.77	3.82	3.87	3.92	3.96	4.01	4.05	4.08	4.12
20	4.16	4.19	4.23	4.26	4.29	4.33	4.36	4.39	4.42	4.45
30	4.48	4.50	4.53	4.56	4.59	4.61	4.64	4.67	4.69	4.72
40	4.75	4.77	4.80	4.82	4.85	4.87	4.90	4.92	4.95	4.97
50	5.00	5.03	5.05	5.08	5.10	5.13	5.15	5.18	5.20	5.23
60	5.25	5.28	5.31	5.33	5.36	5.39	5.41	5.44	5.47	5.50
70	5.52	5.55	5.58	5.61	5.64	5.67	5.71	5.74	5.77	5.81
80	5.84	5.88	5.92	5.95	5.99	6.04	6.08	6.13	6.18	6.23
90	6.28	6.34	6.41	6.48	6.55	6.64	6.75	6.88	7.05	7.33
	0.0	0.1	0.2	0.3	0.4	0.5	0.6	0.7	0.8	0.9
99	7.33	7.37	7.41	7.46	7.51	7.58	7.65	7.75	7.88	8.09

Table 4

Coefficients x_* for computing the responses Y_L and Y_H predicted by least squares at the lowest and highest of k log-doses when these are spaced at equal intervals.

No. of Doses	Predicted End Y	Coefficient x_* for Mean Response $\bar{y}_t$ at Log-Dose 1	2	3	4	5	6	Divisor
3	Y_L	5	2	−1				6
	Y_H	−1	2	5				6
4	Y_L	7	4	1	−2			10
	Y_H	−2	1	4	7			10
5	Y_L	3	2	1	0	−1		5
	Y_H	−1	0	1	2	3		5
6	Y_L	11	8	5	2	−1	−4	21
	Y_H	−4	−1	2	5	8	11	21

in mg of iron per kg of body weight are transformed to x_1 = 2.574, x_2 = 2.699, x_3 = 2.875, and x_4 = 3.000. The probits corresponding to the number of deaths observed in each group of 10 mice are designated y_1, y_2, y_3, and y_4, respectively, and are given in Table 3 for mortalities from 10 to 90 percent. For observed deaths of 0 and 10 adjacent to doses giving an intermediate mortality, use the approximate probits 3.02 and 6.98, respectively; omit the end value (at x_1 or x_4) if not adjacent to an intermediate mortality. Since the information in a probit varies with its expectation, assign each probit an approximate relative weight w for computing the LD_{50} of the Injection, as shown in the accompanying table.

No. of Deaths	0 or 10	1 or 9	2 or 8	3 or 7	4 to 6
Weight, w	0.3	0.7	1.0	1.2	1.3

Calculate the weighted means

$$\bar{x} = \Sigma(wx)/\Sigma w$$

and (2a)

$$\bar{y} = \Sigma(wy)/\Sigma w$$

from the sum of the weights, Σw, of the four (or three) acceptable responses and the corresponding weighted sums of the log-doses, $\Sigma(wx)$, and of the probits, $\Sigma(wy)$. From the sums of the weighted products, $\Sigma(wxy)$, and of the weighted squares, $\Sigma(wx^2)$, compute the slope b of the log-dose-probit line as

$$b = \frac{\Sigma(wxy) - \bar{x}\Sigma(wy)}{\Sigma(wx^2) - \bar{x}\Sigma(wx)}. \quad (2b)$$

The LD_{50} for this safety test, in mg of iron per kg of body weight, is calculated as

$$LD_{50} = \text{antilog}[\bar{x} + (5 - \bar{y})/b]. \quad (2c)$$

In quantal assays not included in this Pharmacopeia, such as the mouse assay for insulin, the calculation with probits involves other adjustments that are omitted here.

When the mean response $\bar{y}_t$ for each dose of Standard plots linearly against the log-dose, and the k doses are spaced at equal intervals on the logarithmic scale, the predicted responses (Y_L and Y_H) at the extreme ends of the line of best fit can be computed directly with the coefficients x_* in Table 4, which correspond to the k successive log-doses, as

$$Y_L = \Sigma(x_*\bar{y}_t)/\text{divisor}$$

and (3)

$$Y_H = \Sigma(x_*\bar{y}_t)/\text{divisor},$$

where Σ stands uniformly for "the sum of" the values that follow it. When Y_L and Y_H are plotted against the low and high log-doses, X_L and X_H, respectively, they may be connected by a straight line with the slope

$$b = (Y_H - Y_L)/(X_H - X_L). \quad (4)$$

At any selected log-dose x of *Standard*, the predicted response is

$$Y = \bar{y} + b(x - \bar{x}), \quad (5)$$

where $\bar{x} = \Sigma x/k$, and $\bar{y} = (Y_L + Y_H)/2$, or, for predictions within a set, $\bar{y}$ is the mean response for the *Standard* within the set.

When the log-dose response relationship is linear, but the k doses (expressed in mL) are spaced substantially in an arithmetic sequence as in Table 5 (which refers to the microbial assays set forth under *Antibiotics—Microbial Assays* ⟨81⟩), the slope b of the straight line of best fit may be computed with the terms in Table 5 and the mean response at each dose $\bar{y}_t$, or $T_t = f\bar{y}_t$ where the number of y's(f) is constant at each dose, as

$$b = \Sigma(x_1\bar{y}_t)/e_b'i = \Sigma(x_1T_t)/fe_b'i. \quad (6)$$

The coefficients x_1 are convenient multiples of the differences $(x - \bar{x})$ about the mean log-dose $\bar{x}$, and $e_b'i$ is the corresponding multiple of $\Sigma(x - \bar{x})^2$. The predicted response Y at a given log-dose x may be computed by substitution of the assay slope b in Equation 5 and of the mean $\bar{y}$ either of all the responses on the *Standard* in the entire assay or of those for each set separately.

Table 5

Coefficients x_1 for computing the slope b of a log-dose response curve when the doses are spaced on an arithmetic scale as shown.

No. of Doses	Coefficients x_1 for Computing b from the Responses y at Doses, in mL, of: 1	1.5	2	3	4	5	Divisor $e_b'i$	Mean Log-dose $\bar{x}$
4	—	−29	−12	12	29	—	14.4663	0.38908
5	−34	—	−9	5	15	23	24.7827	0.41584
5	—	−20	−11	2	11	18	13.3249	0.45105
6	−15	−8	−3	4	9	13	14.1017	0.37588

POTENCIES INTERPOLATED FROM A STANDARD CURVE—Where the log-dose response curve of the *Standard* in a given assay is curvilinear and is fitted graphically to the plotted points, the amount of *Standard* that would be expected to produce each observed response y of an *Unknown* is estimated by interpolation from the curve and then adjusted for the known concentration of its test solution.

When the response to the *Standard* can be plotted linearly against the log-dose, it is fitted numerically by a straight line, as described in the preceding section. For assays in randomized sets, a standard curve is computed with b for the assay and $\bar{y}$ for each set and the response y_U in each tube of a given *Unknown* in that set is converted to an estimated log-relative potency,

$$X = (y_U - Y_S)/b, \qquad (7)$$

where Y_S is the response predicted by the standard curve at the assumed log-dose x of the *Unknown*. The average of the separate estimates from each of f sets, $M' = \Sigma X/f$, is the assayed log-relative potency of the *Unknown*.

Factorial Assays from the Response to Each Treatment—When some function of the response can be plotted linearly against the log-dose, the assayed potency is computed from the total response for each treatment, and its precision is measured in terms of confidence intervals. This requires that (1) in suitable units the response (y) depends linearly upon the log-dose within the dosage range of the assay, and (2) the number (f) of responses be the same at each dosage level of both Standard and Unknown. The y's are totaled at each dosage level of each preparation. In different combinations, these totals, T_t, lead directly to the log-relative potency and to tests of assay validity. The factorial coefficients in Tables 6, 7, and 8 determine how they are combined. In a given row, each T_t is multiplied by the corresponding coefficient and the products summed to obtain T_i. The T_i's in the successive rows carry the same meaning in all assays.

T_a in the first row measures the difference in the average response to the Standard and to the Unknown. T_b in the second row leads directly to the combined slope of the dosage-response curves for both Standard and Unknown. The third to the fifth rows (ab, q, and aq) provide tests for the validity of an assay, as described in a later section. From the totals T_a and T_b, compute the log-relative potency of the Unknown, before adjustment for its assumed potency, as

$$M' = ciT_a/T_b, \qquad (8)$$

where i is the interval in logarithms between successive log-doses of both the Standard and the Unknown, and the constant c is given separately at the bottom of each table. Each M' is corrected to its log-potency M by Equation 2.

When doses are spaced unequally on a log scale, as in Table 8, use instead the constant ci at the bottom of the table.

In a fully balanced assay, such as the assay for corticotropin, compute M' with the coefficients in Table 6. If one preparation has one less dose than the other but the successive log-doses of both Standard and Unknown differ by a constant interval i, use the factorial coefficients in Table 7, correcting for the actual difference between the observed mean log-doses, $\bar{x}_S$ and $\bar{x}_U$, by computing

$$M = \bar{x}_S - \bar{x}_U + M'. \qquad (9)$$

In assays where the successive doses are not spaced at equal log-intervals, the log-relative potency of a single Unknown may be computed by Equation 8 with the factorial coefficients and ci in Table 8.

In an assay of two or more Unknowns against a common Standard, all with dosage-response lines that are parallel within the

Table 6

Factorial coefficients x_1 for analyzing a balanced bioassay, in which successive log-doses of Standard (S_i) and of Unknown (U_i) are spaced equally, each with the same number (f) of responses totaling T_t.

Design	Row	Factorial Coefficients x_1 for Each Dose: S_1	S_2	S_3	S_4	U_1	U_2	U_3	U_4	e_i	T_i
2,2	a	−1	−1			1	1			4	T_a
	b	−1	1			−1	1			4	T_b
	ab	1	−1			−1	1			4	T_{ab}
3,3	a	−1	−1	−1		1	1	1		6	T_a
	b	−1	0	1		−1	0	1		4	T_b
	ab	1	0	−1		−1	0	1		4	T_{ab}
	q	1	−2	1		1	−2	1		12	T_q
	aq	−1	2	−1		1	−2	1		12	T_{aq}
4,4	a	−1	−1	−1	−1	1	1	1	1	8	T_a
	b	−3	−1	1	3	−3	−1	1	3	40	T_b
	ab	3	1	−1	−3	−3	−1	1	3	40	T_{ab}
	q	1	−1	−1	1	1	−1	−1	1	8	T_q
	aq	−1	1	1	−1	1	−1	−1	1	8	T_{aq}

For Computing	Equation No.	Constant	Value of Constant for Design: 2,2	3,3	4,4
M'	8, 10	c	1	4/3	5
L	26, 29	c'	1	8/3	5

Table 7

Factorial coefficients x_1 for analyzing a partially balanced assay, in which successive log-doses of Standard (S_i) and of Unknown (U_i) are spaced equally, each with the same number (f) of responses totaling T_t. If the number of successive doses of the Unknown exceeds by one the number on the Standard, interchange S_i and U_i in the heading and reverse all signs in rows *a*, *ab*, and *aq*.

		Factorial Coefficients x_1 for Each Dose								
Design	Row	S_1	S_2	S_3	S_4	U_1	U_2	U_3	e_i	T_i
2,1	*a*	−1	−1			2			6	T_a
	b	−1	1			0			2	T_b
3,2	*a*	−2	−2	−2		3	3		30	T_a
	b	−2	0	2		−1	1		10	T_b
	ab	1	0	−1		−2	2		10	T_{ab}
	q	1	−2	1		0	0		6	T_q
4,3	*a*	−3	−3	−3	−3	4	4	4	84	T_a
	b	−3	−1	1	3	−2	0	2	28	T_b
	ab	3	1	−1	−3	−5	0	5	70	T_{ab}
	q	3	−3	−3	3	2	−4	2	60	T_q
	aq	−1	1	1	−1	1	−2	1	10	T_{aq}

For Computing	Equation No.	Constant	Value of Constant for Design 2,1	3,2	4,3
M'	8, 10	c	1/2	5/6	7/6
L	26, 29	c'	3/4	25/12	49/12

Table 8

Factorial coefficients x_1 for analyzing assays with a 3- or 4-dose sequence of 1.5, 2.0, 3.0, and 4.0, each dose having the same number (f) of responses.

		Dose of Standard				Dose of Unknown					
Design	Row	1.5	2.0	3.0	4.0	1.5	2.0	3.0	4.0	e_i	T_i
4,4	*a*	−1	−1	−1	−1	1	1	1	1	8	T_a
	b	−29	−12	12	29	−29	−12	12	29	3940	T_b
	ab	29	12	−12	−29	−29	−12	12	29	3940	T_{ab}
	q	1	−1	−1	1	1	−1	−1	1	8	T_q
	aq	−1	1	1	−1	1	−1	−1	1	8	T_{aq}
3,3	*a*	−1	−1	−1		1	1	1		6	T_a
	b	−25	−3	28		−25	−3	28		2836	T_b
	ab	25	3	−28		−25	−3	28		2836	T_{ab}
	q	31	−53	22		31	−53	22		8508	T_q
	aq	−31	53	−22		31	−53	22		8508	T_{aq}
3,3	*a*		−1	−1	−1		1	1	1	6	T_a
	b		−28	3	25		−28	3	25	2836	T_b
	ab		28	−3	−25		−28	3	25	2836	T_{ab}
	q		22	−53	31		22	−53	31	8508	T_q
	aq		−22	53	−31		22	−53	31	8508	T_{aq}

For Computing	Equation No.	Constant	Value of Constant for Design 4,4	3,3
M'	8, 10	ci	7.2332	5.3695
L	26, 29	$c'i^2$	0.10623	0.06100

experimental error, each log-relative potency may be computed with the same assay slope as follows: For each preparation, determine the slope factor $T_b' = \Sigma(x_1 T_t)$ or $\Sigma(x_1 y)$, where the values of x_1 are the factorial coefficients for the Standard in the appropriate row *b* of Table 6 or 8. The log-relative potency of each Unknown is

$$M' = cih'T_a/2\Sigma T_b', \qquad (10)$$

where h' is the number of values of T_b' summed in the denominator.

Assays from Differences in Response—When doses of the Standard and Unknown are paired and the difference in response is computed for each pair, these differences are not affected by variations in the average sensitivity of the paired readings. The paired 2-dose insulin assay corresponds to the first design in Table 6, and requires four equal groups of rabbits each injected twice (see *Insulin Assay* ⟨121⟩). The difference (y) in the blood sugar response of each rabbit to the two treatments leads to the log-relative potency M' (see the first two paragraphs of the section, *Calculation of Potency from a Single Assay*). The Vasopressin Injection assay follows a similar design, substituting two or more randomized sets of four successive pairs of injections into rats for the four treatment groups of rabbits in the insulin assay.

Oxytocin Injection is assayed from blood pressure changes in a single test animal following alternating injections of a single dose of Standard and of one of two doses of the Unknown. The calculation of potency from the differences in the response of the Unknown and to the average of the two adjacent responses to the Standard is equivalent to the first design in Table 7 with *S* and *U* reversed, where *i* is the log-interval between the two dosage levels of the Unknown.

Experimental Error and Tests of Assay Validity

As the term is used here, "experimental error" refers to the residual variation in the response of biological indicators, not to a mistake in procedure or to an outlier that needs replacement. It is measured in terms of the error variance of a single response or other unit, which is designated uniformly as s^2, despite differences in the definition of the unit. It is required in tests of assay validity and in computing the *confidence interval.*

Error Variance of a Threshold Dose—The individual threshold dose is measured directly in some assays. In a Digitalis assay, designate each individual threshold dose by the symbol z, the number or frequency of z's by f, and the total of the z's for each preparation by T, with subscripts S and U for Standard and Unknown, respectively. Compute the error variance of z as

$$s^2 = [\Sigma z^2 - T_S^2/f_S - T_U^2/f_U]/n, \qquad (11)$$

with $n = f_S + f_U - 2$ degrees of freedom. In the assay of Tubocurarine Chloride Injection, each log-threshold dose of the Unknown is subtracted from the corresponding log-dose of the Standard in the same rabbit to obtain an individual difference x. Since each x may be either positive or negative (+ or −), it is essential to carry the correct sign in all sums. Designate the total of the x's for the animals injected with the Standard on the first day as T_1, and for those injected with the Standard on the second day as T_2. Compute the error variance of x with $n = N - 2$ degrees of freedom as

$$s^2 = \{\Sigma x^2 - (T_1^2 + T_2^2)/f\}/n, \qquad (12)$$

where N is the total number of rabbits that complete the assay, excluding any replacement for a missing value to equalize the size of the two groups.

Error Variance of an Individual Response—In the Pharmacopeial assays, differences in dose that modify the mean response are assumed not to affect the variability in the response. The calculation of the error variance depends upon the design of the assay and the form of the adjustment for any missing values. Each response is first converted to the unit y used in computing the potency. Determine a single error variance from the combined deviations of the y's around their respective means for each dosage level, summed over all levels. Doubtful values of y may be tested as described under *Rejection of Outlying or Aberrant Observations*, and proved outliers may be replaced as missing values (see *Replacement of Missing Values*).

In the simplest design, the units of response are assigned at random to each dosage level, as in the assay for corticotropin. If a missing value is replaced by adding the mean of the remaining y's at any given dosage level to their total, the degrees of freedom (n) in the error variance are reduced by one for each replacement but no other change is needed in the calculation. Assuming that f is then the same for all doses or groups, compute the error variance from the variation within doses of all the y's as

$$s^2 = \{\Sigma y^2 - \Sigma T_t^2/f\}/n, \qquad (13)$$

where T_t is the total at each dose of the f values of y, there are k totals T_t and the degrees of freedom $n = \Sigma f - k$, with Σf diminished by 1 for each replacement.

If variations in f are adjusted by subtracting a group mean from its group total, compute the error variance from the observed y's and the *unadjusted* totals T_t as

$$s^2 = \{\Sigma y^2 - \Sigma(T_t^2/f)\}/n, \qquad (14)$$

where $n = \Sigma f - k$.

In the calculation of the result of an assay using the coefficients of Table 6 or 8, s^2 may be computed from the response y for each of the h' preparations, including the h Unknowns and the corresponding dosage levels of the Standard. For each preparation, compute $T' = \Sigma y$ and the slope factor $T_b' = \Sigma(x_1 y)$ where the values of x_1 are the factorial coefficients for the Standard in the appropriate row b of Table 6 or 8. The error variance for the assay is

$$s^2 = \{\Sigma y^2 - \Sigma T'^2/k - 2(\Sigma T_b')^2/h'e_b f\}/n, \qquad (15)$$

where the degrees of freedom $n = h'(k - 1) - 1$, and e_b is the e_i from the same table and row as the coefficients x_1.

The Error Variance in Restricted Designs—In some assays, the individual responses occur in randomized sets of three or more. Examples of sets are litter mates in the assay of vitamin D, the cleared areas within each plate in an antibiotic assay, and the responses following four successive pairs of injections in the vasopressin assay. Arrange the individual y's from these assays in a 2-way table, in which each column represents a different treatment or dose and each row a randomized set. Losses may be replaced as described under *Replacement of Missing Values.* The k column totals are the T_t's required for the analysis of balanced designs. The f row totals (T_r) represent a source of variation that does not affect the estimated potency and hence is excluded from the assay error. Compute the approximate error variance from the squares of the individual y's and of the marginal totals as

$$s^2 = \{\Sigma y^2 - \Sigma T_r^2/k - \Sigma T_t^2/f + T^2/N\}/n, \qquad (16)$$

where $T = \Sigma T_r = \Sigma T_t$, and the $n = (k - 1)(f - 1)$ degrees of freedom must be diminished by one for any gap in the original table that has been filled by computation.

When the order of treatment is an additional potential source of variation, its effect can be corrected by the dose regimen for a series of n' Latin squares with k rows in common, such as that for the two Latin squares in the dose regimens 1 to 4 and 5 to 8 in the assay of Glucagon for Injection. List the observed responses y of each test animal in a separate column in the order of dosing. The responses to each of the k doses then occur equally often in each of the k rows and of the $n'k$ columns, where n' is the number of Latin squares. Total the responses y in each row (T_r) in each column (T_c), and, in a separate listing, for each dose or treatment (T_t). An occasional lost reading may be replaced by Equation 1a as described under *Replacement of Missing Values.* Compute the error variance from the squares of the individual y's and of the marginal and treatment totals as

$$s^2 = \{\Sigma y^2 - \Sigma T_r^2/n'k - \Sigma T_c^2/k - \Sigma T_t^2/n'k + 2T^2/N\}/n, \qquad (16a)$$

where $T = \Sigma y = \Sigma T_r = \Sigma T_c = \Sigma T_t$, $N = n'k^2$, and the $n = (k - 1)(n'k - 2)$ degrees of freedom must be diminished by one for any gap in the original table that has been filled by computation.

In assays where the reactions occur in pairs, the differences between test animals or paired reactions are segregated automatically by calculating the assay with the difference within a pair as the response. With insulin, the response is the difference y in the blood sugar of a single rabbit following two injections (see *Insulin Assay* ⟨121⟩). After adjustment for rabbits lost during the assay, compute the error variance of y from the responses in all four groups and from the group totals $T_i = T_1$ to T_4 as

$$s^2 = \{\Sigma y^2 - \Sigma T_i^2/f\}/n, \qquad (17)$$

where the number of rabbits f is the same in each group and the degrees of freedom, $n = 4(f - 1)$, are reduced by one for each replacement of a rabbit lost during the assay. In the Oxytocin Injection assay, each y represents the difference between the blood pressure response to a dose of the Unknown and the average for the two adjacent doses of Standard. Compute the error variance of y as

$$s^2 = \{\Sigma y^2 - (T_1^2 + T_2^2)/f\}/n \qquad (18)$$

with $n = 2(f - 1)$ degrees of freedom, where T_1 is the total of the y's for the low dose of the Unknown and T_2 that for the high dose.

In a microbial assay calculated by interpolation from a standard curve, convert each difference between two paired responses to units of log-dose, X, by the use of Equation 7. With each difference X as the unit, a composite s^2 is computed from the variation in the f values of X for each *Unknown*, totaled over the h *Unknowns* in the assay, as

$$s^2 = \{\Sigma X^2 - \Sigma(T_x^2/f)\}/n, \qquad (19)$$

where $T_x = \Sigma X$ for a single *Unknown* and the degrees of freedom $n = \Sigma f - h$.

Tests of Assay Validity—In addition to the specific requirements in each monograph and a combined log-dose response curve with a significant slope (see the statistic C in the next section), two conditions determine the validity of an individual factorial assay: (1) the log-dose response curve for the Unknown must parallel that for the Standard within the experimental error, and (2) neither curve may depart significantly from a straight line. When the assay has been completely randomized or consists of randomized sets, the necessary tests are computed with the factorial coefficients for *ab*, *q*, and *aq* from Tables 6 to 8 and the treatment totals T_t. Sum the products of the coefficients in each row by the corresponding T_t's to obtain the product total T_i, where the subscript *i* stands in turn for *ab*, *q*, and *aq*, respectively. Each of the three ratios, $T_i^2/e_i f$, is computed with the corresponding value of e_i from the table and with *f* equal to the number of *y*'s in each T_t. That in row *ab* tests whether the dosage-response lines are parallel, and is the only test available in a 2-dose assay. With three or more doses of both preparations, that in row *q* is a test of combined curvature in the same direction, and in row *aq* of separate curvatures in opposite directions. If any ratio in a 3- or 4-dose assay exceeds s^2 as much as three-fold, compute

$$F_3 = \Sigma(T_i^2/e_i f)/3s^2. \qquad (20)$$

For a 2-dose assay, compute instead

$$F_1 = T_{ab}^2/e_{ab} f s^2, \qquad (21)$$

and for a 3,2 assay (Table 7) determine

$$F_2 = \Sigma(T_i^2/e_i f)/2s^2. \qquad (22)$$

For a valid assay, F_1, F_2, or F_3 does not exceed the value given in Table 9 (at odds of 1 in 20) for the degrees of freedom *n* in s^2.

An assay may fail the test for validity and still provide a contributory estimate of potency that can be combined profitably with the result of a second assay of the same Unknown, as described in a later section. An end dosage level for either the Standard or the Unknown, or both, may fall outside the linear zone. With three or more dosage levels and relatively large values of T_a, T_{ab}, and T_{aq}, the total response T_t at an end dose of one preparation may approach an upper or lower limit and be responsible for the large values of T_{ab} and T_{aq}. This T_t may be omitted and the assay recomputed with the appropriate design in Table 7. If the assay then meets the test in Equation 20, or 22, the resulting potency, *M*, may be combined with that of a second assay in computing the log-potency of the Unknown (see under *Combination of Independent Assays*). If T_a is not significant but T_q shows significant combined curvature, the largest (or smallest) dose of both preparations may be too large (or too small). Their omission may lead to a valid assay with the factorial coefficients for the next smaller design in Table 6 or 8. A statistically significant T_q or $\Sigma T_q'$ may be neglected and all dosage levels retained without biasing the computed log-potency *M'* and its confidence interval by more than 5% when the following inequality is true:

$$T_b^2/e_b > 100T_q^2/e_q$$

or

$$(\Sigma T_b')^2/e_b > 100(\Sigma T_q')^2/e_q, \qquad (23)$$

where each T_b' and T_q' is computed with the T_t's (or *y*'s) for a single preparation multiplied by the coefficients for the Standard in rows *b* and *q*, respectively. If both T_a and T_{ab} are significant in a 2-dose assay, one T_t may be outside the linear zone. Sometimes a preliminary or contributory estimate of potency can be computed from the remaining three values of T_t and the first design in Table 7. In assays of insulin and of other drugs in which the responses are paired, the test for parallelism is so insensitive that it is omitted. If the tubes in each set are arranged systematically instead of at random in a microbial assay, the tests for validity may be subject to bias from positional effects.

The Confidence Interval and Limits of Potency

A bioassay provides an estimate of the true potency of an Unknown. This estimate falls within a confidence interval, which is computed so that the odds are not more than 1 in 20 (P = 0.05) that the true potency either exceeds the upper limit of the confidence interval or is less than its lower limit. Since this interval is determined by a number of factors that may influence the estimate of potency, the required precision for most bioassays is given in the monograph in terms of the confidence interval, related either to the potency directly or to its logarithm.

General Calculation—Despite their many forms, bioassays fall into two general categories: (1) those where the log-potency is computed directly from a mean or a mean difference, and (2) those where it is computed from the ratio of two statistics.

(1) When the log-potency of an assay is computed as the mean of several estimated log-potencies that are approximately equal in precision, the log-confidence interval is

$$L = 2st\sqrt{k}, \qquad (24)$$

where *s* is the standard deviation of a single estimated log-potency, *t* is read from Table 9 with the *n* degrees of freedom in *s*, and *k* is the number of estimates that have been averaged. The same equation holds where the log-potency is computed as the mean $\bar{x}$ of *k* differences *x*, with *s* the standard deviation of a single *x*. In either case, the estimated log-potency *M* is in the center of its confidence interval, so that its confidence limits are

$$X_M = M + ½L \text{ and } M - ½L, \text{ or } X_M = M \pm ½L. \qquad (25)$$

Table 9

Values of *t*, t^2, F_i and χ^2 for different degrees of freedom *n* that will be exceeded with a probability P = 0.05 (or 0.95 for confidence intervals).†

n	t	$t^2 = F_1$	F_2	F_3	χ^2	n	t	$t^2 = F_1$	F_2	F_3	χ^2
1	12.706	161.45	—	—	3.84	19	2.093	4.381	3.52	3.13	30.1
2	4.303	18.51	19.00	19.16	5.99	20	2.086	4.351	3.49	3.10	31.4
3	3.182	10.128	9.55	9.28	7.82	21	2.080	4.325	3.47	3.07	32.7
4	2.776	7.709	6.94	6.59	9.49	22	2.074	4.301	3.44	3.05	33.9
5	2.571	6.608	5.79	5.41	11.07	23	2.069	4.279	3.42	3.03	35.2
6	2.447	5.987	5.14	4.76	12.59	24	2.064	4.260	3.40	3.01	36.4
7	2.365	5.591	4.74	4.35	14.07	25	2.060	4.242	3.38	2.99	37.7
8	2.306	5.318	4.46	4.07	15.51	26	2.056	4.225	3.37	2.98	38.9
9	2.262	5.117	4.26	3.86	16.92	27	2.052	4.210	3.35	2.96	40.1
10	2.228	4.965	4.10	3.71	18.31	28	2.048	4.196	3.34	2.95	41.3
11	2.201	4.844	3.98	3.59	19.68	29	2.045	4.183	3.33	2.93	42.6
12	2.179	4.747	3.89	3.49	21.03	30	2.042	4.171	3.32	2.92	43.8
13	2.160	4.667	3.81	3.41	22.36	40	2.021	4.085	3.23	2.84	55.8
14	2.145	4.600	3.74	3.34	23.68	60	2.000	4.001	3.15	2.76	79.1
15	2.131	4.543	3.68	3.29	25.00	120	1.980	3.920	3.07	2.68	146.6
16	2.120	4.494	3.63	3.24	26.30	∞	1.960	3.841	3.00	2.60	
17	2.110	4.451	3.59	3.20	27.59						
18	2.101	4.414	3.55	3.16	28.87						

† Adapted from portions of Tables III to V of "Statistical Tables for Biological, Agricultural and Medical Research," by R. A. Fisher and F. Yates, published by Oliver and Boyd, Ltd., Edinburgh.

The upper and lower limits are converted to their antilogarithms to obtain the limits as explicit potencies.

(2) More often, the log-potency or potency is computed from a ratio, and in these cases the length of the confidence interval is typified by the log-interval in the equation

$$L = 2\sqrt{(C - 1)(CM'^2 + c'i^2)}, \tag{26}$$

where M' is the log-relative potency as defined (see *Calculation of Potency from a Single Assay*), i is the log-interval between successive doses, and c' is a constant characteristic of the assay procedure. The remaining term C depends upon the precision with which the slope of the dosage-response curve has been determined. (This is sometimes expressed in terms of $g = (C - 1)/C$.) In factorial assays, it is computed as

$$C = T_b^2/(T_b^2 - e_b f s^2 t^2), \tag{27}$$

where s^2 is the error variance of a single observation, t^2 is read from Table 9 with the degrees of freedom in s^2, f is the number of responses in each T_t used in calculating T_b, and T_b and e_b are computed with the factorial coefficients for row b in Tables 6 to 8. The s^2 in Equation 26 depends upon the design of the assay, as indicated for each drug in the next section. In a valid assay, C is a positive number.

In an assay of two or more Unknowns against a common Standard, all with dosage-response curves that are parallel within the experimental error, C may be computed with the error variance s^2 for the assay and with the assay slope as

$$C = (\Sigma T_b')^2/\{(\Sigma T_b')^2 - e_b f h' s^2 t^2/2\}. \tag{28}$$

The slope factor $T_b' = \Sigma(x_1 T_t)$ or $\Sigma(x_1 y)$ for each of the h' preparations, including the Standard, is computed with the factorial coefficients x_1 for the Standard in the appropriate row b of Table 6 or 8. If a treatment total T_t includes one or more replacements for a missing response, replace $e_b f$ in Equation 27, or $e_b fh'/2$ in Equation 28, by $f^2\Sigma(x_1^2/f')$, where each x_1 is a factorial coefficient in row b of Tables 6 to 8, in this chapter, and f' is the number of responses in the corresponding T_t *before* adding the replacement. With this C, compute the confidence interval as

$$L = 2\sqrt{(C - 1)(CM'^2 + c'i^2h'/2)}. \tag{29}$$

In assays computed from a ratio, the most probable log-potency M is not in the exact center of the confidence interval. The upper and lower confidence limits in logarithms are

$$X_M = \log R + CM' + \tfrac{1}{2}L \text{ and } \log R + CM' - \tfrac{1}{2}L. \tag{30}$$

C is often very little larger than unity, and the more precise the assay, the more nearly C approaches 1 exactly. $R = z_S/z_U$ is the ratio of corresponding doses of the Standard and of the Unknown or the assumed potency of the Unknown. The upper and lower confidence limits in log-potencies are converted separately to their antilogarithms to obtain the corresponding potencies.

Confidence Intervals for Individual Assays—Since the confidence interval may vary in detail from the above general patterns, compute it for each assay by the special directions given under the name of the substance in the paragraphs following.

Antibiotic Assays—The confidence interval may be computed by Equations 24 and 25.

Calcium Pantothenate—For log-potencies obtained by interpolation from the Standard curve, the confidence interval may be computed with Equations 19 and 24. For log-potencies calculated with Equation 8 or 10, s^2 may be computed with Equation 15, C with Equation 27 or 28, and the confidence interval L with Equation 26 or 29.

Corticotropin Injection—Compute the log confidence interval by Equations 26 and 27, with the coefficients and constants in Table 6 for a 3-dose assay, and s^2 as determined by Equation 13 or 14.

Digitalis—Compute the confidence interval as

$$L = 2\sqrt{(C - 1)\{C(\bar{z}_S/\bar{z}_U)^2 + f_U/f_S\}}, \tag{31}$$

where f_U and f_S are the number of observations on the Unknown and on the Standard, and

$$C = \bar{z}_U^2/(\bar{z}_U^2 - s^2t^2/f_U) \tag{32}$$

is determined with s^2 from Equation 11. The confidence limits for the potency in USP Units are then

$$X_{P*} = R\{C(\bar{z}_S/\bar{z}_U) \pm \tfrac{1}{2}L\}, \tag{33}$$

in which R is as defined in the *Glossary of Symbols*.

Glucagon for Injection—Compute the error variance s^2 by Equation 15a, C by Equation 27 with $e_b f = 16n'$, and the log confidence interval L by Equation 26 with $c'i^2 = 0.09062$.

Chorionic Gonadotropin—Proceed as directed under *Corticotropin Injection*.

Heparin Sodium—If two independent determinations of the log-potency M differ by more than 0.05, carry out additional assays and compute the error variance among the N values of M as

$$s^2 = \{\Sigma M^2 - (\Sigma M)^2/N\}/n \tag{34}$$

with $n = N - 1$ degrees of freedom. Given this value, determine the confidence interval in logarithms (L) by Equation 24.

Insulin Injection—Compute the error variance (s^2) of y by Equation 16 and C as

$$C = T_b^2/(T_b^2 - s^2t^2N), \tag{35}$$

where t^2 from Table 9 depends upon $n = 4(f - 1)$ degrees of freedom in s^2 and $N = 4f$ is the total number of differences in the four groups. By Equation 26, compute the confidence interval L in logarithms, where $c'i^2 = 0.09062$. The upper and lower confidence limits in USP Units of insulin are given by the antilogarithms of X_M from Equation 30.

Oxytocin Injection—Compute the approximate log confidence interval by Equation 26, in which

$$C = (T_2 - T_1)^2/\{(T_2 - T_1)^2 - 4(f + 1)s^2t^2/3\}, \tag{36}$$

where s^2 is defined by Equation 18, and

$$c' = (4f - 1)/8(f + 1). \tag{37}$$

Tubocurarine Chloride Injection—Compute the error variance by Equation 12, and the confidence interval by Equation 24.

Vasopressin Injection—Compute the error variance s^2 by Equation 16, C by Equation 35, and the log confidence interval by Equation 26, where $c' = 1$ and i is the log-interval separating the two dosage levels.

Vitamin B_{12} Activity—Proceed as directed under *Calcium Pantothenate*.

Combination of Independent Assays

When the method permits, additional animals can be added to an insufficiently precise assay until the combined results reduce the confidence interval within the limits specified in the monograph. Where two or more independent assays are required, each leading to a log-potency M, the M's are combined in determining the *weighted mean potency* of the Unknown. Except in the Heparin Sodium assay, where the log-potencies are weighted equally, the relative precisions of the two or more independent M's determine the weight assigned to each value in computing their mean and its confidence interval.

Before combining two or more separate estimates of M, test their mutual consistency. If the M's are consistent, their respective confidence intervals will overlap. Where the intervals do not overlap or where the overlap is small, compute an approximate χ_M^2. Assign each of the h individual assays a weight w, defined as

$$w = 4t^2/L^2, \tag{38}$$

where the length of the confidence interval L is computed with the appropriate equation from the preceding section, and t^2 is read from Table 9 for the degrees of freedom n in the error variance of the assay. Sum the individual weights to obtain Σw. Then an approximate χ^2 with $h - 1$ degrees of freedom is determined as

$$\text{Approx. } \chi_M^2 = \Sigma(wM^2) - \{\Sigma(wM)\}^2/\Sigma w. \tag{39}$$

For two assays with log-potencies M_1 and M_2 and weights w_1 and w_2, Equation 35 reduces to

$$\text{Approx. } \chi_M{}^2 = \frac{w_1 w_2 (M_1 - M_2)^2}{w_1 + w_2}, \qquad (40)$$

with one degree of freedom. If the approximate $\chi_M{}^2$ is well under the critical value for χ^2 in Table 9, use the weights w in computing the mean log-potency $\overline{M}$ and its confidence interval, L. If $\chi_M{}^2$ approaches or exceeds this critical value, use instead the semi-weights w' (Equation 47) when computing $\overline{M}$.

Compute the mean log-potency $\overline{M}$ of two or more mutually consistent assays as

$$\overline{M} = \Sigma(wM)/\Sigma w. \qquad (41)$$

This is the most probable single value within a combined confidence interval of length L_c, defined as the square root of

$$L_c{}^2 = \frac{4t_L{}^2}{\Sigma w}\left\{1 + \frac{4}{\Sigma^2 w}\Sigma\frac{w(\Sigma w - w)}{n'}\right\}, \qquad (42)$$

where each $n' = n - 4(h - 2)/(h - 1)$ and $t_L{}^2$ is interpolated from Table 9 with the degrees of freedom

$$n_L = \Sigma^2 w/\Sigma(w^2/n).$$

For two assays ($h = 2$) with log-potencies M_1 and M_2 and weights w_1 and w_2, respectively, the above equation may be rewritten as

$$L_c{}^2 = \frac{4t_L{}^2}{\Sigma w}\left\{1 + \frac{4w_1w_2}{\Sigma^2 w}\left[\frac{1}{n_1} + \frac{1}{n_2}\right]\right\}, \qquad (43)$$

where $\Sigma w = w_1 + w_2$. Where L_c, the confidence interval for a combined estimate, does not exceed the requirement in a monograph, upper and lower confidence limits are taken $\frac{1}{2}L_c$ above and below $\overline{M}$, to obtain approximately a 95% confidence interval.

Where the variation in the assayed potency between the h independent determinations, as tested by $\chi_M{}^2$, approaches or exceeds $P = 0.05$, the several estimates are assigned semi-weights w'. From the weight w, compute the variance of each M as

$$V = 1/w = L^2/4t^2. \qquad (44)$$

Calculate the variance of the heterogeneity between assays as

$$v = \frac{\Sigma M^2 - (\Sigma M)^2/h}{h - 1} - \frac{\Sigma V}{h}, \qquad (45)$$

or if $h = 2$,

$$v = \frac{(M_1 - M_2)^2}{2} - \frac{V_1 + V_2}{2}. \qquad (46)$$

Where V varies so markedly that v calculated as above is a negative number, compute instead an approximate v by omitting the term following the minus sign in Equations 45 and 46. A semi-weight is defined as

$$w' = 1/(V + v). \qquad (47)$$

Substitute w' and $\Sigma w'$ for w and Σw in Equation 41 to obtain the semi-weighted mean $\overline{M}$. This falls near the middle of a confidence interval of approximate length L_c', where

$$L_c'^2 = 4t^2/\Sigma w' \qquad (48)$$

and t^2 from Table 9 has Σn degrees of freedom.

Where $\chi_M{}^2$ in Equation 39, from $h = 4$ or more estimates of M, exceeds the critical level in Table 9 by more than 50%, and the weights w differ by less than 30%, the h estimates of M may be checked for a suspected outlier with Table 1. Where significant, the outlying M may be omitted in computing $\overline{M}$ with w'.

Where the potency of a drug is determined repeatedly in a given laboratory by the same bioassay method, successive determinations of both the slope b and the error variance s^2 may scatter randomly within the sampling error about a common value for each parameter. Plotting estimates from successive assays on a quality control chart for each statistic and computing the mid-value and control limits defining the allowable random variation make it possible to check continuously the consistency of an assay technique. Where estimates of b and s^2 from a single assay fall within the control limits, they may be replaced by their laboratory means. Reject any assay in which these statistics fall outside the control limits, or accept it only after close scrutiny with respect to its validity.

Joint Assay of Several Preparations

Each monograph describes the assay of a single Unknown against the Standard. Although not provided explicitly, several different Unknowns are often included in the same assay and each is compared separately with the same responses to the Standard. This fact may warrant increasing the number of observations with the Standard. Given f observations at each dosage level of each of h different Unknowns, the number of observations at each dosage level of the Standard may be increased advantageously, if h is large, to $f\sqrt{h}$. This rule can be applied only approximately where litter differences or their equivalent must be segregated, and in any case is merely suggestive.

If all of several assays conducted concurrently meet the requirements for validity, and have linear log-dose response curves with the same slope b and the same error variance s^2 about these lines, these two statistics may be considered as characteristic of the assay. Combining all of the evidence from the same assay into a single value of the assay slope results in a more stable and reliable estimate of b than if each Unknown were analyzed independently. The degrees of freedom and reliability of the error variance s^2 can be increased similarly. Confidence intervals computed with these composite values for b and s^2 are smaller on the average than if based upon only part of the relevant data. For the calculation or application of such assay estimates, see Equations 10, 15, 16, 19, 28, and 29. The potency estimated with a slope computed from a single Unknown and the Standard agrees within a fraction of the confidence interval with that computed from the combined slope for the entire assay. Since it is based upon more evidence, the latter is considered the better estimate.

GLOSSARY OF SYMBOLS

A absorbance for computing % reduction in bacterial growth from turbidimetric readings.

b slope of the straight line relating response (y) to log-dose (x) [Equations 2b, 4, 5, 6].

c constant for computing M' with Equations 8 and 10.

c' constant for computing L with Equations 26 and 29.

ci constant for computing M' when doses are spaced as in Table 8.

$c'i^2$ constant for computing L when doses are spaced as in Table 8.

C term measuring precision of the slope in a confidence interval [Equations 27, 28, 35, 36].

χ^2 statistical constant for testing significance of a discrepancy [Table 9].

$\chi_M{}^2$ χ^2 testing the disagreement between different estimates of log-potency [Equations 39, 40].

e_b e_i from row b in Tables 6 to 8.

$e_b'i$ multiple of $\Sigma(x - \overline{x})^2$ [Table 5; Equation 6].

e_i sum of squares of the factorial coefficients in each row of Tables 6 to 8.

e_q e_i from row q in Tables 6 to 8.

f number of responses at each dosage level of a preparation; number of replicates or sets.

f_S number of observations on the Standard.

f_U number of observations on the Unknown.

F_1 to F_3 observed variance ratio with 1 to 3 degrees of freedom in numerator [Table 9].

G_1, G_2, and G_3 relative gap in test for outlier [Table 1].

h number of Unknowns in a multiple assay.

h' number of preparations in a multiple assay, including the Standard and h Unknowns; i.e., $h' = h + 1$.

i interval in logarithms between successive log-doses, the same for both *Standard* and *Unknown*.

k number of estimated log-potencies in an average [Equation 24]; number of treatments or doses [Table 4; Equations 1, 13, 15, 16]; number of ranges or groups in a series [Table 2]; number of rows, columns, and doses in a single Latin square [Equations 1a, 16a].

L length of the confidence interval in logarithms [Equations 24, 26, 29, 38], or in terms of a proportion of the relative potency of the dilutions compared [Equations 31, 33].

L_c length of a combined confidence interval [Equations 42, 43].

L_c' length of confidence interval for a semi-weighted mean $\overline{M}$ [Equation 48].

LD_{50} lethal dose killing an expected 50% of the animals under test [Equation 2c].

M log-potency [Equation 2].

M' log-potency of an Unknown, relative to its assumed potency.

$\overline{M}$ mean log-potency.

n degrees of freedom in an estimated variance s^2 or in the statistic t or χ^2.

n' number of Latin squares with rows in common [Equations 1a, 16a].

N number; e.g., of observations in a gap test [Table 1], or of responses y in an assay [Equation 16].

P probability of observing a given result, or of the tabular value of a statistic, usually $P = 0.05$ or 0.95 for confidence intervals [Tables 1, 2, 9].

P_* potency, P_* = antilog M or computed directly.

R ratio of a given dose of the Standard to the corresponding dose of the Unknown, or assumed potency of the Unknown [Equations 2, 30, 33].

R_* ratio of largest of k ranges in a series to their sum [Table 2].

$s = \sqrt{s^2}$ standard deviation of a response unit, also of a single estimated log-potency in a direct assay [Equation 24].

s^2 error variance of a response unit.

S_i a log-dose of Standard [Tables 6, 7].

Σ "the sum of."

t Student's t for n degrees of freedom and probability P = 0.05 [Table 9].

T total of the responses y in an assay [Equation 16].

T' incomplete total for an assay in randomized sets with one missing observation [Equation 1].

T_1 $\Sigma(y)$ for the animals injected with the Standard on the first day [Equations 18, 36].

T_2 $\Sigma(y)$ for the animals injected with the Standard on the second day [Equations 18, 36].

T_a T_i for the difference in the responses to the Standard and to the Unknown [Tables 6 to 8].

T_{ab} T_i for testing the difference in slope between Standard and Unknown [Tables 6 to 8].

T_{aq} T_i for testing opposed curvature in the curves for Standard and Unknown [Tables 6 to 8].

T_b T_i for the combined slope of the dosage-response curves for Standard and Unknown [Tables 6 to 8].

T_b' $\Sigma(x_1T_t)$ or $\Sigma(x_1y)$ for computing the slope of the log-dose response curve [Equations 10, 23, 28].

T_i sum of products of T_t multiplied by the corresponding factorial coefficients in each row of Tables 6 to 8.

T_q T_i for testing similar curvature in the curves for Standard and Unknown [Tables 6 to 8].

T_r row or set total in an assay in randomized sets [Equation 16].

T_r' incomplete total for the randomized set with a missing observation in Equation 1.

T_t total of f responses y for a given dose of a preparation [Tables 6 to 8; Equations 6, 13, 14, 16].

T_t' incomplete total for the treatment with a missing observation in Equation 1.

U_i a log-dose of Unknown [Tables 6 to 8].

v variance for heterogeneity between assays [Equation 45].

$V = 1/w$ variance of an individual M [Equations 44 to 47].

w weight assigned to the M for an individual assay [Equation 38], or to a probit for computing an LD_{50} [Equations 2a, 2b].

w' semi-weight of each M in a series of assays [Equations 47, 48].

x a log-dose of drug in a bioassay [Equation 5]; also the difference between two log-threshold doses in the same animal [Equation 12].

x_* coefficients for computing the lowest and highest expected responses Y_L and Y_H in a log-dose response curve [Table 4; Equation 3].

x_1 a factorial coefficient that is a multiple of $(x - \overline{x})$ for computing the slope of a straight line [Table 5; Equation 6].

$\overline{x}$ mean log-dose [Equation 5].

$\overline{x}_S$ mean log-dose for Standard [Equation 9].

$\overline{x}_U$ mean log-dose for Unknown [Equation 9].

X log-potency from a unit response, as interpolated from a standard curve [Equations 7a, 7b, 19].

X_M confidence limits for an estimated log-potency M [Equations 25, 30].

X_{P_*} confidence limits for a directly estimated potency P_* (see *Digitalis* assay) [Equation 33].

y an observed individual response to a dose of drug in the units used in computing potency and the error variance [Equations 13 to 16]; a unit difference between paired responses in 2-dose assays [Equations 17, 18].

$y_1 \ldots y_N$ observed responses listed in order of magnitude, for computing G_1, G_2, or G_3 in Table 1.

$\underline{y}'$ replacement for a missing value [Equation 1].

$\overline{y}$ mean response in a set or assay [Equation 5].

$\overline{y}_t$ mean response to a given treatment [Equations 3, 6].

Y a response predicted from a dosage-response relationship, often with qualifying subscripts [Equations 3 to 5].

z threshold dose determined directly by titration (see *Digitalis* assay) [Equation 11].

$\overline{z}$ mean threshold dose in a set (see *Digitalis* assay) [Equations 31, 32, 33].

⟨115⟩ DEXPANTHENOL ASSAY

The following procedure is provided for the determination of dexpanthenol as an ingredient of multiple-vitamin preparations. It is applicable also to the determination of the dextrorotatory component of racemic panthenol and of other mixtures containing dextrorotatory panthenol.

Media may be prepared as described hereinafter, or dehydrated mixtures yielding similar formulations may be used provided that, when reconstituted as directed by the manufacturer or distributor, they have growth-promoting properties equal to or superior to those obtained from the formulas given herein.

Reference Standard—*USP Dexpanthenol Reference Standard.*

Standard Stock Solution of Dexpanthenol—Dissolve an accurately weighed quantity of USP Dexpanthenol RS in water, dilute with water to obtain a solution having a known concentration of about 800 µg per mL, and mix. Store in a refrigerator, protected from light, and use within 30 days.

Standard Preparation—On the day of the assay, prepare a water dilution of the *Standard Stock Solution of Dexpanthenol* to contain 1.2 µg of dexpanthenol per mL.

Assay Preparation—Proceed as directed in the individual monograph for preparing a solution expected to contain approximately the equivalent of the dexpanthenol concentration in the *Standard Preparation.*

Modified Pantothenate Medium—

Acid-hydrolyzed Casein Solution	25	mL
Cystine-Tryptophane Solution	25	mL
Polysorbate 80 Solution	0.25	mL
Dextrose, Anhydrous	10	g
Sodium Acetate, Anhydrous	5	g
Adenine-Guanine-Uracil Solution	5	mL
Riboflavin–Thiamine Hydrochloride–Biotin Solution	5	mL
Para-aminobenzoic Acid–Niacin–Pyridoxine Hydrochloride Solution	5	mL
Salt Solution A	5	mL
Salt Solution B	5	mL
Pyridoxal–Calcium Pantothenate Solution	5	mL
Polysorbate 40–Oleic Acid Solution	5	mL

Dissolve the anhydrous dextrose and sodium acetate in the solutions previously mixed, and adjust with 1 *N* sodium hydroxide to a pH of 6.8. Finally, dilute with water to 250 mL, and mix.

Double-strength Modified Pantothenate Medium—Prepare as directed under *Modified Pantothenate Medium,* but make the final dilution to 125 mL instead of 250 mL. Prepare fresh.

Acid-hydrolyzed Casein Solution—Mix 100 g of vitamin-free casein with 500 mL of 6 *N* hydrochloric acid, and reflux the mixture for 8 to 12 hours. Remove the hydrochloric acid from the mixture by distillation under reduced pressure until a thick paste remains. Redissolve the resulting paste in about 500 mL of water, adjust the solution with 1 *N* sodium hydroxide to a pH of 3.5 ± 0.1, and add water to make 1000 mL. Add 20 g of activated charcoal, stir for 1 hour, and filter. Repeat the treatment with activated charcoal. Store under toluene in a refrigerator at a temperature not below 10°. Filter the solution if a precipitate forms during storage.

Cystine-Tryptophane Solution—Suspend 4.0 g of L-cystine and 1.0 g of L-tryptophane (or 2.0 g of D,L-tryptophane) in 700 mL to 800 mL of water, heat to 75 ± 5°, and add dilute hydrochloric acid (1 in 2) dropwise, with stirring, until the solids are dissolved. Cool, add water to make 1000 mL, and mix. Store under toluene in a refrigerator at a temperature not below 10°.

Adenine-Guanine-Uracil Solution—Dissolve 200 mg each of adenine sulfate, guanine hydrochloride, and uracil, with the aid of heat, in 10 mL of 4 *N* hydrochloric acid, cool, add water to make 200 mL, and mix. Store under toluene in a refrigerator.

Polysorbate 80 Solution—Dissolve 25 g of polysorbate 80 in alcohol to make 250 mL, and mix.

Riboflavin–Thiamine Hydrochloride–Biotin Solution—Prepare a solution containing, in each mL, 20 µg of riboflavin, 10 µg of thiamine hydrochloride, and 0.04 µg of biotin, by dissolving riboflavin, thiamine hydrochloride, and biotin in 0.02 *N* acetic acid. Store, protected from light, under toluene in a refrigerator.

Para-aminobenzoic Acid–Niacin–Pyridoxine Hydrochloride Solution—Prepare a solution in neutral 25 percent alcohol to contain 10 µg of para-aminobenzoic acid, 50 µg of niacin, and 40 µg of pyridoxine hydrochloride in each mL. Store in a refrigerator.

Salt Solution A—Dissolve 25 g of monobasic potassium phosphate and 25 g of dibasic potassium phosphate in water to make 500 mL. Add 5 drops of hydrochloric acid, mix, and store under toluene.

Salt Solution B—Dissolve 10 g of magnesium sulfate, 0.5 g of sodium chloride, 0.5 g of ferrous sulfate, and 0.5 g of manganese sulfate in water to make 500 mL. Add 5 drops of hydrochloric acid, mix, and store under toluene.

Pyridoxal–Calcium Pantothenate Solution—Dissolve 40 mg of pyridoxal hydrochloride and 375 µg of calcium pantothenate in 10 percent alcohol to make 2000 mL, and mix. Store in a refrigerator, and use within 30 days.

Polysorbate 40–Oleic Acid Solution—Dissolve 25 g of polysorbate 40 and 0.25 g of oleic acid in 20 percent alcohol to make 500 mL, and mix. Store in a refrigerator, and use within 30 days.

Stock Culture of *Pediococcus acidilactici*—Dissolve in about 800 mL of water, with the aid of heat, 6.0 g of peptone, 4.0 g of pancreatic digest of casein, 3.0 g of yeast extract, 1.5 g of beef extract, 1.0 g of dextrose, and 15.0 g of agar. Adjust with 0.1 *N* sodium hydroxide or 0.1 *N* hydrochloric acid to a pH between 6.5 and 6.6, adjust the volume with water to 1000 mL, and mix. Add approximately 10-mL portions of the solution to culture tubes, place caps on the tubes, and sterilize at 121° for 15 minutes. Cool on a slant, and store in a refrigerator. Prepare a stock culture of *Pediococcus acidilactici** on a slant of this medium. Incubate at 35° for 20 to 24 hours, and store in a refrigerator. Maintain the stock culture by monthly transfer onto fresh slants.

Inoculum—Inoculate three 250-mL portions of *Modified Pantothenate Medium* from a stock culture slant, and incubate at 35° for 20 to 24 hours. Centrifuge the suspension from the combined portions, and wash the cells with *Modified Pantothenate Medium.* Resuspend the cells in sufficient *Modified Pantothenate Medium* so that a 1:50 dilution, when tested in a 13-mm diameter test tube, gives 80% light transmission at 530 nm. Transfer 1.2-mL portions of this stock suspension to glass ampuls, seal, freeze in liquid nitrogen, and store in a freezer. On the day of the assay, allow the ampuls to reach room temperature, mix the contents, and dilute 1 mL of thawed culture with sterile saline TS to 150 mL. [NOTE—This dilution may be altered, when necessary, to obtain the desired test response.]

Procedure—Prepare in triplicate a series of eight culture tubes by adding the following quantities of water to the tubes within a set: 5.0 mL, 4.5 mL, 4.0 mL, 3.5 mL, 3.0 mL, 2.0 mL, 1.0 mL, and 0.0 mL. To these same tubes, and in the same order, add 0.0 mL, 0.5 mL, 1.0 mL, 1.5 mL, 2.0 mL, 3.0 mL, 4.0 mL, and 5.0 mL of the *Standard Preparation.*

Prepare in duplicate a series of five culture tubes by adding the following quantities of water to the tubes within a set: 4.0 mL, 3.5 mL, 3.0 mL, 2.0 mL, and 1.0 mL. To these same tubes, and in the same order, add 1.0 mL, 1.5 mL, 2.0 mL, 3.0 mL, and 4.0 mL of the *Assay Preparation.*

Add 5.0 mL of *Double-strength Modified Pantothenate Medium* to each tube, and mix. Cover the tubes with metal caps, and sterilize in an autoclave at 121° for 5 minutes. Cool to room temperature in a chilled water bath, and inoculate each tube with 0.5 mL of the *Inoculum.* Allow to incubate at 37° for 16 hours. Terminate growth by heating to a temperature not below 80°, such as by steaming at atmospheric pressure in a suitable sterilizer, for 5 to 10 minutes. Cool, and concomitantly determine the percentage transmittance of the suspensions, in cells of equal pathlength, on a suitable spectrophotometer, at 530 nm.

Calculation—Draw a dose-response curve on arithmetic graph paper by plotting the average response, in percent transmittance, for each set of tubes of the standard curve against the standard level concentrations. The curve is drawn by connecting each adjacent pair of points with a straight line. From this standard curve, determine by interpolation the potency, in terms of dexpanthenol, of each tube containing portions of the *Assay Preparation.* Divide the potency of each tube by the amount of *Assay Preparation* added to it, to obtain the individual responses. Calculate the mean response by averaging the individual responses that vary from their mean by not more than 15%, using not less than half the total number of tubes. Calculate the potency of the portion of the material taken for assay, in terms of dexpanthenol, by multiplying the mean response by the appropriate dilution factor.

* American Type Culture Collection No. 8042 is suitable.

⟨121⟩ INSULIN ASSAY

The most prominent manifestation of insulin activity, an abrupt decrease in blood glucose, was the basis for biologic assay from the time of the first clinical use of insulin. The procedure, although relatively cumbersome, has the great merit of accurately reflecting the effect on the diabetic patient. Another attribute of insulin, that of reacting under in-vitro conditions to specific antibodies, the amounts of which are rendered measurable by means of radioactive isotopes, is the basis for a procedure that makes possible the rapid measurement of minute amounts of insulin.

Rabbit Blood-sugar Method

Reference Standard—*USP Insulin Reference Standard*—Preserve in a refrigerator, and after opening the ampul, store in a tight container.

Standard Solution—Dissolve a suitable quantity of USP Insulin RS, accurately weighed, in sufficient water, containing 0.1% to 0.25% (w/v) of either phenol or cresol, 1.4% to 1.8% (w/v) of glycerin, and sufficient hydrochloric acid to make a *Standard Solution* containing 40 USP Insulin Units per mL and having a pH between 2.5 and 3.5, unless otherwise directed in the individual monograph. Store in a cold place, protected from freezing, and use within 6 months.

Standard Dilutions—Dilute portions of the *Standard Solution* to make two solutions, one to contain 1.0 USP Insulin Unit per mL (*Standard dilution 1*), and the other to contain 2.0 USP Insulin Units per mL (*Standard dilution 2*). Use as a diluent a solution containing 0.1% to 0.25% (w/v) of either cresol or phenol, 1.4% to 1.8% (w/v) of glycerin, and sufficient hydrochloric acid to produce a pH between 2.5 and 3.5, unless otherwise directed in the individual monograph.

Assay Dilutions—Employing the same diluent used in preparing the *Standard Dilutions*, make two dilutions of the preparation to be assayed, one of which may be expected, on the basis of the assumed potency, to contain 1.0 USP Insulin Unit per mL (*Assay dilution 1*), and the other to contain 2.0 USP Insulin Units per mL (*Assay dilution 2*). In the case of neutral Insulin Injection, adjust to a pH of 2.5 to 3.5 prior to making the dilutions.

Doses of the Dilutions To Be Injected—Select on the basis of trial or experience the dose of the dilutions to be injected, the volume of which usually will be between 0.30 mL and 0.50 mL. For each animal the volume of the *Standard dilution* shall be the same as that of the *Assay dilution.*

The Animals—Select suitable, healthy rabbits each weighing not less than 1.8 kg. Keep the rabbits in the laboratory for not less than 1 week before use in the assay, maintaining them on an adequate uniform diet, with water available at all times except during the assay.

Procedure—Divide the rabbits into four equal groups of preferably not less than six rabbits each. On the preceding day, approximately 20 hours before the assay, provide each rabbit with an amount of food that will be consumed within 6 hours. Follow the same feeding schedule before each test day. During the assay, withhold all food and water until after the final blood specimen is taken. Handle the rabbits with care in order to avoid undue excitement, and inject subcutaneously the doses indicated in the following design, the Second Injection being made on the day after the First Injection or not more than 1 week later.

At 1 hour and 2½ hours after the time of injection obtain from each rabbit a suitable blood specimen from a marginal ear vein.

Determine the dextrose (glucose) content of the blood specimens by a suitable procedure, preferably one that depends upon the reduction of ferricyanide and is adapted to automatic handling. The following method may be used, all steps being carried out in a pre-determined sequence accurately reproduced for each blood specimen and for prepared solutions of dextrose of known concentration.

Pipet into separate, suitable vessels 0.8 mL of each blood specimen and 0.8 mL each of standard solutions containing, respectively, the following concentrations of dextrose: 0.25, 0.50, 0.75, 1.0, and 1.25 mg per mL. Into each vessel pipet 2.4 mL of saline TS, and mix. Place each vessel in a water bath maintained at 38°, and subject the diluted blood to dialysis across a semipermeable membrane for a sufficient time for a definite proportion of the dextrose to pass through the membrane into a solution, in saline TS, of potassium ferricyanide (1 in 1670) and sodium carbonate (1 in 50). Add a measured portion of potassium cyanide solution (1 in 200). Heat at a temperature of 95° for 5 minutes, cool to 40°, and determine the absorbance, at 420 nm, in a recording colorimeter. In a similar manner, determine the absorbances of solutions obtained from one or more standard solutions of dextrose of known concentration in the range of 25 mg to 125 mg per 100 mL.

Calculation—Calculate the response of each rabbit to each injection from the sum of the two blood-sugar values, and subtract its response to *Dilution 1* from that to *Dilution 2*, disregarding the chronological order in which the responses were observed, to obtain the individual differences, *y*, shown in the accompanying table.

Group	First Injection	Second Injection
1	Standard dilution 2	Assay dilution 1
2	Standard dilution 1	Assay dilution 2
3	Assay dilution 2	Standard dilution 1
4	Assay dilution 1	Standard dilution 2

Group	Differences	Individual Response	Total Response
		(y)	(T)
1	Standard 2 − Assay 1	y_1	T_1
2	Assay 2 − Standard 1	y_2	T_2
3	Assay 2 − Standard 1	y_3	T_3
4	Standard 2 − Assay 1	y_4	T_4

When the data for one or more rabbits are missing in an assay, allow for differences in the sizes of the groups by suitable means (see *Replacement of Missing Values* ⟨111⟩).

When the number of rabbits, *f*, carried through the assay is the same in each group, total the *y*'s in each group and compute $T_a = -T_1 + T_2 + T_3 - T_4$ and $T_b = T_1 + T_2 + T_3 + T_4$. The logarithm of the relative potency of the test dilutions is $M' = 0.301T_a/T_b$. The potency of the Injection in USP Units per mL equals the antilog ($\log R + M'$), where $R = v_S/v_U$, in which v_S is the number of USP Units per mL of the Standard dilution and v_U is the number of mL of Injection per mL of the Assay dilution.

Determine the confidence interval of the log-relative potency *M′* (see *Confidence Intervals for Independent Assays* ⟨111⟩). If the confidence interval is more than 0.082, which corresponds at $P = 0.95$ to confidence limits of about ±10% of the computed potency, repeat the assay until the combined data of the two or more assays, re-determined as described under *Combination of Independent Assays* ⟨111⟩, meet this acceptable limit.

⟨141⟩ PROTEIN—BIOLOGICAL ADEQUACY TEST

This test is intended for the evaluation of the biological adequacy, as an index to the completeness of the mixture of amino acids contained, of Protein Hydrolysate Injection.

Depletion Diet—

	Parts by Weight
Dextrin	83.9
Corn Oil	9.0
Salt Mixture	4.0
Agar	2.0
Cod Liver Oil	1.0
Choline Chloride	0.15
Inositol	0.10
Calcium Pantothenate	0.002
Niacinamide	0.0015
Riboflavin	0.0003
Pyridoxine	0.00025
Thiamine	0.0002
p-Aminobenzoic Acid	0.0002
Folic Acid	0.0002
Menadione	0.0002
Biotin	0.00002

Salt Mixture—Prepare the salt mixture specified in the *Depletion Diet* as follows:

Sodium Chloride	139.3 g
Potassium Biphosphate	389.0 g
Magnesium Sulfate, Anhydrous	57.3 g
Calcium Carbonate	381.4 g
Ferrous Sulfate	27.0 g
Manganese Sulfate	4.01 g
Potassium Iodide	0.79 g
Zinc Sulfate	0.548 g
Cupric Sulfate	0.477 g
Cobaltous Chloride	0.023 g

Place a portion of the weighed quantity of sodium chloride in a suitable mortar and add, with grinding, the potassium iodide. Set aside the mixture, and mix in a similar manner all the other salts with the remainder of the sodium chloride, adding finally the previously mixed sodium chloride and potassium iodide. Reduce the entire mixture to a fine powder (see *Powder Fineness* ⟨811⟩).

Control Nitrogen Supplement Mixture—Place 50 g of calcium caseinate and 46 g of anhydrous dextrose in a beaker, add sufficient water to make a paste, and finally add 1000 mL of water. Heat the solution between 70° and 82° for 5 minutes with stirring, and cool. Determine nitrogen on an aliquot using *Nitrogen Determination—Method I* or *Method II* ⟨461⟩. Store in a refrigerator. Mix before removing portions for analysis or use.

Depletion and Control Periods—Select a group of not less than six male rats 2 to 4 months of age and each weighing between 190 g and 225 g. Place the rats in individual cages with free access to water and the *Depletion Diet* for 12 days. Weigh the depleted rats, and discard any rat that weighs more than 90% of its starting weight.

For the next 3 days substitute as drinking water the *Control Nitrogen Supplement Mixture* in a quantity equivalent to 0.12 g of nitrogen per rat per day, diluted with water to 20 mL, and offered at the same time each morning either in a dish suitable for preventing spillage or in a reservoir fitted with a drinking tube. Remove all drinking water from the cages of the depleted rats during each feeding, and return it after the supplement has been consumed or is removed. On the third day, weigh each rat. Discard any rats that have not consumed all of the *Control Nitrogen Supplement Mixture*.

For the next 3 days, replace the *Control Nitrogen Supplement Mixture* with water ad libitum, and continue the rats on the *Depletion Diet*. Weigh the rats, and discard any that have not lost weight since the previous weighing.

Procedure—Assemble not less than six rats that have completed the depletion and control periods. For 5 days maintain the assembled rats on the *Depletion Diet* with a daily supplement of 20 mL, accurately measured, of a solution containing the Protein Hydrolysate Injection in an amount equivalent to 0.12 g of nitrogen offered each morning in the same way as the *Control Nitrogen Supplement Mixture* was offered previously. Withhold water for at least 2 hours prior to offering the supplement and for 4 hours afterward. Then if the supplement has been consumed, offer water ad libitum.

On the afternoon of the fifth day, weigh each rat, and compare the respective final and starting weights. Not fewer than 80% of the group of rats used gain weight or maintain their weight during the test.

⟨151⟩ PYROGEN TEST

The pyrogen test is designed to limit to an acceptable level the risks of febrile reaction in the patient to the administration, by injection, of the product concerned. The test involves measuring the rise in temperature of rabbits following the intravenous injection of a test solution and is designed for products that can be tolerated by the test rabbit in a dose not to exceed 10 mL per kg injected intravenously within a period of not more than 10 minutes. For products that require preliminary preparation or are subject to special conditions of administration, follow the additional directions given in the individual monograph or, in the case of antibiotics or biologics, the additional directions given in the federal regulations (see *Biologics* ⟨1041⟩).

Apparatus and Diluents—Render the syringes, needles, and glassware free from pyrogens by heating at 250° for not less than 30 minutes or by any other suitable method. Treat all diluents and solutions for washing and rinsing of devices or parenteral injection assemblies in a manner that will assure that they are sterile and pyrogen-free. Periodically perform control pyrogen tests on representative portions of the diluents and solutions for washing or rinsing of the apparatus. Where Sodium Chloride Injection is specified as a diluent, use Injection containing 0.9 percent of NaCl.

Temperature Recording—Use an accurate temperature-sensing device such as a clinical thermometer, or thermistor probes or similar probes that have been calibrated to assure an accuracy of ±0.1° and have been tested to determine that a maximum reading is reached in less than 5 minutes. Insert the temperature-sensing probe into the rectum of the test rabbit to a depth of not less than 7.5 cm, and, after a period of time not less than that previously determined as sufficient, record the rabbit's body temperature.

Test Animals—Use healthy, mature rabbits. House the rabbits individually in an area of uniform temperature between 20° and 23° and free from disturbances likely to excite them. The temperature varies not more than ±3° from the selected temperature. Before using a rabbit for the first time in a pyrogen test, condition it not more than seven days before use by a sham test that includes all of the steps as directed under *Procedure* except injection. Do not use a rabbit for pyrogen testing more frequently than once every 48 hours, nor prior to 2 weeks following a maximum rise of its temperature of 0.6° or more while being subjected to the pyrogen test, or following its having been given a test specimen that was adjudged pyrogenic.

Procedure—Perform the test in a separate area designated solely for pyrogen testing and under environmental conditions similar to those under which the animals are housed and free from disturbances likely to excite them. Withhold all food from the rabbits used during the period of the test. Access to water is allowed at all times, but may be restricted during the test. If rectal temperature-measuring probes remain inserted throughout the testing period, restrain the rabbits with light-fitting neck stocks that allow the rabbits to assume a natural resting posture. Not more than 30 minutes prior to the injection of the test dose, determine the "control temperature" of each rabbit: this is the base for the determination of any temperature increase resulting from the injection of a test solution. In any one group of test rabbits, use only those rabbits whose control temperatures do not vary by more than 1° from each other, and do not use any rabbit having a temperature exceeding 39.8°.

Unless otherwise specified in the individual monograph, inject into an ear vein of each of three rabbits 10 mL of the test solution per kg of body weight, completing each injection within 10 minutes after start of administration. The test solution is *either* the product, constituted if necessary as directed in the labeling, *or* the material under test treated as directed in the individual monograph and injected in the dose specified therein. For pyrogen testing of devices or injection assemblies, use washings or rinsings of the surfaces that come in contact with the parenterally administered material or with the injection site or internal tissues of the patient. Assure that all test solutions are protected from contamination. Perform the injection after warming the test solution to a temperature of 37 ± 2°. Record the temperature at 1, 2, and 3 hours subsequent to the injection.

Test Interpretation and Continuation—Consider any temperature decreases as zero rise. If no rabbit shows an individual rise in temperature of 0.6° or more above its respective control temperature, and if the sum of the three individual maximum temperature rises does not exceed 1.4°, the product meets the requirements for the absence of pyrogens. If any rabbit shows an individual temperature rise of 0.6° or more, or if the sum of the three individual maximum temperature rises exceeds 1.4°, continue the test using five other rabbits. If not more than three of the eight rabbits show individual rises in temperature of 0.6° or more, and if the sum of the eight individual maximum temperature rises does not exceed 3.7°, the material under examination meets the requirements for the absence of pyrogens.

RADIOACTIVE PHARMACEUTICALS

Test Dose for Preformulated, Ready-to-use Products Labeled with Radioactivity

AGGREGATED ALBUMIN AND OTHER PARTICLE-CONTAINING PRODUCTS

For the rabbit pyrogen test, dilute the product with Sodium Chloride Injection to not less than 100 μCi per mL, and inject a dose of 3 mL per kg of body weight into each rabbit.

OTHER PRODUCTS

Where Physical Half-life of Radionuclide Is Greater Than 1 Day—Calculate the maximum volume of the product that might be injected into a human subject. This calculation takes into account the maximum recommended radioactive dose of the product, in μCi, and the radioactive assay, in μCi per mL, of the product at its expiration date or time. Using this information, calculate the maximum volume dose per kg to a 70-kg human subject.

For the rabbit pyrogen test, inject a minimum of 10 times this dose per kg of body weight into each rabbit. If necessary, dilute with Sodium Chloride Injection. The total injected volume per rabbit is not less than 1 mL and not more than 10 mL of solution.

Where Physical Half-life of Radionuclide is Less Than 1 Day—For products labeled with radionuclides having a half-life of less than 1 day, the dosage calculations are identical to those described in the first paragraph under *Other Products*. These products may be released for distribution prior to completion of the rabbit pyrogen test, but such test shall be initiated at not more than 36 hours after release.

Test Dose for Pharmaceutical Constituents or Reagents to Be Labeled

The following test dose requirements pertain to reagents that are to be labeled or constituted prior to use by the direct addition of radioactive solutions such as Sodium Pertechnetate Tc 99m Injection, i.e., "cold kits."

Assume that the entire contents of the vial of nonradioactive reagent will be injected into a 70-kg human subject, or that 1/70 of the total contents per kg will be injected. If the contents are dry, constitute with a measured volume of Sodium Chloride Injection.

For the rabbit pyrogen test, inject 1/7 of the vial contents per kg of body weight into each rabbit. The maximum dose per rabbit is the entire contents of a single vial. The total injected volume per rabbit is not less than 1 mL and not more than 10 mL of solution.

⟨161⟩ TRANSFUSION AND INFUSION ASSEMBLIES

The requirements apply to medical devices that are labeled nonpyrogenic in contact directly or indirectly with the cardiovascular system or other soft body tissues. This includes, but is not limited to, solution administration sets, extension sets, transfer sets, blood administration sets, intravenous catheters, implants extracorporeal oxygenator tubings and accessories, dialysers and dialysis tubing and accessories, heart valves, vascular grafts, intramuscular drug delivery catheters, and transfusion and infusion assemblies. These requirements do not apply to orthopedic products, latex gloves, or wound dressings.

Sterility—Proceed as directed for *Sterilized Devices* under *Sterility Tests* ⟨71⟩.

Bacterial Endotoxins (see *Bacterial Endotoxins Test* ⟨85⟩)—*Transfusion and Infusion Assemblies and Similar Devices*—Through the tubing of each of 10 assemblies, pass a separate 40-mL portion of Limulus Amoebocyte Lysate (LAL)-negative *Sterile Water for Injection* at a temperature between 37° and 40°, at a flow rate of approximately 10 mL per minute. The pooled effluent meets the requirements of the *Bacterial Endotoxins Test* ⟨85⟩ with an endotoxin limit not to exceed 0.5 USP Endotoxin Unit per mL.

Implants—For implants as well as other medical devices that are intended to be sterile internally as well as externally, select not less than 3 implants and not more than 10 implants. Cut in small pieces or, if modular, separate the pieces, and soak for not less than 40 minutes and not more than 60 minutes in 400 mL of LAL-negative *Sterile Water for Injection* at a temperature between 37° and 40°, with swirling at 10-minute intervals. The supernatant liquid meets the requirements of the *Bacterial Endotoxins Test* ⟨85⟩ with an endotoxin limit not to exceed 0.5 USP Endotoxin Unit per mL.

Other Medical Devices—Pass equal portions of 400 mL of LAL-negative *Sterile Water for Injection* at a temperature between 37° and 40°, at a flow rate of approximately 10 mL per minute through a number of devices representative of a lot (not less than 3 devices and not more than 10). The 400 mL effluent meets the requirements of the *Bacterial Endotoxins Test* ⟨85⟩ with an endotoxin limit not to exceed 0.5 USP Endotoxin Unit per mL.

Medical Devices in Contact with Cerebrospinal Fluid—Proceed as directed under *Transfusion and Infusion Assemblies, Other Medical Devices*, or *Implants*, whichever is applicable. The 400 mL pooled effluent meets the requirements of the *Bacterial Endotoxins Test* ⟨85⟩ with an endotoxin limit of not more than 0.06 USP Endotoxin Unit per mL.

A *Bacterial Endotoxins Test* failure can be retested once by another *Bacterial Endotoxins Test*. If the cause of the initial failure is due to a level of endotoxin of less than 0.5 USP Endotoxin Unit per mL, the *Pyrogen Test* ⟨151⟩ can be used to retest.

Pyrogen—For samples that meet the requirements of the *Bacterial Endotoxins Test* retest and for samples that cannot be tested by the *Bacterial Endotoxins Test* because of nonremovable inhibition or enhancement of the test, the *Pyrogen Test* ⟨151⟩ is applied. A pooled effluent is obtained by passing a separate 40-mL portion of sterile pyrogen-free saline TS at a flow rate of approximately 10 mL per minute through the tubing of each of 10 Transfusion and Infusion Assemblies, or through 3 to 10 Medical Devices, or by soaking the cut or separated pieces of *Implants* in 400 mL of the Saline TS, whichever is appropriate. The requirements of the *Pyrogen Test* ⟨151⟩ are met.

Safety—It meets the requirements of the *Safety Test* under *Safety Tests—General* in chapter ⟨88⟩, *Biological Reactivity Tests, In-vivo,* or in chapter ⟨87⟩, *Biological Reactivity Tests, In-Vitro.*

Other Requirements—The portions of medical devices that are made of plastics or other polymers meet the requirements under *Containers—Plastics* ⟨661⟩; those made of elastomers meet the requirements under *Elastomeric Closures for Injection* ⟨381⟩.

⟨171⟩ VITAMIN B_{12} ACTIVITY ASSAY

Reference Standard—*USP Cyanocobalamin Reference Standard*—Keep container tightly closed and protected from light. Dry over silica gel for 4 hours before using.

Assay Preparation—Place a suitable quantity of the material to be assayed, previously reduced to a fine powder if necessary and accurately measured or weighed, in an appropriate vessel containing, for each g or mL of material taken, 25 mL of an aqueous extracting solution prepared just prior to use to contain, in each 100 mL, 1.29 g of disodium phosphate, 1.1 g of anhydrous citric acid, and 1.0 g of sodium metabisulfite. Autoclave the mixture at 121° for 10 minutes. Allow any undissolved particles of the extract to settle, and filter or centrifuge, if necessary. Dilute an aliquot of the clear solution with water so that the final test solution contains vitamin B_{12} activity approximately equivalent to that of the *Standard Cyanocobalamin Solution* which is added to the assay tubes.

Standard Cyanocobalamin Stock Solution—To a suitable quantity of USP Cyanocobalamin RS, accurately weighed, add sufficient 25 percent alcohol to make a solution having a known concentration of 1.0 µg of cyanocobalamin per mL. Store in a refrigerator.

Standard Cyanocobalamin Solution—Dilute a suitable volume of *Standard Cyanocobalamin Stock Solution* with water to a measured volume such that after the incubation period as described under *Procedure*, the difference in transmittance between the inoculated blank and the 5.0-mL level of the *Standard Cyanocobalamin Solution* is not less than that which corresponds to a difference of 1.25 mg in dried cell weight. This concentration usually falls between 0.01 ng and 0.04 ng per mL of *Standard Cyanocobalamin Solution.* Prepare a fresh standard solution for each assay.

Basal Medium Stock Solution—Prepare the medium according to the following formula and directions. A dehydrated mixture containing the same ingredients may be used provided that, when constituted as directed in the labeling, it yields a medium comparable to that obtained from the formula given herein.

Add the ingredients in the order listed, carefully dissolving the cystine and tryptophane in the hydrochloric acid before adding the next eight solutions in the resulting solution. Add 100 mL of water, mix, and dissolve the dextrose, sodium acetate, and ascorbic acid. Filter, if necessary, add the polysorbate 80 solution, adjust the solution to a pH between 5.5 and 6.0 with 1 *N* sodium hydroxide, and add purified water to make 250 mL.

L-Cystine	0.1	g
L-Tryptophane	0.05	g
1 *N* Hydrochloric Acid	10	mL
Adenine-Guanine-Uracil Solution	5	mL
Xanthine Solution	5	mL
Vitamin Solution I	10	mL
Vitamin Solution II	10	mL
Salt Solution A	5	mL
Salt Solution B	5	mL
Asparagine Solution	5	mL
Acid-hydrolyzed Casein Solution	25	mL
Dextrose, Anhydrous	10	g
Sodium Acetate, Anhydrous	5	g
Ascorbic Acid	1	g
Polysorbate 80 Solution	5	mL

Acid-Hydrolyzed Casein Solution—Prepare as directed under *Calcium Pantothenate Assay* ⟨91⟩.

Asparagine Solution—Dissolve 2.0 of *l*-asparagine in water to make 200 mL. Store under toluene in a refrigerator.

Adenine-Guanine-Uracil Solution—Prepare as directed under *Calcium Pantothenate Assay* ⟨91⟩.

Xanthine Solution—Suspend 0.20 g of xanthine in 30 mL to 40 mL of water, heat to about 70°, add 6.0 mL of 6 *N* ammonium hydroxide, and stir until the solid is dissolved. Cool, and add water to make 200 mL. Store under toluene in a refrigerator.

Salt Solution A—Dissolve 10 g of monobasic potassium phosphate and 10 g of dibasic potassium phosphate in water to make 200 mL. Add 2 drops of hydrochloric acid, and store under toluene.

Salt Solution B—Dissolve 4.0 g of magnesium sulfate, 0.20 g of sodium chloride, 0.20 g of ferrous sulfate, and 0.20 g of manganese sulfate in water to make 200 mL. Add 2 drops of hydrochloric acid, and store under toluene.

Polysorbate 80 Solution—Dissolve 20 g of polysorbate 80 in alcohol to make 200 mL. Store in a refrigerator.

Vitamin Solution I—Dissolve 10 mg of riboflavin, 10 mg of thiamine hydrochloride, 100 µg of biotin, and 20 mg of niacin in 0.02 *N* glacial acetic acid to make 400 mL. Store, protected from light, under toluene in a refrigerator.

Vitamin Solution II—Dissolve 20 mg of para-aminobenzoic acid, 10 mg of calcium pantothenate, 40 mg of pyridoxine hydrochloride, 40 mg of pyridoxal hydrochloride, 8 mg of pyridoxamine dihydrochloride, and 2 mg of folic acid in dilute neutralized alcohol (1 in 4) to make 400 mL. Store, protected from light, in a refrigerator.

Tomato Juice Preparation—Centrifuge commercially canned tomato juice so that most of the pulp is removed. Suspend about 5 g per liter of analytical filter-aid in the supernatant liquid, and filter, with the aid of reduced pressure, through a layer of the filter-aid. Repeat, if necessary, until a clear, straw-colored filtrate is obtained. Store under toluene in a refrigerator.

Culture Medium—[NOTE—A dehydrated mixture containing the same ingredients may be used provided that, when constituted as directed in the labeling, it yields a medium equivalent to that obtained from the formula given herein.] Dissolve 0.75 g of water-soluble yeast extract, 0.75 g of dried peptone, 1.0 g of anhydrous dextrose, and 0.20 g of potassium biphosphate in 60 mL to 70 mL of water. Add 10 mL of *Tomato Juice Preparation* and 1 mL of *Polysorbate 80 Solution*. Adjust the solution with 1 *N* sodium hydroxide to a pH of 6.8, and add water to make 100 mL. Place 10-mL portions of the solution in test tubes, and plug with cotton. Sterilize the tubes and contents in an autoclave at 121° for 15 minutes. Cool as rapidly as possible to avoid color formation resulting from overheating the medium.

Suspension Medium—Dilute a measured volume of *Basal Medium Stock Solution* with an equal volume of water. Place 10-mL portions of the diluted medium in test tubes. Sterilize, and cool as directed above for the *Culture Medium*.

Stock Culture of *Lactobacillus leichmannii*—To 100 mL of *Culture Medium* add 1.0 g to 1.5 g of agar, and heat the mixture, with stirring, on a steam bath, until the agar dissolves. Place approximately 10-mL portions of the hot solution in test tubes, cover the tubes suitably, sterilize at 121° for 15 minutes in an autoclave (exhaust line temperature), and allow the tubes to cool in an upright position. Inoculate three or more of the tubes, by stab transfer of a pure culture of *Lactobacillus leichmannii*.* (Before first using a fresh culture in this assay, make not fewer than 10 successive transfers of the culture in a 2-week period.) Incubate 16 to 24 hours at any selected temperature between 30° and 40° but held constant to within ±0.5°, and finally store in a refrigerator.

Prepare fresh stab cultures at least three times each week, and do not use them for preparing the inoculum if more than 4 days old. The activity of the microorganism can be increased by daily or twice-daily transfer of the stab culture, to the point where definite turbidity in the liquid inoculum can be observed 2 to 4 hours after inoculation. A slow-growing culture seldom gives a suitable response curve, and may lead to erratic results.

Inoculum—[NOTE—A frozen suspension of *Lactobacillus leichmannii* may be used as the stock culture, provided it yields an inoculum comparable to a fresh culture.] Make a transfer of cells from the *Stock Culture of Lactobacillus leichmannii* to 2 sterile tubes containing 10 mL of the *Culture Medium* each. Incubate these cultures for 16 to 24 hours at any selected temperature between 30° and 40° but held constant to within ±0.5°. Under aseptic conditions, centrifuge the cultures, and decant the supernatant liquid. Suspend the cells from the culture in 5 mL of sterile *Suspension Medium*, and combine. Using sterile *Suspension Medium*, adjust the volume so that a 1 in 20 dilution in saline TS produces 70% transmittance when read on a suitable spectrophotometer that has been set at a wavelength of 530 nm, equipped with a 10-mm cell, and read against saline TS set at 100% transmittance. Prepare a 1 in 400 dilution of the adjusted suspension using *Basal Medium Stock Solution*, and use it for the test inoculum. (This dilution may be altered, when necessary, to obtain the desired test response.)

Calibration of Spectrophotometer—Check the wavelength of the spectrophotometer periodically, using a standard wavelength cell or other suitable device. Before reading any tests, calibrate the spectrophotometer for 0% and 100% transmittance, using water and with the wavelength set at 530 nm.

Procedure—Cleanse meticulously by suitable means, followed preferably by heating at 250° for 2 hours, hard-glass test tubes, about 20 mm × 150 mm in size, and other necessary glassware because of the high sensitivity of the test organism to minute amounts of vitamin B_{12} activity and to traces of many cleansing agents.

To test tubes add, in duplicate, 1.0 mL, 1.5 mL, 2.0 mL, 3.0 mL, 4.0 mL, and 5.0 mL, respectively, of the *Standard Cyanocobalamin Solution*. To each of these tubes and to four similar empty tubes add 5.0 mL of *Basal Medium Stock Solution* and water to make 10 mL.

To similar test tubes add, in duplicate, respectively, 1.0 mL, 1.5 mL, 2.0 mL, 3.0 mL, and 4.0 mL of the *Assay Preparation*. To each tube add 5.0 mL of *Basal Medium Stock Solution* and water to make 10 mL. Place one complete set of standard and assay tubes together in one tube rack and the duplicate set in a second rack or section of a rack, preferably in random order.

Cover the tubes suitably to prevent bacterial contamination, and sterilize the tubes and contents in an autoclave at 121° for 5 minutes, arranging to reach this temperature in not more than 10 minutes by preheating the autoclave, if necessary. Cool as rapidly as practicable to avoid color formation resulting from overheating the medium. Take precautions to maintain uniformity of sterilizing and cooling conditions throughout the assay, since packing tubes too closely in the autoclave, or overloading it, may cause variation in the heating rate.

Aseptically add 0.5 mL of *Inoculum* to each tube so prepared, except two of the four containing no *Standard Cyanocobalamin Solution* (the uninoculated blanks). Incubate the tubes at a temperature between 30° and 40° held constant to within ±0.5°, for 16 to 24 hours.

Terminate growth by heating to a temperature not lower than 80° for 5 minutes. Cool to room temperature. After agitating its contents, place the container in a spectrophotometer that has been set at a wavelength of 530 nm, and read the transmittance when a steady state is reached. This steady state is observed a few seconds after agitation when the reading remains constant for 30 seconds or more. Allow approximately the same time interval for the reading on each tube.

With the transmittance set at 100% for the uninoculated blank, read the transmittance of the inoculated blank. If the difference is greater than 5% or if there is evidence of contamination with a foreign microorganism, disregard the results of the assay.

With the transmittance set at 100% for the uninoculated blank, read the transmittance of each of the remaining tubes. Disregard the results of the assay if the slope of the standard curve indicates a problem with sensitivity.

Calculation—Prepare a standard concentration–response curve by the following procedure. Test for and replace any aberrant individual transmittances. For each level of the standard, calculate the response from the sum of the duplicate values of the transmittances (Σ) as the difference, $y = 2.00 - \Sigma$. Plot this response on the ordinate of cross-section paper against the logarithm of the mL of *Standard Cyanocobalamin Solution* per tub[e]
on the abscissa, using for the ordinate either an ar
logarithmic scale, whichever gives the better appro
straight line. Draw the straight line or smooth cu
fits the plotted points.

Calculate the response, y, adding together the t
tances for each level of the *Assay Preparation*. Re
standard curve the logarithm of the volume of th

* Pure cultures of *Lactobacillus leichmannii* may be obtained as No. 7830 from the American Type Culture Collection, 12301 Parklawn Drive, Rockville, MD 20852.

Preparation corresponding to each of those values of y that falls within the range of the lowest and highest points plotted for the standard. Subtract from each logarithm so obtained the logarithm of the volume, in mL, of the *Assay Preparation* to obtain the difference, x, for each dosage level. Average the values of x for each of three or more dosage levels to obtain $\bar{x} = M'$, the log-relative potency of the *Assay Preparation.* Determine the quantity, in µg, of USP Cyanocobalamin RS corresponding to the cyanocobalamin in the portion of material taken for assay by the equation antilog M = antilog (M' + log R), in which R is the number of µg of cyanocobalamin that was assumed to be present in each mg (or capsule or tablet) of the material taken for assay.

Replication—Repeat the entire determination at least once, using separately prepared *Assay Preparations.* If the difference between the two log potencies M is not greater than 0.08, their mean, $\bar{M}$, is the assayed log-potency of the test material (see *Vitamin B_{12} Activity Assay* under *Design and Analysis of Biological Assays* ⟨111⟩). If the two determinations differ by more than 0.08, conduct one or more additional determinations. From the mean of two or more values of M that do not differ by more than 0.15, compute the mean potency of the preparation under assay.

Chemical Tests and Assays

IDENTIFICATION TESTS

⟨181⟩ IDENTIFICATION—ORGANIC NITROGENOUS BASES

This test is for the identification of tertiary amine compounds.

Dissolve 50 mg of the substance under test, if in bulk, in 25 mL of 0.01 N hydrochloric acid, or shake a quantity of powdered tablets or the contents of capsules equivalent to 50 mg of the substance with 25 mL of 0.01 N hydrochloric acid for 10 minutes. Transfer the liquid to a separator, if necessary filtering it and washing the filter and the residue with several small portions of water. In a second separator dissolve 50 mg of the corresponding USP Reference Standard in 25 mL of 0.01 N hydrochloric acid. Treat each solution as follows: Add 2 mL of 1 N sodium hydroxide and 4 mL of carbon disulfide, and shake for 2 minutes. Centrifuge if necessary to clarify the lower phase, and filter it through a dry filter, collecting the filtrate in a small flask provided with a glass stopper.

Determine the absorption spectra of the filtered solutions of both standard and sample without delay, in 1-mm cells between 7 µm and 15 µm, with a suitable infrared spectrophotometer, using carbon disulfide in a matched cell as the blank. The spectrum of the solution prepared from the sample shows all of the significant absorption bands present in the spectrum of the solution prepared from the Reference Standard.

⟨191⟩ IDENTIFICATION TESTS—GENERAL

Under this heading are placed tests that are frequently referred to in the Pharmacopeia for the identification of official articles.

NOTE—The tests are not intended to be applicable to mixtures of substances unless so specified.

Acetate—When acetic acid or an acetate is warmed with sulfuric acid and alcohol, ethyl acetate, recognizable by its characteristic odor, is evolved. With neutral solutions of acetates, ferric chloride TS produces a deep red color which is destroyed v the addition of mineral acids.

luminum—Solutions of aluminum salts yield with 6 N am- m hydroxide a gelatinous, white precipitate that is insoluble in an excess of 6 N ammonium hydroxide. 1 N sodium hydroxide or sodium sulfide TS produces the same precipitate, which dissolves in an excess of either of these reagents.

Ammonium—Ammonium salts are decomposed by the addition of an excess of 1 N sodium hydroxide, with the evolution of ammonia, recognizable by its odor and by its alkaline effect upon moistened red litmus paper exposed to the vapor. Warming the solution accelerates the decomposition.

Antimony—Solutions of antimony (III) compounds, strongly acidified with hydrochloric acid, yield with hydrogen sulfide an orange precipitate of antimony sulfide which is insoluble in 6 N ammonium hydroxide, but is soluble in ammonium sulfide TS.

Barium—Solutions of barium salts yield a white precipitate with 2 N sulfuric acid. This precipitate is insoluble in hydrochloric and in nitric acid. Barium salts impart a yellowish green color to a nonluminous flame, which appears blue when viewed through green glass.

Benzoate—In neutral solutions, benzoates yield a salmon-colored precipitate with ferric chloride TS. In moderately concentrated solutions, benzoates yield a precipitate of benzoic acid upon acidification with 2 N sulfuric acid. This precipitate is readily soluble in ether.

Bicarbonate—See *Carbonate.*

Bismuth—When dissolved in a slight excess of nitric or hydrochloric acid, bismuth salts yield a white precipitate upon dilution with water. This precipitate is colored brown by hydrogen sulfide, and the resulting compound dissolves in a warm mixture of equal parts of nitric acid and water.

Bisulfite—See *Sulfite.*

Borate—To 1 mL of a borate solution, acidified with hydrochloric acid to litmus, add 3 or 4 drops of a saturated solution of iodine and 3 or 4 drops of polyvinyl alcohol solution (1 in 50): an intense blue color is produced. When a borate is treated with sulfuric acid, methanol is added, and the mixture is ignited, it burns with a green-bordered flame.

Bromide—Solutions of bromides, upon the addition of chlorine TS, dropwise, liberate bromine, which is dissolved by shaking with chloroform, coloring the chloroform red to reddish brown. Silver nitrate TS produces in solutions of bromides a yellowish white precipitate, which is insoluble in nitric acid and is slightly soluble in 6 N ammonium hydroxide.

Calcium—Solutions of calcium salts form insoluble oxalates when treated as follows: To a solution of the calcium salt (1 in 20) add 2 drops of methyl red TS, and neutralize with 6 N ammonium hydroxide. Add 3 N hydrochloric acid, dropwise, until the solution is acid to the indicator. Upon the addition of ammonium oxalate TS, a white precipitate is formed. This precipitate is insoluble in 6 N acetic acid but dissolves in hydrochloric acid. Calcium salts moistened with hydrochloric acid impart a transient yellowish red color to a nonluminous flame.

Carbonate—Carbonates and bicarbonates effervesce with acids, evolving a colorless gas, which when passed into calcium hydroxide TS produces a white precipitate immediately. A cold solution of a soluble carbonate is colored red by phenolphthalein TS, while a similar solution of a bicarbonate remains unchanged or is only slightly colored.

Chlorate—Solutions of chlorates yield no precipitate with silver nitrate TS. The addition of sulfurous acid to this mixture produces a white precipitate which is insoluble in nitric acid, but is soluble in 6 N ammonium hydroxide. Upon ignition, chlorates yield chlorides, recognizable by appropriate tests. When sulfuric acid is added to a dry chlorate, decrepitation occurs and a greenish yellow gas is evolved. [*Caution—Use only a small amount of chlorate for this test and exercise extreme caution in performing it.*]

Chloride—Solutions of chlorides yield with silver nitrate TS a white, curdy precipitate, which is insoluble in nitric acid, but is soluble in a slight excess of 6 N ammonium hydroxide. When testing alkaloidal hydrochlorides, add 6 N ammonium hydroxide, filter, acidify the filtrate with nitric acid, and proceed as directed above. Dry chlorides, when mixed with an equal weight of manganese dioxide, moistened with sulfuric acid, and gently heated, evolve chlorine which is recognizable by the production of a blue color with moistened starch iodide paper.

Citrate—To 15 mL of pyridine add a few mg of a citrate salt, dissolved or suspended in 1 mL of water, and shake. To this mixture add 5 mL of acetic anhydride, and shake: a light red color is produced.

Cobalt—Solutions of cobalt salts (1 in 20) in 3 *N* hydrochloric acid yield a red precipitate when heated on a steam bath with an equal volume of a hot, freshly prepared solution of 1-nitroso-2-naphthol (1 in 10) in 9 *N* acetic acid. Solutions of cobalt salts, when saturated with potassium chloride and treated with potassium nitrite and acetic acid, yield a yellow precipitate.

Copper—Solutions of cupric compounds, acidified with hydrochloric acid, deposit a red film of metallic copper upon a bright, untarnished surface of metallic iron. An excess of 6 *N* ammonium hydroxide, added to a solution of a cupric salt, produces first a bluish precipitate and then a deep blue-colored solution. With potassium ferrocyanide TS, solutions of cupric salts yield a reddish brown precipitate, insoluble in diluted acids.

Hypophosphite—When strongly heated, hypophosphites evolve spontaneously flammable phosphine. Hypophosphites in solution yield a white precipitate with mercuric chloride TS. This precipitate becomes gray when an excess of hypophosphite is present. Solutions of hypophosphites, acidified with sulfuric acid, and warmed with cupric sulfate TS yield a red precipitate.

Iodide—Solutions of iodides, upon the addition of chlorine TS, dropwise, liberate iodine, which colors the solution yellow to red. When the solution is shaken with chloroform, the latter is colored violet. The iodine thus liberated gives a blue color with starch TS. Silver nitrate TS produces in solutions of iodides a yellow, curdy precipitate, which is insoluble in nitric acid and in 6 *N* ammonium hydroxide.

Iron—Ferrous and ferric compounds in solution yield a black precipitate with ammonium sulfide TS. This precipitate is dissolved by cold 3 *N* hydrochloric acid with the evolution of hydrogen sulfide.

FERRIC SALTS—Acid solutions of ferric salts yield a dark blue precipitate with potassium ferrocyanide TS. With an excess of 1 *N* sodium hydroxide, a reddish brown precipitate is formed. Solutions of ferric salts produce with ammonium thiocyanate TS a deep red color which is not destroyed by dilute mineral acids.

FERROUS SALTS—Solutions of ferrous salts yield a dark blue precipitate with potassium ferricyanide TS. This precipitate is insoluble in 3 *N* hydrochloric acid, but is decomposed by 1 *N* sodium hydroxide. Solutions of ferrous salts yield with 1 *N* sodium hydroxide a greenish white precipitate, the color rapidly changing to green and then to brown when shaken.

Lactate—When solutions of lactates are acidified with sulfuric acid, potassium permanganate TS is added, and the mixture is heated, acetaldehyde, recognizable by its distinctive odor, is evolved.

Lead—Solutions of lead salts yield with 2 *N* sulfuric acid a white precipitate which is insoluble in 3 *N* hydrochloric or 2 *N* nitric acid, but is soluble in warm 1 *N* sodium hydroxide and in ammonium acetate TS. With potassium chromate TS, solutions of lead salts, free or nearly free from mineral acids, yield a yellow precipitate which is insoluble in 6 *N* acetic acid, but is soluble in 1 *N* sodium hydroxide.

Lithium—Moderately concentrated solutions of lithium salts, made alkaline with sodium hydroxide, yield with sodium carbonate TS a white precipitate on boiling. The precipitate is soluble in ammonium chloride TS. Lithium salts moistened with hydrochloric acid impart an intense crimson color to a nonluminous flame. Solutions of lithium salts are not precipitated by 2 *N* sulfuric acid or soluble sulfates (*distinction from strontium*).

Magnesium—Solutions of magnesium salts in the presence of ammonium chloride yield no precipitate when neutralized with ammonium carbonate TS, but on the subsequent addition of dibasic sodium phosphate TS, a white, crystalline precipitate, which is insoluble in 6 *N* ammonium hydroxide, is formed.

Manganese—Solutions of manganous salts yield with ammonium sulfide TS a salmon-colored precipitate, which dissolves in acetic acid.

Mercury—Solutions of mercury salts, free from an excess of nitric acid, when applied to bright copper foil, yield a deposit, which, upon rubbing, becomes bright and silvery in appearance. With hydrogen sulfide, solutions of mercury compounds yield a black precipitate, which is insoluble in ammonium sulfide TS and in boiling 2 *N* nitric acid.

MERCURIC SALTS—Solutions of mercuric salts yield a yellow precipitate with 1 *N* sodium hydroxide. They yield also, in neutral solutions with potassium iodide TS, a scarlet precipitate, which is very soluble in an excess of the reagent.

MERCUROUS SALTS—Mercurous compounds are decomposed by 1 *N* sodium hydroxide, producing a black color. Solutions of mercurous salts yield with hydrochloric acid a white precipitate which is blackened by 6 *N* ammonium hydroxide. With potassium iodide TS, a yellow precipitate, which may become green upon standing, is formed.

Nitrate—When a solution of a nitrate is mixed with an equal volume of sulfuric acid, the mixture is cooled, and a solution of ferrous sulfate is superimposed, a brown color is produced at the junction of the two liquids. When a nitrate is heated with sulfuric acid and metallic copper, brownish red fumes are evolved. Nitrates do not decolorize acidified potassium permanganate TS (*distinction from nitrites*).

Nitrite—When treated with dilute mineral acids or with 6 *N* acetic acid, nitrites evolve brownish red fumes. The solution colors starch-iodide paper blue.

Oxalate—Neutral and alkaline solutions of oxalates yield a white precipitate with calcium chloride TS. This precipitate is insoluble in 6 *N* acetic acid, but is dissolved by hydrochloric acid. Hot acidified solutions of oxalates decolorize potassium permanganate TS.

Permanganate—Solutions of permanganates acidified with sulfuric acid are decolorized by hydrogen peroxide TS and by sodium bisulfite TS, in the cold, and by oxalic acid TS, in hot solution.

Peroxide—Solutions of peroxides slightly acidified with sulfuric acid yield a deep blue color upon the addition of potassium dichromate TS. On shaking the mixture with an equal volume of ether and allowing the liquids to separate, the blue color is found in the ether layer.

Phosphate—Neutral solutions of orthophosphates yield with silver nitrate TS a yellow precipitate, which is soluble in 2 *N* nitric acid and in 6 *N* ammonium hydroxide. With ammonium molybdate TS, a yellow precipitate, which is soluble in 6 *N* ammonium hydroxide, is formed. Pyrophosphates obtained by ignition yield with silver nitrate TS a white precipitate, which is soluble in 2 *N* nitric acid and in 6 *N* ammonium hydroxide. With ammonium molybdate TS, a yellow precipitate, which is soluble in 6 *N* ammonium hydroxide, is formed.

Potassium—Potassium compounds impart a violet color to a nonluminous flame, but the presence of small quantities of sodium masks the color unless the yellow color produced by sodium is screened out by viewing through cobalt glass. In neutral, concentrated or moderately concentrated solutions of potassium salts (depending upon the solubility and the potassium content), sodium bitartrate TS produces a white crystalline precipitate, which is soluble in 6 *N* ammonium hydroxide and in solutions of alkali hydroxides and carbonates. The formation of the precipitate, which is usually slow, is accelerated by stirring or rubbing the inside of the test tube with a glass rod. The addition of a small amount of glacial acetic acid or alcohol also promotes the precipitation.

Salicylate—In moderately dilute solutions of salicylates, ferric chloride TS produces a violet color. The addition of acids to moderately concentrated solutions of salicylates produces a white, crystalline precipitate of salicylic acid, which melts between 158° and 161°.

Silver—Solutions of silver salts yield with hydrochloric acid a white, curdy precipitate, which is insoluble in nitric acid, but is readily soluble in 6 *N* ammonium hydroxide. A solution of a silver salt to which 6 *N* ammonium hydroxide and a small quantity of formaldehyde TS are added deposits, upon warming, a mirror of metallic silver upon the sides of the container.

Sodium—Solutions of sodium compounds after conversion to chloride or nitrate yield, with five times their volume of cobalt–uranyl acetate TS, a golden yellow precipitate, which forms after agitation for several minutes. Sodium compounds impart an intense yellow color to a nonluminous flame.

Sulfate—Solutions of sulfates yield with barium chloride TS a white precipitate, which is insoluble in hydrochloric acid and in nitric acid. With lead acetate TS, sulfates yield a white precipitate, which is soluble in ammonium acetate solution. Hydrochloric acid produces no precipitate when added to solutions of sulfates (*distinction from thiosulfates*).

Sulfite—When treated with 3 *N* hydrochloric acid, sulfites and bisulfites yield sulfur dioxide, recognizable by its characteristic, pungent odor. This gas blackens filter paper moistened with mercurous nitrate TS.

Tartrate—Dissolve a few mg of a tartrate salt in 2 drops of sodium periodate solution (1 in 20). Add a drop of 1 *N* sulfuric acid, and after 5 minutes add a few drops of sulfurous acid followed by a few drops of fuchsin–sulfurous acid TS: a reddish pink color is produced within 15 minutes.

Thiocyanate—Solutions of thiocyanates yield with ferric chloride TS a red color, which is not destroyed by moderately concentrated mineral acids.

Thiosulfate—Solutions of thiosulfates yield with hydrochloric acid a white precipitate which soon turns yellow, and sulfur dioxide, recognizable by its odor, is liberated. The addition of ferric chloride TS to solutions of thiosulfates produces a dark violet color which quickly disappears.

Zinc—In the presence of sodium acetate, solutions of zinc salts yield a white precipitate with hydrogen sulfide. This precipitate is insoluble in acetic acid, but is dissolved by 3 *N* hydrochloric acid. Ammonium sulfide TS produces a similar precipitate in neutral and in alkaline solutions. Zinc salts in solution yield with potassium ferrocyanide TS a white precipitate, which is insoluble in 3 *N* hydrochloric acid.

⟨193⟩ IDENTIFICATION—TETRACYCLINES

The following chromatographic procedures are provided to confirm the identity of Pharmacopeial drug substances that are of the tetracycline type, such as doxycycline, oxytetracycline, and tetracycline, and to confirm the identity of such compounds in their respective Pharmacopeial dosage forms. Two procedures are provided, one based on paper chromatography (*Method I*) and the other on thin-layer chromatography (*Method II*). *Method I* is to be used unless otherwise directed in the individual monograph.

Standard Solution—Unless otherwise directed in the individual monograph, dissolve the USP Reference Standard for the drug substance being identified in the same solvent and at the same concentration as for the *Test solution*.

Test Solution—Prepare as directed in the individual monograph.

Method I

pH 3.5 Buffer—Dissolve 13.4 g of anhydrous citric acid and 16.3 g of dibasic sodium phosphate in 1000 mL of water, and mix.

Developing Solvent—On the day of use, mix 10 volumes of chloroform, 20 volumes of nitromethane, and 3 volumes of pyridine.

Mixed Test Solution—Mix equal volumes of the *Standard solution* and the *Test solution*.

Chromatographic Sheet—Draw a spotting line 2.5 cm from one edge of a 20-cm × 20-cm sheet of filter paper (Whatman No. 1, or equivalent). Impregnate the sheet with *pH 3.5 Buffer* by passing it through a trough filled with *pH 3.5 Buffer*, and remove the excess solvent by firmly pressing the sheet between non-fluorescent blotting papers.

Procedure—To a suitable chromatographic chamber, prepared for ascending chromatography (see *Chromatography* ⟨621⟩), add *Developing Solvent* to a depth of 0.6 cm. On the spotting line of the *Chromatographic Sheet* apply at 1.5-cm intervals 2 µL each of the *Standard Solution*, the *Test Solution*, and the *Mixed Test Solution*. Allow the sheet to dry partially, and while still damp place it in the chromatographic chamber with the bottom edge touching the *Developing Solvent*. When the solvent front has risen about 10 cm, remove the sheet from the chamber, and expose the sheet to ammonia vapor. Examine the chromatogram under long-wavelength ultraviolet light. Record the positions of the major yellow fluorescent spots: the R_f value of the principal spot obtained from the *Test Solution* and from the *Mixed Test Solution* corresponds to that obtained from the *Standard Solution*.

Method II

Resolution Solution—Prepare as directed in the individual monograph.

Developing Solvent—Prepare a mixture of 0.5 *M* oxalic acid, previously adjusted with ammonium hydroxide to a pH of 2.0, acetonitrile, and methanol (80:20:20).

Chromatographic Plate—Use a suitable thin-layer chromatographic plate (see *Thin-layer Chromatography* under *Chromatography* ⟨621⟩), coated with a 0.25-mm layer of octylsilanized chromatographic silica gel mixture. Activate the plate by heating it at 130° for 20 minutes, allow to cool, and use while still warm.

Procedure—On the *Chromatographic plate* separately apply 1 µL each of the *Standard Solution*, the *Test Solution*, and the *Resolution Solution*. Allow the spots to dry, and develop the chromatogram in the *Developing Solvent* until the solvent front has moved about three-fourths of the length of the plate. Remove the plate from the developing chamber, mark the solvent front, and allow to air-dry. Expose the plate to ammonia vapors, for 5 minutes, and promptly locate the spots on the plate by viewing under long-wavelength ultraviolet light: the chromatogram of the *Resolution Solution* shows clearly separated spots, and the principal spot obtained from the *Test Solution* corresponds in R_f value, intensity, and appearance to that obtained from the *Standard Solution*.

⟨201⟩ THIN-LAYER CHROMATOGRAPHIC IDENTIFICATION TEST

The following procedure is applicable as an aid in verifying the identities of many compendial drug substances as such and in their respective dosage forms.

Prepare a test solution as directed in the individual monograph. On a line parallel to and about 2 cm from the edge of a suitable thin-layer chromatographic plate, coated with a 0.25-mm layer of chromatographic silica gel with a suitable fluorescing substance (see *Chromatography* ⟨621⟩), apply 10 µL of this solution and 10 µL of a Standard solution prepared from the USP Reference Standard for the drug substance being identified, in the same solvent and at the same concentration as the test solution, unless otherwise directed in the individual monograph. Allow the spots to dry, and develop the chromatogram in a solvent system consisting of a mixture of chloroform, methanol, and water (180:15:1), unless otherwise directed in the individual monograph, until the solvent front has moved about three-fourths of the length of the plate. Remove the plate from the developing chamber, mark the solvent front, and allow the solvent to evaporate. Unless otherwise directed in the individual monograph, locate the spots on the plate by examination under short-wavelength ultraviolet light. The R_f value of the principal spot obtained from the test solution corresponds to that obtained from the Standard solution.

LIMIT TESTS

⟨211⟩ ARSENIC

This procedure is designed to determine the presence of trace amounts of arsenic (As) by converting the arsenic in a substance under test to arsine, which is then passed through a solution of silver diethyldithiocarbamate to form a red complex. The red color so produced is compared, either visually or spectrophotometrically, to the color produced similarly in a control containing an amount of arsenic equivalent to the limit given in the individual monograph. Limits are stated in terms of arsenic (As). The content of arsenic does not exceed the limit given in the individual monograph.

Two methods are provided, the methods differing only in the preliminary treatment of the test substance and the standard. Generally, *Method I* is used for inorganic materials, while *Method II* is used for organic materials.

Apparatus—The apparatus (see illustration)* consists of an arsine generator (*a*) fitted with a scrubber unit (*c*) and an absorber tube (*e*) with standard-taper or ground glass ball-and-socket joints (*b* and *d*) between the units. However, any other suitable apparatus, embodying the principle of the assembly described and illustrated, may be used.

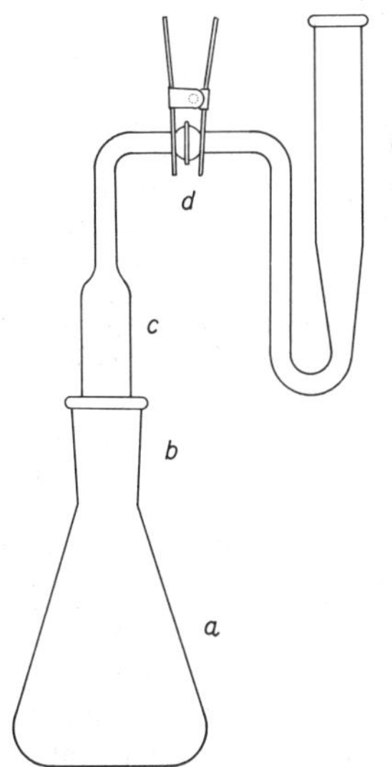

Arsenic Test Apparatus

Arsenic Trioxide Stock Solution—Dissolve 132.0 mg of arsenic trioxide, previously dried at 105° for 1 hour and accurately weighed, in 5 mL of sodium hydroxide solution (1 in 5) in a 1000-mL volumetric flask. Neutralize the solution with 2 *N* sulfuric acid, add 10 mL more of 2 *N* sulfuric acid, then add recently boiled and cooled water to volume, and mix.

Standard Arsenic Solution—Transfer 10.0 mL of *Arsenic Trioxide Stock Solution* to a 1000-mL volumetric flask, add 10 mL of 2 *N* sulfuric acid, then add recently boiled and cooled water to volume, and mix. Each mL of *Standard Arsenic Solution* contains the equivalent of 1 µg of arsenic (As). Keep this solution in an all-glass container, and use within 3 days.

Method I

Standard Preparation—Pipet 3.0 mL of *Standard Arsenic Solution* into a generator flask, and dilute with water to 35 mL.

Test Preparation—Unless otherwise directed in the individual monograph, transfer to the generator flask the quantity, in g, of the test substance calculated by the formula:

$$3.0/L,$$

in which L is the arsenic limit in ppm, dissolve in water, and dilute with water to 35 mL.

Procedure—Treat the *Standard Preparation* and the *Test Preparation* similarly as follows: Add 20 mL of 7 *N* sulfuric acid, 2 mL of potassium iodide TS, 0.5 mL of stronger acid stannous chloride TS, and 1 mL of isopropyl alcohol, and mix. Allow to stand at room temperature for 30 minutes. Pack the scrubber tube (*c*) with two pledgets of cotton that have been soaked in saturated lead acetate solution, freed from excess solution by expression, and dried in vacuum at room temperature, leaving a 2-mm space between the two pledgets. Lubricate the joints (*b* and *d*) with a suitable stopcock grease designed for use with organic solvents, and connect the scrubber unit to the absorber tube (*e*). Transfer 3.0 mL of silver diethyldithiocarbamate TS to the absorber tube. Add 3.0 g of granular zinc (No. 20 mesh) to the mixture in the flask, immediately connect the assembled scrubber unit, place the generator flask (*a*) in a water bath maintained at a temperature of 25 ± 3°, and allow the evolution of hydrogen and the color development to proceed for 45 minutes, swirling the flask gently at 10-minute intervals. Disconnect the absorber tube from the generator and scrubber units, and transfer the absorbing solution to a 1-cm absorption cell. Any red color produced by the *Test Preparation* does not exceed that produced by the *Standard Preparation.* If necessary or desirable, determine the absorbance at the wavelength of maximum absorbance between 535 nm and 540 nm, with a suitable spectrophotometer or colorimeter, using silver diethyldithiocarbamate TS as the blank.

* A suitable apparatus is obtainable from Fisher Scientific Co., 711 Forbes Ave., Pittsburgh, PA 15219.

Method II

NOTES—

(1) *Caution—Some substances may react with explosive violence when digested with hydrogen peroxide. Exercise safety precautions at all times.*

(2) If halogen-containing compounds are present, use a lower temperature while heating the test specimen with sulfuric acid, avoid boiling the mixture, and add the hydrogen peroxide with caution, before charring begins, to prevent loss of trivalent arsenic.

(3) If the test substance reacts too rapidly and begins charring with 5 mL of sulfuric acid before heating, use instead 10 mL of cooled dilute sulfuric acid (1 in 2), and add a few drops of the hydrogen peroxide before heating.

Standard Preparation—Pipet 3.0 mL of *Standard Arsenic Solution* into a generator flask, add 2 mL of sulfuric acid, mix, and add the total amount of 30 percent hydrogen peroxide used in preparing the *Test Preparation.* Heat the mixture to strong fuming, cool, add cautiously 10 mL of water, and again heat to strong fumes. Repeat this procedure with another 10 mL of water to remove any traces of hydrogen peroxide. Cool, and dilute with water to 35 mL.

Test Preparation—Unless otherwise directed in the individual monograph, transfer to a generator flask the quantity, in g, of the test substance calculated by the formula:

$$3.0/L,$$

in which L is the arsenic limit in ppm. Add 5 mL of sulfuric acid and a few glass beads, and digest in a fume hood, preferably on a hot plate and at a temperature not exceeding 120°, until charring begins. (Additional sulfuric acid may be necessary to wet some specimens completely, but the total volume added should not exceed 10 mL.) Cautiously add, dropwise, 30 percent hydrogen peroxide, allowing the reaction to subside and again heating between drops. Add the first few drops very slowly with sufficient mixing, in order to prevent a rapid reaction. Discontinue heating if foaming becomes excessive. When the reaction has abated, heat cautiously, rotating the flask occasionally to prevent the specimen from caking on glass exposed to the heating unit. *Maintain oxidizing conditions at all times during the digestion by adding small quantities of the hydrogen peroxide solution whenever the mixture turns brown or darkens.* Continue the digestion until the organic matter is destroyed, gradually raising the temperature of the hot plate until fumes of sulfur trioxide are copiously evolved, and the solution becomes colorless or retains only a light straw color. Cool, add cautiously 10 mL of water, mix, and again evaporate to strong fuming, repeating this procedure to remove any trace of hydrogen peroxide. Cool, add cautiously 10 mL of water, wash the sides of the flask with a few mL of water, and dilute with water to 35 mL.

Procedure—Proceed as directed for *Procedure* under *Method I.*

Interfering Chemicals—Metals or salts of metals such as chromium, cobalt, copper, mercury, molybdenum, nickel, palladium, and silver may interfere with the evolution of arsine. *Antimony*, which forms stibine, produces a positive interference in the color development with silver diethyldithiocarbamate TS; when the presence of antimony is suspected, the red colors produced in the two silver diethyldithiocarbamate solutions may be compared at the wavelength of maximum absorbance between 535 nm and 540 nm, with a suitable colorimeter, since at this wavelength the interference due to stibine is negligible.

⟨216⟩ CALCIUM, POTASSIUM, AND SODIUM

The flame photometer characteristically is equipped with a photomultiplier phototube detector for determination of calcium or sodium, a red-sensitive phototube detector for the determination of potassium, a monochromator, an adjustable exit slit, sensitivity controls, and an oxyacetylene burner. An oxyhydrogen burner is necessary for the determination of potassium in the presence of large amounts of calcium.

Standard Calcium Ion Solution—Transfer 249.7 mg of calcium carbonate, previously dried at 300° for 3 hours and cooled in a desiccator for 2 hours, to a 100-mL volumetric flask, dissolve in a mixture of 20 mL of water and 5 mL of 3 *N* hydrochloric acid, dilute with water to volume, and mix. Each mL contains 1.00 mg of calcium ion (Ca).

Standard Potassium Ion Solution—Transfer 190.7 mg of potassium chloride, previously dried at 105° for 2 hours, to a 100-mL volumetric flask, dissolve in water, dilute with water to volume, and mix. Each mL contains 1.00 mg of potassium ion (K).

Standard Sodium Ion Solution—Transfer 254.2 mg of sodium chloride, previously dried at 105° for 2 hours, to a 100-mL volumetric flask, dissolve in water, dilute with water to volume, and mix. Each mL contains 1.00 mg of sodium ion (Na).

Standard Preparation—Transfer a 50-mL aliquot of the *Test Preparation* to a 100-mL volumetric flask, add the volume(s) of *Standard Ion Solution(s)* specified in the individual monograph, dilute with water to volume, and mix. Quantitatively dilute aliquots of this solution with water as necessary to bring the concentration of the ion to be determined into the proper range for the flame photometer used.

Test Preparation—Unless otherwise directed in the individual monograph, transfer 2.000 g of test specimen to a 100-mL volumetric flask, chill in an ice bath, add 5 mL of nitric acid, swirl to dissolve, and allow to warm to room temperature. Heat gently, if necessary, to obtain a clear or just slightly turbid mixture. Cool to room temperature, if necessary, dilute with water to volume, and mix. Filter or centrifuge if necessary, to obtain a clear solution.

Adjust the flame photometer to give a reading as near as possible to 100 percent transmittance with the *Standard Preparation* at the wavelength setting giving maximum emission corresponding to the designated characteristic wavelength as shown in the accompanying table. Use an exit slit width corresponding as nearly as possible to the designated bandwidth. Record the transmittance reading, labeling it as *S*.

Dilute aliquots of the *Test Preparation* with water as necessary to prepare a solution in which the concentration is similar to that in the *Standard Preparation*. Without changing any of the adjustments of the flame photometer, determine the emission of the solution as percent transmittance, and record the reading, labeling it as *T*. Readjust only the monochromator to the designated wavelength for background determination. Determine the emission of the solution at this wavelength as percent transmittance, and record the reading, labeling it as *B*.

The requirements of the test are met if the value of *T* minus *B* is less than or equal to the value of *S* minus *T*.

Ion	Wavelength (nm)		Bandwidth (nm)
	Characteristic	Background	
Calcium	422.7	430	0.8
Potassium	766.5	750	12
Sodium	589	580	0.8

⟨221⟩ CHLORIDE AND SULFATE

The following limit tests are provided as general procedures for use where limits for chloride and sulfate are specified in the individual monographs.

Perform the tests and the controls in glass cylinders of the same diameter and matched as closely as practicable in other respects (see *Visual Comparison* ⟨851⟩). Use the same quantities of the same reagents for both the solution under test and the control solution containing the specified volume of chloride or sulfate. If, after acidification, the solution is not perfectly clear, filter it through a filter paper that gives negative tests for chloride and sulfate. Add the precipitant, silver nitrate TS or barium chloride TS as required, to both the test solution and the control solution in immediate sequence.

Where the individual monograph calls for applying the test to a specific volume of a solution of the substance, and the limit for chloride or sulfate corresponds to 0.20 mL or less of 0.020 *N* hydrochloric acid or sulfuric acid, respectively, apply the test to the solution without further dilution. In such cases maintain the same volume relationships for the control solution as specified for the solution under test. In applying the test to the salts of heavy metals, which normally show an acid reaction, omit the acidification and do not neutralize the solution. Dissolve bismuth salts in a few mL of water and 2 mL of nitric acid before treating with the precipitant.

Chloride—Dissolve the specified quantity of the substance under test in 30 to 40 mL of water, or, where the substance is already in solution, add water to make a total volume of 30 to 40 mL, and, if necessary, neutralize the solution with nitric acid to litmus. Add 1 mL each of nitric acid and of silver nitrate TS, and sufficient water to make 50 mL. Mix, allow to stand for 5 minutes protected from direct sunlight, and compare the turbidity, if any, with that produced in a solution containing the volume of 0.020 *N* hydrochloric acid specified in the monograph.

Sulfate—Dissolve the specified quantity of the substance under test in 30 to 40 mL of water, or, where the substance is already in solution, add water to make a total volume of 30 to 40 mL, and, if necessary, neutralize the solution with hydrochloric acid to litmus. Add 1 mL of 3 *N* hydrochloric acid, 3 mL of barium chloride TS, and sufficient water to make 50 mL. Mix, allow to stand for 10 minutes, and compare the turbidity, if any, with that produced in a solution containing the volume of 0.020 *N* sulfuric acid specified in the monograph.

⟨224⟩ DIOXANE

The following limit test is provided as a general procedure, where specified in the individual monograph, for the gas chromatographic determination of traces of 1,4-dioxane that may be separated from compendial articles.

Apparatus—Assemble a closed-system vacuum distillation apparatus, employing glass vacuum stopcocks (*A*, *B*, and *C*), as shown in the accompanying diagram. The concentrator tube (*D*)* is made of borosilicate or quartz (not flint) glass, graduated precisely enough to measure the 0.9 mL or more of distillate collected and marked so that the analyst can dilute accurately to 2.0 mL.

Standard preparation—Prepare a solution of dioxane in water having a known concentration of 100 µg per mL. Use a freshly prepared solution.

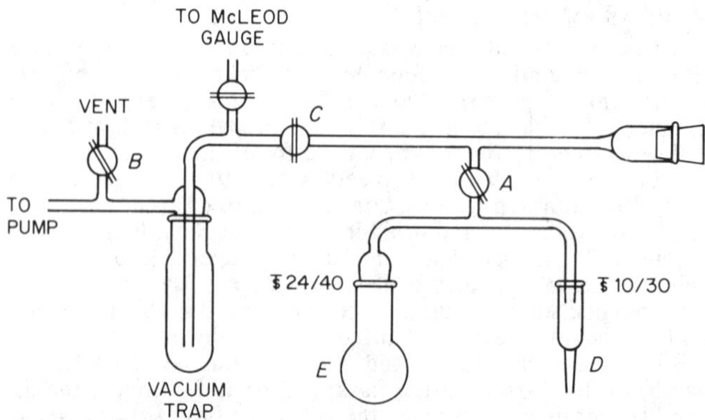

Closed-system Vacuum Distillation Apparatus for Dioxane

* A suitable tube is available as Chromaflex concentrator tube, Kontes Glass Co., Vineland, NJ (Catalog No. K42560-0000).

Test preparation—Unless otherwise specified in the individual monograph, transfer 20 g of the substance to be tested, accurately weighed, to a 50-mL round-bottom flask (*E*) having a 24/40 ground-glass neck joint. Liquefy semisolid or waxy test specimens by heating on a steam bath before making the transfer. Add 2.0 mL of water to the flask for crystalline specimens, and 1.0 mL for liquid, semisolid, or waxy specimens. Place a small polytef-covered stirring bar in the flask, insert the stopper, and stir to mix. Immerse the flask in an ice bath, and chill for about 1 minute.

Wrap heating tape around the tube connecting the concentrator tube (*D*) and the round-bottom flask (*E*), and apply about 10 volts to the tape. Apply a light coating of high-vacuum silicone grease to the ground-glass joints, and connect the tube *D* to the 10/30 joint and the round-bottom flask *E* to the 24/40 joint. Immerse the vacuum trap in a Dewar flask filled with liquid nitrogen, close stopcocks *A* and *B*, open stopcock *C*, and begin evacuating the system with a vacuum pump. Prepare a slurry bath from powdered dry ice and methanol, and raise the bath to the neck of the round-bottom flask. After freezing the contents of the flask for about 10 minutes, and when the vacuum system is operating at 0.05-mm pressure or lower, open stopcock *A* for 20 seconds, then close it. Remove the slurry bath, and allow the flask to warm in air for about 1 minute. Immerse the flask in a water bath maintained at a temperature of 20° to 25°, and after about 5 minutes warm the water bath to 35° to 40° (sufficient to liquefy most specimens) while stirring slowly but constantly with the magnetic bar. Cool the water in the bath by adding ice, and chill for about 2 minutes. Replace the water bath with the slurry bath, freeze the contents of the flask for about 10 minutes, then open stopcock *A* for 20 seconds, and close it. Remove the slurry bath, and repeat the heating steps as before, this time reaching a final temperature of 45° to 50° or a temperature necessary to melt the specimen completely. If there is any condensation in the tube connecting the round-bottom flask to the concentrator tube *D*, slowly increase the voltage to the heating tape, and heat until condensation disappears.

Stir with the magnetic stirrer throughout the following steps: Very slowly immerse the tube *D* in a Dewar flask containing liquid nitrogen. [*Caution*—When there is liquid distillate in tube *D*, immerse the tube in the liquid nitrogen *very slowly* or the tube will break.]

Water will begin to distil into the concentrator tube. As ice forms in the tube, raise the Dewar flask to keep the liquid nitrogen level only slightly below the level of ice in the tube. When water begins to freeze in the neck of the 10/30 joint, or when liquid nitrogen reaches the 2.0-mL graduation mark on the tube *D*, remove the Dewar flask, and allow the ice to melt without heating. After the ice has melted, check the volume of water that has distilled, and repeat the sequence of chilling and thawing until not less than 0.9 mL of water has been collected. Freeze the tube once again for about 2 minutes, and release the vacuum first by opening stopcock *B*, followed by opening stopcock *A*. Remove the tube *D* from the apparatus, close it with a greased stopper, and allow the ice to melt without heating. Mix the contents of the tube by swirling, note the volume of distillate, and dilute with water to 2.0 mL, if necessary. Use this *Test preparation* as directed under *Procedure*.

Procedure—Use a gas chromatograph equipped with a flame-ionization detector. Under typical conditions, the instrument contains a 2-mm × 1.8-m glass column packed with 80- to 100- or 100- to 120-mesh support S10. The column is maintained isothermally at a temperature of about 140°, the injection port at 200°, and the detector at 250°. Nitrogen or helium is the carrier gas, flowing at a rate of about 35 mL per minute. Install an oxygen scrubber between the carrier gas line and the column. Condition the column for about 72 hours at 230° with 30 to 40 mL per minute carrier flow. [NOTE—Support S10 is oxygen-sensitive. Flush both new and used columns with carrier gas for 30 to 60 minutes before heating each time they are installed in the gas chromatograph.]

Inject a volume of the *Standard preparation*, accurately measured, to produce about 20% of maximum recorder response. Where possible, keep the injection volume in the range of 2 µL to 4 µL, and use the solvent-flush technique to minimize errors associated with injection volumes. In the same manner, inject an equal volume of the *Test preparation*. The height of the peak produced by the *Test preparation* is not greater than that produced by the *Standard preparation*. The limit is 10 ppm, unless otherwise specified in the individual monograph.

⟨226⟩ 4-EPIANHYDROTETRACYCLINE

This chromatographic procedure is provided to demonstrate that the content of 4-epianhydrotetracycline, a degradation product of tetracycline, does not exceed the limit given in the individual monograph.

EDTA Buffer—Dissolve 37.2 g of disodium ethylenediaminetetraacetate in 800 mL of water, adjust with ammonium hydroxide to a pH of 7.8, dilute with water to 1000 mL, and mix.

Support Phase—Add 5 mL of *EDTA buffer* to 10 g of acid-washed chromatographic siliceous earth for column chromatography, and mix until the siliceous earth is uniformly moistened.

Test Solution—Prepare as directed in the individual monograph.

Procedure—Prepare a 15-mm × 170-mm chromatographic tube with a 4-mm × 50-mm outlet by packing it, in increments, with *Support phase*, firmly tamping down each increment, until the tube is filled to a height of about 10 cm. In a beaker, prepare a mixture of 1 g of acid-washed chromatographic siliceous earth for column chromatography and 1 mL of *Test solution*. Transfer the mixture to the top of the column. Dry-wash the beaker with *Support phase*, and transfer to the column to provide an additional 1-cm layer on top of the mixture containing the *Test solution*. Within 30 minutes, pass chloroform through the column, and collect successive fractions of 5.0 mL, 5.0 mL, 10.0 mL, 10.0 mL, and 5.0 mL. Observe the column during elution, and note the appearance of two separate yellow bands. The fraction or fractions containing the first yellow band contain the anhydrotetracyclines. Discard these fractions. The fractions after the first yellow band contain the 4-epianhydrotetracycline. Determine the absorbance of each 4-epianhydrotetracycline fraction at the wavelength of maximum absorbance at about 438 nm, with a suitable spectrophotometer, diluting each fraction, if necessary, with chloroform, and using chloroform as the blank. Calculate the quantity, in mg, of 4-epianhydrotetracycline in each fraction by the formula:

$$AVD/20.08,$$

in which A is the absorbance, V is the volume, in mL, of the fraction taken, D is the dilution factor, if the fraction was diluted, and 20.08 is the absorptivity of 4-epianhydrotetracycline at 438 nm. From the sum of the quantities of 4-epianhydrotetracycline found in the fractions, calculate the percentage of 4-epianhydrotetracycline in relation to the tetracycline hydrochloride equivalent contained in the *Test solution*.

⟨231⟩ HEAVY METALS

This test is provided to demonstrate that the content of metallic impurities that are colored by sulfide ion, under the specified test conditions, does not exceed the *Heavy metals* limit specified in the individual monograph in terms of the percentage (by weight) of lead in the test substance, as determined by concomitant visual comparison (see *Visual Comparison* in the section *Procedure* under *Spectrophotometry and Light-scattering* ⟨851⟩) with a control prepared from a *Standard Lead Solution*.

Determine the amount of heavy metals by *Method I*, unless otherwise specified in the individual monograph. *Method I* is used for substances that yield clear, colorless preparations under the specified test conditions. *Method II* is used for substances that do not yield clear, colorless preparations under the test conditions specified for *Method I*, or for substances that, by virtue of their complex nature, interfere with the precipitation of metals by sulfide ion, or for fixed and volatile oils. *Method III*, a wet-digestion method, is used only in those cases where neither *Method I* nor *Method II* can be utilized.

Special Reagents

Lead Nitrate Stock Solution—Dissolve 159.8 mg of lead nitrate in 100 mL of water to which has been added 1 mL of nitric acid, then dilute with water to 1000 mL. Prepare and store this solution in glass containers free from soluble lead salts.

Standard Lead Solution—On the day of use, dilute 10.0 mL of *Lead Nitrate Stock Solution* with water to 100.0 mL. Each mL of *Standard Lead Solution* contains the equivalent of 10 µg of lead. A comparison solution prepared on the basis of 100 µL of *Standard Lead Solution* per g of substance being tested contains the equivalent of 1 part of lead per million parts of substance being tested.

Method I

Standard Preparation—Into a 50-mL color-comparison tube pipet 2 mL of *Standard Lead Solution* (20 µg of Pb), and dilute with water to 25 mL. Adjust with 1 *N* acetic acid or 6 *N* ammonium hydroxide to a pH between 3.0 and 4.0, using short-range pH indicator paper as external indicator, dilute with water to 40 mL, and mix.

Test Preparation—Into a 50-mL color-comparison tube place 25 mL of the solution prepared for the test as directed in the individual monograph; or, using the designated volume of acid where specified in the individual monograph, dissolve and dilute with water to 25 mL the quantity, in g, of the substance to be tested, as calculated by the formula:

$$2.0/(1000L),$$

in which L is the *Heavy metals* limit, in percentage. Adjust with 1 *N* acetic acid or 6 *N* ammonium hydroxide to a pH between 3.0 and 4.0, using short-range pH indicator paper as external indicator, dilute with water to 40 mL, and mix.

Monitor Preparation—Into a third 50-mL color-comparison tube place 25 mL of a solution prepared as directed for *Test Preparation*, and add 2.0 mL of *Standard Lead Solution*. Adjust with 1 *N* acetic acid or 6 *N* ammonium hydroxide to a pH between 3.0 and 4.0, using short-range pH indicator paper as external indicator, dilute with water to 40 mL, and mix.

Procedure—To each of the three tubes containing the *Standard Preparation*, the *Test Preparation*, and the *Monitor Preparation*, respectively, add 10 mL of freshly prepared hydrogen sulfide TS, mix, allow to stand for 5 minutes, and view downward over a white surface: the color of the solution from the *Test Preparation* is not darker than that of the solution from the *Standard Preparation*, and the intensity of the color of the *Monitor Preparation* is equal to or greater than that of the *Standard Preparation*. [NOTE—If the color of the *Monitor Preparation* is lighter than that of the *Standard Preparation*, use *Method II* instead of *Method I* for the substance being tested.]

Method II

Standard Preparation—Prepare as directed under *Method I*.

Test Preparation—Use a quantity, in g, of the substance to be tested as calculated by the formula:

$$2.0/(1000L),$$

in which L is the *Heavy metals* limit, in percentage. Transfer the weighed quantity of the substance to a suitable crucible, add sufficient sulfuric acid to wet the substance, and carefully ignite at a low temperature until thoroughly charred. (The crucible may be loosely covered with a suitable lid during the charring.) Add to the carbonized mass 2 mL of nitric acid and 5 drops of sulfuric acid, and heat cautiously until white fumes no longer are evolved. Ignite, preferably in a muffle furnace, at 500° to 600°, until the carbon is completely burned off. Cool, add 4 mL of 6 *N* hydrochloric acid, cover, digest on a steam bath for 15 minutes, uncover, and slowly evaporate on a steam bath to dryness. Moisten the residue with 1 drop of hydrochloric acid, add 10 mL of hot water, and digest for 2 minutes. Add 6 *N* ammonium hydroxide dropwise, until the solution is just alkaline to litmus paper, dilute with water to 25 mL, and adjust with 1 *N* acetic acid to a pH between 3.0 and 4.0, using short-range pH indicator paper as external indicator. Filter if necessary, rinse the crucible and the filter with 10 mL of water, combine the filtrate and rinsing in a 50-mL color-comparison tube, dilute with water to 40 mL, and mix.

Procedure—To each of the tubes containing the *Standard Preparation* and the *Test Preparation*, respectively, add 10 mL of freshly prepared hydrogen sulfide TS, mix, allow to stand for 5 minutes, and view downward over a white surface: the color of the solution from the *Test Preparation* is not darker than that of the solution from the *Standard Preparation*.

Method III

Standard Preparation—Transfer a mixture of 8 mL of sulfuric acid and 10 mL of nitric acid to a clean, dry, 100-mL Kjeldahl flask, and add a further volume of nitric acid equal to the incremental volume of nitric acid added to the *Test Preparation*. Heat the solution to the production of dense, white fumes, cool, cautiously add 10 mL of water and, if hydrogen peroxide was used in treating the *Test Preparation*, add a volume of 30 percent hydrogen peroxide equal to that used for the substance being tested, and boil gently to the production of dense, white fumes. Again cool, cautiously add 5 mL of water, mix, and boil gently to the production of dense, white fumes and to a volume of 2 to 3 mL. Cool, dilute cautiously with a few mL of water, add 2.0 mL of *Standard Lead Solution* (20 µg of Pb), and mix. Transfer to a 50-mL color-comparison tube, rinse the flask with water, adding the rinsing to the tube until the volume is 25 mL, and mix.

Test Preparation—

If the substance is a solid—Transfer the quantity of the test substance specified in the individual monograph to a clean, dry, 100-mL Kjeldahl flask [NOTE—A 300-mL flask may be used if the reaction foams excessively], clamp the flask at an angle of 45°, and add sufficient of a mixture of 8 mL of sulfuric acid and 10 mL of nitric acid to moisten the substance thoroughly. Warm gently until the reaction commences, allow the reaction to subside, and add additional portions of the same acid mixture, heating after each addition, until a total of 18 mL of the acid mixture has been added. Increase the amount of heat, and boil gently until the solution darkens. Cool, add 2 mL of nitric acid, and again heat until the solution darkens. Continue the heating, followed by addition of nitric acid until no further darkening occurs, then heat strongly to the production of dense, white fumes. Cool, cautiously add 5 mL of water, boil gently to the production of dense, white fumes, and continue heating until the volume is reduced to a few mL. Cool, cautiously add 5 mL of water, and examine the color of the solution. If the color is yellow, cautiously add 1 mL of 30 percent hydrogen peroxide, and again evaporate to the production of dense, white fumes and a volume of 2 to 3 mL. If the solution is still yellow in color, repeat the addition of 5 mL of water and the peroxide treatment. Cool, dilute cautiously with a few mL of water, and rinse into a 50-mL color-comparison tube, taking care that the combined volume does not exceed 25 mL.

If the substance is a liquid—Transfer the quantity of the test substance specified in the individual monograph to a clean, dry, 100-mL Kjeldahl flask [NOTE—A 300-mL flask may be used if the reaction foams excessively], clamp the flask at an angle of 45°, and cautiously add a few mL of a mixture of 8 mL of sulfuric acid and 10 mL of nitric acid. Warm gently until the reaction commences, allow the reaction to subside, and proceed as directed under *If the substance is a solid*, beginning with "add additional portions of the same acid mixture."

Procedure—Treat the *Test Preparation* and the *Standard Preparation* as follows: Adjust the solution to a pH between 3.0 and 4.0, using short-range pH indicator paper as external indicator, with ammonium hydroxide (a dilute ammonia solution may be used, if desired, as the specified range is approached), dilute with water to 40 mL, and mix.

To each tube add 10 mL of freshly prepared hydrogen sulfide TS, mix, allow to stand for 5 minutes, and view downward over a white surface: the color of the *Test Preparation* is not darker than that of the *Standard Preparation*.

⟨241⟩ IRON

This limit test is provided to demonstrate that the content of iron, in either the ferric or the ferrous form, does not exceed the limit for iron specified in the individual monograph. The determination is made by concomitant visual comparison with a control prepared from a standard iron solution.

Special Reagents—

STANDARD IRON SOLUTION—Dissolve 863.4 mg of ferric ammonium sulfate [$FeNH_4(SO_4)_2 . 12H_2O$] in water, add 10 mL of 2 *N* sulfuric acid, and dilute with water to 100.0 mL. Pipet 10 mL of this solution into a 1000-mL volumetric flask, add 10 mL of 2 *N* sulfuric acid, dilute with water to volume, and mix. This solution contains the equivalent of 0.01 mg (10 µg) of iron per mL.

AMMONIUM THIOCYANATE SOLUTION—Dissolve 30 g of ammonium thiocyanate in water to make 100 mL.

Standard Preparation—Into a 50-mL color-comparison tube pipet 1 mL of *Standard Iron Solution* (10 µg of Fe), dilute with water to 45 mL, add 2 mL of hydrochloric acid, and mix.

Test Preparation—Into a 50-mL color comparison tube place the solution prepared for the test as directed in the individual monograph and if necessary dilute with water to 45 mL; or, dissolve in water, and dilute with water to 45 mL the quantity, in g, of the substance to be tested, as calculated by the formula:

$$1.0/(1000L),$$

in which *L* is the *Iron* limit in percentage. Add 2 mL of hydrochloric acid, and mix.

Procedure—To each of the tubes containing the *Standard Preparation* and the *Test Preparation* add 50 mg of ammonium peroxydisulfate crystals and 3 mL of *Ammonium Thiocyanate Solution*, and mix: the color of the solution from the *Test Preparation* is not darker than that of the solution from the *Standard Preparation.*

⟨251⟩ LEAD

The imposition of stringent limits on the amounts of lead that may be present in pharmaceutical products has resulted in the use of two methods, of which the one set forth following depends upon extraction of lead by solutions of dithizone. For determination of the content of heavy metals generally, expressed as a lead equivalent, see *Heavy Metals* ⟨231⟩.

Select all reagents for this test to have as low a content of lead as practicable, and store all reagent solutions in containers of borosilicate glass. Rinse thoroughly all glassware with warm dilute nitric acid (1 in 2), followed by water.

Special Reagents—

AMMONIA-CYANIDE SOLUTION—Dissolve 2 g of potassium cyanide in 15 mL of ammonium hydroxide, and dilute with water to 100 mL.

AMMONIUM CITRATE SOLUTION—Dissolve 40 g of citric acid in 90 mL of water. Add 2 or 3 drops of phenol red TS, then cautiously add ammonium hydroxide until the solution acquires a reddish color. Remove any lead that may be present by extracting the solution with 20-mL portions of *Dithizone Extraction Solution* (see below), until the dithizone solution retains its orange-green color.

DILUTED STANDARD LEAD SOLUTION—Dilute an accurately measured volume of *Standard Lead Solution* (see *Heavy Metals* ⟨231⟩) [containing 10 µg of lead per mL], with 9 volumes of dilute nitric acid (1 in 100) to obtain a solution that contains 1 µg of lead per mL.

DITHIZONE EXTRACTION SOLUTION—Dissolve 30 mg of dithizone in 1000 mL of chloroform, and add 5 mL of alcohol. Store the solution in a refrigerator.

Before use, shake a suitable volume of the dithizone extraction solution with about half its volume of dilute nitric acid (1 in 100), discarding the nitric acid.

HYDROXYLAMINE HYDROCHLORIDE SOLUTION—Dissolve 20 g of hydroxylamine hydrochloride in sufficient water to make approximately 65 mL. Transfer to a separator, add 5 drops of thymol blue TS, then add ammonium hydroxide until the solution assumes a yellow color. Add 10 mL of sodium diethyldithiocarbamate solution (1 in 25), mix, and allow to stand for 5 minutes. Extract this solution with successive 10- to 15-mL portions of chloroform until a 5-mL portion of the chloroform extract does not assume a yellow color when shaken with cupric sulfate TS. Add 3 *N* hydrochloric acid until the solution is pink (if necessary, add 1 or 2 drops more of thymol blue TS), and then dilute with water to 100 mL.

POTASSIUM CYANIDE SOLUTION—Dissolve 50 g of potassium cyanide in sufficient water to make 100 mL. Remove the lead from this solution by extraction with successive portions of *Dithizone Extraction Solution*, as described under *Ammonium Citrate Solution* above, then extract any dithizone remaining in the cyanide solution by shaking with chloroform. Finally dilute the cyanide solution with sufficient water so that each 100 mL contains 10 g of potassium cyanide.

STANDARD DITHIZONE SOLUTION—Dissolve 10 mg of dithizone in 1000 mL of chloroform. Keep the solution in a glass-stoppered, lead-free bottle, suitably wrapped to protect it from light, and store in a refrigerator.

NOTE—The following special reagents are called for in the test for *Lead* under *Ferrous Sulfate.*

CITRATE-CYANIDE WASH SOLUTION—To 50 mL of water add 50 mL of *Ammonium Citrate Solution* and 4 mL of *Potassium Cyanide Solution*, mix, and adjust the pH, if necessary, with ammonium hydroxide to 9.0.

pH 2.5 BUFFER SOLUTION—To 25.0 mL of 0.2 *M* potassium biphthalate add 37.0 mL of 0.1 *N* hydrochloric acid, and dilute with water to 100.0 mL.

DITHIZONE–CARBON TETRACHLORIDE SOLUTION—Dissolve 10 mg of dithizone in 1 liter of carbon tetrachloride. Prepare this solution on the day of use.

pH 2.5 WASH SOLUTION—To 500 mL of dilute nitric acid (1 in 100) add 6 *N* ammonium hydroxide until the pH of the mixture is 2.5, then add 10 mL of *pH 2.5 Buffer Solution*, and mix.

AMMONIA-CYANIDE WASH SOLUTION—To 35 mL of *pH 2.5 Wash Solution* add 4 mL of *Ammonia-Cyanide Solution*, and mix.

Test Preparation—[NOTE—If, in the following preparation, the substance under test reacts too rapidly and begins charring with 5 mL of sulfuric acid before heating, use instead 10 mL of cooled dilute sulfuric acid (1 in 2), and add a few drops of the hydrogen peroxide before heating]. Where the monograph does not specify preparation of a solution, prepare a *Test Preparation* as follows:

Caution—Exercise safety precautions in this procedure, as some substances may react with explosive violence when digested with hydrogen peroxide. Transfer 1.0 g of the substance under test to a suitable flask, add 5 mL of sulfuric acid and a few glass beads, and digest on a hot plate in a hood until charring begins. Other suitable means of heating may be substituted. (Add additional sulfuric acid, if necessary, to wet the substance completely, but do not add more than a total of 10 mL.) Add, dropwise and with caution, 30 percent hydrogen peroxide, allowing the reaction to subside and again heating between drops. Add the first few drops very slowly, mix carefully to prevent a rapid reaction, and discontinue heating if foaming becomes excessive. Swirl the solution in the flask to prevent unreacted substance from caking on the walls of the flask. [*NOTE—Add peroxide whenever the mixture turns brown or darkens.*] Continue the digestion until the substance is completely destroyed, copious fumes of sulfur trioxide are evolved, and the solution is colorless. Cool, cautiously add 10 mL of water, evaporate until sulfur trioxide again is evolved, and cool. Repeat this procedure with another 10 mL of water to remove any traces of hydrogen peroxide. Cautiously dilute with 10 mL of water, and cool.

Procedure—Transfer the *Test Preparation*, rinsing with 10 mL of water, or the volume of the prepared sample specified in the monograph to a separator, and, unless otherwise directed in the monograph, add 6 mL of *Ammonium Citrate Solution* and 2 mL of *Hydroxylamine Hydrochloride Solution.* (For the determination of lead in iron salts use 10 mL of *Ammonium Citrate Solution.*) Add 2 drops of phenol red TS, and make the solution just alkaline (red in color) by the addition of ammonium hydroxide. Cool the solution if necessary, and add 2 mL of *Potassium Cyanide Solution.* Immediately extract the solution with 5-mL portions of *Dithizone Extraction Solution*, draining off each extract into another separator, until the dithizone solution retains its green color. Shake the combined dithizone solutions for 30 seconds with 20 mL of dilute nitric acid (1 in 100), and discard the chloroform layer. Add to the acid solution 5.0 mL of *Standard Dithizone Solution* and 4 mL of *Ammonia-Cyanide Solution*, and shake for 30 seconds: the color of the chloroform layer is of no deeper shade of violet than that of a control made with a volume of *Diluted Standard Lead Solution* equivalent to the amount of lead permitted in the sample under examination, and the same quantities of the same reagents and in the same manner as in the test with the sample.

⟨261⟩ MERCURY

Method I

[NOTE—Mercuric dithizonate is light-sensitive. Perform this test in subdued light.]

Reagents—

Dithizone Stock Solution—Dissolve 40 mg of dithizone in 1000 mL of chloroform.

Dithizone Titrant—Dilute 30.0 mL of *Dithizone Stock Solution* with chloroform to 100.0 mL. This solution contains approximately 12 mg of dithizone per liter.

Mercury Stock Solution—Transfer 135.4 mg of mercuric chloride to a 100-mL volumetric flask, and dilute with 1 *N* sulfuric acid to volume. This solution contains the equivalent of 100 mg of Hg in 100 mL.

Mercury Solution for Standardizing Dithizone Titrant—Transfer 2.0 mL of *Mercury Stock Solution* to a 100-mL volumetric flask, and dilute with 1 *N* sulfuric acid to volume. Each mL of this solution contains the equivalent of 20 µg of Hg.

The following solutions are called for in the limit test for mercury that is specified in the monographs on Ferrous Fumarate, Ferrous Sulfate, and Dried Ferrous Sulfate.

HYDROXYLAMINE HYDROCHLORIDE SOLUTION—Prepare as directed in the test for *Lead* ⟨251⟩.

STANDARD MERCURY SOLUTION—On the day of use, quantitatively dilute 1.0 mL of *Mercury Stock Solution* with 1 *N* sulfuric acid to 1000 mL. Each mL of the resulting solution contains the equivalent of 1 µg of mercury.

DITHIZONE EXTRACTION SOLUTION—Prepare as directed in the test for *Lead* ⟨251⟩.

DILUTED DITHIZONE EXTRACTION SOLUTION—Just prior to use, dilute 5 mL of *Dithizone Extraction Solution* with 25 mL of chloroform.

Standardization of Dithizone Titrant—Transfer 1.0 mL of *Mercury Solution for Standardizing Dithizone Titrant* to a 250-mL separator, and add 100 mL of 1 *N* sulfuric acid, 90 mL of water, 1 mL of glacial acetic acid, and 10 mL of hydroxylamine hydrochloride solution (1 in 5). Titrate the solution with *Dithizone Titrant* from a 10-mL microburet, shaking the mixture 20 times after each addition and allowing the chloroform layer to separate, then discarding the chloroform layer. Continue until a final addition of *Dithizone Titrant* is green in color after shaking. Calculate the quantity, in µg, of Hg equivalent to each mL of *Dithizone Titrant* by the formula:

$$20/V,$$

in which *V* is the volume, in mL, of *Dithizone Titrant* added.

Test Preparation—Transfer about 2 g of the substance under test, accurately weighed, to a glass-stoppered, 250-mL conical flask, add 20 mL of a mixture of equal volumes of nitric acid and sulfuric acid, attach a suitable condenser, reflux the mixture for 1 hour, cool, cautiously dilute with water, and boil until fumes of nitrous acid no longer are noticeable. Cool the solution, cautiously dilute with water, transfer to a 200-mL volumetric flask, dilute with water to volume, mix, and filter.

Procedure—Transfer 50.0 mL of *Test Preparation* to a 250-mL separator, and extract with successive small portions of chloroform until the last chloroform extract remains colorless. Discard the chloroform extract, and add to the extracted *Test Preparation* 50 mL of 1 *N* sulfuric acid, 90 mL of water, 1 mL of glacial acetic acid, and 10 mL of hydroxylamine hydrochloride solution (1 in 5). Proceed as directed under *Standardization of Dithizone Titrant*, beginning with "Titrate the solution." Calculate the amount of mercury.

Method IIa and Method IIb

Mercury Detection Instrument—Use any suitable atomic absorption spectrophotometer equipped with a fast-response recorder and capable of measuring the radiation absorbed by mercury vapors at the mercury resonance line of 253.6 nm. [NOTE—Wash all glassware associated with the test with nitric acid, and rinse thoroughly with water before use.]

Aeration Apparatus—The apparatus (see accompanying diagram) consists of a flowmeter capable of measuring flow rates from 500 to 1000 mL per minute, connected via a three-way stopcock fitted with a polytef plug to an aeration vessel (250-mL gas washing bottle), followed by a trap, a drying tube packed with magnesium perchlorate, a 10-cm × 25-mm flow-through cell with quartz windows, and terminating with a vent to a fume hood.

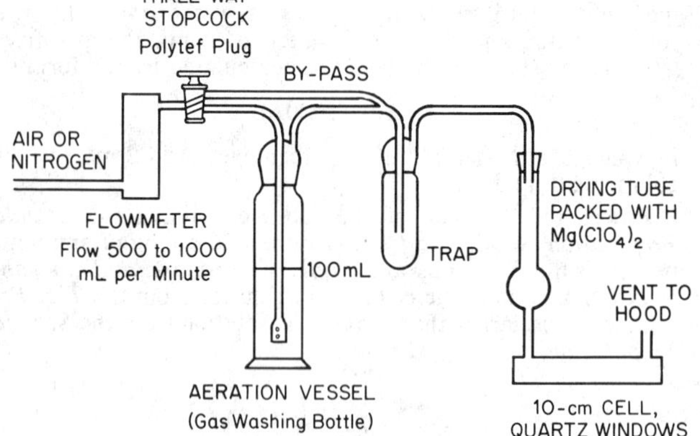

Mercury Aeration Apparatus

Reagents—

Potassium Permanganate Solution—Dissolve 5 g of potassium permanganate in 100 mL of water.

Hydroxylamine Hydrochloride Solution—Dissolve 10 g of hydroxylamine hydrochloride in 100 mL of water.

Stannous Chloride Solution—Dissolve 10 g of $SnCl_2.2H_2O$ in 20 mL of warm hydrochloric acid, and add 80 mL of water. Prepare fresh each week.

Standard Mercury Solution—Prepare from *Mercury Stock Solution* as directed under *Method I*. Each mL of the *Standard Mercury Solution* contains the equivalent of 1 µg of mercury.

Test Preparation—Unless otherwise directed in the individual monograph, use the quantity, in g, of the test substance calculated by the formula:

$$2.0/L,$$

in which *L* is the mercury limit, in ppm.

Method IIa

Standard Preparation—Pipet 2.0 mL of *Standard Mercury Solution* into a 100-mL beaker, and add 35 mL of water, 3 mL of sulfuric acid, and 1 mL of potassium permanganate solution. Cover the beaker with a watch glass, boil for a few seconds, and cool.

Test Preparation—Transfer the calculated amount of the test substance to a 100-mL beaker, and add 35 mL of water. Stir, and warm to assist solution, if necessary. Add 2 drops of phenolphthalein TS, and, as necessary, slowly neutralize with constant stirring, using 1 *N* sodium hydroxide or 1 *N* sulfuric acid. Add 3 mL of sulfuric acid and 1 mL of *Potassium Permanganate Solution*. Cover the beaker with a watch glass, boil for a few seconds, and cool.

Procedure—Assemble the *Aeration apparatus* as shown in the accompanying diagram, with the aeration vessel and the trap empty, and the stopcock in the bypass position. Connect the apparatus to the absorption cell, and adjust the air or nitrogen flow rate so that, in the following procedure, maximum absorption and reproducibility are obtained without excessive foaming in the test solution. Obtain a smooth baseline reading at 253.6 nm, following the manufacturer's instructions for operating the instrument.

Treat the *Standard Preparation* and the *Test Preparation* similarly, as follows: Destroy the excess permanganate by adding *Hydroxylamine Hydrochloride Solution*, dropwise, until the solution is colorless. Immediately wash the solution into the aeration vessel with water, and dilute with water to 100 mL. Add 2 mL of *Stannous Chloride Solution*, and immediately reconnect the aeration vessel to the aeration apparatus. Turn the stopcock from the bypass position to the aerating position, and continue the aeration until the absorption peak has been passed and the recorder pen returns to the baseline. Disconnect the aeration vessel from the apparatus, and wash with water after each use. After correcting for any reagent blank, any absorbance produced by the *Test Preparation* does not exceed that produced by the *Standard Preparation.*

Method IIb

[*Caution—Some substances may react with explosive violence when digested with hydrogen peroxide. Exercise safety precautions at all times.*]

Standard Preparation—Pipet 2.0 mL of *Standard Mercury Solution* into a 125-mL conical flask, add 3 mL each of nitric acid and sulfuric acid, mix, and add an amount of 30 percent hydrogen peroxide equal to the total amount used in preparing the *Test Preparation.* Attach a suitable water-cooled condenser with a standard-taper joint to fit the flask, and reflux the mixture in a fume hood for 1 hour. Turn off the water circulating through the condenser, and heat until white fumes appear in the flask. Cool, and cautiously add 10 mL of water through the condenser, while swirling the flask. Again heat until white fumes appear, cool, and add an additional 15 mL of water. Remove the condenser, and rinse the sides of the flask to obtain a volume of 35 mL. Add 1 mL of *Potassium Permanganate Solution*, boil for a few seconds, and cool.

Test Preparation—Transfer the calculated amount of the test substance to a 125-mL conical flask. Add 5 mL each of nitric acid and sulfuric acid and a few glass beads. Attach a suitable water-cooled condenser with a standard-taper joint to fit the flask, and digest in a fume hood, preferably on a hot plate, and at a temperature not exceeding 120°, until charring begins. (If additional sulfuric acid is necessary to wet the specimen completely, add it carefully through the condenser, but do not allow the total volume added to exceed 10 mL.) After the test substance has been decomposed by the acid, cautiously add, dropwise through the condenser, 30 percent hydrogen peroxide, allowing the reaction to subside and again heating between drops (add the first few drops very slowly with sufficient mixing, in order to prevent a rapid reaction; discontinue heating if foaming becomes excessive). When the reaction has abated, heat cautiously, rotating the flask occasionally to prevent the specimen from caking on glass exposed to the heating unit. Maintain oxidizing conditions at all times during the digestion by adding small quantities of the hydrogen peroxide solution whenever the mixture turns brown or darkens. Continue the digestion until the organic matter is destroyed, and then reflux the mixture for 1 hour. Turn off the water circulating through the condenser, and heat until fumes of sulfur trioxide are copiously evolved and the solution becomes colorless or retains only a light straw color. Cool, and cautiously add 10 mL of water through the condenser, while swirling the flask. Again heat until white fumes appear. Cool, and cautiously add 15 mL of water. Remove the condenser, and rinse the sides of the flask with a few mL of water to obtain a volume of 35 mL. Add 1 mL of *Potassium Permanganate Solution*, boil for a few seconds, and cool.

Procedure—Proceed as directed for *Procedure* under *Method IIa.*

⟨271⟩ READILY CARBONIZABLE SUBSTANCES TEST

In tests for readily carbonizable substances, unless otherwise directed, add the specified quantity of the substance, finely powdered if in solid form, in small portions to the comparison container, which is made of colorless glass resistant to the action of sulfuric acid and contains the specified volume of sulfuric acid TS (see under *Test Solutions*).

Stir the mixture with a glass rod until solution is complete, allow the solution to stand for 15 minutes, unless otherwise directed, and compare the color of the solution with that of the specified matching fluid in a comparison container which also is of colorless glass and has the same internal and cross-section dimensions, viewing the fluids transversely against a background of white porcelain or white glass.

When heat is directed in order to effect solution of the substance in the sulfuric acid TS, mix the sample and the acid in a test tube, heat as directed, and transfer the solution to the comparison container for matching with the designated Matching Fluid (see *Color and Achromicity* ⟨631⟩).

Special attention is directed to the importance of the concentration of sulfuric acid used in this test. The reagent of the required strength, i.e., 95.0 ± 0.5 percent of H_2SO_4, is designated as a "Test Solution."

⟨281⟩ RESIDUE ON IGNITION

Weigh accurately 1 to 2 g of the substance, or the amount specified in the individual monograph, in a suitable crucible that previously has been ignited, cooled, and weighed. Heat, gently at first, until the substance is thoroughly charred, cool, then, unless otherwise directed in the individual monograph, moisten the residue with 1 mL of sulfuric acid, heat gently until white fumes no longer are evolved, and ignite at 800 ± 25° until the carbon is consumed. Cool in a desiccator, weigh, and calculate the percentage of residue. If the amount of the residue so obtained exceeds the limit specified in the individual monograph, again moisten the residue with 1 mL of sulfuric acid, heat and ignite as before, and again calculate the percentage of residue. Continue the ignition until constant weight is attained, unless otherwise specified.

Conduct the ignition in a well-ventilated hood, but protected from air currents, and at as low a temperature as is possible to effect the complete combustion of the carbon. A muffle furnace may be used, if desired, and its use is recommended for the final ignition at 800 ± 25°.

Calibration of the muffle furnace may be carried out using an appropriate digital temperature meter and a working thermocouple probe calibrated against a standard thermocouple traceable to the National Bureau of Standards.

Verify the accuracy of the measuring and controlling circuitry of the muffle furnace by checking the positions in the furnace at the control set point temperature of intended use. Select positions that reflect the eventual method of use with respect to location of the specimen under test. The tolerance is ±25° at each position measured.

⟨291⟩ SELENIUM

Stock Solution—Dissolve 40.0 mg of metallic selenium in 100 mL of dilute nitric acid (1 in 2) in a 1000-mL volumetric flask, warming gently on a steam bath if necessary to effect solution, add water to volume, and mix. Pipet 5 mL of this solution into a 200-mL volumetric flask, add water to volume, and mix. Each mL of the resulting solution contains the equivalent of 1 µg of selenium (Se).

Diaminonaphthalene Solution—Dissolve 100 mg of 2,3-diaminonaphthalene and 500 mg of hydroxylamine hydrochloride in 0.1 *N* hydrochloric acid to make 100 mL. Prepare this solution fresh on the day of use.

Standard Solution—Pipet 6 mL of *Stock Solution* into a 150-mL beaker, and add 25 mL of dilute nitric acid (1 in 30) and 25 mL of water.

Test Solution—Clean combustion of the test material is an important factor in conducting the test. For compounds that burn poorly and produce soot, the addition of magnesium oxide usually results in more thorough combustion and reduces soot formation. Where the need to add magnesium oxide has been identified, it is specified in the individual monograph. Using a 1000-mL com bustion flask and using 25 mL of dilute nitric acid (1 in 30) as the absorbing liquid, proceed as directed under *Oxygen Flask Combustion* ⟨471⟩. Upon completion of the combustion, place a few mL of water in the cup, loosen the stopper, and rinse the stopper, the specimen holder, and the sides of the flask with about 10 mL of water. Transfer the solution with the aid of about 20 mL of water to a 150-mL beaker, and heat gently to the boiling temperature. Boil for 10 minutes, and allow the solution to cool to room temperature.

Procedure—Treat the *Standard Solution*, the *Test Solution*, and the reagent blank consisting of 25 mL of dilute nitric acid (1 in 30) and 25 mL of water, concomitantly and in parallel, as follows: Add ammonium hydroxide solution (1 in 2) to adjust to a pH of 2.0 ± 0.2. Dilute with water to 60.0 mL, and transfer to a low-actinic separator with the aid of 10.0 mL of water, adding the 10.0 mL of rinsings to the separator. Add 200 mg of hydroxylamine hydrochloride, swirl to dissolve, immediately add 5.0 mL of *Diaminonaphthalene Solution*, insert the stopper, and swirl to mix. Allow the solution to stand at room temperature for 100 minutes. Add 5.0 mL of cyclohexane, shake vigorously for 2 minutes, and allow the layers to separate. Discard the aqueous layer, and centrifuge the cyclohexane extract to remove any dispersed water. Determine the absorbances of the cyclohexane extracts of the *Test Solution* and the *Standard Solution* in a 1-cm cell at the wavelength of maximum absorbance at about 380 nm, with a suitable spectrophotometer, using the cyclohexane extract of the reagent blank as the blank, and compare the absorbances: the absorbance of the *Test Solution* is not greater than that of the *Standard Solution* where a 200-mg test specimen has been taken, or is not greater than one-half that of the *Standard Solution* where a 100-mg test specimen has been taken.

OTHER TESTS AND ASSAYS

⟨301⟩ ACID-NEUTRALIZING CAPACITY

NOTE—All tests shall be conducted at a temperature of 37 ± 3°.

Standardization of pH Meter—Standardize a pH meter using the 0.05 *m* potassium biphthalate and 0.05 *m* potassium tetraoxalate standardizing buffers as described under *pH* ⟨791⟩.

Magnetic Stirrer—Transfer 100 mL of water to a 250-mL beaker containing a 40- × 10-mm magnetic stirring bar that is coated with solid perfluorocarbon and has a spin ring at its center. Adjust the power setting of the magnetic stirrer to produce a stirring rate of 300 ± 30 rpm when the stirring bar is centered in the beaker, as determined by a suitable optical tachometer.

Test Preparation—

Powders—Transfer the accurately weighed portion of the substance specified in the individual monograph to a 250-mL beaker, add 70 mL of water, and mix on the *Magnetic Stirrer* for 1 minute.

Effervescent Solids—Transfer an accurately weighed quantity, equivalent to the minimum labeled dosage, to a 250-mL beaker, add 10 mL of water, and swirl the beaker gently while allowing the reaction to subside. Add another 10 mL of water, and swirl gently. Wash the walls of the beaker with 50 mL of water, and mix on the *Magnetic Stirrer* for 1 minute.

Suspensions and Other Liquids—Shake the container until the contents are uniform, and determine the density. Transfer an accurately weighed quantity of the uniform mixture, equivalent to the minimum labeled dosage, to a 250-mL beaker, add water to make a total volume of about 70 mL, and mix on the *Magnetic Stirrer* for 1 minute.

Non-chewable Tablets—Weigh not less than 20 tablets, and determine the average tablet weight. Grind the tablets to a fine powder, mix to obtain a uniform mixture, and transfer an accurately weighed quantity of it, equivalent to the minimum labeled dosage, to a 250-mL beaker. If wetting is desired, add not more than 5 mL of alcohol (neutralized to an apparent pH of 3.5), and mix to wet the specimen thoroughly. Add 70 mL of water, and mix on the *Magnetic Stirrer* for 1 minute.

Chewable Tablets—Prepare as directed for *Non-chewable Tablets.*

Tablets That Are Required To Be Chewed—Transfer 1 Tablet to a 250-mL beaker, add 50 mL of water, and mix on the *Magnetic Stirrer* for 1 minute.

Capsules—Weigh accurately not less than 20 capsules. Remove the capsule contents completely, with the aid of a cotton swab if necessary. Accurately weigh the empty capsules, and determine the average weight of the contents per capsule. Mix the combined capsule contents to obtain a uniform mixture, and proceed as directed for *Non-chewable Tablets*, beginning with "transfer an accurately weighed quantity of it."

Procedure for Powders, Effervescent Solids, Suspensions and Other Liquids, Non-chewable Tablets, Chewable Tablets, and Capsules—Pipet 30.0 mL of 1.0 *N* hydrochloric acid VS into the *Test Preparation* while continuing to stir with the *Magnetic Stirrer*. [NOTE—Where the acid-neutralizing capacity of the specimen under test is greater than 25 mEq, use 60.0 mL of 1.0 *N* hydrochloric acid VS.] Stir for 15 minutes, accurately timed, after the addition of the acid, begin to titrate immediately, and in a period not to exceed an additional 5 minutes, titrate the excess hydrochloric acid with 0.5 *N* sodium hydroxide VS to attain a stable (for 10 to 15 seconds) pH of 3.5. Calculate the number of mEq of acid consumed, and express the result in terms of mEq of acid consumed per g of the substance tested. Each mL of 1.0 *N* hydrochloric acid is equal to 1 mEq of acid consumed.

Procedure for Tablets That Are Required To Be Chewed—Pipet 30.0 mL of 1.0 *N* hydrochloric acid VS into the *Test Preparation* while continuing to stir with the *Magnetic Stirrer* for 10 minutes, accurately timed, after the addition of the acid. Discontinue stirring briefly, and without delay remove any gum base from the beaker using a long needle. Promptly rinse the needle with 20 mL of water, collecting the washing in the beaker, and resume stirring for 5 minutes, accurately timed, then begin to titrate immediately, and in a period not to exceed an additional 5 minutes, titrate the excess hydrochloric acid with 0.5 *N* sodium hydroxide VS to attain a stable (for 10 to 15 seconds) pH of 3.5. Calculate the number of mEq of acid consumed by the Tablet tested. Each mL of 1.0 *N* hydrochloric acid is equal to 1 mEq of acid consumed.

⟨311⟩ ALGINATES ASSAY

Apparatus—The apparatus required is shown in the accompanying diagram. It consists essentially of a soda lime column, *A*, a mercury valve, *B*, connected through a side arm to a reaction flask, *D*, by means of a rubber connection, *C*. Flask *D* is a 100-mL round-bottom, long-neck boiling flask, resting in a suitable heating mantle, *E*.

The reaction flask is provided with a reflux condenser, *F*, to which is fitted a delivery tube, *G*, of 40-mL capacity, having a stopcock, *H*. The reflux condenser terminates in a trap, *I*, containing 25 g of 20-mesh zinc or tin, which can be connected with an absorption tower, *J*.

The absorption tower consists of a 45-cm tube fitted with a medium-porosity, sintered-glass disk sealed to the inner part above the side arm and having a delivery tube sealed to it extending down to the end of the tube. A trap, consisting of a bulb of approximately 100-mL capacity, is blown above the sintered-glass disk, and the outer portion of a ground spherical joint is sealed on above the bulb. A 250-mL conical flask, *K*, is connected to the bottom of the absorption tower. The top of the tower is connected to a soda lime tower, *L*, which is connected to a suitable pump to provide vacuum and air supply, the selection of which is made by a three-way stopcock, *M*. The volume of air or vacuum is controlled by a capillary-tube regulator or needle valve, *N*.

All joints are size $^{35}/_{25}$, ground spherical type.

Procedure—Unless otherwise directed, transfer a specimen of about 250 mg, previously dried in vacuum for 4 hours at 60° and accurately weighed, into the reaction flask, *D*, add 25 mL of dilute hydrochloric acid (1 in 120), insert several boiling chips, and connect the flask to the reflux condenser, *F*, using phosphoric acid as a lubricant. [NOTE—Stopcock grease may be used for the other connections.] Check the system for air leaks by forcing

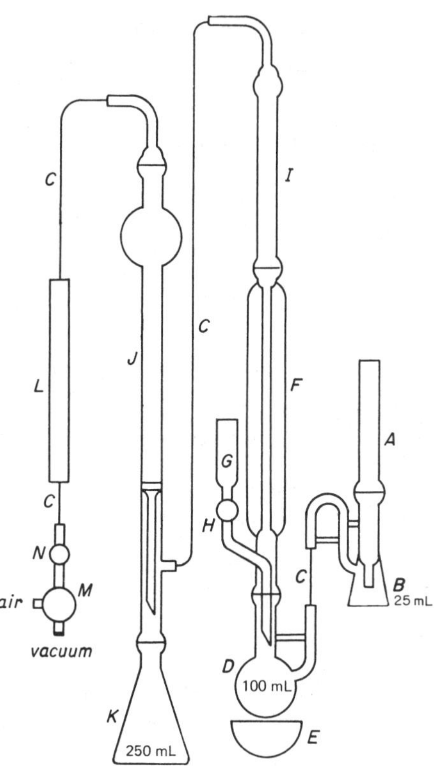

Apparatus for Alginates Assay

mercury up into the inner tube of the mercury valve, *B*, to a height of about 5 cm. Turn off the pressure using the stopcock, *M*. If the mercury level does not fall appreciably after 1 to 2 minutes, the apparatus may be considered to be free from leaks. Draw carbon dioxide–free air through the apparatus at a rate of 3000 to 6000 mL per hour. Raise the heating mantle, *E*, to the flask, heat the specimen to boiling, and boil gently for 2 minutes. Turn off and lower the mantle, and allow the specimen to cool for 15 minutes. Charge the delivery tube, *G*, with 23 mL of hydrochloric acid. Disconnect the absorption tower, *J*, rapidly transfer 25.0 mL of 0.25 *N* sodium hydroxide VS to the tower, add 5 drops of butyl alcohol, and again connect the absorption tower. Draw carbon dioxide–free air through the apparatus at the rate of about 2000 mL per hour, add the hydrochloric acid to the reaction flask through the delivery tube, raise the heating mantle, and heat the reaction mixture to boiling. After 2 hours, discontinue the current of air and heating. Force the sodium hydroxide solution down into the flask, *K*, using gentle air pressure, and then rinse down the absorption tower with three 15-mL portions of water, forcing each washing into the flask with air pressure. Remove the flask, and add to it 10 mL of barium chloride solution (1 in 10). Insert the stopper in the flask, shake gently for about 2 minutes, add phenolphthalein TS, and titrate with 0.1 *N* hydrochloric acid VS. Perform a blank determination (see *Residual Titrations* under *Titrimetry* ⟨541⟩). Each mL of 0.25 *N* sodium hydroxide consumed is equivalent to 5.5 mg of carbon dioxide (CO_2).

⟨321⟩ ALKALOIDAL DRUG ASSAYS; PROXIMATE ASSAYS

Most alkaloids are slightly or very slightly soluble in water, but soluble in certain organic solvents immiscible with water, such as chloroform, ether, amyl alcohol, and benzene, or mixtures of these. Salts of alkaloids, however, are usually soluble in water, but in most cases very slightly soluble or practically insoluble in nearly all of the organic solvents. The process of assay by immiscible solvents, generally known as the "shaking out" process, is based on these partitioning properties of alkaloids. It is carried out by treating the drug, or a concentrated liquid extract of it, with a solvent immiscible with water, in the presence of an excess of alkali which liberates the alkaloid. The free alkaloid is dissolved by the immiscible solvent from which it is subsequently removed by means of an excess of dilute aqueous acid. The acid solutions are then extracted with an immiscible solvent in the presence of an excess of alkali, and the immiscible solvent is evaporated to obtain the alkaloid which is either weighed or determined volumetrically.

Preparation of Drug for Assay—Grind the drug to be extracted to a powder of the fineness designated (see *Powder Fineness* ⟨811⟩). Care should be taken to avoid the loss of water during the powdering of the drug. If it is impossible to avoid this loss, dry the drug at a low temperature before powdering, note the loss of water, and make a correction in the final calculations.

Weighing for Assay—In weighing bulky, crude drugs for the assay, an accuracy to within 10 mg for quantities of 5 g and over is sufficient. Portions of pilular extracts or ointments may be weighed on a tared piece of waxed or parchmentized paper, the surplus paper cut away, and the paper with the specimen dropped into the vessel containing the solvent. In transferring weighed portions to a separator, thoroughly rinse the vessel in which the material to be assayed was weighed, and add the rinsings to the separator.

Extraction of Drugs—The alkaloidal content of alkaloid-bearing drugs is usually extracted by one of the following methods:

A. Maceration—Treat an accurately weighed portion of the ground drug with the specified solvent or mixture of solvents, made alkaline with ammonia TS, and thoroughly mixed. Allow to macerate for 12 to 24 hours with occasional agitation or for a shorter period with continuous agitation. At the end of this period, allow the drug to settle, decant an aliquot of the solvent, and treat as directed for *Purification of Alkaloids*.

B. Percolation—Place an accurately weighed quantity of the ground drug in a suitable container, saturate it with the specified solvent or mixture of solvents, and allow to stand for 5 minutes. Add a quantity of ammonia TS sufficient to make the mixture distinctly alkaline, and mix thoroughly with the drug. Transfer the mixture to a cylindrical percolator, previously prepared by packing the outlet with purified cotton. Use a small amount of the solvent to rinse the container, and add the rinsing to the percolator. Allow the drug to macerate for a suitable period of time (from 1 to 12 hours or overnight, depending upon the drug to be assayed). Then pack the drug firmly, place a pledget of purified cotton above it, and percolate slowly with the solvent until the drug is completely exhausted of its alkaloid content. Determine the completeness of extraction of the alkaloid by evaporating about 4 mL of the last percolate to dryness, dissolving the residue in 500 µL of approximately 0.5 *N* acid, and adding a drop of mercuric iodide TS (Valser's Reagent): not more than a slight turbidity is produced. Treat the percolate as directed for *Purification of Alkaloids*.

C. Continuous Extraction—Place an accurately weighed portion of the ground drug in an extraction thimble, and insert the thimble into a suitable extractor (a Soxhlet extractor of appropriate size is satisfactory). Moisten the drug with the specified solvent, mix by means of a stirring rod, and allow to stand for about 5 minutes. Render the mixture alkaline with the specified quantity of ammonia TS, and mix. Rinse the stirring rod with a small portion of the solvent, and allow the drug to macerate for 6 to 12 hours or overnight. Then pack the drug in the thimble, cover it with a pledget of purified cotton, add a sufficient quantity of solvent, and extract the drug for a specified period of time or until extraction is complete.

Purification of Alkaloids—The alkaloidal solution obtained by any of the extraction methods is usually contaminated with other extractives that interfere with the quantitative determinations of the alkaloids. To effect their purification remove the alkaloids from the immiscible solvent by shaking out with an acid, then render the acid solution alkaline, usually with an alkali hydroxide, and extract with an immiscible solvent.

The volume and strength of the acid to be used are usually left to the discretion of the operator. It is best, however, to keep the total volume as small as possible. For the first extraction, use not less than 10 mL of approximately 1 *N* acid or sufficient to render the mixture distinctly acid. When the drug contains a large amount of fat, use a smaller volume of more concentrated acid to prevent the formation of emulsions in the first extraction.

For succeeding extractions, use a dilution of 5 mL of the acid with 5 mL of water. In all assays, continue the extraction until 500 µL of the last acid washing shows not more than a slight turbidity on the addition of a drop of mercuric iodide TS. The acid extracts, before proceeding with the next step, should be clear or practically so. If not clear, filter or treat as follows: Shake the combined acid extracts with one or more 10-mL portions of the appropriate immiscible solvent until the acid solution is clear or practically so. Then wash the immiscible solvent extracts with one or more 5-mL portions of water acidified with hydrochloric or sulfuric acid, and add these washings to the acid solution.

Render the acid solution alkaline, in most cases with ammonia TS, and extract it with several successive portions of the appropriate immiscible solvent. Use a volume of the latter in each operation not less than half that of the water solution, and repeat the operation as long as any alkaloid is extracted by the immiscible solvent. To determine the completeness of extraction, evaporate 1 mL of the last extraction, and dissolve the residue in 0.5 mL of approximately 0.5 *N* hydrochloric acid: the resulting solution shows not more than a slight turbidity on the addition of a drop of mercuric iodide TS. The number of extractions required depends largely on the partitioning character of the alkaloid. With most alkaloids, extract several times before testing.

Washing—Carefully wash the stems of separators and funnels and the lips of flasks, separators, and graduates from which solvents containing alkaloids have been drawn or poured with some of the solvent to prevent loss and to remove any of the alkaloids left by evaporation. Add these washings to the other extractions containing the alkaloids.

Determination of Alkaloids—Evaporate the solution of the purified alkaloids in the immiscible solvent on a steam bath or with a current of air to dryness. When the alkaloidal residue is to be determined volumetrically, soften it by the addition of about 1 mL of neutralized alcohol or ether, add an accurately measured volume of standard acid, equivalent to about one and one-half to two times the volume estimated for the quantity of alkaloid present, and warm the mixture gently to ensure the complete solution of the alkaloid. If preferred, dissolve the alkaloidal residue in chloroform, add the standard acid, and remove the chloroform completely by evaporation. Then add water to make the volume of the mixture measure not less than 25 mL, and titrate the excess of acid with standard alkali, using the appropriate indicator.

If the alkaloidal residue is to be weighed, dry it at 105° to constant weight. If the final solvent has been chloroform, remove the last traces of that solvent by the addition of a few mL of neutralized ether or alcohol, followed by evaporation. Avoid loss by decrepitation, especially when evaporating chloroform solutions of alkaloids, by the addition of a little alcohol after the solution has been reduced to a volume of 1 or 2 mL, and evaporate at a low temperature, rotating the container during the evaporation.

Indicators—Unless otherwise directed in the individual monograph, use methyl red TS as the indicator in volumetric determinations and for standardizing volumetric solutions.

Aliquots—When using aliquots, measure the solvent and the aliquot at the same temperature. When handling volatile liquids, a lower temperature and a more quickly conducted operation reduce the loss by evaporation.

Adsorbants—In assaying fluidextracts, tinctures, and other preparations of alkaloid-bearing drugs, it is often necessary to evaporate these to dryness and, to avoid loss and to aid in the evaporation, they are usually added to some adsorbent material. For this purpose use paper pulp, previously acid- and alkali-washed, then made neutral by washing with water, and dried before use.

Emulsions—Shake or rotate a water solution with an immiscible solvent in a separator for about 1 minute. Avoid long or violent agitation as emulsions are likely to form, especially in alkaline solutions. Belladonna leaves sometimes contain saponins that cause troublesome emulsions. If emulsions prove persistent, draw off the emulsified portion, and add an excess of either solvent. This usually breaks the emulsion and permits a complete separation. A separated emulsion may sometimes be broken by the addition of a small amount of anhydrous sodium sulfate. If this is done, wash the residue with additional solvent to remove the alkaloid completely.

Emulsification may sometimes be prevented by increasing the volume of the water or of the immiscible solvent. Chloroform and ether solutions of drugs that contain large proportions of fat may form troublesome emulsions. In such cases, add sufficient sulfuric acid to acidify, and evaporate the volatile solvent, while stirring with a glass rod. When the resinous and fatty matter has been agglutinated, cool the acid solution, and filter through a small, wetted filter into a separator. Redissolve the residue in 15 mL of ether, add 5 to 10 mL of 0.1 *N* acid, evaporate the ether as before, with continued stirring, and pour the acid solution through the filter into the separator. Repeat the extraction of the fatty residue with dilute acid two or three times, and finally wash the filter free from alkaloids.

⟨331⟩ AMPHETAMINE ASSAY

Reference Standard—*USP Dextroamphetamine Sulfate Reference Standard*—Keep container tightly closed and protected from light. Dry at 105° for 2 hours before using.

Standard Preparation—Dissolve a suitable quantity of USP Dextroamphetamine Sulfate RS, accurately weighed, in 2 *N* sulfuric acid (saturated with chloroform), and dilute quantitatively with the same solvent to obtain a solution having a known concentration of about 0.5 mg of dextroamphetamine sulfate per mL.

Assay Preparation—Prepare as directed in the individual monograph.

Preparation of Chromatographic Column (see *Chromatography* ⟨621⟩)—Pack a pledget of fine glass wool in the base of a 25- × 300-mm chromatographic tube. Place 2 g of purified siliceous earth in a 100-mL beaker, add 1 mL of 0.1 *N* hydrochloric acid, and mix until a fluffy mixture is obtained. Transfer the mixture to the column, and tamp moderately to compress the material into a uniform mass. Transfer the *Assay Preparation* to the column, dry-rinse the beaker with 1 g of purified siliceous earth, and transfer to the column. Tamp a pledget of fine glass wool into place at the top of the column.

Procedure—Wash the column with 100 mL of chloroform previously saturated with water, and discard the washings. Place under the column, as a receiver, a 125-mL separator containing 10.0 mL of 2 *N* sulfuric acid previously saturated with chloroform. Pass through the column 35 mL of ammoniacal chloroform, prepared by equilibrating 2 mL of ammonium hydroxide and 100 mL of chloroform, and complete the elution with 70 mL of chloroform previously saturated with water. Remove the separator, shake vigorously for 1 minute, allow the layers to separate, discard the chloroform layer, and use the 10.0-mL acid solution of the sulfate salt of the amphetamine as the *Assay Solution.* Concomitantly determine the absorbance of the solution from the *Standard Preparation* and that of the *Assay Solution* in 1-cm cells at 280 nm and at the wavelength of maximum absorbance at about 257 nm, with a suitable spectrophotometer, using 2 *N* sulfuric acid previously saturated with chloroform as the blank. Record the absorbance of the solution from the *Standard Preparation* as A_S and that of the *Assay Solution* as A_U, and calculate as directed in the individual monograph.

⟨341⟩ ANTIMICROBIAL AGENTS—CONTENT

An essential component of Injections preserved in multiple-dose containers is the agent or agents present to reduce the hazard of having introduced, in the course of removing some of the contents, accidental microbial contamination of the contents remaining. It is a Pharmacopeial requirement that the presence and amount added of such agent(s) be declared on the label of the container. The methods provided herein are to be used to demonstrate that the declared agent is present but does not exceed the labeled amount.

The concentration of an antimicrobial preservative added to a multiple-dose and single-dose parenteral, otic, nasal, and ophthalmic preparation may diminish during the shelf-life of the product. The quantitative label statement of the preservative content is not intended to mean that the labeled quantity is retained during the shelf-life of the product; rather, it is a statement of the amount added and which is not exceeded. An example of such a label statement is "____ (unit) added as preservative." [NOTE—"____ (unit)" would be a number followed by the unit of measurement, e.g., 0.015 mg per mL or 0.1%.]

The most commonly used agents include the two mercurials, phenylmercuric nitrate and thimerosal, the four homologous esters of *p*-hydroxybenzoic acid, phenol, benzyl alcohol, and chlorobutanol. The methods for the first two named are polarographic, while quantitative gas chromatography is employed in the determination of the other agents.

GENERAL GAS CHROMATOGRAPHIC METHOD

The general procedure set forth in the following paragraphs is applicable to the quantitative determination of benzyl alcohol, chlorobutanol, phenol, and the methyl, ethyl, propyl, and butyl esters of *p*-hydroxybenzoic acid, the latter being treated as a group, the individual members of which, if present, are capable of separate determination. Prepare the *Internal Standard Solution* and the *Standard Preparation* for each agent as directed individually below. Unless otherwise directed below, prepare the *Test Preparation* from accurately measured portions of the *Internal Standard Solution* and the sample under test, of such size that the concentration of the agent and the composition of the solvent correspond closely to the concentration and composition of the *Standard Preparation*. Suggested operating parameters of the gas chromatograph apparatus are given in the accompanying table, the carrier gas being helium or nitrogen, and the detector being the flame-ionization type.

Suggested Operating Parameters of Gas Chromatograph Apparatus

Agent	Column Size		Column Packing Phases and Support	Flow Rate, mL per min.	Column Temperature
	Length	ID			
Benzyl Alcohol	1.8 m	3 mm	5 percent G16/S1	50	140°
Chlorobutanol	1.2 m	3 mm	5 percent G16/S1	40	110°
Phenol	1.2 m	3 mm	5 percent G16/S1	50	145°
Parabens	1.8 m	2 mm	5 percent G2/S1	20	150°

Benzyl Alcohol

Internal Standard Solution—Dissolve about 380 mg of phenol in 10 mL of methanol contained in a 200-mL volumetric flask. Add water to volume, and mix.

Standard Preparation—Dissolve about 180 mg of benzyl alcohol, accurately weighed, in 20.0 mL of methanol contained in a 100-mL volumetric flask. Add *Internal Standard Solution* to volume, and mix.

Procedure—Using 5-µL portions of the *Standard Preparation* and the *Test Preparation*, record their gas chromatograms with the apparatus adjusted to the parameters set forth in the accompanying table. Measure the areas under the peaks for benzyl alcohol and phenol of the chromatogram for the *Standard Preparation*, designating them P_1 and P_2, respectively. Similarly, determine the corresponding values p_1 and p_2 for the *Test Preparation*. Calculate the content, in mg per mL, of benzyl alcohol (C_7H_8O) in the specimen taken by the formula:

$$100(C/V)(p_1/p_2)(P_2/P_1),$$

in which C is the concentration, in mg per mL, of benzyl alcohol in the *Standard Preparation*, and V is the volume, in mL, of the specimen under test used in preparing each 100 mL of the *Test Preparation*.

Chlorobutanol

Internal Standard Solution—Dissolve about 130 mg of benzaldehyde in 5 mL of methanol contained in a 100-mL volumetric flask. Add water to volume, and mix.

Standard Preparation—Dissolve about 500 mg of anhydrous chlorobutanol, accurately weighed, in 5 mL of methanol contained in a 100-mL volumetric flask. Add water to volume, and mix. Pipet 2 mL of this solution and 2 mL of the *Internal Standard Solution* into a 50-mL volumetric flask, add dilute methanol (1 in 20) to volume, and mix.

Procedure—Using 5-µL portions of the *Standard Preparation* and the *Test Preparation*, record their gas chromatograms with the apparatus adjusted to the parameters set forth in the accompanying table. Measure the areas under the peaks for chlorobutanol and benzaldehyde of the chromatogram for the *Standard Preparation*, designating them P_1 and P_2, respectively. Similarly, determine the corresponding values p_1 and p_2 for the *Test Preparation*. Calculate the content, in mg per mL, of chlorobutanol ($C_4H_7Cl_3O$) in the specimen taken by the formula:

$$100(C/V)(p_1/p_2)(P_2/P_1),$$

in which C is the concentration, in mg per mL, of chlorobutanol in the *Standard Preparation*, and V is the volume, in mL, of the specimen under test used in preparing each 100 mL of the *Test Preparation*.

Phenol

Internal Standard Solution—Pipet 1 mL of benzyl alcohol into a 500-mL volumetric flask, add methanol to volume, and mix.

Standard Preparation—Dissolve about 75 mg of phenol, accurately weighed, in 7.5 mL of methanol contained in a 100-mL volumetric flask. Add 20.0 mL of *Internal Standard Solution*, then add water to volume, and mix.

Procedure—Using 3-µL portions of the *Standard Preparation* and the *Test Preparation*, record their gas chromatograms with the apparatus adjusted to the parameters set forth in the accompanying table. Measure the areas under the peaks for phenol and benzyl alcohol of the chromatogram for the *Standard Preparation*, designating them P_1 and P_2, respectively. Similarly, determine the corresponding values p_1 and p_2 for the *Test Preparation*. Calculate the content, in mg per mL, of phenol (C_6H_6O) in each mL of the specimen taken by the formula:

$$100(C/V)(p_1/p_2)(P_2/P_1),$$

in which C is the concentration, in mg per mL, of phenol in the *Standard Preparation*, and V is the volume, in mL, of the specimen under test used in preparing each 100 mL of the *Test Preparation*.

Methylparaben and Propylparaben

Internal Standard Solution—Place about 200 mg of benzophenone in a 250-mL volumetric flask, add ether to volume, and mix.

Standard Preparation—Place 100 mg of methylparaben and 10 mg of propylparaben, each accurately weighed, in a 200-mL volumetric flask, add *Internal Standard Solution* to volume, and mix. Place 10 mL of this solution in a 25-mL conical flask, and proceed as directed under *Test Preparation*, beginning with "Add 3 mL of pyridine."

Test Preparation—Pipet 10 mL of the specimen under test and 10 mL of the *Internal Standard Solution* into a small separator. Shake vigorously, allow the layers to separate, draw off the aqueous layer into a second separator, and transfer the ether layer into a small flask through a funnel containing anhydrous sodium sulfate. Extract the aqueous layer with two 10-mL portions of ether, also filtering the extracts through the anhydrous sodium sulfate. Evaporate the combined extracts under a stream of dry air until the volume is reduced to about 10 mL, then transfer the residue to a 25-mL conical flask. Add 3 mL of pyridine, complete the evaporation of the ether, and boil on a hot plate until the volume is reduced to about 1 mL. Cool, and add 1.0 mL of a suitable silylation agent, such as hexamethyldisilazane to which has been added trimethylchlorosilane, bis(trimethylsilyl)acetamide, or bis(trimethylsilyl)trifluoroacetamide. Mix, and allow to stand for not less than 15 minutes.

Procedure—Using a 2-μL portion of the silanized solution from the *Standard Preparation*, record the gas chromatogram with the apparatus adjusted to the parameters set forth in the accompanying table. Measure the areas under the peaks for methylparaben, propylparaben, and benzophenone, designating them P_1, P_2, and P_3, respectively. Similarly, measure the corresponding areas for the silanized solution from the *Test Preparation*, designating them p_1, p_2, and p_3, respectively. Calculate the content, in μg per mL, of methylparaben ($C_8H_8O_3$) in the sample under test by the formula:

$$10(C_M/V)(p_1/p_3)(P_3/P_1),$$

in which C_M is the concentration, in μg per mL, of methylparaben in the *Standard Preparation*, and V is the volume, in mL, of the specimen taken. Similarly, calculate the content, in μg per mL, of propylparaben ($C_{10}H_{12}O_3$) in the specimen under test by the formula:

$$10(C_P/V)(p_2/p_3)(P_3/P_2),$$

in which C_P is the concentration, in μg per mL, of propylparaben in the *Standard Preparation*.

Ethylparaben and Butylparaben may be determined in a similar manner.

POLAROGRAPHIC METHOD

Phenylmercuric Nitrate

Standard Preparation—Dissolve about 100 mg of phenylmercuric nitrate, accurately weighed, in sodium hydroxide solution (1 in 250) contained in a 1000-mL volumetric flask, warming if necessary to effect solution, add the sodium hydroxide solution to volume, and mix. Pipet 10 mL of this solution into a 25-mL volumetric flask, and proceed as directed under *Test Preparation*, beginning with "add 2 mL of potassium nitrate solution (1 in 100)."

Test Preparation—Pipet 10 mL of the specimen under test into a 25-mL volumetric flask, add 2 mL of potassium nitrate solution (1 in 100) and 10 mL of pH 9.2 alkaline borate buffer (see under *Buffer Solutions* in the section, *Reagents, Indicators, and Solutions*), and adjust to a pH of 9.2, if necessary, by the addition of 2 *N* nitric acid. Add 1.5 mL of freshly prepared gelatin solution (1 in 1000), then add the pH 9.2 alkaline borate buffer to volume, and mix.

Procedure—Pipet a portion of the *Test Preparation* into the polarographic cell, and deaerate by bubbling nitrogen through the solution for 15 minutes. Insert the dropping mercury electrode of a suitable polarograph (see *Polarography* ⟨801⟩), and record the polarogram from −0.6 to −1.5 volts versus the saturated calomel electrode. Determine the diffusion current of the *Test Preparation*, $(i_d)_U$, as the difference between the residual current and the limiting current. Similarly and concomitantly determine the diffusion current, $(i_d)_S$, of the *Standard Preparation*. Calculate the quantity, in μg, of phenylmercuric nitrate ($C_6H_5HgNO_3$) in each mL of the specimen taken by the formula:

$$2.5C[(i_d)_U/(i_d)_S],$$

in which C is the concentration, in μg per mL, of phenylmercuric nitrate in the *Standard Preparation*.

Thimerosal

Standard Preparation—On the day of use, place about 25 mg of thimerosal, accurately weighed, in a 250-mL volumetric flask, add water to volume, and mix. Protect from light. Pipet 15 mL of this solution into a 25-mL volumetric flask, add 1.5 mL of gelatin solution (1 in 1000), then add potassium nitrate solution (1 in 100) to volume, and mix.

Test Preparation—Pipet 15 mL of the test specimen into a 25-mL volumetric flask, add 1.5 mL of gelatin solution (1 in 1000), add potassium nitrate solution (1 in 100) to volume, and mix.

Procedure—Transfer a portion of the *Test Preparation* to a polarographic cell, and deaerate by bubbling nitrogen through the solution for 15 minutes. Insert the dropping mercury electrode of a suitable polarograph (see *Polarography* ⟨801⟩), and record the polarogram from −0.2 to −1.4 volts versus the saturated calomel electrode. Determine the diffusion current, $(i_d)_U$, as the difference between the residual current and the limiting current. Similarly and concomitantly determine the diffusion current, $(i_d)_S$, of the *Standard Preparation*. Calculate the quantity, in μg, of thimerosal ($C_9H_9HgNaO_2S$) in each mL of the test specimen taken by the formula:

$$1.667C[(i_d)_U/(i_d)_S],$$

in which C is the concentration, in μg per mL, of thimerosal in the *Standard Preparation*.

⟨351⟩ ASSAY FOR STEROIDS

The following procedure is applicable for determination of those Pharmacopeial steroids that possess reducing functional groups such as α-ketols.

Standard Preparation—Dissolve in alcohol a suitable quantity of the USP Reference Standard specified in the individual monograph, previously dried under the conditions specified in the individual monograph and accurately weighed, and dilute quantitatively and stepwise with alcohol to obtain a solution having a concentration of about 10 μg per mL. Pipet 20 mL of this solution into a glass-stoppered, 50-mL conical flask.

Assay Preparation—Prepare as directed in the individual monograph.

Procedure—To each of the two flasks containing the *Assay Preparation* and the *Standard Preparation*, respectively, and to a similar flask containing 20.0 mL of alcohol to serve as the blank, add 2.0 mL of a solution prepared by dissolving 50 mg of blue tetrazolium in 10 mL of methanol, and mix. Then to each flask add 2.0 mL of a mixture of alcohol and tetramethylammonium hydroxide TS (9:1), mix, and allow to stand in the dark for 90 minutes. Without delay, concomitantly determine the absorbances of the solutions from the *Assay Preparation* and the *Standard Preparation* at about 525 nm, with a suitable spectrophotometer, against the blank. Calculate the result by the formula given in the individual monograph, in which C is the concentration, in μg per mL, of the Reference Standard in the *Standard Preparation*, and A_U and A_S are the absorbances of the solutions from the *Assay Preparation* and the *Standard Preparation*, respectively.

⟨361⟩ BARBITURATE ASSAY

Internal Standard, Internal Standard Solution, Standard Preparation, and *Assay Preparation*—Prepare as directed in the individual monograph.

Chromatographic System—Under typical conditions, the gas chromatograph is equipped with a flame-ionization detector and contains a 0.9-m × 4-mm glass column packed with 3 percent liquid phase G10 on support 80- to 100-mesh S1A. The column is maintained at a temperature of 200 ± 10°, and the injection port and detector are maintained at about 225°, the column temperature being varied within the designated tolerance, as necessary, to meet *System Suitability* specifications and provide suitable retention times. Use a suitable carrier gas, such as dry nitrogen, at an appropriate flow rate, such as 60 to 80 mL per minute. Use on-column injection. [NOTE—If the instrument is not equipped for on-column injection, use an injection port lined with glass that has been washed successively with chromic acid cleansing solution, water, methanol, chloroform, a 1 in 10 solution of trimethylchlorosilane in chloroform, and chloroform.]

System Suitability (see *Chromatography* ⟨621⟩)—Chromatograph five replicate injections of the *Standard Preparation*, and record peak responses as directed under *Procedure*. The relative standard deviation for the ratio R_S is not more than 1.5%. In a suitable chromatogram, the resolution, R, between the barbituric acid and the *Internal Standard* is not less than the value given in the individual monograph, and the tailing factor, T, for each of the two peaks is not more than 2.0.

Procedure—Inject a suitable portion (about 5 μL) of the *Standard Preparation* into a suitable gas chromatograph, and record the chromatogram. Similarly inject a suitable portion of the *Assay Preparation*, and record the chromatogram. Calculate the content of the barbiturate or barbituric acid in the assay specimen by the formula given in the individual monograph, in which R_U is the ratio of the peak response of the barbituric acid to that of the *Internal Standard* obtained for the *Assay Preparation*, Q_S is the ratio of the weight of the barbituric acid to that of the *Internal Standard* in the *Standard Preparation*, C_i is the concentration, in mg per mL, of *Internal Standard* in the *Internal Standard Solution*, and R_S is the ratio of the peak response of the barbituric acid to that of the *Internal Standard* in the *Standard Preparation.*

⟨371⟩ COBALAMIN RADIOTRACER ASSAY

All radioactive determinations required by this method should be made with a suitable counting assembly over a period of time optimal for the particular counting assembly used. All procedures should be performed in replicate to obtain the greatest accuracy.

Reference Standard—*USP Cyanocobalamin Reference Standard*—Dry over silica gel for 4 hours before using.

Cyanocobalamin Tracer Reagent—Dilute an accurately measured volume of a solution of radioactive cyanocobalamin* with water to yield a solution having a radioactivity between 500 and 5000 counts per minute per mL. Add 1 drop of cresol per liter of solution prepared, and store in a refrigerator.

Standardization—Prepare a solution of a weighed quantity of USP Cyanocobalamin RS in water to contain 20 to 50 µg per mL. Perform the entire assay on a 10.0-mL portion of this solution, proceeding as directed under *Assay Preparation*, beginning with "Add water to make a measured volume."

Cresol–Carbon Tetrachloride Solution—Mix equal volumes of carbon tetrachloride and freshly distilled cresol.

Phosphate-Cyanide Solution—Dissolve 100 mg of potassium cyanide in 1000 mL of a saturated solution of dibasic sodium phosphate, and mix.

Butanol–Benzalkonium Chloride Solution—Dilute benzalkonium chloride solution (17 in 100) with water (3:1), and mix with 36 volumes of butyl alcohol.

Alumina-Resin Column—Place a pledget of glass wool in the bottom of a constricted glass tube such as a 50-mL buret. With the tube held in an upright position, add a volume of a slurry of ion-exchange resin (see in the section, *Reagents, Indicators, and Solutions*), in water, sufficient to give a column of settled resin 7 cm in height. When the solid has settled somewhat, allow the water to drain so that there is only 1 cm of liquid above the resin column, and tamp the resin lightly. Then add a volume of a slurry of anhydrous alumina (not acid-washed) in water sufficient to increase the height of the settled column to 10 cm, and allow the water to drain to about 1 cm from the top of the alumina. Add a pledget of glass wool, and wash the column, using a total of 50 mL of water, and again drain to within 1 cm of the top of the column. Prepare a fresh column for each determination.

Assay Preparation—Transfer to a beaker a weighed quantity or measured volume of the preparation to be assayed, equivalent in vitamin B_{12} activity to that of 200 to 500 µg of cyanocobalamin. Add water to make a measured volume of not less than 25 mL, then add 5.0 mL of *Cyanocobalamin Tracer Reagent*. Add, while working *under a hood*, 5 mg of sodium nitrite and 2 mg of potassium cyanide for each mL of the resulting solution. Adjust the solution with diluted hydrochloric acid to a pH of approximately 4, and heat on a steam bath for 15 minutes. Cool, and adjust the solution with 1 *N* sodium hydroxide to a pH between 7.6 and 8.0. Centrifuge or filter to remove any undissolved solids.

Procedure—Transfer the *Assay Preparation* to a 250-mL centrifuge bottle, add 10 mL of *Cresol–Carbon Tetrachloride Solution*, suitably close the bottle with a glass, polyethylene, or foil-wrapped rubber stopper, shake vigorously for 2 to 5 minutes, and centrifuge. Remove and save the lower, solvent layer. Repeat the extraction using a 5-mL portion of *Cresol–Carbon Tetrachloride Solution*, and combine the lower, solvent-layer extracts in a centrifuge bottle or separator of 50- to 100-mL capacity.

Wash the combined extracts with successive 10-mL portions of 5 *N* sulfuric acid until the last washing is practically colorless (two washings usually suffice). During each washing, shake for 2 to 5 minutes, allow the layers to separate, centrifuge, if necessary, and discard the acid layer. Wash further with two successive 10-mL portions of *Phosphate-Cyanide Solution*. Finally, wash with 10 mL of water. Discard all of the washings.

To the washed extract add 30 mL of a mixture of *Butanol–Benzalkonium Chloride Solution* and carbon tetrachloride (2:1). Extract with two 5-mL portions of water, each time shaking vigorously for 1 minute, centrifuging, and removing and saving the upper, aqueous layer.

Pass the combined aqueous extracts through the *Alumina-Resin Column* at a rate of about 1 mL per minute, maintaining a 1-cm layer of liquid on the head of the column by adding water as needed. Discard as much of the forerun as is colorless (usually about 5 mL), and collect the colored eluate (usually about 10 mL) in a 50-mL centrifuge tube or separator containing 500 µL of diluted acetic acid. Extract the eluate by shaking for 2 to 5 minutes with 5 mL of *Cresol–Carbon Tetrachloride Solution*, and discard the upper, aqueous layer. To the extract add 5.0 mL of water, 5 mL of carbon tetrachloride, and 10 mL of butyl alcohol. Shake, allow to separate until the upper layer is clear, and remove the upper, aqueous layer.

Determine the absorbances of the aqueous extract, in a 1-cm cell, at 361 nm and 550 nm, with a suitable spectrophotometer, using a tungsten light source. Make the 361-nm reading using a filter capable of reducing stray light. Calculate the ratio A_{361}/A_{550}: the purity of the aqueous extract is acceptable if the ratio is between 3.10 and 3.40. If a ratio outside this range is observed, purify the aqueous extract by repeating the extraction cycle, proceeding as directed in the foregoing paragraph.

If an acceptable absorbance ratio is observed in the aqueous extract, determine the radioactivity, in counts per minute, using a suitable counter over a period optimal for the particular counting assembly used. Average the results, and correct the average for the observed background radioactivity determined over two or more 30-minute periods.

Calculation—Calculate the cobalamin content, expressed in µg of cyanocobalamin, of the portion taken for assay by the formula:

$$R(C_S/C_U)(A_U/A_S),$$

in which R is the quantity, in µg, of cyanocobalamin in the portion of the standard solution taken, C_S and C_U are the corrected average radioactivity values, expressed in counts per minute per mL, of the standard and assay solutions, respectively, and A_U and A_S are the absorbances determined at 361 nm of the assay and standard solutions, respectively.

* A solution of cyanocobalamin made radioactive by the incorporation of ^{60}Co is available from Merck and Co., Inc., Rahway, NJ 07065.

⟨381⟩ ELASTOMERIC CLOSURES FOR INJECTIONS

An elastomeric closure may be of synthetic or natural origin. It is generally a complex mixture of many ingredients. These include the basic polymer, fillers, accelerators, vulcanizing agents, and pigments. The properties of the elastomeric closure are dependent not only upon these ingredients, but also on the processing procedure, such as mixing, milling, dusting agents used, molding, and curing.

Factors such as cleansing procedures, contacting media, and conditions of storage may also affect the suitability of an elastomeric closure for a specific use. Evaluation of such factors should be made by appropriate additional specific tests to determine the suitability of an elastomeric closure for its intended use. Criteria for the selection of an elastomeric closure should also include a careful review of all the ingredients to assure that no known or suspected carcinogens, or other toxic substances are added.

Definition—An *elastomeric closure* is a packaging component that is, or may be, in direct contact with the drug.

Biological Test Procedures

Two stages of testing are indicated. The first stage is the performance of *in-vitro* tests according to the procedures set forth in chapter ⟨87⟩, *Biological Reactivity Tests, In-vitro.* Materials that meet the requirements of the *in-vitro* tests are not required to undergo further testing. Materials that do not meet the requirements of the *in-vitro* tests are subjected to the second stage of testing which is the performance of *in-vivo* tests, i.e., the *Systemic Injection Test* and *Intracutaneous Test*, according to the procedures set forth in chapter ⟨88⟩, *Biological Reactivity Tests, In-vivo.*

Physicochemical Test Procedures

The following tests are designed to determine pertinent physicochemical extraction characteristics of elastomeric closures. Since the tests are based on the extraction of the elastomer, it is essential that the designated amount of surface area of sample be available. In each case, the specified surface area is available for extraction at the designated temperature. The test methods are devised to detect the majority of expected variations.

Extraction Solvents—

A: Purified water.
B: Drug product vehicle (where applicable).
C: Isopropyl alcohol.

Apparatus—

Autoclave—Use an autoclave capable of maintaining a temperature of 121 ± 2°, equipped with a thermometer, a pressure gauge, and a rack adequate to accommodate the test containers above the water level.

Oven—Use an oven, preferably a forced-draft model, that will maintain an operating temperature of 105° ± 2°.

Reflux Apparatus—Use a suitable reflux apparatus having a capacity of about 500 mL.

Procedure—

Preparation of Sample—Place in a suitable extraction container a sufficient number of elastomeric closures to provide 100 cm^2 of exposed surface area. Add 300 mL of purified water to each container, cover with a suitable inverted beaker, and autoclave at 121 ± 0.5° for 30 minutes. [NOTE—Adjust so that the temperature rises rapidly, preferably within 2 to 5 minutes.] Decant, using a stainless steel screen to hold the closures in the containers. Rinse with 100 mL of purified water, gently swirl, and discard the rinsings. Repeat with a second 100-mL portion of purified water. Treat all *blank* containers in a similar manner.

Extracts (with use of *Extraction Solvent A*)—Place a properly prepared *sample*, having an exposed surface area of 100 cm^2, in a suitable container, and add 200 mL of purified water. Cover with a suitable inverted beaker, and extract by heating in an autoclave at 121° for 2 hours, allowing adequate time for the liquid within the container to reach the extraction temperature. Allow the autoclave to cool rapidly, and cool to room temperature. Treat the *blank* container in a similar manner.

Extracts (with use of *Extraction Solvent B* or *C*)—Place a properly prepared sample, having an exposed surface area of 100 cm^2, in a suitable *Reflux Apparatus* containing 200 mL of *Extraction Solvent*, and reflux for 30 minutes. Treat the *blank* in a similar manner.

Turbidity—[NOTE—Use *Extracts* prepared with *Extraction Solvent A, B,* or *C.*] Agitate the container, and transfer a sufficient quantity of *Extract*, diluted with *Extraction Solvent*, if necessary, to a cell. Measure the turbidity in a suitable nephelometer (see *Spectrophotometry and Light-scattering* ⟨851⟩), against fixed reproducible standards.* The turbidity is the difference between the values obtained for the blank and the sample expressed in Nephelos units, an arbitrary linear numerical scale expressing a haze range from absolute clarity to the zone of turbidity.

Reducing Agents—[NOTE—Use *Extracts* prepared with *Extraction Solvent A.*] Agitate the container, transfer 50 mL of *sample* extract to a suitable container, and titrate with 0.01 *N* iodine VS, using 3 mL of starch TS as the indicator. Treat the *blank* extract in a similar manner. The difference between the *blank* and the *sample* titration is expressed in mL of 0.01 *N* iodine.

Heavy Metals ⟨231⟩—[NOTE—Use *Extracts* prepared with *Extraction Solvent A* or *B.*] Transfer 20 mL of the *blank* and the *sample* extracts to separate color-comparison tubes. Transfer 2, 6, and 10 mL of *Standard Lead Solution* into separate color-comparison tubes, add 2 mL of 1 *N* acetic acid to each tube, and adjust the volume to 25 mL with purified water. Add 10 mL of freshly prepared hydrogen sulfide TS to each tube, mix, allow to stand for 5 minutes, and view downward over a white surface. Determine the amount of heavy metals in the *blank* and in the *sample*. The heavy metals content is the difference between the *blank* and the *sample.*

pH Change—[NOTE—Use *Extracts* prepared with *Extraction Solvent A* or *B*, adding to *extracts* obtained with *Solvent A* sufficient potassium chloride to provide a concentration of 0.1%.] Determine the pH of *sample* extracts *A* and *B* potentiometrically, performing blank determinations with *blank* extracts *A* and *B*, and making any necessary corrections. The pH change is the difference between the *blank* and the *sample.*

Total Extractables—[NOTE—Use *Extracts* prepared with *Extraction Solvent A, B,* or *C.*] Agitate the containers, and transfer 100-mL aliquots of the *blank* and the *sample* to separate, tared evaporating dishes. Evaporate on a steam bath to dryness (*Extracts* prepared with *Extraction Solvent C*) or in an oven at 100°, dry at 105° for 1 hour, cool in a desiccator, and weigh. Calculate the total extractables, in mg, by the formula:

$$2(W_U - W_B),$$

in which W_U is the weight, in mg, of residue found in the sample extract aliquot, and W_B is the weight, in mg, of residue found in the blank solution aliquot.

* A suitable Nephelos Standard is available from Coleman Instruments, Inc., Maywood, IL 60153.

⟨391⟩ EPINEPHRINE ASSAY

Reference Standard—*USP Epinephrine Bitartrate Reference Standard*—Keep container tightly closed and protected from light. Dry in vacuum over silica gel for 18 hours before using.

Ferro-citrate Solution—On the day needed, dissolve 1.5 g of ferrous sulfate in 200 mL of water to which have been added 1.0 mL of dilute hydrochloric acid (1 in 12) and 1.0 g of sodium bisulfite. Dissolve 500 mg of sodium citrate in 10 mL of this solution, and mix.

Buffer Solution—In a 50-mL volumetric flask mix 4.2 g of sodium bicarbonate, 5.0 g of potassium bicarbonate, and 18 mL of water (not all of the solids will dissolve at this stage). To another 18 mL of water add 3.75 g of aminoacetic acid and 1.7 mL of 6 *N* ammonium hydroxide, mix to dissolve, and transfer this solution to the 50-mL volumetric flask containing the other mixture. Dilute with water to volume, and mix until solution is complete.

Standard Preparation—Transfer about 18 mg of USP Epinephrine Bitartrate RS, accurately weighed, to a 100-mL volumetric flask with the aid of 20 mL of sodium bisulfite solution (1 in 50), dilute with water to volume, and mix. Transfer 5.0 mL of this solution to a 50-mL volumetric flask, dilute with sodium bisulfite solution (1 in 500) to volume, and mix. [NOTE—Make the final dilution when the assay is carried out.] The concentration of USP Epinephrine Bitartrate RS in the *Standard Preparation* is about 18 µg per mL.

Assay Preparation—Transfer to a 50-mL volumetric flask an accurately measured volume of the Injection under assay, equivalent to about 500 µg of epinephrine, dilute with sodium bisulfite solution (1 in 500) to volume, if necessary, and mix. [NOTE—The final concentration of sodium bisulfite is in the range of 1 to 3 mg per mL, any bisulfite present in the Injection under assay being taken into consideration.]

Procedure—Into three 50-mL glass-stoppered conical flasks transfer, separately, 20.0-mL aliquots of the *Standard Preparation*, the *Assay Preparation*, and sodium bisulfite solution (1 in 500) to provide the blank. To each flask add 200 µL of *Ferro-citrate Solution* and 2.0 mL of *Buffer Solution*, mix, and allow the solutions to stand for 30 minutes. Determine the absorbances of the solutions in 5-cm cells at the wavelength of maximum absorbance at about 530 nm, with a suitable spectrophotometer, using the blank to set the instrument. Calculate the quantity, in mg, of epinephrine ($C_9H_{13}NO_3$) in each mL of the Injection taken by the formula:

$$(183.21/333.29)(0.05C/V)(A_U/A_S),$$

in which 183.21 and 333.29 are the molecular weights of epinephrine and epinephrine bitartrate, respectively, C is the concentration, in µg per mL, of USP Epinephrine Bitartrate RS in the *Standard Preparation*, and V is the volume, in mL, of Injection taken.

⟨401⟩ FATS AND FIXED OILS

The following definitions and general procedures apply to fats, fixed oils, waxes, resins, balsams, and similar substances.

Preparation of Specimen

If a specimen of oil shows turbidity owing to separated stearin, warm the container in a water bath at 50° until the oil is clear, or if the oil does not become clear on warming, filter it through dry filter paper in a funnel contained in a hot-water jacket. Thoroughly mix, and weigh at one time as many portions as are needed for the various determinations, using preferably a bottle having a pipet dropper, or a weighing buret. Keep the specimen melted, if solid at room temperature, until the desired portions of specimen are withdrawn.

Specific Gravity

Determine the specific gravity of a fat or oil as directed under *Specific Gravity* ⟨841⟩.

Melting Temperature

Determine the melting temperature as directed for substances of *Class II* (see *Melting Range or Temperature* ⟨741⟩).

Solidification Temperature of Fatty Acids

Preparation of the Fatty Acids—Heat 75 mL of glycerin–potassium hydroxide solution (made by dissolving 25 g of potassium hydroxide in 100 mL of glycerin) in an 800-mL beaker to 150°, and add 50 mL of the clarified fat, melted if necessary. Heat the mixture for 15 minutes with frequent stirring, but do not allow the temperature to rise above 150°. Saponification is complete when the mixture is homogeneous, with no particles clinging to the beaker at the meniscus. Pour the contents of the beaker into 500 mL of nearly boiling water in an 800-mL beaker or casserole, add slowly 50 mL of dilute sulfuric acid (made by adding water and sulfuric acid (3:1)), and heat the solution, with frequent stirring, until the fatty acids separate cleanly as a transparent layer. Wash the acids with boiling water until free from sulfuric acid, collect them in a small beaker, place on a steam bath until the water has settled and the fatty acids are clear, filter into a dry beaker while hot, and dry at 105° for 20 minutes. Place the warm fatty acids in a suitable container, and cool in an ice bath until they congeal.

Test for Complete Saponification—Place 3 mL of the dry acids in a test tube, and add 15 mL of alcohol. Heat the solution to boiling, and add an equal volume of 6 N ammonium hydroxide. A clear solution results.

Procedure—Using an apparatus similar to the "Congealing Temperature Apparatus" specified therein, proceed as directed for *Procedure* under *Congealing Temperature* ⟨651⟩, reading "solidification temperature" for "congealing point" (the terms are synonymous). The average of not less than four consecutive readings of the highest point to which the temperature rises is the solidification temperature of the fatty acids.

Acid Value (Free Fatty Acids)

The acidity of fats and fixed oils in this Pharmacopeia may be expressed as the number of mL of 0.1 N alkali required to neutralize the free acids in 10.0 g of substance. Acidity is frequently expressed as the *Acid Value*, which is the number of mg of potassium hydroxide required to neutralize the free acids in 1.0 g of the substance.

Procedure—Unless otherwise directed, dissolve about 10.0 g of the substance, accurately weighed, in 50 mL of a mixture of equal volumes of alcohol and ether (which has been neutralized to phenolphthalein with 0.1 N sodium hydroxide) contained in a flask. If the test specimen does not dissolve in the cold solvent, connect the flask with a suitable condenser and warm slowly, with frequent shaking, until the specimen dissolves. Add 1 mL of phenolphthalein TS, and titrate with 0.1 N sodium hydroxide VS until the solution remains faintly pink after shaking for 30 seconds. Calculate either the *Acid Value* or the volume of 0.1 N alkali required to neutralize 10.0 g of specimen (free fatty acids), whichever is appropriate.

If the volume of 0.1 N sodium hydroxide VS required for the titration is less than 2 mL, a more dilute titrant may be used, or the sample size may be adjusted accordingly. The results may be expressed in terms of the volume of titrant used or in terms of the equivalent volume of 0.1 N sodium hydroxide.

If the oil has been saturated with carbon dioxide for the purpose of preservation, gently reflux the alcohol-ether solution for 10 minutes before titration. The oil may be freed from carbon dioxide also by exposing it in a shallow dish in a vacuum desiccator for 24 hours before weighing the test specimens.

Ester Value

The Ester Value is the number of mg of potassium hydroxide required to saponify the esters in 1.0 g of the substance. If the *Saponification Value* and the *Acid Value* have been determined, the difference between these two represents the Ester Value.

Procedure—Place 1.5 g to 2 g of the substance in a tared, 250-mL flask, weigh accurately, add 20 mL to 30 mL of neutralized alcohol, and shake. Add 1 mL of phenolphthalein TS, and titrate with 0.5 N alcoholic potassium hydroxide VS until the free acid is neutralized. Add 25.0 mL of 0.5 N alcoholic potassium hydroxide VS, and proceed as directed under *Saponification Value*, beginning with "Heat the flask" and omitting the further addition of phenolphthalein TS. The difference between the volumes, in mL, of 0.5 N hydrochloric acid consumed in the actual test and in the blank test, multiplied by 28.05 and divided by the weight in g of the specimen taken, is the Ester Value.

Hydroxyl Value

The Hydroxyl Value is the number of mg of potassium hydroxide equivalent to the hydroxyl content of 1.0 g of the substance.

Pyridine–Acetic Anhydride Reagent—Just before use, mix 3 volumes of freshly distilled pyridine with 1 volume of freshly distilled acetic anhydride.

Procedure—Transfer a quantity of the substance, determined by reference to the accompanying table and accurately weighed, to a glass-stoppered, 250-mL conical flask, and add 5.0 mL of *Pyridine–Acetic Anhydride Reagent*. Transfer 5.0 mL of *Pyridine–Acetic Anhydride Reagent* to a second glass-stoppered, 250-mL conical flask to provide the reagent blank. Fit both flasks with suitable glass-jointed reflux condensers, heat on a steam bath for 1 hour, add 10 mL of water through each condenser, and heat on the steam bath for 10 minutes more. Cool, and to each add 25 mL of butyl alcohol, previously neutralized to phenolphthalein TS with 0.5 N alcoholic potassium hydroxide, by pouring 15 mL through each condenser and, after removing the condensers, washing the sides of both flasks with the remaining 10-mL portions. To each flask add 1 mL of phenolphthalein TS, and titrate with 0.5 N alcoholic potassium hydroxide VS, recording the volume, in mL, consumed by the residual acid in the test solution as T and that consumed by the blank as B. In a 125-mL conical flask, mix about 10 g of the substance, accurately weighed, with 10 mL of freshly distilled pyridine, previously neutralized to phenolphthalein TS, add 1 mL of phenolphthalein TS, and titrate with 0.5 N alcoholic potassium hydroxide VS, recording the volume, in mL, consumed by the free acid in the test specimen as

Hydroxyl Value Range	Weight of Test Specimen, g
0 to 20	10
20 to 50	5
50 to 100	3
100 to 150	2
150 to 200	1.5
200 to 250	1.25
250 to 300	1.0
300 to 350	0.75

A, or use the Acid Value to obtain *A*. Calculate the Hydroxyl Value by the formula:

$$(56.11N/W)[B + (WA/C) - T],$$

in which *W* and *C* are the weights, in g, of the substances taken for the acetylation and for the free acid determination, respectively, *N* is the exact normality of the alcoholic potassium hydroxide, and 56.11 is the molecular weight of potassium hydroxide.

Iodine Value

The Iodine Value represents the number of g of iodine absorbed, under the prescribed conditions, by 100 g of the substance. Unless otherwise specified in the individual monograph, determine the Iodine Value by *Method I.*

METHOD I (HANUS METHOD)

Procedure—Introduce about 800 mg of a solid fat or about 200 mg of an oil, accurately weighed, into a 250-mL iodine flask, dissolve it in 10 mL of chloroform, add 25.0 mL of iodobromide TS, insert the stopper in the vessel securely, and allow it to stand for 30 minutes protected from light, with occasional shaking. Then add, in the order named, 30 mL of potassium iodide TS and 100 mL of water, and titrate the liberated iodine with 0.1 *N* sodium thiosulfate VS, shaking thoroughly after each addition of thiosulfate. When the iodine color becomes quite pale, add 3 mL of starch TS, and continue the titration with 0.1 *N* sodium thiosulfate VS until the blue color is discharged. Perform a blank test at the same time with the same quantities of the same reagents and in the same manner (see *Residual Titrations* ⟨541⟩). The difference between the volumes, in mL, of 0.1 *N* sodium thiosulfate VS consumed by the blank test and the actual test, multiplied by 1.269 and divided by the weight in g of the substance taken for test, is the Iodine Value.

NOTE—If more than half of the iodobromide TS is absorbed by the portion of the substance taken, repeat the determination, using a smaller portion of the substance under examination.

METHOD II (WIJS METHOD)

To a 500-mL iodine flask transfer an accurately weighed quantity, in g, of the substance to be tested, about equal to that calculated by the formula $25/I$, in which *I* is the iodine value, except that, for substances having iodine values not greater than 2.5, take about 10 g, accurately weighed, for the test.

Procedure—Dissolve it in 20 mL of carbon tetrachloride, add 25.0 mL of iodochloride TS, insert the stopper securely in the vessel, and allow it to stand at 25 ± 5° for 30 minutes, protected from light, with occasional shaking. Then add, in the order named, 20 mL of potassium iodide TS and 100 mL of recently boiled and cooled water, and titrate the liberated iodine with 0.1 *N* sodium thiosulfate VS, shaking thoroughly after each addition of thiosulfate. When the iodine color becomes quite pale, add 3 mL of starch TS, and continue the titration with 0.1 *N* sodium thiosulfate VS until the blue color is discharged. Perform a blank test at the same time with the same quantities of the same reagents and in the same manner (see *Residual Titrations* ⟨541⟩). The difference between the volumes, in mL, of 0.1 *N* sodium thiosulfate consumed by the blank test and the actual test, multiplied by 1.269 and divided by the weight in g of the sample taken, is the Iodine Value.

Saponification Value

The Saponification Value is the number of mg of potassium hydroxide required to neutralize the free acids and saponify the esters contained in 1.0 g of the substance.

Procedure—Place 1.5 g to 2 g of the substance in a tared, 250-mL flask, weigh accurately, and add to it 25.0 mL of 0.5 *N* alcoholic potassium hydroxide VS. Heat the flask on a steam bath, under a suitable condenser to maintain reflux for 30 minutes, frequently rotating the contents. Then add 1 mL of phenolphthalein TS, and titrate the excess potassium hydroxide with 0.5 *N* hydrochloric acid VS. Perform a blank determination at the same time, using the same amount of 0.5 *N* alcoholic potassium hydroxide VS (see *Residual Titrations* ⟨541⟩). The difference between the volumes, in mL, of 0.5 *N* hydrochloric acid consumed in the actual test and in the blank test, multiplied by 28.05 and divided by the weight in g of specimen taken, is the Saponification Value.

If the oil has been saturated with carbon dioxide for the purpose of preservation, expose it in a shallow dish in a vacuum desiccator for 24 hours before weighing the test specimens.

Unsaponifiable Matter

The term, Unsaponifiable Matter, in oils or fats, refers to those substances that are not saponifiable by alkali hydroxides but are soluble in the ordinary fat solvents, and to products of saponification that are soluble in such solvents.

Procedure—Weigh 5.0 g of the oil or fat into a 250-mL conical flask, add a solution of 2 g of potassium hydroxide in 40 mL of alcohol, and heat the flask on a steam bath under a suitable condenser to maintain reflux for 2 hours. Evaporate the alcohol on a steam bath, dissolve the residue in 50 mL of hot water, and transfer the solution to a separator having a polytetrafluoroethylene stopcock, rinsing the flask with two 25-mL portions of hot water that are added to the separator (do not use grease on stopcock). Cool to room temperature, add a few drops of alcohol to facilitate the separation of the two liquids, and extract with two 50-mL portions of ether, combining the ether extracts in another separator. Wash the combined extracts first with 20 mL of 0.1 *N* sodium hydroxide, then with 20 mL of 0.2 *N* sodium hydroxide, and finally with 15-mL portions of water until the last washing is not reddened by the addition of 2 drops of phenolphthalein TS. Transfer the ether extract to a tared beaker, and rinse the separator with 10 mL of ether, adding the rinsings to the beaker. Evaporate the ether on a steam bath just to dryness, and dry the residue at 100° for 30 minutes. Cool the beaker in a desiccator for 30 minutes, and weigh the residue of Unsaponifiable Matter.

Water and Sediment in Fixed Oils

Apparatus—The preferred centrifuge has a diameter of swing (d = distance from tip to tip of whirling tubes) of 38 to 43 cm and is operated at a speed of about 1500 rpm. If a centrifuge of different dimensions is used, calculate the desired rate of revolution by the formula:

$$\text{rpm} = 1500\sqrt{40.6/d}.$$

The centrifuge tubes are pear-shaped, and are shaped to accept closures. The total capacity of each tube is about 125 mL. The graduations are clear and distinct, reading upward from the bottom of the tube according to the scale shown in the accompanying table.

Volume (mL)	Scale Division (mL)
0 to 3	0.1
3 to 5	0.5
5 to 10	1.0
10 to 25	5.0
25 to 50	25.0
50 to 100	50.0

Procedure—Place 50.0 mL of benzene in each of two centrifuge tubes, and to each tube add 50.0 mL of the oil, warmed if necessary to re-incorporate separated stearin, and thoroughly mixed at 25°. Tightly stopper the tubes, and shake them vigorously until the contents are thoroughly mixed, then immerse the tubes in a water bath at 50° for 10 minutes. Centrifuge for 10 minutes. Read the combined volume of water and sediment at the bottom of each tube. Centrifuge repeatedly for 10-minute periods until the combined volume of water and sediment remains constant for 3 consecutive readings. The sum of the volumes of combined water and sediment in the two tubes represents the percentage, by volume, of water and sediment in the oil.

⟨411⟩ FOLIC ACID ASSAY

The following procedure is provided for the estimation of folic acid as an ingredient of Pharmacopeial preparations containing other active constituents.

Reference Standard—*USP Folic Acid Reference Standard*—Do not dry; determine the water content at the time of use.

Mobile Phase—Place 2.0 g of monobasic potassium phosphate in a 1-liter volumetric flask, and dissolve in about 650 mL of water. Add 12.0 mL of a 1 in 4 solution of tetrabutylammonium hydroxide in methanol, 7.0 mL of 3 *N* phosphoric acid, and 240 mL of methanol. Cool to room temperature, adjust with either 3 *N* phosphoric acid or 6 *N* ammonium hydroxide to a pH of 7.0, dilute with water to volume, and mix. Filter through a 0.45-µm filter, and recheck the pH before use. [NOTE—The methanol-to-water ratio may be varied by up to 3 percent and the pH may be increased up to 7.15 to achieve better separation.]

Diluting Solvent—Prepare as directed under *Mobile Phase*. Adjust to a pH of 7.0, and bubble nitrogen through the solution for 30 minutes before use.

Internal Standard Solution—Dissolve about 25 mg of methylparaben in 2.0 mL of methanol, dilute with *Diluting Solvent* to 50 mL, and mix.

Standard Folic Acid Solution—Transfer about 12 mg of USP Folic Acid RS, accurately weighed, to a low-actinic, 50-mL volumetric flask, dissolve in 2 mL of ammonium hydroxide, dilute with *Diluting Solvent* to volume, and mix.

Standard Preparation—Transfer 2.0 mL of *Standard Folic Acid Solution* to a low-actinic, 25-mL volumetric flask, add 2.0 mL of *Internal Standard Solution*, add *Diluting Solvent* to volume, and mix.

Assay Preparation—Transfer an accurately weighed or measured portion of the preparation to be assayed, containing about 1 mg of folic acid, to a low-actinic, 50-mL volumetric flask, add 4.0 mL of *Internal Standard Solution*, add *Diluting Solvent* to volume, and mix.

Chromatographic System (see *Chromatography* ⟨621⟩)—The liquid chromatograph is equipped with a 280-nm detector and a 15-cm × 3.9-mm column that contains packing L1. The flow rate is about 1.0 mL per minute. Chromatograph the *Standard Preparation*, and record the peak response as directed under *Procedure:* there is baseline separation of folic acid and methylparaben.

Procedure—Separately inject equal volumes (about 10 µL) of *Standard Preparation* and *Assay Preparation* into the chromatograph, record the chromatograms, and measure the responses for the major peaks. The relative retention times are about 0.8 for folic acid and 1.0 for methylparaben. Calculate the quantity, in µg, of $C_{19}H_{19}N_7O_6$ in the portion of the preparation taken by the formula:

$$50C(R_U/R_S),$$

in which C is the concentration, in µg per mL, of USP Folic Acid RS in the *Standard Preparation*, and R_U and R_S are the ratios of the response of the folic acid peak to that of the methylparaben peak obtained from the *Assay Preparation* and the *Standard Preparation*, respectively.

⟨421⟩ HYDROXYPROPOXY DETERMINATION

Reference Standard—*USP Methylcellulose Reference Standard*—Dry at 105° for 2 hours before using. Keep container tightly closed.

Apparatus—The apparatus for hydroxypropoxy group determinations is shown diagrammatically in Figure 1. The boiling or reaction flask, *D*, consisting of a 25-mL conical-bottom micro boiling flask modified to provide a side-arm outlet, is fitted with an aluminum foil–jacketed Vigreaux column, *E*, 95 mm long and with an adapter bleeder tube, *C*, having a 0.25- to 1.25-mm capillary tip through the neck and to the bottom of the flask for the introduction of steam and nitrogen. A steam generator, *B*, consisting of a 25- × 150-mm test tube and a gas inlet tube with a 0.25- to 1.25-mm capillary tip is attached to the bleeder tube, *C*, while a microcondenser with a 100-mm jacket, *F*, is attached to the Vigreaux column, *E*. The reaction flask and the steam generator are immersed in an oil bath, *A*, equipped with an electric heater capable of heating the bath at the desired rate and maintaining the temperature at 155°. The distillate is collected in a 125-mL graduated conical flask, *G*, fitted with a glass stopper.

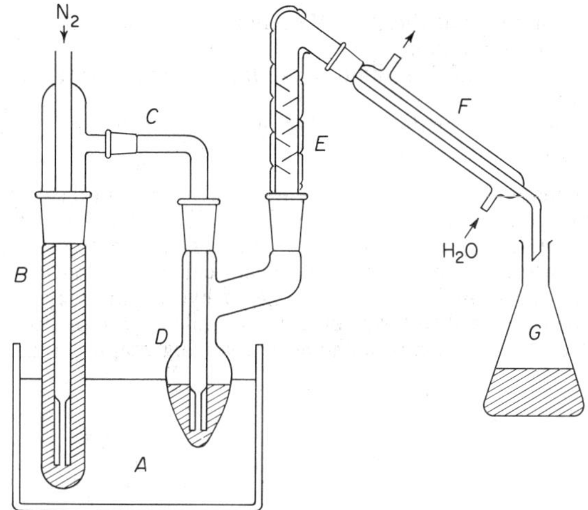

Fig. 1. Apparatus for Hydroxypropoxy Determination.

Procedure—Transfer about 100 mg of Hydroxypropyl Methylcellulose, previously dried at 105° for 2 hours and accurately weighed, into flask *D*, and add 10 mL of chromium trioxide solution (60 g in 140 mL). Fill the steam generator, *B*, with water to the bottom of the standard-taper joint, then assemble the apparatus as shown in the diagram. Immerse the steam generator and sample flask in the oil bath to the level of the chromium trioxide solution. Start the condenser cooling water, and pass nitrogen gas through the flask at a rate of 1 bubble per second. Raise the temperature of the oil bath to 155° during a 30-minute period, and maintain it at this temperature throughout the determination. [NOTE—Too rapid an initial rise in temperature results in high blanks.] Distil until 50 mL of the distillate has been collected. Detach the condenser, *F*, from the Vigreaux column, *E*, and wash with water, collecting the washings in the graduated conical flask containing the distillate. Titrate the solution with 0.02 *N* sodium hydroxide VS to a pH of 7.0 ± 0.1, using an expanded-scale pH meter equipped with glass and calomel electrodes. Record the volume, *V*, of the 0.02 *N* sodium hydroxide used, then add 500 mg of sodium bicarbonate and 10 mL of 2 *N* sulfuric acid. After evolution of carbon dioxide has ceased, add 1 g of potassium iodide, insert the stopper in the flask, shake the mixture, and allow the solution to stand in the dark for 5 minutes. Titrate the liberated iodine with 0.02 *N* sodium thiosulfate VS to the sharp disappearance of the yellow iodine color, adding a few drops of starch TS to confirm the endpoint, and record the volume, *Y*, required. This titration, *Y* mL, multiplied by the empirical factor, *K*, appropriate to the particular apparatus and reagents in use, gives the acid equivalent not caused by acetic acid. The acetic acid equivalent is $(V - KY)$ mL of 0.02 *N* sodium hydroxide.

Empirical Factor, K—Obtain the empirical factor, *K*, for each apparatus by performing a blank determination in which the cellulose ether is omitted. The acidity of the blank for a given apparatus and given reagents is in a fixed ratio to the oxidizing equivalent of the distillate in terms of sodium thiosulfate:

$$K \text{ factor} = (V_b \times N_1)/(Y_b \times N_2), \text{ in which}$$

V_b = mL of 0.02 *N* sodium hydroxide required in blank run,
N_1 = normality of the 0.02 *N* sodium hydroxide,
Y_b = mL of 0.02 *N* sodium thiosulfate required in blank run, and
N_2 = normality of the 0.02 *N* sodium thiosulfate.

Methylcellulose blank—Conduct several determinations using USP Methylcellulose Reference Standard as directed above in the *Procedure*. Calculate the percentage of uncorrected hydroxypropoxy group as follows:

$$OCH_2CHOHCH_3 \text{ percent (uncorrected)} = ([(V_aN_1 - KY_aN_2) \times 0.075]/W) \times 100, \text{ in which}$$

V_a = mL of 0.02 *N* sodium hydroxide required for titration of the sample,

N_1 = normality of the 0.02 N sodium hydroxide,
K = empirical factor,
Y_a = mL of 0.02 N sodium thiosulfate required for titration of the sample,
N_2 = normality of the 0.02 N sodium thiosulfate, and
W = g of sample used.

Calculate the corrected percentage of hydroxypropoxy group by subtracting the percentage of $OCH_2CHOHCH_3$ obtained in the *Methylcellulose blank* determination from the percentage of the uncorrected hydroxypropoxy group calculated above.

The results obtained as percentage of hydroxypropoxy content may be converted to terms of average molecular substitution of glucose units by means of the accompanying graph (Figure 2).

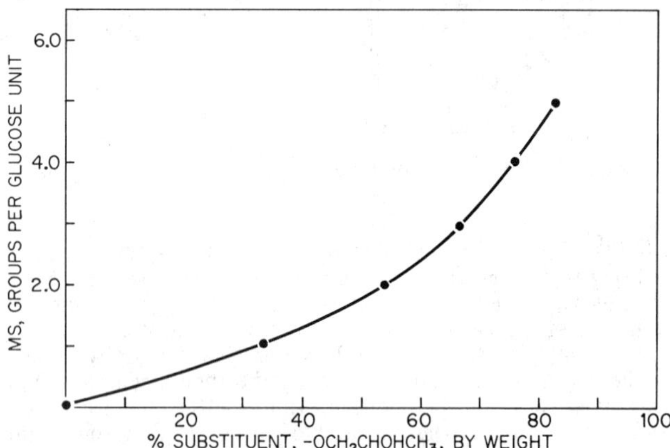

Fig. 2. Graph for Converting Percentage of Substitution, by Weight, of Hydroxypropoxy Groups to Molecular Substitution per Glucose Unit.

⟨425⟩ IODOMETRIC ASSAY—ANTIBIOTICS

The following method is provided for the assay of most of the Pharmacopeial penicillin antibiotic drugs and their dosage forms, for which iodometric titration is particularly suitable.

Standard Preparation—Dissolve in the solvent specified in the table of *Solvents and Final Concentrations* a suitable quantity of the USP Reference Standard specified in the individual monograph, previously dried under the conditions specified in the individual monograph and accurately weighed, and dilute quantitatively and stepwise with the same solvent to obtain a solution having a known concentration of about that specified in the table. Pipet 2.0 mL of this solution into each of two 125-mL glass-stoppered conical flasks.

Solvents and Final Concentrations

Antibiotic	Solvent*	Final concentration
Amoxicillin	Water	1.0 mg per mL
Ampicillin	Water	1.25 mg per mL
Ampicillin Sodium	*Buffer No. 1*	1.25 mg per mL
Cloxacillin Sodium	Water	1.25 mg per mL
Cyclacillin	Water	1.0 mg per mL
Dicloxacillin Sodium	*Buffer No. 1*	1.25 mg per mL
Methicillin Sodium	*Buffer No. 1*	1.25 mg per mL
Nafcillin Sodium	*Buffer No. 1*	1.25 mg per mL
Oxacillin Sodium	*Buffer No. 1*	1.25 mg per mL
Penicillin G Potassium	*Buffer No. 1*	2,000 units per mL
Penicillin G Sodium	*Buffer No. 1*	2,000 units per mL
Penicillin V Potassium	*Buffer No. 1*	2,000 units per mL
Phenethicillin Potassium	*Buffer No. 1*	2,000 units per mL

* Unless otherwise noted, the *Buffers* are the potassium phosphate buffers defined in the section *Media and Diluents* under *Antibiotics—Microbial Assays* ⟨81⟩, except that sterilization is not required before use.

Assay Preparation—Unless otherwise specified in the individual monograph, dissolve in the solvent specified in the table of *Solvents and Final Concentrations* a suitable quantity, accurately weighed, of the specimen under test, and dilute quantitatively with the same solvent to obtain a solution having a known final concentration of about that specified in the table. Pipet 2 mL of this solution into each of two 125-mL glass-stoppered conical flasks.

Procedure—

Inactivation and titration—To 2.0 mL of the *Standard Preparation* and of the *Assay Preparation*, in respective flasks, add 2.0 mL of 1.0 N sodium hydroxide, mix by swirling, and allow to stand for 15 minutes. To each flask add 2.0 mL of 1.2 N hydrochloric acid, add 10.0 mL of 0.01 N iodine VS, immediately insert the stopper, and allow to stand for 15 minutes. Titrate with 0.01 N sodium thiosulfate VS. As the end-point is approached, add 1 drop of starch iodide paste TS, and continue the titration to the discharge of the blue color.

Blank determination—To a flask containing 2.0 mL of the *Standard Preparation* add 10.0 mL of 0.01 N iodine VS. If the *Standard Preparation* contains amoxicillin or ampicillin, immediately add 0.1 mL of 1.2 N hydrochloric acid. Immediately titrate with 0.01 N sodium thiosulfate VS. As the end-point is approached, add 1 drop of starch iodide paste TS, and continue the titration to the discharge of the blue color. Similarly treat a flask containing 2.0 mL of the *Assay Preparation.*

Calculations—Calculate the microgram (or unit) equivalent (F) of each mL of 0.01 N sodium thiosulfate consumed by the *Standard Preparation* by the formula:

$$(2CP)/(B - I),$$

in which C is the concentration, in mg per mL, of Reference Standard in the *Standard Preparation*, P is the potency, in µg (or units) per mg, of the Reference Standard, B is the volume, in mL, of 0.01 N sodium thiosulfate consumed in the *Blank determination*, and I is the volume, in mL, of 0.01 N sodium thiosulfate consumed in the *Inactivation and titration.* Calculate the potency of the specimen under test by the formula given in the individual monograph.

⟨431⟩ METHOXY DETERMINATION

Apparatus—The apparatus for methoxy determination is shown diagrammatically in the accompanying figure. The boiling flask, *A*, is fitted with a capillary side-arm for the introduction of carbon dioxide or nitrogen and is connected to a column, *B*, which serves to separate aqueous hydriodic acid from the more volatile methyl iodide. The methyl iodide passes through water in a scrubber trap, *C*, and is finally absorbed in the bromine–acetic acid solution in absorption tube *D*. The carbon dioxide or nitrogen is introduced through a pressure-regulating device and connected to the apparatus by a small capillary containing a small cotton pledget. [NOTE—Avoid the use of organic solvents in cleaning this apparatus, since traces remaining may interfere with the determination. This test is used also for ethoxy determination with an 80-minute reaction time and a titrant equivalent of 0.751 mg of (OC_2H_5).]

For greater convenience in use and cleaning, a ground-glass ball joint connects the two upright columns of the apparatus. The top of the scrubber *C* consists of a 35/20 ball joint, the upper half of which is connected to the side-arm leading into tube *D*. This permits taking the apparatus apart and facilitates adding the water to the trap. Also, it allows access to the loose inverted (10-mm) test tube that serves as the trap over the inner tube of the scrubber *C*.

Reagents—

BROMINE–ACETIC ACID SOLUTION—Dissolve 100 g of potassium acetate in 1000 mL of a solution consisting of 900 mL of glacial acetic acid and 100 mL of acetic anhydride. On the day of use, to 145 mL of this solution add 5 mL of bromine.

HYDRIODIC ACID—A colorless, or nearly colorless, constant-boiling reagent solution, prepared for this purpose, is available commercially. If not obtained commercially, it may be prepared

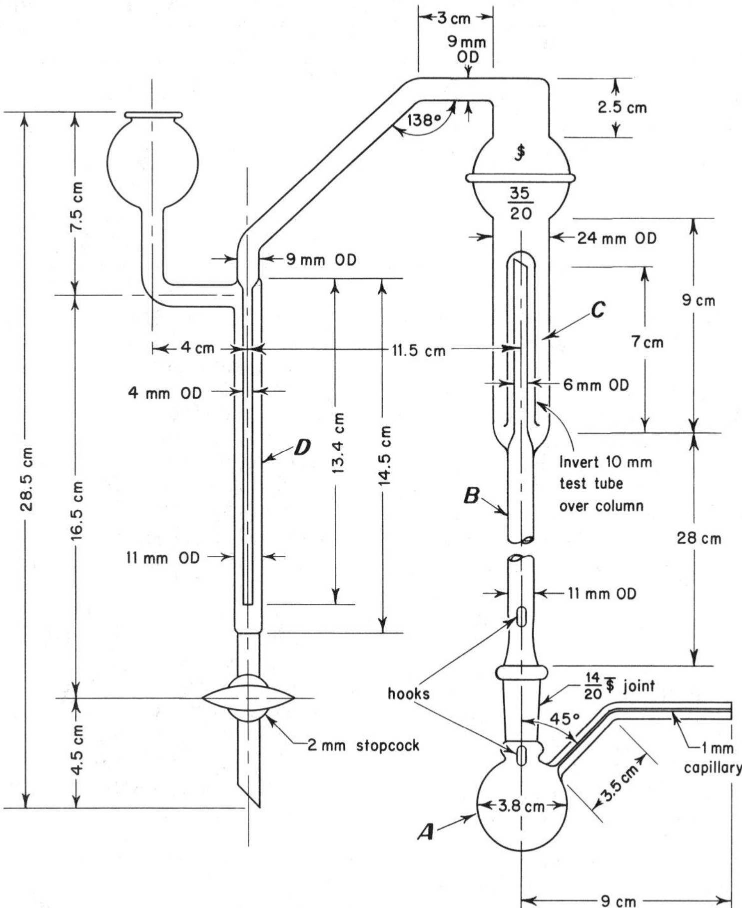

Apparatus for Methoxy Determination

by distilling hydriodic acid over red phosphorus, passing carbon dioxide or nitrogen through the apparatus during the distillation. Use the constant-boiling mixture (between 55% and 58% of HI) distilling between 126° and 127°, which is colorless or nearly colorless. (*Caution—Exercise safety precautions when distilling hydriodic acid.*) Place the acid in small, amber, glass-stoppered bottles previously flushed with carbon dioxide, or nitrogen, seal with paraffin, and store in a cool, dark place.

Procedure—Prepare the apparatus by disconnecting the ball joint and pouring water into trap *C* until it is half-full. Connect the two parts, using a minimal amount of a suitable silicone grease to seal the ball joint. Add 7 mL of *Bromine–Acetic Acid Solution* to absorption tube *D*. Weigh the sample in a tared gelatin capsule, and add it to the boiling flask along with a few boiling chips or pieces of porous plate. Finally add 6 mL of *Hydriodic Acid* and attach the flask to the column, using a minimal amount of a suitable silicone grease to seal the junction. Bubble the carbon dioxide or nitrogen through the apparatus at the rate of 2 bubbles per second, place the boiling flask in an oil bath or heating mantle heated to 150°, and continue the reaction for 40 minutes for methoxy determination, or 80 minutes for ethoxy determination. Drain the contents of the absorption tube into a 500-mL conical flask containing 10 mL of sodium acetate solution (1 in 4). Rinse the tube with water, adding the rinsings to the flask, and finally dilute with water to about 125 mL. Add formic acid, dropwise, with swirling, until the reddish brown color of the bromine is discharged, then add 3 additional drops. A total of 12 to 15 drops usually is required. Allow to stand for 3 minutes, and add 15 mL of diluted sulfuric acid and 3 g of potassium iodide, and titrate immediately with 0.1 *N* sodium thiosulfate VS, using 3 mL of starch TS as the indicator. Perform a blank determination, including also a gelatin capsule, and make any necessary correction. Each mL of 0.1 *N* sodium thiosulfate is equivalent to 0.517 mg of (OCH_3).

⟨441⟩ NIACIN OR NIACINAMIDE ASSAY

Reference Standards—*USP Niacin Reference Standard*—Dry at 105° for 1 hour before using. *USP Niacinamide Reference Standard*—Dry over silica gel for 4 hours before using. [NOTE—The previously dried Reference Standards may be stored in a desiccator over silica gel, protected from light.]

Chemical Method

NOTE—Determine from the labeling if the vitamin in the assay specimen is niacin or niacinamide, and use the corresponding standard preparation (either *Standard Niacin Preparation* or *Standard Niacinamide Preparation*) as directed in the *Procedure*.

Cyanogen Bromide Solution—Dissolve 5 g of cyanogen bromide in water to make 50 mL. (*Caution—Prepare this solution under a hood, as cyanogen bromide volatilizes at room temperature, and the vapor is highly irritating and poisonous.*)

Sulfanilic Acid Solution—To 2.5 g of sulfanilic acid add 15 mL of water and 3 mL of 6 *N* ammonium hydroxide. Mix, add, with stirring, more 6 *N* ammonium hydroxide, if necessary, until the acid dissolves, adjust the solution with 3 *N* hydrochloric acid to a pH of about 4.5, using bromocresol green TS as an external indicator, and dilute with water to 25 mL.

Standard Niacin Stock Solution—Transfer 25.0 mg of USP Niacin RS to a 500-mL volumetric flask, dissolve in alcohol solution (1 in 4), dilute with alcohol solution (1 in 4) to volume, and mix. Store in a refrigerator. Each mL of this solution contains 50 µg of USP Niacin RS.

Standard Niacin Preparation—Transfer 10.0 mL of *Standard*

Reaction Mixtures for Niacin or Niacinamide Assay—Chemical Method

Constituent	Tube 1, mL	Tube 2, mL	Tube 3, mL	Tube 4, mL
Standard Preparation	1.0	1.0	—	—
Assay Preparation	—	—	1.0	1.0
Ammonia Dilution (ammonium hydroxide, diluted to 1 in 50)	0.5	0.5	0.5	0.5
Water	6.5	1.5	6.5	1.5
Cyanogen Bromide Solution	—	5.0	—	5.0
Sulfanilic Acid Solution	2.0	2.0	2.0	2.0
Hydrochloric Acid	1 drop	—	1 drop	—

Niacin Stock Solution to a 100-mL volumetric flask, dilute with water to volume, and mix. Each mL of this solution contains 5 µg of USP Niacin RS.

Standard Niacinamide Stock Solution—Transfer 50.0 mg of USP Niacinamide RS to a 500-mL volumetric flask, dissolve in alcohol solution (1 in 4), dilute with alcohol solution (1 in 4) to volume, and mix. Store in a refrigerator. Each mL of this solution contains 100 µg of USP Niacinamide RS.

Standard Niacinamide Preparation—Transfer 10.0 mL of *Standard Niacinamide Stock Solution* to a 100-mL volumetric flask, dilute with water to volume, and mix. Each mL of this solution contains 10 µg of USP Niacinamide RS.

Assay Preparation—Prepare as directed in the individual monograph.

Procedure—Pipet into four marked tubes the quantities of the appropriate *Standard Preparation*, the *Assay Preparation*, the ammonia dilution, and water indicated in the accompanying table. Then add the other constituents, respectively, as listed in the table, according to the directions given herein.

To Tube 1 add the *Sulfanilic Acid Solution*, shake well, add the hydrochloric acid, mix, place in a suitable spectrophotometer, and adjust to zero absorbance at 450 nm. To Tube 2 add the *Cyanogen Bromide Solution*, mix, and 30 seconds, accurately timed, after completion of the addition of the cyanogen bromide add the *Sulfanilic Acid Solution*, with swirling. Close the tube, place it in the spectrophotometer, and after 2 minutes measure its absorbance at 450 nm against Tube 1 as a blank, designating the absorbance as A_S. Repeat the procedure with Tubes 3 (as blank) and 4, designating the absorbance of Tube 4 as A_U. Calculate the quantity of niacin or niacinamide in the sample as directed in the individual monograph.

Microbiological Method

Test Solution of Material to be Assayed—Place the prescribed amount of the material to be assayed in a flask of suitable size, and proceed by one of the methods given below. The concentrations of the sulfuric acid and sodium hydroxide solutions used are not stated in each instance because these concentrations may be varied depending upon the amount of material taken for assay, volume of test solution, and buffering effect of material.

(*a*) *For Dry or Semidry Materials that Contain No Appreciable Amount of Basic Substances*—Add a volume of dilute sulfuric acid (1 in 35) equal, in mL, to not less than 10 times the dry weight of the material, in g, but the resulting solution shall contain not more than 5.0 mg of niacin in each mL. If the material is not readily soluble, comminute it so that it may be evenly dispersed in the liquid, then agitate vigorously, and wash down the sides of the flask with dilute sulfuric acid (1 in 35).

Heat the mixture in an autoclave at 121° to 123° for 30 minutes, and cool. If lumping occurs, agitate the mixture until the particles are evenly dispersed. Adjust the mixture with sodium hydroxide solution to a pH of 6.8, dilute with water to make a final measured volume that has a concentration of niacin equivalent to that of *Standard Niacin Solution*, and filter.

(*b*) *For Dry or Semidry Materials that Contain Appreciable Amounts of Basic Substances*—Add sufficient sulfuric acid solution to bring the pH of the mixture to between 5.0 and 6.0. Add such an amount of water that the total volume of liquid shall be equal in mL to not less than ten times the dry weight of the assay specimen, in g, but the resulting solution shall contain not more than 5.0 mg of niacin in each mL. Then add the equivalent of 10 mL of dilute sulfuric acid (2 in 7) for each 100 mL of liquid, and proceed as directed under (*a*), beginning with the second paragraph.

(*c*) *For Liquid Materials*—Adjust the material with either sulfuric acid solution or sodium hydroxide solution to a pH of 5.0 to 6.0. Add such an amount of water that the total volume of liquid shall be equal, in mL, to not less than 10 times the volume of the specimen, in mL, but the resulting solution shall contain not more than 5.0 mg of niacin in each mL. Then add the equivalent of 10 mL of dilute sulfuric acid (2 in 7) for each mL of liquid, and proceed as directed under (*a*), beginning with the second paragraph.

Standard Niacin Stock Solution I—Transfer 50.0 mg of USP Niacin RS to a 500-mL volumetric flask, dissolve in alcohol, dilute with alcohol to volume, and mix. Store in a refrigerator. Each mL of this solution contains 100 µg of USP Niacin RS.

Standard Niacin Stock Solution II—To 100.0 mL of *Niacin Stock Solution I* add water to make 1000.0 mL. Store under toluene in a refrigerator. Each mL of this solution contains 10 µg of USP Niacin RS.

Standard Niacin Solution—Dilute a suitable volume of *Niacin Stock Solution II* with water to such a measured volume so that after incubation as described in the *Assay Procedure* the transmittance of the 5.0-mL level of *Standard Niacin Solution* is equivalent to that of a dried cell weight of not less than 1.25 mg, when the inoculated blank is set at 100 percent transmittance. This concentration is usually between 10 ng and 40 ng of niacin per mL. Prepare a fresh *Standard Niacin Solution* for each assay.

Basal Medium Stock Solution—

Acid-hydrolyzed Casein Solution	25 mL
Cystine-Tryptophan Solution	25 mL
Dextrose Anhydrous	10 g
Sodium Acetate Anhydrous	5 g
Adenine-Guanine-Uracil Solution	5 mL
Riboflavin–Thiamine Hydrochloride–Biotin Solution	5 mL
Aminobenzoic Acid–Calcium Pantothenate–Pyridoxine Hydrochloride Solution	5 mL
Salt Solution A	5 mL
Salt Solution B	5 mL

Dissolve the anhydrous dextrose and sodium acetate in the solutions previously mixed, and adjust with 1 *N* sodium hydroxide to a pH of 6.8. Finally, add water to make 250 mL.

Acid-Hydrolyzed Casein Solution—Mix 100 g of vitamin-free casein with 500 mL of constant-boiling hydrochloric acid [approximately 20 percent (w/w) HCl], and reflux the mixture for 24 hours. Remove the hydrochloric acid from the mixture by distillation under reduced pressure until a thick paste remains. Redissolve the resulting paste in water, adjust the solution with 1 *N* sodium hydroxide to a pH of 3.5 (±0.1), and add water to make 1000 mL. Add 20 g of activated charcoal, stir for 1 hour, and filter. Repeat the treatment with activated charcoal if the filtrate does not appear straw-colored to colorless. Store under toluene in a refrigerator. Filter the solution if a precipitate forms upon storage.

Cystine-Tryptophan Solution—Suspend 4.0 g of *l*-cystine and 1.0 g of *l*-tryptophan (or 2.0 g of *dl*-tryptophan) in 700 to 800 mL of water, heat to 70° to 80°, and add the 20 percent (w/w) hydrochloric acid, dropwise, with stirring, until the solids are dissolved. Cool, and add water to make 1000 mL. Store under toluene in a refrigerator at a temperature not below 10°.

Adenine-Guanine-Uracil Solution—Dissolve 100 mg each of adenine sulfate, guanine hydrochloride, and uracil, with the aid of heat, in 5.0 mL of the 20 percent (w/w) hydrochloric acid, cool, and add water to make 100 mL. Store under toluene in a refrigerator.

Riboflavin–Thiamine Hydrochloride–Biotin Solution—Prepare a solution containing, in each mL, 20 µg of riboflavin, 10 µg of thiamine hydrochloride, and 0.04 µg of biotin by dissolving crystalline riboflavin, crystalline thiamine hydrochloride, and crystalline biotin (free acid) in dilute glacial acetic acid (1 in 850). Store, protected from light, under toluene in a refrigerator.

Aminobenzoic Acid–Calcium Pantothenate–Pyridoxine Hydrochloride Solution—Prepare a solution of neutral 25 percent alcohol having a concentration of 10 µg of aminobenzoic acid, 20 µg of calcium pantothenate, and 40 µg of pyridoxine hydrochloride per mL. Store in a refrigerator.

Salt Solution A—Dissolve 25 g of monobasic potassium phosphate and 25 g of dibasic potassium phosphate in water to make 500 mL. Add 5 drops of hydrochloric acid, and store under toluene.

Salt Solution B—Dissolve 10 g of magnesium sulfate, 500 mg of sodium chloride, 500 mg of ferrous sulfate, and 500 mg of manganese sulfate in water to make 500 mL. Add 5 drops of hydrochloric acid, and store under toluene.

Stock Culture of *Lactobacillus plantarum*—Dissolve 2.0 g of water-soluble yeast extract in 100 mL of water, add 500 mg of anhydrous dextrose, 500 mg of anhydrous sodium acetate, and 1.5 g of agar, and heat the mixture with stirring, on a steam bath, until the agar dissolves. Add approximately 10-mL portions of the hot solution to test tubes, plug the tubes with cotton, sterilize for 15 minutes in an autoclave at 121° to 123°, and allow the tubes to cool in an upright position. Prepare stab cultures in three or more of the tubes, using a pure culture of *Lactobacillus plantarum,** incubating for 16 to 24 hours at any selected temperature between 30° and 37°, but held constant to within ±0.5°, and finally store in a refrigerator. Prepare a fresh stab of the stock culture every week, and do not use for inoculum if the culture is more than 1 week old.

Culture Medium—To each of a series of test tubes containing 5.0 mL of the *Basal Medium Stock Solution* add 5.0 mL of water containing 1.0 µg of niacin. Plug the tubes with cotton, sterilize for 15 minutes in an autoclave at 121° to 123°, and cool.

Inoculum—Make a transfer of cells from the stock culture of *Lactobacillus plantarum* to a sterile tube containing 10 mL of culture medium. Incubate this culture for 16 to 24 hours at any selected temperature between 30° and 37°, but held constant to within ±0.5°. The cell suspension so obtained is the inoculum.

Calibration of Spectrophotometer—Add aseptically 1 mL of *Inoculum* to approximately 300 mL of *Culture Medium* containing 1 mL of *Standard Niacin Solution.* Incubate the inoculated medium for the same period and at the same temperature to be employed in the *Assay Procedure.*

Following the incubation period, centrifuge and wash the cells three times with approximately 50-mL portions of saline TS, and then resuspend the cells in about 25 mL of the saline solution.

Dry to constant weight a 10-mL portion, accurately measured, using a steam bath and completing the drying in vacuum at 100°, and calculate the dry weight of the cells, in mg per mL, corrected for the amount of sodium chloride present.

Dilute a second portion, accurately measured, of the saline cell suspension with the saline solution so that each mL contains a known quantity of cells equivalent to 500 µg on a dried basis. To test tubes add, in triplicate, 0.5 mL, 1.0 mL, 1.5 mL, 2.0 mL, 2.5 mL, 3.0 mL, 4.0 mL, and 5.0 mL, respectively, of this diluted cell suspension and 5.0 mL of *Basal Medium Stock Solution,* and make the volume in each tube to 10.0 mL with saline solution. Using as the blanks three similar tubes containing no cell suspension, measure the light transmittance of each tube under the same conditions to be employed in the assay. Plot the observations as the ordinate on cross-section paper against the cell content, expressed as mg of dry weight, as the abscissa.

Repeat this procedure at least twice for the spectrophotometer to be used in the assay. Draw the composite curve best representing the three or more individual curves relating transmittance to cell density for the spectrophotometer under the conditions of the assay.

Assay Procedure—Prepare standard niacin tubes as follows: To test tubes add, in duplicate, 0.0 mL, 0.5 mL, 1.0 mL, 1.5 mL, 2.0 mL, 2.5 mL, 3.0 mL, 3.5 mL, 4.0 mL, 4.5 mL, and 5.0 mL, respectively, of *Standard Niacin Solution.* To each tube add 5.0 mL of *Basal Medium Stock Solution* and water to make 10.0 mL.

Prepare tubes containing the material to be assayed as follows: To test tubes add, in duplicate, 1.0 mL, 2.0 mL, 3.0 mL, and 4.0 mL, respectively, of the test solution of the material to be assayed. To each tube add 5.0 mL of *Basal Medium Stock Solution* and water to make 10.0 mL. After mixing, plug the tubes with cotton or cover with caps, and sterilize in an autoclave at 121° to 123°. (Overheating the assay tubes may produce unsatisfactory results.) Cool, aseptically inoculate each tube with 1 drop of *Inoculum,* and incubate for 16 to 24 hours at any selected temperature between 30° and 37°, but held constant to within ±0.5°. Contamination of the assay tubes with any foreign organism invalidates the assay.

Determine the transmittance of the tubes in the following manner. Mix the contents of each tube, to which 1 drop of a suitable antifoam agent solution may be added, and transfer to an optical container. After agitating its contents, place the container in a spectrophotometer that has been set at a specific wavelength between 540 nm and 660 nm, and read the transmittance when a steady state is reached. This steady state is observed a few seconds after agitation when the reading remains constant for 30 seconds or more. Allow approximately the same time interval for the reading on each tube.

With the transmittance set at 1.00 for the uninoculated blank, read the transmittance of the inoculated blank. If this transmittance reading corresponds to a dried cell weight greater than 600 µg per tube, or if there is evidence of contamination with a foreign microorganism, disregard the results of the assay.

Then with the transmittance set at 1.00 for the inoculated blank, read the transmittance for each of the remaining tubes. Disregard the results of the assay if the difference between the transmittance observed at the highest level of the standard and that of the inoculated blank is less than the difference corresponding to a dried cell weight of 1.25 mg per tube.

Calculation—Prepare a standard curve of the niacin standard transmittances for each level of *Standard Niacin Solution* plotted against µg of niacin contained in the respective tubes. From this standard curve, determine by interpolation the niacin content of the test solution in each tube. Disregard transmittance values equivalent to less than 0.5 mL or more than 4.5 mL of *Standard Niacin Solution.* The niacin content of the test material is calculated from the average values obtained from not less than six tubes that do not vary by more than ±10 percent from the average. If the transmittance values of less than six tubes containing the test solution are within the range of the 0.5- to 4.5-mL levels of the niacin standard tubes, the data are insufficient to permit calculation of the concentration of niacin in the test material. Transmittance values of inoculated blank exceeding readings corresponding to dried cell weights of more than 600 µg per tube indicate the presence of an excessive amount of niacin in the *Basal Medium Stock Solution* and invalidate the assay.

Multiply the values obtained by 0.992 if the results are to be expressed as niacinamide.

* Pure cultures of *Lactobacillus plantarum* may be obtained, as number 8014, from the the American Type Culture Collection, 12301 Parklawn Drive, Rockville, MD 20852.

⟨451⟩ NITRITE TITRATION

The following general method is provided for the determination of most of the Pharmacopeial sulfonamide drugs and their dosage forms, as well as of other Pharmacopeial drugs for which nitrite titration is particularly suitable.

Reference Standard—*USP Sulfanilamide Reference Standard*—Dry at 105° for 3 hours before using. Keep container tightly closed and protected from light.

Procedure—Weigh accurately about 500 mg in the case of a sulfonamide, or otherwise the quantity specified in the individual monograph, and transfer to a suitable open vessel. Add 20 mL of hydrochloric acid and 50 mL of water, stir until dissolved, cool to about 15°, and slowly titrate with 0.1 *M* sodium nitrite VS that previously has been standardized against USP Sulfanilamide RS.

Determine the end-point electrometrically, using suitable electrodes (platinum-calomel or platinum-platinum). Place the buret tip below the surface of the solution to eliminate air oxidation of the sodium nitrite, and stir the solution gently, using a magnetic stirrer, without pulling a vortex of air under the surface, maintaining the temperature at about 15°. The titration may be carried out manually, or by means of an automatic titrator. In performing it manually, add the titrant until the titration is within 1 mL of the end-point, and then add it in 0.1-mL portions, allowing not less than 1 minute between additions. (The instrument needle deflects and then returns to approximately its original position until the end-point is reached.)

The weight, in mg, of the substance to which each mL of 0.1 *M* sodium nitrite VS is equivalent is as stated in the individual monograph.

For the assay of Tablets of the sulfonamides or other drugs, reduce not less than 20 tablets to a fine powder, weigh accurately a portion of the powder, equivalent to about 500 mg if a sulfonamide, or the quantity of drug specified in the individual monograph, and proceed as directed in the foregoing, beginning with "transfer to a suitable open vessel."

For the assay of Injections and other liquid forms where the nitrite titration is specified, pipet a portion, equivalent to about 500 mg if a sulfonamide, or the quantity of drug specified in the individual monograph, into a suitable open vessel, and proceed as directed in the foregoing, beginning with "Add 20 mL of hydrochloric acid."

⟨461⟩ NITROGEN DETERMINATION

Some alkaloids and other nitrogen-containing organic compounds fail to yield all of their nitrogen upon digestion with sulfuric acid; therefore these methods cannot be used for the determination of nitrogen in all organic compounds.

Method I

Nitrates and Nitrites Absent—Place about 1 g of the substance, accurately weighed, in a 500-mL Kjeldahl flask of hard borosilicate glass. The material to be tested, if solid or semisolid, may be wrapped in a sheet of nitrogen-free filter paper for convenience in transferring it to the flask. Add 10 g of powdered potassium sulfate or anhydrous sodium sulfate, 500 mg of powdered cupric sulfate, and 20 mL of sulfuric acid. Incline the flask at an angle of about 45°, and gently heat the mixture, keeping the temperature below the boiling point until frothing has ceased. Increase the heat until the acid boils briskly, and continue the heating until the solution has been clear green in color or almost colorless for 30 minutes. Allow to cool, add 150 mL of water, mix the contents of the flask, and again cool. Add cautiously 100 mL of sodium hydroxide solution (2 in 5), in such manner as to cause the solution to flow down the inner side of the flask to form a layer under the acid solution. Immediately add a few pieces of granulated zinc, and without delay connect the flask to a Kjeldahl connecting bulb (trap), previously attached to a condenser, the delivery tube from which dips beneath the surface of 100 mL of boric acid solution (1 in 25) contained in a conical flask or a wide-mouth bottle of about 500-mL capacity. Mix the contents of the Kjeldahl flask by gentle rotation, and distil until about four-fifths of the contents of the flask has distilled over. Add not less than 3 drops of methyl red–methylene blue TS to the contents of the receiving vessel, and determine the ammonia by titration with 0.5 *N* sulfuric acid VS. Perform a blank determination, and make any necessary correction. Each mL of 0.5 *N* sulfuric acid VS is equivalent to 7.003 mg of nitrogen.

When the nitrogen content of the substance is known to be low, the 0.5 *N* sulfuric acid VS may be replaced by 0.1 *N* sulfuric acid VS. Each mL of 0.1 *N* sulfuric acid VS is equivalent to 1.401 mg of nitrogen.

Nitrates and Nitrites Present—Place a quantity of the substance, accurately weighed, corresponding to about 150 mg of nitrogen, in a 500-mL Kjeldahl flask of hard borosilicate glass, and add 25 mL of sulfuric acid in which 1 g of salicylic acid previously has been dissolved. Mix the contents of the flask, and allow the mixture to stand for 30 minutes with frequent shaking. To the mixture add 5 g of powdered sodium thiosulfate, again mix, then add 500 mg of powdered cupric sulfate, and proceed as directed under *Nitrates and Nitrites Absent*, beginning with "Incline the flask at an angle of about 45°."

When the nitrogen content of the substance is known to exceed 10%, 500 mg to 1 g of benzoic acid may be added, prior to digestion, to facilitate the decomposition of the substance.

Method II

Apparatus—Select a unit of the general type known as a semimicro Kjeldahl apparatus, by which the nitrogen is first liberated by acid digestion and then transferred quantitatively to the titration vessel by steam distillation.

Procedure—Place an accurately weighed or measured quantity of the material, equivalent to 2 to 3 mg of nitrogen, in the digestion flask of the apparatus. Add 1 g of a powdered mixture of potassium sulfate and cupric sulfate (10:1), and wash down any adhering material from the neck of the flask with a fine jet of water. Add 7 mL of sulfuric acid, allowing it to rinse down the wall of the flask, then, while swirling the flask, add 1 mL of 30 percent hydrogen peroxide cautiously down the side of the flask. (Do not add hydrogen peroxide during the digestion.)

Heat the flask over a free flame or an electric heater until the solution has a clear blue color and the sides of the flask are free from carbonaceous material. Cautiously add to the digestion mixture 20 mL of water, cool the solution, and arrange for steam distillation. Add through a funnel 30 mL of sodium hydroxide solution (2 in 5), rinse the funnel with 10 mL of water, tightly close the apparatus, and begin the distillation with steam immediately. Receive the distillate in 15 mL of boric acid solution (1 in 25), to which has been added 3 drops of methyl red–methylene blue TS and sufficient water to cover the end of the condensing tube. Continue the distillation until the distillate measures 80 to 100 mL. Remove the absorption flask, rinse the end of the condensing tube with a small quantity of water, and titrate the distillate with 0.01 *N* sulfuric acid VS. Perform a blank determination, and make any necessary correction. Each mL of 0.01 *N* acid VS is equivalent to 140.1 μg of nitrogen.

When a quantity of material containing more than 2 to 3 mg of nitrogen is taken, 0.02 *N* or 0.1 *N* sulfuric acid may be employed, provided that at least 15 mL is required for the titration. If the total dry weight of material taken is greater than 100 mg, increase proportionately the quantities of sulfuric acid and sodium hydroxide.

⟨466⟩ ORDINARY IMPURITIES

This test, where called for in the individual monograph, is provided to evaluate the impurity profile of an article. See *Chromatography* ⟨621⟩ for a general discussion of the thin-layer chromatographic technique. Unless otherwise specified in the individual monograph, use the following method.

Test Solution—Prepare, in the solvent specified in the monograph, a solution of the substance under test having an accurately known final concentration of about 10 mg per mL. [NOTE—Heat or sonication may be used to dissolve the drug substance where use of such does not adversely affect the compound.]

Standard Solutions—Prepare, in the solvent specified in the monograph, solutions of the USP Reference Standard or designated substance having accurately known concentrations of 0.01 mg per mL, 0.05 mg per mL, 0.1 mg per mL, and 0.2 mg per mL. [NOTE—Heat or sonication may be used to dissolve the drug substance where use of such does not adversely affect the compound.]

Procedure—Use a thin-layer chromatographic plate coated with a 0.25-mm layer of chromatographic silica gel mixture, and the *Eluant* specified in the monograph. Apply equal volumes (20 μL) of the *Test Solution* and *Standard Solutions* to the plate, using a stream of nitrogen to dry the spots.

Allow the chromatogram to develop in a pre-equilibrated chamber until the solvent front has moved about three-fourths of the length of the plate. Remove the plate from the chamber, and air-dry. View the plate using the visualization technique(s) specified. Locate any spots other than the principal spot, in the chromatogram of the *Test Solution*, and determine their relative intensities by comparison with the chromatograms of the appropriate *Standard Solutions*. The total of any ordinary impurities observed does not exceed 2.0%, unless otherwise specified in the individual monograph.

KEY FOR VISUALIZATION TECHNIQUES

(1) Use ultraviolet light at 254 nm and 366 nm.
(2) Use Iodoplatinate TS.
(3) Solution A—Mix 850 mg of bismuth subnitrate with 40 mL of water and 10 mL of glacial acetic acid.
Solution B—Dissolve 8 g of potassium iodide in 20 mL of water. Mix A and B together to obtain a Stock Solution which can be stored for several months in a dark bottle. Mix 10 mL of the Stock Solution with 20 mL of glacial acetic acid, and dilute with water to make 100 mL, to prepare the spray reagent.
(4) *Ninhydrin Spray*—Dissolve 200 mg of ninhydrin in 100 mL of alcohol. Heat the plate after spraying.
(5) *Acid Spray*—In an ice bath, add slowly and cautiously, with stirring, 10 mL of sulfuric acid to 90 mL of alcohol. Spray the plate, and heat until charred.
(6) *Acid–Dichromate Spray*—Add sufficient potassium dichromate to 100 mL of sulfuric acid to make a saturated solution. Spray the plate, and heat until charred.
(7) *Vanillin*—Dissolve 1 g of vanillin in 100 mL of sulfuric acid.
(8) *Chloramine T–Trichloroacetic Acid*—Mix 10 mL of a 3% aqueous solution of chloramine T with 40 mL of a 25% alcoholic solution of trichloroacetic acid. Prepare immediately before use.
(9) *Folin-C*—Add 10 g of sodium tungstate and 2.5 g of sodium molybdate to 70 mL of water, add 5 mL of 85% phosphoric acid and 10 mL of 36% hydrochloric acid, and reflux this solution for 10 hours.
(10) *$KMnO_4$*—Dissolve 100 mg of Potassium Permanganate in 100 mL of water.
(11) *DAB*—Mix 1 g of *p*-dimethylaminobenzaldehyde in 100 mL of 0.6 *N* hydrochloric acid.
(12) *DAC*—Mix 100 mg of *p*-dimethylaminocinnamaldehyde in 100 mL of 1 *N* hydrochloric acid.
(13) *Ferricyanide*—Mix equal volumes of a 1% ferric chloride solution and a 1% potassium ferricyanide solution. Use immediately.
(14) *Fast Blue B*—Reagent A—Dissolve 500 mg of Fast Blue B Salt in 100 mL of water.
Reagent B—0.1 *N* sodium hydroxide.
Spray first with A, then with B.
(15) *Alkaline Ferric Cyanide*—Dilute 1.5 mL of a 1% potassium ferricyanide solution with water to 20 mL, and add 10 mL of 15% sodium hydroxide solution.
(16) *Iodine Spray*—Prepare a 0.5% solution of iodine in chloroform.
(17) Expose the plate for 10 minutes to iodine vapors in a pre-equilibrated closed chamber, on the bottom of which there are iodine crystals.
(18) Solution A—Dissolve 0.5 g of potassium iodide in 50 mL of water.
Solution B—Prepare a solution of 0.5 g of soluble starch in 50 mL of hot water.
Just prior to use, mix equal volumes of Solution A and Solution B.
(19) *PTSS*—Dissolve 20 g of *p*-toluenesulfonic acid in 100 mL of alcohol, spray the plate, dry for 15 minutes at 110°, and view under ultraviolet light at 366 nm.
(20) *o-Tolidine Spray*—Dissolve 160 mg of *o*-tolidine in 30 mL of glacial acetic acid, dilute with water to make 500 mL, add 1 g of potassium iodide, and mix until the potassium iodide has dissolved.
(21) Mix 3 mL of chloroplatinic acid solution (1 in 10) with 97 mL of water, followed by the addition of 100 mL of potassium iodide solution (6 in 100) to prepare the spray reagent.
(22) *Iodine-Methanol Spray*—Prepare a mixture of iodine TS and methanol (1:1).

⟨468⟩ OXYGEN DETERMINATION

For the measurement of oxygen concentrations in air or mixtures of oxygen with air or inert gas diluents, employ an instrument utilizing the variations of electric current produced by the interaction of oxygen with an electrochemical cell to display the oxygen strength of a confined sample or an in-line flow of the gas. This current generates a signal proportional to the oxygen concentration which is displayed on a meter.

The instrument is basically maintenance-free but is to be periodically calibrated. When the monitor can no longer be calibrated with the sensor adjustment, replace or regenerate the cell.

⟨471⟩ OXYGEN FLASK COMBUSTION

The oxygen flask combustion procedure is provided as the preparatory step in the determination of bromine, chlorine, iodine, selenium, and sulfur in some Pharmacopeial articles. Combustion of the material under test (usually organic) yields water-soluble inorganic products, which are analyzed for specific elements as directed in the individual monograph or general chapter.

The caution statement given under *Procedure* covers minimum safety precautions only, and serves to emphasize the need for exceptional care throughout.

Apparatus—The apparatus[1] consists of a heavy-walled conical, deeply lipped or cupped 500-mL flask (unless a larger flask is specified), fitted with a ground-glass stopper to which is fused a test specimen carrier consisting of heavy-gauge platinum wire and a piece of welded platinum gauze measuring about 1.5 × 2 cm.

Procedure—[CAUTION—The analyst should wear safety glasses and use a suitable safety shield between himself and the apparatus. Exercise care to ensure that the flask is scrupulously clean and free from even traces of organic solvents.] Weigh the substance, if a solid, on a piece of halide-free filter paper measuring about 4 cm square, and fold the paper to enclose it. Liquid substances are weighed in tared capsules, cellulose acetate capsules[2] being used for liquids in volumes not exceeding 200 µL, and gelatin capsules being satisfactory for use for larger volumes. [NOTE—Gelatin capsules may contain significant amounts of combined halide or sulfur. If such capsules are used, perform a blank determination, and make any necessary correction.] Place the specimen, together with a filter paper fuse-strip, in the platinum gauze specimen holder. Place the absorbing liquid specified in the individual monograph or general chapter in the flask, moisten the joint of the stopper with water, and flush the air from the flask with a stream of rapidly flowing oxygen, swirling the liquid to favor its taking up oxygen. [NOTE—Saturation of the liquid with oxygen is essential for the successful performance of the combustion procedure.] Ignite the fuse-strip by suitable means. If the strip is ignited outside the flask, immediately plunge the specimen holder into the flask, invert the flask so that the absorption solution makes a seal around the stopper, and hold the stopper firmly in place. If the ignition is carried out in a closed system, the inversion of the flask may be omitted. After combustion is complete, shake the flask vigor-

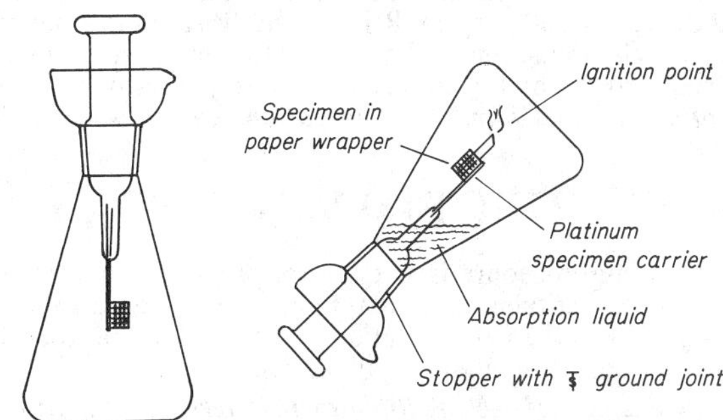

Apparatus for Oxygen Flask Combustion

[1] A suitable apparatus [Catalog Nos. 6513-C20 (500-mL capacity) and 6513-C30 (1000-mL capacity)] and suitable capsules [Catalog Nos. 6513-C80 (100 capsules) and 6513-C82 (1000 capsules)] are obtainable from Arthur H. Thomas Co., P. O. Box 779, Philadelphia, PA 19105.

[2] A suitable apparatus [Catalog Nos. 6513-C20 (500-mL capacity) and 6513-C30 (1000-mL capacity)] and suitable capsules [Catalog Nos. 6513-C80 (100 capsules) and 6513-C82 (1000 capsules)] are obtainable from Arthur H. Thomas Co., P. O. Box 779, Philadelphia, PA 19105.

ously, and allow to stand for not less than 10 minutes with intermittent shaking. Then proceed as directed in the individual monograph or general chapter.

⟨475⟩ PENICILLIN G DETERMINATION

The following procedure is used to determine the content of penicillin G moiety in an antibiotic drug substance when such a requirement is specified in the individual monograph.

0.05 *M* Phosphate Buffer, pH 6—Dissolve 6.8 g of monobasic potassium phosphate in 900 mL of water, adjust with 1 *N* sodium hydroxide to a pH of 6.00, dilute with water to 1000 mL, and mix.

Mobile Phase—Prepare a mixture of *0.05 M Phosphate Buffer, pH 6* and acetonitrile (4:1), filter through a membrane filter of 5-µm or finer porosity, and degas.

Standard Preparation—Transfer about 80 mg of USP Penicillin G Potassium RS, accurately weighed, to a 100-mL volumetric flask, add about 50 mL of *Mobile Phase*, swirl to dissolve, dilute with *Mobile Phase* to volume, and mix.

Test Preparation—Unless otherwise directed in the individual monograph, proceed with the specimen under test as directed under *Standard Preparation.*

System Suitability Preparation—Prepare a solution of penicillin V potassium in *Mobile Phase* containing about 1 mg per mL. Mix equal volumes of this solution and the *Standard Preparation.*

Chromatographic System (see *Chromatography* ⟨621⟩)—The liquid chromatograph is equipped with a 225-nm detector and a 4-mm × 30-cm column that contains 10-µm packing L1. The flow rate is about 2 mL per minute. Chromatograph the *Standard Preparation* and the *System Suitability Preparation*, and record the peak responses as directed under *Procedure:* the column efficiency determined from the analyte peak is not less than 600 theoretical plates, the resolution, *R*, between the penicillin G and penicillin V peaks is not less than 2.0, and the relative standard deviation for replicate injections of the *Standard Preparation* is not more than 1.0%.

Procedure—Separately inject equal volumes (about 10 µL) of the *Standard Preparation*, the *Test Preparation*, and the *System Suitability Preparation* into the chromatograph, record the chromatograms, and measure the responses for the major peaks. The relative retention times are about 0.7 for penicillin G and 1.0 for penicillin V. Calculate the percentage of penicillin G ($C_{16}H_{18}N_2O_4S$) in the specimen under test by the formula:

$$(G_S W_S / W_U)(r_U / r_S),$$

in which G_S is the designated penicillin G content, in percentage, of USP Penicillin G Potassium RS, W_S and W_U are the amounts, in mg, of USP Penicillin G Potassium RS and test specimen taken, respectively, and r_U and r_S are the responses of the *Test Preparation* and the *Standard Preparation* peaks, respectively.

⟨481⟩ RIBOFLAVIN ASSAY

The following procedure is suitable for preparations in which riboflavin is a constituent of a mixture of several ingredients. In employing it, keep the pH of solutions below 7.0, and protect the solutions from direct sunlight at all stages.

Reference Standard—*USP Riboflavin Reference Standard*—Dry at 105° for 2 hours before using.

Standard Riboflavin Stock Solution—To 50.0 mg of USP Riboflavin RS, previously dried and stored protected from light in a desiccator over phosphorus pentoxide, add about 300 mL of 0.02 *N* acetic acid, and heat the mixture on a steam bath, with frequent agitation, until the riboflavin has dissolved. Then cool, add 0.02 *N* acetic acid to make 500 mL, and mix. Store under toluene in a refrigerator.

Dilute an accurately measured portion of this solution, using 0.02 *N* acetic acid, to a concentration of 10.0 µg of the dried USP Riboflavin RS per mL, to obtain the *Standard Riboflavin Stock Solution.* Store under toluene in a refrigerator.

Standard Preparation—Dilute 10.0 mL of *Standard Riboflavin Stock Solution* with water in a 100-mL volumetric flask to volume, and mix. Each mL represents 1.0 µg of USP Riboflavin RS. Prepare fresh *Standard Preparation* for each assay.

Assay Preparation—Place an amount of the material to be assayed in a flask of suitable size, and add a volume of 0.1 *N* hydrochloric acid equal in mL to not less than 10 times the dry weight of the material in g, but the resulting solution shall contain not more than 100 µg of riboflavin per mL. If the material is not readily soluble, comminute it so that it may be evenly dispersed in the liquid. Then agitate vigorously, and wash down the sides of the flask with 0.1 *N* hydrochloric acid.

Heat the mixture in an autoclave at 121° to 123° for 30 minutes, and cool. If clumping occurs, agitate the mixture until the particles are evenly dispersed. Adjust the mixture, with vigorous agitation, to a pH of 6.0 to 6.5 with sodium hydroxide solution,* then add hydrochloric acid solution* immediately until no further precipitation occurs (usually at a pH of approximately 4.5, the isoelectric point of many of the proteins present). Dilute the mixture with water to make a measured volume that contains about 0.11 µg of riboflavin in each mL, and filter through paper known not to adsorb riboflavin. To an aliquot of the filtrate add, with vigorous agitation, sodium hydroxide solution* to produce a pH of 6.6 to 6.8, dilute the solution with water to make a final measured volume that contains approximately 0.1 µg of riboflavin in each mL, and if cloudiness occurs, filter again.

Procedure—To each of four or more tubes (or reaction vessels) add 10.0 mL of the *Assay Preparation.* To each of two or more of these tubes add 1.0 mL of the *Standard Preparation*, and mix, and to each of two or more of the remaining tubes add 1.0 mL of water, and mix. To each tube add 1.0 mL of glacial acetic acid, mix, then add, with mixing, 0.50 mL of potassium permanganate solution (1 in 25), and allow to stand for 2 minutes. To each tube add, with mixing, 0.50 mL of hydrogen peroxide solution, whereupon the permanganate color is destroyed within 10 seconds. Shake the tubes vigorously until excess oxygen is expelled. Remove any gas bubbles remaining on the sides of the tubes after foaming has ceased, by tipping the tubes so that the solution flows slowly from end to end.

In a suitable fluorophotometer, having an input filter of narrow transmittance range with a maximum at about 440 nm and an output filter of narrow transmittance range with a maximum at about 530 nm, measure the fluorescence of all tubes, designating the average reading from the tubes containing only the *Assay Preparation* as I_U and the average from the tubes containing both the *Assay Preparation* and the *Standard Preparation* as I_S. Then to each of one or more tubes of each kind add, with mixing, 20 mg of sodium hydrosulfite, and within 5 seconds again measure the fluorescence, designating the average reading as I_B.

Calculation—Calculate the quantity, in mg, of $C_{17}H_{20}N_4O_6$ in each mL of the *Assay Preparation* taken by the formula:

$$0.0001(I_U - I_B)/(I_S - I_U).$$

Calculate the quantity, in mg, of $C_{17}H_{20}N_4O_6$ in each capsule or tablet.

⟨501⟩ SALTS OF ORGANIC NITROGENOUS BASES

Standard Preparation—Unless otherwise directed, prepare a solution in dilute sulfuric acid (1 in 70) containing, in each mL, about 500 µg of the specified USP Reference Standard, calculated on the anhydrous basis, and accurately weighed.

Assay Preparation—If the dosage form is a tablet, weigh and finely powder not less than 20 tablets, weigh accurately a portion of the powder, equivalent to about 25 mg of the active ingredient, and transfer to a 125-mL separator; or, if the dosage form is a liquid, transfer a volume of it, equivalent to about 25 mg of the active ingredient and accurately measured, to a 125-mL separator. Then to the separator add 20 mL of dilute sulfuric acid (1 in 350), and shake vigorously for 5 minutes. Add 20 mL of

* The concentrations of the hydrochloric acid and sodium hydroxide solutions used are not stated in each instance because these concentrations may be varied depending upon the amount of material taken for assay, volume of test solution, and buffering effect of material.

ether, shake carefully, and filter the acid phase into a second 125-mL separator. Shake the ether phase with two 10-mL portions of dilute sulfuric acid (1 in 350), filter each portion of acid into the second separator, and discard the ether. To the acid extract add 10 mL of sodium hydroxide TS and 50 mL of ether, shake carefully, and transfer the aqueous phase to a third 125-mL separator containing 50 mL of ether. Shake the third separator carefully, and discard the aqueous phase. Wash the two ether solutions, in succession, with a single 20-mL portion of water, and discard the water. Extract each of the two ether solutions with 20-, 20-, and 5-mL portions of dilute sulfuric acid (1 in 70), in the order listed, but each time extract first the ether solution in the third separator and then that in the second separator. Combine the acid extracts in a 50-mL volumetric flask, dilute with the acid to volume, and mix.

NOTE—Hexane or heptane may be substituted for ether if the distribution ratio of the nitrogenous base between water and hexane, or between water and heptane, favors complete extraction by the organic phase.

Procedure—Unless otherwise directed, dilute 5.0 mL each of the *Standard Preparation* and the *Assay Preparation* with dilute sulfuric acid (1 in 70) to 100.0 mL, and determine the absorbance of each solution at the specified wavelength, using dilute sulfuric acid (1 in 70) as the blank. Designate the absorbance of the solution from the *Standard Preparation* as A_S and that from the *Assay Preparation* as A_U, and calculate the result of the assay as directed in the individual monograph.

⟨511⟩ SINGLE-STEROID ASSAY

In the following procedure, the steroid to be assayed is separated from related foreign steroids and excipients by thin-layer chromatography and determined following recovery from the chromatogram.

Preparation of the Plate—Prepare a slurry from 30 g of chromatographic silica gel with a suitable fluorescing substance by the gradual addition, with mixing, of about 65 mL of a mixture of water and alcohol (5:2). Transfer the slurry to a clean, 20- × 20-cm plate, spread to make a uniform layer 250 µm thick, and allow to dry at room temperature for 15 minutes. Heat the plate at 105° for 1 hour, and store in a desiccator.

Solvent A—Mix methylene chloride with methanol (180:16).

Solvent B—Mix chloroform with acetone (4:1).

Standard Preparation—Dissolve in a mixture of equal volumes of chloroform and alcohol a suitable quantity of the USP Reference Standard specified in the individual monograph, previously dried as directed in the individual monograph, and accurately weighed, to obtain a solution having a known concentration of about 2 mg per mL.

Assay Preparation—Prepare as directed in the individual monograph.

Procedure—Divide the area of the chromatographic plate into three equal sections, the left and right sections to be used for the *Assay Preparation* and the *Standard Preparation*, respectively, and the center section for the blank. Apply 200 µL each of the *Assay Preparation* and the *Standard Preparation* as streaks 2.5 cm from the bottom of the appropriate section of the plate. Dry the solution as it is being applied, with the aid of a stream of air. Using the *Solvent* specified in the individual monograph, develop the chromatogram in a suitable chamber, previously equilibrated and lined with absorbent paper, until the solvent front has moved 15 cm above the initial streaks.

Remove the plate, evaporate the solvent, and locate the principal band occupied by the *Standard Preparation* by viewing under ultraviolet light. Mark this band, as well as corresponding bands in the *Assay Preparation* and blank sections of the plate. Remove the silica gel from each band separately, either by scraping onto glazed weighing papers or by using a suitable vacuum collecting device, and transfer it to a glass-stoppered, 50-mL centrifuge tube. To each tube add 25.0 mL of alcohol, and shake for not less than 2 minutes. Centrifuge the tubes for 5 minutes, pipet 20 mL of the supernatant liquid from each tube into a glass-stoppered, 50-mL conical flask, add 2.0 mL of a solution prepared by dissolving 50 mg of blue tetrazolium in 10 mL of methanol, and mix. Proceed as directed for *Procedure* under *Assay for Steroids* ⟨351⟩, beginning with "Then to each flask."

⟨521⟩ SULFONAMIDES

Identification of Individual Sulfonamides in Mixed Sulfonamides

NOTE—The following instructions for preparations and procedure are applicable to all sulfonamides except sulfadiazine. When testing for sulfadiazine proceed in the same manner, except to use sulfadiazine preparations having one-half the designated concentration, and apply twice the designated volumes of sulfadiazine preparations to the chromatographic plates.

Standard Preparation—Transfer a quantity of the pertinent USP Reference Standard to a suitable glass-stoppered, conical flask, dissolve in methanol to obtain a solution having a concentration of about 2 mg per mL, and mix. A separate *Standard Preparation* is required for each sulfonamide present in mixed sulfonamides.

Test Preparation—Transfer a portion of the thoroughly mixed suspension or finely powdered tablets, equivalent to about 100 mg of each sulfonamide, to a 50-mL volumetric flask containing 10 mL of ammonia TS, and swirl. Add methanol to volume, mix, filter, and use the filtrate in the *Procedure*.

Preparation of Chromatographic Plates—Prepare three identical chromatographic plates according to the following directions. On a suitable thin-layer chromatographic plate (see *Chromatography* ⟨621⟩), coated with a 0.25-mm layer of chromatographic silica gel mixture, apply separately, and 2 cm apart along a spotting line 1.5 cm from the bottom of the plate and parallel to it, 2 µL of each *Standard Preparation* and 2 µL of the *Test Preparation*. On another spot, 2 cm along the spotting line from the application of the *Test Preparation*, apply, successively, 2 µL of each *Standard Preparation* to obtain a mixed standard. Dry the spots immediately with the aid of a stream of nitrogen.

Procedure—Prepare a chromatographic chamber lined with filter paper and containing a solvent system consisting of ethyl acetate, methanol, and a 1 in 4 aqueous solution of ammonium hydroxide (17:6:5), and allow to equilibrate for 1 hour. Similarly prepare a second chamber to contain a solvent system consisting of solvent hexane, chloroform, and butyl alcohol (1:1:1), and a third chamber to contain a solvent system consisting of chloroform and methanol (95:5). Place one prepared chromatographic plate in each equilibrated chamber, and develop the chromatograms until the solvent front has moved about three-fourths of the length of each plate. Remove each plate from its developing chamber, mark the solvent front, and allow the solvent to evaporate. Locate the spots on the plates by viewing under short-wavelength ultraviolet light. Spray the plates with a 1 in 100 solution of *p*-dimethylaminobenzaldehyde in dilute hydrochloric acid (1 in 20), and heat at 110° for 5 minutes or until bright yellow spots become visible. The R_f values of the yellow spots obtained from each *Test Preparation* correspond to those obtained from the mixed *Standard Preparations* on the respective plates. The individual sulfonamides may be identified by comparison of the R_f values of the yellow spots obtained from the *Test Preparations* and individual *Standard Preparations* on the respective plates.

Determination of Individual Sulfonamides in Mixed Sulfonamides

Standard Preparation—A separate *Standard Preparation* is required for each sulfonamide being determined. Transfer about 50 mg, accurately weighed, of the pertinent USP Reference Standard to a 50-mL volumetric flask containing 1.5 mL of ammonium hydroxide, add methanol, dissolve in methanol, dilute with methanol to volume, and mix. Transfer 1.0 mL of this solution to a 100-mL volumetric flask, add dilute hydrochloric acid (1 in 100) to volume, and mix. [NOTE—Retain the methanol solutions for the *Mixed Standard Preparation.* The methanol solutions are stable for at least 1 week, and the acid solutions for at least 1 month.]

Mixed Standard Preparation—Transfer 1.0 mL of each methanol solution, prepared as required for each *Standard Preparation*, to a small glass-stoppered flask, and mix. [NOTE—This Standard is used to identify the components of the *Assay Preparation* on the chromatogram.]

Assay Preparation—Prepare as directed in the individual monograph.

Procedure—Prepare the necessary number of chromatographic sheets (Whatman No. 1 filter paper, or equivalent), about 20 × 20 cm in size, by drawing a pencil line parallel to and 2.5 cm from one edge of the paper. Mark the line at points 2.5 and 5 cm from each edge of the paper. Impregnate the paper by dipping it in the immobile solvent (prepared fresh by dissolving 30 mL of redistilled formamide in 70 mL of acetone) for 30 seconds. Remove the paper, drain for 10 seconds, and blot between filter paper. Place the impregnated paper on dry filter paper, and air-dry for 3 to 5 minutes. With a micropipet, and with repeated applications, streak 100 µL of the *Assay Preparation* along the starting line, applying the volume in five streaks of about 20 µL each and evaporating the solvent with a gentle stream of nitrogen between applications. [NOTE—Make the streak as narrow as possible along the starting line, and keep within the 5-cm border.] Rinse the tip of the pipet with a drop of methanol–ammonia TS mixture (9:1), and then streak the rinse along the starting line between the 5- and 2.5-cm points at the right edge. Repeat the rinsing with two additional drops, and then blow out the pipet.

Apply 10 µL of the *Mixed Standard Preparation* at the mark 2.5 cm from the left edge.

Place 50 mL of methylene chloride (mobile solvent) in a tray in a 23- × 23- × 7.5-cm chromatographic chamber arranged for ascending chromatography (see *Chromatography* ⟨621⟩), and allow the chamber to equilibrate for about 15 minutes. Remove the cover, place from 7 to 10 mL of water in a second tray, and without delay, suspend the prepared chromatographic paper sheet so that it dips into the mobile solvent. Cover and seal the chamber, and allow the chromatogram to develop for 1 hour. Remove the paper from the chamber, and allow to air-dry for 5 minutes. Place the chromatogram on a dry sheet of filter paper, and view it under short-wavelength ultraviolet light. [NOTE—Conduct the following identification and marking without delay to avoid excessive exposure of the sulfonamide spots to ultraviolet irradiation.] Identify and mark the respective spots by matching R_f values with those of the spots produced by the *Mixed Standard Preparation.* [NOTE—Sulfadiazine and sulfamerazine are chromatographed with increasing R_f, respectively.]

Cut the marked zones from the paper, cut each zone into five or six pieces, and place the pieces from each spot in separate, glass-stoppered, 50-mL flasks. Add 20.0 mL of dilute hydrochloric acid (1 in 100) to each flask, and allow to stand for about 30 minutes, swirling each flask at least five times during this period. Filter the solutions through dry glass wool into separate test tubes, discarding the first 5 mL of filtrate. Transfer 5.0 mL of the subsequent filtrate from each solution into separate 10-mL volumetric flasks. Transfer 3.0 mL of each required *Standard Preparation* into separate, 10-mL volumetric flasks. To each flask, and to a blank flask containing 5 mL of dilute hydrochloric acid (1 in 100), add 1.0 mL of sodium nitrite solution (1 in 1000) and 0.10 mL of hydrochloric acid, and allow to stand for 5 minutes with frequent swirling. To each flask add 1.0 mL of ammonium sulfamate solution (1 in 200), and allow to stand for 5 minutes, swirling frequently. Finally, to each flask add 1.0 mL of freshly prepared *N*-(1-naphthyl)ethylenediamine dihydrochloride solution (1 in 1000), mix, dilute with water to volume, and mix. Allow each solution to stand between 15 and 60 minutes, and then concomitantly determine the absorbances of the solutions, in 1-cm cells, recording the spectra from 440 to 700 nm, with a suitable spectrophotometer, using the blank to set the instrument. Draw a baseline, and determine the corrected absorbance for each solution at the wavelength of maximum absorbance at about 545 nm.

Calculate the concentration, in mg per mL, of each sulfonamide in the *Assay Preparation* by the formula:

$$0.12C(A_U/A_S),$$

in which C is the concentration, in µg per mL, of the pertinent USP Reference Standard in the *Standard Preparation,* A_U is the corrected absorbance of the *Assay Preparation,* and A_S is the corrected absorbance of the pertinent *Standard Preparation.* From the concentration of the *Assay Preparation* thus determined, and applying appropriate dilution factors, calculate the percentage of sulfonamide in the specimen taken.

⟨531⟩ THIAMINE ASSAY

Reference standard—*USP Thiamine Hydrochloride Reference Standard*—Do not dry; determine the water content titrimetrically at the time of use.

The following procedure is provided for the determination of thiamine as an ingredient of Pharmacopeial preparations containing other active constituents.

Special Solutions and Solvents—

POTASSIUM FERRICYANIDE SOLUTION—Dissolve 1.0 g of potassium ferricyanide in water to make 100 mL. Prepare fresh on the day of use.

OXIDIZING REAGENT—Mix 4.0 mL of *Potassium Ferricyanide Solution* with sufficient 3.5 *N* sodium hydroxide to make 100 mL. Use this solution within 4 hours.

QUININE SULFATE STOCK SOLUTION—Dissolve 10 mg of quinine sulfate in 0.1 *N* sulfuric acid to make 1000 mL. Preserve this solution, protected from light, in a refrigerator.

QUININE SULFATE STANDARD SOLUTION—Dilute 0.1 *N* sulfuric acid with *Quinine Sulfate Stock Solution* (39:1). This solution fluoresces to approximately the same degree as the thiochrome obtained from 1 µg of thiamine hydrochloride and is used to correct the fluorometer at frequent intervals for variation in sensitivity from reading to reading within an assay. Prepare this solution fresh on the day of use.

Standard Thiamine Hydrochloride Stock Solution—Transfer about 25 mg of USP Thiamine Hydrochloride RS, accurately weighed, to a 1000-mL volumetric flask. Dissolve the weighed Standard in about 300 mL of dilute alcohol solution (1 in 5) adjusted with 3 *N* hydrochloric acid to a pH of 4.0, and add the acidified, dilute alcohol to volume. Store in a light-resistant container, in a refrigerator. Prepare this stock solution fresh each month.

Standard Preparation—Dilute a portion of *Standard Thiamine Hydrochloride Stock Solution* quantitatively and stepwise with 0.2 *N* hydrochloric acid to obtain the *Standard Preparation,* each mL of which represents 0.2 µg of USP Thiamine Hydrochloride RS.

Assay Preparation—Place in a suitable volumetric flask sufficient of the material to be assayed, accurately weighed or measured by volume as directed, such that when diluted to volume with 0.2 *N* hydrochloric acid, the resulting solution will contain about 100 µg of thiamine hydrochloride (or mononitrate) per mL. If the sample is difficultly soluble, the solution may be heated on a steam bath, and then cooled and diluted with the acid to volume. Dilute 5 mL of this solution, quantitatively and stepwise, using 0.2 *N* hydrochloric acid, to an estimated concentration of 0.2 µg of thiamine hydrochloride (or mononitrate) per mL.

Procedure—Into each of three or more tubes (or other suitable vessels), of about 40-mL capacity, pipet 5 mL of *Standard Preparation.* To each of two of these tubes add rapidly (within 1 to 2 seconds), with mixing, 3.0 mL of *Oxidizing Reagent,* and within 30 seconds add 20.0 mL of isobutyl alcohol, then mix vigorously for 90 seconds by shaking the capped tubes manually, or by bubbling a stream of air through the mixture. Prepare a blank in the remaining tube of the standard by substituting for the *Oxidizing Reagent* an equal volume of 3.5 *N* sodium hydroxide and proceeding in the same manner.

Into each of three or more similar tubes pipet 5 mL of the *Assay Preparation.* Treat these tubes in the same manner as directed for the tubes containing the *Standard Preparation.*

Into each of the six tubes pipet 2 mL of dehydrated alcohol, swirl for a few seconds, allow the phases to separate, and decant or draw off about 10 mL of the clear, supernatant isobutyl alcohol solution into standardized cells, then measure the fluorescence in a suitable fluorometer, having an input filter of narrow transmittance range with a maximum at about 365 nm and an output filter of narrow transmittance range with a maximum at about 435 nm.

Calculation—The number of µg of $C_{12}H_{17}ClN_4OS.HCl$ in each 5 mL of the *Assay Preparation* is given by the formula:

$$(A - b)/(S - d),$$

in which A and S are the average fluorometer readings of the portions of the *Assay Preparation* and the *Standard Preparation* treated with *Oxidizing Reagent*, respectively, and *b* and *d* are the readings for the blanks of the *Assay Preparation* and the *Standard Preparation*, respectively.

Calculate the quantity, in mg, of thiamine hydrochloride ($C_{12}H_{17}ClN_4OS.HCl$) in the assay material on the basis of the aliquots taken. Where indicated, the quantity, in mg, of thiamine mononitrate ($C_{12}H_{17}N_5O_4S$) may be calculated by multiplying the quantity of $C_{12}H_{17}ClN_4OS.HCl$ found by 0.9706.

⟨541⟩ TITRIMETRY

Direct Titrations—Direct titration is the treatment of a soluble substance, contained in solution in a suitable vessel (the titrate), with an appropriate standardized solution (the titrant), the end-point being determined instrumentally or visually with the aid of a suitable indicator.

The titrant is added from a suitable buret and is so chosen, with respect to its strength (normality), that the volume added is between 30% and 100% of the rated capacity of the buret. [NOTE—Where less than 10 mL of titrant is required, a suitable microburet is to be used.] The end-point is approached directly but cautiously, and finally the titrant is added dropwise from the buret in order that the final drop added will not over run the end-point. The quantity of the substance being titrated may be calculated from the volume and the normality or molarity factor of the titrant and the equivalence factor for the substance given in the individual monograph.

Residual Titrations—Some Pharmacopeial assays require the addition of a measured volume of a volumetric solution, in excess of the amount actually needed to react with the substance being assayed, the excess of this solution then being titrated with a second volumetric solution. This constitutes a residual titration and is known also as a "back titration." The quantity of the substance being titrated may be calculated from the difference between the volume of the volumetric solution originally added and that consumed by the titrant in the back titration, due allowance being made for the respective normality or molarity factors of the two solutions, and the equivalence factor for the substance given in the individual monograph.

Chelometric Titrations—Simple, direct titrations of some polyvalent cations are possible by the use of reagents with which the cations form complexes. The titration of the calcium ion by this means is particularly advantageous, in comparison to the oxalate precipitation method previously used for Pharmacopeial purposes. The success of complexometry depends in large measure upon the indicator chosen. Often the color change of an indicator can be improved by the addition of a screening agent.

Titrations in Nonaqueous Solvents—Acids and bases have long been defined as substances that furnish, when dissolved in water, hydrogen, and hydroxyl ions, respectively. This definition, introduced by Arrhenius, fails to recognize the fact that properties characteristic of acids or bases may be developed also in other solvents. A more generalized definition is that of Brönsted, who defined an acid as a substance that furnishes protons, and a base as a substance that combines with protons. Even broader is the definition of Lewis, who defined an acid as any material that will accept an electron pair, a base as any material that will donate an electron pair, and neutralization as the formation of a coordination bond between an acid and a base.

The apparent strength of an acid or a base is determined by the extent of its reaction with a solvent. In water solution all strong acids appear equally strong because they react with the solvent to undergo almost complete conversion to oxonium ion and the acid anion (leveling effect). In a weakly protophilic solvent such as acetic acid the extent of formation of the acetate acidium ion shows that the order of decreasing strength for acids is perchloric, hydrobromic, sulfuric, hydrochloric, and nitric (differentiating effect).

Acetic acid reacts incompletely with water to form oxonium ion and is, therefore, a weak acid. In contrast, it dissolves in a base such as ethylenediamine, and reacts so completely with the solvent that it behaves as a strong acid. The same holds for perchloric acid.

This leveling effect is observed also for bases. In sulfuric acid almost all bases appear to be of the same strength. As the acid properties of the solvent decrease in the series sulfuric acid, acetic acid, phenol, water, pyridine, and butylamine, the bases become progressively weaker until all but the strongest have lost their basic properties. In order of decreasing strength, the strong bases are sodium 2-aminoethoxide, potassium methoxide, sodium methoxide, and lithium methoxide.

Many water-insoluble compounds acquire enhanced acidic or basic properties when dissolved in organic solvents. Thus the choice of the appropriate solvent permits the determination of a variety of such materials by nonaqueous titration. Furthermore, depending upon which part of a compound is the physiologically active moiety, it is often possible to titrate that part by proper selection of solvent and titrant. Pure compounds can be titrated directly, but it is often necessary to isolate the active ingredient in pharmaceutical preparations from interfering excipients and carriers.

The types of compounds that may be titrated as acids include acid halides, acid anhydrides, carboxylic acids, amino acids, enols such as barbiturates and xanthines, imides, phenols, pyrroles, and sulfonamides. The types of compounds that may be titrated as bases include amines, nitrogen-containing heterocyclic compounds, oxazolines, quaternary ammonium compounds, alkali salts of organic acids, alkali salts of weak inorganic acids, and some salts of amines. Many salts of halogen acids may be titrated in acetic acid or acetic anhydride after the addition of mercuric acetate, which removes halide ion as the un-ionized mercuric halide complex and introduces the acetate ion.

For the titration of a basic compound, a volumetric solution of perchloric acid in glacial acetic acid is preferred, although perchloric acid in dioxane is used in special cases. The calomel-glass electrode system is useful in this case. In acetic acid solvent this electrode system functions as predicted by theory.

For the titration of an acidic compound, two classes of titrant are available: the alkali metal alkoxides and the tetraalkylammonium hydroxides. A volumetric solution of sodium methoxide in a mixture of methanol and toluene is used frequently, although lithium methoxide in methanol-benzene solvent is used for those compounds yielding a gelatinous precipitate on titration with sodium methoxide.

The alkali error limits the use of the glass electrode as an indicating electrode in conjunction with alkali metal alkoxide titrants, particularly in basic solvents. Thus, the antimony-indicating electrode, though somewhat erratic, is used in such titrations. The use of quaternary ammonium hydroxide compounds, e.g., tetra-*n*-butylammonium hydroxide and trimethylhexadecylammonium hydroxide (in benzene-methanol or isopropyl alcohol), has two advantages over the other titrants in that (a) the tetraalkylammonium salt of the titrated acid is soluble in the titration medium, and (b) the convenient and well-behaved calomel-glass electrode may be used to conduct potentiometric titrations.

Because of interference by carbon dioxide, solvents for acidic compounds need to be protected from excessive exposure to the atmosphere by a suitable cover or by an inert atmosphere during the titration. Absorption of carbon dioxide may be determined by performing a blank titration. The blank should not exceed 0.01 mL of 0.1 *N* sodium methoxide VS per mL of solvent.

The end-point may be determined visually by color change, or potentiometrically, as indicated in the individual monograph. If the calomel reference electrode is used, it is advantageous to replace the aqueous potassium chloride salt bridge with 0.1 *N* lithium perchlorate in glacial acetic acid for titrations in acidic solvents or potassium chloride in methanol for titrations in basic solvents.

Where these or other mixtures are specified in individual monographs, the calomel reference electrode is modified by first removing the aqueous potassium chloride solution and residual potassium chloride, if any, by rinsing with water, then eliminating residual water by rinsing with the required nonaqueous solvent, and finally filling the electrode with the designated nonaqueous mixture.

The more useful systems for titration in nonaqueous solvents are listed in Table 1.

Indicator and Potentiometric End-point Detection—The simplest and most convenient method by which the equivalence point, i.e., the point at which the stoichiometric analytical reaction is complete, may be determined is with the use of indicators. These chemical substances, usually colored, respond to changes in solution conditions before and after the equivalence point by exhibiting color changes that may be taken visually as the end-point, a reliable estimate of the equivalence point.

Table 1. Systems for Nonaqueous Titrations.

Type of Solvent	*Acidic* (for titration of bases and their salts)	*Relatively Neutral* (for differential titration of bases)	*Basic* (for titration of acids)	*Relatively Neutral* (for differential titration of acids)
Solvent[1]	Glacial Acetic Acid Acetic Anhydride Formic Acid Propionic Acid Sulfuryl Chloride	Acetonitrile Alcohols Chloroform Benzene Toluene Chlorobenzene Ethyl Acetate Dioxane	Dimethylformamide *n*-Butylamine Pyridine Ethylenediamine Morpholine	Acetone Acetonitrile Methyl Ethyl Ketone Methyl Isobutyl Ketone *tert*-Butyl Alcohol
Indicator	Crystal Violet Quinaldine Red *p*-Naphtholbenzein Alphazurine 2-G Malachite green	Methyl Red Methyl Orange *p*-Naphtholbenzein	Thymol Blue Thymolphthalein Azo Violet *o*-Nitroaniline *p*-Hydroxyazobenzene	Azo Violet Bromothymol Blue *p*-Hydroxyazobenzene Thymol Blue
Electrodes	Glass-calomel Glass-silver–silver chloride Mercury–mercuric acetate	Glass-calomel Calomel-silver–silver chloride	Antimony-calomel Antimony-glass Antimony-antimony[2] Platinum-calomel Glass-calomel	Antimony-calomel Glass-calomel Glass-platinum[2]

[1] Relatively neutral solvents of low dielectric constant such as benzene, toluene, chloroform or dioxane may be used in conjunction with any acidic or basic solvent in order to increase the sensitivity of the titration end-points.
[2] In titrant.

A useful method of end-point determination results from the use of electrochemical measurements. If an indicator electrode, sensitive to the concentration of the species undergoing titrimetric reaction, and a reference electrode, whose potential is insensitive to any dissolved species, are immersed in the titrate to form a galvanic cell, the potential difference between the electrodes may be sensed by a pH meter and used to follow the course of the reaction. Where such a series of measurements is plotted correctly (i.e., for an acid-base titration, pH versus mL of titrant added; for a precipitimetric, complexometric, or oxidation-reduction titration, mV versus mL of titrant added), a sigmoid curve results with a rapidly changing portion (the "break") in the vicinity of the equivalence point. The mid-point of this linear vertical portion or the inflection point may be taken as the end-point. However, it should be noted that in asymmetrical reactions, which are reactions in which the number of anions reacting is not the same as the number of cations reacting, the end-point as defined by the inflection of the titration curve does not occur exactly at the stoichiometric equivalence point. Thus, potentiometric end-point detection by this method is not suitable in the case of asymmetric reactions, examples of which are the precipitation reaction,

$$2Ag^+ + CrO_4^{-2}$$

and the oxidation-reduction reaction,

$$5Fe^{+2} + MnO_4^-.$$

All acid-base reactions, however, are symmetrical. Thus, potentiometric end-point detection may be employed in acid-base titrations and in other titrations involving symmetrical reversible reactions where an indicator is specified, unless otherwise directed in the individual monograph.

Two types of automatic electrometric titrators are available. The first is one that carries out titrant addition automatically and records the electrode potential differences during the course of titration as the expected sigmoid curve. In the second type, titrant addition is performed automatically until a preset potential or pH, representing the end-point, is reached, at which point the titrant addition ceases.

Several acceptable electrode systems for potentiometric titrations are summarized in Table 2.

Blank Corrections—As previously noted, the end-point determined in a titrimetric assay is an estimate of the reaction equivalence point. The validity of this estimate depends upon, among other factors, the nature of the titrate constituents and the concentration of the titrant. An appropriate *blank correction* is em-

Table 2. Potentiometric Titration Electrode Systems.

Titration	Indicating Electrode	Equation[1]	Reference Electrode	Applicability[2]
Acid-base	Glass	$E = k + 0.0591\ \mathrm{pH}$	Calomel or silver–silver chloride	Titration of acids and bases
Precipitimetric (silver)	Silver	$E = E° + 0.0591 \log [Ag^+]$	Calomel (with potassium nitrate salt bridge)	Titration with or of silver involving halides or thiocyanate
Chelometric	Mercury-mercury(II)	$E = E° + 0.0296(\log k' - \mathrm{pM})$	Calomel	Titration of various metals (M), e.g., Mg^{+2}, Ca^{+2}, Al^{+3}, Bi^{+3}, with EDTA
Oxidation-reduction	Platinum	$E = E° + \frac{0.0591}{n} \log \frac{[ox]}{[red]}$	Calomel or silver–silver chloride	Titrations with arsenite, bromine, cerate, dichromate, hexacyanoferrate(III), iodate, nitrite, permanganate, thiosulfate

[1] Appropriate form of Nernst equation describing the indicating electrode system: k = glass electrode constant; k' = constant derived from Hg-Hg(II)-EDTA equilibrium; M = any metal undergoing EDTA titration; [ox] and [red] from the equation, ox + $ne \rightleftarrows$ red.
[2] Listing is representative but not exhaustive.

ployed in titrimetric assays to enhance the reliability of the end-point determination. Such a blank correction is usually obtained by means of a *residual blank titration*, wherein the required procedure is repeated in every detail except that the substance being assayed is omitted. In such instances, the actual volume of titrant equivalent to the substance being assayed is the difference between the volume consumed in the residual blank titration and that consumed in the titration with the substance present. The corrected volume so obtained is used in calculating the quantity of the substance being titrated, in the same manner as prescribed under *Residual Titrations*. Where potentiometric end-point detection is employed, the blank correction is usually negligible.

⟨551⟩ ALPHA TOCOPHEROL ASSAY

The following procedure is provided for the determination of tocopherol as an ingredient of *Decavitamin Capsules* and *Decavitamin Tablets*.

Hydrogenator—A suitable device for low-pressure hydrogenation may be assembled as follows: Arrange in a rack or in clamps two 50-mL conical centrifuge tubes, connected in series by means of glass and inert plastic tubing and suitable stoppers of glass, polymer, or cork (avoiding all use of rubber). Use one tube for the blank and the other for the assay specimen. Arrange a gas-dispersion tube so that the hydrogen issues as bubbles at the bottom of each tube. Pass the hydrogen first through the blank tube and then through the specimen tube.

Procedure—Pipet into a suitable vessel 25 mL of the final washed ether solution of the unsaponifiable fraction obtained as directed for *When Tocopherol Is Present* under *Procedure* in the *Vitamin A Assay* ⟨571⟩, and evaporate to about 5 mL. *Without applying heat*, remove the remaining ether in a stream of inert gas or by vacuum. Dissolve the residue in sufficient alcohol to give an expected concentration of about 0.15 mg of alpha tocopherol per mL. Pipet 15 mL into a 50-mL centrifuge tube, add about 200 mg of palladium catalyst, stir with a glass rod, and hydrogenate for 10 minutes in the *Hydrogenator*, using hydrogen that has been passed through alcohol in a blank tube. Add about 300 mg of chromatographic siliceous earth, stir with a glass rod, and immediately centrifuge until the solution is clear.

Test a 1-mL aliquot of the solution by removing the solvent by evaporation, dissolving the residue in 1 mL of chloroform, and adding 10 mL of antimony trichloride TS: no detectable blue color appears. [NOTE—If a blue color appears, repeat the hydrogenation for a longer time period, or with a new lot of catalyst.]

Pipet 2 mL of the supernatant solution into a glass-stoppered, opaque flask, add 1.0 mL of a 1 in 500 solution of ferric chloride in dehydrated alcohol,* and begin timing the reaction, preferably with a stop watch. Add immediately 1.0 mL of a 1 in 200 solution of 2,2′-bipyridine in dehydrated alcohol, mix with swirling, add 21.0 mL of dehydrated alcohol, close the tube, and shake vigorously to ensure complete mixing. When about 9½ minutes have elapsed from the beginning of the reaction, transfer part of the mixture to one of a pair of matched 1-cm spectrophotometer cells. After 10 minutes, accurately timed, following the addition of the ferric chloride–dehydrated alcohol solution, determine the absorbance at 520 nm, with a suitable spectrophotometer, using dehydrated alcohol as the blank. Perform a blank determination with the same quantities of the same reagents and in the same manner, but using 2 mL of dehydrated alcohol in place of the 2 mL of the hydrogenated solution. Subtract the absorbance determined for the blank from that determined for the assay specimen, and designate the difference as A_D.

Calculate the alpha tocopherol content, in mg, in the assay specimen taken by the formula:

$$30.2\, A_D/(LC_D),$$

in which A_D is the corrected absorbance, L is the length, in cm, of the absorption cell, and C_D is the content of the assay specimen in the alcohol solution employed for the measurement of absorbance, expressed as g, capsules, or tablets per 100 mL.

* NOTE—The absorbance of the blank may be reduced, and the precision of the determination thereby improved, by purification of the dehydrated alcohol that is used throughout the assay. Purification may be accomplished by the addition of a few crystals (about 0.02%) of potassium permanganate and of a few pellets of potassium hydroxide to the dehydrated alcohol, and subsequent redistillation.

⟨561⟩ VEGETABLE DRUGS—SAMPLING AND METHODS OF ANALYSIS

Sampling

In order to reduce the effect of sampling bias in qualitative and quantitative results, it is necessary to ensure that the composition of the sample used be representative of the batch of drugs being examined. The following sampling procedures are the minimum considered applicable to vegetable drugs. Some articles, or some tests, may require more rigorous procedures involving more containers being sampled and/or more samples per container.

Gross Sample—Where external examination of containers, markings, and labels indicates that the batch can be considered to be homogeneous, take individual samples from the number of randomly selected containers indicated below. Where the batch cannot be considered to be homogeneous, divide it into sub-batches that are as homogeneous as possible, then sample each one as a homogeneous batch.

No. of Containers in Batch (N)	No. of Containers to be Sampled (n)
1 to 10	all
11 to 19	11
>19	$n = 10 + (N/10)$

(Round calculated "n" to next highest whole number.)

Samples shall be taken from the upper, middle, and lower sections of each container. If the crude material consists of component parts which are 1 cm or less in any dimension, and in the case of all powdered or ground materials, withdraw the sample by means of a sampling device that removes a core from the top to the bottom of the container, not less than two cores being taken in opposite directions. For materials with component parts over 1 cm in any dimension, withdraw samples by hand. In the case of large bales or packs, samples should be taken from a depth of 10 cm because the moisture content of the surface layer may be different from that of the inner layers.

Prepare the gross sample by combining and mixing the individual samples taken from each opened container, taking care not to increase the degree of fragmentation or significantly affect the moisture content.

Laboratory Sample—Prepare the laboratory sample by repeated quartering of the gross sample. (NOTE—Quartering consists of placing the sample, adequately mixed, as an even and square-shaped heap and dividing it diagonally into four equal parts. The two opposite parts are then taken and carefully mixed. The process is repeated as necessary until the required quantity is obtained.)

The laboratory sample should be of a size sufficient for performing all the necessary tests.

Test Sample—Unless otherwise directed in the individual monograph or test procedure below, prepare the test sample as follows:

Decrease the size of the laboratory sample by quartering, taking care that each withdrawn portion remains representative. In the case of unground or unpowdered drugs, grind the withdrawn sample so that it will pass through a No. 20 standard-mesh sieve, and mix the resulting powder well. If the material cannot be ground, reduce it to as fine a state as possible, mix by rolling it on paper or sampling cloth, spread it out in a thin layer and withdraw the portion for analysis.

Foreign Organic Matter

Test Sample—Unless otherwise specified in the individual monograph, weigh the following quantities of the laboratory sample, taking care that the withdrawn portion is representative (quartering if necessary):

Roots, rhizomes, bark, and herbs	500 g
Leaves, flowers, seeds, and fruit	250 g
Cut vegetable drugs (average weight of the pieces is less than 0.5 g)	50 g

Spread the sample out in a thin layer, and separate the foreign organic matter by hand as completely as possible. Weigh it, and determine the percentage of foreign organic matter in the weight of drug taken.

Total Ash

Accurately weigh a quantity of the *test sample*, representing 2 to 4 g of the air-dried material, in a tared crucible, and incinerate, gently at first, and gradually increase the temperature to 675 ± 25°, until free from carbon, and determine the weight of the ash. If a carbon-free ash cannot be obtained in this way, extract the charred mass with hot water, collect the insoluble residue on an ashless filter paper, incinerate the residue and filter paper until the ash is white or nearly so, then add the filtrate, evaporate it to dryness, and heat the whole to a temperature of 675 ± 25°. If a carbon-free ash cannot be obtained in this way, cool the crucible, add 15 mL of alcohol, break up the ash with a glass rod, burn off the alcohol, and again heat the whole to a temperature of 675 ± 25°. Cool in a desiccator, weigh the ash, and calculate the percentage of total ash from the weight of the drug taken.

Acid-insoluble Ash

Boil the ash obtained as directed under *Total Ash*, above, with 25 mL of 3 *N* hydrochloric acid for 5 minutes, collect the insoluble matter on a tared filtering crucible or ashless filter, wash with hot water, ignite, and weigh. Determine the percentage of acid-insoluble ash calculated from the weight of drug taken.

Crude Fiber

Exhaust a weighed quantity of the *test sample*, representing about 2 g of the drug, with ether. Add 200 mL of boiling dilute sulfuric acid (1 in 78) to the ether-exhausted marc, in a 500-mL flask, and connect the flask to a reflux condenser. Reflux the mixture for 30 minutes, accurately timed, then filter through a linen or hardened-paper filter, and wash the residue on the filter with boiling water until the effluent washing is no longer acid. Rinse the residue back into the flask with 200 mL of boiling sodium hydroxide solution, adjusted to 1.25 percent by titration and free from sodium carbonate. Again reflux the mixture for 30 minutes, accurately timed, then rapidly filter through a tared filter, wash the residue with boiling water until the last washing is neutral, and dry it at 110° to constant weight. Incinerate the dried residue, ignite to constant weight, cool in a desiccator, and weigh the ash: the difference between the weight obtained by drying at 110° and that of the ash represents the weight of the crude fiber.

NOTE—The boiling with acid and alkali should continue for 30 minutes, accurately timed, from the time that the liquid (which is cooled below the boiling point by being added to the cold flask) again boils. After the solution has been brought to boiling, the heat should be turned low enough just to maintain boiling. During the boiling, the flask should be gently rotated from time to time to wash down any particles that may adhere to the walls of the flask. A slow current of air introduced into the flask during the boiling operation aids in preventing excessive frothing.

Volatile Oil Determination

Set up a round-bottom, shortneck, 1-liter flask in a heating mantle set over a magnetic stirrer. Insert an egg-shaped stirring bar magnet in the flask, and attach a cold-finger condenser and an appropriate volatile oil trap of the type illustrated.

Coarsely comminute a sufficient quantity of the drug to yield from 1 to 3 mL of volatile oil. Small seeds, fruits, or broken leaves of herbs ordinarily do not need comminution. Very fine powders are to be avoided. If this is not possible, it may be necessary to mix them with purified sawdust or purified sand. Place a suitable quantity of the drug, accurately weighed, in the flask, and fill it one-half with water. Attach the condenser and the proper separator. Boil the contents of the flask, using a suit-

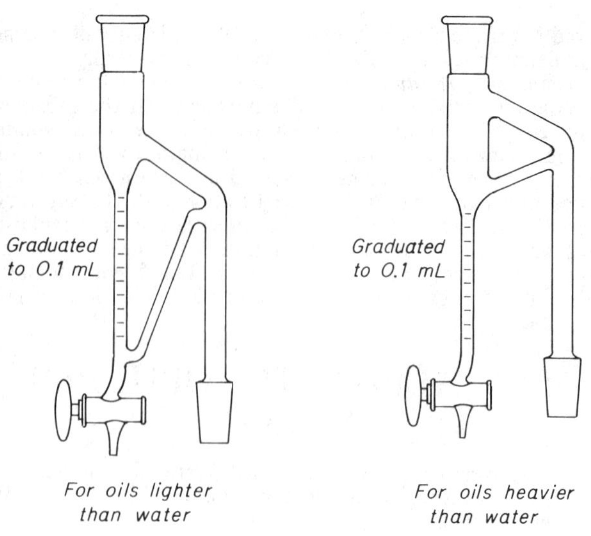

Traps for Volatile Oil Apparatus

able amount of heat to maintain gentle boiling for 2 hours, or until the volatile oil has been completely separated from the drug and no longer collects in the graduated tube of the separator.

If a proper quantity of the volatile oil has been obtained in the graduated tube of the separator, it can be read to tenths of 1 mL, and the volume of volatile oil from each 100 g of drug can be calculated from the weight of the drug taken. The graduations on the separator "for oils heavier than water" are so placed that oil remains below the aqueous condensate that automatically flows back into the flask.

Water

For unground or unpowdered drugs, prepare about 10 g of the *laboratory sample* by cutting, granulating, or shredding, so that the parts are about 3 mm in thickness. Seeds or fruits smaller than 3 mm should be cracked. Avoid the use of high-speed mills in preparing the sample, and exercise care that no appreciable amount of moisture is lost during the preparation and that the portion taken is representative of the *laboratory sample*. Determine the water content as directed under *Procedure for Vegetable Drugs* in the test for *Water Determination—Gravimetric Method* ⟨921⟩.

⟨571⟩ VITAMIN A ASSAY

The following procedure is provided for the determination of vitamin A as an ingredient of Pharmacopeial preparations. It conforms to that which was adopted in 1956 for international use by the International Union of Pure and Applied Chemistry.

Complete the assay promptly, and exercise care throughout the procedure to keep to a minimum the exposure to actinic light and to atmospheric oxygen and other oxidizing agents, preferably, by the use of non-actinic glassware and an atmosphere of an inert gas.

Special Reagents—

ETHER—Use ethyl ether, and use it within 24 hours after opening the container.

ISOPROPYL ALCOHOL—Use spectrophotometric-grade isopropyl alcohol (see *Isopropyl Alcohol* under *Reagent Specifications* in the section, *Reagents, Indicators, and Solutions*).

Procedure—Accurately weigh, count, or measure a portion of the test specimen expected to contain the equivalent of not less than 0.15 mg of retinol but containing not more than 1 g of fat. If in the form of capsules, tablets, or other solid, so that it cannot be saponified efficiently by the ensuing instructions, reflux the portion taken in 10 mL of water on a steam bath for about 10 minutes, crush the remaining solid with a blunt glass rod, and warm for about 5 minutes longer.

Transfer to a suitable borosilicate glass flask, and add 30 mL of alcohol, followed by 3 mL of potassium hydroxide solution (9 in 10). Reflux in an all-borosilicate glass apparatus for 30 minutes. Cool the solution, add 30 mL of water, and transfer to a conical separator. Add 4 g of finely powdered sodium sulfate decahydrate. Extract by shaking with one 150-mL portion of ether for 2 minutes, and then, if an emulsion forms, with three 25-mL portions of ether. Combine the ether extracts, if necessary, and wash by swirling gently with 50 mL of water. Repeat the washing more vigorously with three additional 50-mL portions of water. Transfer the washed ether extract to a 250-mL volumetric flask, add ether to volume, and mix.

Evaporate a 25.0-mL portion of the ether extract to about 5 mL. *Without applying heat and with the aid of a stream of inert gas or vacuum*, continue the evaporation to about 3 mL. Dissolve the residue in sufficient isopropyl alcohol to give an expected concentration of the equivalent to 3 µg to 5 µg of vitamin A per mL or such that it will give an absorbance in the range 0.5 to 0.8 at 325 nm. Determine the absorbances of the resulting solution at the wavelengths 310 nm, 325 nm, and 334 nm, with a suitable spectrophotometer fitted with matched quartz cells, using isopropyl alcohol as the blank.

WHEN TOCOPHEROL IS PRESENT—Transfer to a suitable borosilicate glass flask a test specimen, accurately measured, or not less than 5 previously crushed Decavitamin Capsules or Decavitamin Tablets. Reflux in an all-borosilicate glass apparatus with 30 mL of alcohol and 3 mL of potassium hydroxide solution (9 in 10) for 30 minutes. Add through the condenser 2.0 g of citric acid monohydrate, washing the walls of the condenser with 10 mL of water. Cool, and transfer the solution to a conical separator with the aid of 20 mL of water. Add 4 g of finely powdered sodium sulfate decahydrate. Extract with one 150-mL portion of ether and then, if an emulsion forms, with three 25-mL portions of ether. Combine the ether extracts, if necessary, and wash by swirling gently with 50 mL of water. Repeat the washing more vigorously with three additional 50-mL portions of water. Transfer the washed ether extract to a 250-mL volumetric flask, and add ether to volume. Transfer a 100.0-mL aliquot of the resulting ether solution to a conical separator, and wash once with 50 mL of potassium hydroxide solution (1 in 33), using alcohol, if necessary, to break any emulsion that forms. Wash by swirling gently with 50 mL of water. Repeat the washing more vigorously with three additional 50-mL portions of water. Transfer the washed ether extract to a 100-mL volumetric flask, add ether to volume, and mix.

Evaporate a 50.0-mL aliquot of the ether solution of the unsaponifiable extract to about 5 mL. *Without applying heat and with the aid of a stream of inert gas or vacuum*, remove the residual ether. Dissolve the residue in 50.0 mL of isopropyl alcohol.

Hydrogenated portion—Pipet 15.0 mL of the isopropyl alcohol solution into a 50-mL centrifuge tube, add approximately 200 mg of palladium catalyst, stir with a glass rod, and hydrogenate for 10 minutes in a *Hydrogenator* such as is described in the *Alpha Tocopherol Assay* ⟨551⟩, using isopropyl alcohol in the blank tube. Add about 300 mg of chromatographic siliceous earth, stir with a glass rod, and immediately centrifuge until the solution is clear.

Test a 1-mL aliquot of the solution by removing the solvent by evaporation, dissolving the residue in 1 mL of chloroform, and adding 10 mL of phosphomolybdic acid TS: no detectable blue-green color appears. [NOTE—If a blue-green color appears, repeat the hydrogenation for a longer time period, or with a new lot of catalyst.]

Into two separate flasks pipet equal volumes of the *Hydrogenated portion* and the untreated isopropyl alcohol solution, respectively, and add sufficient isopropyl alcohol to give an expected concentration of vitamin A equivalent to 3 µg to 5 µg per mL. Determine the absorbances of the untreated solution against the solution from the *Hydrogenated portion* as a blank, at the wavelengths 310 nm, 325 nm, and 334 nm, with a suitable spectrophotometer fitted with matched quartz cells.

Calculation—Calculate the vitamin A content as follows:

$$\text{Content (in mg)} = 0.549A_{325}/LC,$$

in which A_{325} is the observed absorbance at 325 nm, L is the length, in cm, of the absorption cell, and C is the amount of test specimen expressed as g, capsule, or tablet in each 100 mL of the final isopropyl alcohol solution, provided that A_{325} has a value not less than $[A_{325}]/1.030$ and not more than $[A_{325}]/0.970$, where $[A_{325}]$ is the *corrected* absorbance at 325 nm and is given by the equation:

$$[A_{325}] = 6.815A_{325} - 2.555A_{310} - 4.260A_{334},$$

in which A designates the absorbance at the wavelength indicated by the subscript.

Where $[A_{325}]$ has a value less than $A_{325}/1.030$, apply the following equation:

$$\text{Content (in mg)} = 0.549[A_{325}]/LC,$$

in which the values are as defined herein. Each mg of vitamin A (alcohol) represents 3333 USP Units of vitamin A.

Confidence Interval—The range of the limits of error, indicating the extent of discrepancy to be expected in the results of different laboratories at $P = 0.05$, is approximately ±8%.

⟨581⟩ VITAMIN D ASSAY

Chromatographic Method

The following pressurized liquid chromatographic procedure is provided for the determination of vitamin D, as cholecalciferol or as ergocalciferol, as an ingredient of Pharmacopeial multiple-vitamin preparations.

Throughout this assay, protect solutions containing, and derived from, the test specimen and the Reference Standard from the atmosphere and light, preferably by the use of a blanket of inert gas and low-actinic glassware.

Reference Standard—[NOTE—Use USP Ergocalciferol RS, or USP Cholecalciferol RS, for assaying pharmaceutical dosage forms that are labeled to contain vitamin D as ergocalciferol, or as cholecalciferol, respectively.] *USP Ergocalciferol Reference Standard*—Store in a cold place, protected from light. Allow it to attain room temperature before opening ampul. Use the material promptly, and discard the unused portion.

USP Cholecalciferol Reference Standard—Store in a cold place, protected from light. Allow it to attain room temperature before opening ampul. Use the material promptly, and discard the unused portion.

USP Vitamin D Assay System Suitability Reference Standard—Store in a cool place, protected from light. Allow it to attain room temperature before opening ampul. Do not dry. Transfer unused contents of ampul to a tightly closed container, and store under nitrogen, in the dark, in a cool place.

USP Δ4,6-Cholestadienol Reference Standard—Store in a cool place, protected from light. Allow it to attain room temperature before opening ampul. Do not dry. Transfer unused contents of ampul to a tightly closed container, and store under nitrogen, in the dark, in a cool place.

Special Reagents and Solutions—

ETHER—Use ethyl ether. Use within 24 hours after opening container.

DEHYDRATED HEXANE—Prepare a chromatographic column by packing a chromatographic tube, 60 cm × 8 cm in diameter, with 500 g of 50- to 250-µm chromatographic siliceous earth, activated by drying at 150° for 4 hours (see *Column adsorption chromatography* under *Chromatography* ⟨621⟩). Pass 500 mL of hexanes through the column, and collect the eluate in a glass-stoppered flask.

BUTYLATED HYDROXYTOLUENE SOLUTION—Dissolve a quantity of butylated hydroxytoluene in chromatographic hexane to obtain a solution containing 10 mg per mL.

AQUEOUS POTASSIUM HYDROXIDE SOLUTION—Dissolve 500 g of potassium hydroxide in 500 mL of freshly boiled water, mix, and cool. Prepare this solution fresh daily.

ALCOHOLIC POTASSIUM HYDROXIDE SOLUTION—Dissolve 3 g of potassium hydroxide in 50 mL of freshly boiled water, add 10 mL of alcohol, dilute with freshly boiled water to 100 mL, and mix. Prepare this solution fresh daily.

SODIUM ASCORBATE SOLUTION—Dissolve 3.5 g of ascorbic acid in 20 mL of 1 N sodium hydroxide. Prepare this solution fresh daily.

SODIUM SULFIDE SOLUTION—Dissolve 12 g of sodium sulfide in 20 mL of water, dilute with glycerin to 100 mL, and mix.

Mobile Phase A—Prepare a mixture of acetonitrile, methanol, and water (25:25:1). The amount of water and the flow rate may be varied to meet system suitability requirements.

Mobile Phase B—Prepare a 3 in 1000 mixture of *n*-amyl alcohol in *Dehydrated hexane*. The ratio of components and the flow rate may be varied to meet system suitability requirements.

Internal Standard Solution—Transfer 15 mg of USP Δ4,6-Cholestadienol RS, accurately weighed, to a 200-mL volumetric flask, add a 1 in 10 mixture of toluene and *Mobile Phase B* to volume, and mix.

Standard Preparation—Transfer about 25 mg of USP Ergocalciferol RS or Cholecalciferol RS, accurately weighed, to a 50-mL volumetric flask, dissolve without heat in toluene, add toluene to volume, and mix. Pipet 10 mL of this stock solution into a 100-mL volumetric flask, dilute with toluene to volume, and mix. Prepare stock solution fresh daily.

Assay Preparation—

For oily solutions—Accurately weigh a portion of the specimen to be assayed, preferably more than 0.5 g and equivalent to about 125 µg of cholecalciferol or ergocalciferol (5000 USP Units). Add 1 mL of *Sodium Ascorbate Solution*, 25 mL of alcohol, and 2 mL of *Aqueous Potassium Hydroxide Solution*, and mix.

For capsules or tablets—Reflux not less than 10 capsules or tablets with a mixture of 10 mL of *Sodium Ascorbate Solution* and 2 drops of *Sodium Sulfide Solution* on a steam bath for 10 minutes, crush any remaining solids with a blunt glass rod, and continue heating for 5 minutes. Cool, add 25 mL of alcohol and 3 mL of *Aqueous Potassium Hydroxide Solution*, and mix.

For dry preparations and aqueous dispersions—Accurately weigh a portion of the specimen to be assayed, preferably more than 0.5 g and equivalent to about 125 µg of cholecalciferol or ergocalciferol (5000 USP Units). Add, in small quantities and with gentle swirling, 25 mL of alcohol, 5 mL of *Sodium Ascorbate Solution*, and 3 mL of *Aqueous Potassium Hydroxide Solution*.

SAPONIFICATION AND EXTRACTION—Reflux the mixture prepared from the specimen to be assayed on a steam bath for 30 minutes. Cool rapidly under running water, and transfer the saponified mixture to a conical separator, rinsing the saponification flask with two 15-mL portions of water, 10 mL of alcohol, and two 50-mL portions of ether. Shake the combined saponified mixture and rinsings vigorously for 30 seconds, and allow to stand until both layers are clear. Transfer the aqueous phase to a second conical separator, add a mixture of 10 mL of alcohol and 50 mL of solvent hexane, and shake vigorously. Allow to separate, transfer the aqueous phase to a third conical separator, and transfer the hexane phase to the first separator, rinsing the second separator with two 10-mL portions of solvent hexane, adding the rinsings to the first separator. Shake the aqueous phase in the third separator with 50 mL of solvent hexane, and add the hexane phase to the first separator. Wash the combined ether-hexane extracts by shaking vigorously with three 50-mL portions of *Alcoholic Potassium Hydroxide Solution*, and wash with 50-mL portions of water vigorously until the last washing is neutral to phenolphthalein. Drain any remaining drops of water from the combined ether-hexane extracts, add 2 sheets of 9-cm filter paper, in strips, to the separator, and shake. Transfer the washed ether-hexane extracts to a round-bottom flask, rinsing the separator and paper with solvent hexane. Combine the hexane rinsings with the ether-hexane extracts, add 5.0 mL of *Internal Standard Solution* and 100 µL of *Butylated Hydroxytoluene Solution*, and mix. Evaporate to dryness in vacuum by swirling in a water bath maintained at a temperature not higher than 40°. Cool under running water, and introduce nitrogen sufficient to restore atmospheric pressure. Without delay, dissolve the residue in 5.0 mL of a mixture of equal volumes of acetonitrile and methanol, or in a measured portion of the acetonitrile-methanol mixture until the concentration of vitamin D is about 25 µg per mL, to obtain the *Assay Preparation.*

Chromatographic System—Use a chromatograph, operated at room temperature, fitted with an ultraviolet detector that monitors absorption at 254 nm, a 30-cm × 4.6-mm stainless steel cleanup column packed with column packing L7 and using *Mobile Phase A*, and a 25-cm × 4.6-mm stainless steel analytical column packed with column packing L3 and using *Mobile Phase B*.

CLEANUP COLUMN SYSTEM SUITABILITY TEST—Pipet 5 mL of the *Standard Preparation* into a round-bottom flask fitted with a reflux condenser, and add 2 or 3 crystals of butylated hydroxytoluene. Displace the air with nitrogen, and heat in a water bath maintained at a temperature of 90° in subdued light under an atmosphere of nitrogen for 45 minutes, to obtain a solution containing vitamin D and pre-vitamin D. Cool, add 10.0 mL of *Internal Standard Solution*, mix, and evaporate in vacuum to dryness by swirling in a water bath maintained at a temperature not higher than 40°. Cool under running water, and introduce nitrogen sufficient to restore atmospheric pressure. Without delay, dissolve the residue in 10.0 mL of a mixture of equal volumes of acetonitrile and methanol, and mix. Inject 500 µL of this solution into the cleanup column, and record the chromatogram as directed under *Procedure*. The chromatogram exhibits a peak exhibiting a retention time between 5 and 9 minutes, corresponding to the separation under a single peak of the mixture of vitamin D, pre-vitamin D, and Δ4,6-cholestadienol from other substances. Adjust the water content or other operating parameters, if necessary (see *Mobile Phase A*).

ANALYTICAL COLUMN SYSTEM SUITABILITY TEST—Transfer about 100 mg of USP Vitamin D Assay System Suitability RS to a 100-mL volumetric flask, add a 1 in 20 mixture of toluene and *Mobile Phase B* to volume, and mix. Heat a portion of this solution, under reflux, at 90° for 45 minutes, and cool. Chromatograph five injections of the resulting solution, and measure the peak responses as directed under *Procedure*. The relative standard deviation for the cholecalciferol peak response does not exceed 2.0%, and the resolution between *trans*-cholecalciferol and pre-cholecalciferol is not less than 1.0. [NOTE—Chromatograms obtained as directed for this test exhibit relative retention times of approximately 0.4 for pre-cholecalciferol, 0.5 for *trans*-cholecalciferol, and 1.0 for cholecalciferol.]

Calibration—

VITAMIN D RESPONSE FACTOR—Transfer 4.0 mL of the *Standard Preparation* and 10.0 mL of *Internal Standard Solution* to a 100-mL volumetric flask, dilute with *Mobile Phase B* to volume, and mix to obtain the *Working Standard Preparation.* Store this *Working Standard Preparation* at a temperature not above 0°, retaining the unused portion for the *Procedure*. Inject 200 µL of the *Working Standard Preparation* into the analytical column, and measure the peak responses for vitamin D and for Δ4,6-cholestadienol. The relative retention time of Δ4,6-cholestadienol is about 1.3. Calculate the response factor, F_D, by the formula:

$$C_s/(R_sC_r),$$

in which C_s and C_r are the concentrations, in µg per mL, of vitamin D and Δ4,6-cholestadienol, respectively, in the *Working Standard Preparation*, and R_s is the ratio of the peak response of vitamin D to that of Δ4,6-cholestadienol.

PRE-VITAMIN D RESPONSE FACTOR—Pipet 4 mL of the *Standard Preparation* into a round-bottom flask fitted with a reflux condenser, and add 2 or 3 crystals of butylated hydroxytoluene. Displace the air with nitrogen, and heat in a water bath maintained at a temperature of 90° in subdued light under a nitrogen atmosphere for 45 minutes, to obtain a solution containing vitamin D and pre-vitamin D. Cool, transfer with the aid of several portions of *Mobile Phase B* to a 100-mL volumetric flask containing 10.0 mL of *Internal Standard Solution*, dilute with *Mobile Phase B* to volume, and mix to obtain the *Working Mixture.* Inject 200 µL of this *Working Mixture* into the analytical column, and measure the peak responses for vitamin D, pre-vitamin D, and Δ4,6-cholestadienol. Calculate the concentration, C'_s, in µg per mL, of vitamin D in the (heated) *Working Mixture* by the formula:

$$F_DC_rR'_s,$$

in which C_r is the concentration, in µg per mL, of Δ4,6-cholestadienol, and R'_s is the ratio of the peak response for vitamin D to that for Δ4,6-cholestadienol. Calculate the concentration, C'_{pre}, in µg per mL, of pre-vitamin D, in the *Working Mixture* by the formula:

$$C'_{pre} = C_s - C'_s.$$

Calculate the response factor, F_{pre}, for pre-vitamin D by the formula:

$$(F_D R'_s C'_{pre})/(R'_{pre} C'_s),$$

in which R'_{pre} is the ratio of the peak response of pre-vitamin D to that of Δ4,6-cholestadienol. [NOTE—Value of F_{pre} determined in duplicate, on different days, can be used during the whole procedure.]

Procedure—Inject 500 µL of the *Assay Preparation* into the cleanup column, and collect the fraction representing 0.7 to 1.3 relative to the retention time of the mixed vitamin D peak (see *Cleanup column system suitability test*) in a round-bottom flask. Add 50 µL of *Butylated Hydroxytoluene Solution*, mix, and evaporate in vacuum to dryness by swirling in a water bath maintained at a temperature not higher than 40°. Cool under running water, and introduce nitrogen sufficient to restore atmospheric pressure. Without delay, dissolve the residue in 5.0 mL of a 1 in 20 mixture of toluene and *Mobile Phase B*, and mix. Inject 200 µL of this solution into the analytical column, and measure the peak responses for vitamin D, pre-vitamin D, and Δ4,6-cholestadienol. Calculate the concentration, in µg per mL, of cholecalciferol ($C_{27}H_{44}O$) or ergocalciferol ($C_{28}H_{44}O$) in the *Assay Preparation* by the formula:

$$(R''_D F_D + R''_{pre} F_{pre}) C''_r,$$

in which R''_D is the ratio of the peak response of vitamin D to that of Δ4,6-cholestadienol, R''_{pre} is the ratio of the peak response of pre-vitamin D to that of Δ4,6-cholestadienol, and C''_r is the concentration, in µg per mL, of Δ4,6-cholestadienol in the *Assay Preparation.*

Chemical Method

The following procedure is provided for the determination of vitamin D as an ingredient of Pharmacopeial preparations.

Complete the assay promptly, and exercise care throughout the procedure to keep to a minimum the exposure to air and to actinic light, preferably by the use of a blanket of inert gas and low-actinic glassware.

Reference Standard—[NOTE—Use USP Ergocalciferol RS, or USP Cholecalciferol RS, for assaying pharmaceutical dosage forms that are labeled to contain vitamin D as ergocalciferol, or as cholecalciferol, respectively.] *USP Ergocalciferol Reference Standard*—Store in a cold place, protected from light. Allow it to attain room temperature before opening ampul. Use the material promptly, and discard the unused portion. *USP Cholecalciferol Reference Standard*—Store in a cold place, protected from light. Allow it to attain room temperature before opening ampul. Use the material promptly, and discard the unused portion.

Special Reagents and Solutions—

CHROMATOGRAPHIC FULLER'S EARTH—Use chromatographic fuller's earth having a water content corresponding to between 8.5% and 9.0% of loss on drying.

SOLVENT HEXANE—Use solvent hexane (see under *Reagents*), redistilling if necessary so that it meets the following additional specification:

Spectral purity—Measure in a 1-cm cell at 300 nm, with a suitable spectrophotometer, against air as the blank: the absorbance is not more than 0.070.

ETHYLENE DICHLORIDE—Purify by passage through a column of granular (20 to 200 mesh) silica gel.

POTASSIUM HYDROXIDE SOLUTION—Dissolve 500 g of potassium hydroxide in water to make 1000 mL.

BUTYLATED HYDROXYTOLUENE SOLUTION—Dissolve 10 mg of butylated hydroxytoluene in 100 mL of alcohol. Prepare this solution fresh daily.

ETHER—Use freshly distilled ether, discarding the first and last 10% portions of the distillate.

COLOR REAGENT—Prepare two stock solutions as follows.

Solution A—Empty, without weighing, the entire contents of a previously unopened 113-g bottle of dry, crystalline antimony trichloride into a flask containing about 400 mL of *Ethylene Dichloride.* Add about 2 g of anhydrous alumina, mix, and filter through filter paper into a clear-glass, glass-stoppered container calibrated at 500 mL. Add *Ethylene Dichloride* to make 500 mL, and mix: the absorbance of the solution, measured in a 20-mm cell at 500 nm, with a suitable spectrophotometer, against *Ethylene Dichloride*, does not exceed 0.070.

Solution B—Mix, under a hood, 100 mL of acetyl chloride and 400 mL of *Ethylene Dichloride.*

Mix 45 mL of *Solution A* and 5 mL of *Solution B* to obtain the *Color Reagent.* Store in a tight container, and use within 7 days, but discard any reagent in which a color develops.

Chromatographic Tubes—

FIRST COLUMN—Arrange for descending column chromatography a tube of 2.5-cm (inside) diameter, about 25 cm long, and constricted to 8-mm diameter for a distance of 5 cm at the lower end, by inserting at the point of constriction a coarse-porosity, sintered-glass disk or a small plug of glass wool. The constricted portion may be fitted with an inert, plastic stopcock.

SECOND COLUMN—Select a tube that is made up of three sections: (1) a flared top section, 18 mm in (inside) diameter and approximately 14 cm long, (2) a middle section, 6 mm in (inside) diameter and approximately 25 cm long, and (3) a tapered, constricted lower exit tube approximately 5 cm long. Insert a small plug of glass wool in the upper 1-cm portion of the constricted section.

Chromatographic Columns—

FIRST COLUMN—To about 125 mL of isooctane contained in a screw-capped, wide-mouth bottle add 25 g of chromatographic siliceous earth, and shake until a slurry is formed. Add, dropwise and with vigorous mixing, 10 mL of polyethylene glycol 600. Replace the bottle cover, and shake vigorously for 2 minutes. Pour about half of the resulting slurry into the chromatographic tube, and allow it to settle by gravity. Then apply gentle suction, and add the remainder of the slurry in small portions, packing each portion with a 20-mm disk plunger. When a solid surface has formed, remove the vacuum, and add about 2 mL of isooctane.

SECOND COLUMN—Pack the midsection of the tube with 3 g of moderately coarse *Chromatographic Fuller's Earth* with the aid of gentle suction (about 125 mm of mercury).

Standard Preparation—Dissolve about 25 mg of *Reference Standard*, accurately weighed, in isooctane to give a known concentration of about 250 µg per mL. Store in a refrigerator.

On the day of assay, pipet 1 mL of the standard solution into a 50-mL volumetric flask, remove the solvent with a stream of nitrogen, and dissolve the residue in, and dilute to volume with, *Ethylene Dichloride*, and mix.

Sample Preparation—Accurately weigh or measure a portion of the sample to be assayed, equivalent to not less than 125 µg but preferably about 250 µg of ergocalciferol (10,000 USP Units). If little or no vitamin A is present in the sample, add about 1.5 mg (the equivalent of 3000 USP Units) of vitamin A acetate to provide the needed pilot bands in the subsequent chromatography.

For capsules or tablets, reflux not less than 10 of them in 10 mL of water on a steam bath for about 10 minutes, crush the remaining solid with a blunt glass rod, and warm for 5 minutes longer.

Add a volume of *Potassium Hydroxide Solution* representing 2.5 mL for each g of the total weight of the sample, but not less than a total of 3.0 mL. Add 10 mL of *Butylated Hydroxytoluene Solution* and 20 mL of alcohol. Reflux vigorously on a steam bath for 30 minutes. Cool, and transfer the saponified mixture to a conical separator, rinsing the saponification flask with three 10-mL portions of water and three 50-mL portions of *Ether*, adding each rinse to the separator. Add about 4 g of sodium sulfate decahydrate to the separator, and extract by shaking for 2 minutes. If an emulsion forms, extract with three 25-mL portions of *Ether.* Combine the ether extracts, if necessary, and wash by swirling gently with 50 mL of water. Repeat the washing more vigorously with additional 50-mL portions of water until the last portion shows no pink color on the addition of phenolphthalein TS. Transfer the washed ether extract to a 250-mL volumetric flask, add *Ether* to volume, and mix. Transfer the entire sample or an accurately measured aliquot containing about 250 µg to a tall-form, 400-mL beaker containing about 5 g of anhydrous sodium sulfate. Stir for 2 minutes, then decant the solution into a second 400-mL beaker. Rinse the sodium sulfate with three 25-mL portions of *Ether*, adding each rinse to the main portion. Reduce the total volume to about 30 mL by evaporation on a steam bath, and transfer the concentrate to a small, round-bottom evaporation flask. Rinse the beaker with three 10-mL portions of *Ether*, adding the rinsings to the flask. With the aid of vacuum in a water bath at a temperature not exceeding 40°, or with a stream of nitrogen at room temperature, remove the remaining solvent completely. Dissolve the residue in a small amount of *Solvent Hexane*, transfer to a 10-mL volumetric flask, dilute with *Solvent Hexane* to volume, and mix, to obtain the *Sample Preparation.*

Procedure—

FIRST COLUMN CHROMATOGRAPHY—Just as the 2 mL of isooctane disappears into the surface of the prepared *First Column*, pipet 2 mL of the *Sample Preparation* onto the column. As the meniscus of the *Sample Preparation* reaches the column surface, add the first of three 2-mL portions of *Solvent Hexane*, adding each succeeding portion as the preceding portion disappears into the column. Continue adding *Solvent Hexane* in portions of 5 to 10 mL until 100 mL has been added. If necessary, adjust the flow rate to between 3 and 6 mL per minute, by application of gentle pressure at the top of the chromatographic tube.

Discard the first 20 mL of effluent, and collect the remainder. Examine the column under ultraviolet light at intervals during the chromatography, and stop the flow when the front of the fluorescent band representing vitamin A is about 5 mm from the bottom of the column. (The ultraviolet lamp should provide *weak* radiation in the 300-nm region. It is frequently necessary to use a narrow aperture or screen with commercial lamps to reduce the amount of radiation to the minimum required to visualize the vitamin A on the column.)

Transfer the eluate to a suitable evaporation flask, and remove the solvent hexane completely under vacuum at a temperature not above 40° or with a stream of nitrogen at room temperature. Dissolve the residue in about 10 mL of *Solvent Hexane.*

SECOND COLUMN CHROMATOGRAPHY—Add the solvent hexane solution obtained as directed under *First Column Chromatography* onto the *Second Column.* Rinse the evaporation flask with a total of 10 mL of *Solvent Hexane* in small portions, adding each portion to the *Second Column* and allowing it to flow through the column, and discard the effluent. When about 1 mL of the hexane remains above the surface of the column, add 75 mL of benzene, and elute with the aid of gentle suction (about 125 mm of mercury), collecting the eluate. Evaporate the benzene under vacuum at a temperature not above 40°, or with a stream of nitrogen at room temperature.

ASSAY PREPARATION—Dissolve the residue obtained as directed under *Second Column Chromatography* in a small amount of *Ethylene Dichloride*, transfer to a 10-mL volumetric flask, dilute with *Ethylene Dichloride* to volume, and mix, to obtain the *Assay Preparation.*

COLOR DEVELOPMENT—Into each of three suitable, matched colorimeter tubes of about 20-mm (inside) diameter, and designated *1*, *2*, and *3*, respectively, pipet 1 mL of the *Assay Preparation.* Into tube *1* pipet 1 mL of the *Standard Preparation*, into tube *2*, 1 mL of *Ethylene Dichloride*, and into tube *3*, 1 mL of a mixture of equal volumes of acetic anhydride and *Ethylene Dichloride.* To each tube add quickly, and preferably from an automatic pipet, 5.0 mL of *Color Reagent*, and mix. After 45 seconds, accurately timed, following the addition of the color reagent, determine the absorbances of the three solutions at 500 nm, with a suitable spectrophotometer, using *Ethylene Dichloride* as the blank. Similarly, 45 seconds after making the first reading on each solution, determine the absorbances of the solutions in tubes *2* and *3* at 550 nm, in a similar manner. Designate the absorbances as $A^1{}_{500}$, $A^2{}_{500}$, $A^3{}_{500}$, $A^2{}_{550}$, and $A^3{}_{550}$, respectively, in which the superscript indicates the number of the tube and the subscript the wavelength.

Calculation—Calculate the quantity, in µg, of vitamin D in the portion of the sample taken by the formula:

$$(C_S/C)(A_U/A_S),$$

in which C_S is the concentration of vitamin D, in µg per mL, of the *Standard Preparation*, C is the concentration of the sample (as g, capsules, tablets, etc.) in each mL of the final solution, A_U has the value of $(A^2{}_{500} - A^3{}_{500}) - 0.67(A^2{}_{550} - A^3{}_{550})$ determined from the absorbances observed on the solution from the *Assay Preparation*, and A_S has the value of $A^1{}_{500} - A^2{}_{500}$ determined on the solutions from the *Standard Preparation.*

Biological Method

The biological assay of vitamin D comprises the recording and interpretation of observations on groups of rats maintained on specified dietary regimens throughout specified periods of their lives whereby the biological response to the preparation under assay is compared with the response to USP Vitamin D Capsules RS.

Reference Standard—*USP Cholecalciferol Reference Standard*—Store in a cold place, protected from light. Allow it to attain room temperature before opening ampul. Use the material promptly, and discard the unused portion.

Preliminary Period—Throughout the preliminary period in the life of a rat, which is not longer than 30 days and extends from birth to the first day of the depletion period, maintain litters of rats under the immediate supervision of, or according to the directions of, the individual responsible for the assay. During the preliminary period, use a dietary regimen that provides for normal development but is limited in its content of vitamin D, so that when placed upon the *Rachitogenic Diet* in the depletion period the rats develop rickets. At the end of the preliminary period, reject any rat that weighs less than 44 g or more than 60 g, or that shows evidence of injury, disease, or anatomical abnormality.

Depletion Period—Through the depletion period, which extends from the end of the preliminary period to the first day of the assay period, provide each rat ad libitum with the *Rachitogenic Diet* and water, and allow access to no other food or dietary supplement.

Rachitogenic Diet—The *Rachitogenic Diet* consists of a uniform mixture of the following ingredients in the proportions shown in the accompanying table.

Rachitogenic Diet

Ingredient	Parts by weight
Whole yellow corn, ground	76
Wheat gluten, ground	20
Calcium carbonate	3
Sodium chloride	1

When a chemical analysis of the entire ration shows a Ca:P ratio of less than 4:1 or more than 5:1, the proportion of calcium carbonate may be varied to bring the adjusted ratio to a uniform level within this range.

Assigning Rats to Groups for Assay Period—Consider a litter suitable for the assay period when individual rats in the litter show evidence of rickets such as enlarged joints and a distinctive wobbly, rachitic gait, provided that the depletion period is not less than 19 or more than 25 days. The presence of rickets may be established also from the width of the rachitic metaphysis upon X-ray examination or by applying the *Line Test* (described below) to a leg bone of one member of each litter.

Record the weight of each rat, and assign it to a group, in which each rat will be fed a specified dose of the Reference Standard or of an assay sample that is under examination for its vitamin D potency. For each assay sample provide one or more assay groups and not less than two standard groups. The two standard groups may be used for the concurrent assay of more than one assay sample. Within an interval not exceeding 30 days, complete the assignment of rats to groups according to a design that divides litters among the groups, to achieve a complete balance.

For complete balance, whereby each litter is represented equally in every group, use 7 or more litters containing at least as many depleted rats as there are groups. From a given litter, assign one rat, selected at random, to each group on the same day. If a litter contains twice as many rats as there are groups, assign a second series of rats similarly. The last one or two litters to be assigned may be allotted to groups so that at the start of the assay period the average body weight of any completed groups will not differ by more than 8 g from that of any other group.

Assay Doses—Select two dosage levels of the USP Cholecalciferol RS, spaced so that the ratio of the larger to the smaller dose is not less than 1.5 or more than 2.5. Select one or two dosage levels based upon a single assumed potency for each sample. The dosage levels of the sample are equivalent to those of the standard or to a mid-level equal to the square root of the product of the two dosage levels of the standard.

Select dosage levels such that, when fed to rachitic rats, they are expected to produce degrees of calcification within the range specified under the test of data acceptability. Before feeding, the Reference Standard and/or sample may be diluted with cottonseed oil, provided that not more than 0.2 mL is fed on any one day. Store the oil solutions in well-closed bottles, protected from light, at a temperature not exceeding 10°, and use within 5 weeks.

Assign one group of rats to each dosage level of the standard and of the one or more samples.

Assay Period—During the assay period, which extends from the end of the depletion period for a fixed interval of 7 to 10 days, cage each rat individually and provide it ad libitum with the *Rachitogenic Diet* and water. Supply a *Rachitogenic Diet* prepared from the same lots of ingredients to all rats. On the first and on the third (or fourth) day of the assay period, feed each rat one-half of its total assigned dose.

Throughout the assay period, maintain as uniform environmental conditions as possible for all rats, and exclude exposure to antirachitic radiations. At the end of a fixed period of 7 to 10 days, weigh and kill each rat. From those rats that do not weigh less at the end than at the start of the assay period and that have consumed each assigned dose within 24 hours of the time it was fed, dissect out one or more leg bones for examination by the *Line Test*.

Line Test—Remove the proximal end of a tibia or the distal end of a radius, and clean adhering tissue from it, in any one assay using the same bone from all animals. With a clean, sharp blade cut a median, longitudinal section through the juncture of the epiphysis and diaphysis at the same place on each bone. Rinse both sections in purified water, immerse immediately in silver nitrate solution (1 in 50) for 1 minute, and rinse again in purified water. Expose the cut surface of bone, in water, to daylight or another source of actinic light until the calcified areas develop a clearly defined stain without marked discoloration of the uncalcified areas. The staining procedure may be modified to differentiate more clearly between calcified and uncalcified areas.

Score the degree of calcification of the rachitic metaphysis in each rat, according to a scale that allows the average response to be plotted as a straight line against the logarithm of the dose.

Acceptability—Observations are acceptable for use in calculation of the potency only from those groups in which two-thirds or more but not less than 7 rats show calcification at least as great as the lowest level and not greater than the highest level illustrated in the figure. If the average score of the standard group on the high dosage level is not greater than the average score of the standard group on the low dosage level, discard the results, and repeat the assay. If an assay sample is represented solely by assay groups that are not acceptable for measuring vitamin D potency and in each of which the average score is less than the average score of the standard group on the low dosage level or more than the average score of the standard group on the high dosage level, its assayed content of vitamin D is respectively less than that represented by the low dose or more than that represented by the high dose of the Reference Standard.

Calculation—Tabulate the scores (y), listing each litter in a separate row with treatment groups in columns. Omit any groups that do not meet the test for *Acceptability*. Equalize the number of observations in the acceptable groups by disregarding the results on all litters not equally represented in the groups or by other suitable means (see *Design and Analysis of Biological Assays* ⟨111⟩). Total the f scores for each of the treatment groups, where f is the number of litters, and designate each total as T with subscripts 1 and 2 for the low and high dosage levels, respectively. Compute the slope b from the sums of T_1, i.e., ΣT_1, and of T_2, i.e., ΣT_2, for the standard and sample, provided the latter is represented at both dosage levels, from the equation:

$$b = (\Sigma T_2 - \Sigma T_1)/ifh',$$

in which i is the logarithm of the ratio of the high dose to the low dose and is the same for each preparation, and h' is the number of preparations represented by two dosage levels and included in the calculation of the value of b.

Compute the logarithm of the relative potency of each specimen under assay from the equation:

$$\begin{aligned}\log (\text{relative potency}) &= M' \\ &= (\bar{y}_U - \bar{y}_S)/b \\ &= ih'T_a/2\Sigma T_b,\end{aligned}$$

in which each mean score, $\bar{y}_U$ for the assay sample and $\bar{y}_S$ for the Reference Standard, is the average of the individual scores for an intermediate dosage level or of the two means for the high and the low dosage levels and where $T_b = \Sigma T_2 - \Sigma T_1$ and T_a is as defined (see *Design and Analysis of Biological Assays* ⟨111⟩). Convert each observed M' to its antilogarithm to obtain the relative potency of the sample. Multiply the relative potency by the assumed potency of the assay oil in Units per g, adopted at the start of the assay, to obtain its assayed content of vitamin D in USP Units per g.

⟨591⟩ ZINC DETERMINATION

The need for a quantitative determination of zinc in the Pharmacopeial insulin preparations reflects the fact that the element is an essential component of zinc-insulin crystals. In common with lead, zinc may be determined either by the dithizone method or by atomic absorption.

Dithizone Method

Select all reagents for this test to have as low a content of heavy metals as practicable. If necessary, distil water and other solvents into hard or borosilicate glass apparatus. Rinse thoroughly all glassware with warm dilute nitric acid (1 in 2) followed by water. Avoid using on the separator any lubricants that dissolve in chloroform.

Special Solutions and Solvents—

ALKALINE AMMONIUM CITRATE SOLUTION—Dissolve 50 g of dibasic ammonium citrate in water to make 100 mL. Add 100 mL of ammonium hydroxide. Remove any heavy metals that may be present by extracting the solution with 20-mL portions of *Dithizone Extraction Solution* (see *Lead* ⟨251⟩) until the dithizone solution retains a clear green color, then extract any dithizone remaining in the citrate solution by shaking with chloroform.

CHLOROFORM—Distil chloroform in hard or borosilicate glass apparatus, receiving the distillate in sufficient dehydrated alcohol to make the final concentration 1 mL of alcohol for each 100 mL of distillate.

DITHIZONE SOLUTION—Use *Standard Dithizone Solution* (see *Lead* ⟨251⟩), prepared with the distilled *Chloroform*.

STANDARD ZINC SOLUTION—Dissolve 625 mg of zinc oxide, accurately weighed, and previously gently ignited to constant weight, in 10 mL of nitric acid, and add water to make 500.0 mL. This solution contains 1.0 mg of zinc per mL.

DILUTED STANDARD ZINC SOLUTION—Dilute 1 mL of *Standard Zinc Solution*, accurately measured, with 2 drops of nitric acid and sufficient water to make 100.0 mL. This solution contains 10 µg of zinc per mL. Use this solution within 2 weeks.

TRICHLOROACETIC ACID SOLUTION—Dissolve 100 g of trichloroacetic acid in water to make 1000 mL.

Procedure—Transfer 1 to 5 mL of the preparation to be tested, accurately measured, to a centrifuge tube graduated at 40 mL. If necessary, add 0.25 N hydrochloric acid, dropwise, to obtain a clear solution. Add 5 mL of *Trichloroacetic Acid Solution* and sufficient water to make 40.0 mL. Mix, and centrifuge.

Transfer to a hard-glass separator an accurately measured volume of the supernatant liquid believed to contain from 5 to 20 µg of zinc, and add water to make about 20 mL. Add 1.5 mL of *Alkaline Ammonium Citrate Solution* and 35 mL of *Dithizone Solution*. Shake vigorously 100 times. Allow the chloroform phase to separate. Insert a cotton plug in the stem of the separator to remove any water emulsified with the chloroform. Collect the chloroform extract (discarding the first portion that comes through) in a test tube, and determine the absorbance at 530 nm, with a suitable spectrophotometer.

Calculate the amount of zinc present by reference to a standard absorbance-concentration curve obtained by using 0.5 mL, 1.0 mL, 1.5 mL, and, if the zinc content of the sample extracted exceeds 15 µg, 2.0 mL of the *Diluted Standard Zinc Solution*, corrected as indicated by a blank determination run concomitantly, using all of the reagents but no added zinc.

Physical Tests and Determinations

⟨601⟩ AEROSOLS

Delivery Rate—Select not less than four aerosol containers, shake, if the label includes this directive, remove the caps and covers, and actuate each valve for 2 to 3 seconds. Weigh each container accurately, and immerse in a constant-temperature bath until the internal pressure is equilibrated at a temperature of 25 ± 1° as determined by constancy of internal pressure as directed under *Pressure Testing*. Remove the containers from the bath, remove the excess moisture by blotting with a paper towel, shake, if the label includes this directive, actuate each valve for 5.0 seconds (accurately timed by use of a stopwatch), and weigh each container again. Return the containers to the constant-temperature bath, and repeat the foregoing procedure three times for each container. Calculate the average delivery rate, in g per second, for each container.

Leak Testing—Select 12 aerosol containers, and record the date and time to the nearest half-hour. Weigh each container to the nearest mg, and record the weight, in mg, of each as W_1. Allow the containers to stand in an upright position at room temperature for not less than 3 days, and again weigh each container, recording the weight, in mg, of each as W_2 and recording the date and time to the nearest half-hour. Determine the time, T, in hours, during which the containers were under test. Calculate the leakage rate, in mg per year, of each container by the formula:

$$(365)(24/T)(W_1 - W_2).$$

Where plastic-coated glass aerosol containers are tested, dry the containers in a desiccator for 12 to 18 hours, and allow them to stand in a constant-humidity environment for 24 hours prior to determining the initial weight as indicated above. Conduct the test under the same humidity conditions.

Empty the contents of each container tested by employing any safe technique; e.g., chill to reduce the internal pressure, remove the valve, and pour. Remove any residual contents by rinsing with suitable solvents, then rinse with a few portions of methanol. Retain as a unit the container, the valve, and all associated parts, and heat them at 100° for 5 minutes. Cool, weigh, record the weight as W_3, and determine the net fill weight ($W_1 - W_3$) for each container tested.

The requirements are met if the average leakage rate of the 12 containers is not more than 3.5% of the net fill weight per year, and none of the containers leaks more than 5.0% of the net fill weight per year. If 1 container leaks more than 5.0% per year, and if none of the containers leaks more than 7.0% per year, determine the leakage rate of an additional 24 containers as directed herein. Not more than 2 of the 36 containers leak more than 5.0% of the net fill weight per year, and none of the 36 containers leaks more than 7.0% of the net fill weight per year.

Where the net fill weight is less than 15 g and the label bears an expiration date, the requirements are met if the average leakage rate of the 12 containers is not more than 525 mg per year, and none of the containers leaks more than 750 mg per year. If 1 container leaks more than 750 mg per year, but not more than 1.1 g per year, determine the leakage rate of an additional 24 containers as directed herein. Not more than 2 of the 36 containers leak more than 750 mg per year, and none of the 36 containers leaks more than 1.1 g per year.

This test is in addition to the customary in-line leak testing of each container.

Pressure Testing—Select not less than four aerosol containers, remove the caps and covers, and immerse in a constant-temperature bath until the internal pressure is constant at a temperature of 25 ± 1°. Remove the containers from the bath, shake well, and remove the actuator and water, if any, from the valve stem. Place each container in an upright position, and determine the pressure in each container by placing a pre-pressurized gauge on the valve stem, holding firmly, and actuating the valve so that it is fully open. The gauge is of a calibration approximating the expected pressure and is fitted with an adapter appropriate for the particular valve stem dimensions. Read the pressure direct from the gauge.

Unit Spray Sampling Apparatus—The apparatus described herein is employed, where indicated in the individual monograph, to obtain a test specimen from metered-dose aerosols through the inhalation actuators provided.

The apparatus consists of an intake system comprising the inhalation actuator (*A*), intake adapter (*B*), and intake tube (*C*, approximately 5 cm × 15 cm, drawn to 8 mm at one end); a delivery tube to which is attached a coarse-porosity, sintered-glass dispersion bubbler (*D*); a collection chamber (*E*, gas-washing bottle) that contains an absorbing solution; and a vacuum system comprising a vacuum source, a flow-regulator, and a flowmeter. The intake adapter is constructed to provide the necessary coupling to the inhalation actuator provided with the aerosol under test. To avoid loss of the drug into the atmosphere when the aerosol is discharged, air is drawn continuously at the rate of 12 ± 1 liters per minute through the intake system into the collection chamber and absorbing solution by means of a suitable vacuum system. Alternatively, an apparatus embodying the principle of the assembly described and illustrated may be used.

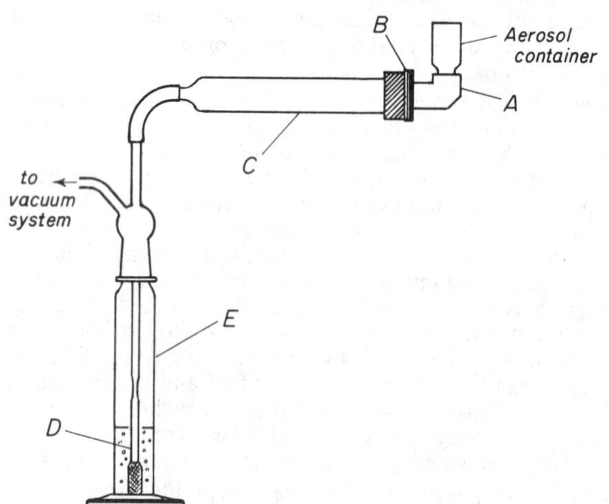

Apparatus for Metered-dose Aerosols

Propellants—

Caution—Hydrocarbon propellants are highly flammable and explosive. Observe precautions and perform sampling and analytical operations in a well-ventilated fume hood.

General Sampling Procedure—This procedure is used to obtain test specimens for those propellants that occur as gases at about 25° and that are stored in pressurized cylinders. Use a stainless steel specimen cylinder, equipped with a stainless steel valve, having a capacity of not less than 200 mL and a pressure rating of 240 psi or more. Dry the cylinder with the valve open at 110° for 2 hours, and evacuate the hot cylinder to less than 1 mm of mercury. Close the valve, cool, and weigh. Connect one end of a charging line tightly to the propellant container and the other end loosely to the specimen cylinder. Carefully open the propellant container, and allow the propellant to flush out the charging line through the loose connection. Avoid excessive flushing that causes moisture to freeze in the charging line and connections. Tighten the fitting on the specimen cylinder, and open the specimen cylinder valve, allowing the propellant to flow into the evacuated cylinder. Continue sampling until the desired amount of specimen is obtained, then close the propellant container valve, and finally close the specimen cylinder valve. [*Caution—Do not overload the specimen cylinder.*] Again weigh the charged specimen cylinder, and calculate the specimen weight.

Approximate Boiling Temperature—Transfer a 100-mL specimen to a tared, pear-shaped, 100-mL centrifuge tube containing a few boiling stones, and weigh. Suspend a suitable thermometer in the liquid, and place the tube in a medium maintained at a temperature 32 ± 1° above the expected boiling temperature. When the thermometer reading becomes constant, record as the boiling temperature the thermometer reading after 5% of the specimen has distilled. Retain the remainder of the specimen for the determination of *High-boiling Residues*.

High-boiling Residues, Method I—Allow 85 mL of the specimen to distil as directed in the test for *Approximate Boiling Temperature*, and transfer the centrifuge tube containing the remaining 15 mL of specimen to a medium maintained at a temperature 10 ± 1° above the boiling temperature. After 30 minutes, remove the tube from the water bath, blot dry, and weigh. Calculate the weight of the residue.

High-boiling Residue, Method II—Prepare a cooling coil from copper tubing (about 6 mm outside diameter × about 6.1 m long) to fit into a suitable vacuum-jacketed flask. Immerse the cooling coil in a mixture of dry ice and acetone in a vacuum-jacketed flask, and connect one end of the tubing to the propellant specimen cylinder. Carefully open the specimen cylinder valve, flush the cooling coil with about 50 mL of the propellant, and discard this portion of liquefied propellant. Continue delivering liquefied propellant from the cooling coil, and collect it in a previously chilled 1000-mL sedimentation cone until the cone is filled to the 1000-mL mark. Allow the propellant to evaporate, using a warm water bath maintained at about 40° to reduce evaporating time. When all of the liquid has evaporated, rinse the sedimentation cone with two 50-mL portions of pentane, and combine the rinsings in a tared 150-mL evaporating dish. Transfer 100 mL of the pentane solvent to a second tared 150-mL evaporating dish, place both evaporating dishes on a water bath, evaporate to dryness, and heat the dishes in an oven at 100° for 60 minutes. Cool the dishes in a desiccator, and weigh. Repeat the heating for 15-minute periods until successive weighings are within 0.1 mg, and calculate the weight of the residue obtained from the propellant as the difference between the weights of the residues in the two evaporating dishes.

Water Content—Proceed as directed under *Water Determination* ⟨921⟩, with the following modifications: (a) Provide the closed-system titrating vessel with an opening through which passes a coarse-porosity gas dispersion tube connected to a sampling cylinder. (b) Dilute the *Reagent* with anhydrous methanol to give a water equivalence factor of between 0.2 and 1.0 mg per mL; age this diluted solution not less than 16 hours before standardization. (c) Obtain a 100-g specimen as directed under *General Sampling Procedure*, and introduce the specimen into the titration vessel through the gas dispersion tube at a rate of about 100 mL of gas per minute; if necessary, heat the specimen cylinder gently to maintain this flow rate.

⟨611⟩ ALCOHOL DETERMINATION

Method I—Distillation Method

Method I is to be used for the determination of alcohol, unless otherwise specified in the individual monograph. It is suitable for examining most fluidextracts and tinctures, provided the capacity of the distilling flask is sufficient (commonly two to four times the volume of the liquid to be heated) and the rate of distillation is such that clear distillates are produced. Cloudy distillates may be clarified by agitation with talc, or with precipitated calcium carbonate, and filtered, after which the temperature of the filtrate is adjusted and the alcohol content determined from the specific gravity. During all manipulations, take precautions to minimize the loss of alcohol by evaporation.

FROTHING—Treat liquids that froth to a troublesome extent during distillation by rendering them strongly acid with phosphoric, sulfuric, or tannic acid, or treat with a slight excess of calcium chloride solution, or with a small amount of paraffin or silicone oil before starting the distillation.

BUMPING—Prevent bumping during distillation by adding porous chips of insoluble material such as silicon carbide, or beads.

For liquids presumed to contain 30% of alcohol or less—By means of a pipet, transfer to a suitable distilling apparatus not less than 25 mL of the liquid in which the alcohol is to be determined, and note the temperature at which the volume was measured. Add an equal volume of water, distil, and collect a volume of distillate about 2 mL less than the volume taken of the original test liquid, adjust to the temperature at which the original test liquid was measured, add sufficient water to measure exactly the original volume of the test liquid, and mix. The distillate is clear or not more than slightly cloudy, and does not contain more than traces of volatile substances other than alcohol and water. Determine the specific gravity of the liquid at 25°, as directed under *Specific Gravity* ⟨841⟩, using this result to ascertain the percentage, by volume, of C_2H_5OH contained in the liquid examined by reference to the *Alcoholometric Table* in the section, *Reference Tables*.

For liquids presumed to contain more than 30% of alcohol—Proceed as directed in the foregoing paragraph, except: dilute the specimen with about twice its volume of water, and collect a volume of distillate about 2 mL less than twice the volume of the original test liquid, bring to the temperature at which the original liquid was measured, add sufficient water to measure exactly twice the original volume of the test liquid, mix, and determine its specific gravity. The proportion of C_2H_5OH, by volume, in this distillate, as ascertained from its specific gravity, equals one-half that in the liquid examined.

Special Treatment—

VOLATILE ACIDS AND BASES—Render preparations containing volatile bases slightly acidic with diluted sulfuric acid before distilling. If volatile acids are present, render the preparation slightly alkaline with sodium hydroxide TS.

GLYCERIN—To liquids that contain glycerin add sufficient water so that the residue, after distillation, contains not less than 50% of water.

IODINE—Treat all solutions containing free iodine with powdered zinc before the distillation, or decolorize with just sufficient sodium thiosulfate solution (1 in 10), followed by a few drops of sodium hydroxide TS.

OTHER VOLATILE SUBSTANCES—Spirits, elixirs, tinctures, and similar preparations that contain appreciable proportions of volatile materials other than alcohol and water, such as volatile oils, chloroform, ether, camphor, etc., require special treatment, as follows:

For liquids presumed to contain 50% of alcohol or less—Mix 25 mL of the specimen under examination, accurately measured, with about an equal volume of water in a separator. Saturate this mixture with sodium chloride, then add 25 mL of solvent hexane, and shake the mixture to extract the interfering volatile ingredients. Draw off the separated, lower layer into a second separator, and repeat the extraction twice with two further 25-mL portions of solvent hexane. Extract the combined solvent hexane solutions with three 10-mL portions of a saturated solution of sodium chloride. Combine the saline solutions, and distil in the usual manner, collecting a volume of distillate having a simple ratio to the volume of the original specimen.

For liquids presumed to contain more than 50% of alcohol—Adjust the specimen under examination to a concentration of approximately 25% of alcohol by diluting it with water, then proceed as directed in the preceding paragraph, beginning with "Saturate this mixture with sodium chloride."

In preparing *Collodion* or *Flexible Collodion* for distillation, use water in place of the saturated solution of sodium chloride directed above.

If volatile oils are present in small proportions only, and a cloudy distillate is obtained, the solvent hexane treatment not having been employed, the distillate may be clarified and rendered suitable for the specific gravity determination by shaking it with about one-fifth its volume of solvent hexane, or by filtering it through a thin layer of talc.

Method II—Gas-liquid Chromatographic Method

Method II is to be used where specified in the individual monograph. For a discussion of the principles upon which it is based, see *Gas Chromatography* under *Chromatography* ⟨621⟩.

Apparatus—Under typical conditions, use a gas chromatograph equipped with a flame-ionization detector and a 1.8-m × 4-mm (ID) glass column packed with 100- to 120-mesh chromatographic column packing No. S3, using nitrogen or helium as the carrier. Prior to use, condition the column overnight at 235° with a slow flow of carrier gas. Adjust the carrier flow and temperature (about 120°) so that acetonitrile, the internal standard, elutes in 5 to 10 minutes.

Solutions—

Standard Solution—Dilute 5.0 mL of dehydrated alcohol with water to 250 mL.

Internal Standard Solution—Dilute 5.0 mL of acetonitrile with water to 250 mL.

Test Solution—Dilute the specimen under examination stepwise with water to obtain a solution containing approximately 2% (v/v) of alcohol.

Test Preparation—Pipet 10 mL each of the *Test Solution* and the *Internal Standard Solution* into a 100-mL volumetric flask, and dilute with water to volume.

Standard Preparation—Pipet 10 mL each of the *Standard Solution* and the *Internal Standard Solution* into a 100-mL volumetric flask, dilute with water to volume, and mix.

Procedure—Inject about 5 µL each of *Test Preparation* and *Standard Preparation*, in duplicate, into the gas chromatograph, record the chromatograms, and determine the peak response ratios. Calculate the percentage of alcohol (v/v) in the specimen under test according to the formula:

$$2R_U/R_SD,$$

in which D is the dilution factor, expressed as a fraction, used in preparing the *Test Solution*, and R_U and R_S are the peak response ratios obtained for the *Test Preparation* and the *Standard Preparation*, respectively.

System Suitability Test—In a suitable chromatogram, the resolution factor, R, is not less than 2, and six replicate injections of the *Standard Preparation* show a relative standard deviation of not more than 4.0% in the ratio of the peak of alcohol to the peak of the internal standard, and the tailing factor of the alcohol peak is not greater than 1.5.

⟨621⟩ CHROMATOGRAPHY

This chapter defines the terms and procedures used in chromatography and provides general information. Specific requirements for chromatographic tests and assays of drug substances and dosage forms, including adsorbant and developing solvents, are given in the individual monographs.

Chromatography is defined as a procedure by which solutes are separated by a dynamic differential migration process in a system consisting of two or more phases, one of which moves continuously in a given direction and in which the individual substances exhibit different mobilities by reason of differences in adsorption, partition, solubility, vapor pressure, molecular size, or ionic charge density. The individual substances thus obtained can be identified or determined by analytical methods.

The general chromatographic technique requires that a solute undergo distribution between two phases, one of them fixed (stationary phase), the other moving (mobile phase). It is the mobile phase that transfers the solute through the medium until it eventually emerges separated from other solutes that are eluted earlier or later. Generally, the solute is transported through the separation medium by means of a flowing stream of a liquid or a gaseous solvent known as the "eluant." The stationary phase may act through adsorption, as in the case of adsorbants such as activated alumina, silica gel, and ion-exchange resins, or it may act by dissolving the solute, thus partitioning the latter between the stationary and mobile phases. In the latter process, a liquid coating held on an inert support serves as the stationary phase. Partitioning is the predominant mechanism of separation in gas-liquid chromatography, paper chromatography, and forms of column chromatography designated as liquid-liquid chromatography. In practice, separations frequently result from a combination of adsorption and partitioning effects.

The types of chromatography useful in qualitative and quantitative analysis that are employed in the USP assays and tests are Column, Gas, Paper, Thin-layer, and Pressurized Liquid Chromatography (commonly called high-pressure or high-performance liquid chromatography). Paper and thin-layer chromatography are ordinarily more useful for purposes of identification, because of their convenience and simplicity. Column chromatography offers a wider choice of stationary phases and is useful for the separation of individual compounds, in quantity, from mixtures. Both gas chromatography and pressurized liquid chromatography require more elaborate apparatus and usually provide high-resolution methods that will identify and quantitate very small amounts of material.

Use of Reference Substances in Identity Tests—In paper and thin-layer chromatography, the ratio of the distance (this distance being measured to the point of maximum intensity of the spot) traveled on the medium by a given compound to the distance traveled by the front of the mobile phase, from the point of application of the test substance, is designated as the R_f value of the compound. The ratio between the distances traveled by a given compound and a reference substance is the R_r value. R_f values vary with the experimental conditions, and thus identification is best accomplished where an authentic specimen of the compound in question is used as a reference substance on the same chromatogram.

For this purpose, chromatograms are prepared by applying on the thin-layer adsorbant or on the paper in a straight line, parallel to the edge of the chromatographic plate or paper, solutions of the substance to be identified, the authentic specimen, and a mixture of nearly equal amounts of the substance to be identified and the authentic specimen. Each sample application contains approximately the same quantity by weight of material to be chromatographed. If the substance to be identified and the authentic specimen are identical, all chromatograms agree in color and R_f value and the mixed chromatogram yields a single spot; i.e., R_r is 1.0.

Location of Components—The spots produced by paper or thin-layer chromatography may be located by: (1) direct inspection if the compounds are visible under white or either short- (254 nm) or long-wavelength (360 nm) ultraviolet light, (2) inspection in white or ultraviolet light after treatment with reagents that will make the spots visible (reagents are most conveniently applied with an atomizer), (3) use of a Geiger-Müller counter or autoradiographic techniques in the case of the presence of radioactive substances, or (4) evidence resulting from stimulation or inhibition of bacterial growth by the placing of removed portions of the adsorbant and substance on inoculated media.

In open-column chromatography, in pressurized liquid chromatography performed under conditions of constant flow rate, and in gas chromatography, the retention time, t, defined as the time elapsed between sample injection and appearance of the peak concentration of the eluted sample zone, may be used as a parameter of identification. Solutions of the substance to be identified or derivatives thereof, of the reference compound, and of a mixture of equal amounts of these two are chromatographed successively on the same column under the same chromatographic conditions. Only one peak should be observed for the mixture. The ratio of the retention times of the test substance, the reference compound, and a mixture of these, to the retention time of an internal standard is called the relative retention time R_r and is also used frequently as a parameter of identification.

The deviations of R_r, R_f, or t values measured for the test substance from the values obtained for the reference compound and mixture should not exceed the reliability estimates determined statistically from replicate assays of the reference compound.

Chromatographic identification by these methods under given conditions strongly indicates identity but does not constitute definitive identification. Coincidence of identity parameters under 3 to 6 different sets of chromatographic conditions (temperatures, column packings, adsorbants, eluants, developing solvents, various chemical derivatives, etc.) increases the probability that the test and reference substances are identical. However, many isomeric compounds cannot be separated. Specific and pertinent chemical, spectroscopic, or physicochemical identification of the eluted component combined with chromatographic identity is the most valid criterion of identification. For this purpose, the individual components separated by chromatography may be collected for further identification.

Paper Chromatography

In paper chromatography the adsorbant is a sheet of paper of suitable texture and thickness. Chromatographic separation may proceed through the action of a single liquid phase in a process analogous to adsorption chromatography in columns. Since the natural water content of the paper, or selective imbibition of a hydrophilic component of the liquid phase by the paper fibers, may be regarded as a stationary phase, a partitioning mechanism may contribute significantly to the separation.

Alternatively, a two-phase system may be used. The paper is impregnated with one of the phases, which then remains stationary (usually the more polar phase in the case of unmodified paper). The chromatogram is developed by slow passage of the other, mobile, phase over the sheet. Development may be ascending, in which case the solvent is carried up the paper by capillary forces, or descending, in which case the solvent flow is also assisted by gravitational force.

Differences in the value of R_f have been reported where chromatograms developed in the direction of the paper grain (machine direction) are compared with others developed at right angles to the grain. Therefore, the orientation of paper grain with respect to solvent flow should be maintained constant in a series of chromatograms. (The machine direction is usually designated by the manufacturer on packages of chromatography paper.)

DESCENDING CHROMATOGRAPHY

In descending chromatography, the mobile phase flows downward on the chromatographic sheet.

Apparatus—The essential equipment for descending chromatography consists of the following.

A *vapor-tight chamber* provided with inlets for addition of solvent or for releasing internal pressure. The chamber is constructed preferably of glass, stainless steel, or porcelain and is so designed as to permit observation of the progress of the chromatographic run without opening of the chamber. Tall glass cylinders are convenient if they are made vapor-tight with suitable covers and a sealing compound.

A *rack of corrosion-resistant material* about 5 cm shorter than the inside height of the chamber. The rack serves as a support for solvent troughs and for antisiphoning rods which, in turn, hold up the chromatographic sheets.

One or more *glass troughs* capable of holding a volume of solvent greater than that needed for one chromatographic run. The troughs must also be longer than the width of the chromatographic sheets.

Heavy glass anti-siphoning rods to be supported by the rack and running outside of, parallel to, and slightly above the edge of the glass trough.

Chromatographic sheets of special filter paper at least 2.5 cm wide and not wider than the length of the troughs are cut to a length approximately equal to the height of the chamber. A fine pencil line is drawn horizontally across the filter paper at a distance from one end such that, when the sheet is suspended from the anti-siphoning rods with the upper end of the paper resting in the trough and the lower portion hanging free into the chamber, the line is located a few centimeters below the rods. Care is necessary to avoid contaminating the filter paper by excessive handling or by contact with dirty surfaces.

Procedure—The substance or substances to be analyzed are dissolved in a suitable solvent. Convenient volumes, delivered from suitable micropipets, of the resulting solution, normally containing 1 to 20 µg of the compound, are placed in 6- to 10-mm spots along the pencil line not less than 3 cm apart. If the total volume to be applied would produce spots of a diameter greater than 6 to 10 mm, it is applied in separate portions to the same spot, each portion being allowed to dry before the next is added.

The spotted chromatographic sheet is suspended in the chamber by use of the anti-siphoning rod, which holds the upper end of the sheet in the solvent trough. The bottom of the chamber is covered with the prescribed solvent system. Saturation of the chamber with solvent vapor is facilitated by lining the inside walls with paper that is wetted with the prescribed solvent system. It is important to ensure that the portion of the sheet hanging below the rods is freely suspended in the chamber without touching the rack or the chamber walls or the fluid in the chamber. The chamber is sealed to allow equilibration (saturation) of the chamber and the paper with the solvent vapor. Any excess pressure is released as necessary. For large chambers, equilibration overnight may be necessary.

A volume of the mobile phase in excess of the volume required for complete development of the chromatogram is saturated with the immobile phase by shaking. After equilibration of the chamber, the prepared mobile solvent is introduced into the trough through the inlet. The inlet is closed and the mobile solvent phase is allowed to travel down the paper the desired distance. Precautions must be taken against allowing the solvent to run down the sheet when opening the chamber and removing the chromatogram. The location of the solvent front is quickly marked, and the sheets are dried.

The chromatogram is observed and measured directly or after suitable development to reveal the location of the spots of the isolated drug or drugs. The paper section(s) predetermined to contain the isolated drug(s) may be cut out and eluted by an appropriate solvent, and the solutions may be made up to a known volume and quantitatively analyzed by appropriate chemical or instrumental techniques. Similar procedures should be conducted with various amounts of similarly spotted reference standard on the same paper in the concentration range appropriate to prepare a valid calibration curve.

ASCENDING CHROMATOGRAPHY

In ascending chromatography the lower edge of the sheet (or strip) is dipped into the mobile phase, to permit the mobile phase to rise on the chromatographic sheet by capillary action.

Apparatus—The essential equipment for ascending chromatography is substantially the same as that described under *Descending Chromatography*.

Procedure—The test materials are applied to the chromatographic sheets as directed under *Descending Chromatography*, and above the level to which the paper is dipped into the developing solvent. The bottom of the developing chamber is covered with the developing solvent system. If a two-phase system is used, both phases are added. It is also desirable to line the walls of the chamber with paper and to saturate this lining with the solvent system. Empty solvent troughs are placed on the bottom of the chamber, and the chromatographic sheets are suspended so that the end on which the spots have been added hangs free inside the empty trough.

The chamber is sealed, and equilibration is allowed to proceed as described under *Descending Chromatography*. Then the developing solvent (mobile phase) is added through the inlet to the trough in excess of the solvent required for complete moistening of the chromatographic sheet. The chamber is re-sealed. When the solvent front has reached the desired height, the chamber is opened and the sheet is removed and dried.

Quantitative analyses of the spots may be conducted as described under *Descending Chromatography*.

Thin-layer Chromatography

In thin-layer chromatography, the adsorbant is a relatively thin, uniform layer of dry, finely powdered material applied to a glass, plastic, or metal sheet or plate, glass plates being most commonly employed. The coated plate can be considered an "open chromatographic column" and the separations achieved may be based upon adsorption, partition, or a combination of both effects, depending on the particular type of support, its preparation, and its use with different solvents. Thin-layer chromatography on ion-exchange films can be used for the fractionation of polar compounds. Presumptive identification can be effected by observation of spots of identical R_f value and about equal magnitude obtained, respectively, with an unknown and a reference sample chromatographed on the same plate. A visual comparison of the size of the spots may serve for semiquantitative estimation. Quantitative measurements are possible by means of densitometry, fluorescence, and fluorescence quenching; or, the spots may be carefully removed from the plate, followed by elution with a suitable solvent and spectrophotometric measurement. For two-dimensional thin-layer chromatography, the chromatographed plate is turned at a right angle and again chromatographed, usually in another chamber equilibrated with a different solvent system.

Apparatus—Acceptable apparatus and materials for thin-layer chromatography consist of the following.

Flat *glass plates* of convenient size, typically 20 cm × 20 cm.[1]

An *aligning tray* or a flat surface upon which to align and rest the plates during the application of the adsorbant.

[1] Commercially prepared plates may be substituted for plates prepared as directed herein.

A *storage rack* to hold the prepared plates during drying and transportation. The rack holding the plates should be kept in a desiccator or be capable of being sealed in order to protect the plates from the environment after removal from the drying oven.

The *adsorbant* consists of finely divided adsorbent materials, normally 5 µm to 40 µm in diameter, suitable for chromatography. It can be applied directly to the glass plate or can be bonded to the plate by means of plaster of Paris (hydrated calcium sulfate) [at a ratio of 5 to 15%] or with starch paste or other binders. The former will not yield as hard a surface as will the starch, but it is not affected by strongly oxidizing spray reagents. The adsorbant may contain fluorescing material to aid in the visualization of spots that absorb ultraviolet light.

A *spreader*, which, when moved over the glass plate, will apply a uniform layer of adsorbant of desired thickness over the entire surface of the plate.

A *developing chamber* that can accommodate one or more plates and can be properly closed and sealed as described under *Ascending Chromatography*. The chamber is fitted with a *plate-support rack* that supports the plates, back to back, with the lid of the chamber in place.

A *template* (generally made of plastic) to aid in placing the test spots at definite intervals, to mark distances as needed, and to aid in labeling the plates.

A graduated *micropipet* capable of delivering 10-µL quantities. Total quantities of test and standard solutions are specified in the individual monograph.

A *reagent sprayer* that will emit a fine spray and will not itself be attacked by the reagent.

An *ultraviolet light source* suitable for observations with short (254 nm) and long (360 nm) ultraviolet wavelengths.

Procedure—[NOTE—In this procedure, use purified water that is obtained by distillation.] Clean the plates scrupulously, as by immersion in chromic acid cleansing mixture, rinsing them with copious quantities of water until the water runs off the plates without leaving any visible water or oily spots, then dry. It is important that the plates be completely free from lint and dust when the adsorbant is applied.

Arrange the plate or plates on the aligning tray, place a 5- × 20-cm plate adjacent to the front edge of the first square plate and another 5- × 20-cm plate adjacent to the rear edge of the last square, and secure all of the plates so that they will not slip during the application of the adsorbant. Position the spreader on the end plate opposite to the raised end of the aligning tray. Mix 1 part of adsorbant with 2 parts of water (or in the ratio suggested by the supplier) by shaking vigorously for 30 seconds in a glass-stoppered conical flask, and transfer the slurry to the spreader. Usually 30 g of adsorbant and 60 mL of water are sufficient for five 20- × 20-cm plates. Complete the application of adsorbants using plaster of Paris binder within 2 minutes of addition of the water, since thereafter the mixture begins to harden. Draw the spreader smoothly over the plates toward the raised end of the aligning tray, and remove the spreader when it is on the end plate next to the raised end of the aligning tray. (Wash away all traces of adsorbant from the spreader immediately after use.) Allow the plates to remain undisturbed for 5 minutes, then transfer the square plates, layer side up, to the storage rack, and dry at 105° for 30 minutes. Preferably place the rack at an angle in the drying oven to prevent the condensation of moisture on the back side of plates in the rack. When the plates are dry, allow them to cool to room temperature, and inspect the uniformity of the distribution and the texture of the adsorbant layer; transmitted light will show uniformity of distribution, and reflected light will show uniformity of texture. Store the satisfactory plates over silica gel in a suitable chamber.

Place two filter-paper wicks, 18 cm in height and as wide as the length of the developing chamber, into the chamber, add about 100 mL of the solvent (sufficient to have a depth of 5 mm to 10 mm at the bottom of the chamber), seal the cover to the top of the chamber, and allow the system to equilibrate; it is essential that the wicks become completely wet. Alternatively, the chamber may be completely lined with filter paper. In either case, assure that the filter paper dips into the solvent at the bottom of the chamber. Where vapor saturation of the chamber by these methods is undesirable, it is so indicated in the individual monograph.

Apply the *Test solution* and the *Standard solution*, as directed in the individual monograph, at points about 1.5 cm apart and about 2 cm from the lower edge of the plate (the lower edge is the first part over which the spreader moved in the application of the adsorbant layer), and allow to dry. Avoid physical disturbance of the adsorbant during the spotting procedure (by the pipet or other applicator) or when handling the plates. The template will aid in determining the spot points and the 10- to 15-cm distance through which the solvent front should pass.

Place a mark 10 cm to 15 cm above the spot point. Arrange the plate on the supporting rack (test spots toward the bottom), and introduce the rack into the developing chamber. Allow the solvent in the chamber to reach the lower edge of the adsorbant, but do not allow the spot points to be immersed. Put the cover in place, and maintain the system until the solvent ascends to a point 10 cm to 15 cm above the initial spots, this usually requiring about 15 minutes to 1 hour. Remove the plate from the developing chamber, mark the solvent front, air-dry the plates, and observe first under short-wavelength ultraviolet light (254 nm) and then under long-wavelength ultraviolet light (360 nm). Measure and record the distance of each spot from the point of origin, and indicate for each spot the wavelength under which it was observed. Determine the R_f values for the principal spots (see *Glossary of Symbols*). If further directed, spray the spots with the reagent specified, observe, and compare the test chromatogram with the standard chromatogram.

Continuous Development Thin-layer Chromatography

In contrast to conventional thin-layer chromatography, which is carried out in a closed tank, the continuous development or continuous flow technique allows the upper end of the plate to project through a slot in the cover of the developing chamber. When the developing solvent reaches the slot, continuous evaporation occurs, producing a steady flow of solvent over the plate. In conventional thin-layer chromatography, spot migration ceases when the solvent reaches the top of the plate, after which the spots simply enlarge by diffusion. In the continuous flow process, spot migration continues as long as the plate remains in the tank and the developing solvent is not exhausted.

Development may be continued for several hours after the solvent reaches the top of the plate, to provide adequate migration of the spots. Usually spots of a standard solution, a test solution, and a mixture of equal amounts of test and standard solutions, are initially applied at a standard distance from the base of the plate. Identity of the standard and test substances is confirmed by their migrating equal distances from the origin and by the observation that the two substances applied as a mixture show no tendency to separate.

A major advantage of continuous development thin-layer chromatography stems from the greater solvent selectivity for solvents of low solvent strength. Solvent strength refers to the property of a developing solvent that causes solutes to migrate, and it is strongly influenced by the polarity of the solvent. Increasing the solvent strength by adding a more polar solvent causes the R_f value to increase. Solvent selectivity refers to the ability of a solvent system to produce different R_f values for closely related substances. In conventional thin-layer chromatography, a solvent system giving an R_f value in the range of 0.3 to 0.7, but with adequate selectivity to permit separation of the substances being examined is usually selected. It is much easier to find solvent systems producing adequate migration than to find those affording adequate selectivity.

Solvent systems of lower strength generally exhibit higher selectivity, but are difficult to employ in conventional thin-layer chromatography because they result in very little migration before the solvent reaches the top of the plate. Migration may be increased, however, by repeated drying and redevelopment of the plate or, more conveniently, by providing means for evaporation of solvent at the top of the plate, which results in continuous development. Two techniques are used: continuous development and short-bed continuous development thin-layer chromatography.

An R_f value cannot be measured in continuous development thin-layer chromatography. Substances may be compared either by their migration distance over a fixed period of time or by comparison with the migration of a standard substance applied to the plate. The comparison may be expressed as a relative retention, R_r (see *Glossary of Symbols*).

Continuous development—*Continuous development* thin-layer chromatography as required for antibiotics is described in the Appendix to USP XX, *Antibiotic Regulations*.

APPARATUS—Acceptable apparatus and materials for continuous development thin-layer chromatography are the same as those described under conventional *Thin-layer Chromatography*, except as follows:

A *developing chamber* is used that consists of a rectangular tank, approximately 23 cm × 23 cm × 9 cm, equipped with a glass solvent trough and a platform about 3.75 cm high to elevate the solvent trough above the base of the tank. The chamber is fitted with a cover having a 2 cm × 6 cm slot in the front edge.

PROCEDURE—On a line about 2 cm from the base of the plate, apply the *Standard solution*, the *Test solution*, and a mixture of equal amounts of the *Standard* and *Test solutions*. Place the plate in the elevated empty solvent trough with the adsorbant on the underside of the leaning plate. The adsorbant rests against a piece of heavy (about 1 mm thick)[2] filter paper measuring 20 cm × 3 cm, folded lengthwise and placed over the front edge of the tank. Place the developing solvent in the trough; set the cover in place and seal all openings except where the adsorbant contacts the paper wick. The plate extends about 1 cm beyond the top of the tank. After the solvent reaches the top of the plate, allow development to continue for an appropriate time. Then remove and dry the plate, and detect the spots by suitable means.

Short-bed continuous development—A major advantage of the short-bed technique derives from the fact that solvent velocity is inversely related to bed length. Since spot migration depends upon the total amount of solvent passing over the plate, the short-bed permits useful migration to be obtained in a reasonable time with solvent having very low solvent strength. Lower diffusion in solvents of low solvent strength produces smaller and more dense spots, which enhances both detectability and discernment of small differences in migration distance.

APPARATUS—Acceptable apparatus and materials for short-bed continuous development thin-layer chromatography are the same as those described under conventional *Thin-layer Chromatography*, except as follows:

A shallow *developing chamber*[3] approximately 22 cm × 9 cm × 3 cm, equipped with a cover plate and tight-fitting polytef wings that enable the chamber to be sealed against the plate, is used. The inside bottom of the chamber contains ridges that support the plate and allow it to be inserted at different angles, thereby varying the length of the plate contained within the tank.

PROCEDURE—On a line about 2 cm from the base of the plate, apply the *Standard solution*, the *Test solution*, and a mixture of equal parts of the *Standard* and *Test solutions*. Place the plate in the developing chamber (adsorbant side up), and add the developing solvent to the chamber. No paper wick is employed. After the solvent reaches the top of the plate, allow development to continue for an appropriate time. Then remove and dry the plate, and detect the spots by suitable means.

Column Chromatography

Apparatus—The apparatus required for column chromatographic procedures is simple, consisting only of the chromatographic tube itself and a tamping rod which may be needed to pack a pledget of glass wool or cotton, if needed, in the base of the tube and compress the adsorbant or slurry uniformly within the tube. In some cases a porous glass disk is sealed at the base of the tube in order to support the contents. The tube is cylindrical and is made of glass, unless another material is specified in the individual monograph. A smaller-diameter delivery tube is fused or otherwise attached by a leak-proof joint to the lower end of the main tube. Column dimensions are variable; the dimensions of those commonly used in pharmaceutical analysis range from 10 mm to 30 mm in uniform inside diameter and 150 mm to 400 mm in length, exclusive of the delivery tube. The delivery tube, usually 3 mm to 6 mm in inside diameter, may include a stopcock for accurate control of the flow rate of solvents through the column. The tamping rod, a cylindrical ram firmly attached to a shaft, may be constructed of plastic, glass, stainless steel, or aluminum, unless another material is specified in the individual monograph. The shaft of the rod is substantially smaller in diameter than the column and is not less than 5 cm longer than the effective length of the column. The ram has a diameter about 1 mm smaller than the inside diameter of the column.

[2] Whatman No. 3MM filter paper or equivalent.

[3] Suitable equipment is available from Regis Chemical Company, Morton Grove, IL.

COLUMN ADSORPTION CHROMATOGRAPHY

The adsorbant (such as activated alumina or silica gel, calcined diatomaceous silica, or chromatographic purified siliceous earth) as a dry solid or as a slurry is packed into a glass or quartz chromatographic tube. A solution of the drug in a small amount of solvent is added to the top of the column and allowed to flow into the adsorbant. The drug principles are quantitatively removed from the solution and are adsorbed in a narrow transverse band at the top of the column. As additional solvent is allowed to flow through the column, either by gravity or by application of air pressure, each substance progresses down the column at a characteristic rate resulting in a spatial separation to give what is known as the *chromatogram*. The rate of movement for a given substance is affected by several variables, including the adsorptive power of the adsorbant and its particle size and surface area; the nature and polarity of the solvent; the hydrostatic head or applied pressure; and the temperature of the chromatographic system.

If the separated compounds are colored or if they fluoresce under ultraviolet light, the adsorbant column may be extruded and, by transverse cuts, the appropriate segments may then be isolated. The desired compounds are then extracted from each segment with a suitable solvent. If the compounds are colorless, they may be located by means of painting or spraying the extruded column with color-forming reagents. Chromatographed radioactive substances may be located by means of Geiger-Müller detectors or similar sensing and recording instruments. Clear plastic tubing made of a material such as nylon, which is inert to most solvents and transparent to short-wavelength ultraviolet light, may be packed with adsorbant and used as a chromatographic column. Such a column may be sliced with a sharp knife without removing the packing from the tubing. If a fluorescent adsorbant is used, the column may be marked under ultraviolet light in preparation for slicing.

A "flowing" chromatogram, which is extensively used, is obtained by a procedure in which solvents are allowed to flow through the column until the separated drug appears in the effluent solution, known as the "eluate." The drug may be determined in the eluate by titration or by a spectrophotometric or colorimetric method, or the solvent may be evaporated, leaving the drug in more or less pure form. If a second drug principle is involved, it is eluted by continuing the first solvent or by passing a solvent of stronger eluting power through the column. The efficiency of the separation may be checked by obtaining a thin-layer chromatogram on the individual fractions.

A modified procedure for adding the mixture to the column is sometimes employed. The drug, in a solid form, and, as in the case of a powdered tablet, without separation from the excipients, is mixed with some of the adsorbant and added to the top of a column. The subsequent flow of solvent moves the drug down the column in the manner described.

COLUMN PARTITION CHROMATOGRAPHY

In partition chromatography the substances to be separated are partitioned between two immiscible liquids one of which, the immobile phase, is adsorbed on a *Solid Support*, thereby presenting a very large surface area to the flowing solvent or mobile phase. The exceedingly high number of successive liquid-liquid contacts allows an efficiency of separation not achieved in ordinary liquid-liquid extraction.

The *Solid Support* is usually polar, and the adsorbed immobile phase more polar than the mobile phase. The *Solid Support* that is most widely used is chromatographic siliceous earth having a particle size suitable to permit proper flow of eluant.[4] In *Reverse-phase* partition chromatography the adsorbed immobile phase is less polar than the mobile phase and the solid adsorbant is rendered nonpolar by suitable treatment with a silanizing agent, such as dichlorodimethylsilane, to give silanized chromatographic siliceous earth.

[4] A suitable grade is acid-washed Celite 545, available from Johns-Manville Corp., 22 East 40th St., New York, NY 10016.

The sample to be chromatographed is usually introduced into the chromatographic system in one of two ways: (a) a solution of the sample in a small volume of the mobile phase is added to the top of the column; or, (b) a solution of the sample in a small volume of the immobile phase is mixed with the *Solid Support* and transferred to the column as a layer above a bed of a mixture of immobile phase with adsorbant.

Development and elution are accomplished with flowing solvent as before. The mobile solvent usually is saturated with the immobile solvent before use.

In conventional liquid-liquid partition chromatography, the degree of partition of a given compound between the two liquid phases is expressed by its partition or distribution coefficient. In the case of compounds that dissociate, distribution can be controlled by modifying the pH, dielectric constant, ionic strength, and other properties of the two phases. Selective elution of the components of a mixture can be achieved by successively changing the mobile phase to one that provides a more favorable partition coefficient, or by changing the pH of the immobile phase *in situ* with a mobile phase consisting of a solution of an appropriate acid or base in an organic solvent.

Unless otherwise specified in the individual monograph, assays and tests that employ column partition chromatography are performed according to the following general method.

Solid Support—Use purified siliceous earth. Use silanized chromatographic siliceous earth for reverse-phase partition chromatography.

Stationary Phase—Use the solvent or solution specified in the individual monograph. If a mixture of liquids is to be used as the *Stationary Phase*, mix them prior to the introduction of the *Solid Support*.

Mobile Phase—Use the solvent or solution specified in the individual monograph. Equilibrate it with water if the *Stationary Phase* is an aqueous solution; if the *Stationary Phase* is a polar organic fluid, equilibrate with that fluid.

Preparation of Chromatographic Column—Unless otherwise specified in the individual monograph, the chromatographic tube is about 22 mm in inside diameter and 200 mm to 300 mm in length, without porous glass disk, to which is attached a delivery tube, without stopcock, about 4 mm in inside diameter and about 50 mm in length. Pack a pledget of fine glass wool in the base of the tube. Place the specified volume of *Stationary Phase* in a 100- to 250-mL beaker, add the specified amount of *Solid Support*, and mix to produce a homogeneous, fluffy mixture. Transfer this mixture to the chromatographic tube, and tamp, using gentle pressure, to obtain a uniform mass. If the specified amount of *Solid Support* is more than 3 g, transfer the mixture to the column in portions of approximately 2 g, and tamp each portion. If the assay or test requires a multi-segment column, with a different *Stationary Phase* specified for each segment, tamp after the addition of each segment, and add each succeeding segment directly to the previous one.

If a solution of the analyte is incorporated in the *Stationary Phase*, complete the quantitative transfer to the chromatographic tube by scrubbing the beaker used for the preparation of the test mixture with a mixture of about 1 g of *Solid Support* and several drops of the solvent used to prepare the test solution.

Pack a pledget of fine glass wool above the completed column packing. The *Mobile Phase* flows through a properly packed column as a moderate stream or, if reverse-phase chromatography is applied, as a slow trickle.

Procedure—Transfer the *Mobile Phase* to the column space above the column packing, and allow it to flow through the column under the influence of gravity. Rinse the tip of the chromatographic column with about 1 mL of *Mobile Phase* before each change in composition of *Mobile Phase* and after completion of the elution. If the analyte is introduced into the column as a solution in the *Mobile Phase*, allow it to pass completely into the column packing, then add *Mobile Phase* in several small portions, allowing each to drain completely, before adding the bulk of the *Mobile Phase*. Where the assay or test requires the use of multiple chromatographic columns mounted in series and the addition of *Mobile Phase* in divided portions is specified, allow each portion to drain completely through each column, and rinse the tip of each with *Mobile Phase* prior to the addition of each succeeding portion.

Pressurized Liquid Chromatography

Advances in column technology, high-pressure pumping systems, and sensitive detectors have transformed liquid column chromatography into a high-speed, high-efficiency method of separation. This method is sometimes referred to as HPLC, which is alternatively expressed as either high-performance liquid chromatography or high-pressure liquid chromatography. The column technology is based upon the use of small-bore (2- to 5-mm ID) columns and small-particle (3- to 50-µm) packings that allow fast equilibrium between mobile and stationary phases. This small-particle column technology requires high-pressure pumping systems capable of delivering the mobile phase at high pressure, as much as 300 atmospheres, to achieve flow rates of several mL per minute. Since it is often necessary to use small amounts of analyte (usually less than 20 µg) with the column packings, sensitive detectors are needed. With this technology, liquid chromatography can give high-speed separations comparable in many cases to those achieved by gas chromatography, with the advantage that nonvolatile or thermally unstable materials can be chromatographed without decomposition or the necessity for making volatile derivatives.

One type of stationary phase support used in these packings consists of micro-particles 30 µm to 50 µm in diameter, having a solid center and a thin porous crust. Some of these pellicular support materials can be pre-activated to give them adsorptive properties, while others can be covered with a thin film of stationary phase for partition or ion-exchange separations. The stationary phase can be either a liquid or a polymer, either coated or chemically bonded to the surface of the support in a thin film that reduces mass transfer resistances so that fast equilibrium between mobile and stationary phases can be attained. A liquid stationary phase must be largely immiscible in the mobile phase solvent; it is usually necessary to pre-saturate the mobile phase solvent with the stationary phase liquid to prevent stripping of the stationary phase from the column. Polymer stationary phases coated on the support are more durable. Stationary phases that have been chemically bonded to the support provide greater convenience for use with a variety of solvents and at elevated temperatures.

Other, smaller-diameter packing materials of 3- to 10-µm diameter are almost completely porous and give much more efficient separations than the 30- to 50-µm particle packings. The particles can also be made adsorptive or covered with a stationary phase. It is essential that these packings be slurry-packed in order to obtain high-efficiency columns, in contrast to the 30- to 50-µm particles which can be dry-packed.

The three forms of high-performance liquid chromatography most often used are ion-exchange, partition, and adsorption. Ion-exchange chromatography is used mainly for separation of water-soluble ionic or ionizable materials of molecular weight less than 1500. The stationary phases of ion-exchange chromatography are usually synthetic organic resins having different active groups present. Cation-exchange resins contain negatively charged active sites and are used to separate basic substances such as amines, while anion-exchange resins have positively charged active sites, which will attract substances such as those carrying phosphate, sulfonate, or carboxylate groups which are negatively charged. Water-soluble ionic or ionizable compounds are attracted to the resins, and differences in affinity bring about the chromatographic separation. The pH of the mobile phase, temperature, ion type, ionic concentration, and organic modifiers affect the attraction of the solute, and these variables can be adjusted to obtain the desired degree of separation.

In partition chromatography, mobile and stationary phases of different polarity are used. If the mobile phase is polar and the stationary phase nonpolar, referred to as reverse-phase chromatography, then nonpolar, hydrocarbon-soluble compounds of molecular weight less than 1000, such as fat-soluble vitamins and anthraquinones, can be separated by their affinity for the stationary phase. Modification of the polar mobile phase solvent with a less polar solvent causes a decrease in affinity and the retention of the compounds on the column. If the mobile phase is nonpolar and the stationary phase polar, then polar material such as alcohols and amines can be chromatographed. The nonpolar mobile phase can then be modified with a more polar solvent to decrease retention and change the separation.

A wide range of nonionic compounds can be chromatographed by adsorption chromatography with the choice of the proper stationary and mobile phases. (For further details of both the partition and absorption techniques, see the preceding sections.)

Apparatus—The liquid chromatograph consists basically of a pumping system, analyte injection device, chromatographic column, detector, amplifier, and recorder. The high-pressure pumping system delivers the mobile phase solvent from the solvent reservoir to the column through high-pressure tubing and fittings. Two means for analyte introduction onto the column are injection into a flowing stream and a "stop-flow" injection. These techniques can be used with either a syringe or an injection valve. For the syringe injection technique, a septum may be used where the column head pressures are less than 70 atmospheres (about 1000 psi). At higher pressures, an injection valve may be used to introduce the test specimen. Some valve systems incorporate a calibrated loop that is loaded with a test specimen and then transferred by the valve system to the flowing stream of mobile phase. Other valve systems permit introduction of the analyte into a cavity using a syringe. The loaded cavity is then switched into the high-pressure stream by the valve system. In the "stop-flow" technique, column flow is stopped, and after the injection port pressure drops to zero, the port is opened and the analyte injected with a syringe. The injection port is then closed and pumping is resumed. High pressure is rapidly re-established and little zone-spreading is incurred in the process. The "stop-flow" technique allows better injection reproducibility at higher pressures than does the use of a septum, and the problem of septum deterioration with many solvents is averted.

The columns normally used for analytical separations have small (2- to 4-mm) internal diameters; larger-diameter columns are used for preparative steps. Columns may be heated to give more efficient separations, but only rarely are they used at temperatures above 60° because of potential difficulties resulting from stationary phase degradation or mobile phase volatility at elevated temperatures. Unless otherwise specified in the individual monograph, the column is maintained at ambient temperature.

The detectors commonly used include the ultraviolet photometer, the differential refractometer, and the fluorometer. The low-pressure mercury ultraviolet photometer is a common and stable detector, but its use is limited to the detection of materials that absorb radiation at a wavelength of 254 nm. Its limit of sensitivity to compounds that absorb ultraviolet light strongly may be about 1 ng. Compounds that do not absorb light at 254 nm appreciably may often be converted to suitable derivatives that absorb at this wavelength, thereby increasing the range of applicability of the single-wavelength detector. The introduction of spectrophotometers equipped with micro cells and detectors capable of operating at additional wavelengths has extended the scope of ultraviolet detection.

The differential refractometer detects differences between the refractive indexes of the pure solvent and of a solution of the chromatographed test substance in the solvent. While more generally applicable, it is a less sensitive detector having a lower limit of about 1 µg and is responsive to small changes in solvent composition, flow rate, and temperature, so that a reference column and flow of mobile phase may be required to give a satisfactory baseline.

The fluorometer is a sensitive detector for compounds that are inherently fluorescent or that can be converted to fluorescent derivatives either by chemical transformation of the compound or by coupling with fluorescent reagents at specific functional groups.

With some reagents, derivatization is carried out prior to chromatographic separation of the derivatives. In another approach, the reagent is introduced into the eluent stream and reacts with the analyte *in situ*, and the derivative is then exposed to the detector.

An electrochemical detector employing carbon-paste electrodes mounted in a thin-film cell of very small volume may advantageously be employed to measure very small amounts (1 ng) of easily oxidized compounds, particularly phenols and catechols.

In general, the signal from the detector is amplified before being fed to a suitable automatic recording device, usually a strip-chart potentiometric recorder, where the signal is plotted versus time. The signal may go also to an electronic digital integrator for the automatic measurement of chromatogram peak areas.

The mobile phase composition significantly influences chromatographic performance and should be controlled carefully. The composition can have a far greater effect on capacity factors (k or ratio of amount of time spent in stationary phase to time spent in the mobile phase; see under *Gas Chromatography*) than the temperature.

In partition and adsorption chromatography, the mobile phase may be modified with another solvent, while in ion-exchange chromatography, both pH and ionic strength as well as modification of the solvent can change capacity factors. The technique of continuously changing the solvent composition during the chromatographic run is called gradient elution, or solvent programming, and is sometimes used to chromatograph complex samples having components of greatly differing capacity factors. Detectors that are sensitive to change in solvent composition, such as the differential refractometer, are more difficult to use with the gradient elution technique.

Procedure—The procedures for compound identification and the techniques of calibration and data reduction used in pressurized liquid chromatography are essentially the same as those for gas chromatography (see *Procedure* under *Gas Chromatography*). For accurate quantitative work, it is necessary that the detector have a large linear dynamic range and that the components to be measured be resolved from any interfering substances. The linear dynamic range is defined as the sample size range from the minimum detectable to maximum sample size over which the detector signal response, i.e., the peak responses on the recorder chart, is linearly proportional to the concentration of test substance. For maximum flexibility in quantitative work, this range should be about three orders of magnitude.

Both peak height and area can be related to sample concentration. The term "peak response," designated r_U for sample response and r_S for standard response, has been adopted for use in specifying measurements in chromatography. This encompasses peak areas, peak heights, and other electronic measurements. Peak heights are easy to measure but are greatly influenced by changes in retention time caused by variations in temperature and solvent composition. For these reasons, peak areas are considered to be a more accurate parameter for quantitation. The detector response may be calibrated by relating peak responses to a known concentration of reference standard using either an external or an internal standardization procedure.

One drawback to the method of external calibration, i.e., direct comparison of the peak responses obtained on chromatographing the test specimen and a known concentration of the corresponding Reference Standard, is that the accuracy and precision are dependent upon the reproducibility of analyte injection. Since the reproducibility of injection at high pressure may vary considerably, the better quantitative results usually are obtained when the method for internal calibration is used. An internal standard at known concentration is added both to the test solution and to a solution of the Reference Standard of known concentration, and the ratios of peak responses of drug and internal standard are compared. Because of normal variations in equipment, supplies, and techniques, a system suitability test (see *System Suitability*) may be useful to ensure that a given operating system may be generally applicable.

Gas Chromatography

In gas chromatography, the moving phase is a *gas*. The stationary phase is usually a liquid but may be a solid or a combination of solid and liquid.

In gas-liquid chromatography (GLC), the stationary liquid phase is immobilized as a thin film on a finely divided, inert solid support, such as chromatographic siliceous earth, crushed firebrick, glass beads, or even the inner wall of a small-diameter tube. If the tube is filled with liquid-covered, finely divided solid, it is called a packed column. If the inner wall of a small-diameter tube is coated with the liquid, it is called an open tubular or capillary column. If the inner wall of the open tubular column is treated so as to deposit a porous or irregular support on its surface before coating with the liquid phase, it is called a support coated open tubular (SCOT) column. In gas-solid chromatography (GSC), the identical situation holds except that the liquid phase is absent and the solid is an active adsorbant, such as alumina, silica gel, or carbon. In either case, the mobile phase continuously moves over the stationary phase.

When a vaporized substance is introduced into the gas stream at the head of the column, it is swept into the column and undergoes distribution between the gas and liquid or solid phases. The distribution process reaches a dynamic equilibration, that is adequately described by an extension of the mathematical treatment of the stepwise process of countercurrent distribution. The behavior of a solute in such a partition process is conveniently defined by a dimensionless partition ratio, k', called the capacity factor, which may be defined alternatively in terms of the relative amounts, or relative residence times, of the substance in the respective phases. The gas phase simply serves to move the substance down the column between excursions into the stationary phase, and *all* substances spend the *same time in the gas phase* in any particular column. The value of the capacity factor, and, therefore, the time in a gas-liquid chromatographic column, depend upon the following considerations: (a) the specific solute; (b) the specific liquid phase; (c) the amount of liquid phase; (d) the temperature; and (e) the gas flow rate. Therefore, a partition ratio exists for each column, solute, and temperature, and in order to reproduce the behavior of a particular solute, every parameter must be carefully reproduced.

Apparatus—The basic apparatus required for gas chromatography is relatively simple. The *carrier gas*, usually available in compressed form in a cylinder fitted with a suitable pressure-reducing valve, is conducted into a flow meter, which is used to reproduce the particular flow found to be satisfactory for the resolution of a particular mixture. Helium, nitrogen, and other inert gases are suitable carriers. The actual carrier gas used is often determined by the characteristics of the detector being used. Since solutes to be chromatographed must be in the vapor phase, the injection port is heated to a temperature high enough to ensure rapid vaporization but not high enough to cause thermal degradation. Most test specimens are injected by syringe through a silicone rubber septum in the injection port. Preferably the test specimen is injected directly into the column packing. Alternatively, the test specimen vapor is mixed with the flowing carrier gas and then swept into the *column*. In pyrolysis gas chromatography, nonvolatile solids are decomposed by heating to several hundred degrees Celsius, and the volatile products produced are passed directly onto the column. It is in the column that the different components of the vaporized test specimen are separated by virtue of their different interactions with the stationary column packing. The tubing that contains the packing usually is made of glass or metal, and is located in a controlled-temperature oven maintained at a selected temperature, which determines the retention time and, to a degree, the resolution and efficiency obtained. Temperature-programmable components allow efficient elution of compounds over a wide range of vapor pressure. As the components emerge individually from the column, they pass through a differential-type *detector*, which indicates the amount of each component leaving the column. The detector temperature is controlled to prevent condensation. The choice of detector is specified in the individual monograph.

Signals from the detector are passed to an *amplifier* or electrometer, which is coupled to an automatic recording device. The resulting record is a signal-time plot, the chromatogram, which is to be used to determine the identities and concentrations of the components. The usual detectors emit a signal proportional to the concentration of the solute in the carrier as it leaves the column, so that the chromatogram for each drug appears as a bell-shaped peak on a time axis. The resulting curves accurately represent the distribution process as it has occurred during the residence time of the solutes in column. Malfunctions in any of these components can degrade the accuracy and precision of measurements.

Detectors—Detectors commonly used for gas chromatography include those that depend upon thermal conductivity, flame-ionization, alkali flame-ionization, electron-capture, and conductivity. For accurate quantitative work, the detector should have a large linear dynamic range. Helium, because of its high thermal conductivity, is the carrier gas of choice for use with a thermal conductivity detector. The thermal conductivity detector is applicable to all organic compounds but has a lower sensitivity and lower dynamic range than some other detectors. Unless otherwise specified in the individual monograph, the use of a flame-ionization detector with either helium or nitrogen carrier gas is assumed. This detector is sensitive to all carbon compounds and has a wide dynamic range. Nitrogen, by virtue of its higher viscosity, reduces zone spreading in the gas phase and may yield higher efficiencies than helium, but the lower viscosity of helium leads to higher carrier gas flow rates at optimum efficiencies and, therefore, to shorter elution times and faster analyses. The alkali flame-ionization detector contains an alkali-metal salt or a glass element containing rubidium or other metal that results in the suppression of the response to carbon, thereby increasing the relative response to nitrogen, sulfur, and phosphorus several fold. It is, therefore, a selective detector which shows little response to hydrocarbons. The electron-capture detector is also selective, showing little response to hydrocarbons and extremely high response to some compounds, such as those containing halogens or some ketones. Depending on the mode of operation, nitrogen or argon containing a small percentage of methane is used as the carrier gas for electron-capture detection. The electron-capture detector is the most sensitive detector available for those compounds to which it responds. The conductivity detection system includes a heated reaction chamber in which compounds are reacted with a reagent gas such as oxygen or hydrogen that converts some compounds to electrically conductive species such as hydrochloric acid or ammonia while simultaneously removing carbon. The conductive species is then trapped in an electrolyte, and the observed change in electrical conductivity is continuously monitored. The conductivity detector can be made selectively responsive to halogens, sulfur, nitrogen, or phosphorus, and provides very high sensitivity.

Combustion of chlorinated solvents, such as chloroform, in the flame-ionization detector produces hydrochloric acid, which in time may damage detector components. Combustion of silicone derivatives produces a deposit of silica within the detector. Frequent inspection and maintenance of the detector is required to obtain optimum performance.

The specified carrier gas flow rate is the flow rate of the gas exiting the column and is usually expressed in mL per minute at atmospheric pressure and room temperature. It is commonly measured, with the column at operating temperature, by the use of a soap-bubble flow-meter attached to the exit of the column. The gas quickly cools and is essentially at room temperature within the flow-meter. It is usually necessary to disconnect the column from the detector to make this measurement. For a given flow rate, the linear flow rate through the column is related to the square of the column diameter. Thus a flow rate of 60 mL per minute for a 4-mm column and a flow rate of 15 mL per minute for a 2-mm column produce comparable linear flow rates in the respective columns and thus give comparable retention times. Unless otherwise specified in the individual monograph, a 30- to 60-mL flow rate is to be used.

Columns—Pharmaceutical analyses usually employ packed columns and, ideally, only the packing influences the relative movement of solutes through the system. Columns should be of glass unless otherwise specified. Columns of various dimensions are used, but normally they are 0.6 m to 1.8 m in length and 2 mm to 4 mm in internal diameter. Low-capacity columns, having about 5% (w/w) or less of liquid phase on the solid support, are preferred for analytical use. High-capacity columns, such as those having 20% (w/w) liquid loadings, may be used for large test specimens and for the determination of low molecular weight compounds such as water. The desired capacity influences the choice of solid supports.

Support materials are available in various mesh-size ranges, with 80- to 100- and 100- to 120-mesh being the most commonly used with 2- to 4-mm diameter columns. The support material should be as inert as possible, particularly for polar drugs being chromatographed on low-capacity, low-polarity liquid phase columns. Reactive supports can result in decomposition, rearrangement, or peak tailing of the solute. Acid-washed, flux-calcined diatomaceous earth is often used for drug analysis. Reactivity of the support is reduced by treatment with a silanizing reagent prior to coating with liquid phase. Supports receiving an additional, alkaline wash should be used with care, since residual alkali decomposes some liquid phases. Polyaromatic porous resins are sometimes specified. They do not require coating with a liquid phase.

Liquid phases are drawn from a wide range of chemical classes, such as polyethylene glycols, high molecular weight esters and amides, hydrocarbons, and silicone gums and fluids (polysiloxanes substituted normally by methyl, phenyl, nitrile, vinyl, or fluoroalkyl groups or mixtures of these). In all cases, batches must be carefully selected for use in gas chromatography. At operating temperatures, even these materials have sufficient vapor pressure

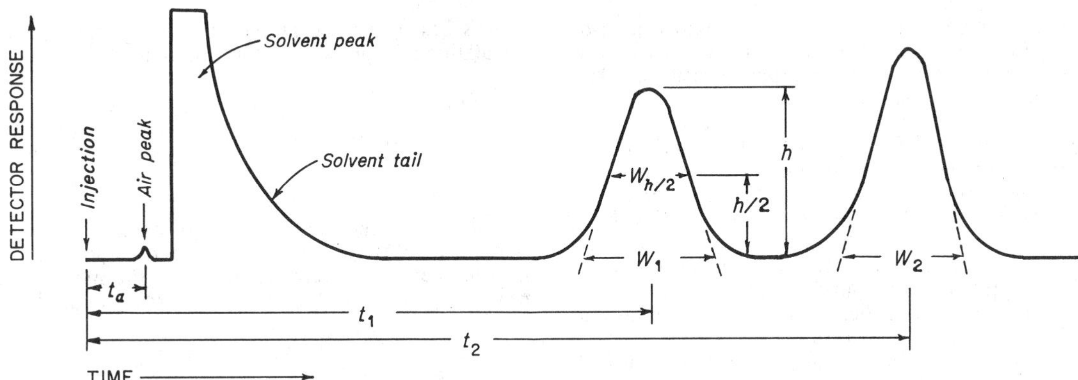

Fig. 1. Chromatographic Separation of Two Substances.

to result in gradual loss of liquid phase by bleeding. Some phases are characterized by quite low bleed rates at operating temperatures and, in such cases, columns may be rejuvenated by repacking the first 10 cm to 15 cm to remove injection residues. Silanized glass wool inserts may be used in the injection ports of some gas chromatographs to trap nonvolatile residues. These should be as small as practicable, to avoid analyte decomposition resulting from the highly active surface.

Figure 1 represents a typical elution chromatographic separation of two substances where t_1 and t_2 are the retention times of substances 1 and 2, h and $h/2$ are the height and half-height of peak 1, $W_{h/2}$ is the width-at-half-height for peak 1, and W_1 and W_2 are the base-widths of peaks 1 and 2, respectively. The air peak is characteristic of gas chromatograms obtained with the thermal conductivity detector and may just precede or even coincide with the front of the solvent peak. A deflection corresponding to the air peak is usually not seen with other gas chromatographic or liquid chromatographic detectors. The interval t_a is the dead time or holdup time and corresponds to the retention time of a nonretarded substance.

Procedure—Since gas chromatography is primarily a separation method, it cannot be used to identify compounds without comparison to a Reference Standard. For qualitative analysis, the retention time for a peak in the chromatogram obtained for a test specimen is "the same as," or "corresponding to," that obtained for a standard preparation under the conditions specified in the individual monograph. The differences in retention times from chromatogram to chromatogram are small, usually less than one-tenth of a minute, as compared with differences between neighboring peaks in the same chromatogram. When a peak appears at the same time or volume under the same experimental conditions, the probability of correct identification is quite high. Alternatively, the individual components may be collected in a cold trap as they emerge from the column for independent analysis by other instrumental or chemical methods, such as mass spectrometry or infrared absorption spectrometry. The retention time or volume for air is an important quantity, since it is used to obtain absolute and relative retention values for characterization of compounds. Drugs may be identified by means of their relative retention, α, determined by the equation

$$\alpha = \frac{t_2 - t_a}{t_1 - t_a}, \qquad (1)$$

in which t_2 is the retention time measured from the point of injection of the desired drug and t_1 is the same for a reference standard material determined with the same column and temperature, and t_a is the retention time for an inert component, such as air, which is not retarded in its passage through the column.

In this and the following expressions written in terms of retention times, the corresponding retention volumes or distances on the chromatogram, both of which are directly proportional to retention time, may be substituted in the equations.

With the flame-ionization detector, which responds to neither air nor water, the retention of a nonretarded compound such as methane, for which natural gas is a convenient source, may be used to estimate t_a. Where t_a is small, α may be estimated from the retentions from the point of injection alone (t_2/t_1).

The capacity factor is related to retention by the equation

$$k'_2 = \frac{t_2}{t_a} - 1. \qquad (2)$$

A measure of the efficiency of a particular column is provided by calculating the number of theoretical plates, n, in the column with the equation

$$n = 16\left(\frac{t}{W}\right)^2, \qquad (3)$$

in which t is the retention time of the substance and W is the width of the base of the peak obtained by extrapolating the relatively straight sides of the peak to the baseline. The value of n is dependent upon the substance being chromatographed as well as the operating conditions such as flow rate and temperature, the quality of the packing, and the uniformity of the packing within the column.

As a measure of efficiency of the separation of two components in a mixture, the resolution, R, is determined by the equation

$$R = \frac{2(t_2 - t_1)}{W_2 + W_1}, \qquad (4)$$

in which t_2 and t_1 are the retention times of the two components and W_2 and W_1 are the corresponding widths of the bases of the peaks, obtained by extrapolating the relatively straight sides of the peaks to the baseline.

Quantitative data can be obtained from the areas under the peaks, determined graphically or by means of an automatic electronic integrator or planimeter. Peak areas are less accurate for small peaks and those having short retention times. The product of peak width at half height and peak height may be substituted for peak areas to minimize graphical error for symmetrical peaks. The chart should be run faster than usual or a comparator should be used to measure the width-at-half-height, so as to minimize error in this measurement. For accurate quantitative work, the components to be measured should be separated from any interfering components. Peak measurements on solvent tails are to be avoided.

Area percentages, % A_i, of species within a chromatogram are used in purity analysis and are equal to 100 times the ratio of the peak areas of the species, A_i, to the sum, ΣA_i, of all of the area of the peaks in the chromatogram. [NOTE—Where used in the individual monograph, the expression "percentage (a/a)" represents 100 times the ratio ($A_i/\Sigma A_i$).] Where the individual components and response factors are known, measure the area of each peak and convert this to mass of that component by multiplying the peak area by the response factor. Where the identity of other components is known, calibration curves may be based on area percentage alone, such as in water content tests for solvents, provided the detector type is specified.

Assays require quantitative comparison of one chromatogram with another, and lack of control of the specimen size injected is a major source of error. Addition of an internal standard to the test specimen minimizes this error. The ratio of peak response of the species of interest to the internal standard is compared from one chromatogram to another. Where the internal standard is chemically similar to the substance being determined, minor variations in column and detector parameters are controlled also. In some cases, the internal standard may be carried through the assay procedure prior to gas chromatography to control other quantitative aspects of the procedure.

A quantity of solute may be adsorbed within the system, and this would be reflected in the failure of the calibration curve to pass through zero, terminating instead along the abscissa. This effect may contribute error, particularly for the measurement of small specimens and the use of a single reference point. At high test-specimen concentrations, the solute may overload the liquid phase, leading to relative loss of peak height or symmetry. Before any column is accepted for assay purposes, a calibration curve should be constructed to control these errors prior to the use of the single sample size used in most assays.

Special-grade solvents are available for use especially where extensive evaporative concentration of test specimen is necessary prior to chromatography. Such chromatographic-grade reagents may be specified in the individual monograph. Since most drugs are polar molecules having reactive groups, successful chromatography may require the conversion of the drug to a more volatile or less polar derivative by treatment of reactive groups with appropriate reagents.

Columns should be conditioned by being operated until stable at a temperature higher than that specified for use in the individual monograph. In the case of thermally stable methyl- and phenyl-substituted polysiloxanes, a special sequence increases inertness and efficiency: maintain the column at a temperature of 250° for 1 hour with helium flowing to remove oxygen and solvents, stop the flow of helium, heat at about 340° for 4 hours, then reduce the heating so as to attain a temperature of 250°, and condition with helium flowing until stable. A suitable test for support inertness, which is necessary with low-polarity liquid phases, is the appearance of a single, symmetric peak for injected cholesterol with no evidence of decomposition. A column may occasionally be conditioned by repeated injections of the compound or mixture to be chromatographed.

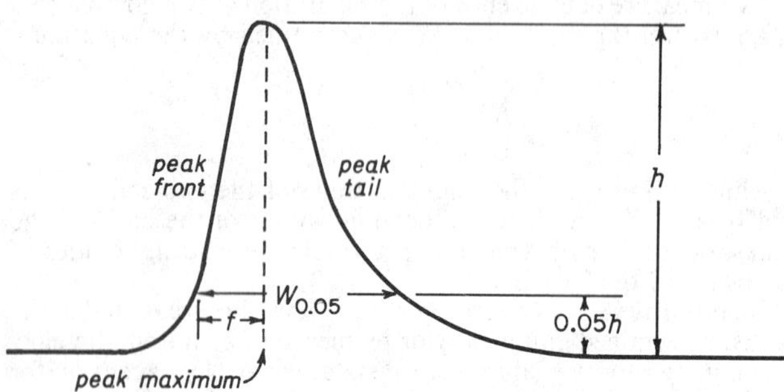

Fig. 2. Asymmetrical Chromatographic Peak.

System Suitability

It is generally desirable to ascertain the suitability and effectiveness of the operating system when employing chromatographic methods such as pressurized liquid chromatography or gas chromatography. It should be noted that the specification of definitive parameters in a monograph does not preclude the use of other suitable operating conditions (see *Procedures* in the *General Notices*). Adjustment of operating conditions to obtain acceptable operation and chromatograms may be required.

To ascertain the effectiveness of the final operating system, it should be subjected to a suitability test prior to use. The essence of such a test is the concept that the electronics, the equipment, the specimens, and the analytical operations constitute a single analytical system, which is amenable to an overall test of system function. Specific data are collected from replicate injections of the assay preparation or standard preparation. These are matched to specified maximum and minimum values, such as efficiency, internal precision, tailing factor, resolution, retention time, nature of the calibration curve, response, and recovery, as specified in the individual monograph.

A useful parameter is the reproducibility of replicate injections of the analytical solution. The reproducibility of replicate injections is best expressed as the relative standard deviation. The calculation is expressed by the equation

$$S_R\,(\%) = \frac{100}{\bar{X}}\left[\frac{\sum_{i=1}^{N}(X_i - \bar{X})^2}{N-1}\right]^{1/2}, \qquad (1)$$

in which S_R is the relative standard deviation in percentage, $\bar{X}$ is the mean of the set of N measurements, and X_i is an individual measurement. The term X_i refers to the measurement of the peak response ratio, R_S (where an internal standard is employed),

$$X_i = R_S = \frac{r_S}{r_i}, \qquad (2)$$

in which r_S is the peak response corresponding to the reference standard, and r_i is the peak response corresponding to the internal standard, or to the peak response, r_S, for an external standard method.

Replicate injections of the *Standard preparations* are usually specified in the individual monograph, and the resulting measurements are compared to ascertain whether requirements for precision are met. Unless otherwise specified in the individual monograph, data from five replicate chromatograms are used for calculation if the stated limit for relative standard deviation is 2.0% or less, and data from six replicate chromatograms are used for calculation if the stated relative standard deviation limit is more than 2.0%.

It is useful also to specify a tailing factor to limit the maximum permissible asymmetry of the peak. For Pharmacopeial purposes, the tailing factor, T, is defined as the ratio of the distance from the leading edge to the trailing edge of the peak, $W_{0.05}$, divided by twice the distance, f, from the peak maximum to the leading edge of the peak, the distances being measured at a point 5% of the peak height from the baseline. For a symmetrical peak, the tailing factor, T, is unity, and the value of T increases as tailing becomes more pronounced.

Resolution, R, is specified to ensure separation of closely eluting components, to establish the general separatory efficiency of the system, or where an internal standard is used.

GLOSSARY OF SYMBOLS

To promote uniformity of interpretation, the following symbols and definitions are employed where applicable in presenting formulas in the individual monographs. [NOTE—Where the terms W and t both appear in the same equation they must be expressed in the same units.]

α relative retention,

$$\alpha = \frac{t_2 - t_a}{t_1 - t_a}.$$

c_r, c_i, c_u concentrations of Reference Standard, internal standard, and analyte in a particular solution.

C_A concentration ratio of analyte and internal standard in test solution or *Assay preparation*,

$$C_A = \frac{q_u}{q_i}.$$

C_S concentration ratio of Reference Standard and internal standard in Standard solution,

$$C_S = \frac{c_r}{c_i}.$$

k' capacity factor,

$$k' = \frac{\text{amount of substance in stationary phase}}{\text{amount of substance in mobile phase}}$$

$$k' = \frac{\text{time spent by substance in stationary phase}}{\text{time spent by substance in mobile phase}} = \frac{t}{t_a} - 1.$$

n number of theoretical plates in a chromatographic column,

$$n = 16\left(\frac{t}{W}\right)^2.$$

q_r, q_i, q_u total quantities (weights) of Reference Standard, internal standard, and analyte in a particular solution.

Q_A quantity ratio of analyte and internal standard in test solution or *Assay preparation*,

$$Q_A = \frac{q_u}{q_i}.$$

Q_S quantity ratio of Reference Standard and internal standard in Standard solution,

$$Q_S = \frac{q_r}{q_i}.$$

r_S peak response of the Reference Standard obtained from a chromatogram.

r_U peak response of the analyte obtained from a chromatogram.

R resolution between two chromatographic peaks,

$$R = \frac{2(t_2 - t_1)}{W_1 + W_2}.$$

R_f chromatographic retardation factor equal to the ratio of the distance from the origin to the center of a zone divided by the distance from the origin to the solvent front.

R_r relative retention

$$R_r = \frac{\text{distance traveled by test substance}}{\text{distance traveled by standard}}.$$

R_S peak response ratio for *Standard preparation* containing Reference Standard and internal standard,

$$R_S = \frac{r_S}{r_i}.$$

R_U peak response ratio for *Assay preparation* containing the analyte and internal standard,

$$R_U = \frac{r_U}{r_i}.$$

S_R (%) relative standard deviation in percentage,

$$S_R\,(\%) = \frac{100}{\overline{X}}\left[\frac{\sum_{i=1}^{N}(X_i - \overline{X})^2}{N - 1}\right]^{1/2},$$

where X_i is an individual measurement in a set of N measurements and $\overline{X}$ is the arithmetic mean of the set.

T tailing factor,

$$T = \frac{W_{0.05}}{2f}.$$

t retention time measured from time of injection to time of elution of peak maximum.

t_a retention time of nonretarded component, air with thermal conductivity detection.

W width of peak measured by extrapolating the relatively straight sides to the baseline.

$W_{h/2}$ width of peak at half height.

$W_{0.05}$ width of peak at 5% height.

CHROMATOGRAPHIC REAGENTS

The following list of packings (L), phases (G), and supports (S) is intended to be a convenient reference for the chromatographer. These materials are listed also under *Reagent Specifications* in the section, *Reagents, Indicators, and Solutions*, as *Packings for High-pressure Liquid Chromatography* (L); *Phases for Gas Chromatography* (G); and *Supports for Gas Chromatography* (S). [NOTE—Particle sizes given in this listing are those generally provided. Where other, usually finer, sizes are required, the individual monograph specifies the desired particle size.]

Packings

L1—Octadecyl silane chemically bonded to porous silica or ceramic micro-particles, 5 to 10 µm in diameter.

L2—Octadecyl silane chemically bonded to silica gel of a controlled surface porosity that has been bonded to a solid spherical core, 30 to 50 µm in diameter.

L3—Porous silica particles, 5 to 10 µm in diameter.

L4—Silica gel of controlled surface porosity bonded to a solid spherical core, 30 to 50 µm in diameter.

L5—Alumina of controlled surface porosity bonded to a solid spherical core, 30 to 50 µm in diameter.

L6—Strong cation-exchange packing–sulfonated fluorocarbon polymer coated on a solid spherical core, 30 to 50 µm in diameter.

L7—Octylsilane chemically bonded to totally porous silica particles, 5 to 10 µm in diameter.

L8—An essentially monomolecular layer of aminopropylsilane chemically bonded to totally porous silica gel support, 10 µm in diameter.

L9—10-µm irregular, totally porous silica gel having a chemically bonded, strongly acidic cation-exchange coating.

L10—Nitrile groups chemically bonded to porous silica particles, 5 to 10 µm in diameter.

L11—Phenyl groups chemically bonded to porous silica particles, 5 to 10 µm in diameter.

L12—A strong anion-exchange packing made by chemically bonding a quaternary amine to a solid silica spherical core, 30 to 50 µm in diameter.

L13—Trimethylsilane chemically bonded to porous silica particles, 5 to 10 µm in diameter.

L14—Silica gel 10 µm in diameter having a chemically bonded, strongly basic quaternary ammonium anion-exchange coating.

L15—Hexylsilane chemically bonded to totally porous silica particles, 3 to 10 µm in diameter.

L16—Dimethylsilane chemically bonded to porous silica particles, 5 to 10 µm in diameter.

L17—Strong cation-exchange resin consisting of sulfonated cross-linked styrene-divinylbenzene copolymer in the hydrogen form, 7 to 11 µm in diameter.

L18—Amino and cyano groups chemically bonded to porous silica particles, 5 to 10 µm in diameter.

L19—Strong cation-exchange resin consisting of sulfonated cross-linked styrene-divinylbenzene copolymer in the calcium form, about 9 µm in diameter.

L20—Dihydroxypropane groups chemically bonded to porous silica particles, 5 µm to 10 µm in diameter.

L21—A rigid, spherical styrene-divinylbenzene copolymer, 5 to 10 µm in diameter.

L22—A cation exchange resin made of porous polystyrene gel with sulfonic acid groups, about 10 µm in size.

L23—An anion exchange resin made of porous polymethacrylate or polyacrylate gel with quaternary ammonium groups, about 10 µm in size.

L24—A semi-rigid hydrophilic gel consisting of vinyl polymers with numerous hydroxyl groups on the matrix surface, 32 to 63 µm in diameter.[5]

Phases

G1—Dimethylpolysiloxane oil.

G2—Dimethylpolysiloxane gum.

G3—50% Phenyl-50% methylpolysiloxane.

G4—Diethylene glycol succinate polyester.

G5—3-Cyanopropylpolysiloxane.

[5] Available as Fractogel TSK-HW-40F and distributed by Merck and Co.

G6—Trifluoropropylmethylpolysiloxane.
G7—50% 3-Cyanopropyl–50% phenylmethylsilicone.
G8—90% 3-Cyanopropyl–10% phenylmethylsilicone.
G9—Methylvinylpolysiloxane.
G10—Polyamide.
G11—Bis(2-ethylhexyl) sebacate polyester.
G12—Phenyldiethanolamine succinate polyester.
G13—Sorbitol.
G14—Polyethylene glycol (av. mol. wt. of 950 to 1050).
G15—Polyethylene glycol (av. mol. wt. of 3000 to 3700).
G16—Polyethylene glycol compound (av. mol. wt. about 15,000). A high molecular weight compound of polyethylene glycol and a diepoxide.
G17—75% Phenyl–25% methylpolysiloxane.
G18—Polyalkylene glycol.
G19—25% Phenyl–25% cyanopropyl–50% methylsilicone.
G20—Polyethylene glycol (av. mol. wt. of 380 to 420).
G21—Neopentyl glycol succinate.
G22—Bis(2-ethylhexyl) phthalate.
G23—Polyethylene glycol adipate.
G24—Diisodecyl phthalate.
G25—Polyethylene glycol compound TPA. A high molecular weight compound of a polyethylene glycol and a diepoxide that is esterified with terephthalic acid.
G26—25% 2-Cyanoethyl–75% methylpolysiloxane.
G27—5% Phenyl–95% methylpolysiloxane.
G28—25% Phenyl–75% methylpolysiloxane.
G29—β-β'-Thiodipropionitrile.
G30—Tetraethylene glycol dimethyl ether.
G31—Nonylphenoxypoly(ethyleneoxy)ethanol (av. ethyleneoxy chain length is 30); Nonoxynol 30.
G32—20% Phenylmethyl–80% dimethylpolysiloxane.
G33—20% Carborane–80% methylsilicone.
G34—Diethylene glycol succinate polyester stabilized with phosphoric acid.
G35—A high molecular weight compound of a polyethylene glycol and a diepoxide that is esterified with nitro-terephthalic acid.
G36—1% Vinyl–5% phenylmethylpolysiloxane.
G37—Polyimide.

Supports

[NOTE—Unless otherwise specified, mesh sizes of 80 to 100 or, alternatively, 100 to 120 are intended.]

S1A—Siliceous earth for gas chromatography has been flux-calcined by mixing diatomite with Na_2CO_3 flux and calcining above 900°. The siliceous earth is acid-washed, then water-washed until neutral, but not base-washed. The siliceous earth may be silanized by treating with an agent such as dimethyldichlorosilane[6] to mask surface silanol groups.

S1AB—The siliceous earth as described above is both acid- and base-washed.[6]

S1C—A support prepared from crushed firebrick and calcined or burned with a clay binder above 900° with subsequent acid-wash. It may be silanized.

S1NS—The siliceous earth is untreated.

S2—Styrene-divinylbenzene copolymer having a nominal surface area of less than 50 m^2 per g and an average pore diameter of 0.3 to 0.4 μm.

S3—Copolymer of ethylvinylbenzene and divinylbenzene having a nominal surface area of 500 to 600 m^2 per g and an average pore diameter of 0.0075 μm.

S4—Styrene-divinylbenzene copolymer with aromatic –O and –N groups, having a nominal surface area of 400 to 600 m^2 per g and an average pore diameter of 0.0076 μm.

S5—40- to 60-mesh, high-molecular weight tetrafluorethylene polymer.

S6—Styrene-divinylbenzene copolymer having a nominal surface area of 250 to 350 m^2 per g and an average pore diameter of 0.0091 μm.

S7—Graphitized carbon having a nominal surface area of 12 m^2 per g.

S8—Copolymer of 4-vinyl-pyridine and styrene-divinylbenzene.

S9—A porous polymer based on 2,6-diphenyl-*p*-phenylene oxide.

[6] Unless otherwise specified in the individual monograph, silanized support is intended.

S10—A highly polar cross-linked copolymer of acrylonitrite and divinylbenzene.

S11—Graphitized carbon having a nominal surface area of 9 m^2 per gram modified with small amounts of petrolatum and polyethylene glycol compound.[7]

⟨631⟩ COLOR AND ACHROMICITY

Definition—For the purposes of this chapter, color may be defined as the perception or subjective response by an observer to the objective stimulus of radiant energy in the visible spectrum extending over the range 400 nm to 700 nm in wavelength. Perceived color is a function of three variables: spectral properties of the object, both absorptive and reflective; spectral properties of the source of illumination; and visual characteristics of the observer.

Two objects are said to have a color match for a particular source of illumination when an observer cannot detect a color difference. Where a pair of objects exhibit a color match for one source of illumination and not another, they constitute a metameric pair. Color matches of two objects occur for all sources of illumination if the absorption and reflectance spectra of the two objects are identical.

Achromicity or colorlessness is one extreme of any color scale for transmission of light. It implies the complete absence of color, and therefore the visible spectrum of the object lacks absorbances. For practical purposes, the observer in this case perceives little if any absorption taking place in the visible spectrum.

Color Attributes—Because the sensation of color has both a subjective and an objective part, color cannot be described solely in spectrophotometric terms. The common attributes of color therefore cannot be given a one-to-one correspondence with spectral terminology.

Three attributes are commonly used to identify a color: (1) hue, or the quality by which one color family is distinguished from another, such as red, yellow, blue, green, and intermediate terms; (2) value, or the quality that distinguishes a light color from a dark one; and (3) chroma, or the quality that distinguishes a strong color from a weak one, or the extent to which a color differs from a gray of the same value.

The three attributes of color may be used to define a three-dimensional color space in which any color is located by its coordinates. The color space chosen is a visually uniform one if the geometric distance between two colors in the color space is directly a measure of the color distance between them. Cylindrical coordinates are often conveniently chosen:

Points along the long axis represent value from dark to light or black to white and have indeterminate hue and no chroma. Focusing on a cross-section perpendicular to the value axis, hue is determined by the angle about the long axis and chroma is determined by the distance from the long axis. Red, yellow, green, blue, purple, and intermediate hues are given by different angles. Colors along a radius of a cross-section have the same hue, which become more intense farther out. For example, colorless or achromic water has indeterminate hue, high value, and no chroma. If a colored solute is added, the water takes on a particular hue. As more is added, the color becomes darker, more intense, or deeper; i.e., the chroma generally increases and value decreases. If, however, the solute is a neutral color, i.e., gray, the value decreases, no increase in chroma is observed, and the hue remains indeterminate.

Laboratory spectroscopic measurements can be converted to measurements of the three color attributes. Spectroscopic results for three chosen lights or stimuli are weighted by three distribution functions to yield the tristimulus values, X, Y, Z (see *Color—Instrumental Measurement* ⟨1061⟩). The distribution functions were determined in color matching experiments with human subjects.

The tristimulus values are not coordinates in a visually uniform color space; however, several transformations have been proposed that are close to being uniform, one of which is given in the chapter cited ⟨1061⟩. The value is often a function of only the Y value. Obtaining uniformity in the chroma-hue subspace has

[7] Commercially available as SP1500 on Carbopack C from Supelco.

been less satisfactory. In a practical sense, this means in visual color comparison that if two objects differ significantly in hue, deciding which has a higher chroma becomes difficult. This points out the importance of matching standard to sample color as closely as possible, especially for the attributes of hue and chroma.

Color Determination and Standards—The perception of color and color matches is dependent on conditions of viewing and illumination. Determinations should be made using diffuse, uniform illumination under conditions that reduce shadows and nonspectral reflectance to a minimum. The surface of powders should be smoothed with gentle pressure so that a planar surface free from irregularities is presented. Liquids should be compared in matched color-comparison tubes, against a white background. If results are found to vary with illumination, those obtained in natural or artificial daylight are to be considered correct. Instead of visual determination, a suitable instrumental method may be used.

Colors of standards should be as close as possible to those of test specimens for quantifying color differences. Standards for opaque materials are available as sets of color chips that are arranged in a visually uniform space.* Standards identified by a letter for matching the colors of fluids can be prepared according to the accompanying table. To prepare the matching fluid required, pipet the prescribed volumes of the colorimetric test solutions [see under *Colorimetric Solutions* (CS) in the section, *Reagents, Indicators, and Solutions*] and water into one of the matching containers, and mix the solution in the container. Make the comparison as directed in the individual monograph, under the viewing conditions previously described. The matching fluids, or other combinations of the colorimetric solutions, may be used in very low concentrations to measure deviation from achromicity.

Matching Fluids

Matching Fluid	Parts of Cobaltous Chloride CS	Parts of Ferric Chloride CS	Parts of Cupric Sulfate CS	Parts of Water
A	0.1	0.4	0.1	4.4
B	0.3	0.9	0.3	8.5
C	0.1	0.6	0.1	4.2
D	0.3	0.6	0.4	3.7
E	0.4	1.2	0.3	3.1
F	0.3	1.2	0.0	3.5
G	0.5	1.2	0.2	3.1
H	0.2	1.5	0.0	3.3
I	0.4	2.2	0.1	2.3
J	0.4	3.5	0.1	1.0
K	0.5	4.5	0.0	0.0
L	0.8	3.8	0.1	0.3
M	0.1	2.0	0.1	2.8
N	0.0	4.9	0.1	0.0
O	0.1	4.8	0.1	0.0
P	0.2	0.4	0.1	4.3
Q	0.2	0.3	0.1	4.4
R	0.3	0.4	0.2	4.1
S	0.2	0.1	0.0	4.7
T	0.5	0.5	0.4	3.6

⟨641⟩ COMPLETENESS OF SOLUTION

Place the quantity of the substance specified in the individual monograph in a meticulously cleansed, glass-stoppered, 10-mL glass cylinder approximately 13 mm × 125 mm in size. Using the solvent that is specified in the monograph or on the label of the product, fill the cylinder almost to the constriction at the neck. Shake gently to effect solution: the solution is not less clear than an equal volume of the same solvent contained in a similar vessel and examined similarly.

* Collections of color chips, arranged according to hue, value, and chroma in a visually uniform space and suitable for use in color designation of specimens by visual matching are available from Munsell Color, Macbeth Division of Kollmorgen Corp., 2441 N. Calvert St., Baltimore, MD 21218.

⟨651⟩ CONGEALING TEMPERATURE

The temperature at which a substance passes from the liquid to the solid state upon cooling is a useful index to purity if heat is liberated when the solidification takes place, provided that any impurities present dissolve in the liquid only, and not in the solid. Pure substances have a well-defined freezing point, but mixtures generally freeze over a range of temperatures. For many mixtures, the congealing temperature, as determined by strict adherence to the following empirical methods, is a useful index of purity. The method for determining congealing temperatures set forth here is applicable to substances that melt between −20° and 150°, the range of the thermometer used in the bath. The congealing temperature is the maximum point (or lacking a maximum, the point of inflection) in the temperature-time curve.

Apparatus—Assemble an apparatus similar to that illustrated, in which the container for the substance is a 25- × 100-mm test tube. This is provided with a suitable, short-range thermometer suspended in the center, and a wire stirrer, about 30 cm long, bent at its lower end into a horizontal loop around the thermometer.

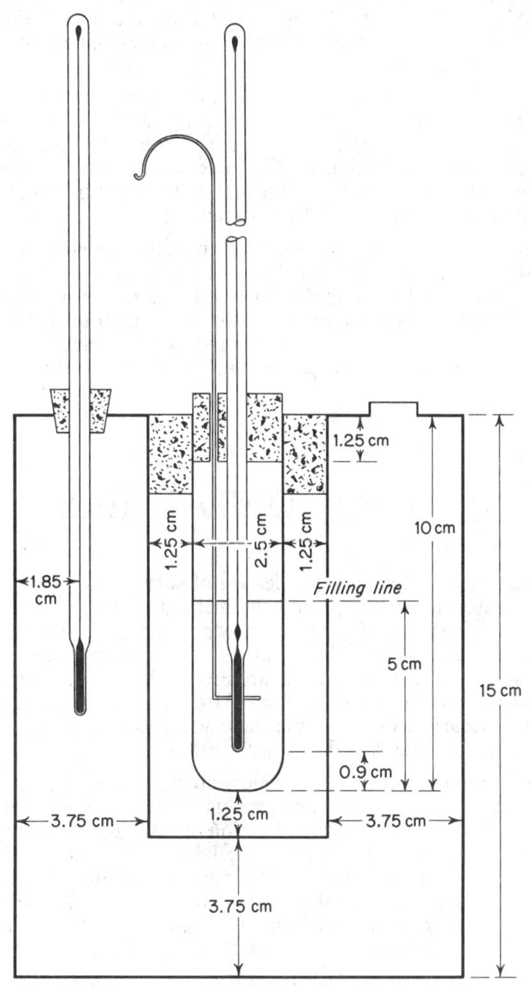

Congealing Temperature Apparatus

The specimen container is supported, by means of a cork, in a suitable water-tight cylinder about 50 mm in internal diameter and 11 cm in length. The cylinder, in turn, is supported in a suitable bath sufficient to provide not less than a 37-mm layer surrounding the sides and bottom of the cylinder. The outside bath is provided with a suitable thermometer.

Procedure—Use a thermometer having a range not exceeding 30°, graduated in 0.1° divisions, and calibrated for, but not used at, 76-mm immersion. A suitable series of thermometers, covering a range from −20° to +150°, is available as the ASTM

E-1 series 89C through 96C (see *Thermometers* ⟨21⟩). Melt the substance, if a solid, at a temperature not exceeding 20° above its expected congealing point, and pour it into the test tube to a height of 50 mm to 57 mm. Assemble the apparatus with the bulb of the test tube thermometer immersed halfway between the top and bottom of the specimen in the test tube. Fill the bath to about 12 mm from the top of the tube with suitable fluid at a temperature 4° to 5° below the expected congealing point.

In case the substance is a liquid at room temperature, carry out the determination using a bath temperature about 15° below the expected congealing point.

When the test specimen has cooled to about 5° above its expected congealing point, adjust the bath to a temperature 7° to 8° below the expected congealing point. Stir the specimen continuously during the remainder of the test by moving the loop up and down between the top and bottom of the specimen, at a regular rate of 20 complete cycles per minute.

Congelation frequently may be induced by rubbing the inner walls of the test tube with the thermometer, or by introducing a small fragment of the previously congealed substance. Pronounced supercooling may cause deviation from the normal pattern of temperature changes. If the latter occurs, repeat the test, introducing small particles of the material under test in solid form at 1° intervals as the temperature approaches the expected congealing point.

Record the reading of the test tube thermometer every 30 seconds. Continue stirring only so long as the temperature is gradually falling, stopping when the temperature becomes constant or starts to rise slightly. Continue recording the temperature in the test tube every 30 seconds for at least 3 minutes after the temperature again begins to fall after remaining constant.

The average of not less than four consecutive readings that lie within a range of 0.2° constitutes the congealing temperature. These readings lie about a point of inflection or a maximum, in the temperature-time curve, that occurs after the temperature becomes constant or starts to rise and before it again begins to fall. The average to the nearest 0.1° is the congealing temperature.

⟨661⟩ CONTAINERS

Many Pharmacopeial articles are of such nature as to require the greatest attention to the containers in which they are stored or maintained even for short periods of time. While the needs vary widely and some of them are not fully met by the containers available, objective standards are essential. It is the purpose of this chapter to provide such standards as have been developed for the materials of which pharmaceutical containers principally are made, i.e., glass and plastic.

A container intended to provide protection from light or offered as a "light-resistant" container meets the requirements for *Light Transmission*, where such protection or resistance is by virtue of the specific properties of the material of which the container is composed, including any coating applied thereto. A clear and colorless or a translucent container that is made light-resistant by means of an opaque enclosure (see *General Notices*) is exempt from the requirements for *Light Transmission.*

Containers composed of glass meet the requirements for *Chemical Resistance—Glass Containers*, and containers composed of plastic and intended for packaging products prepared for parenteral use meet the requirements under *Biological Tests—Plastics* and *Physicochemical Tests—Plastics.*

Where dry oral dosage forms, not meant for constitution into solution, are intended to be packaged in a container defined in the section *Polyethylene Containers*, the requirements given in that section are to be met.

Guidelines and requirements under *Single-unit Containers and Unit-dose Containers for Nonsterile Solid and Liquid Dosage Forms* apply to official dosage forms that are repackaged into single-unit or unit-dose containers or mnemonic packs for dispensing pursuant to prescription.

LIGHT TRANSMISSION

Apparatus[1]—Use a spectrophotometer of suitable sensitivity and accuracy, adapted for measuring the amount of light transmitted by either transparent or translucent glass or plastic materials used for pharmaceutical containers. For glass containers of nominal capacity up to 5 mL, use a suitable spectrophotometer having an aperture not larger than 2 mm × 1 cm. For containers made of translucent materials other than glass, use a suitable spectrophotometer equipped with an attachment that is capable of measuring and recording light transmitted in diffused as well as parallel rays.

Preparation of Specimen—

GLASS—Break the container or cut it with a circular saw fitted with a wet abrasive wheel, such as a carborundum or a bonded diamond wheel. Select sections to represent the average wall thickness in the case of blown glass containers, and trim them as necessary to give segments of a size convenient for mounting in the spectrophotometer. After cutting, wash and dry each specimen, taking care to avoid scratching the surfaces. If the specimen is too small to cover the opening in the specimen holder, mask the uncovered portion of the opening with opaque paper or masking tape, provided that the length of the specimen is greater than that of the slit in the spectrophotometer. Immediately before mounting in the specimen holder, wipe the specimen with lens tissue. Mount the specimen with the aid of a tacky wax, or by other convenient means, taking care to avoid leaving fingerprints or other marks on the surfaces through which light must pass.

PLASTIC—Cut circular sections from two or more areas of the container, and wash and dry them, taking care to avoid scratching the surfaces. Mount in the apparatus as described for *Glass.*

Procedure—Place the section in the spectrophotometer with its cylindrical axis parallel to the plane of the slit and approximately centered with respect to the slit. When properly placed, the light beam is normal to the surface of the section and reflection losses are at a minimum.

Measure the transmittance of the section with reference to air in the spectral region of interest, continuously with a recording instrument or at intervals of about 20 nm with a manual instrument, in the region of 290 nm to 450 nm.

Limits—The observed light transmission does not exceed the limits given in Table 1 for containers intended for parenteral use.

The observed light transmission for containers of Type NP glass and for plastic containers for products intended for oral or topical administration does not exceed 10% at any wavelength in the range from 290 nm to 450 nm.

Table 1. Limits for Glass Types I, II, and III and Plastic Classes I–VI.

Nominal Size (in mL)	Maximum Percentage of Light Transmission at Any Wavelength Between 290 nm and 450 nm	
	Flame-sealed Containers	Closure-sealed Containers
1	50	25
2	45	20
5	40	15
10	35	13
20	30	12
50	15	10

NOTE—Any container of a size intermediate to those listed above exhibits a transmission not greater than that of the next larger size container listed in the table. For containers larger than 50 mL, the limits for 50 mL apply.

[1] For further detail regarding apparatus and procedures, reference may be made to the following publications of the American Society for Testing and Materials, 1916 Race St., Philadelphia, PA 19103: "Standard Method of Test for Haze and Luminous Transmittance of Transparent Plastics," ASTM Designation D-1003-61; "Tentative Method of Test for Luminous Reflectance, Transmittance, and Color of Materials," ASTM E 308-66.

CHEMICAL RESISTANCE—GLASS CONTAINERS

The following tests are designed to determine the resistance to water attack of new (not previously used) glass containers. The degree of attack is determined by the amount of alkali released from the glass under the influence of the attacking medium under the conditions specified. This quantity of alkali is extremely small in the case of the more resistant glasses, thus calling for particular attention to all details of the tests and the use of apparatus of high quality and precision. The tests should be conducted in an area relatively free from fumes and excessive dust.

Glass Types—Glass containers suitable for packaging Pharmacopeial preparations may be classified as in Table 2 on the basis of the tests set forth in this section. Containers of Type I borosilicate glass are generally used for preparations that are intended for parenteral administration. Containers of Type I glass, or of Type II glass, i.e., soda-lime glass that is suitably de-alkalized, are usually used for packaging acidic and neutral parenteral preparations. Type I glass containers, or Type II glass containers (where stability data demonstrate their suitability), are used for alkaline parenteral preparations. Type III soda-lime glass containers usually are not used for parenteral preparations, except where suitable stability test data indicate that Type III glass is satisfactory for the parenteral preparations that are packaged therein. Containers of Type NP glass are intended for packaging nonparenteral articles; i.e., those intended for oral or topical use.

Table 2. Glass Types and Test Limits.

			Limits	
Type	General Description[a]	Type of Test	Size,[b] mL	mL of 0.020 *N* Acid
I	Highly resistant, borosilicate glass	*Powdered Glass*	All	1.0
II	Treated soda-lime glass	*Water Attack*	100 or less	0.7
			Over 100	0.2
III	Soda-lime glass	*Powdered Glass*	All	8.5
NP	General-purpose soda-lime glass	*Powdered Glass*	All	15.0

[a] The description applies to containers of this type of glass usually available.

[b] Size indicates the overflow capacity of the container.

Apparatus—

AUTOCLAVE—For these tests, use an autoclave capable of maintaining a temperature of 121 ± 2.0°, equipped with a thermometer, a pressure gauge, a vent cock, and a rack adequate to accommodate at least 12 test containers above the water level.

MORTAR AND PESTLE—Use a hardened-steel mortar and pestle, made according to the specifications in the accompanying illustration.

OTHER EQUIPMENT—Also required are 20.3-cm (8-inch) sieves made of stainless steel including the No. 20, No. 40, and No. 50 sieves along with the pan and cover (see *Openings of Standard Sieves* ⟨811⟩), 250-mL conical flasks made of resistant glass aged as specified, a 900-g (2-lb) hammer, a permanent magnet, a desiccator, and adequate volumetric apparatus.

Reagents—

HIGH-PURITY WATER—The water used in these tests has a conductivity at 25°, as measured in an in-line cell just prior to dispensing, of not greater than 0.15 μmho per cm. It meets the requirements of the test for *Heavy metals* under *Purified Water*, and is free from copper. The water may be prepared by passing distilled water through a deionizer cartridge packed with a mixed bed of nuclear-grade resin, then through a cellulose ester membrane having openings not exceeding 0.45 μm.[3] Do not use copper tubing. Flush the discharge lines before water is dispensed into test vessels. When the low conductivity specification can no longer be met, replace the deionizer cartridge.

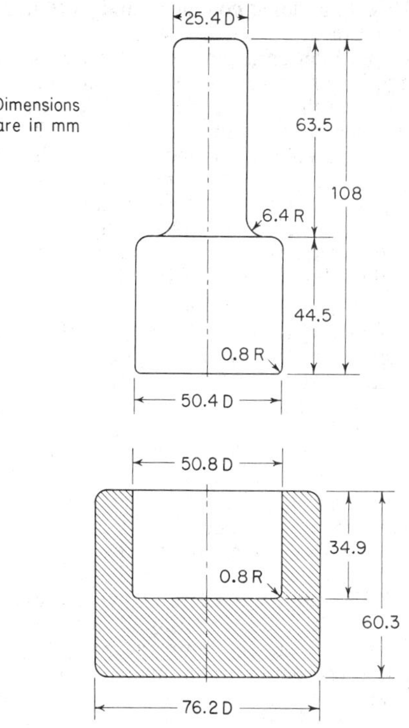

Special Mortar and Pestle for Pulverizing Glass[2]

METHYL RED SOLUTION—Dissolve 24 mg of methyl red sodium in purified water to make 100 mL. If necessary, neutralize the solution with 0.02 *N* sodium hydroxide or acidify it with 0.02 *N* sulfuric acid so that the titration of 100 mL of *High-purity Water*, containing 5 drops of indicator, does not require more than 0.020 mL of 0.020 *N* sodium hydroxide to effect the color change of the indicator, which should occur at a pH of 5.6.

Powdered Glass Test

Rinse thoroughly with Purified Water 6 or more containers selected at random, and dry them with a current of clean, dry air. Crush the containers into fragments about 25 mm in size, divide about 100 g of the coarsely crushed glass into three approximately equal portions, and place one of the portions in the special mortar. With the pestle in place, crush the glass further by striking 3 or 4 blows with the hammer. Nest the sieves, and empty the mortar into the No. 20 sieve. Repeat the operation on each of the two remaining portions of glass, emptying the mortar each time into the No. 20 sieve. Shake the sieves for a short time, then remove the glass from the No. 20 and No. 40 sieves, and again crush and sieve as before. Repeat again this crushing and sieving operation. Empty the receiving pan, reassemble the nest of sieves, and shake on a mechanical sieve shaker for 5 minutes or by hand for an equivalent length of time. Transfer the portion retained on the No. 50 sieve, which should weigh

[2] A suitable mortar and pestle is available (catalog No. H-17280) from Humboldt Manufacturing Co., 7300 West Agatite, Norridge, Chicago, IL 60656.

[3] A suitable nuclear-grade resin mixture of the strong acid cation exchanger in the hydrogen form and the strong base anion exchanger in the hydroxide form, with a one-to-one cation to anion equivalence ratio, is available from the Millipore Corp., Bedford, MA 01730; Barnstead Co., 225 Rivermoor St., Boston, MA 02132; Illinois Water Treatment Co., 840 Cedar St., Rockford, IL 61105; and Vaponics, Inc., 200 Cordage Park, Plymouth, MA 02360.

A suitable in-line filter is available from the Millipore Corp.; Gelman Instrument Co., 600 S. Wagner Rd., Ann Arbor, MI 48106; and Schleicher and Schuell, Inc., 540 Washington St., Keene, NH 10003.

in excess of 10 g, to a closed container, and store in a desiccator until used for the test.

Spread the specimen on a piece of glazed paper, and pass a magnet through it to remove particles of iron that may be introduced during the crushing. Transfer the specimen to a 250-mL conical flask of resistant glass, and wash it with six 30-mL portions of acetone, swirling each time for about 30 seconds and carefully decanting the acetone. After washing, the specimen should be free from agglomerations of glass powder, and the surface of the grains should be practically free from adhering fine particles. Dry the flask and contents for 20 minutes at 140°, transfer the grains to a weighing bottle, and cool in a desiccator. Use the test specimen within 48 hours after drying.

Procedure—Transfer 10.00 g of the prepared specimen, accurately weighed, to a 250-mL conical flask that has been digested (aged) previously with *High-purity Water* in a bath at 90° for at least 24 hours or at 121° for 1 hour. Add 50.0 mL of *High-purity Water* to this flask and to one similarly prepared to provide a blank. Cap all flasks with borosilicate glass beakers that previously have been treated as described for the flasks and that are of such size that the bottoms of the beakers fit snugly down on the top rims of the containers. Place the containers in the autoclave, and close it securely, leaving the vent cock open. Heat until steam issues vigorously from the vent cock, and continue heating for 10 minutes. Close the vent cock, and adjust the temperature to 121°, taking 19 to 23 minutes to reach the desired temperature. Hold the temperature at 121 ± 2.0° for 30 minutes, counting from the time this temperature is reached. Reduce the heat so that the autoclave cools and comes to atmospheric pressure in 38 to 46 minutes, being vented as necessary to prevent the formation of a vacuum. Cool the flask at once in running water, decant the water from the flask into a suitably cleansed vessel, and wash the residual powdered glass with four 15-mL portions of *High-purity Water*, adding the decanted washings to the main portion. Add 5 drops of *Methyl Red Solution*, and titrate immediately with 0.020 *N* sulfuric acid. If the volume of titrating solution is expected to be less than 10 mL, use a microburet. Record the volume of 0.020 *N* sulfuric acid used to neutralize the extract from 10 g of the prepared specimen of glass, corrected for a blank. The volume does not exceed that indicated in Table 2 for the type of glass concerned.

Water Attack at 121°

Rinse thoroughly 3 or more containers, selected at random, twice with *High-purity Water*.

Procedure—Fill each container to 90% of its overflow capacity with *High-purity Water*, and proceed as directed for *Procedure* under *Powdered Glass Test*, beginning with "Cap all flasks," except that the time of autoclaving shall be 60 minutes instead of 30 minutes, and ending with "to prevent the formation of a vacuum." Empty the contents from 1 or more containers into a 100-mL graduated cylinder, combining, in the case of smaller containers, the contents of several containers to obtain a volume of 100 mL. Place the pooled specimen in a 250-mL conical flask of resistant glass, add 5 drops of *Methyl Red Solution*, and titrate, while warm, with 0.020 *N* sulfuric acid. Complete the titration within 60 minutes after opening the autoclave. Record the volume of 0.020 *N* sulfuric acid used, corrected for a blank obtained by titrating 100 mL of *High-purity Water* at the same temperature and with the same amount of indicator. The volume does not exceed that indicated in Table 2 for the type of glass concerned.

Arsenic

Arsenic ⟨211⟩—Use as the *Test Preparation* 35 mL of the water from one Type I glass container or, in the case of smaller containers, 35 mL of the combined contents of several Type I glass containers, prepared as directed for *Procedure* under *Water Attack at 121°*: the limit is 0.1 ppm.

BIOLOGICAL TESTS—PLASTICS AND OTHER POLYMERS

Two stages of testing are indicated. The first stage is the performance of *in-vitro* biological tests according to the procedures set forth in chapter ⟨87⟩, *Biological Reactivity Tests, In-vitro*. Materials that meet the requirements of the *in-vitro* tests are not required to undergo further testing. No plastic class designation is assigned to these materials. Materials that do not meet the requirements of the *in-vitro* tests are subjected to the second stage of testing which is the performance of *in-vivo* tests, i.e., the *Systemic Injection Test*, *Intracutaneous Test*, and *Implantation Test* according to the procedures set forth in chapter ⟨88⟩, *Biological Reactivity Tests, In-vivo*.

PHYSICOCHEMICAL TESTS—PLASTICS

The following tests, designed to determine physical and chemical properties of plastics and their extracts, are based on the extraction of the plastic material, and it is essential that the designated amount of the plastic be used. Also, the specified surface area must be available for extraction at the designated temperature.

Extracting Medium—Unless otherwise directed in a specific test below, use *Purified Water* (see monograph) as the extracting medium, maintained at a temperature of 70° during the extraction of the prepared *Sample*.

Apparatus—Use a water bath and the *Extraction Containers* as described under Chapter ⟨88⟩ *Biological Reactivity Tests, In-vivo*.

Preparation of Apparatus—Proceed as directed in the first paragraph under *Preparation of Apparatus* in Chapter ⟨88⟩ *Biological Reactivity Tests, In-vivo*. [NOTE—The containers and equipment need not be sterile.]

Procedure—

Preparation of Sample—From a homogeneous plastic specimen, use a portion, for each 20.0 mL of extracting medium, equivalent to 120 cm^2 total surface area (both sides combined), and subdivide into strips approximately 3 mm in width and as near to 5 cm in length as is practical. Transfer the subdivided *Sample* to a glass-stoppered, 250-mL graduated cylinder of Type I glass, and add about 150 mL of Purified Water. Agitate for about 30 seconds, drain off and discard the liquid, and repeat with a second washing.

Transfer the prepared *Sample* to a suitable extraction flask, and add the required amount of *Extracting Medium*. Extract by heating in a water bath at the temperature specified for the *Extracting Medium* for 24 hours. Cool, but not below 20°. Pipet 20 mL of the extract of the prepared *Sample* into a suitable container. Use this portion in the test for *Buffering Capacity*. Immediately decant the remaining extract into a suitably cleansed container, and seal.

Blank—Use *Purified Water* where a blank is specified in the following tests.

NONVOLATILE RESIDUE—Transfer, in suitable portions, 50.0 mL of the extract of the prepared *Sample* to a suitable, tared crucible (preferably a fused silica crucible that has been acid-cleaned), and evaporate the volatile matter on a steam bath. Similarly evaporate 50.0 mL of the *Blank* in a second crucible. [NOTE—If an oily residue is expected, inspect the crucible repeatedly during the evaporation and drying period, and reduce the amount of heat if the oil tends to creep along the walls of the crucible.] Dry at 105° for 1 hour: the difference between the amounts obtained from the *Sample* and the *Blank* does not exceed 15 mg.

RESIDUE ON IGNITION ⟨281⟩—Proceed with the *Nonvolatile Residue* obtained from the *Sample* and from the *Blank*, using, if necessary, additional sulfuric acid but adding the same amount of sulfuric acid to each crucible: the difference between the amounts of residue on ignition obtained from the *Sample* and the *Blank* does not exceed 5 mg.

HEAVY METALS—Pipet 20 mL of the extract of the prepared *Sample*, filtered if necessary, into one of two matched 50-mL color-comparison tubes. Adjust with 1 *N* acetic acid or 6 *N* ammonium hydroxide to a pH between 3.0 and 4.0, using short-range pH paper as external indicator, dilute with water to about 35 mL, and mix.

Into the second color-comparison tube pipet 2 mL of *Standard Lead Solution* (see *Heavy Metals* ⟨231⟩), and add 20 mL of the *Blank*. Adjust with 1 *N* acetic acid or 6 *N* ammonium hydroxide to a pH between 3.0 and 4.0, using short-range pH paper as external indicator, dilute with water to about 35 mL, and mix.

Add 10 mL of freshly prepared hydrogen sulfide TS to each tube, dilute with water to 50 mL, and mix: any brown color produced within 10 minutes in the tube containing the extract of the prepared *Sample* does not exceed that in the tube containing the *Standard Lead Solution*, both tubes being viewed downward over a white surface (1 ppm in extract).

BUFFERING CAPACITY—Titrate the previously collected 20-mL portion of the extract of the prepared *Sample* potentiometrically to a pH of 7.0, using either 0.010 *N* hydrochloric acid or 0.010 *N* sodium hydroxide, as required. Treat a 20.0-mL portion of the *Blank* similarly: if the same titrant was required for both *Sample* and *Blank*, the difference between the two volumes is not greater than 10.0 mL; and if acid was required for either the *Sample* or the *Blank* and alkali for the other, the total of the two volumes required is not greater than 10.0 mL.

CONTAINERS FOR OPHTHALMICS—PLASTICS

Plastics for ophthalmics are composed of a mixture of homologous compounds, having a range of molecular weights. Such plastics frequently contain other substances such as residues from the polymerization process, plasticizers, stabilizers, antioxidants, pigments, and lubricants. Factors such as plastic composition, processing and cleaning procedures, contacting media, inks, adhesives, absorption, adsorption and permeability of preservatives, and conditions of storage may also affect the suitability of a plastic for a specific use. Evaluation of such factors should be made by appropriate additional specific tests to determine the suitability of a plastic for use as a container for ophthalmics.

Definition—For the purposes of this chapter, a *container* is that which holds the drug and is or may be in direct contact with the drug.

Biological Tests—Plastics and Other Polymers—Two stages of testing are indicated. The first stage is the performance of *in-vitro* biological tests according to the procedures set forth in chapter ⟨87⟩, *Biological Reactivity Tests, In-vitro.* Materials that meet the requirements of these tests are tested according to the procedures set forth for the *Systemic Injection Test*, *Intracutaneous Test*, and *Eye Irritation Test* in chapter ⟨88⟩, *Biological Reactivity Tests, In-vivo.* Materials that do not meet the requirements of the *in-vitro* tests cannot be used for ophthalmics.

POLYETHYLENE CONTAINERS

The standards and tests provided in this section characterize high-density and low-density polyethylene containers that are interchangeably suitable for packaging dry oral dosage forms not meant for constitution into solution.

Where stability studies have been performed to establish the expiration date of a particular dry oral dosage form not meant for constitution into solution in a container meeting the requirements set forth herein for either high- or low-density polyethylene containers, then any other polyethylene container meeting the same sections of these requirements may be similarly used to package such dosage form, provided that the appropriate stability programs are expanded to include the alternative container, in order to assure that the identity, strength, quality, and purity of the dosage form are maintained throughout the expiration period.

Both high- and low-density polyethylene are long-chain polymers synthesized under controlled conditions of heat and pressure, with the aid of catalysts from not less than 85.0 percent ethylene and not less than 95.0 percent total olefins. The other olefin ingredients most frequently used are butene, hexene, and propylene. The ingredients used to manufacture the polyethylene, and those used in the fabrication of the containers, conform to the requirements in the applicable sections of the Code of Federal Regulations, Title 21.

High-density polyethylene and low-density polyethylene both have an infrared absorption spectrum that is distinctive for polyethylene, and each possesses characteristic thermal properties. High-density polyethylene has a density between 0.941 and 0.965 g per cm^3. Low-density polyethylene has a density between 0.850 and 0.940 g per cm^3. The permeation properties of molded polyethylene containers may be altered when re-ground polymer is incorporated, depending upon the proportion of re-ground material in the final product. Other properties that may affect the suitability of polyethylene used in containers for packaging drugs are: oxygen and moisture permeability, modulus of elasticity, melt index, environmental stress crack resistance, and degree of crystallinity after molding. The requirements in this section are to be met when dry oral dosage forms, not meant for constitution into solution, are intended to be packaged in a container defined by this section.

Multiple Internal Reflectance—

APPARATUS—Use a double-beam infrared spectrophotometer equipped with a multiple internal reflectance accessory and a KRS-5 internal reflection plate.[4] A KRS-5 crystal 2 mm thick having an angle of incidence of 45° provides a sufficient number of reflections.

PREPARATION OF SPECIMEN—Cut 2 flat sections representative of the average wall thickness of the container, and trim them as necessary to obtain segments that are convenient for mounting in the multiple internal reflectance accessory. Taking care to avoid scratching the surfaces, wipe the specimens with dry paper or, if necessary, clean them with a soft cloth dampened with methanol, and permit them to dry. Securely mount the specimens on both sides of the KRS-5 internal reflection plate, ensuring adequate surface contact. Prior to mounting the specimens on the plate, they may be compressed to thin uniform films by exposing them to temperatures of about 177° under high pressures (15,000 psi or more).

PROCEDURE—Place the mounted specimen sections within the multiple internal reflectance accessory, and place the assembly in the specimen beam of the infrared spectrophotometer. Adjust the specimen position and mirrors within the accessory to permit maximum light transmission of the unattenuated reference beam. Upon completing the adjustments in the accessory, attenuate the reference beam to permit full-scale deflection during the scanning of the specimen. Determine the infrared spectrum from 3500 to 600 cm^{-1}: the spectrum of the specimen exhibits major absorption bands only at the same wavelengths as the spectrum of USP High-density Polyethylene RS or USP Low-density Polyethylene RS, similarly determined.

Thermal Analysis—Cut a section weighing about 12 mg, and place it in the test-specimen pan. Determine the thermogram under nitrogen, using equipment capable of performing the determinations described under *Thermal Analysis* ⟨891⟩.

High-density Polyethylene—The thermogram of the specimen is similar to the thermogram of USP High-density Polyethylene RS, similarly determined, and the temperatures of the endotherms and exotherms in the thermogram of the specimen do not differ from those of the standard by more than 6.0°.

Low-density Polyethylene—The thermogram of the specimen is similar to the thermogram of USP Low-density Polyethylene RS, similarly determined, and the temperatures of the endotherms and exotherms in the thermogram of the specimen do not differ from those of the standard by more than 8.0°.

Light Transmission—Polyethylene containers intended to provide protection from light meet the requirements under *Light Transmission.*

Water Vapor Permeation—Fit the containers with impervious seals obtained by heat-sealing the bottles with an aluminum foil–polyethylene laminate or other suitable seal.[5] Test the containers as described under *Containers—Permeation* ⟨671⟩: the high-density polyethylene containers so tested meet the requirements if the moisture permeability exceeds 10 mg per day per liter in not more than 1 of the 10 test containers and exceeds 25 mg per day per liter in none of them. The low-density polyethylene containers so tested meet the requirements if the moisture permeability exceeds 20 mg per day per liter in not more than 1 of the 10 test containers and exceeds 30 mg per day per liter in none of them.

Heavy Metals and Nonvolatile Residue—Prepare extracts of specimens for these tests as directed for *Preparation of Sample* in the *Procedure* under *Physicochemical Tests—Plastics*, except

[4] The multiple internal reflectance accessory and KRS-5 plate are available from several sources, including Beckman Instruments, Inc., 2500 Harbor Blvd., Fullerton, CA 92634, and from Perkin Elmer Corp., Main Ave., Norwalk, CT 06856.

[5] A suitable laminate for sealing has as the container contact layer polyethylene of not less than 0.025 mm (0.001 inch) and a second layer of aluminum foil of not less than 0.018 mm (0.0007 inch), with additional layers of suitable backing materials. A suitable seal can be obtained also by using glass plates and a sealing wax consisting of 60% of refined amorphous wax and 40% of refined crystalline paraffin wax.

that for each 20.0 mL of *Extracting Medium* the portion shall be 60 cm^2, regardless of thickness.

HEAVY METALS—Containers meet the requirements for *Heavy Metals* under *Physicochemical Tests—Plastics.*

NONVOLATILE RESIDUE—Proceed as directed for *Nonvolatile Residue* under *Physicochemical Tests—Plastics*, except that the blank shall be the same solvent used in each of the tests set forth below. The difference between the amounts obtained from the specimen and the blank does not exceed 12.0 mg when water maintained at a temperature of 70° is used as the extracting medium; does not exceed 75.0 mg when alcohol maintained at a temperature of 70° is used as the extracting medium; and does not exceed 100.0 mg for high-density polyethylene and does not exceed 350.0 mg for low-density polyethylene when hexanes maintained at a temperature of 50° are used as the extracting medium. Containers meet these requirements for *Nonvolatile Residue* for all of the above extracting media. [NOTE—Hexanes and alcohol are flammable. When evaporating these solvents, use a stream of air with the water bath; when drying the residue, use an explosion-proof oven.]

SINGLE-UNIT CONTAINERS AND UNIT-DOSE CONTAINERS FOR NON-STERILE SOLID AND LIQUID DOSAGE FORMS

An official dosage form is required to bear on its label an expiration date assigned for the particular formulation and package of the article. This date limits the time during which the product may be dispensed or used. Because the expiration date stated on the manufacturer's or distributor's package has been determined for the drug in that particular package and may not be applicable to the product where it has been repackaged in a different container, repackaged drugs dispensed pursuant to a prescription are exempt from this label requirement. It is necessary, therefore, that other precautions be taken by the dispenser to preserve the strength, quality, and purity of drugs that are repackaged for ultimate distribution or sale to patients.

The following guidelines and requirements are applicable where official dosage forms are repackaged into single-unit or unit-dose containers or mnemonic packs for dispensing pursuant to prescription.

Labeling—It is the responsibility of the dispenser, taking into account the nature of the drug repackaged, the characteristics of the containers, and the storage conditions to which the article may be subjected, to determine a suitable beyond-use date to be placed on the label. Such date is not later than the expiration date of the original package. In the absence of stability data to the contrary, such date should not exceed (1) 25% of the remaining time between the date of repackaging and the expiration date on the original manufacturer's bulk container, or (2) a six-month period from the date the drug is repackaged, whichever is earlier. Each single-unit or unit-dose container bears a separate label, unless the device holding the unit-dose form does not allow for the removal or separation of the intact single-unit or unit-dose container therefrom.

Storage—Store the repackaged article in a humidity-controlled environment and at the temperature specified in the individual monograph or in the product labeling. Where no temperature or humidity is specified in the monograph or in the labeling of the product, controlled room temperature and a relative humidity corresponding to 75% at 23° are not to be exceeded during repackaging or storage.

A refrigerator or freezer shall not be considered to be a humidity-controlled environment, and drugs that are to be stored at a cold temperature in a refrigerator or freezer shall be placed within an outer container that meets the monograph requirements for the drug contained therein.

CUSTOMIZED PATIENT MEDICATION PACKAGES

In lieu of dispensing two or more prescribed drug products in separate containers, a pharmacist may, with the consent of the patient, the patient's caregiver, or a prescriber, provide a customized patient medication package (patient med pak).[6]

A patient med pak is a package prepared by a pharmacist for a specific patient comprising a series of containers and containing two or more prescribed solid oral dosage forms. The patient med pak is so designed or each container is so labeled as to indicate the day and time, or period of time, that the contents within each container are to be taken.

Label—(A) The patient med pak shall bear a label stating:

(1) the name of the patient;

(2) a serial number for the patient med pak itself and a separate identifying serial number for each of the prescription orders for each of the drug products contained therein;

(3) the name, strength, physical description or identification, and total quantity of each drug product contained therein;

(4) the directions for use and cautionary statements, if any, contained in the prescription order for each drug product therein;

(5) any storage instructions or cautionary statements required by the official compendia;

(6) the name of the prescriber of each drug product;

(7) the date of preparation of the patient med pak and the beyond-use date assigned to the patient med pak (such beyond-use date shall be not later than 60 days from the date of preparation);

(8) the name, address, and telephone number of the dispenser and the dispenser's registration number where necessary; and

(9) any other information, statements, or warnings required for any of the drug products contained therein.

(B) If the patient med pak allows for the removal or separation of the intact containers therefrom, each individual container shall bear a label identifying each of the drug products contained therein.

Labeling—The patient med pak shall be accompanied by a patient package insert, in the event that any medication therein is required to be dispensed with such insert as accompanying labeling. Alternatively, such required information may be incorporated into a single, overall educational insert provided by the pharmacist for the total patient med pak.

Packaging—In the absence of more stringent packaging requirements for any of the drug products contained therein, each container of the patient med pak shall comply with the moisture permeation requirements for a Class B single-unit or unit-dose container (see *Containers—Permeation* ⟨671⟩). Each container shall be either not reclosable or so designed as to show evidence of having been opened.

Guidelines—It is the responsibility of the dispenser, when preparing a patient med pak, to take into account any applicable compendial requirements or guidelines and the physical and chemical compatibility of the dosage forms placed within each container, as well as any therapeutic incompatibilities that may attend the simultaneous administration of the medications. In this regard, pharmacists are encouraged to report to USP headquarters any observed or reported incompatibilities.

Record keeping—In addition to any individual prescription filing requirements, a record of each patient med pak shall be made and filed. Each record shall contain, as a minimum:

(1) the name and address of the patient;

(2) the serial number of the prescription order for each drug product contained therein;

(3) the name of the manufacturer or labeler and lot number for each drug product contained therein;

(4) information identifying or describing the design, characteristics, or specifications of the patient med pak sufficient to allow subsequent preparation of an identical patient med pak for the patient;

(5) the date of preparation of the patient med pak and the beyond-use date that was assigned;

(6) any special labeling instructions; and

(7) the name or initials of the pharmacist who prepared the patient med pak.

[6] It should be noted that there is no special exemption for patient med paks from the requirements of the Poison Prevention Packaging Act. Thus the patient med pak, if it does not meet child-resistant standards, shall be placed in an outer package that does comply, or the necessary consent of the purchaser or physician, to dispense in a container not intended to be child-resistant, shall be obtained.

⟨671⟩ CONTAINERS—PERMEATION

This test is provided to determine the moisture permeability of a container utilized for a drug when dispensed on prescription where packaging and storage in a *tight container* or a *well-closed container* is specified in the individual monograph. It is applicable to multiple-unit containers (see *Preservation, Packaging, Storage, and Labeling* under *General Notices*). As used herein, the term "container" refers to the entire system comprising, usually, the container itself, the liner (if used), and the closure. Where the manufacturer's container, previously unopened, is utilized for dispensing the drug, such container is exempt from the requirements of this test.

Desiccant—Place a quantity of 4-mesh, anhydrous calcium chloride[1] in a shallow container, taking care to exclude any fine powder, then dry at 110° for 1 hour, and cool in a desiccator.

Procedure—Select 12 containers of a uniform size and type, clean the sealing surfaces with a lint-free cloth, and close and open each container 30 times. Apply the closure firmly and uniformly each time the container is closed. Close screw-capped containers with a torque that is within the range of tightness specified in the accompanying table. Add *Desiccant* to 10 of the containers, designated *test containers*, filling each to within 13 mm of the closure if the container volume is 20 mL or more, or filling each to two-thirds of capacity if the container volume is less than 20 mL. If the interior of the container is more than 63 mm in depth, an inert filler or spacer may be placed in the bottom to minimize the total weight of the container and *Desiccant;* the layer of *Desiccant* in such a container shall be not less than 5 cm in depth. Close each immediately after adding *Desiccant*, applying the torque designated in the accompanying table when closing screw-capped containers. To each of the remaining 2 containers, designated *controls*, add a sufficient number of glass beads to attain a weight approximately equal to that of each of the *test containers*, and close, applying the torque designated in the accompanying table when closing screw-capped containers. Record the weight of the individual containers so prepared to the nearest 0.1 mg if the container volume is less than 20 mL; to the nearest mg if the container volume is 20 mL or more but less than 200 mL; or to the nearest centigram (10 mg) if the container volume is 200 mL or more; and store at 75 ± 3% relative humidity and a temperature of 20 ± 2°. [NOTE—A saturated system of 35 g of sodium chloride with each 100 mL of water placed in the bottom of a desiccator maintains the specified humidity. Other methods may be employed to maintain these conditions.] After 336 ± 1 hours (14 days), record the weight of the individual containers in the same manner. Completely fill 5 empty containers of the same size and type as the containers under test with water or a non-compressible, free-flowing solid such as well-tamped fine glass beads, to the level indicated by the closure surface when in place. Transfer the contents of each to a graduated cylinder, and determine the average container volume, in mL. Calculate the rate of moisture permeability, in mg per day per liter, by the formula:

$$(1000/14V)[(T_f - T_i) - (C_f - C_i)],$$

in which V is the volume, in mL, of the container, $(T_f - T_i)$ is the difference, in mg, between the final and initial weights of each *test container*, and $(C_f - C_i)$ is the average of the differences, in mg, between the final and initial weights of the 2 *controls*. The containers so tested are *tight containers* if not more than one of the 10 *test containers* exceeds 100 mg per day per liter in moisture permeability, and none exceeds 200 mg per day per liter.

The containers are *well-closed containers* if not more than one of the 10 *test containers* exceeds 2000 mg per day per liter in moisture permeability, and none exceeds 3000 mg per day per liter.

[1] Suitable 4-mesh, anhydrous calcium chloride is available commercially as Item SC11204-5006M from Sargent-Welch Scientific Co., P. O. Box 1026, Skokie, IL 60077.

SINGLE-UNIT CONTAINERS AND UNIT-DOSE CONTAINERS FOR CAPSULES AND TABLETS

To permit an informed judgment regarding the suitability of the packaging for a particular type of product, the following procedure and classification scheme are provided for evaluating the moisture-permeation characteristics of single-unit and unit-dose containers. Inasmuch as equipment and operator performance may affect the moisture permeation of a container formed or closed, the moisture-permeation characteristics of the packaging system being utilized shall be determined.

Desiccant—Dry suitable desiccant pellets[2] at 110° for 1 hour prior to use. Use pellets weighing approximately 400 mg each and having a diameter of approximately 8 mm.

Procedure—

Method I—Seal not less than 10 unit-dose containers with 1 pellet in each, and seal 10 additional, empty unit-dose containers to provide the controls, using finger cots or padded forceps to handle the sealed containers. Number the containers, and record the individual weights[3] to the nearest mg. Weigh the controls as a unit, and divide the total weight by the number of controls to obtain the average. Store all of the containers at 75 ± 3% relative humidity and at a temperature of 20 ± 2°. [NOTE—A saturated system of 35 g of sodium chloride with each 100 mL of water placed in the bottom of a desiccator maintains the specified humidity. Other methods may be employed to maintain these conditions.] After a 24-hour interval, or a multiple thereof (see *Results*), remove the containers from the chamber, and allow them to equilibrate for 15 to 60 minutes in the weighing area. Again record the weight of the individual containers and the combined controls in the same manner. [NOTE—If any indicating pellets turn pink during this procedure, or if the pellet weight increase exceeds 10%, terminate the test, and regard only earlier determinations as valid.] Return the containers to the humidity chamber. Calculate the rate of moisture permeation, in mg per day, of each container by the formula:

$$(1/N)[(W_f - W_i) - (C_f - C_i)],$$

in which N is the number of days expired in the test period, $(W_f - W_i)$ is the difference, in mg, between the final and initial weights of each test container, and $(C_f - C_i)$ is the average of the difference, in mg, between the final and initial weights of the controls, the data being calculated to two significant figures. [NOTE—Where the permeations measured are less than 5 mg per day, and where the controls are observed to reach a steady state in 7 days, the individual permeations may be determined more accurately after an initial 7 days of equilibration by using that weight as W_i, zero time, in the calculation.]

Method II—Use this procedure for packs (e.g., punch-out cards) that incorporate a number of separately sealed unit-dose containers or blisters. Seal a sufficient number of packs, such that not less than 4 packs and a total of not less than 10 unit-dose containers or blisters filled with 1 pellet in each unit are tested. Seal a corresponding number of empty packs, each pack containing the same number of unit-dose containers or blisters as used in the test packs, to provide the controls. Store all of the containers at 75 ± 3% relative humidity and at a temperature of 20 ± 2°. [See *Note* under *Method I.*] After 24 hours, and at multiples thereof (see *Results*), remove the packs from the chamber, and allow them to equilibrate for approximately 45 minutes. Record the weights of the individual packs, and return them to the chamber. Weigh the control packs as a unit, and divide the total weight by the number of control packs to obtain the average empty pack weight. [NOTE—If any indicating pel-

[2] Suitable moisture-indicating desiccant pellets are available commercially from sources such as Medical Packaging, Inc., 525 White Horse Pike, Atco, NJ 08004 [Telephone 800-257-5282; in N. J., 609-767-3604], as Indicating Desiccant Pellets, item No. TK-1002.

[3] Accurate comparisons of *Class A* containers may require test periods in excess of 28 days if weighings are performed on a *Class A* prescription balance (see *Prescription Balances and Volumetric Apparatus* ⟨1176⟩). The use of an analytical balance on which weights can be recorded to 4 or 5 decimal places may permit more precise characterization between containers and/or shorter test periods.

lets turn pink during the procedure, or if the average pellet weight increase in any pack exceeds 10%, terminate the test, and regard only earlier determinations as valid.] Calculate the average rate of moisture permeation, in mg per day, for each unit-dose container or blister in each pack by the formula:

$$(1/NX)[(W_f - W_i) - (C_f - C_i)],$$

in which N is the number of days expired in the test period (beginning after the initial 24 hour equilibration period), X is the number of separately sealed units per pack, $(W_f - W_i)$ is the difference, in mg, between the final and initial weights of each test pack, and $(C_f - C_i)$ is the average of the difference, in mg, between the final and initial weights of the control packs, the rates being calculated to two significant figures.

Results—The individual unit-dose containers as tested in *Method I* are designated *Class A* if not more than 1 of 10 containers tested exceeds 0.5 mg per day in moisture permeation rate and none exceeds 1 mg per day; they are designated *Class B* if not more than 1 of 10 containers tested exceeds 5 mg per day and none exceeds 10 mg per day; they are designated *Class C* if not more than 1 of 10 containers tested exceeds 20 mg per day and none exceeds 40 mg per day; and they are designated *Class D* if the containers tested meet none of the moisture permeation rate requirements.

The packs as tested in *Method II* are designated *Class A* if no pack tested exceeds 0.5 mg per day in average blister moisture permeation rate; they are designated *Class B* if no pack tested exceeds 5 mg per day in average blister moisture permeation rate; they are designated *Class C* if no pack tested exceeds 20 mg per day in average blister moisture permeation rate; and they are designated *Class D* if the packs tested meet none of the above average blister moisture permeation rate requirements.

With the use of the *Desiccant* described herein, suitable test intervals for the final weighings, W_f, are: 24 hours for *Class D;* 48 hours for *Class C;* 7 days for *Class B;* and not less than 28 days for *Class A.*

Torque Applicable to Screw-Type Container

Closure Diameter[1] (mm)	Suggested Tightness Range with Manually Applied Torque[2] (inch-pounds)
15	6 to 9
18	7 to 11
20	8 to 12
22	9 to 13
24	10 to 15
28	11 to 17
33	13 to 20
38	15 to 23
43	17 to 26
48	19 to 29
53	21 to 32
58	23 to 35
63	25 to 38
70	28 to 42
83	34 to 49
86	35 to 51
89	36 to 53
100	40 to 60
110	45 to 65
120	48 to 72
132	53 to 79

[1] The torque designated for the next larger closure diameter is to be applied in testing containers having a closure diameter intermediate to the diameters listed.

[2] A suitable apparatus is available from Owens-Illinois, Toledo, Ohio 43666. (Model 25 torque tester is used for testing between 0 and 25; Model 50 for testing between 0 and 50; and Model 100 for testing between 0 and 100 inch-pounds of torque.) The torque values refer to application, not removal, of the closure. For further detail regarding instructions, reference may be made to "Standard Method of Measuring Application and Removal Torque of Threaded Closures," ASTM Designation D 3198-73, published by the American Society for Testing and Materials, 1916 Race St., Philadelphia, PA 19103.

⟨691⟩ COTTON

Preparatory to the determination of absorbency and of fiber length, remove the Cotton from its wrappings, and condition it for not less than 4 hours in a standard atmosphere of 65 ± 2% relative humidity at 21 ± 1.1° (70 ± 2°F).

Absorbency Test

Procedure—Prepare a test basket, weighing not more than 3 g, from copper wire approximately 0.4 mm in diameter (No. 26 B. & S.) in the form of a cylinder approximately 5 cm in diameter and 8 cm deep, with spaces of about 2 cm between the wires. Take portions of purified cotton weighing 1 ± 0.05 g from five different parts of the package by pulling, not cutting, the specimens, place the combined portions in the basket, and weigh. Hold the basket on its side approximately 12 mm above the surface of water at 25 ± 1°, and drop it into the water. Determine, preferably by use of a stop watch, the time in seconds required for complete submersion.

Remove the basket from the water, allow it to drain for 10 seconds in the same horizontal position, then place it immediately in a tared, covered vessel and weigh, deducting the weight of the test basket and of the purified cotton to find the weight of water absorbed.

Fiber Length

For the determination of the length and of the length distribution of cotton fibers in purified cotton use the following method:

Carry out all operations associated with the determination of fiber length of purified cotton in an atmosphere maintained at 65 ± 2% relative humidity at 21 ± 1.1° (70 ± 2°F).

These directions describe the mode of procedure that is well adapted to the sorter* most extensively used in the United States at the present time.

Apparatus—The sorter (see illustration) consists of two banks of combs rigidly mounted side by side on a common base. Each bank of combs consists of at least 12 individual combs spaced 3.2 mm apart, one behind the other, and mounted in grooves so that as they are approached during the fractionating process and no longer needed, they may be dropped below the working plane. Each individual comb has a single row of accurately aligned and sharply pointed teeth, 12 mm long, consisting of needles 0.38 mm in diameter. The teeth are spaced 62 to 25 mm over an extent of approximately 50 mm.

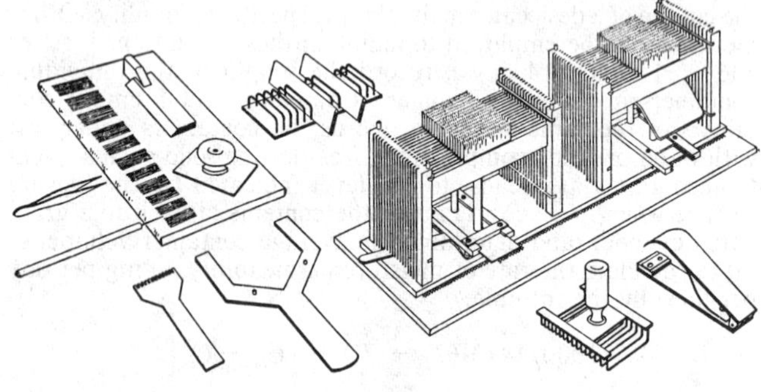

Duplex Cotton Fiber Sorter

Accessory equipment consists of fiber-sorter forceps, fiber-depressing grid, fiber-depressing smooth plate, and velvet-covered plates. The sorter forceps consist of two brass pieces approximately 75 mm long, hinged on one end and slightly curved to present a beaked aspect at the gripping end for gripping the protruding fibers close to the surfaces of the combs. Usually, one of the gripping edges has a leather or other fibrous padding. The gripping edge is approximately 19 mm wide.

The fiber-depressing grid consists of a series of brass rods spaced 3.2 mm apart so that they may be placed between the combs to

*NOTE—The method here described is especially adapted to the Suter-Webb Duplex Cotton Fiber sorting apparatus, but with more or less obvious alteration in procedure, may be carried out with two Baer sorters in tandem arrangement, or with a Johannsen or other similar apparatus.

press the fibers down between the teeth. The fiber-depressing smooth plate consists of a polished brass plate approximately 25 × 50 mm, with a knob or handle on the upper surface whereby the plate may be smoothed over the fibers as they are laid on the velvet surface of the array plates. The velvet-covered plates, upon which the fibers may be arrayed, are aluminum sheets approximately 100 mm × 225 mm × 2.4 mm thick, covered on both sides with high-grade velvet, preferably black.

Selection of Cotton—After unrolling the cotton, prepare a representative laboratory test specimen by taking from a package containing from 8 to 16 ounces, 32 pinches (about 75 mg each) well distributed throughout the bulk of the lap, 16 representative pinches being taken from each longitudinal half of the lap. Avoid the cut ends of the lap, and take particular care to secure portions throughout the thickness of the lap. To avoid biased selection of long or short fibers, remove all fibers of the group pinched and do not allow them to slip from between the fingers.

From packages of not more than 4 ounces in weight, take 8 pinches, and from packages weighing more than 4 ounces and not more than 8 ounces, take 16 pinches, all well distributed.

Mix the pinches in pairs promiscuously, and combine each pair by gently drawing and lapping them in the fingers. Then divide each combined pair by splitting longitudinally into two approximately equal parts and utilize one part in the further mixing. (The other part may be discarded or reserved for any further tests or checks.)

Repeat the process described in the preceding paragraph with the successive halves of the bifurcated series until only 1 pinch, the final composite test portion, results. Gently parallel and straighten the fibers of the final composite test portion by drawing and lapping them in the fingers. Take care to retain all of the fibers, including as far as possible those of the neps (specks of entangled fibers) and naps (matted masses of fibers), discarding only motes (immature seed fragments with fibers) and nonfiber foreign material such as stem, leaf, and fragments of seedcoats.

From the final composite portion described in the preceding paragraph, separate longitudinally a test portion of 75 ± 2 mg, accurately weighed. Retain the residue for any check test necessary.

Procedure—With the fiber-depressing grid carefully insert the weighed test portion into one bank of combs of the cotton sorter, so that it extends across the combs at approximately right angles.

With the sorter forceps, grip by the free ends a small portion of the fibers extending through the teeth of the comb nearest to the operator; gently and smoothly draw them forward out of the combs, and transfer them to the tips of the teeth in the second bank of combs, laying them parallel to themselves, straight, and approximately at right angles to the faces of the combs, releasing the gripped ends as near to the face of the front comb as possible. With the depressor grid carefully press the transferred fibers down into the teeth of the combs. Continue the operation until all of the fibers are transferred to the second bank of combs. During this transfer of the fibers, drop the combs of the first bank in succession when and as all of the protruding fibers have been removed.

Turn the machine through 180°, and transfer the cotton fibers back to the *first bank* of combs in the manner described in the preceding paragraph.

Take great care in evening up the ends of the fibers during both of the above transfers, arranging them as closely as possible to the front surface of the proximal comb. Such evening out of the ends of the protruding fibers may involve drawing out straggling fibers from both the front and rear aspects of the banks of combs, and re-depositing them into and over the main bundle in the combs.

Turn the machine again through 180°. Drop successive combs if necessary to expose the ends of the longest fibers. It may be necessary to re-deposit some straggling fibers. With the forceps withdraw the few most protuberant fibers. In this way continue to withdraw successively the remaining protuberant fibers back to the front face of the proximal comb. Drop this comb and repeat the series of operations in the same manner until all of the fibers have been drawn out. In order not to disturb seriously the portion being tested, and thereby vitiate the length fractionation into length groups, make several pulls (as many as 8 to 10) between each pair of combs.

Lay the pulls on the velvet-covered plates alongside each other, as straight as possible, with the ends as clearly defined as possible, and with the distal ends arranged in a straight line, pressing them down gently and smoothly with the fiber-depressing smooth plate before releasing the pull from the forceps. Employ not less than 50 and not more than 100 pulls to fractionate the test portion.

Group together all of the fibers measuring 12.5 mm (about ½ inch) or more in length, and weigh the group to the nearest 0.3 mg. In the same manner, group together all fibers 6.25 mm (about ¼ inch) or less in length, and weigh in the same manner. Finally, group the remaining fibers of intermediate lengths together and weigh. The sum of the three weights does not differ from the initial weight of the test portion by more than 3 mg. Divide the weight of each of the first two groups by the weight of the test portion to obtain the percentage by weight of fiber in the two ranges of length.

⟨695⟩ CRYSTALLINITY

This test is provided to determine compliance with the crystallinity requirement where stated in the individual monograph for a drug substance.

Procedure—Unless otherwise specified in the individual monograph, mount a few particles of the specimen in mineral oil on a clean glass slide. Examine the mixture using a polarizing microscope: the particles show birefringence (interference colors) and extinction positions when the microscope stage is revolved.

⟨701⟩ DISINTEGRATION

This test is provided to determine compliance with the limits on *Disintegration* stated in the individual monographs except where the label states that the tablets or capsules are intended for use as troches, or are to be chewed, or are designed to liberate the drug content gradually over a period of time or to release the drug over two or more separate periods with a distinct time interval between such release periods. Determine the type of units under test from the labeling and from observation, and apply the appropriate procedure to 6 or more dosage units.

For the purposes of this test, disintegration does not imply complete solution of the unit or even of its active constituent. Complete disintegration is defined as that state in which any residue of the unit, except fragments of insoluble coating or capsule shell, remaining on the screen of the test apparatus is a soft mass having no palpably firm core.

Apparatus

The apparatus[1] consists of a basket-rack assembly, a 1000-mL, low-form beaker for the immersion fluid, a thermostatic arrangement for heating the fluid between 35° and 39°, and a device for raising and lowering the basket in the immersion fluid at a constant frequency rate between 29 and 32 cycles per minute through a distance of not less than 5.3 cm and not more than 5.7 cm. The volume of the fluid in the vessel is such that at the highest point of the upward stroke the wire mesh remains at least 2.5 cm below the surface of the fluid and descends to not less than 2.5 cm from the bottom of the vessel on the downward stroke. The time required for the upward stroke is equal to the time required for the downward stroke, and the change in stroke direction is a smooth transition, rather than an abrupt reversal of motion. The basket-rack assembly moves vertically along its axis. There is no appreciable horizontal motion or movement of the axis from the vertical.

Basket-rack Assembly—The basket-rack assembly consists of six open-ended transparent tubes, each 7.75 ± 0.25 cm long and having an inside diameter of approximately 21.5 mm and a wall approximately 2 mm thick; the tubes are held in a vertical position by two plastic plates, each about 9 cm in diameter and 6 mm in thickness, with six holes, each about 24 mm in diameter, equidistant from the center of the plate and equally spaced from one another. Attached to the under surface of the lower plate is 10-mesh No. 23 (0.025-inch) W. and M. gauge woven stainless-steel

[1] A suitable apparatus, meeting these specifications, is available from laboratory supply houses, from Van-Kel Industries, Inc., 36 Meridian Rd., Edison, NJ 08820, or from Hanson Research Corp., P. O. Box 35, Northridge, CA 91324.

wire cloth having a plain square weave. The parts of the apparatus are assembled and rigidly held by means of three bolts passing through the two plastic plates. A suitable means is provided to suspend the basket-rack assembly from the raising and lowering device using a point on its axis.

The design of the basket-rack assembly may be varied somewhat provided the specifications for the glass tubes and the screen mesh size are maintained.

Disks[2]—Each tube is provided with a slotted and perforated cylindrical disk 9.5 ± 0.15 mm thick and 20.7 ± 0.15 mm in diameter. The disk is made of a suitable, transparent plastic material having a specific gravity of between 1.18 and 1.20. Five 2-mm holes extend between the ends of the cylinder, one of the holes being through the cylinder axis and the others parallel with it equally spaced on a 6-mm radius from it. Equally spaced on the sides of the cylinder are four notches that form V-shaped planes that are perpendicular to the ends of the cylinder. The dimensions of each notch are such that the openings on the bottom of the cylinder are 1.60 mm square and those on the top are 9.5 mm wide and 2.55 mm deep. All surfaces of the disk are smooth.

Procedure

Uncoated Tablets—Place 1 tablet in each of the six tubes of the basket, add a disk to each tube, and operate the apparatus, using water maintained at 37 ± 2° as the immersion fluid unless another fluid is specified in the individual monograph. At the end of the time limit specified in the monograph, lift the basket from the fluid, and observe the tablets: all of the tablets have disintegrated completely. If 1 or 2 tablets fail to disintegrate completely, repeat the test on 12 additional tablets: not less than 16 of the total of 18 tablets tested disintegrate completely.

Plain Coated Tablets—Place 1 tablet in each of the six tubes of the basket and, if the tablet has a soluble external coating, immerse the basket in water at room temperature for 5 minutes. Then add a disk to each tube, and operate the apparatus, using simulated gastric fluid TS maintained at 37 ± 2° as the immersion fluid. After 30 minutes of operation in simulated gastric fluid TS, lift the basket from the fluid, and observe the tablets. If the tablets have not disintegrated completely, substitute simulated intestinal fluid TS maintained at 37 ± 2° as the immersion fluid, and continue the test for a total period of time, including previous exposure to water and simulated gastric fluid TS, equal to the time limit specified in the individual monograph plus 30 minutes, lift the basket from the fluid, and observe the tablets: all of the tablets have disintegrated completely. If 1 or 2 tablets fail to disintegrate completely, repeat the test on 12 additional tablets: not less than 16 of the total of 18 tablets tested disintegrate completely.

Enteric-coated Tablets—Place 1 tablet in each of the six tubes of the basket and, if the tablet has a soluble external coating, immerse the basket in water at room temperature for 5 minutes. Then operate the apparatus, without adding the disks, using simulated gastric fluid TS maintained at 37 ± 2° as the immersion fluid. After 1 hour of operation in simulated gastric fluid TS, lift the basket from the fluid, and observe the tablets: the tablets show no evidence of disintegration, cracking, or softening. Then add a disk to each tube, and operate the apparatus, using simulated intestinal fluid TS maintained at 37 ± 2° as the immersion fluid, for a period of time equal to 2 hours plus the time limit specified in the individual monograph, or, where only an enteric-coated tablet is recognized, for only the time limit specified in the monograph. Lift the basket from the fluid, and observe the tablets: all of the tablets disintegrate completely. If 1 or 2 tablets fail to disintegrate completely, repeat the test on 12 additional tablets: not less than 16 of the total of 18 tablets tested disintegrate completely.

Buccal Tablets—Apply the test for *Uncoated Tablets*, but omit the use of the disks. After 4 hours, lift the basket from the fluid, and observe the tablets: all of the tablets have disintegrated. If 1 or 2 tablets fail to disintegrate completely, repeat the test on 12 additional tablets: not less than 16 of the total of 18 tablets tested disintegrate completely.

Sublingual Tablets—Apply the test for *Uncoated Tablets*, but omit the use of the disks. Observe the tablets within the time limit specified in the individual monograph: all of the tablets have disintegrated. If 1 or 2 tablets fail to disintegrate completely, repeat the test on 12 additional tablets: not less than 16 of the total of 18 tablets tested disintegrate completely.

Hard Gelatin Capsules—Apply the test for *Uncoated Tablets*, but omit the use of disks. In place of disks attach a removable 10-mesh wire cloth,[3] as described under *Basket-rack Assembly*, to the surface of the upper plate of the basket-rack assembly. Observe the capsules within the time limit specified in the individual monograph: all of the capsules have disintegrated except for fragments from the capsule shell. If 1 or 2 capsules fail to disintegrate completely, repeat the test on 12 additional capsules: not less than 16 of the total of 18 capsules tested disintegrate completely.

Soft Gelatin Capsules—Proceed as directed under *Hard Gelatin Capsules*.

[2] Disks meeting these specifications are obtainable from Van-Kel Industries, Inc.

[3] A suitable wire cloth cover is available as Van-Kel Industries Part TT-1030.

⟨711⟩ DISSOLUTION

This test is provided to determine compliance with the dissolution requirements where stated in the individual monograph for a tablet or capsule dosage form, except where the label states that the tablets are to be chewed. Requirements for *Dissolution* do not apply to soft gelatin capsules unless specified in the individual monograph. Where the label states that an article is enteric-coated, and a dissolution or disintegration test that does not specifically state that it applied to enteric-coated articles is included in the individual monograph, the test for *Delayed-release Articles* under *Drug Release* ⟨724⟩ is applied unless otherwise specified in the individual monograph. Of the types of apparatus described herein, use the one specified in the individual monograph.

Apparatus 1—The assembly consists of the following: a covered vessel made of glass or other inert, transparent material[1]; a motor; a metallic drive shaft; and a cylindrical basket. The vessel is partially immersed in a suitable water bath of any convenient size that permits holding the temperature inside the vessel at 37 ± 0.5° during the test and keeping the bath fluid in constant, smooth motion. No part of the assembly, including the environment in which the assembly is placed, contributes significant motion, agitation, or vibration beyond that due to the smoothly rotating stirring element. Apparatus that permits observation of the specimen and stirring element during the test is preferable. The vessel is cylindrical, with a hemispherical bottom. It is 160 mm to 175 mm high, its inside diameter is 98 mm to 106 mm, and its nominal capacity is 1000 mL. Its sides are flanged at the top. A fitted cover may be used to retard evaporation.[2] The shaft is positioned so that its axis is not more than 2 mm at any point from the vertical axis of the vessel and rotates smoothly and without significant wobble. A speed-regulating device is used that allows the shaft rotation speed to be selected and maintained at the rate specified in the individual monograph, within ±4%.

Shaft and basket components of the stirring element are fabricated of stainless steel, type 316 or equivalent, to the specifications shown in Figure 1. Unless otherwise specified in the individual monograph, use 40-mesh cloth. A basket having a gold coating 0.0001 inch (2.5 μm) thick may be used. The dosage unit is placed in a dry basket at the beginning of each test. The distance between the inside bottom of the vessel and the basket is maintained at 25 ± 2 mm during the test.

Apparatus 2—Use the assembly from *Apparatus 1*, except that a paddle formed from a blade and a shaft is used as the stirring element. The shaft is positioned so that its axis is not more than 2 mm at any point from the vertical axis of the vessel, and rotates smoothly without significant wobble. The blade passes through the diameter of the shaft so that the bottom of the blade is flush with the bottom of the shaft. The paddle conforms to the specifications shown in Figure 2. The distance of 25 ± 2 mm between the blade and the inside bottom of the vessel is maintained during

[1] The materials should not sorb, react, or interfere with the specimen being tested.

[2] If a cover is used, it provides sufficient openings to allow ready insertion of the thermometer and withdrawal of specimens.

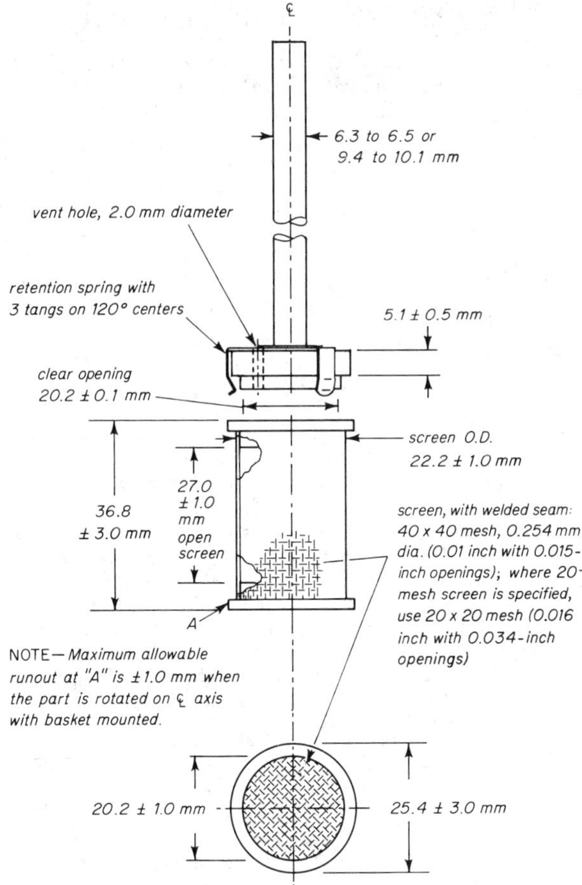

Fig. 1. Basket Stirring Element.

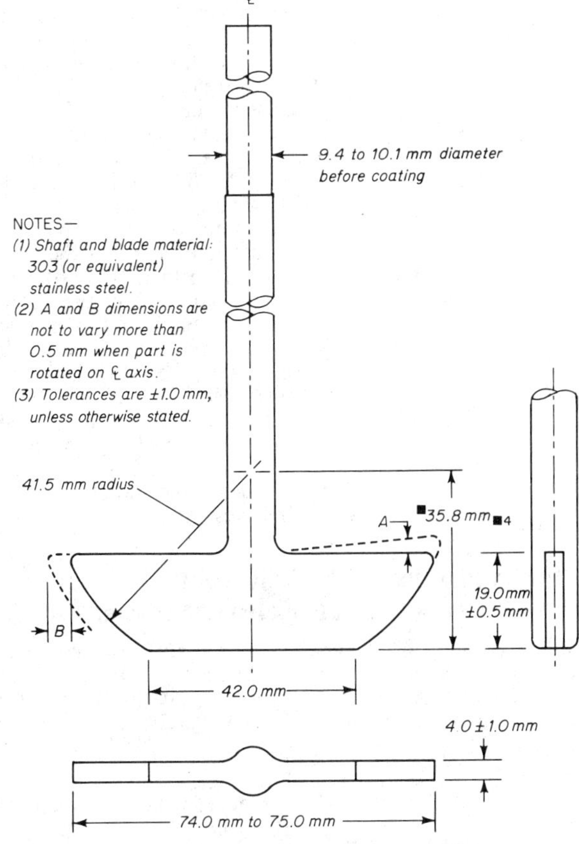

Fig. 2. Paddle Stirring Element.

the test. The metallic blade and shaft comprise a single entity that may be coated with a suitable inert coating. The dosage unit is allowed to sink to the bottom of the vessel before rotation of the blade is started. A small, loose piece of nonreactive material such as not more than a few turns of wire helix may be attached to dosage units that would otherwise float.

Apparatus Suitability Test—Individually test 1 tablet of the *USP Dissolution Calibrator, Disintegrating Type*[3] and 1 tablet of *USP Dissolution Calibrator, Nondisintegrating Type*,[3] according to the operating conditions specified. The apparatus is suitable if the results obtained are within the acceptable range stated in the certificate for that calibrator in the apparatus tested.

Dissolution Medium—Use the solvent specified in the individual monograph. If the *Dissolution Medium* is a buffered solution, adjust the solution so that its pH is within 0.05 unit of the pH specified in the individual monograph. [NOTE—Dissolved gases can cause bubbles to form which may change the results of the test. In such cases, dissolved gases should be removed prior to testing.]

Time—Where a single time specification is given, the test may be concluded in a shorter period if the requirement for minimum amount dissolved is met. If two or more times are specified, specimens are to be withdrawn only at the stated times, within a tolerance of ±2%.

Procedure for Capsules, Uncoated Tablets, and Plain Coated Tablets—Place the stated volume of the *Dissolution Medium* in the vessel of the apparatus specified in the individual monograph, assemble the apparatus, equilibrate the *Dissolution Medium* to 37 ± 0.5°, and remove the thermometer. Place 1 tablet or 1 capsule in the apparatus, taking care to exclude air bubbles from the surface of the dosage-form unit, and immediately operate the apparatus at the rate specified in the individual monograph. Within the time interval specified, or at each of the times stated, withdraw a specimen from a zone midway between the surface of the *Dissolution Medium* and the top of the rotating basket or blade, not less than 1 cm from the vessel wall. Perform the analysis as directed in the individual monograph. Repeat the test with additional dosage form units.

Where capsule shells interfere with the analysis, remove the contents of not less than 6 capsules as completely as possible, and dissolve the empty capsule shells in the specified volume of *Dissolution Medium*. Perform the analysis as directed in the individual monograph. Make any necessary correction. Correction factors greater than 25% of the labeled content are unacceptable.

Interpretation—Unless otherwise specified in the individual monograph, the requirements are met if the quantities of active ingredient dissolved from the units tested conform to the accompanying acceptance table. Continue testing through the three stages unless the results conform at either S_1 or S_2. The quantity, Q, is the amount of dissolved active ingredient specified in the individual monograph, expressed as a percentage of the labeled content; both the 5% and 15% values in the acceptance table are percentages of the labeled content so that these values and Q are in the same terms.

Acceptance Table

Stage	Number Tested	Acceptance Criteria
S_1	6	Each unit is not less than Q + 5%.
S_2	6	Average of 12 units ($S_1 + S_2$) is equal to or greater than Q, and no unit is less than Q − 15%.
S_3	12	Average of 24 units ($S_1 + S_2 + S_3$) is equal to or greater than Q, not more than 2 units are less than Q − 15%, and no unit is less than Q − 25%.

⟨721⟩ DISTILLING RANGE

To determine the range of temperatures within which an official liquid distils, or the percentage of the material that distils between two specified temperatures, use Method I or Method II

[3] Available from USP-NF Reference Standards, 12601 Twinbrook Parkway, Rockville, MD 20852.

as directed in the individual monograph. The *lower limit* of the range is the temperature indicated by the thermometer when the first drop of condensate leaves the tip of the condenser, and the *upper limit* is the Dry Point, i.e., the temperature at which the last drop of liquid evaporates from the lowest point in the distillation flask, without regard to any liquid remaining on the side of the flask, or the temperature observed when the proportion specified in the individual monograph has been collected.

[NOTE—Cool all liquids that distil below 80° to between 10° and 15° before measuring the sample to be distilled.]

Method I

Apparatus—Use apparatus similar to that specified for *Method II*, except that the distilling flask is of 50- to 60-mL capacity, and the neck of the flask is 10 to 12 cm long and 14 to 16 mm in internal diameter. The perforation in the upper asbestos board, if one is used, should be such that when the flask is set into it, the portion of the flask below the upper surface of the asbestos has a capacity of 3 to 4 mL.

Procedure—Proceed as directed for *Method II*, but place in the flask only 25 mL of the liquid to be tested.

Method II

Apparatus—Use an apparatus consisting of the following parts:

Distilling Flask—A round-bottom distilling flask, of heat-resistant glass, of 200-mL capacity, and having a total length of 17 to 19 cm and an inside neck diameter of 20 to 22 mm. Attached about midway on the neck, approximately 12 cm from the bottom of the flask, is a side-arm 10 to 12 cm long and 5 mm in internal diameter, which forms an angle of 70° to 75° with the lower portion of the neck.

Condenser—A straight glass condenser 55 to 60 cm in length with a water jacket about 40 cm in length, or a condenser of other design having equivalent condensing capacity. The lower end of the condenser may be bent to provide a delivery tube, or it may be connected to a bent adapter that serves as a delivery tube.

Asbestos Boards—Two pieces of asbestos board, 5 to 7 mm thick and 14 to 16 cm square, suitable for confining the heat to the lower part of the flask. Each board has a hole in its center, and the two boards differ only with respect to the diameter of the hole, i.e., the diameters are 4 and 10 cm, respectively. In use, the boards are placed one upon the other, and resting on a tripod or other suitable support, with the board having the larger hole on top.

Receiver—A 100-mL cylinder graduated in 1-mL subdivisions.

Thermometer—In order to avoid the necessity for an emergent stem correction, an accurately standardized, partial-immersion thermometer having the smallest practical subdivisions (not greater than 0.2°) is recommended. Suitable thermometers are available as the ASTM E-1 series 37C through 41C, and 102C through 107C (see *Thermometers* ⟨21⟩). When placed in position, the stem is located in the center of the neck and the top of the contraction chamber (or bulb, if 37C or 38C is used) is level with the bottom of the outlet to the side-arm.

Heat Source—A small Bunsen burner or an electric heater or mantle capable of adjustment comparable to that possible with a Bunsen burner.

Procedure—Assemble the apparatus, and place in the flask 100 mL of the liquid to be tested, taking care not to allow any of the liquid to enter the side-arm. Insert the thermometer, shield the entire burner and flask assembly from external air currents, and apply heat, regulating it so that between 5 and 10 minutes elapse before the first drop of distillate falls from the condenser. Continue the distillation at a rate of 4 to 5 mL of distillate per minute, collecting the distillate in the receiver. Note the temperature when the first drop of distillate falls from the condenser, and again when the last drop of liquid evaporates from the bottom of the flask or when the specified percentage has distilled over. Correct the observed temperature readings for any variation in the barometric pressure from the normal (760 mm), adding if the pressure is lower or subtracting if the pressure is higher than 760 mm, and apply the emergent stem correction where necessary. Unless otherwise specified in the individual monograph, allow 0.1° for each 2.7 mm (0.037° per mm) of variation.

⟨724⟩ DRUG RELEASE

This test is provided to determine compliance with drug-release requirements where specified in individual monographs. Use the apparatus specified in the individual monograph.

Apparatus 1, Apparatus 2, Apparatus Suitability Test, Dissolution Medium, and **Procedure**—Proceed as directed under *Dissolution* ⟨711⟩. [NOTE—Replace the aliquots withdrawn for analysis with equal volumes of fresh *Dissolution Medium* at 37° or, where it can be shown that replacement of the medium is not necessary, correct for the volume change in the calculation. Keep the vessel covered for the duration of the test, and verify the temperature of the mixture under test at suitable times.]

Extended-release Articles—General Drug Release Standard

Time—The test-time points, generally three, are expressed in terms of the labeled dosing interval, *D*, expressed in hours. Specimens are to be withdrawn within a tolerance of ±2% of the stated time.

Interpretation—Unless otherwise specified in the individual monograph, the requirements are met if the quantities of active ingredient dissolved from the units tested conform to *Acceptance Table 1*. Continue testing through the three levels unless the results conform at either L_1 or L_2. Limits on the amounts of active ingredient dissolved are expressed in terms of the percentage of labeled content. The limits embrace each value of Q_i, the amount dissolved at each specified fractional dosing interval.

Acceptance Table 1

Level	Number Tested	Criteria
L_1	6	No individual value lies outside each of the stated ranges and no individual value is less than the stated amount at the final test time.
L_2	6	The average value of the 12 units (L_1 + L_2) lies within each of the stated ranges and is not less than the stated amount at the final test time; none is more than 10% of labeled content outside each of the stated ranges; and none is more than 10% of labeled content below the stated amount at the final test time.
L_3	12	The average value of the 24 units (L_1 + L_2 + L_3) lies within each of the stated ranges, and is not less than the stated amount at the final test time; not more than 2 of the 24 units are more than 10% of labeled content outside each of the stated ranges; not more than 2 of the 24 units are more than 10% of labeled content below the stated amount at the final test time; and none of the units is more than 20% of labeled content outside each of the stated ranges or more than 20% of labeled content below the stated amount at the final test time.

Delayed-release (Enteric-coated) Articles—General Drug Release Standard

Use *Method A* or *Method B* and the apparatus specified in the individual monograph. Conduct the *Apparatus Suitability Test* as directed under *Dissolution* ⟨711⟩. All test times stated are to be observed within a tolerance of ±2%, unless otherwise specified.

Method A:

Procedure (unless otherwise directed in the individual monograph)—

Acid stage—Place 750 mL of 0.1 *N* hydrochloric acid in the vessel, and assemble the apparatus. Allow the medium to equilibrate to a temperature of 37 ± 0.5°. Place 1 tablet or 1 capsule in the apparatus, cover the vessel, and operate the apparatus for 2 hours at the rate specified in the monograph.

After 2 hours of operation in 0.1 *N* hydrochloric acid, withdraw an aliquot of the fluid, and proceed immediately as directed under *Buffer stage*.

Perform an analysis of the aliquot using the *Procedure* specified in the test for *Drug release* in the individual monograph.

Unless otherwise specified in the individual monograph, the requirements of this portion of the test are met if the quantities, based on the percentage of the labeled content, of active ingredient dissolved from the units tested conform to *Acceptance Table 2*. Continue testing through all levels unless the results of both acid and buffer stages conform at an earlier level.

Acceptance Table 2

Level	Number Tested	Criteria
A_1	6	No individual value exceeds 10% dissolved.
A_2	6	Average of the 12 units ($A_1 + A_2$) is not more than 10% dissolved, and no individual unit is greater than 25% dissolved.
A_3	12	Average of the 24 units ($A_1 + A_2 + A_3$) is not more than 10% dissolved, and no individual unit is greater than 25% dissolved.

Buffer stage—[NOTE—Complete the operations of adding the buffer, and adjusting the pH within 5 minutes.] With the apparatus operating at the rate specified in the monograph, add to the fluid in the vessel 250 mL of 0.20 *M* tribasic sodium phosphate that has been equilibrated to 37 ± 0.5°. Adjust, if necessary, with 2 *N* hydrochloric acid or 2 *N* sodium hydroxide to a pH of 6.8 ± 0.05. Continue to operate the apparatus for 45 minutes, or for the time specified in the individual monograph. At the end of the time period, withdraw an aliquot of the fluid, and perform the analysis using the *Procedure* specified in the test for *Drug release* in the individual monograph. The test may be concluded in a shorter time period than that specified for the *Buffer stage* if the requirement for minimum amount dissolved is met at an earlier time.

Interpretation—Unless otherwise specified in the individual monograph, the requirements are met if the quantities of active ingredient dissolved from the units tested conform to *Acceptance Table 3*. Continue testing through the three levels unless the results of both stages conform at an earlier level. The value of *Q* in *Acceptance Table 3* is 75% dissolved unless otherwise specified in the individual monograph. The quantity, *Q*, specified in the individual monograph, is the total amount of active ingredient dissolved in both the acid and buffer stages, expressed as a percentage of the labeled content. The 5% and 15% values in *Acceptance Table 3* are percentages of the labeled content so that these values and *Q* are in the same terms.

Acceptance Table 3

Level	Number Tested	Criteria
B_1	6	Each unit is not less than Q + 5%.
B_2	6	Average of 12 units ($B_1 + B_2$) is equal to or greater than Q, and no unit is less than Q − 15%.
B_3	12	Average of 24 units ($B_1 + B_2 + B_3$) is equal to or greater than Q, not more than 2 units are less than Q − 15%, and no unit is less than Q − 25%.

Method B:

Procedure (unless otherwise directed in the individual monograph)—

Acid stage—Place 1000 mL of 0.1 *N* hydrochloric acid in the vessel, and assemble the apparatus. Allow the medium to equilibrate to a temperature of 37 ± 0.5°. Place 1 tablet or 1 capsule in the apparatus, cover the vessel, and operate the apparatus for 2 hours at the rate specified in the monograph. After 2 hours of operation in 0.1 *N* hydrochloric acid, withdraw an aliquot of the fluid, and proceed immediately as directed under *Buffer stage*.

Perform an analysis of the aliquot using the *Procedure* specified in the test for *Drug release* in the individual monograph.

Unless otherwise specified in the individual monograph, the requirements of this portion of the test are met if the quantities, based on the percentage of the labeled content, of active ingredient dissolved from the units tested conform to *Acceptance Table 2* under *Method A*. Continue testing through all levels unless the results of both acid and buffer stages conform at an earlier level.

Buffer stage—[NOTE—For this stage of the procedure, use buffer that previously has been equilibrated to a temperature of 37 ± 0.5°.] Drain the acid from the vessel, and add to the vessel 1000 mL of pH 6.8 phosphate buffer, prepared by mixing 0.1 *N* hydrochloric acid with 0.20 *M* tribasic sodium phosphate (3:1) and adjusting, if necessary, with 2 *N* hydrochloric acid or 2 *N* sodium hydroxide to a pH of 6.8 ± 0.05. [NOTE—This may be accomplished also by removing from the apparatus the vessel containing the acid and replacing it with another vessel containing the buffer and transferring the dosage unit to the vessel containing the buffer.] Continue to operate the apparatus for 45 minutes, or for the time specified in the individual monograph. At the end of the time period, withdraw an aliquot of the fluid, and perform the analysis using the *Procedure* specified in the test for *Drug release* in the individual monograph. The test may be concluded in a shorter time period than that specified for the *Buffer stage* if the requirement for minimum amount dissolved is met at an earlier time.

Interpretation—Proceed as directed for *Interpretation* under *Method A*.

Transdermal Delivery Systems—General Drug Release Standards

Time—The test-time points, generally three, are expressed in terms of the labeled dosing interval, *D*, expressed in hours. Specimens are to be withdrawn within a tolerance of ±15 minutes or ±2% of the stated time, the tolerance that results in the narrowest time interval being selected.

Apparatus 3—

PADDLE OVER DISK—

APPARATUS—Use the paddle and vessel assembly from *Apparatus 2* as described under *Dissolution* ⟨711⟩, with the addition of a stainless steel disk assembly[1] designed for holding the transdermal system at the bottom of the vessel. The temperature is maintained at 32 ± 0.5°. A distance of 25 ± 2 mm between the paddle blade and the surface of the disk assembly is maintained during the test. The vessel may be covered during the test to minimize evaporation. The disk assembly for holding the transdermal system is designed to minimize any "dead" volume between the disk assembly and the bottom of the vessel. The disk assembly holds the system flat and is positioned such that the release surface is parallel with the bottom of the paddle blade (see Fig. 1).

Apparatus Suitability Test and Dissolution Medium—Proceed as directed for *Apparatus 2* under *Dissolution* ⟨711⟩.

Procedure—Place the stated volume of the *Dissolution Medium* in the vessel, assemble the apparatus without the disk assembly, and equilibrate the medium to 32 ± 0.5°. Apply the transdermal system to the disk assembly, assuring that the release surface of the system is as flat as possible. The system may be attached to the disk by applying a suitable adhesive[2] to the disk assembly. Dry for 1 minute. Press the system, release surface side up, onto the adhesive-coated side of the disk assembly. If

[1] Disk assembly (stainless support disk) may be obtained from Millipore Corp., Ashley Rd., Bedford, MA 01730.

Other appropriate devices may be used, provided they do not sorb, react with, or interfere with the specimen being tested.

[2] Use Dow Corning, 355 Medical Adhesive 18.5% in Freon 113, or the equivalent.

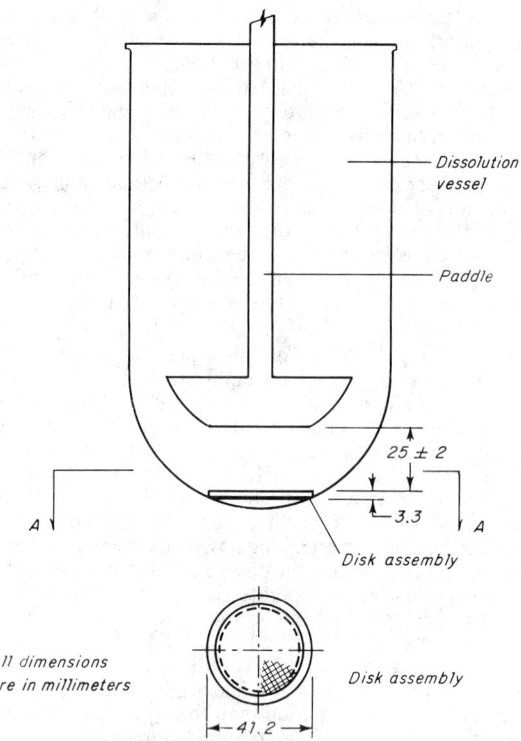

Fig. 1. Paddle Over Disk.

a membrane[3] is used to support the system, it is applied so that no air bubbles occur between the membrane and the release surface. Place the disk assembly flat at the bottom of the vessel with the release surface facing up and parallel to the edge of the paddle blade and surface of the *Dissolution Medium.* The bottom edge of the paddle is 25 ± 2 mm from the surface of the disk assembly. Immediately operate the apparatus at the rate specified in the monograph. At each sampling time interval, withdraw a specimen from a zone midway between the surface of the *Dissolution Medium* and the top of the blade, not less than 1 cm from the vessel wall. Perform the analysis on each sampled aliquot as directed in the individual monograph, correcting for any volume losses, as necessary. Repeat the test with additional transdermal systems.

Interpretation—Unless otherwise specified in the individual monograph, the requirements are met if the quantities of active ingredient released from the system conform to *Acceptance Table 4* for transdermal drug delivery systems. Continue testing through the three levels unless the results conform at either L_1 or L_2.

Acceptance Table 4

Level	Number Tested	Criteria
L_1	6	No individual values lies outside the stated range.
L_2	6	The average value of the 12 units (L_1 + L_2) lies within the stated range. No individual value is outside the stated range by more than 10% of the average of the stated range.
L_3	12	The average value of the 24 units (L_1 + L_2 + L_3) lies within the stated range. Not more than 2 of the 24 units are outside the stated range by more than 10% of the average of the stated range; and none of the units is outside the stated range by more than 20% of the average of the stated range.

[3] Use Cuprophan, Type 150 pm, 11 ± 0.5-μm thick, an inert, porous cellulosic material, which is available from ENKA AG, 1601 Castle Cove Circle, Corona DelMar, CA 92625, or LifeMed Corp., 2107 Delano Blvd., Compton, CA 90220.

Apparatus 4—Cylinder

APPARATUS—Use the vessel assembly from *Apparatus 1* as described under *Dissolution* ⟨711⟩, except to replace the basket and shaft with a stainless steel cylinder stirring element and to maintain the temperature at 32 ± 0.5° during the test. The shaft and cylinder components of the stirring element are fabricated of stainless steel to the specifications shown in Fig. 2. The dosage unit is placed on the cylinder at the beginning of each test. The distance between the inside bottom of the vessel and the cylinder is maintained at 25 ± 2 mm during the test.

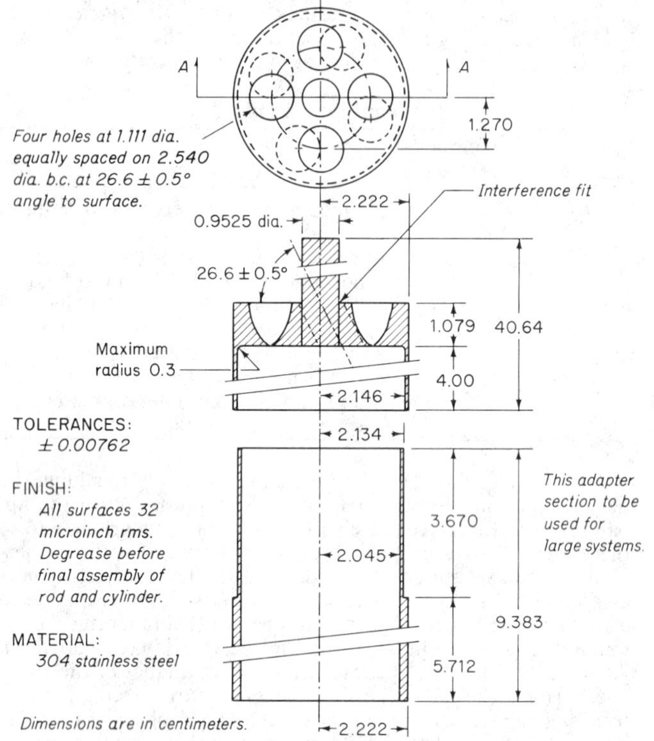

Fig. 2. Cylinder Stirring Element.[4]

Dissolution Medium—Use the medium specified in the individual monograph (see *Dissolution* ⟨711⟩).

Procedure—Place the stated volume of the *Dissolution Medium* in the vessel of the apparatus specified in the individual monograph, assemble the apparatus, and equilibrate the dissolution medium to 32 ± 0.5°. Unless otherwise directed in the individual monograph, prepare the test system prior to test as follows: Remove the protective liner from the system, and place the adhesive side on a piece of Cuprophan[3] that is not less than 1 cm larger on all sides than the system. Place the system, Cuprophan covered side down, on a clean surface, and apply a suitable adhesive[2] to the exposed Cuprophan borders. If necessary, apply additional adhesive to the back of the system. Dry for 1 minute. Carefully apply the adhesive-coated side of the system to the exterior of the cylinder such that the long axis of the system fits around the circumference of the cylinder. Press the Cuprophan covering to remove trapped air bubbles. Place the cylinder in the apparatus, and immediately rotate at the rate specified in the individual monograph. Within the time interval specified, or at each of the times stated, withdraw a quantity of *Dissolution Medium* for analysis from a zone midway between the surface of the *Dissolution Medium* and the top of the rotating cylinder, not less than 1 cm from the vessel wall. Perform the analysis as directed in the individual monograph, correcting for any volume losses as necessary. Repeat the test with additional transdermal drug delivery systems.

Interpretation—Unless otherwise specified in the individual monograph, the requirements are met if the quantities of active ingredient released from the system conform to *Acceptance Table*

[4] The cylinder stirring element is available from Accurate Tool, Inc., 25 Diaz St., Stamford, CT 06907, or from Van-Kel Industries, Inc., 36 Meridian Rd., Edison, NJ 08820.

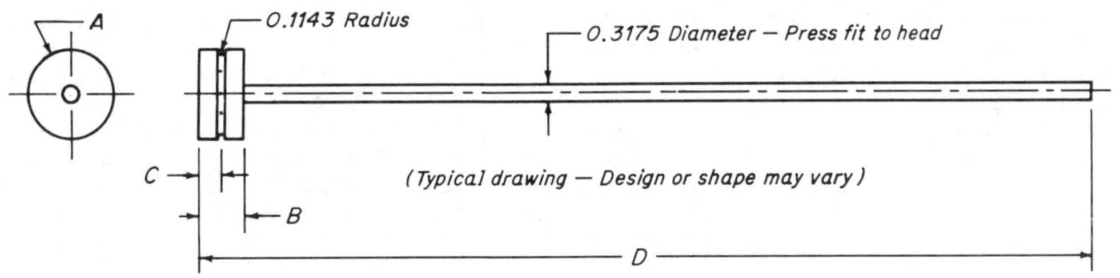

	HEAD				ROD		O-RING
System[a]	A (Diameter)	B	C	Material[b]	D	Material[c]	(not shown)
1.6 cm²	1.428	0.9525	0.4750	SS/VT	30.48	SS/P	Parker 2-113-V884-75
2.5 cm²	1.778	0.9525	0.4750	SS/VT	30.48	SS/P	Parker 2-016-V884-75
5 cm²	2.6924	0.7620	0.3810	SS/VT	8.890	SS/P	Parker 2-022-V884-75
7 cm²	3.1750	0.7620	0.3810	SS/VT	30.48	SS/P	Parker 2-124-V884-75
10 cm²	5.0292	0.6350	0.3505	SS/VT	31.01	SS/P	Parker 2-225-V884-75

[a] Typical system sizes.
[b] SS/VT = Either stainless steel or virgin Teflon.
[c] SS/P = Either stainless steel or Plexiglas.

Fig. 3. Reciprocating Disk Sample Holder.[6]

4 for transdermal drug delivery systems. Continue testing through the three levels unless the results conform at either L_1 or L_2.

Apparatus 5—Reciprocating Disk—

APPARATUS—The assembly consists of a set of volumetrically calibrated or tared solution containers made of glass or other suitable inert material,[5] a motor and drive assembly to reciprocate the system vertically and to index the system horizontally to a different row of vessels automatically if desired, and a set of disk-shaped sample holders (see Fig. 3). The solution containers are partially immersed in a suitable water bath of any convenient size that permits maintaining the temperature inside the containers at 32 ± 0.5° during the test. No part of the assembly, including the environment in which the assembly is placed, contributes significant motion, agitation, or vibration beyond that due to the smooth, vertically reciprocating sample holder. Apparatus that permits observation of the system and holder during the test is preferable. Use the size container and sample holder as specified in the individual monograph.

Dissolution Medium—Use the dissolution medium specified in the individual monograph (see *Dissolution* ⟨711⟩).

Procedure—Remove the transdermal system from its backing. Press the system onto a dry, unused piece of Cuprophan[3] or equivalent with the adhesive side against the Cuprophan, taking care to eliminate air bubbles between the Cuprophan and the release surface. Attach the system to a suitable size sample holder with a suitable O-ring such that the back of the system is adjacent to and centered on the bottom of the sample holder. Trim the excess Cuprophan with a sharp blade. Suspend each sample holder from a vertically reciprocating shaker such that each system is continuously immersed in an accurately measured volume of *Dissolution Medium* within a calibrated container pre-equilibrated to 32 ± 0.5°. Reciprocate at a frequency of about 30 cycles per minute with an amplitude of about 1.9 cm for the specified time in the medium specified for each time point. Perform the analysis as directed in the individual monograph. Repeat the test with additional transdermal drug delivery systems.

Interpretation—Unless otherwise specified in the individual monograph, the requirements are met if the quantities of active ingredient released from the system conform to *Acceptance Table 4* for transdermal drug delivery systems. Continue testing through the three levels unless the results conform at either L_1 or L_2.

[5] The materials should not sorb, react with, or interfere with the specimen being tested.
[6] The reciprocating disk sample holder may be purchased from ALZA Corp., 950 Page Mill Rd., Palo Alto, CA 94304 or Van-Kel Industries, Inc.

⟨726⟩ ELECTROPHORESIS

Electrophoresis refers to the migration of electrically charged proteins, colloids, molecules, or other particles when dissolved or suspended in an electrolyte through which an electric current is passed.

Based upon the type of apparatus used, electrophoretic methods may be divided into two categories, one called *free solution* or moving boundary electrophoresis and the other called *zone electrophoresis*.

In the *free solution* method, a buffered solution of proteins in a U-shaped cell is subjected to an electric current which causes the proteins to form a series of layers in order of decreasing mobility, which are separated by boundaries. Only a part of the fastest moving protein is physically separated from the other proteins, but examination of the moving boundaries using a schlieren optical system provides data for calculation of mobilities and information on the qualitative and quantitative composition of the protein mixture.

In *zone electrophoresis*, the sample is introduced as a narrow zone or spot in a column, slab, or film of buffer. Migration of the components as narrow zones permits their complete separation. Remixing of the separated zones by thermal convection is prevented by stabilizing the electrolyte in a porous matrix such as a powdered solid, or a fibrous material such as paper, or a gel such as starch, agar, or polyacrylamide.

Various methods of zone electrophoresis are widely employed. *Gel electrophoresis*, particularly the variant called *disk electrophoresis*, is especially useful for protein separation because of its high resolving power.

Gel electrophoresis, which is employed by the compendium, is discussed in more detail following the presentation of some theoretical principles and methodological practices, which are shared in varying degrees by all electrophoretic methods.

The electrophoretic migration observed for particles of a particular substance depends on characteristics of the particle, primarily its electrical charge, its size or molecular weight, and its shape, as well as characteristics and operating parameters of the system. These latter include the pH, ionic strength, viscosity and temperature of the electrolyte, density or cross-linking of any stabilizing matrix such as gel, and the voltage gradient employed.

Effect of charge, particle size, electrolyte viscosity, and voltage gradient—Electrically charged particles migrate toward the electrode of opposite charge, and molecules with both positive and negative charges move in a direction dependent on the net charge. The rate of migration is directly related to the magnitude of the net charge on the particle and is inversely related to the size of the particle, which in turn is directly related to its molecular weight.

Very large spherical particles, for which Stokes' law is valid, exhibit an electrophoretic mobility, u_0, which is inversely related to the first power of the radius as depicted in the equation

$$u_0 = \frac{v}{E} = \frac{Q}{6\pi r\eta},$$

where v is the velocity of the particle, E is the voltage gradient imposed on the electrolyte, Q is the charge on the particle, r is the particle radius, and η is the viscosity of the electrolyte. This idealized expression is strictly valid only at infinite dilution and in the absence of a stabilizing matrix such as paper or a gel.

Ions, and peptides up to molecular weights of at least 5000, particularly in the presence of stabilizing media, do not obey Stokes' law, and their electrophoretic behavior is best described by an equation of the type

$$u_0 = \frac{Q}{A\pi r^2\eta}$$

where A is a shape factor generally in the range of 4 to 6, which shows an inverse dependence of the mobility on the square of the radius. In terms of molecular weight, this implies an inverse dependence of mobility on the ⅔ power of the molecular weight.

Effect of pH—The direction and rate of migration of molecules containing a variety of ionizable functional groups, such as amino acids and proteins, depends upon the pH of the electrolyte. For instance, the mobility of a simple amino acid such as glycine varies with pH approximately as shown in Figure 1. The pK_a values of 2.2 and 9.9 coincide with the inflection points of the sigmoid portions of the plot. Since the respective functional groups are 50% ionized at the pH values where pH = pK_a, the electrophoretic mobilities at these points are half of the value observed for the fully ionized cation and anion obtained at very low and very high pH, respectively. The zwitterion that exists at the intermediate pH range is electrically neutral and has zero mobility.

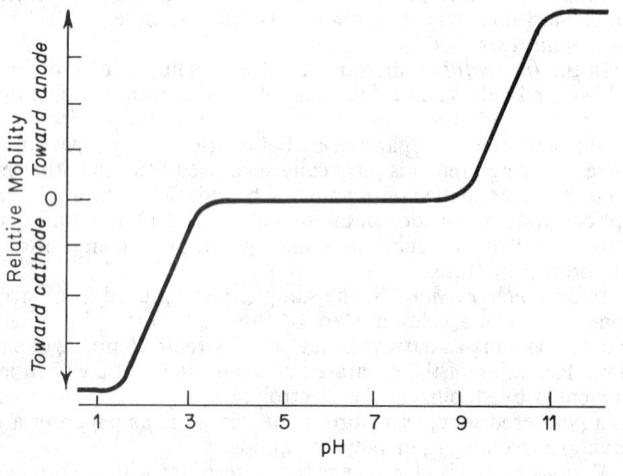

Fig. 1.

Effect of ionic strength and temperature—Electrophoretic mobility decreases with increasing ionic strength of the supporting electrolyte. Ionic strength, μ, is defined as

$$\mu = 0.5\Sigma C_i Z_i^2,$$

where C_i is the concentration of an ion in moles per liter and Z_i is its valence, and the sum is calculated for all ions in the solution. For buffers in which both the anion and cation are univalent, ionic strength is identical with molarity.

Ionic strengths of electrolytes employed in electrophoresis commonly range from about 0.01 to 0.10. A suitable strength is somewhat dependent on the sample composition, since the buffer capacity must be great enough to maintain a constant pH over the area of the component zones. Zones become sharper or more compact as ionic strength is increased.

Temperature affects mobility indirectly, since the viscosity, η, of the supporting electrolyte is temperature-dependent. The viscosity of water decreases at a rate of about 3% per °C in the range of 0° to 5° and at a slightly lower rate in the vicinity of room temperature. Mobility, therefore, increases with increasing electrolyte temperature.

Considerable heat is evolved as a result of current passing through the supporting electrolyte. This heat increases with the applied voltage and with increasing ionic strength. Particularly in larger apparatus, despite the circulation of a coolant, this heat produces a temperature gradient across the bed which may lead to distortion of the separated zones. Therefore, practical considerations and the design of the particular apparatus dictate the choice of ionic strength and operating voltage.

Effect of a stabilizing medium, electroosmosis—When an electrical current is passed through an electrolyte contained in a glass tube or contained between plates of glass or plastic, a bulk flow of the electrolyte toward one of the electrodes is observed. This flow is called electroosmosis. It results from the surface charge on the walls of the apparatus, which arises either from ionizable functional groups inherent in the structural material or from ions adsorbed on the cell walls from the electrolyte contacting them. The effect is usually increased when the cell is filled with a bed of porous substance, such as a gel, used to stabilize the supporting electrolyte and prevent remixing of separated zones by thermal convection or diffusion. The solution immediately adjacent to the surface builds up an electrical charge, equal but opposite to the surface charge, and the electrical field traversing the cell produces a movement of solution toward the electrode of opposite charge.

The substances commonly used as stabilizing media in zone electrophoresis develop a negative surface charge, and therefore electroosmotic flow of the electrolyte is toward the cathode. As a result, all zones, including neutral substances, are carried toward the cathode during the electrophoretic run.

The degree of electroosmosis observed varies with the stabilizing substance. It is appreciable with agar gel, while it is negligibly small with polyacrylamide gel.

Molecular sieving—In the absence of a stabilizing medium or in cases where the medium is very porous, electrophoretic separation of molecules results from differences in the ratio of their electrical charge to their size. In the presence of a stabilizing medium, differences in adsorptive or other affinity of molecules for the medium introduces a chromatographic effect that may enhance the separation.

If the stabilizing medium is a highly cross-linked gel such that the size of the resultant pores is of the order of the dimensions of the molecules being separated, a molecular sieving effect is obtained. This effect is analogous to that obtained in separations based on gel permeation or molecular exclusion chromatography, but in gel electrophoresis the effect is superimposed on the electrophoretic separation. Molecular sieving may be visualized to result from a steric barrier to the passage of larger molecules. Small molecules pass through pores of a wide size range, and therefore their electrophoretic passage through the gel will not be impeded. As size increases, fewer pores will permit passage of the molecules, causing a retardation of the migration of substances of large molecular weight.

Gel Electrophoresis

Processes employing a gel such as agar, starch, or polyacrylamide as a stabilizing medium are broadly termed gel electrophoresis. The method is particularly advantageous for protein separations. The separation obtained depends upon the electrical charge to size ratio coupled with a molecular sieving effect dependent primarily on the molecular weight.

Polyacrylamide gel has several advantages that account for its extensive use. It has minimal adsorptive properties and produces a negligible electroosmotic effect. Gels of a wide range of pore size can be reproducibly prepared by varying the total gel concentration (based on monomer plus cross-linking agent) and the percentage of cross-linking agent used to form the gel. These quantities are conveniently expressed as

$$T(\%) = \frac{a + b}{V} 100,$$

$$C(\%) = \frac{b}{a + b} 100,$$

where T is the total gel concentration in %, C is the percentage of cross-linking agent used to prepare the gel, V is the volume, in mL, of buffer used in preparing the gel, and a and b are the weights, in g, of monomer (acrylamide) and cross-linking agent (usually *N,N′*-methylenebisacrylamide) used to prepare the gel. Satisfactory gels ranging in concentration (T) from about 3% to 30% have been prepared. The amount of cross-linking agent is usually about one-tenth to one-twentieth of the quantity of monomer (C = 10% to 5%), a smaller percentage being used for higher values of T.

In the preparation of the gel, the bed of the electrophoresis apparatus is filled with an aqueous solution of monomer and cross-linking agent, usually buffered to the pH desired in the later run, and polymerized in place by a free radical process. Polymerization may be initiated by a chemical process, frequently using ammonium persulfate plus *N,N,N′,N′*-tetramethylenediamine or photochemically using a mixture of riboflavin and *N,N,N′,N′*-tetramethylenediamine. Polymerization is inhibited by molecular oxygen and by acidic conditions. The gel composition and polymerization conditions chosen must be adhered to rigorously to ensure reproducible qualities of the gel.

Apparatus for Gel Electrophoresis—In general, the bed or medium in which electrophoresis is carried out may be supported horizontally or vertically, depending upon the design of the apparatus. A series of separations to be compared may also be carried out in several individual tubes or by placing different samples in adjacent wells, cast or cut into a single slab of gel. A vertical slab assembly such as that depicted schematically in Figure 2 is convenient for direct comparison of several samples. A particular advantage derives from the comparison of the samples in a single bed of gel which is likely to be more uniform in composition than gels cast in a series of chambers.

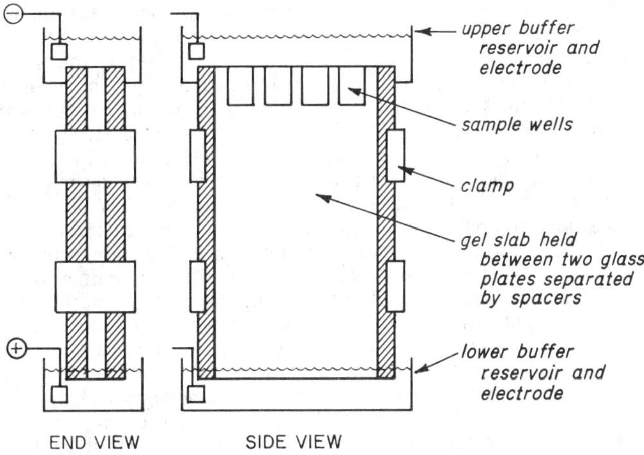

Fig. 2. Vertical Slab Gel Electrophoresis Apparatus.

A feature of many types of apparatus, not illustrated in the schematic view, seals the lower buffer chamber to the base of the bed and allows the level of the buffer in the lower chamber to be made equal to that in the upper chamber, thereby eliminating hydrostatic pressure on the gel. In addition, some units provide for the circulation of coolant on one or both sides of the gel bed.

In the preparation of the gel, the base of the gel chamber is closed with a suitable device and the unit is filled with the solution of monomer, cross-linking agent, and catalyst. A comb, having teeth of an appropriate size, is inserted in the top, and polymerization is allowed to proceed to completion. Removal of the comb leaves a series of sample wells in the polymerized gel.

In simple gel electrophoresis, an identical buffer is used to fill the upper and lower buffer chambers as well as in the solution used to prepare the gel. After filling the chambers, the samples, dissolved in sucrose or other dense and somewhat viscous solution to prevent diffusion, are introduced with a syringe or micropipet into the bottoms of the sample wells, and the electrophoresis is begun immediately thereafter.

DISK ELECTROPHORESIS

An important variant of polyacrylamide gel electrophoresis, which employs a discontinuous series of buffers and often also a discontinuous series of gel layers, is called disk electrophoresis. The name is derived from the discoid shape of the very narrow zones that result from the technique. As a result of the narrow zones produced, this technique exhibits an extremely high resolving power and is to be recommended for the characterization of protein mixtures and for the detection of contaminants that may have mobilities close to that of the major component.

The basis of disk electrophoresis is outlined in the following paragraphs with reference to an anionic system suitable for separating proteins bearing a net negative charge. To understand disk electrophoresis, it is essential to have a knowledge of the general aspects of electrophoresis and the apparatus already described.

Basis of Disk Electrophoresis—The high resolution obtained in disk electrophoresis depends on the use of a buffer system that is discontinuous with respect to both pH and composition. This is usually combined with a discontinuous series of two or three gels that differ in density.

A typical system is illustrated schematically in Figure 3.

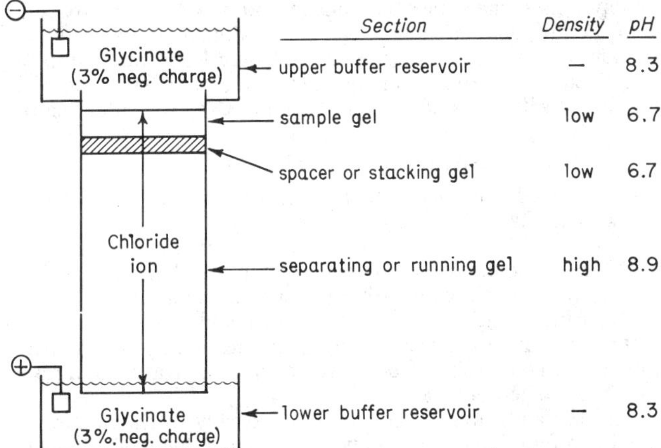

Fig. 3. Terminology, Buffer pH, and Buffer Composition for Acrylamide Gel Disk Electrophoresis.

A high density (T = 10% to 30%) separating gel several centimeters high is polymerized in a tris-chloride buffer in the bed of the apparatus. During polymerization the buffer is overlayered with a thin layer of water to prevent fixation of a meniscus in the top of the gel. The overlayer of water is then removed and a thin layer, 3 mm to 10 mm thick, of low density (T = 3%) gel, called the spacer or stacking gel, is polymerized in a tris-chloride buffer on top of the separating gel. An overlayer of water is again used to ensure a flat surface. The sample is mixed with a small amount of the spacer gel monomer solution which is applied on top of the spacer gel and allowed to polymerize. The pH of the separating gel is typically 8.9, while that of the spacer and sample gels is 6.7. All three gels are prepared using chloride as the anion.

The upper and lower buffer reservoirs are filled with a pH 8.3 buffer prepared from tris and glycine. At this pH about 3% of the glycine molecules bear a net negative charge.

When a voltage is applied across the system, the glycinate-chloride interface moves downward toward the anode. It was initially positioned at the junction of the buffer in the upper reservoir and the top of the sample gel layer. The chloride anion, by virtue of its small size, migrates faster than any of the proteins present in the sample. The pH of the sample and spacer layers was chosen to be about 3 units below the higher pK_a of glycine. Therefore, in traversing these layers, only about 0.1% of the glycine molecules bear a net negative charge. Consequently, glycine migrates more slowly than chloride. The tendency for the faster-moving chloride to move away from glycinate lowers the concentration at the interface, producing a greater voltage drop at the interface, which in turn causes the glycinate to catch up to the chloride. Under these conditions, a very sharp interface is maintained, and as it moves through the sample and spacer layers, the proteins in the sample tend to stack themselves at the interface in very thin layers in order of mobility. The process is called stacking and is the source of the disks which are separated.

When the stacked proteins reach the high-density separating gel, they are slowed down by a molecular sieving process. The higher pH encountered in the running gel also causes the glycinate to migrate faster, so that the discontinuous buffer interface overtakes the proteins and eventually reaches the bottom of the separating gel. During this period, the disks of protein continue to separate by electrophoresis and molecular sieving in the separating gel. At the end of the run, the pH of the separating gel will have risen above its original value of 8.9 to a value of about pH 9.5.

Relative Mobility—Bromophenol blue is often used as a standard for calculating the relative mobility of separated zones and to judge visually the progress of a run. It may be added to one of the sample wells, or mixed with the sample itself, or simply added to the buffer in the upper sample reservoir.

Relative mobility, M_B, is calculated as

$$M_B = \frac{\text{distance from origin to sample zone}}{\text{distance from origin to bromophenol blue zone}}.$$

Visualization of Zones—Since polyacrylamide is transparent, protein bands may be located by scanning in a densitometer with ultraviolet light. The zones may be fixed by immersing in protein precipitants such as phosphotungstic acid or 10% trichloroacetic acid. A variety of staining reagents including naphthalene black (amido black) and Coomassie brilliant blue R250 may be used. The fixed or stained zones may be conveniently viewed and photographed with transmitted light from an X-ray film illuminator.

SAFETY PRECAUTIONS

Voltages used in electrophoresis can readily deliver a lethal shock. The hazard is increased by the use of aqueous buffer solutions and the possibility of working in damp environments.

The equipment, with the possible exception of the power supply, should be enclosed in either a grounded metal case or a case made of insulating material. The case should have an interlock that deenergizes the power supply when the case is opened, after which reactivation should be prevented until activation of a reset switch is carried out.

High-voltage cables from the power supply to the apparatus should preferably be a type in which a braided metal shield completely encloses the insulated central conductor, and the shield should be grounded. The base of the apparatus should be grounded metal or contain a grounded metal rim which is constructed in such a way that any leakage of electrolyte will produce a short which will deenergize the power supply before the electrolyte can flow beyond the protective enclosure.

If the power supply contains capacitors as part of a filter circuit, it should also contain a bleeder resistor to ensure discharge of the capacitors before the protective case is opened. A shorting bar that is activated by opening the case may be considered as an added precaution.

Because of the potential hazard associated with electrophoresis, laboratory personnel should be completely familiar with electrophoresis equipment before using it.

⟨731⟩ LOSS ON DRYING

The procedure set forth in this chapter determines the amount of volatile matter of any kind that is driven off under the conditions specified. For substances appearing to contain water as the only volatile constituent, the procedure given in the chapter, *Water Determination* ⟨921⟩, is appropriate, and is specified in the individual monograph.

Mix and accurately weigh the substance to be tested, and, unless otherwise directed in the individual monograph, conduct the determination on 1 to 2 g. If the test specimen is in the form of large crystals, reduce the particle size to about 2 mm by quickly crushing. Tare a glass-stoppered, shallow weighing bottle that has been dried for 30 minutes under the same conditions to be employed in the determination. Put the test specimen in the bottle, replace the cover, and accurately weigh the bottle and the contents. By gentle, sidewise shaking, distribute the test specimen as evenly as practicable to a depth of about 5 mm generally, and not more than 10 mm in the case of bulky materials. Place the loaded bottle in the drying chamber, removing the stopper and leaving it also in the chamber. Dry the test specimen at the temperature and for the time specified in the monograph. [NOTE—The temperature specified in the monograph is to be regarded as being within the range of ±2° of the stated figure.] Upon opening the chamber, close the bottle promptly, and allow it to come to room temperature in a desiccator before weighing.

If the substance melts at a lower temperature than that specified for the determination of *Loss on drying*, maintain the bottle with its contents for 1 to 2 hours at a temperature 5° to 10° below the melting temperature, then dry at the specified temperature.

Where the specimen under test is Capsules, use a portion of the mixed contents of not less than 4 capsules.

Where the specimen under test is Tablets, use powder from not less than 4 tablets ground to a fine powder.

Where the individual monograph directs that loss on drying be determined by thermogravimetric analysis, a sensitive electrobalance is to be used.

Where drying in vacuum over a desiccant is directed in the individual monograph, a vacuum desiccator or a vacuum drying pistol, or other suitable vacuum drying apparatus, is to be used.

Where drying in a desiccator is specified, exercise particular care to ensure that the desiccant is kept fully effective by frequent replacement.

Where drying in a capillary-stoppered bottle in vacuum is directed in the individual monograph, use a bottle or tube fitted with a stopper having a 225 ± 25 μm diameter capillary, and maintain the heating chamber at a pressure of 5 mm or less of mercury. At the end of the heating period, admit dry air to the heating chamber, remove the bottle, and with the capillary stopper still in place allow it to cool in a desiccator before weighing.

⟨733⟩ LOSS ON IGNITION

This procedure is provided for the purpose of determining the percentage of test material that is volatilized and driven off under the conditions specified. The procedure, as generally applied, is nondestructive to the substance under test; however, the substance may be converted to another form such as an anhydride.

Perform the test on finely powdered material, and break up lumps, if necessary, with the aid of a mortar and pestle before weighing the specimen. Weigh the specimen to be tested without further treatment, unless a preliminary drying at a lower temperature, or other special pretreatment, is specified in the individual monograph. Unless other equipment is designated in the individual monograph, conduct the ignition in a suitable muffle furnace or oven that is capable of maintaining a temperature within 25° of that required for the test, and use a suitable crucible, complete with cover, previously ignited for 1 hour at the temperature specified for the test, cooled in a desiccator, and accurately weighed.

Unless otherwise directed in the individual monograph, transfer to the tared crucible an accurately weighed quantity, in g, of the substance to be tested, about equal to that calculated by the formula:

$$10/L,$$

in which L is the limit (or the mean value of the limits) for *Loss on ignition*, in percentage. Ignite the loaded uncovered crucible, and cover at the temperature (±25°) and for the period of time designated in the individual monograph. Ignite for successive 1-hour periods where ignition to constant weight is indicated. Upon completion of each ignition, cover the crucible, and allow it to cool in a desiccator to room temperature before weighing.

⟨736⟩ MASS SPECTROMETRY

Mass spectrometers can be used for the measurement of ionic mass-to-charge ratio, for the determination of ionic abundance, and for the study of the ionization process. In addition, studies of ionic reactions in the gas phase such as unimolecular decomposition processes, and ion molecule reactions, are also possible.

A mass spectrometer is an instrument that produces a beam of ions from a substance under investigation, sorts these ions into a spectrum according to their mass-to-charge ratio (m/z), and records the relative abundance of each ionic species present. Traditionally only the positive ions have been studied, principally because the negative ion yield from electron impact (EI) sources is normally low. With the introduction of the chemical ionization (CI) and fast atom bombardment (FAB) techniques, both of which can produce a high negative ion yield, interest in the analysis of negative ions has increased.

In general, a mass spectrometer consists of three major components, as shown in the accompanying figure: an ion source for producing gaseous ions from the substance(s) being studied; an analyzer for resolving the ions into their characteristic mass components according to the mass-to-charge ratios of the ions present; and a detector system for recording the relative abundance or intensity of each of the resolved ionic species present. In addition, a sample introduction system is necessary in order to admit the samples to be studied to the ion source while still maintaining the high vacuum requirements ($\sim 10^{-6}$ to 10^{-8} torr) of the technique. As the accompanying figure indicates, most commercial instruments include a computer for conveniently handling the large amounts of data produced by these instruments.

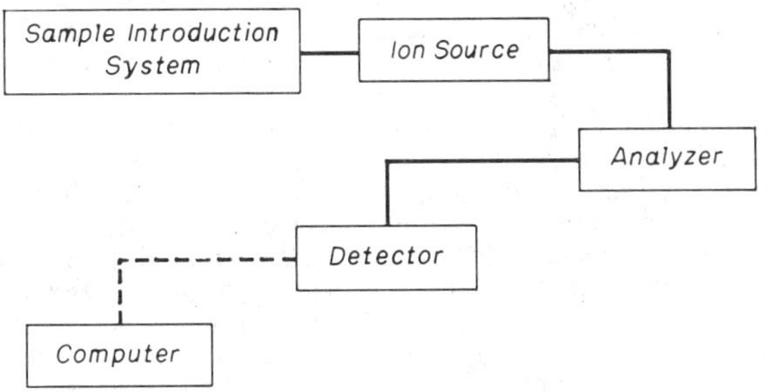

Analyzers—The mass analyzer sorts the different masses present in the ionized sample, and this allows one to determine the mass and ultimately the abundance or relative intensity of each ionic species present. Four of the several methods that are commonly used for analysis are (1) the quadrupole, (2) the magnetic analyzer, (3) the time-of-flight analyzer, and (4) the fourier transform analyzer. Electrostatic analyzers are often used in conjunction with other mass analyzers.

In the quadrupole, mass separation may be achieved in an instrument composed of four coaxial rods; ideally each rod possesses a hyperbolic cross section, but in practice circular rods are commonly used. Two opposite rods have a fixed electric potential (U), the other two have a radio-frequency alternating potential (V,Ω). Under the action of these electric fields, all of the ions (except one selected mass) in the ion beam are deflected to the sides and lost; the selected mass (determined by the settings of U,V,Ω) is allowed through the rods. Thus, because all other m/z values but the selected one are rejected, the analyzer is sometimes called a *mass filter*. The theory is that as high mass ions take longer to traverse the analyzer and consequently have more opportunity both to fragment and to collide with residual gas molecules, sensitivity decreases with increasing mass: this phenomenon is called mass discrimination, and it is a fundamental characteristic of quadrupole analyzers.

In the presence of a magnetic field perpendicular to the motion of the positive ion beam, each ion experiences a force at right angles to both its direction of motion and the direction of the magnetic field, thereby deflecting the beam of ions. The following equation of motion applies:

$$m/z = H^2r^2/2V,$$

in which m is the mass in atomic mass units; z is the number of electronic charges; H is the magnetic field strength in gauss; r is the ion trajectory radius in centimeters; and V is the accelerating voltage. The mass spectrum is scanned by varying the strength of the magnetic field and detecting those ions passing through the exit slit as they come into "focus."

In the time-of-flight analyzer, separation of ions of different masses is based on all ions being given equal energy; therefore, ions of different masses have different velocities. If there is a fixed distance for the ions to travel, the time of their travel will vary with their mass, the lighter masses traveling more rapidly and thereby reaching the detector in a shorter period of time. The time of flight is given by:

$$t(f) = k\sqrt{m/z},$$

in which $t(f)$ is the time of flight in seconds, and m and z are the same as defined previously. Thus, the time-of-flight of the various ions is simply proportional to the square root of the mass-to-charge ratio of the ions.

Fourier transform mass spectrometry (FT–MS) is a technique based on the cyclotron motion of ions in a uniform magnetic field. In such a field of flux density B, ions are constrained to move in circular (cyclotron) orbits. The angular frequency, ω, of the cyclotron motion is given by the equation:

$$\omega = \frac{z \times B}{m}.$$

In the cyclotron resonance mass spectrometer, the cyclotron orbits can be expanded by subjecting the ions to an alternating electric field. When the frequency of the signal generator matches the cyclotron frequency, the ions are steadily accelerated to larger and larger radii leading to a coherent motion (of an ensemble of ions) corresponding to a significant amount of kinetic energy. After excitation is turned off, the cyclotronic ions give rise to an alternating image current on the electrodes, which is amplified. A frequency analysis of the corresponding receiver signal yields the mass of the ions involved with high precision. Thus, the Fourier transform of the time domain transient signal yields the corresponding frequency spectrum from which the mass spectrum is computed.

Ionization Techniques—Positive ions may be produced by passing a beam of electrons through a gas at pressures of about 10^{-4} to 10^{-6} mm (Hg). Pressures other than these may be employed, but this range is the most common. The energy of the electron beam is usually controlled. If the energy is greater than the ionization potential of the gas, the electrons may cause ionization and/or fragmentation of the gas molecules, represented as follows:

$$e^- + M \rightarrow M^{+\bullet} + 2e^-.$$

Sources of this type are called electron bombardment (or electron impact, EI) sources. The electrons are usually emitted from a heated tungsten or rhenium filament.

Ions formed in the ionization chamber are accelerated through the source exit slit toward the analyzer region by a repeller/draw out field determined partly by field penetration through the source exit slit, and partly by a small potential applied to an ion repeller plate in the ionization chamber. The ions are further accelerated by the much larger field existing between the ionization chamber and the source exit slits, the final slits being at ground potential.

Conversion of a mass spectrometer to the negative ion mode is straightforward, and modern equipment is designed to execute the procedure automatically on selection of a single parameter. All that is required, in theory, is to reverse all operating voltages and fields. While negative ions are also formed in the various ionization processes discussed, the introduction of a sample with a high electron capture cross section leads to the formation of abundant negative ions. For this reason, multi-halide derivatives of compounds to be studied are often prepared.

Negative ion MS studies have been successfully applied to pesticide analyses, since their structures are favorable for the technique.

In the field ionization (FI) source ions are formed in the strong electrostatic field set up at the tip of a wire electrode to which a high voltage is applied. Ions formed from molecules present on the tip of the wire are almost all parent ions. The source is not widely used but is of considerable value in studying very unstable molecules or very complex mixtures and in surface reaction studies.

Field desorption (FD) may be considered as an extension of field ionization; the main difference is that the sample is coated on the field ion emitter tip and ionization occurs from the solid phase. The technique requires experience to obtain reliable results. Mass spectra consisting mainly of molecular ions may be recorded from highly nonvolatile and thermally labile compounds.

Chemical ionization (CI) is a popular secondary ionization technique, and most new instruments are purchased with this capability.

In chemical ionization, a reagent gas at a pressure of about 0.1 to 10 torr is admitted to the source and ionized. At this pressure, ion-molecule reactions occur and the primary reagent gas ions react further. The most commonly used reagent gases are methane, isobutane, and ammonia. Typical reactions for methane are shown in the following equations:

$$CH_4 + e^- \rightarrow CH_4^{+\bullet} + CH_3^{+\bullet} + CH_2^{+\bullet}$$

$$CH_4^{+\bullet} + CH_4 \rightarrow CH_5^+ + CH_3^\bullet$$

$$CH_3^+ + CH_4 \rightarrow C_2H_5^+ + H_2$$

The species CH_5^+ is a strong Bronsted acid and can transfer a proton to most organic compounds, as follows:

$$CH_5^+ + M \rightarrow MH^+ + CH_4$$

With methane, the protonated molecule ion $(MH)^+$ formed initially may be sufficiently energetic to dissociate further.

The fast atom bombardment (FAB) method uses a beam of fast (neutral) atoms to bombard the sample. Thus, the first requirement of this technique is a beam of fast-moving atoms, properly aimed at the target sample, which is dissolved in a nonvolatile liquid matrix. This is relatively easy to achieve, and methods for producing such beams are well developed. Essentially a fast atom gun consists of an ion gun with a collision cell in front of it. The ion gun is used to produce a xenon ion beam, which is then charge exchanged in the collision cell with xenon gas to produce the required beam of fast xenon atoms. The process is summarized in the following equations:

$$Xe \rightarrow Xe^+ + e^-$$

$$\underset{\rightarrow}{Xe^+} + Xe \rightarrow \underset{\rightarrow}{Xe} + Xe^+$$

in which the subscript arrow indicates the fast-moving particle.

Since FAB is a surface analysis technique, the preparation of the sample, in order to optimize the surface conditions, is of paramount importance. When the sample is coated on the probe by evaporation of a solution, the resultant sample ion beam is often transitory. Adduct ions are frequently produced. The preferential formation of $(M + \mathrm{Na})$ and $(M + \mathrm{K})$ adduct ions has some parallels in FD, especially in the ionization of sugars. This phenomenon can be used to good effect to assist in the ionization of these classes of compounds. Frequently, the sample surface is treated with sodium chloride solution to enhance the yield of the adduct ions. Heating the sample during analysis can sometimes increase the ion yield.

The suppression of sample ion is probably due to destruction of the sample surface, and a means of continuously replenishing the sample surface during the analysis is required. Dissolving the sample in a suitable, nonvolatile liquid and coating this mixture onto the probe tip achieves this. Using this approach, sample lifetimes of greater than 1 hour have been realized in the ion source, and the range of compounds amenable to FAB has expanded dramatically. These long sample lifetimes and higher sensitivity make FAB an important mass spectral technique for producing mass spectra from novel, difficult-to-handle, biochemicals, and also allow unequivocal identification of the elemental formula of the material through accurate mass determination. A further advantage of FAB, useful in structural determination, is the presence of fragment ions within the spectra.

Sample Introduction—The sample is to be admitted to the ionization chamber in the gaseous or vapor state. Since many samples are gases or liquids at room temperature and atmospheric pressure, a sample handling system and a leak arrangement to the ion source are all that are required.

To produce from solids a molecular beam directly within the vacuum system can be a simple matter of heating a solid sample in a crucible to a sufficiently high temperature. Several commercially available probes, or cartridges, are used, depending upon the particular instrument and applications involved.

Other analytical instruments are used as inlets into the mass spectrometer. The most popular and most successful early development in mass spectrometry was the combination of a gas chromatograph and mass spectrometer (GC/MS). This combined instrument was a ready success, since the effluent of the gas chromatograph was in the vapor state and the primary problem of the combined instrument was the task of selectively removing the unwanted carrier gas.

Combining the liquid chromatograph with the mass spectrometer (LC/MS) was a far more challenging problem. While the liquid chromatograph is a powerful separative instrument, the widely used eluting solvents are often quite polar, complex, and relatively nonvolatile. Nevertheless, the coupling of the two instruments has been achieved and LC/MS instruments are commercially available.

Finally, nearly all of the various combinations of one mass spectrometer being an inlet system with another mass spectrometer (MS/MS) have been developed and studied (e.g., TOF with Magnetic sector, two magnetic sectors, quadrupole with a magnetic sector, etc.). Application of this technique has been most successful for mixture analysis. It has been applicable also to structure analysis where it was necessary to ionize the molecule of interest by a technique yielding mostly parent ions, then introducing these parent ions into a second mass spectrometer in order to study fragmentation patterns.

Data Analysis and Interpretation—Although molecules are normally electrically neutral, if one electron is taken away or added, a molecular ion results. The mass of this ion is the molecular weight of the molecule under study. Furthermore, it is often possible to determine the accurate mass of this ion with sufficient precision to enable the calculation of the empirical formula of the compound. Accurate masses may be determined at high resolution by either scanning or by peak-matching measurements.

Fragment ions are those produced from the molecular ion by various bond cleavage processes. Numerous papers in the literature relate the bond cleavage patterns (fragmentation patterns) to molecular structure. Correlations of mass spectra and molecular structure are discussed for steroids, aromatics, aliphatics, and, recently, complex compounds arising from biotechnology.

The mass spectrum is often very complex and not all of the ions may be separated by the mass spectrometer. The limit of the ability of the instrument to separate two ions very close in mass is called the resolving power of the instrument. The most common definition of the resolving power of a mass spectrometer is the "10% valley" definition. This states that the resolving power of a mass spectrometer is the highest mass number at which peaks of adjacent molecular weight and equal heights have a valley between them of 10% of the peak height. In mass spectrometry, low resolution covers the range of about 100 to 2000, medium resolution 2000 to 10,000, and high resolution greater than 10,000.

Quantitative analysis in mass spectrometry is usually performed in one of two ways. The first is selective ion monitoring. In this technique the ions, or group of ions of interest, are individually focused on the detector and measured. Both sensitivity and selectivity are enhanced by this technique.

The second most popular quantitative technique is isotope dilution. This method may be applied through the use of either radioactive or stable isotopes, the latter is most popular for mass spectrometry. The isotope dilution technique has the unique advantage that it is not necessary to recover all of the original material being analyzed to obtain the quantitative information desired. The technique has been successfully applied in biological studies, often in combination with GC/MS or LC/MS.

⟨741⟩ MELTING RANGE OR TEMPERATURE

For Pharmacopeial purposes, the melting range or temperature of a solid is defined as those points of temperature within which, or the point at which, the solid coalesces and is completely melted, except as defined otherwise for *Classes II* and *III* below. Any apparatus or method capable of equal accuracy may be used. The accuracy should be checked frequently by the use of one or more of the six USP Melting Point Reference Standards, preferably the one that melts nearest the melting temperature of the compound to be tested (see *USP Reference Standards* ⟨11⟩).

Five procedures for the determination of melting range or temperature are given herein, varying in accordance with the nature of the substance. When no class is designated in the monograph, use the procedure for *Class Ia*.

The procedure known as the mixed-melting point determination, whereby the melting range of a solid under test is compared with that of an intimate mixture of equal parts of the solid and an authentic specimen of it, e.g., the corresponding USP Reference Standard, if available, may be used as a confirmatory identification test. Agreement of the observations on the original and the mixture constitutes reliable evidence of chemical identity.

Apparatus—An example of a suitable melting range apparatus consists of a glass container for a bath of transparent fluid, a suitable stirring device, an accurate thermometer (see *Thermometers* ⟨21⟩),* and a controlled source of heat. The bath fluid is selected with a view to the temperature required, but light paraffin is used generally and certain liquid silicones are well adapted to the higher temperature ranges. The fluid is deep enough to permit immersion of the thermometer to its specified immersion depth so that the bulb is still about 2 cm above the bottom of the bath. The heat may be supplied by an open flame or electrically. The capillary tube is about 10 cm long and 0.8 to 1.2 mm in internal diameter with walls 0.2 to 0.3 mm in thickness.

Procedure for Class I—Reduce the substance under test to a very fine powder, and, unless otherwise directed, render it anhydrous when it contains water of hydration by drying it at the temperature specified in the monograph, or, when the substance contains no water of hydration, dry it over a suitable desiccant for not less than 16 hours.

Charge a capillary glass tube, one end of which is sealed, with sufficient of the dry powder to form a column in the bottom of the tube 2.5 to 3.5 mm high when packed down as closely as possible by moderate tapping on a solid surface.

Heat the bath until the temperature is about 30° below the expected melting point. Remove the thermometer, and quickly attach the capillary tube to the thermometer by wetting both with a drop of the liquid of the bath or otherwise, and adjust its height so that the material in the capillary is level with the thermometer bulb. Replace the thermometer, and continue the heating, with constant stirring, sufficiently to cause the temperature to rise at a rate of about 3° per minute. When the temperature is about 3° below the lower limit of the expected melting range, reduce the heating so that the temperature rises at a rate of about 1° to 2° per minute. Continue heating until melting is complete.

The temperature at which the column of the substance under test is observed to collapse definitely against the side of the tube at any point is defined as the beginning of melting, and the temperature at which the test substance becomes liquid throughout is defined as the end of melting or the "melting point." The two temperatures fall within the limits of the melting range.

Procedure for Class Ia—Prepare the test substance and charge the capillary as directed for *Class I*. Heat the bath until the temperature is about 10° below the expected melting point and is rising at a rate of 1 ± 0.5° per minute. Insert the capillary as directed under *Class I* when the temperature is about 5° below the lower limit of the expected melting range, and continue heating until melting is complete. Record the melting range as directed for *Class I*.

Procedure for Class Ib—Place the test substance in a closed container and cool to 10°, or lower, for at least 2 hours. Without previous powdering, charge the cooled material into the capillary tube as directed for *Class I*, then immediately place the charged tube in a vacuum desiccator and dry at a pressure not exceeding 20 mm of mercury for 3 hours. Immediately upon removal from the desiccator, fire-seal the open end of the tube, and as soon as practicable proceed with the determination of the melting range as follows: Heat the bath until a temperature 10 ± 1° below the expected melting range is reached, then introduce the charged tube, and heat at a rate of rise of 3 ± 0.5° per minute until melting is complete. Record the melting range as directed for *Class I*.

If the particle size of the material is too large for the capillary, pre-cool the test substance as above directed, then with as little pressure as possible gently crush the particles to fit the capillary, and immediately charge the tube.

Procedure for Class II—Carefully melt the material to be tested at as low a temperature as possible, and draw it into a capillary tube, which is left open at both ends, to a depth of about 10 mm. Cool the charged tube at 10°, or lower, for 24 hours, or in contact with ice for at least 2 hours. Then attach the tube to the thermometer by suitable means, adjust it in a water bath so that the upper edge of the material is 10 mm below the water level, and heat as directed for *Class I* except, within 5° of the expected melting temperature, to regulate the rate of rise of temperature to 0.5° to 1.0° per minute. The temperature at which the material is observed to rise in the capillary tube is the melting temperature.

Procedure for Class III—Melt a quantity of the test substance slowly, while stirring, until it reaches a temperature of 90° to 92°. Remove the source of the heat and allow the molten substance to cool to a temperature of 8° to 10° above the expected melting point. Chill the bulb of a suitable thermometer (see *Thermometers* ⟨21⟩) to 5°, wipe it dry, and while it is still cold dip it into the molten substance so that approximately the lower half of the bulb is submerged. Withdraw it immediately, and hold it vertically away from the heat until the wax surface dulls, then dip it for 5 minutes into a water bath having a temperature not higher than 16°.

Fix the thermometer securely in a test tube so that the lower point is 15 mm above the bottom of the test tube. Suspend the test tube in a water bath adjusted to about 16°, and raise the temperature of the bath at the rate of 2° per minute to 30°, then change to a rate of 1° per minute, and note the temperature at which the first drop of melted substance leaves the thermometer. Repeat the determination twice on a freshly melted portion of the test substance. If the variation of three determinations is less than 1°, take the average of the three as the melting point. If the variation of three determinations is greater than 1°, make two additional determinations and take the average of the five.

* ASTM Method E77 deal with "Verification and Calibration of Liquid-in-glass Thermometers."

⟨751⟩ METAL PARTICLES IN OPHTHALMIC OINTMENTS

The following test is designed to limit to a level considered to be unobjectionable the number and size of discrete metal particles that may occur in ophthalmic ointments.

Procedure—Extrude, as completely as practicable, the contents of 10 tubes individually into separate, clear, flat-bottom, 60-mm petri dishes that are free from scratches. Cover the dishes, and heat at 85° for 2 hours, increasing the temperature slightly if necessary to ensure that a fully fluid state is obtained. Taking precautions against disturbing the melted sample, allow each to cool to room temperature and to solidify.

Remove the covers, and invert each petri dish on the stage of a suitable microscope adjusted to furnish 30 times magnification and equipped with an eye-piece micrometer disk that has been calibrated at the magnification being used. In addition to the usual source of light, direct an illuminator from above the ointment at a 45° angle. Examine the entire bottom of the petri dish for metal particles. Varying the intensity of the illuminator from above allows such metal particles to be recognized by their characteristic reflection of light.

Count the number of metal particles that are 50 µm or larger in any dimension: the requirements are met if the total number of such particles in all 10 tubes does not exceed 50, and if not more than 1 tube is found to contain more than 8 such particles. If these results are not obtained, repeat the test on 20 additional tubes: the requirements are met if the total number of metal particles that are 50 µm or larger in any dimension does not exceed 150 in all 30 tubes tested, and if not more than 3 of the tubes are found to contain more than 8 such particles each.

⟨755⟩ MINIMUM FILL

The following tests and specifications apply to articles such as creams, gels, jellies, ointments, pastes, and powders that are packaged in containers in which the labeled net weight is not more than 150 g.

Select a sample of 10 filled containers, and remove any labeling that might be altered in weight during the removal of the container contents. Thoroughly cleanse and dry the outside of the containers by a suitable means, and weigh individually. Quantitatively remove the contents from each container, cutting the latter open and washing with a suitable solvent, if necessary, taking care to retain the closure and other parts of each container.

Dry, and again weigh each empty container together with its corresponding parts. The difference between the two weights is the net weight of the contents of the container. The average net weight of the contents of the 10 containers is not less than the labeled amount, and the net weight of the contents of any single container is not less than 90% of the labeled amount where the labeled amount is 60 g or less, or not less than 95% of the labeled amount where the labeled amount is more than 60 g but not more than 150 g. If this requirement is not met, determine the net weight of the contents of 20 additional containers. The average net weight of the contents of the 30 containers is not less than the labeled amount, and the net weight of the contents of not more than 1 of the 30 containers is less than 90% of the labeled amount where the labeled amount is 60 g or less, or less than 95% of the labeled amount where the labeled amount is more than 60 g but not more than 150 g.

⟨761⟩ NUCLEAR MAGNETIC RESONANCE

Nuclear magnetic resonance (NMR) spectroscopy is a useful analytical procedure because it is specific: every drug substance possesses a unique, characteristic NMR spectrum.

Since atomic nuclei are charged and may spin on their nuclear axes, the spinning nuclei create a magnetic dipole having a magnetic moment, μ, along this axis. The angular momentum of the spinning nucleus is characterized by a spin quantum number (I). If the mass number is odd, I is ½ or an integral multiple of ½. In contrast to those nuclei having an I of ½ and a spherical nuclear charge distribution, nuclei having $I \geq 1$ exhibit a nonspherical nuclear charge distribution that is characterized by a nuclear quadrupole moment and that results in demonstrable spectral perturbations.

Nuclei having a spin quantum number, I, when placed in an external uniform static magnetic field of strength, H_0, tend to be oriented similarly to a bar magnet in $(2I + 1)$ possible orientations. Thus, for nuclei with I = ½ there are two possible orientations, namely, +½ or −½, a lower or an upper energy state. Since two energy states exist, transitions from the lower to the higher state should be possible if the proper amount of energy is introduced. In a static magnetic field the nuclear magnetic axis spins or precesses (Larmor precession) about the external field axis. The precessional angular velocity, ω_0, is related to the external magnetic field strength through the equation:

$$\omega_0 = \gamma H_0,$$

in which γ is the magnetogyric ratio, a parametric constant for each nucleus. In addition, if energy from an oscillating radiofrequency field is introduced, the absorption of radiation takes place according to the relationship:

$$\Delta E = h\nu_0 = \mu H_0/I,$$

and

$$\nu_0 = \omega_0/2\pi = \gamma H_0/2\pi.$$

Thus, when the frequency (ν_0) of the external energy ($E = h\nu$) is the same as the precessional angular velocity, resonance is achieved, and the nucleus attains the upper state. The existence of two energy states (I = ½) introduces the question of relative population of the two states under usual conditions. With the use of the Boltzmann distribution law, it has been found that the lower level population exceeds that of the upper level by only a few parts per million because the separation of the energy levels is about 0.01 calorie at 10 to 15 kilogauss field strength—a small quantity compared to the Boltzmann energy unit kT. Because of this small lower energy level excess, any useful measurement scheme must have the capability to sense and to amplify weak signals.

NMR involves energy absorption by nuclei in the radiofrequency range, i.e., wavelengths of 1 to 100 m or frequencies of 3×10^8 to 3×10^6 Hz. In practice, NMR measurements are generally useful in studies of nuclei such as ^{1}H, ^{13}C, ^{19}F, ^{31}P, and ^{11}B, which have odd mass numbers and I values of ½ or integral multiples of ½. The proton, ^{1}H, is most often studied and most frequently used in quantitative analysis. A list of common elements studied and their significant NMR properties is presented in Table 1.

Table 1. Properties of Some Nuclei Amenable to NMR Study.

Nucleus	I	Natural Abundance, %	Sensitivity	Resonance Frequency (MHz) at 14.1 kilogauss
^{1}H	½	99.98	1.000	60.00
^{13}C	½	1.11	0.016	15.086
^{19}F	½	100	0.834	56.446
^{31}P	½	100	0.066	24.288
^{11}B	$^3/_2$	80.42	0.165	19.252

The Spectrum

The magnitude of the separation of the frequency of resonance of a proton from that of some standard (tetramethylsilane or sodium 2,2-dimethyl-2-silapentane-5-sulfonate) is called the chemical shift, which is proportional to the strength of the applied field. The latter is a composite of the external field and the field caused by the circulation of surrounding electrons about the protons. The conventional NMR spectrum is shown with the magnetic field strength increasing in the direction left to right, and a proton that resonates at a high magnetic field strength (near tetramethylsilane) is said to be more shielded (greater electron density) than a proton that resonates at a lower magnetic field strength and thus is said to be de-shielded (lower electron density).

Figure 1 shows the proton NMR spectrum of 2,3-dimethyl-2-butenyl methyl ether. This compound contains protons in a methylene group (marked d in the graphic formula) and in three methyl groups (a, b, and c). Because the three methyl groups are situated in different molecular environments, three different modes of methyl proton resonance are observed as spectral peaks in addition to the peak corresponding to methylene proton resonance. In some NMR spectra (see a and c in Figure 2), the observed signal corresponding to a particular proton is split into a set or multiplet of peaks. Such peak splitting is ascribed to the influence of nuclei that possess magnetic moments and are within several valence bonds of the nucleus being studied. This intramolecular effect is called coupling.

The coupling between two nuclei may be described in terms of the coupling constant, J, which is the separation (in Hz) between the individual peaks of the multiplet. Where two nuclei interact and cause reciprocal splitting, the measured coupling constants in the two resulting multiplets are equal. Furthermore, J is independent of magnetic field strength.

In a comparatively non-complex spin system, the number of individual peaks present in a multiplet and the relative peak in-

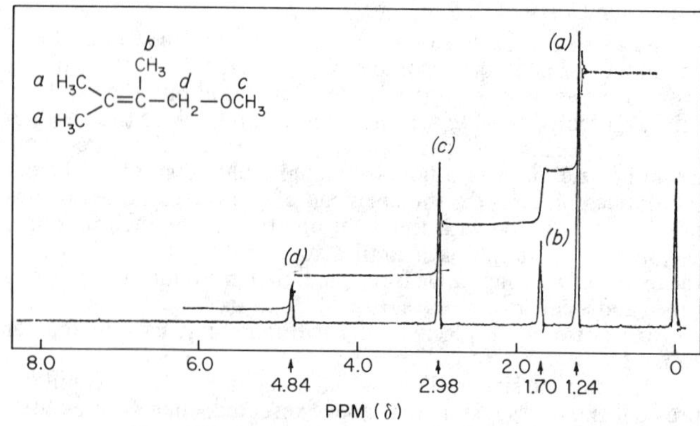

Fig. 1. NMR spectrum of 2,3-dimethyl-2-butenyl methyl ether (15% in CCl_4) showing four nonequivalent, uncoupled types of protons with a normal integral trace (peak area ratio from low H_0 to high H_0 of 2:3:3:6). (Tetramethylsilane, the NMR Reference, appears at 0 ppm.) The system of units represented by δ is defined under *Spectral Standards and Units of Measurement*, in this chapter.

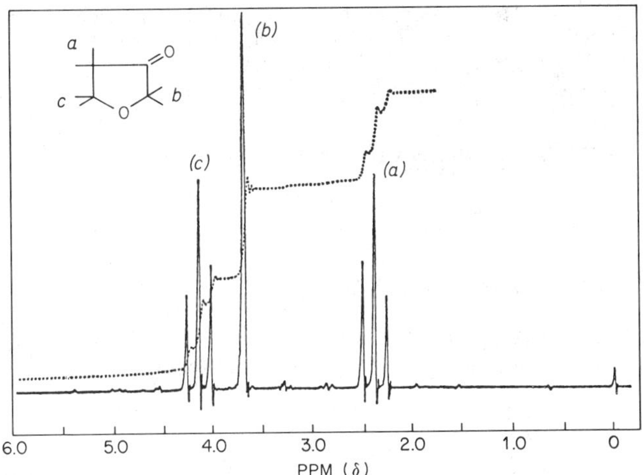

Fig. 2. NMR spectrum of 3-keto-tetrahydrofuran (10% in CCl_4) showing three nonequivalent types of protons with a normal integral trace (peak area ratio from low H_0 to high H_0 of 1:1:1). Note two sets of methylene groups coupled to each other at 4.2 and 2.4 ppm. (Tetramethylsilane, the NMR Reference, appears at 0 ppm.)

tensities are predictable. The peak number is determined by n, the number of protons on adjacent groups that are active in splitting. The total number of observed peaks is $(n + 1)$. Also, the relative intensity of each separate peak follows the coefficient of the binomial expansion $(a + b)^n$; these coefficients may be conveniently found by use of the Pascal triangle, which produces the following relative areas for the specified multiplets: doublet, 1:1; triplet, 1:2:1; quartet, 1:3:3:1; quintet, 1:4:6:4:1; sextet, 1:5:10:10:5:1; and septet, 1:6:15:20:15:6:1. This orderly arrangement, generally referred to as first-order behavior, may be expected when the ratio of $\Delta\nu$ to J is greater than about 10; $\Delta\nu$ is the chemical shift difference between two nuclei or two groups of equivalent nuclei, and J is the spin–spin coupling constant. Two examples of idealized spectra arising from first-order coupling are shown in Figure 3.

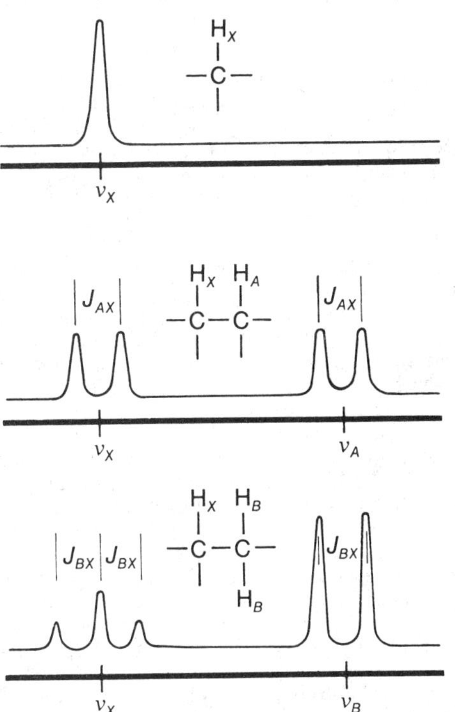

Fig. 3. Diagrammatic Representation of Simple First-order Coupling of Adjacent Protons.

The introduction of extra peaks not ascribable to NMR phenomena may complicate the spectrum. In an attempt to negate magnetic field inhomogeneity and increase resolution, the tube containing the test substance is spun. If this spinning frequency is too low, the desired field averaging is not complete, and an absorption signal is accompanied by aberrant signals of lower intensity called spinning side bands. These side bands are located symmetrically around the main signal, and the separation is equal to the spinning frequency or some integral multiple of that frequency. Thus, spinning side bands are readily identifiable, since their location changes with spinning frequency. Side bands can result also from uneven spinning.

Double resonance or spin decoupling is a useful means of simplifying spectra. This technique removes spin coupling between nuclei or groups of nuclei. For example, in a simple two-proton system, generally designated an AX system (see Figure 3), which is manifested as a pair of doublets, where a strong radiofrequency field at the frequency of X is introduced while the usual radiofrequency field is present to provide resonance possibilities for A, the coupling of X with A is removed, and A is no longer split but instead is manifested as a singlet. This technique provides a convenient way of establishing coupling relationships.

Another valuable property of the recorded spectrum is the peak area. During a single instrumental scan, the intensity of energy absorption is constant for all protons regardless of their nature. Thus the area of a single peak or a multiplet is directly proportional to the number of protons giving rise to that peak or multiplet. As a result, it is possible to determine the relative ratio of different kinds of protons in a molecule (note the integral traces in Figures 1 and 2). Moreover, if the solution of the analyte is prepared quantitatively and an Internal Standard is used, the peak areas, measured by instrumental integration, may be used for quantitative analysis.

Apparatus

Various instrument configurations are possible; the arrangement of a typical double-coil spectrometer is illustrated in Figure 4.

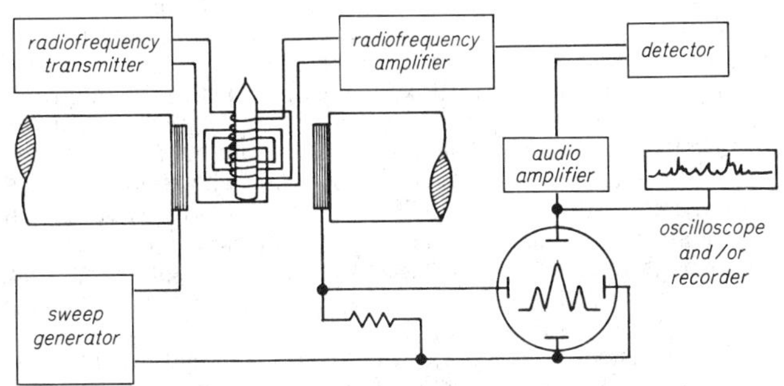

Fig. 4. Block Diagram of Typical NMR Spectrometer.

To detect variations in signal as a function of test substance constitution, NMR spectrometers can be operated either at fixed magnetic fields or at fixed oscillator (transmitter) frequency. In this manner, at a fixed magnetic field strength the test substance is scanned by varying the frequency over a relatively narrow region, or, if the frequency is fixed, the test substance is studied by varying the magnetic field.

The spectrometer must be able to reproduce line widths of a few tenths of a Hz at field strengths of the order of 100 MHz, and thus a stability in the range of 1 part in 10^8 is required. Some spectrometers maintain H_0 sufficiently constant so that line shape distortions do not occur when the instrument is operated at a fixed ν_0. However, problems in magnet stability make this difficult. As may be noted from $\nu_0 = \omega_0/2\pi = \gamma H_0/2\pi$, the ratio H_0/ν_0 is a constant for a particular nuclear resonance. This fact provides the foundation for maintaining reproducible spectra in terms of a combined field-frequency lock control system.

The field-frequency (H_0/ν_0) lock system requires a reference nucleus that is continuously irradiated at its exact resonance frequency with concomitant monitoring of the NMR signal. If the irradiation frequency does not match the resonance frequency (e.g., because of magnetic field changes) at any time, an error signal of an appropriate polarity is produced and taken into a feed-back loop. In this way the magnet coils may react to correct the field, or, if it is appropriate, the modulation frequency may be adjusted in accordance with the manufacturer's instructions.

The reference nucleus may be part of the instrumental arrangement (external lock), or it may be placed in the test solution (internal lock). Spectrometers should be equipped with both external and internal lock systems. The external lock system, the most commonly used, requires a control compound (e.g., water) whose side-band frequency is used by the instrument feed-back loop to maintain a system stability of 1 part in more than 1×10^8/hour, in terms of field strength to frequency ratios. It is used with field sweep for easy operation, stable integration of peak areas, and easy scale expansion. The proton internal lock system utilizes two side-band frequencies, separated by at least a few Hz, derived from a sharp signal within the test substance, usually the internal reference. It is essential that internal lock stability be 1 part in more than 1×10^9/hour in order to be useful for accurate measurements and spin-decoupling (double resonance). The proton internal lock is not suitable for integration over very wide ranges on some instruments.

Some internal lock systems, using a separate radiofrequency oscillator, lock on the deuterium resonance of deuterated solvent. These systems allow easy scale expansion and integration over the entire spectral range.

The integrator in an NMR instrument is an electronic integrator generally present as part of the spectrometer. The integrator determines the relative areas of resonance peaks through connection with the spectrometer output stage and presents the areas as stepping horizontal lines when a tracing is made of the test specimen with the spectrometer in the integration mode (see Figures 1 and 2).

General Method

The method and procedures discussed herein refer specifically to ^1H (proton) and ^{19}F NMR. They are applicable, with modification, to other nuclei.

Useful interpretation of an NMR spectrum requires a relatively narrow Lorentzian peak. Line broadening severely hampers interpretation and may be caused by local inhomogeneities in the magnetic field. To decrease the effect of small variations in peak shape of nuclear signals, the tube containing the test substance is rotated about the vertical axis, perpendicular to the magnetic field, so that the nuclei in the effective field of the magnet are uniformly exposed to an average field strength. This tends to increase resolution and reduce line widths. The spinning speed (average spin speed is about 20 to 40 rps) should be sufficient to produce averaging and to remove spinning side bands, but not fast enough to produce an extended vortex in the specimen tube. A vortex extended near the region exposed to the radiofrequency coils decreases resolution.

The presence of paramagnetic substances tends to increase broadening. One of the most common paramagnetic species, dissolved oxygen, causes interferences where oxygenated solvents are used for dissolution and where extremely high resolution is required. To remove dissolved oxygen, immerse the test substance contained in a preconstricted, heavy-walled (about 0.5-mm) tube in liquid nitrogen until it is frozen. Evacuate the tube by applying a vacuum of about 10 mm of mercury for several minutes. Allow the tube to stand until thawed, then repeat the procedure twice. After the third operation, while the test substance is frozen, seal the tube at the constriction prior to inserting it into the spectrometer.

Line broadening is also often observed for single protons attached to nitrogen because of the quadrupole moment of the nitrogen nucleus.

Selection of Solvent—Since narrow Lorentzian peaks are possible only in solution NMR spectroscopy, the test substance must be either a liquid or a solution of the solid in a suitable solvent. Choice of solvent is critical for solubility purposes and for making use of the apparent resonance peaks, since solvent peaks, if present, tend to obscure structural features of the test substance. To avoid interferences, for proton NMR determinations, special deuterated solvents are employed, with at least 99.5% isotopic purity. Deuterium ($I = 1$) does not exhibit resonance under ^1H conditions (e.g., 14.1 kilogauss, 60.0 MHz), but a small amount of ^1H present causes a small resonance to be shown. Some solvents (e.g., D_2O) enter into fast exchange reactions with protons and may abolish resonance signals from —COOH, —OH, and —NH_2 structural groups present. The protons from alcohols and amines do not take part in rapid exchange, unless catalyzed by small concentrations of acid and base, except in the presence of D_2O. The most commonly used solvents for proton NMR are listed in Table 2.

Table 2. Solvents Commonly Used for Proton NMR.

Solvent	Residual Proton Signal, δ[a]
CCl_4[b]	—
CS_2[b]	—
SO_2 (liquid)	—
$(CF_3)_2CO$	—
$CDCl_3$	7.27
CD_3OD	3.35, 4.8[c]
$(CD_3)_2CO$	2.05
D_2O	4.7[c]
DMSO-d_6[d]	2.50[e]
C_6D_6	7.20
p-Dioxane-d_8	3.55
CD_3CO_2D	2.05, 8.5[c]
DMF-d_7[f]	2.77, 2.93, 8.05

[a] δ in ppm relative to tetramethylsilane arbitrarily taken as 0δ.
[b] Spectrophotometric grade.
[c] Highly variable; depends on solute and temperature.
[d] Dimethyl sulfoxide-d_6.
[e] Poor grades show an additional peak of water associate at about 3.7 ppm.
[f] Dimethylformamide-d_7.

For ^{19}F NMR, most solvents used in proton NMR may be employed, the most common ones being $CHCl_3$, CCl_4, H_2O, CS_2, aqueous acids and bases, and dimethylacetamide. In general, any nonfluorinated solvent may be used, provided that it is of spectral quality.

Spectral Standards and Units of Measurement—For uniformity, and to avoid unnecessary conversions of units, spectrometer outputs to recorders for proton NMR are designed and calibrated for spectral sweep in terms of ν and δ values relative to tetramethylsilane NMR Reference which has been given a spectral line value of 0 Hz (ν) or 0 ppm (δ). This appears at the far right end of the recorder or oscilloscope where the spectrometer is operated in the normal range and is in the diamagnetic (upfield) direction. The relationship of ν to δ is given by the equation:

$$\delta = (\nu_s - \nu_r)/\nu_0 ,$$

in which ν_s is the test substance line frequency, ν_r is the reference line frequency, and ν_0 is the instrument oscillator frequency.

Since ν_r generally is at arbitrary zero, and ν_0 is the specific resonance frequency, either δ_s or ν_s can be determined from the instrument recorder calibration of the other. For example, on a 60-MHz spectrometer, a resonance line signal of 420 Hz shows a δ value of 7 ppm, i.e., $\delta = [(420 - 0)(\text{Hz})]/60 \text{ MHz} = 420 \text{ Hz}/60{,}000{,}000 \text{ Hz} = 7 \times 10^{-6}$ parts $= 7$ ppm. Where the spectral scale is defined in terms of δ values, numerical values decrease from left to right (as does frequency). In order to have a scale that increases from left to right, a more conventional arrangement, a scale of τ values has been defined in which the tetramethylsilane resonance line has an assigned value of 10.00. The relationship between τ and δ is: $\tau = 10.00 - \delta$ (ppm). Thus, in the foregoing example, where $\delta = 7$ ppm, $\tau = 3.00$ (ppm).

The chemical shift value is obtained from the recorder trace and is specific for types of nuclei in a particular chemical environment. The NMR Reference tetramethylsilane (1% concentration) is mixed with the test solution where nonaqueous solvents are used or where chemical interaction between tetramethylsilane and the compound studied is not possible. For aqueous systems, an NMR Reference of sodium 2,2-dimethyl-2-silapentane-5-sulfonate (1% concentration) is used because it has an absolute magnetic resonance value for its methyl groups closely approximating that of tetramethylsilane. The sodium 2,2-dimethyl-2-silapentane-5-sulfonate has the disadvantage, however, of showing a number of CH_2 multiplets at high gain and internal lock that may interfere with signals from the test substance. Where the use of NMR Reference tetramethylsilane is not possible because of solubility difficulties or interaction, an *external reference* is used.

For ^{19}F lock calibration, the most widely used NMR Reference is trichloromonofluoromethane, which, like tetramethylsilane, has an arbitrary shift value of 0.00 ppm (Φ^* scale). [NOTE—Φ is fluorine chemical shift in trichloromonofluoromethane extrapolated to infinite dilution; Φ^* is a nonextrapolated value. The difference $\Phi - \Phi^*$ is measurable, but generally not large.] Where operation is at subambient temperatures, trichloromonofluoromethane may be employed as both the reference and the solvent. Hexafluorobenzene, perfluorocyclobutane, and 1,4-dibromotetrafluorobenzene have been used also. Trifluoroacetic acid is the usual choice if an external reference is required. ^{19}F chemical shifts are much larger than ^{1}H values (−70 to about 250 ppm) and, as a result, instruments capable of use for ^{19}F have a wide sweep accessory in addition to a crystal oscillator, tunable to the appropriate fluorine (56.446 MHz at 14.1 kilogauss) resonance frequency.

The greatest difference in the recorder presentation of ^{19}F spectra with regard to ^{1}H is that the wide spectral range covered by ^{19}F scans makes it desirable to calibrate "zero" reference at some point paramagnetic (downfield) to the maximum field position. In this way, all the usable ^{19}F resonance signals are "on-chart" positive (upfield) to "lock-calibration zero."

Qualitative and Quantitative Analysis

NMR measurements are useful for a variety of analytical purposes. The various types of protons, fluorine atoms, carbon atoms, etc., each with a different environment, appear as different resonance signals with respect to their different chemical environments. The spectrum thus affords information about the molecular structure. The general multiplicity of each individual resonance (e.g., singlet, doublet, triplet) adds more structural information, and the combination of chemical shift and spin-coupling pattern enables the determination of (a) the number of the atoms being measured, (b) the chemical environment of each atom, (c) the structural and/or isomeric relationships, and (d) the presence of impurities. The integration of peak areas is an important step in interpretation since the ratio of areas yields the relative ratios of the various kinds of resonant nuclei. In addition, the integration may be extended to quantitative analysis.

The quantitative analysis of a compound by NMR exemplifies the use of a specific intimate property for measurement purposes. Once the position of a definite structural unit is known, the area of its resonance peak(s) can be related to that of other peaks, to obtain ratios of the various atoms represented in the spectrum. Quantitative estimation is limited largely by the accuracy and reproducibility of the built-in integrator that is usually part of the instrument recorder. By employing multiple integration tracings across the entire spectrum, as well as several independent analyses, a relative accuracy of $\pm 2\%$ can be achieved. If small parts of the scan yield the quantitative information of interest, these partial scans can be integrated at higher gain to improve the accuracy. The principal advantages of quantitative NMR are: (a) the intensity of a signal for a given nuclear isotope is proportional to the number of nuclei contributing to the signal but independent of its chemical nature; (b) a compound being analyzed in a mixture need not be available in pure form for use as a standard; and (c) resonance lines are narrow in relation to chemical shift differences (line positions). The likelihood of appreciable overlap of signals of different components of a mixture is thereby reduced.

For first-order interpretive qualitative usefulness, the spectrum should first be analyzed for total number of protons by area estimate with the integrator. [NOTE—Neighboring nonequivalent protons interact with the resonance field of the measured proton, producing multiple lines.]

Simple multiplet patterns may be recognized by their relative intensities, by a set of matched or coupled lines in a different part of the spectrum, or by observation of the spectrum on an instrument having a different field strength. In the last-mentioned case, the coupling constant (J) has the same value in Hz, but the field-dependent expression of the chemical shift (ν) changes. The easiest, most direct way is to search for a matched (equal J values) set of multiplets. Coupling usually occurs only over a distance of three chemical bonds, although in some instances long-range coupling (transmitted through four or five bonds) may occur. The coupling and intensity for first-order coupling follow the coefficients of the binomial law, as discussed previously.

Once the total number of protons has been ascertained, their field position may then be matched against the range of positions of known types of groups for the structure(s) present, e.g., —CH_3 types, —CH_2 types, and —$COCH_2$ types. In general, the closer the protons are to an electronegative group, the more de-shielded they are and thus they resonate at lower H_0 values (higher δ). Groups having significant alkyl character resonate at high H_0 values (0.9 to 2.4 ppm).

Valuable structural information may be obtained for test substances having —OH, —NH_2, or —COOH groups by taking advantage of the rapid exchange of these with D_2O. To determine the presence and position of these groups, scan the test substance in $CDCl_3$ or CCl_4, then add a few drops of D_2O to the sample tube, and scan again. Resonance peaks from —OH, —NH_2, and —COOH collapse in the second scan and are replaced by a singlet at 4.7 ppm from HDO.

More structural information is obtained from an inspection of the J values when coupling has occurred. The magnitude of J in the aromatic proton region (6.5 to 8.0 ppm) gives valuable data about the nature of aryl ring substituents, since *ortho*-, *meta*-, and *para*-protons show different coupling (8, 2, and about 0.5 Hz, respectively). A similar inspection of coupling in unsaturated systems gives data on isomer content (*cis*-HC═CH, 6 to 12 Hz; *trans*-HC═CH, 9 to 18 Hz), and conformational isomers have been successfully identified (ax–ax, 8 to 10 Hz; ax–eq, 2 to 5 Hz; and eq–eq, 1 to 4 Hz).

If alcoholic —OH is possibly present and has been noted by D_2O exchange, —OH coupling may be observed by scanning a test substance in DMSO-d_6. In this procedure, —CH_2OH shows a triplet, —CHRO*H* shows a doublet, and —CR_2OH shows a singlet.

If a complex spectrum is not interpretable by usual first-order rules, double resonance, as previously discussed, may be employed. Instruments used at 60 MHz or above should be equipped with homonuclear spin decouplers.

The spectra of compounds containing ethers, esters, ketones, etc., often may have resonance lines grouped in a narrow region of the spectrum (1.0 to 3.0 ppm) and are difficult to analyze because of peak overlap. In these cases it is often advantageous to add a shift reagent such as the dipyridine adduct of tris(2, 2,-6, 6-tetramethyl-3, 5-heptanedionato)europium(III) [$Eu(dpm)_3$], or tris(1,1,1,2,2,3,3-heptafluoro-7,7-dimethyl-4,6-octanedionato)europium(III) [$Eu(fod)_3$], or their corresponding praseodymium compounds. [$Eu(fod)_3$] is the most useful, and in its presence, coordination occurs between the functional group and the transition metal complex causing a downfield shift of the resonance lines. The relative shift indicates the type of groups proximal to the protons. In order to improve the spectrum, increments of reagent, not to exceed the weight of the test substance, may be added to the test substance.

Quantitative analysis can be done either in a relative manner or by use of an internal standard. In the former method, a resonance peak area attributed to the main absorbing group is compared proton for proton with a different area from a second compound or impurity, and a percentage value, in mole percent, is determined. For example, the partial spectrum (see Figure 5) shows the impurity A (one proton of type —CH═) in a specimen of B (one proton of type —CH<). The amount of A present is calculated in terms of the total areas of A and B. For the greatest accuracy, quantitative evaluation of this type should be done usually by comparing areas of equal numbers of protons. The areas should always be obtained not less than five times to average integrator variations.

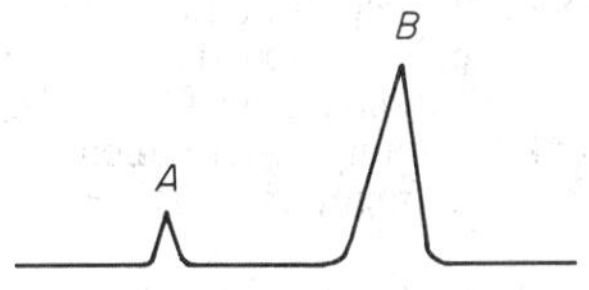

Fig. 5.

In the internal standard method a resonance peak area arising from the test substance is compared with a resonance peak area ascribable to part of an internal standard molecule. If both test substance and internal standard are accurately weighed, the absolute purity of the substance may be calculated. A good internal standard has the following properties: (a) it presents a reference resonance peak, preferably a singlet, at a field position at least

30 Hz removed from a sample peak; (b) it is soluble in the analytical solvent; (c) its proton equivalent weight, i.e., the molecular weight divided by the number of protons giving rise to the reference peak, is as small as possible; and (d) it does not interact with the compound being tested. Typical examples of useful standards are benzene and benzyl benzoate for comparing nonaromatic areas (—CH_2 of $C_6H_5CO_2CH_2C_6H_5$ at about 5.2 ppm) and maleic acid for nonalkene types. The utility and disadvantages of this method are the same as in gas chromatography, except that a standard of the measured unknown is not needed if some other pure reference standard is available.

Quantitative analysis by the internally relative method or the internal standard method, as well as detection of trace impurities, can be measurably improved by the use of a time-averaging accessory. This is a signal-to-noise booster that additively accumulates signals while averaging out randomly occurring noise level. An enhanced trace spectrum is obtained along with a digitalized count area, thus measurably improving accuracy.

For either qualitative or quantitative scans, coupling may occur between ^1H and other nuclei, especially ^{19}F, ^{31}P, and ^{199}Hg. In some cases, e.g., ^{31}P and ^{199}Hg, the coupling constants may be large enough so that part of the multiplet is off the chart at either the upfield or downfield end. Couplings of these types can occur over the normal "three-bond distance," as for ^1H-^1H coupling.

Qualitative Scans—Proceed as directed in the individual monograph, or weigh 60 to 90 mg of test substance into a 15- × 40-mm amber vial, and add about 500 µL of solvent, using a pipet or tuberculin syringe. For general organic solvents (such as $CDCl_3$, CCl_4, DMSO-d_6, and DMF-d_7), add about 5 µL of tetramethylsilane, mix, and transfer 400 to 500 µL of the solution to a 5-mm × 17.5-cm tube suitable for spinning. Take precautions to prevent evaporation of the tetramethylsilane prior to mixing with the solution. Place the cap on the tube. Using field sweep and external lock, scan the test substance from 0 ppm to about 8 ppm at a scan rate of about 1.5 to 4 minutes for a full-scale recording, adjusting amplification so that all peaks remain on scale. Adjust the spin rate so that no spinning side bands interfere with the peaks of interest. If the response is low at reasonable amplitude, increase the radiofrequency power to obtain the highest peak response without peak broadening.

In adjusting resolution, take care to ensure that the tetramethylsilane peak shows definite ringing. The phenomenon of "ringing" is the repeated excursion or "wiggling" of the recorder trace after the magnetic field has passed through a resonance value and the peak value has been recorded. The ringing noted on a number of the peaks in Figure 2 is seen during rapid scans and decays exponentially, finally reaching the baseline value. Ringing is a good indication of a homogeneous field.

After the initial scan, quickly check for peaks downfield of 8 ppm by offsetting the instrument response by 5 ppm. Record any peaks from 13 ppm to 5 ppm so that peak positions of 5 ppm to 8 ppm may be matched, since the tetramethylsilane position (0 ppm) may shift slightly in the higher range.

After scanning, set the recorder sweep time to about 1 minute, reset phase, and record the total integral using an integrator amplitude set to keep the total integrator response on chart scale. Compare the area of each set of peaks to the area of an assigned peak, and determine the number of nuclei contributing to each set.

Note peak positions, in ppm, from the tetramethylsilane peak, and measure the associated integrals. Check the spectrum for coupled peaks either visually or by spin decoupling. For $CDCl_3$, C_6D_6, $(CD_3)_2CO$, and CCl_4 solutions, add 2 drops of D_2O to the tube, shake for about 30 seconds, and again scan after resetting the 0-ppm position. Check the second scan for any disappearance of peaks, noting the additional peak for HDO.

For accurate peak matching, where accuracy of 0.5 to 1.0 Hz is needed, reset the instrument calibration on internal lock, and again scan.

Where shift reagents are specified, prepare a solution of the dipyridine adduct of [Eu(dpm)$_3$], or [Eu(fod)$_3$], or their corresponding praseodymium analogs in $CDCl_3$ or CCl_4 as directed in the individual monograph. In an amber vial, dilute the test solution with a specific volume of the shift solution, containing reagent equal to 0.2 times the weight of test substance, and reduce the volume to about 500 µL by using a thin stream of cold, dry nitrogen. Filter the solution through a fine-porosity, sintered-glass filter into the sample tube. Again scan the test substance, noting shifts in major peaks. Repeat using 0.4-, 0.6-, and 0.8-weight equivalent of the shift reagent. Assign the peak position by the nature of the shift. [NOTE—With shift reagent present, an additional peak results from the reagent and some degree of broadening occurs.]

Absolute Method of Quantitation—Where the individual monograph directs that the *Absolute Method of Quantitation* be employed, proceed as follows.

Solvent, Internal Standard, and *NMR Reference*—Use as directed in the individual monograph.

Test Preparation—Transfer a quantity of the article, containing about 4.5 proton mEq of the analyte, accurately weighed, to a glass-stoppered, graduated centrifuge tube. Transfer to the tube about 4.5 proton mEq of *Internal Standard,* accurately weighed, add 3.0 mL of *Solvent,* insert the stopper, and shake. When dissolution has been completed, add about 30 µL (30 mg if a solid) of *NMR Reference* if it will not interfere with subsequent measurements, and shake.

Procedure—Transfer about 0.4 mL of *Test Preparation* to a standard 5-mm NMR spinning tube, and record the spectrum, adjusting the spin-rate so that no spinning side bands interfere with the peaks of interest. Measure the area under each of the peaks specified in the individual monograph by integrating not fewer than five times. Record the average area of the *Internal Standard* peak as A_S and that of the *Test Preparation* peak as A_U.

Calculate the quantity, in mg, of the analyte in the *Test Preparation* by the formula:

$$W_S(A_U/A_S)(E_U/E_S),$$

in which W_S is the weight, in mg, of *Internal Standard* taken, and E_U and E_S are the proton equivalent weights (i.e., the molecular weights divided by the number of protons giving rise to the reference peak) of the analyte and the *Internal Standard,* respectively.

Relative Method of Quantitation—Where the individual monograph directs that the *Relative Method of Quantitation* be employed, proceed as follows.

Solvent, NMR Reference, and *Test Preparation*—Use as directed under *Absolute Method of Quantitation.*

Procedure—Transfer about 0.4 mL of *Test Preparation* to a standard 5-mm NMR spinning tube, and record the spectrum, adjusting the spin-rate so that no spinning side bands interfere with the peaks of interest. Measure the area under each of the peaks specified in the individual monograph by integrating not fewer than five times. Record the average areas resulting from the resonances of the groups designated in the individual monograph as A_1 and A_2.

Calculate the quantity, in mole percent, of the analyte in the *Test Preparation* by the formula:

$$100(A_1/n_1)/[(A_1/n_1) + (A_2/n_2)],$$

in which n_1 and n_2 are, respectively, the numbers of protons in the designated groups.

⟨771⟩ OPHTHALMIC OINTMENTS

Added Substances—Suitable substances may be added to ophthalmic ointments to increase stability or usefulness, unless proscribed in the individual monograph, provided they are harmless in the amounts administered and do not interfere with the therapeutic efficacy or with the responses to the specified assays and tests. No coloring agent may be added, solely for the purpose of coloring the finished preparation, to an article intended for ophthalmic use (see also *Added Substances* under *General Notices,* and *Antimicrobial Preservatives—Effectiveness* ⟨51⟩).

A suitable substance or mixture of substances to prevent the growth of microorganisms must be added to ophthalmic ointments that are packaged in multiple-use containers, regardless of the method of sterilization employed, unless otherwise directed in the individual monograph, or unless the formula itself is bacteriostatic. Such substances are used in concentrations that will prevent the growth of or kill microorganisms in the ophthalmic ointments (see also *Antimicrobial Preservatives—Effectiveness* ⟨51⟩ and *Antimicrobial Agents—Content* ⟨341⟩). Sterilization processes are employed for the finished ointment or for all ingredients, if the ointment is manufactured under rigidly aseptic conditions, even though such substances are used (see also *Parenteral and Topical Preparations* in the section, *Added Substances*, under *General Notices*, and *Sterilization and Sterility Assurance of Compendial Articles* ⟨1211⟩). Ophthalmic ointments that are packaged in single-use containers are not required to contain antibacterial agents; however, they meet the requirements for *Sterility Tests* ⟨71⟩.

Containers—Containers, including the closures, for ophthalmic ointments do not interact physically or chemically with the preparation in any manner to alter the strength, quality, or purity beyond the official requirements under the ordinary or customary conditions of handling, shipment, storage, sale, and use (see also *Containers for Articles Intended for Ophthalmic Use*, under *General Notices*).

Metal Particles—Follow the *Procedure* set forth under *Metal Particles in Ophthalmic Ointments* ⟨751⟩.

Leakage—Select 10 tubes of the Ointment, with seals applied when specified. Thoroughly clean and dry the exterior surfaces of each tube with an absorbent cloth. Place the tubes in a horizontal position on a sheet of absorbent blotting paper in an oven maintained at a temperature of 60 ± 3° for 8 hours. No significant leakage occurs during or at the completion of the test (disregard traces of ointment presumed to originate externally from within the crimp of the tube or from the thread of the cap). If leakage is observed from one, but not more than one, of the tubes, repeat the test with 20 additional tubes of the Ointment. The requirement is met if no leakage is observed from the first 10 tubes tested, or if leakage is observed from not more than one of 30 tubes tested.

⟨781⟩ OPTICAL ROTATION

Many drugs, in a pure state or in a solution, are optically active in the sense that they cause incident plane polarized light to emerge in a continuum of planes at different intensities such that the plane of maximum intensity forms a measurable angle with the plane of the incident light. Where this effect is large enough for precise measurement, it may serve as the basis for an assay or an identity test. The optical rotation is expressed in degrees, as either *angular rotation* (observed) or *specific rotation* (calculated with reference to the specific concentration of 1 g of solute in 1 mL of solution, measured under stated conditions). Substances that cause the light plane to rotate clockwise, as viewed toward the light source, are termed dextrorotatory and the *angular rotation* is designated (+); those that cause counter-clockwise rotation are termed levorotatory and the *angular rotation* is designated (−).

Specific rotation usually is expressed by the term:

$$[\alpha]_x^t,$$

in which *t* represents, in degrees centigrade (Celsius), the temperature at which the rotation is determined, and *x* represents the characteristic spectral line or wavelength of the light used. Unless otherwise specified, the values cited in this Pharmacopeia relate to measurements at 25° with the use of the D line of sodium (a doublet at 589.0 nm and 589.6 nm).

The accuracy and precision of optical rotatory measurements will be increased if they are carried out with due regard for the following general considerations.

The instrument itself must be in good condition. Optical elements must be brilliantly clean and in exact alignment. The match point should lie close to the normal zero mark. The light source should be rigidly set and well aligned with respect to the optical bench. It should be supplemented by a filtering system capable of transmitting light of a sufficiently monochromatic nature. Precision polarimeters generally are designed to accommodate interchangeable disks to isolate the D line from sodium light or the 546.1-mm line from the mercury spectrum. With polarimeters not thus designed, cells containing suitably colored liquids may be employed as filters. Temperature control of the solution and of the polarimeter requires special attention, since rotatory power varies appreciably with temperature. The temperature specified for the determination applies to the solution, and is maintained within 0.2° of the stated value. Accuracy is assured by calibration of the instrument with suitable standards.* Generally a polarimeter capable of giving replicate readings within 0.020° suffices for Pharmacopeial purposes.

Polarimeter tubes should be filled in such a way as to avoid creating or leaving air bubbles that interfere with the passage of the beam of light. Interference from bubbles is minimized with tubes in which the bore is expanded at one end. However, with tubes of uniform bore, such as semimicro or micro tubes, care is required for proper filling. Non-metallic tubes are recommended for use in testing corrosive articles or solutions of articles in corrosive solvents.

In closing tubes having removable end-plates fitted with gaskets and caps, the latter should be tightened only enough to ensure a leak-proof seal between the end-plate and the body of the tube. Excessive pressure on the end-plate may set up strains that result in interference with the measurement. In determining the specific rotation of a substance of low rotatory power, it is desirable to loosen the caps and tighten them again between successive readings in the measurement of both the rotation and the zero point. Differences arising from end-plate strain thus generally will be revealed, and appropriate adjustments to eliminate the cause may be made.

Procedure—Where the substance is a liquid, adjust its temperature to 25°, and transfer it to the polarimeter tube. Proceed as directed below, beginning with "Make at least five readings," but carry out the blank determination on the empty, dry tube.

Where the substance is a solid, accurately weigh a suitable portion and transfer it to a volumetric flask by means of water, or other solvent if specified, reserving a portion of the solvent for the blank determination. Add enough solvent to bring the meniscus close to but still below the mark, and adjust the flask contents to 25° as by suspending the flask in a constant-temperature bath. Add solvent to the mark, and mix. Transfer the solution to the polarimeter tube within 30 minutes from the time the substance was dissolved, taking care to standardize the elapsed time in the case of substances known to undergo racemization or mutarotation. During the elapsed time interval, maintain the solution at a temperature of 25°.

Make at least five readings, at 25°, of the observed rotation. Substitute the reserved solvent for the solution, and make an equal number of readings on it. The zero correction is the average of the blank readings, and is subtracted from the average observed rotation. It is necessary in this calculation to use the observed signs of rotation, whether positive or negative, to give the corrected observed rotation.

Where an automatic photoelectric polarimeter that possesses the necessary degree of accuracy and precision is employed, the need for five or more repetitive readings is obviated.

Calculation—Calculate the specific rotation of a liquid substance, or of a solid in solution, by application of one of the following formulas:

I. For liquid substances, $[\alpha]_x^t = \frac{a}{ld}$;

II. For solutions, $[\alpha]_x^t = \frac{100a}{lpd} = \frac{100a}{lc}$;

in which *a* is the corrected observed rotation, in degrees, at temperature *t* at wavelength *x*; *l* is the length of the polarimeter tube, in decimeters; *d* is the specific gravity of the liquid or solution at the temperature of observation; *p* is the concentration of the solution expressed as the number of g of substance in 100 g of solution; and *c* is the concentration of the solution expressed as the number of g of substance in 100 mL of solution.

⟨785⟩ OSMOLARITY

Osmotic pressure is fundamentally related to all biological processes that involve diffusion of solutes or transfer of fluids through membranes. Thus, knowledge of the osmolar concentrations of parenteral fluids is essential. The labels of Pharmacopeial solutions that provide intravenous replenishment of fluid, nu-

* Suitable calibrators are available from the Office of Standard Reference Materials, National Institute of Science and Technology, Washington, DC 20234, as Standard Reference Material 41b, Dextrose, and Standard Reference Material 17a, Sucrose.

trient(s), or electrolyte(s), as well as of the osmotic diuretic Mannitol Injection, are required to state the osmolar concentration.

The declaration of osmolar concentration on the label of a parenteral solution serves primarily to inform the practitioner whether the solution is hypo-osmotic, iso-osmotic, or hyper-osmotic. A quantitative statement facilitates calculation of the dilution required to render a hyper-osmotic solution iso-osmotic. It also simplifies many calculations involved in peritoneal dialysis and hemodialysis procedures. The osmolar concentration of an extemporaneously compounded intravenous solution prepared in the pharmacy (e.g., a hyperalimentation solution) from osmolar-labeled solutions also can be obtained simply by summing the osmoles contributed by each constituent.

The units of osmolar concentration are usually expressed as milliosmoles (abbreviation: mOsmol) of solute per liter of solution. In general terms, the weight of an osmole is the gram molecular weight of a substance divided by the number of ions or chemical species (n) formed upon dissolution. In ideal solutions, for example, $n = 1$ for glucose, $n = 2$ for sodium chloride or magnesium sulfate, $n = 3$ for calcium chloride, and $n = 4$ for sodium citrate.

The ideal osmolar concentration may be determined according to the formula:

$$\text{osmolar concentration (mOsmol/liter)} = \text{mOsM}$$

$$= \frac{\text{wt. of substance (g/liter)}}{\text{mol. wt. (g)}} \times \text{number of species} \times 1000.$$

As the concentration of the solute increases, interaction among solute particles increases, and actual osmolar values decrease when compared to ideal values. Deviation from ideal conditions is usually slight in solutions within the physiologic range and for more dilute solutions, but for highly concentrated solutions the actual osmolarities may be appreciably lower than ideal values. For example, the ideal osmolarity of 0.9% Sodium Chloride Injection is $9/58.4 \times 2 \times 1000 = 308$ milliosmoles per liter. In fact, however, n is slightly less than 2 for solutions of sodium chloride at this concentration, and the actual measured osmolarity of 0.9% Sodium Chloride Injection is about 286 milliosmoles per liter.

The theoretical osmolarity of a complex mixture, such as Protein Hydrolysate Injection, cannot be readily calculated. In such instances, actual values of osmolar concentration are to be used to meet the labeling requirement set forth in the individual monograph. They are determined by calculating the osmolarity from measured values of osmolal concentration and water content. Each osmole of solute added to 1 kg of water lowers the freezing point approximately 1.86° and lowers the vapor pressure approximately 0.3 mm of mercury (at 25°). These physical changes are measurable, and they permit accurate estimations of osmolal concentrations.

Where osmometers that measure the freezing-point depression are employed, a measured volume of solution (usually 2 mL) is placed in a glass tube immersed in a temperature-controlled bath. A thermistor and a vibrator are lowered into the mixture, and the temperature of the bath is decreased until the mixture is super-cooled. The vibrator is activated to induce crystallization of the water in the test solution, and the released heat of fusion raises the temperature of the mixture to its freezing point. By means of a Wheatstone bridge, the recorded freezing point is converted to a measurement in terms of milliosmolality, or its near equivalent for dilute solutions, milliosmolarity. The instrument is calibrated by using two standard solutions of sodium chloride that span the expected range of osmolarities.

Osmometers that measure the vapor pressures of solutions are less frequently employed. They require a smaller volume of specimen (generally about 5 µL), but the accuracy and precision of the resulting osmolality determination are comparable to those obtained by the use of osmometers that depend upon the observed freezing points of solutions.

Labeling—Where an osmolarity declaration is required in the individual monograph, the label states the total osmolar concentration in milliosmoles per liter. Where the contents are less than 100 mL, or where the label states that the article is not for direct injection but is to be diluted before use, the label alternatively may state the total osmolar concentration in milliosmoles per milliliter.

⟨788⟩ PARTICULATE MATTER IN INJECTIONS

Particulate Matter—Particulate matter consists of extraneous, mobile, undissolved substances, other than gas bubbles, unintentionally present in parenteral solutions. Injectable solutions, including solutions constituted from sterile solids intended for parenteral use, are essentially free from particles that can be observed on visual inspection. In the following tests, for large-volume and small-volume Injections, the results obtained in examining a discrete unit or group of units for particulate matter cannot be extrapolated with certainty to other units that remain untested. Statistically sound sampling plans based upon a known set of given operational factors must be elaborated if valid inferences are to be drawn from observed data to characterize the level of particulate matter in a large group of units. Two procedures for the determination of particulate matter are given herein, differing in accordance with the labeled volume of an article in a container. All large-volume Injections for single-dose infusion, and those small-volume Injections for which the monographs specify such requirements, are subject to the particulate matter limits set forth for the test being applied.

LARGE-VOLUME INJECTIONS FOR SINGLE-DOSE INFUSION

Limits for particulate matter are prescribed herein for individual articles in containers that are labeled as containing more than 100 mL of a single-dose large-volume Injection intended for administration by intravenous infusion. The limits do not apply to multiple-dose Injections, to single-dose, small-volume Injections, nor to injectable solutions constituted from sterile solids.

This test for particulate matter is suitable for revealing the presence of particles whose longest axis, or effective linear dimension, is 10 µm or greater. Alternative procedures or procedural details may be employed to measure particulate matter, provided the results obtained are of equivalent reliability. However, where a difference appears, or in the event of a dispute, only the result obtained by the procedure given in this Pharmacopeia is conclusive.

PROCEDURE—[NOTE—Throughout this procedure, use suitable, nonpowdered gloves and scrupulously clean glassware and equipment that have been rinsed successively with a warm solution of detergent, hot water, water, and isopropyl alcohol. Apply the water as a strong jet back and forth across the surface of the vertically held object, working slowly from top to bottom. Perform the rinsing with isopropyl alcohol under a laminar flow hood equipped with ultra-HEPA (high-efficiency particulate air) filters. Permit the objects to dry under the hood upstream of all other operations. Preferably, locate the hood in a separate room that is supplied with filtered, air-conditioned air, and maintained under positive pressure with respect to the surrounding area. Prior to conducting the test, clean the laminar flow hood (except the surfaces of the filter media) with an appropriate solvent. Maintain airflow velocity at 90 ± 20 feet per minute.]

Membrane Filter and Assembly—Using forceps, remove a color contrast grid membrane filter from its container. Wash both sides of the membrane with a stream of water that has been further purified by filtration through a suitable membrane to remove particulate matter having an effective linear dimension greater than 5 µm, by holding the filter in a vertical position, and, starting at the top of the non-gridded side, sweeping the stream back and forth across the surface, working slowly from top to bottom so that particles will be rinsed downward off the filter, and repeating the process on the gridded side. Place the membrane (grid side up) on the filter holder base, and install the filtering funnel on the base without sliding the funnel over the membrane filter. Invert the assembled unit, and wash the inside of the funnel for about 10 seconds with a jet of filtered water. Allow the water to drain, and place the unit on the filter flask.

Test Preparation—Mix the solution by inverting the container 20 times. Thoroughly clean the outer surface of the container with a jet of water, and remove the closure carefully, avoiding contamination of the contents. Transfer 25 mL of the well-mixed solution to the funnel, allow to stand for 1 minute, and apply the vacuum and filter. Release the vacuum gently, and wash the inner walls of the funnel with a jet of 25 mL of the filtered water. Direct the jet of filtered water in such manner as to wash the walls of the funnel free from any particles that may have become lodged on the walls, but avoid directing the stream onto the filter surface. After turbulence has dissipated, vacuum-filter the rinsing. Carefully remove the upper section of the filter assembly while maintaining vacuum. Release the vacuum, and remove the membrane filter with forceps. Place the filter in a plastic petri slide, using a very thin film of stopcock grease as pre-coating, if necessary to hold the filter flat and in place. Allow the filter to dry with the cover of the petri slide slightly ajar. Cover the slide carefully on the micrometer stage of the microscope, and count the particles on the filter as described below.

Determination—Examine the entire membrane filter in a suitable microscope under 100× magnification with the incident light at an angle of 10° to 20° with the horizontal. Count the number of particles having effective linear dimensions equal to or larger than 10 µm and equal to or larger than 25 µm. Perform a blank determination, using a *Membrane filter and assembly*, as directed under *Test preparation*, beginning with "wash the inner walls of the funnel with a jet." Subtract the total counts obtained in the blank determination from the uncorrected total counts for the *Test preparation*. [NOTE—For Dextrose-containing solutions, do not enumerate morphologically indistinct material showing little or no surface relief and presenting a gelatinous or film-like appearance. Since in solution this material consists of units of the order of 1 µm or less and is liable to be counted only after aggregation and/or deformation on the membrane, interpretation of enumeration may be aided by testing a specimen of the solution with a suitable electronic particle counter.]

Interpretation—Duplicate *Test preparations* and blanks may be examined as directed. If the blank determination yields more than 5 particles having effective linear dimension of 25 µm or greater, the operational environment is unsatisfactory and the test is invalid.

The large-volume Injection for single-dose infusion meets the requirements of the test if it contains not more than 50 particles per mL that are equal to or larger than 10 µm and not more than 5 particles per mL that are equal to or larger than 25 µm in effective linear dimension.

SMALL-VOLUME INJECTIONS

This test is applicable to all small-volume Injections, in containers that are labeled as containing 100 mL or less, single- or multiple-dose, either in solution or in solution constituted from sterile solids, wherever a requirement for a limit for particulate matter appears in the individual monograph. Injections packaged in prefilled syringes and cartridges are exempt from these requirements unless an individual monograph states specifically that prefilled syringes and cartridges are to be included. The requirement does not apply where the monograph specifies that the label shall state that the product is to be used with a final filter.

The test calls for the use of an electronic liquid-borne particle counter system utilizing a light-obscuration based sensor with a suitable sample feeding device. [NOTE—see *Tests and Assays, Apparatus*, page 5 under *General Notices and Requirements*].[1,2]

Acceptable resolution of the sensor, and accuracy of the sampling apparatus used, are critical to this test. The following two methods are intended to aid in assuring system suitability.

Determination of Sensor Resolution—The particle size resolution of the instrumental particle counter is dependent upon the particle sensor used. Determine the resolution of the particle counter for 10-µm particles. Use the monosized particle size standard. The relative standard deviation of the standard is less than 5%. Two acceptable methods of determining particle size resolution are: (1) manually generating a particle count versus particle size response curve, and (2) using an electronic method of measuring and sorting particle sensor voltage output with a multichannel analyzer.

Sensor Flow Rate—Verify that the flow rate is within the manufacturer's specifications for the sensor used.

MANUAL METHOD—Sample successive aliquots of the 10-µm particle size standard suspension at various particle size threshold settings (typically from 5 µm to 15 µm). The breadth of the particle size response range depends on the particle sensor resolution and the distribution of the particle size standard. Plot the particle counts versus the corresponding particle size thresholds to determine the observed size distribution (a Gaussian distribution results). Calculate the percentage resolution by the formula:

$$\% \text{ resolution} = (100/D)(\text{Var}_{\text{Obs}} - \text{Var}_{\text{Std}})^{1/2},$$

in which D is the mean particle diameter, and Var_{Obs} and Var_{Std} are the variances of the observed size distribution and the labeled distribution of the particle standard, respectively. The resolution is not greater than 10%.

ELECTRONIC METHOD—Record the voltage output distribution of the particle sensor, using a multi-channel analyzer while sampling a suspension of the 10-µm particle size standard. Proceed with the calculations as directed for the *Manual Method*. The resolution is not greater than 10%.

Sample Volume Accuracy—Since particle count varies directly with the volume of fluid sampled, it is important that the sampling accuracy be known (or known to be within a certain range). Fill all of the dead volume in the feeder with *Water for Injection*. Take 10 mL of *Water for Injection* in a tared container. Withdraw 5 mL through the sample feeding device, and again weigh the container. The accuracy of the volume sampled is ±5%.

PROCEDURE—[NOTE—Prepare the sample, glassware, closures, and other required equipment in an environment protected by HEPA (high-efficiency particulate air) filters. Particle-free garments and nonpowdered gloves preferably are worn throughout the preparation. Preferably locate the hood in a separate room supplied with HEPA-filtered, air-conditioned air, maintained under positive pressure with respect to the surrounding area.]

Use a pressure vessel capable of withstanding a pressure of 100 psi, having pressure tubing which does not shed particles and a hand-held spray nozzle with filter holder, for filtering water for cleaning and sample preparation. Use non-gridded filters of 5.0-µm pore size or less.

For standardization and sample preparation, use hardened, non-particle shedding, glass containers having openings of minimum size to reduce inadvertent contamination. Where closures are used, they are to have non-shedding liners such as polytef.

Glassware and Closure Cleaning—Cleanse glassware, closures, and other required equipment by immersing and scrubbing in warm, nonionic detergent solution, then rinsing in flowing warm tap water, followed by rinsing in flowing filtered water. Organic solvents may be used to facilitate cleaning. Finally, pressure-rinse in filtered water, using a hand-held pressure nozzle with final filter, or other appropriate equipment.

Particulate Control Test—Conduct this test to determine that the environment is suitable for the analysis and that the glassware is properly cleaned, and to assure the water to be used for analysis is particle-free.

Using filtered water and cleaned glassware, take 5 consecutive water samples of 5 mL each. Invert each sample 20 times. Degas by ultrasonication for 30 seconds, or by allowing to stand for 2 minutes. Stir each water sample by mechanical means at a speed sufficient to maintain a slight vortex throughout the analysis. If 5 particles of 25-µm or 25 particles of 10-µm or greater size are observed for the combined 25 mL, either the environment is not suitable for particulate analysis or the filtered water and glassware have not been properly prepared. Repeat the preparatory steps until environment, water, and glassware are suitable for this test.

Calibration—Calibrate the instrument with 3 standards, each consisting of monosized polystyrene spheres, approximately 10 µm, 20 µm, and 30 µm, in an aqueous vehicle.[3] When using particulate reference standards, take care to reduce particle agglomeration and assure particle purity. Suitable methods are

[1] Equivalency may be determined by procedures outlined in ASTM F-660-83, available from the American Society for Testing and Materials, 1916 Race St., Philadelphia, PA 19103.

[2] Suitable apparatus is available from Pacific Scientific, Instrument Div., 2431 Linden Lane, Silver Spring, MD 20910.

Suitable sensors are available also from Russell Laboratories, 3314 Rubio Crest Drive, Altadena, CA 91001.

[3] See ASTM F322-80.

available for examining commercial spheres where desired.[4] Use the procedure in ASTM 658-87 (defining counting and sizing accuracy of a liquid-borne particle counter using near-mono disperse spherical particulate materials) to calibrate automated particle counters.

Test Preparation—Prepare the test specimens in the following sequence:

Remove outer closures, sealing bands, and any loose or shedding paper labels, wash the exterior of containers as described under *Glassware and Closure Cleaning,* and dry in a particle-free airflow. Withdraw the contents of the containers in the normal or customary manner of use, or as instructed in the package labeling, except that containers with removable stoppers may be sampled by removing the closure and emptying the contents into a clean container.

Determination—

(*A*) *Liquid Products*

(1) Mix by inverting 25 times within 10 seconds.

NOTE—Because of the small volume of some products, it is necessary to agitate the solution more vigorously in order to suspend the particles properly.

(2) Open and combine the contents of not less than 10 containers, to obtain a volume of not less than 20 mL, in a cleaned container.

(3) Degas by ultrasonication for 30 seconds or by allowing to stand for 2 minutes.

(4) Gently stir contents of containers by hand swirling or by mechanical means, taking care not to introduce air bubbles or contamination. Stir continuously throughout the analysis.

(5) Take 3 consecutive portions, each not less than 5 mL. Discard the data from the first portion.

(*B*) *Dry or Lyophilized Products*

(1) Open the container, taking care not to contaminate the opening or cover.

(2) Constitute with a suitable volume of filtered water, or with the appropriate filtered diluent if water is not suitable.

(3) Replace cover, and agitate as in (*A*).

(4) Analyze as in (*A*).

(*C*) For products packaged in containers that are constructed to hold the drug product and a solvent in separate compartments, mix each unit as directed in the labeling. Analyze the solutions as in (*A*).

(*D*) For products labeled as "Pharmacy Bulk Package—Not for direct infusion," proceed as directed under (A) or (B) above, performing the test on a portion taken from pooled units that is equal to the maximum dose given in the labeling. For the calculations below, consider this portion to be the equivalent of the contents of one full container.

Calculations—Average the counts resulting from the 2 portions of the sample analyzed. Calculate the number of particles in each container, P_C, by the equation:

$$P_C = \overline{C}V_T/V_PN,$$

in which $\overline{C}$ is the average particle count obtained from the portions analyzed, V_T is the volume, in mL, of pooled sample, V_P is the volume, in mL, of each portion analyzed, and N is the number of containers pooled.

Interpretation—The small-volume Injection meets the requirements of the test if the average number of particles it contains is not more than 10,000 per container that are equal to or greater than 10 µm in effective spherical diameter and not more than 1000 per container equal to or greater than 25 µm in effective spherical diameter.

⟨791⟩ pH

For compendial purposes, pH is defined as the value given by a suitable, properly standardized, potentiometric instrument (pH meter) capable of reproducing pH values to 0.02 pH unit using an indicator electrode sensitive to hydrogen-ion activity, the glass electrode, and a suitable reference electrode such as calomel or silver–silver chloride. The instrument should be capable of sensing the potential across the electrode pair and, for pH standardization purposes, applying an adjustable potential to the circuit by manipulation of "standardization," "zero," "asymmetry," or "calibration" control, and should be able to control the change in millivolts per unit change in pH reading through a "temperature" and/or "slope" control. Measurements are made at 25 ± 2°, unless otherwise specified in the individual monograph or herein.

The pH scale is defined by the equation:

$$\mathrm{pH} = \mathrm{pHs} + (E - E_S)/k,$$

in which E and E_S are the measured potentials where the galvanic cell contains the solution under test, represented by pH, and the appropriate *Buffer Solution for Standardization,* represented by pHs, respectively. The value of k is the change in potential per unit change in pH and is theoretically $[0.05916 + 0.000198(t - 25°)]$ volts at any temperature t. This operational pH scale is established by assigning rounded pH values to the *Buffer Solutions for Standardization* from the corresponding National Bureau of Standards *molal* solutions.

It should be emphasized that the definitions of pH, the pH scale, and the values assigned to the *Buffer Solutions for Standardization* are for the purpose of establishing a practical, operational system so that results may be compared between laboratories. The pH values thus measured do not correspond exactly to those obtained by the classical definition, $\mathrm{pH} = -\log [H^+(aq)]$. So long as the solution being measured is sufficiently similar in composition to the buffer used for standardization, the operational pH corresponds fairly closely to the theoretical pH. Although no claim is made with respect to the suitability of the system for measuring hydrogen-ion activity or concentration, the values obtained are closely related to the activity of the hydrogen ion in aqueous solutions.

Where a pH meter is standardized by use of an aqueous buffer and then used to measure the "pH" of a nonaqueous solution or suspension, the ionization constant of the acid or base, the dielectric constant of the medium, the liquid-junction potential (which may give rise to errors of approximately 1 pH unit), and the hydrogen-ion response of the glass electrode are all changed. For these reasons, the values so obtained with solutions that are only partially aqueous in character can be regarded only as apparent pH values. However, acidity may be accurately measured with the proper use of electrodes and instrument standardization.

Buffer Solutions for Standardization of the pH Meter—

Buffer Solutions for Standardization are to be prepared as directed in the accompanying table.* Buffer salts of requisite purity can be obtained from the National Institute of Science and Technology. Solutions may be stored in chemically resistant, tight containers, such as Type I glass bottles. Fresh solutions should be prepared at intervals not to exceed 3 months. The table indicates the pH of the buffer solutions as a function of temperature. The instructions presented here are for the preparation of solutions having the designated molal (*m*) concentrations. For convenience, and to facilitate their preparation, however, instructions are given in terms of dilution to a 1000-mL volume rather than specifying the use of 1000 g of solvent, which is the basis of the molality system of solution concentration. The indicated quantities cannot be computed simply without additional information.

Potassium Tetraoxalate, 0.05 *m*—Dissolve 12.61 g of $KH_3(C_2O_4)_2 \cdot 2H_2O$ in water to make 1000 mL.

Potassium Biphthalate, 0.05 *m*—Dissolve 10.12 g of $KHC_8H_4O_4$, previously dried at 110° for 1 hour, in water to make 1000 mL.

Equimolal Phosphate, 0.05 *m*—Dissolve 3.53 g of Na_2HPO_4 and 3.39 g of KH_2PO_4, each previously dried at 120° for 2 hours, in water to make 1000 mL.

[4] For example, ASTM F322-80. The National Bureau of Standards SRM 1960, approximately 10-µm spheres, is useful in this regard.

* Commercially available buffer solutions for pH meter standardization, standardized by methods traceable to the National Institute of Science and Technology (NIST), labeled with a pH value accurate to 0.01 pH unit, and provided with a table showing the pH values at various temperatures, may be used. Solutions prepared from ACS reagent grade materials or other suitable materials, in the stated quantities, may be used provided the pH of the resultant solution is the same as that of the solution prepared from the NBS certified material.

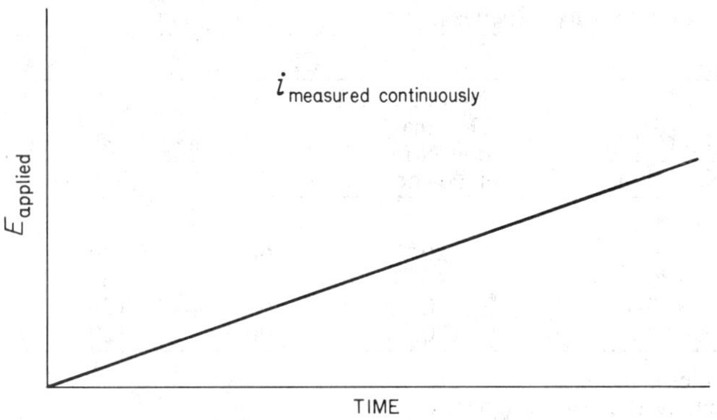

Fig. 2. Direct Current (dc) Polarography.

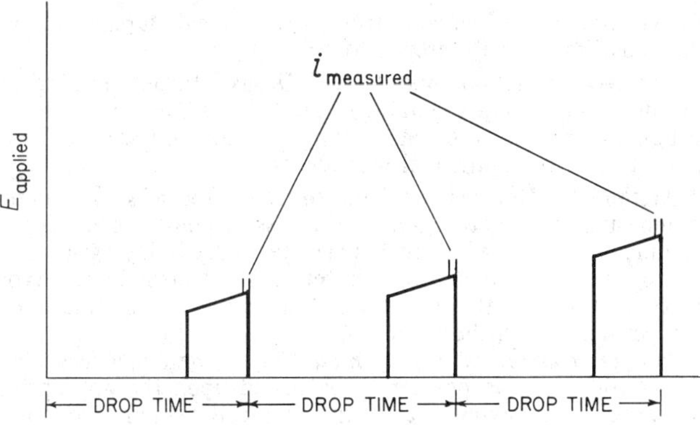

Fig. 3. Pulse Polarography.

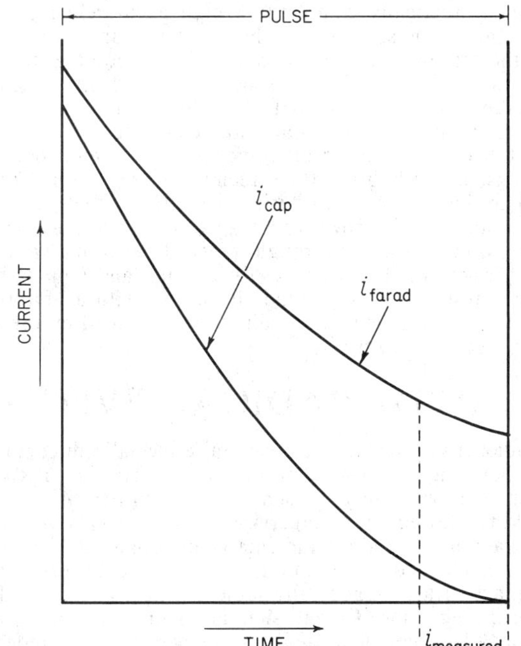

Fig. 4. Plot of Current Versus Time in Pulse Polarography.

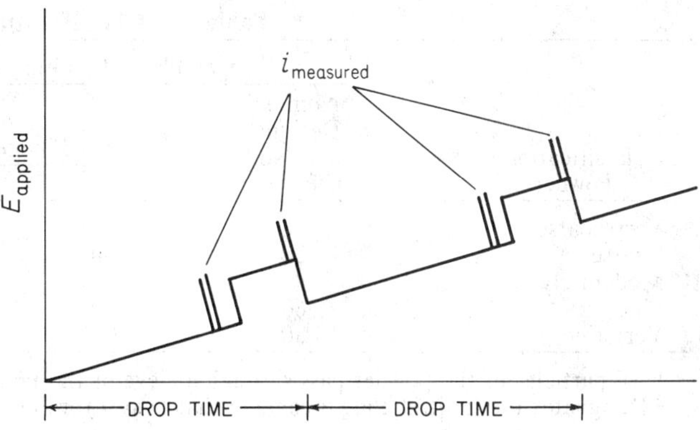

Fig. 5. Differential Pulse Polarography.

Limiting current values are more easily measured, since the waves are free from oscillations.

Differential pulse polarography is a technique whereby a fixed-height pulse applied at the end of the life of each drop is superimposed on a linear increasing dc ramp (see Figure 5). Current flow is measured just before application of the pulse and again at the end of the pulse. The difference between these two currents is measured and presented to the recorder. Such a differential signal provides a curve approximating the derivative of the polarographic wave, and gives a peak presentation. The peak potential is equivalent to:

$$E_{1/2} - \Delta E/2,$$

where ΔE is the pulse height. The peak height is directly proportional to concentration at constant scan rates and constant pulse heights. This technique is especially sensitive (levels of 10^{-7} *M* may be determined) and affords improved resolution between closely spaced waves.

Anodic Stripping Voltammetry—Anodic stripping voltammetry is an electrochemical technique whereby trace amounts of substances in solution are concentrated (reduced) onto an electrode and then stripped (oxidized) back into solution by scanning the applied voltage anodically. The measurement of the current flow as a function of this voltage and scanning rate provides qualitative and quantitative information on such substances. The concentration step permits analyses at 10^{-7} *M* to 10^{-9} *M* levels.

Basic instrumentation includes a voltage ramp generator; current-measuring circuitry; a cell with working, reference, and counter electrodes; and a recorder or other read-out device. Instruments having dc or pulse-polarographic capabilities are generally quite adequate for stripping application. The working electrode commonly used is the hanging mercury drop electrode (HMDE), although the mercury thin-film electrode (MTFE) has acquired acceptance. For analysis of metals such as silver, platinum, and gold, whose oxidation potentials are more positive than mercury, and mercury itself, the use of solid electrodes such as platinum, gold, or carbon is required. A saturated calomel electrode or a silver–silver chloride electrode serves as the reference except for the analysis of mercury or silver. A platinum wire is commonly employed as the counter electrode.

Test specimens containing suitable electrolyte are pipeted into the cell. Dissolved oxygen is removed by bubbling nitrogen through the cell for 5 to 10 minutes.

Generally, an electrolysis potential equivalent to 200 to 300 mV more negative than the half-wave potential of the material to be analyzed is applied (although this potential is to be determined experimentally), with stirring for 1 to 10 minutes. For reproducible results, maintain constant conditions (i.e., deposition time, stirring rate, temperature, specimen volume, and drop size if HMDE is used).

After deposition, the stirring is discontinued and the solution and electrode are allowed to equilibrate for a short period. The potential is then rapidly scanned anodically (10 mV/second or greater in dc polarography and 5 mV/second in differential pulse polarography). As in polarography, the limiting current is proportional to concentration of the species (wave height in dc and pulse; peak height in differential pulse), while the half-wave potential (dc, pulse) or peak potential (differential pulse) identifies the species. It is imperative that the choice of supporting electrolyte be made carefully in order to obtain satisfactory behavior. Quantitation is usually achieved by a standard addition or calibration method.

This technique is appropriate for trace-metal analysis, but has limited use in organic determinations, since many of these reactions are irreversible. In analyzing substances such as chloride, cathodic stripping voltammetry may be used. The technique is the same as anodic stripping voltammetry, except that the substance is deposited anodically and then stripped by a cathodic voltage scan.

Table 1. Classification of Powders by Fineness.

Classification of Powder	Vegetable and Animal Drugs: Nominal Designation No.[1] of Powder	Fineness Limit[2] %	Fineness Limit[2] Sieve No.	Chemicals: Nominal Designation No.[1] of Powder	Fineness Limit[2] %	Fineness Limit[2] Sieve No.
Very coarse	8	20	60			
Coarse	20	40	60	20	60	40
Moderately coarse	40	40	80	40	60	60
Fine	60	40	100	80	60	120
Very fine	80	100	80	120	100	120

[1] All particles of the powder pass through a sieve of the nominal designation.
[2] Designates the limit of the percentage that passes through a sieve of the size designated.

⟨811⟩ POWDER FINENESS

The fineness of powders in this Pharmacopeia, expressed in descriptive terms, is related to the number assigned to a standard (U. S. Series) sieve, as indicated in Table 1.

For practical reasons, sieves are the preferred means of measuring powder fineness for most pharmaceutical purposes; however, their applicability does not extend into the range of particle size that is of increasing interest with respect to the attainment of prompt and complete gastrointestinal absorption of administered drugs. For the measurement of particles less than 100 µm in nominal size, devices other than sieves may be more useful.

The efficiency and speed of particle separation by sieves vary inversely with the number of particles in the charge. The effectiveness of separation falls off rapidly when the depth of the charge exceeds a layer of 6 to 8 particles.

Sieves for Pharmacopeial Testing—Sieves for Pharmacopeial testing are of wire cloth woven, not twilled, except the cloth for the sizes Nos. 230, 270, 325, and 400, from brass, bronze, stainless steel, or other suitable wire, and are not coated or plated. Table 2 gives the average dimensions of the openings of woven wire cloth standard sieves.

For details on the standardization of sieves, reference may be made to Specification E11-70 of the American Society for Testing and Materials. For use in the evaluation of the effective opening of test sieves in the size range of No. 20 through No. 70, Standard Glass Spheres are available from the National Bureau of Standards as Standard Reference Material 1018.

Table 2. Openings of Standard Sieves.

Sieve Designation: Nominal Designation No.	Sieve Opening
2[1]	9.5 mm
3.5	5.6 mm
4	4.75 mm
8	2.36 mm
10	2.00 mm
14	1.40 mm
16	1.18 mm
18	1.00 mm
20	850 µm
25	710 µm
30	600 µm
35	500 µm
40	425 µm
45	355 µm
50	300 µm
60	250 µm
70	212 µm
80	180 µm
100	150 µm
120	125 µm
200	75 µm
230	63 µm
270	53 µm
325	45 µm
400	38 µm

[1] Designated as ⅜ inch in ASTM Specification E11-70.

Powdered Vegetable and Animal Drugs—In determining the powder fineness of a vegetable or animal drug, no portion of the drug may be rejected during milling or sifting unless specifically permitted in the individual monograph.

Method for Determining Uniformity of Fineness—For determining uniformity of degree of fineness of powdered drugs and chemicals, the following process may be used, employing standard testing sieves that meet the requirements set forth above. Avoid prolonged shaking that would result in increasing the fineness of the powder during the testing.

For *very coarse*, *coarse*, and *moderately coarse powders*, place 25 to 100 g of the powder to be tested upon the appropriate standard sieve having a close-fitting receiving pan and cover. Shake the sieve in a rotary horizontal direction and vertically by tapping on a hard surface for not less than 20 minutes or until sifting is practically complete. Weigh accurately the amount remaining on the sieve and in the receiving pan.

In the case of *fine* or *very fine powders*, proceed as for *coarse powders*, except that the test specimen should not exceed 25 g, and except that the sieve is to be shaken for not less than 30 minutes or until sifting is practically complete.

In the case of oily or other powders that tend to clog the openings, carefully brush the screen at intervals during the test. Break up lumps that form during the sifting.

The fineness of a powdered drug or chemical may be determined also by screening through standard sieves in a *mechanical sieve shaker*, which reproduces the circular and tapping motion given to testing sieves in hand sifting but with a uniform mechanical action, following the directions provided by the manufacturer of the shaker.

⟨821⟩ RADIOACTIVITY

Radioactive pharmaceuticals require specialized techniques in their handling and testing in order that correct results may be obtained and hazards to personnel be minimized. All operations should be carried out or supervised by personnel having had expert training in handling radioactive materials.

The facilities for the production, use, and storage of radioactive pharmaceuticals are generally subject to licensing by the federal Nuclear Regulatory Commission, although in certain cases this authority has been delegated to state agencies. The federal Department of Transportation regulates the conditions of shipment of radioactive materials. State and local agencies often have additional special regulations. Each producer or user must be thoroughly cognizant of the applicable regulations of the federal Food, Drug, and Cosmetic Act, and any additional requirements of the U. S. Public Health Service and of state and local agencies pertaining to the articles concerned.

Definitions, special considerations, and procedures with respect to the Pharmacopeial monographs on radioactive drugs are set forth in this chapter.

GENERAL CONSIDERATIONS

Fundamental Decay Law

The decay of a radioactive source is described by the equation:

$$N_t = N_o e^{-\lambda t},$$

in which N_t is the number of atoms of a radioactive substance at elapsed time t, N_o is the number of those atoms when $t = 0$, and λ is the transformation or decay constant, which has a characteristic value for each radionuclide. The *half-life*, $T_{1/2}$, is the time interval required for a given activity of a radionuclide to decay to one-half of its initial value, and is related to the decay constant by the equation:

$$T_{1/2} = \frac{0.69315}{\lambda}.$$

The activity of a radioactive source (A) is related to the number of radioactive atoms present by the equation:

$$A = \lambda N,$$

from which the number of radioactive atoms at time t can be computed, and hence the mass of the radioactive material can be determined.

The activity of a pure radioactive substance as a function of time can be obtained from the exponential equation or from decay tables, or by graphical means based on the half-life (see Normalized Decay Chart, Figure 1).

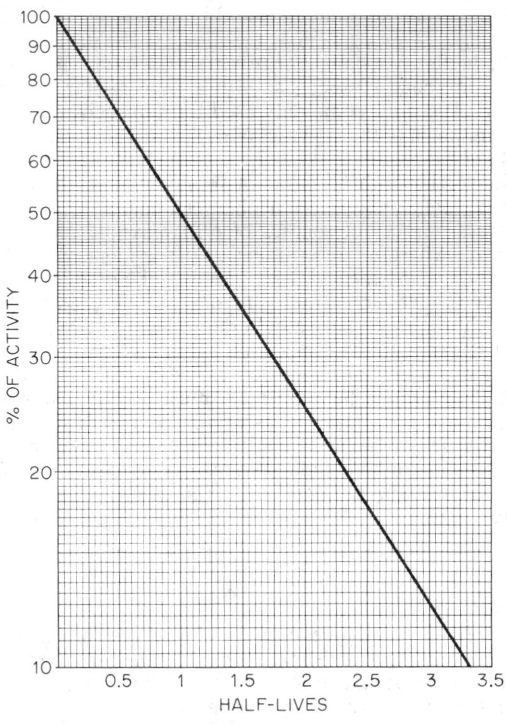

Fig. 1. Normalized Decay Chart.

The activity of a radioactive material is expressed as the number of nuclear transformations per unit time. The fundamental unit of radioactivity, the *curie* (Ci), is defined as 3.700×10^{10} nuclear transformations per second. The *millicurie* (mCi) and *microcurie* (μCi) are commonly used subunits. The "number of nuclear transformations per unit time" is the sum of rates of decay from all competing modes of disintegration of the parent nuclide. Before the activity of any given radionuclide in a measured specimen can be expressed in curies, it is often necessary to know the abundance(s) of the emitted radiation(s) measured.

Geometry

The validity of relative calibration and measurement of radionuclides is dependent upon the reproducibility of the relationship of the source to the detector and its surroundings. Appropriate allowance must be made for source configuration.

Background

Cosmic rays, radioactivity present in the detector and shielding materials, and radiation from nearby radioactive sources not properly shielded from the measuring equipment, all contribute to the background count rate. All radioactivity measurements must be corrected by subtracting the background count rate from the gross count rate in the test specimen.

Statistics of Counting

Since the process of radioactive decay is a random phenomenon, the events being counted form a random sequence in time. Therefore, counting for any finite time can yield only an estimate of the true counting rate. The precision of this estimate, being subject to statistical fluctuations, is dependent upon the number of counts accumulated in a given measurement and can be expressed in terms of the standard deviation σ. An *estimate* for σ is $\sqrt{n}$, where n is the number of counts accumulated in a given measurement. The probability of a single measurement falling within $\pm 100/\sqrt{n}\%$ of the mean of a great many measurements is 0.68. That is, if many measurements of n counts each were to be made, approximately two-thirds of the observations would lie within $\pm 100/\sqrt{n}\%$ of the mean, and the remainder outside.

Because of the statistical nature of radioactive decay, repeated counting of an undisturbed source in a counting assembly will yield count-rate values in accordance with the frequency of a normal distribution. Deviations in these values from the normal distribution conform to the χ^2 test. For this reason, the χ^2 test is frequently applied to determine the performance and correct operation of a counting assembly. In the selection of instruments and conditions for assay of radioactive sources, the figure of merit ϵ^2/B should be maximized (where ϵ = counter efficiency = observed count rate/sample disintegration rate, and B = background count rate).

Counting Losses

The minimum time interval that is required for the counter to resolve two consecutive signal pulses is known as the dead time. The dead time varies typically from the order of microseconds for proportional and scintillation counters, to hundreds of microseconds for Geiger-Müller counters. Nuclear events occurring within the dead time of the counter will not be registered. To obtain the corrected count rate, R, from the observed count rate, r, it is necessary to use the formula:

$$R = \frac{r}{1 - r_\tau},$$

in which τ is the dead time. The foregoing correction formula assumes a nonextendable dead time. Thus, for general validity, the value of r_τ should not exceed 0.1. The observed count rate, r, refers to the gross specimen count rate and is not to be corrected for background before use in the foregoing equation.

Calibration Standards

Perform all radioactivity assays using measurement systems calibrated with appropriately certified radioactivity standards. Such calibration standards may be purchased either direct from the National Institute of Standards and Technology or from other sources that have established traceability to the National Institute of Standards and Technology through participation in a program of inter-comparative measurements. Where such calibration standards are unavailable, the Pharmacopeia provides the nuclear decay data required for calibration. These data, as well as half-life values, are obtained from the Evaluated Nuclear Structure Data File of the Oak Ridge Nuclear Data Project, and reflect the most recent values at the time of publication.

Carrier

The total mass of radioactive atoms or molecules in any given radioactive source is directly proportional to the activity of the radionuclide for a given half-life, and the amount present in radiopharmaceuticals is usually too small to be measured by ordinary chemical or physical methods. For example, the mass of ^{131}I having an activity of 100 mCi is 8×10^{-7} g. Since suc' small amounts of material behave chemically in an anom manner, carriers in the form of nonradioactive isotopes of the same radionuclide may be added during processing to permit ready handling. In many cases, adsorption can be prevented merely by increasing the hydrogen-ion concentration of the solution. Amounts of such material, however, must be sufficiently small that undesirable physiological effects are not produced. The term "carrier-free" refers only to radioactive preparations in which nonradioactive isotopes of the radionuclide are absent. This implies that radioactive pharmaceuticals produced by means of (n, γ) reactions cannot be considered carrier-free.

The activity per unit volume or weight of a medium or vehicle containing a radionuclide either in the carrier-free state or in the presence of carrier is referred to as the radioactive concentration, whereas the term specific activity is used to express the activity of a radionuclide per gram of its element.

Radiochemical Purity

Radiochemical purity of a radiopharmaceutical preparation refers to the fraction of the stated radionuclide present in the stated chemical form. Radiochemical impurities in radiopharmaceuticals may result from decomposition and from improper preparative procedures. Radiation causes decomposition of water, a main ingredient of most radiopharmaceuticals, leading to the production of reactive hydrogen atoms and hydroxyl radicals, hydrated electrons, hydrogen, hydrogen ions, and hydrogen peroxide. The last-mentioned is formed in the presence of oxygen radicals, originating from the radiolytic decomposition of dissolved oxygen. Many radiopharmaceuticals show improved stability if oxygen is excluded. Radiation may also affect the radiopharmaceutical itself, giving rise to ions, radicals, and excited states. These species may combine with one another and/or with the active species formed from water. Radiation decomposition may be minimized by the use of chemical agents that act as electron or radical scavengers. Electrons trapped in solids cause discoloration due to formation of F-centers and the darkening of glass containers for radiopharmaceuticals, a situation that typifies the case. The radiochemical purity of radiopharmaceuticals is determined by column, paper, and thin-layer chromatography or other suitable analytical separation techniques as specified in the individual monograph.

Radionuclidic Purity

Radionuclidic purity of a radiopharmaceutical preparation refers to the proportion of radioactivity due to the desired radionuclide in the total radioactivity measured. Radionuclidic purity is important in the estimation of the radiation dose received by the patient when the preparation is administered. Radionuclidic impurities may arise from impurities in the target materials, differences in the values of various competing production cross-sections, and excitation functions at the energy or energies of the bombarding particles during production.

Terms and Definitions

The *date of manufacture* is the date on which the manufacturing cycle for the finished product is completed.

The *date of assay* is the date (and time, if appropriate) when the actual assay for radioactivity is performed.

The *date of calibration* is an arbitrary assigned date and time to which the radioactivity of the product is calculated for the convenience of the user.

The *expiration date* is the date that establishes a limit for the use of the product. The expiration period (i.e., the period of time between the date of manufacture and the expiration date) is based on a knowledge of the radioactive properties of the product and the results of stability studies on the finished dosage form.

Labeling

Individual radiopharmaceutical monographs indicate the expiration date, the calibration date, and the statement, "Caution—Radioactive Material." The labeling indicates that in making dosage calculations, correction is to be made for radioactive decay, and also indicates the radioactive half-life of the radionuclide. Articles that are Injections comply with the requirements for *Labeling* under *Injections* ⟨1⟩, and those that are Biologics comply with the requirements for *Labeling* under *Biologics* ⟨1041⟩.

IDENTIFICATION AND ASSAY OF RADIONUCLIDES

Instrumentation

IONIZATION CHAMBERS

An ionization chamber is an instrument in which an electric field is applied across a volume of gas for the purpose of collecting ions produced by a radiation field. The positive ions and negative electrons drift along the lines of force of the electric field, and are collected on electrodes, producing an ionization current. In a properly designed well-type ionization chamber, the ionization current should not be too dependent on the position of the radioactive specimen, and the value of the current per unit activity, known as the calibration factor, is characteristic of each gamma-ray-emitting radionuclide.

The ionization current produced in an ionization chamber is related to the mean energy of the emitted radiation and is proportional to the intensity of the radiation. If standard sources of known disintegration rates are used for efficiency calibration, the ionization chamber may then be used for activity determinations between several microcuries and several hundred millicuries or more. The upper limit of activity that may be measured in an ionization chamber usually is not sharply defined and may be limited by saturation considerations, range of the amplifier, and design of the chamber itself. The data supplied with or obtained from a particular instrument should be reviewed to ascertain the useful ranges of energies and intensities of the device.

Reproducibility within approximately 5% or less can be readily obtained in about 10 seconds, with a deep re-entrant well-type chamber. The most commonly used form of ionization chamber for measurement of the activities of radiopharmaceuticals is known as a dose calibrator.

Although the calibration factor for a radionuclide may be interpolated from an ionization chamber energy-response curve, there are a number of sources of error possible in such a procedure. It is therefore recommended that all ionization chamber calibrations be performed with the use of authentic reference sources of the individual radionuclides, as described hereinafter.

The calibration of a dose calibrator should be maintained by relating the measured response of a standard to that of a long-lived performance standard, such as radium 226 in equilibrium with its daughters. The instrument must be checked daily with the ^{226}Ra or other source to ascertain the stability over a long period of time. This check should include performance standard readings at all radionuclide settings employed. To obtain the activity (A_x) of the radionuclide being measured, use the relationship:

$$A_x = \frac{R_c R}{R_n},$$

in which R_n is the new reading for the radium or other source, R_c is the reading for the same source obtained during the initial calibration procedure, and R is the observed reading for the radionuclide specimen. Obviously, any necessary corrections for radioactive decay of the reference source must first be applied. Use of this procedure should minimize any effects due to drift in the response of the instrument. The recommended activity of the ^{226}Ra or other monitor used in the procedure described above is 75 to 150 µCi. It is recommended also that the reproducibility and/or stability of multirange instruments be checked for all ranges with the use of appropriate standards.

The size and shape of a radioactive source may affect the response of a dose calibrator, and it is often necessary to apply a small correction when measuring a bulky specimen.

SCINTILLATION AND SEMICONDUCTOR DETECTORS

When all or part of the energy of beta or gamma radiation is dissipated within scintillators, photons of intensity proportional to the amount of dissipated energy are produced. These pulses are detected by an electron multiplier phototube and converted to electrical pulses, which are subsequently analyzed with a pulse-height analyzer to yield a pulse-height spectrum related to the energy spectrum of the radiation emitted by the source. In general, a beta-particle scintillation pulse-height spectrum approximates the true beta-energy spectrum, provided that the beta-particle source is prepared in such a manner that self-absorption is minimized. Beta-ray spectra may be obtained by using calcium fluoride or anthracene as the scintillator, whereas gamma-ray spectra are usually obtained with a thallium-activated sodium iodide crystal or a large-volume lithium-drifted germanium semiconductor detector. The spectra of charged particles also may be obtained using silicon semiconductor detectors and/or gas proportional counters. Semiconductor detectors are in essence solid-state ionization chambers, but the energy required to create an electron-hole pair or to promote an electron from the valence band to the conduction band in the semiconductor is about one-tenth the energy required for creation of an ion-pair in a gas-filled ionization chamber or proportional counter and is far less than the energy needed to produce a photon in a NaI(Tl) scintillation crystal. In gamma-ray spectrometry, a Ge(Li) detector can yield an energy resolution of 0.33% for 1.33 MeV gamma-rays from ^{60}Co, while a 3- × 3-inch NaI(Tl) crystal can give a value of 5.9% for the same gamma-ray energy. The energy resolution is a measure of the ability to distinguish the presence of two gamma rays closely spaced in energy and is defined by convention as the full width of the photopeak at its half maximum (FWHM), expressed in percentage of the photopeak energy.

Gamma-ray spectra exhibit one or more sharp, characteristic photopeaks, or full-energy peaks, as a result of total absorption in the detector of the full energy of gamma radiations from the source; these photopeaks are useful for identification purposes. Other secondary peaks are observed as a consequence of backscatter, annihilation radiation, coincidence summing, fluorescent X-rays, etc., accompanied by a broad band known as the Compton continuum arising from scattering of the photons in the detector and from surrounding materials. Since the photopeak response varies with gamma-ray energy, calibration of a gamma-ray spectrometer should be achieved with radionuclide standards having well-known gamma-ray energies and emission rates. The shape of the gamma-ray spectrum is dependent upon the shape and size of the detector and the types of shielding materials used.

When confirming the identity of a radionuclide by gamma-ray spectrometry, it is necessary to make a comparison of the specimen spectrum with that of a specimen of known purity of the same radionuclide obtained under *identical instrument parameters and specimen geometry*. Where the radionuclides emit coincident X- or gamma-radiations, the character of the pulse-height distribution often changes quite dramatically because of the summing effect of these coincident radiations in the detector as the efficiency of detection is increased (e.g., by bringing the source closer to the detector). Such an effect is particularly evident in the case of iodine 125. Among the more useful applications of gamma-ray spectrometry are those for the identification of radionuclides and the determination of radionuclidic impurities.

Where confirmation of the identity of a given radionuclide by means of a direct comparison with the spectrum of a specimen of the same radionuclide of known purity is not possible, the identity of the radionuclide in question must then be established by the following method. Two or more of the following nuclear decay scheme parameters of the radionuclide specimen to be identified shall be measured, and agreement shall be within ± 10%: (1) half-life, (2) energy of each gamma- or X-ray emitted, (3) the abundance of each emission, and (4) E_{max} for those radionuclides that decay with beta-particle emissions. Such measurements are to be performed as directed in the *Identification* and *Assay* sections of this chapter. Agreement of two or more of the measured parameters with the corresponding published nuclear decay scheme data constitutes confirmation of the identity of the radionuclide.

LIQUID-SCINTILLATION COUNTERS

Alpha- and beta-emitting radionuclides may be assayed with the use of a liquid-scintillation detector system. In the liquid scintillator, the radiation energy is ultimately converted into light quanta that are usually detected by two multiplier phototubes so arranged as to count only coincidence radiation. The liquid scintillator is a solution consisting of a solvent, primary and secondary solutes, and additives. The charged particle dissipates its energy in the solvent, and a fraction of this energy is converted into fluorescence in the primary solute. The function of the secondary solute is to shift the fluorescence radiation to longer wavelengths that are more efficiently detected by the multiplier phototubes. Frequently used solvents are toluene and *p*-xylene; primary solutes are 2,5-diphenyloxazole (PPO) and 2-(4′-*tert*-butylphenyl)-5-(4-biphenylyl)-1,3,4-oxadiazole (butyl-PBD); and secondary solutes are 2,2′-*p*-phenylenebis[4-methyl-5-phenyloxazole] (dimethyl-POPOP) and *p*-bis(*o*-methylstrylyl)benzene (bis-MSB). As a means of attaining compatibility and miscibility with aqueous specimens to be assayed, many additives, such as surfactants and solubilizing agents, are also incorporated into the scintillator. For an accurate determination of radioactivity of the specimen, care must be exercised to prepare a specimen that is truly homogeneous. The presence of impurities or color in solution causes a decrease in photon output of the scintillator; such a decrease is known as quenching. Accurate radioactivity measurement requires correcting for count-rate loss due to quenching.

The disintegration rate of a beta-particle source may be determined by a procedure in which the integral count rate of the specimen is measured as a function of the pulse-height discriminator bias, and the emission rate is then obtained by extrapolation to zero bias. Energetic alpha-particle emitters may be similarly measured by this method.

Identification

A radionuclide can be identified by its mode of decay, its half-life, and the energies of its nuclear emissions.

The radioactive half-life is readily determined by successive counting of a given source of the radionuclide over a period of time that is long compared to its half-life. The response of the counting assembly when employed for the decay measurement of long-lived radionuclides should be monitored with an even longer-lived reference source to assess and compensate for errors arising from electronic drift. In the case of short-lived radionuclides, when the counting period constitutes a significant fraction of the half-life of the radionuclide, the recorded count rate must be corrected to the time when the count is initiated, as follows:

$$R_t = \frac{r\lambda t}{1 - e^{-\lambda t}},$$

in which R_t is the count rate at the beginning of a counting period, r is the count rate observed over the entire counting period, t is the duration of the counting period, λ is the decay constant of the radionuclide, and e is the base of the natural logarithm. When t is small compared to the half-life of the radionuclide under study so that $\lambda t < 0.05$, then $(1 - e^{-\lambda t})$ approaches λt, and no such correction is necessary.

The energy of nuclear emissions is often determined by the maximum range of penetration of the radiation in matter (in the case of alpha- and beta-particles) and by the full-energy peak or photopeak in the gamma-ray spectrum (in the case of X- and gamma-rays). Since beta-particles are emitted with a continuous energy spectrum, the maximum beta-energy, E_{max}, is a unique index for each beta-emitting radionuclide. In addition to the maximum range and energy spectrum of the beta-particles, the absorption coefficient, when obtained under reproducible counting conditions, can serve as a reliable index for identification of a beta-emitter. Fortuitously, beta-particles are absorbed in matter in an approximately exponential manner, and a plot of the logarithm of the beta-particle count rate as a function of the absorber thickness is known as the absorption curve. The initial portion of the absorption curve shows linearity from which the absorption coefficient can be obtained. The maximum range is determined by the use of absorbers of varying thickness, and the energy spectrum is measured by beta-ray scintillation spectrometry.

The absorption of gamma-rays in matter is strictly exponential, but the half-value layers of attenuation have not been very useful for the purpose of radionuclide characterization. Gamma-rays from each isomeric transition are mono-energetic; their energy can be directly measured by gamma-ray spectrometry. Because of their high energy resolution, solid-state detectors [Ge(Li)] are vastly superior to scintillation detectors [NaI(Tl)] in gamma-ray spectrometry.

The activities of radiopharmaceutical solutions are frequently in the range of millicuries per mL. Such solutions usually must be extensively diluted before they can be accurately assayed. The diluent should be compatible with the radiopharmaceutical with respect to factors such as pH and redox potentials, so that no hydrolysis or change in oxidation state occurs upon dilution, which could lead to adsorption and separation of the radionuclide from solution.

BETA-EMITTING RADIONUCLIDES

Mass Absorption Coefficient Procedure—Deposit and dry an aliquot of the radioactive phosphorus 32 solution on a thin plastic film to minimize backscattering, and place it under a suitable counter. Determine the counting rates successively, using not less than six different "thicknesses" of aluminum each between 20 and 50 mg/cm^2 and a single absorber thicker than 800 mg/cm^2, which is used to measure the background. (The absorbers are inserted between the test specimen and the counter but are placed nearer the counter window to minimize scattering.) Net beta-particle count rates are obtained after subtraction of the count rate found with the absorber having a thickness of 800 mg/cm^2 or greater. Plot the logarithm of the net beta-particle count rate as a function of the total absorber "thickness." The total absorber "thickness" is the "thickness" of the aluminum absorbers plus the "thickness" of the counter window (as stated by the manufacturer) plus the air-equivalent "thickness" (the distance in centimeters of the specimen from the counter window multiplied by 1.205 mg/cm^3 at 20° and 76 cm of mercury), all expressed in mg/cm^2. An approximately straight line results.

Choose two total absorber "thicknesses" that differ by 20 mg/cm^2 or more and that fall on the linear plot, and calculate the mass absorption coefficient, μ, by the equation:

$$\mu = \frac{1}{t_2 - t_1} \cdot \ln\left(\frac{N_{t_1}}{N_{t_2}}\right) = \frac{2.303}{t_2 - t_1}(\log N_{t_1} - \log N_{t_2})$$

in which t_1 and t_2 represent the total absorber "thicknesses," in mg/cm^2, t_2 being the thicker absorber, and N_{t_1} and N_{t_2} being the net beta-particle rates with the t_1 and t_2 absorbers, respectively.

For characterization of the radionuclide, the mass absorption coefficient should be within ±5% of the value found for a pure specimen of the same radionuclide when determined under identical counting conditions and geometry.

Other Methods of Identification—Other methods for determining the identity of a beta emitter also rely upon the determination of E_{max}. This may be accomplished in several ways. For example, (1) utilization of the range energy relationships of beta particles in an absorber, or (2) determination of E_{max} from a beta-particle spectrum obtained on an energy-calibrated beta-spectrometer using a thin source of the radionuclide (see *Scintillation and Semiconductor Detectors* in this chapter).

GAMMA-EMITTING RADIONUCLIDES

The gamma-ray spectrum of a radionuclide is a valuable tool for the qualitative identification of gamma-ray emitting radionuclides. The full-energy peak, or the photopeak, is identified with the gamma-ray transition energy that is given in the decay scheme of the radionuclide.

In determining radionuclidic identity and purity, the gamma-ray spectrum of a radioactive substance is obtained with either a NaI(Tl) crystal or a semiconductor Ge(Li) detector. The latter has an energy resolution more than an order of magnitude better than the former and is highly preferred for analytical purposes. The spectrum obtained shall be identical in shape to that of a specimen of the pure radionuclide, measured with the same detection system and in the same geometry. The gamma-ray spectrum of the radiopharmaceutical shall contain only photopeaks identifiable with the gamma-ray transition energies found in the decay scheme of the same radionuclide. For low geometrical efficiencies, the areas under the photopeaks, after correction for the measured detector efficiency, shall be proportional to the abundances or emission rates of the respective gamma-rays in the radionuclide.

RADIONUCLIDIC IMPURITIES

Because they are extremely toxic, alpha-emitting nuclides must be strictly limited in radiopharmaceutical preparations. Procedures for identifying beta- and gamma-active radionuclides as given in the foregoing text are applicable to the detection of gamma and usually beta contaminants.

The gross alpha-particle activity in radiopharmaceutical preparations can be measured by the use of a windowless proportional counter or a scintillation detector employing a silver-activated zinc-sulfide phosphor or by the techniques of liquid-scintillation counting.

The heavy ionization caused by alpha particles allows the measurement of alpha-emitting radionuclides in the presence of large quantities of beta- and gamma-active nuclides by the use of appropriate techniques for discriminating the amplitudes of signal pulses. In proportional counting, the operating voltage region for counting alpha particles, referred to as the "alpha plateau," is considerably lower than the "beta plateau" for counting beta and gamma radiations. Typical "alpha plateau" and "beta plateau" voltage settings with P-10 counting gas are 900 to 1300 and 1600 to 2000 volts, respectively.

When silver-activated zinc-sulfide phosphor is employed for alpha-particle detection, the alpha particles can be distinguished from other interfering radiation by pulse-height discrimination. Care must be exercised to minimize self-absorption at the source whenever specimens are prepared for alpha-particle counting.

Assay

BETA-EMITTING RADIONUCLIDES

Procedure—The disintegration rate (A) of a beta-particle-emitting specimen is obtained by counting a quantitatively deposited aliquot in a fixed geometry according to the formula:

$$A = \frac{R}{\epsilon \times f_r \times f_b \times f_s},$$

in which ϵ is the counting efficiency of the counter; f_r is the correction factor for counter dead time; f_b is the correction factor for backscatter; and f_s is the correction factor for self-absorption. The count rate for zero absorber is obtained by extrapolation of the initial linear portion of the absorption curve to zero absorber "thickness," taking into consideration the mg/cm^2 "thickness" of specimen coverings, counter window, and the intervening air space between specimen and the counter window. The counter efficiency, ϵ, is determined by use of a long-lived secondary standard with similar spectral characteristics. RaD + E has frequently been used for efficiency calibration of counters for phosphorus 32. By the use of identical measurement conditions for the specimen and the standard (and extrapolation to zero absorber), the ratio of the values of f_r, f_b, and f_s for the standard and the specimen approaches unity.

The previous relationship is valid also when the counter has been calibrated with a standard of the radionuclide to be assayed. In this case, however, the extrapolations to zero absorber "thickness" for the specimen and standard are not required, as the two absorption corrections cancel for a given geometry.

Another useful and frequently employed method for the determination of the disintegration rate of beta-emitting radionuclides is liquid-scintillation counting, which also utilizes an extrapolation of the specimen count rate to zero pulse-height discriminator bias.

GAMMA-EMITTING RADIONUCLIDES

For the assay of gamma-emitting radionuclides, three methods are provided. The selection of the preferred method is dictated by the availability of a calibration standard of the radionuclide to be assayed and the radionuclidic purity of the article itself.

Direct comparison with a calibration standard is required if a calibration standard of the radionuclide to be assayed is available and if the upper limit of conceivable error in the activity determination arising from the presence of radionuclidic impurities has been determined to be less than 3%. If the required calibration standard is not routinely available, as would probably be the case for a short-lived radionuclide, but was available at some time prior to the performance of the assay for determination of efficiency of the counting system for the radionuclide to be assayed, use a calibrated counting system, provided the radionuclidic impurity content of the specimen meets the requirements stated for the direct comparison method. If the requirements for either of the first two methods cannot be met, use the method for determination of activity from a calibration curve.

With the exception of the first method, the counting systems used are monitored for stability. This requirement is met by daily checks with a long-lived performance check source and weekly checks with at least three sources covering a broad range of gamma-ray emission energies (e.g., ^{57}Co, ^{137}Cs, and ^{60}Co). If a discrepancy for any of the aforementioned measurements is found, either completely recalibrate or repair and recalibrate the system prior to further use.

Assay by Direct Comparison with a Calibration Standard—An energy selective measurement system (e.g., pulse-height analyzer) is not required for this procedure. Use either an ionization chamber or an integral counting system with a NaI(Tl) detector. A consistently reproducible geometrical factor from specimen to specimen is essential for accurate results. With proper precautions, the accuracy of this method approaches the accuracy with which the disintegration rate of the calibration standard is known.

Determine the counting rate of the detector system for a calibration standard of the radionuclide to be assayed (e.g., active enough to give good measurement statistics in a reasonable time, but not so active as to cause serious dead-time problems), selecting such a standard as to provide optimum accuracy with the particular assembly used. Place an accurately measured aliquot of the unknown assay specimen (diluted, if necessary) in a container identical to that used for the standard, and measure this specimen at approximately the same time and under the same geometrical conditions as for the standard. If the elapsed time between the measurements of the calibration standard and the specimen exceeds 12 hours, check the stability of the measurement system within 8 hours of the specimen measurement time with a long-lived performance check source. Record the system response with respect to the same check source at the time of calibration, and if subsequent checks exceed the original recorded response by more than ±3%, recalibration is required. Correct both activity determinations for background, and calculate the activity, in µCi per mL, by the formula:

$$SD(g/b),$$

in which S is the µCi strength of the standard, D is the dilution factor, and g and b are the measured values of counting rate for the specimen and the standard, respectively.

Assay with a Calibrated Integral Counting System—The procedure and precautions given for the preceding direct-comparison

Table of Nuclear Properties[1,2]

Principal Photon Emissions	Energy (keV)	Photons per 100 Disintegrations
^{129}I ($T_{1/2}$ = 1.57 × 10^7 years)		
$K_{\alpha 1}$[3]	29.8	37.0
$K_{\alpha 2}$	29.5	20.0
K_β	33.6	13.2
γ_1	39.6	7.52
Weighted Mean[4]	(31.3)	(77.80)
^{241}Am ($T_{1/2}$ = 432.2 years)		
ΣX_L	13.9	38.2
γ_1	26.4	2.5
γ_2	59.5	35.9
^{109}Cd ($T_{1/2}$ = 464 days)		
$K_{\alpha 1}$	22.2	55.1
$K_{\alpha 2}$	22.0	29.1
K_β	24.9	17.8
Weighted Mean[4]	(22.6)	(102.0)
γ_1	88.0	3.72
^{195}Au ($T_{1/2}$ = 183 days)		
$K_{\alpha 1}$	66.83	50
$K_{\alpha 2}$	65.12	29.0
K_β	75.7	21.7
Weighted Mean[4]	(68.25)	(100.7)
γ_1	30.88	0.83
γ_2	98.86	10.9
γ_3	129.7	0.89
^{57}Co ($T_{1/2}$ = 270.9 days)		
ΣX_K	6.5	56.0
γ_1	14.4	9.5
γ_2	122.1	85.6
γ_3	136.4	10.6
Weighted Mean ($\gamma_2 + \gamma_3$)[4]	(125.0)	(96.2)
^{139}Ce ($T_{1/2}$ = 137.7 days)		
$K_{\alpha 1}$	33.4	42.6
$K_{\alpha 2}$	33.0	23.1
K_β	37.8	15.6
Weighted Mean[4]	(34.1)	(81.3)
γ_1	165.8	80.0

Principal Photon Emissions	Energy (keV)	Photons per 100 Disintegrations
^{203}Hg ($T_{1/2}$ = 46.6 days)		
ΣX_L	10.3	5.6
$K_{\alpha 1}$	72.87	6.27
$K_{\alpha 2}$	70.83	3.72
K_β	82.6	2.79
Weighted Mean[4]	(74.6)	(12.8)
γ_1	279.2	81.5
^{113}Sn-^{113m}In ($T_{1/2}$ = 115.1 days)		
ΣX_L	3.3	9.0
$K_{\alpha 1}$	24.2	51.5
$K_{\alpha 2}$	24.0	27.5
K_β	27.3	17.3
Weighted Mean[4]	(24.7)	(96.7)
γ_1	255.1	1.9
γ_2	391.7	64.6
^{85}Kr ($T_{1/2}$ = 10.72 years)		
γ_1	514.0	0.43
^{137}Cs-^{137m}Ba ($T_{1/2}$ = 30.0 years)		
$K_{\alpha 1}$	32.2	3.90
$K_{\alpha 2}$	31.8	2.11
K_β	36.4	1.42
Weighted Mean[4]	(32.9)	(7.43)
γ_1	661.6	89.9
^{94}Nb ($T_{1/2}$ = 2 × 10^4 years)		
γ_1	702	100.0
γ_2	871	100.0
^{22}Na ($T_{1/2}$ = 2.60 years)		
$h\nu$	511	179.78[5]
γ_1	1274.5	99.95
^{60}Co ($T_{1/2}$ = 5.27 years)		
γ_1	1173.2[6]	99.88
γ_2	1332.5[6]	99.98

(1) In measurements for gamma- (or X-)ray assay purposes, fluorescent radiation from lead shielding (specifically, lead K X-rays ~76 keV) may interfere with quantitative results. Allowance must be made for these effects, or the radiation suppressed; a satisfactory means of absorbing this radiation is covering the exposed lead with cadmium sheet 0.06 to 0.08 inch thick, and then covering the cadmium with copper 0.02 to 0.04 inch thick.

(2) Only those photon emissions having an abundance ⩾1% are normally included.

(3) The K notation refers to X-ray emissions.

(4) The weighted mean energies and total intensities are given for groups of photons that would not be resolved by a NaI(Tl) detector.

(5) For this photon intensity to be usable, all emitted positrons must be annihilated in the source material.

(6) Cascade.

method apply, except that the efficiency of the detector system is determined and recorded for each radionuclide to be assayed, rather than simply recording the counting rate of the standard. Thus, the efficiency for a given radionuclide, x, is determined by $\epsilon_x = b_x/s_x$, in which b_x is the counting rate, corrected for background and dead-time, for the calibration standard of the radionuclide, x, and s_x is the corresponding activity of the certified calibration standard in nuclear transformations per second. For subsequent specimen assays, the activity is given by the formula:

$$A_x = Dg_x/\epsilon_x,$$

in which D is the dilution factor, g_x is the specimen counting rate (corrected for background and dead-time), and ϵ_x is the corresponding efficiency for the radionuclide.

Determination of Activity from a Calibration Curve—Versatility in absolute gamma-ray intensity measurements can be achieved by employing multi-channel pulse-height analysis. The photopeak efficiency of a detector system can be determined as a function of gamma-ray energy by means of a series of gamma-ray emission rate standard specimens, and the gamma-ray emission rate of any radionuclide for which no standard is available can be determined by interpolation from this efficiency curve. However, exercise care to ensure that the efficiency curve for the detector system is adequately defined over the entire region of interest by using a sufficient number of calibration points along the photopeak-energy axis.

Selection of a Counting Assembly—A gamma-ray spectrometer is used for the identification of radionuclides that emit X-rays or gamma rays in their decay. Requirements for an assembly suitable for identification and assay of the radionuclides used in radiopharmaceuticals are that (a) the resolution of the detector based on the 662-keV photopeak of ^{137}Cs-^{137m}Ba must be 8.0% or better, (b) the detector must be equipped with a specimen holder designed to facilitate exact duplication of counting geometry, and (c) the pulse-height analyzer must have enough channels to delineate clearly the photopeak being observed.

Procedure—Minimal requirements for the maintenance of instrument calibrations shall consist of weekly performance checks with a suitable reference source and a complete recalibration semi-annually. Should the weekly performance check deviate from the value determined at the time of calibration by more than 4.0%, a complete recalibration of the instrument is required at that time.

This method involves three basic steps, namely photopeak integration, determination of the photopeak efficiency curve, and calculation of the activity of the specimen.

PHOTOPEAK INTEGRATION—The method for the determination of the required photopeak area utilizes a Gaussian approximation for fitting the photopeak. A fixed fraction of the total number of photopeak counts can be obtained by taking the peak width, a, at some fraction of the maximum, where the shape has been experimentally found to be very close to Gaussian, and multiplying by the counting rate of the peak channel, P, after correction for any Compton and background contributions to the peak channel count rate. This background usually can be adequately determined by linear interpolation. This is illustrated in Figure 2.

The photopeak-curve shape is closest to a straight line at $0.606P$, and the contribution of the fractional channels to a can be accurately estimated by interpolation. Calculate a by the equation:

$$a = D' - D + \frac{d - 0.606P}{d - c} + \frac{d' - 0.606P}{d' - c'},$$

in which c and d and also c' and d' are the single *channel counting* rates on either side of $0.606P$, and D and D' are the channel *numbers* (locations) of d and d', respectively. The location of the required variables on the photopeak is illustrated in Figure 3.

From the known values for the counting rate in the peak channel of the photopeak, P, and the width of the peak at $0.606P$, a, a calibrated fraction of the photopeak area is then obtained from the product, (aP).

To summarize the procedures involved in obtaining a calibrated fraction of a photopeak area using this method, the necessary steps or calculations are presented below in a stepwise manner:

(1) Subtract any Compton and background contributions from the photopeak to be measured.

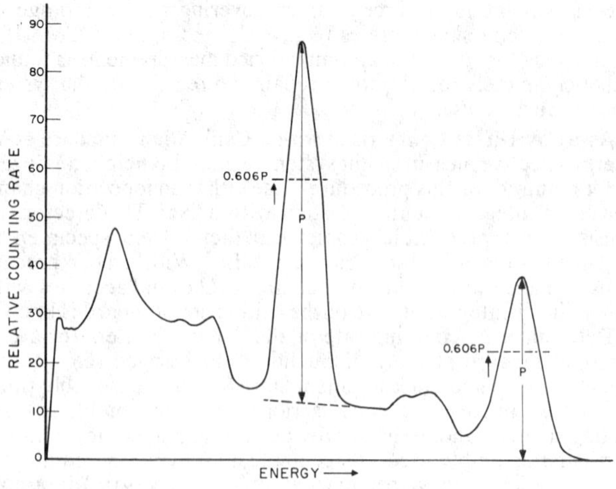

Fig. 2. Typical Gamma-ray Spectrum Showing the Selection of the Peak Channel Counting Rate, P, After Correction for Compton and Background Contributions.

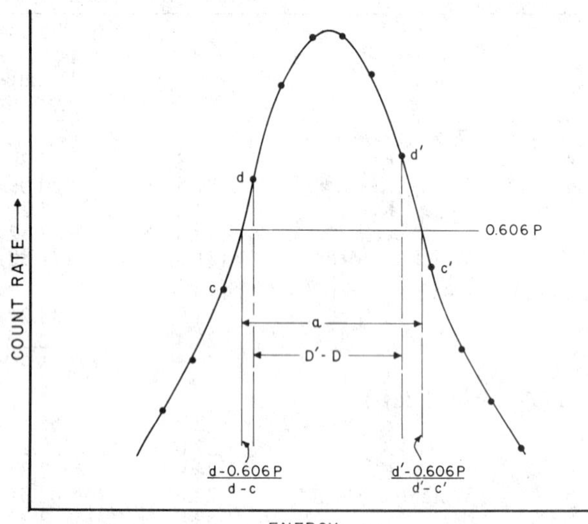

Fig. 3. Location of the Variables Required for the Determination of the Peak Width, a, at $0.606P$.

(2) Determine the counting rate of the peak channel (maximum channel counting rate after subtracting Compton and background), P.

(3) Multiply P by 0.606, and locate the horizontal line corresponding to the peak width, a.

(4) Obtain the peak width, a, by inserting the values of variables (obtained as shown in the preceding figure) into the equation defining a.

(5) The desired calibrated fraction of the peak area is then equal to the product of a times P or $F = aP$, where F is a fractional area of the peak proportional to the emission rate of the source.

This method provides a quick and accurate means of determining the gamma-ray emission rate of sources while avoiding, to a large extent, subjective estimates of the detailed shape of the tails of the peaks. The error due to using the maximum channel counting rate, rather than the theoretical maximum or peak channel rate, is of the order of 1.0% if a is 6 or greater.

PHOTOPEAK EFFICIENCY CALIBRATION—Radionuclides such as those listed in the accompanying table together with some of their nuclear decay data are available as certified reference standards.* A sufficient number of radioactive standard reference sources should be selected in order to obtain the calibration curve over the desired range. Where possible, standard sources of those radionuclides that are to be assayed should be included.

* These certified reference standards are obtainable from the National Institute of Standards and Technology, Washington, DC 20234.

Calculate the gamma-ray emission rate from the equation:

$$\Gamma = A_s b,$$

in which A_s is the activity, in disintegrations per second, of the standard used, and b is the number of gamma rays per disintegration at that energy. Accurately measure quantities of standard solutions of each radionuclide into identical containers, and determine the fractional photopeak area (F) for each of the standards.

Using the equation $\epsilon_p = F/\Gamma$, calculate the photopeak efficiency, ϵ_p, and construct a log–log plot of ϵ_p versus the gamma-ray energy as shown in Figure 4.

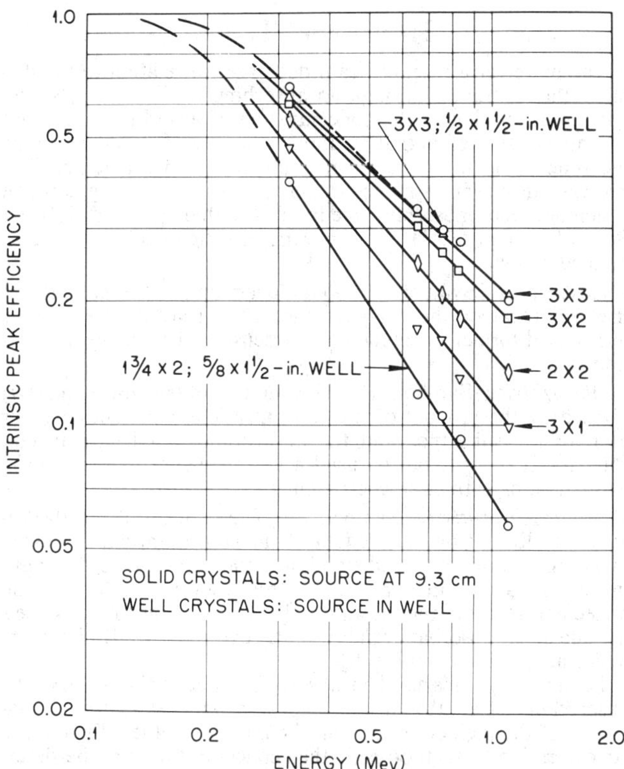

Fig. 4. Typical Photopeak Efficiency Calibration Curves for Various NaI(Tl) Detectors.

DETERMINATION OF SPECIMEN ACTIVITY—In the same manner as in the preparation of the calibration curve, determine the fractional area (F) of the principal photopeak of the specimen under assay or an accurately measured aliquot adjusted to the same volume in an identical container as used for the standards. From the calibration curve, find the value of ϵ_p for this radionuclide. Using the equation $\Gamma = F/\epsilon_p$, calculate the gamma-ray emission rate (Γ). Calculate the activity (A), in disintegrations per second, of the specimen using the equation $A = (\Gamma/b)(D)$, in which b is the number of gamma rays per disintegration and D is the dilution factor. To obtain the activity, in μCi or mCi, divide A by 3.7×10^4 or 3.7×10^7, respectively. The above relationship is equally valid for obtaining the activity of an undiluted specimen or capsule; in this case, the dilution factor, D, is unity.

⟨831⟩ REFRACTIVE INDEX

The refractive index (n) of a substance is the ratio of the velocity of light in air to the velocity of light in the substance. It is valuable in the identification of substances and the detection of impurities.

Although the standard temperature for Pharmacopeial measurements is 25°, many of the refractive index specifications in the individual monographs call for determining this value at 20°. The temperature should be carefully adjusted and maintained, since the refractive index varies significantly with temperature.

The values for refractive index given in this Pharmacopeia are for the D line of sodium (doublet at 589.0 nm and 589.6 nm). Most instruments available are designed for use with white light but are calibrated to give the refractive index in terms of the D line of sodium light.

The Abbé refractometer measures the range of refractive index for those Pharmacopeial materials for which such values are given. Other refractometers of equal or greater accuracy may be employed.

To achieve the theoretical accuracy of ±0.0001, it is necessary to calibrate the instrument against a standard provided by the manufacturer and to check frequently the temperature control and cleanliness of the instrument by determining the refractive index of distilled water, which is 1.3330 at 20° and 1.3325 at 25°.

⟨841⟩ SPECIFIC GRAVITY

Unless otherwise stated in the individual monograph, the specific gravity determination is applicable only to liquids, and, unless otherwise stated, is based on the ratio of the weight of a substance in air at 25° to that of an equal volume of water at the same temperature. Where a temperature is specified in the individual monograph, the specific gravity is the ratio of the weight of the substance in air at the specified temperature to that of an equal volume of water at the same temperature. When the substance is a solid at 25°, determine the specific gravity at the temperature directed in the individual monograph, and refer to water at 25°.

Procedure—Select a scrupulously clean, dry pycnometer that previously has been calibrated by determining its weight and the weight of recently boiled water contained in it at 25°. Adjust the temperature of the substance to about 20°, and fill the pycnometer with it. Adjust the temperature of the filled pycnometer to 25°, remove any excess of the substance, and weigh. Subtract the tare weight of the pycnometer from the filled weight of the pycnometer.

The specific gravity of the substance is the quotient obtained by dividing the weight of the substance contained in the pycnometer by the weight of water contained, both determined at 25° unless otherwise directed in the individual monograph.

⟨851⟩ SPECTROPHOTOMETRY AND LIGHT-SCATTERING

ULTRAVIOLET, VISIBLE, INFRARED, ATOMIC ABSORPTION, FLUORESCENCE, TURBIDIMETRY, NEPHELOMETRY, AND RAMAN MEASUREMENT

Absorption spectrophotometry is the measurement of an interaction between electromagnetic radiation and the molecules, or atoms, of a chemical substance. Techniques frequently employed in pharmaceutical analysis include ultraviolet, visible, infrared, and atomic absorption spectroscopy. Spectrophotometric measurement in the visible region was formerly referred to as *colorimetry;* however, it is more precise to use the term "colorimetry" only when considering human perception of color.

Fluorescence spectrophotometry is the measurement of the emission of light from a chemical substance while it is being exposed to ultraviolet, visible, or other electromagnetic radiation. In general, the light emitted by a fluorescent solution is of maximum intensity at a wavelength longer than that of the exciting radiation, usually by some 20 nm to 30 nm.

Light-scattering involves measurement of the light scattered because of submicroscopic optical density inhomogeneities of solutions and is useful in the determination of weight-average molecular weights of polydisperse systems in the molecular weight range from 1000 to several hundred million. Two such techniques utilized in pharmaceutical analysis are *turbidimetry* and *nephelometry.*

Raman spectroscopy (inelastic light-scattering) is a light-scattering process in which the specimen under examination is irradiated with intense monochromatic light (usually laser light) and the light scattered from the specimen is analyzed for frequency shifts.

The wavelength range available for these measurements extends from the short wavelengths of the ultraviolet through the infrared. For convenience of reference, this spectral range is roughly divided into the ultraviolet (190 nm to 380 nm), the visible (380 nm to 780 nm), the near-infrared (780 nm to 3000 nm), and the infrared (2.5 μm to 40 μm or 4000 cm^{-1} to 250 cm^{-1}).

Comparative Utility of Spectral Ranges

For many pharmaceutical substances, measurements can be made in the ultraviolet and visible regions of the spectrum with greater accuracy and sensitivity than in the near-infrared and infrared. When solutions are observed in 1-cm cells, concentrations of about 10 μg of the specimen per mL often will produce absorbances of 0.2 to 0.8 in the ultraviolet or the visible region. In the infrared and near-infrared, concentrations of 1 to 10 mg per mL and up to 100 mg per mL, respectively, may be needed to produce sufficient absorption; for these spectral ranges, cell lengths of from 0.01 mm to upwards of 3 mm are commonly used.

The ultraviolet and visible spectra of substances generally do not have a high degree of specificity. Nevertheless, they are highly suitable for quantitative assays, and for many substances they are useful as additional means of identification.

The near-infrared region is especially suitable for the determination of —OH and —NH groups, such as water in alcohol, —OH in the presence of amines, alcohols in hydrocarbons, and primary and secondary amines in the presence of tertiary amines.

The infrared spectrum is unique for any given chemical compound with the exception of optical isomers, which have identical spectra. However, polymorphism may occasionally be responsible for a difference in the infrared spectrum of a given compound in the solid state. Frequently, small differences in structure result in significant differences in the spectra. Because of the large number of maxima in an infrared absorption spectrum, it is sometimes possible to measure quantitatively the individual components of a mixture of known qualitative composition without prior separation.

The Raman spectrum and the infrared spectrum provide similar data, although the intensities of the spectra are governed by different molecular properties. Raman and infrared spectroscopy exhibit different relative sensitivities for different functional groups, e.g., Raman spectroscopy is particularly sensitive to C—S and C—C multiple bonds, and some aromatic compounds are more easily identified by means of their Raman spectra. Water has a highly intense infrared absorption spectrum, but a particularly weak Raman spectrum. Therefore, water has only limited infrared "windows" that can be used to examine aqueous solutes, while its Raman spectrum is almost completely transparent and useful for solute identification. The two major limitations of Raman spectroscopy are that the minimum detectable concentration of specimen is typically 10^{-1} *M* to 10^{-2} *M* and that the impurities in many substances fluoresce and interfere with the detection of the Raman scattered signal.

Optical reflectance measurements provide spectral information similar to that obtained by transmission measurements. Since reflectance measurements probe only the surface composition of the specimen, difficulties associated with the optical thickness and the light-scattering properties of the substance are eliminated. Thus, reflectance measurements are frequently more simple to perform on intensely absorbing materials. A particularly common technique used for infrared reflectance measurements is termed attenuated total reflectance (ATR), also known as multiple internal reflectance (MIR). In the ATR technique, the beam of the infrared spectrometer is passed through an appropriate infrared window material (e.g., KRS-5, a TlBr-TlI eutectic mixture), which is cut at such an angle that the infrared beam enters the first (front) surface of the window, but is totally reflected when it impinges on the second (back) surface (i.e., the angle of incidence of the radiation upon the second surface of the window exceeds the critical angle for that material). By appropriate window construction, it is possible to have many internal reflections of the infrared beam before it is transmitted out of the window. If a specimen is placed in close contact with the window along the sides that totally reflect the infrared beam, the intensity of reflected radiation is reduced at each wavelength (frequency) that the specimen absorbs. Thus, the ATR technique provides a reflectance spectrum that has been increased in intensity, when compared to a simple reflectance measurement, by the number of times that the infrared beam is reflected within the window. The ATR technique provides excellent sensitivity, but it yields poor reproducibility, and is not a reliable quantitative technique unless an internal standard is intimately mixed with each test specimen.

Fluorescence spectrophotometry is often more sensitive than absorption spectrophotometry. In absorption measurements, the specimen transmittance is compared to that of a blank; and at low concentrations, both solutions give high signals. Conversely, in fluorescence spectrophotometry, the solvent blank has low rather than high output, so that the background radiation that may interfere with determinations at low concentrations is much less. Whereas few compounds can be determined conveniently at concentrations below 10^{-5} *M* by light absorption, it is not unusual to employ concentrations of 10^{-7} *M* to 10^{-8} *M* in fluorescence spectrophotometry.

Theory and Terms

The power of a radiant beam decreases in relation to the distance that it travels through an absorbing medium. It also decreases in relation to the concentration of absorbing molecules or ions encountered in that medium. These two factors determine the proportion of the total incident energy that emerge. The decrease in power of monochromatic radiation passing through a homogeneous absorbing medium is stated quantitatively by Beer's law, $\log_{10}(1/T) = A = abc$, in which the terms are as defined below.

Absorbance [Symbol: *A*]—The logarithm, to the base 10, of the reciprocal of the transmittance (*T*). [NOTE—Descriptive terms used formerly include optical density; absorbancy; and extinction.]

Absorptivity [Symbol: *a*]—The quotient of the absorbance (*A*) divided by the product of the concentration, expressed in *g* per liter, of the substance and the absorption path length in cm. [NOTE—It is not to be confused with absorbancy index; specific extinction; or extinction coefficient.]

Molar Absorptivity [Symbol: ϵ]—The quotient of the absorbance (*A*) divided by the product of the concentration, expressed in *moles* per liter, of the substance and the absorption path length in cm. It is also the product of the absorptivity (*a*) and the molecular weight of the substance. [NOTE—Terms formerly used include molar absorbancy index; molar extinction coefficient; and molar absorption coefficient.]

For most systems used in absorption spectrophotometry, the absorptivity of a substance is a constant independent of the intensity of the incident radiation, the internal cell length, and the concentration, with the result that concentration may be determined photometrically.

Beer's law gives no indication of the effect of temperature, wavelength, or the type of solvent. For most analytical work the effects of normal variation in temperature are negligible.

Deviations from Beer's law may be caused by either chemical or instrumental variables. Apparent failure of Beer's law may result from a concentration change in solute molecules because of association between solute molecules or between solute and solvent molecules, or dissociation or ionization. Other deviations might be caused by instrumental effects such as polychromatic radiation, slit-width effects, or stray light.

Even at a fixed temperature in a given solvent, the absorptivity may not be truly constant. However, in the case of specimens having only one absorbing component, it is not necessary that the absorbing system conform to Beer's law for use in quantitative analysis. The concentration of an unknown may be found by comparison with an experimentally determined standard curve.

Although, in the strictest sense, Beer's law does not hold in atomic absorption spectrophotometry because of the lack of quantitative properties of the cell length and the concentration, the absorption processes taking place in the flame under conditions of reproducible aspiration do follow the Beer relationship in principle. Specifically, the negative log of the transmittance, or the absorbance, is directly proportional to the absorption coefficient, and, consequently, is proportional to the number of absorbing atoms. On this basis, calibration curves may be constructed to permit evaluation of unknown absorption values in terms of concentration of the element in solution.

Absorption Spectrum—A graphic representation of absorbance, or any function of absorbance, plotted against wavelength or function of wavelength.

Transmittance [Symbol: *T*]—The quotient of the radiant power transmitted by a specimen divided by the radiant power incident upon the specimen. [NOTE—Terms formerly used include transmittancy and transmission.]

Fluorescence Intensity [Symbol: I]—An empirical expression of fluorescence activity, commonly given in terms of arbitrary units proportional to detector response. The *fluorescence emission spectrum* is a graphical presentation of the spectral distribution of radiation emitted by an activated substance, showing intensity of emitted radiation as ordinate, and wavelength as abscissa. The *fluorescence excitation spectrum* is a graphical presentation of the activation spectrum, showing intensity of radiation emitted by an activated substance as ordinate, and wavelength of the incident (activating) radiation as abscissa. As in absorption spectrophotometry, the important regions of the electromagnetic spectrum encompassed by the fluorescence of organic compounds are the ultraviolet, visible, and near-infrared, i.e., the region from 250 nm to 800 nm. After a molecule has absorbed radiation, the energy can be lost as heat or released in the form of radiation of the same or longer wavelength as the absorbed radiation. Both absorption and emission of radiation are due to the transitions of electrons between different energy levels, or orbitals, of the molecule. There is a time delay between the absorption and emission of light; this interval, the duration of the excited state, has been measured to be about 10^{-9} second to 10^{-8} second for most organic fluorescent solutions. The short lifetime of fluorescence distinguishes this type of luminescence from phosphorescence, which is a long-lived afterglow having a lifetime of 10^{-3} second up to several minutes.

Turbidance [Symbol: S]—The light-scattering effect of suspended particles. The amount of suspended matter may be measured by observation of either the transmitted light (turbidimetry) or the scattered light (nephelometry).

Turbidity [Symbol: τ]—In light-scattering measurements, the turbidity is the measure of the decrease in incident beam intensity per unit length of a given suspension.

Raman scattering activity—The molecular property (in units of cm^4 per g) governing the intensity of an observed Raman band for a randomly oriented specimen. The scattering activity is determined from the derivative of the molecular polarizability with respect to the molecular motion giving rise to the Raman shifted band. In general, the Raman band intensity is linearly proportional to the concentration of the analyte.

Use of Reference Standards

With few exceptions, the Pharmacopeial spectrophotometric tests and assays call for comparison against a USP Reference Standard. This is to ensure measurement under conditions identical for the test specimen and the reference substance. These conditions include wavelength setting, slit-width adjustment, cell placement and correction, and transmittance levels. It should be noted that cells exhibiting identical transmittance at a given wavelength may differ considerably in transmittance at other wavelengths. Appropriate cell corrections should be established and used where required.

The expressions, "similar preparation" and "similar solution," as used in tests and assays involving spectrophotometry, indicate that the reference specimen, generally a USP Reference Standard, is to be prepared and observed in a manner identical for all practical purposes to that used for the test specimen. Usually in making up the solution of the specified Reference Standard, a solution of about (i.e., within 10%) the desired concentration is prepared and the absorptivity is calculated on the basis of the exact amount weighed out; if a previously dried specimen of the Reference Standard has not been used, the absorptivity is calculated on the anhydrous basis.

The expressions, "concomitantly determine" and "concomitantly measured," as used in tests and assays involving spectrophotometry, indicate that the absorbances of both the solution containing the test specimen and the solution containing the reference specimen, relative to the specified test blank, are to be measured in immediate succession.

Apparatus

Many types of spectrophotometers are available. Fundamentally, most types, except those used for infrared spectrophotometry, provide for passing essentially monochromatic radiant energy through a specimen in suitable form, and measuring the intensity of the fraction that is transmitted. Fourier transform infrared spectrophotometers use an interferometric technique whereby polychromatic radiation passes through the analyte and onto a detector on an intensity and time basis. Ultraviolet, visible, and dispersive infrared spectrophotometers comprise an energy source, a dispersing device (e.g., a prism or grating), slits for selecting the wavelength band, a cell or holder for the test specimen, a detector of radiant energy, and associated amplifiers and measuring devices. Fourier transform infrared systems utilize an interferometer instead of a dispersing device and a digital computer to process the spectral data. Some instruments are manually operated, whereas others are equipped for automatic and continuous recording. Instruments that are interfaced to a digital computer have the capabilities also of co-adding and storing spectra, performing spectral comparisons, and performing difference spectroscopy (accomplished with the use of a digital absorbance subtraction method).

Instruments are available for use in the visible; in the visible and ultraviolet; in the visible, ultraviolet, and near-infrared; and in the infrared regions of the spectrum. Choice of the type of spectrophotometric analysis and of the instrument to be used depends upon factors such as the composition and amount of available test specimen, the degree of accuracy, sensitivity, and selectivity desired, and the manner in which the specimen is handled.

The apparatus used in atomic absorption spectrophotometry has several unique features. For each element to be determined, a specific source that emits the spectral line to be absorbed should be selected. The source is usually a hollow-cathode lamp, the cathode of which is designed to emit the desired radiation when excited. Since the radiation to be absorbed by the test specimen element is usually of the same wavelength as that of its emission line, the element in the hollow-cathode lamp is the same as the element to be determined. The apparatus is equipped with an aspirator for introducing the test specimen into a flame, which is usually provided by air-acetylene, air-hydrogen, or, for refractory cases, nitrous oxide–acetylene. The flame, in effect, is a heated specimen chamber. A detector is used to read the signal from the chamber. Interfering radiation produced by the flame during combustion may be negated by the use of a chopped source lamp signal of a definite frequency. The detector should be tuned to this alternating current frequency so that the direct current signal arising from the flame is ignored. The detecting system, therefore, reads only the change in signal from the hollow-cathode source, which is directly proportional to the number of atoms to be determined in the test specimen. For Pharmacopeial purposes, apparatus that provides the readings directly in absorbance units is usually required. However, instruments providing readings in percent transmission, percent absorption, or concentration may be used if the calculation formulas provided in the individual monographs are revised as necessary to yield the required quantitative results. Percent absorption or percent transmittance may be converted to absorbance, A, by the following two equations:

$$A = 2 - \log_{10} (100 - \% \text{ absorption})$$

or:

$$A = 2 - \log_{10} (\% \text{ transmittance}).$$

Depending upon the type of apparatus used, the readout device may be a meter, digital counter, recorder, or printer. Both single-beam and double-beam instruments are commercially available, and either type is suitable.

Measurement of fluorescence intensity can be made with a simple *filter fluorometer*. Such an instrument consists of a radiation source, a primary filter, a specimen chamber, a secondary filter, and a fluorescence detection system. In most such fluorometers, the detector is placed on an axis at 90° from that of the exciting beam. This right-angle geometry permits the exciting radiation to pass through the test specimen and not contaminate the output signal received by the fluorescence detector. However, the detector unavoidably receives some of the exciting radiation as a result of the inherent scattering properties of the solutions themselves, or if dust or other solids are present. Filters are used to eliminate this residual scatter. The primary filter selects short-wavelength radiation capable of exciting the test specimen, while the secondary filter is normally a sharp cut-off filter that allows the longer-wavelength fluorescence to be transmitted but blocks the scattered excitation.

Most fluorometers use photomultiplier tubes as detectors, many types of which are available, each having special characteristics with respect to spectral region of maximum sensitivity, gain, and electrical noise. The photocurrent is amplified and read out on a meter or recorder.

A *spectrofluorometer* differs from a filter fluorometer in that filters are replaced by monochromators, of either the prism or the grating type. For analytical purposes, the spectrofluorometer is superior to the filter fluorometer in wavelength selectivity, flexibility, and convenience, in the same way in which a spectrophotometer is superior to a filter photometer.

Many radiation sources are available. Mercury lamps are relatively stable and emit energy mainly at discrete wavelengths. Tungsten lamps provide an energy continuum in the visible region. The high-pressure xenon arc lamp is often used in spectrofluorometers because it is a high-intensity source that emits an energy continuum extending from the ultraviolet into the infrared.

In spectrofluorometers, the monochromators are equipped with slits. A narrow slit provides high resolution and spectral purity, while a large slit sacrifices these for high sensitivity. Choice of slit size is determined by the separation between exciting and emitting wavelengths as well as the degree of sensitivity needed.

Specimen cells used in fluorescence measurements may be round tubes or rectangular cells similar to those used in absorption spectrophotometry, except that they are polished on all four vertical sides. A convenient test specimen size is 2 to 3 mL, but some instruments can be fitted with small cells holding 100 to 300 μL, or with a capillary holder requiring an even smaller amount of specimen.

Light-scattering instruments are available and consist in general of a mercury lamp, with filters for the strong green or blue lines, a shutter, a set of neutral filters with known transmittance, and a sensitive photomultiplier to be mounted on an arm that can be rotated around the solution cell and set at any angle from $-135°$ to $0°$ to $+135°$ by a dial outside the light-tight housing. Solution cells are of various shapes, such as square for measuring 90° scattering; semioctagonal for 45°, 90°, and 135° scattering; and cylindrical for scattering at all angles. Since the determination of molecular weight requires a precise measure of the difference in refractive index between the solution and solvent, $[(n - n_0)/c]$, a second instrument, a differential refractometer, is needed to measure this small difference.

Raman spectrometers include the following major components: a source of intense monochromatic radiation (invariably a laser); optics to collect the light scattered by the test specimen; a (double) monochromator to disperse the scattered light and reject the intense incident frequency; and a suitable light-detection and amplification system. Raman measurement is simple in that most specimens are examined directly in melting-point capillaries. Because the laser source can be focused sharply, only a few microliters of the specimen is required.

Procedure

Spectrophotometry—Detailed instructions for operating spectrophotometers are supplied by the manufacturers. To achieve significant and valid results, the operator of a spectrophotometer should be aware of its limitations and of potential sources of error and variation. The instruction manual should be followed closely on such matters as care, cleaning, and calibration of the instrument, and techniques of handling absorption cells, as well as instructions for operation. The following points require special emphasis.

Check the instrument for accuracy of calibration. Where a continuous source of radiant energy is used, attention should be paid to both the wavelength and photometric scales; where a spectral line source is used, only the photometric scale need be checked. A number of sources of radiant energy have spectral lines of suitable intensity, adequately spaced throughout the spectral range selected. The best single source of ultraviolet and visible calibration spectra is the quartz-mercury arc, of which the lines at 253.7, 302.25, 313.16, 334.15, 365.48, 404.66, and 435.83 nm may be used. The glass-mercury arc is equally useful above 300 nm. The 486.13-nm and 656.28-nm lines of a hydrogen discharge lamp may be used also. The wavelength scale may be calibrated also by means of suitable glass filters, which have useful absorption bands through the visible and ultraviolet regions. Standard glasses containing didymium (a mixture of praseodymium and neodymium) have been used widely. Glass containing holmium[1] is considered superior. The wavelength scales of near-infrared and infrared spectrophotometers are readily checked by the use of absorption bands provided by polystyrene films, carbon dioxide, water vapor, or ammonia gas.

For checking the photometric scale, a number of standard inorganic glass filters as well as standard solutions of known transmittances such as potassium chromate or potassium dichromate are available.[2]

Quantitative absorbance measurements usually are made on solutions of the substance in liquid-holding cells. Since both the solvent and the cell window absorb light, compensation must be made for their contribution to the measured absorbance. Matched cells are available commercially for ultraviolet and visible spectrophotometry for which no cell correction is necessary. In infrared spectrophotometry, however, corrections for cell differences usually must be made. In such cases, pairs of cells are filled with the selected solvent and the difference in their absorbances at the chosen wavelength is determined. The cell exhibiting the greater absorbance is used for the solution of the test specimen and the measured absorbance is corrected by subtraction of the cell difference.

With the use of a computerized Fourier transform infrared system, this correction need not be made, since the same cell can be used for both the solvent blank and the test solution. However, it must be ascertained that the transmission properties of the cell are constant.

Comparisons of a test specimen with a Reference Standard are best made at a peak of spectral absorption for the compound concerned. Assays prescribing spectrophotometry give the commonly accepted wavelength for peak spectral absorption of the substance in question. It is known that different spectrophotometers may show minor variation in the apparent wavelength of this peak. Good practice demands that comparisons be made at the wavelength at which peak absorption occurs. Should this differ by more than ± 1 nm from the wavelength specified in the individual monograph, re-calibration of the instrument may be indicated.

Test Preparation—For determinations utilizing ultraviolet or visible spectrophotometry, the specimen generally is dissolved in a solvent. Many solvents are suitable for these ranges, including water, alcohols, chloroform, lower hydrocarbons, ethers, and dilute solutions of strong acids and alkalies. Precautions should be taken to utilize solvents free from contaminants absorbing in the spectral region being used. It is usually advisable to use water-free methanol or alcohol, or alcohol denatured by the addition of methanol but not containing benzene or other interfering impurities, as the solvent. Solvents of special spectrophotometric quality, guaranteed to be free from contaminants, are available commercially from several sources. Some other analytical reagent-grade organic solvents may contain traces of impurities that absorb strongly in the ultraviolet region. New lots of these solvents should be checked for their transparency, and care should be taken to use the same lot of solvent for preparation of the test solution and the standard solution and for the blank.

No solvent in appreciable thickness is completely transparent throughout the near-infrared and infrared spectrum. Carbon tetrachloride (up to 5 mm in thickness) is practically transparent to 6 μm (1666 cm^{-1}). Carbon disulfide (1 mm in thickness) is

[1] Certified holmium glass filters are obtainable from the National Institute of Science and Technology (212.141g, 212.141h, and 212.141i). The same type of filter, without certification, is obtainable from Corning Glass Works, Corning, NY 14830 (Cat. No. 3130). The performance of an uncertified filter should be checked against one properly certified.

[2] For further detail regarding checks on both the wavelength and the photometric scales of a spectrophotometer, reference may be made to the following publications of the National Institute of Science and Technology, Department of Commerce, Washington, DC 20234: "Standards for Checking the Calibration of Spectrophotometers (200 to 1000 mμ)," Letter Circular LC-1017, January, 1955; "Spectrophotometry (200 to 1000 mμ)," Circular 484, September 15, 1949; "Calibrating Wave Lengths in the Region from 0.6 to 2.6 μ," *J. Research Natl. Bur. Standards* **49**, 13 (1952), RP 2338 and "Reference Wave Lengths for Calibrating Prism Spectrometers," *J. Research Natl. Bur. Standards* **58**, 195 (1957), RP 2752; "Liquid Absorbance Standards for UV-VIS," SRM #931.

suitable as a solvent to 40 μm (250 cm^{-1}) with the exception of the 4.2-μm to 5.0-μm (2381-cm^{-1} to 2000-cm^{-1}) and the 5.5-μm to 7.5-μm (1819-cm^{-1} to 1333-cm^{-1}) regions, where it has strong absorption. Other solvents have relatively narrow regions of transparency. For infrared spectrophotometry, an additional qualification for a suitable solvent is that it must not affect the material, usually sodium chloride, of which the cell is made. Where such a suitable solvent is not available, alternative methods for preparation of the test specimen include dispersing the finely ground solid specimen in mineral oil, or incorporating it in a transparent disk or pellet obtained by mixing it intimately with previously dried alkali halide salt (usually potassium bromide) and pressing the mixture in a die. A mineral oil dispersion is preferable where disproportionation between the alkali halide and the test specimen is encountered. For suitable materials the test specimen may be suspended neat as a thin film. For Raman spectrometry, most common solvents are suitable, and ordinary (nonfluorescing) glass specimen cells can be used. While the infrared region of the spectrum extends from about 2.5 μm to 40 μm, comparative spectra, obtained to ascertain compliance with a monograph specification for infrared absorption, are generally determined in the range from about 2.6 μm to 15 μm (3800 cm^{-1} to 650 cm^{-1}), unless otherwise specified in the individual monograph.

Where values for infrared line spectra are given in an individual monograph, the letters *s*, *m*, and *w* signify strong, medium, and weak absorption, respectively; *sh* signifies a shoulder, *bd* signifies a band, and *v* means very. The values may vary as much as 0.1 μm or 10 cm^{-1}, depending upon the particular instrument used. Polymorphism gives rise to variations in the infrared spectra of many compounds in the solid state. Therefore, when conducting infrared absorption tests, if a difference appears in the infrared spectra of the analyte and the standard, dissolve equal portions of the test substance and the standard in equal volumes of a suitable solvent, evaporate the solutions to dryness in similar containers under identical conditions, and repeat the test on the residues.

In atomic absorption spectrophotometry, the nature of the solvent and the concentration of solids must be given special consideration. An ideal solvent is one that interferes to a minimal extent in the absorption or emission processes and one that produces neutral atoms in the flame. If there is a significant difference between the surface tension or viscosity of the test solution and standard solution, the solutions are aspirated or atomized at a different rate, causing significant differences in the signals generated. The acid concentration of the solutions also affects the absorption processes. Thus, the solvents used in preparing the test specimen and the standard should be the same or as much alike in these respects as possible, and should yield solutions that are easily aspirated via the specimen tube of the burner-aspirator. Since undissolved solids present in the solutions may give rise to matrix or bulk interferences, the total undissolved solids content in all solutions should be kept below 2% wherever possible.

Calculations—The application of absorption spectrophotometry in an assay or a test generally requires the use of a Reference Standard. Where such a measurement is specified in an assay, a formula is provided in order to permit calculation of the desired result. A numerical constant is frequently included in the formula. The following derivation is provided to introduce a logical approach to the deduction of the constants appearing in formulas in the assays in many monographs.

The Beer's law relationship is valid for the solutions of both the Reference Standard (*S*) and the test specimen (*U*):

(1) $A_S = abC_S$

(2) $A_U = abC_U$

in which A_S is the absorbance of the Standard solution of concentration C_S, and A_U is the absorbance of the test specimen solution of concentration C_U. If C_S and C_U are expressed in the same units and the absorbances of both solutions are measured in matching cells having the same dimensions, the absorptivity, *a*, and the cell thickness, *b*, are the same; consequently, the two equations may be combined and rewritten to solve for C_U:

(3) $C_U = C_S(A_U/A_S)$.

Quantities of solid test specimens to be taken for analysis are generally specified in mg. Instructions for dilution are given in the assay and, since dilute solutions are used for absorbance measurements, concentrations are usually expressed for convenience in units of μg per mL. Taking a quantity, in mg, of a test specimen of a drug substance or solid dosage form for analysis, it therefore follows that a volume (V_U), in liters, of solution of concentration C_U may be prepared from the amount of test specimen that contains a quantity W_U, in mg, of the drug substance [NOTE—C_U is numerically the same whether expressed as μg per mL or mg per liter], such that:

(4) $W_U = V_UC_U$.

The form in which the formula appears in the assay in a monograph for a solid article may be derived by substituting C_U of equation (3) into equation (4):

(5) $W_U = V_UC_S(A_U/A_S)$.

In summary, the use of equation (4), with due consideration for any unit conversions necessary to achieve equality in equation (5), permits the calculation of the constant factor (V_U) occurring in the final formula.

The same derivation is applicable to formulas that appear in monographs for liquid articles that are assayed by absorption spectrophotometry. For liquid dosage forms, results of calculations are generally expressed in terms of the quantity, in mg, of drug substance in each mL of the article. Thus it is necessary to include in the denominator an additional term, the volume (*V*), in mL, of the test preparation taken.

Spectrophotometry in the Visible Region—Assays in the visible region usually call for comparing concomitantly the absorbance produced by the *Assay preparation* with that produced by a *Standard preparation* containing approximately an equal quantity of a USP Reference Standard. In some situations, it is permissible to omit the use of a Reference Standard. This is true where spectrophotometric assays are made with routine frequency, and where a suitable standard curve is available, prepared with the respective USP Reference Standard, and where the substance assayed conforms to Beer's law within the range of about 75% to 125% of the final concentration used in the assay. Under these circumstances, the absorbance found in the assay may be interpolated on the standard curve, and the assay result calculated therefrom.

Such standard curves should be confirmed frequently, and always when a new spectrophotometer or new lots of reagents are put into use.

In spectrophotometric assays that direct the preparation and use of a standard curve, it is permissible and preferable, when the assay is employed infrequently, not to use the standard curve but to make the comparison directly against a quantity of the Reference Standard approximately equal to that taken of the specimen, and similarly treated.

NOTE—See *Calculations* under *Spectrophotometry*.

Visual Comparison—Where a color or a turbidity comparison is directed, color-comparison tubes that are matched as closely as possible in internal diameter and in all other respects should be used. For color comparison, the tubes should be viewed downward, against a white background, with the aid of a light source directed from beneath the bottoms of the tubes, while for turbidity comparison the tubes should be viewed horizontally, against a dark background, with the aid of a light source directed from the sides of the tubes.

In conducting limit tests that involve a comparison of colors in two like containers (e.g., matched color-comparison tubes), a suitable instrument, rather than the unaided eye, may be used.

Fluorescence Spectrophotometry—The measurement of fluorescence is a useful analytical technique. *Fluorescence* is light emitted from a substance in an excited state that has been reached by the absorption of radiant energy. A substance is said to be *fluorescent* if it can be made to fluoresce. Many compounds can be assayed by procedures utilizing either their inherent fluorescence or the fluorescence of suitable derivatives.

Test specimens prepared for fluorescence spectrophotometry are usually one-tenth to one-hundredth as concentrated as those used in absorption spectrophotometry, for the following reason. In analytical applications, it is preferable that the fluorescence signal be linearly related to the concentration; but if a test specimen is too concentrated, a significant part of the incoming light is absorbed by the specimen near the cell surface, and the light reaching the center is reduced. That is, the specimen itself acts as an "inner filter." However, fluorescence spectrophotometry is inherently a highly sensitive technique, and concentrations of 10^{-5} *M* to 10^{-7} *M* frequently are used. It is necessary in any analytical procedure to make a working curve of fluorescence intensity versus concentration in order to establish a linear relationship. All readings should be corrected for a solvent blank.

Fluorescence measurements are sensitive to the presence of dust and other solid particles in the test specimen. Such impurities may reduce the intensity of the exciting beam or give misleading high readings because of multiple reflections in the specimen cell. It is, therefore, wise to eliminate solid particles by centrifugation; filtration also may be used, but some filter papers contain fluorescent impurities.

Temperature regulation is often important in fluorescence spectrophotometry. For some substances, fluorescence efficiency may be reduced by as much as 1% to 2% per degree of temperature rise. In such cases, if maximum precision is desired, temperature-controlled specimen cells are useful. For routine analysis, it may be sufficient to make measurements rapidly enough so that the specimen does not heat up appreciably from exposure to the intense light source. Many fluorescent compounds are light-sensitive. Exposed in a fluorometer, they may be photo-degraded into more or less fluorescent products. Such effects may be detected by observing the detector response in relationship to time, and may be reduced by attenuating the light source with filters or screens.

Change of solvent may markedly affect the intensity and spectral distribution of fluorescence. It is inadvisable, therefore, to alter the solvent specified in established methods without careful preliminary investigation. Many compounds are fluorescent in organic solvents but virtually nonfluorescent in water; thus, a number of solvents should be tried before it is decided whether or not a compound is fluorescent. In many organic solvents, the intensity of fluorescence is increased by elimination of dissolved oxygen, which has a strong quenching effect. Oxygen may be removed by bubbling an inert gas such as nitrogen or helium through the test specimen.

A semiquantitative measure of the strength of fluorescence is given by the ratio of the fluorescence intensity of a test specimen and that of a standard obtained with the same instrumental settings. Frequently, a solution of stated concentration of quinine in 0.1 *N* sulfuric acid or fluorescein in 0.1 *N* sodium hydroxide is used as a reference standard.

Light-scattering, Turbidimetry, and Nephelometry—Turbidity can be measured with a standard photoelectric filter photometer or spectrophotometer, preferably with illumination in the blue portion of the spectrum. Nephelometric measurements require an instrument with a photocell placed so as to receive scattered rather than transmitted light; this geometry applies also to fluorometers, so that, in general, fluorometers can be used as nephelometers, by proper selection of filters.

In practice, it is advisable to ensure that settling of the particles being measured is negligible. This is usually accomplished by including a protective colloid in the liquid suspending medium. It is important that results be interpreted by comparison of readings with those representing known concentrations of suspended matter, produced under precisely the same conditions.

Turbidimetry or nephelometry may be useful for the measurement of precipitates formed by the interaction of highly dilute solutions of reagents, or other particulate matter, such as suspensions of bacterial cells. In order that consistent results may be achieved, all variables must be carefully controlled. Where such control is possible, extremely dilute suspensions may be measured.

Light-scattering—The specimen solute is dissolved in the solvent at several different accurately known concentrations, the choice of concentrations being dependent on the molecular weight of the solute and ranging from 1% for M_w = 10,000 to 0.01% for M_w = 1,000,000. Each solution must be very carefully cleaned before measurement by repeated filtration through fine filters. A dust particle in the solution vitiates the intensity of the scattered light measured. A criterion for a clear solution is that the dissymmetry, 45°/135° scattered intensity ratio, has attained a minimum.

The turbidity and refractive index of the solutions are measured. From the general 90° light-scattering equation, a plot of HC/τ versus C is made and extrapolated to infinite dilution, and the weight-average molecular weight, *M*, is calculated from the intercept, $1/M$.

⟨861⟩ SUTURES—DIAMETER

The gauge for determining the diameter of sutures is of the dead-weight type, mechanical or electrical, and equipped with a direct-reading dial, a digital readout, or a printed readout. Use a gauge graduated to 0.002 mm (to 0.0001 inch if the gauge is in English units) or smaller. The anvil of the gauge is about 50 mm in diameter, and the presser foot is 12.70 ± 0.02 mm in diameter. The presser foot and moving parts connected therewith are weighted so as to apply a total load of 210 ± 3 g to the specimen. The presser foot and anvil surfaces are plane to within 0.005 mm and parallel to each other to within 0.005 mm. For measuring the diameter of sutures of metric size 0.4 and smaller, remove the additional weight from the presser foot so that the total load on the suture does not exceed 60 g.

Collagen Absorbable Surgical Suture—Determine the diameter immediately after removal from the immediate container and without stretching. Lay the strand across the center of the anvil and presser foot, and gently lower the foot until its entire weight rests upon the suture. Measure the diameter of each strand at three points corresponding roughly to one-fourth, one-half, and three-fourths of its length.

Synthetic Absorbable Surgical Suture—Proceed as directed for *Nonabsorbable Surgical Suture.*

Nonabsorbable Surgical Suture—Lay the strand across the center of the anvil and presser foot, and gently lower the foot until its entire weight rests upon the suture. Measure nonabsorbable sutures, whether packaged in dry form or in fluid, immediately after removal from the container, without prior drying or conditioning.

Measure the diameter of the suture at three points corresponding roughly to one-fourth, one-half, and three-fourths of its length. In the case of braided suture of sizes larger than 3–0 (metric size 2), make two measurements at each point at right angles to each other, and use the average as the observed diameter at that point.

In measuring multifilament sutures, attach a portion of the designated section of the strand in a fixed clamp in such a way that the strand lies across the center of the anvil. While holding the strand in the same plane as the surface of the anvil, place the strand under tension by suitable means, such as by passing the free end of the strand around a cylinder or a pulley and attaching to the free end a weight of about one-half of the knot-pull limit for the non-sterilized Class I suture of the size concerned, taking care not to permit the strand, if twisted, to untwist. Measure the diameter at the designated points on the strand, and calculate the average diameter likewise as directed.

⟨871⟩ SUTURES—NEEDLE ATTACHMENT

Absorbable (collagen) surgical sutures and nonabsorbable surgical sutures with *Standard Needle Attachment* are such that the needles are firmly attached and are not intended to be separated. Sutures supplied with eyeless needles attached fall into either the category of *Standard Needle Attachment* or the category of *Removable Needle Attachment. Removable Needle Attachment* of both absorbable and nonabsorbable surgical sutures is such that the needle may be deliberately separated from the suture by means of a quick tug. Both types of attachments are tested on equipment as specified under *Tensile Strength* ⟨881⟩.

Procedure—Clamp each of 5 sutures in the tensilometer so that the needle is in the fixed clamp with all of the swaged portion exposed and in line with the direction of force applied to the suture by the moving clamp. Determine the force required to detach the suture from the needle. In the case of *Standard Needle Attachment*, the suture may break without needle detachment.

STANDARD NEEDLE ATTACHMENT—The requirements are met if neither the average of the 5 values nor any individual value is less than the limit given for the designated size in Table 1.

Table 1. Standard Needle Attachment for Absorbable and Nonabsorbable Sutures.

Metric Size (Gauge No.)			Limits on Needle Attachment	
Absorbable (Collagen) Suture	Nonabsorbable and Synthetic Absorbable Suture	USP Size	Average (kgf) (Min.)	Individual (kgf) (Min.)
	0.1	11–0	0.007	0.005
	0.2	10–0	0.014	0.010
0.4	0.3	9–0	0.021	0.015
0.5	0.4	8–0	0.050	0.025
0.7	0.5	7–0	0.080	0.040
1	0.7	6–0	0.17	0.08
1.5	1	5–0	0.23	0.11
2	1.5	4–0	0.45	0.23
3	2	3–0	0.68	0.34
3.5	3	2–0	1.10	0.45
4	3.5	0	1.50	0.45
5	4	1	1.80	0.60
6 and larger	5 and larger	2 and larger	1.80	0.70

Table 2. Removable Needle Attachment for Absorbable and Nonabsorbable Sutures.

Metric Size			Attachment Limits	
Absorbable (Collagen)	Nonabsorbable and Synthetic Absorbable	USP Size	Minimum (kgf)	Maximum (kgf)
1.5	1	5–0		
2	1.5	4–0		
3	2	3–0		
3.5	3	2–0	0.028	1.59
4	3.5	0		
5	4	1		
6	5	2		

REMOVABLE NEEDLE ATTACHMENT—The requirements are met if the individual values of the 5 sutures are within the limits shown in Table 2.

NOTE—For either type of attachment, if not more than 1 of the individual values falls outside the prescribed limits, repeat the test on an additional 10 sutures: the requirements of the test are met if none of the 10 additional values falls outside the individual limit requirements.

⟨881⟩ TENSILE STRENGTH

Devices for measurement of tensile strength used in the United States may be calibrated in the English units of measure. The following directions are given in metric units with the understanding that the corresponding English equivalents may be used.

Surgical Sutures

Equipment—Determine the tensile strength of surgical sutures on a motor-driven tensile strength testing machine using the principle of the constant specimen-rate-of-load, having suitable clamps for holding the specimen firmly.

This description applies specifically to that known as the Incline Plane Tester.

The carriage used in any test is of a weight such that when the break occurs, the position of the recording pen on the chart is between 20% and 80% of the capacity that may be recorded on the chart. The friction in the carriage is low enough to permit the recording pen to depart from the zero line of the chart at a point not exceeding 2.5% of the capacity of the chart when no specimen is held in the clamps.

For surgical sutures of intermediate and larger sizes, the clamp for holding the specimen is of the roll type, with a flat gripping surface. The roll has a diameter of 19 mm and the flat gripping surface is not less than 25 mm in length. The length of the specimen, when inserted in the clamps, is at least 127 mm from nip to nip. The speed of inclination of the plane of the tester is such that it reaches its full inclination of 30° from the horizontal in 20 ± 1 seconds from the start of the test.

For surgical sutures of small sizes, the suitable clamp has a flat gripping surface of not less than 13 mm in length. The length of the specimen, when inserted in the clamps, is at least 127 mm from clamp to clamp or 35 mm less than the labeled length, whichever distance is shorter. In the event that a labeled length is less than 47 mm, use a clamp to clamp distance of 12 mm. The speed of inclination of the plane is such that it reaches its full inclination of 30° from the horizontal in 60 ± 5 seconds from the start of the test.

Procedure—Determine the tensile strength of sutures, whether packaged in dry form or in fluid, promptly after removal from the container, without prior drying or conditioning.

Except where straight pull (no knot required) is indicated in the suture monograph, tie the test suture into a surgeon's knot with one turn of suture around flexible rubber tubing of 6.5-mm inside diameter and 1.6-mm wall thickness. The surgeon's knot is a square knot in which the free end is first passed twice, instead of once, through the loop, and pulled taut, then passed once through a second loop, and the ends are drawn taut so that a single knot is superimposed upon a compound knot. Start the first knot with the left end over the right end, exerting sufficient tension to tie the knot securely. Where the test specimen includes a knot, place the specimen in the testing device with the knot approximately midway between the clamps.

Attach one end of the suture in the clamp at the load end of the machine, pass the other end through the opposite clamp, with sufficient tension so that the specimen is taut between the clamps, and close the second clamp. Perform as many breaks as are specified in the individual monograph. If the break occurs outside the central 80% of the specimen length, discard the reading on the specimen. If the labeled length of the strand exceeds 7 meters (25 feet), take 2 meters from each of five strands, selected at random from the lot, rejecting the first 30 cm (12 inches), and make at least two breaks on each strand, about 60 cm to 100 cm apart.

Textile Fabrics and Films

Equipment—Determine the tensile strength of textile fabrics, including adhesive tape, on a constant-speed or pendulum type of testing machine, of the following general description.

The clamps for holding the specimen are smooth, flat, parallel jaws that are not less than 25 mm in length in the dimension parallel to the direction of application of the load. When the width of the strip being tested does not exceed 19 mm, the jaws of the clamp should be at least 25 mm wide. If the width of the strip is greater than 19 mm and not greater than 44 mm, the width of the jaws of the clamp should be at least 50 mm. If the width of the specimen is greater than 44 mm, cut a 25-mm strip, and use a clamp with jaws not less than 50 mm wide. Round all edges that might have a cutting action on the specimen to a radius of 0.4 mm (1/64 inch). The jaws are 76.2 mm apart at the beginning of the test, and they separate at the rate of 30.5 cm ± 13 mm per minute. The machine is of such capacity that when the break occurs, the deviation of the pendulum from the vertical is between 9° and 45°.

⟨891⟩ THERMAL ANALYSIS

Precisely determined thermodynamic events, such as a change of state, can indicate the identity and purity of drugs. Compendial standards have long been established for the melting or boiling temperatures of substances. These transitions occur at characteristic temperatures, and the compendial standards therefore contribute to the identification of the substances. Because impurities affect these changes in predictable ways, the same compendial standards contribute to the control of the purity of the substances.

Thermal analysis in the broadest sense is the measurement of physical-chemical properties of materials as a function of temperature. Instrumental methods have largely supplanted older methods dependent on visual inspection and on measurements under fixed or arbitrary conditions, because they are objective, they provide more information, they afford permanent records, and they are generally more sensitive, more precise, and more accurate. Furthermore, they may provide information on crystal perfection, polymorphism, melting temperature, sublimation, glass transitions, dehydration, evaporation, pyrolysis, solid-solid interactions, and purity. Such data are useful in the characterization of substances with respect to compatibility, stability, packaging, and quality control. The measurements used most often in thermal analysis, i.e., transition temperature, thermogravimetry, and impurity analysis, are described here.

Transition Temperature—As a specimen is heated, its uptake (or evolution) of heat can be measured [differential scanning calorimetry (DSC)] or the resulting difference in temperature from that of an inert reference heated identically [differential thermal analysis (DTA)] can be measured. Either technique provides a record of the temperature at which phase changes, glass transitions, or chemical reactions occur. In the case of melting, both an "onset" and a "peak" temperature can be determined objectively and reproducibly, often to within a few tenths of a degree. While these temperatures are useful for characterizing substances, and the difference between the two temperatures is indicative of purity, the values cannot be correlated with subjective, visual "melting-range" values or with constants such as the triple point of the pure material.

A complete description of the conditions employed should accompany each thermogram, including make and model of instrument; record of last calibration; specimen size and identification (including previous thermal history); container; identity, flow rate, and pressure of gaseous atmosphere; direction and rate of temperature change; and instrument and recorder sensitivity.

It is appropriate to make a preliminary examination over a wide range (room temperature to decomposition temperature) at a high (10° to 20° per minute) heating rate in order to reveal unusual effects, then to make replicate examinations over a narrow range bracketing the transition of interest at lower heating rates (about 2° per minute). As the reliability of the measurements varies from one substance to another, statements of the number of significant figures to be used in the reporting, of intralaboratory repeatability, and of interlaboratory reproducibility cannot be given here, but should be included in the individual monograph.

Thermogravimetric Analysis—Thermogravimetric analysis involves the determination of the mass of a specimen as a function of temperature, or time of heating, or both, and when properly applied, provides more useful information than does loss on drying at fixed temperature, often for a fixed time and in what is usually an ill-defined atmosphere. Usually, loss of surface-absorbed solvent can be distinguished from solvent in the crystal lattice and from degradation losses. The measurements can be carried out in atmospheres having controlled humidity and oxygen concentration to reveal interactions with the drug substance, between drug substances, and between active substances and excipients or packaging materials.

While the details depend on the manufacturer, the essential features of the equipment are a recording balance and a programmable heat source. Equipment differs in the ability to handle specimens of various sizes, the means of sensing specimen temperature, and the range of atmosphere control. Calibration is required with all systems, i.e., the mass scale is calibrated by the use of standard weights; calibration of the temperature scale, which is more difficult, involving either variations in positioning of thermocouples and their calibration; or in other systems, calibration involves the use of standard materials because it is assumed that the specimen temperature is the furnace temperature.

Procedural details are specified in order to provide for valid interlaboratory comparison of results. The specimen weight, source, and thermal history are noted. The equipment description covers dimensions and geometry, the materials of the test specimen holder, and the location of the temperature transducer. Alternatively, the make and model number of commercial equipment are specified. In all cases, the calibration record is specified. Data on the temperature environment include the initial and final temperatures and the rate of change or other details if non-linear. The test atmosphere is critical; the volume, pressure, composition, whether static or dynamic, and if the latter, the flow rate and temperature are specified.

Eutectic Impurity Analysis—The basis of any calorimetric purity method is the relationship between the melting and freezing point depression, and the level of impurity. The melting of a compound is characterized by the absorption of latent heat of fusion, ΔH_f, at a specific temperature, T_o. In theory, a melting transition for an absolutely pure crystalline compound should occur within an infinitely narrow range. A broadening of the melting range, due to impurities, provides a sensitive criterion of purity. The effect is apparent visually by examination of thermograms of specimens differing by a few tenths percent in impurity content. A material that is 99% pure is about 20% molten at 3° below the melting point of the pure material (see accompanying figure).

The parameters of melting (melting range, ΔH_f, and calculated eutectic purity) are readily obtained from the thermogram of a single melting event using a small test specimen, and the method does not require multiple, precise actual temperature measurements. Thermogram units are directly convertible to heat transfer, millicalories per second.

The lowering of the freezing point in *dilute solutions* by molecules of nearly equal size is expressed by a modified van't Hoff equation:

$$\frac{dT}{dX_2} = \frac{RT^2}{\Delta H_f} \cdot (K - 1), \tag{1}$$

in which T = absolute temperature in degrees Kelvin (°K), X_2 = mole fraction of minor component (solute; impurity), ΔH_f = molar heat of fusion of the major component, R = gas constant, and K = distribution ratio of solute between the solid and liquid phases.

Assuming that the temperature range is small and that no solid solutions are formed ($K = 0$), integration of the van't Hoff equation yields the following relationship between mole fraction of impurity and the melting-point depression:

$$X_2 = \frac{(T_o - T_m)\Delta H_f}{RT_o^2}, \tag{2}$$

in which T_o = melting point of the pure compound, in °K, and T_m = melting point of the test specimen, in °K.

With no solid solution formation, the concentration of impurity in the liquid phase at any temperature during the melting is inversely proportional to the fraction melted at that temperature, and the melting-point depression is directly proportional to the mole fraction of impurity. A plot of the observed test specimen temperature, T_s, versus the reciprocal of the fraction melted, $1/F$, at temperature T_s, should yield a straight line with the slope equal to the melting-point depression ($T_o - T_m$). The theoretical melting point of the pure compound is obtained by extrapolation to $1/F = 0$:

$$T_s = T_o - \frac{RT_o^2X_2(1/F)}{\Delta H_f}. \tag{3}$$

Substituting the experimentally obtained values for $T_o - T_m$, ΔH_f, and T_o in Equation 2 yields the mole fraction of the total eutectic impurity, which, when multiplied by 100, gives the mole percentage of total eutectic impurities.

Deviations from the theoretical linear plot also may be due to solid solution formation ($K \neq 0$), so that care must be taken in interpreting the data.

To observe the linear effect of the impurity concentration on the melting-point depression, the impurity must be *soluble* in the liquid phase or melt of the compound, but *insoluble* in the solid phase, i.e., no solid solutions are formed. Some chemical similarities are necessary for solubility in the melt. For example, the presence of ionic compounds in neutral organic compounds and the occurrence of thermal decomposition may not be reflected in purity estimates. The extent of these theoretical limitations has been only partially explored.

Impurities present from the synthetic route often are similar to the end product, hence there usually is no problem of solubility in the melt. Impurities consisting of molecules of the same shape, size, and character as those of the major component can fit into the matrix of the major component without disruption of the lattice, forming solid solutions or inclusions; such impurities are not detectable by DSC. Purity estimates are too high in such cases. This is more common with less-ordered crystals as indicated by low heats of fusion.

Impurity levels calculated from thermograms are reproducible and probably reliable within 0.1% for ideal compounds. Melting-point determinations by scanning calorimetry have a reproducibility with a standard deviation of about 0.2°. Calibration against standards may allow about 1° accuracy for the melting point, so that this technique is comparable to other procedures.

Compounds that exist in polymorphic form cannot be used in purity determination unless the compound is completely converted to one form. On the other hand, DSC and DTA are inherently useful for detecting, and therefore monitoring, polymorphism.

Procedure—The actual procedure and the calculations to be employed are dependent on the particular instrument used. Consult the manufacturer's literature and/or the thermal analysis literature for the most appropriate technique for a given instrument. In any event, it is imperative to keep in mind the limitations of solid solution formation, insolubility in the melt, polymorphism, and decomposition during the analysis.

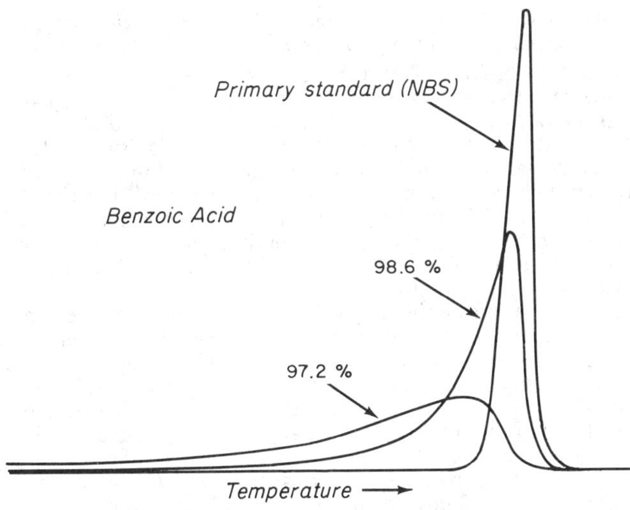

Superimposed Thermograms Illustrating the Effect of Impurities on DSC Melting Peak Shape

⟨901⟩ ULTRAVIOLET ABSORBANCE OF CITRUS OILS

Place about 250 mg, accurately weighed, of the oil in a 100-mL volumetric flask, add alcohol to volume, and mix. Determine the ultraviolet absorption spectrum of this solution in the range from 260 nm to 400 nm in a 1-cm cell, with a suitable recording or manual spectrophotometer, using alcohol as the blank. If a manual instrument is used, read the absorbances at 5-nm intervals from 260 nm to a point about 12 nm from the expected maximum absorbance, then at 3-nm intervals for 3 readings, and at 1-nm intervals to a point about 5 nm beyond the maximum, and then at 10-nm intervals to 400 nm; and from these data, plot the absorbances as ordinates against wavelength on the abscissa, and draw the absorption spectrum or spectrogram. Draw a base-line tangent to the latter, as shown in the accompanying figure (which is typical of Lemon Oil), between points *A* and *B*. Locate the point of maximum absorbance (*C*), and from it drop a vertical line, perpendicular to the abscissa, that intersects line *AB* at *D*. Read from the ordinate the absorbances corresponding to points *D* and *C*, subtract the former from the latter, and calculate the difference on the basis of a 250-mg sample.

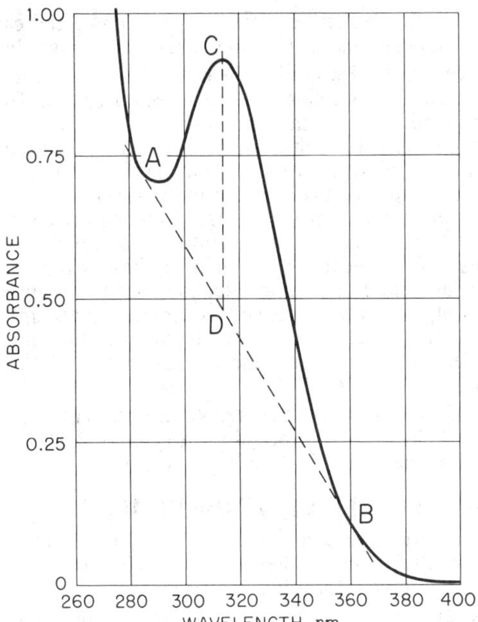

Typical Spectrum of Lemon Oil

⟨905⟩ UNIFORMITY OF DOSAGE UNITS

The uniformity of dosage units can be demonstrated by either of two methods, weight variation or content uniformity. The requirements of this chapter apply both to dosage forms containing a single active ingredient and to dosage forms containing two or more active ingredients.

Weight Variation requirements may be applied where the product is a liquid-filled soft capsule, or where the product to be tested contains 50 mg or more of an active ingredient comprising 50% or more, by weight, of the dosage-form unit. Uniformity with respect to other active ingredients, if present in lesser proportions, is demonstrated by *Content Uniformity* requirements. *Weight Variation* requirements may be applied to solids (including sterile solids) that contain no inactive or active added substances.

Weight Variation requirements may be applied to solids (including sterile solids), with or without inactive or active added substances, that have been prepared from true solutions and freeze-dried in the final containers, and labeled to indicate this method of preparation.

Content Uniformity requirements may be applied in all cases. The test for *Content Uniformity* is required for all coated tablets, including film-coated tablets, for transdermal systems, and for suspensions in single-unit containers or in soft capsules. The test for *Content Uniformity* is required for solids (including sterile solids) that contain inactive or active added substances, except that the test for *Weight Variation* may be applied for special situations as stated above.

WEIGHT VARIATION

For the determination of dosage-form uniformity by weight variation, select not less than 30 units, and proceed as follows for the dosage form designated. [NOTE—Specimens other than these test units may be drawn from the same batch for *Assay* determinations.]

UNCOATED TABLETS—Weigh accurately 10 tablets individually, and calculate the average weight. From the result of the *Assay*, obtained as directed in the individual monograph, calculate the content of active ingredient in each of 10 tablets, assuming homogeneous distribution of the active ingredient.

HARD CAPSULES—Weigh accurately 10 capsules individually, taking care to preserve the identity of each capsule. Remove the contents of each capsule by a suitable means. Weigh accurately the emptied shells individually, and calculate for each capsule the net weight of its contents by subtracting the weight of the shell from the respective gross weight. From the results of the *Assay*, obtained as directed in the individual monograph, calculate the content of active ingredient in each of the capsules, assuming homogeneous distribution of the active ingredient.

SOFT CAPSULES—Determine the net weight of the contents of individual capsules as follows: Weigh accurately the 10 intact capsules individually to obtain their gross weights, taking care to preserve the identity of each capsule. Then cut open the capsules by means of a suitable clean, dry cutting instrument such as scissors or a sharp open blade, and remove the contents by washing with a suitable solvent. Allow the occluded solvent to evaporate from the shells at room temperature over a period of about 30 minutes, taking precautions to avoid uptake or loss of moisture. Weigh the individual shells, and calculate the net contents. From the results of the *Assay*, obtained as directed in the individual monograph, calculate the content of active ingredient in each of the capsules, assuming homogeneous distribution of the active ingredient.

SOLIDS IN SINGLE-UNIT CONTAINERS and STERILE SOLIDS FOR PARENTERAL USE—Proceed as directed under *Hard Capsules*, treating each unit as described therein.

CONTENT UNIFORMITY

For the determination of dosage-form uniformity by assay of individual units, select not less than 30 units, and proceed as follows for the dosage form designated.

UNCOATED AND COATED TABLETS, HARD AND SOFT CAPSULES, TRANSDERMAL SYSTEMS, SUSPENSIONS IN SINGLE-UNIT CONTAINERS, INHALATIONS IN SINGLE-UNIT CONTAINERS, and SOLIDS (INCLUDING STERILE SOLIDS) IN SINGLE-UNIT CONTAINERS—Assay 10 units individually as directed in the *Assay* in the individual monograph, unless otherwise specified in the test for *Content uniformity*. Where the amount of active ingredient in a single dose unit is less than required in the *Assay*, adjust the degree of dilution of the solutions and/or the volume of aliquots so that the concentration of the active ingredients in the final solution is of the same order as that obtained in the *Assay* procedure; or, in the case of a titrimetric assay, use a more dilute titrant, if necessary, so that an adequate volume of titrant is required (see *Titrimetry* ⟨541⟩); see also *Procedures* under *Tests and Assays* in the *General Notices*. If any such modifications are made in the *Assay* procedure set forth in the individual monograph, make the appropriate corresponding changes in the calculation formula and titration factor.

Where a special procedure is specified in the test for *Content uniformity* in the individual monograph, make any necessary correction of the results obtained as follows.

(1) Prepare a composite specimen of a sufficient number of dosage units to provide the amount of specimen called for in the *Assay* in the individual monograph plus the amount required for the special procedure given in the test for *Content uniformity* in the monograph by finely powdering tablets or mixing the contents of capsules or suspensions or solids in single-unit containers to obtain a homogeneous mixture. If a homogeneous mixture cannot be obtained in this manner, use suitable solvents or other procedures to prepare a solution containing all of the active ingredient, and use appropriate aliquot portions of this solution for the specified procedures.

(2) Assay separate, accurately measured portions of the composite specimen of capsules or tablets or suspensions or inhalations or solids in single-unit containers, both (a) as directed in the *Assay*, and (b) using the special procedure given in the test for *Content uniformity* in the monograph.

(3) Calculate the weight of active ingredient equivalent to 1 average dosage unit, by (a) using the results obtained by the *Assay* procedure, and by (b) using the results obtained by the special procedure.

(4) Calculate the correction factor, *F*, by the formula:

$$F = A/P,$$

in which *A* is the weight of active ingredient equivalent to 1 average dosage unit obtained by the *Assay* procedure, and *P* is the weight of active ingredient equivalent to 1 average dosage unit obtained by the special procedure.

(5) If *F* is between 0.970 and 1.030 no correction is required.

(6) If *F* is not within 0.970 and 1.030, calculate the weight of active ingredient in each dosage unit by multiplying each of the weights found using the special procedure by *F*.

Calculation of the Relative Standard Deviation

The use of pre-programmed calculators or computers is acceptable. A manual mathematical method is as follows:

s = sample standard deviation.

RSD = relative standard deviation (the sample standard deviation expressed as a percentage of the mean).

$\overline{X}$ = mean of the values obtained from the units tested, expressed as a percentage of the label claim.

n = number of units tested.

$x_1, x_2, x_3 \ldots x_n$ = individual values (x_i) of the units tested, expressed as a percentage of the label claim.

$$s = \left[\frac{\Sigma(x_i - \overline{X})^2}{n - 1}\right]^{1/2}$$

$$RSD = \frac{100s}{\overline{X}}.$$

Criteria

Apply the following criteria, unless otherwise specified in the individual monograph.

(*A*) *If the Average of the Limits Specified in the Potency Definition in the Individual Monograph is 100.0 Percent or Less*—

COMPRESSED TABLETS (COATED OR UNCOATED), SUSPENSIONS IN SINGLE-UNIT CONTAINERS, SOLIDS (INCLUDING STERILE SOLIDS) IN SINGLE-UNIT CONTAINERS, and STERILE SOLIDS FOR PARENTERAL USE—Unless otherwise specified in the individual monograph, the requirements for dose uniformity are met if the amount of the active ingredient in each of the 10 dosage units as determined from the *Weight variation* or the *Content uniformity* method lies within the range of 85.0 percent to 115.0 percent of the label claim and the *Relative standard deviation* is less than or equal to 6.0 percent.

If 1 unit is outside the range of 85.0 percent to 115.0 percent of label claim and no unit is outside the range of 75.0 percent to 125.0 percent of label claim, or if the *Relative standard deviation* is greater than 6.0 percent, or if both conditions prevail, test 20 additional units. The requirements are met if not more than 1 unit of the 30 is outside the range of 85.0 percent to 115.0 percent of label claim and no unit is outside the range of 75.0 percent to 125.0 percent of label claim and the *Relative standard deviation* of the 30 dosage units does not exceed 7.8 percent.

CAPSULES, TRANSDERMAL SYSTEMS, INHALATIONS, AND MOLDED TABLETS—Unless otherwise specified in the individual monograph, the requirements for dose uniformity are met if the amount of the active ingredient in not less than 9 of the 10 dosage units as determined from the *Weight variation* or the *Content uniformity* method lies within the range of 85.0 percent to 115.0 percent of label claim and no unit is outside the range of 75.0 percent to 125.0 percent of label claim and the *Relative standard deviation* of the 10 dosage units is less than or equal to 6.0 percent.

If 2 or 3 dosage units are outside the range of 85.0 percent to 115.0 percent of label claim, but not outside the range of 75.0 percent to 125.0 percent of label claim, or if the *Relative standard deviation* is greater than 6.0 percent or if both conditions prevail, test 20 additional units. The requirements are met if not more than 3 units of the 30 are outside the range of 85.0 percent to 115.0 percent of label claim and no unit is outside the range of 75.0 percent to 125.0 percent of label claim, and the *Relative standard deviation* of the 30 dosage units does not exceed 7.8 percent.

(*B*) *If the Average of the Limits Specified in the Potency Definition in the Individual Monograph is Greater than 100.0 Percent*—

(1) If the average value of the dosage units tested is 100.0 percent or less, the requirements are as in (*A*).

(2) If the average value of the dosage units tested is greater than or equal to the average of the limits specified in the potency definition in the individual monograph, the requirements are as in (*A*), except that the words "label claim" are replaced by the words "label claim multiplied by the average of the limits specified in the potency definition in the monograph divided by 100."

(3) If the average value of the dosage units tested is between 100 percent and the average of the limits specified in the potency definition in the individual monograph, the requirements are as

in (*A*), except that the words "label claim" are replaced by the words "label claim multiplied by the average value of the dosage units tested (expressed as a percent of label claim) divided by 100."

⟨911⟩ VISCOSITY

Viscosity is a property of liquids that is closely related to the resistance to flow. It is defined in terms of the force required to move one plane surface continuously past another under specified steady-state conditions when the space between is filled by the liquid in question. It is defined as the shear stress divided by the rate of shear strain. The basic unit is the *poise;* however, viscosities commonly encountered represent fractions of the poise, so that the *centipoise* (1 poise = 100 centipoises) proves to be the more convenient unit. The specifying of temperature is important because viscosity changes with temperature; in general, viscosity decreases as temperature is raised. While on the absolute scale viscosity is measured in poises or centipoises, for convenience the kinematic scale, in which the units are *stokes* and *centistokes* (1 stoke = 100 centistokes) commonly is used. To obtain the kinematic viscosity from the absolute viscosity, the latter is divided by the density of the liquid at the same temperature, i.e., kinematic viscosity = (absolute viscosity)/(density). The sizes of the units are such that viscosities in the ordinary ranges are conveniently expressed in centistokes. The approximate viscosity in centistokes at room temperature of ether is 0.2; of water, 1; of kerosene, 2.5; of mineral oil, 20 to 70; and of honey, 10,000.

Absolute viscosity can be measured directly if accurate dimensions of the measuring instruments are known, but it is more common practice to calibrate the instrument with a liquid of known viscosity and to determine the viscosity of the unknown fluid by comparison with that of the known.

Many substances, such as the gums employed in pharmacy, have variable viscosity, and most of them are less resistant to flow at higher flow rates. In such cases, a given set of conditions is selected for measurement, and the measurement obtained is considered to be an apparent viscosity. Since a change in the conditions of measurement would yield a different value for the apparent viscosity of such substances, the instrument dimensions and conditions for measurement must be closely adhered to by the operator.

Measurement of Viscosity—The usual method for measurement of viscosity involves the determination of the time required for a given volume of liquid to flow through a capillary. Many capillary-tube viscosimeters have been devised, but Ostwald and Ubbelohde viscosimeters are among the most frequently used. Several types are described, with directions for their use, by the American Society for Testing and Materials (ASTM, D-445). The viscosity of oils is expressed on arbitrary scales that vary from one country to another, there being several corresponding instruments. The most widely used are the Redwood No. I and No. II, the Engler, the Saybolt Universal, and the Saybolt Furol. Each of these instruments uses arbitrary units that bear the name of the instrument. Standard temperatures are adopted as a matter of convenience with these instruments. For the Saybolt instruments, measurements usually are made at 100°F and 210°F; Redwood instruments may be used at several temperatures up to 250°F; and values obtained on the Engler instrument usually are reported at 20°C and 50°C. A particularly convenient and rapid type of instrument is a rotational viscosimeter, which utilizes a bob or spindle immersed in the test specimen and measures the resistance to movement of the rotating part. Different spindles are available for given viscosity ranges, and several rotational speeds generally are available. Other rotational instruments may have a stationary bob and a rotating cup. The Brookfield, Rotouisco, and Stormer viscosimeters are examples of rotating-bob instruments, and the MacMichael is an example of the rotating-cup instrument. Numerous other rotational instruments of advanced design with special devices for reading or recording, and with wide ranges of rotational speed, have been devised.

Where only a particular type of instrument is suitable, the individual monograph so indicates.

For measurement of viscosity or apparent viscosity, the temperature of the substance being measured must be accurately controlled, since small temperature changes may lead to marked changes in viscosity. For usual pharmaceutical purposes, the temperature should be held to within ±0.1°.

Procedure for Cellulose Derivatives—Measurement of the viscosity of solutions of the high-viscosity types of methylcellulose is a special case, since they are too viscous for the commonly available viscosimeters. The Ubbelohde viscosimeter may be adapted (cf. ASTM, D-1347) to the measurement of the ranges of viscosity encountered in methylcellulose solutions.

Calibration of Capillary-Type Viscosimeters—Determine the viscosimeter constant, *k*, for each viscosimeter by the use of an oil of known viscosity.*

Ostwald-Type Viscosimeter—Fill the tube with the exact amount of oil (adjusted to 20.0 ± 0.1°) as specified by the manufacturer. Adjust the meniscus of the column of liquid in the capillary tube to the level of the top graduation line with the aid of either pressure or suction. Open both the filling and capillary tubes in order to permit the liquid to flow into the reservoir against atmospheric pressure. [NOTE—Failure to open either of these tubes will yield false values.] Record the time, in seconds, for liquid to flow from the upper mark to the lower mark in the capillary tube.

Ubbelohde-Type Viscosimeter—Place a quantity of the oil (adjusted to 20.0 ± 0.1°) in the filling tube, and transfer to the capillary tube by gentle suction, taking care to prevent bubble formation in the liquid by keeping the air vent tube closed. Adjust the meniscus of the column of liquid in the capillary tube to the level of the top graduation line. Open both the vent and capillary tubes in order to permit the liquid to flow into the reservoir against atmospheric pressure. [NOTE—Failure to open the vent tube before releasing the capillary tube will yield false values.] Record the time, in seconds, for the liquid to flow from the upper mark to the lower mark in the capillary tube.

Calculations—

Calculate the viscosimeter constant, *k*, from the equation:

$$k = v/dt,$$

in which v is the known viscosity of the liquid in centipoises, d is the specific gravity of the liquid tested at 20°/20°, and t is the time in seconds for the liquid to pass from the upper mark to the lower mark.

If a viscosimeter is repaired, it must be recalibrated, since even minor repairs frequently cause significant changes in the value of its constant, *k*.

⟨921⟩ WATER DETERMINATION

Many Pharmacopeial articles either are hydrates or contain water in adsorbed form. As a result, the determination of the water content is important in demonstrating compliance with the Pharmacopeial standards. Generally one of the methods given below is called for in the individual monograph, depending upon the nature of the article. In rare cases, a choice is allowed between two methods. When the article contains water of hydration, the Titrimetric Method, the Azeotropic Method, or the Gravimetric Method is employed, as directed in the individual monograph, and the requirement is given under the heading, *Water.*

The heading, *Loss on drying* (see ⟨731⟩), is used in those cases where the loss sustained on heating may be not entirely water.

I—Titrimetric Method

Principle—The titrimetric determination of water is based upon the quantitative reaction of water with an anhydrous solution of sulfur dioxide and iodine in the presence of a buffer which reacts with hydrogen ions.

In the original titrimetric solution, known as Karl Fischer Reagent, the sulfur dioxide and iodine are dissolved in pyridine and methanol. The test specimen may be titrated with the *Reagent* directly, or the analysis may be carried out by a residual titration

* Oils of known viscosities may be obtained from the Cannon Instrument Co., Box 16, State College, PA 16801. For methylcellulose, choose an oil the viscosity of which is as close as possible to that of the type of methylcellulose to be determined.

procedure. In the residual titration, excess *Reagent* is added to the test specimen, sufficient time is allowed for the reaction to reach completion, and the unconsumed *Reagent* is titrated with a standard solution of water in a solvent such as methanol. The residual titration procedure is applicable generally and avoids the difficulties that may be encountered in the direct titration of substances from which the bound water is released slowly.

The stoichiometry of the reaction is not exact, and the reproducibility of a determination depends upon such factors as the relative concentrations of the *Reagent* ingredients, the nature of the inert solvent used to dissolve the test specimen, and the technique used in the particular determination. Therefore, an empirically standardized technique is used in order to achieve the desired accuracy. Precision in the method is governed largely by the extent to which atmospheric moisture is excluded from the system. The titration of water is usually carried out with the use of anhydrous methanol as the solvent for the test specimen; however, other suitable solvents may be used for special or unusual test specimens.

Apparatus—Any apparatus may be used that provides for adequate exclusion of atmospheric moisture and determination of the end-point. In the case of a colorless solution that is titrated directly, the end-point may be observed visually as a change in color from canary yellow to amber. The reverse is observed in the case of a test specimen that is titrated residually. More commonly, however, the end-point is determined electrometrically with an apparatus employing a simple electrical circuit that serves to impress about 200 mV of applied potential between a pair of platinum electrodes (about 5 square mm in area and about 2.5 cm apart) immersed in the solution to be titrated. At the end-point of the titration a slight excess of the reagent increases the flow of current to between 50 and 150 microamperes for 30 seconds to 30 minutes, depending upon the solution being titrated. The time is shortest for substances that dissolve in the reagent. With some automatic titrators, the abrupt change in current or potential at the end-point serves to close a solenoid-operated valve that controls the buret delivering the titrant. Commercially available apparatus generally comprises a closed system consisting of one or two automatic burets and a tightly covered titration vessel fitted with the necessary electrodes and a magnetic stirrer. The air in the system is kept dry with a suitable desiccant such as phosphorus pentoxide, and the titration vessel may be purged by means of a stream of dry nitrogen or current of dry air.

Reagent—

Prepare the Karl Fischer Reagent as follows:

Add 125 g of iodine to a solution containing 670 mL of methanol and 170 mL of pyridine, and cool. Place 100 mL of pyridine in a 250-mL graduated cylinder and, keeping the pyridine cold in an ice bath, pass in dry sulfur dioxide until the volume reaches 200 mL. Slowly add this solution, with shaking, to the cooled iodine mixture. Shake well to dissolve the iodine, transfer the solution to the apparatus, and allow to stand overnight before standardizing. One mL of this solution when freshly prepared is equivalent to approximately 5 mg of water, but it deteriorates gradually; therefore, standardize it within 1 hour before use, or daily if in continuous use. Protect from light while in use. Store any bulk stock of the reagent in a suitably sealed, glass-stoppered container, fully protected from light, and under refrigeration.

A commercially available, stabilized solution of Karl Fischer type reagent may be used. Commercially available reagents containing solvents or bases other than pyridine and/or alcohols other than methanol may be used also. These may be single solutions or reagents formed in situ by combining the components of the reagents present in two discrete solutions. The diluted *Reagent* called for in some monographs should be diluted as directed by the manufacturer. Either methanol or other suitable solvent, such as ethylene glycol monomethyl ether, may be used as the diluent.

Test Preparation—Unless otherwise specified in the individual monograph, use an accurately weighed or measured amount of the specimen under test estimated to contain 10 to 250 mg of water.

Where the specimen under test is an aerosol with propellant, store it in a freezer for not less than 2 hours, open the container, and test 10.0 mL of the well-mixed specimen. In titrating the specimen, determine the end-point at a temperature of 10° or higher.

Where the specimen under test is capsules, use a portion of the mixed contents of not less than 4 capsules.

Where the specimen under test is tablets, use powder from not less than 4 tablets ground to a fine powder in an atmosphere of about 10% relative humidity.

Where the monograph specifies that the specimen under test is hygroscopic, using a dry syringe inject an appropriate volume of methanol, or other suitable solvent, accurately measured, into the container, previously accurately weighed, and shake to dissolve the specimen. Using the same syringe, remove the solution from the container and transfer it to a titration vessel prepared as directed under *Procedure*. Repeat the procedure with a second portion of methanol, or other suitable solvent, accurately measured, add this washing to the titration vessel, and immediately titrate. Determine the water content, in mg, of a portion of solvent of the same total volume as that used to dissolve the specimen and to wash the container and syringe, as directed under *Standardization of Water Solution for Residual Titrations*, and subtract this value from the water content, in mg, obtained in the titration of the specimen under test. Dry the container and its closure at 100° for 3 hours, allow to cool in a desiccator, and weigh. Determine the weight of specimen tested from the difference in weight from the initial weight of the container.

Standardization of the Reagent—Place enough methanol or other suitable solvent in the titration vessel to cover the electrodes, and add sufficient *Reagent* to give the characteristic end-point color, or 100 ± 50 microamperes of direct current at about 200 mV of applied potential.

For determination of trace amounts of water (less than 1%), sodium tartrate may be used as a convenient water reference substance. Quickly add 150 to 350 mg of sodium tartrate ($C_4H_4Na_2O_6 \cdot 2H_2O$), accurately weighed by difference, and titrate to the end-point. The water equivalence factor F, in mg of water per mL of reagent, is given by the formula:

$$2(18.02/230.08)(W/V),$$

in which 18.02 and 230.08 are the molecular weights of water and sodium tartrate dihydrate, respectively. W is the weight, in mg, of sodium tartrate dihydrate, and V is the volume, in mL, of the *Reagent* consumed in the second titration.

For the precise determination of significant amounts of water (more than 1%), use purified water obtained by distillation as the reference substance. Quickly add between 25 mg and 250 mg of water, accurately weighed by difference, from a weighing pipet or from a pre-calibrated syringe or micropipet, the amount taken being governed by the reagent strength and the buret size, as referred to under *Volumetric Apparatus* ⟨31⟩. Titrate to the end-point. Calculate the water equivalence factor, F, in mg of water per mL of reagent, by the formula:

$$W/V,$$

in which W is the weight, in mg, of the water, and V is the volume, in mL, of the reagent required.

Standardization of Water Solution for Residual Titration—Prepare a *Water Solution* by diluting 2 mL of water with methanol or other suitable solvent to 1000 mL. Standardize this solution by titrating 25.0 mL with the *Reagent*, previously standardized as directed under *Standardization of the Reagent*. Calculate the water content, in mg per mL, of the *Water Solution* by the formula:

$$V'F/25,$$

in which V' is the volume of the *Reagent* consumed, and F is the water equivalence factor of the *Reagent*. Determine the water content of the *Water Solution* weekly, and standardize the *Reagent* against it periodically as needed.

Procedure—Determine the water by *Method Ia*, unless otherwise specified in the individual monograph.

Method Ia (*Direct Titration*)—Unless otherwise specified, transfer 35 to 40 mL of methanol or other suitable solvent to the titration vessel, and titrate with the *Reagent* to the electrometric or visual end-point to consume any moisture that may be present. (Disregard the volume consumed, since it does not enter into the calculations.) Quickly add the *Test Preparation*, mix, and again titrate with the *Reagent* to the electrometric or visual end-point. Calculate the water content of the specimen, in mg, by the formula:

$$SF,$$

in which S is the volume, in mL, of the *Reagent* consumed in the second titration, and F is the water equivalence factor of the *Reagent*.

Method Ib (*Residual Titration*)—Where the individual monograph specifies that the water content is to be determined by *Method Ib*, the *residual titration* procedure, transfer 35 to 40 mL of methanol or other suitable solvent to the titration vessel, and titrate with the *Reagent* to the electrometric or visual end-point. Quickly add the *Test Preparation*, mix, and add an accurately measured excess of the *Reagent*. Allow sufficient time for the reaction to reach completion, and titrate the unconsumed *Reagent* with standardized *Water Solution* to the electrometric or visual end-point. Calculate the water content of the specimen, in mg, by the formula:

$$F(X' - XR),$$

in which F is the water equivalence factor of the *Reagent*, X' is the volume, in mL, of the *Reagent* added after introduction of the specimen, X is the volume, in mL, of standardized *Water Solution* required to neutralize the unconsumed *Reagent*, and R is the ratio, $V'/25$ (mL *Reagent*/mL *Water Solution*), determined from the *Standardization of Water Solution for Residual Titration*.

II—Azeotropic (Toluene Distillation) Method

Apparatus—Use a 500-mL glass flask A connected by means of a trap B to a reflux condenser C by ground glass joints (see illustration).

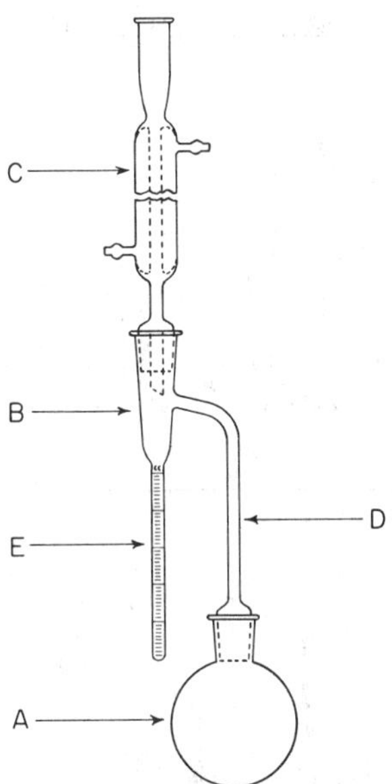

Toluene Moisture Apparatus

The critical dimensions of the parts of the apparatus are as follows: The connecting tube D is 9 to 11 mm in internal diameter. The trap is 235 to 240 mm in length. The condenser, if of the straight-tube type, is approximately 400 mm in length and not less than 8 mm in bore diameter. The receiving tube E has a 5-mL capacity and its cylindrical portion, 146 to 156 mm in length, is graduated in 0.1-mL subdivisions, so that the error of reading is not greater than 0.05 mL for any indicated volume. The source of heat is preferably an electric heater with rheostat control or an oil bath. The upper portion of the flask and the connecting tube may be insulated with asbestos.

Clean the receiving tube and the condenser with chromic acid cleansing mixture, thoroughly rinse with water, and dry in an oven. Prepare the toluene to be used by first shaking with a small quantity of water, separating the excess water, and distilling the toluene.

Procedure—Place in the dry flask a quantity of the substance, weighed accurately to the nearest centigram, which is expected to yield 2 to 4 mL of water. If the substance is of a pasty character, weigh it in a boat of metal foil of a size that will just pass through the neck of the flask. If the substance is likely to cause bumping, add enough dry, washed sand to cover the bottom of the flask, or a number of capillary melting-point tubes, about 100 mm in length, sealed at the upper end. Place about 200 mL of toluene in the flask, connect the apparatus, and fill the receiving tube E with toluene poured through the top of the condenser. Heat the flask gently for 15 minutes and, when the toluene begins to boil, distil at the rate of about 2 drops per second until most of the water has passed over, then increase the rate of distillation to about 4 drops per second. When the water has apparently all distilled over, rinse the inside of the condenser tube with toluene while brushing down the tube with a tube brush attached to a copper wire and saturated with toluene. Continue the distillation for 5 minutes, then remove the heat, and allow the receiving tube to cool to room temperature. If any droplets of water adhere to the walls of the receiving tube, scrub them down with a brush consisting of a rubber band wrapped around a copper wire and wetted with toluene. When the water and toluene have separated completely, read the volume of water, and calculate the percentage that was present in the substance.

III—Gravimetric Method

Procedure for Chemicals—Proceed as directed in the individual monograph preparing the chemical as directed under *Loss on Drying* ⟨731⟩.

Procedure for Biologics—Proceed as directed in the individual monograph.

Procedure for Vegetable Drugs—Place about 10 g of the drug, prepared as directed (see *Vegetable Drugs—Methods of Analysis* ⟨561⟩) and accurately weighed, in a tared evaporating dish. Dry at 105° for 5 hours, and weigh. Continue the drying and weighing at 1-hour intervals until the difference between two successive weighings corresponds to not more than 0.25%.

⟨941⟩ X-RAY DIFFRACTION

Every crystal form of a compound produces its own characteristic X-ray diffraction pattern. These diffraction patterns can be derived either from a single crystal or from a powdered specimen (containing numerous crystals) of the material. The spacings between and the relative intensities of the diffracted maxima can be used for qualitative and quantitative analysis of crystalline materials. Powder diffraction techniques are most commonly employed for routine identification and the determination of relative purity of crystalline materials. Small amounts of impurity, however, are not normally detectable by the X-ray diffraction method, and for quantitative measurements it is necessary to prepare the sample carefully to avoid preferred orientation effects.

The powder methods provide an advantage over other means of analysis in that they are usually nondestructive in nature (specimen preparation is usually limited to grinding to ensure a randomly oriented sample, and deleterious effects of X-rays on solid pharmaceutical compounds are not commonly encountered). The principal use of single-crystal diffraction data is for the determination of molecular weights and analysis of crystal structures at the atomic level. However, diffraction established for a single crystal can be used to support a specific powder pattern as being truly representative of a single phase.

Solids—A solid substance can be classified as being crystalline, noncrystalline, or a mixture of the two forms. In crystalline materials, the molecular or atomic species are ordered in a three-dimensional array, called a lattice, within the solid particles. This ordering of molecular components is lacking in noncrystalline material. Noncrystalline solids sometimes are referred to as glasses or amorphous solids when repetitive order is nonexistent in all three dimensions. It is also possible for order to exist in only one or two dimensions, resulting in mesomorphic phases (liquid crystals). Although crystalline materials are usually considered to have well-defined visible external morphologies (their habits), this is not a necessity for X-ray diffraction analysis.

The relatively random arrangement of molecules in noncrystalline substances makes them poor coherent scatterers of X-rays, resulting in broad, diffuse maxima in diffraction patterns. Their X-ray patterns are quite distinguishable from crystalline specimens, which give sharply defined diffraction patterns.

Many compounds are capable of crystallizing in more than one type of crystal lattice. At any particular temperature and pressure, only one crystalline form (polymorph) is thermodynamically stable. Since the rate of phase transformation of a metastable polymorph to the stable one can be quite slow, it is not uncommon to find several polymorphs of crystalline pharmaceutical compounds existing under normal handling conditions.

In addition to exhibiting polymorphism, many compounds form crystalline solvates in which the solvent molecule is an integral part of the crystal structure. Just as every polymorph has its own characteristic X-ray patterns, so does every solvate. Sometimes the differences in the diffraction patterns of different polymorphs are relatively minor, and must be very carefully evaluated before a definitive conclusion is reached. In some instances, these polymorphs and/or solvates show varying dissolution rates. Therefore, on the time scale of pharmaceutical bioavailability, different total amounts of drug are dissolved, resulting in potential bioinequivalence of the several forms of the drug.

Fundamental Principles—A collimated beam of monochromatic X-rays is diffracted in various directions when it impinges upon a rotating crystal or randomly oriented powdered crystal. The crystal acts as a three-dimensional diffraction grating to this radiation. This phenomenon is described by Bragg's law, which states that diffraction (constructive interference) can occur only when waves that are scattered from different regions of the crystal, in a specific direction, travel distances differing by integral numbers (n) of the wavelength (λ). Under such circumstances, the waves are in phase. This condition is described by the Bragg equation:

$$\frac{n\lambda}{2 \sin \theta} = d_{hkl},$$

in which d_{hkl} denotes the interplanar spacings and θ is the angle of diffraction.

A family of planes in space can be indexed by three whole numbers, usually referred to as Miller indices. These indices are the reciprocals, reduced to smallest integers, of the intercepts that a plane makes along the axes corresponding to three nonparallel edges of the unit cell (basic crystallographic unit). The unit cell dimensions are given by the lengths of the spacings along the three axes, a, b, c, and the angles between them, α, β, and γ. The interplanar spacing for a specific set of parallel planes hkl is denoted by d_{hkl}. Each such family of planes may show higher orders of diffraction where the d values for the related families of planes nh, nk, nl are diminished by the factor $1/n$ (n being an integer: 2, 3, 4, etc.). Every set of planes throughout a crystal has a corresponding Bragg diffraction angle associated with it (for a specific λ).

The amplitude of a diffracted X-ray beam from any set of planes is dependent upon the following atomic properties of the crystal: (1) position of each atom in the unit cell; (2) the respective atomic scattering factors; and (3) the individual thermal motions. Other factors that directly influence the intensities of the diffracted beam are: (1) the intensity and wavelength of the incident radiation; (2) the volume of crystalline specimen; (3) the absorption of the X-radiation by the specimen; and (4) the experimental arrangement utilized to record the intensity data. Thus, the experimental conditions are especially important for measurement of diffraction intensities.

Only a limited number of Bragg planes are in a position to diffract when monochromatized X-rays pass through a single crystal. Techniques of recording the intensities of all of the possible diffracting hkl planes involve motion of the single crystal and the recording media. Recording of these data is accomplished by photographic techniques (film) or with radiation detectors.

A beam passing through a very large number of small, randomly oriented crystals produces continuous cones of diffracted rays from each set of lattice planes. Each cone corresponds to the diffraction from various planes having a similar interplanar spacing. The intensities of these Bragg reflections are recorded by either film or radiation detectors. The Bragg angle can be measured easily from a film, but the advent of radiation detectors has made possible the construction of diffractometers that read this angle directly. The intensities and d spacings are more conveniently determined with powder diffractometers employing radiation detectors than by film methods. Microphotometers are frequently used for precise intensity measurements of films.

An example of the type of powder patterns obtained for four different solid phases of ampicillin are shown in the accompanying figure. These diffraction patterns were derived from a powder diffractometer equipped with a Geiger-Müller detector; nickel-filtered Cu $K\alpha$ radiation was used.

Radiation—The principal radiation sources utilized for X-ray diffraction are vacuum tubes utilizing copper, molybdenum, iron, and chromium as anodes; copper X-rays are employed most commonly for organic substances. For each of these radiations there is an element that will filter off the $K\beta$ radiation and permit the $K\alpha$ radiation to pass (nickel is used, in the case of copper radiation). In this manner the radiation is practically monochromatized. The choice of radiation to be used depends upon the absorption characteristics of the material and possible fluorescence by atoms present in the specimen.

Caution—Care must be taken in the use of such radiation. Those not familiar with the use of X-ray equipment should seek expert advice. Improper use can result in harmful effects to the operator.

Test Preparation—In an attempt to improve randomness in the orientation of crystallites (and, for film techniques, to avoid a grainy pattern), the specimen may be ground in a mortar to a fine powder. Grinding pressure has been known to induce phase

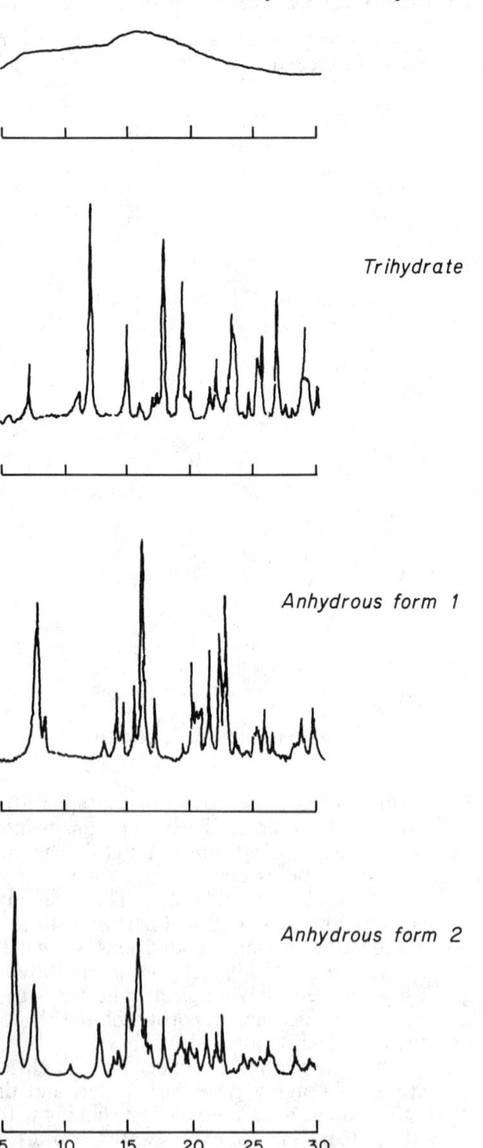

Typical Powder Patterns Obtained for Four Solid Phases of Ampicillin

transformations; therefore, it is advisable to check the diffraction pattern of the unground sample.

In general, the shapes of many crystalline particles tend to give a specimen that exhibits some degree of preferred orientation in the specimen holder. This is especially evident for needle-like or plate-like crystals where size reduction yields finer needles or platelets. Preferred orientation in the specimen influences the relative intensities of various reflections.

Several specialized handling techniques may be employed to minimize preferred orientation, but further reduction of particle size is often the best approach.

Where very accurate measurement of the Bragg angles is necessary, a small amount of an internal standard can be mixed into the specimen. This enables the film or recorder tracing to be calibrated. If comparisons to literature values (including compendial limits) of *d* are being made, calibrate the diffractometer. NBS standards are available covering to a *d*-value of 0.998 nm. Tetradecanol[1] may be used (*d* is 3.963 nm) for larger spacing.

The absorption of the radiation by any specimen is determined by the number and kinds of atoms through which the X-ray beam passes. An organic matrix usually absorbs less of the diffracted radiation than does an inorganic matrix. Therefore, it is important in quantitative studies that standard curves relating amount of material to the intensity of certain *d* spacings for that substance be determined in a matrix similar to that in which the substance will be analyzed.

In quantitative analyses of materials, a known amount of standard usually is added to a weighed amount of specimen to be analyzed. This enables the amount of the substance to be determined relative to the amount of standard added. The standard used should have approximately the same density as the specimen and similar absorption characteristics. More important, its diffraction pattern should not overlap to any extent with that of the material to be analyzed. Under these conditions a linear relationship between line intensity and concentration exists. In favorable cases, amounts of crystalline materials as small as 10% may be determined in solid matrices.

Identification of crystalline materials can be accomplished by comparison of X-ray powder diffraction patterns obtained for known[2] materials with those of the unknown. The intensity ratio (ratio of the peak intensity of a particular *d* spacing to the intensity of the strongest maxima in the diffraction pattern) and the *d* spacing are used in the comparison. If a reference material (e.g., USP Reference Standard) is available, it is preferable to generate a primary reference pattern on the same equipment used for running the unknown sample, and under the same conditions. For most organic crystals, it is appropriate to record the diffraction pattern to include values for 2θ that range from as near zero degrees as possible to 40 degrees. Agreement between sample and reference should be within the calibrated precision of the diffractometer for diffraction angle (2θ values should typically be reproducible to ± 0.10 or 0.20 degrees), while relative intensities between sample and reference may vary up to 20 percent. For other types of samples (e.g., inorganic salts), it may be necessary to extend the 2θ region scanned to well beyond 40 degrees. It is generally sufficient to scan past the ten strongest reflections identified in the JCPDS[2] file.

[1] Brindley, GW and Brown, G, eds., *Crystal Structures of Clay Minerals and their X-ray Identification*, Mineralogical Society Monograph No. 5, London, 1980, pp. 318 ff.

[2] The Joint Committee on Powder Diffraction Standards, 1601 Park Lane, Swarthmore, PA 19081, maintains a file on more than 21,000 crystalline materials, both organic and inorganic, suitable for such comparisons.

GENERAL CHAPTERS

General Information

The chapters in this section are primarily informational, and, aside from excerpts given herein from federal Acts and regulations that may be applicable, they contain no standards, tests, or assays, nor other mandatory specifications, with respect to any Pharmacopeial article. The excerpts from pertinent federal Acts and regulations included in this section are placed here inasmuch as they are not of Pharmacopeial authorship. Revisions of the federal requirements that affect these excerpts will be included in USP Supplements as promptly as practicable. The official requirements for Pharmacopeial articles are set forth in the *General Notices*, the individual monographs, and the *General Tests and Assays* chapters of this Pharmacopeia.

⟨1001⟩ ANTACID EFFECTIVENESS

Presented here are the tablet disintegration test and the acid-neutralizing capacity test, as prescribed in the *Code of Federal Regulations*, that are part of the federal Food and Drug Administration requirements for assuring that over-the-counter (OTC) antacid preparations shall be generally recognized as safe and effective and that they shall not be misbranded. These tests are general in nature, and while it is obligatory that OTC antacid products comply with these requirements for safety and efficacy, for purposes of determining the strength, quality, and purity of official antacids, the Pharmacopeial standards and tests set forth in the individual monographs are determinative.

§ 331.23 Temperature standardization.

All tests shall be conducted at 25° C ± 3°, or 37° C ± 3°.

§ 331.24 Tablet disintegration test.

A tablet disintegration test shall be performed on tablets that are not to be chewed following the procedures described in the United States Pharmacopeia XVIII (page 932). If the label states the tablet may be swallowed, it must disintegrate within a 10-minute time limit pursuant to the test procedure using simulated gastric fluid test solution without enzymes, the United States Pharmacopeia XVIII page 1026, rather than water as the immersion fluid.

§ 331.25 Preliminary antacid test.

(a) *pH meter.* Standardize the pH meter at pH 4.0 with the standardizing buffer and check for proper operation at pH 1 with 0.1 *N* HCl.

(b) *Dosage form testing*—(1) *Liquid sample.* Place an accurately weighed (calculate density) and well mixed amount of the antacid product equivalent to the minimum labeled dosage; e.g., 5 mL, into a 100-mL beaker. Add sufficient water to obtain a total volume of about 40 mL and mix on magnetic stirrer at 300 ± 30 r.p.m. for about 1 minute. Analyze the sample according to the procedure set forth in § 331.25.

(2) *Chewable and non-chewable tablet sample.* Place an accurately weighed amount of a tablet composite equivalent to the minimum labeled dosage into a 100-mL beaker. (The composite shall be prepared by determining the average weight of not less than 20 tablets and then comminuting the tablets sufficiently to pass through a number 20 U. S. standard mesh sieve and held by a number 100 U. S. standard mesh sieve.) Mix the sieved material to obtain a uniform sample. If wetting is desired, add not more than 5 mL of 95 percent ethanol and mix to wet the sample thoroughly (ethanol may effect the acid neutralizing capacity). Add water to a volume of 40 mL and mix on magnetic stirrer at 300 ± 30 r.p.m. for about 1 minute. (Capsules should be tested in the same manner using the sieved capsule powder as the sample.) Analyze the sample according to the procedure set forth in § 331.25.

(3) *Effervescent sample.* Place an amount equivalent to the minimum labeled dosage into a 100-mL beaker. Add 10 mL water and swirl the beaker gently while allowing the reaction to subside. Add another 10 mL of water and swirl the beaker gently. Wash down the walls of the beaker with 20 mL of water and mix on magnetic stirrer at 300 ± 30 r.p.m. for about 1 minute. Analyze the sample according to the procedure set forth in § 331.25.

(4) *Chewing gum samples with antacid in coating.* Place the number of pieces of gum equivalent to the minimum labeled dosage in a 100-mL beaker. Add 40 mL of water and mix on magnetic stirrer at 300 ± 30 r.p.m. for about 2 to 3 minutes. Analyze the sample according to the procedure set forth in § 331.25.

(c) *Test procedure*—(1) Add 10.0 mL of 0.5 *N* HCl to the test solution while stirring on the magnetic stirrer at 300 ± 30 r.p.m.

(2) Stir for exactly 10 minutes after addition of acid.

(3) Read and record pH.

(4) If pH is below 3.5, the product shall not be labeled as

NOTE—Volume 21, part 331, of the *Code of Federal Regulations*, revised as of April 1, 1978, covers regulations pertaining to antacid products for over-the-counter (OTC) human use.

an antacid. If the pH is 3.5 or greater, determine the acid neutralizing capacity according to the procedure set forth in § 331.26.

§ 331.26 Acid neutralizing capacity test.

(a) *pH meter.* Standardize the pH meter at a pH of 4.0 with the standardizing buffer and check for proper operation at pH 1 with 0.1 *N* HCl.

(b) *Dosage form testing*—(1) *Liquid sample.* Place an accurately weighed (calculate density) and well mixed amount of product equivalent to the minimum labeled dosage (e.g., 5 mL, etc.) into a 250 mL beaker. Add sufficient water to obtain a total volume of about 70 mL and mix on the magnetic stirrer at 300 ± 30 r.p.m. for about 1 minute. Analyze the sample according to the procedure set forth in § 331.26.

(2) *Chewable and non-chewable tablet sample.* Place an accurately weighed amount of a tablet composite equivalent to the minimum labeled dosage into a 250 mL beaker. (The composite shall be prepared by determining the average weight of not less than 20 tablets and then comminuting the tablets sufficiently to pass through a number 20 U. S. standard mesh sieve and held by a number 100 U. S. standard mesh sieve. Mix the sieved material to obtain a uniform sample.) If wetting is desired, add not more than 5 mL of 95 percent ethanol and mix to wet the sample thoroughly (ethanol may effect the acid neutralizing capacity). Add water to a volume of 70 mL and mix on magnetic stirrer at 300 ± 30 r.p.m. for about 1 minute. (Capsules should be tested in the same manner using the sieved capsule powder as the sample.) Analyze the sample according to the procedure set forth in § 331.26.

(3) *Effervescent sample.* Place an amount equivalent to the minimum labeled dosage into a 250 mL beaker. Add 10 mL water and swirl the beaker gently while allowing the reaction to subside. Add another 10 mL of water and swirl the beaker gently. Wash down the walls of the beaker with 50 mL of water and mix on magnetic stirrer at 300 ± 30 r.p.m. for about 1 minute. Analyze the sample according to the procedure set forth in § 331.26.

(4) *Sample and test procedure for chewing gum with antacid in coating.* Assay six pieces of gum individually in the following manner.

(i) Place one piece of gum in a 250-mL beaker and add 50 mL of water.

(ii) Pipette in 30.0 mL of 1.0 *N* HCl and stir on magnetic stirrer at 300 ± 30 r.p.m.

(iii) Stir for exactly 10 minutes after addition of acid.

(iv) Stop the stirrer and remove the gum using a long needle or similar utensil.

(v) Rinse the long needle or utensil and the gum with 20 mL of water into the sample beaker.

(vi) Stir for exactly 5 additional minutes.

(vii) Begin titrating immediately and in a period of time not to exceed 5 minutes titrate the excess 1.0 *N* HCl with 0.5 *N* NaOH to stable pH of 3.5.

(viii) Check sample solution 10 to 15 seconds after obtaining a pH of 3.5 to determine that the pH is stable.

(ix) Average the results of the six individual assays and calculate the total mEq based on the minimum labeled dosage as follows:

mEq/piece of gum = (30.0 mL)(normality of HCl) − (mL of NaOH)(normality of NaOH). Total mEq per labeled minimum dose = (number of pieces of gum in minimum dosage) × (mEq/piece of gum).

(c) *Acid neutralizing capacity test procedure (except chewing gum)*—(1) Pipette 30.0 mL of 1.0 *N* HCl into the sample solution while stirring on the magnetic stirrer at 300 ± 30 r.p.m.

(2) Stir for exactly 15 minutes after addition of acid.

(3) Begin titrating immediately and in a period not to exceed an additional 5 minutes titrate the excess 1.0 *N* HCl with 0.5 *N* NaOH to stable pH of 3.5.

(4) Check the sample solution 10 to 15 seconds after obtaining a pH of 3.5 to make sure the pH is stable.

(5) Calculate the number of mEq of acid neutralized by the sample as follows:

Total mEq = (30.0 mL)(normality of HCl) − (mL of NaOH)(*N* of NaOH).

Use appropriate factors, i.e., density, average tablet weight, etc., to calculate the total mEq of acid neutralized per minimum labeled dosage.

⟨1035⟩ BIOLOGICAL INDICATORS

A biological indicator is a characterized preparation of specific microorganisms resistant to a particular sterilization process. It is used to assist in the qualification of the physical operation of sterilization apparatus in the development and establishment of a validated sterilization process for a particular article, and the sterilization of equipment, materials, and packaging components for aseptic processing. It may also be used to monitor a sterilization cycle, once established, and periodically in the program to revalidate previously established and documented sterilization cycles. It is in one of two main forms, each of which incorporates a viable culture of a known species of microorganism. In one, the spores are added to a carrier (disk or strip of filter paper, glass, or plastic) and packaged so as to maintain the integrity of the inoculated carrier but, when used appropriately in the individual immediate package, so as to allow the sterilizing agent to exert its effect. In the other, the spores are added to representative units of the lot to be sterilized (inoculated product) or to similar units (inoculated similar product). An inoculated product should not adversely affect the performance characteristics of the viable spores. If the material to be sterilized is a liquid, and if it is not practicable to add a biological indicator to selected units of the lot, viable spores may be added to a simulated product but in such a way that the resistance of the simulated product to the sterilization process does not differ from the resistance to sterilization of the product to be sterilized.

A biological indicator used for monitoring of a sterilization process may not be suitable, and may even be unsatisfactory for validation of sterilization cycles which may differ in their needs for particular applications.

The effective use of biological indicators for the monitoring of a sterilization process requires a thorough knowledge of the product being sterilized and its component parts (materials and packaging) and, for an overkill approach (see *Sterilization and Sterility Assurance of Compendial Articles* ⟨1211⟩), at least a general idea of the probable types and numbers of microorganisms constituting the microbial burden in the product immediately prior to sterilization.

For an overkill sterilization mode using a biological indicator containing 5×10^5 to 5×10^6 spores of the specified strain per carrier, the following characteristics are obtained (see accompanying table).

Where such a biological indicator is labeled with a particular spore count outside the range mentioned above, the D value, survival time, and kill time may differ from the ranges given, but would bear the same relationship to the actual spore count, specified by the formulas in the individual monographs. Biological indicators with indeterminate labeled spore counts or without such labeled information at all, or with a vague description of the sterilization mode for which the biological indicator is to be used (e.g., "for steam sterilization cycles") are unsatisfactory and cannot be used rationally, unless the user determines the required resistance characteristics and the total viable spore count per carrier, with the necessary precision under the user's sterilization conditions.

Where a product is likely to be damaged by the sterilization process, it is necessary to design a suitable cycle, and a more extensive evaluation of the microbial burden must be made for validation, as detailed in that informational chapter. The selection of a biological indicator is critical, and requires that due weight be given to a knowledge of the resistance of the biological indicator to the specific sterilization process, so that when it is used within its performance characteristics, it provides a challenge to the sterilization process that exceeds the challenge of the natural microbial burden in or on the product.

In some situations, the user may wish to employ a biological indicator containing a low total viable spore count with a high D value or one with a high count and a smaller D value. Thus if a resistance window (survival time/kill time) of approximately 5 minutes and 15 minutes is desired, a preparation with a D value of about 1.5 minutes containing between 5×10^5 and 5×10^6

Typical Biological Indicator Characteristics[a]

Sterilization mode[b]	*Example of a typical D value*[c] (minutes)	*Range of D values for selecting a suitable BI* (minutes)	*Limits for a suitable resistance* (Depending on the particular D value) (minutes)	
			Survival time	*Kill time*
Dry heat				
160°	1.9	Min. 1.0 Max. 3.0	Min. 4.0 Max. 12.0	10.0 30.0
121°	5.0	Min. 2.0 Max. 15.0	Min. 8.0 Max. 60.0	20.0 150
Ethylene oxide				
600 mg/1 54° 60 RH%	3.0	Min. 2.6 Max. 5.8	Min. 10.4 Max. 23.2	26.0 58.0
Steam				
121°	1.9	Min. 1.5 Max. 3.0	Min. 6.0 Max. 12.0	15.0 30.0

[a] For 5×10^5 to 5×10^6 spores per carrier.
[b] For an acceptable range for each condition, see the individual monograph.
[c] A prototype or basic biological indicator for each mode of sterilization, dry heat, ethylene oxide, or steam, under specified conditions which is the form with extensive experience in use for an overkill approach. This form of each has been described repeatedly.

spores per carrier should be selected. If however, the D value is smaller, say 0.8 minutes, more spores per carrier would be required, perhaps more than 5×10^8 spores. A particular strain of microbial spores selected for use as a biological indicator and resistant to one sterilization process may not necessarily be suitable for other sterilization processes or even differing sterilizing conditions of the same mode of sterilization.

The characterization of each form of biological indicator which provides the basis for label claims generally would be done using specialized and standardized apparatus under precisely defined conditions. It is unlikely that many users of a particular biological indicator would have, or have access to, such specialized sterilizing equipment. The resistance characteristics of the same biological indicator, determined under conditions of use, may not necessarily be identical with the original determinations for basing label claims owing to differences in the sterilizing apparatus and conditions. The individual monographs have labeling requirements for a statement to that effect. Detailed descriptions have been published of apparatus which has been claimed to provide consistent results, applicable to characterization of biological indicators for ethylene oxide and steam sterilization. Some information is also available for dry-heat sterilizers. But even with such standardized apparatus, under apparently identical sterilization cycles, the results of exposure of the same biological indicator may be found to be different. Differing results under the user's sterilization conditions, compared to the label claims, are not necessarily due to variance from those claims. They may rather be due to the difference between the user's conditions and the sterilization conditions used for characterizing the biological indicators. Even, e.g., between user's sterilization autoclaves claimed to provide particular steam sterilization cycles, the results of graded exposure of the same biological indicator may be found to be different. Provided there is reasonable consistency between user's cycles in the *same* sterilizer, the biological indicator label information can be used for determining whatever consistent difference would be operative, e.g., the margin of safety for the survival time and kill time under the user's sterilizing conditions. The user could, for example, make a series of determinations of the D value under those conditions, and verify the relationship of the resistance window. A suitable modification of the method for the D value test in the individual monograph could be used for this purpose, but which is not intended to replace the compendial test.

It would be the responsibility of the user to have ascertained the effect of such performance variations and to use the biological indicator properly for the required purpose (sterilization validation or monitoring of sterilization cycles), i.e., to provide acceptable evidence that sterilization did occur and with adequate lethality.

In steam sterilization at certain temperatures, spores of suitable strains of *Bacillus stearothermophilus* are commonly employed because of their resistance to this mode of sterilization. In dry-heat or ethylene oxide sterilization, spores of a subspecies of *Bacillus subtilis* are commonly used. Spores of suitable strains of *Bacillus pumilus* have been utilized as biological indicators to monitor sterilization processes using ionizing radiation.

The preparation of stock spore suspensions of selected microorganisms requires the development of appropriate procedures, including mass culturing, harvesting, and maintaining of the spore suspensions. The stock suspension should contain predominantly dormant (non-germinating) spores that have been held in a non-nutritive liquid. Where the biological indicator is in a self-contained package that includes an available culture medium, the package design offers minimal resistance to the sterilizing substrate penetration, particularly where the sterilization control indicator is used in conjunction with rapid steam sterilization cycles. In placing an indicator in any particular selected location, pay attention to its positioning, e.g., vertical, sideways, to assure maximum penetration of the sterilizing substrate. The performance of the biological indicator is a function of both its initial viable spore count and the resistance of the viable spores to the sterilization process. It is important therefore that the biological indicator maintains its numbers of viable spores and resistance characteristics throughout the dating period.

A biological indicator made from spores of a particular microbial strain and intended to be used for a variety of cycles of even the same mode of sterilization may be available in different forms. One, containing a relatively larger number of spores per inoculum or per carrier, e.g., 10^6, 10^7, requires that the extent of sterilization exposure, e.g., number of D values applied (see *Sterilization and Sterility Assurance of Compendial Articles* ⟨1211⟩), be determined by the degree of reduction of viable spores through an actual count. Another contains only the particular number of spores required to indicate that the validated cycle has been applied, e.g., 10^4. The requisite end result would be the presence or absence of microbial growth when the biological indicator preparation is cultured; no count is made. Still other forms may have particularly high or particularly low numbers of spores and/or resistance characteristics for particular special applications, e.g., to monitor rapid steam sterilization of operating room surgical instruments for emergency use. Such a biological indicator may have 10^9 spores per carrier with a small D value (0.1 to 0.2 minutes).

Each form of biological indicator should have been validated for such application.

Where a biological indicator is used outside of the label claim basis and recommended conditions of use, minimal verification of its resistance parameters would be insufficient; it should be validated for the actual purpose and conditions under which it is applied.

⟨1041⟩ BIOLOGICS

Products such as antitoxins, antivenins, blood, blood derivatives, immune serums, immunologic diagnostic aids, toxoids, vaccines, and related articles that are produced under license in accordance with the terms of the federal Public Health Service Act (58 Stat. 682) approved July 1, 1944, as amended, have long been known as "biologics." However, in Table III, Part F, of the Act, the term "biological products" is applied to the group of licensed products as a whole. For Pharmacopeial purposes, the term "biologics" refers to those products that must be licensed under the Act and comply with Food and Drug Regulations—Code of Federal Regulations, Title 21 Parts 600–680, pertaining to federal control of these products (other than certain diagnostic aids), as administered by the Center for Biologics Evaluation and Research or, in the case of the relevant diagnostic aids, by the Center for Devices and Radiological Health of the federal Food and Drug Administration.

Each lot of a licensed biologic is approved for distribution when it has been determined that the lot meets the specific control requirements for that product as set forth by the Office. Licensing includes approval of a specific series of production steps and in-process control tests as well as end-product specifications that must be met on a lot-by-lot basis. These can be altered only upon approval by the Center for Biologics Evaluation and Research and with the support of appropriate data demonstrating that the change will yield a final product having equal or superior safety, purity, potency, and efficacy. No lot of any licensed biological product is to be distributed by the manufacturer prior to the completion of the specified tests. Provisions generally applicable to biologic products include tests for potency, general safety, sterility, purity, water (residual moisture), pyrogens, identity, and constituent materials (Sections 610.10 to 610.15 and see *Safety Tests-General* under *Biological Reactivity Tests, In-vivo* ⟨88⟩, *Sterility Tests* ⟨71⟩, *Water Determination* ⟨921⟩, and *Pyrogen Test* ⟨151⟩, as well as *Bacterial Endotoxins Test* ⟨157⟩). Constituent materials include ingredients, preservatives, diluents and adjuvants (which generally should meet compendial standards), extraneous protein in cell-culture produced vaccines (which, if other than serum-originating, is excluded) and antibiotics other than penicillin added to the production substrate of viral vaccines (for which compendial monographs on antibiotics and antibiotic substances are available). Additional specific safety tests are also required to be performed on live vaccines and certain other items. Where standard preparations are made available by the Center for Biologics Evaluation and Research (Section 610.20), such preparations are specified for comparison in potency or virulence testing. The U. S. Opacity Standard is used in estimating the bacterial concentration of certain bacterial vaccines and/or evaluating challenge cultures used in tests of them. (See also *Units of Potency*, in the *General Notices.*)

The Pharmacopeial monographs conform to the Food and Drug Regulations in covering those aspects of identity, quality, purity, potency, and packaging and storage that are of particular interest to pharmacists and physicians responsible for the purchase, storage, and use of biologics. Revisions of the federal requirements affecting the USP monographs will be made the subjects of USP Supplements as promptly as practicable.

Vehicles and Added Substances—Vehicles and added substances suitable for biologics are those named in the Food and Drug Regulations.

Containers for Injections—Containers for biologics intended to be administered by injection meet the requirements for *Containers for Injections* under *Injections* ⟨1⟩.

Volume in Container—The volumes in containers of biologics intended to be administered by injection meet the requirements for *Volume in Container* under *Injections* ⟨1⟩.

Labeling—Biologics intended to be administered by injection comply with the requirements for *Labeling* under *Injections* ⟨1⟩. In addition, the label on the final container for each biologic states: the title or proper name (the name under which the product is licensed under the Public Health Service Act); the name, address, and license number of the manufacturer; the lot number; the expiration date; and the recommended individual dose for multiple-dose containers. The package label includes all of the above, with the addition of: the preservative used and its amount; the number of containers, if more than one; the amount of product in the container; the recommended storage temperature; a statement, if necessary, that freezing is to be avoided; and such other information as the Food and Drug regulations may require.

Packaging and Storage—The labeling gives the recommended storage temperature (see *General Notices*). Precautions should be taken where products labeled to be stored at a temperature between 2° and 8° are stored in a refrigerator, in order to assure that they will not be frozen. Diluents packaged with biologics should not be frozen. Some products (as defined in Section 600.15) are to be maintained during shipment at specified temperatures.

Expiration Date—For compendial articles the expiration date identifies the time during which the article may be expected to meet the requirements of the Pharmacopeial monograph, provided it is kept under the prescribed storage conditions. This date limits the time during which the product may be dispensed or used (see *General Notices*, page 1). However, for biological products the stated date on each lot determines the dating period, which begins on the date of manufacture (Section 610.50) and beyond which the product cannot be expected beyond reasonable doubt to yield its specific results and to retain the required safety, purity, and potency (Section 300.3 (1) and (m)). Such a dating period may comprise an in-house storage period during which it is permitted to be held under prescribed conditions in the manufacturer's storage, followed by a period after issue therefrom. The individual monographs usually indicate both, the latter period and (in parentheses) the permissible in-house storage period. If the product is held for a longer period in the manufacturer's storage than that indicated (in parentheses), the expiration date is set so as to reduce the dating period after issue from the manufacturer's storage by a corresponding amount.

⟨1051⟩ CLEANING GLASS APPARATUS

Success in conducting many Pharmacopeial assays depends upon the utmost cleanliness of the apparatus used. To be mentioned especially in this connection are the assays of heparin sodium and of vitamin B_{12} activity, and the pyrogen test.

For cleansing glassware, one of the most effective agents is hot nitric acid. An effective method of removing organic matter from glass without heating is by treatment with a chromic acid cleansing mixture as described herein. Crystalline chromic acid tends to separate from the mixture on standing, and may be removed by decantation.

Glass tends to adsorb the chromic acid, which makes long rinsing imperative. Alkaline cleansing agents, such as trisodium phosphate and the synthetic detergents, are highly useful but also require prolonging rinsing. Special care is required for cleaning containers used for optical measurements, and the use of both chromic acid and highly alkaline solutions should be avoided.

Chromic Acid Cleansing Mixture may be prepared as follows.

Chromic Acid Cleansing Mixture

Sodium Dichromate	200 g
Water	100 mL
Sulfuric Acid	1500 mL

Dissolve the sodium dichromate in the water, and slowly and cautiously add the sulfuric acid, with stirring.

Caution—Wear safety goggles. Prepare this mixture in a hard, borosilicate-glass, 2000-mL beaker, since the heat produced may cause soft-glass containers to break. Chromic Acid Cleansing Mixture is extremely corrosive and hygroscopic, and should be stored in glass-stoppered bottles in a safe place. When the mixture acquires a green color, it should not be returned to the storage bottle, but should be discarded under continuously flowing water.

⟨1061⟩ COLOR—INSTRUMENTAL MEASUREMENT

The observed color (see *Color and Achromicity* ⟨631⟩) of an object depends on the spectral energy of the illumination, the absorbing characteristics of the object, and the visual sensitivity of the observer over the visible range. Similarly, it is essential that any instrumental method that is widely applicable take these same factors into account.

Instrumental methods for measurement of color provide more objective data than the subjective viewing of colors by a small number of individuals. With adequate maintenance and calibration, instrumental methods can provide accurate and precise measurements of color and color differences that do not drift with time. The basis of any instrumental measurement of color is that the human eye has been shown to detect color via three "receptors." Hence, all colors can be broken down into a mixture of three radiant stimuli that are suitably chosen to excite all three receptors in the eye. Although no single set of real light sources can be used to match all colors (i.e., for any three lights chosen, some colors require a negative amount of one or more of the lights), three arbitrary stimuli have been defined, with which it is possible to define all real colors. Through extensive color-matching experiments with human subjects having normal color vision, distributing coefficients have been measured for each visible wavelength (400 nm to 700 nm) giving the relative amount of stimulation of each receptor caused by light of that wavelength. These distribution coefficients $\bar{x}$, $\bar{y}$, $\bar{z}$, are shown below. Similarly, for any color the amount of stimulation of each receptor in the eye is defined by the set of *Tristimulus values* (X, Y, and Z) for that color.

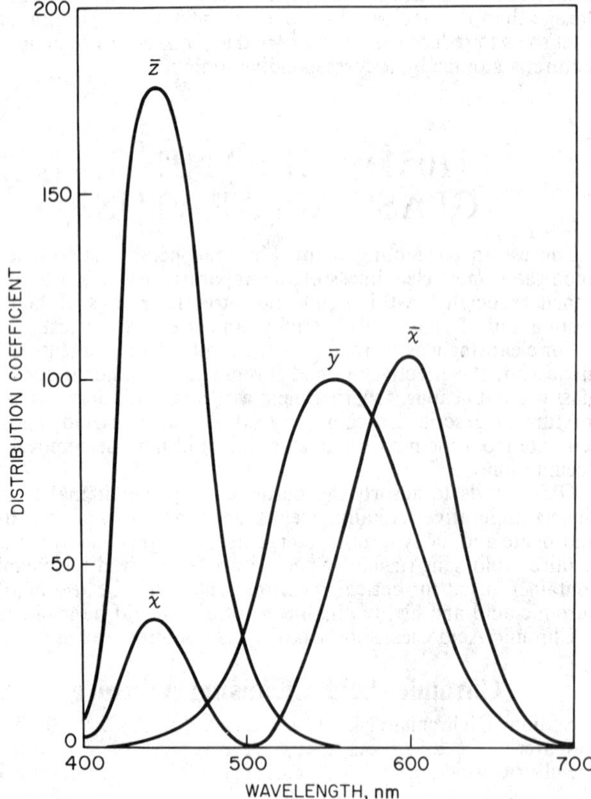

Distribution Coefficients from 400 to 700 nm

The relationships between the distribution coefficient (see accompanying figure) and the tristimulus values are given in the equations

$$X = \int_0^\infty f_\lambda \bar{x}_\lambda P_\lambda d\lambda / Y',$$

$$Y = \int_0^\infty f_\lambda \bar{y}_\lambda P_\lambda d\lambda / Y', \text{ and}$$

$$Z = \int_0^\infty f_\lambda \bar{z}_\lambda P_\lambda d\lambda / Y',$$

in which $Y' = \int_0^\infty \bar{y}_\lambda P_\lambda d\lambda$, P_λ is the spectral power of the illuminant, and f_λ is either the spectral reflectance (ρ_λ) or spectral transmittance (τ_λ) of the material.

Once the tristimulus values of a color have been determined, they may be used to calculate the coordinates of the color in an idealized three-dimensional color space referred to as a *visually uniform color space.* Many sets of color equations have been developed in an attempt to define such a space. The equations given in this chapter represent a compromise between simplicity of calculation and conformance with ideality.

The coordinates of a color in a visually uniform color space may be used to calculate the deviation of a color from a chosen reference point. Where the instrumental method is used to determine the result of a test requiring color comparison of a test preparation with that of a standard or matching fluid, the parameter to be compared is the difference, in visually uniform color space, between the color of the blank and the color of the test specimen or standard.

Procedure

The considerations discussed under *Spectrophotometry and Light-scattering* ⟨851⟩ apply to instrumental color measurement as well. In the spectrophotometric method, reflectance or transmittance values are obtained at discrete wavelengths throughout the visible spectrum, a band width of 10 nm or less being used. These values are then used to calculate the tristimulus values through the use of weighting factors.[1] In the colorimetric method, the weighting is performed through the use of filters.

In the measurement of the spectral reflectance of opaque solids, the angle of viewing is separated from the angle of illumination in such a manner that only rays reflected diffusely from the test specimen enter the receptor. Specular reflection and stray light are excluded.

For the measurement of the spectral transmittance of clear liquids, the specimen is irradiated from within 5 degrees of the normal to its surface, and the transmitted energy measured is that confined within 5 degrees from the normal. The color of solutions changes with the thickness of the layer measured. Unless special considerations dictate otherwise, a layer 1 cm thick should be used.

The methods described here are not applicable to hazy liquids or translucent solids.

CALIBRATION

For purposes of calibration, one of the following reference materials may be used, as required by instrument geometry. For transmittance measurements, purified water may be used as a white standard and assigned a transmittance of 1.000 at all wavelengths. Then the tristimulus values X, Y, and Z for CIE source C are 98.0, 100.0, and 118.1, respectively. For reflectance measurements, opaque porcelain plaques, whose calibration base is the perfect diffuse reflector and whose reflectance characteristics have been determined for the appropriate instrumental geometry, may be used.[2] If the geometry of sample presentation precludes the use of such plaques, pressed barium sulfate, white reflectance standard grade, may be used.[3]

After calibration with the above-mentioned materials, it is desirable whenever possible to measure a reference material as close to the color of the sample as possible. If a sample of the material being tested is not suitable for use as a long-term standard, color chips are available[4] which span the entire visually uniform color space in small increments. The use of such a reference standard is encouraged as a means of monitoring instrument performance even for absolute color determinations.

[1] Typical weighting factors are given by ASTM Z58.7.1-1951 as reported in the Journal of the Optical Society of America, Vol. 41, 1951, pages 431–439.

[2] Suitable items are available from Pacific Scientific Co., Gardner Instrument Div., 2431 Linden Lane, Silver Spring, MD 20910, or from Hunter Associates Laboratory, Inc., 11491 Sunset Hills Road, Reston, VA 22090.

[3] Suitable material is available from Eastman Kodak Company, Rochester, NY 14650, as "White Reflectance Standard."

[4] National Instititute of Standards and Technology, Standard Reference Material #2107, ISCC-NBS Centroid Color Charts available from U. S. Dept. of Commerce, National Institute of Standards and Technology, Wash., DC 20234.

SPECTROPHOTOMETRIC METHOD

Determine the reflectance or transmittance from 380 to 770 nm at intervals of 10 nm. Express the result as a percentage, the maximum being 100.0. Calculate the tristimulus values X, Y, and Z as follows.

Reflecting Materials—For reflecting materials the quantities X, Y, and Z are

$$X = \sum_{380}^{770} \rho_\lambda \bar{x}_\lambda P_\lambda \Delta\lambda / Y',$$

$$Y = \sum_{380}^{770} \rho_\lambda \bar{y}_\lambda P_\lambda \Delta\lambda / Y', \text{ and}$$

$$Z = \sum_{380}^{770} \rho_\lambda \bar{z}_\lambda P_\lambda \Delta\lambda / Y',$$

in which $Y' = \sum_{380}^{770} \bar{y}_\lambda P_\lambda \Delta\lambda$, ρ_λ is the spectral reflectance of the material, $\bar{x}_\lambda P_\lambda$, $\bar{y}_\lambda P_\lambda$ and $\bar{z}_\lambda P_\lambda$ are known values associated with each Standard Source,[1,2] and $\Delta\lambda$ is expressed in nm.

Transmitting Materials—For transmitting materials, the quantities X, Y, and Z are calculated as above, τ_λ (spectral transmittance) being substituted for ρ_λ.

COLORIMETRIC METHOD

Operate a suitable colorimeter[5] to obtain values equivalent to the tristimulus values, X, Y, and Z. The accuracy with which the results obtained from the filter colorimeter match the tristimulus values may be indicated by determining the tristimulus values of plaques of strongly saturated colors and comparing these values with those computed from spectral measurements on a spectrophotometer.

Interpretation

COLOR COORDINATES

The Color Coordinates, L^*, a^*, and b^* are defined by

$$L^* = 116\,(Y/Y_o)^{1/3} - 16,$$
$$a^* = 500\,[(X/X_o)^{1/3} - (Y/Y_o)^{1/3}], \text{ and}$$
$$b^* = 200\,[(Y/Y_o)^{1/3} - (Z/Z_o)^{1/3}],$$

in which X_o, Y_o, and Z_o are the tristimulus values of the nominally white or colorless standard, and $Y/Y_o > 0.01$. Usually they are equal to the tristimulus values of the standard illuminant, with Y_o set equal to 100.0. In this case $X_o = 98.0$ and $Z_o = 118.1$.

COLOR DIFFERENCE

The total Color Difference ΔE is

$$\Delta E = [(\Delta L^*)^2 + (\Delta a^*)^2 + (\Delta b^*)^2]^{1/2},$$

in which ΔL^*, Δa^*, and Δb^* are the differences in color coordinates of the specimens being compared.

Instrumental variables can influence results. Although reliable comparisons can be made between similar colors measured concomitantly, results obtained on different instruments or under different operating conditions should be compared with caution. If it is necessary to compare data obtained from different instruments or taken at different times, etc., it is very helpful to have concomitant data obtained on a standard reference material such as the NIST #2107 color chips for opaque materials. Comparison of the readings on the reference material helps to identify variations caused by instrument performance.

⟨1071⟩ CONTROLLED SUBSTANCES ACT REGULATIONS

Selected portions of the regulations promulgated under the Controlled Substances Act that are believed to be of most concern to practitioners and students of pharmacy and medicine are presented here as a service to the professions and at the suggestion of the Drug Enforcement Administration in accordance with its registrant information and self-regulation program.

The publication of these regulations in The United States Pharmacopeia is for purposes of information and does not impart to them any legal effect.

The Drug Enforcement Administration was established July 1, 1973, through the merging of various Bureaus and offices of the federal government with the Bureau of Narcotics and Dangerous Drugs.

INDEX TO THE PORTIONS OF CONTROLLED SUBSTANCES ACT REGULATIONS PRESENTED HEREIN

NOTE—A complete set of the regulations appears in Volume 21, part 1300 to end, of the *Code of Federal Regulations*, revised annually, and covers regulations pertaining to manufacturers, distributors, researchers, exporters, and importers.

Information regarding procedures under these regulations and instructions implementing them are obtainable from the Drug Enforcement Administration, Department of Justice, 1405 I Street, N.W., Washington, DC 20537.

---

[5]A suitable tristimulus colorimeter is available from Pacific Scientific Co., Gardner Instrument Div., 2431 Linden Lane, Silver Spring, MD 20910, or from Hunter Associates Laboratory, Inc., 11491 Sunset Hills Road, Reston, VA 22090.

General Provisions

§ 290.05 Drugs; statement of required warning.

The label of any drug listed as a "controlled substance" in Schedule II, III, or IV of the Federal Controlled Substances Act shall, when dispensed to or for a patient, contain the following warning: "Caution: Federal law prohibits the transfer of this drug to any person other than the patient for whom it was prescribed." This statement is not required to appear on the label of a controlled substance dispensed for use in clinical investigations which are "blind."

§ 290.06 Spanish-language version of required warning.

By direction of section 305(c) of the Federal Controlled Substances Act, § 290.05, promulgated under section 503(b) of the Federal Food, Drug, and Cosmetic Act, requires the following warning on the label of certain drugs when dispensed to or for a patient: "Caution: Federal law prohibits the transfer of this drug to any person other than the patient for whom it was prescribed." The Spanish version of this is: "Precaucion: La ley Federal prohibe el transferir de esta droga a otra persona que no sea el paciente para quien fue recetada." (Secs. 502, 503; 53 Stat. 854, 65 Stat. 648; 21 U.S.C. 352, 353)

§ 290.10 Definition of emergency situation.

For the purposes of authorizing an oral prescription of a controlled substance listed in Schedule II of the Federal Controlled Substances Act, the term "emergency situation" means those situations in which the prescribing practitioner determines:

(a) That immediate administration of the controlled substance is necessary, for proper treatment of the intended ultimate user; and

(b) That no appropriate alternative treatment is available, including administration of a drug which is not a controlled substance under Schedule II of the Act, and

(c) That it is not reasonably possible for the prescribing practitioner to provide a written prescription to be presented to the person dispensing the substance, prior to the dispensing.

General Information

§ 1301.01 Scope of Part 1301.

Procedures governing the registration of manufacturers, distributors, and dispensers of controlled substances pursuant to sections 1301 through 1304 of the Act (21 U.S.C. 821–824) are set forth generally by those sections and specifically by the sections of this part.

§ 1301.02 Definitions.

As used in this part, the following terms shall have the meanings specified:

(a) The term "Act" means the Controlled Substances Act (84 Stat. 1242; 21 U.S.C. 801) and/or the Controlled Substances Import and Export Act (84 Stat. 1285; 21 U.S.C. 951).

(b) The term "basic class" means, as to controlled substances listed in Schedules I and II:

(1) Each of the opiates, including its isomers, esters, ethers, salts, and salts of isomers, esters, and ethers whenever the existence of such isomers, esters, ethers, and salts is possible within the specific chemical designation, listed in § 1308.11(b) of this chapter;

(2) Each of the opium derivatives, including its salts, isomers, and salts of isomers whenever the existence of such salts, isomers, and salts of isomers is possible within the specific chemical designation, listed in § 1308.11(c) of this chapter;

(3) Each of the hallucinogenic substances, including its salts, isomers, and salts of isomers whenever the existence of such salts, isomers, and salts of isomers is possible within the specific chemical designation, listed in § 1308.11(d) of this chapter;

(4) Each of the following substances, whether produced directly or indirectly by extraction from substances of vegetable origin, or independently by means of chemical synthesis, or by a combination of extraction and chemical synthesis:

(i) Opium, including raw opium, opium extracts, opium fluid extracts, powdered opium, granulated opium, deodorized opium and tincture of opium;
(ii) Apomorphine;
(iii) Codeine;
(iv) Etorphine hydrochloride;
(v) Ethylmorphine;
(vi) Hydrocodone;
(vii) Hydromorphone;
(viii) Metopon;
(ix) Morphine;
(x) Oxycodone;
(xi) Oxymorphone;
(xii) Thebaine;
(xiii) Mixed alkaloids of opium listed in § 1308.12(b)(2) of this chapter;
(xiv) Cocaine; and
(xv) Ecgonine.

(5) Each of the opiates, including its isomers, esters, ethers, salts, and salts of isomers, esters, and ethers whenever the existence of such isomers, esters, ethers, and salts is possible within the specific chemical designation, listed in § 1308.12(c) of this chapter; and

(6) Methamphetamine, its salts, isomers, and salts of its isomers;

(7) Amphetamine, its salts, optical isomers, and salts of its optical isomers;

(8) Phenmetrazine and its salts;

(9) Methylphenidate;

(10) Each of the substances having a depressant effect on the central nervous system, including its salts, isomers, and salts of isomers whenever the existence of such salts, isomers, and salts of isomers is possible within the specific chemical designation, listed in § 1308.12(e) of this chapter.

(c) The term "Administration" means the Drug Enforcement Administration.

(d) The term "compounder" means any person engaging in maintenance or detoxification treatment who also mixes, prepares, packages or changes the dosage form of a narcotic drug listed in Schedules II, III, IV or V for use in maintenance or detoxification treatment by another narcotic treatment program.

(e) The term "detoxification treatment" means the dispensing for a period not in excess of twenty-one days, of a narcotic drug or narcotic drugs in decreasing doses to an individual in order to alleviate adverse physiological or psychological effects incident to withdrawal from the continuous or sustained use of a narcotic drug and as a method of bringing the individual to a narcotic drug-free state within such period of time.

(f) The term "Administrator" means the Administrator of the Drug Enforcement Administration. The Administrator has been delegated authority under the Act by the Attorney General (28 CFR 0.100).

(g) The term "hearing" means any hearing held pursuant to this part for the granting, denial, revocation, or suspension of a registration pursuant to sections 303 and 304 of the Act (21 U.S.C. 823–824).

(h) The term "maintenance treatment" means the dispensing for a period in excess of twenty-one days, of a narcotic drug or narcotic drugs in the treatment of an individual for dependence upon heroin or other morphine-like drug.

(i) The term "narcotic treatment program" means a program engaged in maintenance and/or detoxification treatment with narcotic drugs.

(j) The term "person" includes any individual, corporation, government or governmental subdivision or agency, business trust, partnership, association, or other legal entity.

(k) The terms "register" and "registration" refer only to registration required and permitted by section 303 of the Act (21 U.S.C. 823).

(l) The term "registrant" means any person who is registered pursuant to either section 303 or section 1008 of the Act (21 U.S.C. 823 or 958).

(m) Any term not defined in this section shall have the definition set forth in section 102 of the Act (21 U.S.C. 802).

§ 1301.03 Information; special instructions.

Information regarding procedures under these rules and instructions supplementing these rules will be furnished upon request by writing to the Registration Unit, Drug Enforcement Administration, Department of Justice, Post Office Box 28083, Central Station, Washington, D. C. 20005.

Fees for Registration and Reregistration

§ 1301.11 Fee amounts.

(c) For each registration or reregistration to dispense, or to conduct instructional activities with, controlled substances listed in Schedules II through V, the registrant shall pay an application fee of $60 for a three-year registration.

(d) For each registration to conduct research or instructional activities with a controlled substance listed in Schedule I, or to conduct research with a controlled substance in Schedules II through V, the registrant shall pay an application fee of $20.

(f) For each registration or reregistration to engage in a narcotic treatment program, including a compounder, the registrant shall pay an application fee of $5.

§ 1301.12 Time and method of payment; refund.

Application fees shall be paid at the time when the application for registration or reregistration is submitted for filing. Payments should be made in the form of a personal, certified, or cashier's check or money order made payable to "Drug Enforcement Administration." Payments made in the form of stamps, foreign currency, or third party endorsed checks will not be accepted. These application fees are not refundable.

Requirements for Registration

§ 1301.21 Persons required to register.

Every person who manufactures, distributes, or dispenses any controlled substance or who proposes to engage in the manufacture, distribution, or dispensing of any controlled substance shall obtain annually a registration unless exempted by law or pursuant to §§ 1301.24–1301.29. Only persons actually engaged in such activities are required to obtain a registration; related or affiliated persons who are not engaged in such activities are not required to be registered. (For example, a stockholder or parent corporation of a corporation manufacturing controlled substances is not required to obtain a registration.)

§ 1301.22 Separate registration for independent activities.

(a) The following groups of activities are deemed to be independent of each other:

(1) Manufacturing controlled substances;

(2) Distributing controlled substances;

(3) Dispensing controlled substances listed in Schedules II through V;

(4) Conducting research with controlled substances listed in Schedules II through V;

(5) Conducting instructional activities with controlled substances listed in Schedules II through V;

(6) Conducting a narcotic treatment program using any narcotic drug listed in Schedules II, III, IV or V, however, pursuant to § 1301.24, employees, agents, or affiliated practitioners, in programs, need not register separately. Each program site located away from the principal location and at which place narcotic drugs are stored or dispensed must be separately registered and obtain narcotic drugs by use of order forms pursuant to § 1305.03;

(7) Conducting research and instructional activities with controlled substances listed in Schedule I;
(8) Conducting chemical analysis with controlled substances listed in any schedule;
(9) Importing controlled substances;
(10) Exporting controlled substances; and
(11) A compounder as defined by § 1301.02(d).

(b) Every person who engages in more than one group of independent activities shall obtain a separate registration for each group of activities, except as provided in this paragraph. Any person, when registered to engage in the group of activities described in each subparagraph in this paragraph, shall be authorized to engage in the coincident activities described in that subparagraph without obtaining a registration to engage in such coincident activities, provided that, unless specifically exempted, he complies with all requirements and duties prescribed by law for persons registered to engage in such coincident activities;

(6) A person registered to dispense controlled substances listed in Schedules II through V shall be authorized to conduct research and to conduct instructional activities with those substances.

(c) A single registration to engage in any group of independent activities may include one or more controlled substances listed in the schedules authorized in that group of independent activities. A person registered to conduct research with controlled substances listed in Schedule I may conduct research with any substance listed in Schedule I for which he has filed and had approved a research protocol.

§ 1301.23 Separate registrations for separate locations.

(a) A separate registration is required for each principal place of business or professional practice at one general physical location where controlled substances are manufactured, distributed, or dispensed by a person.

(b) The following locations shall be deemed not to be places where controlled substances are manufactured, distributed, or dispensed:

(1) A warehouse where controlled substances are stored by or on behalf of a registered person, unless such substances are distributed directly from such warehouse to registered locations other than the registered location from which the substances were delivered or to persons not required to register by virtue of subsection 302(c)(2) of the Act (21 U.S.C. 822(c)(2));

(2) An office used by agents of a registrant where sales of controlled substances are solicited, made, or supervised but which neither contains such substances (other than substances for display purposes or lawful distribution as samples only) nor serves as a distribution point for filling sales orders; and

(3) An office used by a practitioner (who is registered at another location) where controlled substances are prescribed but neither administered nor otherwise dispensed as a regular part of the professional practice of the practitioner at such office, and where no supplies of controlled substances are maintained.

§ 1301.24 Exemption of agents and employees; affiliated practitioners.

(a) The requirement of registration is waived for any agent or employee of a person who is registered to engage in any group of independent activities, if such agent or employee is acting in the usual course of his business or employment.

(b) An individual practitioner, as defined in § 1304.02 of this chapter (other than an intern, resident, foreign-trained physician, or physician on the staff of a Veterans Administration facility or physician who is an agent or employee of the Health Bureau of the Canal Zone Government), who is an agent or employee of another practitioner registered to dispense controlled substances may, when acting in the usual course of his employment, administer and dispense (other than by issuance of prescription) controlled substances if and to the extent that such individual practitioner is authorized or permitted to do so by the jurisdiction in which he practices, under the registration of the employer or principal practitioner in lieu of being registered himself. (For example, a staff physician employed by a hospital need not be registered individually to administer and dispense, other than by prescribing, controlled substances within the hospital.)

(c) An individual practitioner, as defined in § 1304.02 of this chapter, who is an intern, resident, or foreign-trained physician or physician on the staff of a Veterans Administration facility or physician who is an agent or employee of the Health Bureau of the Canal Zone Government, may dispense, administer and prescribe controlled substances under the registration of the hospital or other institution which is registered and by whom he is employed in lieu of being registered himself, provided that:

(1) Such dispensing, administering, or prescribing is done in the usual course of his professional practice;

(2) Such individual practitioner is authorized or permitted to do so by the jurisdiction in which he is practicing;

(3) The hospital or other institution by whom he is employed has verified that the individual practitioner is so permitted to dispense, administer, or prescribe drugs within the jurisdiction;

(4) Such individual practitioner is acting only within the scope of his employment in the hospital or institution;

(5) The hospital or other institution authorizes the intern, resident, or foreign-trained physician to dispense or prescribe under the hospital registration and designates a specific internal code number for each intern, resident, or foreign-trained physician so authorized. The code number shall consist of numbers, letters, or a combination thereof and shall be a suffix to the institution's DEA registration number, preceded by a hyphen (e.g., APO 123456-10 or APO 123456-A12); and

(6) A current list of internal codes and the corresponding individual practitioners is kept by the hospital or other institution and is made available at all times to other registrants and law enforcement agencies upon request for the purpose of verifying the authority of the prescribing individual practitioner.

§ 1301.25 Exemption of certain military and other personnel.

(a) The requirement of registration is waived for any official of the U. S. Army, Navy, Marine Corps, Air Force, Coast Guard, Public Health Service, or Bureau of Prisons who is authorized to prescribe, dispense, or administer, but not to procure or purchase, controlled substances in the course of his official duties. Such officials shall follow procedures set forth in Part 1306 of this chapter regarding prescriptions, but shall state the branch of service or agency (e.g., "U. S. Army" or "Public Health Service") and the service identification number of the issuing official in lieu of the registration number required on prescription forms. The service identification number for a Public Health Service employee is his Social Security identification number.

(b) If any official exempted by this section also engages as a private individual in any activity or group of activities for which registration is required, such official shall obtain a registration for such private activities.

§ 1301.28 Registration regarding ocean vessels.

(a) If acquired by and dispensed under the general supervision of a medical officer described in paragraph (b) of this section, or the master or first officer of the vessel under the circumstances described in paragraph (d) of this section, controlled substances may be held for stocking, be maintained in, and dispensed from medicine chests, first aid packets, or dispensaries:

(1) On board any vessel engaged in international trade or in trade between ports of the United States and any merchant vessel belonging to the U. S. Government;

(2) On board any aircraft operated by an air carrier under a certificate of permit issued pursuant to the Federal Aviation Act of 1958 (49 U.S.C. 1301); and

(3) In any other entity of fixed or transient location approved by the Administrator as appropriate for application of this section (e.g., emergency kits at field sites of an industrial firm).

(b) A medical officer shall be:

(1) Licensed in a state as a physician;

(2) Employed by the owner or operator of the vessel, aircraft or other entity; and

(3) Registered under the Act at either of the following locations:

(i) The principal office of the owner or operator of the vessel, aircraft or other entity or

(ii) At any other location provided that the name, address, registration number, and expiration date as they appear on his Certificate of Registration (DEA Form 223) for this location are maintained for inspection at said principal office in a readily retrievable manner.

(c) A registered medical officer may serve as medical officer for more than one vessel, aircraft, or other entity under a single registration, unless he serves as medical officer for more than one owner or operator, in which case he shall either maintain a separate registration at the location of the principal office of each such owner or operator or utilize one or more registrations pursuant to paragraph (b)(3)(ii) of this section.

(d) If no medical officer is employed by the owner or operator of a vessel, or in the event such medical officer is not accessible and the acquisition of controlled substances is required, the master or first officer of the vessel, who shall not be registered under the Act, may purchase controlled substances from a registered manufacturer or distributor, or from an authorized pharmacy as described in paragraph (f) of this section, by following the procedure outlined below:

(1) The master or first officer of the vessel must personally appear at the vendor's place of business, present proper identification (e.g., Seaman's photographic identification card) and a written requisition for the controlled substances.

(2) The written requisition must be on the vessel's official stationery or purchase order form and must include the name and address of the vendor, the name of the controlled substance, description of the controlled substance (dosage form, strength and number or volume per container), number of containers ordered, the name of the vessel, the vessel's official number and country of registry, the owner or operator of the vessel, the port at which the vessel is located, signature of the vessel's officer who is ordering the controlled substances and the date of the requisition.

(3) The vendor may, after verifying the identification of the vessel's officer requisitioning the controlled substances, deliver the control substances to that officer. The transaction shall be documented, in triplicate, on a record of sale in a format similar to that outlined in paragraph (d)(4) of this section. The vessel's requisition shall be attached to copy 1 of the record of sale and filed with the controlled substances records of the vendor, copy 2 of the record of sale shall be furnished to the officer of the vessel and retained aboard the vessel, copy 3 of the record of sale shall be forwarded to the nearest DEA Division Office within 15 days after the end of the month in which the sale is made.

(4) The vendor's record of sale should be similar to, and must include all the information contained in, the below listed format.

Sale of Controlled Substances to Vessels

(Name of registrant) ______________
(Address of registrant) ______________
(DEA registration number) ______________

Line No.	Number of packages ordered	Size of packages	Name of product	Packages distributed	Date distributed
1					
2					
3					

Line numbers may be continued according to needs of the vendor.

Number of lines completed ______________
Name of vessel ______________
Vessel's official number ______________
Vessel's country of registry ______________
Owner or operator of the vessel ______________
Name and title of vessel's officer who presented the requisition ______________
Signature of vessel's officer who presented the requisition ______________

(e) Any medical officer described in paragraph (b) of this section shall, in addition to complying with all requirements and duties prescribed for registrants generally, prepare an annual report as of the date on which his registration expires, which shall give in detail an accounting for each vessel, aircraft, or other entity, and a summary accounting for all vessels, aircraft, or other entities under his supervision for all controlled substances purchased, dispensed or disposed of during the year. The medical officer shall maintain this report with other records required to be kept under the Act and, upon request, deliver a copy of the report to the Administration. The medical officer need not be present when controlled substances are dispensed, if the person who actually dispensed the controlled substances is responsible to the medical officer to justify his actions.

(f) Any registered pharmacy which wishes to distribute controlled substances pursuant to this section shall be authorized to do so, provided that:

(1) The registered pharmacy notifies the nearest Division Office of the Administration of its intention to so distribute controlled substances prior to the initiation of such activity. This notification shall be by registered mail and shall contain the name, address, and registration number of the pharmacy as well as the date upon which such activity will commence; and

(2) Such activity is authorized by State law; and

(3) The total number of dosage units of all controlled substances distributed by the pharmacy during any calendar year in which the pharmacy is registered to dispense does not exceed the limitations imposed upon such distribution by § 1307.11(a)(4) and (b) of this chapter.

(g) Owners or operators of vessels, aircraft, or other entities described in this section shall not be deemed to possess or dispense any controlled substance acquired, stored and dispensed in accordance with this section.

(h) The Master of a vessel shall prepare a report for each calendar year which shall give in detail an accounting for all controlled substances purchased, dispensed, or disposed of during the year. The Master shall file this report with the medical officer employed by the owner or operator of his vessel, if any, or, if not, he shall maintain this report with other records required to be kept under the Act and, upon request, deliver a copy of the report to the Administration.

(i) Controlled substances acquired and possessed in accordance with this section shall not be distributed to persons not under the general supervision of the medical officer employed by the owner or operator of the vessel, aircraft, or other entity, except in accordance with § 1307.21 of this chapter.

§ 1301.29 Provisional registration of narcotic treatment programs; compounders.

(a) All persons currently approved by the Food and Drug Administration under § 310.505 (formerly § 130.44) of this title to conduct a methadone treatment program and who are registered by the Drug Enforcement Administration under this section will be granted a Provisional Narcotic Treatment Program Registration.

(b) The provisions of §§ 1301.45–1301.57 relating to revocation and suspension of registration, shall apply to a provisional registration.

(c) Unless sooner revoked or suspended under paragraph (b) of this section, a provisional registration shall remain in effect until (1) the date on which such person has registered under this section or has had his registration denied, or (2) such date as may be prescribed by written notification to the person from the Drug Enforcement Administration for the person to become registered to conduct a narcotic treatment program, whichever occurs first.

Applications for Registration

§ 1301.31 Time for application for registration; expiration date.

(a) Any person who is required to be registered and who is not so registered may apply for registration at any time. No person required to be registered shall engage in any activity for which registration is required until the application for registration is granted and a Certificate of Registration is issued by the Administrator to such person.

(b) Any person who is registered may apply to be reregistered not more than 60 days before the expiration date of his registration.

(c) At the time a manufacturer, distributor, researcher, analytical lab, importer, exporter or narcotic treatment program is first registered, that business activity shall be assigned to one of twelve groups, which shall correspond to the months of the year. The expiration date of the registrations of all registrants within any group will be the last date of the month designated for that group. In assigning any of the above business activities to a group, the Administration may select a group the expiration date of which is less than one year from the date such business activity was registered. If the business activity is assigned to a group which has an expiration date less than three months from the date on which the business activity is registered, the registration shall not expire until one year from that expiration date; in all other cases, the registration shall expire on the expiration date following the date on which the business activity is registered.

(d) At the time a retail pharmacy, hospital/clinic, practitioner or teaching institution is first registered, that business activity shall be assigned to one of twelve groups, which shall correspond to the months of the year. The expiration date of the registrations of all registrants within any group will be the last day of the month designated for that group. In assigning any of the above business activities to a group, the Administration may select a group the expiration date of which is not less than 28 months nor more than 39 months from the date such business activity was registered. After the initial registration period, the registration shall expire 36 months from the initial expiration date.

§ 1301.32 Application forms; contents; signature.

(a) If any person is required to be registered, and is not so registered and is applying for registration:

(1) To manufacture or distribute controlled substances, he shall apply on DEA Form 225;

(2) To dispense controlled substances listed in Schedules II through V, he shall apply on DEA Form 224;

(9) To conduct a narcotic treatment program, including a compounder, he shall apply on DEA Form 363.

(b) If any person is registered and is applying for reregistration:

(1) To manufacture or distribute controlled substances, he shall apply on DEA Form 225a;

(2) To dispense controlled substances listed in Schedules II through V, he shall apply on DEA Form 224A;

(9) To conduct a narcotic treatment program, including a compounder, he shall apply on DEA Form 363a (Renewal Form).

(c) DEA (or BND) Forms 224 and 225 may be obtained at any regional office of the Administration or by writing to the Registration Unit, Drug Enforcement Administration, Department of Justice, Post Office Box 28083, Central Station, Washington, D. C. 20005. DEA Forms 224a, 225a and 363a will be mailed, as applicable, to each registered person approximately 60 days before the expiration date of his registration; if any registered person does not receive such forms within 45 days before the expiration date of his registration, he must promptly give notice of such fact and request such forms by writing to the Registration Unit of the Administration at the foregoing address.

(d) Each application for registration to handle any basic class of controlled substance listed in Schedule I (except to conduct chemical analysis with such classes) and each application for registration to manufacture a basic class of controlled substance listed in Schedule II shall include the Administration Controlled Substances Code Number, as set forth in Part 1308 of this chapter, for each basic class to be covered by such registration.

(e) Each application for registration to conduct research with any basic class of controlled substance listed in Schedule II shall include the Administration Controlled Substances Code Number, as set forth in Part 1308 of this chapter, for each such basic class to be manufactured or imported as a coincident activity of that registration. A statement listing the quantity of each such basic class or controlled substance to be imported or manufactured during the registration period for which application is being made shall be included with each such application. For purposes of this paragraph only, manufacturing is defined as the production of a controlled substance by synthesis, extraction or by agricultural/horticultural means.

(f) Each application shall include all information called for in the form, unless the item is not applicable, in which case this fact shall be indicated.

(g) Each application, attachment, or other document filed as part of an application, shall be signed by the applicant, if an individual; by a partner of the applicant, if a partnership; or by an officer of the applicant, if a corporation, corporate division, association, trust or other entity. An applicant may authorize one or more individuals, who would not otherwise be authorized to do so, to sign applications for the applicant by filing with the Registration Unit of the Administration a power of attorney for each such individual. The power of attorney shall be signed by a person who is authorized to sign applications under this paragraph and shall contain the signature of the individual being authorized to sign applications. The power of attorney shall be valid until revoked by the applicant.

§ 1301.34 Filing of application; joint filings.

(a) All applications for registration shall be submitted for filing to the Registration Unit, Drug Enforcement Administration, Department of Justice, Post Office Box 28083, Central Station, Washington, D. C. 20005. The appropriate registration fee and any required attachments must accompany the application.

(b) Any person required to obtain more than one registration may submit all applications in one package. Each application must be complete and should not refer to any accompanying application for required information.

§ 1301.35 Acceptance for filing; defective applications.

(a) Applications submitted for filing are dated upon receipt. If found to be complete, the application will be accepted for filing. Applications failing to comply with the requirements of this part will not generally be accepted for filing. In the case of minor defects as to completeness, the Administrator may accept the application for filing with a request to the applicant for additional information. A defective application will be returned to the applicant within 10 days following its receipt with a statement of the reason for not accepting the application for filing. A defective application may be corrected and resubmitted for filing at any time; the Administrator shall accept for filing any application upon resubmission by the applicant, whether complete or not.

(b) Accepting an application for filing does not preclude any subsequent request for additional information pursuant to § 1301.36 and has no bearing on whether the application will be granted.

§ 1301.36 Additional information.

The Administrator may require an applicant to submit such documents or written statements of fact relevant to the application as he deems necessary to determine whether the application should be granted. The failure of the applicant to provide such documents or statements within a reasonable time after being requested to do so shall be deemed to be a waiver by the applicant of an opportunity to present such documents or facts for consideration by the Administrator in granting or denying the application.

§ 1301.37 Amendments to and withdrawal of applications.

(a) An application may be amended or withdrawn without permission of the Administrator at any time before the date on which the applicant receives an order to show cause pursuant to § 1301.48, or before the date on which a notice of hearing on the application is published pursuant to § 1301.43, whichever is sooner. An application may be amended or withdrawn with permission of the Administrator at any time where good cause is shown by the applicant or where the amendment or withdrawal is in the public interest.

(b) After an application has been accepted for filing, the request by the applicant that it be returned or the failure of the applicant to respond to official correspondence regarding the application, when sent by registered or certified mail, return receipt requested, shall be deemed to be a withdrawal of the application.

§ 1301.38 Special procedures for certain applications.

(a) If, at the time of application for registration of a new pharmacy, the pharmacy has been issued a license from the appropriate State licensing agency, the applicant may include with his application an affidavit as to the existence of the State license in the following form:

AFFIDAVIT FOR NEW PHARMACY

I, ______________, the ______________________
(Title of officer, official, partner,

______________ of ________________________
or other position) (Corporation, partnership, or sole

__________, doing business as ____________ at __________
proprietor) (Store name) (Number

____,
and Street) (City) (State) (Zip code)

hereby certify that said store was issued a pharmacy permit No. ________ by the ________________________________
(Board of Pharmacy or Licensing Agency)
of the State of ________________ on ________________ .
(Date)

This statement is submitted in order to obtain a Drug Enforcement Administration registration number. I understand that if any information is false, the Administration may immediately suspend the registration for this store and commence proceedings to revoke under 21 U.S.C. 824(a) because of the danger to public health and safety. I further understand that any false information contained in this affidavit may subject me personally and the above-named corporation/partnership/business to prosecution under 21 U.S.C. 843, the penalties for conviction of which include imprisonment for up to 4 years, a fine of not more than $30,000 or both.

____
Signature (Person who Signs Application for Registration)

State of ____________
County of ____________
Subscribed to and sworn before me this ____________ day of ________________, 19 _____ .

____
Notary Public

(b) Whenever the ownership of a pharmacy is being transferred from one person to another, if the transferee owns at least one other pharmacy licensed in the same State as the one the ownership of which is being transferred, the transferee may apply for registration prior to the date of transfer. The Administrator may register the applicant and authorize him to obtain controlled substances at the time of transfer. Such registration shall not authorize the transferee to dispense controlled substances until the pharmacy has been issued a valid State license. The transferee shall include with his application the following affidavit:

AFFIDAVIT FOR TRANSFER OF PHARMACY

I, ______________, the ______________________________
(Title of officer, official, partner or other position)
______________ of ______________________________
(Corporation, partnership, or sole proprietor)
__________ , doing business as ________________________ ,
(Store name)
hereby certify:

(1) That said company was issued a pharmacy permit No. ________ by the ________________________________
(Board of Pharmacy or Licensing Agency)
of the State of ______________ and a DEA Registration Number ________ for a pharmacy located at ________________________
(Number and Street)
____ ; and
(City) (State) (Zip code)

(2) That said company is acquiring the pharmacy business of ______________ doing business as ______________
(Name of Seller)
with DEA Registration Number ______________ on or about ______________ and that said company has applied (or will
(Date of Transfer)
apply) on ______________ for a pharmacy permit from the
(Date)
board of pharmacy (or licensing agency) of the State of ______________ to do business as ________________________
(Store name)
at ____ .
(Number and Street) (City) (State) (Zip code)

This statement is submitted in order to obtain a Drug Enforcement Administration registration number.

I understand that if a DEA registration number is issued, the pharmacy may acquire controlled substances but may not dispense them until a pharmacy permit or license is issued by the State board of pharmacy or licensing agency.

I understand that if any information is false, the Administration may immediately suspend the registration for this store and commence proceedings to revoke under 21 U.S.C. 824(a) because of the danger to public health and safety. I further understand that any false information contained in this affidavit may subject me personally to prosecution under 21 U.S.C. 843, the penalties for conviction of which include imprisonment for up to 4 years, a fine of not more than $30,000 or both.

____
Signature (Person who Signs Application for Registration)

State of ____________
County of ____________
Subscribed to and sworn before me this ______________ day of ______________, 19 ______ .

____
Notary Public

(c) The Administrator shall follow the normal procedures for approving an application to verify the statements in the affidavit. If the statements prove to be false, the Administrator may revoke the registration on the basis of section 1304(a)(1) of the Act (21 U.S.C. 824(a)(1)) and suspend the registration immediately by pending revocation on the basis of section 1304(d) of the Act (21 U.S.C. 824(d)). At the same time, the Administrator may seize and place under seal all controlled substances possessed by the applicant under section 1304(f) of the Act (21 U.S.C. 824(f)). Intentional misuse of the affidavit procedure may subject the applicant to prosecution for fraud under section 403(a)(4) of the Act (21 U.S.C. 843(a)(4)), and obtaining controlled substances under a registration fraudulently gotten may subject the applicant to prosecution under section 403(a)(3) of the Act (21 U.S.C. 843(a)(3)). The penalties for conviction of either offense include imprisonment for up to 4 years, a fine not exceeding $30,000 or both.

Action on Applications for Registration: Revocation or Suspension of Registration

§ 1301.41 Administrative review generally.

The Administrator may inspect, or cause to be inspected, the establishment of an applicant or registrant, pursuant to Subpart A of Part 1316 of this chapter. The Administrator shall review the application for registration and other information gathered by the Administrator regarding an applicant in order to determine whether the applicable standards of section 1303 of the Act (21 U.S.C. 823) have been met by the applicant.

§ 1301.44 Certificate of registration; denial of registration.

(a) The Administrator shall issue a Certificate of Registration (DEA Form 223) to an applicant if the issuance of registration or reregistration is required under the applicable provisions of section 303 of the Act (21 U.S.C. 823). In the event that the issuance of registration or reregistration is not required, the Administrator shall deny the application. Before denying any application, the Administrator shall issue an order to show cause pursuant to § 1301.48 and, if requested by the applicant, shall hold a hearing on the application pursuant to § 1301.51.

(b) The Certificate of Registration (DEA Form 223) shall contain the name, address, and registration number of the registrant, the activity authorized by the registration, the schedules and/or Administration Controlled Substances Code Number (as set forth in Part 1308 of this chapter) of the controlled substances which the registrant is authorized to handle, the amount of fee paid (or exemption), and the expiration date of the registration. The registrant shall maintain the certificate of registration at the registered location in a readily retrievable manner and shall permit inspection of the certificate by any official, agent or employee of the Administration or of any Federal, State, or local agency engaged in enforcement of laws relating to controlled substances.

§ 1301.47 Extension of registration pending final order.

In the event that an applicant for reregistration (who is doing business under a registration previously granted and not revoked or suspended) has applied for reregistration at least 45 days before the date on which the existing registration is due to expire, and the Administrator has issued no order on the application on the date on which the existing registration is due to expire, the existing registration of the applicant shall automatically be extended and continue in effect until the date on which the Administrator so issues his order. The Administrator may extend any other existing registration under the circumstances contemplated in this section even though the registrant failed to apply for reregistration at least 45 days before expiration of the existing registration, with or without request by the registrant, if the Administrator finds that such extension is not inconsistent with the public health and safety.

Modification, Transfer and Termination of Registration

§ 1301.61 Modification in registration.

Any registrant may apply to modify his registration to authorize the handling of additional controlled substances or to change his name or address, by submitting a letter of request to the Registration Unit, Drug Enforcement Administration, Department of Justice, Post Office Box 28083, Central Station, Washington, DC 20005. The letter shall contain the registrant's name, address, and registration number as printed on the certificate of registration, and the substances and/or schedules to be added to his registration or the new name or address and shall be signed in accordance with § 1301.32(f). If the registrant is seeking to handle additional controlled substances listed in Schedule I for the purpose of research or instructional activities, he shall attach three copies of a research protocol describing each research project involving the additional substances, or two copies of a statement describing the nature, extent, and duration of such instructional activities, as appropriate. No fee shall be required to be paid for the modification. The request for modification shall be handled in the same manner as an application for registration. If the modification in registration is approved, the Administrator shall issue a new certificate of registration (DEA Form 223) to the registrant, who shall maintain it with the old certificate of registration until expiration.

§ 1301.62 Termination of registration.

The registration of any person shall terminate if and when such person dies, ceases legal existence, or discontinues business or professional practice. Any registrant who ceases legal existence or discontinues business or professional practice shall notify the Administrator promptly of such fact.

§ 1301.63 Transfer of registration.

No registration or any authority conferred thereby shall be assigned or otherwise transferred except upon such conditions as the Administrator may specifically designate and then only pursuant to his written consent.

Security Requirements

§ 1301.71 Security requirements generally.

(a) All applicants and registrants shall provide effective controls and procedures to guard against theft and diversion of controlled substances. In order to determine whether a registrant has provided effective controls against diversion, the Administrator shall use the security requirements set forth in §§ 1301.72–1301.76 as standards for the physical security controls and operating procedures necessary to prevent diversion. Materials and construction which will provide a structural equivalent to the physical security controls set forth in §§ 1301.72, 1301.73, and 1301.75 may be used in lieu of the materials and construction described in those sections.

§ 1301.72 Physical security controls for non-practitioners; narcotic treatment programs and compounders for narcotic treatment programs; storage areas.

(a) *Schedules I* and *II*. Raw materials, bulk materials awaiting further processing, and finished products which are controlled substances listed in Schedule I or II shall be stored in one of the following secure storage areas:

(1) Where small quantities permit, a safe or steel cabinet:

(i) Which safe or steel cabinet shall have the following specifications or the equivalent: 30 man-minutes against surreptitious entry, 10 man-minutes against forced entry, 20 man-hours against lock manipulation, and 20 man-hours against radiological techniques;

(ii) Which safe or steel cabinet, if it weighs less than 750 pounds, is bolted or cemented to the floor or wall in such a way that it cannot be readily removed; and

(iii) Which safe or steel cabinet, if necessary, depending upon the quantities and type of controlled substances stored, is equipped with an alarm system which, upon attempted unauthorized entry, shall transmit a signal directly to a central protection company or a local or State police agency which has a legal duty to respond, or a 24-hour control station operated by the registrant, or such other protection as the Administrator may approve.

(2) A vault constructed before, or under construction on, September 1, 1971, which is of substantial construction with a steel door, combination or key lock, and an alarm system; or

(3) A vault constructed after September 1, 1971:

(i) The walls, floors, and ceilings of which vault are constructed of at least 8 inches of reinforced concrete or other substantial masonry, reinforced vertically and horizontally with ½-inch steel rods tied 6 inches on center, or the structural equivalent to such reinforced walls, floors, and ceilings;

(ii) The door and frame unit of which vault shall conform to the following specifications or the equivalent: 30 man-minutes against surreptitious entry, 10 man-minutes against forced entry, 20 man-hours against lock manipulation, and 20 man-hours against radiological techniques;

(iii) Which vault, if operations require it to remain open for frequent access, is equipped with a "day-gate" which is self-closing and self-locking, or the equivalent, for use during the hours of operation in which the vault door is open;

(iv) The walls or perimeter of which vault are equipped with an alarm, which upon unauthorized entry shall transmit a signal directly to a central station protection company, or a local or State police agency which has a legal duty to respond, or a 24-hour control station operated by the registrant, or such other protection as the Administrator may approve, and, if necessary, holdup buttons at strategic points of entry to the perimeter area of the vault;

(v) The door of which vault is equipped with contact switches; and

(vi) Which vault has one of the following: complete electrical lacing of the walls, floor and ceilings; sensitive ultrasonic equipment within the vault; a sensitive sound accumulator system; or such other device designed to detect illegal entry as may be approved by the Administration.

(b) *Schedules III, IV,* and *V*. Raw materials, bulk materials awaiting further processing, and finished products which are controlled substances listed in Schedules III, IV, and V shall be stored in the following secure storage areas:

(1) A safe or steel cabinet as described in paragraph (a)(1) of this section;

(2) A vault as described in paragraph (a)(2) or (3) of this section equipped with an alarm system as described in paragraph (b)(4)(v) of this section;

(3) A building used for storage of Schedules III through V controlled substances with perimeter security which limits access during working hours and provides security after working hours and meets the following specifications:

(i) Has an electronic alarm system as described in paragraph (b)(4)(v) of this section,

(ii) Is equipped with self-closing, self-locking doors constructed of substantial material commensurate with the type of building construction, provided, however, a door which is kept closed and locked at all times when not in use and when in use is kept under direct observation of a responsible employee or agent of the registrant is permitted in lieu of a self-closing, self-locking door. Doors may be sliding or hinged. Regarding hinged doors, where hinges are mounted on the outside, such hinges shall be sealed, welded or otherwise constructed to inhibit removal. Locking devices for such doors shall be either of the multiple-position combination or key lock type and:

(a) In the case of key locks, shall require key control which limits access to a limited number of employees, or;

(b) In the case of combination locks, the combination shall be limited to a minimum number of employees and can be changed upon termination of employment of an employee having knowledge of the combination;

(4) A cage, located within a building on the premises, meeting the following specifications:

(i) Having walls constructed of not less than No. 10 gauge steel fabric mounted on steel posts, which posts are:

(a) At least one inch in diameter;

(b) Set in concrete or installed with lay bolts that are pinned or brazed; and

(c) Which are placed no more than ten feet apart with horizontal one and one-half inch reinforcements every sixty inches.

(ii) Having a mesh construction with openings of not more than two and one-half inches across the square,

(iii) Having a ceiling constructed of the same material, or in the alternative, a cage shall be erected which reaches and is securely attached to the structural ceiling of the building. A lighter gauge mesh may be used for the ceilings of large enclosed areas if walls are at least 14 feet in height,

(iv) Is equipped with a door constructed of No. 10 gauge steel fabric on a metal door frame in a metal door flange, and in all other respects conforms to all the requirements of 21 CFR 1301.72(b)(3)(ii), and

(v) Is equipped with an alarm system which upon unauthorized entry shall transmit a signal directly to a central station protection agency or a local or State police agency, each having a legal duty to respond, or to a 24-hour control station operated by the registrant, or to such other source of protection as the Administrator may approve;

(5) An enclosure of masonry or other material, approved in writing by the Administrator as providing security comparable to a cage;

(6) A building or enclosure within a building which has been inspected and approved by DEA or its predecessor agency, BNDD, and continues to provide adequate security against the diversion of Schedule III through V controlled substances, of which fact written acknowledgment has been made by the Special Agent in Charge of DEA for the area in which such building or enclosure is situated;

(7) Such other secure storage areas as may be approved by the Administrator after considering the factors listed in § 1301.71(b), (1) through (14);

(8)(i) Schedule III through V controlled substances may be stored with Schedules I and II controlled substances under security measures provided by 21 CFR 1301.72(a);

(ii) Noncontrolled drugs, substances and other materials may be stored with Schedule III through V controlled substances in any of the secure storage areas required by 21 CFR 1301.72(b), provided that permission for such storage of noncontrolled items is obtained in advance, in writing, from the Special Agent in Charge of DEA for the area in which such storage area is situated. Any such permission tendered must be upon the Special Agent in Charge's written determination that such nonsegregated storage does not diminish security effectiveness for Schedule III through V controlled substances.

(c) *Multiple storage areas.* Where several types or classes of controlled substances are handled separately by the registrant or applicant for different purposes (e.g., returned goods or goods in process), the controlled substances may be stored separately, provided that each storage area complies with the requirements set forth in this section.

(d) *Accessibility to storage areas.* The controlled substances storage areas shall be accessible only to an absolute minimum number of specifically authorized employees. When it is necessary for employee maintenance personnel, nonemployee maintenance personnel, business guests, or visitors to be present in or pass through controlled substances storage areas, the registrant shall provide for adequate observation of the area by an employee specifically authorized in writing.

§ 1301.73 Physical security controls for nonpractitioners; compounders for narcotic treatment programs; manufacturing and compounding areas.

All manufacturing activities (including processing, packaging and labeling) involving controlled substances listed in any schedule and all activities of compounders shall be conducted in accordance with the following:

(a) All in-process substances shall be returned to the controlled substances storage area at the termination of the process. If the process is not terminated at the end of a workday (except where a continuous process or other normal manufacturing operation should not be interrupted), the processing area or tanks, vessels, bins, or bulk containers containing such substances shall be securely locked, with adequate security for the area or building. If such security requires an alarm, such alarm, upon unauthorized entry, shall transmit a signal directly to a central station protection company, or local or State police agency which has a legal duty to respond, or a 24-hour control station operated by the registrant.

(b) Manufacturing activities with controlled substances shall be conducted in an area or areas of clearly defined limited access which is under surveillance by an employee or employees designated in writing as responsible for the area. "Limited access" may be provided, in the absence of physical dividers such as walls or partitions, by traffic control lines or restricted space designation. The employee designated as responsible for the area may be engaged in the particular manufacturing operation being conducted: *Provided*, That he is able to provide continuous surveillance of the area in order that unauthorized persons may not enter or leave the area without his knowledge.

(c) During the production of controlled substances, the manufacturing areas shall be accessible to only those employees required for efficient operation. When it is necessary for employee maintenance personnel, nonemployee maintenance personnel, business guests, or visitors to be present in or pass through manufacturing areas during production of controlled substances, the registrant shall provide for adequate observation of the area by an employee specifically authorized in writing.

§ 1301.74 Other security controls for nonpractitioners; narcotic treatment programs and compounders for narcotic treatment programs.

(c) The registrant shall notify the Field Division Office of the Administration in his area of any theft or significant loss of any controlled substances upon discovery of such theft or loss. The supplier shall be responsible for reporting in-transit losses of controlled substances by the common or contract carrier selected pursuant to § 1301.74(e), upon discovery of such theft or loss. The registrant shall also complete DEA Form 106 regarding such theft or loss. Thefts must be reported whether or not the controlled substances are subsequently recovered and/or the responsible parties are identified and action taken against them.

(h) The acceptance of delivery of narcotic substances by a narcotic treatment program shall be made only by a licensed practitioner employed at the facility or other authorized individuals designated in writing. At the time of delivery, the licensed practitioner or other authorized individual designated in writing (excluding persons currently or previously dependent on narcotic drugs), shall sign for the narcotics and place his specific title (if any) on any invoice. Copies of these signed invoices shall be kept by the distributor.

(i) Narcotics dispensed or administered at a narcotic treatment program will be dispensed or administered directly to the patient by either (1) the licensed practitioner, (2) a registered nurse under the direction of the licensed practitioner, (3) a licensed practical nurse under the direction of the licensed practitioner, or (4) a pharmacist under the direction of the licensed practitioner.

(j) Persons enrolled in a narcotic treatment program will be required to wait in an area physically separated from the narcotic storage and dispensing area. This requirement will be enforced by the program physician and employees.

(k) All narcotic treatment programs must comply with standards established by the Secretary of Health and Human Services (after consultation with the Administration) respecting the quantities of narcotic drugs which may be provided to persons enrolled in a narcotic treatment program for unsupervised use.

(l) DEA may exercise discretion regarding the degree of security required in narcotic treatment programs based on such factors as the location of a program, the number of patients enrolled in a program and the number of physicians, staff members and security guards. Similarly, such factors will be taken into consideration when evaluating existing security or requiring new security at a narcotic treatment program.

§ 1301.75 Physical security controls for practitioners.

(a) Controlled substances listed in Schedule I shall be stored in a securely locked, substantially constructed cabinet.

(b) Controlled substances listed in Schedules II, III, IV, and V shall be stored in a securely locked, substantially constructed cabinet. However, pharmacies and institutional practitioners (as defined in § 1304.02(e) of this chapter) may disperse such substances throughout the stock of noncontrolled substances in such a manner as to obstruct the theft or diversion of the controlled substances.

(d) Etorphine hydrochloride and diprenorphine shall be stored in a safe or steel cabinet equivalent to a U. S. Government Class V security container.

§ 1301.76 Other security controls for practitioners.

(a) The registrant shall not employ as an agent or employee who has access to controlled substances any person who has had an application for registration denied, or has had his registration revoked, at any time.

(b) The registrant shall notify the Field Division Office of the Administration in his area of the theft or significant loss of any controlled substances upon discovery of such loss or theft. The registrant shall also complete DEA (or BND) Form 106 regarding such loss or theft.

(c) Whenever the registrant distributes a controlled substance (without being registered as a distributor, as permitted in § 1301.22(b) and/or §§ 1307.11–1307.14), he shall comply with the requirements imposed on nonpractitioners in § 1301.74(a), (b), and (e).

Labeling and Packaging Requirements for Controlled Substances

§ 1302.02 Definitions.

As used in this part, the following terms shall have the meanings specified:

(a) The term "commercial container" means any bottle, jar, tube, ampule, or other receptacle in which a substance is held for distribution or dispensing to an ultimate user, and in addition, any box or package in which the receptacle is held for distribution or dispensing to an ultimate user. The term "commercial container" does not include any package liner, package insert or other material kept with or within a commercial container, nor any carton, crate, drum, or other package in which commercial containers are stored or are used for shipment of controlled substances.

(b) The term "label" means any display of written, printed, or graphic matter placed upon the commercial container of any controlled substance by any manufacturer of such substance.

(c) The term "labeling" means all labels and other written, printed, or graphic matter (1) upon any controlled substance or any of its commercial containers or wrappers, or (2) accompanying such controlled substance.

(d) The term "manufacture" means the producing, preparation, propagation, compounding, or processing of a drug or other substance or the packaging or repackaging of such substance, or the labeling or relabeling of the commercial container of such substance, but does not include the activities of a practitioner who, as an incident to his administration or dispensing such substance in the course of his professional practice, prepares, compounds, packages or labels such substance. The term "manufacturer" means a person who manufactures a drug or other substance, whether under a registration as a manufacturer or under authority of registration as a researcher or chemical analyst.

§ 1302.03 Symbol required; exceptions.

(a) Each commercial container of a controlled substance (except for a controlled substance excepted by the Administrator pursuant to § 1308.31 of this chapter) shall have printed on the label the symbol designating the schedule in which such controlled substance is listed. Each such commercial container, if it otherwise has no label, must bear a label complying with the requirement of this part.

(b) Each manufacturer shall print upon the labeling of each controlled substance distributed by him the symbol designating the schedule in which such controlled substance is listed.

(c) The following symbols shall designate the schedule corresponding thereto:

Schedule	*Symbol*
Schedule I	CI or C-I.
Schedule II	CII or C-II.
Schedule III	CIII or C-III.
Schedule IV	CIV or C-IV.
Schedule V	CV or C-V.

The word "schedule" need not be used. No distinction need be made between narcotic and nonnarcotic substances.

(d) The symbol is not required on a carton or wrapper in which a commercial container is held if the symbol is easily legible through such carton or wrapper.

(e) The symbol is not required on a commercial container too small or otherwise unable to accommodate a label, if the symbol is printed on the box or package from which the commercial container is removed upon dispensing to an ultimate user.

(f) The symbol is not required on a commercial container containing, or on the labeling of, a controlled substance being utilized in clinical research involving blind and double blind studies.

§ 1302.04 Location and size of symbol on label.

(a) The symbol shall be prominently located on the right upper corner of the principal panel of the label of the commercial container and/or the panel of the commercial container normally displayed to dispensers of any controlled substance listed in Schedules I through V. The symbol must be at least two times as large as the largest type otherwise printed on the label.

(b) In lieu of locating the symbol in the corner of the label, as prescribed in paragraph (a) of this section, the symbol may be overprinted on the label, in which case the symbol must be printed at least one-half the height of the label and in a contrasting color providing clear visibility against the background color of the label.

(c) In all cases the symbol shall be clear and large enough to afford easy identification of the schedule of the controlled substance upon inspection without removal from the dispenser's shelf.

§ 1302.05 Location and size of symbol on labeling.

The symbol shall be prominently located on all labeling other than labels covered by § 1302.04. In all cases the symbol shall be clear and large enough to afford prompt identification of the controlled substance upon inspection of the labeling.

§ 1302.07 Sealing of controlled substances.

(a) On each bottle, multiple-dose vial, or other commercial container of any controlled substance listed in Schedule I or II or of any narcotic controlled substance listed in Schedule III or IV, there shall be securely affixed to the stopper, cap, lid, covering, or wrapper of such container a seal to disclose upon inspection any tampering or opening of the container.

Records and Reports of Registrants

§ 1304.02 Definitions.

As used in this part, the following terms shall have the meanings specified:

(a) The term "Act" means the Controlled Substances Act (84 Stat. 1242; 21 U.S.C. 801) and/or the Controlled Substances Import and Export Act (84 Stat. 1285; 21 U.S.C. 951).

(b) The term "commercial container" means any bottle, jar, tube, ampule, or other receptacle in which a substance is held for distribution or dispensing to an ultimate user, and in addition, any box or package in which the receptacle is held for distribution or dispensing to an ultimate user. The term "commercial container" does not include any package liner, package insert or other material kept with or within a commercial container, nor any carton, crate, drum, or other package in which commercial containers are stored or are used for shipment of controlled substances.

(c) The term "dispenser" means an individual practitioner, institutional practitioner, pharmacy or pharmacist who dispenses a controlled substance.

(d) The term "individual practitioner" means a physician, dentist, veterinarian, or other individual licensed, registered, or otherwise permitted, by the United States or the jurisdiction in which he practices, to dispense a controlled substance in the course of professional practice, but does not include a pharmacist, a pharmacy, or an institutional practitioner.

(e) The term "institutional practitioner" means a hospital or other person (other than an individual) licensed, registered, or otherwise permitted, by the United States or the jurisdiction in which it practices, to dispense a controlled substance in the course of professional practice, but does not include a pharmacy.

(f) The term "name" means the official name, common or usual name, chemical name, or brand name of a substance.

(g) The term "pharmacist" means any pharmacist licensed by a State to dispense controlled substances, and shall include any other person (e.g., pharmacist intern) authorized by a State to dispense controlled substances under the supervision of a pharmacist licensed by such State.

(h) The term "readily retrievable" means that certain records are kept by automatic data processing systems or other electronic or mechanized recordkeeping systems in such a manner that they can be separated out from all other records in a reasonable time and/or records are kept on which certain items are asterisked, redlined, or in some other manner visually identifiable apart from other items appearing on the records.

§ 1304.03 Persons required to keep records and file reports.

(a) Each registrant shall maintain the records and inventories and shall file the reports required by this part, except as exempted by this section. Any registrant who is authorized to conduct other activities without being registered to conduct those activities, either pursuant to § 1301.22(b) of this chapter or pursuant to §§ 1307.11–1307.15 of this chapter, shall maintain the records and inventories and shall file the reports required by this part for persons registered to conduct such activities. This latter requirement should not be construed as requiring stocks of controlled substances being used in various activities under one registration to be stored separately, nor that separate records are required for each activity. The intent of the Administration is to permit the registrant to keep one set of records which are adapted by the registrant to account for controlled substances used in any activity. Also, the Administration does not wish to require separate stocks of the same substance to be purchased and stored for separate activities. Otherwise, there is no advantage gained by permitting several activities under one registration. Thus, when a researcher manufactures a controlled item, he must keep a record of the quantity manufactured; when he distributes a quantity of the item, he must use and keep invoices or order forms to document the transfer; when he imports a substance, he keeps as part of his records the documentation required of an importer; and when substances are used in chemical analysis, he need not keep a record of this because such a record would not be required of him under a registration to do chemical analysis. All of these records may be maintained in one consolidated record system. Similarly, the researcher may store all of his controlled items in one place, and every two years take inventory of all items on hand, regardless of whether the substances were manufactured by him, imported by him, or purchased domestically by him, or whether the substances will be administered to subjects, distributed to other researchers, or destroyed during chemical analysis.

(b) A registered individual practitioner is required to keep records, as described in § 1304.04 of controlled substances in Schedules II, III, IV, and V which are dispensed, other than by prescribing or administering in the lawful course of professional practice.

(c) A registered individual practitioner is not required to keep records of controlled substances in Schedules II, III, IV, and V which are prescribed in the lawful course of professional practice, unless such substances are prescribed in the course of maintenance or detoxification treatment of an individual.

(d) A registered individual practitioner is not required to keep records of controlled substances listed in Schedules II, III, IV and V which are administered in the lawful course of professional practice unless the practitioner regularly engages in the dispensing or administering of controlled substances and charges patients, either separately or together with charges for other professional services, for substances so dispensed or administered. Records are required to be kept for controlled substances administered in the course of maintenance or detoxification treatment of an individual.

§ 1304.04 Maintenance of records and inventories.

(a) Every inventory and other records required to be kept under this Part shall be kept by the registrant and be available, for at least 2 years from the date of such inventory or records, for inspection and copying by authorized employees of the Administration, except that financial and shipping records (such as invoices and packing slips but not executed order forms subject to paragraph 1305.13 of this chapter) may be kept at a central location, rather than at the registered location, if the registrant has notified the Administration of his intention to keep central records. Written notification must be submitted by registered or certified mail, return receipt requested, in triplicate, to the Special Agent in Charge of the Administration in the area in which the registrant is located. Unless the registrant is informed by the Special Agent in Charge that permission to keep central records is denied, the registrant may maintain central records commencing 14 days after receipt of his notification by the Special Agent in Charge.

All notifications must include:

(1) The nature of the records to be kept centrally.

(2) The exact location where the records will be kept.

(3) The name, address, DEA registration number and type of DEA registration of the registrant whose records are being maintained centrally.

(4) Whether central records will be maintained in a manual, or computer readable form.

(b) All registrants that are authorized to maintain a central recordkeeping system shall be subject to the following conditions:

(1) The records to be maintained at the central record location shall not include executed order forms, prescriptions and/or inventories which shall be maintained at each registered location.

(2) If the records are kept on microfilm, computer media or in any form requiring special equipment to render the records easily readable, the registrant shall provide access to such equipment with the records. If any code system is used (other than pricing information), a key to the code shall be provided to make the records understandable.

(3) The registrant agrees to deliver all or any part of such records to the registered location within two business days upon receipt of a written request from the Administration for such records, and if the Administration chooses to do so in lieu of requiring delivery of such records to the registered location, to allow authorized employees of the Administration to inspect such records at the central location upon request by such employees without a warrant of any kind.

(4) In the event that a registrant fails to comply with these conditions, the Special Agent in Charge may cancel such central recordkeeping authorization, and all other central recordkeeping authorizations held by the registrant without a hearing or other procedures. In the event of a cancellation of central recordkeeping authorizations under this sub-paragraph the registrant shall, within the time specified by the Special Agent in Charge, comply with the requirements of this section that all records be kept at the registered location.

(c) Registrants need not notify the Special Agent in Charge or obtain central recordkeeping approval in order to maintain records on an in-house computer system.

(d) ARCOS participants who desire authorization to report from other than their registered locations must obtain a separate central reporting identifier. Request for central reporting identifiers will be submitted to: ARCOS Unit, P. O. Box 28293, Central Station, Washington, D. C. 20005.

(e) All central recordkeeping permits previously issued by the Administration will expire on September 30, 1980. Registrants who desire to continue maintaining central records will make notification to the local Special Agent in Charge as provided in (a) above.

(f) Each registered manufacturer, distributor, importer, exporter, narcotic treatment program and compounder for narcotic treatment program shall maintain inventories and records of controlled substances as follows:

(1) Inventories and records of controlled substances listed in Schedules I and II shall be maintained separately from all of the records of the registrant; and

(2) Inventories and records of controlled substances listed in Schedules III, IV, and V shall be maintained either separately from all other records of the registrant or in such form that the information required is readily retrievable from the ordinary business records of the registrant.

(g) Each registered individual practitioner required to keep records and institutional practitioner shall maintain inventories and records of controlled substances in the manner prescribed in paragraph (f) of this section.

(h) Each registered pharmacy shall maintain the inventories and records of controlled substances as follows:

(1) Inventories and records of all controlled substances listed in Schedules I and II shall be maintained separately from all other records of the pharmacy, and prescriptions for such substances shall be maintained in a separate prescription file; and

(2) Inventories and records of controlled substances listed in Schedules III, IV, and V shall be maintained either separately from all other records of the pharmacy or in such form that the information required is readily retrievable from ordinary business records of the pharmacy, and prescriptions for such substances shall be maintained either in a separate prescription file for controlled substances listed in Schedules III, IV, and V only or in such form that they are readily retrievable from the other prescription records of the pharmacy. Prescriptions will be deemed readily retrievable if, at the time they are initially filed, the face of the prescription is stamped in red ink in the lower right corner with the letter "C" no less than 1-inch high and filed either in the prescription file for controlled substances listed in Schedules I and II or in the usual consecutively numbered prescription file for noncontrolled substances.

Inventory Requirements

§ 1304.11 General requirements for inventories.

(a) Each inventory shall contain a complete and accurate record of all controlled substances on hand on the date the inventory is taken. Controlled substances shall be deemed to be "on hand" if they are in the possession of or under the control of the registrant, including substances returned by a customer, substances ordered by a customer but not yet invoiced, substances stored in a warehouse on behalf of the registrant, and substances in the possession of employees of the registrant and intended for distribution as complimentary samples.

(b) A separate inventory shall be made by a registrant for each registered location. In the event controlled substances are in the possession or under the control of the registrant at a location for which he is not registered, the substances shall be included in the inventory of the registered location to which they are subject to control or to which the person possessing the substance is responsible. Each inventory for a registered location shall be kept at the registered location.

(c) A separate inventory shall be made by a registrant for each independent activity for which he is registered, except as provided in § 1304.18.

(d) A registrant may take an inventory on a date that is within 4 days of his biennial inventory date pursuant to § 1304.13 if he notifies in advance the Special Agent in Charge of the Administration in his area of the date on which he will take the inventory. A registrant may take an inventory either as of the opening of business or as of the close of business on the inventory date. The registrant shall indicate on the inventory records whether the inventory is taken as of the opening or as of the close of business and the date the inventory is taken.

(e) An inventory must be maintained in a written, typewritten or printed form. An inventory taken by use of an oral recording device must be promptly transcribed.

§ 1304.12 Initial inventory date.

(b) Every person required to keep records who is registered after May 1, 1971, and who was not provisionally registered on that date, shall take an inventory of all stocks of controlled substances on hand on the date he first engages in the manufacture, distribution, or dispensing of controlled substances, in accordance with §§ 1304.15–1304.19, as applicable. In the event a person commences business with no controlled substances on hand, he shall record this fact as his initial inventory.

§ 1304.13 Biennial inventory date.

Every 2 years following the date on which the initial inventory is taken by a registrant pursuant to § 1304.12, the registrant shall take a new inventory of all stocks of controlled substances on hand. The biennial inventory may be taken (a) on the day of the year on which the initial inventory was taken or (b) on the registrant's regular general physical inventory date, if any, which is nearest to and does not vary by more than 6 months from the biennial date that would otherwise apply or (c) on any other fixed date which does not vary by more than 6 months from the biennial date that would otherwise apply. If the registrant elects to take the biennial inventory on his regular general physical inventory date or another fixed date, he shall notify the Administration of this election and of the date on which the biennial inventory will be taken.

§ 1304.14 Inventory date for newly controlled substances.

On the effective date of a rule by the Administrator pursuant to §§ 1308.48–1308.49, or 1308.50 of this chapter adding a substance to any schedule of controlled substances, which substance was, immediately prior to that date, not listed on any such schedule, every registrant required to keep records who possesses that substance shall take an inventory of all stocks of the substance on hand. Thereafter such substance shall be included in each inventory made by the registrant pursuant to § 1304.13.

§ 1304.15 Inventories of manufacturers.

Each person registered or authorized (by § 1301.22(b), § 1307.12, or § 1307.15 of this chapter) to manufacture controlled substances shall include the following information in his inventory:

(a) For each controlled substance in bulk form to be used in (or capable of use in) the manufacture of the same or other controlled or noncontrolled substances in finished form:

(1) The name of the substance; and

(2) The total quantity of the substance to the nearest metric unit weight consistent with unit size (except that for inventories made in 1971, avoirdupois weights may be utilized where metric weights are not readily available).

(b) For each controlled substance in the process of manufacture on the inventory date:

(1) The name of the substance;

(2) The quantity of the substance in each batch and/or stage of manufacture, identified by the batch number or other appropriate identifying number;

(3) The physical form which the substance is to take upon completion of the manufacturing process (e.g., granulations, tablets, capsules, or solutions), identified by the batch number or other appropriate identifying number, and if possible the finished form of the substance (e.g., 10-milligram tablet or 10-milligram concentration per fluid ounce or milliliter) and the number or volume thereof; and

(c) For each controlled substance in finished form:

(1) The name of the substance;

(2) Each finished form of the substance (e.g., 10-milligram tablet or 10-milligram concentration per fluid ounce or milliliter);

(3) The number of units or volume of each finished form in each commercial container (e.g., 100-tablet bottle or 3-milliliter vial); and

(4) The number of commercial containers of each such finished form (e.g., four 100-tablet bottles or six 3-milliliter vials).

(d) For each controlled substance not included in paragraphs (a), (b) or (c) of this section (e.g., damaged, defective, or impure substances awaiting disposal, substances held for quality control purposes, or substances maintained for extemporaneous compoundings):

(1) The name of the substance;

(2) The total quantity of the substance to the nearest metric unit weight or the total number of units of finished form; and

(3) The reason for the substance being maintained by the registrant and whether such substance is capable of use in the manufacture of any controlled substance in finished form.

§ 1304.17 Inventories of dispensers and researchers.

Each person registered or authorized (by § 1301.22(b) of this chapter) to dispense or conduct research with controlled substances and required to keep records pursuant to § 1304.03 shall include in his inventory the same information required of manufacturers pursuant to § 1304.15(c) and (d). In determining the number of units of each finished form of a controlled substance in a commercial container which has been opened, the dispenser shall do as follows:

(a) If the substance is listed in Schedule I or II, he shall make an exact count or measure of the contents; and

(b) If the substance is listed in Schedule III, IV, or V, he shall make an estimated count or measure of the contents, unless the container holds more than 1,000 tablets or capsules in which case he must make an exact count of the contents.

Continuing Records

§ 1304.21 General requirements for continuing records.

(a) On and after May 1, 1971, every registrant required to keep records pursuant to § 1304.03 shall maintain on a current basis a complete and accurate record of each such substance manufactured, imported, received, sold, delivered, exported, or otherwise disposed of by him, except that no registrant shall be required to maintain a perpetual inventory.

(b) Separate records shall be maintained by a registrant for each registered location except as provided in § 1304.04(a). In the event controlled substances are in the possession or under the control of a registrant at a location for which he is not registered, the substances shall be included in the records of the registered location to which they are subject to control or to which the person possessing the substance is responsible.

(c) Separate records shall be maintained by a registrant for each independent activity for which he is registered, except as provided in §§ 1304.25 and 1304.26.

(d) In recording dates of receipt, importation, distribution, exportation, or other transfers, the date on which the controlled substances are actually received, imported, distributed, exported, or otherwise transferred shall be used as the date of receipt or distribution of any documents of transfer (e.g., invoices or packing slips).

§ 1304.23 Records for distributors.

Each person registered or authorized (by § 1301.22(b) or §§ 1307.11–1307.14 of this chapter) to distribute controlled substances shall maintain records with the following information for each controlled substance:

(a) The name of the substance;

(b) Each finished form (e.g., 10-milligram tablet or 10-milligram concentration per fluid ounce or milliliter) and the number of units or volume of finished form in each commercial container (e.g., 100-tablet bottle or 3-milliliter vial);

(c) The number of commercial containers of each such finished form received from other persons, including the date of and number of containers in each receipt and the name, address, and registration number of the person from whom the containers were received;

(d) The number of commercial containers or each such finished form imported directly by the person (under a registration or authorization to import), including the date of, the number of commercial containers in, and the import permit or declaration number for, each importation;

(e) The number of commercial containers of each such finished form distributed to other persons, including the date of and number of containers in each distribution and the name, address, and registration number of the person to whom the containers were distributed;

(f) The number of commercial containers of each such finished form exported directly by the person (under a registration or authorization to export), including the date of, the number of commercial containers in, and the export permit or declaration number for, each exportation; and

(g) The number of units or volume of finished forms and/or commercial containers distributed or disposed of in any other manner by the person (e.g., by distribution as complimentary samples or by destruction) including the date and manner of distribution or disposal, the name, address, and registration number of the person to whom distributed, and the quantity of the substance in finished form distributed or disposed.

§ 1304.24 Records for dispensers and researchers.

Each person registered or authorized (by § 1301.22(b) of this chapter) to dispense or conduct research with controlled substances and required to keep records pursuant to § 1304.03 shall maintain records with the following information for each controlled substance:

(a) The name of the substance;

(b) Each finished form (e.g., 10-milligram tablet or 10-milligram concentration per fluid ounce or milliliter) and the number of units or volume of finished form in each commercial container (e.g., 100-tablet bottle or 3-milliliter vial);

(c) The number of commercial containers of each such finished form received from other persons, including the date of and number of containers in each receipt and the name, address, and registration number of the person from whom the containers were received;

(d) The number of units or volume of such finished form dispensed, including the name and address of the person to whom it was dispensed, the date of dispensing, the number of units or volume dispensed, and the written or typewritten name or initials of the individual who dispensed or administered the substance on behalf of the dispenser; and

(e) The number of units or volume of such finished forms and/or commercial containers disposed of in any other manner by the registrant, including the date and manner of disposal and the quantity of the substance in finished form disposed.

§ 1304.28 Records for maintenance treatment programs and detoxification treatment programs.

(a) Each person registered or authorized (by § 1301.22 of this chapter) to maintain and/or detoxify controlled substance users in a narcotic treatment program shall maintain records with the following information for each narcotic controlled substance:

(1) Name of substance;
(2) Strength of substance;
(3) Dosage form;
(4) Date dispensed;
(5) Adequate identification of patient (consumer);
(6) Amount consumed;
(7) Amount and dosage form taken home by patient; and
(8) Dispenser's initials.

(b) The records required by paragraph (a) of this section will be maintained in a dispensing log at the narcotic treatment program site and will be maintained in compliance with § 1304.24 without reference to § 1304.03.

(c) All sites which compound a bulk narcotic solution from bulk narcotic powder to liquid for on-site use must keep a separate batch record of the compounding.

(d) Records of identity, diagnosis, prognosis, or treatment of any patients which are maintained in connection with the performance of a narcotic treatment program shall be confidential, except that such records may be disclosed for purposes and under the circumstances authorized by Part 310 and Part 1401 of this title.

§ 1304.29 Records for treatment programs which compound narcotics for treatment programs and other locations.

Each person registered or authorized by § 1301.22 of this chapter to compound narcotic drugs for off-site use in a narcotic treatment program shall maintain records which include the following information for each narcotic drug:

(a) For each narcotic controlled substance in bulk form to be used in, or capable of use in, or being used in, the compounding of the same or other noncontrolled substances in finished form:

(1) The name of the substance;

(2) The quantity compounded in bulk form by the registrant, including the date, quantity and batch or other identifying number of each batch compounded;

(3) The quantity received from other persons, including the date and quantity of each receipt and the name, address, and registration number of the other person from whom the substance was received;

(4) The quantity imported directly by the registrant (under a registration as an importer) for use in compounding by him, including the date, quantity and import permit or declaration number of each importation;

(5) The quantity used to compound the same substance in finished form, including:

(i) The date and batch or other identifying number of each compounding;
(ii) The quantity used in the compound;
(iii) The finished form (e.g., 10-milligram tablets or 10-milligram concentration per fluid ounce or milliliter);
(iv) The number of units of finished form compounded;
(v) The quantity used in quality control;
(vi) The quantity lost during compounding and the causes therefore, if known;
(vii) The total quantity of the substance contained in the finished form;
(viii) The theoretical and actual yields; and
(ix) Such other information as is necessary to account for all controlled substances used in the compounding process;

(6) The quantity used to manufacture other controlled and noncontrolled substances; including the name of each substance

manufactured and the information required in paragraph (a)(5) of this section;

(7) The quantity distributed in bulk form to other programs, including the date and quantity of each distribution and the name, address and registration number of each program to whom a distribution was made;

(8) The quantity exported directly by the registrant (under a registration as an exporter), including the date, quantity, and export permit or declaration number of each exportation; and

(9) The quantity disposed of by destruction, including the reason, date and manner of destruction. All other destruction of narcotic controlled substances will comply with § 1307.22.

(b) For each narcotic controlled substance in finished form:

(1) The name of the substance;

(2) Each finished form (e.g., 10-milligram tablet or 10-milligram concentration per fluid ounce or milliliter) and the number of units or volume or finished form in each commercial container (e.g., 100-tablet bottle or 3-milliliter vial);

(3) The number of containers of each such commercial finished form compounded from bulk form by the registrant, including the information required pursuant to paragraph (a)(5) of this section;

(4) The number of units of finished forms and/or commercial containers received from other persons, including the date of and number of units and/or commercial containers in each receipt and the name, address and registration number of the person from whom the units were received;

(5) The number of units of finished forms and/or commercial containers imported directly by the person (under a registration or authorization to import), including the date of, the number of units and/or commercial containers in, and the import permit or declaration number for, each importation;

(6) The number of units and/or commercial containers compounded by the registrant from units in finished form received from others or imported, including:

(i) The date and batch or other identifying number of each compounding;

(ii) The operation performed (e.g., repackaging or relabeling);

(iii) The number of units of finished form used in the compound, the number compounded and the number lost during compounding, with the causes for such losses, if known; and

(iv) Such other information as is necessary to account for all controlled substances used in the compounding process;

(7) The number of containers distributed to other programs, including the date, the number of containers in each distribution, and the name, address and registration number of the program to whom the containers were distributed;

(8) The number of commercial containers exported directly by the registrant (under a registration as an exporter), including the date, number of containers and export permit or declaration number for each exportation; and

(9) The number of units of finished forms and/or commercial containers destroyed in any manner by the registrant, including the reason, the date and manner of destruction. All other destruction of narcotic controlled substances will comply with §1307.22.

Order Forms

§ 1305.02 Definitions.

As used in this part, the following terms shall have the meanings specified:

(b) The term "purchaser" means any registered person entitled to obtain and execute order forms pursuant to § 1305.04 and § 1305.06.

(c) The term "supplier" means any registered person entitled to fill order forms pursuant to § 1305.08.

§ 1305.03 Distributions requiring order forms.

An order form (DEA Form 222) is required for each distribution of a controlled substance listed in Schedule I or II, except for the following:

(a) The exportation of such substances from the United States in conformity with the Act;

(b) The delivery of such substances to or by a common or contract carrier for carriage in the lawful and usual course of its business, or to or by a warehouseman for storage in the lawful and usual course of its business (but excluding such carriage or storage by the owner of the substance in connection with the distribution to a third person);

(c) The procurement of a sample of such substances by an exempt law enforcement official pursuant to § 1301.26(b) of this chapter, provided that the receipt required by that section is used and is preserved in the manner prescribed in this part for order forms;

(d) The procurement of such substances by a civil defense or disaster relief organization, pursuant to § 1301.27 of this chapter, provided that the Civil Defense Emergency Order Form required by that section is used and is preserved with other records of the registrant; and

(e) The purchase of such substances by the master or first officer of a vessel pursuant to § 1301.28 of this chapter: Provided, that copies of the record of sale are generated, distributed, and preserved by the vendor according to that section.

(f) The delivery of such substances to a registered analytical laboratory, or its agent approved by DEA, from an anonymous source for the analysis of the drug sample, provided the laboratory has obtained a written waiver of the order form requirement from the Regional Director[1] of the Region in which the laboratory is located, which waiver may be granted upon agreement of the laboratory to conduct its activities in accordance with Administration guidelines.

§ 1305.04 Persons entitled to obtain and execute order forms.

(a) Order forms may be obtained only by persons who are registered under section 303 of the Act (21 U.S.C. 823) to handle controlled substances listed in Schedules I and II, and by persons who are registered under section 1008 of the Act (21 U.S.C. 958) to export such substances. Persons not registered to handle controlled substances listed in Schedule I or II and persons registered only to import controlled substances listed in any schedule are not entitled to obtain order forms.

(b) An order form may be executed only on behalf of the registrant named thereon and only if his registration as to the substances being purchased has not expired or been revoked or suspended.

§ 1305.05 Procedure for obtaining order forms.

(a) Order Forms are issued in mailing envelopes containing either seven or fourteen forms, each form containing an original duplicate and triplicate copy (respectively, Copy 1, Copy 2, and Copy 3). A limit, which is based on the business activity of the registrant, will be imposed on the number of order forms which will be furnished on any requisition unless additional forms are specifically requested and a reasonable need for such additional forms is shown.

(b) Any person applying for a registration which would entitle him to obtain order forms may requisition such forms by so indicating on the application form; order forms will be supplied upon the registration of the applicant. Any person holding a registration entitling him to obtain order forms may requisition such forms for the first time by contacting any Division Office or the Registration Unit of the Administration. Any person already holding order forms may requisition additional forms on DEA Form 222a which is mailed to a registrant approximately 30 days after each shipment of order forms to that registrant or by contacting any Division Office or the Registration Unit of the Administration. All requisition forms (DEA Form 222a) shall be submitted to the Registration Unit, Drug Enforcement Administration, Department of Justice, Post Office Box 28083, Central Station, Washington, DC 20005.

(c) Each requisition shall show the name, address, and registration number of the registrant and the number of books of order forms desired. Each requisition shall be signed and dated by the same person who signed the most recent application for registration or for reregistration, or by any person authorized to obtain and execute order forms by a power of attorney pursuant to § 1305.07.

(d) Order forms will be serially numbered and issued with the name, address and registration number of the registrant, the authorized activity and schedules of the registrant. This information cannot be altered or changed by the registrant; any errors must be corrected by the Registration Unit of the Administration by returning the forms with notification of the error.

---

[1] Special Agent in Charge.

§ 1305.06 Procedure for executing order forms.

(a) Order forms shall be prepared and executed by the purchaser simultaneously in triplicate by means of interleaved carbon sheets which are part of the DEA Form 222 . Order forms shall be prepared by use of a typewriter, pen, or indelible pencil.

(b) Only one item shall be entered on each numbered line. There are ten lines on each order form. If one order form is not sufficient to include all items in an order, additional forms shall be used. Order forms for etorphine hydrochloride and diprenorphine shall contain only these substances. The total number of items ordered shall be noted on that form in the space provided.

(c) An item shall consist of one or more commercial or bulk containers of the same finished or bulk form and quantity of the same substance; a separate item shall be made for each commercial or bulk container of different finished or bulk form, quantity or substance. For each item the form shall show the name of the article ordered, the finished or bulk form of the article (e.g., 10-milligram tablet, 10-milligram concentration per fluid ounce or milliliter, or USP), the number of units or volume in each commercial or bulk container (e.g., 100-tablet bottle or 3-milliliter vial) or the quantity or volume of each bulk container (e.g., 10 kilograms), the number of commercial or bulk containers ordered, and the name and quantity per unit of the controlled substance or substances contained in the article if not in pure form. The catalogue number of the article may be included at the discretion of the purchaser.

(d) The name and address of the supplier from whom the controlled substances are being ordered shall be entered on the form. Only one supplier may be listed on any one form.

(e) Each order form shall be signed and dated by a person authorized to sign a requisition for order forms on behalf of the purchaser pursuant to § 1305.05(c). The name of the purchaser, if different from the individual signing the order form, shall also be inserted in the signature space. Unexecuted order forms may be kept and may be executed at a location other than the registered location printed on the form, provided that all unexecuted forms are delivered promptly to the registered location upon an inspection of such location by any officer authorized to make inspections, or to enforce, any Federal, State, or local law regarding controlled substances.

§ 1305.07 Power of attorney.

Any purchaser may authorize one or more individuals, whether or not located at the registered location of the purchaser, to obtain and execute order forms on his behalf by executing a power of attorney for each such individual. The power of attorney shall be signed by the same person who signed (or was authorized to sign, pursuant to § 1301.32(f) of this chapter or § 1311.32(f) of this chapter) the most recent application for registration or reregistration and by the individual being authorized to obtain and execute order forms. The power of attorney shall be filed with the executed order forms of the purchaser, and shall be retained for the same period as any order form bearing the signature of the attorney. The power of attorney shall be available for inspection together with other order form records. Any power of attorney may be revoked at any time by executing a notice of revocation, signed by the person who signed (or was authorized to sign) the power of attorney or by a successor, whoever signed the most recent application for registration or reregistration, and filing it with the power of attorney being revoked. The form for the power of attorney and notice of revocation shall be similar to the following:

POWER OF ATTORNEY FOR DEA ORDER FORMS

____
(Name of registrant)

____
(Address of registrant)

____
(DEA registration number)

I, ________________________ , the undersigned,
(Name of person granting power)

who is authorized to sign the current application for registration of the above-named registrant under the Controlled Substances Act or Controlled Substances Import and Export Act, have made, constituted, and appointed, and by these presents, do make, constitute, and appoint ________________________ ,
(Name of attorney-in-fact)

my true and lawful attorney for me in my name, place, and stead, to execute applications for books of official order forms and to sign such order forms in requisition for Schedule I and II controlled substances, in accordance with section 308 of the Controlled Substances Act (21 U.S.C. 828) and Part 305 of Title 21 of the Code of Federal Regulations. I hereby ratify and confirm all that said attorney shall lawfully do or cause to be done by virtue hereof.

____
(Signature of person granting power)

I, ________________________ , hereby affirm that
(Name of attorney-in-fact)

I am the person named herein as attorney-in-fact and that the signature affixed hereto is my signature.

____
(Signature of attorney-in-fact)

Witnesses:

1. ________________ .
2. ________________ .

Signed and dated on the ______ day of ________ , 19___ , at ________________ .

NOTICE OF REVOCATION

The foregoing power of attorney is hereby revoked by the undersigned, who is authorized to sign the current application for registration of the above-named registrant under the Controlled Substances Act or the Controlled Substances Import and Export Act. Written notice of this revocation has been given to the attorney-in-fact ________________________ this same day.

____
(Signature of person revoking power)

Witnesses:

1. ________________ .
2. ________________ .

Signed and dated on the ______ day of ____________ , 19___ , at ____________ .

§ 1305.08 Persons entitled to fill order forms.

An order form may be filled only by a person registered as a manufacturer or distributor of controlled substances listed in Schedule I or II under section 303 of the Act (21 U.S.C. 823) or as an importer of such substances under section 1008 of the Act (21 U.S.C. 958), except for the following:

(a) A person registered to dispense such substances under section 303 of the Act, or to export such substances under section 1008 of the Act, if he is discontinuing business or if his registration is expiring without reregistration, may dispose of any controlled substances listed in Schedule I or II in his possession pursuant to order forms in accordance with § 1307.14 of this chapter;

(b) A person who has obtained any controlled substance in Schedule I or II by order form may return such substance, or portion thereof, to the person from whom he obtained the substance or the manufacturer of the substance pursuant to the order form of the latter person;

(c) A person registered to dispense such substances may distribute such substances to another dispenser pursuant to, and only in the circumstances described in, § 1307.11 of this chapter; and

(d) A person registered or authorized to conduct chemical analysis or research with controlled substances may distribute a controlled substance listed in Schedule I or II to another person registered or authorized to conduct chemical analysis, instructional activities, or research with such substances pursuant to the order form of the latter person, if such distribution is for the purpose of furthering such chemical analysis, instructional activities, or research.

(e) A person registered as a compounder of narcotic substances for use at off-site locations in conjunction with a narcotic treatment program at the compounding location, who is authorized to handle Schedule II narcotics, is authorized to fill order forms for distribution of narcotic drugs to off-site narcotic treatment programs only.

§ 1305.09 Procedure for filling order forms.

(a) The purchaser shall submit Copy 1 and Copy 2 of the order form to the supplier, and retain Copy 3 in his own files.

(b) The supplier shall fill the order, if possible and if he desires to do so, and record on Copies 1 and 2 the number of commercial or bulk containers furnished on each item and the date on which such containers are shipped to the purchaser. If an order cannot be filled in its entirety, it may be filled in part and the balance supplied by additional shipments within 60 days following the date of the order form. No order form shall be valid more than 60 days after its execution by the purchaser, except as specified in paragraph (f) of this section.

(c) The controlled substances shall only be shipped to the purchaser and at the location printed by the Administration on the order form, except as specified in paragraph (f) of this section.

(d) The supplier shall retain Copy 1 of the order form for his own files and forward Copy 2 to the Special Agent in Charge of the Drug Enforcement Administration in the area in which the supplier is located. Copy 2 shall be forwarded at the close of the month during which the order is filled; if an order is filled by partial shipments, Copy 2 shall be forwarded at the close of the month during which the final shipment is made or during which the 60-day validity period expires.

(e) The purchaser shall record on Copy 3 of the order form the number of commercial or bulk containers furnished on each item and the dates on which such containers are received by the purchaser.

(f) Order forms submitted by registered procurement officers of the Defense Personnel Support Center of Defense Supply Agency for delivery to armed services establishments within the United States may be shipped to locations other than the location printed on the order form, and in partial shipments at different times not to exceed six months from the date of the order, as designated by the procurement officer when submitting the order.

§ 1305.10 Procedure for endorsing order forms.

(a) An order form made out to any supplier who cannot fill all or a part of the order within the time limitation set forth in § 1305.09 may be endorsed to another supplier for filling. The endorsement shall be made only by the supplier to whom the order form was first made, shall state (in the spaces provided on the reverse sides of Copies 1 and 2 of the order form) the name and address of the second supplier, and shall be signed by a person authorized to obtain and execute order forms on behalf of the first supplier. The first supplier may not fill any part of an order on an endorsed form. The second supplier shall fill the order, if possible and if he desires to do so, in accordance with § 1305.09 (b), (c), and (d), including shipping all substances directly to the purchaser.

(b) Distributions made on endorsed order forms shall be reported by the second supplier in the same manner as all other distributions except that where the name of the supplier is requested on the reporting form, the second supplier shall record the name, address and registration number of the first supplier.

§ 1305.11 Unaccepted and defective order forms.

(a) No order form shall be filled if it:

(1) Is not complete, legible, or properly prepared, executed, or endorsed; or

(2) Shows any alteration, erasure, or change of any description.

(b) If an order form cannot be filled for any reason under this section, the supplier shall return Copies 1 and 2 to the purchaser with a statement as to the reason (e.g., illegible or altered). A supplier may for any reason refuse to accept any order and if a supplier refuses to accept the order, a statement that the order is not accepted shall be sufficient for purposes of this paragraph.

(c) When received by the purchaser, Copies 1 and 2 of the order form and the statement shall be attached to Copy 3 and retained in the files of the purchaser in accordance with § 1305.13. A defective order form may not be corrected; it must be replaced by a new order form in order for the order to be filled.

§ 1305.12 Lost and stolen order forms.

(a) If a purchaser ascertains that an unfilled order form has been lost, he shall execute another in triplicate and a statement containing the serial number and date of the lost form, and stating that the goods covered by the first order form were not received through loss of that order form. Copy 3 of the second form and a copy of the statement shall be retained with Copy 3 of the order form first executed. A copy of the statement shall be attached to Copies 1 and 2 of the second order form sent to the supplier. If the first order form is subsequently received by the supplier to whom it was directed, the supplier shall mark upon the face thereof "Not accepted" and return Copies 1 and 2 to the purchaser, who shall attach it to Copy 3 and the statement.

(b) Whenever any used or unused order forms are stolen from or lost (otherwise than in the course of transmission) by any purchaser or supplier, he shall immediately upon discovery of such theft or loss, report the same to the Registration Unit, Drug Enforcement Administration, Department of Justice, Post Office Box 28083, Central Station, Washington, DC 20005, stating the serial number of each form stolen or lost. If the theft or loss includes any original order forms received from purchasers and the supplier is unable to state the serial numbers of such order forms, he shall report the date or approximate date of receipt thereof and the names and addresses of the purchasers. If an entire book of order forms is lost or stolen, and the purchaser is unable to state the serial numbers of the order forms contained therein, he shall report, in lieu of the numbers of the forms contained in such book, the date or approximate date of issuance thereof. If any unused order form reported stolen or lost is subsequently recovered or found, the Registration Unit of the Administration shall immediately be notified.

§ 1305.13 Preservation of order forms.

(a) The purchaser shall retain Copy 3 of each order form which has been filled. He shall also retain in his files all copies of each unaccepted or defective order form and each statement attached thereto.

(b) The supplier shall retain Copy 1 of each order form which he has filled.

(c) Order forms must be maintained separately from all other records of the registrant. Order forms are required to be kept available for inspection for a period of 2 years. If a purchaser has several registered locations, he must retain Copy 3 of the executed order forms and any attached statements or other related documents (not including unexecuted order forms which may be kept elsewhere pursuant to § 1305.06(e)) at the registered location printed on the order form.

(d) The supplier of etorphine hydrochloride and diprenorphine shall maintain order forms for these substances separately from all other order forms and records required to be maintained by the registrant.

§ 1305.14 Return of unused order forms.

If the registration of any purchaser terminates (because the purchaser dies, ceases legal existence, discontinues business or professional practice, or changes his name or address as shown on his registration) or is suspended or revoked pursuant to §§ 1301.45 or 1301.46 of this chapter as to all controlled substances listed in Schedules I and II for which he is registered, he shall return all unused order forms for such substance to the nearest office of the Administration.

§ 1305.15 Cancellation and voiding of order forms.

(a) A purchaser may cancel part or all of an order on an order form by notifying the supplier in writing of such cancellation. The supplier shall indicate the cancellation on Copies 1 and 2 of the order form by drawing a line through the canceled items and printing "canceled" in the space provided for number of items shipped.

(b) A supplier may void part or all of an order on an order form by notifying the purchaser in writing of such voiding. The supplier shall indicate the voiding in the manner prescribed for cancellation in paragraph (a) of this section.

(c) No cancellation or voiding permitted by this section shall affect in any way contract rights of either the purchaser or the supplier.

Prescriptions

§ 1306.02 Definitions.

As used in this part, the following terms shall have the meanings specified:

(b) The term "individual practitioner" means a physician, dentist, veterinarian, or other individual licensed, registered, or otherwise permitted, by the United States or the jurisdiction in which he practices, to dispense a controlled substance in the course of professional practice, but does not include a pharmacist, a pharmacy, or an institutional practitioner.

(c) The term "institutional practitioner" means a hospital or other person (other than an individual) licensed, registered, or otherwise permitted, by the United States or the jurisdiction in which it practices, to dispense a controlled substance in the course of professional practice, but does not include a pharmacy.

(d) The term "pharmacist" means any pharmacist licensed by a State to dispense controlled substances, and shall include any other person (e.g., a pharmacist intern) authorized by a State to dispense controlled substances under the supervision of a pharmacist licensed by such State.

(e) A "Long Term Care Facility" (LTCF) means a nursing home, retirement care, mental care or other facility or institution which provides extended health care to resident patients.

(f) The term "prescription" means an order for medication which is dispensed to or for an ultimate user but does not include an order for medication which is dispensed for immediate administration to the ultimate user. (E.g., an order to dispense a drug to a bed patient for immediate administration in a hospital is not a prescription.)

(g) The terms "register" and "registered" refer to registration required and permitted by section 303 of the Act (21 U.S.C. 823).

§ 1306.03 Persons entitled to issue prescriptions.

(a) A prescription for a controlled substance may be issued only by an individual practitioner who is:

(1) Authorized to prescribe controlled substances by the jurisdiction in which he is licensed to practice his profession and

(2) Either registered or exempted from registration pursuant to §§ 1301.24(c) and 1301.25 of this chapter.

(b) A prescription issued by an individual practitioner may be communicated to a pharmacist by an employee or agent of the individual practitioner.

§ 1306.04 Purpose of issue of prescription.

(a) A prescription for a controlled substance to be effective must be issued for a legitimate medical purpose by an individual practitioner acting in the usual course of his professional practice. The responsibility for the proper prescribing and dispensing of controlled substances is upon the prescribing practitioner, but a corresponding responsibility rests with the pharmacist who fills the prescription. An order purporting to be a prescription issued not in the usual course of professional treatment or in legitimate and authorized research is not a prescription within the meaning and intent of section 309 of the Act (21 U.S.C. 829) and the person knowingly filling such a purported prescription, as well as the person issuing it, shall be subject to the penalties provided for violations of the provisions of law relating to controlled substances.

(b) A prescription may not be issued in order for an individual practitioner to obtain controlled substances for supplying the individual practitioner for the purpose of general dispensing to patients.

(c) A prescription may not be issued for the dispensing of narcotic drugs listed in any schedule for "detoxification treatment" or "maintenance treatment" as defined in Section 102 of the Act (21 U.S.C. 802).

§ 1306.05 Manner of issuance of prescriptions.

(a) All prescriptions for controlled substances shall be dated as of, and signed on, the day when issued and shall bear the full name and address of the patient, and the name, address, and registration number of the practitioner. A practitioner may sign a prescription in the same manner as he would sign a check or legal document (e.g., J. H. Smith or John H. Smith). Where an oral order is not permitted, prescriptions shall be written with ink or indelible pencil or typewriter and shall be manually signed by the practitioner. The prescriptions may be prepared by a secretary or agent for the signature of a practitioner, but the prescribing practitioner is responsible in case the prescription does not conform in all essential respects to the law and regulations. A corresponding liability rests upon the pharmacist who fills a prescription not prepared in the form prescribed by these regulations.

(b) An intern, resident, or foreign-trained physician, or physician on the staff of a Veterans Administration facility, exempted from registration under § 1301.24(c) shall include on all prescriptions issued by him the registration number of the hospital or other institution and the special internal code number assigned to him by the hospital or other institution as provided in § 1301.24(c), in lieu of the registration number of the practitioner required by this section. Each written prescription shall have the name of the physician stamped, typed, or handprinted on it, as well as the signature of the physician.

(c) An official exempted from registration under § 1301.25 shall include on all prescriptions issued by him his branch of service or agency (e.g., "U. S. Army" or "Public Health Service") and his service identification number, in lieu of the registration number of the practitioner required by this section. The service identification number for a Public Health Service employee is his Social Security identification number. Each prescription shall have the name of the officer stamped, typed, or handprinted on it, as well as the signature of the officer.

§ 1306.06 Persons entitled to fill prescriptions.

A prescription for controlled substances may only be filled by a pharmacist acting in the usual course of his professional practice and either registered individually or employed in a registered pharmacy or registered institutional practitioner.

§ 1306.07 Administering or dispensing of narcotic drugs.

(a) The administering or dispensing directly (but not prescribing) of narcotic drugs listed in any schedule to a narcotic drug dependent person for "detoxification treatment" or "maintenance treatment" as defined in section 102 of the Act (21 U.S.C. 802) shall be deemed to be within the meaning of the term "in the course of his professional practice or research" in section 308(e) and section 102(20) of the Act (21 U.S.C. 828(e)): *Provided*, That the practitioner is separately registered with the Attorney General as required by section 303(g) of the Act (21 U.S.C. 823(g)) and then thereafter complies with the regulatory standards imposed relative to treatment qualification, security, records and unsupervised use of drugs pursuant to such Act.

(b) Nothing in this section shall prohibit a physician who is not specifically registered to conduct a narcotic treatment program from administering (but not prescribing) narcotic drugs to a person for the purpose of relieving acute withdrawal symptoms when necessary while arrangements are being made for referral for treatment. Not more than one day's medication may be administered to the person or for the person's use at one time. Such emergency treatment may be carried out for not more than three days and may not be renewed or extended.

(c) This section is not intended to impose any limitations on a physician or authorized hospital staff to administer or dispense narcotic drugs in a hospital to maintain or detoxify a person as an incidental adjunct to medical or surgical treatment of conditions other than addiction, or to administer or dispense narcotic drugs to persons with intractable pain in which no relief or cure is possible or none has been found after reasonable efforts.

Controlled Substances Listed in Schedule II

§ 1306.11 Requirement of prescription.

(a) A pharmacist may dispense directly a controlled substance listed in Schedule II, which is a prescription drug as determined under the Federal Food, Drug, and Cosmetic Act, only pursuant to a written prescription signed by the prescribing individual practitioner, except as provided in paragraph (d) of this section.

(b) An individual practitioner may administer or dispense directly a controlled substance listed in Schedule II in the course of his professional practice without a prescription, subject to § 1306.07.

(c) An institutional practitioner may administer or dispense directly (but not prescribe) a controlled substance listed in Schedule II only pursuant to a written prescription signed by the prescribing individual practitioner or to an order for medication made by an individual practitioner which is dispensed for immediate administration to the ultimate user.

(d) In the case of an emergency situation, as defined by the

Secretary in § 290.10 of this title, a pharmacist may dispense a controlled substance listed in Schedule II upon receiving oral authorization of a prescribing individual practitioner, provided that:

(1) The quantity prescribed and dispensed is limited to the amount adequate to treat the patient during the emergency period (dispensing beyond the emergency period must be pursuant to a written prescription signed by the prescribing individual practitioner);

(2) The prescription shall be immediately reduced to writing by the pharmacist and shall contain all information required in § 1306.05, except for the signature of the prescribing individual practitioner;

(3) If the prescribing individual practitioner is not known to the pharmacist, he must make a reasonable effort to determine that the oral authorization came from a registered individual practitioner, which may include a callback to the prescribing individual practitioner using his phone number as listed in the telephone directory and/or other good faith efforts to ensure his identity; and

(4) Within 72 hours after authorizing an emergency oral prescription, the prescribing individual practitioner shall cause a written prescription for the emergency quantity prescribed to be delivered to the dispensing pharmacist. In addition to conforming to the requirements of § 1306.05, the prescription shall have written on its face "Authorization for Emergency Dispensing," and the date of the oral order. The written prescription may be delivered to the pharmacist in person or by mail, but if delivered by mail it must be postmarked within the 72-hour period. Upon receipt, the dispensing pharmacist shall attach this prescription to the oral emergency prescription which had earlier been reduced to writing. The pharmacist shall notify the nearest office of the Administration if the prescribing individual practitioner fails to deliver a written prescription to him; failure of the pharmacist to do so shall void the authority conferred by this paragraph to dispense without a written prescription of a prescribing individual practitioner.

§ 1306.12 Refilling prescriptions.

The refilling of a prescription for a controlled substance listed in Schedule II is prohibited.

§ 1306.13 Partial filling of prescriptions.

(a) The partial filling of a prescription for a controlled substance listed in Schedule II is permissible, if the pharmacist is unable to supply the full quantity called for in a written or emergency oral prescription and he makes a notation of the quantity supplied on the face of the written prescription (or written record of the emergency oral prescription). The remaining portion of the prescription may be filled within 72 hours of the first partial filling; however, if the remaining portion is not or cannot be filled within the 72-hour period, the pharmacist shall so notify the prescribing individual practitioner. No further quantity may be supplied beyond 72 hours without a new prescription.

(b) Prescriptions for Schedule II controlled substances written for patients in Long Term Care Facilities (LTCF) may be filled in partial quantities, to include individual dosage units. For each partial filling, the dispensing pharmacist shall record on the back of the prescription (or on another appropriate record, uniformly maintained, and readily retrievable) the date of the partial filling, quantity dispensed, remaining quantity authorized to be dispensed and the identification of the dispensing pharmacist. The total quantity of Schedule II controlled substances dispensed in all partial fillings must not exceed the total quantity prescribed. Schedule II prescriptions, for patients in a LTCF, shall be valid for a period not to exceed 60 days from the issue date unless sooner terminated by the discontinuance of medication.

(c) Information pertaining to current Schedule II prescriptions for patients in a LTCF may be maintained in a computerized system if this system has the capability to permit:

(1) Output (display or printout) of the original prescription number, date of issue, identification of prescribing individual practitioner, identification of patient, identification of LTCF, identification of medication authorized (to include dosage form strength and quantity), listing of partial fillings that have been dispensed under each prescription and the information required in § 1306.13(b).

(2) Immediate (real time) updating of the prescription record each time a partial filling of the prescription is conducted.

(3) Retrieval of partially filled Schedule II prescription information is the same as required by § 1306.22(b)(4) and (5) for Schedule III and IV prescription refill information.

§ 1306.14 Labeling of substances.

(a) The pharmacist filling a written or emergency oral prescription for a controlled substance listed in Schedule II shall affix to the package a label showing date of filling, the pharmacy name and address, the serial number of the prescription, the name of the patient, the name of the prescribing practitioner, and directions for use and cautionary statements, if any, contained in such prescription or required by law.

(b) The requirements of paragraph (a) of this section do not apply when a controlled substance listed in Schedule II is prescribed for administration to an ultimate user who is institutionalized: *Provided*, That:

(1) Not more than a 7-day supply of the controlled substance listed in Schedule II is dispensed at one time;

(2) The controlled substance listed in Schedule II is not in the possession of the ultimate user prior to the administration;

(3) The institution maintains appropriate safeguards and records regarding the proper administration, control, dispensing, and storage of the controlled substance listed in Schedule II; and

(4) The system employed by the pharmacist in filling a prescription is adequate to identify the supplier, the product, and the patient, and to set forth the directions for use and cautionary statements, if any, contained in the prescription or required by law.

§ 1306.15 Filing of prescriptions.

All written prescriptions and written records of emergency oral prescriptions shall be kept in accordance with requirements of § 1304.04(h) of this chapter.

Controlled Substances Listed in Schedules III and IV

§ 1306.21 Requirement of prescription.

(a) A pharmacist may dispense directly a controlled substance listed in Schedule III or IV, which is a prescription drug as determined under the Federal Food, Drug, and Cosmetic Act, only pursuant to either a written prescription signed by a prescribing individual practitioner or an oral prescription made by a prescribing individual practitioner and promptly reduced to writing by the pharmacist containing all information required in § 1306.05, except for the signature of the prescribing individual practitioner.

(b) An individual practitioner may administer or dispense directly a controlled substance listed in Schedule III or IV in the course of his professional practice without a prescription, subject to § 1306.07.

(c) An institutional practitioner may administer or dispense directly (but not prescribe) a controlled substance listed in Schedule III or IV pursuant to a written prescription signed by a prescribing individual practitioner, or pursuant to an oral prescription made by a prescribing individual practitioner and promptly reduced to writing by the pharmacist (containing all information required in § 1306.05 except for the signature of the prescribing individual practitioner), or pursuant to an order for medication made by an individual practitioner which is dispensed for immediate administration to the ultimate user, subject to § 1306.07.

§ 1306.22 Refilling of prescriptions.

(a) No prescription for a controlled substance listed in Schedule III or IV shall be filled or refilled more than 6 months after the date on which such prescription was issued and no such prescription authorized to be refilled may be refilled more than five times. Each refilling of a prescription shall be entered on the back of the prescription or on another appropriate document. If entered on another document, such as a medication record, the document must be uniformly maintained and readily retrievable. The following information must be retrievable by the prescription number consisting of the name and dosage form of the controlled substance, the date filled or refilled, the quantity dispensed, initials of the dispensing pharmacist for each refill, and the total number of refills for that prescription. If the pharmacist merely initials and dates the back of the prescription it shall be deemed that the full face amount of the prescription has been dispensed. The prescribing practitioner may authorize additional refills of

Schedule III or IV controlled substances on the original prescription through an oral refill authorization transmitted to the pharmacist provided the following conditions are met:

(1) The total quantity authorized, including the amount of the original prescription, does not exceed five refills nor extend beyond six months from the date of issue of the original prescription.

(2) The pharmacist obtaining the oral authorization records on the reverse of the original prescription the date, quantity of refill, number of additional refills authorized, and initials the prescription showing who received the authorization from the prescribing practitioner who issued the original prescription.

(3) The quantity of each additional refill authorized is equal to or less than the quantity authorized for the initial filling of the original prescription.

(4) The prescribing practitioner must execute a new and separate prescription for any additional quantities beyond the five refill, six-month limitation.

(b) As an alternative to the procedures provided by subsection (a), an automated data processing system may be used for the storage and retrieval of refill information for prescription orders for controlled substances in Schedules III and IV, subject to the following conditions:

(1) Any such proposed computerized system must provide on-line retrieval (via CRT display or hard-copy printout) of original prescription order information for those prescription orders which are currently authorized for refilling. This shall include, but is not limited to, data such as the original prescription number, date of issuance of the original prescription order by the practitioner, full name and address of the patient, name, address, and DEA registration number of the practitioner, and the name, strength, dosage form, quantity of the controlled substance prescribed (and quantity dispensed if different from the quantity prescribed), and the total number of refills authorized by the prescribing practitioner.

(2) Any such proposed computerized system must also provide on-line retrieval (via CRT display or hard-copy printout) of the current refill history for Schedule III or IV controlled substance prescription orders (those authorized for refill during the past six months). This refill history shall include, but is not limited to, the name of the controlled substance, the date of refill, the quantity dispensed, the identification code, or name or initials of the dispensing pharmacist for each refill and the total number of refills dispensed to date for that prescription order.

(3) Documentation of the fact that the refill information entered into the computer each time a pharmacist refills an original prescription order for a Schedule III or IV controlled substance is correct must be provided by the individual pharmacist who makes use of such a system. If such a system provides a hard-copy printout of each day's controlled substance prescription order refill data, that printout shall be verified, dated, and signed by the individual pharmacist who refilled such a prescription order. The individual pharmacist must verify that the data indicated is correct and then sign this document in the same manner as he would sign a check or legal document (e.g., J. H. Smith, or John H. Smith). This document shall be maintained in a separate file at that pharmacy for a period of two years from the dispensing date. This printout of the day's controlled substance prescription order refill data must be provided to each pharmacy using such a computerized system within 72 hours of the date on which the refill was dispensed. It must be verified and signed by each pharmacist who is involved with such dispensing. In lieu of such a printout, the pharmacy shall maintain a bound log book, or separate file, in which each individual pharmacist involved in such dispensing shall sign a statement (in the manner previously described) each day, attesting to the fact that the refill information entered into the computer that day has been reviewed by him and is correct as shown. Such a book or file must be maintained at the pharmacy employing such a system for a period of two years after the date of dispensing the appropriately authorized refill.

(4) Any such computerized system shall have the capability of producing a printout of any refill data which the user pharmacy is responsible for maintaining under the Act and its implementing regulations. For example, this would include a refill-by-refill audit trail for any specified strength and dosage form of any controlled substance (by either brand or generic name or both). Such a printout must indicate name of the prescribing practitioner, name and address of the patient, quantity dispensed on each refill, date of dispensing for each refill, name or identification code of the dispensing pharmacist, and the number of the original prescription order. In any computerized system employed by a user pharmacy the central recordkeeping location must be capable of sending the printout to the pharmacy within 48 hours, and if a DEA Special Agent or Compliance Investigator requests a copy of such printout from the user pharmacy, it must, if requested to do so by the Agent or Investigator, verify the printout transmittal capability of its system by documentation (e.g., postmark).

(5) In the event that a pharmacy which employs such a computerized system experiences system down-time, the pharmacy must have an auxiliary procedure which will be used for documentation of refills of Schedule III and IV controlled substance prescription orders. This auxiliary procedure must ensure that refills are authorized by the original prescription order, that the maximum number of refills has not been exceeded, and that all of the appropriate data are retained for on-line data entry as soon as the computer system is available for use again.

(c) When filing refill information for original prescription orders for Schedule III or IV controlled substances, a pharmacy may use only one of the two systems described in paragraph (a) or (b) of this section.

§ 1306.23 Partial filling of prescriptions.

The partial filling of a prescription for a controlled substance listed in Schedule III or IV is permissible, provided that:

(a) Each partial filling is recorded in the same manner as a refilling,

(b) The total quantity dispensed in all partial fillings does not exceed the total quantity prescribed, and

(c) No dispensing occurs after 6 months after the date on which the prescription was issued.

§ 1306.24 Labeling of substances.

(a) The pharmacist filling a prescription for a controlled substance listed in Schedule III or IV shall affix to the package a label showing the pharmacy name and address, the serial number and date of initial filling, the name of the patient, the name of the practitioner issuing the prescription, and directions for use and cautionary statements, if any, contained in such prescription as required by law.

(b) The requirements of paragraph (a) of this section do not apply when a controlled substance listed in Schedule III or IV is prescribed for administration to an ultimate user who is institutionalized: *Provided*, That:

(1) Not more than a 34-day supply or 100 dosage units, whichever is less, of the controlled substance listed in Schedule III or IV is dispensed at one time;

(2) The controlled substance listed in Schedule III or IV is not in the possession of the ultimate user prior to administration;

(3) The institution maintains appropriate safeguards and records the proper administration, control, dispensing, and storage of the controlled substance listed in Schedule III or IV; and

(4) The system employed by the pharmacist in filling a prescription is adequate to identify the supplier, the product and the patient, and to set forth the directions for use and cautionary statements, if any, contained in the prescription or required by law.

§ 1306.25 Filing prescriptions.

All prescriptions for controlled substances listed in Schedules III and IV shall be kept in accordance with § 1304.04(h) of this chapter.

§ 1306.26 Transfer between pharmacies of prescription information for Schedules III, IV, and V controlled substances for refill purposes.

(a) The transfer of original prescription information for a controlled substance listed in Schedules III, IV or V for the purpose of refill dispensing is permissible between pharmacies on a one time basis subject to the following requirements:

(1) The transfer is communicated directly between two licensed pharmacists and the transferring pharmacist records the following information:

(i) Write the word "VOID" on the face of the invalidated prescription.

(ii) Record on the reverse of the invalidated prescription the name, address and DEA registration number of the pharmacy to which it was transferred and the name of the pharmacist receiving the prescription information.

(iii) Record the date of the transfer and the name of the pharmacist transferring the information.

(b) The pharmacist receiving the transferred prescription information shall reduce to writing the following:

(1) Write the word "transfer" on the face of the transferred prescription.

(2) Provide all information required to be on a prescription pursuant to 21 CFR 1306.05 and include:

(i) Date of issuance of original prescription;

(ii) Original number of refills authorized on original prescription;

(iii) Date of original dispensing;

(iv) Number of valid refills remaining and date of last refill;

(v) Pharmacy's name, address, DEA registration number and original prescription number from which the prescription information was transferred;

(vi) Name of transferor pharmacist.

(3) Both the original and transferred prescription must be maintained for a period of two years from the date of last refill.

(c) Pharmacies electronically accessing the same prescription record must satisfy all information requirements of a manual mode for prescription transferral.

(d) The procedure allowing the transfer of prescription information for refill purposes is permissible only if allowable under existing state or other applicable law.

Controlled Substances Listed in Schedule V

§ 1306.31 Requirement of prescription.

(a) A pharmacist may dispense directly a controlled substance listed in Schedule V pursuant to a prescription as required for controlled substances listed in Schedules III and IV in § 1306.21. A prescription for a controlled substance listed in Schedule V may be refilled only as expressly authorized by the prescribing individual practitioner on the prescription; if no such authorization is given, the prescription may not be refilled. A pharmacist dispensing such substance pursuant to a prescription shall label the substance in accordance with § 1306.24 and file the prescription in accordance with § 1306.25.

(b) An individual practitioner may administer or dispense directly a controlled substance listed in Schedule V in the course of his professional practice without a prescription, subject to § 1306.07.

(c) An institutional practitioner may administer or dispense directly (but not prescribe) a controlled substance listed in Schedule V only pursuant to a written prescription signed by the prescribing individual practitioner, or pursuant to an oral prescription made by a prescribing individual practitioner and promptly reduced to writing by the pharmacist (containing all information required in § 1306.05 except for the signature of the prescribing individual practitioner), or pursuant to an order for medication made by an individual practitioner which is dispensed for immediate administration to the ultimate user, subject to § 1306.07.

§ 1306.32 Dispensing without prescription.

A controlled substance listed in Schedule V, and a controlled substance listed in Schedule II, III, or IV which is not a prescription drug as determined under the Federal Food, Drug, and Cosmetic Act, may be dispensed by a pharmacist without a prescription to a purchaser at retail, provided that:

(a) Such dispensing is made only by a pharmacist (as defined in § 1306.02(d)), and not by a nonpharmacist employee even if under the supervision of a pharmacist (although after the pharmacist has fulfilled his professional and legal responsibilities set forth in this section, the actual cash, credit transaction, or delivery, may be completed by a nonpharmacist);

(b) Not more than 240 cc. (8 ounces) of any such controlled substance containing opium, nor more than 120 cc. (4 ounces) of any other such controlled substance nor more than 48 dosage units of any such controlled substance containing opium, nor more than 24 dosage units of any other such controlled substance may be dispensed at retail to the same purchaser in any given 48-hour period;

(c) The purchaser is at least 18 years of age;

(d) The pharmacist requires every purchaser of a controlled substance under this section not known to him to furnish suitable identification (including proof of age where appropriate);

(e) A bound record book for dispensing of controlled substances under this section is maintained by the pharmacist, which book shall contain the name and address of the purchaser, the name and quantity of controlled substance purchased, the date of each purchase, and the name or initials of the pharmacist who dispensed the substance to the purchaser (the book shall be maintained in accordance with the recordkeeping requirement of § 1304.04 of this chapter); and

(f) A prescription is not required for distribution or dispensing of the substance pursuant to any other Federal, State, or local law.

Miscellaneous

§ 1307.02 Application of State law and other Federal law.

Nothing in Parts 1301–1308, 1311, 1312, or 1316 of this chapter shall be construed as authorizing or permitting any person to do any act which such person is not authorized or permitted to do under other Federal laws or obligations under international treaties, conventions or protocols, or under the law of the State in which he desires to do such act nor shall compliance with such Parts be construed as compliance with other Federal or State laws unless expressly provided in such other laws.

§ 1307.03 Exceptions to regulations.

Any person may apply for an exception to the application of any provision of Parts 1301–1308, 1311, 1312, or 1316 of this chapter by filing a written request stating the reasons for such exception. Requests shall be filed with the Administrator, Drug Enforcement Administration, Department of Justice, Washington, DC 20537. The Administrator may grant an exception in his discretion, but in no case shall he be required to grant an exception to any person which is not otherwise required by law or the regulations cited in this section.

Special Exceptions for Manufacture and Distribution of Controlled Substances

§ 1307.11 Distribution by dispenser to another practitioner.

(a) A practitioner who is registered to dispense a controlled substance may distribute (without being registered to distribute) a quantity of such substance to another practitioner for the purpose of general dispensing by the practitioner to his or its patients: *Provided*, That:

(1) The practitioner to whom the controlled substance is to be distributed is registered under the Act to dispense that controlled substance;

(2) The distribution is recorded by the distributing practitioner in accordance with § 1304.24(e) of this chapter and by the receiving practitioner in accordance with § 1304.24(c) of this chapter;

(3) If the substance is listed in Schedule I or II, an order form is used as required in Part 1305 of this chapter;

(4) The total number of dosage units of all controlled substances distributed by the practitioner pursuant to this section and § 1301.28 of this chapter during each calendar year in which the practitioner is registered to dispense does not exceed 5 percent of the total number of dosage units of all controlled substances distributed and dispensed by the practitioner during the same calendar year.

(b) If, during any calendar year in which the practitioner is registered to dispense, the practitioner has reason to believe that the total number of dosage units of all controlled substances which will be distributed by him pursuant to this section and § 1301.28 of this chapter will exceed 5 percent of the total number of dosage units of all controlled substances distributed and dispensed by him during that calendar year, the practitioner shall obtain a registration to distribute controlled substances.

§ 1307.12 Manufacture and distribution of narcotic solutions and compounds by a pharmacist.

As an incident to a distribution under § 1307.11, a pharmacist may manufacture (without being registered to manufacture) an aqueous or oleaginous solution or solid dosage form containing a narcotic controlled substance in a proportion not exceeding 20 percent of the complete solution, compound, or mixture.

§ 1307.13 Distribution to supplier.

Any person lawfully in possession of a controlled substance

listed in any schedule may distribute (without being registered to distribute) that substance to the person from whom he obtained it or to the manufacturer of the substance, provided that a written record is maintained which indicates the date of the transaction, the name, form and quantity of the substance, the name, address, and registration number, if any, of the person making the distribution, and the name, address, and registration number, if known, of the supplier or manufacturer. In the case of returning a controlled substance listed in Schedule I or II, an order form shall be used in the manner prescribed in Part 1305 of this chapter and be maintained as the written record of the transaction. Any person not required to register pursuant to sections 302(c) or 1007(b)(1) of the Act (21 U.S.C. 823(c) or 957(b)(1)) shall be exempt from maintaining the records required by this section.

§ 1307.14 Distribution upon discontinuance or transfer of business.

(a) Any registrant desiring to discontinue business activities altogether or with respect to controlled substances (without transferring such business activities to another person) shall return for cancellation his certificate of registration, and any unexecuted order forms in his possession, to the Registration Unit, Drug Enforcement Administration, Department of Justice, Post Office Box 28083, Central Station, Washington, DC 20005. Any controlled substances in his possession may be disposed of in accordance with § 1307.21.

(b) Any registrant desiring to discontinue business activities altogether or with respect to controlled substances (by transferring such business activities to another person) shall submit in person or by registered or certified mail, return receipt requested, to the Special Agent in Charge in his area, at least 14 days in advance of the date of the proposed transfer (unless the Special Agent in Charge waives this time limitation in individual instances), the following information:

(1) The name, address, registration number, and authorized business activity of the registrant discontinuing the business (registrant-transferor);

(2) The name, address, registration number, and authorized business activity of the person acquiring the business (registrant-transferee);

(3) Whether the business activities will be continued at the location registered by the person discontinuing business, or moved to another location (if the latter, the address of the new location should be listed);

(4) Whether the registrant-transferor has a quota to manufacture or procure any controlled substance listed in Schedule I or II (if so, the basic class or class of the substance should be indicated); and

(5) The date on which the transfer of controlled substances will occur.

(c) Unless the registrant-transferor is informed by the Regional Administrator,[1] before the date on which the transfer was stated to occur, that the transfer may not occur, the registrant-transferor may distribute (without being registered to distribute) controlled substances in his possession to the registrant-transferee in accordance with the following:

(1) On the date of transfer of the controlled substances, a complete inventory of all controlled substances being transferred shall be taken in accordance with §§ 1304.11–1304.19 of this chapter. This inventory shall serve as the final inventory of the registrant-transferor and the initial inventory of the registrant-transferee, and a copy of the inventory shall be included in the records of each person. It shall not be necessary to file a copy of the inventory with the Administration unless requested by the Regional Administrator.[1] Transfers of any substances listed in Schedule I or II shall require the use of order forms in accordance with Part 1305 of this chapter.

(2) On the date of transfer of the controlled substances, all records required to be kept by the registrant-transferor with reference to the controlled substances being transferred, under Part 1304 of this chapter, shall be transferred to the registrant-transferee. Responsibility for the accuracy of records prior to the date of transfer remains with the transferor, but responsibility for custody and maintenance shall be upon the transferee.

(3) In the case of registrants required to make reports pursuant to Part 1304 of this chapter, a report marked "Final" will be prepared and submitted by the registrant-transferor showing the disposition of all the controlled substances for which a report is required; no additional report will be required from him, if no further transactions involving controlled substances are consummated by him. The initial report of the registrant-transferee shall account for transactions beginning with the day next succeeding the date of discontinuance or transfer of business by the transferor-registrant and the substances transferred to him shall be reported as receipts in his initial report.

[1] Special Agent in Charge.

Disposal of Controlled Substances

§ 1307.21 Procedure for disposing of controlled substances.

(a) Any person in possession of any controlled substance and desiring or required to disposed of such substance may request the Special Agent in Charge of the Administration in the area in which the person is located for authority and instructions to dispose of such substance. The request should be made as follows:

(1) If the person is a registrant required to make reports pursuant to Part 1304 of this chapter, he shall list the controlled substance or substances which he desires to dispose of on the "b" subpart of the report normally filed by him, and submit three copies of that subpart to the Special Agent in Charge of the Administration in his area;

(2) If the person is a registrant not required to make reports pursuant to Part 1304 of this chapter, he shall list the controlled substance or substances which he desires to dispose of on DEA Form 41, and submit three copies of that form to the Special Agent in Charge in his area; and

(3) If the person is not a registrant, he shall submit to the Special Agent in Charge a letter stating:

(i) The name and address of the person;

(ii) The name and quantity of each controlled substance to be disposed of;

(iii) How the applicant obtained the substance, if known; and

(iv) The name, address, and registration number, if known, of the person who possessed the controlled substances prior to the applicant, if known.

(b) The Special Agent in Charge shall authorize and instruct the applicant to dispose of the controlled substance in one of the following manners:

(1) By transfer to person registered under the Act and authorized to possess the substance;

(2) By delivery to an agent of the Administration or to the nearest office of the Administration;

(3) By destruction in the presence of an agent of the Administration or other authorized person; or

(4) By such other means as the Special Agent in Charge may determine to assure that the substance does not become available to unauthorized persons.

(c) In the event that a registrant is required regularly to dispose of controlled substances, the Special Agent in Charge may authorize the registrant to dispose of such substances, in accordance with paragraph (b) of this section, without prior approval of the Administration in each instance, on the condition that the registrant keep records of such disposals and file periodic reports with the Special Agent in Charge summarizing the disposals made by the registrant. In granting such authority, the Special Agent in Charge may place such conditions as he deems proper on the disposal of controlled substances, including the method of disposal and the frequency and detail of reports.

(d) This section shall not be construed as affecting or altering in any way the disposal of controlled substances through procedures provided in laws and regulations adopted by any State.

Schedules of Controlled Substances

§ 1308.11 Schedule I.

(a) Schedule I shall consist of the drugs and other substances, by whatever official name, common or usual name, chemical name, or brand name designated, listed in this section. Each drug or substance has been assigned the DEA Controlled Substances Code Number set forth opposite it.

(b) *Opiates.* Unless specifically excepted or unless listed in another schedule, any of the following opiates, including their isomers, esters, ethers, salts, and salts of isomers, esters and ethers, whenever the existence of such isomers, esters, ethers, and salts is possible within the specific chemical designation (for purposes of paragraph (b)(34) only, the term isomer includes the optical and geometric isomers):

(1) Acetyl-alpha-methylfentanyl (*N*-[1-(1-methyl-2-phenethyl)-4-piperidinyl]-*N*-phenylacetamide) 9815
(2) Acetylmethadol 9601
(3) Allylprodine 9602
(4) Alphacetylmethadol 9603
(5) Alphameprodine 9604
(6) Alphamethadol 9605
(7) Alpha-methylfentanyl (*N*-[1-(alpha-methylbeta-phenyl)ethyl-4-piperidyl] propionanilide; 1-(1-methyl-2-phenylethyl)-4-(N-propanilido) piperidine) 9814
(8) Alpha-methylthiofentanyl (*N*-[1-methyl-2-(2-thienyl)ethyl-4-piperidinyl]-*N*-phenylpropanamide) 9832
(9) Benzethidine 9606
(10) Betacetylmethadol 9607
(11) Beta-hydroxyfentanyl (*N*-[1-(2-hydroxy-2-phenethyl)-4-piperidinyl]-*N*-phenylpropanamide) 9830
(12) Beta-hydroxy-3-methylfentanyl (other name: *N*-[1-(2-hydroxy-2-phenethyl)-3-methyl-4-piperidinyl]-*N*-phenylpropanamide) 9831
(13) Betameprodine 9608
(14) Betamethadol 9609
(15) Betaprodine 9611
(16) Clonitazene 9612
(17) Dextromoramide 9613
(18) Diampromide 9615
(19) Diethylthiambutene 9616
(20) Difenoxin 9168
(21) Dimenoxadol 9617
(22) Dimepheptanol 9618
(23) Dimethylthiambutene 9619
(24) Dioxaphetyl butyrate 9621
(25) Dipipanone 9622
(26) Ethylmethylthiambutene 9623
(27) Etonitazene 9624
(28) Etoxeridine 9625
(29) Furethidine 9626
(30) Hydroxypethidine 9627
(31) Ketobemidone 9628
(32) Levomoramide 9629
(33) Levophenacylmorphan 9631
(34) 3-Methylfentanyl(*N*-[3-methyl-1-(2-phenylethyl)-4-piperidyl]-*N*-phenylpropanamide) 9813
(35) 3-Methylthiofentanyl (*N*-[(3-methyl-1-(2-thienyl)ethyl-4-piperidinyl]-*N*-phenylpropanamide) 9833
(36) Morpheridine 9632
(37) MPPP (1-methyl-4-phenyl-4-propionoxypiperidine) 9661
(38) Noracymethadol 9633
(39) Norlevorphanol 9634
(40) Normethadone 9635
(41) Norpipanone 9636
(42) Para-fluorofentanyl (*N*-(4-fluorophenyl)-*N*-[1-(2-phenethyl)-4-piperidinyl]-propanamide 9812
(43) Phenadoxone 9637
(44) PEPAP (1-(-2-phenethyl)-4-phenyl-4-acetoxypiperidine 9663
(45) Phenampromide 9638
(46) Phenomorphan 9647
(47) Phenoperidine 9641
(48) Piritramide 9642
(49) Proheptazine 9643
(50) Properidine 9644
(51) Propiram 9649
(52) Racemoramide 9645
(53) Thiofentanyl (*N*-phenyl-*N*-[1-(2-thienyl)-ethyl-4-piperidinyl]-propanamide .. 9835
(54) Tilidine 9750
(55) Trimeperidine 9646

(c) *Opium derivatives.* Unless specifically excepted or unless listed in another schedule, any of the following opium derivatives, its salts, isomers, and salts of isomers whenever the existence of such salts, isomers, and salts of isomers is possible within the specific chemical designation:

(1) Acetorphine 9319
(2) Acetyldihydrocodeine 9051
(3) Benzylmorphine 9052
(4) Codeine methylbromide 9070
(5) Codeine-N-Oxide 9053
(6) Cyprenorphine 9054
(7) Desomorphine 9055
(8) Dihydromorphine 9145
(9) Drotebanol 9335
(10) Etorphine (except hydrochloride salt) .. 9056
(11) Heroin 9200
(12) Hydromorphinol 9301
(13) Methyldesorphine 9302
(14) Methyldihydromorphine 9304
(15) Morphine methylbromide 9305
(16) Morphine methylsulfonate 9306
(17) Morphine-N-Oxide 9307
(18) Myrophine 9308
(19) Nicocodeine 9309
(20) Nicomorphine 9312
(21) Normorphine 9313
(22) Pholcodine 9314
(23) Thebacon 9315

(d) *Hallucinogenic substances.* Unless specifically excepted or unless listed in another schedule, any material, compound, mixture, or preparation, which contains any quantity of the following hallucinogenic substances, or which contains any of its salts, isomers, and salts of isomers whenever the existence of such salts, isomers, and salts of isomers is possible within the specific chemical designation (for purposes of this paragraph only, the term "isomer" includes the optical, position and geometric isomers):

(1) 4-bromo-2,5-dimethoxyamphetamine ... 7391
Some trade or other names: 4-bromo-2,5-dimethoxy-α-methylphenethylamine; 4-bromo-2,5-DMA.
(2) 2,5-dimethoxyamphetamine 7396
Some trade or other names: 2,5-dimethoxy-α-methylphenethylamine; 2,5-DMA.
(3) 4-methoxyamphetamine 7411
Some trade or other names: 4-methoxy-α-methylphenethylamine; para-methoxy-amphetamine; PMA.
(4) 5-methoxy-3,4-methylenedioxyamphetamine 7401
(5) 4-methyl-2,5-dimethoxyamphetamine ... 7395
Some trade and other names: 4-methyl-2,5-dimethoxy-α-methylphenethylamine; "DOM"; and "STP."
(6) 3,4-methylenedioxy amphetamine 7400
(7) 3,4-methylenedioxymethamphetamine (MDMA) 7405
(8) 3,4,5-trimethoxy amphetamine 7390
(9) Bufotenine 7433
Some trade and other names: 3-(β-Dimethylaminoethyl)-5-hydroxyindole; 3-(2-dimethylaminoethyl)-5-indolol; *N,N*-dimethylserotonin; 5-hydroxy-*N,N*-dimethyltryptamine; mappine.
(10) Diethyltryptamine 7434
Some trade and other names: *N,N*-Diethyltryptamine; DET.
(11) Dimethyltryptamine 7435
Some trade or other names: DMT.
(12) Ibogaine 7260
Some trade and other names: 7-Ethyl-6,6β,7,8,9,10,12,13-octahydro-2-methoxy-6,9-methano-5*H*-pyrido [1′,2′:1,2]azepino [5,4-b] indole; tabernanthe iboga.

(13) Lysergic acid diethylamide 7315
(14) Marihuana 7360
(15) Mescaline 7381
(16) Parahexyl 7374
Some trade or other names: 3-Hexyl-1-hydroxy-7,8,9,10-tetrahydro-6,6,9-trimethyl-6*H*-dibenzo[b,d]pyran; Synhexyl.
(17) Peyote 7415
Meaning all parts of the plant presently classified botanically as *Lophophora Williamsii Lemaire*, whether growing or not, the seeds thereof, any extract from any part of such plant, and every compound, manufacture, salt, derivative, mixture, or preparation of such plant, its seeds or extracts.

(Interprets 21 U.S.C 812(c), Schedule I(c)(12))
(18) *N*-ethyl-3-piperidyl benzilate 7482
(19) *N*-methyl-3-piperidyl benzilate 7484
(20) Psilocybin 7437
(21) Psilocyn 7438
(22) Tetrahydrocannabinols 7370
Synthetic equivalents of the substances contained in the plant, or in the resinous extractives of Cannabis, sp. and/or synthetic substances, derivatives, and their isomers with similar chemical structure and pharmacological activity such as the following:
Δ1 cis or trans tetrahydrocannabinol, and their optical isomers.
Δ6 cis or trans tetrahydrocannabinol, and their optical isomers.
Δ3,4 cis or trans tetrahydrocannabinol, and its optical isomers.
(Since nomenclature of these substances is not internationally standardized, compounds of these structures, regardless of numerical designation of atomic positions covered.)
(23) Ethylamine analog of phencyclidine 7455
Some trade or other names: *N*-ethyl-1-phenylcyclohexylamine, (1-phenylcyclohexyl)-ethylamine, *N*-(1-phenylcyclohexyl)-ethylamine, cyclohexamine, PCE.
(24) Pyrrolidine analog of phencyclidine 7458
Some trade or other names: 1-(1-phenylcyclohexyl)-pyrrolidine, PCPy, PHP.
(25) Thiophene analog of phencyclidine 7470
Some trade or other names: 1-[1-(2-thienyl)-cyclohexyl]-piperidine, 2-thienyl analog of phencyclidine, TPCP, TCP.

(e) *Depressants.* Unless specifically excepted or unless listed in another schedule, any material compound, mixture, or preparation which contains any quantity of the following substances having a depressant effect on the central nervous system, including its salts, isomers, and salts of isomers whenever the existence of such salts, isomers, and salts of isomers is possible within the specific chemical designation:

(1) Mecloqualone 2572
(2) Methaqualone 2565

(f) *Stimulants.* Unless specifically excepted or unless listed in another schedule, any material, compound, mixture, or preparation which contains any quantity of the following substances having a stimulant effect on the central nervous system, including its salts, isomers, and salts of isomers:

(1) Fenethylline 1503
(2) *N*-ethylamphetamine 1475

(g) Temporary listing of substances subject to emergency scheduling. Any material, compound, mixture, or preparation which contains any quantity of the following substances:

(1) *N*-[1-benzyl-4-piperidyl]-*N*-phenylpropanamide (benzylfentanyl), its optical isomers, salts, and salts of isomers 9818
(2) *N*-[1-(2-thienyl)methyl-4-piperidyl]-*N*-phenylpropanamide (thenylfentanyl), its optical isomers, salts, and salts of isomers 9834
(3) 3,4-methylenedioxy-*N*-ethylamphetamine (also known as N-ethyl-alpha-methyl-3,4(methylenedioxy)phenethylamine, *N*-ethyl MDA, MDE, and MDEA) 7404
(4) *N*-hydroxy-3,4-methylenedioxyamphetamine (also known as *N*-hydroxy-alpha-methyl-3,4-(methylenedioxy)-phenethylamine, and *N*-hydroxy MDA) 7402
(5) 4-methylaminorex (also known as 2-amino-4-methyl-5-phenyl-2-oxazoline) 1590
(6) *N,N*-dimethylamphetamine (Some other names: *N,N*,alpha-trimethylbenzeneethaneamine; *N,N*,alpha-trimethylphenethylamine), its salts, optical isomers, and salts of optical isomers 1480

§ 1308.12 Schedule II.

(a) Schedule II shall consist of the drugs and other substances, by whatever official name, common or usual name, chemical name, or brand name designated, listed in this section. Each drug or substance has been assigned the Controlled Substances Code Number set forth opposite it.

(b) *Substances, vegetable origin, or chemical synthesis.* Unless specifically excepted or unless listed in another schedule, any of the following substances whether produced directly or indirectly by extraction from substances of vegetable origin, or independently by means of chemical synthesis, or by a combination of extraction and chemical synthesis:

(1) Opium and opiate, and any salt, compound, derivative, or preparation of opium or opiate, excluding apomorphine, dextrorphan, nalbuphine, nalmefene, naloxone, and naltrexone, and their respective salts, but including the following:

(1) Raw opium 9600
(2) Opium extracts 9610
(3) Opium fluid extracts 9620
(4) Powdered opium 9639
(5) Granulated opium 9640
(6) Tincture of opium 9630
(7) Codeine 9050
(8) Ethylmorphine 9190
(9) Etorphine hydrochloride 9059
(10) Hydrocodone 9193
(11) Hydromorphone 9150
(12) Metopon 9260
(13) Morphine 9300
(14) Oxycodone 9143
(15) Oxymorphone 9652
(16) Thebaine 9333

(2) Any salt, compound, derivative, or preparation thereof which is chemically equivalent or identical with any of the substances referred to in paragraph (b)(1) of this section, except that these substances shall not include the isoquinoline alkaloids of opium.

(3) Opium poppy and poppy straw.

(4) Coca leaves (9040) and any salt, compound, derivative or preparation of coca leaves (including cocaine (9041) and ecgonine (9180) and their salts, isomers, derivatives and salts of isomers and derivatives), and any salt, compound, derivative, or preparation thereof which is chemically equivalent or identical with any of these substances, except that the substances shall not include decocainized coca leaves or extraction of coca leaves, which extractions do not contain cocaine or ecgonine.

(5) Concentrate of poppy straw (the crude extract of poppy straw in either liquid, solid or powder form which contains the phenanthrine alkaloids of the opium poppy), 9670.

(c) *Opiates.* Unless specifically excepted or unless in another schedule any of the following opiates, including its isomers, esters, ethers, salts and salts of isomers, esters and ethers whenever the existence of such isomers, esters, ethers, and salts is possible within the specific chemical designation, dextrorphan and levopropoxyphene excepted:

(1)	Alfentanil	9737
(2)	Alphaprodine	9010
(3)	Anileridine	9020
(4)	Benzitramide	9800
(5)	Bulk Dextropropoxyphene (nondosage forms)	9273
(6)	Carfentanil	9743
(7)	Dihydrocodeine	9120
(8)	Diphenoxylate	9170
(9)	Fentanyl	9801
(10)	Isomethadone	9226
(11)	Levomethorphan	9210
(12)	Levorphanol	9220
(13)	Metazocine	9240
(14)	Methadone	9250
(15)	Methadone-Intermediate, 4-cyano-2-dimethyl amino-4,4-diphenyl butane	9254
(16)	Moramide-Intermediate, 2-methyl-3-morpholino-1,1-diphenylpropane-carboxylic acid	9802
(17)	Pethidine (meperidine)	9230
(18)	Pethidine-Intermediate-A, 4-cyano-1-methyl-4-phenylpiperidine	9232
(19)	Pethidine-Intermediate-B, ethyl-4-phenyl-piperidine-4-carboxylate	9233
(20)	Pethidine-Intermediate-C, 1-methyl-4-phenyl-piperidine-4-carboxylic acid	9234
(21)	Phenazocine	9715
(22)	Piminodine	9730
(23)	Racemethorphan	9732
(24)	Racemorphan	9733
(25)	Sufentanil	9740

(d) *Stimulants.* Unless specifically excepted or unless listed in another schedule, any material, compound, mixture, or preparation which contains any quantity of the following substances having a stimulant effect on the central nervous system:

(1)	Amphetamine, its salts, optical isomers, and salts of its optical isomers	1100
(2)	Methamphetamine, its salts, isomers, and salts of its isomers	1105
(3)	Phenmetrazine and its salts	1631
(4)	Methylphenidate	1724

(e) *Depressants.* Unless specifically excepted or unless listed in another schedule, any material, compound, mixture, or preparation which contains any quantity of the following substances having a depressant effect on the central nervous system, including its salts, isomers, and salts of isomers whenever the existence of such salts, isomers, and salts of isomers is possible within the specific chemical designation:

(1)	Amobarbital	2125
(2)	Pentobarbital	2270
(3)	Phencyclidine	7471
(4)	Secobarbital	2315

(f) *Hallucinogenic substances.*

(1)	Dronabinol (synthetic) in sesame oil and encapsulated in a soft gelatin capsule in a U.S. Food and Drug Administration approved drug product[2]	7369
	Some other names for dronabinol: (6a*R*-*trans*)-6a,7,8,10a-tetrahydro-6,6,9-trimethyl-3-pentyl-6*H*-dibenzo-[*b,d*]pyran-1-ol, or (−)-delta-9-(trans)-tetrahydrocannabinol	
(2)	Nabilone	7379
	Another name for nabilone: (±)-*trans*-3-(1,1-dimethylheptyl)-6,6a,7,8,10,10a-hexahydro-1-hydroxy-6,6-dimethyl-9H-dibenzo[b,d]pyran-9-one.	

[2] DEA Statement of Policy: *Any person registered by DEA to distribute, prescribe, administer, or dispense controlled substances in Schedule II who engages in the distribution or dispensing of dronabinol for medical indications outside the approved use associated with cancer treatment, except within the confines of a structured and recognized research program, may subject his or her controlled substances registration to review under the provisions of 21 U.S.C. 823(f) and 824(a)(4) as being inconsistent with the public interest. DEA will take action to revoke that such distribution or dispensing constitutes a threat to the public health and safety, and in addition will pursue any criminal sanctions which may be warranted under 21 U.S.C. 841(a)(1).* See *United States* v. *Moore*, 423 U.S. 122 (1975).

(g) *Immediate precursors.* Unless specifically excepted or unless listed in another schedule, any material, compound, mixture, or preparation which contains any quantity of the following substances:

(1)	Immediate precursor to amphetamine and methamphetamine:	
	(i) Phenylacetone	8501
	Some trade or other names: phenyl-2-propanone; P2P; benzyl methyl ketone; methyl benzyl ketone;	
(2)	Immediate precursors to phencyclidine (PCP):	
	(i) 1-phenylcyclohexylamine	7460
	(ii) 1-piperidinocyclohexanecarbonitrile (PCC)	8603

§ 1308.13 Schedule III.

(a) Schedule III shall consist of the drugs and other substances, by whatever official name, common or usual name, chemical name, or brand name designated, listed in this section. Each drug or substance has been assigned the DEA Controlled Substances Code Number set forth opposite it.

(b) *Stimulants.* Unless specifically excepted or unless listed in another schedule, any material, compound, mixture, or preparation which contains any quantity of the following substances having a stimulant effect on the central nervous system, including its salts, isomers (whether optical, position, or geometric), and salts of such isomers whenever the existence of such salts, isomers, and salts of isomers is possible within the specific chemical designation:

(1)	Those compounds, mixtures, or preparations in dosage unit form containing any stimulant substances listed in Schedule II which compounds, mixtures, or preparations were listed on August 25, 1971, as excepted compounds under § 308.32, and any other drug of the quantitative composition shown in that list for those drugs or which is the same except that it contains a lesser quantity of controlled substances	1405
(2)	Benzphetamine	1228
(3)	Chlorphentermine	1645
(4)	Clortermine	1647
(5)	Phendimetrazine	1615

(c) *Depressants.* Unless specifically excepted or unless listed in another schedule, any material, compound, mixture, or preparation which contains any quantity of the following substances having a depressant effect on the central nervous system:

(1)	Any compound, mixture, or preparation containing:	
	(i) Amobarbital	2128
	(ii) Secobarbital	2316
	(iii) Pentobarbital	2271
	or any salt thereof and one or more other active medicinal ingredients which are not listed in any schedule.	
(2)	Any suppository dosage form containing:	
	(i) Amobarbital	2126
	(ii) Secobarbital	2316
	(iii) Pentobarbital	2271
	or any salt of any of these drugs and approved by the Food and Drug Administration for marketing only as a suppository.	
(3)	Any substance which contains any quantity of a derivative of barbituric acid or any salt thereof	2100

(4) Chlorhexadol 2510
(5) Glutethimide 2550
(6) Lysergic acid 7300
(7) Lysergic acid amide 7310
(8) Methyprylon 2575
(9) Sulfondiethylmethane 2600
(10) Sulfonethylmethane 2605
(11) Sulfonmethane 2610
(12) Tiletamine and zolazepam or any salt thereof 7295
Some trade or other names for a tiletamine-zolazepam combination product: Telazol.
Some trade or other names for tiletamine: 2-(ethylamino)-2-(2-thienyl)-cyclohexanone.
Some trade or other names for zolazepam: 4-(2-fluorophenyl)-6,8-dihydro-1,3,8-trimethylpyrazolo-[3,4-*e*][1,4]-diazepin-7(1*H*)-one, flupyrazapon.

(d) *Nalorphine* 9400

(e) *Narcotic drugs.* Unless specifically excepted or unless listed in another schedule, any material, compound, mixture, or preparation containing any of the following narcotic drugs, or their salts calculated as the free anhydrous base or alkaloid, in limited quantities as set forth below:

(1) Not more than 1.8 grams of codeine per 100 milliliters or not more than 90 milligrams per dosage unit, with an equal or greater quantity of an isoquinoline alkaloid of opium 9803
(2) Not more than 1.8 grams of codeine per 100 milliliters or not more than 90 milligrams per dosage unit, with one or more active, nonnarcotic ingredients in recognized therapeutic amounts 9804
(3) Not more than 300 milligrams of dihydrocodeinone (hydrocodone) per 100 milliliters or not more than 15 milligrams per dosage unit, with a fourfold or greater quantity of an isoquinoline alkaloid of opium 9805
(4) Not more than 300 milligrams of dihydrocodeinone (hydrocodone) per 100 milliliters or not more than 15 milligrams per dosage unit, with one or more active nonnarcotic ingredients in recognized therapeutic amounts 9806
(5) Not more than 1.8 grams of dihydrocodeine per 100 milliliters or not more than 90 milligrams per dosage unit, with one or more active nonnarcotic ingredients in recognized therapeutic amounts 9807
(6) Not more than 300 milligrams of ethylmorphine per 100 milliliters or not more than 15 milligrams per dosage unit, with one or more active, nonnarcotic ingredients in recognized therapeutic amounts 9808
(7) Not more than 500 milligrams of opium per 100 milliliters or per 100 grams or not more than 25 milligrams per dosage unit, with one or more active, nonnarcotic ingredients in recognized therapeutic amounts 9809
(8) Not more than 50 milligrams of morphine per 100 milliliters or per 100 grams, with one or more active, nonnarcotic ingredients in recognized therapeutic amounts 9810

§ 1308.14 Schedule IV.

(a) Schedule IV shall consist of the drugs and other substances, by whatever official name, common or usual name, chemical name, or brand name designated, listed in this section. Each drug or substance has been assigned the DEA Controlled Substances Code Number set forth opposite it.

(b) *Narcotic drugs.* Unless specifically excepted or unless listed in another schedule, any material, compound, mixture, or preparation containing any of the following narcotic drugs, or their salts calculated as the free anhydrous base or alkaloid, in limited quantities as set forth below:

(1) Not more than 1 milligram of difenoxin (DEA Drug Code No. 9618) and not less than 25 micrograms of atropine sulfate per dosage unit.
(2) Dextropropoxyphene (alpha-(+)-4-dimethylamino-1,2-diphenyl-3-methyl-2-propionoxybutane) 9273

(c) *Depressants.* Unless specifically excepted or unless listed in another schedule, any material, compound, mixture, or preparation which contains any quantity of the following substances, including its salts, isomers, and salts of isomers whenever the existence of such salts, isomers, and salts of isomers is possible within the specific chemical designation:

(1) Alprazolam 2882
(2) Barbital 2145
(3) Bromazepam 2748
(4) Camazepam 2749
(5) Chloral betaine 2460
(6) Chloral hydrate 2465
(7) Chlordiazepoxide 2744
(8) Clobazam 2751
(9) Clonazepam 2737
(10) Clorazepate 2768
(11) Clotiazepam 2752
(12) Cloxazolam 2753
(13) Delorazepam 2754
(14) Diazepam 2765
(15) Estazolam 2756
(16) Ethchlorvynol 2540
(17) Ethinamate 2545
(18) Ethyl loflazepate 2758
(19) Fludiazepam 2759
(20) Flunitrazepam 2763
(21) Flurazepam 2767
(22) Halazepam 2762
(23) Haloxazolam 2771
(24) Ketazolam 2772
(25) Loprazolam 2773
(26) Lorazepam 2885
(27) Lormetazepam 2774
(28) Mebutamate 2800
(29) Medazepam 2836
(30) Meprobamate 2820
(31) Methohexital 2264
(32) Methylphenobarbital (mephobarbital) .. 2250
(33) Midazolam 2884
(34) Nimetazepam 2837
(35) Nitrazepam 2834
(36) Nordiazepam 2838
(37) Oxazepam 2835
(38) Oxazolam 2839
(39) Paraldehyde 2585
(40) Petrichloral 2591
(41) Phenobarbital 2285
(42) Pinazepam 2883
(43) Prazepam 2764
(44) Quazepam 2881
(45) Temazepam 2925
(46) Tetrazepam 2886
(47) Triazolam 2887

(d) *Fenfluramine.* Any material, compound, mixture, or preparation which contains any quantity of the following substances, including its salts, isomers (whether optical, position, or geometric), and salts of such isomers whenever the existence of such salts, isomers, and salts of isomers is possible:

(1)	Fenfluramine	1670

(e) *Stimulants.* Unless specifically excepted or unless listed in another schedule, any material, compound, mixture, or preparation which contains any quantity of the following substances having a stimulant effect on the central nervous system, including its salts, isomers and salts of isomers:

(1)	Cathine ((+)-norpseudoephedrine)	1230
(2)	Diethylpropion	1610
(3)	Fencamfamin	1760
(4)	Fenproporex	1575
(5)	Mazindol	1605
(6)	Mefenorex	1580
(7)	Pemoline (including organometallic complexes and chelates thereof)	1530
(8)	Phentermine	1640
(9)	Pipradrol	1750
(10)	SPA ((−)-1-dimethylamino-1,2-diphenylethane)	1635

(f) *Other substances.* Unless specifically excepted or unless listed in another schedule, any material, compound, mixture, or preparation which contains any quantity of the following substances, including its salts:

(1)	Pentazocine	9709

§ 1308.15 Schedule V.

(a) Schedule V shall consist of the drugs and other substances, by whatever official name, common or usual name, chemical name, or brand name designated, listed in this section.

(b) *Narcotic drugs.* Unless specifically excepted or unless listed in another schedule, any material, compound, mixture, or preparation containing any of the following narcotic drugs and their salts, as set forth below:

(1)	Buprenorphine	9064

(c) Narcotic drugs containing nonnarcotic active medicinal ingredients. Any compound, mixture, or preparation containing any of the following narcotic drugs, or their salts calculated as the free anhydrous base or alkaloid, in limited quantities as set forth below, which shall include one or more nonnarcotic active medicinal ingredients in sufficient proportion to confer upon the compound, mixture, or preparation valuable medicinal qualities other than those possessed by narcotic drugs alone:

(1) Not more than 200 milligrams of codeine per 100 milliliters or per 100 grams.

(2) Not more than 100 milligrams of dihydrocodeine per 100 milliliters or per 100 grams.

(3) Not more than 100 milligrams of ethylmorphine per 100 milliliters or per 100 grams.

(4) Not more than 2.5 milligrams of diphenoxylate and not less than 25 micrograms of atropine sulfate per dosage unit.

(5) Not more than 100 milligrams of opium per 100 milliliters or per 100 grams.

(6) Not more than 0.5 milligram of difenoxin (DEA Drug Code No. 9618) and not less than 25 micrograms of atropine sulfate per dosage unit.

(d) *Stimulants.* Unless specifically exempted or excluded or unless listed in another schedule, any material, compound, mixture, or preparation which contains any quantity of the following substances having a stimulant effect on the central nervous system, including its salts, isomers and salts of isomers:

(1)	Propylhexedrine	8161
(2)	Pyrovalerone	1485

Inspections

§ 1316.02 Definitions.

As used in this Subpart, the following terms shall have the meanings specified:

(c) The term "controlled premises" means—(1) Places where original or other records or documents required under the Act are kept or required to be kept, and

(2) Places, including factories, warehouses, or other establishments, and conveyances, where persons registered under the Act or exempted from registration under the Act may lawfully hold, manufacture, or distribute, dispense, administer, or otherwise dispose of controlled substances.

(e) The term "inspector" means an officer or employee of the Administration authorized by the Administrator to make inspections under the Act.

(f) The term "register" and "registration" refer to registration required and permitted by sections 303 and 1008 of the Act (21 U.S.C. 823 and 958).

§ 1316.03 Authority to make inspections.

In carrying out his functions under the Act, the Administrator, through his inspectors, is authorized in accordance with sections 510 and 1015 of the Act (21 U.S.C. 880 and 965) to enter controlled premises and conduct administrative inspections thereof, for the purpose of:

(a) Inspecting, copying, and verifying the correctness of records, reports, or other documents required to be kept or made under the Act and the regulations promulgated under the Act, including, but not limited to, inventory and other records required to be kept pursuant to Part 1304 of this chapter, order form records required to be kept pursuant to Part 1305 of this chapter, prescription and distribution records required to be kept pursuant to Part 1306 of this chapter, shipping records identifying the name of each carrier used and the date and quantity of each shipment, and storage records identifying the name of each warehouse used and the date and quantity of each storage;

(b) Inspecting within reasonable limits and in a reasonable manner all pertinent equipment, finished and unfinished controlled substances and other substances or materials, containers, and labeling found at the controlled premises relating to this Act;

(c) Making a physical inventory of all controlled substances on-hand at the premises;

(d) Collecting samples of controlled substances or precursors (in the event any samples are collected during an inspection, the inspector shall issue a receipt for such samples on DEA Form 84 to the owner, operator, or agent in charge of the premises);

(e) Checking of records and information on distribution of controlled substances by the registrant as they relate to total distribution of the registrant (i.e., has the distribution in controlled substances increased markedly within the past year, and if so why); and

(f) Except as provided in § 1316.04, all other things therein (including records, files, papers, processes, controls and facilities) appropriate for verification of the records, reports, documents referred to above or otherwise bearing on the provisions of the Act and the regulations thereunder.

§ 1316.04 Exclusion from inspection.

(a) Unless the owner, operator or agent in charge of the controlled premises so consents in writing, no inspection authorized by these regulations shall extend to:

(1) Financial data;
(2) Sales data other than shipping data; or
(3) Pricing data.

§ 1316.05 Entry.

An inspection shall be carried out by an inspector. Any such inspector, upon (a) stating his purpose and (b) presenting to the owner, operator or agent in charge of the premises to be inspected (1) appropriate credentials, and (2) written notice of his inspection authority under § 1314.06 of this chapter, and (c) receiving informed consent under § 1316.08 or through the use of administrative warrant issued under §§ 1316.09–1316.14, shall have the right to enter such premises and conduct inspections at reasonable times and in a reasonable manner.

§ 1316.06 Notice of inspection.

The notice of inspection (DEA (or DNB) Form 82) shall contain:

(a) The name and title of the owner, operator, or agent in charge of the controlled premises;
(b) The controlled premises name;
(c) The address of the controlled premises to be inspected;
(d) The date and time of the inspection;

(e) A statement that a notice of inspection is given pursuant to section 510 of the Act (21 U.S.C. 880);

(f) A reproduction of the pertinent parts of section 510 of the Act; and

(g) The signature of the inspector.

§ 1316.07 Requirement for administrative inspection warrant; exceptions.

In all cases where an inspection is contemplated, an administrative inspection warrant is required pursuant to section 510 of the Act (21 U.S.C. 880), except that such warrant shall not be required for establishments applying for initial registration under the Act, for the inspection of books and records pursuant to an administrative subpoena issued in accordance with section 506 of the Act (21 U.S.C. 876) nor for entries in administrative inspections (including seizures of property):

(a) With the consent of the owner, operator, or agent in charge of the controlled premises as set forth in § 1316.08;

(b) In situations presenting imminent danger to health or safety;

(c) In situations involving inspection of conveyances where there is reasonable cause to obtain a warrant;

(d) In any other exceptional or emergency circumstance or time or opportunity to apply for a warrant is lacking; or

(e) In any other situations where a warrant is not constitutionally required.

§ 1316.08 Consent to inspection.

(a) An administrative inspection warrant shall not be required if informed consent is obtained from the owner, operator, or agent in charge of the controlled premises to be inspected;

(b) Wherever possible, informed consent shall consist of a written statement signed by the owner, operator, or agent in charge of the premises to be inspected and witnessed by two persons. The written consent shall contain the following information:

(1) That he (the owner, operator, or agent in charge of the premises) has been informed of his constitutional right not to have an administrative inspection made without an administrative inspection warrant;

(2) That he has right to refuse to consent to such an inspection;

(3) That anything of an incriminating nature which may be found may be seized and used against him in a criminal prosecution;

(4) That he has been presented with a notice of inspection as set forth in § 1316.06;

(5) That the consent given by him is voluntary and without threats of any kind; and

(6) That he may withdraw his consent at any time during the course of inspection.

(c) The written consent shall be produced in duplicate and be distributed as follows:

(1) The original will be retained by the inspector; and

(2) The duplicate will be given to the person inspected.

§ 1316.11 Execution of warrants.

An administrative inspection warrant shall be executed and returned as required by, and any inventory or seizure made shall comply with the requirements of, section 510(d)(3) of the Act (21 U.S.C. 880(d)(3)). The inspection shall begin as soon as is practicable after the issuance of the administrative inspection warrant and shall be completed with reasonable promptness. The inspection shall be conducted during regular business hours and shall be completed in a reasonable manner.

§ 1316.12 Refusal to allow inspection with an administrative warrant.

If a registrant or any person subject to the Act refuses to permit execution of an administrative warrant or impedes the inspector in the execution of that warrant, he shall be advised that such refusal or action constitutes a violation of section 402(a)(6) of the Act (21 U.S.C. (a)(6)). If he persists and the circumstances warrant, he shall be arrested and the inspection shall commence or continue.

§ 1316.13 Frequency of administrative inspections.

Except where circumstances otherwise dictate, it is the intent of the Administration to inspect all manufacturers of controlled substances listed in Schedules I and II and distributors of controlled substances listed in Schedule I once each year; and to inspect all distributors of controlled substances listed in Schedules II through V and manufacturers of controlled substances listed in Schedules III through V once every 3 years.

⟨1076⟩ FEDERAL FOOD, DRUG, AND COSMETIC ACT REQUIREMENTS RELATING TO DRUGS FOR HUMAN USE

Selected portions of the Federal Food, Drug, and Cosmetic Act as it relates to the regulation of drugs for human use are presented here as a service to practitioners and students of pharmacy and medicine. The complete text of the Act can be found in Title 21 of the United States Code. The corresponding section number of the code appear in brackets after the section number of the Act.

In addition to federal requirements, statutes governing drugs and their quality have been enacted by various states. In many cases, state requirements parallel those of the federal law. However, this should not be assumed to be the case and individual state laws and requirements should be consulted also.

Publication of these sections in the United States Pharmacopeia is for purposes of information and does not impart to them any legal effect.

INDEX TO SELECTED PORTIONS OF THE FEDERAL FOOD, DRUG, AND COSMETIC ACT RELATED TO DRUGS FOR HUMAN USE PRESENTED HEREIN

failure to conform to good manufacturing practices
poisonous or deleterious container
unsafe color additive
(b) failure to comply with compendial standards
(c) failure to comply with purported strength, quality, or purity
(d) other substance mixed with or substituted therefor

§ 502 Misbranded Drugs
(a) false and misleading labeling
(b) name and place of manufacturer, packer, or distributor quantity statement
(c) conspicuousness of statements
(d) label statement for certain narcotic and hypnotic substances
(e) established name and quantity requirements
(f) directions for use and adequate warnings
(g) compendial packaging and labeling requirements
(h) packaging requirements for drugs subject to deterioration
(i) misleading containers and imitations
(j) dangerous to health as labeled
(k) uncertified insulin
(l) uncertified antibiotic
(m) nonconforming color additive
(n) advertising requirements
(o) unregistered establishment or failure to bear identification symbol
(p) failure to comply with Poison Prevention Packaging Act

§ 503 Exemptions from Labeling and Packaging Requirements; Sales Restrictions of Drug Samples
(a) exemptions for repacking
(b) (1) prescription drug classification
(b) (2) dispensed prescription drug labeling requirements
(b) (4) federal caution statement required
(c) (1) drug sample marketing prohibited
(c) (2) coupon marketing prohibited
(c) (3) hospital or charity resale restrictions
(d) (1) restrictions on distribution of drug samples
(d) (2) requirements for sampling by mail
(d) (3) other distribution requirements
(e) (1) wholesale distributor restrictions
(e) (2) state licensure requirements
(e) (3) distributor and wholesale distribution definitions

§ 505 New Drugs

§ 506 Certification of Drugs Containing Insulin

§ 507 Certification of Antibiotics

§ 508 Authority to Designate Official Names

§ 510 Registration Requirements
(g) nonapplicability to pharmacies and practitioners

§ 525 Recommendations for Investigations of Drugs for Rare Diseases or Conditions

§ 526 Designation of Drugs for Rare Diseases or Conditions

§ 527 Protection of Unpatented Drugs for Rare Diseases or Conditions

§ 528 Open Protocols for Investigations of Drugs for Rare Diseases or Conditions

§ 703 Records of Interstate Shipment

§ 704 Inspections

§ 707 Cooperation in Revision of USP and Development of Analytical Methods

Short Title

§ 1 This Act may be cited as the Federal Food, Drug, and Cosmetic Act.

Definitions

§ 201 [321] For the purposes of this Act—

(e) The term "person" includes individual, partnership, corporation, and association.

(g) (1) The term "drug" means (A) articles recognized in the official United States Pharmacopeia, official Homeopathic Pharmacopeia of the United States, or official National Formulary, or any supplement to any of them; and (B) articles intended for use in the diagnosis, cure, mitigation, treatment, or prevention of disease in man or other animals; and (C) articles (other than food) intended to affect the structure or any function of the body of man or other animals; and (D) articles intended for use as a component of any articles specified in clause (A), (B), or (C); but does not include devices or their components, parts, or accessories.

(2) The term "counterfeit drug" means a drug which, or the container or labeling of which, without authorization, bears the trademark, trade name, or other identifying mark, imprint, or device, or any likeness thereof, of a drug manufacturer, processor, packer, or distributor other than the person or persons who in fact manufactured, processed, packed, or distributed such drug and which thereby falsely purports or is represented to be the product of, or to have been packed or distributed by, such other drug manufacturer, processor, packer, or distributor.

(j) The term "official compendium" means the official United States Pharmacopeia, official Homeopathic Pharmacopeia of the United States, official National Formulary, or any supplement to any of them.

(k) The term "label" means a display of written, printed, or graphic matter upon the immediate container of any article; and a requirement made by or under authority of this Act that any word, statement, or other information appear on the label shall not be considered to be complied with unless such word, statement, or other information also appears on the outside container or wrapper, if any there be, of the retail package of such article, or is easily legible through the outside container or wrapper.

(l) The term "immediate container" does not include package liners.

(m) The term "labeling" means all labels and other written, printed, or graphic matter (1) upon any article or any of its containers or wrappers, or (2) accompanying such article.

(n) If an article is alleged to be misbranded because the labeling or advertising is misleading, then in determining whether the labeling or advertising is misleading there shall be taken into account (among other things) not only representations made or suggested by statement, word, design, device, or any combination thereof, but also the extent to which the labeling or advertising fails to reveal facts material in the light of such representations or material with respect to consequences which may result from the use of the article to which the labeling or advertising relates under the conditions of use prescribed in the labeling or advertising thereof or under such conditions of use as are customary or usual.

(o) The representation of a drug, in its labeling, as an antiseptic shall be considered to be a representation that it is a germicide, except in the case of a drug purporting to be, or represented as, an antiseptic for inhibitory use as a wet dressing, ointment, dusting powder, or such other use as involves prolonged contact with the body.

(p) The term "new drug" means—

(1) Any drug (except a new animal drug or an animal feed bearing or containing a new animal drug) the composition of which is such that such drug is not generally recognized, among experts qualified by scientific training and experience to evaluate the safety and effectiveness of drugs, as safe and effective for use under the conditions prescribed, recommended, or suggested in the labeling thereof, except that such a drug not so recognized shall not be deemed to be a "new drug" if at any time prior to the enactment of this Act it was subject to the Food and Drugs Act of June 30, 1906, as amended, and if at such time its labeling contained the same representations concerning the conditions of its use; or

(2) Any drug (except a new animal drug or an animal feed bearing or containing a new animal drug) the composition of which is such that such drug, as a result of investigations to determine its safety and effectiveness for use under such conditions, has become so recognized, but which has not, otherwise than in such investigations, been used to a material extent or for a material time under such conditions.

(t) (1) The term "color additive" means a material which—

(A) is a dye, pigment, or other substance made by a process of synthesis or similar artifice, or extracted, isolated, or otherwise derived, with or without intermediate or final change of identity, from a vegetable, animal, mineral, or other source, and

(B) when added or applied to a food, drug, or cosmetic, or

to the human body or any part thereof, is capable (alone or through reaction with another substance) of imparting color thereto:

except that such term does not include any material which the Secretary, by regulation, determines is used (or intended to be used) solely for a purpose or purposes other than coloring.

(2) The term "color" includes black, white, and intermediate grays.

(u) The term "safe," as used in paragraph(s) of this section and in sections 409, 512, and 706, has reference to the health of man or animal.

Prohibited Acts

§ **301** [331] The following acts and the causing thereof are hereby prohibited:

(a) The introduction or delivery for introduction into interstate commerce of any food, drug, device, or cosmetic that is adulterated or misbranded.

(b) The adulteration or misbranding of any food, drug, device, or cosmetic in interstate commerce.

(c) The receipt in interstate commerce of any food, drug, device, or cosmetic that is adulterated or misbranded, and the delivery or proffered delivery thereof for pay or otherwise.

(d) The introduction or delivery for introduction into interstate commerce of any article in violation of section 404 or 505.

(e) The refusal to permit access to or copying of any record as required by section 703; or the failure to establish or maintain any record, or make any report, required under section 505 (i) or (j), 507 (d) or (g), 512 (j), (l) or (m), 515 (f) or 519, or the refusal to permit access to or verification or copying of any such required record.

(f) The refusal to permit entry or inspection as authorized by section 704.

(g) The manufacture within any Territory of any food, drug, device, or cosmetic that is adulterated or misbranded.

(h) The giving of a guaranty or undertaking referred to in section 303 (c) (2), which guaranty or undertaking is false, except by a person who relied upon a guaranty or undertaking to the same effect signed by, and containing the name and address of, the person residing in the United States from whom he received in good faith the food, drug, device, or cosmetic; or the giving of a guaranty or undertaking referred to in section 303 (c) (3), which guaranty or undertaking is false.

(i) (1) Forging, counterfeiting, simulating, or falsely representing, or without proper authority using any mark, stamp, tag, label, or other identification device authorized or required by regulations promulgated under the provisions of section 404, 506, 507, or 706.

(2) Making, selling, disposing of, or keeping in possession, control, or custody, or concealing any punch, die, plate, stone, or other thing designed to print, imprint, or reproduce the trademark, trade name, or other identifying mark, imprint, or device of another or any likeness of any of the foregoing upon any drug or container or labeling thereof so as to render such drug a counterfeit drug.

(3) The doing of any act which causes a drug to be a counterfeit drug, or the sale or dispensing, or the holding for sale or dispensing, of a counterfeit drug.

(k) The alteration, mutilation, destruction, obliteration, or removal of the whole or any part of the labeling of, or the doing of any other act with respect to, a food, drug, device, or cosmetic, if such act is done while such article is held for sale (whether or not the first sale) after shipment in interstate commerce and results in such article being adulterated or misbranded.

(l) The using, on the labeling of any drug or device or in any advertising relating to such drug or device, of any representation or suggestion that approval of an application with respect to such drug or device is in effect under section 505, 515, or 520 (g), as the case may be, or that such drug or device complies with the provisions of such action.

(n) The using, in labeling, advertising or other sales promotion of any reference to any report or analysis furnished in compliance with scction 704.

(p) The failure to register in accordance with section 510, the failure to provide any information required by section 510 (j) or 510 (k), or the failure to provide a notice required by section 510 (j) (2).

(t) The importation of a drug in violation of section 801 (d) (1), the sale, purchase, or trade of a drug or drug sample or the offer to sell, purchase, or trade a drug or drug sample in violation of section 503 (c), the sale, purchase, or trade of a coupon, the offer to sell, purchase, or trade such a coupon, or the counterfeiting of such a coupon in violation of section 503 (c) (2), the distribution of a drug sample in violation of section 503 (d) or the failure to otherwise comply with the requirements of section 503 (d), or the distribution of drugs in violation of section 503 (e) or the failure to otherwise comply with the requirements of section 503 (e).

Penalties

§ **303** [333] (a) (1) Any person who violates a provision of section 301 shall be imprisoned for not more than one year or fined not more than $1,000, or both.

(2) Notwithstanding the provisions of paragraph (1) of this section, if any person commits such a violation after a conviction of him under this section has become final, or commits such a violation with the intent to defraud or mislead, such person shall be imprisoned for not more than three years or fined not more than $10,000 or both.

(b) (1) Notwithstanding subsection (a), any person who violates section 301 (t) because of an importation of a drug in violation of section 801 (d) (1), because of a sale, purchase, or trade of a drug or drug sample or the offer to sell, purchase, or trade a drug or drug sample in violation of section 503 (c), because of the sale, purchase, or trade of a coupon, the offer to sell, purchase, or trade such a coupon, or the counterfeiting of such a coupon in violation of section 503 (c) (2), or the distribution of drugs in violation of section 503 (e) (2) (A) shall be imprisoned for not more than 10 years or fined not more than $250,000, or both.

(2) Any manufacturer or distributor who distributes drug samples by means other than the mail or common carrier whose representative, during the course of the representative's employment or association with the manufacturer or distributor, violated section 301 (t) because of a violation of section 503 (c) (1) or violated any State law prohibiting the sale, purchase, or trade of a drug sample subject to section 503 (b) or the offer to sell, purchase, or trade such a drug sample shall, upon conviction of the representative for such violation, be subject to the following civil penalties:

(A) A civil penalty of not more than $50,000 for each of the first two such violations resulting in a conviction of any representative of the manufacturer or distributor in any 10-year period.

(B) A civil penalty of not more than $1,000,000 for each violation resulting in a conviction of any representative after the second conviction in any 10-year period.

For the purposes of this paragraph, multiple convictions of one or more persons arising out of the same event or transaction, or a related series of events or transactions, shall be considered as one violation.

(3) Any manufacturer or distributor who violates section 301 (t) because of a failure to make a report required by section 503 (d) (3) (E) shall be subject to a civil penalty of not more than $100,000.

(4) (A) If a manufacturer or distributor or any representative of such manufacturer or distributor provides information leading to the arrest and conviction of any representative of that manufacturer or distributor for a violation of section 301 (t) because of a sale, purchase, or trade or offer to purchase, sell, or trade a drug sample in violation of section 503 (c) (1) or for a violation of State law prohibiting the sale, purchase, or trade or offer to sell, purchase, or trade a drug sample, the conviction of such representative shall not be considered as a violation for purposes of paragraph (2).

(B) If, in an action brought under paragraph (2) against a manufacturer or distributor relating to the conviction of a representative of such manufacturer or distributor for the sale, purchase, or trade of a drug or the offer to sell, purchase, or trade a drug, it is shown, by clear and convicting evidence—

(i) that the manufacturer or distributor conducted, before the arrest of such representative for the violation which resulted in such conviction, an investigation of events or transactions which would have led to the reporting of information leading to the arrest and conviction of such representative for such purchase, sale, or trade or offer to purchase, sell, or trade, or

(ii) that, except in the case of the conviction of a representative employed in a supervisory function, despite diligent implementation by the manufacturer or distributor of an independent audit and security system designed to detect such a violation, the manufacturer or distributor could not reasonably have been expected to have detected such violation, the conviction of such representative shall not be considered as a conviction for purpose of paragraph (2).

(5) If a person provides information leading to the arrest and conviction of a person for a violation of section 301 (t) because of the sale, purchase, or trade of a drug sample or the offer to sell, purchase, or trade a drug sample in violation of section 503 (c) (1), such person shall be entitled to one-half of the criminal fine imposed and collected for such violation but not more than $125,000.

(c) No person shall be subject to the penalties of subsection (a) of this section, (1) for having received in interstate commerce any article and delivered it or proffered delivery of it, if such delivery or proffer was made in good faith, unless he refuses to furnish on request of an officer or employee duly designated by the Secretary the name and address of the person from whom he purchased or received such article and copies of all documents, if any there be, pertaining to the delivery of the article to him; or (2) for having violated section 301 (a) or (d), if he establishes a guaranty or undertaking signed by, and containing the name and address of, the person residing in the United States from whom he received in good faith the article, to the effect, in case of an alleged violation of section 301 (a), that such article is not adulterated or misbranded, within the meaning of this Act, designating this Act, or to the effect, in case of an alleged violation of section 301 (d), that such article is not an article which may not, under the provisions of section 404 or 505, be introduced into interstate commerce; or (3) for having violated section 301 (a), where the violation exists because the article is adulterated by reason of containing a color additive not from a batch certified in accordance with regulations promulgated by the Secretary under this Act, if such person establishes a guaranty or undertaking signed by, and containing the name and address of, the manufacturer of the color additive, to the effect that such color additive was from a batch certified in accordance with the applicable regulations promulgated by the Secretary under this Act; or (4) for having violated section 301 (b), (c), or (k) by failure to comply with section 502 (f) in respect to an article received in interstate commerce to which neither section 503 (a) nor section 503 (b) (1) is applicable if the delivery or proffered delivery was made in good faith and the labeling at the time thereof contained the same directions for use and warning statements as were contained in the labeling at the time of such receipt of such article; or (5) for having violated section 301 (i) (2) if such person acted in good faith and had no reason to believe that use of the punch, die, plate, stone, or other thing involved would result in a drug being a counterfeit drug, or for having violated section 301 (i) (3) if the person doing the act or causing it to be done acted in good faith and had no reason to believe that the drug was a counterfeit drug.

Adulterated Drugs

§ **501** [351] A drug or device shall be deemed to be adulterated—

(a) (1) if it consists in whole or in part of any filthy, putrid, or decomposed substance; or (2) (A) if it has been prepared, packed, or held under insanitary conditions whereby it may have been contaminated with filth, or whereby it may have been rendered injurious to health; or (B) if it is a drug and the methods used in, or the facilities or controls used for, its manufacture, processing, packing, or holding do not conform to or are not operated or administered in conformity with current good manufacturing practice to assure that such drug meets the requirements of this Act as to safety and has the identity and strength, and meets the quality and purity characteristics, which it purports or is represented to possess; or (3) if its container is composed, in whole or in part, of any poisonous or deleterious substance which may render the contents injurious to health; or (4) if (A) it bears or contains, for purposes of coloring only, a color additive which is unsafe within the meaning of section 706 (a), or (B) it is a color additive the intended use of which in or on drugs or devices is for purposes of coloring only and is unsafe within the meaning of section 706 (a);

(b) If it purports to be or is represented as a drug the name of which is recognized in an official compendium, and its strength differs from, or its quality or purity falls below, the standards set forth in such compendium. Such determination as to strength, quality, or purity shall be made in accordance with the tests or methods of assay set forth in such compendium, except that whenever tests or methods of assays have not been prescribed in such compendium, or such tests or methods of assay as are prescribed are, in the judgment of the Secretary, insufficient for the making of such determination, the Secretary shall bring such fact to the attention of the appropriate body charged with the revision of such compendium, and if such body fails within a reasonable time to prescribe tests or methods of assay which, in the judgment of the Secretary, are sufficient for purposes of this paragraph, then the Secretary shall promulgate regulations prescribing appropriate tests or methods of assay in accordance with which such determination as to strength, quality, or purity shall be made. No drug defined in an official compendium shall be deemed to be adulterated under this paragraph because it differs from the standard of strength, quality, or purity therefor set forth in such compendium, if its difference in strength, quality, or purity from such standards is plainly stated on its label. Whenever a drug is recognized in both the United States Pharmacopeia and the Homeopathic Pharmacopeia of the United States it shall be subject to the requirements of the United States Pharmacopeia unless it is labeled and offered for sale as a homeopathic drug, in which case it shall be subject to the provisions of the Homeopathic Pharmacopeia of the United States and not to those of the United States Pharmacopeia.

(c) If it is not subject to the provisions of paragraph (b) of this section and its strength differs from, or its purity or quality falls below, that which it purports or is represented to possess.

(d) If it is a drug and any substance has been (1) mixed or packed therewith so as to reduce its quality or strength or (2) substituted wholly or in part therefor.

Misbranded Drugs

§ **502** [352] A drug or device shall be deemed to be misbranded—

(a) If its labeling is false or misleading in any particular.

(b) If in a package form unless it bears a label containing (1) the name and place of business of the manufacturer, packer, or distributor; and (2) an accurate statement of the quantity of the contents in terms of weight, measure, or numerical count: *Provided*, That under clause (2) of this paragraph reasonable variations shall be permitted, and exemptions as to small packages shall be established, by regulations prescribed by the Secretary.

(c) If any word, statement, or other information required by or under authority of this Act to appear on the label or labeling is not prominently placed thereon with such conspicuousness (as compared with other words, statements, designs, or devices, in the labeling) and in such terms as to render it likely to be read and understood by the ordinary individual under customary conditions of purchase and use.

(d) If it is for use by man and contains any quantity of the narcotic or hypnotic substance alpha-eucaine, barbituric acid, beta-eucaine, bromal, cannabis, carbromal, chloral, coca, cocaine, codeine, heroin, marijuana, morphine, opium, paraldehyde, peyote, or sulfonmethane; or any chemical derivative of such substance, which derivative has been by the Secretary, after investigation, found to be, and by regulations designated as, habit forming; unless its label bears the name, and quantity or proportion of such substance or derivative and in juxtaposition therewith the statement "Warning—May be habit forming."

(e) (1) If it is a drug, unless (A) its label bears, to the exclusion of any other nonproprietary name (except the applicable systematic chemical name or the chemical formula), (i) the established name (as defined in subparagraph (3)) of the drug, if such there be, and (ii) in case it is fabricated from two or more ingredients, the established name and quantity of each active ingredient, including the quantity, kind, and proportion of any alcohol, and also including whether active or not, the established name and quantity or proportion of any bromides, ether, chloroform, acetanilide, acetophenetidin, amidopyrine, antipyrine, atropine, hyoscine, hyoscyamine, arsenic, digitalis, digitalis glucosides, mercury, ouabain, strophanthin, strychnine, thyroid, or any derivative or preparation of any such substances, contained therein: *Provided*, That the requirement for stating the quantity of the active ingredients, other than the quantity of those specifically named in this paragraph, shall apply only to prescription drugs; and (B) for any prescription drug the established name of such drug or ingredient, as the case may be, on such label (and on any labeling on which a name for such drug or ingredient is used) is printed prominently and in type at least half as large as that used thereon for any proprietary name or designation for such drug or ingredient: and *Provided*, That to the extent that compliance with the requirements of clause (A) (ii) or clause (B) of this subparagraph is impracticable, exemptions shall be established by regulations promulgated by the Secretary.

(3) As used in paragraph (l) the term "established name," with respect to a drug or ingredient thereof, means (A) the applicable official name designated pursuant to section 508, or (B) if there is no such name and such drug, or such ingredient, is an article recognized in an official compendium, then the official title thereof in such compendium or (C) if neither clause (A) nor clause (B) of this subparagraph applies, then the common or usual name, if any, of such drug or of such ingredient: *Provided further,* That where clause (B) of this subparagraph applies to an article recognized in the United States Pharmacopeia and in the Homeopathic Pharmacopeia under different official titles, the official title used in the United States Pharmacopeia shall apply unless it is labeled and offered for sale as a homeopathic drug, in which case the official title used in the Homeopathic Pharmacopeia shall apply.

(f) Unless its labeling bears (1) adequate directions for use; and (2) such adequate warnings against use in those pathological conditions or by children where its use may be dangerous to health, or against unsafe dosage or methods or duration of administration or application, in such manner and form, as are necessary for the protection of users: *Provided,* That where any requirement of clause (1) of this paragraph, as applied to any drug or device, is not necessary for the protection of the public health, the Secretary shall promulgate regulations exempting such drug or device from such requirement.

(g) If it purports to be a drug the name of which is recognized in an official compendium, unless it is packaged and labeled as prescribed therein: *Provided*, That the method of packing may be modified with the consent of the Secretary. Whenever a drug is recognized in both the United States Pharmacopeia and the Homeopathic Pharmacopeia of the United States, it shall be subject to the requirements of the United States Pharmacopeia with respect to packaging, and labeling unless it is labeled and offered for sale as a homeopathic drug, in which case it shall be subject to the provisions of the Homeopathic Pharmacopeia of the United States, and not to those of the United States Pharmacopeia: *Provided further*, That, in the event of inconsistency between the requirements of this paragraph and those of paragraph (e) as to the name by which the drug or its ingredients shall be designated, the requirements of paragraph (e) shall prevail.

(h) If it has been found by the Secretary to be a drug liable to deterioration, unless it is packaged in such form and manner, and its label bears a statement of such precautions, as the Secretary shall by regulations require as necessary for the protection of the public health. No such regulation shall be established for any drug recognized in an official compendium until the Secretary shall have informed the appropriate body charged with the revision of such compendium of the need for such packaging or labeling requirements and such body shall have failed within a reasonable time to prescribe such requirements.

(i) (1) If it is a drug and its container is so made, formed, or filled as to be misleading; or (2) if it is an imitation of another drug; or (3) if it is offered for sale under the name of another drug.

(j) If it is dangerous to health when used in the dosage or manner, or with the frequency or duration prescribed, recommended, or suggested in the labeling thereof.

(k) If it is, or purports to be, or is represented as a drug composed wholly or partly of insulin, unless (1) it is from a batch with respect to which a certificate or release has been issued pursuant to section 506, and (2) such certificate or release is in effect with respect to such drug.

(l) If it is, or purports to be, or is represented as a drug (except a drug for use in animals other than man) composed wholly or partly of any kind of penicillin, streptomycin, chlortetracycline, chloramphenicol, bacitracin, or any other antibiotic drug, or any derivative thereof, unless (1) it is from a batch with respect to which a certificate or release has been issued pursuant to section 507, and (2) such certificate or release is in effect with respect to such drug; *Provided*, That this paragraph shall not apply to any drug or class of drugs exempted by regulations promulgated under section 507 (c) or (d).

(m) If it is a color additive the intended use of which is for the purpose of coloring only, unless its packaging and labeling are in conformity with such packaging and labeling requirements applicable to such color additive, as may be contained in regulations issued under section 706.

(n) In the case of any prescription drug distributed or offered for sale in any State, unless the manufacturer, packer, or distributor thereof includes in all advertisements and other descriptive printed matter issued or caused to be issued by the manufacturer, packer, or distributor with respect to that drug a true statement of (1) the established name as defined in section 502 (e), printed prominently and in type at least half as large as that used for any trade or brand name thereof, (2) the formula showing quantitatively each ingredient of such drug to the extent required for labels under section 502 (e), and (3) such other information in brief summary relating to side effects, contraindications, and effectiveness as shall be required in regulations which shall be issued by the Secretary in accordance with the procedure specified in section 701 (e) of this Act: *Provided*, That (A) except in extraordinary circumstances, no regulation issued under this paragraph shall require prior approval by the Secretary of the content of any advertisement, and (B) no advertisement of a prescription drug, published after the effective date of regulations issued under this paragraph applicable to advertisements of prescription drugs, shall, with respect to the matters specified in this paragraph or covered by such regulations, be subject to the provisions of sections 12 through 17 of the Federal Trade Commission Act, as amended (15 U.S.C. 52–57). This paragraph (n) shall not be applicable to any printed matter which the Secretary determines to be labeling as defined in section 201 (m) of this Act. Nothing in the Convention on Psychotropic Substances, signed at Vienna, Austria, on February 21, 1971, shall be construed to prevent drug price communications to consumers.

(o) If it was manufactured, prepared, propagated, compounded, or processed in an establishment in any State not duly registered under section 510, if it was not included in a list required by section 510 (j), if a notice or other information respecting it was not provided as required by such section or section 510 (k), or if it does not bear such symbols from the uniform system for identification of devices prescribed under section 510 (e) as the Secretary by regulation requires.

(p) If it is a drug and its packaging or labeling is in violation of an applicable regulation issued pursuant to section 3 or 4 of the Poison Prevention Packing Act of 1970.

Exemptions in Case of Drugs

§ **503** [353] (a) The Secretary is hereby directed to promulgate regulations exempting from any labeling or packaging requirement of this Act drugs and devices which are, in accordance with the practice of the trade, to be processed, labeled, or repacked in substantial quantities at establishments other than those where originally processed or packed, on condition that such drugs and devices are not adulterated or misbranded under the provisions of this Act upon removal from such processing, labeling, or repacking establishment.

(b) (1) A drug intended for use by man which—

(A) is a habit-forming drug to which section 502 (d) applies; or

(B) because of its toxicity or other potentiality for harmful effect, or the method of its use, or the collateral measures necessary to its use, is not safe for use except under the supervision of a practitioner licensed by law to administer such drug; or

(C) is limited by an approved application under section 505 to use under the professional supervision of a practitioner licensed by law to administer such drug;

shall be dispensed only (i) upon a written prescription of a practitioner licensed by law to administer such drug, or (ii) upon an oral prescription of such practitioner which is reduced promptly to writing and filed by the pharmacist, or (iii) by refilling any such written or oral prescription if such refilling is authorized by the prescriber either in the original prescription or by oral order which is reduced promptly to writing and filed by the pharmacist. The act of dispensing a drug contrary to the provisions of this paragraph shall be deemed to be an act which results in the drug being misbranded while held for sale.

(2) Any drug dispensed by filling or refilling a written or oral prescription of a practitioner licensed by law to administer such drug shall be exempt from the requirements of section 502, except paragraphs (a), (i) (2) and (3), (k), and (l), and the packaging requirements of paragraphs (g), (h), and (p), if the drug bears a label containing the name and address of the dispenser, the serial number and date of the prescription or of its filling, the name of the prescriber, and, if stated in the prescription, the name of the patient, and the directions for use and cautionary statements, if any, contained in such prescription. This exemption shall not apply to any drug dispensed in the course of the conduct of a business of dispensing drugs pursuant to diagnosis by mail, or to a drug dispensed in violation of paragraph (1) of this subsection.

(3) The Secretary may by regulation remove drugs subject to section 502 (d) and section 505 from the requirements of paragraph (1) of this subsection when such requirements are not necessary for the protection of the public health.

(4) A drug which is subject to paragraph (1) of this subsection shall be deemed to be misbranded if at any time prior to dispensing its label fails to bear the statement "Caution: Federal law prohibits dispensing without prescription." A drug to which paragraph (1) of this subsection does not apply shall be deemed to be misbranded if at any time prior to dispensing its label bears the caution statement quoted in the preceding sentence.

(5) Nothing in this subsection shall be construed to relieve any person from any requirement prescribed by or under authority of law with respect to drugs now included or which may hereafter be included within the classifications stated in section 3220 of the Internal Revenue Code (26 U.S.C. 3220), or to marijuana as defined in section 3238 (b) of the Internal Revenue Code (26 U.S.C. 3238 (b)).

(c) (1) No person may sell, purchase, or trade or offer to sell, purchase, or trade any drug sample. For purpose of this paragraph and subsection (d), the term 'drug sample' means a unit of a drug, subject to subsection (b), which is not intended to be sold and is intended to promote the sale of the drug. Nothing in this paragraph shall subject an officer or executive of a drug manufacturer or distributor to criminal liability solely because of a sale, purchase, trade, or offer to sell, purchase, or trade in violation of this paragraph by other employees of the manufacturer or distributor.

(2) No person may sell, purchase, or trade, offer to sell, purchase, or trade, or counterfeit any coupon. For purposes of this paragraph, the term 'coupon' means a form which may be redeemed, at no cost or at a reducec cost, for a drug which is prescribed in accordance with section 503 (b).

(3) (A) No person may sell, purchase, or trade, or offer to sell, purchase, or trade, any drug—

(i) which is subject to subsection (b), and

(ii) (I) which was purchased by a public or private hospital or other health care entity, or

(II) which was donated or supplied at a reduced price to a charitable organization described in section 501 (c) (3) of the Internal Revenue Code of 1954.

(B) Subparagraph (A) does not apply to—

(i) the purchase or other acquisition by a hospital or other health care entity which is a member of a group purchasing organization of a drug for its own use from the group purchasing organization or from other hospitals or health care entities which are members of such organization,

(ii) the sale, purchase, or trade of a drug or an offer to sell, purchase, or trade a drug by an organization described in subparagraph (A) (ii) (II) to a nonprofit affiliate of the organization to the extent otherwise permitted by law,

(iii) a sale, purchase, or trade of a drug or an offer to sell, purchase, or trade a drug among hospitals or other health care entities which are under common control,

(iv) a sale, purchase, or trade of a drug or an offer to sell, purchase, or trade a drug for emergency medical reasons, or

(v) a sale, purchase, or trade of a drug, an offer to sell, purchase, or trade a drug, or the dispensing of a drug pursuant to a prescription executed in accordance with section 503 (b).

For purposes of this paragraph, the term 'entity' does not include a wholesale distributor of drugs or a retail pharmacy licensed under State law and the term 'emergency medical reasons' includes transfers of a drug between health care entities or from a health care entity to a retail pharmacy undertaken to alleviate temporary shortages of the drug arising from delays in or interruptions of regular distribution schedules.

(d) (1) Except as provided in paragraphs (2) and (3), no representative of drug manufacturer or distributor may distribute any drug sample.

(2) (A) The manufacturer or distributor of a drug subject to subsection (b) may, in accordance with this paragraph, distribute drug samples by mail or common carrier to practitioners licensed to prescribe such drugs, or, at the request of a licensed practitioner, to pharmacies of hospitals or other health care entities. Such a distribution of drug samples may only be made—

(i) in response to a written request for drug samples made on a form which meets the requirements of subparagraph (B), and

(ii) under a system which requires the recipient of the drug sample to execute a written receipt for the drug sample upon its delivery and the return of the receipt to the manufacturer or distributor.

(B) A written request for a drug sample required by subparagraph (A) (i) shall contain—

(i) the name, address, professional designation, and signature of the practitioner making the request,

(ii) the identity of the drug sample requested and the quantity requested,

(iii) the name of the manufacturer of the drug sample requested, and

(iv) the date of the request.

(C) Each drug manufacturer or distributor which makes distributions by mail or common carrier under this paragraph shall maintain, for a period of 3 years, the request forms submitted for such distributions and shall maintain a record of distributions of drug samples which identifies the drugs distributed and the recipients of the distributions. Forms, receipts, and records required to be maintained under this subparagraph shall be made available by the drug manufacturer or distributor to Federal and State officials engaged in the regulation of drugs and in the enforcement of laws applicable to drugs.

(3) The manufacturer or distributor of a drug subject to subsection (b) may, by means other than mail or common carrier, distribute drug samples only if the manufacturer or distributor makes the distributions in accordance with subparagraph (A) and carries out the activities described in subparagraphs (B) through (F) as follows:

(A) Drug samples may only be distributed—

(i) to practitioners licensed to prescribe such drugs if they make a written request for the drug samples, or

(ii) at the written request of such a licensed practitioner, to pharmacies of hospitals or other health care entities.

A written request for drug samples shall be made on a form which contains the practitioner's name, address, and professional designation, the identity of the drug sample requested, the quantity of drug samples requested, the name of the manufacturer or distributor of the drug sample, the date of the request and signature of the practitioner making the request.

(B) Drug manufacturers or distributors shall store drug samples under conditions that will maintain their stability, integrity, and effectiveness and will assure that the drug samples will be free of contamination, deterioration, and adulteration.

(C) Drug manufacturers or distributors shall conduct, at least annually, a complete and accurate inventory of all drug samples in the possession of representatives of the manufacturer or distributor. Drug manufacturers or distributors shall maintain lists of the names and addresses of each of their representatives who distribute drug samples and of the sites where drug samples are stored. Drug manufacturers or distributors shall maintain records for at least 3 years of all drug samples distributed, destroyed, or returned to the manufacturer or distributor, of all inventories maintained under this subparagraph, of all thefts or significant losses of drug samples, and of all requests made under subparagraph (A) for drug samples. Records and lists maintained under this subparagraph shall be made available by the drug manufacturer or distributor to the Secretary upon request.

(D) Drug manufacturers or distributors shall notify the

Secretary of any significant loss of drug samples and any known theft of drug samples.

(E) Drug manufacturers or distributors shall report to the Secretary any conviction of their representatives for violations of section 503 (c) (1) or a State law because of the sale, purchase, or trade of a drug sample or the offer to sell, purchase, or trade a drug sample.

(F) Drug manufacturers or distributors shall provide to the Secretary the name and telephone number of the individual responsible for responding to a request for information respecting drug samples.

(e) (1) Each person who is engaged in the wholesale distribution of drugs subject to subsection (b) and who is not an authorized distributor of record of such drugs shall provide to each wholesale distributor of such drugs a statement identifying each sale of the drug (including the date of the sale) before the sale to such wholesale distributor. Each manufacturer shall maintain at its corporate offices a curent list of such authorized distributors.

(2) (A) No person may engage in the wholesale distribution in interstate commerce of drugs subject to subsection (b) in a State unless such person is licensed by the State in accordance with the guidelines issued under subparagraph (B).

(B) The Secretary shall by regulation issue guidelines establishing minimum standards, terms, and conditions for the licensing of persons to make wholesale distributions in interstate commerce of drugs subject to subsection (b). Such guidelines shall prescribe requirements for the storage and handling of such drugs and for the establishment and maintenance of records of the distributions of such drugs.

(3) For the purposes of this subsection—

(A) the term 'authorized distributors of record' means those distributors with whom a manufacturer has established an ongoing relationship to distribute such manufacturer's products, and

(B) the term 'wholesale distribution' means distribution of drugs subject to subsection (b) to other than the consumer or patient but does not include intracompany sales and does not include distributions of drugs described in subsection (c) (3) (B).

New Drugs

§ **505** [355] (a) No person shall introduce or deliver for introduction into interstate commerce any new drug, unless an approval of an application filed pursuant to subsection (b) or (j) is effective with respect to such drug.

(b) (1) Any person may file with the Secretary an application with respect to any drug subject to the provisions of subsection (a). Such persons shall submit to the Secretary as a part of the application (A) full reports of investigations which have been made to show whether or not such drug is safe for use and whether such drug is effective in use; (B) a full list of the articles used as components of such drug; (C) a full statement of the composition of such drug; (D) a full description of the methods used in, and the facilities and controls used for, the manufacture, processing, and packing of such drug; (E) such samples of such drug and of the articles used as components thereof as the Secretary may require; and (F) specimens of the labeling proposed to be used for such drug. The applicant shall file with the application the patent number and the expiration date of any patent which claims the drug for which the applicant submitted the application or which claims a method of using such drug and with respect to which a claim of patent infringement could reasonably be asserted if a person not licensed by the owner engaged in the manufacture, use, or sale of the drug. If an application is filed under this subsection for a drug and a patent which claims such drug or a method of using such drug is issued after the filing date but before approval of the application, the applicant shall amend the application to include the information required by the preceding sentence. Upon approval of the application, the Secretary shall publish information submitted under the two preceding sentences.

(2) An application submitted under paragraph (1) for a drug for which the investigations described in clause (A) of such paragraph and relied upon by the applicant for approval of the application were not conducted by or for the applicant and for which the applicant has not obtained a right of reference or use from the person by or for whom the investigations were conducted shall also include—

(A) a certification, in the opinion of the applicant and to the best of his knowledge, with respect to each patent which claims the drug for which such investigations were conducted or which claims a use for such drug for which the applicant is seeking approval under this subsection and for which information is required to be filed under paragraph (1) or subsection (c)—

(i) that such patent information has not been filed,

(ii) that such patent has expired,

(iii) of the date on which such patent will expire, or

(iv) that such patent is invalid or will not be infringed by the manufacture, use, or sale of the new drug for which the application is submitted; and

(B) if with respect to the drug for which investigations described in paragraph (1) (A) were conducted information was filed under paragraph (1) or subsection (c) for a method of use patent which does not claim a use for which the applicant is seeking approval under this subsection, a statement that the method of use patent does not claim such a use.

(3) (A) An applicant who makes a certification described in paragraph (2) (A) (iv) shall include in the application a statement that the applicant will give the notice required by subparagraph (B) to—

(i) each owner of the patent which is the subject of the certification or the representative of such owner designated to receive such notice, and

(ii) the holder of the approved application under subsection (b) for the drug which is claimed by the patent or a use of which is claimed by the patent or the representative of such holder designated to receive such notice.

(B) The notice referred to in subparagraph (A) shall state that an application has been submitted under this subsection for the drug with respect to which the certification is made to obtain approval to engage in the commercial manufacture, use, or sale of the drug before the expiration of the patent referred to in the certification. Such notice shall include a detailed statement of the factual and legal basis of the applicant's opinion that the patent is not valid or will not be infringed.

(C) If an application is amended to include a certification described in paragraph (2) (A) (iv), the notice required by subparagraph (B) shall be given when the amended application is submitted.

(c)(1) Within one hundred and eighty days after the filing of an application under subsection (b), or such additional period as may be agreed upon by the Secretary and the applicant, the Secretary shall either—

(A) approve the application if he then finds that none of the grounds for denying approval specified in subsection (d) applies, or

(B) give the applicant notice of an opportunity for a hearing before the Secretary under subsection (d) on the question whether such application is approvable. If the applicant elects to accept the opportunity for hearing by written request within thirty days after such notice, such hearing shall commence not more than ninety days after the expiration of such thirty days unless the Secretary and the applicant otherwise agree. Any such hearing shall thereafter be conducted on an expedited basis and the Secretary's order thereon shall be issued within ninety days after the date fixed by the Secretary for filing final briefs.

(2) If the patent information described in subsection (b) could not be filed with the submission of an application under subsection (b) because the application was filed before the patent information was required under subsection (b) or a patent was issued after the application was approved under such subsection, the holder of an approved application shall file with the Secretary the patent number and the expiration date of any patent which claims the drug for which the application was submitted or which claims a method of using such drug and with respect to which a claim of patent infringement could reasonably be asserted if a person not licensed by the owner engaged in the manufacture, use, or sale of the drug. If the holder of an approved application could not file patent information under subsection (b) because it was not required at the time the application was approved, the holder shall file such information under this subsection not later than thirty days after the date of the enactment of this sentence, and if the holder of an approved application could not file patent information under subsection (b) because no patent had been issued when an application was filed or approved, the holder shall file such information under this subsection not later than thirty

days after the date the patent involved is issued. Upon the submission of patent information under this subsection, the Secretary shall publish it.

(3) The approval of an application filed under subsection (b) which contains a certification required by paragraph (2) of such subsection shall be made effective on the last applicable date determined under the following:

(A) If the applicant only made a certification described in clause (i) or (ii) of subsection (b) (2) (A) or in both such clauses, the approval may be made effective immediately.

(B) If the applicant made a certification described in clause (iii) of subsection (b) (2) (A), the approval may be made effective on the date certified under clause (iii).

(C) If the applicant made a certification described in clause (iv) of subsection (b) (2) (A), the approval shall be made effective immediately unless an action is brought for infringement of a patent which is the subject of the certification before the expiration of forty-five days from the date the notice provided under paragraph (3) (B) is received. If such an action is brought before the expiration of such days, the approval may be made effective upon the expiration of the thirty-month period beginning on the date of the receipt of the notice provided under paragraph (3) (B) or such shorter or longer period as the court may order because either party to the action failed to reasonably cooperate in expediting the action, except that—

(i) if before the expiration of such period the court decides that such patent is invalid or not infringed, the approval may be made effective on the date of the court decision,

(ii) if before the expiration of such period the court decides that such patent has been infringed, the approval may be made effective on such date as the court orders under section 271 (e) (4) (A) of title 35, United States Code, or

(iii) if before the expiration of such period the court grants a preliminary injunction prohibiting the applicant from engaging in the commercial manufacture or sale of the drug until the court decides the issues of patent validity and infringement and if the court decides that such patent is invalid or not infringed, the approval shall be made effective on the date of such court decision.

In such an action, each of the parties shall reasonably cooperate in expediting the action. Until the expiration of forty-five days from the date the notice made under paragraph (3) (B) is received, no action may be brought under section 2201 of title 28, United States Code, for a declaratory judgment with respect to the patent. Any action brought under such section 2201 shall be brought in the judicial district where the defendent has its principal place of business or a regular and established place of business.

(D) (i) If an application (other than an abbreviated new drug application) submitted under subsection (b) for a drug, no active ingredient (including any ester or salt of the active ingredient) of which has been approved in any other application under subsection (b), was approved during the period beginning January 1, 1982, and ending on the date of the enactment of this subsection, the Secretary may not make the approval of another application for a drug for which the investigations described in clause (A) of subsection (b) (1) and relied upon by the applicant for approval of the application were not conducted by or for the applicant and for which the applicant has not obtained a right of reference or use from the person by or for whom the investigations were conducted effective before the expiration of ten years from the date of the approval of the application previously approved under subsection (b).

(ii) If an application submitted under subsection (b) for a drug, no active ingredient (including any ester or salt of the active ingredient) of which has been approved in any other application under subsection (b), is approved after the date of the enactment of this clause, no application which refers to the drug for which the subsection (b) application was submitted and for which the investigations described in clause (A) of subsection (b) (1) and relied upon by the applicant for approval of the application were not conducted by or for the applicant and for which the applicant has not obtained a right of reference or use from the person by or for whom the investigations were conducted may be submitted under subsection (b) before the expiration of five years from the date of the approval of the application under subsection (b), except that such an application may be submitted under subsection (b) after the expiration of four years from the date of the approval of the subsection (b) application if it contains a certification of patent invalidity or noninfringement described in clause (iv) of subsection (b) (2) (A). The approval of such an application shall be made effective in accordance with this paragraph except that, if an action for patent infringement is commenced during the one-year period beginning forty-eight months after the date of the approval of the subsection (b) application, the thirty-month period referred to in subparagraph (C) shall be extended by such amount of time (if any) which is required for seven and one-half years to have elapsed from the date of approval of the subsection (b) application.

(iii) If an application submitted under subsection (b) for a drug, which includes an active ingredient (including any ester or salt of the active ingredient) that has been approved in another application approved under subsection (b), is approved after the date of the enactment of this clause and if such application contains reports of new clinical investigations (other than bioavailability studies) essential to the approval of the application and conducted or sponsored by the applicant, the Secretary may not make the approval of an application submitted under subsection (b) for the conditions of approval of such drug in the approved subsection (b) application effective before the expiration of three years from the date of the approval of the application under subsection (b) if the investigations described in clause (A) of subsection (b) (1) and relied upon by the applicant for approval of the application were not conducted by or for the applicant and if the applicant has not obtained a right of reference or use from the person by or for whom the investigations were conducted.

(iv) If a supplement to an application approved under subsection (b) is approved after the date of enactment of this clause and the supplement contains reports of new clinical investigations (other than bioavailability studies) essential to the approval of the supplement and conducted or sponsored by the person submitting the supplement, the Secretary may not make the approval of an application submitted under subsection (b) for a change approved in the supplement effective before the expiration of three years from the date of the approval of the supplement under subsection (b) if the investigations described in clause (A) of subsection (b) (1) and relied upon by the applicant for approval of the application were not conducted by or for the applicant and if the applicant has not obtained a right of reference or use from the person by or for whom the investigations were conducted.

(v) If an application (or supplement to an application) submitted under subsection (b) for a drug, which includes an active ingredient (including any ester or salt of the active ingredient) that has been approved in another application under subsection (b), was approved during the period beginning January 1, 1982, and ending on the date of the enactment of this clause, the Secretary may not make the approval of an application submitted under this subsection and for which the investigations described in clause (A) of subsection (b) (1) and relied upon by the applicant for approval of the application were not conducted by or for the applicant and for which the applicant has not obtained a right of reference or use from the person by or for whom the investigations were conducted and which refers to the drug for which the subsection (b) application was submitted effective before the expiration of two years from the date of enactment of this clause.

(d) If the Secretary finds, after due notice to the applicant in accordance with subsection (c) and giving him an opportunity for a hearing, in accordance with said subsection, that (1) the investigations, reports of which are required to be submitted to the Secretary pursuant to subsection (b), do not include adequate tests by all methods reasonably applicable to show whether or not such drug is safe for use under the conditions prescribed, recommended, or suggested in the proposed labeling thereof; (2) the results of such tests show that such drug is unsafe for use under such conditions or do not show that such drug is safe for use under such conditions; (3) the methods used in, and the facilities and controls used for, the manufacture, processing, and packing of such drug are inadequate to preserve its identity, strength, quality, and purity; (4) upon the basis of the information submitted to him as part of the application, or upon the basis of any other information before him with respect to such drug, he has insufficient information to determine whether such drug is safe for use under such conditions; or (5) evaluated on the basis of the information submitted to him as part of the application and any other information before him with respect to such drug, there is a lack of substantial evidence that the drug will have the effect it purports or is represented to have under the conditions of use prescribed, recommended, or suggested in the proposed labeling thereof; or

(6) the application failed to contain the patent information prescribed by subsection (b); or

(7) based on a fair evaluation of all material facts, such labeling is false or misleading in any particular; he shall issue an order refusing to approve the application. If, after such notice and opportunity for hearing, the Secretary finds that clauses (1) through (6) do not apply, he shall issue an order approving the application. As used in this subsection and subsection (e), the term "substantial evidence" means evidence consisting of adequate and well-controlled investigations, including clinical investigations, by experts qualified by scientific training and experience to evaluate the effectiveness of the drug involved, on the basis of which it could fairly and responsibly be concluded by such experts that the drug will have the effect it purports or is represented to have under the conditions of use prescribed, recommended, or suggested in the labeling or proposed labeling thereof.

(e) The Secretary shall, after due notice and opportunity for hearing to the applicant, withdraw approval of an application with respect to any drug under this section if the Secretary finds (1) that clinical or other experience, tests, or other scientific data show that such drug is unsafe for use under the conditions of use upon the basis of which the application was approved; (2) that new evidence of clinical experience, not contained in such application or not available to the Secretary until after such application was approved, or tests by new methods, or tests by methods not deemed reasonably applicable when such application was approved, evaluated together with the evidence available to the Secretary when the application was approved, shows that such drug is not shown to be safe for use under the conditions of use upon the basis of which the application was approved; or (3) on the basis of new information before him with respect to such drug, evaluated together with the evidence available to him when the application was approved, that there is a lack of substantial evidence that the drug will have the effect it purports or is represented to have under the conditions of use prescribed, recommended, or suggested in the labeling thereof; or

(4) the patent information prescribed by subsection (c) was not filed within thirty days after the receipt of written notice from the Secretary specifying the failure to file such information; or

(5) that the application contains any untrue statement of a material fact: *Provided*, That if the Secretary (or in his absence the officer acting as Secretary) finds that there is an imminent hazard to the public health, he may suspend the approval of such application immediately, and give the applicant prompt notice of his action and afford the applicant the opportunity for an expedited hearing under this subsection; but the authority conferred by this proviso to suspend the approval of an application shall not be delegated. The Secretary may also, after due notice and opportunity for hearing to the applicant, withdraw the approval of an application submitted under subsection (b) or (j) with respect to any drug under this section if the Secretary finds (1) that the applicant has failed to establish a system for maintaining required records, or has repeatedly or deliberately failed to maintain such records or to make required reports, in accordance with a regulation or order under subsection (k) or to comply with the notice requirements of section 510 (k) (2), or the applicant has refused to permit access to, or copying or verification of, such records as required by paragraph (2) of such subsection; or (2) that on the basis of new information before him, evaluated together with the evidence before him when the application was approved, the methods used in, or the facilities and controls used for, the manufacture, processing, and packing of such drug are inadequate to assure and preserve its identity, strength, quality, and purity and were not made adequate within a reasonable time after receipt of written notice from the Secretary specifying the matter complained of; or (3) that on the basis of new information before him, evaluated together with the evidence before him when the application was approved, the labeling of such drug, based on a fair evaluation of all material facts, is false or misleading in any particular and was not corrected within a reasonable time after receipt of written notice from the Secretary specifying the matter complained of. Any order under this subsection shall state the findings upon which it is based.

(f) Whenever the Secretary finds that the facts so require, he shall revoke any previous order under subsection (d) or (e) refusing, withdrawing, or suspending approval of an application and shall approve such application or reinstate such approval, as may be appropriate.

(g) Orders of the Secretary issued under this section shall be served (1) in person by any officer or employee of the Department designated by the Secretary or (2) by mailing the order by registered mail or by certified mail addressed to the applicant or respondent at his last-known address in the records of the Secretary.

(h) An appeal may be taken by the applicant from an order of the Secretary refusing or withdrawing approval of an application under this section. Such appeal shall be taken by filing in the United States court of appeals for the circuit wherein such applicant resides or has his principal place of business, or in the United States Court of Appeals for the District of Columbia Circuit, within sixty days after the entry of such order, a written petition praying that the order of the Secretary be set aside. A copy of such petition shall be forthwith transmitted by the clerk of the court to the Secretary, or any officer designated by him for that purpose, and thereupon the Secretary shall certify and file in the court the record upon which the order complained of was entered, as provided in section 2112 of title 28, United States Code. Upon the filing of such petition such court shall have exclusive jurisdiction to affirm or set aside such order, except that until the filing of the record the Secretary may modify or set aside his order. No objection to the order of the Secretary shall be considered by the court unless such objection shall have been urged before the Secretary or unless there were reasonable grounds for failure so to do. The finding of the Secretary as to the facts, if supported by substantial evidence, shall be conclusive. If any person shall apply to the court for leave to adduce additional evidence, and shall show to the satisfaction of the court that such additional evidence is material and that there were reasonable grounds for failure to adduce such evidence in the proceeding before the Secretary, the court may order such additional evidence to be taken before the Secretary and to be adduced upon the hearing in such manner and upon such terms and conditions as to the court may seem proper. The Secretary may modify his findings as to the facts by reason of the additional evidence so taken, and he shall file with the court such modified findings which, if supported by substantial evidence, shall be conclusive, and his recommendation, if any, for the setting aside of the original order. The judgment of the court affirming or setting aside any such order of the Secretary shall be final, subject to review by the Supreme Court of the United States upon certiorari or certification as provided in section 1254 of title 28 of the United States Code. The commencement of proceedings under this subsection shall not, unless specifically ordered by the court to the contrary, operate as a stay of the Secretary's order.

(i) The Secretary shall promulgate regulations for exempting from the operation of the foregoing subsections of this section drugs intended solely for investigational use by experts qualified by scientific training and experience to investigate the safety and effectiveness of drugs. Such regulations may, within the discretion of the Secretary, among other conditions relating to the protection of the public health, provide for conditioning such exemption upon—

(1) the submission to the Secretary, before any clinical testing of a new drug is undertaken, of reports, by the manufacturer or the sponsor of the investigation of such drug, or preclinical tests (including tests on animals) of such drug adequate to justify the proposed clinical testing;

(2) the manufacturer or the sponsor of the investigation of a new drug proposed to be distributed to investigators for clinical testing obtaining a signed agreement from each of such investigators that patients to whom the drug is administered will be under his personal supervision, or under the supervision of investigators responsible to him, and that he will not supply such drug to any other investigator, or to clinics, for administration to human beings; and

(3) the establishment and maintenance of such records, and the making of such reports to the Secretary, by the manufacturer or the sponsor of the investigation of such drug, of data (including but not limited to analytical reports by investigators) obtained as

the result of such investigational use of such drug, as the Secretary finds will enable him to evaluate the safety and effectiveness of such drug in the event of the filing of an application pursuant to subsection (b).
Such regulations shall provide that such exemption shall be conditioned upon the manufacturer, or the sponsor of the investigation, requiring that experts using such drugs for investigational purposes certify to such manufacturer or sponsor that they will inform any human beings to whom such drugs, or any controls used in connection therewith, are being administered, or their representatives, that such drugs are being used for investigational purposes and will obtain the consent of such human beings or their representatives, except where they deem it not feasible or, in their professional judgment, contrary to the best interests of such human beings. Nothing in this subsection shall be construed to require any clinical investigator to submit directly to the Secretary reports on the investigational use of drugs.

(j)(1) Any person may file with the Secretary an abbreviated application for the approval of a new drug.

(2) (A) An abbreviated application for a new drug shall contain—

(i) information to show that the conditions of use prescribed, recommended, or suggested in the labeling proposed for the new drug have been previously approved for a drug listed under paragraph (6) (hereinafter in this subsection referred to as a listed drug);

(ii) (I) if the listed drug referred to in clause (i) has only one active ingredient, information to show that the active ingredient of the new drug is the same as that of the listed drug;

(II) if the listed drug referred to in clause (i) has more than one active ingredient, information to show that the active ingredients of the new drug are the same as those of the listed drug, or

(III) if the listed drug referred to in clause (i) has more than one active ingredient and if one of the active ingredients of the new drug is different and the application is filed pursuant to the approval of a petition filed under subparagraph (C), information to show that the other active ingredients of the new drug are the same as the active ingredients of the listed drug, information to show that the different active ingredient is an active ingredient of a listed drug or of a drug which does not meet the requirements of section 201 (p), and such other information respecting the different active ingredient with respect to which the petition was filed as the Secretary may require;

(iii) information to show that the route of administration, the dosage form, and the strength of the new drug are the same as those of the listed drug referred to in clause (i) or, if the route of administration, the dosage form, or the strength of the new drug is different and the application is filed pursuant to the approval of a petition filed under subparagraph (C), such information respecting the route of administration, dosage form, or strength with respect to which the petition was filed as the Secretary may require;

(iv) information to show that the new drug is bioequivalent to the listed drug referred to in clause (i), except that if the application is filed pursuant to the approval of a petition filed under subparagraph (C), information to show that the active ingredients of the new drug are of the same pharmacological or therapeutic class as those of the listed drug referred to in clause (i) and the new drug can be expected to have the same therapeutic effect as the listed drug when administered to patients for a condition of use referred to in clause (i);

(v) information to show that the labeling proposed for the new drug is the same as the labeling approved for the listed drug referred to in clause (i) except for changes required because of differences approved under a petition filed under subparagraph (C) or because the new drug and the listed drug are produced or distributed by different manufacturers;

(vi) the items specified in clauses (B) through (F) of subsection (b) (1);

(vii) a certification, in the opinion of the applicant and to the best of his knowledge, with respect to each patent which claims the listed drug referred to in clause (i) or which claims a use for such listed drug for which the applicant is seeking approval under this subsection and for which information is required to be filed under subsection (b) or (c)—

(I) that such patent information has not been filed,

(II) that such patent has expired,

(III) of the date on which such patent will expire, or

(IV) that such patent is invalid or will not be infringed by the manufacture, use, or sale of the new drug for which the application is submitted; and

(viii) if with respect to the listed drug referred to in clause (i) information was filed under subsection (b) or (c) for a method of use patent which does not claim a use for which the applicant is seeking approval under this subsection, a statement that the method of use patent does not claim such a use.

The Secretary may not require that an abbreviated application contain information in addition to that required by clauses (i) through (viii).

(B) (i) An applicant who makes a certification described in subparagraph (A) (vii) (IV) shall include in the application a statement that the applicant will give the notice required by clause (ii) to—

(I) each owner of the patent which is the subject of the certification or the representative of such owner designated to receive such notice, and

(II) the holder of the approved application under subsection (b) for the drug which is claimed by the patent or a use of which is claimed by the patent or the representative of such holder designated to receive such notice.

(ii) The notice referred to in clause (i) shall state that an application, which contains data from bioavailability or bioequivalence studies, has been submitted under this subsection for the drug with respect to which the certification is made to obtain approval to engage in the commercial manufacture, use, or sale of such drug before the expiration of the patent referred to in the certification. Such notice shall include a detailed statement of the factual and legal basis of the applicant's opinion that the patent is not valid or will not be infringed.

(iii) If an application is amended to include a certification described in subparagraph (A) (vii) (IV), the notice required by clause (ii) shall be given when the amended application is submitted.

(C) If a person wants to submit an abbreviated application for a new drug which has a different active ingredient or whose route of administration, dosage form, or strength differ from that of a listed drug, such person shall submit a petition to the Secretary seeking permission to file such an application. The Secretary shall approve or disapprove a petition submitted under this subparagraph within ninety days of the date the petition is submitted. The Secretary shall approve such a petition unless the Secretary finds—

(i) that investigations must be conducted to show the safety and effectiveness of the drug or of any of its active ingredients, the route of administration, the dosage form, or the strength, which differs from the listed drug; or

(ii) that any drug with a different active ingredient may not be adequately evaluated for approval as safe and effective on the basis of the information required to be submitted in an abbreviated application.

(3) Subject to paragraph (4), the Secretary shall approve an application for a drug unless the Secretary finds that—

(A) the methods used in, or the facilities and controls used for, the manufacture, processing, and packing of the drug are inadequate to assure and preserve its identity, strength, quality, and purity;

(B) the information submitted with the application is insufficient to show that each of the proposed conditions of use have been previously approved for the listed drug referred to in the application;

(C) (i) if the listed drug has only one active ingredient, information submitted with the application is insufficient to show that the active ingredient is the same as that of the listed drug;

(ii) if the listed drug has more than one active ingredient, information submitted with the application is insufficient to show that the active ingredients are the same as the active ingredients of the listed drug, or

(iii) if the listed drug has more than one active ingredient and if the application is for a drug which has an active ingredient different from the listed drug, information submitted with the application is insufficient to show—

(I) that the other active ingredients are the same as the active ingredients of the listed drug, or

(II) that the different active ingredient is an active ingredient of a listed drug or a drug which does not meet the requirements of section 201 (p), or no petition to file an application for the drug with the different ingredient was approved under paragraph (2) (C);

(D) (i) if the application is for a drug whose route of administration, dosage form, or strength of the drug is the same as the route of administration, dosage form, or strength of the listed drug referred to in the application, information submitted in the application is insufficient to show that the route of administration, dosage form, or strength is the same as that of the listed drug, or

(ii) if the application is for a drug whose route of administration, dosage form, or strength of the drug is different from that of the listed drug referred to in the application, no petition to file an application for the drug with the different route of administration, dosage form, or strength was approved under paragraph (2) (C);

(E) if the application was filed pursuant to the approval of a petition under paragraph (2) (C), the application did not contain the information required by the Secretary respecting the active ingredient, route of administration, dosage form, or strength which is not the same;

(F) information submitted in the application is insufficient to show that the drug is bioequivalent to the listed drug referred to in the application or, if the application was filed pursuant to a petition approved under paragraph (2) (C), information submitted in the application is insufficient to show that the active ingredients of the new drug are of the same pharmacological or therapeutic class as those of the listed drug referred to in paragraph (2) (A) (i) and that the new drug can be expected to have the same therapeutic effect as the listed drug when administered to patients for a condition of use referred to in such paragraph;

(G) information submitted in the application is insufficient to show that the labeling proposed for the drug is the same as the labeling approved for the listed drug referred to in the application except for changes required because of differences approved under a petition filed under paragraph (2) (C) or because the drug and the listed drug are produced or distributed by different manufacturers;

(H) information submitted in the application or any other information available to the Secretary shows that (i) the inactive ingredients of the drug are unsafe for use under the conditions prescribed, recommended, or suggested in the labeling proposed for the drug, or (ii) the composition of the drug is unsafe under such conditions because of the type or quantity of inactive ingredients included or the manner in which the inactive ingredients are included;

(I) the approval under subsection (c) of the listed drug referred to in the application under this subsection has been withdrawn or suspended for grounds described in the first sentence of subsection (e), the Secretary has published a notice of opportunity for hearing to withdraw approval of the listed drug under subsection (c) for grounds described in the first sentence of subsection (e), the approval under this subsection of the listed drug referred to in the application under this subsection has been withdrawn or suspended under paragraph (5), or the Secretary has determined that the listed drug has been withdrawn from sale for safety or effectiveness reasons;

(J) the application does not meet any other requirement of paragraph (2) (A); or

(K) the application contains an untrue statement of material fact.

(4) (A) Within one hundred and eighty days of the initial receipt of an application under paragraph (2) or within such additional period as may be agreed upon by the Secretary and the applicant, the Secretary shall approve or disapprove the application.

(B) The approval of an application submitted under paragraph (2) shall be made effective on the last applicable date determined under the following:

(i) If the applicant made a certification described in subclause (I) or (II) of paragraph (2) (A) (vii) or in both such subclauses, the approval may be made effective immediately.

(ii) If the applicant made a certification described in subclause (III) of paragraph (2) (A) (vii), the approval may be made effective on the date certified under subclause (III).

(iii) If the applicant made a certification described in subclause (IV) of paragraph (2) (A) (vii), the approval shall be made effective immediately unless an action is brought for infringement of a patent which is the subject of the certification before the expiration of forty-five days from the date the notice provided under paragraph (2) (B) (i) is received. If such an action is brought before the expiration of such days, the approval shall be made effective upon the expiration of the thirty-month period beginning on the date of the receipt of the notice provided under paragraph (2) (B) (i) or such shorter or longer period as the court may order because either party to the action failed to reasonably cooperate in expediting the action, except that—

(I) if before the expiration of such period the court decides that such patent is invalid or not infringed, the approval shall be made effective on the date of the court decision,

(II) if before the expiration of such period the court decides that such patent has been infringed, the approval shall be made effective on such date as the court orders under section 271 (e) (4) (A) of title 35, United States Code, or

(III) if before the expiration of such period, the court grants a preliminary injunction prohibiting the applicant from engaging in the commercial manufacture or sale of the drug until the court decides the issues of patent validity and infringement, and if the court decides that such patent is invalid or not infringed, the approval shall be made effective on the date of such court decision.

In such an action, each of the parties shall reasonably cooperate in expediting the action. Until the expiration of forty-five days from the date the notice made under paragraph (2) (B) (i) is received, no action may be brought under section 2201 of title 28, United States Code, for a declaratory judgment with respect to the patent. Any action brought under section 2201 shall be brought in the judicial district where the defendant has its principal place of business or a regular and established place of business.

(iv) If the application contains a certification described in subclause (IV) of paragraph (2) (A) (vii) and is for a drug for which a previous application has been submitted under this subsection continuing such a certification, the application shall be made effective not earlier than one hundred and eighty days after—

(I) the date the Secretary receives notice from the applicant under the previous application of the first commercial marketing of the drug under the previous application, or

(II) the date of a decision of a court in an action described in clause (iii) holding the patent which is the subject of the certification to be invalid or not infringed, whichever is earlier.

(C) If the Secretary decides to disapprove an application, the Secretary shall give the applicant notice of an opportunity for a hearing before the Secretary on the question of whether such application is approvable. If the applicant elects to accept the opportunity for hearing by written request within thirty days after such notice, such hearing shall commence not more than ninety days after the expiration of such thirty days unless the Secretary and the applicant otherwise agree. Any such hearing shall thereafter be conducted on an expedited basis and the Secretary's order thereon shall be issued within ninety days after the date fixed by the Secretary for filing final briefs.

(D) (i) If an application (other than an abbreviated new drug application) submitted under subsection (b) for a drug, no active ingredient (including any ester or salt of the active ingredient) of which has been approved in any other application under subsection (b), was approved during the period beginning January 1, 1982, and ending on the date of the enactment of this subsection, the Secretary may not make the approval of an application submitted under this subsection which refers to the drug for which the subsection (b) application was submitted effective before the expiration of ten years from the date of the approval of the application under subsection (b).

(ii) If an application submitted under subsection (b) for a drug, no active ingredient (including any ester or salt of the active ingredient) of which has been approved in any other application under subsection (b), is approved after the date of the enactment of this subsection, no application may be submitted under this subsection which refers to the drug for which the subsection (b) application was submitted before the expiration of five years from the date of the approval of the application under subsection (b), except that such an application may be submitted under this subsection after the expiration of four years from the date of the approval of the subsection (b) application if it contains a certification of patent invalidity or noninfringement described in subclause (IV) of paragraph (2) (A) (vii). The approval of such an application shall be made effective in accordance with subparagraph (B) except that, if an action for patent infringement is commenced during the one-year period beginning forty-eight months after the date of the approval of the subsection (b) application, the thirty-month period referred to in subparagraph (B) (iii) shall be extended by such amount of time (if any) which is required for seven and one-half years to have elapsed from the date of approval of the subsection (b) application.

(iii) If an application submitted under subsection (b) for a drug, which includes an active ingredient (including any ester or salt of the active ingredient) that has been approved in another application approved under subsection (b), is approved after the date of enactment of this subsection and if such application contains reports of new clinical investigations (other than bioavailability studies) essential to the approval of the application and conducted or sponsored by the applicant, the Secretary may not make the approval of an application submitted under this subsection for the conditions of approval of such drug in the subsection (b) application effective before the expiration of three years from the date of the approval of the application under subsection (b) for such drug.

(iv) If a supplement to an application approved under subsection (b) is approved after the date of enactment of this subsection and the supplement contains reports of new clinical investigations (other than bioavailability studies) essential to the approval of the supplement and conducted or sponsored by the person submitting the supplement, the Secretary may not make the approval of an application submitted under this subsection for a change approved in the supplement effective before the expiration of three years from the date of the approval of the supplement under subsection (b).

(v) If an application (or supplement to an application) submitted under subsection (b) for a drug, which includes an active ingredient (including any ester or salt of the active ingredient) that has been approved in another application under subsection (b), was approved during the period beginning January 1, 1982, and ending on the date of the enactment of this subsection, the Secretary may not make the approval of an application submitted under this subsection which refers to the drug for which the subsection (b) application was submitted or which refers to a change approved in a supplement to the subsection (b) application effective before the expiration of two years from the date of enactment of this subsection.

(5) If a drug approved under this subsection refers in its approved application to a drug the approval of which was withdrawn or suspended for grounds described in the first sentence of subsection (e) or was withdrawn or suspended under this paragraph or which, as determined by the Secretary, has been withdrawn from sale for safety or effectiveness reasons, the approval of the drug under this subsection shall be withdrawn or suspended—

(A) for the same period as the withdrawal or suspension under subsection (e), or this paragraph, or

(B) if the listed drug has been withdrawn from sale, for the period of withdrawal from sale or, if earlier, the period ending on the date the Secretary determines that the withdrawal from sale is not for safety or effectiveness reasons.

(6) (A) (i) Within sixty days of the date of the enactment of this subsection, the Secretary shall publish and make available to the public—

(I) a list in alphabetical order of the official and proprietary name of each drug which has been approved for safety and effectiveness under subsection (c) before the date of the enactment of this subsection;

(II) the date of approval if the drug is approved after 1981 and the number of the application which was approved; and

(III) whether in vitro or in vivo bioequivalence studies, or both such studies, are required for applications filed under this subsection which will refer to the drug published.

(ii) Every thirty days after the publication of the first list under clause (i) the Secretary shall revise the list to include each drug which has been approved for safety and effectiveness under subsection (c) or approved under this subsection during the thirty-day period.

(iii) When patent information submitted under subsection (b) or (c) respecting a drug included on the list is to be published by the Secretary the Secretary shall, in revisions made under clause (ii), include such information for such drug.

(B) A drug approved for safety and effectiveness under subsection (c) or approved under this subsection shall, for purposes of this subsection, be considered to have been published under subparagraph (A) on the date of its approval or the date of enactment, whichever is later.

(C) If the approval of a drug was withdrawn or suspended for grounds described in the first sentence of subsection (e) or was withdrawn or suspended under paragraph (5) or if the Secretary determines that a drug has been withdrawn from sale for safety or effectiveness reasons, it may not be published in the list under subparagraph (A) or, if the withdrawal or suspension occurred after its publication in such list, it shall be immediately removed from such list—

(i) for the same period as the withdrawal or suspension under subsection (e) or paragraph (5), or

(ii) if the listed drug has been withdrawn from sale, for the period of withdrawal from sale or, if earlier, the period ending on the date the Secretary determines that the withdrawal from sale is not for safety or effectiveness reasons.

A notice of the removal shall be published in the Federal Register.

(7) For purposes of this subsection:

(A) The term 'bioavailability' means the rate and extent to which the active ingredient or therapeutic ingredient is absorbed from a drug and becomes available at the site of drug action.

(B) A drug shall be considered to be bioequivalent to a listed drug if—

(i) the rate and extent of absorption of the drug do not show a significant difference from the rate and extent of absorption of the listed drug when administered at the same molar dose of the therapeutic ingredient under similar experimental conditions in either a single dose or multiple doses; or

(ii) the extent of absorption of the drug does not show a significant difference from the extent of absorption of the listed drug when administered at the same molar dose of the therapeutic ingredient under similar experimental conditions in either a single dose or multiple doses and the difference from the listed drug in the rate of absorption of the drug is intentional, is reflected in its proposed labeling, is not essential to the attainment of effective body drug concentrations on chronic use, and is considered medically insignificant for the drug.

(k)(1) In the case of any drug for which an approval of an application filed under subsection (b) or (j) is in effect, the applicant shall establish and maintain such records, and make such reports to the Secretary, of data relating to clinical experience and other data or information, received or otherwise obtained by such applicant with respect to such drug, as the Secretary may by general regulation, or by order with respect to such application, prescribe on the basis of a finding that such records and reports are necessary in order to enable the Secretary to determine, or facilitate a determination, whether there is or may be ground for invoking subsection (e) of this section: *Provided, however,* That regulations and orders issued under subsection and under subsection (i) shall have due regard for the professional ethics of the medical profession and the interests of patients and shall provide, where the Secretary deems it to be appropriate, for the examination, upon request, by the persons to whom such regulations or orders are applicable, of similar information received or otherwise obtained by the Secretary.

(2) Every person required under this section to maintain records, and every person in charge or custody thereof, shall, upon request of an officer or employee designated by the Secretary, permit such officer or employee at all reasonable times to have access to and copy and verify such records.

(l) Safety and effectiveness data and information which has been submitted in an application under subsection (b) for a drug and which has not previously been disclosed to the public shall be made available to the public, upon request, unless extraordinary circumstances are shown—

(1) if no work is being or will be undertaken to have the application approved,

(2) if the Secretary has determined that the application is not approvable and all legal appeals have been exhausted,

(3) if approval of the application under subsection (c) is withdrawn and all legal appeals have been exhausted,

(4) if the Secretary has determined that such drug is not a new drug, or

(5) upon the effective date of the approval of the first application under subsection (j) which refers to such drug or upon the date upon which the approval of an application under subsection (j) which refers to such drug could be made effective if such an application had been submitted.

(m) For purposes of this section, the term 'patent' means a patent issued by the Patent and Trademark Office of the Department of Commerce.

Certification of Drugs Containing Insulin

§ **506** [356] (a) The Secretary, pursuant to regulations promulgated by him, shall provide for the certification of batches of drugs composed wholly or partly of insulin. A batch of any such drug shall be certified if such drug has such characteristics of identity and such batch has such characteristics of strength, quality, and purity, as the Secretary prescribes in such regulations as necessary to adequately insure safety and efficacy of use, but shall not otherwise be certified. Prior to the effective date of such regulations the Secretary, in lieu of certification, shall issue a release for any batch which, in his judgment, may be released without risk as to the safety and efficacy of its use. Such release shall prescribe the date of its expiration and other conditions under which it shall cease to be effective as to such batch and as to portions thereof.

(b) Regulations providing for such certification shall contain such provisions as are necessary to carry out the purposes of this section, including provisions prescribing (1) standards of identity and of strength, quality, and purity; (2) tests and methods of assay to determine compliance with such standards; (3) effective periods for certificates, and other conditions under which they shall cease to be effective as to certified batches and as to portions thereof; (4) administration and procedure; and (5) such fees, specified in such regulations, as are necessary to provide, equip, and maintain an adequate certification service. Such regulations shall prescribe no standard of identity or of strength, quality, or purity for any drug different from the standard of identity, strength, quality, or purity set forth for such drug in an official compendium.

(c) Such regulations, insofar as they prescribe tests or methods of assay to determine strength, quality, or purity of any drug, different from the tests or methods of assay set forth for such drug in an official compendium, shall be prescribed, after notice and opportunity for revision of such compendium, in the manner provided in the second sentence of section 501 (b). The provisions of subsections (e), (f), and (g) of section 701 shall be applicable to such portion of any regulation as prescribes any such different test or method, but shall not be applicable to any other portion of any such regulation.

Certification of Antibiotics

§ **507** [357] (a) The Secretary, pursuant to regulations promulgated by him, shall provide for the certification of batches of drugs (except drugs for use in animals other than man) composed wholly or partly of any kind of penicillin, streptomycin, chlortetracycline, chloramphenicol, bacitracin, or any other antibiotic drug, or any derivative thereof. A batch of any such drug shall be certified if such drug has such characteristics of identity and such batch has such characteristics of strength, quality, and purity, as the Secretary prescribes in such regulations as necessary to adequately insure safety and efficacy of use, but shall not otherwise be certified. Prior to the effective date of such regulations the Secretary, in lieu of certification, shall issue a release for any batch which, in his judgment, may be released without risk as to the safety and efficacy of its use. Such release shall prescribe the date of its expiration and other conditions under which it shall cease to be effective as to such batch and as to portions thereof. For purposes of this section and of section 502 (l), the term "antibiotic drug" means any drug intended for use by man containing any quantity of any chemical substance which is produced by a microorganism and which has the capacity to inhibit or destroy microorganisms in dilute solution (including the chemically synthesized equivalent of any such substance).

(b) Regulations providing for such certification shall contain such provisions as are necessary to carry out the purposes of this section, including provisions prescribing (1) standards of identity and of strength, quality, and purity; (2) tests and methods of assay to determine compliance with such standards; (3) effective periods for certificates, and other conditions under which they shall cease to be effective as to certified batches and as to portions thereof; (4) administration and procedure; and (5) such fees, specified in such regulations, as are necessary to provide, equip, and maintain an adequate certification service. Such regulations shall prescribe only such tests and methods of assay as will provide for certification or rejection within the shortest time consistent with the purposes of this section.

(c) Whenever in the judgment of the Secretary, the requirements of this section and of section 502 (l) with respect to any drug or class of drugs are not necessary to insure safety and efficacy of use, the Secretary shall promulgate regulations exempting such drug or class of drugs from such requirements. In deciding whether an antibiotic drug, or class of antibiotic drugs, is to be exempted from the requirement of certification the Secretary shall give consideration, among other relevant factors, to—

(1) whether such drug or class of drugs is manufactured by a person who has, or hereafter shall have, produced fifty consecutive batches of such drug or class of drugs in compliance with the regulations for the certification thereof within a period of not more than eighteen calendar months, upon the application by such person to the Secretary; or

(2) whether such drug or class of drugs is manufactured by any person who has otherwise demonstrated such consistency in the production of such drug or class of drugs, in compliance with the regulations for the certification thereof, as in the judgment of the Secretary is adequate to insure the safety and efficacy of use thereof. When an antibiotic drug or a drug manufacturer has been exempted from the requirement of certification, the manufacturer may still obtain certification of a batch or batches of that drug if he applies for and meets the requirements for certification. Nothing in this Act shall be deemed to prevent a manufacturer or distributor of an antibiotic drug from making a truthful statement in labeling or advertising of the product as to whether it has been certified or exempted from the requirement of certification.

(d) The Secretary shall promulgate regulations exempting from any requirement of this section and of section 502 (l), (1) drugs which are to be stored, processed, labeled, or repacked at establishments other than those where manufactured, on condition that such drugs comply with all such requirements upon removal from such establishments; (2) drugs which conform to applicable standards of identity, strength, quality, and purity prescribed by these regulations and are intended for use in manufacturing other drugs; and (3) drugs which are intended solely for investigational use by experts qualified by scientific training and experience to investigate the safety and efficacy of drugs. Such regulations may, within the discretion of the Secretary, among other conditions relating to the protection of the public health, provide for conditioning the exemption under clause (3) upon—

(1) the submission to the Secretary, before any clinical testing of a new drug is undertaken, of reports, by the manufacturer or the sponsor of the investigation of such drug, of preclinical tests (including tests on animals) of such drug adequate to justify the proposed clinical testing;

(2) the manufacturer or the sponsor of the investigation of a new drug proposed to be distributed to investigators for clinical testing obtaining a signed agreement from each of such investigators that patients to whom the drug is administered will be under his personal supervision, or under the supervision of investigators responsible to him, and that he will not apply such drug to any other investigator, or to clinics, for administration to human beings; and

(3) the establishment and maintenance of such records, and the making of such reports to the Secretary, by the manufacturer or the sponsor of the investigation of such drug, of data (including but not limited to analytical reports by investigators) obtained as the result of such investigational use of such drug, as the Secretary finds will enable him to evaluate the safety and effectiveness of such drug in the event of the filing of an application for certification or release pursuant to subsection (a).

Such regulations shall provide that such exemption shall be conditioned upon the manufacturer, or the sponsor of the investigation, requiring that experts using such drugs for investigational purposes certify to such manufacturer or sponsor that they will inform any human beings to whom such drugs, or any controls used in connection therewith, are being administered, or their representatives, that such drugs are being used for investigational purposes and will obtain the consent of such human beings or their representatives, except where they deem it not feasible or, in their professional judgment, contrary to the best interests of such human beings. Nothing in this subsection shall be construed to require any clinical investigator to submit directly to the Secretary reports on the investigational use of drugs.

(e) No drug which is subject to section 507 shall be deemed to be subject to any provision of section 505 except a new drug exempted from the requirements of this section and of section 502 (l) pursuant to regulations promulgated by the Secretary: *Provided*, That, for purposes of section 505, the initial request for certification, as thereafter duly amended, pursuant to section 507, of a new drug so exempted shall be considered a part of the application filed pursuant to section 505 (b) with respect to the person filing such request and to such drug as of the date of the exemption. Compliance of any drug subject to section 502 (l) or 507 with section 501 (b) and 502 (g) shall be determined by the application of the standards of strength, quality, and purity, the tests and methods of assay, and the requirements of packaging, and labeling, respectively, prescribed by regulations promulgated under section 507.

(f) Any interested person may file with the Secretary a petition proposing the issuance, amendment, or repeal of any regulation contemplated by this section. The petition shall set forth the proposal in general terms and shall state reasonable grounds therefor. The Secretary shall give public notice of the proposal and an opportunity for all interested persons to present their views thereon, orally or in writing, and as soon as practicable thereafter shall make public his action upon such proposal. At any time prior to the thirtieth day after such action is made public any interested person may file objections to such action, specifying with particularity the changes desired, stating reasonable grounds therefor, and requesting a public hearing upon such objections. The Secretary shall thereupon, after due notice, hold such public hearing. As soon as practicable after completion of the hearing, the Secretary shall by order make public his action on such objections. The Secretary shall base his order only on substantial evidence of record at the hearing and shall set forth as part of the order detailed findings of fact on which the order is based. The order shall be subject to the provision of section 701 (f) and (g).

(g) (1) Every person engaged in manufacturing, compounding, or processing any drug within the purview of this section with respect to which a certificate or release has been issued pursuant to this section shall establish and maintain such records, and make such reports to the Secretary, of data relating to clinical experience and other data or information, received or otherwise obtained by such person with respect to such drug, as the Secretary may by general regulation, or by order with respect to such certification or release, prescribe on the basis of a finding that such records and reports are necessary in order to enable the Secretary to make, or to facilitate, a determination as to whether such certification or release should be rescinded or whether any regulation issued under this section should be amended or repealed: *Provided, however*, That regulations and orders issued under this subsection and under clause (3) of subsection (d) shall have due regard for the professional ethics of the medical profession and the interests of patients and shall provide, where the Secretary deems it to be appropriate, for the examination, upon request, by the persons to whom such regulations or orders are applicable, of similar information received or otherwise obtained by the Secretary.

(2) Every person required under this section to maintain records, and every person having charge or custody thereof, shall, upon request of an officer or employee designated by the Secretary, permit such officer or employee at all reasonable times to have access to and copy and verify such records.

(h) In the case of a drug for which, on the day immediately preceding the effective date of this subsection, a prior approval of an application under section 505 had not been withdrawn under section 505 (e), the initial issuance of regulations providing for certification or exemption of such drug under this section 507 shall, with respect to the conditions of use prescribed, recommended, or suggested in the labeling covered by such application, not be conditioned upon an affirmative finding of the efficacy of such drug. Any subsequent amendment or repeal of such regulations so as no longer to provide for such certification or exemption on the ground of a lack of efficacy of such drug for use under such conditions of use may be effected only on or after that effective date of clause (3) of the first sentence of section 505 (e) which would be applicable to such drug under such conditions of use if such drug were subject to section 505 (e), and then only if (1) such amendment or repeal is made in accordance with the procedure specified in subsection (f) of this section (except that such amendment or repeal may be initiated either by a proposal of the Secretary or by a petition of any interested person) and (2) the Secretary finds, on the basis of new information with respect to such drug evaluated together with the information before him when the application under section 505 became effective or was approved, that there is a lack of substantial evidence (as defined in section 505 (d)) that the drug has the effect it purports or is represented to have under such conditions of use.

Authority to Designate Official Names

§ **508** [358] (a) The Secretary may designate an official name for any drug or device if he determines that such action is necessary or desirable in the interest of usefulness and simplicity. Any official name designated under this section for any drug or device shall be the only official name of that drug or device used in any official compendium published after such name has been prescribed or for any other purpose of this Act. In no event, however, shall the Secretary establish an official name so as to infringe a valid trademark.

(b) Within a reasonable time after the effective date of this section, and at such other times as he may deem necessary, the Secretary shall cause a review to be made of the official names by which drugs are identified in the official United States Pharmacopeia, the official Homeopathic Pharmacopeia of the United States, and the official National Formulary, and all supplements thereto and at such times as he may deem necessary shall cause a review to be made of the official names by which devices are identified in any official compendium (and all supplements thereto), to determine whether revision of any of those names is necessary or desirable in the interest of usefulness and simplicity.

(c) Whenever he determines after any such review that (1) any such official name is unduly complex or is not useful for any other reason, (2) two or more official names have been applied to a single drug or device, or to two or more drugs which are identical in chemical structure and pharmacological action and which are substantially identical in strength, quality, and purity or to two or more devices which are substantially equivalent in design and purpose, or (3) no official name has been applied to a medically useful drug or device, he shall transmit in writing to the compiler of each official compendium in which that drug or drugs or device are identified and recognized his request for the recommendation of a single official name for such drug or drugs or device which will have usefulness and simplicity. Whenever such a single official name has not been recommended within one hundred and eighty days after such request, or the Secretary determines that any name so recommended is not useful for any reason, he shall designate a single official name for such drug or drugs or device. Whenever he determines that the name so recommended is useful, he shall designate that name as the official name of such drug or drugs or device. Such designation shall be made as a regulation upon public notice and in accordance with the procedure set forth in section 4 of the Administrative Procedure Act (5 U.S.C. 1003).

(d) After each such review, and at such other times as the Secretary may determine to be necessary or desirable, the Secretary shall cause to be compiled, published, and publicly distributed a list which shall list all revised official names of drugs or devices designated under this section and shall contain such descriptive and explanatory matter as the Secretary may determine to be required for the effective use of those names.

(e) Upon a request in writing by any compiler of an official compendium that the Secretary exercise the authority granted to him under section 508 (a), he shall upon public notice and in accordance with the procedure set forth in section 4 of the Administrative Procedure Act (5 U.S.C. 1003) designate the official name of the drug or device for which the request is made.

Registration Requirements

§ **510** [360] (a) As used in this section—

(1) the term "manufacture, preparation, propagation, compounding, or processing" shall include repackaging or otherwise changing the container, wrapper, or labeling of any drug package or device package in furtherance of the distribution of the drug or device from the original place of manufacture to the person who makes final delivery or sale to the ultimate consumer or user; and

(2) the term "name" shall include in the case of a partnership the name of each partner and, in the case of a corporation, the name of each corporate officer and director, and the State of incorporation.

(b) On or before December 31 of each year every person who owns or operates any establishment in any State engaged in the manufacture, preparation, propagation, compounding, or processing of a drug or drugs or a device or devices shall register with the Secretary his name, places of business, and all such establishments.

(c) Every person upon first engaging in the manufacture, preparation, propagation, compounding, or processing of a drug or drugs or a device or devices in any establishment which he owns or operates in any State shall immediately register with the Secretary his name, place of business, and such establishment.

(d) Every person duly registered in accordance with the foregoing subsections of this section shall immediately register with the Secretary any additional establishment which he owns or operates in any State and in which he begins the manufacture, preparation, propagation, compounding, or processing of a drug or drugs or a device or devices.

(e) The Secretary may assign a registration number to any person or any establishment registered in accordance with this section. The Secretary may also assign a listing number to each drug or class of drugs listed under subsection (j). Any number assigned pursuant to the preceding sentence shall be the same as that assigned pursuant to the National Drug Code. The Secretary may by regulation prescribe a uniform system for the identification of devices intended for human use and may require that persons who are required to list such devices pursuant to subsection (j) shall list such devices in accordance with such system.

(f) The Secretary shall make available for inspection, to any person so requesting, any registration filed pursuant to this section, except that any list submitted pursuant to paragraph (3) of subsection (j) and the information accompanying any list or notice filed under paragraph (1) or (2) of that subsection shall be exempt from such inspection unless the Secretary finds that such an exemption would be inconsistent with protection of the public health.

(g) The foregoing subsections of this section shall not apply to—

(1) pharmacies which maintain establishments in conformance with any applicable local laws regulating the practice of pharmacy and medicine and which are regularly engaged in dispensing prescription drugs or devices, upon prescriptions of practitioners licensed to administer such drugs or devices to patients under the care of such practitioners in the course of their professional practice, and which do not manufacture, prepare, propagate, compound, or process drugs or devices for sale other than in the regular course of their business of dispensing or selling drugs or devices at retail;

(2) practitioners licensed by law to prescribe or administer drugs or devices and who manufacture, prepare, propagate, compound, or process drugs or devices solely for use in the course of their professional practice;

(3) persons who manufacture, prepare, propagate, compound, or process drugs or devices solely for use in research, teaching, or chemical analysis and not for sale;

(4) such other classes of persons as the Secretary may by regulation exempt from the application of this section upon a finding that registration by such classes of persons in accordance with this section is not necessary for the protection of the public health.

(h) Every establishment in any State registered with the Secretary pursuant to this section shall be subject to inspection pursuant to section 704 and every such establishment engaged in the manufacture, propagation, compounding, or processing of a drug or drugs or of a device or devices classified in class II or III shall be so inspected by one or more officers or employees duly designated by the Secretary at least once in the two-year period beginning with the date of registration of such establishment pursuant to this section and at least once in every successive two-year period thereafter.

(i) Any establishment within any foreign country engaged in the manufacture, preparation, propagation, compounding, or processing of a drug or drugs or a device or devices shall be permitted to register under this section pursuant to regulations promulgated by the Secretary. Such regulations shall require such establishment to provide the information required by subsection (j) and shall require such establishment to provide the information required by subsection (j) in the case of a device or devices and shall include provisions for registration of any such establishment upon condition that adequate and effective means are available, by arrangement with the government of such foreign country or otherwise, to enable the Secretary to determine from time to time whether drugs or devices manufactured, prepared, propagated, compounded or processed in such establishment, if imported or offered for import into the United States, shall be refused admission on any of the grounds set forth in section 801 (a) of this Act.

(j) (1) Every person who registers with the Secretary under subsection (b), (c), or (d) shall, at the time of registration under any such subsection, file with the Secretary a list of all drugs and a list of all devices and a brief statement of the basis for believing that each device included in the list is a device rather than a drug (with each drug and device in each list listed by its established name as defined in section 502 (e) and by any proprietary name) which is being manufactured, prepared, propagated, compounded, or processed by him for commercial distribution and which he has not included in any list of drugs or devices filed by him with the Secretary under this paragraph or paragraph (2) before such time of registration. Such list shall be prepared in such form and manner as the Secretary may prescribe and shall be accompanied by—

(A) in the case of a drug contained in the applicable list and subject to section 505, 506, 507, or 523, or a device intended for human use contained in the applicable list with respect to which a performance standard has been established under section 514 or which is subject to section 515, a reference to the authority for the marketing of such drug or device and a copy of all labeling for such drug or device;

(B) in the case of any other drug or device contained in an applicable list—

(i) which drug is subject to section 503 (b) (1), or which device is a restricted device, a copy of all labeling for such drug or device, a representative sampling of advertisements for such drug or device, and, upon request made by the Secretary for good cause, a copy of all advertisements for a particular drug product or device, or

(ii) which drug is not subject to section 503 (b) (1) or which device is not a restricted device, the label and package insert for such drug or device and a representative sampling of any other labeling for such drug or device;

(C) in the case of any drug contained in an applicable list which is described in subparagraph (B), a quantitative listing of its active ingredient or ingredients, except that with respect to a particular drug product the Secretary may require the submission of a quantitative listing of all ingredients if he finds that such submission is necessary to carry out the purposes of this Act; and

(D) if the registrant filing a list has determined that a particular drug product or device contained in such list is not subject to section 505, 506, 507, or 512, or the particular device contained in such list is not subject to a performance standard established under section 514 or to section 515 or is not a restricted device, a brief statement of the basis on which the registrant made such determination if the Secretary requests such a statement with respect to that particular drug product or device.

(2) Each person who registers with the Secretary under this subsection shall report to the Secretary once during the month of June of each year and once during the month of December of each year the following information:

(A) A list of each drug or device introduced by the registrant for commercial distribution which has not been included in any list previously filed by him with the Secretary under this subparagraph or paragraph (1) of this subsection. A list under this subparagraph shall list a drug or device by its established name (as defined in section 502 (e)) and by any proprietary name it may have and shall be accompanied by the other information required by paragraph (1).

(B) If since the date the registrant last made a report under this paragraph (or if he has not made a report under this paragraph, since the effective date of this subsection) he has discontinued the manufacture, preparation, propagation, compounding, or processing for commercial distribution of a drug or device included in a list filed by him under subparagraph (A) or paragraph (1); notice of such discontinuance, the date of such discontinuance, and the identity (by established name as defined in section 502 (e) and by any proprietary name) of such drug or device.

(C) If since the date the registrant reported pursuant to subparagraph (B) a notice of discontinuance he has resumed the manufacture, preparation, propagation, compounding, or processing for commercial distribution of the drug or device with respect to which such notice of discontinuance was reported; notice of such resumption, the date of such resumption, the identity of such drug or device (each by established name (as defined in section 502 (e)) and by any proprietary name), and the other information required by paragraph (1), unless the registrant has previously reported such resumption to the Secretary pursuant to this subparagraph.

(D) Any material change in any information previously submitted pursuant to this paragraph or paragraph (1).

(3) The Secretary may also require each registrant under this section to submit a list of each drug product which (A) the registrant is manufacturing, preparing, propagating, compounding, or processing for commercial distribution, and (B) contains a particular ingredient. The Secretary may not require the submission of such a list unless he has made a finding that the submission of such a list is necessary to carry out the purposes of this Act.

Drugs for Rare Diseases or Conditions

RECOMMENDATIONS FOR INVESTIGATIONS OF DRUGS FOR RARE DISEASES OR CONDITIONS

§ **525** [360aa] (a) The sponsor of a drug for a disease or condition which is rare in the States may request the Secretary to provide written recommendations for the non-clinical and clinical investigations which must be conducted with the drug before—

(1) it may be approved for such disease or condition under section 505, or

(2) if the drug is a biological product, before it may be licensed for such disease or condition under section 351 of the Public Health Service Act.

If the Secretary has reason to believe that a drug for which a request is made under this section is a drug for a disease or condition which is rare in the States, the Secretary shall provide the person making the request written recommendations for the non-clinical and clinical investigations which the Secretary believes, on the basis of information available to the Secretary at the time of the request under this section, would be necessary for approval of such drug for such disease or condition under section 505 or licensing under section 351 of the Public Health Service Act for such disease or condition.

(b) The Secretary shall by regulation promulgate procedures for the implementation of subsection (a).

DESIGNATION OF DRUGS FOR RARE DISEASES OR CONDITIONS

§ **526** [360bb] (a) (1) The manufacturer or the sponsor of a drug may request the Secretary to designate the drug as a drug for a rare disease or condition. If the Secretary finds that a drug for which a request is submitted under this subsection is being or will be investigated for a rare disease or condition and—

(A) if an application for such drug is approved under section 505, or

(B) if the drug is a biological product, a license is issued under section 351 of the Public Health Service Act,

the approval or license would be for use for such disease or condition, the Secretary shall designate the drug as a drug for such disease or condition. A request for a designation of a drug under this subsection shall contain the consent of the applicant to notice being given by the Secretary under subsection (b) respecting the designation of the drug.

(2) For purposes of paragraph (1), the term 'rare disease or condition' means any disease or condition *which* (*A*) *affects less than 200,000 persons in the U.S. or* (*B*) *affects more than 200,000 persons in the U.S. and for which* there is no reasonable expectation that the cost of developing and making available in the United States a drug for such disease or condition will be recovered from sales in the United States of such drug. Determinations under the preceding sentence with respect to any drug shall be made on the basis of the facts and circumstances as of the date the request for designation of the drug under this subsection is made.

(b) Notice respecting the designation of a drug under subsection (a) shall be made available to the public.

(c) The Secretary shall by regulation promulgate procedures for the implementation of subsection (a).

PROTECTION FOR UNPATENTED DRUGS FOR RARE DISEASES OR CONDITIONS

§ **527** [360cc] (a) Except as provided in subsection (b), if the Secretary—

(1) approves an application filed pursuant to section 505 (b), or

(2) issues a license under section 351 of the Public Health Service Act for a drug designated under section 526 for a rare disease or condition and for which a United States Letter of Patent may not be issued, the Secretary may not approve another application under section 505 (b) or issue another license under section 351 of the Public Health Service Act for such drug for such disease or condition for a person who is not the holder of such approved application or of such license until the expiration of seven years from the date of the approval of the approved application or the issuance of the license. Section 505 (c) (2) does not apply to the refusal to approve an application under the preceding sentence.

(b) If an application filed pursuant to section 505 (b) is approved for a drug designated under section 526 for a rare disease or condition or a license is issued under section 351 of the Public Health Service Act for such a drug and if a United States Letter of Patent may not be issued for the drug, the Secretary may, during the seven-year period beginning on the date of the application approval or of the issuance of the license, approve another application under section 505 (b), or, if the drug is a biological product, issue a license under section 351 of the Public Health Service Act, for such drug for such disease or condition for a person who is not the holder of such approved application or of such license if—

(1) The Secretary finds, after providing the holder notice and opportunity for the submission of views, that in such period the holder of the approved application or of the license cannot assure the availability of sufficient quantities of the drug to meet the needs of persons with the disease or condition for which the drug was designated; or

(2) such holder provides the Secretary in writing the consent of such holder for the approval of other applications or the issuance of other licenses before the expiration of such seven-year period.

OPEN PROTOCOLS FOR INVESTIGATIONS OF DRUGS FOR RARE DISEASES OR CONDITIONS

§ **528** [360dd] If a drug is designated under section 526 as a drug for a rare disease or condition and if notice of a claimed exemption under section 505 (i) or regulations issued thereunder is filed for such drug, the Secretary shall encourage the sponsor of such drug to design protocols for clinical investigations of the drug which may be conducted under the exemption to permit the addition to the investigations of persons with the disease or condition who need the drug to treat the disease or condition and who cannot be satisfactorily treated by available alternative drugs.

Records of Interstate Shipment

§ **703** [373] For the purpose of enforcing the provisions of this Act, carriers engaged in interstate commerce, and persons receiving foods, drugs, devices, or cosmetics in interstate commerce or holding such articles so received, shall, upon the request of an officer or employee duly designated by the Secretary, permit such officer or employee, at reasonable times, to have access to and to copy all records showing the movement in interstate commerce of any food, drug, device, or cosmetic, or the holding thereof during or after such movement, and the quantity, shipper, and consignee thereof; and it shall be unlawful for any such carrier or person to fail to permit such access to and copying of any such record so requested when such request is accompanied by a statement in writing specifying the nature or kind of food, drug, device, or cosmetic to which such request relates: *Provided*, That evidence obtained under this section, or any evidence which is directly or indirectly derived from such evidence, shall not be used in a criminal prosecution of the person from whom obtained: *Provided further*, That carriers shall not be subject to the other provisions of this Act by reason of their receipt, carriage, holding, or delivery of food, drugs, devices, or cosmetics in the usual course of business as carriers.

Inspections

§ **704** [374] (a) (1) For purposes of enforcement of this Chapter, officers or employees duly designated by the Secretary, upon presenting appropriate credentials and a written notice to the owner, operator, or agent in charge, are authorized (A) to enter, at reasonable times, any factory, warehouse, or establishment in which food, drugs, devices, or cosmetics are manufactured, processed, packed, or held, for introduction into interstate commerce or after such introduction, or to enter any vehicle being used to transport or hold such food, drugs, devices or cosmetics in interstate commerce; and (B) to inspect, at reasonable times and within reasonable limits and in a reasonable manner, such factory, warehouse, establishment, or vehicle and all pertinent equipment, finished and unfinished materials, containers, and labeling therein. In the case of any factory, warehouse, establishment, or consulting laboratory in which prescription drugs or restricted devices are manufactured, processed, packed, or held, the inspection shall extend to all things therein (including records, files, papers, processes, controls, and facilities) bearing on whether prescription drugs or restricted devices which are adulterated or misbranded within the meaning of this Chapter, or which may not be manufactured, introduced into interstate commerce, or sold, or offered for sale by reason of any provision of this Chapter, have been or are being manufactured, processed, packed, transported, or held in any such place, or otherwise bearing on violation of this Chapter. No inspection authorized by the preceding sentence or by paragraph (3) shall extend to financial data, sales data other than shipment data, pricing data, personnel data (other than data as to qualifications of technical and professional personnel performing functions subject to this Chapter, and research data (other than data relating to new drugs, antibiotic drugs, and devices and subject to reporting and inspection under regulations lawfully issued pursuant to section 505 (i) or (j), section 507 (d) or (g), section 519, or 520 (g), and data relating to other drugs or devices which in the case of a new drug would be subject to reporting or inspection under lawful regulations issued pursuant to section 505 (j) of the title). A separate notice shall be given for each such inspection, but a notice shall not be given for each such inspection, but a notice shall not be required for each entry made during the period covered by the inspection. Each such inspection shall be commenced and completed with reasonable promptness.

(2) The provisions of the second sentence of this subsection shall not apply to—

(A) pharmacies which maintain establishments in conformance with any applicable local laws regulating the practice of pharmacy and medicine and which are regularly engaged in dispensing prescription drugs, or devices upon prescriptions of practitioners licensed to administer such drugs or devices to patients under the care of such practitioners in the course of their professional practice, and which do not, either through a subsidiary or otherwise, manufacture, prepare, propagate, compound, or process drugs or devices for sale other than in the regular course of their business of dispensing or selling drugs or devices at retail;

(B) practitioners licensed by law to prescribe or administer drugs or prescribe or use devices, as the case may be, and who manufacture, prepare, propagate, compound, or process drugs or manufacture or process devices solely for use in the course of their professional practice;

(C) persons who manufacture, prepare, propagate, compound, or process drugs or manufacture or process devices solely for use in research, teaching, or chemical analysis and not for sale;

(D) such other classes of persons as the Secretary may by regulation exempt from the application of this section upon a finding that inspection as applied to such classes of persons in accordance with this section is not necessary for the protection of the public health.

(b) Upon completion of any such inspection of a factory, warehouse, consulting laboratory, or other establishment, and prior to leaving the premises, the officer or employee making the inspection shall give to the owner, operator, or agent in charge a report in writing setting forth any conditions or practices observed by him which, in his judgment, indicate that any food, drug, device, or cosmetic in such establishment (1) consists in whole or in part of any filthy, putrid, or decomposed substance, or (2) has been prepared, packed, or held under insanitary conditions whereby it may have become contaminated with filth, or whereby it may have been rendered injurious to health. A copy of such report shall be sent promptly to the Secretary.

(c) If the officer or employee making any such inspection of a factory, warehouse, or other establishment has obtained any sample in the course of the inspection, upon completion of the inspection and prior to leaving the premises he shall give to the owner, operator, or agent in charge a receipt describing the samples obtained.

Revision of United States Pharmacopeia; Development of Analysis and Mechanical and Physical Tests

§ **707** [377] The Secretary, in carrying into effect the provisions of this chapter, is authorized hereafter to cooperate with associations and scientific societies in the revision of the United States Pharmacopeia and in the development of methods of analysis and mechanical and physical tests necessary to carry out the work of the Food and Drug Administration.

Inquiries regarding these requirements should be directed to the U.S. Food and Drug Administration, 5600 Fishers Lane, Rockville, MD 20857.

⟨1077⟩ GOOD MANUFACTURING PRACTICE FOR FINISHED PHARMACEUTICALS

As is indicated in the General Notices, tolerances stated in the United States Pharmacopeia and in the National Formulary are based upon the consideration that the article is produced under recognized principles of good manufacturing practice. In the United States, a drug not produced in accordance with current good manufacturing practices may be considered to be adulterated. The U. S. Food and Drug Administration has published regulations setting forth minimum current good manufacturing practices for the preparation of drug products. While the regulations are directed primarily to drug manufacturers, the principles embodied therein may be helpful to those engaged in the practice of pharmacy and it is for this reason that these regulations are reproduced here.

Publication of these regulations in this Pharmacopeia is for purposes of information and does not impart to them any legal effect under the Federal Food, Drug, and Cosmetic Act.

Part 210—Current Good Manufacturing Practices in Manufacturing, Processing, Packing, or Holding of Drugs: General

§ 210.1 Status of current good manufacturing practice regulations.

(a) The regulations set forth in this part and in Parts 211 through 229 of this chapter contain the minimum current good manufacturing practice for methods to be used in, and the facilities or controls to be used for, the manufacture, processing, packing, or holding of a drug to assure that such drug meets the requirements of the act as to safety, and has the identity and strength and meets the quality and purity characteristics that it purports or is represented to possess.

(b) The failure to comply with any regulation set forth in this part and in Parts 211 through 229 of this chapter in the manufacture, processing, packing, or holding of a drug shall render such drug to be adulterated under section 501(a)(2)(B) of the act and such drug, as well as the person who is responsible for the failure to comply, shall be subject to regulatory action.

§ 210.2 Applicability of current good manufacturing practice regulations.

(a) The regulations in this part and in Parts 211 through 229 of this chapter as they may pertain to a drug and in Parts 600 through 680 of this chapter as they may pertain to a biological product for human use, shall be considered to supplement, not supersede, each other, unless the regulations explicitly provide otherwise. In the event that it is impossible to comply with all applicable regulations in these parts, the regulations specifically applicable to the drug in question shall supersede the more general.

(b) If a person engages in only some operations subject to the regulations in this part and in Parts 211 through 229 and Parts 600 through 680 of this chapter, and not in others, that person need only comply with those regulations applicable to the operations in which he or she is engaged.

§ 210.3 Definitions.

(a) The definitions and interpretations contained in section 201 of the act shall be applicable to such terms when used in this part and in Parts 211 through 229 of this chapter.

(b) The following definitions of terms apply to this part and to Parts 211 through 229 of this chapter.

(1) "Act" means the Federal Food, Drug, and Cosmetic Act, as amended (21 U.S.C. 301 et seq.).

(2) "Batch" means a specific quantity of a drug or other material that is intended to have uniform character and quality, within specified limits, and is produced according to a single manufacturing order during the same cycle of manufacture.

(3) "Component" means any ingredient intended for use in the manufacture of a drug product, including those that may not appear in such drug product.

(4) "Drug product" means a finished dosage form, for example, tablet, capsule, solution, etc., that contains an active drug ingredient generally, but not necessarily, in association with inactive ingredients. The term also includes a finished dosage form that does not contain an active ingredient but is intended to be used as a placebo.

(5) "Fiber" means any particulate contaminant with a length at least three times greater than its width.

(6) "Non-fiber-releasing filter" means any filter, which after any appropriate pretreatment such as washing or flushing, will not release fibers into the component or drug product that is being filtered. All filters composed of asbestos are deemed to be fiber-releasing filters.

(7) "Active ingredient" means any component that is intended to furnish pharmacological activity or other direct effect in the diagnosis, cure, mitigation, treatment, or prevention of disease, or to affect the structure of any function of the body of man or other animals. The term includes those components that may undergo chemical change in the manufacture of the drug product and be present in the drug product in a modified form intended to furnish the specified activity or effect.

(8) "Inactive ingredient" means any component other than an "active ingredient."

(9) "In-process material" means any material fabricated, compounded, blended, or derived by chemical reaction that is produced for, and used in the preparation of the drug product.

(10) "Lot" means a batch, or a specific identified portion of a batch, having uniform character and quality within specified limits; or, in the case of a drug product produced by continuous process, it is a specific identified amount produced in a unit of time or quantity in a manner that assures its having uniform character and quality within specified limits.

(11) "Lot number, control number, or batch number" means any distinctive combination of letters, numbers, or symbols, or any combination of them, from which the complete history of the manufacture, processing, packing, holding, and distribution of a batch or lot of drug product or other material can be determined.

(12) "Manufacture, processing, packing, or holding of a drug product" includes packaging and labeling operations, testing, and quality control of drug products.

(15) "Quality control unit" means any person or organizational element designated by the firm to be responsible for the duties relating to quality control.

(16) "Strength" means:

(i) The concentration of the drug substance (for example, weight/weight, weight/volume, or unit dose/volume basis), and/or

(ii) The potency, that is, the therapeutic activity of the drug product as indicated by appropriate laboratory tests or by adequately developed and controlled clinical data (expressed, for example, in terms of units by reference to a standard).

(17) "Theoretical yield" means the quantity that would be produced at any appropriate phase of manufacture, processing, or packing of a particular drug product, based upon the quantity of components to be used, in the absence of any loss or error in actual production.

(18) "Actual yield" means the quantity that is actually produced at any appropriate phase of manufacture, processing, or packing of a particular drug product.

(19) "Percentage of theoretical yield" means the ratio of the actual yield (at any appropriate phase of manufacture, processing, or packing of a particular drug product) to the theoretical yield (at the same phase), stated as a percentage.

(20) "Acceptance criteria" means the product specifications and acceptance/rejection criteria, such as acceptable quality level and unacceptable quality level, with an associated sampling plan, that are necessary for making a decision to accept or reject a lot or batch (or any other convenient subgroups of manufactured units).

(21) "Representative sample" means a sample that consists of a number of units that are drawn based on rational criteria such as random sampling and intended to assure that the sample accurately portrays the material being sampled.

Part 211—Current Good Manufacturing Practice for Finished Pharmaceuticals

Subpart A—General Provisions

Subpart B—Organization and Personnel

Subpart C—Buildings and Facilities

Subpart D—Equipment

Subpart E—Control of Components and Drug Product Containers and Closures

Subpart A—General Provisions

§ 211.1 Scope.

(a) The regulations in this part contain the minimum current good manufacturing practice for preparation of drug products for administration to humans or animals.

(b) The current good manufacturing practice regulations in this chapter, as they pertain to drug products, and in Parts 600 through 680 of this chapter, as they pertain to biological products for human use, shall be considered to supplement, not supersede, the regulations in this part unless the regulations explicitly provide otherwise. In the event it is impossible to comply with applicable regulations both in this part and in other parts of this chapter or in Parts 600 through 680 of this chapter, the regulation specifically applicable to the drug product in question shall supersede the regulation in this part.

§ 211.3 Definitions.

The definitions set forth in § 210.3 of this chapter apply in this part.

Subpart B—Organization and Personnel

§ 211.22 Responsibilities of quality control unit.

(a) There shall be a quality control unit that shall have the responsibility and authority to approve or reject all components, drug product containers, closures, in-process materials, packaging material, labeling, and drug products, and the authority to review production records to assure that no errors have occurred or, if errors have occurred, that they have been fully investigated. The quality control unit shall be responsible for approving or rejecting drug products manufactured, processed, packed, or held under contract by another company.

(b) Adequate laboratory facilities for the testing and approval (or rejection) of components, drug product containers, closures, packaging materials, in-process materials, and drug products shall be available to the quality control unit.

(c) The quality control unit shall have the responsibility for approving or rejecting all procedures or specifications impacting on the identity, strength, quality, and purity of the drug product.

(d) The responsibilities and procedures applicable to the quality control unit shall be in writing; such written procedures shall be followed.

§ 211.25 Personnel qualifications.

(a) Each person engaged in the manufacture, processing, packing, or holding of a drug product shall have education, training, and experience, or any combination thereof, to enable that person to perform the assigned functions. Training shall be in the particular operations that the employee performs and in current good manufacturing practice (including the current good manufacturing practice regulations in this chapter and written procedures required by these regulations) as they relate to the employee's functions. Training in current good manufacturing practice shall be conducted by qualified individuals on a continuing basis and with sufficient frequency to assure that employees remain familiar with CGMP requirements applicable to them.

(b) Each person responsible for supervising the manufacture, processing, packing, or holding of a drug product shall have the education, training, and experience, or any combination thereof, to perform assigned functions in such a manner as to provide assurance that the drug product has the safety, identity, strength, quality, and purity that it purports or is represented to possess.

(c) There shall be an adequate number of qualified personnel to perform and supervise the manufacture, processing, packing, or holding of each drug product.

§ 211.28 Personnel responsibilities.

(a) Personnel engaged in the manufacture, processing, packing, or holding of a drug product shall wear clean clothing appropriate for the duties they perform. Protective apparel, such as head, face, hand, and arm coverings, shall be worn as necessary to protect drug products from contamination.

(b) Personnel shall practice good sanitation and health habits.

(c) Only personnel authorized by supervisory personnel shall enter those areas of the buildings and facilities designated as limited-access areas.

(d) Any person shown at any time (either by medical examination or supervisory observation) to have an apparent illness or open lesions that may adversely affect the safety or quality of drug products shall be excluded from direct contact with components, drug product containers, closures, in-process materials, and drug products until the condition is corrected or determined by competent medical personnel not to jeopardize the safety or quality of drug products. All personnel shall be instructed to report to supervisory personnel any health conditions that may have an adverse effect on drug products.

§ 211.34 Consultants.

Consultants advising on the manufacture, processing, packing, or holding of drug products shall have sufficient education, training, and experience, or any combination thereof, to advise on the subject for which they are retained. Records shall be maintained stating the name, address, and qualifications of any consultants and the type of service they provide.

Subpart C—Buildings and Facilities

§ 211.42 Design and construction features.

(a) Any building or buildings used in the manufacture, processing, packing, or holding of a drug product shall be of suitable size, construction and location to facilitate cleaning, maintenance, and proper operations.

(b) Any such building shall have adequate space for the orderly placement of equipment and materials to prevent mixups between different components, drug product containers, closures, labeling, in-process materials, or drug products, and to prevent contamination. The flow of components, drug product containers, closures, labeling, in-process materials, and drug products through the building or buildings shall be designed to prevent contamination.

(c) Operations shall be performed within specifically defined areas of adequate size. There shall be separate or defined areas for the firm's operations to prevent contamination or mixups as follows:

(1) Receipt, identification, storage, and withholding from use of components, drug product containers, closures, and labeling, pending the appropriate sampling, testing, or examination by the quality control unit before release for manufacturing or packaging;

(2) Holding rejected components, drug product containers, closures, and labeling before disposition;

(3) Storage of released components, drug product containers, closures, and labeling;

(4) Storage of in-process materials;

(5) Manufacturing and processing operations;

(6) Packaging and labeling operations;

(7) Quarantine storage before release of drug products;

(8) Storage of drug products after release;

(9) Control and laboratory operations;

(10) Aseptic processing, which includes as appropriate:

(i) Floors, walls, and ceilings of smooth, hard surfaces that are easily cleanable;

(ii) Temperature and humidity controls;

(iii) An air supply filtered through high-efficiency particulate air filters under positive pressure, regardless of whether flow is laminar or nonlaminar;

(iv) A system for monitoring environmental conditions;

(v) A system for cleaning and disinfecting the room and equipment to produce aseptic conditions;

(vi) A system for maintaining any equipment used to control the aseptic conditions.

(d) Operations relating to the manufacture, processing, and packing of penicillin shall be performed in facilities separate from those used for other drug products for human use.

§ 211.44 Lighting.

Adequate lighting shall be provided in all areas.

§ 211.46 Ventilation, air filtration, air heating and cooling.

(a) Adequate ventilation shall be provided.

(b) Equipment for adequate control over air pressure, microorganisms, dust, humidity, and temperature shall be provided when appropriate for the manufacture, processing, packing, or holding of a drug product.

(c) Air filtration systems, including prefilters and particulate matter air filters, shall be used when appropriate on air supplies to production areas. If air is recirculated to production areas, measures shall be taken to control recirculation of dust from production. In areas where air contamination occurs during production, there shall be adequate exhaust systems or other systems adequate to control contaminants.

(d) Air-handling systems for the manufacture, processing, and packing of penicillin shall be completely separate from those for other drug products for human use.

§ 211.48 Plumbing.

(a) Potable water shall be supplied under continuous positive pressure in a plumbing system free of defects that could contribute contamination to any drug product. Potable water shall meet the standards prescribed in the Environmental Protection Agency's Primary Drinking Water Regulations set forth in 40 CFR Part 141. Water not meeting such standards shall not be permitted in the potable water system.

(b) Drains shall be of adequate size and, where connected directly to a sewer, shall be provided with an air break or other mechanical device to prevent back-siphonage.

§ 211.50 Sewage and refuse.

Sewage, trash, and other refuse in and from the building and immediate premises shall be disposed of in a safe and sanitary manner.

§ 211.52 Washing and toilet facilities.

Adequate washing facilities shall be provided, including hot and cold water, soap or detergent, air driers or single-service towels, and clean toilet facilities easily accessible to working areas.

§ 211.56 Sanitation.

(a) Any building used in the manufacture, processing, packing, or holding of a drug product shall be maintained in a clean and sanitary condition. Any such building shall be free of infestation by rodents, birds, insects, and other vermin (other than laboratory animals). Trash and organic waste matter shall be held and disposed of in a timely and sanitary manner.

(b) There shall be written procedures assigning responsibility for sanitation and describing in sufficient detail the cleaning schedules, methods, equipment, and materials to be used in cleaning the buildings and facilities; such written procedures shall be followed.

(c) There shall be written procedures for use of suitable rodenticides, insecticides, fungicides, fumigating agents, and cleaning and sanitizing agents. Such written procedures shall be designed to prevent the contamination of equipment, components, drug product containers, closures, packaging, labeling materials, or drug products and shall be followed. Rodenticides, insecticides, and fungicides shall not be used unless registered and used in accordance with the Federal Insecticide, Fungicide, and Rodenticide Act (7 U.S.C. 135).

(d) Sanitation procedures shall apply to work performed by contractors or temporary employees as well as work performed by full-time employees during the ordinary course of operations.

§ 211.58 Maintenance.

Any building used in the manufacture, processing, packing, or holding of a drug product shall be maintained in a good state of repair.

Subpart D—Equipment

§ 211.63 Equipment design, size, and location.

Equipment used in the manufacture, processing, packing, or holding of a drug product shall be of appropriate design, adequate size, and suitably located to facilitate operations for its intended use and for its cleaning and maintenance.

§ 211.65 Equipment construction.

(a) Equipment shall be constructed so that surfaces that contact components, in-process materials, or drug products shall not be reactive, additive, or absorptive so as to alter the safety, identity, strength, quality, or purity of the drug product beyond the official or other established requirements.

(b) Any substances required for operation, such as lubricants or coolants, shall not come into contact with components, drug product containers, closures, in-process materials, or drug products so as to alter the safety, identity, strength, quality, or purity of the drug product beyond the official or other established requirements.

§ 211.67 Equipment cleaning and maintenance.

(a) Equipment and utensils shall be cleaned, maintained, and sanitized at appropriate intervals to prevent malfunctions or contamination that would alter the safety, identity, strength, quality, or purity of the drug product beyond the official or other established requirements.

(b) Written procedures shall be established and followed for cleaning and maintenance of equipment, including utensils, used in the manufacture, processing, packing, or holding of a drug product. These procedures shall include, but are not necessarily limited to, the following:

(1) Assignment of responsibility for cleaning and maintaining equipment;

(2) Maintenance and cleaning schedules, including, where appropriate, sanitizing schedules;

(3) A description in sufficient detail of the methods, equipment, and materials used in cleaning and maintenance operations, and the methods of disassembling and reassembling equipment as necessary to assure proper cleaning and maintenance;

(4) Removal or obliteration of previous batch identification;

(5) Protection of clean equipment from contamination prior to use;

(6) Inspection of equipment for cleanliness immediately before use.

(c) Records shall be kept of maintenance, cleaning, sanitizing, and inspection as specified in §§ 211.180 and 211.182.

§ 211.68 Automatic, mechanical, and electronic equipment.

(a) Automatic, mechanical, or electronic equipment or other types of equipment, including computers, or related systems that will perform a function satisfactorily, may be used in the manufacture, processing, packing, and holding of a drug product. If such equipment is so used, it shall be routinely calibrated, inspected, or checked according to a written program designed to assure proper performance. Written records of those calibration checks and inspections shall be maintained.

(b) Appropriate controls shall be exercised over computer or related systems to assure that changes in master production and control records or other records are instituted only by authorized personnel. Input to and output from the computer or related system of formulas or other records or data shall be checked for accuracy. A backup file of data entered into the computer or related system shall be maintained except where certain data, such as calculations performed in connection with laboratory analysis, are eliminated by computerization or other automated processes. In such instances a written record of the program shall be maintained along with appropriate validation data. Hard copy or alternative systems, such as duplicates, tapes, or microfilm, designed to assure that backup data are exact and complete and that it is secure from alteration, inadvertent erasures, or loss shall be maintained.

§ 211.72 Filters.

Filters for liquid filtration used in the manufacture, processing, or packing of injectable drug products intended for human use shall not release fibers into such products. Fiber-releasing filters may not be used in the manufacture, processing, or packing of these injectable drug products unless it is not possible to manufacture such drug products without the use of such filters. If use of a fiber-releasing filter is necessary, an additional non-fiber-releasing filter of 0.22 micron maximum mean porosity (0.45 micron if the manufacturing conditions so dictate) shall subsequently be used to reduce the content of particles in the injectable drug product. Use of an asbestos-containing filter, with or without subsequent use of a specific non-fiber-releasing filter, is permissible only upon submission of proof to the appropriate bureau of the Food and Drug Administration that use of a non-fiber-releasing filter will, or is likely to, compromise the safety or effectiveness of the injectable drug product.

Subpart E—Control of Components and Drug Product Containers and Closures

§ 211.80 General requirements.

(a) There shall be written procedures describing in sufficient detail the receipt, identification, storage, handling, sampling, testing, and approval or rejection of components and drug product containers and closures; such written procedures shall be followed.

(b) Components and drug product containers and closures shall at all times be handled and stored in a manner to prevent contamination.

(c) Bagged or boxed components of drug product containers, or closures shall be stored off the floor and suitably spaced to permit cleaning and inspection.

(d) Each container or grouping of containers for components or drug product containers, or closures shall be identified with a distinctive code for each lot in each shipment received. This code shall be used in recording the disposition of each lot. Each lot shall be appropriately identified as to its status (i.e., quarantined, approved, or rejected).

§ 211.82 Receipt and storage of untested components, drug product containers, and closures.

(a) Upon receipt and before acceptance, each container or grouping of containers of components, drug product containers, and closures shall be examined visually for appropriate labeling as to contents, container damage or broken seals, and contamination.

(b) Components, drug product containers, and closures shall be stored under quarantine until they have been tested or examined, as appropriate, and released. Storage within the area shall conform to the requirements of § 211.80.

§ 211.84 Testing and approval or rejection of components, drug product containers, and closures.

(a) Each lot of components, drug product containers, and closures shall be withheld from use until the lot has been sampled, tested, or examined, as appropriate, and released for use by the quality control unit.

(b) Representative samples of each shipment of each lot shall be collected for testing or examination. The number of containers to be sampled, and the amount of material to be taken from each container, shall be based upon appropriate criteria such as statistical criteria for component variability, confidence levels, and degree of precision desired, the past quality history of the supplier, and the quantity needed for analysis and reserve where required by § 211.170.

(c) Samples shall be collected in accordance with the following procedures:

(1) The containers of components selected shall be cleaned where necessary, by appropriate means.

(2) The containers shall be opened, sampled, and resealed in a manner designed to prevent contamination of their contents and contamination of other components, drug product containers, or closures.

(3) Sterile equipment and aseptic sampling techniques shall be used when necessary.

(4) If it is necessary to sample a component from the top, middle, and bottom of its container, such sample subdivisions shall not be composited for testing.

(5) Sample containers shall be identified so that the following information can be determined: name of the material sampled, the lot number, the container from which the sample was taken, the data on which the sample was taken, and the name of the person who collected the sample.

(6) Containers from which samples have been taken shall be marked to show that samples have been removed from them.

(d) Samples shall be examined and tested as follows:

(1) At least one test shall be conducted to verify the identity of each component of a drug product. Specific identity tests, if they exist, shall be used.

(2) Each component shall be tested for conformity with all appropriate written specifications for purity, strength, and quality. In lieu of such testing by the manufacturer, a report of analysis may be accepted from the supplier of a component, provided that at least one specific identity test is conducted on such component by the manufacturer, and provided that the manufacturer establishes the reliability of the supplier's analyses through appropriate validation of the supplier's test results at appropriate intervals.

(3) Containers and closures shall be tested for conformance with all appropriate written procedures. In lieu of such testing by the manufacturer, a certificate of testing may be accepted from the supplier, provided that at least a visual identification is conducted on such containers/closures by the manufacturer and provided that the manufacturer establishes the reliability of the supplier's test results through appropriate validation of the supplier's test results at appropriate intervals.

(4) When appropriate, components shall be microscopically examined.

(5) Each lot of a component, drug product container, or closure that is liable to contamination with filth, insect infestation, or other extraneous adulterant shall be examined against established specifications for such contamination.

(6) Each lot of a component, drug product container, or closure that is liable to microbiological contamination that is objectionable in view of its intended use shall be subjected to microbiological tests before use.

(e) Any lot of components, drug product containers, or closures that meets the appropriate written specifications of identity, strength, quality, and purity and related tests under paragraph (d) of this section may be approved and released for use. Any lot of such material that does not meet such specifications shall be rejected.

§ 211.86 Use of approved components, drug product containers, and closures.

Components, drug product containers, and closures approved for use shall be rotated so that the oldest approved stock is used first. Deviation from this requirement is permitted if such deviation is temporary and appropriate.

§ 211.87 Retesting of approved components, drug product containers, and closures.

Components, drug product containers, and closures shall be retested or reexamined, as appropriate, for identity, strength, quality, and purity and approved or rejected by the quality control unit in accordance with § 211.84 as necessary, e.g., after storage for long periods or after exposure to air, heat or other conditions that might adversely affect the component, drug product container, or closure.

§ 211.89 Rejected components, drug product containers, and closures.

Rejected components, drug product containers, and closures shall be identified and controlled under a quarantine system designed to prevent their use in manufacturing or processing operations for which they are unsuitable.

§ 211.94 Drug product containers and closures.

(a) Drug product containers and closures shall not be reactive, additive, or absorptive so as to alter the safety, identity, strength, quality, or purity of the drug beyond the official or established requirements.

(b) Container closure systems shall provide adequate protection against foreseeable external factors in storage and use that can cause deterioration or contamination of the drug product.

(c) Drug product containers and closures shall be clean and, where indicated by the nature of the drug, sterilized and processed to remove pyrogenic properties to assure that they are suitable for their intended use.

(d) Standards or specifications, methods of testing, and, where indicated, methods of cleaning, sterilizing, and processing to remove pyrogenic properties shall be written and followed for drug product containers and closures.

Subpart F—Production and Process Controls

§ 211.100 Written procedures; deviations.

(a) There shall be written procedures for production and process control designed to assure that the drug products have the identity, strength, quality, and purity they purport or are represented to possess. Such procedures shall include all requirements in this subpart. These written procedures, including any changes, shall be drafted, reviewed, and approved by the appropriate organizational units and reviewed and approved by the quality control unit.

(b) Written production and process control procedures shall be followed in the execution of the various production and process control functions and shall be documented at the time of performance. Any deviation from the written procedures shall be recorded and justified.

§ 211.101 Charge-in of components.

Written production and control procedures shall include the following, which are designed to assure that the drug products produced have the identity, strength, quality, and purity they purport or are represented to possess:

(a) The batch shall be formulated with the intent to provide not less than 100 percent of the labeled or established amount of active ingredient.

(b) Components for drug product manufacturing shall be weighed, measured, or subdivided as appropriate. If a component is removed from the original container to another, the new container shall be identified with the following information:

(1) Component name or item code;
(2) Receiving or control number;
(3) Weight or measure in new container;
(4) Batch for which component was dispensed, including its product name, strength, and lot number.

(c) Weighing, measuring, or subdividing operations for components shall be adequately supervised. Each container of component dispensed to manufacturing shall be examined by a second person to assure that:

(1) The component was released by the quality control unit;
(2) The weight or measure is correct as stated in the batch production records;
(3) The containers are properly identified.

(d) Each component shall be added to the batch by one person and verified by a second person.

§ 211.103 Calculation of yield.

Actual yields and percentages of theoretical yield shall be determined at the conclusion of each appropriate phase of manufacturing, processing, packaging, or holding of the drug product. Such calculations shall be performed by one person and independently verified by a second person.

§ 211.105 Equipment identification.

(a) All compounding and storage containers, processing lines, and major equipment used using the production of a batch of a drug product shall be properly identified at all times to indicate their contents and, when necessary, the phase of processing of the batch.

(b) Major equipment shall be identified by a distinctive identification number or code that shall be recorded in the batch production record to show the specific equipment used in the manufacture of each batch of a drug product. In cases where only one of a particular type of equipment exists in a manufacturing facility, the name of the equipment may be used in lieu of a distinctive identification number or code.

§ 211.110 Sampling and testing of in-process materials and drug products.

(a) To assure batch uniformity and integrity of drug products, written procedures shall be established and followed that describe the in-process controls, and tests, or examinations to be conducted on appropriate samples of in-process materials of each batch. Such control procedures shall be established to monitor the output and to validate the performance of those manufacturing processes that may be responsible for causing variability in the characteristics of in-process material and the drug product. Such control procedures shall include, but are not limited to, the following, where appropriate:

(1) Tablet or capsule weight variation;
(2) Disintegration time;
(3) Adequacy of mixing to assure uniformity and homogeneity;
(4) Dissolution time and rate;
(5) Clarity, completeness, or pH of solutions.

(b) Valid in-process specifications for such characteristics shall be consistent with drug product final specifications and shall be derived from previous acceptable process average and process variability estimates where possible and determined by the application of suitable statistical procedures where appropriate. Examination and testing of samples shall assure that the drug product and in-process material conform to specifications.

(c) In-process materials shall be tested for identity, strength, quality, and purity as appropriate, and approved or rejected by the quality control unit, during the production process, e.g., at commencement or completion of significant phases or after storage for long periods.

(d) Rejected in-process materials shall be identified and controlled under a quarantine system designed to prevent their use in manufacturing or processing operations for which they are unsuitable.

§ 211.111 Time limitations on production.

When appropriate, time limits for the completion of each phase of production shall be established to assure the quality of the drug product. Deviation from established time limits may be acceptable if such deviation does not compromise the quality of the drug product. Such deviation shall be justified and documented.

§ 211.113 Control of microbiological contamination.

(a) Appropriate written procedures, designed to prevent objectionable microorganisms in drug products not required to be sterile, shall be established and followed.

(b) Appropriate written procedures, designed to prevent microbiological contamination of drug products purporting to be sterile, shall be established and followed. Such procedures shall include validation of any sterilization process.

§ 211.115 Reprocessing.

(a) Written procedures shall be established and followed prescribing a system for reprocessing batches that do not conform to standards or specifications and the steps to be taken to insure that the reprocessed batches will conform with all established standards, specifications, and characteristics.

(b) Reprocessing shall not be performed without the review and approval of the quality control unit.

Subpart G—Packaging and Labeling Control

§ 211.122 Materials examination and usage criteria.

(a) There shall be written procedures describing in sufficient detail the receipt, identification, storage, handling, sampling, examination, and/or testing of labeling and packaging materials; such written procedures shall be followed. Labeling and packaging materials shall be representatively sampled, and examined or tested upon receipt and before use in packaging or labeling of a drug product.

(b) Any labeling or packaging materials meeting appropriate written specifications may be approved and released for use. Any labeling or packaging materials that do not meet such specifications shall be rejected to prevent their use in operations for which they are unsuitable.

(c) Records shall be maintained for each shipment received of each different labeling and packaging material indicating receipt, examination or testing, and whether accepted or rejected.

(d) Labels and other labeling materials for each different drug product, strength, dosage form, or quantity of contents shall be stored separately with suitable identification. Access to the storage area shall be limited to authorized personnel.

(e) Obsolete and outdated labels, labeling, and other packaging materials shall be destroyed.

(f) Gang printing of labeling to be used for different drug products or different strengths of the same drug product (or labeling of the same size and identical or similar format and/or color schemes) shall be minimized. If gang printing is employed, packaging and labeling operations shall provide for special control procedures, taking into consideration sheet layout, stacking, cutting, and handling during and after printing.

(g) Printing devices on, or associated with, manufacturing lines used to imprint labeling upon the drug product unit label or case shall be monitored to assure that all imprinting conforms to the print specified in the batch production record.

§ 211.125 Labeling issuance.

(a) Strict control shall be exercised over labeling issued for use in drug product labeling operations.

(b) Labeling materials issued for a batch shall be carefully examined for identity and conformity to the labeling specified in the master or batch production records.

(c) Procedures shall be utilized to reconcile the quantities of labeling issued, used, and returned, and shall require evaluation of discrepancies found between the quantity of drug product finished and the quantity of labeling issued when such discrepancies are outside narrow preset limits based on historical operating data. Such discrepancies shall be investigated in accordance with § 211.192.

(d) All excess labeling bearing lot or control numbers shall be destroyed.

(e) Returned labeling shall be maintained and stored in a manner to prevent mixups and provide proper identification.

(f) Procedures shall be written describing in sufficient detail the control procedures employed for the issuance of labeling; such written procedures shall be followed.

§ 211.130 Packaging and labeling operations.

There shall be written procedures designed to assure that correct labels, labeling, and packaging materials are used for drug products; such written procedures shall be followed. These procedures shall incorporate the following features:

(a) Prevention of mixups and cross-contamination by physical or spatial separation from operations on other drug products.

(b) Identification of the drug product with a lot or control number that permits determination of the history of the manufacture and control of the batch.

(c) Examination of packaging and labeling materials for suitability and correctness before packaging operations, and documentation of such examination in the batch production record.

(d) Inspection of the packaging and labeling facilities immediately before use to assure that all drug products have been removed from previous operations. Inspection shall also be made to assure that packaging and labeling materials not suitable for subsequent operations have been removed. Results of inspection shall be documented in the batch production records.

§ 211.132 Tamper-resistant packaging requirements for over-the-counter human drug products.

(a) *General.* Because most over-the-counter (OTC) human drug products are not now packaged in tamper-resistant retail packages, there is the opportunity for the malicious adulteration of OTC drug products with health risks to individuals who unknowingly purchase adulterated products and with loss of consumer confidence in the security of OTC drug product packages. The Food and Drug Administration has the authority and responsibility under the Federal Food, Drug, and Cosmetic Act (the act) to establish a uniform national requirement for tamper-resistant packaging of OTC drug products that will improve the security of OTC drug packaging and help assure the safety and effectiveness of OTC drug products. An OTC drug product (except a dermatological, dentifrice, insulin, or lozenge product) for retail sale that is not packaged in a tamper-resistant package or that is not properly labeled under this section is adulterated under section 501 of the act or misbranded under section 502 of the act, or both.

(b) *Requirement for tamper-resistant package.* Each manufacturer and packer who packages an OTC drug product (except a dermatological, dentifrice, insulin, or lozenge product) for retail sale, shall package the product in a tamper-resistant package, if this product is accessible to the public while held for sale. A tamper-resistant package is one having an indicator or barrier to entry which, if breached or missing, can reasonably be expected to provide visible evidence to consumers that tampering has occurred. To reduce the likelihood of substitution of a tamper-resistant feature after tampering, the indicator or barrier to entry is required to be distinctive by design (e.g., an aerosol product container) or by the use of an identifying characteristic (e.g., a pattern, name, registered trademark, logo, or picture). For purposes of this section, the term "distinctive by design" means the packaging cannot be duplicated with commonly available materials or through commonly available processes. For purposes of this section, the term "aerosol product" means a product which depends upon the power of a liquified or compressed gas to expel the contents from the container. A tamper-resistant package may involve an immediate-container and closure system or secondary-container or carton system or any combination of systems intended to provide a visual indication of package integrity. The tamper-resistant feature shall be designed to and shall remain intact when handled in a reasonable manner during manufacture, distribution, and retail display.

(c) *Labeling.* Each retail package of an OTC drug product covered by this section, except ammonia inhalant in crushable glass ampules, aerosol products as defined in paragraph (b) of this section, or containers of compressed medical oxygen, is required to bear a statement that is prominently placed so that consumers are alerted to the specific tamper-resistant feature of the package. The labeling statement is also required to be so placed that it will be unaffected if the tamper-resistant feature of the package is breached or missing. If the tamper-resistant feature chosen to meet the requirement in paragraph (b) of this section is one that uses an identifying characteristic, that characteristic is required to be referred to in the labeling statement. For example, the labeling statement on a bottle with a shrink band could say "For your protection, this bottle has an imprinted seal around the neck."

(d) *Requests for exemptions from packaging and labeling requirements.* A manufacturer or packer may request an exemption from the packaging and labeling requirements of this section. A request for an exemption is required to be submitted in the form of a citizen petition under § 10.30 of this chapter and should be clearly identified on the envelope as a "Request for Exemption from Tamper-resistant Rule." The petition is required to contain the following:

(1) The name of the drug product or, if the petition seeks an exemption for a drug class, the name of the drug class, and a list of products within that class.

(2) The reasons that the drug product's compliance with the tamper-resistant packaging or labeling requirements of this section is unnecessary or cannot be achieved.

(3) A description of alternative steps that are available, or that the petitioner has already taken, to reduce the likelihood that the product or drug class will be the subject of malicious adulteration.

(4) Other information justifying an exemption.

This information collection requirement has been approved by the Office of Management and Budget under number 0910–0149.

(e) *OTC drug products subject to approved new drug applications.* Holders of approved new drug applications for OTC drug products are required under § 314.8 (a) (4) (vi), (5) (xi), or (d) (5) of this chapter to provide for changes in packaging, and under § 314.8 (a) (5) (xii) to provide for changes in labeling to comply with the requirements of this section.

(f) *Poison Prevention Packaging Act of 1970.* This section does not affect any requirements for "special packaging" as defined under § 310.3 (l) of this chapter and required under the Poison Prevention Packaging Act of 1970.

(g) *Effective date.* OTC drug products, except dermatological, dentifrice, insulin, and lozenge products, are required to comply with the requirements of this section on the dates listed below except to the extent that a product's manufacturer or packer has obtained an exemption from a packaging or labeling requirement.

(1) *Initial effective date for packaging requirements.*

(i) The packaging requirement in paragraph (b) of this section is effective on February 7, 1983 for each affected OTC drug product (except oral and vaginal tablets, vaginal and rectal suppositories, and one-piece soft gelatin capsules) packaged for retail sale on or after that date, except for the requirement in paragraph (b) of this section for a distinctive indicator or barrier to entry.

(ii) The packaging requirement in paragraph (b) of this section is effective on May 5, 1983 for each OTC drug product that is an oral or vaginal tablet, a vaginal or rectal suppository, or one-piece soft gelatin capsules packaged for retail sale on or after that date.

(2) *Initial effective date for labeling requirements.* The requirement in paragraph (b) of this section that the indicator or barrier to entry be distinctive by design and the requirement in paragraph (c) of this section for a labeling statement are effective on May 5, 1983 for each affected OTC drug product packaged on or after that date.

(3) *Retail level effective date.* The tamper-resistant packaging requirement of paragraph (b) of this section is effective on February 6, 1984 for each affected OTC drug product held for sale on or after that date that was packaged for retail sale before May 5, 1983. This does not include the requirement in paragraph (b) of this section that the indicator or barrier to entry be distinctive by design. Products packaged for retail sale after May 5, 1983, are required to be in compliance with all aspects of the regulations without regard to the retail level effective date.

§ 211.134 Drug product inspection.

(a) Packaged and labeled products shall be examined during finishing operations to provide assurance that containers and packages in the lot have the correct label.

(b) A representative sample of units shall be collected at the completion of finishing operations and shall be visually examined for correct labeling.

(c) Results of these examinations shall be recorded in the batch production or control records.

§ 211.137 Expiration dating.

(a) To assure that a drug product meets applicable standards of identity, strength, quality, and purity at the time of use, it shall bear an expiration date determined by appropriate stability testing described in § 211.166.

(b) Expiration dates shall be related to any storage conditions stated on the labeling, as determined by stability studies described in § 211.166.

(c) If the drug product is to be reconstituted at the time of dispensing, its labeling shall bear expiration information for both the reconstituted and unreconstituted drug products.

(d) Expiration dates shall appear on labeling in accordance with the requirements of § 201.17 of this chapter.

(e) Homeopathic drug products shall be exempt from the requirements of this section.

(f) Allergenic extracts that are labeled "No U.S. Standard of Potency" are exempt from the requirements of this section.

(g) Pending consideration of a proposed exemption, published in the FEDERAL REGISTER of September 29, 1978, the requirements in this section shall not be enforced for human OTC drug products if their labeling does not bear dosage limitations and they are stable for at least 3 years as supported by appropriate stability data.

Subpart H—Holding and Distribution

§ 211.142 Warehousing procedures.

Written procedures describing the warehousing of drug products shall be established and followed. They shall include:

(a) Quarantine of drug products before release by the quality control unit.

(b) Storage of drug products under appropriate conditions of temperature, humidity, and light so that the identity, strength, quality, and purity of the drug products are not affected.

§ 211.150 Distribution procedures.

Written procedures shall be established, and followed, describing the distribution of drug products. They shall include:

(a) A procedure whereby the oldest approved stock of a drug product is distributed first. Deviation from this requirement is permitted if such deviation is temporary and appropriate.

(b) A system by which the distribution of each lot of drug product can be readily determined to facilitate its recall if necessary.

Subpart I—Laboratory Controls

§ 211.160 General requirements.

(a) The establishment of any specifications, standards, sampling plans, test procedures, or other laboratory control mechanisms required by this subpart, including any change in such specifications, standards, sampling plans, test procedures, or other laboratory control mechanisms, shall be drafted by the appropriate organizational unit and reviewed and approved by the quality control unit. The requirements in this subpart shall be followed and shall be documented at the time of performance. Any deviation from the written specifications, standards, sampling plans, test procedures, or other laboratory control mechanisms shall be recorded and justified.

(b) Laboratory controls shall include the establishment of scientifically sound and appropriate specifications, standards, sampling plans, and test procedures designed to assure that components, drug product containers, closures, in-process materials, labeling, and drug products conform to appropriate standards of identity, strength, quality, and purity. Laboratory controls shall include:

(1) Determination of conformance to appropriate written specifications for the acceptance of each lot within each shipment of components, drug product containers, closures, and labeling used in the manufacture, processing, packing, or holding of drug products. The specifications shall include a description of the sampling and testing procedures used. Samples shall be representative and adequately identified. Such procedures shall also require appropriate retesting of any component, drug product container, or closure that is subject to deterioration.

(2) Determination of conformance to written specifications and a description of sampling and testing procedures for in-process materials. Such samples shall be representative and properly identified.

(3) Determination of conformance to written descriptions of sampling procedures and appropriate specifications for drug products. Such samples shall be representative and properly identified.

(4) The calibration of instruments, apparatus, gauges, and recording devices at suitable intervals in accordance with an established written program containing specific directions, schedules, limits for accuracy and precision, and provisions for remedial action in the event accuracy and/or precision limits are not met. Instruments, apparatus, gauges, and recording devices not meeting established specifications shall not be used.

§ 211.165 Testing and release for distribution.

(a) For each batch of drug product, there shall be appropriate laboratory determination of satisfactory conformance to final specifications for the drug product, including the identity and strength of each active ingredient, prior to release. Where sterility and/or pyrogen testing are conducted on specific batches of short-lived radiopharmaceuticals, such batches may be released prior to completion of sterility and/or pyrogen testing, provided such testing is completed as soon as possible.

(b) There shall be appropriate laboratory testing, as necessary, of each batch of drug product required to be free of objectionable microorganisms.

(c) Any sampling and testing plans shall be described in written procedures that shall include the method of sampling and the number of units per batch to be tested; such written procedure shall be followed.

(d) Acceptance criteria for the sampling and testing conducted by the quality control unit shall be adequate to assure that batches of drug products meet each appropriate specification and appropriate statistical quality control criteria as a condition for their approval and release. The statistical quality control criteria shall include appropriate acceptance levels and/or appropriate rejection levels.

(e) The accuracy, sensitivity, specificity, and reproducibility of test methods employed by the firm shall be established and documented. Such validation and documentation may be accomplished in accordance with § 211.194 (a) (2).

(f) Drug products failing to meet established standards or specifications and any other relevant quality control criteria shall be rejected. Reprocessing may be performed. Prior to acceptance and use, reprocessed material must meet appropriate standards, specifications, and any other relevant criteria.

§ 211.166 Stability testing.

(a) There shall be a written testing program designed to assess the stability characteristics of drug products. The results of such stability testing shall be used in determining appropriate storage conditions and expiration dates. The written program shall be followed and shall include:

(1) Sample size and test intervals based on statistical criteria for each attribute examined to assure valid estimates of stability;

(2) Storage conditions for samples retained for testing;

(3) Reliable, meaningful, and specific test methods;

(4) Testing of the drug product in the same container-closure system as that in which the drug product is marketed;

(5) Testing of drug products for reconstitution at the time of dispensing (as directed in the labeling) as well as after they are reconstituted.

(b) An adequate number of batches of each drug product shall be tested to determine an appropriate expiration date and a record of such data shall be maintained. Accelerated studies, combined with basic stability information on the components, drug products, and container-closure system, may be used to support tentative expiration dates provided full shelf life studies are not available and are being conducted. Where data from accelerated studies are used to project a tentative expiration date that is beyond a date supported by actual shelf life studies, there must be stability studies conducted, including drug product testing at appropriate intervals, until the tentative expiration date is verified or the appropriate expiration date determined.

(c) For homeopathic drug products, the requirements of this section are as follows:

(1) There shall be a written assessment of stability based at least on testing or examination of the drug product for compatibility of the ingredients, and based on marketing experience with the drug product to indicate that there is no degradation of the product for the normal or expected period of use.

(2) Evaluation of stability shall be based on the same container-closure system in which the drug product is being marketed.

(d) Allergenic extracts that are labeled "No U.S. Standard of Potency" are exempt from the requirements of this section.

§ 211.167 Special testing requirements.

(a) For each batch of drug product purporting to be sterile and/or pyrogen-free, there shall be appropriate laboratory testing to determine conformance to such requirements. The test procedures shall be in writing and shall be followed.

(b) For each batch of ophthalmic ointment, there shall be appropriate testing to determine conformance to specifications regarding the presence of foreign particles and harsh or abrasive substances. The test procedures shall be in writing and shall be followed.

(c) For each batch of controlled-release dosage form, there shall be appropriate laboratory testing to determine conformance to the specifications for the rate of release of each active ingredient. The test procedures shall be in writing and shall be followed.

§ 211.170 Reserve samples.

(a) An appropriately identified reserve sample that is representative of each lot in each shipment of each active ingredient shall be retained. The reserve sample consists of at least twice the quantity necessary for all tests required to determine whether the active ingredient meets its established specifications, except for sterility and pyrogen testing. The retention time is as follows:

(1) For an active ingredient in a drug product other than those described in paragraphs (a) (2) and (3) of this section, the reserve sample shall be retained for 1 year after the expiration date of the last lot of the drug product containing the active ingredient.

(2) For an active ingredient in a radioactive drug product, except for nonradioactive reagent kits, the reserve sample shall be retained for:

(i) Three months after the expiration date of the last lot of the drug product containing the active ingredient if the expiration dating period of the drug product is 30 days or less; or

(ii) Six months after the expiration date of the last lot of the drug product containing the active ingredient if the expiration dating period of the drug product is more than 30 days.

(3) For an active ingredient in an OTC drug product that is exempt from bearing an expiration date under § 211.137, the reserve sample shall be retained for 3 years after distribution of the last lot of the drug product containing the active ingredient.

(b) An appropriately identified reserve sample that is representative of each lot or batch of drug product shall be retained and stored under conditions consistent with product labeling. The reserve sample shall be stored in the same immediate container-closure system in which the drug product is marketed or in one that has essentially the same characteristics. The reserve sample consists of at least twice the quantity necessary to perform all the required tests, except those for sterility and pyrogens. Reserve samples, except those drug products described in paragraph (b) (2), shall be examined visually at least once a year for evidence of deterioration unless visual examination would affect the integrity of the reserve samples. Any evidence of reserve sample deterioration shall be investigated in accordance with § 211.192. The results of the examination shall be recorded and maintained with other stability data on the drug product. Reserve samples of compressed medical gases need not be retained. The retention time is as follows:

(1) For a drug product other than those described in paragraphs (b) (2) and (3) of this section, the reserve sample shall be retained for 1 year after the expiration date of the drug product.

(2) For a radioactive drug product, except for nonradioactive reagent kits, the reserve sample shall be retained for:

(i) Three months after the expiration date of the drug product if the expiration dating period of the drug product is 30 days or less; or

(ii) Six months after the expiration date of the drug product if the expiration dating period of the drug product is more than 30 days.

(3) For an OTC drug product that is exempt from bearing an expiration date under § 211.137, the reserve sample must be retained for 3 years after the lot or batch of drug product is distributed.

§ 211.173 Laboratory animals.

Animals used in testing components, in-process materials, or drug products for compliance with established specifications shall be maintained and controlled in a manner that assures their suitability for their intended use. They shall be identified, and adequate records shall be maintained showing the history of their use.

§ 211.176 Penicillin contamination.

If a reasonable possibility exists that a non-penicillin drug product has been exposed to cross-contamination with penicillin, the non-penicillin drug product shall be tested for the presence of penicillin. Such drug product shall not be marketed if detectable levels are found when tested according to procedures specified in 'Procedures for Detecting and Measuring Penicillin Contamination in Drugs,' which is incorporated by reference. Copies are available from the Bureau of Drugs (HFD-430), Food and Drug Administration, 200 C St., SW., Washington, DC 20204, or available for inspection at the Office of the Federal Register, 1100 L St. NW., Washington, DC 20408. [Ed. Note—The Bureau of Drugs (HFD-430) is now designated as National Center for Drugs and Biologics (HFN-416).]

Subpart J—Records and Reports

§ 211.180 General requirements.

(a) Any production, control, or distribution record that is required to be maintained in compliance with this part and is specifically associated with a batch of a drug product shall be retained for at least 1 year after the expiration date of the batch or, in the case of certain OTC drug products lacking expiration dating because they meet the criteria for exemption under § 211.137, 3 years after distribution of the batch.

(b) Records shall be maintained for all components, drug product containers, closures, and labeling for at least 1 year after the expiration date or, in the case of certain OTC drug products lacking expiration dating because they meet the criteria for exemption under § 211.137, 3 years after distribution of the last lot of drug product incorporating the component or using the container, closure, or labeling.

(c) All records required under this part, or copies of such records, shall be readily available for authorized inspection during the retention period at the establishment where the activities described in such records occurred. These records or copies thereof shall be subject to photocopying or other means of reproduction as part of such inspection. Records that can be immediately retrieved from another location by computer or other electronic means shall be considered as meeting the requirements of this paragraph.

(d) Records required under this part may be retained either as original records or as true copies such as photocopies, microfilm, microfiche, or other accurate reproductions of the original records. Where reduction techniques, such as microfilming, are used, suitable reader and photocopying equipment shall be readily available.

(e) Written records required by this part shall be maintained so that data therein can be used for evaluating, at least annually, the quality standards of each drug product to determine the need for changes in drug product specifications or manufacturing or control procedures. Written procedures shall be established and followed for such evaluations and shall include provisions for:

(1) A review of every batch, whether approved or rejected, and, where applicable, records associated with the batch.

(2) A review of complaints, recalls, returned or salvaged drug products, and investigations conducted under § 211.192 for each drug product.

(f) Procedures shall be established to assure that the responsible officials of the firm, if they are not personally involved in or immediately aware of such actions, are notified in writing of any investigations conducted under §§ 211.198, 211.204, or 211.208 of these regulations, any recalls, reports of inspectional observations issued by the Food and Drug Administration, or any regulatory actions relating to good manufacturing practices brought by the Food and Drug Administration.

§ 211.182 Equipment cleaning and use log.

A written record of major equipment cleaning, maintenance (except routine maintenance such as lubrication and adjustments), and use shall be included in individual equipment logs that show the date, time, product, and lot number of each batch processed. If equipment is dedicated to manufacture of one product, then individual equipment logs are not required, provided that lots or batches of such product follow in numerical order and are manufactured in numerical sequence. In cases where dedicated equipment is employed, the records of cleaning, maintenance, and use shall be part of the batch record. The persons performing and double-checking the cleaning and maintenance shall date and sign or initial the log indicating that the work was performed. Entries in the log shall be in chronological order.

§ 211.184 Component, drug product container, closure, and labeling records.

These records shall include the following:

(a) The identity and quantity of each shipment of each lot of components, drug product containers, closures, and labeling; the name of the supplier; the supplier's lot number(s) if known; the receiving code as specified in § 211.80; and the date of receipt. The name and location of the prime manufacturer, if different from the supplier, shall be listed if known.

(b) The results of any test or examination performed (including those performed as required by § 211.82 (a), § 211.84 (d), or § 211.122 (a)) and the conclusions derived therefrom.

(c) An individual inventory record of each component, drug product container, and closure and, for each component, a reconciliation of the use of each lot of such component. The inventory record shall contain sufficient information to allow determination of any batch or lot of drug product associated with the use of each component, drug product container, and closure.

(d) Documentation of the examination and review of labels and labeling for conformity with established specifications in accord with §§ 211.122 (c) and 211.130 (c).

(e) The disposition of rejected components, drug product containers, closure, and labeling.

§ 211.186 Master production and control records.

(a) To assure uniformity from batch to batch, master production and control records for each drug product, including each batch size thereof, shall be prepared, dated, and signed (full signature, handwritten) by one person and independently checked, dated, and signed by a second person. The preparation of master production and control records shall be described in a written procedure and such written procedure shall be followed.

(b) Master production and control records shall include:

(1) The name and strength of the product and a description of the dosage form;

(2) The name and weight or measure of each active ingredient per dosage unit or per unit of weight or measure of the drug product, and a statement of the total weight or measure of any dosage unit;

(3) A complete list of components designated by names or codes sufficiently specific to indicate any special quality characteristic;

(4) An accurate statement of the weight or measure of each component, using the same weight system (metric, avoirdupois, or apothecary) for each component. Reasonable variations may be permitted, however, in the amount of components necessary for the preparation in the dosage form, provided they are justified in the master production and control records;

(5) A statement concerning any calculated excess of component;

(6) A statement of theoretical weight or measure at appropriate phases of processing;

(7) A statement of theoretical yield, including the maximum and minimum percentages of theoretical yield beyond which investigation according to § z211.192 is required;

(8) A description of the drug product containers, closures, and packaging materials, including a specimen or copy of each label and all other labeling signed and dated by the person or persons responsible for approval of such labeling;

(9) Complete manufacturing and control instructions, sampling and testing procedures, specifications, special notations, and precautions to be followed.

§ 211.188 Batch production and control records.

Batch production and control records shall be prepared for each batch of drug product produced and shall include complete information relating to the production and control of each batch. These records shall include:

(a) An accurate reproduction of the appropriate master production or control record, checked for accuracy, dated, and signed;

(b) Documentation that each significant step in the manufacture, processing, packing, or holding of the batch was accomplished, including:

(1) Dates;

(2) Identity of individual major equipment and lines used;

(3) Specific identification of each batch of component or in-process material used;

(4) Weights and measures of components used in the course of processing;

(5) In-process and laboratory control results;

(6) Inspection of the packaging and labeling area before and after use;

(7) A statement of the actual yield and a statement of the percentage of theoretical yield at appropriate phases of processing;

(8) Complete labeling control records, including specimens or copies of all labeling used;

(9) Description of drug product containers and closures;

(10) Any sampling performed;

(11) Identification of the persons performing and directly supervising or checking each significant step in the operation;

(12) Any investigation made according to § 211.192;

(13) Results of examinations made in accordance with § 211.134.

§ 211.192 Production record review.

All drug product production and control records, including those for packaging and labeling, shall be reviewed and approved by the quality control unit to determine compliance with all established, approved written procedures before a batch is released or distributed. Any unexplained discrepancy (including a percentage of theoretical yield exceeding the maximum or minimum percentages established in master production and control records) or the failure of a batch or any of its components to meet any of its specifications shall be thoroughly investigated, whether or not the batch has already been distributed. The investigation shall extend to other batches of the same drug product and other drug products that may have been associated with the specific failure or discrepancy. A written record of the investigation shall be made and shall include the conclusions and followup.

§ 211.194 Laboratory records.

(a) Laboratory records shall include complete data derived from all tests necessary to assure compliance with established specifications and standards, including examinations and assays, as follows:

(1) A description of the sample received for testing with identification of source (that is, location from where sample was obtained), quantity, lot number or other distinctive code, date sample was taken, and date sample was received for testing.

(2) A statement of each method used in the testing of the sample. The statement shall indicate the location of data that establish that the methods used in the testing of the sample meet proper standards of accuracy and reliability as applied to the product tested. (If the method employed is in the current revision of the United States Pharmacopeia, National Formulary, Association of Official Analytical Chemists, Book of Methods,* or in other recognized standard references, or is detailed in an approved new drug application and the referenced method is not modified, a statement indicating the method and reference will suffice.) The suitability of all testing methods used shall be verified under actual conditions of use.

(3) A statement of the weight or measure of sample used for each test, where appropriate.

(4) A complete record of all data secured in the course of each test, including all graphs, charts, and spectra from laboratory instrumentation, properly identified to show the specific component, drug product container, closure, in-process material, or drug product, and lot tested.

(5) A record of all calculations performed in connection with the test, including units of measure, conversion factors, and equivalency factors.

(6) A statement of the results of tests and how the results compare with established standards of identity, strength, quality, and purity for the component, drug product container, closure, in-process material, or drug product tested.

(7) The initials or signature of the person who performs each test and the date(s) the tests were performed.

(8) The initials or signature of a second person showing that the original records have been reviewed for accuracy, completeness, and compliance with established standards.

(b) Complete records shall be maintained of any modification of an established method employed in testing. Such records shall include the reason for the modification and data to verify that the modification produced results that are at least as accurate and reliable for the material being tested as the established method.

(c) Complete records shall be maintained of any testing and standardization of laboratory reference standards, reagents, and standard solutions.

(d) Complete records shall be maintained of the periodic calibration of laboratory instruments, apparatus, gauges, and recording devices required by § 211.160 (b) (4).

(e) Complete records shall be maintained of all stability testing performed in accordance with § 211.166.

§ 211.196 Distribution records.

Distribution records shall contain the name and strength of the product and a description of the dosage form, name and address of the consignee, date and quantity shipped, and lot or control number of the drug product.

* Copies may be obtained from: Association of Official Analytical Chemists, P. O. Box 540, Benjamin Franklin Station, Washington, DC 20204.

§ 211.198 Complaint files.

(a) Written procedures describing the handling of all written and oral complaints regarding a drug product shall be established and followed. Such procedures shall include provisions for review by the quality control unit, of any complaint involving the possible failure of a drug product to meet any of its specifications and, for such drug products, a determination as to the need for an investigation in accordance with § 211.192.

(b) A written record of each complaint shall be maintained in a file designated for drug product complaints. The file regarding such drug product complaints shall be maintained at the establishment where the drug product involved was manufactured, processed, or packed, or such file may be maintained at another facility if the written records in such files are readily available for inspection at that other facility. Written records involving a drug product shall be maintained until at least 1 year after the expiration date of the drug product, or 1 year after the date that the complaint was received, whichever is longer. In the case of certain OTC drug products lacking expiration dating because they meet the criteria for exemption under § 211.137, such written records shall be maintained for 3 years after distribution of the drug product.

(1) The written record shall include the following information, where known: the name and strength of the drug product, lot number, name of complainant, nature of complaint, and reply to complainant.

(2) Where an investigation under § 211.192 is conducted, the written record shall include the findings of the investigation and followup. The record or copy of the record of the investigation shall be maintained at the establishment where the investigation occurred in accordance with § 211.180 (c).

(3) Where an investigation under § 211.192 is not conducted, the written record shall include the reason that an investigation was found not to be necessary and the name of the responsible person making such a determination.

Subpart K—Returned and Salvaged Drug Products

§ 211.204 Returned drug products.

Returned drug products shall be identified as such and held. If the conditions under which returned drug products have been held, stored, or shipped before or during their return, or if the condition of the drug product, its container, carton, or labeling, as a result of storage or shipping, casts doubt on the safety, identity, strength, quality or purity of the drug product, the returned drug product shall be destroyed unless examination, testing, or other investigations prove the drug product meets appropriate standards of safety, identity, strength, quality, or purity. A drug product may be reprocessed provided the subsequent drug product meets appropriate standards, specifications, and characteristics. Records of returned drug products shall be maintained and shall include the name and label potency of the drug product dosage form, lot number (or control number or batch number), reason for the return, quantity returned, date of disposition, and ultimate disposition of the returned drug product. If the reason for a drug product being returned implicates associated batches, an appropriate investigation shall be conducted in accordance with the requirements of § 211.192. Procedures for the holding, testing, and reprocessing of returned drug products shall be in writing and shall be followed.

§ 211.208 Drug product salvaging.

Drug products that have been subjected to improper storage conditions including extremes in temperature, humidity, smoke, fumes, pressure, age, or radiation due to natural disasters, fires, accidents, or equipment failures shall not be salvaged and returned to the marketplace. Whenever there is a question whether drug products have been subjected to such conditions, salvaging operations may be conducted only if there is (a) evidence from laboratory tests and assays (including animal feeding studies where applicable) that the drug products meet all applicable standards of identity, strength, quality, and purity and (b) evidence from inspection of the premises that the drug products and their associated packaging were not subjected to improper storage conditions as a result of the disaster or accident. Organoleptic examinations shall be acceptable only as supplemental evidence that the drug products meet appropriate standards of identity, strength, quality, and purity. Records including name, lot number, and disposition shall be maintained for drug products subject to this section.

⟨1081⟩ GEL STRENGTH OF GELATIN

Pipet 105 mL of water at 10° to 15° into a standard Bloom bottle, add 7.5 g of Gelatin, and stir. Allow to stand for 1 hour, then bring to a temperature of 62° in 15 minutes by placing in a water bath regulated at 65° (the substance may be swirled several times to aid solution). Finally mix by inversion, allow to stand for 15 minutes, and place in a water bath at 10 ± 0.1°. Chill, without disturbance, for 17 hours. Determine the gel strength in a Bloom Gelometer (a device developed to make this determination under standardized conditions) adjusted for 4-mm depression and to deliver 200 ± 5 g of shot per 5 seconds, using the 12.7-mm diameter (nonbeveled) plunger.

⟨1086⟩ IMPURITIES IN OFFICIAL ARTICLES

Concepts about purity change with time and are inseparable from developments in analytical chemistry. If a material previously considered to be pure can be resolved into more than one component, that material can be redefined into new terms of purity and impurity. Inorganic, organic, biochemical, isomeric, or polymeric components can all be considered impurities. Microbiological species or strains are sometimes described in similar terms of resolving into more than one component.

Communications about compendial articles may be improved by including in this Pharmacopeia the definitions of terms and the contexts in which these terms are used. (See *Definitions* below.) There has been much activity and discussion in recent years about term definition. Certain industry-wide concerns about terminology and context deserve widespread publication and ready retrievability and are included here. (See *Industrial Concepts* below.) See *Foreign Substances and Impurities*, in the section *Tests and Assays*, under *General Notices* and *Requirements*, as well as the recently adopted general chapter, *Ordinary Impurities* ⟨466⟩. Some other general chapters added over the years have also addressed topics of purity or impurity as these have come into focus or as analytical methodology has become available. Analytical aspects are enlarged upon in the chapter *Validation of Compendial Methods* ⟨1225⟩.

Monographs on bulk pharmaceutical chemicals usually cite one of three types of purity tests: (1) a chromatographic purity test coupled with a nonspecific assay; (2) a chromatographic purity-indicating method that serves as the assay; or (3) a specific test and limit for a known impurity, an approach that usually requires a reference standard for that impurity. Modern separation methods clearly play a dominant role in scientific research today because these methods simultaneously separate and measure components and fulfill the analytical ideal of making measurements only on purified specimens. Nevertheless, the more classical methods based on titrimetry, colorimetry, spectrophotometry, single or multiple partitions, or changes in physical constants (or any other tests or assays) lose none of their previous validities. The *purity profile* of a specimen that is constructed from the results of experiments using a number of analytical methods is the ultimate goal.

Purity or impurity measurements on finished preparations present a challenge to Pharmacopeial standard setting. Where degradation of a preparation over time is at issue, the same analytical methods that are stability-indicating are also purity-indicating. Resolution of the active ingredient(s) from the excipients necessary to the preparation presents the same qualitative problem. Thus, many monographs for Pharmacopeial preparations feature chromatographic assays. Where more significant impurities are known, some monographs set forth specific limit tests. In general, however, this Pharmacopeia does not repeat impurity tests in subsequent preparations where these appear in the monographs of bulk pharmaceutical chemicals and where these impurities are not expected to increase. There is consistency between compendial standards and *Good Manufacturing Practice for Finished Pharmaceuticals* ⟨1077⟩, and it is presumed that adequate retention specimens are in storage for the exact batch of bulk chemicals used in any specific lot of a preparation. Whenever analysis of an official preparation raises a question of the official attributes of any of the bulks used, subsequent analysis of retention specimens is in order.

Definitions—

Foreign Substances—Foreign substances, which are introduced by contamination or adulteration, are not consequences of the synthesis or preparation of compendial articles and thus cannot be anticipated when monograph tests and assays are selected. The presence of objectionable foreign substances not revealed by monograph tests and assays constitutes a variance from the official standard. Examples of foreign substances include ephedrine in Ipecac or a pesticide in an oral liquid analgesic. Allowance is made in this Pharmacopeia for the detection of foreign substances by unofficial methods. (See *Foreign Substances and Impurities*, in the section *Tests and Assays*, under *General Notices and Requirements.*)

Toxic Impurities—Toxic impurities have significant undesirable biological activity, even as minor components, and require individual identification and quantitation by specific tests. These impurities may arise out of the synthesis, preparation, or degradation of compendial articles. Based on validation data, individualized tests and specifications are selected. These feature comparison to a reference standard of the impurity, if available. It is incumbent on the manufacturer to provide data that would support the classification of such impurities as toxic impurities.

Concomitant Components—Concomitant components are characteristic of many bulk pharmaceutical chemicals and are not considered to be impurities in the pharmacopeial sense. Limits on contents, or specified ranges, or defined mixtures are set forth for concomitant components in this Pharmacopeia. Examples of concomitant components are geometric and optical isomers (or racemates) and antibiotics that are mixtures. Any component that can be considered a toxic impurity because of significant undesirable biological effect is not considered to be a concomitant component.

Signal Impurities—Signal impurities are distinct from ordinary impurities in that they require individual identification and quantitation by specific tests. Based on validation data, individualized tests and specifications are selected. These feature a comparison to a reference standard of the impurity, if available.

Signal impurities may include some process-related impurities or degradation products that provide key information about the process, such as diazotizable substances in thiazides. It is incumbent on the manufacturer to provide data that would support the classification of such impurities as signal impurities rather than ordinary impurities.

Ordinary Impurities—Ordinary impurities are those species in bulk pharmaceutical chemicals that are innocuous by virtue of having no significant, undesirable biological activity in the amounts present. These impurities may arise out of the synthesis, preparation, or degradation of compendial articles. Selections of tests and assays allow for anticipated amounts of impurities that are unobjectionable for the customary use of the article. The presence of ordinary impurities is controlled in monographs in this Pharmacopeia by including tests for *Ordinary Impurities* ⟨466⟩. Tests for *related substances* or *chromatographic purity* may also control the presence of ordinary impurities.

Unless otherwise specified in an individual monograph, estimation of the amount and number of ordinary impurities is made by relative methods rather than by strict comparison to individual Reference Standards. Nonspecific detection of ordinary impurities is also consistent with this classification.

The value of 2.0% was selected as the general limit on ordinary impurities in monographs where documentation did not support adoption of other values. This value represents the maximum allowable impact from this source of variation, when taken with the variation allowed by the composite of other Pharmacopeial tests and assays for both the bulk pharmaceutical chemical and the preparations.

Where a monograph sets limits on concomitant components, signal impurities, and/or toxic impurities, these species are not to be included in the estimation of ordinary impurities unless so stated in the individual monograph.

Industrial Concepts—Pharmaceutical manufacturers interact with regulatory agencies in developing new drug substances and new drug products, and cooperate with the compendia in writing official monographs for the compendial articles the manufacturers produce. Establishment of impurity limits in drug substances should proceed on a rational basis so that everyone involved in the development and approval phases can carry on their work in a predictable fashion. Although drug development in the United States is the primary focus of this section of the chapter, the subject also has broad applicability across national boundaries.

Manufacturers share with regulatory agencies and with the compendia the goal of making available to the public high-quality products that are both safe and efficacious. This goal continues to be achieved through rational approaches to the complex process of drug development. Tests used at all stages of drug development and marketing should not be interpreted individually but as a whole. Controls on raw materials and on manufacturing as well as those on drug substances, along with toxicological and clinical studies performed, ensure the safety and efficacy of drug products. It has been suggested that impurities should be identified when they exceed some set amount, e.g., 0.1, 0.3, or 0.5%. It is more rational to identify impurities and to set limits based on the factors detailed here, relying on the scientific judgments of manufacturers, the compendia, and regulators to arrive at sets of acceptable limits for identified and unidentified impurities.

Limits are set for impurity levels as one of the steps in ensuring the identity, strength, quality, and chemical purity of drug substances. The ultimate goal is to produce a final drug product of high quality and at a reasonable cost that is safe and efficacious and remains so throughout its shelf life. The setting of limits for impurities in bulk drug substances is a complex process that considers a number of factors:

(1) the toxicology of a drug substance containing typical levels of impurities and/or the toxicology of impurities relative to a drug substance;

(2) the route of administration, e.g., oral, topical, parenteral, or intrathecal;

(3) the daily dose, i.e., frequency and amount (micrograms or grams) administered of a drug substance;

(4) the target population (age and disease state), e.g., neonates, children, or senior citizens;

(5) the pharmacology of an impurity, when appropriate;

(6) the source of a drug substance, e.g., synthetic, natural product, or biotechnology;

(7) the duration of therapy, i.e., administration over a long period (treatment of chronic conditions) versus administration intended for a short duration (treatment of acute conditions); and

(8) the capability of a manufacturer to produce high-quality material at a reasonable cost to consumers.

Concepts for setting impurity limits in bulk drug substances are the concerns of the regulatory and compendial agencies as well as the pharmaceutical industry. The basic tenet for setting limits is that levels of impurities in a drug substance must be controlled to ensure its safety and quality throughout its development into and use as a drug product. The concepts are derived from issues and experiences with drug substances from traditional sources and technologies. Issues arising from biotechnologically produced drug substances, e.g., recombinant DNA and hybridomas, are still being defined and so are not necessarily covered by these concepts. However, the concepts can serve as a general foundation to address specific issues arising from biotechnology.

The setting of limits on impurities in drug substances is an evolutionary process, beginning in the United States before an investigational new drug (IND) is filed and continuing until well after the approval of a new drug application (NDA). Therefore, it is appropriate to address different stages in drug development as separate issues. There are three points in the drug development process where the setting of limits may be significantly different: (1) at the initial IND application, (2) at the filing of the NDA, and (3) after NDA approval. The filing of an abbreviated new drug application (ANDA) is another activity in which limits are set on impurities. Since the approach may vary from that of filing an NDA, it is addressed as a separate issue. The underlying assumption is that the analytical methods used to evaluate impurities in a drug substance are suitable for their intended purpose at each stage in the development.

An impurity is any component of a drug substance (excluding water) that is not the chemical entity defined as the drug substance. The impurity profile of a drug substance is a description of the impurities present in a typical lot of a drug substance produced by a given manufacturing process. The description includes the identity or some qualitative analytical designation (if unidentified), the range of each impurity observed, and the classification of each identified impurity.

Following are two more terms that enlarge upon those given under *Definitions.*

Related Substances—Related substances are structurally related to a drug substance. These substances may be identified or unidentified degradation products or impurities arising from a manufacturing process or during storage of a material.

Process Contaminants—Process contaminants are identified or unidentified substances (excluding related substances and water), including reagents, inorganics (e.g., heavy metals, chloride, or sulfate), raw materials, and solvents. These substances may be introduced during manufacturing or handling procedures.

Initial IND Filing—At the initial IND filing, the chemical nature of a bulk substance has generally been defined. The manufacturing process normally is in an early stage of development, and materials may be produced on a laboratory scale. Usually few batches have been made and, therefore, little historical data are available. The reference materials of a drug substance may be relatively impure. Limits for the purity of a drug substance are set to indicate drug quality. The setting of limits on related substances and process contaminants can be characterized as follows.

(1) Limits are set on total impurities, and an upper limit may be set on any single impurity. The limit for total impurities should maintain, if possible, a nominal composition material balance.

(2) Impurity profiles are documented. These are profiles of the lots of drug substances used in clinical studies and in toxicological studies that establish the safety of drug substances. The lots used in these studies should be typical products of the manufacturing process in use at that time.

(3) Limits for residual solvents are based on the known toxicology of the solvents and on the manufacturing capabilities and dosing regimens.

(4) General inorganic contaminants are monitored by appropriate tests such as a heavy metals limit test and/or a test for residue on ignition. Traditional compendial limits are applied unless otherwise indicated. Specific metal contaminants that appear during manufacturing should be monitored by appropriate analytical techniques, and limits should be set based on the toxicological properties of these metals.

(5) Appropriate limits are set for impurities known to be toxic.

(6) If appropriate, enantiomeric purity is controlled.

Although water is not classified as an impurity, limits for water content may be needed to ensure the stability or ease of processing a drug substance.

NDA Filing—During the IND phases of drug development, the manufacturing process for a drug substance may undergo a number of revisions. Generally, the scale will have changed from laboratory size and will approach or reach full production batch size. A number of batches will normally have been produced, and a historical data base of the results of testing for impurities will exist. When significant changes in a manufacturing process are made, the impurity profile should be reviewed to determine if the toxicological studies are still supportive.

At the NDA stage a reference standard of defined purity is available, analytical methods have been validated, impurity and degradation profiles are known, and enantiomeric purity has been evaluated. The setting of limits on related substances and process contaminants can be characterized as follows.

(1) Consistency of the impurity profile of a drug substance has been established.

(2) IND limits for total and individual impurities (identified and unidentified) are reviewed and adjusted based on manufacturing experience and toxicological data.

(3) Impurities present in significant amounts are identified and individual limits are set. However, it is not always possible to identify and/or prepare authentic substances for impurities. The labile nature of some impurities precludes this possibility. Limits may be set on these substances based on comparison of lots produced and used in toxicological and clinical studies.

(4) The impurity profiles of the lots designated for marketing

should not be significantly different from those of the lot(s) used for toxicological and clinical studies.

(5) The composition material balance should be used, if possible, to evaluate the adequacy of the controls.

(6) Limits for residual solvents are based on the known toxicology of the solvents and on the manufacturing capabilities and dosing regimens.

(7) Limits are set for inorganic contaminants by appropriate tests such as a heavy metals limit test and/or by a test for residue on ignition. Traditional compendial limits are applied unless otherwise indicated. Based on toxicological properties, limits may be set for specific metal contaminants that appear during manufacturing.

Post NDA Approval—After approval and marketing of a pharmaceutical product, significant changes may be made in manufacturing the bulk drug substance. There may be technological, ecological, economic, or safety reasons for these changes. If they occur, the Pharmacopeial and NDA impurity limits and rationale should be reviewed; the limits should be revised when indicated to ensure similar or improved quality of the drug substance.

ANDA Filing—The drug substance for a pharmaceutical product eligible for ANDA status normally is an official article and should be well characterized analytically. Drug substances are typically available from multiple sources, and each source may have a different manufacturing process. Therefore, it is essential that the dosage-form manufacturer evaluate each supplier's drug substance impurity profiles. Limits can then be set based on the more detailed concepts described for NDA filing, including review of compendial monographs for appropriateness.

⟨1091⟩ LABELING OF INACTIVE INGREDIENTS

This informational chapter provides guidelines for labeling of inactive ingredients present in dosage forms.

Within the past few years a number of trade associations representing pharmaceutical manufacturers have adopted voluntary guidelines for the disclosure and labeling of inactive ingredients. This is helpful to individuals who are sensitive to particular substances and who wish to identify the presence or confirm the absence of such substances in drug products. Because of the actions of these associations, the labeling of therapeutically inactive ingredients currently is deemed to constitute good pharmaceutical practice.

Although the manufacturers represented by these associations produce most of the products sold in this country, not all manufacturers, repackagers, or labelers here or abroad are members of these associations. Further, there are some differences in association guidelines. The guidelines presented here are designed to help promote consistency in labeling.

In accordance with good pharmaceutical practice, all dosage forms [NOTE—for requirements on parenteral and topical preparations, see the General Notices] should be labeled to state the identity of all added substances (therapeutically inactive ingredients) present therein, including colors, except that flavors and fragrances may be listed by the general term "flavor" or "fragrance." Such listing should be in alphabetical order by name and be distinguished from the identification statement of the active ingredient(s).

The name of an inactive ingredient should be taken from the current edition of one of the following reference works (in the following order of precedence): (1) the *United States Pharmacopeia* or the *National Formulary;* (2) *USAN and the USP Dictionary of Drug Names;* (3) CTFA *Cosmetic Ingredient Dictionary;* (4) *Food Chemicals Codex*. An ingredient not listed in any of the aforementioned reference works should be identified by its common or usual name (the name generally recognized by consumers or health-care professionals) or, if no common or usual name is available, by its chemical or other technical name.

An ingredient that may be, but not always is, present in a product should be qualified by words such as "or" or "may also contain."

The name of an ingredient whose identity is a trade secret may be omitted from the list if the list states "and other ingredients." For the purposes of this guideline, an ingredient is considered to be a trade secret only if its presence confers a significant competitive advantage upon its manufacturer and if its identity cannot be ascertained by the use of modern analytical technology.

An incidental trace ingredient having no functional or technical effect on the product need not be listed unless it has been demonstrated to cause sensitivity reactions or allergic responses.

Inactive ingredients should be listed on the label of a container of a product intended for sale without prescription, except that in the case of a container too small, such information may be contained in other labeling on or within the package.

⟨1101⟩ MEDICINE DROPPER

The Pharmacopeial medicine dropper consists of a tube made of glass or other suitable transparent material that generally is fitted with a collapsible bulb and, while varying in capacity, is constricted at the delivery end to a round opening having an external diameter of about 3 mm. The dropper, when held vertically, delivers water in drops each of which weighs between 45 mg and 55 mg.

In using a medicine dropper, one should keep in mind that few medicinal liquids have the same surface and flow characteristics as water, and therefore the size of drops varies materially from one preparation to another.

Where accuracy of dosage is important, a dropper that has been calibrated especially for the preparation with which it is supplied should be employed. The volume error incurred in measuring any liquid by means of a calibrated dropper should not exceed 15%, under normal use conditions.

⟨1111⟩ MICROBIOLOGICAL ATTRIBUTES OF NON-STERILE PHARMACEUTICAL PRODUCTS

Few raw materials used in making pharmaceutical products are sterile as received, and special treatment may be required to render them microbiologically acceptable for use. Strict adherence to effective environmental control and sanitation, equipment cleaning practices, and good personal hygiene practices in pharmaceutical manufacture is vital in minimizing both the type and the number of microorganisms.

Monitoring, in the form of regular surveillance, should include an examination of the microbiological attributes of Pharmacopeial articles and a determination of compliance with such microbiological standards as are set forth in the individual monographs. It may be necessary also to monitor the early and intermediate stages of production, with emphasis being placed on raw materials, especially those of animal or botanical origin, or from natural mineral sources, which may harbor objectionable microorganisms not destroyed during subsequent processing. It is essential that ingredients and components be stored under conditions designed to deter microbial proliferation.

The nature and frequency of testing vary according to the product. Monographs for some articles require freedom from one or more species of selected indicator microorganisms such as *Salmonella* species, *Escherichia coli*, *Staphylococcus aureus*, and *Pseudomonas aeruginosa*. For some articles, a specific limit on the total aerobic count of viable microorganisms and/or the total combined molds and yeasts count is set forth in the individual monograph; in these cases a requirement for freedom from specified indicator microorganisms may also be included. The significance of microorganisms in non-sterile pharmaceutical products should be evaluated in terms of the use of the product, the nature of the product, and the potential hazard to the user. Also taken into account is the processing of the product in relation to an acceptable quality for pharmaceutical purposes.

It is suggested that certain categories of products should be tested routinely for total microbial count and for specified indicator microbial contaminants, e.g., natural plant, animal, and some mineral products for *Salmonella* species; oral solutions and suspensions for *E. coli;* articles applied topically for *P. aeruginosa* and *S. aureus;* and articles intended for rectal, urethral, or vaginal administration for yeasts and molds.

Definitive microbial limits (stipulated microorganisms and/or counts) are incorporated into specific monographs on the basis of a major criterion, i.e., the potential of the stipulated microorganisms and/or counts, and of any others that they may reflect, to constitute a hazard in the end product. Such considerations also take into account the processing to which the product components are subjected, the current technology for testing, and the availability of desired quality material. Any of these may preclude the items from specific requirements under *Microbial Limit Tests* ⟨61⟩. Regardless of such preclusion, it remains essential to apply strict good manufacturing practices to assure a lowest possible load of microorganisms.

The relevant tests for determining the total count of viable aerobic microorganisms and the total combined molds and yeasts count, and for detection and identification of designated species are given under *Microbial Limit Tests* ⟨61⟩. For reliable results, the personnel responsible for the conduct of the test should have specialized training in microbiology and in the interpretation of microbiological data.

⟨1121⟩ NOMENCLATURE

The USP (or NF) titles are legally recognized as the designations for use in labeling the articles to which they apply.

The value of designating each drug by one and only one nonproprietary[1] name is obvious, in terms of achieving simplicity and uniformity in drug nomenclature. In support of the U. S. Adopted Names program (see *Preface*), of which the U. S. Pharmacopeial Convention is a co-sponsor, the USP Committee of Revision gives consideration to the adoption of the U. S. Adopted Name, if any, as the official title for any compound that attains compendial recognition.

A compilation of the U. S. Adopted Names (USAN) published from the start of the USAN program in 1961, as well as other names for drugs, both current and retrospective, is provided in *USAN and the USP Dictionary of Drug Names*. This publication is intended to serve as a book of names useful for identifying and distinguishing all kinds of names for drugs, whether public or proprietary or chemical or code-designated names.[2]

A nonproprietary name of a drug serves numerous and varied purposes, its principal function being to identify the substance to which it applies by means of a designation that may be used by the professional and lay public free from the restrictions associated with registered trademarks. Teaching in pharmacy and medicine requires a common designation, especially for a drug that is available from several sources or is incorporated into a combination drug product; nonproprietary names facilitate communication among physicians; nonproprietary names must be used as the titles of the articles recognized by official drug compendia; a nonproprietary name is essential to the pharmaceutical manufacturer as a means of protecting trademark rights in the brand name for the article concerned; and, finally, the manufacturer is obligated by federal law to include the established nonproprietary name in advertising and labeling.

Under the terms of the Drug Amendments of 1962 to the Federal Food, Drug, and Cosmetic Act, which became law October 10, 1962, the Secretary of Health and Human Services is authorized to designate an official name for any drug wherever deemed "necessary or desirable in the interest of usefulness and simplicity." [3]

The Commissioner of Food and Drugs and the Secretary of Health and Human Services published in the *Federal Register* regulations effective November 26, 1984, which state, in part:

Sec. 299.4 Established names of drugs.

(e) "The Food and Drug Administration will not routinely designate official names under section 508 of the act. As a result, the established name under section 502(e) of the act will ordinarily be either the compendial name of the drug or, if there is no compendial name, the common or usual name of the drug. Interested persons, in the absence of the designation by the Food and Drug Administration of an official name, may rely on as the established name for any drug the current compendial name or the USAN adopted name listed in *USAN and the USP Dictionary of Drug Names*. . . ." [4]

[1] The term "generic" has been widely used in place of the more accurate and descriptive term "nonproprietary," with reference to drug nomenclature.

[2] *USAN and the USP Dictionary of Drug Names* is obtainable on order from the USAN Division, USP Convention, Inc., 12601 Twinbrook Parkway, Rockville, MD 20852.

[3] F.D.&C. Act, Sec. 508 [358].

[4] 53 Fed. Reg. 5369 (1988) amending 21 CFR § 299.4.

It will be noted that the monographs on the biologics, which are produced under licenses issued by the Secretary of the U. S. Department of Health and Human Services, represent a special case. Although efforts continue toward achieving uniformity, there may be a difference between the respective title required by federal law and the USP title. Such differences are fewer than in past revisions of the Pharmacopeia. The USP title, where different from the FDA Bureau of Biologics title, does not constitute a synonym for labeling purposes; the conditions of licensing the biologic concerned require that each such article be designated by the name appearing in the product license issued to the manufacturer. Where a USP title differs from the title in the federal regulations, the former has been adopted with a view to usefulness and simplicity and conformity with the principles governing the selection of monograph titles generally.

⟨1141⟩ PACKAGING—CHILD-SAFETY

The Poison Prevention Packaging Act is administered and enforced by the Consumer Product Safety Commission of the federal government. The purpose of the law is to decrease the chance that children may obtain access to poisons. The act applies not only to drugs, but also to other household substances. The Commission is authorized to promulgate regulations providing standards for special packaging of any household substance where that special packaging will help to protect children from serious injury or illness resulting from the handling, using, or ingesting of such substance. Special packaging of this type is not necessarily packaging that all children under 5 years of age cannot open, but is packaging that makes it difficult for most children under the age of 5 to open or to obtain a harmful amount of the contents. On the other hand, the packaging should not be difficult for normal adults to open.

Not all hazardous household substances are subject to special packaging requirements. However, once a special packaging requirement is published, special packaging becomes the rule, and conventional packaging the exception. There are basically four ways by which exceptions are allowed for drugs:

(1) The substance is specifically exempted by regulation;

(2) If the drug is dispensed on prescription, the prescriber directs in the prescription order that it be dispensed in a noncomplying package;

(3) The purchaser of a prescription drug so requests;

(4) The manufacturer (or packer) of a drug for sale without prescription may provide a single-size, noncomplying package provided it bears conspicuous labeling stating, "This package for households without young children," or, if the package is small, stating, "Package not child-resistant," and provided the manufacturer also supplies the substance in complying packages.

The Commission has taken the view that in the case of prescription drugs, the manufacturer has the primary responsibility to provide special packaging where the manufacturer places the drug in a container clearly intended to be utilized in dispensing the drug for use in the home. If the pharmacist transfers a drug from the manufacturer's container to a dispensing container, the responsibility shifts to him. The fact that a manufacturer fails to provide suitable dispensing packaging does not relieve the pharmacist of this responsibility. It should be noted also that there is no special exemption for single-unit or unit-dose packages in the Act. Therefore, in order to comply, such unit packages, if they do not meet child-resistant standards, are to be placed in an outer package that does comply. Drugs placed in containers having dual-purpose closures must be dispensed in the child-resistant mode unless the physician has specified, or the patient has requested, the conventional mode.

Although special packaging should not be difficult for normal adults to open, it is recognized that some individuals having physical limitations, e.g., some elderly persons or arthritics, may have difficulties. Where pharmacists are aware of such infirmities, the proper procedure is to make the patient aware that noncomplying packages are available for his or her prescription medication and that the choice is his or hers, with regard to the need for taking into account the likelihood of young children's gaining access to the medication.

Failure to dispense a drug in a child-resistant container where such is required may be a violation of the Federal Food, Drug,

(3) *Methyl salicylate.* Liquid preparations containing more than 5 percent by weight of methyl salicylate, other than those packaged in pressurized spray containers, shall be packaged in accordance with the provisions of § 1700.15(a), (b), and (c).

(4) *Controlled drugs.* Any preparation for human use that consists in whole or in part of any substance subject to control under the Comprehensive Drug Abuse Prevention and Control Act of 1970 (21 U.S.C. 801 et seq.) and that is in a dosage form intended for oral administration shall be packaged in accordance with the provisions of § 1700.15(a), (b), and (c).

(6) *Turpentine.* Household substances in liquid form containing 10 percent or more by weight of turpentine shall be packaged in accordance with the provisions of § 1700.15(a) and (b).

(10) *Prescription drugs.* Any drug for human use that is in a dosage form intended for oral administration and that is required by Federal law to be dispensed only by or upon an oral or written prescription of a practitioner licensed by law to administer such drug shall be packaged in accordance with the provisions of § 1700.15(a), (b), and (c), except for the following:

(i) Sublingual dosage forms of nitroglycerin.

(ii) Sublingual and chewable forms of isosorbide dinitrate in dosage strengths of 10 milligrams or less.

(iii) Erythromycin ethylsuccinate granules for oral suspension and oral suspensions in packages containing not more than 8 grams of the equivalent of erythromycin.

(iv) Cyclically administered oral contraceptives in manufacturers' mnemonic (memory-aid) dispenser packages that rely solely upon the activity of one or more progestogen or estrogen substances.

(v) Anhydrous cholestyramine in powder form.

(vi) All unit-dose forms of potassium supplements, including individually wrapped effervescent tablets, unit-dose vials of liquid potassium, and powdered potassium in unit-dose packets, containing not more than 50 milliequivalents of potassium per unit dose.

(vii) Sodium fluoride drug preparations, including liquid and tablet forms, containing no more than 264 milligrams of sodium fluoride per package and containing no other substances subject to this § 1700.14(a)(10).

(viii) Betamethasone tablets packaged in manufacturers' dispenser packages, containing no more than 12.6 milligrams betamethasone.

(ix) Pancrelipase preparations in tablet, capsule, or powder form and containing no other substances subject to this § 1700.14(a)(10).

(x) Prednisone in tablet form, when dispensed in packages containing no more than 105 mg of the drug, and containing no other substances subject to this § 1700.14(a)(10).

(xiii) Mebendazole in tablet form in packages containing not more than 600 mg of the drug, and containing no other substance subject to the provisions of this section.

(xiv) Methylprednisolone in tablet form in packages containing not more than 84 mg of the drug and containing no other substance subject to the provisions of this section.

(xv) Colestipol in powder form in packages containing not more than 5 grams of the drug and containing no other substance subject to the provisions of this section.

(xvi) Erythromycin ethylsuccinate tablets in packages containing no more than the equivalent of 16 grams erythromycin.

(xvii) Conjugated Estrogen Tablets, USP, when dispensed in mnemonic packages containing not more than 32.0 mg of the drug and containing no other substances subject to this § 1700.14(a)(10).

(xviii) Norethindrone Acetate Tablets, USP, when dispensed in mnemonic packages containing not more than 50 mg of the drug and containing no other substances subject to this § 1700.14(a)(10).

(12) *Iron-containing drugs.* With the exception of: (i) Animal feeds used as vehicles for the administration of drugs, and (ii) those preparations in which iron is present solely as a colorant, noninjectable animal and human drugs providing iron for therapeutic or prophylactic purposes, and containing a total amount of elemental iron, from any source, in a single package, equivalent to 250 mg or more elemental iron in a concentration of 0.025 percent or more on a weight to volume basis for liquids and 0.05 percent or more on a weight-to-weight basis for nonliquids (e.g., powders, granules, tablets, capsules, wafers, gels, viscous products, such as pastes and ointments, etc.) shall be packaged in accordance with the provisions of § 1700.15(a), (b), and (c).

(13) *Dietary supplements containing iron.* With the exception of those preparations in which iron is present solely as a colorant, dietary supplements, as defined in § 1700.1(a)(3), that contain an equivalent of 250 mg or more of elemental iron, from any source, in a single package in concentrations of 0.025 percent or more on a weight to volume basis for liquids and 0.05 percent or more on a weight-to-weight basis for nonliquids (e.g., powders, granules, tablets, capsules, wafers, gels, viscous products, such as pastes and ointments, etc.) shall be packaged in accordance with the provisions of § 1700.15(a), (b), and (c).

(16) *Acetaminophen.* Preparations for human use in a dosage form intended for oral administration and containing in a single package a total of more than one gram acetaminophen shall be packaged in accordance with the provisions of § 1700.15(a), (b), and (c), except the following—

(i) Effervescent tablets or granules containing acetaminophen, provided the dry tablet or granules contain less than 10 percent acetaminophen, the tablet or granules have an oral LD-50 of greater than 5 grams per kilogram of body weight, the measured dosage of the product, when placed in water, releases at least 85 milliliters of carbon dioxide per grain of acetaminophen in the dry form when measured stoichiometrically at standard conditions (0°C, 760 mm Hg) and the tablets or granules contain no other substance subject to this § 1700.14(a).

(ii) Unflavored acetaminophen-containing preparations in powder form (other than those intended for pediatric use) that are packaged in unit doses providing not more than 13 grains of acetaminophen per unit dose and that contain no other substance subject to this § 1700.14(a).

(17) *Diphenhydramine.* Preparations for human use in a dosage form intended for oral administration and containing more than the equivalent of 66 mg diphenhydramine base in a single package shall be packaged in accordance with the provisions of § 1700.15(a), (b), and (c), if packaged on or after February 11, 1985.

⟨1151⟩ PHARMACEUTICAL DOSAGE FORMS

Dosage forms are provided for most of the Pharmacopeial drug substances, but the processes for the preparation of many of them are, in general, beyond the scope of the Pharmacopeia. In addition to defining the dosage forms, this section presents the general principles involved in the manufacture of some of them, particularly on a small scale. Other information that is given bears on the use of the Pharmacopeial substances in extemporaneous compounding of dosage forms.

BIOAVAILABILITY

Bioavailability, or the extent to which the therapeutic constituent of a pharmaceutical dosage form intended for oral or topical use is available for absorption is influenced by a variety of factors. Among the inherent factors known to affect absorption are the method of manufacture or method of compounding; the particle size and crystal form or polymorph of the drug substance; and the diluents and excipients used in formulating the dosage form, including fillers, binders, disintegrating agents, lubricants, coatings, solvents, suspending agents, and dyes. Lubricants and coatings are foremost among these. The maintenance of a demonstrably high degree of bioavailability requires particular attention to all aspects of production and quality control that may affect the nature of the finished dosage form.

STABILITY

The term "stability," with respect to a drug dosage form, refers to the time lapse from initial preparation and packaging during which the dosage form continues to fulfill the specifications presented in the monograph on identity, strength, quality, and purity. The monograph specifications apply throughout the shelf-life of the product. Stability of manufactured dosage forms should be demonstrated by the manufacturer by the use of methods adequate for the purpose. Monograph assays may be used for stability testing if they are stability-indicating, i.e., if they accurately differentiate between the intact drug molecules and their degradation products. Stability considerations should include not only the specific compendial requirements, but also changes in physical appearance of the product that would warn users of the product that its continued integrity is questionable.

Expiration dating is a valuable quality attribute and is required for all Pharmacopeial dosage forms. The expiration date preferably should be accompanied by specific storage conditions as provided in the Pharmacopeia for this purpose (see *Preservation, Packaging, Storage, and Labeling* under *General Notices*). Adequate stability data acquired by the manufacturer should be available to support the expiration date and storage conditions specified.

The stability parameters of a drug dosage form can be influenced by environmental conditions of storage (temperature, light, air, and humidity), as well as the package components. Pharmacopeial articles preferably should include required storage conditions on their labeling. These are the conditions under which the expiration date shall apply. The storage requirements must be observed throughout the distribution of the article, i.e., beyond the time it leaves the manufacturer up to and including its handling by the dispenser or seller of the article to the consumer. Although labeling for the consumer should indicate proper storage conditions, it is recognized that control beyond the dispenser or seller is difficult.

A discussion of aspects of drug product stability that are of primary concern to the pharmacist in the dispensing of medications may be found under *Stability Considerations in Dispensing Practice* ⟨1191⟩.

Inasmuch as this chapter is for purposes of general information only, no statement herein is intended to modify or supplant any of the specific requirements pertinent to pharmaceutical preparations, which are given elsewhere in this Pharmacopeia.

Terminology

Occasionally it is necessary to add solvent to the contents of a container just prior to use, usually because of instability of some drugs in the diluted form. Thus, a solid diluted to yield a suspension is called "_______ for Suspension;" a solid dissolved and diluted to yield a solution is called "_______ for Solution;" and a solution or suspension diluted to yield a more dilute form of the drug is called "_______ Oral Concentrate" or "_______ Injection Concentrate" as appropriate for the route of administration of the drug. After dilution, it is important that the drug be homogeneously dispersed before administration.

Aerosols

Pharmaceutical aerosols are products that are packaged under pressure and contain therapeutically active ingredients that are released upon activation of an appropriate valve system. They are intended for topical application to the skin as well as local application into the nose (nasal aerosols), mouth (lingual aerosols), or lungs (inhalation aerosols).

The term "aerosol" refers to the fine mist of spray that results from most pressurized systems. However, the term has been broadly misapplied to all self-contained pressurized products, some of which deliver foams or semisolid fluids. In the case of *Inhalation Aerosols*, the particle size of the delivered medication must be carefully controlled and the average size of the particles should be under 10 μm. These products are also known as metered dose inhalers (MDIs). (See *Inhalations*.) Other aerosol sprays may contain particles up to several hundred micrometers in diameter.

The basic components of an aerosol system are the container, the propellant, the concentrate containing the active ingredient(s), the valve, and the actuator. The nature of these components determines such characteristics as particle size distribution, uniformity of valve delivery for metered valves, delivery rate, wetness and temperature of the spray, foam density, or fluid viscosity.

TYPES OF AEROSOLS

Aerosols consist of two-phase (gas and liquid) or three-phase (gas, liquid, and solid or liquid) systems. The two-phase aerosol consists of a solution of active ingredients in liquefied propellant and the vaporized propellant. The solvent is composed of the propellant or a mixture of the propellant and co-solvents such as alcohol, propylene glycol, and polyethylene glycols which are often used to enhance the solubility of the active ingredients.

Three-phase systems consist of a suspension or emulsion of the active ingredient(s) in addition to the vaporized propellants. A suspension consists of the active ingredient(s) which may be dispersed in the propellant system with the aid of suitable excipients such as wetting agents and/or solid carriers such as talc or colloidal silicas.

A foam aerosol is an emulsion containing one or more active ingredients, surfactants, aqueous or nonaqueous liquids, and the propellants. If the propellant is in the internal (discontinuous) phase (i.e., of the oil-in-water type), a stable foam is discharged, and if the propellant is in the external (continuous) phase (i.e., of the water-in-oil type), a spray or a quick-breaking foam is discharged.

PROPELLANTS

The propellant supplies the necessary pressure within an aerosol system to expel material from the container and, in combination with other components, to convert the material into the desired physical form. Propellants may be broadly classified as liquefied or compressed gases having vapor pressures generally exceeding atmospheric pressure. Propellants within this definition include various hydrocarbons, especially fluorochloro-derivatives of methane and ethane, low molecular weight hydrocarbons such as the butanes and pentanes, and compressed gases such as carbon dioxide, nitrogen, and nitrous oxide. Mixtures of propellants are frequently used to obtain desirable pressure, delivery, and spray characteristics. A good propellant system should have the proper vapor pressure characteristics consistent with the other aerosol components.

VALVES

The primary function of the valve is to regulate the flow of the therapeutic agent and propellant from the container. The spray characteristics of the aerosol are influenced by orifice dimension, number, and location. Most aerosol valves provide for continuous spray operation and are used on most topical products. However, pharmaceutical products for oral or nasal inhalation often utilize metered dose valves that must deliver a uniform quantity of spray upon each valve activation. The accuracy and reproducibility of the doses delivered from metering valves are generally good, comparing favorably to the uniformity of solid dosage forms such as tablets and capsules. However, when aerosol packages are stored improperly, or when they have not been used for long periods of time, valves must be primed before use. Materials used for the manufacture of valves should be inert to the formulations used. Plastic, rubber, aluminum, and stainless steel valve components are commonly used. Metered-dose valves must deliver an accurate dose within specified tolerances.

ACTUATORS

An actuator is the fitting attached to an aerosol valve stem which, when depressed or moved, opens the valve, and directs the spray containing the drug preparation to the desired area. The actuator usually indicates the direction in which the preparation is dispensed and protects the hand or finger from the refrigerant effects of the propellant. Actuators incorporate an orifice which may vary widely in size and shape. The size of this orifice, the expansion chamber design, and the nature of the propellant and formulation influence the physical characteristics of the spray, foam, or stream of solid particles dispensed. For inhalation or oral dose aerosols, an actuator capable of delivering the medication in the proper particle size range is utilized.

CONTAINERS

Aerosol containers usually are made of glass, plastic, or metal, or a combination of these materials. Glass containers must be precisely engineered to provide the maximum in pressure safety and impact resistance. Plastics may be employed to coat glass containers for improved safety characteristics, or to coat metal containers to improve corrosion resistance and enhance stability of the formulation. Suitable metals include stainless steel, aluminum, and tin-plated steel.

MANUFACTURE

Aerosols are usually prepared by one of two general processes. In the "cold-fill" process, the concentrate (generally cooled to a temperature below 0°) and the refrigerated propellant are measured into open containers (usually chilled). The valve-actuator assembly is then crimped onto the container to form a pressure-tight seal. During the interval between propellant addition and crimping, sufficient volatilization of propellant occurs to displace air from the container. In the "pressure-fill" method, the concentrate is placed in the container, and either the propellant is forced under pressure through the valve orifice after the valve is sealed, or the propellant is allowed to flow under the valve cap and then the valve assembly is sealed ("under-the-cap" filling). In both cases of the "pressure-fill" method, provision must be made for evacuation of air by means of vacuum or displacement with a small amount of propellant. Manufacturing process controls usually include monitoring of proper formulation and propellant fill weight, and pressure testing and leak testing of the finished aerosol.

LABELING

Medicinal aerosols should contain at least the following warning information on the label as in accordance with appropriate regulations.

Warning—Avoid inhaling. Keep away from eyes or other mucous membranes.

The statement "Avoid inhaling" is not necessary for preparations specifically designed for use by inhalation.

The phrase "or other mucous membranes" is not necessary for preparations specifically designed for use on mucous membranes.

Warning—Contents under pressure. Do not puncture or incinerate container. Do not expose to heat or store at temperatures above 120°F (49°C). Keep out of reach of children.

In addition to the aforementioned warnings, the label of a drug packaged in an aerosol container in which the propellant consists in whole or in part of a halocarbon or hydrocarbon shall, where required under regulations of the Food and Drug Administration, bear the following warning:

Warning—Do not inhale directly; deliberate inhalation of contents can cause death.

or

Warning—Use only as directed; intentional misuse by deliberately concentrating and inhaling the contents can be harmful or fatal.

Capsules

Capsules are solid dosage forms in which the drug is enclosed within either a hard or soft soluble container or "shell" made from a suitable form of gelatin. Hard gelatin capsule sizes range from No. 5, the smallest, to No. 000, which is the largest, except for veterinary sizes. However, size No. 0 generally is the largest size acceptable to patients. Hard shell capsules consist of two, telescoping cap and body pieces. Generally, there are unique grooves or indentations molded into the cap and body portions to provide a positive closure when fully engaged which helps prevent the accidental separation of the filled capsules during shipping and handling. Positive closure also may be effected by spot fusion ("welding") of the cap and body pieces together through direct thermal means or by application of ultrasonic energy. Factory-filled hard shell capsules may be completely sealed by banding, a process in which one or more layers of gelatin are applied over the seam of the cap and body, or by a liquid fusion process wherein the filled capsules are wetted with a hydroalcoholic solution which penetrates into the space where the cap overlaps the body, and then dried. The latter two methods in particular may improve the stability of contents by limiting oxygen penetration as well as enhancing the consumer safety of over-the-counter capsules by making them difficult to open without causing visible, obvious damage. Industrially filled capsules also are often of distinctive color and shape or are otherwise marked to identify them with the manufacturer.

In extemporaneous prescription practice, hard gelatin capsules may be hand-filled; this permits the prescriber a latitude of choice in selecting either a single drug or a combination of drugs at the exact dosage level considered best for the individual patient. This flexibility gives hard shell capsules an advantage over tablets and soft gelatin capsules as a dosage form.

Hard gelatin capsules are made from gelatins having relatively high gel strength. Either type may be used, but blends of pork skin and bone gelatin are often used to optimize shell clarity and toughness. Hard gelatin capsules may also contain colorants, such as D & C and FD & C dyes or the various iron oxides, opaquing agents such as titanium dioxide, dispersing agents, hardening agents such as sucrose, and preservatives. They normally contain between 10 and 15% water.

Hard shell capsules are made by a process that involves dipping shaped pins into gelatin solutions, after which the gelatin films are dried, trimmed, and removed from the pins, and the body and cap pieces are joined. The empty capsules should be stored in tight containers until they are filled. Since they are of animal origin, they should be protected from potential sources or microbial contamination.

Hard shell capsules typically are filled with powder, beads, or granules. Semisolids or liquids also may be filled into hard shell capsules; however, when the latter are encapsulated, one of the sealing techniques must be employed to prevent leakage.

In all filling operations, the body and cap of the shell are separated prior to dosing. Machines employing various dosing principles may be employed to fill powders into hard shell capsules; however, most fully automatic machines form powder plugs by compression and eject them into empty capsule bodies. Accessories to these machines generally are available for the other types of fills. Powder formulations often require adding fillers, lubricants, and glidants to the active ingredients to facilitate encapsulation. The formulation, as well as the method of filling, particularly the degree of compaction, may influence the rate of drug release. The addition of wetting agents to the powder mass is common where the active ingredient is hydrophobic. Disintegrants also may be included in powder formulations to facilitate deaggregation and dispersal of capsule plugs in the gut.

Powder mixtures which tend to liquefy may be dispensed in hard shell capsules if an absorbent such as magnesium carbonate, colloidal silicon dioxide or other suitable substance is used. Potent drugs are often mixed with an inert diluent before being filled into capsules. Where two mutually incompatible drugs are prescribed together, it is sometimes possible to place one in a small capsule and then enclose it with the second drug in a larger capsule. Incompatible drugs also can be separated by placing coated pellets or tablets, or soft gelatin capsules of one drug into the capsule shell before adding the second drug.

Thixotropic semisolids may be formed by gelling liquid drugs or vehicles with colloidal silicas or powdered high molecular weight polyethylene glycols. Various waxy or fatty compounds may be used to prepare semisolid matrices by fusion.

Soft gelatin capsules (sometimes called softgels) require large-scale production methods. The gelatin shell is somewhat thicker than that of hard shell capsules and is plasticized by the addition of a polyol such as sorbitol or glycerin. The ratio of dry plasticizer to dry gelatin determines the "hardness" of the shell and may be varied to accommodate environmental conditions as well as the nature of the contents. Like hard shells, the shell composition may include approved dyes and pigments, opaquing agents such as titanium dioxide, and preservatives. Flavors may be added and up to 5% sucrose may be included for its sweetness and to produce a chewable shell. Soft gelatin shells normally contain 6% to 13% water.

In most cases, soft gelatin capsules are filled with liquid contents. Typically, active ingredients are dissolved or suspended in a liquid vehicle. Classically, an oleaginous vehicle such as a vegetable oil was used; however, nonaqueous, water-miscible liquid vehicles such as the lower molecular weight polyethylene glycols are more common today due to fewer bioavailability problems.

Available in a wide variety of sizes and shapes, soft gelatin capsules are both formed, filled, and sealed in the same machine; typically, this is a rotary die process, although a plate process or reciprocating die process also may be employed. Soft gelatin capsules also may be manufactured in a bubble process which forms seamless spherical capsules. With suitable equipment, powders, and other dry solids also may be filled into soft gelatin capsules.

Liquid-filled capsules of either type involve similar formulation technology and offer similar advantages and limitations. For instance, both may offer advantages over dry-filled capsules and tablets in content uniformity and drug dissolution. Greater homogeneity is possible in liquid systems, and liquids can be metered more accurately. Drug dissolution may be a benefit because the drug may already be in solution or at least suspended in a hydrophilic vehicle. However, the contact between the hard or soft shell and its liquid content is more intimate than exists with dry-filled capsules, and this may enhance the chances for undesired interactions. The liquid nature of capsule contents presents different technological problems than dry-filled capsules in regard to disintegration and dissolution testing. From formulation, technological, and biopharmaceutical points of view, liquid-filled capsules of either type have more in common than liquid-filled and dry-filled capsules having the same shell composition. Thus, for compendial purposes, standards and methods should be established based on capsule contents rather than on whether the contents are filled into hard or soft gelatin capsules.

Enteric-coated Capsules—Capsules may be coated, or, more commonly, encapsulated granules may be coated to resist releasing the drug in the gastric fluid of the stomach where a delay is important to alleviate potential problems of drug inactivation or gastric mucosal irritation. The term "delayed-release" is used for Pharmacopeial monographs on enteric-coated capsules that are intended to delay the release of medicament until the capsule has passed through the stomach, and the individual monographs include tests and specifications for *Drug release* (see *Drug Release* ⟨724⟩).

Extended-release Capsules—Extended-release capsules are formulated in such manner as to make the contained medicament available over an extended period of time following ingestion. Expressions such as "prolonged-action," "repeat-action," and "sustained-release" have been used also to describe such dosage forms. However, the term "extended-release" is used for Pharmacopeial purposes and requirements for *Drug release* ⟨724⟩ typically are specified in the individual monographs.

Creams

Creams are semisolid dosage forms containing one or more drug substances dissolved or dispersed in a suitable base. This term has traditionally been applied to semisolids that possess a relatively fluid consistency formulated as either water-in-oil (*Cold Cream*) or oil-in-water (e.g., *Fluocinolone Acetonide Cream*) emulsions. However, more recently the term has been restricted to products consisting of oil-in-water emulsions or aqueous microcrystalline dispersions of long chain fatty acids or alcohols that are water washable and more cosmetically and aesthetically acceptable. Creams can be used for administering drugs via the vaginal route (e.g., *Triple Sulfa Vaginal Cream*).

Elixirs

See *Solutions*.

Emulsions

Emulsions are two-phase systems in which one liquid is dispersed throughout another liquid in the form of small droplets. Where oil is the dispersed phase and an aqueous solution is the continuous phase, the system is designated as an oil-in-water emulsion. Conversely, where water or an aqueous solution is the dispersed phase and oil or oleaginous material is the continuous phase, the system is designated as a water-in-oil emulsion. Emulsions are stabilized by emulsifying agents which prevent coalescence, the merging of small droplets into larger droplets and, ultimately, into a single separated phase. Emulsifying agents (surfactants) do this by concentrating in the interface between the droplet and external phase, and by providing a physical barrier around the particle to coalescence. Surfactants also reduce the interfacial tension between the phases, thus increasing the ease of emulsification upon mixing.

Natural, semisynthetic, and synthetic hydrophilic polymers may be used in conjunction with surfactants in oil-in-water emulsions as they accumulate at interfaces and also increase the viscosity of the aqueous phase, thereby decreasing the rate of formation of aggregates of droplets. Aggregation is generally accompanied by a relatively rapid separation of an emulsion into a droplet-rich and droplet-poor phase. Normally the density of an oil is lower than that of water, in which case the oil droplets and droplet aggregates rise, a process referred to as creaming. The greater the rate of aggregation, the greater the droplet size and the greater the rate of creaming. The water droplets in a water-in-oil emulsion generally sediment because of their greater density.

The consistency of emulsions varies widely, ranging from easily pourable liquids to semisolid creams. Generally oil-in-water creams are prepared at high temperature, where they are fluid, and cooled to room temperature, whereupon they solidify as a result of solidification of the internal phase. When this is the case, a high internal-phase volume to external-phase volume ratio is not necessary for semisolid character, and, for example, stearic acid creams or vanishing creams are semisolid with as little as 15% internal phase. Any semisolid character with water-in-oil emulsions generally is attributable to a semisolid external phase.

All emulsions require an antimicrobial agent because the aqueous phase is favorable to the growth of microorganisms. The presence of a preservative is particularly critical in oil-in-water emulsions where contamination of the external phase occurs readily. Since fungi and yeasts are found with greater frequency than bacteria, fungistatic as well as bacteriostatic properties are desirable. Bacteria have been shown to degrade nonionic and anionic emulsifying agents, glycerin, and many natural stabilizers such as tragacanth and guar gum.

Complications arise in preserving emulsion systems, as a result of partitioning of the antimicrobial agent out of the aqueous phase where it is most needed, or of complexation with emulsion ingredients that reduce effectiveness. Therefore, the effectiveness of the preservative system should always be tested in the final product. Preservatives commonly used in emulsions include methyl-, ethyl-, propyl-, and butyl-parabens, benzoic acid, and quaternary ammonium compounds.

(See also *Creams* and *Ointments*.)

Extracts and Fluidextracts

Extracts are concentrated preparations of vegetable or animal drugs obtained by removal of the active constituents of the respective drugs with suitable menstrua, evaporation of all or nearly all of the solvent, and adjustment of the residual masses or powders to the prescribed standards.

In the manufacture of most extracts, the drugs are extracted by percolation. The entire percolates are concentrated, generally by distillation under reduced pressure, in order to subject the drug principles to as little heat as possible.

Fluidextracts are liquid preparations of vegetable drugs, containing alcohol as a solvent or as a preservative, or both, and so made that, unless otherwise specified in an individual monograph, each mL contains the therapeutic constituents of 1 g of the standard drug that it represents.

A fluidextract that tends to deposit sediment may be aged and filtered or the clear portion decanted, provided the resulting clear liquid conforms to the Pharmacopeial standards.

Fluidextracts may be prepared from suitable extracts.

Gels

Gels (sometimes called Jellies) are semisolid systems consisting of either suspensions made up of small inorganic particles or large organic molecules interpenetrated by a liquid. Where the gel mass consists of a network of small discrete particles, the gel is classified as a two-phase system (e.g., *Aluminum Hydroxide Gel*). In a two-phase system, if the particle size of the dispersed phase is relatively large, the gel mass is sometimes referred to as a magma (e.g., *Bentonite Magma*). Both gels and magmas may be thixotropic, forming semisolids on standing and becoming liquid on agitation. They should be shaken before use to ensure homogeneity and should be labeled to that effect. (See *Suspensions*.)

Single-phase gels consist of organic macromolecules uniformly distributed throughout a liquid in such a manner that no apparent boundaries exist between the dispersed macromolecules and the liquid. Single-phase gels may be made from synthetic macromolecules (e.g., *Carbomer*) or from natural gums (e.g., *Tragacanth*). The latter preparations are also called mucilages. Although these gels are commonly aqueous, alcohols and oils may be used as the continuous phase. For example, mineral oil can be combined with a polyethylene resin to form an oleaginous ointment base.

Gels can be used to administer drugs topically or into body cavities (e.g., *Phenylephrine Hydrochloride Nasal Jelly*).

Implants (Pellets)

Implants or pellets are small sterile solid masses consisting of a highly purified drug (with or without excipients) made by compression or molding. They are intended for implantation in the body (usually subcutaneously) for the purpose of providing continuous release of the drug over long periods of time. Implants are administered by means of a suitable special injector or surgical incision. This dosage form has been used to administer hormones such as testosterone or estradiol. They are packaged individually in sterile vials or foil strips.

Infusions, Intramammary

Intramammary infusions are suspensions of drugs in suitable oil vehicles. These preparations are intended for veterinary use only, and are administered by instillation via the teat canals into the udders of milk-producing animals.

Inhalations

Inhalations are drugs or solutions or suspensions of one or more drug substances administered by the nasal or oral respiratory route for local or systemic effect.

Solutions of drug substances in sterile water for inhalation or in sodium chloride inhalation solution may be nebulized by use of inert gases. Nebulizers are suitable for the administration of inhalation solutions only if they give droplets sufficiently fine and uniform in size so that the mist reaches the bronchioles. Nebulized solutions may be breathed directly from the nebulizer or the nebulizer may be attached to a plastic face mask, tent, or intermittent positive pressure breathing (IPPB) machine.

Another group of products, also known as metered dose inhalers (MDIs) are propellant driven drug suspensions or solution in liquified gas propellant with or without a cosolvent and are intended for delivering metered doses of the drug to the respiratory tract. An MDI contains multiple doses, often exceeding several hundred. The most common single-dose volumes delivered are from 25 to 100 μl (also expressed as mg) per actuation.

Examples of MDIs containing drug solutions and suspension in this pharmacopeia are *Epinephrine Inhalation Aerosol* and *Isoproterenol Hydrochloride and Phenylephrine Bitartrate Inhalation Aerosol*, respectively.

Powders may also be administered by mechanical devices that require manually produced pressure or a deep inhalation by the patient, e.g., *Cromolyn Sodium for Inhalation.*

A special class of inhalations termed inhalants consists of drugs or combination of drugs, that by virtue of their high vapor pressure, can be carried by an air current into the nasal passage where they exert their effect. The container from which the inhalant generally is administered is known as an inhaler.

Injections

See *Injections* ⟨1⟩.

Irrigations

Irrigations are sterile solutions intended to bathe or flush open wounds or body cavities. They are used topically, never parenterally. They differ from large volume parenteral solutions in the way they are packaged (screw cap containers).

Lotions

See *Solutions* or *Suspensions.*

Lozenges

Lozenges are solid preparations containing one or more medicaments, usually in a flavored, sweetened base which are intended to dissolve or disintegrate slowly in the mouth. They can be prepared by molding (gelatin and/or fused sucrose or sorbitol base) or by compression of sugar based tablets. Molded lozenges are sometimes referred to as pastilles while compressed lozenges are often referred to as troches. They are usually intended for treatment of local irritation or infections of the mouth or throat but may contain active ingredients intended for systemic absorption after swallowing.

Ointments

Ointments are semisolid preparations intended for external application to the skin or mucous membranes.

Ointment bases recognized for use as vehicles fall into four general classes: the hydrocarbon bases, the absorption bases, the water-removable bases, and the water-soluble bases. Each therapeutic ointment possesses as its base a representative of one of these four general classes.

Hydrocarbon Bases—These bases, which are known also as "oleaginous ointment bases," are represented by *White Petrolatum* and *White Ointment*. Only small amounts of an aqueous component can be incorporated into them. They serve to keep medicaments in prolonged contact with the skin and act as occlusive dressings. Hydrocarbon bases are used chiefly for their emollient effects, and are difficult to wash off. They do not "dry out" or change noticeably on aging.

Absorption Bases—This class of bases may be divided into two groups, the first group consisting of bases that permit the incorporation of aqueous solutions with the formation of a water-in-oil emulsion (*Hydrophilic Petrolatum* and *Anhydrous Lanolin*), and the second group consisting of water-in-oil emulsions that permit the incorporation of additional quantities of aqueous solutions (*Lanolin* and *Cold Cream*). Absorption bases are useful also as emollients.

Water-removable Bases—Such bases are oil-in-water emulsions, e.g., *Hydrophilic Ointment*, and are more correctly called "creams." (See *Creams.*) They are described also as "water-washable," since they may be readily washed from the skin or clothing with water, an attribute that makes them more acceptable for cosmetic reasons. Some medicaments may be more effective in these bases than in hydrocarbon bases. Other advantages of the water-removable bases are that they may be diluted with water and that they favor the absorption of serous discharges in dermatological conditions.

Water-soluble Bases—This group of so-called "greaseless ointment bases" is comprised of water-soluble constituents. *Polyethylene Glycol Ointment* is the only Pharmacopeial preparation in this group. Bases of this type offer many of the advantages of the water-removable bases and, in addition, contain no water-insoluble substances such as petrolatum, anhydrous lanolin, or waxes. They are more correctly called "Gels." (See *Gels.*)

Choice of Base—The choice of an ointment base depends upon many factors, such as the action desired, the nature of the medicament to be incorporated and its bioavailability and stability, and the requisite shelf-life of the finished product. In some cases, it is necessary to use a base that is less than ideal in order to achieve the stability required. Drugs that hydrolyze rapidly, for example, are more stable in hydrocarbon bases than in bases containing water, even though they may be more effective in the latter.

Ophthalmic Preparations

Drugs are administered to the eyes in a wide variety of dosage forms, some of which require special consideration. They are discussed in the following paragraphs.

Ointments—Ophthalmic ointments are ointments for application to the eye. Special precautions must be taken in the preparation of ophthalmic ointments. They are manufactured from sterilized ingredients under rigidly aseptic conditions and meet the requirements under *Sterility Tests* ⟨71⟩. If the specific ingredients used in the formulation do not lend themselves to routine sterilization techniques, ingredients that meet the sterility requirements described under *Sterility Tests* ⟨71⟩, along with aseptic manufacture, may be employed. Ophthalmic ointments must contain a suitable substance or mixture of substances to prevent growth of, or to destroy, microorganisms accidentally introduced when the container is opened during use, unless otherwise directed in the individual monograph, or unless the formula itself is bacteriostatic (see *Added Substances* under *Ophthalmic Ointments* ⟨771⟩). The medicinal agent is added to the ointment base either as a solution or as a micronized powder. The finished ointment must be free from large particles and must meet the requirements for *Leakage* and for *Metal Particles* under *Ophthalmic Ointments* ⟨771⟩. The immediate containers for ophthalmic ointments shall be sterile at the time of filling and closing. It is mandatory that the immediate containers for ophthalmic ointments be sealed and tamper-proof so that sterility is assured at time of first use.

The ointment base that is selected must be non-irritating to the eye, permit diffusion of the drug throughout the secretions bathing the eye, and retain the activity of the medicament for a reasonable period under proper storage conditions.

Petrolatum is mainly used as a base for ophthalmic drugs. Some absorption bases, water-removable bases, and water-soluble bases may be desirable for water-soluble drugs. Such bases allow for better dispersion of water-soluble medicaments, but they must be non-irritating themselves to the eye.

Solutions—Ophthalmic solutions are sterile solutions, essentially free from foreign particles, suitably compounded and packaged for instillation into the eye. Preparation of an ophthalmic solution requires careful consideration of such factors as the inherent toxicity of the drug itself, isotonicity value, the need for buffering agents, the need for a preservative (and, if needed, its selection), sterilization, and proper packaging. Similar considerations are also made for nasal and otic products.

Isotonicity Value—Lacrimal fluid is isotonic with blood, having an isotonicity value corresponding to that of a 0.9% sodium chloride solution. Ideally, an ophthalmic solution should have this isotonicity value; but the eye can tolerate isotonicity values as low as that of a 0.6% sodium chloride solution and as high as that of a 2.0% sodium chloride solution without marked discomfort.

Some ophthalmic solutions are necessarily hypertonic in order to enhance absorption and provide a concentration of the active ingredient(s) strong enough to exert a prompt and effective action. Where the amount of such solutions used is small, dilution with lacrimal fluid takes place rapidly so that discomfort from the hypertonicity is only temporary. However, any adjustment toward isotonicity by dilution with tears is negligible where large volumes of hypertonic solutions are used as collyria to wash the eyes; it is therefore important that solutions used for this purpose be approximately isotonic.

Buffering—Many drugs, notably alkaloidal salts, are most effective at pH levels that favor the undissociated free bases. At such pH levels, however, the drug may be unstable so that compromise levels must be found and held by means of buffers. One purpose of buffering some ophthalmic solutions is to prevent an increase in pH caused by the slow release of hydroxyl ions by glass. Such a rise in pH can affect both the solubility and the stability of the drug. The decision whether or not buffering agents should be added in preparing an ophthalmic solution must be based on several considerations. Normal tears have a pH of about 7.4 and possess some buffer capacity. The application of a solution to the eye stimulates the flow of tears and the rapid neutralization of any excess hydrogen or hydroxyl ions within the buffer capacity of the tears. Many ophthalmic drugs, such as alkaloidal salts, are weakly acidic and have only weak buffer capacity. Where only 1 or 2 drops of a solution containing them are added to the eye, the buffering action of the tears is usually adequate to raise the pH and prevent marked discomfort. In some cases pH may vary between 3.5 and 8.5. Some drugs, notably pilocarpine hydrochloride and epinephrine bitartrate, are more acid and overtax the buffer capacity of the lacrimal fluid. Ideally, an ophthalmic solution should have the same pH, as well as the same isotonicity value, as lacrimal fluid. This is not usually possible since, at pH 7.4, many drugs are not appreciably soluble in water. Most alkaloidal salts precipitate as the free alkaloid at this pH. Additionally, many drugs are chemically unstable at pH levels approaching 7.4. This instability is more marked at the high temperatures employed in heat sterilization. For this reason, the buffer system should be selected that is nearest to the physiological pH of 7.4 and does not cause precipitation of the drug or its rapid deterioration.

An ophthalmic preparation with a buffer system approaching the physiological pH can be obtained by mixing a sterile solution of the drug with a sterile buffer solution using aseptic technique. Even so, the possibility of a shorter shelf-life at the higher pH must be taken into consideration, and attention must be directed toward the attainment and maintenance of sterility throughout the manipulations.

Many drugs, when buffered to a therapeutically acceptable pH, would not be stable in solution for long periods of time. These products are lyophilized and are intended for reconstitution immediately before use (e.g., *Acetylcholine Chloride for Ophthalmic Solution*).

Sterilization—The sterility of solutions applied to an injured eye is of the greatest importance. Sterile preparations in special containers for individual use on one patient should be available in every hospital, office, or other installation where accidentally or surgically traumatized eyes are treated. The method of attaining sterility is determined primarily by the character of the particular product (see *Sterilization and Sterility Assurance of Compendial Articles* ⟨1211⟩).

Whenever possible, sterile membrane filtration under aseptic conditions is the preferred method. If it can be shown that product stability is not adversely affected, sterilization by autoclaving in the final container is also a preferred method.

Buffering certain drugs near the physiological pH range makes them quite unstable at high temperature.

Avoiding the use of heat by employing a bacteria-retaining filter is a valuable technique, provided caution is exercised in the selection, assembly, and use of the equipment. Single-filtration, presterilized disposable units are available and should be utilized wherever possible.

Preservation—Ophthalmic solutions may be packaged in multiple-dose containers when intended for the individual use of one patient and where the ocular surfaces are intact. It is mandatory that the immediate containers for ophthalmic solutions be sealed and tamper-proof so that sterility is assured at time of first use. Each solution must contain a suitable substance or mixture of substances to prevent the growth of, or to destroy, microorganisms accidentally introduced when the container is opened during use.

Where intended for use in surgical procedures, ophthalmic solutions, although they must be sterile, should not contain antibacterial agents, since they may be irritating to the ocular tissues.

Thickening agent—A pharmaceutical grade of methylcellulose (e.g., 1% if the viscosity is 25 centipoises, or 0.25% if 4000 centipoises) or other suitable thickening agents such as hydroxypropyl methylcellulose or polyvinyl alcohol occasionally are added to ophthalmic solutions to increase the viscosity and prolong contact of the drug with the tissue. The thickened ophthalmic solution must be free from visible particles.

Suspensions—Ophthalmic suspensions are sterile liquid preparations containing solid particles dispersed in a liquid vehicle intended for application to the eye (see *Suspensions*). It is imperative that such suspensions contain the drug in a micronized form to prevent irritation and/or scratching of the cornea. Ophthalmic suspensions should never be dispensed if there is evidence of caking or aggregation.

Strips—Fluorescein sodium solution should be dispensed in a sterile, single-use container or in the form of a sterile, impregnated paper strip. The strip releases a sufficient amount of the drug for diagnostic purposes when touched to the eye being examined for a foreign body or a corneal abrasion. Contact of the paper with the eye may be avoided by leaching the drug from the strip onto the eye with the aid of sterile water or sterile sodium chloride solution.

Pastes

Pastes are semisolid dosage forms that contain one or more drug substances intended for topical application. One class is made from a single phase aqueous gel, e.g., *Carboxymethylcellulose Sodium Paste*. The other class, the fatty pastes, e.g., *Zinc Oxide Paste*, consists of thick, stiff ointments that do not ordinarily flow at body temperature, and therefore serve as protective coatings over the areas to which they are applied.

The fatty pastes appear less greasy and more absorptive than ointments by reason of a high proportion of drug substance(s) having an affinity for water. These pastes tend to absorb serous secretions, and are less penetrating and less macerating than ointments, so that they are preferred for acute lesions that have a tendency towards crusting, vesiculation or oozing.

A dental paste is intended for adhesion to the mucous membrane for local effect (e.g., *Triamcinolone Acetonide Dental Paste*).

Pellets

See *Implants*

Powders

Powders are intimate mixtures of dry, finely divided drugs and/or chemicals that may be intended for internal (Oral Powders) or external (Topical Powders) use. Because of their greater specific surface area, powders disperse and dissolve more readily than compacted dosage forms. Children and those adults who experience difficulty in swallowing tablets or capsules may find powders more acceptable. Drugs that are too bulky to be formed into tablets or capsules of convenient size may be administered as powders. Immediately prior to use, oral powders are mixed in a beverage or apple sauce.

Often, stability problems encountered in liquid dosage forms are avoided in powdered dosage forms. Drugs that are unstable in aqueous suspensions or solutions may be prepared in the form of granules or powders. These are intended to be constituted by the pharmacist by the addition of a specified quantity of water just prior to dispensing. Because these constituted products have limited stability, they are required to have a specified expiration date after constitution and may require storage in a refrigerator.

Oral powders may be dispensed in doses pre-measured by the pharmacist, i.e., divided powders, or in bulk. Traditionally, divided powders have been wrapped in materials such as bond paper and parchment. However, the pharmacist may provide greater protection from the environment by sealing individual doses in small cellophane or polyethylene envelopes.

Bulk oral powders are limited to relatively non-potent drugs such as laxatives, antacids, dietary supplements, and certain analgesics that the patient may safely measure by the teaspoonful or capful. Other bulky powders include douche powders, tooth powders, and dusting powders. Bulk powders are best dispensed in tight, wide-mouth glass containers to afford maximum protection from the atmosphere and to prevent the loss of volatile constituents.

Dusting powders are impalpable powders intended for topical application. They may be dispensed in sifter-top containers to facilitate dusting onto the skin. In general, dusting powders should be passed through at least a 100-mesh sieve to assure freedom from grit that could irritate traumatized areas (see *Powder Fineness* ⟨811⟩).

Solutions

Solutions are liquid preparations that contain one or more chemical substances dissolved, i.e., molecularly dispersed, in a suitable solvent or mixture of mutually miscible solvents. Since molecules in solutions are uniformly dispersed, the use of solutions as dosage forms generally provides for the assurance of uniform dosage upon administration, and good accuracy when diluting or otherwise mixing solutions.

Substances in solutions, however, are more susceptible to chemical instability than the solid state and dose for dose, generally require more bulk and weight in packaging relative to solid dosage forms. For all solutions, but particularly those containing volatile solvents, tight containers, stored away from excessive heat, should be used. Consideration should also be given to the use of light-resistant containers when photolytic chemical degradation is a potential stability problem. Dosage forms categorized as "Solutions" are classified according to route of administration, such as "Oral Solutions" and "Topical Solutions," or by their solute and solvent systems, such as "Spirits," "Tinctures," and "Waters." Solutions intended for parenteral administration are officially entitled, "Injections" (see *Injections* ⟨1⟩).

Oral Solutions—Oral Solutions are liquid preparations, intended for oral administration, that contain one or more substances with or without flavoring, sweetening, or coloring agents dissolved in water or cosolvent-water mixtures. Oral Solutions may be formulated for direct oral administration to the patient or they may be dispensed in a more concentrated form that must be diluted prior to administration. It is important to recognize that dilution with water of Oral Solutions containing cosolvents, such as alcohol, could lead to precipitation of some ingredients. Hence, great care must be taken in diluting concentrated solutions when cosolvents are present. Preparations dispensed as soluble solids or soluble mixtures of solids, with the intent of dissolving them in a solvent and administering them orally, are designated "for Oral Solution," e.g., *Potassium Chloride for Oral Solution.*

Oral Solutions containing high concentrations of sucrose or other sugars traditionally have been designated as Syrups. A near-saturated solution of sucrose in purified water, for example, is known as Syrup or "Simple Syrup." Through common usage the term, syrup, also has been used to include any other liquid dosage form prepared in a sweet and viscid vehicle, including oral suspensions.

In addition to sucrose and other sugars, certain polyols such as sorbitol or glycerin may be present in Oral Solutions to inhibit crystallization and to modify solubility, taste, mouth-feel, and other vehicle properties. Antimicrobial agents to prevent the growth of bacteria, yeasts, and molds are generally also present. Some sugarless Oral Solutions contain sweetening agents such as sorbitol or aspartame, as well as thickening agents such as the cellulose gums. Such viscid sweetened solutions, containing no sugars, are occasionally prepared as vehicles for administration of drugs to diabetic patients.

Many oral solutions, which contain alcohol as a cosolvent, have been traditionally designated as Elixirs. Many others, however, designated as Oral Solutions, also contain significant amounts of alcohol. Since high concentrations of alcohol can produce a pharmacologic effect when administered orally, other cosolvents, such as glycerin and propylene glycol, should be used to minimize the amount of alcohol required. To be designated as an Elixir, however, the solution must contain alcohol.

Topical Solutions—Topical Solutions are solutions, usually aqueous but often containing other solvents, such as alcohol and polyols, intended for topical application to the skin, or as in the case of Lidocaine Oral Topical Solution, to the oral mucosal surface. The term "lotion" is applied to solutions or suspensions applied topically.

Otic Solutions—Otic Solutions, intended for instillation in the outer ear, are aqueous, or they are solutions prepared with glycerin or other solvents and dispersing agents, e.g., *Antipyrine and Benzocaine Otic Solution* and *Neomycin and Polymyxin B Sulfates and Hydrocortisone Otic Solution.*

Ophthalmic Solutions (See *Ophthalmic Preparations.*)

Spirits—Spirits are alcoholic or hydroalcoholic solutions of volatile substances prepared usually by simple solution or by admixture of the ingredients. Some spirits serve as flavoring agents while others have medicinal value. Reduction of the high alcoholic content of spirits by admixture with aqueous preparations often causes turbidity.

Spirits require storage in tight, light-resistant containers to prevent loss by evaporation and to limit oxidative changes.

Tinctures—Tinctures are alcoholic or hydroalcoholic solutions prepared from vegetable materials or from chemical substances.

The proportion of drug represented in the different chemical tinctures is not uniform but varies according to the established standards for each. Traditionally, tinctures of potent vegetable drugs essentially represent the activity of 10 g of the drug in each 100 mL of tincture, the potency being adjusted following assay. Most other vegetable tinctures represent 20 g of the respective vegetable material in each 100 mL of tincture.

Process P—Carefully mix the ground drug or mixture of drugs with a sufficient quantity of the prescribed solvent or solvent mixture to render it evenly and distinctly damp, allow it to stand for 15 minutes, transfer it to a suitable percolator, and pack the drug firmly. Pour on enough of the prescribed solvent or solvent mixture to saturate the drug, cover the top of the percolator and, when the liquid is about to drip from the percolator, close the lower orifice, and allow the drug to macerate for 24 hours or for the time specified in the monograph. If no assay is directed, allow the percolation to proceed slowly, or at the specified rate, gradually adding sufficient solvent or solvent mixture to produce 1000 mL of tincture, and mix (for definitions of flow rates, see under *Fluidextracts*). If an assay is directed, collect only 950 mL of percolate, mix this, and assay a portion of it as directed. Dilute the remainder with such quantity of the prescribed solvent or solvent mixture as calculation from the assay indicates is necessary to produce a tincture that conforms to the prescribed standard, and mix.

Process M—Macerate the drug with 750 mL of the prescribed solvent or solvent mixture in a container that can be closed, and put in a warm place. Agitate it frequently during 3 days or until the soluble matter is dissolved. Transfer the mixture to a filter, and when most of the liquid has drained away, wash the residue on the filter with a sufficient quantity of the prescribed solvent or solvent mixture, combining the filtrates, to produce 1000 mL of tincture, and mix.

Tinctures require storage in tight, light-resistant containers, away from direct sunlight and excessive heat.

Waters, Aromatic—Aromatic waters are clear, saturated aqueous solutions (unless otherwise specified) of volatile oils or other aromatic or volatile substances. Their odors and tastes are similar, respectively, to those of the drugs or volatile substances from which they are prepared, and they are free from empyreumatic and other foreign odors. Aromatic waters may be prepared by distillation or solution of the aromatic substance, with or without the use of a dispersing agent.

Aromatic waters require protection from intense light and excessive heat.

Suppositories

Suppositories are solid bodies of various weights and shapes, adapted for introduction into the rectal, vaginal, or urethral orifice of the human body. They usually melt, soften, or dissolve at body temperature. A suppository may act as a protectant or palliative to the local tissues at the point of introduction or as a carrier of therapeutic agents for systemic or local action. Suppository bases usually employed are cocoa butter, glycerinated gelatin, hydrogenated vegetable oils, mixtures of polyethylene glycols of various molecular weights, and fatty acid esters of polyethylene glycol.

The suppository base employed has a marked influence on the release of the active ingredient incorporated in it. While cocoa butter melts quickly at body temperature, it is immiscible with body fluids and this inhibits the diffusion of fat-soluble drugs to the affected sites. Polyethylene glycol is a suitable base for some antiseptics. In cases where systemic action is expected, it is preferable to incorporate the ionized rather than the nonionized form of the drug, in order to maximize bioavailability. Although unionized drugs partition out of water-miscible bases such as glycerinated gelatin and polyethylene glycol more readily, the bases themselves tend to dissolve very slowly and thus retard release in this manner. Oleaginous vehicles such as cocoa butter are seldom used in vaginal preparations because of the nonabsorbable residue formed, while glycerinated gelatin is seldom used rectally because of its slow dissolution. Cocoa butter is superior for allaying irritation, as in preparations intended for treating internal hemorrhoids.

Cocoa Butter Suppositories—Suppositories having cocoa butter as the base may be made by means of incorporating the finely divided medicinal substance into the solid oil at room temperature and suitably shaping the resulting mass, or by working with the oil in the melted state and allowing the resulting suspension to cool in molds. A suitable quantity of hardening agents may be added to counteract the tendency of some medicaments such as chloral hydrate and phenol to soften the base. It is important that the finished suppository melt at body temperature.

The approximate weights of suppositories prepared with cocoa butter are given below. Suppositories prepared from other bases vary in weight and generally are heavier than the weights indicated here.

Rectal Suppositories for adults are tapered at one or both ends and usually weigh about 2 g each.

Vaginal Suppositories are usually globular or oviform and weigh about 5 g each. They are made from water soluble or water miscible vehicles such as polyethylene glycol or glycerinated gelatin.

Suppositories with cocoa butter base require storage in well-closed containers, preferably at a temperature below 30° (controlled room temperature).

Cocoa Butter Substitutes—Fat-type suppository bases can be produced from a variety of vegetable oils, such as coconut or palm kernel, which are modified by esterification, hydrogenation, and fractionation to obtain products of varying composition and melting temperatures (e.g., *Hydrogenated Vegetable Oil* and *Hard Fat*). These products can be so designed as to reduce rancidity. At the same time, desired characteristics such as narrow intervals between melting and solidification temperatures, and melting ranges to accommodate various formulation and climatic conditions, can be built in.

Glycerinated Gelatin Suppositories—Medicinal substances may be incorporated into glycerinated gelatin bases by addition of the prescribed quantities to a vehicle consisting of about 70 parts of glycerin, 20 parts of gelatin, and 10 parts of water.

Glycerinated gelatin suppositories require storage in tight containers, preferably at a temperature below 35°.

Polyethylene Glycol–Base Suppositories—Several combinations of polyethylene glycols having melting temperatures that are above body temperature have been used as suppository bases. Inasmuch as release from these bases depends on dissolution rather than on melting, there are significantly fewer problems in preparation and storage than exist with melting-type vehicles. However, high concentrations of higher molecular weight polyethylene glycols may lengthen dissolution time, resulting in problems with retention. Labels on polyethylene glycol suppositories should contain directions that they be moistened with water before inserting. Although they can be stored without refrigeration, they should be packaged in tightly closed containers.

Surfactant Suppository Bases—Several nonionic surface-active agents closely related chemically to the polyethylene glycols can be used as suppository vehicles. Examples of such surfactants are polyoxyethylene sorbitan fatty acid esters and the polyoxyethylene stearates. These surfactants are used alone or in combination with other suppository vehicles to yield a wide range of melting temperatures and consistencies. One of the major advantages of such vehicles is their water-dispersibility. However, care must be taken with the use of surfactants, because they may either increase the rate of drug absorption or interact with drug molecules, causing a decrease in therapeutic activity.

Tableted Suppositories or Inserts—Vaginal suppositories occasionally are prepared by the compression of powdered materials into a suitable shape. They are prepared also by encapsulation in soft gelatin.

Suspensions

Suspensions are liquid preparations which consist of solid particles dispersed throughout a liquid phase in which the particles are not soluble. Dosage forms officially categorized as Suspensions are designated as such if they are not included in other more specific categories of suspensions, such as Oral Suspensions, Topical Suspensions, etc. (see these other categories). Some suspensions are prepared and ready for use, while others are prepared as solid mixtures intended for constitution just before use with an appropriate vehicle. Such products are designated "for Oral Suspension," etc. The term, Milk, is sometimes used for suspensions in aqueous vehicles intended for oral administration, e.g., *Milk of Magnesia*. The term, Magma, is often used to describe suspensions of inorganic solids such as clays, in water, where there is a tendency for strong hydration and aggregation of the solid, giving rise to gel-like consistency and thixotropic rheological behavior, e.g., *Bentonite Magma*. The term "Lotion" has been used to categorize many topical suspensions and emulsions intended for application to the skin, e.g., *Calamine Lotion*. Some suspensions are prepared in sterile form and are used as Injectables, as well as for ophthalmic and otic administration. These may be of two types, ready to use or intended for constitution with a prescribed amount of Water for Injection or other suitable diluent before use by the designated route. Suspensions should not be injected intravenously or intrathecally.

Suspensions intended for any route of administration should contain suitable antimicrobial agents to protect against bacteria, yeast, and mold contamination (see *Emulsions* for some consideration of antimicrobial preservative properties which apply also to Suspensions). By its very nature, the particular matter in a suspension may settle or sediment to the bottom of the container upon standing. Such sedimentation may also lead to caking and solidification of the sediment with a resulting difficulty in redispersing the suspension upon agitation. To prevent such problems, suitable ingredients which increase viscosity and the gel state of the suspension, such as clays, surfactants, polyols, polymers, or sugars, should be added. It is important that suspensions always be shaken well before use to ensure uniform distribution of the solid in the vehicle, thereby ensuring uniform and proper dosage. Suspensions require storage in tight containers.

Oral Suspensions—Oral Suspensions are liquid preparations containing solid particles dispersed in a liquid vehicle, with suitable flavoring agents, intended for oral administration. Some suspensions labeled as Milks or Magmas fall into this category.

Topical Suspensions—Topical Suspensions are liquid preparations containing solid particles dispersed in a liquid vehicle, intended for application to the skin. Some suspensions labeled as Lotions fall into this category.

Otic Suspensions—Otic Suspensions are liquid preparations containing micronized particles intended for instillation in the outer ear.

Ophthalmic Suspensions—(See *Ophthalmic Preparations*).

Syrups

See *Solutions*

Systems

In recent years, a number of dosage forms have been developed using modern technology that allows for the uniform release or targeting of drugs to the body. These products are commonly called delivery systems. The most widely used of these are Transdermal Systems.

Transdermal Systems—Transdermal drug delivery systems are self-contained, discrete dosage forms which, when applied to intact skin are designed to deliver the drug(s) through the skin to the systemic circulation. Systems typically comprise an outer covering (barrier), a drug reservoir which may have a rate controlling membrane, a contact adhesive applied to some or all parts of the system and the system/skin interface, and a protective liner which is removed before applying the system. The activity of these systems is defined in terms of the release rate of the drug(s) from the system. The total duration of drug release from the system and the system surface area may also be stated.

Transdermal drug delivery systems work by diffusion: the drug diffuses from the drug reservoir, directly or through the rate controlling membrane and/or contact adhesive if present, and then through the skin into the general circulation. Typically, modified-release systems are designed to provide drug delivery at a constant rate, such that a true steady state blood concentration is achieved and maintained until the system is removed. At that time, blood concentration declines at a rate consistent with the pharmacokinetics of the drug.

Transdermal drug delivery systems are applied to body areas consistent with the labeling for the product(s). As long as drug concentration at the system/skin interface remains constant, the amount of drug in the dosage form does not influence plasma concentrations. The functional lifetime of the system is defined by the initial amount of drug in the reservoir and the release rate from the reservoir.

[NOTE—Drugs for local rather than systemic effect are commonly applied to the skin embedded in glue on a cloth or plastic backing. These products are defined traditionally as plasters or tapes.]

Ocular System—Another type of system is the ocular system which is intended for placement in the lower conjunctival formix from which the drug diffuses through a membrane at a constant rate over a seven-day period (e.g., *Pilocarpine Ocular System*).

Intrauterine System—An intrauterine system, based on a similar principle but intended for release of drug over a much longer period of time, i.e., one year, is also available (e.g., *Progesterone Intrauterine Contraceptive System*).

Tablets

Tablets are solid dosage forms containing medicinal substances with or without suitable diluents. They may be classed, according to the method of manufacture, as molded tablets or compressed tablets.

The vast majority of all tablets manufactured are made by compression, and compressed tablets are the most widely used dosage form in this country. Compressed tablets are prepared by the application of high pressures, utilizing steel punches and dies, to powders or granulations. Tablets can be produced in a wide variety of sizes, shapes, and surface markings, depending upon the design of the punches and dies. Capsule-shaped tablets are commonly referred to as caplets. Boluses are large tablets intended for veterinary use, usually for large animals.

Molded tablets are prepared by forcing dampened powders under low pressure into die cavities. Solidification depends upon crystal bridges built up during the subsequent drying process, and not upon the compaction force.

Tablet triturates are small, usually cylindrical, molded or compressed tablets. Tablet triturates were traditionally used as dispensing tablets in order to provide a convenient, measured quantity of a potent drug for compounding purposes. Such tablets are rarely used today. Hypodermic tablets are molded tablets made from completely and readily water-soluble ingredients and formerly were intended for use in making preparations for hypodermic injection. They are employed orally, or where rapid drug availability is required such as in the case of *Nitroglycerin Tablets*, sublingually.

Buccal tablets are intended to be inserted in the buccal pouch, and sublingual tablets are intended to be inserted beneath the tongue, where the active ingredient is absorbed directly through the oral mucosa. Few drugs are readily absorbed in this way, but for those that are (such as nitroglycerin and certain steroid hormones), a number of advantages may result.

Soluble, effervescent tablets are prepared by compression and contain, in addition to active ingredients, mixtures of acids (citric acid, tartaric acid) and sodium bicarbonate, which release carbon dioxide when dissolved in water. They are intended to be dissolved or dispersed in water before administration. Effervescent tablets should be stored in tightly closed containers, or moisture-proof packs, labeled to indicate that they are not to be swallowed directly.

CHEWABLE TABLETS

Chewable tablets are intended to be chewed, producing a pleasant tasting residue in the oral cavity that is easily swallowed and does not leave a bitter or unpleasant after-taste. These tablets have been used in tablet formulations for children, especially multivitamin formulations, and for the administration of antacids and selected antibiotics. Chewable tablets are prepared by compression, usually utilizing mannitol, sorbitol, or sucrose as binders and fillers, and containing colors and flavors to enhance their appearance and taste.

PREPARATION OF MOLDED TABLETS

Molded tablets are prepared from mixtures of medicinal substances and a diluent usually consisting of lactose and powdered sucrose in varying proportions. The powders are dampened with solutions containing high percentages of alcohol. The concentration of alcohol depends upon the solubility of the active ingredients and fillers in the solvent system and the desired degree of hardness of the finished tablets. The dampened powders are pressed into molds, removed, and allowed to dry. Molded tablets are quite friable and care must be taken in packaging and dispensing.

FORMULATION OF COMPRESSED TABLETS

Most compressed tablets consist of the active ingredient and a diluent (filler), binder, disintegrating agent, and lubricant. Approved FD&C and D&C dyes or lakes (dyes adsorbed onto insoluble aluminum hydroxide), flavors, and sweetening agents may be present also. Diluents are added where the quantity of active ingredient is small or difficult to compress. Common tablet fillers include lactose, starch, dibasic calcium phosphate, and microcrystalline cellulose. Chewable tablets often contain sucrose, mannitol, or sorbitol as a filler. Where the amount of active ingredient is small, the overall tableting properties are in large measure determined by the filler. Because of problems encountered with bioavailability of hydrophobic drugs of low water-solubility, water-soluble diluents are used as fillers for these tablets.

Binders give adhesiveness to the powder during the preliminary granulation and to the compressed tablet. They add to the cohesive strength already available in the diluent. While binders may be added dry, they are more effective when added out of solution. Common binders include acacia, gelatin, sucrose, povidone, methylcellulose, carboxymethylcellulose, and hydrolyzed starch pastes. The most effective dry binder is microcrystalline cellulose, which is commonly used for this purpose in tablets prepared by direct compression.

A disintegrating agent serves to assist in the fragmentation of the tablet after administration. The most widely used tablet disintegrating agent is starch. Chemically modified starches and cellulose, alginic acid, microcrystalline cellulose, and cross-linked povidone, are also used for this purpose. Effervescent mixtures are used in soluble tablet systems as disintegrating agents. The concentration of the disintegrating agent, method of addition, and degree of compaction play a role in effectiveness.

Lubricants reduce friction during the compression and ejection cycle. In addition, they aid in preventing adherence of tablet material to the dies and punches. Metallic stearates, stearic acid, hydrogenated vegetable oils, and talc are used as lubricants. Because of the nature of this function, most lubricants are hydrophobic, and as such tend to reduce the rates of tablet disintegration and dissolution. Consequently, excessive concentrations of lubricant should be avoided. Polyethylene glycols and some lauryl sulfate salts have been used as soluble lubricants, but such agents generally do not possess optimal lubricating properties, and comparatively high concentrations are usually required.

Glidants are agents that improve powder fluidity, and they are commonly employed in direct compression where no granulation step is involved. The most effective glidants are the colloidal pyrogenic silicas.

Colorants are often added to tablet formulations for esthetic value or for product identification. Both D&C and FD&C dyes and lakes are used. Most dyes are photosensitive and they fade when exposed to light. The federal Food and Drug Administration regulates the colorants employed in drugs.

MANUFACTURE

Tablets are prepared by three general methods: wet granulation, dry granulation (roll compaction or slugging), and direct compression. The purpose of both wet and dry granulation is to improve flow of the mixture and/or to enhance its compressibility.

Dry granulation (slugging) involves the compaction of powders at high pressures into large, often poorly formed tablet compacts. These compacts are then milled and screened to form a granulation of the desired particle size. The advantage of dry granulation is the elimination of both heat and moisture in the processing. Dry granulations can be produced also by extruding powders between hydraulically operated rollers to produce thin cakes which are subsequently screened or milled to give the desired granule size.

Excipients are available that allow production of tablets at high speeds without prior granulation steps. These directly compressible excipients consist of special physical forms of substances such as lactose, sucrose, dextrose, or cellulose, which possess the desirable properties of fluidity and compressibility. The most widely used direct-compaction fillers are microcrystalline cellulose, anhydrous lactose, spray-dried lactose, compressible sucrose, and some forms of modified starches. Direct compression avoids many of the problems associated with wet and dry granulations. However, the inherent physical properties of the individual filler materials are highly critical, and minor variations can alter flow and compression characteristics so as to make them unsuitable for direct compression.

Physical evidence of poor tablet quality is discussed under *Stability Considerations in Dispensing Practice* ⟨1191⟩.

Weight Variation and Content Uniformity—Tablets are required to meet a weight variation test (see *Uniformity of Dosage Units* ⟨905⟩) where the active ingredient comprises a major portion of the tablet and where control of weight may be presumed to be an adequate control of drug content uniformity. Weight variation is not an adequate indication of content uniformity where the drug substance comprises a relatively minor portion of the tablet, or where the tablet is sugar-coated. Thus, the Pharmacopeia generally requires that coated tablets and tablets containing 50 mg or less of active ingredient, comprising less than 50% by weight of the dosage-form unit, pass a content uniformity test (see *Uniformity of Dosage Units* ⟨905⟩), wherein individual tablets are assayed for actual drug content.

Disintegration and Dissolution—Disintegration is an essential attribute of tablets intended for administration by mouth, except those intended to be chewed before being swallowed and except some types of extended-release tablets. A disintegration test is provided (see *Disintegration* ⟨701⟩), and limits on the times in which disintegration is to take place, appropriate for the types of tablets concerned, are given in the individual monographs.

For drugs of limited water-solubility, dissolution may be a more meaningful quality attribute than disintegration. A dissolution test (see *Dissolution* ⟨711⟩) is required in a number of monographs on tablets. In many cases, it is possible to correlate dissolution rates with biological availability of the active ingredient. However, such tests are useful mainly as a means of screening preliminary formulations and as a routine quality-control procedure.

COATINGS

Tablets may be coated for a variety of reasons, including protection of the ingredients from air, moisture, or light, masking of unpleasant tastes and odors, improvement of appearance, and control of the site of drug release in the gastrointestinal tract.

Plain Coated Tablets—Classically, tablets have been coated with sugar applied from aqueous suspensions containing insoluble powders such as starch, calcium carbonate, talc, or titanium dioxide, suspended by means of acacia or gelatin. For purposes of identification and esthetic value, the outside coatings may be colored. The finished coated tablets are polished by application of dilute solutions of wax in solvents such as chloroform or powdered mix. Water-protective coatings consisting of substances such as shellac or cellulose acetate phthalate are often applied out of nonaqueous solvents prior to application of sugar coats. Excessive quantities should be avoided. Drawbacks of sugar coating include the lengthy time necessary for application, the need for waterproofing, which also adversely affects dissolution, and the increased bulk of the finished tablet. These factors have resulted in increased acceptance of film coatings. Film coatings consist of water-soluble or dispersible materials such as hydroxypropyl methylcellulose, methylcellulose, hydroxypropylcellulose, carboxymethylcellulose sodium, and mixtures of cellulose acetate phthalate and polyethylene glycols applied out of nonaqueous or aqueous solvents. Evaporation of the solvents leaves a thin film that adheres directly to the tablet and allows it to retain the original shape, including grooves or identification codes.

Enteric-coated Tablets—Where the drug may be destroyed or inactivated by the gastric juice or where it may irritate the gastric mucosa, the use of "enteric" coatings is indicated. Such coatings are intended to delay the release of the medication until the tablet has passed through the stomach. The term "delayed-release" is used for Pharmacopeial purposes, and the individual monographs include tests and specifications for *Drug release* (see *Drug Release* ⟨724⟩).

Extended-release Tablets—Extended-release tablets are formulated in such manner as to make the contained medicament available over an extended period of time following ingestion. Expressions such as "prolonged-action," "repeat-action," and "sustained-release" have been used also to describe such dosage forms. However, the term "extended-release" is used for Pharmacopeial purposes, and requirements for *Drug release* typically are specified in the individual monographs.

⟨1171⟩ PHASE-SOLUBILITY ANALYSIS

Phase-solubility analysis is the quantitative determination of the purity of a substance through the application of precise solubility measurements. At a given temperature, a definite amount of a pure substance is soluble in a definite quantity of solvent. The resulting solution is saturated with respect to the particular substance, but the solution remains unsaturated with respect to other substances, even though such substances may be closely related in chemical structure and physical properties to the particular substance being tested. Constancy of solubility, just as constancy of melting temperature or other physical properties, indicates that a material is pure or is free from foreign admixture except in the unique case where the percentage composition of the substance under test is in direct ratio to solubilities of the respective components. Conversely, variability of solubility indicates the presence of an impurity or impurities.

Phase-solubility analysis is applicable to all species of compounds that are crystalline solids and that form stable solutions. It is not readily applicable to compounds that form solid solutions with impurities.

The standard solubility method consists of six distinct steps: (1) mixing, in a series of separate systems, increasing quantities of material with measured, fixed amounts of a solvent; (2) establishment of equilibrium for each system at identical constant temperature and pressure; (3) separation of the solid phase from the solutions; (4) determination of the concentration of the material dissolved in the various solutions; (5) plotting the concentration of the dissolved materials per unit of solvent (*y*-axis or solution composition) against the weight of material per unit of solvent (*x*-axis or system composition); and (6) extrapolation and calculation.

Solvents

A proper solvent for phase-solubility analysis meets the following criteria: (1) The solvent is of sufficient volatility so that it can be evaporated under vacuum, but is not so volatile that difficulty is experienced in transferring and weighing the solvent and its solutions. Normally, solvents having boiling points between 60° and 150° are suitable. (2) The solvent does not adversely affect the substance being tested. Solvents that cause decomposition or react with the test substance are not to be used. Solvents that solvate or form salts are to be avoided, if possible. (3) The solvent is of known purity and composition. Carefully prepared mixed solvents are permissible. Trace impurities may affect solubility greatly. (4) A solubility of 10 mg to 20 mg per g is optimal, but a wider working range can be utilized.

Apparatus*

Constant-temperature Bath—Use a constant-temperature bath that is capable of maintaining the temperature within ±0.1° and that is equipped with a horizontal shaft capable of rotating at approximately 25 rpm. The shaft is equipped with clamps to hold the *Ampuls*. Alternatively, the bath may contain a suitable vibrator, capable of agitating the ampuls at 100 to 120 vibrations per second, and equipped with a shaft and suitable clamps to hold the ampuls.

Ampuls—Use 15-mL ampuls of the type shown in the accompanying illustration. Other containers may be used provided that they are leakproof and otherwise suitable.

Solubility Flasks—Use solubility flasks of the type shown in the accompanying illustration.

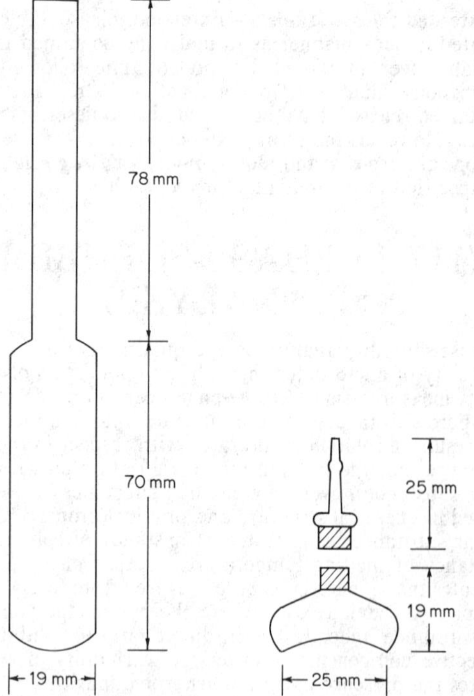

Ampul (left) and Solubility Flask (right) Used in Phase-solubility Analysis

* Available from Hanson Research Corp., 19727 Bahama St., P. O. Box 35, Northridge, CA 91324.

Procedure

NOTE—Make all weighings within ±10 µg.

System Composition—Weigh accurately, in g, not less than 7 scrupulously cleaned 15-mL ampuls. Weigh accurately, in g, increasingly larger amounts of the test substance into each of the ampuls. The weight of the test substance is selected so that the first ampul contains slightly less material than will go into solution in 5 mL of the selected solvent, the second ampul contains slightly more material, and each subsequent ampul contains increasingly more material than meets the indicated solubility. Transfer 5.0 mL of the solvent to each of the ampuls, cool in a dry ice–acetone mixture, and seal, using a doublet-jet air–gas burner and taking care to save all glass. Allow the ampuls and their contents to come to room temperature, and weigh the individual sealed ampuls with the corresponding glass fragments. Calculate the system composition, in mg per g, for each ampul by the formula:

$$1000(W_2 - W_1)/(W_3 - W_2),$$

in which W_2 is the weight of the ampul plus test substance, W_1 is the weight of the empty ampul, and W_3 is the weight of ampul plus test substance, solvent, and separated glass.

Equilibration—The time required for equilibration varies with the substance, the method of mixing (rotation or vibration), and the temperature. Normally, equilibrium is obtained more rapidly by the vibration method (1 to 7 days) than by the rotational method (7 to 14 days). In order to determine whether equilibration has been effected, 1 ampul, i.e., the next to the last in the series, may be warmed to 40° to produce a super-saturated solution. Equilibration is assured if the solubility obtained on the super-saturated solution falls in line with the test specimens that approach equilibrium from an undersaturated solution.

Solution Composition—After equilibration, place the ampuls vertically in a rack in the constant-temperature bath, with the necks of the ampuls above the water level, and allow the contents to settle. Open the ampuls, and remove a portion greater than 2 mL from each by means of a pipet equipped with a small pledget of cotton membrane or other suitable filter. Transfer a 2.0-mL aliquot of clear solution from each ampul to a marked, tared solubility flask, and weigh each flask plus its solution to obtain the weight of the solution. Cool the flasks in a dry ice–acetone bath, and then evaporate the solvent in vacuum. Gradually increase the temperature to a temperature consistent with the stability of the compound, and dry the residue to constant weight. Calculate the solution composition, in mg per g, by the formula:

$$1000(F_3 - F_1)/(F_2 - F_3),$$

in which F_3 is the weight of the flask plus residue, F_1 is the weight of the solubility flask, and F_2 is the weight of the flask plus solution.

Calculation

For each portion of the test substance taken, plot the solution composition as the ordinate and the system composition as the abscissa. As shown in the accompanying diagram, the points for those containers, frequently only one, that represent a true solution fall on a straight line (AB) with a slope of 1, passing through the origin; the points corresponding to saturated solutions fall on another straight line (BC), the slope, *S*, of which represents the

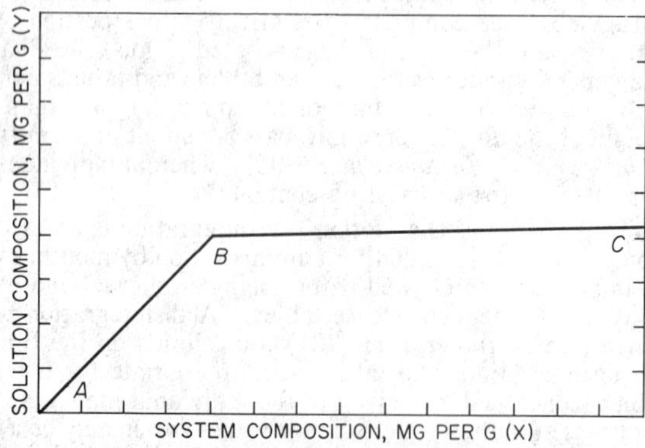

Typical Phase-solubility Diagram

weight fraction of impurity or impurities present in the test substance. Failure of points to fall on a straight line indicates that equilibrium has not been achieved. A curve indicates the material under test may be a solid solution. Calculate the percentage purity of the test substance by the formula:

$$100 - 100S.$$

The slope, S, may be calculated graphically or by least-squares treatment for best fit of the experimental values to a straight line.

The solubility of the main component is obtained by extending the solubility line (BC) through the y-axis. The point of interception on the y-axis is the extrapolated solubility, in mg per g, and is a constant for a given compound.

Purification Technique

Since the solvent phase in all combinations of solvent and solute that are used to construct segment BC of a phase-solubility diagram contains essentially all of the impurities originally present in the substance under analysis, while the solid phase is essentially free from impurities, phase-solubility analysis can be used to prepare pure reference specimens of desired compounds as well as concentrates of impurities from substances otherwise considered pure. A simple modification of this technique can be used to accomplish these purposes with considerably less effort than is usually required for rigorous phase-solubility analysis.

In practice, a weighed amount of test specimen is suspended in a nonreactive solvent of suitable composition and amount so that about 10% of the material is dissolved at equilibrium. The suspension is sealed (a screw-cap vial is usually adequate) and shaken at room temperature until equilibrium is attained (usually 24 hours is sufficient for this purpose). The mother liquor is then drawn off and evaporated at or near room temperature to dryness. Since the mother liquor contained essentially all of the impurities that were present in the specimen, the residue has been concentrated with respect to the impurities roughly in proportion to the ratio of the weight of specimen taken to the weight of solids dissolved in the volume of solvent used.

The undissolved crystals remaining after withdrawal of the mother liquor are usually sufficiently pure to be used as a reference standard after appropriate rinsing and drying.

⟨1176⟩ PRESCRIPTION BALANCES AND VOLUMETRIC APPARATUS

Prescription Balances

NOTE—Balances other than the type described herein may be used provided these afford equivalent or better accuracy. This includes micro-, semimicro-, or electronic single-pan balances (see *Weights and Balances* ⟨41⟩). Some balances offer digital or direct-reading features. All balances should be calibrated and tested frequently using appropriate test weights, both singly and in combination.

Description—A prescription balance is a scale or balance adapted to weighing medicinal and other substances required in prescriptions or in other pharmaceutical compounding. It is constructed so as to support its full capacity without developing undue stresses, and its adjustment is not altered by repeated weighings of the capacity load. The removable pans or weighing vessels should be of equal weight. The balance should have leveling feet or screws. The balance may feature dial-in weights and also a precision spring and dial instead of a weighbeam. A balance that has a graduated weighbeam must have a stop that halts the rider or poise at the "zero" reading. The reading edge of the rider is parallel to the graduations on the weighbeam. The distance from the face of the index plate to the indicator pointer or pointers should be not more than 1.0 mm (0.04 inch), the points should be sharp, and when there are two, their ends should be separated by not more than 1.0 mm (0.04 inch) when the scale is in balance. The indicating elements and the lever system should be protected against drafts, and the balance lid should permit free movement of the loaded weighing pans when the lid is closed. The balance must have a mechanical arresting device.

Definitions—

Capacity—Maximum weight, including the weight of tares, to be placed on one pan. The *N.B.S. Handbook 44*, 4th ed., states: "*In the absence of information to the contrary*, the nominal capacity of a Class *A* balance shall be assumed to be 15.5 g (½ apothecaries' ounce)." Most of the commercially available Class *A* balances have a capacity of 120 g (4 ounces) and bear a statement to that effect.

Weighbeam or Beam—A graduated bar equipped with a movable poise or rider. Metric graduations are in 0.01-g increments up to a maximum of 1.0 g; apothecaries' graduations are in ⅛-grain increments up to a maximum of 15 ⅜-grains. The bar may be graduated in both systems.

Tare Bar—An auxiliary ungraduated weighbeam bar with a movable poise. This can be used to correct for variations in weighing-glasses or papers.

Balance Indicator—A combination of elements, one or both of which will oscillate with respect to the other, to indicate the equilibrium state of the balance during weighing.

Rest Point—The point on the index plate at which the indicator or pointer stops when the oscillations of the balance cease; or the index plate position of the indicator or pointer calculated from recorded consecutive oscillations in both directions past the "zero" of the index plate scale. If the balance has a two-pointer indicating mechanism, the position or the oscillations of only one of the pointers need be recorded or used to determine the rest point.

Sensitivity Requirements (*SR*)—The maximum change in load that will cause a specified change, one subdivision on the index plate, in the position of rest of the indicating element or elements of the balance.

Class A Prescription Balance—A balance that meets the tests for this type of balance has a sensitivity requirement of 6 mg or less with no load and with a load of 10 g on each pan. The Class *A* balance should be used for all of the weighing operations required in prescription compounding.

In order to avoid errors of 5 percent or more that might be due to the limit of sensitivity of the Class *A* prescription balance, do not weigh less than 120 mg (2 grains) of any material. If a smaller weight of dry material is required, mix a larger known weight of the ingredient with a known weight of dry diluent, and weigh an aliquot portion of the mixture for use.

Testing the Prescription Balance—A Class *A* prescription balance meets the following four basic tests. Use a set of test weights, and keep the rider on the weighbeam at zero unless directed to change its position.

1. *Sensitivity Requirement*—Level the balance, determine the rest point, and place a 6-mg weight on one of the empty pans. Repeat the operation with a 10-g weight in the center of each pan. The rest point is shifted not less than one division on the index plate each time the 6-mg weight is added.

2. *Arm Ratio Test*—This test is designed to check the equality of length of both arms of the balance. Determine the rest point of the balance with no weight on the pans. Place in the center of each pan a 30-g test weight, and determine the rest point. If the second rest point is not the same as the first, place a 20-mg weight on the lighter side; the rest point should move back to the original place on the index plate scale or farther.

3. *Shift Tests*—These tests are designed to check the arm and lever components of the balance.

A. Determine the rest point of the indicator without any weights on the pans.

B. Place one of the 10-g weights in the center of the left pan, and place the other 10-g weight successively toward the right, left, front, and back of the right pan, noting the rest point in each case. If in any case the rest point differs from the rest point determined in Step *A*, add a 10-mg weight to the lighter side; this should cause the rest point to shift back to the rest point determined in Step *A* or farther.

C. Place a 10-g weight in the center of the right pan, and place a 10-g weight successively toward the right, left, front, and back of the left pan, noting the rest point in each case. If in any case the rest point is different from that obtained with no weights on the pans, this difference should be overcome by addition of the 10-mg weight to the lighter side.

D. Make a series of observations in which both weights are simultaneously shifted to off-center positions on their pans, both toward the outside, both toward the inside, one toward the outside, and the other toward the inside, both toward the back, and so on until all combinations have been checked. If in any case the rest point differs from that obtained with no weights on the pan, the addition of the 10-mg weight to the lighter side should overcome this difference.

A balance that does not meet the requirements of these tests must be adjusted.

4. *Rider and Graduated Beam Tests*—Determine the rest point for the balance with no weight on the pans. Place on the left pan the 500-mg test weight, move the rider to the 500-mg point on the beam, and determine the rest point. If it is different from the zero rest point, add a 6-mg weight to the lighter side. This should bring the rest point back to its original position or farther. Repeat this test, using the 1-g test weight and moving the rider to the 1-g division on the beam. If the rest point is different, it should be brought back at least to the zero rest point position by addition of 6 mg to the lighter pan. If the balance does not meet this test, the weighbeam graduations or the rider must be corrected.

Metric or apothecaries' weights for use with a prescription balance should be kept in a special rigid and compartmentalized box and handled with plastic or plastic-tipped forceps to prevent scratching or soiling. For prescription use, analytical weights (Class P or better) are recommended. However, Class Q weights have tolerances well within the limits of accuracy of the prescription balance, and they retain their accuracy for a long time with proper care. Apothecaries' weights should have the same general (cylindrical) construction as metric weights. Coin-type (or disk-shaped) weights should not be used.

Test weights consisting of two 20-g or two 30-g, two 10-g, one 1-g, one 500-mg, one 20-mg, one 10-mg, and one 6-mg (or suitable combination totaling 6 mg) weights, adjusted to N.B.S. tolerances for analytical weights (Class P or better) should be used for testing the prescription balances. These weights should be kept in a tightly closed box and should be handled only with plastic or plastic-tipped forceps. The set of test weights should be used only for testing the balance or constantly used weights. If properly cared for, the set lasts indefinitely.

Volumetric Apparatus

Pharmaceutical devices for measuring volumes of liquids, including burets, pipets, and cylinders graduated either in metric or apothecary units meet the standard specifications for glass volumetric apparatus described in NTIS COM-73-10504 of the National Technical Information Service.[1] Conical graduates meet the standard specifications described in Handbook 44, 4th Edition, of the National Institute of Standards and Technology.[2] Graduated medicine droppers meet the specifications (see *Medicine Dropper* ⟨1101⟩). An acceptable ungraduated medicine dropper has a delivery end 3 mm in external diameter and delivers 20 drops of water, weighing 1 g at a temperature of 15°. A tolerance of ±10% of the delivery specification is reasonable.

Selection and Use of Graduates—

Capacity—The capacity of a graduate is the designated volume, at the maximum graduation, that the graduate will contain, or deliver, as indicated, at the specified temperature.

Cylindrical and Conical Graduates—The error in a measured volume caused by a deviation of ±1 mm, in reading the lower meniscus in a graduated cylinder remains constant along the height of the uniform column. The same deviation of ±1 mm causes a progressively larger error in a conical graduate, the extent of the error being further dependent upon the angle of the flared sides to the perpendicular of the upright graduate. A deviation of ±1 mm in the meniscus reading causes an error of approximately 0.5 mL in the measured volume at any mark on the uniform 100-mL cylinder graduate. The same deviation of ±1 mm can cause an error of 1.8 mL at the 100-mL mark on an acceptable conical graduate marked for 125 mL (4 fluidounces).

A general rule for selection of a graduate for use is to use the graduate with a capacity equal to or *just exceeding* the volume to be measured. Measurement of small volumes in large graduates tends to increase errors, because the larger diameter increases the volume error in a deviation of ±1 mm from the mark. The relation of the volume error to the internal diameters of graduated cylinders is based upon the equation $V = \pi r^2 h$. An acceptable 10-mL cylinder having an internal diameter of 1.18 cm holds 109 µL in 1 mm of the column. Reading 4.5 mL in this graduate with a deviation of ±1 mm from the mark causes an error of about ±2.5%, while the same deviation in a volume of 2.2 mL in the same graduate causes an error of about ±5%. Minimum volumes that can be measured within certain limits of error in graduated cylinders of different capacities are incorporated in the design details of graduates in Handbook 44, 4th ed., of the National Bureau of Standards. Conical graduates having a capacity of less than 25 mL should not be used in prescription compounding.

[1] NTIS COM-73-10504 is for sale by the National Technical Information Service, Springfield, VA 22151.

[2] N.B.S. Handbook 44, 4th ed. (1971) is for sale by the Superintendent of Documents, U.S. Government Printing Office, Washington, DC 20402.

⟨1181⟩ SCANNING ELECTRON MICROSCOPY

Scanning electron microscopy (SEM) is an electron optical imaging technique that yields both topographic images and elemental information when used in conjunction with energy-dispersive X-ray analysis (EDX) or wavelength-dispersive X-ray spectrometry (WDS). SEM is useful for characterizing the size and morphology of microscopic specimens. Together, image and X-ray analyses are important for the identification of small particles. Elemental analyses using SEM/EDX or SEM/WDS are useful for qualitative and semiquantitative determination of elemental content. Accurate quantitation is possible only for bulk samples with smooth surfaces and thus is not practical for particle specimens.

Typically, SEM analysis requires a small amount—10^{-10} to 10^{-12} g—of a solid specimen that is coated with a conductive substance to inhibit sample charging. The sample is placed in an evacuated chamber and scanned in a controlled raster pattern by an electron beam. Interaction of the electron beam with the specimen produces a variety of physical phenomena that, when detected, are used to form images and provide elemental information about the specimen. These phenomena include (1) emission of secondary electrons (SE), (2) reflection of backscattered electrons (BSE), (3) characteristic X-ray fluorescence, (4) emission of Auger electrons, (5) cathodoluminescence (CL), (6) conduction of current, (7) charging from induced voltages (IV) or adsorbed electrons (AE), (8) electron transmission, (9) heat generation, and (10) electromotive forces (see Fig. 1). Of these, SE and BSE are the most important for constructing SEM images, and X-ray fluorescence analyses are the most common methods for detecting the presence of particular elements. Use of ancillary instrumentation to detect the variety of phenomena other than (1), (2), or (3) above greatly increases the cost and complexity of the SEM system and will not be addressed here.

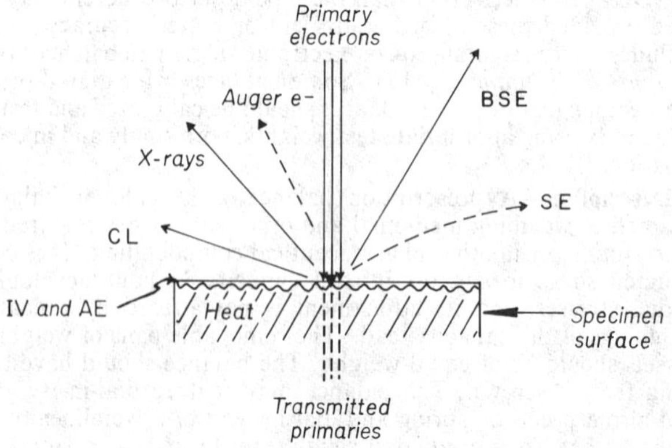

Fig. 1. Interaction Diagram

Electron Beam/Sample Interaction

Imaging—Images are formed in a SEM system by detection and manipulation of electrons. SE are emitted from a specimen surface as the result of inelastic collisions between primary (incident) electrons (PE) and electrons within a specimen. When the energy imparted to a specimen electron exceeds the work function of a sample, that electron is emitted as an SE. Most SE have energies of 5 to 20 eV; electrons in this low-energy range can be efficiently collected, yielding high signal/noise images. Because such low-energy electrons can penetrate only short distances through the specimen, SE originate from within 2 to 30 nm of the surface and generate highly resolved images. The actual PE penetration depth is dependent on PE accelerating voltage, elemental composition, and specimen mounting angle. Excitation volumes of 0.5 to 5 μm in diameter are common.

Backscattered electrons are PE that have been reflected from the sample. The PE can undergo multiple collisions prior to exiting from the specimen; therefore, BSE have energies over a broad range and emerge from relatively deep penetration ($\simeq$0.1 to 5 μm) (see Fig. 2). These high-energy (15 to 25 keV) BSE are collected less efficiently than SE, and they yield images with poorer resolution. The efficiency of BSE reflection is a function of the atomic number (Z) of the specimen atoms; thus, the contrast of BSE images depends on elemental composition. The penetration depth of all electrons is affected by elemental composition, specimen tilt, and incident beam energy (accelerating voltage). For example, the SE images of sodium phosphate and zinc phosphate crystals are quite similar. However, the heavier nuclei of the zinc species produce more efficient BSE reflection and BSE images with higher contrast. BSE images of heavy- versus light-element phases, or mixtures of species, show dramatic contrast differences that are representative of elemental heterogeneity.

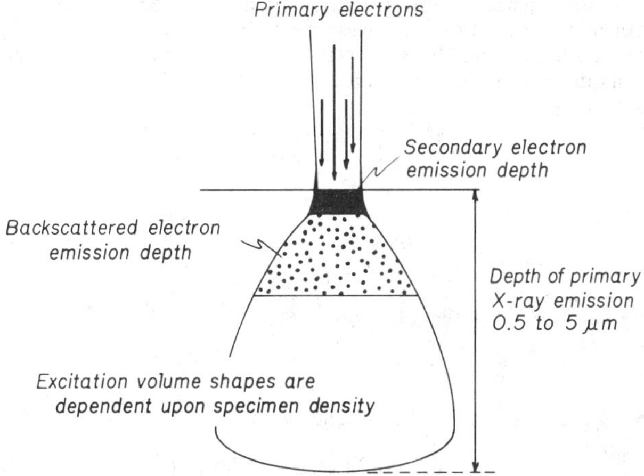

Fig. 2. Bulk Penetration

While single-angstrom resolution is possible, practical SEM image resolution is limited to $\geq$100 Å ($\geq$0.1 μm for X-ray images). These limits depend not only on instrument performance but also on operator acuity. Resolution is optimized under the following conditions: minimum working distance, high accelerating voltage, excellent grounding, excellent mechanical alignment, excellent electronic alignment, minimum incident spot diameter, minimum final aperture diameter, and cleanest column conditions. Sample preparations can be viewed in a variety of orientations and detector modes. Often the examination of a specimen at an oblique angle reveals features unobserved by an electron beam normal to the surface. This is especially true of specimens that have flat, featureless surfaces or that are poor conductors, e.g., glass surfaces. The PE accelerating voltage can be varied to change the PE penetration depth. This procedure is useful for characterizing specimens that are laminated or otherwise heterogeneous between surface and bulk content.

Coating a sample allows observation of a specimen's topography, undisturbed by flare and distortion caused by thermal effects and insufficient grounding. Coatings such as gold, gold-palladium, and carbon are often used because they are highly conductive, easy to apply, and relatively inert. Either evaporation or sputter-coating systems can be used to apply metal films; carbon films must be evaporated. Metal coatings give superior resolution, although their fluorescence can interfere with elemental analysis. Specimen charging affects not only image quality but also X-ray fluorescence yield.

X-ray Fluorescence Analysis—When a PE encounters an orbital electron in an atom, the resultant collision can either promote that orbital electron to a higher energy level or ionize the atom. Stabilization of an atom by relaxation of a higher energy electron to fill a vacancy results in the emission of an X-ray photon. These X-ray energies are discrete and element-specific; they equal the differences between the shell electron energies for the various shells of a given element. For instance, an ejected K-shell electron can be stabilized by a higher energy L-shell electron, yielding a net energy ($E_L - E_K$), which is specific for the X-ray photon energy of the elemental K line. X-ray fluorescence lines are classified according to the electron shell in which the vacancy existed, e.g., K, L, M, etc. The lines are further categorized according to the shell from which the relaxing electron originates. Thus, a $K\alpha$ X-ray line arises from a vacancy in a K-shell that is filled from an L-shell; a $K\beta$ X-ray line arises from a K-shell vacancy filled from an M-shell, and so on (see Fig. 3). Since each shell above K possesses a number of energy levels, electron transitions yield a number of lines, such as $K\alpha_1$, $K\alpha_2$, $K\beta_1$, $K\beta_2$, etc. The existence of several fluorescence lines for each element (a few for $Z \leq 11$ and many for $Z \geq 11$) is useful in overcoming detection problems due to (1) interelement spectral interferences, e.g., titanium $K\alpha$ and barium $L\alpha$; (2) sample matrix effects on energy or intensity; and (3) insufficient PE energy to excite some elemental lines, e.g., lead.

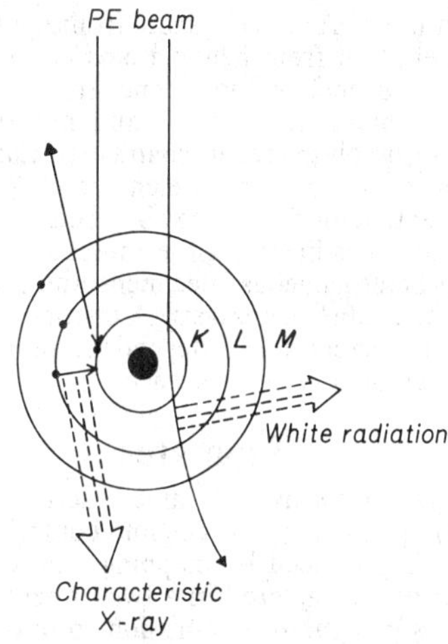

Fig. 3. Atom Model

The energies normally encountered in a SEM/EDX (or WDS) analysis range from 0.28 keV ($\simeq$ 447 nm) for carbon $K\alpha$ to the upper end of the instrument accelerating voltage, typically $\leq$ 40 keV ($\leq$ 1 nm). The natural line width, which is inversely proportional to the lifetime of the upper electronic state, is governed primarily by the transition probabilities for X-ray emission and Auger electron emission. Interaction of X-ray photons with electrons within the specimen can result in Compton scattering to produce a broadened line shifted to lower energy. X-ray photons are also emitted as a result of inelastic acceleration of electrons by atomic nuclei within a specimen. These X-ray photons, termed bremsstrahlung or white radiation, have a broad continuous energy distribution; and their characteristic lines are superimposed on this background signal.

For lighter elements, $Z \leq 11$, the low-energy X-ray photons originating from K-shell transitions can be detected only with wavelength-dispersive spectrometers or specially configured energy-dispersive detectors. All other elements emit easily detectable X-ray photons. Heavier elements, $Z \geq 26$, emit three or more detectable lines corresponding to K-, L-, and M-shell transitions. For a given element, X-ray intensities generally vary as follows: $K\alpha > K\beta > L\alpha > L\beta$, etc. (see Fig. 4).

The elemental content of a sample has a bearing on the selection of conditions for analysis. The most useful range of accelerating voltage is $\simeq$3 to 20 kV; most elements of interest can be ionized by electrons with energies in this range. The energy required to excite fluorescence from a given line is termed its critical excitation potential. The critical excitation potential for a K line can be approximated by the sum of the primary line ener-

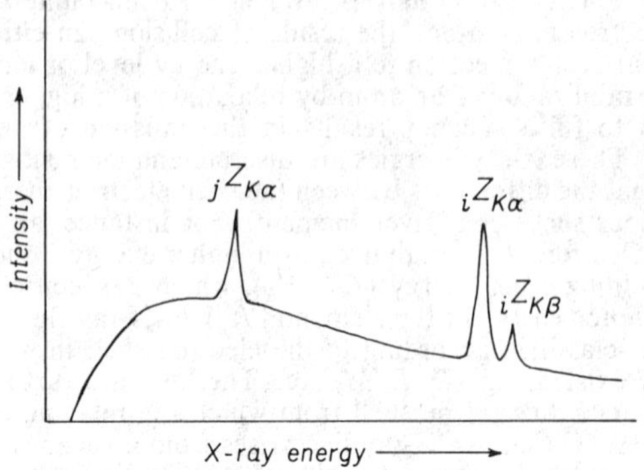

Fig. 4. X-ray Spectrum

gies ($K\alpha + L\alpha + M\alpha$). Selection of an accelerating voltage equal to 1.5 times this sum is usually sufficient for semiquantitative analyses. For example, copper has $K\alpha$ at 8.05 keV + $L\alpha$ at 0.93 keV = 8.98 keV: 1.5 × 8.98 keV = 13.47 keV. Selection of 15 kV accelerating voltage yields sufficient energy to ionize the *K*-shell of copper atoms and generate a useful analytical signal.

Interelement interferences originate from many effects. High-energy X-rays emitted from heavy atoms can ionize lighter elements to produce secondary fluorescence from the lighter species. Lower high-*Z* element fluorescence and higher low-*Z* element fluorescence can be observed, in contrast to that expected from the PE-induced signal of a pure element. Conversely, fluorescence from a light element may be absorbed by a heavier matrix to yield a negative bias in the light-element signal. These effects always exist in heterogeneous specimens and must be corrected for during any quantitative analysis. A common algorithm, ZAF, may be used to correct for *Z*-dependent interferences due to absorption and secondary fluorescence.

Apparatus

The SEM system consists of three electronic groups: (1) illumination, (2) optics, and (3) scanning control-display (see Fig. 5). The image is produced by mapping a specimen with an incident electron beam, rastered in a two-dimensional array. The electron beam is generated by emission from one of three types of sources, listed in order of increasing current density, vacuum requirements, and cost: (1) a tungsten filament cathode, (2) a LaB_6 cathode, or (3) a field emission gun. By far the most common SEM source is the tungsten filament, although the high current density of the LaB_6 source is especially useful where high resolution and/or detection of low-*Z* elements is required.

The optics consist of condenser and objective lenses, in conjunction with selected apertures. The size of the final aperture controls the beam diameter and, accordingly, the image resolution and total current at a specimen. Selection of an objective aperture is an important choice. Small apertures are required for high resolution and large apertures provide high current for optimal X-ray fluorescence intensity. In many systems, the objective aperture can be adjusted during use with a sliding or rotating holder. Flexibility in trading resolution for specimen current is

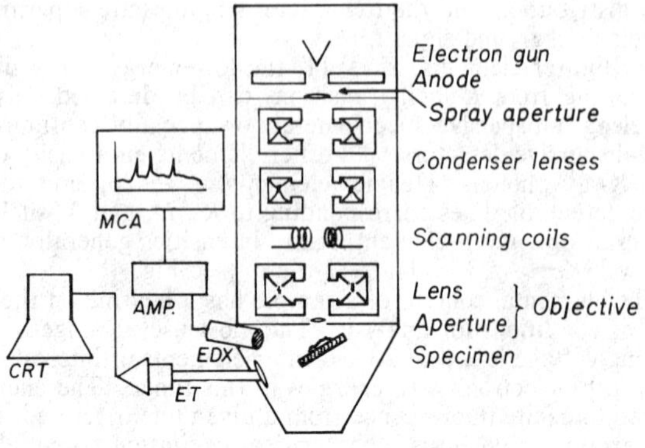

Fig. 5. Optics Diagram

also important because sample characteristics affect these two criteria differently. This feature is beneficial to users requiring high magnification and elemental detection, especially of $Z \leq 11$ elements.

Image magnification is controlled by altering the area of the eletron beam raster; smaller areas yield higher magnification since the cathode-ray tube (CRT) area remains constant. An Everhart-Thornley (ET) detector is used for electron detection; the resultant images are most similar to those of reflected light microscopy. An ET detector consists of a Faraday cage and a scintillator disk connected by a light pipe to a photomultiplier tube. The Faraday cage serves three functions: (1) at positive bias it attracts SE; (2) at negative bias it repels SE to enable the ET to collect BSE signals alone; and (3) it shields the PE beam from the scintillator potential. Various scintillator coatings are used. For example, phosphorus-based coatings yield intense, high-contrast images. Aluminum-based coatings, while less sensitive, can withstand the high SE flux generated during elemental analyses. Solid-state detectors provide up to 10 times greater sensitivity for BSE collection. They can be placed at a variety of positions and distances with respect to a specimen.

Procedure

Preparation—Samples for analysis are easily prepared, especially with an optical aid or a stereomicroscope.

PARTICLES—Place isolated or selected particles onto the SEM sample substrate (e.g., a pyrolytic carbon or metal pedestal). In all cases, the specimen should be attached to an exposed area of the substrate with a suitable liquid cement* (see Fig. 6). Use silver or carbon paint to provide a conductive path between the cement surface and the substrate. A single preparation can accommodate from one to several hundred particles, placed in an identifiable pattern.

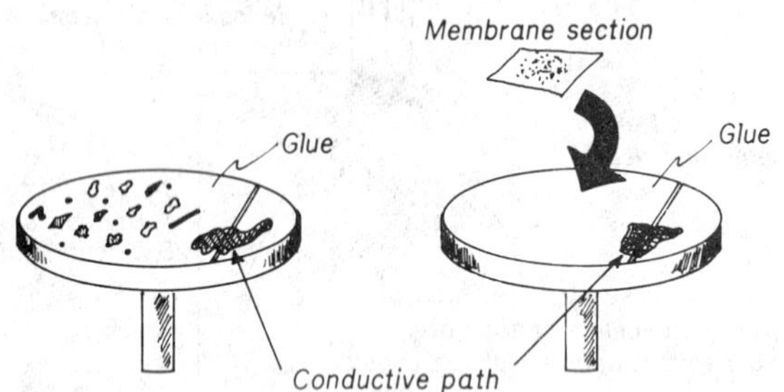

Fig. 6. SEM Pedestal Mount

Particles isolated from liquid samples by membrane filtration can be examined by placing a filter membrane on a sample substrate, as described above (see Fig. 6). This preparation is especially useful for examining a random sample of particles from a liquid. Membrane or film filters rather than depth filters are recommended, since small particles are easily lost in the open pores of a depth filter. Portions of a random isolate can be used for the SEM examination, and the remainder can be saved for other tests. Alternatively, particles can be dispersed in a solvent and concentrated onto a substrate. If a membrane filter is precoated and used to collect particles, a sample can be examined directly without further coating. This procedure alleviates excessive handling of a specimen and lessens exposure of the specimen to coater (sputter or evaporation) environments and to background conductive film signals.

BULK MATERIALS—Scatter a loose powder over double-sided sticky tape on a substrate; excess material can be blown off with a clean air jet. Liquid adhesives can be used instead of tape. However, rapid filming of an adhesive surface can occur and can hinder particle adhesion. Also, particles may sink into an adhesive before it dries. Pack loose powder into small holes cut into the surface of a metal pedestal. This technique is favored for semiquantitative analyses because it produces a relatively flat surface, is, for practical purposes, infinitely thick to the electron

*Any "superglue," copper diallylphthalate, double-sided sticky tape, or other types of mounting resins can be used. The cement with the least residual organic phase will have the longest stability.

beam, provides effective grounding, and requires a minimal amount of sample.

Cement or clamp large materials directly onto the mounting substrate. Nearly any sample that will fit into a vacuum chamber, withstand evacuated conditions, and be effectively grounded is amenable to analysis. For all of the above methods, specimen coating is required. In the absence of coating, a preparation must be viewed at sufficiently low energies, usually ≤ 5 kV, to yield suitable micrograph quality.

Analysis—The method of probing each particle and obtaining a composition is dependent on the level of information required. Several factors must be considered if quantitation is desired. The presence of multiple elements, the type of elements contained in the specimen, and the size and surface characteristics of each particle are a few of the considerations.

Any specimen ≥ 10 µm in diameter can be probed quickly and without subsequent data reduction. Particles from 10 µm to 0.2 µm must be considered more carefully, since their size approaches that of the excitation volume of the probe. Elemental characterization of a particle should be conducted on a specimen sufficiently thick that the particle volume is equal to or greater than the volume of X-ray production. Depth of signal excitation (d_X) determines the volume of X-ray production. Conditions such as accelerating voltage, source current, and spot size that are appropriate for analyzing a given specimen volume can be determined empirically, if particles are mounted on a metal substrate from which fluorescence can be detected. Substrate fluorescence indicates that the excitation volume has exceeded the volume of the particle. Quantitatively, d_X depends on particle characteristics and SEM conditions as follows:

$$d_X = 0.033 \frac{A_{ave}}{dZ_{ave}}(E_O{}^{1.69} - E_C{}^{1.69}),$$

where d_X is the depth of signal excitation, Z_{ave} is the average atomic number, A_{ave} is the average atomic mass, E_O is the energy of incident electrons, E_C is the critical excitation energy of the measured X-ray line, and d is the density of the particle.

Elemental content is estimated directly from elemental line intensities. Use the following procedures.

(1) Mount the specimen(s) by any of the above procedures and ensure good electronic and physical alignment of the SEM.

(2) Tilt or align the specimen toward the detector window at an angle that optimizes collection of the X-rays, i.e., 45° for horizontal detectors.

(3) Select the symmetrical center of the particle and use the raster mode, rather than the point mode, for the analysis. This eliminates topographical variations and generates an average signal for the sample. Choose a magnification at which the raster is $\simeq$ 50% of the particle area to eliminate background signal from the substrate.

(4) Adjust the accelerating voltage to produce adequate signal/noise for detection of the element of choice.

(5) Integrate the signal as long as necessary to achieve statistically significant results. This period can be determined through analysis of reference materials using the SEM conditions of choice (40 seconds per particle is a good working rule).

(6) Ensure sufficient specimen current. A good working rule is 20 to 40% dead time, as measured by the detector multichannel analyzer (MCA).

(7) Subtract a background spectrum that is representative of the coating, chamber, and handling.

(8) Perform an analysis, using exactly the same conditions as above on a reference standard that contains the element of choice.

(9) Perform an analysis, using exactly the same conditions as above on an internal reference (such as the substrate or other metal), for use in normalizing the analytical conditions.

The weight percent of a given element (E) in the specimen can be calculated as

$$\text{Weight \%} = \frac{(I_E/I_{ref})\text{ Sample}}{(I_E/I_{ref})\text{ Standard}} \times C^{E_{std}} \times 100\%,$$

where I_E is the intensity of the element of interest, I_{ref} is the intensity of the internal reference element, and $C^{E_{std}}$ is the concentration of the element of interest in the standard specimen. More accurate quantitative analysis must take into account the effects of matrix type, of interelements, of counting times, and of takeoff angle. The ZAF algorithm used to correct for these effects involves multiplying the measured weight percent by a series of correction factors.

⟨1191⟩ STABILITY CONSIDERATIONS IN DISPENSING PRACTICE

NOTE—Inasmuch as this chapter is for purposes of general information only, no statement in the chapter is intended to modify or supplant any of the specific requirements pertinent to Pharmacopeial articles, which are given elsewhere in this Pharmacopeia.

Aspects of drug product stability that are of primary concern to the pharmacist in the dispensing of medications are discussed herein.

Stability is defined as the extent to which a product retains, within specified limits, and throughout its period of storage and use, i.e., its shelf-life, the same properties and characteristics that it possessed at the time of its manufacture. Five types of stability generally recognized are shown in the accompanying table.

Factors Affecting Product Stability

Each ingredient, whether therapeutically active or inactive, in a dosage form can affect stability. Environmental factors, such as temperature, radiation, light, air (specifically oxygen, carbon dioxide, and water vapor), and humidity also can affect stability. Similarly, such factors as particle size, pH, the properties of water and other solvents employed, the nature of the container, and the presence of other chemicals resulting from contamination or from the intentional mixing of different products can influence stability.

Criteria for Acceptable Levels of Stability

Type of Stability	Conditions Maintained Throughout the Shelf-Life of the Drug Product
Chemical	Each active ingredient retains its chemical integrity and labeled potency, within the specified limits.
Physical	The original physical properties, including appearance, palatability, uniformity, dissolution, and suspendability are retained.
Microbiological	Sterility or resistance to microbial growth is retained according to the specified requirements. Antimicrobial agents that are present retain effectiveness within the specified limits.
Therapeutic	The therapeutic effect remains unchanged.
Toxicological	No significant increase in toxicity occurs.

Stability Studies in Manufacturing

The scope and design of a stability study vary according to the product and the manufacturer concerned. Ordinarily the formulator of a product first determines the effects of temperature, light, air, pH, moisture, and trace metals, and commonly used excipients or solvents on the active ingredient(s). From this information, one or more formulations of each dosage form are prepared, packaged in suitable containers, and stored under a variety of environmental conditions, both exaggerated and normal. At appropriate time intervals, samples of the product are assayed for potency by use of a stability-indicating method, observed for physical changes, and, where applicable, tested for sterility and/or for resistance to microbial growth and for toxicity and bioavailability. Such a study in combination with clinical and toxicological results enables the manufacturer to select the optimum formulation and container, and to assign recommended storage conditions and an expiration date for each dosage form in its package.

Responsibility of the Pharmacist

The pharmacist helps to ensure that the products under his supervision meet acceptable criteria of stability by (1) dispensing oldest stock first and observance of expiration dates; (2) storing products under the environmental conditions stated in the individual monographs and/or in the labeling; (3) observing products for evidence of instability; (4) properly treating and labeling products that are repackaged, diluted, or mixed with other products; (5) dispensing in the proper container with the proper closure; and (6) informing and educating patients concerning the proper storage and use of the products, including the disposition of outdated or excessively aged prescriptions.

Rotating Stock and Observance of Expiration Dates—Proper rotation of stock is necessary to ensure the dispensing of suitable products. A product that is dispensed on an infrequent basis should be closely monitored so that old stocks are given special attention, particularly with regard to expiration dates. The manufacturer can guarantee the quality of a product up to the time designated as its expiration date only if the product has been stored in the original container under recommended storage conditions.

Storage under Recommended Environmental Conditions—In most instances, the recommended storage conditions are stated on the label, in which case it is imperative to adhere to those conditions. They may include a specified temperature range or a designated storage place or condition (e.g., "refrigerator," or "controlled room temperature") as defined in the General Notices. Supplemental instructions, such as a direction to protect the product from light, also should be followed carefully. Where a product is required to be protected from light and is in a clear or translucent container enclosed in an opaque outer covering, such outer covering is not to be removed and discarded until the contents have been used. In the absence of specific instructions, the product should be stored at controlled room temperature (see *Storage Temperature* in the *General Notices*). The product should be stored away from locations where excessive or variable heat, cold, or light prevails, such as near heating pipes or fluorescent lighting.

Observing Products for Evidence of Instability—Loss of potency usually results from a chemical change, the most common reactions being hydrolysis, oxidation-reduction, and photolysis. Chemical changes may occur also through interaction between ingredients within a product, or rarely between product and container. An apparent loss of potency in the active ingredient(s) may result from diffusion of the drug into or its combination with the surface of the container-closure system. An apparent gain in potency usually is caused by solvent evaporation or by leaching of materials from the container-closure system.

The chemical potency of the active ingredient(s) is required to remain within the limits specified in the monograph definition. Potency is determined by means of an assay procedure that differentiates between the intact molecule and its degradation products; and chemical stability data should be available from the manufacturer. Although chemical degradation ordinarily cannot be detected by the pharmacist, excessive chemical degradation sometimes is accompanied by observable physical changes. In addition, some physical changes not necessarily related to chemical potency, such as change in color and odor, or formation of a precipitate, or clouding of solution, may serve to alert the pharmacist to the possibility of a stability problem. It should be assumed that a product that has undergone a physical change not explained in the labeling may also have undergone a chemical change and such a product is never to be dispensed. Excessive microbial growth and/or contamination also may appear as a physical change. A gross change in a physical characteristic such as color or odor is a sign of instability in any product. Other common physical signs of deterioration of dosage forms include the following.

SOLID DOSAGE FORMS—Many solid dosage forms are designed for storage under low-moisture conditions. They require protection from environmental water, and therefore should be stored in tight containers (see *Containers* in the *General Notices*) or in the container supplied by the manufacturer. The appearance of fog or liquid droplets, or clumping of the product, inside the container signifies improper conditions. The presence of a desiccant inside the manufacturer's container indicates that special care should be taken in dispensing. Some degradation products, for example, salicylic acid from aspirin, may sublime and be deposited as crystals on the outside of the dosage form or on the walls of the container.

Hard and Soft Gelatin Capsules—Since the capsule formulation is encased in a gelatin shell, a change in gross physical appearance or consistency, including hardening or softening of the shell, is the primary evidence of instability. Evidence of release of gas, such as a distended paper seal, is another sign of instability.

Uncoated Tablets—Evidence of physical instability in uncoated tablets may be shown by excessive powder and/or pieces (i.e., crumbling as distinct from breakage) of tablet at the bottom of the container (from abraded, crushed, or broken tablets); cracks or chips in tablet surfaces; swelling; mottling; discoloration; fusion between tablets; or the appearance of crystals that obviously are not part of the tablet itself on the container walls or on the tablets.

Coated Tablets—Evidence of physical instability in coated tablets is shown by cracks, mottling, or tackiness in the coating and the clumping of tablets.

Dry Powders and Granules—Dry powders and granules that are not intended for constitution into a liquid form in the original container may cake into hard masses or change color, which may render them unacceptable.

Powders and Granules Intended for Constitution as Solutions or Suspensions—Dry powders and granules intended for constitution into solutions or suspensions require special attention. Usually such forms are those antibiotics or vitamins that are particularly sensitive to moisture. Since they are always dispensed in the original container, they generally are not subject to contamination by moisture. However, an unusual caked appearance necessitates careful evaluation, and the presence of a fog or liquid droplets inside the container generally renders the preparation unfit for use. Presence of an objectionable odor also may be evidence of instability.

Effervescent Tablets, Granules, and Powders—Effervescent products are particularly sensitive to moisture. Swelling of the mass or development of gas pressure is a specific sign of instability, indicating that some of the effervescent action has occurred prematurely.

LIQUID DOSAGE FORMS—Of primary concern with respect to liquid dosage forms are homogeneity and freedom from excessive microbial contamination and growth. Instability may be indicated by cloudiness or precipitation in a solution, breaking of an emulsion, non-resuspendable caking of a suspension, or organoleptic changes. Microbial growth may be accompanied by discoloration, turbidity, or gas formation.

Solutions, Elixirs, and Syrups—Precipitation and evidence of microbial or chemical gas formation are the two major signs of instability.

Emulsions—The breaking of an emulsion, i.e., separation of an oil phase that is not easily dispersed, is a characteristic sign of instability; this is not to be confused with creaming, an easily redispersible separation of the oil phase that is a common occurrence with stable emulsions.

Suspensions—A caked solid phase that cannot be resuspended by a reasonable amount of shaking is a primary indication of instability in a suspension. The presence of relatively large particles may mean that excessive crystal growth has occurred.

Tinctures and Fluidextracts—Tinctures, fluidextracts, and similar preparations usually are dark in color because they are concentrated, and thus they should be scrutinized carefully for evidence of precipitation.

Sterile Liquids—Maintenance of sterility is of course critical for sterile liquids. The presence of microbial contamination in sterile liquids usually cannot be detected visually, but any haze, color change, cloudiness, surface film, particulate or flocculent matter, or gas formation is sufficient reason to suspect possible contamination. Clarity of sterile solutions intended for ophthalmic or parenteral use is of utmost importance. Evidence that the integrity of the seal has been violated on such products should make them suspect.

SEMISOLIDS (CREAMS, OINTMENTS, AND SUPPOSITORIES)—For creams, ointments, and suppositories, the primary indication of instability is often either discoloration or a noticeable change in consistency or odor.

Creams—Unlike ointments, creams usually are emulsions containing water and oil. Indications of instability in creams are emulsion breakage, crystal growth, shrinking due to evaporation of water, and gross microbial contamination.

Ointments—Common signs of instability in ointments are a change in consistency and excessive "bleeding," i.e., separation of excessive amounts of liquid; and formation of granules or grittiness.

Suppositories—Excessive softening is the major indication of instability in suppositories, although some suppositories may dry out and harden or shrivel. Evidence of oil stains on packaging material should warn the pharmacist to examine individual suppositories more closely by removing any foil covering if necessary. As a general rule (although there are exceptions), suppositories should be stored in a refrigerator (see *Storage Temperature* in the *General Notices*).

Proper Treatment of Products Subjected to Additional Manipulations—In repackaging, diluting, or mixing a product with another product, the pharmacist may become responsible for its stability.

REPACKAGING—In general, repackaging is inadvisable. However, if repackaging is necessary, the manufacturer should be consulted concerning potential problems. In the filling of prescriptions, it is essential that suitable containers be used. Appropriate storage conditions and, where appropriate, an expiration date, should be indicated on the label of the prescription container. Single-unit packaging calls for care and judgment, and for strict observance of the following guidelines: (1) use appropriate packaging materials; (2) where stability data on the new package are not available, repackage at any one time only sufficient stock for a limited time; (3) include on the unit-dose label a lot number and an appropriate expiration date; (4) where a sterile product is repackaged from a multiple-dose vial into unit-dose (disposable) syringes, discard the latter if not used within 24 hours, unless data are available to support longer storage; (5) where quantities are repackaged in advance of immediate needs, maintain suitable repackaging records showing name of manufacturer, lot number, date, and designation of persons responsible for repackaging and for checking; (6) where safety closures are required, use container closure systems that ensure compliance with compendial and regulatory standards for storage.

DILUTION OR MIXING—Where a product is diluted, or where two products are mixed, the pharmacist should observe good professional and scientific procedures to guard against incompatibility and instability. For example, tinctures such as those of belladonna and digitalis contain high concentrations of alcohol to dissolve the active ingredient(s), and they may develop a precipitate if they are diluted or mixed with aqueous systems. Pertinent technical literature and labeling should be consulted routinely; it should be current literature, because at times formulas are changed by the manufacturer. If a particular combination is commonly used, consultation with the manufacturer(s) is advisable. Since the chemical stability of extemporaneously prepared mixtures is unknown, the use of such combinations should be discouraged; if such a mixture involved an incompatibility, the pharmacist might be responsible. Oral antibiotic preparations constituted from powder into liquid form should never be mixed with other products.

Combining parenteral products necessitates special care, particularly in the case of intravenous solutions, primarily because of the route of administration. This area of practice demands the utmost in care, aseptic technique, judgment, and diligence. Because of potential unobservable problems with respect to sterility and chemical stability, all extemporaneous parenteral preparations should be used within 24 hours unless data are available to support longer storage.

Informing and Educating the Patient—As a final step in meeting responsibility for the stability of drugs dispensed, the pharmacist is obligated to inform the patient regarding the proper storage conditions (for example, in a cool, dry place—not in the bathroom), for both prescription and nonprescription products, and to suggest a reasonable estimate of the time after which the medication should be discarded. Where expiration dates are applied, the pharmacist should emphasize to the patient that the dates are applicable only when proper storage conditions are used. Patients should be encouraged to clean out their drug storage cabinets periodically.

⟨1211⟩ STERILIZATION AND STERILITY ASSURANCE OF COMPENDIAL ARTICLES

This informational chapter provides a general description of the concepts and principles involved in the quality control of articles that must be sterile. Any modifications or variations in sterility test procedures from those described under *Sterility Tests* ⟨71⟩ should be validated in the context of the entire sterility assurance program and are not intended to be alternative methods to those described in that chapter.

Within the strictest definition of sterility, a specimen would be deemed sterile only when there is complete absence of viable microorganisms from it. However, this absolute definition cannot currently be applied to an entire lot of finished compendial articles because of limitations in testing. Absolute sterility cannot be practically demonstrated without complete destruction of every finished article. The sterility of a lot purported to be sterile is therefore defined in probabalistic terms, where the likelihood of a contaminated unit or article is acceptably remote. Such a state of sterility assurance can be established only through the use of adequate sterilization cycles and subsequent aseptic processing, if any, under appropriate current good manufacturing practice, and not by reliance solely on sterility testing. The basic principles for validation and certification of a sterilizing process are enumerated as follows:

(1) Establish that the process equipment has capability of operating within the required parameters.

(2) Demonstrate that the critical control equipment and instrumentation are capable of operating within the prescribed parameters for the process equipment.

(3) Perform replicate cycles representing the required operational range of the equipment and employing actual or simulated product. Demonstrate that the processes have been carried out within the prescribed protocol limits and finally that the probability of microbial survival in the replicate processes completed is not greater than the prescribed limits.

(4) Monitor the validated process during routine operation. Periodically as needed, requalify and recertify the equipment.

(5) Complete the protocols, and document steps (1) through (4) above.

The principles and implementation of a program to validate an aseptic processing procedure are similar to the validation of a sterilization process. In aseptic processing, the components of the final dosage form are sterilized separately and the finished article is assembled in an aseptic manner.

Proper validation of the sterilization process or the aseptic process requires a high level of knowledge of the field of sterilization and clean room technology. In order to comply with currently acceptable and achievable limits in sterilization parameters, it is necessary to employ appropriate instrumentation and equipment to control the critical parameters such as temperature and time, humidity, and sterilizing gas concentration, or absorbed radiation. An important aspect of the validation program in many sterilization procedures involves the employment of biological indicators. (See *Biological Indicators* ⟨1035⟩.) The validated and certified process should be revalidated periodically; however, the revalidation program need not necessarily be as extensive as the original program.

A typical validation program, as outlined below, is one designed for the steam autoclave, but the principles are applicable to the other sterilization procedures discussed in this informational chapter. The program comprises several stages.

The *installation qualification* stage is intended to establish that controls and other instrumentation are properly designed and calibrated. Documentation should be on file demonstrating the quality of the required utilities such as steam, water, and air. The *operational qualification* stage is intended to confirm that the empty chamber functions within the parameters of temperature at all of the key chamber locations prescribed in the protocol. It is usually appropriate to develop heat profile records, i.e., simultaneous temperatures in the chamber employing multiple temperature-sensing devices. A typical acceptable range of temperature in the empty chamber is $\pm 1°$ when the chamber temperature is not less than 121°. The *confirmatory* stage of the validation program is the actual sterilization of materials or articles. This determination requires the employment of temperature-sensing devices inserted into samples of the articles as well as *either*, samples of the articles to which appropriate concentrations of suitable test microorganisms have been added *or*, separate biological indicators in operationally fully loaded autoclave configurations. The effectiveness of heat delivery or penetration into the actual articles and the time of the exposure are the two main factors that determine the lethality of the sterilization process. The *final* stage of the validation program requires the documentation of the supporting data developed in executing the program.

It is generally accepted that terminally sterilized injectable articles or critical devices purporting to be sterile, when processed in the autoclave, attain a 10^{-6} microbial survivor probability, i.e., assurance of less than one chance in one million that viable microorganisms are present in the sterilized article or dosage form. With heat-stable articles, the approach often is to considerably exceed the critical time necessary to achieve the 10^{-6} microbial survivor probability (overkill). However, with an article where extensive heat exposure may have a damaging effect, it may not be feasible to employ this overkill approach. In this latter instance, the development of the sterilization cycle depends heavily on knowledge of the microbial burden of the product based on examination, over a suitable time period, of a substantial number of lots of the presterilized product.

The D value is the time (in minutes) required to reduce the microbial population by 90% or 1 log cycle (i.e., to a surviving fraction of 1/10), at a specific temperature. Therefore, where the D value of a biological indicator preparation of, for example, *Bacillus stearothermophilus* spores is 1.5 minutes under the total process parameters, e.g. at 121°, if it is treated for 12 minutes under the same conditions, it can be stated that the lethality input is 8D. The effect of applying this input to the product would depend on the initial microbial burden. Assuming that its resistance to sterilization is equivalent to that of the biological indicator, if the microbial burden of the product in question is 10^2 microorganisms, a lethality input of 2D yields a microbial burden of 1 (10^0 theoretical) and a further 6D yields a calculated microbial survivor probability of 10^{-6}. (Under the same conditions, a lethality input of 12D may be used in a typical "overkill" approach.) Generally the survivor probability achieved for the article under the validated sterilization cycle is not completely correlated with what may occur with the biological indicator. For valid use, therefore, it is essential that the resistance of the biological indicator be greater than that of the natural microbial burden of the article sterilized. It is then appropriate to make a worst-case assumption and treat the microbial burden as though its heat resistance were equivalent to that of the biological indicator, although it is not likely that the most resistant of a typical microbial burden isolates will demonstrate a heat resistance of the magnitude shown by this species, frequently employed as a biological indicator for steam sterilization. In the above example, a 12-minute cycle is considered adequate for sterilization if the product had a microbial burden of 10^2 microorganisms. However, if the indicator originally had 10^6 microorganisms content, actually a 10^{-2} probability of survival could be expected; i.e., 1 in 100 biological indicators may yield positive results. This type of situation may be avoided by selection of the appropriate biological indicator. Alternatively, high content indicators may be used on the basis of a predetermined acceptable count reduction.

The D value for the *Bacillus stearothermophilus* preparation determined or verified for these conditions should be re-established when a specific program of validation is changed. Determination of survival curves (see under *Biological Indicators* ⟨1035⟩) or what has been called the fractional cycle approach may be employed to determine the D value of the biological indicator preferred for the specific sterilization procedure. The fractional cycle approach may also be used to evaluate the resistance of the microbial burden. Fractional cycles are studied either for microbial count-reduction or for fraction negative achievement. These numbers may be used to determine the lethality of the process under production conditions. The data can be used in qualified production equipment to establish appropriate sterilization cycles. A suitable biological indicator such as the *Bacillus stearothermophilus* preparation may be employed also during routine sterilization. Any microbial burden method for sterility assurance requires adequate surveillance of the microbial resistance of the article to detect any changes, in addition to periodic surveillance of other attributes.

Methods of Sterilization

In this informational chapter five methods of terminal sterilization, including removal of microorganisms by filtration, and guidelines for aseptic processing are described. Modern technological developments, however, have led to the use of additional procedures. These include blow-molding (at high temperatures), forms of moist heat other than saturated steam and ultraviolet irradiation, as well as on-line continuous filling in aseptic processing. The choice of the appropriate process for a given dosage form or component requires a high level of knowledge of sterilization techniques and information concerning any effects of the process on the material being sterilized.[1]

STEAM STERILIZATION

The process of thermal sterilization employing saturated steam under pressure is carried out in a chamber called an autoclave. It is probably the most widely employed sterilization process.[2] The basic principle of operation is that the air in the sterilizing chamber is displaced by the saturated steam, achieved by employing vents or traps. In order to displace air more effectively from the chamber and from within articles, the sterilization cycle may include air and steam evacuation stages. The design or choice of a cycle for given products or components depends on a number of factors, including the heat lability of the material, knowledge of heat penetration into the articles, and other factors described under the validation program (see above). Apart from that description of sterilization cycle parameters, using a temperature of 121°, the F_0 concept may be appropriate. The F_0, at a particular temperature other than 121°, is the time (in minutes) required to provide the lethality equivalent to that provided at 121° for a stated time. Modern autoclaves generally operate with a control system that is significantly more responsive than the steam reduction valve of older units that have been in service for many years. In order for these older units to achieve the precision and level of control of the cycle discussed in this chapter, it may be necessary to upgrade or modify the control equipment and instrumentation on these units. This modification is warranted only if the chamber and steam jacket are intact for continued safe use and if deposits which interfere with heat distribution can be removed.

DRY-HEAT STERILIZATION

The process of thermal sterilization of Pharmacopeial articles by dry heat is usually carried out by a batch process in an oven designed expressly for that purpose. A modern oven is supplied

[1] A number of guidelines dealing particularly with the development and validation of sterilization cycles and related topics have been published. These include, of the Parenteral Drug Association, Inc. (PDA) *Validation of Steam Sterilization Cycles* (Technical Monograph No. 1), *Validation of Aseptic Filling For Solution Drug Products* (Technical Monograph No. 2) and *Validation of Dry Heat Processes Used for Sterilization and Depyrogenation* (Technical Monograph No. 3), and of the Pharmaceutical Manufacturers Association (PMA) *Validation of Sterilization of Large-Volume Parenterals—Current Concepts* (Science and Technology Publication No. 25). Other series of technical publications on these subjects of the Health Industry Manufacturers Association (HIMA) include *Validation of Sterilization Systems* (Report No. 78-4.1), *Sterilization Cycle Development* (Report No. 78-4.2), *Industrial Sterility: Medical Device Standards and Guidelines* (Document #9, Vol. 1) and *Operator Training* .... for *Ethylene Oxide Sterilization*, for *Steam Sterilization Equipment*, for *Dry Heat Sterilization Equipment* and for *Radiation Sterilization Equipment* Report Nos. 78-4.5 through 4.8). Recommended practice guidelines published by the Association for the Advancement of Medical Instrumentation (AAMI) include *Guideline for Industrial Ethylene Oxide Sterilization of Medical Devices—Process Design, Validation, Routine Sterilization* (No. OPEO-12/81) and *Process Control Guidelines for the Radiation Sterilization of Medical Devices* (No. RS-P 10/82). These detailed publications should be consulted for more extensive treatment of the principles and procedures described in this chapter.

[2] An autoclave cycle, where specified in the compendia for media or reagents, is a period of 15 minutes at 121°, unless otherwise indicated.

with heated, filtered air, distributed uniformly throughout the chamber by convection or radiation and employing a blower system with devices for sensing, monitoring, and controlling the critical parameters. The validation of a dry-heat sterilization facility is carried out in a manner similar to that for a steam sterilizer described earlier. Where the unit is employed for sterilizing components such as containers intended for intravenous solutions, care should be taken to avoid accumulation of particulate matter in the chamber. A typical acceptable range in temperature in the empty chamber is ±15° when the unit is operating at not less than 250°.

In addition to the batch process described above, a continuous process is frequently employed to sterilize and depyrogenate glassware as part of an integrated continuous aseptic filling and sealing system. Heat distribution may be by convection or by direct transfer of heat from an open flame. The continuous system usually requires a much higher temperature than cited above for the batch process because of a much shorter dwell time. However, the total temperature input during the passage of the product should be equivalent to that achieved during the chamber process. The continuous process also usually necessitates a rapid cooling stage prior to the aseptic filling operation. In the qualification and validation program, in view of the short dwell time, parameters for uniformity of the temperature, and particularly the dwell time, should be established.

A microbial survival probability of 10^{-12} is considered achievable for heat-stable articles or components. An example of a biological indicator for validating and monitoring dry-heat sterilization is a preparation of *Bacillus subtilis* spores. Since dry heat is frequently employed to render glassware or containers free from pyrogens as well as viable microbes, a pyrogen challenge, where necessary, should be an integral part of the validation program, e.g., by inoculating one or more of the articles to be treated with 1000 or more USP Units of bacterial endotoxin. The test with *Limulus* lysate could be used to demonstrate that the endotoxic substance has been inactivated to not more than 1/1000 of the original amount (3 log cycle reduction). For the test to be valid, both the original amount and, after acceptable inactivation, the remaining amount of endotoxin should be measured. For additional information on the endotoxin assay, see *Bacterial Endotoxins Test* ⟨85⟩.

GAS STERILIZATION

The choice of gas sterilization as an alternative to heat is frequently made when the material to be sterilized cannot withstand the high temperatures obtained in the steam sterilization or dry-heat sterilization processes. The active agent generally employed in gaseous sterilization is ethylene oxide of acceptable sterilizing quality. Among the disadvantages of this sterilizing agent are its highly flammable nature unless mixed with suitable inert gases, its mutagenic properties, and the possibility of toxic residues in treated materials, particularly those containing chloride ions. The sterilization process is generally carried out in a pressurized chamber designed similarly to a steam autoclave but with the additional features (see below) unique to sterilizers employing this gas. Facilities employing this sterilizing agent should be designed to provide adequate post-sterilization degassing, to enable microbial survivor monitoring, and to minimize exposure of operators to the potentially harmful gas.[3]

Qualification of a sterilizing process employing ethylene oxide gas is accomplished along the lines discussed earlier. However, the program is more comprehensive than for the other sterilization procedures, since in addition to temperature, the humidity, vacuum/positive pressure, and ethylene oxide concentration also require rigid control. An important determination is to demonstrate that all critical process parameters in the chamber are adequate during the entire cycle. Since the sterilization parameters applied to the articles to be sterilized are critical variables, it is frequently advisable to pre-condition the load to achieve the required moisture content, to minimize the time of holding at the required temperature, prior to placement of the load in the ethylene oxide chamber. The validation process is generally made employing product inoculated with appropriate biological indicators such as spore preparations of *Bacillus subtilis*. For validation they may be used in full chamber loads of product, or simulated product. The monitoring of moisture and gas concentration requires the utilization of sophisticated instrumentation that only knowledgeable and experienced individuals can calibrate, operate, and maintain. The biological indicators may be employed also in monitoring routine runs.

As is indicated elsewhere in this chapter, the biological indicator may be employed in a fraction negative mode to establish the ultimate microbiological survivor probability in designing an ethylene oxide sterilization cycle using inoculated product or inoculated simulated product.

One of the principal limitations of the ethylene oxide sterilization process is the limited ability of the gas to diffuse to the innermost product areas that require sterilization. Package design and chamber loading patterns therefore must be determined so that there is minimal resistance to gas diffusion.

STERILIZATION BY IONIZING RADIATION

The rapid proliferation of medical devices unable to withstand heat sterilization and the concerns about the safety of ethylene oxide have resulted in increasing applications of radiation sterilization. It is, however, applicable also to drug substances and final dosage forms. The advantages of sterilization by irradiation include low chemical reactivity, low measurable residues, and the fact that there are fewer variables to control. In fact, radiation sterilization is unique in that the basis of control is essentially that of the absorbed radiation dose, which can be precisely measured. Because of this characteristic, new procedures have been developed to determine the sterilizing dose. These, however, are still under review and appraisal, particularly with regard to the need, or otherwise, for additional controls and safety measures. Irradiation causes only a minimal temperature rise, but can affect certain grades and types of plastics and glass.

The two types of ionizing radiation in use are radioisotope decay (gamma radiation) and electron-beam radiation. In either case the radiation dose to yield the required degree of sterility assurance should be established such that within the range of minimum and maximum doses set, the properties of the article being sterilized are acceptable.

For gamma irradiation, the validation of a procedure includes the establishment of article materials compatibility, establishment of product loading pattern and completion of dose mapping in the sterilization container (including identification of the minimum and maximum dose zones), establishment of timer setting, and demonstration of the delivery of the required sterilization dose. For electron-beam irradiation, in addition the on-line control of voltage, current, conveyor speed, and electron beam scan dimension must be validated.

For gamma radiation sterilization, an effective sterilizing dose which is tolerated without damaging effect should be selected. Although 2.5 megarads (Mrad) of absorbed radiation was historically selected, it is desirable and acceptable in some cases to employ lower doses for devices, drug substances, and finished dosage forms. In other cases, however, higher doses are essential. In order to validate the efficacy particularly of the lower exposure levels, it is necessary to determine the magnitude (number and/or degree) of the natural radiation resistance of the microbial population of the product. Specific product loading patterns must be established and absorbed minimum and maximum dosage distribution must be determined by use of chemical dosimeters. (These dosimeters are usually dyed plastic cylinders, slides or squares that show color intensification based directly on the amount of absorbed radiation energy; they require careful calibration.)

The setting of the preferred absorbed dose has been carried out on the basis of pure cultures of resistant microorganisms and employing inoculated product, e.g., with spores of *Bacillus pumilus* as biological indicators. A fractional experimental cycle approach provides the data to be utilized to determine the D_{10} value of the biological indicator. This information is then applied in extrapolating the amount of absorbed radiation to establish an appropriate microbial survivor probability. The most recent pro-

[3] See *Ethylene Oxide*, Encyclopedia of Industrial Chemical Analysis, 1971, *12*, 317–340, John Wiley & Sons, Inc., and *Use of Ethylene Oxide as a Sterilant in Medical Facilities*, NIOSH Special Occupational Hazard Review with Control Recommendations, August 1977, U. S. Department of Health and Human Services, Public Health Service, Center for Disease Control, National Institute for Occupational Safety and Health, Division of Criteria Documentation and Standards Development, Priorities and Research Analysis Branch, Rockville, MD.

cedures for gamma radiation sterilization, base the dose upon the radiation resistance of the natural heterogeneous microbial burden contained on the product to be sterilized. Such procedures are currently being refined but may provide a more representative assessment of radiation resistance, especially where significant numbers of radiation-resistant organisms are present.[4] These range from inoculation with standard resistant organisms such as *Bacillus pumilus* to subprocess (sublethal) dose exposure of finished product samples taken from production lines. Certain hypotheses are common to all of these methods. While the total microbial population present on an article generally consists of a mixture of microorganisms of differing sensitivity to radiation, the step of subjecting the article to a less than totally lethal sterilization dose eliminates the less resistant microbial fraction. This results in a residual relatively homogeneous population with respect to radiation resistance, and yields consistent and reproducible results of determinations with the residual population. The amount of laboratory manipulation required is dependent upon the particular procedure used.

One such procedure requires the enumeration of the microbial population on representative samples of independently manufactured lots of the article. The resistance of the microbial population is not determined and dose setting is based on a standard arbitrary radiation resistance assigned to the microbial population, derived from data obtained from manufacturers and from the literature. The assumption is made that the distribution of resistances chosen represents a more severe challenge than the natural microbial population on the product to be sterilized. This assumption, however, is verified by experiment. After verification the appropriate radiation sterilization dose is read from a table.

Another, more elaborate, method does not require the enumeration of the microbial population but uses a series of incremental dose exposures to allow a dose to be established such that approximately one out of 100 samples irradiated at that dose will be nonsterile. This is not the ultimate sterilization dose, but provides the basis to determine the sterilization dose by extrapolation from the dose yielding one out of 100 nonsterile samples, using an appropriate resistance factor which characterizes the remaining microorganism-resistant population. A periodic audit is conducted to check that the findings continue to be operative.

More elaborate procedures, requiring more experimentation and including the isolation of microbial cultures, include one where, after determining the substerilization dose (yielding one out of 100 nonsterile samples), the resistance of the surviving microorganisms is used to determine the sterilizing dose. Another is based on different determinations, starting with a substerilization incremental dose which results in not more than 50% of the samples being nonsterile. After irradiation of sufficient samples at this dose, a number of microbial isolates are obtained. The radiation resistance of each of these is determined. The sterilization dose is then calculated using the resistance determinations and the 50% sterilizing dose initially determined. Audit procedures are required for these methods as for the others described.

Where the required minimum radiation dose has been determined and delivery of that dose has been confirmed (by chemical or physical dosimeters), release of the article being sterilized could be effected within the overall validation of sterility assurance (which may include such confirmation of applied dosage, the use of biological indicators, and other means).

STERILIZATION BY FILTRATION

Filtration through microbial retentive materials is frequently employed for the sterilization of heat-labile solutions by physical removal of the contained microorganisms. A filter assembly generally consists of a porous matrix sealed or clamped into an impermeable housing. The effectiveness of a filter medium or substrate depends upon the pore size of the porous material and may depend upon adsorption of bacteria on or in the filter matrix or upon a sieving mechanism. Fiber-shedding filters, particularly those containing asbestos, are to be avoided unless no alternative filtration procedures are possible. Where a fiber-shedding filter is required, it is obligatory that the process include a nonfiber-shedding filter introduced downstream or subsequent to the initial filtration step.

Since the effectiveness of the filtration process is also influenced by the microbial burden of the solution to be filtered, the determination of the microbiological quality of solutions prior to filtration is an important aspect of the validation of the filtration process in addition to establishment of the other parameters of the filtration procedure, such as pressures, flow rates, and filter unit characteristics.

The process of sterilization of solutions by filtration has recently achieved new levels of proficiency, largely as a result of the development and proliferation of membrane filter technology. This class of filter media lends itself to more effective standardization and quality control and also gives the user greater opportunity to confirm the characteristics or properties of the filter assembly before and after use. The fact that membrane filters are thin polymeric films offers many advantages but also some disadvantages when compared to depth filters such as porcelain or sintered material. Since much of the membrane surface is a void or open space, the properly assembled and sterilized filter offers the advantage of a high flow rate. A disadvantage is that since the membrane is usually fragile, it is essential to determine that the assembly was properly made and that the membrane was not ruptured during assembly, sterilization, or use. Additionally, there are other tests to be made by the manufacturer of the membrane filter which are not usually repeated by the user. These include microbiological challenge tests. Results of these tests on each filter batch should be obtained from the manufacturer by the user for his records.

Filtration for sterilization purposes is usually carried out with assemblies having membranes of porosity not greater than 0.22 µm. Membranes of this porosity should retain appropriate suspensions of *Pseudomonas diminuta* (ATCC No. 19146). However, membranes of smaller porosities are also used and may be needed for some products. Membrane filter media which are now available include cellulose acetate, cellulose nitrate, fluorocarbonate, acrylic polymers, polycarbonate, polyester, polyvinyl chloride, vinyl, and even metal membranes, and they may be reinforced or supported by an internal fabric. While it may be possible to mix assemblies and filter membranes produced by different manufacturers, the compatibility of these hybrid assemblies should first be validated. A membrane filter assembly should be tested for integrity of the membrane. This could be for initial integrity prior to use by a nondestructive procedure, if one is available and can be feasibly employed, and should be after the filtration process is completed to demonstrate that the filter assembly maintained its integrity throughout the entire filtration procedure. A typical post-use test is the bubble-point test, whereby it is determined that a prescribed pressure is necessary to force air bubbles through the intact membrane wetted with either product, water or a hydrocarbon liquid.

ASEPTIC PROCESSING

While there is general agreement that sterilization of the final filled container as a dosage form or final packaged device is the preferred process for assuring the minimal risk of microbial contamination in a lot, there is a substantial class of products that are not terminally sterilized but are prepared by a series of aseptic steps. These are designed to prevent the introduction of viable microorganisms into components, where sterile, or once an intermediate process has rendered the bulk product or its components free from viable microorganisms. This section provides a review of the principles involved in producing aseptically processed products with a minimal risk of microbial contamination in the finished lot of final dosage forms.

A product defined as aseptically processed is likely to consist of components that have been sterilized by one of the processes described earlier in this chapter. For example, the bulk product, if a filterable liquid, may have been sterilized by filtration. The final empty container components would probably be sterilized by heat, dry heat being employed for glass vials and an autoclave being employed for rubber closures. The areas of critical concern are the immediate microbial environment where these presterilized components are exposed during assembly to produce the finished dosage form and the aseptic filling operation.

[4] Detailed descriptions of these procedures have been published by the Association for the Advancement of Medical Instrumentation (AAMI) in the document entitled "*Process Control Guidelines for Radiation Sterilization of Medical Devices*" (No. AAMI RS-P 10/82).

The requirements for a properly designed, validated and maintained filling or other aseptic processing facility are mainly directed to (i) an air environment free from viable microorganisms, of a proper design to permit effective maintenance of air supply units and (ii) the provision of trained operating personnel who are adequately equipped and gowned. The desired environment may be achieved through the high level of air filtration[5] technology now available, which contribute[5] to the delivery of air of the requisite microbiological quality. The facilities include both primary (in the vicinity of the exposed article) and secondary (where the aseptic processing is carried out) barrier systems.

For a properly designed aseptic processing facility or aseptic filling area, consideration should be given to such features as nonporous and smooth surfaces, including walls and ceilings that can be sanitized frequently; gowning rooms with adequate space for personnel and storage of sterile garments; adequate separation of preparatory rooms for personnel from final aseptic processing rooms, with the availability where necessary of such devices as airlocks and/or air showers; proper pressure differentials between rooms, the most positive pressure being in the aseptic processing rooms or areas; the employment of laminar (unidirectional) airflow in the immediate vicinity of exposed product or components, and filtered air exposure thereto, with adequate air change frequency; appropriate humidity and temperature environmental controls; and a documented sanitization program. Proper training of personnel in hygienic and gowning techniques should be undertaken so that, for example, gowns, gloves, and other body coverings substantially cover exposed skin surfaces.

Certification and validation of the aseptic process and facility are achieved by establishing the efficiency of the filtration systems, by employing microbiological environmental monitoring procedures, and by processing of sterile culture medium as simulated product.

Monitoring of the aseptic facility should include periodic environmental filter examination as well as routine particulate and microbiological environmental monitoring, and may include periodic sterile culture medium processing.

Sterility Testing of Lots

It should be recognized that the referee sterility test might not detect microbial contamination if present in only a small percentage of the finished articles in the lot because the specified number of units to be taken imposes a significant statistical limitation on the utility of the test results. This inherent limitation, however, has to be accepted since current knowledge offers no nondestructive alternatives for ascertaining the microbiological quality of every finished article in the lot, and it is not a feasible option to increase the number of specimens significantly.

The primary means of supporting the claim that a lot of finished articles purporting to be sterile meets the specifications consist of the documentation of the actual production and sterilization record of the lot and of the additional validation records that the sterilization process possesses the capability of totally inactivating the established product microbial burden or a more resistant challenge. Further, it should be demonstrated that any processing steps involving exposed product following the sterilization procedure are performed in an aseptic manner, to prevent contamination. If data derived from the manufacturing process sterility assurance validation studies and from in-process controls are judged to provide greater assurance that the lot meets the required low probability of containing a contaminated unit (compared to sterility testing results from finished units drawn from that lot), any sterility test procedures adopted may be minimal, or dispensed with on a routine basis. However, assuming that all of the above production criteria have been met, it may still be desirable to perform sterility testing on samples of the lot of finished articles. Such sterility testing is usually carried out directly after the lot is manufactured as a final product quality control test.[6] Sterility tests employed in this way in manufacturing control should not be confused with those described under *Sterility Tests* ⟨71⟩. The procedural details may be the same with regard to media, inocula and handling of specimens, but the number of units and/or incubation time(s) selected for testing may differ. The number should be chosen relative to the purpose to be served, i.e., according to whether greater or lesser reliance is placed on sterility testing in the context of all the measures for sterility assurance in manufacture. Also, longer times of incubation would make the test more sensitive to slow-growing microorganisms. In the growth promotion tests for media, such slow growers, particularly if isolated from the product microbial burden, should be included with the other test stains. Negative or satisfactory sterility test results serve only as further support of the existing evidence concerning the quality of the lot if all of the pertinent production records of the lot are in order and the sterilizing or aseptic process is known to be effective. Unsatisfactory test results, however, in manufacturing quality control indicate a need for further action (see under *Performance, Observation, and Interpretation*).

DEFINITION OF A LOT AND SELECTION OF SPECIMENS FOR STERILITY TEST PURPOSES

Articles may be terminally sterilized either in a chamber or by a continuous process. In the chamber process, a number of articles are sterilized simultaneously under controlled conditions, for example, in a steam autoclave, so that for the purpose of sterility testing, the lot is considered to be the contents of a single chamber. In the continuous process, the articles are sterilized individually and consecutively, for example, by exposure to electron-beam radiation, so that the lot is considered to be not larger than the total number of similar items subjected to uniform sterilization for a period of not more than 24 hours.

For aseptic fills, the term "filling operation" describes a group of final containers, identical in all respects, that have been aseptically filled with the same product from the same bulk within a period of time not longer than 24 consecutive hours without an interruption or change that would affect the integrity of the filling assembly. The items tested should be representative of each filling assembly and should be selected at appropriate intervals throughout the entire filling operation. If more than three filling machines, each with either single or multiple filling stations, are used for filling a single lot, a minimum of 20 filled containers (not less than 10 per medium) should be tested for each filling machine, but the total number generally need not exceed 100 containers.

For small lots, in the case of either aseptic filling or terminal sterilization, if the number of final containers in the lot is between 20 and 200, about 10% of the containers should usually be tested. If the number of final containers in the lot is 20 or less, not fewer than 2 final containers should be tested.

Performance, Observation, and Interpretation

The facility for sterility testing should be such as to offer no greater a microbial challenge to the articles being tested than that of an aseptic processing production facility. The sterility testing procedure should be performed by individuals having a high level of aseptic technique proficiency. The test performance records of these individuals should be documented.

The extensive aseptic manipulations required to perform sterility testing may result in a probability of nonproduct-related contamination of the order of 10^{-3}, a level similar to the overall efficiency of an aseptic operation and comparable to the microbial survivor probability of aseptically processed articles. This level of probability is significantly greater than that usually attributed to a terminal sterilization process, namely one in one million or 10^{-6} microbial survivor probability. Appropriate, known-to-be-sterile, finished articles should be employed periodically as negative controls as a check on the reliability of the test procedure. Preferably the technicians performing the test should be unaware

[5] Available published standards for such controlled work areas include the following: (1) Federal Standard No. 209B, Clean Room and Work Station Requirements for a Controlled Environment, Apr. 24, 1973. (2) NASA Standard for Clean Room and Work Stations for Microbially Controlled Environment, publication NHB5340.2, Aug. 1967. (3) Contamination Control of Aerospace Facilities, U. S. Air Force, T.O. 00-25-203 1 Dec. 1972, change 1–1 Oct. 1974.

[6] *Radioactive Pharmaceutical Products*—Because of rapid radioactive decay, it is not feasible to delay the release of some radioactive pharmaceutical products in order to complete sterility tests on them. In such cases, results of sterility tests provide only retrospective confirmatory evidence for sterility assurance, which therefore depends on the primary means thereto established in the manufacturing and validation/certification procedures.

that they are testing negative controls. Of these tests a false positive frequency not exceeding 2% is desirable.

For aseptically processed articles, these facts support the routine use of the test set forth under *Sterility Tests* ⟨71⟩ or a more elaborate one. The production and validation documentation should be acceptable and complete. For effectively terminally sterilized products, however, the lower microbial survivor probability may direct the use of a less extensive test than the compendial procedure specified under *Sterility Tests* ⟨71⟩, or even preclude the necessity altogether for performing one. This added reliability of sterility assurance of terminal sterilization depends upon a properly validated and documented sterilization process. Sterility testing alone is no substitute.

Interpretation of Quality Control Tests—The overall responsibility for the operation of the test unit and the interpretation of test results in relation to acceptance or rejection of a lot should be in the hands of those who have appropriate formal training in microbiology and have knowledge of industrial sterilization, aseptic processing, and the statistical concepts involved in sampling. These individuals should be knowledgeable also concerning the environmental control program in the test facility to assure that the microbiological quality of the air and critical work surfaces are consistently acceptable.

Quality control sterility tests (either according to the official referee test or modified tests) may be carried out in two separate stages in order to rule out false positive results. *First Stage*. Regardless of the sampling plan used, if no evidence of microbial growth is found, the results of the test may be taken as indicative of absence of intrinsic contamination of the lot.

If microbial growth is found, proceed to the *Second Stage* (unless the *First Stage* test can be invalidated). Evidence for invalidating a *First Stage* test in order to repeat it as a *First Stage* test may be obtained from a review of the testing environment and the relevant records thereto. Finding of microbial growth in negative controls need not be considered the sole grounds for invalidating a *First Stage* test. When proceeding to the *Second Stage*, particularly where depending on the results of the test for lot release, concurrently, initiate and document a complete review of all applicable production and control records. In this review consideration should be paid to the following: (1) a check on monitoring records of the validated sterilization cycle applicable to the product; (2) sterility test history relating to the particular product for both finished and in-process samples, as well as sterilization records of supporting equipment, containers/closures, and sterile components, if any; (3) environmental control data, including those obtained from media fills, exposure plates, filtering records, any sanitization records and microbial monitoring records of operators, gowns, gloves, and garbing practices.

Failing any lead from the above review, the current microbial profile of the product should be checked against the known historical profile for possible change. Records should be checked concomitantly for any changes in source of product components and/or in-processing procedures that might be contributory. Depending on the findings, and in extreme cases, consideration may have to be given to re-validation of the total manufacturing process. For the *Second Stage* it is not possible to specify a particular number of specimens to be taken for testing. It is usual to select double the number specified for the *First Stage* under *Sterility Tests* ⟨71⟩, or other reasonable number. The minimum volumes tested from each specimen, the media, and the incubation periods are the same as those indicated for the *First Stage*.

If no microbial growth is found in the *Second Stage*, and the documented review of appropriate records and the indicated product investigation does not support the possibility of intrinsic contamination, the lot may meet the requirements of a test for sterility. If growth is found, the lot fails to meet the requirements of the test. As was indicated for the *First Stage* test, the *Second Stage* test may similarly be invalidated with appropriate evidence, and, if so done, repeated as a *Second Stage* test.

⟨1221⟩ TEASPOON

For household purposes, an American Standard Teaspoon has been established by the American National Standards Institute* as containing 4.93 ± 0.24 mL. In view of the almost universal practice of employing teaspoons ordinarily available in the household for the administration of medicine, the teaspoon may be regarded as representing 5 mL. Preparations intended for administration by teaspoon should be formulated on the basis of dosage in 5-mL units. Any dropper, syringe, medicine cup, special spoon, or other device used to administer liquids should deliver 5 mL wherever a teaspoon calibration is indicated. Under ideal conditions of use, the volume error incurred in measuring liquids for individual dose administration by means of such calibrated devices should be not greater than 10% of the indicated amount.

Household units are used often to inform the patient of the size of the dose. Fifteen milliliters should be considered 1 standard tablespoonful; 10 mL, 2 standard teaspoonfuls; and 5 mL, 1 standard teaspoonful. Doses of less than 5 mL are frequently stated as fractions of a teaspoonful or in drops.

Because of the difficulties involved in measuring liquids under normal conditions of use, patients should be cautioned that household spoons are not appropriate for measuring medicines. They should be directed to use the standard measures in the cooking-and-baking measuring spoon sets or, preferably, oral dosing devices that may be provided by the practitioner. It must be kept in mind that the actual volume of a spoonful of any given liquid is related to the latter's viscosity and surface tension, among other influencing factors. These factors can also cause variability in the true volumes contained in or delivered by medicine cups. Where accurate dosage is required, a calibrated syringe or dropper should be used.

* American National Standards Institute, 1430 Broadway, New York, NY 10018.

⟨1225⟩ VALIDATION OF COMPENDIAL METHODS

Test procedures for assessment of the quality levels of pharmaceutical products are subject to various requirements. According to Section 501 of the Federal Food, Drug, and Cosmetic Act, assays and specifications in monographs of the United States Pharmacopeia and the National Formulary constitute legal standards. The Current Good Manufacturing Practice regulations [21 CFR 211.194(a)] require that test methods, which are used for assessing compliance of pharmaceutical products with established specifications, must meet proper standards of accuracy and reliability. Also, according to these regulations [21 CFR 211.194(a)(2)], users of analytical methods described in the USP and the NF are not required to validate accuracy and reliability of these methods, but merely verify their suitability under actual conditions of use. Recognizing the legal status of USP and NF standards, it is essential, therefore, that proposals for adoption of new or revised compendial analytical methods be supported by sufficient laboratory data to document the validity of these procedures.

Submissions to the Compendia

Submissions to the compendia for new or revised analytical methods should contain sufficient information to enable members of the USP Committee of Revision to evaluate the relative merit of proposed procedures. In most cases, evaluations involve assessment of the clarity and completeness of the description of the analytical methods, determination of the need for the methods, and documentation that the methods have been appropriately validated. Information may vary depending upon the type of assay involved. However, in most cases a submission will consist of the following sections.

Rationale—This section should identify the need for the assay and describe the capability of the specific method proposed and why it is preferred over other types of determinations. For revised procedures, a comparison should be provided of limitations of the current compendial assay and advantages offered by the proposed method.

Proposed Analytical Method—This section should contain a complete description of the analytical method sufficiently detailed to enable persons "skilled in the art" to replicate it. The write-up should include all important operational parameters and specific instructions such as preparation of reagents, performance of systems suitability tests, precautions, and explicit formulas for calculation of test results.

Data Elements—This section should provide thorough and complete documentation of the validation of the analytical method. It should include summaries of experimental data and calculations substantiating each of the applicable analytical performance parameters. These parameters are described in the following section.

Validation

Validation of an analytical method is the process by which it is established, by laboratory studies, that the performance characteristics of the method meet the requirements for the intended analytical applications. Performance characteristics are expressed in terms of analytical parameters. Typical analytical parameters that should be considered in the validation of the types of assays described in this document are listed in Table 1. Since opinions may differ with respect to terminology and use, each of the parameters is defined in the next section of this chapter along with a delineation of a typical method by which it may be measured.

Table 1. Typical Analytical Parameters Used in Assay Validation.

Parameter
Precision
Accuracy
Limit of Detection
Limit of Quantitation
Selectivity
Range
Linearity
Ruggedness

ANALYTICAL PERFORMANCE PARAMETERS

Precision—

Definition—The precision of an analytical method is the degree of agreement among individual test results when the procedure is applied repeatedly to multiple samplings of a homogeneous sample. The precision of an analytical method is usually expressed as the standard deviation or relative standard deviation (coefficient of variation). Precision is a measure of the degree of reproducibility of the analytical method under normal operating circumstances.

Determination—The precision of an analytical method is determined by assaying a sufficient number of aliquots of a homogeneous sample to be able to calculate statistically valid estimates of standard deviation or relative standard deviation (coefficient of variation). Assays in this context are independent analyses of samples that have been carried through the complete analytical procedure from sample preparation to final test result.

Accuracy—

Definition—The accuracy of an analytical method is the closeness of test results obtained by that method to the true value. Accuracy may often be expressed as percent recovery by the assay of known, added amounts of analyte. Accuracy is a measure of the exactness of the analytical method.

Determination—The accuracy of an analytical method may be determined by applying that method to samples or mixtures of excipients to which known amounts of analyte have been added both above and below the normal levels expected in the samples. The accuracy is then calculated from the test results as the percentage of analyte recovered by the assay.

Limit of Detection—

Definition—The limit of detection is a parameter of limit tests. It is the lowest concentration of analyte in a sample that can be detected, but not necessarily quantitated, under the stated experimental conditions. Thus, limit tests merely substantiate that the analyte concentration is above or below a certain level. The limit of detection is usually expressed as the concentration of analyte (e.g., percentage, parts per billion) in the sample.

Determination—Determination of the limit of detection of an analytical method will vary depending on whether it is an instrumental or a noninstrumental procedure. For instrumental procedures, different techniques may be used. Some investigators determine the signal-to-noise ratio by comparing test results from samples with known concentrations of analyte with those of blank samples and establish the minimum level at which the analyte can be reliably detected. A signal-to-noise ratio of 2:1 or 3:1 is generally accepted. Other investigators measure the magnitude of analytical background response by analyzing a number of blank samples and calculating the standard deviation of this response. The standard deviation multiplied by a factor, usually 2 or 3, provides an estimate of the limit of detection. This limit is subsequently validated by the analysis of a suitable number of samples known to be near or prepared at the limit of detection.

For noninstrumental methods, the limit of detection is generally determined by the analysis of samples with known concentrations of analyte and by establishing the minimum level at which the analyte can be reliably detected.

Limit of Quantitation—

Definition—Limit of quantitation is a parameter of quantitative assays for low levels of compounds in sample matrices, such as impurities in bulk drug substances and degradation products in finished pharmaceuticals. It is the lowest concentration of analyte in a sample that can be determined with acceptable precision and accuracy under the stated experimental conditions. The limit of quantitation is expressed as the concentration of analyte (e.g., percentage, parts per billion) in the sample.

Determination—Determination of the limit of quantitation of an analytical method may vary depending on whether it is an instrumental or a noninstrumental procedure. For instrumental procedures, a common approach is to measure the magnitude of analytical background response by analyzing a number of blank samples and calculating the standard deviation of this response. The standard deviation multiplied by a factor, usually 10, provides an estimate of the limit of quantitation. This limit is subsequently validated by the analysis of a suitable number of samples known to be near or prepared at the limit of quantitation.

For noninstrumental methods, the limit of quantitation is generally determined by the analysis of samples with known concentrations of analyte and by establishing the minimum level at which the analyte can be detected with acceptable accuracy and precision.

Selectivity—

Definition—The selectivity (also termed specificity) of an analytical method is its ability to measure accurately and specifically the analyte in the presence of components that may be expected to be present in the sample matrix. Selectivity may often be expressed as the degree of bias of test results obtained by analysis of samples containing added impurities, degradation products, related chemical compounds, or placebo ingredients when compared to test results from samples without added substances. The bias may be expressed as the difference in assay results between the two groups of samples. Selectivity is a measure of the degree of interference (or absence thereof) in the analysis of complex sample mixtures.

Determination—The selectivity of an analytical method is determined by comparing test results from the analysis of samples containing impurities, degradation products, or placebo ingredients with those obtained from the analysis of samples without impurities, degradation products, or placebo ingredients. The bias of the assay, if any, is the difference in test results between the two groups of samples.

When impurities or degradation products are unidentified or unavailable, selectivity may be demonstrated by analysis by the method in question of samples containing impurities or degradation products and comparing the results to those from additional purity assays (e.g., chromatographic assay, phase solubility, differential scanning calorimetry). The degree of agreement of test results is a measure of the selectivity.

Linearity and Range—

Definition of Linearity—The linearity of an analytical method is its ability to elicit test results that are directly, or by a well-defined mathematical transformation, proportional to the concentration of analyte in samples within a given range. Linearity is usually expressed in terms of the variance around the slope of the regression line calculated according to an established mathematical relationship from test results obtained by the analysis of samples with varying concentrations of analyte.

Definition of Range—The range of an analytical method is the interval between the upper and lower levels of analyte (including these levels) that have been demonstrated to be determined with precision, accuracy, and linearity using the method as written. The range is normally expressed in the same units as test results (e.g., percent, parts per million) obtained by the analytical method.

Determination of Linearity and Range—The linearity of an analytical method is determined by mathematical treatment of test results obtained by analysis of samples with analyte concentrations across the claimed range of the method. The treatment is normally a calculation of a regression line by the method of least squares of test results versus analyte concentrations. In some cases, to obtain proportionality between assays and sample concentrations, the test data may have to be subjected to a mathematical transformation prior to the regression analysis. The slope of the regression line and its variance provide a mathematical measure of linearity; the y-intercept is a measure of the potential assay bias. Plotting the test results graphically as a function of analyte concentration on appropriate graph paper may be an acceptable alternative to the regression line calculation.

The range of the method is validated by verifying that the analytical method provides acceptable precision, accuracy, and linearity when applied to samples containing analyte at the extremes of the range as well as within the range.

Ruggedness—

Definition—The ruggedness of an analytical method is the degree of reproducibility of test results obtained by the analysis of the same samples under a variety of normal test conditions, such as different laboratories, different analysts, different instruments, different lots of reagents, different elapsed assay times, different assay temperatures, different days, etc. Ruggedness is normally expressed as the lack of influence on test results of operational and environmental variables of the analytical method. Ruggedness is a measure of reproducibility of test results under normal, expected operational conditions from laboratory to laboratory and from analyst to analyst.

Determination—The ruggedness of an analytical method is determined by analysis of aliquots from homogeneous lots in different laboratories, by different analysts, using operational and environmental conditions that may differ but are still within the specified parameters of the assay. The degree of reproducibility of test results is then determined as a function of the assay variables. This reproducibility may be compared to the precision of the assay under normal conditions to obtain a measure of the ruggedness of the analytical method.

DATA ELEMENTS REQUIRED FOR ASSAY VALIDATION

Compendial assay procedures vary from highly exacting analytical determinations to subjective evaluation of attributes. Considering this variety of assays, it is only logical that different test methods require different validation schemes. This chapter covers only the most common categories of assays for which validation data should be required. These categories are as follows:

Category I—Analytical methods for quantitation of major components of bulk drug substances or active ingredients (including preservatives) in finished pharmaceutical products fall under Category I.

Category II—Analytical methods for determination of impurities in bulk drug substances or degradation compounds in finished pharmaceutical products are in Category II. These methods include quantitative assays and limit tests.

Category III—Analytical methods for determination of performance characteristics (e.g., dissolution, drug release) are considered Category III.

For each assay category, different analytical information is needed. Listed in Table 2 are data elements that are normally required for each of the categories of assays.

Already established general assays and tests (e.g., titrimetric method of water determination, bacterial endotoxins test) should also be validated to verify their accuracy (and absence of possible interference) when used for a new product or raw material.

The validity of an analytical method can be verified only by laboratory studies. Therefore, documentation of the successful completion of such studies is a basic requirement for determining whether a method is suitable for its intended applications. Appropriate documentation should accompany any proposal for new or revised compendial analytical procedures.

⟨1231⟩ WATER FOR PHARMACEUTICAL PURPOSES

Water is the most copiously and widely used substance in pharmaceutical manufacturing. Control of the chemical and microbiological quality of water for pharmaceutical purposes is difficult because its basic sources—municipal and non-municipal water systems—are influenced by many factors.

Monitoring of quality parameters in source water is necessary to ensure an acceptable water supply.

Water is required for a variety of purposes ranging from the needs of manufacturing processes to the final preparation of therapeutic agents just prior to their administration to patients.

Drinking water, which is subject to federal Environmental Protection Agency regulations and which is delivered by the municipal or other local public system or drawn from a private well or reservoir, is the starting material for most forms of water covered by Pharmacopeial monographs. Water prepared from other starting material may have to be processed to meet drinking water standards. Drinking water may be used in the preparation of USP drug substances but not in the preparation of dosage forms, or in the preparation of reagents or test solutions.

The Pharmacopeia provides several monographs for water. Of these, *Purified Water* and *Water for Injection* represent ingredient materials, while the other monographs for water provide standards for compendial pharmaceutical articles in themselves.

Purified Water (see USP monograph)—This article represents water rendered suitable for pharmaceutical purposes by processes such as distillation, ion-exchange treatment (deionization or demineralization), or reverse osmosis. It meets rigid specifications for chemical purity, the requirements of the federal Environmental Protection Agency with respect to drinking water, and it contains no added substances. However, the various methods of production each present different potential for contaminating products. Purified Water produced by distillation is sterile, provided the production equipment is suitable and is sterile. On the other hand, ion-exchange columns and reverse osmosis units re-

Table 2. Data Elements Required for Assay Validation.

Analytical Performance Parameter	Assay Category I	Assay Category II: Quantitative	Assay Category II: Limit Tests	Assay Category III
Precision	Yes	Yes	No	Yes
Accuracy	Yes	Yes	*	*
Limit of Detection	No	No	Yes	*
Limit of Quantitation	No	Yes	No	*
Selectivity	Yes	Yes	Yes	*
Range	Yes	Yes	*	*
Linearity	Yes	Yes	No	*
Ruggedness	Yes	Yes	Yes	Yes

* May be required, depending on the nature of the specific test.

quire special attention in that they afford sites for microorganisms to foul the system and to contaminate the effluent water. Thus, frequent monitoring may be called for, particularly with the use of these units following periods of shutdown of more than a few hours.

Water for Injection (see USP monograph)—By definition, this article is water purified by distillation or by reverse osmosis, and it meets the purity requirements under *Purified Water*. Although not intended to be sterile, it meets a test for a limit of bacterial endotoxin. It must be produced, stored, and distributed under conditions designed to prevent production of endotoxin.

Sterile Water for Injection (see USP monograph)—As a form in which water is distributed in sterile packages, Sterile Water for Injection is intended mainly for use as a solvent for parenteral products such as sterile solids that must be distributed dry because of limited stability of their solutions. It must be packaged only in single-dose containers of not larger than 1-liter size.

Bacteriostatic Water for Injection (see USP monograph)—Inasmuch as it serves the same purposes as Sterile Water for Injection, it meets the same standards, with the exception that it may be packaged in either single-dose or multiple-dose containers of not larger than 30-mL size.

Sterile Water for Irrigation (see USP monograph)—This form of water meets most, but not all, of the requirements for Sterile Water for Injection. The exceptions are with respect to the following: (1) container size (i.e., the container may contain a volume of more than 1 liter of Sterile Water for Irrigation), (2) container design (i.e., the container may be designed so as to empty rapidly the contents as a single dose), (3) *Particulate matter* requirements (i.e., it need not meet the requirement for particulate matter for Large-volume Injections for single-dose infusions), and (4) *Labeling* requirements (e.g., the designations "For irrigation only" and "Not for injection" appear prominently on the label).

ACTION GUIDELINES FOR THE MICROBIAL CONTROL OF INGREDIENT WATER—Criteria for controlling the microbial quality of Purified Water and Water for Injection may vary according to the method of production, distribution and/or storage and use. The suitability of water systems to produce water of acceptable microbiological quality should be validated prior to production. Suitable microbiological, chemical, and operating controls should be in place. The compendial *Microbial Limit Tests* ⟨61⟩ have not been designed for testing of ingredient waters. Suitable standard methods are found elsewhere.*

A total microbial (aerobic) count that may be used for source drinking water is 500 colony-forming units (cfu) per mL. Since *Purified Water* is used to manufacture a variety of products, the action limit set should be based on the intended use of the water, the nature of the product to be made, and the effect of the manufacturing process on the fate of the microorganisms. A general guideline for *Purified Water* may be 100 cfu/mL. Since ingredient waters are not produced as a lot or batch, these numbers, when exceeded or approached, serve as an alert for corrective action. In practice, the supply of ingredient water and manufacture of pharmaceutical articles are usually carried out concurrently, and the results of the microbial tests made may be available only after some pharmaceutical articles have already been manufactured. The actions to be taken to bring the microbial quality of the ingredient water into desired conformance may include sanitization of the system, for example, flushing with hot water, steam, or suitable disinfectants. Further sampling and monitoring to ensure that the corrective action has been adequate should be conducted.

⟨1241⟩ WATER-SOLID INTERACTIONS IN PHARMACEUTICAL SYSTEMS

Pharmaceutical solids as raw materials or as dosage forms most often come in contact with water during processing and storage. This may occur (a) during crystallization, lyophilization, wet granulation, or spray drying; and (b) because of exposure upon handling and storage to an atmosphere containing water vapor or exposure to other materials in a dosage form that contain water capable of distributing to other ingredients. Some properties known to be altered by the association of solids with water include rates of chemical degradation in the "solid-state," crystal growth and dissolution, dispersibility and wetting, powder flow, lubricity, powder compactibility, and compact hardness.

Water can associate with solids in two ways. It can interact only at the surface (adsorption) and it can penetrate the bulk solid structure (absorption). When both adsorption and absorption occur, the term sorption is often used. Adsorption is particularly critical in affecting the properties of solids when the specific surface area is large. Large values of specific surface area are seen with solids having very small particles, as well as with solids having a high degree of intraparticle porosity.

Absorption is characterized by an association of water per g of solid that is much greater than that which can form a monomolecular layer on the available surface, and an amount that is generally independent of the specific surface area. Most crystalline solids will not absorb water into their bulk structures because of the close packing and high degree of order of the crystal lattice. Indeed, it has been shown that the degree of absorption into solids exhibiting partial crystallinity and partial amorphous structure is often directly proportional to the percent crystallinity. With some crystalline solids, however, crystal hydrates may form. These hydrates may exhibit a stoichiometric relationship, in terms of water molecules bound per solid molecule, or they may be nonstoichiometric. Upon dehydration, crystal hydrates may (a) retain their original crystal structure, (b) lose their crystallinity and become amorphous, or (c) transform to a new anhydrous or less hydrated crystal form.

Amorphous or partially amorphous solids are capable of taking up significant amounts of water when there is sufficient molecular disorder in the solid to permit penetration and dissolution of the water molecule. Such behaviour is observed with most amorphous polymers and with small molecular weight solids rendered amorphous during preparation, e.g., lyophilization, or after milling. The introduction of defects into highly crystalline solids will also produce this behavior. The greater the chemical affinity of water for the solid, the greater the total amount that can be absorbed. When water is absorbed by amorphous solids, the bulk properties of the solid can be significantly altered. It is well established, for example, that amorphous solids, depending on the temperature, can exist in at least one of two states, "glassy" or "fluid"; the temperature at which one state transforms to the other is the glass transition temperature, Tg. Water absorbed into the bulk solid structure, by virtue of its effect on the free volume of the solid, can act as an efficient plasticizer and reduce the value of Tg. Since the rheological properties of "fluid" and "glassy" states are quite different, i.e., the "fluid" state exhibits much less viscosity as one goes increasingly above the glass transition temperature, it is not surprising that a number of important bulk properties dependent on the rheology of the solid are affected by moisture content. Since amorphous solids are metastable relative to the crystalline form of the material, with small molecular weight materials it is possible for absorbed moisture to initiate reversion of the solid to the crystalline form, particularly if the solid is transformed by the sorbed water to a fluid state. This is the basis of "cake collapse" often observed during the lyophilization process. An additional phenomenon noted specifically with water-soluble solids is their tendency to deliquesce, i.e., to dissolve in their own sorbed water, at relative humidities, RH_i, in excess of the relative humidity of a saturated solution of the solid, RH_o. Deliquescence arises because of the high water solubility of the solid and the significant effect it has on the colligative properties of water. It is a dynamic process that continues to occur as long as RH_i is greater than RH_o.

Although precautions can be taken when water is perceived to be a problem, i.e., eliminate all moisture, reduce contact with the atmosphere, or control the relative humidity of the atmosphere, such precautions generally add expense to the process with no guarantee that during the life of the product further problems associated with moisture will be avoided. It is also important to recognize that there are many situations where a certain level of water in a solid is required for proper performance, e.g., powder compaction. It is essential for both reasons, therefore, that as much as possible be known about the effects of moisture on solids before strategies are developed for their handling, storage, and use. Some of the more critical pieces of required information concerning water–solid interactions are (a) total amount of water

* Standard Methods for the Examination of Water and Wastewater—American Public Health Association, Current Edition, Washington, DC 20005.

present; (b) the extent to which adsorption and absorption occur; (c) whether or not crystal hydrates form; (d) specific surface area of the solid, as well as such properties as degree of crystallinity, degree of porosity, and glass transition and melting temperatures; (e) site of water interaction, the extent of binding, and the degree of molecular mobility; (f) effects of temperature and relative humidity; (g) various factors that might influence the rate at which water vapor can be taken up by a solid; and (h) for water-soluble solids capable of being solubilized by the sorbed water, under what conditions dissolution will take place.

Determination of Sorption-Desorption Isotherms—The tendency to take up water vapor is best assessed by measuring sorption or desorption as a function of relative humidity, at constant temperature, and under conditions where sorption or desorption is essentially occurring independently of time, i.e., equilibrium. Relative humidity, RH, is defined as:

$$\mathrm{RH} = \frac{P_c}{P_o} \times 100,$$

where P_c is the pressure of water vapor in the system and P_o is the vapor pressure of pure water under the same conditions. The ratio P_c/P_o is referred to as the relative pressure. It is usually varied by the use of saturated salt solutions in a closed system. Sorption or water uptake is best assessed starting with dried samples and subjecting them to a known relative humidity. Desorption is studied by beginning with a system already containing sorbed water and reducing the relative humidity. Ordinarily, if we are at equilibrium, moisture content at a particular relative humidity should be the same, whether determined from sorption or desorption measurements. However, it is common to see sorption-desorption hysteresis for certain types of systems, particularly those with microporous solids and amorphous solids, both capable of sorbing large amounts of water vapor. Here, the amount of water associated with the solid as relative humidity is decreased is greater than the amount that originally sorbed as the relative humidity is increased.

For microporous solids, vapor adsorption-desorption hysteresis is an equilibrium phenomenon associated with the process of capillary condensation. This takes place because of the high degree of irregular curvature of the micropores and the fact that they "fill" (adsorption) and "empty" (desorption) under different equilibrium conditions. For nonporous solids capable of absorbing water, hysteresis occurs because of a change in the degree of vapor–solid interaction due to a change in the equilibrium state of the solid, e.g., conformation of polymer chains, or because the time scale for structural equilibrium is longer than the time scale for water desorption.

In measuring sorption-desorption isotherms, it is important to establish that indeed something close to an equilibrium state has been reached. Particularly with hydrophilic polymers at high relative humidities, the establishment of water sorption or desorption values independent of time is quite difficult, since one is usually dealing with a polymer plasticized into its "fluid" state, where the solid is undergoing significant change. Storing samples in chambers at various relative humidities and removing them to measure weight gained or lost is most commonly carried out. The major advantage of this method is convenience, while the major disadvantages are the slow rate of reaching constant weight, particularly at high relative humidities, and the error introduced in opening and closing the chamber for weighing. Studies under vacuum in a closed system, using an electrobalance to measure weight change, avoid these problems but reduce the number of samples that can be concurrently run. It is also possible to measure amounts of water uptake not detectable gravimetrically using volumetric techniques. In the case of adsorption, to improve sensitivity one can increase the specific surface area of the sample by reducing particle size or by using larger samples to increase the total area. It is important, however, that such comminution of the solid not alter the surface structure of the solid or render it more amorphous or otherwise less ordered in crystallinity. For absorption, where water uptake is independent of specific surface area, only increasing sample size will help. Increasing sample size, however, will increase the time to establish some type of equilibrium. To establish accurate values, it is important to dry the sample as thoroughly as possible. Higher temperatures and lower pressures (vacuum) facilitate this process; however, one must be aware of any adverse effects this might have on the solid such as chemical degradation or sublimation. Using higher temperatures to induce desorption, as in a thermogravimetric apparatus, likewise must be carefully carried out with these possible pitfalls in mind. In some cases, direct analysis of water content by methods such as Karl Fischer titration or inverse gas chromatography may be advantageous. Sorption is usually expressed as weight of water taken up per unit weight of solid and plotted versus relative humidity. In most cases, the shape of the curve obtained resembles that normally seen for gas adsorption fitted to the Langmuir or Brunauer, Emmett, and Teller equations. Since crystal hydrate formation involving a phase change is usually a distinct first-order phase transition, the plot of water uptake versus pressure or relative humidity will in these cases exhibit a sharp increase in uptake at a particular pressure and the amount of water taken up usually will exhibit a stoichiometric mole: mole ratio of water to solid. In some cases, however, crystal hydrates will not appear to undergo a phase change or the anhydrous form will appear amorphous. Consequently, water sorption or desorption may appear more like that seen with adsorption processes. X-ray crystallographic analysis and thermal analysis are particularly useful for the study of such systems. For situations where water vapor adsorption occurs predominantly, it is very helpful to measure the specific surface area of the solid by an independent method and to express adsorption as weight of water sorbed *per unit area* of solid surface. This can be very useful in assessing the possible importance of water sorption in affecting solid properties. For example, 0.5% w/w uptake of water could hardly cover the bare surface of 100 m^2/g, while for 1.0 m^2/g this amounts to 100 times more surface coverage. Since we generally find that pharmaceutical solids are in the range of 0.01 to 10 m^2/g in specific surface area, what appears to be a low water content could represent a significant amount of water for the surface available.

Since the "dry surface area" is not a factor in absorption, sorption of water with amorphous or partially amorphous solids is best expressed on the basis of *unit mass* corrected for crystallinity when the crystal form does not sorb significant amounts of water relative to the amorphous regions.

Rates of Water Uptake—The rate at which solids exposed to the atmosphere might either sorb or desorb water vapor can be a critical factor in the handling of solids. Even the simple act of weighing out samples of solid on an analytical balance and the exposure, therefore, of a thin layer of powder to the atmosphere for a few minutes can lead to significant error in, for example, the estimation of loss on drying values. It is well established that water-soluble solids exposed to relative humidities above that exhibited by a saturated solution of that solid will spontaneously dissolve via deliquescence and continue to dissolve over a long time period. The rate of water uptake in general depends on a number of parameters not found to be critical in equilibrium measurements because rates of sorption are primarily mass-transfer controlled with some contributions from heat-transfer mechanisms. Thus, factors such as vapor diffusion coefficients in air and in the solid, convective airflow, and the surface area and geometry of the solid bed and surrounding environment, can play an important role. Indeed, the method used to make such measurements can often be the rate-determining factor because of these environmental and geometric factors.

Physical States of Sorbed Water—The key to understanding the effects water can have on the properties of solids, and vice versa, rests with an understanding of the location of the water molecule and its physical state. More specifically, water associated with solids can exist in a highly immobile state, as well as in a state of mobility approaching that of bulk water. This difference in mobility has been observed through such measurements as heats of sorption, freezing point, nuclear magnetic resonance, dielectric properties, and diffusion. Such changes in mobility have been interpreted as arising because of changes in the thermodynamic state of water as more and more water is sorbed. Thus water bound directly to a solid is often thought of as "tightly" bound and unavailable to affect the properties of the solid, whereas larger amounts of sorbed water tend to become more clustered and form water more like that exhibiting solvent properties. In the case of crystal hydrates, the combination of intermolecular forces (hydrogen bonding) and crystal packing can produce very strong water–solid interactions. However, there are reported situations where hydration and dehydration of crystals occur quite easily at low temperatures. More recently the concept of "tightly" bound water in amorphous systems has been questioned. Recognizing that the presence of water in an amorphous solid can affect the glass transition temperature and hence the physical state of the solid, it is argued that at low levels of water most polar amorphous solids are in a highly viscous glassy state because of their high values of Tg. Hence, water is "frozen" into the solid structure and is rendered immobile by the high viscosity, e.g., 10^{14} poise. As the amount of water sorbed increases and Tg decreases and approaches ambient temperatures, the glassy state approaches that of a "fluid" state and water mobility along with the solid itself increases significantly. At high RH the degree of water plasticization of the solid can be sufficiently high so that water and the solid now can assume significant amounts of mobility. In general, therefore, this picture of the nature of sorbed water helps to explain the rather significant effect moisture can have on a number of bulk properties of solids such as chemical reactivity and mechanical deformation. It suggests strongly that methods of evaluating chemical and physical stability of solids and solid dosage forms should take into account the effects water can have on the solid when it is sorbed, particularly when it enters the solid structure and acts as a plasticizer. Much research still remains to be done in assessing the underlying mechanisms involved in water–solid interactions of pharmaceutical importance.

Reagents, Indicators, *and* Solutions

This section deals with the reagents and solutions required in conducting the Pharmacopeial and the National Formulary tests and assays.

As is stated in the General Notices, listing of reagents, indicators, and solutions in the Pharmacopeia in no way implies that they have therapeutic utility; thus, any reference to the USP in their labeling is to include the term "reagent" or "reagent grade."

Reagents required in the tests and assays for the Pharmacopeial and National Formulary articles are listed in this section, generally with specifications appropriate to their intended uses. Exceptions to the latter include those reagents for which corresponding specifications are presented in the current edition of *Reagent Chemicals*, published by the American Chemical Society, and reagents for which specifications could not be drafted in time for inclusion here. Thus, where it is directed to "Use ACS reagent grade," it is intended that a grade meeting the corresponding specifications of the current edition of ACS *Reagent Chemicals* shall be used. Where no such specifications exist, and where it is directed to "Use a suitable grade," the intent is that a suitable reagent grade available commercially shall be used. Occasionally, additional test(s) augment the designation "suitable grade," as indicated in the text. Listed also are some, but not all, reagents that are required only in determining the quality of other reagents. For those reagents that are not listed, satisfactory specifications are available in standard reference works.

In those instances in which a reagent required in a Pharmacopeial or National Formulary test or assay need not be of analytical reagent quality, it suffices to refer to the monograph for that article appearing in this Pharmacopeia or the National Formulary or the current edition of the Food Chemicals Codex (FCC). In such cases it is to be understood that the specifications are minimum requirements and that any substance meeting more rigid specifications for chemical purity is suitable.

Where cross-reference is made to a USP or an NF monograph article, the name is followed by a parenthetic expression "(USP monograph)" or "(NF monograph)." In such cross-references, where no USP or NF designation is given, the substance referred to is a reagent in this section.

Reagents and solutions should be preserved in tight containers made of resistant glass or other suitable material. Directions for storage in light-resistant containers should be carefully observed.

Stoppers and stopcocks brought into contact with substances capable of attacking or penetrating their surfaces may be given a protective coating of a thin film of a suitable lubricant unless specifically interdicted.

Where a particular brand or source of a material or piece of equipment, or the name and address of a manufacturer, is mentioned (ordinarily in a footnote), this identification is furnished solely for informational purposes as a matter of convenience, without implication of approval, endorsement, or certification.

Atomic absorption and flame photometry require the use of a number of metal-ion standard solutions. While the individual monographs usually provide directions for preparation of these solutions, use of commercially prepared standardized solutions of the appropriate ions is permissible, provided that the analyst confirms the suitability of the solutions and has data to support their use.

NOTE—Footnotes for this section are shown at the end of the section, *Reagents, Indicators, and Solutions*.

Reagents are substances used either as such or as constituents of solutions.

Indicators are reagents used to determine the specified endpoint in a chemical reaction, to measure hydrogen-ion concentration (pH), or to indicate that a desired change in pH has been effected. They are listed together with indicator and test papers.

Buffer Solutions are referred to separately.

Colorimetric Solutions, abbreviated "CS," are solutions used in the preparation of colorimetric standards for comparison purposes.

Test Solutions, abbreviated "TS," are solutions of reagents in such solvents and of such definite concentrations as to be suitable for the specified purposes.

Volumetric Solutions, abbreviated "VS" and known also as **Standard Solutions,** are solutions of reagents of known concentration intended primarily for use in quantitative determinations. Concentrations are usually expressed in terms of normality.

Water—As elsewhere in the Pharmacopeia, where "water," without qualification, is mentioned in the tests for reagents or in directions for preparing test solutions, etc., *Purified Water* (USP monograph) is always to be used. "Carbon dioxide–free water" is purified water that has been boiled vigorously for 5 minutes or more and allowed to cool while protected from absorption of carbon dioxide from the atmosphere. "Deaerated water" is purified water that has been treated to reduce the content of dissolved air by suitable means, such as by boiling vigorously for 5 minutes and cooling or by the application of ultrasonic vibration.

Chromatographic Solvents and Carrier Gases—The chromatographic procedures set forth in the Pharmacopeia may require use of solvents and gases that have been especially purified for such use. The purpose may be (a) to exclude certain impurities that interfere with the proper conduct of the test procedure, or (b) to extend the life of a column by reducing the build-up of impurities on the column. Where solvents and gases are called for in chromatographic procedures, it is the responsibility of the analyst to ensure the suitability of the solvent or gas for the specific use. Solvents and gases suitable for specific high-pressure or other chromatographic uses are available as specialty products from various reagent supply houses, although there is no assurance that similar products from different suppliers are of equivalent suitability in any given procedure. The reagent specifications provided herein are for general analytical uses of the solvents and gases and not for chromatographic uses for which the especially purified specialty products may be required.

Reagents

For the purposes of the following specifications, these definitions apply: A *blank* consists of the same quantities of the same reagents treated in the same manner as the specimen under test. A *control* is a blank to which has been added the limiting quantity of the substance being tested for, or is a specified comparison solution prepared as directed in the particular test.

The values given in boldface type following chemical symbols and formulas represent, respectively, atomic and molecular weights of the substances concerned.

Color and turbidity comparisons are to be made in color-comparison tubes that are matched as closely as possible in internal diameter and in all other respects, as directed for *Visual Comparison* under *Spectrophotometry and Light-scattering* ⟨851⟩. Such tubes frequently are called "Nessler tubes."

In making visual comparisons of the densities of turbid fluids, compensate for differences in color, if necessary, by viewing the turbidity through a column of water, the depth of which is determined by the volume specified in the individual reagent specification. Place the water in color-comparison tubes, and hold one of the tubes above the control tube and the other below the specimen tube.

Where an expression such as "Retain the filtrate" appears it is to be understood, unless otherwise indicated, that the washings of the residue are not to be added to the filtrate obtained.

In the test heading, *Calcium, magnesium, and R_2O_3 precipitate*, the expression R_2O_3 is intended to indicate the residue on ignition from compounds precipitated upon the addition of ammonium hydroxide, such as Fe_2O_3 and Al_2O_3.

GENERAL TESTS FOR REAGENTS

The following general test methods are provided for the examination of reagents to determine their compliance with the specifications of the individual reagents and are to be used unless it is otherwise directed in such specifications.

Boiling or Distilling Range for Reagents

Use the following procedure for determining the boiling or distilling range of reagents, unless otherwise directed in the individual specifications:

APPARATUS—Use apparatus similar to that specified for *Distilling Range—Method I* ⟨721⟩, except that the distilling flask is to be of 250-mL capacity, to have a short neck, and to be connected to the condenser by means of a three-way connecting tube fitted with standard-taper ground joints.

PROCEDURE—Place the distilling flask in an upright position in the perforation in the asbestos board, and connect it to the condenser.

Measure 100 mL of the liquid to be tested in a graduated cylinder, and transfer to the boiling flask together with some device to prevent bumping. Use the cylinder as the receiver for the distillate. Insert the thermometer, and heat so as to distil at the rate of 3 mL to 5 mL per minute. Make a preliminary trial, if necessary, to determine the adjustment for the proper rate of heating. Read the thermometer when about 20 drops have distilled and thereafter at volumes of distillate of 5, 10, 40, 50, 60, 90, and 95 mL. Continue the distillation until the dry point is reached.

The *Boiling or Distilling Range* is the interval between the temperatures when 1 mL and 95 mL, respectively, have distilled.

Arsenic in Reagents

Select reagents for this test for a low arsenic content, so that a blank test results in either no stain or one that is barely perceptible.

APPARATUS—Prepare a generator by fitting a 1-hole rubber stopper into a wide-mouth bottle of about 60-mL capacity. Through the perforation insert a vertical exit tube about 12 cm in total length and 1 cm in diameter along the entire upper portion (for about 8 cm) and constricted at its lower extremity to a tube about 4 cm in length and about 5 mm in diameter. The smaller portion of the tube should extend to just slightly below the stopper. Place washed sand or a pledget of purified cotton in the upper portion to about 3 cm from the top of the tube. Moisten the sand or cotton uniformly with lead acetate TS, and remove any excess or adhering droplets of the latter from the walls of the tube. Into the upper end of this tube fit a second glass tube 12 cm in length, having an internal diameter of 2.5 to 3 mm, by means of a rubber stopper. Just before running the test, place a strip of mercuric bromide test paper (see under *Indicator and Test Papers*) in this tube, crimping the upper end of the strip so that it will remain in position about 2 cm above the rubber stopper. Clean and dry the tube thoroughly each time it is used.

STANDARD ARSENIC SOLUTION—Use *Standard Preparation* prepared as directed under *Arsenic* ⟨211⟩.

TEST PREPARATION—Add 1 mL of sulfuric acid to 5 mL of a solution of the chemical substance (1 in 25), unless another quantity is directed in the individual reagent specification. Omit its addition entirely in the case of inorganic acids. Unless especially directed otherwise, add 10 mL of sulfurous acid. Evaporate the liquid in a small beaker, on a steam bath, until it is free from sulfurous acid and has been reduced to about 2 mL in volume. Dilute with water to 5 mL to obtain the *Test Preparation*. Substances subjected to special treatments specified in the individual reagent specification may be used directly as the *Test Preparation*.

NOTE—Solutions prepared by the dissolving of the chemical substances in dilute acids are not considered to have undergone special treatment.

STANDARD STAIN—Place in the generator bottle 5 mL of potassium iodide TS, 2.0 mL of *Standard Arsenic Solution*, 5 mL of acid stannous chloride TS, and 28 mL of water. Add 1.5 g of granulated zinc (in No. 20 powder), and immediately insert the stopper containing the exit tube. Keep the generator bottle immersed in water at 25° during the period of the test to moderate the reaction so that the stain will take the form of a distinctive band to facilitate the comparison of color intensity. When evolution of hydrogen has continued for 1 hour, remove the mercuric bromide test paper for comparison. This stain represents 2 μg of arsenic.

PROCEDURE—Pipet into the generator bottle 5 mL of potassium iodide TS and 5 mL of the *Test Preparation*, and add 5 mL of acid stannous chloride TS. Set the apparatus aside at room temperature for a period of 10 minutes, then add 25 mL of water and 1.5 g of granulated zinc (in No. 20 powder), and proceed as directed under *Standard Stain*. Remove the mercuric bromide test paper, and compare the stain upon it with the *Standard Stain:* the stain produced by the chemical tested does not exceed the standard stain in length or in intensity of color, indicating not more than 10 parts of arsenic per million parts of the substance being tested. Since light, heat, and moisture cause the stain to fade rapidly, place the papers in clean, dry tubes, and make comparisons promptly.

INTERFERING CHEMICALS—*Antimony*, if present in the substance being tested, produces a gray stain. *Sulfites, sulfides, thiosulfates*, and other compounds that liberate hydrogen sulfide or sulfur dioxide when treated with sulfuric acid must be oxidized by means of nitric acid and then reduced by means of sulfur dioxide as directed under *Test Preparation* before they are placed in the apparatus. Certain *sulfur compounds*, as well as *phosphine*, give a bright yellow band on the test paper. If *sulfur compounds* are present, the lead acetate–moistened cotton or sand will darken. In that case, repeat the operation as directed under *Test Preparation* upon a fresh portion of the solution being tested and use greater care in effecting the complete removal of the sulfurous acid. In testing hypophosphites, observe special care to oxidize completely the solution being tested as directed, otherwise the evolution of phosphine may result in a yellow stain which may be confused with the orange-yellow color produced by arsine. The stain produced by phosphine may be differentiated from that given by arsine by means of moistening it with 6 *N* ammonium hydroxide. A stain caused by arsine becomes dark when so treated, but a stain produced by phosphine does not materially change in color.

Chloride in Reagents

STANDARD CHLORIDE SOLUTION—Dissolve 165.0 mg of dried sodium chloride in water to make 1000.0 mL. This solution contains the equivalent of 0.10 mg of chlorine (Cl) in each mL.

PROCEDURE—Neutralize, if alkaline, a solution of the quantity of the reagent indicated in the test in 25 mL of water, or a solution prepared as directed in the test, with nitric acid, litmus paper being used as the indicator, and add 3 mL more of nitric acid. Filter the solution, if necessary, through a filter paper previously washed with water until the paper is free from chloride, and add 1 mL of silver nitrate TS. Mix, and allow to stand for 5 minutes protected from direct sunlight. Compare the turbidity, if any, with that produced in a control made with the same quantities of the same reagents as in the final test and a volume of *Standard Chloride Solution* equivalent to the quantity of chloride (Cl) permitted by the test. Adjust the two solutions with water to the same volume before adding the silver nitrate TS, and compare the turbidities.

In *testing barium salts*, neutralize, if alkaline, the solution containing the reagent, with nitric acid, and add only 3 drops more of nitric acid. Conduct the remainder of the test as described previously.

In *testing salts giving colored solutions*, dissolve 2 g of the reagent in 25 mL of water, and add 3 mL of nitric acid. Filter the solution, if necessary, through a filter paper previously washed with water, and divide the filtrate into two equal portions. Treat one portion with 1 mL of silver nitrate TS, allow to stand for 10 minutes, and, if any turbidity is produced, filter it through a washed filter paper until clear, and use the filtrate as a blank. Treat the other portion with 1 mL of silver nitrate TS, mix, and allow to stand for 5 minutes protected from direct sunlight. Compare the turbidity with that produced in the blank by the addition of a volume of *Standard Chloride Solution* equivalent to the quantity of chloride (Cl) permitted in the test, both solutions being adjusted with water to the same volume.

Flame Photometry for Reagents

The use of flame photometric procedures to determine traces of calcium, potassium, sodium, and strontium is called for in some of the reagent specifications. The suitability of such determinations depends upon the use of adequate apparatus, and several instruments of suitable selectivity are available. The preferred type of flame photometer is one that has a red-sensitive phototube, a multiplier phototube, a monochromator, an adjustable slit-width control, a selector switch, and a sensitivity control. Other types of photometers may be used, provided the operator has proved that the instrument will determine accurately the amount of impurities permitted in the reagent to be tested.

The flame photometric procedures depend upon the use of semi-internal standards, and thus require both a *Sample Solution* and a *Control Solution*. For the *Sample Solution*, a specified weight of specimen is dissolved and diluted to a definite volume. For the *Control Solution*, the same amount of specimen is dissolved, the limiting amounts of the suspected impurities are added, and the solution is then diluted to the same definite volume as the *Sample Solution*. The flame photometer is set as directed in the general procedures and then adjusted to give an emission reading as near 100% transmittance as is possible with the *Control Solution* at the wavelength specified for the particular impurity concerned. With the instrument settings left unchanged, the emission from the *Sample Solution* is read at the same wavelength and at a specified background wavelength. The background reading is then used to correct the observed emission of the *Sample Solution* for the emission due to the specimen and the solvent. The specimen being tested contains less than the specified limit of impurity if the difference between the observed background and total emissions for the *Sample Solution* is less than the difference between the observed emissions for the *Control Solution* and the *Sample Solution* at the wavelength designated for the particular impurity.

CALCIUM IN REAGENTS—

Standard Calcium Solution—Dissolve 250 mg of calcium carbonate in a mixture of 20 mL of water and 5 mL of diluted hydrochloric acid, and when solution is complete, dilute with water to 1 liter. This solution contains 0.10 mg of calcium (Ca) per mL.

Procedure—Use the *Sample Solution* and the *Control Solution* prepared as directed in the individual test procedure.

Set the slit-width control of a suitable flame photometer at 0.03 mm, and set the selector switch at 0.1. Adjust the instrument to give the maximum emission with the *Control Solution* at the 422.7-nm calcium line, and record the transmittance. Without changing any of the instrument settings, record the transmittance for the emission of the *Sample Solution* at 422.7 nm. Change the monochromator to the wavelength specified in the individual test procedure, and record the background transmittance for the background emission of the *Sample Solution:* the difference between the transmittances for the *Sample Solution* at 422.7 nm and at the background wavelength is not greater than the difference between transmittances observed at 422.7 nm for the *Sample Solution* and the *Control Solution.*

POTASSIUM IN REAGENTS—

Standard Potassium Solution—Dissolve 191 mg of potassium chloride in a few mL of water, and dilute with water to 1 liter. Dilute a portion of this solution with water in the ratio of 1 to 10 to obtain a concentration of 0.01 mg of potassium (K) per mL.

Procedure—Use the *Sample Solution* and the *Control Solution* prepared as directed in the individual test procedure. [NOTE—In testing calcium salts, use an oxyhydrogen burner.]

Set the slit-width control of a suitable flame photometer equipped with a red-sensitive detector at 0.1 mm, unless otherwise directed, and set the selector switch at 0.1. Adjust the instrument to give the maximum emission with the *Control Solution* at the 766.5-nm potassium line, and record the transmittance. Without changing any of the instrument settings, record the transmittance for the emission of the *Sample Solution* at 766.5 nm. Change the monochromator to 750 nm, and record the background transmittance for the background emission of the *Sample Solution:* the difference between the transmittances for the *Sample Solution* at 766.5 nm and 750 nm is not greater than the difference between transmittances observed at 766.5 nm for the *Sample Solution* and the *Control Solution.*

SODIUM IN REAGENTS—

Standard Sodium Solution—Dissolve 254 nm of sodium chloride in a few mL of water, and dilute with water to 1 liter. Dilute a portion of this solution with water in the ratio of 1 to 10 to obtain a concentration of 0.01 mg of sodium (Na) per mL.

Procedure—Use the *Sample Solution* and the *Control Solution* prepared as directed in the individual test procedure.

Set the slit-width control of a suitable flame photometer at 0.01 mm, and set the selector switch at 0.1. Adjust the instrument to give the maximum emission with the *Control Solution* at the 589-nm sodium line, and record the transmittance. Without changing any of the instrument settings, record the transmittance for the emission of the *Sample Solution* at 589 nm. Change the monochromator to 580 nm, and record the background transmittance for the background emission of the *Sample Solution:* the difference between the transmittances for the *Sample Solution* at 589 nm and 580 nm is not greater than the difference between transmittances observed at 589 nm for the *Sample Solution* and the *Control Solution.*

STRONTIUM IN REAGENTS—

Standard Strontium Solution—Dissolve 242 mg of strontium nitrate in a few mL of water, and dilute with water to 1 liter. Dilute a portion of this solution with water in the ratio of 1 to 10 to obtain a concentration of 0.01 mg of strontium (Sr) per mL.

Procedure—Use the *Sample Solution* and the *Control Solution* prepared as directed in the individual test procedure.

Set the slit-width control of a suitable flame photometer at 0.03 mm, and set the selector switch at 0.1. Adjust the instrument to give the maximum emission with the *Control Solution* at the 460.7-nm strontium line, and record the transmittance. Without changing any of the instrument settings, record the transmittance for the emission of the *Sample Solution* at 460.7 nm. Change the monochromator to the wavelength specified in the individual test procedure, and record the background transmittance for the background emission of the *Sample Solution:* the difference between the transmittances for the *Sample Solution* at 460.7 nm and at the background wavelength is not greater than the difference between transmittances observed at 460.7 nm for the *Sample Solution* and the *Control Solution.*

Heavy Metals in Reagents

STANDARD LEAD SOLUTION—Use *Standard Lead Solution* (see *Heavy Metals* ⟨231⟩). Each mL of this solution contains the equivalent of 0.01 mg of Pb.

PROCEDURE—Unless otherwise directed, test for heavy metals as follows:

(*a*) If the heavy metals limit is 0.0005% (5 ppm), dissolve 6.0 g of the specimen in water to make 42 mL.

(*b*) If the heavy metals limit is 0.001% (10 ppm) or more, or in the event of limited solubility, use 4 g, and dissolve in water to make 40 mL, warming, if necessary, to aid solution.

For the control, transfer 7 mL of the solution from (*a*) to a color-comparison tube, and add a volume of *Standard Lead Solution* equivalent to the amount of lead permitted in 4 g of the reagent. Dilute with water to 35 mL, and add diluted acetic acid, or ammonia TS, until the pH is about 3.5, determined potentiometrically, then dilute with water to 40 mL, and mix. Transfer the remaining 35 mL of the solution from (*a*) to a color-comparison tube closely matching that used for the control, and add diluted acetic acid, or ammonia TS, until the pH is about 3.5, determined potentiometrically, then dilute with water to 40 mL, and mix. Then to each tube add 10 mL of hydrogen sulfide TS, mix, and compare the colors by viewing through the color-comparison tube downward against a white surface. The color in the test specimen is not darker than that of the control.

If the solution of the reagent is prepared as in (*b*), use for the control 10 mL of the solution, and add to it a volume of *Standard Lead Solution* equivalent to the amount of lead permitted in 2 g of the reagent. Dilute the remaining 30 mL of solution (*b*) with water to 35 mL, and proceed as directed in the preceding paragraph, beginning with "add diluted acetic acid, or ammonia TS," in the second sentence.

If the reagent to be tested for heavy metals is a salt of an aliphatic organic acid, substitute 1 *N* hydrochloric acid for the diluted acetic acid specified in the foregoing method.

Insoluble Matter in Reagents

Dissolve the quantity of reagent specified in the test in 100 mL of water, heat to boiling unless otherwise directed, in a covered beaker, and warm on a steam bath for 1 hour. Filter the hot solution through a suitable, tared crucible with an asbestos mat, or through a tared, sintered-glass crucible of fine porosity. Wash the beaker and the filter thoroughly with hot water, dry at 105°, cool in a desiccator, and weigh.

Loss on Drying for Reagents

Determine as directed under *Loss on Drying* ⟨731⟩.

Nitrate in Reagents

STANDARD NITRATE SOLUTION—Dissolve 163 mg of potassium nitrate in water, add water to make 100 mL, and dilute 10 mL of this solution with water to 1 liter, to obtain a solution containing the equivalent of 0.01 mg of NO_3 per mL.

BRUCINE SULFATE SOLUTION—Dissolve 600 mg of brucine sulfate in 600 mL of nitrate-free, dilute sulfuric acid (2 in 3) that previously has been cooled to room temperature, and dilute with the acid to 1 liter. [NOTE—Prepare the nitrate-free sulfuric acid by adding 4 parts of sulfuric acid to 1 part of water, heating the solution to dense fumes of sulfur trioxide, and cooling. Repeat the dilution and heating three or four times.]

SAMPLE SOLUTION—To the weight of sample specified in the individual reagent specification, dissolved in the designated volume of water, add *Brucine Sulfate Solution* to make 50 mL.

CONTROL SOLUTION—To a volume of *Standard Nitrate Solution* equivalent to the weight of nitrate (NO_3) specified in the individual reagent specification, add the weight of sample specified in the individual reagent specification and then add *Brucine Sulfate Solution* to make 50 mL.

BLANK SOLUTION—Use 50 mL of *Brucine Sulfate Solution.*

PROCEDURE—Heat the *Sample Solution, Control Solution,* and *Blank Solution* in a boiling water bath for 10 minutes, then cool rapidly in an ice bath to room temperature. Adjust a suitable spectrophotometer to zero absorbance at 410 nm with the *Blank Solution.* Determine the absorbance of the *Sample Solution,* note the result, and adjust the instrument to zero absorbance with the *Sample Solution.* Determine the absorbance of the *Control Solution:* the absorbance reading for the *Sample Solution* does not exceed that for the *Control Solution.*

Nitrogen Compounds in Reagents

PROCEDURE—Unless otherwise directed, test for nitrogen compounds as follows: Dissolve the specified quantity of test specimen in 60 mL of ammonia-free water in a Kjeldahl flask connected through a spray trap to a condenser, the end of which dips below the surface of 10 mL of 0.1 *N* hydrochloric acid. Add 10 mL of freshly boiled sodium hydroxide solution (1 in 10) and 500 mg of aluminum wire, in small pieces, to the Kjeldahl flask, and allow to stand for 1 hour, protected from loss of, and exposure to, ammonia. Distil 35 mL, and dilute the distillate with water to 50 mL. Add 2 mL of freshly boiled sodium hydroxide solution (1 in 10), mix, add 2 mL of alkaline mercuric-potassium iodide TS, and again mix: the color produced is not darker than that of a control containing the amount of added N (as ammonium chloride) specified in the individual test procedure.

Phosphate in Reagents

STANDARD PHOSPHATE SOLUTION—Dissolve 143.3 mg of dried monobasic potassium phosphate, KH_2PO_4, in water to make 1000.0 mL. This solution contains the equivalent of 0.10 mg of phosphate (PO_4) in each mL.

PHOSPHATE REAGENT A—Dissolve 5 g of ammonium molybdate in 1 *N* sulfuric acid to make 100 mL.

PHOSPHATE REAGENT B—Dissolve 200 mg of *p*-methylaminophenol sulfate in 100 mL of water, and add 20 g of sodium bisulfite. Store this reagent in well-filled, tightly stoppered bottles, and use within one month.

PROCEDURE—[NOTE—The tests with the specimen and the control are made preferably in matched color-comparison tubes.] Dissolve the quantity of the reagent specified in the test, or the residue obtained after the prescribed treatment, in 20 mL of water, by warming, if necessary, add 2.5 mL of dilute sulfuric acid (1 in 7), and dilute with water to 25 mL. (If preferable, the test specimen or the residue may be dissolved in 25 mL of approximately 0.5 *N* sulfuric acid.) Then add 1 mL each of *Phosphate Reagents A* and *B*, mix, and allow to stand at room temperature for 2 hours. Compare any blue color produced with that produced in a control made with the same quantities of the same reagents as in the test with the specimen, and a volume of *Standard Phosphate Solution* equivalent to the quantity of phosphate (PO_4) designated in the reagent specifications.

Residue on Ignition in Reagents

PROCEDURE—Unless otherwise directed, determine the residue on ignition as follows: Weigh accurately 1 to 2 g of the substance to be tested in a suitable crucible that previously has been ignited, cooled, and weighed. Ignite the substance, gently and slowly at first and then at a more rapid rate, until it is thoroughly charred, if organic in nature, or until it is completely volatilized, if inorganic in nature. If the use of sulfuric acid is specified, cool the crucible, add the specified amount of acid, and ignite the crucible gently until fumes no longer are evolved. Then ignite the crucible at 800 ± 25°, cool in a suitable desiccator, and weigh. If the use of sulfuric acid is not specified, the crucible need not be cooled but can be ignited directly at 800 ± 25° once the charring or volatilization is complete. Continue the ignition until constant weight is attained, unless otherwise specified.

Conduct the ignition in a well-ventilated hood, but protected from air currents, and at as low a temperature as is possible to effect the complete combustion of the carbon. A muffle furnace may be used, if desired, and its use is recommended for the final ignition at 800 ± 25°.

Sulfate in Reagents

STANDARD SULFATE SOLUTION—Dissolve 181.4 mg of dried potassium sulfate in water to make 1000 mL. This solution contains the equivalent of 0.10 mg of sulfate (SO_4) per mL.

PROCEDURE—*Method I*—Neutralize, if necessary, a solution of the quantity of the reagent or residue indicated in the test in 25 mL of water, or a solution prepared as directed in the test, with hydrochloric acid or with ammonia TS, litmus paper being used as the indicator, and add 1 mL of 1 *N* hydrochloric acid. Filter the solution, if necessary, through a filter paper previously washed with water, and add 2 mL of barium chloride TS. Mix, allow to stand for 10 minutes, and compare the turbidity, if any, with that produced in a control containing the same quantities of the same reagents used in the test and a quantity of *Standard Sulfate Solution* equivalent to the quantity of sulfate (SO_4) permitted in the test. Adjust the two solutions with water to the same volume before adding the barium chloride TS.

Method II—Heat to boiling the solution, prepared as directed in the individual test procedure, or the filtrate designated in the procedure, and add 5 mL of barium chloride TS. Then digest the solution on a steam bath for 2 hours, and allow to stand overnight. If any precipitate is formed, filter the solution through paper, wash the residue with hot water, and transfer the paper containing the residue to a tared crucible. Char the paper, without burning, and ignite the crucible and its contents to constant weight. Perform a blank determination concurrently with the test specimen determination, and subtract the weight of residue obtained from that obtained in the test specimen determination to obtain the weight of residue attributable to the sulfate content of the specimen.

REAGENT SPECIFICATIONS

Absolute Ether—See *Ethyl Ether, Anhydrous.*

Absorbent Cotton—Use *Purified Cotton* (USP monograph).

Acacia—Use *Acacia* (NF monograph).

Acetaldehyde, CH_3CHO—**44.05**—Use Acetaldehyde, FCC.

5-Acetamido-3-amino-2,4,6-triiodobenzoic Acid—Use USP 5-Acetamido-3-amino-2,4,6-triiodobenzoic Acid RS.

Acetaminophen—Use *Acetaminophen* (USP monograph).

Acetanilide, C_8H_9NO—**135.17**—White, shiny crystals, usually in scales, or a white, crystalline powder. Is odorless and is stable in air. Slightly soluble in water; freely soluble in alcohol and in chloroform; soluble in boiling water, in ether, and in glycerin.

Melting range ⟨741⟩: between 114° and 116°.

Reaction—Its saturated solution is neutral to litmus.

Loss on drying ⟨731⟩—Dry it over sulfuric acid for 2 hours: it loses not more than 0.5% of its weight.

Residue on ignition (Reagent test): not more than 0.05%.

Acetic Acid (*6 N Acetic Acid*)—Use *Acetic Acid* (NF monograph).

Acetic Acid, Diluted (*1 N Acetic Acid*)—Dilute 60.0 mL of glacial acetic acid with water to make 1000 mL.

Residue on evaporation—Evaporate 50 mL on a steam bath, and dry the residue at 105° for 2 hours: the residue weighs not more than 1 mg (0.002%).

Chloride (Reagent test)—Five mL shows not more than 0.01 mg of Cl (2 ppm).

Sulfate (Reagent test, *Method I*)—Ten mL shows not more than 0.5 mg of SO_4 (5 ppm).

Heavy metals (Reagent test)—Evaporate 20 mL on a steam bath to dryness. Add to the residue 2 mL of the acid, dilute with water to 25 mL, and add 10 mL of hydrogen sulfide TS: any brown color produced is not darker than that of a control containing 0.04 mg of added Pb and 2 mL of the diluted acetic acid (2 ppm).

Acetic Acid, Glacial, CH_3COOH—**60.05**—Use ACS reagent grade, which also meets the requirements of the following test.

Sensitivity—To 20 mL add about 5 mg of crystal violet: the color of the resulting solution is purple.

Acetic Anhydride, $(CH_3CO)_2O$—**102.09**—Use ACS reagent grade.

Acetone, CH_3COCH_3—**58.08**—Use ACS reagent grade.

NOTE—For ultraviolet spectrophotometric determinations, use ACS reagent grade Acetone Suitable for Use in Ultraviolet Spectrophotometry.

Acetone, Anhydrous, CH_3COCH_3—**58.08**—Use ACS reagent grade Acetone.

Acetonitrile (*Methyl Cyanide*), CH_3CN—**41.05**—Use ACS reagent grade.

Acetonitrile, Spectrophotometric—Use ACS reagent grade, which meets also the requirements of the following test.

Spectral purity—Measure in a 1-cm cell between 250 nm and 280 nm, with a suitable spectrophotometer, against air as the blank: its absorbance is not more than 0.01.

Acetylacetone (*2,4-Pentanedione*), $C_5H_8O_2$—**100.11**—Clear, colorless to slightly yellow, flammable liquid. Soluble in water; miscible with alcohol, with chloroform, with acetone, with ether, and with glacial acetic acid.

Assay—Not less than 98% of $C_5H_8O_2$, a suitable gas chromatograph equipped with a flame-ionization detector being used and helium being used as the carrier gas. The following conditions have been found suitable: a 1.83-m × 3-mm stainless steel column containing 10 percent phase G on support S1A; the injection port and detector temperatures are maintained at 250° and 310°, respectively; the column temperature is programmed to rise at 8° per minute, from 50° to 220°.

Refractive index ⟨831⟩: between 1.4505 and 1.4525, at 20°.

Acetyl Chloride, CH_3COCl—**78.50**—Clear, colorless liquid, having a strong, pungent odor. Is decomposed by water and by alcohol. Miscible with benzene and with chloroform. Specific gravity: about 1.1.

Boiling range (Reagent test)—Not less than 94% distils between 49° and 53°.

Residue on evaporation—Evaporate 10 mL on a steam bath, and dry at 105° for 1 hour: the residue weighs not more than 2.5 mg (about 0.02%).

Miscibility with benzene and with chloroform—Separate 5-mL portions give clear solutions with 20 mL of benzene and with 20 mL of chloroform.

Solubility—Place 5 mL in a 50-mL graduated cylinder, and cautiously add, dropwise, about 3 mL of water, shaking after each addition until the reaction is complete, then dilute with water to 50 mL: the solution is clear.

Phosphorus compounds—To 5 mL of the solution obtained in the preceding test add 3 mL of nitric acid, and evaporate on a steam bath to dryness. The residue, dissolved in 20 mL of water, shows not more than 0.03 mg of PO_4 (0.02% as P) [see *Phosphate in Reagents*].

Heavy metals—Dilute 10 mL of the solution obtained in the test for *Solubility* with 30 mL of water, add 10 mL of hydrogen sulfide TS, and render alkaline with ammonia TS: no noticeable change in color is produced.

Acetylcholine Chloride, $[CH_3COOCH_2CH_2N(CH_3)_3]Cl$—**181.66**—White, crystalline, odorless or nearly odorless powder. Very deliquescent; very soluble in water; freely soluble in alcohol.

Melting range ⟨741⟩—When previously dried at 110° in a capillary tube for 1 hour, it melts between 149° and 152°.

Reaction—Its 1 in 10 solution is neutral to litmus.

Residue on ignition (Reagent test): negligible, from 200 mg.

Solubility in alcohol—A solution of 500 mg in 5 mL of alcohol is complete and colorless.

Percent of acetyl (CH_3CO)—Weigh accurately about 400 mg, previously dried at 105° for 3 hours, and dissolve in 15 mL of water in a glass-stoppered conical flask. Add 40.0 mL of 0.1 *N* sodium hydroxide VS, and heat on a steam bath for 30 minutes. Insert the stopper, allow to cool, add phenolphthalein TS, and titrate the excess alkali with 0.1 *N* sulfuric acid VS. Determine the exact normality of the 0.1 *N* sodium hydroxide by titrating 40.0 mL after it has been treated in the same manner as in the test. Each mL of 0.1 *N* sodium hydroxide is equivalent to 4.305 mg of CH_3CO. Between 23.2% and 24.2% is found.

Percent of chlorine (*Cl*)—Weigh accurately about 400 mg, previously dried at 105° for 3 hours, and dissolve in 50 mL of water in a glass-stoppered, 125-mL flask. Add with agitation 30.0

mL of 0.1 *N* silver nitrate VS, then add 5 mL of nitric acid and 5 mL of nitrobenzene, shake, add 2 mL of ferric ammonium sulfate TS, and titrate the excess silver nitrate with 0.1 *N* ammonium thiocyanate VS: each mL of 0.1 *N* silver nitrate is equivalent to 3.545 mg of Cl. Between 19.3% and 19.8% of Cl is found.

***N*-Acetyl-L-tyrosine Ethyl Ester,** $C_{13}H_{17}NO_4$—**251.28**—Determine the suitability of the material as directed in the *Assay* under *Chymotrypsin* (USP monograph).

Acid Orange 5—See *Tropaeolin OO* (USP reagent).

Acid-washed Aluminum Oxide—See *Aluminum Oxide, Acid-washed.*

Activated Alumina—See *Alumina, Activated.*

Activated Charcoal—See *Charcoal, Activated.*

Activated Magnesium Silicate—See *Magnesium Silicate, Activated.*

Adenine Sulfate, $(C_5H_5N_5)_2 . H_2SO_4 . 2H_2O$—**404.36**—White crystals or crystalline powder. Melts, after drying at 110°, at about 200° with some decomposition. One g dissolves in about 160 mL of water; less soluble in alcohol. Soluble in solutions of sodium hydroxide. It is not precipitated from solution by iodine TS or mercuric-potassium iodide TS, but a precipitate is produced with trinitrophenol TS.

Residue on ignition (Reagent test): negligible, from 100 mg.

Water—Dry it at 105° to constant weight: it loses not more than 10.0% of its weight.

Adipic Acid, $C_6H_{10}O_4$—**146.14**—Colorless to white, crystalline powder. Slightly soluble in water and in cyclohexane; soluble in alcohol, in methanol, and in acetone; practically insoluble in benzene and in petroleum benzin.

Assay—Weigh accurately about 0.3 g, and dissolve in 50 mL of alcohol. Add 25 mL of water, mix, and titrate with 0.5 *N* sodium hydroxide VS to a pH of 9.5. Perform a blank determination, and make any necessary correction. Each mL of 0.5 *N* sodium hydroxide is equivalent to 36.54 mg of $C_6H_{10}O_4$. Not less than 98% is found.

Melting range ⟨741⟩: between 151° and 155°, but the range between beginning and end of melting does not exceed 2°.

Agar—Use *Agar* (NF monograph). When used for bacteriological purposes, it is to be dried to a water content of not more than 20%.

Air-Helium Certified Standard—A mixture of 1.0% air in industrial grade helium. It is available from most suppliers of specialty gases.

Alcohol, C_2H_5OH—**46.07**—Use *Alcohol* (USP monograph).

Alcohol, Absolute, C_2H_5OH—**46.07**—Use ACS reagent grade Ethyl Alcohol, Absolute.

Alcohol, 70 percent, 80 percent, and 90 percent—Prepare by mixing alcohol and water in the proportions given, the measurements being made at 25°.

Percent by Volume of C_2H_5OH at 15.56°	Specific Gravity at 25°	Relative Proportions: Alcohol, mL	Relative Proportions: Water, mL	Volume in mL of Alcohol, 94.9% v/v, Required for 100 mL
70	0.884	38.6	15	73.7
80	0.857	45.5	9.5	84.3
90	0.827	51	3	94.8

Alcohol, Aldehyde-free—Dissolve 2.5 g of lead acetate in 5 mL of water, add the solution to 1000 mL of alcohol contained in a glass-stoppered bottle, and mix. Dissolve 5 g of potassium hydroxide in 25 mL of warm alcohol, cool the solution, and add it slowly, without stirring, to the alcohol solution of lead acetate. After 1 hour shake the mixture vigorously, allow it to stand overnight, decant the clear liquid, and recover the alcohol by distillation.

Alcohol, Amyl—See *Amyl Alcohol.*

Alcohol, Dehydrated (*Absolute Alcohol*), C_2H_5OH—**46.07**—Use ACS reagent grade Ethyl Alcohol, Absolute.

Alcohol, Dehydrated Isopropyl—See *Isopropyl Alcohol, Dehydrated.*

Alcohol, Diluted—Use *Diluted Alcohol* (NF monograph).

Alcohol, Isobutyl—See *Isobutyl Alcohol.*

Alcohol, Isopropyl—See *Isopropyl Alcohol.*

Alcohol, Methyl—See *Methanol.*

Alcohol, Neutralized—To a suitable quantity of alcohol add 2 or 3 drops of phenolphthalein TS and just sufficient 0.02 *N* or 0.1 *N* sodium hydroxide to produce a faint pink color. Prepare neutralized alcohol just prior to use.

Alcohol, *n*-Propyl—See *n-Propyl Alcohol.*

Alcohol, Secondary Butyl—See *Butyl Alcohol, Secondary.*

Alcohol, Tertiary Butyl—See *Butyl Alcohol, Tertiary.*

Alkaline Phosphatase Enzyme—See *Phosphatase Enzyme, Alkaline.*

Alkylphenoxypolyethoxyethanol—A nonionic surfactant. Use a suitable grade.[1]

Alphanaphthol—See *1-Naphthol.*

Alum (*Ammonium Alum, Aluminum Ammonium Sulfate*), $AlNH_4(SO_4)_2 . 12H_2O$—**453.32**—Large, colorless crystals or crystalline fragments or a white powder. Soluble in 7 parts of water and in about 0.5 part of boiling water; insoluble in alcohol.

Insoluble matter (Reagent test): not more than 1 mg, from 10 g (0.01%).

Chloride (Reagent test)—Two g shows not more than 0.02 mg of Cl (0.001%).

Alkalies and alkaline earths—Dissolve 2 g in 140 mL of water, add 2 drops of methyl orange TS, then add ammonia TS in small portions until the color just turns yellow. Boil for 2 minutes, dilute with water to 150 mL, and filter. Evaporate 75 mL of the filtrate, and ignite the residue: the ignited residue weighs not more than 2.5 mg (0.25%).

Arsenic (Reagent test)—The stain produced by 1 g does not exceed that produced by 0.002 mg of As (2 ppm as As).

Heavy metals (Reagent test): 0.001%.

Iron ⟨241⟩—Dissolve 1.0 g in 40 mL of water, add 4 mL of hydrochloric acid, mix, and dilute with water to 50 mL. Dilute 25 mL of this solution with water to 47 mL: the solution shows not more than 0.01 mg of Fe (0.002%).

Ammonium Alum—See *Alum.*

Alumina—See *Aluminum Oxide, Acid-washed.*

Alumina, Activated—Use a suitable grade.[2]

Alumina, Anhydrous (*Aluminum Oxide; Alumina specially prepared for use in chromatographic analysis*)—A white or practically white powder, 80- to 200-mesh. It does not soften, swell, or decompose in water. It is not acid-washed. Store it in well-closed containers.

Alumina Mixture, Chromatographic (*Aluminum Oxide, Chromatographic*)—A mixture of alumina with a suitable fluorescing substance.

Aluminum, Al—At. Wt. **26.98154**—Use ACS reagent grade, which also meets the requirements of the following test.

Arsenic—Place 750 mg in a generator bottle (see *Arsenic in Reagents* under *General Tests for Reagents*), omitting the pledget of cotton. Add 10 mL of water and 10 mL of sodium hydroxide solution (3 in 10), and allow the reaction to proceed for 30 minutes: not more than a barely perceptible stain is produced on the mercuric bromide test paper.

Aluminum Chloride—Use *Aluminum Chloride,* NF XIV.

Aluminum Oxide, Acid-washed (*Alumina specially prepared for use in chromatographic analysis*)—White or practically white powder or fine granules. Very hygroscopic. Store in tight containers.

pH of Slurry—The pH of a well-mixed slurry of 5 g in 150 mL of ammonia-free and carbon dioxide–free water, after 10 minutes' standing, is between 3.5 and 4.5.

Loss on ignition—Weigh accurately about 1 g, and ignite, preferably in a muffle furnace at 800° to 825°, to constant weight: it loses not more than 5.0% of its weight.

Silica—Fuse 500 mg with 10 g of potassium bisulfate for 1 hour in a platinum crucible, cool, and dissolve in hot water: not more than a small amount of insoluble matter remains.

Suitability for chromatographic adsorption—Dissolve 50 mg of *o*-nitroaniline in benzene to make 50.0 mL. Dilute 10 mL of the resulting solution with benzene to 100.0 mL, and mix (*Solution A*).

Weigh quickly about 2 (±0.005) g of specimen in a glass-stoppered weighing bottle, and rapidly transfer it to a dry, glass-stoppered test tube. Add 20.0 mL of *Solution A*, insert the stopper, shake vigorously for 3 minutes, and allow to settle.

Pipet 10 mL of the clear, supernatant liquid into a 100-mL volumetric flask, dilute with benzene to volume, and mix (*Solution B*).

Determine the absorbances of *Solutions A* and *B* at 395 nm, with a suitable spectrophotometer, using benzene as the blank. Calculate the quantity, in mg, adsorbed per g of test specimen by the formula:

$$[2(1 - A_B/A_A)]/W,$$

in which A_A and A_B are the absorbances of *Solutions A* and *B*, respectively, and W is the weight, in g, of the aluminum oxide. Not less than 0.3 mg of *o*-nitroaniline is adsorbed for each g of the aluminum oxide.

Aluminum Potassium Sulfate, $AlK(SO_4)_2.12H_2O$—**474.38**—Use ACS reagent grade.

Aminoacetic Acid (*Glycine*), NH_2CH_2COOH—**75.07**—White, crystalline powder. Very soluble in water; slightly soluble in alcohol.

Nitrogen content (Reagent test)—Determine by the Kjeldahl method, using a test specimen previously dried at 105° for 2 hours: between 18.4% and 18.8% of N is found, corresponding to not less than 98.5% of $C_2H_5NO_2$.

Insoluble matter (Reagent test): not more than 1 mg, from 10 g (0.01%).

Residue on ignition (Reagent test): not more than 0.05%.

Chloride (Reagent test)—One g shows not more than 0.1 mg of Cl (0.01%).

Sulfate (Reagent test, *Method I*)—Two g shows not more than 0.1 mg of SO_4 (0.005%).

Heavy metals (Reagent test): 0.001%, 5 mL of 1 *N* hydrochloric acid being used to acidify the solution of the test specimen.

Iron ⟨241⟩—One g, dissolved in 47 mL of water containing 3 mL of hydrochloric acid, shows not more than 0.01 mg of Fe (0.001%).

***p*-Aminoacetophenone,** $H_2N.C_6H_4.CO.CH_3$—**135.17**—Pale yellow crystals or amorphous powder, having a characteristic odor. Slightly soluble in cold water; soluble in hot water, in alcohol, and in dilute mineral acids.

Solubility in alcohol—A solution of 1 g in 20 mL is completely clear.

Solubility in diluted hydrochloric acid—A cooled solution prepared by dissolving 500 mg in 10 mL of diluted hydrochloric acid, with the aid of gentle heat, is completely clear.

Melting range ⟨741⟩: between 104° and 106°.

Residue on ignition (Reagent test): not more than 1.0 mg, from 1 g ignited with 0.5 mL of sulfuric acid (0.10%).

4-Aminoantipyrine, $C_{11}H_{13}N_3O$—**203.24**—Light yellow crystalline powder. A 500-mg portion dissolves completely in 30 mL of water and yields a clear solution.

Melting range ⟨741⟩: between 108° and 110°.

4-Aminoantipyrine Hydrochloride, $C_{11}H_{13}N_3O.HCl$—**239.70**—Pale pink to light orange, crystalline powder. Soluble in water.

Assay—Dissolve about 500 mg, accurately weighed, in 50 mL of water. If necessary, neutralize the solution with 0.1 *N* sodium hydroxide, using litmus paper as the indicator. Add 4 drops of dichlorofluorescein TS, and titrate with 0.1 *N* silver nitrate VS. Each mL of 0.1 *N* silver nitrate is equivalent to 23.97 mg of $C_{11}H_{13}N_3O.HCl$. Between 100.6% and 108.5% is found.

Solubility—A solution of 1 g in 25 mL of water is clear or shows not more than a slight cloudiness.

Melting range ⟨741⟩: between 232° and 238°, with decomposition.

***p*-Aminobenzoic Acid**—See *Para-aminobenzoic Acid.*

4-Amino-6-chloro-1,3-benzenedisulfonamide, $C_6H_8ClN_3O_4S_2$—**285.72**—White, odorless powder. Insoluble in water and in chloroform; soluble in ammonia TS.

Residue on ignition (Reagent test): not more than 2 mg from 2 g (0.1%).

Absorbance—A 1 in 200,000 solution in methanol exhibits absorbance maxima at about 223 nm, 265 nm, and 312 nm. Its absorptivity (see *Spectrophotometry and Light-scattering* ⟨851⟩) at 265 nm is about 64.0.

4-Amino-2-chlorobenzoic Acid, $C_6H_3Cl(NH_2)(COOH)$—**171.58**—White crystals or white, crystalline powder.

Melting range ⟨741⟩: between 208° and 212°.

2-Amino-5-chlorobenzophenone, $C_{13}H_{10}ClNO$—**231.68**—Use USP 2-Amino-5-chlorobenzophenone RS.

1-(2-Aminoethyl)piperazine, $C_6H_{15}N_3$—**129.21**—Viscous, colorless liquid.

Assay—Inject an appropriate specimen into a suitable gas chromatograph (see *Chromatography* ⟨621⟩) equipped with a flame-ionization detector, helium being used as the carrier gas. The following conditions have been found suitable: a 30-m × 0.25-mm capillary column coated with G2. The injection port temperature is maintained at 280°; the column temperature is maintained at 180° and programmed to rise 10° per minute to 280° and held there for 10 minutes. The detector temperature is maintained at 300°. The area of the main peak is not less than 97% of the total peak area.

Refractive index ⟨831⟩: between 1.4978 and 1.5010 at 20°.

***N*-Aminohexamethyleneimine** (*N-Aminohomopiperidine, 1-Aminohomopiperidine*), $C_6H_{14}N_2$—**114.19**—Colorless liquid.

Assay—Inject an appropriate specimen into a suitable gas chromatograph (see *Chromatography* ⟨621⟩) equipped with a flame-ionization detector, helium being used as the carrier gas. The following conditions have been found suitable: a 30-m × 0.25-mm capillary column coated with G2. The injection port temperature is maintained at 180°; the column temperature is maintained at 80° and programmed to rise 10° per minute to 230° and held there for 5 minutes. The detector temperature is maintained at 300°. The area of the main peak is not less than 95% of the total peak area.

Refractive index ⟨831⟩: between 1.4840 and 1.4860 at 20°.

1,2,4-Aminonaphtholsulfonic Acid, $C_{10}H_9NO_4S$—**239.25**—White to slightly brownish pink powder. Sparingly soluble in water.

Sensitiveness—Dissolve 100 mg in 50 mL of freshly prepared sodium bisulfite solution (1 in 5), warming if necessary to effect solution, and filter. Add 1 mL of the filtrate to a solution prepared by adding 2 mL of dilute sulfuric acid (1 in 6) and 1 mL of *Phosphate Reagent A* (see Reagent test) to 20 mL of a 1 in 100 dilution of *Standard Phosphate Solution* (see Reagent test): a distinct blue color develops within 5 minutes.

Solubility in sodium carbonate solution—Dissolve 100 mg in 3 mL of sodium carbonate TS, and add 17 mL of water: not more than a trace remains undissolved.

Residue on ignition (Reagent test)—To 1 g add 0.5 mL of sulfuric acid, and ignite at 800 ± 25° to constant weight: the residue weighs not more than 5 mg (0.5%).

Sulfate (Reagent test, *Method I*)—Heat 500 mg with a mixture of 25 mL of water and 2 drops of hydrochloric acid on a steam bath for 10 minutes. Cool, dilute with water to 200 mL, and filter: 20 mL of the filtrate shows not more than 0.25 mg of SO_4 (0.5%).

***m*-Aminophenol,** C_6H_7NO—**109.13**—Cream-colored to pale yellow flakes. Sparingly soluble in cold water; freely soluble in hot water, in alcohol, and in ether.

Assay—Dissolve about 1.5 g, accurately weighed, in about 400 mL of water in a 500-mL volumetric flask, dilute with water to volume, and mix. Transfer 25.0 mL of this solution to an iodine flask, add 50.0 mL of 0.1 *N* bromine VS, dilute with 50 mL of water, add 5 mL of hydrochloric acid, and immediately insert the stopper in the flask. Shake for 1 minute, allow to stand for 2 minutes, and add 5 mL of potassium iodide TS through the slightly loosened stopper. Shake thoroughly, allow to stand for 5 minutes, remove the stopper, and rinse it and the neck of the flask with 20 mL of water, adding the rinsing to the flask. Titrate the liberated iodine with 0.1 *N* sodium thiosulfate VS, adding 3 mL of starch TS as the end-point is approached. From the volume of 0.1 *N* sodium thiosulfate used, calculate the volume, in mL, of 0.1 *N* bromine consumed by the test specimen. Each mL of 0.1 *N* bromine is equivalent to 1.819 mg of C_6H_7NO: not less than 99.5% is found.

Melting range ⟨741⟩: between 121° and 123°.

Loss on drying ⟨731⟩—Dry it over calcium chloride for 4 hours: the loss in weight is negligible.

Residue on ignition (Reagent test): negligible, from 2 g.

***p*-Aminophenol,** C_6H_7NO—**109.13**—Fine, yellowish, crystalline powder. Slightly soluble in water and in alcohol.

Melting range ⟨741⟩: between 187° and 189°.

Aminosalicylic Acid—Use *Aminosalicyclic Acid* (USP monograph).

3-Amino-2,4,6-triiodobenzoic Acid—Use USP 3-Amino-2,4,6-triiodobenzoic Acid RS.

5-Amino-2,4,6-triiodo-*N*-methylisophthalamic Acid—Use USP 5-Amino-2,4,6-triiodo-*N*-methylisophthalamic Acid RS.

Ammonia Detector Tube—A fuse-sealed glass tube so designed that gas may be passed through it, and containing suitable absorbing filters and support media for the indicator bromophenol blue.

NOTE—A suitable detector tube that conforms to the monograph specification is available from National Draeger, Inc., 401 Parkway View Drive, Pittsburgh, Pa. 15205, as Reference Number CH 20501, Measuring Range 5 to 70 ppm. Tubes having conditions other than those specified in the monograph may be used in accordance with the section entitled *Tests and Assays* in the *General Notices*.

Ammonia Solution, Diluted—Use *Ammonia TS*.

Ammonia Water, Stronger (*Ammonium Hydroxide*)—Use ACS reagent grade Ammonium Hydroxide.

Ammonium Acetate, $NH_4C_2H_3O_2$—**77.08**—Use ACS reagent grade.

Ammonium Bromide, NH_4Br—**91.94**—Use ACS reagent grade.

Ammonium Carbonate—Use ACS reagent grade.

Ammonium Chloride, NH_4Cl—**53.49**—Use ACS reagent grade.

Ammonium Citrate, Dibasic, $(NH_4)_2HC_6H_5O_7$—**226.19**—Use ACS reagent grade.

Ammonium Dihydrogen Phosphate—See *Ammonium Phosphate, Monobasic*.

Ammonium Fluoride, NH_4F—**37.04**—Use ACS reagent grade.

Ammonium Hydroxide—Use ACS reagent grade.

Ammonium Molybdate, $(NH_4)_6Mo_7O_{24}.4H_2O$—**1235.86**—Use ACS reagent grade.

Ammonium Nitrate, NH_4NO_3—**80.04**—Use ACS reagent grade.

Ammonium Oxalate, $(NH_4)_2C_2O_4.H_2O$—**142.11**—Use ACS reagent grade.

Ammonium Persulfate, $(NH_4)_2S_2O_8$—**228.19**—Use ACS reagent grade Ammonium Peroxydisulfate.

Ammonium Phosphate, Dibasic (*Diammonium Hydrogen Phosphate*), $(NH_4)_2HPO_4$—**132.06**—Use ACS reagent grade.

Ammonium Phosphate, Monobasic (*Ammonium Dihydrogen Phosphate*), $NH_4H_2PO_4$—**115.03**—Use ACS reagent grade.

Ammonium Reineckate (*Reinecke Salt*), $NH_4[Cr(NH_3)_2(SCN)_4].H_2O$—**354.42**—Dark red crystals or red, crystalline powder. Moderately soluble in cold water; more soluble in hot water. Gradually decomposes in solution.

Sensitiveness—Dissolve 50 mg in 10 mL of water. Add 0.2 mL of the solution to 1 mL of a solution of 10 mg of choline chloride in 20 mL of water, and shake gently: a distinct precipitate forms within 5 to 10 seconds.

Ammonium Sulfamate, $NH_4OSO_2NH_2$—**114.12**—Use ACS reagent grade.

Ammonium Sulfate, $(NH_4)_2SO_4$—**132.13**—Use ACS reagent grade.

Ammonium Sulfide Solution—Yellow liquid having the odors of ammonia and hydrogen sulfide. Darkens to red on standing. Its sulfide sulfur content is usually equivalent to between 16% and 20% expressed as $(NH_4)_2S$.

For the *Assay* and the test for *Chloride*, prepare a *test solution* by diluting 10 mL with oxygen-free water to 100 mL, quantitatively.

Assay—Pipet 50 mL of 0.1 *N* silver nitrate VS into a 100-mL volumetric flask, and add 7 mL of ammonium hydroxide. Then add 5.0 mL of *test solution*, in small portions, with swirling. Dilute with water to volume, mix, and filter through a dry filter, rejecting the first 10 mL of the filtrate. To 50.0 mL of the retained filtrate add 7 mL of nitric acid and 2 mL of ferric ammonium sulfate TS, and titrate the excess silver nitrate with 0.1 *N* ammonium thiocyanate VS. Each mL of 0.1 *N* silver nitrate is equivalent to 3.407 mg of $(NH_4)_2S$.

Chloride—To 5 mL of *test solution* add 20 mL of water, 5 mL of ammonium hydroxide, and a solution containing 1 g of silver nitrate in 10 mL of water. Filter, and acidify 20 mL of the filtrate with nitric acid: not more than a slight opalescence develops.

Residue on ignition—Evaporate 5 mL to dryness, and ignite the residue at 800 ± 25° for 15 minutes: not more than 1 mg remains.

Arsenic (Reagent test)—Evaporate 1 mL on a steam bath to dryness, and add 1 mL of nitric acid and 2 mL of sulfuric acid to the residue. Evaporate to the appearance of strong fumes, but not to dryness, cool, and dilute with water to 5 mL: the stain produced by this solution does not exceed that produced by 0.0015 mg of As (7 ppm as As).

Carbonate—To 10 mL add 4 mL of calcium chloride TS, and warm the resulting mixture: no precipitate forms.

Ammonium Thiocyanate, NH_4SCN—**76.12**—Use ACS reagent grade.

Ammonium Vanadate (*Ammonium Metavanadate*), NH_4VO_3—**116.98**—White, crystalline powder. Slightly soluble in cold water; soluble in hot water and in dilute ammonia TS.

Assay—Weigh accurately about 500 mg, transfer to a suitable container, add 30 mL of water and 2 mL of dilute sulfuric acid (1 in 4), swirl to dissolve, and pass sulfur dioxide gas through the solution until reduction is complete and the solution is bright blue in color. Boil gently while passing a stream of carbon dioxide through the solution to remove any excess sulfur dioxide, then cool, and titrate with 0.1 *N* potassium permanganate VS. Each mL of 0.1 *N* potassium permanganate consumed is equivalent to 11.7 mg of NH_4VO_3. Not less than 98.0% is found.

Solubility in ammonium hydroxide—Dissolve 1 g in a mixture of 3 mL of ammonium hydroxide and 50 mL of warm water: the solution is clear and colorless.

Carbonate—To 500 mg add 1 mL of water and 2 mL of diluted hydrochloric acid: no effervescence is produced.

Chloride—Dissolve 250 mg in 40 mL of hot water, add 2 mL of nitric acid, and allow to stand for 1 hour. Filter, and to the filtrate add 0.5 mL of silver nitrate TS: any turbidity produced does not exceed that of a blank containing 0.5 mg of added Cl (0.2%).

Sulfate—Dissolve 500 mg in 50 mL of hot water, and add 2 mL of diluted hydrochloric acid and 1.5 g of hydroxylamine hydrochloride. Heat at 60° for 3 minutes, filter, cool, and add to the filtrate 2 mL of barium chloride TS: no turbidity or precipitate is produced within 30 minutes.

Amyl Acetate (*Isoamyl Acetate*), $CH_3CO_2C_5H_{11}$—**130.19**—Clear, colorless liquid with a characteristic, banana oil–like odor. Slightly soluble in water. Miscible with alcohol, with amyl alcohol, with benzene, and with ether. Specific gravity: about 0.87.

Boiling range (Reagent test, *Method I*): not less than 90%, between 137° and 142°.

Solubility in diluted alcohol—A 1.0-mL portion dissolves in 20 mL of diluted alcohol to form a clear solution.

Acidity—Add 5.0 mL to 40 mL of neutralized alcohol, and, if the pink color is discharged, titrate with 0.10 *N* sodium hydroxide: not more than 0.20 mL is required to restore the pink color (about 0.02% as CH_3COOH).

Water—A 5-mL portion gives a clear solution with 5 mL of carbon disulfide.

Amyl Alcohol (*Isoamyl Alcohol*), $C_5H_{11}OH$—**88.15**—Use ACS reagent grade Isopentyl Alcohol.

***tert*-Amyl Alcohol,** $C_5H_{12}O$—**88.15**—Clear, colorless, flammable, volatile liquid with a characteristic odor. Specific gravity: about 0.81.

Boiling range (Reagent test): not less than 95%, between 100° and 103°.

Residue on evaporation—Evaporate 50 mL (40 g) on a steam bath, and dry at 105° for 1 hour: the residue weighs not more than 1.6 mg (0.004%).

Acids and esters—Dilute 20 mL with 20 mL of alcohol, add 5.0 mL of 0.1 *N* sodium hydroxide VS, and reflux gently for 10 minutes. Cool, add 2 drops of phenolphthalein TS, and titrate the excess sodium hydroxide with 0.1 *N* hydrochloric acid VS: not more than 0.75 mL of the 0.10 *N* sodium hydroxide is consumed, correction being made for the amount consumed in a blank (0.06% as amyl acetate).

Aldehydes—Shake 5 mL with 5 mL of potassium hydroxide solution (30 in 100) in a glass-stoppered cylinder for 5 minutes, and allow to separate: no color develops in either layer.

Amylose—Use a suitable grade.

Anhydrous Alumina—See *Alumina, Anhydrous.*

Anhydrous Barium Chloride—See *Barium Chloride, Anhydrous.*

Anhydrous Calcium Chloride—See *Calcium Chloride, Anhydrous.*

Anhydrous Calcium Sulfate—See *Calcium Sulfate, Anhydrous.*

Anhydrous Cupric Sulfate—See *Cupric Sulfate, Anhydrous.*

Anhydrous Dibasic Sodium Phosphate—See *Sodium Phosphate, Dibasic, Anhydrous.*

Anhydrous Magnesium Perchlorate—See *Magnesium Perchlorate, Anhydrous.*

Anhydrous Magnesium Sulfate—See *Magnesium Sulfate, Anhydrous.*

Anhydrous Methanol—See *Methanol, Anhydrous.*

Anhydrous Potassium Carbonate—See *Potassium Carbonate, Anhydrous.*

Anhydrous Sodium Acetate—See *Sodium Acetate, Anhydrous.*

Anhydrous Sodium Carbonate—See *Sodium Carbonate, Anhydrous.*

Anhydrous Sodium Sulfate—See *Sodium Sulfate, Anhydrous.*

Anhydrous Sodium Sulfite—See *Sodium Sulfite, Anhydrous.*

Aniline, $C_6H_5NH_2$—**93.13**—Use ACS reagent grade.

Aniline Blue (*Certified Biological Aniline Blue*)—A water-soluble dye consisting of a mixture of the tri-sulfonates of triphenylpararosaniline and of diphenylrosaniline.

Anion-exchange Resin, Chloromethylated Polystyrene-Divinylbenzene—Strongly basic, cross-linked resin containing quaternary ammonium groups. It consists of small, moist, yellow beads having a characteristic amine odor. It is available in the chloride form which can be converted to the hydroxide form by regeneration with sodium hydroxide solution (1 in 4). For satisfactory regeneration a contact time of about 25 minutes is required, after which it must be washed with water until neutral. Suitable for use in column chromatography.[3]

Anion-exchange Resin, Polystyrene (*Polystyrene Quaternary Ammonium Anion-exchange Resin*). Moist, resinous particles containing quaternary ammonium groups that can exchange anions. The particles are insoluble and infusible, and have a faint, amine-like odor. Before use, rinse the particles with sodium hydroxide solution (1 in 20), using a ratio of 10 mL of the sodium hydroxide solution for each milliliter volume of resin, then wash with water until the pH of the last washing is not greater than 10.

Loss on drying ⟨731⟩—Dry it at 110° for 8 hours: it loses not less than 40.0% and not more than 70.0% of its weight.

Fineness—All of the particles pass through a No. 16 mesh sieve (see *Powder Fineness* ⟨811⟩).

Anion-exchange Resin, Strong, Lightly Cross-Linked, in the Chloride Form—Use a suitable grade.[4]

Anion-exchange Resin, 50- to 100-Mesh, Styrene-Divinylbenzene—Strongly basic, cross-linked resin containing quaternary ammonium groups and about 4% of divinylbenzene. It consists of tan-colored beads that may be relatively free flowing. It is available in the chloride form which can be converted to the hydroxide form by regeneration with a sodium hydroxide solution (5 in 100). For satisfactory regeneration a contact time of at least 30 minutes is required after which it must be washed free of excess alkali. Insoluble in water, in methanol, and in acetonitrile. Suitable for use in column chromatography.[5]

Moisture content of fully regenerated and expanded resin—Transfer 10 to 12 mL of the resin (as received) to a flask, and convert it completely to the chloride form by stirring with 150 mL of hydrochloric acid (5 in 100) for not less than 30 minutes. Decant the acid, and wash the resin in the same manner with distilled water until the wash water is neutral to litmus.

Transfer 5 to 7 mL of the regenerated resin to a glass filtering crucible, and remove only the excess surface water by very careful suction filtration. Transfer the conditioned, dried resin to a tared weighing bottle, and weigh. Dry in a vacuum oven at 100° to 105° and at a pressure of 50 mm of mercury for 16 hours. Transfer from the vacuum oven to a desiccator, and cool to room temperature. Reweigh. The loss in weight is between 50% and 65%.

Total new volume capacity—Transfer 2.5 to 3 mL of the conditioned, undried (see *Moisture content,* above) resin to a 5-mL graduated cylinder, and fill it with water. Remove any air bubbles from the resin bed with a stainless steel wire, and settle the resin to its minimum volume by tapping the graduated cylinder. Record the volume of the resin.

Transfer the resin with 100 mL of water to a 250-mL flask. Add 2 mL of sulfuric acid, heat to 70° to 80°, and hold at that temperature for 5 minutes with occasional stirring (do not boil). Cool to room temperature, and add 2.5 mL of nitric acid (1 in 2), 2 mL of ferric ammonium sulfate TS, and 0.20 mL of 0.1 *N* ammonium thiocyanate. Titrate with 0.1 *N* silver nitrate VS until the solution turns colorless, and add a measured excess (1 to 5 mL). Heat to boiling to coagulate the silver chloride precipitate. Cool to room temperature, add 10 mL of nitrobenzene, shake vigorously, and titrate the excess silver nitrate with 0.1 *N* ammonium thiocyanate VS.

$$\frac{\text{net mL } AgNO_3 \times N}{\text{mL of resin}} = \text{mEq/mL}$$

The total exchange capacity of the regenerated, wet resin is more than 1.0 mEq per mL.

Wet screen analysis—The purpose of this test is to identify properly the mesh size of the resin. To obtain an accurate screen analysis requires special apparatus and technique.

Add 150 mL of resin to 200 mL of distilled water in an appropriate bottle, and allow it to stand at least 4 hours to completely swell the resin.

Transfer by means of a graduated cylinder 100 mL of settled and completely swollen resin to the top screen of a series (20-, 50-, 100-mesh) of 20.3-cm brass screens. Thoroughly wash the resin on each screen with a stream of distilled water until the resin is completely classified, collecting the wash water in a suitable container. Wash the beads remaining on the respective screens back into the 100-mL cylinder, and record the volume of settled resin on each screen: not less than 80% of the resin is between 50- and 100-mesh.

Anisole, $CH_3OC_6H_5$—**108.14**—Colorless liquid.

Assay—Inject an appropriate specimen (about 0.5 µL) into a suitable gas chromatograph (see *Chromatography* ⟨621⟩) equipped with a flame-ionization detector, nitrogen being used as the carrier gas. The following conditions have been found suitable: a 30-m capillary column is coated with phase G3; the injection port and detector are maintained at 140° and 300°, respectively; the column temperature is maintained at 70° and programmed to rise 10° per minute to 170°. The area of the anisole peak is not less than 99% of the total peak area.

Refractive index ⟨831⟩: 1.5160 at 20°.

Anthracene, $C_{14}H_{10}$—**178.23**—White to off-white crystals or platelets. Darkens in sunlight. Insoluble in water; sparingly soluble in alcohol, in benzene, and in chloroform.

Melting range ⟨741⟩: between 215° and 218°.

Anthrone, $C_{14}H_{10}O$—**194.23**—Pale yellow crystals. Insoluble in water; slightly soluble in alcohol and in diluted sulfuric acid.

Solubility in carbon tetrachloride—Add 500 mg to 10 mL of carbon tetrachloride: it dissolves to give a clear, nonfluorescent solution.

Anthranol—Add 100 mg to 10 mL of warm alcohol: a nonfluorescent solution results.

Melting range ⟨741⟩: between 154° and 156°.

Antimony Pentachloride, $SbCl_5$—**299.01**—Clear, reddish yellow, oily, hygroscopic, caustic liquid. Fumes in moist air and solidifies by absorption of one molecule of water. Is decomposed by water, soluble in dilute hydrochloric acid and in chloroform. Boils at about 92° at a pressure of 30 mm of mercury and has a specific gravity of about 2.34 at 25°.

Caution—Antimony pentachloride causes severe burns, and the vapor is hazardous.

Assay (*$SbCl_5$*)—Weigh accurately a glass-stoppered, 125-mL flask, quickly introduce about 0.3 mL of the test specimen, and reweigh. Dissolve with 20 mL of diluted hydrochloric acid (1 in 5), and add 10 mL of potassium iodide solution (1 in 10) and 1 mL of carbon disulfide. Titrate the liberated iodine with 0.1 *N* sodium thiosulfate VS. The brown color will gradually disappear from the solution, and the last traces of free iodine will be collected in the carbon disulfide, giving a pink color. When this pink color disappears the end-point has been reached. Each mL of 0.1 *N* sodium thiosulfate is equivalent to 14.95 mg of $SbCl_5$: not less than 99.0% of $SbCl_5$ is found.

Sulfate (Reagent test, *Method II*)—Dissolve 4.3 mL (10 g) in the minimum volume of hydrochloric acid, dilute with water to 150 mL, neutralize with ammonium hydroxide, and filter. To the filtrate add 2 mL of hydrochloric acid: the solution, 10 mL of barium chloride TS being used, yields not more than 1.3 mg of residue, correction being made for a complete blank test (0.005%).

Arsenic—Add 10 mL of a recently prepared solution of 20 g of stannous chloride in 30 mL of hydrochloric acid to 100 mg of specimen dissolved in 5 mL of hydrochloric acid. Mix, transfer to a color-comparison tube, and allow to stand for 30 minutes. Any color in the solution of the specimen should not be darker than that in a control containing 0.02 mg of arsenic (As), which has been treated in the same manner as the test specimen, when viewed downward over a white surface (0.02% of As).

Iron ⟨241⟩—To the residue from the test for *Substances not precipitated by hydrogen sulfide* add 2 mL of hydrochloric acid and 5 drops of nitric acid, and evaporate on a steam bath to dryness. Take up the residue in 2 mL of hydrochloric acid, and dilute with water to 47 mL: the solution shows not more than 0.01 mg of Fe (0.001%).

Other heavy metals (*as Pb*)—Dissolve the precipitate on the filter paper, from the test for *Substances not precipitated by hydrogen sulfide*, with 75 mL of a solution containing 6 g of sodium sulfide and 4 g of sodium hydroxide dissolved in water and diluted with water to 100 mL. Collect the filtrate in the original flask containing the remainder of the sulfide precipitate. Warm the solution to dissolve the soluble sulfides, and allow the insoluble sulfides to settle. Filter, wash thoroughly with hydrogen sulfide TS, and dissolve any precipitate remaining on the filter paper with 10 mL of hot diluted hydrochloric acid. Dilute the filtrate with water to 50 mL. Neutralize a 25-mL portion of this solution with 1 *N* sodium hydroxide, and add 1 mL of 1 *N* acetic acid and 10 mL of hydrogen sulfide TS. Any brown color should not exceed that produced by 0.05 mg of lead ion in an equal volume of solution containing 1 mL of 1 *N* acetic acid and 10 mL of hydrogen sulfide TS (0.005%).

Substances not precipitated by hydrogen sulfide (*as SO_4*)—Dissolve 0.90 mL (2 g) in 5 mL of hydrochloric acid, and dilute with 95 mL of water. Precipitate the antimony completely with hydrogen sulfide, allow the precipitate to settle, and filter, being careful not to transfer much of the precipitate to the filter paper. (Retain the precipitate.) To 50 mL of the filtrate, add 0.5 mL of sulfuric acid, evaporate in a tared porcelain crucible to dryness, and ignite at 800 ± 25° for 15 minutes. (Retain the residue.) The weight of the ignited residue should not be more than 0.0010 g greater than the weight obtained in a complete blank test (0.10%).

Antimony Potassium Tartrate—Use *Antimony Potassium Tartrate* (USP monograph).

Antimony Trichloride (*Antimonous Chloride*), $SbCl_3$—**228.11**—Use ACS reagent grade.

Antithrombin-III for Anti-Factor X_a test—Antithrombin-III is the serine protease inhibitor derived from bovine plasma which inhibits the enzyme Factor X_a and other blood coagulation factors. It is a glycoprotein having a molecular weight of 58,000. By gel electrophoresis (see *Electrophoresis* ⟨726⟩), the principal protein of interest constitutes not less than 90% of the total protein zones. The specific activity of Antithrombin-III for Anti-Factor X_a test corresponds to not less than 4 USP Antithrombin-III Units[6] per mg of protein in the presence of heparin, the test being made on a solution containing 0.25 mg of protein equivalent and 0.1 USP Heparin Unit per mL. Antithrombin-III for Anti-Factor X_a test contains no detectable heparin when a test is performed with the use of a solution containing 1 USP Antithrombin-III Unit per mL with 1 µL of *Toluidine Blue Solution* (see under *Toluidine Blue*): in the presence of heparin the color changes from blue to purple.

Arginine, $C_6H_{14}N_4O_2$—**174.20**—Use *Arginine* (USP monograph).

Arsanilic Acid, $C_6H_8AsNO_3$—**217.06**—White, odorless, crystalline powder. Soluble in water, in dilute hydrochloric acid, and in sodium hydroxide; practically insoluble in organic solvents.

Loss on drying ⟨731⟩—Dry it at 105° for 16 hours: it loses not more than 2.0% of its weight.

Assay—

HYDROCHLORIC ACID—Boil about 100 mL of hydrochloric acid for 2 minutes, accurately timed, to remove free chlorine, and cool in a glass-stoppered flask under a stream of running water.

SODIUM THIOSULFATE SOLUTION—Dissolve 26 g of sodium thiosulfate and 200 mg of sodium carbonate in 1000 mL of carbon dioxide–free water, and standardize as follows:

Weigh accurately about 150 mg of potassium iodate into a conical flask, dissolve in 25 mL of carbon dioxide–free water, and add 10 mL of the *Hydrochloric acid* and 10 mL of potassium iodide solution (2 in 10). After 2 minutes, titrate the liberated iodine with the sodium thiosulfate solution, adding 3 mL of starch TS as the end-point is approached. Calculate the normality by the formula:

$$28.03(W/V),$$

in which *W* is the weight, in g, of potassium iodate, and *V* is the volume, in mL, of sodium thiosulfate solution.

PROCEDURE—Accurately weigh about 300 mg, and transfer to a dry Kjeldahl flask containing a few glass beads. Wash any particles adhering to the walls of the flask into the bulb with 10 mL of dilute sulfuric acid (2 in 10), add 3 mL of nitric acid, and digest over a small flame for about 15 minutes. Cool, add 10 mL of dilute sulfuric acid (2 in 10) and 3 mL of nitric acid, and again digest over a small flame. Continue the digestion until the solution is clear, adding additional nitric acid as necessary. Cool, add, dropwise and with swirling, 2.5 mL of 30 percent hydrogen peroxide, and digest for 5 minutes over a small flame. Cool, add 10 mL of water, and evaporate to fumes. Cool, *cautiously* add 10 mL of water, and again evaporate to heavy, white fumes. Cool, transfer the solution to a glass-stoppered flask, and rinse the digestion flask with several 5-mL portions of the *Hydrochloric acid*. Add 10 mL of potassium iodide TS, insert the stopper in the flask, and allow to stand for 10 minutes in a dark place. Dilute with water to 150 mL, and titrate the liberated iodine with the sodium thiosulfate solution, adding 3 mL of starch TS as the end-point is approached, to a faint pink end-point. Perform a complete blank determination, and make any necessary correction. Calculate the percentage of $C_6H_8AsNO_3$ by the formula:

$$10850(AB/W),$$

in which *A* is the normality of the sodium thiosulfate solution, *B* is the volume, in mL, of sodium thiosulfate solution consumed, and *W* is the weight, in mg, of test specimen: it contains between 98.5% and 100.8% of $C_6H_8AsNO_3$, on the anhydrous basis.

Arsenic Trioxide, As_2O_3—**197.84**—Use ACS reagent grade.

NOTE—Arsenic trioxide of a quality suitable as a primary standard is available from the National Institute of Standards and Technology, Office of Standard Reference Materials, Washington, DC 20234, as standard sample No. 83b.

Asbestos—The silky, well-matted variety of long-fiber, amphibole asbestos. Digest with diluted hydrochloric acid for 24 hours, and wash with water until free from acid. Then digest with sodium hydroxide TS for 24 hours. Wash until free from alkali, digest with diluted nitric acid for 3 hours, finally wash with water until free from acid, and shake with water to a fine pulp.

Ascorbic Acid—Use *Ascorbic Acid* (USP monograph).

L-Asparagine (*L-2-Aminosuccinamic Acid*), COOHCH(NH_2)$CH_2CONH_2.H_2O$—**150.13**—Colorless, odorless crystals. One g dissolves in 50 mL of water; soluble in acids and in alkalies; insoluble in alcohol and in ether. Its neutral or alkaline solutions are levorotatory; its acid solutions are dextrorotatory.

Specific rotation ⟨781⟩: between +31° and +33°, determined in a solution in diluted hydrochloric acid containing the equivalent of 5 g (on the anhydrous basis, as determined by drying at 105° for 5 hours) in each 100 mL.

Residue on ignition (Reagent test): not more than 0.1%.

Chloride (Reagent test)—One g shows not more than 0.03 mg of Cl (0.003%).

Sulfate (Reagent test, *Method I*)—One g shows not more than 0.05 mg of SO_4 (0.005%).

Heavy metals (Reagent test): 0.002%.

Nitrogen content, Method II ⟨461⟩: between 18.4% and 18.8% of N is found.

Atropine Sulfate—Use *Atropine Sulfate* (USP monograph).

Azobenzene, $C_{12}H_{10}N_2$—**182.22**—Orange, crystalline solid.

Assay—[NOTE—Use low-actinic glassware and spectrophotometric grade methanol.] Transfer about 30 mg of azobenzene, accurately weighed, to a 100-mL volumetric flask. Dilute with methanol to volume, and mix. Dilute 2.0 mL of this solution with methanol to 100.0 mL, and mix. Determine the absorbance of this solution at 316 nm in a 1-cm cell, with a suitable spectrophotometer, using methanol as the blank. Calculate the absorptivity, *a*, of the solution. Calculate the percentage purity by the formula:

$$a/124.0.$$

Not less than 98% is found.

Solubility—500 mg dissolved in 50 mL of hexane yields a clear and complete solution.

Azolitmin—A water-soluble coloring matter obtained from litmus. Occurs in dark violet scales. Soluble with difficulty in water, more easily soluble in hot water; very soluble in dilute alkalies, forming deep blue solutions; insoluble in alcohol and in dilute acid.

Dissolve 1 g of azolitmin in 80 mL of hot water, and add 20 mL of alcohol. Add 0.10 mL of this solution to 50 mL of carbon dioxide–free water. The bluish red color of the liquid is changed to red by the addition of 0.05 mL of dilute hydrochloric acid (1 in 120), and this color is changed to bluish violet by the addition of 0.05 mL of sodium hydroxide solution (1 in 250).

Barium Bromide, $BaBr_2.2H_2O$—**333.17**—Colorless crystals. Readily soluble in water; soluble in methanol and in alcohol.

Insoluble matter—A 10-g portion shows not more than 1.0 mg of insoluble matter (0.01%).

Bromate—Dissolve 500 mg in 10 mL of oxygen-free water, add 2 drops of potassium iodide solution (1 in 10), 1 mL of starch TS, and 5 drops of dilute sulfuric acid (1 in 36), and allow to stand at 25°: no blue color is produced within 10 minutes.

Substances not precipitated by sulfuric acid—Dissolve 5 g in 150 mL of water, add 1 mL of hydrochloric acid, and heat to boiling. Add 25 mL of diluted sulfuric acid, cool, dilute with water to 250 mL, and allow to stand overnight. Decant through a filter, evaporate 50 mL of the filtrate to dryness, and ignite to constant weight: the residue weighs not more than 1.0 mg (0.1% as SO_4).

Heavy metals (Reagent test): 0.0005%.

Chloride—Dissolve about 500 mg, previously dried at 120° to constant weight and accurately weighed, in 50 mL of water. Add slowly and with stirring 40.0 mL of 0.1 *N* silver nitrate VS, then add 2 mL of nitric acid and 2 mL of ferric nitrate solution (1 in 10), and titrate the excess silver nitrate with 0.1 *N* ammonium thiocyanate VS: between 66.8 and 67.8 mL of 0.1 *N* silver nitrate is required for each g of dried specimen taken.

Barium Chloride, $BaCl_2.2H_2O$—**244.27**—Use ACS reagent grade.

Barium Chloride, Anhydrous, $BaCl_2$—**208.24**—This may be made by drying barium chloride in thin layers at 125° until the loss in weight between two successive, 3-hour drying periods does not exceed 1%.

Barium Chloride Dihydrate—Use *Barium Chloride.*

Barium Hydroxide, $Ba(OH)_2.8H_2O$—**315.47**—Use ACS reagent grade.

Barium Nitrate, $Ba(NO_3)_2$—**261.34**—Use ACS reagent grade.

Barium Oxide (*Barium Monoxide*), BaO—**153.33**—White to yellowish white lumps or powder. Absorbs moisture and carbon dioxide on exposure to air. Soluble in water and in dilute acids. Is toxic, as it contains a soluble form of barium. Store in tight containers.

Melting temperature ⟨741⟩: about 1920°.

Basic Fuchsin—See *Fuchsin, Basic.*

Beef Extract—A concentrate from beef broth obtained by extraction from fresh, sound, lean beef by means of cooking with water and evaporating the broth at a low temperature, usually in vacuum, until a thick, pasty residue results. Yellowish brown to dark brown, slightly acid, pasty mass having an agreeable meat-like odor and taste. Store it in tight, light-resistant containers.

For the following tests, prepare a *test solution* by dissolving 25 g in water to make 250 mL of a practically clear and practically sediment-free solution.

Assay for nitrogen content of alcohol-soluble substances—Place a portion of the alcohol filtrate and washings remaining from the test for *Alcohol-insoluble substances*, corresponding to 1 g of the alcohol-soluble solids, in a 500-mL Kjeldahl flask. Add about 10 g of powdered potassium sulfate and 20 mL of sulfuric acid. Heat the mixture at a low temperature until frothing ceases, then raise the temperature and boil until the mixture acquires a pale yellow color or becomes practically colorless. Cool the flask, add about 250 mL of water, and cautiously add sodium hydroxide solution (3 in 10) until the contents are alkaline, then add 5 mL more. Connect the flask at once by means of a spray trap to a condenser, the lower outlet tube of which dips beneath the surface of 50.0 mL of 0.1 *N* sulfuric acid VS contained in a receiving flask. Distil the mixture until about 100 mL of distillate has been collected in the acid. Add methyl red TS, and titrate the excess acid with 0.1 *N* sodium hydroxide VS. Each mL of 0.1 *N* sulfuric acid is equivalent to 1.401 mg of N. Not less than 60 mg of nitrogen is found.

Assay for nitrogen as ammonia—To 100 mL of *test solution*, contained in a 500-mL Kjeldahl flask, add 5 g of barium carbonate and 100 mL of water, and by means of a spray trap, connect the flask to a condenser, the lower outlet tube of which dips beneath the surface of 50.0 mL of 0.1 *N* sulfuric acid VS contained in a receiving flask. Distil the mixture until about 100 mL of distillate has been collected, add methyl red TS, and titrate the excess acid with 0.1 *N* sodium hydroxide VS. Each mL of 0.1 *N* sulfuric acid is equivalent to 1.703 mg of NH_3. The amount of ammonia found does not exceed 0.35% of the total solids in the portion of *test solution* taken.

Total solids—Distribute 10 mL of *test solution* over clean, dry sand or asbestos, tared in a porcelain dish, and dry at 105° for 16 hours: the residue weighs not less than 750 mg (75%).

Residue on ignition—Incinerate the residue obtained in the test for *Total solids* by heating the dish to a dull-red heat: the residue does not exceed 30% of the total solids.

Chlorides calculated as sodium chloride—Dissolve the ash obtained in the test for *Residue on ignition* in about 50 mL of water, and carefully transfer to a 100-mL volumetric flask. Add to the solution a few drops of nitric acid and 10.0 mL of 0.1 *N* silver nitrate VS. Add water to volume, and mix. Filter into a dry flask through a dry filter, rejecting the first 10 mL of the filtrate. To 50.0 mL of the subsequent filtrate add 1 mL of ferric ammonium sulfate TS, and titrate with 0.1 *N* ammonium thiocyanate VS. Each mL of 0.1 *N* silver nitrate is equivalent to 5.844 mg of NaCl. The weight of chlorides calculated as sodium chloride obtained, when multiplied by 2, does not exceed 6% of the total solids.

Alcohol-insoluble substances—Transfer 25 mL of *test solution* to a 100-mL conical flask, add 50 mL of alcohol, and shake thoroughly. Collect the precipitate on a counterpoised filter, wash it three times with a mixture of 2 volumes of alcohol and 1 volume of water, and dry at 105° for 2 hours: the weight of the precipitate, representing the alcohol-insoluble solids, does not exceed 10% of the total solids in the portion of *test solution* taken.

Nitrate—Boil 10 mL of *test solution* for 1 minute with 1.5 g of activated charcoal, add water to replace that lost by evaporation, filter, and add 1 drop of the filtrate to 3 drops of a 1 in 100 solution of diphenylamine in sulfuric acid: no blue color is produced.

Benzaldehyde, C_7H_6O—**106.12**—Colorless, strongly refractive liquid having an odor resembling that of bitter almond oil. Soluble in water; miscible with alcohol, with ether, and with fixed and volatile oils.

Assay—Pipet about 1 mL into a tared, glass-stoppered weighing bottle, and weigh accurately. Loosen the stopper, and transfer both the weighing bottle and its contents to a 250-mL conical flask containing 25 mL of a hydro-alcoholic solution of hydroxylamine hydrochloride (prepared by dissolving 34.7 g of hydroxylamine hydrochloride in 160 mL of water, then adding alcohol to make 1000 mL, and neutralizing to bromophenol blue by the addition of sodium hydroxide TS). Using a graduated cylinder to measure the volume, rinse the sides of the flask with an additional 50 mL of this reagent solution. Allow the solution to stand for 10 minutes, add 1 mL of bromophenol blue TS, and titrate the liberated hydrochloric acid with 1 *N* sodium hydroxide VS. Perform a blank determination with the same quantities of the same reagents, and make any necessary correction. Each mL of 1 *N* sodium hydroxide consumed is equivalent to 106.1 mg of C_7H_6O. Not less than 98% is found.

Specific gravity ⟨841⟩: between 1.041 and 1.046.

Refractive index ⟨831⟩: between 1.5440 and 1.5465 at 20°.

Hydrocyanic acid—Shake 0.5 mL with 5 mL of water, add 0.5 mL of sodium hydroxide TS and 0.1 mL of ferrous sulfate TS, and warm the mixture gently. Add a slight excess of hydrochloric acid: no greenish blue color or blue precipitate is observed within 15 minutes.

Benzalkonium Chloride—Use *Benzalkonium Chloride* (NF monograph).

Benzanilide, $C_{13}H_{11}NO$—**197.24**—Off-white, light gray to grayish green powder. Insoluble in water; sparingly soluble in alcohol; slightly soluble in ether.

Melting range ⟨741⟩: between 162° and 165°.

Solubility in acetone—A 1.0-g portion dissolves completely in 50 mL of acetone to yield a clear solution.

Benzene, C_6H_6—**78.11**—Use ACS reagent grade.

Benzenesulfonyl Chloride, $C_6H_5SO_2Cl$—**176.62**—Colorless, oily liquid. Insoluble in cold water; soluble in alcohol and in ether. Solidifies at 0°.

Melting range ⟨741⟩: between 14° and 17°.

Boiling range (Reagent test): between 251° and 252°.

Benzhydrol (*α-Phenylbenzenemethanol*), $C_{13}H_{12}O$—**184.24**—White to pale yellow crystals. Very slightly soluble in water; soluble in alcohol, in ether, and in chloroform.

Melting range ⟨741⟩: between 65° and 67°, but the range between beginning and end of melting does not exceed 2°.

Benzocaine—Use *Benzocaine* (USP monograph).

Benzoic Acid, C_6H_5COOH—**122.12**—Use ACS reagent grade. NOTE—Benzoic Acid of a quality suitable as a primary standard is available from the National Institute of Standards and Technology, Office of Standard Reference Materials, Washington, DC 20234, as standard sample No. 350.

Benzophenone, $(C_6H_5)_2CO$—**182.22**—White, crystalline powder.

Melting range ⟨741⟩: between 47° and 49°.

***p*-Benzoquinone,** $C_6H_4O_2$—**108.10**—Dark yellow powder having a green cast. Slightly soluble in water; soluble in alcohol, in ether, and in fixed alkali solutions. May darken on standing. Darkened material may be purified by sublimation in vacuum.

Melting range ⟨741⟩: between 113° and 115°.

Benzoyl Chloride, C_6H_5COCl—**140.57**—Use ACS reagent grade.

***N*-Benzoyl-DL-arginine-*p*-nitroanilide Hydrochloride,** $C_{19}H_{22}N_6O_4$—**434.89**—Pale yellow powder. Melts at about 260°, with decomposition.

***N*-Benzoyl-L-arginine Ethyl Ester Hydrochloride,** $C_{15}H_{22}N_4O_3 \cdot HCl$—**342.82**—Determine suitability for use as a substrate as directed under *Crystallized Trypsin* (USP monograph).

Benzphetamine Hydrochloride, $C_{17}H_{21}N \cdot HCl$—**275.82**—White to off-white, odorless, crystalline powder. Freely soluble in water, in alcohol, and in chloroform; slightly soluble in ether.

Assay—Dissolve about 500 mg, accurately weighed, in a mixture of 50 mL of glacial acetic acid and 10 mL of mercuric acetate TS, add 1 drop of crystal violet TS, and titrate with 0.1 *N* perchloric acid VS to a blue-green end-point. Perform a blank determination, and make any necessary correction. Each mL of 0.1 *N* perchloric acid is equivalent to 27.58 mg of $C_{17}H_{21}N \cdot HCl$. Between 98.0% and 101.0%, calculated on the dried basis, is found.

Melting range ⟨741⟩: between 152° and 158°.

Specific rotation ⟨781⟩: between +22° and +26°, determined in a solution containing 200 mg in 10 mL, the specimen having been previously dried in vacuum at 60° for 3 hours.

Loss on drying ⟨731⟩—Dry in vacuum at 60° for 3 hours: it loses not more than 1% of its weight.

Residue on ignition ⟨281⟩: not more than 0.2%.

Benzyl Alcohol—Use *Benzyl Alcohol* (NF monograph).

Benzyltrimethylammonium Chloride, $C_6H_5CH_2N(CH_3)_3Cl$—**185.70**—Available as a 60 percent aqueous solution. Is clear and is colorless or not more than slightly yellow, and has a slight amine-like odor.

Assay—Pipet 2 mL into a 50-mL volumetric flask, and add water to volume. Pipet 20 mL of the solution into a 125-mL conical flask, add about 30 mL of water, then add 0.25 mL of dichlorofluorescein TS, and titrate with 0.1 *N* silver nitrate VS. Each mL of 0.1 *N* silver nitrate is equivalent to 18.57 mg of $C_6H_5CH_2N(CH_3)_3Cl$. Between 59.5% and 60.5% is found.

Betamethasone—Use *Betamethasone* (USP monograph).

Betanaphthol—See *2-Naphthol.*

Bibenzyl (*Dibenzyl*), $C_{14}H_{14}$—**182.27**—Colorless crystals. Freely soluble in chloroform and ether; sparingly soluble in alcohol; practically insoluble in water.

Melting range ⟨741⟩: between 53° and 55°.

Bile Salts—A concentrate of beef bile, the principal constituent of which is sodium desoxycholate, determined as cholic acid. Soluble in water and in alcohol; the solutions foam strongly when shaken.

Insoluble substances—Dissolve 5 g in 100 mL of dilute alcohol (84 in 100), warming if necessary to aid solution. Filter within 15 minutes through a tared filter, and wash with small portions of the dilute alcohol until the last washing is colorless or practically so, then dry the residue at 105° for 1 hour, and weigh: the weight of the residue does not exceed 0.1%.

Assay—

STANDARD CHOLIC ACID SOLUTION—Dissolve 50.0 mg of cholic acid, accurately weighed, in dilute acetic acid (6 in 10) to make 100 mL, and mix. Store in a refrigerator.

PROCEDURE—Dissolve 1.0 g, accurately weighed, in 50 mL of dilute acetic acid (6 in 10). Filter the solution, if necessary, into a 100-mL volumetric flask, wash the original container and the filter with small portions of dilute acetic acid (6 in 10), add the same acetic acid to volume, and mix. Dilute 10 mL of this solution, accurately measured, with dilute acetic acid (6 in 10) to make 100 mL, and mix.

Pipet 1 mL each of the *Standard Cholic Acid Solution* and the solution of the Bile Salts into two matched test tubes. To each tube add 1 mL, accurately measured, of freshly prepared furfural solution (1 in 100), immediately place the tubes in an ice-bath for 5 minutes, then add to each tube 13 mL, accurately measured, of dilute sulfuric acid, made by cautiously mixing 50 mL of sulfuric acid with 65 mL of water. Mix the contents of the tubes, and place them in a water bath maintained at a temperature of 70° for 10 minutes. Immediately transfer the tubes to an ice-bath for 2 minutes, then determine the absorbance of each solution at the wavelength of maximum absorbance at about 670 nm, with a suitable spectrophotometer. Calculate the quantity, in mg, of cholic acid ($C_{24}H_{40}O_5$) in the weight of the Bile Salts taken by the formula:

$$500(A_U/A_S),$$

in which A_U and A_S are the absorbances of the solutions from the Bile Salts and the *Standard Cholic Acid Solution*, respectively. Not less than 45% of cholic acid is found.

Biotin (*cis-Hexahydro-2-oxo-1H-thieno-[3,4]-imidazoline-4-valeric Acid*), $C_{10}H_{16}N_2O_3S$—**244.31**—Use Biotin, FCC II.

Biphenyl, $C_{12}H_{10}$—**154.21**—Colorless to white crystals or crystalline powder, having a pleasant odor. Insoluble in water; soluble in alcohol and in ether. Boils at about 254°.

Melting range ⟨741⟩—between 68° and 72°.

2,2′-Bipyridine (*α,α′-Dipyridyl*), $C_{10}H_8N_2$—**156.19**—White or pink, crystalline powder. Soluble in water and in alcohol. Melts at about 69°, and boils at about 272°.

Sensitiveness—Prepare the following solutions: (*A*)—Dissolve 350 mg of ferrous ammonium sulfate in 50 mL of water containing 1 mL of sulfuric acid, and add 500 mg of hydrazine sulfate, then add water to make 500 mL. For use, dilute this solution with water in the ratio of 1 in 100 mL. (*B*)—Dissolve 8.3 g of sodium acetate and 12 mL of glacial acetic acid in water to make 100 mL. Add 1 mL of a solution of the specimen (1 in 1000) to a mixture of 10 mL of water and 1 mL of each of solutions *A* and *B:* a pink color results immediately.

Solubility—A 100-mg portion dissolves completely in 10 mL of water.

Residue on ignition (Reagent test): not more than 0.2%.

4,4′-Bis(4-amino-1-naphthylazo)-2,2′-stilbenedisulfonic Acid, $C_{34}H_{26}N_6O_6S_2$—**678.74**—Use a suitable grade.[7]

Bis(2-ethylhexyl) Maleate, $C_{20}H_{36}O_4$—**340.50**—Colorless to pale yellow, clear liquid. Miscible with acetone and with alcohol. Specific gravity about 0.945.

Assay—Place about 2.5 g, accurately weighed, in a 250-mL flask, add 50.0 mL of 0.5 *N* alcoholic potassium hydroxide VS, and reflux for 45 minutes. Cool, add 0.5 mL of phenolphthalein TS, and titrate the excess alkali with 0.5 *N* hydrochloric acid VS. Perform a blank determination at the same time, using the same amount of 0.5 *N* alcoholic potassium hydroxide (see *Residual Titrations* under *Titrimetry* ⟨541⟩). The difference, in mL, between the volumes of 0.5 *N* hydrochloric acid consumed in the test titration and blank titration, multiplied by 85.1, represents the quantity, in mg, of bis(2-ethylhexyl) maleate in the portion taken. Not less than 97% is found.

Bis(2-ethylhexyl) Phthalate, C_6H_4-1,2-[$COOCH_2(C_2H_5)CH(CH_2)_3CH_3]_2$—**390.56**—Colorless to light yellow liquid.

Refractive index ⟨831⟩: between 1.4855 and 1.4875, at 20°.

Bis(2-ethylhexyl) Sebacate (*Dioctyl Sebacate*), $C_8H_{17}OOC(CH_2)_8COOC_8H_{17}$—**426.68**—Pale straw-colored liquid. Insoluble in water. Refractive index about 1.448. Suitable for use in gas chromatography.[8]

Specific gravity, 20°/20° ⟨841⟩: between 0.913 and 0.917.

Boiling range: between 243° and 248° at 5 mm of mercury.

Bis(2-ethylhexyl)phosphoric Acid [*Bis-(2-ethylhexyl) Phosphate*], $[CH_3(CH_2)_3CH(C_2H_5)CH_2]_2HPO_4$—**322.42**—Light yellow, viscous liquid. Insoluble in water; freely soluble in chloroform and in ethyl acetate. Refractive index: about 1.443. Specific gravity: about 0.997.

Assay—Dissolve about 250 mg, accurately weighed, in 50 mL of dimethylformamide, add 3 drops of a 1 in 100 solution of thymol blue in dimethylformamide, and titrate with 0.1 *N* sodium methoxide VS to a blue end-point. Perform a blank determination, and make any necessary correction. Each mL of 0.1 *N* sodium methoxide is equivalent to 32.24 mg of $(C_8H_{17})_2HPO_4$. Between 95% and 105% is found.

Solubility—One volume dissolves in 9 volumes of chloroform to yield a clear solution, and 1 volume dissolves in 9 volumes of ethyl acetate to yield a clear solution.

Color—A 1 in 100 solution in chloroform exhibits an absorptivity of not more than 0.03 at 420 nm.

Bismuth Nitrate Pentahydrate, $Bi(NO_3)_3 \cdot 5H_2O$—**485.07**—Use ACS reagent grade.

Bismuth Subnitrate—White, slightly hygroscopic powder having an acid reaction to moistened blue litmus paper. Insoluble in water and alcohol; readily soluble in hydrochloric acid and in nitric acid.

Assay—Weigh accurately in a porcelain crucible about 1 g, previously dried at 105° for 2 hours, and ignite to constant weight. From the weight of Bi_2O_3 so obtained, calculate the percentage in the test specimen. Not less than 79.0% is found.

Loss on drying ⟨731⟩—Dry it at 105° for 2 hours: it loses not more than 3.0% of its weight.

Carbonate—Add 3 g to 3 mL of warm nitric acid: no effervescence occurs. (Retain the solution.)

Alkalies and alkaline earths—Boil 1 g with 20 mL of a mixture of equal volumes of acetic acid and water, cool, and filter. Add 2 mL of diluted hydrochloric acid, precipitate the bismuth by the addition of hydrogen sulfide gas, boil the mixture, and filter. Add 5 drops of sulfuric acid to the filtrate, evaporate to dryness, and ignite to constant weight: the residue weighs not more than 5 mg (0.5%).

Ammonium salts—Boil about 100 mg with 5 mL of sodium hydroxide TS: the vapor does not turn moistened red litmus paper blue.

For the following tests, prepare a *test solution* as follows: Pour the solution retained from the test for *Carbonate* into 100 mL of water. Filter the precipitate away from the solution, and evaporate the filtrate on a steam bath to 30 mL. Filter the concentrated filtrate, and adjust the volume of filtrate, if necessary, to 30 mL.

Chloride (Reagent test)—A 10-mL portion of *test solution* shows not more than 0.35 mg of Cl (0.035%).

Sulfate—To 5 mL of *test solution* add 5 drops of barium nitrate TS: no turbidity is produced immediately.

Copper—To 5 mL of *test solution* add a slight excess of ammonia TS: the mixture does not show a bluish color.

Lead—To 5 mL of *test solution* add 5 mL of diluted sulfuric acid, and mix: the solution does not become cloudy.

Silver—To 5 mL of *test solution* add hydrochloric acid, dropwise: no precipitate insoluble in hydrochloric acid but soluble in ammonia TS is formed.

Bismuth Sulfite Agar—Use a suitable grade.

Bis(trimethylsilyl)acetamide (*N,O-Bis(trimethylsilyl)-acetamide; BSA*), $CH_3CON[Si(CH_3)_3]_2$—**203.43**—Clear, colorless liquid. Readily hydrolyzes when exposed to moist air. Handle under nitrogen, and store in a cool place.

Assay—Not less than 90% of $CH_3CON[Si(CH_3)_3]_2$, a suitable gas chromatograph equipped with a thermal conductivity detector being used. The following conditions are suitable and provide a retention time of approximately 15 minutes.

COLUMN: 1.83-m × 3-mm stainless steel containing 5 percent phase G1 on support S1A.

INJECTION TEMPERATURE: 160°.

COLUMN TEMPERATURE: 90°, programmed to rise 4° per minute to 160°.

CARRIER GAS: Helium.

Refractive index ⟨831⟩: between 1.4150 and 1.4170 at 20°.

Bis(trimethylsilyl)trifluoroacetamide (*BSTFA*), $CF_3CON[Si(CH_3)_3]_2$—**257.40**—Clear, colorless liquid. Readily hydrolyzes when exposed to moist air. Store in a cool place.

Assay—Not less than 98% of $CF_3CON[Si(CH_3)_3]_2$, a suitable gas chromatograph equipped with a thermal conductivity detector being used. The following conditions are suitable and provide a retention time of approximately 15 minutes.

COLUMN: 1.83-m × 3-mm stainless steel containing 5 percent phase G1 on support S1A.

INJECTION TEMPERATURE: 170°.

COLUMN TEMPERATURE: 70°, programmed to rise 4° per minute to 140°.

CARRIER GAS: Helium.

Refractive index ⟨831⟩: between 1.3820 and 1.3860 at 20°.

Bis(trimethylsilyl)trifluoroacetamide with Trimethylchlorosilane—Use a suitable grade.[9]

Blood (for carbon monoxide test in gases)—*Use oxalated* or *defibrinated blood* of dogs, sheep, cattle, or human beings within 24 hours after bleeding. Prepare *oxalated blood* by adding 10 mg of sodium oxalate to each mL of the freshly drawn blood.

Blue Tetrazolium (*3,3′-(3,3′-Dimethoxy[1,1′-biphenyl]-4,4′-diyl)bis[2,5-diphenyl-2H-tetrazolium]dichloride*), $C_{40}H_{32}Cl_2$-

N_8O_2—**727.65**—Lemon-yellow crystals. Slightly soluble in water; freely soluble in chloroform and in methanol; insoluble in acetone and in ether.

Solubility in methanol—Dissolve 1 g in 100 mL of methanol: complete solution results, and the solution is clear.

Color—Transfer a portion of the methanol solution obtained in the preceding test to a 1-cm cell, and determine its absorbance at 525 nm, against water as the blank: the absorbance does not exceed 0.20.

Molar absorptivity ⟨851⟩—Its molar absorptivity in methanol, at 252 nm, is not less than 50,000.

Suitability test—

STANDARD PREPARATION—Dissolve in alcohol a suitable quantity of USP Hydrocortisone RS, previously dried at 105° for 3 hours and accurately weighed, and prepare by stepwise dilution a solution containing about 10 µg per mL.

PROCEDURE—Pipet 10-, 15-, and 20-mL portions of *Standard Preparation* into separate, glass-stoppered, 50-mL conical flasks. Add 10 mL and 5 mL, respectively, of alcohol to the flasks containing the 10- and 15-mL portions of *Standard Preparation*, and swirl to mix. To each of the flasks, and to a fourth flask containing 20 mL of alcohol, add 2.0 mL of a solution prepared by dissolving 50 mg of blue tetrazolium in 10 mL of alcohol, mix, and then add 2.0 mL of a solution prepared by diluting 1 mL of tetramethylammonium hydroxide TS with alcohol to 10 mL. Mix, allow the flasks to stand in the dark for 90 minutes, and determine the absorbances of the three solutions of the steroid standard at 525 nm, with a suitable spectrophotometer, using the solution in the fourth flask as the blank. Plot the absorbances on the abscissa and the amount of hydrocortisone on the ordinate scale of arithmetic coordinate paper, and draw the curve of best fit: the absorbance of each solution is proportional to the concentration, and the absorbance of the solution containing 200 µg of hydrocortisone is not less than 0.50.

Boric Acid, H_3BO_3—**61.83**—Use ACS reagent grade.

Boron Fluoride Ethyl Ether (*Boron Trifluoride Etherate*), $BF_3.(C_2H_5)_2O$—**141.93**—White to light amber liquid. Is hydrolyzed immediately by moisture in air. Density: about 1.125. *Keep tightly closed.*

Caution—Handle with care. It may release toxic fumes of fluorides.

Assay—Place 100 mL of water, 0.5 mL of octyl alcohol, 0.1 mL of methyl orange TS, and several boiling beads in a 500-mL distilling flask equipped with 2 standard-taper glass joints. Cool the flask and its contents in an ice-bath. Using a weight buret, transfer about 2 mL of the test specimen to the distilling flask, immediately insert the stopper in the flask using matching stoppers, and again weigh the buret to obtain the weight of test specimen added. When the gas cloud in the flask has condensed, quickly add 25 g of calcium chloride, and replace the stoppers with a condenser and a buret having its delivery tip sealed to a matching standard-taper joint. Heat the mixture in the flask to boiling, and slowly titrate with 1 *N* sodium hydroxide VS while the mixture is refluxing. Continue the refluxing for 30 minutes while titrating, but do not pass the end-point. Cool the mixture, and continue the titration to the end-point. If the amount of alkali required to reach the end-point of the cooled solution exceeds 1 mL, reflux for an additional 10 minutes, cool, and complete the titration to the exact end-point. Each mL of 1 *N* sodium hydroxide is equivalent to 22.61 mg of BF_3. Not less than 47% of BF_3 is found.

Boron Trifluoride, BF_3—**67.81**—Use a suitable grade.

For use in the *Assay* for *Stearic Acid*, a suitable reagent solution may be prepared as follows: Place 1000 g of methanol in a 2-liter conical flask, and cool in an ice bath for 30 minutes. Weigh 261.6 g of boron fluoride ethyl ether into a 400-mL beaker, using a platform or other suitable balance, and add slowly, with frequent stirring, to the cold methanol. Store the solution in a glass-stoppered bottle in a refrigerator.

Bovine Albumin—Collect blood from beef cattle in a vessel containing sodium citrate solution (8 in 100) in the proportion of 1 volume for each 19 volumes of blood to be collected. Centrifuge the collected blood, siphon the plasma, and promptly cool it to just above its freezing temperature. Maintain the temperature between 0° and −5° during the entire processing to prevent denaturation of the albumin.

Measure the plasma, and slowly add 883 mL of dilute alcohol (53% v/v) for each 1000 mL of plasma. Take special care in adding the alcohol to ensure that heat evolved is removed rapidly enough to prevent a temperature rise. Filter off the precipitate formed, and for each 1000 mL of filtrate slowly add a mixture of 1130 mL of dilute alcohol (53% v/v) and 76 mL of 1.0 *M* sodium acetate that has been buffered with acetic acid to a pH of 5.1 measured when diluted 1 to 20. A dilution of the final mixture (1 in 5) has a pH of 5.8. Filter off the precipitated globulins, and adjust the filtrate to a pH of 5.2 by the addition of 1.0 *M* sodium acetate that has been buffered with acetic acid to a pH of 4.0 measured when diluted 1 in 20. Filter off the precipitated albumin, and remove it from the filter paper, using water if necessary. Freeze the resulting mixture, and dry in vacuum from the frozen state.

Bovine Plasma Albumin, Dried—Use a suitable grade.

Brilliant Green (*Malachite Green G*), $C_{27}H_{34}N_2O_4S$—**482.64**—Glistening, golden-yellow crystals. Soluble in water and in alcohol. Absorption maximum: 623 nm.

Bromine, Br—At. wt. **79.904**—Use ACS reagent grade.

Bromine Detector Tube—A fuse-sealed glass tube so designed that gas may be passed through it, and containing suitable absorbing filters and support media for the indicator *o*-tolidine.

NOTE—A suitable detector tube that conforms to the monograph specification is available from National Draeger, Inc., 401 Parkway View Drive, Pittsburgh, Pa. 15205 as Reference Number CH 24301, Measuring Range 0.2 to 3 ppm. Tubes having conditions other than those specified in the monograph may be used in accordance with the section entitled *Tests and Assays* in the *General Notices.*

α-Bromo-2′-acetonaphthone (*Bromomethyl 2-naphthyl ketone*), $C_{12}H_9BrO$—**249.11**—Tannish pink crystals.

Melting range ⟨741⟩: between 81° and 83°.

***p*-Bromoaniline,** C_6H_6BrN—**172.02**—White to off-white crystals. Insoluble in water; soluble in alcohol and in ether.

Assay—Transfer about 650 mg, accurately weighed, to a suitable container, and dissolve in 50 mL of glacial acetic acid TS. Add crystal violet TS, and titrate with 0.1 *N* perchloric acid VS. Perform a blank determination, and make any necessary correction. Each mL of 0.1 *N* perchloric acid is equivalent to 17.20 mg of C_6H_6BrN. Not less than 98% is found.

Melting range ⟨741⟩: between 60° and 65°, within a 2° range.

***N*-Bromosuccinimide,** $C_4H_4BrNO_2$—**177.98**—White to off-white crystals or powder, having a faint odor. Freely soluble in water, acetone, and glacial acetic acid. [*Caution—Highly irritating to eyes, skin, and mucous membranes.*]

Assay—Transfer 200 mg, accurately weighed, to a conical flask, add 25 mL of 0.5 *N* alcoholic potassium hydroxide, cover with a watch glass, heat to boiling, and boil for 5 minutes. Cool, transfer the solution to a beaker, rinsing the flask with water until the total volume of solution plus rinsings is about 100 mL, and add 10 mL of glacial acetic acid. Insert suitable electrodes, and titrate with 0.1 *N* silver nitrate VS, determining the end-point potentiometrically. Each mL of 0.1 *N* silver nitrate is equivalent to 17.80 mg of $C_4H_4BrNO_2$. Not less than 98% is found.

Brucine Sulfate, $(C_{23}H_{26}N_2O_4)_2.H_2SO_4.7H_2O$—**1013.12**—Use ACS reagent grade.

Buffers—See *Buffer Solutions* under *Solutions.*

Butabarbital Sodium—Use *Butabarbital Sodium* (USP monograph).

1,3-Butanediol (*1,3-Butylene Glycol*), $C_4H_{10}O_2$—**90.12**—Viscous, colorless liquid. Very hygroscopic. Soluble in water, in alcohol, in acetone, and in methyl ethyl ketone; practically insoluble in aliphatic hydrocarbons, in benzene, and in toluene.

Assay—Inject an appropriate specimen into a suitable gas chromatograph (see *Chromatography* ⟨621⟩) equipped with a flame-ionization detector, helium being used as the carrier gas. The following conditions have been found suitable: a 1.8-m × 3-mm stainless steel column containing 20 percent phase G16 on support S1A; the injection port temperature is maintained at 265°; the column temperature is maintained at 150° and programmed to rise 8° per minute to 210°. The area of the butanediol peak is not less than 98% of the total peak area.

Refractive index ⟨831⟩: between 1.4390 and 1.4410 at 20°.

2,3-Butanedione (*Diacetyl*), $CH_3COCOCH_3$—**86.09**—Bright yellow to yellowish green liquid having a strong, pungent odor. Soluble in water. Miscible with alcohol and with ether. Boils at about 88°. Specific gravity: about 0.98.

Assay—

HYDROXYLAMINE HYDROCHLORIDE SOLUTION—Dissolve 20 g of hydroxylamine hydrochloride in 40 mL of water, and dilute with alcohol to 400 mL. Add, with stirring, 300 mL of 0.5 *N* alcoholic potassium hydroxide, and filter. Discard after 2 days.

PROCEDURE—Transfer about 1 g, accurately weighed, to a glass-stoppered, 250-mL flask, add 75.0 mL of *Hydroxylamine hydrochloride solution*, and insert the stopper in the flask. Reflux the mixture for 1 hour, then cool to room temperature. Add bromophenol blue TS, and titrate with 0.5 *N* hydrochloric acid VS to a greenish yellow end-point. [NOTE—Alternatively, the solution may be titrated potentiometrically to a pH of 3.4.] Perform a blank test with the same quantities of reagent used for the test specimen, and make any necessary correction. Each mL of 0.5 *N* hydrochloric acid is equivalent to 43.05 mg of $CH_3COCOCH_3$. Not less than 97% of $CH_3COCOCH_3$ is found.

Refractive index ⟨831⟩: between 1.3935 and 1.3965, at 20°.

Congealing temperature ⟨651⟩: between −2.0° and −5.5°.

Butanol—See *Butyl Alcohol.*

Butyl Acetate, Normal, $CH_3COO(CH_2)_3CH_3$—**116.16**—Clear, colorless liquid having a characteristic odor. Slightly soluble in water; miscible with alcohol. Specific gravity: about 0.88.

Distilling range ⟨721⟩—Not less than 95% distils between 123° and 126°.

Butyl Alcohol (*1-Butanol; Normal Butyl Alcohol*), $CH_3(CH_2)_2CH_2OH$—**74.12**—Use ACS reagent grade 1-Butanol.

Butyl Alcohol, Normal—See *Butyl Alcohol.*

Butyl Alcohol, Secondary (*2-Butanol*), $CH_3CH_2CH(OH)CH_3$—**74.12**—Use ACS reagent grade Isobutyl Alcohol.

Butyl Alcohol, Tertiary, $(CH_3)_3COH$—**74.12**—Colorless crystals, becoming liquid at a temperature above 25.5°. Has a camphoraceous odor. Miscible with water and with common organic solvents.

Miscibility—Mix a 5-mL portion with 15 mL of water, and mix another 5-mL portion with 15 mL of carbon disulfide. Allow each mixture to stand for 15 minutes: neither mixture is more turbid than an equal volume of the diluent.

Specific gravity ⟨841⟩: not less than 0.778 and not more than 0.782.

Boiling range (Reagent test): between 82.5° and 83.5°.

Freezing point: not less than 25°.

Residue on evaporation—Evaporate about 20 g, accurately weighed, in a crucible on a steam bath, and dry at 105° for 1 hour: not more than 0.005% is found.

Acidity—Add 20 mL of it to 20 mL of water previously neutralized to phenolphthalein TS with 0.02 *N* sodium hydroxide, mix, and titrate with 0.020 *N* sodium hydroxide until the pink color is restored: not more than 0.40 mL is required (about 0.003% as CH_3COOH).

Alkalinity—Dilute 10 mL with 20 mL of water, and add 1 drop of methyl red TS: if the solution is yellow, not more than 0.25 mL of 0.020 *N* sulfuric acid is required to change it to pink (about 0.001% as NH_3).

Butyl Benzoate, $C_{11}H_{14}O_2$—**178.23**—Thick, oily, colorless to pale yellow liquid. Practically insoluble in water; soluble in alcohol and in ether.

Assay—When examined by gas-liquid chromatography, it shows a purity of not less than 98%. The following conditions have been found suitable for assaying it: A 1.8-m × 3-mm stainless steel column packed with liquid phase G4 on support S1A. Helium is the carrier gas, the injection port temperature is maintained at 180°, the column temperature is 190°, and the flame-ionization detector is maintained at 280°. The retention time is about 15 minutes.

Refractive index ⟨831⟩: between 1.4980 and 1.5000, at 20°.

Butyl Ether (*n-Dibutyl Ether*), $C_8H_{18}O$—**130.23**—Use a suitable grade.

***n*-Butyl Chloride** (*1-Chlorobutane*), C_4H_9Cl—**92.57**—Clear, colorless, volatile liquid, having a slight, characteristic odor. *Highly flammable.* Practically insoluble in water. Miscible with alcohol and with ether.

Assay—When examined by gas-liquid chromatography, it shows a purity of not less than 98%. The following conditions have been found suitable for assaying the article: A 1.8-m × 3-mm stainless steel column packed with phase G16 on support S1. Helium, flowing at a rate of about 40 mL per minute, is the carrier gas, the detector temperature is about 310°, the injection port temperature is about 230°, and the column temperature is programmed to rise at 10° per minute from 35° to 150°. A flame-ionization detector is employed.

Boiling range ⟨721⟩: between 76° and 80°, within a 2° range.

Refractive index ⟨831⟩: between 1.4015 and 1.4035 at 20°.

Acidity—Add phenolphthalein TS to 75 mL, and titrate with 0.1 *N* potassium hydroxide in methanol to a faint pink color that persists, with shaking, for 1.5 seconds: not more than 0.91 mL is required (about 0.005% as HCl).

Water ⟨921⟩: not more than 0.02%, determined by the *Titrimetric Method.*

Residue after evaporation—Evaporate about 60 mL (50 g), accurately weighed, in a tared platinum dish on a steam bath, and dry at 105° for 1 hour: not more than 0.005% is found.

***tert*-Butyl Hydroperoxide, 70 Percent in Water,** $C_4H_{10}O_2$—**90.12**—Colorless to light yellow liquid.

Assay—Dissolve about 500 mg, accurately weighed, in 25 mL of glacial acetic acid. Add 10 mL of potassium iodide solution (3 in 10) and 0.2 mL of sulfuric acid, and allow to stand in the dark for 2 hours. Titrate with 0.1 *N* sodium thiosulfate VS, using starch TS as the indicator. Perform a blank determination, and make any necessary correction. Each mL of 0.1 *N* sodium thiosulfate is equivalent to 4.506 mg of $C_4H_{10}O_2$: not less than 70% is found.

***n*-Butyl Nitrite,** $CH_3CH_2CH_2CH_2NO_2$—**103.12**—A colorless to yellowish liquid, having a characteristic odor. Is flammable. Slightly soluble in water and is gradually decomposed by it. Miscible with alcohol and with some other organic liquids. Specific gravity: about 0.88. Store it in tight containers, protected from light.

Assay—Shake about 3 mL with 500 mg of anhydrous potassium carbonate, decant carefully, transfer to a 100-mL volumetric flask tared with about 20 mL of alcohol, and weigh accurately. Add alcohol to volume, and mix. Fill a nitrometer with a saturated solution of sodium chloride, pipet into the funnel top 10 mL of the alcohol solution, and, after drawing this into the measuring tube of the nitrometer without the admission of air, follow it with 5 mL of alcohol as a rinse, then with 10 mL of potassium iodide TS, and afterwards with 5 mL of diluted sulfuric acid, introducing each reagent separately into the measuring tube. When the volume of gas has become constant (in about 30 to 60 minutes), note the amount collected, the temperature at the nitrometer, and the barometric pressure. Multiply the volume of gas in mL by 4.8, and divide the product by the weight of specimen taken. At 25° and 760 mm pressure, the quotient represents the percentage of $C_4H_9NO_2$ in the liquid. The temperature correction is $\frac{1}{298}$ of the total percentage just found for each degree, added if the temperature is below 25°, and subtracted if it is above 25°. The barometric correction is $\frac{1}{760}$ of the total percentage just found for each mm, added if it is above 760 mm, and subtracted if it is below 760 mm. Not less than 90% of $C_4H_9NO_2$ is found.

Boiling range (Reagent test)—Not less than 90% distils between 76° and 79°.

***n*-Butylamine,** $CH_3CH_2CH_2CH_2NH_2$—**73.14**—Colorless to pale yellow, flammable liquid. Miscible with water, with alcohol, and with ether. Store it in tight containers. Specific gravity: about 0.740.

Distilling range, Method I ⟨721⟩—Not less than 95% distils between 76° and 78°.

Water, Method I ⟨921⟩: not more than 1.0%, determined by the *Titrimetric Method.*

Chloride (Reagent test)—One g (1.5 mL) shows not more than 0.01 mg of Cl (0.001%).

Acidic impurities—To 50 mL add 5 drops of a saturated solution of azo violet in benzene, and titrate quickly with 0.1 *N* sodium methoxide VS to a deep blue end-point, observing precautions to prevent absorption of atmospheric carbon dioxide as by use of an atmosphere of nitrogen: not more than 1.0 mL of 0.1 *N* sodium methoxide is required for neutralization.

4-(Butylamino)benzoic Acid, $C_{11}H_{15}NO_2$—**193.25**—Off-white powder.

Melting range ⟨741⟩: between 153° and 156°.

Butylated Hydroxyanisole (*BHA*), $C_{11}H_{16}O_2$—**180.25**—Use *Butylated Hydroxyanisole* (NF monograph).

***n*-Butylboronic Acid,** $C_4H_9B(OH)_2$—**101.94**—Use a suitable grade.[10] [NOTE—This reagent is usually shipped and stored under water. Before use, remove any excess water by light vacuum filtration.]

4-*tert*-Butylphenol, $C_{10}H_{14}O$—**150.21**—White, crystalline flakes or needles. Practically insoluble in water; soluble in alcohol and in ether.

Melting range ⟨741⟩: between 98° and 101°.

***n*-Butyrophenone,** $C_{10}H_{12}O$—**148.20**—Colorless liquid.

Assay—Inject an appropriate specimen into a suitable gas chromatograph (see *Chromatography* ⟨621⟩) equipped with a flame-ionization detector, nitrogen being used as the carrier gas. The following conditions have been found suitable: a 30-m × 0.3-mm capillary column coated with dimethylcyanopropylphenylpolysiloxane; the injection port temperature is maintained at 220°; the detector temperature is maintained at 300°; the column temperature is maintained at 120° and programmed to rise 10° per minute to 220°. The area of the main peak is not less than 99% of the total peak area.

Refractive index ⟨831⟩: about 1.5196 at 20°.

Cadmium Acetate, $C_4H_6CdO_4 . 2H_2O$—**266.53**—Colorless, transparent to translucent crystals. Is odorless or has a slight odor of acetic acid. Freely soluble in water; soluble in alcohol.

Insoluble matter (Reagent test): not more than 1 mg, from 20 g (0.005%).

Chloride (Reagent test)—One g shows not more than 0.01 mg of Cl (0.001%).

Sulfate (Reagent test, *Method II*)—Dissolve 10 g in 100 mL of water, add 1 mL of hydrochloric acid, and filter: the residue weighs not more than 1.2 mg more than the residue obtained in a complete blank test (0.005%).

Substances not precipitated by hydrogen sulfide—Dissolve 2 g in a mixture of 135 mL of water and 15 mL of 1 *N* sulfuric acid, heat to boiling, and pass a rapid stream of hydrogen sulfide through the solution as it cools. Filter, and to 75 mL of the clear filtrate add 0.25 mL of sulfuric acid, then evaporate to dryness, and ignite gently: the residue weighs not more than 1 mg (0.1%).

Cadmium Nitrate, $Cd(NO_3)_2 . 4H_2O$—**308.47**—Colorless, hygroscopic crystals. Very soluble in water; soluble in alcohol.

Insoluble matter (Reagent test): not more than 1 mg, from 20 g (0.005%).

Chloride (*Cl*) (Reagent test)—One g shows not more than 0.01 mg of Cl (0.001%).

Sulfate (Reagent test, *Method II*)—Evaporate a mixture of 12 g of specimen and 25 mL of hydrochloric acid on a steam bath to dryness. Add another 15 mL of hydrochloric acid, and again evaporate to dryness. Dissolve the residue in 100 mL of water, filter, and add 1 mL of hydrochloric acid: the residue weighs not more than 1.0 mg more than the residue obtained in a blank test (0.003%).

Copper (*Cu*)—Dissolve 0.5 g in 10 mL of water, add 10 mL of *Ammonium Citrate Solution* (see *Lead* ⟨251⟩), and adjust the reaction of a pH of about 9 by the addition of 1 *N* ammonium hydroxide (about 30 mL). Add 1 mL of sodium diethyldithiocarbamate solution (1 in 1000), and mix. Add 5 mL of amyl alcohol, shake for about 1 minute, and allow the layers to separate: any yellow color in the amyl alcohol layer is not darker than that of a blank to which 0.01 mg of Cu has been added (0.002%).

Iron (*Fe*)—Dissolve 1 g in 15 mL of water, add 2 mL of hydrochloric acid, and boil for 2 minutes. Cool, and add about 30 mg of ammonium persulfate and 15 mL of a solution of potassium thiocyanate in normal butyl alcohol (made by dissolving 10 g of potassium thiocyanate in 10 mL of water, warming the solution to about 30°, diluting with normal butyl alcohol to 100 mL, and shaking until clear). Shake vigorously for 30 seconds, and allow the layers to separate: any red color in the clear alcoholic layer is not darker than that of a blank to which 0.01 mg of Fe has been added (0.001%).

Lead (*Pb*)—Dissolve 1.0 g in 10 mL of water, add 0.2 mL of glacial acetic acid, and filter if necessary. To a 7-mL portion of water add 0.2 mL of glacial acetic acid and 3 mL of *Standard Lead Solution* (see *Lead* ⟨251⟩), and mix, to provide a blank. Then add to each solution 1.0 mL of potassium chromate solution (1 in 10), and mix: after 5 minutes, the test solution is not more turbid than the blank (0.003%).

Substances not precipitated by hydrogen sulfide—Dissolve 2 g in 145 mL of water, add 5 mL of sulfuric acid (1 in 10), heat to boiling, and pass a rapid stream of hydrogen sulfide through the solution as it cools. Filter, and to 75 mL of the clear filtrate add 0.25 mL of sulfuric acid, then evaporate to dryness, and ignite gently: the residue weighs not more than 1 mg (0.1%).

Calcium Acetate, $Ca(C_2H_3O_2)_2 . H_2O$—**176.18**—White, crystalline granules or powder. Soluble in about 3 parts of water; slightly soluble in alcohol.

Insoluble matter (Reagent test): not more than 1.0 mg, from 10 g (0.010%).

Alkalinity and acidity—To a solution of 2.0 g in 25 mL of water add phenolphthalein TS: no pink color is produced. Then add 0.10 *N* sodium hydroxide until a pink color is produced after shaking: not more than 0.70 mL of the alkali is required (0.2% as CH_3COOH).

Chloride (Reagent test)—One g shows not more than 0.05 mg of Cl (0.005%).

Sulfate (Reagent test, *Method I*)—One g shows not more than 0.4 mg of SO_4 (0.04%).

Alkalies and magnesium—Dissolve 1 g in 50 mL of water. Add 2 mL of hydrochloric acid, heat to boiling, and add 35 mL of oxalic acid solution (1 in 20). Slowly neutralize the solution, while it is cooling, with stronger ammonia water, then dilute with water to 100 mL, and allow to stand for 4 hours or overnight. Filter, and to 50 mL of the filtrate add 5 drops of sulfuric acid, evaporate, and ignite to constant weight: not more than 1.5 mg of residue remains (0.3% as SO_4).

Barium—Dissolve 2 g in 15 mL of water, add 2 drops of glacial acetic acid, filter, and add to the filtrate 0.3 mL of potassium dichromate solution (1 in 10): no turbidity is produced within 10 minutes (about 0.01%).

Heavy metals (Reagent test): 0.001%.

Iron ⟨241⟩—Dissolve 500 mg in 47 mL of water containing 2 mL of hydrochloric acid. The solution shows not more than 0.01 mg of Fe (0.002%).

Calcium Biphosphate (*Calcium Phosphate, Monobasic*), $Ca(H_2PO_4)_2 . H_2O$—**252.07**—White, somewhat deliquescent crystals, or crystalline powder. Sparingly soluble in water; insoluble in alcohol.

Insoluble in hydrochloric acid—Dissolve 10 g in 100 mL of dilute hydrochloric acid (1 in 10): the residue, filtered, washed, and dried at 105° for 2 hours, weighs not more than 2 mg (0.02%).

Arsenic (Reagent test)—The stain produced by 1 g does not exceed that produced by 0.002 mg of As (2 ppm).

Barium—Heat 1 g with 10 mL of water, and add hydrochloric acid, dropwise, stirring after each addition, until just dissolved. Filter, and add to the filtrate 5 mL of calcium sulfate TS: no turbidity is produced within 1 hour.

Chloride (Reagent test)—One g shows not more than 0.05 mg of Cl (0.005%).

Dibasic salt or excess acid—Triturate 1 g with 3 mL of water, then dilute with 100 mL of water, and add 1 drop of methyl orange TS: a red color is produced (*absence* of *dibasic salt*), which is changed to yellow by not more than 1 mL of 1 *N* sodium hydroxide (*excess acid*).

Heavy metals—Heat 2 g with 20 mL of water, and add hydrochloric acid, dropwise, stirring after each addition, until dissolved. Dilute with water to 40 mL, and to 20 mL of the dilution add 5 mL of hydrogen sulfide TS: the solution is not darkened.

Nitrate—Mix 500 mg with 5 mL of water, and add just sufficient hydrochloric acid to dissolve. Dilute with water to 10 mL, add 0.1 mL of indigo carmine TS, and follow with 10 mL of sulfuric acid: the blue color persists for 1 minute (about 0.03%).

Sulfate (Reagent test, *Method II*)—The filtrate from the test for *Insoluble in hydrochloric acid* yields not more than 10 mg, no blank correction being made (0.04%).

Calcium Carbonate, $CaCO_3$—**100.09**—Use ACS reagent grade. NOTE—Calcium Carbonate of a quality suitable as a primary standard is available from the National Institute of Standards and Technology, Office of Standard Reference Materials, Washington, DC 20234, as standard sample No. 915.

Calcium Carbonate, Chelometric Standard, $CaCO_3$—**100.09**—Use ACS reagent grade.

Calcium Carbonate, Precipitated—Use *Precipitated Calcium Carbonate* (USP monograph).

Calcium Caseinate—White or slightly yellow, nearly odorless, powder. Insoluble in cold water, but forms a milky solution when suspended in water, stirred, and heated.

Residue on ignition (Reagent test)—Ignite 5 g at 550°: the residue weighs between 150 and 300 mg (3.0% to 6.0%).

Calcium—Treat the residue from the preceding test with 10 mL of diluted hydrochloric acid, filter, and to the clear filtrate add 5 mL of ammonium oxalate TS: it shows a white precipitate upon standing.

Loss on drying ⟨731⟩—Dry it in vacuum at 70° to constant weight: it loses not more than 7.0% of its weight.

Fat—Suspend 1.0 g in 5 mL of alcohol in a Mojonnier flask, add 0.8 mL of stronger ammonia water and 9 mL of water, and shake. Add a second 5-mL portion of alcohol, then add successive portions of 25 mL each of ether and solvent hexane, shaking after each addition by inverting the flask 30 times. Centrifuge, decant the solvent layer, evaporate it at a low temperature, and dry on a steam bath: the residue weighs not more than 20 mg (2.0%).

Nitrogen content, Method I ⟨461⟩—Between 12.5% and 14.3% of N is found, calculated on the anhydrous basis.

Suspensibility in water—Place 2 g in a beaker, and add cool water slowly with stirring to form a thin, smooth paste. Add additional water to make a total of 100 mL. Stir, and heat to 80°: a milky suspension is formed that does not settle after standing for 2 hours.

Calcium Chloride, $CaCl_2 . 2H_2O$—**147.02**—Use ACS reagent grade Calcium Chloride Dihydrate.

Calcium Chloride, Anhydrous (for drying), $CaCl_2$—**110.99**—Use ACS reagent grade.

Calcium Citrate, $Ca_3(C_6H_5O_7)_2 . 4H_2O$—**570.50**—A white, odorless, crystalline powder. Slightly soluble in water; freely soluble in 3 *N* hydrochloric acid and in 2 *N* nitric acid; insoluble in alcohol. To 15 mL of hot 2 *N* sulfuric acid add in small portions and with stirring about 500 mg of calcium citrate. Boil the mixture for 5 minutes, and filter while hot: the cooled filtrate responds to the identification test for *Citrate* ⟨191⟩.

Assay—Weigh accurately about 400 mg of the salt, previously dried at 150° to constant weight, and transfer to a 250-mL beaker. Dissolve the test specimen in 150 mL of water containing 2 mL of 3 *N* hydrochloric acid, add 15 mL of 1 *N* sodium hydroxide and 250 mg of hydroxy naphthol blue, and titrate with 0.05 *M* disodium ethylenediaminetetraacetate VS until the solution turns deep blue in color. Each mL of 0.05 *M* disodium ethylenediaminetetraacetate is equivalent to 8.307 mg of $Ca_3(C_6H_5O_7)_2$: between 97.5% and 101% is found.

Calcium oxide and carbonate—Triturate 1 g of calcium citrate with 5 mL of water for 1 minute: the mixture does not turn red litmus blue. Then add 5 mL of warm 3 *N* hydrochloric acid: only a few isolated bubbles escape.

Hydrochloric acid–insoluble matter—Dissolve 5 g by heating with a mixture of 10 mL of hydrochloric acid and 50 mL of water for 30 minutes: not more than 2.5 mg of insoluble residue remains (0.05%).

Loss on drying ⟨731⟩—Dry it at 150° to constant weight: it loses between 12.2% and 13.3% of its weight.

Arsenic ⟨211⟩—Proceed with 0.50 g as directed for organic compounds (6 ppm of As).

Heavy metals, Method I ⟨231⟩: 0.002%.

Calcium Gluconate Injection—Use *Calcium Gluconate Injection* (USP monograph).

Calcium Hydroxide—Use ACS reagent grade.

Calcium Lactate, $(CH_3CHOHCOO)_2Ca . 5H_2O$—**308.30**—White, almost odorless, granules or powder. Is somewhat efflorescent and at 120° becomes anhydrous. One g dissolves in 20 mL of water; practically insoluble in alcohol. Store it in tight containers.

Assay—Weigh accurately about 500 mg, previously dried at 120° for 4 hours, transfer to a suitable container, and dissolve in 150 mL of water containing 2 mL of diluted hydrochloric acid. Add 15 mL of sodium hydroxide TS and 300 mg of hydroxy naphthol blue indicator, and titrate with 0.05 *M* disodium ethylenediaminetetraacetate VS until the solution is deep blue in color. Each mL of 0.05 *M* disodium ethylenediaminetetraacetate is equivalent to 10.91 mg of $C_6H_{10}CaO_6$. Not less than 98% is found.

Loss on drying ⟨731⟩—Dry it at 120° for 4 hours: it loses between 25.0% and 30.0% of its weight.

Acidity—Add phenolphthalein TS to 20 mL of a 1 in 20 solution, and titrate with 0.10 *N* sodium hydroxide: not more than 0.50 mL is required to produce a pink color.

Heavy metals (Reagent test)—Dissolve 1 g in 2.5 mL of diluted hydrochloric acid, dilute with water to 40 mL, and add 10 mL of hydrogen sulfide TS: any brown color produced is not darker than that of a control containing 0.02 mg of added Pb (0.002%).

Magnesium and alkali salts—Mix 1 g with 40 mL of water, carefully add 5 mL of hydrochloric acid, heat the solution, boil for 1 minute, and add rapidly 40 mL of oxalic acid TS. Add immediately to the warm mixture 2 drops of methyl red TS, then add ammonia TS dropwise, from a buret, until the mixture is just alkaline. Cool to room temperature, transfer to a 100-mL graduated cylinder, dilute with water to 100 mL, mix, and allow to stand for 4 hours or overnight. Filter, and transfer to a platinum dish 50 mL of the clear filtrate, to which has been added 0.5 mL of sulfuric acid. Evaporate the mixture on a steam bath to a small bulk. Carefully heat over a free flame to dryness, and continue heating to complete decomposition and volatilization of ammonium salts. Finally ignite the residue at 800 ± 25° for 15 minutes: the residue weighs not more than 5 mg (1%).

Volatile fatty acid—Stir about 500 mg with 1 mL of sulfuric acid, and warm: the mixture does not emit an odor of volatile fatty acid.

Calcium Nitrate, $Ca(NO_3)_2 . 4H_2O$—**236.15**—Use ACS reagent grade.

Calcium Pantothenate—Use *Calcium Pantothenate* (USP monograph).

Calcium Pantothenate, Dextro—Use *Calcium Pantothenate* (USP monograph).

Calcium Sulfate, $CaSO_4 . 2H_2O$—**172.17**—Use ACS reagent grade.

Calcium Sulfate, Anhydrous, $CaSO_4$—**136.14**—It conforms to the requirements of the ACS tests for chloride, magnesium and alkali salts, and heavy metals under ACS reagent grade Calcium Sulfate.

Loss on ignition—Ignite 1 g, accurately weighed, at 500 ± 25° to constant weight: it loses not more than 2.0% of its weight.

Camphor—Use *Camphor* (USP monograph).

***dl*-10-Camphorsulfonic Acid,** $C_{10}H_{16}O_4S$—**232.29**—White to off-white, crystals or powder. Is optically inactive.

Melting range ⟨741⟩: decomposes about 199°.

***dl*-10-Camphorsulfonic Acid, Sodium Salt,** $C_{10}H_{15}NaO_4S$—**254.28**—White to off-white powder. Soluble in water.

Melting range ⟨741⟩: between 283° and 286°.

Water, Method I ⟨921⟩: not more than 2.7%.

Capryl Alcohol (*2-Octanol*), $CH_3(CH_2)_5CHOHCH_3$—**130.23**—Clear, colorless, oily liquid, having a pungent, aromatic odor. Insoluble in water. Miscible with alcohol, with benzene, and with ether. Specific gravity: about 0.82.

Boiling range (Reagent test)—Not less than 95% distils between 178° and 181°.

Refractive index ⟨831⟩: between 1.424 and 1.426 at 20°.

Residue on evaporation—Evaporate 12 mL on a steam bath, and dry it at 150° for 2 hours: the residue weighs not more than 1.0 mg (0.01%).

Water—Mix 5 mL with 20 mL of solvent hexane: the mixture is clear.

Carbon Dioxide—Use *Carbon Dioxide* (USP monograph).

Carbon Dioxide Detector Tube—A fuse-sealed glass tube so designed that gas may be passed through it, and containing suitable absorbing filters and support media for the indicators hydrazine and crystal violet.

NOTE—A suitable detector tube that conforms to the monograph specification is available from National Draeger, Inc., 401 Parkway View Drive, Pittsburgh, PA 15205 as Reference Number CH 30801, Measuring Range 0.02 to 0.30%. Tubes having conditions other than those specified in the monograph may be used in accordance with the section entitled *Tests and Assays* in the *General Notices*.

Carbon Disulfide, CS_2—**76.13**—Use ACS reagent grade.

Carbon Disulfide, Chromatographic—Use a suitable grade.

Carbon Monoxide Detector Tube—A fuse-sealed glass tube so designed that gas may be passed through it, and containing suitable absorbing filters and support media for the indicators iodine pentoxide and selenium dioxide with chromic acid and phosphoric acid.

NOTE—A suitable detector tube that conforms to the monograph specification is available from National Draeger, Inc., 401 Parkway View Drive, Pittsburgh, PA 15205 as Reference Number CH 25601, Measuring Range 5 to 150 ppm. Tubes having conditions other than those specified in the monograph may be used in accordance with the section entitled *Tests and Assays* in the *General Notices*.

Carbon Tetrachloride, CCl_4—**153.82**—Use ACS reagent grade.

Carborane–Methyl Silicone—Use a suitable grade.[11]

Carboxylate (Sodium Form) Cation-exchange Resin (50- to 100-mesh)—See *Cation-exchange Resin, Carboxylate (Sodium Form) (50- to 100-mesh)*.

Carboxymethyl Cellulase—Use a suitable grade.

Casein[12]—White or slightly yellow, odorless, granular powder. Insoluble in water and in other neutral solvents; readily dissolved by ammonia TS and by solutions of alkali hydroxides, usually forming a cloudy solution.

Residue on ignition (Reagent test)—Ignite 2 g: the residue weighs not more than 20 mg (1.0%).

Loss on drying ⟨731⟩—Dry it at 105° to constant weight: it loses not more than 10.0% of its weight.

Alkalinity—Shake 1 g with 20 mL of water for 10 minutes, and filter: the filtrate is not alkaline to red litmus paper.

Soluble substances—When the filtrate from the *Alkalinity* test is evaporated and dried at 105°, the residue weighs not more than 1 mg (0.1%).

Fats—Dissolve 1 g in a mixture of 10 mL of water and 5 mL of alcoholic ammonia TS, and shake out with two 20-mL portions of solvent hexane. Evaporate the hexane at a low temperature, and dry at 80°: the weight of the residue does not exceed 5 mg (0.5%).

Nitrogen content, Method I ⟨461⟩—Between 15.2% and 16.0% of N is found, on the anhydrous basis.

Where vitamin-free casein is required, use casein that has been rendered free from the fat-soluble vitamins by continuous extraction with hot alcohol for 48 hours followed by air-drying to remove the solvent.

Castor Oil—Use *Castor Oil* (USP monograph).

Catechol (*o-Dihydroxybenzene*), $C_6H_4(OH)_2$—**110.11**—White crystals, which become discolored on exposure to air and light. Readily soluble in water, in alcohol, in benzene, in ether, in chloroform, and in pyridine, forming clear solutions.

Melting range ⟨741⟩: between 104° and 105°.

Residue on ignition (Reagent test)—Ignite 500 mg with 5 drops of sulfuric acid: the residue weighs not more than 1 mg (0.2%).

Cation-exchange Resin—Use a suitable grade.[13]

Cation-exchange Resin, Carboxylate (Sodium Form) (50- to 100-mesh)—Use a suitable grade.[14]

Cation-exchange Resin, Methacrylic Carboxylic Acid—Use a suitable grade.

Cation-exchange Resin, Polystyrene—Use a suitable grade.[15]

Cation-exchange Resin, Styrene-Divinylbenzene—A strongly acidic, cross-linked sulfonated resin containing about 2% of divinylbenzene. It consists of white to light tan–colored beads which may be relatively free flowing. It is available in the hydrogen form in the 25- to 50-, 45- to 100-, and 80- to 270-mesh sizes. It can be regenerated to the hydrogen form by treating with a hydrochloric acid solution (5 in 100). For satisfactory regeneration a contact time of at least 30 minutes is required after which it must be washed free of excess acid. It is insoluble in water, methanol, and acetonitrile. Suitable for use in column chromatography.[16]

Moisture content of fully regenerated and expanded resin—Transfer 10 to 12 mL of the resin (as received) to a flask, and convert it completely to the hydrogen form by stirring with 150 mL of hydrochloric acid solution (5 in 100) for not less than 30 minutes. Decant the acid, and wash the resin in the same manner with water until the wash water is neutral to litmus (pH 3.5).

Transfer 5 to 7 mL of the regenerated resin to a glass filtering crucible, and remove only the excess surface water by very careful suction filtration. Transfer the conditioned resin to a tared weighing bottle, and weigh. Dry in a vacuum oven at a pressure of 50 mm of mercury at 100° to 105° for 16 hours. Transfer from the vacuum oven to a desiccator, and cool to room temperature. Again weigh. The loss in weight is between 75% and 83%.

Total wet volume capacity—Transfer 3 to 5 mL of the regenerated, undried (see *Moisture content* above) resin to a 5-mL graduated cylinder, and fill it with water. Remove any air bubbles from the resin bed with a stainless steel wire, and settle the resin to its minimum volume by tapping the graduated cylinder. Record the volume of the resin.

Transfer the resin to a 400-mL beaker. Add about 5 g of sodium chloride, and titrate, stirring well, with 0.1 *N* sodium hydroxide to the blue end-point of bromothymol blue (pH 7.0).

$$\frac{\text{net mL NaOH} \times N}{\text{mL of resin}} = \text{mEq/mL}$$

The total wet volume capacity of the resin is more than 0.6 mEq per mL.

Wet screen analysis—The purpose of this test is to properly identify the mesh size of the resin. To obtain an accurate screen analysis would require special apparatus and technique.

Add 150 mL of resin to 200 mL of water in an appropriate bottle, and allow it to stand at least 4 hours to completely swell the resin.

Transfer by means of a graduated cylinder 100 mL of settled and completely swollen resin to the top screen of a series of the designated U. S. Standard 20.3-cm brass screens. Thoroughly wash the resin on each screen with a stream of water until the resin is completely classified, collecting the wash water in a suitable container. Wash the beads remaining on the respective screens back into the 100-mL graduate, and record the volume of settled resin on each screen. At least 70% of the resin will be within the specific mesh size.

Cation-exchange Resin, Styrene-Divinylbenzene, Strongly Acidic—Use a suitable grade.[16]

Cation-exchange Resin, Sulfonic Acid—Use a suitable grade.[17]

Cedar Oil (for clearing microscopic sections)—A selected, distilled oil from the wood of the red cedar, *Juniperus virginiana* Linné (Fam. Pinaceae), should be used for this purpose. Refractive index: about 1.504 at 20°. For use with homogeneous immersion lenses, a specially prepared oil having a refractive index of 1.5150 ± 0.0002 at 20° is required.

Cellulase—Use a suitable grade.

Cellulose, Cereal—A non-nutritive dietetic bulking agent, consisting mainly of finely ground vegetable fibers.

Cellulose, Chromatographic—Use a suitable grade.[18]

Cellulose, Microcrystalline—Use Cellulose, Microcrystalline, FCC.

Cellulose Mixture, Chromatographic—Use a suitable grade.[19]

Ceric Ammonium Nitrate, $Ce(NO_3)_4 . 2NH_4NO_3$—**548.23**—Use ACS reagent grade.

Ceric Ammonium Sulfate, $Ce(SO_4)_2 . 2(NH_4)_2SO_4 . 2H_2O$—**632.53**—Yellow to yellowish orange crystals. Dissolves slowly in water, but more rapidly when mineral acids are present.

Assay—Weigh accurately about 1 g, dissolve in 10 mL of dilute sulfuric acid (1 in 10), and add 40 mL of water. Add orthophenanthroline TS, and titrate with freshly standardized 0.1 *N* ferrous ammonium sulfate VS. Each mL of 0.1 *N* ferrous ammonium sulfate is equivalent to 63.26 mg of $Ce(SO_4)_2 \cdot 2(NH_4)_2SO_4 \cdot 2H_2O$: not less than 94% is found.

Iron—Dissolve 100 mg in 30 mL of dilute sulfuric acid (1 in 10), and add hydrogen peroxide TS, dropwise, until the solution is colorless. Add stronger ammonia TS until the pH is between 1 and 3, cool to room temperature, further adjust to a pH of 3.5 (using a glass electrode), and dilute to 50 mL. To 5 mL of this solution add 5 mL of water, mix, and add 6 mL of hydroxylamine hydrochloride solution (1 in 10) and 4 mL of a slightly acidified solution of orthophenanthroline (1 in 1000): any red color produced is not darker than that of a control containing 0.1 mg of added Fe and the volumes of acid and hydrogen peroxide TS used with the test specimen (0.1%).

Phosphate—Dissolve 200 mg in 30 mL of dilute sulfuric acid (1 in 10), add 30 percent hydrogen peroxide until the solution is colorless, and boil to remove the excess peroxide. Cool, and dilute to 100 mL. To 5 mL of the resulting solution add 55 mL of water, and adjust to a pH of 2 to 3 with ammonium hydroxide. [NOTE—Adjust the pH carefully, since the formation of a permanent precipitate will interfere with succeeding operations. Should a permanent precipitate be formed, discard the solution, and start with a fresh aliquot of the test solution.] Add 500 mg of ammonium molybdate, and adjust to a pH of 1.8 (using a glass electrode) with dilute hydrochloric acid (1 in 10). Heat the solution to boiling, cool, add 10 mL of hydrochloric acid, and then dilute to 100 mL. Transfer to a separator, add 35 mL of ether, shake vigorously, allow to separate, and discard the water layer. Wash the ether layer twice by shaking with separate 10-mL portions of dilute hydrochloric acid (1 in 10), discarding the aqueous layer each time. Add 0.20 mL of a freshly prepared solution of 2 g of stannous chloride in 100 mL of hydrochloric acid, shake well, and allow the layers to separate: any blue color in the ether layer is not darker than that of a control prepared by adding the equivalent of 0.01 mg of PO_4 to 5 mL of dilute sulfuric acid (3 in 25) and treating exactly as the 5 mL of test solution (0.1%).

Ceric Sulfate, $Ce(SO_4)_2$ with a variable amount of water—(anhydrous) **332.24**—It may also contain sulfates of other associated rare earth elements. Yellow to orange-yellow crystals or crystalline powder. Practically insoluble in cold water; slowly soluble in cold dilute mineral acids, but more readily soluble when heated with these solvents.

Assay—Weigh accurately about 800 mg, transfer to a flask, add 25 mL of water and 3 mL of sulfuric acid, and warm until dissolved. Cool, and add 60 mL of a mixture of 1 volume of phosphoric acid and 20 volumes of water. Add 25 mL of potassium iodide solution (1 in 10), insert the stopper in the flask, and allow to stand for 15 minutes. Replace the air over the solution with carbon dioxide, and while continuing the flow of carbon dioxide into the flask, titrate the liberated iodine with 0.1 *N* sodium thiosulfate VS, adding 3 mL of starch TS as the end-point is approached. Each mL of 0.1 *N* sodium thiosulfate is equivalent to 33.22 mg of $Ce(SO_4)_2$. Not less than 80.0% is found.

Chloride (Reagent test)—Dissolve 1 g in a mixture of 5 mL of nitric acid and 4 mL of water. Filter, if necessary, and dilute with water to 20 mL. To 10 mL of the dilution add 1 mL of silver nitrate TS, allow to stand for 10 minutes, and filter until clear. To the remaining 10 mL of test solution add 1 mL of silver nitrate TS: any turbidity produced does not exceed that in a control prepared by adding 0.05 mg of Cl to the filtrate obtained from the first 10 mL of test solution (0.01%).

Heavy metals—Heat 500 mg with a mixture of 10 mL of water and 0.5 mL of sulfuric acid until solution is complete. Cool, dilute with water to 50 mL, and bubble hydrogen sulfide gas through the solution until it is saturated: the precipitate that is formed is white or not darker than pale yellow.

Iron—Dissolve 100 mg in a mixture of 5 mL of water and 2 mL of hydrochloric acid, warming if necessary, and cool. Transfer to a glass-stoppered cylinder, dilute with water to 25 mL, and add 5 mL of ammonium thiocyanate TS and 25 mL of ether. Shake gently, but well, and allow the layers to separate: any pink color in the ether layer is not darker than that of a control, similarly prepared, containing 0.02 mg of added Fe (0.02%).

Cetalkonium Chloride (*Benzylhexadecyldimethylammonium Chloride*), $C_{25}H_{46}ClN$—**396.10**—White leaflets. Soluble in water, in acetone, in alcohol, in carbon tetrachloride, in ether, in ethyl acetate, in glycerin, and in propylene glycol. Melting temperature: about 59°. A solution (1 in 100) has a pH of about 5.

Water, Method I ⟨921⟩: not more than 5.0%.

"Cetrimide"—See *Cetyltrimethylammonium Bromide.*

Cetylpyridinium Chloride—Use *Cetylpyridinium Chloride* (USP monograph).

Cetyltrimethylammonium Bromide, $C_{19}H_{42}BrN$—**364.45**—White, crystalline powder. Soluble in water; freely soluble in alcohol; sparingly soluble in acetone; insoluble in benzene and in ether. (Available commercially as "Cetrimide," in the form of an aqueous solution.)

Assay—Weigh accurately about 300 mg, and dissolve in 100 mL of water. Add 1 drop of glacial acetic acid, and titrate with 0.1 *N* silver nitrate VS, determining the end-point potentiometrically, using a silver–silver chloride electrode system. Each mL of 0.1 *N* silver nitrate is equivalent to 36.45 mg of $C_{19}H_{42}BrN$. Not less than 95.0% is found.

Melting range ⟨741⟩: between 232° and 247°.

Charcoal, Activated (*Activated Carbon; Decolorizing Carbon*)—A fine, black, odorless powder, which is the residue from the destructive distillation of various organic materials, treated to increase its high capacity for adsorbing organic coloring substances, as well as nitrogenous bases.

Adsorptive power—Dissolve 100 mg of strychnine sulfate in 50 mL of water, add 1 g of the test specimen, shake during 5 minutes, and filter through a dry filter, rejecting the first 10 mL of the filtrate. To a 10-mL portion of the subsequent filtrate add 1 drop of hydrochloric acid and 5 drops of mercuric iodide TS: no turbidity is produced.

Residue on ignition (Reagent test)—Ignite 500 mg: the residue weighs not more than 20 mg (4.0%).

Reaction—Boil 2 g with 50 mL of water for 5 minutes, allow to cool, restore the original volume by the addition of sufficient water, and filter: the filtrate is colorless and is neutral to litmus paper.

Acid-soluble substances—Boil 1.0 g with 25 mL of dilute hydrochloric acid (1 in 5) for 5 minutes, filter into a tared porcelain crucible, and wash the residue with 10 mL of hot water, adding the washings to the filtrate. To the combined filtrate and washings add 1 mL of sulfuric acid, evaporate to dryness, and ignite to constant weight: the residue weighs not more than 35 mg (3.5%).

Alcohol-soluble substances—Boil 2 g with 40 mL of alcohol for 5 minutes under a reflux condenser, and filter. Evaporate 20 mL of the filtrate on a steam bath, and dry at 105° for 1 hour: the residue weighs not more than 2 mg (0.2%).

Uncarbonized constituents—To 250 mg add 10 mL of sodium hydroxide TS, heat to boiling, and filter: the filtrate is colorless.

Chloride (Reagent test)—A 5-mL portion of the filtrate obtained in the test for *Reaction* shows not more than 0.04 mg of Cl (0.02%).

Sulfate (Reagent test, *Method I*)—A 5-mL portion of the filtrate from the test for *Reaction* shows not more than 0.3 mg of SO_4 (0.15%).

Sulfide—Place 1 g in a small flask with a narrow neck, add 35 mL of water and 5 mL of hydrochloric acid, and boil gently: the escaping vapors do not darken paper moistened with lead acetate TS.

Charcoal, Adsorbant—Use a suitable grade.

Chloral Hydrate—Use *Chloral Hydrate* (USP monograph).

Chloramine T (*Sodium p-Toluenesulfonchloramide*), $C_7H_7ClNNaO_2S \cdot 3H_2O$—**281.69**—White or faintly yellow efflorescent crystals or crystalline powder, having a slight odor of chlorine. Freely soluble in water and in boiling water; soluble in alcohol with decomposition; insoluble in benzene, in chloroform, and in ether. Store in tight containers, protected from light, in a cold place.

Assay—Weigh accurately about 400 mg, and dissolve in 50 mL of water. Add, in the order named, 10 mL of potassium iodide TS and 5 mL of diluted sulfuric acid, allow to stand for 10 minutes, and titrate the liberated iodine with 0.1 *N* sodium thiosulfate VS. Each mL of 0.1 *N* sodium thiosulfate solution is equivalent to 14.1 mg of $C_7H_7ClNNaO_2S \cdot 3H_2O$. Between 98.0% and 103.0% of $C_7H_7ClNNaO_2S \cdot 3H_2O$ is found.

Ortho-compound—Boil 2.0 g with a mixture of 10 mL of water and 1.0 g of sodium metabisulfite, cool in ice, and filter rapidly: the residue, after being washed with three 5-mL portions of ice-cold water and dried in vacuum over phosphorus pentoxide, melts at a temperature not lower than 134°.
Sodium chloride—Weigh accurately about 1 g, shake with 15 mL of dehydrated alcohol, filter, wash the residue with two 5-mL portions of dehydrated alcohol, and dry at 105° to constant weight: the residue represents not more than 1.5% of the weight taken.

Chloramphenicol—Use *Chloramphenicol* (USP monograph).

Chloranil (*Tetrachlorobenzoquinone*), $C_6Cl_4O_2$—**245.88**—Golden yellow leaflets. Insoluble in water; slightly soluble in alcohol; soluble in ether. Dissolves in sodium hydroxide solutions, forming a violet-red solution.
Melting range ⟨741⟩—When determined in a sealed tube in a bath preheated to 270°, it melts at about 295°.
Residue on ignition ⟨281⟩: negligible, from 200 mg.

Chlorinated Lime—A white or grayish white powder, having the odor of chlorine. Rapidly decomposes on exposure to air. Partially soluble in water and in alcohol. Preserve in air-tight containers in a cool, dry place.
Assay—Transfer to a mortar about 4 g, accurately weighed in a tared weighing bottle, using 50 mL of water. Triturate thoroughly, and pour the mixture into a 1000-mL volumetric flask. Rinse the mortar with water into the flask, dilute with water to volume, and mix. Add to 100 mL of the mixture 1 g of potassium iodide and 5 mL of acetate acid, and titrate the liberated iodine with 0.1 *N* sodium thiosulfate VS, adding 3 mL of starch TS as the end-point is approached. Each mL of 0.1 *N* sodium thiosulfate is equivalent to 3.545 mg of available chlorine (Cl). It shows not less than 30% of available chlorine.

***p*-Chloroacetanilide,** C_8H_8ClNO—**169.61**—White or pale yellow, needle-shaped crystals or crystalline powder. Insoluble in water; soluble in alcohol and in ether.
Solubility—One g dissolves in 30 mL of alcohol to form a clear solution.
Melting range ⟨741⟩: between 178° and 181°.
Residue on ignition (Reagent test): not more than 0.1%.

2-Chloro-4-aminobenzoic Acid—See *4-Amino-2-chlorobenzoic Acid.*

5-Chloro-2-aminobenzophenone—See *2-Amino-5-chlorobenzophenone.*

***p*-Chloroaniline,** $ClC_6H_4NH_2$—**127.57**—White or grayish white crystals. Slightly soluble in water; very soluble in alcohol; soluble in most organic solvents.
Melting range ⟨741⟩—It melts between 70° and 72°, and its acetyl derivative melts within a range of 2° between 169° and 178°.
Chloride—To a solution of 50 mg in 5 mL of diluted alcohol add 3 drops each of diluted nitric acid and of silver nitrate TS: no turbidity is produced in 1 minute.
Sensitiveness—Dissolve 10 mg in 100 mL of water. Dilute 0.5 mL of the solution with water to 5 mL. Add 1 drop of hydrochloric acid, mix, and add 1 drop of sodium nitrite solution (1 in 10). Cool to about 15°, add 2 mL of sodium carbonate TS, then add 0.5 mL of 2-naphthol-3,6-sodium disulfonate solution (1 in 10): a deep red color is produced.
Residue on ignition ⟨281⟩: negligible, from 100 mg.

Chlorobenzene, C_6H_5Cl—**112.56**—Clear, colorless liquid having a characteristic odor. Insoluble in water; soluble in alcohol, in benzene, in chloroform, and in ether.
Specific gravity ⟨841⟩: between 1.100 and 1.111.
Boiling range (Reagent test): not less than 95% distils between 129° and 131°.
Acidity—To 200 mL of methanol add methyl red TS, and neutralize with 0.1 *N* sodium hydroxide, disregarding the amount of alkali consumed. Dissolve 23 mL of the test specimen in the neutralized methanol, and titrate with 0.10 *N* sodium hydroxide: not more than 1.0 mL is required to neutralize the specimen (about 0.015% as HCl).
Residue on evaporation—Evaporate 91 mL on a hot plate, and dry at 105° for 30 minutes: the residue weighs not more than 10 mg (about 0.010%).

4-Chlorobenzoic Acid—ClC_6H_4COOH—**156.57**—White, crystalline solid.
Assay—Dissolve about 700 mg, accurately weighed, in a mixture of 100 mL of hot alcohol and 50 mL of water. Titrate with 0.5 *N* sodium hydroxide VS, determining the end-point potentiometrically. Perform a blank determination, and make any necessary correction. Each mL of 0.5 *N* sodium hydroxide is equivalent to 78.28 mg of ClC_6H_4COOH. Not less than 98% is found.
Solubility—One g dissolved in 25 mL of 0.5 *N* sodium hydroxide yields a clear and complete solution.

4-Chlorobenzophenone, $C_{13}H_9ClO$—**216.66**—Use a suitable grade.

1-Chlorobutane—See *n-Butyl Chloride.*

Chlorobutanol—Use *Chlorobutanol* (NF monograph).

7-Chloro-1,3-dihydro-5-phenyl-2*H*-1,4-benzodiazepin-2-one 4-Oxide—Use USP 7-Chloro-1,3-dihydro-5-phenyl-2*H*-1,4-benzodiazepin-2-one 4-Oxide RS.

Chloroform, $CHCl_3$—**119.38**—Use ACS reagent grade.

Chloroform, Alcohol-free—Use a suitable grade.[20]

Chloroform, Methyl—See *Methyl Chloroform.*

Chloromethylated Polystyrene-Divinylbenzene Anion-exchange Resin—See *Anion-exchange Resin, Chloromethylated Polystyrene-Divinylbenzene.*

1-Chloronaphthalene (*alphachloronaphthalene*), $C_{10}H_7Cl$—**162.62**—Colorless to light yellow liquid.
Assay—Not less than 98% of $C_{10}H_7Cl$ is found, a suitable gas chromatograph equipped with a flame-ionization detector being used. The following conditions have been found suitable: a 1.83-m × 3.2-mm stainless steel column is packed with 7 percent phase G2 on support S1A; the injection port is maintained at 250° and the detector at 310°; the column temperature is programmed to rise at 10° per minute from 50° to 250°.
Refractive index ⟨831⟩: between 1.6320 and 1.6340, at 20°.

2-Chloro-4-nitrophenol, $C_6H_4ClNO_3$—**173.56**—Beige powder.
Melting range ⟨741⟩: between 111° and 113°.
Water, Method I ⟨921⟩: not more than 1.5%.

***p*-Chlorophenol,** C_6H_5ClO—**128.56**—White to pale yellow, crystalline solid, having a characteristic odor. Sparingly soluble in water; very soluble in acetone, in benzene, in ether, and in methanol.
Assay—Transfer about 200 mg, accurately weighed, to a 100-mL beaker, add 25 mL of water, swirl to dissolve, and cautiously add, dropwise, sufficient sodium hydroxide solution to ensure complete solution of the specimen. Transfer the solution to a glass-stoppered, 500-mL flask, using water to rinse the beaker, and dilute with water to about 100 mL. Add 25.0 mL of 0.1 *N* potassium bromide–bromate VS and 10 mL of hydrochloric acid, immediately insert the stopper in the flask, and swirl vigorously for 2 to 3 minutes. Remove the stopper, quickly add 5 mL of potassium iodide solution (1 in 5), taking care to avoid loss of bromine, immediately insert the stopper in the flask, and shake thoroughly for about 1 minute. Remove the stopper, rinse it and the neck of the flask with water, and then titrate with 0.1 *N* sodium thiosulfate VS, adding 3 mL of starch TS as the end-point is approached. Each mL of 0.1 *N* potassium bromide–bromate consumed is equivalent to 6.43 mg of C_6H_5OCl. Not less than 99% is found.
Melting range ⟨741⟩: between 42° and 44°.
Boiling range (Reagent test)—Not less than 90% distils between 218.5° and 221.5°.

Chloroplatinic Acid, $H_2PtCl_6 \cdot 6H_2O$—**517.91**—Use ACS reagent grade Chloroplatinic Acid Hexahydrate.

Chlorotrimethylsilane, C_3H_9ClSi—**108.64**—Clear, colorless to light yellow liquid. Fumes when exposed to moist air.
Caution—It reacts vigorously with water, alcohols, and other hydrogen donors. Store in tight glass containers.
Refractive index ⟨831⟩: between 1.3850 and 1.3890 at 20°.

Chlortetracycline Hydrochloride—Use *Chlortetracycline Hydrochloride* (USP monograph).

Cholestane, $C_{27}H_{48}$—**372.68**—Use a suitable grade.

Cholesterol—Use *Cholesterol* (NF monograph).

Cholesteryl Benzoate, $C_{34}H_{50}O_2$—**490.77**—Use a suitable grade.

Cholesteryl Caprylate—Use USP Cholesteryl Caprylate RS.

Cholesteryl *n*-Heptylate—Use a suitable grade.

Cholic Acid, $C_{24}H_{40}O_5$—**408.58**—Colorless plates or white, crystalline powder. Slightly soluble in water; freely soluble in glacial acetic acid; soluble in alcohol and in acetone.

Assay—Weigh accurately about 400 mg, transfer to a 250-mL conical flask, add 20 mL of water and 40 mL of alcohol, cover with a watch glass, and heat gently on a steam bath until dissolution is complete. Cool, add 5 drops of phenolphthalein TS, and titrate with a 0.1 *N* sodium hydroxide VS to the first pink color that persists for 15 seconds. Perform a blank determination, and make any necessary correction. Each mL of 0.1 *N* sodium hydroxide is equivalent to 40.86 mg of $C_{24}H_{40}O_5$. Not less than 98.0% is found.

Melting range ⟨741⟩: between 197° and 202°.

Specific rotation ⟨781⟩: not less than +37°, determined in a 2 in 100 solution in alcohol.

Loss on drying ⟨731⟩—Dry it in vacuum at 140° for 4 hours: it loses not more than 0.5% of its weight.

Residue on ignition (Reagent test): not more than 0.1%.

Choline Chloride, $HOCH_2CH_2N(CH_3)_3Cl$—**139.62**—White crystals or crystalline powder. Very soluble in water. Is hygroscopic. Store in tight containers.

Assay—Transfer about 100 mg, previously dried at 105° for 2 hours and accurately weighed, to a beaker, add 20 mL of water and 1 drop of aluminum chloride solution (1 in 10), and mix. Add, slowly, 20 mL of a freshly prepared, filtered sodium tetraphenylborate solution (1 in 50), and allow the mixture to stand for 30 minutes with occasional swirling. Filter through a medium-porosity, sintered-glass filter, and wash the beaker and the precipitate with four 10-mL portions of water. The weight of the precipitate, determined after drying at 105° for 2 hours, and multiplied by 0.3298, gives the equivalent weight of $C_5H_{14}ClNO$. Not less than 99.5% is found.

Residue on ignition ⟨281⟩: not more than 0.1%.

Chromatographic Fuller's Earth—See *Fuller's Earth, Chromatographic.*

Chromatographic *n*-Heptane—See *n-Heptane, Chromatographic.*

Chromatographic Magnesium Oxide—See *Magnesium Oxide, Chromatographic.*

Chromatographic Reagents—[NOTE—Listings of the numerical designations for phases (G), packings (L), and supports (S), together with corresponding brand names, are published periodically in *Pharmacopeial Forum* as a guide for the chromatographer.]

Chromatographic Silica Gel—See *Silica Gel, Chromatographic.*

Chromatographic Silica Gel Mixture—See *Silica Gel Mixture, Chromatographic.*

Chromatographic Siliceous Earth—See *Siliceous Earth, Chromatographic.*

Chromatographic Siliceous Earth, Silanized—See *Siliceous Earth, Chromatographic, Silanized.*

Chromatographic Solvent Hexane—See *Hexane, Solvent, Chromatographic.*

Chromium Trioxide, CrO_3—**99.99**—Use ACS reagent grade.

Chromophore Substrate (coagulation)—A reaction compound consisting of synthetic molecules of tri- or tetra-peptides coupled to a chromophore, e.g., methane sulfonyl-D-Leu-Gly-Arg-pNA. The arginine group at the terminal amino acid portion is specific for the protease utilized and it has a *p*-nitroaniline end-group covalently attached to the carbonyl group of the arginine. The synthetic peptide molecules mimic the peptide sequence of the site of action of the natural substrate specific for the activated coagulation factor to be measured. Its label states that its molecular weight is within the range of 600 to 750. The complete substrate is colorless, and the coagulation factor to be measured catalyzes the splitting of the chromophore (*p*-nitroaniline) from the peptide. The amount of release is measured directly by the colored chromophore. The substrate, to be usable for the purpose, is soluble to the extent necessary, and for the measurement of Factor X_a is reactive at a concentration based on the molecular weight stated on the label of 2.5 µm per mL or less. Different chromophore substrate preparations differ in sensitivity, and are to be used with optimal incubation periods.

Chromotropic Acid (*1,8-Dihydroxynaphthalene-3,6-disulfonic Acid*), $C_{10}H_8O_8S_2 \cdot 2H_2O$—**356.32**—Use ACS reagent grade.

Cinchonidine, $C_{19}H_{22}N_2O$—**294.40**—White crystals, crystalline or granular powder. Soluble in alcohol and in chloroform; practically insoluble in water.

Assay—Dissolve about 125 mg, accurately weighed, in 50 mL of glacial acetic acid. Add a few drops of *p*-naphtholbenzein TS and titrate with 0.1 *N* perchloric acid VS. Perform a blank determination and make any necessary correction. Each mL of 0.1 *N* perchloric acid is equivalent to 14.72 mg of $C_{19}H_{22}N_2O$. Not less than 99.0% is found.

Loss on drying ⟨731⟩—Dry it at 105° to constant weight: it loses not more than 1.0% of its weight.

Melting range ⟨741⟩: between 200° and 205°.

Specific rotation ⟨781⟩: between −105° and −115°, calculated on the dried basis, determined in a solution in alcohol containing 10 mg per mL.

Cinchonine, $C_{19}H_{22}N_2O$—**294.40**—White crystals, crystalline or granular powder. Slightly soluble in chloroform, sparingly soluble in alcohol and practically insoluble in water.

Assay—Dissolve about 125 mg, accurately weighed, in 50 mL of glacial acetic acid. Add a few drops of *p*-naphtholbenzein TS and titrate with 0.1 *N* perchloric acid VS. Perform a blank determination and make any necessary correction. Each mL of 0.1 *N* perchloric acid is equivalent to 14.72 mg of $C_{19}H_{22}N_2O$. Not less than 99.0% is found.

Loss on drying ⟨731⟩—Dry it at 105° to constant weight: it loses not more than 1.0% of its weight.

Melting range ⟨741⟩: between 255° and 261°.

Specific rotation ⟨781⟩: between +219° and +229°, calculated on the dried basis, determined in a solution in alcohol containing 50 mg per 10 mL.

Citric Acid—Use *Citric Acid* (USP monograph).

Citric Acid, Anhydrous, $C_6H_8O_7$—**192.12**—Use ACS reagent grade Citric Acid, Anhydrous.

Cobalt Chloride (*Cobaltous Chloride*), $CoCl_2 \cdot 6H_2O$—**237.93**—Use ACS reagent grade.

Cobalt Nitrate, $Co(NO_3)_2 \cdot 6H_2O$—**291.03**—Use ACS reagent grade.

Cobaltous Acetate (*Cobalt Acetate*), $Co(C_2H_3O_2)_2 \cdot 4H_2O$—**249.08**—Red, needle-like crystals. Soluble in water and in alcohol.

Insoluble matter (Reagent test): not more than 1 mg, from 5 g, dissolved in 100 mL of water containing 2 mL of glacial acetic acid (0.02%).

Chloride (Reagent test)—One g shows not more than 0.1 mg of Cl (0.01%).

Nitrate—Dissolve 500 mg in 10 mL of water, add, with stirring, 10 mL of sodium hydroxide TS, and heat on a steam bath for 30 minutes. Cool, dilute with water to 20 mL, mix, and filter. To 10 mL of the filtrate add 5 mg of sodium chloride, 0.1 mL of indigo carmine TS, and 10 mL of sulfuric acid: the blue color does not entirely disappear in 1 minute (about 0.02%).

Sulfate (Reagent test, *Method II*)—The filtrate from the test for *Insoluble matter*, exclusive of washings, yields not more than 2.5 mg of residue (0.02%).

Substances not precipitated by hydrogen sulfide—Dissolve 2 g in about 90 mL of water, and add 2 g of ammonium chloride and sufficient ammonia TS to redissolve the precipitate first formed. Pass hydrogen sulfide into this solution until the cobalt is completely precipitated. Dilute with water to 100.0 mL, mix, and filter. Evaporate 50 mL of the filtrate nearly to dryness, add 0.5 mL of sulfuric acid, and ignite at 800 ± 25° to constant weight: the residue weighs not more than 3 mg (0.3% as SO_4).

Copper—Dissolve 500 mg in 30 mL of water, and add 1 mL of hydrochloric acid (A). Dissolve another 500 mg in 20 mL of water, and add 1 mL of hydrochloric acid and 10 mL of hydrogen sulfide TS (B). No difference in color between A and B is noticeable.

Nickel—Dissolve 1 g in 200 mL of water, add 1 g of sodium citrate, heat to boiling, add 100 mL of an alcohol solution of dimethylglyoxime (1 in 100), then add 15 mL of ammonia TS, and allow to stand overnight. Filter through a tared filtering crucible, wash with water, then with diluted alcohol, and dry at 105° to constant weight: the precipitate weighs not more than 25 mg (0.5%).

Cobaltous Chloride—See *Cobalt Chloride.*

Cod Liver Oil—Use *Cod Liver Oil* (USP monograph).

Congo Red, $C_{32}H_{22}N_6Na_2O_6S_2$—**696.66**—A dark red or reddish brown powder. Is odorless and decomposes on exposure to acid fumes. Its solutions have a pH of about 8 to 9.5. One g dissolves in about 30 mL of water. Is slightly soluble in alcohol.

Loss on drying ⟨731⟩—Dry it at 105° for 4 hours: it loses not more than 3.0% of its weight.

Residue on ignition—Weigh accurately about 1 g, previously dried at 105° for 4 hours, and place it in a porcelain dish or crucible. Ignite carefully until well charred, cool, add 2 mL of sulfuric acid, and carefully ignite until the residue is white or practically so. Cool, add 0.5 mL of sulfuric acid and 1 mL of nitric acid, evaporate, and again ignite to constant weight: the weight of the sodium sulfate so obtained is between 20.0% and 24.0% of the weight of the dried specimen taken.

Sensitiveness—To 50 mL of carbon dioxide–free water add 0.1 mL of congo red solution (1 in 1000). The red color of the solution is changed to violet by the addition of 0.05 mL of 0.10 *N* hydrochloric acid and is restored by the subsequent addition of 0.05 mL of 0.10 *N* sodium hydroxide.

Copper, Cu—At. Wt. **63.546**—Use ACS reagent grade.

Corn Oil—Use *Corn Oil* (NF monograph).

Cortisone, $C_{21}H_{28}O_5$—**360.45**—White, crystalline powder. Practically insoluble in water; sparingly soluble in alcohol and in acetone. Melts at about 220°, with decomposition.

Absorption maximum—The ultraviolet absorption spectrum of a 1 in 100,000 solution in alcohol shows a maximum at about 238 nm.

Specific rotation ⟨781⟩: about +209°, determined in a 1 in 100 solution in alcohol.

Cotton, Absorbent—Use *Purified Cotton* (USP monograph).

Cotton, Nonabsorbent—The hairs of the seed of cultivated varieties of *Gossypium hirsutum* Linné, or of other species of *Gossypium* (Fam. Malvaceae), freed from adhering impurities and linters, and sterilized.

Cotton, Purified—Use *Purified Cotton* (USP monograph).

Cottonseed Oil—Use *Cottonseed Oil* (NF monograph).

Cresol—Use *Cresol* (NF monograph).

Pipet 20 mL from each of the two tubes containing *Standard phenol solution* into separate 100-mL volumetric flasks, add 5 mL of nitric acid, dilute with water to volume, and mix. Transfer the solutions to separate burets designated *B1* and *B2*, representing, respectively, the solution not treated and the solution treated with formaldehyde.

Pipet 10 mL from each of the two tubes containing the test specimen into separate 50-mL color comparison tubes designated *N1* and *N2*, representing, respectively, the solution treated with formaldehyde and the solution not treated with formaldehyde.

Add to tube *N1* the solution from buret *B1*, and add to tube *N2* the solution from buret *B2* until the colors in tubes *N1* and *N2* match when compared in a colorimeter. Calculate the percentage of phenol in the test specimen by the formula:

$$5V/W,$$

in which V is the volume, in mL, of solution taken from buret *B1*, and W is the weight, in g, of specimen. Not more than 5.0% is found.

***m*-Cresol Purple,** $C_{21}H_{18}O_5S_2$—**382.43**—Use a suitable grade.

Cupric Acetate, $Cu(C_2H_3O_2)_2 \cdot H_2O$—**199.65**—Use ACS reagent grade.

Cupric Chloride, $CuCl_2 \cdot H_2O$—**170.48**—Bluish green deliquescent crystals. Freely soluble in water; soluble in alcohol; slightly soluble in ether.

Insoluble matter (Reagent test): not more than 1.0 mg, from 10 g (0.010%). Save the combined filtrate and washings for the test for *Sulfate.*

Nitrate—Dissolve 500 mg in 30 mL of dilute sulfuric acid (1 in 30). Slowly add the solution, with constant stirring, to 20 mL of sodium hydroxide solution (1 in 10), and digest on a steam bath for 15 minutes. Cool, with water dilute to 50 mL, and filter. To 10 mL of the clear filtrate, add 0.05 mL of indigo carmine TS followed by 10 mL of sulfuric acid: the blue color does not entirely disappear within 5 minutes (about 0.15%).

Sulfate (Reagent test, *Method II*)—The combined filtrate and washings retained from the test for *Insoluble matter* yield not more than 1.2 mg of residue (0.005%).

Substances not precipitated by hydrogen sulfide—Dissolve 2 g in 100 mL of water, add 1 mL of sulfuric acid, heat the solution to 70°, and pass hydrogen sulfide into the solution until the copper is completely precipitated. Allow the precipitate to settle, and filter without washing. Transfer 50.0 mL of the filtrate to a tared evaporating dish, and evaporate on a steam bath to dryness. Gently ignite the dish over a flame, and then at 800 ± 25° to constant weight. Cool, and weigh: the residue weighs not more than 1.0 mg (0.1%). Retain the residue for the *Iron* test.

Iron ⟨241⟩—To the residue retained from the preceding test, add 2 mL of hydrochloric acid, 2 mL of water, and 0.05 mL of nitric acid. Evaporate slowly on a steam bath to dryness, then take up the residue in 1 mL of hydrochloric acid and 10 mL of water. Dilute with water to 100 mL, and mix. To 20 mL of the dilution add 10 mL of water, and mix: 10 mL of this solution shows not more than 0.01 mg of Fe (0.015%). Retain the residue dilution for use in the test for *Other metals.*

Other metals—To 20 mL of the residue solution retained from the test for *Iron* add a slight excess of ammonium hydroxide, boil the solution for 1 minute, filter, and wash the residue with water until the combined filtrate and washings measure 20 mL. Neutralize the filtrate with diluted hydrochloric acid, dilute with water to 25 mL, and add 0.15 mL of ammonium hydroxide and 1 mL of hydrogen sulfide TS: any color produced is not darker than that of a control containing, in the same volume, 0.15 mL of ammonium hydroxide, 1 mL of hydrogen sulfide TS, and 0.02 mg of added Ni (0.01% as Ni).

Cupric Citrate (*[Citrato(4-)]dicopper*), $Cu_2C_6H_4O_7$—**315.19**—Use a suitable grade.

Cupric Nitrate Hydrate, $Cu(NO_3)_2 \cdot 2.5H_2O$—**232.59** or $Cu(NO_3)_2 \cdot 3H_2O$—**241.60**—Use ACS reagent grade.

Cupric Sulfate, $CuSO_4 \cdot 5H_2O$—**249.68**—Use ACS reagent grade.

Cupric Sulfate, Anhydrous, $CuSO_4$—**159.60**—A white or grayish white powder free from a blue tinge. Upon the addition of a small quantity of water, it becomes blue. Soluble in water. Store in tight containers.

Chloride (Reagent test)—One g shows not more than 0.02 mg of Cl (0.002%).

Substances not precipitated by hydrogen sulfide—Determine as directed for ACS reagent grade of *Cupric Acetate:* the residue weighs not more than 6 mg (0.15%).

Cyanogen Bromide, BrCN—**105.92**—Colorless crystals. Volatilizes at room temperature. Its vapors are highly irritating and *very toxic.* Melts at about 52°. Freely soluble in water and in alcohol. Store in tight containers in a cold place.

Solubility—Separate 1-g portions dissolve completely in 10 mL of water and in 10 mL of alcohol, respectively, to yield colorless solutions.

Cyanopropyl-phenyl Silicone—Use a suitable grade.[21]

Cyclohexane, C_6H_{12}—**84.16**—Use ACS reagent grade.

Cyclohexanol, $C_6H_{12}O$—**100.16**—A clear liquid having a camphoraceous odor. Melting temperature: about 23°. Specific gravity: about 0.962, at 20°. Freely soluble in water. Miscible with alcohol, with ethyl acetate, and with aromatic hydrocarbons.

Assay—When examined by gas-liquid chromatography, using suitable gas chromatographic apparatus and conditions, it shows a purity of not less than 98%.

Cyclohexanone, Chromatographic—Use a suitable grade.

Cycloheximide (*3-[2-(3,5-Dimethyl-2-oxocyclohexyl)-2-hydroxyethyl]glutarimide*), $C_{15}H_{23}NO_4$—**281.35**—Antibiotic substance isolated from the beers of streptomycin-producing strains of *Streptomyces griseus.* Sparingly soluble in water at 2°; soluble in amyl acetate at 2°; soluble in chloroform, in ether, in acetone, in methanol, and in ethanol.

Specific rotation ⟨781⟩: about −6.8°, determined in a solution containing 2 g in each 100 mL.

Melting range ⟨741⟩: between 115° and 119°.

Cyclohexylamine, $\overline{CH_2(CH_2)_4C}HNH_2$—**99.18**—A clear, colorless to slightly yellow liquid, having a strong characteristic amine odor. Strong base. Miscible with water, with alcohol, with ether, with esters, and with ketones. Refractive index: about 1.456. Flash point: about 30°.

Specific gravity ⟨841⟩: between 0.8645 and 0.8655.

pH of solution (1 in 10,000) ⟨791⟩: about 10.5.

Boiling range (Reagent test)—Not less than 90% distils between 134° and 135°.

(1,2-Cyclohexylenedinitrilo)tetraacetic Acid (*trans-1,2-Diaminocyclohexane-N,N,N′,N′-tetraacetic Acid*), $(HOCOCH_2)_2$-$\overline{NCH(CH_2)_4C}HN(CH_2COOH)_2 . H_2O$—**364.35**—Use ACS reagent grade.

L-Cystine, $HOOC(NH_2)CHCH_2S$—$SCH_2CH(NH_2)COOH$—**240.29**—A white, crystalline powder. Very slightly soluble in water; soluble in dilute mineral acids and in solutions of alkali hydroxides; insoluble in alcohol and in other organic solvents.

Specific rotation ⟨781⟩: between −215° and −225°, determined in a 2 in 100 solution of test specimen, previously dried over silica gel for 4 hours, in dilute hydrochloric acid (1 in 10) at a temperature of 20°.

Loss on drying ⟨731⟩—Dry it over silica gel for 4 hours: it loses not more than 0.2% of its weight.

Residue on ignition (Reagent test): not more than 0.1%.

Decane Sodium Sulfonate—Use a suitable grade.[22]

Decanol (*n-Decyl Alcohol*)—$C_{10}H_{22}O$—**158.28**—A clear, viscous liquid. Specific gravity: about 0.83 at 20°. Solidifies at about 6.5°. Insoluble in water; soluble in alcohol and in ether.

Assay—When examined by gas-liquid chromatography, using suitable gas chromatographic apparatus and conditions, it shows a purity of not less than 99%.

Decyl Sodium Sulfate, $C_{10}H_{21}NaO_4S$—**260.32**—White, crystalline solid.

Assay—Transfer about 1 g, accurately weighed, to a suitable, tared crucible, moisten with a few drops of sulfuric acid, and ignite gently to constant weight. Each mg of residue is equivalent to 3.662 mg of $C_{10}H_{21}NaO_4S$. Not less than 95% is found.

Dehydrated Alcohol—See *Alcohol, Dehydrated.*

Desoxycorticosterone Acetate—Use *Desoxycorticosterone Acetate* (USP monograph).

Deuterated Water—See *Deuterium Oxide.*

Deuterium Oxide, D_2O—**20.028**—Use a suitable grade having a minimum isotopic purity of 99.8 atom % of deuterium.

Deuterochloroform, $CDCl_3$—**120.37**—Use a suitable grade.

Devarda's Alloy (*Devarda's Metal*)—A gray powder composed of 50 parts of copper, 45 parts of aluminum, and 5 parts of zinc.

Dextran 80—Use a suitable grade.

Dextran Gel, Chromatographic Cross-linked—Use a suitable grade.[23]

Dextrin, $(C_6H_{10}O_5)_n . xH_2O$—A white amorphous powder. Slowly soluble in cold water; more readily soluble in hot water; insoluble in alcohol.

Insoluble matter—Boil 1 g with 30 mL of water in a small flask: the solution is colorless and clear, or not more than opalescent.

Loss on drying ⟨731⟩—Dry it at 105° to constant weight: it loses not more than 10.0% of its weight.

Residue on ignition (Reagent test)—Ignite 1 g with 0.5 mL of sulfuric acid: the residue weighs not more than 5 mg (0.5%).

Chloride (Reagent test)—Dissolve 3 g in 75 mL of boiling water, cool, dilute with water to 75 mL, and filter if necessary. To 25 mL of the filtrate add 2 mL of nitric acid and 1 mL of silver nitrate TS, and allow to stand for 5 minutes: any turbidity produced is not greater than that of a control containing 0.02 mg of added Cl (0.002%).

Sulfate (Reagent test, *Method I*)—To a 25-mL portion of the filtrate from the preceding test add 0.5 mL of diluted hydrochloric acid and 2 mL of barium chloride TS, and allow to stand for 10 minutes: any turbidity produced is not greater than that of a control containing 0.2 mg of added SO_4 (0.02%).

Alcohol-soluble substances—Boil 1 g with 20 mL of alcohol for 5 minutes under a reflux condenser, and filter while hot. Evaporate 10 mL of the filtrate on a steam bath, and dry at 105°: the residue weighs not more than 5 mg (1%).

Reducing sugars—Shake 2 g with 100 mL of water for 10 minutes, and filter until clear. To 50 mL of the filtrate add 50 mL of alkaline cupric tartrate TS, and boil for 3 minutes. Filter through a tared filtering crucible, wash with water, then with alcohol, and finally with ether, and dry at 105° for 2 hours: the precipitate of cuprous oxide weighs not more than 115 mg (corresponding to about 5% of reducing sugars as dextrose).

Dextro Calcium Pantothenate—Use *Calcium Pantothenate* (USP monograph).

Dextromethorphan Hydrobromide—Use *Dextromethorphan Hydrobromide* (USP monograph).

Dextrose—Use *Dextrose* (USP monograph).

Dextrose, Anhydrous, $C_6H_{12}O_6$—**180.16**—Use ACS reagent grade D-Glucose, Anhydrous.

Diacetyl—See *2,3-Butanedione.*

3,3′-Diaminobenzidine Hydrochloride, $(NH_2)_2C_6H_3C_6H_3$-$(NH_2)_2 . 4HCl$—**360.11**—White to yellowish tan (occasionally purple), needle-shaped crystals. Soluble in water. Stable in organic solvents but unstable in aqueous solution at room temperature. Store aqueous solutions in a refrigerator.

Insoluble matter—Dissolve 2 g in 100 mL of water, without heating, and filter immediately: the insoluble residue does not exceed 1 mg (0.05%).

Residue on ignition (Reagent test): not more than 1 mg, from 2 g (0.05%).

Suitability test for detection of selenium—Dissolve 1.633 g of selenious acid (H_2SeO_3) in water, and dilute with water to 1 liter. Dilute 10 mL of this solution with water to 1 liter, to make a solution containing 0.010 mg of Se per mL. Place 1 mL of the resulting solution in a 100-mL beaker, add 2 mL of formic acid solution (1 in 7), and dilute with water to 50 mL. Add 2 mL of 3,3′-diaminobenzidine hydrochloride solution (1 in 200), and allow to stand for 30 to 50 minutes. Adjust with 6 *N* ammonium hydroxide to a pH between 6 and 7. Transfer to a 125-mL separator, add 10.0 mL of toluene, and shake vigorously for 30 seconds: a distinct yellow color is produced in the toluene layer. A blank containing diaminobenzidine hydrochloride but no selenium standard, treated in the same manner, shows no color in the toluene layer.

2,3-Diaminonaphthalene, $C_{10}H_{10}N_2$—**158.20**—Use a suitable grade.

***o*-Dianisidine Dihydrochloride,** $C_{14}H_{16}N_2O_2 . 2HCl$—**317.21**—Pale yellow crystals. Soluble in water.

Assay—Transfer about 800 mg, accurately weighed, to a beaker, dissolve in 100 mL of water, and add 50 mL of alcohol. When solution is complete, titrate with 0.5 *N* sodium hydroxide VS, determining the end-point potentiometrically. Perform a blank titration, and make any necessary correction. Each mL of 0.5 *N* sodium hydroxide is equivalent to 158.6 mg of $C_{14}H_{16}N_2O_2 . 2HCl$. Not less than 98% is found.

Solubility—Dissolve 500 mg in 25 mL of water: the solution is clear and complete.

Diastatic Enzyme Preparation—Determine suitability for use in preparation of the *Assay Preparation* in the *Thiamine Assay* ⟨531⟩.

Diatomaceous Earth, Flux-calcined—Use a suitable grade.[24]

Diatomaceous Earth, Silanized—Use a suitable grade.[25]

Diatomaceous Silica, Calcined[26]—A form of silica (SiO_2) consisting of fused frustules and fragments of diatoms. It is an amorphous, fine, light pink or white powder. Insoluble in water, in acids, and in dilute solutions of alkali hydroxides.

Loss on ignition—Weigh accurately about 4 g, and ignite to constant weight: it loses not more than 10.0% of its weight.

Organic impurities—It does not darken appreciably upon ignition.

Loss on drying ⟨731⟩—Dry it at 110° for 2 hours: it loses not more than 2.0% of its weight.

Diatomaceous Silica, Flux-calcined.[27]

Dibasic Ammonium Citrate—See *Ammonium Citrate, Dibasic.*

Dibasic Ammonium Phosphate—See *Ammonium Phosphate, Dibasic.*

Dibasic Potassium Phosphate—See *Potassium Phosphate, Dibasic.*

Dibenzyl—See *Bibenzyl.*

2,6-Dibromoquinone-chlorimide (*2,6-Dibromo-N-chloro-p-benzoquinone Imine; DBQ Reagent*), $O{:}C_6H_2Br_2{:}NCl$—**299.35**—A yellow, crystalline powder. Insoluble in water; soluble in alcohol and in dilute alkali hydroxide solutions.

Melting range ⟨741⟩: between 82° and 84°.

Solubility in alcohol—A solution of 100 mg in 10 mL of alcohol is not more than faintly turbid.

Residue on ignition (Reagent test)—Ignite 500 mg with 0.5 mL of sulfuric acid: the residue weighs not more than 1 mg (0.2%).

Sensitiveness—To 10 mL of a water solution containing 0.01 mg of phenol add 0.3 mL of a sodium borate buffer (made by dissolving 2.84 g of crystallized sodium borate in 90 mL of warm water, adding 8.2 mL of 1 *N* sodium hydroxide and diluting with water to 100 mL) and 0.1 mL of a solution of 10 mg of the test specimen in 20 mL of alcohol: a distinct blue color develops within 10 minutes.

Dibutyl Phthalate, $C_{16}H_{22}O_4$—**278.35**—Clear, colorless liquid.

Assay—Weigh accurately about 2 g into a suitable flask, add 25.0 mL of 1 *N* sodium hydroxide and 30 mL of isopropyl alcohol, and mix. Digest the mixture at a temperature near boiling for 30 minutes, then cool in a water bath to room temperature. Add phenolphthalein TS, and titrate with 1 *N* sulfuric acid VS to the disappearance of the pink color. Perform a complete blank determination, and make any necessary correction. Each mL of 1 *N* sulfuric acid consumed is equivalent to 139.2 mg of $C_{16}H_{22}O_4$. Not less than 98% is found.

Refractive index ⟨831⟩: between 1.491 and 1.493 at 20°.

Acid content—Weigh accurately about 10 g, and dissolve in 100 mL of an alcohol-ether mixture (1:1). Add phenolphthalein TS, and titrate immediately with 0.05 *N* alcoholic potassium hydroxide VS. Each mL of 0.05 *N* alcoholic potassium hydroxide is equivalent to 4.15 mg of phthalic acid: not more than 0.02% is found.

2,5-Dichloroaniline, $Cl_2C_6H_3NH_2$—**162.02**—White, needle-like crystals. Slightly soluble in water; soluble in alcohol and in ether.

Melting range, Class I ⟨741⟩: between 49° and 50°.

2,6-Dichloroaniline, $C_6H_5Cl_2N$—**162.02**—Off-white powder.

Melting range ⟨741⟩: between 38° and 41°.

***o*-Dichlorobenzene,** $C_6H_4Cl_2$—**147.00**—Clear liquid, having a light yellowish brown tint (about APHA 20) and an aromatic odor. Practically insoluble in water. Miscible with alcohol and with ether. Boils at about 180°.

Assay—When examined by gas-liquid chromatography, with the use of suitable apparatus and conditions, it shows a purity of not less than 98%.

Density: between 1.299 and 1.301.

Refractive index ⟨831⟩: between 1.548 and 1.550 at 25°.

Residue on evaporation—Evaporate 80 mL on a steam bath, and dry at 105° for 1 hour: the residue weighs not more than 50 mg (0.005%).

Acidity—Add phenolphthalein TS to 25 mL of methanol, and titrate with 0.02 *N* alcoholic potassium hydroxide VS until a faint pink color persists for 15 seconds. Pipet 25 mL of test specimen into the solution, mix, avoiding exposure to the atmosphere, and titrate with 0.02 *N* alcoholic potassium hydroxide VS: not more than 2.2 mL is required to restore the pink color (about 0.005%).

1,2-Dichloroethane—See *Ethylene Dichloride.*

Dichlorofluorescein, $C_{20}H_{10}Cl_2O_5$—**401.20**—[NOTE—This specification covers both the 4,5- and 2,7-isomers of dichlorofluorescein, either of which is suitable for the preparation of dichlorofluorescein TS.] A weak orange-colored, crystalline powder. Sparingly soluble in water; soluble in alcohol and in solutions of alkali hydroxides.

Residue on ignition (Reagent test)—Ignite 200 mg with 5 drops of sulfuric acid: the residue weighs not more than 1 mg (0.5%).

Sensitiveness—Dissolve 100 mg in 60 mL of alcohol, add 2.5 mL of 0.1 *N* sodium hydroxide, and dilute with water to 100 mL. Add 1 mL of this solution to a solution of potassium iodide prepared by dissolving 100 mg of potassium iodide, previously dried at 105° to constant weight and accurately weighed, in 50 mL of water containing 1 mL of glacial acetic acid, and titrate with 0.1 *N* silver nitrate VS until the color of the precipitate changes from pale yellowish orange to pink. The volume of 0.1 *N* silver nitrate consumed is not more than 0.10 mL greater than the calculated volume, the calculated volume being based upon the KI content of the dried specimen as determined in the *Assay* under *Potassium Iodide* (USP monograph).

2,6-Dichloroindophenol Sodium—See *2,6-Dichlorophenolindophenol Sodium.*

Dichloromethane—Use *Methylene Chloride.*

2,4-Dichloro-1-naphthol, $C_{10}H_6OCl_2$—**213.06**—Light tan powder.

Melting range ⟨741⟩: between 103° and 107°, but the range between beginning and end of melting does not exceed 2°.

2,6-Dichlorophenol-indophenol Sodium (*2,6-Dichloro-indophenol Sodium*), $O{:}C_6H_2Cl_2{:}NC_6H_4ONa$ with about $2H_2O$—**290.08** (anhydrous)—Use ACS reagent grade.

2,6-Dichloroquinone-chlorimide (*2,6-Dichloro-N-chloro-p-benzoquinone Imine*), $O{:}C_6H_2Cl_2{:}NCl$—**210.45**—Pale yellow, crystalline powder. Insoluble in water; soluble in alcohol and in dilute alkali hydroxide solutions.

Melting range ⟨741⟩: between 65° and 67°.

Solubility in alcohol—A solution of 100 mg in 10 mL of alcohol is complete and clear.

Residue on ignition—Ignite 500 mg with 0.5 mL of sulfuric acid: the residue weighs not more than 1 mg (0.2%).

Sensitiveness—It meets the requirements of the test for *Sensitiveness* under *2,6-Dibromoquinone-chlorimide.*

Dichlorotetrafluoroethane—Use *Dichlorotetrafluoroethane* (NF monograph).

Dicyclohexylamine, $(C_6H_{11})_2NH$—**181.32**—Clear, strongly alkaline liquid, having a faint, fishy odor. Sparingly soluble in water. Miscible with common organic solvents. Density: 0.9104. Solidifies at −0.1°; melts at about 20°.

Assay—Weigh accurately about 400 mg in a tared, small weighing bottle equipped with a well-fitting closure. Transfer the stoppered bottle to a 250-mL beaker, add sufficient glacial acetic acid TS to cover the bottle, and open the bottle under the surface of the acid. Add crystal violet TS, and titrate with 0.1 *N* perchloric acid VS. Each mL of 0.1 *N* perchloric acid is equivalent to 18.13 mg of $(C_6H_{11})_2NH$. Not less than 98% is found.

Specific gravity ⟨841⟩: between 0.911 and 0.917.

Boiling range (Reagent test): between 255° and 257°.

Water, Method I ⟨921⟩: not more than 0.5%.

Diethylamine, $(C_2H_5)_2NH$—**73.14**—Colorless, flammable, strongly alkaline liquid. Miscible with water and with alcohol. Forms a hydrate with water. *May be irritating to skin and mucous membranes.* Store in well-closed containers.

Assay—To 50 mL of water add 6 to 8 drops of a freshly prepared mixed indicator (prepared by mixing 5 parts of a 1 in 1000 solution of bromocresol green in methanol with 1 part of a 1 in 1000 solution of methyl red in methanol), and neutralize with 0.1 *N* hydrochloric acid to the disappearance of the green color. Weigh accurately about 2 g of specimen in a tared, glass-stoppered, 250-mL conical flask containing a few mL of the neutralized water. Add the remainder of the neutralized water, and titrate with 1 *N* hydrochloric acid VS to the disappearance of the green color. Each mL of 1 *N* hydrochloric acid is equivalent to 73.1 mg of $(C_2H_5)_2NH$. Not less than 99.0% is found.

Specific gravity ⟨841⟩: between 0.700 and 0.705.

Boiling range (Reagent test): between 55° and 58°.

Residue after evaporation—Evaporate 14 mL (10 g) in a tared dish on a steam bath to dryness, dry at 105° for 1 hour, cool, and weigh: the weight of the residue does not exceed 1.0 mg (0.010%).

Water-insoluble substances—Transfer 25 mL to a 125-mL conical flask, and add 25 mL of water in 5-mL portions, shaking the flask well after each addition. Add another 25 mL of specimen to 25 mL of water in the same manner. No cloudiness or turbidity is produced in either instance.

***N,N*-Diethylaniline,** $C_6H_5N(C_2H_5)_2$—**149.24**—Light yellow to amber liquid.

Assay—Inject an appropriate specimen (about 0.2 µL) into a suitable gas chromatograph (see *Chromatography* ⟨621⟩) equipped with a flame-ionization detector, helium being used as the carrier gas flowing at about 40 mL per minute. The following conditions have been found suitable: a 1.8-m × 3-mm stainless steel column containing 20 percent phase G16 on support S1A; the injection port temperature is maintained at 250°; the column temperature is maintained at 140° and programmed to rise 6° per minute to 200°. The detector temperature is maintained at 310°. The area of the *N,N*-diethylaniline peak having a retention time of about 4.9 minutes is not less than 99% of the total peak area.

Refractive index ⟨831⟩: between 1.5405 and 1.5425 at 20°.

Diethylene Glycol, $C_4H_{10}O_3$—**106.12**—A colorless to faintly yellow liquid, having a mild odor. Miscible with water, with alcohol, with ether, and with acetone. Insoluble in benzene and in carbon tetrachloride.

Specific gravity ⟨841⟩: between 1.117 and 1.120 at 20°.

Distilling range ⟨721⟩: between 240° and 250°.

Acidity—Transfer 54 mL (60 g) to a 250-mL conical flask, add phenolphthalein TS, and titrate with 0.02 *N* alcoholic potassium hydroxide VS to the production of a pink color that is stable for at least 15 seconds: not more than 2.5 mL is consumed (0.005% as CH_3COOH).

Water ⟨921⟩: not more than 0.2%.

Residue on ignition ⟨281⟩—Transfer 50 g to a tared platinum dish, heat the dish gently until the vapors ignite, and allow the specimen to burn completely. Ignite the residue at 800 ± 25°, cool, and weigh: the residue weighs not more than 2.5 mg (0.005%).

Diethylene Glycol Succinate Polyester, $(OCH_2CH_2OCH_2CH_2OOCCH_2CH_2COO)_n$—Clear, viscous liquid. Soluble in chloroform. Is stabilized by modification of the polyester, diethylene glycol succinate, to render it suitable for use in gas-liquid chromatography to a temperature of 200°.[28]

Diethylene Succinate—Use a suitable grade.

Diethylenetriamine, $C_4H_{13}N_3$—**103.17**—Colorless liquid.

Assay—Inject an appropriate specimen into a suitable gas chromatograph (see *Chromatography* ⟨621⟩) equipped with a flame-ionization detector, helium being used as the carrier gas. The following conditions have been found suitable: a 30-m × 0.25-mm capillary column coated with G2. The injection port temperature is maintained at 200°; the column temperature is maintained at 100° and programmed to rise 10° per minute to 250° and held there for 5 minutes. The detector temperature is maintained at 300°. The area of the main peak is not less than 95% of the total peak area.

Refractive index ⟨831⟩: between 1.4815 and 1.4845 at 20°.

Di(2-ethylhexyl)phthalate [*Bis*(*2-ethylhexyl*) *phthalate*], $C_{24}H_{38}O_4$—**390.54**—Use a suitable grade.

Digitonin, $C_{56}H_{92}O_{29}$—**1229.33**—White, crystalline powder. Almost insoluble in water; soluble in warm alcohol, and in glacial and in 75 percent acetic acid; insoluble in chloroform and in ether. Melts at about 230°, with decomposition.

Specific rotation ⟨781⟩: between −47° and −49°, determined in a solution in 75 percent acetic acid containing 100 mg per mL.

Solubility in alcohol—A solution of 500 mg in 20 mL of warm alcohol is colorless and complete.

Loss on drying ⟨731⟩—Dry it at 105° to constant weight: it loses not more than 6% of its weight.

Residue on ignition (Reagent test): not more than 0.3%.

Digoxigenin Bisdigitoxoside—Use a suitable grade.[29]

Dihydroquinidine, $C_{20}H_{26}N_2O_2$—**326.44**—Use a suitable grade.[30]

10β,17-Dihydroxy-19-nor-17α-pregn-4-en-20-yn-3-one—Use USP 10β-Hydroxynorethinyltestosterone RS.

Diiodofluorescein, $C_{20}H_{10}I_2O_5$—**584.10**—Orange-red, odorless powder. Slightly soluble in water, soluble in alcohol and in solutions of alkali hydroxides.

Residue on ignition—Ignite 200 mg with 5 drops of sulfuric acid: the weight of the residue does not exceed 1.0 mg (0.5%).

Sensitiveness—Weigh accurately about 100 mg of potassium iodide, previously dried at 105° to constant weight, and dissolve it in 50 mL of water. Add 1 mL of diiodofluorescein TS prepared from the test specimen and 1 mL of glacial acetic acid, and titrate with 0.1 *N* silver nitrate VS until the color of the precipitate changes from brownish red to a bluish red. The volume of 0.1 *N* silver nitrate consumed is not in excess of 0.10 mL over the calculated volume, based on the KI content of the dried potassium iodide determined as follows: Dissolve about 500 mg of potassium iodide, accurately weighed, in about 10 mL of water, and add 35 mL of hydrochloric acid and 5 mL of chloroform. Titrate with 0.05 *M* potassium iodate VS until the purple color of iodine disappears from the chloroform. Add the last portions of the iodate solution dropwise, agitating vigorously and continuously. After the chloroform has been decolorized, allow the mixture to stand for 5 minutes. If the chloroform develops a purple color, titrate further with the iodate solution. Each mL of 0.05 *M* potassium iodate is equivalent to 16.60 mg of KI.

Diiodothyronine (*3,5-Diiodo-4-phenoxy-L-phenylalanine*), $C_{15}H_{13}I_2NO_4$—**525.08**—Light tan, odorless, crystalline powder. Is slightly soluble in water and in alcohol. Melts at about 255°, with decomposition.

Residue on ignition (Reagent test): not more than 0.1%.

Specific rotation ⟨781⟩: between +25° and +27°, determined at 22° in a mixture of 2 parts of 1 *N* hydrochloric acid and 1 part of alcohol.

Diisodecyl Phthalate [*Bis*(*isodecyl*)*phthalate*], $C_{28}H_{46}O_4$—**446.65**—Use a suitable grade.

Diisopropyl Ether (*Isopropyl Ether*) $[(CH_3)_2CH]_2O$—**102.18**—Colorless, mobile liquid. Slightly soluble in water. Miscible with alcohol and with ether. *It is highly flammable. Do not use where it may be ignited. Do not evaporate to the point of near dryness, since it tends to form explosive peroxides.*

Specific gravity: between 0.716 and 0.720.

Distilling range, Method II ⟨721⟩—Not less than 95% distils between 65° and 70°.

Peroxides—To 10 mL, contained in a clean, glass-stoppered cylinder previously rinsed with a portion of the ether under examination, add 1 mL of freshly prepared potassium iodide solution (1 in 10). Shake, and allow to stand for 1 minute: no yellow color is observed in either layer (about 0.001% as H_2O_2).

Residue on evaporation—[*Note—If peroxide is present, do not carry out this procedure.*] Evaporate 14 mL (10 g) from a tared shallow dish, and dry at 105° for 1 hour: the residue weighs not more than 1 mg (0.01%).

Acidity—Add 2 drops of bromothymol blue TS to 10 mL of water in a glass-stoppered, 50-mL flask, and titrate with 0.010 *N* sodium hydroxide until a blue color persists after vigorous shaking. Add 5 mL of Diisopropyl Ether, and titrate with 0.010 *N* sodium hydroxide: not more than 0.30 mL is required to restore the blue color (0.005% as CH_3COOH).

NOTE—For spectrophotometric determinations, use diisopropyl ether that meets the following additional requirement:

Absorbance—Its absorbance at 255 nm, in a 10-mm quartz cell, does not exceed 0.2, water being used as the blank.

Diisopropylamine, $[(CH_3)_2CH]_2NH$—**101.19**—Colorless liquid.

Assay—Not less than 98% of $C_6H_{15}N$ is found, a suitable gas chromatograph equipped with a flame-ionization detector being used. The following conditions have been found suitable: a 1.83-m × 3.2-mm stainless steel column is packed with a cross-linked polystyrene support; the injection port temperature is maintained at 250° and the detector at 310°; the column temperature is programmed to rise at 10° per minute from 50° to 220°.

Refractive index ⟨831⟩: between 1.3915 and 1.3935, at 20°.

Diisopropylethylamine (*N,N-Diisopropylethylamine*), $C_8H_{19}N$—**129.24**—Clear, colorless liquid. Soluble in glacial acetic acid.

Assay—Weigh accurately about 500 mg, dissolve in 50 mL of glacial acetic acid, mix, add crystal violet TS, and titrate with

0.1 *N* perchloric acid VS. Each mL of 0.1 *N* perchloric acid is equivalent to 12.92 mg of $C_8H_{19}N$. Not less than 98% is found.
Refractive index ⟨831⟩: between 1.4125 and 1.4145 at 20°.

Diluted Acetic Acid—See *Acetic Acid, Diluted.*

Diluted Alcohol—Use *Diluted Alcohol* (NF monograph).

Diluted Hydrochloric Acid—See *Hydrochloric Acid, Diluted.*

Diluted Nitric Acid—See *Nitric Acid, Diluted.*

Diluted Sulfuric Acid—See *Sulfuric Acid, Diluted.*

2,5-Dimethoxybenzaldehyde, $C_9H_{10}O_3$—**166.18**—Off-white crystals.
Assay—Inject an appropriate specimen into a suitable gas chromatograph (see *Chromatography* ⟨621⟩) equipped with a flame-ionization detector, nitrogen being used as the carrier gas. The following conditions have been found suitable: a 30-m × 0.3-mm capillary column coated with phase G1; the injection port temperature is maintained at 270°; the detector temperature is maintained at 300°; the column temperature is maintained at 150° and programmed to rise 10° per minute to 270°. The area of the main peak is not less than 97% of the total peak area.
Melting range ⟨741⟩: between 50° and 52°.

1,2-Dimethoxyethane, $C_4H_{10}O_2$—**90.12**—Clear, colorless liquid, having a sharp, ethereal odor. Miscible with water and with alcohol. Soluble in hydrocarbon solvents. *May form peroxides on standing.*
Boiling range (Reagent test)—Not less than 95% distils between 83° and 86°.
Refractive index ⟨831⟩: between 1.379 and 1.381, at 20°.
Acidity—To 20 mL add bromophenol blue TS, and titrate with 0.020 *N* sodium hydroxide: not more than 2.0 mL is consumed (about 0.015% as CH_3COOH).
Water, Method I ⟨921⟩: not more than 0.2%.

(3,4-Dimethoxyphenyl)-acetonitrile (*Homoveratronitrile*), $C_{10}H_{11}NO_2$—**177.20**—Off-white fibers.
Melting range ⟨741⟩: between 65° and 67°.

Dimethyl Polysiloxane Fluid 350—Clear, colorless, essentially odorless, oily silicone liquid, having a viscosity at 25° of about 350 centistokes. Suitable for use as an adsorbant in gas-liquid chromatography.[31]

Dimethyl Sulfate $(CH_3)_2SO_4$—**126.13**—Clear, colorless or pale brownish, liquid having an ethereal odor. Very slightly soluble in water. Miscible with alcohol, with acetone, and with ether. Hydrolyzes gradually in aqueous solution. Boils with decomposition at about 188°. Specific gravity: about 1.33.
Caution—It is toxic.
Solubility in acetone and in ether—To separate 2-mL portions add 10 mL of acetone and 10 mL of ether, respectively: both solutions are clear.
Residue on ignition (Reagent test)—Evaporate 4 mL, and ignite: the residue weighs not more than 1 mg (0.02%).
Acidity—Add 1.0 mL to 20 mL of neutralized alcohol, mix, add phenolphthalein TS, and titrate with 0.10 *N* sodium hydroxide: not more than 2.5 mL is required to produce a pink color.

Dimethyl Sulfone (*Methyl Sulfone*), $(CH_3)_2SO_2$—**94.13**—White crystals.
Melting range ⟨741⟩: between 109° and 111°.

Dimethyl Sulfoxide—See *Methyl Sulfoxide.*

Dimethyl Sulfoxide, Spectrophotometric Grade—Use methyl sulfoxide reagent meeting the following additional specifications:
Assay—Inject an appropriate specimen (about 0.1 µL) into a suitable gas chromatograph (see *Chromatography* ⟨621⟩) equipped with a thermal conductivity detector, and obtain the chromatogram. Under typical conditions, the instrument contains a 2-m × 3-mm glass column packed with 20 percent phase G16 on support S1. The column is maintained at about 95°, and the injection port and the thermal conductivity detector block are maintained at about 180°. Helium is used as the carrier gas at a flow rate of about 50 mL per minute. The area of the symmetric methyl sulfoxide peak is at least 99% of the total peak area.
Ultraviolet absorption—Determine the ultraviolet absorbance of the specimen in a 1-cm cell, from 400 to 262 nm, using water as the blank: the absorbance does not exceed 1.00 at 262 nm, 0.360 at 270 nm, 0.080 at 300 nm, and 0.010 in the range of 340 to 400 nm. The absorbance curve is smooth and does not show extraneous absorbances within the range observed.

***N,N*-Dimethylacetamide,** C_4H_9NO—**87.12**—Clear, colorless liquid. Miscible with water and with many organic solvents.
Assay—When examined by gas-liquid chromatography, with the use of suitable apparatus and conditions, it shows a purity of not less than 99%.
Distilling range ⟨721⟩: between 164.5° and 167.5°.
Residue on evaporation—Evaporate 215 mL on a steam bath, and dry at 105° for 1 hour: the residue weighs not more than 2 mg (0.001%).
pH of 20% solution—Weigh 20 g of it into a 100-mL volumetric flask, and dilute with carbon dioxide–free water to volume: the solution shows a pH between 4.0 and 7.0.
Ultraviolet absorbance—Determine its absorbance throughout the range 270 to 400 nm, using a 1-cm cell, a suitable spectrophotometer, and water to set the instrument: the absorbance does not exceed 1.00 at 270 nm, 0.30 at 280 nm, 0.15 at 290 nm, 0.05 at 310 nm, 0.03 at 320 nm, and 0.01 at 360 to 400 nm.
Water, Method I ⟨921⟩: not more than 0.05%.

***p*-Dimethylaminoazobenzene** (*Methyl Yellow, Butter Yellow*), $C_6H_5N{:}NC_6H_4N(CH_3)_2$—**225.29**—Yellow leaflets or yellow crystalline powder.
Solubility—Insoluble in water; sparingly soluble in chloroform, in ether, or in fatty oils. Dissolve 100 mg in 20 mL of alcohol: the solution is complete or practically so and clear.
Melting range ⟨741⟩: between 115° and 117°.
Residue on ignition ⟨281⟩: not more than 0.1%.
Sensitiveness—Add 0.05 mL of an alcohol solution (1 in 200) and 2 g of ammonium chloride to 25 mL of carbon dioxide–free water: the lemon-yellow color of the solution is changed to orange by the addition of 0.05 mL of 0.1 *N* hydrochloric acid and restored on the subsequent addition of 0.05 mL of 0.1 *N* sodium hydroxide.

***p*-Dimethylaminobenzaldehyde,** $(CH_3)_2NC_6H_4CHO$—**149.19**—Use ACS reagent grade.

***p*-Dimethylaminocinnamaldehyde,** $(CH_3)_2NC_6H_4CH{:}CHCHO$—**175.23**—Orange-yellow powder. Soluble in acetone, in alcohol, and in benzene.
Melting range ⟨741⟩: between 132° and 136°.

Dimethylaminophenol (*meta isomer*), $C_8H_{11}NO$—**137.18**—Purplish black, gray, or tan-colored, crystalline solid.
Melting range ⟨741⟩: between 83° and 85°.

2,6-Dimethylaniline, $C_8H_{11}N$—**121.18**—Yellow liquid.
Refractive index ⟨831⟩: about 1.5609 at 20°.

***N,N*-Dimethylaniline,** $C_6H_5N(CH_3)_2$—**121.18**—Light yellow liquid. Clear, colorless liquid when freshly distilled, but acquiring a reddish to reddish brown color. Specific gravity: about 0.960. Freezing point about 2°. Insoluble in water; soluble in alcohol, in chloroform, in ether, and in dilute mineral acids.
Assay—Inject an appropriate specimen (about 0.2 µL) into a suitable gas chromatograph (see *Chromatography* ⟨621⟩) equipped with a flame-ionization detector, helium being used as the carrier gas flowing at about 40 mL per minute. The following conditions have been found suitable: a 1.8-m × 3-mm stainless steel column containing 20 percent phase G16 on support S1A; the injection port temperature is maintained at 250°; the column temperature is maintained at 50° and programmed to rise 10° per minute to 200°. The detector temperature is maintained at 310°. The area of the *N,N*-dimethylaniline peak having a retention time of about 11.5 minutes is not less than 99% of the total peak area.
Refractive index ⟨831⟩: between 1.5571 and 1.5591 at 20°.
Boiling range (Reagent test)—Distil 100 mL: the difference between the temperatures observed, when 1 mL and 95 mL have distilled, is not more than 2.5°. Its boiling temperature at a pressure of 760 mm of mercury is 194.2°.
Hydrocarbons—Dissolve 5 mL in a mixture of 10 mL of hydrochloric acid and 15 mL of water: a clear solution results and it remains clear on cooling to about 10°.
Aniline or monomethylaniline—Place 5 mL in a glass-stoppered flask, add 5 mL of a solution of acetic anhydride in benzene (1 in 10), mix, and allow to stand for 30 minutes. Add 30.0 mL of 0.5 *N* sodium hydroxide VS, shake the mixture, add phenol-

phthalein TS, and titrate with 0.5 *N* hydrochloric acid VS. Perform a blank determination, and make any necessary correction. Not more than 0.30 mL of 0.5 *N* sodium hydroxide is consumed by the test specimen.

3,4-Dimethylbenzophenone, $C_{15}H_{14}O$—**210.28**—White chunks melting at about 45°.

Assay—Inject an appropriate specimen into a suitable gas chromatograph (see *Chromatography* ⟨621⟩) equipped with a flame-ionization detector, helium being used as the carrier gas. The following conditions have been found suitable: a 30-m × 0.25-mm capillary column coated with phase G1: the detector temperature and the injection port temperature are maintained at 300°; the column temperature is maintained at 180° and programmed to rise at the rate of 10° per minute to 280° and held at that temperature for 10 minutes. The area of the main peak is not less than 99% of the total peak area.

5,5-Dimethyl-1,3-cyclohexanedione, $C_8H_{12}O_2$—**140.18**—White, crystalline solid. Slightly soluble in water; soluble in alcohol, in methanol, in chloroform, and in acetic acid.

Melting range ⟨741⟩: between 148° and 150°.

Dimethylethyl(3-hydroxyphenyl)ammonium Chloride—See *Edrophonium Chloride.*

Dimethylformamide (*N,N-Dimethylformamide*), $HCON(CH_3)_2$—**73.09**—Use ACS reagent grade.

Dimethylglyoxime, $(CH_3)_2C_2(NOH)_2$—**116.12**—Use ACS reagent grade.

***N,N*-Dimethyl-1-naphthylamine,** $C_{12}H_{13}N$—**171.24**—Pale yellow to yellow, aromatic liquid. Soluble in alcohol and in ether.

Assay—Transfer about 250 mg, accurately weighed, to a suitable beaker, add 100 mL of glacial acetic acid, and dissolve by stirring. When solution is complete, titrate with 0.1 *N* perchloric acid VS, determining the end-point potentiometrically. Perform a blank determination, and make any necessary correction. Each mL of 0.1 *N* perchloric acid is equivalent to 17.12 mg of $C_{12}H_{13}N$. Not less than 98% is found.

Refractive index ⟨831⟩: between 1.6210° and 1.6230° at 20°, sodium light being used.

Sulfanilamide test—Dissolve 20 mg of USP Sulfanilamide RS in 100 mL of water to obtain the *Sulfanilamide solution.* Into two 150-mL beakers pipet 1.0 mL and 2.5 mL of the *Sulfanilamide solution*, respectively. Dilute with water to 90 mL. To provide a blank, place 90 mL of water in a third beaker. To each beaker add 8.0 mL of trichloroacetic acid solution (3 in 20) and 1.0 mL of sodium nitrite solution (1 in 1000). Stir the solutions for 5 minutes, then add 10 mL of acetate buffer TS, and 1.0 mL of a 1 in 1000 solution of *N,N*-dimethyl-1-napthylamine in alcohol. The pH is about 5 to 6, using pH paper. Stir for an additional 5 minutes, then add 20 mL of glacial acetic acid. The pH is about 3 to 4, using pH paper. In comparison with the blank, the beaker containing 1.0 mL of the *Sulfanilamide solution* shows a pink color, while the other beaker shows a deep pink to red color.

***N,N*-Dimethyloctylamine,** $C_{10}H_{23}N$—**157.30**—Colorless liquid.

Refractive index ⟨831⟩: 1.4243 at 20°.

2,6-Dimethylphenol, $(CH_3)_2C_6H_3OH$—**122.17**—White to pale yellow crystalline solid.

Assay—Inject a 1 in 3 solution of it in xylene into a suitable gas chromatograph equipped with a flame-ionization detector, helium being used as the carrier gas at a flow rate of about 40 mL per minute. The following conditions have been found suitable: a 1.83-m × 3.2-mm stainless steel column packed with 10 percent phase G25 on support S1A; the injection port is maintained at about 250° and the detector at about 310°; the column temperature is programmed to rise at 8° per minute from 100° to 200°. Similarly inject a specimen of xylene. The area of the $C_8H_{10}O$ peak is not less than 98% of the total peak area corrected for xylene.

Melting range ⟨741⟩: between 44° and 46°.

***N,N*-Dimethyl-*p*-phenylenediamine Dihydrochloride,** $(CH_3)_2NC_6H_4NH_2.2HCl$—**209.12**—Nearly white, fine, crystalline, hygroscopic solid that may have a pinkish cast. Freely soluble in water; soluble in alcohol.

Assay—Transfer about 400 mg, accurately weighed, to a 250-mL beaker, and dissolve in about 75 mL of water. Titrate with 0.1 *N* sodium hydroxide VS, determining the end-point potentiometrically. Each mL of 0.1 *N* sodium hydroxide is equivalent to 10.46 mg of $C_8H_{12}N_2.2HCl$. Not less than 98% is found.

Solubility—A solution of 1 g in 10 mL of water produces not more than a slight haze.

***m*-Dinitrobenzene,** $C_6H_4(NO_2)_2$—**168.11**—Pale yellow crystals or crystalline powder. Almost insoluble in cold water; slightly soluble in hot water. Soluble in chloroform and in benzene; sparingly soluble in alcohol. Is volatile in steam.

Melting range ⟨741⟩: between 89° and 92°.

Residue on ignition (Reagent test): not more than 0.5%.

3,5-Dinitrobenzoyl Chloride, $(NO_2)_2C_6H_3COCl$—**230.56**—Pale yellow, crystalline powder. Freely soluble in dilute sodium hydroxide solutions; soluble in alcohol.

Melting range ⟨741⟩: between 67° and 69°.

Solubility in sodium hydroxide—A solution of 500 mg in 25 mL of 1 *N* sodium hydroxide is clear or not more than faintly turbid.

Residue on ignition—Ignite 1 g with 0.5 mL of sulfuric acid: the residue weighs not more than 1 mg (0.1%).

2,4-Dinitrochlorobenzene, $C_6H_3(NO_2)_2Cl$—**202.55**—Yellow to brownish yellow crystals. Insoluble in water; soluble in hot alcohol, in ether, and in benzene.

Melting range ⟨741⟩: between 51° and 53°.

Residue on ignition—Ignite 500 mg with 5 drops of sulfuric acid: the residue weighs not more than 1 mg (0.2%).

2,4-Dinitrofluorobenzene (*1-Fluoro-2,4-dinitrobenzene*), $C_6H_3FN_2O_4$—**186.10**—Light yellow solid.

Assay—Inject an appropriate specimen (about 0.2 µL) into a suitable gas chromatograph (see *Chromatography* ⟨621⟩) equipped with a thermal conductivity detector, helium being used as the carrier gas. The following conditions have been found suitable: a 1.8-m × 4-mm stainless steel column containing 10 percent phase G1 on support S1A; the injection port and detector are maintained at 290° and 300°, respectively; the column temperature is maintained at 140° and programmed to rise 3° per minute to 190°. The area of the 2,4-dinitrofluorobenzene peak is not less than 99% of the total peak area.

Melting range ⟨741⟩: between 28° and 31°.

2,4-Dinitrophenylhydrazine, 2,4-$C_6H_3(NO_2)_2NHNH_2$—**198.14**—Orange-red crystals, which under the microscope appear individually to be lemon-yellow, lath-like needles. Very slightly soluble in water; slightly soluble in alcohol; moderately soluble in dilute inorganic acids.

Melting range ⟨741⟩: between 197° and 200°.

Solubility in sulfuric acid—Dissolve 500 mg in a mixture of 25 mL of sulfuric acid and 25 mL of water: the solution is clear or not more than slightly turbid.

Residue on ignition (Reagent test): negligible, from 500 mg.

Dioctyl Sodium Sulfosuccinate—See *Docusate Sodium.*

Dioxane (*Diethylene Dioxide; 1,4-Dioxane*), $\overline{OCH_2CH_2OCH_2CH_2}$—**88.11**—Use ACS reagent grade.

Diphenamid (*N,N-Dimethyl-2,2-diphenylacetamide*), $C_{16}H_{17}NO$—**239.32**—Use a suitable grade.[32]

Diphenhydramine Hydrochloride—Use *Diphenhydramine Hydrochloride* (USP monograph).

Diphenyl Ether (*Phenyl Ether*), $(C_6H_5)_2O$—**170.21**—A colorless liquid. Insoluble in water; soluble in glacial acetic acid and in most organic solvents. Boils at about 259°.

Melting range ⟨741⟩: between 26° and 28°.

Diphenylamine, $(C_6H_5)_2NH$—**169.23**—Use ACS reagent grade.

Diphenylcarbazide, $(C_6H_5NHNH)_2CO$—**242.28**—Use ACS reagent grade 1,5-Diphenylcarbohydrazide.

Diphenylcarbazone [*Diphenylcarbazone compd. with s-Diphenylcarbazide* (*1:1*)], $C_6H_5NHNHCON{:}NC_6H_5.C_6H_5NHNHCONHNHC_6H_5$—**482.54**—Use ACS reagent grade Diphenylcarbazone Compound with *s*-Diphenylcarbazide (1:1).

2,2-Diphenylglycine, $C_{14}H_{13}NO_2$—**227.26**—Off-white powder. Melts at about 244°, with decomposition.

Assay—Dissolve about 115 mg, accurately weighed, in 30 mL of methanol. Slowly add about 20 mL of water, heating slightly if necessary for complete solution. Titrate with 0.1 *N* sodium hydroxide VS, determining the end-point potentiometrically. Perform a blank determination and make any necessary correction. Each mL of 0.1 *N* sodium hydroxide is equivalent to 22.73 mg of $C_{14}H_{13}NO_2$. Not less than 98.0% is found.

Dipicrylamine—See *Hexanitrodiphenylamine.*

α,α′-Dipyridyl—See *2,2′-Bipyridine.*

Disodium Chromotropate—See *Chromotropic Acid.*

Disodium Ethylenediaminetetraacetate—Use ACS reagent grade (Ethylenedinitrilo)tetraacetic Acid Disodium Salt.

Disodium Phosphate—See *Sodium Phosphate.*

5,5′-Dithiobis (2-nitrobenzoic Acid) (*3-Carboxy-4-nitrophenyl disulfide; Ellman's reagent*), $C_{14}H_8N_2O_8S_2$—**396.35**—Yellow powder, melting at about 242°. Sparingly soluble in alcohol.

Dithizone (*Diphenylthiocarbazone; Phenylazothioformic Acid 2-Phenylhydrazide*), $C_6H_5N{:}NCSNHNHC_6H_5$—**256.32**—Use ACS reagent grade.

Docosane, $C_{22}H_{46}$—**310.61**—Use a suitable grade.[33]

Docusate Sodium—Use *Docusate Sodium* (USP monograph).

1-Dodecanol (*Dodecyl Alcohol*), $CH_3(CH_2)_{11}OH$—**186.34**—A clear, colorless liquid. Crystallizes as leaflets from dilute alcohol solution.

Melting range ⟨741⟩: between 23° and 25°.

Dodecyl Alcohol—See *1-Dodecanol.*

Dodecyl Sodium Sulfate, $C_{12}H_{25}SO_4Na$—**288.38**—Light yellow, crystalline powder.

Assay—Accurately weigh about 800 mg, and dissolve in 100 mL of water. Pour the solution through a cation-exchange column, collecting the eluate in a suitable container. Wash the column with 400 mL of water, collecting the wash in the same container as for the eluate. Titrate the collected solution with 0.1 *N* sodium hydroxide VS, using phenolphthalein TS as the indicator. Each mL of 0.1 *N* sodium hydroxide is equivalent to 28.84 mg of $C_{12}H_{25}SO_4Na$. Not less than 99.0% is found.

Chloride (Reagent test)—A 0.2-g portion shows no more than 0.02 mg of Cl (0.01%).

Heavy metals, Method II ⟨231⟩: 2 ppm.

Phosphate (Reagent test)—Ignite 10 g in a crucible, and cool. The residue, dissolved in 25 mL of approximately 0.5 *N* sulfuric acid, shows not more than 0.01 mg of PO_4 (1 ppm).

***n*-Dotriacontane,** $C_{32}H_{66}$—**450.87**—White to practically white crystals. Soluble in hot ether and in hot acetic acid; very slightly soluble in alcohol.

Assay—When examined by gas-liquid chromatography, it shows a purity of not less than 99%, determined by use of a suitable chromatographic system, as follows: A 6-inch × ⅛-inch stainless steel column packed with 5 percent phase G2 on support S1 is employed. The carrier gas is helium, the port and column temperatures are 225°, and the detector temperature is 275°. A flame-ionization detector is employed.

Melting range ⟨741⟩: between 68° and 70°.

Dried Peptone—See *Peptone, Dried.*

Dried Yeast—See *Yeast, Dried.*

Dyphylline, $C_{10}H_{14}N_4O_4$—**254.25**—Use *Dyphylline* (USP monograph).

Earth, Chromatographic, Silanized, Acid-base Washed—Use a suitable grade.[34]

Edrophonium Chloride [*Dimethylethyl(3-hydroxyphenyl)ammonium Chloride; 3-Hydroxyphenyl-dimethylethylammonium Chloride*]—Use *Edrophonium Chloride* (USP monograph).

***n*-Eicosane,** $C_{20}H_{42}$—**282.55**—White, crystalline solid.

Melting range ⟨741⟩: between 37° and 39°.

Eosin Y (Eosin Yellowish Y) (*Certified Biological Eosin Y; Sodium Tetrabromofluorescein*), $C_{20}H_6Br_4Na_2O_5$—**691.86**—Red to brownish red pieces or powder. One g dissolves in about 2 mL of water and in 50 mL of alcohol.

Color characteristics—Its 1 in 500 solution is yellowish to purplish red with a greenish fluorescence. Its 1 in 12,000 alcohol solution is pink to purplish red with a greenish yellow fluorescence. The addition of mineral acids to a solution (1 in 100) produces an orange to reddish orange precipitate of tetrabromofluorescein. On the addition of 2 mL of saturated sodium hydroxide solution to 10 mL of a solution of the dye (1 in 100), a red precipitate is formed.

Equilenin, $C_{18}H_{18}O_2$—**266.34**—Colorless or white crystals or crystalline powder. Insoluble in water; soluble in chloroform and in dioxane; moderately soluble in alcohol.

Melting range, Class II ⟨741⟩: between 256° and 260°.

Specific rotation ⟨781⟩: between +85° and +88°, determined in a solution in dioxane containing 75 mg of equilenin in each 10 mL.

Absorption maxima—An alcohol solution exhibits absorption maxima at 231, 282, 325, and 340 nm.

Equilin Reagent—Purify white, crystalline reagent phenol or liquefied phenol free from hypophosphorous stabilizer by distillation through a column, the equivalent of a silvered, vacuum-jacketed Vigreaux tube, at a rate of approximately 20 mL per minute. Discard the first 10% and the last 10% of the distillate. Allow the distillate to recrystallize at room temperature over a period of several hours. Discard any distillate that is colored or shows the presence of fluid or noncrystalline material.

To a given weight of solidified phenol add a volume of sulfuric acid equivalent to 0.62 times the weight of phenol. Stir constantly, without cooling, until all of the phenol is dissolved. Cool, with swirling, to a temperature of about 25° in an ice-and-water bath. Insert the stopper in the flask, and allow to stand in the dark for 12 to 16 hours.

To a given volume of the phenol–sulfuric acid mixture add a volume of dilute sulfuric acid (1 in 2) equivalent to 3.33 times the volume of the mixture. Ensure the complete transfer of the phenol–sulfuric acid mixture by rinsing the container with several portions of the measured amount of dilute sulfuric acid solution.

Mix 20 mL of hydrochloric acid and 8.5 mL of cobalt nitrate hexahydrate solution (1 in 1000), and dilute with water to 100 mL. To 100 volumes of the dilute phenol–sulfuric acid mixture in a round-bottom flask add 7.7 volumes of the hydrochloric acid–cobalt nitrate solution, and mix. Immediately insert the stopper loosely in the flask, heat evenly, and raise the temperature of the mixture quickly to 95°.

Continue to heat for 30 minutes, maintaining a temperature of 95° to 100°. Cool the mixture quickly, with swirling, in an ice-and-water bath. To 100 volumes of this mixture add 0.6 volume of sodium hypochlorite TS, and store the reagent in an amber bottle fitted with a glass or polyethylene stopper. Allow the reagent to stand for 48 hours, and before use ascertain its suitability for the *Assay* for *Sodium equilin sulfate* by the following test.

Suitability test—Proceed as directed in the *Assay* for *Sodium equilin sulfate content* under *Conjugated Estrogens* (USP monograph), using a mixture of 1.0 mL each of the *Standard equilin preparation* and the *Standard estrone preparation.* The Equilin Reagent is satisfactory if the resultant scan exhibits absorbance maxima at about 635 nm and about 527 nm only. Use only reagent that conforms to this *Suitability test.*

Eriochrome Cyanine R, $C_{23}H_{15}Na_3O_9S$—**536.40**—Dark, red-brown powder. Freely soluble in water; insoluble in alcohol.

Solubility—200 mg in 100 mL of water yields a solution that remains clear and free from undissolved matter for 30 minutes.

Loss on drying ⟨731⟩—Dry it in vacuum over silica gel to constant weight: it loses not more than 2% of its weight.

Residue on ignition (Reagent test)—0.5 g, treated with 1 mL of sulfuric acid and 2 mL of nitric acid, yields between 42.0% and 44.0% of the dry weight (theory is 42.9% of Na_2SO_4).

Sensitiveness—Add 2 mL of a solution (1 in 1000) to 1 mL of aluminum sulfate solution (1 in 10,000), heat at 37 ± 3° for 5 minutes, cool, and add 1 mL of sodium acetate TS: a strong red to red-violet color is produced in not more than 5 minutes.

Estradiol—Use *Estradiol* (USP monograph).

Ether—See *Ethyl Ether.*

Ether, Absolute—See *Ethyl Ether, Anhydrous.*

Ether, Diphenyl—See *Diphenyl Ether.*

Ether, Isopropyl—See *Diisopropyl Ether.*

Ether, Nonyl Phenyl Polyethylene Glycol—See (*p-tert-Octylphenoxy*)*nonaethoxyethanol.*

2-(2-Ethoxyethoxy)ethanol (*Diethylene Glycol Monoethyl Ether*), $C_6H_{14}O_3$—**134.17**—Clear, colorless, hygroscopic liquid. Miscible with water, with acetone, with chloroform, with alcohol, and with ether.

Assay—Inject an appropriate specimen into a suitable gas chromatograph (see *Chromatography* ⟨621⟩) equipped with a flame-ionization detector, helium being used as the carrier gas. The following conditions have been found suitable: a 1.8-m × 3-mm stainless steel column containing 20 percent of phase G16 on support S1A; the injection port, detector, and column temperatures are maintained at 180°, 250°, and 120°, respectively. The area of the 2-(2-ethoxyethoxy)ethanol peak is not less than 98% of the total peak area.

Refractive index ⟨831⟩: between 1.4260 and 1.4280 at 20°.

Ethyl Acetate, $CH_3COOC_2H_5$—**88.11**—Use ACS reagent grade.

Ethyl Acrylate—Use a suitable grade.

Ethyl Alcohol (*Alcohol; Ethanol*), C_2H_5OH—**46.07**—Use *Alcohol* (USP monograph).

Ethyl Benzoate, $C_9H_{10}O_2$—**150.18**—Clear, colorless liquid. Has an aromatic odor. Practically insoluble in water; miscible with alcohol, with chloroform, and with ether.

Assay—Inject an appropriate specimen into a suitable gas chromatograph (see *Chromatography* ⟨621⟩), helium being used as the carrier gas. The following conditions have been found suitable: a 2.4-m × 3-mm stainless steel column containing 20 percent phase G16 on support S1A; the injection port, column, and detector temperatures are maintained at 180°, 195°, and 250°, respectively. The area of the ethyl benzoate peak is not less than 98% of the total peak area.

Refractive index ⟨831⟩: between 1.5048 and 1.5058 at 20°.

Ethyl Cyanoacetate, $CNCH_2COOC_2H_5$—**113.12**—Colorless to pale yellow liquid, having a pleasant odor. Slightly soluble in water. Miscible with alcohol and with ether. At atmospheric pressure it boils between 205° and 209° with decomposition. At a pressure of 10 mm of mercury it distils at about 90°.

Specific gravity ⟨841⟩: between 1.057 and 1.062.

Acidity—Dissolve 2 mL in 25 mL of neutralized alcohol, add phenolphthalein TS, and titrate with 0.10 *N* sodium hydroxide: not more than 1.5 mL is required to produce a pink color.

Ethyl Ether (*Diethyl Ether; Ether*), $(C_2H_5)_2O$—**74.12**—Use ACS reagent grade.

Ethyl Ether, Anhydrous (*Diethyl Ether, Anhydrous; Ether, Absolute*), $(C_2H_5)_2O$—**74.12**—Use ACS reagent grade.

Ethyl Iodide, C_2H_5I—**155.97**—Clear liquid, having an ether-like odor. Is colorless when freshly distilled, but on exposure to air and light becomes yellow through liberation of iodine which may be removed by shaking with a globule of mercury or with silver leaf. Slightly soluble in water. Miscible with alcohol and with ether.

Boiling range (Reagent test)—Not less than 90% distils between 71° and 72.5°.

Specific gravity ⟨841⟩: between 1.910 and 1.940.

Residue on evaporation—Evaporate 5 mL on a steam bath, and dry at 105° for 1 hour: the residue weighs not more than 1 mg (about 0.01%).

Ethyl Oxide (*Solvent Ether*), $C_4H_{10}O$—**74.12**—Use ACS reagent grade.

4-Ethylbenzaldehyde, $C_2H_5C_6H_4CHO$—**134.18**—Colorless to pale yellow liquid.

Assay—Dissolve about 600 mg, accurately weighed, in a mixture of 100 mL of alcohol and 25 mL of 1 *M* hydroxylamine hydrochloride in a beaker. Cover the beaker with a watch glass. Heat gently until condensate begins to form on the watch glass. Allow to cool for about 30 minutes. Titrate with 0.5 *N* sodium hydroxide VS, determining the end-point potentiometrically. Perform a blank determination, and make any necessary correction. Each mL of 0.5 *N* sodium hydroxide is equivalent to 67.09 mg of $C_2H_5C_6H_4CHO$. Not less than 98% is found.

Ethylene Dichloride (*1,2-Dichloroethane*), $C_2H_4Cl_2$—**98.96**—Use ACS reagent grade 1,2-Dichloroethane.

Ethylene Glycol, $HOCH_2CH_2OH$—**62.07**—Clear, colorless, slightly viscous, hygroscopic, practically odorless liquid. Slightly soluble in ether; practically insoluble in benzene. Miscible with water and with alcohol. Specific gravity: about 1.11.

Boiling range (Reagent test): between 194° and 200°.

Residue on ignition—Evaporate 100 mL (110 g) in a tared evaporating dish over a flame until the vapors continue to burn after the flame is removed. Allow the vapors to burn until the specimen is consumed. Ignite at 800 ± 25° for 1 hour, cool, and weigh: the residue weighs not more than 5.5 mg (0.005%).

Acidity—Add 0.2 mL of phenol red TS to 50 mL of water, and titrate with 0.1 *N* sodium hydroxide to a red end-point. Add 50 mL (55 g) of ethylene glycol, and titrate with 0.1 *N* sodium hydroxide: not more than 1 mL is required to restore the red color (0.01% as CH_3COOH).

Chloride (Reagent test)—A 4.5-mL (5-g) portion shows not more than 0.025 mg of Cl (5 ppm).

Water, Method I ⟨921⟩: not more than 0.20%.

Ethylene Glycol Monoethyl Ether (*2-Ethoxyethanol*), $C_4H_{10}O_2$—**90.12**—Clear, colorless liquid having a slight, characteristic odor. Miscible with water, with alcohol, with ether, and with acetone. Specific gravity: about 0.93.

Boiling range (Reagent test)—Not less than 95% distils between 133° and 135°.

Ethylenediamine, $C_2H_8N_2$—**60.10**—Use *Ethylenediamine* (USP monograph).

Ethylparaben—Use *Ethylparaben* (NF monograph).

***N*-Ethylpiperidine,** $C_5H_{10}NC_2H_5$—**113.20**—Clear, colorless liquid having a characteristic odor and strongly alkaline reaction. Is hygroscopic and absorbs carbon dioxide from the air. Soluble in about 25 parts of water. Miscible with alcohol and with amyl acetate. Store in tight, light-resistant containers, in a cold place.

Assay—To a tared, glass-stoppered flask containing about 10 mL of water add 0.5 mL of the *N*-ethylpiperidine, insert the stopper immediately, and weigh. Add 50.0 mL of 0.1 *N* hydrochloric acid VS, then add methyl red TS, and titrate the excess acid with 0.1 *N* sodium hydroxide VS: each mL of 0.1 *N* acid corresponds to 11.32 mg of $C_5H_{10}NC_2H_5$. Not less than 99% is found.

Specific gravity ⟨841⟩: between 0.822 and 0.825.

Distilling range, Method II ⟨721⟩—Distil 50 mL: not more than 1% distils below 125°, and not less than 95% distils between 129.5° and 131°.

1-Ethylquinaldinium Iodide, $C_{12}H_{14}IN$—**299.15**—Yellow-green solid. Sparingly soluble in water.

Assay—Dissolve about 290 mg, accurately weighed, in 100 mL of water, and add 10 mL of glacial acetic acid. Titrate with 0.1 *N* silver nitrate VS, determining the end-point potentiometrically, using a silver–ion selective electrode and a calomel reference electrode containing 1 *M* potassium nitrate. Perform a blank determination, and make any necessary correction. Each mL of 0.1 *N* silver nitrate is equivalent to 29.92 mg of $C_{12}H_{14}IN$: not less than 97.0% is found.

Factor X_a (Activated Factor X) for Anti-Factor X_a test—Factor X_a is the proteolytic enzyme derived from bovine plasma, which cleaves prothrombin to form thrombin. It is a glycoprotein having a molecular weight of 40,000. By gel electrophoresis (see *Electrophoresis* ⟨726⟩), the principal protein of interest constitutes not less than 90% of the total protein zones. In Factor X_a for Anti-Factor X_a test, the enzyme is activated by Russel's viper venom, the activating agent is removed by chromatography, and the preparation is stabilized with bovine Albumin and lyophilized. The specific activity of Factor X_a for Anti-Factor X_a test corresponds to not less than 40 USP Factor X_a Units[35] per mg of protein, when tested as follows: Mix 0.1 mL of a saturated solution of cephalin derived from rabbit brain thromboplastin (equivalent to the thromboplastin from approximately 20 mg of rabbit brain–acetone powder per mL) in bovine plasma and 0.1 mL of 0.025 *M* calcium chloride, immediately add 0.1 mL of Factor X_a solution corresponding to a concentration of 0.01 mg of specific protein per mL, and preserve at 37°: it produces a clot in 15 seconds. Factor X_a for Anti-Factor X_a test in a solution containing 0.4 USP Factor X_a Unit per mL in pH 8.4 buffer (see monograph on Heparin Sodium) meets the test for freedom from thrombin; it does not clot excess of pure fibrinogen within a period of 24 hours when preserved at 20° in the absence of calcium ions.

Fast Blue BB Salt, $(C_{17}H_{18}ClN_3O_3)_2 . ZnCl_2$—**831.89**—Yellow powder melting at about 162°, with decomposition. Sparingly soluble in water.

Chloride—Transfer about 80 mg, accurately weighed, to a suitable beaker. Add 25 mL of acetone, 25 mL of water and 500 mg of sodium nitrate. Stir until solution is complete. Titrate with 0.01 *N* silver nitrate VS, determining the end-point potentiometrically. Perform a blank determination and make any necessary correction. Not less than 15.0% of chloride is found.

Fast Blue Salt B, $C_{14}H_{12}N_4O_2 . ZnCl_4$—**475.47**—Green powder.

Loss on drying ⟨731⟩—Dry it in vacuum at 110° for 1 hour: it loses not more than 5.0% of its weight.

Absorbance—Dissolve 50 mg in 100 mL of water. In a second container dissolve 100 mg of 2-naphthol in 100 mL of 2-methoxyethanol. Pipet 5 mL of the test solution and 10 mL of the 2-naphthol solution into a 100-mL volumetric flask, and dilute with acetone to volume. For the blank, pipet 5 mL of water and 10 mL of 2-naphthol solution into a second 100-mL volumetric flask, and dilute with acetone to volume. Determine the absorbance of the test solution in a 1-cm cell at the wavelength of maximum absorbance at about 545 nm, with a suitable spectrophotometer, using the blank to set the instrument: the absorbance is not less than 0.80.

Ferric Ammonium Citrate—Thin, transparent, garnet-red scales or granules or brownish yellow powder, odorless or having a slightly ammoniacal odor, and having a saline, mildly ferruginous taste. Is deliquescent and is affected by light. Very soluble in water; insoluble in alcohol.

Assay—Weigh accurately about 1 g, dissolve in 25 mL of water in a glass-stoppered flask, add 5 mL of hydrochloric acid and 4 g of potassium iodide, insert the stopper in the flask, and allow to stand in the dark for 15 minutes. Add 100 mL of water, and titrate the liberated iodine with 0.1 *N* sodium thiosulfate VS, adding 3 mL of starch TS as the end-point is approached. Perform a blank determination, and make any necessary correction. Each mL of 0.1 *N* sodium thiosulfate is equivalent to 5.585 mg of Fe: between 16.5% and 18.5% is found.

Ferric citrate—To 250 mg dissolved in 25 mL of water add 1 mL of potassium ferrocyanide TS: no blue precipitate is formed.

Tartrate—Dissolve 1 g in 10 mL of water, add 1 mL of potassium hydroxide TS, boil to coagulate the ferric hydroxide, adding more potassium hydroxide TS, if necessary, to precipitate all of the iron, filter, and slightly acidify the filtrate with glacial acetic acid. Add 2 mL of glacial acetic acid, and allow to stand for 24 hours: no crystalline white precipitate is formed.

Lead ⟨251⟩—Dissolve 1.0 g in 30 mL of water, add 5 mL of dilute nitric acid (1 in 21), boil gently for 5 minutes, cool, and dilute with water to 50 mL: 20 mL of the solution shows not more than 0.008 mg of Pb (0.002%).

Ferric Ammonium Sulfate, $FeNH_4(SO_4)_2 . 12H_2O$—**482.18**—Use ACS reagent grade.

Ferric Chloride, $FeCl_3 . 6H_2O$—**270.30**—Use ACS reagent grade.

Ferric Citrate, $FeC_6H_5O_7 . xH_2O$—(anhydrous) **244.95**—Thin, transparent, garnet-red scales, or brown granules. Slowly soluble in water; more readily soluble in hot water, the solubility diminishing with age; insoluble in alcohol.

Assay—Weigh accurately about 1 g, and dissolve in a mixture of 5 mL of hydrochloric acid and 25 mL of water in a glass-stoppered flask, warming if necessary to aid solution. Cool, add 4 g of potassium iodide, insert the stopper in the flask, and allow to stand for 15 minutes. Dilute with 100 mL of water, and titrate the liberated iodine with 0.1 *N* sodium thiosulfate VS, adding 3 mL of starch TS as the end-point is approached: each mL of 0.1 *N* sodium thiosulfate is equivalent to 5.585 mg of Fe. Between 16.5% and 18.5% of Fe is found.

Alkali citrate—Ignite about 500 mg until it is thoroughly charred, cool, and add 2 mL of hot water: the water is neutral or shows only a slight alkaline reaction to litmus.

Ammonium—Heat 500 mg with 5 mL of sodium hydroxide TS: the odor of ammonia is not perceptible.

Chloride—Heat 1 g with 25 mL of water and 2 mL of nitric acid until dissolved. Cool, dilute with water to 100 mL, and mix. To 10 mL of the solution add 1 mL of silver nitrate TS: no turbidity is produced at once.

Sulfate—To 10 mL of the solution obtained in the preceding test add 1 mL of barium nitrate TS: no turbidity is produced within 15 seconds.

Ferric Nitrate, $Fe(NO_3)_3 . 9H_2O$—**404.00**—Use ACS reagent grade.

Ferric Sulfate, $Fe_2(SO_4)_3 . xH_2O$—Grayish white, hygroscopic powder, or fawn-colored pearls, slowly soluble in water.

Assay—Weigh accurately about 700 mg, and dissolve it in a mixture of 50 mL of water and 3 mL of hydrochloric acid in a glass-stoppered flask. Add 3 g of potassium iodide, and allow to stand in the dark for 30 minutes. Then dilute with 100 mL of water, and titrate with 0.1 *N* sodium thiosulfate VS, adding 3 mL of starch TS as the end-point is approached. Each mL of 0.1 *N* sodium thiosulfate is equivalent to 5.585 mg of Fe: not less than 21.0% and not more than 23.0% is found.

Insoluble matter (Reagent test)—A 10-g portion, dissolved in a mixture of 100 mL of water and 5 mL of sulfuric acid, shows not more than 2 mg of insoluble matter (0.02%).

Chloride—Dissolve 1 g by warming with a mixture of 10 mL of water and 1 mL of nitric acid, add 4 mL of additional nitric acid, and dilute with water to 50 mL. To 25 mL add 1 mL of phosphoric acid and 1 mL of silver nitrate TS. Any turbidity does not exceed that produced in a control containing 0.01 mg of chloride ion (Cl), 1 mL of nitric acid, 1 mL of phosphoric acid, and 1 mL of silver nitrate TS (0.002%).

Ferrous iron—Dissolve 4 g by warming with 50 mL of dilute sulfuric acid (1 in 10), cool, and titrate with 0.1 *N* potassium permanganate: not more than 0.16 mL is required to produce a permanent pink color (0.02% as Fe^{+2}).

NOTE—Since the reagents used in the tests for *Copper* and *Zinc* may contain excessive amounts of copper and zinc, they should first be purified by extracting with *Dithizone Extraction Solution* (see *Lead* ⟨251⟩).

Copper—Dissolve 1.2 g in 100 mL of water. To 10 mL add 50 mL of a solution containing 5 g of ammonium tartrate and 5 mL of ammonium hydroxide. Add 10 mL of *Standard Dithizone Solution* (see *Lead* ⟨251⟩), shake for 2 minutes, draw off the dithizone layer, and compare the pink color with that in a control containing 6 µg of copper ion (Cu) and treated exactly as the 10-mL portion of test solution. If the color in the test solution is less than that in the control, then the test specimen contains less than the limit of both *Copper* and *Zinc*. If the color in the test solution is more than that in the control, add 15 mL of dilute hydrochloric acid (1 in 250), and shake for 2 minutes. Draw off the dithizone solution, and shake with a second 15 mL of dilute hydrochloric acid (1 in 250) for 2 minutes. Draw off the dithizone, combine the two acid extracts, and reserve for the *Zinc* test. Any pink color in the dithizone solution is not darker than that in the control solution treated exactly as the test solution (0.005%).

Zinc—To the combined acid extracts saved from the *Copper* test, add 0.5 *M* sodium acetate to bring the pH between 5.0 and 5.5, and then add 1 mL of 0.1 *N* sodium thiosulfate. Add 10 mL of *Standard Dithizone Solution* (see *Lead* ⟨251⟩), shake for 2 minutes, and allow the layers to separate. Draw off the dithizone, and discard the water layer. Any pink color is not greater than that in a control prepared by adding 0.006 mg of zinc ion (Zn) to the combined acid extracts from the control used in the test for *Copper* (0.005%).

Nitrate—Dissolve 10 g in 100 mL of dilute sulfuric acid (1 in 100), heat to boiling, and pour, slowly, into a mixture of 140 mL of water and 50 mL of stronger ammonia TS. Filter through a folded filter while still hot, wash with hot water until the volume of filtrate is 300 mL, mix, and cool. To 15 mL of this solution add 1 mL of sodium chloride solution (1 in 200), 0.10 mL of indigo carmine TS, and 15 mL of sulfuric acid. The blue color is not completely discharged at the end of 5 minutes (0.01%).

Substances not precipitated by ammonia—Evaporate to dryness 30 mL of the filtrate obtained in the test for *Nitrate*, and ignite gently: the weight of residue does not exceed 1 mg (0.01%).

Ferrous Ammonium Sulfate, $Fe(NH_4)_2(SO_4)_2 . 6H_2O$—**392.13**—Use ACS reagent grade.

Ferrous Sulfate, $FeSO_4 . 7H_2O$—**278.01**—Use ACS reagent grade.

Filter Paper, Quantitative—For the *Mercuric Bromide Test Paper* used in testing for arsenic, use Swedish O filter paper or other makes of like surface, quality, and ash.

Fluorene, $C_{13}H_{10}$—**166.22**—White to off-white crystals or powder. Soluble in benzene, in carbon disulfide, in ether, and in hot alcohol; freely soluble in glacial acetic acid.
Solubility test—One g dissolves in 10 mL of acetone to yield a clear and complete solution.
Melting range ⟨741⟩: between 113° and 117°, within a 2° range.

Fluorescamine, $C_{17}H_{10}O_4$—**278.26**—White to off-white powder. Very slightly soluble in water; freely soluble in methylene chloride; soluble in alcohol; slightly soluble in chloroform.
Assay—Dissolve about 600 mg in 75 mL of dimethylformamide, and titrate with 0.1 *N* lithium methoxide to a blue endpoint, using 1% thymol blue in dimethylformamide as the indicator. Perform a blank determination, and make any necessary correction. Each mL of 0.1 *N* lithium methoxide is equivalent to 27.83 mg of $C_{17}H_{10}O_4$. Not less than 99% is found.
Loss on drying ⟨731⟩—Dry it at 105° for 4 hours: it loses not more than 0.5% of its weight.

Fluorosilicone—Use a suitable grade.

Fluoxymesterone—Use *Fluoxymesterone* (USP monograph).

Flurazepam Hydrochloride—Use *Flurazepam Hydrochloride* (USP monograph).

Folic Acid—Use *Folic Acid* (USP monograph).

Formaldehyde Solution, HCHO (**30.03**) and water—Use ACS reagent grade.

Formamide, $HCONH_2$—**45.04**—Use ACS reagent grade.
Preparation for Digitoxin Assay—To ensure freedom from ammonia, treat Formamide as follows: Shake a suitable quantity of formamide with about 10% of its weight of anhydrous potassium carbonate for 15 minutes, and filter. Distil the filtrate in all-glass apparatus under vacuum at a pressure of about 25 mm of mercury or less. Reject the first portion of distillate containing water, and collect the fraction that boils at about 115° at a pressure of 25 mm of mercury or at 101° at a pressure of 12 mm of mercury. Store in tight containers, protected from light.

Formic Acid, HCOOH—**46.03**—Use ACS reagent grade Formic Acid, 88 Percent.

Formic Acid, Anhydrous—Use ACS reagent grade Formic Acid, 96 Percent.

Formic Acid, 96 Percent, HCOOH—**46.03**—Use ACS reagent grade Formic Acid, 96 Percent.

Fructose—Use *Fructose* (USP monograph).

Fuchsin, Basic—A mixture of rosaniline and pararosaniline hydrochlorides. Crystals or crystalline fragments with a glossy, greenish bronze luster. Soluble in water, in alcohol, and in amyl alcohol.
To 10 mL of a solution (1 in 500) add 10 mL of ammonia TS and 500 mg of zinc dust, and agitate the mixture: the solution becomes colorless. Place a few drops of the decolorized solution on filter paper and nearby, on the same paper, place a few drops of diluted hydrochloric acid: a red color develops at the zone of contact.
Loss on drying ⟨731⟩—Dry it at 105° to constant weight: it loses not more than 5.0% of its weight.
Residue on ignition (Reagent test)—Ignite 1 g with 0.5 mL of sulfuric acid: the residue weighs not more than 3 mg (0.3%).

Fuller's Earth, Chromatographic[36]—(*Very Fine and Moderately Coarse*)—Gray or grayish white powder or granules consisting mainly of hydrous aluminum-magnesium silicate.
Powder fineness—See *Powder Fineness* ⟨811⟩.
Soluble matter—Twenty g, treated with 50 mL of cold water and filtered, yields not more than 60 mg of residue upon evaporation of the filtrate (0.3%). A second 20-g portion, treated with 50 mL of cold alcohol and filtered, yields not more than 14 mg upon evaporation of the filtrate (0.07%).
Loss on drying ⟨731⟩—Dry it at 105° for 6 hours: it loses between 7.0% and 10.0% of its weight.
NOTE—Adjust the water content, if necessary, by drying in vacuum at room temperature, restoring the water required, and equilibrating by shaking for 2 hours.

Fumaric Acid—Use *Fumaric Acid* (NF monograph).

Fuming Nitric Acid—See *Nitric Acid, Fuming.*

Fuming Sulfuric Acid—See *Sulfuric Acid, Fuming.*

Furfural, C_4H_3OCHO—**96.09**—A clear, colorless liquid when freshly distilled, but soon turns reddish brown. Soluble in water. Miscible with alcohol. Store in tight, light-resistant containers. Before use it should be freshly distilled.
Boiling range (Reagent test)—Not less than 95% distils between 159° and 162°.

G1; G2; G3; G4; G5; G6; G7; G8; G9; G10; G11; G12; G13; G14; G15; G16; G17; G18; G19; G20; G21; G22; G23; G24; G25; G26; G27; G28; G29; G30; G31; G32; G33; G34; G35; G36; G37—See *Phases, Liquid, for Gas and High-pressure Liquid Chromatography.*

β-Galactosidase Suspension—Use a suitable grade.[37]
Suitability—When used to assay lactulose, determine that a suitable absorbance-versus-concentration slope is obtained using USP Lactulose RS, the reagent blank absorbance being not more than 0.020.

Gallic Acid, $C_6H_2(OH)_3COOH.H_2O$—**188.14**—White, or nearly white, crystals, or powder. Sparingly soluble in cold water; very soluble in boiling water and in alcohol.
Distinction from tannic acid—Its cold, saturated solution neither colors nor precipitates solutions of pure ferrous salts and yields no precipitate with gelatin TS.
Insoluble matter (Reagent test)—A 10-g portion, dissolved in 300 mL of hot water, shows not more than 1 mg of insoluble matter (0.01%).
Residue on ignition—Ignite 10 g, cool, add 1 mL of sulfuric acid, and ignite again: the residue weighs not more than 1 mg (0.01%).
Sulfate—Dissolve 2 g in 50 mL of hot water, cool in ice water while stirring, and filter. Dilute the filtrate to 50 mL, and to 25 mL add 1 mL of 1 *N* hydrochloric acid and 2 mL of barium chloride TS. Any turbidity produced in 10 minutes does not exceed that of a standard containing 0.05 mg of sulfate (SO_4), 1 mL of dilute hydrochloric acid (1 in 12), and 2 mL of barium chloride TS (0.005%).

Gelatin—Use *Gelatin* (NF monograph).

Girard Reagent T—See *Trimethylacethydrazide Ammonium Chloride.*

Gitoxin, $C_{41}H_{64}O_{14}$—**780.95**—White, crystalline powder. Practically insoluble in water, in chloroform, and in ether; slightly soluble in pyridine and in diluted alcohol. Melts at about 250° with decomposition.
Specific rotation ⟨781⟩: between +3.8° and +4.8°, determined in a solution of pyridine containing 10 mg per mL, with the use of a mercury light at 546.1 nm; between +21° and +25°, determined in a solution of equal parts of chloroform and methanol containing 5 mg per mL, with the use of sodium light.
Suitability—Dissolve 10 mg each of USP Digitoxin RS, previously dried, USP Digoxin RS, previously dried, and gitoxin, respectively, in separate 5-mL portions of a mixture of 2 parts of chloroform and 1 part of methanol, and dilute each with additional solvent mixture to 10 mL. Then proceed as directed in the *Identification test* under *Digoxin.* The chromatogram of gitoxin shows one fluorescent spot, located between the digoxin and digitoxin spots.

Glacial Acetic Acid—See *Acetic Acid, Glacial.*

Glass Wool—Fine threads of glass.
Acid-soluble substances—Boil 1 g for 30 minutes with 30 mL of diluted hydrochloric acid, and filter. Evaporate the filtrate, and dry the residue at 105° to constant weight: the residue weighs not more than 5 mg (0.5%).
Heavy metals—Boil 2 g with a mixture of 25 mL each of diluted nitric acid and water for 5 minutes, and filter. Evaporate one-half of the filtrate to dryness, dissolve the residue in 10 mL of water to which 3 drops of hydrochloric acid have been added, filter if necessary, and add an equal volume of hydrogen sulfide TS to the filtrate: no darkening is produced.

D-Gluconic Acid, 50 Percent in Water, $C_6H_{12}O_7$—**196.16**—Pale yellow liquid.

Assay—Dilute about 200 mg of the solution, accurately weighed, with 30 mL of water. Titrate with 0.1 *N* sodium hydroxide VS, determining the end-point potentiometrically. Perform a blank determination and make any necessary correction. Each mL of 0.1 *N* sodium hydroxide is equivalent to 19.62 mg of $C_6H_{12}O_7$. Not less than 49.0% is found.

Refractive index ⟨831⟩: between 1.4160 and 1.4180 at 20°.

Specific rotation ⟨781⟩: between +9.9° and +11.9°, determined as is, at 20°.

Gluten, Ground Wheat—A powder derived from wheat flour by the almost complete removal of starch.

Loss on drying ⟨731⟩—Dry it at 105° to constant weight: it loses not more than 10.0% of its weight.

Nitrogen, Method I ⟨461⟩—Not less than 14.2% of N is found, on the anhydrous basis.

Glycerin (*Glycerol*)—Use ACS reagent grade Glycerol.

Glycine—Use *Glycine* (USP monograph).

Gold Chloride (*Chlorauric Acid*), $HAuCl_4 . 3H_2O$—**393.83**—Use ACS reagent grade.

Ground Wheat Gluten—See *Gluten, Ground Wheat.*

Guaiacol (*o-Methoxyphenol*), $C_7H_8O_2$—**124.14**—Colorless to yellowish, refractive liquid having a characteristic odor. Soluble in about 65 parts of water; soluble in sodium hydroxide solution; miscible with alcohol, with chloroform, with ether, and with glacial acetic acid.

Assay—When examined by gas-liquid chromatography, it shows a purity of not less than 98%. The following conditions have been found suitable for assaying it: A 1.8-m × 3-mm stainless steel column containing liquid phase G16 on 60- to 80-mesh support S1A. Helium is the carrier gas, the injection port temperature is maintained at 180°, the column temperature is 200°, and the flame-ionization detector is maintained at 280°. The retention time is about 8 minutes.

Refractive index ⟨831⟩: between 1.5430 and 1.5450, at 20°.

Guaifenesin—Use *Guaifenesin* (USP monograph).

Guanine Hydrochloride, $C_5H_5N_5O . HCl . H_2O$—**205.60**—White, crystalline powder. Melts above 250°, with decomposition. Slightly soluble in water and in alcohol; soluble in acidulated water and in sodium hydroxide TS. Its solutions are not precipitated by iodine TS or by mercuric-potassium iodide TS, but form a precipitate with trinitrophenol TS.

Residue on ignition (Reagent test): negligible, from 100 mg.

Loss on drying ⟨731⟩—Dry it at 105° to constant weight: it loses not more than 10.0% of its weight.

Helium—Use *Helium* (USP monograph).

Hematein, $C_{16}H_{12}O_6$—**300.27**—Prepared from logwood extract or from hematoxylin by treatment with ammonia and exposure to air. Reddish brown crystals with a yellowish green metallic luster. Very slightly soluble in water (about 1 in 1700); slightly soluble in alcohol and in ether; insoluble in benzene and in chloroform; freely soluble in diluted ammonia solution to form a solution of dusky purplish red color and in an aqueous solution of sodium hydroxide (1 in 50), to form a solution of bright red color, viewed in each case through a layer 1 cm in depth. Melts at a temperature above 200° and tends to decompose at 250°.

Hematin (*Hydroxyheme*), $C_{34}H_{33}FeN_4O_5$—**633.51**—A derivative of the ferriprotoporphyrin fraction of hemoglobin.

Hematoxylin (*Hydroxybrasilin*), $C_{16}H_{14}O_6 . 3H_2O$—**356.33**—A crystalline substance derived from the heart-wood of *Haematoxylon campechianum* Linné (Fam. Leguminosae). Colorless to yellow prisms. Very slightly soluble in cold water and in ether; rapidly soluble in hot water and in hot alcohol. When exposed to light, it acquires a red color and yields a yellow solution. Dissolves in ammonia TS and in solutions of alkali hydroxides and carbonates. When dissolved in solutions of the following salts, it develops the colors indicated: in alum solution a red color; in stannous chloride solution a rose color; in solutions of cupric salts a greenish gray color. It gradually turns black in potassium dichromate solution. Store hematoxylin and its solutions protected from light and air.

Heparin Sodium—Use *Heparin Sodium* (USP monograph).

***n*-Heptane**—Use *n-Heptane, Chromatographic.*

***n*-Heptane, Chromatographic**—Clear, colorless, volatile, flammable liquid consisting essentially of C_7H_{16}. It has a characteristic odor. Practically insoluble in water; soluble in absolute alcohol. Miscible with ether, with chloroform, with benzene, and with most fixed and volatile oils.

NOTE—*n*-Heptane may require purification by passage through a column of silica gel, a ratio of about 25 g of the gel for each 100 mL of *n*-heptane being used, and subsequent fractional distillation.

Boiling range (Reagent test): between 94.5° and 99.0°.

Spectral purity—Measure in a 1-cm cell at 250 nm, with a suitable spectrophotometer, against water as the blank: its absorbance is not more than 0.10.

Residue on evaporation—It meets the requirements of the test for *Residue on evaporation* under *Hexane, Solvent.*

Heptanesulfonic Acid, $C_7H_{16}O_3S$—**180.26**—Use a suitable grade.

Hexacosane, $C_{26}H_{34}$—**366.71**—Use a suitable grade.

Hexadecyl Hexadecanoate (*Hexadecyl Palmitate; Cetyl Palmitate*), $C_{32}H_{64}O_2$—**480.86**—Use a suitable grade.[38]

Hexamethyldisilazane, $C_6H_{19}NSi_2$—**161.39**—Clear, colorless liquid, having a characteristic odor.

Assay—When examined by gas-liquid chromatography, it shows a purity of not less than 95%. The following conditions have been found suitable for assaying the article: A 1.8-m × 2-mm glass column packed with phase G3 on support S1. Helium, flowing at a rate of about 40 mL per minute, is the carrier gas, the detector temperature is about 310°, the injection port temperature is about 100°, and the column temperature is programmed to start at 35°, hold for 5 minutes, then rise at a rate of 8° per minute to 200°. A flame-ionization detector is employed.

Residue after evaporation—Transfer 200 g to a tared dish, and evaporate on a steam bath to dryness. Dry the residue at 105° for 1 hour, cool, and weigh: not more than 0.0025% of residue is found.

Hexamethyleneimine (*Homopiperidine*), $C_6H_{12}NH$—**99.18**—Colorless to nearly colorless liquid.

Refractive index ⟨831⟩: between 1.4640 and 1.4660 at 20°.

***n*-Hexane,** C_6H_{14}—**86.18** (for use in spectrophotometry)—Use *Hexanes.*

Hexane, Solvent (*Petroleum Benzin; Petroleum Ether*)—Clear, volatile liquid, having an ethereal or faint, petroleum-like odor. Practically insoluble in water; soluble in absolute alcohol. Miscible with ether, with chloroform, with benzene, and with most fixed and volatile oils.

Caution—It is dangerously flammable. Keep it away from flames and store in tight containers in a cool place.

Appearance and color—Pour 100 mL, previously well-mixed in its original container, into a 100-mL color-comparison tube, and compare with a standard, in a similar tube, containing 2 mL of platinum-cobalt TS in similar volume: the two liquids are equally clear and free from suspended matter or sediment and when viewed across the columns by transmitted light, the test specimen is not darker in color than the standard.

Odor—Its odor is not disagreeable or suggestive of mercaptans or thiophene.

Distilling range (Reagent test)—Distil 100 mL: none distils below 30° and not less than 100% distils between 30° and 60°.

Residue on evaporation—Evaporate 150 mL (100 g) on a steam bath, and dry at 105° for 30 minutes: the residue weighs not more than 1 mg (0.001%).

Acidity—Shake 10 mL with 5 mL of water for 2 minutes, and allow the layers to separate: the water layer does not turn blue litmus red within 15 seconds.

Heavy oils and fats—Gradually pour 10 mL onto the center of a clean filter paper: there is no disagreeable odor and no greasy stain visible on the paper after it has stood for 30 minutes.

Hexane, Solvent, Chromatographic—It complies with the specifications for *Hexane, Solvent,* and meets the requirements of the following additional test.

Spectral purity—Measure in a 1-cm cell at 300 nm, with a suitable spectrophotometer, against air as the blank: its absorbance is not more than 0.08.

Hexanes (suitable for use in ultraviolet spectrophotometry); usually a mixture of several isomers of hexane (C_6H_{14}), predominantly *n*-hexane, and methylcyclopentane (C_6H_{12})—Use ACS reagent grade.

Hexanitrodiphenylamine (*Dipicrylamine*), $C_{12}H_5N_7O_{12}$—**439.21**—Yellow-gold powder or prisms. *Explosive.* Usually contains about 15% of water as a safety precaution. Insoluble in water, in alcohol, in acetone, and in ether; soluble in glacial acetic acid and in alkalies.

Water, Method I ⟨921⟩: not more than 16%.

Hexokinase and Glucose-6-phosphate Dehydrogenase Suspension—Use a suitable grade.[37]

Suitability—When used in the assay of lactulose, determine that a suitable absorbance-versus-concentration slope is obtained, using USP Lactulose RS, the reagent blank absorbance being not more than 0.020.

Histamine Dihydrochloride, $C_5H_9N_3 . 2HCl$—**184.07**—Use USP Histamine Dihydrochloride RS.

Histamine Phosphate—Use *Histamine Phosphate* (USP monograph).

Homatropine Hydrobromide—Use *Homatropine Hydrobromide* (USP monograph).

Hydrazine Hydrate, 85% in Water, $(NH_2)_2 . H_2O$—**50.06**—Colorless liquid.

Assay—Transfer 600 mg, accurately weighed, to a 100-mL volumetric flask. Dilute with water to volume, and mix. Pipet 10 mL into a suitable beaker, add 1.0 g of sodium bicarbonate and 50.0 mL of 0.1 *N* iodine VS. Titrate the excess iodine with 0.1 *N* sodium thiosulfate VS, using starch TS as the indicator. Perform a blank determination, and make any necessary correction. Each mL of 0.1 *N* iodine is equivalent to 12.52 mg of $(NH_2)_2 . H_2O$. Not less than 83% is found.

Hydrazine Sulfate, $(NH_2)_2 . H_2SO_4$—**130.12**—Use ACS reagent grade.

Hydriodic Acid, HI—**127.91**—Use ACS reagent grade (containing not less than 47.0% of HI).

NOTE—For *Methoxy Determination* (see ⟨431⟩), use hydriodic acid that is labeled "for alkoxyl determination," or that is purified as directed under *Methoxy Determination* ⟨431⟩. Use this grade also for alkoxyl determinations in assays in the individual monographs.

Hydrobromic Acid, 48 Percent, HBr—**80.91**—Use ACS reagent grade.

Hydrochloric Acid, HCl—**36.46**—Use ACS reagent grade.

Hydrochloric Acid, Diluted (10 percent)—Prepare by mixing 226 mL of hydrochloric acid with sufficient water to make 1000 mL.

Hydrocortisone—Use *Hydrocortisone* (USP monograph).

Hydrofluoric Acid, HF—**20.01**—Use ACS reagent grade.

Hydrogen Peroxide, 30 Percent, H_2O_2—**34.01**—Use ACS reagent grade.

Hydrogen Peroxide Solution—Use *Hydrogen Peroxide Topical Solution* (USP monograph).

Hydrogen Sulfide, H_2S—**34.08**—Colorless, poisonous gas, heavier than air. Soluble in water. Is generated by treating ferrous sulfide with diluted sulfuric or diluted hydrochloric acid. Other sulfides yielding hydrogen sulfide with diluted acids may be used. Is also available in compressed form in cylinders.

Hydrogen Sulfide Detector Tube—A fuse-sealed glass tube so designed that gas may be passed through it, and containing suitable absorbing filters and support media for the indicator, the latter consisting of a suitable lead salt.

NOTE—A suitable detector tube that conforms to the monograph specification is available from National Draeger, Inc., 401 Parkway View Drive, Pittsburgh, Pa. 15205 as Reference Number 6719001, Measuring Range 1 to 20 ppm. Tubes having conditions other than those specified in the monograph may be used in accordance with the section entitled *Tests and Assays* in the *General Notices.*

Hydroquinone, $C_6H_4(OH)_2$—**110.11**—Fine, colorless or white, needle crystals. Darkens on exposure to air and light. Soluble in water, in alcohol, and in ether.

Assay—Weigh accurately about 250 mg, and dissolve in a mixture of 100 mL of water and 10 mL of 0.1 *N* sulfuric acid in a 250-mL conical flask. Add 3 drops of a 1 in 100 solution of diphenylamine in sulfuric acid, and titrate with 0.1 *N* ceric sulfate VS until the solution is red-violet in color. Each mL of 0.1 *N* ceric sulfate is equivalent to 5.506 mg of $C_6H_4(OH)_2$. Not less than 99% is found.

Melting range ⟨741⟩: between 172° and 174°.

3′-Hydroxyacetophenone, $C_8H_8O_2$—**136.15**—Light brown powder chips and chunks. Melts at about 96°. Sparingly soluble in chloroform, yielding a clear, light yellow solution.

Assay—Inject an appropriate specimen into a suitable gas chromatograph (see *Chromatography* ⟨621⟩) equipped with a flame-ionization detector, helium being used as the carrier gas. The following conditions have been found suitable: a 0.25-mm × 30-m capillary column coated with G1; the detector and the injection port temperature are maintained at 300°; the column temperature is maintained at 180° and programmed to rise 10° per minute to 280° and held at that temperature for 10 minutes. The area of the main peak is not less than 97% of the total peak area.

4′-Hydroxyacetophenone, $HOC_6H_4COCH_3$—**136.15**—Gray powder, melting at about 109°.

***p*-Hydroxybenzoic Acid,** $C_7H_6O_3$—**138.12**—White crystals.

Assay—Transfer about 700 mg, accurately weighed, to a suitable container, and dissolve in 50 mL of acetone. Add 100 mL of water, mix, and titrate with 0.5 *N* sodium hydroxide VS, determining the end-point potentiometrically. Perform a blank determination, and make any necessary correction. Each mL of 0.5 *N* sodium hydroxide is equivalent to 69.06 mg of $C_7H_6O_3$: not less than 97% is found.

Melting range ⟨741⟩: over a range of 2° that includes 216°.

Hydroxylamine Hydrochloride, $NH_2OH . HCl$—**69.49**—Use ACS reagent grade.

10β-Hydroxynorandrostenedione (*10β-Hydroxy-19-norandrost-4-ene-3,17-dione*), $C_{18}H_{24}O_3$—**288.39**—Use a suitable grade.

10β-Hydroxynorethinyltestosterone—Use USP 10β-Hydroxynorethinyltestosterone Reagent.

3-Hydroxyphenyldimethylethyl Ammonium Chloride [Dimethylethyl(3-hydroxyphenyl)ammonium Chloride]—Use *Edrophonium Chloride* (USP monograph).

8-Hydroxyquinoline (*Oxine*), C_9H_7NO—**145.16**—Use ACS reagent grade 8-Quinolinol.

Hydroxytoluene, Butylated (*2,6-Di-tertiary Butyl-p-cresol*), $C_{15}H_{24}O$—**220.35**—Crystals. Insoluble in water; freely soluble in toluene; soluble in acetone, in alcohol, in benzene, in ether, in isopropyl alcohol, in methanol, in petroleum, and in other hydrocarbon solvents, and in liquid petrolatum, in food oils, and in fats. *Caution—May cause sensitization-type dermatitis.*

Congealing temperature ⟨651⟩: not less than 69.2°, indicating not less than 99.0% of $C_{15}H_{24}O$.

Residue on ignition (Reagent test): not more than 0.002%, determined on a 50-g portion and 1 mL of sulfuric acid being used.

Hypophosphorous Acid, 50 Percent (*Hypophosphorous Acid*), HPH_2O_2—**66.00**—A colorless to faintly yellow liquid. Miscible with water and with alcohol.

Assay—Weigh accurately about 4 mL, dilute with 25 mL of water, add methyl red TS, and titrate with 1 *N* sodium hydroxide VS: each mL of 1 *N* sodium hydroxide is equivalent to 66.00 mg of HPH_2O_2. Not less than 48% is found.

Chloride—Add 0.2 mL to a mixture of 10 mL of silver nitrate TS and 5 mL of nitric acid, and heat until brown fumes are no longer evolved: any white, insoluble residue remaining is negligible.

Phosphate—Dilute 1 mL with water to 50 mL, render alkaline with ammonia TS, filter if a precipitate is formed, and add to the filtrate 5 mL of magnesia mixture TS: not more than a slight precipitate is formed within 5 minutes.

Sulfate (Reagent test, *Method I*)—Dilute 1 mL with water to 50 mL: 20 mL of the solution shows not more than 0.2 mg of SO_4.

Imidazole, $C_3H_4N_2$—**68.08**—White to light yellow crystals. Freely soluble in water.

Assay—Transfer about 0.7 g, accurately weighed, to a 250-mL beaker. Dissolve in 100 mL of glacial acetic acid. Titrate with 0.1 *N* perchloric acid VS, determining the end-point potentiometrically. Perform a blank determination, and make any necessary correction. Each mL of 0.1 *N* perchloric acid is equivalent to 6.808 mg of $C_3H_4N_2$. Not less than 98% is found.

Iminostilbene, $C_{14}H_{11}N$—**193.25**—Yellow-orange powder. Sparingly soluble in ethyl acetate.

Melting range ⟨741⟩: between 197° and 201°.

Indicators—See separate subsection.

Indigo Carmine—Use *Indigotindisulfonate Sodium* (USP monograph).

Inositol (*Hexahydroxycyclohexane*), $C_6H_6(OH)_6$—**180.16**—Fine, white crystals or a white, crystalline powder, odorless, having a sweet taste, and stable in air. Its solutions are neutral to litmus. Optically inactive. One g dissolves in 5.7 mL of water. Slightly soluble in alcohol; insoluble in ether and in chloroform. Store in well-closed containers.

Melting range ⟨741⟩: between 223° and 226°.

Loss on drying ⟨731⟩—Dry it at 105° for 4 hours: it loses not more than 0.5% of its weight.

Residue on ignition (Reagent test): not more than 0.1%.

Insulin Antibody Preparation, Dried—Use a suitable grade.

Iodic Acid, HIO_3—**175.91**—Use ACS reagent grade.

Iodine, I—At. Wt. **126.9045**—Use ACS reagent grade.

Iodine Monochloride, ICl—**162.36**—Use ACS reagent grade.

Iodine Pentoxide (*Iodic Anhydride*), I_2O_5—**333.81**—White, crystalline powder. Soluble in water; insoluble in alcohol and in ether. When heated to about 300°, it decomposes into iodine and oxygen. When it is heated in the presence of sulfur or organic matter, deflagration takes place.

Assay—Dry about 500 mg at 200° to constant weight, weigh accurately, and dissolve in water to make 100.0 mL. Dilute 20.0 mL of this solution with 30 mL of water, add 2 g of potassium iodide and 5 mL of diluted sulfuric acid, allow to stand for 10 minutes in the dark, and titrate the liberated iodine with 0.1 *N* sodium thiosulfate VS, adding 3 mL of starch TS as the end-point is approached. Correct for any iodine liberated in a blank made with the same quantity of the reagents. Each mL of 0.1 *N* sodium thiosulfate is equivalent to 2.782 mg of I_2O_5. Not less than 98.5% is found.

Residue on ignition (Reagent test)—Ignite 4 g in a porcelain crucible: the residue weighs not more than 2 mg (0.05%).

Heavy metals—To the *Residue on ignition* add 1 mL of hydrochloric acid, and evaporate on a steam bath to dryness. Take up in 1 mL of 0.1 *N* hydrochloric acid and water to make 40 mL. Label this *Solution A*. To 20 mL add 5 mL of water, adjust to a pH of about 3.5, determined potentiometrically, by adding diluted acetic acid or ammonia TS, as required. Add water to make 40 mL, mix, then add 10 mL of hydrogen sulfide TS, and mix: any brown color produced is not darker than that of a control containing 0.02 mg of Pb and 0.3 mL of 0.1 *N* hydrochloric acid in a volume of 20 mL and treated with 10 mL of hydrogen sulfide TS (0.001%).

Iron—Five mL of *Solution A* prepared in the *Heavy metals* test shows not more than 0.01 mg of Fe (0.002%).

Iodohydroxyquinolinesulfonic Acid (*Iodoxyquinsulfonic Acid; 7-Iodo-8-hydroxyquinoline-5-sulfonic Acid*), $C_9H_6INO_4S$—**351.12**—Yellow, crystalline powder. Slightly soluble in water and in alcohol. Melts at about 260° to 270°, with decomposition.

Sensitiveness—To 1 mL of a solution containing 0.025 mg of ferric chloride add 2 drops of hydrochloric acid and a drop of hydrogen peroxide TS, and mix. To this mixture add 0.1 mL of a solution of the test specimen (1 in 1000): a green or bluish green color is produced.

Residue on ignition (Reagent test): not more than 0.2%.

Ion-Exchange Resin[39]—An intimate mixture of 4 parts of a strongly acidic cation-exchanger in the hydrogen form (produced by sulfonation of a styrene-divinylbenzene copolymer, representing 8 to 10% divinylbenzene) and 6 parts of a strongly basic anion-exchanger in the hydroxyl form (produced by amination with trimethylamine of a chloromethylated styrene-divinylbenzene copolymer, representing 3 to 5% divinylbenzene).

Iron, Reduced, Fe—At. Wt. **55.847**—Gray, fine powder. Insoluble in water; soluble in dilute acids. Oxidizes on exposure to air and moisture. Store in well-closed containers.

Assay—Weigh accurately about 200 mg, and transfer it to a 300-mL conical flask. Add 50 mL of diluted sulfuric acid, and close the flask with a stopper containing a valve, made by inserting a glass tube connected to a short piece of rubber tubing with a slit on the side and a glass rod inserted in the other end, arranged so that gases can escape, but air cannot enter. Heat on a steam bath until the iron is dissolved. Cool the solution, dilute with 50 mL of freshly boiled and cooled water, add 2 drops of orthophenanthroline TS, and titrate with 0.1 *N* ceric sulfate VS until the red color is changed to weak blue. Each mL of 0.1 *N* ceric sulfate is equivalent to 5.585 mg of Fe: not less than 93% is found.

Sulfuric acid-insoluble matter—To 1 g add 25 mL of diluted sulfuric acid, and warm on a steam bath until the evolution of hydrogen ceases. Collect any undissolved residue on a filter, wash it with 2% sulfuric acid, then with water, and dry at 105°: the weight of the residue does not exceed 5 mg (0.5%).

Nitrogen—Add in small portions 1.5 g to a mixture of 2 mL of sulfuric acid and 30 mL of water contained in a distilling flask. Cool, add 20 mL of water and 50 mL of sodium hydroxide solution (4 in 10), and slowly distil about 40 mL into 5 mL of water containing 1 drop of diluted hydrochloric acid. To the distillate add 2 mL of sodium hydroxide TS and 2 mL of alkaline mercuric–potassium iodide TS: any color produced is not darker than that obtained by treating 0.12 mg of ammonium chloride (0.03 mg N) and 500 mg of the test specimen in the same manner (0.003%).

Sulfide—To 1 g contained in a 150-mL flask add 20 mL of diluted sulfuric acid: the gas evolved does not darken moistened lead acetate paper within 2 minutes.

Iron Wire, Fe—At. Wt. **55.847**—Use ACS reagent grade.

Iron-phenol Reagent—Dissolve 1.054 g of ferrous ammonium sulfate in 20 mL of water, and add 1 mL of sulfuric acid and 1 mL of 30 percent hydrogen peroxide. Mix, heat until effervescence ceases, and add water to make 50 mL. To 3 volumes of this solution contained in a volumetric flask add sulfuric acid, with cooling, to make 100 volumes.

Purify white, crystalline reagent phenol or liquefied phenol free from hydrophosphorous stabilizer by distillation through a column, the equivalent of a silvered, vacuum-jacketed Vigreaux tube, at a rate of approximately 20 mL per minute. Discard the first 10% and the last 10% of the distillate. Allow the distillate to recrystallize at room temperature for several hours. Discard any distillate that is colored or shows the presence of fluid or non-crystalline material. Weigh the flask and its contents, add to the phenol 1.13 times its weight of the iron–sulfuric acid solution prepared as directed in the preceding paragraph, insert the stopper in the flask, and allow to stand, without cooling but with occasional shaking, until the phenol is liquefied, then shake the mixture vigorously until mixed. Allow to stand in the dark for 16 to 24 hours, and again weigh the flask and its contents. To the mixture add 23.5% of its weight of a solution of 100 volumes of sulfuric acid in 110 volumes of water, mix, transfer to dry, glass-stoppered bottles, and store in the dark, protected from atmospheric moisture. Use within 6 months.

Isatin, $C_8H_5NO_2$—**147.13**—Small, yellowish red crystals. Slightly soluble in water; soluble in alcohol and in ether. Water, alcohol, and ether solutions of isatin are reddish brown in color. When isatin is dissolved in a solution of an alkali hydroxide, it yields a violet-colored solution, which, upon heating or long standing, becomes yellow.

Melting range ⟨741⟩: between 198° and 201°.

Residue on ignition (Reagent test)—Ignite 500 mg: the residue weighs not more than 2 mg (0.4%).

Isoamyl Alcohol—Use *Amyl Alcohol*.

Isobutyl Acetate, $C_6H_{12}O_2$—**116.16**—Clear, colorless liquid. Slightly soluble in water. Miscible with alcohol.

Assay—Inject an appropriate specimen into a suitable gas chromatograph (see *Chromatography* ⟨621⟩) equipped with a flame-ionization detector, helium being used as the carrier gas.

The following conditions have been found suitable: a 30-m × 0.25-mm capillary column coated with G2. The injection port temperature is maintained at 130°; the column temperature is maintained at 30° and programmed to rise 10° per minute to 180° and held there for 10 minutes. The detector temperature is maintained at 300°. The area of the main peak is not less than 99% of the total peak area.

Specific gravity ⟨841⟩: between 0.863 and 0.868.

Refractive index ⟨831⟩: between 1.3900 and 1.3920 at 20°.

Isobutyl Alcohol (*2-Methyl-1-propanol*), $(CH_3)_2CHCH_2OH$—**74.12**—Use ACS reagent grade.

Isoflupredone Acetate[40] (*9-α-Fluoroprednisolone Acetate*), $C_{23}H_{29}FO_6$—**420.48**—White to yellowish white powder. Insoluble in water; freely soluble in pyridine; soluble in alcohol and in dioxane; slightly soluble in chloroform. Melts at about 240°, with decomposition.

Loss on drying ⟨731⟩—Dry it at 105° for 4 hours: it loses not more than 1.0% of its weight.

Isoniazid—Use *Isoniazid* (USP monograph).

Isonicotinic Acid Hydrazide—Use *Isoniazid* (USP monograph).

Isooctane—See *2,2,4-Trimethylpentane.*

Isopropyl Acetate, $C_5H_{10}O_2$—**102.13**—Colorless liquid, containing not more than 0.5% alcohol as isopropyl alcohol. Neutral to indicator paper. Slightly soluble in water. Miscible with alcohol and with ether. Specific gravity: about 0.875.

Boiling range (Reagent test): between 86° and 88°.

Refractive index (at 20°) ⟨831⟩: between 1.377 and 1.378.

Isopropyl Alcohol (*2-Propanol*), $(CH_3)_2CHOH$—**60.10**—Use ACS reagent grade.

[NOTE—For use in assays and tests involving ultraviolet spectrophotometry, use ACS reagent grade Isopropyl Alcohol Suitable for Use in Ultraviolet Spectrophotometry.]

Isopropyl Alcohol, Dehydrated—Use Isopropyl Alcohol that previously has been dried by being shaken with a suitable molecular sieve capable of adsorbing water, and filtered.

Isopropyl Ether—See *Diisopropyl Ether.*

Isopropyl Iodide (*2-Iodopropane*), C_3H_7I—**169.99**—Colorless liquid, discoloring upon exposure to air and light. Sparingly soluble in water; miscible with alcohol, with benzene, with chloroform, and with ether.

Density: between 1.696 and 1.704.

Refractive index ⟨831⟩: between 1.4987 and 1.4997 at 20°.

Isopropyl Myristate, $C_{17}H_{34}O_2$—**270.45**—Use *Isopropyl Myristate* (NF monograph). For use as a solvent in sterility test procedures, Isopropyl Myristate conforms to the following additional specification:

pH of water extract—Transfer 100 mL to a 250-mL centrifuge bottle, add 10 mL of twice-distilled water, close the bottle with a suitable closure, and shake vigorously for 60 minutes. Centrifuge the mixture at 1800 rpm for 20 minutes, aspirate the upper (isopropyl myristate) layer, and determine the pH of the residual water layer: the pH is not less than 6.5.

Isopropyl Myristate not conforming to the test for *pH of water extract* may be rendered suitable for use in sterility test procedures as follows:

Using a 20-mm × 20-cm glass column, add activated alumina, and tamp down to a height of 15 cm. Pass 500 mL of the isopropyl myristate through the column, using a slight positive pressure to maintain an even flow, and use the eluate collected directly in the sterility test procedure.

Isopropylamine (*2-Aminopropane*), $C_3H_7NH_2$—**59.11**—Clear, colorless, flammable liquid having a strong odor of ammonia. Miscible with water, with alcohol, and with ether.

Assay—Transfer about 0.2 g, accurately weighed, to a suitable container, add 50 mL of water, and mix. Titrate with 0.1 *N* hydrochloric acid VS, using a mixture of methyl red TS and bromocresol green TS (1:5) as indicator. Each mL of 0.1 *N* hydrochloric acid is equivalent to 59.11 mg of C_3H_9N. Not less than 98% is found.

Boiling range (Reagent test)—Not less than 95% distils between 31° and 33°.

Refractive index ⟨831⟩: between 1.3743 and 1.3753, at 20°.

Kerosene—A mixture of hydrocarbons, chiefly of the methane series. A clear, colorless liquid, possessing a characteristic, but not disagreeable, odor. Specific gravity: about 0.80. Distils between 180° and 300°.

L1; L2; L3; L4; L5; L6; L7; L8; L9; L10; L11; L12; L13; L14; L15; L16; L17; L18; L19; L20; L21; L22; L23; L24—See *Packings for High-pressure Liquid Chromatography.*

Lactic Acid—Use *Lactic Acid* (USP monograph).

Lactose, $C_{12}H_{22}O_{11}$—**342.30**—Use ACS reagent grade.

Lanthanum Alizarin Complexan Mixture—Use a suitable grade.[41]

Lanthanum Chloride, $LaCl_3$—**245.26**—Use ACS reagent grade.

Lead Acetate, $Pb(C_2H_3O_2)_2 \cdot 3H_2O$—**379.33**—Use ACS reagent grade.

Lead Dioxide, PbO_2—**239.20**—Use ACS reagent grade.

Lead Monoxide (*Litharge*), PbO—**223.20**—Heavy, yellowish or reddish yellow powder. Insoluble in water and in alcohol; soluble in acetic acid, in diluted nitric acid, and in warm solutions of the fixed alkali hydroxides.

Assay—Weigh accurately about 300 mg, freshly ignited in a muffle furnace at 600 ± 50°, and dissolve it by warming with 10 mL of water and 1 mL of glacial acetic acid. Dilute with 75 mL of water, heat to boiling, add 50.0 mL of 0.1 *N* potassium dichromate VS, and boil for 2 to 3 minutes. Cool, transfer to a 200-mL volumetric flask with the aid of water, dilute with water to volume, mix, and allow to settle. Withdraw 100.0 mL of the clear liquid, and transfer to a glass-stoppered flask. Add 10 mL of diluted sulfuric acid and 1 g of potassium iodide, insert the stopper, mix gently, and allow to stand for 10 minutes. Then titrate the liberated iodine, representing the excess of dichromate, with 0.1 *N* sodium thiosulfate VS, adding 3 mL of starch TS as the end-point is approached: each mL of 0.1 *N* potassium dichromate is equivalent to 7.440 mg of PbO. Not less than 98% is found.

Insoluble in acetic acid—Dissolve 2 g in 30 mL of dilute glacial acetic acid (1 in 2), boil gently for 5 minutes, filter, wash the residue with diluted acetic acid, and dry at 105° for 2 hours: the residue weighs not more than 10 mg (0.5%).

Substances not precipitated by hydrogen sulfide—Completely precipitate the lead from the filtrate obtained in the test for *Insoluble in acetic acid* by passing hydrogen sulfide into it, filter, and wash the precipitate with 20 mL of water. To one-half of the mixed filtrate and washings add 5 drops of sulfuric acid, evaporate to dryness, and ignite at 800 ± 25° for 15 minutes: the residue weighs not more than 5 mg (0.5%).

Volatile substances—Weigh accurately about 5 g, and heat strongly in a covered porcelain crucible: it loses not more than 2.0% of its weight.

Lead Nitrate, $Pb(NO_3)_2$—**331.21**—Use ACS reagent grade.

Lead Perchlorate, $Pb(ClO_4)_2 \cdot 3H_2O$—**460.15**—White crystals.

Assay—Transfer about 1.8 g, accurately weighed, to a suitable container, and dissolve in 50 mL of water. Pass the solution through a suitable short cation-exchange column, collecting the eluate in a suitable container. Wash the column with water until the eluate is neutral to blue litmus, and combine the washings with the first eluate. Add 5 drops of phenolphthalein TS, and titrate with 0.1 *N* sodium hydroxide VS. Perform a blank determination, and make any necessary correction. Each mL of 0.1 *N* sodium hydroxide is equivalent to 23.07 mg of $Pb(ClO_4)_2 \cdot 3H_2O$.

Leucovorin Calcium—Use *Leucovorin Calcium* (USP monograph).

Light Mineral Oil—Use *Mineral Oil, Light* (NF monograph).

Liquid Petrolatum—Use *Mineral Oil* (USP monograph).

Lithium Chloride, LiCl—**42.39**—White, deliquescent crystals or granules. Freely soluble in water; soluble in acetone, in alcohol, in amyl alcohol, and in ether. Preserve in tight containers.

Assay—Dissolve about 1.3 g, previously dried at 120° for 1 hour and accurately weighed, in water to make 50.0 mL. Pipet 5 mL of the solution into a 250-mL conical flask, and add 5 mL of glacial acetic acid, 50 mL of methanol, and 2 drops of eosin Y TS. Titrate slowly with 0.1 *N* silver nitrate VS, adding it dropwise toward the end, until the color changes to an intense, slightly fluorescent red. Each mL of 0.1 *N* silver nitrate is equivalent to 4.239 mg of LiCl. Not less than 98% is found.

Neutrality—Dissolve 2 g in 20 mL of water, and add 1 drop of methyl red TS: any red color produced is changed to yellow on the addition of not more than 0.30 mL of 0.020 *N* sodium hydroxide. Any yellow color produced is changed to pink on the addition of not more than 0.30 mL of 0.020 *N* hydrochloric acid.

Insoluble matter (Reagent test): not more than 1.0 mg, from 10 g (0.010%).

Nitrate (Reagent test)—A 1-g portion dissolved in 2 mL of water shows no more color than that observed in 1.0 mL of *Standard Nitrate Solution* (0.001%).

Phosphate (Reagent test)—A 2-g portion shows not more than 0.02 mg of PO_4 (0.001%).

Sulfate (Reagent test, *Method I*)—A 1-g portion shows not more than 0.2 mg of SO_4 (0.02%).

Ammonium—

STANDARD AMMONIUM SOLUTION—Dissolve 296 mg of ammonium chloride in water to make 1 liter. This solution contains the equivalent of 0.1 mg of ammonium (NH_4) per mL.

PROCEDURE—To a solution of 900 mg in 50 mL of water add 1 mL of sodium hydroxide solution (1 in 10) and 2 mL of alkaline mercuric-potassium iodide TS: no more color is produced than that produced by 0.3 mL of *Standard Ammonium Solution*, diluted with water to 50 mL and treated similarly (0.003%).

Barium—Dissolve 2 g in 20 mL of water, filter, and divide the filtrate into two equal portions. To one portion add 1 mL of diluted sulfuric acid, and to the other add 1 mL of water: after 2 hours, the two portions are equally clear.

Calcium (Reagent test)—Dissolve 2.50 g in water to make 100 mL (*Test Solution*). Dissolve another 2.50 g in a mixture of 5.00 mL of *Standard Calcium Solution* and water to make 100 mL (*Control Solution*). Determine the calcium as directed under *Flame Photometry for Reagents* (Reagent test) (0.02%).

Heavy metals (Reagent test): 0.001%.

Iron ⟨241⟩—A solution of 500 mg in 47 mL of water containing 2 mL of hydrochloric acid shows not more than 0.01 mg of Fe (0.002%).

Magnesium—

STANDARD MAGNESIUM SOLUTION—Dissolve 1.014 g of clear, non-effloresced crystals of magnesium sulfate in water to make 1 liter. This solution contains the equivalent of 0.1 mg of magnesium (Mg) per mL.

PROCEDURE—To a solution of 1 g in 45 mL of water add 0.5 mL of thiazole yellow solution (1 in 10,000) and 5 mL of sodium hydroxide solution (1 in 10): no more pink color is produced than that produced by 1 mL of *Standard Magnesium Solution*, diluted with water to 45 mL and treated similarly (0.1%).

Potassium (Reagent test)—Dissolve 5.0 g in water to make 100 mL (*Sample Solution*). Dissolve another 5.0 g in a mixture of 1.00 mL of *Standard Potassium Solution* and water to make 100 mL (*Control Solution*). Determine the potassium as directed under *Flame Photometry for Reagents* (Reagent test) (0.01%).

Sodium (Reagent test)—Dissolve 0.20 g in water to make 100 mL (*Test Solution*). Dissolve another 0.20 g in a mixture of 20 mL of *Standard Sodium Solution* and water to make 100 mL (*Control Solution*). Determine the sodium as directed under *Flame Photometry for Reagents* (Reagent test) (0.1%).

Lithium Hydroxide, $LiOH.H_2O$—**41.97**—White crystals. Soluble in water; insoluble in alcohol.

Assay—Dissolve about 160 mg, accurately weighed, in 50 mL of water, add phenolphthalein TS, and titrate with 0.1 *N* hydrochloric acid VS to a colorless end-point. Each mL of 0.1 *N* hydrochloric acid is equivalent to 4.196 mg of $LiOH.H_2O$. Not less than 98% is found.

Insoluble matter (Reagent test): not more than 1.0 mg, from 10 g (0.01%).

Chloride (Reagent test)—A 200-mg portion shows not more than 0.02 mg of Cl (0.01%).

Sulfate (Reagent test, *Method I*)—Dissolve 400 mg in 10 mL of water, and neutralize with 3 *N* hydrochloric acid. Add 0.1 mL of bromine TS, boil to remove the excess bromine, add 2 mL of 3 *N* hydrochloric acid, filter, and dilute with water to 40 mL: 20 mL of this solution shows not more than 0.10 mg of SO_4 (0.05%).

Heavy metals (Reagent test): 0.002%.

Iron ⟨241⟩: One g shows not more than 0.02 mg of Fe (0.002%).

Lithium Metaborate, $LiBO_2$—**49.75**—Use ACS reagent grade.

Lithium Nitrate, $LiNO_3$—**68.95**—Colorless crystals. Use a suitable grade labeled to contain not less than 97.0%.

Lithium Oxalate, $Li_2C_2O_4$—**101.90**—White, crystalline powder. Soluble in about 17 parts of water; insoluble in alcohol.

Assay—Weigh accurately about 1 g, previously dried at 105° to constant weight, dissolve in water, and dilute to 200.0 mL. Dilute 25.0 mL of the resulting solution with 75 mL of water, add 3 mL of sulfuric acid, then add slowly 20 mL of 0.1 *N* potassium permanganate VS. Heat to 70°, and complete the titration with the permanganate until a pale pink color persists for 15 seconds. Each mL of 0.1 *N* potassium permanganate is equivalent to 5.095 mg of $Li_2C_2O_4$: not less than 99% of $Li_2C_2O_4$ is found.

Neutrality—Aqueous solutions are faintly alkaline to litmus, but are not reddened by 1 drop of phenolphthalein TS.

Chloride—Ignite 2 g, and to the residue add 20 mL of water. Neutralize the solution with nitric acid, and add 0.5 mL excess of the acid. Filter, and add to the filtrate 1 mL of silver nitrate TS. Any turbidity produced is not greater than that produced in a blank to which 0.04 mg of chloride has been added.

Sulfate—Ignite 2 g in a porcelain crucible protected from sulfur in the flame. Boil the residue with 20 mL of water and 2 mL of bromine TS, then add 5 mL of hydrochloric acid, and evaporate on a water bath to dryness. Dissolve the residue in 20 mL of water and 1 mL of dilute hydrochloric acid (1 in 12), filter, and add to the filtrate 2 mL of barium chloride TS. Any turbidity is not greater than that in a control made as follows: evaporate 2 mL of bromine TS and 5 mL of hydrochloric acid on a water bath to dryness, dissolve the residue and 0.3 mg of SO_4 in sufficient water to make 20 mL, and then add 1 mL of dilute hydrochloric acid (1 in 12) and 2 mL of barium chloride TS.

Heavy metals—Ignite 4 g gently in a porcelain crucible. Add to the residue 5 mL of water, 5 mL of hydrochloric acid, and 2 mL of nitric acid, and evaporate on a water bath to dryness. Dissolve the residue in 40 mL of water, and filter. To 10 mL of the filtrate add 0.04 mg of lead, as lead acetate, dilute with water to 30 mL, and add 1 mL of 1 *N* acetic acid (A). To the remaining 30 mL of the filtrate add 1 mL of 1 *N* acetic acid (B). Then to each add 10 mL of hydrogen sulfide TS: B is no darker than A.

Sodium—Ignite 1 g in a platinum crucible, dissolve the residue in 10 mL of diluted hydrochloric acid, and filter. The solution tested on a platinum wire in the full-heat flame of a Bunsen burner imparts no distinct yellow color and only a fleeting orange tinge to the red lithium flame (less than 0.05% of sodium oxalate).

Potassium—Ignite 5 g of the salt to carbonate in a platinum crucible, dissolve it in a small amount of water, neutralize the solution with hydrochloric acid, and filter. Evaporate this solution on a water bath to dryness, and redissolve the residue in 15 mL of water. Add 5 mL of sodium cobaltinitrite TS, and allow to stand overnight. Any precipitate formed is not greater than that produced by an amount of potassium chloride equivalent to 0.5 mg of K, dissolved in 15 mL of water, treated with 5 mL of the same sodium cobaltinitrite TS as used in the test, and allowed to stand overnight (0.01% of K).

Lithium Perchlorate, $LiClO_4$—**106.39**—Small, white crystals. Freely soluble in water; sparingly soluble in alcohol, in acetone, in ether, and in ethyl acetate.

Insoluble matter (Reagent test): not more than 1 mg, from 20 g, dissolved in 200 mL of water (0.005%).

pH: between 6.0 and 7.5, in a solution of 10 g in 200 mL of ammonia- and carbon dioxide–free water.

Chloride (Reagent test)—One g shows not more than 0.03 mg of Cl (0.003%).

Sulfate (Reagent test, *Method II*)—Dissolve 40 g in 300 mL of water, add 2 mL of hydrochloric acid, and heat the solution to boiling. Add 5 mL of barium chloride TS, digest on a steam bath for 2 hours, and allow to stand overnight. If any precipitate is formed, filter, wash thoroughly, and ignite: the residue weighs not more than 1 mg (0.001%).

Heavy metals (Reagent test): 5 ppm.

Iron—Dissolve 1 g in water, and dilute with water to 20 mL. Add 1 mL of hydroxylamine hydrochloride solution (1 in 10), 4 mL of a slightly acidified solution of orthophenanthroline, and 1 mL of sodium acetate solution (1 in 10), and allow to stand for

1 hour: any red color produced is not darker than that of a control containing 0.005 mg of added Fe (5 ppm).

Lithium Sulfate, $Li_2SO_4 . H_2O$—**127.95**—Use ACS reagent grade.

Litmus—A blue pigment prepared from various species of *Rocella* DeCandolle, *Lecanora* Acharius, or other lichens (Fam. Parmeliaceae).

Description—Cubes, masses, fragments, or granules, of an indigo blue or deep violet color. Has the combined odor of indigo and violets, tinges the saliva a deep blue, and is somewhat pungent and saline to the taste. The indicator substances it contains are soluble in water and less soluble or insoluble in alcohol.

Ash—It yields not more than 60.0% of ash.

Locust Bean Gum—A gum obtained from the ground endosperms of *Ceratonia siliqua* Linné Taub. (Fam. Leguminosae). Use Locust Bean Gum, FCC.

L-Lysine (*2,6-Diaminohexanoic Acid*), $C_6H_{14}N_2O_2$—**146.19**—Crystalline needles or hexagonal plates. Soluble in water; very slightly soluble in alcohol; insoluble in ether.

Specific rotation ⟨781⟩: between +25.5° and +26.0°, determined in dilute hydrochloric acid (1 in 2).

Nitrogen content, Method I ⟨461⟩: between 18.88% and 19.44% of N is found, corresponding to not less than 98.5% of $C_6H_{14}N_2O_2$, the test specimen previously having been dried at 105° for 2 hours.

Magnesia, Milk of—Use *Milk of Magnesia* (USP monograph).

Magnesium, Mg—**24.305**—Silvery metal in ribbon form. Reacts slowly with water at room temperature. Dissolves readily in dilute acids with the liberation of hydrogen.

Assay—Transfer 1 g, accurately weighed, to a 250-mL volumetric flask, and dissolve in a mixture of 15 mL of hydrochloric acid and 85 mL of water. When solution is complete, dilute with water to volume, and mix. Pipet 25 mL of the dilution into a 400-mL beaker, dilute with water to 250 mL, add 20 mL of ammonia–ammonium chloride TS and a few mg of eriochrome black T trituration, and titrate with 0.1 *M* disodium ethylenediaminetetraacetate VS to a blue end-point. Each mL of 0.1 *M* disodium ethylenediaminetetraacetate VS is equivalent to 2.430 mg of Mg. Not less than 99% is found.

Magnesium Acetate, $Mg(C_2H_3O_2)_2 . 4H_2O$—**214.45**—Use ACS reagent grade.

Magnesium Chloride, $MgCl_2 . 6H_2O$—**203.30**—Use ACS reagent grade.

Magnesium Hydroxide—Use *Magnesium Hydroxide* (USP monograph).

Magnesium Nitrate, $Mg(NO_3)_2 . 6H_2O$—**256.41**—Use ACS reagent grade.

Magnesium Oxide, MgO—**40.30**—Use ACS reagent grade.

Magnesium Oxide, Chromatographic—Use a suitable grade.

Magnesium Perchlorate, Anhydrous, $Mg(ClO_4)_2$—**223.21**—Use ACS reagent grade.

Magnesium Silicate, Activated—Use a suitable grade.[42]

Magnesium Silicate, Chromatographic—Extremely white, hard, powdered (60- to 100-mesh) magnesia–silica gel. Suitable for use as an adsorbant in column chromatography.[43]

Magnesium Sulfate, $MgSO_4 . 7H_2O$—**246.47**—Use ACS reagent grade.

Magnesium Sulfate, Anhydrous, $MgSO_4$—**120.36**—Anhydrous Magnesium Sulfate may be prepared as follows: place a suitable quantity of magnesium sulfate (see above), preferably powdered, in a shallow vessel, and expose to a temperature of about 80° for several hours with occasional stirring. Then heat at 275° to 300° until the weight is practically constant. Transfer the product while still warm to tight containers, as the anhydrous salt is very hygroscopic.

Malachite Green G—See *Brilliant Green.*

Maleic Acid, $C_4H_4O_4$—**116.07**—White, odorless, crystalline powder. Soluble in 1.5 parts of water, in 2 parts of alcohol, and in 12 parts of ether.

Assay—Dissolve about 2 g, accurately weighed, in 100 mL of water and titrate with 1 *N* sodium hydroxide VS, using phenolphthalein TS as the indicator. Each mL of 1 *N* sodium hydroxide is equivalent to 58.04 mg of $C_4H_4O_4$: not less than 99% of $C_4H_4O_4$, calculated on the dried basis, is found.

Loss on drying—Dry it in vacuum over phosphorus pentoxide for 2 hours: it loses not more than 1.5% of its weight.

Residue on ignition ⟨281⟩: not more than 0.1%.

Malt Extract—Obtained by extracting malt with water and then concentrating or drying. A sweet, viscous, light-brown liquid or a powder. Dissolves partially in cold water, but readily in warm water. For use in the preparation of *Malt, Agar Medium*, it must contain no glycerin. It is capable of converting not less than five times its weight of starch into water-soluble sugars.

Assay—Determine the percentage of moisture in potato starch by drying about 500 mg, accurately weighed, at 120° for 4 hours. Mix a quantity equivalent to 5 g of dried starch, in a beaker, with 10 mL of cold water. Add 140 mL of boiling water, and heat on a steam bath, with constant stirring, for 2 minutes or until a translucent, uniform paste is obtained. Cool to 40° in a suitable bath previously adjusted to this temperature. Add 20 mL of a fresh solution prepared by dissolving 5 g of Malt Extract in water to make 100 mL of solution at 40°. Mix, and maintain the same temperature for 30.0 minutes, stirring frequently. A thin, clear liquid is produced. Stir, and add immediately 0.10 mL of this liquid to a previously made mixture of 0.20 mL of 0.1 *N* iodine and 60 mL of water: no blue or reddish color develops.

Solubility and heat stability—Dissolve 3 g in 100 mL of water: the solution is clear and remains so upon standing. Heat a portion of the solution in an autoclave at 121° for 20 minutes: the solution remains clear.

Loss on drying ⟨731⟩—Dry it at 105° for 1 hour: it loses not more than 5.0% of its weight.

Manganese, Mn—**54.94**—Use a suitable grade.

Manganese Dioxide—Use *Manganese Dioxide, Precipitated.*

Manganese Dioxide, Precipitated, MnO_2—**86.94**—A fine, black powder. Insoluble in water. Is decomposed and dissolved by hydrochloric acid with the evolution of chlorine. In the presence of hydrogen peroxide it dissolves in other acids.

Assay—Weigh accurately about 100 mg, transfer to a 300-mL flask with the aid of 20 mL of water, and add 3 mL of sulfuric acid previously mixed with 10 mL of water. In a similar flask place a mixture of 3 mL of sulfuric acid and 30 mL of water. To each flask add 20.0 mL of diluted hydrogen peroxide (made by diluting 30 percent hydrogen peroxide with 40 volumes of water), allow to stand in the dark with occasional agitation until no more black particles are visible, and titrate the contents of each flask with 0.1 *N* potassium permanganate VS: the difference in the volumes of the potassium permanganate consumed, multiplied by 4.347, represents the quantity, in mg, of MnO_2 in the weight of the test specimen and corresponds to not less than 90%.

Insoluble in hydrochloric acid—Dissolve 2 g in 60 mL of dilute hydrochloric acid (1 in 6) by adding sufficient hydrogen peroxide to achieve complete solution: the residue, filtered, washed with hot water, and ignited to constant weight, weighs not more than 2 mg (0.1%).

Manganese Sulfate (*Manganese Sulfate Monohydrate*), $MnSO_4 . H_2O$—**169.01**—Use ACS reagent grade.

D-Mannitol—Use *Mannitol* (USP monograph).

Menadione—Use *Menadione* (USP monograph).

Menthol—Use *Menthol* (USP monograph).

Mercuric Acetate, $Hg(C_2H_3O_2)_2$—**318.68**—Use ACS reagent grade.

Mercuric Bromide, $HgBr_2$—**360.40**—Use ACS reagent grade.

Mercuric Chloride, $HgCl_2$—**271.50**—Use ACS reagent grade.

Mercuric Iodide, Red, HgI_2—**454.40**—Use ACS reagent grade.

Mercuric Nitrate, $Hg(NO_3)_2 . H_2O$—**342.62**—Use ACS reagent grade.

Mercuric Oxide, Yellow, HgO—**216.59**—Use ACS reagent grade.

Mercuric Sulfate, $HgSO_4$—**296.65**—Fine, white, heavy powder. Is odorless. One g dissolves in about 20 mL of sodium chloride solution (1 in 5).

Assay—Weigh accurately about 500 mg, and dissolve in 50 mL of dilute nitric acid (1 in 2). Add 1 mL of ferric nitrate solution (1 in 10), and titrate with 0.1 *N* ammonium thiocyanate VS. Each mL of 0.1 *N* ammonium thiocyanate consumed is equivalent to 10.03 mg of Hg. Between 67% and 67.5% of Hg is found.

Residue on ignition—Ignite 10 g at a rate such that 1 to 2 hours is required to volatilize the test specimen, and ignite at 800 ± 25° for 15 minutes: the residue weighs not more than 1 mg (0.01%).

Chloride—Mix 1 g with 50 mL of water, add 1 mL of formic acid, and add, dropwise, sodium hydroxide solution (1 in 10) until a small amount of permanent precipitate is formed. Reflux the suspension until all of the mercury is reduced to metal and the solution is clear. Cool, filter through a chloride-free paper, wash with two 15-mL portions of water, and dilute with water to 90 mL. To 30 mL add 1 mL of nitric acid and 1 mL of silver nitrate TS, mix, and allow to stand for 5 minutes: any turbidity produced does not exceed that of a control prepared by adding 0.01 mg of Cl to 30 mL of water and treating as the 30 mL of test solution (0.003%).

Iron ⟨241⟩—To the *Residue on ignition* add 3 mL of dilute hydrochloric acid (1 in 2), cover with a watch glass, and digest on a steam bath for 20 minutes. Remove the watch glass, and evaporate to dryness. Take up the residue in a mixture of 1 mL of dilute hydrochloric acid (1 in 2) and 30 mL of water, filter if necessary, and dilute with water to 100 mL. To 10 mL of the solution add 2 mL of hydrochloric acid, and dilute with water to 47 mL: the solution shows not more than 0.01 mg of Fe (0.001%).

Mercurous mercury—Place 5.0 g in a glass-stoppered flask, add 100 mL of potassium iodide solution (15 in 100), 5.0 mL of 0.1 *N* iodine VS, and 3 mL of 1 *N* hydrochloric acid, and allow to stand in the dark, with frequent agitation, for 1 hour. Titrate the excess iodine with 0.1 *N* sodium thiosulfate VS, adding 3 mL of starch TS as the end-point is approached: not more than 0.38 mL of the 0.1 *N* iodine is consumed, correction being made for any iodine consumed in a blank (0.15%).

Nitrate—Disperse 1 g in 9 mL of water, add 1 mL of sodium chloride solution (1 in 200), mix, and add 0.1 mL of indigo carmine TS and 10 mL of sulfuric acid: the blue color of the clear solution is not entirely discharged within 5 minutes (0.005%).

Mercurous Nitrate, [$HgNO_3$ (**280.61**), about $1H_2O$]—Use ACS reagent grade.

Mercury, Hg—At. Wt. **200.59**—Use ACS reagent grade.

Metaphenylenediamine Hydrochloride (*Metaphenylenediamine Dihydrochloride*), $C_6H_4(NH_2)_2.2HCl$—**181.06**—White or slightly reddish white, crystalline powder. Easily soluble in water. On exposure to light it acquires a reddish color. Store it protected from light.

Solubility—A solution of 1 g in 200 mL of water is colorless.

NOTE—Metaphenylenediamine hydrochloride solution can be decolorized by treatment with a small quantity of activated charcoal.

Residue on ignition (Reagent test)—Ignite 1 g with 0.5 mL of sulfuric acid: the residue weighs not more than 1 mg (0.1%).

Metaphosphoric Acid (*Vitreous Sodium Acid Metaphosphate*), HPO_3—**79.98**—Use ACS reagent grade.

Methacrylic Acid—Use a suitable grade.

Methacrylic Carboxylic Acid Cation-exchange Resin—See *Cation-exchange Resin, Methacrylic Carboxylic Acid.*

Methanesulfonic Acid, CH_4O_3S—**96.10**—Use a suitable grade.

Methanol (*Methyl Alcohol*), CH_3OH—**32.04**—Use ACS reagent grade.

Methanol, Anhydrous—Use *Methanol.*

Methanol, Spectrophotometric—Use ACS reagent grade Methanol Suitable for Use in Ultraviolet Spectrophotometry.

Methenamine—Use *Methenamine* (USP monograph).

Methimazole (*1-Methyl-2-mercaptoimidazole*)—Use *Methimazole* (USP monograph).

4-Methoxybenzaldehyde (*p-Anisaldehyde*), $C_8H_8O_2$—**136.15**—Yellow liquid.

Assay—Inject an appropriate specimen into a suitable gas chromatograph (see *Chromatography* ⟨621⟩) equipped with a flame-ionization detector, helium being used as the carrier gas. The following conditions have been found suitable: a 30-m × 0.25-mm capillary column coated with G2. The injection port temperature is maintained at 280°; the column temperature is maintained at 180° and programmed to rise 10° per minute to 280° and held there for 10 minutes. The detector temperature is maintained at 300°. The area of the main peak is not less than 98% of the total peak area.

Refractive index ⟨831⟩: between 1.5724 and 1.5744 at 20°.

Methoxyethanol (*Ethylene Glycol Monomethyl Ether; 2-Methoxyethanol*), $CH_3OCH_2CH_2OH$—**76.09**—Clear, colorless to slightly yellow liquid. Miscible with water, with acetone, with alcohol, with ether, with dimethylformamide, and with glycerin. Refractive index (n_D^{20}): 1.420. [*Caution—Is poisonous; use with adequate ventilation.*]

Specific gravity ⟨841⟩: between 0.960 and 0.964.

Boiling range (Reagent test)—Distil 100 mL: 95% distils between 123° and 126°.

Acidity—To 62 mL (60 g) add phenolphthalein TS, and titrate with 0.1 *N* alcoholic potassium hydroxide: not more than 1 mL is required to produce a pink end-point that persists for not less than 15 seconds (0.01% as CH_3COOH).

Dilution test—Measure 10 mL into a glass-stoppered, 100-mL graduate. Dilute with water to 100 mL, insert the stopper, and mix: no haze or turbidity is observed after the mixture has been allowed to stand at room temperature for 2 hours.

Water, Method I ⟨921⟩: not more than 0.05%.

Methyl Acetate, $C_3H_6O_2$—**74.08**—Colorless liquid, having a characteristic odor. Soluble in water. Miscible with alcohol and with ether. Specific gravity: about 0.933.

Refractive index ⟨831⟩: between 1.3615 and 1.3625 at 20°.

Boiling range (Reagent test)—Not less than 95% distils between 57° and 58°.

Methyl Carbamate, $C_2H_5NO_2$—**75.07**—White crystals. Freely soluble in water.

Melting range ⟨741⟩: between 54° and 56°.

Methyl Chloroform (*Methylchloroform; 1,1,1-Trichloroethane*), CH_3CCl_3—**133.40**—Colorless, heavy liquid. Insoluble in water but is slightly hygroscopic. Miscible with alcohol, with ether, and with chloroform.

Boiling range (Reagent test)—Distil 100 mL: the difference between the temperatures observed when 1 mL and 95 mL have distilled does not exceed 16°. Its boiling temperature at 760 mm of Hg pressure is about 74°.

Specific gravity ⟨841⟩: between 1.312 and 1.321.

Acidity—Pipet 25 mL into 25 mL of alcohol neutralized to bromothymol blue TS with 0.02 *N* sodium hydroxide. Mix gently, and titrate with 0.020 *N* sodium hydroxide VS: not more than 0.50 mL is required to restore the pink color (0.001% as HCl).

Residue on evaporation—Evaporate 76 mL on a steam bath, and dry at 105° for 1 hour: the residue weighs not more than 1 mg (about 0.001%).

Methyl Ethyl Ketone, $CH_3COC_2H_5$—**72.11**—Colorless liquid, having an acetone-like odor. Soluble in water. Miscible with alcohol, with ether, and with benzene.

Boiling range (Reagent test): between 79.0° and 81.0°.

Specific gravity ⟨841⟩: between 0.801 and 0.803.

Residue on evaporation—Evaporate 50 mL on a steam bath, and dry at 105° for 1 hour: the residue weighs not more than 1.0 mg (0.0025%).

Acidity—Add 25 mL to 10 mL of 80 percent alcohol, previously neutralized to phenolphthalein TS with 0.02 *N* sodium hydroxide. Titrate with 0.020 *N* sodium hydroxide VS to the production of a pink color that persists for not less than 15 seconds: not more than 0.50 mL is consumed (0.003% as CH_3COOH).

Solubility in water—Add 5 mL to 40 mL of carbon dioxide-free water, and allow to stand for 20 minutes: the solution remains clear.

Methyl Iodide, CH_3I—**141.94**—Colorless, heavy, transparent liquid. Slightly soluble in water. Miscible with alcohol, with ether, and with solvent hexane. Turns brown on exposure to light as a result of liberation of iodine.

Assay—Add 1 mL to a 100-mL volumetric flask tared with 10 mL of alcohol. Weigh again, add alcohol to volume, and mix. Pipet 20 mL into a glass-stoppered flask, and add 50.0 mL of 0.1 *N* silver nitrate VS and 2 mL of nitric acid. Insert the stopper immediately, shake frequently during 2 hours, and allow to stand in the dark overnight. Shake again during 2 hours, then add 50 mL of water and 3 mL of ferric ammonium sulfate, and titrate the excess silver nitrate with 0.1 *N* ammonium thiocyanate VS. Each mL of 0.1 *N* silver nitrate is equivalent to 14.19 mg of CH_3I: not less than 98.5% is found.

Boiling range (Reagent test)—Distil 50 mL into a chilled, partly closed receiver: not less than 48 mL distils between 41.5° and 43°.

Density: between 2.270 and 2.285.

Residue on evaporation—Evaporate 4 mL (10 g) on a steam bath, and dry the residue at 105° for 1 hour: the residue weighs not more than 1 mg (0.01%).

Acidity—Shake 3 mL with 5 mL of water for 30 seconds, and immediately draw off the lower layer: the aqueous layer is neutral to litmus and when 1 mL of silver nitrate TS is added, it shows not more than a slight opalescence.

Methyl Isobutyl Ketone—See *4-Methyl-2-pentanone.*

Methyl Methacrylate—Use a suitable grade.

***N*-Methyl-*N*-nitroso-*p*-toluenesulfonamide** (*p-Tolylsulfonylmethylnitrosamide*), $C_8H_{10}N_2O_3S$—**214.24**—Light yellow crystals or powder. Insoluble in water; soluble in benzene, in carbon tetrachloride, and in chloroform.

Melting range ⟨741⟩: between 59° and 63°, but the range between beginning and end of melting does not exceed 2°.

Methyl Nonadecanoate, $C_{20}H_{40}O_2$—**312.53**—Use a suitable grade.[44]

Methyl Polysiloxane Gum, 3 Percent—Use a suitable grade.

Methyl Silicone Gum Rubber, 5 Percent—Use a suitable grade.[45]

Methyl Stearate—Use a suitable chromatographic grade.

Methyl Sulfoxide (*Dimethyl Sulfoxide*), $(CH_3)_2SO$—**78.13**—Use ACS reagent grade.

Methylamine, 40 Percent in Water, CH_5N—**31.06**—Colorless liquid.

Assay—Using a syringe, transfer about 0.5 mL of a well-shaken specimen to 100 mL of water at a point below the surface of the water. Determine the weight of the specimen by weighing the syringe before and after the transfer. Mix, and titrate with 0.5 *N* hydrochloric acid VS, determining the end-point potentiometrically, using a silver–silver chloride pH electrode and a calomel reference electrode. Perform a blank determination, and make any necessary correction. Each mL of 0.5 *N* hydrochloric acid is equivalent to 15.53 mg of CH_5N: between 39.0% and 41.0% is found.

Refractive index ⟨831⟩: between 1.3680 and 1.3710, at 20°.

***p*-Methylaminophenol Sulfate,** $(p\text{-}CH_3NHC_6H_4OH)_2 . H_2SO_4$—**344.38**—White, or yellow-white, small crystals or a crystalline powder. Discolors on exposure to air. Soluble in cold water; freely soluble in boiling water; slightly soluble in alcohol; insoluble in ether. Store in well-closed containers, protected from light.

Solubility in HCl—Add 100 mg to 2 mL of hydrochloric acid: it dissolves quickly and completely.

o-Aminophenol—To the solution from the preceding test add 1 drop of ferric chloride TS: no reddish brown color is produced.

Residue on ignition (Reagent test)—The residue from 2 g weighs not more than 2 mg (0.1%).

Chloride—To a solution of 1 g in 20 mL of water add 1 mL of nitric acid and 1 mL of silver nitrate TS: not more than a faint opalescence is produced.

Suitability for phosphate test—Dissolve 2 g in 100 mL of water. To 10 mL of this solution add 90 mL of water and 20 g of sodium bisulfite. Confirm the suitability of the reagent solution by the following test:

Add 1 mL of the reagent solution to each of four solutions containing 25 mL of 0.5 *N* sulfuric acid and 1 mL of a solution of 5 g ammonium molybdate in 100 mL of 1 *N* sulfuric acid. Add 0.005 mg of phosphate (PO_4) to one of the solutions, 0.01 mg to a second, and 0.02 mg to a third. Allow to stand at room temperature for 2 hours: the solutions in the three tubes show readily perceptible differences in blue color corresponding to the relative amounts of phosphate added, and the one to which 0.005 mg of phosphate was added is perceptibly a deeper blue than the blank.

Methylboron Dihydroxide, $CH_3B(OH)_2$—**59.86.**

Melting temperature ⟨741⟩: about 96°.

Methylcellulose—Use *Methylcellulose* (USP monograph).

Methylene Blue, $C_{16}H_{18}ClN_3S . 3H_2O$—**373.90**—Dark green crystals or a crystalline powder, having a bronze-like luster. One g dissolves in about 25 mL of water and in about 65 mL of alcohol. Soluble in chloroform.

Absorptivity ratio—The ratio of its absorptivity (see *Spectrophotometry and Light-scattering* ⟨851⟩) at 635 nm to that at 665 nm, measured in a dilute solution of the dye in diluted alcohol, is between 0.56 and 0.62.

Residue on ignition (Reagent test)—Ignite 1 g with 0.5 mL of sulfuric acid: the residue weighs not more than 10 mg (1%).

Loss on drying ⟨731⟩—Dry it at 105° for 18 hours: it loses not more than 15.0% of its weight.

Methylene Chloride (*Dichloromethane*), CH_2Cl_2—**84.93**—Use ACS reagent grade Dichloromethane.

Methylparaben—Use *Methylparaben* (NF monograph).

4-Methyl-2-pentanone (*Methyl Isobutyl Ketone*), $(CH_3)_2CHCH_2COCH_3$—**100.16**—Use ACS reagent grade.

2-Methyl-2-propyl-1,3-propanediol, $C_7H_{16}O_2$—**132.20**—White crystals, melting at about 58°.

Methylvinyl Silicone—Use a suitable grade.

Milk of Magnesia—Use *Milk of Magnesia* (USP monograph).

Mineral Oil—Use *Mineral Oil* (USP monograph).

Mineral Oil, Light—Use *Mineral Oil, Light* (NF monograph).

Molybdic Acid (*85 Percent Molybdic Acid*)—Use ACS reagent grade.

Monobasic Potassium Phosphate—See *Potassium Phosphate, Monobasic.*

Monobasic Sodium Phosphate—See *Sodium Phosphate, Monobasic.*

Monochloroacetic Acid, $CH_2ClCOOH$—**94.50**—Colorless or white, deliquescent crystals, odorless in the cold. Very soluble in water; soluble in alcohol and in ether. Store in well-closed containers in a cool place.

Assay—Weigh accurately about 3 g, transfer to a suitable container, and dissolve in 50 mL of water. Add phenolphthalein TS, and titrate with 1 *N* sodium hydroxide VS. Each mL of 1 *N* sodium hydroxide is equivalent to 94.50 mg of $CH_2ClCOOH$. Not less than 99.0% is found.

Melting range ⟨741⟩: between 61.0° and 64.0°.

Insoluble matter: not more than 1.0 mg, from 10 g (0.010%).

Residue on ignition (Reagent test)—Ignite 5.0 g: the residue weighs not more than 1.0 mg (0.02%) (retain the residue).

Chloride (Reagent test)—One g shows not more than 0.01 mg of Cl (0.001%).

Sulfate (Reagent test, *Method I*)—One g shows not more than 0.2 mg of SO_4 (0.02%).

Heavy metals (Reagent test)—Test 2.0 g: the limit is 0.001%.

Iron ⟨241⟩—Digest the residue retained from the test for *Residue on ignition* with 6 mL of hydrochloric acid on a steam bath until solution is complete, then dilute with water to 150 mL. To 10 mL of the solution add 1.5 mL of hydrochloric acid, and dilute with water to 47 mL: the solution shows not more than 0.01 mg of Fe (0.003%).

Monoethanolamine, C_2H_7NO—**61.08**—Clear, colorless to faintly yellow, viscous liquid having an ammoniacal odor. Miscible with water, with methanol, and with acetone. Melts at about 9°.

Assay—Weigh accurately a glass-stoppered weighing bottle containing 25 mL of water. Add about 2 g of test specimen, insert the stopper, and again weigh accurately. Add 3 drops of a mixed indicator prepared by adding 5 volumes of bromocresol green TS to 6 parts of methyl red TS (prepared from methyl red hydrochloride), and titrate with 1 *N* hydrochloric acid VS. Each mL of 1 *N* hydrochloric acid VS is equivalent to 61.08 mg of C_2H_7NO. Not less than 99% is found.

Refractive index ⟨831⟩: between 1.453 and 1.455 at 20°.

Residue on ignition ⟨281⟩—Evaporate 20 g on a steam bath to dryness, and ignite the residue at 800 ± 25° for 15 minutes: the residue weighs not more than 1 mg (0.005%).

Morpholine (*Tetrahydro-1,4-oxazine*), $\overline{OCH_2CH_2NHCH_2CH_2}$—**87.12**—Use ACS reagent grade.

Naphazoline Hydrochloride—Use *Naphazoline Hydrochloride* (USP monograph).

Naphthalene, $C_{10}H_8$—**128.17**—Monoclinic prismatic plates, or white scales or powder. A solution in solvent hexane shows a purple fluorescence under light from a mercury-arc lamp. Insoluble in water; very soluble in ether and in fixed and volatile oils; freely soluble in benzene, in carbon disulfide, in carbon tetrachloride, in chloroform, in olive oil, and in toluene; soluble in alcohol and in methanol. Sublimes at temperatures above the melting temperature.

Melting range ⟨741⟩: between 80° and 81°.

Boiling range (Reagent test): between 217° and 219°.

1,3-Naphthalenediol (*Naphthoresorcinol*), $C_{10}H_6(OH)_2$—**160.17**—Grayish white to tan crystals or powder. Freely soluble in methanol; sparingly soluble in water, in alcohol, and in ether.

Melting range ⟨741⟩: between 122° and 127°.

Solubility in methanol—Dissolve 500 mg in 50 mL of methanol: the solution is clear and complete.

2,7-Naphthalenediol (*2,7-Dihydroxynaphthalene*), $C_{10}H_8O_2$—**160.17**—Off-white to yellow crystalline solid or powder. Dissolves in acetone.

Melting range ⟨741⟩: between 187° and 191°.

2-Naphthalenesulfonic Acid, $C_{10}H_8O_3S.H_2O$—**226.25**—Off-white to light gray crystals. Soluble in water.

Assay—Dissolve about 1 g, accurately weighed, in 100 mL of water, add phenolphthalein TS, and titrate with 0.1 *N* sodium hydroxide VS. Perform a blank determination, and make any necessary correction. Each mL of 0.1 *N* sodium hydroxide is equivalent to 22.63 mg of $C_{10}H_8O_3S.H_2O$. Not less than 98.0% is found.

Melting range ⟨741⟩: between 122° and 126°, but the range between beginning and end of melting does not exceed 2°.

1-Naphthol (*Alphanaphthol*), $C_{10}H_7OH$—**144.17**—Colorless or slightly pinkish crystals or crystalline powder, having a characteristic odor. Insoluble in water; soluble in alcohol, in benzene, and in ether.

Melting range ⟨741⟩: between 95° and 97°.

Solubility—Separate 1-g portions dissolve in alcohol and in benzene to yield solutions that are clear and colorless or nearly colorless.

Acidity—Shake 1 g with 50 mL of water occasionally during 15 minutes, and filter: the filtrate is neutral to litmus.

Residue on ignition (Reagent test): not more than 0.05%.

2-Naphthol (*Betanaphthol*), $C_{10}H_7OH$—**144.17**—White leaflets or crystalline powder, having a faint, characteristic odor. Discolors on exposure to light. Very slightly soluble in water; soluble in alcohol, in ether, in chloroform, and in solutions of alkali hydroxides.

Melting range ⟨741⟩: between 121° and 123°.

Solubility in alcohol—A solution of 1 g in 10 mL of alcohol is complete and colorless or practically so.

Residue on ignition (Reagent test): not more than 0.05%.

Acidity—Shake 1 g with 50 mL of water occasionally during 15 minutes, and filter: the filtrate is neutral to litmus.

1-Naphthol—Boil 100 mg with 10 mL of water until dissolved, cool, and filter. Add to the filtrate 0.3 mL of 1 *N* sodium hydroxide and 0.3 mL of 0.1 *N* iodine: no violet color is produced.

Insoluble in ammonia (naphthalene, etc.)—Shake 500 mg with 30 mL of ammonia TS: the 2-naphthol dissolves completely and the solution is not darker than pale yellow.

Naphthol Potassium Disulfonate (*2-Naphthol-6,8-dipotassium Disulfonate*), $C_{10}H_6K_2O_7S_2$—**380.48**—Use a suitable grade.

β-Naphthoquinone-4-sodium Sulfonate, $C_{10}H_5NaO_5S$—**260.20**—Yellow to orange-yellow crystals or crystalline powder. Soluble in about 10 parts of water; insoluble in alcohol.

Loss on drying ⟨731⟩—Dry it in vacuum at about 50°: it loses not more than 2.0% of its weight.

Residue on ignition (Reagent test)—Ignite 1 g of dried sample with 3 mL of sulfuric acid: the residue weighs between 265 and 280 mg (between 26.5% and 28.0%).

Naphthoresorcinol (*1,3-Dihydroresorcinol*), $C_{10}H_8O_2$—**160.17**—Use a suitable grade.

1-Naphthylamine Hydrochloride, $C_{10}H_7NH_2.HCl$—**179.65**—White, crystalline powder that turns bluish upon exposure to light and air. Soluble in water, in alcohol, and in ether.

A 1 in 100 solution, make slightly acid with acetic acid, gives a violet color with 5 drops of ferric chloride TS. A 1 in 40 solution in diluted acetic acid is colorless and not more than slightly opalescent.

Residue on ignition (Reagent test)—Ignite 200 mg with a few drops of sulfuric acid: the weight of the residue is negligible.

***N*-1-Naphthylethylenediamine Dihydrochloride,** $C_{10}H_7NH(CH_2)_2NH_2.2HCl$—**259.18**—Use ACS reagent grade.

Neutralized Alcohol—See *Alcohol, Neutralized.*

Niacin—Use *Niacin* (USP monograph).

Niacinamide—Use *Niacinamide* (USP monograph).

Nickel, Ni—**58.69**—Use a suitable grade.

Nickel-Aluminum Catalyst—Use a suitable grade.[46]

Nickel Ammonium Sulfate, $NiSO_4.(NH_4)_2SO_4.6H_2O$—**394.98**—Bluish green crystals. Soluble in about 10 parts of water; insoluble in alcohol.

Insoluble matter (Reagent test): not more than 1 mg, from 10 g (0.01%).

Alkalies and alkaline earths—Dissolve 2 g in 95 mL of water, add 5 mL of ammonium hydroxide, precipitate the nickel with H_2S, and filter. Boil 50 mL of the filtrate until evaporated to about 10 mL, filter again if necessary, add a few drops of sulfuric acid, evaporate, and ignite at 800 ± 25° for 15 minutes: the residue weighs not more than 2 mg (0.2%).

Cobalt—Dissolve 2 g in 40 mL of water, add 30 mL of potassium hydroxide solution (1 in 20), boil until the NH_3 is removed, and evaporate, or dilute with water, to 30 mL. Add just sufficient glacial acetic acid to dissolve the precipitate, then add 2 mL of excess. Add a solution of 5 g of potassium nitrite in 10 mL of water, stir well, digest on a steam bath for 30 minutes, and allow to stand overnight. Filter through a filtering crucible, and wash with small quantities of cold ammonium nitrate solution (1 in 20) dissolved in acetic acid solution (1 in 100) until free from nitrite. Place the crucible in the same beaker in which the precipitation was made, add 10.0 mL of 0.1 *N* potassium permanganate VS, then add 25 mL of diluted sulfuric acid and sufficient water to cover the crucible. Heat to about 50° until the yellow precipitate has dissolved, then cool, add 1 g of potassium iodide, and titrate the liberated iodine representing the excess of permanganate with 0.1 *N* sodium thiosulfate VS. Determine the value of the permanganate in terms of 0.1 *N* thiosulfate by running a blank with the same volumes of permanganate, sulfuric acid, and potassium iodide in the same manner as with the test specimen. Not more than 2.05 mL of 0.1 *N* potassium permanganate is consumed.

Copper—Dissolve 2 g in a mixture of 2 mL of hydrochloric acid and sufficient water to make 50 mL, add 2 mL of mercuric chloride solution (1 in 100), and saturate with hydrogen sulfide. Filter through a small filter, wash with hydrogen sulfide TS, and ignite the filter and the precipitate in porcelain. Dissolve the residue by warming with 0.5 mL of nitric acid and a few drops of water, dilute with water to 10 mL, and filter if necessary. Dissolve in the solution 1 g of ammonium acetate, and add 5 drops of freshly prepared potassium ferrocyanide TS: any red color produced is not darker than that of a control made with 0.04 mg of Cu and with the same quantities of nitric acid, ammonium acetate, and potassium ferrocyanide, and in the same volume as the test specimen.

Nickel Chloride, $NiCl_2.6H_2O$—**237.70**—Small, odorless, light-green crystals. Soluble in water.

Insoluble matter (Reagent test)—A 20-g portion, dissolved in 150 mL of water containing 1 mL of hydrochloric acid, shows not more than 1 mg of insoluble matter (0.005%).

Cobalt—Dissolve 1 g in 40 mL of water, transfer 10 mL of the solution to a separator, and add 15 mL of water. Into a second separator place 25 mL of water containing 0.5 mg of added cobalt (Co). To each solution add 8 g of ammonium thiocyanate, 2 g of ammonium acetate, a few drops of tartaric acid solution (1 in 10), and 10 drops of acetic acid, and mix to ensure complete solution of the reagents. Add 20 mL of ether and 2 mL of amyl alcohol, shake vigorously, allow the layers to separate, and withdraw the ether-alcohol layers. If the layer from the test solution shows a bluish coloration, repeat the extraction, using fresh portions of ether and amyl alcohol, until the final extract is colorless. Combine the test extracts, mix, and compare the color of the solution with that of the control solution adjusted to a similar volume with the ether-amyl alcohol mixture: the color in the test solution is not darker than that of the control (0.2%).

Copper—Dissolve 2 g in 50 mL of dilute hydrochloric acid (1 in 25), add 2 mL of mercuric chloride solution (1 in 100), and saturate the solution with hydrogen sulfide. Filter through a small filter, and wash with hydrogen sulfide TS until the filtrate is colorless. Ignite the filter in porcelain in a well-ventilated hood, but avoid too high a temperature which might cause fusion of the copper into the glaze of the dish. Take up the residue in a mixture of 0.5 mL of nitric acid and a few drops of water, dilute to 10 mL, and filter, if necessary. Transfer the solution to a separator, and add 5 mL of ammonium acetate solution (1 in 10), 2 mL of ammonium thiocyanate TS, 2 mL of glacial acetic acid, 0.5 mL of pyridine, and 10 mL of chloroform. Shake vigorously, cool in ice water, and allow the layers to separate: any greenish yellow color produced in the chloroform layer is not darker than that of a control containing 0.04 mg of added Cu (0.002%).

Iron—Dissolve 1 g in water, dilute with water to 50 mL, and add 2 mL of hydrochloric acid, 50 mg of ammonium persulfate crystals, and 3 mL of ammonium thiocyanate solution (3 in 10). Add 20 mL of amyl alcohol, shake vigorously, allow the layers to separate, and drain the alcohol layer into a clean, dry comparison tube: any red color in the alcohol extract is not darker than that from a control containing 0.02 mg of added Fe (0.002%).

Lead—Dissolve 2 g in 30 mL of water, and add 1 mL of acetic acid and 1 mL of potassium chromate solution (1 in 10): any turbidity produced does not exceed that of a control containing 0.1 mg of added Pb (0.005%).

Substances not precipitated by ammonium sulfide—Dissolve 2 g in 100 mL of dilute ammonium hydroxide (5 in 100), saturate the solution with hydrogen sulfide, and filter. Evaporate 50 mL of the clear filtrate to a small volume, filter again, if necessary, evaporate to dryness, and ignite at 800 ± 25° for 15 minutes: the residue weighs not more than 2.0 mg (0.20%).

Sulfate (Reagent test, *Method II*)—Ten g, dissolved in 150 mL of water and 1 mL of hydrochloric acid, yields not more than 1.2 mg of residue (0.005%).

Nicotinamide Adenine Dinucleotide Phosphate-adenosine-5′-triphosphate Mixture—Use a suitable grade.

Suitability—When used in the assay of lactulose, determine that a suitable absorbance-versus-concentration slope is obtained, using USP Lactulose RS, the reagent blank absorbance being not more than 0.020. The commercially available reagent[37] contains 64 mg of nicotinamide adenine dinucleotide phosphate and 160 mg of adenosine-5′-triphosphate per vial. The mixture is buffered and stabilized. For use in the Assay of lactulose it is diluted with water to 100 mL.

Nicotinic Acid—Use *Niacin* (USP monograph).

Ninhydrin—See *Triketohydrindene Hydrate*.

Nitranilic Acid (*2,5-Dihydroxy-3,6-dinitro-1,4-benzoquinone*), $C_6H_2N_2O_8$—**230.09**—Yellow plates or prisms. Freely soluble in water and in alcohol; insoluble in ether. Effloresces at about 100°; deflagrates at about 170°. [*Caution—Is unstable in the presence of water, forming toxic oxalic acid and hydrogen cyanide.*]

Nitric Acid, HNO_3—**63.01**—Use ACS reagent grade.

Nitric Acid, Diluted (10 percent HNO_3)—Dilute 105 mL of nitric acid with water to 1000 mL.

Nitric Acid, Fuming (*90 Percent Nitric Acid*), HNO_3—**63.01**—Use ACS reagent grade Nitric Acid, 90 Percent.

Nitric Oxide–Nitrogen Dioxide Detector Tube—A fuse-sealed glass tube so designed that gas may be passed through it, and containing suitable absorbing filters and support media for an oxidizing layer and the indicator diphenyl benzidine.

NOTE—A suitable detector tube that conforms to the monograph specification is available from National Draeger, Inc., 401 Parkway View Drive, Pittsburgh, Pa. 15205 as Reference Number CH 29401, Measuring Range 0.5 to 10 ppm. Tubes having conditions other than those specified in the monograph may be used in accordance with the section entitled *Tests and Assays* in the *General Notices*.

Nitrile Silicone Gum, 5 Percent—Use a suitable grade.

Nitrilotriacetic Acid, $N(CH_2COOH)_3$—**191.14**—Use ACS reagent grade.

4′-Nitroacetophenone (*p′-Nitroacetophenone*), $C_8H_7NO_3$—**165.15**—Yellow crystals.

Assay—Inject an appropriate ether solution of the specimen (about 0.5 µL) into a suitable gas chromatograph (see *Chromatography* ⟨621⟩) equipped with a thermal conductivity detector, helium being used as the carrier gas. The following conditions have been found suitable: a 1.8-m × 4-mm stainless steel column containing 10 percent phase G1 on support S1A; the injection port and detector are maintained at 200° and 300°, respectively; the column temperature is maintained at 170° and programmed to rise 3° per minute to 220°. The area of the 4′-nitroacetophenone peak is not less than 97% of the total peak area.

Melting range ⟨741⟩: between 78° and 80°.

***m*-Nitroaniline,** $C_6H_6N_2O_2$—**138.13**—Yellow crystals, or a yellow crystalline solid. Slightly soluble in water; soluble in alcohol, in ether, and in methanol. [NOTE—May be absorbed through the skin.]

Assay—Dissolve 0.15 g, accurately weighed, in 100 mL of glacial acetic acid TS, and titrate with 0.1 *N* perchloric acid in glacial acetic acid VS, determining the end-point potentiometrically. Perform a blank determination, and make any necessary correction. Each mL of 0.1 *N* perchloric acid is equivalent to 13.81 mg of $C_6H_6N_2O_2$. Not less than 97.5% is found.

Melting range ⟨741⟩: between 112° and 114°.

***o*-Nitroaniline,** $NO_2C_6H_4NH_2$—**138.13**—Orange-yellow crystals. Slightly soluble in cold water; soluble in hot water; freely soluble in alcohol and in chloroform. It forms water-soluble salts with mineral acids.

Melting range ⟨741⟩: between 71° and 72°.

***p*-Nitroaniline,** $NO_2C_6H_4NH_2$—**138.13**—Bright yellow, crystalline powder. Insoluble in water; soluble in alcohol and in ether.

Melting range ⟨741⟩: between 146° and 148°.

Solubility—Separate 1-g portions dissolve in 30 mL of alcohol and in 40 mL of ether, respectively, to yield solutions that are clear or practically so.

Residue on ignition (Reagent test): not more than 0.2%.

***m*-Nitrobenzaldehyde,** $NO_2C_6H_4CHO$—**151.12**—Yellow, crystalline powder. Insoluble in water; soluble in alcohol, in chloroform, and in ether.

Melting range ⟨741⟩: between 57° and 59°.

***o*-Nitrobenzaldehyde,** $NO_2C_6H_4CHO$—**151.12**—Pale yellow or light greenish yellow needles, having an odor resembling that of benzaldehyde. Practically insoluble in water; soluble in alcohol, in benzene, and in ether.

Melting range ⟨741⟩: between 42° and 44°.

Solubility in alcohol—Dissolve 1 g in 20 mL of alcohol: a practically clear, pale yellow or greenish yellow solution is produced.

Residue on ignition (Reagent test): not more than 0.1%.

Nitrobenzene, $C_6H_5NO_2$—**123.11**—Use ACS reagent grade.

***p*-Nitrobenzenediazonium Tetrafluoroborate,** $NO_2C_6H_4N_2BF_4$—**236.92**—Yellow-gold crystals. Soluble in acetonitrile. [*Caution—Shock-sensitive; keep refrigerated.*]

Assay—Transfer about 30 mg, accurately weighed, to a low-actinic, 100-mL volumetric flask. Dissolve in 0.01 *N* hydrochloric acid, dilute with 0.01 *N* hydrochloric acid to volume, and mix. Using low-actinic glassware, dilute 2.0 mL of the resulting solution with spectrophotometric grade methanol to 50.0 mL. Measure the absorbance of this solution in a 1-cm cell at about 255 nm, using methanol as the blank. Calculate the absorptivity of the solution by dividing the measured absorbance by the concentration in g per mL. Calculate the assay value by the formula:

$$100a/59.4,$$

in which a is the absorptivity of the solution: not less than 95.0% is found.

***p*-Nitrobenzoic Acid,** $C_6H_4NO_2COOH$—**167.12**—Yellowish white to yellow crystals or crystalline powder. Very slightly soluble in water; slightly soluble in alcohol and in chloroform; sparingly soluble in ether; soluble in acetone.

Assay—Dissolve about 0.7 g, accurately weighed, in 150 mL of methanol, heating if necessary. Add thymol blue TS, and titrate with 0.1 *N* sodium methoxide VS to a blue end-point. Perform a blank determination, and make any necessary correction. Each mL of 0.1 *N* sodium methoxide is equivalent to 16.71 mg of $C_6H_4NO_2COOH$. Not less than 98% is found.

Melting range ⟨741⟩: between 238° and 241°.

Residue on ignition (Reagent test): not more than 0.01%.

***p*-Nitrobenzyl Bromide,** $NO_2C_6H_4CH_2Br$—**216.03**—Almost white to pale yellow crystals, darkening on exposure to light. Practically insoluble in water; freely soluble in alcohol, in ether, and in glacial acetic acid. Store in tight, light-resistant containers.

Melting range ⟨741⟩: between 98° and 100°.

Solubility—Separate 200-mg portions yield clear solutions in 5 mL of alcohol and in 5 mL of glacial acetic acid.

Residue on ignition (Reagent test): negligible, from 200 mg.

4-(*p*-Nitrobenzyl)pyridine, $C_{12}H_{10}N_2O_2$—**214.22**—Yellow crystals. Soluble in acetone.

Insoluble matter—Dissolve 1 g in 10 mL of acetone: the solution is clear and complete.

Melting range ⟨741⟩: between 71° and 74°.

Nitrogen—Use *Nitrogen* (NF monograph).

Nitromethane, CH_3NO_2—**61.04**—Oily liquid. Soluble in water, in alcohol, in ether, and in dimethylformamide. Specific gravity: about 1.132. Water solutions are acid to litmus.

Refractive index ⟨831⟩: about 1.380 at 22°.

Boiling range: between 101° and 103°.

Residue on evaporation: negligible, from 50 mL.

5-Nitro-1,10-phenanthroline, $C_{12}H_7N_3O_2$—**225.21**—White, odorless powder. Soluble in water.

Melting range ⟨741⟩: between 198° and 200°.

Suitability as redox indicator—Dissolve 25 mg in a minimum volume of diluted sulfuric acid, add 10 mg of ferrous sulfate, and dilute with water to 100 mL: the solution is deep red in color and exhibits an absorption maximum at 510 nm. To 1.0 mL of the solution add 1.0 mL of 0.01 *M* ceric sulfate: the red color is discharged.

1-Nitroso-2-naphthol, $C_{10}H_7NO_2$—**173.17**—Brown to yellowish brown powder. Insoluble in water; soluble in alcohol, in benzene, in ether, in carbon tetrachloride, and in acetic acid.

Assay—Transfer about 250 mg, previously dried over silica gel to constant weight and accurately weighed, to a glass-stoppered flask, and dissolve in 10 mL of sodium hydroxide solution (1 in 10). Cool the solution in an ice bath, add dilute sulfuric acid (1 in 6) until a slight, permanent precipitate is formed and the solution is slightly acid, then add 3 g of potassium iodide, shake to dissolve, add 20 mL of dilute sulfuric acid (1 in 6), immediately insert the stopper in the flask, and allow to stand in the dark for 2 hours. Titrate the liberated iodine with 0.1 *N* sodium thiosulfate VS, adding 3 mL of starch TS as the end-point is approached. Perform a complete blank determination, and make any necessary correction. Each mL of 0.1 *N* sodium thiosulfate is equivalent to 8.66 mg of $C_{10}H_7NO_2$: not less than 95.0% is found.

Melting range ⟨741⟩: between 109° and 111°.

Residue on ignition (Reagent test): not more than 0.2%.

Nitroso R Salt (*1-Nitroso-2-naphthol-3,6-disodium Disulfonate*), $NOC_{10}H_4OH(SO_3Na)_2$—**377.25**—Yellow crystals or crystalline powder. One g dissolves in about 40 mL of water; insoluble in alcohol.

Sensitiveness—Dissolve 500 mg of sodium acetate in a solution of 0.4 mg of cobaltous chloride (0.1 mg of cobalt) in 5 mL of water. Add 1 mL of diluted acetic acid, and follow with 1 mL of a solution of the nitroso R salt (1 in 500): a red color, which is produced at once, persists when the solution is boiled with 1 mL of hydrochloric acid for 1 minute.

Nitroterephthalic Acid, $C_8H_5NO_6$—**211.13**—Use a suitable grade.[47]

Nitrous Oxide Certified Standard—A container of 99.9% nitrous oxide. It is available from most suppliers of specialty gases.

Nonadecane, $C_{19}H_{40}$—**268.53**—White solid.

Assay—Inject an appropriate specimen into a suitable gas chromatograph (see *Chromatography* ⟨621⟩) equipped with a thermal conductivity detector, helium being used as the carrier gas. The following conditions have been found suitable: 1.8-m × 3-mm stainless steel column containing 5 percent phase G2 on support S1AB; the injection port temperature is maintained at 330°; the detector temperature is maintained at 300°; and the oven temperature is held initially at 190° and allowed to rise gradually to 250°. The area of the nonadecane peak is not less than 99% of the total peak area.

Melting range ⟨741⟩: between 31.5° and 33.5°.

1-Nonyl Alcohol (*1-Nonanol*), $CH_3(CH_2)_8OH$ **144.26**—Colorless liquid.

Assay—Not less than 97% of $C_9H_{20}O$ is found, a suitable gas chromatograph equipped with a flame-ionization detector and helium being used as the carrier gas at a flow rate of about 40 mL per minute. The following conditions have been found suitable: a 1.83-m × 3.2-mm stainless steel column packed with 20 percent phase G16 on support S1A; the injection port, column, and detector temperatures are maintained at about 250°, 160°, and 310°, respectively.

Refractive index ⟨831⟩: between 1.432 and 1.434 at 20°.

Nonylphenoxypoly(ethyleneoxy)ethanol—Clear, viscous, pale yellow liquid having an aromatic odor. May exhibit slight solidification on cooling; warming with agitation will restore to original condition. Density: about 1.06. Soluble in alcohol, in xylene, and in water. Suitable for use in gas-liquid chromatography.[48]

***p*-Nonylphenyl Polyethylene Glycol Ether** [(*p-Nonylphenoxy*)*polyethoxyethanol*]—See (*p-tert-Octylphenoxy*)*nonaethoxyethanol.*

19-Norandrostenedione—Use USP 19-Norandrostenedione Reagent.

Norepinephrine Bitartrate—Use *Norepinephrine Bitartrate* (USP monograph).

Norethindrone—Use *Norethindrone* (USP monograph).

Normal Butyl Acetate—See *Butyl Acetate, Normal.*

Normal Butyl Alcohol—See *Butyl Alcohol.*

Normal Butyl Nitrite—See *n-Butyl Nitrite.*

Normal Butylamine—See *n-Butylamine.*

***n*-Octadecane,** $C_{18}H_{38}$—**254.50.**

Melting range ⟨741⟩: between 26° and 30°, but the range between beginning and end of melting does not exceed 2°.

Octadecyl Silane—This reagent is formed *in situ* by reaction of the column support with a suitable silylating agent such as octadecyl trichlorosilane.

Octoxynol 9—Use *Octoxynol 9* (NF monograph).

***n*-Octylamine,** $C_8H_{17}NH_2$—**129.25**—Clear, colorless liquid.

Assay—Dissolve 1.3 g, accurately weighed, in 100 mL of alcohol. When solution is complete, titrate with 0.5 *N* hydrochloric acid VS, determining the end-point potentiometrically. Perform a blank determination, and make any necessary correction. Each mL of 0.5 *N* hydrochloric acid is equivalent to 64.62 mg of $C_8H_{17}NH_2$. Not less than 97% is found.

Refractive index ⟨831⟩: between 1.4289 and 1.4299 at 20°.

(*p-tert*-Octylphenoxy)nonaethoxyethanol, $C_{34}H_{62}O_{11}$—**646.86**—Use a suitable grade.[49]

(*p-tert*-Octylphenoxy)polyethoxyethanol—Use a suitable grade.[50]

Odorless Absorbent Paper—See *Filter Paper, Quantitative.*

Olefin Detector Tube—A fuse-sealed glass tube so designed that gas may be passed through it, and containing suitable absorbing filters and support media for the indicator in a stabilized form of permanganate.

NOTE—A suitable detector tube that conforms to the monograph specification is available from National Draeger, Inc., 401 Parkway View Drive, Pittsburgh, Pa. 15205 as Reference Number CH 31201, Measuring Range 1 to 50 mg per liter. Tubes having conditions other than those specified in the monograph may be used in accordance with the section entitled *Tests and Assays* in the *General Notices.*

Olive Oil—Use *Olive Oil* (NF monograph).

Orange G (the sodium salt of azobenzene-betanaphthol disulfonic acid), $C_6H_5N{:}NC_{10}H_4(OH)(SO_3Na)_2$-2,6,8—**452.36**—Orange to brick-red powder or dark red crystals. Readily soluble in water, yielding an orange-yellow solution; slightly soluble in alcohol; insoluble in ether and in chloroform. The addition of tannic acid TS to its 1 in 500 solution causes no precipitation (*acid color*). The addition of hydrochloric acid to a mixture of 500 mg of zinc dust and 10 mL of its 1 in 500 solution produces decolorization. When filtered, the colorless filtrate, on standing exposed to air, does not regain its original color (*presence of azo-group*). When heated, orange G does not deflagrate (distinction from *nitro colors*). The addition of barium or calcium chloride TS to a concentrated solution of orange G produces a colored, crystalline precipitate. The addition of hydrochloric acid to its 1 in 500 solution produces no change; the addition of sodium hydroxide TS to a similar solution produces a yellowish red to a Bordeaux color but no precipitation. Orange G dissolves in sulfuric acid with an orange to yellowish red color. No change in color results upon diluting the solution cautiously with water.

Orange IV—Use *Tropaeolin OO* (USP reagent).

Orcinol (*5-Methylresorcinol*), $C_7H_8O_2 . H_2O$—**142.15**—White to light tan crystals.

Assay—Transfer about 60 mg, accurately weighed, to a 100-mL volumetric flask, dissolve in methanol, dilute with methanol to volume, and mix. Transfer 5.0 mL of this solution to a 50-mL volumetric flask and dilute with methanol to volume, and mix. Using a suitable spectrophotometer, 1-cm cells, and methanol as the blank, record the absorbance of the solution at the wavelength of maximum absorbance at about 273 nm. From the observed absorbance, calculate the absorptivity (see *Spectrophotometry and Light-scattering* ⟨851⟩): the absorptivity is not less than 13.2, corresponding to not less than 98% of $C_7H_8O_2 . H_2O$.

Melting range ⟨741⟩: between 58° and 61°.

Orthophenanthroline—See *1,10-Phenanthroline.*

Osmium Tetroxide (*Osmic Acid; Perosmic Anhydride*), OsO_4—**254.20**—Colorless or slightly yellow, hygroscopic crystals or crystalline granules. Very pungent odor. Decomposed by light. Slowly soluble in about 20 parts of water; soluble in alcohol and in ether, with decomposition. It softens at about 35°, melts between 40° and 42°, and boils at about 130°.

Caution—Osmium Tetroxide vapors are poisonous and highly irritating to the eyes and to the respiratory membranes.

Solubility—Dissolve 200 mg in 1 mL of carbon tetrachloride: a clear and not more than slightly yellow solution results, and no appreciable amount of insoluble residue remains.

Nonvolatile matter—Evaporate the solution remaining from the test for *Solubility* on a steam bath *in a well-ventilated hood* to dryness, and dry at 105° for 1 hour: the residue weighs not more than 0.4 mg (0.2%).

Heavy metals—To the residue from the test for *Nonvolatile matter* add 2 mL of hydrochloric acid, and evaporate the solution to dryness. Take up the residue in a few mL of water, dilute with water to 25 mL, and add 10 mL of hydrogen sulfide TS: any brown color produced is not darker than that of a control containing 0.01 mg of added Pb (0.005%).

Oxalic Acid, $H_2C_2O_4 . 2H_2O$—**126.07**—Use ACS reagent grade.

Oxprenolol Hydrochloride, $C_{15}H_{23}NO_3 . HCl$—**301.81**—Use a suitable grade.

3,3′-Oxydipropionitrile, $O(CH_2CH_2CN)_2$—**124.14**—Clear, colorless to slightly yellow liquid. Refractive index: about 1.446 at 20°.

Boiling range: between 174° and 176° at 10 mm of mercury.

Oxygen—Use *Oxygen* (USP monograph).

Oxygen-Helium Certified Standard—A mixture of 1.0% oxygen in industrial grade helium. It is available from most suppliers of specialty gases.

Oxygen-Nitrogen Certified Standard—A mixture of 5.0% oxygen in nitrogen, which is available from most suppliers of specialty gases.

Packing, Coated Strong Cation-exchange—Use a suitable grade.[51]

Packings for High-pressure Liquid Chromatography—[NOTE—Unless otherwise specified, the particle diameters of column packings designated in compendial procedures correspond to the nominal sizes or are within the ranges stated herein.]

L1—Octadecyl silane chemically bonded to porous silica or ceramic microparticles, 5 to 10 µm in diameter.

L2—Octadecyl silane chemically bonded to silica gel of a controlled surface porosity that has been bonded to a solid spherical core, 30 to 50 µm in diameter.

L3—Porous silica microparticles, 5 to 10 µm in diameter.

L4—Silica gel of controlled surface porosity bonded to a solid spherical core, 30 to 50 µm in diameter.

L5—Alumina of controlled surface porosity bonded to a solid spherical core, 30 to 50 µm in diameter.

L6—Strong cation-exchange packing—sulfonated fluorocarbon polymer coated on a solid spherical core, 30 to 50 µm in diameter.

L7—Octyl silane chemically bonded to totally porous microsilica particles, 5 to 10 µm in diameter.

L8—An essentially monomolecular layer of aminopropylsilane chemically bonded to totally porous silica gel support, 10 µm in diameter.

L9—10-µm irregular, totally porous silica gel having a chemically bonded, strongly acidic cation-exchange coating.

L10—Nitrile groups chemically bonded to porous silica microparticles, 5 to 10 µm in diameter.

L11—Phenyl groups chemically bonded to porous silica particles, 5 to 10 µm in diameter.

L12—A strong anion-exchange packing-quaternary amine bonded on a solid spherical core, 30 to 50 µm in diameter.

L13—Trimethylsilane chemically bonded to porous silica microparticles, 5 to 10 µm in diameter.

L14—Silica gel, 10 µm in diameter, having a chemically bonded, strongly basic quaternary ammonium anion-exchange coating.

L15—Hexyl silane chemically bonded to totally porous silica particles, 3 to 10 µm in diameter.

L16—Dimethylsilane chemically bonded to porous silica particles, 5 to 10 µm in diameter.

L17—Strong cation-exchange resin consisting of sulfonated cross-linked styrene-divinylbenzene copolymer in the hydrogen form, 7 to 11 µm in diameter.

L18—Amino and cyano groups chemically bonded to porous silica particles, 5 to 10 µm in diameter.

L19—Strong cation-exchange resin consisting of sulfonated cross-linked styrene-divinylbenzene copolymer in the calcium form, about 9 µm in diameter.

L20—Dihydroxypropane groups chemically bonded to porous silica particles, 5 µm to 10 µm in diameter.

L21—A rigid, spherical styrene-divinylbenzene copolymer, 5 to 10 µm in diameter.

L22—A cation exchange resin made of porous polystyrene gel with sulfonic acid groups, about 10 µm in size.

L23—An anion exchange resin made of porous polymethacrylate or polyacrylate gel with quaternary ammonium groups, about 10 µm in size.

L24—A semi-rigid hydrophilic gel consisting of vinyl polymers with numerous hydroxyl groups on the matrix surface, 32 to 63 µm in diameter.[52]

Palladium Catalyst—Use a suitable grade.[53]

Palladium-Charcoal Catalyst—Use a suitable grade.[54]

Palladium Chloride, $PdCl_2$—**177.31**—Brown, crystalline powder. Soluble in water, in alcohol, in acetone, and in diluted hydrochloric acid.

Assay—Dissolve 80 mg, accurately weighed, in 10 mL of diluted hydrochloric acid, dilute with water to 50 mL, and add 25 mL of a 1 in 100 solution of dimethylglyoxime in alcohol. Allow to stand for 1 hour, and filter. Check for complete precipitation with the dimethylglyoxime solution. Ignite the precipitate in a tared platinum crucible at 850° for 2 hours, cool, and weigh the palladium. The weight of the residue is not less than 59.0% of the weight of the test specimen.

Palladous Chloride—See *Palladium Chloride.*

Pancreatic Digest of Casein (a bacteriological peptone; *Tryptone*)—A grayish yellow powder, having a characteristic, but not putrescent, odor. Freely soluble in water; insoluble in alcohol and in ether. The casein used in preparation of this digest meets the following specifications:

Residue on ignition	not more than 2.5%
Loss on drying	not more than 8%
Free acid (as lactic acid)	not more than 0.25%
Fat	not more than 0.5%
Reducing sugars	not more than a trace
Fineness	all passes through a 20-mesh sieve

Degree of digestion—Dissolvc 1 g in 10 mL of water.

(*a*) Overlay 1 mL of the digest solution with 0.5 mL of a solution of 1 mL of glacial acetic acid in 10 mL of diluted alcohol: no ring or precipitate forms at the junction of the two liquids, and when shaken no turbidity results (indicating the absence of undigested casein).

(*b*) Mix 1 mL of the digest solution with 4 mL of a saturated solution of zinc sulfate: a moderate amount of precipitate is formed (indicating the presence of proteoses). Filter, and retain the filtrate.

(*c*) To 1 mL of the filtrate from the preceding test add 3 mL of water, and follow with 1 drop of bromine TS: a violet-red color is produced, indicating the presence of tryptophane.

Nitrogen content (Reagent test)—Determine by the Kjeldahl method, using a test specimen previously dried at 105° to constant weight: not less than 10.0% is found.

Loss on drying ⟨731⟩—Dry it at 100° to constant weight: it loses not more than 7.0% of its weight.

Residue on ignition ⟨281⟩—Ignite 500 mg with 1 mL of sulfuric acid: the residue weighs not more than 75 mg (15%).

Nitrite—To 5 mL of a solution of the digest (1 in 50) add 0.5 mL of sulfanilic-α-naphthylamine TS, mix, and allow to stand for 15 minutes: no pink or red color develops.

Microbial content—Dissolve 1 g in 10 mL of water. Spread 0.01 mL on one square centimeter of a glass slide. Stain by the Gram method, and examine with an oil-immersion lens: not more than a total of 50 microorganisms, or clumps, are visible in 10 consecutive fields.

Bacteriological test—The digest meets the following tests for bacteria-nutrient properties. Prepare media of the following compositions:

(*a*) 2% of digest, in water;

(*b*) 0.1% of digest, in water;

(*c*) 1% of digest, 0.5% of sodium chloride, 0.5% of dextrose, in water;

(*d*) 1% of digest, in water;

(*e*) 2% of digest, 0.5% of sodium chloride, 1.5% of agar, in water.

Adjust all media to a pH of 7.2 to 7.4.

Freedom from fermentable carbohydrate—To medium (*a*) add sufficient phenolsulfonphthalein TS to give a readable color, place in Durham fermentation tubes, and autoclave. Inoculate with a loop of 24-hour culture of *Escherichia coli:* no acid, or only a trace in the inner tube, and no gas are produced during incubation for 48 hours.

Production of indole—Inoculate 5 mL of medium (*b*) with *Escherichia coli*, incubate for 24 hours, and test by addition of about 0.5 mL of *p*-dimethylaminobenzaldehyde TS: it shows a distinct pink or red color which is soluble in chloroform.

Production of acetylmethylcarbinol—Inoculate 5 mL of medium (*c*) with *Aerobacter aerogenes*, and incubate for 24 hours. Test by adding to the culture an equal volume of sodium hydroxide solution (1 in 10), shake, and allow to stand at room temperature for several hours: appearance of a pink color indicates the presence of acetylmethylcarbinol.

Production of hydrogen sulfide—Inoculate 5 mL of medium (*d*) with *Salmonella typhosa.* Hold a strip or loop of lead acetate test paper between the cotton plug and the mouth of the test tube so that it hangs about 5 cm above the medium. After incubation for 24 hours, the lower tip of the lead acetate test paper shows little if any darkening. After 48 hours, it shows an appreciable amount of brownish blackening (*lead sulfide*).

Growth-supporting properties—In the foregoing tests the media support good growth of *Escherichia coli, Aerobacter aerogenes*, and *Salmonella typhosa.* Medium (*e*) stab-inoculated with a stock culture of *Brucella abortus* shows good growth in the line of the stab after incubation for 48 hours. Slants of medium (*e*), inoculated with *Escherichia coli, Aerobacter aerogenes, Salmonella typhosa, Pseudomonas aeruginosa, Staphylococcus aureus*, and *Staphylococcus albus*, show characteristic growth after incubation for 24 hours. Medium (*e*), to which about 5% of sheep blood or of rabbit blood has been added and which has been inoculated and poured into petri dishes, shows characteristic alpha or beta zones about colonies of *pneumococci* and *beta hemolytic streptococci* (serological groups A and B), recognizable within 24 hours and fully developed after 48 hours' incubation. Medium (*e*), to which about 10% of blood has been added and which then has been heated to 80° to 90° until the blood has turned chocolate-brown, permits the growth of *gonococcus* colonies within 48 hours when incubated in an atmosphere containing about 10% of carbon dioxide.

Pancreatic Hydrolysate of Gelatin—Use a suitable grade.

Pancreatin—Use *Pancreatin* (USP monograph).

Papaic Digest of Soybean Meal—A soluble nutrient material prepared by the action of the enzyme papain on soybean meal followed by suitable purification and concentration. It meets the specifications under *Pancreatic Digest of Casein*, except with respect to *Nitrogen content* and except that it shows substantial amounts of reducing sugars. It contains fermentable carbohydrates and gives positive tests for indole, acetylmethylcarbinol, and sulfide upon inoculation and incubation with the specified organisms.

Nitrogen content (Reagent test)—Determine by the Kjeldahl method, using a test specimen previously dried at 105° to constant weight: not less than 8.5% is found.

Papaverine Hydrochloride, $C_{20}H_{21}NO_4.HCl$—**375.85**—Use *Papaverine Hydrochloride* (USP monograph).

Paper, Odorless Absorbent—See *Filter Paper, Quantitative.*

Para-aminobenzoic Acid (*p-Aminobenzoic Acid*), $H_2NC_6H_4COOH$—**137.14**—White or slightly yellow, odorless crystals or crystalline powder, becoming discolored on exposure to air or light. One g dissolves in 170 mL of water, in 9 mL of boiling water, in 8 mL of alcohol, and in 50 mL of ether. Freely soluble in solutions of alkali hydroxides and carbonates; soluble in warm glycerin; sparingly soluble in diluted hydrochloric acid; slightly soluble in chloroform. Store in tight, light-resistant containers.

Assay—Weigh accurately about 300 mg, previously dried at 105° for 2 hours, and transfer to a beaker or casserole. Add 5 mL of hydrochloric acid and 50 mL of water, and stir until dissolved. Cool to about 15°, add about 25 g of crushed ice, and slowly titrate with 0.1 *M* sodium nitrite VS until a glass rod dipped into the titrated solution produces an immediate blue ring when touched to starch iodide paper. When the titration is complete, the end-point is reproducible after the mixture has been allowed to stand for 1 minute. Each mL of 0.1 *M* sodium nitrite is equivalent to 13.71 mg of $C_7H_7NO_2$. Not less than 98.5% is found.

Melting range ⟨741⟩: between 186° and 189°.

Loss on drying ⟨731⟩—Dry it at 105° for 2 hours: it loses not more than 0.2% of its weight.

Residue on ignition (Reagent test): not more than 0.1%.

Paraffin—Use *Paraffin* (NF monograph).

Paraformaldehyde, $(CH_2O)_n$—Fine, white powder, having the characteristic odor of formaldehyde.

Assay—Transfer about 1 g, accurately weighed, to a 250-mL conical flask containing 50.0 mL of 1 *N* sodium hydroxide VS, and mix by swirling. Immediately, and slowly, add 50 mL of hydrogen peroxide TS, previously neutralized to bromothymol blue, through a small funnel placed in the neck of the flask. After the reaction moderates, rinse the funnel and inner wall of the flask with water, allow the solution to stand for 30 minutes, add bromothymol blue TS, and titrate the excess alkali with 1 *N* sulfuric acid VS. Each mL of 1 *N* sodium hydroxide is equivalent to 30.03 mg of HCHO: not less than 95% is found.

Residue on ignition: not more than 0.1%.

Solubility in ammonia—Dissolve 5 g in 50 mL of ammonia TS: a practically clear, colorless solution results.

Reaction—Shake 1 g with 20 mL of water for about 1 minute, and filter: the filtrate is neutral to litmus.

Penicillin V Potassium—Use *Penicillin V Potassium* (USP Monograph).

Penicillinase—Penicillinase is an enzyme produced by a variety of bacteria, but is usually obtained from culture filtrates of a strain of *Bacillus cereus.* It has the specific property of inactivating penicillin by splitting the bond linking the nitrogen of the thiazolidine to the adjacent carbonyl carbon.

It occurs in the form of small, brown, easily pulverizable pieces or granules. Freely soluble in water, forming a slightly opalescent solution that is practically neutral to litmus paper. Is precipitated from its water solutions by acetone, by alcohol, and by dioxane, and is inactivated by contact with these solvents. Is rapidly inactivated by ethyl acetate and is irreversibly destroyed at a temperature of about 80°.

Penicillinase is assayed by a procedure depending upon a determination of the amount of potassium or sodium penicillin destroyed at a pH of 7.0 in a solution of such concentration that the inactivation proceeds as a zero-order reaction.

Pentaerythritol Tetranitrate, Diluted—Use *Diluted Pentaerythritol Tetranitrate* (USP monograph).[55]

Pentane (*n-Pentane*), C_5H_{12}—**72.15**—Clear, colorless, *flammable* liquid. Very slightly soluble in water. Miscible with alcohol, with ether, and with many organic solvents. Specific gravity: about 0.62.

Boiling range (Reagent test)—Not less than 95% distils between 34° and 36°.

1-Pentanesulfonic Acid Sodium Salt—See *Sodium 1-Pentanesulfonate.*

***p*-Pentylphenol,** $CH_3(CH_2)_4C_6H_4OH$—**164.25**—Colorless liquid.

Melting range ⟨741⟩: melts within a 2° range that includes 24°.

Refractive index ⟨831⟩: between 1.5130 and 1.5150 at 20°.

Pepsin—Use *Pepsin* (*Enzyme Preparations*) FCC, having an activity of 1.0 to 1.17 Pepsin units per mg. Pepsin of higher activity may be reduced to this activity by admixture with pepsin of lower activity or with lactose.

Peptic Digest of Animal Tissue (a bacteriological peptone)—Tan powder, having a characteristic, but not putrescent, odor. Soluble in water; insoluble in alcohol and in ether. An autoclaved solution (2 in 100) is clear and is neutral or nearly so in its reaction.

Degree of digestion—Dissolve 1 g in 10 mL of water, and use this solution for the following tests:

(*a*) Overlay 1 mL of the digest solution with 0.5 mL of a solution of 1 mL of glacial acetic acid in 10 mL of diluted alcohol: no ring or precipitate forms at the junction of the two liquids, and on shaking, no turbidity results, indicating the absence of undigested protein.

(*b*) Mix 1 mL of the digest solution with 4 mL of saturated zinc sulfate: a small amount of precipitate is formed, indicating the presence of proteoses. Filter, and retain the filtrate.

(*c*) To 1 mL of the filtrate from the preceding test add 1 drop of bromine TS: the light yellow color changes to a red-brown, indicating the presence of tryptophane.

Nitrogen content, *Loss on drying*, *Residue on ignition*, and *Nitrite*—Proceed as directed under *Pancreatic Digest of Casein.*

Microbial content—Dissolve 1 g in 10 mL of water. Spread 0.01 mL on one square centimeter of a glass slide. Stain by the Gram method, and examine with an oil-immersion lens: not more than a total of 50 microorganisms, or clumps, are visible in 10 consecutive fields.

Bacteriologic test—It meets the following tests for bacteria-nutrient properties. Prepare media of the following compositions:

(*a*) 2% of digest and sufficient phenol red TS to give a perceptible color in water;

(*b*) 0.1% of digest in water;

(*c*) 0.1% of digest and 0.5% of dextrose in water;

(*d*) 1% of digest in water.

Adjust all media to a final pH of 7.2 to 7.4. Place 5 mL of (*a*) in Durham fermentation tubes, and 5 mL each of (*b*), (*c*), and (*d*) in ordinary test tubes. Autoclave the media at 121° for 15 minutes. After autoclaving, and after standing for 24 hours, all media are clear.

Presence of fermentable carbohydrate—Inoculate medium (*a*) with *Escherichia coli* and with *Streptococcus liquefaciens:* acid is produced by *E. coli* but not by *S. liquefaciens* during incubation for 24 hours.

Production of indole—Inoculate medium (*b*) with *Escherichia coli* and with *Aerobacter aerogenes*, and incubate for 24 hours. Test by adding about 0.5 mL of *p*-dimethylaminobenzaldehyde TS: the appearance of a pink or red color (soluble in chloroform) indicates the production of indole by *E. coli.* The *A. aerogenes* culture gives a negative test.

Production of acetylmethylcarbinol—Inoculate medium (*c*) with *Escherichia coli* and with *Aerobacter aerogenes*, and incubate for 24 hours. Test by adding to the culture an equal volume of sodium hydroxide solution (1 in 10), shaking well, and allowing to stand at room temperature for several hours: the appearance of a pink color indicates the production of acetylmethylcarbinol by *A. aerogenes.* The *E. coli* culture gives a negative test.

Production of hydrogen sulfide—Inoculate medium (*d*) with *Salmonella typhosa.* Hold a strip or loop of lead acetate test paper between the cotton plug and the mouth of the test tube so that it hangs about 5 cm above the medium. Then incubate for 24 hours: the lower part of the lead acetate test paper shows an appreciable amount of brownish blackening (*lead sulfide*).

Peptone, Dried (*Meat Peptone*)—Reddish yellow to brown powder, having a characteristic, but not putrescent, odor. Soluble in water, forming a yellowish brown solution having a slight acid reaction; insoluble in alcohol and in ether.

Nitrogen content (Reagent test)—Determine by the Kjeldahl method, using a test specimen previously dried at 105° to constant weight: between 14.2% and 15.5% of N is found, corresponding to not less than 89% of protein.

Residue on ignition (Reagent test)—Ignite 500 mg with 1 mL of sulfuric acid: the residue weighs not more than 25 mg (5.0%).

Loss on drying ⟨731⟩—Dry it at 105° to constant weight: it loses not more than 7.0% of its weight.

Coagulable protein—Heat a filtered solution (1 in 20) to boiling: no precipitate forms.

Proteoses—Mix 5 mL of a filtered solution (1 in 10) with 20 mL of a filtered solution of zinc sulfate (made by dissolving 50 g of the salt in 35 mL of water): not more than a slight, flocculent precipitate is formed.

Perchloric Acid (*70 Percent Perchloric Acid*), $HClO_4$—**100.46**—Use ACS reagent grade (containing between 70.0% and 72.0% of $HClO_4$).

Perchloric Acid, 60 Percent, $HClO_4$—**100.46**—Use ACS reagent grade Perchloric Acid, 60 Percent (containing between 60.0% and 62.0% of $HClO_4$).

Periodic Acid, H_5IO_6—**227.94**—White to pale yellow crystals. Very soluble in water. Undergoes slow decomposition to iodic acid.

Assay—Dissolve about 120 mg, accurately weighed, in water. Add 5 mL of hydrochloric acid and 5 g of potassium iodide, then add 3 mL of starch TS, and titrate the liberated iodine with 0.1 *N* sodium thiosulfate VS. Perform a blank determination, and make any necessary correction. Each mL of 0.1 *N* sodium thiosulfate is equivalent to 2.849 mg of H_5IO_6. Not less than 99% is found.

Insoluble matter (Reagent test): not more than 1.0 mg, from 10.0 g (0.01%).

Residue on ignition (Reagent test): not more than 2 mg, from 10.0 g, ignited for 10 minutes (0.02%). [Retain for *Heavy metals* test.]

Sulfate—Weigh accurately 1 g, add 10 to 20 mg of anhydrous sodium carbonate, and evaporate to dryness three times with 5-mL portions of hydrochloric acid. Dissolve in 25 mL of water, add 1 mL of dilute hydrochloric acid (1 in 10), then add 1 mL of barium chloride TS, and compare the turbidity with that of *Standard Sulfate Solution* (see *Sulfate in Reagents*). One g shows not more than 0.1 mg of SO_4 (0.01%).

Other halogens—Dissolve 1.0 g in 100 mL of water, add 1 mL of phosphoric acid and 5 mL of 30 percent hydrogen peroxide, and boil to expel iodine. Dilute with water to 100 mL. To a 20-mL aliquot add 3 mL of nitric acid and 1 mL of silver nitrate TS. Compare the turbidity with that of a solution similarly prepared from a *Standard Chloride Solution* (see *Chloride in Reagents*) containing 0.02 mg of chloride (0.01%).

Heavy metals—To the *Residue on ignition* add several drops of acetic acid, and warm to dissolve. Transfer to a test tube and add 10 mL of hydrogen sulfide TS. Compare the color with that of *Standard Lead Solution* (see *Heavy Metals* ⟨231⟩) containing 0.5 mg of Pb (0.005%).

Iron—Dissolve 1.0 g in 50 mL of water, add 1 mL of dilute sulfuric acid (1 in 2) and 10 mL of hydroxylamine hydrochloride solution (1 in 5), and evaporate to dryness to expel iodine. Dissolve the residue in water, add 2 mL of 1,10-phenanthroline solution (1 in 1000) and 10 mL of sodium acetate solution (1 in 5), and compare the color with that of a solution containing 0.03 mg of iron, treated similarly (0.003%).

Petrolatum, Liquid—Use *Mineral Oil* (USP monograph).

Petroleum Benzin—See *Hexane, Solvent.*

Phases for Gas Chromatography—

G1—Dimethylpolysiloxane oil.

G2—Dimethylpolysiloxane gum.

G3—50% Phenyl—50% methylpolysiloxane.

G4—Diethylene glycol succinate polyester.

G5—3-Cyanopropylpolysiloxane.

G6—Trifluoropropylmethylpolysiloxane.

G7—50% 3-Cyanopropyl—50% phenylmethylsilicone.

G8—90% 3-Cyanopropyl—10% phenylmethylsilicone.

G9—Methylvinylpolysiloxane.

G10—Polyamide formed by reacting a C_{36} dicarboxylic acid with 1,3-di-4-piperidylpropane and piperidine in the respective mole ratios of 1.00:0.90:0.20.

G11—Bis(2-ethylhexyl) sebacate polyester.

G12—Phenyldiethanolamine succinate polyester.

G13—Sorbitol.

G14—Polyethylene glycol (av. mol. wt. of 950 to 1050).

G15—Polyethylene glycol (av. mol. wt. of 3000 to 3700).

G16—Polyethylene glycol compound (av. mol. wt. about 15,000). A high molecular weight compound of a polyethylene glycol and a diepoxide. Available commercially as Polyethylene Glycol Compound 20M, or as Carbowax 20M, from suppliers of chromatographic reagents.

G17—75% Phenyl—25% methylpolysiloxane.

G18—Polyalkylene glycol.

G19—25% Phenyl—25% cyanopropylmethylsilicone.

G20—Polyethylene glycol (av. mol. wt. of 380 to 420).

G21—Neopentyl glycol succinate.

G22—Bis(2-ethylhexyl) phthalate.

G23—Polyethylene glycol adipate.

G24—Diisodecyl phthalate.

G25—Polyethylene glycol compound TPA. A high molecular weight compound of a polyethylene glycol and a diepoxide that is esterified with terephthalic acid. Available commercially as Carbowax 20M-TPA from suppliers of chromatographic reagents.

G26—25% 2-Cyanoethyl—75% methylpolysiloxane.

G27—5% Phenyl—95% methylpolysiloxane.

G28—25% Phenyl—75% methylpolysiloxane.

G29—3,3′-Thiodipropionitrile.

G30—Tetraethylene glycol dimethyl ether.

G31—Nonylphenoxypoly(ethyleneoxy)ethanol (av. ethyleneoxy chain length is 30); Nonoxynol 30.

G32—20% Phenylmethyl—80% dimethylpolysiloxane.

G33—20% Carborane—80% methylsilicone.

G34—Diethylene glycol succinate polyester stabilized with phosphoric acid.

G35—A high molecular weight compound of a polyethylene glycol and a diepoxide that is esterified with nitroterephthalic acid.

G36—1% Vinyl—5% phenylmethylpolysiloxane.

G37—Polyimide.

G38—Phase G1 containing a small percentage of a tailing inhibitor. Available as SP2100/0.1% Carbowax 1500 from Supelco.

Phenacetin—Use a suitable grade.

1,10-Phenanthroline (*Orthophenanthroline*), $C_{12}H_8N_2 . H_2O$—**198.22**—Use ACS reagent grade.

Phenobarbital Sodium—Use *Phenobarbital Sodium* (USP monograph).

Phenol—Use ACS reagent grade.

Phenol, Liquefied—Use *Liquefied Phenol* (USP monograph).

Phenol, Purified—Use ACS reagent grade Phenol that has been subjected to the following purification procedure.

Transfer a suitable quantity of phenol to an all-glass reflux apparatus so fitted that nitrogen can be bubbled through the molten material. Melt the phenol, and reflux for 20 minutes while passing nitrogen through the molten material. Stop the flow of nitrogen, rearrange the apparatus for distillation, and distil at a reflux ratio of 10 to 1. Determine the suitability of the distilled phenol by warming 3 g with 10 mL of sulfuric acid at 95° for 10 minutes: the color of the mixture is not darker than that of an equal volume of the acid.

Phenolsulfonphthalein—Use *Phenolsulfonphthalein* (USP XXI monograph).

Phenoxybenzamine Hydrochloride [*N-(2-Chloroethyl)-N-(1-methyl-2-phenoxyethyl)benzylamine Hydrochloride*], $C_{18}H_{22}ClNO . HCl$—**340.29**—White, crystalline powder.

Melting range ⟨741⟩: between 137° and 140°.

Absorptivity—Its absorptivity, 1%, 1 cm, in the range of 272 nm to 290 nm, in chloroform solution is about 178.

2-Phenoxyethanol, $C_6H_5OCH_2CH_2OH$—**138.17**—Colorless, slightly viscous liquid. Soluble in water. Miscible with alcohol, with acetone, and with glycerin. Density: about 1.107.

Assay—To 2 g, accurately weighed, add 10 mL of a freshly prepared solution made by dissolving 25 g of acetic anhydride in 100 g of anhydrous pyridine. Swirl to mix the liquids, heat on a steam bath for 45 minutes, add 10 mL of water, heat for 2 additional minutes, and cool. Add 10 mL of normal butyl alcohol, shake vigorously, add phenolphthalein TS, and titrate with 1 *N* sodium hydroxide VS. Perform a blank test using the same quantities of the same reagents, and in the same manner, and make any necessary correction. Each mL of 1 *N* sodium hydroxide is equivalent to 138.2 mg of $C_8H_{10}O_2$. Not less than 99% is found.

Phenol—Add 0.2 mL of it to 20 mL of water, mix, and to 5 mL of the mixture add 0.2 mL of Millon's reagent. Warm the solution at 60° for 90 seconds, and allow to stand: no pink or red color is produced within 1 minute.

Phenyl Ether—See *Diphenyl Ether.*

Phenyl Isocyanate, C_6H_5NCO—**119.12**—Clear, colorless to straw-yellow liquid of medium volatility.

Caution—Phenyl Isocyanate is a violent lacrimator, and the vapor is highly toxic. Handle with care.

Assay—Transfer 250 mg, accurately weighed, to a glass-stoppered, 250-mL flask. Exercise care to avoid loss by volatilization, and avoid breathing the vapor. Add 20 mL of butylamine solution (25 g of butylamine diluted to 1000 mL with dioxane previously dried over potassium hydroxide pellets), insert the stopper in the flask, and allow to stand for 15 minutes. Add a few drops of methyl red TS and 25 mL of water, and titrate the excess amine with 0.1 *N* sulfuric acid VS. Perform a blank titration on 20 mL of the butylamine solution (see *Residual Titrations* ⟨541⟩). Subtract the volume of 0.1 *N* sulfuric acid consumed in the test specimen titration from that consumed in the blank titration. Each mL of 0.1 *N* sulfuric acid, representing this difference, is equivalent to 11.91 mg of C_6H_5NCO: not less than 97.0% of C_6H_5NCO is found.

Phenyl Methyl Silicone—Use a suitable grade.[56]

Phenyl Sulfide, $C_{12}H_{10}S$—**186.27**—Colorless, practically odorless liquid. Refractive index: about 1.635. Insoluble in water; soluble in hot alcohol. Miscible with benzene, with ether, and with carbon disulfide.

Boiling range (Reagent test): between 151° and 153°, at a pressure of 15 mm of mercury.

***dl*-Phenylalanine,** $C_9H_{11}NO_2$—**165.19**—Use a suitable grade.

***p*-Phenylenediamine Dihydrochloride**—See *p-Phenylenediamine Hydrochloride.*

***p*-Phenylenediamine Hydrochloride** (*1,4-Diaminobenzene Dihydrochloride*), $C_6H_8N_2 . 2HCl$—**181.06**—White to pale tan crystals or crystalline powder, turning red on exposure to air. Freely soluble in water; slightly soluble in alcohol and in ether. Preserve in well-closed containers, protected from light.

Insoluble matter—Dissolve 1 g in 10 mL of water: the solution is clear and complete.

Molar absorptivity (see *Spectrophotometry and Light-scattering* ⟨851⟩)—Dissolve 60 mg in 100.0 mL of water, and mix. Pipet 2 mL of this solution into a 50-mL volumetric flask, dilute with pH 7 buffer solution to volume, and mix. The molar absorptivity of this solution, at 239 nm, is not less than 9000.

Phenylethyl Alcohol—Use *Phenylethyl Alcohol* (USP monograph).

Phenylhydrazine, $C_6H_5NHNH_2$—**108.14**—A colorless, or slightly yellowish, highly refractive liquid. [NOTE—Protect from light, and distil under reduced pressure shortly prior to use.]

Congealing temperature ⟨651⟩: not below 16°.

Insoluble matter—Shake 1 mL with 20 mL of diluted acetic acid: the resulting solution is clear or practically so.

Residue on ignition (Reagent test)—Ignite 1 mL with 0.5 mL of sulfuric acid: the residue weighs not more than 1 mg (0.1%).

Phenylhydrazine Hydrochloride, $C_6H_5NHNH_2 . HCl$—**144.60**—White or yellowish crystals or powder. Soluble in water and in alcohol. Store in tight containers, protected from light.

Melting range ⟨741⟩: between 242° and 246°, with slight darkening.

Solubility—Separate 500-mg portions dissolve in 10 mL of water and in 10 mL of alcohol, respectively, to yield solutions that are clear and complete or practically so.

Residue on ignition (Reagent test)—Ignite 1 g with 0.5 mL of sulfuric acid: the residue weighs not more than 1 mg (0.1%).

Phenylmercuric Nitrate—Use *Phenylmercuric Nitrate* (NF monograph).

3-Phenylphenol (*m-Phenylphenol*), $C_6H_5C_6H_4OH$—**170.21**—White to off-white, crystalline powder.

Assay—Inject an appropriate specimen into a suitable gas chromatograph (see *Chromatography* ⟨621⟩) equipped with a flame-ionization detector, helium being used as the carrier gas. The following conditions have been found suitable: a 30-m × 0.25-mm capillary column coated with G1; the injection port temperature is maintained at 250°; the column temperature is maintained at 150° and programmed to rise 15° per minute to 250°; and the detector temperature is maintained at 310°. The area of the 3-phenylphenol peak is not less than 98% of the total peak area.

Melting range ⟨741⟩: between 76° and 79°.

Phloroglucinol, $C_6H_3(OH)_3 . 2H_2O$—**162.14**—White or yellowish white crystals or a crystalline powder. Slightly soluble in water; soluble in alcohol and in ether.

Insoluble in alcohol—Dissolve 1 g in 20 mL of alcohol: a clear and complete solution results.

Melting range, Class Ia ⟨741⟩: between 215° and 219°.

Residue on ignition (Reagent test)—Ignite 1 g with 0.5 mL of sulfuric acid: the residue weighs not more than 1 mg (0.1%).

Diresorcinol—Heat to boiling a solution of 100 mg in 10 mL of acetic anhydride, cool the solution, and superimpose it upon 10 mL of sulfuric acid: no violet color appears at the zone of contact of the liquids.

Phosphatase Enzyme, Alkaline—Use a suitable grade.[57]

Phosphatic Enzyme—An enzyme preparation of microbial origin, high in both phosphatase and amylase activity, the former being the property that renders it suitable for use in the liberation of thiamine from its orthophosphate and pyrophosphate esters. Light cream-colored or slightly gray powder. Freely soluble in water. It hydrolyzes 300 times its weight of starch in 30 minutes.

Amylase activity—Place in a test tube 5 mL of a 1 in 50 solution of soluble starch in 0.2 *M*, pH 5 sodium acetate buffer (containing 1.6 g of anhydrous sodium acetate in each liter and sufficient glacial acetic acid to adjust to a pH of 5), and add 4 mL of water. Mix, and place in a water bath at 40°. Add 1 mL of a solution containing 0.3 mg of the phosphatic enzyme, mix, and note the exact time. After 30 minutes remove 1.0 mL of the mixture, and add it to 5.0 mL of 0.0005 *N* iodine in a 20- × 150-mm test tube: a clear, red color results.

Phosphine Detector Tube[58]—A fuse-sealed glass tube so designed that gas may be passed through it, and containing suitable absorbing filters and support media for the indicator, the latter consisting of a suitable gold salt.

Phosphoglucose Isomerase—Use a suitable grade.[37]

Suitability—When used in the assay of lactulose, determine that a suitable absorbance-versus-concentration slope is obtained using USP Lactulose RS, the reagent blank absorbance being not more than 0.020.

Phosphomolybdic Acid, approximately $20MoO_3 . P_2O_5 . 51H_2O$—**3939.48**—Use ACS reagent grade.

Phosphoric Acid, H_3PO_4—**98.00**—Use ACS reagent grade.

Phosphorus Pentoxide (*Phosphoric Anhydride*), P_2O_5—**141.94**—Use ACS reagent grade.

Phosphorus, Red, P—At. Wt. **30.97376**—A dark red powder. Insoluble in water and in dilute acids; soluble in dehydrated alcohol.

Yellow phosphorus—Shake 20 g with 75 mL of carbon disulfide in a glass-stoppered vessel, and allow to stand in the dark overnight. Filter, and wash the residue with carbon disulfide until the filtrate, collected in a graduated cylinder, measures 100 mL. Evaporate the solvent to 10 mL by immersing the cylinder in hot water. Dip a strip of cupric sulfate test paper in the remaining solvent: no more color is produced than in a similar strip dipped into 10 mL of solution in carbon disulfide containing 3 mg of yellow phosphorus (0.015% as P).

Soluble substances—Digest 2 g with 30 mL of acetic acid on a steam bath for 15 minutes. Cool, dilute with water to 40 mL, and filter. Evaporate 20 mL of the filtrate on a steam bath, and dry at 105° for 2 hours: the residue weighs not more than 6 mg (0.6%).

Phosphotungstic Acid, approximately $24WO_3 . P_2O_5 . 51H_2O$—**6625.08**—White or yellowish green crystals or a crystalline powder. Soluble in water, in alcohol, and in ether.

Insoluble matter (Reagent test): not more than 1 mg, from 5 g (0.02%).

Chloride (Reagent test)—One g shows not more than 0.3 mg of Cl (0.03%).

Nitrate—Dissolve 500 mg in 10 mL of water, and add about 10 mg of sodium chloride, 0.1 mL of indigo carmine TS, and 10 mL of sulfuric acid: the blue color does not disappear within 1 minute (about 0.01%).

Sulfate (Reagent test, *Method I*)—A 500-mg portion shows not more than 0.1 mg of SO_4 (0.02%).

Phthalic Acid, $C_8H_6O_4$—**166.13**—Colorless to white crystalline powder. Soluble in alcohol and in methanol; slightly soluble in water; practically insoluble in chloroform.

Assay—Transfer about 2.8 g, accurately weighed, to a 250-mL conical flask, and add 50.0 mL of 1 *N* sodium hydroxide VS. Add 25 mL of water, and warm on a hot plate until solution is complete. Add phenolphthalein TS, and titrate the excess sodium hydroxide with 1 *N* sulfuric acid VS. Perform a blank determination, and make any necessary correction. Each mL of 1 *N* sodium hydroxide is equivalent to 83.06 mg of $C_8H_6O_4$. Not less than 98% is found.

Melting range ⟨741⟩: between 205° and 209°, with decomposition, a sealed capillary tube being used.

Phthalic Anhydride, $C_8H_4O_3$—**148.12**—Use ACS reagent grade.

Physostigmine Salicylate—Use *Physostigmine Salicylate* (USP monograph).

Picric Acid (*2,4,6-Trinitrophenol; Trinitrophenol*), $C_6H_2(OH)(NO_2)_3$-1,2,4,6—**229.11**—Use ACS reagent grade.

Picrolonic Acid (*3-Methyl-4-nitro-1-(p-nitrophenyl)-5-pyrazolone*), $C_{10}H_8N_4O_5$—**264.20**—Yellow to brownish yellow crystalline powder. Slightly soluble in water; soluble in alcohol, in chloroform, in ether, in benzene, and in solutions of alkali hydroxides.

Melting range ⟨741⟩: between 115° and 117°.

Residue on ignition (Reagent test): negligible, from 200 mg.

Sensitiveness—Dissolve 25 mg in 10 mL of warm water containing 0.1 mL of glacial acetic acid, and filter the solution, if necessary. Dissolve 100 mg of calcium chloride in 250 mL of water, and mix. Heat 1 mL of the calcium chloride solution in a test tube to about 60°, then add to it 1 mL of the picrolonic acid solution: a bulky precipitate forms in 5 minutes or less.

Piperidine, $C_5H_{11}N$—**85.15**—Colorless liquid. Miscible with water and with alcohol. Specific gravity: about 0.860.

Congealing range ⟨651⟩: between 12° and 15°.

Boiling range (Reagent test)—Not less than 95% distils between 104° and 106°.

Refractive index: about 1.454.

Plasma, Mammalian—Use a suitable grade.

Platinic Chloride (*Chloroplatinic Acid*), $H_2PtCl_6 . 6H_2O$—**517.91**—Use ACS reagent grade Chloroplatinic Acid.

Poloxamer 124—Use *Poloxamer 124* (NF Monograph).

Polyamide Resin, 0.5 Percent—Use a suitable grade.[59]

Polyamino Undecanoic Acid—Use a suitable grade.[60]

Polyethylene Glycol 20 M—Use a suitable grade.

Polyethylene Glycol 300—Use *Polyethylene Glycol 300* (NF monograph).

Polyethylene Glycol 400—Use *Polyethylene Glycol 400* (NF monograph).

Polyethylene Glycol 600—A clear, practically colorless, viscous liquid condensation polymer represented by $H(OCH_2CH_2)_nOH$, in which *n* varies from 12 to 14. Its average molecular weight is about 600.

It meets the requirements of all of the tests under *Polyethylene Glycol 400* (NF monograph), except *Limit of ethylene and diethylene glycols.*

Polyethylene Glycol 3350—Use *Polyethylene Glycol 3350* (NF monograph).

Polyethylene Glycol 4000—Use *Polyethylene Glycol 4000* (NF monograph).

Polyethylene Glycol 20,000—Molecular weight range: 15,000–20,000. Hard, white, waxy solid, usually supplied in flake form. Soluble in water with subsequent gel formation.

Viscosity of 25% solution ⟨911⟩—Add 50.0 g of test specimen to a 250-mL wide-mouth, screw-cap jar containing 150.0 g of water. Attach the cap securely to the jar, and roll on a mechanical roller until the test specimen is completely dissolved, in 2 to 4 hours. Allow the solution to stand until all air bubbles have disappeared. Another 2 to 4 hours may be required. Adjust the temperature of the solution to 37.8 ± 0.1°, and determine the kinematic viscosity on a suitable viscosimeter of the Ubbelohde type. The viscosity is not less than 100 centistokes.

pH ⟨791⟩: between 6.5 and 8.0 in a solution (1 in 20). [NOTE—A five-fold dilution of the test solution prepared for the *Viscosity of 25% solution* test may be used.]

Residue on ignition ⟨281⟩: not more than 0.7%, the use of sulfuric acid being omitted.

Polyglycol Nitroterephthalate—Use a suitable grade.[61]

Polyoxyethylated Vegetable Oil Derivative—Use a suitable grade.[62]

Polyoxyethylene (23) Lauryl Ether—Use a suitable grade.[63]

Polysorbate 20—Use *Polysorbate 20* (NF monograph).

Polysorbate 80—Use *Polysorbate 80* (NF monograph).

Polystyrene Anion-exchange Resin—See *Anion-exchange Resin, Polystyrene.*

Polystyrene Cation-exchange Resin—See *Cation-exchange Resin, Polystyrene.*

Polytef—Use *Poly(tetrafluoroethylene).*

Polyvinyl Alcohol,[64] $(C_2H_4O)_n$—White powder. Soluble in water; insoluble in organic solvents.

pH ⟨791⟩: between 5.0 and 8.0, in a solution (1 in 25).

Loss on drying—Dry it at 110° to constant weight: it loses not more than 5% of its weight.

Residue on ignition: not more than 0.75%.

Ponceau S (*3-Hydroxy-4-[2-sulfo-4-(4-sulfophenylazo)phenylazo]2,7-naphthalene disulfonic acid, sodium salt*), $C_{22}H_{12}N_4O_{13}S_4Na_4$—**760.56**—Reddish brown crystals.

Assay—Dissolve about 50 mg, accurately weighed, in 100.0 mL of water. Dilute 2.0 mL of the resulting solution with water to 25.0 mL. Determine the absorbance of this solution in a 1-cm cell at the wavelength of maximum absorbance at about 519 nm, with a suitable spectrophotometer, using water as the blank. Using an absorptivity value of 45.6, determine the purity: not less than 98%, calculated on the anhydrous basis, is found.

Water, Method I ⟨921⟩: not more than 10.0%.

Potassium Acetate, $KC_2H_3O_2$—**98.14**—Use ACS reagent grade.

Potassium Alum—Use *Potassium Alum* [see *Potassium Alum* (USP monograph)].

Potassium Aluminum Sulfate, $AlK(SO_4)_2 . 12H_2O$—**474.38**—Use ACS reagent grade Aluminum Potassium Sulfate Dodecahydrate.

Potassium Bicarbonate, $KHCO_3$—**100.12**—Use ACS reagent grade.

Potassium Biphosphate—See *Potassium Phosphate, Monobasic.*

Potassium Biphthalate (*Acid Potassium Phthalate; Phthalic Acid Monopotassium Salt; Potassium Hydrogen Phthalate Acidimetric Standard*), $KHC_6H_4(COO)_2$—**204.22**—Use ACS reagent grade Potassium Hydrogen Phthalate, Acidimetric Standard.

Potassium Bisulfate, $KHSO_4$—**136.16**—Fused, white, deliquescent masses or granules. Very soluble in water. When ignited, it evolves SO_3 and H_2O, changing first to potassium pyrosulfate, then to sulfate.

Acidity—Dissolve 4 g, accurately weighed, in 50 mL of water, add phenolphthalein TS, and titrate with 1 *N* alkali: it contains between 34% and 36%, calculated as H_2SO_4.

Insoluble matter and ammonium hydroxide precipitate—Dissolve 10 g in 100 mL of water, add methyl red TS, render slightly alkaline with ammonia TS, boil for 1 minute, and digest on a steam bath for 1 hour. Filter through a tared filtering crucible, wash thoroughly, and dry at 105° for 2 hours: the precipitate weighs not more than 1 mg (0.01%).

For the following tests, prepare a *Test solution* as follows: Dissolve 6 g in 45 mL of water, add 2 mL of hydrochloric acid, boil gently for 10 minutes, cool, and dilute with water to 60 mL.

Heavy metals (Reagent test)—To 30 mL of *Test solution* add phenolphthalein TS, and neutralize with ammonia TS. Add 0.5 mL of glacial acetic acid, dilute with water to 40 mL, and add 10 mL of hydrogen sulfide TS: any brown color produced is not darker than that of a control containing 10 mL of *Test solution* and 0.02 mg of added Pb (0.001%).

Iron ⟨241⟩—To 5 mL of *Test solution* add 2 mL of hydrochloric acid, and dilute with water to 47 mL: the solution shows not more than 0.01 mg of Fe (0.002%).

Potassium Bromate, $KBrO_3$—**167.00**—Use ACS reagent grade.

Potassium Bromide, KBr—**119.00**—Use ACS reagent grade.

Potassium Carbonate—See *Potassium Carbonate, Anhydrous.*

Potassium Carbonate, Anhydrous, K_2CO_3—**138.21**—Use ACS reagent grade.

Potassium Chlorate, $KClO_3$—**122.55**—Use ACS reagent grade.

Potassium Chloride, KCl—**74.55**—Use ACS reagent grade.

Potassium Chloroplatinate, K_2PtCl_6—**486.00**—Heavy, yellow powder. Soluble in hydrochloric acid and in nitric acid.

Assay—Weigh accurately about 300 mg, transfer to a 600-mL beaker, add 20 mL of hydrochloric acid, and heat gently if necessary to achieve complete solution. Add zinc granules, slowly, until no more dissolves, then add 2 mL of hydrochloric acid, and digest for 1 hour on a steam bath to coagulate the reduced platinum. Add more acid, if necessary, to ensure that all of the zinc has dissolved. Filter through paper, rinsing the beaker with diluted hydrochloric acid until all of the precipitate is transferred to the filter, then wash with several small portions of water. Ignite the filter in a tared crucible at 800 ± 25° to constant weight. Each mg of residue is equivalent to 1.0 mg of platinum. Not less than 40% is found.

Potassium Chromate, K_2CrO_4—**194.19**—Use ACS reagent grade.

Potassium Cyanate, KOCN—**81.12**—Colorless crystals or crystalline powder. Soluble in water; very slightly soluble in alcohol.

Caution—This reagent, and the gas evolved on treatment with acid, are extremely poisonous. Perform all tests in a hood with a strong draft, and take precautions to avoid inhaling the fumes. Do not use pipets in measuring its solutions.

Insoluble matter (Reagent test): not more than 1 mg from 20 g, 150 mL of water being used as the solvent (0.005%).

For the following tests, prepare *Solution A* as follows: Dilute the filtrate from the test for *Insoluble matter* with water to a volume of 200 mL.

Chloride—Transfer 10 mL (1 g) of *Solution A* to a 100-mL volumetric flask, add 20 mL of water and 25 mL of diluted sulfuric acid, and evaporate to half the original volume. Replace the water lost, and re-evaporate to half the original volume. Cool, filter, dilute the filtrate with water to 50 mL, and add 2 mL of nitric acid and 1 mL of silver nitrate TS: any turbidity does not exceed that produced by 0.05 mg of chloride (Cl) contained in a similarly treated control (0.005%).

Sulfate (Reagent test, *Method II*)—To 50 mL of *Solution A* add 10 mL of hydrochloric acid, and evaporate on a steam bath to dryness. Warm the residue with 50 mL of water, add 1 mL of hydrochloric acid, filter if necessary, and dilute with water to 100 mL: the solution, 10 mL of barium chloride TS being used, yields not more than 6.1 mg of residue, correction being made for a complete blank test (0.05%).

Iron ⟨241⟩—Treat 10 mL (1 g) of *Solution A* with 3 mL of hydrochloric acid and 2 mL of 30 percent hydrogen peroxide in a covered beaker on the steam bath until the reaction ceases. Remove the cover, and evaporate to dryness. Dissolve the residue in 30 mL of water, and to 10 mL of the solution add 2 mL of hydrochloric acid. Dilute with water to 47 mL: the solution shows not more than 0.01 mg of Fe (0.003%).

Potassium Cyanide, KCN—**65.12**—Use ACS reagent grade.

Potassium Dichromate, $K_2Cr_2O_7$—**294.18**—Use ACS reagent grade.

NOTE—Potassium dichromate of a quality suitable as a primary standard is available from the National Institute of Standards and Technology, Washington, DC 20234, as standard sample No. 136a.

Potassium Ferricyanide, $K_3Fe(CN)_6$—**329.25**—Use ACS reagent grade.

Potassium Ferrocyanide, $K_4Fe(CN)_6 \cdot 3H_2O$—**422.39**—Use ACS reagent grade.

Potassium Hyaluronate—White to cream-colored powder. Freely soluble in water. Store in a tight container, in a refrigerator.

Inhibitor content—Prepare as directed in the *Assay* under *Hyaluronidase for Injection* (USP monograph) a quantity of *Standard solution* containing 1 USP Hyaluronidase Unit in each mL, and a similar quantity of acetate-buffered *Standard solution* using as the solvent 0.1 *M*, pH 6 sodium acetate buffer (prepared by diluting the 0.2 *M* buffer prepared as directed below with an equal volume of water). Prepare from the potassium hyaluronate under test 10 mL of *Potassium hyaluronate stock solution*, and dilute 2 mL of it with the specified *Phosphate buffer solution* to make a *Hyaluronate solution.* In the same way, and concurrently, dilute a second 2-mL portion of the stock solution with 0.2 *M*, pH 6 sodium acetate buffer (containing 16.4 g of anhydrous sodium acetate and 0.45 mL of glacial acetic acid in each 1000 mL).

Place 0.50-mL portions of the *Hyaluronate solution* in each of four 16- × 100-mm test tubes, and place 0.50-mL portions of the acetate-buffered *Hyaluronate solution* in two similar tubes. To two of the four tubes containing *Hyaluronate solution* add 0.50 mL of *Diluent for hyaluronidase solutions*, prepared as directed in the *Assay* under *Hyaluronidase for Injection* (USP monograph). To the remaining two tubes, on a rigid schedule, at 30-second intervals, add 0.50 mL of *Standard solution.* Similarly, to the two tubes containing acetate-buffered *Hyaluronate solution* add at 30-second intervals 0.50-mL portions of acetate-buffered *Standard solution.* Then proceed as directed in the second paragraph under *Procedure*, beginning with "Mix the contents," as far as "Plot the average." The reduction in absorbance of acetate-buffered *Hyaluronate solution* is not less than 25% of that observed in the *Hyaluronate solution.*

Turbidity production—The average absorbance of the solutions in the two tubes containing *Hyaluronate solution* and *Diluent for hyaluronidase solutions* prepared in the test for *Inhibitor content* is not less than 0.26 at a wavelength of 640 nm in a suitable spectrophotometer using a 1-cm cell.

Potassium Hydroxide, KOH—**56.11**—Use ACS reagent grade.

Potassium Iodate, KIO_3—**214.00**—Use ACS reagent grade.

Potassium Iodide, KI—**166.00**—Use ACS reagent grade.

Potassium Nitrate, KNO_3—**101.10**—Use ACS reagent grade.

Potassium Nitrite, KNO_2—**85.10**—Use ACS reagent grade.

Potassium Oxalate, $K_2C_2O_4 \cdot H_2O$—**184.23**—Use ACS reagent grade.

Potassium Perchlorate, $KClO_4$—**138.55**—Use ACS reagent grade.

Potassium Periodate (*Potassium meta-Periodate*), KIO_4—**230.00**—Use ACS reagent grade.

Potassium Permanganate, $KMnO_4$—**158.03**—Use ACS reagent grade.

Potassium Persulfate, $K_2S_2O_8$—**270.31**—Use ACS reagent grade Potassium Peroxydisulfate.

Potassium Phosphate, Dibasic (*Dipotassium Hydrogen Phosphate; Dipotassium Phosphate*), K_2HPO_4—**174.18**—Use ACS reagent grade.

Potassium Phosphate, Monobasic (*Potassium Biphosphate; Potassium Dihydrogen Phosphate*), KH_2PO_4—**136.09**—Use ACS reagent grade.

NOTE—Certified Potassium Dihydrogen Phosphate is available from the National Institute of Standards and Technology, Washington, DC 20234, as standard sample No. 186-I.

Potassium Pyrosulfate—Usually available as a mixture of potassium pyrosulfate ($K_2S_2O_7$) and potassium bisulfate ($KHSO_4$)—Use ACS reagent grade.

Potassium Sodium Tartrate, $KNaC_4H_4O_6 \cdot 4H_2O$—**282.22**—Use ACS reagent grade.

Potassium Sulfate, K_2SO_4—**174.25**—Use ACS reagent grade.

Potassium Tellurite (*Potassium Tellurate IV*), K_2TeO_3—**253.79**—White, granular powder. Soluble in water. Its solution is alkaline.

Assay—Weigh accurately about 120 mg, transfer to a beaker, and dissolve in a mixture of 10 mL of nitric acid, 10 mL of sulfuric acid, and 25 mL of water. Heat to boiling, and boil until copious fumes of sulfur trioxide are evolved. Cool, cautiously add 100 mL of water, heat to boiling, add 6 g of sodium fluoride, and titrate the hot solution with 0.1 *N* potassium permanganate VS. Each mL of 0.1 *N* potassium permanganate is equivalent to 12.69 mg of K_2TeO_3. Not less than 98% is found.

Chloride (Reagent test)—One g shows not more than 0.1 mg of Cl (0.01%).

Potassium Thiocyanate, KSCN—**97.18**—Use ACS reagent grade.

Potassium Xanthogenate (*Potassium Xanthate*), C_2H_5O-CSSK—**160.29**—White or pale yellow crystals, or crystalline powder. Very soluble in water; freely soluble in alcohol. Usually contains about 10% of water. Store in tight containers.

Assay—Weigh accurately about 500 mg, and dissolve in 50 mL of water. Add to the solution 50.0 mL of 0.1 *N* iodine VS, and allow to stand for 5 minutes. Then add 2 mL of glacial acetic acid, and titrate the excess iodine with 0.1 *N* sodium thiosulfate VS, adding 3 mL of starch TS as the end-point is approached: each mL of 0.1 *N* iodine is equivalent to 16.03 mg of C_2H_5OCSSK. Not less than 87% is found.

Insoluble matter—A solution of 1 g in 5 mL of water is complete, or practically so.

Alkalinity—Add phenolphthalein TS to 20 mL of a solution (1 in 20), and titrate with 0.1 *N* sulfuric acid: not more than 2 mL is required to discharge the pink color.

Sulfide—To 5 mL of lead acetate TS add sodium hydroxide solution (1 in 10) until the precipitate first formed redissolves. Add 5 drops of this solution to a solution of 1 g of the test specimen in 20 mL of water: no darkening is produced within 2 minutes.

Potato Starch—See *Starch, Potato.*

Precipitated Manganese Dioxide—See *Manganese Dioxide, Precipitated.*

Procainamide Hydrochloride—Use *Procainamide Hydrochloride* (USP monograph).

Procaine Hydrochloride—Use *Procaine Hydrochloride* (USP monograph).

Progesterone—Use *Progesterone* (USP monograph).

Promazine Hydrochloride, $C_{17}H_{20}N_2S.HCl$—**320.88**—Use USP Promazine Hydrochloride RS.

Propionic Anhydride, $C_6H_{10}O_3$—**130.14**—Colorless liquid, having a pungent odor. Is decomposed by water. Soluble in methanol, in alcohol, in ether, and in chloroform.

Assay—Accurately weigh about 350 mg into a tared, glass-stoppered flask containing 50 mL of dimethylformamide previously neutralized to the thymol blue end-point with 0.1 *N* sodium methoxide in methanol VS. Titrate with 0.1 *N* sodium methoxide in methanol VS to the thymol blue end-point. Perform a blank determination, and make any necessary correction. Each mL of 0.1 *N* sodium methoxide is equivalent to 13.014 mg of $C_6H_{10}O_3$. Not less than 97.0% is found.

Refractive index ⟨831⟩: between 1.4035 and 1.4045 at 20°.

***iso*-Propyl Alcohol**—See *Isopropyl Alcohol.*

***n*-Propyl Alcohol** (*1-Propanol*), $CH_3CH_2CH_2OH$—**60.10**—Clear, colorless liquid, having an ethanol-like odor. Miscible with water and with most organic solvents. Specific gravity: about 0.803.

Boiling range (Reagent test)—Not less than 95% distils between 95° and 98°.

Residue on evaporation—Evaporate 25 mL (20 g) on a steam bath, and dry at 105° for 1 hour: the residue weighs not more than 1 mg (0.005%).

Acidity—Add 0.2 mL of phenolphthalein TS to 20 mL of water, and titrate with 0.1 *N* sodium hydroxide until a slight pink color persists after shaking. Add 10 mL of the alcohol, and titrate with 0.10 *N* sodium hydroxide: not more than 0.20 mL is required to restore the pink color (about 0.015% as CH_3COOH).

Alkalinity—Add 2 drops of methyl red TS to a solution of 6 mL in 25 mL of carbon dioxide-free water, and titrate with 0.02 *N* sulfuric acid: not more than 0.3 mL is required to produce a red color (about 0.002% as NH_3).

Propylene Glycol—Use *Propylene Glycol* (USP monograph).

Propylparaben—Use *Propylparaben* (NF monograph).

Pumice—A substance of volcanic origin consisting chiefly of complex silicates of aluminum and alkali metals. Occurs as very light, hard, rough, porous, gray masses, or as a gray-colored powder. Is insoluble in water and is not attacked by diluted acids.

Acid- and water-soluble substances—Boil 2.0 g of powdered pumice with 50 mL of diluted hydrochloric acid under a reflux condenser for 30 minutes. Cool, and filter. To half of the filtrate add 5 drops of sulfuric acid, evaporate to dryness, ignite, and weigh: the residue weighs not more than 60 mg (6.0%).

Purine, $C_5H_4N_4$—**120.11**—White to off-white powder.

Melting range ⟨741⟩: between 214° and 217°.

A single spot is exhibited when it is examined by thin-layer chromatography, with the use of plates coated with chromatographic silica gel mixture and a developing system consisting of butyl alcohol, water, and glacial acetic acid (60:25:15).

Pyrazoline, $C_3H_6N_2$—**70.09**—Liquid having a faint, amine odor. Miscible with water and with alcohol.

Boiling range (Reagent test): between 143° and 145°.

Pyrene, $C_{16}H_{10}$—**202.26**—White to light yellow crystals.

Assay—Transfer about 9 mg, accurately weighed, to a 100-mL volumetric flask, dissolve in methanol, dilute with methanol to volume, and mix. Transfer 2.0 mL of this solution to a 100-mL volumetric flask, dilute with methanol to volume, and mix. Using a suitable spectrophotometer, 1-cm cells, and methanol as the blank, record the absorbance of the solution at the wavelength of maximum absorbance at about 238 nm. From the observed absorbance, calculate the absorptivity (see *Spectrophotometry and Light-scattering* ⟨851⟩): the absorptivity is not less than 432.9, corresponding to not less than 98% of $C_{16}H_{10}$.

Melting range ⟨741⟩: between 149° and 153° over a 2° range.

Pyridine, C_5H_5N—**79.10**—Use ACS reagent grade.

Pyridine, Dried—Use ACS reagent grade.

Pyridoxal Hydrochloride, $C_8H_9NO_3.HCl$—**203.62**—White to slightly yellow crystals or crystalline powder. Gradually darkens on exposure to air or sunlight. One g dissolves in about 2 mL of water and in about 25 mL of alcohol. Insoluble in acetone, in chloroform, and in ether. Its solutions are acid (pH about 3).

Melting range ⟨741⟩: between 171° and 175° with some decomposition.

Residue on ignition (Reagent test): not more than 0.1%.

Loss on drying ⟨731⟩—Dry it at 105° for 2 hours: it loses not more than 0.5% of its weight.

Nitrogen content (Reagent test)—Determine by the Kjeldahl method, using a test specimen previously dried at 105° for 2 hours: between 6.7% and 7.1% of N is found.

Chloride content—Weigh accurately about 500 mg, previously dried at 105° for 2 hours, and dissolve in 50 mL of water. Add 3 mL of nitric acid and 50.0 mL of 0.1 *N* silver nitrate VS, then add 5 mL of nitrobenzene, shake for about 2 minutes, add ferric ammonium sulfate TS, and titrate the excess silver nitrate with 0.1 *N* ammonium thiocyanate VS: each mL of 0.1 *N* silver nitrate is equivalent to 3.545 mg of Cl. Between 17.2% and 17.7% is found.

Pyridoxal 5-phosphate, 4-$CHOC_5HN$-2-CH_3, 3-OH, 5-CH_2-$PO_4H_2.H_2O$—**265.16**—Light yellow powder.

Assay—Transfer about 500 mg, accurately weighed, to a suitable flask. Add 20.0 mL of 0.5 *N* sodium hydroxide VS and 130 mL of water and heat under reflux for 1 hour. Cool, transfer the solution to a 250-mL beaker, rinse the flask with about 30 mL of water and add the rinsing to the beaker. Titrate the solution with 0.5 *N* hydrochloric acid VS, determining the first end-point potentiometrically. Perform a blank determination and make any necessary correction. Each mL of 0.1 *N* sodium hydroxide consumed is equivalent to 88.39 mg of $C_8H_{10}NO_6P.H_2O$: not less than 95% is found.

Melting range ⟨741⟩: between 140° and 143° with decomposition.

Water, Method I ⟨921⟩: between 8.5% and 9.5%.

Pyridoxamine Dihydrochloride, $C_8H_{12}N_2O_2.2HCl$—**241.12**—White to slightly yellow crystals or crystalline powder. Gradually darkens on exposure to air or sunlight. One g dissolves in about 1 mL of water and in about 60 mL of alcohol. Insoluble in chloroform and in ether. Its solutions are acid.

Melting range ⟨741⟩: between 225° and 230°, with some decomposition.

Residue on ignition (Reagent test): not more than 0.15%.

Loss on drying ⟨731⟩—Dry it at 105° for 2 hours: it loses not more than 0.5% of its weight.

Nitrogen content (Reagent test)—Determine by the Kjeldahl method, using a test specimen previously dried at 105° for 2 hours: between 11.3% and 11.8% of N is found.

Chloride content—Determine as directed in the test for *Chloride content* under *Pyridoxal Hydrochloride:* between 29.1% and 29.6% of Cl is found.

Pyridoxine Hydrochloride—Use *Pyridoxine Hydrochloride* (USP monograph).

1-(2-Pyridylazo)-2-naphthol, $C_{15}H_{11}N_3O$—**249.27**—Stable, orange-red crystals. Slightly soluble in water; soluble in alcohol and in hot solutions of dilute alkalies.

Melting range ⟨741⟩: between 140° and 142°.

Sensitiveness—Add 0.1 mL of a 1 in 1000 solution of it in alcohol to a mixture of 10 mL of water and 1 mL of a buffer solution prepared by mixing 80 mL of 0.2 *M* acetic acid and 20 mL of sodium acetate solution (8.2 in 100), and mix. To this solution add 1 mL of a mixture of 1 mL of cupric sulfate TS and 2 mL of water, and mix: the color changes from yellow to red.

Pyrilamine Maleate—Use *Pyrilamine Maleate* (USP monograph).

Pyrocatechol Violet (*Pyrocatecholsulfonphthalein*), $C_{19}H_{14}O_7S$—**386.38**—Reddish brown powder. Freely soluble in water; sparingly soluble in alcohol. In the pH range between 2 and 3, its solution is blue in the presence of bismuth ion and yellow in the presence of excess disodium ethylenediaminetetraacetate.

Pyrogallol, $C_6H_3(OH)_3$—**126.11**—Use ACS reagent grade.

Pyrrole, C_4H_5N—**67.09**—Clear liquid, colorless when freshly distilled, becoming yellow in a few days. Has a characteristic odor. Specific gravity: about 0.94. Insoluble in water; soluble in alcohol, in benzene, and in ether.

Boiling range (Reagent test)—Not less than 90% distils between 128° and 132°.

Pyruvic Acid, $CH_3COCOOH$—**88.06**—Colorless to light yellow liquid. Refractive index: about 1.43 at 20°. Miscible with water, with alcohol, and with ether.

Assay—Weigh accurately about 1 g, transfer to a suitable container, and add 100 mL of water. Mix, add phenolphthalein TS, and titrate with 0.5 *N* sodium hydroxide VS. Each mL of 0.5 *N* sodium hydroxide is equivalent to 44.03 mg of $CH_3COCOOH$: not less than 98.5% of $CH_3COCOOH$ is found.

Quantitative Filter Paper—See *Filter Paper, Quantitative.*

Quinhydrone, $C_6H_4(OH)_2 . C_6H_4O_2$—**218.21**—Green crystals having a metallic luster. Slightly soluble in cold water; soluble in hot water, in alcohol, and in ether.

Assay—Transfer about 450 mg, accurately weighed, to a glass-stoppered flask, add 50 mL of 1 *N* sulfuric acid and 3 g of potassium iodide, insert the stopper in the flask, and shake until dissolved. Titrate the liberated iodine with 0.1 *N* sodium thiosulfate VS, adding 3 mL of starch TS as the end-point is approached. Each mL of 0.1 *N* sodium thiosulfate is equivalent to 5.405 mg of quinone ($C_6H_4O_2$). Between 49.0% and 51.0% is found.

Alcohol-insoluble matter—Dissolve 10 g in 100 mL of hot alcohol, filter through a suitable tared crucible of fine porosity, and wash with hot alcohol until the last washing is colorless. Dry at 105°, cool in a desiccator, and weigh: the residue weighs not more than 1.0 mg (0.010%).

Residue on ignition (Reagent test): not more than 0.050%, a 2.0-g test specimen being used. Save the residue.

Sulfate—Transfer 1 g to a platinum crucible, add 10 mL of hot water and 0.5 g of sodium carbonate, evaporate to dryness, and ignite, protected from the sulfur in the flame, until the residue is nearly white. Cool, add 20 mL of water and 1 mL of 30 percent hydrogen peroxide, boil gently for a few minutes, add 2 mL of hydrochloric acid, and evaporate on a steam bath to dryness. Cool, dissolve the residue in 20 mL of water, filter, and to the filtrate add 1 mL of 1 *N* hydrochloric acid and 3 mL of barium chloride TS: any turbidity produced within 10 minutes does not exceed that in a control containing 0.2 mg of added SO_4 and 0.5 mg of sodium carbonate, 1 mL of 30 percent hydrogen peroxide, and 2 mL of hydrochloric acid previously evaporated on a steam bath to dryness (0.02%).

Heavy metals—To the residue retained from the test for *Residue on ignition* add 2 mL of hydrochloric acid and 0.5 mL of nitric acid, and evaporate on a steam bath to dryness. Dissolve the residue in 30 mL of hot water containing 1 mL of 1 *N* hydrochloric acid, cool, dilute with water to 40 mL, and mix. Dilute 20 mL of this solution (retain the rest of the solution) with water to 25 mL, adjust to a pH between 3.0 and 4.0 by the addition of 1 *N* acetic acid or 6 *N* ammonium hydroxide as necessary, dilute with water to 40 mL, and add 10 mL of freshly prepared hydrogen sulfide TS: any brown color produced does not exceed that in a control containing 0.02 mg of added Pb (0.002%).

Iron ⟨241⟩—To 10 mL of the solution retained from the test for *Heavy metals* add 2 mL of hydrochloric acid, and dilute with water to 47 mL: the solution shows not more than 0.01 mg of Fe (0.002%).

Quinidine Sulfate—Use *Quinidine Sulfate* (USP monograph).

Quinine Sulfate—Use *Quinine Sulfate* (USP monograph).

Quinone—See *p-Benzoquinone.*

Red Mercuric Iodide—See *Mercuric Iodide, Red.*

Red Phosphorus—See *Phosphorus, Red.*

Resazurin (Sodium), $C_{12}H_6NNaO_4$—**251.17**—A brownish purple, crystalline powder. One g dissolves in 100 mL of water, forming a deep violet-colored solution.

Hydrogen sulfide and other compounds containing the thiol group decolorize solutions of resazurin sodium, forming dihydroresorufin. When the decolorized solution is shaken in the presence of air, a rose color develops as a result of the formation of resorufin.

Resorcinol—Use *Resorcinol* (USP monograph).

Rhodamine B (*Tetraethylrhodamine*), $C_{28}H_{31}ClN_2O_3$—**479.02**—Green crystals or a reddish violet powder. Very soluble in water, yielding a bluish red solution that is strongly fluorescent when dilute. Very soluble in alcohol; slightly soluble in dilute acids and in alkali solutions. In strong acid solution, it forms a pink complex with antimony that is soluble in isopropyl ether.

Clarity of solution—Its solution (1 in 200) is complete and clear.

Residue on ignition (Reagent test)—Ignite 1 g with 1 mL of sulfuric acid: the residue weighs not more than 2 mg (0.2%).

Riboflavin—Use *Riboflavin* (USP monograph).

Rose Bengal Sodium (*Disodium Salt of 4,5,6,7-Tetrachloro-2',4',5',7'-tetraiodofluorescein*), $C_{20}H_2Cl_4I_4Na_2O_5$—**1017.64**—Fine, rose-colored crystals or crystalline powder. Is practically odorless. Soluble in water.

NOTE—Render commercially available material suitably pure by the following treatment:

Dissolve 8 g in 200 mL of water, and adjust to a pH between 10 and 11, using short-range pH indicator paper. Add 200 mL of acetone, while stirring gently, then add dilute hydrochloric acid (1 in 10), while continuing to stir, until the pH of the solution reaches 4.0. Add 400 mL more of water, with stirring, and continue the stirring for 5 minutes. Filter the crystals on a filtering funnel, and return the crystals to the beaker used for crystallization. Recrystallize three more times in the same manner, and dry the crystals at 110° for 12 hours. Store in an amber bottle in a refrigerator at a temperature between 2° and 8°. Prepare this reagent fresh monthly.

Chromatographic purity—Dissolve 100 mg of rose bengal sodium, prepared as described above, in 100 mL of water, and apply 10 µL of the solution on suitable chromatographic paper.[65] Develop the chromatogram by ascending chromatography, using a mixture of 1 part of dilute alcohol (1 in 4) and 1 part of dilute stronger ammonia water (1 in 12). Examine the chromatogram in daylight and under ultraviolet light (360 nm): no colored or fluorescent spot is visible other than the rose bengal sodium spot.

Ruthenium Red (*Ruthenium Oxychloride, Ammoniated*), $Ru_2(OH)_2Cl_4 . 7NH_3 . 3H_2O$—**551.32**—A brownish red to dark purple powder. Soluble in water.

S1A; S1AB; S1C; S1NS; S2; S3; S4; S5; S6; S7; S8; S9; S10; S11—See *Supports for Gas Chromatography.*

Saccharose—Use *Sucrose* (NF monograph).

Safranin O—Dark red powder consisting of a mixture of 3,7-diamino-2,8-dimethyl-5-phenylphenazinium chloride, $C_{20}H_{19}ClN_4$—**350.85**, and 3,7-diamino-2,8-dimethyl-5-*o*-tolylphenazinium chloride, $C_{21}H_{21}ClN_4$—**364.88**—Sparingly soluble in 70 percent alcohol yielding a clear red solution with a yellowish red fluorescence.

Identification—

A: To 10 mL of a 0.5% w/v solution add 5 mL of hydrochloric acid: a bluish violet solution is produced.

B: To 10 mL of a 0.5% w/v solution add 5 mL of sodium hydroxide solution (1 in 5): a brownish red precipitate is produced.

C: To 100 mg add 5 mL of sulfuric acid: a green solution is produced, which, on dilution, changes to blue and finally to red.

Absorption characteristics—Dissolve 50 mg in 250 mL of 50 percent alcohol. Dilute 3 mL of this solution with 50 percent alcohol to 200 mL. Determine the absorbance, in a 1-cm cell, with a suitable spectrophotometer. The absorbance maximum is in the range of 530 to 533 nm; the ratio $(P - 15)/(P + 15)$ is between 1.10 and 1.32, in which *P* is the wavelength of maximum absorbance.[66]

Salicylaldehyde, 2-HOC_6H_4CHO—**122.12**—Clear, colorless to yellowish green liquid. Specific gravity: about 1.17. Slightly soluble in water; soluble in alcohol and in ether. May contain a stabilizer.

Congealing temperature ⟨651⟩: between 1.0° and 3.0°.

Refractive index ⟨831⟩: between 1.573 and 1.574 at 20°.

Assay—When examined by gas–liquid chromatography, using suitable apparatus and conditions, it shows a purity of not less than 98%.

Salicylaldazine, $C_{14}H_{12}N_2O_2$—**240.26**—Use a suitable grade or prepare as follows. Dissolve 300 mg of hydrazine sulfate in 5 mL of water, add 1 mL of glacial acetic acid and 2 mL of a freshly prepared 1 in 5 solution of salicylaldehyde in isopropyl alcohol, mix, and allow to stand until a yellow precipitate is formed. Extract the mixture with two 15-mL portions of methylene chloride. Combine the methylene chloride extracts, and dry over anhydrous sodium sulfate. Decant the methylene chloride solution, and evaporate it to dryness. Recrystallize the residue of salicylaldazine from a mixture of warm toluene and methanol (60:40) with cooling. Filter, and dry the crystals in vacuum.

Melting range ⟨741⟩: between 213° and 219°, but the range between beginning and end of melting does not exceed 1°.

Salicylamide—Use *Salicylamide* (USP monograph).

Salicylic Acid—Use *Salicylic Acid* (USP monograph).

Sand, Washed—It may be prepared as follows. Digest clean, hard sand at room temperature with a mixture of 1 part of hydrochloric acid and 2 parts of water (about 13% of HCl) for several days, or at an elevated temperature for several hours. Collect the sand on a filter, wash with water until the washings are neutral and show only a slight reaction for chloride, and finally dry. Washed sand meets the following tests.

Substances soluble in hydrochloric acid—Digest 10 g with a mixture of 10 mL of hydrochloric acid and 40 mL of water on a steam bath for 4 hours, replacing from time to time the water lost by evaporation. Filter, and to 25 mL of the filtrate add 5 drops of sulfuric acid, evaporate, and ignite to constant weight: the residue weighs not more than 8 mg (0.16%).

Chloride (Reagent test)—Shake 1 g with 20 mL of water for 5 minutes, filter, and add to the filtrate 1 mL of nitric acid and 1 mL of silver nitrate TS: any turbidity produced corresponds to not more than 0.03 mg of Cl (0.003%).

Sawdust, Purified—It may be prepared as follows. Extract sawdust in a percolator, first with sodium hydroxide solution (1 in 100), and then with dilute hydrochloric acid (1 in 100) until the acid percolate gives no test for alkaloid with mercuric–potassium iodide TS or with iodine TS. Then wash with water until free from acid and soluble salts, and dry. Purified sawdust meets the following test.

Alkaloids—To 5 g of purified sawdust contained in a flask add 50 mL of a mixture of 2 volumes of ether and 1 volume of chloroform and 10 mL of ammonia TS, and shake frequently for 2 hours. Decant 20 mL of the clear, ether-chloroform liquid, and evaporate to dryness. Dissolve the residue in 2 mL of dilute hydrochloric acid (1 in 12), and divide into two portions. To 1 portion add mercuric–potassium iodide TS, and to the other add iodine TS: no turbidity is produced in either portion.

Secondary Butyl Alcohol—See *Butyl Alcohol, Secondary.*

Selenious Acid (*Selenous Acid*), H_2SeO_3—**128.97**—Colorless or white crystals, efflorescent in dry air and hygroscopic in moist air. Soluble in water and in alcohol.

Assay—Weigh accurately about 100 mg, transfer to a glass-stoppered flask, and dissolve in 50 mL of water. Add 10 mL of potassium iodide solution (3 in 10) and 5 mL of hydrochloric acid, mix, insert the stopper in the flask, and allow to stand for 10 minutes. Dilute with 50 mL of water, add 3 mL of starch TS, and titrate with 0.1 *N* sodium thiosulfate VS until the color is no longer diminished, then titrate with 0.1 *N* iodine VS to a blue color. Subtract the volume of 0.1 *N* iodine solution from the volume of 0.1 *N* sodium thiosulfate to give the volume of 0.1 *N* thiosulfate equivalent to selenious acid. Each mL of 0.1 *N* sodium thiosulfate is equivalent to 3.225 mg of H_2SeO_3: not less than 93% is found.

Insoluble matter—Dissolve 1 g in 5 mL of water: the solution is clear and complete.

Residue on ignition (Reagent test): not more than 1.0 mg (0.01%), from 10 g.

Selenate and sulfate—Dissolve 500 mg in 10 mL of water, and add 0.1 mL of hydrochloric acid and 1 mL of barium chloride TS: no turbidity or precipitate is formed within 10 minutes.

Selenium, Se—At. Wt. **78.96**—Dark-red amorphous, or bluish black crystalline, powder. Insoluble in water; soluble in solutions of sodium and potassium hydroxides or sulfides.

Residue on ignition (Reagent test)—One g yields not more than 2 mg (0.2%).

Heavy metals—To the *Residue on ignition* add 3 mL of hydrochloric acid and 2 mL of nitric acid, evaporate on a steam bath to dryness, take up the residue in a mixture of 2 mL of diluted hydrochloric acid and 50 mL of hot water, cool, filter, and wash the filter with sufficient water to make 100 mL of filtrate (*Test Solution*). To a 30-mL aliquot of the *Test Solution* add 10 mL of water and 10 mL of hydrogen sulfide TS: the color produced is not darker than that of a *Control Solution* prepared from 3 mL of *Standard Lead Solution* (see *Heavy Metals* ⟨231⟩; 0.03 mg of Pb), 0.2 mL of 1 *N* hydrochloric acid, 37 mL of water, and 10 mL of hydrogen sulfide TS (0.01%).

Iron ⟨241⟩—To 20 mL of the *Test Solution* prepared in the test for *Heavy metals* add 2 mL of hydrochloric acid, and dilute with water to 47 mL: the solution shows not more than 0.01 mg of Fe (0.005%).

Nitrogen—

STANDARD NITROGEN SOLUTION—Dissolve 382 mg of ammonium chloride in water to make 1 liter. Each mL of this solution contains the equivalent of 0.1 mg of nitrogen (N).

PROCEDURE—Heat 1.0 g with 10 mL of sulfuric acid in a Kjeldahl flask until the test specimen is dissolved and the volume of acid is reduced to about 5 mL. Cool, cautiously dilute with 100 mL of water, render strongly alkaline with sodium hydroxide solution (3 in 10), and distil about 75 mL of the solution into 5 mL of water containing 2 drops of 1 *N* hydrochloric acid. Dilute the distillate with water to 250 mL. To a 50-mL aliquot of the solution add 1 mL of sodium hydroxide solution (1 in 10) and 2 mL of mercuric-potassium iodide TS: the color produced is not darker than that produced by 0.1 mL of *Standard nitrogen solution* (0.01 mg of N) treated in the same manner as the test specimen (0.005%).

Sulfur—To 1.0 g in a beaker add, successively, 5 mL of nitric acid, then 10 mL of hydrochloric acid, and evaporate on a steam bath to dryness. Add 10 mL of hydrochloric acid, and slowly evaporate again to dryness. Take up the residue in 30 mL of dilute hydrochloric acid (1 in 30), filter, and wash the filter with water to make about 100 mL of filtrate. Heat the filtrate to boiling, and add slowly, with stirring, 5 mL of barium chloride TS. Digest on the steam bath for 4 hours. Filter on a fine-porosity filter paper, wash the precipitate until it is free from chloride, ignite, and weigh. The weight of the barium sulfate residue, multiplied by 0.1374, represents sulfur (S). Not more than 0.5 mg of S is found (0.05%).

Selenomethionine, $C_5H_{11}NO_2Se$—**196.10**—[*Caution—Handle with care, as this reagent is highly toxic.*]

Assay—Weigh accurately about 750 mg, dissolve in 100 mL of methanol, add crystal violet TS, and titrate with 0.1 *N* perchloric acid to a blue-green end-point. Each mL of 0.1 *N* perchloric acid is equivalent to 19.61 mg of $C_5H_{11}NO_2Se$: between 97.0% and 103.0%, calculated on the as-is basis, is found.

Melting range ⟨741⟩: about 260°, with decomposition.

Nitrogen content ⟨461⟩—Determine by the Kjeldahl method: between 6.8% and 7.4%, calculated on the as-is basis, is found.

Semicarbazide Hydrochloride, $NH_2CONHNH_2 . HCl$—**111.53**—White to faintly yellow crystals or crystalline powder. Freely soluble in water; sparingly soluble in alcohol.

Melting range ⟨741⟩: between 181° and 184°.

Solubility—A solution of 1 g in 20 mL of water is clear and colorless.

Residue on ignition (Reagent test): not more than 0.2%.

Sesame Oil—Use *Sesame Oil* (NF monograph).

Silica, Calcined Diatomaceous—See *Diatomaceous Silica, Calcined.*

Silica, Chromatographic, Silanized, Flux-calcined, Acid-washed—Use a suitable grade.[67]

Silica Gel—An amorphous, partly hydrated SiO_2 occurring in glassy granules of varying size. When used as a desiccant, it frequently is coated with a substance that changes color when the capacity to absorb water is exhausted. Such colored products may be regenerated (i.e., may regain their capacity to absorb water) by being heated at 110° until the gel assumes the original color.

NOTE—The following procedures and limits are designed only for use in testing the desiccant grade of silica gel.

Loss on ignition—Ignite 2 g, accurately weighed, at 950 ± 50° to constant weight: it loses not more than 6.0% of its weight.

Water absorption—Place about 10 g in a tared weighing bottle, and weigh. Then place the bottle, with cover removed, for 24 hours in a closed container in which the atmosphere is maintained at 80% relative humidity by being in equilibrium with sulfuric acid having a specific gravity of 1.19. Weigh again: the increase in weight is not less than 31.0% of the weight of test specimen.

Silica Gel, Binder-free—Silica gel for chromatographic use formulated without a binder, since only activated forms of the silica gel are used as the binding agent.[65]

Silica Gel, Chromatographic—Use a suitable grade.[68]

Silica Gel–Impregnated Glass Microfiber Sheet—Use a suitable grade.[69]

Silica Gel Mixture, Chromatographic—A mixture of silica gel with a suitable fluorescing substance.[70]

Silica Gel Mixture, Dimethylsilanized, Chromatographic—Use a suitable grade.[71]

Silica Gel Mixture, Octadecylsilanized Chromatographic—Use a suitable grade.[72]

Silica Gel, Octadecylsilanized Chromatographic—Use a suitable grade.[73]

Silica Gel, Pellicular—Use a suitable grade.[74]

Silica Gel, Porous—Use a grade suitable for high-pressure liquid chromatography.[75]

Silica Gel, Prepurified Chromatographic[76]—A mixture of silica gel with hydrated silicon dioxide as a binder.

Silica Microspheres—Use a suitable grade.[77]

Siliceous Earth, Chromatographic[78]—

For gas chromatography, use a specially prepared grade meeting the following general description: Purified siliceous earth of suitable mesh size that has been acid- and/or base-washed. It may or may not be silanized.[79]

For column partition chromatography,[80] it is essential that the material be free from interfering substances. If such interferences are known or thought to be present, purify the material as follows: Place a pledget of glass wool in the base of a chromatographic column having a diameter of 100 mm or larger, and add *Purified Siliceous Earth* (NF monograph) to a height equal to 5 times the diameter of the column. Add a volume of hydrochloric acid equivalent to one-third the volume of siliceous earth, and allow the acid to percolate into the column. Wash the column with methanol, using small volumes at first to rinse the walls of the column, and continue washing with methanol until the last washing is neutral to moistened litmus paper. Extrude the washed column into shallow dishes, heat on a steam bath to remove the excess methanol, and dry at 105° until the material is powdery and free from traces of methanol. Store the dried material in well-closed containers.

Siliceous Earth, Chromatographic, Silanized—Place about 450 g of purified siliceous earth in a large, open, glass crystallizing dish in a vacuum desiccator containing 30 mL of a suitable silane, e.g., a mixture of 1 volume of dimethyldichlorosilane and 1 volume of trimethylchlorosilane, or a mixture of 1 volume of methyltrichlorosilane and 2 volumes of dimethyldichlorosilane. Apply vacuum intermittently for several hours, until no liquid silane remains. Float the treated purified siliceous earth on water, and gently agitate to allow any uncoated particles to sink. Skim the silanized material off the surface, wash it on a sintered-glass funnel with warm methanol until the filtrate no longer is acidic, and dry at 110°.

Silicic Acid, $SiO_2 . xH_2O$—(anhydrous) **60.08**—White, amorphous powder. Insoluble in water and in acids; soluble in hot solutions of strong alkalies.

Residue on ignition (Reagent test): not less than 80.0%.

Nonvolatile with hydrofluoric acid—Heat 500 mg with 1 mL of sulfuric acid and 10 mL of hydrofluoric acid in a platinum crucible to dryness, and ignite to constant weight: the weight of the residue does not exceed 1.0 mg (0.2%).

Chloride (Reagent test)—One g shows not more than 0.05 mg of Cl (0.005%).

Sulfate (Reagent test)—Boil 2 g with 20 mL of dilute hydrochloric acid (1 in 40), filter, neutralize the filtrate with ammonia TS, and dilute with water to 20.0 mL. A 10-mL aliquot of the solution shows not more than 0.1 mg of SO_4 (0.01%).

Heavy metals (Reagent test)—Boil 2.5 g with 50 mL of dilute hydrochloric acid (1 in 10) for 5 minutes, filter while hot, and evaporate the filtrate on a steam bath to dryness. Take up the residue in 20 mL of dilute hydrochloric acid (1 in 500), digest for 5 minutes, cool, add water to make 100 mL, and filter. To 40 mL of the filtrate add 10 mL of hydrogen sulfide TS: any color produced is not darker than that produced by adding 10 mL of hydrogen sulfide TS to a control containing 0.03 mg of Pb (0.003%).

Iron ⟨241⟩—To 20 mL of the filtrate obtained in the test for *Heavy metals* add 1 mL of hydrochloric acid, and dilute with water to 47 mL: the solution shows not more than 0.015 mg of Fe (0.003%).

Silicic Acid—Impregnated Glass Microfilament Sheets with Fluorescent Indicator—Use a suitable grade.[81]

Silicon Carbide, SiC—**40.10**—In small clean chips, suitable for use in promoting ebullition.

Silicone Fluid—Use a suitable grade.

Silicone Gum—Use a suitable grade.[82]

Silicone Nitrile Gum, 2 Percent—Use a suitable grade.[83]

Silicone (75 Percent Phenyl, Methyl)—Use a suitable grade.[84]

Silver Diethyldithiocarbamate, $(C_2H_5)_2NCS_2Ag$—**256.13**—Use ACS reagent grade.

Silver Nitrate, $AgNO_3$—**169.87**—Use ACS reagent grade.

Silver Oxide, Ag_2O—**231.74**—Brownish black, heavy odorless powder. Slowly decomposes on exposure to light. Absorbs carbon dioxide when moist. Practically insoluble in water; freely soluble in dilute nitric acid and in ammonia; insoluble in alcohol. Store in well-closed containers; do not expose to ammonia fumes or easily oxidizable substances.

Assay—Dissolve about 500 mg, previously dried at 120° for 3 hours and accurately weighed, in a mixture of 20 mL of water and 5 mL of nitric acid. Dilute with 100 mL of water, add 2 mL of ferric ammonium sulfate TS, and titrate with 0.1 *N* ammonium thiocyanate VS to a permanent reddish brown color. Each mL of 0.1 *N* ammonium thiocyanate is equivalent to 11.59 mg of Ag_2O: not less than 99.7% of Ag_2O is found.

Loss on drying—Dry it at 120° for 3 hours: it loses not more than 0.25% of its weight.

Nitrate—To 500 mg add 30 mg of sodium carbonate and 2 mL of phenoldisulfonic acid TS, mix, and heat on a steam bath for 15 minutes. Cool, *cautiously* add 20 mL of water, render alkaline with ammonia TS, and dilute with water to 30 mL: any color produced by the test solution is not darker than that produced in a control containing 0.01 mg of NO_3 (0.002%).

Substances insoluble in nitric acid—Dissolve 5 g in a mixture of 5 mL of nitric acid and 10 mL of water, dilute with water to about 65 mL, and filter any undissolved residue on a tared filtering crucible (retain the filtrate for the test for *Substances not precipitated by hydrochloric acid*). Wash the crucible with water until the last washing shows no opalescence with 1 drop of hydrochloric acid, and dry at 105° to constant weight: the residue weighs not more than 1 mg (0.02%).

Substances not precipitated by hydrochloric acid—Dilute the filtrate obtained in the test for *Substances insoluble in nitric acid* with water to 250 mL, heat to boiling, and add, dropwise, sufficient hydrochloric acid to precipitate all of the silver (about 5 mL), avoiding any great excess. Cool, dilute with water to 300 mL, and allow to stand overnight. Filter, evaporate 200 mL of the filtrate in a suitable tared porcelain dish to dryness, and ignite: the residue weighs not more than 1.7 mg (0.05%).

Alkalinity—Heat 2 g with 40 mL of water on a steam bath for 15 minutes, cool, and dilute with water to 50 mL. Filter, discarding the first 10 mL of the filtrate. To 25 mL of the subsequent filtrate add 2 drops of phenolphthalein TS, and titrate with 0.02 *N* hydrochloric acid VS to the disappearance of any pink color: not more than 0.20 mL is required (0.016% as NaOH).

Soda Lime—Use *Soda Lime* (NF monograph).

Sodium, Na—At. Wt. **22.98977**—Use ACS reagent grade.

Sodium Acetate, $NaC_2H_3O_2 . 3H_2O$—**136.08**—Use ACS reagent grade.

Sodium Acetate, Anhydrous, $NaC_2H_3O_2$—**82.03**—Grayish white masses or powder. Hygroscopic. Freely soluble in water.

Loss on drying ⟨731⟩—Dry it at 120° to constant weight: it loses not more than 3.0% of its weight.

Neutrality—Dissolve 5 g in 100 mL of carbon dioxide-free water, cool to 10°, and add phenolphthalein TS. If a pink color is produced, it is discharged by the addition of not more than 0.50 mL of 0.020 *N* hydrochloric acid. If no pink color is produced, the addition of 0.50 mL of 0.020 *N* sodium hydroxide produces a pink color (about 0.02% of alkali as Na_2CO_3 or about 0.012% of acid as CH_3COOH).

Chloride (Reagent test)—One g shows not more than 0.1 mg of Cl (0.01%).

Heavy metals (Reagent test): 0.0015%.

Sulfate (Reagent test, *Method I*)—One g shows not more than 0.2 mg of SO_4 (0.02%).

Sodium Alizarinsulfonate (*Alizarin Red S; Alizarin Sodium Monosulfonate*), $C_{14}H_7NaO_7S.H_2O$—**360.27**—Yellow-brown or orange-yellow powder. Freely soluble in water, with production of a yellow color; sparingly soluble in alcohol.

Sensitiveness—Add 3 drops of a solution of it (1 in 100) to 100 mL of water, and add 0.25 mL of 0.02 *N* sodium hydroxide: a red color is produced. Add 0.25 mL of 0.02 *N* hydrochloric acid: the original yellow color returns.

Sodium Ammonium Phosphate (*Microcosmic Salt*), $NaNH_4HPO_4.4H_2O$—**209.07**—Colorless crystals or white granules. Freely soluble in water; insoluble in alcohol. Effloresces in air and loses ammonia.

Insoluble matter and ammonium hydroxide precipitate—Dissolve 10 g in 100 mL of water, add 10 mL of ammonia TS, and heat on a steam bath for 1 hour. If any precipitate is formed, filter, wash well with water, and ignite: the ignited precipitate weighs not more than 1 mg (0.01%).

Chloride (Reagent test)—One g shows not more than 0.02 mg of Cl (0.002%).

Heavy metals—Dissolve 3 g in 25 mL of water, add 15 mL of 1 *N* sulfuric acid, then add 10 mL of hydrogen sulfide TS: any brown color developed in 1 minute is not darker than that of a control containing 3 mL of *Standard Lead Solution* (see ⟨231⟩) and 0.5 mL of 1 *N* sulfuric acid (0.001%).

Nitrate—Dissolve 1 g in 10 mL of water, add 0.1 mL of indigo carmine TS, then add, with stirring, 10 mL of sulfuric acid: the blue color persists for 10 minutes (about 0.005%).

Sulfate (Reagent test, *Method II*)—Dissolve 10 g in 100 mL of water, add 5 mL of hydrochloric acid, and filter if necessary: the filtrate yields not more than 5 mg of residue (0.02%).

Sodium Arsenite, $NaAsO_2$—**129.91**—White, crystalline, odorless powder. Soluble in water; slightly soluble in alcohol.

Assay—Transfer about 5.5 g, accurately weighed, to a 500-mL volumetric flask, dissolve in water, dilute with water to volume, and mix. Pipet 25 mL of this solution into a suitable container, add 50 mL of water and 5 g of dibasic sodium phosphate, swirl to dissolve, and titrate with 0.1 *N* iodine VS, adding 3 mL of starch TS as the end-point is approached. Each mL of 0.1 *N* iodine is equivalent to 3.746 mg of As. Between 57.0% and 60.5% is found (equivalent to 98.8% to 104.9% of $NaAsO_2$).

Chloride (Reagent test)—One g shows not more than 0.10 mg of Cl (0.01%).

Heavy metals—Dissolve 200 mg in 8 mL of dilute hydrochloric acid (3 in 8), and evaporate on a steam bath to dryness. Dissolve the residue in 5 mL of dilute hydrochloric acid (2 in 5), and again evaporate to dryness. Dissolve the residue in 10 mL of water, and add 2 mL of diluted acetic acid and 10 mL of hydrogen sulfide TS. Any brown color produced is not darker than that of a control containing 0.01 mg of added Pb (0.005%).

Iron—Dissolve 1 g in 20 mL of dilute hydrochloric acid (1 in 5), and add, dropwise, a slight excess of bromine TS. Boil the solution to remove the excess bromine, cool, dilute with water to 40 mL, and add 10 mL of ammonium thiocyanate solution (3 in 10). Any red color produced is not darker than that of a control containing 0.02 mg of added Fe (0.02%).

Sulfide—Dissolve 1 g in 20 mL of water, and add 5 drops of lead acetate TS: no brown color is produced (about 0.0005%).

Sulfate (Reagent test, *Method II*)—Dissolve 5 g in 100 mL of water, add methyl orange TS, neutralize with 1 *N* hydrochloric acid, add 3 mL of the acid in excess, and filter: the filtrate yields not more than 3 mg of residue (0.02%).

Sodium Bicarbonate, $NaHCO_3$—**84.01**—Use ACS reagent grade.

Sodium Biphenyl, $C_{12}H_9Na$—**176.19**—Supplied as a solution (10 percent to 30 percent, w/w) in a mixture of dimethoxyethane and toluene or xylene. The solution is a viscous, dark green liquid. [NOTE—The solution deteriorates at a rate of about 10% per month. Use only freshly prepared solution.]

Activity—Place 20 mL of dry toluene in a titration flask equipped with a magnetic stirring bar and a stopper having a hole through which the delivery tip of a weight buret may be inserted. Add a quantity of sodium biphenyl sufficient to produce a blue color in the mixture, and titrate with amyl alcohol, contained in a weight buret, to the disappearance of the blue color. (Disregard the amounts of sodium biphenyl and amyl alcohol used in this adjustment.) Weigh accurately the weight buret containing the amyl alcohol. Transfer the contents of a vial of well-mixed test specimen to the titration flask, and titrate quickly with the amyl alcohol to the disappearance of the blue color. Weigh the buret to determine the weight of amyl alcohol consumed, and calculate the activity, in mEq per vial, by the formula:

$$11.25W,$$

in which W is the weight of amyl alcohol consumed. Not less than 10% activity is found.

Iodine content—Add 10 mL to 5 mL of toluene contained in a 125-mL separator fitted with a suitable inert plastic stopcock, and shake vigorously for 2 minutes. Extract gently with three 10-mL portions of dilute phosphoric acid (1 in 3), combining the lower phases in a 125-mL iodine flask. Add sodium hypochlorite TS, dropwise, to the combined extracts until the solution turns brown, then add 0.5 mL in excess. Shake intermittently for 3 minutes, add 5 mL of freshly prepared, saturated phenol solution, mix, and allow to stand for 1 minute, accurately timed. Add 1 g of potassium iodide, shake for 30 seconds, add 3 mL of starch TS, and titrate with 0.1 *N* sodium thiosulfate VS: not more than 0.1 mL of 0.1 *N* sodium thiosulfate is consumed.

Sodium Biphosphate, $NaH_2PO_4.H_2O$—**137.99**—Use ACS reagent grade Sodium Phosphate, Monobasic.

Sodium Bisulfite—Use ACS reagent grade Sodium Metabisulfite.

Sodium Bitartrate, $NaHC_4H_4O_6.H_2O$—**190.08**—White crystals or a crystalline powder. Soluble in cold water.

Assay—Dissolve about 500 mg, accurately weighed, in 30 mL of water, add phenolphthalein TS, and titrate with 0.1 *N* sodium hydroxide VS: each mL of 0.1 *N* sodium hydroxide is equivalent to 19.01 mg of $NaHC_4H_4O_6.H_2O$. Between 99% and 100.5% is found.

Insoluble matter (Reagent test): not more than 1 mg, from 10 g (0.01%).

Chloride (Reagent test)—One g shows not more than 0.2 mg of Cl (0.02%).

Heavy metals (Reagent test)—Dissolve 4 g in 25 mL of water, add 2 drops of phenolphthalein TS, and then add ammonia TS, dropwise, until the solution is slightly pink. Add 4 mL of 1 *N* hydrochloric acid, dilute with water to 40 mL, and add 10 mL of hydrogen sulfide TS: any brown color produced is not darker than that of a control containing 0.04 mg of added Pb (0.001%).

Sulfate (Reagent test, *Method I*)—One g shows not more than 0.2 mg of SO_4 (0.02%).

Sodium Borate (*Borax; Sodium Tetraborate*), $Na_2B_4O_7.10H_2O$—**381.37**—Use ACS reagent grade.

NOTE—Certified Borax is available from the National Bureau of Standards, Washington, DC 20234, as standard sample No. 187.

Sodium Borohydride, $NaBH_4$—**37.83**—White, crystalline solid. Freely soluble in water; soluble (with reaction) in methanol. Its solutions are rapidly decomposed by boiling.

Assay—

POTASSIUM IODATE SOLUTION (0.25 *N*)—Dissolve 8.917 g, previously dried at 110° to constant weight and accurately weighed, in water to make 1000.0 mL.

PROCEDURE—Dissolve about 500 mg, accurately weighed, in 125 mL of sodium hydroxide solution (1 in 25) in a 250-mL volumetric flask, dilute with the sodium hydroxide solution to volume, and mix. Pipet 10 mL of the solution into a 250-mL iodine flask, add 35.0 mL of *Potassium iodate solution*, and mix. Add 2 g of potassium iodide, mix, add 10 mL of dilute sulfuric acid (1 in 10), insert the stopper in the flask, and allow to stand in the dark for 3 minutes. Titrate the solution with 0.1 *N* sodium thiosulfate VS, adding 3 mL of starch TS as the end-point is approached. Perform a blank determination, and make any necessary correction. Calculate the amount, in mg, of $NaBH_4$ in the specimen titrated by the formula:

$$([(35.0)(0.25)] - 0.1\ V)4.729,$$

in which *V* is the volume, in mL, of 0.1 *N* sodium thiosulfate used in the titration. Not less than 98% is found.

Sodium Bromide, NaBr—**102.89**—White, odorless, cubical crystals or granular powder. Soluble in water; slightly soluble in alcohol.

Insoluble matter (Reagent test)—The insoluble matter from 20 g, dissolved in 150 mL of hot water, weighs not more than 1 mg (0.005%).

pH ⟨791⟩: between 5.5 and 7.5, in a solution (1 in 20).

Barium—Dissolve 6 g in 15 mL of water, add 5 mL of acetic acid, 5 mL of 30 percent hydrogen peroxide, and 1 mL of hydrochloric acid, and digest in a covered beaker until the reaction ceases. Remove the cover, and evaporate to dryness. Dissolve the residue in 15 mL of water, filter if necessary, dilute with water to 23 mL, and add 2 mL of potassium dichromate solution (1 in 10). Add ammonium hydroxide until the orange color has been dissipated and the yellow color persists, then add 25 mL of methanol, stir vigorously, and allow to stand for 10 minutes: any turbidity produced does not exceed that of a control containing 1.0 g of test specimen and 100 µg of added barium ion (0.002%).

Bromate—Dissolve 1 g in 10 mL of oxygen-free water, add 100 µL of potassium iodide solution (1 in 10), 1 mL of starch TS, and 25 µL of dilute sulfuric acid (1 in 36), and allow to stand at 25°: no blue or violet color is produced within 10 minutes (about 0.001%).

Calcium, magnesium, and R_2O_3 precipitate—To the filtrate from the test for *Insoluble matter* add 5 mL of ammonium oxalate TS, 2 mL of ammonium phosphate TS, and 10 mL of ammonium hydroxide. Allow to stand for about 16 hours, filter, wash with dilute ammonia TS (1 in 4), ignite, and weigh: the weight of the residue is not more than 1 mg (0.005%).

Chloride—Dissolve 500 mg in 15 mL of dilute nitric acid (1 in 3) in a small conical flask, add 3 mL of 30 percent hydrogen peroxide, and digest on a steam bath until the solution is colorless. Wash down the sides of the flask with a small quantity of water, digest for an additional 15 minutes, cool, and dilute with water to 200 mL. Dilute a 2-mL aliquot with water to 25 mL, and add 1 mL of nitric acid and 1 mL of silver nitrate TS: any turbidity produced does not exceed that of a control containing 10 µg of added chloride ion (0.2%).

Heavy metals (Reagent test): 0.0005%.

Iron ⟨241⟩—Two g, dissolved in 47 mL of water containing 2 mL of hydrochloric acid, shows not more than 0.01 mg of Fe (5 ppm).

Nitrogen compounds (Reagent test)—One g shows not more than 5 µg of N (0.0005%).

Potassium (Reagent test)—

TEST SOLUTION—Dissolve 10 g in water, dilute with water to 100 mL, and mix.

SAMPLE SOLUTION—Dilute 10.0 mL of *Test solution* with water to 100 mL, and mix.

CONTROL SOLUTION—To 10.0 mL of *Test solution* add 50 µg of potassium ion (K), dilute with water to 100 mL, and mix. The limit is 0.005%.

Sulfate—Dissolve 10 g in 100 mL of water, filter if necessary, and add 1 mL of hydrochloric acid: the solution yields not more than 1.2 mg of residue (0.005%).

Sodium Carbonate—Use *Sodium Carbonate, Anhydrous.*

NOTE—Sodium Carbonate of a quality suitable as a primary standard is available from the National Institute of Standards and Technology, Office of Standard Reference Materials, Washington, DC 20234, as standard sample No. 192.

Sodium Carbonate, Anhydrous, Na_2CO_3—**105.99**—Use ACS reagent grade.

Sodium Chlorate, $NaClO_3$—**106.44**—Colorless crystals or white granules. Is odorless. Very soluble in boiling water; freely soluble in cold water and in glycerin; slightly soluble in alcohol. Its water solubility is reduced by the presence of sodium chloride. When heated at 300°, it liberates oxygen; at higher temperatures it is decomposed.

Caution—Keep out of contact with organic matter or other readily oxidizable materials.

Assay—Weigh accurately about 100 mg, previously dried over silica gel for 3 hours, transfer to a 250-mL flask, and dissolve in 10 mL of water. Add 35.0 mL of acid ferrous sulfate TS, close the flask with a safety valve, and boil for 10 minutes. Cool, add 10 mL of manganese sulfate solution (1 in 10), and titrate with 0.1 *N* potassium permanganate VS. Perform a blank determination, and make any necessary correction. Each mL of 0.1 *N* potassium permanganate consumed by the assay specimen is equivalent to 1.774 mg of $NaClO_3$. Not less than 99% is found.

pH: between 5.0 and 7.0, in a solution of 10 g in 200 mL of ammonia- and carbon dioxide–free water.

Insoluble matter (Reagent test): not more than 1 mg, from 20 g (0.005%).

Bromate—Dissolve 9 g in 150 mL of carbon dioxide–free water in a glass-stoppered flask, add 10 mL of dilute hydrochloric acid (8.6 in 100), 10 mL of freshly prepared potassium iodide TS, and 5 mL of freshly prepared starch TS, and swirl to mix. Immediately insert the stopper in the flask, and allow to stand, protected from light, for 1 hour, then titrate with 0.1 *N* sodium thiosulfate VS. Not more than 0.45 mL of thiosulfate is consumed (0.01%).

Chloride—Dissolve 300 mg in 40 mL of water, cool, and add 0.25 mL of nitrogen oxide–free nitric acid and 2 mL of silver nitrate TS: any turbidity produced does not exceed that of a blank containing 0.015 mg of added Cl (0.005%).

Sulfate (Reagent test, *Method II*)—Ten g, dissolved in a mixture of 100 mL of water and 2 mL of diluted hydrochloric acid, yields not more than 2.4 mg of residue (0.001%).

Nitrogen compounds (Reagent test)—One g shows not more than 0.01 mg of N (0.001%).

Calcium, magnesium, and R_2O_3 precipitate—Boil 20 g with 15 mL of hydrochloric acid and 50 mL of water until no more chlorine is evolved. Dilute with water to about 120 mL, filter if necessary, heat to boiling, and add 5 mL of ammonium oxalate TS, 3 mL of dibasic ammonium phosphate TS, and 20 mL of ammonium hydroxide. Allow to stand overnight, filter, wash with dilute ammonia TS (1 in 4), ignite, and weigh. The residue weighs not more than 1 mg (0.005%).

Heavy metals—Dissolve 3 g in 20 mL of dilute hydrochloric acid (1 in 2), and evaporate on a steam bath to dryness. Add 5 mL more of dilute hydrochloric acid (1 in 2), and evaporate again to dryness. Take up the residue in 20 mL of water, add 1 drop of phenolphthalein TS, and neutralize with 0.1 *N* sodium hydroxide. Adjust with diluted acetic acid to a pH between 3 and 4, determined potentiometrically, dilute with water to 40 mL, and add 10 mL of hydrogen sulfide TS: any brown color produced is not darker than that of a control containing 1 g of test specimen and 0.02 mg of added Pb (0.001%).

Iron ⟨241⟩—Dissolve 2 g in a mixture of 10 mL of hot water and 10 mL of hydrochloric acid, and evaporate to dryness. Take up the residue in 5 mL of hydrochloric acid, and evaporate again to dryness: the residue, dissolved in 47 mL of water containing 2 mL of hydrochloric acid, shows not more than 0.01 mg of Fe (5 ppm).

Sodium Chloride, NaCl—**58.44**—Use ACS reagent grade.

Sodium Chloride Injection—Use *Sodium Chloride Injection* (USP monograph).

Sodium Chloride Solution, Isotonic—Use *Saline TS.*

Sodium Chromate, $Na_2CrO_4 \cdot 4H_2O$—**234.03**—Lemon-yellow, odorless crystals. Soluble in water.

Assay—Weigh accurately about 300 mg, and dissolve in 10 mL of water contained in a 500-mL flask. Add 3 g of potassium iodide and 10 mL of diluted sulfuric acid, and dilute with 350 mL of oxygen-free and carbon dioxide–free water. Titrate the liberated iodine with 0.1 *N* sodium thiosulfate VS, adding 3 mL of starch TS as the end-point is approached. Each mL of 0.1 *N* sodium thiosulfate consumed is equivalent to 7.802 mg of $Na_2CrO_4 \cdot 4H_2O$. Not less than 99% is found.

Insoluble matter (Reagent test): not more than 1 mg, from 20 g dissolved in 150 mL of water (0.005%).

Aluminum—Dissolve 20 g in 140 mL of water, filter, and add 5 mL of glacial acetic acid to the filtrate. Add stronger ammonia water until alkaline, and digest for 2 hours on a steam bath. Filter through hardened filter paper, wash thoroughly, ignite, and weigh: the residue weighs not more than 0.8 mg (0.002%).

Calcium—Determine as directed in the test for calcium for ACS reagent grade Potassium Chromate (0.005%).

Chloride—Determine as directed in the test for chloride for ACS reagent grade Potassium Chromate (about 0.005%).

Sulfate—Determine as directed in the test for sulfate for ACS reagent grade Potassium Dichromate, but add 4.5 mL of hydrochloric acid to the water used to dissolve the test specimen: the residue weighs not more than 2.4 mg (0.01%).

Sodium Chromotropate—See *Chromotropic Acid.*

Sodium Citrate—Use *Sodium Citrate* (dihydrate) (USP monograph).

Sodium Cobaltinitrite, $Na_3Co(NO_2)_6$—**403.94**—Use ACS reagent grade.

Sodium Cyanide, NaCN—**49.01**—Use ACS reagent grade.

Sodium Desoxycholate—Use *Bile Salts.*

Sodium Dichromate, $Na_2Cr_2O_7 . 2H_2O$ (for chromic acid cleaning mixture)—**298.00**—Orange-red crystals or granules. Very soluble in water; insoluble in alcohol.

Sodium Diethyldithiocarbamate, $(C_2H_5)_2NCS_2Na . 3H_2O$—**225.30**—Use ACS reagent grade.

Sodium 2,2-dimethyl-2-silapentane-5-sulfonate—See *Sodium 3-(trimethylsilyl)-1-propane sulfonate.*

Sodium Dithionate, $Na_2S_2O_6 . 2H_2O$—**242.16**—Use a suitable grade.

Sodium Dithionite—Use *Sodium Hydrosulfite.*

Sodium Ferrocyanide, $Na_4Fe(CN)_6 . 10H_2O$—**484.06**—Yellow crystals or granules. Freely soluble in water.

Assay—Dissolve 2 g, accurately weighed, in 400 mL of water, add 10 mL of sulfuric acid, and titrate with 0.1 *N* potassium permanganate VS. Each mL of 0.1 *N* potassium permanganate is equivalent to 48.41 mg of $Na_4Fe(CN)_6 . 10H_2O$. Not less than 98% is found.

Insoluble matter (Reagent test): not more than 1 mg, from 10 g (0.01%).

Chloride (Reagent test)—Dissolve 1 g in 75 mL of water, add a solution prepared by dissolving 1.2 g of cupric sulfate in 25 mL of water, mix, and allow to stand for 15 minutes. To 20 mL of the decanted, clear liquid add 2 mL of nitric acid and 1 mL of silver nitrate TS: any turbidity produced does not exceed that of a control containing 0.02 mg of Cl, 2 mL of nitric acid, 1 mL of silver nitrate TS, and sufficient cupric sulfate to match the color of the test solution.

Sulfate—Dissolve 5 g in 100 mL of water without heating, filter, and to the filtrate add 0.25 mL of glacial acetic acid and 5 mL of barium chloride TS: no turbidity is produced in 10 minutes (about 0.01% as SO_4).

Sodium Fluorescein—$C_{20}H_{10}Na_2O_5$—**376.28**—Orange-red, hygroscopic powder. Freely soluble in water; slightly soluble in alcohol. Its water solution is yellowish red in color and exhibits a strong yellowish green fluorescence that disappears when the solution is acidified and reappears when the solution is neutralized or made basic.

Loss on drying ⟨731⟩—Dry it at 120° to constant weight: it loses not more than 7.0% of its weight.

Sodium Fluoride, NaF—**41.99**—Use ACS reagent grade.

Sodium Glycocholate, $C_{26}H_{42}NNaO_6$—**487.61**—White to tan, odorless or practically odorless powder. Is hygroscopic. Freely soluble in water and in alcohol.

Specific rotation ⟨781⟩: between +28° and +31°, calculated on the dried basis (it is rendered anhydrous by drying at 100° for 2 hours), determined at 20° in a solution containing 10 mg per mL.

Nitrogen, Method I ⟨461⟩—Between 2.6% and 3.2% of N is found, calculated on the dried basis.

Sodium Heparin—Use *Heparin Sodium* (USP monograph).

Sodium 1-Heptanesulfonate, $C_7H_{15}NaO_3S$—**202.24**—Use a suitable grade.

Sodium 1-Hexanesulfonate, $C_6H_{13}NaO_3S$—**188.22**—Use a suitable grade.

Sodium Hydrosulfite (*Sodium Dithionite*), $Na_2S_2O_4$—**174.10**—White or grayish white crystalline powder. Soluble in water; slightly soluble in alcohol. Gradually oxidizes in air, more readily when in solution, to bisulfite, acquiring an acid reaction. Is affected by light.

Assay—Weigh accurately about 1 g, dissolve it in a mixture of 10 mL of formaldehyde TS and 10 mL of water contained in a small glass-stoppered flask, and allow to stand for 30 minutes with frequent agitation. Transfer the solution to a 250-mL volumetric flask, add 150 mL of water and 3 drops of methyl orange TS, and then add, dropwise, 1 *N* sulfuric acid to a slightly acid reaction. Dilute with water to 250 mL, and mix. To 50.0 mL of the dilution add 2 drops of phenolphthalein TS and just sufficient 0.1 *N* sodium hydroxide to produce a slight, pink color, then titrate with 0.1 *N* iodine, adding 3 mL of starch TS as the indicator. Then discharge the blue color of the solution with 1 drop of 0.1 *N* sodium thiosulfate, and titrate with 0.1 *N* sodium hydroxide VS to a pink color: each mL of 0.1 *N* sodium hydroxide is equivalent to 3.482 mg of $Na_2S_2O_4$. Not less than 88% is found.

Sulfide—Add sodium hydroxide solution (1 in 10) to lead acetate TS until the precipitate dissolves. Add 5 drops of this solution to a solution of 1 g of the sodium hydrosulfite in 10 mL of water: no immediate darkening is observed.

Heavy metals—Dissolve 1 g in 10 mL of water, add 10 mL of hydrochloric acid, and evaporate on a steam bath to dryness. Dissolve the residue in 20 mL of water and 0.5 mL of diluted hydrochloric acid, filter, and add to the filtrate 10 mL of hydrogen sulfide TS: no darkening is produced. Render the solution alkaline with ammonia TS: a slight, greenish color may be produced, but not a dark or white precipitate.

Suitability for riboflavin assay—To each of 2 or more tubes add 10 mL of water and 1.0 mL of a standard riboflavin solution containing 20 µg of riboflavin in each mL, and mix. To each tube add 1.0 mL of glacial acetic acid, mix, add with mixing, 0.5 mL of potassium permanganate solution (1 in 25), and allow to stand for 2 minutes. Then to each tube add, with mixing, 0.5 mL of hydrogen peroxide TS: the permanganate color is destroyed within 10 seconds. Shake the tubes vigorously until excess oxygen is expelled. If gas bubbles remain on the sides of tubes after foaming has ceased, remove the bubbles by tipping the tubes so that the solution flows slowly from end to end. In a suitable fluorometer, measure the fluorescence of the solution. Then add, with mixing, 8.0 mg of sodium hydrosulfite: the riboflavin is completely reduced in not more than 5 seconds.

Sodium Hydroxide, NaOH—**40.00**—Use ACS reagent grade.

Sodium Hypochlorite Solution—A solution of sodium hypochlorite (NaOCl) in water. Usually yellow to yellowish green in color. Has an odor of chlorine. Is affected by light and gradually deteriorates. Store it in light-resistant containers, preferably below 25°. [*Caution—This solution is corrosive and may evolve gases that are corrosive and toxic. It is a powerful oxidant that can react violently with reducing agents. Is irritating and corrosive to skin and mucous membranes.*]

Assay—Transfer about 3 mL to a tared, glass-stoppered iodine flask, and weigh accurately. Add 50 mL of water, 2 g of potassium iodide, and 10 mL of acetic acid, insert the stopper in the flask, and allow to stand in the dark for 10 minutes. Remove the stopper, rinse the walls of the flask with a few mL of water, and titrate the liberated iodine with 0.1 *N* sodium thiosulfate VS, adding 3 mL of starch TS as the end-point is neared. Each mL of 0.1 *N* sodium thiosulfate consumed is equivalent to 3.723 mg of NaOCl: not less than 5.25% is found. If it is desired to calculate the percentage of available chlorine, note that each mL of 0.1 *N* sodium thiosulfate consumed is equivalent to 3.545 mg of available chlorine.

Calcium—Transfer 10.0 g to a 150-mL beaker, dissolve in 10 mL of water, and add 5 mL of hydrochloric acid and 2 g of potassium iodide. Heat the mixture for 5 minutes, cool, and add 2 mL of 30 percent hydrogen peroxide. Evaporate to dryness, cool, and add 2 mL of hydrochloric acid and 2 mL of 30 percent hydrogen peroxide. Rinse the inner walls of the beaker with a few mL of water, and evaporate to dryness. Take up the residue in 20 mL of water, and filter if necessary. To the filtrate add ammonium hydroxide until the solution is just alkaline, then add 4 drops of ammonium hydroxide and 5 mL of ammonium oxalate TS: any turbidity produced within 15 minutes does not exceed that in a blank containing 0.1 mg of added Ca carried through the entire procedure (0.001%).

Phosphate (Reagent test)—Transfer 2 g to a beaker, and add 5 mL of hydrochloric acid and 2 g of potassium iodide. Heat the solution for 5 minutes, and cool. Add 2 mL of 30 percent hydrogen peroxide, and evaporate the solution to dryness. Rinse the walls of the beaker with a few mL of water, and add 2 mL of hydrochloric acid and 2 mL of 30 percent hydrogen peroxide. Evaporate again to dryness: the residue shows not more than 0.01 mg of PO_4 (5 ppm).

Sodium Indigotindisulfonate—Use *Indigotindisulfonate Sodium* (USP monograph).

Sodium Iodide—Use *Sodium Iodide* (USP monograph).

Sodium Lauryl Sulfate—Use *Sodium Lauryl Sulfate* (NF monograph).

Sodium Metabisulfite, $Na_2S_2O_5$—**190.10**—Use ACS reagent grade.

Sodium Metaperiodate, $NaIO_4$—**213.89**—Use ACS reagent grade Sodium Periodate.

Sodium Methoxide, CH_3ONa—**54.02**—Fine, white powder. Reacts violently with water with evolution of heat. Soluble in alcohol and in methanol.

Assay—Transfer about 220 mg to a tared, glass-stoppered flask, and weigh accurately. Dissolve the test specimen in about 10 mL of methanol, then add 100 mL of water slowly, with stirring. Add phenolphthalein TS, and titrate with 0.1 *N* hydrochloric acid VS to a colorless end-point: each mL of 0.1 *N* hydrochloric acid VS is equivalent to 5.402 mg of CH_3ONa. Not less than 98.0% is found.

Sodium Molybdate, $Na_2MoO_4 . 2H_2O$ **241.95**—Use ACS reagent grade.

Sodium Nitrate, $NaNO_3$—**84.99**—Use ACS reagent grade.

Sodium Nitrite, $NaNO_2$—**69.00**—Use ACS reagent grade.

Sodium Nitroferricyanide (*Sodium Nitroprusside*), $Na_2Fe(NO)(CN)_5 . 2H_2O$—**297.95**—Use ACS reagent grade.

Sodium 1-Octanesulfonate, $C_8H_{17}NaO_3S$—**216.27**—Use a suitable grade.

Sodium Oxalate, $Na_2C_2O_4$—**134.00**—Use ACS reagent grade.

NOTE—Sodium oxalate of a quality suitable as a primary standard is available from the Office of Standard Reference Materials, National Institute of Standards and Technology, Washington, DC 20234, as standard sample No. 40g.

Sodium 3-(trimethylsilyl)-1-propane sulfonate (*Sodium 2,2-dimethyl-2-silapentane-5-sulfonate*), $C_6H_{15}SiNaO_3S$—**218.32**—Use a suitable grade.

Sodium (tri)Pentacyanoamino Ferrate [*Trisodium Aminepentacyanoferrate(3-)*], $Na_3[Fe(CN)_5NH_3]$—**271.94**—Yellow to tan powder. Soluble in water.

Solubility—Dissolve 500 mg in 50 mL of water, and allow to stand for 1 hour: the solution is clear and free from foreign matter.

Sensitivity—

1,1-DIMETHYLHYDRAZINE STANDARD SOLUTION—Place 500 mL of water in a 1-liter volumetric flask, and add from a buret 1.27 mL of anhydrous 1,1-dimethylhydrazine. Dilute with water to volume, and mix. Pipet 10 mL of this solution into a 100-mL volumetric flask, and dilute with water to volume. Each mL of this solution contains the equivalent of 100 μg of 1,1-dimethylhydrazine.

BUFFER SOLUTION—Transfer 4.8 g of citric acid monohydrate to a 1-liter volumetric flask, dissolve in water, add 14.6 g of sodium phosphate, swirl to dissolve, and dilute with water to volume.

TEST PREPARATION—Dissolve 100 mg of sodium (tri)-pentacyanoamino ferrate in 100 mL of water.

PROCEDURE—Into each of five 25-mL volumetric flasks pipet 0, 0.25, 0.50, 1.0, and 1.5 mL, respectively, of *1,1-Dimethylhydrazine standard solution*, to each add 15 mL of *Buffer solution*, and swirl to mix. To each flask, add by pipet 2 mL of *Test preparation*, mix, dilute with *Buffer solution* to volume, and allow to stand for 1 hour. Using a suitable spectrophotometer, 1-cm cells, and the solution containing no *1,1-Dimethylhydrazine standard solution* as the blank, determine the absorbances of the remaining solutions at 500 nm. Plot the observed absorbance as the ordinate versus the concentration of standard as the abscissa on coordinate paper, and draw the curve of best fit. The plot is linear and the absorbance of the 150-μg solution is not less than 0.65.

Sodium 1-Pentanesulfonate, $C_5H_{11}NaO_3S . H_2O$—**192.20**—White, crystalline solid. Soluble in water.

Solubility—One g, dissolved in 25 mL of water, yields a clear and complete solution.

Water, Method I ⟨921⟩: not more than 2.0%.

Sodium Perchlorate, $NaClO_4 . H_2O$—**140.46**—Colorless, deliquescent crystals. Decomposes at about 150°. Soluble in 95% alcohol.

Assay—Dry about 1.5 g in a vacuum desiccator at 80° to constant weight. Mix 750 mg, accurately weighed, of the dried and powdered test specimen with 5 g of powdered sodium nitrite in a nickel crucible, cover the crucible, and heat it over a free flame until the mixture is well melted. Maintain it in this state, without raising the temperature much higher, for 30 minutes. Allow to cool, add 20 mL of hot water, and digest until the melt is dissolved. Filter into a 200-mL volumetric flask, wash any undissolved matter thoroughly with hot water, cool, dilute with water to volume, and mix.

Transfer 50.0 mL of the solution to a 250-mL, glass-stoppered flask, add 25.0 mL of 0.1 *N* silver nitrate VS, then add slowly 6 mL of dilute nitric acid (1 in 6), and heat on the steam bath to expel oxides of nitrogen. Cool, add 3 mL of nitrobenzene, shake well for 1 to 2 minutes, then add 4 mL of ferric ammonium sulfate TS, and titrate the excess silver nitrate with 0.1 *N* ammonium thiocyanate VS. Perform a blank determination, and make any necessary correction. Each mL of 0.1 *N* silver nitrate is equivalent to 12.24 mg of $NaClO_4$: not less than 98.0% of $NaClO_4$ is found.

Insoluble matter—Dissolve 10 g in 50 mL of water, heat to boiling, and filter through a tared sintered-glass crucible. Wash well with water, rinsing the beaker thoroughly. Dry at 105° for 2 hours, and weigh. The weight of the insoluble residue does not exceed 1 mg (0.01%).

Chlorate and chloride (*as Cl*)—Dissolve 1 g in 10 mL of water, add 1 mL of 0.1 *N* ferrous sulfate, and heat on the steam bath for 15 minutes. Cool, dilute with water to 50 mL, and add 1 mL of nitric acid and 1 mL of silver nitrate TS. Any turbidity does not exceed that produced by 0.1 mg of chloride (Cl) contained in a similarly treated blank (0.01% of Cl).

Sulfate—Dissolve 1 g in 10 mL of water, and add 0.05 mL of diluted hydrochloric acid and 1 mL of barium chloride TS. Any turbidity produced in 10 minutes does not exceed that produced in a blank containing 0.05 mg of added SO_4 (0.005%).

Calcium—Dissolve 500 mg in 10 mL of hot water, add 0.25 mL of ammonia TS and 3 mL of ammonium oxalate TS, and keep the solution hot. No turbidity is produced in 5 minutes (about 0.02%).

Heavy metals (Reagent test)—Dissolve 1 g in 25 mL of water: the heavy metals limit is 0.002%.

***p*-Sodium Periodate,** $Na_3H_2IO_6$—**293.89**—Use ACS reagent grade.

Sodium Peroxide, Na_2O_2—**77.98**—Use ACS reagent grade.

Sodium Phenobarbital—Use *Phenobarbital Sodium* (USP monograph).

Sodium Phosphate, Dibasic (*Disodium Phosphate; Disodium Hydrogen Phosphate; Sodium Phosphate, Dibasic, Heptahydrate*), $Na_2HPO_4 . 7H_2O$—**268.07**—Use ACS reagent grade Sodium Phosphate, Dibasic, Heptahydrate.

Sodium Phosphate, Dibasic, Anhydrous (*Anhydrous Disodium Hydrogen Phosphate*) (for buffer solutions), Na_2HPO_4—**141.96**—Use ACS reagent grade Sodium Phosphate, Dibasic, Anhydrous.

Sodium Phosphate, Monobasic, $NaH_2PO_4 . H_2O$—**137.99**—Use ACS reagent grade.

Sodium Phosphate, Tribasic, $Na_3PO_4 . 12H_2O$—**380.12**—Use ACS reagent grade.

Sodium Pyrophosphate, $Na_4P_2O_7$—**265.90**—Use ACS reagent grade.

Sodium Pyruvate, CH_3COCO_2Na—**110.04**—White to practically white powder or crystalline solid. Soluble in water.

Assay—Transfer about 300 mg, accurately weighed, to a high-form titration beaker, add 150 mL of glacial acetic acid, and stir until dissolved. Titrate with 0.1 *N* perchloric acid VS, determining the end-point potentiometrically, using a glass electrode and a calomel electrode modified to use 0.1 *N* tetramethylammonium chloride in methanol as the electrolyte. Perform a blank determination, and make any necessary correction. Each mL of 0.1 *N* perchloric acid is equivalent to 11.00 mg of CH_3COCO_2Na: not less than 98.0% is found.

Solubility—Dissolve 1.5 g in 25 mL of water: the solution is clear and complete.

Free acid—Dissolve 10 g in 150 mL of water, and titrate with 0.5 *N* sodium hydroxide VS, determining the end-point potentiometrically: not more than 2.8 mL of 0.5 *N* sodium hydroxide is consumed (about 1% as $C_3H_4O_3$).

Sodium Salicylate—It complies with the specifications under *Sodium Salicylate* (USP monograph), and in addition meets the requirements of the following test.

Nitrate—Dissolve 100 mg in 5 mL of water, and superimpose the solution upon 5 mL of sulfuric acid: no brownish red color appears at the junction of the two liquids.

Sodium Selenite, Na_2SeO_3—**172.94**—White, odorless, crystalline powder, usually partially hydrated. Freely soluble in water; insoluble in alcohol.

Assay—Weigh accurately about 180 mg, previously dried at 120° to constant weight, and dissolve it in 50 mL of water in a glass-stoppered flask. Add, successively, 3 g of potassium iodide and then 5 mL of hydrochloric acid, insert the stopper, and allow to stand for 10 minutes. Add 50 mL of water, 50.0 mL of 0.1 *N* sodium thiosulfate VS, and 3 mL of starch TS, and immediately titrate with 0.1 *N* iodine VS to a blue color. Perform a blank determination. The difference in volumes of 0.1 *N* iodine required for the blank and the test preparation represents sodium selenite. Each mL of 0.1 *N* iodine is equivalent to 4.323 mg of Na_2SeO_3. Between 98% and 101% is found.

Solubility—One g in 10 mL of water shows not more than a faint haze.

Carbonate—To 500 mg add 1 mL of water and 2 mL of diluted hydrochloric acid: no effervescence is produced.

Chloride (Reagent test)—A 500-mg portion shows not more than 0.05 mg of Cl (0.01%).

Nitrate (Reagent test)—A 200-mg portion dissolved in 3 mL of water shows not more than 0.02 mg of NO_3 (0.01%).

Selenate and sulfate (as SO_4)—To 500 mg in a small evaporating dish add 20 mg of sodium carbonate and 10 mL of hydrochloric acid. Slowly evaporate the solution on a steam bath under a hood to dryness. Wash the sides of the dish with 5 mL of hydrochloric acid, and again evaporate to dryness. Dissolve the residue in a mixture of 15 mL of hot water and 1 mL of hydrochloric acid. Proceed as directed under *Sulfate in Reagents* (Reagent test, *Method I*), beginning with "Filter the solution." The test specimen shows no more turbidity than that produced by 0.15 mg of SO_4 (0.03%).

Sodium Sulfate (*Glauber's Salt*), $Na_2SO_4 . 10H_2O$—**322.19** — Colorless, odorless crystals or white granules. Is efflorescent. Melts at 32.5°. Soluble in 1.5 parts of water; soluble in glycerin; insoluble in alcohol. Store in well-closed containers, protected from heat.

Insoluble matter (Reagent test): not more than 1 mg, from 10 g (0.01%).

pH—The pH of a solution of 10 g in 200 mL of ammonia-free and carbon dioxide–free water is between 5.2 and 8.2.

Arsenic (Reagent test)—The stain produced by 3 g does not exceed that produced by 0.003 mg of As (1 ppm).

Calcium, magnesium, and R_2O_3 precipitate—Dissolve 5 g in 75 mL of water, filter, and add 5 mL of ammonium oxalate TS, 2 mL of ammonium phosphate TS, and 10 mL of ammonium hydroxide. Stir well, and allow to stand overnight. If any precipitate forms, filter, wash with dilute ammonia TS (1 in 4), and ignite: the residue weighs not more than 1 mg (0.02%).

Chloride—Dissolve 1 g in 50 mL of water, and filter if necessary. To 25 mL of the solution add 1 mL of nitric acid and 1 mL of silver nitrate TS: any turbidity produced does not exceed that of a control containing 0.01 mg of Cl (0.002%).

Heavy metals (Reagent test): 5 ppm.

Iron ⟨241⟩—One g, dissolved in 47 mL of water containing 2 mL of hydrochloric acid, shows not more than 0.01 mg of Fe (0.001%).

Nitrogen compounds (Reagent test)—Two g shows not more than 0.01 mg of N (5 ppm).

Sodium Sulfate, Anhydrous, Na_2SO_4—**142.04**—Use ACS reagent grade.

For use in assaying alkaloids by gas-liquid chromatography, it conforms also to the following additional test.

Suitability for alkaloid assays—Transfer about 10 mg of atropine, accurately weighed, to a 25-mL volumetric flask, dissolve in alcohol, and dilute with alcohol to volume. Pipet 3 mL of the solution into each of two 60-mL separators, and add to each 10 mL of water, 1 mL of 1 *N* sodium hydroxide, and 10 mL of chloroform. Shake thoroughly, and allow the layers to separate. Filter the organic phase from one separator through phase-separating paper, previously washed with 5 mL of chloroform, supported in a funnel, and collect the filtrate in a suitable container. Add 10 mL of chloroform to the separator, shake thoroughly, and filter the organic layer through the same phase-separating paper, collecting and combining the filtrates in the same container. Designate the combined filtrates as *Solution A*. Filter the organic phase from the second separator through 30 g of the Anhydrous Sodium Sulfate, supported on a pledget of glass wool in a small funnel, and previously washed with chloroform, and collect the filtrate in a suitable container. Add 10 mL of chloroform to the separator, shake thoroughly, and filter the organic layer through the same portion of anhydrous sodium sulfate, collecting and combining the two filtrates in the same container. Designate the combined filtrates as *Solution B*. Evaporate the two solutions in vacuum to a volume of about 1 mL. Inject an accurately measured volume of *Solution A* into a suitable gas chromatography, and record the peak height. Repeat the determination with a second accurately measured volume of *Solution A*, record the peak height, and obtain the average of the two results. In a similar manner, determine the peak height of two portions of *Solution B*, and obtain the average of the results. The average value obtained for *Solution B* is within 5.0% of the value obtained for *Solution A*.

Under typical conditions, the gas chromatograph contains a 1.2-m × 4-mm glass column packed with 3 percent phase G3 on packing S1A. After curing and conditioning, the column is maintained at 210°, the injector port at 225°, and the detector block at 240° during the determinations. The carrier gas is helium, flowing at a rate of 60 mL per minute.

Sodium Sulfate Decahydrate—Use *Sodium Sulfate.*

Sodium Sulfide, $Na_2S.9H_2O$—**240.18**—Use ACS reagent grade.

Sodium Sulfite—Use *Sodium Sulfite, Anhydrous.*

Sodium Sulfite, Anhydrous (*Exsiccated Sodium Sulfite*), Na_2SO_3—**126.04**—Use ACS reagent grade.

Sodium *p*-Sulfophenylazochromotropate[85]—[*Trisodium Salt of 4,5-Dihydroxy-3-(p-sulfophenylazo)-2,7-naphthalenedisulfonic acid*], $C_{16}H_9N_2Na_3O_{11}S_3 . 3H_2O$—**624.45**—Bright red powder. Very soluble in water; insoluble in alcohol. Combines with zirconium oxychloride to form a soluble pink zirconium lake.

Sodium Tartrate, $Na_2C_4H_4O_6 . 2H_2O$—**230.08**—Use ACS reagent grade.

Sodium Tetraphenylborate (*Sodium Tetraphenylboron*), $(C_6H_5)_4BNa$—**342.22**—Use ACS reagent grade.

Sodium Tetraphenylboron—See *Sodium Tetraphenylborate.*

Sodium Thioglycolate (*Sodium Thioglycollate*), $HSCH_2COONa$—**114.09**—A white, crystalline powder, having a slight, characteristic odor. Very soluble in water; slightly soluble in alcohol. Is hygroscopic, and oxidizes in air. Store in tight, light-resistant containers. It should not be used if it is pale yellow or darker in color.

Assay—Weigh accurately about 250 mg, and dissolve in 50 mL of oxygen-free water. Add 5 mL of diluted hydrochloric acid, boil for 2 minutes, cool, and titrate the solution with 0.1 *N* iodine VS, adding 3 mL of starch TS toward the end: each mL of 0.1

N iodine is equivalent to 11.41 mg of $HSCH_2COONa$. Not less than 75% is found.

Insoluble matter—A solution of 1 g in 10 mL of water is clear, and practically complete.

Sulfide—Dissolve 500 mg in 10 mL of water in a small flask, add 2 mL of hydrochloric acid, then place a strip of filter paper, moistened with lead acetate TS, over the mouth of the flask, and bring the solution to a boil: the lead acetate paper is not darkened.

Sodium Thiosulfate, $Na_2S_2O_3.5H_2O$—**248.17**—Use ACS reagent grade.

Sodium L-Thyroxine—Use *Levothyroxine Sodium* (USP monograph).

Sodium 3-(trimethylsilyl)-1-propane sulfonate (*Sodium 2,2-dimethyl-2-silapentane-5-sulfonate*), $C_6H_{15}SiNaO_3S$—**218.32**—Use a suitable grade.

Sodium Tungstate, $Na_2WO_4.2H_2O$—**329.86**—Use ACS reagent grade.

Soluble Starch—See *Starch, Soluble.*

Solvent Hexane—See *Hexane, Solvent.*

Sorbic Acid—Use *Sorbic Acid* (NF monograph).

Sorbitan Mono-oleate[86]—A partial ester of oleic acid and hexitol anhydrides derived from sorbitol. A viscous, colorless to light straw-colored liquid. Insoluble in water. Miscible with oils.

Sorbitol—Use *Sorbitol* (NF monograph).

Squalane (*2,6,10,15,19,23-Hexamethyltetracosane*), $C_{30}H_{62}$—**422.82**—Oil usually obtained by hydrogenation of shark liver oil. Soluble in ether, in benzene, in chloroform, and in many other oils; slightly soluble in acetone, in alcohol, in glacial acetic acid, and in methanol.

Specific gravity ⟨841⟩: between 0.811 and 0.813 at 20°.

Refractive index ⟨831⟩: between 1.4509 and 1.4525, at 20°.

Stannous Chloride, $SnCl_2.2H_2O$—**225.63**—Use ACS reagent grade.

Starch, Arrowroot—The starch separated from the root of *Maranta arundinacea* Linné (Fam. Marantaceae). A white powder consisting of starch grains of characteristic shape and appearance when examined microscopically.

Sensitiveness—Prepare a paste of 1.0 g with a small amount of cold water, and add to it, with stirring, 200 mL of boiling water. Cool, add 5 mL of the cooled solution to 100 mL of water containing about 50 mg of potassium iodide, mix, and then add 0.05 mL of 0.1 *N* iodine: a deep blue color that is discharged by 0.05 mL of 0.1 *N* sodium thiosulfate is produced.

Starch, Potato—The starch separated from the tubers of *Solanum tuberosum* Linné (Fam. Solanaceae). A more or less finely granular powder, consisting of starch grains of characteristic shape and appearance when examined microscopically.

Starch, Soluble (for iodimetry)—Use ACS reagent grade.

Starch, Soluble, Purified[87]—White, amorphous powder; under microscopic examination it shows the characteristic form of potato starch. Soluble in hot water; very slightly soluble in alcohol.

TEST SOLUTION FOR DETERMINATION OF PH AND SENSITIVITY—Stir 2.0 g in 10 mL of water, add boiling water to make 100 mL, and boil for 2 minutes. The hot solution is almost clear. On cooling, the solution may become opalescent or turbid, but does not gel. Use it as the *Test solution.*

pH ⟨791⟩—The pH of the *Test solution* is between 6.0 and 7.5.

Sensitivity—Mix 2.5 mL of *Test solution,* 97.5 mL of water, and 0.50 mL of 0.010 *N* iodine: a distinct blue color results, and it disappears upon the addition of 0.50 mL of 0.010 *N* sodium thiosulfate.

Absorbance—Prepare a pH 5.3 buffer solution by dissolving 43.5 g of sodium acetate (trihydrate) and 4.5 mL of glacial acetic acid in water, transferring the resultant solution to a 250-mL volumetric flask, adding water to volume, and mixing.

Dissolve 1.00 g of Soluble, Purified Starch in 2.5 mL of the buffer solution by warming, transfer to a 100-mL volumetric flask, add water to volume, and mix. Add 0.50 mL of this solution to a 100-mL volumetric flask containing about 75 mL of water, 1 mL of 1 *N* hydrochloric acid, and 1.5 mL of 0.020 *N* iodine, swirling the flask during the addition. Add water to volume, mix, and allow to stand in the dark for 1 hour. The absorbance of this solution, measured at 575 nm in a 1-cm cell against a blank, is between 0.5 and 0.6.

Reducing substances—Shake 10.0 g with 100 mL of water for 15 minutes, and allow to settle for about 12 hours. Filter a portion of the supernatant solution through fine sintered glass. To 50 mL of the filtrate add 50 mL of alkaline cupric tartrate TS, and boil for one to two minutes. Filter the resulting cuprous oxide, wash it with hot water and then with alcohol, and dry it at 105° for 2 hours: not more than 47 mg is found, corresponding to 0.7% of reducing sugars as maltose.

Loss on drying ⟨731⟩—Dry it at 105° for 2 hours: it loses not more than 10% of its weight.

Residue on ignition ⟨281⟩: not more than 0.5%.

Stearic Acid, $C_{18}H_{36}O_2$—**284.48**—Hard, white crystals or amorphous, white powder. Freely soluble in chloroform and in ether; soluble in alcohol and in solvent hexane.

Congealing temperature ⟨651⟩: between 67° and 69°.

Acid value ⟨401⟩: between 196 and 199.

Iodine value ⟨401⟩: not more than 1.

Saponification value ⟨401⟩: between 197 and 200.

Palmitic acid—Determine as directed in the *Assay* under *Stearic Acid* (NF monograph): not more than 5.0% is found.

Stearyl Alcohol (*1-Octadecanol*), $C_{18}H_{38}O$—**270.50**—White flakes, granules, or crystals. Insoluble in water; soluble in alcohol, in ether, in acetone, and in benzene.

Melting range ⟨741⟩: between 56° and 58°.

Other requirements—It conforms to the tests for *Acid value, Iodine value,* and *Hydroxyl value* under *Stearyl Alcohol* (NF monograph).

Stronger Ammonia Water—See *Ammonia Water, Stronger.*

Strontium Acetate, $Sr(CH_3COO)_2.\frac{1}{2}H_2O$—**214.72**—White, crystalline powder. Soluble in 3 parts of water; slightly soluble in alcohol.

Assay—Ignite about 3 g, accurately weighed, in a platinum crucible, protecting from sulfur in the flame. Cool, transfer the crucible with the residue to a beaker, and add 50 mL of water and 40.0 mL of 1 *N* hydrochloric acid VS. Boil gently for 30 minutes or longer, if necessary, filter, wash with hot water until the washings are neutral, add methyl red TS, and titrate the excess acid with 1 *N* sodium hydroxide VS. Each mL of 1 *N* hydrochloric acid is equivalent to 107.4 mg of $Sr(CH_3COO)_2.\frac{1}{2}H_2O$: not less than 99% is found.

Insoluble matter (Reagent test): not more than 2 mg, from 10 g (0.02%).

Free alkali, or free acid—Dissolve 3 g in 30 mL of water, and add 3 drops of phenolphthalein TS: no pink color is produced. Titrate with 0.1 *N* sodium hydroxide VS to a pink color: not more than 0.30 mL of the 0.1 *N* sodium hydroxide is required.

Barium—Dissolve 1 g in 10 mL of water, and add 1 drop of glacial acetic acid and 5 drops of potassium dichromate solution (1 in 10): no turbidity is produced within 2 minutes (about 0.02%).

Calcium—Ignite 1 g until completely carbonized. Warm the residue with a mixture of 3 mL of nitric acid and 10 mL of water, filter, wash with 5 mL of water, and evaporate the filtrate on a steam bath to dryness. Powder the residue, and dry it at 120° for 3 hours. Reflux the dried powder with 15 mL of dehydrated alcohol for 10 minutes, cool in ice, and filter. Repeat the extraction with 10 mL of dehydrated alcohol. Evaporate the combined filtrates to dryness, add 0.5 mL of sulfuric acid, and ignite: the weight of the residue is not more than 10 mg (0.3% of Ca).

Chloride (Reagent test)—One g shows not more than 0.1 mg of Cl (0.01%).

Heavy metals (Reagent test): 0.001%.

Iron ⟨241⟩—Dissolve 1.0 g in 45 mL of water, and add 2 mL of hydrochloric acid: the solution shows not more than 0.01 mg of Fe (0.001%).

Alkali salts—Dissolve 2 g in 80 mL of water, heat to boiling, add an excess of ammonium carbonate TS, boil for 5 minutes, dilute with water to 100 mL, and filter. Evaporate 50 mL of the filtrate, and ignite: the residue after correcting for the ignition residue from half the volume of the clear ammonium carbonate TS used above, is not more than 3 mg (0.3%).

Nitrate—Dissolve 1 g in 10 mL of water, add 0.10 mL of indigo carmine TS, and then add 10 mL of sulfuric acid: the blue color persists for 5 minutes (about 0.01% of NO_3).

Strontium Hydroxide (*Strontium Hydroxide Octahydrate*), $Sr(OH)_2.8H_2O$—**265.76**—White, crystalline, free-flowing powder. Sparingly soluble in water. May absorb carbon dioxide from the air. Keep tightly closed.

Assay and carbonate—Weigh accurately about 5 g, dissolve in 200 mL of warm carbon dioxide-free water in a glass-stoppered, 500-mL flask, add phenolphthalein TS, and titrate with 1 *N* hydrochloric acid VS to determine the hydroxide alkalinity. Then add methyl orange TS, and titrate with 1 *N* hydrochloric acid VS. Each mL of 1 *N* hydrochloric acid required to reach the phenolphthalein end-point is equivalent to 132.9 mg of $Sr(OH)_2.8H_2O$, and each additional mL of 1 *N* hydrochloric acid VS required to reach the methyl orange end-point is equivalent to 73.8 mg of $SrCO_3$. Not less than 95.0% of $Sr(OH)_2.8H_2O$ and not more than 3.0% of $SrCO_3$ are found.

Chloride (Reagent test)—Dissolve 1.0 g in 100 mL of water, and filter if necessary: 1.0 mL of the solution shows not more than 0.01 mg of Cl (0.1%).

Calcium (Reagent test)—Dissolve 5.0 g in water, and dilute with water to 100 mL, to obtain the *test solution*.

SAMPLE SOLUTION—Dilute 10.0 mL of *test solution* with water to 100 mL.

CONTROL SOLUTION—To 10.0 mL of *test solution* add 0.50 mg of calcium ion (Ca), and dilute with water to 100 mL.

PROCEDURE—Determine the background emission at 416.7 nm: the limit is 0.1%.

Iron—Dissolve 1 g in warm water, and dilute with water to 100 mL. To 20 mL of this solution add 2 mL of hydrochloric acid and 0.1 mL of 0.1 *N* potassium permanganate, allow to stand for 5 minutes, and add 3 mL of ammonium thiocyanate solution (3 in 10). Any red color produced is not darker than that of a control containing 0.03 mg of added Fe (0.015%).

For the following test, prepare a *test solution* as follows: Dissolve 2.0 g in 14 mL of dilute hydrochloric acid (1 in 6), and evaporate on a steam bath to dryness. Take up the residue in 25 mL of water, filter, and dilute with water to 100 mL.

Heavy metals—To 5.0 mL of *test solution* add 0.02 mg of lead (Pb), and dilute with water to 30 mL, to provide the standard. For the test specimen, use 30 mL of *test solution*. Adjust each solution with diluted acetic acid or ammonia TS to a pH between 3.0 and 4.0 (using short-range pH paper), dilute with water to 40 mL, and add 10 mL of freshly prepared hydrogen sulfide TS: any brown color developed in the sample solution is not darker than that in the control solution (0.004%).

Strontium Nitrate, $Sr(NO_3)_2$—**211.63**—Use ACS reagent grade.

Strychnine Sulfate, $(C_{21}H_{22}N_2O_2)_2.H_2SO_4.5H_2O$—**856.98**—Colorless or white crystals, or a white, crystalline powder. Its solutions are levorotatory. One g dissolves in about 35 parts of water, in 85 mL of alcohol, and in about 220 mL of chloroform. Insoluble in ether.

Solubility—A solution of 500 mg in 25 mL of water is complete, clear, and colorless.

Residue on ignition (Reagent test): not more than 0.1%.

Brucine—To 100 mg add 1 mL of dilute nitric acid (1 in 2): a yellow color may be observed, but not a red or reddish brown color.

Styrene-Divinylbenzene Anion-exchange Resin, 50- to 100-Mesh—See *Anion-exchange Resin, 50- to 100-Mesh, Styrene-Divinylbenzene.*

Styrene-Divinylbenzene Cation-exchange Resin, Strongly Acidic—See *Cation-exchange Resin, Styrene-Divinylbenzene, Strongly Acidic.*

Sucrose—Use *Sucrose* (NF monograph).

Sudan IV, $C_{24}H_{20}N_4O$—**380.45**—Brown to reddish brown powder.

Assay—Transfer about 25 mg, accurately weighed, to a 100-mL volumetric flask. Dissolve in chloroform, dilute with chloroform to volume, and mix. Dilute 2.0 mL of the resulting solution with chloroform to 50.0 mL. Determine the absorbance of this solution in 1-cm cells at the wavelength of maximum absorbance at about 520 nm, with a suitable spectrophotometer, using chloroform as the blank. Calculate the percentage of Sudan IV in the test specimen taken by the formula:

$$(100A)/(85C),$$

in which *A* is the absorbance at 520 nm and *C* is the concentration of the test specimen in g per liter. Not less than 90% is found.

Loss on drying ⟨731⟩—Dry it at 105° for 2 hours: it loses not more than 10% of its weight.

Sulfadiazine—Use *Sulfadiazine* (USP monograph).

Sulfamerazine—Use *Sulfamerazine* (USP monograph).

Sulfamethazine—Use *Sulfamethazine* (USP monograph).

Sulfamic Acid, HSO_3NH_2—**97.09**—Colorless or white crystals. Soluble in water; slightly soluble in alcohol.

Assay—Weigh accurately about 400 mg, previously dried over sulfuric acid for 2 hours, and dissolve in 30 mL of water contained in a small flask. Add phenolphthalein TS, and titrate with 0.1 *N* sodium hydroxide VS. Each mL of 0.1 *N* sodium hydroxide consumed is equivalent to 9.709 mg of HSO_3NH_2. Not less than 99.5% is found.

Insoluble matter (Reagent test): not more than 1 mg, from 10 g dissolved in 200 mL of water (0.01%).

Residue on ignition (Reagent test)—Ignite 5 g: the residue weighs not more than 0.5 mg (0.01%).

Chloride (Reagent test)—One g shows not more than 0.01 mg of Cl (0.001%).

Heavy metals—Dissolve 4 g in 30 mL of water, neutralize with stronger ammonia water to litmus, and dilute with water to 40 mL. To 30 mL add 2 mL of diluted acetic acid, dilute with water to 40 mL, and add 10 mL of hydrogen sulfide TS: any brown color produced is not darker than that of a control containing the remaining 10 mL of test solution and 0.02 mg of added Pb (0.001%).

Iron ⟨241⟩—Two g, dissolved in 47 mL of water containing 2 mL of hydrochloric acid, shows not more than 0.01 mg of Fe (5 ppm).

Sulfate (Reagent test, *Method I*)—Dissolve 1 g in 50 mL of water: 20 mL of the solution shows not more than 0.2 mg of SO_4 (0.05%).

Sulfanilamide, $C_6H_8N_2O_2S$—**172.20**—Use USP Sulfanilamide Melting Point RS.

Sulfanilic Acid, *p*-$NH_2C_6H_4SO_3H.H_2O$—**191.20**—Use ACS reagent grade.

Sulfapyridine—Use *Sulfapyridine* (USP monograph).

Sulfatase Enzyme Preparation—Use a suitable grade.[88]

Sulfathiazole, $C_9H_9N_3O_2S_2$—**255.31**—White crystals, granules, or powder. Is odorless or practically so. Very slightly soluble in water; soluble in acetone, in diluted mineral acids, and in solutions of alkali hydroxides; slightly soluble in alcohol.

Assay—Weigh accurately about 500 mg, previously dried at 105° for 2 hours, and transfer to a beaker or casserole. Add 5 mL of hydrochloric acid and 50 mL of water, cool to 15°, add about 25 g of crushed ice, and slowly titrate with 0.1 *M* sodium nitrite VS until a blue color is produced immediately when a glass rod dipped into the titrated solution is streaked on a smear of starch iodide paste TS. When the titration is complete, the end-point is reproducible after the mixture has been allowed to stand for 1 minute. Each mL of 0.1 *M* sodium nitrite is equivalent to 25.53 mg of $C_9H_9N_3O_2S_2$. Not less than 99% is found.

Melting range ⟨741⟩: between 200° and 204°.

Loss on drying ⟨731⟩—Dry about 1 g, accurately weighed, at 105° for 2 hours: it loses not more than 0.5% of its weight.

Residue on ignition (Reagent test): not more than 0.1%.

Sulfonic Acid Cation-exchange Resin—See *Cation-exchange Resin, Sulfonic Acid.*

Sulfosalicylic Acid, $C_6H_3(COOH)(OH)(SO_3H)$-1,2,5.$2H_2O$—**254.21**—Use ACS reagent grade.

Sulfur—Use *Precipitated Sulfur* (USP monograph).

Sulfur Dioxide—Use *Sulfur Dioxide* (USP monograph).

Sulfur Dioxide Detector Tube—A fuse-sealed glass tube so designed that gas may be passed through it, and containing suitable absorbing filters and support media for an iodine-starch indicator.

NOTE—A suitable detector tube that conforms to the monograph specification is available from National Draeger, Inc., 401 Parkway View Drive, Pittsburgh, PA 15205 as Reference Number CH 31701, Measuring Range 1 to 25 ppm. Tubes having conditions other than those specified in the monograph may be used in accordance with the section entitled *Tests and Assays* in the *General Notices*.

Sulfuric Acid, H_2SO_4—**98.07**—Use ACS reagent grade.

Sulfuric Acid, Diluted (10 percent)—Cautiously add 57 mL of sulfuric acid to about 100 mL of water, cool to room temperature, and dilute with water to 1000 mL.

Sulfuric Acid, Fluorometric—Use ACS reagent grade Sulfuric Acid that conforms to the following additional test:

Fluorescence—Using a suitable fluorometer having a sharp cut-off 360-nm excitation filter and a sharp cut-off 415-nm excitation filter, determine the fluorescence of the sulfuric acid in a cuvette previously rinsed with water followed by several portions of the acid under examination: the fluorescence does not exceed that of quinine sulfate solution (1 in 1,600,000,000), similarly measured.

Sulfuric Acid, Fuming, H_2SO_4 plus free SO_3—having a nominal content of 15%, 20%, or 30% of free SO_3—Use ACS reagent grade (containing between 15.0% and 18.0%, between 20.0% and 23.0%, or between 30.0% and 33.0% of free SO_3).

Sulfurous Acid, H_2SO_3—**82.07**—A water solution of sulfur dioxide—Use ACS reagent grade.

Supports for Gas Chromatography—NOTE—Unless otherwise specified, mesh sizes of 80 to 100 or 100 to 120 are intended.

S1A[89]—Siliceous earth for gas chromatography that has been flux-calcined by mixing diatomite with Na_2CO_3 flux and calcining above 900°. The siliceous earth is acid-washed, then water-washed until neutral, but it is not base-washed. The siliceous earth may be silanized by treating with an agent such as dimethyldichlorosilane to mask surface silanol groups.

S1AB[89]—The siliceous earth, as described under *S1A*[89] above, is both acid- and base-washed.

S1C[89]—A support prepared from crushed firebrick and calcined or burned with a clay binder above 900° with subsequent acid-wash. It may be silanized.

S1NS—The siliceous earth is untreated.

S2—Styrene-divinylbenzene copolymer having a nominal surface area of less than 50 m^2 per g and an average pore diameter of 0.3 to 0.4 μm.

S3—Copolymer of ethylvinylbenzene and divinylbenzene having a nominal surface area of 500 to 600 m^2 per g and an average pore diameter of 0.0075 μm.

S4—Styrene-divinylbenzene copolymer with aromatic –O and –N groups, having a nominal surface area of 400 to 600 m^2 per g and an average pore diameter of 0.0076 μm.

S5—40- to 60-mesh, high-molecular weight tetrafluoroethylene polymer.

S6—Styrene-divinylbenzene copolymer having a nominal surface area of 250 to 350 m^2 per g and an average pore diameter of 0.0091 μm.[90]

S7—Graphitized carbon having a nominal surface area of 12 m^2 per g.

S8—Copolymer of 4-vinylpyridine and styrene-divinylbenzene.[91]

S9—A porous polymer based on 2,6-diphenyl-*p*-phenylene oxide.

S10—A highly polar cross-linked copolymer of acrylonitrile and divinylbenzene.

S11—Graphitized carbon having a nominal surface area of 100 m^2 per g modified with small amounts of petrolatum and polyethylene glycol compound.[92]

Talc—Use *Talc* (USP monograph).

Tannic Acid (*Tannin*)—Yellowish to light brown, glistening scales, or an amorphous powder. Is odorless or has a faint, characteristic odor. Very soluble in water and in alcohol; less soluble in dehydrated alcohol. Soluble in acetone; practically insoluble in benzene, in chloroform, and in ether. Store in light-resistant containers.

Solubility—A solution of 2 g in 10 mL of water is clear or practically so.

Residue on ignition (Reagent test)—Ignite 1 g with 1 mL of sulfuric acid: the residue weighs not more than 1 mg (0.1%).

Loss on drying ⟨731⟩—Dry it at 105° for 3 hours: it loses not more than 12% of its weight.

Dextrin, gum, and resinous substances—Dissolve 2 g in 10 mL of warm water: the solution is clear or not more than faintly turbid. Filter if necessary, and divide the filtrate into two equal portions. To one portion add 10 mL of alcohol. To the other portion add 10 mL of water: no turbidity is produced in either solution.

Heavy metals—To the *Residue on ignition* add 1 mL each of hydrochloric acid and nitric acid, and evaporate on a steam bath to dryness. Take up with 1 mL of 1 *N* hydrochloric acid and a few mL of hot water, dilute with water to 40 mL, and add 10 mL of hydrogen sulfide TS: any brown color produced is not darker than that of a control containing 0.02 mg of added Pb (0.002%).

Tartaric Acid, $H_2C_4H_4O_6$—**150.09**—Use ACS reagent grade.

Tertiary Butyl Alcohol—See *Butyl Alcohol, Tertiary*.

Testosterone—Use *Testosterone* (USP monograph).

Testosterone Benzoate, $C_{26}H_{32}O_2$—**376.54**—Use a suitable grade.

Testosterone Propionate—Use *Testosterone Propionate* (USP monograph).

Tetrabromophenolphthalein Ethyl Ester, $C_{22}H_{14}Br_4O_4$—**661.97**—Use ACS reagent grade.

Tetrabutylammonium Hydrogen Sulfate, $C_{16}H_{37}NO_4S$—**339.53**—White crystalline powder. Soluble in alcohol yielding a slightly hazy, colorless solution.

Assay—Dissolve about 170 mg, accurately weighed, in 40 mL of water. Titrate with 0.1 *N* sodium hydroxide VS determining the end-point potentiometrically. Perform a blank determination and make any necessary correction. Each mL of 0.1 *N* sodium hydroxide is equivalent to 33.95 mg of $C_{16}H_{37}NO_4S$. Not less than 97.0% is found.

Melting range ⟨741⟩: between 169° and 173°.

Tetrabutylammonium Hydroxide, 1.0 *M* in Methanol—Use a suitable grade.

Tetrabutylammonium Iodide, $(C_4H_9)_4NI$—**369.37**—White, shiny, crystalline flakes. Soluble in alcohol and in ether; slightly soluble in water.

Assay—Dissolve 200 mg, accurately weighed, in 40 mL of boiling water, with vigorous stirring, and cool to room temperature. Stir the solution by mechanical means, add 5 mL of 2 *N* nitric acid, and titrate with 0.1 *N* silver nitrate VS, determining the end-point potentiometrically, using a glass-silver electrode system and adding the titrant in 0.1-mL increments as the end-point is approached. Perform a complete blank determination, and make any necessary correction. Each mL of 0.1 *N* silver nitrate is equivalent to 36.94 mg of $(C_4H_9)_4NI$: not less than 99.0% is found.

Melting range ⟨741⟩: between 146° and 147°.

Tetrabutylammonium Phosphate, $(C_4H_9)_4NH_2PO_4$—**339.46**—White to off-white powder. Soluble in water.

Assay—Dissolve about 1.5 g, accurately weighed, in 100 mL of water. Without delay, titrate with 0.5 *N* sodium hydroxide VS, determining the end-point potentiometrically. Perform a blank determination, and make any necessary correction. Each mL of 0.5 *N* sodium hydroxide is equivalent to 169.7 mg of $(C_4H_9)_4NH_2PO_4$. Not less than 97.0% is found.

Tetracosane, $C_{24}H_{50}$—**338.66**—White powder.

Melting range ⟨741⟩: between 51° and 53°.

Tetradecane, $C_{14}H_{30}$—**198.39**—Clear, colorless liquid.

Assay—When examined by gas-liquid chromatography, it shows a purity of not less than 98%. The following conditions have been found suitable for assaying the reagent: A 2.4-m × 3-mm stainless steel column packed with phase G16 on support S1. The carrier gas is helium, flowing at 27.5 mL per minute, the column temperature is 250°, the port temperature is 200°, and the detector temperature is 280°. A flame-ionization detector is employed.

Melting range, Class II ⟨741⟩: between 4° and 8°, within a 2° range.

Refractive index ⟨831⟩: between 1.4280 and 1.4300 at 20°.

Tetraethylammonium Hydroxide, $(C_2H_5)_4NOH$—**147.26**—This base is known only in solution or as solid tetrahydrate or hexahydrate. In commerce it is usually available as a 10% or 25%

aqueous solution. This solution is clear and colorless, having a strong ammonia-like odor. Tetraethylammonium hydroxide is a strong base, absorbing CO_2 from the air. Specific gravity of the 10% solution: about 1.01. Preserve in tightly closed containers.

Assay—Accurately weigh a glass-stoppered flask with about 15 mL of water. Add a quantity of the solution equivalent to about 500 mg of tetraethylammonium hydroxide, and again weigh. Titrate the solution with 0.1 *N* hydrochloric acid VS, using methyl red TS as the indicator. Each mL of 0.1 *N* hydrochloric acid is equivalent to 14.73 mg of $(C_2H_5)_4NOH$: not less than 98.0% of the labeled amount is found.

Other amines—Weigh accurately a quantity of the solution, corresponding to about 500 mg of tetraethylammonium hydroxide, in a wide weighing bottle tared with 5 mL of water. Add a slight excess of dilute hydrochloric acid (1 in 12) (about 5 mL), evaporate on a steam bath to dryness, and dry at 105° for 2 hours. The weight of the tetraethylammonium chloride so obtained, multiplied by 0.8883, represents the quantity of $(C_2H_5)_4NOH$ in the weight of the specimen taken for the test, and it corresponds to within 1% of that found in the *Assay*, the percentage found in the *Assay* representing 100%.

Residue on evaporation—Evaporate 5 mL on a steam bath, and dry at 105° for 1 hour. The weight of the residue does not exceed 1.0 mg (0.02%). (Save the residue.)

Heavy metals ⟨231⟩—To the residue obtained in the test for *Residue on evaporation* add 0.25 mL of nitric acid and 1 mL of water, and evaporate on a steam bath to dryness. Take up the residue in 1 mL of dilute glacial acetic acid (1 in 18) and 5 mL of hot water, and dilute with water to 25 mL: the heavy metals limit is 5 ppm.

Tetraethylammonium Perchlorate, $(C_2H_5)_4NClO_4$—**229.70**—White crystals. Soluble in water. Use a suitable grade.

Tetraethylene Glycol, $C_8H_{18}O_5$—**194.23**—Nearly colorless liquid. Refractive index: about 1.46.

Assay—When examined by gas-liquid chromatography, using suitable gas chromatographic apparatus and conditions, it shows a purity of not less than 90%.

Boiling range (Reagent test): between 177° and 187°, at a pressure of 9 mm of Hg.

Tetraheptylammonium Bromide, $(C_7H_{15})_4NBr$—**490.17**—White, flaky powder.

Melting range ⟨741⟩: between 89° and 91°.

Tetrahydrofuran, C_4H_8O—**72.11**—Colorless liquid, having a characteristic, pungent odor. Miscible with water and with common organic solvents. When mixed with water, generates some heat and shrinks in volume; when mixed with chloroform, generates considerable heat. The name and concentration of any suitable preservative, not exceeding 0.1%, added to prevent formation of peroxides, are stated on the label. Preserve in small, well-filled containers, protected from light.

Specific gravity ⟨841⟩: between 0.884 and 0.886.

Distilling range, Method II ⟨721⟩: between 65° and 66°.

Acidity—Mix 5.0 mL with 10 mL of water and 1 drop of methyl red TS: any pink color produced is changed to yellow by addition of not more than 0.25 mL of 0.020 *N* sodium hydroxide.

Water, Method I ⟨921⟩: not more than 0.1%.

Residue on evaporation—Evaporate 10 mL (12 g) on a steam bath to dryness, and dry the residue at 105° for 1 hour: the weight of the residue is not more than 2 mg if a preservative is present, or not more than 1 mg if no preservative is declared on the label.

Tetrahydrofuran, Stabilizer-free—Use a suitable grade.[93]

1,2,3,4-Tetrahydronaphthalene, $C_{10}H_{12}$—**132.21**—Colorless liquid.

Refractive index ⟨831⟩: 1.5401 at 20°.

Tetramethylammonium Bromide, $(CH_3)_4NBr$—**154.05**—Colorless crystals. Soluble in water; sparingly soluble in absolute alcohol; insoluble in chloroform and in ether.

Assay—Transfer about 400 mg, accurately weighed, to a beaker, add 50 mL of water and 10 mL of diluted nitric acid, swirl to dissolve the test specimen, add 50.0 mL of 0.1 *N* silver nitrate VS, and mix. Add 2 mL of ferric ammonium sulfate TS, and titrate the excess silver nitrate with 0.1 *N* ammonium thiocyanate VS: each mL of 0.1 *N* silver nitrate consumed is equivalent to 15.41 mg of $(CH_3)_4NBr$. Not less than 98% is found.

Tetramethylammonium Chloride, $(CH_3)_4NCl$—**109.60**—Colorless crystals. Soluble in water and in alcohol; insoluble in chloroform.

Assay—Transfer about 200 mg, accurately weighed, to a beaker, add 50 mL of water and 10 mL of diluted nitric acid, swirl to dissolve the test specimen, add 50.0 mL of 0.1 *N* silver nitrate VS, and mix. Add 2 mL of ferric ammonium sulfate TS and 5 mL of nitrobenzene, shake, and titrate the excess silver nitrate with 0.1 *N* ammonium thiocyanate VS: each mL of 0.1 *N* silver nitrate is equivalent to 10.96 mg of $(CH_3)_4NCl$. Not less than 98% is found.

Tetramethylammonium Hydroxide, $(CH_3)_4NOH$—**91.15**—Available as an approximately 10% or approximately 25% aqueous solution, or as the crystalline pentahydrate. Is clear and colorless, and has a strong, ammonia-like odor. Tetramethylammonium hydroxide is a stronger base than ammonia, and rapidly absorbs carbon dioxide from the air. Store in tight containers.

Assay—Accurately weigh a glass-stoppered flask containing about 15 mL of water. Add a quantity of a solution of tetramethylammonium hydroxide, equivalent to about 200 mg of $(CH_3)_4NOH$, and again weigh. Add methyl red TS, and titrate the solution with 0.1 *N* hydrochloric acid VS: each mL of 0.1 *N* acid is equivalent to 9.115 mg of $(CH_3)_4NOH$.

Residue on evaporation—Evaporate 5 mL of solution on a steam bath, and dry at 105° for 1 hour: the weight of the residue is equivalent to not more than 0.02% of the weight of the test specimen.

Ammonia and other amines—Weigh accurately a quantity of solution, corresponding to about 300 mg of $(CH_3)_4NOH$, in a low-form weighing bottle tared with 5 mL of water. Add a slight excess of 1 *N* hydrochloric acid (about 4 mL), evaporate on a steam bath to dryness, and dry at 105° for 2 hours: the weight of the tetramethylammonium chloride so obtained, multiplied by 0.8317, represents the quantity, in mg, of $(CH_3)_4NOH$ in the portion of test specimen taken and corresponds to within 0.2% above or below that found in the *Assay*.

Tetramethylammonium Hydroxide, Pentahydrate, $(CH_3)_4NOH.5H_2O$—**181.23**—White to off-white crystals. Is hygroscopic. Strong base. *Keep well-closed.* Soluble in water and in methanol.

Assay—Weigh accurately about 800 mg, dissolve in 100 mL of water, and titrate with 0.1 *N* hydrochloric acid VS, determining the end-point potentiometrically. Perform a blank determination, and make any necessary correction. Each mL of 0.1 *N* hydrochloric acid is equivalent to 18.22 mg of $(CH_3)_4NOH.5H_2O$: not less than 98% is found.

Tetramethylammonium Hydroxide Solution in Methanol—A solution in methanol of tetramethylammonium hydroxide [$(CH_3)_4NOH$—**91.15**]. Is generally available in concentrations of 10% and 25%. The following specifications apply specifically to the 25% concentration; for other concentrations, appropriate adjustments in the procedures may be necessary.

Assay—Weigh accurately about 1 g of the solution, and dilute with water to about 50 mL. Add phenolphthalein TS, and titrate with 0.1 *N* hydrochloric acid VS to the disappearance of the pink color: each mL of 0.1 *N* hydrochloric acid VS is equivalent to 91.15 mg of $(CH_3)_4NOH$. Between 23% and 25% is found.

Clarity—A portion of it in a test tube is clear, or only slightly turbid, when viewed transversely.

Tetramethylammonium Nitrate, $(CH_3)_4NNO_3$—**136.15**—White crystals. Freely soluble in water.

4,4′-Tetramethyldiaminodiphenylmethane [*(4,4′-Methylenebis(N,N-dimethylaniline)*], $[(CH_3)_2NC_6H_4]_2CH_2$—**254.38**—Off-white crystals.

Melting range ⟨741⟩: between 87° and 90°.

Tetramethylsilane, $(CH_3)_4Si$—**88.22**—Use ACS reagent grade.

Tetraphenylcyclopentadienone—$C_{29}H_{20}O$—**384.48**—Dark gray to black crystals.

Melting range ⟨741⟩: between 217° and 220°.

Tetrasodium Ethylenediaminetetraacetate (*(Ethylenedinitrilo)tetraacetic Acid Tetrasodium Salt*), $C_{10}H_{12}N_2Na_4O_8$—**380.17**—Fine, white, crystalline powder. Soluble in water.

Loss on drying ⟨731⟩—Dry it at 105° for 4 hours: it loses not more than 8% of its weight.

Thallous Chloride, TlCl—**239.82**—Fine, white, crystalline powder. Soluble in about 260 parts of cold water and in about 70 parts of boiling water; insoluble in alcohol. *Poisonous; use with adequate ventilation.*

Assay—Dissolve about 500 mg, accurately weighed, in a mixture of 80 mL of water and 0.5 mL of sulfuric acid. When dissolution is complete, add 20 mL of hydrochloric acid. Heat to 60° and maintain this temperature while titrating with 0.1 *N* ceric sulfate VS, determining the end-point potentiometrically, using silver–silver chloride and platinum electrodes. Each mL of 0.1 *N* ceric sulfate is equivalent to 11.99 mg of TlCl. Not less than 99% is found.

Thiamine Hydrochloride—Use *Thiamine Hydrochloride* (USP monograph).

Thiazole Yellow (*CI Direct Yellow 9; Clayton Yellow; Titan Yellow*), $C_{28}H_{19}N_5Na_2O_6S_4$—**695.71**—Yellowish brown powder. Soluble in water and in alcohol to yield in each instance a yellow solution; soluble in dilute alkali to yield a brownish red solution. Protect from light.

Solubility—A 200-mg portion mixed with 50 mL of water shows not more than a faint haze.

Residue on ignition—Weigh accurately about 1.5 g, previously dried at 105° for 2 hours, and ignite until thoroughly charred. Cool, add 2 mL of nitric acid and 2 mL of sulfuric acid, ignite gently to expel excess acids, then at 600° to 800° to constant weight: the residue of sodium sulfate (Na_2SO_4) is between 19.8% and 21.5% of the weight of the test specimen (theory is 20.4%).

Sensitiveness to magnesium—Add 0.2 mL of a solution (1 in 10,000) and 2 mL of 1 *N* sodium hydroxide to a mixture of 9.5 mL of water and 0.5 mL of a solution prepared by dissolving 1.014 g of clear crystals of magnesium sulfate in water, diluting with water to 100 mL, then diluting 10 mL of the resulting solution with water to 1 liter: a distinct pink color is produced within 10 minutes.

Thimerosal, $C_9H_9HgNaO_2S$—**404.81**—Use *Thimerosal* (USP monograph).

2-Thiobarbituric Acid, $C_4H_4N_2O_2S$—**144.15**—White leaflets. Slightly soluble in water.

Melting temperature ⟨741⟩: 236°, with decomposition.

2,2′-Thiodiethanol, $(HOCH_2CH_2)_2S$—**122.18**—Pale yellow to colorless liquid.

Assay—Not less than 98% of $C_4H_{10}O_2S$ is found, a suitable gas chromatograph equipped with a flame-ionization detector being used. The following conditions have been found suitable: a 1.83-m × 4.0-mm glass column is packed with 10 percent phase G25 on support S1A; the column, injection port, and detector temperatures are maintained at 200°, 250°, and 310°, respectively.

Refractive index ⟨831⟩: between 1.4250 and 1.4270, at 20°.

Thioglycolic Acid (*Thioglycolic Acid*), $HSCH_2COOH$—**92.11**—A colorless or nearly colorless liquid, having a strong, unpleasant odor. Miscible with water. Soluble in alcohol.

Residue on ignition (Reagent test): not more than 0.1%.

Solubility—A solution of 1 mL in 10 mL of water is clear and colorless.

Sensitiveness—Mix 1 mL with 2 mL of stronger ammonia water, and dilute with water to 20 mL. Add 1 mL of this solution to a mixture of 20 mL of water and 0.1 mL of dilute ferric chloride TS (1 in 100), then add 5 mL of ammonia TS: a distinct pink color is produced.

Thiourea, $(NH_2)_2CS$—**76.12**—White, odorless crystals or a white, crystalline powder. Soluble in water and in alcohol.

Assay—Weigh accurately 1 g, transfer to a 250-mL volumetric flask, dissolve in 50 mL of water, and dilute with water to volume. Pipet 20 mL of the well-mixed solution into a suitable flask, and add 25.0 mL of 0.1 *N* silver nitrate VS and 10 mL of ammonia TS. Shake vigorously for 2 minutes, heat to boiling, and cool. Add 60 mL of diluted nitric acid, shake vigorously, filter, and wash the residue well with water. Add 2 mL of ferric ammonium sulfate TS to the combined filtrate and washings, and titrate with 0.1 *N* ammonium thiocyanate VS. Each mL of 0.1 *N* silver nitrate consumed is equivalent to 3.806 mg of $(NH_2)_2CS$. Not less than 99% is found.

Solubility—A solution of 1 g in 20 mL of water is complete, clear, and colorless.

Melting range, Method I ⟨741⟩: between 176° and 182°.

Loss on drying ⟨731⟩—Dry it at 105° for 2 hours: it loses not more than 0.5% of its weight.

Residue on ignition (Reagent test)—Ignite 1 g with 0.5 mL of sulfuric acid: the residue weighs not more than 1.5 mg (0.15%).

Sensitiveness—Dissolve 280 mg of bismuth subnitrate in 12 mL of nitric acid, and dilute with water to 200 mL. Dilute 1 mL of this solution with water to 100 mL, and to 10 mL of the dilution add 1 mL of test solution (1 in 5): a distinct yellow color is produced immediately.

Thorium Nitrate, $Th(NO_3)_4 \cdot 4H_2O$—**552.12**—Use ACS reagent grade.

Thrombin—Use *Thrombin* (USP monograph).

Thromboplastin—Buff-colored powder, or opalescent or turbid suspension. It exhibits thrombokinase activity derived from the acetone-extracted brain and/or lung tissue of freshly killed rabbits. It may contain sodium chloride and calcium chloride in suitable proportions, and it may contain a suitable antimicrobial agent. It may have the characteristic odor of dried animal tissue. It is used in suspension form for the determination of the prothrombin time and activity of blood. Its thrombokinase activity is such that it gives a clotting time of 11 to 16 seconds with normal human plasma and the proper concentration of calcium ions. Store in tight containers, preferably at a temperature below 5°.

Loss on drying ⟨731⟩—[NOTE—This test is applicable only to the dry form.] Dry it in vacuum at 60° for 6 hours: it loses not more than 5.0% of its weight.

Thymol, $C_6H_3[CH_3][OH][CH(CH_3)_2]1,3,4$—**150.22**—Colorless, often large, crystals, or a white, crystalline powder, having an aromatic, thyme-like odor and a pungent taste. Is affected by light. Has greater density than water, but when liquefied by fusion is less dense than water. Its alcohol solutions are neutral to litmus. One g dissolves in about 1000 mL of water, in 1 mL of alcohol, in 1 mL of chloroform, in 1.5 mL of ether, and in about 2 mL of olive oil. Soluble in glacial acetic acid and in fixed or volatile oils. Store in tight, light-resistant containers.

Melting range ⟨741⟩: between 48° and 51°, but when melted it remains liquid at a considerably lower temperature.

Nonvolatile matter—Volatilize 2 g on a steam bath, and dry at 105° to constant weight: the residue weighs not more than 1 mg (0.05%).

Tin, Sn—At. Wt. **118.69**—Use ACS reagent grade.

Titan Yellow—See *Thiazole Yellow.*

Titanium Tetrachloride, $TiCl_4$—**189.69**—Clear, colorless liquid. Fumes in air. [*Caution—It reacts violently with water.*]

Assay—Weigh accurately 0.75 g into 100 mL of 2 *N* sulfuric acid contained in a Smith weighing buret. Pour the solution through a zinc-mercury reduction column into 50 mL of 0.1 *N* ferric ammonium sulfate VS. Elute with 100 mL of 2 *N* sulfuric acid and 100 mL of water. Add 10 mL of phosphoric acid, and titrate with 0.1 *N* potassium permanganate VS. Perform a blank determination, and make any necessary correction. Each mL of 0.1 *N* potassium permanganate is equivalent to 18.97 mg of $TiCl_4$. Not less than 99.5% is found.

Boiling range (Reagent test): between 135° and 140°.

Titanium Trichloride (*Titanous Chloride*), $TiCl_3$—**154.24**—Black, hygroscopic powder, unstable in air. Soluble in water, the solution depositing titanic acid on exposure to air. Is available usually as 15% to 20%, dark violet-blue, aqueous solutions. Store the solution in tightly closed, glass-stoppered bottles, protected from light.

***o*-Tolidine** (*4,4′-Diamino-3,3′-dimethylbiphenyl*), $(NH_2)(CH_3)$-$C_6H_3 \cdot C_6H_3(CH_3)(NH_2)$-4,3,3′,4′—**212.29**—White to reddish crystals or crystalline powder. Slightly soluble in water; soluble in alcohol, in ether, and in dilute acids. Preserve in well-closed containers, protected from light.

Caution—Avoid contact with o-tolidine and mixtures containing o-tolidine, and conduct all tests in a well-ventilated fume hood.

Melting range ⟨741⟩: between 129° and 131°.

Tolualdehyde (*o-Tolualdehyde*), C_8H_8O—**120.15**—Use a suitable grade.

***p*-Tolualdehyde,** C_8H_8O—**120.15**—Colorless to yellow, clear liquid.

Assay—When examined by gas-liquid chromatography, it shows a purity of not less than 98%. The following conditions have been found suitable for assaying the article: A 1.8-m × 3-mm stainless steel column packed with a 5 percent phase G4 on support S1. Nitrogen, having a flow rate of about 12 mL per minute, is the carrier gas, the detector and column temperature are about 125°, and the injection port temperature is about 205°. A flame-ionization detector is employed and the specimen is a 5% solution in carbon disulfide.
Refractive index ⟨831⟩: between 1.544 and 1.546, at 20°.

Toluene (*Toluol*), $C_6H_5CH_3$—**92.14**—Use ACS reagent grade.

***p*-Toluenesulfonic Acid,** $CH_3C_6H_4SO_3H.H_2O$—**190.21**—White, hygroscopic crystals or crystalline powder. Soluble in water, in alcohol, and in ether.
Assay—Weigh accurately about 5 g, previously dried over sulfuric acid for 18 hours, and dissolve in about 250 mL of water contained in a 500-mL conical flask. Add 0.15 mL of bromothymol blue TS, and titrate with 1 *N* sodium hydroxide VS to a blue end-point. Each mL of 1 *N* sodium hydroxide is equivalent to 190.2 mg of $CH_3C_6H_4SO_3H.H_2O$. Not less than 99% is found.
Melting range ⟨741⟩: between 104° and 106°, the test specimen having been dried over sulfuric acid for 18 hours.
Loss on drying ⟨731⟩—Dry it over sulfuric acid to constant weight: it loses not more than 1% of its weight.
Solubility—Separate 200-mg portions dissolve completely in 5 mL of alcohol and in 5 mL of ether, respectively.
Residue on ignition ⟨281⟩: negligible, from 200 mg.
Free sulfate—Dissolve 500 mg in 10 mL of water, and add 1 mL of dilute hydrochloric acid (1 in 20) and 1 mL of barium chloride TS: any turbidity produced within 10 minutes does not exceed that of a control containing 0.05 mg of added SO_4 (0.01%).

***p*-Toluenesulfonyl-L-arginine Methyl Ester Hydrochloride,** $C_{14}H_{22}N_4O_4S.HCl$—**378.87**—Determine its suitability as directed in the test for *Trypsin* under *Chymotrypsin* (USP monograph).

***p*-Toluic Acid,** $CH_3C_6H_4COOH$—**136.15**—White, crystalline powder. Sparingly soluble in hot water; very soluble in alcohol, in methanol, and in ether.
Assay—Transfer about 650 mg, accurately weighed, to a suitable container, dissolve in 125 mL of alcohol, add 25 mL of water, and mix. Titrate with 0.5 *N* sodium hydroxide VS, determining the end-point potentiometrically. Perform a blank determination, and make any necessary correction. Each mL of 0.5 *N* sodium hydroxide is equivalent to 68.07 mg of $C_8H_8O_2$: not less than 98% is found.
Melting range ⟨741⟩: over a range of 2° that includes 181°.

***o*-Toluidine** (*2-Aminotoluene; 2-Methylaniline*), $C_6H_4(CH_3)(NH_2)$-1,2—**107.15**—Light yellow liquid becoming reddish brown on exposure to air and light. Slightly soluble in water; soluble in alcohol, in ether, and in dilute acids. Preserve in well-closed containers, protected from light.
Specific gravity ⟨841⟩: 1.008 at 20°.
Boiling range (Reagent test): between 200° and 202°.

Toluidine Blue (*Tolonium Chloride; Toluidine Blue O; Tolazul*), 3-amino-7-dimethylamino-2-methylphenazathionium chloride—**305.83**—Dark green powder, soluble in water and slightly soluble in alcohol.
Toluidine Blue Solution—Dissolve 15 mg of Toluidine Blue in 1 mL of alcohol, and dilute with water to 100 mL.

***p*-Toluidine,** C_7H_9N—**107.15**—White to beige crystals or flakes. Freely soluble in alcohol, in acetone, in methanol, and in dilute acids; slightly soluble in water.
Assay—Dissolve 400 mg, accurately weighed, in 100 mL of glacial acetic acid, and titrate with 0.1 *N* perchloric acid VS, determining the end-point potentiometrically. Perform a blank determination, and make any necessary correction. Each mL of 0.1 *N* perchloric acid is equivalent to 10.72 mg of $CH_3C_6H_4NH_2$. Not less than 98%, calculated on the dried basis, is found.
Loss on drying—Weigh accurately about 1 g, and dry at 30° to constant weight: it loses not more than 2% of its weight.

***n*-Triacontane,** $C_{30}H_{62}$—**422.82**—Use a suitable grade.

Triamcinolone Diacetate—Use *Triamcinolone Diacetate* (USP monograph).

Tributyl Phosphate (*Tri-n-butyl Phosphate*), $(C_4H_9)_3PO_4$—**266.32**—Clear, almost colorless liquid. Slightly soluble in water. Miscible with common organic solvents. Specific gravity: about 0.976.
Refractive index ⟨831⟩: between 1.4205 and 1.4225.

Tributylethylammonium Hydroxide, $C_{14}H_{33}NO$—**231.42**—Use a suitable grade.

Tributyrin (*Glyceryl Tributyrate*), $C_{15}H_{26}O_6$—**302.37**—Colorless, oily liquid. Insoluble in water; very soluble in alcohol and in ether.
Assay—Inject an appropriate specimen into a suitable gas chromatograph (see *Chromatography* ⟨621⟩) equipped with a flame-ionization detector, nitrogen being used as the carrier gas. The following conditions have been found suitable: a 1.8-m × 3-mm stainless steel column containing phase G4 on support S1A; the injection port temperature is maintained at 270°; the detector temperature is maintained at 300°. The area of the tributyrin peak is not less than 98% of the total peak area.
Refractive index ⟨831⟩: between 1.4345 and 1.4365 at 20°.
Acid content—Transfer 1.0 g, accurately weighed, to a beaker, add 75 mL of methanol, and dissolve by stirring. When dissolution is complete, add 25 mL of water, and titrate with 0.05 *N* potassium hydroxide VS, using phenolphthalein TS as indicator. Perform a blank determination, and make any necessary correction. Each mL of 0.05 *N* potassium hydroxide is equivalent to 88.1 mg of butyric acid: not more than 0.5% is found.

Trichloroacetic Acid, CCl_3COOH—**163.39**—Use ACS reagent grade.

Trichloroethane—See *Methyl Chloroform.*

Trichlorotrifluoroethane—Use a suitable grade.[94]

***n*-Tricosane,** $C_{23}H_{48}$—**324.63**—Colorless or white, more or less translucent mass, showing a crystalline structure. Odorless and tasteless, or practically so. Has a slightly greasy feel. Insoluble in water and in alcohol; soluble in chloroform, in ether, in volatile oils, and in most warm fixed oils; slightly soluble in dehydrated alcohol. Boils at about 380°.
Melting range ⟨741⟩: between 47° and 49°.
Suitability—Determine its suitability for use in the test for *Related compounds* under *Propoxyphene Hydrochloride* (USP monograph) as follows. Dissolve a suitable quantity in chloroform to yield a solution containing 20 µg per mL. Following the directions given in the test for *Related compounds* under *Propoxyphene Hydrochloride*, inject a suitable volume of the solution into the chromatograph, and record the chromatogram. Concomitantly record the chromatogram from the *Standard preparation* prepared as directed in the test for *Related compounds:* only one main peak is obtained from the *n*-tricosane solution, and no minor peaks are observed at, or near, the peak positions obtained for propoxyphene, acetoxy, or carbinol in the chromatogram from the *Standard preparation.*

Triethanolamine—See *Trolamine.*

Triethylamine, $(C_2H_5)_3N$—**101.19**—Colorless liquid, having a strong, ammoniacal odor. Slightly soluble in water. Miscible with alcohol, with ether, and with cold water. Store in well-closed containers.
Boiling range (Reagent test): between 89° and 90°.
Absorbance—To 1 mL in a 50-mL volumetric flask add 10 mL of methanol and 1 mL of hydrochloric acid, and add chloroform to volume. The absorbance of this solution, determined at the wavelength of maximum absorbance at about 285 nm, with a suitable spectrophotometer, does not exceed 0.01. [NOTE—If the absorbance exceeds 0.01, purify the triethylamine as follows: Reflux 100 mL with 20 mL of water and 2 g of sodium hydrosulfite for not less than 8 hours, wash with water, dry by refluxing, using a Dean-Stark trap, and distil, collecting only the first 75 mL of filtrate. Store over anhydrous sodium carbonate or anhydrous potassium carbonate.]

Triethylene Glycol, $C_6H_{14}O_4$—**150.17**—Colorless to pale yellow liquid. Is hygroscopic. Miscible with water, with alcohol, and with toluene.
Assay—Inject an appropriate test specimen into a suitable gas chromatograph equipped with a flame-ionization detector (see *Chromatography* ⟨621⟩), helium being used as the carrier gas. The following conditions have been found suitable: a 1.85-m × 3-mm stainless steel column packed with support S2; the injection port, column, and detector temperatures are maintained at 250°,

230°, and 310°, respectively. The area of the $C_6H_{14}O_4$ peak is not less than 97% of the total peak area.

Refractive index ⟨831⟩: between 1.4550 and 1.4570, at 20°.

Trifluoperazine Hydrochloride—Use *Trifluoperazine Hydrochloride* (USP monograph).

Trifluoroacetic Anhydride, $(F_3CCO)_2O$—**210.03**—Colorless liquid. Boils between 40° and 42°. Extremely volatile. Avoid exposure to air or water.

Assay—Transfer about 0.8 g, accurately weighed, to a glass-stoppered flask containing 50 mL of methanol. Add 500 mg of phenolphthalein, and titrate with 0.1 *N* sodium methoxide VS to a pink end-point. Calculate *A* by the formula:

$$V/W,$$

in which *V* is the volume, in mL, of 0.1 *N* sodium methoxide and *W* is the weight, in mg, of test specimen. To a second glass-stoppered flask containing 50 mL of dimethylformamide-water mixture (1:1) transfer 0.4 g, accurately weighed, of the specimen under test, add 500 mg of phenolphthalein, and titrate with 0.1 *N* sodium hydroxide VS to a pink end-point. Calculate *B* by the formula:

$$V^1/W^1,$$

in which V^1 is the volume, in mL, of 0.1 *N* sodium hydroxide and W^1 is the weight, in mg, of test specimen. Calculate the percentage of $(F_3CCO)_2O$ by the formula:

$$21.003(B - A).$$

Not less than 97% is found. If 2*A* is greater than *B*, calculate the percentage of $(F_3CCO)_2O$ by the formula:

$$11.403(2A - B).$$

Triketohydrindene Hydrate (*Ninhydrin*), $C_9H_4O_3 . H_2O$—**178.14**—White to brownish white crystals or crystalline powder. Soluble in water and in alcohol; slightly soluble in ether and in chloroform. When heated above 100°, it becomes red. Store it protected from light.

Melting range ⟨741⟩: between 240° and 245°, with decomposition, determined in a bath preheated at 220°.

Residue on ignition (Reagent test): negligible, from 100 mg.

Sensitiveness—Prepare a solution of 10 mg of aminoacetic acid in 25 mL of water. To 1 mL of this solution add a solution of 50 mg of sodium acetate in 2 mL of water, then add 0.2 mL of a solution of 5 mg of triketohydrindene hydrate in 1 mL of water, and boil the mixture for 1 to 2 minutes: a violet color is produced, and it becomes intense on standing for a few minutes.

Trimethylacethydrazide Ammonium Chloride (*Betaine Hydrazide Chloride; Girard Reagent T*), $[(CH_3)_3N{+}CH_2CONHNH_2]Cl^-$—**167.64**—Colorless or white crystals. Freely soluble in water. One g dissolves in about 25 mL of alcohol. Insoluble in chloroform and in ether. Hygroscopic.

Melting range ⟨741⟩: between 185° and 192°, determined after recrystallization from hot alcohol, if necessary.

Residue on ignition (Reagent test)—Ignite 1 g with 0.5 mL of sulfuric acid: the residue weighs not more than 10 mg (1%).

Trimethylchlorosilane—See *Chlorotrimethylsilane.*

2,2,4-Trimethylpentane (*Isooctane*), C_8H_{18}—**114.23**—Use ACS reagent grade.

2,4,6-Trimethylpyridine (*5-Collidine*), $C_8H_{11}N$—**121.18**—Clear, colorless liquid, having an aromatic odor. Soluble in cold water and less soluble in hot water; soluble in alcohol, in chloroform, and in methanol. Miscible with ether.

Assay—Inject an appropriate test specimen into a suitable gas chromatograph (see *Chromatography* ⟨621⟩), helium being used as a carrier gas. The following conditions have been found suitable: A 1.85-m × 3-mm stainless steel column containing phase G16 on support S1A; the injection port, column, and detector temperatures are maintained at 180°, 165°, and 270°, respectively, and a flame-ionization detector is used. The area of the $C_8H_{11}N$ peak is not less than 98% of the total peak area.

Refractive index ⟨831⟩: between 1.4970 and 1.4990, at 20°.

***N*-(Trimethylsilyl)-imidazole,** $C_6H_{12}N_2Si$—**140.26**—A clear, colorless, to light yellow liquid.

Refractive index ⟨831⟩: between 1.4744 and 1.4764 at 20°.

Trinitrophenol—See *Picric Acid.*

Trioctylphosphine Oxide, $C_{24}H_{51}PO$—**386.64**—White, crystalline powder. Insoluble in water; soluble in organic solvents.

Melting range ⟨741⟩: between 54° and 56°.

Trioxsalen—Use *Trioxsalen* (USP monograph).

Triphenylchloromethane (*Trityl Chloride*), $(C_6H_5)_3CCl$—**278.78**—White to grayish white or yellowish crystals or crystalline powder. Soluble in water with decomposition; soluble in glacial acetic acid. Melts between 112° and 113°. To separate 5-mL portions of a saturated solution of triphenylchloromethane in glacial acetic acid add 1 mL of water and 1 mL of hydrochloric acid: white and yellow precipitates, respectively, are formed.

Triphenylstibine—Use a suitable grade.[7]

Triphenyltetrazolium Chloride, $C_{19}H_{15}ClN_4$—**334.81**—White to yellowish, crystalline powder. Soluble in about 10 parts of water and of alcohol; slightly soluble in acetone; insoluble in ether. Usually contains solvent of crystallization, and when dried at 105° it melts at about 240° with decomposition.

Solubility—Separate 100-mg portions dissolve completely in 10 mL of water and in 10 mL of alcohol, respectively, to yield solutions that are clear, or practically so.

Loss on drying ⟨731⟩—Dry it at 105° to constant weight: it loses not more than 5.0% of its weight.

Residue on ignition (Reagent test): negligible, from 100 mg.

Sensitiveness—Dissolve 10 mg in 10 mL of dehydrated alcohol (*A*). Then dissolve 10 mg of dextrose in 20 mL of dehydrated alcohol (*B*). To 0.2 mL of *B* add 1 mL of dehydrated alcohol and 0.5 mL of dilute tetramethylammonium hydroxide TS (1 volume diluted with 9 volumes of dehydrated alcohol), then add 0.2 mL of *A:* a pronounced red color develops within about 10 minutes.

Tris(2-aminoethyl)amine, $C_6H_{18}N_4$—**146.24**—Yellow liquid. Soluble in methanol.

Assay—Dissolve about 80 mg in 30 mL of methanol. Add 40 mL of water and titrate with 1 *N* hydrochloric acid, determining the end-point potentiometrically. Perform a blank determination and make any necessary correction. Each mL of 1 *N* hydrochloric acid is equivalent to 48.75 mg of $C_6H_{18}N_4$. Not less than 98.0% is found.

Refractive index ⟨831⟩: between 1.4956 and 1.4986 at 20°.

Tris(hydroxymethyl)aminomethane—Use ACS reagent grade—See also *Tromethamine.*

Trolamine—Use *Trolamine* (NF monograph).

Tromethamine [*Tris(hydroxymethyl)aminomethane; THAM; 2-Amino-2-(hydroxymethyl)-1,3-propanediol*], $C_4H_{11}NO_3$—**121.14**—Use ACS reagent grade Tris(hydroxymethyl)aminomethane.

Tropaeolin OO (*Acid Orange 5*), $C_{18}H_{14}N_3NaO_3S$—**375.38**—Orange-yellow scales, or yellow powder. Soluble in water.

pH range: from 1.4 (red) to 2.6 (yellow).

Tryptone—Use *Pancreatic Digest of Casein.*

L-Tryptophane, $C_{11}H_{12}N_2O_2$—**204.23**—White or not more than slightly yellow leaflets or powder. One g dissolves in about 100 mL of water; slightly soluble in alcohol; soluble in dilute acids and in solutions of the alkali hydroxides.

Assay—Weigh accurately about 300 mg, dissolve in a mixture of 3 mL of formic acid and 50 mL of glacial acetic acid, add 2 drops of crystal violet TS, and titrate with 0.1 *N* perchloric acid VS to a green end-point. Each mL of 0.1 *N* perchloric acid is equivalent to 20.42 mg of $C_{11}H_{12}N_2O_2$. Between 98.0% and 102.0%, calculated on the dried basis, is found.

Specific rotation ⟨781⟩: between −30.0° and −33.0°, determined in a solution containing 1.0 g of test specimen, previously dried at 105° for 3 hours, in 100 mL.

Loss on drying ⟨731⟩—Dry it at 105° for 3 hours: it loses not more than 0.3% of its weight.

Residue on ignition (Reagent test): not more than 0.1%.

Tyrosine—Dissolve 100 mg in 3 mL of diluted sulfuric acid, add 10 mL of mercuric sulfate TS, and heat on a steam bath for 10 minutes. Filter, wash with 5 mL of mercuric sulfate TS, and add to the combined filtrate 0.5 mL of sodium nitrite solution (1 in 20): no red color is produced within 15 minutes.

Tyrosine—Use *Tyrosine* (USP monograph).

Uracil, $C_4H_4N_2O_2$—**112.09**—White to cream-colored, crystalline powder. Melts above 300°. One g dissolves in about 500 mL of water; less soluble in alcohol; soluble in ammonia TS and in sodium hydroxide TS. Its solutions yield no precipitate with the usual alkaloidal precipitants.

Residue on ignition (Reagent test): negligible, from 100 mg.

Loss on drying ⟨731⟩—Dry it at 105° for 2 hours: it loses not more than 2% of its weight.

Uranyl Acetate (*Uranium Acetate*), $UO_2(C_2H_3O_2)_2 . 2H_2O$—**424.15**—Use ACS reagent grade.

Urea, NH_2CONH_2—**60.06**—Use ACS reagent grade.

Urethane (*Ethyl carbamate*), $C_3H_7NO_2$—**89.09**—White powder with chunks. Freely soluble in water.

Melting range ⟨741⟩: between 48° and 50°.

Vanadium Pentoxide, V_2O_5—**181.88**—Fine, yellow to orange-yellow powder. Slightly soluble in water; soluble in concentrated acids and in alkalies; insoluble in alcohol.

Assay—Transfer about 400 mg, accurately weighed, to a 500-mL conical flask, and add 150 mL of water and 30 mL of dilute sulfuric acid (1 in 2). Boil the solution on a hot plate for 5 minutes, add 50 mL of water, and continue boiling until a yellow solution is obtained. Transfer the hot plate and the flask to a well-ventilated hood, and bubble sulfur dioxide gas through the solution for 10 minutes, or until the solution is a clear, brilliant blue color. Rinse the gas delivery tube into the flask with a few mL of water, then bubble carbon dioxide gas through the solution for 30 minutes while continuing to boil the solution gently. Cool the solution to about 80°, and titrate with 0.1 *N* potassium permanganate VS to a yellow-orange end-point. Perform a complete blank determination, and make any necessary correction. Each mL of 0.1 *N* potassium permanganate is equivalent to 9.095 mg of V_2O_5. Not less than 99.5% is found.

Vanadyl Sulfate, $VOSO_4 . xH_2O$ (anhydrous)—**163.00**—Blue, hygroscopic crystals. Slowly and usually incompletely soluble in water.

Assay—Weigh accurately about 400 mg of the dried test specimen obtained in the test for *Water*, and transfer with 15 to 20 mL of water into a beaker. Add 3 mL of sulfuric acid, cover the beaker with a watch glass, and heat on a steam bath until all dissolves. Cool, dilute with 125 mL of water, and titrate with 0.1 *N* potassium permanganate VS to the production of a pinkish color that persists for 1 minute: each mL of 0.1 *N* potassium permanganate is equivalent to 16.30 mg of $VOSO_4$. Not less than 97% is found.

Water—Dry about 1 g, accurately weighed, at 220° to constant weight: it loses not more than 50.0% of its weight.

Pentavalent vanadium—Heat 1 g, accurately weighed, with 50 mL of water and 5 mL of hydrochloric acid in a flask until dissolved. Cool, add 2 g of potassium iodide, insert the stopper, and allow to stand for 30 minutes. Add 50 mL of water, and titrate the liberated iodine with 0.1 *N* sodium thiosulfate VS, adding 3 mL of starch TS as the indicator. Correct for the volume of thiosulfate consumed by a blank. Each mL of 0.1 *N* thiosulfate is equivalent to 5.095 mg of vanadium (V). Not more than 0.5% is found, calculated on the dried basis.

Substances not precipitated by ammonia—Dissolve 1.0 g by heating with 20 mL of water and 2 mL of hydrochloric acid. Dilute with water to about 75 mL, and neutralize to litmus paper with ammonia TS. Transfer the solution to a cylinder, slowly add 5 mL of ammonia TS and sufficient water to make 100 mL, and allow to stand overnight. Decant 50 mL of the supernatant liquid through a filter, add 5 drops of sulfuric acid, evaporate to dryness, and ignite: the residue weighs not more than 10 mg (2.0%).

Vanillin—Use *Vanillin* (NF monograph).

Washed Sand—See *Sand, Washed.*

Water, Carbon Dioxide-free—See *Water*, in the introductory section.

Water, Deaerated—See *Water*, in the introductory section.

Water for Injection—Use *Water for Injection* (USP monograph).

Water Vapor Detector Tube—A fuse-sealed glass tube so designed that gas may be passed through it, and containing suitable absorbing filters and support media for the indicator, which consists of a selenium sol in suspension in sulfuric acid.

NOTE—A suitable detector tube that conforms to the monograph specification is available from National Draeger, Inc., 401 Parkway View Drive, Pittsburgh, PA 15205 as Reference Number 67 28531, Measuring Range 5 to 200 mg per cubic meter. Tubes having conditions other than those specified in the monograph may be used in accordance with the section entitled *Tests and Assays* in the *General Notices.*

Xanthine, $C_5H_4N_4O_2$—**152.11**—White, crystalline powder. Decomposes on heating. Slightly soluble in water and in alcohol; soluble in sodium hydroxide TS; sparingly soluble in diluted hydrochloric acid. When subjected to the murexide reaction, a purple color is produced with the ammonia, but on the subsequent addition of fixed alkali hydroxides, the color is not discharged but is changed to violet.

Residue on ignition (Reagent test): negligible, from 100 mg.

Loss on drying ⟨731⟩—Dry it at 105° for 2 hours: it loses not more than 1% of its weight.

Xanthydrol, $C_{13}H_{10}O_2$—**198.22**—Pale yellow, crystalline powder. Insoluble in water; soluble in alcohol, in chloroform, and in ether. Soluble in glacial acetic acid, forming a practically colorless solution; but when the powder is treated with diluted hydrochloric acid, a lemon-yellow color is produced.

Melting range ⟨741⟩: between 121° and 123°.

Residue on ignition—Ignite 500 mg with 0.5 mL of sulfuric acid: the residue weighs not more than 10 mg (2.0%).

Xylene, C_8H_{10}—**106.17**—Use ACS reagent grade.

***o*-Xylene,** C_8H_{10}—**106.17**—Clear, colorless, mobile, flammable liquid. Insoluble in water; miscible with alcohol and with ether.

Assay—When examined by gas-liquid chromatography, it shows a purity of not less than 95%. The following conditions have been found suitable for assaying the substance: A 1.8-m × 3-mm stainless steel column packed with 1.75% hydrated aluminum silicate plus 5.0 percent diisodecylphthalate on support S1. Helium, having a flow rate of about 27.5 mL per minute is the carrier gas, the detector temperature is about 280°, the injection port temperature is about 180°, and the column temperature is 80°. A flame-ionization detector is employed.

Refractive index ⟨831⟩: between 1.5040 and 1.5060, at 20°.

Xylene Cyanole FF, $C_{25}H_{27}N_2NaO_6S_2$—**538.61**—Gray-blue to dark blue powder. Soluble in water.

Assay—Transfer about 50 mg, accurately weighed, to a 100-mL volumetric flask, dissolve in water, dilute with water to volume, and mix. Transfer 2.0 mL of the solution to a 50-mL volumetric flask, dilute with pH 7.0 phosphate buffer to volume (see *Solutions*, in this section), and mix. Using a suitable spectrophotometer, 1-cm cells, and water as the blank, record the absorbance of the solution at the wavelength of maximum absorbance at about 614 nm. From the observed absorbance, calculate the absorptivity (see *Spectrophotometry and Light-scattering* ⟨851⟩): the absorptivity is not less than 55.9, corresponding to about 83% of $C_{25}H_{27}N_2NaO_6S_2$.

Loss on drying ⟨731⟩—Dry it at 110° to constant weight: it loses not more than 6.0% of its weight.

Xylometazoline Hydrochloride—Use *Xylometazoline Hydrochloride* (USP monograph).

Xylose, $C_5H_{10}O_5$—**150.13**—Use a suitable grade.

Yeast Extract—A water-soluble, peptone-like derivative of yeast cells (Saccharomyces) prepared under optimum conditions, clarified, and dried to a reddish yellow to brown powder, having a characteristic but not putrescent odor. Soluble in water, forming a yellow to brown solution, having a slightly acid reaction. Contains no added carbohydrate. One g represents not less than 7.5 g of yeast.

Loss on drying ⟨731⟩—Dry it at 105° to constant weight: it loses not more than 5% of its weight.

Residue on ignition—Ignite 500 mg with 1 mL of sulfuric acid: the residue weighs not more than 75 mg (15%).

Coagulable protein—Heat a filtered solution (1 in 20) to boiling: no precipitate is formed.

Chloride (Reagent test)—It shows not more than 5% of Cl, calculated as sodium chloride.

Nitrogen content (Reagent test)—Determine by the Kjeldahl method, using a test specimen previously dried at 105° to constant weight: between 7.2% and 9.5% of N is found.

Microbial content—It meets the requirements of the test for *Microbial content* under *Pancreatic Digest of Casein.*

Yellow Mercuric Oxide—See *Mercuric Oxide, Yellow.*

Zinc, Zn—At. Wt. **65.38**—Use ACS reagent grade.

Zinc Acetate, $Zn(CH_3COO)_2 . 2H_2O$—**219.50**—Colorless crystals or white, crystalline plates, having a slight odor of acetic acid. A 1-g portion dissolves in about 2.5 mL of water and in about 30 mL of alcohol.

Insoluble matter (Reagent test)—Twenty g, dissolved in 200 mL of water and 2 mL of glacial acetic acid, shows not more than 1.0 mg of insoluble matter (0.005%).

Chloride (Reagent test)—Two g shows not more than 0.01 mg of Cl (5 ppm).

Nitrate—Dissolve 1 g in 10 mL of water and 50 µL of indigo carmine TS, and then add 10 mL of sulfuric acid: the blue color persists for 5 minutes (about 0.005%).

Sulfate (Reagent test, *Method II*)—Dissolve 20 g in 200 mL of water, add 1 mL of hydrochloric acid, and filter: the filtrate, 10 mL of barium chloride TS being used, yields not more than 1.0 mg of residue (0.002% as SO_4).

Alkalies and earths—Dissolve 2 g in 140 mL of water, add 10 mL of stronger ammonia TS, completely precipitate the zinc with hydrogen sulfide, and filter. To 75 mL of the filtrate add 5 drops of sulfuric acid, evaporate, and ignite: the residue weighs not more than 1 mg (0.1%).

Arsenic ⟨211⟩—Dissolve 6 g in water: the limit is 0.5 ppm.

Iron ⟨241⟩—Dissolve 2 g in 45 mL of water, and add 2 mL of hydrochloric acid: the solution shows not more than 0.01 mg (5 ppm).

Lead ⟨251⟩—Dissolve 1 g in 20 mL of water. To 5 mL of the solution add 0.02 mg of Pb and 12 mL of potassium cyanide solution (3 in 20), and dilute with water to 50 mL (*A*). To the remaining 15 mL add 12 mL of the potassium cyanide solution, and dilute to 50 mL with water (*B*). Then to each add 5 drops of sodium sulfide TS: *B* is not darker than *A* (0.004%).

Zinc Amalgam—Add 54 g of mossy or granular zinc to 100 mL of mercury in a beaker. Heat, with stirring, on a hot plate *under a hood* (*Caution—mercury vapor is extremely toxic*) until solution of the zinc is complete or practically so. Allow to cool to room temperature, and if necessary add sufficient mercury to prevent solidification of the amalgam. Transfer the amalgam to a glass-stoppered bottle, and shake a few times with dilute hydrochloric acid (1 in 2), to remove any zinc oxide formed.

Zinc Chloride—Use *Zinc Chloride* (USP monograph).

Zinc Chloride, Anhydrous, Powdered—Use *Zinc Chloride* (USP monograph) that has been dried and powdered.

Zinc Oxide—Use *Zinc Oxide* (USP monograph).

Zinc Sulfate—Use *Zinc Sulfate* (USP monograph).

Zirconium Oxychloride, $ZrOCl_2 . 8H_2O$—**322.25**—White, silky needles. Soluble in water, in alcohol, and in methanol; insoluble in chloroform and in hexane. Its solutions are acid to litmus.

Zirconyl Chloride, Octahydrate, Basic—Use *Zirconium Oxychloride.*

Zirconyl Nitrate, $ZrO(NO_3)_2$—**231.23**—Use a suitable grade.

Indicators and Indicator Test Papers

INDICATORS

Indicators are required in the Pharmacopeial tests and assays either to indicate the completion of a chemical reaction in volumetric analysis or to indicate the hydrogen-ion concentration (pH) of solutions. The necessary solutions of indicators are listed among the Test Solutions, abbreviated "TS."

Solutions of indicators of the basic type and of the phthaleins are prepared by dissolving in alcohol. With indicators containing an acidic group, the acid must first be neutralized with sodium hydroxide as follows:

Triturate 100 mg of the indicator in a smooth-surfaced mortar with the volume of 0.05 *N* sodium hydroxide specified in the directions for preparing its Test Solution, or with the equivalent of 0.02 *N* sodium hydroxide. When the indicator has dissolved, dilute the solution with carbon dioxide–free water to 200 mL (0.05%). Store the solutions in suitably resistant containers, protected from light.

Listed in ascending order of the lower limit of their range, useful pH indicators are: thymol blue, pH 1.2–2.8; methyl yellow, pH 2.9–4.0; bromophenol blue, pH 3.0–4.6; bromocresol green, pH 4.0–5.4; methyl red, pH 4.2–6.2; bromocresol purple, pH 5.2–6.8; bromothymol blue, pH 6.0–7.6; phenol red, pH 6.8–8.2; thymol blue, pH 8.0–9.2; and thymolphthalein, pH 8.6–10.0.

Alphazurine 2G—Use a suitable grade.

Azo Violet [*4-(p-Nitrophenylazo)resorcinol*], $C_{12}H_9N_3O_4$—**259.22**—Red powder. It melts at about 193°, with decomposition.

Bismuth Sulfite—Use a suitable grade.

Brilliant Green—See *Brilliant Green* in the section, *Reagents.*

Brilliant Yellow (*C.I. 24890*), $C_{26}H_{18}N_4Na_2O_8S$—**592.49**—Orange to rust-colored powder. Soluble in water.

Loss on drying ⟨731⟩—Dry it in vacuum at 60° for 1 hour: it loses not more than 5% of its weight.

Bromocresol Blue—Use *Bromocresol Green.*

Bromocresol Green (*Bromocresol Blue; Tetrabromo-m-cresolsulfonphthalein*), $C_{21}H_{14}Br_4O_5S$—**698.01**—White or pale buff-colored powder. Slightly soluble in water; soluble in alcohol and in solutions of alkali hydroxides. Transition interval: from pH 4.0 to 5.4. Color change: from yellow to blue.

Bromocresol Green Sodium Salt—Use a suitable grade.

Bromocresol Purple (*Dibromo-o-cresolsulfonphthalein*), $C_{21}H_{16}Br_2O_5S$—**540.22**—White to pink, crystalline powder. Insoluble in water; soluble in alcohol and in solutions of alkali hydroxides. Transition interval: from pH 5.2 to 6.8. Color change: from yellow to purple.

Bromocresol Purple Sodium Salt, $C_{21}H_{15}Br_2O_5SNa$—**562.20**—Black powder. Soluble in water. Transition interval: from pH 5.0 to 6.8. Color change: from greenish yellow to purple-violet.

Melting range ⟨741⟩: between 261° and 264°.

Bromophenol Blue (*3′,3″,5′,5″-Tetrabromophenolsulfonphthalein*), $C_{19}H_{10}Br_4O_5S$—**669.96**—Pinkish crystals. Insoluble in water; soluble in alcohol and in solutions of alkali hydroxides. Transition interval: from pH 3.0 to 4.6. Color change: from yellow to blue.

Bromophenol Blue Sodium—The sodium salt of 3′,3″,5′,5″ (*Tetrabromophenolsulfonphthalein*), $C_{19}H_9Br_4O_5SNa$—**646.36**—Pinkish crystals. Soluble in water and in alcohol. Transition interval: from pH 3.0 to 4.6. Color change: from yellow to blue.

Bromothymol Blue (*3′,3″-Dibromothymolsulfonphthalein*), $C_{27}H_{28}Br_2O_5S$—**624.38**—Cream-colored powder. Insoluble in water; soluble in alcohol and in solutions of alkali hydroxides. Transition interval: from pH 6.0 to 7.6. Color change: from yellow to blue.

Congo Red—See *Congo Red* in the section, *Reagents.*

Cresol Red (*o-Cresolsulfonphthalein*), $C_{21}H_{18}O_5S$—**382.43**—Red-brown powder. Slightly soluble in water; soluble in alcohol and in dilute solutions of alkali hydroxides. Transition interval: from pH 7.2 to 8.8. Color change: from yellow to red.

Crystal Violet (*Hexamethyl-p-rosaniline Chloride*), $C_{25}H_{30}ClN_3$—**407.99**—Dark-green crystals. Slightly soluble in water;

sparingly soluble in alcohol and in glacial acetic acid. Its solutions are deep violet in color.

Sensitiveness—Dissolve 100 mg in 100 mL of glacial acetic acid, and mix. Pipet 1 mL of the solution into a 100-mL volumetric flask, and dilute with glacial acetic acid to volume: the solution is violet-blue in color and does not show a reddish tint. Pipet 20 mL of the diluted solution into a beaker, and titrate with 0.1 *N* perchloric acid VS, adding the perchloric acid slowly from a microburet: not more than 0.10 mL of 0.1 *N* perchloric acid is required to produce an emerald-green color.

4,5-Dihydroxy-3-(*p*-sulfophenylazo)-2,7-naphthalenedisulfonic Acid, Trisodium Salt—See *2-(4-Sulfophenylazo)-1,8-dihydroxy-3,6-naphthalenedisulfonic Acid, Trisodium Salt.*

Eriochrome Black T [*Sodium 1-(1-Hydroxy-2-naphthylazo)5-nitro-2-naphthol-4-sulfonate*], $C_{20}H_{12}N_3NaO_7S$—**461.38**—Brownish black powder having a faint, metallic sheen. Soluble in alcohol, in methanol, and in hot water.

Sensitiveness—To 10 mL of a 1 in 200,000 solution in a mixture of equal parts of methanol and water add sodium hydroxide solution (1 in 100) until the pH is 10: the solution is pure blue in color and free from cloudiness. Add 0.01 mg of magnesium ion (Mg): the color of the solution changes to red-violet, and with the continued addition of magnesium ion it becomes wine-red.

Eriochrome Black T Trituration—Grind 200 mg of eriochrome black T to a fine powder with 20 g of potassium chloride.

Hydroxy Naphthol Blue Indicator—See *Hydroxy Naphthol Blue Trituration.*

Hydroxy Naphthol Blue Trituration, $C_{20}H_{14}N_2O_{11}S_3$—**554.52**—The disodium salt of 1-(2-naphtholazo-3,6-disulfonic acid)-2-naphthol-4-sulfonic acid deposited on crystals of sodium chloride. Small blue crystals. Freely soluble in water. In the pH range between 12 and 13, its solution is reddish pink in the presence of calcium ion and deep blue in the presence of excess disodium edetate.

Suitability for calcium determination—Dissolve 300 mg in 100 mL of water, add 10 mL of sodium hydroxide TS and 1.0 mL of calcium chloride solution (1 in 200), and dilute with water to 165 mL: the solution is reddish pink in color. Add 1.0 mL of 0.05 *M* disodium ethylenediaminetetraacetate: the solution becomes deep blue in color.

Litmus—Blue powder, cubes, or pieces. Partly soluble in water and in alcohol. Transition interval: from approximately pH 4.5 to 8. Color change: from red to blue. Litmus is unsuitable for determining the pH of alkaloids, carbonates, and bicarbonates.

Malachite Green Oxalate, $[4\text{-}NH(CH_3)_2C_6H_4C(C_6H_5){:}C_6H_4\text{-}4\text{-}{:}N(CH_3)_2(OCOCOOH)]_2(COO)_2$—**927.02**—The oxalate salt, crystallized with oxalic acid, of a triphenylmethane dye. Dark-green powder, having a metallic luster. Sparingly soluble in water; soluble in glacial acetic acid. Transition interval: from pH 0.0 to 2.0. Color change: from yellow to green.

Methyl Orange (*Helianthin* or *Tropaeolin D*), $C_{14}H_{14}N_3NaO_3S$—**327.33**—The sodium salt of dimethylaminoazobenzene sulfonic acid or dimethylaminoazobenzene sodium sulfonate. An orange-yellow powder or crystalline scales. Slightly soluble in cold water; readily soluble in hot water; insoluble in alcohol. Transition interval: from pH 3.2 to 4.4. Color change: from pink to yellow.

Methyl Red (*2-[[4-(Dimethylamino)phenyl]azo]benzoic Acid Hydrochloride*), $2\text{-}[4\text{-}(CH_3)_2NC_6H_4N{:}N]C_6H_4COOH.HCl$—**305.76**—Dark-red powder or violet crystals. Sparingly soluble in water; soluble in alcohol. Transition interval: from pH 4.2 to 6.2. Color change: from red to yellow.

Methyl Red Sodium—The sodium salt of 2-[[4-(dimethylamino)phenyl]azo]benzoic acid. $2\text{-}[4\text{-}(CH_3)_2NC_6H_4N{:}N]C_6H_4COONa$—**291.28**—Orange-brown powder. Freely soluble in cold water and in alcohol. Transition interval: from pH 4.2 to 6.2. Color change: from red to yellow.

Methyl Yellow (*p-Dimethylaminoazobenzene*), $C_{14}H_{15}N_3$—**225.29**—Yellow crystals, melting between 114° and 117°. Insoluble in water; soluble in alcohol, in benzene, in chloroform, in ether, in dilute mineral acids, and in oils. Transition interval: from pH 2.9 to 4.0. Color change: from red to yellow.

***p*-Naphtholbenzein** (*4-[α-(4-Hydroxy-1-naphthyl)benzylidene]-1(4H)-naphthalenone*), $(4\text{-}HOC_{10}H_6)C({:}C_{10}H_6\text{-}4{:}O)(C_6H_5)$—**374.44**—Reddish brown powder. Insoluble in water; soluble in alcohol, in benzene, in ether, and in glacial acetic acid. Transition interval: from pH 8.8 to 10.0. Color change: from orange to green.

Neutral Red (*3-Amino-7-dimethylamino-2-methylphenazine Monohydrochloride*), $C_{15}H_{16}N_4.HCl$—**288.78**—Reddish to olive-green, coarse powder. Sparingly soluble in water and in alcohol. Transition interval: from pH 6.8 to 8.0. Color change: from red to orange.

Nile Blue Hydrochloride (*Nile Blue A, as the hydrochloride; 5-Amino-9-(diethylamino)benzo[a]phenoxazin-7-ium chloride*), $C_{20}H_{20}ClN_3O$—**353.85**—Slightly soluble in alcohol and in glacial acetic acid. Transition interval: from pH 9.0 to 13.0. Color change: from blue to pink.

Oracet Blue B[95] (Solvent blue 19)—A mixture of 1-methylamino-4-anilinoanthraquinone ($C_{21}H_{16}N_2O_2$) and 1-amino-4-anilinoanthraquinine ($C_{20}H_{14}N_2O_2$). Where used for titration in nonaqueous media, it changes from blue (basic) through purple (neutral) to pink (acidic).

Phenol Red [*4,4′-(3H-2,1-Benzoxathiol-3-ylidene)diphenol, S,S-Dioxide*], $C_{19}H_{14}O_5S$—**354.38**—Crystalline powder, varying in color from bright to dark red. Very slightly soluble in water; freely soluble in solutions of alkali carbonates and hydroxides; slightly soluble in alcohol. Transition interval: from pH 6.8 to 8.2. Color change: from yellow to red.

Phenolphthalein [*3,3-Bis(p-hydroxyphenyl)phthalide*], $C_{20}H_{14}O_4$—**318.33**—White or faintly yellowish white, crystalline powder. Insoluble in water; soluble in alcohol. Transition interval: from pH 8.0 to 10.0. Color change: from colorless to red.

Quinaldine Red (*5-Dimethylamino-2-styrylethylquinolinium Iodide*), $C_{21}H_{23}IN_2$—**430.33**—Dark blue-black powder. Sparingly soluble in water; freely soluble in alcohol. Melts at about 260°, with decomposition. Transition interval: from pH 1.4 to 3.2. Color change: from colorless to red.

2-(4-Sulfophenylazo)-1,8-dihydroxy-3,6-naphthalenedisulfonic Acid, Trisodium Salt (*4,5-Dihydroxy-3-(p-sulfophenylazo)-2,7-naphthalenedisulfonic Acid, Trisodium Salt*), $C_{16}H_9N_2O_{11}S_3Na_3$—**570.40**—Red powder. Soluble in water.

Thymol Blue (*Thymolsulfonphthalein*), $C_{27}H_{30}O_5S$—**466.59**—Dark-colored, crystalline powder. Slightly soluble in water; soluble in alcohol and in dilute alkali solutions. *Acid*—Transition interval: from pH 1.2 to 2.8. Color change: from red to yellow. *Alkaline*—Transition interval: from pH 8.0 to 9.2. Color change: from yellow to blue.

Thymolphthalein, $C_{28}H_{30}O_4$—**430.54**—White to slightly yellow, crystalline powder. Insoluble in water; soluble in alcohol and in solutions of alkali hydroxides. Transition interval: from pH 9.3 to 10.5. Color change: from colorless to blue.

Xylenol Orange, (*N,N′-[3H-2,1-Benzoxathiol-3-ylidenebis-[(6-hydroxy-5-methyl-3,1-phenylene)methylene]]bis[N-(carboxymethyl)glycine] S,S-dioxide*), $C_{31}H_{28}N_2Na_4O_{13}S$—**760.59**—Orange powder. Soluble in alcohol and in water. In acid solution, it is lemon-yellow in color, and its metal complexes are intensely red. It yields a distinct end-point where a metal such as bismuth, cadmium, lanthanum, lead, mercury, scandium, thorium, or zinc is titrated with disodium ethylenediaminetetraacetate.

INDICATOR AND TEST PAPERS

Indicator and test papers are strips of paper of suitable dimension and grade (see *Filter Paper, Quantitative*, in the section, *Reagents*) impregnated with an indicator or a reagent that is sufficiently stable to provide a convenient form of the impregnated substance. Some test papers may be obtained from commercial sources of laboratory supplies. Those required in Pharmacopeial tests and assays may be prepared as directed in the following paragraphs, by means of the solutions specified, or to meet the tests set forth herein under the individual titles.

Treat strong, white filter paper with hydrochloric acid, and wash with water until the last washing no longer shows an acid reaction to methyl red. Then treat with ammonia TS, and wash again with water until the last washing is not alkaline to phenolphthalein.

After thorough drying, saturate the paper with the proper strength of indicator solutions, and carefully dry in still air, unless otherwise specified, by suspending it from rods of glass or other inert material in a space free from acid, alkali, and other fumes.

Cut the paper into strips of convenient size, and store the papers in well-closed containers, protected from light and moisture.

Cupric Sulfate Test Paper—Use cupric sulfate TS.

Lead Acetate Test Paper—Usually about 6 × 80 mm in size. Use lead acetate TS, and dry the paper at 100°, avoiding contact with metal.

Litmus Paper, Blue—Usually about 6 × 50 mm in size. It meets the requirements of the following tests.

Phosphate (Reagent test)—Cut 5 strips into small pieces, mix with 500 mg of magnesium nitrate in a porcelain crucible, and ignite. To the residue add 5 mL of nitric acid, and evaporate to dryness: the residue shows not more than 0.02 mg of PO_4.

Residue on ignition—Ignite carefully 10 strips of the paper to constant weight: the weight of the residue corresponds to not more than 0.4 mg per strip of about 3 square cm.

Rosin acids, etc.—Immerse a strip of the blue paper in a solution of 100 mg of silver nitrate in 50 mL of water: the color of the paper does not change in 30 seconds.

Sensitiveness—Drop a 10- to 12-mm strip into 100 mL of 0.0005 *N* acid contained in a beaker, and stir continuously: the color of the paper is changed within 45 seconds. The 0.0005 *N* acid is prepared by diluting 1 mL of 0.1 *N* hydrochloric acid with freshly boiled and cooled purified water to 200 mL.

Litmus Paper, Red—Usually about 6 × 50 mm in size. Red litmus paper meets the requirements of the tests for *Phosphate*, *Residue on ignition*, and *Rosin acids, etc.*, under *Litmus Paper, Blue*.

Sensitiveness—Drop a 10- to 12-mm strip into 100 mL of 0.0005 *N* sodium hydroxide contained in a beaker, and stir continuously: the color of the paper changes within 30 seconds. The 0.0005 *N* sodium hydroxide is prepared by diluting 1 mL of 0.1 *N* sodium hydroxide with freshly boiled and cooled purified water to 200 mL.

Mercuric Bromide Test Paper—Use alcoholic mercuric bromide TS. Store protected from light.

Methyl Yellow Paper—Use a 1 in 2000 solution of methyl yellow in alcohol.

pH Indicator Paper, Short-range—Use a suitable grade.

Phenolphthalein Paper—Use a 1 in 1000 solution of phenolphthalein in diluted alcohol.

Starch Iodate Paper—Use a mixture of equal volumes of starch TS and potassium iodate solution (1 in 20).

Starch Iodide Paper—Use a solution of 500 mg of potassium iodide in 100 mL of freshly prepared starch TS.

Turmeric Paper—Use a solution prepared as follows: Macerate 20 g of powdered turmeric, the dried root of *Curcuma longa* Linné (Fam. Zingiberaceae), with four 100-mL portions of cold water, decanting the clear liquid portion each time and discarding it. Dry the residue at a temperature not over 100°. Macerate with 100 mL of alcohol for several days, and filter.

Sensitiveness—Dip a strip of the paper, of about 1.5-cm length, in a solution of 1.0 mg of boric acid in 5 mL of water, previously mixed with 1 mL of hydrochloric acid. After 1 minute remove the paper from the liquid, and allow it to dry: the yellow color changes to brown. Then moisten the paper with ammonia TS: the color of the paper changes to greenish black.

Solutions

BUFFER SOLUTIONS

The successful completion of many Pharmacopeial tests and assays requires adjustment to or maintenance of a specified pH by the addition of buffer solutions. In pH measurements, standard buffer solutions are required for reference purposes. For convenience, the preparation of these solutions is in some instances described in the sections in which their use is specified; i.e., five separate phosphate buffers are described under *Antibiotics—Microbial Assays* ⟨81⟩, and several miscellaneous single-purpose solutions are described in the individual monographs.

A solution is said to be buffered if it resists changes in the activity of an ion on the addition of substances that are expected to change the activity of that ion. Buffers are substances or combinations of substances that impart this resistance to a solution. Buffered solutions are systems in which the ion is in equilibrium with substances capable of removing or releasing the ion.

Buffer capacity refers to the amount of material that may be added to a solution without causing a significant change in ion activity. It is defined as the ratio of acid or base added (in gram-equivalents per liter) to the change in pH (in pH units). The capacity of a buffered solution is adjusted to the conditions of use, usually by adjustment of the concentrations of buffer substances.

Buffers are used to establish and maintain an ion activity within narrow limits. The most common systems are used (a) to establish hydrogen-ion activity for the calibration of pH meters, (b) in the preparation of dosage forms that approach isotonicity, (c) in analytical procedures, and (d) to maintain stability of various dosage forms. Buffers used in physiological systems are carefully chosen so as not to interfere with pharmacological activity of the medicament or normal function of the organism. It is essential that buffers used in chemical analysis be compatible with the substance determined and the reagents used.

Standard Buffer Solutions—Standard solutions of definite pH are readily available in buffer solutions prepared from the appropriate reagents. In addition, buffer solutions, buffer tablets, and buffer solids may be obtained from commercial sources in convenient prepackaged form. Such preparations are available for the entire working range in pharmaceutical analysis, but are not recommended for pH meter standardization (see *pH* ⟨791⟩).

The required reagents are described in the section, *Reagents*. Previously dry the crystalline reagents, except the boric acid, at 110° to 120° for 1 hour.

NOTE—Where water is specified for solution or dilution of test substances in pH determinations, use carbon dioxide–free water.

Store the prepared solutions in chemically resistant, tight containers such as Type I glass bottles. Use the solutions within 3 months.

Standard buffer solutions for various ranges between pH 1.2 and 10.0 may be prepared by appropriate combinations of 0.2 *M* solutions described herein, used in the proportions shown in the accompanying table. The volumes shown in the table are for 200 mL of buffer solution.

1. *Hydrochloric Acid, 0.2 M*, and *Sodium Hydroxide, 0.2 M*—Prepare and standardize as directed under *Volumetric Solutions*.

2. *Potassium Biphthalate, 0.2 M*—Dissolve 40.85 g of potassium biphthalate [$KHC_6H_4(COO)_2$] in water, and dilute with water to 1000 mL.

3. *Potassium Phosphate, Monobasic 0.2 M*—Dissolve 27.22 g of monobasic potassium phosphate (KH_2PO_4) in water, and dilute with water to 1000 mL.

4. *Boric Acid and Potassium Chloride, 0.2 M*—Dissolve 12.37 g of boric acid (H_3BO_3) and 14.91 g of potassium chloride (KCl) in water, and dilute with water to 1000 mL.

5. *Potassium Chloride, 0.2 M*—Dissolve 14.91 g of potassium chloride (KCl) in water, and dilute with water to 1000 mL.

Composition of Standard Buffer Solutions

Hydrochloric Acid Buffer

Place 50 mL of the potassium chloride solution in a 200-mL volumetric flask, add the specified volume of the hydrochloric acid solution, then add water to volume.

pH	1.2	1.3	1.4	1.5	1.6	1.7	1.8	1.9	2.0	2.1	2.2
HCl, mL	85.0	67.2	53.2	41.4	32.4	26.0	20.4	16.2	13.0	10.2	7.8

Acid Phthalate Buffer

Place 50 mL of the potassium biphthalate solution in a 200-mL volumetric flask, add the specified volume of the hydrochloric acid solution, then add water to volume.

pH	2.2	2.4	2.6	2.8	3.0	3.2	3.4	3.6	3.8	4.0
HCl, mL	49.5	42.2	35.4	28.9	22.3	15.7	10.4	6.3	2.9	0.1

Neutralized Phthalate Buffer

Place 50 mL of the potassium biphthalate solution in a 200-mL volumetric flask, add the specified volume of the sodium hydroxide solution, then add water to volume.

pH	4.2	4.4	4.6	4.8	5.0	5.2	5.4	5.6	5.8
NaOH, mL	3.0	6.6	11.1	16.5	22.6	28.8	34.1	38.8	42.3

Phosphate Buffer

Place 50 mL of the monobasic potassium phosphate solution in a 200-mL volumetric flask, add the specified volume of the sodium hydroxide solution, then add water to volume.

pH	5.8	6.0	6.2	6.4	6.6	6.8	7.0	7.2	7.4	7.6	7.8	8.0
NaOH, mL	3.6	5.6	8.1	11.6	16.4	22.4	29.1	34.7	39.1	42.4	44.5	46.1

Alkaline Borate Buffer

Place 50 mL of the boric acid and potassium chloride solution in a 200-mL volumetric flask, add the specified volume of the sodium hydroxide solution, then add water to volume.

pH	8.0	8.2	8.4	8.6	8.8	9.0	9.2	9.4	9.6	9.8	10.0
NaOH, mL	3.9	6.0	8.6	11.8	15.8	20.8	26.4	32.1	36.9	40.6	43.7

COLORIMETRIC SOLUTIONS (CS)

(For the Preparation of Matching Fluids, see *Color and Achromicity* ⟨631⟩.)

These solutions are used in the preparation of the colorimetric standards for certain drugs, and for the carbonization tests with sulfuric acid that are specified in several monographs. Store the solutions in suitably resistant, tight containers.

Comparison of colors as directed in the Pharmacopeial tests preferably is made in matched color-comparison tubes or in a suitable colorimeter under conditions that ensure that the colorimetric reference solution and that of the specimen under test are treated alike in all respects. The comparison of colors is best made in layers of equal depth, and viewed transversely against a white background (see also *Visual Comparison* ⟨851⟩). It is particularly important that the solutions be compared at the same temperature, preferably 25°.

Cobaltous Chloride CS—Dissolve about 65 g of cobaltous chloride ($CoCl_2.6H_2O$) in enough of a mixture of 25 mL of hydrochloric acid and 975 mL of water to make 1000 mL. Pipet 5 mL of this solution into a 250-mL iodine flask, add 5 mL of hydrogen peroxide TS and 15 mL of sodium hydroxide solution (1 in 5), boil for 10 minutes, cool, and add 2 g of potassium iodide and 20 mL of dilute sulfuric acid (1 in 4). When the precipitate has dissolved, titrate the liberated iodine with 0.1 *N* sodium thiosulfate VS, adding 3 mL of starch TS as the indicator. Perform a blank determination with the same quantities of the same reagents, and make any necessary correction. Each mL of 0.1 *N* sodium thiosulfate is equivalent to 23.79 mg of $CoCl_2.6H_2O$. Adjust the final volume of the solution by the addition of enough of the mixture of hydrochloric acid and water so that each mL contains 59.5 mg of $CoCl_2.6H_2O$.

Cupric Sulfate CS—Dissolve about 65 g of cupric sulfate ($CuSO_4.5H_2O$) in enough of a mixture of 25 mL of hydrochloric acid and 975 mL of water to make 1000 mL. Pipet 10 mL of this solution into a 250-mL iodine flask, add 40 mL of water, 4 mL of acetic acid, 3 g of potassium iodide, and 5 mL of hydrochloric acid, and titrate the liberated iodine with 0.1 *N* sodium thiosulfate VS, adding 3 mL of starch TS as the indicator. Perform a blank determination with the same quantities of the same reagents, and make any necessary correction. Each mL of 0.1 *N* sodium thiosulfate is equivalent to 24.97 mg of $CuSO_4.5H_2O$. Adjust the final volume of the solution by the addition of enough of the mixture of hydrochloric acid and water so that each mL contains 62.4 mg of $CuSO_4.5H_2O$.

Ferric Chloride CS—Dissolve about 55 g of ferric chloride ($FeCl_3.6H_2O$) in enough of a mixture of 25 mL of hydrochloric acid and 975 mL of water to make 1000 mL. Pipet 10 mL of this solution into a 250-mL iodine flask, add 15 mL of water, 3 g of potassium iodide, and 5 mL of hydrochloric acid, and allow the mixture to stand for 15 minutes. Dilute with 100 mL of water, and titrate the liberated iodine with 0.1 *N* sodium thiosulfate VS, adding 3 mL of starch TS as the indicator. Perform a blank determination with the same quantities of the same reagents, and make any necessary correction. Each mL of 0.1 *N* sodium thiosulfate is equivalent to 27.03 mg of $FeCl_3.6H_2O$. Adjust the final volume of the solution by the addition of enough of the mixture of hydrochloric acid and water so that each mL contains 45.0 mg of $FeCl_3.6H_2O$.

INDICATOR SOLUTIONS

SEE *TEST SOLUTIONS*

TEST SOLUTIONS (TS)

Certain of the following test solutions are intended for use as acid-base indicators in volumetric analyses. Such solutions should be so adjusted that when 0.15 mL of the indicator solution is added to 25 mL of carbon dioxide–free water, 0.25 mL of 0.02 *N* acid or alkali, respectively, will produce the characteristic color change. Similar solutions are intended for use in pH measurement. Where no special directions for their preparation are given, the same solution is suitable for both purposes.

Where it is directed that a volumetric solution be used as the test solution, standardization of the solution used as TS is not required.

In general, the directive to prepare a solution "fresh" indicates that the solution is of limited stability and must be prepared on the day of use.

For the preparation of Test Solutions, use reagents of the quality described under *Reagents*.

Acetaldehyde TS—Mix 4 mL of acetaldehyde, 3 mL of alcohol, and 1 mL of water. Prepare this solution fresh.

Acetate Buffer TS—Dissolve 320 g of ammonium acetate in 500 mL of water, add 5 mL of glacial acetic acid, dilute with water to 1000.0 mL, and mix. This solution has a pH between 5.9 and 6.0.

Acetic Acid, Glacial TS—Determine the water content of a specimen of glacial acetic acid by the *Titrimetric Method* (see *Water Determination* ⟨921⟩). If the acid contains more than 0.05% of water, add a few mL of acetic anhydride, mix, allow to stand overnight, and again determine the water content. If the acid contains less than 0.02% of water, add sufficient water to make the final concentration between 0.02% and 0.05%, mix, allow to stand overnight, and again determine the water content. Repeat the adjustment with acetic anhydride or water, as necessary, until the resulting solution shows a water content between 0.02% and 0.05%.

Acetic Acid–Ammonium Acetate Buffer TS—Dissolve 77.1 g of ammonium acetate in water, add 57 mL of glacial acetic acid, and dilute with water to 1000 mL.

Acetone, Buffered, TS—Dissolve 8.15 g of sodium acetate and 42 g of sodium chloride in about 100 mL of water, and add 68 mL of 0.1 *N* hydrochloric acid and 150 mL of acetone. Mix, and dilute with water to 500 mL.

Acid–Ferric Chloride TS—Mix 60 mL of glacial acetic acid with 5 mL of sulfuric acid, add 1 mL of ferric chloride TS, mix, and cool.

Acid Ferrous Sulfate TS—See *Ferrous Sulfate, Acid, TS.*

Acid Stannous Chloride TS—See *Stannous Chloride, Acid, TS.*

Acid Stannous Chloride TS, Stronger—See *Stannous Chloride, Acid, Stronger, TS.*

Albumen TS—Carefully separate the white from the yolk of a strictly fresh hen's egg. Shake the white with 100 mL of water until mixed and all but the chalaza has undergone solution; then filter. Prepare the solution fresh.

Alcohol–Phenol TS—Dissolve 780 mg of phenol in alcohol to make 100 mL.

Alcoholic Ammonia TS—See *Ammonia TS, Alcoholic.*

Alcoholic Mercuric Bromide TS—See *Mercuric Bromide TS, Alcoholic.*

Alcoholic Potassium Hydroxide TS—See *Potassium Hydroxide TS, Alcoholic.*

Alkaline Cupric Citrate TS—See *Cupric Citrate TS, Alkaline.*

Alkaline Cupric Iodide TS—See *Cupric Iodide TS, Alkaline.*

Alkaline Cupric Tartrate TS (*Fehling's Solution*)—See *Cupric Tartrate TS, Alkaline.*

Alkaline Mercuric-Potassium Iodide TS—See *Mercuric-Potassium Iodide TS, Alkaline.*

Alkaline Picrate TS—See *Picrate TS, Alkaline.*

Alkaline Sodium Hydrosulfite TS—See *Sodium Hydrosulfite TS, Alkaline.*

Aminonaphtholsulfonic Acid TS—Weigh accurately 5 g of sodium sulfite, 94.3 g of sodium bisulfite, and 700 mg of 1,2,4-aminonaphtholsulfonic acid, and mix. Prepare aminonaphtholsulfonic acid TS fresh on the day of use by dissolving 1.5 g of the dry mixture in 10 mL of water.

Ammonia–Ammonium Chloride Buffer TS—Dissolve 67.5 g of ammonium chloride in water, add 570 mL of ammonium hydroxide, and dilute with water to 1000 mL.

Ammonia-Cyanide TS—Dissolve 2 g of potassium cyanide in 15 mL of ammonium hydroxide, and dilute with water to 100 mL.

Ammonia TS—It contains between 9.5% and 10.5% of NH_3. Prepare by diluting 400 mL of *Ammonia Water, Stronger* (see in the section, *Reagents*) with water to make 1000 mL.

Ammonia TS, Alcoholic—A solution of ammonia gas in alcohol. Clear, colorless liquid having a strong odor of ammonia. Specific gravity: about 0.80. It contains between 9% and 11% of NH_3. Store it in alkali-resistant containers, in a cold place.

Ammonia TS, Stronger—Use *Ammonia Water, Stronger* (see in the section, *Reagents*).

Ammoniated Cupric Oxide TS—See *Cupric Oxide, Ammoniated, TS.*

Ammonium Acetate TS—Dissolve 10 g of ammonium acetate in water to make 100 mL.

Ammonium Carbonate TS—Dissolve 20 g of ammonium carbonate and 20 mL of ammonia TS in water to make 100 mL.

Ammonium Chloride TS—Dissolve 10.5 g of ammonium chloride in water to make 100 mL.

Ammonium Chloride–Ammonium Hydroxide TS—Mix equal volumes of water and ammonium hydroxide, and saturate with ammonium chloride.

Ammonium Molybdate TS—Dissolve 6.5 g of finely powdered molybdic acid in a mixture of 14 mL of water and 14.5 mL of ammonium hydroxide. Cool the solution, and add it slowly, with stirring, to a well-cooled mixture of 32 mL of nitric acid and 40 mL of water. Allow to stand for 48 hours, and filter through a fine-porosity, sintered-glass crucible. This solution deteriorates upon standing and is unsuitable for use if, upon the addition of 2 mL of dibasic sodium phosphate TS to 5 mL of the solution, an abundant yellow precipitate does not form at once or after slight warming. Store it in the dark. If a precipitate forms during storage, use only the clear, supernatant solution.

Ammonium Oxalate TS—Dissolve 3.5 g of ammonium oxalate in water to make 100 mL.

Ammonium Phosphate, Dibasic, TS (*Ammonium Phosphate TS*)—Dissolve 13 g of dibasic ammonium phosphate in water to make 100 mL.

Ammonium Polysulfide TS—Yellow liquid, made by saturating ammonium sulfide TS with sulfur.

Ammonium Reineckate TS—Shake about 500 mg of ammonium reineckate with 20 mL of water frequently during 1 hour, and filter. Use within 2 days.

Ammonium Sulfide TS—Saturate ammonia TS with hydrogen sulfide, and add two-thirds of its volume of ammonia TS. Residue on ignition: not more than 0.05%. The solution is not rendered turbid either by magnesium sulfate TS or by calcium chloride TS (*carbonate*).

This solution is unsuitable for use if an abundant precipitate of sulfur is present. Store it in small, well-filled, dark amber-colored bottles, in a cold, dark place.

Ammonium Thiocyanate TS—Dissolve 8 g of ammonium thiocyanate in water to make 100 mL.

Ammonium Vanadate TS—Dissolve 2.5 g of ammonium vanadate in 500 mL of boiling water, cool, and add 20 mL of nitric acid. Mix, cool, and add water to make 1 liter. Store in polyethylene containers.

Anthrone TS—Within 12 hours of use, rapidly dissolve 35 mg of anthrone in a hot mixture of 35 mL of water and 65 mL of sulfuric acid. Immediately cool in an ice bath to room temperature, and filter through glass wool. Allow the solution to stand at room temperature for 30 minutes before use.

Antimony Trichloride TS—Dissolve 20 g of antimony trichloride in chloroform to make 100 mL. Filter if necessary.

Barium Chloride TS—Dissolve 12 g of barium chloride in water to make 100 mL.

Barium Hydroxide TS—A saturated solution of barium hydroxide in recently boiled water. Prepare the solution fresh.

Barium Nitrate TS—Dissolve 6.5 g of barium nitrate in water to make 100 mL.

Betanaphthol TS—See *2-Naphthol TS*.

Biuret Reagent TS—Dissolve 1.5 g of cupric sulfate and 6.0 g of potassium sodium tartrate in 500 mL of water in a 1000-mL volumetric flask. Add 300 mL of carbonate-free sodium hydroxide solution (1 in 10), dilute with carbonate-free sodium hydroxide solution (1 in 10) to 1000 mL, and mix.

Blue Tetrazolium TS—Dissolve 500 mg of blue tetrazolium in alcohol to make 100 mL.

Bromine TS (*Bromine Water*)—A saturated solution of bromine, prepared by agitating 2 to 3 mL of bromine with 100 mL of cold water in a glass-stoppered bottle, the stopper of which should be lubricated with petrolatum. Store it in a cold place, protected from light.

Bromine–Sodium Acetate TS—Dissolve 100 g of sodium acetate in 1000 mL of glacial acetic acid, add 50 mL of bromine, and mix.

***p*-Bromoaniline TS**—Add 8 g of *p*-bromoaniline to a mixture of 380 mL of thiourea-saturated glacial acetic acid, 10 mL of sodium chloride solution (1 in 5), 5 mL of oxalic acid solution (1 in 20), and 5 mL of dibasic sodium phosphate solution (1 in 10) in a low-actinic glass bottle. Mix, and allow to stand overnight before using. Protect from light, and use within 7 days.

Bromocresol Blue TS—Use *Bromocresol Green TS*.

Bromocresol Green TS—Dissolve 50 mg of bromocresol green in 100 mL of alcohol, and filter if necessary.

Bromocresol Purple TS—Dissolve 250 mg of bromocresol purple in 20 mL of 0.05 *N* sodium hydroxide, and dilute with water to 250 mL.

Bromophenol Blue TS—Dissolve 100 mg of bromophenol blue in 100 mL of diluted alcohol, and filter if necessary.

Bromothymol Blue TS—Dissolve 100 mg of bromothymol blue in 100 mL of diluted alcohol, and filter if necessary.

Buffered Acetone TS—See *Acetone, Buffered, TS*.

Calcium Chloride TS—Dissolve 7.5 g of calcium chloride in water to make 100 mL.

Calcium Hydroxide TS—Use *Calcium Hydroxide Topical Solution* (USP monograph).

Calcium Sulfate TS—A saturated solution of calcium sulfate in water.

Ceric Ammonium Nitrate TS—Dissolve 6.25 g of ceric ammonium nitrate in 10 mL of 0.25 *N* nitric acid. Use within 3 days.

Chloral Hydrate TS—Dissolve 50 g of chloral hydrate in a mixture of 15 mL of water and 10 mL of glycerin.

Chlorine TS (*Chlorine Water*)—A saturated solution of chlorine in water. Place the solution in small, completely filled, light-resistant containers. Chlorine TS, even when kept from light and air, is apt to deteriorate. Store it in a cold, dark place. For full strength, prepare this solution fresh.

Chromotropic Acid TS—Dissolve 50 mg of chromotropic acid or its sodium salt in 100 mL of 75 percent sulfuric acid, which may be made by cautiously adding 75 mL of sulfuric acid to 33.3 mL of water.

Cobalt-Uranyl Acetate TS—Dissolve, with warming, 40 g of uranyl acetate in a mixture of 30 g of glacial acetic acid and sufficient water to make 500 mL. Similarly, prepare a solution containing 200 g of cobaltous acetate in a mixture of 30 g of glacial acetic acid and sufficient water to make 500 mL. Mix the two solutions while still warm, and cool to 20°. Maintain the temperature at 20° for about 2 hours to separate the excess salts from solution, and then filter through a dry filter.

Cobaltous Chloride TS—Dissolve 2 g of cobaltous chloride in 1 mL of hydrochloric acid and sufficient water to make 100 mL.

Congo Red TS—Dissolve 500 mg of congo red in a mixture of 10 mL of alcohol and 90 mL of water.

***m*-Cresol Purple TS**—Dissolve 0.10 g of metacresol purple in 13 mL of 0.01 *N* sodium hydroxide, dilute with water to 100 mL, and mix.

Cresol Red TS—Triturate 100 mg of cresol red in a mortar with 26.2 mL of 0.01 *N* sodium hydroxide until solution is complete, then dilute the solution with water to 250 mL.

Cresol Red–Thymol Blue TS—Add 15 mL of thymol blue TS to 5 mL of cresol red TS, and mix.

Crystal Violet TS—Dissolve 100 mg of crystal violet in 10 mL of glacial acetic acid.

Cupric Acetate TS—Dissolve 100 mg of cupric acetate in about 5 mL of water to which a few drops of acetic acid has been added. Dilute to 100 mL, and filter, if necessary.

Cupric Acetate TS, Stronger (*Barfoed's Reagent*)—Dissolve 13.3 g of cupric acetate in a mixture of 195 mL of water and 5 mL of acetic acid.

Cupric-Ammonium Sulfate TS—To cupric sulfate TS add ammonia TS, dropwise, until the precipitate at first formed is nearly but not completely dissolved. Allow to settle, and decant the clear solution. Prepare this solution fresh.

Cupric Citrate TS, Alkaline—With the aid of heat, dissolve 173 g of dihydrated sodium citrate and 117 g of monohydrated sodium carbonate in about 700 mL of water, and filter through paper, if necessary, to obtain a clear solution. In a separate container dissolve 17.3 g of cupric sulfate in about 100 mL of water, and slowly add this solution, with constant stirring, to the first solution. Cool the mixture, add water to make 1000 mL, and mix.

Cupric Iodide TS, Alkaline—Dissolve 7.5 g of cupric sulfate ($CuSO_4 . 5H_2O$) in about 100 mL of water. In a separate container dissolve 25 g of anhydrous sodium carbonate, 20 g of sodium bicarbonate, and 25 g of potassium sodium tartrate in about 600 mL of water. With constant stirring, add the cupric sulfate solution to the bottom of the alkaline tartrate solution by means of a funnel that touches the bottom of the container. Add 1.5 g of potassium iodide, 200 g of anhydrous sodium sulfate, 50 to 150 mL of 0.02 *M* potassium iodate, and sufficient water to make 1000 mL.

Cupric Oxide, Ammoniated, TS (*Schweitzer's Reagent*)—Dissolve 10 g of cupric sulfate in 100 mL of water, add sufficient sodium hydroxide solution (1 in 5) to precipitate the copper hydroxide, collect the latter on a filter, and wash free from sulfate with cold water. Dissolve the precipitate, which must be kept wet during the entire process, in the minimum quantity of ammonia TS necessary for complete solution.

Cupric Sulfate TS—Dissolve 12.5 g of cupric sulfate in water to make 100 mL.

Cupric Tartrate TS, Alkaline (*Fehling's Solution*)—*The Copper Solution* (*A*)—Dissolve 34.66 g of carefully selected, small crystals of cupric sulfate, showing no trace of efflorescence of adhering moisture, in water to make 500 mL. Store this solution in small, tight containers.

The Alkaline Tartrate Solution (*B*)—Dissolve 173 g of crystallized potassium sodium tartrate and 50 g of sodium hydroxide in water to make 500 mL. Store this solution in small, alkali-resistant containers.

For use, mix exactly equal volumes of Solutions *A* and *B* at the time required.

Delafield's Hematoxylin TS—Prepare 400 mL of a saturated solution of ammonium alum (*Solution A*). Dissolve 4 g of hematoxylin in 25 mL of alcohol, mix it with *Solution A*, and allow it to stand for 4 days in a flask closed with a pledget of purified cotton, and exposed to light and air (*Solution B*). Then filter *Solution B*, and add to it a *Solution C*, consisting of a mixture of 100 mL of glycerin and 100 mL of methanol. Mix, and allow the mixture to stand in a warm place, exposed to light, for 6 weeks until it becomes dark-colored. Store in tightly stoppered bottles.

For use in staining endocrine tissue, dilute this test solution with an equal volume of water.

Denigès' Reagent—See *Mercuric Sulfate TS.*

Diazobenzenesulfonic Acid TS—Place in a beaker 1.57 g of sulfanilic acid, previously dried at 105° for 3 hours, add 80 mL of water and 10 mL of diluted hydrochloric acid, and warm on a steam bath until dissolved. Cool to 15° (some of the sulfanilic acid may separate but will dissolve later), and add slowly, with constant stirring, 6.5 mL of sodium nitrite solution (1 in 10). Then dilute with water to 100 mL.

Dichlorofluorescein TS—Dissolve 100 mg of dichlorofluorescein in 60 mL of alcohol, add 2.5 mL of 0.1 *N* sodium hydroxide, mix, and dilute with water to 100 mL.

Dicyclohexylamine Acetate TS—Dissolve 50 g of dicyclohexylamine in 150 mL of acetone, cool in an ice bath, and add, with stirring, a solution consisting of 18 mL of glacial acetic acid in 150 mL of acetone. Recrystallize the precipitate that forms, by heating the mixture to boiling and allowing it to cool in an ice bath, then collect the crystals on a filtering funnel, wash with a small volume of acetone, and air-dry. Dissolve 300 mg of the dicyclohexylamine acetate so obtained in 200 mL of a mixture of 6 volumes of chloroform and 4 volumes of water-saturated ether. Use immediately.

2,7-Dihydroxynaphthalene TS—Dissolve 100 mg of 2,7-dihydroxynaphthalene in 1000 mL of sulfuric acid, and allow the solution to stand until the yellow color disappears. If the solution is very dark, discard it and prepare a new solution from a different supply of sulfuric acid. This solution is stable for approximately one month if stored in a dark bottle.

Diiodofluorescein TS—Dissolve 500 mg of diiodofluorescein in a mixture of 75 mL of alcohol and 30 mL of water.

Diluted Lead Subacetate TS—See *Lead Subacetate TS, Diluted.*

***p*-Dimethylaminobenzaldehyde TS**—Dissolve 125 mg of *p*-dimethylaminobenzaldehyde in a cooled mixture of 65 mL of sulfuric acid and 35 mL of water, and add 0.05 mL of ferric chloride TS. Use within 7 days.

Dinitrophenylhydrazine TS—Carefully mix 10 mL of water and 10 mL of sulfuric acid, and cool. To the mixture, contained in a glass-stoppered flask, add 2 g of 2,4-dinitrophenylhydrazine, and shake until dissolved. To the solution add 35 mL of water, mix, cool, and filter.

Diphenylamine TS—Dissolve 1.0 g of diphenylamine in 100 mL of sulfuric acid. The solution should be colorless.

Diphenylcarbazone TS—Dissolve 1 g of crystalline diphenylcarbazone in 75 mL of alcohol, then add alcohol to make 100 mL. Store in a brown bottle.

Disodium Ethylenediaminetetraacetate TS—Dissolve 1 g of disodium ethylenediaminetetraacetate in 950 mL of water, add 50 mL of alcohol, and mix.

Dithizone TS—Dissolve 25.6 mg of dithizone in 100 mL of alcohol. Store in a cold place, and use within 2 months.

Eosin Y TS (adsorption indicator)—Dissolve 50 mg of eosin Y in 10 mL of water.

Eriochrome Black TS—Dissolve 200 mg of eriochrome black T and 2 g of hydroxylamine hydrochloride in methanol to make 50 mL.

Eriochrome Cyanine TS—Dissolve 750 mg of eriochrome cyanine R in 200 mL of water, add 25 g of sodium chloride, 25 g of ammonium nitrate, and 2 mL of nitric acid, and dilute with water to 1000 mL.

Fehling's Solution—See *Cupric Tartrate TS, Alkaline.*

Ferric Ammonium Sulfate TS—Dissolve 8 g of ferric ammonium sulfate in water to make 100 mL.

Ferric Chloride TS—Dissolve 9 g of ferric chloride in water to make 100 mL.

Ferrous Sulfate TS—Dissolve 8 g of clear crystals of ferrous sulfate in about 100 mL of recently boiled and thoroughly cooled water. Prepare this solution fresh.

Ferrous Sulfate, Acid, TS—Dissolve 7 g of ferrous sulfate crystals in 90 mL of recently boiled and thoroughly cooled water, and add sulfuric acid to make 100 mL. Prepare this solution immediately prior to use.

Folin-Ciocalteu Phenol TS—Into a 1500-mL flask introduce 100 g of sodium tungstate, 25 g of sodium molybdate, 700 mL of water, 50 mL of phosphoric acid, and 100 mL of hydrochloric acid. Reflux the mixture gently for about 10 hours, and add 150 g of lithium sulfate, 50 mL of water, and a few drops of bromine. Boil the mixture, without the condenser, for about 15 minutes, or until the excess bromine is expelled. Cool, dilute with water to 1 liter, and filter: the filtrate has no greenish tint. Before use, dilute 1 part of filtrate with 1 part of water.

Formaldehyde TS—Use *Formaldehyde Solution* (see in the section, *Reagents*).

Fuchsin-Pyrogallol TS—Dissolve 100 mg of basic fuchsin in 50 mL of water that previously has been boiled for 15 minutes and allowed to cool slightly. Cool, add 2 mL of a saturated solution of sodium bisulfite, mix, and allow to stand for not less than 3 hours. Add 0.9 mL of hydrochloric acid, mix, and allow to stand overnight. Add 100 mg of pyrogallol, shake until solution is effected, and dilute with water to 100 mL. Store in an amber-glass bottle in a refrigerator.

Fuchsin–Sulfurous Acid TS—Dissolve 200 mg of basic fuchsin in 120 mL of hot water, and allow the solution to cool. Add a solution of 2 g of anhydrous sodium sulfite in 20 mL of water, then add 2 mL of hydrochloric acid. Dilute the solution with water to 200 mL, and allow to stand for at least 1 hour. Prepare this solution fresh.

Gastric Fluid, Simulated, TS—Dissolve 2.0 g of sodium chloride and 3.2 g of pepsin in 7.0 mL of hydrochloric acid and sufficient water to make 1000 mL. This test solution has a pH of about 1.2.

Gelatin TS (for the assay of *Corticotropin Injection*)—Dissolve 340 g of acid-treated precursor gelatin (Type A) in water to make 1000 mL. Heat the solution in an autoclave at 115° for 30 minutes after the exhaust line temperature has reached 115°. Cool the solution, and add 10 g of phenol and 1000 mL of water. Store in tight containers in a refrigerator.

Glacial Acetic Acid TS—See *Acetic Acid, Glacial, TS.*

Glucose oxidase–chromogen TS—A solution containing, in each mL, 0.5 µmol of 4-aminoantipyrine, 22.0 µmol of sodium *p*-hydroxybenzoate, not less than 7.0 units of glucose oxidase, and not less than 0.5 units of peroxidase, and buffered to a pH of 7.0 ± 0.1.[96]

Suitability—When used for determining glucose in Inulin, ascertain that no significant color results by reaction with fructose, and that a suitable absorbance-versus-concentration slope is obtained with glucose.

Gold Chloride TS—Dissolve 1 g of gold chloride in 35 mL of water.

Hydrogen Peroxide TS—Use *Hydrogen Peroxide Topical Solution* (USP monograph).

Hydrogen Sulfide TS—A saturated solution of hydrogen sulfide, made by passing H_2S into cold water. Store it in small, dark amber-colored bottles, filled nearly to the top. It is unsuitable unless it possesses a strong odor of H_2S, and unless it produces at once a copious precipitate of sulfur when added to an equal volume of ferric chloride TS. Store in a cold, dark place.

Hydroxylamine Hydrochloride TS—Dissolve 3.5 g of hydroxylamine hydrochloride in 95 mL of 60 percent alcohol, and add 0.5 mL of bromophenol blue solution (1 in 1000) and 0.5 *N* alcoholic potassium hydroxide until a greenish tint develops in the solution. Then add 60 percent alcohol to make 100 mL.

8-Hydroxyquinoline TS—Dissolve 5 g of 8-hydroxyquinoline in alcohol to make 100 mL.

Indigo Carmine TS (*Sodium Indigotindisulfonate TS*)—Dissolve a quantity of sodium indigotindisulfonate, equivalent to 180 mg of $C_{16}H_8N_2O_2(SO_3Na)_2$, in water to make 100 mL. Use within 60 days.

Indophenol-Acetate TS (for the assay of *Corticotropin Injection*)—To 60 mL of standard dichlorophenol-indophenol solution (see in the section, *Volumetric Solutions*) add water to make 250 mL. Add to the resulting solution an equal volume of sodium acetate solution freshly prepared by dissolving 13.66 g of anhydrous sodium acetate in water to make 500 mL and adjusting with 0.5 *N* acetic acid to a pH of 7. Store in a refrigerator, and use within 2 weeks.

Intestinal Fluid, Simulated, TS—Dissolve 6.8 g of monobasic potassium phosphate in 250 mL of water, mix, and add 190 mL of 0.2 *N* sodium hydroxide and 400 mL of water. Add 10.0 g of pancreatin, mix, and adjust the resulting solution with 0.2 *N* sodium hydroxide to a pH of 7.5 ± 0.1. Dilute with water to 1000 mL.

Iodine TS—Use *0.1 N Iodine* (see in the section, *Volumetric Solutions*).

Iodine Monochloride TS—Dissolve 10 g of potassium iodide and 6.44 g of potassium iodate in 75 mL of water in a glass-stoppered container. Add 75 mL of hydrochloric acid and 5 mL of chloroform, and adjust to a faint iodine color (in the chloroform) by adding dilute potassium iodide or potassium iodate solution. If much iodine is liberated, use a stronger solution of potassium iodate than 0.01 *M* at first, making the final adjustment with the 0.01 *M* potassium iodate. Store in a dark place, and readjust to a faint iodine color as necessary.

Iodine and Potassium Iodide TS—Dissolve 500 mg of iodine and 1.5 g of potassium iodide in 25 mL of water.

Iodobromide TS—Dissolve 13.615 g of iodine, with the aid of heat, in 825 mL of glacial acetic acid that shows no reduction with dichromate and sulfuric acid. Cool, and titrate 25.0 mL of the solution with 0.1 *N* sodium thiosulfate VS, recording the volume consumed as *B*. Prepare another solution containing 3 mL of bromine in 200 mL of glacial acetic acid. To 5.0 mL of this solution add 10 mL of potassium iodide TS, and titrate with the 0.1 *N* sodium thiosulfate VS, recording the volume consumed as *C*. Calculate the quantity, *A*, of the bromine solution needed to double the halogen content of the remaining 800 mL of iodine solution by the formula:

$$800B/5C.$$

Add the calculated volume of bromine solution to the iodine solution, mix, and store in glass containers, protected from light.

Iodochloride TS—Dissolve 16.5 g of iodine monochloride in 1000 mL of glacial acetic acid.

Iodoplatinate TS—Dissolve 300 mg of platinic chloride in 97 mL of water. Immediately prior to use, add 3.5 mL of potassium iodide TS, and mix.

Iron-Phenol TS (*Iron-Kober Reagent*)—Dissolve 1.054 g of ferrous ammonium sulfate in 20 mL of water, and add 1 mL of sulfuric acid and 1 mL of 30 percent hydrogen peroxide. Mix, heat until effervescence ceases, and dilute with water to 50 mL. To 3 volumes of this solution contained in a volumetric flask add sulfuric acid, with cooling, to make 100 volumes. Purify phenol by distillation, discarding the first 10% and the last 5%, collecting the distillate, with exclusion of moisture, in a dry, tared glass-stoppered flask of about twice the volume of the phenol. Solidify the phenol in an ice bath, breaking the top crust with a glass rod to ensure complete crystallization. Weigh the flask and its contents, add to the phenol 1.13 times its weight of the iron–sulfuric acid solution prepared as directed, insert the stopper in the flask, and allow to stand, without cooling but with occasional mixing, until the phenol is liquefied. Shake the mixture vigorously until mixed, allow to stand in the dark for 16 to 24 hours, and again weigh the flask and its contents. To the mixture add 23.5% of its weight of a solution of 100 volumes of sulfuric acid in 110 volumes of water, mix, transfer to dry glass-stoppered bottles, and store in the dark, protected from atmospheric moisture. Use within 6 months. Dispense the reagent from a small-bore buret, arranged to exclude moisture, capable of delivering 1 mL in 30 seconds or less, and having no lubricant, other than reagent, on its stopcock. Wipe the buret tip with tissue before each addition.

Iron Salicylate TS—Dissolve 500 mg of ferric ammonium sulfate in 250 mL of water containing 10 mL of diluted sulfuric acid, and add water to make 500 mL. To 100 mL of the resulting solution add 50 mL of a 1.15% solution of sodium salicylate, 20 mL of diluted acetic acid, and 80 mL of a 13.6% solution of sodium acetate, then add water to make 500 mL. Store in a well-closed container. Protect from light. Use within two weeks.

Lead Acetate TS—Dissolve 9.5 g of clear, transparent crystals of lead acetate in recently boiled water to make 100 mL. Store in well-stoppered bottles.

Lead Acetate TS, Alcoholic—Dissolve 2 g of clear, transparent crystals of lead acetate in alcohol to make 100 mL. Store in tight containers.

Lead Subacetate TS—Triturate 14 g of lead monoxide to a smooth paste with 10 mL of water, and transfer the mixture to a bottle, using an additional 10 mL of water for rinsing. Dissolve 22 g of lead acetate in 70 mL of water, and add the solution to the lead oxide mixture. Shake it vigorously for 5 minutes, then set it aside, shaking it frequently, during 7 days. Finally filter, and add enough recently boiled water through the filter to make 100 mL.

Lead Subacetate TS, Diluted—Dilute 3.25 mL of lead subacetate TS with water, recently boiled and cooled, to make 100 mL. Store in small, well-filled, tight containers.

Litmus TS—Digest 25 g of powdered litmus with three successive, 100-mL portions of boiling alcohol, continuing each extraction for about 1 hour. Filter, wash with alcohol, and discard the alcohol filtrate. Macerate the residue with about 25 mL of cold water for 4 hours, filter, and discard the filtrate. Finally digest the residue with 125 mL of boiling water for 1 hour, cool, and filter.

Locke-Ringer's Solution—See *Locke-Ringer's TS*.

Locke-Ringer's TS (*Locke-Ringer's Solution*)—

Sodium Chloride	9.0 g
Potassium Chloride	0.42 g
Calcium Chloride	0.24 g
Magnesium Chloride	0.2 g
Sodium Bicarbonate	0.5 g
Dextrose	0.5 g
Water, recently distilled from a hard-glass flask, a sufficient quantity, to make	1000 mL

Prepare fresh each day. The constituents (except the dextrose and the sodium bicarbonate) may be made up in stock solutions and diluted as needed.

Magnesia Mixture TS—Dissolve 5.5 g of magnesium chloride and 7 g of ammonium chloride in 65 mL of water, add 35 mL of ammonia TS, set the mixture aside for a few days in a well-stoppered bottle, and filter. If the solution is not perfectly clear, filter it before using.

Magnesium Sulfate TS—Dissolve 12 g of crystals of magnesium sulfate, selected for freedom from efflorescence, in water to make 100 mL.

Malachite Green TS—Dissolve 1 g of malachite green oxalate in 100 mL of glacial acetic acid.

Mallory's Stain—Dissolve 500 mg of water-soluble aniline blue, 2 g of orange G, and 2 g of oxalic acid in 100 mL of water.

Mayer's Reagent—See *Mercuric-Potassium Iodide TS*.

Mercuric Acetate TS—Dissolve 6.0 g of mercuric acetate in glacial acetic acid to make 100 mL. Store in tight containers, protected from direct sunlight.

Mercuric-Ammonium Thiocyanate TS—Dissolve 30 g of ammonium thiocyanate and 27 g of mercuric chloride in water to make 1000 mL.

Mercuric Bromide TS, Alcoholic—Dissolve 5 g of mercuric bromide in 100 mL of alcohol, employing gentle heat to facilitate solution. Store in glass containers, protected from light.

Mercuric Chloride TS—Dissolve 6.5 g of mercuric chloride in water to make 100 mL.

Mercuric Iodide TS (*Valser's Reagent*)—Slowly add potassium iodide solution (1 in 10) to red mercuric iodide until almost all of the latter is dissolved, and filter off the excess. A solution containing 10 g of potassium iodide in 100 mL dissolves approximately 14 g of HgI_2 at 20°.

Mercuric Nitrate TS—Dissolve 40 g of mercuric oxide (red or yellow) in a mixture of 32 mL of nitric acid and 15 mL of water. Store in glass containers, protected from light.

Mercuric-Potassium Iodide TS (*Mayer's Reagent*)—Dissolve 1.358 g of mercuric chloride in 60 mL of water. Dissolve 5 g of potassium iodide in 10 mL of water. Mix the two solutions, and dilute with water to 100 mL.

Mercuric-Potassium Iodide TS, Alkaline (*Nessler's Reagent*)—Dissolve 10 g of potassium iodide in 10 mL of water, and add slowly, with stirring, a saturated solution of mercuric chloride until a slight red precipitate remains undissolved. To this mixture add an ice-cold solution of 30 g of potassium hydroxide in 60 mL of water, then add 1 mL more of the saturated solution of mercuric chloride. Dilute with water to 200 mL. Allow the precipitate to settle, and draw off the clear liquid. A 2-mL portion of this reagent, when added to 100 mL of a 1 in 300,000 solution of ammonium chloride in ammonia-free water, produces at once a yellowish brown color.

Mercuric Sulfate TS (*Denigès' Reagent*)—Mix 5 g of yellow mercuric oxide with 40 mL of water, and while stirring slowly add 20 mL of sulfuric acid, then add another 40 mL of water, and stir until completely dissolved.

Mercurous Nitrate TS—Dissolve 15 g of mercurous nitrate in a mixture of 90 mL of water and 10 mL of diluted nitric acid. Store in dark, amber-colored bottles in which a small globule of mercury has been placed.

Metaphenylenediamine Hydrochloride TS—Dissolve 1 g of metaphenylenediamine hydrochloride in 200 mL of water. The solution must be colorless when used. If necessary, decolorize by heating with activated charcoal.

Metaphosphoric-Acetic Acids TS—Dissolve 15 g of metaphosphoric acid in 40 mL of glacial acetic acid and sufficient water to make 500 mL. Store in a cold place, and use within 2 days.

Methyl Orange TS—Dissolve 100 mg of methyl orange in 100 mL of water, and filter if necessary.

Methyl Purple TS—Use *Methyl Red–Methylene Blue TS*.

Methyl Red TS—Dissolve 100 mg of methyl red in 100 mL of alcohol, and filter if necessary.

Methyl Red TS, Methanolic—Dissolve 1 g of methyl red in 100 mL of methanol, and filter, if necessary. Store protected from light, and use within 21 days.

Methyl Red–Methylene Blue TS—Add 10 mL of methyl red TS to 10 mL of methylene blue TS, and mix.

Methyl Violet TS—Use *Crystal Violet TS*.

Methyl Yellow TS—Dilute with alcohol a commercially available stock solution of methyl yellow in alcohol[97] to obtain a solution having a concentration of 0.10 mg per mL.

Methyl Yellow–Methylene Blue TS—Dissolve 1 g of methyl yellow and 100 mg of methylene blue in 125 mL of methanol.

Methylene Blue TS—Dissolve 125 mg of methylene blue in 100 mL of alcohol, and dilute with alcohol to 250 mL.

Methylthionine Perchlorate TS—To 500 mL of potassium perchlorate solution (1 in 1000) add dropwise, with constant shaking, methylene blue solution (1 in 100) until a slight, permanent turbidity results. Allow the precipitate to settle, decant the supernatant liquid through paper, and use only the clear solution.

Millon's Reagent—To 2 mL of mercury in a conical flask add 20 mL of nitric acid. Shake the flask under a hood to break up the mercury into small globules. After about 10 minutes, add 35 mL of water, and, if a precipitate or crystals appear, add sufficient dilute nitric acid (1 in 5, prepared from nitric acid from which the oxides have been removed by blowing air through it until it is colorless) to dissolve the separated solid. Add sodium hydroxide solution (1 in 10) dropwise, with thorough mixing, until the curdy precipitate that forms after the addition of each drop no longer redissolves but is dispersed to form a suspension. Add 5 mL more of the dilute nitric acid, and mix. Prepare this solution fresh.

Molybdo-phosphotungstate TS (*Folin-Denis Reagent*)—To about 350 mL of water contained in a round-bottom flask add 50 g of sodium tungstate, 12 g of phosphomolybdic acid, and 25 mL of phosphoric acid. Boil the mixture under a reflux condenser for 2 hours, then cool, dilute with water to 500 mL, and mix. Store in tight containers, protected from light, and in a cold place.

1-Naphthol Reagent—Dissolve 1 g of 1-naphthol in 25 mL of methanol. Prepare this solution fresh.

1-Naphthol TS—Use *1-Naphthol Reagent*.

2-Naphthol TS (*Betanaphthol TS*)—Dissolve 1 g of 2-naphthol in 100 mL of sodium hydroxide solution (1 in 100).

***p*-Naphtholbenzein TS**—Dissolve 250 mg of *p*-naphtholbenzein in 100 mL of glacial acetic acid.

***N*-(1-Naphthyl)ethylenediamine Dihydrochloride TS**—Dissolve 100 mg of *N*-(1-naphthyl)ethylenediamine dihydrochloride in 100 mL of a mixture of 7 parts of acetone and 3 parts of water.

Nessler's Reagent—See *Mercuric-Potassium Iodide TS, Alkaline*.

Neutral Red TS—Dissolve 100 mg of neutral red in 100 mL of 50 percent alcohol.

Ninhydrin TS—Use *Triketohydrindene Hydrate TS*.

***p*-Nitroaniline TS**—To 350 mg of *p*-nitroaniline add 1.5 mL of hydrochloric acid, and mix. Dilute with water to 50 mL, mix, and allow to settle. Place 5 mL of the clear, supernatant liquid in a 100-mL volumetric flask, and immerse it in an ice bath. While it is in the ice bath, add 1 mL of hydrochloric acid, then add, in small portions, 2 mL of sodium nitrite solution (1 in 100), dilute with water to volume, and mix.

Nitrophenanthroline TS—Dissolve 150 mg of 5-nitro-1,10-phenanthroline in 15 mL of freshly prepared ferrous sulfate solution (1 in 140).

Oracet Blue B TS—A 1 in 200 solution of *oracet blue B* in *glacial acetic acid*.

Orthophenanthroline TS—Dissolve 150 mg of orthophenanthroline in 10 mL of a solution of ferrous sulfate, prepared by dissolving 700 mg of clear crystals of ferrous sulfate in 100 mL of water. The ferrous sulfate solution must be prepared immediately before dissolving the orthophenanthroline. Store in well-closed containers.

Oxalic Acid TS—Dissolve 6.3 g of oxalic acid in water to make 100 mL.

Palladium Chloride TS, Buffered—Weigh 500 mg of palladium chloride into a 250-mL beaker, add 5 mL of concentrated hydrochloric acid, and warm the mixture on a steam bath. Add 200 mL of hot water in small increments with continued heating until solution is complete. Transfer the solution to a 250-mL volumetric flask, and dilute with water to volume. Transfer 50 mL to a 100-mL volumetric flask. Add 10 mL of 1 *M* sodium acetate and 9.6 mL of 1 *N* hydrochloric acid. Dilute with water to volume.

Phenol Red TS (*Phenolsulfonphthalein TS*)—Dissolve 100 mg of phenolsulfonphthalein in 100 mL of alcohol, and filter if necessary.

Phenoldisulfonic Acid TS—Dissolve 2.5 g of phenol in 15 mL of sulfuric acid in a flask of suitable capacity. Add 7.5 mL of fuming sulfuric acid, stir well, and heat at 100° for 2 hours. Transfer the product, while still fluid, to a glass-stoppered bottle, and, when desired for use, warm in a water bath until liquefied.

Phenolphthalein TS—Dissolve 1 g of phenolphthalein in 100 mL of alcohol.

Phenylhydrazine Acetate TS—Dissolve 10 mL of phenylhydrazine and 5 mL of glacial acetic acid in water to make 100 mL.

Phenylhydrazine–Sulfuric Acid TS—Dissolve 65 mg of phenylhydrazine hydrochloride in 100 mL of a cooled mixture of equal volumes of sulfuric acid and water.

Phloroglucinol TS—Dissolve 500 mg of phloroglucinol in 25 mL of alcohol. Store in tight containers, protected from light.

Phosphatic Enzyme TS—Dissolve 5 g of phosphatic enzyme in water to make 50 mL. Prepare this solution fresh.

Phosphomolybdic Acid TS—Dissolve 20 g of phosphomolybdic acid in alcohol to make 100 mL. Filter the solution, and use only the clear filtrate.

Phosphotungstic Acid TS—Dissolve 1 g of phosphotungstic acid in water to make 100 mL.

Picrate TS, Alkaline—Mix 20 mL of trinitrophenol solution (1 in 100) with 10 mL of sodium hydroxide solution (1 in 20), dilute with water to 100 mL, and mix. Use within 2 days.

Picric Acid TS—See *Trinitrophenol TS*.

Platinic Chloride TS—Dissolve 2.6 g of platinic chloride in water to make 20 mL.

Platinum-Cobalt TS—Dissolve 1.246 g of potassium chloroplatinate (K_2PtCl_6) and 1.000 g of cobalt chloride ($CoCl_2 . 6H_2O$) in water, add 100 mL of hydrochloric acid, and dilute with water to 1 liter.

Potassium Acetate TS—Dissolve 10 g of potassium acetate in water to make 100 mL.

Potassium-Bismuth Iodide TS—Dissolve 12.5 g of tartaric acid in 25 mL of water, then dissolve 1.06 g of bismuth subnitrate in this mixture (*Solution A*). Dissolve 20 g of potassium iodide in 25 mL of water (*Solution B*). Dissolve 100 g of tartaric acid in 450 mL of water (*Solution C*). Add *Solutions A* and *B* to *Solution C*, and mix.

Potassium Carbonate TS—Dissolve 7 g of anhydrous potassium carbonate in water to make 100 mL.

Potassium Chromate TS—Dissolve 10 g of potassium chromate in water to make 100 mL.

Potassium Dichromate TS—Dissolve 7.5 g of potassium dichromate in water to make 100 mL.

Potassium Ferricyanide TS—Dissolve 1 g of potassium ferricyanide in 10 mL of water. Prepare this solution fresh.

Potassium Ferrocyanide TS—Dissolve 1 g of potassium ferrocyanide in 10 mL of water. Prepare this solution fresh.

Potassium Hydroxide TS—Dissolve 6.5 g of potassium hydroxide in water to make 100 mL.

Potassium Hydroxide TS, Alcoholic—Use *0.5 N Potassium Hydroxide, Alcoholic* (see in the section, *Volumetric Solutions*).

Potassium Iodide TS—Dissolve 16.5 g of potassium iodide in water to make 100 mL. Store in light-resistant containers.

Potassium Iodoplatinate TS—Dissolve 200 mg of platinic chloride in 2 mL of water, mix with 25 mL of potassium iodide solution (1 in 25), and add water to make 50 mL.

Potassium Permanganate TS—Use *0.1 N Potassium Permanganate* (see in the section, *Volumetric Solutions*).

Potassium Sulfate TS—Dissolve 1 g of potassium sulfate in water to make 100 mL.

Pyridine-Pyrazolone TS—To 100 mL of a saturated solution of 1-phenyl-3-methyl-2-pyrazoline-5-one add 20 mL of a 1 in 1000 solution of 3,3′-dimethyl-1,1′-diphenyl-[4,4′-bi-2-pyrazoline]-5,5′-dione in pyridine. Store in a dark bottle, and use within 3 days.

Pyrogallol TS, Alkaline—Dissolve 500 mg of pyrogallol in 2 mL of water. Dissolve 12 g of potassium hydroxide in 8 mL of water. The solutions should be freshly prepared and mixed immediately before use.

Quinaldine Red TS—Dissolve 100 mg of quinaldine red in 100 mL of alcohol.

Quinone TS—Dissolve 500 mg of *p*-benzoquinone in 2.5 mL of glacial acetic acid, and dilute with alcohol to 50 mL. Prepare this solution fresh daily.

Resorcinol TS—Dissolve 1 g of resorcinol in hydrochloric acid to make 100 mL.

Ruthenium Red TS—Dissolve 10 g of lead acetate in water, dilute with water to 100 mL, and add 80 mg of ruthenium red. The solution is wine-red in color. [NOTE—If necessary, add additional ruthenium red to obtain a wine-red color.]

Saline TS—Dissolve 9.0 g of sodium chloride in water to make 1000 mL.

NOTE—Where pyrogen-free saline TS is specified in this Pharmacopeia, saline TS that has met the requirements of the *Pyrogen Test* ⟨151⟩ is to be used.

Saline TS, Pyrogen-free—See *Saline TS*.

Schweitzer's Reagent—See *Cupric Oxide, Ammoniated, TS*.

Silver-Ammonia-Nitrate TS—Dissolve 1 g of silver nitrate in 20 mL of water. Add ammonia TS, dropwise, with constant stirring, until the precipitate is almost but not entirely dissolved. Filter, and store in tight, light-resistant containers.

Silver-Ammonium Nitrate TS—See *Silver-Ammonia-Nitrate TS*.

Silver Diethyldithiocarbamate TS—Dissolve 1 g of silver diethyldithiocarbamate in 200 mL of recently distilled pyridine. Store in light-resistant containers, and use within 30 days.

Silver Nitrate TS—Use *0.1 N Silver Nitrate* (see in the section, *Volumetric Solutions*).

Simulated Gastric Fluid TS—See *Gastric Fluid, Simulated, TS*.

Simulated Intestinal Fluid TS—See *Intestinal Fluid, Simulated, TS*.

Sodium Acetate TS—Dissolve 13.6 g of sodium acetate in water to make 100 mL.

Sodium Alizarinsulfonate TS—Dissolve 100 mg of sodium alizarinsulfonate in 100 mL of water, and filter.

Sodium Aminoacetate TS (*Sodium Glycinate TS*)—Dissolve 3.75 g of aminoacetic acid in about 500 mL of water, add 2.1 g of sodium hydroxide, and dilute with water to 1000 mL. Mix 9 mL of the resulting solution with 1 mL of dilute glacial acetic acid (1 in 300). This test solution has a pH between 10.4 and 10.5.

Sodium Bisulfite TS—Dissolve 10 g of sodium bisulfite in water to make 30 mL. Prepare this solution fresh.

Sodium Bitartrate TS—Dissolve 1 g of sodium bitartrate in water to make 10 mL. Prepare this solution fresh.

Sodium Carbonate TS—Dissolve 10.6 g of anhydrous sodium carbonate in water to make 100 mL.

Sodium Chloride TS, Alkaline—Dissolve 2 g of sodium hydroxide in 100 mL of water, saturate the solution with sodium chloride, and filter.

Sodium Cobaltinitrite TS—Dissolve 10 g of sodium cobaltinitrite in water to make 50 mL, and filter if necessary.

Sodium Fluoride TS—Dry about 500 mg of sodium fluoride at 200° for 4 hours. Weigh accurately 222 mg of the dried material, and dissolve in water to make 100.0 mL. Pipet 10 mL of this solution into a 1-liter volumetric flask, and dilute with water to volume. Each mL of this solution corresponds to 0.01 mg of fluorine (F).

Sodium Hydrosulfite TS, Alkaline—Dissolve 25 g of potassium hydroxide in 35 mL of water, and 50 g of sodium hydrosulfite in 250 mL of water. When the test solution is required, mix 40 mL of the hydroxide solution with the 250 mL of the hydrosulfite solution. Prepare this solution fresh.

Sodium Hydroxide TS—Dissolve 4.0 g of sodium hydroxide in water to make 100 mL.

Sodium Hypobromite TS—To a solution of 20 g of sodium hydroxide in 75 mL of water add 5 mL of bromine. After solution has taken place, dilute with water to 100 mL. Prepare this solution fresh.

5V02400

Sodium Hypochlorite TS—Use *Sodium Hypochlorite Solution* (see in the section, *Reagent Specifications*).

Sodium Iodohydroxyquinolinesulfonate TS—Dissolve 8.8 g of iodohydroxyquinoline sulfonic acid in 200 mL of water, and add 6.5 mL of 4 *N* sodium hydroxide. Dilute with water to 250 mL, mix, and filter.

Sodium Nitroferricyanide TS—Dissolve 1 g of sodium nitroferricyanide in water to make 20 mL. Prepare this solution fresh.

Dibasic Sodium Phosphate TS—Dissolve 12 g of clear crystals of dibasic sodium phosphate in water to make 100 mL.

Sodium Phosphotungstate TS—To a solution of 20 g of sodium tungstate in 100 mL of water add sufficient phosphoric acid to impart a strongly acid reaction to litmus, and filter. When required for use, decant the clear solution from any sediment that may be present. Store in tight, light-resistant containers.

Sodium Sulfide TS—Dissolve 1 g of sodium sulfide in water to make 10 mL. Prepare this solution fresh.

Sodium Tartrate TS—Dissolve 11.5 g of sodium tartrate in water to make 100 mL.

Sodium Tetraphenylboron TS—Dissolve 1.2 g of sodium tetraphenylboron in water to make 200 mL. If necessary, stir for 5 minutes with 1 g of freshly prepared hydrous aluminum oxide, and filter to clarify.

Sodium Thioglycolate TS—Dissolve 1.5 g of sodium thioglycolate in 450 mL of water, and add 50 mL of alcohol. Use within 3 days.

Sodium Thiosulfate TS—Use *0.1 N Sodium Thiosulfate* (see in the section, *Volumetric Solutions*).

Standard Lead Solution—See under *Heavy Metals* ⟨231⟩.

Stannous Chloride, Acid, TS—Dissolve 8 g of stannous chloride in 500 mL of hydrochloric acid. Store in glass containers, and use within 3 months.

Stannous Chloride, Acid, Stronger, TS—Dissolve 40 g of stannous chloride in 100 mL of hydrochloric acid. Store in glass containers, and use within 3 months.

Starch Iodide Paste TS—Heat 100 mL of water in a 250-mL beaker to boiling, add a solution of 0.75 g of potassium iodide in 5 mL of water, then add 2 g of zinc chloride dissolved in 10 mL of water, and, while the solution is boiling, add, with stirring, a smooth suspension of 5 g of soluble potato starch in 30 mL of cold water. Continue to boil for 2 minutes, then cool. Store in well-closed containers in a cold place.

Starch iodide paste TS must show a definite blue streak when a glass rod, dipped in a mixture of 1 mL of 0.1 *M* sodium nitrite, 500 mL of water, and 10 mL of hydrochloric acid, is streaked on a smear of the paste.

Starch–Potassium Iodide TS—Dissolve 500 mg of potassium iodide in 100 mL of freshly prepared starch TS. Prepare this solution fresh.

Starch TS—Mix 1 g of soluble starch with 10 mg of red mercuric iodide and sufficient cold water to make a thin paste. Add 200 mL of boiling water, and boil for 1 minute with continuous stirring. Cool, and use only the clear solution. [NOTE—Commercially available, stabilized Starch Indicator solutions may be used.]

Stronger Cupric Acetate TS—See *Cupric Acetate TS, Stronger.*

Sudan IV TS—Dissolve 0.5 g of Sudan IV in chloroform to make 100 mL.

Sulfanilic Acid TS—Dissolve 800 mg of sulfanilic acid in 100 mL of acetic acid. Store in tight containers.

Sulfanilic-α-Naphthylamine TS—See *Sulfanilic-1-Naphthylamine TS.*

Sulfanilic-1-Naphthylamine TS—Dissolve 500 mg of sulfanilic acid in 150 mL of acetic acid. Dissolve 100 mg of 1-naphthylamine hydrochloride in 150 mL of acetic acid, and mix the two solutions. The pink color that may develop on standing can be removed by treatment with zinc.

Sulfuric Acid TS—Add a quantity of sulfuric acid of known concentration to sufficient water to adjust the final concentration to between 94.5% and 95.5% (w/w) of H_2SO_4.

NOTE—Since the acid concentration may change upon standing or upon intermittent use, the concentration should be checked frequently and solutions assaying more than 95.5% or less than 94.5% discarded.

Sulfuric Acid–Formaldehyde TS—Add 1 drop of formaldehyde TS to each mL of sulfuric acid, and mix. Prepare this solution fresh.

Tannic Acid TS—Dissolve 1 g of tannic acid in 1 mL of alcohol, and dilute with water to 10 mL. Prepare this solution fresh.

Tartaric Acid TS—Dissolve 3 g of tartaric acid in water to make 10 mL. Prepare this solution fresh.

Tetrabromophenolphthalein Ethyl Ester TS—Dissolve 100 mg of tetrabromophenolphthalein ethyl ester in 90 mL of glacial acetic acid, and dilute with glacial acetic acid to 100 mL. Prepare this solution fresh.

Tetramethylammonium Hydroxide TS—Use an aqueous solution containing, in each 100 mL, the equivalent of 10 g of anhydrous tetramethylammonium hydroxide.

Thorium Nitrate TS—Dissolve 1 g of thorium nitrate in water to make 100 mL. Filter, if necessary.

Thymol Blue TS—Dissolve 100 mg of thymol blue in 100 mL of alcohol, and filter if necessary.

Thymolphthalein TS—Dissolve 100 mg of thymolphthalein in 100 mL of alcohol, and filter if necessary.

***p*-Toluenesulfonic Acid TS**—Dissolve 2 g of *p*-toluenesulfonic acid in 10 mL of a mixture of 7 parts of acetone and 3 parts of water.

Triketohydrindene Hydrate TS (*Ninhydrin TS*)—Dissolve 200 mg of triketohydrindene hydrate in water to make 10 mL. Prepare this solution fresh.

Trinitrophenol TS (*Picric Acid TS*)—Dissolve the equivalent of 1 g of anhydrous trinitrophenol in 100 mL of hot water. Cool the solution, and filter if necessary.

Triphenyltetrazolium Chloride TS—Dissolve 500 mg of triphenyltetrazolium chloride in dehydrated alcohol to make 100 mL.

Xylenol Orange TS—Dissolve 100 mg of xylenol orange in 100 mL of alcohol.

Zinc Uranyl Acetate TS—Dissolve 50 g of uranyl acetate in a mixture of 15 mL of glacial acetic acid and water to make 500 mL. Then dissolve 150 g of zinc acetate in a mixture of 15 mL of glacial acetic acid and water to make 500 mL. Mix the two solutions, allow to stand overnight, and filter through a dry filter, if necessary.

VOLUMETRIC SOLUTIONS

Normal Solutions—Normal solutions are solutions that contain 1 gram equivalent weight of the active substance in each 1000 mL of solution; that is, an amount equivalent to 1.0079 g of hydrogen or 7.9997 g of oxygen. Normal solutions and solutions bearing a specific relationship to normal solutions, and used in volumetric determinations, are designated as follows: normal, 1 *N;* double-normal, 2 *N;* half-normal, 0.5 *N;* tenth-normal, 0.1 *N;* fiftieth-normal, 0.02 *N;* hundredth-normal, 0.01 *N;* thousandth-normal, 0.001 *N.*

Molar Solutions—Molar solutions are solutions that contain, in 1000 mL, 1 gram-molecule of the reagent. Thus, each liter of a molar solution of sulfuric acid contains 98.07 g of H_2SO_4 and each liter of a molar solution of potassium ferricyanide contains 329.25 g of $K_3Fe(CN)_6$. Solutions containing, in 1000 mL, one-tenth of a gram-molecule of the reagent are designated "tenth-molar," 0.1 *M;* and other molarities are similarly indicated.

Empirical Solutions—It is frequently difficult to prepare standard solutions of a desired theoretical normality, and this is not essential. A solution of approximately the desired normality is prepared and standardized by titration against a primary standard solution. The normality factor so obtained is used in all calculations where such empirical solutions are employed. If desired, an empirically prepared solution may be adjusted downward to a given normality provided it is strong enough to permit dilution.

All volumetric solutions, whether made by direct solution or by dilution of a stronger solution, must be thoroughly mixed by shaking before standardization. As the strength of a standard solution may change upon standing, the factor should be redetermined frequently.

When solutions of a reagent are used in several normalities, the details of the preparation and standardization are usually given for the normality most frequently required. Stronger or weaker solutions are prepared and standardized in the same general manner as described, using proportionate amounts of the reagent. It is possible in many instances to prepare lower normalities accurately by making an exact dilution of a stronger solution. Volumetric solutions prepared by dilution should be restandardized either as directed for the stronger solution or by comparison with another volumetric solution having a known ratio to the stronger solution.

Dilute solutions that are not stable, as, for instance, potassium permanganate 0.01 *N* and more dilute sodium thiosulfate, are preferably prepared by exactly diluting the higher normality with thoroughly boiled and cooled water on the same day they are required for use.

Blank Determinations—Where it is directed that "any necessary correction" be made by a blank determination, the determination is to be conducted with the use of the same quantities of the same reagents treated in the same manner as the solution or mixture containing the portion of the substance under assay or test, but with the substance itself omitted. Appropriate blank corrections are to be made for all Pharmacopeial titrimetric assays (see *Titrimetry* ⟨541⟩).

All Pharmacopeial assays that are volumetric in nature indicate the weight of the substance being assayed to which each mL of the primary volumetric solution is equivalent. In general, these equivalents may be derived by simple calculation from the data given under *Molecular Formulas and Weights*, in the *Reference Tables*.

Preparation and Methods of Standardization of Volumetric Solutions

The following directions give only one method for standardization, but other methods of standardization, capable of yielding at least the same degree of accuracy, may be used. The values obtained in the standardization of volumetric solutions are valid for all Pharmacopeial uses of these solutions, regardless of the instrumental or chemical indicators employed in the individual monographs. Where the apparent normality or molarity of a titrant depends upon the special conditions of its use, the individual monograph sets forth the directions for standardizing the reagent in the specified context. For those salts that usually are available as certified primary standards, or that are available as highly purified salts of primary standard quality, it is permissible to prepare solutions by accurately weighing a suitable quantity of the salt and dissolving it to produce a specific volume of solution of known concentration. Acetic, hydrochloric, and sulfuric acids may be standardized against a sodium hydroxide solution that recently has been standardized against a certified primary standard.

All volumetric solutions, if practicable, are to be prepared, standardized, and used at the standard temperature of 25°. If a titration is carried out with the volumetric solution at a markedly different temperature, standardize the volumetric solution used as the titrant at that different temperature, or make a suitable temperature correction.

Acetic Acid, Double-Normal (2 *N*)

$C_2H_4O_2$, **60.05**
120.10 g in 1000 mL

Add 116 mL of glacial acetic acid to sufficient water to make 1000 mL after cooling to room temperature.

Ammonium Thiocyanate, Tenth-Normal (0.1 *N*)

NH_4SCN, **76.12**
7.612 g in 1000 mL

Dissolve about 8 g of ammonium thiocyanate in 1000 mL of water, and standardize the solution as follows:

Measure accurately about 30 mL of 0.1 *N* silver nitrate VS into a glass-stoppered flask. Dilute with 50 mL of water, then add 2 mL of nitric acid and 2 mL of ferric ammonium sulfate TS, and titrate with the ammonium thiocyanate solution to the first appearance of a red-brown color. Calculate the normality.

If desirable, 0.1 *N* ammonium thiocyanate may be replaced by 0.1 *N* potassium thiocyanate where the former is directed in various tests and assays.

Bromine, Tenth-Normal (0.1 *N*)

Br, **79.90**
7.990 g in 1000 mL

Dissolve 3 g of potassium bromate and 15 g of potassium bromide in water to make 1000 mL, and standardize the solution as follows:

Measure accurately about 25 mL of the solution into a 500-mL iodine flask, and dilute with 120 mL of water. Add 5 mL of hydrochloric acid, insert the stopper in the flask, and shake it gently. Then add 5 mL of potassium iodide TS, again insert the stopper, shake the mixture, allow it to stand for 5 minutes, and titrate the liberated iodine with 0.1 *N* sodium thiosulfate VS, adding 3 mL of starch TS as the end-point is approached. Calculate the normality.

Preserve in dark amber-colored, glass-stoppered bottles.

Ceric Ammonium Nitrate, Twentieth-Normal (0.05 *N*)

$Ce(NO_3)_4 . 2NH_4NO_3$, **548.23**
2.741 g in 100 mL

Dissolve 2.75 g of ceric ammonium nitrate in 1 *N* nitric acid to obtain 100 mL of solution, and filter. Standardize the solution as follows:

Measure accurately 10 mL of freshly standardized 0.1 *N* ferrous ammonium sulfate VS into a flask, and dilute with water to about 100 mL. Add 1 drop of nitrophenanthroline TS, and titrate with the ceric ammonium nitrate solution to a colorless end-point. From the volume of 0.1 *N* ferrous ammonium sulfate VS taken and the volume of ceric ammonium nitrate solution consumed, calculate the normality.

Ceric Sulfate, Tenth-Normal (0.1 *N*)

$Ce(SO_4)_2$, **332.24**
33.22 g in 1000 mL

Transfer 59 g of ceric ammonium nitrate to a beaker, add 31 mL of sulfuric acid, mix, and cautiously add water, in 20-mL portions, until solution is complete. Cover the beaker, allow to stand overnight, filter through a fine-porosity, sintered-glass crucible, dilute with water to 1000 mL, and mix. Standardize the solution as follows:

[NOTE—Prepare the osmium tetroxide solution used in this procedure in a well-ventilated hood, as poisonous vapors are given off by this compound.] Weigh accurately 200 mg of arsenic trioxide, previously dried at 105° for 1 hour, and transfer to a 500-mL conical flask. Wash down the inner walls of the flask with 25 mL of sodium hydroxide solution (2 in 25), swirl to dissolve the substance, and when solution is complete, add 100 mL of water, and mix. Add 10 mL of dilute sulfuric acid (1 in 3), then add 2 drops each of orthophenanthroline TS and a 1 in 400 solution of osmium tetroxide in 0.1 *N* sulfuric acid, and slowly titrate with the ceric sulfate solution until the pink color is changed to a very pale blue. Calculate the normality. Each 4.946 mg of arsenic trioxide is equivalent to 1 mL of 0.1 *N* ceric sulfate.

Cupric Nitrate, Tenth Normal (0.1 *N*)

$Cu(NO_3) . 2.5H_2O$, **232.59**
23.26 g in 1000 mL
$Cu(NO_3) . 3H_2O$, **241.60**
24.16 g in 1000 mL

Dissolve 23.3 g of cupric nitrate 2.5 hydrate, or 24.2 g of the trihydrate, in water to make 1000 mL. Standardize the solution as follows:

Transfer 20.0 mL of the solution to a 250-mL beaker. Add 2 mL of 5 *M* sodium nitrate, 20 mL of ammonium acetate TS, and sufficient water to make 1000 mL. Titrate with 0.05 *M* disodium ethylenediaminetetraacetate VS. Determine the end-point potentiometrically using a cupric ion-double junction reference electrode system. Perform a blank determination, and make any necessary correction. Calculate the normality by the formula:

$$VM/20.0,$$

in which V is the volume, in mL, of disodium ethylenediaminetetraacetate consumed, M is the molarity of the disodium ethylenediaminetetraacetate, and 20.0 is the number of mL of cupric nitrate solution taken.

Standard Dichlorophenol-Indophenol Solution

To 50 mg of 2,6-dichlorophenol-indophenol sodium that has been stored in a desiccator over soda lime add 50 mL of water containing 42 mg of sodium bicarbonate, shake vigorously, and when the dye is dissolved, add water to make 200 mL. Filter into an amber, glass-stoppered bottle. Standardize the solution as follows:

Weigh accurately 50 mg of USP Ascorbic Acid RS, and transfer to a glass-stoppered, 50-mL volumetric flask with the aid of a sufficient volume of metaphosphoric-acetic acids TS to make 50 mL. Immediately transfer 2 mL of the ascorbic acid solution to a 50-mL conical flask containing 5 mL of the metaphosphoric-acetic acids TS, and titrate rapidly with the dichlorophenol-indophenol solution until a distinct rose-pink color persists for at least 5 seconds. Perform a blank titration by titrating 7 mL of the metaphosphoric-acetic acids TS plus a volume of water equal to the volume of the dichlorophenol solution used in titrating the ascorbic acid solution. Express the concentration of the standard solution in terms of its equivalent in mg of ascorbic acid.

Disodium Ethylenediaminetetraacetate, Twentieth-Molar (0.05 *M*)

$C_{10}H_{14}N_2Na_2O_8 . 2H_2O$, **372.24**
18.61 g in 1000 mL

Dissolve 18.6 g of disodium ethylenediaminetetraacetate in water to make 1000 mL, and standardize the solution as follows:

Weigh accurately about 200 mg of chelometric standard calcium carbonate, transfer to a 400-mL beaker, add 10 mL of water, and swirl to form a slurry. Cover the beaker with a watch glass, and introduce 2 mL of diluted hydrochloric acid from a pipet inserted between the lip of the beaker and the edge of the watch glass. Swirl the contents of the beaker to dissolve the calcium carbonate. Wash down the sides of the beaker, the outer surface of the pipet, and the watch glass with water, and dilute with water to about 100 mL. While stirring the solution, preferably with a magnetic stirrer, add about 30 mL of the disodium ethylenediaminetetraacetate solution from a 50-mL buret. Add 15 mL of sodium hydroxide TS and 300 mg of hydroxy naphthol blue indicator, and continue the titration with the disodium ethylenediaminetetraacetate solution to a blue end-point. Calculate the molarity by the formula:

$$W/(100.09V),$$

in which W is the weight, in mg, of $CaCO_3$ in the portion of calcium carbonate taken, and V is the volume, in mL, of disodium ethylenediaminetetraacetate solution consumed.

Ferric Ammonium Sulfate, Tenth-Normal (0.1 *N*)

$FeNH_4(SO_4)_2 . 12H_2O$, **482.18**
48.22 g in 1000 mL

Dissolve 50 g of ferric ammonium sulfate in a mixture of 300 mL of water and 6 mL of sulfuric acid, dilute with water to 1000 mL, and mix. Standardize the solution as follows:

Measure accurately about 40 mL of the solution into a glass-stoppered flask, add 5 mL of hydrochloric acid, mix, and add a solution of 3 g of potassium iodide in 10 mL of water. Insert the stopper, allow to stand for 10 minutes, then titrate the liberated iodine with 0.1 *N* sodium thiosulfate VS, adding 3 mL of starch TS as the end-point is approached. Correct for a blank run on the same quantities of the same reagents, and calculate the normality.

Store in tight containers, protected from light.

Ferrous Ammonium Sulfate, Tenth-Normal (0.1 *N*)

$Fe(NH_4)_2(SO_4)_2 . 6H_2O$, **392.13**
39.21 g in 1000 mL

Dissolve 40 g of ferrous ammonium sulfate in a previously cooled mixture of 40 mL of sulfuric acid and 200 mL of water, dilute with water to 1000 mL, and mix. On the day of use, standardize the solution as follows:

Measure accurately 25 to 30 mL of the solution into a flask, add 2 drops of orthophenanthroline TS, and titrate with 0.1 *N* ceric sulfate VS until the red color is changed to pale blue. From the volume of 0.1 *N* ceric sulfate consumed, calculate the normality.

Hydrochloric Acid, Normal (1 *N*)

HCl, **36.46**
36.46 g in 1000 mL

Dilute 85 mL of hydrochloric acid with water to 1000 mL. Standardize the solution as follows:

Weigh accurately about 1.5 g of primary standard anhydrous sodium carbonate that previously has been heated at a temperature of about 270° for 1 hour. Dissolve it in 100 mL of water, and add 2 drops of methyl red TS. Add the acid slowly from a buret, with constant stirring, until the solution becomes faintly pink. Heat the solution to boiling, cool, and continue the titration. Heat again to boiling, and titrate further as necessary until the faint pink color is no longer affected by continued boiling. Calculate the normality. Each 52.99 mg of anhydrous sodium carbonate is equivalent to 1 mL of 1 *N* hydrochloric acid.

Hydrochloric Acid, Half-Normal (0.5 *N*) in Methanol

HCl, **36.46**
18.23 g in 1000 mL

To a 1000-mL volumetric flask containing 40 mL of water slowly add 43 mL of hydrochloric acid. Cool, and add methanol to volume. Standardize the solution as follows:

Weigh accurately about 800 mg of primary standard anhydrous sodium carbonate that previously has been heated at a temperature of about 270° for 1 hour. Proceed as directed under *Hydrochloric Acid, Normal* (1 *N*), beginning with "Dissolve it in 100 mL of water."

Iodine, Tenth-Normal (0.1 *N*)

I, **126.90**
12.69 g in 1000 mL

Dissolve about 14 g of iodine in a solution of 36 g of potassium iodide in 100 mL of water, add 3 drops of hydrochloric acid, dilute with water to 1000 mL, and standardize the solution as follows:

Weigh accurately about 150 mg of arsenic trioxide, previously dried at 105° for 1 hour, and dissolve in 20 mL of 1 *N* sodium hydroxide by warming if necessary. Dilute with 40 mL of water, add 2 drops of methyl orange TS, and follow with diluted hydrochloric acid until the yellow color is changed to pink. Then add 2 g of sodium bicarbonate, dilute with 50 mL of water, and add 3 mL of starch TS. Slowly add the iodine solution from a buret until a permanent blue color is produced. Calculate the normality. Each 4.946 mg of arsenic trioxide is equivalent to 1 mL of 0.1 *N* iodine.

Preserve in amber-colored, glass-stoppered bottles.

Lithium Methoxide, Fiftieth-Normal (0.02 *N*) in Methanol

CH_3LiO, **37.98**
759.6 mg in 1000 mL

Dissolve 0.12 g of freshly cut lithium metal in 150 mL of methanol, cooling the flask during addition of the metal. When the reaction is complete, add 850 mL of methanol, and mix. Store the solution preferably in the reservoir of an automatic delivery buret suitably protected from carbon dioxide and moisture. Standardize the solution by titration against benzoic acid as described under *Sodium Methoxide, Tenth-Normal* (*0.1 N*) (*in Toluene*), but use only 100 mg of benzoic acid. Each 2.442 mg of benzoic acid is equivalent to 1 mL of 0.02 *N* lithium methoxide.

NOTE—Restandardize the solution frequently.

Lithium Methoxide, Tenth-Normal (0.1 *N*) in Benzene

CH_3OLi, **37.98**
3.798 g in 1000 mL

Dissolve 0.6 g of freshly cut lithium metal in 150 mL of methanol, cooling the flask during addition of the metal. When reaction is complete, add 850 mL of benzene. If cloudiness or precipitation occurs, add sufficient methanol to clarify the solution. Store preferably in the reservoir of an automatic delivery buret suitably protected from carbon dioxide and moisture. Standardize the solution by titration against benzoic acid as described under *Sodium Methoxide, Tenth-Normal* (*0.1 N*) (*in Toluene*).

NOTE—Restandardize the solution frequently.

Lithium Methoxide, Tenth-Normal (0.1 *N*) in Chlorobenzene

CH_3OLi, **37.98**
3.798 g in 1000 mL

Dissolve 0.7 g of freshly cut lithium metal in 150 mL of methanol, cooling the flask during addition of the metal. When reaction is complete, add 850 mL of chlorobenzene. If cloudiness or precipitation occurs, add sufficient methanol to clarify the solution. Store preferably in the reservoir of an automatic delivery buret suitably protected from carbon dioxide and moisture. Standardize the solution by titration against benzoic acid as described under *Sodium Methoxide, Tenth-Normal (0.1 N) (in Toluene).*

NOTE—Restandardize the solution frequently.

Magnesium Acetate, Tenth-Molar (0.1 *M*)

$Mg(C_2H_3O_2)_2 . 4H_2O$, **214.45**
21.45 g in 1000 mL

Dissolve 21.45 g of magnesium acetate, previously dried in a freshly charged calcium chloride–filled desiccator to constant weight, in water to make 1000 mL.

Mercuric Nitrate, Tenth-Molar (0.1 *M*)

$Hg(NO_3)_2$, **324.60**
32.46 g in 1000 mL

Dissolve about 35 g of mercuric nitrate in a mixture of 5 mL of nitric acid and 500 mL of water, and dilute with water to 1000 mL. Standardize the solution as follows:

Transfer an accurately measured volume of about 20 mL of the solution to a conical flask, and add 2 mL of nitric acid and 2 mL of ferric ammonium sulfate TS. Cool to below 20°, and titrate with 0.1 *N* ammonium thiocyanate VS to the first appearance of a permanent brownish color. Calculate the molarity.

Morpholine, Half-Normal (0.5 *N*) in Methanol

C_4H_9NO, **87.12**
43.56 g in 1000 mL

Transfer 44 mL of recently distilled morpholine to a 1-liter reagent bottle, and add methanol to make about 1 liter. Protect from absorption of carbon dioxide during withdrawal of aliquots. It is not necessary to standardize this solution.

Oxalic Acid, Tenth-Normal (0.1 *N*)

$H_2C_2O_4 . 2H_2O$, **126.07**
6.303 g in 1000 mL

Dissolve 6.45 g of oxalic acid in water to make 1000 mL. Standardize by titration against freshly standardized 0.1 *N* potassium permanganate VS as directed under *Potassium Permanganate, Tenth-Normal (0.1 N)*.

Preserve in glass-stoppered bottles, protected from light.

Perchloric Acid, Tenth-Normal (0.1 *N*) (in Glacial Acetic Acid)

$HClO_4$, **100.46**
10.05 g in 1000 mL

NOTE—Where called for in the tests and assays, this volumetric solution is specified as "0.1 *N* perchloric acid." Thus, where 0.1 *N* or other strength of this volumetric solution is specified, the solution in glacial acetic acid is to be used, unless the words "in dioxane" are stated. [See also *Perchloric Acid, Tenth-Normal (0.1 N) in Dioxane.*]

Mix 8.5 mL of perchloric acid with 500 mL of glacial acetic acid and 21 mL of acetic anhydride, cool, and add glacial acetic acid to make 1000 mL. Alternatively, the solution may be prepared as follows. Mix 11 mL of 60 percent perchloric acid with 500 mL of glacial acetic acid and 30 mL of acetic anhydride, cool, and add glacial acetic acid to make 1000 mL.

Allow the prepared solution to stand for 1 day for the excess acetic anhydride to be combined, and determine the water content by the *Titrimetric Method* (see *Water Determination* ⟨921⟩). If the water content exceeds 0.05%, add more acetic anhydride. If the solution contains no titratable water, add sufficient water to obtain a content of between 0.02% and 0.05% of water. Allow the solution to stand for 1 day, and again titrate the water content. The solution so obtained contains between 0.02% and 0.05% of water, indicating freedom from acetic anhydride.

Standardize the solution as follows:

Weigh accurately about 700 mg of potassium biphthalate, previously crushed lightly and dried at 120° for 2 hours, and dissolve it in 50 mL of glacial acetic acid in a 250-mL flask. Add 2 drops of crystal violet TS, and titrate with the perchloric acid solution until the violet color changes to blue-green. Deduct the volume of the perchloric acid consumed by 50 mL of the glacial acetic acid, and calculate the normality. Each 20.42 mg of potassium biphthalate is equivalent to 1 mL of 0.1 *N* perchloric acid.

Perchloric Acid, Tenth-Normal (0.1 *N*) in Dioxane

Mix 8.5 mL of perchloric acid with sufficient dioxane, which has been especially purified by adsorption, to make 1000 mL. Standardize the solution as follows:

Weigh accurately about 700 mg of potassium biphthalate, previously crushed lightly and dried at 120° for 2 hours, and dissolve in 50 mL of glacial acetic acid in a 250-mL flask. Add 2 drops of crystal violet TS, and titrate with the perchloric acid solution until the violet color changes to bluish green. Deduct the volume of the perchloric acid consumed by 50 mL of the glacial acetic acid, and calculate the normality. Each 20.42 mg of potassium biphthalate is equivalent to 1 mL of 0.1 *N* perchloric acid.

Potassium Arsenite, Tenth-Normal (0.1 *N*)

$KAsO_2$, **146.02**
7.301 g in 1000 mL

Dissolve 4.9455 g of arsenic trioxide primary standard, previously dried at 105° for 1 hour, in 75 mL of 1 *N* potassium hydroxide. Add 40 g of potassium bicarbonate, dissolved in about 200 mL of water, and dilute with water to 1000.0 mL.

Potassium Bromate, Tenth-Normal (0.1 *N*)

$KBrO_3$, **167.00**
2.784 g in 1000 mL

Dissolve 2.784 g of potassium bromate in water to make 1000 mL, and standardize the solution as follows:

Transfer an accurately measured volume of about 40 mL of the solution to a glass-stoppered flask, add 3 g of potassium iodide, and follow with 3 mL of hydrochloric acid. Allow to stand for 5 minutes, then titrate the liberated iodine with 0.1 *N* sodium thiosulfate VS, adding 3 mL of starch TS as the end-point is approached. Correct for a blank run on the same quantities of the same reagents, and calculate the normality.

Potassium Bromide–Bromate, Tenth-Normal (0.1 *N*)

Dissolve 2.78 g of potassium bromate ($KBrO_3$) and 12.0 g of potassium bromide (KBr) in water, and dilute with water to 1000 mL. Standardize by the procedure set forth for *Potassium Bromate, Tenth-Normal (0.1 N)*.

Potassium Dichromate, Tenth-Normal (0.1 *N*)

$K_2Cr_2O_7$, **294.18**
4.903 g in 1000 mL

Dissolve about 5 g of potassium dichromate in 1000 mL of water. Standardize the solution as follows:

Transfer 25.0 mL of this solution to a glass-stoppered, 500-mL flask, add 2 g of potassium iodide (free from iodate), dilute with 200 mL of water, add 5 mL of hydrochloric acid, allow to stand for 10 minutes in a dark place, and titrate the liberated iodine with 0.1 *N* sodium thiosulfate VS, adding 3 mL of starch TS as the end-point is approached. Correct for a blank run on the same quantities of the same reagents, and calculate the normality.

Potassium Ferricyanide, Twentieth-Molar (0.05 *M*)

$K_3Fe(CN)_6$, **329.25**
16.46 g in 1000 mL

Dissolve about 17 g of potassium ferricyanide in water to make 1000 mL. Standardize the solution as follows:

Transfer 50.0 mL of this solution to a glass-stoppered, 500-mL flask, dilute with 50 mL of water, add 10 mL of potassium iodide TS and 10 mL of dilute hydrochloric acid, and allow to stand for 1 minute. Then add 15 mL of zinc sulfate solution (1 in 10), and titrate the liberated iodine with 0.1 *N* sodium thiosulfate VS, adding 3 mL of starch TS as the end-point is approached. Calculate the molarity.

Protect from light, and restandardize before use.

Potassium Hydroxide, Normal (1 *N*)

KOH, **56.11**

56.11 g in 1000 mL

Dissolve 68 g of potassium hydroxide in about 950 mL of water. Add a freshly prepared saturated solution of barium hydroxide until no more precipitate forms. Shake the mixture thoroughly, and allow it to stand overnight in a stoppered bottle. Decant the clear liquid, or filter the solution in a tight, polyolefin bottle, and standardize by the procedure set forth for *Sodium Hydroxide, Normal (1 N)*.

Potassium Hydroxide, Alcoholic, Half-Normal (0.5 *N*)

28.06 g in 1000 mL

Dissolve about 34 g of potassium hydroxide in 20 mL of water, and add aldehyde-free alcohol to make 1000 mL. Allow the solution to stand in a tightly stoppered bottle for 24 hours. Then quickly decant the clear supernatant liquid into a suitable, tight container, and standardize the solution as follows:

Measure accurately about 25 mL of 0.5 *N* hydrochloric acid VS. Dilute with 50 mL of water, add 2 drops of phenolphthalein TS, and titrate with the alcoholic potassium hydroxide solution until a permanent, pale pink color is produced. Calculate the normality.

NOTE—Store in tightly stoppered bottles, protected from light.

Potassium Hydroxide, Methanolic, Tenth-Normal (0.1 *N*)

5.612 g in 1000 mL

Dissolve about 6.8 g of potassium hydroxide in 4 mL of water, and add methanol to make 1000 mL. Allow the solution to stand in a tightly stoppered bottle for 24 hours. Then quickly decant the clear supernatant liquid into a suitable, tight container, and standardize the solution as follows:

Measure accurately about 25 mL of 0.1 *N* hydrochloric acid VS. Dilute with 50 mL of water, add 2 drops of phenolphthalein TS, and titrate with the methanolic potassium hydroxide solution until a permanent, pale pink color is produced. Calculate the normality.

NOTE—Store in tightly stoppered bottles, protected from light.

Potassium Iodate, Twentieth-Molar (0.05 *M*)

KIO_3, **214.00**

10.70 g in 1000 mL

Dissolve 10.700 g of potassium iodate, previously dried at 110° to constant weight, in water to make 1000.0 mL.

Potassium Permanganate, Tenth-Normal (0.1 *N*)

$KMnO_4$, **158.03**

3.161 g in 1000 mL

Dissolve about 3.3 g of potassium permanganate in 1000 mL of water in a flask, and boil the solution for about 15 minutes. Insert the stopper in the flask, allow it to stand for at least 2 days, and filter through a fine-porosity, sintered-glass crucible. If necessary, the bottom of the sintered-glass crucible may be lined with a pledget of glass wool. Standardize the solution as follows:

Weigh accurately about 200 mg of sodium oxalate, previously dried at 110° to constant weight, and dissolve it in 250 mL of water. Add 7 mL of sulfuric acid, heat to about 70°, and then slowly add the permanganate solution from a buret, with constant stirring, until a pale pink color, which persists for 15 seconds, is produced. The temperature at the conclusion of the titration should be not less than 60°. Calculate the normality. Each 6.700 mg of sodium oxalate is equivalent to 1 mL of 0.1 *N* potassium permanganate.

Since potassium permanganate is reduced on contact with organic substances such as rubber, the solution must be handled in apparatus entirely of glass or other suitably inert material. It should be frequently restandardized. Store in glass-stoppered, amber-colored bottles.

Silver Nitrate, Tenth-Normal (0.1 *N*)

$AgNO_3$, **169.87**

16.99 g in 1000 mL

Dissolve about 17.5 g of silver nitrate in 1000 mL of water, and standardize the solution as follows:

Transfer about 100 mg, accurately weighed, of reagent-grade sodium chloride, previously dried at 110° for 2 hours, to a 150-mL beaker, dissolve in 5 mL of water, and add 5 mL of acetic acid, 50 mL of methanol, and 3 drops of eosin Y TS. Stir, preferably with a magnetic stirrer, and titrate with the silver nitrate solution. Calculate the normality.

Sodium Hydroxide, Normal (1 *N*)

NaOH, **40.00**

40.00 g in 1000 mL

Dissolve 162 g of sodium hydroxide in 150 mL of carbon dioxide–free water, cool the solution to room temperature, and filter through hardened filter paper. Transfer 54.5 mL of the clear filtrate to a tight, polyolefin container, and dilute with carbon dioxide–free water to 1000 mL.

Weigh accurately about 5 g of potassium biphthalate, previously crushed lightly and dried at 120° for 2 hours, and dissolve in 75 mL of carbon dioxide–free water. Add 2 drops of phenolphthalein TS, and titrate with the sodium hydroxide solution to the production of a permanent pink color. Each 204.2 mg of potassium biphthalate is equivalent to 1 mL of 1 *N* sodium hydroxide.

NOTES—(1) Solutions of alkali hydroxides absorb carbon dioxide when exposed to air. They should be preserved in bottles having well-fitted, suitable stoppers, provided with a tube filled with a mixture of sodium hydroxide and lime (soda-lime tubes) so that air entering the container must pass through this tube, which will absorb the carbon dioxide. (2) Prepare solutions of lower concentration (e.g., 0.1 *N*, 0.01 *N*) by quantitatively diluting accurately measured volumes of the 1 *N* solution with sufficient carbon dioxide–free water to yield the desired concentration.

Restandardize the solution frequently.

Sodium Methoxide, Tenth-Normal (0.1 *N*) (in Toluene)

CH_3ONa, **54.02**

5.402 g in 1000 mL

Cool in ice-water 150 mL of methanol contained in a 1000-mL volumetric flask, and add, in small portions, about 2.5 g of freshly cut sodium metal. When the metal has dissolved, add toluene to make 1000 mL, and mix. Store preferably in the reservoir of an automatic delivery buret suitably protected from carbon dioxide and moisture. Standardize the solution as follows:

Weigh accurately about 400 mg of primary standard benzoic acid, and dissolve in 80 mL of dimethylformamide in a flask. Add 3 drops of a 1 in 100 solution of thymol blue in dimethylformamide, and titrate with the sodium methoxide to a blue endpoint. Correct for the volume of the sodium methoxide solution consumed by 80 mL of the dimethylformamide, and calculate the normality. Each 12.21 mg of benzoic acid is equivalent to 1 mL of 0.1 *N* sodium methoxide.

NOTES—(1) To eliminate any turbidity that may form following dilution with toluene, add methanol (25 to 30 mL usually suffices) until the solution is clear. (2) Restandardize the solution frequently.

Sodium Methoxide, Half-Normal (0.5 *N*) in Methanol

CH_3ONa, **54.02**

27.01 g in 1000 mL

Weigh 11.5 g of freshly cut sodium metal, and cut into small cubes. Place about 0.5 mL of anhydrous methanol in a round-bottom, 250-mL flask equipped with a ground-glass joint, add 1 cube of the sodium metal, and, when the reaction has ceased, add the remaining sodium metal to the flask. Connect a water-jacketed condenser to the flask, and slowly add 250 mL of anhydrous methanol, in small portions, through the top of the condenser. Regulate the addition of the methanol so that the vapors are condensed and do not escape through the top of the condenser. After addition of the methanol is complete, connect a drying tube to the top of the condenser, and allow the solution to cool. Transfer the solution to a 1-liter volumetric flask, dilute with anhydrous methanol to volume, and mix. Standardize the solution as follows:

Measure accurately about 20 mL of freshly standardized 1 *N* hydrochloric acid VS into a 250-mL conical flask, add 0.25 mL of phenolphthalein TS, and titrate with the sodium methoxide solution to the first appearance of a permanent pink color. Calculate the normality.

Sodium Nitrite, Tenth-Molar (0.1 *M*)

$NaNO_2$, **69.00**
6.900 g in 1000 mL

Dissolve 7.5 g of sodium nitrite in water to make 1000 mL, and standardize the solution as follows:

Weigh accurately about 500 mg of USP Sulfanilamide RS, previously dried at 105° for 3 hours, and transfer to a suitable beaker. Add 20 mL of hydrochloric acid and 50 mL of water, stir until dissolved, and cool to 15°. Maintaining the temperature at about 15°, titrate slowly with the sodium nitrite solution, placing the buret tip below the surface of the solution to preclude air oxidation of the sodium nitrite, and stir the solution gently with a magnetic stirrer, but avoid pulling a vortex of air beneath the surface. Use the indicator specified in the individual monograph, or, if a potentiometric procedure is specified, determine the end-point electrometrically, using platinum-calomel or platinum-platinum electrodes. When the titration is within 1 mL of the end-point, add the titrant in 0.1-mL portions, and allow 1 minute between additions. Calculate the molarity. Each 17.22 mg of sulfanilamide is equivalent to 1 mL of 0.1000 *M* sodium nitrite.

Sodium Tetraphenylboron, Fiftieth-Molar (0.02 *M*)

$NaB(C_6H_5)_4$, **342.22**
6.845 g in 1000 mL

Dissolve an amount of sodium tetraphenylboron, equivalent to 6.845 g of $NaB(C_6H_5)_4$, in water to make 1000 mL, and standardize the solution as follows:

Pipet two 75-mL portions of the solution into separate beakers, and to each add 1 mL of acetic acid and 25 mL of water. To each beaker add, slowly and with constant stirring, 25 mL of potassium biphthalate solution (1 in 20), and allow to stand for 2 hours. Filter one of the mixtures through a filtering crucible, and wash the precipitate with cold water. Transfer the precipitate to a container, add 50 mL of water, shake intermittently for 30 minutes, filter, and use the filtrate as the saturated potassium tetraphenylborate solution in the following standardization procedure. Filter the second mixture through a tared filtering crucible, and wash the precipitate with three 5-mL portions of saturated potassium tetraphenylborate solution. Dry the precipitate at 105° for 1 hour. Each g of potassium tetraphenylborate is equivalent to 955.1 mg of sodium tetraphenylboron. From the weight of sodium tetraphenylboron obtained, calculate the molarity of the sodium tetraphenylboron solution.

NOTE—Prepare this solution fresh.

Sodium Thiosulfate, Tenth-Normal (0.1 *N*)

$Na_2S_2O_3 \cdot 5H_2O$, **248.17**
24.82 g in 1000 mL

Dissolve about 26 g of sodium thiosulfate and 200 mg of sodium carbonate in 1000 mL of recently boiled and cooled water. Standardize the solution as follows:

Weigh accurately about 210 mg of primary standard potassium dichromate, previously pulverized and dried at 120° for 4 hours, and dissolve in 100 mL of water in a glass-stoppered, 500-mL flask. Swirl to dissolve the solid, remove the stopper, and quickly add 3 g of potassium iodide, 2 g of sodium bicarbonate, and 5 mL of hydrochloric acid. Insert the stopper gently in the flask, swirl to mix, and allow to stand in the dark for 10 minutes. Rinse the stopper and the inner walls of the flask with water, and titrate the liberated iodine with the sodium thiosulfate solution until the solution is yellowish green in color. Add 3 mL of starch TS, and continue the titration to the discharge of the blue color. Calculate the normality.

Restandardize the solution frequently.

Sulfuric Acid, Half-Normal (0.5 *N*) in Alcohol

H_2SO_4, **98.07**
24.52 g in 1000 mL

Add slowly, with stirring, 13.9 mL of sulfuric acid to a sufficient quantity of dehydrated alcohol to make 1000 mL. Cool, and standardize against anhydrous sodium carbonate as described under *Hydrochloric Acid, Half-Normal (0.5 N) in Methanol.*

Sulfuric Acid, Normal (1 *N*)

H_2SO_4, **98.07**
49.04 g in 1000 mL

Add slowly, with stirring, 30 mL of sulfuric acid to about 1020 mL of water, allow to cool to 25°, and determine the normality by titration against sodium carbonate as described under *Hydrochloric Acid, Normal (1 N).*

Tetrabutylammonium Hydroxide, Tenth-Normal (0.1 *N*)

$(C_4H_9)_4NOH$, **259.48**
25.95 g in 1000 mL

Dissolve 40 g of tetra-*n*-butylammonium iodide in 90 mL of anhydrous methanol in a glass-stoppered flask. Place in an ice bath, add 20 g of powdered silver oxide, insert the stopper in the flask, and agitate vigorously for 60 minutes. Centrifuge a few mL, and test the supernatant liquid for iodide (see *Iodide* ⟨191⟩). If the test is positive, add an additional 2 g of silver oxide, and continue to allow to stand for 30 minutes with intermittent agitation. When all of the iodide has reacted, filter through a fine-porosity, sintered-glass funnel. Rinse the flask and the funnel with three 50-mL portions of anhydrous toluene, adding the rinsings to the filtrate. Dilute with a mixture of three volumes of anhydrous toluene and 1 volume of anhydrous methanol to 1000 mL, and flush the solution for 10 minutes with dry, carbon dioxide–free nitrogen. [NOTE—If necessary to obtain a clear solution, further small quantities of anhydrous methanol may be added.] Store in a reservoir protected from carbon dioxide and moisture, and discard after 60 days. Alternatively, the solution may be prepared by diluting a suitable volume of commercially available tetrabutylammonium hydroxide solution in methanol with a mixture of 4 volumes of anhydrous toluene and 1 volume of anhydrous methanol. [NOTE—If necessary to obtain a clear solution, further small quantities of methanol may be added.]

Standardize the solution on the day of use as follows: Dissolve about 400 mg of primary standard benzoic acid, accurately weighed, in 80 mL of dimethylformamide, add 3 drops of a 1 in 100 solution of thymol blue in dimethylformamide, and titrate to a blue end-point with the tetrabutylammonium hydroxide solution, delivering the titrant from a buret equipped with a carbon dioxide absorption trap. Perform a blank determination, and make any necessary correction. Each mL of 0.1 *N* tetrabutylammonium hydroxide is equivalent to 12.21 mg of benzoic acid.

Tetramethylammonium Bromide, Tenth-Molar (0.1 *M*)

$(CH_3)_4NBr$, **154.05**
15.41 g in 1000 mL

Dissolve 15.41 g of tetramethylammonium bromide in water to make 1000 mL, and standardize the solution as follows:

Transfer an accurately measured volume of about 40 mL of the solution to a beaker, add 10 mL of diluted nitric acid and 50.0 mL of 0.1 *N* silver nitrate VS, and mix. Add 2 mL of ferric ammonium sulfate TS, and titrate the excess silver nitrate with 0.1 *N* ammonium thiocyanate VS. Calculate the molarity.

Tetramethylammonium Chloride, Tenth-Molar (0.1 *M*)

$(CH_3)_4NCl$, **109.60**
10.96 g in 1000 mL

Dissolve 10.96 g of tetramethylammonium chloride in water to make 1000 mL, and standardize the solution as follows:

Transfer an accurately measured volume of about 40 mL of the solution to a flask, add 10 mL of diluted nitric acid and 50.0 mL of 0.1 *N* silver nitrate VS, and mix. Add 5 mL of nitrobenzene and 2 mL of ferric ammonium sulfate TS, shake, and titrate the excess silver nitrate with 0.1 *N* ammonium thiocyanate VS. Calculate the molarity.

Titanium Trichloride, Tenth-Normal (0.1 *N*)

$TiCl_3$, **154.24**
15.42 g in 1000 mL

Add 75 mL of titanium trichloride solution (1 in 5) to 75 mL of hydrochloric acid, dilute to 1000 mL, and mix. Standardize the solution as follows, using the special titration apparatus described.

Apparatus—Store the titanium trichloride solution in the reservoir of a closed-system titration apparatus in an atmosphere of hydrogen.

Use a wide-mouth, 500-mL conical flask as the titration vessel, and connect it by means of a tight-fitting rubber stopper to the titration buret, an inlet tube for carbon dioxide, and an exit tube. Arrange for mechanical stirring. All joints must be air-tight. Arrange to have both the hydrogen and the carbon dioxide pass through wash bottles containing titanium trichloride solution (approximately 1 in 50) to remove any oxygen.

If the solution to be titrated is to be heated before or during titration, connect the titration flask with an upright reflux condenser through the rubber stopper.

Standardization—Place an accurately measured volume of about 40 mL of 0.1 *N* ferric ammonium sulfate VS in the titration flask, and pass in a rapid stream of carbon dioxide until all the air has been removed. Add the titanium trichloride solution from the buret until near the calculated end-point (about 35 mL), then add through the outlet tube 5 mL of ammonium thiocyanate TS, and continue the titration until the solution is colorless. Calculate the normality.

Zinc Sulfate, Twentieth-Molar (0.05 *M*)

$ZnSO_4$, **161.44**
8.072 g in 1000 mL

Dissolve 14.4 g of zinc sulfate in water to make 1 liter. Standardize the solution as follows:

Measure accurately about 10 mL of 0.05 *M* disodium ethylenediaminetetraacetate VS into a 125-mL conical flask, and add, in the order given, 10 mL of acetic acid–ammonium acetate buffer TS, 50 mL of alcohol, and 2 mL of dithizone TS. Titrate with the zinc sulfate solution to a clear, rose-pink color. Calculate the molarity.

NOTE—For many of the reagents mentioned in the foregoing section, the corresponding standards of the 6th edition (1980) of *Reagent Chemicals*, published by the American Chemical Society, should be consulted. For a limited number of other reagents, the standards are adapted from those appearing in *Reagent Chemicals and Standards*, 5th edition, by Joseph Rosin and copyrighted by the publisher, D. Van Nostrand Co., Inc.

Reagent Footnotes

Where a particular brand or source of a material or piece of equipment, or the name and address of a manufacturer, is mentioned (ordinarily in a footnote), this identification is furnished solely for informational purposes as a matter of convenience, without implication of approval, endorsement, or certification.

[1] A suitable grade is available commercially as "Triton X-100" from Rohm and Haas Co., Philadelphia, PA 19105.

[2] A suitable grade is basic aluminum oxide, activity grade I, available from M. Woelm, Eschwege, W. Germany, or from U.S. distributors.

[3] A suitable resin is "Amberlite IRA-400," produced by Rohm and Haas Co., Philadelphia, PA 19105, and available through Mallinckrodt, Inc., P. O. Box 5439, St. Louis, MO 63147.

[4] A suitable resin is "AG 50W-X1," produced by BioRad Laboratories, 32nd and Griffin Sts., Richmond, CA 94804.

[5] A suitable resin is "Dowex 1X4," produced by the Dow Chemical Co., Midland, MI 48640, and available through J. T. Baker Chemical Co., Phillipsburg, NJ 08865.

[6] The USP Antithrombin-III Unit will be defined by the USP Antithrombin-III Reference Standard for Anti-factor X_a test. The USP Unit will be of a size representative of the activity contained in 1.0 mL of pooled human plasma.

[7] A suitable grade is available commercially from Eastman Kodak Co., Rochester, NY 14650.

[8] A suitable grade is "Dioctyl Sebacate," produced by the Harchem Division, Wallace and Tiernan, Inc., 25 Main St., Belleville, NJ 07101.

[9] A suitable grade is available from Regis Chemical Co., 1101 N. Franklin St., Chicago, IL 60610.

[10] A suitable grade is available from Aldrich Chemical Co., 940 W. St. Paul Ave., Milwaukee, WI 53233.

[11] A suitable grade is available commercially as "Dexsil 300 GC," from Applied Science Laboratories, Inc., P. O. Box 440, State College, PA 16801.

[12] Where Hammersten type casein is specified, a suitable grade is available commercially, as Catalog No. 7-E397, from J. T. Baker Chemical Co., 222 Red School Lane, Phillipsburg, NJ 08865.

[13] A suitable grade is available commercially as "Dowex 50-W-X8, 20–50 Mesh" from J. T. Baker Chemical Co., Phillipsburg, NJ 08865.

[14] A suitable grade is available as "Bio-Rex 70" from BioRad Laboratories, 32nd St. and Griffin Ave., Richmond, CA 94804.

[15] A suitable grade is available commercially as ion-exchange resin—analytical grade, 50W-X2, 100–200 mesh, from BioRad Laboratories, 32nd St. and Griffin Ave., Richmond, CA 94804.

[16] A suitable resin is "Dowex 50W-X8," produced by the Dow Chemical Co., Midland, MI 48640, and available through J. T. Baker Chemical Co., Phillipsburg, NJ 08865.

[17] A suitable grade is available commercially as "Amberlyst 15" from Rohm and Haas Co., Philadelphia, PA 19105, and from Mallinckrodt, Inc., Second and Mallinckrodt Sts., St. Louis, MO 63147, or as "Dowex 50-W-X2" from BioRad Laboratories, 32nd St. and Griffin Ave., Richmond, CA 94804.

[18] A suitable grade is available as M-300 from Analtech, Inc., Newark, DE 19711.

[19] A suitable grade is available commercially, in pre-coated plate form, as "Eastman Chromagram Sheet Cellulose with fluorescent indicator," from Eastman Organic Chemicals, Rochester, NY 14650.

[20] A suitable grade is available from Burdick and Jackson Laboratories, Inc., 1953 S. Harvey St., Muskegan, MI 49442.

[21] A suitable grade is available as "Silicone ASI 50 Phenyl 50 Cyanopropyl" from Applied Science Laboratories, Inc., P. O. Box 440, State College, PA 16801.

[22] A suitable grade is available from Research Plus Laboratories, Inc., Bayonne, NJ 07002.

[23] A suitable grade is available commercially as "Sephadex G-25, fine," from Pharmacia Laboratories, Inc., 800 Centennial Avenue, Piscataway, NJ 08854.

[24] A suitable grade is "Chromosorb W, AW-DMCS," available from Applied Science Laboratories, Inc., P. O. Box 440, State College, PA 16801.

[25] Suitable grades are available commercially as "Anachrome Q," "Gas-Chrom Q," and "Varaport 30."

[26] Suitable grades are "Chromosorb P" and "Chromosorb W," available from Johns-Manville Corp., 22 East 40th St., New York, NY 10016.

[27] A suitable grade is "Chromosorb W," available from Johns-Manville Corp., 22 East 40th St., New York, NY 10016.

[28] A suitable grade is available from Analabs, Inc., 80 Republic St., North Haven, CT 06473.

[29] A suitable grade is available from Atomergic Chemetals Corp., 91 Carolyn Boulevard, Farmingdale, NY 11735-1527.

[30] A suitable grade is available as anhydrous Hydroquinidine from Fluka Chemical Corp., 255 Oser Ave., Happauge, NY 11788.

[31] A suitable grade is "Silicone Fluid SF-96 (350)," available from the General Electric Co., Silicone Products Department, Mechanicsville Road, Waterford, NY 12188.

[32] A suitable grade is available from The Upjohn Company, Fine Chemical Division, North Haven, CT.

[33] A suitable grade is available from Lachat Chemical, Inc., Mequon, WI 53092.

[34] A suitable chromatographic grade is "Gas-Chrom Q," available from Applied Science Laboratories, Inc., P. O. Box 440, State College, PA 16801.

[35] The USP Factor X_a Unit will be defined by the USP Factor X_a for Anti-Factor X_a test Reference Standard. The USP Unit

will be of a size representative of the activity contained in 1.0 mL of pooled human plasma when fully activated by optimal concentration of Russel's viper venom in the presence of optimal amount of calcium chloride.

[36] Suitable grades are available commercially under the trade names "Florex XXS" (a moderately coarse powder) and "Florex XXX" (a very fine powder) from the Floridin Co., 3 Penn Center, Pittsburgh, PA 15235, or from laboratory supply houses.

NOTE—For the assay of Sorbitol and Sorbitol Solution, use a grade known as "Florex AARVM," available from the Floridin Co., or the equivalent.

[37] A suitable grade is available from Boehringer Mannheim Biochemicals, P. O. Box 50816, Indianapolis, IN 46250.

[38] A suitable grade is available commercially as Hexadecyl Palmitate, Catalog number LMS-067, from Analabs, Inc., 80 Republic Drive, North Haven, CT 06473.

[39] A suitable resin is "Amberlite MB-1," produced by Rohm and Haas Co., Philadelphia, PA 19105.

[40] A suitable grade is available commercially from The Upjohn Co., Fine Chemicals Marketing, Kalamazoo, MI 49001.

[41] A suitable grade is available as Amadec-F from Burdick and Jackson Laboratories, Inc., Muskegon, MI 49442.

[42] A suitable grade is "Florisil," available from the Floridin Co., 3 Penn Center, Pittsburgh, PA 15235.

[43] A suitable grade is "Florisil, 60–100 mesh," available from the Floridin Co., 3 Penn Center, Pittsburgh, PA 15235.

[44] A suitable grade is available from Applied Science Laboratories, Inc., P. O. Box 440, State College, PA 16801.

[45] A suitable grade is available commercially as "Silicone SE-30."

[46] A suitable grade is Raney Catalyst Powder No. 2813, available from W. R. Grace Co., South Pittsburg, TN.

[47] Ingredient of polyglycol nitroterephthalate, which is available commercially as "FFAP Polyester" from Varian Aerograph, 2700 Mitchell Dr., Walnut Creek, CA 94598.

[48] A suitable grade is "Igepal CO 710," available from General Aniline and Film Corp., 140 West 51st St., New York, NY 10020.

[49] A suitable grade is available commercially as "Tergitol Nonionic NPX," and as "Triton N101," from reagent suppliers.

[50] A suitable grade is available commercially as "Triton 100" from reagent suppliers.

[51] A suitable grade is available commercially as "DuPont SCX," available from E. I. du Pont de Nemours and Co., Wilmington, DE 19898.

[52] Available as Fractogel TSK-HW-40F and distributed by Merck and Co.

[53] A suitable grade is available commercially as "Palladium Catalyst, Type I (5% Palladium on Calcium Carbonate)," from Engelhard Industries, Inc., 113 Astor St., Newark, NJ 07114.

[54] A suitable grade is available commercially as "Palladium on Powdered Charcoal 10%, Catalyst," from EM Science, 2909 Highland Ave., Norwood, OH 45212.

[55] A suitable grade is available from ICI Americas, Specialty Chemical Division, SDM, Wilmington, DE 19897.

[56] A suitable chromatographic grade is "OV-17," available from Applied Science Laboratories, Inc., P. O. Box 440, State College, PA 16801.

[57] A suitable grade is available from Worthington Biochemical Corp., Route 9, Freehold, NJ 07728.

[58] A suitable detector tube is available from Safety First Supply Co., 9430-A Calumet Ave., Munster, IN 46321, and from Scott Aviation Co., 225 Erie St., Lancaster, NY 14086.

[59] Suitable grades are available commercially as "Versamide 900" from Applied Science Laboratories, Inc., P. O. Box 440, State College, PA 16801, and "Versamide 900 Solution" from Supelco, Inc., Supelco Park, Bellefonte, PA 16823.

[60] A suitable grade is available commercially, in pre-coated chromatographic plate form, from Brinkmann Instruments, Inc., Cantiague Rd., Westbury, NY 11590.

[61] A suitable grade is available commercially as "FFAP Polyester," from Varian Aerograph, 2700 Mitchell Dr., Walnut Creek, CA 94598.

[62] A suitable grade is available commercially as "Emulphor EL 620," from Antara Chemicals Div., GAF Corp., 140 West 51st St., New York, NY 10020.

[63] A suitable grade is available commercially as "Brij-35," from ICI Americas Inc., Wilmington, DE 19897.

[64] Suitable grades are available from J. T. Baker Chemical Co., Phillipsburg, NJ 08865 (Catalog No. U 232), or from E. I. du Pont de Nemours and Co., Wilmington, DE 19898 ("Elvanol 51-05").

[65] A suitable grade is available commercially as "Silica Gel H," from Brinkmann Instruments, Inc., Cantiague Rd., Westbury, NY 11590.

[66] From H. J. Conn, "Biological Stains, 7th Edition," 1961, p. 294. Color Index No. 50240, Williams & Wilkins, Baltimore, MD.

[67] Suitable grades are available commercially as "Aeropak 30," "Diatoport S," and "Gas-Chrom Z."

[68] A suitable grade is available commercially as "Silica Gel G," from Brinkmann Instruments, Inc., Cantiague Rd., Westbury, NY 11590.

[69] A suitable grade is available commercially as "Seprachrom" Chamber with Type SG ITLC, Product No. 51923, from Gelman Instrument Co., Ann Arbor, MI 48106.

[70] A suitable grade is available commercially as "Silica Gel GF 254," from Brinkmann Instruments, Inc., Cantiague Rd., Westbury, NY 11590.

[71] A suitable grade is available as "Silica Gel 60 F_{254} silanized, from EM Laboratories, Inc., 500 Executive Blvd., Elmsford, NY 10523.

[72] A suitable grade is available commercially as KC-18F from Whatman Chemical Separation, Inc., 9 Bridewell Place, Clifton, NJ 07014.

[73] A suitable grade is available commercially as "Reversed Phase Uniplates" from Analtech, Inc., Newark, DE 19711.

[74] A suitable grade is available as "Bondapak," from Waters Associates, Inc., Maple St., Milford, MA 01757.

[75] A suitable grade for reverse phase high-pressure liquid chromatography is available as "LiChrosorb SI60, Reverse Phase," from EM Science, 500 Executive Blvd., Elmsford, NY 10523.

[76] A suitable grade is available as "Silica Gel HR," from Brinkmann Instruments, Inc., Cantiague Rd., Westbury, NY 11590.

[77] A suitable grade, in a controlled-diameter, spherical, porous form, is available commercially as "Zorbax Sil," from E. I. du Pont de Nemours and Co., Inc., Instrument Products Div., Wilmington, DE 19898.

[78] A suitable grade is "Chromosorb W-AW," available from Johns-Manville Corp., 22 East 40th St., New York, NY 10016.

[79] Suitable silanized grades for gas chromatography are "Gas Chrom Q," available from Applied Science Labs., Inc., P. O. Box 440, State College, PA 16801, and "Chromosorb W (AW-DMCS-treated)," available from Johns-Manville Corp., 22 East 40th Street, New York, NY 10016.

[80] A suitable grade for column chromatography is acid-washed "Celite 545," available from Johns-Manville Corp., 22 East 40th St., New York, NY 10016.

[81] One example of a suitable grade is "ITLC Type SAF" sheets, available from Gelman Instrument Co., 600 South Wagner Rd., Ann Arbor, MI 48106.

[82] A suitable grade is available commercially as "Silicone Gum XE-60," from Applied Science Laboratories, Inc., P. O. Box 440, State College, PA 16801.

[83] A suitable grade is "Silicone GE, XE-60," available from Applied Science Laboratories, Inc., P. O. Box 440, State College, PA 16801.

[84] A suitable grade is available as "OV-25," from Applied Science Laboratories, Inc., P. O. Box 440, State College, PA 16801.

[85] The reagent is available as Catalog No. 7309 from Distillation Products Industries, Eastman Organic Chemicals Dept., Rochester, NY 14650. A procedure for its preparation is described in *Z. Anal. Chem.*, 146, 417 (1955).

[86] A suitable grade is "SPAN-80," available from ICI Americas Inc., Wilmington, DE 19897.

[87] A suitable grade is "No. 1252 Starch soluble GR meeting FIP standard," available from EM Science, 480 Democrat Road, Gibbstown, NJ 08027.

[88] A suitable grade is available commercially as "Glusulase," from Endo Laboratory, Garden City, NY 11530.

[89] Unless otherwise specified in the individual monograph, silanized support is intended.

[90] Available commercially as Poropak P from suppliers of chromatographic reagents.

[91] Available commercially as Poropak S from suppliers of chromatographic reagents.

[92] Commercially available as SP1500 on Carbopack B from Supelco.

[93] A suitable grade is available from Fisher Scientific Co., 711 Forbes Ave., Pittsburgh, PA 15215, and Burdick and Jackson Laboratories, Inc., Muskegon, MI 49442.

[94] A suitable preparation, listed as "Freon-TF aerosol," is available from E. I. du Pont de Nemours and Co., Wilmington, DE 19898.

[95] A suitable grade is available from Fluka Chemical Corporation, 255 Oser Ave., Hauppause, NY 11788 and from Pfaltz and Bauer, Inc., Div. of Aceto Chemical Corporation, 375 Fairfield Ave., Stamford, CT 06902.

[96] A suitable grade is available, as a concentrate, from Worthington Diagnostics, Division of Millipore Corp., Freehold, NJ 07728.

[97] A suitable stock solution is available commercially as "Topfer Reagent" (0.5% methyl yellow in alcohol), from Anderson Laboratories, Inc., 5901 Fitzhugh Ave., P. O. Box 8429, Fort Worth, TX 76112.

Reference Tables

CONTAINERS FOR DISPENSING CAPSULES AND TABLETS

The following table is provided as a reminder for the pharmacist engaged in the typical dispensing situation who already is acquainted with the *Packaging and storage* requirements set forth in the individual monographs. It lists the capsules and tablets that are official in the United States Pharmacopeia and indicates the relevant tight (T), well-closed (W), and light-resistant (LR) specifications applicable to containers in which the drug that is repackaged should be dispensed.

This table is not intended to replace, nor should it be interpreted as replacing, the definitive requirements stated in the individual monographs.

Container Specifications for Capsules and Tablets

Monograph Title	*Container Specification*
Acetaminophen Capsules	T
Acetaminophen Tablets	T
Acetaminophen and Aspirin Tablets	T
Acetaminophen, Aspirin, and Caffeine Capsules	T
Acetaminophen, Aspirin, and Caffeine Tablets	W
Acetaminophen and Codeine Phosphate Capsules	T, LR
Acetaminophen and Codeine Phosphate Tablets	T, LR
Acetaminophen and Diphenhydramine Citrate Tablets	T
Acetazolamide Tablets	W
Acetohexamide Tablets	W
Acetohydroxamic Acid Tablets	T
Acetophenazine Maleate Tablets	T, LR
Allopurinol Tablets	W
Alprazolam Tablets	T, LR
Alumina and Magnesia Tablets	W
Alumina, Magnesia, and Calcium Carbonate Tablets	W
Alumina, Magnesia, and Simethicone Tablets	W
Alumina and Magnesium Carbonate Tablets	T
Alumina, Magnesium Carbonate, and Magnesium Oxide Tablets	T
Alumina and Magnesium Trisilicate Tablets	W
Aluminum Carbonate Gel, Dried Basic, Capsules	W
Aluminum Carbonate Gel, Dried Basic, Tablets	W
Aluminum Hydroxide Gel, Dried, Capsules	W
Aluminum Hydroxide Gel, Dried, Tablets	W
Amantadine Hydrochloride Capsules	T
Amiloride Hydrochloride Tablets	W
Amiloride Hydrochloride and Hydrochlorothiazide Tablets	W
Aminobenzoate Potassium Capsules	W
Aminobenzoate Potassium Tablets	W
Aminocaproic Acid Tablets	T
Aminoglutethimide Tablets	W
Aminophylline Tablets	T
Aminosalicylate Sodium Tablets	T, LR
Aminosalicylic Acid Tablets	T, LR
Amitriptyline Hydrochloride Tablets	W
Ammonium Chloride Delayed-release Tablets	T
Amobarbital Tablets	W
Amobarbital Sodium Capsules	T
Amodiaquine Hydrochloride Tablets	T
Amoxicillin Capsules	T
Amoxicillin Tablets	T
Amoxicillin and Clavulanate Potassium Tablets	T
Amphetamine Sulfate Tablets	W
Ampicillin Capsules	T
Ampicillin Tablets	T
Ampicillin and Probenecid Capsules	T
Anileridine Hydrochloride Tablets	T, LR
Apomorphine Hydrochloride Tablets	T, LR
Ascorbic Acid Tablets	T, LR
Aspirin Capsules	T
Aspirin Capsules, Delayed-release	T
Aspirin Tablets	T
Aspirin Tablets, Buffered	T
Aspirin Tablets, Delayed-release	T
Aspirin Tablets, Effervescent for Oral Solution	T
Aspirin Tablets, Extended-release	T
Aspirin, Alumina, and Magnesia Tablets	T
Aspirin and Codeine Phosphate Tablets	W, LR
Aspirin, Codeine Phosphate, Alumina, and Magnesia Tablets	W, LR
Aspirin, Codeine Phosphate, and Caffeine Capsules	W, LR
Aspirin, Codeine Phosphate, and Caffeine Tablets	W, LR
Atropine Sulfate Tablets	W
Azatadine Maleate Tablets	W
Azathioprine Tablets	LR
Bacampicillin Hydrochloride Tablets	T
Baclofen Tablets	W
Belladonna Extract Tablets	T, LR
Bendroflumethiazide Tablets	T
Benzonatate Capsules	T, LR
Benzthiazide Tablets	T
Benztropine Mesylate Tablets	W
Beta Carotene Capsules	T, LR
Betamethasone Tablets	W
Bethanechol Chloride Tablets	T
Biperiden Hydrochloride Tablets	T
Bisacodyl Tablets	T
Bromocriptine Mesylate Tablets	T, LR
Bromodiphenhydramine Hydrochloride Capsules	T
Brompheniramine Maleate Tablets	T
Bumetanide Tablets	T, LR
Busulfan Tablets	W
Butabarbital Sodium Capsules	W
Butabarbital Sodium Tablets	W
Butalbital and Aspirin Tablets	T
Calcifediol Capsules	T, LR
Calcium Carbonate Tablets	W

Monograph Title	*Container Specification*
Calcium Carbonate and Magnesia Tablets	W
Calcium and Magnesium Carbonates Tablets	W
Calcium Gluconate Tablets	W
Calcium Lactate Tablets	T
Calcium Pantothenate Tablets	T
Calcium Phosphate, Dibasic, Tablets	W
Candicidin Vaginal Tablets	T
Carbamazepine Tablets	T
Carbenicillin Indanyl Sodium Tablets	T
Carbidopa and Levodopa Tablets	W, LR
Carbinoxamine Maleate Tablets	T, LR
Carboxymethylcellulose Sodium Tablets	T
Carisoprodol Tablets	W
Carisoprodol and Aspirin Tablets	W
Carisoprodol, Aspirin, and Codeine Phosphate Tablets	W
Cascara Tablets	T, W
Castor Oil Capsules	T
Cefaclor Capsules	T
Cefadroxil Capsules	T
Cefadroxil Tablets	T
Cephalexin Capsules	T
Cephalexin Tablets	T
Cephradine Capsules	T
Cephradine Tablets	T
Chloral Hydrate Capsules	T
Chlorambucil Tablets	W; W, LR
Chloramphenicol Capsules	T
Chloramphenicol Tablets	T
Chlordiazepoxide Tablets	T, LR
Chlordiazepoxide and Amitriptyline Hydrochloride Tablets	T, LR
Chlordiazepoxide Hydrochloride Capsules	T, LR
Chlordiazepoxide Hydrochloride and Clidinium Bromide Capsules	T, LR
Chloroquine Phosphate Tablets	W
Chlorothiazide Tablets	W
Chlorotrianisene Capsules	W
Chlorpheniramine Maleate Capsules, Extended-release	T
Chlorpheniramine Maleate Tablets	T
Chlorpromazine Hydrochloride Tablets	W, LR
Chlorpropamide Tablets	W
Chlorprothixene Tablets	W, LR
Chlortetracycline Hydrochloride Capsules	T, LR
Chlortetracycline Hydrochloride Tablets	T, LR
Chlorthalidone Tablets	W
Chlorzoxazone Tablets	T
Chlorzoxazone and Acetaminophen Capsules	T
Cimetidine Tablets	T, LR
Cinoxacin Capsules	W
Clemastine Fumarate Tablets	T, LR
Clidinium Bromide Capsules	T, LR
Clindamycin Hydrochloride Capsules	T
Clofibrate Capsules	W, LR
Clomiphene Citrate Tablets	W
Clonazepam Tablets	T, LR
Clonidine Hydrochloride Tablets	W
Clonidine Hydrochloride and Chlorthalidone Tablets	W
Clotrimazole Vaginal Tablets	W
Cloxacillin Sodium Capsules	T
Cyanocobalamin Co 57 Capsules	W, LR
Cyanocobalamin Co 60 Capsules	W, LR
Cocaine Hydrochloride Tablets for Topical Solution	W, LR
Codeine Phosphate Tablets	W, LR
Codeine Sulfate Tablets	W
Cortisone Acetate Tablets	W
Cromolyn Sodium for Inhalation (in capsules)	T, LR
Cyclacillin Tablets	T
Cyclizine Hydrochloride Tablets	T, LR
Cyclobenzaprine Hydrochloride Tablets	W
Cyclophosphamide Tablets	T

Monograph Title	*Container Specification*
Cycloserine Capsules	T
Cyproheptadine Hydrochloride Tablets	W
Danazol Capsules	W
Dapsone Tablets	W, LR
Dehydrocholic Acid Tablets	W
Demeclocycline Hydrochloride Capsules	T, LR
Demeclocycline Hydrochloride Tablets	T, LR
Demeclocycline Hydrochloride and Nystatin Capsules	T, LR
Demeclocycline Hydrochloride and Nystatin Tablets	T, LR
Desipramine Hydrochloride Capsules	T
Desipramine Hydrochloride Tablets	T
Dexamethasone Tablets	W
Dexchlorpheniramine Maleate Tablets	T
Dextroamphetamine Sulfate Capsules	T
Dextroamphetamine Sulfate Tablets	W
Diazepam Capsules	T, LR
Diazepam Capsules, Extended-release	T, LR
Diazepam Tablets	T, LR
Diazoxide Capsules	W
Dicloxacillin Sodium Capsules	T
Dicumarol Tablets	W
Dicyclomine Hydrochloride Capsules	W
Dicyclomine Hydrochloride Tablets	W
Diethylcarbamazine Citrate Tablets	T
Diethylpropion Hydrochloride Tablets	W
Diethylstilbestrol Tablets	W
Diflunisal Tablets	W
Digitalis Capsules	T
Digitalis Tablets	T
Digitoxin Tablets	W
Digoxin Tablets	T
Dihydrotachysterol Capsules	W, LR
Dihydrotachysterol Tablets	W, LR
Dihydroxyaluminum Aminoacetate Capsules	W
Dihydroxyaluminum Aminoacetate Tablets	W
Dihydroxyaluminum Sodium Carbonate Tablets	W
Diltiazem Tablets	T, LR
Dimenhydrinate Tablets	W
Diphemanil Methylsulfate Tablets	T
Diphenhydramine Hydrochloride Capsules	T
Diphenoxylate Hydrochloride and Atropine Sulfate Tablets	W, LR
Dipyridamole Tablets	T, LR
Disopyramide Phosphate Capsules	W
Disopyramide Phosphate Capsules, Extended-release	W
Disulfiram Tablets	T, LR
Docusate Calcium Capsules	T
Docusate Potassium Capsules	T
Docusate Sodium Capsules	T
Docusate Sodium Tablets	W
Doxepin Hydrochloride Capsules	W
Doxycycline Hyclate Capsules	T, LR
Doxycycline Hyclate Capsules, Delayed-release	T, LR
Doxycycline Hyclate Tablets	T, LR
Doxylamine Succinate Tablets	W, LR
Dydrogesterone Tablets	W
Dyphylline Tablets	T
Dyphylline and Guaifenesin Tablets	T
Enalapril Maleate Tablets	W
Ephedrine Sulfate Capsules	T, LR
Ephedrine Sulfate Tablets	W
Ephedrine Sulfate and Phenobarbital Capsules	W
Ergocalciferol Capsules	T, LR
Ergocalciferol Tablets	T, LR
Ergoloid Mesylates Tablets	T, LR
Ergonovine Maleate Tablets	W
Ergotamine Tartrate Tablets	W, LR
Ergotamine Tartrate and Caffeine Tablets	W, LR

Monograph Title	*Container Specification*
Erythrityl Tetranitrate Tablets	T
Erythromycin Capsules, Delayed-release	T
Erythromycin Tablets	T
Erythromycin Tablets, Delayed-release	T
Erythromycin Estolate Capsules	T
Erythromycin Estolate Tablets	T
Erythromycin Ethylsuccinate Tablets	T
Erythromycin Stearate Tablets	T
Erythrosine Sodium Soluble Tablets	T, LR
Estradiol Tablets	T, LR
Estrogens, Conjugated, Tablets	W
Estrogens, Esterified, Tablets	W
Estropipate Tablets	W
Ethacrynic Acid Tablets	W
Ethambutol Hydrochloride Tablets	W
Ethchlorvynol Capsules	T, LR
Ethinamate Capsules	T
Ethinyl Estradiol Tablets	W
Ethionamide Tablets	W
Ethopropazine Hydrochloride Tablets	W
Ethosuximide Capsules	T
Ethotoin Tablets	T
Ethynodiol Diacetate and Ethinyl Estradiol Tablets	W
Ethynodiol Diacetate and Mestranol Tablets	W
Etidronate Disodium Tablets	T
Famotidine Tablets	W, LR
Fenoprofen Calcium Capsules	W
Fenoprofen Calcium Tablets	W
Ferrous Fumarate Tablets	T
Ferrous Gluconate Capsules	T
Ferrous Gluconate Tablets	T
Ferrous Sulfate Tablets	T
Flucytosine Capsules	T, LR
Fludrocortisone Acetate Tablets	W
Fluoxymesterone Tablets	W
Fluphenazine Hydrochloride Tablets	T, LR
Flurazepam Hydrochloride Capsules	T, LR
Folic Acid Tablets	W
Furazolidone Tablets	T, LR
Furosemide Tablets	W, LR
Gemfibrozil Capsules	T
Glutethimide Capsules	W
Glutethimide Tablets	W
Glycopyrrolate Tablets	T
Griseofulvin Capsules	T
Griseofulvin Tablets	T
Griseofulvin, Ultramicrosize, Tablets	W
Guaifenesin Capsules	T
Guaifenesin Tablets	T
Guanabenz Acetate Tablets	T, LR
Guanadrel Sulfate Tablets	T, LR
Guanethidine Monosulfate Tablets	W
Halazone Tablets for Solution	T, LR
Haloperidol Tablets	T, LR
Hetacillin Tablets	T
Hetacillin Potassium Capsules	T
Hetacillin Potassium Tablets	T
Homatropine Methylbromide Tablets	T, LR
Hydralazine Hydrochloride Tablets	T, LR
Hydrochlorothiazide Tablets	W
Hydrocodone Bitartrate Tablets	T, LR
Hydrocortisone Tablets	W
Hydroflumethiazide Tablets	T
Hydromorphone Hydrochloride Tablets	T, LR
Hydroxychloroquine Sulfate Tablets	W
Hydroxyurea Capsules	T
Hydroxyzine Hydrochloride Tablets	T
Hydroxyzine Pamoate Capsules	W
Hyoscyamine Tablets	W, LR
Hyoscyamine Sulfate Tablets	T, LR
Ibuprofen Tablets	W
Imipramine Hydrochloride Tablets	T
Indomethacin Capsules	W
Indomethacin Capsules, Extended-release	W
Iocetamic Acid Tablets	T
Sodium Iodide I 123 Capsules	W
Sodium Iodide I 125 Capsules	W
Sodium Iodide I 131 Capsules	W
Iodoquinol Tablets	W
Iopanoic Acid Tablets	T, LR
Ipodate Sodium Capsules	T
Isocarboxazid Tablets	W, LR
Isoniazid Tablets	W, LR
Isopropamide Iodide Tablets	W
Isoproterenol Hydrochloride Tablets	W, LR
Isosorbide Dinitrate Capsules, Extended-release	W
Isosorbide Dinitrate Tablets	W
Isosorbide Dinitrate Tablets, Chewable	W
Isosorbide Dinitrate Tablets, Extended-release	W
Isosorbide Dinitrate Tablets, Sublingual	W
Isoxsuprine Hydrochloride Tablets	T
Kanamycin Sulfate Capsules	T
Ketoconazole Tablets	W
Labetalol Hydrochloride Tablets	T, LR
Levodopa Capsules	T, LR
Levodopa Tablets	T, LR
Levonorgestrel and Ethinyl Estradiol Tablets	W
Levopropoxyphene Napsylate Capsules	T
Levorphanol Tartrate Tablets	W
Levothyroxine Sodium Tablets	T, LR
Lincomycin Hydrochloride Capsules	T
Liothyronine Sodium Tablets	T
Liotrix Tablets	T
Lithium Carbonate Capsules	W
Lithium Carbonate Tablets	W
Loperamide Hydrochloride Capsules	W
Lorazepam Tablets	T, LR
Magaldrate Tablets	W
Magaldrate and Simethicone Tablets	W
Magnesia Tablets	W
Magnesia and Alumina Tablets	W
Magnesium Gluconate Tablets	W
Magnesium Oxide Capsules	W
Magnesium Oxide Tablets	W
Magnesium Salicylate Tablets	T
Magnesium Trisilicate Tablets	W
Maprotiline Hydrochloride Tablets	W
Mazindol Tablets	T
Mebendazole Tablets	W
Meclizine Hydrochloride Tablets	W
Meclofenamate Sodium Capsules	T, LR
Medroxyprogesterone Acetate Tablets	W
Megestrol Acetate Tablets	W
Melphalan Tablets	W, LR
Menadiol Sodium Diphosphate Tablets	W, LR
Meperidine Hydrochloride Tablets	W, LR
Mephenytoin Tablets	W
Mephobarbital Tablets	W
Meprobamate Tablets	W
Mercaptopurine Tablets	W
Mesoridazine Besylate Tablets	W, LR
Metaproterenol Sulfate Tablets	W, LR
Methacycline Hydrochloride Capsules	T, LR
Methadone Hydrochloride Tablets	W
Methantheline Bromide Tablets	W
Metharbital Tablets	T
Methazolamide Tablets	W
Methdilazine Tablets	T, LR
Methdilazine Hydrochloride Tablets	T, LR
Methenamine Tablets	W
Methenamine and Sodium Biphosphate Tablets	T
Methenamine Hippurate Tablets	W
Methenamine Mandelate Tablets	W
Methimazole Tablets	W, LR
Methocarbamol Tablets	T

Monograph Title	*Container Specification*
Methotrexate Tablets	W
Methoxsalen Capsules	T, LR
Methscopolamine Bromide Tablets	T
Methsuximide Capsules	T
Methyclothiazide Tablets	W
Methylcellulose Tablets	W
Methyldopa Tablets	W
Methyldopa and Chlorothiazide Tablets	W
Methyldopa and Hydrochlorothiazide Tablets	W
Methylergonovine Maleate Tablets	T, LR
Methylphenidate Hydrochloride Tablets	T
Methylphenidate Hydrochloride Tablets, Extended-release	T
Methylprednisolone Tablets	T
Methyltestosterone Capsules	W
Methyltestosterone Tablets	W
Methyprylon Capsules	T, LR
Methyprylon Tablets	T, LR
Methysergide Maleate Tablets	T
Metoclopramide Tablets	T, LR
Metoprolol Tartrate Tablets	T, LR
Metoprolol Tartrate and Hydrochlorothiazide Tablets	T, LR
Metronidazole Tablets	W, LR
Metyrapone Tablets	T, LR
Metyrosine Capsules	W
Mexiletine Hydrochloride Capsules	T
Minocycline Hydrochloride Capsules	T, LR
Minocycline Hydrochloride Tablets	T, LR
Minoxidil Tablets	T
Mitotane Tablets	T, LR
Nadolol Tablets	T
Nafcillin Sodium Capsules	T
Nafcillin Sodium Tablets	T, LR
Nalidixic Acid Tablets	T
Naproxen Tablets	W
Naproxen Sodium Tablets	W
Neomycin Sulfate Tablets	T
Niacin Tablets	W
Niacinamide Tablets	T
Nifedipine Capsules	T, LR
Nitrofurantoin Capsules	T, LR
Nitrofurantoin Tablets	T, LR
Nitroglycerin Tablets	T
Norethindrone Tablets	W
Norethindrone and Ethinyl Estradiol Tablets	W
Norethindrone and Mestranol Tablets	W
Norethindrone Acetate Tablets	W
Norethindrone Acetate and Ethinyl Estradiol Tablets	W
Norfloxacin Tablets	W
Norgestrel Tablets	W
Norgestrel and Ethinyl Estradiol Tablets	W
Nortriptyline Hydrochloride Capsules	T
Novobiocin Sodium Capsules	T, LR
Nylidrin Hydrochloride Tablets	T
Nystatin Tablets	T, LR
Nystatin Vaginal Tablets	W, LR
Oleovitamin A and D Capsules	T, LR
Oxacillin Sodium Capsules	T
Oxamniquine Capsules	T
Oxandrolone Tablets	T, LR
Oxazepam Capsules	W
Oxazepam Tablets	W
Oxprenolol Tablets	W, LR
Oxtriphylline Tablets, Delayed-release	T
Oxybutynin Chloride Tablets	T, LR
Oxycodone Tablets	T, LR
Oxycodone and Acetaminophen Capsules	T, LR
Oxycodone and Acetaminophen Tablets	T, LR
Oxycodone and Aspirin Tablets	T, LR
Oxymetholone Tablets	W
Oxyphenbutazone Tablets	T
Oxyphencyclimine Hydrochloride Tablets	T
Oxytetracycline Tablets	T, LR
Oxytetracycline and Nystatin Capsules	T, LR
Oxytetracycline Hydrochloride Capsules	T, LR
Oxytetracycline and Phenazopyridine Hydrochlorides and Sulfamethizole Capsules	T, LR
Oxytetracycline Hydrochloride and Polymyxin B Sulfate Vaginal Tablets	W
Pancreatin Capsules	T
Pancreatin Tablets	T
Pancrelipase Capsules	T
Pancrelipase Tablets	T
Papain Tablets for Topical Solution	T, LR
Papaverine Hydrochloride Tablets	T
Paramethadione Capsules	T
Paramethasone Acetate Tablets	W
Pargyline Hydrochloride Tablets	W
Paromomycin Sulfate Capsules	T
Penicillamine Capsules	T
Penicillamine Tablets	T
Penicillin G Benzathine Tablets	T
Penicillin G Potassium Capsules	T
Penicillin G Potassium Tablets	T
Penicillin G Potassium Tablets for Oral Solution	T
Penicillin V Tablets	T
Penicillin V Potassium Tablets	T
Pentaerythritol Tetranitrate Tablets	T
Pentazocine Hydrochloride Tablets	T, LR
Pentazocine Hydrochloride and Aspirin Tablets	T, LR
Pentazocine and Naloxone Hydrochlorides Tablets	T, LR
Pentobarbital Sodium Capsules	T
Perphenazine Tablets	T, LR
Perphenazine and Amitriptyline Hydrochloride Tablets	W
Phenacemide Tablets	W
Phenazopyridine Hydrochloride Tablets	T
Phendimetrazine Tartrate Capsules	T
Phendimetrazine Tartrate Tablets	W
Phenelzine Sulfate Tablets	T
Phenindione Tablets	W
Phenmetrazine Hydrochloride Tablets	T
Phenobarbital Tablets	W
Phenolphthalein Tablets	T
Phenprocoumon Tablets	W
Phensuximide Capsules	T
Phentermine Hydrochloride Capsules	T
Phentermine Hydrochloride Tablets	T
Phenylbutazone Capsules	T
Phenylbutazone Tablets	T
Phenylpropanolamine Hydrochloride Capsules, Extended-release	T, LR
Phenylpropanolamine Hydrochloride Tablets, Extended-release	T, LR
Phenytoin Tablets	W
Phenytoin Sodium Capsules, Extended	T
Phenytoin Sodium Capsules, Prompt	T
Phytonadione Tablets	W, LR
Pimozide Tablets	T, LR
Pindolol Tablets	W, LR
Piperacetazine Tablets	W, LR
Piperazine Citrate Tablets	T
Pipobroman Tablets	W
Piroxicam Capsules	T, LR
Polythiazide Tablets	T, LR
Potassium Bicarbonate Effervescent Tablets for Oral Solution	T
Potassium Bicarbonate and Potassium Chloride Effervescent Tablets for Oral Solution	T
Potassium and Sodium Bicarbonates and Citric Acid Effervescent Tablets for Oral Solution	T

Monograph Title	Container Specification
Potassium Chloride Capsules, Extended-release	T
Potassium Chloride Tablets, Extended-release	T
Potassium Chloride, Potassium Bicarbonate, and Potassium Citrate Effervescent Tablets for Oral Solution	T
Potassium Gluconate Tablets	T
Potassium Iodide Tablets	T
Pralidoxime Chloride Tablets	W
Prazepam Capsules	T, LR
Prazepam Tablets	T, LR
Praziquantel Tablets	T
Prazosin Hydrochloride Capsules	W, LR
Prednisolone Tablets	W
Prednisone Tablets	W
Primaquine Phosphate Tablets	W, LR
Primidone Tablets	W
Probenecid and Colchicine Tablets	W, LR
Procainamide Hydrochloride Capsules	T
Procainamide Hydrochloride Tablets	T
Prochlorperazine Maleate Tablets	W
Procyclidine Hydrochloride Tablets	T
Promazine Hydrochloride Tablets	T, LR
Promethazine Hydrochloride Tablets	T, LR
Propantheline Bromide Tablets	W
Propoxyphene Hydrochloride Capsules	T
Propoxyphene Hydrochloride and Acetaminophen Tablets	T
Propoxyphene Hydrochloride, Aspirin, and Caffeine Capsules	T
Propoxyphene Napsylate Tablets	T
Propoxyphene Napsylate and Acetaminophen Tablets	T
Propoxyphene Napsylate and Aspirin Tablets	T
Propranolol Hydrochloride Capsules, Extended-release	W
Propranolol Hydrochloride Tablets	W, LR
Propranolol Hydrochloride and Hydrochlorothiazide Capsules, Extended-release	W
Propranolol Hydrochloride and Hydrochlorothiazide Tablets	W
Propylthiouracil Tablets	W
Protriptyline Hydrochloride Tablets	T
Pseudoephedrine Hydrochloride Tablets	T
Pyrazinamide Tablets	W
Pyridostigmine Bromide Tablets	T
Pyridoxine Hydrochloride Tablets	W
Pyrilamine Maleate Tablets	W
Pyrimethamine Tablets	T, LR
Pyrvinium Pamoate Tablets	T, LR
Quinacrine Hydrochloride Tablets	T
Quinestrol Tablets	W
Quinethazone Tablets	T
Quinidine Sulfate Capsules	T, LR
Quinidine Sulfate Tablets	W, LR
Quinidine Sulfate Tablets, Extended-release	W, LR
Quinine Sulfate Capsules	T
Quinine Sulfate Tablets	W
Ranitidine Tablets	T, LR
Rauwolfia Serpentina Tablets	T, LR
Reserpine Tablets	T, LR
Reserpine and Chlorothiazide Tablets	T, LR
Reserpine, Hydralazine Hydrochloride, and Hydrochlorothiazide Tablets	T, LR
Reserpine and Hydrochlorothiazide Tablets	T, LR
Riboflavin Tablets	T, LR
Rifampin Capsules	T, LR
Rifampin and Isoniazid Capsules	T, LR
Ritodrine Hydrochloride Tablets	T
Saccharin Sodium Tablets	W
Scopolamine Hydrobromide Tablets	T, LR
Secobarbital Sodium Capsules	T
Secobarbital Sodium and Amobarbital Sodium Capsules	W
Sennosides Tablets	W
Simethicone Tablets	W
Sodium Bicarbonate Tablets	W
Sodium Chloride Tablets	W
Sodium Chloride Tablets for Solution	W
Sodium Chloride and Dextrose Tablets	W
Sodium Fluoride Tablets	T
Sodium Salicylate Tablets	W
Spironolactone Tablets	T, LR
Spironolactone and Hydrochlorothiazide Tablets	T, LR
Stanozolol Tablets	T, LR
Sulfadiazine Tablets	W, LR
Sulfadoxine and Pyrimethamine Tablets	W, LR
Sulfamerazine Tablets	W
Sulfamethizole Tablets	W
Sulfamethoxazole Tablets	W, LR
Sulfamethoxazole and Trimethoprim Tablets	W, LR
Sulfapyridine Tablets	W, LR
Sulfasalazine Tablets	W
Sulfinpyrazone Capsules	W
Sulfinpyrazone Tablets	W
Sulfisoxazole Tablets	W, LR
Sulfoxone Sodium Tablets	T, LR
Sulindac Tablets	W
Talbutal Tablets	T
Tamoxifen Citrate Tablets	W, LR
Terbutaline Sulfate Tablets	T
Testolactone Tablets	T
Tetracycline Hydrochloride Capsules	T, LR
Tetracycline Hydrochloride Tablets	T, LR
Tetracycline Hydrochloride and Novobiocin Sodium Tablets	T
Tetracycline Hydrochloride, Novobiocin Sodium, and Prednisolone Tablets	T
Tetracycline Hydrochloride and Nystatin Capsules	T, LR
Tetracycline Phosphate Complex Capsules	T, LR
Tetracycline Phosphate Complex and Novobiocin Sodium Capsules	T
Theophylline Capsules	W
Theophylline Capsules, Extended-release	W
Theophylline Tablets	W
Theophylline, Ephedrine Hydrochloride, and Phenobarbital Tablets	T
Theophylline and Guaifenisin Capsules	T
Theophylline Sodium Glycinate Tablets	W
Thiabendazole Tablets	T
Thiamine Hydrochloride Tablets	T, LR
Thiethylperazine Maleate Tablets	T, LR
Thioguanine Tablets	T
Thioridazine Hydrochloride Tablets	T, LR
Thiothixene Capsules	W, LR
Thyroglobulin Tablets	T
Thyroid Tablets	T
Timolol Maleate Tablets	W, LR
Timolol Maleate and Hydrochlorothiazide Tablets	W, LR
Tocainide Hydrochloride Tablets	W
Tolazamide Tablets	T
Tolazoline Hydrochloride Tablets	W
Tolbutamide Tablets	W
Tolmetin Sodium Capsules	T
Tolmetin Sodium Tablets	W
Triamcinolone Tablets	W
Triamterene Capsules	T, LR
Triamterene and Hydrochlorothiazide Capsules, Extended-release	T, LR
Triamterene and Hydrochlorothiazide Tablets	T, LR
Triazolam Tablets	T, LR
Trichlormethiazide Tablets	T
Tridihexethyl Chloride Tablets	T
Trientine Hydrochloride Capsules	W
Trifluoperazine Hydrochloride Tablets	W, LR
Triflupromazine Hydrochloride Tablets	W, LR

Monograph Title	*Container Specification*
Trihexyphenidyl Hydrochloride Capsules, Extended-release	T
Trihexyphenidyl Hydrochloride Tablets	T
Trimeprazine Tartrate Tablets	W, LR
Trimethadione Capsules	T
Trimethadione Tablets	T
Trimethobenzamide Hydrochloride Capsules	W
Trimethoprim Tablets	T, LR
Trioxsalen Tablets	W, LR
Tripelennamine Hydrochloride Tablets	W
Triple Sulfa Vaginal Tablets	W, LR
Triprolidine Hydrochloride Tablets	T, LR
Triprolidine and Pseudoephedrine Hydrochlorides Tablets	T, LR
Trisulfapyrimidines Tablets	W
Troleandomycin Tablets	T
Tyropanoate Sodium Capsules	T, LR
Uracil Mustard Capsules	T
Valproic Acid Capsules	T
Vancomycin Hydrochloride Capsules	T
Verapamil Tablets	T, LR
Vitamin A Capsules	T, LR
Vitamin E Capsules	T
Warfarin Sodium Tablets	T, LR

DESCRIPTION AND SOLUBILITY

Description and Relative Solubility of USP and NF Articles

The "description" and "solubility" statements pertaining to an article (formerly included in the individual monograph) are general in nature. The information is provided for those who use, prepare, and dispense drugs, solely to indicate descriptive and solubility properties of an article complying with monograph standards. The properties are not in themselves standards or tests for purity even though they may indirectly assist in the preliminary evaluation of the integrity of an article.

Only where a special, quantitative solubility test is given in the individual monograph, and is designated by a test heading, is it a test for purity.

The approximate solubilities of Pharmacopeial and National Formulary substances are indicated by the descriptive terms in the accompanying table. The term "miscible" as used in this Pharmacopeia pertains to a substance that yields a homogeneous mixture when mixed in any proportion with the designated solvent.

Descriptive Term	Parts of Solvent Required for 1 Part of Solute
Very soluble	Less than 1
Freely soluble	From 1 to 10
Soluble	From 10 to 30
Sparingly soluble	From 30 to 100
Slightly soluble	From 100 to 1000
Very slightly soluble	From 1000 to 10,000
Practically insoluble, or Insoluble	10,000 and over

Soluble Pharmacopeial and National Formulary articles, when brought into solution, may show traces of physical impurities, such as minute fragments of filter paper, fibers, and other particulate matter, unless limited or excluded by definite tests or other specifications in the individual monographs.

Acacia: Is practically odorless and produces a mucilaginous sensation on the tongue. Insoluble in alcohol. *NF category:* Emulsifying and/or solubilizing agent; suspending and/or viscosity-increasing agent; tablet binder.

Acetaminophen: White, odorless, crystalline powder, having a slightly bitter taste. Soluble in boiling water and in 1 *N* sodium hydroxide; freely soluble in alcohol.

Acetazolamide: White to faintly yellowish white, crystalline, odorless powder. Very slightly soluble in water; sparingly soluble in practically boiling water; slightly soluble in alcohol.

Sterile Acetazolamide Sodium: White solid, having the characteristic appearance of freeze-dried products.

Acetic Acid: Clear, colorless liquid, having a strong, characteristic odor, and a sharply acid taste. Specific gravity is about 1.045. Miscible with water, with alcohol, and with glycerin. *NF category:* Acidifying agent; buffering agent

Glacial Acetic Acid: Clear, colorless liquid, having a pungent, characteristic odor and, when well diluted with water, an acid taste. Boils at about 118°. Specific gravity is about 1.05. Miscible with water, with alcohol, and with glycerin. *NF category:* Acidifying agent

Acetohexamide: White, crystalline, practically odorless powder. Practically insoluble in water and in ether; soluble in pyridine and in dilute solutions of alkali hydroxides; slightly soluble in alcohol and in chloroform.

Acetohydroxamic Acid: White, slightly hygroscopic, crystalline powder. Melts, after drying at about 80° for 2 to 4 hours, at about 88°. Freely soluble in water and in alcohol; very slightly soluble in chloroform.

Acetone: Transparent, colorless, mobile, volatile liquid, having a characteristic odor. A solution (1 in 2) is neutral to litmus. Miscible with water, with alcohol, with ether, with chloroform, and with most volatile oils. *NF category:* Solvent.

Acetophenazine Maleate: Fine, yellow powder. Melts at about 165°, with decomposition. Soluble in water; slightly soluble in acetone and in alcohol.

Acetylcholine Chloride: White or off-white crystals or crystalline powder. Very soluble in water; freely soluble in alcohol; insoluble in ether. Is decomposed by hot water and by alkalies.

Acetylcysteine: White, crystalline powder, having a slight acetic odor. Freely soluble in water and in alcohol; practically insoluble in chloroform and in ether.

Acrisorcin: Yellow, odorless powder. Melts at about 190°, with decomposition (see *Melting Range or Temperature* ⟨741⟩). Very slightly soluble in water and in ether; soluble in alcohol; slightly soluble in chloroform.

Adenine: White crystals or crystalline powder. Is odorless and tasteless. Very slightly soluble in water; sparingly soluble in boiling water; slightly soluble in alcohol; practically insoluble in ether and in chloroform.

Agar: Odorless or has a slight odor, and produces a mucilaginous sensation on the tongue. Insoluble in cold water; soluble in boiling water. *NF category:* Suspending and/or viscosity-increasing agent.

Alamic Acid: *NF category:* Suspending and/or viscosity-increasing agent.

Alanine: White, odorless crystals or crystalline powder, having a slightly sweet taste. Freely soluble in water; slightly soluble in 80% alcohol; insoluble in ether.

Albumin Human: Practically odorless, moderately viscous, clear, brownish fluid.

Albuterol: White, crystalline powder. Sparingly soluble in water; soluble in alcohol. Melts at about 156°.

Albuterol Sulfate: White or practically white powder. Freely soluble in water; slightly soluble in alcohol, in chloroform, and in ether.

Alcohol: Clear, colorless, mobile, volatile liquid. Has a characteristic odor and produces a burning sensation on the tongue. Is readily volatilized even at low temperatures, and boils at about 78°. Is flammable. Miscible with water and with practically all organic solvents. *NF category:* Solvent.

Dehydrated Alcohol: Clear, colorless, mobile, volatile liquid. Has a characteristic odor and produces a burning sensation on the tongue. Is readily volatilized even at low temperatures, and boils at about 78°. Is flammable. Miscible with water and with practically all organic solvents.

Diluted Alcohol: Clear, colorless, mobile liquid, having a characteristic odor and producing a burning sensation on the tongue. *NF category:* Solvent.

Rubbing Alcohol: Transparent, colorless, or colored as desired, mobile, volatile liquid. Has an extremely bitter taste and, in the absence of added odorous constituents, a characteristic odor. Is flammable.

Alginic Acid: White to yellowish white, fibrous powder. Is odorless, or practically odorless, and is tasteless. Insoluble in water and in organic solvents; soluble in alkaline solutions. *NF category:* Tablet binder; tablet disintegrant.

Allopurinol: Fluffy white to off-white powder, having only a slight odor. Very slightly soluble in water and in alcohol; soluble in solutions of potassium and sodium hydroxides; practically insoluble in chloroform and in ether.

Almond Oil: Clear, pale straw-colored or colorless, oily liquid, having a bland taste. Remains clear at −10°, and does not congeal until cooled to almost −20°. Slightly soluble in alcohol; miscible with ether, with chloroform, with benzene, and with solvent hexane. *NF category:* Flavors and perfumes.

Aloe: Has a characteristic, somewhat sour and disagreeable, odor.

Alprazolam: A white to off-white crystalline powder. Melts at about 225°. Insoluble in water; slightly soluble in ethyl acetate; sparingly soluble in acetone; soluble in alcohol; freely soluble in chloroform.

Alprostadil: A white to off-white, crystalline powder. Melts at about 110°. Soluble in water; freely soluble in alcohol; soluble in acetone; slightly soluble in ethyl acetate; very slightly soluble in chloroform and in ether.

Ammonium Alum: Large, colorless crystals, crystalline fragments, or white powder. Is odorless, and has a sweetish, strongly astringent taste. Its solutions are acid to litmus. Freely soluble in water; very soluble in boiling water; freely but slowly soluble in glycerin; insoluble in alcohol.

Potassium Alum: Large, colorless crystals, crystalline fragments, or white powder. Is odorless, and has a sweetish, strongly astringent taste. Its solutions are acid to litmus. Freely soluble in water; very soluble in boiling water; freely but slowly soluble in glycerin; insoluble in alcohol.

Aluminum Acetate Topical Solution: Clear, colorless liquid having a faint odor of acetic acid, and a sweetish, astringent taste. Specific gravity is about 1.02.

Aluminum Chloride: White, or yellowish white, deliquescent, crystalline powder. Is practically odorless, and has a sweet, very astringent taste. Its solutions are acid to litmus. Very soluble in water; freely soluble in alcohol; soluble in glycerin.

Aluminum Hydroxide Gel: White, viscous suspension, from which small amounts of clear liquid may separate on standing.

Dried Aluminum Hydroxide Gel: White, odorless, tasteless, amorphous powder. Insoluble in water and in alcohol; soluble in dilute mineral acids and in solutions of fixed alkali hydroxides.

Aluminum Monostearate: Fine, white to yellowish white, bulky powder, having a faint, characteristic odor. Insoluble in water, in alcohol, and in ether. *NF category:* Suspending and/or viscosity-increasing agent.

Aluminum Phosphate Gel: White, viscous suspension from which small amounts of water separate on standing.

Aluminum Subacetate Topical Solution: Clear, colorless or faintly yellow liquid, having an odor of acetic acid and an acid reaction to litmus. Gradually becomes turbid on standing, through separation of a more basic salt.

Aluminum Sulfate: White, crystalline powder, shining plates, or crystalline fragments. Is stable in air. Is odorless, and has a sweet taste, becoming mildly astringent. Freely soluble in water; insoluble in alcohol.

Amantadine Hydrochloride: White or practically white, crystalline powder, having a bitter taste. Freely soluble in water; soluble in alcohol and in chloroform.

Amikacin: White, crystalline powder. Sparingly soluble in water.

Amiloride Hydrochloride: Yellow to greenish yellow, odorless or practically odorless powder. Slightly soluble in water; insoluble in ether, in ethyl acetate, in acetone, and in chloroform; freely soluble in dimethylsulfoxide; sparingly soluble in methanol.

Aminobenzoic Acid: White or slightly yellow, odorless crystals or crystalline powder. Discolors on exposure to air or light. Slightly soluble in water and in chloroform; freely soluble in alcohol and in solutions of alkali hydroxides and carbonates; sparingly soluble in ether.

Aminobenzoic Acid Topical Solution: Straw-colored solution having the odor of alcohol.

Aminocaproic Acid: Fine, white, crystalline powder. Is odorless, or practically odorless. Its solutions are neutral to litmus. Melts at about 205°. Freely soluble in water, in acids, and in alkalies; slightly soluble in methanol and in alcohol; practically insoluble in chloroform and in ether.

Aminoglutethimide: Fine, white, or creamy white, crystalline powder. Very slightly soluble in water; readily soluble in most organic solvents. Forms water-soluble salts with strong acids.

Aminohippuric Acid: White, crystalline powder. Discolors on exposure to light. Melts at about 195°, with decomposition. Sparingly soluble in water and in alcohol; freely soluble in alkaline solutions, with some decomposition, and in diluted hydrochloric acid; very slightly soluble in benzene, in carbon tetrachloride, in chloroform, and in ether.

Aminophylline: White or slightly yellowish granules or powder, having a slight ammoniacal odor and a bitter taste. Upon exposure to air, it gradually loses ethylenediamine and absorbs carbon dioxide with the liberation of free theophylline. Its solutions are alkaline to litmus. 1 g dissolves in 25 mL of water to give a clear solution; 1 g dissolved in 5 mL of water crystallizes upon standing, but redissolves when a small amount of ethylenediamine is added. Insoluble in alcohol and in ether.

Aminophylline Tablets: May have a faint ammoniacal odor.

Aminosalicylate Sodium: White to cream-colored, crystalline powder. Is practically odorless, and has a sweet, saline taste. Its solutions decompose slowly and darken in color. Freely soluble in water; sparingly soluble in alcohol; very slightly soluble in ether and in chloroform.

Aminosalicylic Acid: White or practically white, bulky powder, that darkens on exposure to light and to air. Is odorless, or has a slight acetous odor. Slightly soluble in water and in ether; soluble in alcohol; practically insoluble in benzene.

Amitriptyline Hydrochloride: White or practically white, odorless or practically odorless, crystalline powder or small crystals. Freely soluble in water, in alcohol, in chloroform, and in methanol; insoluble in ether.

Strong Ammonia Solution: Clear, colorless liquid, having an exceedingly pungent, characteristic odor. Specific gravity is about 0.90. *NF category:* Alkalizing agent.

Aromatic Ammonia Spirit: Practically colorless liquid when recently prepared, but gradually acquiring a yellow color on standing. Has the taste of ammonia, has an aromatic and pungent odor, and is affected by light. Specific gravity is about 0.90.

Ammonium Carbonate: White powder, or hard, white or translucent masses, having a strong odor of ammonia, without empyreuma, and a sharp, ammoniacal taste. Its solutions are alkaline to litmus. On exposure to air, it loses ammonia and carbon dioxide, becoming opaque, and is finally converted into friable porous lumps or a white powder of ammonium bicarbonate. Freely soluble in water, but is decomposed by hot water. *NF category:* Alkalizing agent; buffering agent.

Ammonium Chloride: Colorless crystals or white, fine or coarse, crystalline powder. Has a cool, saline taste, and is somewhat hygroscopic. Freely soluble in water and in glycerin, and even more so in boiling water; sparingly soluble in alcohol.

Ammonium Phosphate: Colorless or white granules or powder, having a saline taste. Freely soluble in water; practically insoluble in acetone and in alcohol. *NF category:* Buffering agent.

Amobarbital: White, odorless, crystalline powder, having a bitter taste. Its saturated solution has a pH of about 5.6, determined potentiometrically. Very slightly soluble in water; freely soluble in alcohol and in ether; soluble in chloroform and in solutions of fixed alkali hydroxides and carbonates.

Amobarbital Sodium: White, friable, granular powder. Is odorless, has a bitter taste, and is hygroscopic. Its solutions decompose on standing, heat accelerating the decomposition. Very soluble in water; soluble in alcohol; practically insoluble in ether and in chloroform.

Amodiaquine: Very pale yellow to light tan-yellow, odorless powder. Practically insoluble in water; sparingly soluble in 1.0 *N* hydrochloric acid; slightly soluble in alcohol.

Amodiaquine Hydrochloride: Yellow, crystalline powder. Is odorless and has a bitter taste. Soluble in water; sparingly soluble in alcohol; very slightly soluble in benzene, in chloroform, and in ether.

Amoxicillin: White, practically odorless, crystalline powder. Slightly soluble in water and in methanol; insoluble in benzene, in carbon tetrachloride, and in chloroform.

Amphetamine Sulfate: White, odorless, crystalline powder, having a slightly bitter taste. Its solutions are acid to litmus, having a pH of 5 to 6. Freely soluble in water; slightly soluble in alcohol; practically insoluble in ether.

Amphotericin B: Yellow to orange powder; odorless or practically so. Insoluble in water, in anhydrous alcohol, in ether, in benzene, and in toluene; soluble in dimethylformamide, in dimethyl sulfoxide, and in propylene glycol; slightly soluble in methanol.

Amphotericin B for Injection: It yields a colloidal dispersion in water.

Ampicillin: White, practically odorless, crystalline powder. Slightly soluble in water and in methanol; insoluble in benzene, in carbon tetrachloride, and in chloroform.

Sterile Ampicillin Sodium: White to off-white, odorless or practically odorless, crystalline powder. Is hygroscopic. Very soluble in water and in isotonic sodium chloride and dextrose solutions.

Amprolium ($C_{14}H_{19}ClN_4 . HCl$): White to light yellow powder. Freely soluble in water, in methanol, in alcohol, and in dimethylformamide; sparingly soluble in dehydrated alcohol; practically insoluble in isopropyl alcohol, in butyl alcohol, and in acetone.

Amyl Nitrite: Clear, yellowish liquid, having a peculiar, ethereal, fruity odor. Is volatile even at low temperatures, and is flammable. Boils at about 96°. Practically insoluble in water. Miscible with alcohol and with ether.

Amylene Hydrate: Clear, colorless liquid, having a camphoraceous odor. Its solutions are neutral to litmus. Freely soluble in water. Miscible with alcohol, with chloroform, with ether, and with glycerin. *NF category:* Solvent.

Anethole: Colorless or faintly yellow liquid at or above 23°. Has a sweet taste and the aromatic odor of anise. Is affected by light. Very slightly soluble in water; freely soluble in alcohol. Readily miscible with ether and with chloroform. *NF category:* Flavors and perfumes.

Anileridine: White to yellowish white, odorless to practically odorless, crystalline powder. Is oxidized on exposure to air and light, becoming darker in color. It exhibits polymorphism, and of two crystalline forms observed, one melts at about 80° and the other at about 89°. Very slightly soluble in water; freely soluble in alcohol and in chloroform; soluble in ether although it may show turbidity.

Anileridine Hydrochloride: White or nearly white, odorless, crystalline powder. Is stable in air. Melts at about 270°, with decomposition. Freely soluble in water; sparingly soluble in alcohol; practically insoluble in ether and in chloroform.

Antazoline Phosphate: White to off-white, crystalline powder, having a bitter taste. Soluble in water; sparingly soluble in methanol; practically insoluble in benzene and in ether.

Anthralin: Yellowish brown, crystalline powder. Is odorless and tasteless. Insoluble in water; soluble in chloroform, in acetone, in benzene, and in solutions of alkali hydroxides; slightly soluble in alcohol, in ether, and in glacial acetic acid.

Anticoagulant Citrate Dextrose Solution: Clear, colorless, odorless liquid. Is dextrorotatory.

Anticoagulant Citrate Phosphate Dextrose Solution: Clear, colorless to slightly yellow, odorless liquid. Is dextrorotatory.

Anticoagulant Sodium Citrate Solution: Clear and colorless liquid.

Antihemophilic Factor: White or yellowish powder. On constitution is opalescent with a slight blue tinge or is a yellowish liquid.

Cryoprecipitated Antihemophilic Factor: Yellowish frozen solid. On thawing becomes a very viscous, yellow, gummy liquid.

Antimony Potassium Tartrate: Colorless, odorless, transparent crystals, or white powder. The crystals effloresce upon exposure to air and do not readily rehydrate even on exposure to high humidity. Its solutions are acid to litmus. Freely soluble in boiling water; soluble in water and in glycerin; insoluble in alcohol.

Antimony Sodium Tartrate: Colorless, odorless, transparent crystals, or white powder. The crystals effloresce upon exposure to air. Freely soluble in water; insoluble in alcohol.

Antipyrine: Colorless crystals, or white, crystalline powder. Is odorless and has a slightly bitter taste. Its solutions are neutral to litmus. Very soluble in water; freely soluble in alcohol and in chloroform; sparingly soluble in ether.

Antirabies Serum: Transparent or slightly opalescent liquid, faint brownish, yellowish, or greenish in color, and practically odorless or having a slight odor because of the antimicrobial agent.

Antivenin (Crotalidae) Polyvalent: Solid exhibiting the characteristic structure of a freeze-dried solid; light cream in color.

Antivenin (Micrurus Fulvius): Solid exhibiting the characteristic structure of a freeze-dried solid; light cream in color.

Apomorphine Hydrochloride: Minute, white or grayish white, glistening crystals or white powder. Is odorless. It gradually acquires a green color on exposure to light and to air. Its solutions are neutral to litmus. Sparingly soluble in water and in alcohol; soluble in water at 80°; very slightly soluble in chloroform and in ether.

Arginine: White, practically odorless crystals. Freely soluble in water; sparingly soluble in alcohol; insoluble in ether.

Arginine Hydrochloride: White crystals or crystalline powder, practically odorless. Freely soluble in water.

Aromatic Elixir: *NF category:* Flavored and/or sweetened vehicle.

Ascorbic Acid: White or slightly yellow crystals or powder. On exposure to light it gradually darkens. In the dry state, is reasonably stable in air, but in solution rapidly oxidizes. Melts at about 190°. Freely soluble in water; sparingly soluble in alcohol; insoluble in chloroform, in ether, and in benzene. *NF category:* Antioxidant.

Ascorbyl Palmitate: White to yellowish white powder, having a characteristic odor. Very slightly soluble in water and in vegetable oils; soluble in alcohol. *NF category:* Antioxidant.

Aspartame: White, odorless, crystalline powder, having a sweet taste. Sparingly soluble in water; slightly soluble in alcohol. Melts at about 246°. The pH of an 8 in 1000 solution is about 5. *NF category:* Sweetening agent.

Aspirin: White crystals, commonly tabular or needle-like, or white, crystalline powder. Is odorless or has a faint odor. Is stable in dry air; in the moist air it gradually hydrolyzes to salicylic and acetic acids. Slightly soluble in water; freely soluble in alcohol; soluble in chloroform and in ether; sparingly soluble in absolute ether.

Atropine: White crystals, usually needle-like, or white, crystalline powder. Its saturated solution is alkaline to phenolphthalein TS. Is optically inactive, but usually contains some levorotatory hyoscyamine. Slightly soluble in water, and sparingly soluble in water at 80°; freely soluble in alcohol and in chloroform; soluble in glycerin and in ether.

Atropine Sulfate: Colorless crystals, or white, crystalline powder. Odorless; effloresces in dry air; is slowly affected by light. Very soluble in water; freely soluble in alcohol and even more so in boiling alcohol; freely soluble in glycerin.

Activated Attapulgite: Cream-colored, micronized, non-swelling powder, free from gritty particles. The high heat treatment used in its preparation causes it to yield only moderately viscous aqueous suspensions, its dispersion consisting mainly of particle groups. Insoluble in water.

Colloidal Activated Attapulgite: Cream-colored, micronized, non-swelling powder, free from gritty particles. Yields viscous aqueous suspensions, as a result of dispersion into its constituent ultimate particles. Insoluble in water.

Aurothioglucose: Yellow, odorless or practically odorless powder. Is stable in air. An aqueous solution is unstable on long standing. The pH of its 1 in 100 solution is about 6.3. Freely soluble in water; practically insoluble in acetone, in alcohol, in chloroform, and in ether.

Azatadine Maleate: White to light cream-colored, odorless powder. Melts at about 153°. Freely soluble in water, in alcohol, in chloroform, and in methanol; practically insoluble in benzene and in ether.

Azathioprine: Pale yellow, odorless powder. Insoluble in water; soluble in dilute solutions of alkali hydroxides; sparingly soluble in dilute mineral acids; very slightly soluble in alcohol and in chloroform.

Azathioprine Sodium for Injection: Bright yellow, amorphous mass or cake.

Bacitracin: White to pale buff powder, odorless or having a slight odor. Is hygroscopic. Its solutions deteriorate rapidly at room temperature. Is precipitated from its solutions and is inactivated by salts of many of the heavy metals. Freely soluble in water; soluble in alcohol, in methanol, and in glacial acetic acid, the solution in the organic solvents usually showing some insoluble residue; insoluble in acetone, in chloroform, and in ether.

Sterile Bacitracin: White to pale buff powder, odorless or having a slight odor. Is hygroscopic. Its solutions deteriorate rapidly at room temperature. May be amorphous when prepared by freeze-drying. Is precipitated from its solutions and is inactivated by salts of many of the heavy metals. Freely soluble in water; soluble in alcohol, in methanol, and in glacial acetic acid, the solution in the organic solvents usually showing some insoluble residue; insoluble in acetone, in chloroform, and in ether.

Bacitracin Zinc: White to pale tan powder, odorless or having a slight odor. Is hygroscopic. Sparingly soluble in water.

Baclofen: White to off-white, crystalline powder. Is odorless or practically so. Slightly soluble in water; very slightly soluble in methanol; insoluble in chloroform.

Adhesive Bandage: The compress of Adhesive Bandage is substantially free from loose threads or ravelings. The adhesive strip may be perforated, and the back may be coated with a water-repellent film.

Gauze Bandage: One continuous piece, tightly rolled, in various widths and lengths and substantially free from loose threads and ravelings.

Barium Hydroxide Lime: White or grayish white granules. May have a color if an indicator has been added. *NF category:* Sorbent, carbon dioxide.

Barium Sulfate: Fine, white, odorless, tasteless, bulky powder, free from grittiness. Practically insoluble in water, in organic solvents, and in solutions of acids and of alkalies.

Barium Sulfate for Suspension: White or colored, bulky or granular powder.

BCG Vaccine: White to creamy white, dried mass, having the characteristic texture of material dried in the frozen state.

Beclomethasone Dipropionate: White to cream white, odorless powder. Very slightly soluble in water; very soluble in chloroform; freely soluble in acetone and in alcohol.

Belladonna Leaf: When moistened, its odor is slight, somewhat tobacco-like. Its taste is somewhat bitter and acrid.

Bendroflumethiazide: White to cream-colored, finely divided, crystalline powder. Is odorless, or has a slight odor. Melts at about 220°. Practically insoluble in water; freely soluble in alcohol and in acetone.

Benoxinate Hydrochloride: White, or slightly off-white, crystals or crystalline powder. Is odorless, or has a slight characteristic odor, has a salty taste, and exhibits local anesthetic properties when placed upon the tongue. Its solutions are neutral to litmus, and it melts at about 158°. Very soluble in water; freely soluble in chloroform and in alcohol; insoluble in ether.

Bentonite: Very fine, odorless, pale buff or cream-colored to grayish powder, free from grit. Has a slightly earthy taste. Is hygroscopic. Insoluble in water, but swells to approximately twelve times its volume when added to water; insoluble in, and does not swell in, organic solvents. *NF category:* Suspending and/or viscosity-increasing agent.

Purified Bentonite: Odorless, tasteless, fine (micronized) powder or small flakes that are creamy when viewed on their flat surfaces and tan to brown when viewed on their edges. Insoluble in water and in alcohol. Swells when added to water or glycerin.

Bentonite Magma: *NF category:* Suspending and/or viscosity increasing agent.

Benzaldehyde: Colorless, strongly refractive liquid, having an odor resembling that of bitter almond oil, and having a burning, aromatic taste. Is affected by light. Slightly soluble in water. Miscible with alcohol, with ether, and with fixed and volatile oils. *NF category:* Flavors and perfumes.

Benzaldehyde Elixir, Compound: *NF category:* Flavored and/or sweetened vehicle.

Benzalkonium Chloride: White or yellowish white, thick gel or gelatinous pieces. Usually has a mild, aromatic odor. Its aqueous solution has a bitter taste, foams strongly when shaken, and usually is slightly alkaline. Very soluble in water and in alcohol. Anhydrous form freely soluble in benzene, and slightly soluble in ether. *NF category:* Antimicrobial preservative; wetting and/or solubilizing agent.

Benzalkonium Chloride Solution: Clear liquid; colorless or slightly yellow unless a color has been added. Has an aromatic odor and a bitter taste. *NF category:* Antimicrobial preservative.

Benzethonium Chloride: White crystals, having a mild odor. Its solution (1 in 100) is slightly alkaline to litmus. Soluble in water, in alcohol, and in chloroform; slightly soluble in ether. *NF category:* Antimicrobial preservative; wetting and/or solubilizing agent.

Benzethonium Chloride Solution: Odorless, clear liquid, slightly alkaline to litmus.

Benzethonium Chloride Tincture: Clear liquid, having the characteristic odor of acetone and of alcohol.

Benzocaine: Small, white crystals or white, crystalline powder. Is odorless, is stable in air, and exhibits local anesthetic properties when placed upon the tongue. Very slightly soluble in water; freely soluble in alcohol, in chloroform, and in ether; sparingly soluble in almond oil and olive oil. Dissolves in dilute acids.

Benzoic Acid: White crystals, scales, or needles. Has a slight odor, usually suggesting benzaldehyde or benzoin. Somewhat volatile at moderately warm temperatures. Freely volatile in steam. Slightly soluble in water; freely soluble in alcohol, in chloroform, and in ether. *NF category:* Antimicrobial preservative.

Benzoin: Sumatra Benzoin has an aromatic and balsamic odor. When heated it does not emit a pinaceous odor. When Sumatra Benzoin is digested with boiling water, the odor suggests cinnamates or storax. Its taste is aromatic and slightly acrid. Siam Benzoin has an agreeable, balsamic, vanilla-like odor. Its taste is aromatic and slightly acrid.

Benzonatate: Clear, pale yellow, viscous liquid, having a faint, characteristic odor. Has a bitter taste, and exhibits local anesthetic properties when placed upon the tongue. Miscible with water in all proportions. Freely soluble in chloroform, in alcohol, and in benzene.

Hydrous Benzoyl Peroxide: White, granular powder, having a characteristic odor. Sparingly soluble in water and in alcohol; soluble in acetone, in chloroform, and in ether.

Benzoyl Peroxide Gel: A soft, white gel, having a characteristic odor.

Benzoyl Peroxide Lotion: White, viscous, creamy lotion, having a characteristic odor.

Benzthiazide: White, crystalline powder, having a characteristic odor. Melts at about 240°. Practically insoluble in water; freely soluble in dimethylformamide and in solutions of fixed alkali hydroxides; slightly soluble in acetone; practically insoluble in ether and in chloroform.

Benztropine Mesylate: White, slightly hygroscopic, crystalline powder. Very soluble in water; freely soluble in alcohol; very slightly soluble in ether.

Benzyl Alcohol: Colorless liquid, having a faint, aromatic odor and a sharp, burning taste. Boils at about 206°, without decomposition. Is neutral to litmus. Sparingly soluble in water; freely soluble in 50% alcohol. Miscible with alcohol, with ether, and with chloroform. *NF category:* Antimicrobial preservative.

Benzyl Benzoate: Clear, colorless, oily liquid having a slight aromatic odor and producing a sharp, burning sensation on the tongue. Practically insoluble in water and in glycerin. Miscible with alcohol, with ether, and with chloroform. *NF category:* Solvent.

Beta Carotene: Red or reddish brown to violet-brown crystals or crystalline powder. Insoluble in water and in acids and alkalies; soluble in carbon disulfide, in benzene, and in chloroform; sparingly soluble in ether, in solvent hexane, and in vegetable oils practically insoluble in methanol and in alcohol.

Betamethasone: White to practically white, odorless, crystalline powder. Melts at about 240°, with some decomposition. Insoluble in water; sparingly soluble in acetone, in alcohol, in dioxane, and in methanol; very slightly soluble in chloroform and in ether.

Betamethasone Acetate: White to creamy white, odorless powder. Sinters and resolidifies at about 165°, and remelts at about 200° or 220°, with decomposition (see *Melting Range or Temperature* ⟨741⟩). Practically insoluble in water; freely soluble in acetone; soluble in alcohol and in chloroform.

Betamethasone Benzoate: White to practically white, practically odorless powder. Melts at about 220°, with decomposition. Insoluble in water; soluble in alcohol, in methanol, and in chloroform.

Betamethasone Dipropionate: White to cream-white, odorless powder. Insoluble in water; freely soluble in acetone and in chloroform; sparingly soluble in alcohol.

Betamethasone Sodium Phosphate: White to practically white, odorless powder. Is hygroscopic. Freely soluble in water and in methanol; practically insoluble in acetone and in chloroform.

Betamethasone Valerate: White to practically white, odorless powder. Melts at about 190°, with decomposition. Practically insoluble in water; freely soluble in acetone and in chloroform; soluble in alcohol; slightly soluble in benzene and in ether.

Bethanechol Chloride: Colorless or white crystals or white, crystalline powder, usually having a slight, amine-like odor. Is hygroscopic. Exhibits polymorphism, and of two crystalline forms observed, one melts at about 211° and the other melts at about 219°. Freely soluble in water and in alcohol; insoluble in chloroform and in ether.

Biotin: Practically white, crystalline powder. Very slightly soluble in water and in alcohol; insoluble in other common organic solvents.

Biperiden: White, practically odorless, crystalline powder. Practically insoluble in water; freely soluble in chloroform; sparingly soluble in alcohol.

Biperiden Hydrochloride: White, practically odorless, crystalline powder. Melts at about 275°, with decomposition. Is optically inactive. Slightly soluble in water, in ether, in alcohol, and in chloroform; sparingly soluble in methanol.

Bisacodyl: White to off-white, crystalline powder, in which the number of particles having a longest diameter smaller than 50 µm predominate. Practically insoluble in water; soluble in chloroform and in benzene; sparingly soluble in alcohol and in methanol; slightly soluble in ether.

Milk of Bismuth: Thick, white, opaque suspension that separates on standing. Is odorless and practically tasteless. Miscible with water and with alcohol.

Bismuth Subgallate: Amorphous, bright yellow powder. Odorless and tasteless. Stable in air, but affected by light. Dissolves readily with decomposition in warm, moderately dilute hydrochloric, nitric, or sulfuric acid; readily dissolved by solutions of alkali hydroxides, forming a clear, yellow liquid, which rapidly assumes a deep red color. Practically insoluble in water, in alcohol, in chloroform, and in ether; insoluble in very dilute mineral acids.

Bismuth Subnitrate: White, slightly hygroscopic powder. Practically insoluble in water and in alcohol; readily dissolved by hydrochloric acid or by nitric acid.

Sterile Bleomycin Sulfate: Cream-colored, amorphous powder. Very soluble in water.

Anti-A Blood Grouping Serum: Liquid Serum is a clear or slightly opalescent fluid unless artificially colored blue. Dried Serum is light yellow to deep cream color, unless artificially colored as indicated for liquid Serum. The liquid Serum may develop slight turbidity on storage. The dried Serum may show slight turbidity upon reconstitution for use.

Anti-B Blood Grouping Serum: Liquid Serum is a clear or slightly opalescent fluid unless artificially colored yellow. Dried Serum is light yellow to deep cream color, unless artificially colored as indicated for liquid Serum. The liquid Serum may develop a slight turbidity on storage. The dried Serum may show slight turbidity upon reconstitution for use.

Blood Grouping Serums Anti-D, Anti-C, Anti-E, Anti-c, Anti-e: The liquid Serums are clear, slightly yellowish fluids, that may develop slight turbidity on storage. The dried Serums are light yellow to deep cream color.

Blood Group Specific Substances A, B, and AB: Clear solution that may have a slight odor because of the preservative.

Red Blood Cells: Dark red in color when packed. May show a slight creamy layer on the surface and a small supernatant layer of yellow or opalescent plasma. Also supplied in deep-frozen form with added cryophylactic substance to extend storage time.

Whole Blood: Deep red, opaque liquid from which the corpuscles readily settle upon standing for 24 to 48 hours, leaving a clear, yellowish or pinkish supernatant layer of plasma.

Boric Acid: Colorless, odorless scales of a somewhat pearly luster, or crystals, or white powder that is slightly unctuous to the touch. Is stable in air. Soluble in water and in alcohol; freely soluble in glycerin, in boiling water, and in boiling alcohol. *NF category:* Buffering agent.

Botulism Antitoxin: Transparent or slightly opalescent liquid, practically colorless, and practically odorless or having an odor because of the antimicrobial agent.

Bromocriptine Mesylate: White or slightly colored, fine crystalline powder, odorless or having a weak, characteristic odor.

Bromodiphenhydramine Hydrochloride: White to pale buff, crystalline powder, having no more than a faint odor. Freely soluble in water and in alcohol; soluble in isopropyl alcohol; insoluble in ether and in solvent hexane.

Brompheniramine Maleate: White, odorless, crystalline powder. Freely soluble in water; soluble in alcohol and in chloroform; slightly soluble in ether and in benzene.

Bumetanide: Practically white powder. Slightly soluble in water; soluble in alkaline solutions.

Bupivacaine Hydrochloride: White, odorless, crystalline powder. Melts at about 248°, with decomposition. Freely soluble in water and in alcohol; slightly soluble in chloroform and in acetone.

Bupivacaine Hydrochloride Injection: Clear, colorless solution.

Bupivacaine Hydrochloride and Epinephrine Injection: Clear, colorless solution.

Busulfan: White, crystalline powder. Very slightly soluble in water; sparingly soluble in acetone; slightly soluble in alcohol.

Butabarbital: White, odorless, crystalline powder. Very slightly soluble in water; soluble in alcohol, in chloroform, in ether, and in solutions of alkali hydroxides and carbonates.

Butabarbital Sodium: White powder, having a bitter taste. Freely soluble in water and in alcohol; practically insoluble in absolute ether.

Butalbital: White, crystalline, odorless powder, having a slightly bitter taste. Is stable in air. Its saturated solution is acid to litmus. Freely soluble in alcohol, in ether, and in chloroform; slightly soluble in cold water; soluble in boiling water, and in solutions of fixed alkalies and alkali carbonates.

Butamben: White, crystalline powder. Is odorless and tasteless. Very slightly soluble in water; soluble in dilute acids, in alcohol, in chloroform, in ether, and in fixed oils. Is slowly hydrolyzed when boiled with water.

Butane: Colorless, flammable gas (boiling temperature is about −0.5°). One volume of water dissolves 0.15 volume, and 1 volume of alcohol dissolves 18 volumes at 17° and 770 mm; 1 volume of ether or chloroform at 17° dissolves 25 or 30 volumes, respectively. Vapor pressure at 21° is about 1620 mm of mercury (17 psig). *NF category:* Aerosol propellant.

Butoconazole Nitrate: White to off-white, crystalline powder. Melts at about 160°. Practically insoluble in water; very slightly soluble in ethyl acetate; slightly soluble in acetonitrile, in acetone, in dichloromethane, and in tetrahydrofuran; sparingly soluble in methanol.

Butorphanol Tartrate: White powder. Its solutions are slightly acidic. Melts between 217° and 219°, with decomposition. Sparingly soluble in water; slightly soluble in methanol; insoluble in alcohol, in chloroform, in ethyl acetate, in ethyl ether, and in hexane; soluble in dilute acids.

Butyl Alcohol: Clear, colorless, mobile liquid, having a characteristic, penetrating vinous odor. Soluble in water. Miscible with alcohol, with ether, and with many other organic solvents. *NF category:* Solvent.

Butylated Hydroxyanisole: White or slightly yellow, waxy solid, having a faint, characteristic odor. Insoluble in water; freely soluble in alcohol, in propylene glycol, in chloroform, and in ether. *NF category:* Antioxidant.

Butylated Hydroxytoluene: White, crystalline solid, having a faint, characteristic odor. Insoluble in water and in propylene glycol; freely soluble in alcohol, in chloroform, and in ether. *NF category:* Antioxidant.

Butylparaben: Small, colorless crystals or white powder. Very slightly soluble in water and in glycerin; freely soluble in acetone, in alcohol, in ether, and in propylene glycol. *NF category:* Antimicrobial preservative.

Caffeine: White powder or white, glistening needles, usually matted together. Is odorless and has a bitter taste. Its solutions are neutral to litmus. The hydrate is efflorescent in air. Sparingly soluble in water and in alcohol; freely soluble in chloroform; slightly soluble in ether.

Calamine: Pink, odorless, practically tasteless, fine powder. Insoluble in water; practically completely soluble in mineral acids.

Precipitated Calcium Carbonate: Fine, white, odorless, tasteless, microcrystalline powder. Is stable in air. Practically insoluble in water. Its solubility in water is increased by the presence of any ammonium salt or of carbon dioxide. The presence of any alkali hydroxide reduces its solubility. Insoluble in alcohol. Dissolves with effervescence in 1 *N* acetic acid, in 3 *N* hydrochloric acid, and 2 *N* nitric acid.

Calcium Chloride: White, hard, odorless fragments or granules. Is deliquescent. Freely soluble in water, in alcohol, and in boiling alcohol; very soluble in boiling water. *NF category:* Desiccant.

Calcium Citrate: White, odorless, crystalline powder. Slightly soluble in water; freely soluble in diluted 3 *N* hydrochloric acid and in diluted 2 *N* nitric acid; insoluble in alcohol.

Calcium Gluceptate: White to faintly yellow, amorphous powder. Is stable in air, but the hydrous forms may lose part of their water of hydration on standing. Freely soluble in water; insoluble in alcohol and in many other organic solvents.

Calcium Gluconate: White, crystalline, odorless, tasteless granules or powder. Is stable in air. Its solutions are neutral to litmus. Sparingly (and slowly) soluble in water; freely soluble in boiling water; insoluble in alcohol.

Calcium Hydroxide: White powder. Has an alkaline, slightly bitter taste. Slightly soluble in water; soluble in glycerin and in syrup; very slightly soluble in boiling water; insoluble in alcohol.

Calcium Hydroxide Solution: Clear, colorless liquid having an alkaline taste. Is alkaline to litmus.

Calcium Lactate: White, practically odorless granules or powder. The pentahydrate is somewhat efflorescent and at 120° becomes anhydrous. The pentahydrate is soluble in water; it is practically insoluble in alcohol.

Calcium Levulinate: White, crystalline or amorphous, powder, having a faint odor suggestive of burnt sugar. Has a bitter, salty taste. Freely soluble in water; slightly soluble in alcohol; insoluble in ether and in chloroform.

Calcium Pantothenate: Slightly hygroscopic, white powder. Is odorless and has a bitter taste. Freely soluble in water; soluble in glycerin; practically insoluble in alcohol, in chloroform, and in ether.

Racemic Calcium Pantothenate: White, slightly hygroscopic powder, having a faint, characteristic odor, and a bitter taste. Is stable in air. Its solutions are neutral or alkaline to litmus. Is optically inactive. Freely soluble in water; soluble in glycerin; practically insoluble in alcohol, in chloroform, and in ether.

Dibasic Calcium Phosphate: White, odorless, tasteless powder. Is stable in air. Practically insoluble in water; soluble in 3 *N* hydrochloric acid and in 2 *N* nitric acid; insoluble in alcohol. *NF category:* Tablet and/or capsule diluent.

Tribasic Calcium Phosphate: White, odorless, tasteless powder. Is stable in air. Practically insoluble in water; readily soluble in 3 *N* hydrochloric acid and in 2 *N* nitric acid; insoluble in alcohol. *NF category:* Tablet and/or capsule diluent.

Calcium Polycarbophil: White to creamy white powder. Insoluble in water, in dilute acids, in dilute alkalies, and in common organic solvents.

Calcium Saccharate: White, odorless, tasteless, crystalline powder. Very slightly soluble in cold water; slightly soluble in boiling water; very slightly soluble in alcohol; practically insoluble in ether and in chloroform; soluble in dilute mineral acids and in solutions of calcium gluconate.

Calcium Silicate: White to off-white, free-flowing powder that remains so after absorbing relatively large amounts of water or other liquids. Insoluble in water. Forms a gel with mineral acids. *NF category:* Glidant and/or anticaking agent.

Calcium Stearate: Fine, white to yellowish white, bulky powder having a slight, characteristic odor. Is unctuous, and is free from grittiness. Insoluble in water, in alcohol, and in ether. *NF category:* Tablet and/or capsule lubricant.

Calcium Sulfate: Fine, white to slightly yellow-white, odorless powder. Slightly soluble in water; soluble in 3 *N* hydrochloric acid. *NF category:* Desiccant; tablet and/or capsule diluent.

Camphor: Colorless or white crystals, granules, or crystalline masses; or colorless to white, translucent, tough masses. Has a penetrating, characteristic odor and a pungent, aromatic taste. Specific gravity is about 0.99. Slowly volatilizes at ordinary temperatures. Slightly soluble in water; very soluble in alcohol, in chloroform, and in ether; freely soluble in carbon disulfide, in solvent hexane, and in fixed and volatile oils.

Candicidin: Yellow to brown powder. Sparingly soluble in water; very slightly soluble in alcohol, in acetone, and in butyl alcohol.

Sterile Capreomycin Sulfate: White to practically white, amorphous powder. Freely soluble in water; practically insoluble in most organic solvents.

Captopril: White to off-white, crystalline powder, which may have a characteristic, sulfide-like odor. Melts in the range of 104° to 110°. Freely soluble in water, in methanol, in alcohol, and in chloroform.

Caramel: Thick, dark brown liquid having the characteristic odor of burnt sugar, and a pleasant, bitter taste. One part dissolved in 1000 parts of water yields a clear solution having a distinct yellowish orange color. The color of this solution is not changed and no precipitate is formed after exposure to sunlight for 6 hours. When spread in a thin layer on a glass plate, it appears homogeneous, reddish brown, and transparent. Miscible with water. Soluble in dilute alcohol up to 55% (v/v). Immiscible with ether, with chloroform, with acetone, with benzene, and with solvent hexane. *NF category:* Color.

Carbamazepine: White to off-white powder. Practically insoluble in water; soluble in alcohol and in acetone.

Carbamide Peroxide Topical Solution: Clear, colorless, viscous liquid, having a characteristic odor and taste.

Sterile Carbenicillin Disodium: White to off-white, crystalline powder. Freely soluble in water; soluble in alcohol; practically insoluble in chloroform and in ether.

Carbenicillin Indanyl Sodium: White to off-white powder. Soluble in water and in alcohol.

Carbidopa: White to creamy white, odorless or practically odorless, powder. Slightly soluble in water; freely soluble in 3 *N* hydrochloric acid; slightly soluble in methanol; practically insoluble in alcohol, in acetone, in chloroform, and in ether.

Carbinoxamine Maleate: White, odorless, crystalline powder. Very soluble in water; freely soluble in alcohol and in chloroform; very slightly soluble in ether.

Carbol-Fuchsin Topical Solution: Dark purple liquid, which appears purplish red when spread in a thin film.

Carbomer 934P: White, fluffy powder, having a slight, characteristic odor. Is hygroscopic. The pH of a 1 in 100 dispersion is about 3. When neutralized with alkali hydroxides or with amines, it dissolves in water, in alcohol, and in glycerin. *NF category:* Suspending and/or viscosity-increasing agent.

Carbon Dioxide: Odorless, colorless gas. Its solutions are acid to litmus. One liter at 0° and at a pressure of 760 mm of mercury weighs 1.977 g. One volume dissolves in about 1 volume of water. *NF category:* Air displacement.

Carbon Tetrachloride: Clear, colorless, mobile liquid, having a characteristic, ethereal odor resembling that of chloroform. Practically insoluble in water. Miscible with alcohol, with ether, with chloroform, with benzene, with solvent hexane, and with fixed and volatile oils. *NF category:* Solvent.

Carboxymethylcellulose Calcium: White to yellowish white powder. Is hygroscopic. Practically insoluble in alcohol, in acetone, in ether, in chloroform, and in benzene. It swells with water to form a suspension; the pH of the suspension, obtained by shaking 1 g with 100 mL of water, is between 4.5 and 6.0. *NF category:* Suspending and/or viscosity-increasing agent.

Carboxymethylcellulose Sodium: White to cream-colored powder or granules. The powder is hygroscopic. Is easily dispersed in water to form colloidal solutions. Insoluble in alcohol, in ether, and in most other organic solvents. *NF category:* Coating agent; suspending and/or viscosity-increasing agent; tablet binder.

Carisoprodol: White, crystalline powder, having a mild, characteristic odor and a bitter taste. Very slightly soluble in water; freely soluble in alcohol, in chloroform, and in acetone.

Carphenazine Maleate: Yellow, finely divided powder. Is odorless, or has a slight odor. Slightly soluble in water and in alcohol; practically insoluble in ether.

Carrageenan: Yellowish or tan to white, coarse to fine powder. Is practically odorless and has a mucilaginous taste. Soluble in water at a temperature of about 80°, forming a viscous, clear or slightly opalescent solution that flows readily. Disperses in water more readily if first moistened with alcohol, glycerin, or a saturated solution of sucrose in water. *NF category:* Suspending and/or viscosity-increasing agent.

Cascara Sagrada: Has a distinct odor and a bitter and slightly acrid taste.

Castor Oil: Pale yellowish or almost colorless, transparent, viscid liquid. Has a faint, mild odor; is free from foreign and rancid odor; and has a bland, characteristic taste. Soluble in alcohol. Miscible with dehydrated alcohol, with glacial acetic acid, with chloroform, and with ether. *NF category:* Plasticizer.

Hydrogenated Castor Oil: White, crystalline wax. Insoluble in water and in most common organic solvents. *NF category:* Stiffening agent.

Sterile Cefazolin Sodium: White to off-white, practically odorless, crystalline powder, or white to off-white solid having the characteristic appearance of products prepared by freeze-drying. Freely soluble in water, in saline TS, and in dextrose solutions; very slightly soluble in alcohol; practically insoluble in chloroform and in ether.

Sterile Cefoxitin Sodium: White to off-white, granules or powder, having a slight characteristic odor. Is somewhat hygroscopic. Very soluble in water; soluble in methanol; sparingly soluble in dimethylformamide; slightly soluble in acetone; insoluble in ether and in chloroform.

Cellulose Acetate: Fine, white powder or free-flowing pellets. Available in a range of viscosities and acetyl contents. High viscosity, which reflects high molecular weight, decreases solubility slightly. High acetyl content cellulose acetates generally have more limited solubility in commonly used organic solvents than low acetyl content cellulose acetates, but are more soluble in methylene chloride. All acetyl content cellulose acetates are insoluble in alcohol and in water; soluble in dioxane and in dimethylformamide. *NF category:* Polymer membrane, insoluble.

Microcrystalline Cellulose: Fine, white, odorless, crystalline powder. It consists of free-flowing, nonfibrous particles that may be compressed into self-binding tablets which disintegrate rapidly in water. Insoluble in water, in dilute acids, and in most organic solvents; practically insoluble in sodium hydroxide solution (1 in 20). *NF category:* Tablet binder; tablet disintegrant; tablet and/or capsule diluent.

Microcrystalline Cellulose and Carboxymethylcellulose Sodium: Tasteless, odorless, white to off-white, coarse to fine powder. Swells in water, producing, when dispersed, a white, opaque dispersion or gel. Insoluble in organic solvents and in dilute acids. *NF category:* Suspending and/or viscosity-increasing agent.

Oxidized Cellulose: In the form of gauze or lint. Is slightly off-white in color, is acidic to the taste, and has a slight, charred odor. Insoluble in water and in acids; soluble in dilute alkalies.

Oxidized Regenerated Cellulose: A knit fabric, usually in the form of sterile strips. Slightly off-white, having a slight odor. Insoluble in water and in dilute acids; soluble in dilute alkalies.

Powdered Cellulose: White, odorless substance, consisting of fibrous particles. Exhibits degrees of fineness ranging from a free-flowing dense powder to a coarse, fluffy, nonflowing material. Insoluble in water, in dilute acids, and in nearly all organic solvents; slightly soluble in sodium hydroxide solution (1 in 20). *NF category:* Filtering aid sorbent; tablet and/or capsule diluent.

Cellulose Acetate Phthalate: Free-flowing, white powder. May have a slight odor of acetic acid. Insoluble in water and in alcohol; soluble in acetone and in dioxane. *NF category:* Coating agent.

Cellulose Sodium Phosphate: Free-flowing cream-colored, odorless, tasteless powder. Insoluble in water, in dilute acids, and in most organic solvents.

Cephalexin: White to off-white, crystalline powder. Slightly soluble in water; practically insoluble in alcohol, in chloroform, and in ether.

Sterile Cephalothin Sodium: White to off-white, practically odorless, crystalline powder. Freely soluble in water, in saline TS, and in dextrose solutions; insoluble in most organic solvents.

Sterile Cephapirin Sodium: White to off-white crystalline powder, odorless or having a slight odor. Very soluble in water; insoluble in most organic solvents.

Cephradine: White to off-white, crystalline powder. Sparingly soluble in water; very slightly soluble in alcohol and in chloroform; practically insoluble in ether.

Cetostearyl Alcohol: Unctuous, white flakes or granules, having a faint, characteristic odor, and a bland, mild taste. Insoluble in water; soluble in alcohol and in ether. *NF category:* Stiffening agent.

Cetyl Alcohol: Unctuous, white flakes, granules, cubes, or castings. Has a faint characteristic odor and a bland, mild taste. Insoluble in water; soluble in alcohol and in ether, the solubility increasing with an increase in temperature. *NF category:* Stiffening agent.

Cetyl Esters Wax: White to off-white, somewhat translucent flakes, having a crystalline structure and a pearly luster when caked. Has a faint odor and a bland, mild taste, free from rancidity, and has a specific gravity of about 0.83 at 50°. Insoluble in water; soluble in boiling alcohol, in ether, in chloroform, and in fixed and volatile oils; slightly soluble in cold solvent hexane; practically insoluble in cold alcohol. *NF category:* Stiffening agent.

Cetylpyridinium Chloride: White powder, having a slight, characteristic odor. Very soluble in water, in alcohol, and in chloroform; slightly soluble in benzene and in ether. *NF category:* Antimicrobial preservative; wetting and/or solubilizing agent.

Cetylpyridinium Chloride Topical Solution: Clear liquid. Is colorless unless a color has been added; has an aromatic odor and a bitter taste.

Activated Charcoal: Fine, black, odorless, tasteless powder, free from gritty matter. *NF category:* Sorbent.

Chloral Hydrate: Colorless, transparent, or white crystals having an aromatic, penetrating, and slightly acrid odor, and a slightly bitter, caustic taste. Melts at about 55°, and slowly volatilizes when exposed to air. Very soluble in water and in olive oil; freely soluble in alcohol, in chloroform, and in ether.

Chlorambucil: Off-white, slightly granular powder. Very slightly soluble in water; freely soluble in acetone; soluble in dilute alkali.

Chloramphenicol: Fine, white to grayish white or yellowish white, needle-like crystals or elongated plates. Its solutions are practically neutral to litmus. Is reasonably stable in neutral or moderately acid solutions. Its alcohol solution is dextrorotatory and its ethyl acetate solution is levorotatory. Slightly soluble in water; freely soluble in alcohol, in propylene glycol, in acetone, and in ethyl acetate.

Chloramphenicol Palmitate: Fine, white, unctuous, crystalline powder, having a faint odor and a bland, mild taste. Insoluble in water; freely soluble in acetone and in chloroform; soluble in ether; sparingly soluble in alcohol; very slightly soluble in solvent hexane.

Sterile Chloramphenicol Sodium Succinate: Light yellow powder. Freely soluble in water and in alcohol.

Chlordiazepoxide: Yellow, practically odorless, crystalline powder. Is sensitive to sunlight. Melts at about 240°. Insoluble in water; sparingly soluble in chloroform and in alcohol.

Chlordiazepoxide Hydrochloride: White or practically white, odorless, crystalline powder. Is affected by sunlight. Soluble in water and in alcohol; insoluble in solvent hexane.

Sterile Chlordiazepoxide Hydrochloride: White or practically white, odorless, crystalline powder. Is affected by sunlight. Soluble in water and in alcohol; insoluble in solvent hexane.

Chlorobutanol: Colorless to white crystals, having a characteristic, somewhat camphoraceous, odor and taste. Anhydrous form melts at about 95°, and hydrous form melts at about 76°. Slightly soluble in water; freely soluble in alcohol, in ether, in chloroform, and in volatile oils; soluble in glycerin. *NF category:* Antimicrobial preservative.

Chlorocresol: Colorless or practically colorless crystals or crystalline powder, having a characteristic, non-tarry odor. Volatile in steam. Slightly soluble in water and more soluble in hot water; very soluble in alcohol; soluble in ether, in terpenes, in fixed oils, and in solutions of alkali hydroxides.

Chloroform: Clear, colorless, mobile liquid, having a characteristic, ethereal odor, and a burning, sweet taste. Is not flammable, but its heated vapor burns with a green flame. Boils at about 61°. Is affected by light. Slightly soluble in water. Miscible with alcohol, with ether, with benzene, with solvent hexane, and with fixed and volatile oils. *NF category:* Solvent.

Chloroprocaine Hydrochloride: White, crystalline powder. Is odorless, and is stable in air. Its solutions are acid to litmus. Exhibits local anesthetic properties when placed upon the tongue. Soluble in water; slightly soluble in alcohol; very slightly soluble in chloroform; practically insoluble in ether.

Chloroquine: White or slightly yellow, crystalline powder. Is odorless, and has a bitter taste. Very slightly soluble in water; soluble in dilute acids, in chloroform, and in ether.

Chloroquine Hydrochloride Injection: Colorless liquid.

Chloroquine Phosphate: White, crystalline powder. Is odorless, has a bitter taste, and is discolored slowly on exposure to light. Its solutions have a pH of about 4.5. Exists in two polymorphic forms, one melting between 193° and 195° and the other between 210° and 215° (see *Melting Range or Temperature* ⟨741⟩); mixture of the forms melts between 193° and 215°. Freely soluble in water; practically insoluble in alcohol, in chloroform, and in ether.

Chlorothiazide: White or practically white, crystalline, odorless powder. Melts at about 340°, with decomposition. Very slightly soluble in water; freely soluble in dimethylformamide and in dimethyl sulfoxide; slightly soluble in methanol and in pyridine; practically insoluble in ether, in benzene, and in chloroform.

Chloroxylenol: White crystals or crystalline powder, having a characteristic odor. Volatile in steam. Very slightly soluble in water; freely soluble in alcohol, in ether, in terpenes, in fixed oils, and in solutions of alkali hydroxides.

Chlorpheniramine Maleate: White, odorless, crystalline powder. Its solutions have a pH between 4 and 5. Freely soluble in water; soluble in alcohol and in chloroform; slightly soluble in ether and in benzene.

Chlorpromazine: White, crystalline solid, having an amine-like odor. Darkens on prolonged exposure to light. Melts at about 60°. Practically insoluble in water and in dilute alkali hydroxides; freely soluble in alcohol, in benzene, in chloroform, in ether, and in dilute mineral acids.

Chlorpromazine Hydrochloride: White or slightly creamy white, odorless, crystalline powder. Darkens on prolonged exposure to light. Very soluble in water; freely soluble in alcohol and in chloroform; insoluble in ether and in benzene.

Chlorpropamide: White, crystalline powder, having a slight odor. Practically insoluble in water; soluble in alcohol; sparingly soluble in chloroform.

Chlorprothixene: Yellow, crystalline powder, having a slight amine-like odor. Practically insoluble in water; soluble in alcohol and in ether; freely soluble in chloroform.

Chlortetracycline Hydrochloride: Yellow, crystalline powder. Is odorless, and has a bitter taste. Is stable in air, but is slowly affected by light. Sparingly soluble in water; soluble in solutions of alkali hydroxides and carbonates; slightly soluble in alcohol; practically insoluble in acetone, in chloroform, in dioxane, and in ether.

Chlorthalidone: White to yellowish white, crystalline powder. Melts at a temperature above 215°, with decomposition. Practically insoluble in water, in ether, and in chloroform; soluble in methanol; slightly soluble in alcohol.

Chlorzoxazone: White or practically white, practically odorless, crystalline powder. Slightly soluble in water; sparingly soluble in alcohol, in isopropyl alcohol, and in methanol; soluble in solutions of alkali hydroxides and ammonia.

Cholecalciferol: White, odorless crystals. Is affected by air and by light. Melts at about 85°. Insoluble in water; soluble in alcohol, in chloroform, and in fatty oils.

Cholera Vaccine: Practically water-clear liquid to milky suspension, nearly odorless or having a faint odor because of the antimicrobial agent.

Cholesterol: White or faintly yellow, practically odorless, pearly leaflets needles, powder, or granules. Acquires a yellow to pale tan color on prolonged exposure to light. Insoluble in water; soluble in acetone, in chloroform, in dioxane, in ether, in ethyl acetate, in solvent hexane, and in vegetable oils; sparingly soluble in dehydrated alcohol; slightly (and slowly) soluble in alcohol. *NF category:* Emulsifying and/or solubilizing agent.

Cholestyramine Resin: White to buff-colored, hygroscopic, fine powder. Is odorless or has not more than a slight amine-like odor. Insoluble in water, in alcohol, in chloroform, and in ether.

Sodium Chromate Cr 51 Injection: Clear, slightly yellow solution.

Chromic Chloride: Dark green, odorless, slightly deliquescent crystals. Soluble in water and in alcohol; slightly soluble in acetone; practically insoluble in ether.

Chymotrypsin: White to yellowish white, crystalline or amorphous, odorless, powder. An amount equivalent to 100,000 USP Units is soluble in 10 mL of water and in 10 mL of saline TS.

Cimetidine: White to off-white, crystalline powder; odorless, or having a slight mercaptan odor. Soluble in alcohol and in polyethylene glycol 400; freely soluble in methanol; sparingly soluble in isopropyl alcohol; slightly soluble in water and in chloroform; practically insoluble in ether.

Cinoxacin: White to yellowish white, crystalline solid. Is odorless, and has a bitter taste and a lingering aftertaste. Insoluble in water and in most common organic solvents; soluble in alkaline solution.

Cinoxate: Slightly yellow, practically odorless, viscous liquid. Very slightly soluble in water; slightly soluble in glycerin; soluble in propylene glycol. Miscible with alcohol and with vegetable oils.

Citric Acid: Colorless, translucent crystals, or white, granular to fine, crystalline powder. Is odorless, or practically odorless, and has a strongly acidic taste. The hydrous form is efflorescent in dry air. Very soluble in water; freely soluble in alcohol; sparingly soluble in ether. *NF category:* Acidifying agent; buffering agent.

Clemastine Fumarate: Colorless to faintly yellow, odorless, crystalline powder. Its solutions are acid to litmus. Very slightly soluble in water; slightly soluble in methanol; very slightly soluble in chloroform.

Clidinium Bromide: White to nearly white, practically odorless, crystalline powder. Is optically inactive. Melts at about 242°. Soluble in water and in alcohol; slightly soluble in benzene and in ether.

Clindamycin Hydrochloride: White or practically white, crystalline powder. Is odorless or has a faint mercaptan-like odor. Is stable in the presence of air and light. Its solutions are acidic and are dextrorotatory. Freely soluble in water, in dimethylformamide, and in methanol; soluble in alcohol; practically insoluble in acetone.

Clindamycin Palmitate Hydrochloride: White to off-white amorphous powder, having a characteristic odor. Very soluble in ethyl acetate and in dimethylformamide; freely soluble in water, in benzene, in ether, in chloroform, and in alcohol.

Clindamycin Phosphate: White to off-white, hygroscopic, crystalline powder. Is odorless or practically odorless, and has a bitter taste. Freely soluble in water; slightly soluble in dehydrated alcohol; very slightly soluble in acetone; practically insoluble in chloroform, in benzene, and in ether.

Clioquinol: Voluminous, spongy, yellowish white to brownish yellow powder, having a slight, characteristic odor. Darkens on exposure to light. Melts at about 180°, with decomposition. Practically insoluble in water and in alcohol; soluble in hot ethyl acetate and in hot glacial acetic acid.

Clocortolone Pivalate: White to yellowish white, odorless powder. Melts at about 230°, with decomposition. Freely soluble in chloroform and in dioxane; soluble in acetone; sparingly soluble in alcohol; slightly soluble in benzene and in ether.

Clofibrate: Colorless to pale yellow liquid having a characteristic odor. Insoluble in water; soluble in acetone, in alcohol, in benzene, and in chloroform.

Clomiphene Citrate: White to pale yellow, essentially odorless powder. Slightly soluble in water and in chloroform; freely soluble in methanol; sparingly soluble in alcohol; insoluble in ether.

Clonazepam: Light yellow powder, having a faint odor. Melts at about 239°. Insoluble in water; sparingly soluble in acetone and in chloroform; slightly soluble in alcohol and in ether.

Clorazepate Dipotassium: Light yellow, crystalline powder. Darkens on exposure to light. Soluble in water but, upon standing, may precipitate from the solution; slightly soluble in alcohol and in isopropyl alcohol; practically insoluble in acetone, in benzene, in chloroform, in ether, and in methylene chloride.

Clotrimazole: White to pale yellow, crystalline powder. Melts at about 142°, with decomposition. Practically insoluble in water; freely soluble in methanol, in acetone, in chloroform, and in alcohol.

Cloxacillin Sodium: White, odorless, crystalline powder. Freely soluble in water; soluble in alcohol; slightly soluble in chloroform.

Coal Tar: Nearly black, viscous liquid, heavier than water, having a characteristic, naphthalene-like odor, and producing a sharp, burning sensation on the tongue. Slightly soluble in water, to which it imparts its characteristic odor and taste and a faintly alkaline reaction. Partially soluble in acetone, in alcohol, in carbon disulfide, in chloroform, in ether, in methanol, and in solvent hexane. Is more soluble in benzene, only about 5% remaining undissolved, and is almost completely soluble in nitrobenzene, only a small amount of undissolved matter remaining suspended in the solution.

Cyanocobalamin Co 57 Capsules: May contain a small amount of solid or solids, or may appear empty.

Cyanocobalamin Co 57 Oral Solution: Clear, colorless to pink solution.

Cyanocobalamin Co 60 Capsules: Capsules may contain a small, rectangular solid, or may appear empty.

Cyanocobalamin Co 60 Oral Solution: Clear, colorless to pink solution.

Cocaine: Colorless to white crystals or white, crystalline powder. Is levorotatory in 3 *N* hydrochloric acid solution. Its saturated solution is alkaline to litmus. Slightly soluble in water; very soluble in warm alcohol; freely soluble in alcohol, in chloroform, and in ether; soluble in olive oil; sparingly soluble in mineral oil.

Cocaine Hydrochloride: Colorless crystals or white, crystalline powder. Very soluble in water; freely soluble in alcohol; soluble in chloroform and in glycerin; insoluble in ether.

Coccidioidin: Clear, practically colorless or amber-colored liquid.

Cocoa Butter: Yellowish white solid, having a faint, agreeable odor, and a bland, chocolate-like taste if the cocoa butter is obtained by pressing. If obtained by extraction, the taste is bland. Is usually brittle at temperatures below 25°. Freely soluble in ether and in chloroform; soluble in boiling dehydrated alcohol; slightly soluble in alcohol. *NF category:* Suppository base.

Cod Liver Oil: Thin, oily liquid, having a characteristic, slightly fishy but not rancid odor, and a fishy taste. Slightly soluble in alcohol; freely soluble in ether, in chloroform, in carbon disulfide, and in ethyl acetate.

Codeine: Colorless or white crystals or white, crystalline powder. It effloresces slowly in dry air, and is affected by light. In acid or alcohol solutions it is levorotatory. Its saturated solution is alkaline to litmus. Slightly soluble in water; very soluble in chloroform; freely soluble in alcohol; sparingly soluble in ether. When heated in an amount of water insufficient for complete solution, it melts to oily drops which crystallize on cooling.

Codeine Phosphate: Fine, white, needle-shaped crystals, or white, crystalline powder. Odorless. Is affected by light. Its solutions are acid to litmus. Freely soluble in water; very soluble in hot water; slightly soluble in alcohol but more so in boiling alcohol.

Codeine Sulfate: White crystals, usually needle-like, or white, crystalline powder. Is affected by light. Soluble in water; freely soluble in water at 80°; very slightly soluble in alcohol; insoluble in chloroform and in ether.

Colchicine: Pale yellow to pale greenish yellow, amorphous scales, or powder or crystalline powder. Is odorless or nearly so, and darkens on exposure to light. Soluble in water; freely soluble in alcohol and in chloroform; slightly soluble in ether.

Colestipol Hydrochloride: Yellow to orange beads. Swells but does not dissolve in water or dilute aqueous solutions of acid or alkali. Insoluble in the common organic solvents.

Sterile Colistimethate Sodium: White to slightly yellow, odorless, fine powder. Freely soluble in water; soluble in methanol; insoluble in acetone and in ether.

Colistin Sulfate: White to slightly yellow, odorless, fine powder. Freely soluble in water; slightly soluble in methanol; insoluble in acetone and in ether.

Collodion: Clear, or slightly opalescent, viscous liquid. Is colorless, or slightly yellowish, and has the odor of ether.

Flexible Collodion: Clear, or slightly opalescent, viscous liquid. Is colorless or slightly yellow, and has the odor of ether. The strong odor of camphor becomes noticeable as the ether evaporates.

Corn Oil: Clear, light yellow, oily liquid, having a faint, characteristic odor and taste. Slightly soluble in alcohol. Miscible with ether, with chloroform, with benzene, and with solvent hexane. *NF category:* Solvent; vehicle (oleaginous).

Corticotropin Injection: Colorless or light straw-colored liquid.

Corticotropin for Injection: White or practically white, soluble, amorphous solid having the characteristic appearance of substances prepared by freeze-drying.

Repository Corticotropin Injection: Colorless or light straw-colored liquid, which may be quite viscid at room temperature. Is odorless or has an odor of an antimicrobial agent.

Sterile Corticotropin Zinc Hydroxide Suspension: Flocculent, white, aqueous suspension, free from large particles following moderate shaking.

Cortisone Acetate: White or practically white, odorless, crystalline powder. Is stable in air. Melts at about 240°, with some decomposition (see *Melting Range or Temperature* ⟨741⟩). Insoluble in water; freely soluble in chloroform; soluble in dioxane; sparingly soluble in acetone; slightly soluble in alcohol.

Purified Cotton: White, soft, fine filament-like hairs appearing under the microscope as hollow, flattened, and twisted bands, striate and slightly thickened at the edges. Is practically odorless and practically tasteless. Insoluble in ordinary solvents; soluble in ammoniated cupric oxide TS.

Cottonseed Oil: Pale yellow, oily liquid. Is odorless or nearly so, and has a bland taste. At temperatures below 10° particles of solid fat may separate from the Oil, and at about 0° to −5° the Oil becomes a solid or nearly so. Slightly soluble in alcohol. Miscible with ether, with chloroform, with solvent hexane, and with carbon disulfide. *NF category:* Solvent; vehicle (oleaginous).

Cresol: Colorless, or yellowish to brownish yellow, or pinkish, highly refractive liquid, becoming darker with age and on exposure to light. Has a phenol-like, sometimes empyreumatic odor. A saturated solution of it is neutral or only slightly acid to litmus. Sparingly soluble in water, usually forming a cloudy solution; dissolves in solutions of fixed alkali hydroxides. Miscible with alcohol, with ether, and with glycerin. *NF category:* Antimicrobial preservative.

Cromolyn Sodium: White, odorless, crystalline powder. Is tasteless at first, with a slightly bitter aftertaste. Is hygroscopic. Soluble in water; insoluble in alcohol and in chloroform.

Cromolyn Sodium for Inhalation: White to creamy white, odorless, hygroscopic, and very finely divided powder.

Croscarmellose Sodium: White, free-flowing powder. Partially soluble in water; insoluble in alcohol, in ether, and in other organic solvents. *NF category:* Tablet disintegrant.

Crospovidone: White to creamy-white, hygroscopic powder, having a faint odor. Insoluble in water and in ordinary organic solvents. *NF category:* Tablet disintegrant.

Crotamiton: Colorless to slightly yellowish oil, having a faint amine-like odor. Soluble in alcohol and in methanol.

Cupric Chloride: Bluish green, deliquescent crystals. Freely soluble in water; soluble in alcohol; slightly soluble in ether.

Cupric Sulfate: Deep blue, triclinic crystals or blue, crystalline granules or powder. It effloresces slowly in dry air. Its solutions are acid to litmus. Freely soluble in water and in glycerin; very soluble in boiling water; slightly soluble in alcohol.

Cyanocobalamin: Dark red crystals or amorphous or crystalline red powder. In the anhydrous form, it is very hygroscopic and when exposed to air it may absorb about 12% of water. Sparingly soluble in water; soluble in alcohol; insoluble in acetone, in chloroform, and in ether.

Cyclizine: White, or creamy white, crystalline, practically odorless powder. Slightly soluble in water; soluble in alcohol and in chloroform.

Cyclizine Hydrochloride: White, crystalline powder or small, colorless crystals. Is odorless or nearly so, and has a bitter taste. Melts indistinctly at about 285°, with decomposition. Slightly soluble in water and in alcohol; sparingly soluble in chloroform; insoluble in ether.

Cyclobenzaprine Hydrochloride: White to off-white, odorless, crystalline powder. Freely soluble in water, in alcohol, and in methanol; sparingly soluble in isopropanol; slightly soluble in chloroform and in methylene chloride; insoluble in hydrocarbons.

Cyclopentolate Hydrochloride: White, crystalline powder, which upon standing develops a characteristic odor. Its solutions are acid to litmus. Melts at about 138°, the melt appearing opaque. Very soluble in water; freely soluble in alcohol; insoluble in ether.

Cyclophosphamide: White, crystalline powder. Liquefies upon loss of its water of crystallization. Soluble in water and in alcohol.

Cyclopropane: Colorless gas having a characteristic odor. Has a pungent taste. One liter at a pressure of 760 mm and a temperature of 0° weighs about 1.88 g. One volume dissolves in about 2.7 volumes of water at 15°. Freely soluble in alcohol; soluble in fixed oils.

Cycloserine: White to pale yellow, crystalline powder. Is odorless or has a faint odor. Is hygroscopic and deteriorates upon absorbing water. Its solutions are dextrorotatory. Freely soluble in water.

Cyproheptadine Hydrochloride: White to slightly yellow, odorless or practically odorless, crystalline powder. Slightly soluble in water; freely soluble in methanol; soluble in chloroform; sparingly soluble in alcohol; practically insoluble in ether.

Cysteine Hydrochloride: White crystals or crystalline powder. Soluble in water, in alcohol, and in acetone.

Cytarabine: Odorless, white to off-white, crystalline powder. Freely soluble in water; slightly soluble in alcohol and in chloroform.

Dactinomycin: Bright red, crystalline powder. Is somewhat hygroscopic and is affected by light and heat. Soluble in water at 10° and slightly soluble in water at 37°; freely soluble in alcohol; very slightly soluble in ether.

Danazol: White to pale yellow, crystalline powder. Melts at about 225°, with some decomposition. Practically insoluble or insoluble in water and in hexane; freely soluble in chloroform; soluble in acetone; sparingly soluble in alcohol and in benzene; slightly soluble in ether.

Dapsone: White or creamy white, crystalline powder. Is odorless and has a slightly bitter taste. Very slightly soluble in water; freely soluble in alcohol; soluble in acetone and in dilute mineral acids.

Deferoxamine Mesylate: White to off-white powder. Freely soluble in water; slightly soluble in methanol.

Dehydroacetic Acid: White or nearly white, crystalline powder. Is odorless or practically odorless, and has a faint, acid taste. Very slightly soluble in water; freely soluble in acetone and in benzene; soluble in aqueous solutions of fixed alkalies; sparingly soluble in alcohol. *NF category:* Antimicrobial preservative.

Dehydrocholic Acid: White, fluffy, odorless powder, having a bitter taste. Practically insoluble in water; soluble in glacial acetic acid and in solutions of alkali hydroxides and carbonates; slightly soluble in alcohol and in ether; sparingly soluble in chloroform (the solutions in alcohol and in chloroform usually are slightly turbid).

Demecarium Bromide: White or slightly yellow, slightly hygroscopic, crystalline powder. Freely soluble in water and in alcohol; soluble in ether; sparingly soluble in acetone.

Demeclocycline: Yellow, crystalline odorless powder, having a bitter taste. Sparingly soluble in water; soluble in alcohol. Dissolves readily in 3 *N* hydrochloric acid and in alkaline solutions.

Demeclocycline Hydrochloride: Yellow, crystalline, odorless powder, having a bitter taste. Sparingly soluble in water and in solutions of alkali hydroxides and carbonates; slightly soluble in alcohol; practically insoluble in acetone and in chloroform.

Denatonium Benzoate: Freely soluble in water and in alcohol; very soluble in chloroform and in methanol; very slightly soluble in ether. *NF category:* Alcohol denaturant.

Desipramine Hydrochloride: White to off-white, crystalline powder. Melts at about 213°. Soluble in water and in alcohol; freely soluble in methanol and in chloroform; insoluble in ether.

Desoximetasone: White to practically white, odorless, crystalline powder. Insoluble in water; freely soluble in alcohol, in acetone, and in chloroform.

Desoxycorticosterone Acetate: White or creamy white, crystalline powder. Is odorless, and is stable in air. Practically insoluble in water; sparingly soluble in alcohol, in acetone, and in dioxane; slightly soluble in vegetable oils.

Desoxycorticosterone Pivalate: White or creamy white, crystalline powder. Is odorless, and is stable in air. Practically insoluble in water; soluble in dioxane; sparingly soluble in acetone; slightly soluble in alcohol, in methanol, in ether, and in vegetable oils.

Dexamethasone: White to practically white, odorless, crystalline powder. Is stable in air. Melts at about 250°, with some decomposition. Practically insoluble in water; sparingly soluble in acetone, in alcohol, in dioxane, and in methanol; slightly soluble in chloroform; very slightly soluble in ether.

Dexamethasone Acetate: Clear, white to off-white, odorless powder. Practically insoluble in water; freely soluble in methanol, in acetone, and in dioxane.

Dexamethasone Sodium Phosphate: White or slightly yellow, crystalline powder. Is odorless or has a slight odor of alcohol, and is exceedingly hygroscopic. Freely soluble in water; slightly soluble in alcohol; very slightly soluble in dioxane; insoluble in chloroform and in ether.

Dexbrompheniramine Maleate: White, odorless, crystalline powder. Exists in two polymorphic forms, one melting between 106° and 107° and the other between 112° and 113°. Mixtures of the forms may melt between 105° and 113°. The pH of a solution (1 in 100) is about 5. Freely soluble in water; soluble in alcohol and in chloroform.

Dexchlorpheniramine Maleate: White, odorless, crystalline powder. Freely soluble in water; soluble in alcohol and in chloroform; slightly soluble in benzene and in ether.

Dexpanthenol: Clear, viscous, somewhat hygroscopic liquid, having a slight, characteristic odor. Some crystallization may occur on standing. Freely soluble in water, in alcohol, in methanol, and in propylene glycol; soluble in chloroform and in ether; slightly soluble in glycerin.

Dextrates: Free-flowing, porous, white, odorless, spherical granules consisting of aggregates of microcrystals, having a sweet taste and producing a cooling sensation in the mouth. May be compressed directly into self-binding tablets. Freely soluble in water (heating increases its solubility in water); soluble in dilute acids and alkalies and in basic organic solvents such as pyridine; insoluble in the common organic solvents. *NF category:* Sweetening agent; tablet and/or capsule diluent.

Dextrin: Free-flowing, white, yellow, or brown powder. Its solubility in water varies; it is usually very soluble, but often contains an insoluble portion.

Dextroamphetamine Sulfate: White, odorless, crystalline powder. Soluble in water; slightly soluble in alcohol; insoluble in ether.

Dextromethorphan: Practically white to slightly yellow, odorless, crystalline powder. Eleven mg of Dextromethorphan is equivalent to 15 mg of dextromethorphan hydrobromide monohydrate. Practically insoluble in water; freely soluble in chloroform.

Dextromethorphan Hydrobromide: Practically white crystals or crystalline powder, having a faint odor. Melts at about 126°, with decomposition. Sparingly soluble in water; freely soluble in alcohol and in chloroform; insoluble in ether.

Dextrose: Colorless crystals or white, crystalline or granular powder. Is odorless, and has a sweet taste. Freely soluble in water; very soluble in boiling water; soluble in boiling alcohol; slightly soluble in alcohol. *NF category:* Sweetening agent; tonicity agent.

Dextrose Excipient: Colorless crystals or white, crystalline or granular powder. Is odorless and sweet-tasting. Freely soluble in water; very soluble in boiling water; sparingly soluble in boiling alcohol; slightly soluble in alcohol. *NF category:* Sweetening agent; tablet and/or capsule diluent.

Diacetylated Monoglycerides: Clear liquid. Very soluble in 80 percent (w/w) aqueous alcohol, in vegetable oils, and in mineral oils; sparingly soluble in 70 percent alcohol. *NF category:* Plasticizer.

Diatrizoate Meglumine: White, odorless powder. Freely soluble in water.

Diatrizoate Meglumine Injection: Clear, colorless to pale yellow, slightly viscous liquid.

Diatrizoate Meglumine and Diatrizoate Sodium Injection: Clear, colorless to pale yellow, slightly viscous liquid. May crystallize at room temperature or below.

Diatrizoate Sodium: White, odorless powder. Soluble in water; slightly soluble in alcohol; practically insoluble in acetone and in ether.

Diatrizoate Sodium Injection: Clear, colorless to pale yellow, slightly viscous liquid.

Diatrizoate Sodium Solution: Clear, pale yellow to light brown liquid.

Diatrizoic Acid: White, odorless powder. Very slightly soluble in water and in alcohol; soluble in dimethylformamide and in alkali hydroxide solutions.

Diazepam: Off-white to yellow, practically odorless, crystalline powder. Practically insoluble in water; freely soluble in chloroform; soluble in alcohol.

Diazoxide: White or cream-white crystals or crystalline powder. Practically insoluble to sparingly soluble in water and in most organic solvents; very soluble in strong alkaline solutions; freely soluble in dimethylformamide.

Dibucaine: White to off-white powder, having a slight, characteristic odor. Darkens on exposure to light. Slightly soluble in water; soluble in 1 *N* hydrochloric acid and in ether.

Dibucaine Hydrochloride: Colorless or white to off-white crystals or white to off-white, crystalline powder. Is odorless, is somewhat hygroscopic, and darkens on exposure to light. Its solutions have a pH of about 5.5. Freely soluble in water, in alcohol, in acetone, and in chloroform.

Dichlorodifluoromethane: Clear, colorless gas, having a faint, ethereal odor. Its vapor pressure at 25° is about 4880 mm of mercury (80 psig). *NF category:* Aerosol propellant.

Dichlorotetrafluoroethane: Clear, colorless gas, having a faint, ethereal odor. Its vapor pressure at 25° is about 1620 mm of mercury (17 psig). Usually contains between 6% and 10% of its isomer, $CCl_2F—CF_3$. *NF category:* Aerosol propellant.

Dicloxacillin Sodium: White to off-white, crystalline powder. Freely soluble in water.

Dicumarol: White or creamy white, crystalline powder, having a faint, pleasant odor and a slightly bitter taste. Melts at about 290°. Practically insoluble in water, in alcohol, and in ether; readily soluble in solutions of fixed alkali hydroxides; slightly soluble in chloroform.

Dicyclomine Hydrochloride: Fine, white, crystalline powder. Is practically odorless and has a very bitter taste. Soluble in water; freely soluble in alcohol and in chloroform; very slightly soluble in ether.

Dicyclomine Hydrochloride Injection: Colorless solution, which may have the odor of a preservative.

Dienestrol: Colorless, white or practically white, needle-like crystals, or white or practically white, crystalline powder. Is odorless. Practically insoluble in water; soluble in alcohol, in acetone, in ether, in methanol, in propylene glycol, and in solutions of alkali hydroxides; slightly soluble in chloroform and in fatty oils.

Diethanolamine: White or clear, colorless crystals, deliquescing in moist air; or colorless liquid. Miscible with water, with alcohol, with acetone, with chloroform, and with glycerin. Slightly soluble to insoluble in benzene, in ether, and in petroleum ether. *NF category:* Alkalizing agent; emulsifying and/or solubilizing agent.

Diethylcarbamazine Citrate: White, crystalline powder. Melts at about 136°, with decomposition. Is odorless or has a slight odor; is slightly hygroscopic. Very soluble in water; sparingly soluble in alcohol; practically insoluble in acetone, in chloroform, and in ether.

Diethyl Phthalate: Colorless, practically odorless, oily liquid. Insoluble in water. Miscible with alcohol, with ether, and with other usual organic solvents. *NF category:* Plasticizer.

Diethylpropion Hydrochloride: White to off-white, fine crystalline powder. Is odorless, or has a slight characteristic odor. It melts at about 175°, with decomposition. Freely soluble in water, in chloroform, and in alcohol; practically insoluble in ether.

Diethylstilbestrol: White, odorless, crystalline powder. Practically insoluble in water; soluble in alcohol, in chloroform, in ether, in fatty oils, and in dilute alkali hydroxides.

Diethylstilbestrol Diphosphate: Off-white, odorless, crystalline powder. Sparingly soluble in water; soluble in alcohol and in dilute alkali.

Diethylstilbestrol Diphosphate Injection: Colorless to light, straw-colored liquid.

Diethyltoluamide: Colorless liquid, having a faint, pleasant odor. Boils at about 111° under a pressure of 1 mm of mercury. Practically insoluble in water and in glycerin. Miscible with alcohol, with isopropyl alcohol, with ether, with chloroform, and with carbon disulfide.

Diflorasone Diacetate: White to pale yellow, crystalline powder. Insoluble in water; soluble in methanol and in acetone; sparingly soluble in ethyl acetate; slightly soluble in toluene; very slightly soluble in ether.

Diflunisal: White to off-white, practically odorless powder. Freely soluble in alcohol and in methanol; soluble in acetone and in ethyl acetate; slightly soluble in chloroform, in carbon tetrachloride, and in methylene chloride; insoluble in hexane and in water.

Digitoxin: White or pale buff, odorless, microcrystalline powder. Practically insoluble in water; sparingly soluble in chloroform; slightly soluble in alcohol; very slightly soluble in ether.

Digoxin: Clear to white, odorless crystals or white, odorless crystalline powder. Practically insoluble in water and in ether; freely soluble in pyridine; slightly soluble in diluted alcohol and in chloroform.

Dihydroergotamine Mesylate: White to slightly yellowish powder, or off-white to faintly red powder, having a faint odor. Slightly soluble in water and in chloroform; soluble in alcohol.

Dihydrotachysterol: Colorless or white, odorless crystals, or white, odorless, crystalline powder. Practically insoluble in water; soluble in alcohol; freely soluble in ether and in chloroform; sparingly soluble in vegetable oils.

Dihydroxyaluminum Aminoacetate: White, odorless powder having a faintly sweet taste. Insoluble in water and in organic solvents; soluble in dilute mineral acids and in solutions of fixed alkalies.

Dihydroxyaluminum Aminoacetate Magma: White, viscous suspension, from which small amounts of water may separate on standing.

Dihydroxyaluminum Sodium Carbonate: Fine, white, odorless powder. Practically insoluble in water and in organic solvents; soluble in dilute mineral acids with the evolution of carbon dioxide.

Diisopropanolamine: *NF category:* Alkalizing agent.

Diltiazem Hydrochloride: White, odorless, crystalline powder or small crystals. Freely soluble in chloroform, in formic acid, in methanol, and in water; sparingly soluble in dehydrated alcohol; insoluble in ether. Melts at about 210°, with decomposition.

Dimenhydrinate: White, crystalline, odorless powder. Slightly soluble in water; freely soluble in alcohol and in chloroform; sparingly soluble in ether.

Dimercaprol: Colorless or practically colorless liquid, having a disagreeable, mercaptan-like odor. Soluble in water, in alcohol, in benzyl benzoate, and in methanol.

Dimercaprol Injection: Yellow, viscous solution having a pungent, disagreeable odor. Specific gravity is about 0.978.

Dimethicone: A clear, colorless, and odorless liquid. Insoluble in water, in methanol, in alcohol, and in acetone; very slightly soluble in isopropyl alcohol; soluble in chlorinated hydrocarbons, in benzene, in toluene, in xylene, in *n*-hexane, in petroleum spirits, in ether, and in amyl acetate.

Dinoprost Tromethamine: White to off-white, crystalline powder. Very soluble in water; freely soluble in dimethylformamide; soluble in methanol; slightly soluble in chloroform.

Dioxybenzone: Yellow powder. Practically insoluble in water; freely soluble in alcohol and in toluene.

Diperodon: White to cream-colored powder having a characteristic odor. Insoluble in water.

Diphemanil Methylsulfate: White or nearly white, crystalline solid, having a bitter taste and a faint characteristic odor. Is stable to heat and to light, and is somewhat hygroscopic. Sparingly soluble in water, in alcohol, and in chloroform.

Diphenhydramine Hydrochloride: White, odorless, crystalline powder. Slowly darkens on exposure to light. Its solutions are practically neutral to litmus. Freely soluble in water, in alcohol, and in chloroform; sparingly soluble in acetone; very slightly soluble in benzene and in ether.

Diphenoxylate Hydrochloride: White, odorless, crystalline powder. Its saturated solution has a pH of about 3.3. Slightly soluble in water and in isopropanol; freely soluble in chloroform; soluble in methanol; sparingly soluble in alcohol and in acetone; practically insoluble in ether and in solvent hexane.

Diphtheria Antitoxin: Transparent or slightly opalescent liquid, practically colorless, and practically odorless or having an odor because of the preservative.

Diphtheria Toxin for Schick Test: Transparent liquid.

Diphtheria Toxoid: Clear, brownish yellow, or slightly turbid liquid, free from evident clumps or particles, having a faint, characteristic odor.

Diphtheria Toxoid Adsorbed: White, slightly gray, or slightly pink suspension, free from evident clumps after shaking.

Diphtheria and Tetanus Toxoids: Clear, colorless to brownish yellow or very slightly turbid liquid, free from evident clumps or particles, having a characteristic odor.

Diphtheria and Tetanus Toxoids Adsorbed: Turbid, and white, slightly gray, or slightly pink suspension, free from evident clumps after shaking.

Diphtheria and Tetanus Toxoids and Pertussis Vaccine: More or less turbid, whitish to light yellowish or brownish liquid, free from evident clumps after shaking, having a faint odor because of the toxoid components, the antimicrobial agent, or both.

Diphtheria and Tetanus Toxoids and Pertussis Vaccine Adsorbed: Markedly turbid, whitish liquid, free from evident clumps after shaking; nearly odorless or having a faint odor because of the preservative.

Dipivefrin Hydrochloride: White, crystalline powder or small crystals, having a faint odor. Very soluble in water.

Dipyridamole: Intensely yellow, crystalline powder or needles. Very soluble in methanol, in alcohol, and in chloroform; slightly soluble in water; very slightly soluble in acetone and in ethyl acetate.

Disopyramide Phosphate: White or practically white, odorless powder. Melts at about 205°, with decomposition. Freely soluble in water; slightly soluble in alcohol; practically insoluble in chloroform and in ether.

Disulfiram: White to off-white, odorless, crystalline powder. Very slightly soluble in water; soluble in acetone, in alcohol, in carbon disulfide, and in chloroform.

Docusate Calcium: White, amorphous solid, having the characteristic odor of octyl alcohol. It is free of the odor of other solvents. Very slightly soluble in water; very soluble in alcohol, in polyethylene glycol 400, and in corn oil.

Docusate Potassium: White, amorphous solid, having a characteristic odor suggestive of octyl alcohol. Sparingly soluble in water; very soluble in solvent hexane; soluble in alcohol and in glycerin.

Docusate Sodium: White, wax-like, plastic solid, having a characteristic odor suggestive of octyl alcohol, but no odor of other solvents. Sparingly soluble in water; very soluble in solvent hexane; freely soluble in alcohol and in glycerin. *NF category:* Wetting and/or solubilizing agent.

Dopamine Hydrochloride: White to off-white, crystalline powder. May have a slight odor of hydrochloric acid. Melts at about 240°, with decomposition. Freely soluble in water, in methanol, and in aqueous solutions of alkali hydroxides; insoluble in ether and in chloroform.

Doxapram Hydrochloride: White to off-white, odorless, crystalline powder. Melts at about 220°. Soluble in water and in chloroform; sparingly soluble in alcohol; practically insoluble in ether.

Doxorubicin Hydrochloride: Red-orange, hygroscopic, crystalline powder. Soluble in water, in isotonic sodium chloride solution, and in methanol; practically insoluble in chloroform, in ether, and in other organic solvents.

Doxycycline: Yellow, crystalline powder. Very slightly soluble in water; freely soluble in dilute acid and in alkali hydroxide solutions; sparingly soluble in alcohol; practically insoluble in chloroform and in ether.

Doxycycline Hyclate: Yellow, crystalline powder. Soluble in water and in solutions of alkali hydroxides and carbonates; slightly soluble in alcohol; practically insoluble in chloroform and in ether.

Doxylamine Succinate: White or creamy white powder, having a characteristic odor. Very soluble in water and in alcohol; freely soluble in chloroform; very slightly soluble in ether and in benzene.

Droperidol: White to light tan, amorphous or microcrystalline powder. Practically insoluble in water; freely soluble in chloroform; slightly soluble in alcohol and in ether.

Absorbable Dusting Powder: White, odorless powder.

Dyclonine Hydrochloride: White crystals or white crystalline powder, which may have a slight odor. Exhibits local anesthetic properties when placed upon the tongue. Soluble in water, in acetone, in alcohol, and in chloroform.

Dydrogesterone: White to pale yellow, crystalline powder. Practically insoluble in water; sparingly soluble in alcohol.

Dyphylline: White, odorless, extremely bitter, amorphous or crystalline solid. Freely soluble in water; sparingly soluble in alcohol and in chloroform; practically insoluble in ether.

Echothiophate Iodide: White, crystalline, hygroscopic solid having a slight mercaptan-like odor. Its solutions have a pH of about 4. Freely soluble in water and in methanol; soluble in dehydrated alcohol; practically insoluble in other organic solvents.

Echothiophate Iodide for Ophthalmic Solution: White, amorphous powder.

Edetate Calcium Disodium: White, crystalline granules or white, crystalline powder. Is odorless, is slightly hygroscopic, and has a faint, saline taste. Is stable in air. Freely soluble in water.

Edetate Disodium: White, crystalline powder. Soluble in water. *NF category:* Chelating agent.

Edetic Acid: White, crystalline powder. Melts above 220°, with decomposition. Very slightly soluble in water; soluble in solutions of alkali hydroxides. *NF category:* Chelating agent.

Edrophonium Chloride: White, odorless, crystalline powder. Its solution (1 in 10) is practically colorless. Very soluble in water; freely soluble in alcohol; insoluble in chloroform and in ether.

Emetine Hydrochloride: White or very slightly yellowish, odorless, crystalline powder. Is affected by light. Freely soluble in water and in alcohol.

Enalapril Maleate: Off-white, crystalline powder. Melts at about 144°. Practically insoluble in nonpolar organic solvents; slightly soluble in semipolar organic solvents; sparingly soluble in water; soluble in alcohol; freely soluble in methanol and in dimethylformamide.

Enflurane: Clear, colorless, stable, volatile liquid, having a mild, sweet odor. Is non-flammable. Slightly soluble in water. Miscible with organic solvents, fats, and oils.

Ephedrine: Unctuous, practically colorless solid or white crystals or granules. Gradually decomposes on exposure to light. Melts between 33° and 40°, the variability in the melting point being the result of differences in the moisture content, anhydrous Ephedrine having a lower melting point than the hemihydrate of Ephedrine. Its solutions are alkaline to litmus. Soluble in water, in alcohol, in chloroform, and in ether; moderately and slowly soluble in mineral oil, the solution becoming turbid if the Ephedrine contains more than about 1% of water.

Ephedrine Hydrochloride: Fine, white, odorless crystals or powder. Is affected by light. Freely soluble in water; soluble in alcohol; insoluble in ether.

Ephedrine Sulfate: Fine, white, odorless crystals or powder. Darkens on exposure to light. Freely soluble in water; sparingly soluble in alcohol.

Ephedrine Sulfate Nasal Solution: Clear, colorless solution. Is neutral or slightly acid to litmus.

Epinephrine: White to practically white, odorless, microcrystalline powder or granules, gradually darkening on exposure to light and air. With acids, it forms salts that are readily soluble in water, and the base may be recovered by the addition of ammonia water or alkali carbonates. Its solutions are alkaline to litmus. Very slightly soluble in water and in alcohol; insoluble in ether, in chloroform, and in fixed and volatile oils.

Epinephrine Injection: Practically colorless, slightly acid liquid. Gradually turns dark on exposure to light and to air.

Epinephrine Inhalation Solution: Practically colorless, slightly acid liquid. Gradually turns dark on exposure to light and air.

Epinephrine Nasal Solution: Nearly colorless, slightly acid liquid. Gradually turns dark on exposure to light and air.

Epinephrine Ophthalmic Solution: Colorless to faint yellow solution. Gradually turns dark on exposure to light and air.

Epinephrine Bitartrate: White, or grayish white or light brownish gray, odorless, crystalline powder. Slowly darkens on exposure to air and light. Its solutions are acid to litmus, having a pH of about 3.5. Freely soluble in water; slightly soluble in alcohol; practically insoluble in chloroform and in ether.

Epinephrine Bitartrate for Ophthalmic Solution: White to off-white solid.

Epinephryl Borate Ophthalmic Solution: Clear, pale yellow liquid, gradually darkening on exposure to light and to air.

Ergocalciferol: White, odorless crystals. Is affected by air and by light. Insoluble in water; soluble in alcohol, in chloroform, in ether, and in fatty oils.

Ergocalciferol Oral Solution: Clear liquid having the characteristics of the solvent used in preparing the Solution.

Ergoloid Mesylates: White to off-white, microcrystalline or amorphous, practically odorless powder. Slightly soluble in water; soluble in methanol and in alcohol; sparingly soluble in acetone.

Ergonovine Maleate: White to grayish white or faintly yellow, odorless, microcrystalline powder. Darkens with age and on exposure to light. Sparingly soluble in water; slightly soluble in alcohol; insoluble in ether and in chloroform.

Ergotamine Tartrate: Colorless crystals or white to yellowish white, crystalline powder. Is odorless. Melts at about 180°, with decomposition. One g dissolves in about 3200 mL of water; in the presence of a slight excess of tartaric acid one g dissolves in about 500 mL of water. Slightly soluble in alcohol.

Diluted Erythrityl Tetranitrate: White powder, having a slight odor of nitric oxides. Undiluted erythrityl tetranitrate is practically insoluble in water; soluble in acetone, in acetonitrile, and in alcohol.

Erythromycin: White or slightly yellow, crystalline powder. Is odorless or practically odorless. Slightly soluble in water; soluble in alcohol, in chloroform, and in ether.

Erythromycin Estolate: White, crystalline powder. Is odorless or practically odorless, and is practically tasteless. Soluble in alcohol, in acetone, and in chloroform; practically insoluble in water.

Erythromycin Ethylsuccinate: White or slightly yellow crystalline powder. Is odorless or practically odorless, and is practically tasteless. Very slightly soluble in water; freely soluble in alcohol, in chloroform, and in polyethylene glycol 400.

Sterile Erythromycin Gluceptate: White powder. Is odorless or practically odorless, and is slightly hygroscopic. Its solution (1 in 20) is neutral or slightly acid. Freely soluble in water, in alcohol, and in methanol; slightly soluble in acetone and in chloroform; practically insoluble in ether.

Erythromycin Lactobionate for Injection: White or slightly yellow crystals or powder, having a faint odor. Its solution (1 in 20) is neutral or slightly alkaline. Freely soluble in water, in alcohol, and in methanol; slightly soluble in acetone and in chloroform; practically insoluble in ether.

Erythromycin Stearate: White or slightly yellow crystals or powder. Is odorless or may have a slight, earthy odor, and has a slightly bitter taste. Practically insoluble in water; soluble in alcohol, in chloroform, in methanol, and in ether.

Erythrosine Sodium: Red or brownish red, odorless powder. Dissolves in water to form a bluish red solution that shows no fluorescence in ordinary light. Is hygroscopic. Soluble in water, in glycerin, and in propylene glycol; sparingly soluble in alcohol; insoluble in fats and in oils.

Estradiol: White or creamy white, small crystals or crystalline powder. Is odorless, and is stable in air. Is hygroscopic. Practically insoluble in water; soluble in alcohol, in acetone, in dioxane, in chloroform, and in solutions of fixed alkali hydroxides; sparingly soluble in vegetable oils.

Estradiol Cypionate: White to practically white, crystalline powder. Is odorless or has a slight odor. Insoluble in water; soluble in alcohol, in acetone, in chloroform, and in dioxane; sparingly soluble in vegetable oils.

Estradiol Valerate: White, crystalline powder. Is usually odorless but may have a faint, fatty odor. Practically insoluble in water; soluble in castor oil, in methanol, in benzyl benzoate, and in dioxane; sparingly soluble in sesame oil and in peanut oil.

Estriol: White to practically white, odorless, crystalline powder. Melts at about 280°. Insoluble in water; sparingly soluble in alcohol; soluble in acetone, in chloroform, in dioxane, in ether, and in vegetable oils.

Conjugated Estrogens: Conjugated Estrogens obtained from natural sources is a buff-colored, amorphous powder, odorless or having a slight, characteristic odor. The synthetic form is a white to light buff, crystalline or amorphous powder, odorless or having a slight odor.

Esterified Estrogens: White or buff-colored, amorphous powder, odorless or having a slight, characteristic odor.

Estrone: Small, white crystals or white to creamy white, crystalline powder. Is odorless, and is stable in air. Melts at about 260°. Practically insoluble in water; soluble in alcohol, in acetone, in dioxane, and in vegetable oils; slightly soluble in solutions of fixed alkali hydroxides.

Estropipate: White to yellowish white, fine crystalline powder. Is odorless, or may have a slight odor. Melts at about 190° to a light brown, viscous liquid which solidifies, on further heating, and finally melts at about 245°, with decomposition. Very slightly soluble in water, in alcohol, in chloroform, and in ether; soluble in warm water.

Ethacrynic Acid: White or practically white, odorless or practically odorless, crystalline powder. Very slightly soluble in water; freely soluble in alcohol, in chloroform, and in ether.

Ethambutol Hydrochloride: White, crystalline powder. Freely soluble in water; soluble in alcohol and in methanol; slightly soluble in ether and in chloroform.

Ethchlorvynol: Colorless to yellow, slightly viscous liquid, having a characteristic pungent odor. Darkens on exposure to light and to air. Immiscible with water; miscible with most organic solvents.

Ether: Colorless, mobile, volatile liquid, having a characteristic odor. Is slowly oxidized by the action of air and light, with the formation of peroxides. It boils at about 35°. Soluble in water. Miscible with alcohol, with benzene, with chloroform, with solvent hexane, and with fixed and volatile oils.

Ethinamate: White, essentially odorless powder. Its saturated aqueous solution has a pH of about 6.5. Slightly soluble in water; freely soluble in alcohol, in chloroform, and in ether.

Ethinyl Estradiol: White to creamy white, odorless, crystalline powder. Insoluble in water; soluble in alcohol, in chloroform, in ether, in vegetable oils, and in solutions of fixed alkali hydroxides.

Ethiodized Oil Injection: Straw-colored to amber-colored, oily liquid. It may possess an alliaceous odor. Insoluble in water; soluble in acetone, in chloroform, in ether, and in solvent hexane.

Ethionamide: Bright yellow powder, having a faint to moderate sulfide-like odor. Slightly soluble in water, in chloroform, and in ether; soluble in methanol; sparingly soluble in alcohol and in propylene glycol.

Ethopropazine Hydrochloride: White or slightly off-white, odorless, crystalline powder. Melts at about 210°, with decomposition. Soluble in water at 40°; slightly soluble in water at 20°; soluble in alcohol and in chloroform; sparingly soluble in acetone; insoluble in ether and in benzene.

Ethosuximide: White to off-white, crystalline powder or waxy solid, having a characteristic odor. Freely soluble in water and in chloroform; very soluble in alcohol and in ether; very slightly soluble in solvent hexane.

Ethotoin: White, crystalline powder. Insoluble in water; freely soluble in dehydrated alcohol and in chloroform; soluble in ether.

Ethyl Acetate: Transparent, colorless liquid, having a fragrant, refreshing, slightly acetous odor, and a peculiar, acetous, burning taste. Soluble in water. Miscible with alcohol, with ether, with fixed oils, and with volatile oils. *NF category:* Flavors and perfumes.

Ethyl Chloride: Colorless, mobile, very volatile liquid at low temperatures or under pressure, having a characteristic, ethereal odor. It boils between 12° and 13°, and its specific gravity at 0° is about 0.921. When liberated at room temperature from its sealed container, it vaporizes immediately. It burns with a smoky, greenish flame, producing hydrogen chloride. Slightly soluble in water; freely soluble in alcohol and in ether.

Ethyl Oleate: Mobile, practically colorless liquid, having an agreeable taste. Insoluble in water. Miscible with vegetable oils, with mineral oil, with alcohol, and with most organic solvents. *NF category:* Vehicle.

Ethyl Vanillin: Fine, white or slightly yellowish crystals. Its taste and odor are similar to the taste and odor of vanillin. Affected by light. Its solutions are acid to litmus. Sparingly soluble in water at 50°; freely soluble in alcohol, in chloroform, in ether, and in solutions of alkali hydroxides. *NF category:* Flavors and perfumes.

Ethylcellulose: Free-flowing, white to light tan powder. It forms films that have a refractive index of about 1.47. Its aqueous suspensions are neutral to litmus. Insoluble in water, in glycerin, and in propylene glycol. Ethylcellulose containing less than 46.5 percent of ethoxy groups is freely soluble in tetrahydrofuran, in methyl acetate, in chloroform, and in mixtures of aromatic hydrocarbons with alcohol. Ethylcellulose containing not less than 46.5 percent of ethoxy groups is freely soluble in alcohol, in methanol, in toluene, in chloroform, and in ethyl acetate. *NF category:* Coating agent; tablet binder.

Ethylenediamine: Clear, colorless or only slightly yellow liquid, having an ammonia-like odor and a strong alkaline reaction. Miscible with water and with alcohol.

Ethylnorepinephrine Hydrochloride: White to practically white, crystalline powder, which gradually darkens on exposure to light. Melts at about 190°, with decomposition. Soluble in water and in alcohol; practically insoluble in ether.

Ethylparaben: Small, colorless crystals or white powder. Slightly soluble in water and in glycerin; freely soluble in acetone, in alcohol, in ether, and in propylene glycol. *NF category:* Antimicrobial preservative.

Ethynodiol Diacetate: White, odorless, crystalline powder. Is stable in air. Insoluble in water; very soluble in chloroform; freely soluble in ether; soluble in alcohol; sparingly soluble in fixed oils.

Eucatropine Hydrochloride: White, granular, odorless powder. Its solutions are neutral to litmus. Very soluble in water; freely soluble in alcohol and in chloroform; insoluble in ether.

Eugenol: Colorless or pale yellow liquid, having a strongly aromatic odor of clove and a pungent, spicy taste. Upon exposure to air, it darkens and thickens. Is optically inactive. Slightly soluble in water. Miscible with alcohol, with chloroform, with ether, and with fixed oils.

Evans Blue: Green, bluish green, or brown, odorless powder. Very soluble in water; very slightly soluble in alcohol; practically insoluble in benzene, in carbon tetrachloride, in chloroform, and in ether.

Famotidine: White to pale yellowish-white crystalline powder. Sensitive to light. Freely soluble in dimethylformamide and in glacial acetic acid; slightly soluble in methanol; very slightly soluble in water; practically insoluble in acetone, in alcohol, in chloroform, in ether, and in ethyl acetate.

Hard Fat: White, brittle mass, practically odorless and free from rancid odor; greasy to the touch. Practically insoluble in water; freely soluble in ether; slightly soluble in alcohol.

Fenoprofen Calcium: White, crystalline powder. Slightly soluble in *n*-hexanol, in methanol, and in water; practically insoluble in chloroform.

Fentanyl Citrate: White, crystalline powder or white, glistening crystals. Sparingly soluble in water; soluble in methanol; slightly soluble in chloroform.

Ferric Oxide: Powder exhibiting three basic colors (red, yellow, and black) or other shades produced on blending the basic colors. Insoluble in water and in organic solvents; dissolves in hydrochloric acid upon warming, a small amount of insoluble residue usually remaining. *NF category:* Color.

Ferrous Fumarate: Reddish orange to red-brown, odorless powder. May contain soft lumps that produce a yellow streak when crushed. Slightly soluble in water; very slightly soluble in alcohol. Its solubility in dilute hydrochloric acid is limited by the separation of fumaric acid.

Ferrous Gluconate: Yellowish gray or pale greenish yellow, fine powder or granules, having a slight odor resembling that of burned sugar. Its solution (1 in 20) is acid to litmus. Soluble in water, with slight heating; practically insoluble in alcohol.

Ferrous Sulfate: Pale, bluish green crystals or granules. Is odorless, has a saline, styptic taste, and is efflorescent in dry air. Oxidizes readily in moist air to form brownish yellow basic ferric sulfate. Its solution (1 in 10) is acid to litmus, having a pH of about 3.7. Freely soluble in water; very soluble in boiling water; insoluble in alcohol.

Dried Ferrous Sulfate: Grayish white to buff-colored powder, consisting primarily of $FeSO_4 . H_2O$ with varying amounts of $FeSO_4 . 4H_2O$. Slowly soluble in water; insoluble in alcohol.

Flucytosine: White to off-white, crystalline powder. Is odorless or has a slight odor. Sparingly soluble in water; slightly soluble in alcohol; practically insoluble in chloroform and in ether.

Fludrocortisone Acetate: White to pale yellow crystals or crystalline powder. Is odorless or practically odorless. Is hygroscopic. Insoluble in water; slightly soluble in ether; sparingly soluble in alcohol and in chloroform.

Flumethasone Pivalate: White to off-white, crystalline powder. Insoluble in water; slightly soluble in methanol; very slightly soluble in chloroform and in methylene chloride.

Flunisolide: White to creamy-white, crystalline powder. Melts at about 245°, with decomposition. Practically insoluble in water; soluble in acetone; sparingly soluble in chloroform; slightly soluble in methanol.

Fluocinolone Acetonide: White or practically white, odorless, crystalline powder. Is stable in air. Melts at about 270°, with decomposition. Insoluble in water; soluble in methanol; slightly soluble in ether and in chloroform.

Fluocinonide: White to cream-colored, crystalline powder, having not more than a slight odor. Practically insoluble in water; sparingly soluble in acetone and in chloroform; slightly soluble in alcohol, in methanol, and in dioxane; very slightly soluble in ether.

Fluorescein: Yellowish red to red, odorless powder. Insoluble in water; soluble in dilute alkali hydroxides.

Fluorescein Sodium: Orange-red, hygroscopic, odorless powder. Freely soluble in water; sparingly soluble in alcohol.

Fluorescein Sodium Ophthalmic Strip: Each Strip is a dry, white piece of paper, one end of which is rounded and is uniformly orange-red in color because of the fluorescein sodium impregnated in the paper.

Fluorometholone: White to yellowish white, odorless, crystalline powder. Melts at about 280°, with some decomposition. Practically insoluble in water; slightly soluble in alcohol; very slightly soluble in chloroform and in ether.

Fluorouracil: White to practically white, practically odorless, crystalline powder. Decomposes at about 282°. Sparingly soluble in water; slightly soluble in alcohol; practically insoluble in chloroform and in ether.

Fluoxymesterone: White or practically white, odorless, crystalline powder. Melts at about 240°, with some decomposition. Practically insoluble in water; sparingly soluble in alcohol; slightly soluble in chloroform.

Fluphenazine Enanthate: Pale yellow to yellow-orange, clear to slightly turbid, viscous liquid, having a characteristic odor. Is unstable in strong light, but stable to air at room temperature. Insoluble in water; freely soluble in alcohol, in chloroform, and in ether.

Fluphenazine Hydrochloride: White or nearly white, odorless, crystalline powder. Melts, within a range of 5°, at a temperature above 225°. Freely soluble in water; slightly soluble in acetone, in alcohol, and in chloroform; practically insoluble in benzene and in ether.

Flurandrenolide: White to off-white, fluffy, crystalline powder. Is odorless. Practically insoluble in water and in ether; freely soluble in chloroform; soluble in methanol; sparingly soluble in alcohol.

Flurazepam Hydrochloride: Off-white to yellow, crystalline powder. Is odorless, or has a slight odor, and its solutions are acid to litmus. Melts at about 212°, with decomposition. Freely soluble in water and in alcohol; slightly soluble in isopropyl alcohol and in chloroform.

Folic Acid: Yellow, yellow-brownish, or yellowish orange, odorless, crystalline powder. Very slightly soluble in water; insoluble in alcohol, in acetone, in chloroform, and in ether. It readily dissolves in dilute solutions of alkali hydroxides and carbonates, and is soluble in hot, 3 *N* hydrochloric acid and in hot, 2 *N* sulfuric acid. Soluble in hydrochloric acid and in sulfuric acid, yielding very pale yellow solutions.

Folic Acid Injection: Clear, yellow to orange-yellow, alkaline liquid.

Formaldehyde Solution: Clear, colorless or practically colorless liquid, having a pungent odor. The vapor from it irritates the mucous membrane of the throat and nose. On long standing, especially in the cold, it may become cloudy because of the separation of paraformaldehyde. This cloudiness disappears when the solution is warmed. Miscible with water and with alcohol.

Fructose: Colorless crystals or as a white, crystalline powder. Is odorless, and has a sweet taste. Freely soluble in water; soluble in alcohol and in methanol. *NF category:* Sweetening agent; tablet and/or capsule diluent.

Basic Fuchsin: Dark green powder or greenish glistening crystalline fragments, having a bronze-like luster and not more than a faint odor. Soluble in water, in alcohol, and in amyl alcohol; insoluble in ether.

Fumaric Acid: White, odorless granules or crystalline powder. Soluble in alcohol; slightly soluble in water and in ether; very slightly soluble in chloroform. *NF category:* Acidifying agent.

Furazolidone: Yellow, odorless, crystalline powder. Is tasteless at first, then a bitter aftertaste develops. Practically insoluble in water, in alcohol, and in carbon tetrachloride.

Furosemide: White to slightly yellow, odorless, crystalline powder. Practically insoluble in water; freely soluble in acetone, in dimethylformamide, and in solutions of alkali hydroxides; soluble in methanol; sparingly soluble in alcohol; slightly soluble in ether; very slightly soluble in chloroform.

Furosemide Injection: Clear, colorless solution.

Gallamine Triethiodide: White, odorless, amorphous powder. Is hygroscopic. Very soluble in water; sparingly soluble in alcohol; very slightly soluble in chloroform.

Petrolatum Gauze: The petrolatum recovered by draining in the *Assay* is a white or faintly yellowish, unctuous mass, transparent in thin layers even after cooling to 0°.

Gelatin: Sheets, flakes, or shreds, or coarse to fine powder. Is faintly yellow or amber in color, the color varying in depth according to the particle size. Has a slight, characteristic bouillon-like odor in solution. Is stable in air when dry, but is subject to microbic decomposition when moist or in solution. Gelatin has any suitable strength that is designated by Bloom Gelometer number (see *Gel Strength of Gelatin* ⟨1081⟩). Type A Gelatin exhibits an isoelectric point between pH 7 and pH 9, and Type B Gelatin exhibits an isoelectric point between pH 4.7 and pH 5.2. Insoluble in cold water, but swells and softens when immersed in it, gradually absorbing from 5 to 10 times its own weight of water. Soluble in hot water, in 6 *N* acetic acid, and in a hot mixture of glycerin and water. Insoluble in alcohol, in chloroform, in ether, and in fixed and volatile oils. *NF category:* Coating agent; suspending and/or viscosity-increasing agent; tablet binder.

Absorbable Gelatin Film: Light amber, transparent, pliable film which becomes rubbery when moistened. Insoluble in water.

Absorbable Gelatin Sponge: Light, nearly white, nonelastic, tough, porous, hydrophilic solid. Insoluble in water.

Gemfibrozil: White, waxy, crystalline solid. Very slightly soluble in water; soluble in alcohol, in methanol, and in chloroform.

Gentamicin Sulfate: White to buff powder. Soluble in water; insoluble in alcohol, in acetone, and in benzene.

Gentamicin Sulfate Injection: Clear, slightly yellow solution, having a faint odor.

Gentian Violet: Dark green powder or greenish, glistening pieces having a metallic luster, and having not more than a faint odor. Sparingly soluble in water; soluble in alcohol, in glycerin, and in chloroform; insoluble in ether.

Gentian Violet Cream: Dark purple, water-washable cream.

Gentian Violet Topical Solution: Purple liquid, having a slight odor of alcohol. A dilution (1 in 100), viewed downward through 1 cm of depth, is deep purple in color.

Glaze, Pharmaceutical: *NF category:* Coating agent.

Gentisic Acid Ethanolamide: White to tan powder. Sparingly soluble in water; freely soluble in acetone, in methanol, and in alcohol; very slightly soluble in ether; practically insoluble in chloroform. Melts at about 149°.

Immune Globulin: Transparent or slightly opalescent liquid, either colorless or of a brownish color due to denatured hemoglobin. Is practically odorless. May develop a slight, granular deposit during storage.

Rh_0 (D) Immune Globulin: Transparent or slightly opalescent liquid. Is practically colorless and practically odorless. May develop a slight, granular deposit during storage.

Glucagon: Fine, white or faintly colored, crystalline powder. Is practically odorless and tasteless. Soluble in dilute alkali and acid solutions; insoluble in most organic solvents.

Glucagon for Injection: White, odorless powder.

Liquid Glucose: Colorless or yellowish, thick, syrupy liquid. Odorless or nearly odorless, and has a sweet taste. Miscible with water; sparingly soluble in alcohol. *NF category:* Tablet binder.

Glutaral Concentrate: Clear, colorless or faintly yellow liquid, having a characteristic, irritating odor.

Glutethimide: White, crystalline powder. Its saturated solution is acid to litmus. Practically insoluble in water; freely soluble in ethyl acetate, in acetone, in ether, and in chloroform; soluble in alcohol and in methanol.

Glycerin: Clear, colorless, syrupy liquid, having a sweet taste. Has not more than a slight characteristic odor, which is neither harsh nor disagreeable. Is hygroscopic. Its solutions are neutral to litmus. Miscible with water and with alcohol. Insoluble in chloroform, in ether, and in fixed and volatile oils. *NF category:* Humectant; plasticizer; solvent; tonicity agent.

Glyceryl Behenate: Fine powder, having a faint odor. Melts at about 70°. Practically insoluble in water and in alcohol; soluble in chloroform.

Glyceryl Monostearate: White, wax-like solid or as white, wax-like beads or flakes. Slight, agreeable, fatty odor and taste. Is affected by light. Dissolves in hot organic solvents such as alcohol, minerals or fixed oils, benzene, ether, and acetone. Insoluble in water, but it may be dispersed in hot water with the aid of a small amount of soap or other suitable surface-active agent. *NF category:* Emulsifying and/or solubilizing agent.

Glycine: White, odorless, crystalline powder, having a sweetish taste. Its solutions are acid to litmus. Freely soluble in water; very slightly soluble in alcohol and in ether.

Glycopyrrolate: White, odorless, crystalline powder. Soluble in water and in alcohol; practically insoluble in chloroform and in ether.

Chorionic Gonadotropin: White or practically white, amorphous powder. Freely soluble in water.

Chorionic Gonadotropin for Injection: White or practically white, amorphous solid having the characteristic appearance of substances prepared by freeze-drying.

Gramicidin: White or practically white, odorless, crystalline powder. Insoluble in water; soluble in alcohol.

Green Soap: Soft, unctuous, yellowish white to brownish or greenish yellow, transparent to translucent mass. Has a slight, characteristic odor, often suggesting the oil from which it was prepared. Its solution (1 in 20) is alkaline to bromothymol blue TS.

Griseofulvin: White to creamy white, odorless powder, in which particles of the order of 4 μm in diameter predominate. Very slightly soluble in water; soluble in acetone, in dimethylformamide, and in chloroform; sparingly soluble in alcohol.

Guaifenesin: White to slightly gray, crystalline powder, having a bitter taste. May have a slight characteristic odor. Soluble in water, in alcohol, in chloroform, in glycerin, and in propylene glycol.

Guanabenz Acetate: White or almost white powder having not more than a slight odor. Sparingly soluble in water and in 0.1 *N* hydrochloric acid; soluble in alcohol and in propylene glycol.

Guanadrel Sulfate: White to off-white, crystalline powder. Melts at about 235°, with decomposition. Soluble in water; sparingly soluble in methanol; slightly soluble in alcohol and in acetone.

Guanethidine Monosulfate: White to off-white, crystalline powder. Very soluble in water; sparingly soluble in alcohol; practically insoluble in chloroform.

Guar Gum: White to yellowish white, practically odorless powder. Dispersible in hot or cold water, forming a colloidal solution. *NF category:* Suspending and/or viscosity-increasing agent; tablet binder.

Gutta Percha: Lumps or blocks of variable size; externally brown or grayish brown to grayish white in color; internally reddish yellow or reddish gray and having a laminated or fibrous appearance. Is flexible but only slightly elastic. Has a slight, characteristic odor and a slight taste. Insoluble in water; about 90% soluble in chloroform; partly soluble in benzene, in carbon disulfide, and in turpentine oil.

Halazepam: Fine, white to light cream-colored powder. Melts at about 165°. Freely soluble in chloroform; soluble in methanol; very slightly soluble in water.

Halazone: White, crystalline powder, having a characteristic chlorine-like odor. Is affected by light. Melts at about 194°, with decomposition. Very slightly soluble in water and in chloroform; soluble in glacial acetic acid. Dissolves in solutions of alkali hydroxides and carbonates with the formation of a salt.

Halazone Tablets for Solution: Soluble in water.

Halcinonide: White to off-white, odorless, crystalline powder. Soluble in acetone and in chloroform; slightly soluble in alcohol and in ethyl ether; insoluble in water and in hexanes.

Haloperidol: White to faintly yellowish, amorphous or microcrystalline powder. Its saturated solution is neutral to litmus. Practically insoluble in water; soluble in chloroform; sparingly soluble in alcohol; slightly soluble in ether.

Halothane: Colorless, mobile, nonflammable, heavy liquid, having a characteristic odor resembling that of chloroform. Its taste is sweet and produces a burning sensation. Slightly soluble in water. Miscible with alcohol, with chloroform, with ether, and with fixed oils.

Helium: Colorless, odorless, tasteless gas, which is not combustible and does not support combustion. Very slightly soluble in water. At 0° and at a pressure of 760 mm of mercury, 1000 mL of the gas weighs about 180 mg.

Heparin Sodium: White or pale-colored, amorphous powder. Is odorless or practically so, and is hygroscopic. Soluble in water.

Hetacillin: White to off-white, crystalline powder. Practically insoluble in water and in most organic solvents; soluble in dilute sodium hydroxide solution and in methanol.

Hetacillin Potassium: White to light buff, crystalline powder. Freely soluble in water; soluble in alcohol.

Hexachlorophene: White to light tan, crystalline powder. Is odorless or has only a slight, phenolic odor. Insoluble in water; freely soluble in acetone, in alcohol, and in ether; soluble in chloroform and in dilute solutions of fixed alkali hydroxides.

Hexachlorophene Liquid Soap: Clear, amber-colored liquid, having a slight, characteristic odor. Its solution (1 in 20) is clear and has an alkaline reaction.

Hexylcaine Hydrochloride: White powder, having a bitter taste and not more than a slight aromatic odor. Soluble in water;

freely soluble in alcohol and in chloroform; practically insoluble in ether.

Hexylcaine Hydrochloride Topical Solution: Clear, colorless solution of Hexylcaine Hydrochloride in water.

Hexylene Glycol: Clear, colorless, viscous liquid. Absorbs moisture when exposed to moist air. Miscible with water and with many organic solvents, including alcohol, ether, chloroform, acetone, and hexanes.

Histamine Phosphate: Colorless, odorless, long prismatic crystals. Is stable in air but is affected by light. Its solutions are acid to litmus. Freely soluble in water.

Histidine: White, odorless crystals, having a slightly bitter taste. Soluble in water; very slightly soluble in alcohol; insoluble in ether.

Histoplasmin: Clear, red liquid. Miscible with water.

Homatropine Hydrobromide: White crystals, or white, crystalline powder. Is affected by light. Freely soluble in water; sparingly soluble in alcohol; slightly soluble in chloroform; insoluble in ether.

Homatropine Methylbromide: White, odorless powder. Slowly darkens on exposure to light. Melts at about 190°. Very soluble in water; freely soluble in alcohol and in acetone containing about 20% of water; practically insoluble in ether and in acetone.

Hydralazine Hydrochloride: White to off-white, odorless, crystalline powder. Melts at about 275°, with decomposition. Soluble in water; slightly soluble in alcohol; very slightly soluble in ether.

Hydrochloric Acid: Colorless, fuming liquid having a pungent odor. It ceases to fume when it is diluted with 2 volumes of water. Specific gravity is about 1.18. *NF category:* Acidifying agent.

Diluted Hydrochloric Acid: Colorless, odorless liquid. Specific gravity is about 1.05. *NF category:* Acidifying agent.

Hydrochlorothiazide: White, or practically white, practically odorless, crystalline powder. Slightly soluble in water; freely soluble in sodium hydroxide solution, in *n*-butylamine, and in dimethylformamide; sparingly soluble in methanol; insoluble in ether, in chloroform, and in dilute mineral acids.

Hydrocodone Bitartrate: Fine, white crystals or a crystalline powder. Is affected by light. Soluble in water; slightly soluble in alcohol; insoluble in ether and in chloroform.

Hydrocortisone: White to practically white, odorless, crystalline powder. Melts at about 215°, with decomposition. Very slightly soluble in water and in ether; sparingly soluble in acetone and in alcohol; slightly soluble in chloroform.

Hydrocortisone Acetate: White to practically white, odorless, crystalline powder. Melts at about 200°, with decomposition. Insoluble in water; slightly soluble in alcohol and in chloroform.

Hydrocortisone Butyrate: White to practically white, practically odorless, crystalline powder. Practically insoluble in water; slightly soluble in ether; soluble in methanol, in alcohol, and in acetone; freely soluble in chloroform.

Hydrocortisone Cypionate: White to practically white crystalline powder. Is odorless, or has a slight odor. Insoluble in water; very soluble in chloroform; soluble in alcohol; slightly soluble in ether.

Hydrocortisone Sodium Phosphate: White to light yellow, odorless or practically odorless, powder. Is exceedingly hygroscopic. Freely soluble in water; slightly soluble in alcohol; practically insoluble in chloroform, in dioxane, and in ether.

Hydrocortisone Sodium Succinate: White or nearly white, odorless, hygroscopic, amorphous solid. Very soluble in water and in alcohol; very slightly soluble in acetone; insoluble in chloroform.

Hydroflumethiazide: White to cream-colored, finely divided, odorless, crystalline powder. Very slightly soluble in water; freely soluble in acetone; soluble in alcohol.

Hydrogen Peroxide Concentrate: Clear, colorless liquid. Is acid to litmus. Slowly decomposes, and is affected by light.

Hydrogen Peroxide Solution: Clear, colorless liquid, odorless, or having an odor resembling that of ozone. Is acid to litmus and to the taste and produces a froth in the mouth. Rapidly decomposes when in contact with many oxidizing as well as reducing substances. When rapidly heated, it may decompose suddenly. Is affected by light. Specific gravity is about 1.01.

Hydromorphone Hydrochloride: Fine, white or practically white, odorless, crystalline powder. Is affected by light. Freely soluble in water; sparingly soluble in alcohol; practically insoluble in ether.

Hydroquinone: Fine white needles. Darkens upon exposure to light and to air. Freely soluble in water, in alcohol, and in ether.

Hydroxocobalamin: Dark red crystals or red crystalline powder. Is odorless, or has not more than a slight acetone odor. The anhydrous form is very hygroscopic. Sparingly soluble in water, in alcohol, and in methanol; practically insoluble in acetone, in ether, in chloroform, and in benzene.

Hydroxyamphetamine Hydrobromide: White, crystalline powder. Its solutions are slightly acid to litmus, having a pH of about 5. Freely soluble in water and in alcohol; slightly soluble in chloroform; practically insoluble in ether.

Hydroxychloroquine Sulfate: White or practically white, crystalline powder. Is odorless, and has a bitter taste. Its solutions have a pH of about 4.5. Exists in two forms, the usual form melting at about 240° and the other form melting at about 198°. Freely soluble in water; practically insoluble in alcohol, in chloroform, and in ether.

Hydroxyethyl Cellulose: White to light tan, practically odorless and tasteless, hygroscopic powder. Soluble in hot water and in cold water, giving a colloidal solution; practically insoluble in alcohol and in most organic solvents. *NF category:* Suspending and/or viscosity-increasing agent.

Hydroxyprogesterone Caproate: White or creamy white, crystalline powder. Is odorless or has a slight odor. Insoluble in water; soluble in ether; slightly soluble in benzene.

Hydroxypropyl Cellulose: White to cream-colored, practically odorless and tasteless, granular solid or powder. Is hygroscopic after drying. Soluble in cold water, in alcohol, in chloroform, and in propylene glycol, giving a colloidal solution; insoluble in hot water. *NF category:* Coating agent; suspending and/ or viscosity-increasing agent.

Hydroxypropyl Methylcellulose 2208: White to slightly off-white, fibrous or granular powder. Swells in water and produces a clear to opalescent, viscous, colloidal mixture. Insoluble in anhydrous alcohol, in ether, and in chloroform. *NF category:* Coating agent; suspending and/or viscosity-increasing agent; tablet binder.

Hydroxypropyl Methylcellulose 2906: White to slightly off-white, fibrous or granular powder. Swells in water and produces a clear to opalescent, viscous, colloidal mixture. Insoluble in anhydrous alcohol, in ether, and in chloroform. *NF category:* Coating agent; suspending and/or viscosity-increasing agent; tablet binder.

Hydroxypropyl Methylcellulose 2910: White to slightly off-white, fibrous or granular powder. Swells in water and produces a clear to opalescent, viscous, colloidal mixture. Insoluble in dehydrated alcohol, in ether, and in chloroform. *NF category:* Coating agent; suspending and/or viscosity-increasing agent; tablet binder.

Hydroxypropyl Methylcellulose Phthalate: White powder or granules. Is odorless and tasteless. Practically insoluble in water, in dehydrated alcohol, and in hexane. Produces a viscous solution in a mixture of methanol and dichloromethane (1:1), or in a mixture of dehydrated alcohol and acetone (1:1). Dissolves in 1 *N* sodium hydroxide. *NF category:* Coating agent.

Hydroxystilbamidine Isethionate: Yellow, fine, odorless, crystalline powder. Is stable in air but decomposes upon exposure to light. Melts at about 280°. Soluble in water; slightly soluble in alcohol; insoluble in ether.

Sterile Hydroxystilbamidine Isethionate: Yellow, fine, odorless, crystalline powder. Is stable in air but decomposes upon exposure to light. Melts at about 280°. Soluble in water; slightly soluble in alcohol; insoluble in ether.

Hydroxyurea: White to off-white powder. Is somewhat hygroscopic, decomposing in the presence of moisture. Melts at a temperature exceeding 133°, with decomposition. Freely soluble in water and in hot alcohol.

Hydroxyzine Hydrochloride: White, odorless powder. Melts at about 200°, with decomposition. Very soluble in water; soluble in chloroform; slightly soluble in acetone; practically insoluble in ether.

Hydroxyzine Pamoate: Light yellow, practically odorless powder. Practically insoluble in water and in methanol; freely soluble in dimethylformamide.

Hyoscyamine: White, crystalline powder. Is affected by light. Its solutions are alkaline to litmus. Slightly soluble in water and in benzene; freely soluble in alcohol, in chloroform, and in dilute acids; sparingly soluble in ether.

Hyoscyamine Hydrobromide: White, odorless crystals or crystalline powder. The pH of a solution (1 in 20) is about 5.4. Is affected by light. Freely soluble in water, in alcohol, and in chloroform; very slightly soluble in ether.

Hyoscyamine Sulfate: White, odorless crystals or crystalline powder. Is deliquescent and is affected by light. The pH of a solution (1 in 100) is about 5.3. Very soluble in water; freely soluble in alcohol; practically insoluble in ether.

Hypophosphorous Acid: Colorless or slightly yellow, odorless liquid. Specific gravity is about 1.13. *NF category:* Antioxidant.

Ibuprofen: White to off-white, crystalline powder, having a slight, characteristic odor. Practically insoluble in water; very soluble in alcohol, in methanol, in acetone, and in chloroform; slightly soluble in ethyl acetate.

Ichthammol: Reddish brown to brownish black, viscous fluid, having a strong, characteristic, empyreumatic odor. Miscible with water, with glycerin, and with fixed oils and fats. Partially soluble in alcohol and in ether.

Idoxuridine: White, crystalline, practically odorless powder. Slightly soluble in water and in alcohol; practically insoluble in chloroform and in ether.

Imidurea: White, odorless, tasteless powder. Soluble in water and in glycerin; sparingly soluble in propylene glycol; insoluble in most organic solvents.

Imipramine Hydrochloride: White to off-white, odorless or practically odorless, crystalline powder. Freely soluble in water and in alcohol; soluble in acetone; insoluble in ether and in benzene.

Indigotindisulfonate Sodium: Dusky, purplish blue powder, or blue granules having a coppery luster. Is affected by light. Its solutions have a blue or bluish purple color. Slightly soluble in water and in alcohol; practically insoluble in most other organic solvents.

Indocyanine Green: Olive-brown, dark green, blue-green, dark blue, or black powder. Is odorless or has a slight odor. Its solutions are deep emerald-green in color. The pH of a solution (1 in 200) is about 6. Its aqueous solutions are stable for about 8 hours. Soluble in water and in methanol; practically insoluble in most other organic solvents.

Sterile Indocyanine Green: Olive-brown, dark green, blue-green, dark blue, or black powder. Is odorless or has a slight odor. Its solutions are deep emerald-green in color. The pH of a solution (1 in 200) is about 6. Its aqueous solutions are stable for about 8 hours.

Indomethacin: Pale yellow to yellow-tan, crystalline powder, having not more than a slight odor. Is sensitive to light. Melts at about 162°. Exhibits polymorphism. Practically insoluble in water; sparingly soluble in alcohol, in chloroform, and in ether.

Influenza Virus Vaccine: Slightly turbid liquid or suspension, which may have a slight yellow or reddish tinge and may have an odor because of the preservative.

Insulin: White or practically white crystals. Soluble in solutions of dilute acids and alkalies.

Insulin Injection: The Injection containing, in each mL, not more than 100 USP Units is a clear, colorless or almost colorless liquid; the Injection containing, in each mL, 500 Units may be straw-colored. Contains between 0.1% and 0.25% (w/v) of either phenol or cresol. Contains between 1.4% and 1.8% (w/v) of glycerin.

Insulin Zinc Suspension: Practically colorless suspension of a mixture of characteristic crystals predominantly between 10 μm and 40 μm in maximum dimension and many particles that have no uniform shape and do not exceed 2 μm in maximum dimension. Contains between 0.15% and 0.17% (w/v) of sodium acetate, between 0.65% and 0.75% (w/v) of sodium chloride, and between 0.09% and 0.11% (w/v) of methylparaben.

Isophane Insulin Suspension: White suspension of rod-shaped crystals, free from large aggregates of crystals following moderate agitation. Contains either (1) between 1.4% and 1.8% (w/v) of glycerin, between 0.15% and 0.17% (w/v) of metacresol, and between 0.06% and 0.07% (w/v) of phenol, or (2) between 1.4% and 1.8% (w/v) of glycerin and between 0.20% and 0.25% (w/v) of phenol. Contains between 0.15% and 0.25% (w/v) of dibasic sodium phosphate. When examined microscopically, the insoluble matter in the Suspension is crystalline, and contains not more than traces of amorphous material.

Extended Insulin Zinc Suspension: Practically colorless suspension of a mixture of characteristic crystals the maximum dimension of which is predominantly between 10 μm and 40 μm. Contains between 0.15% and 0.17% (w/v) of sodium acetate, between 0.65% and 0.75% (w/v) of sodium chloride, and between 0.09% and 0.11% (w/v) of methylparaben.

Prompt Insulin Zinc Suspension: Practically colorless suspension of particles that have no uniform shape and the maximum dimension of which does not exceed 2 μm. Contains between 0.15% and 0.17% (w/v) of sodium acetate, between 0.65% and 0.75% (w/v) of sodium chloride, and between 0.09% and 0.11% (w/v) of methylparaben.

Protamine Zinc Insulin Suspension: White or practically white suspension, free from large particles following moderate agitation. Contains between 1.4% and 1.8% (w/v) of glycerin, and either between 0.18% and 0.22% (w/v) of cresol or between 0.22% and 0.28% (w/v) of phenol. Contains between 0.15% and 0.25% (w/v) of dibasic sodium phosphate (Na_2HPO_4), and between 1.0 mg and 1.5 mg of protamine for each 100 USP Insulin Units.

Inulin: White, friable, chalk-like, amorphous, odorless, tasteless powder. Soluble in hot water; slightly soluble in cold water and in organic solvents.

Iodine: Heavy, grayish black plates or granules, having a metallic luster and a characteristic odor. Very slightly soluble in water; freely soluble in carbon disulfide, in chloroform, in carbon tetrachloride, and in ether; soluble in alcohol and in solutions of iodides; sparingly soluble in glycerin.

Iodine Topical Solution: Transparent, reddish brown liquid, having the odor of iodine.

Strong Iodine Solution: Transparent liquid having a deep brown color and having the odor of iodine.

Iodine Tincture: Transparent liquid having a reddish brown color and the odor of iodine and of alcohol.

Sodium Iodide I 123 Capsules: Capsules may contain a small amount of solid or solids, or may appear empty.

Sodium Iodide I 123 Solution: Clear, colorless solution. Upon standing, both the Solution and the glass container may darken as a result of the effects of the radiation.

Iodinated I 125 Albumin Injection: Clear, colorless to slightly yellow solution. Upon standing, both the Albumin and the glass container may darken as a result of the effects of the radiation.

Sodium Iodide I 125 Solution: Clear, colorless solution. Upon standing, both the Solution and the glass container may darken as a result of the effects of the radiation.

Iodinated I 131 Albumin Injection: Clear, colorless to slightly yellow solution. Upon standing, both the Albumin and the glass container may darken as a result of the effects of the radiation.

Iodinated I 131 Albumin Aggregated Injection: Dilute suspension of white to faintly yellow particles, which may settle on standing. The glass container may darken on standing, as a result of the effects of the radiation.

Sodium Rose Bengal I 131 Injection: Clear, deep-red solution.

Iodohippurate Sodium I 131 Injection: Clear, colorless solution. Upon standing, both the Injection and the glass container may darken as a result of the effects of the radiation.

Sodium Iodide I 131 Capsules: May contain a small amount of solid or solids, or may appear empty.

Sodium Iodide I 131 Solution: Clear, colorless solution. Upon standing, both the Solution and the glass container may darken as a result of the effects of the radiation.

Iodipamide: White, practically odorless, crystalline powder. Very slightly soluble in water, in chloroform, and in ether; slightly soluble in alcohol.

Iodipamide Meglumine Injection: Clear, colorless to pale yellow, slightly viscous liquid.

Iodoquinol: Light yellowish to tan, microcrystalline powder not readily wetted by water. Is odorless or has a faint odor; is stable in air. Melts with decomposition. Practically insoluble in water; sparingly soluble in alcohol and in ether.

Iopamidol: Practically odorless, white to off-white powder. Very soluble in water; sparingly soluble in methanol; practically insoluble in alcohol and in chloroform.

Iopanoic Acid: Cream-colored powder. Is tasteless or practically so, and has a faint, characteristic odor. Is affected by light. Insoluble in water; soluble in alcohol, in chloroform, and in ether; soluble in solutions of alkali hydroxides and carbonates.

Iophendylate: Colorless to pale yellow, viscous liquid, the color darkening on long exposure to air. Is odorless or has a faintly ethereal odor. Very slightly soluble in water; freely soluble in alcohol, in benzene, in chloroform, and in ether.

Iophendylate Injection: Colorless to pale yellow, viscous liquid, the color darkening on long exposure to air. Is odorless or has a faintly ethereal odor. Very slightly soluble in water; freely soluble in alcohol, in benzene, in chloroform, and in ether.

Iothalamate Meglumine Injection: Clear, colorless to pale yellow, slightly viscous liquid.

Iothalamate Meglumine and Iothalamate Sodium Injection: Clear, colorless to pale yellow, slightly viscous liquid.

Iothalamate Sodium Injection: Clear, colorless to pale yellow, slightly viscous liquid.

Iothalamic Acid: White, odorless powder. Slightly soluble in water and in alcohol; soluble in solutions of alkali hydroxides.

Powdered Ipecac: Pale brown, weak yellow, or light olive-gray powder.

Ipodate Calcium: White to off-white, odorless, fine, crystalline powder. Slightly soluble in water, in alcohol, in chloroform, and in methanol.

Ipodate Sodium: White to off-white, odorless, fine, crystalline powder. Freely soluble in water, in alcohol, and in methanol; very slightly soluble in chloroform.

Ferrous Citrate Fe 59 Injection: Clear, slightly yellow solution.

Iron Dextran Injection: Dark brown, slightly viscous liquid.

Iron Sorbitex Injection: Clear liquid, having a dark brown color.

Iso-alcoholic elixir: *NF category:* Flavored and/or sweetened vehicle.

Isobutane: Colorless, flammable gas (boiling temperature is about −11°). Vapor pressure at 21° is about 2950 mm of mercury (31 psig). *NF category:* Aerosol propellant.

Isocarboxazid: White or practically white, crystalline powder. Has a slight, characteristic odor. Slightly soluble in water; very soluble in chloroform; soluble in alcohol.

Isoetharine Inhalation Solution: Colorless or slightly yellow, slightly acid liquid, gradually turning dark on exposure to air and light.

Isoetharine Hydrochloride: White to off-white, odorless, crystalline solid. Melts between 196° and 208°, with decomposition. Soluble in water; sparingly soluble in alcohol; practically insoluble in ether.

Isoetharine Mesylate: White or practically white, odorless crystals having a salty, bitter taste. Freely soluble in water; soluble in alcohol; practically insoluble in acetone and in ether.

Isoflurane: Clear, colorless, volatile liquid, having a slight odor. Boils at about 49°. Insoluble in water. Miscible with common organic solvents; miscible with fats and oils.

Isoflurophate: Clear, colorless or faintly yellow, liquid. Its vapor is extremely irritating to the eye and mucous membranes. Is decomposed by moisture, with the formation of hydrogen fluoride. Specific gravity is about 1.05. Sparingly soluble in water; soluble in alcohol and in vegetable oils.

Isoleucine: White, practically odorless crystals, having a slightly bitter taste. Soluble in water; slightly soluble in hot alcohol; insoluble in ether.

Isoniazid: Colorless or white crystals or white, crystalline powder. Is odorless and is slowly affected by exposure to air and to light. Freely soluble in water; sparingly soluble in alcohol; slightly soluble in chloroform and in ether.

Isoniazid Injection: Clear, colorless to faintly greenish yellow liquid. Gradually darkens on exposure to air and to light. Tends to crystallize at low temperatures.

Isopropamide Iodide: White to pale yellow, crystalline powder, having a bitter taste. Sparingly soluble in water; freely soluble in chloroform and in alcohol; very slightly soluble in benzene and in ether.

Isopropyl Alcohol: Transparent, colorless, mobile, volatile liquid, having a characteristic odor and a slightly bitter taste. Is flammable. Miscible with water, with alcohol, with ether, and with chloroform. *NF category:* Solvent.

Azeotropic Isopropyl Alcohol: Transparent, colorless, mobile, volatile liquid, having a characteristic odor and a slightly bitter taste. Is flammable. Miscible with water, with alcohol, with ether, and with chloroform.

Isopropyl Myristate: Clear, practically colorless, oily liquid. Is practically odorless, and congeals at about 5°. Insoluble in water, in glycerin, and in propylene glycol; freely soluble in 90% alcohol. Miscible with most organic solvents and with fixed oils. *NF category:* Vehicle (oleaginous).

Isopropyl Palmitate: Colorless, mobile liquid having a very slight odor. Soluble in acetone, in castor oil, in chloroform, in cottonseed oil, in ethyl acetate, in alcohol, and in mineral oil; insoluble in water, in glycerin, and in propylene glycol. *NF category:* Vehicle (oleaginous).

Isoproterenol Inhalation Solution: Colorless or practically colorless, slightly acid liquid, gradually turning dark on exposure to air and to light.

Isoproterenol Hydrochloride: White to practically white, odorless, crystalline powder, having a slightly bitter taste. Gradually darkens on exposure to air and to light. Its solutions become pink to brownish pink on standing exposed to air, and almost immediately so when rendered alkaline. Its solution (1 in 100) has a pH of about 5. Freely soluble in water; sparingly soluble in alcohol and less soluble in dehydrated alcohol; insoluble in chloroform and in ether.

Isoproterenol Hydrochloride Injection: Colorless or practically colorless liquid, gradually turning dark on exposure to air and to light.

Isoproterenol Sulfate: White to practically white, odorless, crystalline powder. It gradually darkens on exposure to air and to light. Its solutions become pink to brownish pink on standing exposed to air, doing so almost immediately when rendered alkaline. A solution (1 in 100) has a pH of about 5. Freely soluble in water; very slightly soluble in alcohol, in benzene, and in ether.

Isosorbide Concentrate: Colorless to slightly yellow liquid. Soluble in water and in alcohol.

Diluted Isosorbide Dinitrate: Ivory-white, odorless powder. [NOTE—Undiluted isosorbide dinitrate occurs as white, crystalline rosettes.] Undiluted isosorbide dinitrate is very slightly soluble in water; very soluble in acetone; freely soluble in chloroform; sparingly soluble in alcohol.

Isotretinoin: Yellow crystals. Practically insoluble in water; soluble in chloroform; sparingly soluble in alcohol, in isopropyl alcohol, and in polyethylene glycol 400.

Isoxsuprine Hydrochloride: White, odorless, crystalline powder, having a bitter taste. Melts, at about 200° with decomposition. Slightly soluble in water; sparingly soluble in alcohol.

Juniper Tar: Dark brown, clear, thick liquid, having a tarry odor and a faintly aromatic, bitter taste. Very slightly soluble in water; partially soluble in solvent hexane. One volume dissolves in 9 volumes of alcohol. Dissolves in 3 volumes of ether, leaving only a slight, flocculent residue. Miscible with amyl alcohol, with chloroform, and with glacial acetic acid.

Kanamycin Sulfate: White, odorless, crystalline powder. Freely soluble in water; insoluble in acetone, in ethyl acetate, and in benzene.

Kaolin: Soft, white or yellowish white powder or lumps. Has an earthy or clay-like taste and, when moistened with water, assumes a darker color and develops a marked clay-like odor. Insoluble in water, in cold dilute acids, and in solutions of alkali hydroxides. *NF category:* Tablet and/or capsule diluent.

Ketamine Hydrochloride: White, crystalline powder, having a slight, characteristic odor. Freely soluble in water and in methanol; soluble in alcohol; sparingly soluble in chloroform.

Labetalol Hydrochloride: White to off-white powder. Melts at about 180°, with decomposition. Soluble in water and in alcohol; insoluble in ether and in chloroform.

Lactic Acid: Colorless or yellowish, practically odorless, syrupy liquid. Is hygroscopic. When it is concentrated by boiling, lactic acid lactate is formed. Specific gravity is about 1.20. Miscible with water, with alcohol, and with ether. Insoluble in chloroform. *NF category:* Buffering agent.

Lactose: White or creamy white, hard, crystalline masses or powder. Is odorless, and has a faintly sweet taste. Is stable in air, but readily adsorbs odors. Freely (and slowly) soluble in water and even more soluble in boiling water; very slightly soluble in alcohol; insoluble in chloroform and in ether. *NF category:* Tablet and/or capsule diluent.

Lactulose Concentrate: Colorless to amber syrupy liquid, which may exhibit some precipitation and darkening upon standing. Miscible with water.

Lanolin: Yellowish white, ointment-like mass, having a slight, characteristic odor. Upon being heated on a steam bath, it separates at first into two layers; continued heating with frequent stirring drives off the water that makes up the lower layer. The residue is transparent while warm, but cools to form a yellowish, tenacious, unctuous mass completely soluble in ether and in chloroform and only sparingly soluble in alcohol. Insoluble in water; soluble in chloroform and in ether, with the separation of water. *NF category:* Ointment base.

Lanolin Alcohols: Hard, waxy, amber solid, having a characteristic odor. Insoluble in water; slightly soluble in alcohol; freely soluble in chloroform, in ether, and in petroleum ether. *NF category:* Emulsifying and/or solubilizing agent.

Anhydrous Lanolin: Yellow, tenacious, unctuous mass, having a slight, characteristic odor. Insoluble in water, but mixes without separation with about twice its weight of water. Sparingly soluble in cold alcohol; more soluble in hot alcohol; freely soluble in ether and in chloroform. *NF category:* Ointment base.

Lecithin: The consistency of both natural grades and refined grades of lecithin may vary from plastic to fluid, depending upon free fatty acid and oil content, and upon the presence or absence of other diluents. Its color varies from light yellow to brown, depending on the source, on crop variations, and on whether it is bleached or unbleached. Is odorless or has a characteristic, slight nutlike odor and a bland taste. Partially soluble in water, but it readily hydrates to form emulsions. The oil-free phosphatides are soluble in fatty acids, but are practically insoluble in fixed oils. When all phosphatide fractions are present, lecithin is partially soluble in alcohol and practically insoluble in acetone. *NF category:* Emulsifying and/or solubilizing agent.

Leucine: White, practically odorless, tasteless crystals. Sparingly soluble in water; insoluble in ether.

Leucovorin Calcium: Yellowish white or yellow, odorless powder. Very soluble in water; practically insoluble in alcohol.

Leucovorin Calcium Injection: Clear, yellowish solution.

Levocarnitine: White, odorless crystals or crystalline powder. Hygroscopic. Freely soluble in water, and in hot alcohol. Practically insoluble in acetone, in ether, and in benzene.

Levodopa: White to off-white, odorless, crystalline powder. In the presence of moisture, is rapidly oxidized by atmospheric oxygen and darkens. Slightly soluble in water; freely soluble in 3 *N* hydrochloric acid; insoluble in alcohol.

Levonordefrin: White to buff-colored, odorless, crystalline solid. Melts at about 210°. Practically insoluble in water; freely soluble in aqueous solutions of mineral acids; slightly soluble in acetone, in chloroform, in alcohol, and in ether.

Levonorgestrel: White or practically white, odorless powder. Practically insoluble in water; soluble in chloroform; slightly soluble in alcohol.

Levopropoxyphene Napsylate: White powder, having essentially no odor, but having a bitter taste. Very slightly soluble in water; soluble in methanol, in alcohol, in chloroform, and in acetone.

Levorphanol Tartrate: Practically white, odorless, crystalline powder. Sparingly soluble in water; slightly soluble in alcohol; insoluble in chloroform and in ether. Melts, in a sealed tube, at about 110°, with decomposition.

Levothyroxine Sodium: Light yellow to buff-colored, odorless, tasteless, hygroscopic powder. Is stable in dry air but may assume a slight pink color upon exposure to light. The pH of a saturated solution is about 8.9. Very slightly soluble in water; soluble in solutions of alkali hydroxides and in hot solutions of alkali carbonates; slightly soluble in alcohol; insoluble in acetone, in chloroform, and in ether.

Lidocaine: White or slightly yellow, crystalline powder. Has a characteristic odor and is stable in air. Practically insoluble in water; very soluble in alcohol and in chloroform; freely soluble in benzene and in ether. Dissolves in oils.

Lidocaine Hydrochloride: White, odorless, crystalline powder, having a slightly bitter taste. Very soluble in water and in alcohol; soluble in chloroform; insoluble in ether.

Lime: Hard, white or grayish white masses or granules, or white or grayish white powder. Is odorless. Slightly soluble in water; very slightly soluble in boiling water.

Lincomycin Hydrochloride: White or practically white, crystalline powder. Is odorless or has a faint odor. Is stable in the presence of air and light. Its solutions are acid and are dextrorotatory. Freely soluble in water; soluble in dimethylformamide; very slightly soluble in acetone.

Lincomycin Hydrochloride Injection: Clear, colorless to slightly yellow solution, having a slight odor.

Lindane: White, crystalline powder, having a slight, musty odor. Practically insoluble in water; freely soluble in chloroform; soluble in dehydrated alcohol; sparingly soluble in ether; slightly soluble in ethylene glycol.

Liothyronine Sodium: Light tan, odorless, crystalline powder. Very slightly soluble in water; slightly soluble in alcohol; practically insoluble in most other organic solvents.

Lithium Carbonate: White, granular, odorless powder. Sparingly soluble in water; very slightly soluble in alcohol. Dissolves, with effervescence, in dilute mineral acids.

Lithium Citrate: White, odorless, deliquescent powder or granules, having a cooling, faintly alkaline taste. Freely soluble in water; slightly soluble in alcohol.

Loperamide Hydrochloride: White to slightly yellow powder. Melts at about 225°, with some decomposition. Freely soluble in methanol, in isopropyl alcohol, and in chloroform; slightly soluble in water and in dilute acids.

Lorazepam: White or practically white, practically odorless powder. Insoluble in water; sparingly soluble in alcohol; slightly soluble in chloroform.

Lysine Acetate: White, odorless crystals or crystalline powder, having an acid taste. Freely soluble in water.

Lysine Hydrochloride: White, odorless powder. Freely soluble in water.

Mafenide Acetate: White, crystalline powder. Freely soluble in water.

Magaldrate: White, odorless, crystalline powder. Insoluble in water and in alcohol; soluble in dilute solutions of mineral acids.

Milk of Magnesia: White, opaque, more or less viscous suspension from which varying proportions of water usually separate on standing. pH is about 10.

Magnesium Aluminum Silicate: Odorless, tasteless, fine (micronized) powder or small flakes that are creamy when viewed on their flat surfaces and tan to brown when viewed on their edges. Insoluble in water and in alcohol. Swells when added to water or glycerin. *NF category:* Suspending and/or viscosity-increasing agent.

Magnesium Carbonate: Light, white, friable masses or bulky, white powder. Is odorless, and is stable in air. Practically insoluble in water; to which, however, it imparts a slightly alkaline reaction; insoluble in alcohol, but is dissolved by dilute acids with effervescence.

Magnesium Chloride: Colorless, odorless, deliquescent flakes or crystals, which lose water when heated to 100° and lose hydrochloric acid when heated to 110°. Very soluble in water; freely soluble in alcohol.

Magnesium Citrate Oral Solution: Colorless to slightly yellow, clear, effervescent liquid, having a sweet, acidulous taste and a lemon flavor.

Magnesium Gluconate: Colorless crystals or white powder or granules. Is odorless and tasteless. Freely soluble in water; very slightly soluble in alcohol; insoluble in ether.

Magnesium Hydroxide: Bulky, white powder. Practically insoluble in water and in alcohol; soluble in dilute acids.

Magnesium Oxide: Very bulky, white powder known as Light Magnesium Oxide or relatively dense, white powder known as Heavy Magnesium Oxide. Five g of Light Magnesium Oxide occupies a volume of approximately 40 to 50 mL, while 5 g of Heavy Magnesium Oxide occupies a volume of approximately 10 to 20 mL. Practically insoluble in water; soluble in dilute acids; insoluble in alcohol.

Magnesium Phosphate: White, odorless, tasteless powder. Almost insoluble in water; readily soluble in diluted mineral acids.

Magnesium Silicate: Fine, white, odorless, tasteless powder, free from grittiness. Insoluble in water and in alcohol. Is readily decomposed by mineral acids. *NF category:* Glidant and/or anticaking agent.

Magnesium Stearate: Fine, white, bulky powder, having a faint, characteristic odor. Is unctuous, adheres readily to the skin, and is free from grittiness. Insoluble in water, in alcohol, and in ether. *NF category:* Tablet and/or capsule lubricant.

Magnesium Sulfate: Small, colorless crystals, usually needle-like, with a cooling, saline, bitter taste. It effloresces in warm, dry air. Freely soluble in water; freely (and slowly) soluble in glycerin; very soluble in boiling water; sparingly soluble in alcohol.

Magnesium Trisilicate: Fine, white, odorless, tasteless powder, free from grittiness. Insoluble in water and in alcohol. Is readily decomposed by mineral acids.

Malathion: Yellow to deep brown liquid, having a characteristic odor. Congeals at about 2.9°. Slightly soluble in water. Miscible with alcohols, with esters, with ketones, with ethers, with aromatic and alkylated aromatic hydrocarbons, and with vegetable oils.

Malic Acid: White or practically white, crystalline powder or granules, having a strongly acid taste. Melts at about 130°. Very soluble in water; freely soluble in alcohol. *NF category:* Acidifying agent.

Manganese Chloride: Large, irregular, pink, odorless, translucent crystals. Soluble in water and in alcohol; insoluble in ether.

Manganese Sulfate: Pale red, slightly efflorescent crystals, or purple, odorless powder. Soluble in water; insoluble in alcohol.

Mannitol: White, crystalline powder or free-flowing granules. Is odorless and has a sweet taste. Freely soluble in water; soluble in alkaline solutions; slightly soluble in pyridine; very slightly soluble in alcohol; practically insoluble in ether. *NF category:* Sweetening agent; tablet and/or capsule diluent; tonicity agent.

Maprotiline Hydrochloride: Fine, white to off-white, crystalline powder. Is practically odorless. Freely soluble in methanol and in chloroform; slightly soluble in water; practically insoluble in isooctane.

Mazindol: White to off-white, crystalline powder, having not more than a faint odor. Insoluble in water; slightly soluble in methanol and in chloroform.

Measles Virus Vaccine Live: Solid having the characteristic appearance of substances dried from the frozen state. Undergoes loss of potency on exposure to sunlight. The Vaccine is to be constituted with a suitable diluent just prior to use.

Measles and Mumps Virus Vaccine Live: Solid having the characteristic appearance of substances dried from the frozen state. The Vaccine is to be constituted with a suitable diluent just prior to use. Constituted vaccine undergoes loss of potency on exposure to sunlight.

Measles, Mumps, and Rubella Virus Vaccine Live: Solid having the characteristic appearance of substances dried from the frozen state. The Vaccine is to be constituted with a suitable diluent just prior to use. Constituted vaccine undergoes loss of potency on exposure to sunlight.

Measles and Rubella Virus Vaccine Live: Solid having the characteristic appearance of substances dried from the frozen state. The Vaccine is to be constituted with a suitable diluent just prior to use. Constituted vaccine undergoes loss of potency on exposure to sunlight.

Mebendazole: White to slightly yellow powder. Is almost odorless. Melts at about 290°. Practically insoluble in water, in dilute solutions of mineral acids, in alcohol, in ether, and in chloroform; freely soluble in formic acid.

Mechlorethamine Hydrochloride: White, crystalline powder. Is hygroscopic.

Meclizine Hydrochloride: White or slightly yellowish, crystalline powder. Has a slight odor and is tasteless. Practically insoluble in water and in ether; freely soluble in chloroform, in pyridine, and in acid-alcohol-water mixtures; slightly soluble in dilute acids and in alcohol.

Meclofenamate Sodium: A white to creamy white, odorless to almost odorless, crystalline powder. Soluble in methanol; slightly soluble in chloroform; practically insoluble in ether. Freely soluble in water, the solution sometimes being somewhat turbid due to partial hydrolysis and absorption of carbon dioxide; the solution is clear above pH 11.5.

Medroxyprogesterone Acetate: White to off-white, odorless, crystalline powder. Melts at about 205°. Is stable in air. Insoluble in water; freely soluble in chloroform; soluble in acetone and in dioxane; sparingly soluble in alcohol and in methanol; slightly soluble in ether.

Medrysone: White to off-white, crystalline powder. Is odorless or may have a slight odor. Melts at about 158°, with decomposition. Sparingly soluble in water; soluble in methylene chloride and in chloroform.

Megestrol Acetate: White to creamy white, tasteless and essentially odorless, crystalline powder. Insoluble in water, sparingly soluble in alcohol, slightly soluble in ether and in fixed oils, soluble in acetone, very soluble in chloroform. Is unstable under aqueous conditions at pH 7 or above.

Meglumine: White to faintly yellowish white, odorless crystals or powder. Freely soluble in water; sparingly soluble in alcohol.

Melphalan: Off-white to buff powder, having a faint odor. Melts at about 180°, with decomposition. Practically insoluble in water, in chloroform, and in ether; soluble in dilute mineral acids; slightly soluble in alcohol and in methanol.

Menadiol Sodium Diphosphate: White to pink powder, having a characteristic odor. Is hygroscopic. Its solutions are neutral or slightly alkaline to litmus, having a pH of about 8. Very soluble in water; insoluble in alcohol.

Menadione: Bright yellow, crystalline, practically odorless powder. Is affected by sunlight. Practically insoluble in water; soluble in vegetable oils; sparingly soluble in chloroform and in alcohol.

Menthol: Colorless, hexagonal crystals, usually needle-like, or in fused masses, or crystalline powder. Has a pleasant, peppermint-like odor. Slightly soluble in water; very soluble in alcohol, in chloroform, in ether, and in solvent hexane; freely soluble in glacial acetic acid, in mineral oil, and in fixed and volatile oils. *NF category:* Flavors and perfumes.

Meperidine Hydrochloride: Fine, white, crystalline, odorless powder. The pH of a solution (1 in 20) is about 5. Very soluble in water; soluble in alcohol; sparingly soluble in ether.

Mephentermine Sulfate: White, odorless crystals or crystalline powder. Its solutions are slightly acid to litmus, having a pH of about 6. Soluble in water; slightly soluble in alcohol; insoluble in chloroform.

Mephenytoin: White, crystalline powder. Very slightly soluble in water; freely soluble in chloroform; soluble in alcohol and in aqueous solutions of alkali hydroxides; sparingly soluble in ether.

Mephobarbital: White, odorless, crystalline powder, having a bitter taste. Its saturated solution is acid to litmus. Slightly soluble in water, in alcohol, and in ether; soluble in chloroform and in solutions of fixed alkali hydroxides and carbonates.

Mepivacaine Hydrochloride: White, odorless, crystalline solid. The pH of a solution (1 in 50) is about 4.5. Freely soluble in water and in methanol; very slightly soluble in chloroform; practically insoluble in ether.

Meprobamate: White powder, having a characteristic odor and a bitter taste. Slightly soluble in water; freely soluble in acetone and in alcohol; sparingly soluble in ether.

Meprylcaine Hydrochloride: White, odorless, crystalline solid. The pH of a solution (1 in 50) is about 5.7. Freely soluble in water, in alcohol, and in chloroform; slightly soluble in acetone.

Mercaptopurine: Yellow, odorless or practically odorless, crystalline powder. Melts at a temperature exceeding 308°, with decomposition. Insoluble in water, in acetone, and in ether; soluble in hot alcohol and in dilute alkali solutions; slightly soluble in 2 *N* sulfuric acid.

Ammoniated Mercury: White, pulverulent pieces or white, amorphous powder. Is odorless, and is stable in air, but darkens on exposure to light. Insoluble in water, and in alcohol; readily soluble in warm hydrochloric, nitric, and acetic acids.

Mesoridazine Besylate: White to pale yellowish powder, having not more than a faint odor. Melts at about 178°, with decomposition. Freely soluble in water, in chloroform, and in methanol.

Mestranol: White to creamy white, odorless, crystalline powder. Insoluble in water; freely soluble in chloroform; soluble in dioxane; sparingly soluble in dehydrated alcohol; slightly soluble in methanol.

Metaproterenol Sulfate: White to off-white, crystalline powder. Freely soluble in water.

Methacholine Chloride: Colorless or white crystals, or white, crystalline powder. Is odorless or has a slight odor, and is very hygroscopic. Its solutions are neutral to litmus. Very soluble in water; freely soluble in alcohol and in chloroform.

Methacrylic Acid Copolymer: White powder having a faint, characteristic odor. The polymer is insoluble in water, in diluted acids, in simulated gastric fluid TS, and in buffer solutions of up to pH 5; soluble in diluted alkali, in simulated intestinal fluid TS, and in buffer solutions of pH 7 and above. The solubility between pH 5.5 and pH 7 depends on the content of methacrylic acid units in the copolymer. The polymer is soluble to freely soluble in methanol, in alcohol, in isopropyl alcohol, and in acetone, each of which contains not less than 3% of water.

Methacycline Hydrochloride: Yellow to dark yellow, crystalline powder. Soluble in water.

Methadone Hydrochloride: Colorless crystals or white, crystalline, odorless powder. Soluble in water; freely soluble in alcohol and in chloroform; practically insoluble in ether and in glycerin.

Methadone Hydrochloride Oral Concentrate: Clear to slightly hazy, syrupy liquid.

Methantheline Bromide: White or nearly white, practically odorless powder, having a very bitter taste. Its solutions have a pH of about 5. Very soluble in water; freely soluble in alcohol and in chloroform; practically insoluble in ether. Its water solution decomposes on standing.

Sterile Methantheline Bromide: White or nearly white, practically odorless powder, having a very bitter taste. Its solutions have a pH of about 5. Very soluble in water; freely soluble in alcohol and in chloroform; practically insoluble in ether. Its water solution decomposes on standing.

Metharbital: White to nearly white, crystalline powder, having a faint aromatic odor. The pH of a saturated solution is about 6. Slightly soluble in water; soluble in alcohol; sparingly soluble in ether.

Methazolamide: White or faintly yellow, crystalline powder having a slight odor. Melts at about 213°. Very slightly soluble in water and in alcohol; soluble in dimethylformamide; slightly soluble in acetone.

Methdilazine: Light tan, crystalline powder, having a characteristic odor. Practically insoluble in water; freely soluble in 3 *N* hydrochloric acid; soluble in alcohol and in chloroform.

Methdilazine Hydrochloride: Light tan, crystalline powder, having a slight, characteristic odor. Freely soluble in water, in alcohol, and in chloroform.

Methenamine: Colorless, lustrous crystals or white, crystalline powder. Is practically odorless. When brought into contact with fire, it readily ignites, burning with a smokeless flame. It sublimes at about 260°, without melting. Its solutions are alkaline to litmus. Freely soluble in water; soluble in alcohol and in chloroform.

Methenamine Mandelate: White, crystalline powder. Has a sour taste and is practically odorless. Its solutions have a pH of about 4. Melts at about 127°, with decomposition. Very soluble in water; soluble in alcohol and in chloroform; slightly soluble in ether.

Methicillin Sodium for Injection: Fine, white, crystalline powder, odorless or having a slight odor. Freely soluble in water, in methanol, and in pyridine; slightly soluble in propyl and amyl alcohols, in chloroform, and in ethylene chloride; insoluble in acetone, in ether, and in benzene.

Sterile Methicillin Sodium: Fine, white, crystalline powder, odorless or having a slight odor. Freely soluble in water, in methanol, and in pyridine; slightly soluble in propyl and amyl alcohols, in chloroform, and in ethylene chloride; insoluble in acetone, in ether, and in benzene.

Methimazole: White to pale buff, crystalline powder, having a faint, characteristic odor. Its solutions are practically neutral to litmus. Freely soluble in water, in alcohol, and in chloroform; slightly soluble in ether.

Methionine: White crystals, having a characteristic odor and taste. Soluble in water, in warm dilute alcohol, and in dilute mineral acids; insoluble in ether, in absolute alcohol, in benzene, and in acetone (L-form).

Methocarbamol: White powder, odorless, or having a slight characteristic odor. Melts at about 94°, or, if previously ground to a fine powder, melts at about 90°. Sparingly soluble in water and in chloroform; soluble in alcohol only with heating; insoluble in benzene and in *n*-hexane.

Methohexital: White to faintly yellowish white, crystalline, odorless powder. Very slightly soluble in water; slightly soluble in alcohol, in chloroform, and in dilute alkalies.

Methohexital Sodium for Injection: White to off-white hygroscopic powder. Is essentially odorless.

Methotrexate: Orange-brown, or yellow, crystalline powder. Practically insoluble in water, in alcohol, in chloroform, and in ether; freely soluble in dilute solutions of alkali hydroxides and carbonates; slightly soluble in 6 *N* hydrochloric acid.

Methotrimeprazine: Fine, white, practically odorless, crystalline powder. Melts at about 126°. Practically insoluble in water; freely soluble in chloroform and in ether; sparingly soluble in methanol. Is sparingly soluble in alcohol at 25°, but is freely soluble in boiling alcohol.

Methoxamine Hydrochloride: Colorless or white, plate-like crystals or white, crystalline powder. Is odorless or has only a slight odor. Its solutions have a pH of about 5. Freely soluble in water; soluble in alcohol; practically insoluble in chloroform and in ether.

Methoxsalen: White to cream-colored, fluffy, needle-like crystals. Is odorless. Practically insoluble in water; freely soluble in chloroform; soluble in boiling alcohol, in acetone, in acetic acid, in propylene glycol, and in benzene; sparingly soluble in boiling water and in ether.

Methoxsalen Topical Solution: Clear, colorless liquid.

Methoxyflurane: Clear, practically colorless, mobile liquid, having a characteristic odor. Boils at about 105°. Miscible with alcohol, with acetone, with chloroform, with ether, and with fixed oils.

Methscopolamine Bromide: White crystals or white, odorless, crystalline powder. Melts at about 225°, with decomposition (see *Melting Range or Temperature* ⟨741⟩). Freely soluble in water; slightly soluble in alcohol; insoluble in acetone and in chloroform.

Methsuximide: White to grayish white, crystalline powder. Is odorless, or has not more than a slight odor. Slightly soluble in hot water; very soluble in chloroform; freely soluble in alcohol and in ether.

Methyclothiazide: White or practically white, crystalline powder. Is odorless, or has a slight odor. Very slightly soluble in water, in chloroform, and in benzene; freely soluble in acetone and in pyridine; sparingly soluble in methanol; slightly soluble in alcohol.

Methyl Alcohol: Clear, colorless liquid, having a characteristic odor. Is flammable. Miscible with water, with alcohol, with ether, with benzene, and with most other organic solvents. *NF category:* Solvent.

Methyl Isobutyl Ketone: Transparent, colorless, mobile, volatile liquid, having a faint ketonic and camphoraceous odor. Slightly soluble in water; miscible with alcohol, with ether, and with benzene. *NF category:* Alcohol denaturant; solvent.

Methyl Salicylate: Colorless, yellowish, or reddish liquid, having the characteristic odor and taste of wintergreen. It boils between 219° and 224°, with some decomposition. Slightly soluble in water; soluble in alcohol and in glacial acetic acid. *NF category:* Flavors and perfumes.

Methylbenzethonium Chloride: White, hygroscopic crystals, having a mild odor. Its solutions are neutral or slightly alkaline to litmus. Very soluble in water, in alcohol, and in ether; practically insoluble in chloroform.

Methylcellulose: White, fibrous powder or granules. Its aqueous suspensions are neutral to litmus. It swells in water and produces a clear to opalescent, viscous, colloidal suspension. Insoluble in alcohol, in ether, and in chloroform; soluble in glacial acetic acid and in a mixture of equal volumes of alcohol and chloroform. *NF category:* Coating agent; suspending and/or viscosity-increasing agent; tablet binder.

Methyldopa: White to yellowish white, odorless, fine powder, which may contain friable lumps. Sparingly soluble in water; very soluble in 3 *N* hydrochloric acid; slightly soluble in alcohol; practically insoluble in ether.

Methyldopate Hydrochloride: White or practically white, odorless or practically odorless, crystalline powder. Freely soluble in water, in alcohol, and in methanol; slightly soluble in chloroform; practically insoluble in ether.

Methylene Blue: Dark green crystals or crystalline powder having a bronze-like luster. Is odorless or practically so, and is stable in air. Its solutions in water and in alcohol are deep blue in color. Soluble in water and in chloroform; sparingly soluble in alcohol.

Methylene Chloride: Clear, colorless, mobile liquid, having an odor resembling that of chloroform. Miscible with alcohol, with ether, and with fixed and volatile oils. *NF category:* Solvent.

Methylergonovine Maleate: White to pinkish tan, microcrystalline powder. Is odorless. Slightly soluble in water and in alcohol; very slightly soluble in chloroform and in ether.

Methylparaben: Small, colorless crystals, or white, crystalline powder. Is odorless or has a faint, characteristic odor, and has a slight, burning taste. Slightly soluble in water, in benzene, and in carbon tetrachloride; freely soluble in alcohol and in ether. *NF category:* Antimicrobial preservative.

Methylparaben Sodium: White, hygroscopic powder. Freely soluble in water; sparingly soluble in alcohol; insoluble in fixed oils.

Methylphenidate Hydrochloride: White, odorless, fine, crystalline powder. Its solutions are acid to litmus. Freely soluble in water and in methanol; soluble in alcohol; slightly soluble in chloroform and in acetone.

Methylprednisolone: White to practically white, odorless, crystalline powder. Melts at about 240°, with some decomposition (see *Melting Range or Temperature* ⟨741⟩). Practically insoluble in water; sparingly soluble in alcohol, in dioxane, and in methanol; slightly soluble in acetone and in chloroform; very slightly soluble in ether.

Methylprednisolone Acetate: White or practically white, odorless, crystalline powder. Melts at about 225°, with some decomposition (see *Melting Range or Temperature* ⟨741⟩). Practically insoluble in water; soluble in dioxane; sparingly soluble in acetone, in alcohol, in chloroform, and in methanol; slightly soluble in ether.

Methylprednisolone Hemisuccinate: White or nearly white, odorless or nearly odorless, hygroscopic solid. Very slightly soluble in water; freely soluble in alcohol; soluble in acetone.

Methylprednisolone Sodium Succinate: White or nearly white, odorless, hygroscopic, amorphous solid. Very soluble in water and in alcohol; very slightly soluble in acetone; insoluble in chloroform.

Methyltestosterone: White or creamy white crystals or crystalline powder. Is odorless and is stable in air, but is slightly hygroscopic. Is affected by light. Practically insoluble in water; soluble in alcohol, in methanol, in ether, and in other organic solvents; sparingly soluble in vegetable oils.

Methyprylon: White, or practically white, crystalline powder, having a slight, characteristic odor. Soluble in water; very soluble in benzene; freely soluble in alcohol, in chloroform, and in ether.

Methysergide Maleate: White to yellowish white or reddish white, crystalline powder. Is odorless or has not more than a slight odor. Slightly soluble in water and in alcohol; very slightly soluble in chloroform; practically insoluble in ether.

Metoclopramide Hydrochloride: White or practically white, crystalline, odorless or practically odorless powder. Very soluble in water; freely soluble in alcohol; sparingly soluble in chloroform; practically insoluble in ether.

Metocurine Iodide: White or pale yellow, crystalline powder. Slightly soluble in water, in 3 *N* hydrochloric acid, and in dilute solutions of sodium hydroxide. Very slightly soluble in alcohol; practically insoluble in benzene, in chloroform, and in ether.

Metoprolol Tartrate: White, crystalline powder. Very soluble in water; freely soluble in methylene chloride, in chloroform, and in alcohol; slightly soluble in acetone; insoluble in ether.

Metronidazole: White to pale yellow, odorless crystals or crystalline powder. Is stable in air, but darkens on exposure to light. Sparingly soluble in water, in alcohol, and in chloroform; slightly soluble in ether.

Metyrapone: White to light amber, fine, crystalline powder, having a characteristic odor. Darkens on exposure to light. Sparingly soluble in water; soluble in methanol and in chloroform. It forms water-soluble salts with acids.

Miconazole: White to pale cream powder. Melts in the range of 83° to 87°. Insoluble in water; soluble in ether; freely soluble in alcohol, in methanol, in isopropyl alcohol, in acetone, in propylene glycol, in chloroform, and in dimethylformamide.

Miconazole Nitrate: White or practically white, crystalline powder, having not more than a slight odor. Melts in the range of 178° to 183°, with decomposition. Insoluble in ether; very slightly soluble in water and in isopropyl alcohol; slightly soluble in alcohol, in chloroform, and in propylene glycol; sparingly soluble in methanol; soluble in dimethylformamide; freely soluble in dimethylsulfoxide.

Mineral Oil: Colorless, transparent, oily liquid, free or practically free from fluorescence. Is odorless and tasteless when cold, and develops not more than a faint odor of petroleum when heated. Insoluble in water and in alcohol; soluble in volatile oils. Miscible with most fixed oils but not with castor oil. *NF category:* Solvent; vehicle (oleaginous).

Light Mineral Oil: Colorless, transparent, oily liquid, free, or practically free, from fluorescence. Is odorless and tasteless when cold, and develops not more than a faint odor of petroleum when heated. Insoluble in water and in alcohol; soluble in volatile oils. Miscible with most fixed oils, but not with castor oil. *NF category:* Tablet and/or capsule lubricant; vehicle (oleaginous).

Minocycline Hydrochloride: Yellow, crystalline powder. Soluble in water and in solutions of alkali hydroxides and carbonates; slightly soluble in alcohol; practically insoluble in chloroform and in ether.

Minoxidil: White to off-white, crystalline powder. Melts in the approximate range of between 248° and 268°, with decomposition. Soluble in alcohol and in propylene glycol; sparingly soluble in methanol; slightly soluble in water; practically insoluble in chloroform, in acetone, in ethyl acetate, and in hexane.

Mitomycin: Blue-violet, crystalline powder. Soluble in water, in acetone, in methanol, in butyl acetate, and in cyclohexanone.

Mitotane: White, crystalline powder, having a slight, aromatic odor. Practically insoluble in water; soluble in alcohol, in ether, in solvent hexane, and in fixed oils and fats.

Monobenzone: White, odorless, crystalline powder. Practically insoluble in water; soluble in alcohol, in chloroform, in ether, and in acetone.

Monobenzone Ointment: Dispersible with, but not soluble in, water.

Mono- and Di-glycerides: Varies in consistency from yellow liquids through ivory-colored plastics to hard, ivory-colored solids having a bland odor and taste. Insoluble in water; soluble in alcohol, in ethyl acetate, in chloroform, and in other chlorinated hydrocarbons.

Monoethanolamine: Clear, colorless, moderately viscous liquid, having a distinctly ammoniacal odor. Miscible with water, with acetone, with alcohol, with glycerin, and with chloroform. Immiscible with ether, with solvent hexane, and with fixed oils, although it dissolves many essential oils.

Mono- and Di-acetylated Monoglycerides: White to pale yellow, waxy solid, melting at about 45°. Soluble in ether and in chloroform; slightly soluble in carbon disulfide; insoluble in water. *NF category:* Plasticizer.

Monosodium Glutamate: White, practically odorless, free-flowing crystals or crystalline powder. Freely soluble in water; sparingly soluble in alcohol. May have either a slightly sweet or a slightly salty taste. *NF category:* Flavors and perfumes.

Monothioglycerol: Colorless or pale yellow, viscous liquid, having a slight sulfidic odor. Is hygroscopic. Miscible with alcohol. Freely soluble in water; insoluble in ether. *NF category:* Antioxidant.

Morphine Sulfate: White, feathery, silky crystals, cubical masses of crystals, or white, crystalline powder. Is odorless, and when exposed to air it gradually loses water of hydration. Darkens on prolonged exposure to light. Soluble in water; freely soluble in hot water; slightly soluble in alcohol but more so in hot alcohol; insoluble in chloroform and in ether.

Mumps Skin Test Antigen: Slightly turbid liquid.

Mumps Virus Vaccine Live: Solid having the characteristic appearance of substances dried from the frozen state. The Vaccine is to be constituted with a suitable diluent just prior to use. Constituted vaccine undergoes loss of potency on exposure to sunlight.

Myristyl Alcohol: White wax-like mass. Soluble in ether; slightly soluble in alcohol; insoluble in water.

Nadolol: White to off-white, practically odorless, crystalline powder. Freely soluble in water, in alcohol, and in methanol; slightly soluble in chloroform.

Nafcillin Sodium: White to yellowish white powder, having not more than a slight characteristic odor. Freely soluble in water and in chloroform; soluble in alcohol.

Nafcillin Sodium for Injection: White to yellowish white powder, having not more than a slight characteristic odor. Freely soluble in water and in chloroform; soluble in alcohol.

Nalidixic Acid: White to slightly yellow, odorless, crystalline powder. Practically insoluble in water; soluble in chloroform and in solutions of fixed alkali hydroxides and carbonates; slightly soluble in alcohol; very slightly soluble in ether.

Naloxone Hydrochloride: White to slightly off-white powder. Its aqueous solution is acidic. Soluble in water, in dilute acids, and in strong alkali; slightly soluble in alcohol; practically insoluble in ether and in chloroform.

Naloxone Hydrochloride Injection: Clear, colorless liquid.

Nandrolone Decanoate: Fine, white to creamy white, crystalline powder. Is odorless, or may have a slight odor. Practically insoluble in water; soluble in chloroform, in alcohol, in acetone, and in vegetable oils.

Naphazoline Hydrochloride: White, crystalline powder. Is odorless and has a bitter taste. Melts at a temperature of about 255°, with decomposition. Freely soluble in water and in alcohol; very slightly soluble in chloroform; practically insoluble in ether.

Naproxen: White to off-white, practically odorless, crystalline powder. Practically insoluble in water; freely soluble in chloroform and in dehydrated alcohol; soluble in alcohol; sparingly soluble in ether.

Naproxen Sodium: White to creamy crystalline powder. Soluble in water and in methanol; sparingly soluble in alcohol; very slightly soluble in acetone; and practically insoluble in chloroform and in toluene. Melts at about 255°, with decomposition.

Natamycin: Off-white to cream-colored powder, which may contain up to 3 moles of water. Practically insoluble in water; slightly soluble in methanol; soluble in glacial acetic acid and in dimethylformamide.

Neomycin Sulfate: White to slightly yellow powder, or cryodesiccated solid. Is odorless or practically so and is hygroscopic. Its solutions are dextrorotatory. Freely soluble in water; very slightly soluble in alcohol; insoluble in acetone, in chloroform, and in ether.

Niacin: White crystals or crystalline powder. Is odorless, or has a slight odor. Melts at about 235°. Sparingly soluble in water; freely soluble in boiling water, in boiling alcohol, and in solutions of alkali hydroxides and carbonates; practically insoluble in ether.

Niacinamide: White, crystalline powder. Is odorless or practically so, and has a bitter taste. Its solutions are neutral to litmus. Freely soluble in water and in alcohol; soluble in glycerin.

Nifedipine: Yellow powder. Is affected by exposure to light. Practically insoluble in water; freely soluble in acetone.

Nitric Acid: Highly corrosive fuming liquid, having a characteristic, highly irritating odor. Stains animal tissues yellow. Boils at about 120°. Specific gravity is about 1.41. *NF category:* Acidifying agent.

Nitrofurantoin: Lemon-yellow, odorless crystals or fine powder. Has a bitter aftertaste. Very slightly soluble in water and in alcohol; soluble in dimethylformamide.

Nitrofurazone: Lemon yellow, odorless, crystalline powder. Darkens slowly on exposure to light. Melts at about 236°, with decomposition. Very slightly soluble in alcohol and in water; soluble in dimethylformamide; slightly soluble in propylene glycol and in polyethylene glycol mixtures; practically insoluble in chloroform and in ether.

Nitrofurazone Cream: Yellow, opaque, water-miscible cream.

Nitrofurazone Ointment: Yellow, opaque, and water-miscible, and has ointment-like consistency.

Nitrofurazone Topical Solution: Light yellow, clear, somewhat viscous liquid, having a faint characteristic odor. Miscible with water.

Nitrogen: Colorless, odorless, tasteless gas. Is nonflammable and does not support combustion. One liter at 0° and at a pressure of 760 mm of mercury weighs about 1.251 g. One volume dissolves in about 65 volumes of water and in about 9 volumes of alcohol at 20° and at a pressure of 760 mm of mercury. *NF category:* Air displacement.

Diluted Nitroglycerin: When diluted with lactose, it is a white, odorless powder. When diluted with propylene glycol or alcohol, it is a clear, colorless, or pale yellow liquid. [NOTE—Undiluted nitroglycerin occurs as a white to pale yellow, thick, flammable, explosive liquid.] Undiluted nitroglycerin is slightly soluble in water; soluble in methanol, in alcohol, in carbon disulfide, in acetone, in ethyl ether, in ethyl acetate, in glacial acetic acid, in benzene, in toluene, in nitrobenzene, in phenol, in chloroform, and in methylene chloride.

Nitromersol: Brownish yellow to yellow granules or brownish yellow to yellow powder. Is odorless and tasteless and is affected by light. Very slightly soluble in water, in alcohol, in acetone, and in ether; soluble in solutions of alkalies and of ammonia by opening of the anhydride ring and the formation of a salt.

Nitromersol Topical Solution: Clear, reddish orange solution. Is affected by light.

Nitrous Oxide: Colorless gas, without appreciable odor or taste. One liter at 0° and at a pressure of 760 mm of mercury weighs about 1.97 g. One volume dissolves in about 1.4 volumes of water at 20° and at a pressure of 760 mm of mercury. Freely soluble in alcohol; soluble in ether and in oils.

Nonoxynol 9: Clear, colorless to light yellow, viscous liquid. Soluble in water, in alcohol, and in corn oil.

Nonoxynol 10: Colorless to light amber viscous liquid having an aromatic odor. Soluble in polar organic solvents and in water. *NF category:* Wetting and/or solubilizing agent.

Norepinephrine Bitartrate: White or faintly gray, odorless, crystalline powder. Slowly darkens on exposure to air and light. Its solutions are acid to litmus, having a pH of about 3.5. Freely soluble in water; slightly soluble in alcohol; practically insoluble in chloroform and in ether. Melts between 98° and 104°, without previous drying of the specimen, the melt being turbid.

Norepinephrine Bitartrate Injection: Colorless or practically colorless liquid, gradually turning dark on exposure to air and light.

Norethindrone: White to creamy white, odorless, crystalline powder. Is stable in air. Practically insoluble in water; soluble in chloroform and in dioxane; sparingly soluble in alcohol; slightly soluble in ether.

Norethindrone Acetate: White to creamy white, odorless, crystalline powder. Practically insoluble in water; very soluble in chloroform; freely soluble in dioxane; soluble in ether and in alcohol.

Norethynodrel: White or practically white, odorless, crystalline powder. Melts at about 175°, over a range of about 3°. Is stable in air. Very slightly soluble in water and in solvent hexane; freely soluble in chloroform; soluble in acetone; sparingly soluble in alcohol.

Norfloxacin: Very slightly soluble in water, in methanol, and in ethyl acetate; slightly soluble in acetone and in alcohol; freely soluble in acetic acid; sparingly soluble in chloroform; insoluble in ether.

Norgestrel: White or practically white, practically odorless, crystalline powder. Insoluble in water; freely soluble in chloroform; sparingly soluble in alcohol.

Nortriptyline Hydrochloride: White to off-white powder, having a slight, characteristic odor. Its solution (1 in 100) has a pH of about 5. Soluble in water and in chloroform; sparingly soluble in methanol; practically insoluble in ether, in benzene, and in most other organic solvents.

Noscapine: Fine, white or practically white, crystalline powder. Freely soluble in chloroform; soluble in acetone; slightly soluble in alcohol and in ether; practically insoluble in water.

Nylidrin Hydrochloride: White,, odorless, crystalline powder. Sparingly soluble in water and in alcohol; slightly soluble in chloroform and in ether.

Nystatin: Yellow to light tan powder, having an odor suggestive of cereals. Is hygroscopic, and is affected by long exposure to light, heat, and air. Very slightly soluble in water; slightly to sparingly soluble in alcohol, in methanol, in *n*-propyl alcohol, and in *n*-butyl alcohol; insoluble in chloroform, in ether, and in benzene.

Octoxynol 9: Clear, pale yellow, viscous liquid, having a faint odor and a bitter taste. Miscible with water, with alcohol, and with acetone. Soluble in benzene and in toluene; practically insoluble in solvent hexane. *NF category:* Wetting and/or solubilizing agent.

Octyldodecanol: Clear water-white free-flowing liquid. Insoluble in water; soluble in alcohol and in ether.

Hydrophilic Ointment: *NF category:* Ointment base.

White Ointment: *NF category:* Ointment base.

Yellow Ointment: *NF category:* Ointment base.

Oleic Acid: Colorless to pale yellow, oily liquid when freshly prepared, but on exposure to air it gradually absorbs oxygen and darkens. Has a characteristic, lard-like odor and taste. When strongly heated in air, it is decomposed with the production of acrid vapors. Practically insoluble in water. Miscible with alcohol, with chloroform, with ether, with benzene, and with fixed and volatile oils. *NF category:* Emulsifying and/or solubilizing agent.

Oleovitamin A and D: Yellow to red, oily liquid, practically odorless or having a fish-like odor, and having no rancid odor or taste. Is a clear liquid at temperatures exceeding 65°, and may crystallize on cooling. Is unstable in air and in light. Insoluble in water and in glycerin; very soluble in ether and in chloroform; soluble in dehydrated alcohol and in vegetable oils.

Oleovitamin A and D Capsules: The oil contained in Oleovitamin A and D Capsules is a yellow to red, oily liquid, practically odorless or having a fish-like odor, and having no rancid odor or taste. Is a clear liquid at temperatures exceeding 65°, and may crystallize on cooling. Is unstable in air and in light.

Oleyl Alcohol: Clear, colorless to light yellow, oily liquid. Has a faint characteristic odor and a bland taste. Insoluble in water; soluble in alcohol, in ether, in isopropyl alcohol, and in light mineral oil. *NF category:* Emulsifying and/or solubilizing agent.

Olive Oil: Pale yellow, or light greenish yellow, oily liquid, having a slight, characteristic odor and taste, with a faintly acrid aftertaste. Slightly soluble in alcohol. Miscible with ether, with chloroform, and with carbon disulfide. *NF category:* Vehicle (oleaginous).

Opium: Has a very characteristic odor and a very bitter taste.

Powdered Opium: Light brown or moderately yellowish brown powder.

Orange Flower Oil: Pale yellow, slightly fluorescent liquid, which becomes reddish brown on exposure to light and air. Has a distinctive, fragrant odor, similar to that of orange blossoms and an aromatic, at first sweet, then somewhat bitter, taste. May become turbid or solid at low temperatures. Is neutral to litmus. *NF category:* Flavors and perfumes.

Orphenadrine Citrate: White, practically odorless, crystalline powder, having a bitter taste. Sparingly soluble in water; slightly soluble in alcohol; insoluble in chloroform, in benzene, and in ether.

Oxacillin Sodium: Fine, white, crystalline powder, odorless or having a slight odor. Freely soluble in water, in methanol, and in dimethylsulfoxide; slightly soluble in absolute alcohol, in chloroform, in pyridine, and in methyl acetate; insoluble in ethyl acetate, in ether, in benzene, and in ethylene chloride.

Oxacillin Sodium for Injection: Fine, white, crystalline powder, odorless or having a slight odor. Freely soluble in water, in methanol, and in dimethylsulfoxide; slightly soluble in absolute alcohol, in chloroform, in pyridine, and in methyl acetate; insoluble in ethyl acetate, in ether, in benzene, and in ethylene chloride.

Oxamniquine: Yellow-orange crystalline solid. Sparingly soluble in water; soluble in methanol, in chloroform, and in acetone.

Oxandrolone: White, odorless, crystalline powder. Is stable in air, but darkens on exposure to light. Melts at about 225°. Practically insoluble in water; freely soluble in chloroform; sparingly soluble in alcohol and in acetone.

Oxazepam: Creamy white to pale yellow powder. Is practically odorless. Practically insoluble in water; slightly soluble in alcohol and in chloroform; very slightly soluble in ether.

Oxprenolol Hydrochloride: White, crystalline powder. Freely soluble in alcohol, in chloroform, and in water; sparingly soluble in acetone; practically insoluble in ether.

Oxtriphylline: White, crystalline powder, having an amine-like odor. A solution (1 in 100) has a pH of about 10.3. Freely soluble in water and in alcohol; very slightly soluble in chloroform.

Oxybenzone: Pale yellow powder. Practically insoluble in water; freely soluble in alcohol and in toluene.

Oxybutynin Chloride: White, crystalline, practically odorless powder. Freely soluble in water and in alcohol; very soluble in methanol and in chloroform; soluble in acetone; slightly soluble in ether; very slightly soluble in hexane.

Oxygen: Colorless, odorless, tasteless gas, which supports combustion more energetically than does air. One liter at 0° and at a pressure of 760 mm of mercury weighs about 1.429 g. One volume dissolves in about 32 volumes of water and in about 7 volumes of alcohol at 20° and at a pressure of 760 mm of mercury.

Oxymetazoline Hydrochloride: White to practically white, fine crystalline powder. Is hygroscopic. Melts at about 300°, with decomposition. Soluble in water and in alcohol; practically insoluble in benzene, in chloroform, and in ether.

Oxymetholone: White to creamy white, crystalline powder. Is odorless, and is stable in air. Practically insoluble in water; freely soluble in chloroform; soluble in dioxane; sparingly soluble in alcohol; slightly soluble in ether.

Oxymorphone Hydrochloride: White or slightly off-white, odorless powder. Darkens on exposure to light. Its aqueous solutions are slightly acidic. Freely soluble in water; sparingly soluble in alcohol and in ether.

Oxyphenbutazone: White to yellowish white, odorless, crystalline powder. Melts over a wide range between about 85° and 100°. Very slightly soluble in water; soluble in alcohol; freely soluble in acetone and in ether.

Oxyphencyclimine Hydrochloride: White, odorless, crystalline powder, having a characteristic bitter taste. Melts at about 234°. Sparingly soluble in water; soluble in methanol; slightly soluble in chloroform.

Oxyquinoline Sulfate: Yellow powder. Melts at about 185°. Very soluble in water; freely soluble in methanol; slightly soluble in alcohol; practically insoluble in acetone and in ether.

Oxytetracycline: Pale yellow to tan, odorless, crystalline powder. Is stable in air, but exposure to strong sunlight causes it to darken. It loses potency in solutions of pH below 2, and is rapidly destroyed by alkali hydroxide solutions. Very slightly soluble in water; freely soluble in 3 *N* hydrochloric acid and in alkaline solutions; sparingly soluble in alcohol.

Oxytetracycline Calcium: Yellow to light brown, crystalline powder. Insoluble in water.

Oxytetracycline Hydrochloride: Yellow, odorless, crystalline powder, having a bitter taste. Is hygroscopic. Decomposes at a temperature exceeding 180°, and exposure to strong sunlight or to temperatures exceeding 90° in moist air causes it to darken. Its potency is diminished in solutions having a pH below 2, and is rapidly destroyed by alkali hydroxide solutions. Freely soluble in water, but crystals of oxytetracycline base separate as a result of partial hydrolysis of the hydrochloride. Sparingly soluble in alcohol and in methanol, and even less soluble in dehydrated alcohol; insoluble in chloroform and in ether.

Padimate O: A light yellow, mobile liquid having a faint, aromatic odor. Practically insoluble in water; soluble in alcohol, in isopropyl alcohol, and in mineral oil; practically insoluble in glycerin and in propylene glycol.

Pancreatin: Cream-colored, amorphous powder, having a faint, characteristic, but not offensive odor. It hydrolyzes fats to glycerol and fatty acids, changes protein into proteoses and derived substances, and converts starch into dextrins and sugars. Its greatest activities are in neutral or faintly alkaline media; more than traces of mineral acids or large amounts of alkali hydroxides make it inert. An excess of alkali carbonate also inhibits its action.

Pancrelipase: Cream-colored, amorphous powder, having a faint, characteristic, but not offensive odor. Pancrelipase hydrolyzes fats to glycerol and fatty acids, changes protein into proteoses and derived substances, and converts starch into dextrins and sugars. Its greatest activities are in neutral or faintly alkaline media; more than traces of mineral acids or large amounts of alkali hydroxides make it inert. An excess of alkali carbonate also inhibits its action.

Pancrelipase Capsules: The contents of Capsules conform to the *Description* under *Pancrelipase*, except that the odor may vary with the flavoring agent used.

Panthenol: White to creamy white, crystalline powder having a slight, characteristic odor. Freely soluble in water, in alcohol, and in propylene glycol; soluble in chloroform and in ether; slightly soluble in glycerin.

Papain: White to light tan, amorphous powder. Soluble in water, the solution being colorless to light yellow and more or less opalescent; practically insoluble in alcohol, in chloroform, and in ether.

Papaverine Hydrochloride: White crystals or white, crystalline powder. Is odorless, and has a slightly bitter taste. Is optically inactive. Its solutions are acid to litmus. Melts at about 220°, with decomposition. Soluble in water and in chloroform; slightly soluble in alcohol; practically insoluble in ether.

Parachlorophenol: White or pink crystals having a characteristic phenolic odor. When undiluted, it whitens and cauterizes the skin and mucous membranes. Melts at about 42°. Sparingly soluble in water and in liquid petrolatum; very soluble in alcohol, in glycerin, in chloroform, in ether, and in fixed and volatile oils; soluble in petrolatum.

Paraffin: Colorless or white, more or less translucent mass showing a crystalline structure. It is odorless and tasteless, and is slightly greasy to the touch. Insoluble in water and in alcohol; freely soluble in chloroform, in ether, in volatile oils, and in most warm fixed oils; slightly soluble in dehydrated alcohol. *NF category:* Stiffening agent.

Paraldehyde: Colorless, transparent liquid. Has a strong, characteristic but not unpleasant or pungent odor, and a disagreeable taste. Specific gravity is about 0.99. Soluble in water, but less soluble in boiling water. Miscible with alcohol, with chloroform, with ether, and with volatile oils.

Paramethadione: Clear, colorless liquid. May have an aromatic odor. A solution (1 in 40) has a pH of about 6. Sparingly soluble in water; freely soluble in alcohol, in benzene, in chloroform, and in ether.

Paramethasone Acetate: Fluffy, white to creamy white, odorless, crystalline powder. Melts at about 240°, with decomposition. Insoluble in water; soluble in chloroform, in ether, and in methanol.

Pargyline Hydrochloride: White or practically white, crystalline powder, having a slight odor. It sublimes slowly at elevated temperatures. Very soluble in water; freely soluble in alcohol and in chloroform; very slightly soluble in acetone and in benzene.

Paromomycin Sulfate: Creamy white to light yellow powder. Is odorless or practically odorless, and is very hygroscopic. Very soluble in water; insoluble in alcohol, in chloroform, and in ether.

Peanut Oil: Colorless or pale yellow, oily liquid with a bland taste. May have a characteristic, nutty odor. Very slightly soluble in alcohol. Miscible with ether, with chloroform, and with carbon disulfide. *NF category:* Solvent; vehicle (oleaginous).

Pectin: Coarse or fine powder, yellowish white in color, almost odorless, and having a mucilaginous taste. Almost completely soluble in 20 parts of water, forming a viscous, opalescent, colloidal solution that flows readily and is acid to litmus. Is practically insoluble in alcohol or in diluted alcohol and in other organic solvents. Pectin dissolves in water more readily if first moistened with alcohol, glycerin, or simple syrup, or if first mixed with 3 or more parts of sucrose. *NF category:* Suspending and/or viscosity-increasing agent.

Penicillamine: White or practically white, crystalline powder, having a slight, characteristic odor. Freely soluble in water; slightly soluble in alcohol; insoluble in chloroform and in ether.

Penicillin G Benzathine: White, odorless, crystalline powder. Very slightly soluble in water; sparingly soluble in alcohol.

Sterile Penicillin G Potassium: Colorless or white crystals, or white, crystalline powder. Is odorless or practically so, and is moderately hygroscopic. Its solutions are dextrorotatory. Its solutions retain substantially full potency for several days at temperatures below 15°, but are rapidly inactivated by acids, alkali hydroxides, glycerin, and oxidizing agents. Very soluble in water, in saline TS, and in dextrose solutions; sparingly soluble in alcohol.

Sterile Penicillin G Procaine: White crystals or white, very fine, microcrystalline powder. Is odorless or practically odorless, and is relatively stable in air. Its solutions are dextrorotatory. Is rapidly inactivated by acids, by alkali hydroxides, and by oxidizing agents. Slightly soluble in water; soluble in alcohol and in chloroform.

Sterile Penicillin G Sodium: Colorless or white crystals or white to slightly yellow, crystalline powder. Is odorless or practically odorless, and is moderately hygroscopic. Its solutions are dextrorotatory. Is relatively stable in air, but is inactivated by prolonged heating at about 100°, especially in the presence of moisture. Its solutions lose potency fairly rapidly at room temperature, but retain substantially full potency for several days at temperatures below 15°. Its solutions are rapidly inactivated by acids, alkali hydroxides, oxidizing agents, and penicillinase.

Penicillin V: White, odorless, crystalline powder. Very slightly soluble in water; freely soluble in alcohol and in acetone; insoluble in fixed oils.

Penicillin V Benzathine: Practically white powder, having a characteristic odor. Very slightly soluble in water; slightly soluble in alcohol and in ether; sparingly soluble in chloroform.

Penicillin V Potassium: White, odorless, crystalline powder. Very soluble in water; slightly soluble in alcohol; insoluble in acetone.

Diluted Pentaerythritol Tetranitrate: White to ivory-colored powder, having a faint, mild odor. Undiluted pentaerythritol tetranitrate is soluble in acetone; slightly soluble in alcohol and in ether; practically insoluble in water.

Pentazocine: White or very pale, tan-colored powder. Practically insoluble in water; freely soluble in chloroform; soluble in alcohol, in acetone, and in ether; sparingly soluble in benzene and in ethyl acetate.

Pentazocine Hydrochloride: White, crystalline powder. It exhibits polymorphism, one form melting at about 254° and the other at about 218°. Freely soluble in chloroform; soluble in alcohol; sparingly soluble in water; very slightly soluble in acetone and in ether; practically insoluble in benzene.

Pentobarbital: White to practically white, fine, practically odorless powder. May occur in a polymorphic form that melts at about 116°. This form gradually reverts to the more stable higher-melting form upon being heated at about 110°. Very slightly soluble in water and in carbon tetrachloride; very soluble in alcohol, in methanol, in ether, in chloroform, and in acetone; soluble in benzene.

Pentobarbital Sodium: White, crystalline granules or white powder. Is odorless or has a slight characteristic odor, and has a slightly bitter taste. Its solutions decompose on standing, heat accelerating the decomposition. Very soluble in water; freely soluble in alcohol; practically insoluble in ether.

Peppermint: Has an aromatic, characteristic odor and a pungent taste, and produces a cooling sensation in the mouth. *NF category:* Flavors and perfumes.

Peppermint Oil: Colorless or pale yellow liquid, having a strong, penetrating, characteristic odor and a pungent taste, followed by a sensation of cold when air is drawn into the mouth. *NF category:* Flavors and perfumes.

Peppermint Water: *NF category:* Flavored and/or sweetened vehicle.

Perphenazine: White to creamy white, odorless powder. Practically insoluble in water; freely soluble in alcohol and in chloroform; soluble in acetone.

Persic Oil: Clear, pale straw-colored or colorless, oily liquid. Is almost odorless, and has a bland taste. Is not turbid at temperatures exceeding 15°. Slightly soluble in alcohol. Miscible with ether, with chloroform, with benzene, and with solvent hexane. *NF category:* Vehicle (oleaginous).

Pertussis Immune Globulin: Transparent or slightly opalescent liquid, practically colorless, free from turbidity or particles, and practically odorless. May develop a slight, granular deposit during storage. Is standardized for agglutinating activity with the U. S. Standard Antipertussis Serum.

Pertussis Vaccine: More or less turbid, whitish liquid. Practically odorless, or having a faint odor because of the antimicrobial agent.

Pertussis Vaccine Adsorbed: Markedly turbid, whitish liquid. Is substantially odorless, or has a faint odor because of the antimicrobial agent.

Petrolatum: Unctuous yellowish to light amber mass, having not more than a slight fluorescence even after being melted. Is transparent in thin layers. Is free or practically free from odor and taste. Insoluble in water; freely soluble in benzene, in carbon disulfide, in chloroform, and in turpentine oil; soluble in ether, in solvent hexane, and in most fixed and volatile oils; practically insoluble in cold alcohol and hot alcohol and in cold dehydrated alcohol. *NF category:* Ointment base.

Hydrophilic Petrolatum: *NF category:* Ointment base.

White Petrolatum: White or faintly yellowish, unctuous mass, transparent in thin layers even after cooling to 0°. Insoluble in water; slightly soluble in cold or hot alcohol, and in cold dehydrated alcohol; freely soluble in benzene, in carbon disulfide, and in chloroform; soluble in ether, in solvent hexane, and in most fixed and volatile oils. *NF category:* Ointment base.

Phenacemide: White to practically white, fine crystalline powder. Is odorless, or practically odorless, and melts at about 213°. Very slightly soluble in water, in alcohol, in benzene, in chloroform, and in ether; slightly soluble in acetone and in methanol.

Phenazopyridine Hydrochloride: Light or dark red to dark violet, crystalline powder. Is odorless, or has a slight odor. Melts at about 235°, with decomposition. Slightly soluble in water, in alcohol, and in chloroform.

Phendimetrazine Tartrate: White, odorless, crystalline powder. Freely soluble in water; sparingly soluble in warm alcohol; insoluble in chloroform, in acetone, in ether, and in benzene. Phendimetrazine base is extracted by organic solvents from alkaline solution.

Phenelzine Sulfate: White to yellowish white powder, having a characteristic odor. Freely soluble in water; practically insoluble in alcohol, in chloroform, and in ether.

Phenindione: Creamy white to pale yellow, almost odorless crystals or crystalline powder. Very slightly soluble in water; freely soluble in chloroform; slightly soluble in alcohol and in ether.

Phenmetrazine Hydrochloride: White to off-white, crystalline powder. Very soluble in water; freely soluble in alcohol and in chloroform.

Phenobarbital: White, odorless, glistening, small crystals, or white, crystalline powder, which may exhibit polymorphism. Is stable in air. Its saturated solution has a pH of about 5. Very slightly soluble in water; soluble in alcohol, in ether, and in solutions of fixed alkali hydroxides and carbonates; sparingly soluble in chloroform.

Phenobarbital Sodium: Flaky crystals, or white, crystalline granules, or white powder. Is odorless, has a bitter taste, and is hygroscopic. Its solutions are alkaline to phenolphthalein TS, and decompose on standing. Very soluble in water; soluble in alcohol; practically insoluble in ether and in chloroform.

Phenol: Colorless to light pink, interlaced or separate, needle-shaped crystals, or white to light pink, crystalline mass. Has a characteristic odor. Is liquefied by warming and by the addition of 10% of water. Boils at about 182°, and its vapor is flammable. Gradually darkens on exposure to light and air. Soluble in water; very soluble in alcohol, in glycerin, in chloroform, in ether, and in fixed and volatile oils; sparingly soluble in mineral oil. *NF category:* Antimicrobial preservative.

Liquefied Phenol: Colorless to pink liquid, which may develop a red tint upon exposure to air or light. Has a characteristic, somewhat aromatic odor. It whitens and cauterizes the skin and mucous membranes. Specific gravity is about 1.065. Miscible with alcohol, with ether, and with glycerin. A mixture of equal volumes of Liquefied Phenol and glycerin is miscible with water.

Phenolphthalein: White or faintly yellowish white, crystalline powder. Is odorless, and is stable in air. Practically insoluble in water; soluble in alcohol; sparingly soluble in ether.

Phenprocoumon: Fine, white, crystalline powder. Is odorless, or has a slight odor. Practically insoluble in water; soluble in chloroform, in methanol, and in solutions of alkali hydroxides.

Phensuximide: White to off-white, crystalline powder. Is odorless, or has not more than a slight odor. Slightly soluble in water; very soluble in chloroform; soluble in alcohol.

Phentermine Hydrochloride: White, odorless, hygroscopic, crystalline powder. Soluble in water and in the lower alcohols; slightly soluble in chloroform; insoluble in ether.

Phentolamine Mesylate: White or off-white, odorless, crystalline powder. Its solutions are acid to litmus, having a pH of about 5, and slowly deteriorate. Melts at about 178°. Freely soluble in water and in alcohol; slightly soluble in chloroform.

Phenylalanine: White, odorless crystals, having a slightly bitter taste. Sparingly soluble in water; very slightly soluble in methanol, in alcohol, and in dilute mineral acids.

Phenylbutazone: White to off-white, odorless, crystalline powder. Very slightly soluble in water; freely soluble in acetone and in ether; soluble in alcohol.

Phenylephrine Hydrochloride: White or practically white, odorless crystals, having a bitter taste. Freely soluble in water and in alcohol.

Phenylephrine Hydrochloride Nasal Solution: Clear, colorless or slightly yellow, odorless liquid. Is neutral or acid to litmus.

Phenylephrine Hydrochloride Ophthalmic Solution: Clear, colorless or slightly yellow liquid, depending on the concentration.

Phenylethyl Alcohol: Colorless liquid, having a rose-like odor and a sharp, burning taste. Sparingly soluble in water; very soluble in alcohol, in fixed oils, in glycerin, and in propylene glycol; slightly soluble in mineral oil. *NF category:* Antimicrobial preservative.

Phenylmercuric Acetate: White to creamy white, crystalline powder, or small white prisms or leaflets. Is odorless. Slightly soluble in water; soluble in alcohol and in acetone. *NF category:* Antimicrobial preservative.

Phenylmercuric Nitrate: White, crystalline powder. Is affected by light. Its saturated solution is acid to litmus. Very slightly soluble in water; slightly soluble in alcohol and in glycerin. It is more soluble in the presence of either nitric acid or alkali hydroxides. *NF category:* Antimicrobial preservative.

Phenylpropanolamine Hydrochloride: White, crystalline powder, having a slight aromatic odor. Is affected by light. Freely soluble in water and in alcohol; insoluble in ether.

Phenytoin: White, odorless powder. Melts at about 295°. Practically insoluble in water; soluble in hot alcohol; slightly soluble in cold alcohol, in chloroform, and in ether.

Phenytoin Sodium: White, odorless powder. Is somewhat hygroscopic and on exposure to air gradually absorbs carbon dioxide. Freely soluble in water, the solution usually being somewhat turbid due to partial hydrolysis and absorption of carbon dioxide; soluble in alcohol; practically insoluble in ether and in chloroform.

Sodium Phosphate P 32 Solution: Clear, colorless solution. Upon standing, both the Solution and the glass container may darken as a result of the effects of the radiation.

Phosphoric Acid: Colorless, odorless liquid of syrupy consistency. Specific gravity is about 1.71. Miscible with water and with alcohol. *NF category:* Acidifying agent; buffering agent.

Diluted Phosphoric Acid: Clear, colorless, odorless liquid. Specific gravity is about 1.057. *NF category:* Acidifying agent.

Physostigmine: White, odorless, microcrystalline powder. Acquires a red tint when exposed to heat, light, air, or contact with traces of metals. Melts at a temperature not lower than 103°. Slightly soluble in water; very soluble in chloroform and in dichloromethane; freely soluble in alcohol; soluble in benzene and in fixed oils.

Physostigmine Salicylate: White, shining, odorless crystals or white powder. Acquires a red tint when exposed to heat, light, air, or contact with traces of metals for long periods. Melts at about 184°. Sparingly soluble in water; freely soluble in chloroform; soluble in alcohol; slightly soluble in ether.

Physostigmine Sulfate: White, odorless, microcrystalline powder. Is deliquescent in moist air and acquires a red tint when exposed to heat, light, air, or contact with traces of metals for long periods. Melts at about 143°. Freely soluble in water; very soluble in alcohol; very slightly soluble in ether.

Phytonadione: Clear, yellow to amber, very viscous, odorless or practically odorless liquid, having a specific gravity of about 0.967. Is stable in air, but decomposes on exposure to sunlight. Insoluble in water; soluble in dehydrated alcohol, in benzene, in chloroform, in ether, and in vegetable oils; slightly soluble in alcohol.

Pilocarpine: A viscous, oily liquid, or crystals melting at about 34°. Exceedingly hygroscopic. Soluble in water, in alcohol, and in chloroform; practically insoluble in petroleum ether; sparingly soluble in ether and in benzene.

Pilocarpine Hydrochloride: Colorless, translucent, odorless, faintly bitter crystals. Is hygroscopic and is affected by light. Its solutions are acid to litmus. Very soluble in water; freely soluble in alcohol; slightly soluble in chloroform; insoluble in ether.

Pilocarpine Nitrate: Shining, white crystals. Is stable in air but is affected by light. Its solutions are acid to litmus. Freely soluble in water; sparingly soluble in alcohol; insoluble in chloroform and in ether.

Pimozide: White, crystalline powder. Insoluble in water; slightly soluble in ether and in alcohol; freely soluble in chloroform.

Pindolol: White to off-white, crystalline powder, having a faint odor. Practically insoluble in water; slightly soluble in methanol; very slightly soluble in chloroform.

Piperacetazine: Yellow, granular powder. Practically insoluble in water; freely soluble in chloroform; soluble in alcohol and in dilute hydrochloric acid.

Piperazine: White to slightly off-white lumps or flakes, having an ammoniacal odor. Soluble in water and in alcohol; insoluble in ether.

Piperazine Citrate: White, crystalline powder, having not more than a slight odor. Its solution (1 in 10) has a pH of about 5. Soluble in water; insoluble in alcohol and in ether.

Pipobroman: White or practically white, crystalline powder, having a slightly sharp, fruity odor. Slightly soluble in water; freely soluble in chloroform; soluble in acetone; sparingly soluble in alcohol; very slightly soluble in ether.

Piroxicam: Off-white to light tan or light yellow, odorless powder. Forms a monohydrate that is yellow. Very slightly soluble in water, in dilute acids, and in most organic solvents; slightly soluble in alcohol and in aqueous alkaline solutions.

Plague Vaccine: Turbid, whitish liquid, practically odorless or having a faint odor because of the preservative.

Plantago Seed: All varieties are practically odorless and have a bland, mucilaginous taste.

Plicamycin: Yellow, odorless, hygroscopic, crystalline powder. Slightly soluble in water and in methanol; very slightly soluble in alcohol; freely soluble in ethyl acetate.

Podophyllum: Has a slight odor and a disagreeably bitter and acrid taste.

Podophyllum Resin: Amorphous powder, varying in color from light brown to greenish yellow, turning darker when subjected to a temperature exceeding 25° or when exposed to light. Has a slight, peculiar, faintly bitter taste. Its alcohol solution is acid to moistened litmus paper. Soluble in alcohol with a slight opalescence; partially soluble in ether and in chloroform.

Polacrilin Potassium: White to off-white, free-flowing powder. Has a faint odor or is odorless. Insoluble in water and in most liquids. *NF category:* Tablet disintegrant.

Poliovirus Vaccine Inactivated: Clear, reddish-tinged or yellowish liquid, that may have a slight odor because of the preservative.

Poliovirus Vaccine Live Oral: Is generally frozen but, in liquid form, is clear and colorless, or may have a yellow or red tinge.

Poloxamer: *NF category:* Emulsifying and/or solubilizing agent. *Poloxamer 124:* Colorless liquid, having a mild odor. When solidified, it melts at about 16°. Freely soluble in water, in alcohol, in isopropyl alcohol, in propylene glycol, and in xylene.

Poloxamer 188: White, prilled or cast solid. Is odorless, or has a very mild odor. Melts at about 52°. Freely soluble in water and in alcohol.

Poloxamer 237: White, prilled or cast solid. Is odorless, or has a very mild odor. Melts at about 49°. Freely soluble in water and in alcohol; sparingly soluble in isopropyl alcohol and in xylene.

Poloxamer 338: White, prilled or cast solid. Is odorless, or has a very mild odor. Melts at about 57°. Freely soluble in water and in alcohol; sparingly soluble in propylene glycol.

Poloxamer 407: White, prilled or cast solid. Is odorless, or has a very mild odor. Melts at about 56°. Freely soluble in water, in alcohol, and in isopropyl alcohol.

Polyethylene Excipient: White, translucent, partially crystalline and partially amorphous resin. Available in various grades and types, differing from one another in molecular weight, molecular weight distribution, degree of chain branching, and extent of crystallinity. Insoluble in water; soluble in hot benzene.

Polyethylene Glycol: Polyethylene Glycol is usually designated by a number that corresponds approximately to its average molecular weight. As the average molecular weight increases, the water solubility, vapor pressure, hygroscopicity, and solubility in organic solvents decrease, while congealing temperature, specific gravity, flash point, and viscosity increase. Liquid grades occur as clear to slightly hazy, colorless or practically colorless, slightly hygroscopic, viscous liquids, having a slight, characteristic odor, and a specific gravity at 25° of about 1.12. Solid grades occur as practically odorless and tasteless, white, waxy, plastic material having a consistency similar to beeswax, or as creamy white flakes, beads, or powders. The accompanying table states the approximate congealing temperatures that are characteristic of commonly available grades. Liquid grades are miscible with water; solid grades are freely soluble in water; and all are soluble in acetone, in alcohol, in chloroform, in ethylene glycol monoethyl ether, in ethyl acetate, and in toluene; all are insoluble in ether and in hexane. *NF category:* Coating agent; plasticizer; solvent; suppository base; tablet and/or capsule lubricant.

Nominal Molecular Weight Polyethylene Glycol	Approximate Congealing Temperature (°)
300	−11
400	6
600	20
900	34
1000	38
1450	44
3350	56
4500	58
8000	60

Polyethylene Oxide: Polyethylene oxide resins are high molecular weight polymers having the common structure:

$$(\text{–O–}CH_2CH_2\text{–})_n$$

in which n, the degree of polymerization, varies from about 2000 to over 100,000. Polyethylene oxide, being a polyether, strongly hydrogen, bonds with water. It is nonionic and undergoes salting-out effects associated with neutral molecules in solutions of high dielectric media. Salting-out effects manifest themselves in depressing the upper temperature limit of solubility, and in reducing the viscosity of both dilute and concentrated solutions of the polymers. All molecular weight grades are powdered or granular solids. They are soluble in water but, because of the high solution viscosities obtained (see table), solutions over 1% in water may be difficult to prepare.

Approximate Molecular Weight	Typical Solution Viscosity (cps), 25°	
	5% Solution	1% Solution
100,000	40	
200,000	100	
300,000	800	
400,000	3000	
600,000	6000	
900,000	15000	
4,000,000		3500
5,000,000		5500

The water solubility, hygroscopicity, solubility in organic solvents, and melting point do not vary in the specified molecular weight range. At room temperature polyethylene oxide is miscible with water in all proportions. At concentrations of about 20% polymer in water the solutions are nontacky, reversible, elastic gels. At higher concentrations, the solutions are tough, elastic materials with the water acting as a plasticizer. Polyethylene oxide is also freely soluble in acetonitrile, in ethylene dichloride, in trichloroethylene, and in methylene chloride. Heating may be required to obtain solutions in many other organic solvents. It is insoluble in aliphatic hydrocarbons, in ethylene glycol, in diethylene glycol, and in glycerol.

Polyethylene 50 Stearate: *NF category:* Emulsifying and/or solubilizing agent.

Polymyxin B Sulfate: White to buff-colored powder. Is odorless or has a faint odor. Freely soluble in water; slightly soluble in alcohol.

Sterile Polymyxin B Sulfate: White to buff-colored powder. Is odorless or has a faint odor. Freely soluble in water; slightly soluble in alcohol.

Polyoxyl 10 Oleyl Ether: White, soft semisolid, or pale yellow liquid, having a bland odor. Soluble in water and in alcohol; dispersible in mineral oil and in propylene glycol, with possible separation on standing. *NF category:* Emulsifying and/or solubilizing agent; wetting and/or solubilizing agent.

Polyoxyl 20 Cetostearyl Ether: Cream-colored, waxy, unctuous mass, melting, when heated, to a clear brownish yellow liquid. Soluble in water, in alcohol, and in acetone; insoluble in solvent hexane. *NF category:* Emulsifying and/or solubilizing agent; wetting and/or solubilizing agent.

Polyoxyl 35 Castor Oil: Yellow oily liquid, having a faint, characteristic odor and a somewhat bitter taste. Very soluble in water, producing a practically odorless and colorless solution; soluble in alcohol and in ethyl acetate; insoluble in mineral oils. *NF category:* Emulsifying and/or solubilizing agent; wetting and/or solubilizing agent.

Polyoxyl 40 Hydrogenated Castor Oil: White to yellowish paste or pasty liquid, having a faint odor and slight taste. Very soluble in water, producing a practically tasteless, odorless, and colorless solution; soluble in alcohol and in ethyl acetate; insoluble in mineral oils. *NF category:* Emulsifying and/or solubilizing agent; wetting and/or solubilizing agent.

Polyoxyl 40 Stearate: Waxy, white to light tan solid. Is odorless or has a faint, fat-like odor. Soluble in water, in alcohol, in ether, and in acetone; insoluble in mineral oil and in vegetable oils. *NF category:* Emulsifying and/or solubilizing agent; wetting and/or solubilizing agent.

Polyoxyl 50 Stearate: Soft, cream-colored, waxy solid, having a faint, fat-like odor. Melts at about 45°. Soluble in water and in isopropyl alcohol. *NF category:* Wetting and/or solubilizing agent.

Polysorbate 20: Lemon to amber liquid having a faint characteristic odor. Soluble in water, in alcohol, in ethyl acetate, in methanol, and in dioxane; insoluble in mineral oil. *NF category:* Emulsifying and/or solubilizing agent; wetting and/or solubilizing agent.

Polysorbate 40: Yellow liquid having a faint, characteristic odor. Soluble in water and in alcohol; insoluble in mineral oil and in vegetable oils. *NF category:* Emulsifying and/or solubilizing agent; wetting and/or solubilizing agent.

Polysorbate 60: Lemon- to orange-colored, oily liquid or semi-gel having a faint, characteristic odor. Soluble in water, in ethyl acetate, and in toluene; insoluble in mineral oil and in vegetable oils. *NF category:* Emulsifying and/or solubilizing agent; wetting and/or solubilizing agent.

Polysorbate 80: Lemon- to amber-colored, oily liquid having a faint, characteristic odor and a warm, somewhat bitter taste. Very soluble in water, producing an odorless and practically colorless solution; soluble in alcohol and in ethyl acetate; insoluble in mineral oil. *NF category:* Emulsifying and/or solubilizing agent; wetting and/or solubilizing agent.

Polythiazide: White, crystalline powder, having a characteristic odor. Practically insoluble in water and in chloroform; soluble in methanol and in acetone.

Polyvinyl Acetate Phthalate: Free-flowing white powder. May have a slight odor of acetic acid. Insoluble in water, in methylene chloride, and in chloroform; soluble in methanol and in alcohol. *NF category:* Coating agent.

Polyvinyl Alcohol: White to cream-colored granules, or white to cream-colored powder. Is odorless. Freely soluble in water at room temperature. Solution may be effected more rapidly at somewhat higher temperatures. *NF category:* Suspending and/or viscosity-increasing agent.

Sulfurated Potash: Irregular, liver-brown pieces when freshly made, changing to a greenish yellow. Has an odor of hydrogen sulfide and a bitter, acrid, and alkaline taste, and decomposes on exposure to air. A solution (1 in 10) is light brown in color and is alkaline to litmus. Freely soluble in water, usually leaving a slight residue. Alcohol dissolves only the sulfides.

Potassium Acetate: Colorless, monoclinic crystals or white, crystalline powder having a saline and slightly alkaline taste. Is odorless, or has a faint acetous odor. Deliquesces on exposure to moist air. Very soluble in water; freely soluble in alcohol.

Potassium Benzoate: White, odorless, or practically odorless, granular or crystalline powder. Is stable in air. Freely soluble in water; sparingly soluble in alcohol and somewhat more soluble in 90 percent alcohol.

Potassium Bicarbonate: Colorless, transparent, monoclinic prisms or as a white, granular powder. Is odorless, and is stable in air. Its solutions are neutral or alkaline to phenolphthalein TS. Freely soluble in water; practically insoluble in alcohol.

Potassium Chloride: Colorless, elongated, prismatic, or cubical crystals, or white, granular powder. Is odorless, has a saline taste, and is stable in air. Its solutions are neutral to litmus. Freely soluble in water and even more soluble in boiling water; insoluble in alcohol. *NF category:* Tonicity agent.

Potassium Citrate: Transparent crystals or white, granular powder. Is odorless, has a cooling, saline taste, and is deliquescent when exposed to moist air. Freely soluble in water; almost insoluble in alcohol.

Potassium Gluconate: White to yellowish white, crystalline powder or granules. Is odorless, has a slightly bitter taste, and is stable in air. Its solutions are slightly alkaline to litmus. Freely soluble in water; practically insoluble in dehydrated alcohol, in ether, in benzene, and in chloroform.

Potassium Hydroxide: White or practically white, fused masses, or small pellets, or flakes, or sticks, or other forms. Is hard and brittle and shows a crystalline fracture. Exposed to air, it rapidly absorbs carbon dioxide and moisture, and deliquesces. Freely soluble in water, in alcohol, and in glycerin; very soluble in boiling alcohol. *NF category:* Alkalizing agent.

Potassium Iodide: Hexahedral crystals, either transparent and colorless or somewhat opaque and white, or a white, granular powder. Is slightly hygroscopic. Its solutions are neutral or alkaline to litmus. Very soluble in water and even more soluble in boiling water; freely soluble in glycerin; soluble in alcohol.

Potassium Iodide Oral Solution: Clear, colorless, odorless liquid, having a characteristic, strongly salty taste. Is neutral or alkaline to litmus. Specific gravity is about 1.70.

Potassium Metaphosphate: White, odorless powder. Insoluble in water; soluble in dilute solutions of sodium salts. *NF category:* Buffering agent.

Potassium Permanganate: Dark purple crystals, almost opaque by transmitted light and of a blue metallic luster by reflected light. Its color is sometimes modified by a dark bronze-like appearance. Is stable in air. Soluble in water; freely soluble in boiling water.

Monobasic Potassium Phosphate: Colorless crystals or white, granular or crystalline powder. Is odorless, and is stable in air. The pH of a solution (1 in 100) is about 4.5. Freely soluble in water; practically insoluble in alcohol. *NF category:* Buffering agent.

Potassium Sodium Tartrate: Colorless crystals or white, crystalline powder, having a cooling, saline taste. As it effloresces slightly in warm, dry air, the crystals are often coated with a white powder. Freely soluble in water; practically insoluble in alcohol.

Potassium Sorbate: White crystals or powder, having a characteristic odor. Melts at about 270°, with decomposition. Freely soluble in water; soluble in alcohol. *NF category:* Antimicrobial preservative.

Povidone: White to creamy white powder, having a faint odor. Is hygroscopic. Soluble in water, in alcohol, and in chloroform; insoluble in ether. *NF category:* Suspending and/or viscosity-increasing agent; tablet binder.

Povidone-Iodine: Yellowish brown, amorphous powder, having a slight, characteristic odor. Its solution is acid to litmus. Soluble in water and in alcohol; practically insoluble in chloroform, in carbon tetrachloride, in ether, in solvent hexane, and in acetone.

Povidone-Iodine Topical Aerosol Solution: The liquid obtained from Povidone-Iodine Topical Aerosol Solution is transparent, having a reddish brown color.

Pralidoxime Chloride: White to pale-yellow, crystalline powder. Is odorless and is stable in air. Freely soluble in water.

Sterile Pralidoxime Chloride: White to pale-yellow, crystalline powder. Is odorless and is stable in air. Freely soluble in water.

Pramoxine Hydrochloride: White to practically white, crystalline powder, having a numbing taste. May have a slight aromatic odor. The pH of a solution (1 in 100) is about 4.5. Freely soluble in water and in alcohol; soluble in chloroform; very slightly soluble in ether.

Prazepam: White to off-white crystalline powder. Freely soluble in acetone; soluble in dilute mineral acids, in alcohol, and in chloroform.

Praziquantel: White or practically white, crystalline powder; odorless or having a faint characteristic odor. Very slightly soluble in water; freely soluble in alcohol and in chloroform.

Prazosin Hydrochloride: White to tan powder. Slightly soluble in water, in methanol, in dimethylformamide, and in dimethylacetamide; very slightly soluble in alcohol; practically insoluble in chloroform and in acetone.

Prednisolone: White to practically white, odorless, crystalline powder. Melts at about 235°, with some decomposition (see *Melting Range or Temperature* ⟨741⟩). Very slightly soluble in water; soluble in methanol and in dioxane; sparingly soluble in acetone and in alcohol; slightly soluble in chloroform.

Prednisolone Acetate: White to practically white, odorless, crystalline powder. Melts at about 235°, with some decomposition (see *Melting Range or Temperature* ⟨741⟩). Practically insoluble in water; slightly soluble in acetone, in alcohol, and in chloroform.

Prednisolone Hemisuccinate: Fine, creamy white powder with friable lumps; practically odorless. Melts at about 205°, with decomposition. Very slightly soluble in water; freely soluble in alcohol; soluble in acetone.

Prednisolone Sodium Phosphate: White or slightly yellow, friable granules or powder. Is odorless or has a slight odor. Is slightly hygroscopic. Freely soluble in water; soluble in methanol; slightly soluble in alcohol and in chloroform; very slightly soluble in acetone and in dioxane.

Prednisolone Sodium Succinate for Injection: Creamy white powder with friable lumps, having a slight odor.

Prednisolone Tebutate: White to slightly yellow, free-flowing powder, which may show some soft lumps. Is odorless or has not more than a moderate, characteristic odor. Is hygroscopic. Very slightly soluble in water; freely soluble in chloroform and in dioxane; soluble in acetone; sparingly soluble in alcohol and in methanol.

Prednisone: White to practically white, odorless, crystalline powder. Melts at about 230°, with some decomposition (see *Melting Range or Temperature* ⟨741⟩). Very slightly soluble in water; slightly soluble in alcohol, in chloroform, in dioxane, and in methanol.

Prilocaine Hydrochloride: White, odorless, crystalline powder, having a bitter taste. Freely soluble in water and in alcohol; slightly soluble in chloroform; very slightly soluble in acetone; practically insoluble in ether.

Primaquine Phosphate: Orange-red, crystalline powder. Is odorless and has a bitter taste. Its solutions are acid to litmus. Melts at about 200°. Soluble in water; insoluble in chloroform and in ether.

Primidone: White, crystalline powder. Is odorless and has a slightly bitter taste. Very slightly soluble in water and in most organic solvents; slightly soluble in alcohol.

Probucol: White to off-white, crystalline powder. Insoluble in water; freely soluble in chloroform and in *n*-propyl alcohol; soluble in alcohol and in solvent hexane.

Probenecid: White or practically white, fine, crystalline powder. Is practically odorless. Practically insoluble in water and in dilute acids; soluble in dilute alkali, in chloroform, in alcohol, and in acetone.

Procainamide Hydrochloride: White to tan, crystalline powder. Is odorless. Its solution (1 in 10) has a pH between 5 and 6.5. Very soluble in water; soluble in alcohol; slightly soluble in chloroform; very slightly soluble in benzene and in ether.

Procainamide Hydrochloride Injection: Colorless, or having not more than a slight yellow color.

Procaine Hydrochloride: Small, white crystals or white, crystalline powder. Is odorless. Exhibits local anesthetic properties when placed on the tongue. Freely soluble in water; soluble in alcohol; slightly soluble in chloroform; practically insoluble in ether.

Procaine Hydrochloride Injection: Clear, colorless liquid.

Sterile Procaine Hydrochloride: Small, white crystals or white, crystalline powder. Is odorless. Exhibits local anesthetic properties when placed on the tongue. Freely soluble in water; soluble in alcohol; slightly soluble in chloroform; practically insoluble in ether.

Prochlorperazine: Clear, pale yellow, viscous liquid. Is sensitive to light. Very slightly soluble in water; freely soluble in alcohol, in chloroform, and in ether.

Prochlorperazine Edisylate: White to very light yellow, odorless, crystalline powder. Its solutions are acid to litmus. Freely soluble in water; very slightly soluble in alcohol; insoluble in ether and in chloroform.

Prochlorperazine Maleate: White or pale yellow, practically odorless, crystalline powder. Its saturated solution is acid to litmus. Practically insoluble in water and in alcohol; slightly soluble in warm chloroform.

Procyclidine Hydrochloride: White, crystalline powder, having a moderate, characteristic odor. Melts at about 225°, with decomposition. Soluble in water and in alcohol; insoluble in ether and in acetone.

Progesterone: White or creamy white, odorless, crystalline powder. Is stable in air. Practically insoluble in water; soluble in alcohol, in acetone, and in dioxane; sparingly soluble in vegetable oils.

Proline: White, odorless crystals, having a slightly sweet taste. Freely soluble in water and in absolute alcohol; insoluble in ether, in butanol, and in isopropanol.

Promazine Hydrochloride: White to slightly yellow, practically odorless, crystalline powder. It oxidizes upon prolonged exposure to air and acquires a blue or pink color. Freely soluble in water and in chloroform.

Promethazine Hydrochloride: White to faint yellow, practically odorless, crystalline powder. Slowly oxidizes, and acquires a blue color, on prolonged exposure to air. Very soluble in water, in hot dehydrated alcohol, and in chloroform; practically insoluble in ether, in acetone, and in ethyl acetate.

Propane: Colorless, flammable gas (boiling temperature is about −42°). One hundred volumes of water dissolves 6.5 volumes at 17.8° and 753 mm pressure; 100 volumes of anhydrous alcohol dissolves 790 volumes at 16.6° and 754 mm pressure; 100 volumes of ether dissolves 926 volumes at 16.6° and 757 mm pressure; 100 volumes of chloroform dissolves 1299 volumes at 21.6° and 757 mm pressure. Vapor pressure at 21° is about 10290 mm of mercury (108 psig). *NF category:* Aerosol propellant.

Propantheline Bromide: White or practically white crystals. Is odorless and has a bitter taste. Melts at about 160°, with decomposition. Very soluble in water, in alcohol, and in chloroform; practically insoluble in ether and in benzene.

Sterile Propantheline Bromide: White or practically white crystals. Is odorless and has a bitter taste. Very soluble in water, in alcohol, and in chloroform; practically insoluble in ether and in benzene.

Proparacaine Hydrochloride: White to off-white, or faintly buff-colored, odorless, crystalline powder. Its solutions are neutral to litmus. Soluble in water, in warm alcohol, and in methanol; insoluble in ether and in benzene.

Proparacaine Hydrochloride Ophthalmic Solution: Colorless or faint yellow solution.

Propiomazine Hydrochloride: Yellow, practically odorless powder. Very soluble in water; freely soluble in alcohol; insoluble in benzene.

Propionic Acid: Oily liquid having a slight pungent, rancid odor. Miscible with water and with alcohol and various other organic solvents.

Propoxycaine Hydrochloride: White, odorless, crystalline solid which discolors on prolonged exposure to light and to air. The pH of a solution (1 in 50) is about 5.4. Freely soluble in water; soluble in alcohol; sparingly soluble in ether; practically insoluble in acetone and in chloroform.

Propoxyphene Hydrochloride: White, crystalline powder. Is odorless, and has a bitter taste. Freely soluble in water; soluble in alcohol, in chloroform, and in acetone; practically insoluble in benzene and in ether.

Propoxyphene Napsylate: White powder, having essentially no odor, but having a bitter taste. Very slightly soluble in water; soluble in methanol, in alcohol, in chloroform, and in acetone.

Propranolol Hydrochloride: White to off-white, crystalline powder. Is odorless and has a bitter taste. Melts at about 164°. Soluble in water and in alcohol; slightly soluble in chloroform; practically insoluble in ether.

Propyl Gallate: White, crystalline powder having a very slight, characteristic odor. Slightly soluble in water; freely soluble in alcohol. *NF category:* Antioxidant.

Propylene Carbonate: Clear, colorless, mobile liquid. Miscible with alcohol and with chloroform. Freely soluble in water; insoluble in hexane.

Propylene Glycol: Clear, colorless, viscous liquid having a slight, characteristic taste. Is practically odorless. Absorbs moisture when exposed to moist air. Miscible with water, with acetone, and with chloroform. Soluble in ether and will dissolve many essential oils, but is immiscible with fixed oils. *NF category:* Humectant; plasticizer; solvent.

Propylene Glycol Alginate: White to yellowish fibrous or granular powder. Practically odorless and tasteless. Soluble in water, in solutions of dilute organic acids, and, depending on the degree of esterification, in hydroalcoholic mixture containing up to 60% by weight of alcohol to form stable, viscous colloidal solutions at a pH of 3.

Propylene Glycol Diacetate: Clear, colorless liquid having a mild, fruity odor. Soluble in water. *NF category:* Emulsifying and/or solubilizing agent.

Propylene Glycol Monostearate: White, wax-like solid or as white, wax-like beads or flakes. Has a slight, agreeable, fatty odor and taste. Insoluble in water, but may be dispersed in hot water with the aid of a small amount of soap or other suitable surface-active agent; soluble in organic solvents such as alcohol, mineral or fixed oils, benzene, ether, and acetone. *NF category:* Emulsifying and/or solubilizing agent.

Propylhexedrine: Clear, colorless liquid, having a characteristic, amine-like odor. Volatilizes slowly at room temperature. Absorbs carbon dioxide from the air, and its solutions are alkaline to litmus. Boils at about 205°. Very slightly soluble in water. Miscible with alcohol, with chloroform, and with ether.

Propyliodone: White or almost white, crystalline powder. Is odorless or has a faint odor. Practically insoluble in water; soluble in acetone, in alcohol, and in ether.

Propylparaben: Small, colorless crystals or white powder. Very slightly soluble in water; freely soluble in alcohol and in ether; slightly soluble in boiling water. *NF category:* Antimicrobial preservative.

Propylparaben Sodium: White powder. Is odorless and hygroscopic. Freely soluble in water; sparingly soluble in alcohol; insoluble in fixed oils.

Propylthiouracil: White, powdery, crystalline substance. Is starch-like in appearance and to the touch, and has a bitter taste. Slightly soluble in water; sparingly soluble in alcohol; slightly soluble in chloroform and in ether; soluble in ammonium hydroxide and in alkali hydroxides.

Protamine Sulfate Injection: Colorless solution, which may have the odor of a preservative.

Protamine Sulfate for Injection: White, odorless powder, having the characteristic appearance of solids dried from the frozen state.

Protein Hydrolysate Injection: Yellowish to reddish amber, transparent liquid.

Protriptyline Hydrochloride: White to yellowish powder. Is odorless, or has not more than a slight odor. Melts at about 168°. Freely soluble in water, in alcohol, and in chloroform; practically insoluble in ether.

Pseudoephedrine Hydrochloride: Fine, white to off-white crystals or powder, having a faint characteristic odor. Very soluble in water; freely soluble in alcohol; sparingly soluble in chloroform.

Pseudoephedrine Sulfate: White crystals or crystalline powder. Is odorless. Freely soluble in alcohol.

Pumice: Very light, hard, rough, porous, grayish masses or gritty, grayish powder. Is odorless and tasteless, and is stable in air. Practically insoluble in water; is not attacked by acids.

Pyrantel Pamoate: Yellow to tan solid. Practically insoluble in water and in methanol; soluble in dimethylsulfoxide; slightly soluble in dimethylformamide.

Pyrazinamide: White to practically white, odorless or practically odorless, crystalline powder. Sparingly soluble in water; slightly soluble in alcohol, in ether, and in chloroform.

Pyridostigmine Bromide: White or practically white, crystalline powder, having an agreeable, characteristic odor. Is hygroscopic. Freely soluble in water, in alcohol, and in chloroform; slightly soluble in solvent hexane; practically insoluble in ether.

Pyridoxine Hydrochloride: White to practically white crystals or crystalline powder. Is stable in air, and is slowly affected by sunlight. Its solutions have a pH of about 3. Freely soluble in water; slightly soluble in alcohol; insoluble in ether.

Pyrilamine Maleate: White, crystalline powder, usually having a faint odor. Its solutions are acid to litmus. Very soluble in water; freely soluble in alcohol and in chloroform; slightly soluble in ether and in benzene.

Pyrimethamine: White, odorless, crystalline powder. Practically insoluble in water; slightly soluble in acetone, in alcohol, and in chloroform.

Pyrvinium Pamoate: Bright orange or orange-red to practically black, crystalline powder. Practically insoluble in water and in ether; freely soluble in glacial acetic acid; slightly soluble in chloroform and in methoxyethanol; very slightly soluble in methanol.

Pyrvinium Pamoate Oral Suspension: Dark red, opaque suspension of essentially very fine, amorphous particles or aggregates, usually less than 10 μm in size. Larger particles, some of which may be crystals, up to 100 μm in size also may be present.

Quinacrine Hydrochloride: Bright yellow, crystalline powder. Is odorless and has a bitter taste. Its solution (1 in 100) has a pH of about 4.5. Melts at about 250°, with decomposition. Sparingly soluble in water; soluble in alcohol.

Quinestrol: White, practically odorless powder. Insoluble in water; soluble in alcohol, in chloroform, and in ether.

Quinethazone: White to yellowish white, crystalline powder. Very slightly soluble in water; freely soluble in solutions of alkali hydroxides and carbonates; sparingly soluble in pyridine; slightly soluble in alcohol.

Quinidine Gluconate: White powder. Is odorless and has a very bitter taste. Freely soluble in water; slightly soluble in alcohol.

Quinidine Sulfate: Fine, needle-like, white crystals, frequently cohering in masses, or fine, white powder. Is odorless, has a very bitter taste, and darkens on exposure to light. Its solutions are neutral or alkaline to litmus. Slightly soluble in water; soluble in alcohol and in chloroform; insoluble in ether.

Quinine Sulfate: White, fine, needle-like crystals, usually lusterless, making a light and readily compressible mass. Is odorless and has a persistent, very bitter taste. It darkens on exposure to light. Its saturated solution is neutral or alkaline to litmus. Slightly soluble in water, in alcohol, in chloroform, and in ether; freely soluble in alcohol at 80°, and in a mixture of 2 volumes of chloroform and 1 volume of dehydrated alcohol; sparingly soluble in water at 100°.

Rabies Immune Globulin: Transparent or slightly opalescent liquid, practically colorless and practically odorless. May develop a slight, granular deposit during storage.

Rabies Vaccine: White to straw-colored, amorphous pellet, which may or may not become fragmented when shaken.

Racepinephrine: White to nearly white, crystalline, odorless powder, gradually darkening on exposure to light and air. With acids, it forms salts that are readily soluble in water, and the base may be recovered by the addition of ammonium hydroxide. Very slightly soluble in water and in alcohol; insoluble in ether, in chloroform, and in fixed and volatile oils.

Racepinephrine Hydrochloride: Fine, white, odorless powder. Darkens on exposure to light and air. Its solutions are acid to litmus. Freely soluble in water; sparingly soluble in alcohol.

Ranitidine Hydrochloride: White to pale yellow, crystalline, practically odorless powder. Is sensitive to light and moisture. Melts at about 140°, with decomposition. Very soluble in water; moderately soluble in alcohol; and sparingly soluble in chloroform.

Purified Rayon: White, lustrous or dull, fine, soft, filamentous fibers, appearing under the microscope as round, oval, or slightly flattened translucent rods, straight or crimped, striate and with serrate cross-sectional edges. Is practically odorless and practically tasteless. Very soluble in ammoniated cupric oxide TS and in dilute sulfuric acid (3 in 5); insoluble in ordinary solvents.

Reserpine: White or pale buff to slightly yellowish, odorless, crystalline powder. Darkens slowly on exposure to light, but more rapidly when in solution. Insoluble in water; freely soluble in acetic acid and in chloroform; slightly soluble in benzene; very slightly soluble in alcohol and in ether.

Resorcinol: White, or practically white, needle-shaped crystals or powder. Has a faint, characteristic odor and a sweetish, followed by a bitter, taste. Acquires a pink tint on exposure to light and air. Its solution (1 in 20) is neutral or acid to litmus. Freely soluble in water, in alcohol, in glycerin, and in ether; slightly soluble in chloroform.

Resorcinol Monoacetate: Viscous, pale yellow or amber liquid, having a faint characteristic odor and a burning taste. Boils at about 283°, with decomposition. Its saturated solution is acid to litmus. Sparingly soluble in water; soluble in alcohol and in most organic solvents.

Riboflavin: Yellow to orange-yellow, crystalline powder having a slight odor. Melts at about 280°. Its saturated solution is neutral to litmus. When dry, it is not appreciably affected by diffused light, but when in solution, light induces quite rapid deterioration, especially in the presence of alkalies. Very slightly soluble in water, in alcohol, and in isotonic sodium chloride solution; very soluble in dilute solutions of alkalies; insoluble in ether and in chloroform.

Riboflavin 5′-Phosphate Sodium: Fine, orange-yellow, crystalline powder, having a slight odor. Sparingly soluble in water. When dry, it is not affected by diffused light, but when in solution, light induces deterioration rapidly. Is hygroscopic.

Rifampin: Red-brown, crystalline powder. Very slightly soluble in water; freely soluble in chloroform; soluble in ethyl acetate and in methanol.

Ritodrine Hydrochloride: White to nearly white, odorless or practically odorless, crystalline powder. Melts at about 200°. Freely soluble in water and in alcohol; soluble in *n*-propyl alcohol; practically insoluble in ether.

Sterile Rolitetracycline: Light yellow, crystalline powder, having a characteristic, musty, amine-like odor. Soluble in water and in acetone; slightly soluble in dehydrated alcohol; very slightly soluble in ether.

Rose Oil: Colorless or yellow liquid, having the characteristic odor and taste of rose. At 25° is a viscous liquid. Upon gradual cooling, changes to a translucent, crystalline mass, easily liquefied by warming. *NF category:* Flavors and perfumes.

Rose Water Ointment: *NF category:* Ointment base.

Stronger Rose Water: Practically colorless and clear, having the pleasant odor and taste of fresh rose blossoms. Is free from empyreuma, mustiness, and fungal growths. *NF category:* Flavors and perfumes.

Rubella Virus Vaccine Live: Solid having the characteristic appearance of substances dried from the frozen state. Undergoes loss of potency on exposure to sunlight. The Vaccine is to be constituted with a suitable diluent just prior to use.

Rubella and Mumps Virus Vaccine Live: Solid having the characteristic appearance of substances dried from the frozen state. The Vaccine is to be constituted with a suitable diluent just prior to use. Constituted vaccine undergoes loss of potency on exposure to sunlight.

Saccharin: White crystals or white, crystalline powder. Is odorless or has a faint, aromatic odor. In dilute solution, it is intensely sweet. Its solutions are acid to litmus. Slightly soluble in water, in chloroform, and in ether; soluble in boiling water; sparingly soluble in alcohol. Is readily dissolved by dilute solutions of ammonia, by solutions of alkali hydroxides, and by solutions of alkali carbonates with the evolution of carbon dioxide. *NF category:* Sweetening agent.

Saccharin Calcium: White crystals or white, crystalline powder. Is odorless, or has a faint, aromatic odor, and has an intensely sweet taste even in dilute solutions. Its dilute solution is about 300 times as sweet as sucrose. Freely soluble in water. *NF category:* Sweetening agent.

Saccharin Sodium: White crystals or white, crystalline powder. Is odorless, or has a faint, aromatic odor, and has an intensely sweet taste even in dilute solutions. Its dilute solution is about 300 times as sweet as sucrose. When in powdered form it usually contains about one-third the theoretical amount of water of hydration as a result of efflorescence. Freely soluble in water; sparingly soluble in alcohol. *NF category:* Sweetening agent.

Saccharin Sodium Oral Solution: Clear, colorless, odorless liquid, having a sweet taste.

Salicylamide: White, practically odorless, crystalline powder. Slightly soluble in water and in chloroform; soluble in alcohol and in propylene glycol; freely soluble in ether and in solutions of alkalies.

Salicylic Acid: White crystals, usually in fine needles, or fluffy, white, crystalline powder. Has a sweetish, followed by an acrid, taste and is stable in air. The synthetic form is white and odorless. When prepared from natural methyl salicylate, it may have a slightly yellow or pink tint, and a faint, mint-like odor. Slightly soluble in water and in benzene; freely soluble in alcohol and in ether; soluble in boiling water; sparingly soluble in chloroform.

Schick Test Control: Transparent liquid.

Scopolamine Hydrobromide: Colorless or white crystals or white, granular powder. Is odorless, and slightly efflorescent in dry air. Freely soluble in water; soluble in alcohol; slightly soluble in chloroform; insoluble in ether.

Secobarbital: White, amorphous or crystalline, odorless powder, having a slightly bitter taste. Its saturated solution has a pH of about 5.6. Very slightly soluble in water; freely soluble in alcohol, in ether, and in solutions of fixed alkali hydroxides and carbonates; soluble in chloroform.

Secobarbital Sodium: White powder. Is odorless, has a bitter taste, and is hygroscopic. Its solutions decompose on standing, heat accelerating the decomposition. Very soluble in water; soluble in alcohol; practically insoluble in ether.

Selenomethionine Se 75 Injection: Clear, colorless to pale yellow liquid.

Selenium Sulfide: Reddish brown to bright orange powder, having not more than a faint odor. Practically insoluble in water and in organic solvents.

Sennosides: Brownish powder.

Serine: White, odorless crystals, having a sweet taste. Soluble in water; practically insoluble in absolute alcohol and in ether.

Sesame Oil: Pale yellow, oily liquid. Is practically odorless, and has a bland taste. Slightly soluble in alcohol. Miscible with ether, with chloroform, with solvent hexane, and with carbon disulfide. *NF category:* Solvent, vehicle (oleaginous).

Shellac: *Orange Shellac*—Thin, hard, brittle, transparent, pale lemon-yellow to brownish orange flakes, having little or no odor; *Bleached Shellac*—Opaque, amorphous cream to yellow granules or coarse powder, having little or no odor. Insoluble in water; soluble (very slowly) in alcohol, 85% to 95% (w/w); in ether, 13% to 15%; in benzene, 10% to 20%; in petroleum ether, 2% to 6%; soluble in aqueous solutions of ethanolamines, alkalies, and borax; sparingly soluble in oil of turpentine. *NF category:* Coating agent.

Silica Gel: Fine, white, hygroscopic, odorless, amorphous powder, in which the diameter of the average particles ranges between 2 μm and 10 μm. Insoluble in water, in alcohol, and in other organic solvents; soluble in hot solutions of alkali hydroxides. *NF category:* Desiccant; suspending and/or viscosity-increasing agent.

Purified Siliceous Earth: Very fine, white, light gray, or pale buff mixture of amorphous powder and lesser amounts of crystalline polymorphs, including quartz and cristobalite. Is gritty, readily absorbs moisture, and retains about four times its weight of water without becoming fluid. Insoluble in water, in acids, and in dilute solutions of the alkali hydroxides. *NF category:* Filtering aid; sorbent.

Colloidal Silicon Dioxide: Light, white, nongritty powder of extremely fine particle size (about 15 nm). Insoluble in water and in acid (except hydrofluoric); soluble in hot solutions of alkali hydroxides. *NF category:* Glidant and/or anticaking agent; suspending and/or viscosity-increasing agent.

Silver Nitrate: Colorless or white crystals. The pH of its solutions is about 5.5. On exposure to light in the presence of organic matter, it becomes gray or grayish black. Very soluble in water and even more so in boiling water; sparingly soluble in alcohol; freely soluble in boiling alcohol; slightly soluble in ether.

Toughened Silver Nitrate: White, crystalline masses generally molded as pencils or cones. It breaks with a fibrous fracture. Its solutions are neutral to litmus. It becomes gray or grayish black upon exposure to light. Soluble in water to the extent of its nitrate content (there is always a residue of silver chloride). Partially soluble in alcohol; slightly soluble in ether.

Simethicone: Translucent, gray, viscous fluid. Insoluble in water and in alcohol. The liquid phase is soluble in chloroform, in ether, and in benzene, but silicon dioxide remains as a residue in these solvents.

Smallpox Vaccine: Liquid Vaccine is a turbid, whitish to greenish suspension, which may have a slight odor due to the antimicrobial agent. Dried Vaccine is a yellow to grayish pellet, which may or may not become fragmented when shaken.

Soda Lime: White or grayish white granules. May have a color if an indicator has been added. *NF category:* Sorbent, carbon dioxide.

Sodium Acetate: Colorless, transparent crystals, or white, granular crystalline powder, or white flakes. Is odorless or has a faint acetous odor, and has a slightly bitter, saline taste. Is efflorescent in warm, dry air. Very soluble in water; soluble in alcohol. *NF category:* Buffering agent.

Sodium Alginate: Practically odorless and tasteless, coarse or fine powder, yellowish white in color. Soluble in water, forming a viscous, colloidal solution; insoluble in alcohol and in hydroalcoholic solutions in which the alcohol content is greater than about 30% by weight; insoluble in chloroform, in ether, and in acids when the pH of the resulting solution becomes lower than about 3. *NF category:* Suspending and/or viscosity-increasing agent.

Sodium Ascorbate: White or very faintly yellow crystals or crystalline powder. Is odorless or practically odorless. Is relatively stable in air. On exposure to light it gradually darkens. Freely soluble in water; very slightly soluble in alcohol; insoluble in chloroform and in ether.

Sodium Benzoate: White, odorless or practically odorless, granular or crystalline powder. Is stable in air. Freely soluble in water; sparingly soluble in alcohol and somewhat more soluble in 90 percent alcohol. *NF category:* Antimicrobial preservative.

Sodium Bicarbonate: White, crystalline powder. Is stable in dry air, but slowly decomposes in moist air. Its solutions, when freshly prepared with cold water, without shaking, are alkaline to litmus. The alkalinity increases as the solutions stand, as they are agitated, or as they are heated. Soluble in water; insoluble in alcohol. *NF category:* Alkalizing agent.

Sodium Bisulfite: *NF category:* Antioxidant.

Sodium Borate: Colorless, transparent crystals or white, crystalline powder. Is odorless. Its solutions are alkaline to phenolphthalein TS. As it effloresces in warm, dry air, the crystals are often coated with white powder. Soluble in water; freely soluble in boiling water and in glycerin; insoluble in alcohol. *NF category:* Alkalizing agent.

Sodium Carbonate: Colorless crystals, or white, crystalline powder or granules. Is stable in air under ordinary conditions. When exposed to dry air above 50°, the hydrous salt effloresces and, at 100°, becomes anhydrous. Freely soluble in water, but still more soluble in boiling water. *NF category:* Alkalizing agent.

Sodium Chloride: Colorless, cubic crystals or white crystalline powder. Has a saline taste. Freely soluble in water; and slightly more soluble in boiling water; soluble in glycerin; slightly soluble in alcohol. *NF category:* Tonicity agent.

Sodium Chloride Inhalation Solution: Clear, colorless solution.

Bacteriostatic Sodium Chloride Injection: Clear, colorless solution, odorless or having the odor of the bacteriostatic substance. *NF category:* Vehicle (sterile).

Sodium Chloride Irrigation: Clear, colorless solution.

Sodium Citrate: Colorless crystals, or white, crystalline powder. Hydrous form freely soluble in water and very soluble in boiling water. Insoluble in alcohol. *NF category:* Buffering agent.

Sodium Citrate and Citric Acid Oral Solution: Clear solution having the color of any added preservative or flavoring agents.

Sodium Dehydroacetate: White or practically white, odorless powder, having a slight characteristic taste. Freely soluble in water, in propylene glycol, and in glycerin. *NF category:* Antimicrobial preservative.

Sodium Fluoride: White, odorless powder. Soluble in water; insoluble in alcohol.

Sodium Formaldehyde Sulfoxylate: White crystals or hard, white masses, having the characteristic odor of garlic. Freely soluble in water; slightly soluble in alcohol, in ether, in chloroform, and in benzene. *NF category:* Antioxidant.

Sodium Hydroxide: White, or practically white, fused masses, in small pellets, in flakes, or sticks, and in other forms. Is hard and brittle and shows a crystalline fracture. Exposed to the air, it rapidly absorbs carbon dioxide and moisture. Freely soluble in water and in alcohol. *NF category:* Alkalizing agent.

Sodium Hypochlorite Solution: Clear, pale greenish yellow liquid, having the odor of chlorine. Is affected by light.

Sodium Iodide: Colorless, odorless crystals, or white, crystalline powder. Is deliquescent in moist air, and develops a brown tint upon decomposition. Very soluble in water; freely soluble in alcohol and in glycerin.

Sodium Lactate Solution: Clear, colorless or practically colorless, slightly viscous liquid, odorless or having a slight, not unpleasant odor. Miscible with water. *NF category:* Buffering agent.

Sodium Lauryl Sulfate: Small, white or light yellow crystals having a slight, characteristic odor. Freely soluble in water, forming an opalescent solution. *NF category:* Emulsifying and/or solubilizing agent; wetting and/or solubilizing agent.

Sodium Metabisulfite: White crystals or white to yellowish crystalline powder, having the odor of sulfur dioxide. Freely soluble in water and in glycerin; slightly soluble in alcohol. *NF category:* Antioxidant.

Sodium Monofluorophosphate: White to slightly gray, odorless powder. Freely soluble in water.

Sodium Nitrite: White to slightly yellow, granular powder, or white or practically white, opaque, fused masses or sticks. Has a mild, saline taste and is deliquescent in air. Its solutions are alkaline to litmus. Freely soluble in water; sparingly soluble in alcohol.

Sodium Nitrite Injection: Clear, colorless liquid.

Sodium Nitroprusside: Reddish brown, practically odorless, crystals or powder. Freely soluble in water; slightly soluble in alcohol; very slightly soluble in chloroform; insoluble in benzene.

Sterile Sodium Nitroprusside: Reddish brown, practically odorless, crystals or powder. Freely soluble in water; slightly soluble in alcohol; very slightly soluble in chloroform; insoluble in benzene.

Dibasic Sodium Phosphate (*dried*): White powder that readily absorbs moisture. Freely soluble in water; insoluble in alcohol. *NF category:* Buffering agent.

Dibasic Sodium Phosphate (*heptahydrate*): Colorless or white, granular or caked salt. Effloresces in warm, dry air. Its solutions are alkaline to phenolphthalein TS, a 0.1 *M* solution having a pH of about 9. Freely soluble in water; very slightly soluble in alcohol. *NF category:* Buffering agent.

Monobasic Sodium Phosphate: Colorless crystals or white, crystalline powder. Is odorless and is slightly deliquescent. Its solutions are acid to litmus and effervesce with sodium carbonate. Freely soluble in water; practically insoluble in alcohol. *NF category:* Buffering agent.

Sodium Polystyrene Sulfonate: Golden brown, fine powder. Is odorless and tasteless. Insoluble in water.

Sodium Propionate: Colorless, transparent crystals or granular, crystalline powder. Is odorless, or has a faint acetic-butyric odor and is deliquescent in moist air. Very soluble in water; soluble in alcohol. *NF category:* Antimicrobial preservative.

Sodium Salicylate: Amorphous or microcrystalline powder or scales. Is colorless, or has not more than a faint, pink tinge. Is odorless, or has a faint, characteristic odor, and is affected by light. A freshly made solution (1 in 10) is neutral or acid to litmus. Freely (and slowly) soluble in water and in glycerin; very soluble in boiling water and in boiling alcohol; slowly soluble in alcohol.

Sodium Starch Glycolate: White, tasteless, odorless, relatively free-flowing powder; available in several different viscosity grades. A 2% (w/v) dispersion in cold water settles, on standing, in the form of a highly hydrated layer. *NF category:* Tablet disintegrant.

Sodium Stearate: Fine, white powder, soapy to the touch, usually having a slight, tallow-like odor. Is affected by light. Its solutions are alkaline to phenolphthalein TS. Slowly soluble in cold water and in cold alcohol; readily soluble in hot water and in hot alcohol. *NF category:* Emulsifying and/or solubilizing agent.

Sodium Stearyl Fumarate: Fine, white powder. Slightly soluble in methanol; practically insoluble in water.

Sodium Sulfate: Large, colorless, odorless, transparent crystals, or a granular powder. Effloresces rapidly in air, liquefies in its water of hydration at about 33°, and loses all of its water of hydration at about 100°. Freely soluble in water; soluble in glycerin; insoluble in alcohol.

Sodium Thiosulfate: Large, colorless crystals or coarse, crystalline powder. Is deliquescent in moist air and effloresces in dry air at temperatures exceeding 33°. Its solutions are neutral or faintly alkaline to litmus. Very soluble in water; insoluble in alcohol. *NF category:* Antioxidant.

Sorbic Acid: Free-flowing, white, crystalline powder, having a characteristic odor. Slightly soluble in water; soluble in alcohol and in ether. *NF category:* Antimicrobial preservative.

Sorbitan Monolaurate: Yellow to amber oily liquid, having a bland, characteristic odor. Insoluble in water; soluble in mineral oil; slightly soluble in cottonseed oil and in ethyl acetate. *NF category:* Emulsifying and/or solubilizing agent; wetting and/or solubilizing agent.

Sorbitan Monooleate: Viscous, yellow to amber-colored, oily liquid, having a bland, characteristic odor. Insoluble in water and in propylene glycol. Miscible with mineral and vegetable oils. *NF category:* Emulsifying and/or solubilizing agent; wetting and/or solubilizing agent.

Sorbitan Monopalmitate: Cream-colored, waxy solid having a faint fatty odor. Insoluble in water; soluble in warm absolute alcohol; soluble, with haze, in warm peanut oil and in warm mineral oil. *NF category:* Emulsifying and/or solubilizing agent; wetting and/or solubilizing agent.

Sorbitan Monostearate: Cream-colored to tan, hard, waxy solid, having a bland odor and taste. Insoluble in cold water and in acetone; dispersible in warm water; soluble, with haze, above 50° in mineral oil and in ethyl acetate. *NF category:* Emulsifying and/or solubilizing agent; wetting and/or solubilizing agent.

Sorbitol: White, hygroscopic powder, granules, or flakes, having a sweet taste. Very soluble in water; slightly soluble in alcohol, in methanol, and in acetic acid. *NF category:* Humectant; sweetening agent; tablet and/or capsule diluent.

Sorbitol Solution: Clear, colorless, syrupy liquid, having a sweet taste. Is neutral to litmus. *NF category:* Sweetening agent; vehicle (flavored and/or sweetened).

Soybean Oil: Clear, pale yellow, oily liquid having a characteristic odor and taste. Insoluble in water. Miscible with ether and with chloroform. *NF category:* Vehicle (oleaginous).

Sterile Spectinomycin Hydrochloride: White to pale-buff crystalline powder. Freely soluble in water; practically insoluble in alcohol, in chloroform, and in ether.

Spironolactone: Light cream-colored to light tan, crystalline powder. Has a faint to mild mercaptan-like odor; is stable in air. Practically insoluble in water; freely soluble in benzene and in chloroform; soluble in ethyl acetate and in alcohol; slightly soluble in methanol and in fixed oils.

Squalane: Colorless, practically odorless transparent oil. Insoluble in water; very slightly soluble in absolute alcohol; slightly soluble in acetone. Miscible with ether and with chloroform. *NF category:* Ointment base; vehicle (oleaginous).

Stannous Fluoride: White, crystalline powder, having a bitter, salty taste. Melts at about 213°. Freely soluble in water; practically insoluble in alcohol, in ether, and in chloroform.

Stanozolol: Odorless, crystalline powder, occurring in two forms: as needles, melting at about 155°, and as prisms, melting at about 235°. Insoluble in water; soluble in dimethylformamide; sparingly soluble in alcohol and in chloroform; slightly soluble in ethyl acetate and in acetone; very slightly soluble in benzene.

Starch: Irregular, angular, white masses or fine powder. Is odorless, and has a slight, characteristic taste. Insoluble in cold water and in alcohol. *NF category:* Tablet and/or capsule diluent; tablet disintegrant.

Pregelatinized Starch: Moderately coarse to fine, white to off-white powder. Is odorless and has a slight, characteristic taste. Slightly soluble to soluble in cold water; insoluble in alcohol. *NF category:* Tablet binder; tablet and/or capsule diluent; tablet disintegrant.

Stearic Acid: Hard, white or faintly yellowish, somewhat glossy and crystalline solid, or white or yellowish white powder. Its odor and taste are slight, suggesting tallow. Practically insoluble in water; freely soluble in chloroform and in ether; soluble in alcohol. *NF category:* Emulsifying and/or solubilizing agent; tablet and/or capsule lubricant.

Purified Stearic Acid: Hard, white or faintly yellowish, somewhat glossy and crystalline solid, or white or yellowish white powder. Its odor and taste are slight, suggesting tallow. Practically insoluble in water; freely soluble in chloroform and in ether; soluble in alcohol.

Stearyl Alcohol: Unctuous, white flakes or granules. Has a faint, characteristic odor and a bland, mild taste. Insoluble in water; soluble in alcohol and in ether. *NF category:* Stiffening agent.

Storax: Semiliquid, grayish to grayish brown, sticky, opaque mass depositing on standing a heavy dark brown layer (Levant Storax); or semisolid, sometimes a solid mass, softened by gently warming (American Storax). Is transparent in thin layers, has a characteristic odor and taste, and is more dense than water. Insoluble in water; soluble, usually incompletely, in an equal weight of warm alcohol; soluble in acetone, in carbon disulfide, and in ether, some insoluble residue usually remaining.

Streptomycin Sulfate Injection: Clear, colorless to yellow, viscous liquid. Is odorless or has a slight odor.

Sterile Streptomycin Sulfate: White or practically white powder. Is odorless or has not more than a faint odor. Is hygroscopic, but is stable in air and on exposure to light. Its solutions are acid to practically neutral to litmus. Freely soluble in water; very slightly soluble in alcohol; practically insoluble in chloroform.

Succinylcholine Chloride: White, odorless, crystalline powder. Its solutions have a pH of about 4. The dihydrate form melts at about 160°; the anhydrous form melts at about 190°, and is hygroscopic. Freely soluble in water; slightly soluble in alcohol and in chloroform; practically insoluble in ether.

Sterile Succinylcholine Chloride: White, odorless, crystalline powder. Its solutions have a pH of about 4. The dihydrate form melts at about 160°; the anhydrous form melts at about 190°, and is hygroscopic. Freely soluble in water; slightly soluble in alcohol and in chloroform; practically insoluble in ether.

Sucrose: Colorless or white crystals, crystalline masses or blocks, or white, crystalline powder. Is odorless, has a sweet taste, and is stable in air. Its solutions are neutral to litmus. Very soluble in water, and even more soluble in boiling water; slightly soluble in alcohol; insoluble in chloroform and in ether. *NF category:* Coating agent; tablet and/or capsule diluent.

Sucrose Octaacetate: White, practically odorless powder, having an intensely bitter taste. Is hygroscopic. Very slightly soluble in water; very soluble in methanol and in chloroform; soluble in alcohol and in ether. *NF category:* Alcohol denaturant.

Compressible Sugar: Practically white, crystalline, odorless powder, having a sweet taste. Is stable in air. The sucrose portion of Compressible Sugar is very soluble in water. *NF category:* Sweetening agent; tablet and/or capsule diluent.

Confectioner's Sugar: Fine, white, odorless powder, having a sweet taste. Is stable in air. The sucrose portion of Confectioner's Sugar is soluble in cold water. Confectioner's Sugar is freely soluble in boiling water. *NF category:* Sweetening agent; tablet and/or capsule diluent.

Sugar Spheres: Hard, brittle, free-flowing, spherical masses ranging generally in size from 10- to 60-mesh. Usually white, but may be colored. Solubility in water varies according to the sugar-to-starch ratio.

Sulfabenzamide: Fine, white, practically odorless powder. Insoluble in water and in ether; soluble in alcohol, in acetone, and in sodium hydroxide TS.

Sulfacetamide: White, crystalline powder, odorless and having a characteristic sour taste. Its aqueous solutions are sensitive to light, and are unstable when acidic or strongly alkaline. Slightly soluble in water and in ether; freely soluble in dilute mineral acids and in solutions of potassium and sodium hydroxides; soluble in alcohol; very slightly soluble in chloroform; practically insoluble in benzene.

Sulfacetamide Sodium: White, crystalline powder. Is odorless and has a bitter taste. Freely soluble in water; sparingly soluble in alcohol; practically insoluble in chloroform and in ether.

Sulfadiazine: White or slightly yellow powder. Is odorless or nearly odorless and is stable in air, but slowly darkens on exposure to light. Practically insoluble in water; freely soluble in dilute mineral acids, in solutions of potassium and sodium hydroxides, and in ammonia TS; sparingly soluble in alcohol and in acetone; slightly soluble in human serum at 37°.

Sulfadiazine Sodium: White powder. On prolonged exposure to humid air it absorbs carbon dioxide with the liberation of sulfadiazine and becomes incompletely soluble in water. Its solutions are alkaline to phenolphthalein. Is affected by light. Freely soluble in water; slightly soluble in alcohol.

Sulfamerazine: White or faintly yellowish white crystals or powder. Has a slightly bitter taste and is odorless or practically odorless. Is stable in air, but slowly darkens on exposure to light. Very slightly soluble in water; sparingly soluble in acetone; slightly soluble in alcohol; very slightly soluble in ether and in chloroform.

Sulfamethazine: White to yellowish white powder, which may darken on exposure to light. Has a slightly bitter taste and is practically odorless. Very slightly soluble in water and in ether; soluble in acetone; slightly soluble in alcohol.

Sulfamethizole: White crystals or powder, having a slightly bitter taste. Is practically odorless, and has no odor of hydrogen sulfide. Very slightly soluble in water, in chloroform, and in ether; freely soluble in solutions of ammonium, potassium, and sodium hydroxides; soluble in dilute mineral acids and in acetone; sparingly soluble in alcohol; practically insoluble in benzene.

Sulfamethoxazole: White to off-white, practically odorless, crystalline powder. Practically insoluble in water, in ether, and in chloroform; freely soluble in acetone and in dilute solutions of sodium hydroxide; sparingly soluble in alcohol.

Sulfapyridine: White or faintly yellowish white crystals, granules, or powder. Is odorless or practically odorless, and is stable in air, but slowly darkens on exposure to light. Very slightly soluble in water; freely soluble in dilute mineral acids and in solutions of potassium and sodium hydroxides; sparingly soluble in acetone; slightly soluble in alcohol.

Sulfasalazine: Bright yellow or brownish yellow, odorless, fine powder. Melts at about 255°, with decomposition. Very slightly soluble in alcohol; practically insoluble in water, in ether, in chloroform, and in benzene; soluble in aqueous solutions of alkali hydroxides.

Sulfathiazole: Fine, white or faintly yellowish white, practically odorless powder. Very slightly soluble in water; soluble in acetone, in dilute mineral acids, in solutions of alkali hydroxides, and in 6 *N* ammonium hydroxide; slightly soluble in alcohol.

Sulfinpyrazone: White to off-white powder. Practically insoluble in water and in solvent hexane; soluble in alcohol and in acetone; sparingly soluble in dilute alkali.

Sulfisoxazole: White to slightly yellowish, odorless, crystalline powder. Very slightly soluble in water; soluble in boiling alcohol and in 3 *N* hydrochloric acid.

Sulfisoxazole Acetyl: White or slightly yellow, crystalline powder. Practically insoluble in water; sparingly soluble in chloroform; slightly soluble in alcohol.

Sulfisoxazole Diolamine: White to off-white, fine crystalline, odorless powder. Freely soluble in water; soluble in alcohol.

Sulfobromophthalein Sodium: White, crystalline powder. Is odorless and has a bitter taste. Is hygroscopic. Soluble in water; insoluble in acetone and in alcohol.

Sulfoxone Sodium: White to pale yellow powder, having a characteristic odor. Soluble in water, yielding a clear to hazy, pale yellow solution; very slightly soluble in alcohol.

Precipitated Sulfur: Very fine, pale yellow, amorphous or microcrystalline powder. Is odorless and tasteless. Practically insoluble in water; very soluble in carbon disulfide; slightly soluble in olive oil; very slightly soluble in alcohol.

Sublimed Sulfur: Fine, yellow, crystalline powder, having a faint odor and taste. Practically insoluble in water; sparingly soluble in olive oil; practically insoluble in alcohol.

Sulfur Dioxide: Colorless, nonflammable gas, possessing a strong, suffocating odor characteristic of burning sulfur. Under pressure, it condenses readily to a colorless liquid that boils at −10° and has a density of approximately 1.5. At 20° and at standard pressure, approximately 36 volumes dissolve in 1 volume of water and approximately 114 volumes dissolve in 1 volume of alcohol. Soluble also in ether and in chloroform. *NF category:* Antioxidant.

Sulfuric Acid: Clear, colorless, oily liquid. Miscible with water and with alcohol with the generation of much heat. Is very caustic and corrosive. Specific gravity is about 1.84. *NF category:* Acidifying agent.

Sulindac: Yellow, crystalline powder, which is odorless or practically so. Slightly soluble in methanol, in alcohol, in acetone, and in chloroform; very slightly soluble in isopropanol and in ethyl acetate; practically insoluble in hexane and in water.

Sutilains: Cream-colored powder.

Syrup: *NF category:* Sweetening agent; tablet binder; flavored and/or sweetened vehicle.

Talbutal: White, crystalline powder, which may have a slight odor of caramel. Melts at about 108°, or at about 111° in the polymorphic form. Slightly soluble in water; freely soluble in alcohol and in chloroform; soluble in glacial acetic acid and in aqueous solutions of sodium hydroxide and sodium carbonate; sparingly soluble in ether.

Talc: Very fine, white or grayish white, crystalline powder. Is unctuous, adheres readily to the skin, and is free from grittiness. *NF category:* Glidant and/or anticaking agent; tablet and/or capsule lubricant.

Tamoxifen Citrate: White, fine, crystalline powder. Very slightly soluble in water, in acetone, in chloroform, and in alcohol; soluble in methanol.

Tannic Acid: Amorphous powder, glistening scales, or spongy masses, varying in color from yellowish white to light brown. Is odorless or has a faint, characteristic odor, and has a strongly astringent taste. Very soluble in water, in acetone, and in alcohol; is freely soluble in diluted alcohol, and only slightly soluble in dehydrated alcohol; practically insoluble in benzene, in chloroform, in ether, and in solvent hexane; 1 g dissolves in about 1 mL of warm glycerin.

Tartaric Acid: Colorless or translucent crystals or white, fine to granular, crystalline powder. Is odorless, has an acid taste, and is stable in air. Very soluble in water; freely soluble in alcohol. *NF category:* Acidifying agent.

Technetium Tc 99m Aggregated Albumin Injection: Milky suspension, from which particles settle upon standing.

Technetium Tc 99m Ferpentetate Injection: Clear, light brown to yellow solution.

Technetium Tc 99m Pentetate Injection: Clear, colorless solution.

Sodium Pertechnetate Tc 99m Injection: Clear, colorless solution.

Technetium Tc 99m (Pyro- and trimeta-) Phosphates Injection: Clear solution.

Technetium Tc 99m Sulfur Colloid Injection: Colloidal dispersion. Slightly opalescent, colorless to light tan liquid.

Terbutaline Sulfate: White to gray-white, crystalline powder. Is odorless or has a faint odor of acetic acid. Soluble in water and in 0.1 *N* hydrochloric acid; slightly soluble in methanol; insoluble in chloroform.

Terpin Hydrate: Colorless, lustrous crystals or white powder. Has a slight odor, and effloresces in dry air. A hot solution (1 in 100) is neutral to litmus. When dried in vacuum at 60° for 2 hours, it melts at about 103°. Slightly soluble in water, in chloroform, and in ether; very soluble in boiling alcohol; soluble in alcohol; sparingly soluble in boiling water.

Testolactone: White to off-white, practically odorless, crystalline powder. Melts at about 218°. Slightly soluble in water and in benzyl alcohol; soluble in alcohol and in chloroform; insoluble in ether and in solvent hexane.

Testosterone: White or slightly creamy white crystals or crystalline powder. Is odorless, and is stable in air. Practically insoluble in water; freely soluble in dehydrated alcohol and in chloroform; soluble in dioxane and in vegetable oils; slightly soluble in ether.

Testosterone Cypionate: White or creamy white, crystalline powder. Is odorless or has a slight odor, and is stable in air. Insoluble in water; freely soluble in alcohol, in chloroform, in dioxane, and in ether; soluble in vegetable oils.

Testosterone Enanthate: White or creamy white, crystalline powder. Is odorless or has a faint odor characteristic of heptanoic acid. Insoluble in water; very soluble in ether; soluble in vegetable oils.

Testosterone Propionate: White or creamy white crystals or crystalline powder. Is odorless and is stable in air. Insoluble in water; freely soluble in alcohol, in dioxane, in ether, and in other organic solvents; soluble in vegetable oils.

Tetanus Antitoxin: Transparent or slightly opalescent liquid, faint brownish, yellowish, or greenish in color and practically odorless or having an odor because of the antimicrobial agent.

Tetanus Immune Globulin: Transparent or slightly opalescent liquid, practically colorless and practically odorless. May develop a slight granular deposit during storage.

Tetanus Toxoid: Clear, colorless to brownish yellow, or slightly turbid liquid, free from evident clumps or particles, having a characteristic odor or an odor of formaldehyde.

Tetanus Toxoid Adsorbed: Turbid, white, slightly gray, or slightly pink suspension, free from evident clumps after shaking.

Tetanus and Diphtheria Toxoids Adsorbed for Adult Use: Turbid, white, slightly gray, or cream-colored suspension, free from evident clumps after shaking.

Tetracaine: White or light yellow, waxy solid. Very slightly soluble in water; soluble in alcohol, in ether, in benzene, and in chloroform.

Tetracaine Hydrochloride: Fine, white, crystalline, odorless powder. Has a slightly bitter taste followed by a sense of numbness. Its solutions are neutral to litmus. Melts at about 148°, or may occur in either of two other polymorphic modifications that melt at about 134° and 139°, respectively. Mixtures of the forms may melt within the range of 134° to 147°. Is hygroscopic. Very soluble in water; soluble in alcohol; insoluble in ether and in benzene.

Sterile Tetracaine Hydrochloride: Fine, white, crystalline, odorless powder. Has a slightly bitter taste followed by a sense of numbness. Its solutions are neutral to litmus. Melts at about 148°, or may occur in either of two other polymorphic modifications that melt at about 134° and 139°, respectively. Mixtures of the forms may melt within the range of 134° to 147°. Is hygroscopic. Very soluble in water; soluble in alcohol; insoluble in ether and in benzene.

Tetracycline: Yellow, odorless, crystalline powder. Is stable in air, but exposure to strong sunlight causes it to darken. It loses potency in solutions of pH below 2, and is rapidly destroyed by alkali hydroxide solutions. Very slightly soluble in water; freely soluble in dilute acid and in alkali hydroxide solutions; sparingly soluble in alcohol; practically insoluble in chloroform and in ether.

Tetracycline Hydrochloride: Yellow, odorless, crystalline powder. Is moderately hygroscopic. Is stable in air, but exposure to strong sunlight in moist air causes it to darken. It loses potency in solution at a pH below 2, and is rapidly destroyed by alkali hydroxide solutions. Soluble in water and in solutions of alkali hydroxides and carbonates; slightly soluble in alcohol; practically insoluble in chloroform and in ether.

Tetracycline Phosphate Complex: Yellow, crystalline powder, having a faint, characteristic odor. Sparingly soluble in water; slightly soluble in methanol; very slightly soluble in acetone.

Tetrahydrozoline Hydrochloride: White, odorless solid. Melts at about 256°, with decomposition. Freely soluble in water and in alcohol; very slightly soluble in chloroform; practically insoluble in ether.

Theophylline: White, odorless, crystalline powder, having a bitter taste. Is stable in air. Slightly soluble in water, but more soluble in hot water; freely soluble in solutions of alkali hydroxides and in ammonia; sparingly soluble in alcohol, in chloroform, and in ether.

Theophylline Sodium Glycinate: White, crystalline powder having a slight ammoniacal odor and a bitter taste. Freely soluble in water; very slightly soluble in alcohol; practically insoluble in chloroform.

Thiabendazole: White to practically white, odorless or practically odorless powder. Practically insoluble in water; slightly soluble in acetone and in alcohol; very slightly soluble in chloroform and in ether.

Thiamine Hydrochloride: White crystals or crystalline powder, usually having a slight, characteristic odor. When exposed to air, the anhydrous product rapidly absorbs about 4% of water. Melts at about 248°, with some decomposition. Freely soluble in water; soluble in glycerin; slightly soluble in alcohol; insoluble in ether and in benzene.

Thiamine Mononitrate: White crystals or crystalline powder, usually having a slight, characteristic odor. Sparingly soluble in water; slightly soluble in alcohol and in chloroform.

Thiamylal Sodium for Injection: Pale yellow, hygroscopic powder, having a disagreeable odor.

Thiethylperazine Maleate: Yellowish, granular powder. Is odorless or has not more than a slight odor. Melts at about 183°, with decomposition. Practically insoluble in water; slightly soluble in methanol; practically insoluble in chloroform.

Thimerosal: Light cream-colored, crystalline powder, having a slight characteristic odor. Is affected by light. The pH of a solution (1 in 100) is about 6.7. Freely soluble in water; soluble in alcohol; practically insoluble in ether. *NF category:* Antimicrobial preservative.

Thimerosal Topical Solution: Clear liquid, having a slight characteristic odor. Is affected by light.

Thimerosal Tincture: Transparent, mobile liquid, having the characteristic odor of alcohol and acetone. Is affected by light.

Thioguanine: Pale yellow, odorless or practically odorless, crystalline powder. Insoluble in water, in alcohol, and in chloroform; freely soluble in dilute solutions of alkali hydroxides.

Thiopental Sodium: White to off-white, crystalline powder, or yellowish white to pale greenish yellow, hygroscopic powder. May have a disagreeable odor. Its solutions are alkaline to litmus. Its solutions decompose on standing, and on boiling precipitation occurs. Soluble in water and in alcohol; insoluble in benzene, in absolute ether, and in solvent hexane.

Thiopental Sodium for Injection: White to off-white, crystalline powder, or yellowish white to pale greenish yellow, hygroscopic powder. May have a disagreeable odor. Its solutions are alkaline to litmus. Its solutions decompose on standing, and on boiling precipitation occurs.

Thioridazine: White to slightly yellow, crystalline or micronized powder, odorless or having a faint odor. Practically insoluble in water; freely soluble in dehydrated alcohol and in ether; very soluble in chloroform.

Thioridazine Hydrochloride: White to slightly yellow, granular powder, having a faint odor and a very bitter taste. Freely soluble in water, in methanol, and in chloroform; insoluble in ether.

Thiotepa: Fine, white, crystalline flakes, having a faint odor. Freely soluble in water, in alcohol, in chloroform, and in ether.

Thiotepa for Injection: White powder.

Thiothixene: White to tan, practically odorless crystals. Is affected by light. Practically insoluble in water; very soluble in chloroform; slightly soluble in methanol and in acetone.

Thiothixene Hydrochloride: White, or practically white, crystalline powder, having a slight odor. Is affected by light. Soluble in water; slightly soluble in chloroform; practically insoluble in benzene, in acetone, and in ether.

Threonine: White, odorless crystals, having a slightly sweet taste. Freely soluble in water; insoluble in absolute alcohol, in ether, and in chloroform.

Thrombin: White to grayish, amorphous substance dried from the frozen state.

Thymol: Colorless, often large, crystals, or white, crystalline powder, having an aromatic, thyme-like odor and a pungent taste. Is affected by light. Its alcohol solution is neutral to litmus. Very slightly soluble in water; freely soluble in alcohol, in chloroform, in ether, and in olive oil; soluble in glacial acetic acid and in fixed and volatile oils. *NF category:* Antimicrobial preservative; flavors and perfumes.

Thyroglobulin: Cream to tan-colored, free-flowing powder, having a slight, characteristic odor. Insoluble in water, in dimethylformamide, in alcohol, in hydrochloric acid, in chloroform, and in carbon tetrachloride.

Thyroid: Yellowish to buff-colored, amorphous powder, having a slight, characteristic, meat-like odor and a saline taste.

Sterile Ticarcillin Disodium: White to pale yellow powder, or white to pale yellow solid having the characteristic appearance of products prepared by freeze-drying. Freely soluble in water.

Timolol Maleate: White to practically white, odorless or practically odorless, powder. Freely soluble in water; soluble in alcohol and in methanol; sparingly soluble in chloroform and in propylene glycol; insoluble in ether and in cyclohexane.

Titanium Dioxide: White, odorless, tasteless powder. Its 1 in 10 suspension in water is neutral to litmus. Insoluble in water, in hydrochloric acid, in nitric acid, and in 2 *N* sulfuric acid. Dissolves in hydrofluoric acid and in hot sulfuric acid. Is rendered soluble by fusion with potassium bisulfate or with alkali carbonates or hydroxides. *NF category:* Coating agent.

Tobramycin: White to off-white, hygroscopic powder. Freely soluble in water; very slightly soluble in alcohol; practically insoluble in chloroform and in ether.

Tobramycin Sulfate Injection: Clear, colorless solution.

Tocainide Hydrochloride: Fine, white, odorless powder. Freely soluble in water and in alcohol; practically insoluble in chloroform and in ether.

Tocopherol: *NF category:* Antioxidant.

Tocopherols Excipient: Brownish red to red, clear, viscous oil, having a mild, characteristic odor and taste. May show a slight separation of waxlike constituents in microcrystalline form. Oxidizes and darkens slowly in air and on exposure to light, particularly in alkaline media. Insoluble in water; soluble in alcohol; miscible with acetone, with chloroform, with ether, and with vegetable oils. *NF category:* Antioxidant.

Tolazamide: White to off-white, crystalline powder, odorless or having a slight odor. Melts with decomposition in the approximate range of 161° to 173°. Very slightly soluble in water; freely soluble in chloroform; soluble in acetone; slightly soluble in alcohol.

Tolazoline Hydrochloride: White to off-white, crystalline powder. Its solutions are slightly acid to litmus. Freely soluble in water and in alcohol.

Tolbutamide: White, or practically white, crystalline powder. Is slightly bitter and practically odorless. Practically insoluble in water; soluble in alcohol and in chloroform.

Sterile Tolbutamide Sodium: White to off-white, practically odorless, crystalline powder, having a slightly bitter taste. Freely soluble in water; soluble in alcohol and in chloroform; very slightly soluble in ether.

Tolmetin Sodium: Light yellow to light orange, crystalline powder. Freely soluble in water and in methanol; slightly soluble in alcohol; very slightly soluble in chloroform.

Tolnaftate: White to creamy white, fine powder, having a slight odor. Practically insoluble in water; freely soluble in acetone and in chloroform; sparingly soluble in ether; slightly soluble in alcohol.

Tolu Balsam: Brown or yellowish brown, plastic solid, transparent in thin layers and brittle when old, dried, or exposed to cold temperatures. Has a pleasant, aromatic odor resembling that of vanilla, and a mild, aromatic taste. Practically insoluble in water and in solvent hexane; soluble in alcohol, in chloroform, and in ether, sometimes with slight residue or turbidity. *NF category:* Flavors and perfumes.

Tolu Balsam Syrup: *NF category:* Flavored and/or sweetened vehicle.

Tragacanth: Is odorless, and has an insipid, mucilaginous taste. *NF category:* Suspending and/or viscosity-increasing agent.

Tretinoin: Yellow to light-orange, crystalline powder. Insoluble in water; slightly soluble in alcohol and in chloroform.

Triacetin: Colorless, somewhat oily liquid having a slight, fatty odor and a bitter taste. Soluble in water; slightly soluble in carbon disulfide. Miscible with alcohol, with ether, and with chloroform. *NF category:* Plasticizer.

Triamcinolone: White or practically white, odorless, crystalline powder. Very slightly soluble in water, in chloroform, and in ether; slightly soluble in alcohol and in methanol.

Triamcinolone Acetonide: White to cream-colored, crystalline powder, having not more than a slight odor. Practically insoluble in water; very soluble in dehydrated alcohol, in chloroform, and in methanol.

Triamcinolone Diacetate: Fine, white to off-white, crystalline powder, having not more than a slight odor. Practically insoluble in water; soluble in chloroform; sparingly soluble in alcohol and in methanol; slightly soluble in ether.

Triamcinolone Hexacetonide: White to cream-colored powder. Practically insoluble in water; soluble in chloroform; slightly soluble in methanol.

Triamterene: Yellow, odorless, crystalline powder. Practically insoluble in water, in benzene, in chloroform, in ether, and in dilute alkali hydroxides; soluble in formic acid; sparingly soluble in methoxyethanol; very slightly soluble in acetic acid, in alcohol, and in dilute mineral acids.

Triazolam: White to off-white, practically odorless, crystalline powder. Soluble in chloroform; slightly soluble in alcohol; practically insoluble in ether and in water.

Trichlormethiazide: White or practically white, crystalline powder. Is odorless, or has a slight characteristic odor. Melts at about 274°, with decomposition. Very slightly soluble in water, in ether, and in chloroform; freely soluble in acetone; soluble in methanol; sparingly soluble in alcohol.

Trichloromonofluoromethane: Clear, colorless gas, having a faint, ethereal odor. Its vapor pressure at 25° is about 796 mm of mercury (1 psig). *NF category:* Aerosol propellant.

Tridihexethyl Chloride: White, odorless, crystalline powder. Freely soluble in water, in methanol, and in chloroform; practically insoluble in ether and in acetone.

Trientine Hydrochloride: White to pale yellow, crystalline powder. Melts at about 117°. Insoluble in chloroform and in ether; slightly soluble in alcohol; soluble in methanol; freely soluble in water.

Triethyl Citrate: Odorless, practically colorless, oily liquid. Slightly soluble in water; miscible with alcohol and with ether.

Trifluoperazine Hydrochloride: White to pale yellow, crystalline powder. Is practically odorless, and has a bitter taste. Melts at about 242°, with decomposition. Freely soluble in water; soluble in alcohol; sparingly soluble in chloroform; insoluble in ether and in benzene.

Triflupromazine: Viscous, light amber-colored, oily liquid, which crystallizes on prolonged standing into large, irregular crystals. Practically insoluble in water.

Triflupromazine Hydrochloride: White to pale tan, crystalline powder, having a slight, characteristic odor. Melts between 170° and 178°. Soluble in water, in alcohol, and in acetone; insoluble in ether.

Trihexyphenidyl Hydrochloride: White or slightly off-white, crystalline powder, having not more than a very faint odor. Melts at about 250°. Slightly soluble in water; soluble in alcohol and in chloroform.

Trimeprazine Tartrate: White to off-white, odorless, crystalline powder. Freely soluble in water and in chloroform; soluble in alcohol; very slightly soluble in ether and in benzene.

Trimethadione: White, crystalline granules. Has a slight, camphor-like odor. Soluble in water; freely soluble in alcohol, in ether, and in chloroform.

Trimethaphan Camsylate: White crystals or white, crystalline powder. Is odorless or has a slight odor. Its solution (1 in 10) is clear and practically colorless. Melts at about 232°, with decomposition. Freely soluble in water, in alcohol, and in chloroform; insoluble in ether.

Trimethobenzamide Hydrochloride: White, crystalline powder having a slight phenolic odor. Soluble in water and in warm alcohol; insoluble in ether and in benzene.

Trimethoprim: White to cream-colored, odorless crystals, or crystalline powder. Very slightly soluble in water; soluble in benzyl alcohol; sparingly soluble in chloroform and in methanol; slightly soluble in alcohol and in acetone; practically insoluble in ether and in carbon tetrachloride.

Trioxsalen: White to off-white or grayish, odorless, crystalline solid. Melts at about 230°. Practically insoluble in water; sparingly soluble in chloroform; slightly soluble in alcohol.

Tripelennamine Citrate: White, crystalline powder. Its solutions are acid to litmus. Melts at about 107°. Freely soluble in water and in alcohol; very slightly soluble in ether; practically insoluble in benzene and in chloroform.

Tripelennamine Hydrochloride: White, crystalline powder. Slowly darkens on exposure to light. Its solutions are practically neutral to litmus. Freely soluble in water, in alcohol, and in chloroform; slightly soluble in acetone; insoluble in benzene, in ether, and in ethyl acetate.

Triprolidine Hydrochloride: White, crystalline powder, having no more than a slight, but unpleasant, odor. Its solutions are alkaline to litmus, and it melts at about 115°. Soluble in water, in alcohol, and in chloroform; insoluble in ether.

Trolamine: Colorless to pale yellow, viscous, hygroscopic liquid having a slight, ammoniacal odor. Miscible with water and with alcohol. Soluble in chloroform. *NF category:* Alkalizing agent; emulsifying and/or solubilizing agent.

Tromethamine: White, crystalline powder, having a slight, characteristic odor. Freely soluble in water and in low molecular weight aliphatic alcohols; practically insoluble in chloroform, in benzene, and in carbon tetrachloride.

Tropicamide: White or practically white, crystalline powder, odorless or having not more than a slight odor. Slightly soluble in water; freely soluble in chloroform and in solutions of strong acids.

Crystallized Trypsin: White to yellowish white, odorless, crystalline or amorphous powder.

Crystallized Trypsin for Inhalation Aerosol: White to yellowish white, crystalline or amorphous powder.

Tryptophan: White to slightly yellowish white crystals or crystalline powder, having a slightly bitter taste. Soluble in hot alcohol and in dilute hydrochloric acid.

Tuaminoheptane: Volatile, colorless to pale yellow liquid, having an amine-like odor. On exposure to air it may absorb carbon dioxide with the formation of a white precipitate of tuaminoheptane carbonate. Sparingly soluble in water; freely soluble in alcohol, in benzene, in chloroform, and in ether.

Tuberculin: Old Tuberculin is a clear, brownish liquid, which is readily miscible with water and has a characteristic odor. Purified Protein Derivative (PPD) of Tuberculin is a very slightly opalescent, colorless solution. Old Tuberculin and PPD concentrates contain 50% of glycerin for use with various application devices. Old Tuberculin and PPD are also dried on the tines of multiple-puncture devices.

Tubocurarine Chloride: White or yellowish white to grayish white, crystalline powder. Melts at about 270°, with decomposition. Soluble in water; sparingly soluble in alcohol.

Tyloxapol: Viscous, amber liquid, having a slight, aromatic odor. May exhibit a slight turbidity. Slowly but freely miscible with water. Soluble in glacial acetic acid, in benzene, in toluene, in carbon tetrachloride, in chloroform, and in carbon disulfide. *NF category:* Wetting and/or solubilizing agent.

Typhoid Vaccine: More or less turbid, milky fluid, practically odorless or having a faint odor due to the antimicrobial agent, or white solid having the characteristic appearance of freeze-dried products.

Tyropanoate Sodium: White, hygroscopic, odorless powder, having a bitter taste. Soluble in water, in alcohol, and in dimethylformamide; very slightly soluble in acetone and in ether.

Tyrosine: White, odorless, tasteless crystals or crystalline powder. Very slightly soluble in water; insoluble in alcohol and in ether.

Undecylenic Acid: Clear, colorless to pale yellow liquid having a characteristic odor. Practically insoluble in water; miscible with alcohol, with chloroform, with ether, with benzene, and with fixed and volatile oils.

Uracil Mustard: Off-white, odorless, crystalline powder. Melts at about 200°, with decomposition (see *Melting Range or Temperature* ⟨741⟩). Very slightly soluble in water; slightly soluble in acetone and in alcohol; practically insoluble in chloroform.

Urea: Colorless to white, prismatic crystals, or white, crystalline powder, or small white pellets. Is practically odorless, but may gradually develop a slight odor of ammonia upon long standing. Its solutions are neutral to litmus. Freely soluble in water and in boiling alcohol; practically insoluble in chloroform and in ether.

Sterile Urea: Colorless to white, prismatic crystals, or white, crystalline powder, or small white pellets. Is practically odorless, but may gradually develop a slight odor of ammonia upon long standing. Its solutions are neutral to litmus.

Vaccinia Immune Globulin: Transparent or slightly opalescent liquid. Is practically colorless and practically odorless. May develop a slight, granular deposit during storage.

Valine: White, odorless, tasteless crystals. Soluble in water; practically insoluble in ether, in alcohol, and in acetone.

Valproic Acid: Colorless to pale yellow, slightly viscous, clear liquid, having a characteristic odor. Refractive index: about 1.423 at 20°. Slightly soluble in water; freely soluble in 1 *N* sodium hydroxide, in methanol, in alcohol, in acetone, in chloroform, in benzene, in ether, and in *n*-heptane; slightly soluble in 0.1 *N* hydrochloric acid.

Vancomycin Hydrochloride: Tan to brown, free-flowing powder, odorless and having a bitter taste. Freely soluble in water; insoluble in ether and in chloroform.

Sterile Vancomycin Hydrochloride: Tan to brown, free-flowing powder, odorless and having a bitter taste. Freely soluble in water; insoluble in ether and in chloroform.

Vanilla: Has a characteristic, agreeably fragrant odor and taste. *NF category:* Flavors and perfumes.

Vanillin: Fine, white to slightly yellow crystals, usually needle-like, having an odor and taste suggestive of vanilla. Is affected by light. Its solutions are acid to litmus. Slightly soluble in water; freely soluble in alcohol, in chloroform, in ether, and in solutions of the fixed alkali hydroxides; soluble in glycerin and in hot water. *NF category:* Flavors and perfumes.

Vasopressin Injection: Clear, colorless or practically colorless liquid, having a faint, characteristic odor.

Hydrogenated Vegetable Oil: Fine, white powder at room temperature, and a pale yellow, oily liquid above its melting temperature. Insoluble in water; soluble in hot isopropyl alcohol, in hexane, and in chloroform. *NF category:* Tablet and/or capsule lubricant.

Verapamil Hydrochloride: White or practically white, crystalline powder. Is practically odorless and has a bitter taste. Soluble in water; freely soluble in chloroform; sparingly soluble in alcohol; practically insoluble in ether.

Sterile Vidarabine: White to off-white powder. Very slightly soluble in water; slightly soluble in dimethylformamide.

Vinblastine Sulfate: White or slightly yellow, odorless, amorphous or crystalline powder. Is hygroscopic. Freely soluble in water.

Sterile Vinblastine Sulfate: Yellowish white solid, having the characteristic appearance of products prepared by freeze-drying.

Vincristine Sulfate: White to slightly yellow, odorless, amorphous or crystalline powder. Is hygroscopic. Freely soluble in water; soluble in methanol; slightly soluble in alcohol.

Vincristine Sulfate for Injection: Yellowish white solid, having the characteristic appearance of products prepared by freeze-drying.

Vitamin A: In liquid form, a light-yellow to red oil that may solidify upon refrigeration. In solid form, has the appearance of any diluent that has been added. May be practically odorless or may have a mild fishy odor, but has no rancid odor or taste. Is unstable to air and light. In liquid form, insoluble in water and in glycerin; very soluble in chloroform and in ether; soluble in absolute alcohol and in vegetable oils. In solid form, may be dispersible in water.

Vitamin E: Practically odorless and tasteless. The alpha tocopherols and alpha tocopheryl acetates occur as clear, yellow, or greenish yellow, viscous oils. *d*-Alpha tocopheryl acetate may solidify in the cold. Alpha tocopheryl acid succinate occurs as a white powder; the *d*-isomer melts at about 75°, and the *dl*-form melts at about 70°. The alpha tocopherols are unstable to air and to light, particularly when in alkaline media. The esters are stable to air and to light, but are unstable to alkali; the acid succinate is also unstable when held molten. Alpha tocopheryl acid succinate is insoluble in water; slightly soluble in alkaline solutions; soluble in alcohol, in ether, in acetone, and in vegetable oils; very soluble in chloroform. The other forms of Vitamin E are insoluble in water; soluble in alcohol; miscible with ether, with acetone, with vegetable oils, and with chloroform.

Vitamin E Preparation: The liquid forms are clear, yellow to brownish red, viscous oils. The solid forms are white to tan-white granular powders. The liquid forms are insoluble in water; soluble in alcohol; miscible with ether, with acetone, with vegetable oils, and with chloroform. The solid forms disperse in water to give cloudy suspensions.

Warfarin Sodium: White, odorless, amorphous or crystalline powder, having a slightly bitter taste. Is discolored by light. Very soluble in water; freely soluble in alcohol; very slightly soluble in chloroform and in ether.

Water for Injection: Clear, colorless, odorless liquid. *NF category:* Solvent.

Bacteriostatic Water for Injection: Clear, colorless liquid, odorless or having the odor of the antimicrobial substance. *NF category:* Vehicle (sterile).

Sterile Water for Inhalation: Clear, colorless solution.

Sterile Water for Injection: Clear, colorless, odorless liquid. *NF category:* Solvent.

Sterile Water for Irrigation: Clear, colorless, odorless liquid. *NF category:* Solvent.

Purified Water: Clear, colorless, odorless liquid. *NF category:* Solvent.

Carnauba Wax: Light brown to pale yellow, moderately coarse powder or flakes, possessing a characteristic bland odor, and free from rancidity. Specific gravity is about 0.99. Insoluble in water; freely soluble in warm benzene; soluble in warm chloroform and in toluene; slightly soluble in boiling alcohol. *NF category:* Coating agent.

Emulsifying Wax: Creamy white, wax-like solid, having a mild, characteristic odor. Insoluble in water; freely soluble in ether, in chloroform, in most hydrocarbon solvents, and in aerosol propellants; soluble in alcohol. *NF category:* Emulsifying and/or solubilizing agent; stiffening agent.

Microcrystalline Wax: White or cream-colored, odorless, waxy solid. Insoluble in water; sparingly soluble in dehydrated alcohol; soluble in chloroform, in ether, in volatile oils, and in most warm fixed oils. *NF category:* Coating agent.

White Wax: Yellowish white solid, somewhat translucent in thin layers. Has a faint, characteristic odor, and is free from rancidity. Specific gravity is about 0.95. Insoluble in water; sparingly soluble in cold alcohol. Boiling alcohol dissolves the cerotic acid and a portion of the myricin, which are constituents of White Wax. Completely soluble in chloroform, in ether, and in fixed and volatile oils. Partly soluble in cold benzene and in cold carbon disulfide; completely soluble in these liquids at about 30°. *NF category:* Stiffening agent.

Yellow Wax: Solid varying in color from yellow to grayish brown. Has an agreeable, honey-like odor. Is somewhat brittle when cold, and presents a dull, granular, noncrystalline fracture when broken. It becomes pliable from the heat of the hand. Specific gravity is about 0.95. Insoluble in water; sparingly soluble in cold alcohol. Boiling alcohol dissolves the cerotic acid and a portion of the myricin, that are constituents of Yellow Wax. Completely soluble in chloroform, in ether, in fixed oils, and in volatile oils; partly soluble in cold benzene and in cold carbon disulfide; completely soluble in these liquids at about 30°. *NF category:* Stiffening agent.

Xanthan Gum: Cream-colored powder. Its solutions in water are neutral to litmus. Soluble in hot or cold water. *NF category:* Suspending and/or viscosity-increasing agent.

Xenon Xe 127: Clear, colorless gas.

Xenon Xe 133 Injection: Clear, colorless solution.

Xylometazoline Hydrochloride: White to off-white, odorless, crystalline powder. Melts above 300°, with decomposition. Soluble in water; freely soluble in alcohol; sparingly soluble in chloroform; practically insoluble in benzene and in ether.

Xylose: Colorless needles or white, crystalline powder. Is odorless, and has a slightly sweet taste. Very soluble in water; slightly soluble in alcohol.

Yellow Fever Vaccine: Slightly dull, light-orange colored, flaky or crustlike, desiccated mass.

Ytterbium Yb 169 Pentetate Injection: Clear, colorless to light tan, solution.

Zein: White to yellow powder. Soluble in aqueous alcohols, in glycols, in ethylene glycol ethyl ether, in furfuryl alcohol, in tetrahydrofurfuryl alcohol, and in aqueous alkaline solutions of pH 11.5 or greater. Insoluble in water and in acetone; readily soluble in acetone-water mixtures between the limits of 60% and 80% of acetone by volume; insoluble in all anhydrous alcohols except methanol.

Zinc Acetate: White crystals or granules, having a slight acetous odor and an astringent taste. Is slightly efflorescent. Freely soluble in water and in boiling alcohol; slightly soluble in alcohol.

Zinc Chloride: White or practically white, odorless, crystalline powder, or white or practically white crystalline granules. May also be in porcelain-like masses or molded into cylinders. Is very deliquescent. A solution (1 in 10) is acid to litmus. Very soluble in water; freely soluble in alcohol and in glycerin. Its solution in water or in alcohol is usually slightly turbid, but the turbidity disappears when a small quantity of hydrochloric acid is added.

Zinc Gluconate: White or practically white powder or granules. Soluble in water; very slightly soluble in alcohol.

Zinc Oxide: Very fine, odorless, amorphous, white or yellowish white powder, free from gritty particles. It gradually absorbs carbon dioxide from air. Insoluble in water and in alcohol; soluble in dilute acids.

Zinc Stearate: Fine, white, bulky powder, free from grittiness. Has a faint, characteristic odor. Is neutral to moistened litmus paper. Insoluble in water, in alcohol, and in ether. *NF category:* Tablet and/or capsule lubricant.

Zinc Sulfate: Colorless, transparent prisms, or small needles. May occur as a granular, crystalline powder. Is odorless and is efflorescent in dry air. Its solutions are acid to litmus. Very soluble in water; freely soluble in glycerin; insoluble in alcohol.

Zinc Undecylenate: Fine, white powder. Practically insoluble in water and in alcohol.

SOLUBILITIES

Approximate Solubilities of USP and NF Articles

Solute (1 g)	Water	Boiling Water	Alcohol	Chloroform	Ether	Other
	Name and Volume, in mL, of Solvent					
Acenocoumarol	67,000		280	130	1800	
Acetaminophen		20	10			1 *N* sodium hydroxide, 15
Acetohexamide			230	210		
Acetophenazine Maleate	10		260	2850	6000	acetone, 370; propylene glycol, 11
Acetylcysteine	5		4			
Acetyldigitoxin	6100		62.5	12	>10,000	
Acrisorcin	1000		18	320		acetone, 55; dimethylformamide, 3
Ammonium Alum	7	0.5				
Aluminum Chloride	0.9		4			
Aluminum Sulfate	1					
Amantadine Hydrochloride	2.5		5.1	18		polyethylene glycol 400, 70
Amaranth	15					
Aminocaproic Acid	3					methanol, 450
Aminohippuric Acid	45		50			3 *N* hydrochloric acid, 5
Aminosalicylate Sodium	2					
Ammonium Carbonate	4					
Amodiaquine Hydrochloride	25		78	>10,000	>10,000	benzene, 10,000
Anethole[1]			2			
Anileridine	>10,000		2	1		
Anileridine Hydrochloride	5		80	>10,000	>10,000	
Antimony Potassium Tartrate	12	3				glycerin, 15
Apomorphine Hydrochloride	50		50			water at 80°, 20
Ascorbic Acid	3		40			
Ascorbyl Palmitate	>1000		125	>1000	>1000	
Aspirin	300		5	17	10 to 15	
Atropine	460		2	1	25	water at 80°, 90
Atropine Sulfate	0.5	2.5	5			glycerin, 2.5
Bendroflumethiazide			23		200	
Benoxinate Hydrochloride	0.8		2.6	2.5	>10,000	
Benzalkonium Chloride (anhydrous)					100	benzene, 6
Gamma Benzene Hexachloride				3.5	40	dehydrated alcohol, 20
Benzethonium Chloride	<1		<1	<1	6000	
Benzocaine	2500		5	2	4	almond oil or olive oil, 30–50
Benzoic Acid	300		3	5	3	
Benzonatate	<1		<1	<1	<1	
Benzthiazide	41,000		480	24,000	2900	
Betamethasone	5300		65	325		warm alcohol, 15; methanol, 3
Betamethasone Acetate	2000		9	16		
Betamethasone Sodium Phosphate	2		470	>10,000	>10,000	
Betamethasone Valerate	10,000		16	<10	400	
Bisacodyl	>10,000		210	2.5	275	
Boric Acid	18	4	18			boiling alcohol, 6; glycerin, 4
Bromodiphenhydramine Hydrochloride	<1		2	2	3500	isopropyl alcohol, 31
Brompheniramine Maleate	5		15	15		
Busulfan						acetone, 45
Butabarbital Sodium	2		7	7000	10,000	
Butamben	7000					
Butylated Hydroxyanisole			4.0	2.0	1.2	
Butylated Hydroxytoluene			4.0	1.1	1.1	
Caffeine (hydrous)	50		75	6	600	
Calcium Chloride	0.7	0.2	4			boiling alcohol, 2
Calcium Gluconate	30 (slowly)	5				
Calcium Hydroxide	630	1300				
Calcium Lactate	20					
Calcium Pantothenate	3					
Calcium Sulfate	375	485				
Camphor	800		1	0.5	1	
Candicidin	75		260	10,000	33,000	dimethyl sulfoxide, 50
Carbinoxamine Maleate	<1		1.5	1.5	8300	
Carisoprodol	2083		2.5	2.3		acetone, 2.5
Carphenazine Maleate	600		400	>10,000	>10,000	
Cephaloridine	5		1000	10,000	10,000	
Cetylpyridinium Chloride	4.5		2.5	4.5		
Chloral Betaine	1		4	>10,000	>10,000	0.1 *N* hydrochloric acid, 1; 0.1 *N* sodium hydroxide, 1
Chloral Hydrate	0.25		1.3	2	1.5	
Chlorambucil						acetone, 2
Chloramphenicol	400					

[1] Solubility data for compounds that ordinarily are liquids at 25° are expressed in terms of the ratio of the *volume* of solute to the *volume* of solvent; i.e., 1 mL dissolved in ____ mL of solvent.

Approximate Solubilities of USP and NF Articles—*Continued*

Solute (1 g)	Name and Volume, in mL, of Solvent					
	Water	Boiling Water	Alcohol	Chloroform	Ether	Other
Chlordiazepoxide	>10,000		50	6250	130	
Chlorobutanol	125		1			glycerin, 10
Chlorocresol	260		0.4			
Chloroprocaine Hydrochloride	20		100			
Chlorpheniramine Maleate	4		10	10		
Chlorpromazine			3	2	3	benzene, 2
Chlorpromazine Hydrochloride	1		1.5	1.5		
Chlorprothixene	1700		29	2	14	acetone, 18
Chlortetracycline Hydrochloride	75		560			
Cholesterol			100 (slowly)			dehydrated alcohol, 50
Citric Acid	0.5		2		30	
Clindamycin Palmitate Hydrochloride			3			ethyl acetate, 9
Clindamycin Phosphate	2.5		>1000	>1000	>1000	
Clioquinol	>100,000		3500	120	4500	
Cocaine	600		7	1	3.5	olive oil, 12; liquid petrolatum, 80–100
Cocaine Hydrochloride	0.5		3.5	15		
Codeine	120		2	0.5	50	
Codeine Phosphate	2.5		325			boiling alcohol, 125; water at 80°, 0.5
Codeine Sulfate	30		1300			water at 80°, 6.5
Colchicine	25				220	
Cortisone Acetate			350	4		acetone, 75; dioxane, 30
Cupric Sulfate	3	0.5	500			glycerin, 3
Cyanocobalamin	80					
Cyclizine	11,000		6	0.9	6	
Cyclizine Hydrochloride	115		115	75		
Cyproheptadine Hydrochloride	275		35	26		methanol, 1.5
Dehydrocholic Acid			100	35	2200 (15°)	acetic acid at 15°, 135; acetone at 15°, 130; benzene at 15°, 960; ethyl acetate at 15°, 135
Demeclocycline			200			methanol, 40
Demeclocycline Hydrochloride	60					methanol, 50
Denatonium Benzoate	20		2.4	2.9	5000	
Desipramine Hydrochloride	12		14	3.5	>10,000	
Desoxycorticosterone Pivalate	>10,000		500	3	640	
Dexamethasone Sodium Phosphate	2					
Dexbrompheniramine Maleate	1.2		2.5	2	3000	
Dexchlorpheniramine Maleate	1.1		2	1.7	2500	
Dextroamphetamine Sulfate	10		800			
Dextromethorphan Hydrobromide	65					
Dextrose	1		100			
Diazepam	333		16	2	39	
Dibucaine	4600		0.7	0.5	1.4	
Dicyclomine Hydrochloride	13		5	2	770	glacial acetic acid, 2
Diethylpropion Hydrochloride	0.5		3	3		
Digitoxin			150	40		
Dihydroergotamine Mesylate	125		90	175	2600	
Dimercaprol	20					
Dimethisterone			3	0.7		
Diperodon	15,000		3	10	4	methanol, 1
Diphemanil Methylsulfate	33		33	33		
Diphenhydramine Hydrochloride	1		2	2		acetone, 50
Disulfiram	>5000		30		15	
Docusate Calcium	3300		>1	>1	>1	
Docusate Sodium	70 (slowly)					
Doxapram Hydrochloride	50					
Doxylamine Succinate	1		2	2	370	
Droperidol	10,000		140	4	500	
Dyclonine Hydrochloride	60		24	2.3	>10,000	hexane, >10,000
Dydrogesterone	>10,000		40	2	200	
Echothiophate Iodide	1					dehydrated alcohol, 25; methanol, 3
Edrophonium Chloride	0.5		5			
Ephedrine	20		0.2			
Ephedrine Hydrochloride	3		14			
Ephedrine Sulfate	1.3		90			
Epinephrine Bitartrate	3					
Ergotamine Tartrate	500		500			
Erythromycin	1000					
Erythromycin Estolate			20	10		acetone, 15
Estradiol			28	435	150	
Estradiol Cypionate	>10,000		40	7	2800	

Approximate Solubilities of USP and NF Articles—*Continued*

	Name and Volume, in mL, of Solvent					
Solute (1 g)	Water	Boiling Water	Alcohol	Chloroform	Ether	Other
Estrone			250 (15°)	110 (15°)		boiling alcohol, 50; boiling chloroform, 80; acetone at 50°, 50; boiling acetone, 33; boiling benzene, 145
Estropipate	>2000		>2000	>2000	>2000	warm alcohol, 500
Ethacrynic Acid			1.6	6	3.5	
Ether[1]	12					
Ethinamate	400		2.9			hexane, 50; sesame oil, 140; 1,2-propylene glycol, 4.6
Ethyl Vanillin	100 (50°)		2			
Ferrous Gluconate	5					
Ferrous Sulfate	1.5	0.5				
Flumethasone Pivalate	>10,000		89	350	2800	
Fluocinolone Acetonide	>1000		45	25	350	
Fluorometholone	>10,000		200	2200	>10,000	
Fluphenazine Enanthate			<1	<1	2	
Fluphenazine Hydrochloride	1.4		6.7			
Flurandrenolide			72	10		methanol, 25
Flurazepam Hydrochloride	2		4	90	5000	methanol, 3; isopropanol, 69; benzene, 2500; petroleum ether, 5000
Fluroxene	220					
Fructose			15			methanol, 14
Gentian Violet			10			glycerin, 15
Glyceryl Monostearate				10.0	100.0	methanol, 100; isopropyl alcohol, 33
Glycine	4		1254			water at 50°, 2.6; at 75°, 1.9; at 100°, 1.5
Glycopyrrolate	4.2		30	260	35,000	
Guaifenesin	60–70					
Halazone	>1000		140	>1000	>2000	
Haloperidol	>10,000		60	15	200	
Heparin Sodium	20					
Hexylcaine Hydrochloride	17					
Histamine Phosphate	4					
Homatropine Hydrobromide	6		40	420		
Hydralazine Hydrochloride	25		500			
Hydrocortisone			40			acetone, 80
Hydrocortisone Acetate			230	200		
Hydrocortisone Sodium Phosphate	1.5					
Hydroflumethiazide	>5000		39	>5000	2500	
Hydromorphone Hydrochloride	3					
Hydroquinone	17		4	51	16.5	
Hydroxocobalamin	50		100	10,000	10,000	
Hydroxyzine Hydrochloride	1		4.5	13	>1000	
Hydroxyzine Pamoate	>1000		700	>1000	>1000	dimethylformamide, 10; 10 *N* sodium hydroxide, 3.5
Hyoscyamine Hydrobromide			2.5	1.7	2300	
Hyoscyamine Sulfate	0.5		5			
Ichthammol	10					
Indigotindisulfonate Sodium	100					
Indomethacin			50	30	40	
Iodine	3000		13			carbon disulfide, 4; glycerin, 80
Ipodate Calcium				2.6		
Ipodate Sodium	<1		2			dimethylacetamide, 2; dimethylformamide, 3.5; dimethylsulfoxide, 3.5
Isocarboxazid	2000		83	2	58	
Isoniazid	8		50			
Isopropamide Iodide	50		10	5		
Isopropyl Alcohol	<1		<1	<1	<1	
Isoproterenol Hydrochloride	3		50			
Isoproterenol Sulfate	4		>2000	>2000	>2000	
Isoxsuprine Hydrochloride	500		100	>10,000	>10,000	0.1 *N* hydrochloric acid, 2500; 0.1 *N* sodium hydroxide, 100
Ketamine Hydrochloride	4		14	60	>10,000	methanol, 6; absolute alcohol, 60
Lactose	5 (slowly)	2.6				
Levopropoxyphene Napsylate			17	2		
Levorphanol Tartrate	50		120			
Levothyroxine Sodium	700		300			

[1] Solubility data for compounds that ordinarily are liquids at 25° are expressed in terms of the ratio of the *volume* of solute to the *volume* of solvent; i.e., 1 mL dissolved in ____ mL of solvent.

Approximate Solubilities of USP and NF Articles—*Continued*

Solute (1 g)	Name and Volume, in mL, of Solvent					
	Water	Boiling Water	Alcohol	Chloroform	Ether	Other
Magnesium Hydroxide	>10,000		>10,000	>10,000	>10,000	
Magnesium Sulfate	0.8	0.5				glycerin, 1 (slowly)
Mannitol	5.5					
Menadione			60			benzene, 10; vegetable oils, 50
Mephentermine Sulfate	18		220	>1000	>10,000	
Mephenytoin	1400		15	3	90	
Mephobarbital	>1000		>1000	50	>1000	
Meprylcaine Hydrochloride	6		5	3	12	
Mesoridazine Besylate	1		11	3	6300	
Methacholine Chloride	1.2		1.7	2.1		
Methacrylic Acid Copolymer						water in methanol (≥3 in 100), 10; water in alcohol (≥3 in 100), 10; water in isopropyl alcohol (≥3 in 100), 10; water in acetone (≥3 in 100), 10
Methacycline Hydrochloride	100		300	>1000	>1000	0.1 *N* sodium hydroxide, 25
Methantheline Bromide	<5		<5	<5	>10,000	acetone, 390
Metharbital	830		23		40	
Methdilazine	>10,000		2	1	8	0.1 *N* hydrochloric acid, 40; 0.1 *N* sodium hydroxide, >10,000
Methdilazine Hydrochloride	2		2	6	>10,000	0.1 *N* hydrochloric acid, 1; 0.1 *N* sodium hydroxide, 1
Methenamine	1.5		12.5	10	320	
Methenamine Mandelate			10	20	350	
Methimazole	5		5	4.5	125	
Methocarbamol	40 (20°)					
Methotrimeprazine	10		10	2		methanol, 10
Methoxamine Hydrochloride	2.5		12			
Methoxyflurane	500		<1	<1	<1	
Methsuximide	350		3	<1	2	
Methyclothiazide	>10,000		92.5	>10,000	2700	
Methylbenzethonium Chloride	0.8		0.9	>10,000	0.7	
Methylene Blue	25		65			
Methylergonovine Maleate	100		175	1900	8400	
Methylparaben	400		3		10	water at 80°, 50
Methylprednisolone	10,000		100	800	800	
Methylprednisolone Acetate	1500		400	250	1500	
Methylprednisolone Sodium Succinate	1.5		12	>10,000	>10,000	
Methyprylon	11		2	2	2	
Methysergide Maleate	200		165	3400	>10,000	
Metocurine Iodide	400		10,000	10,000	10,000	
Miconazole	>100,000		9.5	2	15	isopropyl alcohol, 4; propylene glycol, 9; methanol, 5.3
Miconazole Nitrate	6250		312	525	50,000	isopropyl alcohol, 1408; propylene glycol, 119; methanol, 75
Monobenzone	>10,000		14.5	29	14	
Morphine Sulfate	16		570			alcohol at 60°, 240; water at 80°, 1
Nalidixic Acid	>1000		910	29	>1000	
Neomycin Sulfate	1					
Niacin	60					
Niacinamide	1.5	10	5.5			
Nifedipine	>10,000					acetone, 10
Nitrofurazone	4200		590			propylene glycol, 350
Nitromersol	>2000		>2000	>2000	>2000	
Norepinephrine Bitartrate	2.5		300			
Norethindrone Acetate	>10,000		10	<1	18	dioxane, 2
Nortriptyline Hydrochloride	90		30	20		methanol, 10
Nylidrin Hydrochloride	65		40			
Oxandrolone	5200		57	<5	860	acetone, 69
Oxazepam	>10,000		220	270	2200	
Oxtriphylline	1					
Oxymetazoline Hydrochloride	6.7		3.6	862		
Oxymetholone	>10,000		40	5	82	dioxane, 14
Oxymorphone Hydrochloride	4		100	>1000	>1000	methanol, 25
Oxyphenbutazone	>10,000		1.5	4.0	15	
Oxyphencyclimine Hydrochloride	100		75	500	>1000	
Oxytetracycline	4150			>10,000	6250	absolute alcohol, 66
Oxytetracycline Calcium	>1000		>1000	>1000	>1000	0.1 *N* sodium hydroxide, 15
Papaverine Hydrochloride	30		120			
Paraldehyde[1]	10	17				

[1] Solubility data for compounds that ordinarily are liquids at 25° are expressed in terms of the ratio of the *volume* of solute to the *volume* of solvent; i.e., 1 mL dissolved in ____ mL of solvent.

Approximate Solubilities of USP and NF Articles—*Continued*

	Name and Volume, in mL, of Solvent					
Solute (1 g)	Water	Boiling Water	Alcohol	Chloroform	Ether	Other
Paramethasone Acetate				50		methanol, 40
Pargyline Hydrochloride	0.6		5	7.2		
Paromomycin Sulfate	<1		>10,000	>10,000	>10,000	
Pectin	20					
Penicillin G Benzathine	5000		65			
Sterile Penicillin G Procaine	250		30	60		
Sterile Penicillin G Sodium	40					
Penicillin V Benzathine	3200		330	42	910	acetone, 37
Penicillin V Potassium			150			
Pentazocine	>1000		11	2	42	
Pentazocine Hydrochloride	30		20	4	>10,000	
Pentobarbital	>2000		4.5	4.0	10	
Pentolinium Tartrate	0.5		475	>1000	>2000	
Perphenazine			7			acetone, 13
Phenacemide	>2000		>2000	>2000	>2000	warm alcohol, 500; methanol, 300
Phenazopyridine Hydrochloride	<10	20	59	331	>5000	cold water, 300; glycerin, 100
Phenindamine Tartrate	40		350			
Phenindione	5000		100	7	110	0.1 *N* sodium hydroxide, 100
Phenmetrazine Hydrochloride	0.4		2	2		
Phenobarbital	1000		10			
Phenol	15					mineral oil, 70
Phenolphthalein			15		100	
Phensuximide	210		11	<1	19	
Phentolamine Mesylate	1		4	700		
Phenylethyl Alcohol	60		<1	<1	<1	alcohol solution (1 in 2), 2; diethyl phthalate, <1; benzyl benzoate, <1
Phenylmercuric Acetate	180		225	6.8	200	
Phenylmercuric Nitrate	600					
Phenylpropanolamine Hydrochloride	1.1		7.4	4100		
Physostigmine Salicylate	75		16	6	250	
Physostigmine Sulfate	4		0.4		1200	
Pilocarpine Hydrochloride	0.3		3	360		
Pilocarpine Nitrate	4		75			
Piminodine Esylate	>1000		6	2	>1000	
Pimozide	>10,000		1000	10	1000	acetone, 100; methanol, 1000; 0.1 *N* hydrochloride acid, >1000
Piperacetazine	10,000		11	1.3	1200	1 *N* hydrochloric acid, 25
Pipobroman	230		35	4.8	530	
Polyethylene Glycol 1540	1			3		absolute alcohol, 100
Polyethylene Glycol 4000	4		2.5	2		
Polyoxyl 50 Stearate	0.7			0.45	14,000	absolute alcohol, 13,000
Polythiazide	>1000		150	175	>1000	
Potash, Sulfurated	2					
Potassium Acetate	0.5	0.2	3			
Potassium Alum	7	0.3				
Potassium Benzoate	2		75			alcohol solution (9 in 10), 50
Potassium Chloride	2.8	2				
Potassium Citrate	1					glycerin, 2.5
Potassium Gluconate	3					
Potassium Hydroxide	1		3			glycerin, 2.5
Potassium Iodide	0.7	0.5	22			glycerin, 2
Potassium Permanganate	15	3.5				
Potassium Sodium Tartrate	1					
Potassium Sorbate	4.5		35	>1000	>1000	
Pramoxine Hydrochloride				35		
Prednisolone			30	180		acetone, 50
Prednisolone Acetate			120			
Prednisolone Hemisuccinate	4170		6.3	1064	248	
Prednisolone Sodium Phosphate	4					methanol, 13
Prednisone			150	200		
Prilocaine Hydrochloride	3.5		4.2	175		
Primaquine Phosphate	15					
Primidone	2000		200			
Procaine Hydrochloride	1		15			
Prochlorperazine Edisylate	2		1500			
Prochlorperazine Maleate			1200			
Procyclidine Hydrochloride	35		9	6	11,000	
Promazine Hydrochloride	3					
Propiomazine Hydrochloride	<1		6	2	>10,000	
Propoxycaine Hydrochloride	2		10	>10,000	80	
Propoxyphene Napsylate	10,000		15	10		
Propylhexedrine	>500		0.4	0.2	0.1	
Propylparaben	2500	400	1.5		3	
Protriptyline Hydrochloride	2		3.5	2.5	>10,000	

Approximate Solubilities of USP and NF Articles—*Continued*

Solute (1 g)	Water	Boiling Water	Alcohol	Chloroform	Ether	Other
	Name and Volume, in mL, of Solvent					
Pseudoephedrine Hydrochloride	0.5		3.6	91	7000	
Pyrazinamide	67			135	1000	absolute alcohol, 175; methanol, 72
Pyridoxine Hydrochloride	5		115			
Pyrilamine Maleate	0.5		3	2		absolute alcohol, 15
Pyrimethamine			200	125		
Pyrrocaine Hydrochloride	1.5		12	8		
Quinacrine Hydrochloride	35					
Quinidine Sulfate	100		10	15		
Quinine Sulfate	500		120			
Reserpine			1800	6		
Resorcinol	1		1			
Sterile Rolitetracycline	1.1		200			
Rotoxamine Tartrate	10		100	>10,000	>10,000	
Saccharin	290	25	31			
Saccharin Calcium	2.6		4.7			
Saccharin Sodium	1.5		50			
Salicylic Acid	460	15	3	45	3	benzene, 135
Scopolamine Hydrobromide	1.5		20			
Secobarbital						0.5 *N* sodium hydroxide, 8.5
Selenium Sulfide				161	1667	
Sennosides	35		2100	3700	6100	
Silver Nitrate	0.4	0.1	30			boiling alcohol, 6.5
Simethicone[2]	>10,000			10^2	10^2	benzene, 10^2; absolute alcohol, >10,000
Sodium Acetate	0.8	0.6	19			
Sodium Ascorbate	1.3					
Sodium Benzoate	2		75			90 percent alcohol, 50
Sodium Bicarbonate	12					
Sodium Bisulfite	4					
Sodium Borate	16	1				glycerin, 1
Sodium Carbonate	3	1.8				
Sodium Chloride	2.8	2.7				glycerin, 10
Sodium Citrate (hydrous)	1.5	0.6				
Sodium Fluoride	25					
Sodium Formaldehyde Sulfoxylate	3.4		510	175	180	
Sodium Hydroxide	1					
Sodium Iodide	0.6		2			glycerin, 1
Sodium Lauryl Sulfate	10					
Sodium Nitrite	1.5					
Sodium Phosphate, Dried	8					
Sodium Propionate	1	0.65	24	>10,000	>10,000	
Sodium Thiosulfate	0.5					
Sorbic Acid	1000		10	15	30	absolute alcohol, 8; methanol, 8; propylene glycol, 19
Sorbitol	0.45					
Stanozolol	>1000		41	74	370	
Stearic Acid			20	2	3	
Stibophen	1		>10,000	>10,000	10,000	
Succinylcholine Chloride	1		350			
Sucrose	0.5	0.2	170			
Sucrose Octaacetate	1100		11			acetone, 0.3; benzene, 0.6; toluene, 0.5
Sulfacetamide Sodium	2.5					
Sulfadiazine	13,000					human serum at 37°, 620
Sulfadiazine Sodium	2					
Sulfadimethoxine			200			2 *N* hydrochloric acid, 50
Sulfaethidole	>3000		75	1300	1700	methanol, 51; acetone, 13; benzene, 2277
Sulfamerazine						water at 20°, 6250; water at 37°, 3300
Sulfamethizole	2000		38	1900	1900	acetone, 13
Sulfamethoxazole	3400		50	1000	1000	carbon disulfide, 2 (slowly and usually incompletely)
Sulfapyridine	3500		440			acetone, 65
Sulfasalazine	>10,000		2900	>10,000	>10,000	methanol, 1500
Sulfisoxazole	7700					boiling alcohol, 10
Sulfisoxazole Acetyl			176	35	1064	methanol, 203
Sulfisoxazole Diolamine	2		16	1000	>10,000	methanol, 4; isopropanol, 250; benzene, >10,000; petroleum ether, >10,000
Sulfoxone Sodium	13.5		>1000	>1000	>2000	
Precipitated Sulfur						carbon disulfide, 2 (slowly and usually incompletely) olive oil, 100
Talbutal	500		1	2	40	
Tartaric Acid	0.8	0.5	3		250	methanol, 1.7
Terpin Hydrate	200	35	13	140	140	boiling alcohol, 3

[2] Liquid phase only; silicon dioxide remains as a residue in these solvents.

Approximate Solubilities of USP and NF Articles—*Continued*

Solute (1 g)	Name and Volume, in mL, of Solvent					
	Water	Boiling Water	Alcohol	Chloroform	Ether	Other
Testolactone	4050					
Testosterone				2	100	absolute alcohol, 6
Tetracaine	>1000		5	2	2	
Tetracycline	2500		50			
Tetracycline Hydrochloride	10		100			
Tetracycline Phosphate Complex	31		130			
Tetrahydrozoline Hydrochloride	3.5		7.5	>1000	>1000	
Theophylline Sodium Glycinate	6					
Thiamine Hydrochloride	1		170			
Thiamine Mononitrate	44					
Thiethylperazine Maleate	1700		530	>10,000	>10,000	
Thimerosal	1		12			
Thioguanine			7700			
Thiotepa	13		8.3	1.9	4.1	
Thiothixene	>10,000			2	120	absolute alcohol, 110
Thiothixene Hydrochloride	8			280	>10,000	absolute alcohol, 270
Thymol	1000		1	1	1.5	olive oil, 2
Tolazoline Hydrochloride	<1		2	3	>10,000	
Triamcinolone Diacetate			13	80		methanol, 40
Triamterene						formic acid, 30; 2-methoxyethanol, 85
Triazolam	>10,000		1000	25	>10,000	0.1 *N* hydrochloric acid, 600
Trichlormethiazide	1100		48	5000	1400	dioxane, 9.1; dimethylformamide, 4.35
Trichloroethylene	>10,000					
Tridihexethyl Chloride	3		3	2		
Triethylenemelamine	2.5		13	3.6		methanol, 8; acetone, 9.5; benzene, 18
Trifluoperazine Hydrochloride	3.5		11	100		
Triflupromazine Hydrochloride	<1		<1	1.7		
Trimeprazine Tartrate	2		20	5	1800	
Trimethobenzamide Hydrochloride	2		59	67	720	
Trioxsalen			1150	84		methylenedichloride, 43; 4-methyl-2-pentanone, 100
Tripelennamine Citrate	1					
Tripelennamine Hydrochloride	1		6	6		acetone, 350
Triprolidine Hydrochloride	2.1		1.8	1	2000	
Tromethamine	1.8		45.5	>10,000		
Tuaminoheptane	100		25	20		
Tubocurarine Chloride	20		45			
Uracil Mustard	>1000		150	>1000		
Urea	1.5		10			boiling alcohol, 1
Vanillin	100					glycerin, 20; water at 80°, 20
Xylometazoline Hydrochloride	35					
Zinc Acetate	2.5		30			
Zinc Chloride	0.5		1.5			glycerin, 2
Zinc Sulfate	0.6					glycerin, 2.5

PHARMACEUTIC INGREDIENTS

USP and NF Pharmaceutic Ingredients, Listed by Categories

Acidifying Agent
Acetic Acid
Acetic Acid, Glacial
Citric Acid
Fumaric Acid
Hydrochloric Acid
Hydrochloric Acid, Diluted
Malic Acid
Nitric Acid
Phosphoric Acid
Phosphoric Acid, Diluted
Sulfuric Acid
Tartaric Acid

Aerosol Propellant
Butane
Dichlorodifluoromethane
Dichlorotetrafluoroethane
Isobutane
Propane
Trichloromonofluoromethane

Air Displacement
Carbon Dioxide
Nitrogen

Alcohol Denaturant
Denatonium Benzoate
Methyl Isobutyl Ketone
Sucrose Octaacetate

Alkalizing Agent
Ammonia Solution, Strong
Ammonium Carbonate
Diethanolamine
Diisopropanolamine
Potassium Hydroxide
Sodium Bicarbonate
Sodium Borate
Sodium Carbonate
Sodium Hydroxide
Trolamine

Anticaking Agent (See *Glidant*)

Antifoaming Agent
Dimethicone
Simethicone

Antimicrobial Preservative
Benzalkonium Chloride
Benzalkonium Chloride Solution
Benzethonium Chloride
Benzoic Acid
Benzyl Alcohol
Butylparaben
Cetylpyridinium Chloride
Chlorobutanol
Chlorocresol
Cresol
Dehydroacetic Acid
Ethylparaben
Methylparaben
Methylparaben Sodium
Phenol
Phenylethyl Alcohol
Phenylmercuric Acetate
Phenylmercuric Nitrate
Potassium Benzoate
Potassium Sorbate
Propylparaben
Propylparaben Sodium
Sodium Benzoate
Sodium Dehydroacetate
Sodium Propionate
Sorbic Acid
Thimerosal
Thymol

Antioxidant
Ascorbic Acid
Ascorbyl Palmitate
Butylated Hydroxyanisole
Butylated Hydroxytoluene
Hypophosphorous Acid
Monothioglycerol
Propyl Gallate
Sodium Formaldehyde Sulfoxylate
Sodium Metabisulfite
Sodium Thiosulfate
Sulfur Dioxide
Tocopherol
Tocopherols Excipient

Buffering Agent
Acetic Acid
Ammonium Carbonate
Ammonium Phosphate
Boric Acid
Citric Acid
Lactic Acid
Phosphoric Acid
Potassium Citrate
Potassium Metaphosphate
Potassium Phosphate, Monobasic
Sodium Acetate
Sodium Citrate
Sodium Lactate Solution
Sodium Phosphate, Dibasic
Sodium Phosphate, Monobasic

Capsule Lubricant (See *Tablet and/or Capsule Lubricant*)

Chelating Agent
Edetate Disodium
Edetic Acid

Coating Agent
Carboxymethylcellulose, Sodium
Cellulose Acetate
Cellulose Acetate Phthalate
Ethylcellulose
Gelatin
Glaze, Pharmaceutical
Hydroxypropyl Cellulose
Hydroxypropyl Methylcellulose
Hydroxypropyl Methylcellulose Phthalate
Methacrylic Acid Copolymer
Methylcellulose
Polyethylene Glycol
Polyvinyl Acetate Phthalate
Shellac
Sucrose
Titanium Dioxide
Wax, Carnauba
Wax, Microcrystalline
Zein

Color
Caramel
Ferric Oxide, red
yellow, black, or blends

Complexing Agent
Edetate Disodium
Edetic Acid
Gentisic Acid Ethanolamide
Oxyquinoline Sulfate

Desiccant
Calcium Chloride
Calcium Sulfate
Silicon Dioxide

Emulsifying and/or Solubilizing Agent
Acacia
Cholesterol
Diethanolamine (Adjunct)
Glyceryl Monostearate
Lanolin Alcohols
Lecithin
Mono- and Di-glycerides
Monoethanolamine (Adjunct)
Oleic Acid (Adjunct)
Oleyl Alcohol (Stabilizer)
Poloxamer
Polyoxyethylene 50 Stearate
Polyoxyl 35 Castor Oil
Polyoxyl 40 Hydrogenated Castor Oil
Polyoxyl 10 Oleyl Ether
Polyoxyl 20 Cetostearyl Ether
Polyoxyl 40 Stearate
Polysorbate 20
Polysorbate 40
Polysorbate 60
Polysorbate 80
Propylene Glycol Diacetate
Propylene Glycol Monostearate
Sodium Lauryl Sulfate
Sodium Stearate
Sorbitan Monolaurate
Sorbitan Monooleate
Sorbitan Monopalmitate
Sorbitan Monostearate
Stearic Acid
Trolamine
Wax, Emulsifying

Filtering Aid
Cellulose, Powdered
Siliceous Earth, Purified

Flavors and Perfumes
Anethole
Benzaldehyde
Ethyl Vanillin
Menthol
Methyl Salicylate
Monosodium Glutamate
Orange Flower Oil
Peppermint
Peppermint Oil
Peppermint Spirit
Rose Oil
Rose Water, Stronger
Thymol
Tolu Balsam Tincture
Vanilla
Vanilla Tincture
Vanillin

Glidant and/or Anticaking Agent
Calcium Silicate
Magnesium Silicate
Silicon Dioxide, Colloidal
Talc

Humectant
Glycerin
Hexylene Glycol
Propylene Glycol
Sorbitol

Ointment Base
Lanolin
Lanolin, Anhydrous
Ointment, Hydrophilic
Ointment, White
Ointment, Yellow
Polyethylene Glycol Ointment
Petrolatum
Petrolatum, Hydrophilic
Petrolatum, White
Rose Water Ointment
Squalane

Plasticizer
Castor Oil
Diacetylated Monoglycerides
Diethyl Phthalate
Glycerin
Mono- and Di-acetylated Monoglycerides
Polyethylene Glycol
Propylene Glycol
Triacetin
Triethyl Citrate

Polymer Membrane
Cellulose Acetate

Solvent
Acetone
Alcohol
Alcohol, Diluted
Amylene Hydrate
Benzyl Benzoate
Butyl Alcohol
Carbon Tetrachloride
Chloroform
Corn Oil
Cottonseed Oil
Ethyl Acetate
Glycerin
Hexylene Glycol
Isopropyl Alcohol
Methyl Alcohol
Methylene Chloride
Methyl Isobutyl Ketone
Mineral Oil
Peanut Oil
Polyethylene Glycol
Propylene Carbonate
Propylene Glycol
Sesame Oil
Water for Injection
Water for Injection, Sterile
Water for Irrigation, Sterile
Water, Purified

Sorbent
Cellulose, Powdered
Charcoal
Siliceous Earth, Purified

Sorbent, Carbon Dioxide
Barium Hydroxide Lime
Soda Lime

Stiffening Agent
Castor Oil, Hydrogenated
Cetostearyl Alcohol
Cetyl Alcohol
Cetyl Esters Wax
Hard Fat
Paraffin
Polyethylene Excipient
Stearyl Alcohol
Wax, Emulsifying
Wax, White
Wax, Yellow

Suppository Base
Cocoa Butter
Hard Fat
Polyethylene Glycol

Suspending and/or Viscosity-increasing Agent
Acacia
Agar
Alginic Acid
Aluminum Monostearate
Bentonite
Bentonite, Purified
Bentonite Magma
Carbomer 934P
Carboxymethylcellulose Calcium
Carboxymethylcellulose Sodium
Carboxymethylcellulose Sodium 12
Carrageenan
Cellulose, Microcrystalline, and Carboxymethylcellulose Sodium
Dextrin
Gelatin
Guar Gum
Hydroxyethyl Cellulose
Hydroxypropyl Cellulose
Hydroxypropyl Methylcellulose
Magnesium Aluminum Silicate
Methylcellulose
Pectin
Polyethylene Oxide
Polyvinyl Alcohol
Povidone
Propylene Glycol Alginate
Silicon Dioxide
Silicon Dioxide, Colloidal
Sodium Alginate
Tragacanth
Xanthan Gum

Sweetening Agent
Aspartame
Dextrates
Dextrose
Dextrose Excipient
Fructose
Mannitol
Saccharin
Saccharin Calcium
Saccharin Sodium
Sorbitol
Sorbitol Solution
Sucrose
Sugar, Compressible
Sugar, Confectioner's
Syrup

Tablet Binder
Acacia
Alginic Acid
Carboxymethylcellulose, Sodium
Cellulose, Microcrystalline
Dextrin
Ethylcellulose
Gelatin
Glucose, Liquid
Guar Gum
Hydroxypropyl Methylcellulose
Methylcellulose
Polyethylene Oxide
Povidone
Starch, Pregelatinized
Syrup

Tablet and/or Capsule Diluent
Calcium Carbonate
Calcium Phosphate, Dibasic
Calcium Phosphate, Tribasic
Calcium Sulfate
Cellulose, Microcrystalline
Cellulose, Powdered
Dextrates
Dextrin
Dextrose Excipient
Fructose
Kaolin
Lactose
Mannitol
Sorbitol
Starch
Starch, Pregelatinized
Sucrose
Sugar, Compressible
Sugar, Confectioner's

Tablet Disintegrant
Alginic Acid
Cellulose, Microcrystalline
Croscarmellose Sodium
Crospovidone
Polacrilin Potassium
Sodium Starch Glycolate
Starch
Starch, Pregelatinized

Tablet and/or Capsule Lubricant
Calcium Stearate
Glyceryl Behenate
Magnesium Stearate
Mineral Oil, Light
Polyethylene Glycol
Sodium Stearyl Fumarate
Stearic Acid
Stearic Acid, Purified
Talc
Vegetable Oil, Hydrogenated
Zinc Stearate

Tonicity Agent
Dextrose
Glycerin
Mannitol
Potassium Chloride
Sodium Chloride

Vehicle

FLAVORED AND/OR SWEETENED
Aromatic Elixir
Benzaldehyde Elixir, Compound
Iso-alcoholic Elixir
Peppermint Water
Sorbitol Solution
Syrup
Tolu Balsam Syrup

OLEAGINOUS
Almond Oil
Corn Oil
Cottonseed Oil
Ethyl Oleate
Isopropyl Myristate
Isopropyl Palmitate
Mineral Oil
Mineral Oil, Light
Myristyl Alcohol
Octyldodecanol
Olive Oil
Peanut Oil
Persic Oil
Sesame Oil
Soybean Oil
Squalane

SOLID CARRIER
Sugar Spheres

STERILE
Sodium Chloride Injection, Bacteriostatic
Water for Injection, Bacteriostatic

Viscosity-Increasing (See *Suspending Agent*)

Water Repelling Agent
Cyclomethicone
Dimethicone
Simethicone

Wetting and/or Solubilizing Agent

Benzalkonium Chloride
Benzethonium Chloride
Cetylpyridinium Chloride
Docusate Sodium
Nonoxynol 9
Nonoxynol 10
Octoxynol 9
Poloxamer
Polyoxyl 35 Castor Oil
Polyoxyl 40 Hydrogenated Castor Oil
Polyoxyl 50 Stearate
Polyoxyl 10 Oleyl Ether
Polyoxyl 20 Cetostearyl Ether
Polyoxyl 40 Stearate
Polysorbate 20
Polysorbate 40
Polysorbate 60
Polysorbate 80
Sodium Lauryl Sulfate
Sorbitan Monolaurate
Sorbitan Monooleate
Sorbitan Monopalmitate
Sorbitan Monostearate
Tyloxapol

ATOMIC WEIGHTS

Atomic Weights, Recommended by the Commission on Atomic Weights of the International Union of Pure and Applied Chemistry (1981)

Name	Symbol	Atomic Number	Atomic Weight†
Actinium	Ac	89	227.0278[z]
Aluminum	Al	13	26.98154
Americium	Am	95	(243)
Antimony	Sb	51	121.75*
Argon	Ar	18	39.948*[w,x]
Arsenic	As	33	74.9216
Astatine	At	85	(210)
Barium	Ba	56	137.33[x]
Berkelium	Bk	97	(247)
Beryllium	Be	4	9.01218
Bismuth	Bi	83	208.9804
Boron	B	5	10.81[w,y]
Bromine	Br	35	79.904
Cadmium	Cd	48	112.41[x]
Calcium	Ca	20	40.08[x]
Californium	Cf	98	(251)
Carbon	C	6	12.011[w]
Cerium	Ce	58	140.12[x]
Cesium	Cs	55	132.9054
Chlorine	Cl	17	35.453
Chromium	Cr	24	51.996
Cobalt	Co	27	58.9332
Copper	Cu	29	63.546*[w]
Curium	Cm	96	(247)
Dysprosium	Dy	66	162.50*
Einsteinium	Es	99	(252)
Erbium	Er	68	167.26*
Europium	Eu	63	151.96[x]
Fermium	Fm	100	(257)
Fluorine	F	9	18.998403
Francium	Fr	87	(223)
Gadolinium	Gd	64	157.25*[x]
Gallium	Ga	31	69.72
Germanium	Ge	32	72.59*
Gold	Au	79	196.9665
Hafnium	Hf	72	178.49*
Helium	He	2	4.00260[x]
Holmium	Ho	67	164.9304
Hydrogen	H	1	1.00794[w,x,y]
Indium	In	49	114.82[x]
Iodine	I	53	126.9045
Iridium	Ir	77	192.22*
Iron	Fe	26	55.847*
Krypton	Kr	36	83.80[x,y]
Lanthanum	La	57	138.9055*[x]
Lawrencium	Lr	103	(260)
Lead	Pb	82	207.2[w,x]
Lithium	Li	3	6.941*[w,x,y]
Lutetium	Lu	71	174.967
Magnesium	Mg	12	24.305[x]
Manganese	Mn	25	54.9380
Mendelevium	Md	101	(258)
Mercury	Hg	80	200.59*
Molybdenum	Mo	42	95.94[x]
Neodymium	Nd	60	144.24*[x]
Neon	Ne	10	20.179[x,y]
Neptunium	Np	93	237.0482[z]
Nickel	Ni	28	58.69
Niobium	Nb	41	92.9064
Nitrogen	N	7	14.0067
Nobelium	No	102	(259)
Osmium	Os	76	190.2[x]
Oxygen	O	8	15.9994*[w,x]
Palladium	Pd	46	106.42[x]
Phosphorus	P	15	30.97376
Platinum	Pt	78	195.08*
Plutonium	Pu	94	(244)
Polonium	Po	84	(209)
Potassium	K	19	39.0983
Praseodymium	Pr	59	140.9077
Promethium	Pm	61	(145)
Protactinium	Pa	91	231.0359[z]
Radium	Ra	88	226.0254[x,z]
Radon	Rn	86	(222)
Rhenium	Re	75	186.207
Rhodium	Rh	45	102.9055
Rubidium	Rb	37	85.4678*[x]
Ruthenium	Ru	44	101.07*[x]
Samarium	Sm	62	150.36*[x]
Scandium	Sc	21	44.9559
Selenium	Se	34	78.96*
Silicon	Si	14	28.0855*
Silver	Ag	47	107.8682*[x]
Sodium	Na	11	22.98977
Strontium	Sr	38	87.62[x]
Sulfur	S	16	32.06[w]
Tantalum	Ta	73	180.9479
Technetium	Tc	43	(98)
Tellurium	Te	52	127.60*[x]
Terbium	Tb	65	158.9254
Thallium	Tl	81	204.383
Thorium	Th	90	232.0381[x,z]
Thulium	Tm	69	168.9342
Tin	Sn	50	118.69*
Titanium	Ti	22	47.88*
Tungsten	W	74	183.85*
(Unnilhexium)	(Unh)	106	(263)
(Unnilpentium)	(Unp)	105	(262)
(Unnilquadium)	(Unq)	104	(261)
Uranium	U	92	238.0289[x,y]
Vanadium	V	23	50.9415
Xenon	Xe	54	131.29*[x,y]
Ytterbium	Yb	70	173.04*
Yttrium	Y	39	88.9059
Zinc	Zn	30	65.38
Zirconium	Zr	40	91.22[x]

† Atomic weight is an alternative term for "relative atomic mass of an element," $A_r(E)$. Values given here are scaled to $A_r(^{12}C) = 12$ and apply to elements as they exist in materials of terrestrial origin and to certain artificial elements. When used with due regard to the footnotes, they are considered reliable to ±1 in the last digit, or ±3 where followed by an asterisk.* With elements for which a precise atomic weight cannot be included in the table, a value in parentheses is given; this value is the atomic mass number of the isotope of longest known half-life.

[w] Element for which known variations in isotopic composition in normal terrestrial material prevent a more precise atomic weight being given; $A_r(E)$ values should be applicable to any "normal" material.

[x] Element for which geological specimens are known in which the element has an anomalous isotopic composition, such that the difference in atomic weight of the element in such specimens from that given in the table may exceed considerably the implied uncertainty.

[y] Element for which substantial variations in A_r from the value given can occur in commercially available material because of inadvertent or undisclosed change of isotopic composition.

[z] Element for which the value of A_r is that of the radioisotope of longest half-life.

MOLECULAR WEIGHTS

Molecular Formulas and Weights

This table presents, in general, the molecular formulas and molecular weights of the following classes of items, for both the Pharmacopeia and the National Formulary: (1) the chemicals covered by the titles of the individual monographs; (2) specific chemical compounds listed among the Reference Standards; and (3) chemical entities listed as *Reagents*, with the exception of indicators and biological stains. In addition, the table includes the molecular formulas and molecular weights of a number of other items, some of which, although not listed in the main volume, may be included by supplement.

The International Commission on Atomic Weights and Isotopic Abundances has quoted uncertainties in its atomic weight values in recent years. It has not been decided to what extent the information on such uncertainties ought to be utilized in the table of molecular weights in this compendium. At the present time, the molecular weights in the following table should be considered accurate to 4 parts in 10,000 (0.04%) or higher.

Acenocoumarol $C_{19}H_{15}NO_6$ 353.33
Acetaldehyde C_2H_4O 44.05
5-Acetamido-3-amino-2,4,6-triiodobenzoic Acid $C_9H_7I_3N_2O_3$ 571.88
5-Acetamido-2,4,6-triiodo-*N*-methylisophthalamic Acid $C_{11}H_9I_3N_2O_4$ 613.92
Acetaminophen $C_8H_9NO_2$ 151.16
Acetanilide C_8H_9NO 135.17
Acetazolamide $C_4H_6N_4O_3S_2$ 222.24
Sodium $C_4H_5N_4NaO_3S_2$ 244.22
Acetic Acid $C_2H_4O_2$ 60.05
Anhydride $C_4H_6O_3$ 102.09
Acetohexamide $C_{15}H_{20}N_2O_4S$ .. 324.39
Acetohydroxamic Acid $C_2H_5NO_2$ 75.07
Acetone C_3H_6O 58.08
Acetonitrile C_2H_3N 41.05
Acetophenazine $C_{23}H_{29}N_3O_2S$ .. 411.56
Maleate $C_{23}H_{29}N_3O_2S.2C_4H_4O_4$ 643.71
Acetophenetidin (see *Phenacetin*)
Acetosulfone Sodium $C_{14}H_{14}N_3NaO_5S_2$ 391.39
2-Acetoxy-4-dimethylamino-1,2-diphenyl-3-methylbutane (see *Propoxyphene Acetoxy Analog*)
Acetrizoate Sodium $C_9H_5I_3NNaO_3$ 578.85
Acetrizoic Acid $C_9H_6I_3NO_3$ 556.86
Acetyl $C_2H_3O^+$ 43.05
Chloride C_2H_3ClO 78.50
Sulfisoxazole $C_{13}H_{15}N_3O_4S$ .. 309.34
Acetylacetone $C_5H_8O_2$ 100.11
Acetylcholine Chloride $C_7H_{16}ClNO_2$ 181.66
Acetylcysteine $C_5H_9NO_3S$ 163.19
Acetyldigitoxin $C_{43}H_{66}O_{14}$ 806.99
Acetyl-L-tyrosine Ethyl Ester $C_{13}H_{17}NO_4$ 251.28
Acrisorcin $C_{12}H_{18}O_2.C_{13}H_{10}N_2$ 388.51
Acrylic Acid $C_3H_4O_2$ 72.06
Adenine $C_5H_5N_5$ 135.13
Sulfate $(C_5H_5N_5)_2.H_2SO_4.2H_2O$ 404.36
Sulfate, anhydrous $(C_5H_5N_5)_2.H_2SO_4$ 368.33
Adipic Acid $C_6H_{10}O_4$ 146.14
3,3′-(Adipoyldiimino)bis-(2,4,6-triiodobenzoic Acid) $C_{20}H_{14}I_6N_2O_6$ 1139.77
Alanine $C_3H_7NO_2$ 89.09
Albuterol $C_{13}H_{21}NO_3$ 239.31
Sulfate $(C_{13}H_{21}NO_3)_2.H_2SO_4$ 576.70

Alclometasone Dipropionate $C_{28}H_{37}ClO_7$ 521.05
Alcohol C_2H_6O 46.07
Allopurinol $C_5H_4N_4O$ 136.11
Alpha Tocopherol $C_{29}H_{50}O_2$ 430.71
Tocopheryl Acetate $C_{31}H_{52}O_3$ 472.75
Tocopheryl Acid Succinate $C_{33}H_{54}O_5$ 530.79
Alprostadil $C_{20}H_{34}O_5$ 354.49
Alum (see *Ammonium Alum*, and *Potassium Alum*)
Alumina (see *Aluminum Oxide*)
Aluminum Acetate $C_6H_9AlO_6$ 204.12
Chloride $AlCl_3.6H_2O$ 241.43
Chloride, anhydrous $AlCl_3$ ... 133.34
Hydroxide $Al(OH)_3$ 78.00
Monopalmitate $C_{16}H_{33}AlO_4$ .. 316.42
Monostearate $C_{18}H_{37}AlO_4$ 344.47
Nicotinate $C_{18}H_{12}AlN_3O_6$ 393.29
Oxide Al_2O_3 .... 101.96
Phosphate $AlPO_4$ 121.95
Subacetate $C_4H_7AlO_5$ 162.08
Sulfate $Al_2(SO_4)_3.14H_2O$ 594.35
Sulfate, anhydrous $Al_2(SO_4)_3$ 342.14
Alverine $C_{20}H_{27}N$ 281.44
Citrate $C_{20}H_{27}N.C_6H_8O_7$ 473.56
Amantadine $C_{10}H_{17}N$ 151.25
Hydrochloride $C_{10}H_{17}N.HCl$ 187.71
Amaranth $C_{20}H_{11}N_2Na_3O_{10}S_3$ .. 604.46
Amcinonide $C_{28}H_{35}FO_7$ 502.58
Amdinocillin $C_{15}H_{23}N_3O_3S$ 325.43
Amikacin $C_{22}H_{43}N_5O_{13}$ 585.61
Sulfate $C_{22}H_{43}N_5O_{13}.2H_2SO_4$ 781.75
Amiloride Hydrochloride $C_6H_8ClN_7O.HCl.2H_2O$ 302.12
Aminoacetophenone C_8H_9NO .. 135.17
Aminoantipyrine $C_{11}H_{13}N_3O$ ... 203.24
Hydrochloride $C_{11}H_{13}N_3O.HCl$ 239.70
Aminobenzoate Potassium $C_7H_6KNO_2$ 175.23
Aminobenzoic Acid $C_7H_7NO_2$ .. 137.14
Aminobutanol $C_4H_{11}NO$ 89.14
Aminocaproic Acid $C_6H_{13}NO_2$ 131.17
3-Amino-4-carboxamido-pyrazole Hemisulfate $(C_4H_6N_4O)_2.H_2SO_4$ 350.31
4-Amino-6-chloro-1,3-benzene-disulfonamide $C_6H_8ClN_3O_4S_2$ 285.72
4-Amino-2-chlorobenzoic Acid $C_7H_6ClNO_2$ 171.58

2-Amino-5-chlorobenzophenone $C_{13}H_{10}ClNO$ 231.68
3-Amino-6-chloro-1-(2-diethylaminoethyl)-4-(2-fluorophenyl)carbostyril Hydrochloride $C_{21}H_{23}ClFN_3O.HCl$ 424.34
4-Amino-6-chloro-N^3-methyl-*m*-benzenedisulfonamide $C_7H_{10}ClN_3O_4S_2$ 299.75
3-Amino-6-chloro-1-methyl-4-phenylcarbostyril $C_{16}H_{15}ClN_2O$ 286.76
3-Amino-2′-chloro-5-nitrobenzophenone $C_{13}H_9ClN_2O_3$ 276.68
3-Amino-4-(2-chlorophenyl)-6-nitrocarbostyril $C_{15}H_{10}ClN_3O_3$ 315.72
2-Amino-2′,5-dichlorobenzophenone $C_{13}H_9Cl_2NO$ 266.13
4-Amino-6,7-dimethoxy-2-(1-piperazinyl)quinazoline $C_{14}H_{19}N_5O_2$ 289.34
3-Amino-7-(dimethylamino)-phenothiazin-5-ium Chloride $C_{14}H_{14}ClN_3S$ 291.80
Aminoglutethimide $C_{13}H_{16}N_2O_2$ 232.26
Aminohippurate Sodium $C_9H_9N_2NaO_3$ 216.17
Aminohippuric Acid $C_9H_{10}N_2O_3$ 194.19
5-Aminoimidazole-4-carboxamide $C_4H_6N_4O$ 126.12
5-Aminoimidazole-4-carboxamide Hydrochloride $C_4H_6N_4O.HCl$ 162.58
2-[3-Amino-5-(*N*-methylacetamido)-2,4,6-triiodobenzamido]-2-deoxy-D-glucose $C_{16}H_{20}I_3N_3O_7$ 747.06
1-Amino-2-naphthol-4-sulfonic Acid $C_{10}H_9NO_4S$ 239.25
Aminophenol C_6H_7NO 109.13
3-Amino-4-phenoxy-5-sulfamoylbenzoic Acid $C_{13}H_{12}N_2O_5S$ 308.31
Aminophylline, anhydrous $C_{16}H_{24}N_{10}O_4$ 420.43
dihydrate $C_{16}H_{24}N_{10}O_4.2H_2O$ 456.46
Aminopropiophenone $C_9H_{11}NO$ 149.19
Hydrochloride $C_9H_{11}NO.HCl$ 185.65
Aminosalicylate Sodium (see *Sodium Aminosalicylate*)

Aminosalicylic Acid $C_7H_7NO_3$ 153.14
3-Amino-2,4,6-triiodobenzoic Acid $C_7H_4I_3NO_2$ 514.83
5-Amino-2,4,6-triiodo-*N*-methylisophthalamic Acid $C_9H_7I_3N_2O_3$ 571.88
Amitriptyline $C_{20}H_{23}N$ 277.41
  Hydrochloride $C_{20}H_{23}N.HCl$ 313.87
Ammonia NH_3 17.03
Ammonium Acetate $NH_4C_2H_3O_2$ 77.08
  Alum $AlNH_4(SO_4)_2.12H_2O$ 453.32
  Alum, anhydrous $AlNH_4(SO_4)_2$ 237.14
  Bicarbonate NH_4HCO_3 79.06
  Bromide NH_4Br 91.94
  Carbamate $NH_4NH_2CO_2$ 78.07
  Carbonate, normal $(NH_4)_2CO_3$ 96.09
  Chloride NH_4Cl 53.49
  Citrate, Dibasic $(NH_4)_2HC_6H_5O_7$ 226.19
  Fluoride NH_4F 37.04
  Hydroxide NH_4OH 35.05
  Molybdate $(NH_4)_6Mo_7O_{24}.4H_2O$ 1235.86
  Nitrate NH_4NO_3 80.04
  Oxalate $(NH_4)_2C_2O_4.H_2O$ 142.11
  Oxalate, anhydrous $(NH_4)_2C_2O_4$ 124.10
  Persulfate $(NH_4)_2S_2O_8$ 228.19
  Phosphate (see *Ammonium Phosphate, Dibasic*)
  Phosphate, Dibasic $(NH_4)_2HPO_4$ 132.06
  Phosphate, Monobasic $NH_4H_2PO_4$ 115.03
  Reineckate $NH_4[Cr(NH_3)_2(SCN)_4].H_2O$ 354.42
  Sulfamate $NH_4NH_2SO_3$ 114.12
  Sulfate $(NH_4)_2SO_4$ 132.13
  Sulfide $(NH_4)_2S$ 68.14
  Thiocyanate NH_4SCN 76.12
  Vanadate NH_4VO_3 116.98
Amobarbital $C_{11}H_{18}N_2O_3$ 226.27
  Sodium $C_{11}H_{17}N_2NaO_3$ 248.26
Amodiaquine $C_{20}H_{22}ClN_3O$ 355.87
  Hydrochloride $C_{20}H_{22}ClN_3O.2HCl.2H_2O$ 464.82
  Hydrochloride, anhydrous $C_{20}H_{22}ClN_3O.2HCl$ 428.79
Amoxicillin, anhydrous $C_{16}H_{19}N_3O_5S$ 365.40
  trihydrate $C_{16}H_{19}N_3O_5S.3H_2O$ 419.45
Amphetamine $C_9H_{13}N$ 135.21
  Hydrochloride $C_9H_{13}N.HCl$ 171.67
  Phosphate $C_9H_{13}N.H_3PO_4$ 233.20
  Sulfate $(C_9H_{13}N)_2.H_2SO_4$ 368.49
Amphotericin B $C_{47}H_{73}NO_{17}$ 924.09
Ampicillin, anhydrous $C_{16}H_{19}N_3O_4S$ 349.40
  trihydrate $C_{16}H_{19}N_3O_4S.3H_2O$ 403.45
  Sodium $C_{16}H_{18}N_3NaO_4S$ 371.39
Amprolium $C_{14}H_{19}ClN_4.HCl$ 315.25
Amyl Acetate $C_7H_{14}O_2$ 130.19
  Alcohol $C_5H_{12}O$ 88.15
  Nitrite $C_5H_{11}NO_2$ 117.15
Amylene Hydrate $C_5H_{12}O$ 88.15
Anethole $C_{10}H_{12}O$ 148.20
Angiotensin Amide $C_{49}H_{70}N_{14}O_{11}$ 1031.18
Anhydroprednisolone Tebutate $C_{27}H_{36}O_5$ 440.58

Anileridine $C_{22}H_{28}N_2O_2$ 352.48
  Hydrochloride $C_{22}H_{28}N_2O_2.2HCl$ 425.40
Aniline C_6H_7N 93.13
3-Anilino-2-(3,4,5-trimethoxybenzyl)acrylonitrile $C_{19}H_{20}N_2O_3$ 324.37
Anisindione $C_{16}H_{12}O_3$ 252.27
Anisole $CH_3OC_6H_5$ 108.14
Antazoline $C_{17}H_{19}N_3$ 265.36
  Phosphate $C_{17}H_{19}N_3.H_3PO_4$ 363.35
Anthracene $C_{14}H_{10}$ 178.23
Anthralin $C_{14}H_{10}O_3$ 226.23
Anthrone $C_{14}H_{10}O$ 194.23
Antimony Pentachloride $SbCl_5$ 299.01
  Potassium Tartrate $C_8H_4K_2O_{12}Sb_2.3H_2O$ 667.85
  Potassium Tartrate, anhydrous $C_8H_4K_2O_{12}Sb_2$ 613.81
  Sodium Tartrate $C_8H_4Na_2O_{12}Sb_2$ 581.59
  Trichloride $SbCl_3$ 228.11
Antipyrine $C_{11}H_{12}N_2O$ 188.23
Apomorphine Hydrochloride $C_{17}H_{17}NO_2.HCl.½H_2O$ 312.80
  Hydrochloride, anhydrous $C_{17}H_{17}NO_2.HCl$ 303.79
Aprobarbital $C_{10}H_{14}N_2O_3$ 210.23
Arginine $C_6H_{14}N_4O_2$ 174.20
  Hydrochloride $C_6H_{14}N_4O_2.HCl$ 210.66
Arsanilic Acid $C_6H_8AsNO_3$ 217.06
Arsenic Trioxide As_2O_3 197.84
Ascorbic Acid $C_6H_8O_6$ 176.13
Ascorbyl Palmitate $C_{22}H_{38}O_7$ 414.54
Asparagine $C_4H_8N_2O_3.H_2O$ 150.13
Aspartame $C_{14}H_{18}N_2O_5$ 294.31
Aspirin $C_9H_8O_4$ 180.16
  Aluminum $C_{18}H_{15}AlO_9$ 402.29
  Sodium $C_9H_7NaO_4$ 202.14
Atropine $C_{17}H_{23}NO_3$ 289.37
  Sulfate $(C_{17}H_{23}NO_3)_2.H_2SO_4.H_2O$ 694.84
  Sulfate, anhydrous $(C_{17}H_{23}NO_3)_2.H_2SO_4$ 676.82
Aurothioglucose $C_6H_{11}AuO_5S$ 392.18
2-Azahyoxanthine $C_4H_3N_5O$ 137.10
Azapetine $C_{17}H_{17}N$ 235.33
  Phosphate $C_{17}H_{17}N.H_3PO_4$ 333.32
Azatadine $C_{20}H_{22}N_2$ 290.41
  Maleate $C_{20}H_{22}N_2.2C_4H_4O_4$ 522.55
Azathioprine $C_9H_7N_7O_2S$ 277.26
Azobenzene $C_{12}H_{10}N_2$ 182.22
Azure A (see *3-Amino-7-(dimethylamino)phenothiazin-5-ium Chloride*)

Bacampicillin Hydrochloride $C_{21}H_{27}N_3O_7S.HCl$ 501.98
Baclofen $C_{10}H_{12}ClNO_2$ 213.66
Barium Bromide $BaBr_2.2H_2O$ 333.17
  Bromide, anhydrous $BaBr_2$ 297.14
  Carbonate $BaCO_3$ 197.34
  Chloride $BaCl_2.2H_2O$ 244.27
  Chloride, anhydrous $BaCl_2$ 208.24
  Chromate $BaCrO_4$ 253.32
  Hydroxide $Ba(OH)_2.8H_2O$ 315.47
  Hydroxide, anhydrous $Ba(OH)_2$ 171.34
  Nitrate $Ba(NO_3)_2$ 261.34
  Oxide BaO 153.33
  Sulfate $BaSO_4$ 233.39
  Sulfide BaS 169.39
  Sulfite $BaSO_3$ 217.39

Beclomethasone Dipropionate $C_{28}H_{37}ClO_7$ 521.05
Bendroflumethiazide $C_{15}H_{14}F_3N_3O_4S_2$ 421.41
Benoxinate $C_{17}H_{28}N_2O_3$ 308.42
  Hydrochloride $C_{17}H_{28}N_2O_3.HCl$ 344.88
Benzaldehyde C_7H_6O 106.12
Benzanilide $C_{13}H_{11}NO$ 197.24
Benzathine Penicillin G (see *Penicillin G Benzathine*)
Benzene C_6H_6 78.11
  Hexachloride, Delta $C_6H_6Cl_6$ 290.83
  Hexachloride, Gamma (see *Lindane*)
Benzenesulfonyl Chloride $C_6H_5ClO_2S$ 176.62
Benzethonium Chloride $C_{27}H_{42}ClNO_2$ 448.09
Benzhydrol $C_{13}H_{12}O$ 184.24
Benzidine $C_{12}H_{12}N_2$ 184.24
Benzocaine $C_9H_{11}NO_2$ 165.19
Benzoic Acid $C_7H_6O_2$ 122.12
Benzonatate (Av.) $C_{30}H_{53}NO_{11}$ 603
Benzophenone $C_{13}H_{10}O$ 182.22
Benzoquinone $C_6H_4O_2$ 108.10
Benzoyl Chloride C_7H_5ClO 140.57
  Peroxide $C_{14}H_{10}O_4$ 242.23
N-Benzoyl-L-arginine Ethyl Ester Hydrochloride $C_{15}H_{22}N_4O_3.HCl$ 342.82
N-Benzoyl-DL-arginine-*p*-nitroanilide hydrochloride $C_{19}H_{22}N_6O_4$ 434.89
2-Benzoyl-1,2,3,6,7,11b-hexahydro-4*H*-pyrazino[2,1-*a*]isoquinolin-4-one $C_{19}H_{18}N_2O_2$ 306.36
Benzphetamine $C_{17}H_{21}N$ 239.36
  Hydrochloride $C_{17}H_{21}N.HCl$ 275.82
Benzthiazide $C_{15}H_{14}ClN_3O_4S_3$ 431.93
Benztropine $C_{21}H_{25}NO$ 307.43
  Mesylate $C_{21}H_{25}NO.CH_4O_3S$ 403.54
Benzyl Alcohol C_7H_8O 108.14
  Benzoate $C_{14}H_{12}O_2$ 212.25
Benzyldioxopiperazineacetic Acid $C_{13}H_{14}N_2O_4$ 262.26
  Acid Hydrochloride $C_{13}H_{14}N_2O_4.HCl$ 298.73
1-Benzyl-3-methyl-5-aminopyrazole Hydrochloride $C_{11}H_{13}N_3.HCl$ 223.70
Benzyltrimethylammonium Chloride $C_{10}H_{16}ClN$ 185.70
Beta Carotene $C_{40}H_{56}$ 536.88
Betahistine $C_8H_{12}N_2$ 136.20
  Hydrochloride $C_8H_{12}N_2.2HCl$ 209.12
Betaine $C_5H_{11}NO_2$ 117.15
  Hydrochloride $C_5H_{11}NO_2.HCl$ 153.61
Betamethasone $C_{22}H_{29}FO_5$ 392.47
  Acetate $C_{24}H_{31}FO_6$ 434.50
  Benzoate $C_{29}H_{33}FO_6$ 496.57
  Dipropionate $C_{28}H_{37}FO_7$ 504.59
  Sodium Phosphate $C_{22}H_{28}FNa_2O_8P$ 516.41
  Valerate $C_{27}H_{37}FO_6$ 476.58
Betazole $C_5H_9N_3$ 111.15
Bethanechol Chloride $C_7H_{17}ClN_2O_2$ 196.68
  Tetraphenylboron $C_{31}H_{37}BN_2O_2$ 480.46
Bibenzyl $C_{14}H_{14}$ 182.27

Name and Formula	Mol. Wt.
Bichloroacetic Acid $C_2H_2Cl_2O_2$	128.94
Biotin $C_{10}H_{16}N_2O_3S$	244.31
Biperiden $C_{21}H_{29}NO$	311.47
Hydrochloride $C_{21}H_{29}NO.HCl$	347.93
Lactate $C_{21}H_{29}NO.C_3H_6O_3$	401.54
Biphenyl $C_{12}H_{10}$	154.21
Bipyridine $C_{10}H_8N_2$	156.19
Bisacodyl $C_{22}H_{19}NO_4$	361.40
1,4-Bis(4-amino-6,7-dimethoxy-2-quinazolinyl)piperazine $C_{24}H_{28}N_8O_4$	492.54
4,4′-Bis(4-amino-1-naphthylazo)-2,2′-stilbenedisulfonic Acid $C_{34}H_{26}N_6O_6S_2$	678.74
2,5-Bis(D-*arabino*-1,2,3,4-tetrahydroxybutyl)pyrazine $C_{12}H_{20}N_2O_8$	320.30
4,4′-Bis[4-(*p*-chlorophenyl)-4-hydroxypiperidino]butyrophenone $C_{32}H_{36}Cl_2N_2O_3$	567.55
Bis(2-ethylhexyl) Maleate $C_{20}H_{36}O_4$	340.50
Phthalate C_6H_4-1,2-$[COOCH_2(C_2H_5)CH(CH_2)_3CH_3]_2$	390.56
Sebacate $C_{26}H_{50}O_4$	426.68
Bis(2-ethylhexyl)phosphoric Acid $C_{16}H_{35}O_4P$	322.42
1,4-Bis(2-furoyl)piperazine $C_{14}H_{14}N_2O_4$	274.28
Bishydroxycoumarin (see *Dicumarol*)	
Bismuth Hydroxide $Bi(OH)_3$	260.00
Bismuth Subgallate $C_7H_5BiO_6$	394.09
Nitrate Pentahydrate $Bi(NO_3)_3.5H_2O$	485.07
Phosphate $BiPO_4$	303.95
Subnitrate (approx.) $4Bi(OH)_2NO_3.BiOOH$	1461.99
Sulfide Bi_2S_3	514.14
Trioxide Bi_2O_3	465.96
4,4′-Bis[1,2,3,6-tetrahydro-4-(2-oxo-1-benzimidazolinyl)-1-pyridyl]butyrophenone $C_{34}H_{34}N_6O_3$	574.68
Bis(trimethylsilyl)acetamide $C_8H_{21}NOSi_2$	203.43
Bis(trimethylsilyl)trifluoroacetamide $C_8H_{18}F_3NOSi_2$	257.40
Bithionol $C_{12}H_6Cl_4O_2S$	356.05
Blue Tetrazolium $C_{40}H_{32}Cl_2N_8O_2$	727.65
Boric Acid H_3BO_3	61.83
Borneol $C_{10}H_{18}O$	154.25
Bornyl Acetate $C_{12}H_{20}O_2$	196.29
Boron Trifluoride BF_3	67.81
Trifluoride Etherate $C_4H_{10}BF_3O$	141.93
Trioxide B_2O_3	69.62
Brilliant Green $C_{27}H_{34}N_2O_4S$	482.64
Yellow $C_{26}H_{18}N_4Na_2O_8S$	592.49
α-Bromo-2′-acetonaphthone $C_{12}H_9BrO$	249.11
Bromoaniline C_6H_6BrN	172.02
Bromocresol Purple $C_{21}H_{16}Br_2O_5S$	540.22
Purple, Sodium Salt $C_{21}H_{15}Br_2O_5SNa$	562.20
Bromocriptine $C_{32}H_{40}BrN_5O_5$	654.60
Mesylate $C_{32}H_{40}BrN_5O_5.CH_4O_3S$	750.70
Bromodiphenhydramine $C_{17}H_{20}BrNO$	334.26
Hydrochloride $C_{17}H_{20}BrNO.HCl$	370.72
Bromophenol Blue $C_{19}H_{10}Br_4O_5S$	669.96
Bromosuccinimide $C_4H_4BrNO_2$	177.98
Bromothymol Blue $C_{27}H_{28}Br_2O_5S$	624.38
Brompheniramine $C_{16}H_{19}BrN_2$	319.24
Maleate $C_{16}H_{19}BrN_2.C_4H_4O_4$	435.32
Brucine Sulfate $(C_{23}H_{26}N_2O_4)_2.H_2SO_4.7H_2O$	1013.12
Bucrylate $C_8H_{11}NO_2$	153.18
Bumetanide $C_{17}H_{20}N_2O_5S$	364.42
Bupivacaine $C_{18}H_{28}N_2O$	288.43
Hydrochloride, anhydrous $C_{18}H_{28}N_2O.HCl$	324.89
Hydrochloride, monohydrate $C_{18}H_{28}N_2O.HCl.H_2O$	342.91
Busulfan $C_6H_{14}O_6S_2$	246.29
Butabarbital $C_{10}H_{16}N_2O_3$	212.25
Sodium $C_{10}H_{15}N_2NaO_3$	234.23
Butacaine $C_{18}H_{30}N_2O_2$	306.45
Butalbital $C_{11}H_{16}N_2O_3$	224.26
Butamben (see *Butyl Aminobenzoate*)	
Butane C_4H_{10}	58.12
1,3-Butanediol $C_4H_{10}O_2$	90.12
Butanedione $C_4H_6O_2$	86.09
Butaperazine $C_{24}H_{31}N_3OS$	409.59
Maleate $C_{24}H_{31}N_3OS.2C_4H_4O_4$	641.73
Butethamine $C_{13}H_{20}N_2O_2$	236.31
Hydrochloride $C_{13}H_{20}N_2O_2.HCl$	272.77
Butoconazole Nitrate $C_{19}H_{17}Cl_3N_2S.HNO_3$	474.79
Butorphanol $C_{21}H_{29}NO_2$	327.47
Tartrate $C_{21}H_{29}NO_2.C_4H_6O_6$	477.55
Butyl Acetate $C_6H_{12}O_2$	116.16
Alcohol $C_4H_{10}O$	74.12
Aminobenzoate $C_{11}H_{15}NO_2$	193.24
Chloride C_4H_9Cl	92.57
Ether $C_8H_{18}O$	130.23
Nitrite $C_4H_9NO_2$	103.12
Butylamine $C_4H_{11}N$	73.14
4-(Butylamino)benzoic Acid $C_{11}H_{15}NO_2$	193.25
tert-Butyl Hydroperoxide, 70 Percent in Water $C_4H_{10}O_2$	90.12
Butylated Hydroxyanisole $C_{11}H_{16}O_2$	180.25
Hydroxytoluene $C_{15}H_{24}O$	220.35
Butylboronic Acid $C_4H_{11}BO_2$	101.94
Butyl Benzoate $C_{11}H_{14}O_2$	178.23
Butyl 3-(butylamino)-4-phenoxy-5-sulfamoylbenzoate $C_{21}H_{28}N_2O_5S$	420.52
Butylhydroxyanisole $C_{11}H_{16}O_2$	180.25
Butylparaben $C_{11}H_{14}O_3$	194.23
4-*tert*-Butylphenol $C_{10}H_{14}O$	150.21
n-Butyrophenone $C_{10}H_{12}O$	148.20
Cadmium Acetate $C_4H_6CdO_4.2H_2O$	266.53
Cadmium Nitrate $Cd(NO_3)_2.4H_2O$	308.48
Nitrate, anhydrous $Cd(NO_3)_2$	236.42
Caffeine $C_8H_{10}N_4O_2.H_2O$	212.21
anhydrous $C_8H_{10}N_4O_2$	194.19
Calcifediol $C_{27}H_{44}O_2.H_2O$	418.66
Calcitonin $C_{159}H_{232}N_{46}O_{45}S_3$	3604.04
Calcium Acetate $C_4H_6CaO_4.H_2O$	176.18
Acetate, anhydrous $C_4H_6CaO_4$	158.17
Acetic acid, calcium salt $C_4H_6CaO_4$	158.17
Aminosalicylate, anhydrous $C_{14}H_{12}CaN_2O_6$	344.34
Benzoylpas (see *Benzoylpas Calcium*)	
Biphosphate $Ca(H_2PO_4)_2.H_2O$	252.07
Biphosphate, anhydrous $Ca(H_2PO_4)_2$	234.05
Carbonate $CaCO_3$	100.09
Chloride $CaCl_2.2H_2O$	147.02
Chloride, anhydrous $CaCl_2$	110.99
Citrate $C_{12}H_{10}Ca_3O_{14}.4H_2O$	570.50
Citrate, anhydrous $C_{12}H_{10}Ca_3O_{14}$	498.44
Formyltetrahydrofolate (see *Leucovorin Calcium*)	
Glubionate $C_{18}H_{32}CaO_{19}.H_2O$	610.53
Glubionate, anhydrous $C_{18}H_{32}CaO_{19}$	592.52
Gluceptate $C_{14}H_{26}CaO_{16}$	490.43
Gluconate $C_{12}H_{22}CaO_{14}$	430.38
Hydroxide $Ca(OH)_2$	74.09
Hydroxide Phosphate $Ca_5(OH)(PO_4)_3$	502.32
Hypochlorite $Ca(ClO)_2$	142.98
Ipodate (see *Ipodate Calcium*)	
Lactate $C_6H_{10}CaO_6.5H_2O$	308.30
Lactate, anhydrous $C_6H_{10}CaO_6$	218.22
Lactobionate $C_{24}H_{42}CaO_{24}.2H_2O$	790.69
Levulinate $C_{10}H_{14}CaO_6.2H_2O$	306.33
Levulinate, anhydrous $C_{10}H_{14}CaO_6$	270.30
Nitrate $Ca(NO_3)_2.4H_2O$	236.15
Nitrate, anhydrous $Ca(NO_3)_2$	164.09
Orotate $C_{10}H_6CaN_4O_8$	350.26
Oxalate C_2CaO_4	128.10
Oxide CaO	56.08
Palmitate $C_{32}H_{62}CaO_4$	550.92
Pantothenate $C_{18}H_{32}CaN_2O_{10}$	476.54
Phosphate, Dibasic $CaHPO_4.2H_2O$	172.09
Phosphate, Dibasic, anhydrous $CaHPO_4$	136.06
Phosphate, Tribasic $Ca_3(PO_4)_2$	310.18
Phosphate, Tribasic (approx.) $10CaO.3P_2O_5.H_2O$	1004.64
Pyrophosphate $Ca_2P_2O_7$	254.10
Saccharate $C_6H_8CaO_8.4H_2O$	320.27
Saccharate, anhydrous $C_6H_8CaO_8$	248.20
Stearate $C_{36}H_{70}CaO_4$	607.03
Sulfate $CaSO_4.2H_2O$	172.17
Sulfate, anhydrous $CaSO_4$	136.14
Camphor $C_{10}H_{16}O$	152.24
Dinitrophenylhydrazone $C_{16}H_{20}N_4O_4$	332.36
Camphorsulfonic Acid $C_{10}H_{16}O_4S$	232.29
dl-10-Camphorsulfonic Acid, Sodium Salt $C_{10}H_{15}NaO_4S$	254.28
Cannabidiol $C_{21}H_{30}O_2$	314.47
Cannabinol $C_{21}H_{26}O_2$	310.44
Capryl Alcohol $C_8H_{18}O$	130.23

Captopril $C_9H_{15}NO_3S$ 217.28
Carbamazepine $C_{15}H_{12}N_2O$ 236.27
Carbamazine $C_{10}H_{21}N_3O$ 199.30
Carbamide Peroxide
$CH_6N_2O_3$ 94.07
Carbaspirin Calcium
$C_{19}H_{18}CaN_2O_9$ 458.44
Carbenicillin $C_{17}H_{18}N_2O_6S$ 378.40
Indanyl Sodium
$C_{26}H_{25}N_2NaO_6S$ 516.54
Disodium $C_{17}H_{16}N_2Na_2O_6S$ .. 422.36
Monosodium, anhydrous
$C_{17}H_{17}N_2NaO_6S$ 400.38
Monosodium monohydrate
$C_{17}H_{17}N_2NaO_6S.H_2O$ 418.40
Carbetapentane $C_{20}H_{31}NO_3$ 333.47
Citrate $C_{20}H_{31}NO_3.C_6H_8O_7$ .. 525.59
Carbidopa $C_{10}H_{14}N_2O_4.H_2O$ ... 244.25
anhydrous $C_{10}H_{14}N_2O_4$ 226.23
Carbinoxamine $C_{16}H_{19}ClN_2O$ .. 290.79
Maleate
$C_{16}H_{19}ClN_2O.C_4H_4O_4$ 406.87
Carbon Dioxide CO_2 .... 44.01
Disulfide CS_2 .... 76.13
Monoxide CO 28.01
Tetrachloride CCl_4 .... 153.82
Carbonic Acid H_2CO_3 .... 62.02
Carboxyl (carboxy) $-CHO_2$ 45.02
(Carboxymethyl)trimethyl-
ammonium Chloride,
hydrazide $C_5H_{14}ClN_3O$ 167.64
Carisoprodol $C_{12}H_{24}N_2O_4$ 260.33
Carmustine $C_5H_9Cl_2N_3O_2$ 214.05
Carotene $C_{40}H_{56}$ 536.88
Carphenazine $C_{24}H_{31}N_3O_2S$ 425.59
Maleate
$C_{24}H_{31}N_3O_2S.2C_4H_4O_4$ 657.73
Carvone $C_{10}H_{14}O$ 150.22
Catechol $C_6H_6O_2$ 110.11
Cefadroxil $C_{16}H_{17}N_3O_5S$ 363.39
Cefamandole Nafate
$C_{19}H_{17}N_6NaO_6S_2$ 512.50
Cefamandole Sodium
$C_{18}H_{17}N_6NaO_5S_2$ 484.48
Cefazolin $C_{14}H_{14}N_8O_4S_3$ 454.50
Sodium $C_{14}H_{13}N_8NaO_4S_3$ 476.48
Cefonicid $C_{18}H_{18}N_6O_8S_3$ 542.56
Sodium $C_{18}H_{16}N_6Na_2O_8S_3$ ... 586.52
Cefoperazone Sodium
$C_{25}H_{26}N_9NaO_8S_2$ 667.65
Sterile Cefoperazone Sodium
$C_{25}H_{26}N_9NaO_8S_2$ 667.65
Ceforanide $C_{20}H_{21}N_7O_6S_2$ 519.55
Cefotaxime Sodium
$C_{16}H_{16}N_5NaO_7S_2$ 477.44
Sterile Cefotetan Disodium
$C_{17}H_{15}N_7Na_2O_8S_4$ 619.57
Cefoxitin $C_{16}H_{17}N_3O_7S_2$ 427.45
Sodium $C_{16}H_{16}N_3NaO_7S_2$ 449.43
Sterile Ceftazidime
$C_{22}H_{22}N_6O_7S_2.5H_2O$ 636.65
Ceftizoxime Sodium
$C_{13}H_{12}N_5NaO_5S_2$ 405.38
Ceftriaxone $C_{18}H_{18}N_8O_7S_3$ 554.57
Sodium $C_{18}H_{16}N_8Na_2O_7S_3$.-
3½H_2O 661.59
Sodium, anhydrous
$C_{18}H_{16}N_8Na_2O_7S_3$ 598.53
Cefuroxime $C_{16}H_{16}N_4O_8S$ 424.39
Sodium $C_{16}H_{15}N_4NaO_8S$ 446.37
Cephaeline $C_{28}H_{38}N_2O_4$ 466.62
Hydrobromide anhydrous
$C_{28}H_{38}N_2O_4.2HBr$ 628.44
Hydrobromide heptahydrate
$C_{28}H_{38}N_2O_4.2HBr.7H_2O$ ... 754.55
Hydrochloride
$C_{28}H_{38}N_2O_4.2HCl.7H_2O$ ... 665.65
Hydrochloride, anhydrous
$C_{28}H_{38}N_2O_4.2HCl$ 539.54

Cephalexin
$C_{16}H_{17}N_3O_4S.H_2O$ 365.40
anhydrous $C_{16}H_{17}N_3O_4S$ 347.39
Cephaloridine
$C_{19}H_{17}N_3O_4S_2$ 415.48
Cephalothin $C_{16}H_{16}N_2O_6S_2$ 396.43
Sodium $C_{16}H_{15}N_2NaO_6S_2$ 418.41
Cephapirin $C_{17}H_{17}N_3O_6S_2$ 423.46
Sodium $C_{17}H_{16}N_3NaO_6S_2$ 445.44
Cephradine, anhydrous
$C_{16}H_{19}N_3O_4S$ 349.40
dihydrate
$C_{16}H_{19}N_3O_4S.2H_2O$ 385.43
monohydrate
$C_{16}H_{19}N_3O_4S.H_2O$ 367.42
Ceric Ammonium Nitrate
$Ce(NO_3)_4.2NH_4NO_3$ 548.23
Ammonium Sulfate
$Ce(SO_4)_2.2(NH_4)_2SO_4$.-
$2H_2O$ 632.53
Ammonium Sulfate, anhydrous
$Ce(SO_4)_2.2(NH_4)_2SO_4$ 596.50
Sulfate $Ce(SO_4)_2.4H_2O$ 404.30
Sulfate, anhydrous
$Ce(SO_4)_2$ 332.24
Cetalkonium Chloride
$C_{25}H_{46}ClN$ 396.10
Cetyl Alcohol $C_{16}H_{34}O$ 242.44
Cetyl Palmitate (see *Hexadecyl Hexadecanoate*)
Cetylpyridinium Chloride
$C_{21}H_{38}ClN.H_2O$ 358.01
Chloride, anhydrous
$C_{21}H_{38}ClN$ 339.99
Cetyltrimethylammonium
Bromide $C_{19}H_{42}BrN$ 364.45
Chloral Betaine
$C_2H_3Cl_3O_2.C_5H_{11}NO_2$ 282.55
Hydrate $C_2H_3Cl_3O_2$ 165.40
Chlorambucil $C_{14}H_{19}Cl_2NO_2$ ... 304.22
Chloramine T
$C_7H_7ClNNaO_2S.3H_2O$ 281.69
Chloramphenicol
$C_{11}H_{12}Cl_2N_2O_5$ 323.13
Palmitate $C_{27}H_{42}Cl_2N_2O_6$ 561.54
Sodium Succinate
$C_{15}H_{15}Cl_2N_2NaO_8$ 445.19
Chloranil $C_6Cl_4O_2$ 245.88
Chlorauric Acid $HAuCl_4$.-
$3H_2O$ 393.83
Acid, anhydrous $HAuCl_4$ 339.79
Chlordantoin $C_{11}H_{17}Cl_3N_2O_2S$ .. 347.69
Chlordiazepoxide
$C_{16}H_{14}ClN_3O$ 299.76
Hydrochloride
$C_{16}H_{14}ClN_3O.HCl$ 336.22
Chlorhexidine $C_{22}H_{30}Cl_2N_{10}$ 505.45
Gluconate
$C_{22}H_{30}Cl_2N_{10}.2C_6H_{12}O_7$ 897.77
Chloride Cl^- 35.45
Chlormadinone Acetate
$C_{23}H_{29}ClO_4$ 404.93
Chlormerodrin
$C_5H_{11}ClHgN_2O_2$ 367.20
Chloroacetanilide C_8H_8ClNO ... 169.61
2-Chloro-4-amino-6,7-
dimethoxyquinazoline
$C_{10}H_{10}ClN_3O_2$ 239.66
2-Chloro-4-nitrophenol
$C_6H_4ClNO_3$ 173.56
Chloroaniline C_6H_6ClN 127.57
Chlorobenzene C_6H_5Cl 112.56
Chlorobenzoic Acid $C_7H_5ClO_2$. 156.57
4-Chlorobenzophenone
$C_{13}H_9OCl$ 216.66
Chlorobutanol $C_4H_7Cl_3O$ 177.46
7-Chloro-5-(*o*-chlorophenyl)-
1,3-dihydro-3-acetoxy-2*H*-1,4-
benzodiazepin-2-one
$C_{17}H_{12}Cl_2N_2O_3$ 327.75
6-Chloro-4-(*o*-chlorophenyl)-
2-quinazolinecarboxaldehyde
$C_{15}H_8Cl_2N_2O$ 303.15
6-Chloro-4-(*o*-chlorophenyl)-
2-quinazolinecarboxylic Acid
$C_{15}H_8Cl_2N_2O_2$ 319.15
6-Chloro-4-(*o*-chlorophenyl)-
2-quinazolinemethanol
$C_{15}H_{10}Cl_2N_2O$ 305.16
Chlorocresol C_7H_7ClO 142.58
2-Chloro-10-(2-diethylami-
noethyl)-9-acridanone
Hydrochloride
$C_{19}H_{21}ClN_2O.HCl$ 365.30
5-Chloro-2-[[2-(diethylamino)-
ethyl]amino]-2′-fluoro-
benzophenone
$C_{19}H_{22}ClFN_2O$ 348.85
5-Chloro-2-(2-diethylamino-
ethylamino)-2′-fluoro-
benzophenone Hydrochloride
$C_{19}H_{22}ClFN_2O.HCl$ 385.31
7-Chloro-1-(2-diethylamino-
ethyl)-5-(2-fluorophenyl)-
1,3,4,5-tetrahydro-2*H*-1,4-
benzodiazepin-2-one
Hydrochloride
$C_{19}H_{25}ClFN_3O.HCl$ 402.34
7-Chloro-1,3-dihydro-1-(2-
diethylaminoethyl)-5-(2-
fluorophenyl)-2*H*-1,4-
benzodiazepin-2-one
Hydrochloride
$C_{19}H_{23}ClFN_3O.HCl$ 400.32
7-Chloro-1,3-dihydro-5-phenyl-
2*H*-1,4-benzodiazepin-2-one
$C_{15}H_{11}ClN_2O$ 270.72
7-Chloro-1,3-dihydro-5-phenyl-
2*H*-1,4-benzodiazepin-2-one
4-Oxide $C_{15}H_{11}ClN_2O_2$ 286.72
2-Chloro-3,5-dimethylphenol
C_8H_9ClO 156.61
7-Chloro-5-(2-fluorophenyl)-
1,3-dihydro-2*H*-1,4-benzo-
diazepin-2-one
$C_{15}H_{10}ClFN_2O$ 288.71
Chloroform $CHCl_3$ 119.38
Chloroguanide $C_{11}H_{16}ClN_5$ 253.73
Hydrochloride
$C_{11}H_{16}ClN_5.HCl$ 290.19
1-Chloronaphthalene
$C_{10}H_7Cl$ 162.62
p-Chlorophenol (see *Parachlorophenol*)
Chlorophenothane $C_{14}H_9Cl_5$ 354.49
Chlorophenyldiphenylmeth-
anol $C_{19}H_{15}ClO$ 294.78
Chloroplatinic Acid
$H_2PtCl_6.6H_2O$ 517.91
Acid, anhydrous H_2PtCl_6 .... 409.82
Chloroprocaine
$C_{13}H_{19}ClN_2O_2$ 270.76
Hydrochloride
$C_{13}H_{19}ClN_2O_2.HCl$ 307.22
Chloroquine $C_{18}H_{26}ClN_3$ 319.88
Hydrochloride
$C_{18}H_{26}ClN_3.2HCl$ 392.80
Phosphate
$C_{18}H_{26}ClN_3.2H_3PO_4$ 515.87
Chlorosulfamoylanthranilic
4-Chloro-5-sulfamoylanthranilic
Acid $C_7H_7ClN_2O_4S$ 250.66
Chlorothen $C_{14}H_{18}ClN_3S$ 295.83
Citrate
$C_{14}H_{18}ClN_3S.C_6H_8O_7$ 487.95
Chlorotheophylline
$C_7H_7ClN_4O_2$ 214.61
Chlorothiazide

$C_7H_6ClN_3O_4S_2$ 295.72
Sodium $C_7H_5ClN_3NaO_4S_2$ 317.70
Chlorotrimethylsilane C_3H_9ClSi 108.64
Chloroxine $C_9H_5Cl_2NO$ 214.05
Chloroxylenol C_8H_9ClO 156.61
Chlorpheniramine $C_{16}H_{19}ClN_2$ 274.79
Maleate $C_{16}H_{19}ClN_2.C_4H_4O_4$ 390.87
Chlorpromazine $C_{17}H_{19}ClN_2S$ 318.86
Hydrochloride $C_{17}H_{19}ClN_2S.HCl$ 355.32
Chlorpropamide $C_{10}H_{13}ClN_2O_3S$ 276.74
Chlorprothixene $C_{18}H_{18}ClNS$ 315.86
Chlortetracycline $C_{22}H_{23}ClN_2O_8$ 478.89
Hydrochloride $C_{22}H_{23}ClN_2O_8.HCl$ 515.35
Chlorthalidone $C_{14}H_{11}ClN_2O_4S$ 338.76
Chlorzoxazone $C_7H_4ClNO_2$ 169.57
Cholecalciferol $C_{27}H_{44}O$ 384.64
Cholestane $C_{27}H_{48}$ 372.68
Cholesterol $C_{27}H_{46}O$ 386.66
Cholesteryl Benzoate $C_{34}H_{50}O_2$ 490.77
Caprylate $C_{35}H_{60}O_2$ 512.86
Heptylate $C_{34}H_{58}O_2$ 498.83
Cholesteryl Caprylate $C_{35}H_{60}O_2$ 512.86
Heptylate $C_{54}H_{58}O_2$ 498.83
Cholic Acid $C_{24}H_{40}O_5$ 408.58
Choline $C_5H_{15}NO_2$ 121.18
Chloride $C_5H_{14}ClNO$ 139.62
Salicylate $C_{12}H_{19}NO_4$ 241.29
Chromic Chloride $CrCl_3.6H_2O$ 266.48
Chromium Trioxide CrO_3 99.99
Chromotropic Acid $C_{10}H_8O_8S_2.2H_2O$ 356.32
Ciclopirox Olamine $C_{12}H_{17}NO_2.C_2H_7NO$ 268.36
Cimetidine $C_{10}H_{16}N_6S$ 252.34
Cinchonine $C_{19}H_{22}N_2O$ 294.40
Cinchonidine $C_{19}H_{22}N_2O$ 294.40
Cinoxacin $C_{12}H_{10}N_2O_5$ 262.22
Cinoxate $C_{14}H_{18}O_4$ 250.29
Cisplatin $Cl_2H_6N_2Pt$ 300.06
Citral $C_{10}H_{16}O$ 152.24
[Citrato(4-)]dicopper $C_6H_4Cu_2O_7$ 315.19
Citric Acid, anhydrous $C_6H_8O_7$ 192.12
Acid, monohydrate $C_6H_8O_7.H_2O$ 210.14
Clavulanate Potassium $C_8H_8KNO_5$ 237.25
Clemastine $C_{21}H_{26}ClNO$ 343.90
Fumarate $C_{21}H_{26}ClNO.C_4H_4O_4$ 459.97
Clemizole $C_{19}H_{20}ClN_3$ 325.84
Hydrochloride $C_{19}H_{20}ClN_3.HCl$ 362.30
Clidinium Bromide $C_{22}H_{26}BrNO_3$ 432.36
Clindamycin $C_{18}H_{33}ClN_2O_5S$ 424.98
Hydrochloride, anhydrous $C_{18}H_{33}ClN_2O_5S.HCl$ 461.44
Hydrochloride monohydrate $C_{18}H_{35}ClN_2O_5S.HCl.H_2O$ 479.46
Palmitate $C_{34}H_{63}ClN_2O_6S$ 663.39
Palmitate Hydrochloride $C_{34}H_{63}ClN_2O_6S.HCl$ 699.86
Phosphate $C_{18}H_{34}ClN_2O_8PS$ 504.96

Clioquinol C_9H_5ClINO 305.50
Clocortolone Pivalate $C_{27}H_{36}ClFO_5$ 495.03
Clofibrate $C_{12}H_{15}ClO_3$ 242.70
Clomiphene $C_{26}H_{28}ClNO$ 405.97
Citrate $C_{26}H_{28}ClNO.C_6H_8O_7$ 598.09
Clonazepam $C_{15}H_{10}ClN_3O_3$ 315.72
Clonidine $C_9H_9Cl_2N_3$ 230.10
Hydrochloride $C_9H_9Cl_2N_3.HCl$ 266.56
Clorazepate Dipotassium $C_{16}H_{11}ClK_2N_2O_4$ 408.92
Potassium $C_{16}H_{10}ClKN_2O_3$ 352.82
Clotrimazole $C_{22}H_{17}ClN_2$ 344.84
Cloxacillin Benzathine $(C_{19}H_{18}ClN_3O_5S)_2.-C_{16}H_{20}N_2$ 1112.11
Sodium $C_{19}H_{17}ClN_3NaO_5S.H_2O$ 475.88
Sodium, anhydrous $C_{19}H_{17}ClN_3NaO_5S$ 457.86
Cobaltous Acetate $C_4H_6CoO_4.4H_2O$ 249.08
Acetate, anhydrous $C_4H_6CoO_4$ 177.02
Chloride $CoCl_2.6H_2O$ 237.93
Chloride, anhydrous $CoCl_2$ 129.84
Nitrate $Co(NO_3)_2.6H_2O$ 291.03
Nitrate, anhydrous $Co(NO_3)_2$ 182.94
Cocaine $C_{17}H_{21}NO_4$ 303.36
Hydrochloride $C_{17}H_{21}NO_4.HCl$ 339.82
Codeine $C_{18}H_{21}NO_3.H_2O$ 317.38
anhydrous $C_{18}H_{21}NO_3$ 299.37
N-Oxide $C_{18}H_{21}NO_4$ 315.37
Phosphate $C_{18}H_{21}NO_3.-H_3PO_4.½H_2O$ 406.37
Phosphate, anhydrous $C_{18}H_{21}NO_3.H_3PO_4$ 397.36
Sulfate $(C_{18}H_{21}NO_3)_2.-H_2SO_4.3H_2O$ 750.86
Sulfate anhydrous $(C_{18}H_{21}NO_3)_2.H_2SO_4$ 696.81
Colchicine $C_{22}H_{25}NO_6$ 399.44
Colistimethate Sodium
Colistin A component $C_{58}H_{105}N_{16}Na_5O_{28}S_5$ 1749.81
Colistin B component $C_{57}H_{103}N_{16}Na_5O_{28}S_5$ 1735.78
Colistin A $C_{53}H_{100}N_{16}O_{13}$ 1169.47
B $C_{52}H_{98}N_{16}O_{13}$ 1155.45
Sulfate, Colistin A component $C_{53}H_{100}N_{16}O_{13}.2½H_2SO_4$ 1414.65
Sulfate, Colistin B component $C_{52}H_{98}N_{16}O_{13}.2½H_2SO_4$ 1400.63
Congo Red $C_{32}H_{22}N_6Na_2O_6S_2$ 696.66
Copper Compounds (see under *Cupric* and *Cuprous*)
Gluconate $C_{12}H_{22}CuO_{14}$ 453.84
Cortisone $C_{21}H_{28}O_5$ 360.45
Acetate $C_{23}H_{30}O_6$ 402.49
Cosyntropin $C_{136}H_{210}N_{40}O_{31}S$ 2933.46
Coumarin $C_9H_6O_2$ 146.15
Cresol C_7H_8O 108.14
m-Cresol Purple $C_{21}H_{18}O_5S$ 382.43
Cromolyn Sodium $C_{23}H_{14}Na_2O_{11}$ 512.34
Crotamiton $C_{13}H_{17}NO$ 203.28
Crystal Violet (see *Hexamethylpararosaniline Chloride*)
Cupric Acetate $C_4H_6CuO_4.H_2O$ 199.65
Acetate, anhydrous $C_4H_6CuO_4$ 181.63
Chloride $CuCl_2.2H_2O$ 170.48
Chloride, anhydrous $CuCl_2$ 134.45

Citrate (see [*Citrato(4-)*]-*dicopper*)
Nitrate $Cu(NO_3)_2.3H_2O$ 241.60
Oxide CuO 79.55
Sulfate $CuSO_4.5H_2O$ 249.68
Sulfate, anhydrous $CuSO_4$ 159.60
Cuprous Oxide Cu_2O 143.09
Cyanocobalamin $C_{63}H_{88}CoN_{14}O_{14}P$ 1355.38
Cyanogen Bromide $BrCN$ 105.92
Cyclizine $C_{18}H_{22}N_2$ 266.39
Hydrochloride $C_{18}H_{22}N_2.HCl$ 302.85
Lactate $C_{18}H_{22}N_2.C_3H_6O_3$ 356.46
Cyclobenzaprine $C_{20}H_{21}N$ 275.39
Hydrochloride $C_{20}H_{21}N.HCl$ 311.85
Cyclohexane C_6H_{12} 84.16
Cyclohexanol $C_6H_{12}O$ 100.16
Cyclohexanone $C_6H_{10}O$ 98.14
Cycloheximide $C_{15}H_{23}NO_4$ 281.35
Cyclohexylamine $C_6H_{13}N$ 99.18
2-(Cyclohexylcarbonyl)-2,3,6,7-tetrahydro-4*H*-pyrazino[2,1-*a*]isoquinolin-4-one $C_{19}H_{22}N_2O_2$ 310.40
(1,2-Cyclohexylenedinitrilo)-tetraacetic Acid $(HOCO-CH_2)_2NCH(CH_2)_4CHN-(CH_2COOH)_2.H_2O$ 364.35
Cyclomethycaine $C_{22}H_{33}NO_3$ 359.51
Hydrochloride $C_{22}H_{33}NO_3.HCl$ 395.97
Cyclopentolate $C_{17}H_{25}NO_3$ 291.39
Hydrochloride $C_{17}H_{25}NO_3.HCl$ 327.85
Cyclophosphamide $C_7H_{15}Cl_2N_2O_2P.H_2O$ 279.10
anhydrous $C_7H_{15}Cl_2N_2O_2P$ 261.09
Cyclopropane C_3H_6 42.08
Cycloserine $C_3H_6N_2O_2$ 102.09
Cyclosporine $C_{62}H_{111}N_{11}O_{12}$ 1202.63
Cycrimine $C_{19}H_{29}NO$ 287.44
Cyproheptadine $C_{21}H_{21}N$ 287.40
Hydrochloride $C_{21}H_{21}N.HCl.1½H_2O$ 350.89
Hydrochloride, anhydrous $C_{21}H_{21}N.HCl$ 323.86
Cysteine $C_3H_7NO_2S$ 121.15
Hydrochloride $C_3H_7NO_2S.HCl$ 157.61
Hydrochloride monohydrate $C_3H_7NO_2S.HCl.H_2O$ 175.63
Cystine $C_6H_{12}N_2O_4S_2$ 240.29
Cytarabine $C_9H_{13}N_3O_5$ 243.22

Dacarbazine $C_6H_{10}N_6O$ 182.18
Dactinomycin $C_{62}H_{86}N_{12}O_{16}$ 1255.43
Danazol $C_{22}H_{27}NO_2$ 337.46
Dapsone $C_{12}H_{12}N_2O_2S$ 248.30
Decanal $C_{10}H_{20}O$ 156.27
Decanol $C_{10}H_{22}O$ 158.28
Decyl Sodium Sulfate $C_{10}H_{21}NaO_4S$ 206.32
Deferoxamine $C_{25}H_{48}N_6O_8$ 560.69
Mesylate $C_{25}H_{48}N_6O_8.CH_4O_3S$ 656.79
Dehydroacetic Acid $C_8H_8O_4$ 168.15
Dehydrocholic Acid $C_{24}H_{34}O_5$ 402.53
Dehydrocholesterol $C_{27}H_{44}O$ 384.64
Dehydroemetine $C_{29}H_{38}N_2O_4$ 478.63
Hydrochloride $C_{29}H_{38}N_2O_4.2HCl$ 551.55

Substance	Mol. Wt.
Demecarium Bromide $C_{32}H_{52}Br_2N_4O_4$	716.60
Demeclocycline, anhydrous $C_{21}H_{21}ClN_2O_8$	464.86
sesquihydrate $C_{21}H_{21}ClN_2O_8.1\frac{1}{2}H_2O$	491.88
Hydrochloride $C_{21}H_{21}ClN_2O_8.HCl$	501.32
Denatonium Benzoate $C_{28}H_{34}N_2O_3.H_2O$	464.61
Desipramine $C_{18}H_{22}N_2$	266.39
Hydrochloride $C_{18}H_{22}N_2.HCl$	302.85
Desonide $C_{24}H_{32}O_6$	416.51
Desoximetasone $C_{22}H_{29}FO_4$	376.47
Desoxycorticosterone Acetate $C_{23}H_{32}O_4$	372.50
Pivalate $C_{26}H_{38}O_4$	414.58
Deuterium Oxide D_2O	20.028
Deuterochloroform $CDCl_3$	120.37
Dexamethasone $C_{22}H_{29}FO_5$	392.47
Acetate $C_{24}H_{31}FO_6.H_2O$	452.52
Acetate, anhydrous $C_{24}H_{31}FO_6$	434.50
Phosphate $C_{22}H_{30}FO_8P$	472.45
Sodium Phosphate $C_{22}H_{28}FNa_2O_8P$	516.41
Dexbrompheniramine $C_{16}H_{19}BrN_2$	319.24
Maleate $C_{16}H_{19}BrN_2.C_4H_4O_4$	435.32
Dexchlorpheniramine $C_{16}H_{19}ClN_2$	274.79
Maleate $C_{16}H_{19}ClN_2.C_4H_4O_4$	390.87
Dexpanthenol $C_9H_{19}NO_4$	205.25
Dextroamphetamine (see *Amphetamine*)	
Dextromethorphan $C_{18}H_{25}NO$	271.40
Hydrobromide $C_{18}H_{25}NO.HBr.H_2O$	370.33
Hydrobromide, anhydrous $C_{18}H_{25}NO.HBr$	352.31
Dextrose, anhydrous $C_6H_{12}O_6$	180.16
monohydrate $C_6H_{12}O_6.H_2O$	198.17
Diacetylfluorescein $C_{24}H_{16}O_7$	416.39
Diacetylhydroxyamphetamine $C_{13}H_{17}NO_3$	235.28
Diacetylmorphine $C_{21}H_{23}NO_5$	369.42
Hydrochloride $C_{21}H_{23}NO_5.HCl.H_2O$	423.89
Hydrochloride, anhydrous $C_{21}H_{23}NO_5.HCl$	405.88
Diaminobenzidine $C_{12}H_{14}N_4$	214.27
Tetrahydrochloride $C_{12}H_{14}N_4.4HCl$	360.11
Diaminodiphenylsulfone (see *Dapsone*)	
Diaminonaphthalene $C_{10}H_{10}N_2$	158.20
Dianisidine $C_{14}H_{16}N_2O_2$	244.29
Dihydrochloride $C_{14}H_{16}N_2O_2.2HCl$	317.21
Diatrizoate Meglumine $C_{11}H_9I_3N_2O_4.C_7H_{17}NO_5$	809.13
Sodium $C_{11}H_8I_3N_2NaO_4$	635.90
Diatrizoic Acid $C_{11}H_9I_3N_2O_4$	613.92
Diazepam $C_{16}H_{13}ClN_2O$	284.75
Diazoxide $C_8H_7ClN_2O_2S$	230.67
Dibenzyl (see *Bibenzyl*)	
Dibromoparachlorophenol $C_6H_3Br_2ClO$	286.35
Dibromoquinone-chlorimide $C_6H_2Br_2ClNO$	299.35
Dibucaine $C_{20}H_{29}N_3O_2$	343.47
Hydrochloride $C_{20}H_{29}N_3O_2.HCl$	379.93
Dichloroaniline $C_6H_5Cl_2N$	162.02
Dichlorobenzene $C_6H_4Cl_2$	147.00
1-[2,4-Dichloro-β-[(5-bromo-2-chloro-3-thenyl)-oxy]-phenethyl]imidazole hydrochloride $C_{16}H_{13}BrCl_2N_2OS.HCl$	468.62
1-[2,4-Dichloro-β-[(2,5-dichloro-3-thenyl)-oxy]phenethyl]-imidazole hydrochloride $C_{16}H_{12}Cl_4N_2OS.HCl$	458.62
Dichlorodifluoromethane CCl_2F_2	120.91
Dichlorofluorescein $C_{20}H_{10}Cl_2O_5$	401.20
Dichloroindophenol Sodium $C_{12}H_6Cl_2NNaO_2.2H_2O$	326.11
anhydrous $C_{12}H_6Cl_2NNaO_2$	290.08
Dichloromethane (see *Methylene Chloride*)	
Dichloronaphthol $C_{10}H_6Cl_2O$	213.06
2,4-Dichloro-1-naphthol $C_{10}H_6OCl_2$	213.06
Dichlorophenol-indophenol (see *Dichloroindophenol*)	
Dichloroquinone-chlorimide $C_6H_2Cl_3NO$	210.45
Dichlorotetrafluoroethane $C_2Cl_2F_4$	170.92
1-[2,4-Dichloro-β-[(3-thenyl)-oxy]phenethyl]imidazole hydrochloride $C_{16}H_{14}Cl_2N_2OS.HCl$	389.73
Dichlorvos $C_4H_7Cl_2O_4P$	220.98
Dicloxacillin $C_{19}H_{17}Cl_2N_3O_5S$	470.33
Sodium $C_{19}H_{16}Cl_2N_3NaO_5S.H_2O$	510.32
Sodium, anhydrous $C_{19}H_{16}Cl_2N_3NaO_5S$	492.31
Dicumarol $C_{19}H_{12}O_6$	336.30
Dicyclohexylamine $C_{12}H_{23}N$	181.32
Dicyclomine $C_{19}H_{35}NO_2$	309.49
Hydrochloride $C_{19}H_{35}NO_2.HCl$	345.95
Dienestrol $C_{18}H_{18}O_2$	266.34
Diethanolamine $C_4H_{11}NO_2$	105.14
Diethyl Phthalate $C_{12}H_{14}O_4$	222.24
Diethylamine $C_4H_{11}N$	73.14
Diethylaniline $C_{10}H_{15}N$	149.24
Diethylcarbamazine $C_{10}H_{21}N_3O$	199.30
Citrate $C_{10}H_{21}N_3O.C_6H_8O_7$	391.42
Diethylene Glycol $C_4H_{10}O_3$	106.12
Di(2-ethylhexyl)phthalate $C_{24}H_{38}O_4$	390.54
Diethylpropion $C_{13}H_{19}NO$	205.30
Hydrochloride $C_{13}H_{19}NO.HCl$	241.76
Diethylstilbestrol $C_{18}H_{20}O_2$	268.35
Diphosphate $C_{18}H_{22}O_8P_2$	428.31
Diethyltoluamide $C_{12}H_{17}NO$	191.27
Diflorasone Diacetate $C_{26}H_{32}F_2O_7$	494.53
Diflunisal $C_{13}H_8F_2O_3$	250.20
Digitonin $C_{56}H_{92}O_{29}$	1229.33
Digitoxin $C_{41}H_{64}O_{13}$	764.95
Digoxin $C_{41}H_{64}O_{14}$	780.95
Dihydroequilin $C_{18}H_{22}O_2$	270.37
Dihydroergocornine Mesylate $C_{31}H_{41}N_5O_5.CH_4O_3S$	659.80
Dihydroergocristine Mesylate $C_{35}H_{41}N_5O_5.CH_4O_3S$	707.84
Dihydro-α-ergocryptine Mesylate $C_{32}H_{43}N_5O_5.CH_4O_3S$	673.82
Dihydro-β-ergocryptine Mesylate $C_{32}H_{43}N_5O_5.CH_4O_3S$	673.82
Dihydroergotamine $C_{33}H_{37}N_5O_5$	583.69
Mesylate $C_{33}H_{37}N_5O_5.CH_4O_3S$	679.79
Dihydroquinidine $C_{20}H_{26}N_2O_2$	326.44
Dihydroquinine $C_{20}H_{26}N_2O_2$	326.44
Dihydrostreptomycin $C_{21}N_{41}N_7O_{12}$	583.59
Sulfate $(C_{21}H_{41}N_7O_{12})_2.3H_2SO_4$	1461.41
Dihydrotachysterol $C_{28}H_{46}O$	398.67
Dihydroxyaluminum Aminoacetate, anhydrous $C_2H_6AlNO_4$	135.06
Sodium Carbonate, anhydrous $NaAl(OH)_2CO_3$	144.00
Dihydroxyanthraquinone $C_{14}H_8O_4$	240.21
2,7-Dihydroxynaphthalene $C_{10}H_8O_2$	160.17
Dihydroxynorpregnenynone $C_{20}H_{26}O_3$	314.42
4,5-Dihydroxy-3-(*p*-sulfophenylazo)-2,7-naphthalenedisulfonic Acid, Trisodium Salt—(see *2-(4-Sulfophenylazo)-1,8-dihydroxy-3,6-naphthalenedisulfonic Acid, Trisodium Salt*)	
Diiodofluorescein $C_{20}H_{10}I_2O_5$	584.10
Diiodohydroxyquin (see *Iodoquinol*)	
Diiodothyronine $C_{15}H_{13}I_2NO_4$	525.08
Diisodecyl Phthalate $C_{28}H_{46}O_4$	446.65
Diisopropanolamine $C_6H_{15}NO_2$	133.19
Diisopropylamine $[(CH_3)_2CH]_2NH$	101.19
Diisopropyl Ether $C_6H_{14}O$	102.18
Diisopropylethylamine $C_8H_{19}N$	129.24
Diloxanide Furoate $C_{14}H_{11}Cl_2NO_4$	328.15
Diltiazem Hydrochloride $C_{22}H_{26}N_2O_4S.HCl$	450.98
Dimenhydrinate $C_{17}H_{21}NO.C_7H_7ClN_4O_2$	469.97
Dimercaprol $C_3H_8OS_2$	124.22
Dimethindene $C_{20}H_{24}N_2$	292.42
Dimethisoquin $C_{17}H_{24}N_2O$	272.39
Dimethisterone $C_{23}H_{32}O_2.H_2O$	358.52
anhydrous $C_{23}H_{32}O_2$	340.50
2,5-Dimethoxybenzaldehyde $C_9H_{10}O_3$	166.18
Dimethoxyethane $C_4H_{10}O_2$	90.12
3,4-Dimethoxyphenylacetonitrile $C_{10}H_{11}NO_2$	177.20
Dimethylacetamide C_4H_9NO	87.12
Dimethylaminoazobenzene $C_{14}H_{15}N_3$	225.29
Dimethylaminobenzaldehyde $C_9H_{11}NO$	149.19
Dimethylaminocinnamaldehyde $C_{11}H_{13}NO$	175.23
4-Dimethylamino-1,2-diphenyl-3-methyl-2-butanol Hydrochloride (see *Propoxyphene Carbinol Hydrochloride*)	
Dimethylaminophenol $C_8H_{11}NO$	137.18
Dimethylaniline $C_8H_{11}N$	121.18
3,4-Dimethylbenzophenone $C_{15}H_{14}O$	210.28
Dimethylcyclohexanedione $C_8H_{12}O_2$	140.18
N,N-Dimethyl-2,2-diphenylacetamide (see *Diphenamid*)	
3,3′-Dimethyl-1,1′-diphenyl-[4,4′-bi-2-pyrazolin]-5,5′-dione $C_{20}H_{18}N_4O_2$	346.39
Dimethylethyl(3-hydroxyphenyl)-	

ammonium Chloride (see *Edrophonium Chloride*)
Dimethylformamide C_3H_7NO .. 73.09
Dimethylglyoxime $C_4H_8N_2O_2$ ... 116.12
N,N-Dimethyl-1-naphthylamine $C_{12}H_{13}N$ ... 171.24
Dimethyl 4-(2-nitrophenyl)-2,6-dimethylpyridine-3,5-dicarboxylate $C_{17}H_{16}N_2O_6$ ... 344.32
4-(2-nitrosophenyl)-2,6-dimethylpyridine-3,5-dicarboxylate $C_{17}H_{16}N_2O_5$ ... 328.32
N,N-Dimethyloctylamine $C_{10}H_{23}N$ ... 157.30
2,6-Dimethylphenol $(CH_3)_2C_6H_3OH$ ... 122.17
Dimethylphenylenediamine $C_8H_{12}N_2$ ... 136.20
Dihydrochloride $C_8H_{12}N_2.2HCl$ ... 209.12
Dimethylsiloxane (monomer) C_2H_6OSi ... 74.15
Dimethyl Sulfate $C_2H_6O_4S$ ... 126.13
Tubocurarine Iodide (see *Metocurine Iodide*)
Dimethyl Sulfone $C_2H_6O_2S$ ... 94.13
Dimethyl Sulfoxide C_2H_6OS ... 78.13
Dinitrobenzene $C_6H_4N_2O_4$ ... 168.11
Dinitrobenzoyl Chloride $C_7H_3ClN_2O_5$ ... 230.56
Dinitrochlorobenzene $C_6H_3ClN_2O_4$ ... 202.55
2,4-Dinitrofluorobenzene $C_6H_3FN_2O_4$ ... 186.10
Dinitrophenylhydrazine $C_6H_6N_4O_4$ ... 198.14
Dinoprost $C_{20}H_{34}O_5$ ... 354.49
Tromethamine $C_{24}H_{45}NO_8$ ... 475.62
Dinoprostone $C_{20}H_{32}O_5$ ... 352.47
Dioctyl Calcium Sulfosuccinate $C_{40}H_{74}CaO_{14}S_2$ ... 883.22
Potassium Sulfosuccinate $C_{20}H_{37}KO_7S$ ... 460.67
Sodium Sulfosuccinate $C_{20}H_{37}NaO_7S$ ... 444.56
Dioxane $C_4H_8O_2$ ... 88.11
Dioxybenzone $C_{14}H_{12}O_4$ ... 244.25
Diperodon $C_{22}H_{27}N_3O_4.H_2O$ ... 415.49
anhydrous $C_{22}H_{27}N_3O_4$ ... 397.47
Hydrochloride $C_{22}H_{27}N_3O_4.HCl$ ... 433.93
Diphemanil Methylsulfate $C_{21}H_{27}NO_4S$ ... 389.51
Reineckate $C_{20}H_{24}N.[Cr(NH_3)_2(SCN)_4].H_2O$ ... 614.80
Diphenamid $C_{16}H_{17}NO$ ... 239.32
Diphenhydramine $C_{17}H_{21}NO$ ... 255.36
Citrate $C_{17}H_{21}NO.C_6H_8O_7$ ... 447.49
Hydrochloride $C_{17}H_{21}NO.HCl$ ... 291.82
Diphenidol $C_{21}H_{27}NO$ ... 309.45
Hydrochloride $C_{21}H_{27}NO.HCl$ ... 345.91
Diphenoxylate $C_{30}H_{32}N_2O_2$ ... 452.59
Hydrochloride $C_{30}H_{32}N_2O_2.HCl$ ... 489.06
Diphenyl Ether $C_{12}H_{10}O$ ... 170.21
Diphenylamine $C_{12}H_{11}N$ ... 169.23
Diphenylcarbazide $C_{13}H_{14}N_4O$ ... 242.28
Diphenylcarbazone comp'd with Diphenylcarbazide (1:1) $C_{13}H_{12}N_4O.C_{13}H_{14}N_4O$ ... 482.54
2,2-Diphenylglycine $C_{14}H_{13}NO_2$ ... 227.26
Diphenylhydantoin (see *Phenytoin*)
Dipivefrin Hydrochloride $C_{19}H_{29}NO_5.HCl$ ... 387.90
Dipyridamole $C_{24}H_{40}N_8O_4$ ... 504.63
Disodium Ethylenediaminetetraacetate (see *Edetate Disodium*)
Guanylate, anhydrous $C_{10}H_{12}N_5Na_2O_8P$ ... 407.19
Inosinate $C_{10}H_{11}N_4Na_2O_8P.2H_2O$ ... 428.20
Inosinate, anhydrous $C_{10}H_{11}N_4Na_2O_8P$ ... 392.17
Disopyramide $C_{21}H_{29}N_3O$ ... 339.48
Phosphate $C_{21}H_{29}N_3O.H_3PO_4$ ... 437.47
2,4-Disulfamoyl-5-trifluoromethylaniline $C_7H_8F_3N_3O_4S_2$ ... 319.27
2,4-Disulfamoyl-5-trifluoromethylaniline Hydrochloride $C_7H_8F_3N_3O_4S_2.HCl$ ... 355.73
Di-*tert*-butyl-*p*-cresol (see *Butylated Hydroxytoluene*)
4-[(3,5-Di-*tert*-butyl-2-hydroxyphenylthio)isopropylidenethio]-2,6-di-*tert*-butylphenol $C_{31}H_{48}O_2S_2$ ... 516.84
4,4′-(Dithio)bis(2,6-di-*tert*-butylphenol) $C_{28}H_{42}O_2S_2$ ... 474.76
Disulfiram $C_{10}H_{20}N_2S_4$ ... 296.52
5,5′-Dithiobis (2-nitrobenzoic Acid) $C_{14}H_8N_2O_8S_2$ ... 396.35
Dithizone $C_{13}H_{12}N_4S$ ... 256.32
Divinyl Glycol $C_6H_{10}O_2$ ... 114.14
Docosane $C_{22}H_{46}$ ... 310.61
Docusate Calcium (see *Dioctyl Calcium Sulfosuccinate*)
Potassium (see *Dioctyl Potassium Sulfosuccinate*)
Sodium (see *Dioctyl Sodium Sulfosuccinate*)
Dodecanol $C_{12}H_{26}O$ ... 186.34
Dodecyl Sodium Sulfate $C_{12}H_{25}SO_4Na$ ... 288.38
Dopamine $C_8H_{11}NO_2$ ... 153.18
Hydrochloride $C_8H_{11}NO_2.HCl$ ... 189.64
Dotriacontane $C_{32}H_{66}$ ... 450.87
Doxapram $C_{24}H_{30}N_2O_2$ ... 378.51
Hydrochloride $C_{24}H_{30}N_2O_2.HCl.H_2O$ ... 432.99
Hydrochloride, anhydrous $C_{24}H_{30}N_2O_2.HCl$ ... 414.97
Doxepin $C_{19}H_{21}NO$ ... 279.38
Hydrochloride $C_{19}H_{21}NO.HCl$ ... 315.84
Doxorubicin $C_{27}H_{29}NO_{11}$ ... 543.53
Hydrochloride $C_{27}H_{29}NO_{11}.HCl$ ... 579.99
Doxycycline $C_{22}H_{24}N_2O_8.H_2O$. 462.46
anhydrous $C_{22}H_{24}N_2O_8$ ... 444.44
Hyclate $(C_{22}H_{24}N_2O_8.HCl)_2.C_2H_6O.H_2O$ ... 1025.89
Doxylamine $C_{17}H_{22}N_2O$ ... 270.37
Succinate $C_{17}H_{22}N_2O.C_4H_6O_4$ ... 388.46
Droperidol $C_{22}H_{22}FN_3O_2$ ... 379.43
Dyclonine $C_{18}H_{27}NO_2$ ... 289.42
Hydrochloride $C_{18}H_{27}NO_2.HCl$ ... 325.88
Dydrogesterone $C_{21}H_{28}O_2$ ... 312.45
Dyphylline $C_{10}H_{14}N_4O_4$ ... 254.25

Ecgonine, anhydrous $C_9H_{15}NO_3$ ... 185.22
Echothiophate Iodide $C_9H_{23}INO_3PS$ ... 383.22
Edetate Calcium Disodium, anhydrous $C_{10}H_{12}CaN_2Na_2O_8$ ... 374.27
Disodium $C_{10}H_{14}N_2Na_2O_8.2H_2O$ ... 372.24
Disodium, anhydrous $C_{10}H_{14}N_2Na_2O_8$ ... 336.21
Edetic Acid $C_{10}H_{16}N_2O_8$ ... 292.24
Edrophonium Chloride $C_{10}H_{16}ClNO$ ... 201.70
n-Eicosane $C_{20}H_{42}$ ... 282.55
Emetine $C_{29}H_{40}N_2O_4$ ... 480.65
Hydrochloride, anhydrous $C_{29}H_{40}N_2O_4.2HCl$ ... 553.57
Enalapril Maleate $C_{20}H_{28}N_2O_5.C_4H_4O_4$ ... 492.52
Enflurane $C_3H_2ClF_5O$ ... 184.49
Eosin Y $C_{20}H_6Br_4Na_2O_5$ ... 691.86
Ephedrine, anhydrous $C_{10}H_{15}NO$ ... 165.23
hemihydrate $C_{10}H_{15}NO.½H_2O$ ... 174.24
Hydrochloride $C_{10}H_{15}NO.HCl$ ... 201.70
Sulfate $(C_{10}H_{15}NO)_2.H_2SO_4$ ... 428.54
Epitetracycline Hydrochloride ... 480.90
Epinephrine $C_9H_{13}NO_3$ ... 183.21
Bitartrate $C_9H_{13}NO_3.C_4H_6O_6$ ... 333.29
Epinephryl Borate $C_9H_{12}BNO_4$ ... 209.01
Equilenin $C_{18}H_{18}O_2$ ... 266.34
Equilin $C_{18}H_{20}O_2$ ... 268.35
Ergocalciferol $C_{28}H_{44}O$ ... 396.65
Ergonovine $C_{19}H_{23}N_3O_2$ ... 325.41
Maleate $C_{19}H_{23}N_3O_2.C_4H_4O_4$ ... 441.48
Ergosterol $C_{28}H_{44}O$ ... 396.65
Ergotamine $C_{33}H_{35}N_5O_5$ ... 581.67
Tartrate $(C_{33}H_{35}N_5O_5)_2.C_4H_6O_6$ ... 1313.43
Eriochrome Cyanine R $C_{23}H_{15}Na_3O_9S$ ... 536.40
Erythrityl Tetranitrate $C_4H_6N_4O_{12}$ ... 302.11
Erythromycin $C_{37}H_{67}NO_{13}$ ... 733.94
Estolate $C_{40}H_{71}NO_{14}.C_{12}H_{26}O_4S$ ... 1056.39
Ethylcarbonate $C_{40}H_{71}NO_{15}$ ... 806.00
Ethylsuccinate $C_{43}H_{75}NO_{16}$ ... 862.06
Gluceptate $C_{37}H_{67}NO_{13}.C_7H_{14}O_8$ ... 960.12
Lactobionate $C_{37}H_{67}NO_{13}.C_{12}H_{22}O_{12}$ ... 1092.23
Stearate $C_{37}H_{67}NO_{13}.C_{18}H_{36}O_2$ ... 1018.42
Erythrosine Sodium $C_{20}H_6I_4Na_2O_5.H_2O$ ... 897.88
Sodium, anhydrous $C_{20}H_6I_4Na_2O_5$ ... 879.86
Estradiol $C_{18}H_{24}O_2$ ... 272.39
Cypionate $C_{26}H_{36}O_3$ ... 396.57
Dipropionate $C_{24}H_{32}O_4$ ... 384.51
Valerate $C_{23}H_{32}O_3$ ... 356.50
Estriol $C_{18}H_{24}O_3$ ... 288.39
Estrone $C_{18}H_{22}O_2$ ... 270.37
Estropipate $C_4H_{10}N_2.C_{18}H_{22}O_5S$ ... 436.56
Ethacrynate Sodium $C_{13}H_{11}Cl_2NaO_4$ ... 325.12
Ethacrynic Acid $C_{13}H_{12}Cl_2O_4$ .. 303.14
Ethambutol $C_{10}H_{24}N_2O_2$ ... 204.31
Hydrochloride

Compound	Mol. Wt.
$C_{10}H_{24}N_2O_2.2HCl$	277.33
Ethanol C_2H_6O	46.07
Ethchlorvynol C_7H_9ClO	144.60
Ether $C_4H_{10}O$	74.12
Ethinamate $C_9H_{13}NO_2$	167.21
Ethinyl Estradiol $C_{20}H_{24}O_2$	296.41
Ethionamide $C_8H_{10}N_2S$	166.24
Ethisterone $C_{21}H_{28}O_2$	312.45
Ethopropazine $C_{19}H_{24}N_2S$	312.47
Hydrochloride $C_{19}H_{24}N_2S.HCl$	348.93
Ethosuximide $C_7H_{11}NO_2$	141.17
Ethotoin $C_{11}H_{12}N_2O_2$	204.23
Ethoxy C_2H_5O	45.06
2-(2-Ethoxyethoxy)ethanol $C_6H_{14}O_3$	134.17
Ethyl Acetate $C_4H_8O_2$	88.11
Alcohol (see *Alcohol*)	
Benzoate $C_9H_{10}O_2$	150.18
Biscoumacetate $C_{22}H_{16}O_8$	408.36
Chloride C_2H_5Cl	64.51
Cyanoacetate $C_5H_7NO_2$	113.12
Ether $C_4H_{10}O$	74.12
Iodide C_2H_5I	155.97
Iodophenylundecanoate (see *Iophendylate*)	
Maltol $C_7H_8O_3$	140.14
Oleate $C_{20}H_{38}O_2$	310.52
Oxide $C_4H_{10}O$	74.12
Vanillin $C_9H_{10}O$	166.18
4-Ethylbenzaldehyde $C_9H_{10}O$	134.18
Ethylene C_2H_4	28.05
Dichloride $C_2H_4Cl_2$	98.96
Glycol $C_2H_6O_2$	62.07
Glycol Monoethyl Ether $C_4H_{10}O_2$	90.12
Oxide C_2H_4O	44.05
Ethylenediamine $C_2H_8N_2$	60.10
Ethylnorepinephrine $C_{10}H_{15}NO_3$	197.23
Hydrochloride $C_{10}H_{15}NO_3.HCl$	233.69
Ethylparaben $C_9H_{10}O_3$	166.18
N-Ethylpiperidine $C_7H_{15}N$	113.20
Penicillin G $C_{23}H_{33}N_3O_4S$	447.59
1-Ethylquinaldinium Iodide $C_{12}H_{14}IN$	299.15
Ethynodiol Diacetate $C_{24}H_{32}O_4$	384.51
Ethynyl C_2H-	25.03
Etidocaine $C_{17}H_{28}N_2O$	276.42
Hydrochloride $C_{17}H_{28}N_2O.HCl$	312.88
Etidronate Disodium $C_2H_6Na_2O_7P_2$	249.99
Etidronic Acid, monohydrate $C_2H_8O_7P_2.H_2O$	224.04
Eucalyptol $C_{10}H_{18}O$	154.25
Eucatropine $C_{17}H_{25}NO_3$	291.39
Hydrochloride $C_{17}H_{25}NO_3.HCl$	327.85
Eugenol $C_{10}H_{12}O_2$	164.20
Evans Blue $C_{34}H_{24}N_6Na_4O_{14}S_4$	960.79
Famotidine $C_8H_{15}N_7O_2S_3$	337.43
Fast Blue BB Salt $(C_{17}H_{18}ClN_3O_3)_2.ZnCl_2$	831.89
Fenoprofen $C_{15}H_{14}O_3$	242.27
Calcium $C_{30}H_{26}CaO_6.2H_2O$	558.64
Calcium, anhydrous $C_{30}H_{26}CaO_6$	522.61
Sodium, anhydrous $C_{15}H_{13}NaO_3$	264.26
Fentanyl $C_{22}H_{28}N_2O$	336.48
Citrate $C_{22}H_{28}N_2O.C_6H_8O_7$	528.60
Ferric Ammonium Sulfate $FeNH_4(SO_4)_2.12H_2O$	482.18
Ammonium Sulfate, anhydrous $FeNH_4(SO_4)_2$	266.00
Chloride $FeCl_3.6H_2O$	270.30
Chloride, anhydrous $FeCl_3$	162.21
Citrate, anhydrous $C_6H_5FeO_7$	244.95
Hydroxide $Fe(OH)_3$	106.87
Nitrate $Fe(NO_3)_3.9H_2O$	404.00
Nitrate, anhydrous $Fe(NO_3)_3$	241.86
Oxide Fe_2O_3	159.69
Sulfate, anhydrous $Fe_2(SO_4)_3$	399.87
Ferrous Ammonium Sulfate $Fe(NH_4)_2(SO_4)_2.6H_2O$	392.13
Ammonium Sulfate, anhydrous $Fe(NH_4)_2(SO_4)_2$	284.04
Fumarate $C_4H_2FeO_4$	169.90
Gluconate $C_{12}H_{22}FeO_{14}.2H_2O$	482.17
Gluconate, anhydrous $C_{12}H_{22}FeO_{14}$	446.14
Sulfate $FeSO_4.7H_2O$	278.01
Sulfate, anhydrous $FeSO_4$	151.90
Flucytosine $C_4H_4FN_3O$	129.09
Fludrocortisone $C_{21}H_{29}FO_5$	380.46
Acetate $C_{23}H_{31}FO_6$	422.49
Flumethasone Pivalate $C_{27}H_{36}F_2O_6$	494.57
Flunisolide $C_{24}H_{31}FO_6.\frac{1}{2}H_2O$	443.51
Fluocinolone Acetonide $C_{24}H_{30}F_2O_6$	452.49
Fluocinonide $C_{26}H_{32}F_2O_7$	494.53
Fluorene $C_{13}H_{10}$	166.22
Fluorescamine $C_{17}H_{10}O_4$	278.26
Fluorescein $C_{20}H_{12}O_5$	332.31
Sodium $C_{20}H_{10}Na_2O_5$	376.28
1-Fluoro-2,4-dinitrobenzene $C_6H_3FN_2O_4$	186.10
Fluorometholone $C_{22}H_{29}FO_4$	376.47
Fluorouracil $C_4H_3FN_2O_2$	130.08
Fluoxymesterone $C_{20}H_{29}FO_3$	336.45
Fluphenazine $C_{22}H_{26}F_3N_3OS$	437.52
Decanoate $C_{32}H_{44}F_3N_3O_2S$	591.77
Enanthate $C_{29}H_{38}F_3N_3O_2S$	549.69
Enanthate Dihydrochloride $C_{29}H_{38}F_3N_3O_2S.2HCl$	622.62
Hydrochloride $C_{22}H_{26}F_3N_3OS.2HCl$	510.44
Fluprednisolone $C_{21}H_{27}FO_5$	378.44
Flurandrenolide $C_{24}H_{33}FO_6$	436.52
Flurazepam $C_{21}H_{23}ClFN_3O$	387.88
Hydrochloride $C_{21}H_{23}ClFN_3O.2HCl$	460.81
Flurazepam Related Compound A (see *3-Amino-6-chloro-1-(2-diethylaminoethyl)-4-(2-fluorophenyl)carbostyril Hydrochloride*)	
Related Compound B (see *2-Chloro-10-(2-diethylaminoethyl)-9-acridanone Hydrochloride*)	
Related Compound C (see *5-Chloro-2-(2-diethylaminoethylamino)-2′-fluorobenzophenone Hydrochloride*)	
Related Compound D (see *7-Chloro-1,3-dihydro-1-(2-diethylaminoethyl)-5-(2-fluorophenyl)-2H-1,4-benzodiazepin-2-one Hydrochloride*)	
Related Compound E (see *7-Chloro-1-(2-diethylaminoethyl)-5-(2-fluorophenyl)-1,3,4,5-tetrahydro-2H-1,4-benzodiazepin-2-one Hydrochloride*)	
Related Compound F (see *7-Chloro-5-(2-fluorophenyl)-1,3-dihydro-2H-1,4-benzodiazepin-2-one*)	
Flurbiprofen Sodium $C_{15}H_{12}FNaO_2.2H_2O$	302.28
Folic Acid $C_{19}H_{19}N_7O_6$	441.40
Formaldehyde CH_2O	30.03
Formamide CH_3NO	45.04
Formic Acid CH_2O_2	46.03
2-(*N*-Formylhexahydrohippuroyl)-1,2,3,4-tetrahydroisoquinolin-1-one $C_{19}H_{22}N_2O_4$	340.38
Fructose $C_6H_{12}O_6$	180.16
Fumaric Acid $C_4H_4O_4$	116.07
Furazolidone $C_8H_7N_3O_5$	225.16
Furfural $C_5H_4O_2$	96.09
Furosemide $C_{12}H_{11}ClN_2O_5S$	330.74
1-(2-Furoyl)piperazine $C_9H_{12}N_2O_2$	180.21
Galacturonic Acid $C_6H_{10}O_7$	194.14
Gallamine Triethiodide $C_{30}H_{60}I_3N_3O_3$	891.54
Gallic Acid $C_7H_6O_5.H_2O$	188.14
Acid, anhydrous $C_7H_6O_5$	170.12
Gemfibrozil $C_{15}H_{22}O_3$	250.34
Gentian Violet $C_{25}H_{30}ClN_3$	407.99
Gentisic Acid Ethanolamide $C_9H_{11}NO_4$	197.19
Gibberellic Acid $C_{19}H_{22}O_6$	346.38
Girard Reagent T [see *(Carboxymethyl)trimethylammonium Chloride, hydrazide*]	
Gitaloxin $C_{42}H_{64}O_{15}$	808.96
Gitoxin $C_{41}H_{64}O_{14}$	780.95
Glucagon $C_{153}H_{225}N_{43}O_{49}S$	3482.78
D-Gluconic Acid, 50 Percent in Water $C_6H_{12}O_7$	196.16
D-Glucose (see *Dextrose*)	
Glutamic Acid $C_5H_9NO_4$	147.13
Glutaral $C_5H_8O_2$	100.12
Glutethimide $C_{13}H_{15}NO_2$	217.27
Glycerin $C_3H_8O_3$	92.09
Glyceryl Guaiacolate (see *Guaifenesin*)	
Monopalmitate $C_{19}H_{38}O_4$	330.51
Monostearate $C_{21}H_{42}O_4$	358.56
Triacetate $C_9H_{14}O_6$	218.21
Trinitrate $C_3H_5N_3O_9$	227.09
Glycine $C_2H_5NO_2$	75.07
Glycopyrrolate $C_{19}H_{28}BrNO_3$	398.34
Gold Chloride (see *Chlorauric Acid*)	
Griseofulvin $C_{17}H_{17}ClO_6$	352.77
Guaiacol $C_7H_8O_2$	124.14
Guaifenesin $C_{10}H_{14}O_4$	198.22
Guanabenz Acetate $C_{10}H_{12}N_4O_2Cl_2$	291.14
Guanadrel Sulfate $(C_{10}H_{19}N_3O_2)_2.H_2SO_4$	524.63

Name	Formula	Mol. Wt.
Guanethidine	$C_{10}H_{22}N_4$	198.31
Monosulfate	$C_{10}H_{22}N_4.H_2SO_4$	296.38
Guanidine	CH_5N_3	59.07
Hydrochloride	$CH_5N_3.HCl$	95.53
Guanine Hydrochloride	$C_5H_5N_5O.HCl.H_2O$	205.60
Hydrochloride, anhydrous	$C_5H_5N_5O.HCl$	187.59
Halazepam	$C_{17}H_{12}ClF_3N_2O$	352.74
Halazone	$C_7H_5Cl_2NO_4S$	270.09
Halcinonide	$C_{24}H_{32}ClFO_5$	454.97
Haloperidol	$C_{21}H_{23}ClFNO_2$	375.87
Haloprogin	$C_9H_4Cl_3IO$	361.39
Halothane	$C_2HBrClF_3$	197.38
Hematein	$C_{16}H_{12}O_6$	300.27
Hematin	$C_{34}H_{33}FeN_4O_5$	633.51
Hematoxylin	$C_{16}H_{14}O_6.3H_2O$	356.33
anhydrous	$C_{16}H_{14}O_6$	302.28
Heptane	C_7H_{16}	100.20
Heptanesulfonic Acid	$C_7H_{16}O_3S$	180.26
Heroin (see *Diacetylmorphine*)		
Hetacillin	$C_{19}H_{23}N_3O_4S$	389.47
Potassium	$C_{19}H_{22}KN_3O_4S$	427.56
Hexachlorophene	$C_{13}H_6Cl_6O_2$	406.91
Hexacosane	$C_{26}H_{54}$	366.71
Hexadecyl Hexadecanoate	$C_{32}H_{64}O_2$	480.86
Hexamethyldisilazane	$C_6H_{19}NSi_2$	161.39
Hexamethyleneimine	$C_6H_{13}N$	99.18
Hexamethylpararosaniline Chloride	$C_{25}H_{30}ClN_3$	407.99
Hexane	C_6H_{14}	86.18
Hexanitrodiphenylamine	$C_{12}H_5N_7O_{12}$	439.21
Hexocyclium Methylsulfate	$C_{21}H_{36}N_2O_5S$	428.59
Hexylcaine	$C_{16}H_{23}NO_2$	261.36
Hydrochloride	$C_{16}H_{23}NO_2.HCl$	297.82
Hexylene Glycol	$C_6H_{14}O_2$	118.18
Histamine	$C_5H_9N_3$	111.15
Dihydrochloride	$C_5H_9N_3.2HCl$	184.07
Phosphate	$C_5H_9N_3.2H_3PO_4$	307.14
Histidine	$C_6H_9N_3O_2$	155.16
Homatropine	$C_{16}H_{21}NO_3$	275.35
Hydrobromide	$C_{16}H_{21}NO_3.HBr$	356.26
Methylbromide	$C_{17}H_{24}BrNO_3$	370.29
Hycanthone	$C_{20}H_{24}N_2O_2S$	356.48
Mesylate	$C_{20}H_{24}N_2O_2S.CH_4O_3S$	452.58
Hydralazine	$C_8H_8N_4$	160.18
Hydrochloride	$C_8H_8N_4.HCl$	196.64
Hydrastine	$C_{21}H_{21}NO_6$	383.40
Hydrochloride	$C_{21}H_{21}NO_6.HCl$	419.86
Hydrazine Hydrate, 85% in Water	$(NH_2)_2.H_2O$	50.06
Sulfate	$N_2H_4.H_2SO_4$	130.12
Hydriodic Acid	HI	127.91
Hydrobromic Acid	HBr	80.91
Hydrochloric Acid	HCl	36.46
Hydrochlorothiazide	$C_7H_8ClN_3O_4S_2$	297.73
Hydrocodone	$C_{18}H_{21}NO_3$	299.37
Bitartrate	$C_{18}H_{21}NO_3.-C_4H_6O_6.2\frac{1}{2}H_2O$	494.50
Bitartrate, anhydrous	$C_{18}H_{21}NO_3.C_4H_6O_6$	449.46
Hydrocortamate	$C_{27}H_{41}NO_6$	475.62
Hydrochloride	$C_{27}H_{41}NO_6.HCl$	512.08
Hydrocortisone	$C_{21}H_{30}O_5$	362.47
Acetate	$C_{23}H_{32}O_6$	404.50
Butyrate	$C_{25}H_{36}O_6$	432.56
Cypionate	$C_{29}H_{42}O_6$	486.65
Hemisuccinate	$C_{25}H_{34}O_8$	462.54
Phosphate Triethylamine	$C_{21}H_{31}O_8P.C_6H_{15}N$	543.64
Sodium Phosphate	$C_{21}H_{29}Na_2O_8P$	486.41
Sodium Succinate	$C_{25}H_{33}NaO_8$	484.52
Tebutate	$C_{27}H_{40}O_6$	460.61
Valerate	$C_{26}H_{38}O_6$	446.58
Hydroflumethiazide	$C_8H_8F_3N_3O_4S_2$	331.28
Hydrofluoric Acid	HF	20.01
Hydrogen Peroxide	H_2O_2	34.01
Sulfide	H_2S	34.08
Hydromorphone	$C_{17}H_{19}NO_3$	285.34
Hydrochloride	$C_{17}H_{19}NO_3.HCl$	321.80
Sulfate	$(C_{17}H_{19}NO_3)_2.H_2SO_4$	668.76
Hydroquinone	$C_6H_6O_2$	110.11
Hydroxocobalamin	$C_{62}H_{89}CoN_{13}O_{15}P$	1346.37
4′-Hydroxyacetophenone	$C_8H_8O_2$	136.15
Hydroxyamphetamine	$C_9H_{13}NO$	151.21
Hydrobromide	$C_9H_{13}NO.HBr$	232.12
Hydroxybenzamide	$C_7H_7NO_2$	137.14
p-Hydroxybenzoic Acid	$C_7H_6O_3$	138.12
Hydroxychloroquine	$C_{18}H_{26}ClN_3O$	335.88
Sulfate	$C_{18}H_{26}ClN_3O.H_2SO_4$	433.95
Hydroxyl	$-OH$	17.01
Hydroxylamine	NH_3O	33.03
Hydrochloride	$NH_3O.HCl$	69.49
3-Hydroxy-1-methylquinuclidinium Bromide	$C_8H_{16}BrNO$	222.12
Hydroxynorandrostenedione	$C_{18}H_{24}O_3$	288.39
Hydroxynorethynyltestosterone	$C_{20}H_{26}O_3$	314.42
Hydroxyprogesterone Caproate	$C_{27}H_{40}O_4$	428.61
Hydroxypropoxy	$C_3H_7O_2-$	75.09
Hydroxyquinoline	C_9H_7NO	145.16
Hydroxystilbamidine	$C_{16}H_{16}N_4O$	280.33
Isethionate	$C_{16}H_{16}N_4O.2C_2H_6O_4S$	532.58
Hydroxyurea	$CH_4N_2O_2$	76.05
Hydroxyzine	$C_{21}H_{27}ClN_2O_2$	374.91
Hydrochloride, anhydrous	$C_{21}H_{27}ClN_2O_2.2HCl$	447.83
Pamoate	$C_{21}H_{27}ClN_2O_2.C_{23}H_{16}O_6$	763.29
Hyoscyamine	$C_{17}H_{23}NO_3$	289.37
Hydrobromide	$C_{17}H_{23}NO_3.HBr$	370.29
Sulfate	$(C_{17}H_{23}NO_3)_2.-H_2SO_4.2H_2O$	712.85
Sulfate, anhydrous	$(C_{17}H_{23}NO_3)_2.H_2SO_4$	676.82
Hypophosphorous Acid	HPH_2O_2	66.00
Ibuprofen	$C_{13}H_{18}O_2$	206.28
Imidazole	$C_3H_4N_2$	68.08
Iminodibenzyl	$C_{14}H_{15}N$	197.28
Iminostilbene	$C_{14}H_{11}N$	193.25
Imipramine	$C_{19}H_{24}N_2$	280.41
Hydrochloride	$C_{19}H_{24}N_2.HCl$	316.87
Indigo Carmine (see *Indigotindisulfonate Sodium*)		
Indigotindisulfonate Sodium	$C_{16}H_8N_2Na_2O_8S_2$	466.35
Indocyanine Green	$C_{43}H_{47}N_2NaO_6S_2$	774.96
Indomethacin	$C_{19}H_{16}ClNO_4$	357.79
Inositol	$C_6H_{12}O_6$	180.16
Insulin Human	$C_{257}H_{383}N_{65}O_{77}S_6$	5807.62
Insulin (beef)		5777.59
Insulin (pork)		5733.54
Iocarmate Meglumine	$(C_7H_{17}NO_5)_2.-C_{24}H_{20}I_6N_4O_8$	1644.30
Iocetamic Acid	$C_{12}H_{13}I_3N_2O_3$	613.96
Iodic Acid	HIO_3	175.91
Iodine Monobromide	IBr	206.81
Monochloride	ICl	162.36
Pentoxide	I_2O_5	333.81
Iodipamide	$C_{20}H_{14}I_6N_2O_6$	1139.77
Meglumine	$C_{20}H_{14}I_6N_2O_6.-2C_7H_{17}NO_5$	1530.20
Sodium	$C_{20}H_{12}I_6N_2Na_2O_6$	1183.73
Iodohippurate Sodium	$C_9H_7INNaO_3$	327.05
Iodohippuric Acid	$C_9H_8INO_3$	305.07
Iodohydroxyquinolinesulfonic Acid	$C_9H_6INO_4S$	351.12
Iodophthalein	$C_{20}H_{10}I_4O_4$	821.91
Iodoquinol	$C_9H_5I_2NO$	396.95
Iodoxuridine	$C_9H_{11}IN_2O_5$	354.10
Iopamidol	$C_{17}H_{22}I_3N_3O_8$	777.09
Iopanoic Acid	$C_{11}H_{12}I_3NO_2$	570.93
Iophendylate	$C_{19}H_{29}IO_2$	416.34
Iothalamate Meglumine	$C_{11}H_9I_3N_2O_4.C_7H_{17}NO_5$	809.13
Sodium	$C_{11}H_8I_3N_2NaO_4$	635.90
Iothalamic Acid (see *5-Acetamido-2,4,6-triiodo-N-methylisophthalamic Acid*)		
Ipodate Calcium	$C_{24}H_{24}CaI_6N_4O_4$	1233.98
Sodium	$C_{12}H_{12}I_3N_2NaO_2$	619.94
Iron Compounds (see under *Ferric* and *Ferrous*)		
Isatin	$C_8H_5NO_2$	147.13
Isoamyl (see *Amyl*)		
Isobutane	C_4H_{10}	58.12
p-Isobutylacetophenone	$C_{12}H_{16}O$	176.26
Isobutyl Alcohol	$C_4H_{10}O$	74.12
Isocarboxazid	$C_{12}H_{13}N_3O_2$	231.25
Isoetharine	$C_{13}H_{21}NO_3$	239.31
Hydrochloride	$C_{13}H_{21}NO_3.HCl$	275.77
Mesylate	$C_{13}H_{21}NO_3.CH_4O_3S$	335.41
Isoflupredone Acetate	$C_{23}H_{29}FO_6$	420.48
Isoflurane	$C_3H_2ClF_5O$	184.49
Isoflurophate	$C_6H_{14}FO_3P$	184.15

Compound	Molecular Weight
Isoleucine $C_6H_{13}NO_2$	131.17
Isomalathion $C_{10}H_{19}O_6PS_2$	330.35
Isoniazid $C_6H_7N_3O$	137.14
Isooctane C_8H_{18}	114.23
Isopropamide $C_{23}H_{33}N_2O^+$	353.53
Iodide $C_{23}H_{33}IN_2O$	480.43
Isopropyl Acetate $C_5H_{10}O_2$	102.13
Alcohol C_3H_8O	60.10
Ether (see *Diisopropyl Ether*)	
Iodide C_3H_7I	169.99
Myristate $C_{17}H_{34}O_2$	270.45
Palmitate $C_{19}H_{38}O_2$	298.51
Phosphorofluoridate (see *Isoflurophate*)	
Isopropylamine C_3H_9N	59.11
Isoproterenol $C_{11}H_{17}NO_3$	211.26
Hydrochloride $C_{11}H_{17}NO_3.HCl$	247.72
Sulfate $(C_{11}H_{17}NO_3)_2.H_2SO_4.2H_2O$	556.62
Sulfate, anhydrous $(C_{11}H_{17}NO_3)_2.H_2SO_4$	520.59
Isosorbide $C_6H_{10}O_4$	146.14
Dinitrate $C_6H_8N_2O_8$	236.14
Isoxsuprine $C_{18}H_{23}NO_3$	301.38
Hydrochloride $C_{18}H_{23}NO_3.HCl$	337.85
Kanamycin $C_{18}H_{36}N_4O_{11}$	484.50
Sulfate $C_{18}H_{36}N_4O_{11}.H_2SO_4$	582.58
B Sulfate $C_{18}H_{37}N_5O_{10}.H_2SO_4$	581.59
Ketamine $C_{13}H_{16}ClNO$	237.73
Hydrochloride $C_{13}H_{16}ClNO.HCl$	274.19
Ketoconazole $C_{26}H_{28}Cl_2N_4O_4$	531.44
Labetalol Hydrochloride $C_{19}H_{24}N_2O_3.HCl$	364.87
Lactate $C_3H_5O_3^-$	89.07
Lactic Acid $C_3H_6O_3$	90.08
Acid Lactate $C_6H_{10}O_5$	162.14
Lactobionic Acid $C_{12}H_{22}O_{12}$	358.30
Lactose, anhydrous $C_{12}H_{22}O_{11}$	342.30
monohydrate $C_{12}H_{22}O_{11}.H_2O$	360.31
Lactulose $C_{12}H_{22}O_{11}$	342.30
Lanatoside C $C_{49}H_{76}O_{20}$	985.13
Lanthanum Chloride $LaCl_3$	245.26
Lead Acetate $C_4H_6PbO_4.3H_2O$	379.33
Acetate, anhydrous $C_4H_6PbO_4$	325.29
Dioxide (Lead Peroxide) PbO_2	239.20
Monoxide PbO	223.20
Nitrate $Pb(NO_3)_2$	331.21
Perchlorate $Pb(ClO_4)_2.3H_2O$	460.15
Leucine $C_6H_{13}NO_2$	131.17
Leucovorin $C_{20}H_{23}N_7O_7$	473.44
Calcium $C_{20}H_{21}CaN_7O_7$	511.51
Levarterenol (see *Norepinephrine*)	
Bitartrate (see *Norepinephrine Bitartrate*)	
Levo-α-acetylmethadol $C_{23}H_{31}NO_2$	353.50
Hydrochloride $C_{23}H_{31}NO_2.HCl$	389.96
Levobunolol Hydrochloride $C_{17}H_{25}NO_3.HCl$	327.85
Levocarnitine $C_7H_{15}NO_3$	161.20
Levodopa $C_9H_{11}NO_4$	197.19
Levonordephrin $C_9H_{13}NO_3$	183.21
Levonorgestrel $C_{21}H_{28}O_2$	312.45
Levopropoxyphene $C_{22}H_{29}NO_2$	339.48
Napsylate $C_{22}H_{29}NO_2.C_{10}H_8O_3S.H_2O$	565.72
Napsylate, anhydrous $C_{22}H_{29}NO_2.C_{10}H_8O_3S$	547.71
Levorphanol $C_{17}H_{23}NO$	257.37
Tartrate $C_{17}H_{23}NO.C_4H_6O_6.2H_2O$	443.49
Tartrate, anhydrous $C_{17}H_{23}NO.C_4H_6O_6$	407.46
Levothyroxine $C_{15}H_{11}I_4NO_4$	776.87
Sodium, anhydrous $C_{15}H_{10}I_4NNaO_4$	798.86
Lidocaine $C_{14}H_{22}N_2O$	234.34
Hydrochloride $C_{14}H_{22}N_2O.HCl.H_2O$	288.82
Hydrochloride, anhydrous $C_{14}H_{22}N_2O.HCl$	270.80
Linalyl Acetate $C_{12}H_{20}O_2$	196.29
Lincomycin $C_{18}H_{34}N_2O_6S$	406.54
Hydrochloride $C_{18}H_{34}N_2O_6S.HCl.H_2O$	461.01
Hydrochloride, anhydrous $C_{18}H_{34}N_2O_6S.HCl$	443.00
Lindane $C_6H_6Cl_6$	290.83
Liothyronine $C_{15}H_{12}I_3NO_4$	650.98
Sodium $C_{15}H_{11}I_3NNaO_4$	672.96
Lithium Carbonate Li_2CO_3	73.89
Chloride $LiCl$	42.39
Citrate, anhydrous $C_6H_5Li_3O_7$	209.92
Hydroxide, anhydrous $LiOH$	23.95
Metaborate $LiBO_2$	49.75
Methoxide CH_3LiO	37.98
Nitrate $LiNO_3$	68.95
Oxalate $C_2Li_2O_4$	101.90
Perchlorate $LiClO_4$	106.39
Sulfate $Li_2SO_4.H_2O$	127.95
Sulfate, anhydrous Li_2SO_4	109.94
Lomustine $C_9H_{16}ClN_3O_2$	233.70
Loperamide $C_{29}H_{33}ClN_2O_2$	477.04
Hydrochloride $C_{29}H_{33}ClN_2O_2.HCl$	513.51
Lorazepam $C_{15}H_{10}Cl_2N_2O_2$	321.16
Loxapine $C_{18}H_{18}ClN_3O$	327.81
Hydrochloride $C_{18}H_{18}ClN_3O.HCl$	364.27
Succinate $C_{18}H_{18}ClN_3O.C_4H_6O_4$	445.90
Lucanthone $C_{20}H_{24}N_2OS$	340.48
Hydrochloride $C_{20}H_{24}N_2OS.HCl$	376.94
Lypressin $C_{46}H_{65}N_{13}O_{12}S_2$	1056.22
Lysergic Acid Diethylamide $C_{20}H_{25}N_3O$	323.44
Acid Diethylamide Tartrate $(C_{20}H_{25}N_3O)_2.C_4H_6O_6$	796.96
Lysine $C_6H_{14}N_2O_2$	146.19
Acetate $C_6H_{14}N_2O_2.C_2H_4O_2$	206.24
Hydrochloride $C_6H_{14}N_2O_2.HCl$	182.65
Mafenide $C_7H_{10}N_2O_2S$	186.23
Acetate $C_7H_{10}N_2O_2S.C_2H_4O_2$	246.28
Hydrochloride $C_7H_{10}N_2O_2S.HCl$	222.69
Magaldrate anhydrous (approx.) $Al_5Mg_{10}(OH)_{31}(SO_4)_2$	1097.38
Magnesium Acetate $C_4H_6MgO_4.4H_2O$	214.45
Acetate, anhydrous $C_4H_6MgO_4$	142.39
Carbonate, basic (approx.) $(MgCO_3)_4.Mg(OH)_2.5H_2O$	485.65
Carbonate, normal $MgCO_3.H_2O$	102.33
Carbonate, normal, anhydrous $MgCO_3$	84.31
Chloride $MgCl_2.6H_2O$	203.30
Chloride, anhydrous $MgCl_2$	95.21
Citrate $C_{12}H_{10}Mg_3O_{14}.14H_2O$	703.33
Citrate, anhydrous $C_{12}H_{10}Mg_3O_{14}$	451.12
Gluconate $C_{12}H_{22}MgO_{14}.2H_2O$	450.63
Gluconate, anhydrous $C_{12}H_{22}MgO_{14}$	414.60
Hydroxide $Mg(OH)_2$	58.32
Hydroxyquinolate $C_{18}H_{12}MgN_2O_2.2H_2O$	348.64
Nitrate $Mg(NO_3)_2.6H_2O$	256.41
Nitrate, anhydrous $Mg(NO_3)_2$	148.31
Orotate $C_{10}H_6MgN_4O_8$	334.48
Oxide MgO	40.30
Palmitate $C_{32}H_{62}MgO_4$	535.14
Perchlorate $Mg(ClO_4)_2$	223.21
Phosphate, Tribasic $Mg_3(PO_4)_2.5H_2O$	352.93
Phosphate, anhydrous $Mg_3(PO_4)_2$	262.86
Pyrophosphate $Mg_2P_2O_7$	222.55
Salicylate $C_{14}H_{10}MgO_6.4H_2O$	370.60
Salicylate, anhydrous $C_{14}H_{10}MgO_6$	298.53
Stearate $C_{36}H_{70}MgO_4$	591.25
Sulfate $MgSO_4.7H_2O$	246.47
Sulfate, anhydrous $MgSO_4$	120.36
Trisilicate, anhydrous $Mg_2Si_3O_8$	260.86
Malathion $C_{10}H_{19}O_6PS_2$	330.35
Maleic Acid $C_4H_4O_4$	116.07
Malic Acid $C_4H_6O_5$	134.09
Maltol $C_6H_6O_3$	126.11
Maltose $C_{12}H_{22}O_{11}.H_2O$	360.31
anhydrous $C_{12}H_{22}O_{11}$	342.30
Mandelic Acid $C_8H_8O_3$	152.15
Manganese Chloride $MnCl_2.4H_2O$	197.91
Dioxide MnO_2	86.94
Sulfate $MnSO_4.H_2O$	169.01
Sulfate, anhydrous $MnSO_4$	151.00
Mannitol $C_6H_{14}O_6$	182.17
Maprotiline Hydrochloride $C_{20}H_{23}N.HCl$	313.87
Mazindol $C_{16}H_{13}ClN_2O$	284.74
Mebendazole $C_{16}H_{13}N_3O_3$	295.30
Mebutamate $C_{10}H_{20}N_2O_4$	232.28
Mechlorethamine $C_5H_{11}Cl_2N$	156.05
Hydrochloride $C_5H_{11}Cl_2N.HCl$	192.52
Meclizine $C_{25}H_{27}ClN_2$	390.95
Hydrochloride $C_{25}H_{27}ClN_2.2HCl.H_2O$	481.89
Hydrochloride, anhydrous $C_{25}H_{27}ClN_2.2HCl$	463.88
Meclocycline Sulfosalicylate $C_{22}H_{21}ClN_2O_8.C_7H_6O_6S$	695.05
Meclofenamate Sodium $C_{14}H_{10}Cl_2NNaO_2.H_2O$	336.15
Medroxyprogesterone Acetate $C_{24}H_{34}O_4$	386.53

Compound	Mol. Wt.
Medrysone $C_{22}H_{32}O_3$	344.49
Megestrol Acetate $C_{24}H_{32}O_4$	384.52
Meglumine $C_7H_{17}NO_5$	195.21
Melarsoprol $C_{12}H_{15}AsN_6OS_2$	398.33
Melphalan $C_{13}H_{18}Cl_2N_2O_2$	305.20
Hydrochloride $C_{13}H_{18}Cl_2N_2O_2.HCl$	341.67
Menadiol Sodium Diphosphate $C_{11}H_8Na_4O_8P_2.6H_2O$	530.18
Sodium Diphosphate, anhydrous $C_{11}H_8Na_4O_8P_2$	422.09
Menadione $C_{11}H_8O_2$	172.18
Sodium Bisulfite $C_{11}H_8O_2.NaHSO_3.3H_2O$	330.28
Sodium Bisulfite, anhydrous $C_{11}H_8O_2.NaHSO_3$	276.24
Menthol $C_{10}H_{20}O$	156.27
Menthyl Acetate $C_{12}H_{22}O_2$	198.30
Meperidine $C_{15}H_{21}NO_2$	247.34
Hydrochloride $C_{15}H_{21}NO_2.HCl$	283.80
Mephenesin $C_{10}H_{14}O_3$	182.22
Mephentermine $C_{11}H_{17}N$	163.26
Sulfate $(C_{11}H_{17}N)_2.H_2SO_4.2H_2O$	460.63
Sulfate, anhydrous $(C_{11}H_{17}N)_2.H_2SO_4$	424.60
Mephenytoin $C_{12}H_{14}N_2O_2$	218.25
Mephobarbital $C_{13}H_{14}N_2O_3$	246.27
Mepivacaine $C_{15}H_{22}N_2O$	246.35
Hydrochloride $C_{15}H_{22}N_2O.HCl$	282.81
Meprednisone $C_{22}H_{28}O_5$	372.46
Meprobamate $C_9H_{18}N_2O_4$	218.25
Meprylcaine $C_{14}H_{21}NO_2$	235.33
Hydrochloride $C_{14}H_{21}NO_2.HCl$	271.79
Meralluride $C_{16}H_{22}HgN_6O_7$	610.98
3-Mercapto-2-methylpropionic Acid $C_4H_8O_2S$	120.17
3-Mercapto-2-methylpropionic Acid 1,2-Diphenylethylamine Salt $C_4H_7O_2S.C_{14}H_{16}N$	317.45
Mercaptopurine $C_5H_4N_4S.H_2O$	170.19
anhydrous $C_5H_4N_4S$	152.17
Mercuric Acetate $C_4H_6HgO_4$	318.68
Bromide $HgBr_2$	360.40
Chloride $HgCl_2$	271.50
Iodide HgI_2	454.40
Nitrate $Hg(NO_3)_2.H_2O$	342.62
Nitrate, anhydrous $Hg(NO_3)_2$	324.60
Oxide HgO	216.59
Sulfate $HgSO_4$	296.65
Sulfide HgS	232.65
Mercurous Chloride $HgCl$	236.04
Nitrate $HgNO_3.H_2O$	280.61
Nitrate, anhydrous $HgNO_3$	262.59
Mercury, ammoniated $Hg(NH_2)Cl$	252.07
Mesoridazine $C_{21}H_{26}N_2OS_2$	386.57
Besylate $C_{21}H_{26}N_2OS_2.C_6H_6O_3S$	544.74
Mestranol $C_{21}H_{26}O_2$	310.44
Metabutethamine $C_{13}H_{20}N_2O_2$	236.31
Hydrochloride $C_{13}H_{20}N_2O_2.HCl$	272.77
Metaphenylenediamine Hydrochloride (see *Phenylenediamine Dihydrochloride*)	
Metaphosphoric Acid HPO_3	79.98
Metaproterenol $C_{11}H_{17}NO_3$	211.26
Sulfate $(C_{11}H_{17}NO_3)_2.H_2SO_4$	520.59
Methacholine Chloride $C_8H_{18}ClNO_2$	195.69
Methacycline $C_{22}H_{22}N_2O_8$	442.42
Hydrochloride $C_{22}H_{22}N_2O_8.HCl$	478.89
Methadone $C_{21}H_{27}NO$	309.45
Hydrochloride $C_{21}H_{27}NO.HCl$	345.91
Methamphetamine $C_{10}H_{15}N$	149.24
Hydrochloride $C_{10}H_{15}N.HCl$	185.70
Methanesulfonic Acid CH_4O_3S	96.10
Methanol (see *Methyl Alcohol*)	
Methantheline Bromide $C_{21}H_{26}BrNO_3$	420.35
Methapyrilene $C_{14}H_{19}N_3S$	261.38
Metharbital $C_9H_{14}N_2O_3$	198.22
Methazolamide $C_5H_8N_4O_3S_2$	236.26
Methdilazine $C_{18}H_{20}N_2S$	296.43
Hydrochloride $C_{18}H_{20}N_2S.HCl$	332.89
Methenamine $C_6H_{12}N_4$	140.19
Hippurate $C_6H_{12}N_4.C_9H_9NO_3$	319.36
Mandelate $C_6H_{12}N_4.C_8H_8O_3$	292.34
Methicillin Sodium $C_{17}H_{19}N_2NaO_6S.H_2O$	420.41
Sodium, anhydrous $C_{17}H_{19}N_2NaO_6S$	402.40
Methimazole $C_4H_6N_2S$	114.16
Methisazone $C_{10}H_{10}N_4OS$	234.28
Methocarbamol $C_{11}H_{15}NO_5$	241.24
Methohexital $C_{14}H_{18}N_2O_3$	262.31
Sodium $C_{14}H_{17}N_2NaO_3$	284.29
Methotrexate $C_{20}H_{22}N_8O_5$	454.44
Methotrimeprazine $C_{19}H_{24}N_2OS$	328.47
Methoxamine $C_{11}H_{17}NO_3$	211.26
Hydrochloride $C_{11}H_{17}NO_3.HCl$	247.72
Methoxsalen $C_{12}H_8O_4$	216.19
Methoxy (Methoxyl) $CH_3O–$	31.03
p-Methoxybenzaldehyde $C_8H_8O_2$	136.15
Methoxyethanol $C_3H_8O_2$	76.09
Methoxyflurane $C_3H_4Cl_2F_2O$	164.97
Methoxytyrosine $C_{10}H_{13}NO_4$	211.22
Methscopolamine Bromide $C_{18}H_{24}BrNO_4$	398.30
Methsuximide $C_{12}H_{13}NO_2$	203.24
Methyclothiazide $C_9H_{11}Cl_2N_3O_4S_2$	360.23
Methyl Acetate $C_3H_6O_2$	74.08
Alcohol CH_4O	32.04
Ethyl Ketone C_4H_8O	72.11
Iodide CH_3I	141.94
Isobutyl Ketone $C_6H_{12}O$	100.16
5-Methyl-3-isoxazolecarboxylate $C_6H_7NO_3$	141.13
Nonadecanoate $C_{20}H_{40}O_2$	312.53
Red $C_{15}H_{15}N_3O_2$	269.30
Salicylate $C_8H_8O_3$	152.15
Stearate $C_{19}H_{38}O_2$	298.51
Sulfoxide C_2H_6OS	78.13
Methylamine CH_5N	31.06
2-Methylamino-5-chlorobenzophenone $C_{14}H_{12}ClNO$	245.71
Methylaminophenol Sulfate $(C_7H_9NO)_2.H_2SO_4$	344.38
Methylbenzethonium Chloride $C_{28}H_{44}ClNO_2.H_2O$	480.13
Chloride, anhydrous $C_{28}H_{44}ClNO_2$	462.11
Methylbenzylamine	121.18
Methylboron Dihydroxide CH_5BO_2	59.86
3-*O*-Methylcarbidopa $C_{11}H_{16}N_2O_4$	240.26
Methyl Carbamate $C_2H_5NO_2$	75.07
Methylchloroform (see *Trichloroethane*)	
Methyldimethoxyamphetamine $C_{12}H_{19}NO_2$	209.29
Hydrochloride $C_{12}H_{19}NO_2.HCl$	245.75
Methyldopa $C_{10}H_{13}NO_4.1½H_2O$	238.24
anhydrous $C_{10}H_{13}NO_4$	211.22
Methyldopate $C_{12}H_{17}NO_4$	239.27
Hydrochloride $C_{12}H_{17}NO_4.HCl$	275.73
Methylene Blue $C_{16}H_{18}ClN_3S.3H_2O$	373.90
Blue, anhydrous $C_{16}H_{18}ClN_3S$	319.85
Chloride CH_2Cl_2	84.93
Methylenedioxyamphetamine $C_{10}H_{13}NO_2$	179.22
Hydrochloride $C_{10}H_{13}NO_2.HCl$	215.68
Methylergonovine $C_{20}H_{25}N_3O_2$	339.44
Maleate $C_{20}H_{25}N_3O_2.C_4H_4O_4$	455.51
Methylglucamine (see *Meglumine*)	
3-*O*-Methylmethyldopa $C_{11}H_{15}NO_4$	225.24
N-Methyl-*N*-nitroso-*p*-toluenesulfonamide $C_8H_{10}N_2O_3S$	214.24
Methylparaben $C_8H_8O_3$	152.15
Methylpentanone (see *Methyl Isobutyl Ketone*)	
Methylphenidate $C_{14}H_{19}NO_2$	233.31
Hydrochloride $C_{14}H_{19}NO_2.HCl$	269.77
Methylprednisolone $C_{22}H_{30}O_5$	374.48
Acetate $C_{24}H_{32}O_6$	416.51
Hemisuccinate $C_{26}H_{34}O_8$	474.55
Sodium Succinate $C_{26}H_{33}NaO_8$	496.63
Succinate (see *Methylprednisolone Hemisuccinate*)	
2-Methyl-2-propyl-1,3-propanediol $C_7H_{16}O_2$	132.20
Methyltestosterone $C_{20}H_{30}O_2$	302.46
Methylthionine Perchlorate $C_{16}H_{18}ClN_3O_4S$	383.85
Methyprylon $C_{10}H_{17}NO_2$	183.25
Methysergide $C_{21}H_{27}N_3O_2$	353.46
Maleate $C_{21}H_{27}N_3O_2.C_4H_4O_4$	469.54
Metoclopramide $C_{14}H_{22}ClN_3O_2$	299.80
Hydrochloride $C_{14}H_{22}ClN_3O_2.HCl.H_2O$	354.28
Hydrochloride, anhydrous $C_{14}H_{22}ClN_3O_2.HCl$	336.26
Metocurine Iodide $C_{40}H_{48}I_2N_2O_6$	906.64
Metolazone $C_{16}H_{16}ClN_3O_3S$	365.83
Metoprolol Tartrate $(C_{15}H_{25}NO_3)_2.C_4H_6O_6$	684.82
Metrizamide $C_{18}H_{22}I_3N_3O_8$	789.10
Metronidazole $C_6H_9N_3O_3$	171.16
Metyrapone $C_{14}H_{14}N_2O$	226.28
Tartrate $C_{14}H_{14}N_2O.2C_4H_6O_6$	526.45
Metyrosine $C_{10}H_{13}NO_3$	195.22
Mexiletine Hydrochloride $C_{11}H_{17}NO.HCl$	215.72
Mezlocillin Sodium $C_{21}H_{24}NaN_5O_8S_2$	561.56
Miconazole $C_{18}H_{14}Cl_4N_2O$	416.13

  - Nitrate $C_{18}H_{14}Cl_4N_2O.HNO_3$ 479.15
- Minocycline $C_{23}H_{27}N_3O_7$ 457.48
  - Hydrochloride $C_{23}H_{27}N_3O_7.HCl$ 493.94
- Minoxidil $C_9H_{15}N_5O$ 209.25
- Mitomycin $C_{15}H_{18}N_4O_5$ 334.33
- Mitotane $C_{14}H_{10}Cl_4$ 320.04
- Molybdenum Trioxide MoO_3 143.94
- Monobenzone $C_{13}H_{12}O_2$ 200.24
  - Acetate $C_{15}H_{14}O_3$ 242.27
- Monochloroacetic Acid $C_2H_3ClO_2$ 94.50
- Monoethanolamine C_2H_7NO 61.08
- Monosodium Glutamate $C_5H_8NNaO_4.H_2O$ 187.13
- Monothioglycerol $C_3H_8O_2S$ 108.15
- Morphine $C_{17}H_{19}NO_3.H_2O$ 303.36
  - anhydrous $C_{17}H_{19}NO_3$ 285.34
  - Sulfate $(C_{17}H_{19}NO_3)_2.H_2SO_4.5H_2O$ 758.83
  - Sulfate, anhydrous $(C_{17}H_{19}NO_3)_2.H_2SO_4$ 668.76
- Morpholine C_4H_9NO 87.12
- Moxalactam Disodium $C_{20}H_{18}N_6Na_2O_9S$ 564.44
- Murexide, anhydrous $C_8H_8N_6O_6$ 284.19
- Myristalkonium Chloride $C_{23}H_{42}ClN$ 368.04
- Myristyl Alcohol $C_{14}H_{30}O$ 214.39

- Nadolol $C_{17}H_{27}NO_4$ 309.41
- Nafcillin $C_{21}H_{22}N_2O_5S$ 414.48
  - Sodium $C_{21}H_{21}N_2NaO_5S.H_2O$ 454.47
  - Sodium, anhydrous $C_{21}H_{21}N_2NaO_5S$ 436.46
- Nalidixic Acid $C_{12}H_{12}N_2O_3$ 232.24
- Nalorphine $C_{19}H_{21}NO_3$ 311.38
  - Hydrochloride $C_{19}H_{21}NO_3.HCl$ 347.84
- Naloxone $C_{19}H_{21}NO_4$ 327.38
  - Hydrochloride $C_{19}H_{21}NO_4.HCl.2H_2O$ 399.87
  - Hydrochloride, anhydrous $C_{19}H_{21}NO_4.HCl$ 363.84
- Nandrolone $C_{18}H_{26}O_2$ 274.40
  - Decanoate $C_{28}H_{44}O_3$ 428.65
- Naphazoline $C_{14}H_{14}N_2$ 210.28
  - Hydrochloride $C_{14}H_{14}N_2.HCl$ 246.74
- Naphthalene $C_{10}H_8$ 128.17
- Naphthalenediol $C_{10}H_8O_2$ 160.17
- 2-Naphthalenesulfonic Acid $C_{10}H_8O_3S.H_2O$ 226.25
- Naphthol $C_{10}H_8O$ 144.17
- 2-Naphthol-6,8-dipotassium Disulfonate $C_{10}H_6K_2O_7S_2$ 380.47
- 2-Naphthol-6,8-disodium disulfonate $C_{10}H_6Na_2O_7S_2$ 348.25
- Naphthol Potassium Disulfonate (see *2-Naphthol-6,8-dipotassium Disulfonate*)
- β-Naphthoquinone-4-sodium sulfonate $C_{10}H_5NaO_5S$ 260.20
- Naphthoresorcinol $C_{10}H_8O_2$ 160.17
- 1-Naphthylamine Hydrochloride $C_{10}H_9N.HCl$ 179.65
- *N*-1-Naphthylethylenediamine Dihydrochloride $C_{12}H_{14}N_2.2HCl$ 259.18
- Naproxen $C_{14}H_{14}O_3$ 230.26
  - Sodium $C_{14}H_{13}NaO_3$ 252.24
- Natamycin $C_{33}H_{47}NO_{13}$ 665.73
- Netilmicin Sulfate $(C_{21}H_{41}N_5O_7)_2.5H_2SO_4$ 1441.54
- Niacin $C_6H_5NO_2$ 123.11
- Niacinamide $C_6H_6N_2O$ 122.13
- Nialamide $C_{16}H_{18}N_4O_2$ 298.34
- Nickel Ammonium Sulfate $NiSO_4.(NH_4)_2SO_4.6H_2O$ 394.98
  - Ammonium Sulfate, anhydrous $NiSO_4.(NH_4)_2SO_4$ 286.89
  - Chloride $NiCl_2.6H_2O$ 237.70
  - Chloride, anhydrous $NiCl_2$ 129.61
- Niclosamide $C_{13}H_8Cl_2N_2O_4$ 327.12
- Nicotinic Acid $C_6H_5NO_2$ 123.11
- Nifedipine $C_{17}H_{18}N_2O_6$ 346.34
- Nifuroxime $C_5H_4N_2O_4$ 156.10
- Nifurtimox $C_{10}H_{13}N_3O_5S$ 287.29
- Nile Blue Hydrochloride $C_{20}H_{20}ClN_3O$ 353.85
- Ninhydrin (see *Triketohydrindene Hydrate*)
- Niridazole $C_6H_6N_4O_3S$ 214.20
- Nitranilic Acid $C_6H_2N_2O_8$ 230.09
- Nitric Acid HNO_3 63.01
  - Oxide NO 30.01
- Nitrilotriacetic Acid $C_6H_9NO_6$ 191.14
- 4′-Nitroacetophenone $C_8H_7NO_3$ 165.15
- Nitroaniline $C_6H_6N_2O_2$ 138.13
- Nitrobenzaldehyde $C_7H_5NO_3$ 151.12
- Nitrobenzene $C_6H_5NO_2$ 123.11
- *p*-Nitrobenzenediazonium Tetrafluoroborate $C_6H_4BF_4N_3O_2$ 236.92
- *p*-Nitrobenzoic Acid $C_6H_4NO_2COOH$ 167.12
- Nitrobenzyl Bromide $C_7H_6BrNO_2$ 216.03
- Nitrobenzylpyridine $C_{12}H_{10}N_2O_2$ 214.22
- Nitrofurantoin $C_8H_6N_4O_5.H_2O$ 256.17
  - anhydrous $C_8H_6N_4O_5$ 238.16
  - Sodium $C_8H_5N_4NaO_5$ 260.14
- Nitrofurazone $C_6H_6N_4O_4$ 198.14
- Nitrofurfural Diacetate $C_9H_9NO_7$ 243.17
- 5-Nitro-2-furfuraldazine $C_8H_6N_2O_6$ 226.15
- Nitrogen N_2 28.01
  - Trioxide N_2O_3 76.01
- Nitroglycerin (see *Glyceryl Trinitrate*)
- Nitromersol $C_7H_5HgNO_3$ 351.71
- Nitromethane CH_3NO_2 61.04
- Nitrophenanthroline $C_{12}H_7N_3O_2$ 225.21
- 3-Nitro-4-phenoxy-5-sulfamoylbenzoic Acid $C_{13}H_{10}N_2O_7S$ 338.29
- Nitroso R Salt $C_{10}H_5NNa_2O_8S_2$ 377.25
- Nitrosonaphthol $C_{10}H_7NO_2$ 173.17
- Nitroterephthalic Acid $C_8H_5NO_6$ 211.13
- Nitrous Oxide N_2O 44.01
- Nonadecane $C_{19}H_{40}$ 268.53
- 1-Nonyl Alcohol $CH_3(CH_2)_8OH$ 144.26
- Norandrostenedione $C_{18}H_{24}O_2$ 272.39
- Norepinephrine $C_8H_{11}NO_3$ 169.18
  - Bitartrate $C_8H_{11}NO_3.C_4H_6O_6.H_2O$ 337.28
  - Bitartrate, anhydrous $C_8H_{11}NO_3.C_4H_6O_6$ 319.27
- Norethandrolone $C_{20}H_{30}O_2$ 302.46
- Norethindrone $C_{20}H_{26}O_2$ 298.42
  - Acetate $C_{22}H_{28}O_3$ 340.46
- Norethynodrel $C_{20}H_{26}O_2$ 298.42
- Norfloxacin $C_{16}H_{18}FN_3O_3$ 319.34
- Norgestrel $C_{21}H_{28}O_2$ 312.45
- Noroxymorphone $C_{16}H_{17}NO_4$ 287.31
  - Hydrochloride $C_{16}H_{17}NO_4.HCl$ 323.78
- Nortriptyline $C_{19}H_{21}N$ 263.38
  - Hydrochloride $C_{19}H_{21}N.HCl$ 299.84
- Noscapine $C_{22}H_{23}NO_7$ 413.43
- Novobiocin $C_{31}H_{36}N_2O_{11}$ 612.63
  - Calcium $C_{62}H_{70}CaN_4O_{22}$ 1263.33
  - Sodium $C_{31}H_{35}N_2NaO_{11}$ 634.62
- Nylidrin $C_{19}H_{25}NO_2$ 299.41
  - Hydrochloride $C_{19}H_{25}NO_2.HCl$ 335.87

- *n*-Octadecane $C_{18}H_{38}$ 254.50
- Octadecenoic Acid (see *Oleic Acid*)
- Octoxynol (Av.) $C_{34}H_{62}O_{11}$ 647
- *n*-Octylamine $C_8H_{17}NH_2$ 129.25
- Octyldimethylaminobenzoic Acid $C_{17}H_{27}NO_2$ 277.40
- Octyldodecanol $C_{20}H_{42}O$ 298.55
- Octylphenoxynonaethoxyethanol $C_{34}H_{62}O_{11}$ 646.86
- Oleandomycin $C_{35}H_{61}NO_{12}$ 687.87
  - Chloroform Adduct $C_{35}H_{61}NO_{12}.CHCl_3$ 807.24
- Oleic Acid $C_{18}H_{34}O_2$ 282.47
- Oleyl Alcohol $C_{18}H_{36}O$ 268.48
- Orange G $C_{16}H_{10}N_2Na_2O_7S_2$ 452.36
- Orcinol $C_7H_8O_2$ 124.14
- Orphenadrine $C_{18}H_{23}NO$ 269.39
  - Citrate $C_{18}H_{23}NO.C_6H_8O_7$ 461.51
- Orthophenanthroline $C_{12}H_8N_2.H_2O$ 198.22
- Orthophosphoric Acid (see *Phosphoric Acid*)
- Orthotolidine $C_{14}H_{16}N_2$ 212.29
- Osmium Tetroxide (Osmic Acid) OsO_4 254.20
- Oxacillin $C_{19}H_{19}N_3O_5S$ 401.44
  - Sodium $C_{19}H_{18}N_3NaO_5S.H_2O$ 441.43
  - Sodium, anhydrous $C_{19}H_{18}N_3NaO_5S$ 423.42
- Oxalic Acid $C_2H_2O_4.2H_2O$ 126.07
  - Acid, anhydrous $C_2H_2O_4$ 90.04
- Oxamniquine $C_{14}H_{21}N_3O_3$ 279.34
- Oxandrolone $C_{19}H_{30}O_3$ 306.44
- Oxazepam $C_{15}H_{11}ClN_2O_2$ 286.72
- Oxolinic Acid $C_{13}H_{11}NO_5$ 261.23
- Oxprenolol Hydrochloride $C_{15}H_{23}NO_3.HCl$ 301.81
- Oxtriphylline $C_{12}H_{21}N_5O_3$ 283.33
- Oxybenzone $C_{14}H_{12}O_3$ 228.25
- Oxybutynin $C_{22}H_{31}NO_3$ 357.49
  - Chloride $C_{22}H_{31}NO_3.HCl$ 393.95
- Oxycholesterin $C_{27}H_{46}O_2$ 402.66
- Oxycodone $C_{18}H_{21}NO_4$ 315.37
  - Hydrochloride $C_{18}H_{21}NO_4.HCl$ 351.83
  - Terephthalate $(C_{18}H_{21}NO_4)_2.C_8H_6O_4$ 796.87
- Oxy(dimethylsilylene) (see *Dimethylsiloxane*)
- Oxydipropionitrile $C_6H_8N_2O$ 124.14
- Oxygen O_2 32.00
- Oxymetazoline $C_{16}H_{24}N_2O$ 260.38
  - Hydrochloride $C_{16}H_{24}N_2O.HCl$ 296.84
- Oxymetholone $C_{21}H_{32}O_3$ 332.48
- Oxymorphone $C_{17}H_{19}NO_4$ 301.34

Hydrochloride $C_{17}H_{19}NO_4.HCl$ 337.80
Oxyphenbutazone $C_{19}H_{20}N_2O_3.H_2O$ 342.39
anhydrous $C_{19}H_{20}N_2O_3$ 324.38
Oxyphencyclimine $C_{20}H_{28}N_2O_3$ 344.45
Hydrochloride $C_{20}H_{28}N_2O_3.HCl$ 380.91
Oxyquinoline Sulfate $(C_9H_7NO)_2.H_2SO_4$ 388.39
Oxytetracycline $C_{22}H_{24}N_2O_9.2H_2O$ 496.47
anhydrous $C_{22}H_{24}N_2O_9$ 460.44
Calcium $C_{44}H_{46}CaN_4O_{18}$ 958.94
Hydrochloride $C_{22}H_{24}N_2O_9.HCl$ 496.90
Oxytocin $C_{43}H_{66}N_{12}O_{12}S_2$ 1007.19

Padimate A $C_{14}H_{21}NO_2$ 235.33
O $C_{17}H_{27}NO_2$ 277.41
Palladous Chloride, anhydrous $PdCl_2$ 177.31
Sulfate $PdSO_4$ 202.46
Palmitic Acid $C_{16}H_{32}O_2$ 256.43
Pamoic Acid $C_{23}H_{16}O_6$ 388.38
Pancuronium Bromide $C_{35}H_{60}Br_2N_2O_4$ 732.68
Panthenol $C_9H_{19}NO_4$ 205.25
Pantolactone $C_6H_{10}O_3$ 130.14
Pantothenic Acid $C_9H_{17}NO_5$ 219.24
Papaverine $C_{20}H_{21}NO_4$ 339.39
Hydrochloride $C_{20}H_{21}NO_4.HCl$ 375.85
Parabens (see *Butylparaben; Ethylparaben; Methylparaben; Propylparaben*)
Parachlorophenol C_6H_5ClO 128.56
Paraldehyde $C_6H_{12}O_3$ 132.16
Paramethadione $C_7H_{11}NO_3$ 157.17
Paramethasone Acetate $C_{24}H_{31}FO_6$ 434.50
Pararosaniline Chloride $C_{19}H_{18}ClN_3$ 323.82
Hydrochloride (see *Pararosaniline Chloride*)
Parathion $C_{10}H_{14}NO_5PS$ 291.26
Pargyline $C_{11}H_{13}N$ 159.23
Hydrochloride $C_{11}H_{13}N.HCl$ 195.69
Paromomycin $C_{23}H_{45}N_5O_{14}$ 615.63
Pemoline $C_9H_8N_2O_2$ 176.17
Penicillamine $C_5H_{11}NO_2S$ 149.21
Penicillin Phenoxymethyl (see *Penicillin V*)
Potassium Phenoxymethyl (see *Penicillin V Potassium*)
G Benzathine $C_{16}H_{20}N_2.2C_{16}H_{18}N_2O_4S.4H_2O$ 981.19
G Benzathine, anhydrous $C_{16}H_{20}N_2.2C_{16}H_{18}N_2O_4S$ 909.13
G Potassium $C_{16}H_{17}KN_2O_4S$ 372.48
G Procaine $C_{16}H_{18}N_2O_4S.C_{13}H_{20}N_2O_2.H_2O$ 588.72
G Procaine, anhydrous $C_{16}H_{18}N_2O_4S.C_{13}H_{20}N_2O_2$ 570.70
G Sodium $C_{16}H_{17}N_2NaO_4S$ 356.37
V $C_{16}H_{18}N_2O_5S$ 350.39
V Benzathine, anhydrous $C_{16}H_{20}N_2.2C_{16}H_{18}N_2O_5S$ 941.12
V Benzathine tetrahydrate $C_{16}H_{20}N_2.2C_{16}H_{18}N_2O_5S.4H_2O$ 1013.19
V Potassium $C_{16}H_{17}KN_2O_5S$ 388.48

Pentaerythritol Tetranitrate $C_5H_8N_4O_{12}$ 316.14
Pentagastrin $C_{37}H_{49}N_7O_9S$ 767.90
Pentamethylpararosaniline Chloride $C_{24}H_{28}ClN_3$ 393.96
Pentamidine $C_{19}H_{24}N_4O_2$ 340.42
Isethionate $C_{19}H_{24}N_4O_2.2C_2H_6O_4S$ 592.68
Pentane C_5H_{12} 72.15
Pentazocine $C_{19}H_{27}NO$ 285.43
Hydrochloride $C_{19}H_{27}NO.HCl$ 321.89
Lactate $C_{19}H_{27}NO.C_3H_6O_3$ 375.51
Penthienate Bromide $C_{18}H_{30}BrNO_3S$ 420.40
Pentobarbital $C_{11}H_{18}N_2O_3$ 226.27
Sodium $C_{11}H_{17}N_2NaO_3$ 248.26
Pentolinium Tartrate $C_{23}H_{42}N_2O_{12}$ 538.59
Pentose $C_5H_{10}O_5$ 150.13
Pentylenetetrazol $C_6H_{10}N_4$ 138.17
p-Pentylphenol $CH_3(CH_2)_4C_6H_4OH$ 164.25
Pentylphenol $C_{11}H_{16}O$ 164.25
Perchloric Acid $HClO_4$ 100.46
Periodic Acid H_5IO_6 227.94
Perphenazine $C_{21}H_{26}ClN_3OS$ 403.97
Hydrochloride, anhydrous $C_{18}H_{22}N_2O_2.HCl$ 334.84
Phenacemide $C_9H_{10}N_2O_2$ 178.19
Phenacetin $C_{10}H_{13}NO_2$ 179.22
Phenanthroline (see *Orthophenanthroline*)
Phenaphthazine $C_{18}H_8N_4Na_2O_{11}S_2$ 566.38
Phenazocine $C_{22}H_{27}NO$ 321.46
Hydrobromide $C_{22}H_{27}NO.HBr$ 402.37
Phenazopyridine $C_{11}H_{11}N_5$ 213.24
Hydrochloride $C_{11}H_{11}N_5.HCl$ 249.70
Phencyclidine $C_{17}H_{25}N$ 243.39
Hydrochloride $C_{17}H_{25}N.HCl.$ 279.85
Phendimetrazine Tartrate $C_{12}H_{17}NO.C_4H_6O_6$ 341.36
Phenelzine $C_8H_{12}N_2$ 136.20
Sulfate $C_8H_{12}N_2.H_2SO_4$ 234.27
Phenformin $C_{10}H_{15}N_5$ 205.26
Hydrochloride $C_{10}H_{15}N_5.HCl$ 241.72
Phenindamine $C_{19}H_{19}N$ 261.37
Tartrate $C_{19}H_{19}N.C_4H_6O_6$ 411.45
Phenindione $C_{15}H_{10}O_2$ 222.24
Phenmetrazine $C_{11}H_{15}NO$ 177.25
Hydrochloride $C_{11}H_{15}NO.HCl$ 213.71
Phenobarbital $C_{12}H_{12}N_2O_3$ 232.24
Sodium $C_{12}H_{11}N_2NaO_3$ 254.22
Phenol C_6H_6O 94.11
Phenolphthalein $C_{20}H_{14}O_4$ 318.33
Phenoxyethanol $C_8H_{10}O_2$ 138.17
Phenoxymethyl Penicillin (see *Penicillin V*)
Phenprocoumon $C_{18}H_{16}O_3$ 280.32
Phensuximide $C_{11}H_{11}NO_2$ 189.21
Phentermine $C_{10}H_{15}N$ 149.24
Hydrochloride $C_{10}H_{15}N.HCl$ 185.70
Phentolamine $C_{17}H_{19}N_3O$ 281.36
Mesylate $C_{17}H_{19}N_3O.CH_4SO_3$ 377.46
Trichloroacetate $C_{17}H_{19}N_3O.C_2HCl_3O_2$ 444.74
Phenyl Ether (see *Diphenyl Ether*)
Isocyanate C_6H_5NCO 119.12
Isocyanide C_7H_5N 103.12
Sulfide $C_{12}H_{10}S$ 186.27

Phenylalanine $C_9H_{11}NO_2$ 165.19
Phenylbutazone $C_{19}H_{20}N_2O_2$ 308.38
Phenylcyclohexylglycolic Acid $C_{14}H_{18}O_3$ 234.30
2-Phenylcyclopropylamine $C_9H_{11}N$ 133.19
Phenylcyclopropylamine Sulfate (see *Tranylcypromine Sulfate*)
Phenylenediamine Dihydrochloride $C_6H_8N_2.2HCl$ 181.06
Phenylephrine $C_9H_{13}NO_2$ 167.21
Bitartrate $C_9H_{13}NO_2.C_4H_6O_6$ 317.29
Hydrochloride $C_9H_{13}NO_2.HCl$ 203.67
Phenylethyl Alcohol $C_8H_{10}O$ 122.17
2-Phenylethylamine Hydrochloride $C_8H_{11}N.HCl$ 157.64
Phenylhydrazine $C_6H_8N_2$ 108.14
Hydrochloride $C_6H_8N_2.HCl$ 144.60
Phenylmercuric Acetate $C_8H_8HgO_2$ 336.74
Hydroxide C_6H_5HgOH 294.70
Ion $C_6H_5Hg^+$ 277.70
Nitrate $C_6H_5HgNO_3$ 339.70
1-Phenyl-3-methyl-2-pyrazolin-5-one $C_{10}H_{10}N_2O$ 174.20
3-Phenylphenol $C_{12}H_{10}O$ 170.21
Phenylpiperidineacetic Acid $C_{13}H_{17}NO_2$ 219.28
Acid Hydrochloride $C_{13}H_{17}NO_2.HCl$ 255.74
Phenylpropanolamine $C_9H_{13}NO$ 151.21
Hydrochloride $C_9H_{13}NO.HCl$ 187.67
Phenytoin $C_{15}H_{12}N_2O_2$ 252.27
Sodium $C_{15}H_{11}N_2NaO_2$ 274.25
Phloroglucinol $C_6H_6O_3.2H_2O$ 162.14
anhydrous $C_6H_6O_3$ 126.11
Phosphate $PO_4^{\equiv}$ 94.97
Phosphomolybdic Acid (approx.) $20MoO_3.P_2O_5.51H_2O$ 3939.48
Phosphoric Acid H_3PO_4 98.00
Phosphorus Pentoxide P_2O_5 141.94
Phosphotungstic Acid (approx.) $24WO_3.P_2O_5.51H_2O$ 6625.08
o-Phthalaldehyde $C_8H_6O_2$ 134.13
Phthalic Acid $C_8H_6O_4$ 166.13
Anhydride $C_8H_4O_3$ 148.12
Phthalyl (*o*-carboxybenzoyl) $C_8H_5O_3^+$ 149.13
Physostigmine $C_{15}H_{21}N_3O_2$ 275.35
Salicylate $C_{15}H_{21}N_3O_2.C_7H_6O_3$ 413.47
Sulfate $(C_{15}H_{21}N_3O_2)_2.H_2SO_4$ 648.77
Phytonadione $C_{31}H_{46}O_2$ 450.70
Picric Acid (see *Trinitrophenol*)
Picrolonic Acid $C_{10}H_8N_4O_5$ 264.20
Pilocarpine $C_{11}H_{16}N_2O_2$ 208.26
Hydrochloride $C_{11}H_{16}N_2O_2.HCl$ 244.72
Nitrate $C_{11}H_{16}N_2O_2.HNO_3$ 271.27
Piminodine $C_{23}H_{30}N_2O_2$ 366.50
Esylate $C_{23}H_{30}N_2O_2.C_2H_6O_3S$ 476.63
Pimozide $C_{28}H_{29}F_2N_3O$ 461.55
Pindolol $C_{14}H_{20}N_2O_2$ 248.32
Piperacetazine $C_{24}H_{30}N_2O_2S$ 410.57
Piperacillin Sodium $C_{23}H_{26}N_5NaO_7S$ 539.54
Piperazine, anhydrous $C_4H_{10}N_2$ 86.14
hexahydrate $C_4H_{10}N_2.6H_2O$ 194.23

Substance	Mol. Wt.
Calcium Edetate $C_4H_{10}N_2 \cdot C_{10}H_{14}CaN_2O_8 \cdot 2H_2O$	452.48
Calcium Edetate, anhydrous $C_4H_{10}N_2 \cdot C_{10}H_{14}CaN_2O_8$	416.45
Citrate, anhydrous $3C_4H_{10}N_2 \cdot 2C_6H_8O_7$	642.66
Dipicrate $C_4H_{10}N_2 \cdot 2C_6H_3N_3O_7$	544.35
Phosphate, anhydrous $C_4H_{10}N_2 \cdot H_3PO_4$	184.13
Piperidine $C_5H_{11}N$	85.15
Piperidolate $C_{21}H_{25}NO_2$	323.43
Pipobroman $C_{10}H_{16}Br_2N_2O_2$	356.06
Piroxicam $C_{15}H_{13}N_3O_4S$	331.35
Platinic Chloride (see *Chloroplatinic Acid*)	
Plicamycin $C_{52}H_{76}O_{24}$	1085.16
Polythiazide $C_{11}H_{13}ClF_3N_3O_4S_3$	439.87
Ponceau S $C_{22}H_{12}N_4O_{13}S_4Na_4$	760.56
Potassium Acetate $C_2H_3KO_2$	98.14
Alum $AlK(SO_4)_2 \cdot 12H_2O$	474.38
Alum, anhydrous $AlK(SO_4)_2$	258.19
Aluminum Sulfate (see *Aluminum and Potassium Sulfate*)	
Aminobenzoate $C_7H_6KNO_2$	175.23
Aminosalicylate $C_7H_6KNO_3$	191.23
Arsenite (see *Potassium Metarsenite*)	
Benzoate $C_7H_5KO_2$	160.21
Bicarbonate $KHCO_3$	100.12
Biphosphate (see *Potassium Phosphate, Monobasic*)	
Biphthalate $C_8H_5KO_4$	204.22
Bisulfate $KHSO_4$	136.16
Bromate $KBrO_3$	167.00
Bromide KBr	119.00
Carbonate $K_2CO_3 \cdot 1\frac{1}{2}H_2O$	165.23
Carbonate, anhydrous K_2CO_3	138.21
Chlorate $KClO_3$	122.55
Chloride KCl	74.55
Chloroplatinate K_2PtCl_6	486.00
Chromate K_2CrO_4	194.19
Citrate $C_6H_5K_3O_7 \cdot H_2O$	324.41
Citrate, anhydrous $C_6H_5K_3O_7$	306.40
Cyanate $KOCN$	81.12
Cyanide KCN	65.12
Dichromate $K_2Cr_2O_7$	294.18
Ferricyanide $K_3Fe(CN)_6$	329.25
Ferrocyanide $K_4Fe(CN)_6 \cdot 3H_2O$	422.39
Ferrocyanide, anhydrous $K_4Fe(CN)_6$	368.35
Gluconate $C_6H_{11}KO_7$	234.25
Guaiacolsulfonate $C_7H_7KO_5S \cdot \frac{1}{2}H_2O$	251.29
Guaiacolsulfonate, anhydrous $C_7H_7KO_5S$	242.29
Hydroxide KOH	56.11
Iodate KIO_3	214.00
Iodide KI	166.00
Metaphosphate KPO_3	118.07
Metarsenite $KAsO_2$	146.02
Nitrate KNO_3	101.10
Nitrite KNO_2	85.10
Oxalate $C_2K_2O_4 \cdot H_2O$	184.23
Oxalate, anhydrous $C_2K_2O_4$	166.22
Perchlorate $KClO_4$	138.55
Periodate (Potassium *meta*-Periodate) KIO_4	230.00
Permanganate $KMnO_4$	158.03
Persulfate $K_2S_2O_8$	270.31
Phosphate, Dibasic K_2HPO_4	174.18
Phosphate, Monobasic KH_2PO_4	136.09
Pyrosulfate $K_2S_2O_7$	254.31
Salicylate $C_7H_5KO_3$	176.21
Sodium Tartrate $C_4H_4KNaO_6 \cdot 4H_2O$	282.22
Sodium Tartrate, anhydrous $C_4H_4KNaO_6$	210.16
Sorbate $C_6H_7KO_2$	150.22
Sulfate K_2SO_4	174.25
Tellurite K_2TeO_3	253.79
Thiocyanate $KSCN$	97.18
Thiosulfate $K_2S_2O_3$	190.31
Xanthogenate $C_3H_5KOS_2$	160.29
Pralidoxime Chloride $C_7H_9ClN_2O$	172.61
Pramoxine $C_{17}H_{27}NO_3$	293.41
Hydrochloride $C_{17}H_{27}NO_3 \cdot HCl$	329.87
Prazepam $C_{19}H_{17}ClN_2O$	324.81
Praziquantel $C_{19}H_{24}N_2O_2$	312.41
Related Compound A (see *2-Benzoyl-1,2,3,6,7,11b-hexahydro-4H-pyrazino[2,1-a]isoquinolin-4-one*)	
Related Compound B (see *2-(Cyclohexylcarbonyl)-2,3,6,7-tetrahydro-4H-pyrazino[2,1-a]isoquinolin-4-one*)	
Related Compound C (see *2-(N-Formylhexahydrohippuroyl)-1,2,3,4-tetrahydroisoquinolin-1-one*)	
Prazosin $C_{19}H_{21}N_5O_4$	383.41
Hydrochloride $C_{19}H_{21}N_5O_4 \cdot HCl$	419.87
Prednisolone, anhydrous $C_{21}H_{28}O_5$	360.45
sesquihydrate $C_{21}H_{28}O_5 \cdot 1\frac{1}{2}H_2O$	387.47
Acetate $C_{23}H_{30}O_6$	402.49
Hemisuccinate $C_{25}H_{32}O_8$	460.52
Phosphate $C_{21}H_{29}O_8P$	440.43
Sodium Phosphate $C_{21}H_{27}Na_2O_8P$	484.39
Sodium Succinate $C_{25}H_{31}NaO_8$	482.50
Tebutate, anhydrous $C_{27}H_{38}O_6$	458.59
Tebutate monohydrate $C_{27}H_{38}O_6 \cdot H_2O$	476.61
Prednisone $C_{21}H_{26}O_5$	358.43
Tebutate $C_{27}H_{36}O_6$	456.58
Prilocaine $C_{13}H_{20}N_2O$	220.31
Hydrochloride $C_{13}H_{20}N_2O \cdot HCl$	256.77
Primaquine $C_{15}H_{21}N_3O$	259.35
Phosphate $C_{15}H_{21}N_3O \cdot 2H_3PO_4$	455.34
Primidone $C_{12}H_{14}N_2O_2$	218.25
Probenecid $C_{13}H_{19}NO_4S$	285.36
Probucol $C_{31}H_{48}O_2S_2$	516.84
Related Compound A (see *2,2',6,6'-Tetra-tert-butyldiphenoquinone*)	
Related Compound B (see *4,4'-[Dithio]bis[2,6-di-tert-butylphenol]*)	
Related Compound C (see *4-[(3,5-Di-tert-butyl-2-hydroxyphenylthio)isopropylidenethio]-2,6-di-tert-butylphenol*)	
Procainamide $C_{13}H_{21}N_3O$	235.33
Hydrochloride $C_{13}H_{21}N_3O \cdot HCl$	271.79
Procaine (base) $C_{13}H_{20}N_2O_2$	236.31
Hydrochloride $C_{13}H_{20}N_2O_2 \cdot HCl$	272.77
Prochlorperazine $C_{20}H_{24}ClN_3S$	373.94
Edisylate $C_{20}H_{24}ClN_3S \cdot C_2H_6O_6S_2$	564.13
Maleate $C_{20}H_{24}ClN_3S \cdot 2C_4H_4O_4$	606.09
Procyclidine $C_{19}H_{29}NO$	287.44
Hydrochloride $C_{19}H_{29}NO \cdot HCl$	323.91
Progesterone $C_{21}H_{30}O_2$	314.47
Progesteronebis(dinitrophenylhydrazone) $C_{33}H_{38}N_8O_8$	674.71
Proline $C_5H_9NO_2$	115.13
Promazine $C_{17}H_{20}N_2S$	284.42
Hydrochloride $C_{17}H_{20}N_2S \cdot HCl$	320.88
Promethazine $C_{17}H_{20}N_2S$	284.42
Hydrochloride $C_{17}H_{20}N_2S \cdot HCl$	320.88
Propane C_3H_8	44.10
Propantheline Bromide $C_{23}H_{30}BrNO_3$	448.40
Proparacaine $C_{16}H_{26}N_2O_3$	294.39
Hydrochloride $C_{16}H_{26}N_2O_3 \cdot HCl$	330.85
Propiolactone $C_3H_4O_2$	72.06
Propiomazine $C_{20}H_{24}N_2OS$	340.48
Hydrochloride $C_{20}H_{24}N_2OS \cdot HCl$	376.94
Propionic Anhydride $C_6H_{10}O_3$	130.14
Propoxycaine $C_{16}H_{26}N_2O_3$	294.39
Hydrochloride $C_{16}H_{26}N_2O_3 \cdot HCl$	330.85
Propoxyphene $C_{22}H_{29}NO_2$	339.48
Acetoxy Analog $C_{21}H_{27}NO_2$	325.45
Acetoxy Analog Hydrochloride $C_{21}H_{27}NO_2 \cdot HCl$	361.91
Acetoxy Analog Napsylate, anhydrous $C_{21}H_{27}NO_2 \cdot C_{10}H_8O_3S$	533.68
Acetoxy Analog Napsylate monohydrate $C_{21}H_{27}NO_2 \cdot C_{10}H_8O_3S \cdot H_2O$	551.70
Carbinol Hydrochloride $C_{19}H_{25}NO \cdot HCl$	319.87
Carbinol Napsylate, anhydrous $C_{19}H_{25}NO \cdot C_{10}H_8O_3S$	491.64
Carbinol Napsylate monohydrate $C_{19}H_{25}NO \cdot C_{10}H_8O_3S \cdot H_2O$	509.66
Hydrochloride $C_{22}H_{29}NO_2 \cdot HCl$	375.94
Napsylate, anhydrous $C_{22}H_{29}NO_2 \cdot C_{10}H_8O_3S$	547.71
Napsylate monohydrate $C_{22}H_{29}NO_2 \cdot C_{10}H_8O_3S \cdot H_2O$	565.72
Propranolol $C_{16}H_{21}NO_2$	259.35
Hydrochloride $C_{16}H_{21}NO_2 \cdot HCl$	295.81
Propyl Alcohol C_3H_8O	60.10
Gallate $C_{10}H_{12}O_5$	212.20
Propylene Carbonate $C_4H_6O_3$	102.09
Propylene Glycol $C_3H_8O_2$	76.09
Glycol Monopalmitate $C_{19}H_{38}O_3$	314.51
Glycol Monostearate $C_{21}H_{42}O_3$	342.56
Propylhexedrine $C_{10}H_{21}N$	155.28
Propyliodone $C_{10}H_{11}I_2NO_3$	447.01
Propylparaben $C_{10}H_{12}O_3$	180.20
Propylthiouracil $C_7H_{10}N_2OS$	170.23
Protirelin $C_{16}H_{22}N_6O_4$	362.39
Protriptyline $C_{19}H_{21}N$	263.38
Hydrochloride $C_{19}H_{21}N \cdot HCl$	299.84
Pseudoephedrine $C_{10}H_{15}NO$	165.23

Hydrochloride $C_{10}H_{15}NO.HCl$ 201.70
Sulfate $(C_{10}H_{15}NO)_2.H_2SO_4$ 428.54
Psilocin $C_{12}H_{16}N_2O$ 204.27
Psilocybin $C_{12}H_{17}N_2O_4P$ 284.25
Purine $C_5H_4N_4$ 120.11
Pyrantel $C_{11}H_{14}N_2S$ 206.30
Pamoate $C_{11}H_{14}N_2S.C_{23}H_{16}O_6$ 594.68
Pyrazinamide $C_5H_5N_3O$ 123.11
Pyrazoline $C_3H_6N_2$ 70.09
Pyrene $C_{16}H_{10}$ 202.26
Pyridine C_5H_5N 79.10
Pyridostigmine Bromide $C_9H_{13}BrN_2O_2$ 261.12
Pyridoxal Hydrochloride $C_8H_9NO_3.HCl$ 203.62
5-phosphate 4-$CHOC_5HN$-2-CH_3, 3-OH, 5-$CH_2PO_4H_2.H_2O$ 265.16
Pyridoxamine Dihydrochloride $C_8H_{12}N_2O_2.2HCl$ 241.12
Pyridoxine $C_8H_{11}NO_3$ 169.18
Hydrochloride $C_8H_{11}NO_3.HCl$ 205.64
Pyridylazonaphthol $C_{15}H_{11}N_3O$ 249.27
Pyrilamine $C_{17}H_{23}N_3O$ 285.39
Maleate $C_{17}H_{23}N_3O.C_4H_4O_4$ 401.46
Pyrimethamine $C_{12}H_{13}ClN_4$ 248.71
Pyrogallol $C_6H_6O_3$ 126.11
Pyrrocaine $C_{14}H_{20}N_2O$ 232.32
Hydrochloride $C_{14}H_{20}N_2O.HCl$ 268.79
Pyrrole C_4H_5N 67.09
Pyruvic Acid $C_3H_4O_3$ 88.06
Pyrvinium $C_{26}H_{28}N_3^+$ 382.53
Pamoate $C_{75}H_{70}N_6O_6$ 1151.41

Quinacrine $C_{23}H_{30}ClN_3O$ 399.96
Hydrochloride $C_{23}H_{30}ClN_3O.2HCl.2H_2O$ 508.91
Hydrochloride, anhydrous $C_{23}H_{30}ClN_3O.2HCl$ 472.88
Quinaldine Red $C_{21}H_{23}IN_2$ 430.33
Quinestrol $C_{25}H_{32}O_2$ 364.53
Quinethazone $C_{10}H_{12}ClN_3O_3S$ 289.74
Quinhydrone $C_{12}H_{10}O_4$ 218.21
Quinidine, anhydrous $C_{20}H_{24}N_2O_2$ 324.42
Gluconate $C_{20}H_{24}N_2O_2.C_6H_{12}O_7$ 520.58
Sulfate $(C_{20}H_{24}N_2O_2)_2.H_2SO_4.2H_2O$ 782.95
Sulfate, anhydrous $(C_{20}H_{24}N_2O_2)_2.H_2SO_4$ 746.92
Quinine $C_{20}H_{24}N_2O_2.3H_2O$ 378.47
anhydrous $C_{20}H_{24}N_2O_2$ 324.42
Dihydrochloride $C_{20}H_{24}N_2O_2.2HCl$ 397.34
Sulfate $(C_{20}H_{24}N_2O_2)_2.H_2SO_4.2H_2O$ 782.95
Sulfate, anhydrous $(C_{20}H_{24}N_2O_2)_2.H_2SO_4$ 746.92
Quininone $C_{20}H_{22}N_2O_2$ 322.41
Quinuclidinyl Benzilate $C_{21}H_{23}NO_3$ 337.42

Racemic Calcium Pantothenate (see *Calcium Pantothenate*)
Ranitidine Hydrochloride $C_{13}H_{22}N_4O_3S.HCl$ 350.87

Resazurin Sodium $C_{12}H_6NNaO_4$ 251.17
Rescinnamine $C_{35}H_{42}N_2O_9$ 634.72
Reserpine $C_{33}H_{40}N_2O_9$ 608.69
Resorcinol $C_6H_6O_2$ 110.11
Monoacetate $C_8H_8O_3$ 152.15
Retinol $C_{20}H_{30}O$ 286.46
Acetate $C_{22}H_{32}O_2$ 328.49
Palmitate $C_{36}H_{60}O_2$ 524.87
Rhodamine B $C_{28}H_{31}ClN_2O_3$ 479.02
Riboflavin $C_{17}H_{20}N_4O_6$ 376.37
5-Phosphate Sodium $C_{17}H_{20}N_4NaO_9P$ 478.33
Rifampin $C_{43}H_{58}N_4O_{12}$ 822.95
Ritodrine Hydrochloride $C_{17}H_{21}NO_3.HCl$ 323.82
Rolitetracycline $C_{27}H_{33}N_3O_8$ 527.57
Rosaniline Chloride $C_{20}H_{20}ClN_3$ 337.85
Hydrochloride (see *Rosaniline Chloride*)
Rose Bengal $C_{20}H_4Cl_4I_4O_5$ 973.63
Bengal Sodium $C_{20}H_2Cl_4I_4Na_2O_5$ 1017.64
Rotoxamine $C_{16}H_{19}ClN_2O$ 290.79
Tartrate $C_{16}H_{19}ClN_2O.C_4H_6O_6$ 440.88
Ruthenium Red $Ru_2(OH)_2Cl_4.7NH_3.3H_2O$ 551.32

Saccharin $C_7H_5NO_3S$ 183.18
Calcium $C_{14}H_8CaN_2O_6S_2.3\frac{1}{2}H_2O$ 467.48
Calcium, anhydrous $C_{14}H_8CaN_2O_6S_2$ 404.43
Sodium $C_7H_4NNaO_3S.2H_2O$ 241.19
Sodium, anhydrous $C_7H_4NNaO_3S$ 205.16
Salicylaldazine $C_{14}H_{12}N_2O_2$ 240.26
Salicylaldehyde $C_7H_6O_2$ 122.12
Salicylamide $C_7H_7NO_2$ 137.14
Salicylic Acid $C_7H_6O_3$ 138.12
Scopolamine $C_{17}H_{21}NO_4$ 303.36
Hydrobromide $C_{17}H_{21}NO_4.HBr.3H_2O$ 438.31
Hydrobromide, anhydrous $C_{17}H_{21}NO_4.HBr$ 384.27
Secobarbital $C_{12}H_{18}N_2O_3$ 238.29
Sodium $C_{12}H_{17}N_2NaO_3$ 260.27
Secretin (pig) $C_{130}H_{220}N_{44}O_{41}$ 3055.44
Selenious Acid H_2SeO_3 128.97
Selenium Sulfide SeS_2 143.08
Semicarbazide Hydrochloride $CH_5N_3O.HCl$ 111.53
Seractide $C_{207}H_{308}N_{56}O_{58}S$ 4541.11
Serine $C_3H_7NO_3$ 105.09
Silica (see *Silicon Dioxide*)
Silicon Carbide SiC 40.10
Dioxide SiO_2 60.08
Silver Bromide $AgBr$ 187.77
Chloride $AgCl$ 143.32
Diethyldithiocarbamate $C_5H_{10}AgNS_2$ 256.13
3,5-Diiodo-4-oxo-1(4*H*) pyridineacetate $C_7H_4AgI_2NO_3$ 511.79
Iodide AgI 234.77
Nitrate $AgNO_3$ 169.87
Oxide Ag_2O 231.74
Sulfadiazine $C_{10}H_9AgN_4O_2S$ 357.13
Sincalide $C_{49}H_{62}N_{10}O_{16}S_3$ 1143.27
β-Sitosterol $C_{29}H_{50}O$ 414.71

Sodium Acetate $C_2H_3NaO_2.3H_2O$ 136.08
Acetate, anhydrous $C_2H_3NaO_2$ 82.03
Acetyltryptophanate $C_{13}H_{13}N_2NaO_3$ 268.25
Alizarinsulfonate $C_{14}H_7NaO_7S.H_2O$ 360.27
Aminosalicylate $C_7H_6NNaO_3.2H_2O$ 211.15
Aminosalicylate, anhydrous $C_7H_6NNaO_3$ 175.12
Ammonium Phosphate $NaNH_4HPO_4.4H_2O$ 209.07
Ammonium Phosphate, anhydrous $NaNH_4HPO_4$ 137.01
Arsenate Dibasic, anhydrous Na_2HAsO_4 185.91
Arsenate Dibasic, heptahydrate $Na_2HAsO_4.7H_2O$ 312.01
Arsenite $NaAsO_2$ 129.91
Ascorbate $C_6H_7NaO_6$ 198.11
Benzoate $C_7H_5NaO_2$ 144.11
Bicarbonate $NaHCO_3$ 84.01
Biphenyl $C_{12}H_9Na$ 176.19
Bisulfite $NaHSO_3$ 104.06
Bitartrate $C_4H_5NaO_6.H_2O$ 190.08
Bitartrate, anhydrous $C_4H_5NaO_6$ 172.07
Borate $Na_2B_4O_7.10H_2O$ 381.37
Borate, anhydrous $Na_2B_4O_7$ 201.22
Borohydride $NaBH_4$ 37.83
Bromide $NaBr$ 102.89
Butabarbital (see *Butabarbital Sodium*)
Carbonate, anhydrous Na_2CO_3 105.99
Carbonate monohydrate $Na_2CO_3.H_2O$ 124.00
Cephalothin (see *Cephalothin Sodium*)
Chlorate $NaClO_3$ 106.44
Chloride $NaCl$ 58.44
Cholate $C_{24}H_{39}NaO_5$ 430.56
Chromate $Na_2CrO_4.4H_2O$ 234.03
Chromate, anhydrous Na_2CrO_4 161.97
Chromotropate $C_{10}H_6Na_2O_8S_2.2H_2O$ 400.28
Citrate $C_6H_5Na_3O_7.2H_2O$ 294.10
Citrate, anhydrous $C_6H_5Na_3O_7$ 258.07
Cloxacillin (see *Cloxacillin Sodium*)
Cobaltinitrite $Na_3Co(NO_2)_6$ 403.94
Colistimethate (see *Colistimethate Sodium*)
Cyanide $NaCN$ 49.01
Dehydroacetate $C_8H_7NaO_4$ 190.13
Dehydrocholate (see *Dehydrocholate Sodium*)
Desoxycholate $C_{24}H_{39}NaO_4$ 414.56
Diatrizoate (see *Diatrizoate Sodium*)
Dichromate $Na_2Cr_2O_7.2H_2O$ 298.00
Dichromate, anhydrous $Na_2Cr_2O_7$ 261.97
Diethyldithiocarbamate $C_5H_{10}NNaS_2.3H_2O$ 225.30
Diethyldithiocarbamate, anhydrous $C_5H_{10}NNaS_2$ 171.25
Dithionate $Na_2S_2O_6.2H_2O$ 242.16
Equilin Sulfate

$C_{18}H_{19}NaO_5S$ 370.39
Estrone Sulfate
$C_{18}H_{21}NaO_5S$ 372.41
Ethacrynate (see *Ethacrynate Sodium*)
Ferrocyanide
$Na_4Fe(CN)_6.10H_2O$ 484.06
Ferrocyanide, anhydrous
$Na_4Fe(CN)_6$ 303.91
Fluorescein (see *Fluorescein Sodium*)
Fluoride NaF 41.99
Folate $C_{19}H_{18}N_7NaO_6$ 463.38
Formaldehyde Sulfoxylate
CH_3NaO_3S 118.08
Gluconate $C_6H_{11}NaO_7$ 218.14
Glycocholate $C_{26}H_{42}NNaO_6$ .. 487.61
1-Heptanesulfonate
$C_7H_{15}NaO_3S$ 202.24
1-Hexanesulfonate
$C_6H_{13}NaO_3S$ 188.22
Hydrosulfite $Na_2S_2O_4$ 174.10
Hydroxide NaOH 40.00
Hypochlorite NaClO 74.44
Indigotindisulfonate
$C_{16}H_8N_2Na_2O_8S_2$ 466.35
Iodide NaI 149.89
Iodohippurate (see *Iodohippurate Sodium*)
Ipodate (see *Ipodate Sodium*)
Lactate $C_3H_5NaO_3$ 112.06
Lauryl Sulfate
$C_{12}H_{25}NaO_4S$ 288.38
Metabisulfite $Na_2S_2O_5$ 190.10
Metaperiodate $NaIO_4$ 213.89
Methicillin (see *Methicillin Sodium*)
Methoxide CH_3NaO 54.02
Molybdate $Na_2MoO_4.2H_2O$ .. 241.95
Molybdate, anhydrous
Na_2MoO_4 .... 205.92
Monofluorophosphate
Na_2PFO_3 .... 143.95
Nafcillin (see *Nafcillin Sodium*)
Nitrate $NaNO_3$ 84.99
Nitrite $NaNO_2$ 69.00
Nitroferricyanide (see *Sodium Nitroprusside*)
Nitroprusside
$Na_2Fe(CN)_5NO.2H_2O$ 297.95
Nitroprusside, anhydrous
$Na_2Fe(CN)_5NO$ 261.92
1-Octanesulfonate
$C_8H_{17}NaO_3S$ 216.27
Oxacillin (see *Oxacillin Sodium*)
Oxalate $C_2Na_2O_4$ 134.00
Palmitate $C_{16}H_{31}NaO_2$ 278.41
Pantothenate $C_9H_{16}NNaO_5$ 241.22
Penicillin G (see *Penicillin G Sodium*)
1-Pentanesulfonate $C_5H_{11}Na$-$O_3S.H_2O$ 192.21
1-Pentanesulfonate, anhydrous
$C_5H_{11}NaO_3S$ 174.19
Perchlorate $NaClO_4.H_2O$ 140.46
Perchlorate, anhydrous
$NaClO_4$ 122.44
p-Periodate $Na_3H_2IO_6$ 293.89
Peroxide Na_2O_2 .... 77.98
Phosphate, Dibasic
$Na_2HPO_4.7H_2O$ 268.07
Phosphate, Dibasic, anhydrous Na_2HPO_4 .... 141.96
Phosphate, Monobasic
$NaH_2PO_4.H_2O$ 137.99
Phosphate, Monobasic, anhydrous NaH_2PO_4 .... 119.98
Phosphate, Tribasic
$Na_3PO_4.12H_2O$ 380.12
Phosphate, Tribasic, anhydrous
Na_3PO_4 .... 163.94
Propionate, anhydrous
$C_3H_5NaO_2$ 96.06
Pyrophosphate
$Na_4P_2O_7.10H_2O$ 446.05
Pyrophosphate, anhydrous
$Na_4P_2O_7$ 265.90
Pyruvate $C_3H_3NaO_3$ 110.04
Salicylate $C_7H_5NaO_3$ 160.10
Selenite, anhydrous
Na_2SeO_3 .... 172.94
Stearate $C_{18}H_{35}NaO_2$ 306.46
Sulfate $Na_2SO_4.10H_2O$ 322.19
Sulfate, anhydrous Na_2SO_4 .. 142.04
Sulfide $Na_2S.9H_2O$ 240.18
Sulfide, anhydrous Na_2S 78.04
Sulfite, anhydrous Na_2SO_3 ... 126.04
Sulfobromophthalein (see *Sulfobromophthalein Sodium*)
Sulfophenylazochromotropate
$C_{16}H_9N_2Na_3O_{11}S_3.3H_2O$ 624.45
Sulfophenylazochromotropate, anhydrous
$C_{16}H_9N_2Na_3O_{11}S_3$ 570.40
Tartrate $C_4H_4Na_2O_6.2H_2O$ .. 230.08
Tartrate, anhydrous
$C_4H_4Na_2O_6$ 194.05
Taurocholate, anhydrous
$C_{26}H_{44}NNaO_7S$ 537.69
Tetraborate (see *Sodium Borate*)
Tetradecyl Sulfate
$C_{14}H_{29}NaO_4S$ 316.43
Tetraphenylboron
$C_{24}H_{20}BNa$ 342.22
Thioglycollate $C_2H_3NaO_2S$ ... 114.09
Thiosulfate $Na_2S_2O_3.5H_2O$ .. 248.17
Thiosulfate, anhydrous
$Na_2S_2O_3$ 158.10
p-Toluenesulfonchloramide (see *Chloramine T*)
(tri)Pentacyanoamino Ferrate [see *Trisodium Amine-pentacyanoferrate(3-)*]
3-(trimethylsilyl)-1-propane sulfonate $C_6H_{15}SiNaO_3S$ .. 218.32
Tungstate $Na_2WO_4.2H_2O$ ... 329.86
Tungstate, anhydrous
Na_2WO_4 .... 293.83
Somatropin
$C_{990}H_{1529}N_{263}O_{299}S_7$ 22123.99
Sorbic Acid $C_6H_8O_2$ 112.13
1,4-Sorbitan $C_6H_{12}O_5$ 164.16
Sorbitol $C_6H_{14}O_6$ 182.17
Spectinomycin $C_{14}H_{24}N_2O_7$ 332.35
Hydrochloride
$C_{14}H_{24}N_2O_7.2HCl.5H_2O$ ... 495.35
Hydrochloride, anhydrous
$C_{14}H_{24}N_2O_7.2HCl$ 405.27
Spironolactone $C_{24}H_{32}O_4S$ 416.57
Squalane $C_{30}H_{62}$ 422.82
Stannous Chloride $SnCl_2$.-$2H_2O$ 225.63
Chloride, anhydrous $SnCl_2$ ... 189.60
Fluoride SnF_2 .... 156.69
Stanozolol $C_{21}H_{32}N_2O$ 328.50
Stearic Acid $C_{18}H_{36}O_2$ 284.48
Stearyl Alcohol $C_{18}H_{38}O$ 270.50
Stibocaptate
$C_{12}H_6Na_6O_{12}S_6Sb_2$ 915.97
Stibogluconate Sodium
$C_{12}H_{17}Na_3O_{17}Sb_2.9H_2O$ 907.86
Sodium, anhydrous
$C_{12}H_{17}Na_3O_{17}Sb_2$ 745.73
Stibophen
$C_{12}H_4Na_5O_{16}S_4Sb.7H_2O$ 895.20
anhydrous
$C_{12}H_4Na_5O_{16}S_4Sb$ 769.09
Streptomycin $C_{21}H_{39}N_7O_{12}$ 581.58
Sulfate $(C_{21}H_{39}N_7O_{12})_2$.-$3H_2SO_4$ 1457.38
Strontium Acetate
$C_4H_6SrO_4.\frac{1}{2}H_2O$ 214.72
Acetate, anhydrous
$C_4H_6SrO_4$ 205.71
Hydroxide octahydrate
$Sr(OH)_2.8H_2O$ 265.76
Nitrate, anhydrous
$Sr(NO_3)_2$ 211.63
Strychnine $C_{21}H_{22}N_2O_2$ 334.42
Sulfate $(C_{21}H_{22}N_2O_2)_2$.-$H_2SO_4.5H_2O$ 856.98
Sulfate, anhydrous
$(C_{21}H_{22}N_2O_2)_2.H_2SO_4$ 766.91
Succinylcholine Chloride
$C_{14}H_{30}Cl_2N_2O_4.2H_2O$ 397.34
Chloride, anhydrous
$C_{14}H_{30}Cl_2N_2O_4$ 361.31
Sucrose $C_{12}H_{22}O_{11}$ 342.30
Octaacetate $C_{28}H_{38}O_{19}$ 678.60
Sudan IV $C_{24}H_{20}N_4O$ 380.45
Sterile Sulbactam Sodium
$C_8H_{10}NNaO_5S$ 255.22
Sulfabenzamide
$C_{13}H_{12}N_2O_3S$ 276.31
Sulfacetamide $C_8H_{10}N_2O_3S$ 214.24
Sodium $C_8H_9N_2NaO_3S$.-H_2O 254.24
Sodium, anhydrous
$C_8H_9N_2NaO_3S$ 236.22
Sulfacytine $C_{12}H_{14}N_4O_3S$ 294.33
Sulfadiazine $C_{10}H_{10}N_4O_2S$ 250.27
Sodium $C_{10}H_9N_4NaO_2S$ 272.26
Sulfadimethoxine
$C_{12}H_{14}N_4O_4S$ 310.33
Sulfadoxine $C_{12}H_{14}N_4O_4S$ 310.33
Sulfaethidole $C_{10}H_{12}N_4O_2S_2$ 284.35
Sulfamerazine $C_{11}H_{12}N_4O_2S$ ... 264.30
Sodium $C_{11}H_{11}N_4NaO_2S$ 286.28
Sulfamethazine $C_{12}H_{14}N_4O_2S$ .. 278.33
Sulfamethizole $C_9H_{10}N_4O_2S_2$ ... 270.32
Sulfamethoxazole
$C_{10}H_{11}N_3O_3S$ 253.28
N_4-glucoside $C_{16}H_{21}N_3O_8S$ ... 415.42
Sulfamic Acid HSO_3NH_2 .... 97.09
Sulfanilamide $C_6H_8N_2O_2S$ 172.20
Sulfanilic Acid
$C_6H_7NO_3S.H_2O$ 191.20
Acid, anhydrous $C_6H_7NO_3S$ .. 173.19
Sulfapyridine $C_{11}H_{11}N_3O_2S$ 249.29
Sulfasalazine $C_{18}H_{14}N_4O_5S$ 398.39
Sulfate $SO_4^{=}$ 96.06
Sulfathiazole $C_9H_9N_3O_2S_2$ 255.31
Sulfinpyrazone $C_{23}H_{20}N_2O_3S$ ... 404.48
Sulfisoxazole $C_{11}H_{13}N_3O_3S$ 267.30
Acetyl $C_{13}H_{15}N_3O_4S$ 309.34
Diolamine
$C_{11}H_{13}N_3O_3S.C_4H_{11}NO_2$ ... 372.44
Sulfobromophthalein
$C_{20}H_{10}Br_4O_{10}S_2$ 794.03
Sodium $C_{20}H_8Br_4Na_2O_{10}S_2$ ... 837.99
2(4-Sulfophenylazo)-1,8-dihydroxy-3,6-naphthalenedisulfonic Acid, Trisodium Salt
$C_{16}H_9N_2O_{11}S_3Na_3$ 570.40
Sulfosalicylic Acid
$C_7H_6O_6S.2H_2O$ 254.21
Acid, anhydrous $C_7H_6O_6S$ ... 218.18
Sulfoxone Sodium
$C_{14}H_{14}N_2Na_2O_6S_3$ 448.43
Sulfur Dioxide SO_2 .... 64.06

Compound	Mol. Wt.
Trioxide SO_3	80.06
Sulfuric Acid H_2SO_4	98.07
Sulfurous Acid H_2SO_3	82.07
Sulindac $C_{20}H_{17}FO_3S$	356.41
Sulisobenzone $C_{14}H_{12}O_6S$	308.31
Suramin Sodium $C_{51}H_{34}N_6Na_6O_{23}S_6$	1429.15
Syrosingopine $C_{35}H_{42}N_2O_{11}$	666.72
Talbutal $C_{11}H_{16}N_2O_3$	224.26
Tamoxifen $C_{26}H_{29}NO$	371.52
Citrate $C_{26}H_{29}NO.C_6H_8O_7$	563.65
Tartaric Acid $C_4H_6O_6$	150.09
Terbutaline $C_{12}H_{19}NO_3$	225.29
Sulfate $(C_{12}H_{19}NO_3)_2.H_2SO_4$	548.65
Terpin Hydrate $C_{10}H_{20}O_2.H_2O$	190.28
Hydrate, anhydrous $C_{10}H_{20}O_2$	172.27
Testolactone $C_{19}H_{24}O_3$	300.40
Testosterone $C_{19}H_{28}O_2$	288.43
Benzoate $C_{26}H_{32}O_2$	376.54
Cypionate $C_{27}H_{40}O_3$	412.61
Enanthate $C_{26}H_{40}O_3$	400.60
Propionate $C_{22}H_{32}O_3$	344.49
Propionate Semicarbazone $C_{23}H_{35}N_3O_3$	401.55
Tetrabromophenolphthalein, Ethyl Ester $C_{22}H_{14}Br_4O_4$	661.97
Tetrabromphenol Blue $C_{19}H_{10}Br_4O_5S$	669.96
Iodide $C_{16}H_{36}IN$	369.37
Phosphate $C_{16}H_{38}NO_4P$	339.46
Tetrabutylammonium Hydrogen Sulfate $C_{16}H_{37}NO_4S$	339.53
Tetracaine $C_{15}H_{24}N_2O_2$	264.37
Hydrochloride $C_{15}H_{24}N_2O_2.HCl$	300.83
Tetrachlorobenzoquinone (see *Chloranil*)	
2,3,7,8-Tetrachlorodibenzo-*p*-dioxin $C_{12}H_4Cl_4O_2$	321.97
Tetracosane $C_{24}H_{50}$	338.66
Tetracycline, anhydrous $C_{22}H_{24}N_2O_8$	444.44
trihydrate $C_{22}H_{24}N_2O_8.3H_2O$	498.49
Hydrochloride $C_{22}H_{24}N_2O_8.HCl$	480.90
Tetradecane $C_{14}H_{30}$	198.39
Tetraethylammonium Hydroxide $C_8H_{21}NO$	147.26
Perchlorate $C_8H_{20}ClNO_4$	229.70
Tetraethylene Glycol $C_8H_{18}O_5$	194.23
Tetraheptylammonium Bromide $(C_7H_{15})_4NBr$	490.17
Tetrahydrocannabinol $C_{21}H_{30}O_2$	314.47
Tetrahydrofuran C_4H_8O	72.11
1,2,3,4-Tetrahydro-2-[(isopropylamino)methyl]-7-nitro-6-methylquinoline $C_{14}H_{21}N_3O_2$	263.34
methanesulfonate (salt) $C_{14}H_{21}N_3O_2.CH_4O_3S$	359.44
1,2,3,4-Tetrahydro-2-[(isopropylamino)methyl]-5-nitro-6-quinolinemethanol $C_{14}H_{21}N_3O_3$	279.34
1,2,3,4-Tetrahydronaphthalene $C_{10}H_{12}$	132.21
Tetrahydrozoline $C_{13}H_{16}N_2$	200.28
Hydrochloride $C_{13}H_{16}N_2.HCl$	236.74
Tetraiodophenolphthalein (see *Iodophthalein*)	
Tetramethylammonium Bromide $C_4H_{12}BrN$	154.05
Chloride $C_4H_{12}ClN$	109.60
Hydroxide $C_4H_{13}NO$	91.15
Hydroxide, Pentahydrate $(CH_3)_4NOH.5H_2O$	181.23
Nitrate $(CH_3)_4NNO_3$	136.15
4,4′-Tetramethyldiaminodiphenylmethane $[(CH_3)_2NC_6H_4]_2CH_2$	254.38
Tetramethylpararosaniline Chloride $C_{23}H_{26}ClN_3$	379.93
Tetramethylsilane $C_4H_{12}Si$	88.22
Tetraphenylcyclopentadienone $C_{29}H_{20}O$	384.48
Tetrasodium Ethylenediaminetetraacetate $C_{10}H_{12}N_2Na_4O_8$	380.17
2,2′,6,6′-Tetra-*tert*-butyldiphenoquinone $C_{28}H_{40}O_2$	408.62
Thallous Chloride $TlCl$	239.82
Theobromine $C_7H_8N_4O_2$	180.17
Theophylline $C_7H_8N_4O_2.H_2O$	198.18
anhydrous $C_7H_8N_4O_2$	180.17
Sodium $C_7H_7N_4NaO_2$	202.15
Thiabendazole $C_{10}H_7N_3S$	201.25
Thiamine Hydrochloride $C_{12}H_{17}ClN_4OS.HCl$	337.27
Mononitrate $C_{12}H_{17}N_5O_4S$	327.36
Thiamylal Sodium $C_{12}H_{17}N_2NaO_2S$	276.33
Thiazole Yellow $C_{28}H_{19}N_5Na_2O_6S_4$	695.71
Thiethylperazine $C_{22}H_{29}N_3S_2$	399.61
Maleate $C_{22}H_{29}N_3S_2.2C_4H_4O_4$	631.76
Thihexinol Methylbromide $C_{18}H_{26}BrNOS_2$	416.43
Thimerosal $C_9H_9HgNaO_2S$	404.81
Thiobarbituric Acid $C_4H_4N_2O_2S$	144.15
2,2′-Thiodiethanol $(HOCH_2CH_2)_2S$	122.18
Thioglycolic Acid $C_2H_4O_2S$	92.11
Thioguanine $C_5H_5N_5S.½H_2O$	176.20
anhydrous $C_5H_5N_5S$	167.19
Thiopental $C_{11}H_{18}N_2O_2S$	242.34
Sodium $C_{11}H_{17}N_2NaO_2S$	264.32
Thiopropazate $C_{23}H_{28}ClN_3O_2S$	446.01
Hydrochloride $C_{23}H_{28}ClN_3O_2S.2HCl$	518.93
Thioridazine $C_{21}H_{26}N_2S_2$	370.57
Hydrochloride $C_{21}H_{26}N_2S_2.HCl$	407.03
Thiotepa $C_6H_{12}N_3PS$	189.21
Thiothixene $C_{23}H_{29}N_3O_2S_2$	443.62
Hydrochloride $C_{23}H_{29}N_3O_2S_2.2HCl.2H_2O$	552.57
Hydrochloride, anhydrous $C_{23}H_{29}N_3O_2S_2.2HCl$	516.54
Thiourea CH_4N_2S	76.12
Thonzonium Bromide $C_{32}H_{55}BrN_4O$	591.72
Thorium Nitrate $Th(NO_3)_4.4H_2O$	552.12
Nitrate, anhydrous $Th(NO_3)_4$	480.06
Threonine $C_4H_9NO_3$	119.12
Thymol $C_{10}H_{14}O$	150.22
Ticarcillin $C_{15}H_{16}N_2O_6S_2$	384.42
Disodium $C_{15}H_{14}N_2Na_2O_6S_2$	428.38
Monosodium Monohydrate $C_{15}H_{15}N_2NaO_6S_2.H_2O$	424.42
Timolol $C_{13}H_{24}N_4O_3S$	316.42
Maleate $C_{13}H_{24}N_4O_3S.C_4H_4O_4$	432.49
Tin Compounds (see under *Stannic* and *Stannous*)	
Tioconazole $C_{16}H_{13}Cl_3N_2OS$	387.71
Titanium Dioxide TiO_2	79.88
Tetrachloride $TiCl_4$	189.69
Trichloride $TiCl_3$	154.24
Tobramycin $C_{18}H_{37}N_5O_9$	467.52
Sulfate, Sterile $(C_{18}H_{37}N_5O_9)_2.5H_2SO_4$	1425.39
Tocainide Hydrochloride $C_{11}H_{16}N_2O.HCl$	228.72
Tocopherol (see *Alpha Tocopherol*)	
Tolazamide $C_{14}H_{21}N_3O_3S$	311.40
Tolazoline $C_{10}H_{12}N_2$	160.22
Hydrochloride $C_{10}H_{12}N_2.HCl$	196.68
Tolbutamide $C_{12}H_{18}N_2O_3S$	270.35
Tolidine $C_{14}H_{16}N_2$	212.29
Tolmetin $C_{15}H_{15}NO_3$	257.29
Sodium $C_{15}H_{14}NNaO_3.2H_2O$	315.30
Sodium, anhydrous $C_{15}H_{14}NNaO_3$	279.27
Tolnaftate $C_{19}H_{17}NOS$	307.41
Tolualdehyde C_8H_8O	120.15
Toluene C_7H_8	92.14
Toluenesulfonamide $C_7H_9NO_2S$	171.21
Toluenesulfonic Acid $C_7H_8O_3S.H_2O$	190.21
p-Toluenesulfonyl-L-arginine Methyl Ester Hydrochloride $C_{14}H_{22}N_4O_4S.HCl$	378.87
p-Toluic Acid $C_8H_8O_2$	136.15
Toluidine C_7H_9N	107.15
Tretinoin $C_{20}H_{28}O_2$	300.44
Triacetin $C_9H_{14}O_6$	218.21
Triacetylepinephrine $C_{15}H_{19}NO_6$	309.32
Triacetyl-levarterenol $C_{14}H_{17}NO_6$	295.29
n-Triacontane $C_{30}H_{62}$	422.82
Triamcinolone $C_{21}H_{27}FO_6$	394.44
Acetonide $C_{24}H_{31}FO_6$	434.50
Diacetate $C_{25}H_{31}FO_8$	478.51
Hexacetonide $C_{30}H_{41}FO_7$	532.65
Triamterene $C_{12}H_{11}N_7$	253.27
Tributyl Phosphate $C_{12}H_{27}O_4P$	266.32
Tributylethylammonium Hydroxide $C_{14}H_{33}NO$	231.42
Tributyrin $C_{15}H_{26}O_6$	302.37
Trichlormethiazide $C_8H_8Cl_3N_3O_4S_2$	380.65
Trichloroethane $C_2H_3Cl_3$	133.40
Trichloroethylene C_2HCl_3	131.39
Trichloromonofluoromethane CCl_3F	137.37
Trichlorotrifluoroethane $C_2Cl_3F_3$	187.38
Triclobisonium Chloride $C_{36}H_{74}Cl_2N_2$	605.90
Tricosane $C_{23}H_{48}$	324.63
Tridihexethyl Chloride $C_{21}H_{36}ClNO$	353.97
Trientine Hydrochloride $C_6H_{18}N_4.2HCl$	219.16
Triethanolamine (see *Trolamine*)	
Triethyl Citrate $C_{12}H_{20}O_7$	276.29
Triethylamine $C_6H_{15}N$	101.19
Triethylene Glycol $C_6H_{14}O_4$	150.17

Triethylenemelamine $C_9H_{12}N_6$ .. 204.23
Trifluoperazine $C_{21}H_{24}F_3N_3S$ ... 407.50
Hydrochloride $C_{21}H_{24}F_3N_3S.2HCl$ 480.42
Trifluoroacetic Anhydride $(F_3CCO)_2O$ 210.03
2-[*N*-(2,2,2,-Trifluoroethyl)amino]-5-chlorobenzophenone $C_{15}H_{11}ClF_3NO$ 313.71
Triflupromazine $C_{18}H_{19}F_3N_2S$ 352.42
Hydrochloride $C_{18}H_{19}F_3N_2S.HCl$ 388.88
Trihexyphenidyl $C_{20}H_{31}NO$ 301.47
Hydrochloride $C_{20}H_{31}NO.HCl$ 337.93
Trihydroxyphenylalanine $C_9H_{11}NO_5$ 213.19
Triiodothyronine $C_{15}H_{12}I_3NO_4$ .. 650.98
Triisopropanolamine $C_9H_{21}NO_3$ 191.27
Triketohydrindene Hydrate $C_9H_4O_3.H_2O$ 178.14
Trimeprazine $C_{18}H_{22}N_2S$ 298.45
Tartrate $(C_{18}H_{22}N_2S)_2.C_4H_6O_6$ 746.98
Trimercaptopropane $C_3H_8S_3$ 140.28
Trimethadione $C_6H_9NO_3$ 143.14
Trimethaphan Camsylate $C_{32}H_{40}N_2O_5S_2$ 596.80
Trimethobenzamide $C_{21}H_{28}N_2O_5$ 388.46
Hydrochloride $C_{21}H_{28}N_2O_5.HCl$ 424.92
Trimethoprim $C_{14}H_{18}N_4O_3$ 290.32
Trimethylacethydrazide Ammonium Chloride [see (*Carboxymethyl*)*trimethylammonium Chloride, hydrazide*]
Trimethylpentane C_8H_{18} 114.23
2,4,6-Trimethylpyridine $C_8H_{11}N$ 121.18
N-(Trimethylsilyl)-imidazole $C_6H_{12}N_2Si$ 140.26
Trinitrophenol $C_6H_3N_3O_7$ 229.11
Trioctylphosphine Oxide $C_{24}H_{51}OP$ 386.64
Trioxsalen $C_{14}H_{12}O_3$ 228.25
Tripelennamine $C_{16}H_{21}N_3$ 255.36
Citrate $C_{16}H_{21}N_3.C_6H_8O_7$ ... 447.49
Hydrochloride $C_{16}H_{21}N_3.HCl$ 291.82
Triphenylchloromethane $C_{19}H_{15}Cl$ 278.78
Triphenyltetrazolium Chloride $C_{19}H_{15}ClN_4$ 334.81
Triprolidine $C_{19}H_{22}N_2$ 278.40
Hydrochloride $C_{19}H_{22}N_2.HCl.H_2O$ 332.87
Hydrochloride, anhydrous $C_{19}H_{22}N_2.HCl$ 314.86
Tris(2-aminoethyl)amine $C_6H_{18}N_4$ 146.24
Tris(hydroxymethyl)aminomethane $C_4H_{11}NO_3$ 121.14
Trisodium Aminepentacyanoferrate(3−) $Na_3[Fe(CN)_5NH_3]$ 271.94
Trolamine $C_6H_{15}NO_3$ 149.19
Stearate $C_6H_{15}NO_3.C_{18}H_{36}O_2$ 433.67
Troleandomycin $C_{41}H_{67}NO_{15}$ ... 813.98
Tromethamine $C_4H_{11}NO_3$ 121.14
Tropaeolin OO $C_{18}H_{14}N_3NaO_3S$ 375.38
Tropicamide $C_{17}H_{20}N_2O_2$ 284.36
Tryparsamide $C_8H_{10}AsN_2NaO_4.½H_2O$ 305.10
anhydrous $C_8H_{10}AsN_2NaO_4$ .. 296.09
Tryptophan $C_{11}H_{12}N_2O_2$ 204.23
Tuaminoheptane $C_7H_{17}N$ 115.22
Tubocurarine Chloride $C_{37}H_{41}ClN_2O_6.HCl.5H_2O$ .. 771.73
Chloride, anhydrous $C_{37}H_{41}ClN_2O_6.HCl$ 681.65
Tybamate $C_{13}H_{26}N_2O_4$ 274.36
Tyropanoate Sodium $C_{15}H_{17}I_3NNaO_3$ 663.01
Tyrosine $C_9H_{11}NO_3$ 181.19

Undecylenic Acid $C_{11}H_{20}O_2$ 184.28
Uracil $C_4H_4N_2O_2$ 112.09
Mustard $C_8H_{11}Cl_2N_3O_2$ 252.10
Uranyl Acetate $C_4H_6O_6U.2H_2O$ 424.15
Acetate, anhydrous $C_4H_6O_6U$ 388.12
Urea CH_4N_2O 60.06
Urethan $C_3H_7NO_2$ 89.09
Urethane $C_3H_7NO_2$ 89.09

Valeric Acid $C_5H_{10}O_2$ 102.13
Valerophenone $C_{11}H_{14}O$ 162.23
Valethamate Bromide $C_{19}H_{32}BrNO_2$ 386.37
Valine $C_5H_{11}NO_2$ 117.15
Valproic Acid $C_8H_{16}O_2$ 144.21
Vanadium Pentoxide V_2O_5 ..... 181.88
Vanadyl Sulfate, anhydrous $VOSO_4$ 163.00
Vancomycin Hydrochloride $C_{66}H_{75}Cl_2N_9O_{24}.HCl$ 1485.73
Vanillin $C_8H_8O_3$ 152.15
Verapamil Hydrochloride $C_{27}H_{38}N_2O_4.HCl$ 491.07
Vidarabine $C_{10}H_{13}N_5O_4.H_2O$ .. 285.26
anhydrous $C_{10}H_{13}N_5O_4$ 267.24
Vinbarbital $C_{11}H_{16}N_2O_3$ 224.26
Sodium $C_{11}H_{15}N_2NaO_3$ 246.24
Vinblastine $C_{46}H_{58}N_4O_9$ 810.99
Sulfate $C_{46}H_{58}N_4O_9.H_2SO_4$ .. 909.06
Vincristine $C_{46}H_{56}N_4O_{10}$ 824.97
Sulfate $C_{46}H_{56}N_4O_{10}.H_2SO_4$ 923.04
Vinylpyrrolidinone C_6H_9NO 111.14
Viomycin $C_{25}H_{43}N_{13}O_{10}$ 685.70
Vitamin A Acetate (see *Retinol Acetate*)
Alcohol (see *Retinol*)

Warfarin $C_{19}H_{16}O_4$ 308.33
Sodium $C_{19}H_{15}NaO_4$ 330.31
Water H_2O 18.02

Xanthanoic Acid $C_{14}H_{10}O_3$ 226.23
Xanthanone (Xanthone) $C_{13}H_8O_2$ 196.21
Xanthine $C_5H_4N_4O_2$ 152.11
Xanthydrol $C_{13}H_{10}O_2$ 198.22
Xylene (Xylol) C_8H_{10} 106.17
Cyanole $C_{25}H_{27}N_2NaO_6S_2$ ... 538.61
Xylene Cyanol FF $C_{25}H_{27}N_2NaO_6S_2$ 538.61
Xylenol Orange $C_{31}H_{28}N_2Na_4O_{13}S$ 760.59
Xylometazoline $C_{16}H_{24}N_2$ 244.38
Hydrochloride $C_{16}H_{24}N_2.HCl$ 280.84
Xylose $C_5H_{10}O_5$ 150.13

Zinc Acetate $C_4H_6O_4Zn.2H_2O$ 219.50
Acetate, anhydrous $C_4H_6O_4Zn$ 183.47
Carbonate (approx.) $5ZnO.2CO_2.4H_2O$ 566.98
Chloride $ZnCl_2$ 136.29
Gluconate $C_{12}H_{22}O_{14}Zn$ 455.68
Hydroxide $Zn(OH)_2$ 99.39
Hydroxyquinolinate $C_{18}H_{12}N_2O_2Zn$ 353.68
Oxide ZnO 81.38
Palmitate $C_{32}H_{62}O_4Zn$ 576.22
Phosphate $Zn_3(PO_4)_2$ 386.08
Stearate $C_{36}H_{70}O_4Zn$ 632.33
Sulfate $ZnSO_4.7H_2O$ 287.54
Sulfate, anhydrous $ZnSO_4$ ... 161.44
Sulfide ZnS 97.44
Undecylenate $C_{22}H_{38}O_4Zn$ ... 431.92
Zirconium Oxychloride $ZrOCl_2.8H_2O$ 322.25
Oxychloride, anhydrous $ZrOCl_2$ 178.13
Zirconyl Chloride (see *Zirconium Oxychloride*)
Nitrate anhydrous $ZrO(NO_3)_2$ 231.23

ALCOHOLOMETRIC TABLE

Based on data appearing in the National Bureau of Standards Bulletin, vol. 9, pp. 424–5 (publication of the National Institute of Standards and Technology).

Percentage of C_2H_5OH		Specific gravity in air		Percentage of C_2H_5OH		Specific gravity in air	
By volume at 15.56° C (1)	By weight (2)	at $\frac{25°}{25°}$ (3)	at $\frac{15.56°}{15.56°}$ (4)	By weight (5)	By volume at 15.56°C (6)	at $\frac{25°}{25°}$ (7)	at $\frac{15.56°}{15.56°}$ (8)
0	0.00	1.0000	1.0000	0	0.00	1.0000	1.0000
1	0.80	0.9985	0.9985	1	1.26	0.9981	0.9981
2	1.59	0.9970	0.9970	2	2.51	0.9963	0.9963
3	2.39	0.9956	0.9956	3	3.76	0.9945	0.9945
4	3.19	0.9941	0.9942	4	5.00	0.9927	0.9928
5	4.00	0.9927	0.9928	5	6.24	0.9911	0.9912
6	4.80	0.9914	0.9915	6	7.48	0.9894	0.9896
7	5.61	0.9901	0.9902	7	8.71	0.9879	0.9881
8	6.42	0.9888	0.9890	8	9.94	0.9863	0.9867
9	7.23	0.9875	0.9878	9	11.17	0.0948	0.9852
10	8.05	0.9862	0.9866	10	12.39	0.9833	0.9839
11	8.86	0.9850	0.9854	11	13.61	0.9818	0.9825
12	9.68	0.9838	0.9843	12	14.83	0.9804	0.9812
13	10.50	0.9826	0.9832	13	16.05	0.9789	0.9799
14	11.32	0.9814	0.9821	14	17.26	0.9776	0.9787
15	12.14	0.9802	0.9810	15	18.47	0.9762	0.9774
16	12.96	0.9790	0.9800	16	19.68	0.9748	0.9763
17	13.79	0.9778	0.9789	17	20.88	0.9734	0.9751
18	14.61	0.9767	0.9779	18	22.08	0.9720	0.9738
19	15.44	0.9756	0.9769	19	23.28	0.9706	0.9726
20	16.27	0.9744	0.9759	20	24.47	0.9692	0.9714
21	17.10	0.9733	0.9749	21	25.66	0.9677	0.9701
22	17.93	0.9721	0.9739	22	26.85	0.9663	0.9688
23	18.77	0.9710	0.9729	23	28.03	0.9648	0.9675
24	19.60	0.9698	0.9719	24	29.21	0.9633	0.9662
25	20.44	0.9685	0.9708	25	30.39	0.9617	0.9648
26	21.29	0.9673	0.9697	26	31.56	0.9601	0.9653
27	22.13	0.9661	0.9687	27	32.72	0.9585	0.9620
28	22.97	0.9648	0.9676	28	33.88	0.9568	0.9605
29	23.82	0.9635	0.9664	29	35.03	0.9551	0.9590
30	24.67	0.9622	0.9653	30	36.18	0.9534	0.9574
31	25.52	0.9609	0.9641	31	37.32	0.9516	0.9558
32	26.38	0.9595	0.9629	32	38.46	0.9498	0.9541
33	27.24	0.9581	0.9617	33	39.59	0.9480	0.9524
34	28.10	0.9567	0.9604	34	40.72	0.9561	0.9506
35	28.97	0.9552	0.9590	35	41.83	0.9442	0.9488
36	29.84	0.9537	0.9576	36	42.94	0.9422	0.9470
37	30.72	0.9521	0.9562	37	44.05	0.9402	0.9451
38	31.60	0.9506	0.9548	38	45.15	0.9382	0.9432
39	32.48	0.9489	0.9533	39	46.24	0.9362	0.9412
40	33.36	0.9473	0.9517	40	47.33	0.9341	0.9392
41	34.25	0.9456	0.9501	41	48.41	0.9320	0.9372
42	35.15	0.9439	0.9485	42	49.48	0.9299	0.9352
43	36.05	0.9421	0.9469	43	50.55	0.9278	0.9331
44	36.96	0.9403	0.9452	44	51.61	0.9256	0.9310
45	37.87	0.9385	0.9434	45	52.66	0.9235	0.9289
46	38.78	0.9366	0.9417	46	53.71	0.9213	0.9268
47	39.70	0.9348	0.9399	47	54.75	0.9191	0.9246
48	40.62	0.9328	0.9380	48	55.78	0.9169	0.9225
49	41.55	0.9309	0.9361	49	56.81	0.9147	0.9203
50	42.49	0.9289	0.9342	50	57.83	0.9124	0.9181
51	43.43	0.9269	0.9322	51	58.84	0.9102	0.9159
52	44.37	0.9248	0.9302	52	59.85	0.9079	0.9137
53	45.33	0.9228	0.9282	53	60.85	0.9056	0.9114
54	46.28	0.9207	0.9262	54	61.85	0.9033	0.9092
55	47.25	0.9185	0.9241	55	62.84	0.9010	0.9069
56	48.21	0.9164	0.9220	56	63.82	0.8987	0.9046
57	49.19	0.9142	0.9199	57	64.80	0.8964	0.9024
58	50.17	0.9120	0.9177	58	65.77	0.8941	0.9001
59	51.15	0.9098	0.9155	59	66.73	0.8918	0.8978
60	52.15	0.9076	0.9133	60	67.79	0.8895	0.8955
61	53.15	0.9053	0.9111	61	68.64	0.8871	0.8932
62	54.15	0.9030	0.9088	62	69.59	0.8848	0.8909
63	55.17	0.9006	0.9065	63	70.52	0.8824	0.8886
64	56.18	0.8983	0.9042	64	71.46	0.8801	0.8862
65	57.21	0.8959	0.9019	65	72.38	0.8777	0.8839
66	58.24	0.8936	0.8995	66	73.30	0.8753	0.8815
67	59.28	0.8911	0.8972	67	74.21	0.8729	0.8792
68	60.33	0.8887	0.8948	68	75.12	0.8706	0.8768
69	61.38	0.8862	0.8923	69	76.02	0.8682	0.8745
70	62.44	0.8837	0.8899	70	76.91	0.8658	0.8721
71	63.51	0.8812	0.8874	71	77.79	0.8634	0.8697
72	64.59	0.8787	0.8848	72	78.67	0.8609	0.8673
73	65.67	0.8761	0.8823	73	79.54	0.8585	0.8649
74	66.77	0.8735	0.8797	74	80.41	0.8561	0.8625
75	67.87	0.8709	0.8771	75	81.27	0.8537	0.8601
76	68.98	0.8682	0.8745	76	82.12	0.8512	0.8576
77	70.10	0.8655	0.8718	77	82.97	0.8488	0.8552
78	71.23	0.8628	0.8691	78	83.81	0.8463	0.8528
79	72.38	0.8600	0.8664	79	84.64	0.8439	0.8503
80	73.53	0.8572	0.8636	80	85.46	0.8414	0.8479
81	74.69	0.8544	0.8608	81	86.28	0.8389	0.8454
82	75.86	0.8516	0.8580	82	87.08	0.8364	0.8429
83	77.04	0.8487	0.8551	83	87.89	0.8339	0.8404
84	78.23	0.8458	0.8522	84	88.68	0.8314	0.8379
85	79.44	0.8428	0.8493	85	89.46	0.8288	0.8354
86	80.66	0.8397	0.8462	86	90.24	0.8263	0.8328
87	81.90	0.8367	0.8432	87	91.01	0.8237	0.8303
88	83.14	0.8335	0.8401	88	91.77	0.8211	0.8276
89	84.41	0.8303	0.8369	89	92.52	0.8184	0.8250
90	85.69	0.8271	0.8336	90	93.25	0.8158	0.8224
91	86.99	0.8237	0.8303	91	93.98	0.8131	0.8197
92	88.31	0.8202	0.8268	92	94.70	0.8104	0.8170
93	89.65	0.8167	0.8233	93	95.41	0.8076	0.8142
94	91.03	0.8130	0.8196	94	96.10	0.8048	0.8114
95	92.42	0.8092	0.8158	95	96.79	0.8020	0.8086
96	93.85	0.8053	0.8118	96	97.46	0.7992	0.8057
97	95.32	0.8011	0.8077	97	98.12	0.7962	0.8028
98	96.82	0.7968	0.8033	98	98.76	0.7932	0.7988
99	98.38	0.7921	0.7986	99	99.39	0.7902	0.7967
100	100.00	0.7871	0.7936	100	100.00	0.7871	0.7936

THERMOMETRIC EQUIVALENTS

Fahrenheit to Centigrade (Celsius) Scale

$$(°F - 32) \times \frac{5}{9} = °C$$

°F	°C	°F	°C	°F	°C	°F	°C	°F	°C
0	−17.78	51	10.56	101	38.33	151	66.11	201	93.89
1	−17.22	52	11.11	102	38.89	152	66.67	202	94.44
2	−16.67	53	11.67	103	39.44	153	67.22	203	95.
3	−16.11	54	12.22	104	40.	154	67.78	204	95.56
4	−15.56	55	12.78	105	40.56	155	68.33	205	96.11
5	−15.	56	13.33	106	41.11	156	68.89	206	96.67
6	−14.44	57	13.89	107	41.67	157	69.44	207	97.22
7	−13.89	58	14.44	108	42.22	158	70.	208	97.78
8	−13.33	59	15.	109	42.78	159	70.56	209	98.33
9	−12.78	60	15.56	110	43.33	160	71.11	210	98.89
10	−12.22	61	16.11	111	43.89	161	71.67	211	99.44
11	−11.67	62	16.67	112	44.44	162	72.22	212	100.
12	−11.11	63	17.22	113	45.	163	72.78	213	100.56
13	−10.56	64	17.78	114	45.56	164	73.33	214	101.11
14	−10.	65	18.33	115	46.11	165	73.89	215	101.67
15	−9.44	66	18.89	116	46.67	166	74.44	216	102.22
16	−8.89	67	19.44	117	47.22	167	75.	217	102.78
17	−8.33	68	20.	118	47.78	168	75.56	218	103.33
18	−7.78	69	20.56	119	48.33	169	76.11	219	103.89
19	−7.22	70	21.11	120	48.89	170	76.67	220	104.44
20	−6.67	71	21.67	121	49.44	171	77.22	221	105.
21	−6.11	72	22.22	122	50.	172	77.78	222	105.56
22	−5.56	73	22.78	123	50.56	173	78.33	223	106.11
23	−5.	74	23.33	124	51.11	174	78.89	224	106.67
24	−4.44	75	23.89	125	51.67	175	79.44	225	107.22
25	−3.89	76	24.44	126	52.22	176	80.	226	107.78
26	−3.33	77	25.	127	52.78	177	80.56	227	108.33
27	−2.78	78	25.56	128	53.33	178	81.11	228	108.89
28	−2.22	79	26.11	129	53.89	179	81.67	229	109.44
29	−1.67	80	26.67	130	54.44	180	82.22	230	110.
30	−1.11	81	27.22	131	55.	181	82.78	231	110.56
31	−0.56	82	27.78	132	55.56	182	83.33	232	111.11
32	0.	83	28.33	133	56.11	183	83.89	233	111.67
33	0.56	84	28.89	134	56.67	184	84.44	234	112.22
34	1.11	85	29.44	135	57.22	185	85.	235	112.78
35	1.67	86	30.	136	57.78	186	85.56	236	113.33
36	2.22	87	30.56	137	58.33	187	86.11	237	113.89
37	2.78	88	31.11	138	58.89	188	86.67	238	114.44
38	3.33	89	31.67	139	59.44	189	87.22	239	115.
39	3.89	90	32.22	140	60.	190	87.78	240	115.56
40	4.44	91	32.78	141	60.56	191	88.33	241	116.11
41	5.	92	33.33	142	61.11	192	88.89	242	116.67
42	5.56	93	33.89	143	61.67	193	89.44	243	117.22
43	6.11	94	34.44	144	62.22	194	90.	244	117.78
44	6.67	95	35.	145	62.78	195	90.56	245	118.33
45	7.22	96	35.56	146	63.33	196	91.11	246	118.89
46	7.78	97	36.11	147	63.89	197	91.67	247	119.44
47	8.33	98	36.67	148	64.44	198	92.22	248	120.
48	8.89	99	37.22	149	65.	199	92.78	249	120.56
49	9.44	100	37.78	150	65.56	200	93.33	250	121.11
50	10.								

Centigrade (Celsius) to Fahrenheit Scale

$$\frac{9}{5}°C + 32 = °F$$

°C	°F	°C	°F	°C	°F	°C	°F	°C	°F
−20	−4.0	21	69.8	61	141.8	101	213.8	141	285.8
−19	−2.2	22	71.6	62	143.6	102	215.6	142	287.6
−18	−0.4	23	73.4	63	145.4	103	217.4	143	289.4
−17	1.4	24	75.2	64	147.2	104	219.2	144	291.2
−16	3.2	25	77.	65	149.	105	221.	145	293.
−15	5.	26	78.8	66	150.8	106	222.8	146	294.8
−14	6.8	27	80.6	67	152.6	107	224.6	147	296.6
−13	8.6	28	82.4	68	154.4	108	226.4	148	298.4
−12	10.4	29	84.2	69	156.2	109	228.2	149	300.2
−11	12.2	30	86.	70	158.	110	230.	150	302.
−10	14.	31	87.8	71	159.8	111	231.8	151	303.8
−9	15.8	32	89.6	72	161.6	112	233.6	152	305.6
−8	17.6	33	91.4	73	163.4	113	235.4	153	307.4
−7	19.4	34	93.2	74	165.2	114	237.2	154	309.2
−6	21.2	35	95.	75	167.	115	239.	155	311.
−5	23.	36	96.8	76	168.8	116	240.8	156	312.8
−4	24.8	37	98.6	77	170.6	117	242.6	157	314.6
−3	26.6	38	100.4	78	172.4	118	244.4	158	316.4
−2	28.4	39	102.2	79	174.2	119	246.2	159	318.2
−1	30.2	40	104.	80	176.	120	248.	160	320.
0	32.	41	105.8	81	177.8	121	249.8	161	321.8
1	33.8	42	107.6	82	179.6	122	251.6	162	323.6
2	35.6	43	109.4	83	181.4	123	253.4	163	325.4
3	37.4	44	111.2	84	183.2	124	255.2	164	327.2
4	39.2	45	113.	85	185.	125	257.	165	329.
5	41.	46	114.8	86	186.8	126	258.8	166	330.8
6	42.8	47	116.6	87	188.6	127	260.6	167	332.6
7	44.6	48	118.4	88	190.4	128	262.4	168	334.4
8	46.4	49	120.2	89	192.2	129	264.2	169	336.2
9	48.2	50	122.	90	194.	130	266.	170	338.
10	50.	51	123.8	91	195.8	131	267.8	171	339.8
11	51.8	52	125.6	92	197.6	132	269.6	172	341.6
12	53.6	53	127.4	93	199.4	133	271.4	173	343.4
13	55.4	54	129.2	94	201.2	134	273.2	174	345.2
14	57.2	55	131.	95	203.	135	275.	175	347.
15	59.	56	132.8	96	204.8	136	276.8	176	348.8
16	60.8	57	134.6	97	206.6	137	278.6	177	350.6
17	62.6	58	136.4	98	208.4	138	280.4	178	352.4
18	64.4	59	138.2	99	210.2	139	282.2	179	354.2
19	66.2	60	140.	100	212.	140	284.	180	356.
20	68.								

EQUIVALENTS OF WEIGHTS AND MEASURES

Metric, Avoirdupois, and Apothecaries

NOTE—These values are for water at the temperature of 4°C (39.2°F) in vacuum. For practical purposes the values may be used without correction.

This table of exact equivalents should not be confused with the table of *approximate* dose equivalents which appears on the inside of the back cover. The latter table is provided only as a convenience to physicians for prescribing.

For the conversion of specific quantities in pharmaceutical formulas, use the exact equivalents. For prescription compounding, use the exact equivalents rounded to three significant figures.

Weights					**Metric Equivalents g or mL**	**Measures**		
Apothecaries		Avoirdupois				Fluid		Decimal
oz.	grains	lb	oz	grains		ounces	minims	Equivalent
32	72.4	2	3	119.9	**1000**	33	391.1	33.815
30	204.1	2	1	166.6	946.333	**32**	....	32
29	80.0	**2**	..	....	907.185	30	324.6	30.676
16	36.2	1	1	278.7	**500**	16	435.6	16.907
15	102.1	1	..	302.1	473.167	**16**	....	16
14	280.0	**1**	..	....	453.592	15	162.3	15.338
12	....	..	13	72.5	373.242	12	298.1	12.621
8	....	..	8	340.0	248.828	8	198.7	8.414
7	291.0	..	8	151.0	236.583	8	....	8
7	140.0	..	8	....	226.796	7	321.1	7.669
6	206.5	..	7	24.0	**200**	6	366.2	6.763
4	....	..	4	170.0	124.414	4	99.4	4.207
3	385.5	..	4	75.5	118.292	**4**	....	4
3	310.0	..	**4**	....	113.398	3	400.6	3.835
3	103.2	..	3	230.7	**100**	3	183.1	3.381
2	....	..	2	85.0	62.207	2	49.7	2.104
1	432.8	..	2	37.8	59.146	**2**	....	2
1	395.0	..	**2**	....	56.699	1	440.3	1.917
1	291.6	..	1	334.1	**50**	1	331.5	1.691
1	....	..	1	42.5	31.1035	1	24.9	1.052
..	456.380	..	1	18.88	29.5729	**1**	....	1
..	437.5	..	**1**	....	28.350	..	460.15	0.959
..	385.8	..	..	....	**25**	..	405.78	0.845
..	308.6	..	..	....	**20**	..	324.62	0.676
..	154.3	..	..	....	**10**	..	162.31	0.338
..	19.02	..	..	....	1.232	..	**20**	....
..	15.4324	..	..	....	**1**	..	16.23	....
..	9.51	..	..	....	0.616	..	**10**	....
..	**5**	..	..	....	0.324	..	....	....
..	4.75	..	..	....	0.308	..	**5**	....
..	**1**	..	..	....	0.06480	..	1.0517	....

Approximate Metric Equivalents of Grains (5 Grains and Less)

Grains	Milligrams	Grains	Milligrams	Grains	Milligrams
5	324	1/2	32.4	1/30	2.2
4	259	2/5	25.9	1/40	1.6
3	194	1/4	16.2	1/50	1.3
2	130	1/8	8.1	1/60	1.1
1 1/2	97	1/10	6.5	1/64	1.0
1	65	1/16	4.0	1/100	0.6
3/4	48.6	1/20	3.2	1/128	0.5

See Inside Back Cover for
Table of Metric–Apothecary Approximate Dose Equivalents

NF XVII

THE NATIONAL FORMULARY

Official from January 1, 1990

SEVENTEENTH
EDITION

Contents

NF XVII

History of the National Formulary

Formularies and pharmacopeias from very early times have had an important influence upon the establishment and observance of drug standards. The early standards for drugs, however, were not all comparable to those of today. Standards for drugs of any period are always limited by the stage of development of the collateral sciences upon which they depend. The development of new standards for established drugs, and the discovery of new drugs, have closely paralleled the advancements in science. On the other hand, the search for a better knowledge of drugs has frequently contributed to the advance of science in general, and to pharmacy, chemistry, and pharmacology in particular.

Early pharmacopeias and formularies could do little more than provide uniformity of titles for botanical drugs and the few available medicinal chemicals, methods for preparation of the latter, and uniformity of formulas for dosage forms of these drugs.

The events leading to the publication of the first edition of the National Formulary and its progress since its appearance in 1888 illustrate a phase in the evolution of drug standards. When the American Pharmaceutical Association was organized in 1852, the only authoritative and generally recognized book of drug standards available was the third revision of the United States Pharmacopeia. It had been started in 1820 by physicians to serve as "a therapeutic guide to the medical profession." As such, its scope was restricted to drugs selected by representatives of the medical profession, and believed by them to possess the greatest therapeutic merit. This selectivity had, until 1975, always prevented inclusion by the United States Pharmacopeia of a large number of drugs that enjoy a substantial degree of acceptance by physicians and are judged to have significant therapeutic value. This pharmacopeial selectivity was largely responsible for the origin and development of the National Formulary, and it is one of the main reasons why there are these two official drug compendia in the United States.

The idea of a National Formulary for use in the United States is nearly as old as the American Pharmaceutical Association. This association of pharmacists organized in 1852 included among its original objectives the following:

"(a) To improve and regulate the drug market by preventing the importation of inferior, adulterated, or deteriorated drugs and by detecting and exposing home adulterations. (b) To improve the science and art of pharmacy by diffusing scientific knowledge among pharmacists and druggists, fostering pharmaceutical literature, developing talent, stimulating discovery and invention, encouraging home production and manufacture in the several departments of the drug business. (c) To uphold standards of authority in the education, theory, and practice of pharmacy. (d) To create and maintain a standard of professional honesty equal to the amount of our professional knowledge with a view to the highest good and greatest protection to the public."

One means of partially achieving these objectives was to promote the standardization of names and formulas for extensively used dosage forms of drugs not described in the United States Pharmacopeia. It is, therefore, not surprising to find that as early as 1856 an American Pharmaceutical Association committee was appointed to develop plans for the compilation of standards for dosage forms not included in the United States Pharmacopeia of that period. During several ensuing years such a committee was continued, but plans for a formulary entirely acceptable to the Association were not presented. In the meantime, a book on elixirs by John Uri Lloyd, and the New York and Brooklyn Formulary for local use were published in 1883. Pharmaceutical interest in these books stimulated the Association to intensify its efforts to produce a formulary, and in 1888 the first edition of the National Formulary was published under the title, "National Formulary of Unofficial Preparations."

The designation "Unofficial Preparations" in the title was used because the United States Pharmacopeia had earlier adopted the term "official" as applying to the drugs for which it provided standards. The title of the publication was changed to "National Formulary" when, by the terms of the Federal Food and Drugs Act of 1906, both the United States Pharmacopeia and the National Formulary were designated as official compendia.

The First Three Editions—The first edition of the National Formulary was a book of formulas for widely used dosage forms of that period such as elixirs, emulsions, fluidextracts, tinctures, solutions, syrups, and many others. At the time of its publication, there were no federal or state food and drug laws requiring adherence to National Formulary standards of nomenclature and composition of dosage forms. Despite the lack of legal compulsion, the National Formulary—like the United States Pharmacopeia—did much to promote uniformity in many drugs and preparations distributed under distinctive names. When the Federal Food and Drugs Act was passed in 1906, the National Formulary was well established, and it was natural that its standards should be recognized by the terms of that Act as those with which drugs described therein shall comply. New editions representing minor revisions were published in 1896 and in 1906.

The Fourth and Fifth Editions—A fourth edition, the first after the passage of the 1906 Federal Food and Drugs Act, was published in 1916. This edition marked a turning point in the scope of the National Formulary, and established a pattern for its future policies that resulted in its development into a book of modern standards for drugs.

In the 1916 edition, standards of identity, strength, quality, and purity, in addition to distinctive titles and formulas, were introduced into many of the monographs. In that edition, specifications for basic drugs as well as dosage forms were included on a much larger scale than previously. Official formulas for a limited number of parenteral solutions under the class title "Ampuls" were published for the first time in the National Formulary. The fifth edition, published in 1926, did not differ materially from the preceding edition.

The Sixth and Seventh Editions—In 1929, the Council of the American Pharmaceutical Association elected a new Committee on National Formulary. Under the capable direction of the Chairman of this Committee, Professor E. N. Gathercoal, a program was inaugurated and carried out that resulted in more far-reaching changes and greater progress in the technique of revising the National Formulary than in any previous period. A series of prescription surveys, published by the American Pharmaceutical Association in book form under the title, "The Prescription Ingredient Survey" was used as a basis for determining admissions to the sixth edition of the National Formulary, published in 1936. An unusually large number of obsolete drugs were discontinued, and numerous additional chemical, biological, and proximate assays were developed and introduced. One of the outstanding features of the sixth edition was the inclusion of monographs on a large number of ampuls and tablets with adequate standards of identity, strength, quality, and purity, whereby modern trends in dosage forms were recognized that previously had been neglected by the official compendia. The members of the Combined Contact Committee of the American Drug Manufacturers Association and the American Pharmaceutical Manufacturers' Association made outstanding contributions in the development of standards for ampuls and tablets.

By the time of the appearance of the sixth edition, it was evident that a new Federal Food, Drug, and Cosmetic Act, replacing the 1906 law, would be enacted. Plans in connection with the legislation indicated that a greatly strengthened and more comprehensive Act would place greater significance and importance upon official drug standards than obtained under the 1906 law.

Prior to 1936, the work incidental to the establishment of National Formulary standards was done voluntarily and gratuitously by the members of the Committee. The problems that evolved in connection with the sixth edition were so extensive that it was necessary to inaugurate a new and more efficient procedure whereby inadequate standards could be investigated and corrected more promptly than had been possible previously. Provision was made by the American Pharmaceutical Association in 1936 for the establishment of a well-equipped laboratory to serve as a clearing house for the research necessary in the establishment of adequate drug standards. This laboratory was installed in the headquarters building of the Association in 1938.

Other changes in policy relating to the National Formulary, adopted or inaugurated by the Council of the American Pharmaceutical Association after the publication of the sixth edition, included provision for the issuance of interim revision supplements, provision for issuance of the National Formulary at five-year instead of ten-year intervals, and a change in the organization of the Committee.

By means of Interim Revision Announcements, necessary corrections in the sixth edition monographs were issued as promptly as possible. The Bulletin of the National Formulary Committee, issued as a periodical upon a subscription basis, was used for the dissemination of all information concerning the National Formulary.

The decision to issue revisions of the National Formulary at five-year instead of ten-year intervals was necessary because of the rapid advancements and changes in the types of drugs employed. When the principle of more frequent revisions was adopted, the revision program was placed on a continuous basis and activated by provision for an Executive Committee of ten members with a chairman to serve as a full-time employee of the American Pharmaceutical Association. The reorganization of the Committee on National Formulary also provided for designation of the chairman as Director of Revision. This small Executive Committee with power to act made it possible to issue Interim Revision Announcements promptly, and promoted the smooth functioning of a continuous program.

A Committee was elected under the new design of organization in 1939. When this Committee began to function early in 1940, it found that plans for the seventh edition had been well advanced by the previous Committee. The seventh edition, published in 1942, was for the most part an extension of the sixth edition, but with many new specifications added and unsatisfactory standards corrected. The seventh edition also included a much enlarged and thoroughly revised section on materials and preparations for use in the clinical laboratory. In the process of revision, several sixth edition National Formulary monographs were transferred to the United States Pharmacopeia, and the majority of the USP XI monographs not admitted to USP XII were incorporated in the seventh edition of the National Formulary in order to provide for the continuance of official standards. In connection with the seventh edition it was necessary to issue five supplements. Their issuance provided mainly for replacements of drugs and pharmaceutical necessities employed in National Formulary preparations that became unavailable during World War II.

The Eighth and Ninth Editions—Immediately after the publication of the seventh edition in May 1942, work was started on its revision. Surveys of limited scope were conducted to aid the Committee in reaching decisions on deletions and new admissions for the eighth edition. When final decisions on these questions were made during the early part of 1944, it became apparent that again the National Formulary was to undergo a more extensive revision than at any time in the past. It was decided to discontinue official standards for a number of drugs, mostly of botanical origin, that had become obsolete. Monographs on several drugs were also discontinued because of their admission to USP XIII. Approval of the admission of approximately 100 new drugs was given by the Committee. The staff of the American Pharmaceutical Association Laboratory, assisting the members of the Committee and other collaborators, then developed tentative specifications that were published in the Bulletin of the National Formulary Committee. Since all specifications were based upon the results of extensive laboratory investigation, few adverse criticisms were received, this again illustrating the advantage of available laboratory facilities for any drug standardization project. In addition to admitting previously unofficial drugs, the Committee also voted to continue official recognition of a large number of pharmacopeial drugs not admitted to USP XIII.

The eighth edition of the National Formulary, the second published under the plan adopted in 1938 for more frequent revisions, was published in November 1946. The provisions of the new standards became effective on April 1, 1947, thus giving pharmacists time to familiarize themselves with the new edition, and pharmaceutical manufacturers an opportunity to make necessary changes in labels before compliance with the standards was compulsory.

Admissions of drugs to the eighth edition were based upon extent of use and therapeutic value. In addition to extensive changes in content of the National Formulary, numerous editorial changes were made, including the transposition of Latin and English titles, and the use of metric weights and measures was emphasized.

The ninth edition of the National Formulary was published in July 1950, and became official on November 1 of the same year. This edition included specifications for 717 drugs and dosage forms of which 155 had not been in the previous editions.

Essentially the same procedures as those employed in the revision of NF VIII were followed. Admissions to this edition were based, in part, on extent of use as determined by surveys designed to furnish the background information required, but therapeutic value was considered to a greater extent than in previous editions.

One Supplement to NF IX was issued in 1950, and one Interim Revision Announcement was published in 1952 for the correction of errors and the revision of a limited number of specifications.

The Tenth and Eleventh Editions—The tenth edition of the National Formulary was published in August 1955, and became official on December 15 of the same year. Two Interim Revision Announcements were issued, of which one became effective December 15, 1955, and the other on October 1, 1956.

During the revision of the ninth edition of the National Formulary, leading to the publication of the tenth, substantial changes in the content of the text were made. Of the 717 monographs in NF IX, 242 were not admitted to NF X. Titles and standards for 259 drugs, for which official standards would not otherwise have been provided, were added to NF X during the revision period. In the process of revision, many individual specifications for strength, quality, purity, and identity were revised and materially improved.

In 1951 the Bulletin of the National Formulary Committee was retitled Drug Standards, and many of the new specifications were published therein for review and comment.

Several new features were added to NF X. Among these were

specifications for the disintegration time of coated, buccal, and sublingual tablets; weight variation standards for the content of dry-filled capsules; and specifications for solubility of hypodermic tablets. Through the efforts of a joint USP-NF Panel on Sterility Requirements, the general chapters on injections and sterility tests were revised and a new chapter on general sterilization procedures was prepared. A new section on general information was added to NF X. This section included specifications for prescription balances, weights, and measuring devices; formulas for clinical laboratory reagents; a selected list of drugs and formulas for preparations used by chiropodists; basic information relating to dyes for use in coloring pharmaceutical preparations; a table comparing the names and standards of strength of drugs covered in the International Pharmacopoeia with those appearing in NF X; and a description of general sterilization procedures.

The section on clinical laboratory reagents and the table of optical crystallographic characteristics of NF medicinal chemicals underwent considerable revision and were transferred to the general information section.

The eleventh edition of the National Formulary—the last of five editions issued under the direction of Dr. J. L. Powers—was published in July 1960, and became official October 1, 1960. Subsequently, four Interim Revision Announcements were issued which became official January 1, 1961; May 1, 1962; February 1, 1963; and May 1, 1963. As in the case of the preceding volume, NF XI was marked by a large number of new admissions, 285, which together with the 531 monographs admitted from NF X provided a total of 816 monographs. At the same time, 208 of the articles in the previous edition were no longer accorded NF recognition.

NF XI presented a marked change in typography and style, with use of a two-column format for the first time. In addition, the tests and assays reflected somewhat greater use of the new analytical techniques that were beginning to gain widespread acceptance for both quality control and enforcement purposes. This required a substantial increase in the number of Reference Standards, from 10 in NF X to 42 in NF XI.

The Twelfth Edition—Probably the most distinguishing feature of the twelfth edition—particularly from a historical standpoint—was adoption of a sweeping new criterion for determining admission of drugs to the National Formulary. As heralded in the preface to that edition, the philosophy of *extent of use*, which had been the traditional criterion for recognition of articles in the NF, was struck down and eliminated. In its place, the concept of *therapeutic value* was established as the sole basis for inclusion of a drug in the compendium. This action was expected to elevate the importance of the compendium as a guide to rational medical practice, and to raise the stature of the NF within the medically related community.

The great impact of instrumentation in pharmaceutical analysis for the first time was reflected in the individual NF monographs. Ultraviolet spectrophotometry had been employed to some extent in the eleventh edition, but it was in NF XII that instrumentation was fully accepted and embraced as a general approach to the establishment of drug standards.

However, with the general adoption of instrumentation, the need for suitable samples of reference substances to be used for comparative testing purposes also mushroomed. From a mere handful—10 in the case of NF X—the number of NF Reference Standards rose to almost 100. While still ancillary to the compendium itself, supplying this number of reference substances required a major expansion of this phase of the overall program to the point where in itself the Reference Standards activity became a significant facet of the total operation of establishing specifications.

The individual monographs, as presented in NF XII, were also affected in other ways as the result of the 1962 Drug Amendments. Previously, synonyms had been in wide use and were included by name in the pertinent monographs. This practice was specifically prohibited by the directive from Congress. Furthermore, batch certification—with the concomitant power for FDA to set specifications—was extended to all antibiotics. Taking this development into account, the NF deleted all standards and specifications from all of the antibiotic monographs, in order to avoid any confusion from dual and possibly conflicting standards. Abbreviated monographs for antibiotics were still included in the NF to indicate their accepted therapeutic value as determined by the Committee on Admissions and to give related information of interest to pharmaceutical and medical practitioners.

Early during the revision period leading to publication of NF XII, the American Medical Association and the United States Pharmacopeial Convention joined with the American Pharmaceutical Association in sponsorship of the analytical laboratory facilities housed in the Association's building and operated under the auspices of the APhA Foundation. Through this mutual support and cooperation, the Drug Standards Laboratory was revitalized and by the close of the revision period was making a significant contribution to the NF program by providing direct, convenient laboratory services.

A thorough overhauling of all graphic formulas and all systematic chemical names was also instituted in NF XII to bring these into accord with the latest approved practices of recognized authoritative groups such as Chemical Abstracts Service and the International Union of Pure and Applied Chemistry (IUPAC). Furthermore, all molecular weights and assay calculation factors were recalculated on the basis of carbon-12 rather than oxygen-16, as previously employed. This change was in line with the 1961 revision of International Atomic Weights, which changed the element of reference.

NF XII was published in early January 1965, and became official on September 1, 1965. Subsequently, two Supplements were issued that became official on September 1, 1965, and on March 1, 1967. Issuance of Supplements revived a practice that had not been utilized during the several previous revision periods. One Interim Revision Announcement was also issued, which became official on April 1, 1968. Continuing the trend of the last two editions, there were many new admissions to NF XII (248) to provide a total of 783 articles in the volume itself—another 19 articles were admitted by way of the supplements. At the same time, 280 monographs in NF XI were not voted admission to the new edition.

The Thirteenth and Fourteenth Editions—"With ingenuity which rivals that seen in any field of science or technology, pharmaceutical scientists—particularly during the past ten years—have utilized the recent developments in analytical methodology in the testing of pharmaceuticals. The net result permits inspection of infinitely smaller samples, enormously greater selectivity, and drastically improved precision and accuracy." So read the preface of the thirteenth edition, and it was just that ingenuity which made it possible for NF XIII to provide the most advanced and authoritative means yet available for assessing drug quality.

A very significant indication of the progress made in utilizing the most advanced knowledge and sophisticated analytical techniques as applied to drug standards is evident from the following diverse but partial list of new General Tests chapters in NF XIII: Selenium Limit Test, Salmonella Test, Dissolution Test, Steroid Chromatographic Purity, Rubber Closures for Injections, Coulometry, Phase Solubility Analysis, and X-ray Diffraction. Moreover, many of the General Tests chapters carried over from NF XII underwent extensive revision for NF XIII. For example, the chapter on spectrophotometry was expanded to include atomic absorptions spectrophotometry as well as fluorometry and nephelometry.

Formalization of the complete National Formulary revision and publication process so as to render it highly methodical and systematic was a major achievement in the NF XIII revision period. As a result of the growth of the NF, the increased sophistication and complexity of pharmaceutical manufacturing and investigation, and increased legal responsibilities, the procedure by which this compendium was revised and prepared grew vastly more complex. While the suggestion was made in 1959 by Justin L.

Powers, then retiring as NF Director, that consideration be given to describing the NF operation in greater detail, it was not until 1965 that efforts were initiated to reorganize the entire committee and staff structure.

With the adoption in 1967 of revisions in the Constitution of the American Pharmaceutical Association, a new Bylaws chapter pertaining exclusively to NF codification was added with a provision for an appeals route included. This latter aspect greatly enhanced the NF legal foundation.

At the same time, the Committee on National Formulary was renamed the National Formulary Board. Moreover, the new Bylaws chapter provided for members of this Board to be elected by the Board of Trustees of the American Pharmaceutical Association, rather than being appointed as previously. The new name more adequately reflected the responsibilities and functions of this group, including formulating the general principles and policies under which the NF is compiled; considering and acting upon all proposals and recommendations made by NF committees and advisory panels, and other parties; evaluating the safety, effectiveness, quality, and purity of drugs; and prescribing standards and specifications for drugs and other articles that may be used in connection with drugs, and promulgating such revisions as are deemed necessary to ensure a current compendium.

The NF Board emphatically reaffirmed for NF XIII the admissions policy initially employed for NF XII—namely, that therapeutic value serve as the sole criterion in considering articles to be admitted to the NF for medicinal purposes. The Board also defined its policy with regard to the admission of drug combinations—namely, that NF monographs for combination dosage forms should be limited to those combinations for which there is a therapeutic advantage to the patient as contrasted with administration of the separate individual therapeutically active ingredients. This policy was in accord with the philosophy underlying the drug efficacy study subsequently undertaken by the National Academy of Sciences–National Research Council and implemented by FDA.

A broader policy was adopted relative to the admission of pharmaceutical aids or adjuncts, by not limiting such admissions to those ingredients specified in NF monograph formulas. Providing NF monographs for a broader group of such articles served to give greater assurance of quality and purity suitable for pharmaceutical purposes.

NF XIII reflected clear recognition of those factors identified as potentially affecting biological availability—the extent to which the active ingredient in a drug product can be taken up in a form that is physiologically active. The NF Board approached the problem by establishing appropriate specifications both for active ingredients and for final dosage forms, to serve as an index of drug quality from the standpoint of expected biological availability. As a consequence, X-ray diffraction specifications were established for a number of drug substances, where the existence of different polymorphs might have a marked influence on the absorption profile. Moreover, two dissolution test methods were adopted, and dissolution requirements were included as an index of drug release in six individual monographs for tablets and capsules.

Extension of content uniformity specifications was a major advance in NF XIII. The specification was applied in all monographs for tablets containing an active ingredient in relatively low quantities (usually 50 mg or less of any strength available, where test considerations and similar considerations indicated this was feasible). Furthermore, the same concept was applied to capsules and sterile solids.

Automated analytical methods came to prominence in quality control laboratories, as they faced the need for analyzing more and more samples such as in content uniformity testing. The official compendia, with their test procedures being the common legal ground between plaintiff and defendant, had been reluctant to adopt automated procedures, because this would become the only way in which the analysis could be performed in demonstrating compliance or lack of compliance with official standards.

The NF Board formulated a major policy and interpretation on the use of automated procedures which, at the same time, fulfilled pragmatic and practical aspects while still satisfying the legal obligations and responsibilities of NF as an enforcement device. The Board concluded that the NF intent was to recommend equipment that would be suitable, but to require that—in order to demonstrate compliance—the exact chain of chemical and physical steps specified in an official procedure be conducted. A statement of the Board's position was included in the NF XIII General Notices. Equally important, a statement was included also providing for the use of alternative methods of equivalent accuracy, with the proviso that authoritative judgment be by the NF method in the event of doubt or dispute.

Among other noteworthy new features of NF XIII was the official recognition of several aerosols, the first such recognition of this dosage form. A uniform set of category statements to indicate the pharmacological basis for recognition of articles was developed and included in this edition. In the General Information section, a major new chapter was added providing a tabulation of approximate solubility data on NF articles. In matters of nomenclature, a review of NF monograph titles from the standpoint of usefulness and simplicity was continued and a number of titles were changed.

NF XIII became official on September 1, 1970, having been published early in January of that year. A two-column format was maintained but improved by the presentation of titles, preliminary chemical and related information, and the definition in single-column format with the definition distinctively set off from the rest of the monograph. One of the major advances made in NF XIII was implementation of a continuous supplemental revision program. Five supplements were published, the first official concurrently with the main NF XIII volume and the other four issued annually throughout the revision period.

The number of recognized articles jumped to 992 in the main NF XIII volume (additional monographs were added via supplements), 411 being newly admitted. Of the NF XII articles, 221 were not included in NF XIII. The NF Reference Standards program mushroomed (to almost 250 reference materials), reflecting the importance of suitable samples for comparative testing purposes in modern analytical techniques and the reliance placed on these techniques in developing official drug standards and specifications.

During the period covered by the first 14 editions of the National Formulary, many individuals served as members of the revision bodies. The names may be found by reference to previous editions of the National Formulary. The names of the Chairmen of the National Formulary Boards (or, as previously referred to, Committees on National Formulary), the years served by each, and the editions published during each corresponding period are shown in the accompanying table.

Chairman	*NF Edition(s)*	*Years*
Charles Rice	I	1883–1888
C. Lewis Diehl	II, III, IV	1889–1917
Wilbur L. Scoville	V	1919–1929
Edmund N. Gathercoal	VI	1930–1940
Justin L. Powers	VII, VIII, IX, X, XI	1940–1960
Edward G. Feldmann	XII, XIII	1960–1969
John V. Bergen	XIV	1970–1974

The Fifteenth and Sixteenth Editions—Pursuant to the acquisition of the National Formulary by the U.S. Pharmacopeial Convention on January 2, 1975, the responsibility for setting standards for drugs was consolidated within USPC. The two official compendia were combined within a single volume; however, USP and NF continued as two distinct official compendia. In accordance with a resolution adopted by the Pharmacopeial Convention in 1975, to the effect that public standards should be set for all drugs, the scope of both the NF and the USP was changed drastically. The Pharmacopeia was limited to drug substances and dosage forms and the National Formulary was limited to pharmaceutic ingredients. Both USP XIX and NF XIV had individually observed these divisions, and division between the compendia along these lines was expected to minimize label changes in the future.

Although it became the goal of the National Formulary to provide monographs eventually for all pharmaceutic ingredients used in drug dosage forms, precedence was given to those used in formulations for which there were Approved New Drug Applications and indications of widespread use, and which were administered systemically.

Where an article was used as both a therapeutic agent and a pharmaceutic ingredient, it was included in the USP, with a cross-reference from the NF to that USP monograph, solely for the convenience of the user of the compendia.

In the single USP-NF volume, a large amount of duplication of text was eliminated, and it was feasible and convenient to make extensive use of cross-references from the NF to the USP with regard to General Notices and Requirements; General Chapters; Reagents, Indicators, and Solutions; and Reference Tables. In addition, a combined index to the USP and the NF facilitated reference to the contents of the respective compendia.

The Seventeenth Edition—Over the years it had become evident that the compounding activities of pharmacists had practically disappeared and the USP and NF were no longer being used as books of recipes, chiefly because large-scale manufacturing could produce products less expensively and more reliably. Since methods of preparing tablets and other dosage forms produced by large-scale methods were not defined in monographs, there was an inconsistency in requiring definitive preparative methods for some elixirs. For USP XXII, most of the remaining recipes were deleted or revised; stress thus was placed instead on the assay of a finished preparation. Therefore, it became unnecessary to provide tests and standards for 33 flavors and related articles. The monographs for these articles are not included in this edition of NF. The publication in 1986 of the *Handbook of Pharmaceutical Excipients*, a joint effort by the American Pharmaceutical Association and The Pharmaceutical Society of Great Britain, provided a great deal of useful, unofficial, auxiliary information regarding many articles of the NF.

Preface to NF XVII

The *National Formulary* has been published with the *United States Pharmacopeia* in a single volume since 1980 as a consequence of the acquisition, by the USP Convention, of the *National Formulary* in 1975. From its initial publication in 1888 as a sister compendium to the Pharmacopeia (see *History of the National Formulary*, page 1886) through the years to its 1975 edition, the NF was published by the American Pharmaceutical Association.

In the 1980 publications the scope of both USP and NF was changed, the *United States Pharmacopeia* being limited to drug substances and dosage forms, and the *National Formulary* being limited to pharmaceutic ingredients. Where an article was used as both a therapeutic agent and a pharmaceutic ingredient, it was included in USP, with a cross-reference in NF to that USP monograph solely for the convenience of the user of the compendia.

While USP and NF have continued as two distinct official compendia, the consolidation has resulted in the elimination of a large amount of duplication of text. It is feasible and convenient to make extensive use of cross-references in NF to the USP *General Notices and Requirements; General Chapters; Reagents, Indicators,* and *Solutions;* and *Reference Tables.* In addition, a combined index to USP XXII and NF XVII facilitates reference to the contents of both compendia.

A tabulation of pharmaceutic ingredients by categories is included in the *Reference Tables* (see USP XXII, page 1857). Most of these articles are recognized in NF XVII, although a number of them are the subjects of USP XXII monographs. Where an article is recognized in the USP and also is classified as a pharmaceutic ingredient, the NF includes a cross-reference to that title for the convenience of users. Also, USP XXII includes cross-references to all monographs in NF XVII.

Standards and Tests—The goal of the *National Formulary* is to eventually provide monographs for all pharmaceutic ingredients used in drug dosage forms. Precedence is given to those used in formulations for which there are Approved New Drug Applications that have indications of widespread use and that are administered systemically.

In general, NF articles are defined to characterize their composition as closely as possible by providing tests and stating limits, either in the monograph definition explicitly or by reference to required labeling.

Many articles used in foods, cosmetics, organic syntheses, and even various commercial operations are also used as pharmaceutic excipients. Specifications for such articles developed in nonpharmaceutical industrial laboratories reflect the history of those materials and thus may differ from the usual pharmaceutical standards and tests. In some cases these specifications have been approved for adaptation in NF monographs.

Of increasing importance is the proliferation of polymeric products that are useful as excipients. Because the need for further monographs is expected for such articles, the chain length of which may vary, some titles (e.g., Nonoxynol 10) include a number to denote more definitely the chemical structure of the article. The use of family monographs is valuable because such monographs define closely related polymers of various kinds.

Credits—As with any group effort, the responsibility for success in publishing the *National Formulary* falls more heavily on some than on others. Thus, while the Revision Committee as a whole (see page vi) served as the elected agent of the USP Convention in producing this compendium, much credit should be given to the people not on the Revision Committee whose contributions, given out of a sense of public service, greatly enhanced the Committee's effectiveness (see page xii).

Foremost recognition belongs to Dr. Ralph F. Shangraw, who served as Chairman of the USP Subcommittee on Dosage Forms and Excipients, which was directly responsible for most of the contents of the *National Formulary.* Dr. Shangraw's thorough examination of the issues concerning excipients and his careful attention to the details of tests and specifications have led to many additions and improvements.

For further information regarding contributors to the compendial revision program, see the *Preface to USP XXII.*

Admissions

Articles Official in NF XVII

Acacia
Acetic Acid
Acetone
Agar
Diluted Alcohol
Alginic Acid
Almond Oil
Aluminum Monostearate
Strong Ammonia Solution
Ammonium Carbonate
Ammonium Phosphate
Amylene Hydrate
Anethole
Aromatic Elixir
Ascorbyl Palmitate
Aspartame

Bentonite
Purified Bentonite
Bentonite Magma
Benzaldehyde
Compound Benzaldehyde Elixir
Benzalkonium Chloride
Benzalkonium Chloride Solution
Benzyl Alcohol
Boric Acid
Butane
Butyl Alcohol
Butylated Hydroxyanisole
Butylated Hydroxytoluene
Butylparaben

Tribasic Calcium Phosphate
Calcium Silicate
Calcium Stearate
Calcium Sulfate
Caramel
Carbomer 910
Carbomer 934
Carbomer 934P
Carbomer 940
Carbomer 941
Carbomer 1342
Carbon Tetrachloride
Carboxymethylcellulose Calcium
Carboxymethylcellulose Sodium 12
Carrageenan
Hydrogenated Castor Oil
Microcrystalline Cellulose
Microcrystalline Cellulose and Carboxymethylcellulose Sodium
Powdered Cellulose
Cellulose Acetate
Cellulose Acetate Phthalate
Cetostearyl Alcohol
Cetyl Alcohol
Cetyl Esters Wax
Chlorobutanol
Chlorocresol
Chloroform
Cholesterol
Cocoa Butter
Corn Oil
Cottonseed Oil

Cresol
Croscarmellose Sodium
Crospovidone
Cyclomethicone

Dehydroacetic Acid
Denatonium Benzoate
Dextrates
Dextrin
Dextrose Excipient
Diacetylated Monoglycerides
Dichlorodifluoromethane
Dichlorotetrafluoroethane
Diethanolamine
Diethyl Phthalate
Diisopropanolamine
Dimethicone

Edetic Acid
Ethyl Acetate
Ethyl Oleate
Ethyl Vanillin
Ethylcellulose
Ethylcellulose Aqueous Dipersion
Ethylparaben

Hard Fat
Ferric Oxide
Fumaric Acid

Gelatin
Gentisic Acid Ethanolamide
Pharmaceutical Glaze
Liquid Glucose
Glutaral Disinfectant Solution
Glyceryl Behenate
Glyceryl Monostearate
Guar Gum

Hexylene Glycol
Hydrochloric Acid
Diluted Hydrochloric Acid
Hydroxyethyl Cellulose
Hydroxypropyl Cellulose
Hydroxypropyl Methylcellulose Phthalate
Hydroxypropyl Methylcellulose Phthalate 200731
Hydroxypropyl Methylcellulose Phthalate 220824
Hypophosphorous Acid

Imidurea
Isobutane
Isopropyl Myristate
Isopropyl Palmitate

Lactose
Lanolin Alcohols
Lecithin

Magnesium Aluminum Silicate
Magnesium Silicate
Magnesium Stearate

Malic Acid
Methacrylic Acid Copolymer
Methyl Alcohol
Methyl Isobutyl Ketone
Methyl Salicylate
Methylene Chloride
Methylparaben
Methylparaben Sodium
Light Mineral Oil
Mono- and Di-glycerides
Mono- and Di-acetylated Monoglycerides
Monoethanolamine
Monosodium Glutamate
Monothioglycerol
Myristyl Alcohol

Nitric Acid
Nitrogen
Nonoxynol 10

Octoxynol 9
Octyldodecanol
Oleic Acid
Oleyl Alcohol
Olive Oil
Orange Flower Oil
Oxyquinoline Sulfate

Paraffin
Peanut Oil
Peppermint
Peppermint Oil
Peppermint Water
Persic Oil
Phenylmercuric Acetate
Phenylmercuric Nitrate
Phosphoric Acid
Diluted Phosphoric Acid
Polacrilin Potassium
Poloxamer
Polyethylene Excipient
Polyethylene Glycol
Polyethylene Glycol Ointment
Polyethylene Oxide
Polyoxyl 10 Oleyl Ether
Polyoxyl 20 Cetostearyl Ether
Polyoxyl 35 Castor Oil
Polyoxyl 40 Hydrogenated Castor Oil
Polyoxyl 40 Stearate
Polyoxyl 50 Stearate
Polysorbate 20
Polysorbate 40
Polysorbate 60
Polysorbate 80
Polyvinyl Acetate Phthalate
Potassium Benzoate
Potassium Hydroxide
Potassium Metabisulfite
Potassium Metaphosphate
Monobasic Potassium Phosphate
Potassium Sorbate
Propane
Propionic Acid
Propyl Gallate

Propylene Carbonate
Propylene Glycol Alginate
Propylene Glycol Diacetate
Propylene Glycol Monostearate
Propylparaben
Propylparaben Sodium

Rose Oil
Stronger Rose Water

Sesame Oil
Shellac
Purified Siliceous Earth
Silicon Dioxide
Colloidal Silicon Dioxide
Soda Lime
Sodium Alginate
Sodium Benzoate
Sodium Borate
Sodium Carbonate
Sodium Dehydroacetate
Sodium Formaldehyde Sulfoxylate
Sodium Hydroxide
Sodium Lauryl Sulfate
Sodium Metabisulfite
Sodium Propionate
Sodium Starch Glycolate
Sodium Stearate
Sodium Stearyl Fumarate
Sorbic Acid
Sorbitan Monolaurate
Sorbitan Monooleate
Sorbitan Monopalmitate
Sorbitan Monostearate
Sorbitol
Squalane
Starch
Pregelatinized Starch
Stearic Acid
Purified Stearic Acid
Stearyl Alcohol
Sucrose
Sucrose Octaacetate
Compressible Sugar
Confectioner's Sugar
Sugar Spheres
Sulfur Dioxide
Sulfuric Acid
Syrup

Tartaric Acid
Thymol
Tocopherols Excipient
Tolu Balsam Syrup
Tolu Balsam Tincture
Tragacanth
Trichloromonofluoromethane
Triethyl Citrate
Trolamine

Vanilla
Vanilla Tincture
Vanillin
Hydrogenated Vegetable Oil

Carnauba Wax
Emulsifying Wax
Microcrystalline Wax
White Wax
Yellow Wax

Xanthan Gum

Zein

General Notices and Requirements

Applying to Standards, Tests, Assays, and Other Specifications of the National Formulary

The General Notices and Requirements (hereinafter referred to as the "General Notices") provide in summary form the basic guidelines for the interpretation and application of the standards, tests, assays, and other specifications of The National Formulary and obviate the need to repeat throughout the book those requirements that are pertinent in numerous instances.

Where, exceptions to the General Notices are made, the wording in the individual monograph or general test chapter takes precedence and specifically indicates the directions or the intent. To emphasize that such exceptions do exist, the General Notices employ where indicated a qualifying expression such as "unless otherwise specified." Thus, it is understood that the specific wording of standards, tests, assays, and other specifications is binding wherever deviations from the General Notices exist. By the same token, where no language is given specifically to the contrary, the General Notices apply.

TITLE

The full title of this book, including its supplements, is The National Formulary, Seventeenth Edition. This title may be abbreviated to National Formulary XVII or to NF XVII. Where the term NF is used, without further qualification, during the period in which this National Formulary is official, it refers to NF XVII and to any supplement(s) thereto.

"OFFICIAL" AND "OFFICIAL ARTICLES"

The word "official," as used in this National Formulary or with reference hereto, is synonymous with "National Formulary," with "NF," and with "compendial."

The designation NF in conjunction with the official title on the label of an article is a reminder that the article purports to comply with NF standards; such specific designation on the label does not constitute a representation, endorsement, or incorporation by the manufacturer's labeling of the informational material contained in the NF monograph, nor does it constitute assurance by NF that the article is known to comply with NF standards. The standards apply equally to articles bearing the official titles or names derived by transposition of the definitive words of official titles or transposition in the order of the names of two or more ingredients in official titles, whether or not the added designation "NF" is used. Names considered to be synonyms of the official titles may not be used for official titles.

Where an article differs from the standards of strength, quality, and purity as determined by the application of the assays and tests, set forth for it in the National Formulary, its difference shall be plainly stated on its label. Where an article fails to comply in identity with the identity prescribed in the NF, or contains an added substance that interferes with the prescribed assays and tests, such article shall be designated by a name that is clearly distinguishing and differentiating from any name recognized in the National Formulary.

Articles listed herein are official and the standards set forth in the monographs apply to them only when the articles are intended or labeled for use as drugs or as medical devices and when bought, sold, or dispensed for these purposes or when labeled as conforming to this National Formulary.

An article is deemed to be recognized in this National Formulary when a monograph for the article is published in it, including its supplements, addenda, or other interim revisions, and an official date is generally or specifically assigned to it.

Because of differing characteristics not standardized by this Formulary, all sources or types of some excipients may not have identical properties with respect to use in a specific preparation. To assure interchangeability in such instances, users may wish to ascertain final performance equivalency or determine such characteristics prior to use.

OTHER GENERAL NOTICES

The General Notices of USP XXII, beginning with "The following terminology is used" under "*Official*" *and* "*Official Articles*," apply equally to the standards, tests, assays, and other specifications of this National Formulary, the terms "National Formulary" and "NF" being read for "Pharmacopeial" and "USP," respectively.

Similarly, the General Chapters, the section on Reagents, Indicators, and Solutions, and the Reference Tables of USP XXII apply also to this National Formulary.

Official Monographs for NF XVII

Acacia

» Acacia is the dried gummy exudate from the stems and branches of *Acacia senegal* (Linné) Willdenow or of other related African species of *Acacia* (Fam. Leguminosae).

Packaging and storage—Preserve in tight containers.

Solubility and reaction—Dissolve 1 g in 2 mL of water; the resulting solution flows readily and is acid to litmus.

Botanic characteristics—

Acacia—Spheroidal tears up to 32 mm in diameter or in angular fragments of white to yellowish white color. Is translucent or somewhat opaque from the presence of numerous minute fissures; very brittle, the fractured surface glassy and occasionally iridescent. Is almost odorless and produces a mucilaginous sensation on the tongue.

Flake Acacia—White to yellowish white, thin flakes, appearing under the microscope as colorless, striated fragments.

Powdered Acacia—White to yellowish white, angular microscopic fragments with only traces of starch or vegetable tissues present.

Granular Acacia—White to pale yellowish white, fine granules. Under the microscope it appears as colorless, glassy, irregularly angular fragments up to 100 µm in thickness, some of which exhibit parallel linear streaks.

Spray-dried Acacia—White to off-white compacted microscopic fragments or whole spheres.

Identification—To 10 mL of a cold solution (1 in 50) add 0.2 mL of diluted lead subacetate TS: a flocculent, or curdy, white precipitate is formed immediately.

Microbial limit—It meets the requirements of the test for absence of *Salmonella* species under *Microbial Limit Tests* ⟨61⟩.

Water, *Method III* ⟨921⟩—Dry it at 105° for 5 hours: it loses not more than 15.0% of its weight. For unground Acacia, crush it in a mortar until it passes through a No. 40 sieve, and mix the ground material before weighing the test specimen.

Total ash ⟨561⟩: not more than 4.0%.

Acid-insoluble ash ⟨561⟩: not more than 0.5%.

Insoluble residue—Dissolve 5.0 g of powdered or finely ground Acacia in about 100 mL of water in a 250-mL conical flask, add 10 mL of 3 *N* hydrochloric acid, and boil gently for 15 minutes. Filter by suction, while hot, through a tared filtering crucible, wash thoroughly with hot water, dry at 105° for 1 hour, and weigh. The weight of the residue thus obtained does not exceed 50 mg.

Arsenic, *Method II* ⟨211⟩: 3 ppm.

Lead ⟨251⟩: 0.001%.

Heavy metals, *Method II* ⟨231⟩: 0.004%.

Starch or dextrin—Boil a solution (1 in 50), cool, and add iodine TS: no bluish or reddish color is produced.

Tannin-bearing gums—To 10 mL of a solution (1 in 50) add 0.1 mL of ferric chloride TS: no blackish coloration or blackish precipitate is produced.

Acetic Acid

Acetic acid.
Acetic acid [*64-19-7*].

» Acetic Acid is a solution containing not less than 36.0 percent and not more than 37.0 percent, by weight, of $C_2H_4O_2$.

Packaging and storage—Preserve in tight containers.

Identification—It responds to the tests for *Acetate* ⟨191⟩.

Nonvolatile residue—Evaporate 20 mL in a tared porcelain dish on a steam bath, and dry at 105° for 1 hour: the weight of the residue does not exceed 1.0 mg (0.005%).

Chloride—Add 5 drops of silver nitrate TS to 10 mL of a solution (1 in 10): no opalescence is produced.

Sulfate—Add 5 drops of barium chloride TS to 10 mL of a solution (1 in 10): no turbidity is produced.

Heavy metals ⟨231⟩—To the residue obtained in the test for *Nonvolatile residue* add 8 mL of 0.1 *N* hydrochloric acid, warm gently until solution is complete, dilute with water to 100 mL, and use 10 mL of the solution: the limit is 0.001%.

Readily oxidizable substances—Dilute 4.0 mL in a glass-stoppered vessel with 20 mL of water, and add 0.30 mL of 0.10 *N* potassium permanganate: the pink color is not changed to brown at once, and the liquid does not become entirely brown or free from a pink tint in less than 30 seconds.

Assay—Place about 6 mL of Acetic Acid in a tared, glass-stoppered flask, and weigh accurately. Add 40 mL of water, then add phenolphthalein TS, and titrate with 1 *N* sodium hydroxide VS. Each mL of 1 *N* sodium hydroxide is equivalent to 60.05 mg of $C_2H_4O_2$.

Acetic Acid, Glacial—*see* Acetic Acid, Glacial USP

Acetone

CH$_3$COCH$_3$

C_3H_6O 58.08
2-Propanone.
Acetone [*67-64-1*].

» Acetone contains not less than 99.0 percent of C_3H_6O, calculated on the anhydrous basis.

Caution—Acetone is very flammable. Do not use where it may be ignited.

Packaging and storage—Preserve in tight containers, remote from fire.

Identification—Using 0.1-mm cells, obtain the infrared absorption spectrum of a 1 in 10 solution of it in carbon tetrachloride, with carbon tetrachloride in the reference beam. It exhibits a strong band at about 5.8 µm; a strong region of absorption between 6.8 µm and 7.5 µm, with maxima at about 7.0 µm and 7.3 µm; a strong maximum at about 8.2 µm; and weak maxima at about 9.2 µm and 11.0 µm.

Specific gravity ⟨841⟩: not more than 0.789.

Water—

Standard preparation—Transfer 0.50 mL of water to a dry 100-mL volumetric flask, dilute with dehydrated isopropyl alcohol to volume, and mix.

Chromatographic system—Under typical conditions, the gas chromatograph is equipped with a thermal conductivity detector, and contains a 1.5-m × 6-mm stainless steel column containing packing S4. The column is maintained at a temperature of 180°, and helium is used as the carrier gas.

Procedure—Inject 5.0 µL of *Standard preparation* into a suitable gas chromatograph, and determine the area under the water peak. Similarly inject 5.0 µL of the same dehydrated isopropyl alcohol to provide a blank, and determine the area under the water peak. The retention time is about 1 minute. Similarly inject 5.0 µL of Acetone. The area under the water peak for Acetone is not greater than that from the *Standard preparation* corrected for the area under the water peak in the blank chromatogram (0.5%).

Nonvolatile residue—Evaporate 50 mL in a tared porcelain dish on a steam bath, and dry at 105° for 1 hour: the weight of the residue does not exceed 2 mg (0.004%).

Readily oxidizable substances—Mix 20 mL with 0.10 mL of 0.10 *N* potassium permanganate in a glass-stoppered bottle: the permanganate color of the mixture does not completely disappear within 15 minutes.

Assay—

Chromatographic system—Under typical conditions, the instrument is equipped with a flame-ionization detector, and contains a 1.8-m × 3-mm column containing packing S4. The column temperature is programmed at a rate of about 8° per minute from 110° to 220°, and helium is used as the carrier gas.

Procedure—Inject a suitable volume, typically about 0.5 µL, of Acetone into a suitable gas chromatograph, and record the chromatogram. Calculate the percentage of C_3H_6O in the Acetone, on the anhydrous basis, by dividing the area under the acetone peak by the sum of the areas under all of the peaks and multiplying by 100. [NOTE—No separate correction is applied for water content, since water does not respond to the flame-ionization detector.]

Acids—*see complete list in index*
Activated Charcoal—*see* Charcoal, Activated USP

Agar

» Agar is the dried, hydrophilic, colloidal substance extracted from *Gelidium cartilagineum* (Linné) Gaillon (Fam. Gelidiaceae), *Gracilaria confervoides* (Linné) Greville (Fam. Sphaerococcaceae), and related red algae (Class Rhodophyceae).

Botanic characteristics—

Agar—Usually in bundles consisting of thin, membranous, agglutinated strips or in cut, flaked, or granulated forms. May be weak yellowish orange, yellowish gray to pale yellow, or colorless. Is tough when damp, brittle when dry.

Histology—In water mounts Agar appears granular and somewhat filamentous; a few fragments of the spicules of sponges and a few frustules of diatoms may be present; in Japanese Agar, the frustules of *Arachnoidiscus ehrenbergii* Baillon often occur, being disk-shaped and from 100 µm to 300 µm in diameter.

Powdered Agar—White to yellowish white or pale yellow; in chloral hydrate TS its fragments are transparent, more or less granular, striated, and angular, and occasionally they contain frustules of diatoms.

Identification—

A: Iodine TS colors some of the fragments of Agar bluish black, with some areas reddish to violet.

B: When boiled with 65 times its weight of water for 10 minutes, with constant stirring, and adjusted to a concentration of 1.5%, by weight, with hot water, Agar forms a clear liquid which congeals at 32° to 39° to form a firm resilient gel, which does not melt below 85°.

Microbial limit—It meets the requirements of the test for absence of *Salmonella* species under *Microbial Limit Tests* ⟨61⟩.

Water, *Method III* ⟨921⟩—If necessary, cut it into pieces from 2 mm to 5 mm square, and dry at 105° for 5 hours: it loses not more than 20.0% of its weight.

Total ash ⟨561⟩: not more than 6.5%, on a dry-weight basis.

Acid-insoluble ash ⟨561⟩: not more than 0.5%, on a dry-weight basis.

Foreign organic matter ⟨561⟩: not more than 1.0%.

Foreign insoluble matter—To 7.5 g add sufficient water to make 500 g, boil for 15 minutes, and readjust to the original 500 g. To 100 g of the uniformly mixed material add hot water to make 200 mL, heat almost to boiling, filter while hot through a tared filtering crucible, rinse the container with several portions of hot water, and pass these rinsings through the crucible. Dry the crucible and its contents at 105° to constant weight: not more than 15 mg (1.0%) remains.

Arsenic, *Method II* ⟨211⟩: 3 ppm.

Lead ⟨251⟩: 0.001%.

Heavy metals, *Method II* ⟨231⟩: 0.004%.

Foreign starch—A solution made by boiling 0.10 g of it in 100 mL of water does not, upon cooling, produce a blue color upon the addition of iodine TS.

Gelatin—Dissolve about 1 g in 100 mL of boiling water, and allow to cool to about 50°. To 5 mL of the solution add 5 mL of trinitrophenol TS: no turbidity appears within 10 minutes.

Water absorption—Place 5.0 g in a 100-mL graduated cylinder, fill to the mark with water, mix, and allow to stand at 25° for 24 hours. Pour the contents of the cylinder through moistened glass wool, allowing the water to drain into a second 100-mL graduated cylinder: not more than 75 mL of water is obtained.

Alcohol—*see* Alcohol USP
Alcohol, Butyl—*see* Butyl Alcohol
Alcohol, Cetostearyl—*see* Cetostearyl Alcohol

Diluted Alcohol

» Diluted Alcohol is a mixture of Alcohol and water containing not less than 41.0 percent and not more than 42.0 percent, by weight, corresponding to not less than 48.4 percent and not more than 49.5 percent, by volume, at 15.56°, of C_2H_5OH.

Diluted Alcohol may be prepared as follows:

Alcohol	500 mL
Purified Water	500 mL

Measure the Alcohol and the Purified Water separately at the same temperature, and mix. If the water and the Alcohol and the resulting mixture are measured at 25°, the volume of the mixture will be about 970 mL.

Packaging and storage—Preserve in tight containers, remote from fire.

Specific gravity ⟨841⟩: between 0.935 and 0.937 at 15.56°, indicating between 41.0% and 42.0%, by weight, or between 48.4% and 49.5%, by volume, of C_2H_5OH.

Other tests—In other respects it responds to the tests under *Alcohol*, allowance being made for the difference in alcohol concentration.

Alcohols—*see complete list in index*

Alginic Acid

Alginic acid.
Alginic acid [*9005-32-7*].

» Alginic Acid is a hydrophilic colloidal carbohydrate extracted with dilute alkali from various species of brown seaweeds (Phaeophyceae).

Packaging and storage—Preserve in well-closed containers.

Identification—

A: To 5 mL of a 1 in 150 solution in 0.1 *N* sodium hydroxide add 1 mL of calcium chloride TS: a voluminous, gelatinous precipitate is formed.

B: To 5 mL of a 1 in 150 solution in 0.1 *N* sodium hydroxide add 1 mL of 4 *N* sulfuric acid: a heavy, gelatinous precipitate is formed.

C: To about 5 mg in a test tube add 5 mL of water, 1 mL of a freshly prepared 1 in 100 solution of 1,3-naphthalenediol in alcohol, and 5 mL of hydrochloric acid. Heat the mixture to boiling, boil gently for 3 minutes, then cool to about 15°. Transfer the contents of the test tube to a 30-mL separator with the aid of 5 mL of water, and extract with 15 mL of isopropyl ether: the isopropyl ether extract exhibits a deeper purplish hue than that from a blank, similarly prepared.

Microbial limits ⟨61⟩—The total bacterial count does not exceed 200 per g, and the tests for *Salmonella* species and *Escherichia coli* are negative.

pH ⟨791⟩: between 1.5 and 3.5, in a 3 in 100 dispersion in water.

Loss on drying ⟨731⟩—Dry it at 105° for 4 hours: it loses not more than 15.0% of its weight.

Ash—Proceed as directed for *Total Ash* under *Vegetable Drugs—Methods of Analysis* ⟨561⟩, carefully igniting about 4 g of Alginic Acid, accurately weighed, in a tared platinum dish, until the residue is thoroughly carbonized (about 5 minutes), and then igniting in a muffle furnace at a temperature of 800 ± 25° until the carbon is completely burned off (20 to 35 minutes): not more than 4.0% of ash is found.

Arsenic, *Method II* ⟨211⟩: 3 ppm.

Lead—Add 1.0 g to 20 mL of nitric acid in a 250-mL conical flask, mix, and heat carefully until the Alginic Acid is dissolved. Continue heating until the volume is reduced to about 7 mL. Cool rapidly to room temperature, transfer to a 100-mL volumetric flask, and dilute with water to volume. A 50.0-mL portion of this solution contains not more than 5 µg of lead (corresponding to not more than 0.001%) when tested according to the test for *Lead* ⟨251⟩, 15 mL of ammonium citrate solution, 3 mL of potassium cyanide solution, and 500 µL of hydroxylamine hydrochloride solution being used. After the first dithizone extraction, wash the combined chloroform layers with 5 mL of water, discarding the water layer and continuing in the usual manner by extracting with 20 mL of 0.2 *N* nitric acid.

Heavy metals, *Method II* ⟨231⟩: 0.004%, a platinum crucible being used for the ignition and nitric acid being used in place of sulfuric acid to wet the test specimen.

Acid value ⟨401⟩—Suspend about 1 g of Alginic Acid, accurately weighed, in a mixture of 50 mL of water and 30.0 mL of calcium acetate solution (11 in 250). Shake thoroughly, allow the mixture to stand for 1 hour, add phenolphthalein TS, and then titrate the liberated acetic acid with 0.1 *N* sodium hydroxide VS. Perform a blank determination, and calculate the acid value by the formula:

$$5.611(A - B)/W,$$

in which 5.611 is one-tenth the molecular weight of potassium hydroxide, *A* and *B* are the volumes, in mL, of 0.100 *N* sodium hydroxide consumed in the titrations of the test preparation and blank, respectively, and *W* is the weight, in g, of Alginic Acid taken. The acid value of Alginic Acid, calculated on the dried basis, is not less than 230.

Almond Oil

» Almond Oil is the fixed oil obtained by expression from the kernels of varieties of *Prunus amygdalus* Batsch (Fam. Rosaceae).

Packaging and storage—Preserve in tight containers.

Specific gravity ⟨841⟩: between 0.910 and 0.915.

Foreign kernel oils—Shake vigorously 2 mL with a mixture of 1 mL of fuming nitric acid and 1 mL of water for 5 minutes: the mixture is not more than slightly colored.

Cottonseed oil—Mix 5 mL in a test tube with 5 mL of a mixture of equal volumes of amyl alcohol and a 1 in 100 solution of sulfur in carbon disulfide, warm the mixture carefully until the carbon disulfide is expelled, and immerse the test tube to one-third of its length in a boiling, saturated solution of sodium chloride: the mixture develops no reddish color within 2 hours.

Sesame oil—Mix 10 mL with 10 mL of hydrochloric acid, add 0.1 mL of a 1 in 50 solution of furfural in alcohol, and shake the mixture vigorously for 15 seconds: no pink to crimson color appears in the acid layer when separate layers form. Should any color appear in the acid layer, add 10 mL of water, and again shake the mixture vigorously. In the absence of sesame oil the pink color is fugitive.

Mineral oil and foreign fatty oils—Heat 10 mL on a steam bath with 15 mL of 4 *N* sodium hydroxide and 30 mL of alcohol in a flask that has a small, short-stem funnel inserted in the neck, and occasionally agitate the mixture until it becomes clear. Transfer the solution to a shallow dish, evaporate the alcohol on a steam bath, and mix the residue with 100 mL of water: a clear solution results (*absence of mineral oil*). Add an excess of hydrochloric acid to this solution, remove the layer of fatty acids that rises to the surface, wash it with warm water, clarify it by heating on a steam bath, and allow it to cool to 15° without stirring: the fatty acids remain clear for 30 minutes at this temperature (*absence of foreign fatty oils*).

Foreign oils—One volume of the mixed fatty acids obtained in the test for *Mineral oil and foreign fatty oils*, when mixed with 1 volume of alcohol, yields a clear solution, which at 15° does not deposit any fatty acid or become turbid upon the further addition of 1 volume of alcohol (*absence of olive, peanut, or other fixed oils*).

Free fatty acids ⟨401⟩—The free fatty acids in 10 g require for neutralization not more than 5.0 mL of 0.10 *N* sodium hydroxide.

Iodine value ⟨401⟩: between 95 and 105.

Saponification value ⟨401⟩: between 190 and 200.

Aluminum Monostearate

Aluminum, dihydroxy(octadecanoato-*O*-)-.
Dihydroxy(stearato)aluminum [*7047-84-9*].

» Aluminum Monostearate is a compound of aluminum with a mixture of solid organic acids obtained from fats, and consists chiefly of variable proportions of aluminum monostearate and aluminum monopalmitate. It contains the equivalent of not less than 14.5 percent and not more than 16.5 percent of Al_2O_3, calculated on the dried basis.

Packaging and storage—Preserve in well-closed containers.

Identification—

A: Heat 1 g with a mixture of 25 mL of water and 5 mL of hydrochloric acid for 1 hour, replacing the water as it evaporates: fatty acids are liberated, floating as an oily layer on the surface of the liquid, and the water layer responds to the tests for *Aluminum* ⟨191⟩.

B: Mix intimately 25 g with 100 mL of ether in a 500-mL flask, add 150 mL of 3 *N* hydrochloric acid, attach a water-cooled condenser, and heat on a steam bath under reflux for 15 minutes. Cool, and transfer both layers to a separator with the aid of an additional 100 mL of ether. Shake vigorously, and allow the layers to separate. Remove the water layer, and wash the ether layer with three 30-mL portions of water. Transfer the ether layer to a small beaker, warm on a steam bath until the ether has evaporated and the fatty acids are clear, and dry the acids at 105° for 20 minutes: the solidification temperature (see *Fats and Fixed Oils* ⟨401⟩) of the fatty acids is not below 54°.

Loss on drying ⟨731⟩—Dry it at 80° for 16 hours: it loses not more than 2.0% of its weight.

Arsenic, *Method I* ⟨211⟩—Prepare the *Test Preparation* as follows. To 3.75 g add 12.5 mL of hydrochloric acid and 0.5 mL of bromine TS, and heat on a steam bath until a transparent layer of melted fatty acid forms. Add 50 mL of water, heat on a hot plate until the volume is about 25 mL, and filter while hot. Cool, dilute the filtrate with water to 50 mL, and to a 10-mL aliquot of this solution add 2.5 mL of hydrochloric acid, then dilute with water to 55 mL: the resulting solution meets the requirements of the test, the addition of 20 mL of 7 *N* sulfuric acid specified under *Procedure* being omitted. The limit is 4 ppm.

Heavy metals—To 2 g contained in a 250-mL flask add 20 mL of water and 10 mL of hydrochloric acid. Place a small funnel in the neck of the flask, and boil gently, replacing the water as it evaporates, until the fatty acids separate in a clear layer. Cool rapidly by rotating under a stream of cold water until the fatty acids solidify. Decant through a filter previously washed with 3 *N* hydrochloric acid, wash until the combined filtrate and washings measure 50 mL, and mix. To 20 mL of the pooled filtrate add 6 *N* ammonium hydroxide, dropwise, until a permanent turbidity forms. Add 1 *N* acetic acid until the precipitate just dissolves, then add 2 mL in excess, and add water to make 40 mL. Add 10 mL of hydrogen sulfide TS, and allow to stand for 10 minutes: any brown color produced is not darker than that of a control solution prepared from 10 mL of the pooled filtrate and 2 mL of *Standard Lead Solution* (see *Heavy Metals* ⟨231⟩) containing 10 µg of lead per mL, then diluted with water to 20 mL and treated as directed in the foregoing (0.005%).

Assay—Accurately weigh 5 g of Aluminum Monostearate in a covered platinum crucible that previously has been ignited for 20 minutes, cooled over anhydrous magnesium perchlorate, and weighed. Heat the open crucible gently, without allowing the specimen to burst into flame, and gradually increase the heat until the ash is white. Ignite the ash for 20 minutes after the organic matter is removed, and cool. Add 15 mL of water, cover the crucible with a small watch glass, and boil gently for 5 minutes, using a small stirring rod to break up any large lumps of ash. Decant the solution through ashless filter paper, retaining most of the ash in the crucible. Repeat the extraction with water twice, passing the solutions through the same filter. Transfer the ash to the filter by means of a fine stream of water, and wash the crucible and the residue three times with warm water. Transfer the filter paper and the residue to the crucible, dry, and ignite for 20 minutes after the filter paper has burned away. Following the ignition period, cover the crucible, cool over anhydrous magnesium perchlorate for 15 minutes, and weigh the residue of Al_2O_3 rapidly. Repeat the ignition until constant weight is attained, using 20-minute ignition periods and 15-minute cooling periods. From the weight of the residue remaining in the crucible, calculate the content of Al_2O_3.

Strong Ammonia Solution

NH_3 17.03
Ammonia.
Ammonia [*7664-41-7*].

» Strong Ammonia Solution is a solution of NH_3, containing not less than 27.0 percent and not more than 31.0 percent (w/w) of NH_3. On exposure to air it loses ammonia rapidly.

Caution—Use care in handling Strong Ammonia Solution because of the caustic nature of the Solution and the irritating properties of its vapor. Cool the container well before opening, and cover the closure with a cloth or similar material while opening. Do not taste Strong Ammonia Solution, and avoid inhalation of its vapor.

Packaging and storage—Preserve in tight containers, at a temperature not above 25°.

Identification—Hold a glass rod moistened with hydrochloric acid near the surface of the Solution: dense, white fumes are produced.

Nonvolatile residue—Evaporate a 10-mL portion in a tared platinum or porcelain dish to dryness, and dry at 105° for 1 hour: not more than 5 mg of residue remains (0.05%).

Heavy metals, *Method I* ⟨231⟩—Evaporate 1.7 mL on a steam bath to dryness, add to the residue 1 mL of 3 *N* hydrochloric acid, and evaporate to dryness. Dissolve the residue in 2 mL of 1 *N* acetic acid, and dilute with water to 25 mL: the limit is 0.0013%.

Readily oxidizable substances—To a mixture of 4.0 mL of Solution and 6 mL of water add a slight excess of 2 *N* sulfuric acid and 0.10 mL of 0.1 *N* potassium permanganate: the pink color does not completely disappear within 10 minutes.

Assay—Transfer quickly a portion of Strong Ammonia Solution to a stoppered, thick-walled container (a pressure bottle is suitable) to obtain a column height of about 20 cm, insert the stopper, and cool the container and contents to 10° or lower. Weigh accurately a glass-stoppered, 125-mL conical flask containing 35.0 mL of 1 *N* sulfuric acid VS. Insert a graduated 10-mL measuring pipet into the cooled Solution, allow the liquid to rise in the pipet without vacuum, remove the pipet, wipe off adhering liquid, and discard the first mL of the Solution permitted to run from the pipet. Hold the pipet just above the surface of the 1 *N* sulfuric acid VS in the conical flask, and transfer about 2 mL of the Solution into the flask. Insert the stopper, mix, and again weigh to obtain the weight of the specimen. Titrate the excess acid with 1 *N* sodium hydroxide VS, using methyl red TS as the indicator. Perform a blank determination (see *Residual Titrations* under *Titrimetry* ⟨541⟩). Each mL of 1 *N* sulfuric acid is equivalent to 17.03 mg of NH_3.

Ammonium Carbonate

Carbonic acid, monoammonium salt, mixt. with ammonium carbamate.
Monoammonium carbonate mixture with ammonium carbamate [*8000-73-5*].

» Ammonium Carbonate consists of ammonium bicarbonate (NH_4HCO_3) and ammonium carbamate (NH_2COONH_4) in varying proportions. It yields not less than 30.0 percent and not more than 34.0 percent of NH_3.

Packaging and storage—Preserve in tight, light-resistant containers, at a temperature not above 30°.

Identification—When heated, it is volatilized without charring, and the vapor is alkaline to moistened litmus paper. A solution (1 in 20) effervesces with acids.

Residue on ignition ⟨281⟩: not more than 0.1%.

Chloride ⟨221⟩—A 2.0-g portion shows no more chloride than corresponds to 0.10 mL of 0.020 *N* hydrochloric acid (0.0035%).

Sulfate ⟨221⟩—A 2.0-g portion shows no more sulfate than corresponds to 0.10 mL of 0.020 *N* sulfuric acid (0.005%).

Heavy metals, *Method I* ⟨231⟩—Reduce to a coarse powder and volatilize 2 g of it on a steam bath. Add to the residue 1 mL of 3 *N* hydrochloric acid, and evaporate to dryness. Dissolve the residue in 2 mL of 1 *N* acetic acid, and add water to make 25 mL: the limit is 0.001%.

Assay—Place in a weighing bottle about 10 mL of water, tare the bottle and its contents, add about 2 g of Ammonium Carbonate, and weigh accurately. Transfer the contents of the bottle to a 250-mL flask, add 50.0 mL of 1 *N* sulfuric acid VS, and, when solution has been effected, add methyl orange TS, and titrate the excess acid with 1 *N* sodium hydroxide VS. Perform a blank determination (see *Residual Titrations* under *Titrimetry* ⟨541⟩). Each mL of 1 *N* sulfuric acid is equivalent to 17.03 mg of NH_3.

Ammonium Phosphate

$(NH_4)_2HPO_4$ 132.06
Phosphoric acid, diammonium salt.
Diammonium phosphate [*7783-28-0*].

» Ammonium Phosphate contains not less than 96.0 percent and not more than 102.0 percent of $(NH_4)_2HPO_4$.

Packaging and storage—Preserve in tight containers.

Identification—A solution (1 in 20) responds to the tests for *Ammonium* ⟨191⟩ and for *Phosphate* ⟨191⟩.

pH ⟨791⟩: between 7.6 and 8.2, in a solution (1 in 100).

Chloride ⟨221⟩—A 1.0-g portion shows no more chloride than corresponds to 0.40 mL of 0.020 *N* hydrochloric acid (0.03%).

Sulfate ⟨221⟩—A 0.20-g portion shows no more sulfate than corresponds to 0.30 mL of 0.020 *N* sulfuric acid (0.15%).

Arsenic, *Method I* ⟨211⟩: 3 ppm.

Heavy metals ⟨231⟩—Dissolve 2.0 g in 25 mL of water: the limit is 0.001%.

Assay—Dissolve about 600 mg of Ammonium Phosphate, accurately weighed, in 40 mL of water, and titrate with 0.1 *N* sulfuric acid VS to a pH of 4.6, determined potentiometrically. Each mL of 0.1 *N* sulfuric acid is equivalent to 13.21 mg of $(NH_4)_2HPO_4$.

Amylene Hydrate

$CH_3CH_2C(CH_3)(OH)CH_3$

$C_5H_{12}O$ 88.15
2-Butanol, 2-methyl-.
tert-Pentyl alcohol [*75-85-4*].

» Amylene Hydrate contains not less than 99.0 percent and not more than 100.0 percent of $C_5H_{12}O$.

Packaging and storage—Preserve in tight containers.

Identification—

A: Mix 2 mL with 15 mL of water, 5 mL of sulfuric acid, and 10 g of potassium dichromate, heat the mixture under a reflux condenser for 2 hours, and then distil, collecting and reserving the first 2 mL of the distillate. Continue the distillation until most of the water has distilled over, render this distillate alkaline with 1 *N* sodium hydroxide, add 2 *N* sulfuric acid dropwise until the solution is neutral to litmus, and carefully evaporate to dryness: the residue responds to the tests for *Acetate* ⟨191⟩.

B: To 1 mL of the reserved distillate obtained in *Identification test A* add 5 drops of sodium nitroferricyanide TS and 2 mL of 1 *N* sodium hydroxide, then add a slight excess of 6 *N* acetic acid: a deep red liquid, which develops a violet tint when diluted with several volumes of water, is produced.

C: To 10 mL of a solution (1 in 10) quickly add 5 mL of a 1 in 100 solution of vanillin in sulfuric acid: a violet-red color is produced.

Specific gravity ⟨841⟩: between 0.803 and 0.807.

Distilling range, *Method I* ⟨721⟩—It distils completely between 97° and 103°, a correction factor of 0.037° per mm being applied as necessary.

Water—The relative retention time of water on the gas chromatographic column used in the *Assay* is approximately 0.2 times that of Amylene Hydrate. From the area under the curve obtained in the *Assay*, calculate the percentage (a/a) of water in the Amylene Hydrate taken: not more than 0.5% is found.

Nonvolatile residue—Evaporate 10 mL in a tared porcelain dish on a steam bath to a volume of about 1 mL, and allow it to evaporate spontaneously to dryness while protected from dust: the residue, if any, is colorless, and, when dried at 105° for 1 hour, weighs not more than 2 mg (0.02%).

Heavy metals, *Method I* ⟨231⟩—Evaporate 5.0 mL (4 g) on a steam bath to dryness, warm the residue gently with 1 mL of dilute hydrochloric acid (1 in 120), add water to make 25 mL, and filter, if necessary: the limit is 0.0005%.

Readily oxidizable substances—To 10 mL of a solution (1 in 20) add 0.10 mL of 0.10 *N* potassium permanganate: the pink color does not completely disappear within 10 minutes.

Aldehyde—To 10 mL of a solution (1 in 20) add 1 mL of silver–ammonium nitrate TS, and heat the mixture on a water bath at 60° for 10 minutes: no darkening occurs.

Assay—Inject 0.4 µL of Amylene Hydrate into a suitable gas chromatograph (see *Chromatography* ⟨621⟩), equipped with a thermal conductivity detector. Under typical conditions, the instrument contains a 2-m × 4-mm glass column packed with chromatographic support S2. The column is maintained at a temperature of 190°, the injection port is maintained at about 200°, the detector block is maintained at about 200°, and helium is used as a carrier gas at a flow rate of about 50 mL per minute. From the area under the curve, calculate the percentage (a/a) of $C_5H_{12}O$ in the Amylene Hydrate taken.

Anethole

$CH_3O-C_6H_4-CH=CH-CH_3$

$C_{10}H_{12}O$ 148.20

Benzene, 1-methoxy-4-(1-propenyl)-, (*E*)-.
(*E*)-*p*-Propenylanisole [*4180-23-8*].

» Anethole is obtained from Anise Oil and other sources, or is prepared synthetically.

Packaging and storage—Preserve in tight, light-resistant containers.

Labeling—Label it to indicate whether it is obtained from natural sources or is prepared synthetically.

Specific gravity ⟨841⟩: between 0.983 and 0.988.

Congealing temperature ⟨651⟩: not lower than 20°.

Distilling range, *Method I* ⟨721⟩: between 231° and 237°, a correction factor of 0.063° per mm being applied as necessary.

Angular rotation ⟨781⟩: between −0.15° and +0.15° in a 100-mm tube.

Refractive index ⟨831⟩: between 1.557 and 1.561.

Heavy metals, *Method II* ⟨231⟩: 0.004%.

Aldehydes and ketones—Shake 10 mL with 50 mL of a saturated solution of sodium bisulfite in a graduated cylinder, and allow the mixture to stand for 6 hours: no appreciable diminution in the volume of Anethole occurs, and no crystalline deposit separates.

Phenols—Shake 1 mL with 20 mL of water, and allow the liquids to separate. Filter the water layer through a filter paper previously moistened with water, and to 10 mL of the filtrate add 3 drops of ferric chloride TS: no purple or purplish color is produced.

Anhydrous Lanolin—*see* Lanolin, Anhydrous USP

Aromatic Elixir

» Prepare Aromatic Elixir as follows:

Suitable essential oil(s)		
Syrup	375	mL
Talc	30	g
Alcohol, Purified Water, each a sufficient quantity, to make	1000	mL

Dissolve the oil(s) in Alcohol to make 250 mL. To this solution add the Syrup in several portions, agitating vigorously after each addition, and afterwards add, in the same manner, the required quantity of Purified Water. Mix the Talc with the liquid, and filter through a filter wetted with Diluted Alcohol, returning the filtrate until a clear liquid is obtained.

Packaging and storage—Preserve in tight containers.

Alcohol content, *Method II* ⟨611⟩: between 21.0% and 23.0% of C_2H_5OH, acetone being used as the internal standard.

Ascorbic Acid—*see* Ascorbic Acid USP

Ascorbyl Palmitate

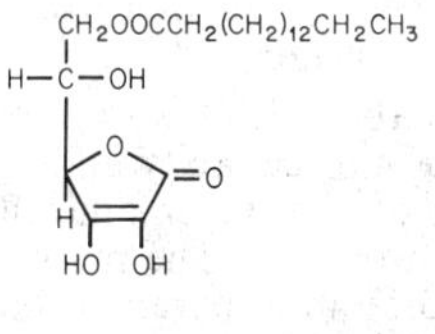

$CH_2OOCCH_2(CH_2)_{12}CH_2CH_3$
H—C—OH
HO OH

$C_{22}H_{38}O_7$ 414.54

L-Ascorbic acid, 6-hexadecanoate.
L-Ascorbic acid 6-palmitate [*137-66-6*].

» Ascorbyl Palmitate contains not less than 95.0 percent and not more than 100.5 percent of $C_{22}H_{38}O_7$, calculated on the dried basis.

Packaging and storage—Preserve in tight containers, in a cool, dry place.

Identification—Dissolve 1 g in 25 mL of alcohol, and add 1 mL of a 1 in 1000 solution of 2,6-dichlorophenol-indophenol sodium in alcohol: the blue color of the 2,6-dichlorophenol-indophenol sodium solution is discharged immediately.

Melting range ⟨741⟩: between 107° and 117°.

Specific rotation ⟨781⟩: between +21° and +24°, calculated on the dried basis, determined in a solution in methanol containing 1 g of Ascorbyl Palmitate in each 10 mL.

Loss on drying ⟨731⟩—Dry it in vacuum at 60° for 1 hour: it loses not more than 2.0% of its weight.

Residue on ignition ⟨281⟩: not more than 0.1%.

Heavy metals, *Method II* ⟨231⟩: 0.001%.

Assay—Dissolve about 300 mg of Ascorbyl Palmitate, accurately weighed, in 50 mL of alcohol, add 30 mL of water, and immediately titrate with 0.1 *N* iodine VS to a yellow color that persists for not less than 30 seconds. Each mL of 0.1 *N* iodine is equivalent to 20.73 mg of $C_{22}H_{38}O_7$.

Aspartame

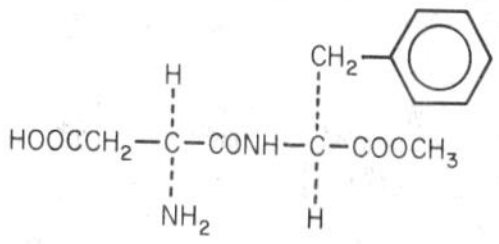

$HOOCCH_2$—C(H)(NH$_2$)—CONH—C(H)(CH_2—C$_6$H$_5$)—$COOCH_3$

$C_{14}H_{18}N_2O_5$ 294.31

L-Phenylalanine, *N*-L-α-aspartyl-, 1-methyl ester.
3-Amino-*N*-(α-carboxyphenethyl)succinamic acid *N*-methyl ester [*22839-47-0*].

» Aspartame contains not less than 98.0 percent and not more than 102.0 percent of $C_{14}H_{18}N_2O_5$, calculated on the dried basis.

Packaging and storage—Preserve in well-closed containers.

Reference standards—*USP Aspartame Reference Standard*—Dry at 105° for 4 hours before using. *USP 5-Benzyl-3,6-dioxo-2-piperazineacetic Acid Reference Standard.*

Identification—The infrared absorption spectrum of a potassium bromide dispersion of it, previously dried, exhibits maxima only at the same wavelengths as that of a similar preparation of USP Aspartame RS.

Transmittance—The transmittance of a 1 in 100 solution of it in 2 *N* hydrochloric acid, prepared by means of sonication, determined in a 1-cm cell at 430 nm with a suitable spectrophotometer, is not less than 0.95, corresponding to an absorbance of not more than about 0.022.

Specific rotation ⟨781⟩: between +14.5° and +16.5°, calculated on the dried basis, determined at 20° in a solution in 15 *N* formic acid containing 40 mg per mL within 30 minutes after preparation of the solution.

Loss on drying ⟨731⟩—Dry it at 105° for 4 hours: it loses not more than 4.5% of its weight.

Residue on ignition ⟨281⟩: not more than 0.2%.

Arsenic, *Method II* ⟨211⟩: 3 ppm.

Heavy metals, *Method II* ⟨231⟩: 0.001%.

Other related substances—In the chromatogram obtained from the *Test preparation* in the test for *5-Benzyl-3,6-dioxo-2-piperazineacetic acid*, the elution being continued to twice the retention time of the aspartame peak, the sum of the response of any peaks detected, other than the peaks from aspartame and 5-benzyl-3,6-dioxo-2-piperazineacetic acid, is not more than 2.0% of the response of the aspartame peak.

5-Benzyl-3,6-dioxo-2-piperazineacetic acid—

Mobile phase—In a 1000-mL flask, dissolve 9.6 g of anhydrous citric acid in about 800 mL of water, adjust with 1 *N* sodium hydroxide to a pH of 4.7, add water to volume, and mix. Filter the solution, and mix 670 mL of it with 330 mL of methanol. Degas the mixture with vacuum and stirring.

Standard preparation—Dissolve an accurately weighed quantity of USP 5-Benzyl-3,6-dioxo-2-piperazineacetic Acid RS in *Mobile phase*, and dilute quantitatively with *Mobile phase* to obtain a solution having a known concentration of 45 µg per mL.

Test preparation—Transfer 300 mg of Aspartame, accurately weighed, to a 100-mL volumetric flask, dissolve in *Mobile phase*, dilute with *Mobile phase* to volume, and mix by inversion.

Chromatographic system (see *Chromatography* ⟨621⟩)—The liquid chromatograph is equipped with a 254-nm detector and a 4-mm × 30-cm column containing packing L1. Chromatograph the *Standard preparation*, and record the peak responses as directed under *Procedure:* the relative standard deviation for replicate injections is not more than 4.0%. The tailing factor, *T*, is not more than 2.0.

Procedure—Separately inject equal volumes (about 25 µL) of the *Standard preparation* and the *Test preparation* into the chromatograph, and record the chromatograms. The flow rate is about 1.5 mL per minute, and the relative retention times for 5-benzyl-3,6-dioxo-2-piperazineacetic acid and aspartame are about 1.0 and 1.9, respectively. The response obtained for any peak observed in the chromatogram of the *Test preparation* at a retention time corresponding to that of the Reference Standard in the *Standard preparation* is not greater than the response obtained for the peak in the chromatogram of the *Standard preparation*, corresponding to not more than 1.5% of 5-benzyl-3,6-dioxo-2-piperazineacetic acid.

Assay—

0.1 N Perchloric acid—Use perchloric acid, tenth-normal (in glacial acetic acid) as specified under *Volumetric Solutions*, in the section, *Reagents, Indicators, and Solutions*, but in the standardization, titrate to a green end-point.

Procedure—Transfer about 300 mg of Aspartame, accurately weighed, to a 150-mL beaker, dissolve in 1.5 mL of anhydrous formic acid, and add 60 mL of glacial acetic acid. Add crystal violet TS, and immediately titrate with *0.1 N Perchloric acid* to a green end-point. Perform a blank determination, and make any necessary correction. Each mL of 0.1 *N* perchloric acid is equivalent to 29.43 mg of $C_{14}H_{18}N_2O_5$. [NOTE—A blank titration exceeding 0.1 mL may be due to excessive water content, and may cause loss of visual end-point sensitivity.]

Bacteriostatic Sodium Chloride Injection—*see* Sodium Chloride Injection, Bacteriostatic USP

Barium Hydroxide Lime—*see* Barium Hydroxide Lime USP

Bentonite

Bentonite.
Bentonite [*1302-78-9*].

» Bentonite is a native, colloidal, hydrated aluminum silicate.

Packaging and storage—Preserve in tight containers.

Labeling—Label it to indicate that absorption of atmospheric moisture should be avoided following the opening of the original package, preferably by storage of the remainder of the contents in a tight container.

Identification—Add 2 g in small portions to 100 mL of water, with intense agitation. Allow to stand for 12 hours to ensure complete hydration. Place 2 mL of the mixture so obtained on a suitable glass slide, and allow to air-dry at room temperature to produce an oriented film. Place the slide in a vacuum desiccator over a free surface of ethylene glycol. Evacuate the desiccator, and close the stopcock so that the ethylene glycol saturates the desiccator chamber. Allow to stand for 12 hours. Record the X-ray diffraction pattern (see *X-ray Diffraction* ⟨941⟩), and calculate the *d* values: the largest peak corresponds to a *d* value between 15.0 and 17.2 angstrom units. Prepare a random powder specimen of Bentonite, record the X-ray diffraction pattern, and determine the *d* values in the region between 1.48 and 1.54 angstrom units: the peak is between 1.492 and 1.504 angstrom units.

Microbial limit—It meets the requirements of the test for absence of *Escherichia coli* under *Microbial Limit Tests* ⟨61⟩.

pH ⟨791⟩—Disperse 4.0 g in 200 mL of water, mixing vigorously to facilitate wetting. The pH of this suspension is between 9.5 and 10.5.

Loss on drying ⟨731⟩—Dry it at 105° for 2 hours: it loses between 5.0% and 8.0% of its weight.

Arsenic ⟨211⟩—

Test preparation—Transfer 8.0 g to a 250-mL beaker containing 100 mL of dilute hydrochloric acid (1 in 25), mix, cover with a watch glass, and boil gently, with occasional stirring, for 15 minutes without allowing excessive foaming. Filter the hot supernatant liquid through a rapid-flow filter paper into a 200-mL volumetric flask, and wash with four 25-mL portions of hot dilute hydrochloric acid (1 in 25), collecting the washings in the volumetric flask. Cool the combined filtrates to room temperature, add dilute hydrochloric acid (1 in 25) to volume, and mix.

Procedure—Use a 25-mL aliquot of *Test preparation* for the *Procedure*. The absorbance due to any red color from the *Test preparation* does not exceed that produced by 5.0 mL of *Standard preparation* (5 µg of As) when treated with the same quantities of the same reagents and in the same manner: the limit is 5 ppm.

Lead—[NOTE—The *Standard preparation* and the *Test preparation* may be modified, if necessary, to obtain solutions, of suitable concentrations, adaptable to the linear or working range of the instrument. It may be necessary to correct the value obtained for the *Test preparation* for interference from the test specimen matrix.]

Standard preparation—On the day of use, dilute 3.0 mL of *Lead Nitrate Stock Solution* (see *Heavy Metals* ⟨231⟩) with water to 100 mL. Each mL of the *Standard preparation* contains the equivalent of 3 µg of lead.

Test preparation—Transfer 3.75 g to a 250-mL beaker containing 100 mL of dilute hydrochloric acid (1 in 25), stir, cover with a watch glass, and boil for 15 minutes. Cool to room temperature, and filter through a rapid-flow filter paper into a 400-mL beaker. Wash the filter with four 25-mL portions of hot water, collecting the washings in the 400-mL beaker. Concentrate the combined extracts by gentle boiling to approximately 20 mL. If a precipitate appears, add 2 to 3 drops of nitric acid, heat to boiling, and cool to room temperature. Filter the concentrated extracts through a rapid-flow filter paper into a 50-mL volumetric flask. Transfer the remaining contents of the 400-mL beaker through the filter paper and into the flask with water. Dilute with water to volume, and mix.

Procedure—Determine the absorbances of the *Test preparation* and the *Standard preparation* at 284 nm in a suitable atomic absorption spectrophotometer equipped with a lead hollow-cathode lamp deuterium arc background correction, and a single-slot burner, using an oxidizing flame of air and acetylene. The absorbance of the *Test preparation* is not greater than that of the *Standard preparation* (0.004%).

Gel formation—Mix 6 g with 300 mg of magnesium oxide. Add the mixture, in several divided portions, to 200 mL of water contained in a blender of approximately 500-mL capacity. Blend thoroughly for 5 minutes at high speed, transfer 100 mL of the mixture to a 100-mL graduated cylinder, and allow to remain undisturbed for 24 hours: not more than 2 mL of supernatant liquid appears on the surface.

Swelling power—To 100 mL of water contained in a glass-stoppered cylinder of 100-mL capacity add 2 g in portions, dropping it upon the surface of the water and allowing each portion to settle before adding the next. The mass at the bottom gradually swells until it occupies an apparent volume of not less than 24 mL at the end of a 2-hour period.

Fineness of powder—Sprinkle 2 g upon 20 mL of water contained in a mortar. Allow to swell, disperse evenly with a pestle, and dilute with water to 100 mL. Pour the suspension through a No. 200 standard sieve, and wash the sieve thoroughly with water. No grit is felt when the fingers are rubbed over the wire mesh of the sieve.

Purified Bentonite

» Purified Bentonite is a colloidal montmorillonite that has been processed to remove grit and non-swellable ore components.

Packaging and storage—Preserve in tight containers.

Identification—It responds to the *Identification test* under *Bentonite*.

Viscosity—After determining the *Loss on drying*, weigh a quantity of Purified Bentonite test specimen equivalent to 25.0 g on the dried basis. Over a period of a few seconds, transfer the undried test specimen to a suitable 1-liter blender jar containing an amount of water, maintained at a temperature of 25 ± 2°, that is sufficient to produce a mixture weighing 500 g. Blend for 3 minutes, accurately timed, at 14,000 to 15,000 rpm (high speed). [NOTE—Heat generated during blending causes a temperature rise to above 30°.] Transfer the contents of the blender to a 600-mL beaker, allow to stand for 5 minutes, and adjust, if necessary, to a temperature of 33 ± 3°. Using a suitable rotational viscosimeter equipped with a spindle having a cylinder 1.87 cm in diameter and 0.69 cm high attached to a shaft 0.32 cm in diameter, the distance from the top of the cylinder to the lower tip of the shaft being 2.54 cm, and the immersion depth being 5.00 cm (No. 2 spindle), operate the viscosimeter at 60 rpm for 6 minutes, accurately timed, and record the scale reading: the viscosity is between 40 centipoises and 200 centipoises.

Microbial limits ⟨61⟩—Its total aerobic microbial count does not exceed 1000 per g, and it meets the requirements of the test for absence of *Escherichia coli*.

pH ⟨791⟩: between 9.0 and 10.0, in a suspension (5 in 100) in water.

Acid demand—After determining the *Loss on drying*, weigh a quantity of Purified Bentonite equivalent to 5.00 g, and disperse in 500 mL of water with the aid of a suitable blender fitted with a 1-liter jar. Using a stopwatch, designate zero time. With constant mixing, add 3.0-mL portions of 0.100 *N* hydrochloric acid at 5, 65, 125, 185, 245, 305, 365, 425, 485, 545, 605, 665, and 725 seconds, and add a 1.0-mL portion at 785 seconds. Determine the pH potentiometrically at 840 seconds: the pH is not more than 4.0.

Loss on drying ⟨731⟩—Dry it at 110° to constant weight: it loses not more than 8.0% of its weight.

Arsenic, *Method I* ⟨211⟩—Prepare the *Test Preparation* as follows. Transfer 13.3 g to a 250-mL beaker containing 100 mL of dilute hydrochloric acid (1 in 25), mix, cover with a watch glass, and boil gently, with occasional stirring, for 15 minutes without allowing excessive foaming. Allow the insoluble material to settle, and decant the hot supernatant liquid through a rapid-flow filter paper into a 200-mL volumetric flask, retaining as much sediment as possible in the beaker. Add 25 mL of hot dilute hydrochloric acid (1 in 25) to the residue in the beaker, stir, heat to boiling, allow the insoluble material to settle, and decant the supernatant liquid through the filter into the 200-mL volumetric flask. Repeat the extraction with four additional 25-mL portions of hot dilute hydrochloric acid (1 in 25), decanting each hot supernatant liquid through the filter into the volumetric flask. At the last extraction, transfer as much of the insoluble material as possible onto the filter. Cool the combined filtrates to room temperature, add dilute hydrochloric acid (1 in 25) to volume, and mix.

Procedure—Use a 25-mL aliquot of *Test Preparation* for the *Procedure*. The absorbance due to any red color from the *Test Preparation* does not exceed that produced by 5.0 mL of *Standard Preparation* (5 µg of As) when treated with the same quantities of the same reagents and in the same manner. The limit is 3 ppm.

Lead—[NOTE—The *Standard preparation* and the *Test preparation* may be modified, if necessary, to obtain solutions, of suitable concentrations, adaptable to the linear or working range of the instrument.]

Standard preparation—On the day of use, dilute 3.0 mL of *Lead Nitrate Stock Solution* (see *Heavy Metals* ⟨231⟩) with water to 100 mL. Each mL of the *Standard preparation* contains the equivalent of 3 µg of lead.

Test preparation—Transfer 10.0 g of Purified Bentonite to a 250-mL beaker containing 100 mL of dilute hydrochloric acid (1 in 25), stir, cover with a watch glass, and boil for 15 minutes. Cool to room temperature, and allow the insoluble matter to settle. Decant the supernatant liquid through a rapid-flow filter paper into a 400-mL beaker. Add 25 mL of hot water to the insoluble matter in the 250-mL beaker, stir, allow the insoluble matter to settle, and decant the supernatant liquid through the filter into the 400-mL beaker. Repeat the extraction with two additional 25-mL portions of water, decanting each supernatant liquid portion through the filter into the 400-mL beaker. Wash the filter with 25 mL of hot water, collecting this filtrate in the 400-mL beaker. Concentrate the combined extracts by gentle boiling to approximately 20 mL. If a precipitate appears, add 2 to 3 drops of nitric acid, heat to boiling, and cool to room temperature. Filter the concentrated extracts through a rapid-flow filter paper into a 50-mL volumetric flask. Transfer the remaining contents of the 400-mL beaker through the filter paper and into the flask with water. Dilute with water to volume, and mix.

Procedure—Determine the absorbances of the *Test preparation* and the *Standard preparation* at 284 nm in a suitable atomic absorption spectrophotometer equipped with a lead hollow-cathode lamp, deuterium arc background correction, and a single-slot burner, using an oxidizing flame of air and acetylene. The absorbance of the *Test preparation* is not greater than that of the *Standard preparation* (0.0015%).

Assay for aluminum and magnesium content—[NOTE—The *Standard preparations* and *Assay preparations* may be diluted quantitatively with water, if necessary, to obtain solutions, of suitable concentrations, adaptable to the linear or working range of the instrument.]

Lanthanum solution—Stir 88.30 g of lanthanum chloride ($LaCl_3$) with 500 mL of 6 *N* hydrochloric acid until solution is complete, transfer with the aid of water to a 1000-mL volumetric flask, and dilute with water to volume.

Specimen preparation—Transfer 0.200 g of Purified Bentonite to a 25-mL platinum crucible containing 1.0 g of lithium metaborate, and mix. Using a muffle furnace or a suitable burner, heat slowly at first, and ignite at 1000° to 1200° for 15 minutes. Cool, place the crucible in a 100-mL beaker containing 25 mL of dilute nitric acid (1 in 20), and add an additional 50 mL of the dilute acid, filling and submerging the upright crucible. Place a polyfluorocarbon-coated magnetic stirring bar into the crucible, and stir gently with a magnetic stirrer until solution is complete. Pour the contents into a 250-mL beaker, and remove the crucible. Warm the solution, and transfer through a rapid-flow filter paper with the aid of water into a 200-mL volumetric flask, dilute with water to volume, and mix.

Aluminum standard preparations—Dissolve 1.000 g of aluminum in a mixture of 10 mL of hydrochloric acid and 10 mL of water by gentle heating. Transfer the solution to a 1000-mL volumetric flask, dilute with water to volume, and mix. This solution contains the equivalent of 1 mg of aluminum per mL. Transfer 2-, 5-, and 10-mL aliquots to separate 100-mL volumetric flasks containing 200 mg of sodium chloride, dilute each with water to volume, and mix.

Aluminum assay preparation—Dissolve 100 mg of sodium chloride in a 50-mL aliquot of *Specimen preparation.*

Procedure for aluminum—In a suitable atomic absorption spectrophotometer equipped with an aluminum hollow-cathode lamp and a single-slot burner, using an oxidizing air-acetylene–nitrous oxide flame, determine the absorbances of the *Aluminum assay preparation* and each of the *Aluminum standard preparations* at 309 nm. From a linear regression equation calculated from the absorbances and concentrations of the *Aluminum standard preparations*, determine the aluminum content of the Purified Bentonite.

Magnesium standard preparations—Place 1.000 g of magnesium in a 250-mL beaker containing 20 mL of water, and carefully add 20 mL of hydrochloric acid, warming, if necessary, to complete the reaction. Transfer the solution to a 1000-mL volumetric flask, dilute with water to volume, and mix. This solution contains the equivalent of 1 mg of magnesium per mL. Transfer 5-, 10-, 15-, and 20-mL aliquots to separate 100-mL volumetric flasks. To each flask add 20.0 mL of *Lanthanum solution*, dilute with water to volume, and mix.

Magnesium assay preparation—Transfer a 25-mL aliquot of *Specimen preparation* to a 50-mL volumetric flask, dilute with water to volume, and mix. Transfer a 5.0-mL aliquot of this dilution to a 100-mL volumetric flask, add 20.0 mL of *Lanthanum solution*, dilute with water to volume, and mix.

Procedure for magnesium—In a suitable atomic absorption spectrophotometer equipped with a magnesium hollow-cathode lamp and a single-slot burner, using a reducing acetylene-air flame, determine the absorbances of the *Magnesium assay preparation* and each of the *Magnesium standard preparations* at 285 nm. From a linear regression equation calculated from the absorbances and concentrations of the *Magnesium standard preparations*, determine the magnesium content of the Purified Bentonite: the ratio of the aluminum content to the magnesium content is between 3.5 and 5.5.

Bentonite Magma

» Prepare Bentonite Magma as follows:

Bentonite	50 g
Purified Water, a sufficient quantity, to make	1000 g

Sprinkle the Bentonite, in portions, upon 800 g of hot purified water, allowing each portion to become thoroughly wetted without stirring. Allow it to stand with occasional stirring for 24 hours. Stir until a uniform magma is obtained, add Purified Water to make 1000 g, and mix.

The Magma may be prepared also by mechanical means such as by use of a blender, as follows. Place about 500 g of Purified Water in the blender, and while the machine is running, add the Bentonite. Add Purified Water to make up to about 1000 g or up to the operating capacity of the blender. Blend the mixture for 5 to 10 minutes, add Purified Water to make 1000 g, and mix.

Packaging and storage—Preserve in tight containers.

Microbial limit—It meets the requirements of the test for absence of *Escherichia coli* under *Microbial Limit Tests* ⟨61⟩.

Benzaldehyde

C_6H_5-CHO

C_7H_6O 106.12

Benzaldehyde.

Benzaldehyde [*100-52-7*].

» Benzaldehyde contains not less than 98.0 percent and not more than 100.5 percent of C_7H_6O.

Packaging and storage—Preserve in well-filled, tight, light-resistant containers.

Specific gravity ⟨841⟩: between 1.041 and 1.046 at 25°.

Refractive index ⟨831⟩: between 1.544 and 1.546 at 20°.

Hydrocyanic acid—Shake 0.5 mL of it with 5 mL of water, add 0.5 mL of 1 *N* sodium hydroxide and 0.1 mL of ferrous sulfate TS, and warm the mixture gently. Upon the addition of a slight excess of hydrochloric acid, no greenish blue color or blue precipitate is produced within 15 minutes.

Chlorinated compounds—Wind a strip of 20-mesh copper gauze 1.5 cm wide and 5 cm long around the end of a copper wire. Heat the gauze in the nonluminous flame of a Bunsen burner until it glows without coloring the flame green. Permit the gauze to cool, and heat several times until a thick coat of oxide has formed. With a medicine dropper, apply 2 drops of Benzaldehyde to the cooled gauze, ignite, and permit it to burn freely in the air. Again cool the gauze, add 2 more drops of it, and burn as before. Repeat this process until a total of 6 drops has been added and ignited. Then hold the gauze in the outer edge of the Bunsen flame, adjusted to a height of about 4 cm: not even a transient green color is imparted to the flame.

Nitrobenzene—Dissolve 1 mL of it in 20 mL of alcohol, and mix with 10 mL of water. Add 1-g portions of zinc and 1-mL portions of 2 *N* sulfuric acid, as needed, to maintain a brisk evolution of hydrogen for 1 hour. Filter, evaporate the liquid to about 20 mL, and boil 10 mL of the concentrated liquid with 1 drop of potassium dichromate TS: no purplish color is produced.

Assay—Transfer about 1 mL of Benzaldehyde to a tared, glass-stoppered weighing bottle, and weigh accurately. Loosen the stopper, and transfer the weighing bottle and contents to a 250-mL conical flask containing 25 mL of a reagent prepared by dissolving 34.7 g of hydroxylamine hydrochloride in 160 mL of water and diluting this solution with alcohol to 1000 mL. Rinse down the sides of the flask with 50 mL of the reagent. Allow to stand for 10 minutes, add 1 mL of bromophenol blue TS, and titrate the liberated hydrochloric acid with 1 *N* sodium hydroxide VS to a light-green end-point. Perform a blank determination (see *Residual Titrations* under *Titrimetry* ⟨541⟩). Each mL of 1 *N* sodium hydroxide is equivalent to 106.1 mg of C_7H_6O.

Compound Benzaldehyde Elixir

» Prepare Compound Benzaldehyde Elixir as follows:

Benzaldehyde	0.5 mL
Vanillin	1 g
Orange Flower Water	150 mL
Alcohol	50 mL
Syrup	400 mL
Purified Water, a sufficient quantity, to make	1000 mL

Dissolve the Benzaldehyde and the Vanillin in the Alcohol; add the Syrup, the Orange Flower Water, and sufficient Purified Water, in several portions, shaking the mixture thoroughly after each addition, to make the product measure 1000 mL; then filter, if necessary, until the product is clear.

Packaging and storage—Preserve in tight, light-resistant containers.

Alcohol content, *Method I* ⟨611⟩: between 3.0% and 5.0% of C_2H_5OH.

Benzalkonium Chloride

Ammonium, alkyldimethyl(phenylmethyl)-, chloride.

Alkylbenzyldimethylammonium chloride [*8001-54-5*].

» Benzalkonium Chloride is a mixture of alkylbenzyldimethylammonium chlorides of the general formula:

$$[C_6H_5CH_2N(CH_3)_2R]Cl,$$

in which R represents a mixture of alkyls, including all or some of the group beginning with n-C_8H_{17} and extending through higher homologs, with n-$C_{12}H_{25}$, n-$C_{14}H_{29}$, and n-$C_{16}H_{33}$ comprising the major portion. On the anhydrous basis, the content of the n-$C_{12}H_{25}$ homolog is not less than 40.0 percent, and the content of the n-$C_{14}H_{29}$ homolog is not less than 20.0 percent, of the total alkylbenzyldimethylammonium chloride content. The amounts of the n-$C_{12}H_{25}$ and n-$C_{14}H_{29}$ homolog components comprise together not less than 70.0 percent of the total alkylbenzyldimethylammonium chloride content. The total alkylbenzyldimethylammonium chloride content, calculated on the anhydrous basis, allowance being made for the amount of residue on ignition, is not less than 97.0 percent and not more than 103.0 percent of $[C_6H_5CH_2N(CH_3)_2R]Cl$ of average molecular weight 360.

Packaging and storage—Preserve in tight containers.

Reference standard—*USP Benzalkonium Chloride Reference Standard*—After opening ampul, store in a tightly closed container.

Identification—

A: To a solution (1 in 100) add 2 *N* nitric acid or mercuric chloride TS: a white precipitate is formed, and it is soluble in alcohol.

B: Dissolve about 200 mg in 1 mL of sulfuric acid, add 100 mg of sodium nitrate, and heat on a steam bath for 5 minutes. Cool, dilute with water to 10 mL, add 500 mg of zinc dust, and warm for 5 minutes on a steam bath. To 2 mL of the clear supernatant liquid add 1 mL of sodium nitrite solution (1 in 20), cool in ice water, then add 3 mL of a solution of 500 mg of 2-naphthol in 10 mL of 6 *N* ammonium hydroxide: an orange-red color is produced.

C: A solution of it in a mixture of equal volumes of water and alcohol responds to the tests for *Chloride* ⟨191⟩.

Water, *Method I* ⟨921⟩: not more than 15.0%.

Residue on ignition ⟨281⟩: not more than 2.0%.

Water-insoluble matter—A solution (1 in 10) is free from turbidity and insoluble matter.

Foreign amines—To 5 mL of a solution (1 in 50) add 3 mL of 1 *N* sodium hydroxide: no precipitate is formed. Heat to boiling: the odor of amines is not perceptible.

Ratio of alkyl components—

Mobile phase—Adjust a 0.1 *M* solution of sodium acetate with glacial acetic acid to a pH of 5.0. Mix 55 parts of this solution with 45 parts of acetonitrile, filter, and degas. The acetonitrile concentration may be varied from 40 parts to 60 parts to meet system suitability requirements.

Standard preparation—Dilute USP Benzalkonium Chloride RS with water to obtain a solution having a concentration of about 4 mg per mL.

Test preparation—Dissolve about 1 g of Benzalkonium Chloride, accurately weighed, in water in a 50-mL volumetric flask, dilute with water to volume, and mix. Pipet 5.0 mL of this solution into a 25-mL volumetric flask, dilute with water to volume, and mix.

Chromatographic system (see *Chromatography* ⟨621⟩)—The liquid chromatograph is equipped with a 254-nm detector and a column that contains packing L10. The flow rate is about 2 mL per minute. Chromatograph the *Standard preparation*, and record the peak areas as directed under *Procedure:* the column efficiency, based on the C_{12} peak, is not less than 1000 theoretical plates, the resolution between the C_{12} and C_{14} peaks is not less than 1.5, and the relative standard deviation for replicate injections is not more than 2.0% for the C_{12} peak.

Procedure—Inject about 20 µL of the *Test preparation* into the chromatograph, record the chromatogram, and measure the areas for the peaks. Identify the homolog peaks by comparison of the retention times with those from the *Standard preparation*, similarly chromatographed. Calculate the percentage of each quaternary ammonium homolog by the formula:

$$100A/B,$$

in which A is the product of the area obtained from the homolog multiplied by its molecular weight, and B is the sum of all of these products. The molecular weights of the C_{10}, C_{12}, C_{14}, and C_{16} homologs are 312, 340, 368, and 396, respectively.

Assay for total alkylbenzyldimethylammonium chlorides—Weigh accurately a quantity of Benzalkonium Chloride equivalent to about 500 mg of anhydrous benzalkonium chloride, and transfer, with the aid of 35 mL of water, to a glass-stoppered, 250-mL conical separator containing 25 mL of chloroform. Add 10.0 mL of freshly prepared potassium iodide solution (1 in 20), insert the stopper in the separator, shake, allow the layers to separate, and discard the chloroform layer. Wash the aqueous layer with three 10-mL portions of chloroform, and discard the washings. Transfer the aqueous layer to a glass-stoppered, 250-mL conical flask, and rinse the separator with three 5-mL portions of water, adding the washings to the flask. Add 40 mL of cold hydrochloric acid to the flask, mix, and titrate with 0.05 *M* potassium iodate VS until the solution becomes light brown in color. Add 5 mL of chloroform, insert the stopper into the flask, and shake vigorously. Continue the titration, dropwise, with shaking after each addition, until the chloroform layer becomes colorless and the aqueous layer is clear yellow. Perform a blank determination, using 20 mL of water as the sample. The difference between the two titrations represents the amount of potassium iodate equivalent to the weight of benzalkonium chloride in the sample. Each mL of 0.05 *M* potassium iodate is equivalent to 36.0 mg of benzalkonium chloride.

Benzalkonium Chloride Solution

» Benzalkonium Chloride Solution contains not less than 95.0 percent and not more than 105.0 percent of the labeled amount of benzalkonium chloride in concentrations of 1.0 percent or more; and not less than 93.0 percent and not more than 107.0 percent of the labeled amount in concentrations of less than 1.0 percent. It may contain a suitable coloring agent and may contain not more than 10 percent of alcohol.

Caution—Mixing Benzalkonium Chloride Solution with ordinary soaps and with anionic detergents may decrease or destroy the bacteriostatic activity of the Solution.

Packaging and storage—Preserve in tight containers, and prevent contact with metals.

Identification—

A: It responds to *Identification tests A* and *C* under *Benzalkonium Chloride.*

B: The residue obtained by evaporating, on a steam bath, a volume of Solution equivalent to about 200 mg of benzalkonium chloride responds to *Identification test B* under *Benzalkonium Chloride.*

Microbial limit—Solution containing less than 5.0% of benzalkonium chloride meets the requirements of the test for absence of *Pseudomonas aeruginosa* under *Microbial Limit Tests* ⟨61⟩.

Alcohol (*if present*): between 95.0% and 105.0% of the labeled amount, determined as directed under *Alcohol Determination* ⟨611⟩.

Foreign amines—A volume of Solution, equivalent to 100 mg of benzalkonium chloride, and adjusted to a concentration of 1 in 50, meets the requirements of the test for *Foreign amines* under *Benzalkonium Chloride.*

Assay—Evaporate or dilute with water to about 30 mL an accurately measured volume of Benzalkonium Chloride Solution, equivalent to about 275 mg of benzalkonium chloride. Transfer the solution with the aid of a minimum quantity of water to a glass-stoppered, 250-mL conical separator. Add 25 mL of chloroform, and proceed as directed in the *Assay for total alkylbenzyldimethylammonium chlorides* under *Benzalkonium Chloride*, beginning with "Add 10.0 mL of freshly prepared potassium iodide solution (1 in 20)."

Benzethonium Chloride—*see* Benzethonium Chloride USP

Benzoic Acid—*see* Benzoic Acid USP

Benzyl Alcohol

C_7H_8O 108.14
Benzenemethanol.
Benzyl alcohol [*100-51-6*].

Packaging and storage—Preserve in tight containers, and prevent exposure to excessive heat.

Identification—Add 2 or 3 drops of it to 5 mL of potassium permanganate solution (1 in 20), and acidify with 2 *N* sulfuric acid: the odor of benzaldehyde is perceptible.

Specific gravity ⟨841⟩: between 1.042 and 1.047.

Distilling range, *Method II* ⟨721⟩—Not less than 94.0%, by volume, distils between 202.5° and 206.5°.

Refractive index ⟨831⟩: between 1.539 and 1.541 at 20°.

Residue on ignition ⟨281⟩—Evaporate 25 mL in a suitable crucible, and ignite to constant weight: not more than 0.005% is found.

Chlorinated compounds—Wind a strip of 20-mesh copper gauze 1.5 cm in width and 5 cm in length around the end of a copper wire. Heat the gauze in the nonluminous flame of a Bunsen burner until it glows without coloring the flame green. Permit the gauze to cool, and heat again several times until a considerable coat of oxide has formed. Apply 2 drops of Benzyl Alcohol from a medicine dropper to the cooled gauze, ignite, and permit it to burn freely in the air. Again cool the gauze, add 2 more drops of Benzyl Alcohol, and burn as before. Continue this process until a total of 6 drops has been added and ignited. Then hold the gauze in the outer edge of the Bunsen burner flame, adjusted to a height of about 4 cm: not even a transient green color is imparted to the flame.

Aldehyde—Transfer 20.0 mL to a 250-mL conical flask containing 5 mL of a reagent prepared by dissolving 3.5 g of hydroxylamine hydrochloride in 100 mL of 60% alcohol, add 50 mL of 60% alcohol, and mix. Allow to stand for 10 minutes, add 1 mL of bromophenol blue TS, and titrate with 0.1 *N* sodium hydroxide VS to a light green end-point. Perform a blank determination (see *Residual Titrations* under *Titrimetry* ⟨541⟩). Match the color of the end-point with that of the titrated test solution: the net volume of 0.1 *N* sodium hydroxide consumed does not exceed 4.0 mL (0.2% as benzaldehyde).

Benzyl Benzoate—*see* Benzyl Benzoate USP

Boric Acid

H_3BO_3 61.83
Boric acid (H_3BO_3).
Boric acid (H_3BO_3) [*10043-35-3*].

» Boric Acid contains not less than 99.5 percent and not more than 100.5 percent of H_3BO_3, calculated on the dried basis.

Packaging and storage—Preserve in well-closed containers.

Labeling—Label the container with a warning that it is not for internal use.

Solubility in alcohol—One g dissolves completely in 10 mL of boiling alcohol.

Completeness of solution—One g dissolves in 25 mL of water, producing a clear solution.

Identification—It responds to the tests for *Borate* ⟨191⟩.

Loss on drying ⟨731⟩—Dry it over silica gel for 5 hours: it loses not more than 0.5% of its weight.

Arsenic, *Method I* ⟨211⟩: 8 ppm.

Heavy metals, *Method I* ⟨231⟩—Dissolve 1 g in 23 mL of water, and add 2 mL of 1 *N* acetic acid: the limit is 0.002%.

Assay—Dissolve about 2 g of Boric Acid, accurately weighed, in 100 mL of a mixture of equal volumes of glycerin and water, previously neutralized to phenolphthalein TS, and titrate with 1 *N* sodium hydroxide VS. Discharge the pink color by the addition of 50 mL of glycerin, neutralized to phenolphthalein TS, and continue the titration until the pink color reappears. Each mL of 1 *N* sodium hydroxide is equivalent to 61.83 mg of H_3BO_3.

Butane

C_4H_{10} 58.12
n-Butane [*106-97-8*].

» Butane contains not less than 97.0 percent of *n*-butane (C_4H_{10}).

Caution—Butane is highly flammable and explosive.

Packaging and storage—Preserve in tight cylinders, and prevent exposure to excessive heat.

Identification—

A: The infrared absorption spectrum of it exhibits maxima, among others, at about the following wavelengths, in μm: 3.4 (*vs*), 6.8 (*s*), 7.2 (*m*), and 10.4 (*m*).

B: The vapor pressure of a test specimen, obtained as directed in the *General Sampling Procedure* under *Propellants* (see *Aerosols* ⟨601⟩), determined at 21° by means of a suitable pressure gauge, is between 205 and 235 kPa absolute (30 and 34 psia).

Water: not more than 0.001%, determined as directed for *Water Content* under *Aerosols* ⟨601⟩.

High-boiling residues: not more than 5 ppm, determined as directed for *High-boiling Residues, Method II*, under *Aerosols* ⟨601⟩.

Acidity of residue—Add 10 mL of water to the residue obtained in the test for *High-boiling residues*, mix by swirling for about 30 seconds, add 2 drops of methyl orange TS, insert the stopper in the tube, and shake vigorously: no pink or red color appears in the aqueous layer.

Sulfur compounds—Carefully open the container valve to produce a moderate flow of gas. Do not direct the gas stream toward the face, but deflect a portion of the stream toward the nose: the odor is free from the characteristic odor of sulfur compounds.

Assay—

Chromatographic system—Under typical conditions, the gas chromatograph is equipped with a thermal-conductivity detector and contains a 6-m × 3-mm aluminum column packed with 10 weight percent of liquid phase G30 on non-acid-washed support S1C, helium is used as the carrier gas at a flow rate of 50 mL per minute, and the temperature of the column is maintained at 33°.

System suitability—The peak responses obtained for *n*-butane in the chromatograms from duplicate determinations agree within 1%.

Procedure—Connect 1 Butane cylinder to the chromatograph through a suitable sampling valve and a flow control valve downstream from the sampling valve. Flush the liquid specimen through the sampling valve, taking care to avoid entrapment of gas or air in the sampling valve. Inject a suitable volume, typically 2 µL, of Butane into the chromatograph, and record the chromatogram. Calculate the percentage purity by dividing 100 times the *n*-butane response by the sum of all of the responses in the chromatogram.

Butyl Alcohol

» Butyl Alcohol is *n*-butyl alcohol.

Packaging and storage—Preserve in tight containers, and prevent exposure to excessive heat.

Specific gravity ⟨841⟩: between 0.807 and 0.809.

Distilling range, *Method II* ⟨721⟩—It distils within a range of 1.5°, including 117.7°.

Acidity—Titrate 74 mL (60 g) of it with 0.020 *N* alcoholic potassium hydroxide, using phenolphthalein TS as the indicator, until a pink color persists for not less than 15 seconds: not more than 2.5 mL is consumed.

Water, *Method I* ⟨921⟩: not more than 0.1%.

Nonvolatile residue—Evaporate 100 mL in a tared porcelain dish on a steam bath, and dry at 105° for 30 minutes: the weight of the residue does not exceed 4 mg (0.004%).

Aldehydes—To 10 mL of silver-ammonia-nitrate TS in a test tube add 10 mL of Butyl Alcohol, and mix. Allow the mixture to stand in a dark place for 30 minutes: no color is produced, although a slight precipitate may form at the interface of the two layers.

Butyl ether—Under typical conditions, the gas chromatograph (see *Chromatography* ⟨621⟩) is equipped with a thermal conductivity detector, and contains a 2-m × 6-mm stainless steel column packed with 25 percent liquid phase G29 3,3′-thiodipropionitrile on 30- to 40-mesh support S1C, maintained at a temperature of 85°. Helium is used as the carrier gas at a flow rate of about 75 mL per minute. Inject a suitable volume (about 10 µL) of Butyl Alcohol, and measure the peak responses. Using these conditions, retention times are approximately 6.0 minutes for butyl ether; 12.0 minutes for 2-butanol; 17.0 minutes for water; 18.0 minutes for isobutyl alcohol; and 25.0 minutes for butyl alcohol: the response due to butyl ether is not more than 0.2% of the sum of all of the responses.

Butylated Hydroxyanisole

OH
C(CH3)3
OCH3

$C_{11}H_{16}O_2$ 180.25
Phenol, (1,1-dimethylethyl)-4-methoxy-.
tert-Butyl-4-methoxyphenol [*25013-16-5*].

» Butylated Hydroxyanisole contains not less than 98.5 percent of $C_{11}H_{16}O_2$.

Packaging and storage—Preserve in well-closed containers.

Reference standards—*USP 3-tert-Butyl-4-hydroxyanisole Reference Standard*—Keep container tightly closed. Do not dry before using. *USP 2-tert-Butyl-4-hydroxyanisole Reference Standard*—Keep container tightly closed. Do not dry before using.

Identification—To 5 mL of a 1 in 10,000 solution in 72 percent alcohol add 2 mL of sodium borate solution (1 in 50) and 1 mL of a 1 in 10,000 solution of 2,6-dichloroquinone-chlorimide in dehydrated alcohol, and mix: a blue color is produced.

Residue on ignition ⟨281⟩: not more than 0.01%, determined on a 10-g specimen.

Arsenic, *Method II* ⟨211⟩: 3 ppm.

Heavy metals, *Method II* ⟨231⟩: 0.001%.

Assay—

Internal standard solution—Dissolve about 500 mg of 4-*tert*-butylphenol, accurately weighed, in acetone in a 100-mL volumetric flask, add acetone to volume, and mix.

Standard preparation—Dissolve accurately weighed quantities of USP 3-*tert*-Butyl-4-hydroxyanisole RS and USP 2-*tert*-Butyl-4-hydroxyanisole RS, respectively, in *Internal standard solution* to obtain a solution having known concentrations of about 9 mg of USP 3-*tert*-Butyl-4-hydroxyanisole RS and about 1 mg of USP 2-*tert*-Butyl-4-hydroxyanisole RS per mL.

Assay preparation—Dissolve about 100 mg of Butylated Hydroxyanisole, accurately weighed, in *Internal standard solution* in a 10-mL volumetric flask, dilute with *Internal standard solution* to volume, and mix.

Chromatographic system (see *Chromatography* ⟨621⟩)—The gas chromatograph is equipped with a flame-ionization detector, and contains a 1.8-m × 2-mm stainless steel column packed with 10 percent liquid phase G26 on support S1A, the column is maintained isothermally at a temperature between 175° and 185°, and helium is used as the carrier gas. Chromatograph a sufficient number of injections of the *Standard preparation*, and record the areas as directed under *Procedure*, to ensure that the relative standard deviation does not exceed 2.0% for the 3-*tert*-butyl-4-hydroxyanisole isomer and 6.0% for the 2-*tert*-butyl-4-hydroxyanisole isomer. The resolution between the isomers is not less than 1.3, and the tailing factor does not exceed 2.0.

Procedure—Separately inject on-column suitable portions (about 5 µL) of the *Standard preparation* and the *Assay preparation* into the gas chromatograph, and record the chromatograms. Measure the areas under the peaks for each isomer and the internal standard in each chromatogram, and calculate the quantity, in mg, of each isomer in the Butylated Hydroxyanisole taken by the formula:

$$10C_S(R_U / R_S),$$

in which C_S is the concentration, in mg per mL, of the isomer in the *Standard preparation*, R_S is the ratio of the area of each isomer to that of the internal standard in the chromatogram from the *Standard preparation*, and R_U is the ratio of the area of each isomer to that of the internal standard in the chromatogram from the *Assay preparation.* Calculate the weight, in mg, of $C_{11}H_{16}O_2$ in the Butylated Hydroxyanisole taken by adding the quantities of the two isomers.

Butylated Hydroxytoluene

OH
(CH3)3C C(CH3)3
CH3

$C_{15}H_{24}O$ 220.35
Phenol, 2,6-bis(1,1-dimethylethyl)-4-methyl-.
2,6-Di-*tert*-butyl-*p*-cresol [*128-37-0*].

» Butylated Hydroxytoluene contains not less than 99.0 percent of $C_{15}H_{24}O$.

Packaging and storage—Preserve in well-closed containers.

Identification—To 10 mL of a 1 in 100,000 solution in methanol add 10 mL of water, 2 mL of sodium nitrite solution (3 in 1000), and 5 mL of dianisidine dihydrochloride solution (1 in 500), prepared by dissolving 200 mg of dianisidine dihydrochloride in a mixture of 40 mL of methanol and 60 mL of 1 *N* hydrochloric acid: an orange-red color develops within 3 minutes. Add 5 mL of chloroform, and shake: the chloroform layer exhibits a purple or magenta color, which fades when exposed to light.

Congealing temperature ⟨651⟩: not lower than 69.2°, corresponding to not less than 99.0% of $C_{15}H_{24}O$.

Residue on ignition ⟨281⟩—Transfer about 50 g, accurately weighed, to a tared crucible, ignite until thoroughly charred, and cool. Moisten the ash with 1 mL of sulfuric acid, and complete the ignition by heating at 800 ± 25° for 15-minute periods to constant weight: the limit is 0.002%.

Arsenic, *Method II* ⟨211⟩: 3 ppm.

Heavy metals, *Method II* ⟨231⟩: 0.001%.

Butylparaben

HO–C6H4–COOCH2(CH2)2CH3

$C_{11}H_{14}O_3$ 194.23

Benzoic acid, 4-hydroxy-, butyl ester.

Butyl *p*-hydroxybenzoate [*94-26-8*].

» Butylparaben contains not less than 99.0 percent and not more than 100.5 percent of $C_{11}H_{14}O_3$, calculated on the dried basis.

Packaging and storage—Preserve in well-closed containers.

Reference standard—*USP Butylparaben Reference Standard*—Dry over silica gel for 5 hours before using.

Identification—The infrared absorption spectrum of a mineral oil dispersion of it, previously dried, exhibits maxima only at the same wavelengths as that of a similar preparation of USP Butylparaben RS.

Melting range ⟨741⟩: between 68° and 72°.

Acidity—Heat 0.75 g in 15 mL of water at 80° for 1 minute, cool, and filter: the filtrate is neutral or acid to litmus. To 10 mL of the filtrate add 0.20 mL of 0.10 *N* sodium hydroxide and 2 drops of methyl red TS: the solution is yellow.

Loss on drying ⟨731⟩—Dry it over silica gel for 5 hours: it loses not more than 0.5% of its weight.

Residue on ignition ⟨281⟩: not more than 0.05%.

Assay—Transfer to a flask equipped for refluxing under a water-cooled condenser about 2 g of Butylparaben, accurately weighed, add 40.0 mL of 1 *N* sodium hydroxide VS, and rinse the sides of the flask with water. Reflux for 1 hour, and cool. Add 5 drops of bromothymol blue TS, and titrate the excess sodium hydroxide with 1 *N* sulfuric acid VS to a pH of 6.6 by matching the color of pH 6.6 phosphate buffer (see *Buffer Solutions* in the section, *Reagents, Indicators, and Solutions*) containing the same proportion of indicator. Perform a blank determination (see *Residual Titrations* under *Titrimetry* ⟨541⟩). Each mL of 1 *N* sodium hydroxide is equivalent to 194.2 mg of $C_{11}H_{14}O_3$.

Calcium Carbonate, Precipitated—*see* Calcium Carbonate, Precipitated USP

Calcium Chloride—*see* Calcium Chloride USP

Calcium Phosphate, Dibasic—*see* Calcium Phosphate, Dibasic USP

Tribasic Calcium Phosphate

$Ca_5(OH)(PO_4)_3$ 502.32

Calcium hydroxide phosphate ($Ca_5(OH)(PO_4)_3$).

Calcium hydroxide phosphate ($Ca_5(OH)(PO_4)_3$) [*12167-74-7*].

» Tribasic Calcium Phosphate consists of a variable mixture of calcium phosphates having the approximate composition $10CaO.3P_2O_5.H_2O$. It contains not less than 34.0 percent and not more than 40.0 percent of calcium (Ca).

Packaging and storage—Preserve in well-closed containers.

Identification—

A: To a warm solution in a slight excess of nitric acid add ammonium molybdate TS: a yellow precipitate is formed.

B: It responds to the flame test for *Calcium* ⟨191⟩.

Loss on ignition ⟨733⟩—Ignite it at 800° for 30 minutes: it loses not more than 8.0% of its weight.

Water-soluble substances—Digest 2 g with 100 mL of water on a steam bath for 30 minutes, cool, add sufficient water to restore the original volume, stir well, and filter. Evaporate 50 mL of the filtrate in a tared porcelain dish on a steam bath to dryness, and dry the residue at 120° to constant weight: the weight of the residue does not exceed 5 mg (0.5%).

Acid-insoluble substances—If an insoluble residue remains in the test for *Carbonate*, boil the solution, filter, wash the residue well with hot water until the last washing is free from chloride, and ignite the residue to constant weight: the weight of the residue does not exceed 4 mg (0.2%).

Carbonate—Mix 2 g with 20 mL of water, and add 3 *N* hydrochloric acid, dropwise, to effect solution: no effervescence is produced.

Chloride ⟨221⟩—Dissolve 500 mg in 25 mL of 2 *N* nitric acid, and add 1 mL of silver nitrate TS: the turbidity does not exceed that produced by 1.0 mL of 0.020 *N* hydrochloric acid (0.14%).

Fluoride—[NOTE—Prepare and store all solutions in plastic containers.]

Buffer solution, Standard solution, and *Electrode system*—Proceed as directed in the test for *Fluoride* under *Dibasic Calcium Phosphate*.

Standard response line—Proceed as directed in the test for *Fluoride* under *Dibasic Calcium Phosphate*, except to use 3.0 mL of hydrochloric acid, instead of 2.0 mL.

Procedure—Proceed as directed in the test for *Fluoride* under *Dibasic Calcium Phosphate*, except to use 3.0 mL of hydrochloric acid, instead of 2.0 mL. The limit is 0.0075%.

Nitrate—Mix 200 mg with 5 mL of water, and add just sufficient hydrochloric acid to effect solution. Dilute with water to 10 mL, add 0.10 mL of indigo carmine TS, then add, with stirring, 10 mL of sulfuric acid: the blue color persists for not less than 5 minutes.

Sulfate ⟨221⟩—Dissolve 500 mg in the smallest possible amount of 3 *N* hydrochloric acid, dilute with water to 100 mL, filter, if necessary, and to 25 mL of the filtrate add 1 mL of barium chloride TS: the turbidity does not exceed that produced by 1.0 mL of 0.020 *N* sulfuric acid (0.8%).

Arsenic, *Method I* ⟨211⟩—Prepare the *Test Preparation* by dissolving 1.0 g in just sufficient 3 *N* hydrochloric acid to dissolve the test specimen. The limit is 3 ppm.

Barium—Mix 500 mg with 10 mL of water, heat, add hydrochloric acid, dropwise, until solution is effected, and then add 2 drops of the acid in excess. Filter, and add to the filtrate 1 mL of potassium sulfate TS: no turbidity appears within 15 minutes.

Dibasic salt and calcium oxide—Weigh accurately about 1.5 g, and dissolve by warming with 25.0 mL of 1 *N* hydrochloric acid VS. Cool, and slowly titrate the excess of 1 *N* hydrochloric acid, while agitating constantly, with 0.1 *N* sodium hydroxide VS to a pH of 4.0, determined potentiometrically. Not less than 13.0 mL and not more than 14.3 mL of 1 *N* hydrochloric acid is consumed for each g of salt, calculated on the ignited basis.

Heavy metals, *Method I* ⟨231⟩—Mix 1.3 g with 9 mL of 3 *N* hydrochloric acid, dilute with water to 50 mL, and heat to boiling. Cool to room temperature, and filter [NOTE—Filter the mixture after the pH adjustment]: the limit is 0.003%.

Assay—Weigh accurately about 150 mg of Tribasic Calcium Phosphate, and proceed as directed in the *Assay* under *Dibasic Calcium Phosphate*, beginning with "dissolve, with the aid of gentle heat." Each mL of 0.05 *M* disodium ethylenediaminetetraacetate is equivalent to 2.004 mg of Ca.

Calcium Silicate

» Calcium Silicate is a compound of calcium oxide and silicon dioxide. It contains not less than 25.0 percent of calcium oxide and not less than 45.0 percent of silicon dioxide.

Packaging and storage—Preserve in well-closed containers.

Identification—

A: Mix 0.5 g with 10 mL of 3 *N* hydrochloric acid, filter, and neutralize the filtrate to litmus paper with 6 *N* ammonium hydroxide: the neutralized filtrate so obtained responds to the tests for *Calcium* ⟨191⟩.

B: Prepare a bead by fusing a few crystals of sodium ammonium phosphate on a platinum loop in the flame of a gas burner. Place the hot, transparent bead in contact with the specimen of Calcium Silicate, and again fuse. Silica floats about in the bead, producing, upon cooling, an opaque bead having a web-like structure.

pH ⟨791⟩: between 8.4 and 10.2, determined in a well-mixed aqueous suspension (1 in 20).

Loss on ignition ⟨733⟩—Transfer about 1 g, accurately weighed, to a suitable tared crucible, dry at 105° for 2 hours, and ignite at 900° to constant weight: it loses not more than 20.0% of its weight.

Fluoride—

Indicator solution—Prepare a solution containing 100 mg of lanthanum alizarin complexan mixture per mL in 60 percent isopropyl alcohol. Filter the solution, if it is not clear.

Test preparation—Prepare a slurry consisting of 5.0 g of Calcium Silicate and 45 mL of 0.1 *N* hydrochloric acid, stir at room temperature for 15 minutes, and filter through a 0.45-μm filter into a 50-mL volumetric flask. Wash the filter with five 1-mL portions of 0.1 *N* hydrochloric acid, collecting the washings in the flask, dilute with 0.1 *N* hydrochloric acid to volume, and mix.

Procedure—Transfer 5.0 mL of the *Test preparation* to a 25-mL volumetric flask, add 5.0 mL of *Indicator solution*, dilute with water to volume, mix, and allow to stand in diffuse light at ambient temperature for 1 hour. Determine the absorbance of this solution in a 1-cm cell at the wavelength of maximum absorbance at about 620 nm, with a suitable spectrophotometer, against a blank consisting of 5.0 mL of 0.1 *N* hydrochloric acid, 5.0 mL of *Indicator solution*, and 15.0 mL of water: the absorbance is not greater than that produced by 5.0 mL of a solution containing 2.21 μg of sodium fluoride per mL of 0.1 *N* hydrochloric acid, when treated in the same manner as the *Test preparation* (10 ppm).

Arsenic, *Method I* ⟨211⟩: 3 ppm.

Lead ⟨251⟩—Dissolve 1.0 g in 20 mL of 3 *N* hydrochloric acid, evaporate on a steam bath to about 10 mL, dilute with water to about 20 mL, and cool: the limit is 0.001%.

Heavy metals ⟨231⟩—Boil 2.0 g with a mixture of 50 mL of water and 5 mL of hydrochloric acid for 20 minutes, adding water to maintain the volume during the boiling. Add ammonium hydroxide until the mixture is only slightly acid to litmus paper. Filter with the aid of suction, and wash with 15 mL to 20 mL of water, combining the washing with the original filtrate. Add 2 drops of phenolphthalein TS, then add a slight excess of 6 *N* ammonium hydroxide. Discharge the pink color with dilute hydrochloric acid (1 in 100). Dilute with water to 100 mL, and use 25 mL of the solution for the test: the limit is 0.004%.

Assay for silicon dioxide—Transfer about 400 mg of Calcium Silicate, accurately weighed, to a beaker, add 5 mL of water and 10 mL of perchloric acid, and heat until dense white fumes of perchloric acid are evolved. Cover the beaker with a watch glass, and continue to heat for 15 minutes longer. Allow to cool, add 30 mL of water, filter, and wash the precipitate with 200 mL of hot water. (Retain the combined filtrate and washings for use in the *Assay for calcium oxide*.) Transfer the filter paper and its contents to a platinum crucible, heat slowly to dryness, then heat sufficiently to char the filter paper. After cooling, add a few drops of sulfuric acid, and ignite at about 1300° to constant weight. Moisten the residue with 5 drops of sulfuric acid, add 15 mL of hydrofluoric acid, heat cautiously on a hot plate until all of the acid is driven off, and ignite at a temperature not lower than 1000° to constant weight. Cool in a desiccator, and weigh: the loss in weight represents the weight of SiO_2.

Assay for calcium oxide—To the combined filtrate and washings retained from the *Assay for silicon dioxide*, add, while stirring, about 30 mL of 0.05 *M* disodium ethylenediaminetetraacetate VS from a 50-mL buret, then add 15 mL of 1 *N* sodium hydroxide and 300 mg of hydroxy naphthol blue trituration, and continue the titration to a blue end-point. Each mL of 0.05 *M* disodium ethylenediaminetetraacetate is equivalent to 2.804 mg of CaO.

Ratio of CaO to SiO_2—Divide the percentage of SiO_2 obtained in the *Assay for silicon dioxide* by the percentage of CaO obtained in the *Assay for calcium oxide:* the quotient obtained is between 1.65 and 2.65.

Sum of CaO, SiO_2, and Loss on ignition—The sum of the percentages obtained in the three tests is not less than 90.0%.

Calcium Stearate

Octadecanoic acid, calcium salt.
Calcium stearate [*1592-23-0*].

» Calcium Stearate is a compound of calcium with a mixture of solid organic acids obtained from fats and consists chiefly of variable proportions of calcium stearate and calcium palmitate. It contains the equivalent of not less than 9.0 percent and not more than 10.5 percent of calcium oxide (CaO).

Packaging and storage—Preserve in well-closed containers.

Identification—

A: Heat 1 g with a mixture of 25 mL of water and 5 mL of hydrochloric acid: fatty acids are liberated and appear as an oily layer floating on the surface of the liquid. The water layer responds to the tests for *Calcium* ⟨191⟩.

B: Mix 25 g with 200 mL of hot water, add 60 mL of 2 *N* sulfuric acid, and heat the mixture, with frequent stirring, until the separated fatty acid layer is clear. Wash the fatty acids with boiling water until free from sulfate, collect them in a small beaker, and warm on a steam bath until the water has separated and the fatty acids are clear. Allow the acids to cool, pour off the water layer, melt the acids, filter into a dry beaker, and dry at 105° for 20 minutes: the fatty acids so obtained congeal at a temperature not below 54° (see *Congealing Temperature* ⟨651⟩).

Loss on drying ⟨731⟩—Dry it at 105° to constant weight, using 2-hour increments of heating: it loses not more than 4.0% of its weight.

Arsenic, *Method I* ⟨211⟩—Prepare the *Test Preparation* as follows. Mix 1.0 g with 25 mL of hydrochloric acid and 1 mL of bromine TS, and heat on a steam bath until the separated fatty acid layer is clear. Add 50 mL of water, evaporate to about 25 mL, and filter while hot. Cool, neutralize with 12 *N* sodium hydroxide, and dilute with water to 35 mL. The limit is 3 ppm.

Heavy metals ⟨231⟩—Place 2.5 g in a porcelain dish, place a 500-mg portion in a second dish to provide the control, and to each add 5 mL of a 1 in 4 solution of magnesium nitrate in alcohol. Cover the dishes with 7.5-cm short-stem funnels so that the stems are straight up. Heat on a hot plate at low heat for 30 minutes, then heat at medium heat for 30 minutes, and cool. Remove the funnels, add 2 mL of *Standard Lead Solution* (20 μg of Pb) to the control, and heat each dish over a suitable burner until most of the carbon is burned off. Cool, add 10 mL of nitric acid, and transfer the solutions into 250-mL beakers. Add 5 mL of 70 percent perchloric acid, cautiously evaporate to dryness, add 2 mL of hydrochloric acid to the residues, and wash down the insides of the beakers with water. Evaporate carefully to dryness again, swirling near the dry point to avoid spattering. Repeat the hydrochloric acid treatment, then cool, and dissolve the residues in about 10 mL of water. To each solution add 1 drop of phenolphthalein TS and add sodium hydroxide TS until the solutions just turn pink, then add 3 *N* hydrochloric acid until the solutions become colorless. Add 1 mL of 1 *N* acetic acid and a small amount of charcoal to each solution, and filter through filter paper into 50-mL color-comparison tubes. Wash with water, dilute with water to 40 mL, and add 10 mL of hydrogen sulfide TS to each tube: the color of the test solution does not exceed that of the control (0.001%).

Assay—Boil about 1.2 g of Calcium Stearate, accurately weighed, with 50 mL of 0.1 *N* hydrochloric acid for 10 minutes, or until the separated fatty acid layer is clear, adding water, if necessary, to maintain the original volume. Cool, filter, and wash the filter and the flask thoroughly with water until the last washing is not acid to litmus. Neutralize the filtrate with 1 *N* sodium hydroxide to litmus. While stirring, preferably with a magnetic stirrer, titrate with 0.05 *M* disodium ethylenediaminetetraacetate VS as follows. Add about 30 mL from a 50-mL buret, then add 15 mL of 1 *N* sodium hydroxide and 300 mg of hydroxy naphthol blue, and continue the titration to a blue end-point. Each mL of 0.05 *M* disodium ethylenediaminetetraacetate is equivalent to 2.804 mg of CaO.

Calcium Sulfate

$CaSO_4$ 136.14
Sulfuric acid, calcium salt (1:1).
Calcium sulfate (1:1) 136.14 [*7778-18-9*].
Dihydrate 172.17 [*10101-41-4*].

» Calcium Sulfate is anhydrous or contains two molecules of water of hydration. It contains not less than 98.0 percent and not more than 101.0 percent of $CaSO_4$, calculated on the dried basis.

Packaging and storage—Preserve in well-closed containers.

Labeling—Label it to indicate whether it is anhydrous or the dihydrate.

Identification—Dissolve about 200 mg by warming in a mixture of 4 mL of 3 *N* hydrochloric acid and 16 mL of water. This solution responds to the tests for *Calcium* ⟨191⟩ and for *Sulfate* ⟨191⟩.

Loss on drying ⟨731⟩—Dry it at a temperature not lower than 250° to constant weight: the anhydrous form loses not more than 1.5% of its weight, and the dihydrate loses between 19.0% and 23.0% of its weight.

Iron ⟨241⟩—Dissolve 100 mg in 8 mL of 3 *N* hydrochloric acid, and dilute with water to 47 mL: the limit is 0.01%.

Heavy metals, *Method I* ⟨231⟩—Mix 2.0 g with 20 mL of water, add 25 mL of 3 *N* hydrochloric acid, and heat to boiling to dissolve the test specimen. Cool, and add ammonium hydroxide to a pH of 7. Filter, evaporate to a volume of about 25 mL, and refilter, if necessary, to obtain a clear solution: the limit is 0.001%.

Assay—Dissolve about 300 mg of Calcium Sulfate, accurately weighed, in 100 mL of water and 4 mL of 3 *N* hydrochloric acid. Boil, if necessary, to effect solution, and cool before titrating. While stirring, preferably with a magnetic stirrer, add, in the order named, 0.5 mL of triethanolamine, 300 mg of hydroxy naphthol blue trituration, and, from a 50-mL buret, about 30 mL of 0.05 *M* disodium ethylenediaminetetraacetate VS. Add sodium hydroxide solution (45 in 100) until the initial red color changes to clear blue, then continue to add it dropwise until the color changes to violet, then add an additional 0.5 mL. The pH is between 12.3 and 12.5. Continue the titration dropwise with 0.05 *M* disodium ethylenediaminetetraacetate VS to the appearance of a clear-blue end-point that persists for not less than 60 seconds. Each mL of 0.05 *M* disodium ethylenediaminetetraacetate is equivalent to 6.807 mg of $CaSO_4$.

Caramel

» Caramel is a concentrated solution of the product obtained by heating sugar or glucose until the sweet taste is destroyed and a uniform dark brown mass results, a small amount of alkali or of alkaline carbonate or a trace of mineral acid being added while heating.

NOTE—Where included in articles for coloring purposes, Caramel complies with the regulations of the federal Food and Drug Administration concerning color additives (21 CFR 73.85, caramel).

Packaging and storage—Preserve in tight containers.

Specific gravity ⟨841⟩: not less than 1.30.

Purity—The addition of 0.5 mL of phosphoric acid to 20 mL of a solution (1 in 20) produces no precipitate.

Microbial limits—It meets the requirements of the tests for absence of *Salmonella* species and *Escherichia coli* under *Microbial Limit Tests* ⟨61⟩.

Ash—It swells when incinerated, and forms a coke-like charcoal that burns off only after prolonged heating at a high temperature. It yields not more than 8.0% of ash.

Arsenic, *Method II* ⟨211⟩: 3 ppm.

Lead ⟨251⟩: 10 ppm.

Carbomer 910

» Carbomer 910 is a high molecular weight polymer of acrylic acid cross-linked with allyl ethers of pentaerythritol. Carbomer 910, previously dried in vacuum at 80° for 1 hour, contains not less than 56.0 percent and not more than 68.0 percent of carboxylic acid (–COOH) groups. The viscosity of a neutralized 1.0 percent aqueous dispersion of Carbomer 910 is between 3000 and 7000 centipoises.

Packaging and storage—Preserve in tight containers.

Labeling—Label it to indicate that it is not intended for internal use.

Reference standard—*USP Carbomer Homopolymer Reference Standard.*

Viscosity—Proceed as directed in the test for *Viscosity* under *Carbomer 934P*, except to perform the test on a 1.0 percent aqueous dispersion (prepared by using 5.00 g instead of 2.50 g) and to use a spindle having a disk 1.3 cm in diameter and 0.2 cm high attached to a shaft 0.32 cm in diameter, the distance from the top of the disk to the lower tip of the shaft being 2.4 cm and the immersion depth being 5 cm (No. 3 spindle). The viscosity is between 3000 and 7000 centipoises.

Benzene—Proceed as directed in the test for *Benzene* under *Carbomer 934P* except dilute the *Test preparation* with *p-xylene* to one-fiftieth of its concentration before use. The limit is 0.5%.

Other requirements—It meets the requirements for *Identification, Loss on drying, Heavy metals,* and *Assay for carboxylic acid content* under *Carbomer 934P*.

Carbomer 934

» Carbomer 934 is a high molecular weight polymer of acrylic acid cross-linked with allyl ethers of sucrose. Carbomer 934, previously dried in vacuum at 80° for 1 hour, contains not less than 56.0 and not more than 68.0 percent of carboxylic acid (–COOH) groups. The viscosity of a neutralized 0.5 percent aqueous dispersion of Carbomer 934 is between 30,500 and 39,400 centipoises.

Packaging and storage—Preserve in tight containers.

Labeling—Label it to indicate that it is not intended for internal use.

Reference standard—*USP Carbomer Homopolymer Reference Standard.*

Viscosity—Proceed as directed in the test for *Viscosity* under *Carbomer 934P*. It is between 30,500 and 39,400 centipoises.

Benzene—Proceed as directed in the test for *Benzene* under *Carbomer 934P* except dilute the *Test preparation* with *p-xylene* to one-fiftieth of its concentration before use. The limit is 0.5%.

Other requirements—It meets the requirements for *Identification, Loss on drying, Heavy metals,* and *Assay for carboxylic acid content* under *Carbomer 934P.*

Carbomer 934P

» Carbomer 934P is a high molecular weight polymer of acrylic acid cross-linked with allyl ethers of sucrose or pentaerythritol. Carbomer 934P, previously dried in vacuum at 80° for 1 hour, contains not less than 56.0 percent and not more than 68.0 percent of carboxylic acid (–COOH) groups. The viscosity of a neutralized 0.5 percent aqueous dispersion of Carbomer 934P is between 29,400 and 39,400 centipoises.

Packaging and storage—Preserve in tight containers.

Reference standard—*USP Carbomer Homopolymer Reference Standard.*

Identification—The infrared absorption spectrum of a potassium bromide dispersion of it exhibits maxima only at the same wavelengths as that of a similar preparation of USP Carbomer Homopolymer RS.

Viscosity ⟨911⟩—Carefully add 2.50 g to 500 mL of water in a 1000-mL beaker, while stirring continuously at 1000 ± 10 rpm, with the stirrer shaft set at an angle of 60 degrees and to one side of the beaker, with the propeller positioned near the bottom of the beaker. Allow 45 to 90 seconds for addition of the test preparation at a uniform rate, being sure that loose aggregates of powder are broken up, and continue stirring at 1000 ± 10 rpm for 15 minutes. Remove the stirrer, and place the beaker containing the dispersion in a 25 ± 0.2° water bath for 30 minutes. Insert the stirrer to a depth necessary to ensure that air is not drawn into the dispersion, and, while stirring at 300 ± 10 rpm, titrate (see *Titrimetry* ⟨541⟩) with a calomel-glass electrode system to a pH of between 7.3 and 7.8 by adding sodium hydroxide solution (18 in 100) below the surface, determining the end-point potentiometrically. The total volume of sodium hydroxide solution (18 in 100) used is about 6.2 mL. Allow 2 to 3 minutes before final pH determination. [NOTE—If the final pH exceeds 7.8, discard the mucilage, and prepare another using a smaller amount of sodium hydroxide for titration.] Return the neutralized mucilage to the 25° water bath for 1 hour, then perform the viscosity determination without delay to avoid slight viscosity changes that occur 75 minutes after neutralization. Equip a suitable rotational viscosimeter with a spindle having a cylinder 1.47 cm in diameter and 0.16 cm high attached to a shaft 0.32 cm in diameter, the distance from the top of the cylinder to the lower tip of the shaft being 3.02 cm, and the immersion depth being 4.92 cm (No. 6 spindle). With the spindle rotating at 20 rpm, observe and record the scale reading. Calculate the viscosity, in centipoises, by multiplying the scale reading by the constant for the spindle used at 20 rpm.

Loss on drying ⟨731⟩—Dry it in vacuum at 80° for 1 hour: it loses not more than 2.0% of its weight.

Heavy metals, *Method II* ⟨231⟩: 0.002%.

Benzene—

p-Xylene—A suitable grade produces a benzene peak response not greater than one-half that of the most dilute *Standard preparation,* when chromatographed as directed.

Standard preparations—Using *p-Xylene* as solvent, prepare a solution containing a known concentration of benzene of about 0.115 µL per mL, equivalent to 0.1 mg per mL. Pipet 5-, 10-, and 15-mL portions into separate 100-mL volumetric flasks, dilute with *p-Xylene* to volume, and mix.

Test preparation—Transfer about 1 g, accurately weighed, to a tared 35-mL centrifuge tube. Add 10.0 mL of *p-Xylene,* and swirl to disperse the solid. Add 10.0 mL of 0.1 *N* sodium hydroxide, affix a polytef-lined-screw cap, and shake by mechanical means for 1 hour. Centrifuge for 5 minutes at high speed. The upper layer is the *Test preparation.*

Chromatographic system (see *Chromatography* ⟨621⟩)—The gas chromatograph is equipped with a flame-ionization detector and a 2-mm × 1.8-m column packed with 30 percent liquid phase G1 on 60- to 80-mesh support S1AB. The column temperature is maintained at about 100°, and nitrogen or helium is used as the carrier gas. Inject the middle strength *Standard preparation,* and measure the peak response for benzene as directed under *Procedure.* The relative standard deviation for replicate injections is not more than 5.0%. None of the points in the *Standard preparation* plot obtained in the *Procedure* deviate from a straight line by more than 10 percent.

Procedure—Separately inject equal volumes (about 1.0 µL) of the *Standard preparations* and the *Test preparation* into the chromatograph, and measure the peak responses for benzene. Plot the responses from the *Standard preparations* against their respective concentrations, in µg per mL. Read the concentration, *C*, of the *Test preparation* from the plot. Calculate the quantity, in µg, of benzene in the weight of Carbomer 934P taken, by multiplying by the appropriate dilution factor. The limit is 0.010%.

Assay for carboxylic acid content—Slowly add about 400 mg previously dried and accurately weighed, to 400 mL of water in a 1000-mL beaker, while stirring continuously at about 1000 rpm, with the stirrer shaft set at an angle of 60° and at the side of the beaker, with the propeller positioned near the bottom of the beaker, and continue stirring for 15 minutes. Reduce the stirring speed, and titrate potentiometrically with 0.25 *N* sodium hydroxide VS, using a calomel-glass electrode system. Allow 1 minute for mixing, after each addition of 0.25 *N* sodium hydroxide VS, before recording the pH. Calculate the carboxylic acid content, as a percentage of carboxylic acid groups, by the formula:

$$100(45.02VN/W),$$

in which V is the volume, in mL, of sodium hydroxide consumed, N is the normality of the sodium hydroxide solution, W is the weight, in mg, of specimen taken, and 45.02 is the molecular weight of the carboxylic acid (–COOH) group.

Carbomer 940

» Carbomer 940 is a high molecular weight polymer of acrylic acid cross-linked with allyl ethers of pentaerythritol. Carbomer 940, previously dried in vacuum at 80° for 1 hour, contains not less than 56.0 percent and not more than 68.0 percent of carboxylic acid (–COOH) groups. The viscosity of a neutralized 0.5 percent aqueous dispersion of Carbomer 940 is between 40,000 and 60,000 centipoises.

Packaging and storage—Preserve in tight containers.

Labeling—Label it to indicate that it is not intended for internal use.

Reference standard—*USP Carbomer Homopolymer Reference Standard.*

Viscosity—Proceed as directed in the test for *Viscosity* under *Carbomer 934P,* except to use a spindle having a shaft 0.32 cm in diameter, the distance from the top of the shaft to the lower tip of the shaft being 5.04 cm, and the immersion depth being 0.95 cm (No. 7 spindle). The viscosity is between 40,000 and 60,000 centipoises.

Benzene—Proceed as directed in the test for *Benzene* under *Carbomer 934P* except dilute the *Test preparation* with *p-xylene* to one-fiftieth of its concentration before use. The limit is 0.5%.

Other requirements—It meets the requirements for *Identification, Loss on drying, Heavy metals,* and *Assay for carboxylic acid content* under *Carbomer 934P.*

Carbomer 941

» Carbomer 941 is a high molecular weight polymer of acrylic acid cross-linked with allyl ethers of pentaerythritol. Carbomer 941, previously dried in vacuum at 80° for 1 hour, contains not less than 56.0 percent and not more than 68.0 percent of carboxylic acid (–COOH) groups. The viscosity of a neutralized 0.5 percent aqueous dispersion of Carbomer 941 is between 4,000 and 11,000 centipoises.

Packaging and storage—Preserve in tight containers.

Labeling—Label it to indicate that it is not intended for internal use.

Reference standard—*USP Carbomer Homopolymer Reference Standard.*

Viscosity—Proceed as directed in the test for *Viscosity* under *Carbomer 934P*, except to use a spindle having a disk 2.08 cm in diameter and 0.16 cm high attached to a shaft 0.32 cm in diameter, the distance from the top of the disk to the lower tip of the shaft being 2.7 cm, and the immersion depth being 3.73 cm (No. 5 spindle). The viscosity is between 4,000 and 11,000 centipoises.

Benzene—Proceed as directed in the test for *Benzene* under *Carbomer 934P* except dilute the *Test preparation* with *p-xylene* to one-fiftieth of its concentration before use. The limit is 0.5%.

Other requirements—It meets the requirements for *Identification, Loss on drying, Heavy metals,* and *Assay for carboxylic acid content* under *Carbomer 934P.*

Carbomer 1342

» Carbomer 1342 is a high molecular weight copolymer of acrylic acid and a long chain alkyl methacrylate cross-linked with allyl ethers of pentaerythritol. Carbomer 1342, previously dried in vacuum at 80° for 1 hour, contains not less than 52.0 percent and not more than 62.0 percent of carboxylic acid (–COOH) groups. The viscosity of a neutralized 1.0 percent aqueous dispersion of Carbomer 1342 is between 9,500 and 26,500 centipoises.

Packaging and storage—Preserve in tight containers.

Labeling—Label it to indicate that it is not intended for internal use.

Reference standard—*USP Carbomer Copolymer Reference Standard.*

Identification—The infrared absorption spectrum of a potassium bromide dispersion of it exhibits maxima only at the same wavelengths as that of a similar preparation of USP Carbomer Copolymer RS.

Viscosity—Proceed as directed in the test for *Viscosity* under *Carbomer 934P* except to perform the test on a 1.0 percent aqueous dispersion (prepared by using 5.00 g instead of 2.50 g) and omit the 30-minute incubation of the solution before the titration. The viscosity is between 9,500 and 26,500 centipoises.

Carboxylic acid content—Proceed as directed in the *Assay for carboxylic acid content* under *Carbomer 934P.*

Benzene—Proceed as directed in the test for *Benzene* under *Carbomer 934P*, except dilute the *Test preparation* with *p-xylene* to one-twentieth of its concentration before use. The limit is 0.2%.

Other requirements—It meets the requirements for *Loss on drying* and *Heavy metals* under *Carbomer 934P.*

Carbon Dioxide—*see* Carbon Dioxide USP

Carbon Tetrachloride

$$Cl-\underset{Cl}{\overset{Cl}{C}}-Cl$$

CCl_4 153.82
Methane, tetrachloro-.
Carbon tetrachloride [*56-23-5*].

» Carbon Tetrachloride contains not less than 99.0 percent and not more than 100.5 percent of CCl_4.

Caution—Avoid contact; vapor and liquid are poisonous. Care should be taken not to vaporize Carbon Tetrachloride in the presence of a flame because of the production of harmful gases (mainly phosgene).

Packaging and storage—Preserve in tight, light-resistant containers, at a temperature not exceeding 30°.

Specific gravity ⟨841⟩: between 1.588 and 1.590, indicating between 99.0% and 100.5% of CCl_4.

Distilling range, *Method I* ⟨721⟩: between 76.0° and 78.0°, a correction factor of 0.043° per mm being applied as necessary.

Acidity—Shake 15 mL with 25 mL of oxygen-free and carbon dioxide–free water for 5 minutes, and allow the layers to separate completely: a 5-mL portion of the aqueous layer is neutral to litmus. Retain the remainder of the aqueous layer for the *Chloride and free chlorine test.*

Nonvolatile residue—Evaporate 50 mL in a platinum or porcelain dish on a steam bath until the volume is reduced to about 1 mL, and allow it to evaporate spontaneously to dryness. Dry at 105° for 1 hour, and weigh: the weight of the residue does not exceed 1 mg (0.002%).

Chloride and free chlorine—A 10-mL portion of the aqueous layer retained from the *Acidity test* exhibits no opalescence upon the addition of a few drops of silver nitrate TS (*chloride*), and a 10-mL portion of the same aqueous solution is not colored blue upon the addition of a few drops of potassium iodide TS and 3 mL of starch TS (*free chlorine*).

Readily carbonizable substances ⟨271⟩—In a glass-stoppered separator, previously rinsed with sulfuric acid TS, combine 40 mL of it and 5 mL of sulfuric acid TS, and shake vigorously for 5 minutes. Allow the layers to separate completely: the acid has no more color than Matching Fluid A.

Carbon disulfide—Mix 10 mL of it with 10 mL of alcohol potassium hydroxide solution (1 in 10), allow the mixture to stand for 1 hour, then add 5 mL of 6 *N* acetic acid followed by 1 mL of cupric sulfate TS: no yellow precipitate is formed within 2 hours.

Carboxymethylcellulose Calcium

Cellulose, carboxymethyl ether, calcium salt.
Cellulose carboxymethyl ether calcium salt [*9050-04-8*].

» Carboxymethylcellulose Calcium is the calcium salt of a polycarboxymethyl ether of cellulose.

Packaging and storage—Preserve in tight containers.

Identification—

A: Shake thoroughly 0.1 g with 10 mL of water, followed by 2 mL of 1 *N* sodium hydroxide, allow to stand for 10 minutes, and use 1 mL of this solution as the test solution, retaining the remainder of it for *Identification tests B* and *C*. To 1 mL of the test solution add water to make 5 mL, then to 1 drop of the resulting solution add 0.5 mL of chromotropic acid TS, and heat in a water bath for 10 minutes: a red-purple color develops.

B: Shake 5 mL of the test solution obtained in *Identification test A* with 10 mL of acetone: a white, flocculent precipitate is formed.

C: Shake 5 mL of the test solution obtained in *Identification test A* with 1 mL of ferric chloride TS: a brown, flocculent precipitate is formed.

D: Ignite 1 g to ash, dissolve the residue in 10 mL of water and 5 mL of 6 *N* acetic acid, and filter, if necessary. Boil the filtrate, cool, and neutralize with 6 *N* ammonium hydroxide: the solution responds to the tests for *Calcium* ⟨191⟩.

Alkalinity—Shake thoroughly 1.0 g with 50 mL of freshly boiled and cooled water, and add 2 drops of phenolphthalein TS: no red color develops.

Loss on drying ⟨731⟩—Dry it at 105° for 4 hours: it loses not more than 10.0% of its weight.

Residue on ignition ⟨281⟩: between 10.0% and 20.0%, about 0.5 g, previously dried, being used for the test, and an ignition temperature of 450° to 550° being used.

Chloride ⟨221⟩—Shake thoroughly 0.80 g with 50 mL of water, dissolve in 10 mL of 1 *N* sodium hydroxide, add water to make 100 mL, and use 20 mL of this solution as the test solution, retaining the remainder of it for the test for *Sulfate*. Heat 20 mL of the test solution with 10 mL of 2 *N* nitric acid in a water bath until a flocculent precipitate is formed, cool, centrifuge, and remove the supernatant liquid. Wash the precipitate with three 10-mL portions of water by centrifuging each time, combine the supernatant liquid and the washings, add water to make 100 mL, and mix: a 25-mL portion of this solution shows no more chloride than is contained in 0.20 mL of 0.020 *N* hydrochloric acid (0.36%).

Silicate—Weigh accurately about 1 g, ignite in a platinum dish, add 20 mL of 3 *N* hydrochloric acid, cover with a watch glass, and boil gently for 30 minutes. Remove the watch glass, and evaporate in a water bath, with the aid of a current of air, to dryness. Continue heating for 1 hour, add 10 mL of hot water, stir well, and filter through quantitative filter paper. Wash the residue with hot water, dry it together with the filter paper after no turbidity is produced on the addition of silver nitrate TS to the last washing, then ignite to constant weight: not more than 1.5% of residue is obtained.

Sulfate ⟨221⟩—Heat 10 mL of the test solution obtained in the test for *Chloride* with 1 mL of hydrochloric acid in a water bath until a flocculent precipitate is formed, cool, centrifuge, and remove the supernatant liquid. Wash the precipitate with three 10-mL portions of water by centrifuging each time, combine the supernatant liquid and the washings, add water to make 100 mL, and mix: a 25-mL portion of this solution shows no more sulfate than is contained in 0.20 mL of 0.020 *N* sulfuric acid (0.96%).

Arsenic, *Method I* ⟨211⟩—Heat cautiously 2.0 g with 10 mL of sulfuric acid and 10 mL of nitric acid in a Kjeldahl flask. Continue heating with occasional addition of 2-mL portions of nitric acid until the liquid becomes colorless to light yellow. Cool, add 15 mL of ammonium oxalate TS, heat to the production of white fumes, cool, mix with 30 mL of water, and filter. Wash the residue with water, combine the washings with the filtrate, add water to make 50 mL, and mix. Transfer a 7.5-mL portion of this solution to the arsine generator, and dilute with water to 55 mL. The resulting solution meets the requirements of the test, the 20 mL of 7 *N* sulfuric acid specified under *Procedure* being added to the *Standard Preparation* only. The limit is 0.001%.

Heavy metals ⟨231⟩—Determine as directed in the test for *Heavy metals* under *Methylcellulose*, except to use only 1 g of Carboxymethylcellulose Calcium. The limit is 0.002%.

Starch—Heat 0.10 g with 10 mL of water, cool, and add 2 drops of iodine TS: no blue color develops.

Carboxymethylcellulose Sodium—*see* Carboxymethylcellulose Sodium USP

Carboxymethylcellulose Sodium, Microcrystalline Cellulose and—*see* Cellulose, Microcrystalline, and Carboxymethylcellulose Sodium

Carboxymethylcellulose Sodium 12

» Carboxymethylcellulose Sodium 12 is the sodium salt of a polycarboxymethyl ether of cellulose. Its degree of substitution is not less than 1.15 and not more than 1.45, corresponding to a sodium (Na) content of not less than 10.5 percent and not more than 12.0 percent, calculated on the dried basis.

Packaging and storage—Preserve in tight containers.

Labeling—Label it to indicate the viscosity in solutions of stated concentrations of either 1% (w/w) or 2% (w/w).

Identification—Add about 1 g of powdered Carboxymethylcellulose Sodium 12 to 50 mL of water, while stirring to produce a uniform dispersion. Continue the stirring until a clear solution is produced, and use the solution for the following tests.

A: To 1 mL of the solution, diluted with an equal volume of water, in a small test tube, add 5 drops of 1-naphthol TS. Incline the test tube, and carefully introduce down the side of the tube 2 mL of sulfuric acid so that it forms a lower layer: a red-purple color develops at the interface.

B: To 5 mL of the solution add an equal volume of barium chloride TS: a fine, white precipitate is formed.

C: A portion of the solution responds to the tests for *Sodium* ⟨191⟩.

Viscosity ⟨911⟩—Determine the viscosity in a water solution at the concentration stated on the label. Using undried Carboxymethylcellulose Sodium 12, weigh accurately the amount which, on the dried basis, will provide 200 g of solution of the stated concentration. Add the substance in small amounts to about 180 mL of stirred water contained in a tared, wide-mouth bottle, continue stirring rapidly until the powder is well wetted, add sufficient water to make the mixture weigh 200 g, and allow to stand, with occasional stirring, until solution is complete. Adjust the temperature to 25 ± 0.2°, and determine the viscosity, using a rotational type of viscosimeter, making certain that the system reaches equilibrium before taking the final reading. The viscosity of solutions of 2 percent concentration is not less than 80.0% and not more than 120.0% of that stated on the label; the viscosity of solutions of 1 percent concentration is not less than 75.0% and not more than 140.0% of that stated on the label.

pH ⟨791⟩: between 6.5 and 8.5, in a solution (1 in 100).

Loss on drying ⟨731⟩—Dry it at 105° for 3 hours: it loses not more than 10.0% of its weight.

Heavy metals—Determine as directed in the test for *Heavy metals* under *Methylcellulose*, using a 500-mg specimen: the limit is 0.004%.

Sodium chloride and Sodium glycolate—

SODIUM CHLORIDE—Weigh accurately about 5 g of it into a 250-mL beaker, add 50 mL of water and 5 mL of 30 percent hydrogen peroxide, and heat on a steam bath for 20 minutes, stirring occasionally to ensure hydration. Cool, add 100 mL of water and 10 mL of nitric acid, and titrate with 0.05 *N* silver nitrate VS, determining the end-point potentiometrically, using a silver electrode and a mercurous sulfate electrode having a potassium sulfate bridge, and stirring constantly. Calculate the percentage of sodium chloride in the specimen by the formula,

$$584.4VN/[(100 - b)W],$$

in which V and N represent the volume, in mL, and the normality, respectively, of the silver nitrate, b is the percentage of *Loss on drying*, determined separately, W is the weight, in g, of the specimen, and 584.4 is an equivalence factor for sodium chloride.

SODIUM GLYCOLATE—Transfer about 500 mg of it, accurately weighed, into a 100-mL beaker, moisten thoroughly with 5 mL of glacial acetic acid, followed by 5 mL of water, and stir with a glass rod to ensure proper hydration (usually about 15 minutes). Slowly add 50 mL of acetone, with stirring, then add 1 g of sodium chloride, and stir for several minutes to ensure complete precipitation of the carboxymethylcellulose. Filter through a soft, open-textured paper, previously wetted with a small amount of acetone, and collect the filtrate in a 100-mL volumetric flask. Use an additional 30 mL of acetone to facilitate the transfer of the solids and to wash the filter cake, then dilute with acetone to volume, and mix.

Prepare a series of standard solutions as follows. Transfer 100 mg of glycolic acid, previously dried in a desiccator at room temperature overnight and accurately weighed, to a 100-mL volumetric flask, dissolve in water, dilute with water to volume, and mix. Use this solution within 30 days. Transfer 1.0 mL, 2.0 mL, 3.0 mL, and 4.0 mL portions of the solution, respectively, to separate 100-mL volumetric flasks, add water to each flask to make 5 mL, then add 5 mL of glacial acetic acid, dilute with acetone to volume, and mix.

Transfer 2.0 mL of the test solution and 2.0 mL of each standard solution to separate 25-mL volumetric flasks, and prepare a blank flask containing 2.0 mL of a solution containing 5% each of glacial acetic acid and water in acetone. Place the uncovered flasks in a boiling water bath for 20 minutes, accurately timed, to remove the acetone, remove from the bath, and cool. Add to each flask 5.0 mL of 2,7-dihydroxynaphthalene TS, mix, add an additional 15 mL, and again mix. Cover the mouth of each flask with a small piece of aluminum foil. Place the flasks upright in a boiling water bath for 20 minutes, then remove from the bath, cool, dilute with sulfuric acid to volume, and mix.

Determine the absorbance of each solution at 540 nm, with a suitable spectrophotometer, against the blank, and prepare a standard curve using the absorbances obtained from the standard solutions. From the standard curve and the absorbance of the test specimen, determine the weight (w), in mg, of glycolic acid in the specimen, and calculate the percentage of sodium glycolate in the specimen by the formula,

$$12.9w/[(100 - b)W],$$

in which 12.9 is a factor converting glycolic acid to sodium glycolate, b is the percentage of *Loss on drying*, determined separately, and W is the weight, in g, of the specimen. The sum of the percentages of sodium chloride and sodium glycolate is not more than 0.5%.

Degree of substitution—Weigh accurately about 200 mg, previously dried at 105° for 3 hours, and transfer to a glass-stoppered, 250-mL conical flask. Add 75 mL of glacial acetic acid, connect the flask with a water-cooled condenser, and reflux gently on a hot plate for 2 hours. Cool, transfer the solution to a 250-mL beaker with the aid of 50 mL of glacial acetic acid, and titrate with 0.1 N perchloric acid in dioxane VS while stirring with a magnetic stirrer. Determine the end-point potentiometrically with a pH meter equipped with a standard glass electrode and a calomel electrode modified as follows. Discard the aqueous potassium chloride solution contained in the electrode, rinse and fill with the supernatant liquid obtained by shaking thoroughly 2 g each of potassium chloride and silver chloride (or silver oxide) with 100 mL of methanol, then add a few crystals of potassium chloride and silver chloride (or silver oxide) to the electrode.

Record the amount, in mL, of 0.1 N perchloric acid versus mV (0- to 700-mV range), and continue the titration to a few mL beyond the end-point. Plot the titration curve, and read the volume, v, in mL, of 0.1 N perchloric acid at the inflection point. Calculate the degree of substitution by the formula:

$$16.2v/\{G(1.000 - [8.0v/G])\},$$

in which G is the weight, in mg, of Carboxymethylcellulose Sodium 12 taken, 16.2 is one-tenth of the molecular weight of 1 anhydroglucose unit, and 8.0 is one-tenth of the molecular weight of 1 sodium carboxymethyl group.

Carnauba Wax—*see* Wax, Carnauba

Carrageenan

Carrageenan.
Carrageenan [*9000-07-1*].

» Carrageenan is the hydrocolloid obtained by extraction with water or aqueous alkali, from some members of the class *Rhodophyceae* (red seaweeds). Carrageenan consists chiefly of potassium, sodium, calcium, magnesium, and ammonium sulfate esters of galactose and 3,6-anhydrogalactose copolymers. These hexoses are alternately linked α-1,3 and β-1,4 in the polymer. The prevalent copolymers in the hydrocolloid are designated *kappa*-, *iota*-, and *lambda*-carrageenan. *Kappa*-carrageenan is mostly the alternating polymer of D-galactose-4-sulfate and 3,6-anhydro-D-galactose. *Iota*-carrageenan is similar, except that the 3,6-anhydrogalactose is sulfated at carbon 2. Between *kappa*-carrageenan and *iota*-carrageenan there is a continuum of intermediate compositions differing in degree of sulfation at carbon 2. In *lambda*-carrageenan, the alternating monomeric units are mostly D-galactose-2-sulfate (1,3-linked) and D-galactose-2,6-disulfate (1,4-linked). The ester sulfate content for Carrageenan ranges from 18 percent to 40 percent. In addition, it contains inorganic salts that originate from the seaweed and from the process of recovery from the extract.

Carrageenan is recovered by alcohol precipitation, by drum drying, or by freezing. The alcohols used during recovery and purification are restricted to methanol, alcohol, and isopropyl alcohol. Carrageenan that is recovered by drum-roll drying may contain mono- and di-glycerides or up to 5 percent of polysorbate 80 used as roll-stripping agents.

Packaging and storage—Preserve in tight containers, preferably in a cool place.

Solubility in water—Not more than 30 mL of water is required to dissolve 1 g at a temperature of 80°.

Identification—

A: A solution (1 in 50) prepared by heating a uniform dispersion in a hot water bath to 80° (*Solution A*) becomes more viscous upon cooling and may form a gel.

B: To 10 mL of *Solution A*, while still hot, add 4 drops of potassium chloride solution (1 in 10), mix, and cool. A short-textured ("brittle") gel indicates a carrageenan of a predominantly *kappa* type; a compliant ("elastic") gel indicates a predominantly *iota* type. If the solution does not gel, the carrageenan is of a predominantly *lambda* type.

C: Dilute a portion of *Solution A* with about 4 parts of water, and add 2 to 3 drops of methylene blue TS: a blue, stringy precipitate is formed (also positive for furcellaran, a similar colloid).

D: Obtain infrared absorption spectra on the gelling and nongelling fractions of the specimen by the following procedure. Disperse 2 g in 200 mL of potassium chloride solution (1 in 40), and stir for 1 hour. Allow to stand for 18 hours, stir again for 1 hour, and transfer to a centrifuge tube. (If the transfer cannot be made because the dispersion is too viscous, dilute with up to 200 mL of the potassium chloride solution.) Centrifuge at approximately 1000 g for 15 minutes.

Remove the clear supernatant liquid, resuspend the residue in 200 mL of potassium chloride solution (1 in 40), and centrifuge again. Coagulate the combined supernatant liquids by adding 2 volumes of dilute alcohol (9 in 10). (Retain the sediment for use subsequently as directed.) Recover the coagulum, and wash with 250 mL of the dilute alcohol. Press the excess liquid from the

Wave Number cm^{-1}	Molecular Assignment	Absorbance Relative to 1050 cm^{-1}		
		Kappa	Iota	Lambda
1220 to 1260	Ester sulfate	0.7 to 1.2	1.2 to 1.6	1.4 to 2.0
928 to 933	3,6-anhydrogalactose	0.3 to 0.6	0.2 to 0.4	0 to 0.2
840 to 850	Galactose-4-sulfate	0.3 to 0.5	0.2 to 0.4	—
825 to 830	Galactose-2-sulfate	—	—	0.2 to 0.4
810 to 820	Galactose-6-sulfate	—	—	0.1 to 0.3
800 to 805	3,6-anhydrogalactose-2-sulfate	0 to 0.2	0.2 to 0.4	—

coagulum, and dry it at 60° for 2 hours: the material so obtained is the nongelling fraction (*lambda* carrageenan).

Disperse the sediment retained from the foregoing procedure in 250 mL of cold water, heat at 90° for 10 minutes, and cool to 60°. Coagulate the mixture, then recover, wash, and dry the coagulum as described above: the material so obtained is the gelling fraction (*kappa*- and *iota*-carrageenan).

Prepare a solution (1 in 500) of each fraction, cast films 5 µm thick (when dry) on a suitable nonsticking surface, and obtain the infrared absorption spectrum of each film. Carrageenan has strong, broad absorption bands, typical of all polysaccharides, in the 1000 to 1100 cm^{-1} region. Absorption maxima are 1065 cm^{-1} and 1020 cm^{-1} for gelling and nongelling types, respectively. Other characteristic absorption bands and their intensities relative to the absorbance at 1050 cm^{-1} are as shown in the accompanying table.

Viscosity ⟨911⟩—Transfer 7.5 g to a tared, tall-form, 600-mL beaker, add 450 mL of water, and disperse with agitation for about 15 minutes. Add water to bring the weight to 500 g, and heat in a water bath, with continuous agitation, until a temperature of 80° is reached. Add water to adjust for loss by evaporation, cool to between 76° and 77°, and place in a constant-temperature bath maintained at 75°. Provide a suitable rotational viscosimeter with a spindle 1.88 cm in diameter and 6.51 cm high, using an immersion depth of 8.10 cm (No. 1 spindle). Allow the spindle to rotate in the solution at 30 rpm for 6 revolutions, then observe the scale reading. Convert the scale reading to centipoises by multiplying by the constant for the spindle and speed employed. The viscosity at 75° is not less than 5 centipoises.

Microbial limits ⟨61⟩—The total bacterial count does not exceed 200 per g, and the tests for *Salmonella* species and *Escherichia coli* are negative.

Loss on drying ⟨731⟩—Dry it at a pressure not exceeding 10 mm of mercury at 70° for 18 hours, cool in a desiccator, and weigh: it loses not more than 12.5% of its weight.

Acid-insoluble matter—Transfer about 2 g, accurately weighed, to a 250-mL beaker containing 150 mL of water and 1.5 mL of sulfuric acid. Cover with a watch glass, and heat on a steam bath for 6 hours, rubbing down the wall of the beaker frequently with a rubber-tipped stirring rod, and replacing any water lost by evaporation. Transfer about 500 mg of a suitable filter aid, accurately weighed, to the beaker, and filter through a tared filtering crucible provided with a 2.4-cm glass fiber filter. Wash the residue several times with hot water, dry at 105° for 3 hours, cool in a desiccator, and weigh. The difference between the total weight and the sum of the weights of the filter aid, crucible, and glass fiber filter is the weight of the acid-insoluble matter. It is not more than 2.0% of the weight of Carrageenan taken.

Total ash ⟨561⟩: not more than 35.0%.

Arsenic ⟨211⟩: 3 ppm.

Lead ⟨251⟩: 0.001%.

Heavy metals, *Method II* ⟨231⟩: 0.004%.

Castor Oil—*see* Castor Oil USP

Hydrogenated Castor Oil

» Hydrogenated Castor Oil is refined, bleached, hydrogenated, and deodorized Castor Oil, consisting mainly of the triglyceride of hydroxystearic acid.

Packaging and storage—Preserve in tight containers, and avoid exposure to excessive heat.

Melting range, *Class II* ⟨741⟩: between 85° and 88°.

Heavy metals, *Method II* ⟨231⟩: 0.001%.

Free fatty acids ⟨401⟩—Weigh accurately 20 g into a conical flask, melt on a steam bath, add 75 mL of hot alcohol that previously has been neutralized with 0.1 *N* sodium hydroxide to phenolphthalein TS, swirl, and add 1 mL of phenolphthalein TS. Titrate with 0.1 *N* sodium hydroxide VS, swirling vigorously, until the solution remains faintly pink after being shaken for 60 seconds: not more than 11.0 mL of 0.1 *N* sodium hydroxide is required.

Hydroxyl value—The hydroxyl value is between 154 and 162, determined as directed in the test for *Hydroxyl value* under *Castor Oil.*

Iodine value ⟨401⟩: not more than 5.

Saponification value ⟨401⟩: between 176 and 182.

Castor Oil, Polyoxyl 35—*see* Polyoxyl 35 Castor Oil

Castor Oil, Polyoxyl 40 Hydrogenated—*see* Polyoxyl 40 Hydrogenated Castor Oil

Microcrystalline Cellulose

Cellulose.
Cellulose [*9004-34-6*].

» Microcrystalline Cellulose is purified, partially depolymerized cellulose prepared by treating alpha cellulose, obtained as a pulp from fibrous plant material, with mineral acids. It contains not less than 97.0 percent and not more than 102.0 percent of cellulose, calculated on the dried basis.

Packaging and storage—Preserve in tight containers.

Identification—Determine the particle size by sieving 20 g for 5 minutes on an air jet sieve equipped with a screen having 37-µm openings. If more than 5% is retained on the screen, mix 30 g of Microcrystalline Cellulose with 270 mL of water; otherwise, mix 45 g with 255 mL of water. Perform the mixing in a high-speed (18,000 rpm) power blender for 5 minutes. Transfer 100 mL of the dispersion to a 100-mL graduated cylinder, and allow to stand for 3 hours: a white, opaque, bubble-free dispersion, which does not form a supernatant liquid at the surface, is obtained.

pH ⟨791⟩—Shake about 5 g with 40 mL of water for 20 minutes, and centrifuge: the pH of the supernatant liquid is between 5.5

and 7.0 for the grades of Microcrystalline Cellulose that have a sieve fraction greater than 5% retained on the 37-µm screen, and between 5.0 and 7.0 for the grades with less than 5% retained on the 37-µm screen.

Loss on drying ⟨731⟩—Dry it at 105° for 3 hours: it loses not more than 5.0% of its weight.

Residue on ignition ⟨281⟩: not more than 0.05%.

Water-soluble substances—Shake 5.0 g with about 80 mL of water for 10 minutes, filter through filter paper (Whatman No. 42 or equivalent) into a tared beaker, evaporate on a steam bath to dryness, and dry at 105° for 1 hour: not more than 8.0 mg (0.16%) of residue is obtained for the grades that have a sieve fraction more than 5% retention on the 37-µm screen, while not more than 12.0 mg (0.24%) of residue is obtained for the grades that have a sieve fraction not more than 5% retention on the 37-µm screen.

Heavy metals, *Method II* ⟨231⟩: 0.001%.

Starch—To 20 mL of the dispersion of Microcrystalline Cellulose obtained in the *Identification test* add a few drops of iodine TS, and mix: no purplish blue or blue color is produced.

Assay—Transfer about 125 mg of Microcrystalline Cellulose, accurately weighed, to a 300-mL conical flask with the aid of about 25 mL of water. Add 50.0 mL of 0.5 *N* potassium dichromate, mix, carefully add 100 mL of sulfuric acid, and heat rapidly to incipient boiling. Remove from heat, allow to stand at room temperature for 15 minutes, cool in a water bath, and transfer to a 250-mL volumetric flask. Dilute with water almost to volume, cool to 25°, dilute with water to volume, and mix. Titrate a 50-mL aliquot with 0.1 *N* ferrous ammonium sulfate VS, using 2 or 3 drops of orthophenanthroline TS as the indicator. Perform a blank determination, and note the difference in volumes required. Each mL of the difference in volumes of 0.1 *N* ferrous ammonium sulfate consumed is equivalent to 0.675 mg of cellulose in the portion titrated (i.e., one-fifth of the specimen weighed for the *Assay*).

Microcrystalline Cellulose and Carboxymethylcellulose Sodium

» Microcrystalline Cellulose and Carboxymethylcellulose Sodium is a colloid-forming, attrited mixture of Microcrystalline Cellulose and Carboxymethylcellulose Sodium. It contains not less than 75.0 percent and not more than 125.0 percent of the labeled amount of carboxymethylcellulose sodium, calculated on the dried basis. The viscosity of its aqueous dispersion of percent by weight stated on the label is not less than 60.0 percent and not more than 140.0 percent of that stated on the label in centipoises.

Packaging and storage—Preserve in tight containers, store in a dry place, and avoid exposure to excessive heat.

Labeling—Label it to indicate the percentage content of carboxymethylcellulose sodium and the viscosity of the dispersion in water of the designated weight percentage composition.

Identification—

A: Mix 6 g with 300 mL of water in a blender at 18,000 rpm for 5 minutes: a white, opaque, dispersion is produced which does not settle on standing.

B: Add several drops of the dispersion obtained in *Identification test A* to a solution of aluminum chloride (1 in 10): each drop forms a white, opaque globule which does not disperse on standing.

C: Add 3 mL of iodine TS to the dispersion obtained in *Identification test A:* no blue or purplish blue color is produced.

Viscosity ⟨911⟩—Determine the amounts of Microcrystalline Cellulose and Carboxymethylcellulose Sodium needed to prepare 600 g of a suitable dispersion, calculated on the dried basis. Transfer an accurately weighed amount of water to a 1000-mL blender bowl. Begin stirring with an 18,000 rpm blender at a reduced speed obtained by adjusting the voltage to 30 volts by means of a suitable transformer, and immediately add the accurately weighed portion of Microcrystalline Cellulose and Carboxymethylcellulose Sodium, taking care to avoid contacting the sides of the bowl with the powder. Continue stirring at this speed for 15 seconds following the addition of the powder, then increase the transformer setting to 115 volts, and mix for 2 minutes, accurately timed, at 18,000 rpm. Stop the blender, and lower the appropriate spindle of a suitable rotational viscosimeter into the dispersion. Thirty seconds after cessation of mixing, start the viscosimeter, and determine the viscosity using the appropriate spindle to obtain a scale reading between 10% and 90% of full-scale at a speed of 20 rpm. Determine the scale reading after 30 seconds of rotation, and calculate the viscosity, in centipoises, by multiplying the scale reading by the constant for the spindle used at 20 rpm.

pH ⟨791⟩: between 6.0 and 8.0, determined on the dispersion prepared in the test for *Viscosity*.

Loss on drying ⟨731⟩—Dry it at 105° to constant weight: it loses not more than 8.0% of its weight.

Residue on ignition ⟨281⟩: not more than 5.0%.

Heavy metals, *Method II* ⟨231⟩: 0.001%.

Assay for carboxymethylcellulose sodium—Transfer about 2 g of Microcrystalline Cellulose and Carboxymethylcellulose Sodium, accurately weighed, to a glass-stoppered, 250-mL conical flask. Add 75 mL of glacial acetic acid, attach a condenser, and reflux for 2 hours. Cool, transfer the mixture to a 250-mL beaker with the aid of small volumes of glacial acetic acid, and titrate with 0.1 *N* perchloric acid in dioxane VS, determining the end-point potentiometrically. Each mL of 0.1 *N* perchloric acid is equivalent to 29.6 mg of carboxymethylcellulose sodium.

Cellulose, Oxidized—*see* Cellulose, Oxidized USP

Cellulose, Oxidized Regenerated—*see* Cellulose, Oxidized Regenerated USP

Powdered Cellulose

» Powdered Cellulose is purified, mechanically disintegrated cellulose prepared by processing alpha cellulose obtained as a pulp from fibrous plant materials. It contains not less than 97.0 percent and not more than 102.0 percent of cellulose, calculated on the dried basis.

Packaging and storage—Preserve in well-closed containers.

Identification—Mix 30 g with 270 mL of water in a high-speed (12,000 rpm) power blender for 5 minutes. The mixture will be either a heavy, lumpy suspension that flows poorly, if at all, settles only slightly, and contains many trapped air bubbles, or a free-flowing suspension. If a free-flowing suspension is obtained, transfer 100 mL of it to a 100-mL graduated cylinder, and allow to stand for 1 hour: the Powdered Cellulose settles in the cylinder, and a supernatant liquid appears above the layer of the cellulose.

pH ⟨791⟩—Mix 10 g with 90 mL of water, and allow to stand with occasional stirring for 1 hour: the pH of the supernatant liquid is between 5.0 and 7.5.

Loss on drying ⟨731⟩—Dry it at 105° for 3 hours: it loses not more than 7.0% of its weight.

Residue on ignition ⟨281⟩: not more than 0.3%, calculated on the dried basis, the addition of sulfuric acid being omitted from the procedure.

Water-soluble substances—Mix 6.0 g with 90 mL of recently boiled and cooled water, and allow to stand with occasional stirring for 10 minutes. Filter, discard the first 10 mL of the filtrate, and pass the filtrate through the same filter a second time, if necessary, to obtain a clear filtrate. Evaporate a 15.0-mL portion of the filtrate in a tared evaporating dish on a steam bath to dryness, dry at 105° for 1 hour, cool in a desiccator, and weigh: not more than 15 mg of residue is obtained (1.5%).

Heavy metals, *Method II* ⟨231⟩: 0.001%.

Starch—To 10 g add 90 mL of water, and boil for 5 minutes. Filter hot. To the filtrate add 2 drops of iodine TS: no change in color from the yellow-red is produced.

Assay—Transfer about 125 mg of Powdered Cellulose, accurately weighed, to a 300-mL conical flask with the aid of about 25 mL of water. Add 50.0 mL of 0.5 *N* potassium dichromate, mix, carefully add 100 mL of sulfuric acid, and heat rapidly to incipient boiling. Remove from heat, allow to stand at room temperature for 15 minutes, cool in a water bath, and transfer to a 250-mL volumetric flask. Dilute with water almost to volume, cool to 25°, dilute with water to volume, and mix. Titrate a 50-mL aliquot with 0.1 *N* ferrous ammonium sulfate VS, using 2 or 3 drops of orthophenanthroline TS as the indicator. Perform a blank determination, and note the difference in volumes required. Each mL of the difference in volumes of 0.1 *N* ferrous ammonium sulfate consumed is equivalent to 0.675 mg of cellulose in the portion titrated (i.e., one-fifth of the specimen weighed for the *Assay*).

Cellulose Acetate

Cellulose acetate.
Cellulose, acetate [*9004-35-7*].
Cellulose, diacetate [*9035-69-2*].
Cellulose, triacetate [*9012-09-3*].

» Cellulose Acetate is partially or completely acetylated cellulose. It contains not less than 29.0 percent and not more than 44.8 percent, by weight, of acetyl (C_2H_3O) groups. Its acetyl content is not less than 90.0 percent and not more than 110.0 percent of that indicated on the label.

Packaging and storage—Preserve in well-closed containers.

Labeling—Label it to indicate the percentage content of acetyl.

Reference standard—*USP Cellulose Acetate Reference Standard*—Dry at 105° for 3 hours before using.

Identification—Prepare a 1 in 10 solution of Cellulose Acetate, previously dried, in dioxane. Spread 1 drop of the solution on a sodium chloride plate, place a second sodium chloride plate over it, and spread the specimen between the plates. Separate the plates, heat them both at 105° for 1 hour, and reassemble the dried plates: the infrared absorption spectrum exhibits maxima only at the same wavelengths as that of a similar preparation of USP Cellulose Acetate RS, treated in the same manner.

Loss on drying ⟨731⟩—Dry it at 105° for 3 hours: it loses not more than 5.0% of its weight.

Residue on ignition ⟨281⟩: not more than 0.1%.

Heavy metals, *Method II* ⟨231⟩: 0.001%.

Free acid—Transfer about 5 g, previously dried and accurately weighed, to a 250-mL flask. Add 150 mL of freshly boiled, cooled water, insert the stopper in the flask, swirl the suspension gently, and allow it to stand for 3 hours. Filter through paper, and wash the flask and the filter with water, adding these washings to the filtrate. Add phenolphthalein TS, and titrate the combined filtrate and washings with 0.01 *N* sodium hydroxide VS. Each mL of 0.01 *N* sodium hydroxide is equivalent to 0.6005 mg of acetic acid: not more than 0.1% is found, on the dried basis.

Acetyl content—

FOR CELLULOSE ACETATE LABELED TO CONTAIN NOT MORE THAN 42.0% OF ACETYL GROUPS—Transfer about 2 g, previously dried and accurately weighed, to a 500-mL flask. Add 100 mL of acetone and 5 mL to 10 mL of water to the flask, insert the stopper in the flask, and stir with a magnetic stirrer until solution is complete. Add 30 mL, accurately measured, of 1.0 *N* sodium hydroxide VS to the solution, with constant stirring. A finely divided precipitate of regenerated cellulose, free from lumps, is obtained. Insert the stopper in the flask, and stir with a magnetic stirrer for 30 minutes. Add 100 mL of hot water, washing down the sides of the flask, and stir for 1 to 2 minutes. Titrate the excess sodium hydroxide solution with 1.0 *N* sulfuric acid VS to a phenolphthalein end-point. Treat a blank in the same manner. Calculate the percentage of acetyl by the formula:

$$4.305(B - A)/W,$$

in which B and A are the volumes, in mL, of 1.0 *N* sulfuric acid consumed by the blank and the Cellulose Acetate, respectively, and W is the weight, in g, of Cellulose Acetate taken.

FOR CELLULOSE ACETATE LABELED TO CONTAIN MORE THAN 42.0% OF ACETYL GROUPS—Transfer about 2 g, previously dried and accurately weighed, to a 500-mL conical flask. Add 30.0 mL of dimethylsulfoxide and 100 mL of acetone, and stir for 16 hours with the aid of a magnetic stirrer. Pipet 30 mL of 1 *N* sodium hydroxide VS slowly into the flask, with constant stirring. Insert the stopper in the flask, and stir for 6 minutes. Allow to stand without stirring for 60 minutes. Resume stirring, and add 100 mL of water that has been pre-heated to 80°, washing down the sides of the flask. Stir for 2 minutes, and cool to room temperature. Add 4 to 5 drops of phenolphthalein TS, and titrate the excess sodium hydroxide with 0.5 *N* hydrochloric acid VS. Add an accurately measured excess of about 0.5 mL of 0.5 *N* hydrochloric acid VS. Stir for 5 minutes. Allow to stand for 30 minutes. Titrate with 0.5 *N* sodium hydroxide VS to a persistent pink end-point, using a magnetic stirrer for agitation. Calculate the net number of milliequivalents of sodium hydroxide consumed, and correct this value by use of the average of two blank determinations run concomitantly through the entire procedure. Calculate the percentage of acetyl by the formula:

$$4.305n/W,$$

in which n is the corrected value of the net number of milliequivalents of sodium hydroxide consumed, and W is the weight, in g, of Cellulose Acetate taken.

Cellulose Acetate Phthalate

Cellulose, acetate, 1,2-benzenedicarboxylate.
Cellulose acetate phthalate [*9004-38-0*].

» Cellulose Acetate Phthalate is a reaction product of phthalic anhydride and a partial acetate ester of cellulose. It contains not less than 21.5 percent and not more than 26.0 percent of acetyl (C_2H_3O) groups and not less than 30.0 percent and not more than 36.0 percent of phthalyl(*o*-carboxy-benzoyl, $C_8H_5O_3$) groups, calculated on the anhydrous, acid-free basis.

Packaging and storage—Preserve in tight containers.

Reference standard—*USP Cellulose Acetate Phthalate Reference Standard.*

Identification—

A: The infrared absorption spectrum of a potassium bromide dispersion of it exhibits maxima only at the same wavelengths as that of a similar preparation of USP Cellulose Acetate Phthalate RS.

B: Dissolve about 150 mg in 1 mL of acetone, and pour onto a clear glass plate in an area of good airflow: a glossy, clear film is deposited as the acetone evaporates.

Viscosity ⟨911⟩—Dissolve 15 g, calculated on the anhydrous basis, in 85 g of a mixture of 249 parts of anhydrous acetone and 1 part of water: the apparent viscosity (see *Procedure for Methylcellulose* under *Viscosity* ⟨911⟩) is between 45 and 90 centipoises, determined at 25 ± 0.2°.

Water, *Method I* ⟨921⟩: not more than 5.0%, a mixture of dehydrated alcohol and methylene chloride (3:2) being used instead of methanol as the solvent.

Residue on ignition ⟨281⟩: not more than 0.1%.

Free acid—Transfer 3.0 g, accurately weighed, to a glass-stoppered flask, add 100 mL of dilute methanol (1 in 2), insert the stopper in the flask, and shake for 2 hours. Filter and wash the flask and the filter with two 10-mL portions of the methanol solution, adding the washings to the filtrate. Titrate the combined filtrate and washings with 0.1 *N* sodium hydroxide VS to a phenolphthalein end-point. Perform a blank determination on 120 mL of the dilute methanol (1 in 2). Calculate the percentage of free acid, *B*, by the formula:

$$0.8306A/W,$$

in which *A* is the volume, in mL, of 0.1 *N* sodium hydroxide consumed, corrected for the blank, and *W* is the weight, in g, of the Cellulose Acetate Phthalate, calculated on the anhydrous basis. Not more than 6.0%, calculated as phthalic acid, is found.

Phthalyl content—Transfer about 1 g, accurately weighed, to a conical flask, dissolve in 50 mL of a mixture of alcohol and dichloromethane (3:2), then add 25 mL of alcohol, add phenolphthalein TS, and titrate with 0.1 *N* sodium hydroxide VS. Perform a blank determination on a mixture of 55 mL of alcohol and 20 mL of dichloromethane. Calculate the percentage of phthalyl, on the acid-free basis, by the formula:

$$100[(1.491A/W) - 1.795B]/(100 - B),$$

in which *A* is the volume, in mL, of 0.1 *N* sodium hydroxide consumed after correction for the blank, *W* is the weight, in g, of Cellulose Acetate Phthalate taken, calculated on the anhydrous basis, and *B* is the percentage of acid found in the test for *Free acid*.

Acetyl content—Transfer about 500 mg, accurately weighed, to a glass-stoppered flask, and add 50 mL of water and 50.0 mL of 0.5 *N* sodium hydroxide VS. Connect the flask to a reflux condenser, and reflux for 60 minutes. Cool, add 5 drops of phenolphthalein TS, and titrate with 0.5 *N* hydrochloric acid VS. Perform a blank determination. Calculate the free and combined acids, as acetyl, by the formula:

$$2.152(A/W),$$

in which *A* is the volume, in mL, of 0.5 *N* sodium hydroxide consumed after correction for the blank, and *W* is the weight, in g, of Cellulose Acetate Phthalate taken, on the anhydrous basis. Calculate the percentage of acetyl, on the acid-free basis, by the formula:

$$[100(P - 0.5182B)/(100 - B)] - 0.5772C,$$

in which *P* is the free and combined acids, as acetyl, *B* is the percentage of acid found in the test for *Free acid*, and *C* is the percentage of phthalyl found in the test for *Phthalyl content*.

Cellulose, Hydroxyethyl—*see* Hydroxyethyl Cellulose

Cellulose, Hydroxypropyl—*see* Hydroxypropyl Cellulose

Cetostearyl Alcohol

» Cetostearyl Alcohol contains not less than 40.0 percent of stearyl alcohol ($C_{18}H_{38}O$), and the sum of the stearyl alcohol content and the cetyl alcohol ($C_{16}H_{34}O$) content is not less than 90.0 percent.

Packaging and storage—Preserve in well-closed containers.

Reference standards—*USP Stearyl Alcohol Reference Standard*—Do not dry before using. *USP Cetyl Alcohol Reference Standard*—Do not dry before using.

Identification—The retention times of the major peaks in the chromatogram of the Assay solution correspond to those of the Standard solution, as obtained in the *Assay*.

Melting range ⟨741⟩: between 48° and 55°.

Acid value ⟨401⟩: not more than 2.

Iodine value ⟨401⟩: not more than 4.

Hydroxyl value ⟨401⟩—Place about 2 g, accurately weighed, in a dry, glass-stoppered, 250-mL flask, add 2 mL of pyridine, then add 10 mL of toluene. To the mixture add 10.0 mL of a solution of acetyl chloride prepared by mixing 10 mL of acetyl chloride with 90 mL of toluene. Insert the stopper in the flask, and immerse in a water bath heated at 60° to 65° for 20 minutes. Add 25 mL of water, again insert the stopper in the flask, and shake vigorously for several minutes to decompose the excess acetyl chloride. Add 0.5 mL of phenolphthalein TS, and titrate with 1 *N* sodium hydroxide VS to a permanent pink end-point, shaking the flask vigorously toward the end of the titration in order to maintain the contents in an emulsified condition. Perform a blank determination with the same quantities of the same reagents and in the same manner. The difference between the number of mL of 1 *N* sodium hydroxide consumed in the test with the specimen under test and that consumed in the blank test, multiplied by 56.1, and the result divided by the weight, in g, of the Cetostearyl Alcohol used, represents the hydroxyl value of the Cetostearyl Alcohol. The hydroxyl value is between 208 and 228.

Assay—Dissolve 100 mg of Cetostearyl Alcohol in 10.0 mL of dehydrated alcohol, and mix. Introduce a 2-µL portion of this Assay solution into a suitable gas chromatograph equipped with a flame-ionization detector. The column is 2 m × 3 mm, and is packed with 10 percent liquid phase G2 on support S1. The carrier gas is helium, flowing at the rate of 100 mL per minute, as measured at the column exit. The temperatures of the injection port and the detector are maintained at about 275° and 250°, respectively, and the column temperature is about 205°. In a suitable chromatogram, the resolution factor, *R* (see *Chromatography* ⟨621⟩), is not less than 4.0 between the peaks due to cetyl alcohol and stearyl alcohol, both peaks being identified by comparison with a chromatogram concomitantly obtained on a portion of a Standard solution prepared by dissolving about 50 mg of USP Stearyl Alcohol RS and 50 mg of USP Cetyl Alcohol RS in 10.0 mL of alcohol, and five replicate injections of a single sample show a relative standard deviation of not more than 1.5% in the percentages of $C_{18}H_{38}O$ and $C_{16}H_{34}O$, respectively. Measure the peak areas of the long-chain alcohol components in the chromatogram of the Assay solution, and determine the percentage of $C_{18}H_{38}O$ of Cetostearyl Alcohol taken by the formula:

$$100A/B,$$

in which *A* is the area due to the stearyl alcohol peak and *B* is the sum of the areas of all the long-chain alcohol peaks in the chromatogram. Determine the percentage of $C_{16}H_{34}O$ in the portion of Cetostearyl Alcohol taken by the formula:

$$100C/B,$$

in which *C* is the area due to the cetyl alcohol peak and *B* is as previously defined.

Cetyl Alcohol

$CH_3(CH_2)_{14}CH_2OH$

$C_{16}H_{34}O$ 242.44

1-Hexadecanol.

1-Hexadecanol [*124-29-8; 36653-82-4*].

» Cetyl Alcohol contains not less than 90.0 percent of cetyl alcohol ($C_{16}H_{34}O$), the remainder consisting chiefly of related alcohols.

Packaging and storage—Preserve in well-closed containers.

Reference standards—*USP Stearyl Alcohol Reference Standard*—Do not dry before using. *USP Cetyl Alcohol Reference Standard*—Do not dry before using.

Identification—The retention time of the major peak obtained in the *Assay* is the same as that obtained from the Standard solution employed in the *Assay*.

Melting range ⟨741⟩: between 45° and 50°, determined by the procedure for *Class I*, except that the specimen is inserted into the bath at about room temperature.

Acid value ⟨401⟩: not more than 2.

Iodine value ⟨401⟩: not more than 5.

Hydroxyl value ⟨401⟩—Place about 2 g, accurately weighed, in a dry, glass-stoppered, 250-mL flask, and add 2 mL of pyridine, followed by 10 mL of toluene. To the mixture add 10.0 mL of a solution of acetyl chloride, prepared by mixing 10 mL of acetyl chloride with 90 mL of toluene. Insert the stopper in the flask, and immerse in a water bath heated at 60° to 65° for 20 minutes. Add 25 mL of water, again insert the stopper in the flask, and shake vigorously for several minutes to decompose the excess acetyl chloride. Add 0.5 mL of phenolphthalein TS, and titrate to a permanent pink end-point with 1 *N* sodium hydroxide VS, shaking the flask vigorously toward the end of the titration to maintain the contents in an emulsified condition. Perform a blank test with the same quantities of the same reagents and in the same manner. The difference between the number of mL of 1 *N* sodium hydroxide consumed in the test with the sample and that consumed in the blank test, multiplied by 56.1, and the result divided by the weight, in g, of the Cetyl Alcohol used, represents the hydroxyl value of the Cetyl Alcohol, which is between 218 and 238.

Assay—

Chromatographic system (see *Chromatography* ⟨621⟩)—The gas chromatograph is equipped with a flame-ionization detector, and it contains a 2-m × 3-mm column that is packed with 10 percent phase G2 on support S1A. Dry helium is used as the carrier gas. The column is maintained at a temperature of about 205°, and the injection port and the detector are maintained at temperatures of about 275° and 250°, respectively.

System suitability—Prepare a system suitability solution by dissolving about 90 mg of USP Cetyl Alcohol RS and 10 mg of USP Stearyl Alcohol RS in 10.0 mL of alcohol. Inject 2 µL of this solution, and calculate the resolution factor, *R* (see *Chromatography* ⟨621⟩), for cetyl alcohol and stearyl alcohol. A suitable resolution is not less than 4.0. Inject 2-µL portions of the suitability standard solution until the area ratio of cetyl alcohol to stearyl alcohol for each of five sequential injections is within 1.5% of the average of the area ratios for the five injections.

Procedure—Inject a suitable volume (about 2 µL) of a solution of Cetyl Alcohol in dehydrated alcohol (1 in 100) into the gas chromatograph. Measure the peak areas of the long-chain alcohol components in the chromatogram and determine the percentage of $C_{16}H_{34}O$ in the portion of Cetyl Alcohol taken by the formula:

$$100C/B,$$

in which C is the area due to the cetyl alcohol peak and B is the sum of the areas of all the long-chain alcohol peaks in the chromatogram.

Cetyl Esters Wax

» Cetyl Esters Wax is a mixture consisting primarily of esters of saturated fatty alcohols (C_{14} to C_{18}) and saturated fatty acids (C_{14} to C_{18}).

Packaging and storage—Preserve in well-closed containers, in a dry place, and prevent exposure to excessive heat.

Melting range, *Class II* ⟨741⟩: between 43° and 47°.

Acid value ⟨401⟩: not more than 5.

Iodine value ⟨401⟩: not more than 1.

Saponification value ⟨401⟩: between 109 and 120.

Paraffin and free acids—A 1-g portion dissolves completely in 50 mL of boiling alcohol and the solution in neutral or acid to moistened litmus paper.

Cetylpyridinium Chloride—*see* Cetylpyridinium Chloride USP

Charcoal, Activated—*see* Charcoal, Activated USP

Chlorobutanol

$Cl_3C.C(CH_3)_2.OH$

$C_4H_7Cl_3O$ (anhydrous) 177.46
2-Propanol, 1,1,1-trichloro-2-methyl-.
1,1,1-Trichloro-2-methyl-2-propanol [*57-15-8*].
Hemihydrate 186.46 [*6001-64-5*].

» Chlorobutanol is anhydrous or contains not more than one-half molecule of water of hydration. It contains not less than 98.0 percent and not more than 100.5 percent of $C_4H_7Cl_3O$, calculated on the anhydrous basis.

Packaging and storage—Preserve in tight containers.

Labeling—Label it to indicate whether it is anhydrous or hydrous.

Reference standard—*USP Chlorobutanol Reference Standard*—Do not dry before using.

Identification—

A: The infrared absorption spectrum, determined in a 1-mm cell, of a 1 in 50 solution in carbon disulfide, passed through phase-separating paper (Whatman No. 1PS), exhibits maxima only at the same wavelengths as that of a similar preparation of USP Chlorobutanol RS.

B: To 5 mL of a freshly prepared solution (1 in 200) add 1 mL of 1 *N* sodium hydroxide, then slowly add 3 mL of iodine TS: a yellow precipitate of iodoform, recognizable by its odor, appears.

Reaction—Shake thoroughly 0.5 g with 25 mL of water: the water remains neutral to litmus.

Water, *Method I* ⟨921⟩: not more than 1.0% anhydrous form) and not more than 6.0% (hydrous form).

Chloride—To a solution of 0.50 g in a mixture of 25 mL of diluted alcohol and 1 mL of nitric acid add 2 mL of silver nitrate TS: any turbidity produced is not greater than that produced from a control solution containing 0.50 mL of 0.020 *N* hydrochloric acid in place of the Chlorobutanol (0.07%).

Assay—Transfer about 200 mg of Chlorobutanol, accurately weighed, to a 250-mL conical flask, add 10 mL of a 3 in 10 solution of potassium hydroxide in alcohol, and attach the flask to a reflux condenser. Reflux for 15 minutes, cool, and rinse the condenser and the tip with water. Remove the flask, and transfer the contents to a suitable titration vessel with the aid of several small portions of a 50-mL portion of water. Add the remaining water to the titration vessel. Add 10 mL of nitric acid and 50.0 mL of 0.1 *N* silver nitrate VS. Stir, and titrate the excess silver nitrate with 0.1 *N* ammonium thiocyanate VS, determining the end-point potentiometrically. Perform a blank determination (see *Residual Titrations* under *Titrimetry* ⟨541⟩). Each mL of 0.1 *N* silver nitrate is equivalent to 5.915 mg of $C_4H_7Cl_3O$.

Chlorocresol

OH, CH_3, Cl (structural formula)

C_7H_7ClO 142.58
Phenol, 4-chloro-3-methyl-.
4-Chloro-*m*-cresol [*59-50-7*].

» Chlorocresol contains not less than 99.0 percent and not more than 101.0 percent of C_7H_7ClO (4-chloro-3-methylphenol).

Packaging and storage—Preserve in tight, light-resistant containers.

Completeness of solution—Transfer 1 g to a test tube, add 0.4 mL of alcohol, and shake: solution is complete.

Identification—

A: Add 40 mg to 10 mL of water, mix, and add 1 drop of ferric chloride TS: a blue color develops.

B: Transfer 50 mg to a crucible, add 500 mg of anhydrous sodium carbonate, and mix. Heat the mixture until fused. Cool, add 5 mL of water, and boil. Acidify with 1 mL of nitric acid, filter, and add 1 mL of silver nitrate TS to the filtrate: a white precipitate is formed.

Melting range ⟨741⟩: between 63° and 66°.

Nonvolatile residue—Heat about 1.0 g, accurately weighed, in a tared crucible on a steam bath until it has evaporated, and dry the residue at 105° for 1 hour: not more than 0.1% of residue remains.

Assay—Transfer about 70 mg of Chlorocresol, accurately weighed, to an iodine flask, add 30 mL of glacial acetic acid, 25.0 mL of 0.1 *N* bromine VS, 10 mL of potassium bromide solution (3 in 20), and 10 mL of hydrochloric acid. Immediately insert the stopper, mix, and allow to stand for 15 minutes, protected from light. Quickly add 10 mL of potassium iodide solution (1 in 10) and 100 mL of water, taking precautions against the escape of bromine vapor, at once insert the stopper, and shake the mixture thoroughly. Remove the stopper, and rinse it and the neck of the flask with a small quantity of water so that the washing flows into the flask. Add 1 mL of chloroform, shake the mixture thoroughly, and titrate the liberated iodine with 0.1 *N* sodium thiosulfate VS, adding 3 mL of starch TS as the end-point is approached. Perform a blank determination (see *Residual Titrations* under *Titrimetry* ⟨541⟩). Each mL of 0.1 *N* bromine is equivalent to 3.565 mg of C_7H_7ClO.

Chloroform

$CHCl_3$ 119.38
Methane, trichloro-.
Trichloromethane [*67-66-3*].

» Chloroform contains not less than 99.0 percent and not more than 99.5 percent of $CHCl_3$, the remainder consisting of alcohol.

Caution—Care should be taken not to vaporize Chloroform in the presence of a flame, because of the production of harmful gases.

Packaging and storage—Preserve in tight, light-resistant containers, at a temperature not exceeding 30°.

Specific gravity ⟨841⟩: between 1.476 and 1.480, indicating between 99.0% and 99.5% of $CHCl_3$.

Nonvolatile residue—Evaporate 50 mL in a platinum or porcelain dish on a steam bath, and dry at 105° for 1 hour: the weight of the residue does not exceed 1 mg (0.002%).

Free chlorine—To 10 mL add 10 mL of water and 0.1 mL of potassium iodide TS, shake for 2 minutes, and allow the liquids to separate: the lower layer does not show a violet tint.

Readily carbonizable substances ⟨271⟩—Transfer 40 mL to a glass-stoppered cylinder, previously rinsed with sulfuric acid TS and allowed to drain for 10 minutes. Add 5 mL of sulfuric acid TS, and shake the mixture vigorously for 5 minutes. Allow the liquids to separate completely: the Chloroform remains colorless, and the acid has no more color than Matching Fluid A.

Chlorinated decomposition products and chloride—Dilute 2 mL of the sulfuric acid separated from the Chloroform in the test for *Readily carbonizable substances* with 5 mL of water: the liquid is colorless and clear. When further diluted with 10 mL of water, it remains clear and is not affected within 1 minute by the addition of 3 drops of silver nitrate TS (*chlorinated decomposition products and chloride*).

Acid and phosgene—Into each of two glass-stoppered, 50-mL color-comparison cylinders having an internal diameter of 20 mm, place 10 mL of water, 2 drops of phenolphthalein TS, and enough 0.010 *N* sodium hydroxide to produce, after vigorous shaking, pink tints of equal intensity. Into one of the cylinders measure 20.0 mL of Chloroform, and again shake the mixture. Add 0.010 *N* sodium hydroxide dropwise from a microburet, shaking the mixture after each addition, until the pink color is reproduced in an intensity equal to that in the cylinder without the Chloroform. Not more than 0.20 mL of 0.010 *N* sodium hydroxide is required to produce a pink color that persists for 15 minutes.

Aldehyde and ketone—Shake 3.0 mL of it with 10 mL of ammonia-free water in a glass-stoppered cylinder for 5 minutes. After the liquids separate, transfer 5 mL of the water extract to another glass-stoppered cylinder containing 40 mL of ammonia-free water, and add 5 mL of alkaline mercuric–potassium iodide TS: no turbidity or precipitate develops within 1 minute.

Cholesterol

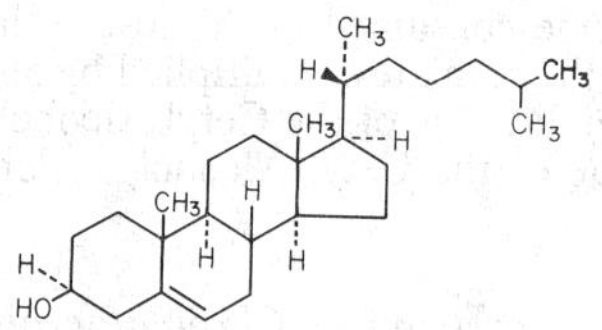

$C_{27}H_{46}O$ 386.67
Cholest-5-en-3-ol, (3β)-.
Cholest-5-en-3β-ol [*57-88-5*].

» Cholesterol is a steroid alcohol used as an emulsifying agent.

Packaging and storage—Preserve in well-closed, light-resistant containers.

Solubility in alcohol—Dissolve 500 mg in 50 mL of warm alcohol in a stoppered flask or cylinder, and allow to stand at room temperature for 2 hours: no deposit or turbidity is formed.

Identification—

A: To a solution of 10 mg in 1 mL of chloroform add 1 mL of sulfuric acid: the chloroform acquires a blood-red color and the sulfuric acid shows a green fluorescence.

B: Dissolve about 5 mg in 2 mL of chloroform, add 1 mL of acetic anhydride, and follow with 1 drop of sulfuric acid: a pink color is produced, and it rapidly changes to red, then to blue, and finally to a brilliant green.

Melting range ⟨741⟩: between 147° and 150°.

Specific rotation ⟨781⟩: between −34° and −38°, determined in a solution in dioxane containing 200 mg in each 10 mL.

Acidity—Dissolve 1.0 g in 10 mL of ether in a small flask, add 10.0 mL of 0.10 *N* sodium hydroxide, and shake for about 1 minute. Heat gently to expel the ether, and then boil for 5 minutes. Cool, dilute with 10 mL of water, add phenolphthalein TS, and titrate with 0.10 *N* sulfuric acid until the pink color just disappears, stirring the solution vigorously throughout the titration. Perform a blank determination (see *Residual Titrations* under *Titrimetry* ⟨541⟩). The difference between the number of mL of 0.10 *N* sulfuric acid consumed in the blank and the number of mL consumed in the test with the Cholesterol is not more than 0.3 mL.

Loss on drying ⟨731⟩—Dry it in vacuum at 60° for 4 hours: it loses not more than 0.3% of its weight.

Residue on ignition ⟨281⟩: not more than 0.1%.

Citric Acid—*see* Citric Acid USP

Cocoa Butter

» Cocoa Butter is the fat obtained from the roasted seed of *Theobroma cacao* Linné (Fam. Sterculiaceae).

Packaging and storage—Preserve in well-closed containers.

Specific gravity ⟨841⟩: between 0.858 and 0.864 at 100°/25°.

Refractive index ⟨831⟩: between 1.454 and 1.458 at 40°.

Heavy metals, *Method II* ⟨231⟩: 0.001%.

Wax, stearin, and tallow—Dissolve 1 g in 3 mL of ether in a test tube at a temperature of 17°, and immerse the tube in a mixture of ice and water: the solution does not become turbid nor deposit white flakes in less than 3 minutes. After the solution has congealed, raise the temperature to 15°: a clear liquid is gradually formed.

Solidification range of the fatty acids ⟨401⟩—The dry, mixed fatty acids of it solidify between 45° and 50°.

Free fatty acids ⟨401⟩—The free fatty acids in 10.0 g of it require for neutralization with not more than 5.0 mL of 0.10 *N* sodium hydroxide.

Iodine value ⟨401⟩: between 35 and 43.

Saponification value ⟨401⟩: between 188 and 195.

Colloidal Silicon Dioxide—*see* Silicon Dioxide, Colloidal

Compound Benzaldehyde Elixir—*see* Benzaldehyde Elixir, Compound

Compressible Sugar—*see* Sugar, Compressible

Confectioner's Sugar—*see* Sugar, Confectioner's

Corn Oil

» Corn Oil is the refined fixed oil obtained from the embryo of *Zea mays* Linné (Fam. Gramineae).

Packaging and storage—Preserve in tight, light-resistant containers, and avoid exposure to excessive heat.

Specific gravity ⟨841⟩: between 0.914 and 0.921.

Heavy metals, *Method II* ⟨231⟩: 0.001%.

Cottonseed oil—Mix 5 mL in a test tube with 5 mL of a mixture of equal volumes of amyl alcohol and a 1 in 100 solution of sulfur in carbon disulfide. Warm the mixture gently until the carbon disulfide is expelled, then immerse the tube to one-third of its depth in a boiling, saturated solution of sodium chloride: no reddish color develops within 15 minutes.

Fatty acid composition—Place about 1 g of Corn Oil in a small conical flask fitted with a reflux attachment. Add 10 mL of methanol and 0.5 mL of 1 *N* methanolic potassium hydroxide solution prepared by dissolving 34 g of potassium hydroxide in sufficient methanol to produce 500 mL, allow it to settle for 24 hours, and decant the clear solution. Reflux the mixture for 10 minutes, cool, transfer to a separator with the aid of 15 mL of *n*-heptane, shake with 10 mL of saturated sodium chloride solution, and allow to separate. Transfer the lower layer to another separator, and shake it with 10 mL of *n*-heptane. Wash the combined organic layers with 10 mL of water, dry over anhydrous sodium sulfate, and filter. Introduce a suitable portion of the filtrate into a gas chromatograph equipped with a flame-ionization detector and a column, preferably glass, 1.8 m in length and 4 mm in internal diameter packed with 10 percent liquid phase G4 on support S1A, maintained at a temperature of 175°. The carrier gas is nitrogen. Measure the 5 main peak areas of the methyl esters of the fatty acids. The order of elution is palmitate, stearate, oleate, linoleate, and linolenate, and their relative areas, expressed as percentages of the total area of the 5 main peaks, are in the ranges 8 to 19, 1 to 4, 19 to 50, 34 to 62, and 1 to 2, respectively.

Free fatty acids ⟨401⟩—The free fatty acids in 10.0 g require for neutralization not more than 2.0 mL of 0.020 *N* sodium hydroxide.

Iodine value ⟨401⟩: between 102 and 130.

Saponification value ⟨401⟩: between 187 and 193.

Unsaponifiable matter ⟨401⟩: not more than 1.5%.

Cottonseed Oil

» Cottonseed Oil is the refined fixed oil obtained from the seed of cultivated plants of various varieties of *Gossypium hirsutum* Linné or of other species of *Gossypium* (Fam. Malvaceae).

Packaging and storage—Preserve in tight, light-resistant containers, and avoid exposure to excessive heat.

Identification—Mix 2 mL in a test tube with 2 mL of mixture of equal volumes of amyl alcohol and a 1 in 100 solution of sulfur in carbon disulfide. Warm the mixture carefully until the carbon disulfide is expelled, and immerse the test tube to one-third of its length in a boiling, saturated solution of sodium chloride: a red color develops in the mixture within 5 to 15 minutes.

Specific gravity ⟨841⟩: between 0.915 and 0.921.

Heavy metals, *Method II* ⟨231⟩: 0.001%.

Trichloroethylene—Add 2 mL of pyridine to 2 mL of sodium hydroxide solution (1 in 10) contained in a small test tube, and heat in a water bath at 90° for 5 minutes. Remove the tube, and immediately add 1 mL of Cottonseed Oil without mixing the layers: no pink color develops in the pyridine layer within 20 minutes.

Solidification range of the fatty acids ⟨401⟩—The dry, mixed fatty acids of it solidify between 31° and 35°.

Free fatty acids ⟨401⟩—The free fatty acids in 10.0 g require for neutralization not more than 2.0 mL of 0.020 *N* sodium hydroxide.

Iodine value ⟨401⟩: between 109 and 120.

Saponification value ⟨401⟩: between 190 and 198.

Cresol

C_7H_8O 108.14

Phenol, methyl-.

Cresol [*1319-77-3*].

» Cresol is a mixture of isomeric cresols obtained from coal tar or from petroleum.

Packaging and storage—Preserve in tight, light-resistant containers.

Identification—To a saturated solution of it add a few drops of ferric chloride TS: a bluish violet color is produced.

Specific gravity ⟨841⟩: between 1.030 and 1.038.

Distilling range ⟨721⟩: not less than 90% distils between 195° and 205°.

Hydrocarbons—A solution (1 in 60) shows no more turbidity than that produced in 58 mL of water by the addition of 1.5 mL of 0.02 *N* sulfuric acid and 1 mL of barium chloride solution (1 in 10), the comparison being made after the control has been shaken and allowed to stand for 5 minutes.

Phenol—

Dilute nitric acid—Bubble air through nitric acid until the acid is colorless, then mix 1 volume of the acid with 4 volumes of water.

Standard phenol solution—Dissolve about 1 g of phenol in about 100 mL of water, and determine the actual C_6H_6O concentration as follows: Pipet 4 mL of the solution into an iodine flask, add 30.0 mL of 0.1 *N* bromine VS, then add 5 mL of hydrochloric acid, and immediately insert the stopper. Shake the flask repeatedly during 30 minutes, allow to stand for 15 minutes, add quickly 5 mL of potassium iodide solution (1 in 5), taking precautions to prevent the escape of bromine vapor, and at once insert the stopper into the flask. Shake thoroughly, remove the stopper, and rinse it and the neck of the flask with a small quantity of water so that the washings flow into the flask. Add 1 mL of chloroform, shake the mixture, and titrate the liberated iodine with 0.1 *N* sodium thiosulfate VS, adding 3 mL of starch TS as the end-point is approached. Perform a blank determination. Each mL of 0.1 *N* bromine is equivalent to 1.569 mg of C_6H_6O. Dilute a suitable volume of the solution with water to obtain a concentration of 250 µg of C_6H_6O per mL.

Procedure—Place about 2.5 g of Cresol, accurately weighed, in a 250-mL volumetric flask, add 10 mL of sodium hydroxide solution (1 in 10), dilute with water to volume, and mix. Pipet 5 mL of this solution into a 200-mL volumetric flask, add 45 mL of water and 1 drop of methyl orange TS, neutralize with *Dilute nitric acid*, added dropwise, then dilute with water to volume, and mix. Pipet 5 mL of the neutralized solution into each of two 20- × 180-mm test tubes, graduated at the 25-mL mark, and pipet 5.0 mL of *Standard phenol solution* into each of two similar test tubes. To the contents of each tube add 5 mL of Millon's Reagent, allowing it to flow down the inner wall of the tube, mix, place the tubes simultaneously in a boiling water bath provided with a rack so that the tubes do not touch the bottom of the bath, and maintain the bath at boiling temperature for 30 minutes, accurately timed. At once remove the tubes from the bath, cool them immediately and thoroughly by placing them in a bath of cold water for not less than 10 minutes, add 5 mL of *Dilute nitric acid* to each tube, and mix. Add 3 mL of a mixture of 1 volume of formaldehyde solution and 50 volumes of water to one of each pair of tubes, add water to fill all tubes to volume, shake thoroughly, and allow to stand for 16 hours, during which time the added formaldehyde imparts a yellow color while the contents of the other 2 tubes acquire an orange-red color.

Pipet 20 mL from each of the two tubes containing *Standard phenol solution* into separate 100-mL volumetric flasks, add 5 mL of *Dilute nitric acid*, then add water to volume, and mix. Transfer the solutions to burets marked *B*1 and *B*2, representing, respectively, the solution not treated and the solution treated with formaldehyde.

Pipet 10 mL of the contents of the tube of formaldehyde-treated Cresol into a 50-mL color-comparison tube marked *N*1, and similarly add 10.0 mL of the contents not treated with formaldehyde to a similar tube marked *N*2.

Add to tube *N*1 the orange-red colored solution from buret *B*1, and add to tube *N*2 an equal volume of the yellow-colored solution from buret *B*2, until the colors in tubes *N*1 and *N*2 match when observed in a colorimeter. Calculate the percentage of phenol by the formula:

$$5V/W,$$

in which *V* is the volume, in mL, of *Standard phenol solution* taken from buret *B*1 and *W* is the weight, in g, of Cresol taken: not more than 5.0% of phenol (C_6H_6O) is found.

Croscarmellose Sodium

» Croscarmellose Sodium is a cross-linked polymer of carboxymethylcellulose sodium.

Packaging and storage—Preserve in tight containers.

Identification—

A: Mix 1 g of it with 100 mL of methylene blue solution (1 in 250,000), stir the mixture, and allow it to settle: the Croscarmellose Sodium absorbs the methylene blue and settles as a blue, fibrous mass.

B: Mix 1 g of it with 50 mL of water. Transfer 1 mL of the mixture to a small test tube, and add 1 mL of water and 5 drops of 1-naphthol TS. Incline the test tube, and carefully add 2 mL of sulfuric acid down the side so that it forms a lower layer: a red-purple color develops at the interface.

C: A portion of the mixture of it with water, prepared as directed in *Identification test B*, responds to the tests for *Sodium* ⟨191⟩.

pH ⟨791⟩—Mix 1 g of it with 99 mL of water for 1 hour: the pH of the dispersion is between 5.0 and 7.0.

Loss on drying ⟨731⟩—Dry it at 105° to constant weight: it loses not more than 10.0% of its weight.

Heavy metals, *Method II* ⟨231⟩: 0.001%.

Sodium chloride and Sodium glycolate—

SODIUM CHLORIDE—Weigh accurately about 5 g of it into a 250-mL beaker, add 50 mL of water and 5 mL of 30 percent hydrogen peroxide, and heat on a steam bath for 20 minutes, stirring occasionally to ensure hydration. Cool, add 100 mL of water and 10 mL of nitric acid, and titrate with 0.05 *N* silver nitrate VS, determining the end-point potentiometrically, using a silver electrode and a mercurous sulfate electrode having a potassium sulfate bridge, and stirring constantly. Calculate the percentage of sodium chloride in the specimen by the formula:

$$584.4VN/[(100 - b)W],$$

in which *V* and *N* represent the volume, in mL, and the normality, respectively, of the silver nitrate, *b* is the percentage of *Loss on drying*, determined separately, *W* is the weight, in g, of the specimen, and 584.4 is an equivalence factor for sodium chloride.

SODIUM GLYCOLATE—Transfer about 500 mg of it, accurately weighed, into a 100-mL beaker, moisten thoroughly with 5 mL of glacial acetic acid, followed by 5 mL of water, and stir with a glass rod to ensure proper hydration (usually about 15 minutes). Slowly add 50 mL of acetone, with stirring, then add 1 g of sodium chloride, and stir for several minutes to ensure complete precipitation of the carboxymethylcellulose. Filter through a soft, open-textured paper, previously wetted with a small amount of acetone, and collect the filtrate in a 100-mL volumetric flask. Use an additional 30 mL of acetone to facilitate the transfer of the solids and to wash the filter cake, then dilute with acetone to volume, and mix.

Prepare a series of standard solutions as follows. Transfer 100 mg of glycolic acid, previously dried in a desiccator at room temperature overnight and accurately weighed, to a 100-mL volumetric flask, dissolve in water, dilute with water to volume, and mix. Use this solution within 30 days. Transfer 1.0 mL, 2.0 mL, 3.0 mL, and 4.0 mL portions of the solution, respectively, to separate 100-mL volumetric flasks, add water to each flask to make 5 mL, then add 5 mL of glacial acetic acid, dilute with acetone to volume, and mix.

Transfer 2.0 mL of the test solution and 2.0 mL of each standard solution to separate 25-mL volumetric flasks, and prepare a blank flask containing 2.0 mL of a solution containing 5% each of glacial acetic acid and water in acetone. Place the uncovered flasks in a boiling water bath for 20 minutes, accurately timed, to remove the acetone, remove from the bath, and cool. Add to each flask 5.0 mL of 2,7-dihydroxynaphthalene TS, mix, add an additional 15 mL, and again mix. Cover the mouth of each flask with a small piece of aluminum foil. Place the flasks upright in a boiling water bath for 20 minutes, then remove from the bath, cool, dilute with sulfuric acid to volume, and mix.

Determine the absorbance of each solution at 540 nm, with a suitable spectrophotometer, against the blank, and prepare a standard curve using the absorbances obtained from the standard solutions. From the standard curve and the absorbance of the test specimen, determine the weight (*w*), in mg, of glycolic acid in the specimen, and calculate the percentage of sodium glycolate in the specimen by the formula:

$$12.9w/[(100 - b)W],$$

in which 12.9 is a factor converting glycolic acid to sodium glycolate, *b* is the percentage of *Loss on drying*, determined separately, and *W* is the weight, in g, of the specimen. The sum of the percentages of sodium chloride and sodium glycolate is not more than 0.5%.

Degree of substitution—Transfer about 1 g of it, accurately weighed, to a glass-stoppered, 500-mL conical flask, add 300 mL of sodium chloride solution (1 in 10), then add 25.0 mL of 0.1 *N* sodium hydroxide VS. Insert the stopper, and allow to stand for 5 minutes with intermittent shaking. Add 5 drops of *m*-cresol purple TS, and from a buret add about 15 mL of 0.1 *N* hydrochloric acid VS. Insert the stopper in the flask, and shake. If the solution is purple, add 0.1 *N* hydrochloric acid VS in 1-mL portions until the solution becomes yellow, shaking after each addition. Titrate with 0.1 *N* sodium hydroxide VS to a purple end-point. Calculate the net number of milliequivalents, *M*, of base required for the neutralization of 1 g of Croscarmellose Sodium, on the dried basis. Determine the percentage of residue on ignition, C, of the Croscarmellose Sodium on the dried basis as directed under *Residue on Ignition* ⟨281⟩, using sufficient sulfuric acid to moisten the entire residue after the initial charring step, and additional sulfuric acid if an excessive amount of carbonaceous material remains after the initial complete volatilization of white fumes.

Calculate the degree of acid carboxymethyl substitution, A, by the formula:

$$1150M/(7102 - 412M - 80C).$$

Calculate the degree of sodium carboxymethyl substitution, S, by the formula:

$$(162 + 58A)C/(7102 - 80C).$$

The degree of substitution is the sum of $A + S$. It is between 0.60 and 0.85, calculated on the dried basis.

Content of water-soluble material—Disperse about 10 g, accurately weighed, in 800 mL of water, accurately measured, and stir for 1 minute every 10 minutes during the first 30 minutes. Allow to stand for an additional hour, or centrifuge, if necessary. Decant about 200 mL of the aqueous slurry onto a rapid-filtering filter paper in a vacuum filtration funnel, apply vacuum, and collect about 150 mL of the filtrate. Pour the filtrate into a tared 250-mL beaker, weigh accurately, and calculate the weight, in g, of the filtrate, W_3, by difference. Concentrate on a hot plate to a small volume, but not to dryness, dry at 105° for 4 hours, again weigh, and calculate the weight, in g, of residue W_1, by difference. Calculate the percentage of water-soluble material in the specimen, on the dried basis, by the formula:

$$100W_1(800 + W_2)/[W_2W_3(1 - 0.01b)],$$

in which W_2 is the weight, in g, of the specimen taken, and b is the percentage *Loss on drying* of the specimen taken. It is between 1.0% and 10.0%.

Settling volume—To 75 mL of water in a 100-mL graduated cylinder add 1.5 g of it in 0.5-g portions, shaking vigorously after each addition. Add water to make 100 mL, shake again until all of the powder is homogeneously distributed, and allow to stand for 4 hours. Note the volume of the settled mass. It is between 10.0 mL and 30.0 mL.

Crospovidone

$$\left[\mathrm{-CHCH_2-}\ (\text{N-linked 2-pyrrolidinone})\right]_n$$

$(C_6H_9NO)_n$
1-Ethenyl-2-pyrrolidinone homopolymer.
1-Vinyl-2-pyrrolidinone homopolymer [*9003-39-8*].

» Crospovidone is a water-insoluble synthetic cross-linked homopolymer of *N*-vinyl-2-pyrrolidinone. It contains not less than 11.0 percent and not more than 12.8 percent of nitrogen (N), calculated on the anhydrous basis.

Packaging and storage—Preserve in tight containers.

Reference standard—*USP Crospovidone Reference Standard*—Dry in vacuum at 105° for 1 hour before using.

Identification—

A: The infrared absorption spectrum of a potassium bromide dispersion of it, previously dried in vacuum at 105° for 1 hour, exhibits maxima only at the same wavelengths as that of a similar preparation of USP Crospovidone RS.

B: Suspend 1 g in 10 mL of water, add 0.1 mL of 0.1 N iodine, and shake for 30 seconds. Add 1 mL of starch TS, and shake: no blue color develops.

pH ⟨791⟩: between 5.0 and 8.0, in an aqueous suspension (1 in 100).

Water, *Method I* ⟨921⟩: not more than 5.0%.

Residue on ignition ⟨281⟩: not more than 0.4%, a 2-g specimen being used.

Water-soluble substances—Transfer 25.0 g to a 400-mL beaker, add 200 mL of water, and stir on a magnetic stirrer, using a 5-cm stirring bar, for 1 hour. Transfer to a 250-mL volumetric flask with the aid of about 25 mL of water, add water to volume, and mix. Allow the bulk of the solids to settle. Filter about 100 mL of the relatively clear supernatant liquid through a 0.45-μm membrane filter, protected against clogging by superimposing a 3-μm membrane filter. While filtering, stir the solution above the filter manually or by means of a mechanical stirrer, taking care not to damage the membrane filter physically. Transfer 50.0 mL of the clear filtrate to a tared 100-mL beaker, evaporate to dryness, and dry at 110° for 3 hours: the weight of the residue does not exceed 75 mg (1.5%).

Heavy metals, *Method II* ⟨231⟩: 0.001%.

Vinylpyrrolidinone—Suspend 4.0 g in 30 mL of water, stir for 15 minutes, centrifuge the suspension, and filter the slightly turbid upper layer through a sintered-glass, 10-μm filter. Stir the lower layer with 50 mL of water, centrifuge, and filter the upper layer through the same filter. Again stir the lower layer with 50 mL of water, and filter similarly. Add 0.5 g of sodium acetate to the combined filtrates, and titrate with 0.1 N iodine VS until the color of iodine no longer fades. Add 3.0 mL of 0.1 N iodine VS, allow to stand for 10 minutes, and titrate the excess iodine with 0.1 N sodium thiosulfate VS, adding 3 mL of starch TS as the end-point is approached. Perform a blank determination (see *Residual Titrations* under *Titrimetry* ⟨541⟩), using the same total volume of the same 0.1 N iodine VS, accurately measured, as was used for titrating the specimen. Before titrating the blank, adjust with acetic acid to the same pH as that of the specimen: not more than 0.72 mL of 0.1 N iodine is consumed, corresponding to not more than 0.1% of vinylpyrrolidinone.

Nitrogen content—Proceed as directed under *Nitrogen Determination, Method II* ⟨461⟩, using about 0.1 g, accurately weighed, of Crospovidone, and repeating the addition of 1 mL of 30 percent hydrogen peroxide (usually 3 to 6 times) until a clear, light green solution is obtained on heating the mixture. Heat for an additional 4 hours, and proceed as directed for *Procedure*, beginning with "Cautiously add to the digestion mixture 20 mL of water."

Cyclomethicone

$$\left[(CH_3)_2SiO\right]_n$$

$(C_2H_6OSi)_n$
Cyclopolydimethylsiloxane.
Cyclomethicone [*69430-24-6*].

» Cyclomethicone is a fully methylated cyclic siloxane containing repeating units of the formula:

$$[-(CH_3)_2SiO-]_n,$$

in which n is 4, 5, or 6, or a mixture of them. It contains not less than 98.0 percent of $(C_2H_6OSi)_n$, calculated as the sum of cyclomethicone 4, cyclomethicone 5, and cyclomethicone 6, and not less than 95.0 percent and not more than 105.0 percent of the labeled amount of any one or more of the individual cyclomethicone components.

Packaging and storage—Preserve in tight containers.

Labeling—Label it to state, as part of the official title, the n-value of the Cyclomethicone. Where it is a mixture of 2 or 3 such cyclic siloxanes, the label states the n-value and percentage of each in the mixture.

Reference standards—*USP Cyclomethicone 4 Reference Standard*—After opening, store in a tightly closed container. *USP Cyclomethicone 5 Reference Standard*—After opening, store in a tightly closed container. *USP Cyclomethicone 6 Reference Standard*—After opening, store in a tightly closed container.

Identification—The infrared absorption spectrum, determined in a 0.1-mm cell, exhibits maxima only at the same wavelengths as that of a similar preparation of USP Cyclomethicone 4 RS, USP Cyclomethicone 5 RS, or USP Cyclomethicone 6 RS.

Nonvolatile residue—Evaporate 2.0 g in an open, tared aluminum dish in a circulating air oven at 150° for 2 hours, allow to cool in a desiccator, and weigh: the weight of the residue so obtained does not exceed 3.0 mg, corresponding to not more than 0.15% (w/w).

Assay—The gas chromatograph is equipped with a thermal conductivity detector and a suitable recorder, and contains a 3.66-m × 3-mm column packed with 20 percent liquid phase G1 on 60- to 80-mesh packing S1A (see *Gas Chromatography* under *Chromatography* ⟨621⟩). The column is temperature-programmed at a rate of about 8° per minute from 125° to 320°, the injection port is maintained at a temperature of about 300°, and the detector block is maintained at a temperature of about 350°. Helium is used as the carrier gas, flowing at a rate of about 20 mL per minute. Separately inject about 1 µL of USP Cyclomethicone 4 RS, USP Cyclomethicone 5 RS, and USP Cyclomethicone 6 RS into the gas chromatograph, record the chromatograms, and note the retention times of the peaks. Similarly inject about 1 µL of Cyclomethicone, record the chromatogram, and measure the responses of the major peaks. Calculate the percentage of cyclomethicone 4, cyclomethicone 5, and cyclomethicone 6 by dividing 100 times the response of each peak at the retention time of the corresponding reference standard by the sum of all of the responses in the chromatogram. The percentages obtained from duplicate injections agree to within 1.0%. Calculate the percentage purity by adding the percentages of cyclomethicone 4, cyclomethicone 5, and cyclomethicone 6.

Dehydroacetic Acid

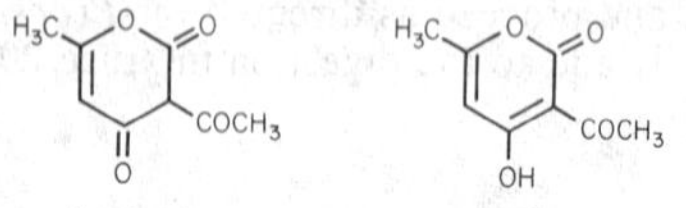

(keto form) (enol form)

$C_8H_8O_4$ 168.15

Keto form:

2*H*-Pyran-2,4(3*H*)-dione, 3-acetyl-6-methyl-.

3-Acetyl-6-methyl-2*H*-pyran-2,4(3*H*)-dione [*520-45-6*].

Enol form:

2*H*-Pyran-2-one, 3-acetyl-4-hydroxy-6-methyl-.

3-Acetyl-4-hydroxy-6-methyl-2*H*-pyran-2-one [*771-03-9*].

» Dehydroacetic Acid contains not less than 98.0 percent and not more than 100.5 percent of $C_8H_8O_4$, calculated on the anhydrous basis.

Packaging and storage—Preserve in well-closed containers.

Reference standard—*USP Dehydroacetic Acid Reference Standard*—Dry at 80° for 4 hours before using.

Identification—The infrared absorption spectrum of a potassium bromide dispersion of it, previously dried at 80° for 4 hours, exhibits maxima only at the same wavelengths as that of a similar preparation of USP Dehydroacetic Acid RS.

Melting range, *Class I* ⟨741⟩: between 109° and 111°.

Water, *Method I* ⟨921⟩: not more than 1.0%.

Residue on ignition ⟨281⟩: not more than 0.1%.

Arsenic, *Method II* ⟨211⟩: 3 ppm.

Heavy metals, *Method II* ⟨231⟩: 0.001%.

Assay—Transfer about 500 mg of Dehydroacetic Acid, accurately weighed, to a 250-mL conical flask, dissolve in 75 mL of neutralized alcohol, add phenolphthalein TS, and titrate with 0.1 *N* sodium hydroxide VS to a pink end-point that persists for not less than 30 seconds. Each mL of 0.1 *N* sodium hydroxide is equivalent to 16.82 mg of $C_8H_8O_4$.

Denatonium Benzoate

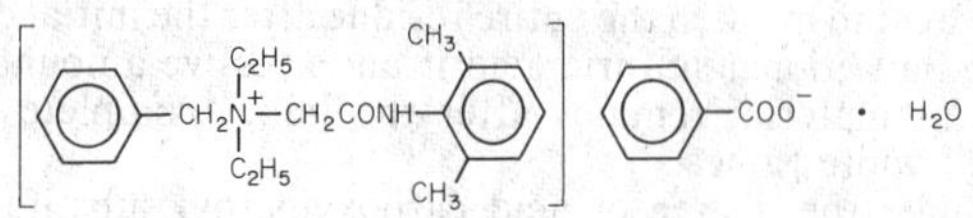

$C_{28}H_{34}N_2O_3 . H_2O$ 464.60

Benzenemethanaminium, *N*-[2-[(2,6-dimethylphenyl)amino]-2-oxoethyl]-*N*,*N*-diethyl-, benzoate, monohydrate.

Benzyldiethyl[(2,6-xylylcarbamoyl)methyl]ammonium benzoate monohydrate [*86398-53-0*].

Anhydrous 446.59 [*3734-33-6*].

» Denatonium Benzoate, dried at 105° for 2 hours, contains one molecule of water of hydration or is anhydrous. When dried at 105° for 2 hours, it contains not less than 99.5 percent and not more than 101.0 percent of $C_{28}H_{34}N_2O_3$.

Packaging and storage—Preserve in tight containers.

Labeling—Label it to indicate whether it is hydrous or anhydrous.

Reference standard—*USP Denatonium Benzoate Reference Standard*—Dry at 105° for 2 hours before using.

Identification—

A: The infrared absorption spectrum of a potassium bromide dispersion of it, previously dried, exhibits maxima only at the same wavelengths as that of a similar preparation of USP Denatonium Benzoate RS.

B: The ultraviolet absorption spectrum of a solution (1 in 10,000) exhibits maxima and minima at the same wavelengths as that of a similar solution of USP Denatonium Benzoate RS, concomitantly measured, and the respective absorptivities, calculated on the dried basis, at the wavelength of maximum absorbance at about 263 nm do not differ by more than 3.0%.

C: Dissolve about 150 mg in 10 mL of water, and add 15 mL of trinitrophenol TS: a yellow precipitate is formed.

D: Dissolve about 100 mg in 10 mL of water, and add 20 mL of 2 *N* sulfuric acid and 15 mL of ammonium reineckate TS. Mix, filter through a sintered-glass crucible using gentle suction, and wash thoroughly with water. Remove as much water as possible with suction, and then dry in an oven at 105° for 1 hour: the denatonium reineckate so obtained melts at about 170° (see *Melting Range or Temperature* ⟨741⟩).

Melting range ⟨741⟩: between 163° and 170°, on a dried specimen.

pH ⟨791⟩: between 6.5 and 7.5, in a solution (3 in 100).

Loss on drying ⟨731⟩—Dry it at 105° for 2 hours: the monohydrate loses between 3.5% and 4.5% of its weight, and the anhydrous form loses not more than 1.0% of its weight.

Residue on ignition ⟨281⟩: not more than 0.1%.

Chloride ⟨221⟩—Dissolve 350 mg in 9 mL of water, add 1 mL of nitric acid, and filter. A 1.0-mL portion of the filtrate shows no more chloride than corresponds to 0.10 mL of 0.020 *N* hydrochloric acid (0.2%).

Assay—Dissolve about 900 mg of Denatonium Benzoate, previously dried and accurately weighed, in 50 mL of glacial acetic acid, add 1 drop of crystal violet TS, and titrate with 0.1 *N* perchloric acid VS to a green end-point. Perform a blank determination, and make any necessary correction. Each mL of 0.1 *N* perchloric acid is equivalent to 44.66 mg of $C_{28}H_{34}N_2O_3$.

Dextrates

» Dextrates is a purified mixture of saccharides resulting from the controlled enzymatic hydrolysis of starch. It is either anhydrous or hydrated. Its dextrose equivalent is not less than 93.0 percent and not more than 99.0 percent, calculated on the dried basis.

Packaging and storage—Preserve in well-closed containers, in a cool, dry place.

Labeling—Label it to state whether it is anhydrous or hydrated.

Reference standard—*USP Dextrose Reference Standard*—Dry at 105° for 16 hours before using.

pH ⟨791⟩: between 3.8 and 5.8, determined in a 1 in 5 solution in carbon dioxide-free water.

Loss on drying ⟨731⟩—Dry it at 105° for 16 hours: the anhydrous form loses not more than 2.0% of its weight; the hydrated form loses between 7.8% and 9.2% of its weight.

Residue on ignition ⟨281⟩: not more than 0.1%.

Heavy metals, *Method II* ⟨231⟩: 5 ppm.

Dextrose equivalent—

Standard solution—Dissolve an accurately weighed quantity of USP Dextrose RS in water, and dilute quantitatively with water to obtain a solution having a known concentration of about 10 mg per mL.

Test solution—Transfer about 5 g of Dextrates, accurately weighed, with the aid of hot water to a 500-mL volumetric flask, cool, add water to volume, and mix.

Procedure—Transfer 25.0-mL portions of alkaline cupric tartrate TS to each of two boiling flasks. Bring the contents of one of the flasks to boiling, and titrate with *Standard solution* to within 0.5 mL of the anticipated end-point. Again heat the flask, with swirling, boil moderately for 2 minutes, add 2 drops of methylene blue solution (1 in 100), immediately add about 2 drops of the *Standard solution* from the buret, and again bring to a boil. Allow the cuprous oxide to settle slightly, and observe the color of the supernatant liquid. Complete the titration within 3 minutes by adding *Standard solution* dropwise, and boiling after each addition, to the disappearance of the blue color, determined by viewing against a white background in daylight or under equivalent illumination. If more than 0.5 mL of the titrant was required after the addition of the indicator, repeat the titration, adding the necessary volume of titrant before adding the indicator. Bring the contents of the second flask to boiling, and similarly titrate with *Test solution.* Calculate the *Dextrose equivalent*, on the dried basis, by the formula:

$$[100/(1 - 0.01A)](C_S/C_U)(V_S/V_U),$$

in which A is the percentage *Loss on drying* of the Dextrates taken, C_U is the concentration, in mg per mL, of Dextrates in the *Test solution*, C_S is the concentration, in mg per mL, of USP Dextrose RS in the *Standard solution*, and V_U and V_S are the titrated volumes, in mL, of *Test solution* and *Standard solution*, respectively.

Dextrin

» Dextrin is starch, or partially hydrolyzed starch, modified by heating in a dry state, with or without acids, alkalies, or pH control agents. During heating, moisture may be added.

Packaging and storage—Preserve in well-closed containers.

Botanic characteristics—

Microscopic—Granules similar in appearance to the starch from which the Dextrin has been prepared, except that when prepared from cornstarch many of the granules show concentric striations and when prepared from potato starch concentric striations are not clearly visible, the hilum is frequently bicleft and a small proportion of the granules are distorted.

Identification—

A: Suspend about 1 g in 20 mL of water, and add a few drops of iodine TS: a blue to reddish brown color results.

B: Dextrin is very soluble in boiling water, forming a mucilaginous solution (*difference from starch*).

Loss on drying ⟨731⟩—Dry it at a pressure not exceeding 100 mm of mercury at 120° for 4 hours: it loses not more than 13.0% of its weight.

Acidity—Add 10.0 g to 100 mL of 70 percent alcohol, previously neutralized to phenolphthalein, shake for 1 hour, filter, and titrate 50 mL of the filtrate with 0.10 *N* sodium hydroxide: not more than 3.0 mL is required.

Residue on ignition ⟨281⟩: not more than 0.5%.

Chloride ⟨221⟩—Dissolve 1.0 g in 75 mL of boiling water, cool, dilute with water to 100 mL, and filter if necessary. To 1.0 mL of this solution add 24 mL of water, 2 mL of nitric acid, and 1 mL of silver nitrate TS: any turbidity produced is not greater than that of a control containing 2.8 mL of 0.020 *N* hydrochloric acid (0.2%).

Arsenic, *Method II* ⟨211⟩: 3 ppm.

Heavy metals, *Method II* ⟨231⟩: 0.004%.

Protein: not more than 1.0%, when determined as directed under *Nitrogen Determination* ⟨461⟩, using a 10-g specimen instead of 1 g, and 60 mL of sulfuric acid instead of 20 mL, and multiplying the percentage of nitrogen found by 6.25.

Reducing sugars—To a quantity of Dextrin equivalent to 2.0 g on the dried basis add 100 mL of water, shake for 30 minutes, dilute with water to 200 mL, accurately measured, and filter. To 10.0 mL of alkaline cupric tartrate TS add 20.0 mL of the filtrate, mix, and heat on a hot plate adjusted to bring the solution to a boil in 3 minutes. Boil for 2 minutes, and cool quickly. Add 5 mL of potassium iodide solution (3 in 10) and 10 mL of 2 *N* sulfuric acid, mix, and titrate immediately with 0.1 *N* sodium thiosulfate VS, using starch TS, added towards the end of the titration, as indicator. Repeat the procedure beginning with "To 10 mL of," using, in place of the filtrate, 20 mL of a 1 in 1000 solution of anhydrous dextrose, accurately prepared. Perform a blank titration. $(V_B - V_U)$ is not greater than $(V_B - V_S)$, in which V_B, V_U, and V_S are the number of mL of 0.1 *N* sodium thiosulfate consumed in the titrations of the blank, the Dextrin and the dextrose, respectively (10 percent, calculated as dextrose, $C_6H_{12}O_6$).

Dextrose—*see* Dextrose USP

Dextrose Excipient

» Dextrose Excipient is a sugar usually obtained by hydrolysis of starch. It contains one molecule of water of hydration.

Packaging and storage—Preserve in well-closed containers.

Labeling—Label it to indicate that it is not intended for parenteral use.

Specific rotation ⟨781⟩: between +52.5° and +53.5°, calculated on the anhydrous basis, determined in a solution containing 10 g of Dextrose Excipient and 0.2 mL of 6 *N* ammonium hydroxide in each 100 mL.

Water ⟨921⟩—Dry it at 105° for 16 hours: it loses between 7.5% and 9.5% of its weight, determined by the *Gravimetric Method*.

Other requirements—It responds to the *Identification test* and meets the requirements of the tests for *Color of solution, Acidity, Residue on ignition, Chloride, Sulfate, Arsenic, Heavy metals, Dextrin,* and *Soluble starch, sulfites* under *Dextrose*.

Diacetylated Monoglycerides

» Diacetylated Monoglycerides is glycerin esterified with edible fat-forming fatty acids and acetic acid. It may be prepared by the interesterification of edible oils with triacetin in the presence of catalytic agents, followed by molecular distillation, or by the direct acetylation of edible monoglycerides with acetic anhydride without the use of catalyst or molecular distillation.

Packaging and storage—Preserve in tight, light-resistant containers.

Reference standard—*USP Diacetylated Monoglycerides Reference Standard.*

Identification—The infrared absorption spectrum of a thin film of it exhibits maxima only at the same wavelengths as that of a similar preparation of USP Diacetylated Monoglycerides RS.

Residue on ignition ⟨281⟩: not more than 0.1%.

Arsenic, *Method II* ⟨211⟩: 3 ppm.

Heavy metals, *Method II* ⟨231⟩: 0.001%.

Acid value ⟨401⟩: not more than 3.

Hydroxyl value ⟨401⟩: not more than 15.

Saponification value ⟨401⟩: between 365 and 385.

Dibasic Calcium Phosphate—*see* Calcium Phosphate, Dibasic USP

Dibasic Sodium Phosphate—*see* Sodium Phosphate, Dibasic USP

Dichlorodifluoromethane

CCl_2F_2

CCl_2F_2 120.91
Methane, dichlorodifluoro-.
Dichlorodifluoromethane [*75-71-8*].

Packaging and storage—Preserve in tight cylinders, and avoid exposure to excessive heat.

Identification—The infrared absorption spectrum of it, determined in a 10-cm cell with sodium chloride windows, at atmospheric pressure, exhibits maxima, among others, at the following wavelengths, in µm: 4.33 (*m*), 4.46 (*m*), 4.56 (*m*), 6.29 (*m*), 7.25 (*m*), 8.05 (*s*), 8.63 (*s*), 9.1 (*vs*), 10.7 (*vs*), 10.8 (*vs*), 11.2 (*m*), 11.3 (*m*), 13.1 (*w*), and 13.9 (*w*). The stronger maxima are best obtained at pressures less than 10 mm of mercury.

Boiling temperature: approximately −30°, determined as directed under *Approximate Boiling Temperature* (see *Aerosols* ⟨601⟩).

Water: not more than 0.001%, determined as directed under *Water Content* (see *Aerosols* ⟨601⟩).

High-boiling residues: not more than 0.01%, determined as directed for *High-boiling Residues, Method I,* under *Aerosols* ⟨601⟩.

Inorganic chlorides—Place 5 mL of anhydrous methanol in a test tube, add 3 drops of a saturated solution of silver nitrate in anhydrous methanol, shake, and add 7 g of Dichlorodifluoromethane: no opalescence or turbidity is produced.

Dichlorotetrafluoroethane

$CClF_2—CClF_2$

$C_2Cl_2F_4$ 170.92
Ethane, 1,2-dichloro-1,1,2,2-tetrafluoro-.
1,2-Dichlorotetrafluoroethane [*76-14-2*].

Packaging and storage—Preserve in tight cylinders, and avoid exposure to excessive heat.

Identification—The infrared absorption spectrum of it, determined in a 10-cm cell with sodium chloride windows, at atmospheric pressure, exhibits maxima, among others, at the following wavelengths, in µm: 4.34 (*s*), 4.48 (*m*), 5.28 (*m*), 5.95 (*m*), 7.36 (*s*), 9.5 (*vs*), 10.9 (*vs*), 11.8 (*vs*), 13.6 (*s*), and 14.8 (*s*). The stronger maxima are best obtained at pressures less than 10 mm of mercury.

Boiling temperature: approximately 4°, determined as directed under *Approximate Boiling Temperature* (see *Aerosols* ⟨601⟩).

Water: not more than 0.001%, determined as directed under *Water Content* (see *Aerosols* ⟨601⟩).

High-boiling residues: not more than 0.01%, determined as directed for *High-boiling Residues, Method I,* under *Aerosols* ⟨601⟩.

Inorganic chlorides—Place 5 mL of anhydrous methanol in a test tube, add 3 drops of a saturated solution of silver nitrate in anhydrous methanol, shake, and add 7 g of Dichlorotetrafluoroethane: no opalescence or turbidity is produced.

Diethanolamine

$NH(CH_2CH_2OH)_2$

$C_4H_{11}NO_2$ 105.14
Ethanol, 2,2′-iminobis-.
2,2′-Iminodiethanol [*111-42-2*].

» Diethanolamine is a mixture of ethanolamines, consisting largely of diethanolamine. It contains not less than 98.5 percent and not more than 101.0 percent of ethanolamines, calculated on the anhydrous basis as $NH(C_2H_4OH)_2$.

Packaging and storage—Preserve in tight, light-resistant containers.

Identification—The infrared absorption spectrum, between 6.6 µm and 12.5 µm, of a thin film of it exhibits maxima at about 6.8 µm, 7.3 µm, 8.1 µm, 8.3 µm, 8.9 µm, 9.4 µm, 10.6 µm, and 11.6 µm.

Refractive index ⟨831⟩: between 1.473 and 1.476, at 30°.

Water, *Method I* ⟨921⟩: not more than 0.15%, a 20-g test specimen being used, and a mixture of 25 mL of glacial acetic acid and 40 mL of methanol being used as the solvent.

Triethanolamine—

Mixed indicator—Dissolve 0.15 g of methyl orange and 0.08 g of xylene cyanole FF in 100 mL of water, and mix.

Procedure—Place 100 mL of methanol and 6 to 8 drops of *Mixed indicator* in a 500-mL glass-stoppered conical flask, and neutralize with 0.1 *N* alcoholic sulfuric acid or 0.1 *N* alcoholic potassium hydroxide. The neutral solution is amber when viewed by transmitted light and is red-brown when viewed by reflected light. Add about 20 g of test specimen, accurately weighed. Cautiously add 75 mL of acetic anhydride, and swirl to effect complete solution. Allow to stand at room temperature for 30 minutes. Cool to room temperature, if necessary. Titrate with 0.5 *N* alcoholic sulfuric acid VS. Perform a blank determination, and make any necessary correction. Each mL of 0.5 *N* alcoholic sulfuric acid is equivalent to 74.6 mg of triethanolamine: the limit is 1.0%, by weight.

Assay—Transfer about 2 g of Diethanolamine, accurately weighed, to a 250-mL conical flask. Add 50 mL of water and bromocresol green TS, and titrate with 0.5 *N* hydrochloric acid VS. Perform a blank determination, and make any necessary correction. Each mL of 0.5 *N* hydrochloric acid is equivalent to 52.57 mg of diethanolamine, expressed as $NH(C_2H_4OH)_2$.

Diethyl Phthalate

$C_6H_4(COOCH_2CH_3)_2$

$C_{12}H_{14}O_4$ 222.24
1,2-Benzenedicarboxylic acid, diethyl ester.
Diethyl phthalate [*84-66-2*].

» Diethyl Phthalate contains not less than 98.0 percent and not more than 102.0 percent of $C_{12}H_{14}O_4$, calculated on the anhydrous basis.

Caution—Avoid contact.

Packaging and storage—Preserve in tight containers.

Reference standard—*USP Diethyl Phthalate Reference Standard*—Do not dry before using.

Identification—The infrared absorption spectrum of a thin film of it exhibits maxima only at the same wavelengths as that of a similar preparation of USP Diethyl Phthalate RS.

Specific gravity ⟨841⟩: between 1.118 and 1.122, at 20°.

Refractive index ⟨831⟩: between 1.500 and 1.505, at 20°.

Acidity—Mix 20.0 g with 50 mL of alcohol that previously has been neutralized to a phenolphthalein end-point, and titrate with 0.10 *N* sodium hydroxide to a phenolphthalein end-point: not more than 0.50 mL is required for neutralization.

Water, *Method I* ⟨921⟩: not more than 0.2%.

Residue on ignition ⟨281⟩—Heat gently about 10 g, accurately weighed, until the liquid has evaporated, and ignite the residue to constant weight. Not more than 0.02% is found.

Assay—Transfer about 1.5 g of Diethyl Phthalate, accurately weighed, to a flask, add 50.0 mL of 0.5 *N* alcoholic potassium hydroxide VS, attach a reflux condenser to the flask, and boil on a water bath for 1 hour. Add 20 mL of water, then add phenolphthalein TS, and titrate the excess potassium hydroxide with 0.5 *N* hydrochloric acid VS. Perform a blank determination (see *Residual Titrations* under *Titrimetry* ⟨541⟩). Each mL of 0.5 *N* potassium hydroxide is equivalent to 55.56 mg of $C_{12}H_{14}O_4$.

Diisopropanolamine

$NH(CH_2CH(OH)CH_3)_2$

$C_6H_{15}NO_2$ 133.19

2-Propanol, 1,1′-iminobis-.

1,1′-Iminodi-2-propanol [*110-97-4*].

» Diisopropanolamine is a mixture of isopropanolamines, consisting largely of diisopropanolamine. It contains not less than 98.0 percent and not more than 102.0 percent of isopropanolamines, calculated on the anhydrous basis as $NH(C_3H_6OH)_2$.

Packaging and storage—Preserve in tight, light-resistant containers.

Identification—The infrared absorption spectrum of a thin film of it exhibits regions of absorption between 2.8 µm and 4.0 µm, between 6.7 µm and 7.1 µm, and between 8.5 µm and 9.4 µm; and several characteristic peaks, the most pronounced being at about 7.3 µm, 7.5 µm, 8.3 µm, 9.6 µm, 10.4 µm, and 10.7 µm.

Water, *Method I* ⟨921⟩: not more than 0.50%, a mixture of 5.0 mL of glacial acetic acid and 25 mL of methanol being used as the solvent.

Triisopropanolamine—

Mixed indicator—Dissolve 0.15 g of methyl orange and 0.08 g of xylene cyanole FF in 100 mL of water, and mix.

Procedure—Place 100 mL of methanol and 6 to 8 drops of *Mixed indicator* in a glass-stoppered, 500-mL conical flask, and neutralize with 0.1 *N* alcoholic sulfuric acid or 0.1 *N* alcoholic potassium hydroxide. The neutral solution is amber when viewed by transmitted light and is red-brown when viewed by reflected light. Add about 20 g of Diisopropanolamine, accurately weighed. Cautiously add 75 mL of acetic anhydride, and swirl to effect complete solution. Allow to stand at room temperature for 30 minutes. Cool to room temperature, if necessary. Titrate with 0.5 *N* alcoholic sulfuric acid VS. Perform a blank determination, and make any necessary correction. Each mL of 0.5 *N* alcoholic sulfuric acid is equivalent to 95.7 mg of triisopropanolamine: the limit is 1.0%, by weight.

Assay—Transfer about 2 g of Diisopropanolamine, accurately weighed, to a 250-mL conical flask, add 50 mL of water and bromocresol green TS, and titrate with 0.5 *N* hydrochloric acid VS. Perform a blank determination, and make any necessary correction. Each mL of 0.5 *N* hydrochloric acid is equivalent to 66.60 mg of isopropanolamines, expressed as $NH(C_3H_6OH)_2$.

Diluted Alcohol—*see* Alcohol, Diluted

Diluted Hydrochloric Acid—*see* Hydrochloric Acid, Diluted

Diluted Phosphoric Acid—*see* Phosphoric Acid, Diluted

Dimethicone

$(CH_3)_3Si{-}[OSi(CH_3)_2]_n{-}CH_3$

Dimethicone.

α-(Trimethylsilyl)-ω-methylpoly[oxy(dimethylsilylene)] [*9006-65-9*].

» Dimethicone is a mixture of fully methylated linear siloxane polymers containing repeating units of the formula:

$$[-(CH_3)_2SiO-]_n,$$

stabilized with trimethylsiloxy end-blocking units of the formula:

$$[(CH_3)_3SiO-],$$

wherein *n* has an average value such that the corresponding nominal viscosity is in a discrete range between 20 and 12,500 centistokes. It contains not less than 97.0 percent and not more than 103.0 percent of polydimethylsiloxane ($[-(CH_3)_2SiO]_n$).

The requirements for viscosity, specific gravity, refractive index, and loss on heating differ for the several types of Dimethicone, as set forth in the accompanying table.

Packaging and storage—Preserve in tight containers.

Labeling—Label it to indicate its nominal viscosity value. Dimethicone intended for use in coating containers that come in contact with articles for parenteral use is so labeled.

Reference standard—*USP Polydimethylsiloxane Reference Standard*—Keep container tightly closed. Do not dry before using.

Identification—The infrared absorption spectrum, determined in a 0.5-mm cell, of the solution of Dimethicone prepared as directed in the *Assay* exhibits maxima only at the same wavelengths as that of the Standard solution prepared as directed in the *Assay*.

Specific gravity ⟨841⟩: within the limits specified in the accompanying table.

Viscosity ⟨911⟩—Determine its viscosity at 25 ± 0.1°, using a capillary viscosimeter. The viscosity is within the limits specified in the accompanying table.

Refractive index ⟨831⟩: within the limits specified in the accompanying table.

Acidity—Dissolve 15.0 g in a mixture of 15 mL of toluene and 15 mL of butyl alcohol, previously neutralized to bromophenol blue TS, and titrate with 0.050 *N* alcoholic potassium hydroxide to a bromophenol blue end-point: not more than 0.10 mL is required.

Loss on heating—Heat about 15 g, accurately weighed, in an open, tared, aluminum vessel at 200° in a circulating air oven

Nominal viscosity (centistokes)	Viscosity (centistokes)		Specific gravity		Refractive index		Loss on heating
	Min.	Max.	Min.	Max.	Min.	Max.	Max.
20	18	22	0.946	0.954	1.3980	1.4020	20.0
100	95	105	0.962	0.970	1.4005	1.4045	2.0
200	190	220	0.964	0.972	1.4013	1.4053	2.0
350	332.5	367.5	0.965	0.973	1.4013	1.4053	2.0
500	475	525	0.967	0.975	1.4013	1.4053	2.0
1000	950	1050	0.967	0.975	1.4013	1.4053	2.0
12,500	11,250	13,750	—	—	1.4015	1.4055	2.0

for 4 hours, and allow to come to room temperature in a desiccator before weighing: it loses not more than the maximum percentage of its weight specified in the accompanying table.

Heavy metals, *Method II* ⟨231⟩: 0.001%.

Other requirements—Dimethicone intended for use in coating containers that come in contact with articles for parenteral use meets the following additional requirements.

PYROGEN—To 20.0 g in a suitable flask add 400 mL of pyrogen-free saline TS, and heat at 85° for 1 hour. Using a pyrogen-free pipet, remove a suitable volume of the saline extract, and allow to cool. The saline extract meets the requirements of the *Pyrogen Test* ⟨151⟩.

BIOLOGICAL SUITABILITY—

A: It meets the requirements of the *Systemic Injection Test* in the section, *Biological Tests—Plastics*, under *Containers* ⟨661⟩. Where the test specimen is unextracted Dimethicone of viscosity of 1000 centistokes or less, inject by the intraperitoneal route unextracted Dimethicone and use *Cottonseed Oil* as the blank. Where the specimen is Dimethicone of viscosity greater than 1000 centistokes use extracts in *Sodium Chloride Injection* and in *Cottonseed Oil* prepared by mixing 4-g portions of *Sample* with 20 mL of *Sodium Chloride Injection* and with 20 mL of *Cottonseed Oil* in an oven at 70° for 24 hours with occasional swirling, and preparing one 20-mL blank of each extracting medium for parallel injections and comparisons.

B: It meets the requirements of the *Intracutaneous Test* in the section, *Biological Tests—Plastics*, under *Containers* ⟨661⟩, the *Sample* and *Blanks* being as described in *Biological suitability test A*.

Assay—Transfer about 250 mg of Dimethicone, accurately weighed, to a 25-mL volumetric flask, dissolve in carbon tetrachloride, dilute with carbon tetrachloride to volume, and mix. Concomitantly determine the absorbances of this solution and a Standard solution of USP Polydimethylsiloxane RS in carbon tetrachloride having a known concentration of about 10 mg per mL, in 0.1-mm cells at the wavelength of maximum absorbance at about 7.9 μm, with a suitable infrared spectrophotometer, using carbon tetrachloride as the blank. Calculate the quantity, in mg, of $[-(CH_3)_2SiO-]_n$ in the Dimethicone taken by the formula:

$$25C(A_U/A_S),$$

in which C is the concentration, in mg per mL, of USP Polydimethylsiloxane RS in the Standard solution, and A_U and A_S are the absorbances of the solution of Dimethicone and the Standard solution, respectively.

Docusate Sodium—*see* Docusate Sodium USP

Edetate Disodium—*see* Edetate Disodium USP

Edetic Acid

$(HOOCCH_2)_2NCH_2CH_2N(CH_2COOH)_2$

$C_{10}H_{16}N_2O_8$ 292.24

Glycine, *N,N'*-1,2-ethanediylbis[*N*-(carboxymethyl)-.

(Ethylenedinitrilo)tetraacetic acid [*60-00-4*].

» Edetic Acid contains not less than 98.0 percent and not more than 100.5 percent of $C_{10}H_{16}N_2O_8$.

Packaging and storage—Preserve in well-closed containers.

Reference standard—*USP Edetic Acid Reference Standard*—Do not dry before using.

Identification—The infrared absorption spectrum of a potassium bromide dispersion of it exhibits maxima only at the same wavelengths as that of a similar preparation of USP Edetic Acid RS.

Residue on ignition ⟨281⟩: not more than 0.2%.

Heavy metals, *Method II* ⟨231⟩: 0.003%.

Nitrilotriacetic acid—

Standard stock solution—Dissolve 1.0 g of nitrilotriacetic acid in 10 mL of potassium hydroxide solution (1 in 10) in a 100-mL volumetric flask, dilute with water to volume, and mix.

Test solution—Dissolve 10.0 g in 87 mL of potassium hydroxide solution (1 in 10) in a 100-mL volumetric flask, dilute with water to volume, and mix.

Standard preparation—Mix 10.0 mL of *Test solution* with 1.0 mL of the *Standard stock solution* in a 100-mL volumetric flask, dilute with water to volume, and mix.

Test preparation—Transfer 10.0 mL of *Test solution* to a 100-mL volumetric flask, dilute with water to volume, and mix.

Procedure—Proceed with the *Standard preparation* and with the *Test preparation*, respectively, through the following steps, as far as the completion of the polarographic determination. To 20.0 mL of the *preparation* in a 150-mL beaker add 1 mL of potassium hydroxide solution (1 in 10), 2 mL of ammonium nitrate solution (1 in 10), and eriochrome black T trituration in an amount (about 50 mg) sufficient to impart color to the solution. Titrate with cadmium nitrate solution (3 in 100) to a red endpoint, record the titer, and discard the solution. To another 20.0-mL aliquot of the *preparation* in a 100-mL volumetric flask add a volume of the same cadmium nitrate solution equal to the volume titrated, and add 0.05 mL in excess. Add 1.5 mL of potassium hydroxide solution (1 in 10), 10 mL of ammonium nitrate solution (1 in 10), and 0.5 mL of methyl red TS, dilute with water to volume, and mix. Transfer a portion of this solution to a polarographic cell and deaerate by bubbling nitrogen through the solution for 10 minutes. Insert the dropping mercury electrode of a suitable polarograph, and record the polarogram from −0.6 to −1.2 volts at a sensitivity of 0.006 microampere per mm, using a saturated calomel electrode as the reference electrode. The diffusion current for the *Test preparation* is not greater than 30.0% of the difference between the diffusion currents for the *Standard preparation* and the *Test preparation* (0.3% nitrilotriacetic acid).

Iron—Char 3.0 g thoroughly, and heat in an oven at 500° until most of the carbon is consumed. Cool, add 0.15 mL of nitric acid, and heat at 500° until all of the carbon is consumed. Dissolve the residue in 2 mL of a mixture of equal volumes of hydrochloric acid and water, digest in a covered dish on a steam bath for 10 minutes, remove the cover, and evaporate to dryness. Dissolve the residue in 1 mL of 1 *N* acetic acid and 20 mL of hot water, digest for 5 minutes on a steam bath, cool, and dilute with water to 30 mL. To 2.0 mL of this solution add 2 mL of hydrochloric acid, and dilute with water to 50 mL. Add about 50 mg of ammonium persulfate and 3 mL of ammonium thiocyanate solution (3 in 10), mix, and transfer to a color-comparison tube. Treat in the same manner 2.0 mL of a solution of ferric ammonium sulfate, prepared by dissolving 43.2 mg of ferric ammonium sulfate in 10 mL of 2 *N* sulfuric acid and adding water to make 1000 mL, each mL representing 5 μg of Fe. The color of the test solution is not deeper than that of the solution containing the standard iron solution (0.005%).

Assay—

Assay preparation—Transfer about 1.4 g of Edetic Acid, accurately weighed, to a 100-mL volumetric flask, dissolve in 11 mL of 1 *N* sodium hydroxide, dilute with water to volume, with cooling, if necessary, and mix.

Procedure—Transfer to a 400-mL beaker about 200 mg of chelometric standard calcium carbonate, previously dried at 300° for 3 hours, cooled in a desiccator for 2 hours, and accurately weighed. Add 10 mL of water, swirl to form a slurry, and cover the beaker with a watch glass. Without removing the watch glass, add 2 mL of 3 *N* hydrochloric acid from a pipet, and swirl to dissolve. Wash down the sides of the beaker, the outer surface of the pipet, and the watch glass with water, and dilute with water to about 100 mL. While stirring with a magnetic stirrer, add about 30 mL of the *Assay preparation* from a 50-mL buret. Add 10 mL of 1 *N* sodium hydroxide and 300 mg of hydroxy naphthol blue, and continue the titration with the *Assay preparation* to a blue end-point. Calculate the weight, in g, of $C_{10}H_{16}N_2O_8$ in the portion of Edetic Acid taken by the formula:

$$(292.24/100.09)(0.1W/V),$$

in which 292.24 and 100.09 are the molecular weights of edetic acid and calcium carbonate, respectively, *W* is the weight, in mg, of calcium carbonate, and *V* is the volume, in mL, of the *Assay preparation* consumed in the titration.

Elixirs—*see complete list in index*

Emulsifying Wax—*see* Wax, Emulsifying

Ether, Polyoxyl 10 Oleyl—*see* Polyoxyl 10 Oleyl Ether

Ether, Polyoxyl 20 Cetostearyl—*see* Polyoxyl 20 Cetostearyl Ether

Ethyl Acetate

$CH_3COOC_2H_5$

$C_4H_8O_2$ 88.11

Acetic acid, ethyl ester.

Ethyl acetate [*141-78-6*].

» Ethyl Acetate contains not less than 99.0 percent and not more than 100.5 percent of $C_4H_8O_2$.

Packaging and storage—Preserve in tight containers, and avoid exposure to excessive heat.

Identification—It is readily volatilized even at low temperatures and is flammable; when burned, a yellow flame and an acetous odor are produced.

Specific gravity ⟨841⟩: between 0.894 and 0.898.

Acidity—A solution of 2.0 mL in 10 mL of neutralized alcohol requires not more than 0.10 mL of 0.10 *N* sodium hydroxide for neutralization, 2 drops of phenolphthalein TS being used as the indicator.

Nonvolatile residue—Evaporate it in a tared porcelain dish on a steam bath, and dry at 105° for 1 hour: not more than 0.02% of residue remains.

Readily carbonizable substances ⟨271⟩—Pour 2 mL carefully upon 10 mL of sulfuric acid TS so as to form separate layers: no dark zone is developed within 15 minutes.

Methyl compounds—Place 20 mL in a 500-mL separator, add a solution of 20 g of sodium hydroxide in 50 mL of water, insert the stopper in the separator, and wrap it securely in a towel for protection against the heat of the reaction. Shake the mixture vigorously for about 5 minutes, cautiously opening the stopcock from time to time to permit the escape of air. Continue shaking vigorously until a homogeneous liquid results, then distil, and collect about 25 mL of the distillate. To 0.05 mL of the distillate add 1 drop of dilute phosphoric acid (1 in 20) and 1 drop of potassium permanganate solution (1 in 20). Mix, allow to stand 1 minute, and add sodium bisulfite solution (1 in 20), dropwise, until the permanganate color is discharged. If a brown color remains, add 1 drop of the dilute phosphoric acid. To the colorless solution add 5 mL of freshly prepared chromotropic acid TS, and heat on a steam bath at 60° for 10 minutes: no violet color appears.

Chromatographic purity—

Chromatographic system (see *Chromatography* ⟨621⟩)—The gas chromatograph is equipped with a flame-ionization detector and a 1.8-m × 4-mm column that contains support S11. The column temperature is maintained at 115° for 6 minutes, then programmed to rise at 16° per minute to 200°, and held at 200° for 15 minutes. Prepare a mixture of chloroform, ethyl acetate, isobutyl acetate, and *n*-butyl acetate (3:1:1:1), and inject 0.1 µL, using a 1-µL syringe, into the chromatograph. In the resulting chromatogram, the tailing factor, *T*, for the ethyl acetate peak is not more than 1.5, the resolution, *R*, between the chloroform and ethyl acetate peaks is not less than 1.3, and the resolution, *R*, between the isobutyl acetate and *n*-butyl acetate peaks is not less than 1.5. The retention times, relative to ethyl acetate as 1.0, are about 0.9 for chloroform, 2.7 for isobutyl acetate and 2.8 for *n*-butyl acetate.

Procedure—Using a 1-µL syringe, inject a suitable volume (about 0.06 µL) of Ethyl Acetate into the chromatograph, record the chromatogram, and measure the areas of all the peaks. The area of the ethyl acetate peak is not less than 99.5% of the sum of the areas of all the peaks.

Assay—Transfer about 1.5 g of Ethyl Acetate, accurately weighed in a tared, stoppered weighing bottle, to a suitable flask, add 50.0 mL of 0.5 *N* sodium hydroxide VS, and heat on a steam bath under a reflux condenser for 1 hour. Allow to cool, add phenolphthalein TS, and titrate the excess sodium hydroxide with 0.5 *N* hydrochloric acid VS. Perform a blank determination (see *Residual Titrations* under *Titrimetry* ⟨541⟩). Each mL of 0.5 *N* sodium hydroxide is equivalent to 44.05 mg of $C_4H_8O_2$.

Ethyl Oleate

$HC-CH_2(CH_2)_6COOC_2H_5$
$\|$
$HC-CH_2(CH_2)_6CH_3$

$C_{20}H_{38}O_2$ 310.52

9-Octadecenoic acid, (*Z*)-, ethyl ester.

Ethyl oleate [*111-62-6*].

» Ethyl Oleate consists of esters of ethyl alcohol and high molecular weight fatty acids, principally oleic acid.

Packaging and storage—Preserve in tight, light-resistant containers.

Specific gravity ⟨841⟩: between 0.866 and 0.874 at 20°.

Viscosity ⟨911⟩: not less than 5.15 centipoises.

Refractive index ⟨831⟩: between 1.443 and 1.450.

Acid value ⟨401⟩: not more than 0.5.

Iodine value ⟨401⟩: between 75 and 85.

Saponification value ⟨401⟩: between 177 and 188.

Ethyl Vanillin

OC_2H_5

HO— —CHO

$C_9H_{10}O_3$ 166.18
Benzaldehyde, 3-ethoxy-4-hydroxy-.
3-Ethoxy-4-hydroxybenzaldehyde [*121-32-4*].

» Ethyl Vanillin, dried over phosphorus pentoxide for 4 hours, contains not less than 98.0 percent and not more than 101.0 percent of $C_9H_{10}O_3$.

Packaging and storage—Preserve in tight, light-resistant containers.

Reference standard—*USP Ethyl Vanillin Reference Standard*—Dry over phosphorus pentoxide for 4 hours before using.

Identification—

A: The infrared absorption spectrum of a potassium bromide dispersion of it exhibits maxima only at the same wavelengths as that of a similar preparation of USP Ethyl Vanillin RS.

B: The ultraviolet absorption spectrum of a 1 in 125,000 solution in methanol exhibits maxima and minima at the same wavelengths as that of a similar solution of USP Ethyl Vanillin RS, concomitantly measured.

Melting range ⟨741⟩: between 76° and 78°.

Loss on drying ⟨731⟩—Dry it over phosphorus pentoxide for 4 hours: it loses not more than 1.0% of its weight.

Residue on ignition ⟨281⟩: not more than 0.1%.

Assay—Dissolve about 300 mg of Ethyl Vanillin, previously dried and accurately weighed, in 50 mL of dimethylformamide contained in a 125-mL conical flask. Add thymol blue TS, and titrate with 0.1 *N* sodium methoxide VS, using a magnetic stirrer and taking precautions against the absorption of atmospheric carbon dioxide. Perform a blank determination, and make any necessary correction. Each mL of 0.1 *N* sodium methoxide is equivalent to 16.62 mg of $C_9H_{10}O_3$.

Ethylcellulose

Cellulose, ethyl ether.
Cellulose ethyl ether [*9004-57-3*].

» Ethylcellulose is an ethyl ether of cellulose. When dried at 105° for 2 hours, it contains not less than 44.0 percent and not more than 51.0 percent of ethoxy ($-OC_2H_5$) groups.

Packaging and storage—Preserve in well-closed containers.

Labeling—Label it to indicate its viscosity (under the conditions specified herein), and its ethoxy content.

Reference standard—*USP Ethylcellulose Reference Standard*—Do not dry before using. Keep container tightly closed.

Identification—Dissolve 5 g in 95 g of a mixture of 80 parts of toluene and 20 parts of alcohol, by weight: a clear, stable, slightly yellow solution results. Pour a few mL of this solution onto a sodium chloride plate, and allow the solvent to evaporate: a thin, tough, continuous, clear film remains. The infrared absorption spectrum of the film so obtained exhibits maxima only at the same wavelengths as that of a similar preparation of USP Ethylcellulose RS. The film, prepared from the test specimen and removed from the plate, is flammable.

Viscosity—

Solvent systems—For Ethylcellulose containing less than 46.5 percent of ethoxy groups, prepare a solvent system consisting of 60 parts of toluene and 40 parts of alcohol, by weight. Otherwise, prepare a solvent system consisting of 80 parts of toluene and 20 parts of alcohol, by weight.

Procedure—Place a quantity of undried Ethylcellulose, accurately weighed and equivalent to 5.0 g of solids on the dried basis, in a bottle containing 95 ± 0.05 g of the appropriate solvent system. Shake or tumble the bottle until the sample is completely dissolved. Adjust the temperature of the solution to 25 ± 0.1°, and determine the viscosity as described in the section, *Procedure for Cellulose Derivatives*, under *Viscosity* ⟨911⟩, but make all determinations at 25°/25° instead of 20°/20° as directed therein. The viscosity is not less than 90.0% and not more than 110.0% of that stated on the label for a labeled viscosity of 10 centipoises or more; not less than 80.0% and not more than 120.0% of that stated on the label for a labeled viscosity of less than 10 centipoises but more than 6 centipoises; and not less than 75.0% and not more than 140.0% of that stated on the label for a labeled viscosity of 6 centipoises or less.

Loss on drying ⟨731⟩—Dry it at 105° for 2 hours: it loses not more than 3.0% of its weight.

Residue on ignition ⟨281⟩: not more than 0.4%.

Arsenic, *Method II* ⟨211⟩: 3 ppm.

Lead ⟨251⟩: 10 ppm.

Heavy metals, *Method II* ⟨231⟩: 40 ppm.

Assay—Proceed as directed under *Methoxy Determination* ⟨431⟩, using about 50 mg of Ethylcellulose, previously dried and accurately weighed. Each mL of 0.1 *N* sodium thiosulfate is equivalent to 0.7510 mg of ($-OC_2H_5$).

Ethylcellulose Aqueous Dispersion

» Ethylcellulose Aqueous Dispersion is a colloidal dispersion of Ethylcellulose in water. It contains not less than 90.0 percent and not more than 110.0 percent of the labeled amount of Ethylcellulose. It contains suitable amounts of Cetyl Alcohol and Sodium Lauryl Sulfate, which assist in the formation and stabilization of the dispersion. It may contain suitable antifoaming and antimicrobial agents.

Packaging and storage—Preserve in tight containers, and protect from freezing.

Labeling—The labeling states the ethoxy content of the Ethylcellulose and the percentage of Ethylcellulose.

Reference standards—*USP Ethylcellulose Reference Standard*—Do not dry before using. *USP Cetyl Alcohol Reference Standard*—Do not dry before using.

Identification—

A: Transfer a small quantity to a silver chloride plate, and allow the water to evaporate: the infrared absorption spectrum of the residue in the 3600 to 2600 cm^{-1} and 1500 to 800 cm^{-1} regions exhibits maxima only at the same wave numbers as that of a film of USP Ethylcellulose RS prepared as directed in the test for *Identification* under *Ethylcellulose*.

B: Transfer about 2 mL to a 100-mm diameter petri dish so that the bottom of the dish is covered uniformly. Place the dish in an oven or on a hot plate to evaporate the water: a transparent film results.

C: Dissolve the film formed in *Identification test B* in 20 mL of chloroform. Inject 2 µL of this solution into a gas chromatograph (see *Chromatography* ⟨621⟩) equipped with a 1.8-meter column that contains liquid phase G1 on support S1A maintained at a temperature of 220° and a flame-ionization detector: the retention time of the major peak following the solvent peak in the resulting chromatogram corresponds to that obtained from a similar solution of USP Cetyl Alcohol RS.

D: *Methylene blue solution*—To a 150-mL graduated beaker containing 0.7 mL of sulfuric acid and 5 g of anhydrous sodium sulfate slowly add water to the 90-mL mark. Add methylene blue solution (3 in 1000) to the 100-mL mark, and mix.

Procedure—To 1 mL of Aqueous Dispersion in a 100-mL graduated mixing cylinder add 9 mL of water followed by 25 mL of *Methylene blue solution*, and mix. Add 15 mL of chloroform, and shake vigorously. Allow the two phases to separate: the lower phase is blue, indicating the presence of sodium lauryl sulfate.

Viscosity ⟨911⟩—Use a rotational viscosimeter equipped with a low-viscosity adapter. Mix the Aqueous Dispersion, and pipet 20 mL of it into the low-viscosity small sample adapter. Start the viscosimeter, and take readings after 60, 90, and 120 seconds at a temperature of 25 ± 2° and at a spindle speed that results in readings between 10% and 90% of full-scale. Multiply the average of the three readings by the factor specified for the spindle speed selected to obtain the viscosity in centipoises. The viscosity is not more than 150 centipoises.

pH ⟨791⟩: between 4.0 and 7.0.

Loss on drying—Place about 10 g of previously dried Ottawa sand (20- to 30-mesh) into a tared petri dish. Add 5 mL of Aqueous Dispersion, and again weigh. Dry at about 60° to constant weight: it loses not more than 71.0% of its weight.

Heavy metals, *Method II* ⟨231⟩: 0.001%.

Assay—Determine the ethoxy content, as directed under ⟨431⟩ *Methoxy Determination*, using an accurately weighed portion of Ethylcellulose Aqueous Dispersion equivalent to about 25 mg of ethylcellulose. Calculate the ethylcellulose content from the ethoxy content found and the ethoxy content of the Ethylcellulose as declared in the labeling. Each mL of 0.1 *N* sodium thiosulfate is equivalent to 0.7510 mg of ($-OC_2H_5$).

Ethylparaben

HO–C_6H_4–$COOC_2H_5$

$C_9H_{10}O_3$ 166.18
Benzoic acid, 4-hydroxy-, ethyl ester.
Ethyl *p*-hydroxybenzoate [*120-47-8*].

» Ethylparaben contains not less than 99.0 percent and not more than 100.5 percent of $C_9H_{10}O_3$, calculated on the dried basis.

Packaging and storage—Preserve in well-closed containers.

Reference standard—*USP Ethylparaben Reference Standard*—Dry over silica gel for 5 hours before using.

Identification—The infrared absorption spectrum of a mineral oil dispersion of it, previously dried, exhibits maxima only at the same wavelengths as that of a similar preparation of USP Ethylparaben RS.

Melting range ⟨741⟩: between 115° and 118°.

Other requirements—It meets the requirements for *Acidity*, *Loss on drying*, and *Residue on ignition* under *Butylparaben*.

Assay—Proceed with Ethylparaben as directed in the *Assay* under *Butylparaben*. Each mL of 1 *N* sodium hydroxide is equivalent to 166.2 mg of $C_9H_{10}O_3$.

Hard Fat

» Hard Fat is a mixture of glycerides of saturated fatty acids.

Packaging and storage—Preserve in tight containers at a temperature that is 5° or more below the melting range stated in the labeling.

Labeling—The labeling includes a melting range, which is not greater than 4° and which is between 27° and 44°.

Melting range ⟨741⟩: conforms to the requirements given under *Labeling*.

Residue on ignition ⟨281⟩: not more than 0.05%.

Acid value ⟨401⟩: not more than 1.0.

Iodine value ⟨401⟩: not more than 7.0.

Saponification value ⟨401⟩: between 215 and 255.

Hydroxyl value ⟨401⟩: not more than 70.

Unsaponifiable matter ⟨401⟩: not more than 3.0%.

Alkaline impurities—Dissolve 2.0 g in a mixture of 1.5 mL of alcohol and 3.0 mL of ether. Add 0.05 mL of bromophenol blue TS, and titrate with 0.01 *N* hydrochloric acid to a yellow endpoint: not more than 0.15 mL of 0.01 *N* hydrochloric acid is required.

Ferric Oxide

» Ferric Oxide contains not less than 97.0 percent and not more than 100.5 percent of Fe_2O_3, calculated on the ignited basis. It complies also with the regulation of the federal Food and Drug Administration (21 CFR 73.2250 iron oxides) that permits not more than 3 ppm arsenic, not more than 0.001% lead, and not more than 3 ppm mercury, all on an "as is" basis.

Packaging and storage—Preserve in well-closed containers.

Identification—Dissolve 0.5 g in 50 mL of hydrochloric acid, and dilute with water to 200 mL: the solution so obtained responds to the test for *Ferric Salts* under *Iron* ⟨191⟩.

Particle size—Triturate it with water to obtain a smooth suspension, and wash it through a 200-mesh sieve (see *Powder Fineness* ⟨811⟩) with water. Dry, and weigh the residue: not more than 1.0% remains.

Water-soluble substances—Digest 2.0 g in 100 mL of water on a boiling water bath for 2 hours, filter, and wash the filter with water. Evaporate the filtrate and washings, and dry the residue at 105° for 1 hour: the weight of the dried residue is not more than 20 mg (1.0%).

Acid-insoluble substances—Digest 2.0 g in 25 mL of hydrochloric acid on a boiling water bath for 20 minutes. Add 100 mL of water and filter quantitatively through a tared filtering crucible, with the aid of wash water. Dry the crucible and contents at 105° for 1 hour: the residue weighs not more than 2 mg (0.1%).

Organic colors and lakes—Place 1.0 g in each of 3 beakers, and add 25 mL of each of the following reagents, respectively: 1-alphachloronaphthalene, alcohol, and chloroform. Heat the beakers containing alcohol and chloroform just to boiling. Heat the other beaker on a boiling water bath for 15 minutes, with occasional swirling. Filter the contents of the beakers through retentive, solvent-resistant filter paper. If any of the filtrates shows visible turbidity, centrifuge for 15 minutes. Record the spectra against respective solvent blanks in 1-cm cells from 350 to 750 nm. No peak, above the noise level, with a slope greater than +0.001 absorbance unit per nm is found.

Assay—Digest about 1.5 g of Ferric Oxide, accurately weighed, in 25 mL of hydrochloric acid on a water bath until dissolved. Add 10 mL of hydrogen peroxide TS, and evaporate on a water bath almost to dryness. Dissolve the residue by warming with 5 mL of hydrochloric acid, add 25 mL of water, filter into a 250-mL volumetric flask, washing the filter well with water, and add water to volume. Transfer a 50-mL aliquot to a glass-stoppered flask, add 3 g of potassium iodide and 5 mL of hydrochloric acid, and insert the stopper in the flask. Allow the mixture to stand for 15 minutes, add 50 mL of water, and titrate the liberated iodine with 0.1 *N* sodium thiosulfate VS, using starch TS as the indicator. Perform a blank test with the same quantities of reagents and in the same manner, and make any necessary corrections. Each mL of 0.1 *N* sodium thiosulfate is equivalent to 7.985 mg of Fe_2O_3. Ignite about 2 g of Ferric Oxide at 800 ± 25° to constant weight as directed under *Loss on Ignition* ⟨733⟩, to enable calculation of percentage Fe_2O_3 on the ignited basis.

Fructose—*see* Fructose USP

Fumaric Acid

HC—COOH
‖
HOOC—CH

$C_4H_4O_4$ 116.07
2-Butenedioic acid, [*E*]-.
Fumaric acid [*110-17-8*].

» Fumaric Acid contains not less than 99.5 percent and not more than 100.5 percent of $C_4H_4O_4$, calculated on the anhydrous basis.

Packaging and storage—Preserve in well-closed containers.

Reference standard—*USP Maleic Acid Reference Standard.*

Identification—Dissolve about 10 mg in 25 mL of water, and to this solution add 1 mL of a solution prepared by mixing 20 mL of copper sulfate solution (1 in 5) and 8 mL of pyridine: a precipitate is formed in the blue solution within 1 minute.

Water, *Method I* ⟨921⟩: 0.5%.

Residue on ignition ⟨281⟩: not more than 0.1%.

Heavy metals, *Method II* ⟨231⟩: 0.001%.

Maleic acid—

Mobile phase—Prepare 0.005 *N* sulfuric acid that has been suitably filtered and degassed.

Standard preparation—Using *Mobile phase* as the solvent, prepare a solution containing an accurately known concentration of about 0.001 mg of USP Maleic Acid RS per mL.

Test preparation—Transfer about 100 mg of Fumaric Acid, accurately weighed, to a 100-mL volumetric flask, dissolve in *Mobile phase*, dilute with *Mobile phase* to volume, and mix.

Resolution solution—Using *Mobile phase* as the solvent, prepare a solution containing about 10 µg of USP Fumaric Acid RS per mL, and about 5 µg of USP Maleic Acid RS per mL.

Chromatographic system (see *Chromatography* ⟨621⟩)—The liquid chromatograph is equipped with a 210-nm detector and a 4.6-mm × 22-cm column that contains packing L17. The flow rate is about 0.3 mL per minute. Chromatograph the *Resolution solution*, and record the peak responses: the resolution, *R*, of the maleic acid and fumaric acid peaks is not less than 2.5, and the relative standard deviation of the maleic acid peak for replicate injections is not more than 2.0%.

Procedure—Separately inject equal volumes (about 5 µL) of the *Standard preparation* and the *Test preparation* into the chromatograph, record the chromatograms, and measure the peak responses. The relative retention times are about 0.5 for maleic acid and 1.0 for fumaric acid. Calculate the quantity, in mg, of maleic acid in the total weight of Fumaric Acid taken by the formula:

$$100C(r_U/r_S),$$

in which *C* is the concentration, in mg per mL, of USP Maleic Acid RS in the *Standard preparation*, and r_U and r_S are the responses of the maleic acid peaks obtained from the *Test preparation* and the *Standard preparation*, respectively. Not more than 0.1% of maleic acid is found.

Assay—Transfer about 1 g of Fumaric Acid, accurately weighed, to a conical flask, add 50 mL of methanol, and warm gently on a steam bath to effect solution. Cool, add phenolphthalein TS, and titrate with 0.5 *N* sodium hydroxide VS to the first appearance of a pink color that persists for not less than 30 seconds. Perform a blank titration, and make any necessary correction. Each mL of 0.5 *N* sodium hydroxide is equivalent to 29.02 mg of $C_4H_4O_4$.

Gel, Silica—*see* Silica Gel

Gelatin

» Gelatin is a product obtained by the partial hydrolysis of collagen derived from the skin, white connective tissue, and bones of animals. Gelatin derived from an acid-treated precursor is known as Type A, and Gelatin derived from an alkali-treated precursor is known as Type B.

Gelatin, where being used in the manufacture of capsules, or for the coating of tablets, may be colored with a certified color, may contain not more than 0.15 percent of sulfur dioxide, and may contain a suitable concentration of sodium lauryl sulfate and suitable antimicrobial agents.

Packaging and storage—Preserve in well-closed containers in a dry place.

Identification—

A: To a solution (1 in 100) add trinitrophenol TS or a solution of potassium dichromate (1 in 15) previously mixed with about one-fourth its volume of 3 *N* hydrochloric acid: a yellow precipitate is formed.

B: To a solution (1 in 5000) add tannic acid TS: turbidity is produced.

Microbial limits ⟨61⟩—The total bacterial count does not exceed 1000 per g, and the tests for *Salmonella* species and *Escherichia coli* are negative.

Residue on ignition ⟨281⟩: Incinerate 5.0 g without the use of sulfuric acid, but with the addition of 1.5 to 2.0 g of paraffin to avoid loss due to swelling, then finish ashing in a muffle furnace at 550° for 15 to 20 hours: the weight of the residue does not exceed 2.0%.

Odor and water-insoluble substances—A hot solution (1 in 40) is free from any disagreeable odor, and when viewed in a layer 2 cm thick is only slightly opalescent.

Sulfur dioxide—Dissolve 20.0 g in 150 mL of hot water in a flask having a round bottom and a long neck, add 5 mL of phosphoric acid and 1 g of sodium bicarbonate, and at once connect the flask with a condenser. [NOTE—Excessive foaming can be alleviated by the addition of a few drops of a suitable antifoaming agent.] Distil 50 mL, receiving the distillate under the surface of 50 mL of 0.1 *N* iodine. Acidify the distillate with a few drops of hydrochloric acid, add 2 mL of barium chloride TS, and heat on a steam bath until the liquid is nearly colorless. The precipitate of barium sulfate, if any, when filtered, washed, and ignited, weighs not more than 3 mg, corresponding to not more than 0.004% of sulfur dioxide, correction being made for any sulfate that may be present in 50 mL of the 0.1 *N* iodine. Gelatin used in the manufacture of capsules or for the coating of tablets yields not more than 109.3 mg of barium sulfate, corresponding to not more than 0.15% of sulfur dioxide.

Arsenic, *Method I* ⟨211⟩—Prepare the *Test Preparation* as follows. Mix 3.75 g with 10 mL of water in the arsine generator flask. Add 10 mL of nitric acid and 10 mL of perchloric acid, mix, and heat cautiously to the production of strong fumes of perchloric acid. Cool, wash down the sides of the generator with water, add 10 mL of nitric acid, and again heat to strong fumes. Cool, wash down the sides of the generator with water, and again heat to fumes. Repeat the digestion, if necessary, until a clear solution results. Cool, dilute with water to 52 mL, add 3 mL of hydrochloric acid: the resulting solution meets the requirements of the test, the addition of 20 mL of 7 *N* sulfuric acid specified under *Procedure* being omitted. The limit is 0.8 ppm.

Heavy metals ⟨231⟩—To the residue obtained in the test for *Residue on ignition* add 2 mL of hydrochloric acid and 0.5 mL of nitric acid, and evaporate on a steam bath to dryness. To the residue add 1 mL of 1 *N* hydrochloric acid and 15 mL of water, and warm for a few minutes. Filter, and wash with water to make the filtrate measure 100 mL. Dilute 8 mL of the solution with water to 25 mL: the limit is 0.005%.

Gentisic Acid Ethanolamide

» Gentisic Acid Ethanolamide contains not less than 99.0 percent and not more than 100.5 percent of $C_9H_{11}NO_4$, calculated on the dried basis.

Packaging and storage—Preserve in well-closed containers.

Reference standard—*USP Gentisic Acid Ethanolamide Reference Standard*—Dry at 105° for 3 hours before using.

Identification—The infrared absorption spectrum of a potassium bromide dispersion of it, previously dried, exhibits maxima only at the same wavelengths as that of a similar preparation of USP Gentisic Acid Ethanolamide RS.

Loss on drying ⟨731⟩—Dry it at 105° for 3 hours: it loses not more than 0.5% of its weight.

Residue on ignition ⟨281⟩: not more than 0.1%, a 5-g test specimen being used.

Chloride—

Standard solution—Mix 14.05 mL of 0.1 *N* hydrochloric acid with 1000 mL of water. Mix 0.4 mL of this solution with 9.6 mL of water.

Test solution—Dissolve 1.0 g of Gentisic Acid Ethanolamide in 10 mL of water, and mix. Mix 2.0 mL of this solution with 8.0 mL of water.

Procedure—To the *Standard solution* and the *Test solution* concomitantly add 0.5 mL of 2 *M* nitric acid and 0.5 mL of silver nitrate TS, mix, and compare the mixtures (see *Visual Comparison* ⟨851⟩). The turbidity resulting from the *Test solution* is not greater than that resulting from the *Standard solution* (0.01%).

Sulfate ⟨221⟩—A 1.6-g portion shows no more sulfate than corresponds to 0.50 mL of 0.020 *N* sulfuric acid (0.02%).

Heavy metals, *Method II* ⟨231⟩: 0.001%.

Chromatographic impurities—Prepare a solution containing 50.0 mg of Gentisic Acid Ethanolamide per mL in methanol. Apply 2-, 4-, and 10-µL portions of this solution to a suitable thin-layer chromatographic plate (see *Chromatography* ⟨621⟩) coated with a 0.25-mm layer of chromatographic silica gel mixture. Apply to the same plate 2-, 4-, and 8-µL portions of a methanol solution containing 0.25 mg of gentisic acid per mL, 0.25 mg of hydroquinone per mL, and 0.25 mg of USP Gentisic Acid Ethanolamide RS per mL. Allow the spots to dry, and develop the chromatogram in a solvent system consisting of a mixture of chloroform, glacial acetic acid, and water (70:30:2) until the solvent front has moved about 12 cm. Remove the plate from the developing chamber, mark the solvent front, and dry the plate with a current of air. Locate the spots on the plate by viewing under short-wavelength ultraviolet light. Spray the plate with a solution containing 5 mg of 2,6-dichloroquinone-chlorimide per mL in methanol, and then with a solution containing 0.2 g of sodium carbonate per mL in water. Compare the spots of any gentisic acid and hydroquinone that may be present in the specimen with the reference spots of these substances. Also compare the sum of the spots obtained at any other R_f values with the spot obtained from the USP Gentisic Acid Ethanolamide RS. The limits are 0.5% of each.

Assay—Transfer about 150 mg of Gentisic Acid Ethanolamide, accurately weighed, to a conical flask, dissolve in 70 mL of isopropyl alcohol, and titrate with 0.1 *N* tetrabutylammonium hydroxide VS, a stream of nitrogen being directed toward the surface of the solution during the titration, and the end-point being determined potentiometrically. Perform a blank determination, and make any necessary correction. Each mL of 0.1 *N* tetrabutylammonium hydroxide is equivalent to 19.72 mg of $C_9H_{11}NO_4$.

Glacial Acetic Acid—*see* Acetic Acid, Glacial USP

Pharmaceutical Glaze

» Pharmaceutical Glaze is a specially denatured alcoholic solution of Shellac containing between 20.0 and 57.0 percent of anhydrous shellac, and is made with either dehydrated alcohol or alcohol containing 5 percent of water by volume. The solvent is a specially denatured alcohol approved for glaze manufacturing by the Internal Revenue Service. It contains not less than 90.0 percent and not more than 110.0 percent of the labeled amount of shellac. It may contain waxes, and it may contain Titanium Dioxide as an opaquing agent.

Packaging and storage—Preserve in tight, lined metal or plastic containers, protected from excessive heat, preferably at a temperature below 25°.

Labeling—Label it to indicate the shellac type (see under *Shellac*) and concentration, the composition of the solvent, and the quantity of titanium dioxide, if present. Where titanium dioxide or waxes are present, the label states that the Glaze requires mixing before use.

Identification, Arsenic, Heavy metals, and Rosin—Pour the remainder of the solution in the volumetric flask, retained from the *Assay*, onto a clean glass plate, and place the plate in a nearly vertical position. After drainage is complete, allow the resulting film to dry in a well-ventilated place at 20° for 1 hour, then place the plate in an oven at a temperature of 43° for 16 to 24 hours. Cool, and scrape the film from the plate with a sharp blade, discarding the thick edges: it responds to the *Identification* test and meets the requirements of the tests for *Arsenic*, *Heavy metals*, and *Rosin* under *Shellac*.

Acid value—Accurately weigh, by difference, a quantity of Glaze containing about 2 g of shellac, dissolve in 50 mL of alcohol that has been neutralized to phenolphthalein with 0.1 *N* sodium hydroxide, add additional phenolphthalein TS, if necessary, and titrate with 0.1 *N* sodium hydroxide VS to a pink end-point. [NOTE—For Glaze containing orange shellac, titrate slowly, stirring vigorously, until a glass rod dipped into the titrated solution produces a color change when touched to a drop of thymol blue TS on a spot plate.] Express the acid value in terms of the number of mg of potassium hydroxide required per g of dried shellac. It meets the requirement for *Acid value* under *Shellac*.

Wax—Accurately weigh, by difference, a quantity of Glaze containing about 10 g of shellac into a 200-mL tall-form beaker. Add, with stirring, 150 mL of hot water containing 2.5 g of sodium carbonate, and proceed as directed in the test for *Wax* under *Shellac*, beginning with "immerse the beaker." It meets the requirement for *Wax* under *Shellac*.

Assay—When testing Glaze that does not contain titanium dioxide, transfer an accurately weighed quantity of Pharmaceutical Glaze, containing about 17 g of shellac, to a 100-mL volumetric flask, add alcohol to volume, and mix. Pipet 3 mL into a tared dish containing about 10 g of washed sand and a small glass rod. [NOTE—The tare weight includes the combined weights of the dish, the washed sand, and the glass rod. Retain the remaining solution in the volumetric flask for the tests for *Identification*, *Arsenic*, *Heavy metals*, and *Rosin*.] Stir until a uniform mixture is obtained, allow the glass rod to remain in the dish, dry at 105° for 1 hour in an explosion-proof oven, cool, and weigh: the weight of shellac in the quantity of Pharmaceutical Glaze taken is obtained by subtracting the tare weight from the final weight of the dried dish and contents. When testing Glaze that contains titanium dioxide, transfer an accurately weighed quantity, containing about 10 g of solids, to a beaker, and add about 10 mL of alcohol. Filter off the pigment with the aid of vacuum. Wash the filter with alcohol, and transfer the combined filtrate and washing, with the aid of alcohol, to a 200-mL volumetric flask, add alcohol to volume, and mix. Pipet 6 mL into a tared dish containing about 10 g of washed sand and a small glass rod. Proceed as directed above, beginning with the *Note*.

Liquid Glucose

» Liquid Glucose is a product obtained by the incomplete hydrolysis of starch. It consists chiefly of dextrose, dextrins, maltose, and water.

Packaging and storage—Preserve in tight containers.

Identification—Add a few drops of a solution (1 in 20) to 5 mL of hot, alkaline cupric tartrate TS: a copious, red precipitate of cuprous oxide is formed (distinction from *sucrose*).

Acidity—To a solution of 5.0 g in 15 mL of water add 5 drops of phenolphthalein TS: not more than 0.60 mL of 0.10 *N* sodium hydroxide is required to produce a pink color.

Water, *Method Ia* ⟨921⟩: not more than 21.0%, determined on a 100-mg specimen dissolved in a mixture of 20 mL of methanol and 20 mL of pyridine, the titration volume being corrected for the volume of titrant consumed by the methanol and pyridine, determined by a blank titration.

Residue on ignition ⟨281⟩: not more than 0.5%.

Sulfite—Dissolve 5 g in 50 mL of water, add 0.2 mL of 0.1 *N* iodine, then add 0.5 mL of starch TS: a blue color is produced.

Arsenic, *Method I* ⟨211⟩: 1 ppm.

Heavy metals ⟨231⟩—Mix 2.0 g with water to make 25 mL: the limit is 0.001%.

Starch—Dissolve 5 g in 50 mL of water, boil the solution for 1 minute, cool, and add 0.2 mL of 0.1 *N* iodine: no blue color is produced.

Glutaral Disinfectant Solution

» Glutaral Disinfectant Solution contains not less than 100.0 percent and not more than 110.0 percent, by weight, of the labeled amount of $C_5H_8O_2$.

Packaging and storage—Preserve in tight, light-resistant containers, and avoid exposure to excessive heat.

Identification—

2,4-Dinitrophenylhydrazine reagent—Prepare as directed in the *Identification* test under *Glutaral Concentrate*.

Procedure—Add 5 mL of Glutaral Disinfectant Solution to 20 mL of *2,4-Dinitrophenylhydrazine reagent*, mix by swirling, allow to stand for 5 minutes, and proceed as directed for *Procedure* in the *Identification* test under *Glutaral Concentrate*, beginning with "Collect the precipitate on a filter."

pH ⟨791⟩: between 2.7 and 3.7.

Assay—

Buffer—Dissolve 2.59 g of monobasic potassium phosphate and 6.77 g of anhydrous dibasic sodium phosphate in 500 mL of water in a 1000-mL volumetric flask, dilute with water to volume, and mix.

Hydroxylamine hydrochloride solution—Prepare a solution of hydroxylamine hydrochloride in *Buffer* containing 70 µg per mL.

Standard preparation—Transfer an accurately weighed quantity of Glutaral Concentrate, previously assayed as directed in the *Assay* under *Glutaral Concentrate*, equivalent to about 2.5 g of glutaraldehyde, to a 100-mL volumetric flask, dilute with water to volume, and mix. Dilute an accurately measured volume of this solution quantitatively and stepwise with water to obtain a solution having a known concentration of about 50 µg per mL.

Standard preparation blank solution—Transfer 10.0 mL of *Standard preparation* and 10.0 mL of *Buffer* to a 50-mL volumetric flask, dilute with water to volume, and mix.

Assay preparation—Transfer an accurately weighed portion of Glutaral Disinfectant Solution, equivalent to about 100 mg of glutaraldehyde, to a 100-mL volumetric flask, dilute with water to volume, and mix. Transfer 5.0 mL of this solution to a second 100-mL volumetric flask, dilute with water to volume, and mix.

Assay preparation blank solution—Transfer 10.0 mL of *Assay preparation* and 10.0 mL of *Buffer* to a 50-mL volumetric flask, dilute with water to volume, and mix.

Reagent blank solution—Transfer 10.0 mL of *Buffer* and 10.0 mL of *Hydroxylamine hydrochloride solution* to a 50-mL volumetric flask, dilute with water to volume, and mix.

Procedure—Transfer 10.0 mL each of the *Standard preparation* and the *Assay preparation* to separate 50-mL volumetric flasks, to each add 10.0 mL of *Hydroxylamine hydrochloride solution*, dilute with water to volume, mix, and allow each to stand for 25 minutes. Concomitantly determine the absorbances of both solutions and of the respective blank solutions at the wavelength of maximum absorption at about 238 nm, with a suitable spectrophotometer, using the *Reagent blank solution* to set the instrument to zero. Calculate the percentage, by weight, of $C_5H_8O_2$ in the Disinfectant Solution taken by the formula:

$$0.2(C/W)(A_U - A_{Ub})/(A_S - A_{Sb}),$$

in which C is the concentration, in µg per mL, of $C_5H_8O_2$ in the *Standard preparation*, W is the weight, in g, of Disinfectant Solution taken, A_U and A_S are the absorbances of the solutions from the *Assay preparation* and the *Standard preparation*, respectively, and A_{Ub} and A_{Sb} are the absorbances of the *Assay preparation blank solution* and the *Standard preparation blank solution*, respectively.

Glycerin—*see* Glycerin USP

Glyceryl Behenate

» Glyceryl Behenate is a mixture of glycerides of fatty acids, mainly behenic acid.

Packaging and storage—Preserve in tight containers, at a temperature not higher than 35°.

Reference standard—*USP Glyceryl Behenate Reference Standard*—Do not dry before using.

Identification—

A: *Solvent mixture*—Mix 96 volumes of chloroform and 4 volumes of acetone.

Chromatographic plates—Use suitable thin-layer chromatographic plates (see *Chromatography* ⟨621⟩) coated with a 0.25-mm layer of chromatographic silica gel. Pre-develop the plates by placing in a chromatographic chamber saturated with ether. Remove the plates from the chamber, allow the ether to evaporate, and immerse the plates in a 2.5% solution of boric acid in alcohol. After about 1 minute, withdraw the plates, allow them to dry at ambient temperature, and activate them at 110° for 30 minutes. Keep the plates in a desiccator.

Procedure—Apply 10 µL of a 6% solution of Glyceryl Behenate in chloroform and 10 µL of a 6% solution of USP Glyceryl Behenate RS in chloroform on a *Chromatographic plate*. Develop the chromatogram in *Solvent mixture* until the solvent front has moved about 12 cm. Remove the plate from the developing chamber, and allow the solvent to evaporate. Spray the chromatogram with a 0.02% solution of dichlorofluorescein in alcohol. Examine the spots under short-wavelength ultraviolet light: the R_f values of the spots obtained from the test solution correspond to those obtained from the Standard solution.

B: Dissolve about 22 mg of Glyceryl Behenate in 1 mL of toluene in a screw-cap vial having a polytef-lined septum. Add about 0.4 mL of 0.2 *N* methanolic (*m*-trifluoromethylphenyl) trimethylammonium hydroxide, attach the cap, and mix. Allow the vial to stand at room temperature for not less than 30 minutes. Introduce a suitable volume into a gas chromatograph equipped with a flame-ionization detector and a column 1.8 m in length and 4 mm in internal diameter packed with 10 percent liquid phase G7 on support S1A, maintained at a temperature of about 225°: the retention time of the main peak in the resulting chromatogram corresponds to that of the main peak in a similar preparation of USP Glyceryl Behenate RS concomitantly chromatographed. The ratio of the response of the main peak to the sum of all the responses is not less than 0.83.

Residue on ignition ⟨281⟩: not more than 0.1%.

Heavy metals, *Method II* ⟨231⟩: 0.001%.

Acid value ⟨401⟩: not more than 4.

Iodine value, *Method II* ⟨401⟩: not more than 3.

Saponification value ⟨401⟩: between 145 and 165.

1-Monoglycerides—

Periodic acid solution—Dissolve 5.4 g of periodic acid in 100 mL of water, add 1900 mL of glacial acetic acid, and mix. Store in a light-resistant, glass-stoppered bottle or in a clear glass-stoppered bottle protected from light.

Chloroform—Use chloroform that meets the following test. To each of three 500-mL flasks add 50.0 mL of *Periodic acid solution.* Add 50 mL of chloroform and 10 mL of water to two of the flasks, and add 50 mL of water to the third flask. To each flask add 20 mL of potassium iodide TS, mix gently, and proceed as directed under *Procedure*, beginning with "allow to stand for not less than 1 minute." The difference between the volumes of 0.1 *N* sodium thiosulfate required in the titrations with and without the chloroform is not greater than 0.5 mL.

Procedure—Melt the Glyceryl Behenate at a temperature not higher than 80°, and mix. Transfer about 1 g, accurately weighed, to a 100-mL beaker, and dissolve in 25 mL of *Chloroform.* Transfer the solution, with the aid of an additional 25 mL of *Chloroform*, to a separator, wash the beaker with 25 mL of water, and add the washing to the separator. Place the stopper in the separator tightly, shake vigorously for 30 to 60 seconds, and allow the layers to separate. Add 1 mL to 2 mL of glacial acetic acid to break any emulsion formed. Collect the aqueous layer in a glass-stoppered, 500-mL conical flask, and extract the nonaqueous layer again, using two 25-mL portions of water. Retain the combined aqueous extracts for the test for *Free glycerin.* Transfer the nonaqueous layer to a glass-stoppered, 500-mL conical flask, and add 50.0 mL of *Periodic acid solution* to this flask and to a blank flask containing 50 mL of chloroform and 10 mL of water, swirling the flasks during the addition. Allow to stand for not less than 30 minutes but not longer than 90 minutes. To each flask add 20 mL of potassium iodide TS, and allow to stand for not less than 1 minute but not longer than 5 minutes before titrating. Add 100 mL of water, and titrate with 0.1 *N* sodium thiosulfate VS, using a magnetic stirrer to keep the solution mixed, to the disappearance of the brown iodine color. Then add 2 mL of starch TS, and continue the titration to the disappearance of the blue color. [NOTE—If the Glyceryl Behenate titration is less than 0.8 of the blank titration, discard, and repeat using a smaller weight of Glyceryl Behenate.] Calculate the percentage of 1-monoglycerides, as glyceryl monobehenate, by the formula:

$$20.73N(B - S)/W,$$

in which 20.73 is one-twentieth of the molecular weight of glyceryl monobehenate, N is the normality of the sodium thiosulfate, B and S are the volumes, in mL, of 0.1 *N* sodium thiosulfate consumed by the blank and the Glyceryl Behenate, respectively, and W is the weight, in g, of Glyceryl Behenate taken: between 12.0% and 18.0% is found.

Free glycerin—

Periodic acid solution and *Chloroform*—Prepare as directed in the test for *1-Monoglycerides.*

Procedure—To the combined aqueous extracts obtained as directed in the test for *1-Monoglycerides* add 50.0 mL of *Periodic acid solution.* Prepare a blank by adding 50.0 mL of *Periodic acid solution* to a glass-stoppered conical flask containing 75 mL of water. Proceed as directed for *Procedure* in the test for *1-Monoglycerides*, beginning with "Allow to stand for not less than 30 minutes." Calculate the percentage of free glycerin by the formula:

$$2.30N(b - s)/W,$$

in which 2.30 is one-fortieth of the molecular weight of glycerin, N is the normality of the sodium thiosulfate, b and s are the volumes, in mL, of 0.1 *N* sodium thiosulfate consumed by the blank and the Glyceryl Behenate, respectively, and W is the weight, in g, of Glyceryl Behenate taken. Not more than 1.0% is found.

Glyceryl Monostearate

Octadecanoic acid, monoester with 1,2,3-propanetriol.
Monostearin [*31566-31-1*].

» Glyceryl Monostearate contains not less than 90.0 percent of monoglycerides of saturated fatty acids, chiefly glyceryl monostearate ($C_{21}H_{42}O_4$) and glyceryl monopalmitate ($C_{19}H_{38}O_4$). It may contain a suitable antioxidant.

Packaging and storage—Preserve in tight, light-resistant containers.

Reference standards—*USP Mono-glycerides Reference Standard*—Do not dry before using. Keep container tightly closed and protected from light. *USP Stearic Acid Reference Standard*—Keep container tightly closed. Do not dry before using. *USP Palmitic Acid Reference Standard*—Keep container tightly closed. Do not dry before using.

Melting range, *Class I* ⟨741⟩: does not melt below 55°.

Residue on ignition ⟨281⟩: not more than 0.5%.

Arsenic, *Method II* ⟨211⟩: 3 ppm.

Heavy metals, *Method II* ⟨231⟩: 0.001%.

Acid value ⟨401⟩: not more than 6.

Iodine value ⟨401⟩: not more than 3.

Saponification value ⟨401⟩: between 155 and 165.

Hydroxyl value ⟨401⟩: between 300 and 330.

Free glycerin—

Propionating reagent—Mix 10 mL of pyridine with 20 mL of propionic anhydride.

Internal standard solution—Dissolve a suitable quantity of tributyrin, accurately weighed, in chloroform, and dilute quantitatively with chloroform to obtain a solution having a concentration of about 0.2 mg per mL.

Standard preparation—Transfer about 15 mg of glycerin and about 50 mg of tributyrin, both accurately weighed, to a glass-stoppered, 25-mL conical flask, add 3 mL of *Propionating reagent*, and heat at 75° for 30 minutes. Volatilize the reagents with the aid of a stream of nitrogen at room temperature, add about 12 mL of chloroform, and mix. Dilute about 1 mL of this mixture with chloroform to about 20 mL, and mix.

Test preparation—Transfer about 50 mg of Glyceryl Monostearate, accurately weighed, to a glass-stoppered, 25-mL conical flask, add by pipet 5 mL of *Internal standard solution*, and mix to dissolve. Immerse the flask in a water bath maintained at a temperature between 45° and 50°, and volatilize the chloroform with the aid of a stream of nitrogen. Add 3 mL of *Propionating reagent*, and heat at 75° for 30 minutes. Volatilize the reagents with the aid of a stream of nitrogen at room temperature, add about 5 mL of chloroform, and mix.

Chromatographic system—Under typical conditions, the instrument is equipped with a flame-ionization detector, and contains a 2.4-m × 4-mm borosilicate glass column packed with 2 percent liquid phase G16 on 80- to 100-mesh support S1A. The column is maintained isothermally at a temperature between 190° and 200° and the injection port and detector block are maintained at about 300° and 310°, respectively, and helium is used as the carrier gas at a flow rate of about 70 mL per minute.

System suitability—Chromatograph six to ten injections of the *Standard preparation* as directed under *Procedure:* the resolution factor, R, between the peaks for the derivatized glycerin and the tributyrin is not less than 4.0, and the relative standard deviation of the ratio of their peak areas is not more than 2.0%.

Procedure—Inject a suitable portion of the *Standard preparation* into a suitable gas chromatograph, and record the chromatogram. Measure the areas under the peaks, and record the values of the areas under the tripropionin and tributyrin peaks as A_S and A_D, respectively. Calculate the response factor, F, by the formula:

$$(A_D/A_S)(W_S/W_D),$$

in which W_S and W_D are the weights, in mg, of glycerin and tributyrin, respectively, in the *Standard preparation.* Similarly inject a suitable portion of the *Test preparation*, and record the chromatogram. Measure the areas under the peaks, and record the values of the areas under the tripropionin and tributyrin peaks as a_U and a_D, respectively. Calculate the percentage of glycerin by the formula:

$$100F(a_U/a_D)(w_D/w_U),$$

in which w_D is the weight, in mg, of tributyrin in 5 mL of *Internal standard solution*, and w_U is the weight, in mg, of Glyceryl Monostearate in the *Test preparation:* the limit is 1.2%.

Assay for mono-glycerides—

Propionating reagent—Mix 10 mL of pyridine and 20 mL of propionic anhydride.

Internal standard solution—Transfer about 400 mg of hexadecyl hexadecanoate, accurately weighed, to a 100-mL volumetric flask, dissolve in chloroform, dilute with chloroform to volume, and mix.

Standard preparation—Transfer about 50 mg of USP Mono-glycerides RS, accurately weighed, to a 25-mL conical flask, add by pipet 5 mL of *Internal standard solution*, and mix. When solution is complete, immerse the flask in a water bath maintained at a temperature between 45° and 50°, and volatilize the chloroform with the aid of a stream of nitrogen. Add 3.0 mL of *Propionating reagent*, and heat on a hot plate at 75° for 30 minutes. Evaporate the reagents with the aid of a stream of nitrogen and gentle steam heat. Add 15 mL of chloroform, and swirl to dissolve the residue.

Assay preparation—Transfer about 50 mg of Glyceryl Monostearate, accurately weighed, to a 25-mL conical flask, and proceed as directed for *Standard preparation*, beginning with "add by pipet 5 mL of *Internal standard solution*."

Chromatographic system—Under typical conditions the instrument is equipped with a flame-ionization detector, and contains a 2.4-m × 4-mm borosilicate glass column packed with 2 percent liquid phase 5 percent phenyl methyl silicone on 80- to 100-mesh support S1. The column is maintained isothermally at a temperature between 270° and 280°, the injection port and detector block are maintained at about 310°, and helium is used as the carrier gas at a flow rate of about 70 mL per minute.

System suitability—Chromatograph six to ten injections of the *Standard preparation* as directed under *Procedure:* the resolution factor, *R*, between the peaks for the derivatized glyceryl hexadecanoate and glyceryl octadecanoate is not less than 2.0, and the relative standard deviation of the ratio of the peak area of the derivatized glyceryl octadecanoate to that of the hexadecyl hexadecanoate is not more than 2.0%.

Procedure—Inject a suitable portion of the *Standard preparation* into a suitable gas chromatograph, and record the chromatogram. Measure the areas under the peaks, and record the values of the sum of the areas under the derivatized mono-glyceride peaks and of the area under the hexadecyl hexadecanoate peak as A_S and A_D, respectively. Calculate the response factor, *F*, taken by the formula:

$$(A_S/A_D)(W_D/W_S),$$

in which W_D and W_S are the weights, in mg, of hexadecyl hexadecanoate and USP Mono-glycerides RS, respectively, in the *Standard preparation.* Similarly inject a suitable portion of the *Assay preparation*, and record the chromatogram. Measure the areas under the peaks, and record the values of the sum of the areas under the derivatized mono-glyceride peaks and of the area under the hexadecyl hexadecanoate peak as a_U and a_D, respectively. Calculate the quantity, in mg, of mono-glycerides in the amount of Glyceryl Monostearate taken by the formula:

$$(W_D/F)(a_U/a_D).$$

Guar Gum

» Guar Gum is a gum obtained from the ground endosperms of *Cyamopsis tetragonolobus* (Linné) Taub. (Fam. Leguminosae). It consists chiefly of a high molecular weight hydrocolloidal polysaccharide, composed of galactan and mannan units combined through glycosidic linkages, which may be described chemically as a galactomannan.

Packaging and storage—Preserve in well-closed containers.

Identification—

A: Place about 2 g in a 400-mL beaker, moisten it with about 4 mL of isopropyl alcohol, add 200 mL of cold water with vigorous stirring, and continue stirring until the gum is completely and uniformly dispersed: an opalescent, viscous solution results.

B: Place about 100 mL of the solution prepared in *Identification test A* in a 400-mL beaker, heat in a boiling water bath for about 10 minutes, and cool: no appreciable increase in viscosity is produced (*distinction from locust bean gum*—see *Reagents* in the section, *Reagents, Indicators, and Solutions*).

Loss on drying ⟨731⟩—Dry it at 105° for 5 hours: it loses not more than 15.0% of its weight.

Ash—Proceed as directed for *Total Ash* (see *Vegetable Drugs—Methods of Analysis* ⟨561⟩). Guar Gum yields not more than 1.5% of ash.

Acid-insoluble matter—Transfer about 1.5 g of Guar Gum, accurately weighed, to a 250-mL beaker containing 150 mL of water and 1.5 mL of sulfuric acid. Cover the beaker with a watch glass, and heat the mixture on a steam bath for 6 hours, rubbing down the wall of the beaker frequently with a rubber-tipped stirring rod, and replacing any water lost by evaporation. At the end of the 6-hour heating period add about 500 mg of a suitable filter aid, accurately weighed, and filter through a suitable tared, ashless filter. Wash the residue several times with hot water, dry the filter and its contents at 105° for 3 hours, cool in a desiccator, and weigh: the amount of acid-insoluble matter, determined by subtracting the weight of filter aid from that of the residue, is not more than 7.0% of the Guar Gum taken.

Arsenic, *Method II* ⟨211⟩: 3 ppm.

Lead ⟨251⟩—Prepare a *Test Preparation* as directed for organic compounds, and use 10 mL of *Diluted Standard Lead Solution* (10 µg of Pb) for the test: the limit is 0.001%.

Heavy metals, *Method II* ⟨231⟩: 0.002%.

Protein—Transfer about 3.5 g of Guar Gum, accurately weighed, to a 500-mL Kjeldahl flask, and proceed as directed for *Method I* under *Nitrogen Determination* ⟨461⟩: the amount of protein, obtained by multiplying by 6.25 the percentage of nitrogen determined, is not more than 10.0% of the Guar Gum taken.

Starch—To a 1 in 10 solution of Guar Gum add a few drops of iodine TS: no blue color is produced.

Galactomannans—Subtract from 100.0 the total percentages of *Loss on drying, Ash, Acid-insoluble matter,* and *Protein:* the content of galactomannans is not less than 66.0%.

Gum, Xanthan—*see* Xanthan Gum
Hard Fat—*see* Fat, Hard

Hexylene Glycol

$$CH_3CH(OH)CH_2C(CH_3)(OH)CH_3$$

$C_6H_{14}O_2$ 118.18

2,4-Pentanediol, 2-methyl-.

2-Methyl-2,4-pentanediol [*107-41-5*].

» Hexylene Glycol is 2-methyl-2,4-pentanediol.

Packaging and storage—Preserve in tight containers.

Reference standard—*USP Hexylene Glycol Reference Standard*—Keep container tightly closed. Do not dry before using.

Identification—The infrared absorption spectrum of a thin film of it exhibits maxima only at the same wavelengths as that of a similar preparation of USP Hexylene Glycol RS.

Specific gravity ⟨841⟩: between 0.917 and 0.923.

Refractive index ⟨831⟩: between 1.424 and 1.430.

Acidity—Mix 1 mL of phenolphthalein TS with 50 mL of water, and add 0.1 *N* sodium hydroxide until the solution remains pink for 30 seconds. Add 10 mL of Hexylene Glycol, accurately measured, and titrate with 0.10 *N* sodium hydroxide until the original pink color returns and remains for 30 seconds: not more than 0.20 mL of 0.10 *N* sodium hydroxide is required.

Water, *Method I* ⟨921⟩: not more than 0.5%.

Hydrochloric Acid

HCl 36.46
Hydrochloric acid.
Hydrochloric acid [*7647-01-0*].

» Hydrochloric Acid contains not less than 36.5 percent and not more than 38.0 percent, by weight, of HCl.

Packaging and storage—Preserve in tight containers.

Identification—It responds to the tests for *Chloride* ⟨191⟩.

Residue on ignition ⟨281⟩—To 20 mL add 2 drops of sulfuric acid, evaporate to dryness, and ignite: not more than 2 mg of residue remains (about 0.008%).

Bromide or iodide, Free bromine or chlorine, Sulfate, and Sulfite—Dilute it with 2 volumes of water, to perform the following tests.

Bromide or iodide—Add 1 mL of chloroform to 10 mL of the dilution, and cautiously add, dropwise, with constant agitation, chlorine TS that has been diluted with an equal volume of water: the chloroform remains free from even a transient yellow, orange, or violet color.

Free bromine or chlorine—Add 1 mL of potassium iodide TS and 1 mL of chloroform to 10 mL of the dilution, and agitate the mixture: the chloroform remains free from any violet coloration for at least 1 minute.

Sulfate—To a mixture of 3 mL of the dilution and 5 mL of water add 5 drops of barium chloride TS: neither turbidity nor precipitate appears within 1 hour.

Sulfite—On the completion of the test for *Sulfate*, the addition to the liquid of 2 drops of 0.1 *N* iodine produces neither turbidity nor decolorization of the iodine.

Arsenic, *Method I* ⟨211⟩—Prepare the *Test Preparation* as follows. To 2.5 mL (3 g) add 2.5 mL of hydrochloric acid (see under *Reagents* in the section, *Reagents, Indicators, and Solutions*), and dilute with water to 55 mL: the resulting solution meets the requirements of the test, the addition of 20 mL of 7 *N* sulfuric acid specified under *Procedure* being omitted. The limit is 1 ppm.

Heavy metals ⟨231⟩—Evaporate 3.4 mL (4 g) on a steam bath to dryness, add 2 mL of 1 *N* acetic acid to the residue, then dilute with water to 25 mL: the limit is 5 ppm.

Assay—Place about 3 mL of Hydrochloric Acid in a glass-stoppered flask, previously tared while containing about 20 mL of water, and weigh again to obtain the weight of the substance under assay. Dilute with about 25 mL of water, add methyl red TS, and titrate with 1 *N* sodium hydroxide VS. Each mL of 1 *N* sodium hydroxide is equivalent to 36.46 mg of HCl.

Diluted Hydrochloric Acid

» Diluted Hydrochloric Acid contains, in each 100 mL, not less than 9.5 g and not more than 10.5 g of HCl.

Diluted Hydrochloric Acid may be prepared as follows:

Hydrochloric Acid	226 mL
Purified Water, a sufficient quantity, to make	1000 mL

Mix the ingredients.

Packaging and storage—Preserve in tight containers.

Identification—It responds to the tests for *Chloride* ⟨191⟩.

Residue on ignition ⟨281⟩—To 20 mL add 2 drops of sulfuric acid, evaporate to dryness, and ignite: the weight of the residue does not exceed 2 mg.

Sulfate—To a mixture of 3 mL of it and 5 mL of water add 5 drops of barium chloride TS: neither turbidity nor precipitate appears within 1 hour.

Sulfite—On the completion of the test for *Sulfate*, the further addition to the liquid of 2 drops of 0.1 *N* iodine produces neither turbidity nor decolorization of the iodine.

Arsenic, *Method I* ⟨211⟩—A 5.0-mL portion meets the requirements of the test. The limit is 0.6 ppm.

Heavy metals ⟨231⟩—To 3.8 mL (4 g) add 5 mL of water and 1 drop of phenolphthalein TS. Add 6 *N* ammonium hydroxide until the solution assumes a faint pink color. Add 2 mL of 1 *N* acetic acid, then add water to make 25 mL: the limit is 5 ppm.

Free bromine or chlorine—To 10 mL add 1 mL of potassium iodide TS and 1 mL of chloroform, and agitate the mixture: the chloroform remains free from any violet coloration for at least 1 minute.

Assay—Transfer 10.0 mL of Diluted Hydrochloric Acid to a conical flask, and add about 20 mL of water. Add 3 drops of methyl red TS, and titrate with 1 *N* sodium hydroxide VS. Each mL of 1 *N* sodium hydroxide is equivalent to 36.46 mg of HCl.

Hydrogenated Castor Oil—*see* Castor Oil, Hydrogenated

Hydrogenated Vegetable Oil—*see* Vegetable Oil, Hydrogenated

Hydrophilic Ointment—*see* Ointment, Hydrophilic USP

Hydrophilic Petrolatum—*see* Petrolatum, Hydrophilic USP

Hydroxyanisole, Butylated—*see* Butylated Hydroxyanisole

Hydroxyethyl Cellulose

Cellulose, 2-hydroxyethyl ether [*9004-62-0*].

» Hydroxyethyl Cellulose is a partially substituted poly(hydroxyethyl) ether of cellulose. It is available in several grades, varying in viscosity and degree of substitution, and some grades are modified to improve their dispersion in water. It may contain suitable anticaking agents.

Packaging and storage—Preserve in well-closed containers.

Labeling—The labeling indicates its viscosity, under specified conditions, in aqueous solution. The indicated viscosity may be in the form of a range encompassing 50% to 150% of the average value.

Identification—

A: Stir about 1 g into 100 mL of water: it is dissolved completely to produce a colloidal solution that remains clear when heated to 60°.

B: Place 1 mL of the solution from *Identification test A* on a glass plate, and allow the water to evaporate: a thin, self-sustaining film is formed.

C: To 1 mL of a solution (1 in 2000) add 1 mL of phenol solution (1 in 20). Add 5 mL of sulfuric acid, shake, and allow to cool: the color of the solution so obtained becomes red.

Viscosity ⟨911⟩—When determined at the concentration and under the conditions specified in the labeling, its viscosity is not less than 50% and not more than 150% of the labeled viscosity, where stated as a single value, or it is between the maximum and minimum values, where stated as a range of viscosities.

pH ⟨791⟩: between 6.0 and 8.5, in a solution (1 in 100).

Loss on drying ⟨731⟩—Dry it at 105° for 3 hours: it loses not more than 10.0% of its weight.

Residue on ignition ⟨281⟩: not more than 5.0%.

Lead ⟨251⟩: 0.001%.

Arsenic, *Method II* ⟨211⟩: 3 ppm.

Heavy metals, *Method II* ⟨231⟩: 0.004%.

Hydroxypropyl Cellulose

Cellulose, 2-hydroxypropyl ether [*9004-64-2*].

» Hydroxypropyl Cellulose is a partially substituted poly(hydroxypropyl) ether of cellulose. It may contain not more than 0.60 percent of silica, or other suitable anticaking agents. When dried at 105° for 3 hours, it contains not more than 80.5 percent of hydroxypropoxy groups.

Packaging and storage—Store in well-closed containers.

Labeling—Label it to indicate the viscosity in an aqueous solution of stated concentration and temperature. The indicated viscosity may be in the form of a range encompassing 50% to 150% of the average value.

Identification—

A: Add about 1 g to 100 mL of water, previously heated to 60°, and stir: a slurry is formed that swells and disperses on cooling to form a colloidal solution.

B: Heat 10 mL of the solution prepared in *Identification test A* on a water bath while stirring: at a temperature of 45° the solution becomes cloudy, or a flocculant precipitate is formed, which disappears on cooling.

C: Place 1 mL of the solution prepared in *Identification test A* on a glass plate, and allow the water to evaporate: a thin, self-sustaining film is formed.

Apparent viscosity (see *Viscosity* ⟨911⟩)—Determine the apparent viscosity at the concentration and temperature specified on the label with a suitable viscosimeter of the rotational type (see *Labeling*).

pH ⟨791⟩: between 5.0 and 8.0, in a solution (1 in 100).

Loss on drying ⟨731⟩—Dry it at 105° for 3 hours: it loses not more than 5.0% of its weight.

Residue on ignition—[*Caution—Perform the mixing and heating of the mixtures containing hydrofluoric acid in a well-ventilated hood.*] Proceed as directed for *Residue on Ignition* ⟨281⟩, using a platinum crucible if silica may be present. If more than 0.2% residue is found, and silica is present, moisten the residue with water, and add about 5 mL of hydrofluoric acid, in small portions. Evaporate on steam bath to dryness, and cool. Add about 5 mL of hydrofluoric acid and about 0.5 mL of sulfuric acid, and evaporate to dryness. Slowly increase the temperature until all of the acids have been volatilized, and ignite at 1000 ± 25°. Cool in a desiccator, and weigh: the difference between the final weight and the weight of the initially ignited portion represents the weight of silica, and the final weight is not more than 0.2% of the weight of test specimen taken for the ignition.

Arsenic, *Method II* ⟨211⟩: 3 ppm.

Lead ⟨251⟩: 0.001%.

Heavy metals, *Method II* ⟨231⟩: 0.004%.

Assay for hydroxypropoxy groups—Proceed as directed under *Hydroxypropoxy Determination* ⟨421⟩, but use only about 65 mg of Hydroxypropyl Cellulose, previously dried and accurately weighed, add 5 mL of water, and swirl slowly for 5 minutes, before adding the chromium trioxide solution, and omit the *Methylcellulose blank*. Each mL of 0.02 *N* sodium hydroxide is equivalent to 1.502 mg of hydroxypropoxy groups ($-OCH_2CHOHCH_3$).

Hydroxypropyl Methylcellulose—*see* Hydroxypropyl Methylcellulose USP

Hydroxypropyl Methylcellulose 2208—*see* Hydroxypropyl Methylcellulose 2208 USP

Hydroxypropyl Methylcellulose 2906—*see* Hydroxypropyl Methylcellulose 2906 USP

Hydroxypropyl Methylcellulose 2910—*see* Hydroxypropyl Methylcellulose 2910 USP

Hydroxypropyl Methylcellulose Phthalate

» Hydroxypropyl Methylcellulose Phthalate is a monophthalic acid ester of hydroxypropyl methylcellulose. When dried at 105° for 1 hour, it contains methoxy ($-OCH_3$), hydroxypropoxy ($-OCH_2CHOHCH_3$), and phthalyl (*o*-carboxybenzoyl, $C_8H_5O_3$) groups, conforming to the limits for the types of Hydroxypropyl Methylcellulose Phthalate set forth in the accompanying table.

Substitution Type	Methoxy (percent)		Hydroxypropoxy (percent)		Phthalyl (percent)	
	Min.	Max.	Min.	Max.	Min.	Max.
200731	18.0	22.0	5.0	9.0	27.0	35.0
220824	20.0	24.0	6.0	10.0	21.0	27.0

Packaging and storage—Preserve in well-closed containers.

Labeling—Label it to indicate its substitution type and its viscosity.

Clarity and color of solution—Dissolve 1.0 g in 10 mL of a mixture of equal volumes of dehydrated alcohol and acetone: the solution is not more turbid than a solution made by mixing, in the order named, 0.30 mL of 0.01 *N* hydrochloric acid, 6 mL of 2 *N* nitric acid, water to make 50 mL, and 1 mL of silver nitrate TS. Also, the solution has no more color than a diluted solution of *Matching Fluid A* (1 in 3) (see *Color and Achromicity* ⟨631⟩).

Identification—

A: Shake about 10 mg with 1 mL of water and 2 mL of a 0.035 in 100 solution of anthrone in sulfuric acid: a green color develops, and it gradually changes to a dark, greenish brown color.

B: Dissolve 1 g in 40 mL of 1 *N* sodium hydroxide, add water to make 100 mL, and mix. To 0.1 mL of this solution add 9 mL of 32 *N* sulfuric acid, and shake. Heat in a water bath for 3 minutes, accurately timed, and immediately cool in an ice bath. While the mixture is cold, carefully add 0.6 mL of ninhydrin TS, and mix. Allow to stand at room temperature: the red color that appears immediately turns to violet within 100 minutes.

C: Place 5 mg in a small test tube, add 2 drops of a 1 in 10 solution of benzoyl peroxide in acetone, evaporate on a water bath to dryness, and fit the test tube with a cork stopper bearing a glass rod extending downward and moistened at its tip with chromotropic acid TS. Heat in a liquid bath at 125° for 5 to 6 minutes: the color of the chromotropic acid TS becomes purple.

D: Shake about 10 mg with 2 mg to 3 mg of resorcinol and 1 mL of sulfuric acid, and heat in a liquid bath at 125° to 130° for 5 minutes. Cool, cautiously add 5 mL of water, add dropwise, with cooling, 10 *N* sodium hydroxide until the solution is alkaline, and add an additional 10 mL of water: the solution fluoresces green-yellow under ultraviolet light.

Viscosity ⟨911⟩—Dissolve 10 g, previously dried, in 90 g of a mixture of equal weights of methanol and methylene chloride by mixing and shaking: the viscosity (see *Procedure for Cellulose Derivatives* under *Viscosity* ⟨911⟩) is not less than 80% and not more than 120% of that indicated by the label.

Loss on drying ⟨731⟩—Dry it at 105° for 1 hour: it loses not more than 5.0% of its weight.

Residue on ignition ⟨281⟩: not more than 0.20%.

Chloride ⟨221⟩—Dissolve 1.0 g in 40 mL of 0.2 *N* sodium hydroxide, add 1 drop of phenolphthalein TS, and add 2 *N* nitric acid dropwise, with stirring, until the red color is discharged. Add an additional 20 mL of 2 *N* nitric acid with stirring. Heat on a water bath, with stirring, until the gel-like precipitate formed becomes granular. Cool the mixture, and centrifuge. Separate the liquid phase, and wash the residue with three successive 20-mL portions of water, separating the washings by centrifuging. Dilute the combined liquids with water to 200 mL, mix, and filter. A 50-mL portion of the filtrate so obtained shows no more chloride than a control solution made by treating 0.50 mL of 0.01 *N* hydrochloric acid with 10 mL of 0.2 *N* sodium hydroxide, adding 7 mL of 2 *N* nitric acid, and diluting with water to 50 mL (0.07%).

Arsenic, *Method II* ⟨211⟩: 2 ppm.

Heavy metals, *Method II* ⟨231⟩: 0.001%.

Free phthalic acid—Transfer about 1.5 g, previously dried and accurately weighed, to a separator, dissolve in 50 mL of a mixture of dehydrated alcohol and methylene chloride (3:2), add 75 mL of water with shaking, shake thoroughly with 50 mL of hexane, and draw off the aqueous layer. Wash the remaining liquid with 50 mL of water, combine the washings with the aqueous layer, and titrate with 0.1 *N* sodium hydroxide VS to a phenolphthalein end-point. Perform a blank determination, and make any necessary correction. Each mL of 0.1 *N* sodium hydroxide is equivalent to 8.307 mg of phthalic acid. Not more than 1.0% is found.

Phthalyl content—Transfer about 1 g, previously dried and accurately weighed, to a conical flask, dissolve in 50 mL of a mixture of alcohol, acetone, and water (2:2:1), add phenolphthalein TS, and titrate with 0.1 *N* sodium hydroxide VS. Perform a blank determination, and make any necessary correction. Calculate the percentage of phthalyl by the formula:

$$0.01(149.1)(V/W) - 2(149.1/166.1)(P),$$

in which 149.1 and 166.1 are the molecular weights of the phthalyl group and phthalic acid, respectively, V is the volume, in mL, of 0.1 *N* sodium hydroxide consumed after correction for the blank, W is the weight, in g, of Hydroxypropyl Methylcellulose Phthalate taken, and P is the percentage of free phthalic acid found as directed in the test for *Free phthalic acid*.

Methoxy and hydroxypropoxy contents—[*Caution—Hydriodic acid and its reaction byproducts are highly toxic. Perform all steps of the Assay preparation and Standard preparation in a properly functioning hood. Specific safety practices to be followed are to be identified to the analyst performing this test.*]

Internal standard solution—Transfer about 2.5 g of toluene, accurately weighed, to a 100-mL volumetric flask containing 10 mL of *o*-xylene, dilute with *o*-xylene to volume, and mix.

Standard preparation—Into a suitable serum vial weigh about 135 mg of adipic acid and 4.0 mL of hydriodic acid, pipet 4 mL of *Internal standard solution* into the vial, and close the vial securely with a suitable septum stopper. Weigh the vial and contents accurately, add 30 µL of isopropyl iodide through the septum with a syringe, again weigh, and calculate the weight of isopropyl iodide added, by difference. Add 90 µL of methyl iodide similarly, again weigh, and calculate the weight of methyl iodide added, by difference. Shake, and allow the layers to separate.

Assay preparation—Transfer about 0.065 g of dried Hydroxypropyl Methylcellulose Phthalate, accurately weighed, to a 5-mL thick-walled reaction vial equipped with a pressure-tight septum-type closure, add an amount of adipic acid equal to the weight of the test specimen, and pipet 2 mL of *Internal standard solution* into the vial. Cautiously pipet 2 mL of hydriodic acid into the mixture, immediately cap the vial tightly, and weigh accurately. Mix the contents of the vial continuously while heating at 150°, for 60 minutes. Allow the vial to cool for about 45 minutes, and again weigh. If the weight loss is greater than 10 mg, discard the mixture, and prepare another *Assay preparation*.

Chromatographic system—Use a gas chromatograph equipped with a thermal conductivity detector. Under typical conditions, the instrument contains a 1.8-m × 4-mm glass column packed with 20 percent liquid phase G28 on 100- to 120-mesh support S1C that is not silanized, the column is maintained at 130°, and helium is used as the carrier gas. In a suitable system, the resolution, R (see *Chromatography* ⟨621⟩), between the toluene and isopropyl iodide peaks is not less than 2.0.

Calibration—Inject about 2 µL of the upper layer of the *Standard preparation* into the gas chromatograph, and record the chromatogram. Under the conditions described above, the relative retention times of methyl iodide, isopropyl iodide, toluene, and *o*-xylene are approximately 1.0, 2.2, 3.6, and 8.0, respectively. Calculate the relative response factor, F_{mi}, of equal weights of toluene and methyl iodide by the formula:

$$Q_{smi}/A_{smi},$$

in which Q_{smi} is the quantity ratio of methyl iodide to toluene in the *Standard preparation*, and A_{smi} is the peak area ratio of methyl iodide to toluene obtained from the *Standard preparation*. Similarly, calculate the relative response factor, F_{ii}, of equal weights of toluene and isopropyl iodide by the formula:

$$Q_{sii}/A_{sii},$$

in which Q_{sii} is the quantity ratio of isopropyl iodide to toluene in the *Standard preparation*, and A_{sii} is the peak area ratio of isopropyl iodide to toluene obtained from the *Standard preparation.*

Procedure—Inject about 2 µL of the upper layer of the *Assay preparation* into the gas chromatograph, and record the chromatogram. Calculate the percentage of methoxy in the Hydroxypropyl Methylcellulose Phthalate by the formula:

$$2(31/142)F_{mi}A_{umi}(W_t/W_u),$$

in which 31/142 is the ratio of the formula weights of methoxy and methyl iodide, F_{mi} is defined under *Calibration*, A_{umi} is the ratio of the area of the methyl iodide peak to that of the toluene peak obtained from the *Assay preparation*, W_t is the weight, in g, of toluene in the *Internal standard solution*, and W_u is the weight, in g, of Hydroxypropyl Methylcellulose Phthalate taken for the *Assay*. Similarly, calculate the percentage of hydroxypropoxy in the Hydroxypropyl Methylcellulose Phthalate by the formula:

$$2(75/170)F_{ii}A_{uii}(W_t/W_u),$$

in which 75/170 is the ratio of the formula weights of hydroxypropoxy and isopropyl iodide, F_{ii} is defined under *Calibration*, A_{uii} is the ratio of the area of the isopropyl iodide peak to that of the toluene peak obtained from the *Assay preparation*, W_t is the weight, in g, of toluene in the *Internal standard solution*, and W_u is the weight, in g, of Hydroxypropyl Methylcellulose Phthalate taken for the *Assay*.

Hydroxypropyl Methylcellulose Phthalate 200731

Where this monograph is specified or referred to in this Pharmacopeia, see the monograph, *Hydroxypropyl Methylcellulose Phthalate*.

Hydroxypropyl Methylcellulose Phthalate 220824

Where this monograph is specified or referred to in this Pharmacopeia, see the monograph, *Hydroxypropyl Methylcellulose Phthalate.*

Hydroxytoluene, Butylated—*see* Butylated Hydroxytoluene

Hypophosphorous Acid

H_3PO_2 66.00
Phosphinic acid.
Hypophosphorous acid [*6303-21-5*].

» Hypophosphorous Acid contains not less than 30.0 percent and not more than 32.0 percent of H_3PO_2.

Packaging and storage—Preserve in tight containers.

Identification—It responds to the tests for *Hypophosphite* ⟨191⟩.

Arsenic, Barium, and Oxalate—

Test solution—Dilute Hypophosphorous Acid with 3 volumes of water, and use this solution for the following tests.

Arsenic, Method I ⟨211⟩—Prepare the *Test Preparation* as follows. Mix 8 mL of *Test solution* with 5 mL of 30 percent hydrogen peroxide solution, and evaporate on a steam bath to dryness. Cool, add 10 mL of 2 *N* sulfuric acid, and evaporate to dense fumes of sulfur trioxide, cool, add cautiously 10 mL of water, and again evaporate to dense fumes, repeating if necessary to remove any trace of hydrogen peroxide. Cool, and dilute cautiously with water to 35 mL. The limit is 1.5 ppm.

Barium—Neutralize 30 mL of *Test solution* with 6 *N* ammonium hydroxide: the mixture exhibits little or no precipitation. Filter, acidify 10 mL of the filtrate with hydrochloric acid, and add 2 mL of potassium sulfate TS: no turbidity is produced.

Oxalate—Another 10-mL portion of the filtrate obtained in the test for *Barium* shows no turbidity upon the addition of 1 mL of calcium chloride TS.

Heavy metals, *Method I* ⟨231⟩—Place 0.90 mL (1 g) in a small beaker, and add 3 mL of water. Add 1 mL of nitric acid, and evaporate on a steam bath to about 1 mL. Again add 1 mL of nitric acid, and evaporate on a steam bath. Dissolve the residue in 3 mL of water, add 6 *N* ammonium hydroxide until the solution is distinctly alkaline to litmus, then boil gently until the odor of ammonia disappears. Add 2 mL of 1 *N* acetic acid and 15 mL of warm water, filter, and dilute the filtrate with water to 25 mL: the limit is 0.002%.

Assay—Pour about 7 mL of Hypophosphorous Acid into a tared, glass-stoppered flask, and weigh accurately. Dilute with about 25 mL of water, add phenolphthalein TS, and titrate with 1 *N* sodium hydroxide VS. Each mL of 1 *N* sodium hydroxide is equivalent to 66.00 mg of H_3PO_2.

Imidurea

[Structural formula: CH2OH, O=C–N–CH–NH–C(=O)–NH–CH2–NH–C(=O)–NH–HC–N–C=O, HN–C=O, O=C–NH, CH2OH]

$C_{11}H_{16}N_8O_8$ 388.30
N,N″-Methylenebis[*N′*-[3-(hydroxymethyl)-2,5-dioxo-4-imidazolidinyl]urea].
1,1′-Methylenebis[3-[3-(hydroxymethyl)-2,5-dioxo-4-imidazolidinyl]urea] [*39236-46-9*].

» Imidurea contains not less than 26.0 percent and not more than 28.0 percent of nitrogen (N), calculated on the dried basis.

Packaging and storage—Preserve in tight containers.

Reference standard—*USP Imidurea Reference Standard*—Dry in vacuum over phosphorus pentoxide for 48 hours before using.

Color and clarity of solution—Dissolve 3.0 g in 7.0 mL of water in a test tube: the solution is clear and colorless.

Identification—The infrared absorption spectrum of a potassium bromide dispersion of it exhibits maxima only at the same wavelengths as that of a similar preparation of USP Imidurea RS.

pH ⟨791⟩: between 6.0 and 7.5, in a solution (1 in 100).

Loss on drying ⟨731⟩—Dry in vacuum over phosphorus pentoxide for 48 hours: it loses not more than 3.0% of its weight.

Residue on ignition ⟨281⟩: not more than 3.0%.

Heavy metals, *Method II* ⟨231⟩: 0.001%.

Nitrogen content—Place about 150 mg of Imidurea, accurately weighed, in a 500-mL Kjeldahl flask, and add 8.0 g of anhydrous sodium sulfate, 0.5 g of anhydrous cupric sulfate, 0.1 g of yellow mercuric oxide, and 11 mL of sulfuric acid. Incline the flask at an angle of about 45°, and gently heat the mixture, keeping the temperature below the boiling point until frothing has ceased. Increase the heat until the acid boils briskly, and continue the heating until the solution has become clear green in color or practically colorless for 30 minutes. Allow to cool, cautiously add 225 mL of water, mix the contents of the flask, and again cool. Add cautiously 1.0 g of zinc metal dust, 2.0 g of sodium thiosulfate, and 18.0 g of sodium hydroxide pellets, and without delay connect the flask to a Kjeldahl connecting bulb (trap), previously attached to a condenser, the delivery tube of which dips beneath the surface of 50 mL of boric acid solution (1 in 25) contained in a 300-mL conical flask. Mix the contents of the Kjeldahl flask by gentle rotation, and distil about 125 mL into the receiver. Add not less than 3 drops of methyl red–methylene blue TS to the contents of the receiving vessel, and determine the ammonia by titration with 0.1 *N* hydrochloric acid VS. Perform a blank determination, and make any necessary correction. Each mL of 0.1 *N* hydrochloric acid is equivalent to 1.401 mg of nitrogen.

Isobutane

$CH_3CH(CH_3)CH_3$

C_4H_{10} 58.12

» Isobutane contains not less than 95.0 percent of isobutane (C_4H_{10}).

Caution—Isobutane is highly flammable and explosive.

Packaging and storage—Preserve in tight cylinders, and prevent exposure to excessive heat.

Identification—

A: The infrared absorption spectrum of it exhibits maxima, among others, at about the following wavelengths, in μm: 3.4 (*vs*), 6.8 (*s*), 7.2 (*m*), 8.5 (*m*), and 10.9 (*m*).

B: The vapor pressure of a test specimen, obtained as directed in the *General Sampling Procedure* under *Propellants* (see *Aerosols* ⟨601⟩), determined at 21° by means of a suitable pressure gauge, is between 303 and 331 kPa absolute (44 and 48 psia).

Water: not more than 0.001%, determined as directed for *Water Content* under *Aerosols* ⟨601⟩.

High-boiling residues: not more than 5 ppm, determined as directed for *High-boiling Residues, Method II*, under *Aerosols* ⟨601⟩.

Acidity of residue—Add 10 mL of water to the residue obtained in the test for *High-boiling residues*, mix by swirling for about 30 seconds, add 2 drops of methyl orange TS, insert the stopper in the tube, and shake vigorously: no pink or red color appears in the aqueous layer.

Sulfur compounds—Carefully open the container valve to produce a moderate flow of gas. Do not direct the gas stream toward the face, but deflect a portion of the stream toward the nose: the odor is free from the characteristic odor of sulfur compounds.

Assay—

Chromatographic system—Under typical conditions, the gas chromatograph is equipped with a thermal-conductivity detector and contains a 6-m × 3-mm aluminum column packed with 10 weight percent of liquid phase G30 on non-acid-washed support S1C, helium is used as the carrier gas at a flow rate of 50 mL per minute, and the temperature of the column is maintained at 33°.

System suitability—The peak responses obtained for Isobutane in the chromatograms from duplicate determinations agree within 1%.

Procedure—Connect 1 Isobutane cylinder to the chromatograph through a suitable sampling valve and a flow control valve downstream from the sampling valve. Flush the liquid specimen through the sampling valve, taking care to avoid entrapment of gas or air in the sampling valve. Inject a suitable volume, typically 2 µL, of Isobutane into the chromatograph, and record the chromatogram. Calculate the percentage purity by dividing 100 times the Isobutane response by the sum of all of the responses in the chromatogram.

Isopropyl Alcohol—*see* Isopropyl Alcohol USP

Isopropyl Myristate

$CH_3(CH_2)_{12}COOCH(CH_3)_2$

$C_{17}H_{34}O_2$ 270.45
Tetradecanoic acid, 1-methylethyl ester.
Isopropyl myristate [*110-27-0*].

» Isopropyl Myristate consists of esters of isopropyl alcohol and saturated high molecular weight fatty acids, principally myristic acid. It contains not less than 90.0 percent of $C_{17}H_{34}O_2$.

Packaging and storage—Preserve in tight, light-resistant containers.

Reference standards—*USP Isopropyl Myristate Reference Standard*—Do not dry before using. *USP Isopropyl Palmitate Reference Standard*—Do not dry before using.

Identification—The retention time of the major peak obtained in the *Assay* is the same as that of the corresponding peak obtained from the Standard solution employed in the *Assay*.

Specific gravity ⟨841⟩: between 0.846 and 0.854.

Refractive index ⟨831⟩: between 1.432 and 1.436 at 20°.

Residue on ignition ⟨281⟩: not more than 0.1%.

Acid value ⟨401⟩: not more than 1.

Saponification value ⟨401⟩: between 202 and 212.

Iodine value ⟨401⟩: not more than 1.

Assay—Dissolve 125 mg of Isopropyl Myristate in 25.0 mL of *n*-hexane, and mix. Introduce a 5-µL portion of this solution in to a suitable temperature-programmable gas chromatograph equipped with a flame-ionization detector. The column is 1.8-m × 4-mm, and is packed with 10 percent liquid phase G8 on 100- to 120-mesh support S1. The carrier gas is nitrogen, flowing at the rate of 45 mL per minute. The column temperature is programmed from 90° to 210° at 2° per minute, with an 8-minute upper limit interval. In a suitable chromatogram, the resolution factor, *R*, is not less than 6.0 between the peaks due to isopropyl myristate and isopropyl palmitate, their relative retention times being about 1 and 1.3, respectively, and both peaks being identified by comparison with a chromatogram concomitantly obtained on a portion of a Standard solution prepared by dissolving about 45 mg of USP Isopropyl Myristate RS and 5 mg of USP Isopropyl Palmitate RS in 10.0 mL of *n*-hexane. The relative standard deviation of five single injections of the same solution does not exceed 2.0%, and the tailing factor for the isopropyl myristate peak does not exceed 2. Measure the peak responses in the chromatograms, and determine the percentage of $C_{17}H_{34}O_2$ in the portion of Isopropyl Myristate taken by the formula:

$$100A/B,$$

in which *A* is the isopropyl myristate peak response, and *B* is the sum of the responses of all the peaks in the chromatogram except the solvent peak.

Isopropyl Palmitate

$CH_3(CH_2)_{14}COOCH(CH_3)_2$

$C_{19}H_{38}O_2$ 298.51
Hexadecanoic acid, 1-methylethyl ester.
Isopropyl palmitate [*142-91-6*].

» Isopropyl Palmitate consists of esters of isopropyl alcohol and saturated high molecular weight fatty acids. It contains not less than 90.0 percent of $C_{19}H_{38}O_2$.

Packaging and storage—Preserve in tight, light-resistant containers.

Reference standards—*USP Isopropyl Myristate Reference Standard*—Do not dry before using. *USP Isopropyl Palmitate Reference Standard*—Do not dry before using.

Identification—The retention time of the major peak obtained in the *Assay* is the same as that of the corresponding peak obtained from the Standard solution employed in the *Assay*.

Specific gravity ⟨841⟩: between 0.850 and 0.855.

Refractive index ⟨831⟩: between 1.435 and 1.438.

Residue on ignition ⟨281⟩: not more than 0.1%.

Acid value ⟨401⟩: not more than 1.

Iodine value ⟨401⟩: not more than 1.

Saponification value ⟨401⟩: between 183 and 193.

Assay—Dissolve 125 mg of Isopropyl Palmitate in 25.0 mL of *n*-hexane, and mix. Introduce a 5-µL portion of this solution into a suitable temperature-programmable gas chromatograph equipped with a flame-ionization detector. The column is 1.8-m × 4-mm, and is packed with 10 percent liquid phase G8 on 100- to 120-mesh support S1. The carrier gas is nitrogen, flowing at the rate of 45 mL per minute, as measured at the column exit. The column temperature is programmed from 90° to 210° at 2° per minute, with an 8-minute upper limit interval. In a suitable chromatogram, the resolution factor, *R*, is not less than 6.0 between the peaks due to isopropyl myristate and isopropyl palmitate, their relative retention times being about 1 and 1.3, respectively, and both peaks being identified by comparison with a chromatogram concomitantly obtained on a portion of a Standard solution prepared by dissolving about 45 mg of USP Isopropyl Palmitate RS and 5 mg of USP Isopropyl Myristate RS in 10.0 mL of *n*-hexane. The relative standard deviation of five single injections of the same solution does not exceed 2.0%, and the tailing factor for the isopropyl palmitate peak does not exceed 2. Measure the peak responses in the chromatograms, and determine the percentage of $C_{19}H_{38}O_2$ in the portion of Isopropyl Palmitate taken by the formula:

$$100A/B,$$

in which *A* is the isopropyl palmitate peak response, and *B* is the sum of the responses of all the peaks in the chromatogram except the solvent peak.

Kaolin—*see* Kaolin USP

Lactic Acid—*see* Lactic Acid USP

Lactose

(α-Lactose)

$C_{12}H_{22}O_{11}$ (anhydrous) 342.30
D-Glucose, 4-*O*-β-D-galactopyranosyl-.
Lactose [*63-42-3*].
Monohydrate 360.31 [*64044-51-5*].

» Lactose is a sugar obtained from milk. It is anhydrous or contains one molecule of water of hydration.

Packaging and storage—Preserve in well-closed containers.

Labeling—Label it to indicate whether it is anhydrous or hydrous. If it has been prepared by a spray-dried process, the label so indicates.

Clarity and color of solution—A solution of 3 g in 10 mL of boiling water is clear, colorless or nearly colorless, and odorless.

Identification—Add 5 mL of 1 *N* sodium hydroxide to 5 mL of a hot, saturated solution of Lactose, and gently warm the mixture: the liquid becomes yellow and finally brownish red. Cool to room temperature, and add a few drops of alkaline cupric tartrate TS: a red precipitate of cuprous oxide is formed.

Specific rotation ⟨781⟩: between +54.8° and +55.5°, calculated on the anhydrous basis, determined at 20° in a solution containing 10 g of Lactose and 0.2 mL of 6 *N* ammonium hydroxide in each 100 mL.

Microbial limits ⟨61⟩—The total bacterial count does not exceed 100 per g, and the tests for *Salmonella* species and *Escherichia coli* are negative.

Acidity or alkalinity—Dissolve 30 g by heating in 100 mL of carbon dioxide–free water, and add 10 drops of phenolphthalein TS: the solution is colorless, and not more than 1.5 mL of 0.1 *N* sodium hydroxide is required to produce a red color.

Water, *Method I* ⟨921⟩: not more than 1.0% for the anhydrous form, and not more than 5.5% for the hydrous form.

Residue on ignition ⟨281⟩: not more than 0.1%.

Alcohol-soluble residue—Add 10 g of very finely powdered Lactose to 40 mL of alcohol, and shake for 10 minutes. Filter, evaporate 10 mL of the filtrate to dryness, and dry at 100° for 10 minutes: the weight of the residue does not exceed 20 mg.

Heavy metals ⟨231⟩—Dissolve 4 g in 20 mL of warm water, add 1 mL of 0.1 *N* hydrochloric acid, and dilute with water to 25 mL: the limit is 5 ppm.

Lanolin—*see* Lanolin USP

Lanolin Alcohols

» Lanolin Alcohols is a mixture of aliphatic alcohols, triterpenoid alcohols, and sterols, obtained by the hydrolysis of Lanolin. It contains not less than 30.0 percent of cholesterol. It may contain not more than 0.1 percent of a suitable antioxidant.

Packaging and storage—Preserve in well-closed, light-resistant containers, preferably at controlled room temperature.

Identification—Dissolve 0.5 g in 5 mL of chloroform, and add 1 mL of acetic anhydride and 2 drops of sulfuric acid: a green color is produced.

Melting range, *Class II* ⟨741⟩: not below 56°.

Acidity and alkalinity—Boil 10 g with 100 mL of water for 5 minutes, with frequent stirring. Remove the source of heat, add 0.5 mL of phenolphthalein TS, and stir: no pink color is produced. Add 0.5 mL of methyl orange TS, and stir: no red color is produced.

Loss on drying ⟨731⟩—Dry it at 105° for 1 hour: it loses not more than 0.5% of its weight.

Residue on ignition ⟨281⟩: not more than 0.15%.

Copper—Heat 5.0 g over a small flame until charred, ignite the residue at about 550°, and dissolve the ash in 5 mL of hydrochloric acid, with the aid of heat. Cool, dilute with water, render alkaline with ammonium hydroxide, boil to remove the excess of ammonia, add a few drops of bromine TS, again boil, and filter. To the filtrate add 1 mL of sodium diethyldithiocarbamate solution (1 in 1000), a few drops of 6 *N* ammonium hydroxide, and sufficient water to bring the volume to 50 mL. The resulting color is not darker than that produced by adding 1 mL of the sodium diethyldithiocarbamate solution and a few drops of 6 *N* ammonium hydroxide to 2.5 mL of a 0.00393% solution of cupric sulfate, and diluting with water to 50 mL (5 ppm).

Acid value ⟨401⟩: not more than 2.

Saponification value ⟨401⟩: not more than 12, 5 g of Lanolin Alcohols being refluxed with the alcoholic potassium hydroxide for 4 hours.

Assay—Melt 20 g of Lanolin Alcohols on a water bath, mix, and allow to cool. Dissolve about 1 g, accurately weighed, in 25 mL of warm (about 60°) 90 percent alcohol, filter while still warm through a medium-porosity, sintered-glass filter, and wash the residue with 50 mL of the warm 90 percent alcohol. Cool the combined filtrate and washings, transfer to a 100-mL volumetric flask with the aid of 90 percent alcohol, dilute with 90 percent alcohol to volume, and mix. Pipet 10 mL of the resulting solution into a 150-mL beaker, add 40 mL of a 1 in 200 solution of digitonin in 90 percent alcohol, warm to 60°, and allow to stand for 18 hours. Filter through a medium-porosity, sintered-glass filter with the aid of gentle vacuum, wash the residue successively with 15 mL of 90 percent alcohol, 15 mL of acetone, and 15 mL of hot carbon tetrachloride, and dry at 105° to constant weight. Each g of residue is equivalent to 0.239 g of cholesterol.

Lanolin, Anhydrous—*see* Lanolin, Anhydrous USP

Lecithin

» Lecithin is a complex mixture of acetone-insoluble phosphatides, which consist chiefly of phosphatidyl choline, phosphatidyl ethanolamine, phosphatidyl serine, and phosphatidyl inositol, combined with various amounts of other substances such as triglycerides, fatty acids, and carbohydrates, as separated from the crude vegetable oil source. It contains not less than 50.0 percent of *Acetone-insoluble matter.*

Packaging and storage—Preserve in well-closed containers.

Water, *Method I* ⟨921⟩: not more than 1.5%.

Arsenic ⟨211⟩: 3 ppm.

Lead ⟨251⟩: 0.001%.

Heavy metals, *Method II* ⟨231⟩: 0.004%.

Acid value ⟨401⟩: not more than 36.

Hexane-insoluble matter—If the substance under test is plastic or semisolid, soften the Lecithin by warming it at a temperature not exceeding 60°, and then mix. Weigh 10.0 g into a 250-mL conical flask, add 100 mL of solvent hexane, and shake until solution is apparently complete or until no more of any residue seems to be dissolving. Filter through a coarse-porosity filtering funnel that previously has been heated at 105° for 1 hour, cooled, and weighed, wash the flask with two 25-mL portions of solvent hexane, and pour both washings through the funnel. Dry the funnel at 105° for 1 hour. [NOTE—Exercise due precautions, since hexane is flammable.] Cool to room temperature, and determine the gain in weight: not more than 0.3% is found.

Acetone-insoluble matter—If the substance under test is plastic or semisolid, soften the Lecithin by warming it briefly at a temperature not exceeding 60°, and then mix. Transfer about 2 g, accurately weighed, to a 40-mL centrifuge tube that previously has been tared with a stirring rod. Add 15.0 mL of acetone, melt in a water bath without evaporating the acetone, but with stirring to aid complete disintegration, and place in an ice-water bath for 5 minutes. Add acetone that previously has been chilled to 0° to 5° to the 40-mL mark on the tube, stirring during the addition. Cool in an ice-water bath for 15 minutes, stir, remove the rod, clarify by centrifuging at about 2000 rpm for 5 minutes, and decant. Break up the residue with the stirring rod, and refill the centrifuge tube to the 40-mL mark with chilled acetone, while stirring, cool in an ice-water bath for 15 minutes, stir, remove the rod, centrifuge, and decant. Break up the residue with the stirring rod. Place the tube in a horizontal position until most of the acetone has evaporated, mix again, and heat the tube and the stirring rod at 105° to constant weight. [NOTE—Exercise due precautions, since acetone is flammable.] Determine the gain in weight, and calculate the percentage of acetone-insoluble matter.

Liquid Glucose—*see* Glucose, Liquid
Magma, Bentonite—*see* Bentonite Magma

Magnesium Aluminum Silicate

» Magnesium Aluminum Silicate is a blend of colloidal montmorillonite and saponite that has been processed to remove grit and non-swellable ore components.

The requirements for viscosity and ratio of aluminum content to magnesium content differ for the several types of Magnesium Aluminum Silicate, as set forth in the accompanying table.

	Viscosity (cps)		Al content/ Mg content	
Type	Min.	Max.	Min.	Max.
IA	225	600	0.5	1.2
IB	150	450	0.5	1.2
IC	800	2200	0.5	1.2
IIA	100	300	1.4	2.8

Packaging and storage—Preserve in tight containers.

Labeling—Label it to indicate its type.

Identification—Add 2 g in small portions to 100 mL of water, with intense agitation. Allow to stand for 12 hours to ensure complete hydration. Place 2 mL of the resulting mixture on a suitable glass slide, and allow to air-dry at room temperature to produce an oriented film. Place the slide in a vacuum desiccator over a free surface of ethylene glycol. Evacuate the desiccator, and close the stopcock so that the ethylene glycol saturates the desiccator chamber. Allow to stand for 12 hours. Record the X-ray diffraction pattern (see *X-ray Diffraction* ⟨941⟩), and calculate the *d* values: the largest peak corresponds to a *d* value between 15.0 and 17.2 angstrom units. Prepare a random powder specimen of Magnesium Aluminum Silicate, record the X-ray diffraction pattern, and determine the *d* values in the region between 1.48 and 1.54 angstrom units: peaks are found between 1.492 and 1.504 angstrom units and between 1.515 and 1.545 angstrom units.

Viscosity—After determining the *Loss on drying*, weigh a quantity of Magnesium Aluminum Silicate test specimen equivalent to 25.0 g on the dried basis. Over a period of a few seconds, transfer the undried test specimen to a suitable 1-liter blender jar containing an amount of water, maintained at a temperature of 25 ± 2°, that is sufficient to produce a mixture weighing 500 g. Blend for 3 minutes, accurately timed, at 14,000 to 15,000 rpm (high speed). [NOTE—Heat generated during blending causes a temperature rise to above 30°.] Transfer the contents of the blender to a 600-mL beaker, allow to stand for 5 minutes, and adjust, if necessary, to a temperature of 33 ± 3°. Using a suitable rotational viscosimeter equipped with a spindle as specified below, operate the viscosimeter at 60 rpm for 6 minutes, accurately timed, and record the scale reading. For Type IA, use a spindle having a cylinder 1.87 cm in diameter and 0.69 cm high attached to a shaft 0.32 cm in diameter, the distance from the top of the cylinder to the lower tip of the shaft being 2.54 cm, and the immersion depth being 5.00 cm (No. 2 spindle); if the scale reading is greater than 90% of full-scale, repeat the measurement, using a spindle similar to the No. 2 spindle but having the cylinder 1.27 cm in diameter and 0.16 cm high instead (No. 3 spindle). For Type IC, use a No. 3 spindle; if the scale reading is greater than 90% of full-scale, repeat the measurement using a spindle consisting of a cylindrical shaft 0.32 cm in diameter and having an immersion depth of 4.05 cm (No. 4 spindle). For Types IB, and IIA, use a No. 2 spindle.

Microbial limits ⟨61⟩—Its total aerobic microbial count does not exceed 1000 per g, and it meets the requirements of the test for absence of *Escherichia coli*.

pH ⟨791⟩: between 9.0 and 10.0, in a suspension (5 in 100) in water.

Acid demand—After determining the *Loss on drying*, weigh a quantity of Magnesium Aluminum Silicate equivalent to 5.00 g, and disperse in 500 mL of water with the aid of a suitable blender fitted with a 1-liter jar. Using a stopwatch, designate zero time. With constant mixing, add 3.0-mL portions of 0.100 *N* hydrochloric acid at 5, 65, 125, 185, 245, 305, 365, 425, 485, 545, 605, 665, and 725 seconds, and add a 1.0-mL portion at 785 seconds. Determine the pH potentiometrically at 840 seconds: the pH is not more than 4.0.

Loss on drying ⟨731⟩—Dry it at 110° to constant weight: it loses not more than 8.0% of its weight.

Arsenic, *Method I* ⟨211⟩—Prepare the *Test Preparation* as follows. Transfer 13.3 g to a 250-mL beaker containing 100 mL of dilute hydrochloric acid (1 in 25), mix, cover with a watch glass, and boil gently, with occasional stirring, for 15 minutes without allowing excessive foaming. Allow the insoluble material to settle, and decant the hot supernatant liquid through a rapid-flow filter paper into a 200-mL volumetric flask, retaining as much sediment as possible in the beaker. Add 25 mL of hot dilute hydrochloric acid (1 in 25) to the residue in the beaker, stir, heat to boiling, allow the insoluble material to settle, and decant the supernatant liquid through the filter into the 200-mL volumetric flask. Repeat the extraction with four additional 25-mL portions of hot dilute hydrochloric acid (1 in 25), decanting each hot supernatant liquid through the filter into the volumetric flask. At the last extraction, transfer as much of the insoluble material as possible onto the filter. Cool the combined filtrates to room temperature, add dilute hydrochloric acid (1 in 25) to volume, and mix.

Procedure—Use a 25-mL aliquot of *Test Preparation* for the *Procedure.* The absorbance due to any red color from the *Test Preparation* does not exceed that produced by 5.0 mL of *Standard Preparation* (5 µg of As) when treated with the same quantities of the same reagents and in the same manner. The limit is 3 ppm.

Lead—[NOTE—The *Standard preparation* and *Test preparation* may be modified if necessary, to obtain solutions, of suitable concentrations, adaptable to the linear or working range of the instrument.]

Standard preparation—On the day of use, dilute 3.0 mL of *Lead Nitrate Stock Solution* (see *Heavy Metals* ⟨231⟩) with water to 100 mL. Each mL of the *Standard preparation* contains the equivalent of 3 µg of lead.

Test preparation—Transfer 10.0 g of Magnesium Aluminum Silicate to a 250-mL beaker containing 100 mL of dilute hydrochloric acid (1 in 25), stir, cover with a watch glass, and boil for 15 minutes. Cool to room temperature, and allow the insoluble matter to settle. Decant the supernatant liquid through a rapid-flow filter paper into a 400-mL beaker. Add 25 mL of hot water to the insoluble matter in the 250-mL beaker, stir, allow the insoluble matter to settle, and decant the supernatant liquid through the filter into the 400-mL beaker. Repeat the extraction with two additional 25-mL portions of water, decanting each supernatant liquid portion through the filter into the 400-mL beaker. Wash the filter with 25 mL of hot water, collecting this filtrate in the 400-mL beaker. Concentrate the combined extracts by gentle boiling to approximately 20 mL. If a precipitate appears, add 2 to 3 drops of nitric acid, heat to boiling, and cool to room temperature. Filter the concentrated extracts through a rapid-flow filter paper into a 50-mL volumetric flask. Transfer the remaining contents of the 400-mL beaker through the filter paper and into the flask with water. Dilute with water to volume, and mix.

Procedure—Determine the absorbances of the *Test preparation* and the *Standard preparation* at 284 nm in a suitable atomic absorption spectrophotometer equipped with a lead hollow-cathode lamp, deuterium arc background correction, and a single-slot burner, using an oxidizing flame of air and acetylene. The absorbance of the *Test preparation* is not greater than that of the *Standard preparation* (0.0015%).

Assay for aluminum and magnesium content—[NOTE—The *Standard preparations* and *Assay preparations* may be diluted quantitatively with water, if necessary, to obtain solutions, of suitable concentrations, adaptable to the linear or working range of the instrument.]

Lanthanum solution—Stir 88.30 g of lanthanum chloride ($LaCl_3$) with 500 mL of 6 *N* hydrochloric acid until solution is complete, transfer with the aid of water to a 1000-mL volumetric flask, and dilute with water to volume.

Specimen preparation—Transfer 0.200 g of Magnesium Aluminum Silicate to a 25-mL platinum crucible containing 1.0 g of lithium metaborate, and mix. Using a muffle furnace or a suitable burner, heat slowly at first, and ignite at 1000° to 1200° for 15 minutes. Cool, place the crucible in a 100-mL beaker containing 25 mL of dilute nitric acid (1 in 20), and add an additional 50 mL of the dilute acid, filling and submerging the upright crucible. Place a polyfluorocarbon-coated magnetic stirring bar into the crucible, and stir gently with a magnetic stirrer until solution is complete. Pour the contents into a 250-mL beaker, and remove the crucible. Warm the solution, and transfer through a rapid-flow filter paper with the aid of water into a 200-mL volumetric flask, dilute with water to volume, and mix.

Aluminum standard preparations—Dissolve 1.000 g of aluminum in a mixture of 10 mL of hydrochloric acid and 10 mL of water by gentle heating. Transfer the solution to a 1000-mL volumetric flask, dilute with water to volume, and mix. This solution contains the equivalent of 1 mg of aluminum per mL. Transfer 2-, 5-, and 10-mL aliquots to separate 100-mL volumetric flasks containing 200 mg of sodium chloride, dilute each with water to volume, and mix.

Aluminum assay preparation—Dissolve 100 mg of sodium chloride in a 50-mL aliquot of *Specimen preparation.*

Procedure for aluminum—In a suitable atomic absorption spectrophotometer equipped with an aluminum hollow-cathode lamp and a single-slot burner, using an oxidizing acetylene-air-nitrous oxide flame, determine the absorbances of the *Aluminum assay preparation* and each of the *Aluminum standard preparations* at 309 nm. From a linear regression equation calculated from the absorbances and concentrations of the *Aluminum standard preparations*, determine the aluminum content of the Magnesium Aluminum Silicate.

Magnesium standard preparations—Place 1.000 g of magnesium in a 250-mL beaker containing 20 mL of water, and carefully add 20 mL of hydrochloric acid, warming, if necessary, to complete the reaction. Transfer the solution to a 1000-mL volumetric flask, dilute with water to volume, and mix. This solution contains the equivalent of 1 mg of magnesium per mL. Transfer 5-, 10-, 15-, and 20-mL aliquots to separate 100-mL volumetric flasks. To each flask add 20.0 mL of *Lanthanum solution*, dilute with water to volume, and mix.

Magnesium assay preparation—Transfer a 25-mL aliquot of *Specimen preparation* to a 50-mL volumetric flask, dilute with water to volume, and mix. Transfer a 5.0-mL aliquot of this dilution to a 100-mL volumetric flask, add 20.0 mL of *Lanthanum solution*, dilute with water to volume, and mix.

Procedure for magnesium—In a suitable atomic absorption spectrophotometer equipped with a magnesium hollow-cathode lamp and a single-slot burner, using a reducing acetylene-air flame, determine the absorbances of the *Magnesium assay preparation* and each of the *Magnesium standard preparations* at 285 nm. From a linear regression equation calculated from the absorbances and concentrations of the *Magnesium standard preparations*, determine the magnesium content of the Magnesium Aluminum Silicate.

Magnesium Silicate

» Magnesium Silicate is a compound of magnesium oxide and silicon dioxide. It contains not less than 15.0 percent of magnesium oxide (MgO) and not less than 67.0 percent of silicon dioxide (SiO_2), calculated on the ignited basis.

Packaging and storage—Preserve in well-closed containers.

Identification—

A: Mix about 500 mg with 10 mL of 3 *N* hydrochloric acid, filter, and neutralize the filtrate to litmus paper with 6 *N* ammonium hydroxide: the neutralized filtrate responds to the tests for *Magnesium* ⟨191⟩.

B: Prepare a bead by fusing a few crystals of sodium ammonium phosphate on a platinum loop in the flame of a Bunsen burner. Place the hot, transparent bead in contact with Magnesium Silicate, and again fuse: silica floats about in the bead, producing, upon cooling, an opaque bead with a web-like structure.

pH ⟨791⟩: between 7.0 and 10.8, determined in a well-mixed aqueous suspension (1 in 10).

Loss on drying ⟨731⟩—Dry it at 105° for 2 hours: it loses not more than 15.0% of its weight. (Retain the dried specimen for the test for *Loss on ignition.*)

Loss on ignition ⟨733⟩—Ignite the specimen retained from the test for *Loss on drying* at 900° to 1000° for 20 minutes: the previously dried specimen loses not more than 15% of its weight.

Soluble salts—Boil 10.0 g with 150 mL of water for 15 minutes. Cool to room temperature, allow the mixture to stand for 15 minutes, filter with the aid of suction, transfer the filtrate to a 200-mL volumetric flask, dilute with water to volume, and mix. Evaporate 50.0 mL of this solution, representing 2.5 g of the Silicate, in a tared platinum dish to dryness, and ignite gently to constant weight: the weight of the residue does not exceed 75.0 mg (3.0%).

Fluoride—

Indicator solution—Prepare a solution containing 100 mg of lanthanum alizarin complexan mixture per mL in 60 percent isopropyl alcohol. Filter the solution, if it is not clear.

Test preparation—Prepare a slurry consisting of 5.0 g of Magnesium Silicate and 45 mL of 0.1 *N* hydrochloric acid, stir at room temperature for 15 minutes, and filter through a 0.45-μm filter into a 50-mL volumetric flask. Wash the filter with five 1-mL portions of 0.1 *N* hydrochloric acid, collecting the washings in the flask, dilute with 0.1 *N* hydrochloric acid to volume, and mix.

Procedure—Transfer 5.0 mL of the *Test preparation* to a 25-mL volumetric flask, add 5.0 mL of *Indicator solution*, dilute with water to volume, mix, and allow to stand for 1 hour in diffuse light at ambient temperature. Determine the absorbance of this solution in a 1-cm cell with a suitable spectrophotometer, at the wavelength of maximum absorbance at about 620 nm, against a blank consisting of 5.0 mL of 0.1 *N* hydrochloric acid, 5.0 mL of *Indicator solution*, and 15.0 mL of water. The absorbance is not greater than that produced by 5.0 mL of a solution containing 2.21 μg of sodium fluoride per mL of 0.1 *N* hydrochloric acid, when treated in the same manner as the *Test preparation* (10 ppm).

Free alkali—Add 2 drops of phenolphthalein TS to 20 mL of the diluted filtrate prepared in the test for *Soluble salts*, representing 1 g of Magnesium Silicate: if a pink color is produced, not more than 2.5 mL of 0.1 *N* hydrochloric acid is required to discharge it.

Arsenic, *Method I* ⟨211⟩: 3 ppm.

Lead ⟨251⟩—Dissolve 1.0 g in 20 mL of 3 *N* hydrochloric acid, evaporate on a steam bath to about 10 mL, dilute with water to about 20 mL, and cool: the limit is 0.001%.

Ratio of SiO_2 to MgO—Divide the percentage of SiO_2 obtained in the *Assay for silicon dioxide* by the percentage of MgO obtained in the *Assay for magnesium oxide:* the quotient obtained is between 2.50 and 4.50.

Heavy metals ⟨231⟩—Boil 2.0 g with a mixture of 50 mL of water and 5 mL of hydrochloric acid for 20 minutes, adding water to maintain the volume during the boiling. Add ammonium hydroxide until the mixture is only slightly acid to litmus paper. Filter with the aid of suction, and wash with 15 to 20 mL of water, combining the washings with the original filtrate. Add 2 drops of phenolphthalein TS, then add a slight excess of 6 *N* ammonium hydroxide. Discharge the pink color with dilute hydrochloric acid (1 in 100), then add 8 mL of dilute hydrochloric acid (1 in 100). Dilute with water to 100 mL, and use 25 mL of the solution for the test: the limit is 0.004%.

Assay for magnesium oxide—Weigh accurately about 1.5 g, and transfer to a 250-mL conical flask. Add 50.0 mL of 1 *N* sulfuric acid VS, and digest on a steam bath for 1 hour. Cool to room temperature, add methyl orange TS, and titrate the excess acid with 1 *N* sodium hydroxide VS. Each mL of 1 *N* sulfuric acid is equivalent to 20.15 mg of MgO.

Assay for silicon dioxide—Transfer about 700 mg of Magnesium Silicate, accurately weighed, to a small platinum dish. Add 10 mL of 1 *N* sulfuric acid, and heat on a steam bath to dryness, leaving the dish uncovered. Treat the residue with 25 mL of water, and digest on a steam bath for 15 minutes. Decant the supernatant liquid through an ashless filter paper, with the aid of suction, and wash the residue, by decantation, three times with hot water, passing the washings through the filter paper. Finally transfer the residue to the filter, and wash thoroughly with hot water. Transfer the filter paper and its contents to the platinum dish previously used. Heat to dryness, incinerate, ignite strongly for 30 minutes, cool, and weigh. Moisten the residue with water, and add 6 mL of hydrofluoric acid and 3 drops of sulfuric acid. Evaporate to dryness, ignite for 5 minutes, cool, and weigh: the loss in weight represents the weight of SiO_2.

Magnesium Stearate

Octadecanoic acid, magnesium salt.
Magnesium stearate   [*557-04-0*].

» Magnesium Stearate is a compound of magnesium with a mixture of solid organic acids obtained from fats, and consists chiefly of variable proportions of magnesium stearate and magnesium palmitate. It contains the equivalent of not less than 6.8 percent and not more than 8.3 percent of MgO.

Packaging and storage—Preserve in well-closed containers.

Identification—

A: Mix 25 g with 200 mL of hot water, then add 60 mL of 2 *N* sulfuric acid, and heat the mixture, with frequent stirring, until the fatty acids separate cleanly as a transparent layer. Separate the aqueous layer, and retain it for *Identification test B*. Wash the fatty acids with boiling water until free from sulfate, collect them in a small beaker, and warm on a steam bath until the water has separated and the fatty acids are clear. Allow to cool, and discard the water layer. Then melt the acids, filter into a dry beaker while hot, and dry at 100° for 20 minutes: the solidification temperature of the fatty acids is not below 54°.

B: The aqueous layer obtained from the separated fatty acids in *Identification test A* responds to the test for *Magnesium* ⟨191⟩.

Microbial limits ⟨61⟩—The total bacterial count does not exceed 1000 per g and the test for *Escherichia coli* is negative.

Loss on drying ⟨731⟩—Dry it at 105° to constant weight: it loses not more than 4.0% of its weight.

Lead ⟨251⟩—Ignite 0.50 g in a silica crucible in a muffle furnace at 475° to 500° for 15 to 20 minutes. Cool, add 3 drops of nitric acid, evaporate over a low flame to dryness, and again ignite at 475° to 500° for 30 minutes. Dissolve the residue in 1 mL of a mixture of equal parts by volume of nitric acid and water, and wash into a separator with several successive portions of water. Add 3 mL of *Ammonium citrate solution* and 0.5 mL of *Hydroxylamine hydrochloride solution*, and render alkaline to phenol red TS with ammonium hydroxide. Add 10 mL of *Potassium cyanide solution.* Immediately extract the solution with successive 5-mL portions of *Dithizone extraction solution*, draining off each extract into another separator, until the last portion of dithizone solution retains its green color. Shake the combined extracts for 30 seconds with 20 mL of 0.2 *N* nitric acid, and discard the chloroform layer. Add to the acid solution 4.0 mL of the *Ammonia-cyanide solution* and 2 drops of *Hydroxylamine hydrochloride solution.* Add 10.0 mL of *Standard dithizone solution*, and shake the mixture for 30 seconds. Filter the chloroform layer through an acid-washed filter paper into a color-comparison tube, and compare the color with that of a standard solution prepared as follows. To 20 mL of 0.2 *N* nitric acid add 5 µg of lead, 4 mL of *Ammonia-cyanide solution* and 2 drops of *Hydroxylamine hydrochloride solution*, and shake with 10.0 mL of *Standard dithizone solution* for 30 seconds. Filter through an acid-washed filter paper into a color-comparison tube. The color of the sample solution does not exceed that in the control (0.001%).

Assay—Boil about 1 g of Magnesium Stearate, accurately weighed, with 50 mL of 0.1 *N* sulfuric acid for about 30 minutes, or until the separated fatty acid layer is clear, adding water, if necessary, to maintain the original volume. Cool, filter, and wash the filter and the flask thoroughly with water until the last washing is not acid to litmus. Neutralize the filtrate with 1 *N* sodium hydroxide to litmus. While stirring, preferably with a magnetic stirrer, titrate with 0.05 *M* disodium ethylenediaminetetraacetate VS as follows: Add about 30 mL from a 50-mL buret, then add 5 mL of ammonia–ammonium chloride buffer TS and 0.15 mL of eriochrome black TS, and continue the titration to a blue endpoint. Each mL of 0.05 *M* disodium ethylenediaminetetraacetate is equivalent to 2.015 mg of MgO.

Malic Acid

HOOC—CHOH—CH_2—COOH

$C_4H_6O_5$ 134.09
Hydroxybutanedioic acid.
Malic acid.
Hydroxysuccinic acid   [*6915-15-7*].

» Malic Acid contains not less than 99.0 percent and not more than 100.5 percent of $C_4H_6O_5$.

Packaging and storage—Preserve in well-closed containers.

Reference standards—*USP Malic Acid Reference Standard*—Do not dry before using. Keep container tightly closed and protected from light. *USP Fumaric Acid Reference Standard*—Do not dry before using. Determine the water content titrimetrically at time of use. Keep container tightly closed. *USP Maleic Acid Reference Standard*—Do not dry. Keep container tightly closed and protected from light.

Identification—The infrared absorption spectrum of a potassium bromide dispersion of it exhibits maxima only at the same wavelengths as that of a similar preparation of USP Malic Acid RS.

Specific rotation ⟨781⟩: between −0.10° and +0.10°, determined in a solution in water containing 85 mg per mL.

Residue on ignition ⟨281⟩: not more than 0.1%.

Water-insoluble substances—Dissolve 25 g in 100 mL of water, filter the solution through a tared filtering crucible, wash the filter with hot water, and dry at 100° to constant weight: the increase in weight is not more than 25 mg (0.1%).

Heavy metals, *Method II* ⟨231⟩: 0.002%.

Fumaric and maleic acids—

Mobile phase—Prepare a suitable filtered and degassed 0.01 *N* sulfuric acid in water solution.

Standard preparation—Using *Mobile phase* as a solvent, prepare a solution containing accurately known concentrations of about 0.005 mg of USP Fumaric Acid RS per mL and about 0.002 mg of USP Maleic Acid RS per mL.

Test preparation—Transfer about 100 mg, accurately weighed, of Malic Acid to a 100-mL volumetric flask, dissolve in *Mobile phase*, dilute with *Mobile phase* to volume, and mix.

Resolution solution—Using *Mobile phase* as a solvent, prepare a solution containing about 1 mg of Malic Acid per mL, about 10 µg of USP Fumaric Acid RS per mL, and about 4 µg of USP Maleic Acid RS per mL.

Chromatographic system (see *Chromatography* ⟨621⟩)—The liquid chromatograph is equipped with a 210-nm detector and a 6.5-mm × 30-cm column that contains packing L17. Maintain the temperature of the column at 37 ± 1°. The flow rate is about 0.6 mL per minute. Chromatograph the *Resolution solution*, and record the peak responses: the resolution, *R*, of the maleic acid and malic acid peaks is not less than 2.5, the resolution, *R*, of the malic acid and fumaric acid peaks is not less than 7.0, and the relative standard deviation of the maleic acid peak for replicate injections is not more than 2.0%.

Procedure—Separately inject equal volumes (about 20 µL) of the *Standard preparation* and the *Test preparation* into the chromatograph, record the chromatograms, and measure the peak responses. The relative retention times are about 0.6 for maleic acid, 1.0 for malic acid, and about 1.5 for fumaric acid. Calculate the quantities, in mg, of maleic acid and of fumaric acid in the portion of Malic Acid taken by the formula:

$$100C(r_U/r_S),$$

in which C is the concentration, in mg per mL, of the corresponding reference standard in the *Standard preparation*, and r_U and r_S are the responses of the corresponding peaks from the *Test preparation* and the *Standard preparation*, respectively. Not more than 1.0% of fumaric acid and not more than 0.05% of maleic acid are found.

Assay—Transfer about 2 g of Malic Acid, accurately weighed, to a conical flask, dissolve in 40 mL of recently boiled and cooled water, add phenolphthalein TS, and titrate with 1 *N* sodium hydroxide VS to the first appearance of a faint pink color that persists for not less than 30 seconds. Each mL of 1 *N* sodium hydroxide is equivalent to 67.04 mg of $C_4H_6O_5$.

Mannitol—*see* Mannitol USP

Menthol—*see* Menthol USP

Methacrylic Acid Copolymer

» Methacrylic Acid Copolymer is a fully polymerized copolymer of methacrylic acid and an acrylic or methacrylic ester. The assay and viscosity requirements differ for the several types, as set forth in the accompanying table.

Type	Methacrylic acid units, dried basis (%)		Viscosity (cps)	
	Min.	Max.	Min.	Max.
A	46.0	50.6	50	200
B	27.6	30.7	50	200
C	46.0	50.6	100	200

Packaging and storage—Preserve in tight containers.

Labeling—Label it to state whether it is Type A, B, or C.

Reference standards—*USP Methacrylic Acid Copolymer, Type A Reference Standard*—Dry at 110° for 6 hours before using. *USP Methacrylic Acid Copolymer, Type B Reference Standard*—Dry at 110° for 6 hours before using. *USP Methacrylic Acid Copolymer, Type C Reference Standard*—Dry at 110° for 6 hours before using.

Identification—

A: The infrared absorption spectrum of a potassium bromide dispersion of it, previously dried, exhibits maxima only at the same wavelengths as that of a similar preparation of the relevant type of USP Methacrylic Acid Copolymer RS.

B: Pour a few mL of the solution prepared for the *Viscosity* test onto a glass plate, and allow the solvent to evaporate: a clear, brittle film results.

Viscosity ⟨911⟩—Place 254.6 g of isopropyl alcohol and 7.9 g of water in a conical flask having a ground-glass joint. Add a quantity of Methacrylic Acid Copolymer, accurately weighed and equivalent to 37.5 g of solids on the dried basis, while stirring by means of a magnetic stirrer. Close the flask, and continue stirring until the polymer has dissolved completely. Adjust the temperature to 20 ± 0.1°. Equip a suitable rotational viscosimeter with a spindle having a cylinder 1.88 cm in diameter and 6.25 cm high attached to a shaft 0.32 cm in diameter, the distance from the top of the cylinder to the lower tip of the shaft being 0.75 cm, and the immersion depth being 8.15 cm (No. 1 spindle). With the spindle rotating at 30 rpm, immediately observe and record the scale reading. Convert the scale reading to centipoises by multiplying the reading by the constant for the viscosimeter spindle and speed employed.

Loss on drying ⟨731⟩—Dry it at 110° for 6 hours: it loses not more than 5.0% of its weight.

Residue on ignition ⟨281⟩: not more than 0.1%.

Arsenic, *Method II* ⟨211⟩: 2 ppm.

Heavy metals, *Method II* ⟨231⟩: 0.002%.

Monomers—

pH 2.0 phosphate buffer, fortieth-molar—Prepare an aqueous solution containing 3.550 g of dibasic sodium phosphate (Na_2HPO_4) and 3.400 g of monobasic potassium phosphate (KH_2PO_4) per liter. Adjust by the addition of phosphoric acid to a pH of 2.0.

Mobile phase—Prepare a solution in methanol to contain 700 mL of *pH 2.0 phosphate buffer, fortieth-molar* per liter.

Standard preparation—Prepare a solution in methanol to contain an accurately known concentration of about 2.4 µg per mL each of methacrylic acid and either methyl methacrylate (for Type A and Type B) or ethyl acrylate (for Type C). To 50.0 mL of this solution, add 25.0 mL of water, and mix.

Test preparation—Dissolve about 40 mg of Methacrylic Acid Copolymer, accurately weighed, in 50.0 mL of methanol, add 25.0 mL of water, and mix.

Chromatographic system (see *Chromatography* ⟨621⟩)—The liquid chromatograph is equipped with a 202-nm detector and a 4-mm × 10-cm column that contains 10 µm packing L1. The flow rate is about 2.5 mL per minute. Chromatograph the *Standard preparation*, and record the peak responses as directed under *Procedure:* the resolution, *R*, of each pair of analytes is not less than 2.0, and the relative standard deviation for replicate injections is not more than 2% for each analyte.

Procedure—Separately inject equal volumes (about 50 µL) of the *Standard preparation* and the *Test preparation* into the chromatograph, record the chromatograms, and measure the responses for the major peaks. The capacity factors, *k′*, for methacrylic acid, ethyl acrylate, and methyl methacrylate are 1.7, 4.3, and 4.8, respectively. Calculate the quantity, in μg, of each monomer in the portion of Methacrylic Acid Copolymer taken by the formula:

$$75C(r_U/r_S),$$

in which C is the concentration, in μg per mL, of the monomer in the *Standard preparation*, and r_U and r_S are the peak responses of the monomer obtained from the *Test preparation* and the *Standard preparation*, respectively. The total amount of monomers found is not more than 0.3%.

Assay—Dissolve about 1 g of Methacrylic Acid Copolymer, previously dried and accurately weighed, in 100 mL of neutral acetone, add 1 drop of phenolphthalein TS, and titrate with 0.1 *N* sodium hydroxide VS until a pink color persists for 15 seconds. Each mL of 0.1 *N* sodium hydroxide is equivalent to 8.609 mg of methacrylic acid ($C_4H_6O_2$) units.

Methyl Alcohol

CH_3OH

CH_4O 32.04
Methanol.
Methanol [*67-56-1*].

» Methyl Alcohol contains not less than 99.5 percent of CH_3OH.
Caution—Methyl Alcohol is poisonous.

Packaging and storage—Preserve in tight containers, remote from heat, sparks, and open flames.

Identification—The infrared absorption spectrum of a 1 in 50 solution of it in carbon tetrachloride, determined in 0.1-mm cells, with carbon tetrachloride in the reference beam, exhibits a broad, strong band at 2.7 μm to 3.2 μm, a strong maximum at about 3.4 μm, a medium strong maximum at about 3.5 μm, a weak region of absorption between 6.6 μm and 7.6 μm, and a very strong maximum at about 9.7 μm.

Acidity —Mix 25 mL of water with 10 mL of alcohol and 0.5 mL of phenolphthalein TS, and add 0.02 *N* sodium hydroxide until a slight pink color persists after shaking for 30 seconds. Taking precautions to avoid absorption of carbon dioxide, add 19 mL (15 g) of Methyl Alcohol, mix, and titrate with 0.020 *N* sodium hydroxide: not more than 0.45 mL is required to produce a pink color.

Alkalinity (as ammonia)—Mix 28.6 mL (22.6 g) of it with 25 mL of water, add 1 drop of methyl red TS, and titrate with 0.020 *N* sulfuric acid: not more than 0.20 mL is required to produce a pink color (3 ppm).

Water, *Method I* ⟨921⟩: not more than 0.1%.

Nonvolatile residue—Evaporate 250 mL of it in a 600-mL beaker on a steam bath, in a well-ventilated hood, until the volume is reduced to about 100 mL. Cool, transfer a portion of the liquid to a suitable, tared 50-mL platinum dish on a steam bath, evaporate, repeat the process until all of the liquid has been transferred, and evaporate to dryness. Dry at 105° for 30 minutes, cool, and weigh: the weight of the residue does not exceed 2 mg, corresponding to not more than 0.001% (w/w).

Readily carbonizable substances ⟨271⟩—Cool 5 mL of sulfuric acid TS, contained in a small conical flask, to 10°, and add 5 mL of Methyl Alcohol dropwise with constant mixing, maintaining the temperature below 20° throughout the test: no discoloration develops.

Readily oxidizable substances—Cool 20 mL of it to 15°, add 0.1 mL of 0.1 *N* potassium permanganate, and allow to stand at 15°: the pink color does not completely disappear within 5 minutes.

Acetone and aldehydes (as acetone)—
Standard preparation—Dilute 1.9 mL (1.5 g) of acetone with water to 1000 mL, then dilute 1.0 mL of this solution with water to 100 mL, and mix. Dilute 2 mL of the resulting solution with water to 5 mL, and mix. The *Standard preparation* contains 30 μg of acetone, and is freshly prepared.

Test preparation—Dilute 1.25 mL (1 g) of it with water to 5 mL, and mix.

Procedure—Adjust and maintain each solution at 20°. Add 5 mL of alkaline mercuric–potassium iodide TS to the *Standard preparation* and to the *Test preparation*, and mix: any turbidity produced in the *Test preparation* is not greater than that produced in the *Standard preparation* (0.003%).

Assay—Under typical conditions, the instrument is equipped with a flame-ionization detector and contains a 2-m × 3-mm stainless steel column packed with 50- to 80-mesh S4. The temperatures of the column, the injection port, and the detector are maintained at 140°, 220°, and 250°, respectively; and dry nitrogen is used as the carrier gas, at a flow rate of 20 mL per minute. Inject about 1 μL of Methyl Alcohol, and determine the peak responses by a convenient means. The retention time of methyl alcohol is about 2.5 minutes and that of acetone is about 7 minutes. Calculate the percentage of CH_4O in the Methyl Alcohol by dividing the response due to the methyl alcohol by the sum of the responses for all the peaks, and multiplying by 100.

Methyl Isobutyl Ketone

$CH_3CH(CH_3)CH_2COCH_3$

$C_6H_{12}O$ 100.16
2-Pentanone, 4-methyl-.
4-Methyl-2-pentanone [*108-10-1*].

» Methyl Isobutyl Ketone contains not less than 99.0 percent of $C_6H_{12}O$.

Packaging and storage—Preserve in tight containers.

Identification—The infrared absorption spectrum of a thin film of it between sodium chloride crystals exhibits maxima, among others, at the following wavelengths, in μm: 5.81 (*vs*), 6.80 (*m*), 7.00 (*m*), 7.09 (*m*), 7.29 (*vs*), 7.72 (*m*), 8.06 (*m*), 8.31 (*sh*), 8.53 (*s*), and 8.91 (*m*).

Specific gravity ⟨841⟩: not more than 0.799, indicating not less than 99.0% of $C_6H_{12}O$.

Distilling range, *Method I* ⟨721⟩: between 114° and 117°, a correction factor of 0.046° per mm being applied as necessary.

Acidity—Mix 15.0 mL with 15 mL of neutralized alcohol, add phenolphthalein TS, and titrate with 0.050 *N* sodium hydroxide: not more than 0.40 mL is required for neutralization.

Nonvolatile residue—Evaporate 50 mL in a tared porcelain dish on a steam bath, and dry at 105° for 1 hour: the weight of the residue does not exceed 4 mg (0.008%).

Methyl Salicylate

COOCH3
OH

$C_8H_8O_3$ 152.15
Benzoic acid, 2-hydroxy-, methyl ester.
Methyl salicylate [*119-36-8*].

» Methyl Salicylate is produced synthetically or is obtained by maceration and subsequent distillation with steam from the leaves of *Gaultheria procumbens* Linné (Fam. Ericaceae) or from the bark of *Betula lenta* Linné (Fam. Betulaceae). It contains not less than 98.0 percent and not more than 100.5 percent of $C_8H_8O_3$.

Packaging and storage—Preserve in tight containers.

Labeling—Label it to indicate whether it was made synthetically or distilled from either of the plants mentioned above.

Solubility in 70 percent alcohol—One volume of synthetic Methyl Salicylate dissolves in 7 volumes of 70 percent alcohol. One volume of natural Methyl Salicylate dissolves in 7 volumes of 70 percent alcohol, the solution having not more than a slight cloudiness.

Identification—Shake 1 drop with about 5 mL of water, and add 1 drop of ferric chloride TS: the resulting mixture has a deep violet color.

Specific gravity ⟨841⟩: between 1.180 and 1.185 for the synthetic variety; between 1.176 and 1.182 for the natural variety.

Angular rotation ⟨781⟩—Synthetic Methyl Salicylate and that from betula are optically inactive. Methyl Salicylate from gaultheria is slightly levorotatory, the angular rotation not exceeding −1.5° in a 100-mm tube.

Refractive index ⟨831⟩: between 1.535 and 1.538 at 20°.

Heavy metals, *Method II* ⟨231⟩: 0.004%.

Assay—Place about 2 g of Methyl Salicylate, accurately weighed, in a flask, add 40.0 mL of 1 *N* sodium hydroxide VS, and boil gently under a reflux condenser for 2 hours. Cool, rinse the condenser and the sides of the flask with a few mL of water, add phenolphthalein TS, and titrate the excess alkali with 1 *N* sulfuric acid VS. Perform a blank determination (see *Residual Titrations* under *Titrimetry* ⟨541⟩). Each mL of 1 *N* sodium hydroxide corresponds to 152.2 mg of $C_8H_8O_3$.

Methylcellulose—*see* Methylcellulose USP

Methylene Chloride

CH_2Cl_2 84.93
Methane, dichloro-.
Dichloromethane [*75-09-2*].

» Methylene Chloride contains not less than 99.0 percent of CH_2Cl_2.

Caution—Perform all steps involving evaporation of methylene chloride in a well-ventilated fume hood.

Packaging and storage—Preserve in tight containers.

Identification—Place about 5 mL into a glass-stoppered, 10-mL conical flask, and shake for several minutes. Remove the stopper, quickly withdraw a portion of the vapor into a 50-mL syringe that is not fitted with a needle, and inject the vapor into a suitable evacuated gas cell: the infrared absorption spectrum of the vapor shows strong doublet peaks at 7.8 μm and 7.9 μm and at 13.2 μm and 13.4 μm, and relatively few minor peaks.

Specific gravity ⟨841⟩: between 1.318 and 1.322.

Distilling range, *Method I* ⟨721⟩: between 39.5° and 40.5°.

Water, *Method I* ⟨921⟩: not more than 0.02%.

Hydrogen chloride—Into each of 2 glass-stoppered, 50-mL color-comparison cylinders having an internal diameter of 20 mm place 10 mL of water, 2 drops of phenolphthalein TS, and sufficient 0.010 *N* sodium hydroxide to produce a pink color that persists after vigorous shaking for 30 seconds, and is of equal intensity in each cylinder. [NOTE—In the following steps, take special care to avoid contamination with carbon dioxide.] Into one of the cylinders place 20.0 mL of Methylene Chloride and 0.70 mL of 0.010 *N* sodium hydroxide, and again shake. The pink color in the test cylinder is at least as intense as that in the comparison cylinder, and the color persists for not less than 15 minutes (0.001%).

Nonvolatile residue—Evaporate 50 g in a platinum or porcelain dish on a steam bath, and dry at 105° for 30 minutes: the weight of the residue does not exceed 1 mg (0.002%).

Heavy metals, *Method I* ⟨231⟩—Evaporate 15 mL (20 g) in a glass evaporating dish on a steam bath to dryness. Cool, add 2 mL of hydrochloric acid, and slowly evaporate again on a steam bath to dryness. Dissolve the residue in 1 mL of 1 *N* acetic acid, add 24 mL of water, and mix: the limit is 1 ppm.

Free chlorine—To 10 mL add 10 mL of water and 0.1 mL of potassium iodide TS, shake for 2 minutes, and allow the liquids to separate: the lower layer does not show a violet tint.

Assay—

Chromatographic system—Under typical conditions, the instrument is equipped with a thermal conductivity detector, and contains a 1.8-m × 4-mm column packed with 15 percent liquid phase G18 on 30- to 60-mesh S1C unsilanized support. The temperatures of the column, the injection port, and the detector are maintained at 60°, 200°, and 250°, respectively; and helium is used as the carrier gas, at a flow rate of about 20 mL per minute.

System suitability—Chromatograph five 1-μL injections of a mixture of 3 mL of methylene chloride with 7 mL of chloroform. The relative standard deviation of the peak response ratio does not exceed 2%, the resolution factor is not less than 4.0, and the tailing factor is not more than 1.4 (see *Chromatography* ⟨621⟩).

Procedure—Inject about 1 μL of Methylene Chloride, and determine the peak responses by any convenient means. The order of elution is amylenes (5 or 6 peaks), if present, and then methylene chloride. Calculate the percentage of CH_2Cl_2 in the Methylene Chloride by dividing the response due to the methylene chloride by the sum of the responses for all the peaks and multiplying by 100.

Methylparaben

HO–C_6H_4–$COOCH_3$

$C_8H_8O_3$ 152.15
Benzoic acid, 4-hydroxy-, methyl ester.
Methyl *p*-hydroxybenzoate [*99-76-3*].

» Methylparaben contains not less than 99.0 percent and not more than 100.5 percent of $C_8H_8O_3$, calculated on the dried basis.

Packaging and storage—Preserve in well-closed containers.

Reference standard—*USP Methylparaben Reference Standard*—Dry over silica gel for 5 hours before using.

Identification—The infrared absorption spectrum of a mineral oil dispersion of it, previously dried, exhibits maxima only at the same wavelengths as that of a similar preparation of USP Methylparaben RS.

Melting range ⟨741⟩: between 125° and 128°.

Other requirements—It meets the requirements of the tests for *Acidity, Loss on drying,* and *Residue on ignition* under *Butylparaben.*

Assay—Proceed with Methylparaben as directed in the *Assay* under *Butylparaben.* Each mL of 1 *N* sodium hydroxide is equivalent to 152.2 mg of $C_8H_8O_3$.

Methylparaben Sodium

» Methylparaben Sodium contains not less than 98.5 percent and not more than 101.5 percent of $C_8H_7NaO_3$, calculated on the anhydrous basis.

Packaging and storage—Preserve in tight containers.

Reference standard—*USP Methylparaben Reference Standard*—Dry over silica gel for 5 hours before using.

Completeness of solution ⟨641⟩—One g of it, dissolved in water, meets the requirement.

Identification—

A: Dissolve 0.5 g in 5 mL of water, acidify with hydrochloric acid, and filter the resulting precipitate. Wash the precipitate with water, and dry it over silica gel for 5 hours: the infrared absorption spectrum of a mineral oil dispersion of it exhibits maxima only at the same wavelengths as that of a similar preparation of USP Methylparaben RS.

B: Ignite about 0.3 g, cool, and dissolve the residue in about 3 mL of 3 *N* hydrochloric acid. A platinum wire dipped in this solution imparts an intense, persistent yellow color to a nonluminous flame.

pH ⟨791⟩: between 9.5 and 10.5, in a solution (1 in 1000).

Water, *Method I* ⟨921⟩: not more than 5.0%.

Chloride ⟨221⟩—A 0.2-g portion shows no more chloride than corresponds to 0.10 mL of 0.020 *N* hydrochloric acid (0.035%).

Sulfate ⟨221⟩—A 0.25-g portion shows no more sulfate than corresponds to 0.30 mL of 0.020 *N* sulfuric acid (0.12%).

Assay—Gently reflux about 100 mg of Methylparaben Sodium, accurately weighed, with 30 mL of 1 *N* sodium hydroxide for 30 minutes. Cool, add 25.0 mL of potassium bromate solution (2.78 in 500), 5 mL of potassium bromide solution (1 in 8), and 10 mL of hydrochloric acid, and immediately insert the stopper into the flask. Cool, shake for 15 minutes, and allow to stand for 15 minutes. Quickly add 15 mL of potassium iodide TS, taking care to avoid the escape of bromine vapor, at once replace the stopper in the flask, and shake vigorously. Rinse the stopper and the neck of the flask with a small quantity of water, and titrate the liberated iodine with 0.1 *N* sodium thiosulfate VS, adding 3 mL of starch TS as the end-point is approached. [NOTE—About 15 mL is needed.] Perform a blank determination (see *Residual Titrations* under *Titrimetry* ⟨541⟩), and note the difference in volumes required. Each mL of the difference in volume of 0.1 *N* sodium thiosulfate is equivalent to 2.902 mg of $C_8H_7NaO_3$.

Microcrystalline Cellulose—*see* Cellulose, Microcrystalline

Microcrystalline Wax—*see* Wax, Microcrystalline

Mineral Oil—*see* Mineral Oil USP

Light Mineral Oil

» Light Mineral Oil is a mixture of liquid hydrocarbons obtained from petroleum. It may contain a suitable stabilizer.

Packaging and storage—Preserve in tight containers.

Labeling—Label it to indicate the name of any substance added as a stabilizer, and label packages intended for direct use by the public to indicate that it is not intended for internal use.

Specific gravity ⟨841⟩: between 0.818 and 0.880.

Viscosity ⟨911⟩—It has a kinematic viscosity of not more than 33.5 centistokes at 40°.

Neutrality, Readily carbonizable substances, Limit of polynuclear compounds, and Solid paraffin—It meets the requirements of the tests for *Neutrality*, *Readily carbonizable substances*, *Limit of polynuclear compounds*, and *Solid paraffin* under *Mineral Oil*.

Monobasic Potassium Phosphate—*see* Potassium Phosphate, Monobasic

Monobasic Sodium Phosphate—*see* Sodium Phosphate, Monobasic USP

Mono- and Di-glycerides

» Mono- and Di-glycerides is a mixture of glycerol mono- and di-esters, with minor amounts of tri-esters, of fatty acids from edible oils. It contains not less than 40.0 percent of monoglycerides. The monoglyceride content, is not less than 90.0 percent and not more than 110.0 percent of the value indicated in the labeling. It may contain suitable stabilizers.

Packaging and storage—Preserve in tight, light-resistant containers.

Labeling—The labeling indicates the monoglyceride content, hydroxyl value, iodine value, saponification value, and the name and quantity of any stabilizers.

Reference standards—*USP Monoglycerides Reference Standard*—Do not dry before using. Keep container tightly closed and protected from light. *USP Stearic Acid Reference Standard*—Keep container tightly closed. Do not dry before using. *USP Palmitic Acid Reference Standard*—Keep container tightly closed. Do not dry before using.

Residue on ignition ⟨281⟩: not more than 0.1%.

Arsenic, *Method II* ⟨211⟩: 3 ppm.

Heavy metals, *Method II* ⟨231⟩: 0.001%.

Acid value ⟨401⟩: not more than 4.

Hydroxyl value ⟨401⟩: not less than 90.0% and not more than 110.0% of the value indicated in the labeling.

Iodine value ⟨401⟩: not less than 90.0% and not more than 110.0% of the value indicated in the labeling.

Saponification value ⟨401⟩: not less than 90.0% and not more than 110.0% of the value indicated in the labeling.

Free glycerin—

Propionating reagent—Mix 10 mL of pyridine with 20 mL of propionic anhydride.

Internal standard solution—Dissolve a suitable quantity of tributyrin, accurately weighed, in chloroform, and dilute quantitatively with chloroform to obtain a solution having a concentration of about 2.0 mg per mL.

Standard preparation—Transfer about 15 mg of glycerin and about 50 mg of tributyrin, both accurately weighed, to a glass-stoppered, 25-mL conical flask, add 3 mL of *Propionating reagent*, and heat at 75° for 30 minutes. Evaporate most of the unreacted *Propionating reagent* without heat, but with the aid of a stream of nitrogen, add about 12 mL of chloroform, and mix.

Test preparation—Transfer about 50 mg of Mono- and Diglycerides, accurately weighed, to a glass-stoppered, 25-mL conical flask, add by pipet 5 mL of *Internal standard solution*, and mix to dissolve. Volatilize the chloroform on a steam bath with the aid of a stream of nitrogen. Add 3 mL of *Propionating reagent*, and heat at 75° for 30 minutes. Evaporate most of the unreacted *Propionating reagent* with the aid of a stream of nitrogen at room temperature, add about 5 mL of chloroform, and mix.

Chromatographic system (see *Chromatography* ⟨621⟩)—Under typical conditions, the instrument is equipped with a flame-ionization detector, and contains a 2.4-m × 4-mm borosilicate glass column packed with 2 percent liquid phase G16 on 80- to 100-mesh support S1A. The column is maintained isothermally at a temperature between 190° and 200° and the injection port and detector block are maintained at about 300° and 310°, respectively, and helium is used as the carrier gas at a flow rate of about 70 mL per minute.

System suitability—Chromatograph a suitable number of injections of the *Standard preparation* as directed under *Procedure:* the resolution factor, *R*, between the peaks for the derivatized glycerin and the tributyrin is not less than 4.0, and the relative standard deviation of the ratio of their peak areas is not more than 2.0%.

Procedure—Inject a suitable portion of the *Standard preparation* into a suitable gas chromatograph, and record the chromatogram. Measure the areas under the peaks, and record the values of the areas under the tripropionin and tributyrin peaks as A_S and A_D, respectively. Calculate the response factor, F, by the formula:

$$(A_D/A_S)(W_S/W_D),$$

in which W_S and W_D are the weights, in mg, of glycerin and tributyrin, respectively, in the *Standard preparation.* Similarly inject a suitable portion of the *Test preparation*, and record the chromatogram. Measure the areas under the peaks, and record the values of the areas under the tripropionin and tributyrin peaks as a_U and a_D, respectively. Calculate the percentage of glycerin by the formula:

$$100F(a_U/a_D)(w_D/w_U),$$

in which w_D is the weight, in mg, of tributyrin in 5 mL of *Internal standard solution*, and w_U is the weight, in mg, of Mono- and Di-glycerides in the *Test preparation:* the limit is 7.0%.

Assay for monoglycerides—

Propionating reagent—Mix 10 mL of pyridine and 20 mL of propionic anhydride.

Internal standard solution—Transfer about 400 mg of hexadecyl hexadecanoate, accurately weighed, to a 100-mL volumetric flask, dissolve in chloroform, dilute with chloroform to volume, and mix.

Standard preparation—Transfer about 50 mg of USP Monoglycerides RS, accurately weighed, to a 25-mL conical flask, add by pipet 5 mL of *Internal standard solution*, and mix. When solution is complete, volatilize the chloroform on a steam bath with the aid of a stream of nitrogen. Add 3.0 mL of *Propionating reagent*, and heat at 75° for 30 minutes. Evaporate most of the unreacted *Propionating reagent* on a steam bath with the aid of a stream of nitrogen. Add 15 mL of chloroform, and swirl to dissolve the residue.

Assay preparation—Transfer about 50 mg of Mono- and Diglycerides, accurately weighed, to a 25-mL conical flask, and proceed as directed for *Standard preparation*, beginning with "add by pipet 5 mL of *Internal standard solution*."

Chromatographic system—Under typical conditions the instrument is equipped with a flame-ionization detector, and contains a 2.4-m × 4-mm borosilicate glass column packed with 2 percent phase G27 on 80- to 100-mesh support S1. The column is maintained isothermally at a temperature between 270° and 280°, the injection port and detector block are maintained at about 310°, and helium is used as the carrier gas at a flow rate of about 70 mL per minute.

System suitability—Chromatograph a suitable number of injections of the *Standard preparation* as directed under *Procedure:* the resolution, R, between the peaks for the derivatized glyceryl hexadecanoate and glyceryl octadecanoate, the two main components of the Reference Standard, is not less than 2.0, and the relative standard deviation of the ratio of the peak area of the derivatized glyceryl octadecanoate to that of the hexadecyl hexadecanoate is not more than 2.0%.

Procedure—Inject a suitable portion of the *Standard preparation* into a suitable gas chromatograph, and record the chromatogram. The relative retention times for the derivatized glyceryl hexadecanoate, the derivatized glyceryl octadecanoate, and the internal standard are about 1.0, 1.4, and 4.0, respectively. Measure the areas under the peaks, and record the values of the sum of the areas under the derivatized monoglyceride peaks and of the area under the hexadecyl hexadecanoate peak as A_S and A_D, respectively. Calculate the response factor, F, by the formula:

$$(A_S/A_D)(W_D/W_S),$$

in which W_D and W_S are the weights, in mg, of hexadecyl hexadecanoate and USP Monoglycerides RS, respectively, in the *Standard preparation.* Similarly inject a suitable portion of the *Assay preparation*, and record the chromatogram. Measure the areas under the peaks, and record the values of the sum of the areas under the two main derivatized monoglyceride peaks and of the area under the hexadecyl hexadecanoate peak as a_U and a_D, respectively. Calculate the content, in mg, of monoglycerides in the amount of Mono- and Di-glycerides taken by the formula:

$$(W_D/F)(a_U/a_D),$$

in which the terms are as defined therein.

Mono- and Di-acetylated Monoglycerides

» Mono- and Di-acetylated Monoglycerides is glycerin esterified with edible fat-forming fatty acids and acetic acid. It may be prepared by the inter-esterification of edible oils with triacetin or a mixture of triacetin and glycerin in the presence of catalytic agents, followed by molecular distillation, or by direct acetylation of edible monoglycerides and diglycerides with acetic anhydride with or without the use of catalysts or molecular distillation.

Packaging and storage—Preserve in tight, light-resistant containers.

Reference standard—*USP Mono- and Di-acetylated Monoglycerides Reference Standard*—Do not dry before using. Keep container tightly closed and protected from light.

Identification—Place about 50 mg, previously melted on a steam bath, in a 10-mL volumetric flask. Add carbon disulfide to volume, and shake vigorously: the infrared absorption spectrum of the solution so obtained, determined in a suitable double-beam spectrophotometer with carbon disulfide in the reference beam, exhibits maxima only at the same wavelengths as that of a similar preparation of USP Mono- and Di-acetylated Monoglycerides RS.

Residue on ignition ⟨281⟩: not more than 0.5%.

Arsenic ⟨211⟩: 3 ppm.

Heavy metals, *Method II* ⟨231⟩: 0.001%.

Acid value ⟨401⟩: not more than 3.

Hydroxyl value ⟨401⟩: between 133 and 152.

Saponification value ⟨401⟩: between 279 and 292.

Free glycerin—

Propionating reagent—Mix 10 mL of pyridine with 20 mL of propionic anhydride.

Internal standard solution—Dissolve a suitable quantity of tributyrin, accurately weighed, in chloroform, and dilute quantitatively with chloroform to obtain a solution having a concentration of about 0.5 mg per mL.

Standard preparation—Transfer about 15 mg of glycerin and about 50 mg of tributyrin, both accurately weighed, to a glass-stoppered, 25-mL conical flask, add 3 mL of *Propionating reagent*, and heat at 75° for 30 minutes in a well-ventilated fume hood. Volatilize the reagents with the aid of a stream of nitrogen, at room temperature, add about 12 mL of chloroform, and mix. Dilute about 2 mL of this mixture with chloroform to about 20 mL, and mix.

Test preparation—Transfer about 50 mg of Mono- and Diacetylated Monoglycerides, accurately weighed, to a glass-stoppered, 25-mL conical flask, add by pipet 5 mL of *Internal standard solution*, and mix to dissolve. Immerse the flask in a water bath maintained at a temperature between 45° and 50°, and volatilize the chloroform with the aid of a stream of nitrogen. Add 3 mL of *Propionating reagent*, and heat at 75° for 30 minutes in a well-ventilated fume hood. Volatilize the reagents with the aid of a stream of nitrogen, at room temperature, add about 5 mL of chloroform, and mix.

Chromatographic system (see *Chromatography* ⟨621⟩)—Under typical conditions, the instrument is equipped with a flame-ionization detector, and contains a 2.4-m × 4-mm borosilicate glass column packed with 2 percent liquid phase G16 on 80- to 100-mesh support S1A. The column is maintained isothermally at a temperature between 190° and 200° and the injection port and detector block at about 300° and 310°, respectively, and helium is used as the carrier gas at a flow rate of about 70 mL per minute.

System suitability—Chromatograph a suitable number of injections of the *Standard preparation* as directed under *Procedure:* the resolution factor, R, between the peaks for the derivatized glycerin and the tributyrin is not less than 4.0, and the relative standard deviation of the ratio of their peak areas is not more than 2.0%.

Procedure—Inject a suitable portion of the *Standard preparation* into a suitable gas chromatograph, and record the chromatogram. Measure the areas under the peaks, and record the values of the areas under the tripropionin and tributyrin peaks as P_S and P_D, respectively. Calculate the response factor, F, by the formula:

$$(P_D/P_S)(W_S/W_D),$$

in which W_S and W_D are the weights, in mg, of glycerin and of tributyrin, respectively, in the *Standard preparation.* Similarly inject a suitable portion of the *Test preparation*, and record the chromatogram. Measure the areas under the peaks, and record the values of the areas under the tripropionin and tributyrin peaks as p_U and p_D, respectively. Calculate the percentage of glycerin by the formula:

$$100F(p_U/p_D)(w_D/w_U),$$

in which w_D is the weight, in mg, of tributyrin in 5 mL of *Internal standard solution*, and w_U is the weight, in mg, of Mono- and Di-acetylated Monoglycerides in the *Test preparation.* The limit is 1.5%.

Monoethanolamine

$HOCH_2CH_2NH_2$

C_2H_7NO 61.08
Ethanol, 2-amino-.
2-Aminoethanol [*141-43-5*].

» Monoethanolamine contains not less than 98.0 percent and not more than 100.5 percent, by weight, of C_2H_7NO.

Packaging and storage—Preserve in tight, light-resistant containers.

Specific gravity ⟨841⟩: between 1.013 and 1.016.

Distilling range, *Method II* ⟨721⟩—Not less than 95% of it distils between 167° and 173°, a correction factor of 0.052° per mm being applied as necessary.

Residue on ignition ⟨281⟩: not more than 0.1%.

Assay—Accurately weigh a glass-stoppered weighing bottle containing 25 mL of water. Add 1 g of Monoethanolamine, and weigh. Transfer to a suitable flask, add a mixed indicator of bromocresol green TS and methyl red TS (5 in 6), mix, and titrate with 0.5 *N* hydrochloric acid VS. Each mL of 0.5 *N* hydrochloric acid is equivalent to 30.54 mg of C_2H_7NO.

Monosodium Glutamate

» Monosodium Glutamate contains not less than 99.0 percent and not more than 100.5 percent of $C_5H_8NNaO_4 . H_2O$.

Packaging and storage—Preserve in tight containers.

Clarity and color of solution—Dissolve 1.0 g in 10 mL of water. The solution is colorless and has no more turbidity (see *Visual Comparison* under *Spectrophotometry and Light-scattering* ⟨851⟩) than a standard mixture prepared as follows: To 0.2 mL of a solution of sodium chloride containing 10 µg of chloride ion (Cl) per mL, add 20 mL of water, and mix, then add 1 mL of 5 *N* nitric acid, 0.2 mL of dextrin solution (1 in 50), and 1 mL of silver nitrate TS, and allow to stand for 15 minutes.

Identification—

A: To 1 mL of a solution (1 in 30) add 1 mL of triketohydrindene hydrate TS and 100 mg of sodium acetate, and heat in a boiling water bath for 10 minutes: an intense, violet blue color is formed.

B: To 10 mL of a solution (1 in 10) add 5.6 mL of 1 *N* hydrochloric acid: a white, crystalline precipitate of glutamic acid is formed on standing. Precipitation is promoted by agitation. When 6 mL of 1 *N* hydrochloric acid is added to the turbid solution, the glutamic acid dissolves on stirring.

C: Prepare a 1 in 10 solution of it in 1 *N* hydrochloric acid. To 1 mL of this solution add 5 mL of cobalt-uranyl acetate TS, and agitate on a vortex mixer for 3 minutes: a golden yellow precipitate is formed, indicating the presence of sodium.

Specific rotation ⟨781⟩: between +24.8° and +25.3°, determined at 20° in a solution containing 10 g in sufficient 2 *N* hydrochloric acid to make 100 mL.

pH ⟨791⟩: between 6.7 and 7.2, in a solution (1 in 20).

Loss on drying ⟨731⟩—Dry it at 100° for 5 hours: it loses not more than 0.5% of its weight.

Chloride ⟨221⟩—A 280-mg portion shows no more chloride than corresponds to 1.0 mL of 0.020 *N* hydrochloric acid (0.25%).

Arsenic, *Method II* ⟨211⟩: 3 ppm.

Lead ⟨251⟩: 10 ppm.

Heavy metals, *Method II* ⟨231⟩: 0.002%.

Assay—Dissolve about 250 mg of Monosodium Glutamate, accurately weighed, then wetted with a few drops of water, in 100 mL of glacial acetic acid. Titrate with 0.1 *N* perchloric acid VS, determining the end-point potentiometrically. Each mL of 0.1 *N* perchloric acid is equivalent to 9.356 mg of $C_5H_8NNaO_4 . H_2O$.

Monostearate, Glyceryl—*see* Glyceryl Monostearate

Monothioglycerol

$HSCH_2CH(OH)CH_2OH$

$C_3H_8O_2S$ 108.15
1,2-Propanediol, 3-mercapto-.
3-Mercapto-1,2-propanediol [*96-27-5*].

» Monothioglycerol contains not less than 97.0 percent and not more than 101.0 percent of $C_3H_8O_2S$, calculated on the anhydrous basis.

Packaging and storage—Preserve in tight containers.

Specific gravity ⟨841⟩: between 1.241 and 1.250.

Refractive index ⟨831⟩: between 1.521 and 1.526.

pH ⟨791⟩: between 3.5 and 7.0, in a solution (1 in 10).

Water, *Method II* ⟨921⟩: not more than 5.0%.

Residue on ignition ⟨281⟩: not more than 0.1%.

Selenium ⟨291⟩: 0.003%, a 200-µL test specimen being used.

Heavy metals, *Method II* ⟨231⟩: 0.002%.

Assay—Dissolve about 400 mg of Monothioglycerol, accurately weighed, in 50 mL of water, and titrate with 0.1 *N* iodine VS, adding 3 mL of starch TS as the end-point is approached. Each mL of 0.1 *N* iodine is equivalent to 10.82 mg of $C_3H_8O_2S$.

Myristyl Alcohol

» Myristyl Alcohol contains not less than 90.0 percent of myristyl alcohol ($C_{14}H_{30}O$), the remainder consisting chiefly of related alcohols.

Packaging and storage—Preserve in well-closed containers.

Reference standards—*USP Myristyl Alcohol Reference Standard*—Do not dry before using. *USP Cetyl Alcohol Reference Standard*—Do not dry before using.

Identification—The retention time of the major peak obtained in the *Assay* is the same as that obtained from the Standard solution employed in the *Assay*.

Melting range, *Class II* ⟨741⟩: between 36° and 40°.

Acid value ⟨401⟩: not more than 2.

Iodine value ⟨401⟩: not more than 1.

Hydroxyl value ⟨401⟩—Place about 2 g, accurately weighed, in a dry, glass-stoppered, 250-mL flask, and add 2 mL of pyridine, followed by 10 mL of toluene. To the mixture add 10.0 mL of a solution of acetyl chloride, prepared by mixing 10 mL of acetyl chloride with 90 mL of toluene. Insert the stopper in the flask, and immerse in a water bath heated at 60° to 65° for 20 minutes. Add 25 mL of water, again insert the stopper in the flask, and shake vigorously for several minutes to decompose the excess acetyl chloride. Add 0.5 mL of phenolphthalein TS, and titrate to a permanent pink end-point with 1 *N* sodium hydroxide VS, shaking the flask vigorously toward the end of the titration to maintain the contents in an emulsified condition. Perform a blank test with the same quantities of the same reagents and in the same manner. The difference between the number of mL of 1 *N* sodium hydroxide consumed in the test with the sample and that consumed in the blank test, multiplied by 56.1, and the result divided by the weight, in g, of the Myristyl Alcohol used, represents the hydroxyl value of the Myristyl Alcohol, which is between 250 and 267.

Assay—

Chromatographic system (see *Chromatography* ⟨621⟩)—The gas chromatograph is equipped with a flame-ionization detector and contains a 2-m × 3-mm column packed with 10 percent phase G2 on support S1A. Dry helium is used as the carrier gas. The column is maintained at a temperature of about 205°, and the injection port and the detector are maintained at temperatures of 275° and 250°, respectively.

System suitability—Prepare a system suitability solution by dissolving about 90 mg of USP Myristyl Alcohol RS and 10 mg of USP Cetyl Alcohol RS in 10.0 mL of alcohol. Inject 2 µL of this solution, and calculate the resolution factor, *R* (see *Chromatography* ⟨621⟩), for cetyl alcohol and myristyl alcohol. A suitable resolution is not less than 4.0. Inject 2-µL portions of the suitability standard solution until the area ratio of cetyl alcohol to myristyl alcohol for each of five sequential injections is within 1.5% of the average of the area ratios for the five injections.

Procedure—Inject a suitable volume (about 2 µL) of a solution of Myristyl Alcohol in dehydrated alcohol (1 in 100) into the gas chromatograph. Measure the peak areas of the long-chain alcohol components in the chromatogram and determine the percentage of $C_{14}H_{30}O$ in the portion of Myristyl Alcohol taken by the formula:

$$100A/B,$$

in which A is the area due to the myristyl alcohol peak and B is the sum of the areas of all the long-chain alcohol peaks in the chromatogram.

Nitric Acid

HNO_3 63.01
Nitric acid.
Nitric acid [*7697-37-2*].

» Nitric Acid contains not less than 69.0 percent and not more than 71.0 percent, by weight, of HNO_3.

Caution—Avoid contact since Nitric Acid rapidly destroys tissues.

Packaging and storage—Preserve in tight containers.

Clarity and color—Mix it in its original container, and transfer 10 mL to a 20- × 150-mm test tube. Compare with water in a similar test tube: the liquids are equally clear and free from suspended matter, and, when viewed transversely by transmitted light, exhibit no apparent difference in color.

Identification—It responds to the tests for *Nitrate* ⟨191⟩.

Residue on ignition ⟨281⟩—To 70 mL (100 g) in a tared crucible add 2 drops of sulfuric acid, and evaporate to dryness. Ignite for 15 minutes: the weight of the residue does not exceed 0.5 mg (5 ppm).

Chloride ⟨221⟩—A 35-mL (50-g) portion shows no more chloride than corresponds to 35 µL of 0.020 *N* hydrochloric acid (0.5 ppm).

Sulfate ⟨221⟩—Add about 10 mg of sodium carbonate to 28 mL of Nitric Acid. Evaporate to dryness, dissolve in a mixture of 4 mL of water and 1 mL of dilute hydrochloric acid (1 in 20), and filter if necessary. Wash with two 2-mL portions of water, dilute with water to 10 mL, and add 1 mL of barium chloride reagent solution. Observe 10 minutes after adding the barium chloride: any turbidity produced is not greater than that produced by 40 µL of 0.020 *N* sulfuric acid in an equal volume of solution containing the quantities of reagents used in the test (1 ppm).

Arsenic, *Method I* ⟨211⟩—Prepare the *Test Preparation* as follows. To 210 mL (300 g) in a 1000-mL beaker, add 5 mL of sulfuric acid, and evaporate to dense fumes of sulfur trioxide. Cool, cautiously add 500 mL of water, again evaporate to dense fumes of sulfur trioxide, and repeat the dilution and evaporation, if necessary, to remove all of the nitric acid. Cautiously dilute with water to 35 mL. The limit is 0.01 ppm.

Iron ⟨241⟩—Evaporate 35 mL (50 g) to dryness, dissolve the residue in 2 mL of hydrochloric acid, and dilute with water to 47 mL: the limit is 0.2 ppm.

Heavy metals, *Method I* ⟨231⟩—To 70 mL (100 g) in a 250-mL beaker add about 10 mg of sodium carbonate, and evaporate on a steam bath to dryness. Add 25 mL of water: the limit is 0.2 ppm.

Assay—Weigh accurately about 2 mL of Nitric Acid in a tared, glass-stoppered conical flask, and add 25 mL of water. Add methyl red TS, and titrate with 1 *N* sodium hydroxide VS. Each mL of 1 *N* sodium hydroxide is equivalent to 63.01 mg of HNO_3.

Nitrogen

N_2 28.01
Nitrogen.
Nitrogen [*7727-37-9*].

» Nitrogen contains not less than 99.0 percent, by volume, of N_2.

Packaging and storage—Preserve in cylinders.

Identification—The flame of a burning wood splinter is extinguished when inserted into a test tube filled with Nitrogen. [NOTE—Exercise caution.]

Odor—Carefully open the container valve to produce a moderate flow of gas. Do not direct the gas stream toward the face, but deflect a portion of the stream toward the nose: no appreciable odor is discernible.

Note—Reduce the container pressure by means of a regulator. Measure the gases with a gas volume meter downstream from the detector tube in order to minimize contamination or change of the specimens.

Carbon monoxide—Pass 1050 ± 50 mL through a carbon monoxide detector tube (see under *Reagents* in the section, *Reagents, Indicators, and Solutions*) at the rate specified for the tube: the indicator change corresponds to not more than 0.001%.

Oxygen—Not more than 1.0% of oxygen is present, determined as directed in the *Assay*.

Assay—Introduce a specimen of Nitrogen into a gas chromatograph by means of a gas sampling valve. Select the operating conditions of the gas chromatograph such that the standard peak signal resulting from the following procedure corresponds to not less than 70% of the full-scale reading. Preferably, use an apparatus corresponding to the general type in which the column is 3 m in length and 4 mm in inside diameter and is packed with a molecular sieve prepared from a synthetic alkali-metal aluminosilicate capable of absorbing molecules having diameters of up to 0.5 nm, which permit complete separation of oxygen from nitrogen. Use industrial grade helium (99.99%) as the carrier gas, with a thermal-conductivity detector, and control the column temperature: the peak response produced by the assay specimen exhibits a retention time corresponding to that produced by an oxygen-helium certified standard (see under *Reagents* in the section, *Reagents, Indicators, and Solutions*), and is equivalent to not more than 1.0% of oxygen when compared to the peak response of the oxygen-helium certified standard, indicating not less than 99.0%, by volume, of N_2.

Nonoxynol 9—*see* Nonoxynol 9 USP

Nonoxynol 10

C_9H_{19}—C_6H_4—$(OCH_2CH_2)_nOH$

Poly(oxy-1,2-ethanediyl), α-(4-nonylphenyl)-ω-hydroxy-.
Polyethylene glycol mono(*p*-nonylphenyl) ether [*26027-38-3*].

» Nonoxynol 10 is an anhydrous liquid mixture consisting chiefly of monononylphenyl ethers of polyethylene glycols corresponding to the formula C_9H_{19}-$C_6H_4(OCH_2CH_2)_nOH$, in which the average value of n is about 10.

Packaging and storage—Preserve in tight containers.

Labeling—The labeling includes a cloud point range which is not greater than 6° and which is between 52° and 67°.

Reference standard—*USP Nonoxynol 10 Reference Standard*—Do not dry before using. Keep container tightly closed.

Identification—Its infrared absorption spectrum, obtained by spreading a capillary film of it between sodium chloride plates, exhibits maxima only at the same wavelengths as that of a similar preparation of USP Nonoxynol 10 RS.

Water, *Method I* ⟨921⟩: not more than 0.5%.

Residue on ignition ⟨281⟩: not more than 0.4%.

Arsenic, *Method II* ⟨211⟩: 2 ppm.

Heavy metals ⟨231⟩: 0.002%.

Hydroxyl value ⟨401⟩: between 81 and 97.

Cloud point—Weigh 1.00 g into a 250-mL beaker, and add 99 g of water. Dissolve completely by careful heating, while stirring at a constant slow speed with a small-propeller blade stirrer. Center a thermometer vertically in the solution, heat rapidly until the entire solution becomes cloudy, then raise the temperature 10°. Remove the source of heat, continue stirring, and record the temperature at which the solution becomes sufficiently clear that the entire thermometer bulb is seen plainly: the cloud point is within the range stated in the labeling.

Octoxynol 9

C_8H_{17}—C_6H_4—$(OCH_2CH_2)_nOH$

$C_{34}H_{62}O_{11}$ (av.) 647 (av.)
Poly(oxy-1,2-ethanediyl), α-(octylphenyl)-ω-hydroxy-.
Polyethylene glycol mono(octylphenyl) ether [*9002-93-1*].

» Octoxynol 9 is an anhydrous liquid mixture consisting chiefly of monooctylphenyl ethers of polyethylene glycols, corresponding to the formula:

$$C_8H_{17}C_6H_4(OCH_2CH_2)_nOH,$$

in which the average value of n is about 9.

Packaging and storage—Preserve in tight containers.

Reference standard—*USP Octoxynol 9 Reference Standard*—Do not dry before using. Keep container tightly closed.

Identification—Its infrared absorption spectrum, obtained by spreading a capillary film of it between sodium chloride plates, exhibits maxima only at the same wavelengths as that of a similar preparation of USP Octoxynol 9 RS.

Water, *Method I* ⟨921⟩: not more than 0.5%.

Residue on ignition ⟨281⟩: not more than 0.4%.

Arsenic, *Method II* ⟨211⟩: 2 ppm.

Heavy metals ⟨231⟩: 0.002%.

Hydroxyl value ⟨401⟩: between 85 and 101.

Cloud point—Weigh 1.00 g into a 250-mL beaker, and add 99 g of water. Dissolve completely by careful heating, while stirring at a constant slow speed with a small-propeller-blade stirrer. Center a thermometer vertically in the solution, and heat rapidly until the entire solution becomes cloudy, then raise the temperature 10°. Remove the source of heat, continue stirring, and record the temperature at which the solution becomes sufficiently clear to permit seeing the entire thermometer bulb plainly: the cloud point is between 63° and 69°.

Free ethylene oxide—Octoxynol 9 meets the requirements for *Free ethylene oxide* under *Nonoxynol 9* (see USP monograph). In performing the test, substitute the term "Octoxynol 9" for "Nonoxynol 9" wherever it occurs. The limit is 5 ppm.

Dioxane—It meets the requirements under *Dioxane* ⟨224⟩.

Octyldodecanol

» Octyldodecanol contains not less than 90.0 percent of 2-octyldodecanol, the remainder consisting chiefly of related alcohols.

Packaging and storage—Preserve in tight containers.

Reference standards—*USP Octyldodecanol Reference Standard*—Do not dry before using. *USP Stearyl Alcohol Reference Standard*—Do not dry before using.

Identification—The retention time of the major peak obtained in the *Assay* is the same as that obtained from the Standard solution employed in the *Assay*.

Acid value ⟨401⟩: not more than 0.5.

Iodine value ⟨401⟩: not more than 8.

Hydroxyl value ⟨401⟩—Place about 2 g, accurately weighed, in a dry, glass-stoppered, 250-mL flask, and add 2 mL of pyridine, followed by 10 mL of toluene. To the mixture add 10.0 mL of a solution of acetyl chloride, prepared by mixing 10 mL of acetyl chloride with 90 mL of toluene. Insert the stopper in the flask, and immerse in a water bath heated at 60° to 65° for 20 minutes. Add 25 mL of water, again insert the stopper in the flask, and shake vigorously for several minutes to decompose the excess acetyl chloride. Add 0.5 mL of phenolphthalein TS, and titrate to a permanent pink end-point with 1 *N* sodium hydroxide VS, shaking the flask vigorously toward the end of the titration to maintain the contents in an emulsified condition. Perform a blank test with the same quantities of the same reagents and in the same manner. The difference between the number of mL of 1 *N* sodium hydroxide consumed in the test with the sample and that consumed in the blank test, multiplied by 56.1, and the result divided by the weight, in g, of the Octyldodecanol used, represents the hydroxyl value of the Octyldodecanol, which is between 170 and 185.

Saponification value ⟨401⟩: not more than 5.

Assay—

Chromatographic system (see *Chromatography* ⟨621⟩)—The gas chromatograph is equipped with a flame-ionization detector, and contains a 2-m × 2-mm column packed with 3 percent phase G2 on support S1A. Dry nitrogen is used as the carrier gas. The column temperature is programmed at a rate of 6° per minute from 80° to 300° and the injection port and the detector are maintained at a temperature of about 280°.

System suitability—Prepare a system suitability solution by dissolving about 90 mg of USP Octyldodecanol RS and 10 mg of USP Stearyl Alcohol RS in 10.0 mL of alcohol. Inject 2 µL of this solution, and calculate the resolution factor, *R* (see *Chromatography* ⟨621⟩), for octyldodecanol and stearyl alcohol. A suitable resolution is not less than 4.0. Inject 2-µL portions of the suitability standard solution until the area ratio of octyldodecanol to stearyl alcohol for each of five sequential injections is within 1.5% of the average of the area ratios for the five injections.

Procedure—Inject a suitable volume (about 2 µL) of a solution of 90 mg of Octyldodecanol in 10 mL of alcohol into the gas chromatograph. Measure the peak areas in the chromatogram and determine the percentage of $C_{20}H_{42}O$ in the portion of Octyldodecanol taken by the formula:

$$100A/B,$$

in which A is the area due to the octyldodecanol peak and B is the sum of the areas of all the peaks in the chromatogram except the solvent peak.

Oils—*see complete list in index*

Ointment, Hydrophilic—*see* Ointment, Hydrophilic USP

Ointment, White—*see* Ointment, White USP

Ointment, Yellow—*see* Ointment, Yellow USP

Oleic Acid

$C_{18}H_{34}O_2$ 282.47
9-Octadecenoic acid, (*Z*)-.
Oleic acid [*112-80-1*].

» Oleic Acid is manufactured from fats and oils derived from edible sources, and consists chiefly of (*Z*)-9-octadecenoic acid [$CH_3(CH_2)_7CH{:}CH(CH_2)_7COOH$].

NOTE—Oleic Acid labeled solely for external use is exempt from the requirement that it be prepared from edible sources.

Packaging and storage—Preserve in tight containers.

Labeling—If it is for external use only, the labeling so indicates.

Specific gravity ⟨841⟩: between 0.889 and 0.895.

Congealing temperature ⟨651⟩: not above 10°.

Residue on ignition ⟨281⟩: not more than 1 mg (about 0.01%), from a 10-mL portion.

Mineral acids—Shake 5 mL with an equal volume of water at a temperature of about 25° for 2 minutes, allow the liquids to separate, and filter the water layer through a paper filter previously moistened with water: the filtrate is not reddened by the addition of 1 drop of methyl orange TS.

Neutral fat or mineral oil—Boil 1 mL with about 500 mg of sodium carbonate and 30 mL of water in a 250-mL flask: the resulting solution, while hot, is clear or, at most, opalescent.

Acid value ⟨401⟩: between 196 and 204, about 2 g, accurately weighed, being used.

Iodine value ⟨401⟩: between 85 and 95.

Oleyl Alcohol

$HC{-}CH_2(CH_2)_7OH$
$\|$
$HC{-}CH_2(CH_2)_6CH_3$

$C_{18}H_{36}O$ 268.48
9-Octadecen-1-ol, (*Z*)-.
(*Z*)-9-Octadecen-1-ol [*143-28-2*].

» Oleyl Alcohol is a mixture of unsaturated and saturated high molecular weight fatty alcohols consisting chiefly of oleyl alcohol.

Packaging and storage—Preserve in well-filled, tight containers, and store at controlled room temperature.

Cloud point—Place about 60 g in a 150-mL beaker, heat to 30°, cool, and immerse the beaker in an ice-water bath with the surfaces of the water and the test specimen at the same level. Insert a thermometer, and, using it as a stirring rod, begin stirring rapidly and steadily when the temperature falls below 20°. Keep the thermometer immersed throughout the test, remove and inspect the beaker containing the test specimen at regular intervals, and record the temperature at which the immersed portion of the thermometer, positioned vertically in the center of the beaker, is no longer visible when viewed horizontally through the beaker and test specimen: the cloud point is not above 10°.

Refractive index ⟨831⟩: between 1.458 and 1.460.

Acid value ⟨401⟩: not more than 1.

Hydroxyl value ⟨401⟩: between 205 and 215.

Iodine value ⟨401⟩: between 85 and 95.

Olive Oil

» Olive Oil is the fixed oil obtained from the ripe fruit of *Olea europaea* Linné (Fam. Oleaceae).

Packaging and storage—Preserve in tight containers, and prevent exposure to excessive heat.

Specific gravity ⟨841⟩: between 0.910 and 0.915.

Heavy metals, *Method II* ⟨231⟩: 0.001%.

Cottonseed oil—Mix 5 mL in a test tube with 5 mL of a mixture of equal volumes of amyl alcohol and a 1 in 100 solution of sulfur in carbon disulfide, warm the mixture carefully to expel the carbon disulfide, and immerse the test tube to one-third of its length in a boiling, saturated solution of sodium chloride for 2 hours: the mixture develops no reddish color.

Peanut oil—Saponify 10 g by heating for 1 hour under a reflux condenser with 80 mL of alcoholic potassium hydroxide TS. Add phenolphthalein TS, neutralize with 1 *N* acetic acid, and wash the solution into 120 mL of boiling lead acetate TS contained in a conical flask. Boil the mixture for 1 minute, and cool by immersing the flask in cold water, rotating the contents occasionally to cause the precipitate to adhere to the walls of the flask. Decant the liquid, wash the precipitate with cold water to remove the excess lead acetate, and then wash with 90 percent (by volume) alcohol. Add 100 mL of ether, stopper the flask, and allow to stand until the precipitate is disintegrated. Connect the flask to a reflux condenser, boil for 5 minutes, cool to about 15°, and allow to stand overnight. Filter, and thoroughly wash the precipitate with ether. With the use of a jet of ether, transfer the precipitate to a 500-mL separator, alternating the jet of ether with 3 *N* hydrochloric acid at the end if any of the precipitate adheres to the filter paper. Add sufficient 3 *N* hydrochloric acid to make the total acid layer measure about 100 mL, and add sufficient ether to make the ether layer measure about 100 mL. Shake the mixture vigorously for several minutes, allow the layers to separate, draw off the acid layer, and wash the ether once by shaking with 50 mL of 3 *N* hydrochloric acid and finally with several portions of water until the last washing is not acid to methyl orange TS. Transfer the ether solution to a dry flask, evaporate the ether, add a small amount of dehydrated alcohol, and evaporate on a steam bath to dryness. Dissolve the residue of dry fatty acids by warming with 60 mL of 90 percent (by volume) alcohol, slowly cool the solution to 15° while shaking frequently, and allow the solution to stand at 15° for 30 minutes: no crystals separate from the solution.

Sesame oil—Mix 10 mL with 10 mL of hydrochloric acid, add 0.1 mL of a 1 in 50 solution of furfural in alcohol, and shake the mixture vigorously for 15 seconds: no pink to crimson color appears in the acid layer when the emulsion breaks. If any color appears in the acid layer, add 10 mL of water, and again shake the mixture vigorously: in the absence of sesame oil any pink color is evanescent.

Teaseed oil—In a dry, 18- × 150-mm test tube place 0.8 mL of acetic anhydride, 1.5 mL of chloroform, and 0.2 mL of sulfuric acid, mix, and cool in a water bath to 25°. Add about 200 mg of Olive Oil (about 7 drops), mix, and cool to 25°. If the solution is cloudy, add acetic anhydride, dropwise, shaking after each addition, until the solution suddenly clears. Allow the mixture to remain in the water bath for 5 minutes: it shows a green color by both reflected and transmitted light. Add 10 mL of absolute ether, and mix by inverting the tube: the initial green color fades to a brownish gray. (Before the dilution with ether, the presence of *teaseed oil* will cause a brown color to appear by transmitted light, and after the dilution, a transient red color.)

Solidification range of fatty acids ⟨401⟩—The dry, mixed fatty acids of it solidify between 17° and 26°.

Free fatty acids ⟨401⟩—The free fatty acids in 10 g require for neutralization not more than 5.0 mL of 0.10 *N* sodium hydroxide.

Iodine value ⟨401⟩: between 79 and 88.

Saponification value ⟨401⟩: between 190 and 195.

Orange Flower Oil

» Orange Flower Oil is the volatile oil distilled from the fresh flowers of *Citrus aurantium* Linné (Fam. Rutaceae).

Packaging and storage—Preserve in tight, light-resistant containers.

Solubility in alcohol—It is miscible with an equal volume of alcohol and with about 2 volumes of 80 percent alcohol, the solution becoming cloudy on the further addition of alcohol of the same percentage.

Identification—An alcohol solution of it has a violet fluorescence.

Specific gravity ⟨841⟩: between 0.863 and 0.880.

Angular rotation ⟨781⟩: between +1.5° and +9.1°, when determined in a 100-mm tube.

Heavy metals, *Method II* ⟨231⟩: 0.004%.

Oxyquinoline Sulfate

OH N [ ]$_2$ · H_2SO_4

$(C_9H_7NO)_2 \cdot H_2SO_4$ 388.39
8-Quinolinol sulfate (2:1) (salt).
8-Quinolinol sulfate (2:1) (salt) [*134-31-6*].

» Oxyquinoline Sulfate is 8-hydroxyquinoline sulfate. It contains not less than 97.0 percent and not more than 101.0 percent of $(C_9H_7NO)_2 \cdot H_2SO_4$, calculated on the anhydrous basis.

Packaging and storage—Preserve in well-closed containers.

Reference standard—*USP Oxyquinoline Sulfate Reference Standard.*

Identification—

A: The infrared absorption spectrum of a potassium bromide dispersion of it exhibits maxima only at the same wavelengths as that of a similar preparation of USP Oxyquinoline Sulfate RS.

B: A solution (1 in 10) responds to the tests for *Sulfate* ⟨191⟩.

Water, *Method I* ⟨921⟩: between 4.0 and 6.0%.

Residue on ignition ⟨281⟩: not more than 0.3%.

Arsenic, *Method II* ⟨211⟩: 3 ppm.

Heavy metals, *Method II* ⟨231⟩: 0.004%.

Assay—Transfer about 100 mg of Oxyquinoline Sulfate, accurately weighed, to an iodine flask, add 30 mL of glacial acetic acid, 25.0 mL of 0.1 *N* bromine VS, 10 mL of potassium bromide solution (3 in 20), and 10 mL of hydrochloric acid, immediately insert the stopper, mix, and allow to stand for 15 minutes, protected from light. Quickly add 10 mL of potassium iodide solution (1 in 10) and 100 mL of water, taking precautions against the escape of bromine vapor, at once insert the stopper, and shake the mixture thoroughly. Remove the stopper, and rinse it and the neck of the flask with a small quantity of water so that the washing flows into the flask. Add 1 mL of chloroform, shake the mixture thoroughly, and titrate the liberated iodine with 0.1 *N* sodium thiosulfate VS, adding 3 mL of starch TS as the endpoint is approached. Perform a blank determination (see *Residual Titrations* under *Titrimetry* ⟨541⟩). Each mL of 0.1 *N* bromine is equivalent to 4.855 mg of $(C_9H_7NO)_2 \cdot H_2SO_4$.

Paraffin

» Paraffin is a purified mixture of solid hydrocarbons obtained from petroleum.

Packaging and storage—Preserve in well-closed containers, and avoid exposure to excessive heat.

Identification—

A: When strongly heated, it ignites with a luminous flame and deposits carbon.

B: Heat about 500 mg in a dry test tube with an equal weight of sulfur: the mixture evolves hydrogen sulfide and becomes black as a result of the liberation of carbon.

Congealing range ⟨651⟩: between 47° and 65°.

Reaction—Shake melted Paraffin with an equal volume of hot alcohol previously neutralized to litmus: the separated alcohol is neutral to litmus.

Readily carbonizable substances ⟨271⟩—Use a clean, dry, heat-resistant, glass-stoppered test tube, 140 ± 3 mm in length and 14 ± 1 mm in diameter, with a capacity of 16 ± 1 mL when the stopper is inserted, and calibrated at the 5- and 10-mL liquid levels. Place in the test tube 5 mL of Paraffin, at a temperature just above the melting point, add 5 mL of sulfuric acid containing 94.5% to 94.9% of H_2SO_4, and heat in a water bath at 70° for 10 minutes. When 5 minutes has elapsed, and at each successive minute thereafter, remove the tube from the bath, place a finger over the stopper, and give the tube three vigorous vertical shakes over an amplitude of about 12 cm, returning the tube to the bath within 3 seconds after the time when it was removed therefrom. At the end of 10 minutes from the time the tube was placed in the bath, the acid has no more color than a mixture of 3 mL of ferric chloride CS, 1.5 mL of cobaltous chloride CS, and 0.50 mL of cupric sulfate CS, overlaid with 5 mL of mineral oil. If the sulfuric acid remains dispersed in the molten paraffin, the color of the emulsion is not darker than that of the standard mixture when shaken vigorously.

Peanut Oil

» Peanut Oil is the refined fixed oil obtained from the seed kernels of one or more of the cultivated varieties of *Arachis hypogaea* Linné (Fam. Leguminosae).

Packaging and storage—Preserve in tight, light-resistant containers, and prevent exposure to excessive heat.

Identification—Saponify 5 g by boiling with 2.5 mL of 7.5 *N* sodium hydroxide and 12.5 mL of alcohol. Evaporate the alcohol, dissolve the soap in 50 mL of hot water, and add hydrochloric acid until the free fatty acids separate as an oily layer. Cool the mixture, and remove the separated fatty acids and dissolve them in 75 mL of ether. To the ether solution add a hot solution of 4 g of lead acetate in 40 mL of alcohol, and allow the mixture to stand for 18 hours. Filter the supernatant liquid, and transfer the precipitate to the filter with the aid of ether. Place the precipitate in a mixture of 40 mL of 3 *N* hydrochloric acid and 20 mL of water, and heat until the oily layer is entirely clear. Cool, decant the water solution, and boil the fatty acids with water that has been acidified with hydrochloric acid, until free from lead. (The fatty acids are free from lead when 100 mg, dissolved in 10 mL of alcohol, is not darkened by the addition of 2 drops of sodium sulfide TS.) Allow the fatty acids to solidify, and press them dry between filter papers on a cold surface. Dissolve the solid fatty acids in 25 mL of 90 percent alcohol, by heating gently, then cool to 15° and maintain at that temperature until the fatty acids have crystallized. Filter the separated fatty acids, recrystallize them from hot 90 percent alcohol, and dry in a vacuum desiccator for 4 hours: the arachidic acid so obtained melts between 73° and 76°.

Specific gravity ⟨841⟩: between 0.912 and 0.920.

Refractive index ⟨831⟩: between 1.462 and 1.464 at 40°.

Heavy metals, *Method II* ⟨231⟩: 0.001%.

Cottonseed oil—Mix 5 mL in a test tube with 5 mL of a mixture of equal volumes of amyl alcohol and a 1 in 100 solution of sulfur in carbon disulfide, warm the mixture carefully until the carbon disulfide is expelled, and immerse the test tube to one-third of its length in a boiling, saturated solution of sodium chloride: no reddish color develops within 15 minutes.

Rancidity—Shake 1 mL of a 1 in 10 solution of Peanut Oil in ether with 1 mL of hydrochloric acid, and add 1 mL of a 1 in 1000 solution of phloroglucinol in ether: no red or pink color develops.

Solidification range of fatty acids ⟨401⟩—The dry mixed fatty acids of it solidify between 26° and 33°.

Free fatty acids ⟨401⟩—The free fatty acids in 10 g require for neutralization not more than 2.0 mL of 0.020 *N* sodium hydroxide.

Iodine value ⟨401⟩: between 84 and 100.

Saponification value ⟨401⟩: between 185 and 195.

Unsaponifiable matter ⟨401⟩: not more than 1.5%.

Pectin—*see* Pectin USP

Peppermint

» Peppermint consists of the dried leaf and flowering top of *Mentha piperita* Linné (Fam. Labiatae).

Stems and other foreign organic matter ⟨561⟩: not more than 2.0% of stems more than 3 mm in diameter and other foreign organic matter.

Botanic characteristics—

Unground peppermint—Leaves, slender stems, and flowering tops. The leaves are opposite, usually more or less crumpled, and frequently detached from the stem. The petiole is from 4 mm to 15 mm in length, slightly pubescent; the blade, when entire, is ovate-oblong to oblong-lanceolate, from 1.5 cm to 9 cm in length with an acute apex, a narrowed or rounded base, and a sharply serrate margin; light green to dark green in color; its upper surface is nearly glabrous, its lower surface has a few hairs on the veins and many amber-colored glandular hairs. The stem is quadrangular, from 1 mm to 3 mm in diameter, glabrous except for a few scattered deflexed hairs, green to dark purple. The flowers occur as a compact, oblong or oval spike of verticillasters, from 1 cm to 1.5 cm in breadth, rounded at the summit, and in fruit attaining a length of from 3 cm to 7 cm. The bracts are oblong-lanceolate, from 4 mm to 7 mm in length; the calyx, tubular-campanulate, equally five-toothed, pubescent, and glandular-punctate, green to dark purple; the corolla is glabrous, light purple, tubular-campanulate, four-cleft, about 3 mm in length; stamens, four, short and equal; style two- or rarely three-cleft at the summit. The nutlets are ellipsoidal, about 500 µm in diameter. Peppermint has an aromatic, characteristic odor and a pungent taste, and produces a cooling sensation in the mouth.

Histology—Leaf: The lamina is dorsiventral. Both the upper and the lower epidermis consist of epidermal cells with wavy, anticlinal walls and stomata, the latter enclosed by a pair of subsidiary cells whose common wall is at right angles to the guard cells. Many of the epidermal cells, especially over the veins and midrib bear nonglandular and glandular hairs. The nonglandular hairs, also numerous along the margin, are uniseriate with longitudinally striate and papillose cuticle, up to eight cells in length and tapered to a pointed apex. The glandular hairs occur in two types. The larger of these are sunken in depressions of the epidermis and consist of a one- to two-celled stalk and a glandular head of eight radiating cells beneath the raised cuticle of which volatile oil is secreted. The smaller type of glandular hair consists of a one- to two-celled stalk and a one-celled glandular head containing volatile oil. Beneath the upper epidermis occurs a single layer of palisade parenchyma up to 80 µm in length and, directly underneath it, spongy parenchyma of three or four layers of chloroplastid-containing cells, through which zone course the fibrovascular tissues of the veins.

Stem—The stem is quadrangular. It shows a layer of epidermis bearing hairs similar to those of the leaf and possessing cuticularized outer convex walls, a narrow cortex of chlorenchyma, a clear endodermis of tangentially elongated, thin-walled cells with colorless contents, a narrow phloem, a cambium, and a xylem broadest in the regions beneath the stem angles and containing narrow wood-wedges separated by xylem rays one cell in width. The wood-wedges consist chiefly of simple pitted and spiral vessels, tracheids, and wood-fibers. Beneath each of the four angles of the stem occurs an elliptic to ovate zone of collenchyma. A large pith composed of thin-walled parenchyma occupies the center.

Powdered Peppermint—Green to light olive green. Shows fragments of leaf epidermis with wavy vertical walls and, if from the lower surface of the leaf, with numerous stomata and glandular and nonglandular hairs, the latter especially numerous along the veins; glandular hairs with a one- to two-celled stalk and one- to eight-celled head, usually set in a depression in the leaf and containing volatile oil and frequently yellowish or brownish crystals which are birefringent; nonglandular hairs with thin, papillose walls and frequently with short, longitudinal striations of one to eight cells and up to 1.4 mm in length, the terminal cell pointed or sometimes globular; fragments of chlorenchyma with vascular tissue, the vessels spiral or with simple pits and but slightly lignified; fragments of collenchyma and of thin-walled, nonlignified fibers associated with parenchyma. The pollen grains are spheroidal and smooth.

Peppermint Oil

» Peppermint Oil is the volatile oil distilled with steam from the fresh overground parts of the flowering plant of *Mentha piperita* Linné (Fam. Labiatae), rectified by distillation and neither partially nor wholly dementholized. It yields not less than 5.0 percent of esters, calculated as menthyl acetate ($C_{12}H_{22}O_2$), and not less than 50.0 percent of total menthol ($C_{10}H_{20}O$), free and as esters.

Packaging and storage—Preserve in tight containers, and prevent exposure to excessive heat.

Solubility in 70 percent alcohol: One volume dissolves in 3 volumes of 70 percent alcohol, with not more than slight opalescence.

Identification—Mix in a dry test tube 6 drops of Peppermint Oil with 5 mL of a 1 in 300 solution of nitric acid in glacial acetic acid, and place the tube in a beaker of boiling water: within 5 minutes the liquid develops a blue color which, on continued heating, deepens and shows a copper-colored fluorescence, and then fades, leaving a golden-yellow solution.

Specific gravity ⟨841⟩: between 0.896 and 0.908.

Angular rotation ⟨781⟩: between −18° and −32° in a 100-mm tube.

Refractive index ⟨831⟩: between 1.459 and 1.465 at 20°.

Heavy metals, *Method II* ⟨231⟩: 0.004%.

Dimethyl sulfide—Distil 1 mL from 25 mL of Peppermint Oil, and carefully superimpose the distillate on 5 mL of mercuric chloride TS in a test tube: a white film does not form at the zone of contact within 1 minute.

Assay for total esters—Place about 10 g of Peppermint Oil, accurately weighed, in a 250-mL conical flask, add 10 mL of neutralized alcohol and 2 drops of phenolphthalein TS, then add, dropwise, 0.1 *N* sodium hydroxide until a faint pink color appears. Add 25.0 mL of 0.5 *N* alcoholic potassium hydroxide VS, connect the flask to a reflux condenser, and heat on a boiling water bath for 1 hour. Allow the mixture to cool, add 20 mL of water, add phenolphthalein TS, and titrate the excess alkali with 0.5 *N* hydrochloric acid VS. Perform a blank determination, disregarding the 0.1 *N* sodium hydroxide (see *Residual Titrations* under *Titrimetry* ⟨541⟩). Each mL of 0.5000 *N* alcoholic potassium hydroxide consumed in the saponification is equivalent to 99.15 mg of total esters calculated as menthyl acetate ($C_{12}H_{22}O_2$).

Assay for total menthol—Place 10 mL of Peppermint Oil in an acetylation flask of 100-mL capacity, and add 10 mL of acetic anhydride and 1 g of anhydrous sodium acetate. Boil the mixture gently for 1 hour, accurately timed, cool, disconnect the flask from the condenser, transfer the mixture to a small separator, rinsing the acetylation flask with three 5-mL portions of warm water, and add the rinsings to the separator. When the liquids have completely separated, discard the water layer, and wash the remaining oil with successive portions of sodium carbonate TS, diluted with an equal volume of water, until the last washing is alkaline to phenolphthalein TS. Dry the resulting oil with anhydrous sodium sulfate, and filter. Transfer 5 mL of the dry acetylated oil to a tared, 100-mL conical flask, and weigh. Add 50.0 mL of 0.5 *N* alcoholic potassium hydroxide VS, connect the flask to a reflux condenser, and boil the mixture on a steam bath for 1 hour, accurately timed. Allow the mixture to cool, add 10 drops of phenolphthalein TS, and titrate the excess alkali with 0.5 *N* hydrochloric acid VS. Perform a blank determination (see *Residual Titrations* under *Titrimetry* ⟨541⟩). Calculate the percentage of total menthol in the Peppermint Oil tested by the formula:

$$7.813A(1 - 0.0021E)/(B - 0.021A),$$

in which A is the result obtained by subtracting the number of mL of 0.5 *N* hydrochloric acid required in the above titration from the number of mL of 0.5 *N* hydrochloric acid required in the residual titration blank, E is the percentage of esters calculated as menthyl acetate ($C_{12}H_{22}O_2$), and B is the weight of acetylated oil taken.

Peppermint Spirit—*see* Peppermint Spirit USP

Peppermint Water

» Peppermint Water is a clear, saturated solution of Peppermint Oil in Purified Water, prepared by one of the processes described under *Aromatic Waters* (see *Pharmaceutical Dosage Forms* ⟨1151⟩).

Packaging and storage—Preserve in tight containers.

Persic Oil

» Persic Oil is the oil expressed from the kernels of varieties of *Prunus armeniaca* Linné (Apricot Kernel Oil), or from the kernels of varieties of *Prunus persica* Sieb. et Zucc. (Peach Kernel Oil) (Fam. Rosaceae).

Packaging and storage—Preserve in tight containers.

Labeling—Label it to indicate whether it was derived from apricot kernels or from peach kernels.

Specific gravity ⟨841⟩: between 0.910 and 0.923.

Heavy metals, *Method II* ⟨231⟩: 0.004%.

Mineral oil—Heat 10 mL of it on a steam bath with 15 mL of 4 *N* sodium hydroxide and 30 mL of alcohol in a flask having a small, short-stem funnel inserted in the neck, and occasionally agitate the mixture until it becomes clear. Transfer the solution to a shallow dish, evaporate the alcohol on a steam bath, and mix the residue with 100 mL of water: a clear solution results.

Cottonseed oil—Mix 5 mL in a test tube with 5 mL of a mixture of equal volumes of amyl alcohol and of a 1% solution of sulfur in carbon disulfide, warm the mixture carefully until the carbon disulfide is expelled, and immerse the test tube to one-third of its length in a boiling, saturated solution of sodium chloride: the mixture develops no reddish color within 2 hours.

Sesame oil—Mix 10 mL of it with 10 mL of hydrochloric acid, add 0.1 mL of a 1 in 50 solution of furfural in alcohol, and shake the mixture vigorously for 15 seconds: no pink to crimson color appears in the acid layer when separate layers form. Should any color appear in the acid layer, add 10 mL of water, and again shake the mixture vigorously. In the absence of sesame oil the pink color is fugitive.

Free fatty acids ⟨401⟩—The free fatty acids in 10 g of it require for neutralization not more than 2.0 mL of 0.020 *N* sodium hydroxide.

Iodine value ⟨401⟩: between 90 and 108.

Saponification value ⟨401⟩: between 185 and 195.

Petrolatum—*see* Petrolatum USP

Petrolatum, Hydrophilic—*see* Petrolatum, Hydrophilic USP

Petrolatum, White—*see* Petrolatum, White USP

Pharmaceutical Glaze—*see* Glaze, Pharmaceutical

Phenol—*see* Phenol USP

Phenylethyl Alcohol—*see* Phenylethyl Alcohol USP

Phenylmercuric Acetate

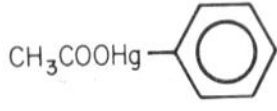

$C_8H_8HgO_2$ 336.74

Mercury, (acetato-*O*)phenyl-.

(Acetato)phenylmercury [*62-38-4*].

» Phenylmercuric Acetate contains not less than 98.0 percent and not more than 100.5 percent of $C_8H_8HgO_2$.

Packaging and storage—Preserve in tight, light-resistant containers.

Identification—

A: Add 0.5 mL of nitric acid to 0.1 g of it, warm gently until a dark brown color is produced, and dilute with water to 10 mL: the characteristic odor of nitrobenzene is evolved.

B: To 0.1 g of it add 0.5 mL of sulfuric acid and 1 mL of alcohol, and warm: the characteristic odor of ethyl acetate is evolved.

C: To 5 mL of a saturated solution in water add a few drops of sodium sulfide TS: a white precipitate is formed, which turns black when the mixture is boiled and then allowed to stand.

Melting range ⟨741⟩: between 149° and 153°.

Residue on ignition ⟨281⟩: not more than 0.2%.

Mercuric salts and Heavy metals—Heat about 100 mg with 15 mL of water, cool, and filter. To the filtrate add a few drops of sodium sulfide TS: the resulting precipitate shows no immediate color.

Polymercurated benzene compounds—Shake 2.0 g with 100 mL of acetone, and filter. Wash the residue with successive portions of acetone until a total of 50 mL is used, then dry the residue at 105° for 1 hour, and weigh: the weight of the residue does not exceed 30 mg (1.5%).

Assay—Transfer about 500 mg of Phenylmercuric Acetate, accurately weighed, to a 100-mL flask, add 15 mL of water, 5 mL of formic acid, and 1 g of zinc dust, and reflux for 30 minutes. Cool, filter, and wash the filter paper and the amalgam with water until the washings are no longer acid to litmus. Dissolve the amalgam in 40 mL of 8 *N* nitric acid. Heat on a steam bath for 3 minutes, and then add 500 mg of urea and enough potassium permanganate TS to produce a permanent pink color. Cool, decolorize the solution with hydrogen peroxide TS, add 1 mL of ferric ammonium sulfate TS, and titrate with 0.1 *N* ammonium thiocyanate VS. Each mL of 0.1 *N* ammonium thiocyanate is equivalent to 16.84 mg of $C_8H_8HgO_2$.

Phenylmercuric Nitrate

Mercury, (nitrato-*O*)phenyl-.
Nitratophenylmercury [*55-68-5*].

» Phenylmercuric Nitrate is a mixture of phenylmercuric nitrate and phenylmercuric hydroxide containing not less than 87.0 percent and not more than 87.9 percent of phenylmercuric ion ($C_6H_5Hg^+$), and not less than 62.75 percent and not more than 63.50 percent of mercury (Hg).

Packaging and storage—Preserve in tight, light-resistant containers.

Identification—

A: Add 3 mL of sulfuric acid to 0.1 g of it: the mixture becomes yellow, and the characteristic odor of nitrobenzene is evolved.

B: To 5 mL of a saturated solution of it add 1 mL of 3 *N* hydrochloric acid: a white precipitate is formed.

C: To 5 mL of a saturated solution of it add 5 mL of ammonium sulfide TS: there is no reaction in the cold, but upon heating in a boiling water bath for 10 minutes a black precipitate is formed.

Melting range ⟨741⟩: between 175° and 185°.

Residue on ignition ⟨281⟩: not more than 0.1%.

Mercury ions—To 5 mL of a saturated solution add 5 mL of 1 *N* sodium hydroxide: no yellow precipitate is formed (*mercuric ions*) and the solution does not darken (*mercurous ions*).

Assay for phenylmercuric ions—Transfer about 200 mg, accurately weighed, to a conical flask, and dissolve in 90 mL of water and 10 mL of nitric acid. Add 2 mL of ferric ammonium sulfate TS, and titrate with 0.05 *N* ammonium thiocyanate VS. Each mL of 0.05 *N* ammonium thiocyanate is equivalent to 13.88 mg of phenylmercuric ion ($C_6H_5Hg^+$).

Assay for mercury—Transfer about 400 mg of Phenylmercuric Nitrate, accurately weighed, to a 100-mL flask, add 15 mL of water, 5 mL of formic acid, and 1 g of zinc dust, and reflux for 30 minutes. Cool, filter, and wash the filter paper and the amalgam with water until the washings are no longer acid to litmus. Dissolve the amalgam in 40 mL of 8 *N* nitric acid. Heat on a steam bath for 3 minutes, then add 0.5 g of urea and enough potassium permanganate TS to produce a permanent pink color. Cool, decolorize the solution with hydrogen peroxide TS, add 1 mL of ferric ammonium sulfate TS, and titrate with 0.1 *N* ammonium thiocyanate VS. Each mL of 0.1 *N* ammonium thiocyanate is equivalent to 10.03 mg of Hg.

Phosphoric Acid

H_3PO_4 98.00
Phosphoric acid.
Phosphoric acid [*7664-38-2*].

» Phosphoric Acid contains not less than 85.0 percent and not more than 88.0 percent, by weight, of H_3PO_4.

Caution—Avoid contact, as Phosphoric Acid rapidly destroys tissues.

Packaging and storage—Preserve in tight containers.

Identification—When carefully neutralized with 1 *N* sodium hydroxide, phenolphthalein TS being used as the indicator, it responds to the tests for *Phosphate* ⟨191⟩.

Nitrate—Dilute it with 14 volumes of water, mix 5 mL of the dilution with about 0.1 mL of indigo carmine TS, then add 5 mL of sulfuric acid: the blue color is not discharged within 1 minute.

Phosphorous or hypophosphorous acid—Dilute it with 14 volumes of water. Gently warm 5 mL of the dilution, and add 2 mL of silver nitrate TS: the mixture does not become brownish.

Sulfate ⟨221⟩—Dilute it with 90 volumes of water, and add 1 mL of barium chloride TS: no precipitate is formed immediately.

Arsenic, *Method I* ⟨211⟩: 3 ppm.

Alkali phosphates—Transfer 1 mL of it to a graduated cylinder, and add 6 mL of ether and 2 mL of alcohol: no turbidity is produced.

Heavy metals, *Method I* ⟨231⟩: 0.001%.

Assay—Weigh accurately about 1 g of Phosphoric Acid in a tared, glass-stoppered flask, and dilute it with water to about 120 mL. Add 0.5 mL of thymolphthalein TS, and titrate with 1 *N* sodium hydroxide VS to the first appearance of a blue color. Perform a blank determination, and make any necessary correction. Each mL of 1 *N* sodium hydroxide is equivalent to 49.00 mg of H_3PO_4.

Diluted Phosphoric Acid

» Diluted Phosphoric Acid contains, in each 100 mL, not less than 9.5 g and not more than 10.5 g of H_3PO_4.

Diluted Phosphoric Acid may be prepared as follows:

Phosphoric Acid	69 mL
Purified Water, a sufficient quantity, to make	1000 mL

Mix the ingredients.

Packaging and storage—Preserve in tight containers.

Alkali phosphates—Evaporate 20 mL on a steam bath to a weight of about 5 g. Cool, transfer 2 mL to a graduated cylinder, and add 6 mL of ether and 2 mL of alcohol: no turbidity is produced.

Arsenic, *Method I* ⟨211⟩: 1.5 ppm.

Heavy metals, *Method I* ⟨231⟩—Dilute 10 g (9.5 mL) with 10 mL of water, add 6 mL of 1 *N* sodium hydroxide, and dilute with water to 50 mL. Dilute 20 mL of this solution with water to 25 mL: the limit is 5 ppm.

Other tests—Diluted Phosphoric Acid, without further dilution, responds to the *Identification test* and meets the requirements of the tests for *Nitrate*, *Phosphorous or hypophosphorous acid*, and *Sulfate* under *Phosphoric Acid*.

Assay—Transfer 10.0 mL of Diluted Phosphoric Acid to a flask, and dilute it with water to about 50 mL. Add 0.5 mL of thymolphthalein TS, and titrate with 1 *N* sodium hydroxide VS to the first appearance of a blue color. Perform a blank determination, and make any necessary correction. Each mL of 1 *N* sodium hydroxide is equivalent to 49.00 mg of H_3PO_4.

Polacrilin Potassium

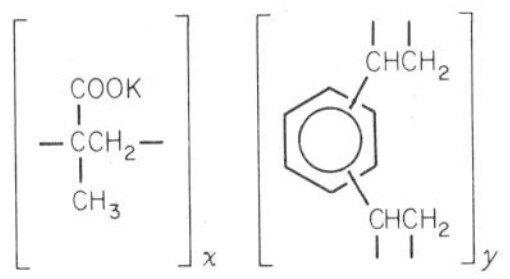

2-Propenoic acid, 2-methyl-, polymer with divinylbenzene, potassium salt.
Methacrylic acid polymer with divinylbenzene, potassium salt [*39394-76-5*].

» Polacrilin Potassium is the potassium salt of a unifunctional low-cross-linked carboxylic cation-exchange resin prepared from methacrylic acid and divinylbenzene. When previously dried at 105° for 6 hours, it contains not less than 20.6 percent and not more than 25.1 percent of potassium.

Packaging and storage—Preserve in well-closed containers.

Reference standard—*USP Polacrilin Potassium Reference Standard*—Dry at 105° for 6 hours before using.

Identification—

A: The infrared absorption spectrum of a potassium bromide dispersion of it, previously dried, exhibits maxima only at the same wavelengths as that of a similar preparation of USP Polacrilin Potassium RS.

B: Shake about 1 g with 10 mL of water: the aqueous phase does not respond to the tests for *Potassium* ⟨191⟩. Shake about 1 g with 10 mL of 0.1 *N* hydrochloric acid: the aqueous phase responds to the tests for *Potassium* ⟨191⟩.

Loss on drying ⟨731⟩—Dry it at 105° for 6 hours: it loses not more than 10.0% of its weight.

Powder fineness ⟨811⟩—Transfer about 4 g, accurately weighed, to a No. 100 standard sieve placed on top of a No. 200 standard sieve and pan. Using a soft 2-cm brush, brush the sample lightly across the No. 100 sieve until no more particles pass through. By brushing and tapping, dust off the particles on the underside of the No. 100 sieve into the No. 200 sieve. Obtain the weight of the material retained on the No. 100 sieve. Similarly, determine the weight of material retained by the No. 200 sieve; not more than 1.0% is retained on the No. 100 sieve, and not more than 30.0% is retained on the No. 200 sieve.

Arsenic, *Method II* ⟨211⟩: not more than 3 ppm.

Iron ⟨241⟩—Transfer 0.10 g to a suitable crucible, and ignite at a low heat until thoroughly ashed. Add to the carbonized mass 2 mL of nitric acid and 5 drops of sulfuric acid, and heat cautiously until white fumes are no longer evolved. Ignite, preferably in a muffle furnace, at 500° to 600°, until the carbon is completely burned off. Cool, add 4 mL of 6 *N* hydrochloric acid, cover, digest on a steam bath for 15 minutes, uncover, and slowly evaporate on a steam bath to dryness. Moisten the residue with 1 drop of hydrochloric acid, add 10 mL of hot water, and digest for 2 minutes. Dilute with water to about 25 mL. Filter, if necessary, rinse the crucible and the filter with 10 mL of water, combining the filtrate and rinsing in a 50-mL color-comparison tube, add 2 mL of hydrochloric acid, dilute with water to 45 mL, and mix. The limit is 0.01%.

Sodium—

Test solution—Transfer about 2 g, accurately weighed, to a 400-mL borosilicate beaker, add 20 mL of sulfuric acid, cover with a borosilicate watch glass, and heat until charring is complete. While continuing to heat the beaker, add 20 mL of nitric acid dropwise. Continue to heat and add nitric acid until all of the organic material has been destroyed as indicated by the contents of the beaker turning from brown to a very pale straw-colored or colorless solution. Continue to evaporate the solution, and if it turns brown during the evaporation, add nitric acid dropwise until the brown color disappears. Evaporate just to dryness, cool, and dissolve the residue in 40 mL of water and 10 mL of 6 *N* hydrochloric acid. Heat to boiling, cool, transfer to a 100-mL volumetric flask, dilute with water to volume, and mix.

Procedure—To three separate 100-mL volumetric flasks add, respectively, 1.00 mL, 2.00 mL, and 3.00 mL of a solution containing 254.2 mg of sodium chloride in 1000 mL of water. Add water to volume, and mix to obtain sodium chloride solutions having concentrations equivalent to 1 µg of Na per mL, 2 µg of Na per mL, and 3 µg of Na per mL, respectively. Adjust the settings of a suitable flame photometer so that the emission of the solution containing 3.00 mL of the sodium chloride solution reads close to 100% at 589 nm. Determine the readings of the three solutions at 589 nm. Readjust the wavelength setting to 580 nm, and determine the background emission reading for one of these standards. Pipet 5 mL of the *Test solution* into a 100-mL volumetric flask, add water to volume, and mix. Observe the emission reading of this solution at 589 nm, using the same instrument settings, then readjust the wavelength setting to 580 nm, and observe the background emission reading. Subtract the corresponding background readings from the standard and test specimen readings. Prepare a standard curve by plotting the corrected standard readings versus the square root of the sodium concentration. From this standard curve, determine the sodium content of the test specimen. It is not greater than 0.20%.

Heavy metals, *Method III* ⟨231⟩: 0.002%.

Assay for potassium—

Sodium stock solution—Transfer 7.306 g of sodium chloride, previously dried at 125° for 30 minutes and accurately weighed, to a 500-mL volumetric flask, add water to volume, and mix. This solution contains 5.76 g of Na per 1000 mL.

Potassium stock solution—Transfer 745.5 mg of potassium chloride, previously dried at 125° for 30 minutes and accurately weighed, to a 1000-mL volumetric flask, add water to volume, and mix. This solution contains 391 mg of K per 1000 mL.

Surfactant solution—Transfer 5.0 g of a suitable nonionic surfactant to a 250-mL beaker, add 200 mL of water, and stir to dissolve. Transfer this solution to a 500-mL volumetric flask, dilute with water to volume, and mix. [NOTE—To prevent foaming when using this solution, gently run the solution down the sides of the vessel, and use gentle action when mixing.]

Diluted sodium solution—Transfer 50.0 mL of *Sodium stock solution* and 10.0 mL of *Surfactant solution* to a 100-mL volumetric flask, dilute with water to volume, and mix gently to prevent foaming.

Standard preparations—To three separate 500-mL volumetric flasks transfer, respectively, 3.0-, 4.0-, and 5.0-mL portions of *Potassium stock solution.* To each flask add 50.0 mL of *Sodium stock solution* and 10.0 mL of *Surfactant solution*, dilute with water to volume, and mix gently to prevent foaming. Each mL of these solutions contains 2.346, 3.128, and 3.910 µg of K, respectively.

Assay preparation—Transfer about 1.4 g of Polacrilin Potassium, previously dried and accurately weighed, to a 50-mL silica crucible, moisten with 4 mL of sulfuric acid, heat over a small flame until the acid has fumed off, moisten the residue with a few drops of sulfuric acid, and ignite strongly. Allow to cool, transfer, with the aid of water, to a 1000-mL volumetric flask, dilute with water to volume, and mix. Transfer 1.00 mL of this solution to a 100-mL volumetric flask, add 20.0 mL of *Diluted sodium solution*, dilute with water to volume, and mix gently to prevent foaming.

Procedure—Concomitantly determine the emittances of the *Standard preparations* and the *Assay preparation* at 766 nm, with a flame photometer, adjusting the instrument so that the most concentrated *Standard preparation* gives a reading near 100%. Prepare a standard curve by plotting the standard readings versus the square root of the potassium concentrations. From the curve, determine the concentration, C_U, in µg per mL, of potassium in the *Assay preparation.* Calculate the weight, in mg, of potassium in the portion of Polacrilin Potassium taken by the formula:

$$100C_U.$$

Poloxamer

$$HO - [C_2H_4O]_a - [C_3H_6O]_b - [C_2H_4O]_a - H$$

$HO(C_2H_4O)_a(C_3H_6O)_b(C_2H_4O)_aH$
Oxirane, methyl-, polymer with oxirane.
α-Hydro-ω-hydroxypoly(oxyethylene)$_a$-poly(oxypropylene)$_b$-poly(oxyethylene)$_a$ block copolymer, in which a and b have the following values:

Poloxamer	a	b
124	12	20
188	79	28
237	64	37
338	141	44
407	101	56

Polyethylene-polypropylene glycol [*9003-11-6*].

» Poloxamer is a synthetic block copolymer of ethylene oxide and propylene oxide. It is available in several types, conforming to the following requirements:

Poloxamer	Physical Form	Average Molecular Weight	Weight % Oxyethylene	Unsaturation, mEq/g
124	Liquid	2090 to 2360	46.7 ± 1.9	0.020 ± 0.008
188	Solid	7680 to 9510	81.1 ± 0.63	0.026 ± 0.004
237	Solid	6840 to 8830	72.4 ± 1.9	0.034 ± 0.008
338	Solid	12700 to 17400	83.1 ± 1.7	0.031 ± 0.008
407	Solid	9840 to 14600	73.2 ± 1.7	0.048 ± 0.017

Packaging and storage—Preserve in tight containers.

Labeling—Label it to state, as part of the official title, the Poloxamer number.

Average molecular weight—

Phthalic anhydride–pyridine solution—Add 144 g of phthalic anhydride to 1000 mL of freshly distilled pyridine containing less than 0.1% of water. Protect from light. Shake vigorously until solution is effected, and allow to stand overnight. To verify that the *Phthalic anhydride–pyridine solution* has adequate strength, pipet 10 mL into a 250-mL conical flask, add 25 mL of pyridine and 50 mL of water, and after about 15 minutes add 0.5 mL of a 1 in 100 solution of phenolphthalein in pyridine, then titrate with 0.5 N sodium hydroxide VS: it consumes between 37.6 mL and 40.0 mL of 0.5 N sodium hydroxide.

Procedure—Accurately weigh a suitable quantity, in g, of Poloxamer, calculated by multiplying the average molecular weight by 0.004, into a glass-stoppered, 250-mL boiling flask. Carefully pipet 25 mL of *Phthalic anhydride–pyridine solution* into the flask, touching the tip of the drained pipet to the protrusion in the flask. Add a few glass beads, and swirl to dissolve the specimen. Pipet 25 mL of *Phthalic anhydride–pyridine solution* into a second glass-stoppered conical flask, add a few glass beads, and use as the reagent blank. Heat both flasks, fitted with suitable reflux condensers, and allow to reflux for 1 hour. Allow to cool, and pour two 10-mL portions of pyridine through each condenser. Remove the flasks from the condensers, add 10 mL of water to each, insert the stopper, swirl, and allow to stand for 10 minutes. To each flask add 50.0 mL of 0.66 N sodium hydroxide and 0.5 mL of a 1 in 100 solution of phenolphthalein in pyridine. Titrate with 0.5 N sodium hydroxide VS to a light pink end-point that persists for not less than 15 seconds, recording the volume, in mL, consumed by the residual acid in the test solution as S, and that consumed by the blank as B. Calculate the average molecular weight by the formula:

$$2000W/[(B - S)(N)],$$

in which W is the weight, in g, of the test specimen taken, and N is the exact normality of the 0.5 N sodium hydroxide VS.

Weight percent oxyethylene—

Solvent—Use deuterated water or deuterochloroform.

NMR reference—Use sodium 2,2-dimethyl-2-silapentane-5-sulfonate (for deuterated water) or tetramethylsilane (for deuterochloroform).

Test preparation—Dissolve 0.1 g to 0.2 g of Poloxamer in deuterated water containing 1% of sodium 2,2-dimethyl-2-silapentane-5-sulfonate to obtain 1 mL of solution, or, if the Poloxamer does not dissolve in water, use deuterochloroform containing 1% of tetramethylsilane as the solvent.

Procedure—Transfer 0.5 mL to 1.0 mL of the *Test preparation* to a standard 5-mm NMR spinning tube, and if deuterochloroform is the solvent, add 1 drop of deuterated water, and shake the tube. Proceed as directed for *Relative Method of Quantitation* under *Nuclear Magnetic Resonance* ⟨761⟩, using the *Test preparation* volumes specified here, scanning the region from 0 ppm to 5 ppm, and using the calculation formulas specified here. Record as A_1 the average area of the doublet appearing at about 1.08 ppm, representing the methyl groups of the oxypropylene units, and record as A_2 the average area of the composite band from 3.2 to 3.8 ppm, due to the CH_2O groups of both the oxyethylene and oxypropylene units and also the CHO groups of the oxypropylene units, with reference to the sodium 2,2-dimethyl-2-silapentane-5-sulfonate or tetramethylsilane singlet at 0 ppm. Calculate, the percentage of oxyethylene, by weight, in the Poloxamer, by the formula:

$$3300\alpha/(33\alpha + 58),$$

in which α is $(A_2/A_1) - 1$.

pH ⟨791⟩: between 5.0 and 7.5, in a solution (1 in 40).

Unsaturation—

Mercuric acetate solution—Place 50 g of mercuric acetate in a 1000-mL volumetric flask, and dissolve with about 900 mL of methanol to which 0.5 mL of glacial acetic acid has been added. Dilute with methanol to volume, and mix. Discard the solution if it is yellow. If it is turbid, filter it. Discard it if it is still turbid. Use fresh reagents if it is necessary to repeat the preparation of the solution. Protect the solution from light by storing it in an amber bottle in the dark.

Procedure—Transfer about 30 g of Poloxamer, accurately weighed, to a 250-mL conical flask. Pipet 50 mL of *Mercuric acetate solution* into the flask, and mix on a magnetic stirrer until solution is complete. Allow to stand for 30 minutes with occasional swirling. Add 10 g of sodium bromide crystals, and stir on a magnetic stirrer for about 2 minutes. Without delay, add about 1 mL of phenolphthalein TS, and titrate the liberated acetic acid with 0.1 N methanolic potassium hydroxide VS. Perform a blank determination and make any necessary correction. Subtract from this corrected titer also the number of milliliters needed to neutralize any acidity of the Poloxamer, determined by dissolving in neutralized methanol under a nitrogen sweep, and titrating with 0.1 N methanolic potassium hydroxide VS under nitrogen. Each milliliter of 0.1 N methanolic potassium hydroxide is equivalent to 0.1 N mEq of unsaturation in the amount of Poloxamer taken.

Arsenic, *Method II* ⟨211⟩—Use 1.0 g: the limit is 3 ppm.

Heavy metals, *Method I* ⟨231⟩: 0.002%.

Free ethylene oxide, propylene oxide, and 1,4-dioxane—

Stripped poloxamer—Into a 5000-mL 4-neck, round-bottom flask, equipped with a stirrer, a thermometer, a gas dispersion tube, a dry ice trap, a vacuum outlet, and a heating mantle, place 3000 g of Poloxamer 124. At room temperature, evacuate the flask carefully to a pressure of less than 1 mm of mercury, applying the vacuum slowly while observing for excessive foaming due to entrapped gases. After any foaming has subsided, sparge with nitrogen, allowing the pressure to rise to 10 mm of mercury. Heat the flask to 130° while increasing the pressure to about 60 mm of mercury. Continue stripping for 4 hours; then cool to room temperature. Shut off the vacuum pump, and bring the flask pressure back to atmospheric while maintaining nitrogen sparging. Remove the sparging tube with the gas still flowing, then turn off the gas flow. Transfer the *Stripped poloxamer* to a suitable nitrogen-filled container.

Standard preparations—[*Caution—Ethylene oxide, propylene oxide, and 1,4-dioxane are toxic and flammable. Prepare these solutions in a well-ventilated fume hood.*] To a known weight of *Stripped poloxamer* add suitable quantities of propylene oxide and 1,4-dioxane. Determine the amounts added by weight difference after each addition. Using the special handling described in the following sentences, complete the preparation. Ethylene oxide is a gas at room temperature. It is usually stored in a lecture-type gas cylinder or small metal pressure bomb. Chill the cylinder in a refrigerator before use. Transfer about 5 mL of the liquid ethylene oxide to a 100-mL beaker chilled in wet ice. Using a gas-tight gas chromatographic syringe that has been chilled in a refrigerator, transfer a suitable amount of the liquid ethylene oxide into the mixture. Immediately reseal the vial, and shake. Determine the amount added by weight difference. By appropriate dilution with *Stripped poloxamer*, prepare four solutions, covering the range from 1 to 20 ppm for the three components added to the matrix (e.g., 5, 10, 15, and 20 ppm). Transfer 1.0 mL of each of these solutions to separate 22-mL pressure headspace vials, seal each with a silicone septum, star spring, and pressure relief safety aluminum sealing cap, and crimp the cap closed with a cap-sealing tool.

Test preparation—Transfer 1 ± 0.01 g of Poloxamer to a 22-mL pressure headspace vial, and seal, cap, and crimp as directed for the *Standard preparations*.

Chromatographic system (see *Chromatography* ⟨621⟩)—The gas chromatograph is equipped with a balanced pressure automatic headspace sampler and a flame-ionization detector and contains a 50-m × 0.32-mm fused silica capillary column containing bonded phase G27 in a 5-µm film thickness. The column temperature is programmed from 70° to 250° at 10° per minute, with the transfer line at 140° and the detector at 250°. The carrier gas is helium, flowing at a rate of about 0.8 mL per minute. The resolution, *R*, of ethylene oxide and propylene oxide, when the *Standard preparations* are chromatographed as directed under *Calibration*, is not less than 2.0. On the three *Calibration* plots, no point digresses from its line by more than 10%.

Calibration—Place the vials containing the *Standard preparations* in the automated sampler, and start the sequence so that each vial is heated at a temperature of 110° for 30 minutes before a suitable portion of its headspace is injected into the chromatograph. Set the automatic sampler for a needle withdrawal time of 0.3 minute, a pressurization time of 1 minute, an injection time of 0.08 minute, and a vial pressure of 22 psig with the vial vent off. Obtain the peak areas for ethylene oxide, propylene oxide, and 1,4-dioxane, which have relative retention times of about 1.0, 1.3, and 3.1, respectively. Plot the area versus parts per million on linear graph paper, and draw the best straight line through the points.

Procedure—Place the vial containing the *Test preparation* in the automatic sampler, and chromatograph its headspace as done for the *Standard preparations*. Obtain the peak areas of each of the components and read the concentrations directly from the *Calibration* plots. Not more than 5 ppm of ethylene oxide, propylene oxide, or 1,4-dioxane is found.

Polyethylene Excipient

» Polyethylene Excipient is a homopolymer produced by the direct polymerization of ethylene.

Packaging and storage—Preserve in well-closed containers.

Reference standards—*USP High-density Polyethylene Reference Standard*—Do not dry before using. *USP Low-density Polyethylene Reference Standard*—Do not dry before using.

Identification—The infrared absorption spectrum of a film of it, prepared by dissolving it in hot toluene and evaporating the solution on a cesium bromide plate, exhibits maxima only at the same wavelengths as that of a similar preparation of USP High-density Polyethylene RS or USP Low-density Polyethylene RS.

Intrinsic viscosity in 1,2,3,4-tetrahydronaphthalene (see *Viscosity* ⟨911⟩)—[*Caution—Avoid contact with 1,2,3,4-tetrahydronaphthalene when performing this test.*] Dissolve about 1 g of Polyethylene Excipient in 95 mL of 1,2,3,4-tetrahydronaphthalene, filter into a 100-mL volumetric flask, dilute with 1,2,3,4-tetrahydronaphthalene to volume, and mix (*Test Solution 1*). Transfer 50.0 mL of *Test Solution 1* to a tared dish, evaporate on a steam bath in a well-ventilated hood for about 1 hour, then evaporate in a vacuum oven at 70° to constant weight. Calculate the concentration, in g per 100 mL, of *Test Solution 1*. Dilute with 1,2,3,4-tetrahydronaphthalene 5.0-mL and 10.0-mL portions of *Test Solution 1* to 50.0 mL to obtain *Test Solution 2* and *Test Solution 3*, and calculate the concentration of each. Determine the flow time, t_0, in seconds, of the solvent, using a capillary viscosimeter immersed in a constant-temperature bath maintained at 130°. Similarly, determine the flow time, t, of each *Test Solution*. Calculate the reduced viscosity of each *Test Solution* by the formula:

$$[(t/t_0) - 1]/C,$$

in which C is the concentration corresponding to the *Test Solution* with flow time t. Plot the reduced viscosities against the respective concentrations, and extrapolate to zero concentration to obtain the intrinsic viscosity. The intrinsic viscosity is not less than 0.126.

Heavy metals, *Method II* ⟨231⟩: 0.004%.

Volatile substances—Proceed as directed under *Loss on drying* ⟨731⟩, heating 4 g at 105° for 45 minutes: it loses not more than 0.5% of its weight. [*Caution—To reduce the explosion hazard, pass carbon dioxide or nitrogen into the lower part of the oven at a rate of about 100 mL per minute, or use an explosion-proof oven.*]

Polyethylene Glycol

$H(OCH_2CH_2)_nOH$

Poly(oxy-1,2-ethanediyl), α-hydro-ω-hydroxy-.
Polyethylene glycol [*25322-68-3*].

» Polyethylene Glycol is an addition polymer of ethylene oxide and water, represented by the formula:

$$H(OCH_2CH_2)_nOH,$$

in which n represents the average number of oxyethylene groups. The average molecular weight is not less than 95.0 percent and not more than 105.0 percent of the labeled nominal value if the labeled nominal value is below 1000; it is not less than 90.0 percent and not more than 110.0 percent of the labeled nominal value if the labeled nominal value is between 1000 and 7000; it is not less than 87.5 percent and not more than 112.5 percent of the labeled nominal value if the labeled nominal value is above 7000.

Packaging and storage—Preserve in tight containers.

Labeling—Label it to state, as part of the official title, the average nominal molecular weight of the Polyethylene Glycol.

Completeness and color of solution—A solution of 5 g of Polyethylene Glycol in 50 mL of water is colorless; it is clear for liquid grades and not more than slightly hazy for solid grades.

Viscosity ⟨911⟩—Determine its viscosity, using a capillary viscosimeter giving a flow time of not less than 200 seconds, and a liquid bath maintained at 98.9 ± 0.3° C (210° F). The viscosity is within the limits specified in the accompanying table. For a Polyethylene Glycol not listed in the table, calculate the limits by interpolation.

Average molecular weight—

Phthalic anhydride solution—Place 49.0 g of phthalic anhydride into an amber bottle, and dissolve in 300 mL of pyridine that has been freshly distilled over phthalic anhydride. Shake the bottle vigorously until solution is effected. Add 7 g of imidazole, swirl carefully to dissolve, and allow to stand for 16 hours before using.

Test preparation for liquid Polyethylene Glycols—Carefully introduce 25.0 mL of the *Phthalic anhydride solution* into a dry, heat-resistant pressure bottle. To the bottle add an accurately weighed amount of the specimen equivalent to its expected average molecular weight divided by 160. Place the stopper in the bottle, and wrap it securely in a cloth bag.

Test preparation for solid Polyethylene Glycols—Carefully introduce 25.0 mL of *Phthalic anhydride solution* into a dry, heat-resistant pressure bottle. To the bottle add an accurately weighed amount of the specimen, equivalent to its expected molecular weight divided by 160; because of limited solubility, however, do not use more than 25 g. Add 25 mL of pyridine, freshly distilled over phthalic anhydride, swirl to effect solution, place the stopper in the bottle, and wrap it securely in a cloth bag.

Procedure—Immerse the bottle in a water bath maintained at a temperature between 96° and 100°, to the same depth as that of the mixture in the bottle. Remove the bottles from the bath after 5 minutes, and, without unwrapping, swirl for 30 seconds to ensure homogeneity. Heat in the water bath for 30 minutes (60 minutes for Polyethylene Glycols having molecular weights of 3000 or higher), then remove the bottle from the bath, and allow it to cool to room temperature. Uncap the bottle carefully to release any pressure, remove the bottle from the bag, add 10 mL of water, and swirl thoroughly. Wait 2 minutes, add 0.5 mL of a 1 in 100 solution of phenolphthalein in pyridine, and titrate with 0.5 *N* sodium hydroxide to the first pink color that persists for 15 seconds, recording the volume, in mL, of 0.5 *N* sodium hydroxide required as *S*. Perform a blank determination on 25.0 mL of *Phthalic anhydride solution* plus any additional pyridine added to the bottle, and record the volume, in mL, of 0.5 *N* sodium hydroxide required as *B*. Calculate the average molecular weight by the formula:

$$[2000W]/[(B - S)(N)],$$

in which W is the weight, in g, of the Polyethylene Glycol, $(B - S)$ is the difference between the volumes of 0.5 *N* sodium hydroxide consumed by the blank and by the specimen, and N is the normality of the sodium hydroxide solution.

Nominal Average Molecular Weight	Viscosity Range, Centistokes	Nominal Average Molecular Weight	Viscosity Range, Centistokes
300	5.4 to 6.4	2400	49 to 65
400	6.8 to 8.0	2500	51 to 70
500	8.3 to 9.6	2600	54 to 74
600	9.9 to 11.3	2700	57 to 78
700	11.5 to 13.0	2800	60 to 83
800	12.5 to 14.5	2900	64 to 88
900	15.0 to 17.0	3000	67 to 93
1000	16.0 to 19.0	3250	73 to 105
1100	18.0 to 22.0	3350	76 to 110
1200	20.0 to 24.5	3500	87 to 123
1300	22.0 to 27.5	3750	99 to 140
1400	24 to 30	4000	110 to 158
1450	25 to 32	4250	123 to 177
1500	26 to 33	4500	140 to 200
1600	28 to 36	4750	155 to 228
1700	31 to 39	5000	170 to 250
1800	33 to 42	5500	206 to 315
1900	35 to 45	6000	250 to 390
2000	38 to 49	6500	295 to 480
2100	40 to 53	7000	350 to 590
2200	43 to 56	7500	405 to 735
2300	46 to 60	8000	470 to 900

pH ⟨791⟩: between 4.5 and 7.5, determined potentiometrically, in a solution prepared by dissolving 5.0 g of Polyethylene Glycol in 100 mL of carbon dioxide–free water and adding 0.30 mL of saturated potassium chloride solution.

Residue on ignition ⟨281⟩: not more than 0.1%, a 25-g specimen and a tared platinum dish being used, and the residue being moistened with 2 mL of sulfuric acid.

Arsenic, *Method II* ⟨211⟩: 3 ppm.

Limit of ethylene glycol and diethylene glycol (for Polyethylene Glycol having a nominal molecular weight less than 450)—

Standard preparation—Prepare an aqueous solution containing 500 µg of ethylene glycol and 500 µg of diethylene glycol in each mL.

Test preparation—Transfer about 4 g of Polyethylene Glycol, accurately weighed, to a 10-mL volumetric flask, dissolve in water, dilute with water to volume, and mix.

Chromatographic system (see *Chromatography* ⟨621⟩)—Under typical conditions, the instrument contains a 1.5-m × 3-mm stainless steel column packed with 12 percent G13 on S1NS. The column is maintained at about 165°, the injection port is maintained at about 260°, and nitrogen or another suitable inert gas is used as the carrier gas at a flow rate of 70 mL per minute.

Procedure—Inject 2.0 µL of the *Standard preparation* into a suitable gas chromatograph equipped with a flame-ionization detector, and record the chromatogram, adjusting the operational conditions to obtain peaks not less than 10 cm in height. Measure the heights of the first (ethylene glycol) and second (diethylene glycol) peaks, and record the values as P_1 and P_2, respectively. Inject 2.0 µL of the *Test preparation* into the chromatograph, and record the chromatogram under the same conditions as those employed for the *Standard preparation*. Measure the heights of the first (ethylene glycol) and second (diethylene glycol) peaks, and record the values as p_1 and p_2, respectively. Calculate the percentage of ethylene glycol by the formula:

$$(C_1p_1)/(P_1W),$$

in which C_1 is the concentration, in µg per mL, of ethylene glycol in the *Standard preparation*, and W is the weight, in mg, of Polyethylene Glycol taken. Calculate the percentage of diethylene glycol by the formula:

$$(C_2p_2)/(P_2W),$$

in which C_2 is the concentration, in µg per mL, of diethylene glycol in the *Standard preparation:* not more than 0.25% of combined ethylene glycol and diethylene glycol is found.

Limit of ethylene glycol and diethylene glycol (for Polyethylene Glycol having a nominal molecular weight 450 or above but not more than 1000)—

Ceric ammonium nitrate solution—Dissolve 6.25 g of ceric ammonium nitrate in 100 mL of 0.25 *N* nitric acid. Use within 3 days.

Standard preparation—Transfer 62.5 mg of diethylene glycol to a 25-mL volumetric flask, dissolve in a mixture of equal volumes of freshly distilled acetonitrile and water, dilute with the same mixture to volume, and mix.

Test preparation—Dissolve 50.0 g of Polyethylene Glycol in 75 mL of diphenyl ether, previously warmed, if necessary, just to melt the crystals, in a 250-mL distilling flask. Slowly distil at a pressure of 1 mm to 2 mm of mercury, into a receiver graduated to 100 mL in 1-mL subdivisions, until 25 mL of distillate has been collected. Add 20.0 mL of water to the distillate, shake vigorously, and allow the layers to separate. Cool in an ice bath to solidify the diphenyl ether and facilitate its removal. Filter the separated aqueous layer, wash the diphenyl ether with 5.0 mL of ice-cold water, pass the washings through the filter, and collect the filtrate and washings in a 25-mL volumetric flask. Warm to room temperature, dilute with water to volume, if necessary, and mix. Mix this solution with 25.0 mL of freshly distilled acetonitrile in a glass-stoppered, 125-mL conical flask.

Procedure—Transfer 10.0 mL each of the *Standard preparation* and the *Test preparation* to separate 50-mL flasks, each containing 15.0 mL of *Ceric ammonium nitrate solution*, and mix. Within 2 to 5 minutes, concomitantly determine the absorbances of the solutions in 1-cm cells at the wavelength of maximum absorbance at about 450 nm, with a suitable spectrophotometer, using a blank consisting of a mixture of 15.0 mL of *Ceric ammonium nitrate solution* and 10.0 mL of a mixture of equal volumes of freshly distilled acetonitrile and water: the absorbance of the solution from the *Test preparation* does not exceed that of the solution from the *Standard preparation*, corresponding to not more than 0.25% of combined ethylene glycol and diethylene glycol.

Ethylene oxide—

Morpholine solution—Mix 1 part of recently distilled morpholine with 9 parts of anhydrous methanol.

Mixed indicator—Weigh 0.050 g of 4,4′-bis(4-amino-1-naphthylazo)-2,2′-stilbenedisulfonic acid and 0.010 g of brilliant yellow into a 60-mL vial. Pipet 1.5 mL of 0.1 *N* sodium hydroxide into the vial, and mix. Add 3.5 mL of water, and mix. Transfer the mixture to a storage bottle with the aid of 45 mL of methanol as a rinse, and mix.

Standard methanolic hydrochloric acid—Mix 8.5 mL of hydrochloric acid and 1 liter of anhydrous methanol, and standardize by titrating 9.00 mL with 0.1 *N* sodium hydroxide VS to a phenolphthalein end-point. Restandardize if this solution is used more than 48 hours after standardization.

Procedure—Place 50 mL of anhydrous methanol into a 250-mL conical flask. Add 4 to 6 drops of *Mixed indicator*, and titrate with *Standard methanolic hydrochloric acid* to a clear blue color. Transfer to the flask about 25 g of Polyethylene Glycol, accurately weighed, to provide the specimen blank, and swirl to effect complete solution. Titrate with *Standard methanolic hydrochloric acid* to a clear blue color, approaching the end-point carefully, using small increments of titrant. Place 50 mL of *Morpholine solution* into a heat-resistant pressure bottle, and place an equal amount into a similar bottle to provide the reagent blank. To the first bottle add about 25 g of Polyethylene Glycol, accurately weighed, and swirl to effect complete solution. Wrap the bottles securely in a cloth bag, and place them close together in a water bath maintained at 98 ± 2° for 30 minutes, keeping the water level in the bath just above the liquid level in the bottles. Remove the bottles from the bath, and allow them to cool in air to room temperature. When the bottles have cooled, loosen the wrappers, uncap to release any pressure, and remove the wrappers. Slowly add 20 mL of acetic anhydride to each bottle, and swirl to effect complete solution. Allow to stand at room temperature for 15 minutes. If the bottles are still warm, cool them to room temperature. To each bottle add 4 to 6 drops of *Mixed indicator*, and titrate with *Standard methanolic hydrochloric acid* to a clear blue color, adding very small increments when approaching the end-point. Calculate the percentage of ethylene oxide by the formula:

$$4.41N[(A - B)/W_1 - C/W_2],$$

in which N is the normality of the *Standard methanolic hydrochloric acid*, A, B, and C are the volumes, in mL, required in the titration of the specimen, the reagent blank, and the specimen blank, respectively, and W_1 and W_2 are the weights, in g, of Polyethylene Glycol taken for the reaction and the specimen blank, respectively: the limit is 0.02%.

Heavy metals ⟨231⟩—Mix 4.0 g with 5.0 mL of 0.1 *N* hydrochloric acid, and dilute with water to 25 mL: the limit is 5 ppm.

Polyethylene Glycol Ointment

» Prepare Polyethylene Glycol Ointment as follows:

Polyethylene Glycol 3350	400 g
Polyethylene Glycol 400	600 g
To make	1000 g

Heat the two ingredients on a water bath to 65°. Allow to cool, and stir until congealed. If a firmer preparation is desired, replace up to 100 g of the polyethylene glycol 400 with an equal amount of polyethylene glycol 3350.

NOTE—If 6 percent to 25 percent of an aqueous solution is to be incorporated in Polyethylene Glycol Ointment, replace 50 g of the polyethylene glycol 3350 with an equal amount of stearyl alcohol.

Packaging and storage—Preserve in well-closed containers.

Polyethylene Oxide

» Polyethylene Oxide is a nonionic homopolymer of ethylene oxide, represented by the formula:

$$(OCH_2CH_2)_n,$$

in which n represents the average number of oxyethylene groups. It is a white powder obtainable in several grades, varying in viscosity profile when dissolved in water. It may contain not more than 3.0 percent of silicon dioxide.

Packaging and storage—Preserve in tight, light-resistant containers.

Labeling—The labeling indicates its viscosity profile in aqueous solution.

Reference standard—*USP Polyethylene Oxide Reference Standard*—Dry in vacuum at room temperature to constant weight before using.

Identification—

A: The infrared absorption spectrum of a potassium bromide dispersion of it, previously dried, exhibits maxima only at the same wavelengths as that of a similar preparation of USP Polyethylene Oxide RS.

B: The aqueous viscosity, determined at 25° and in a concentration as directed in the labeling, falls within the viscosity range indicated by the labeling.

Loss on drying ⟨731⟩—Dry about 4 g of it at 105° for 45 minutes: it loses not more than 1.0% of its weight.

Silicon dioxide and Non-silicon dioxide residue on ignition—Weigh accurately about 1 g into a previously ignited, tared 50-mL platinum crucible. Add 4 drops of sulfuric acid. Heat carefully on a hot plate until the specimen is thoroughly charred and fumes no longer are evolved. Ignite the crucible at 700 ± 25° (see *Residue on Ignition* ⟨281⟩) to constant weight. Wet the residue carefully with 1 mL of water, and slowly add 20 drops of hydrofluoric acid. [*Caution—Hydrofluoric acid is an extremely hazardous chemical. When handling it, wear a face shield, arm protection, and rubber gloves, and perform the operation in a hood.*] Evaporate slowly on a hot plate to dryness, then ignite at 700 ± 25° for 10 minutes, cool to room temperature in a desiccator, and weigh accurately. Repeat the addition of hydrofluoric acid, evaporation, and ignition, to constant weight. Calculate the percentage of silicon dioxide from the difference between the net weights before and after the hydrofluoric acid treatment. Calculate the percentage of Non-silicon dioxide residue on ignition from the final net weight: it is not more than 2.0%.

Arsenic, *Method II* ⟨211⟩: 3 ppm.

Heavy metals, *Method II* ⟨231⟩: 0.001%.

Free ethylene oxide—[NOTE (1)—This section requires the use of flammable, toxic, and reactive chemicals. Handle them safely, with gloves, in an approved hood. (2)—Several 30-mL septum-seal vials with polytef-lined septa and seals are needed in this section.]

Reactant solution—Into a 30-mL vial weigh 3.0 g of 4-chlorothiophenol. Add about 25 mL of dimethylformamide, and swirl to effect solution. When solution is complete, proceed as follows. Add 1.0 g of diazabicyclononene, and swirl. [NOTE—A yellow color develops.] Affix the stopper, crimp-seal the vial, and weigh it accurately. Add 50 µL of heptadecane through the septum with the aid of a syringe, again weigh accurately, and calculate the weight of internal standard added by difference. Calculate the percentage, by weight, of internal standard present in the solution. [NOTE—This solution may be stored for not longer than one week.]

Spiking solution—[NOTE—Ethylene oxide is extremely flammable, reactive, and harmful to human beings. Exercise great care in handling it, until the final solution is sealed and refrigerated.] Obtain liquid ethylene oxide by venting the gas from a cylinder through a cooled trap in a hood. Place 2 mL in a cooled 30-mL vial. Crimp-seal the vial, and store in a refrigerator. Place 25 mL of dimethylformamide in another 30-mL vial, affix the septum-stopper, crimp-seal, and weigh it accurately. Add 50 µL of liquid ethylene oxide through the septum with the aid of a cooled syringe. Again weigh the vial accurately, and calculate the weight of ethylene oxide added by difference. Store under refrigeration. Calculate the percentage, by weight, of ethylene oxide present. [NOTE—This solution may be stored for not longer than one week.]

Retention time solution—Place about 10 mL of dimethylformamide in a 30-mL vial, affix the stopper, and crimp-seal. Add 100 µL of *Reactant solution* through the septum, using a 100-µL syringe. Using another, cooled 100-µL syringe, add 100 µL of *Spiking solution* through the septum. Shake the vial for 30 seconds to ensure mixing. Allow to stand for not less than 4 hours at room temperature for adduct to form.

Test preparation—Accurately weigh about 3 g of cooled Polyethylene Oxide into a 30-mL vial. Add about 10 mL of dimethylformamide, affix the septum, and crimp-seal. Weigh the vial accurately, add 100 µL of *Reactant solution* with the aid of a syringe through the septum, again weigh accurately, and calculate the weight of *Reactant solution* added by difference. Shake the vial for 30 seconds to ensure mixing. Allow to stand overnight at room temperature for adduct to form fully. [NOTE—Do not warm the vial, since warming may cause a gel to form.] Calculate the weight of internal standard in the vial from the weight of *Reactant solution* added and the percentage, by weight, of internal standard in the *Reactant solution.*

Low-level spiked test preparation—Proceed as directed for *Test preparation*, except, after crimp-sealing the vial and weighing, first to add, through the septum, 20 µL of *Spiking solution* with the aid of a cooled 100-µL syringe, then weigh accurately, and calculate the weight of ethylene oxide in the vial from the weight of *Spiking solution* added and the weight percentage of ethylene oxide in the *Spiking solution.* Calculate the percentage, by weight, of added ethylene oxide in the *Spiked test preparation.*

High-level spiked test preparation—Proceed as directed for *Low-level spiked test preparation*, except to add 40 µL of *Spiking solution* instead of 20 µL.

Chromatographic system (see *Chromatography* ⟨621⟩)—The gas chromatograph is equipped with a flame-ionization detector maintained at about 300°, a 30-m × 0.32-mm fused silica capillary column containing liquid phase G1 in the form of a 1-µm layer bonded to the wall, and a split injection system, maintained at about 250°. The column temperature is maintained at 150° for 5 minutes after injection, then programmed to 290° at 10° per minute, and then held at 290° for 11 minutes. The carrier gas and make-up gases are helium, at 20 psi head pressure and 30 mL per minute split flow. Ascertain the relevant retention times by injecting 1 µL of the *Retention time solution.* Retention times are constant within ±0.02 minute. [NOTE—Several injections of *Retention time solution* may be needed to passivate fully the active sites on the column. Incomplete passivation is evidenced by humps and/or broad features for the adduct.] Relative retention times for the adduct and the internal standard are about 1.0 and 1.2, respectively. The two values for *Free ethylene oxide*, determined as directed under *Procedure*, agree to within 10%.

Procedure—Inject 1.0-µL portions of the *Test preparation*, the *Low-level spiked test preparation*, and the *High-level spiked test preparation* into the chromatograph, and record the peak areas for the adduct and the internal standard in each chromatogram. Calculate the percentage of adduct in the *Test preparation* by the formula:

$$100(W_i/W_t)(r_a/r_i),$$

in which W_i and W_t are the weights, in mg, of internal standard and Polyethylene Oxide, respectively, in the vial of *Test preparation*, and r_a and r_i are the areas of the peaks for the adduct and the internal standard, respectively, in the chromatogram of the *Test preparation.* Similarly calculate the percentage of adduct in the *Low-level spiked test preparation* and the *High-level spiked test preparation*, by the same formula, using appropriate values of W_i, W_t, r_a, and r_i from each *Spiked test preparation* and its chromatogram. Calculate the percentage of free ethylene oxide in the Polyethylene Oxide taken by the formula:

$$C_1C_3/(C_2 - C_1),$$

in which C_1 and C_2 are the percentages of adduct found in the analysis of the *Test preparation* and the *Low-level spiked test preparation*, respectively, and C_3 is the percentage of added ethylene oxide in the *Low-level spiked test preparation.* Similarly, calculate the percentage of free ethylene oxide, using C_2 and C_3 values from the *High-level spiked test preparation.* The average of the two values is not more than 0.001%.

Polyoxyethylene 50 Stearate—*see* Polyoxyl 50 Stearate

Polyoxyl 10 Oleyl Ether

Polyoxy-1,2-ethanediyl, α-[(*Z*)-9-octadecenyl-ω-hydroxy-.
Polyethylene glycol monooleyl ether [*9004-98-2*].

» Polyoxyl 10 Oleyl Ether is a mixture of the monooleyl ethers of mixed polyoxyethylene diols, the average polymer length being equivalent to not less than 8.6 and not more than 10.4 oxyethylene units. It may contain suitable stabilizers.

Packaging and storage—Preserve in tight containers, in a cool place.

Labeling—Label it to indicate the names and proportions of any added stabilizers.

Reference standard—*USP Polyoxyl 10 Oleyl Ether Reference Standard.*

Identification—Its infrared absorption spectrum, obtained by spreading a capillary film of it between sodium chloride plates, exhibits maxima only at the same wavelengths as that of a similar preparation of USP Polyoxyl 10 Oleyl Ether RS.

Water, *Method I* ⟨921⟩: not more than 3.0%.

Residue on ignition—Weigh accurately about 25 g into a tared 40-mL porcelain crucible, and heat in contact with air until it ignites spontaneously, or can be ignited with a glowing splint. Allow the flame to go out, place the crucible in a muffle furnace with the door partly open until the carbon is consumed, close the door, and heat at about 700 ± 100° for 1 hour. Cool in a desiccator, weigh, and calculate the percentage of residue. If the amount so obtained exceeds 0.4%, again heat until constant weight is attained: the limit is 0.4%.

Arsenic, *Method II* ⟨211⟩: 2 ppm.

Heavy metals, *Method II* ⟨231⟩: 0.002%.

Acid value ⟨401⟩: not more than 1.0.

Hydroxyl value ⟨401⟩: between 75 and 95.

Iodine value, *Method I* ⟨401⟩: between 23 and 40, about 550 mg of Polyoxyl 10 Oleyl Ether, accurately weighed, being used, and the reaction time being extended to 60 minutes.

Saponification value ⟨401⟩: not more than 3.

Free polyethylene glycols—Transfer about 12 g, accurately weighed, to a 500-mL separator containing 50 mL of ethyl acetate. Add 50 mL of sodium chloride solution (29 in 100), shake vigorously for 2 minutes, and allow to separate for 15 minutes. Drain the lower, aqueous phase into a second 500-mL separator, and extract the upper layer with a second 50-mL portion of sodium chloride solution (29 in 100). To the combined aqueous layers add 50 mL of ethyl acetate, shake vigorously for 2 minutes, and allow to separate as before. Drain the lower, aqueous phase into a third 500-mL separator, and extract with two 50-mL portions of chloroform by shaking for 2 minutes each time. Evaporate the combined chloroform extracts in a 150-mL beaker on a steam bath, with the aid of a stream of nitrogen, to apparent dryness. Redissolve in about 15 mL of chloroform, and transfer to a filter, collecting the filtrate in a 150-mL beaker. Rinse the funnel with several small portions of chloroform, and evaporate the combined filtrate and rinsings, as described above, until no odor of chloroform or ethyl acetate is perceptible. Cool in a desiccator, and weigh: the limit is 7.5%.

Free ethylene oxide—

Internal standard solution—Prepare a solution containing 100 mg of *n*-butyl chloride in each mL of chlorobenzene. Store in a tightly closed container. Prepare fresh weekly.

Standard solution—[NOTE—Ethylene oxide is toxic and flammable. Prepare this solution in a well-ventilated hood, using great care.] Place 250 mL of chlorobenzene in a glass-stoppered, 500-mL conical flask. Bubble ethylene oxide through the chlorobenzene at a moderate rate for about 30 minutes, insert the stopper, and store with protection from heat. Pipet 25 mL of a 0.5 *N* alcoholic hydrochloric acid solution, prepared by mixing 45 mL of hydrochloric acid with 1 liter of alcohol, into a 500-mL conical flask containing 40 g of magnesium chloride hexahydrate. Shake the mixture to effect saturation. Pipet 10 mL of the ethylene oxide solution into the flask, and add 20 drops of bromocresol green TS. If the solution is not yellow (acid) at this point, add an additional volume, accurately measured, of 0.5 *N* alcoholic hydrochloric acid to give an excess of about 10 mL. Record the total volume of 0.5 *N* alcoholic hydrochloric acid added. Insert the stopper in the flask, and allow to stand for 30 minutes. Titrate the excess acid with 0.5 *N* alcoholic potassium hydroxide VS. Perform a blank titration, using 10.0 mL of chlorobenzene instead of ethylene oxide solution, adding the same total volume of 0.5 *N* alcoholic hydrochloric acid, and note the difference in volumes required. Each mL of the difference in volumes of 0.5 *N* alcoholic potassium hydroxide consumed is equivalent to 22.02 mg of ethylene oxide. Calculate the concentration, in mg per mL, of ethylene oxide in the *Standard solution.* Standardize daily.

Standard preparation—Transfer about 5 g of USP Polyoxyl 10 Oleyl Ether RS to a suitable glass bottle of about 60-mL capacity, and add 10 mL of chlorobenzene, exactly 50 µL of *Internal standard solution*, and an accurately measured volume of *Standard solution* containing about 0.5 mg of ethylene oxide. Insert a magnetic stirring bar, cap the bottle tightly, and stir until homogeneity is attained.

Test preparation—Transfer about 5 g of Polyoxyl 10 Oleyl Ether, accurately weighed, to a suitable glass bottle of about 60-mL capacity, and add 10 mL of chlorobenzene and 50 µL, accurately measured, of *Internal standard solution.* Insert a magnetic stirring bar, cap the bottle tightly, and stir until homogeneity is attained.

Chromatographic system—Under typical conditions, the instrument is equipped with a flame-ionization detector, and contains a 1.8-m × 3-mm (OD) stainless steel column packed with S3. The injector port and detector block are maintained at about 210° and 230°, respectively, and the column at about 160°. Helium is used as the carrier gas at a flow rate of 66 mL per minute.

Interference check—Inject a suitable volume of chlorobenzene into the gas chromatograph, and allow the chromatogram to run until the solvent has eluted. Similarly inject and chromatograph the *Internal standard solution*, the *Standard solution*, and a solution prepared according to the directions for the *Test preparation*, but omitting the internal standard. No interfering peaks are observed.

Procedure—Inject about 2 µL of the *Standard preparation* into a suitable gas chromatograph, and record the chromatogram. Similarly, inject about 2 µL of the *Test preparation*, and record the chromatogram. Calculate the quantity, in mg, of ethylene oxide in the portion of Polyoxyl 10 Oleyl Ether taken by the formula:

$$W_S(R_U/R_S),$$

in which W_S is the weight, in mg, of ethylene oxide in the portion of *Standard solution* taken, and R_U and R_S are the area ratios of ethylene oxide to internal standard in the chromatograms for the *Test preparation* and the *Standard preparation*, respectively. The limit is 0.01%.

Average polymer length—If solid material is present, place the Polyoxyl 10 Oleyl Ether in a 60° water bath overnight. Shake vigorously to eliminate any possibility of molecular weight gradients within it. Add about 1 mL of the melt to 1 mL of carbon tetrachloride in a test tube, and shake the test tube until dissolution is complete. Transfer about 0.5 mL to an NMR tube, and add 5 drops of tetramethylsilane as an internal reference standard. Cap the tube tightly, and shake thoroughly. Place the tube in the NMR spectrometer, and record the NMR spectrum at an appropriate RF power level and a sweep time of 250 seconds per 500 Hz (see *Qualitative scans* under *Nuclear Magnetic Resonance* ⟨761⟩). Adjust the spectrum amplitude so that the signal at 1.1 ppm is at least 80% of full scale. Record the integral areas from 0.4 ppm to 2.35 ppm (A_1), and from 2.35 ppm to 4.9 ppm (A_2) at a sweep time of 50 seconds per 500 Hz at an integral power level such that the integral of the largest peak is at least 80% of full chart height. Do not change the power level during the sweep. Record the integral of each peak several times, and calculate the average integral area. Calculate the number of oxyethylene units, *n*, per molecule by the formula:

$$n = (31A_2/A_1 - 3)/4,$$

in which 31 is the total number of protons in the molecule not activated by either oxygen or a double bond, 3 is the number of oxygen-activated protons not included in the oxyethylene unit count, and 4 is the number of protons in each oxyethylene unit.

Polyoxyl 20 Cetostearyl Ether

» Polyoxyl 20 Cetostearyl Ether is a mixture of monocetostearyl (mixed hexadecyl and octadecyl) ethers of mixed polyoxyethylene diols, the average polymer length being equivalent to not less than 17.2 and not more than 25.0 oxyethylene units.

Packaging and storage—Preserve in tight containers, in a cool place.

Reference standard—*USP Polyoxyl 20 Cetostearyl Ether Reference Standard.*

Identification—Its infrared absorption spectrum, obtained by spreading a capillary film of it between sodium chloride plates, exhibits maxima only at the same wavelengths as that of a similar preparation of USP Polyoxyl 20 Cetostearyl Ether RS.

pH ⟨791⟩: between 4.5 and 7.5, determined in a solution (1 in 10).

Water, *Method I* ⟨921⟩: not more than 1.0%.

Residue on ignition—Weigh accurately about 25 g into a tared 40-mL porcelain crucible, and heat in contact with air until it ignites spontaneously, or can be ignited with a glowing splint. Allow the flame to go out, place the crucible in a muffle furnace with the door partly open until the carbon is consumed, close the door, and heat at 700 ± 100° for 1 hour. Cool in a desiccator, weigh, and calculate the percentage of residue. If the amount so obtained exceeds 0.4%, again heat until constant weight is attained: the limit is 0.4%.

Arsenic, *Method II* ⟨211⟩: 2 ppm.

Heavy metals, *Method II* ⟨231⟩: 0.002%.

Acid value ⟨401⟩: not more than 0.5.

Hydroxyl value ⟨401⟩: between 42 and 60.

Saponification value ⟨401⟩: not more than 2.

Free polyethylene glycols—Transfer about 12 g, accurately weighed, to a 500-mL separator containing 50 mL of ethyl acetate. Add 50 mL of sodium chloride solution (29 in 100), shake vigorously for 2 minutes, and allow to separate for 15 minutes. Drain the lower, aqueous phase into a second 500-mL separator, and extract the upper layer with a second 50-mL portion of sodium chloride solution (29 in 100). To the combined aqueous layers add 50 mL of ethyl acetate, shake vigorously for 2 minutes, and allow to separate as before. Drain the lower, aqueous phase into a third 500-mL separator, and extract with two 50-mL portions of chloroform, by shaking for 2 minutes each time. Evaporate the combined chloroform extracts in a 150-mL beaker on a steam bath, with the aid of a stream of nitrogen, to apparent dryness. Redissolve in about 15 mL of chloroform, and transfer to a filter, collecting the filtrate in a 150-mL beaker. Rinse the funnel with several small portions of chloroform, and evaporate the combined filtrate and rinsings, as described above, until no odor of chloroform or ethyl acetate is perceptible. Cool in a desiccator, and weigh: the limit is 7.5%.

Free ethylene oxide—

Internal standard solution—Prepare a solution containing 100 mg of *n*-butyl chloride in each mL of chlorobenzene. Store in a tightly closed container. Prepare fresh weekly.

Standard solution—[NOTE—Ethylene oxide is toxic and flammable. Prepare this solution in a well-ventilated hood, using great care.] Place 250 mL of chlorobenzene in a glass-stoppered, 500-mL conical flask. Bubble ethylene oxide through the chlorobenzene at a moderate rate for about 30 minutes, insert the stopper, and store protected from heat. Pipet 25 mL of 0.5 *N* alcoholic hydrochloric acid solution, prepared by mixing 45 mL of hydrochloric acid with 1 liter of alcohol, into a 500-mL conical flask containing 40 g of magnesium chloride hexahydrate. Shake the mixture to effect saturation. Pipet 10 mL of the ethylene oxide solution into the flask, and add 20 drops of bromocresol green TS. If the solution is not yellow (acid) at this point, add an additional volume, accurately measured, of 0.5 *N* alcoholic hydrochloric acid to give an excess of about 10 mL. Record the total volume of 0.5 *N* alcoholic hydrochloric acid added. Insert the stopper in the flask, and allow to stand for 30 minutes. Titrate the excess acid with 0.5 *N* alcoholic potassium hydroxide VS. Perform a blank titration, using 10.0 mL of chlorobenzene instead of ethylene oxide solution, adding the same total volume of 0.5 *N* alcoholic hydrochloric acid, and note the difference in volumes required. Each mL of the difference in volumes of 0.5 *N* alcoholic potassium hydroxide consumed is equivalent to 22.02 mg of ethylene oxide. Calculate the concentration, in mg per mL, of ethylene oxide in the *Standard solution*. Standardize daily.

Standard preparation—Transfer about 5 g of USP Polyoxyl 20 Cetostearyl Ether RS to a suitable glass bottle of about 60-mL capacity, and add 10 mL of chlorobenzene, exactly 50 µL of *Internal standard solution*, and an accurately measured volume of *Standard solution* containing about 0.5 mg of ethylene oxide. Insert a magnetic stirring bar, cap the bottle tightly, and stir until homogeneity is attained.

Test preparation—Transfer about 5 g of Polyoxyl 20 Cetostearyl Ether, accurately weighed, to a suitable glass bottle of about 60-mL capacity, and add 10 mL of chlorobenzene and 50 µL, accurately measured, of *Internal standard solution*. Insert a magnetic stirring bar, cap the bottle tightly, and stir until homogeneity is attained.

Chromatographic system—Under typical conditions, the instrument is equipped with a flame-ionization detector, and contains a 1.8-m × 3-mm (OD) stainless steel column packed with S3. The injector port and detector block are maintained at about 210° and 230°, respectively, and the column at about 160°. Helium is used as the carrier gas at a flow rate of 66 mL per minute.

Interference check—Inject a suitable volume of chlorobenzene into the gas chromatograph, and allow the chromatogram to run until the solvent has eluted. Similarly inject and chromatograph the *Internal standard solution*, the *Standard solution*, and a solution prepared according to the directions for the *Test preparation*, but omitting the internal standard. No interfering peaks are observed.

Procedure—Inject about 2 µL of the *Standard preparation* into a suitable gas chromatograph, and record the chromatogram. Similarly, inject about 2 µL of the *Test preparation*, and record the chromatogram. Calculate the quantity, in mg, of ethylene oxide in the portion of Polyoxyl 20 Cetostearyl Ether taken by the formula:

$$W_S(R_U/R_S),$$

in which W_S is the weight, in mg, of ethylene oxide in the portion of *Standard solution* taken, and R_U and R_S are the area ratios of ethylene oxide to internal standard in the chromatograms for the *Test preparation* and the *Standard preparation*, respectively. The limit is 0.01%.

Average polymer length—Place the Polyoxyl 20 Cetostearyl Ether in a 50° water bath overnight, in order to melt it completely. Shake vigorously to eliminate any possibility of molecular weight gradients within it, and transfer 200 µL to a 5- × 180-mm high-resolution NMR sample tube. Add 200 µL of carbon tetrachloride by means of a separate microsyringe. Add 5 drops of tetramethylsilane as an internal reference standard. Cap the tube tightly, and shake thoroughly. Place the tube in the NMR spectrometer, and record the NMR spectrum at an appropriate RF power level and a sweep time of 250 seconds per 500 Hz (see *Qualitative scans* under *Nuclear Magnetic Resonance* ⟨761⟩). Adjust the spectrum amplitude so that the signal at 1.1 ppm is at least 80% of full scale. Record the integral areas from 0.4 ppm to 2.35 ppm (A_1), and from 2.35 ppm to 4.9 ppm (A_2) at a sweep time of 50 seconds per 500 Hz at an integral power level such that the integral of the ethylene oxide peak at 3.5 ppm is at least 80% of full chart height. Do not change the power level during the sweep. Record the integral of each peak several times, and calculate the average integral area. Calculate the number of oxyethylene units, *n*, per molecule by the formula:

$$n = (32A_2/A_1 - 3)/4,$$

in which 32 is the total number of protons in the molecule not activated by oxygen, averaged for the cetyl and stearyl radicals, 3 is the number of oxygen-activated protons not included in the oxyethylene unit count, and 4 is the number of protons in each oxyethylene unit.

Polyoxyl 35 Castor Oil

» Polyoxyl 35 Castor Oil contains mainly the triricinoleate ester of ethoxylated glycerol, with smaller amounts of polyethylene glycol ricinoleate and the corresponding free glycols. It results from the reaction of glycerol ricinoleate with about 35 moles of ethylene oxide.

Packaging and storage—Preserve in tight containers.

Identification—

A: Dissolve about 0.1 g in 1 mL of water, add 9 mL of sodium chloride solution (1 in 20), and heat in a water bath: the solution becomes turbid at a temperature between 65° and 85°, and becomes clear on cooling to 40°.

B: Dissolve about 0.1 g in 10 mL of alcoholic potassium hydroxide TS, boil for about 3 minutes, and evaporate to dryness. Mix the residue with 5 mL of water: it dissolves, yielding a clear solution. Add a few drops of glacial acetic acid: a white precipitate is formed.

C: To a solution (1 in 20) add bromine TS, dropwise: the bromine is decolorized.

Specific gravity ⟨841⟩: between 1.05 and 1.06.

Viscosity ⟨911⟩: between 650 and 850 centipoises at 25°, a capillary viscosimeter being used.

Water, *Method I* ⟨921⟩: not more than 3.0%.

Residue on ignition ⟨281⟩: not more than 0.3%.

Heavy metals, *Method II* ⟨231⟩: 0.001%.

Acid value ⟨401⟩: not more than 2.0.

Hydroxyl value ⟨401⟩: between 65 and 80.

Iodine value ⟨401⟩: between 25 and 35.

Saponification value ⟨401⟩: between 60 and 75.

Polyoxyl 40 Hydrogenated Castor Oil

» Polyoxyl 40 Hydrogenated Castor Oil contains mainly the tri-hydroxystearate ester of ethoxylated glycerol, with smaller amounts of polyethylene glycol tri-hydroxystearate and of the corresponding free glycols. It results from the reaction of glycerol tri-hydroxystearate with about 40 to 45 moles of ethylene oxide.

Packaging and storage—Preserve in tight containers.

Identification—

A: Dissolve about 0.1 g in 1 mL of water, add 9 mL of sodium chloride solution (1 in 20), and heat in a water bath: the solution becomes turbid at a temperature between 70° and 85°.

B: Dissolve about 0.1 g in 10 mL of alcoholic potassium hydroxide TS, boil for about 3 minutes, and evaporate to dryness. Mix the residue with 5 mL of water: it dissolves, yielding a clear solution. Add a few drops of glacial acetic acid: a white precipitate is formed.

Congealing temperature ⟨651⟩: between 20° and 30°.

Water, *Method I* ⟨921⟩: not more than 3.0%.

Residue on ignition ⟨281⟩: not more than 0.3%.

Heavy metals, *Method II* ⟨231⟩: 0.001%.

Acid value ⟨401⟩: not more than 2.0.

Hydroxyl value ⟨401⟩: between 60 and 80.

Iodine value ⟨401⟩: not more than 2.0.

Saponification value ⟨401⟩: between 45 and 69.

Polyoxyl 40 Stearate

Poly(oxy-1,2-ethanediyl), α-hydro-ω-hydroxy-, octadecanoate.
Polyethylene glycol monostearate [*9004-99-3*].

» Polyoxyl 40 Stearate is a mixture of the monoesters and di-esters of Stearic Acid or Purified Stearic Acid with mixed polyoxyethylene diols, the average polymer length being equivalent to about 40 oxyethylene units.

Packaging and storage—Preserve in tight containers.

Reference standard—*USP Polyoxyl 40 Stearate Reference Standard*—Do not dry before using. Keep container tightly closed.

Identification—The infrared absorption spectrum of a mineral oil dispersion of it exhibits maxima only at the same wavelengths as that of a similar preparation of USP Polyoxyl 40 Stearate RS.

Congealing temperature ⟨651⟩: between 37° and 47°.

Water, *Method I* ⟨921⟩: not more than 3.0%.

Arsenic, *Method I* ⟨211⟩—Prepare the *Test Preparation* as follows. Mix 1.0 g of it with 3 g of a mixture of equal parts of powdered potassium nitrate and sodium carbonate. Add the mixture, in small portions, to a platinum crucible at red heat, and allow the reaction to cease. Extract the cooled residue by boiling it with 20 mL of 2 *N* sulfuric acid, cool the solution, and filter. Wash the residue with 20 mL of water, and evaporate the combined filtrate and washing to fumes of sulfur trioxide: the residue, dissolved in 5 mL of water, meets the requirements of the test. The limit is 3 ppm.

Heavy metals, *Method II* ⟨231⟩: 0.001%.

Acid value ⟨401⟩: not more than 2.

Hydroxyl value ⟨401⟩: between 25 and 40.

Saponification value ⟨401⟩: between 25 and 35.

Free polyethylene glycols—Transfer about 6 g, accurately weighed, to a 500-mL separator containing 50 mL of ethyl acetate. Dissolve completely, then add 50 mL of sodium chloride solution (29 in 100), shake vigorously for 2 minutes, and allow to separate for 15 minutes. If separation is incomplete, carefully insert the separator into the well of a steam bath for short time intervals. Repeat this technique as many times as necessary to ensure the complete separation of the two phases. Cool, and drain the lower, aqueous, phase into a second 500-mL separator, and extract the upper layer with a second 50-mL portion of sodium chloride solution (29 in 100). Repeat the separation as before, including the steam bath technique to enhance complete separation. To the combined aqueous layers add 50 mL of ethyl acetate, shake vigorously for 2 minutes, and allow to separate as before. Drain the lower, aqueous phase into a third 500-mL separator, and extract it with two 50-mL portions of chloroform, by shaking for 2 minutes each time. Repeat the steam bath technique to ensure complete separation. Evaporate the combined chloroform extracts in a 150-mL beaker on a steam bath, with the aid of a stream of nitrogen, to apparent dryness. Redissolve in about 15 mL of chloroform, and transfer to a filter, collecting the filtrate in a 150-mL beaker. Rinse the funnel with several small portions of chloroform, and evaporate the combined filtrate and rinsings, as described above, until no odor of chloroform or ethyl acetate is perceptible. Dry in vacuum at 60° for 1 hour. Cool in a desiccator, and weigh: not less than 17% and not more than 27% of free polyethylene glycols is found.

Polyoxyl 50 Stearate

Poly(oxy-1,2-ethanediyl), α-(1-oxooctadecyl)-ω-hydroxy-.
Polyethylene glycol monostearate [*9004-99-3*].

» Polyoxyl 50 Stearate is a mixture of the monostearate and distearate esters of mixed polyoxyethylene diols and the corresponding free diols. The average polymer length is equivalent to about 50 oxyethylene units.

Packaging and storage—Preserve in tight containers.

Reference standard—*USP Polyoxyl 50 Stearate Reference Standard*—Do not dry before using. Keep container tightly closed.

Identification—The infrared absorption spectrum of a mineral oil dispersion of it exhibits maxima only at the same wavelengths as that of a similar preparation of USP Polyoxyl 50 Stearate RS.

Free polyethylene glycols—Proceed as directed for *Free polyethylene glycols* under *Polyoxyl 40 Stearate:* not less than 17.0% and not more than 27.0% of free polyethylene glycols is found.

Acid value ⟨401⟩: not more than 2.

Hydroxyl value ⟨401⟩: between 23 and 35.

Saponification value ⟨401⟩: between 20 and 28.

Other requirements—It meets the requirements for *Water*, *Arsenic*, and *Heavy metals* under *Polyoxyl 40 Stearate.*

Polysorbate 20

$HO(C_2H_4O)_w$ ... $(OC_2H_4)_xOH$; $C(OC_2H_4)_yOH$; $H_2C(OC_2H_4)_zR$

[Sum of w, x, y, and z is 20;
R is $(C_{11}H_{23})COO$]

Sorbitan, monododecanoate, poly(oxy-1,2-ethanediyl) derivs.
Polyoxyethylene 20 sorbitan monolaurate. [*9005-64-5*].

» Polysorbate 20 is a laurate ester of sorbitol and its anhydrides copolymerized with approximately 20 moles of ethylene oxide for each mole of sorbitol and sorbitol anhydrides.

Packaging and storage—Preserve in tight containers.

Identification—

A: It responds to *Identification tests A* and *C* under *Polysorbate 80.*

B: To 2 mL of a solution (1 in 20) add 0.5 mL of bromine TS dropwise: the bromine is not decolorized (*distinction from Polysorbate 80*).

Hydroxyl value ⟨401⟩: between 96 and 108.

Saponification value ⟨401⟩: between 40 and 50.

Other requirements—It meets the requirements for *Water*, *Residue on ignition*, *Arsenic*, *Heavy metals*, and *Acid value* under *Polysorbate 80.*

Polysorbate 40

Sorbitan, monohexadecanoate, poly(oxy-1,2-ethanediyl) derivs.
Polyoxyethylene 20 sorbitan monopalmitate. [Compound usually contains also associated fatty acids.] [*9005-66-7*].

» Polysorbate 40 is a palmitate ester of sorbitol and its anhydrides copolymerized with approximately 20 moles of ethylene oxide for each mole of sorbitol and sorbitol anhydrides.

Packaging and storage—Preserve in tight containers.

Identification—

A: It responds to *Identification tests A* and *C* under *Polysorbate 80.*

B: To 2 mL of a solution (1 in 20) add 0.5 mL of bromine TS, dropwise: the bromine is not decolorized (*distinction from Polysorbate 80*).

Hydroxyl value ⟨401⟩: between 89 and 105.

Saponification value ⟨401⟩: between 41 and 52.

Other requirements—It meets the requirements for *Water, Residue on ignition, Arsenic, Heavy metals,* and *Acid value* under *Polysorbate 80.*

Polysorbate 60

Sorbitan, monooctadecanoate, poly(oxy-1,2-ethanediyl) derivs.
Polyoxyethylene 20 sorbitan monostearate. [Compound usually contains also associated fatty acids.] [*9005-67-8*].

» Polysorbate 60 is a mixture of stearate and palmitate esters of sorbitol and its anhydrides copolymerized with approximately 20 moles of ethylene oxide for each mole of sorbitol and sorbitol anhydrides.

Packaging and storage—Preserve in tight containers.

Identification—

A: It responds to *Identification tests A* and *C* under *Polysorbate 80.*

B: To 2 mL of a solution (1 in 20) add 0.5 mL of bromine TS, dropwise: the bromine is not decolorized (*distinction from Polysorbate 80*).

Hydroxyl value ⟨401⟩: between 81 and 96.

Saponification value ⟨401⟩: between 45 and 55.

Other requirements—It meets the requirements for *Water, Residue on ignition, Arsenic, Heavy metals,* and *Acid value* under *Polysorbate 80.*

Polysorbate 80

Sorbitan, mono-9-octadecenoate, poly(oxy-1,2-ethanediyl) derivs., (*Z*)-.
Polyoxyethylene 20 sorbitan monooleate [*9005-65-6*].

» Polysorbate 80 is an oleate ester of sorbitol and its anhydrides copolymerized with approximately 20 moles of ethylene oxide for each mole of sorbitol and sorbitol anhydrides.

Packaging and storage—Preserve in tight containers.

Identification—

A: To 5 mL of a solution (1 in 20) add 5 mL of sodium hydroxide TS. Boil for a few minutes, cool, and acidify with 3 *N* hydrochloric acid: the solution is strongly opalescent.

B: To 2 mL of a solution (1 in 20) add 0.5 mL of bromine TS, dropwise: the bromine is decolorized.

C: A mixture of 60 volumes of it and 40 volumes of water yields a gelatinous mass at normal and lower than normal room temperatures.

Specific gravity ⟨841⟩: between 1.06 and 1.09.

Viscosity ⟨911⟩: between 300 and 500 centistokes when determined at 25°.

Water, *Method I* ⟨921⟩: not more than 3.0%.

Residue on ignition ⟨281⟩: not more than 0.25%.

Arsenic, *Method I* ⟨211⟩: 1 ppm.

Heavy metals, *Method II* ⟨231⟩: 0.001%.

Acid value—Weigh 10.0 g into a wide-mouth, 250-mL conical flask, and add 50 mL of neutralized alcohol. Heat on a steam bath nearly to boiling, thoroughly shaking occasionally while heating. Invert a beaker over the mouth of the flask, cool under running water, add 5 drops of phenolphthalein TS, and titrate with 0.1 *N* sodium hydroxide VS: not more than 4 mL of 0.100 *N* sodium hydroxide is required, corresponding to an acid value of 2.2.

Hydroxyl value ⟨401⟩: between 65 and 80.

Saponification value ⟨401⟩: between 45 and 55.

Polyvinyl Acetate Phthalate

» Polyvinyl Acetate Phthalate is a reaction product of phthalic anhydride and a partially hydrolyzed polyvinyl acetate. It contains not less than 55.0 percent and not more than 62.0 percent of phthalyl (*o*-carboxybenzoyl, $C_8H_5O_3$) groups, calculated on an anhydrous acid-free basis.

Packaging and storage—Prepare in tight containers.

Identification—

A: The solution prepared for measurement of absorbance in the *Assay* exhibits a maximum at 277 ± 3 nm.

B: To about 10 mg contained in a small test tube add 10 mg of resorcinol, mix, add 0.5 mL of sulfuric acid, and heat in a liquid bath at 160° for 3 minutes. Cool, and pour the solution into a mixture of 25 mL of 1 *N* sodium hydroxide and 200 mL of water: the solution shows a vivid green fluorescence.

C: Dissolve about 2 g in 20 mL of methanol, and pour 1 mL of the solution onto a clear glass plate: a film is deposited as the methanol evaporates.

Viscosity ⟨911⟩—Dissolve a quantity, equivalent to 15 g on the anhydrous basis, in 85 g of methanol. Determine the viscosity of this solution, using a capillary viscosimeter. The apparent viscosity is between 9 and 11 centipoises, determined at 25 ± 0.2°.

Water, *Method I* ⟨921⟩: not more than 5.0%.

Residue on ignition ⟨281⟩: not more than 1.0%.

Free phthalic acid—

Standard preparation—Transfer about 50 mg of phthalic anhydride, accurately weighed, to a 100-mL volumetric flask, dissolve in water, dilute with water to volume, and mix.

Test preparation—Dissolve about 1500 mg of Polyvinyl Acetate Phthalate, accurately weighed, in 50 mL of a mixture of methylene chloride and methanol (4:1). Transfer the solution to a separator with the aid of 75 mL of water, and swirl, taking care not to shake. Add 100 mL of hexanes, shake, and allow the mixture to stand until it separates into 2 layers. Transfer the water layer to a 250-mL volumetric flask. Add 100 ml of water to the separator, shake, and allow to stand until the layers separate. Transfer the water layer to the same volumetric flask, dilute with water to volume, and mix. If the solution is cloudy, centrifuge a portion until clear.

Procedure—Concomitantly determine the absorbances of the *Standard preparation* and the *Test preparation* in 1-cm cells at the wavelength of maximum absorbance at about 277 nm. Calculate the percentage of free phthalic acid by the formula:

$$25\,(166.13/148.12)(W_S/W_U)(A_U/A_S),$$

in which 166.13 and 148.12 are the molecular weights of phthalic acid and phthalic anhydride, respectively, W_S and W_U are the weights, in mg, of phthalic anhydride and Polyvinyl Acetate Phthalate, respectively, and A_U and A_S are the absorbances of the *Test preparation* and the *Standard preparation*, respectively. Not more than 0.6%, on the anhydrous basis, is found.

Free acid other than phthalic—Proceed as directed for *Test preparation* under *Free phthalic acid*, but instead of transferring the water extracts to a volumetric flask, transfer them to a 400-mL beaker, and titrate with 0.1 *N* sodium hydroxide VS to a phenolphthalein end-point. Calculate the volume, V_p, in mL, of 0.1 *N* sodium hydroxide consumed by the free phthalic acid in the specimen by the formula:

$$(1/830.6)pW,$$

in which *p* is the percentage of free phthalic acid, previously determined, and *W* is the weight, in mg, of the specimen, on the anhydrous basis. Calculate the percentage of free acid other than phthalic, as acetic acid, by the formula:

$$60.05(10/W)(V—Vp),$$

in which *V* is the total volume, in mL, of 0.1 *N* sodium hydroxide used. Not more than 0.6%, on the anhydrous basis, is found.

Phthalyl content—

Standard preparation—Transfer 50 mg of phthalic anhydride, accurately weighed, to a 1000-mL volumetric flask, dissolve with heat in 100 mL of alcohol, dilute with alcohol to volume, and mix.

Assay preparation—Transfer about 100 mg of Polyvinyl Acetate Phthalate, accurately weighed, to a 1000-mL volumetric flask, dissolve in alcohol, dilute with alcohol to volume, and mix.

Procedure—Concomitantly determine the absorbances of the *Standard preparation* and the *Assay preparation* in-cm cells at the wavelength of maximum absorbance at about 275 nm. Calculate the percentage of phthalyl, on the acid-free basis, by the formula:

$$100(W_S/W_U)(A_U/A_S),$$

in which W_S and W_U are the weights, in mg, of phthalic anhydride and Polyvinyl Acetate Phthalate on the anhydrous basis, respectively, and A_U and A_S are the absorbances of the *Assay preparation* and the *Standard preparation*, respectively.

Polyvinyl Alcohol—*see* Polyvinyl Alcohol USP

Potassium Benzoate

C_6H_5COOK

$C_7H_5KO_2$ 160.21
Benzoic acid, potassium salt.
Potassium benzoate [*582-25-2*].

» Potassium Benzoate contains not less than 99.0 percent and not more than 100.5 percent of $C_7H_5KO_2$, calculated on the anhydrous basis.

Packaging and storage—Preserve in well-closed containers.

Identification—It responds to the flame test for *Potassium* ⟨191⟩ and to the tests for *Benzoate* ⟨191⟩.

Alkalinity—Dissolve 2 g in 20 mL of hot water, and add 2 drops of phenolphthalein TS: the pink color produced, if any, is discharged by the addition of 0.20 mL of 0.10 *N* sulfuric acid.

Water, *Method I* ⟨921⟩: not more than 1.5%.

Arsenic, *Method II* ⟨211⟩: 3 ppm.

Heavy metals ⟨231⟩—Dissolve 4.0 g in 40 mL of water, add, dropwise, with vigorous stirring, 10 mL of 3 *N* hydrochloric acid, and filter. Use 25 mL of the filtrate: the limit is 0.001%.

Assay—Transfer about 600 mg of Potassium Benzoate, accurately weighed, to a 250-mL beaker. Add 100 mL of glacial acetic acid, stir until the assay specimen is completely dissolved, add 2 drops of crystal violet TS, and titrate with 0.1 *N* perchloric acid VS. Perform a blank determination, and make any necessary correction. Each mL of 0.1 *N* perchloric acid is equivalent to 16.02 mg of $C_7H_5KO_2$.

Potassium Chloride—*see* Potassium Chloride USP

Potassium Citrate—*see* Potassium Citrate USP

Potassium Hydroxide

KOH 56.11
Potassium hydroxide.
Potassium hydroxide [*1310-58-3*].

» Potassium Hydroxide contains not less than 85.0 percent of total alkali, calculated as KOH, including not more than 3.5 percent of K_2CO_3.

Caution—Exercise great care in handling Potassium Hydroxide, as it rapidly destroys tissues.

Packaging and storage—Preserve in tight containers.

Identification—A solution (1 in 25) responds to the tests for *Potassium* ⟨191⟩.

Insoluble substances—Dissolve 1 g in 20 mL of water: the solution is complete, clear, and colorless.

Heavy metals ⟨231⟩—Dissolve 0.67 g in a mixture of 5 mL of water and 7 mL of 3 *N* hydrochloric acid. Heat to boiling, cool, and dilute with water to 25 mL: the limit is 0.003%.

Assay—Dissolve about 1.5 g of Potassium Hydroxide, accurately weighed, in 40 mL of carbon dioxide–free water. Cool the solution to 15°, add phenolphthalein TS, and titrate with 1 *N* sulfuric acid VS. At the discharge of the pink color of the indicator, record the volume of acid solution required, then add methyl orange TS, and continue the titration to a persistent pink color. Each mL of 1 *N* sulfuric acid is equivalent to 56.11 mg of total alkali, calculated as KOH, and each mL of acid consumed in the titration with methyl orange is equivalent to 138.2 mg of K_2CO_3.

Potassium Metabisulfite

$K_2S_2O_5$ 222.32
Disulfurous acid, dipotassium salt.
Dipotassium pyrosulfite [*16731-55-8*].

» Potassium Metabisulfite contains an amount of $K_2S_2O_5$ equivalent to not less than 51.8 percent and not more than 57.6 percent of SO_2.

Packaging and storage—Preserve in well-fitted, tight containers, and avoid exposure to excessive heat.

Identification—A solution (1 in 20) responds to the tests for *Potassium* ⟨191⟩ and for *Sulfite* ⟨191⟩.

Arsenic, *Method I* ⟨211⟩—Prepare the *Test Preparation* as follows. Dissolve 1.0 g in 10 mL of water in a 150-mL beaker. Cautiously add 10 mL of nitric acid and 5 mL of sulfuric acid, and evaporate on a steam bath to a volume of about 5 mL. Place the beaker on a hot plate, and heat until dense white fumes of sulfur trioxide first appear. Cool, cautiously wash down the inner wall of the beaker with about 10 mL of water, and again heat to the appearance of dense white fumes of sulfur trioxide. Cool, repeat the washing and fuming, and cool again. This solution meets the requirements of the test, the addition of 20 mL of 7 *N* sulfuric acid specified under *Procedure* being omitted. The limit is 3 ppm.

Iron ⟨241⟩—Dissolve 1.00 g in 14 mL of dilute hydrochloric acid (2 in 7), and evaporate on a steam bath to dryness. Dissolve the residue in 7 mL of dilute hydrochloric acid (2 in 7), and again evaporate to dryness. Dissolve the resulting residue in a mixture of 2 mL of hydrochloric acid and 20 mL of water, add 3 drops of bromine TS, and boil to expel the bromine. Cool, and dilute with water to 47 mL: the limit is 0.001%.

Heavy metals, *Method I* ⟨231⟩—Dissolve 2 g in 20 mL of water, add 5 mL of hydrochloric acid, evaporate on a steam bath to about 1 mL, and dissolve the residue in 25 mL of water: the limit is 0.001%.

Assay—Transfer about 250 mg of Potassium Metabisulfite, accurately weighed, to a glass-stoppered conical flask containing 50.0 mL of 0.1 *N* iodine VS, and swirl to dissolve. Allow to stand for 5 minutes, protected from light, add 1 mL of hydrochloric acid, and titrate the excess iodine with 0.1 *N* sodium thiosulfate VS, adding 3 mL of starch TS as the end-point is approached. Each mL of 0.1 *N* iodine is equivalent to 3.203 mg of SO_2.

Potassium Metaphosphate

KPO_3 118.07
Metaphosphoric acid (HPO_3), potassium salt.
Potassium metaphosphate [*7790-53-6*].

» Potassium Metaphosphate is a straight-chain polyphosphate, having a high degree of polymerization. It contains the equivalent of not less than 59.0 percent and not more than 61.0 percent of P_2O_5.

Packaging and storage—Preserve in well-closed containers.

Identification—

A: Add 1 g of finely powdered Potassium Metaphosphate, slowly and with vigorous stirring, to 100 mL of sodium chloride solution (1 in 50): a gelatinous mass is formed.

B: Boil a mixture of 0.5 g of Potassium Metaphosphate, 10 mL of nitric acid, and 50 mL of water for 30 minutes, and cool: the resulting solution responds to the tests for *Potassium* ⟨191⟩ and for *Phosphate* ⟨191⟩.

Viscosity ⟨911⟩—Mix 300 mg with 200 mL of sodium pyrophosphate solution (3.5 in 1000), using a magnetic stirrer. Determine the viscosity of the clear solution obtained, or of the liquid phase of the mixture obtained after 30 minutes of continuous stirring: the viscosity is between 6.5 and 15 centipoises.

Fluoride—Place 5.0 g of Potassium Metaphosphate, 25 mL of water, 50 mL of perchloric acid, 5 drops of silver nitrate solution (1 in 2), and a few glass beads in a 250-mL distilling flask connected with a condenser and carrying a thermometer and a capillary tube, both of which extend into the liquid. Connect a small dropping funnel, filled with water, or a steam generator to the capillary tube. Support the flask on a distillation shield with a hole that exposes about one-third of the bottom of the flask to the flame. Distil into a 250-mL volumetric flask until the temperature reaches 135°. Add water from the funnel or introduce steam through the capillary to maintain the temperature between 135° and 140°. Continue the distillation until 225 mL to 240 mL has been collected, then dilute with water to volume, and mix. Transfer 50.0 mL of this solution to a 100-mL color-comparison tube, and transfer 50.0 mL of water to a similar tube to serve as a control. Add to each tube 0.1 mL of a filtered solution of sodium alizarin-sulfonate TS and 1 mL of freshly prepared hydroxylamine hydrochloride solution (1 in 4000), and mix. Add, dropwise, and with stirring, 0.05 *N* sodium hydroxide to the tube containing the distillate until its color just matches that of the control, which is faintly pink. Then add to each tube 1.0 mL of 0.1 *N* hydrochloric acid, and mix. From a buret, graduated in 0.05-mL increments, add slowly to the tube containing the distillate enough thorium nitrate solution (1 in 4000) so that, after mixing, the color of the liquid just changes to a faint pink. Note the volume of the solution added, add the same volume, accurately measured, to the control, and mix. Then add to the control sodium fluoride TS (10 µg of F per mL) from a buret to make the colors of the two tubes match after dilution to the same volume. Mix, and allow all air bubbles to escape before making the final color comparison. Check the end-point by adding 1 or 2 drops of sodium fluoride TS to the control. A distinct change in color appears. The volume of sodium fluoride TS required for the control solution does not exceed 1.0 mL (0.001%).

Arsenic, *Method I* ⟨211⟩—Prepare the *Test Preparation* by dissolving 1.0 g in 15 mL of 3 *N* hydrochloric acid. The limit is 3 ppm.

Lead—A solution of 1 g in 10 mL of 3 *N* hydrochloric acid contains not more than 5 µg of lead (corresponding to not more than 5 ppm of Pb) when tested as directed in the test for *Lead* ⟨251⟩.

Heavy metals, *Method I* ⟨231⟩—Warm 1 g with 10 mL of 3 *N* hydrochloric acid until no more dissolves. Add 15 mL of water, mix, and filter: the limit is 0.002%.

Assay—Mix about 200 mg of Potassium Metaphosphate, accurately weighed, with 15 mL of nitric acid and 30 mL of water, boil for 30 minutes, cool, and dilute with water to about 100 mL. Heat to 60°, add an excess of ammonium molybdate TS, and heat at 50° for 30 minutes. Filter, and wash the precipitate, first with 0.5 *N* nitric acid, and then potassium nitrate solution (1 in 100) until the filtrate is no longer acid to litmus. Add 25 mL of water to the precipitate, dissolve it in 50.0 mL of 1 *N* sodium hydroxide VS, add phenolphthalein TS, and titrate the excess sodium hydroxide with 1 *N* sulfuric acid VS. Each mL of 1 *N* sodium hydroxide is equivalent to 3.086 mg of P_2O_5.

Monobasic Potassium Phosphate

KH_2PO_4 136.09
Phosphoric acid, monopotassium salt.
Monopotassium phosphate [*7778-77-0*].

» Monobasic Potassium Phosphate, dried at 105° for 4 hours, contains not less than 98.0 percent and not more than 100.5 percent of KH_2PO_4.

Packaging and storage—Preserve in tight containers.

Identification—A solution (1 in 20) responds to the tests for *Potassium* ⟨191⟩ and for *Phosphate* ⟨191⟩.

Loss on drying ⟨731⟩—Dry it at 105° for 4 hours: it loses not more than 1.0% of its weight.

Insoluble substances—Dissolve 10 g in 100 mL of hot water, filter through a tared filtering crucible, wash the insoluble residue with hot water, and dry at 105° for 2 hours: the residue does not exceed 20 mg (0.2%).

Fluoride—Proceed as directed in the test for *Fluoride* under *Dibasic Calcium Phosphate.* The limit is 0.001%.

Arsenic, *Method I* ⟨211⟩: 3 ppm.

Heavy metals, *Method I* ⟨231⟩—Dissolve 1 g in 25 mL of water: the limit is 0.002%.

Lead—A solution of 1 g in 20 mL of water contains not more than 5 µg of lead (corresponding to not more than 5 ppm of Pb) when tested as directed in the test for *Lead* ⟨251⟩.

Assay—Transfer about 5 g of Monobasic Potassium Phosphate, previously dried and accurately weighed, to a 250-mL beaker, add 100 mL of water and 5.0 mL of 1 *N* hydrochloric acid VS, and stir until the assay specimen is completely dissolved. Place the electrodes of a suitable pH meter in the solution, and slowly titrate the excess acid, stirring constantly, with 1 *N* sodium hydroxide VS to the inflection point occurring at about pH 4. Record the buret reading, and calculate the volume (*A*), if any, of 1 *N* hydrochloric acid consumed by the assay specimen. Continue the titration with 1 *N* sodium hydroxide VS until the inflection point occurring at about pH 8.8 is reached, record the buret reading, and calculate the volume (*B*) of 1 *N* sodium hydroxide required in the titration between the two inflection points (pH 4 and 8.8). Each mL of the volume ($B - A$) of 1 *N* sodium hydroxide is equivalent to 136.1 mg of KH_2PO_4.

Potassium, Polacrilin—*see* Polacrilin Potassium

Potassium Sorbate

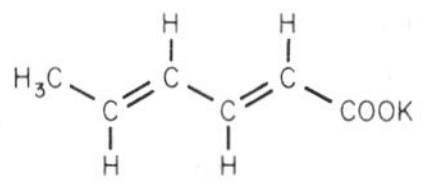

$C_6H_7KO_2$ 150.22
2,4-Hexadienoic acid, (*E,E*)-, potassium salt; 2,4-Hexadienoic acid, potassium salt.
Potassium (*E,E*)-sorbate; Potassium sorbate [*590-00-1*] and [*24634-61-5*].

» Potassium Sorbate contains not less than 98.0 percent and not more than 101.0 percent of $C_6H_7KO_2$, calculated on the dried basis.

Packaging and storage—Preserve in tight containers, protected from light, and avoid exposure to excessive heat.

Identification—

A: Dissolve 1 g in 10 mL of water: the solution responds to the test for *Potassium* ⟨191⟩.

B: Dissolve 0.2 g in 2 mL of water, and add a few drops of bromine TS: the color is discharged.

Acidity or alkalinity—Dissolve 1.1 g in 20 mL of water, and add phenolphthalein TS. If the solution is colorless, titrate with 0.10 *N* sodium hydroxide to a pink color that persists for 15 seconds: not more than 1.1 mL is required. If the solution is pink in color, titrate with 0.10 *N* hydrochloric acid: not more than 0.80 mL is required to discharge the pink color.

Loss on drying ⟨731⟩—Dry it at 105° for 3 hours: it loses not more than 1.0% of its weight.

Heavy metals, *Method II* ⟨231⟩: 0.001%.

Assay—Dissolve about 300 mg of Potassium Sorbate, accurately weighed, in 40 mL of glacial acetic acid, warming, if necessary, to effect solution. Cool to room temperature, add 1 drop of crystal violet TS, and titrate with 0.1 *N* perchloric acid VS to a blue-green end-point. Perform a blank determination, and make any necessary correction. Each mL of 0.1 *N* perchloric acid is equivalent to 15.02 mg of $C_6H_7KO_2$.

Povidone—*see* Povidone USP
Powdered Cellulose—*see* Cellulose, Powdered
Precipitated Calcium Carbonate—*see* Calcium Carbonate, Precipitated USP
Pregelatinized Starch—*see* Starch, Pregelatinized

Propane

C_3H_8 44.10
[*74-98-6*].

» Propane contains not less than 98.0 percent of C_3H_8.

Caution—Propane is highly flammable and explosive.

Packaging and storage—Preserve in tight cylinders, and prevent exposure to excessive heat.

Identification—

A: The infrared absorption spectrum of it exhibits maxima, among others, at about the following wavelengths, in μm: 3.4 (*vs*), 6.8 (*s*), and 7.2 (*m*).

B: The vapor pressure of a test specimen, obtained as directed in the *General Sampling Procedure* under *Propellants* (see *Aerosols* ⟨601⟩), determined at 21° by means of a suitable pressure gauge, is between 820 and 875 kPa absolute (119 and 127 psia).

Water: not more than 0.001%, determined as directed for *Water Content* under *Aerosols* ⟨601⟩.

High-boiling residues: not more than 5 ppm, determined as directed for *High-boiling Residues, Method II*, under *Aerosols* ⟨601⟩.

Acidity of residue—Add 10 mL of water to the residue obtained in the test for *High-boiling residues*, mix by swirling for about 30 seconds, add 2 drops of methyl orange TS, insert the stopper in the tube, and shake vigorously: no pink or red color appears in the aqueous layer.

Sulfur compounds—Carefully open the container valve to produce a moderate flow of gas. Do not direct the gas stream toward the face, but deflect a portion of the stream toward the nose: the odor is free from the characteristic odor of sulfur compounds.

Assay—

Chromatographic system—Under typical conditions, the gas chromatograph is equipped with a thermal-conductivity detector and contains a 6-m × 3-mm aluminum column packed with 10 weight percent of liquid phase G30 on non-acid-washed support S1C, helium is used as the carrier gas at a flow rate of 50 mL per minute, and the temperature of the column is maintained at 33°.

System suitability—The peak responses obtained for Propane in the chromatograms from duplicate determinations agree within 1%.

Procedure—Connect 1 Propane cylinder to the chromatograph through a suitable sampling valve and a flow control valve downstream from the sampling valve. Flush the liquid specimen through the sampling valve, taking care to avoid entrapment of gas or air in the sampling valve. Inject a suitable volume, typically 2 μL, of Propane into the chromatograph, and record the chromatogram. Calculate the percentage purity by dividing 100 times the propane response by the sum of all of the responses in the chromatogram.

Propionic Acid

» Propionic Acid contains not less than 99.5 percent and not more than 100.5 percent, by weight, of $C_3H_6O_2$.

Packaging and storage—Preserve in tight containers.

Specific gravity ⟨841⟩: between 0.988 and 0.993.

Distilling range, *Method I* ⟨721⟩: between 138.5° and 142.5°.

Nonvolatile residue—Evaporate 20 g in a tared dish, and dry at 105° for 1 hour: the weight of the residue does not exceed 2.0 mg.

Arsenic, *Method II* ⟨211⟩: 3 ppm.

Heavy metals ⟨231⟩—To the residue obtained in the test for *Nonvolatile residue* add 8 mL of 0.1 *N* hydrochloric acid, warm gently until solution is complete, dilute with water to 100 mL, and use 20 mL of the solution: the limit is 0.001%.

Readily oxidizable substances—Dissolve 15 g of sodium hydroxide in 50 mL of water, cool, add 6 mL of bromine, stirring to effect complete solution, and dilute with water to 2000 mL. Transfer 25.0 mL of this solution to a glass-stoppered, 250-mL conical flask containing 100 mL of water, and add 10 mL of sodium acetate solution (1 in 5) and 10.0 mL of Propionic Acid. Allow to stand for 15 minutes, add 5 mL of potassium iodide solution (1 in 4) and 10 mL of hydrochloric acid, and titrate with 0.1 *N* sodium thiosulfate VS just to the disappearance of the brown color. Perform a blank determination. The difference between the volume of 0.1 *N* sodium thiosulfate required for the blank and that required for the test specimen is not more than 2.2 mL.

Aldehydes—Transfer 10.0 mL to a glass-stoppered, 250-mL conical flask containing 50 mL of water and 10.0 mL of sodium bisulfite solution (1 in 8), insert the stopper, and shake vigorously. Allow the mixture to stand for 30 minutes, then titrate with 0.1 *N* iodine VS to the same brownish yellow end-point obtained with a blank treated with the same quantities of the same reagents. The difference between the volume of 0.1 *N* iodine required for the blank and that required for the test specimen is not more than 1.75 mL.

Assay—Mix about 1.5 g of Propionic Acid, accurately weighed, with 100 mL of recently boiled and cooled water in a 250-mL conical flask, add phenolphthalein TS, and titrate with 0.5 *N* sodium hydroxide VS to the first appearance of a faint pink endpoint that persists for not less than 30 seconds. Each mL of 0.5 *N* sodium hydroxide is equivalent to 37.04 mg of $C_3H_6O_2$.

Propyl Gallate

OH
HO—⟨benzene ring⟩—COOCH$_2$CH$_2$CH$_3$
OH

$C_{10}H_{12}O_5$ 212.20
Benzoic acid, 3,4,5-trihydroxy-, propyl ester.
Propyl gallate [*121-79-9*].

» Propyl Gallate contains not less than 98.0 percent and not more than 102.0 percent of $C_{10}H_{12}O_5$, calculated on the dried basis.

Packaging and storage—Preserve in tight containers, protected from light, and avoid contact with metals.

Reference standard—*USP Propyl Gallate Reference Standard*—Dry at 105° for 4 hours before using.

Identification—

A: The infrared absorption spectrum of a mineral oil dispersion of it exhibits maxima only at the same wavelengths as that of a similar preparation of USP Propyl Gallate RS.

B: The ultraviolet absorption spectrum of the solution prepared for measurement of absorbance in the *Assay* exhibits maxima and minima at the same wavelengths as that of a 1 in 100,000 solution of USP Propyl Gallate RS in methanol, concomitantly measured.

Melting range ⟨741⟩: between 146° and 150°.

Loss on drying ⟨731⟩—Dry it at 105° for 4 hours: it loses not more than 0.5% of its weight.

Residue on ignition ⟨281⟩: not more than 0.1%.

Heavy metals, *Method II* ⟨231⟩: 0.001%.

Assay—Transfer about 100 mg of Propyl Gallate, accurately weighed, to a 250-mL volumetric flask, dissolve in about 100 mL of methanol, dilute with methanol to volume, and mix. Transfer 5.0 mL of this solution to a 200-mL volumetric flask, dilute with methanol to volume, and mix. Concomitantly determine the absorbances of this solution and a Standard solution of USP Propyl Gallate RS in methanol having a known concentration of about 10 µg per mL, in 1-cm cells at the wavelength of maximum absorbance at about 273 nm, with a suitable spectrophotometer, using methanol as the blank. Calculate the quantity, in mg, of $C_{10}H_{12}O_5$ in the portion of Propyl Gallate taken by the formula:

$$10C(A_U/A_S),$$

in which *C* is the concentration, in µg per mL, of USP Propyl Gallate RS in the Standard solution, and A_U and A_S are the absorbances of the solution of Propyl Gallate and the Standard solution, respectively.

Propylene Carbonate

H$_3$C—(1,3-dioxolan-2-one ring, =O)

$C_4H_6O_3$ 102.09
4-Methyl-1,3-dioxolan-2-one.
Cyclic propylene carbonate [*108-32-7*].

» Propylene Carbonate contains not less than 99.0 percent and not more than 100.5 percent of $C_4H_6O_3$.

Packaging and storage—Preserve in tight containers.

Reference standard—*USP Propylene Carbonate Reference Standard*—Do not dry before using.

Identification—The infrared absorption spectrum of a thin film of it exhibits maxima only at the same wavelengths as that of a similar preparation of USP Propylene Carbonate RS.

Specific gravity ⟨841⟩: between 1.203 and 1.210 at 20°.

pH ⟨791⟩—Gently but throughly mix 10 mL with 0.3 mL of saturated potassium chloride solution in a 100-mL borosilicate volumetric flask. Dilute with carbon dioxide–free water having a pH of 7.0 ± 0.5 to volume. Completely purge the solution by vigorous nitrogen bubbling, and continue the bubbling during the pH measurement. Determine the pH potentiometrically when the reading stabilizes: it is between 6.0 and 7.5.

Residue on ignition ⟨281⟩: not more than 0.01%.

Assay—

Barium hydroxide solution—Dissolve 75 g of barium hydroxide (octahydrate) in 1 liter of water. Filter before using.

Procedure—Flush a 250-mL iodine flask with nitrogen to expel the air, and insert the stopper in the flask to exclude carbon dioxide. Transfer 50.0 mL of *Barium hydroxide solution* and about 600 mg of Propylene Carbonate, accurately weighed, to the flask, and loosely reinsert the stopper. Moisten the stopper with 3 drops of water, and heat the flask on a steam bath for 10 minutes. Remove the flask from the steam bath, add 6 drops of phenolphthalein TS, and titrate while hot with 0.5 *N* hydrochloric acid VS until only a trace of pink color remains. Perform a blank determination, using the same *Barium hydroxide solution.* Each mL of 0.5 *N* hydrochloric acid consumed is equivalent to 25.52 mg of $C_4H_6O_3$.

Propylene Glycol—*see* Propylene Glycol USP

Propylene Glycol Alginate

» Propylene Glycol Alginate is a propylene glycol ester of alginic acid. Each g yields not less than 0.16 and not more than 0.20 g of carbon dioxide, calculated on the dried basis.

Packaging and storage—Preserve in well-closed containers.

Identification—Place 20 mL of the saponified solution obtained in the determination of *Esterified carboxyl groups* in a 250-mL conical flask, add 50 mL of a solution of periodic acid (1 in 50), swirl, and allow to stand for 30 minutes. Add 2 g of potassium iodide, titrate with sodium thiosulfate TS to a faint yellow color, dilute the mixture with water to 200 mL, and mix to obtain the Test solution for tests *A* and *B*.

A: To 10 mL of the Test solution add 5 mL of hydrochloric acid and 10 mL of fuchsin-sulfurous acid TS: a blue to blue-violet color, due to formaldehyde, develops in about 20 minutes.

B: To 10 mL of the Test solution add 1 mL of a saturated solution of piperazine and 0.5 mL of sodium nitroferricyanide TS: a green color, due to acetaldehyde, develops.

Microbial limits ⟨61⟩—The total bacterial count does not exceed 200 per g, and the tests for *Salmonella* species and *Escherichia coli* are negative.

Loss on drying ⟨731⟩—Dry it at 105° for 4 hours: it loses not more than 20.0% of its weight.

Ash—Weigh accurately about 3 g in a tared crucible, and incinerate at 650 ± 25° until free from carbon. Cool in a desiccator, weigh, and determine the weight of the ash: not more than 10.0%, calculated on the dried basis, is found.

Arsenic, *Method II* ⟨211⟩: 3 ppm.

Lead ⟨251⟩—Add 1.0 g to 20 mL of nitric acid in a 250-mL conical flask, mix, and heat carefully until the specimen is dissolved. Continue the heating until the volume is reduced to about 7 mL. Cool rapidly to room temperature, transfer to a 100-mL volumetric flask, dilute with water to volume, and mix. A 50.0-

mL portion of this solution contains not more than 5 µg of lead (corresponding to not more than 0.001% of Pb), 15 mL of ammonium citrate solution, 3 mL of potassium cyanide solution, and 0.5 mL of hydroxylamine hydrochloride solution being used for the test. After the first dithizone extractions, wash the combined chloroform layers with 5 mL of water, discarding the water layer and continuing in the usual manner by extracting with 20 mL of 0.2 *N* nitric acid.

Heavy metals, *Method II* ⟨231⟩: 0.004%, a platinum crucible being used for the ignition, and nitric acid being used in place of sulfuric acid to wet the test specimen.

Free carboxyl groups—Transfer about 1 g, accurately weighed, to a 600-mL beaker. Dissolve in 200 mL of water, stirring by mechanical means for not less than 30 minutes, and titrate with 0.1 *N* sodium hydroxide VS to a pH of 7.0, determining the end-point potentiometrically. Calculate the weight, in g, of free carboxyl groups in the weight, *W*, in g, of specimen taken by the formula:

$$0.0044V/W,$$

in which *V* is the volume, in mL, of 0.1 *N* sodium hydroxide consumed: the weight of free carboxyl groups found, calculated on the dried basis, is not more than 35% of the weight of carbon dioxide yielded by an equal weight of specimen in the *Assay*.

Esterified carboxyl groups—Transfer the solution obtained in the test for *Free carboxyl groups* with the aid of water to a 1000-mL conical flask, add phenolphthalein TS and 50.0 mL of 0.1 *N* sodium hydroxide VS, insert a stopper in the flask, mix, and allow to stand for 30 minutes at ambient temperature. Titrate the excess sodium hydroxide with 0.1 *N* hydrochloric acid VS to a faint pink end-point. Transfer the solution with the aid of water to a 600-mL beaker, and complete the titration to a pH of 7.0, determining the end-point potentiometrically. Calculate the weight, in g, of esterified carboxyl groups in the weight, *W*, in g, of specimen by the formula:

$$0.0044v/W,$$

in which *v* is the volume, in mL, of 0.1 *N* sodium hydroxide consumed: the weight of esterified carboxyl groups found, calculated on the dried basis, is between 40% and 85% of the weight of carbon dioxide yielded by an equal weight of specimen in the *Assay*.

Assay—Proceed as directed for *Procedure* under *Alginates Assay* ⟨311⟩, without preliminary drying of the Propylene Glycol Alginate.

Propylene Glycol Diacetate

» Propylene Glycol Diacetate contains not less than 98.0 percent and not more than 102.0 percent of $C_7H_{12}O_4$.

Packaging and storage—Preserve in tight containers, and avoid contact with metal.

Reference standard—*USP Propylene Glycol Diacetate Reference Standard*—Do not dry before using.

Identification—The infrared absorption spectrum of a thin film of it exhibits maxima and minima only at the same wavelengths as that of a similar preparation of USP Propylene Glycol Diacetate RS.

Specific gravity ⟨841⟩: between 1.040 and 1.060.

Refractive index ⟨831⟩: between 1.4130 and 1.4150, at 20°.

pH ⟨791⟩: between 4.0 and 6.0, in a solution (1 in 20).

Acetic acid—Dissolve about 10 g, accurately weighed, in 30 mL of a mixture of alcohol and water (1:1), previously neutralized to phenolphthalein TS, and titrate with 0.1 *N* sodium hydroxide VS to a phenolphthalein end-point. Each mL of 0.1 *N* sodium hydroxide is equivalent to 6.005 mg of $C_2H_4O_2$: the limit is 0.2%.

Chromatographic purity—

Chromatographic system (see *Chromatography* ⟨621⟩)—The gas chromatograph is equipped with a thermal conductivity detector and a 3-mm × 1.8-m stainless steel column packed with 20 percent liquid phase G4 on support S1A (unsilanized). Maintain the column at a temperature of 145°, and use helium as the carrier gas.

Procedure—Inject a suitable volume, typically about 1 µL, of Propylene Glycol Diacetate into the gas chromatograph, and record the chromatogram. The response of the main peak, which elutes at about 3.7 minutes, is not less than 98.0 percent of the total responses obtained.

Assay—Transfer about 150 mg, accurately weighed, to a 50-mL flask, add 6.0 mL of 0.5 *N* potassium hydroxide VS, connect the flask to a water-jacketed condenser, and reflux for 30 minutes. Cool, and titrate the excess alkali with 0.1 *N* hydrochloric acid VS to a phenolphthalein end-point. Perform a blank determination (see *Residual Titrations* under *Titrimetry* ⟨541⟩). Each mL of 0.5 *N* potassium hydroxide is equivalent to 40.04 mg of $C_7H_{12}O_4$.

Propylene Glycol Monostearate

Octadecanoic acid, monoester with 1,2-propanediol.
1,2-Propanediol monostearate [*1323-39-3*].

» Propylene Glycol Monostearate is a mixture of the propylene glycol mono- and di-esters of stearic and palmitic acids. It contains not less than 90.0 percent of monoesters of saturated fatty acids, chiefly propylene glycol monostearate ($C_{21}H_{42}O_3$) and propylene glycol monopalmitate ($C_{19}H_{38}O_3$).

Packaging and storage—Preserve in well-closed containers.

Congealing temperature ⟨651⟩: not lower than 45°.

Residue on ignition ⟨281⟩: not more than 0.5%.

Acid value ⟨401⟩: not more than 4.

Saponification value ⟨401⟩: between 155 and 165.

Hydroxyl value ⟨401⟩: between 160 and 175.

Iodine value ⟨401⟩: not more than 3.

Free glycerin and propylene glycol—

Periodic acid solution—Dissolve 5.4 g of periodic acid in 100 mL of water, and add 1900 mL of glacial acetic acid. Store in a glass-stoppered bottle, protected from light.

Chloroform—Use chloroform that meets the following additional requirement. To each of three glass-stoppered, 500-mL conical flasks add 50.0 mL of *Periodic acid solution*, then add 50 mL of chloroform and 10 mL of water to two of the flasks and 50 mL of water to the third flask. To each flask add 20 mL of potassium iodide TS, mix gently, and proceed as directed for *Procedure*, beginning with "allow to stand for 1 to 5 minutes." The difference between the volumes of 0.1 *N* sodium thiosulfate required in the titrations with and without the chloroform does not exceed 100 µL.

Procedure—Melt the Propylene Glycol Monostearate at a temperature not above 55°, and mix. Transfer a 3-g portion, accurately weighed, to a 100-mL beaker, and dissolve in 25 mL of *Chloroform*. Transfer this solution, with the aid of another 25-mL portion of *Chloroform*, to a separator, wash the beaker with 25 mL of water, and add the washing to the separator. Insert the stopper, shake vigorously for 30 to 60 seconds, and allow the layers to separate, adding 1 mL to 2 mL of glacial acetic acid, if necessary, to break any emulsion. Transfer the aqueous layer to a glass-stoppered, 500-mL conical flask, wash the chloroform layer with two 25-mL portions of water, combining the washings with the aqueous layer, and discard the chloroform layer. Add, with swirling, 50.0 mL of *Periodic acid solution* to the solution and to another glass-stoppered, 500-mL conical flask containing 75 mL of water to provide the blank. Allow to stand for 30 to 90 minutes. To each flask add 20 mL of potassium iodide TS, mix gently, and allow to stand for 1 to 5 minutes before titrating. Add 100 mL of water, and titrate with 0.1 *N* sodium thiosulfate VS until the brown iodine color fades to pale yellow, add 3 mL of starch TS, and continue the titration to the disappearance of the blue color. Propylene Glycol Monostearate contains not more than 1.0% of free glycerin and propylene glycol, calculated as propylene glycol by the formula:

$$[3.805N(B - T)]/W,$$

in which 3.805 is the molecular weight of propylene glycol divided by 20, N is the exact normality of the sodium thiosulfate solution, B and T are the volumes, in mL, of sodium thiosulfate consumed in the titrations of the blank solution and test solution, respectively, and W is the weight, in g, of Propylene Glycol Monostearate taken.

Propylene glycol monoesters—Transfer about 25 g to a 500-mL, round-bottom flask, add 250 mL of alcohol and 7.5 g of potassium hydroxide, and mix. Connect a suitable condenser to the flask, reflux the mixture for 2 hours, cool, and transfer to an 800-mL beaker, rinsing the flask with about 100 mL of water and combining the rinsing with the mixture in the beaker. Heat on a steam bath to evaporate the alcohol, adding water occasionally to replace the alcohol, and continue the evaporation until the odor of alcohol can no longer be detected. Adjust the volume, with hot water, to about 250 mL, neutralize with a mixture of equal volumes of sulfuric acid and water, noting the volume used, and add a 10% excess of the dilute acid. Heat with stirring until the fatty acid layer separates, and transfer the fatty acids to a 500-mL separator. Wash the fatty acids with four 200-mL portions of hot water, and discard the washings. Dry the fatty acids at 105° for 1 hour, cool, and determine the *Acid value*, A, on a 1-g portion, accurately weighed, as directed for *Acid Value* (*Free Fatty Acids*) under *Fats and Fixed Oils* ⟨401⟩. Calculate the percentage of propylene glycol monoesters by the formula:

$$[M(H - F)]/561.1;$$

calculate M, the average molecular weight of the monoesters, by the formula:

$$(56{,}110/A) + 76.10 - 18.02;$$

and calculate F by the formula:

$$561.1G/38.05,$$

in which H is the *Hydroxyl value* of Propylene Glycol Monostearate, G is the content, in percentage, of glycerin and propylene glycol in Propylene Glycol Monostearate, 561.1 and 56,110 are 10 and 1000 times the molecular weight of potassium hydroxide, respectively, 38.05 is one-half of the molecular weight of propylene glycol (76.10), and 18.02 is the molecular weight of water.

Propylparaben

HO–C_6H_4–COO(CH$_2$)$_2$CH$_3$

$C_{10}H_{12}O_3$ 180.20
Benzoic acid, 4-hydroxy-, propyl ester.
Propyl *p*-hydroxybenzoate [*94-13-3*].

» Propylparaben contains not less than 99.0 percent and not more than 100.5 percent of $C_{10}H_{12}O_3$, calculated on the dried basis.

Packaging and storage—Preserve in well-closed containers.

Reference standard—*USP Propylparaben Reference Standard*—Dry over silica gel for 5 hours before using.

Identification—The infrared absorption spectrum of a mineral oil dispersion of it, previously dried, exhibits maxima only at the same wavelengths as that of a similar preparation of USP Propylparaben RS.

Melting range ⟨741⟩: between 95° and 98°.

Other requirements—It meets the requirements of the tests for *Acidity*, *Loss on drying*, and *Residue on ignition* under *Butylparaben.*

Assay—Proceed with Propylparaben as directed in the *Assay* under *Butylparaben.* Each mL of 1 N sodium hydroxide is equivalent to 180.2 mg of $C_{10}H_{12}O_3$.

Propylparaben Sodium

» Propylparaben Sodium contains not less than 98.5 percent and not more than 101.5 percent of $C_{10}H_{11}NaO_3$, calculated on the anhydrous basis.

Packaging and storage—Preserve in tight containers.

Reference standard—*USP Propylparaben Reference Standard*—Dry over silica gel for 5 hours before using.

Completeness of solution ⟨641⟩—One g of it, dissolved in water, meets the requirement.

Identification—

A: Dissolve 0.5 g in 5 mL of water, acidify with hydrochloric acid, and filter the resulting precipitate. Wash the precipitate with water, and dry it over silica gel for 5 hours: the infrared absorption spectrum of a mineral oil dispersion of it exhibits maxima only at the same wavelengths as that of a similar preparation of USP Propylparaben RS.

B: Ignite about 0.3 g, cool, and dissolve the residue in about 3 mL of 3 N hydrochloric acid. A platinum wire dipped in this solution imparts an intense, persistent yellow color to a nonluminous flame.

pH ⟨791⟩: between 9.5 and 10.5, in a solution (1 in 1000).

Water, *Method I* ⟨921⟩: not more than 5.0%.

Chloride ⟨221⟩—A 0.2-g portion shows no more chloride than corresponds to 0.10 mL of 0.020 N hydrochloric acid (0.035%).

Sulfate ⟨221⟩—A 0.25-g portion shows no more sulfate than corresponds to 0.30 mL of 0.020 N sulfuric acid (0.12%).

Assay—Gently reflux about 100 mg of Propylparaben Sodium, accurately weighed, with 30 mL of 1 N sodium hydroxide for 30 minutes. Cool, add 25.0 mL of potassium bromate solution (2.78 in 500), 5 mL of potassium bromide solution (1 in 8), and 10 mL of hydrochloric acid, and immediately insert the stopper into the flask. Cool, shake for 15 minutes, and allow to stand for 15 minutes. Quickly add 15 mL of potassium iodide TS, taking care to avoid the escape of bromine vapor, at once replace the stopper in the flask, and shake vigorously. Rinse the stopper and the neck of the flask with a small quantity of water, and titrate the liberated iodine with 0.1 N sodium thiosulfate VS, adding 3 mL of starch TS as the end-point is approached. [NOTE—About 15 mL is needed.] Perform a blank determination (see *Residual Titrations* under *Titrimetry* ⟨541⟩), and note the difference in volumes required. Each mL of the difference in volume of 0.1 N sodium thiosulfate is equivalent to 3.37 mg of $C_{10}H_{11}NaO_3$.

Purified Bentonite—*see* Bentonite, Purified
Purified Siliceous Earth—*see* Siliceous Earth, Purified
Purified Stearic Acid—*see* Stearic Acid, Purified
Purified Water—*see* Water, Purified USP

Rose Oil

» Rose Oil is volatile oil distilled with steam from the fresh flowers of *Rosa gallica* Linné, *Rosa damascena* Miller, *Rosa alba* Linné, *Rosa centifolia* Linné, and varieties of these species (Fam. Rosaceae).

Packaging and storage—Preserve in well-filled, tight containers.

Solubility test—One mL is miscible with 1 mL of chloroform without turbidity. Add 20 mL of 90 percent alcohol to this mixture: the resulting liquid is neutral or acid to moistened litmus paper and, upon standing at 20°, deposits crystals within 5 minutes.

Specific gravity ⟨841⟩: between 0.848 and 0.863 at 30° compared with water at 15°.

Angular rotation ⟨781⟩: between −1° and −4° when determined in a 100-mm tube.

Refractive index ⟨831⟩: between 1.457 and 1.463 at 30°.

Rose Water Ointment—*see* Rose Water Ointment USP

Stronger Rose Water

» Stronger Rose Water is a saturated solution of the odoriferous principles of the flowers of *Rosa centifolia* Linné (Fam. Rosaceae) prepared by distilling the fresh flowers with water and separating the excess volatile oil from the clear, water portion of the distillate.

NOTE—Stronger Rose Water, diluted with an equal volume of purified water, may be supplied when "Rose Water" is required.

Packaging and storage—The odor of Stronger Rose Water is best preserved by allowing a limited access of fresh air to the container.

Reaction: neutral or acid to litmus.

Residue on evaporation—Evaporate 100 mL on a steam bath, and dry the residue at 105° for 1 hour: not more than 15 mg of residue remains (0.015%).

Heavy metals, *Method I* ⟨231⟩—Add 2 mL of 1 *N* acetic acid to 10 mL of Stronger Rose Water, and add water to make 25 mL: the limit is 2 ppm.

Saccharin

$C_7H_5NO_3S$ 183.18

1,2-Benzisothiazol-3(2*H*)-one, 1,1-dioxide.

1,2-Benzisothiazolin-3-one 1,1-dioxide [*81-07-2*].

» Saccharin contains not less than 98.0 percent and not more than 101.0 percent of $C_7H_5NO_3S$, calculated on the dried basis.

Packaging and storage—Preserve in well-closed containers.

Reference standards—*USP o-Toluenesulfonamide Reference Standard*—Do not dry before using. Keep container tightly closed. *USP p-Toluenesulfonamide Reference Standard*—Do not dry before using. Keep container tightly closed.

Identification—

A: Dissolve about 100 mg in 5 mL of sodium hydroxide solution (1 in 20), evaporate the solution to dryness, and gently fuse the residue over a small flame until it no longer evolves ammonia. Allow the residue to cool, dissolve it in 20 mL of water, neutralize the solution with 3 *N* hydrochloric acid, and filter: the addition of a drop of ferric chloride TS to the filtrate produces a violet color.

B: Mix 20 mg with 40 mg of resorcinol, add 10 drops of sulfuric acid, and heat the mixture in a suitable liquid bath at 200° for 3 minutes. Allow it to cool, and add 10 mL of water and an excess of 1 *N* sodium hydroxide: a fluorescent green liquid results.

Melting range ⟨741⟩: between 226° and 230°.

Loss on drying ⟨731⟩—Dry it at 105° for 2 hours: it loses not more than 1.0% of its weight.

Residue on ignition ⟨281⟩: not more than 0.2%.

Toluenesulfonamides—

Internal standard solution—Place 10 mg of *n*-tricosane in a 10-mL volumetric flask, dissolve in *n*-heptane, dilute with *n*-heptane to volume, and mix.

Standard stock solution—Transfer 20 mg each, accurately weighed, of USP *o*-Toluenesulfonamide RS and of USP *p*-Toluenesulfonamide RS to a 10-mL volumetric flask, dissolve in methylene chloride, dilute with methylene chloride to volume, and mix.

Standard preparations—Transfer 100, 150, 200, and 250 µL, respectively, of *Standard stock solution* to each of four 10-mL volumetric flasks. Add 250 µL, accurately measured, of *Internal standard solution* to each flask, dilute each with methylene chloride to volume, and mix. These preparations contain, in each mL, 25 µg of *n*-tricosane and, respectively, 20, 30, 40, and 50 µg of each toluenesulfonamide isomer.

Test preparation—Prepare as directed under *Column Partition Chromatography* (see *Chromatography* ⟨621⟩), employing a chromatographic tube fitted with a porous glass disk in its base, a plastic stopcock on the delivery tube, and a reservoir at the top. Add a mixture consisting of 12 g of *Solid Support* and a solution of 2.0 g, accurately weighed, of Saccharin with 12 mL of filtered sodium bicarbonate solution (1 in 11). Add about 200 mg of sodium bicarbonate to effect complete solution of the saccharin. Pack the contents of the tube by tapping the column on a padded surface, and then by tamping firmly from the top. Place 100 mL of methylene chloride in the reservoir, and adjust the stopcock so that 50 mL of eluate is collected in 20 to 30 minutes. To the eluate add 25 µL of *Internal standard solution*, mix, and concentrate the solution, by suitable means, to a volume of 1.0 mL.

Chromatographic system (see *Chromatography* ⟨621⟩)—Under typical conditions, the instrument is equipped with a flame-ionization detector, and contains a 3-m × 2-mm glass column packed with 3 percent liquid phase G3 on 100- to 120-mesh support S1AB, utilizing a glass-lined sample introduction system or on-column injection. The injector port, column, and detector block are maintained at temperatures of about 225°, 180°, and 250°, respectively, and dry helium is used as the carrier gas at a flow rate of 30 mL per minute.

Procedure—Inject portions (about 2.5 µL) of the *Standard preparations*, successively, into a suitable gas chromatograph, and record each chromatogram so as to obtain at least 50% of maximum recorder response. Measure the areas under the first (*o*-toluenesulfonamide), second (*p*-toluenesulfonamide), and third (*n*-tricosane) peaks, and for each chromatogram record the values as A_o, A_p, and A_N, respectively. Calculate the ratios R_o and R_p by the equations:

$$R_o = A_o/A_N \text{ and } R_p = A_p/A_N,$$

and prepare standard curves by plotting the concentrations, in µg per mL, of USP *o*-Toluenesulfonamide RS and of USP *p*-Toluenesulfonamide RS in the *Standard preparations* versus R_o and R_p, respectively. [NOTE—Relative retention times are, approximately, 5 for *o*-toluenesulfonamide, 6 for *p*-toluenesulfonamide, and 15 for *n*-tricosane.] Similarly inject a portion (about 2.5 µL) of the *Test preparation*, and record the chromatogram. Measure the areas under the first (*o*-toluenesulfonamide), second (*p*-toluenesulfonamide), and third (*n*-tricosane) peaks, and record the values as a_o, a_p, and a_N, respectively. Calculate the ratios r_o and r_p by the equations:

$$r_o = a_o/a_N \text{ and } r_p = a_p/a_N,$$

and, from the standard curve, determine the concentration, in µg per mL, of each toluenesulfonamide isomer in the *Test preparation:* the total amount of toluenesulfonamides in the specimen taken is not more than 0.0025%.

Arsenic, *Method II* ⟨211⟩: 3 ppm.

Selenium ⟨291⟩: 0.003%, a 100-mg specimen, mixed with 100 mg of magnesium oxide, being used.

Heavy metals, *Method II* ⟨231⟩: 0.001%.

Readily carbonizable substances ⟨271⟩—Dissolve 200 mg in 5 mL of sulfuric acid TS, and keep at a temperature of 48° to 50° for 10 minutes: the solution has no more color than *Matching Fluid A*.

Benzoic and salicylic acids—To 10 mL of a hot, saturated solution of it add ferric chloride TS, dropwise: no precipitate or violet color appears in the liquid.

Assay—Accurately weigh about 500 mg of Saccharin, dissolve in 40 mL of alcohol, add 40 mL of water, mix, add phenolphthalein TS, and titrate with 0.1 *N* sodium hydroxide VS. Perform a blank determination, and make any necessary correction. Each mL of 0.1 *N* sodium hydroxide is equivalent to 18.32 mg of $C_7H_5NO_3S$.

Saccharin Calcium—*see* Saccharin Calcium USP

Saccharin Sodium—*see* Saccharin Sodium USP

Sesame Oil

» Sesame Oil is the refined fixed oil obtained from the seed of one or more cultivated varieties of *Sesamum indicum* Linné (Fam. Pedaliaceae).

Packaging and storage—Preserve in tight, light-resistant containers, and prevent exposure to excessive heat.

Identification—Shake 1 mL for 30 seconds with a solution of 100 mg of sucrose in 10 mL of hydrochloric acid: the acid layer becomes pink and changes to red on standing (*distinction from most other fixed oils*).

Specific gravity ⟨841⟩: between 0.916 and 0.921.

Heavy metals, *Method II* ⟨231⟩: 0.001%.

Cottonseed oil—Mix 5 mL in a test tube with 5 mL of a mixture of equal volumes of amyl alcohol and a 1 in 100 solution of sulfur in carbon disulfide, warm the mixture carefully until the carbon disulfide is expelled, and immerse the tube to one-third of its depth in a boiling saturated solution of sodium chloride: no reddish color develops within 15 minutes.

Solidification range of fatty acids ⟨401⟩—The dry mixed fatty acids of it solidify between 20° and 25°.

Free fatty acids ⟨401⟩—The free fatty acids in 10 g require for neutralization not more than 2.0 mL of 0.020 *N* sodium hydroxide.

Iodine value ⟨401⟩: between 103 and 116.

Saponification value ⟨401⟩: between 188 and 195.

Unsaponifiable matter ⟨401⟩: not more than 1.5%.

Shellac

» Shellac is obtained by the purification of Lac, the resinuous secretion of the insect *Laccifer Lacca Kerr* (Fam. Coccidae). Orange Shellac is produced either by a process of filtration in the molten state, or by hot solvent process, or both. Orange Shellac may retain most of its wax or be dewaxed, and may contain lesser amounts of the natural color than originally present. Bleached (White) Shellac is prepared by dissolving the Lac in aqueous sodium carbonate, bleaching the solution with sodium hypochlorite and precipitating the Bleached Shellac with 2 *N* sulfuric acid. Removal of the wax, by filtration, during the process results in Refined Bleached Shellac. Shellac conforms to the specifications in the accompanying table.

Packaging and storage—Preserve in well-closed containers, preferably in a cold place.

Labeling—Label it to indicate whether it is bleached or is orange, and whether it is dewaxed or wax-containing.

Identification—To 50 mg of Shellac add a few drops of a mixture of 1 g of ammonium molybdate and 3 mL of sulfuric acid: a green color is produced, and it becomes lilac on standing for 5 minutes.

Loss on drying ⟨731⟩—Dry it at 41 ± 2°, in a well-ventilated oven, for 24 hours.

Arsenic, *Method II* ⟨211⟩: 1.5 ppm.

Heavy metals, *Method II* ⟨231⟩: 0.001%.

Acid value—Dissolve about 2 g of finely ground Shellac, accurately weighed in 50 mL of alcohol that has been neutralized to phenolphthalein with 0.1 *N* sodium hydroxide, add additional phenolphthalein TS, if necessary, and titrate with 0.1 *N* sodium hydroxide VS to a pink end-point. [Note—For orange Shellac, titrate slowly, stirring vigorously, until a glass rod dipped into the titrated solution produces a color change when touched to a drop of thymol blue TS on a spot plate.] Express the acid value in terms of the number of mg of potassium hydroxide required per g of dried Shellac.

Wax—Transfer about 10 g of finely ground Shellac, accurately weighed, and 2.50 g of sodium carbonate to a 200-mL, tall-form beaker. Add 150 mL of hot water, immerse the beaker in a boiling water bath, and stir until the solid is dissolved. Cover the beaker with a watch glass, and maintain the heat for 3 hours more, without agitation. Remove the beaker to a cold water bath. When the wax has floated to the surface, filter the solution through medium-speed quantitative ashless filter paper, transferring the wax to the paper, and wash the filter with water. Pour 5 to 10 mL of alcohol onto the filter to facilitate drying. Wrap the paper loosely in a larger piece of filter paper, bind with a piece of fine wire, and dry with the aid of gentle heat. Extract with chloroform in a suitable continuous extraction apparatus for 2 hours, using a weighed flask to receive the extracted wax and solvent. Evaporate the solvent, and dry the wax at 105° to constant weight.

Rosin—Dissolve 2 g by shaking with 10 mL of dehydrated alcohol, add slowly, with shaking, 50 mL of solvent hexane, wash with two successive 50-mL portions of water, filter the washed alcohol–solvent hexane solution, and evaporate to dryness. To the residue add 2 mL of a mixture of 1 volume of liquified phenol and 2 volumes of carbon tetrachloride, stir, and transfer a portion of the solution to a cavity of a color-reaction plate. Fill an adjacent cavity with a mixture of 1 volume of bromine and 4 volumes of carbon tetrachloride, and cover both cavities with an inverted watch glass: no purple or deep indigo-blue color is produced in or above the liquid containing the residue.

	Acid value (on dried basis)	Loss on drying	Wax
Orange Shellac	between 68 and 76	not more than 2.0%	not more than 5.5%
Dewaxed Orange Shellac	between 71 and 79	not more than 2.0%	not more than 0.2%
Regular Bleached Shellac	between 73 and 89	not more than 6.0%	not more than 5.5%
Refined Bleached Shellac	between 75 and 91	not more than 6.0%	not more than 0.2%

Purified Siliceous Earth

» Purified Siliceous Earth is a form of silica (SiO_2) consisting of the frustules and fragments of diatoms, purified by calcining.

Packaging and storage—Preserve in well-closed containers.

Loss on drying ⟨731⟩—Dry it at 105° for 2 hours: it loses not more than 0.5% of its weight.

Loss on ignition ⟨733⟩—Transfer about 1 g, previously dried and accurately weighed, to a tared platinum or porcelain crucible, and ignite at 980 ± 25° for 1 hour: it loses not more than 2.0% of its weight.

Acid-soluble substances—Digest 10.0 g with 50 mL of 0.5 *N* hydrochloric acid at 70° for 15 minutes, and filter. Wash the residue, adding the washings to the filtrate, to obtain a total volume of 100 mL. Evaporate at 110° in a tared porcelain dish to dryness: the weight of the dried residue so obtained does not exceed 200 mg (2.0%).

Water-soluble substances—Place 12.5 g in a 500-mL conical flask, add 250 mL of water, and shake for 2 hours at room temperature. Filter with the aid of vacuum, and again filter if necessary to obtain a clear filtrate. Evaporate in a tared platinum or porcelain dish, and dry at 110°: the weight of the residue does not exceed 25 mg (0.2%).

Leachable arsenic—Transfer 10.0 g to a 250-mL beaker, add 50 mL of 0.5 *N* hydrochloric acid, cover with a watch glass, and heat at 70° for 15 minutes. Cool, and decant through a Whatman No. 3 filter paper into a 100-mL volumetric flask. Wash the slurry with three 10-mL portions of water, preheated to 70°, dilute with water to volume, and mix: a 3.0-mL portion of this solution meets the requirements under *Arsenic, Method I* ⟨211⟩. The limit is 0.001%.

Leachable lead—A 10.0-mL portion of the solution prepared in the test for *Leachable arsenic* meets the requirements under *Lead* ⟨251⟩, 10 mL of *Diluted Standard Lead Solution* being used for the control. The limit is 0.001%.

Nonsiliceous substances—Transfer about 200 mg, accurately weighed, to a tared platinum crucible, add 5 mL of hydrofluoric acid and 2 drops of dilute sulfuric acid (1 in 2), and evaporate gently to dryness. Cool, add 5 mL of hydrofluoric acid, evaporate again to dryness, and ignite to constant weight: the weight of the residue so obtained does not exceed 50 mg.

Silicon Dioxide

$SiO_2 \cdot xH_2O$
Anhydrous 60.08

» Silicon Dioxide is obtained by insolubilizing the dissolved silica in sodium silicate solution. Where obtained by the addition of sodium silicate to a mineral acid, the product is termed silica gel; where obtained by the destabilization of a solution of sodium silicate in such manner as to yield very fine particles, the product is termed precipitated silica. After ignition at 1000° for not less than 1 hour, it contains not less than 99.0 percent of SiO_2.

Packaging and storage—Preserve in tight containers, protected from moisture.

Labeling—Label it to state whether it is silica gel or precipitated silica.

Identification—Transfer about 5 mg to a platinum crucible, mix with about 200 mg of anhydrous potassium carbonate, ignite at a red heat over a burner for 10 minutes, and cool. Dissolve the melt in 2 mL of recently distilled water, warming if necessary, and slowly add 2 mL of ammonium molybdate TS: a deep yellow color is produced.

pH ⟨791⟩: between 4 and 8, in a slurry (1 in 20).

Loss on drying ⟨731⟩—Dry it at 145° for 4 hours: it loses not more than 5.0% of its weight.

Loss on ignition ⟨733⟩—Ignite about 1 g of it, previously dried and accurately weighed, at 1000° for not less than 1 hour: it loses not more than 8.5% of its weight.

Chloride ⟨221⟩—Boil 5 g in 50 mL of water under a reflux condenser for 2 hours, cool, and filter. A 7-mL portion of the filtrate shows no more chloride than corresponds to 1.0 mL of 0.020 *N* hydrochloric acid (0.1%).

Sulfate ⟨221⟩—A 10-mL portion of the filtrate obtained in the test for *Chloride* shows no more sulfate than corresponds to 5.0 mL of 0.020 *N* sulfuric acid (0.5%).

Arsenic, *Method I* ⟨211⟩—Prepare the *Test Preparation* as follows: Transfer 4.0 g to a platinum dish, add 5 mL of nitric acid and 35 mL of hydrofluoric acid, and evaporate on a steam bath. Cool, add 5 mL of perchloric acid, 10 mL of hydrofluoric acid, and 10 mL of sulfuric acid, and evaporate on a hot plate to the production of heavy fumes. Cool, cautiously transfer to a 100-mL beaker with the aid of a few mL of hydrochloric acid, and evaporate to dryness. Cool, add 5 mL of hydrochloric acid, dilute with water to about 40 mL, and heat to dissolve any residue. Cool, transfer to a 100-mL volumetric flask, dilute with water to volume, and mix. A 25.0-mL portion of this solution meets the requirements of the test. The limit is 3 ppm.

Heavy metals, *Method I* ⟨231⟩—Transfer 16.7 mL of the solution prepared for the test for *Arsenic* into a 100-mL beaker, and neutralize with ammonium hydroxide to litmus paper. Adjust with 6 *N* acetic acid to a pH of between 3 and 4. Filter, using medium-speed filter paper, wash with water until the filtrate and washings measure 40 mL, and mix. The limit is 0.003%.

Assay—Transfer about 1 g of Silica Gel to a tared platinum dish, ignite at 1000° for 1 hour, cool in a desiccator, and weigh. Carefully wet with water, and add about 10 mL of hydrofluoric acid, in small increments. Evaporate on a steam bath to dryness, and cool. Add about 10 mL of hydrofluoric acid and about 0.5 mL of sulfuric acid, and evaporate to dryness. Slowly increase the temperature until all of the acids have been volatilized, and ignite at 1000°. Cool in a desiccator, and weigh. The difference between the final weight and the weight of the initially ignited portion represents the weight of SiO_2.

Colloidal Silicon Dioxide

SiO_2 60.08
Silica.
Silica [*7631-86-9*].

» Colloidal Silicon Dioxide is a submicroscopic fumed silica prepared by the vapor-phase hydrolysis of a silicon compound. When ignited at 1000° for 2 hours, it contains not less than 99.0 percent and not more than 100.5 percent of SiO_2.

Packaging and storage—Preserve in well-closed containers.

Identification—

A: Transfer about 5 mg to a platinum crucible, and mix with about 200 mg of anhydrous potassium carbonate. Ignite at a red heat over a burner for about 10 minutes, and cool. Dissolve the melt in 2 mL of freshly distilled water, warming if necessary, and slowly add 2 mL of ammonium molybdate TS to the solution: a deep yellow color is produced.

B: [*Caution—Avoid contact with o-tolidine when performing this test, and conduct the test in a well-ventilated hood.*] Place 1 drop of the yellow silicomolybdate solution obtained in *Identification test A* on a filter paper, and evaporate the solvent. Add 1 drop of a saturated solution of *o*-tolidine in glacial acetic acid to reduce the silicomolybdate to molybdenum blue, and place the paper over ammonium hydroxide: a greenish blue spot is produced.

pH ⟨791⟩: between 3.5 and 4.4, in a 1 in 25 dispersion.

Loss on drying ⟨731⟩—Dry it in a tared platinum crucible at 105° for 2 hours: it loses not more than 2.5% of its weight. Retain the dried specimen, in the crucible, for the test for *Loss on ignition.*

Loss on ignition ⟨733⟩—Ignite the portion of Colloidal Silicon Dioxide, retained from the test for *Loss on drying*, at 1000 ± 25° to constant weight: the previously dried Colloidal Silicon Dioxide loses not more than 2.0% of its weight.

Arsenic, *Method I* ⟨211⟩—Prepare the *Test Preparation* as follows: Transfer 2.5 g to a flask, add 50 mL of 3 *N* hydrochloric acid, and reflux for 30 minutes using a water condenser. Cool, filter with the aid of suction, and transfer the filtrate to a 100-mL volumetric flask. Wash the filter and flask with several portions of hot water, and add the washings to the flask. Cool, dilute with water to volume, and mix: a 15.0-mL portion of this solution, to which 3 mL of hydrochloric acid has been added, meets the requirements of the test, the addition of the 7 *N* sulfuric acid being omitted. The limit is 8 ppm.

Assay—Transfer about 500 mg of Colloidal Silicon Dioxide to a tared platinum crucible, ignite at 1000 ± 25° for 2 hours, cool in a desiccator, and weigh. Add 3 drops of sulfuric acid, and add enough alcohol to just moisten the sample completely. Add 15 mL of hydrofluoric acid, and in a well-ventilated hood evaporate on a hot plate to dryness, using medium heat (95° to 105°) and taking care that the sample does not spatter as dryness is approached. Heat the crucible to a red color with the aid of a Bunsen burner. Ignite the residue at 1000 ± 25° for 30 minutes, cool in a desiccator, and weigh. If a residue remains, repeat the procedure, beginning with "add 15 mL of hydrofluoric acid." The weight lost by the assay specimen, previously ignited at 1000 ± 25°, represents the weight of SiO_2 in the portion taken.

Soda Lime

» Soda Lime is a mixture of Calcium Hydroxide and Sodium or Potassium Hydroxide or both.

It may contain an indicator that is inert toward anesthetic gases such as Ether, Cyclopropane, and Nitrous Oxide, and that changes color when the Soda Lime no longer can absorb Carbon Dioxide.

Identification—

A: Place a granule of it on a piece of moistened red litmus paper: the paper turns blue immediately.

B: A solution in 6 *N* acetic acid responds to the tests for *Calcium* ⟨191⟩. It also imparts a yellow color to a nonluminous flame that, when viewed through cobalt glass, may show a violet color.

Loss on drying ⟨731⟩—Weigh accurately, in a tared weighing bottle, about 10 g, and dry at 105° for 2 hours: it loses between 12.0% and 19.0% of its weight.

Moisture absorption—Place about 10 g in a tared, 50-mL weighing bottle, having a diameter of 50 mm and a height of 30 mm, and weigh. Then place the bottle, with cover removed, for 24 hours in a closed container in which the atmosphere is maintained at 85% relative humidity by being in equilibrium with sulfuric acid having a specific gravity of 1.16. Weigh again: the increase in weight is not more than 7.5%.

Hardness—Screen 200 g on a mechanical sieve shaker (see *Powder Fineness* ⟨811⟩) having a frequency of oscillation of 285 ± 3 cycles per minute, for 3 minutes, to remove granules both coarser and finer than the labeled particle size. Proceed as directed in the test for *Hardness* under *Barium Hydroxide Lime*, beginning with "Weigh 50 g of the granules." The percentage of Soda Lime retained on the screen is not less than 75.0, and represents the hardness.

Carbon dioxide absorbency—Proceed as directed in the test for *Carbon dioxide absorbency* under *Barium Hydroxide Lime*. The increase in weight is not less than 19.0% of the weight of Soda Lime used for the test.

Other requirements—It meets the requirements for *Packaging and storage*, *Labeling*, and *Size of granules* under *Barium Hydroxide Lime.*

Sodium Acetate—*see* Sodium Acetate USP

Sodium Alginate

Alginic acid, sodium salt.
Sodium alginate [*9005-38-3*].

» Sodium Alginate is the purified carbohydrate product extracted from brown seaweeds by the use of dilute alkali. It consists chiefly of the sodium salt of Alginic Acid, a polyuronic acid composed of β-D-mannuronic acid residues linked so that the carboxyl group of each unit is free while the aldehyde group is shielded by a glycosidic linkage. It contains not less than 90.8 percent and not more than 106.0 percent of sodium alginate of average equivalent weight 222.00, calculated on the dried basis.

Packaging and storage—Preserve in tight containers.

Identification—

A: To 5 mL of a solution (1 in 100) add 1 mL of calcium chloride TS: a voluminous, gelatinous precipitate is formed immediately.

B: To 10 mL of a solution (1 in 100) add 1 mL of 4 *N* sulfuric acid: a heavy, gelatinous precipitate is formed.

Microbial limits ⟨61⟩—The total bacterial count does not exceed 200 per g, and the tests for *Salmonella* species and *Escherichia coli* are negative.

Loss on drying ⟨731⟩—Dry it at 105° for 4 hours: it loses not more than 15.0% of its weight.

Ash—Proceed as directed for *Total Ash* under *Vegetable Drugs—Methods of Analysis* ⟨561⟩, carefully igniting about 4 g, accurately weighed, in a tared platinum dish, until the residue is thoroughly carbonized (about 5 minutes), and then igniting in a muffle furnace at a temperature of 800 ± 25° until the carbon is completely burned off (20 to 35 minutes): between 18.0% and 24.0% of ash is found.

Arsenic, *Method II* ⟨211⟩: 1.5 ppm.

Lead ⟨251⟩—Add 1.0 g to 20 mL of nitric acid in a 250-mL conical flask, mix, and heat carefully until the Sodium Alginate is dissolved. Continue the heating until the volume is reduced to about 7 mL. Cool rapidly to room temperature, transfer to a 100-mL volumetric flask, and dilute with water to volume. A 50.0-mL portion of this solution contains not more than 5 µg of lead (corresponding to not more than 0.001% of Pb), 15 mL of ammonium citrate solution, 3 mL of potassium cyanide solution, and 0.5 mL of hydroxylamine hydrochloride solution being used for the test. After the first dithizone extractions, wash the combined chloroform layers with 5 mL of water, discarding the water layer and continuing in the usual manner by extracting with 20 mL of 0.2 *N* nitric acid.

Heavy metals, *Method II* ⟨231⟩—Conduct the ignition in a platinum crucible, and use nitric acid in place of sulfuric acid to wet the test specimen: the limit is 0.004%.

Assay—Proceed as directed for *Procedure* under *Alginates Assay* ⟨311⟩, using about 250 mg of Sodium Alginate, accurately weighed. Each mL of 0.2500 *N* sodium hydroxide consumed is equivalent to 27.75 mg of sodium alginate.

Sodium Benzoate

C_6H_5COONa

$C_7H_5NaO_2$ 144.11
Benzoic acid, sodium salt.
Sodium benzoate [*532-32-1*].

» Sodium Benzoate contains not less than 99.0 percent and not more than 100.5 percent of $C_7H_5NaO_2$, calculated on the anhydrous basis.

Packaging and storage—Preserve in well-closed containers.

Identification—It responds to the tests for *Sodium* ⟨191⟩ and for *Benzoate* ⟨191⟩.

Alkalinity—Dissolve 2 g in 20 mL of hot water, and add 2 drops of phenolphthalein TS: the pink color produced, if any, is discharged by the addition of 0.20 mL of 0.10 *N* sulfuric acid.

Water, *Method I* ⟨921⟩: not more than 1.5%.

Arsenic, *Method II* ⟨211⟩: 3 ppm.

Heavy metals ⟨231⟩—Dissolve 4.0 g in 40 mL of water, add, dropwise, with vigorous stirring, 10 mL of 3 *N* hydrochloric acid, and filter. Use 25 mL of the filtrate: the limit is 0.001%.

Assay—Transfer about 600 mg of Sodium Benzoate, accurately weighed, to a 250-mL beaker. Add 100 mL of glacial acetic acid, stir until the assay specimen is completely dissolved, add 2 drops of crystal violet TS, and titrate with 0.1 *N* perchloric acid VS to a green end-point. Perform a blank determination, and make any necessary correction. Each mL of 0.1 *N* perchloric acid is equivalent to 14.41 mg of $C_7H_5NaO_2$.

Sodium Bicarbonate—*see* Sodium Bicarbonate USP

Sodium Borate

$Na_2B_4O_7 . 10H_2O$ 381.37
Borax.
Borax [*1303-96-4*].
Anhydrous 201.22 [*1330-43-4*].

» Sodium Borate contains an amount of $Na_2B_4O_7$ equivalent to not less than 99.0 percent and not more than 105.0 percent of $Na_2B_4O_7 . 10H_2O$.

Packaging and storage—Preserve in tight containers.

Identification—A solution (1 in 20) responds to the tests for *Sodium* ⟨191⟩ and for *Borate* ⟨191⟩.

Carbonate and bicarbonate—To 5 mL of a solution (1 in 20), contained in a test tube, add 1 mL of 3 *N* hydrochloric acid: no effervescence is observed.

Arsenic, *Method I* ⟨211⟩: 8 ppm.

Heavy metals ⟨231⟩—Dissolve 1 g in 16 mL of water and 6 mL of 1 *N* hydrochloric acid, and dilute with water to 25 mL: the limit is 0.002%.

Assay—Dissolve about 3 g of Sodium Borate, accurately weighed, in 50 mL of water, add methyl red TS, and titrate with 0.5 *N* hydrochloric acid VS. [Note—Heating on a steam bath may be required initially to effect solution.] Each mL of 0.5 *N* hydrochloric acid is equivalent to 95.34 mg of $Na_2B_4O_7 . 10H_2O$.

Sodium Carbonate

Na_2CO_3 (anhydrous) 105.99
Carbonic acid, disodium salt.
Disodium carbonate [*497-19-8*].
Monohydrate 124.00 [*5968-11-6*].

» Sodium Carbonate is anhydrous or contains one molecule of water of hydration. It contains not less than 99.5 percent and not more than 100.5 percent of Na_2CO_3, calculated on the anhydrous basis.

Packaging and storage—Preserve in well-closed containers.

Labeling—Label it to indicate whether it is anhydrous or hydrous.

Identification—

A: A solution (1 in 10) is strongly alkaline to phenolphthalein TS.

B: It responds to the tests for *Sodium* ⟨191⟩, and for *Carbonate* ⟨191⟩.

Water, *Method III* ⟨921⟩—Dry about 2 g, accurately weighed, at 105° for 4 hours: the anhydrous form loses not more than 0.5%, and the hydrous form between 12.0% and 15.0%, of its weight.

Arsenic, *Method I* ⟨211⟩—Prepare the *Test Preparation* by dissolving 1.0 g in 25 mL of 7 *N* sulfuric acid, and adding 35 mL of water: the resulting solution meets the requirements of the test, the addition of 20 mL of 7 *N* sulfuric acid specified under *Procedure* being omitted. The limit is 3 ppm.

Heavy metals ⟨231⟩—Dissolve 2.0 g in 10 mL of water, add 1 drop of phenolphthalein TS, and neutralize the solution with hydrochloric acid, added dropwise. Heat the solution to boiling, and again neutralize by the dropwise addition of hydrochloric acid. Cool, and dilute with water to 25 mL: the limit is 0.001%.

Assay—Transfer the anhydrous sodium carbonate obtained in the test for *Water* to a flask with the aid of 50 mL of water, add methyl red TS, and titrate with 1 *N* sulfuric acid VS. Add the acid slowly, with constant stirring, until the solution becomes faintly pink. Heat the solution to boiling, cool, and continue the titration. Heat again to boiling, and titrate further as necessary until the faint pink color is no longer affected by continued boiling. Each mL of 1 *N* sulfuric acid is equivalent to 52.99 mg of Na_2CO_3.

Sodium Chloride—*see* Sodium Chloride USP

Sodium Chloride Injection, Bacteriostatic—*see* Sodium Chloride Injection, Bacteriostatic USP

Sodium Citrate—*see* Sodium Citrate USP

Sodium, Croscarmellose—*see* Croscarmellose Sodium

Sodium Dehydroacetate

$C_8H_7NaO_4$ 190.13
2*H*-Pyran-2,4(3*H*)-dione, 3-acetyl-6-methyl-, monosodium salt [*4418-26-2*].

» Sodium Dehydroacetate contains not less than 98.0 percent and not more than 100.5 percent of $C_8H_7NaO_4$, calculated on the anhydrous basis.

Packaging and storage—Preserve in well-closed containers.

Identification—

A: Dissolve about 1.5 g in 10 mL of water, add 5 mL of 3 *N* hydrochloric acid, collect the crystals by filtration with suction, wash with 10 mL of water, and dry at 80° for 4 hours: the crystals so obtained melt between 109° and 111° (see *Melting Range or Temperature* ⟨741⟩).

B: A solution (1 in 20) responds to the tests for *Sodium* ⟨191⟩.

Water, *Method I* ⟨921⟩: between 8.5% and 10.0%.

Arsenic, *Method II* ⟨211⟩: 3 ppm.

Heavy metals, *Method II* ⟨231⟩: 0.001%.

Assay—Transfer to a 125-mL conical flask about 500 mg of Sodium Dehydroacetate, accurately weighed, dissolve it in 25 mL of glacial acetic acid containing *p*-naphtholbenzein TS which has been previously neutralized to a blue color, and titrate with 0.1 *N* perchloric acid VS to the original blue color. Each mL of 0.1 *N* perchloric acid is equivalent to 19.01 mg of $C_8H_7NaO_4$.

Sodium Formaldehyde Sulfoxylate

$$HOCH_2-\overset{\overset{\displaystyle O}{\|}}{S}-ONa$$

CH_3NaO_3S 118.08
Methanesulfinic acid, hydroxy-, monosodium salt.
Monosodium hydroxymethanesulfinate [*149-44-0*].
Dihydrate 154.11 [*6035-47-8*].

» Sodium Formaldehyde Sulfoxylate contains an amount of CH_3NaO_3S equivalent to not less than 45.5 percent and not more than 54.5 percent of SO_2, calculated on the dried basis. It may contain a suitable stabilizer, such as sodium carbonate.

Packaging and storage—Preserve in well-closed, light-resistant containers, and store at controlled room temperature.

Clarity and color of solution—Dissolve 1 g in 20 mL of water, and transfer 10 mL to a 20- × 150-mm test tube. Compare with water in a similar test tube: the liquids are equally clear and, when viewed transversely by transmitted light, exhibit no apparent difference in color.

Identification—

A: Dissolve about 4 g in 10 mL of water in a test tube, and add 1 mL of silver-ammonia-nitrate TS: metallic silver is produced, either as a finely divided, gray precipitate or as a bright metallic mirror on the inner surface of the tube.

B: Dissolve about 40 mg of salicylic acid in 5 mL of sulfuric acid, add about 50 mg of Sodium Formaldehyde Sulfoxylate, and warm very gently: a permanent, deep red color appears.

Alkalinity—Dissolve 1.0 g in 50 mL of water, add phenolphthalein TS, and titrate with 0.10 *N* sulfuric acid: not more than 3.5 mL is required for neutralization.

pH ⟨791⟩: between 9.5 and 10.5, in a solution (1 in 50).

Loss on drying ⟨731⟩—Dry it at 105° for 3 hours: it loses not more than 27.0% of its weight.

Sulfide—Dissolve 6 g in 14 mL of water in a test tube, and wet a strip of lead acetate test paper with the clear solution: no discoloration is evident within 5 minutes.

Iron—Transfer 1.0 g to a suitable crucible, and carefully ignite, initially at a low temperature until thoroughly charred, and finally, preferably in a muffle furnace, at 500° to 600° until the carbon is all burned off. Cool, dissolve the residue in 2 mL of hydrochloric acid, and dilute with water to 50 mL. Add about 50 mg of ammonium persulfate and 5 mL of ammonium thiocyanate TS, mix, and transfer to a color-comparison tube. Treat in the same manner 5.0 mL of a solution of ferric ammonium sulfate, prepared by dissolving 43.2 mg of ferric ammonium sulfate in 10 mL of 2 *N* sulfuric acid and adding water to make 1000 mL, each mL representing 5 µg of Fe. The color of the test solution is not deeper than that of the solution containing the standard iron solution (0.0025%).

Sodium sulfite—Transfer 4.0 mL of the solution prepared for the *Assay* to a conical flask containing 100 mL of water. Add 2 mL of formaldehyde TS, and titrate with the same 0.1 *N* iodine VS that is used for the *Assay*, adding 3 mL of starch TS as the end-point is approached. Calculate the percentage of Na_2SO_3 in the Sodium Formaldehyde Sulfoxylate by the formula:

$$(1.25)(63.02)(V_2 - V_1)(N/W),$$

in which 63.02 is the equivalent weight of sodium sulfite, V_1 and V_2 are the volumes, in mL, of 0.1 *N* iodine VS consumed in this titration and in the titration performed in the *Assay*, respectively, N is the exact normality of the iodine solution, and W is the weight, in g, of Sodium Formaldehyde Sulfoxylate taken for the *Assay:* not more than 5.0% of Na_2SO_3, calculated on the dried basis, is found.

Assay—Transfer about 1 g of Sodium Formaldehyde Sulfoxylate, accurately weighed, to a 50-mL volumetric flask, dissolve in about 25 mL of water, dilute with water to volume, and mix. Reserve a portion of this solution for the test for *Sodium sulfite.* Transfer 4.0 mL of this solution to a conical flask containing 100 mL of water, and titrate with 0.1 *N* iodine VS, adding 3 mL of starch TS as the end-point is approached. Each mL of 0.1 *N* iodine is equivalent to 1.602 mg of SO_2.

Sodium Hydroxide

NaOH 40.00
Sodium hydroxide.
Sodium hydroxide [*1310-73-2*].

» Sodium Hydroxide contains not less than 95.0 percent and not more than 100.5 percent of total alkali, calculated as NaOH, including not more than 3.0 percent of Na_2CO_3.

Caution—Exercise great care in handling Sodium Hydroxide, as it rapidly destroys tissues.

Packaging and storage—Preserve in tight containers.

Identification—A solution (1 in 25) responds to the tests for *Sodium* ⟨191⟩.

Insoluble substances and organic matter—A solution (1 in 20) is complete, clear, and colorless to slightly colored.

Potassium—Acidify 5 mL of a solution (1 in 20) with 6 *N* acetic acid, then add 5 drops of sodium cobaltinitrite TS: no precipitate is formed.

Heavy metals ⟨231⟩—Dissolve 0.67 g in a mixture of 5 mL of water and 7 mL of 3 *N* hydrochloric acid. Heat to boiling, cool, and dilute with water to 25 mL: the limit is 0.003%.

Assay—Dissolve about 1.5 g of Sodium Hydroxide, accurately weighed, in about 40 mL of carbon dioxide–free water. Cool the solution to room temperature, add phenolphthalein TS, and titrate with 1 *N* sulfuric acid VS. At the discharge of the pink color of the indicator, record the volume of acid solution required, add methyl orange TS, and continue the titration until a persistent pink color is produced. Each mL of 1 *N* sulfuric acid is equivalent to 40.00 mg of total alkali, calculated as NaOH, and each mL of acid consumed in the titration with methyl orange is equivalent to 106.0 mg of Na_2CO_3.

Sodium Lauryl Sulfate

Sulfuric acid monododecyl ester sodium salt.
Sodium monododecyl sulfate [*151-21-3*].

» Sodium Lauryl Sulfate is a mixture of sodium alkyl sulfates consisting chiefly of sodium lauryl sulfate [$CH_3(CH_2)_{10}CH_2OSO_3Na$]. The combined content of sodium chloride and sodium sulfate is not more than 8.0 percent.

Packaging and storage—Preserve in well-closed containers.

Identification—A solution (1 in 10) responds to the tests for *Sodium* ⟨191⟩, and, after acidification with hydrochloric acid and gentle boiling for 20 minutes, responds to the tests for *Sulfate* ⟨191⟩.

Alkalinity—Dissolve 1.0 g in 100 mL of water, add phenol red TS, and titrate with 0.10 *N* hydrochloric acid: not more than 0.60 mL is required for neutralization.

Arsenic, *Method II* ⟨211⟩: 3 ppm.

Heavy metals, *Method II* ⟨231⟩: 0.002%.

Sodium chloride—Dissolve about 5 g, accurately weighed, in about 50 mL of water. Neutralize the solution with 0.8 *N* nitric acid, using litmus paper as the indicator, add 2 mL of potassium chromate TS, and titrate with 0.1 *N* silver nitrate VS. Each mL of 0.1 *N* silver nitrate is equivalent to 5.844 mg of NaCl.

Sodium sulfate—

Lead nitrate solution—Dissolve 33.1 g of lead nitrate in water to make 1000 mL.

Procedure—Transfer about 1 g of Sodium Lauryl Sulfate, accurately weighed, to a 250-mL beaker, add 35 mL of water, and warm to dissolve. To the warm solution add 2.0 mL of 1 *N* nitric acid, mix, and add 50 mL of alcohol. Heat the solution to boiling, and slowly add 10 mL of *Lead nitrate solution*, with stirring. Cover the beaker, simmer for 5 minutes, and allow to settle. If the supernatant liquid is hazy, allow to stand for 10 minutes, heat to boiling, and allow to settle. When the solution is almost to a boiling point, decant as much liquid as possible through a 9-cm filter paper (Whatman No. 41 or equivalent). Wash four times by decantation, each time using 50 mL of 50 percent alcohol, and bring the mixture to a boil. Finally, transfer the filter paper to the original beaker, and immediately add 30 mL of water, 20.0 mL of 0.05 *M* disodium ethylenediaminetetraacetate VS, and 1 mL of ammonia–ammonium chloride buffer TS. Warm to dissolve the precipitate, add 0.2 mL of eriochrome black TS and titrate with 0.05 *M* zinc sulfate VS. Each mL of 0.05 *M* disodium ethylenediaminetetraacetate is equivalent to 7.102 mg of Na_2SO_4.

Unsulfated alcohols—Dissolve about 10 g, accurately weighed, in 100 mL of water, and add 100 mL of alcohol. Transfer the solution to a separator, and extract with three 50-mL portions of solvent hexane. If an emulsion forms, sodium chloride may be added to promote separation of the two layers. Wash the combined solvent hexane extracts with three 50-mL portions of water, and dry with anhydrous sodium sulfate. Filter the solvent hexane extract into a tared beaker, evaporate on a steam bath until the odor of solvent hexane no longer is perceptible, dry the residue at 105° for 30 minutes, cool, and weigh. The weight of the residue is not more than 4.0% of the weight of the Sodium Lauryl Sulfate taken.

Total alcohols—Transfer about 5 g, accurately weighed, to an 800-mL Kjeldahl flask, and add 150 mL of water, 50 mL of hydrochloric acid, and a few boiling chips. Attach a reflux condenser to the Kjeldahl flask, heat carefully to avoid excessive frothing, and then boil for about 4 hours. Cool the flask, rinse the condenser with ether, collecting the ether in the flask, and transfer the contents to a 500-mL separator, rinsing the flask twice with ether and adding the washings to the separator. Extract the solution with two 75-mL portions of ether, evaporate the combined ether extracts in a tared beaker on a steam bath, dry the residue at 105° for 30 minutes, cool, and weigh. The residue represents the total alcohols, and is not less than 59.0% of the weight of Sodium Lauryl Sulfate taken.

Sodium Metabisulfite

$Na_2S_2O_5$ 190.10
Disulfurous acid, disodium salt.
Disodium pyrosulfite [*7681-57-4*].

» Sodium Metabisulfite contains an amount of $Na_2S_2O_5$ equivalent to not less than 65.0 percent and not more than 67.4 percent of SO_2.

Packaging and storage—Preserve in well-filled, tight containers, and avoid exposure to excessive heat.

Identification—A solution (1 in 20) responds to the tests for *Sodium* ⟨191⟩ and for *Sulfite* ⟨191⟩.

Chloride—Dissolve 1.0 g in 10 mL of water, filter, if necessary, through a small chloride-free filter, and add 6 mL of 30 percent hydrogen peroxide. Add 1 *N* sodium hydroxide until the solution is slightly alkaline to phenolphthalein, dilute with water to 100 mL, and mix. Dilute 2.0 mL of this solution with water to 20 mL, and add 1 mL of nitric acid and 1 mL of silver nitrate TS. Mix, allow to stand for 5 minutes protected from direct sunlight, and compare the turbidity, if any, with that produced from a 2-mL aliquot of a control (prepared by diluting 0.71 mL of 0.020 *N* hydrochloric acid to 100 mL) in an equal volume of total solution containing the quantities of reagents used in the test (see *Visual Comparison* in the section, *Procedure*, under *Spectrophotometry and Light-scattering* ⟨851⟩): any turbidity produced by the test specimen does not exceed that of the control preparation (0.05%).

Thiosulfate—Mix 2.2 g with 10 mL of 1 *N* hydrochloric acid in a 50-mL beaker. Boil gently for 5 minutes, cool, and transfer to a small test tube. Any turbidity is not greater than that produced by 0.10 mL of 0.10 *N* sodium thiosulfate, similarly treated (0.05%).

Arsenic, *Method I* ⟨211⟩—Prepare the *Test Preparation* as follows. Dissolve 1.0 g in 10 mL of water in a 150-mL beaker. Cautiously add 10 mL of nitric acid and 5 mL of sulfuric acid, and evaporate on a steam bath to a volume of about 5 mL. Place the beaker on a hot plate, and heat until dense white fumes of sulfur trioxide first appear. Cool, cautiously wash down the inner wall of the beaker with about 10 mL of water, and again heat to the appearance of dense white fumes of sulfur trioxide. Cool, repeat the washing and fuming, and cool again. This solution meets the requirements of the test, the addition of 20 mL of 7 *N* sulfuric acid specified under *Procedure* being omitted. The limit is 3 ppm.

Iron ⟨241⟩—Dissolve 500 mg in 14 mL of dilute hydrochloric acid (2 in 7), and evaporate on a steam bath to dryness. Dissolve the residue in 7 mL of dilute hydrochloric acid (2 in 7), and again evaporate to dryness. Dissolve the resulting residue in a mixture of 2 mL of hydrochloric acid and 20 mL of water, add 3 drops of bromine TS, and boil to expel the bromine. Cool, and dilute with water to 47 mL: the limit is 0.002%.

Heavy metals, *Method I* ⟨231⟩—Dissolve 1 g in 10 mL of water, add 5 mL of hydrochloric acid, evaporate on a steam bath to dryness, and dissolve the residue in 25 mL of water: the limit is 0.002%.

Assay—Transfer about 200 mg of Sodium Metabisulfite, accurately weighed, to a glass-stoppered conical flask containing 50.0 mL of 0.1 *N* iodine VS, and swirl to dissolve. Allow to stand for 5 minutes, protected from light, add 1 mL of hydrochloric acid, and titrate the excess iodine with 0.1 *N* sodium thiosulfate VS, adding 3 mL of starch TS as the end-point is approached. Each mL of 0.1 *N* iodine is equivalent to 3.203 mg of SO_2.

Sodium Phosphate, Dibasic—*see* Sodium Phosphate, Dibasic USP

Sodium Phosphate, Monobasic—*see* Sodium Phosphate, Monobasic USP

Sodium Propionate

$CH_3CH_2COONa \cdot xH_2O$

$C_3H_5NaO_2 \cdot xH_2O$
Propanoic acid, sodium salt, hydrate.
Sodium propionate hydrate [*6700-17-0*].
Anhydrous 96.06 [*137-40-6*].

» Sodium Propionate, dried at 105° for 2 hours, contains not less than 99.0 percent and not more than 100.5 percent of $C_3H_5NaO_2$.

Packaging and storage—Preserve in tight containers.

Reference standard—*USP Sodium Propionate Reference Standard.*

Identification—

A: The infrared absorption spectrum of a potassium bromide dispersion of it exhibits maxima only at the same wavelengths as that of a similar preparation of USP Sodium Propionate RS.

B: A solution (1 in 20) responds to the tests for *Sodium* ⟨191⟩.

Alkalinity—Dissolve 2.0 g in 20 mL of water, and add phenolphthalein TS: if a pink color is produced, it is discharged by 0.60 mL of 0.10 *N* sulfuric acid.

Water, *Method I* ⟨921⟩: not more than 1.0%.

Arsenic, *Method II* ⟨211⟩: 3 ppm.

Heavy metals, *Method I* ⟨231⟩—Dissolve 2 g in 1 mL of 1 *N* acetic acid and sufficient water to make 25 mL: the limit is 0.001%.

Assay—Dissolve about 200 mg of Sodium Propionate, previously dried at 105° for 2 hours and accurately weighed, in 50 mL of glacial acetic acid. Add 1 drop of crystal violet TS, and titrate with 0.1 *N* perchloric acid VS to a green end-point. Perform a blank determination, and make any necessary correction. Each mL of 0.1 *N* perchloric acid is equivalent to 9.606 mg of $C_3H_5NaO_2$.

Sodium Starch Glycolate

Starch carboxymethyl ether, sodium salt.

» Sodium Starch Glycolate is the sodium salt of a carboxymethyl ether of starch. It contains not less than 2.8 percent and not more than 4.2 percent of sodium (Na) on the dried, alcohol-washed basis. It may contain not more than 10.0 percent of Sodium Chloride.

Packaging and storage—Preserve in well-closed containers, preferably protected from wide variations in temperature and humidity, which may cause caking.

Labeling—The labeling indicates the pH range.

Identification—A slightly acidified solution of it is colored blue by iodine and potassium iodide TS.

Microbial limits ⟨61⟩—It meets the requirements of the tests for absence of *Salmonella* species and *Escherichia coli.*

pH ⟨791⟩—Disperse 1 g in 30 mL of water: the pH of the resulting suspension is either between 3.0 and 5.0 or between 5.5 and 7.5.

Loss on drying ⟨731⟩—Dry it at 130° for 90 minutes: it loses not more than 10.0% of its weight.

Iron ⟨241⟩: 0.002%, the *Test preparation* being prepared as directed for *Test preparation* under *Heavy metals, Method III* ⟨231⟩, a 0.5-g test specimen being used and the final solution being diluted with water to 47 mL.

Heavy metals, *Method II* ⟨231⟩: 0.002%.

Sodium chloride—Weigh accurately about 1 g, transfer to a conical flask, add 20 mL of 80 percent alcohol, stir for 10 minutes, and filter. Repeat the extraction until chloride has been completely extracted, as shown by a test with silver nitrate TS. Dry the insoluble portion at 105° to constant weight, and reserve it for the *Assay.* Evaporate the combined filtrates, and dry the residue at 105° to constant weight. The weight of the dried residue is not greater than 15% of the weight of Sodium Starch Glycolate taken. Transfer it with the aid of water to a 200-mL volumetric flask, add 5 mL of nitric acid and 40.0 mL of 0.1 *N* silver nitrate VS, mix, and dilute with water to volume. Allow it to stand in the dark for 30 minutes, and filter. To 100.0 mL of the filtrate add 5 mL of ferric ammonium sulfate TS, and titrate with 0.1 *N* ammonium thiocyanate VS. Perform a blank determination, and make any necessary correction (see *Residual Titrations* under *Titrimetry* ⟨541⟩). Each mL of 0.1 *N* silver nitrate is equivalent to 5.844 mg of NaCl in the 100-mL aliquot taken, and to 11.69 mg of NaCl in the amount of Sodium Starch Glycolate weighed.

Assay—Transfer about 700 mg of the dried 80 percent alcohol-insoluble portion obtained in the test for *Sodium chloride,* accurately weighed, to a suitable flask, add 80 mL of glacial acetic acid, heat the mixture under reflux, on a boiling water bath, for 2 hours, cool to room temperature, and titrate with 0.1 *N* perchloric acid VS, determining the end-point potentiometrically. Each mL of 0.1 *N* perchloric acid is equivalent to 2.299 mg of Na in the form of sodium starch glycolate.

Sodium Stearate

Octadecanoic acid, sodium salt.
Sodium stearate [*822-16-2*].

» Sodium Stearate is a mixture of sodium stearate ($C_{18}H_{35}NaO_2$) and sodium palmitate ($C_{16}H_{31}NaO_2$), which together constitute not less than 90.0 percent of the total content. The content of $C_{18}H_{35}NaO_2$ is not less than 40.0 percent of the total. Sodium stearate contains small amounts of the sodium salts of other fatty acids.

Packaging and storage—Preserve in well-closed, light-resistant containers.

Reference standards—*USP Palmitic Acid Reference Standard*—Dry over silica gel for 4 hours before using. *USP Stearic Acid Reference Standard*—Dry over silica gel for 4 hours before using.

Identification—

A: When heated, it fuses. At a high temperature it decomposes, emitting flammable vapors and the odor of burning fat, finally leaving a residue that, when moistened with water, is alkaline to litmus paper, effervesces with acids, and colors a nonluminous flame intensely yellow.

B: Dissolve 25 g in 300 mL of hot water, add 60 mL of 2 *N* sulfuric acid, and heat the solution, with frequent stirring, until the separated fatty acid layer is clear. Wash the fatty acids with boiling water until they are free from sulfate, collect in a small beaker, and warm on a steam bath until the water has settled and the fatty acids are clear. Allow the acids to cool, pour off the water layer, then melt the acids, filter into a dry beaker while hot, and dry at 105° for 20 minutes: the solidification temperature of the fatty acids is not less than 54°.

Acidity—Heat 50 mL of alcohol to the same temperature, ±5°, as that attained when the pink end-point is reached in the titration of the test specimen. Add 3 drops of phenolphthalein TS and sufficient 0.020 *N* sodium hydroxide to produce a faint pink color. Add 2.00 g of Sodium Stearate, and dissolve with the aid of a small amount of heat: no pink color is produced. Titrate the solution with 0.020 *N* sodium hydroxide until a pink color is produced: between 1.00 mL and 4.25 mL of 0.020 *N* sodium hydroxide is required (between 0.28% and 1.2% as stearic acid).

Loss on drying ⟨731⟩—Tare a beaker containing about 1 g of washed sand, previously dried at 105°, add about 500 mg of Sodium Stearate, and again weigh. Add 10 mL of alcohol, evaporate the mixture at about 80° to dryness, and dry at 105° for 4 hours: it loses not more than 5.0% of its weight.

Alcohol-insoluble substances—Reflux 1.0 g with 25 mL of alcohol: it dissolves completely, and the resulting solution is clear or not more than slightly opalescent.

Iodine value of fatty acids ⟨401⟩: not more than 4.0, determined on the fatty acids obtained in *Identification test B.*

Acid value of fatty acids ⟨401⟩: between 196 and 211, determined on 1 g of the fatty acids obtained in *Identification test B.*

Assay—Proceed with Sodium Stearate as directed in the *Assay* under *Stearic Acid,* using about 100 mg of Sodium Stearate, accurately weighed. Determine the percentage of $C_{18}H_{35}NaO_2$ in the portion of Sodium Stearate taken by the formula:

$$100(A/B),$$

in which A is the area due to the methyl stearate peak and B is the sum of the areas of all fatty acid ester peaks in the chromatogram. Similarly, determine the percentage of $C_{16}H_{31}NaO_2$.

Sodium Stearyl Fumarate

» Sodium Stearyl Fumarate contains not less than 99.0 percent and not more than 101.5 percent of $C_{22}H_{39}NaO_4$, calculated on the anhydrous basis.

Packaging and storage—Preserve in well-closed containers.

Reference standards—*USP Sodium Stearyl Fumarate Reference Standard*—Do not dry before using. Keep container tightly closed. *USP Monostearyl Maleate Reference Standard*—Do not dry before using. Keep container tightly closed. *USP Stearyl Alcohol Reference Standard*—Do not dry before using.

Identification—The infrared absorption spectrum of a potassium bromide dispersion of it (1 in 300) exhibits maxima only at the same wavelengths as that of a similar preparation of USP Sodium Stearyl Fumarate RS.

Water, *Method I* ⟨921⟩: not more than 5.0%.

Arsenic, *Method II* ⟨211⟩: 3 ppm.

Lead ⟨251⟩: 0.001%.

Heavy metals, *Method II* ⟨231⟩: 0.002%.

Limit of sodium stearyl maleate and stearyl alcohol—

Standard monostearyl maleate preparation—Using a mixture of chloroform and glacial acetic acid (4:1) as solvent, prepare a solution of USP Monostearyl Maleate RS containing 1 mg per mL. Pipet 5 mL of this solution into a 50-mL volumetric flask, dilute with chloroform to volume, and mix.

Standard stearyl alcohol preparation—Using a mixture of chloroform and glacial acetic acid (4:1) as solvent, prepare a solution of USP Stearyl Alcohol RS containing 1 mg per mL. Pipet 5 mL of this solution into a 50-mL volumetric flask, dilute with chloroform to volume, and mix.

Test preparation—Weigh 200 mg of Sodium Stearyl Fumarate into a small, glass-stoppered conical flask. Add 10.0 mL of a mixture of chloroform and glacial acetic acid (4:1). Dissolve by placing the flask in an ultrasonic bath for about 10 minutes.

Procedure—Apply 5 µL of *Standard monostearyl maleate preparation* and 10 µL each of *Standard stearyl alcohol preparation* and *Test preparation* separately to a suitable thin-layer chromatographic plate (see *Chromatography* ⟨621⟩) coated with a 0.25-mm layer of chromatographic silica gel. Immerse the plate in a tank containing a layer of about 10 mm of chloroform on the bottom. Allow the solvent front to reach the upper edge of the spots. Withdraw the plate, and dry in a current of cold air. Repeat the immersion, development, and drying. This results in spots having a linear shape. Develop the chromatograph in a saturated chamber containing a solvent system consisting of a mixture of hexanes, toluene, and glacial acetic acid (50:50:10) until the solvent front has moved 15 cm, and remove the plate from the chamber. Allow to dry for 10 minutes, and heat in an oven at 90° for 2 minutes. Allow to cool to room temperature. Replace the plate into the chamber for another 15-cm development, remove the plate, and allow to dry at room temperature for 15 minutes. Spray the plate carefully with a solution prepared by adding 50 mL of sulfuric acid cautiously to 50 mL of methanol. Dry the plate in an oven at 150° for 15 minutes. Dark spots appear on a light background. Allow to cool. Approximate R_f values for sodium stearyl fumarate, monostearyl maleate, and stearyl alcohol are 0.5, 0.4, and 0.35, respectively. Faint spots at an R_f value of about 0.9 may result from traces of distearyl maleate and distearyl fumarate. The intensity of any spot from the *Test preparation* is not greater than that from the corresponding spot from the *Standard preparation* (0.25% sodium stearyl maleate, 0.5% stearyl alcohol).

Saponification value—

Ethanolic potassium hydroxide—Dissolve about 5.5 g of potassium hydroxide in absolute alcohol, heating if necessary to effect solution, and dilute with absolute alcohol to about 1000 mL. Prepare fresh daily, and filter if necessary to remove carbonate.

Procedure—Transfer about 450 mg of Sodium Stearyl Fumarate, accurately weighed, to a 300-mL conical flask, and add 50.0 mL of *Ethanolic potassium hydroxide*, rinsing down the inside of the flask during the addition. Reflux the mixture gently on a steam bath for not less than 2 hours, swirling gently occasionally, but avoid splashing the mixture up into the condenser. Rinse the condenser with 10 mL of 70 percent alcohol, followed by three 10-mL portions of water, collecting the rinsings in the flask. Cool, rinse the sides of the flask with two 10-mL portions of 70 percent alcohol, add phenolphthalein TS, and titrate with 0.1 *N* hydrochloric acid VS to the disappearance of any pink color. Perform a blank determination, using the same volume of *Ethanolic potassium hydroxide*. The difference between the volumes, in mL, of 0.1 *N* hydrochloric acid consumed in the actual test and in the blank test, multiplied by 5.61 and divided by the weight, in g, of specimen taken, is the *Saponification value*. It is between 142.2 and 146.0, calculated on the anhydrous basis.

Assay—Transfer about 250 mg of Sodium Stearyl Fumarate, accurately weighed, to a 50-mL conical flask, mix with 1 mL of chloroform, and add 20 mL of glacial acetic acid to dissolve. Add quinaldine red TS, and titrate with 0.1 *N* perchloric acid VS. Each mL of 0.1 *N* perchloric acid is equivalent to 39.05 mg of $C_{22}H_{39}NaO_4$.

Sodium Thiosulfate—*see* Sodium Thiosulfate USP

Solutions—*see complete list in index*

Sorbic Acid

$C_6H_8O_2$ 112.13

2,4-Hexadienoic acid, (*E,E*)-; 2,4-Hexadienoic acid.

(*E,E*)-Sorbic acid; Sorbic acid [*22500-92-1*] and [*110-44-1*].

» Sorbic Acid contains not less than 99.0 percent and not more than 101.0 percent of $C_6H_8O_2$, calculated on the anhydrous basis.

Packaging and storage—Preserve in tight containers, protected from light, and avoid exposure to excessive heat.

Identification—

A: To a solution of 200 mg in 2 mL of alcohol add a few drops of bromine TS: the color is discharged.

B: A 1 in 400,000 solution in isopropyl alcohol exhibits an absorbance maximum at 254 ± 2 nm.

Melting range ⟨741⟩: between 132° and 135°.

Water, *Method I* ⟨921⟩: not more than 0.5%.

Residue on ignition ⟨281⟩: not more than 0.2%.

Heavy metals, *Method II* ⟨231⟩: 0.001%.

Assay—Dissolve about 250 mg of Sorbic Acid, accurately weighed, in 50 mL of methanol that previously has been neutralized with 0.1 *N* sodium hydroxide VS. Add phenolphthalein TS, and titrate with 0.1 *N* sodium hydroxide VS to the first pink color that persists for at least 30 seconds. Each mL of 0.1 *N* sodium hydroxide is equivalent to 11.21 mg of $C_6H_8O_2$.

Sorbitan Monolaurate

[*R* is $(C_{11}H_{23})COO$]

Sorbitan, esters, monododecanoate.

Sorbitan monolaurate [*1338-39-2*].

» Sorbitan Monolaurate is a partial ester of lauric acid with Sorbitol and its mono- and dianhydrides. It yields, upon saponification, not less than 55.0 percent and not more than 63.0 percent of fatty acids, and not less than 39.0 percent and not more than 45.0 percent of polyols (w/w).

Packaging and storage—Preserve in tight containers.

Reference standards—*USP Isosorbide Reference Standard*—After opening the ampul, store in a tightly closed container. *USP 1,4-Sorbitan Reference Standard*—Keep container tightly closed. Do not dry before using.

Identification—

A: The residue of lauric acid obtained in the *Assay for fatty acids* has an acid value (see *Fats and Fixed Oils* ⟨401⟩) between 260 and 280, about 1 g of the residue, accurately weighed, being used, and an iodine value (see *Fats and Fixed Oils* ⟨401⟩) of not more than 5.

B: It responds to *Identification test B* under *Sorbitan Monooleate*.

Water, *Method I* ⟨921⟩: not more than 1.5%.

Residue on ignition ⟨281⟩: not more than 0.5%.

Heavy metals, *Method II* ⟨231⟩: not more than 0.001%.

Acid value ⟨401⟩: not more than 8.

Hydroxyl value ⟨401⟩: between 330 and 358.

Saponification value ⟨401⟩: between 158 and 170.

Assay for fatty acids—Transfer about 10 g of Sorbitan Monolaurate, accurately weighed, to a 500-mL conical flask, cautiously add 100 mL of alcohol and 3.0 g of potassium hydroxide, and mix. Proceed as directed in the *Assay for fatty acids* under *Sorbitan Monooleate*, beginning with "Connect a suitable condenser."

Assay for polyols—Proceed with Sorbitan Monolaurate as directed for *Assay for polyols* under *Sorbitan Monooleate.*

Sorbitan Monooleate

[Graphic formula same as for Sorbitan Monolaurate, except that *R* is $(C_{17}H_{33})COO$.]
Sorbitan, esters, mono(*Z*)-9-octadecenoate.
Sorbitan monooleate [*1338-43-8*].

» Sorbitan Monooleate is a partial oleate ester of Sorbitol and its mono- and dianhydrides. It yields, upon saponification, not less than 72.0 percent and not more than 78.0 percent of fatty acids, and not less than 25.0 percent and not more than 31.0 percent of polyols (w/w).

Packaging and storage—Preserve in tight containers.

Reference standards—*USP Isosorbide Reference Standard*—After opening the ampul, store in a tightly closed container. *USP 1,4-Sorbitan Reference Standard*—Keep container tightly closed. Do not dry before using.

Identification—

A: The residue of oleic acid obtained in the *Assay for fatty acids* has an acid value (see *Fats and Fixed Oils* ⟨401⟩) between 192 and 204, about 1 g of the residue, accurately weighed, being used, and an iodine value (see *Fats and Fixed Oils* ⟨401⟩) between 75 and 95.

B: *Standard preparation*—Transfer 33 mg of USP Isosorbide RS, 25 mg of USP 1,4-Sorbitan RS, and 25 mg of sorbitol to a 1-mL volumetric flask, dilute with water to volume, and mix to dissolve.

Test preparation—Transfer 500 mg of the polyols obtained in the *Assay for polyols* to a 2-mL volumetric flask, dilute with water to volume, and mix to dissolve.

Procedure—On a suitable thin-layer chromatographic plate (see *Chromatography* ⟨621⟩), coated with a 0.25-mm layer of chromatographic silica gel, apply 2 µL each of the *Standard preparation* and of the *Test preparation.* Allow the spots to dry, and develop the chromatogram in a solvent system consisting of a mixture of acetone and glacial acetic acid (50 : 1) until the solvent front has moved about three-fourths of the length of the plate. Remove the plate from the developing chamber, mark the solvent front, and allow the solvent to evaporate. Spray evenly with a mixture of equal volumes of sulfuric acid and water until the surface is uniformly wet (NOTE—Do not overspray.), and immediately place the sprayed plate on a 200° hot plate *in a well-ventilated hood.* Char until white fumes of sulfur trioxide cease, and cool: the R_f values of the spots obtained from the *Test preparation* correspond to those of the spots obtained from the *Standard preparation.*

Water, *Method I* ⟨921⟩: not more than 1.0%.

Residue on ignition ⟨281⟩: not more than 0.5%.

Heavy metals, *Method II* ⟨231⟩: 0.001%.

Acid value ⟨401⟩: not more than 8.

Hydroxyl value ⟨401⟩: between 190 and 215.

Iodine value ⟨401⟩: between 62 and 76.

Saponification value ⟨401⟩: between 145 and 160.

Assay for fatty acids—Transfer about 10 g of Sorbitan Monooleate, accurately weighed, to a 500-mL conical flask, cautiously add 100 mL of alcohol and 3.5 g of potassium hydroxide, then add a few glass beads, and mix. Connect a suitable condenser to the flask, reflux the mixture on a hot plate for 2 hours, add about 100 mL of water, and heat on a steam bath to evaporate the alcohol, adding water occasionally to replace the alcohol. Continue the evaporation until the odor of alcohol can no longer be detected, and transfer the saponification mixture, with the aid of about 100 mL of hot water, to a 500-mL separator. Using extreme caution, neutralize to litmus with a mixture of equal volumes of sulfuric acid and water, noting the volume used, and add a 10% excess of the dilute acid. Allow the solution to cool. If salts appear, add sufficient water to produce a clear solution. Cautiously add 100 mL of solvent hexane, shake thoroughly, and withdraw the lower layer into a second 500-mL separator. Similarly extract with 2 more 100-mL portions of solvent hexane. Extract the combined hexane layers with 50-mL portions of water until neutral to litmus paper, and combine the extracts with the original aqueous phase, for the *Assay for polyols.* Evaporate the solvent hexane in a tared beaker on a steam bath nearly to dryness, dry in vacuum at 60° for 1 hour, cool in a desiccator, and weigh the fatty acids.

Assay for polyols—Neutralize the aqueous solution of polyols retained from the *Assay for fatty acids* with potassium hydroxide solution (1 in 10) to a pH of 7, using a suitable pH meter. Evaporate on a steam bath to a moist residue, extract the polyols from the salts with three 150-mL portions of dehydrated alcohol, boiling the salt residue for 3 minutes, and crushing it, as necessary, with the flattened end of a stirring rod, during each extraction, filtering each extract, while hot, through a medium-porosity, sintered-glass funnel, provided with a sheet of retentive filter paper on which a layer of purified siliceous earth has been superimposed, and receiving the filtrates in a 1-liter suction flask. Transfer the clear alcoholic polyols solution to a tared beaker, evaporate the alcohol on a steam bath, dry in vacuum at 60° for 1 hour, cool in a desiccator, and weigh the polyols.

Sorbitan Monopalmitate

[Graphic formula same as for Sorbitan Monolaurate, except that *R* is $(C_{15}H_{31})COO$.]
Sorbitan, esters, monohexadecanoate.
Sorbitan monopalmitate [*26266-57-9*].

» Sorbitan Monopalmitate is a partial ester of palmitic acid with Sorbitol and its mono- and dianhydrides. It yields, upon saponification, not less than 63.0 percent and not more than 71.0 percent of fatty acids, and not less than 32.0 percent and not more than 38.0 percent of polyols (w/w).

Packaging and storage—Preserve in well-closed containers.

Reference standards—*USP Isosorbide Reference Standard*—After opening the ampul, store in a tightly closed container. *USP 1,4-Sorbitan Reference Standard*—Keep container tightly closed. Do not dry before using.

Identification—

A: The residue of palmitic acid obtained in the *Assay for fatty acids* has an acid value (see *Fats and Fixed Oils* ⟨401⟩) between 210 and 225, about 1 g of the residue, accurately weighed, being used, and an iodine value (see *Fats and Fixed Oils* ⟨401⟩) of not more than 4.

B: It responds to *Identification test B* under *Sorbitan Monooleate.*

Water, *Method I* ⟨921⟩: not more than 1.5%.

Residue on ignition ⟨281⟩: not more than 0.5%.

Heavy metals, *Method II* ⟨231⟩: not more than 0.001%.

Acid value ⟨401⟩: not more than 8.

Hydroxyl value ⟨401⟩: between 275 and 305.

Saponification value ⟨401⟩: between 140 and 150.

Assay for fatty acids—Transfer about 10 g of Sorbitan Monopalmitate, accurately weighed, to a 500-mL conical flask, cautiously add 100 mL of alcohol and 3.0 g of potassium hydroxide, and mix. Proceed as directed in the *Assay for fatty acids* under *Sorbitan Monooleate*, beginning with "Connect a suitable condenser."

Assay for polyols—Proceed with Sorbitan Monopalmitate as directed for *Assay for polyols* under *Sorbitan Monooleate.*

Sorbitan Monostearate

[Graphic formula same as for Sorbitan Monolaurate, except that *R* is $(C_{17}H_{35})COO$.]
Sorbitan, esters, monooctadecanoate.
Sorbitan monostearate [*1338-41-6*].

» Sorbitan Monostearate is a partial ester of Stearic Acid with Sorbitol and its mono- and dianhydrides. It yields, upon saponification, not less than 68.0 percent and not more than 76.0 percent of fatty acids, and not less than 27.0 percent and not more than 34.0 percent of polyols (w/w).

Packaging and storage—Preserve in well-closed containers.

Reference standards—*USP Isosorbide Reference Standard*—After opening the ampul, store in a tightly closed container. *USP 1,4-Sorbitan Reference Standard*—Keep container tightly closed. Do not dry before using.

Identification—

A: The residue of stearic acid obtained in the *Assay for fatty acids* has an acid value (see *Fats and Fixed Oils* ⟨401⟩) between 200 and 215, about 1 g of the residue, accurately weighed, being used, and an iodine value (see *Fats and Fixed Oils* ⟨401⟩) of not more than 4.

B: It responds to *Identification test B* under *Sorbitan Monooleate.*

Water, *Method I* ⟨921⟩: not more than 1.5%.

Residue on ignition ⟨281⟩: not more than 0.5%.

Heavy metals, *Method II* ⟨231⟩: not more than 0.001%.

Acid value ⟨401⟩: not more than 10.

Hydroxyl value ⟨401⟩: between 235 and 260.

Saponification value ⟨401⟩: between 147 and 157.

Assay for fatty acids—Transfer about 10 g of Sorbitan Monostearate, accurately weighed, to a 500-mL conical flask, cautiously add 100 mL of alcohol and 3.0 g of potassium hydroxide, and mix. Proceed as directed in the *Assay for fatty acids* under *Sorbitan Monooleate*, beginning with "Connect a suitable condenser."

Assay for polyols—Proceed with Sorbitan Monostearate as directed for *Assay for polyols* under *Sorbitan Monooleate.*

Sorbitol

```
    H  OH H  OH OH H
    |  |  |  |  |  |
HO—C—C—C—C—C—C—OH
    |  |  |  |  |  |
    H  H  OH H  H  H
```

$C_6H_{14}O_6$ 182.17
D-Glucitol.
D-Glucitol [*50-70-4*].

» Sorbitol contains not less than 91.0 percent and not more than 100.5 percent of $C_6H_{14}O_6$, calculated on the anhydrous basis. It may contain small amounts of other polyhydric alcohols.

Packaging and storage—Preserve in tight containers.

Reference standard—*USP Sorbitol Reference Standard*—Use without drying.

Identification—A solution of 5 g in about 4 mL of water responds to *Identification test B* under *Sorbitol Solution* (USP monograph).

Water, *Method I* ⟨921⟩: not more than 1.0%.

Residue on ignition ⟨281⟩: not more than 0.1%.

Chloride ⟨221⟩—A 1.5-g portion shows no more chloride than corresponds to 0.10 mL of 0.020 *N* hydrochloric acid (0.0050%).

Sulfate ⟨221⟩—A 1.0-g portion shows no more sulfate than corresponds to 0.10 mL of 0.020 *N* sulfuric acid (0.010%).

Arsenic, *Method II* ⟨211⟩: 3 ppm.

Heavy metals ⟨231⟩—Dissolve 2.0 g in 25 mL of water: the limit is 0.001%.

Reducing sugars—Transfer 7 g, accurately weighed, to a 400-mL beaker with the aid of 35 mL of water, and mix. Add 50 mL of alkaline cupric tartrate TS, cover the beaker, heat the mixture at such a rate that it comes to a boil in approximately 4 minutes, and boil for 2 minutes, accurately timed. Collect the precipitated cuprous oxide in a tared filtering crucible previously washed successively with hot water, with alcohol, and with ether and then dried at 105° for 30 minutes. Thoroughly wash the collected cuprous oxide on the filter with hot water, then with 10 mL of alcohol, and finally with 10 mL of ether, and dry at 105° for 30 minutes: the weight of the cuprous oxide does not exceed 50 mg.

Total sugars—Place 2.1 g in a 250-mL flask fitted with a ground-glass joint, add 40 mL of approximately 0.1 *N* hydrochloric acid, attach a reflux condenser, and reflux for 4 hours. Transfer the solution to a 400-mL beaker, rinsing the flask with about 10 mL of water, neutralize with 6 *N* sodium hydroxide, and proceed as directed in the test for *Reducing sugars*, beginning with "Add 50 mL of alkaline cupric tartrate TS": the weight of the cuprous oxide does not exceed 50 mg.

Assay—

Mobile phase—Use degassed water.

Resolution solution—Dissolve mannitol and USP Sorbitol RS in water to obtain a solution having concentrations of about 4.8 mg per mL of each.

Standard preparation—Dissolve an accurately weighed quantity of USP Sorbitol RS in water, and dilute quantitatively with water to obtain a solution having a known concentration of about 4.8 mg per mL.

Assay preparation—Transfer about 0.24 g of Sorbitol, accurately weighed, to a 50-mL volumetric flask, dissolve in 10 mL of water, dilute with water to volume, and mix.

Chromatographic system (see *Chromatography* ⟨621⟩)—The liquid chromatograph is equipped with a refractive index detector that is maintained at a constant temperature and a 7.8-mm × 30-cm column that contains packing L19. The column temperature is maintained at 30 ± 2° and the flow rate is about 0.2 mL per minute. Chromatograph the *Standard preparation*, and record the peak responses as directed under *Procedure:* the relative standard deviation for replicate injections is not more than 2.0%. In a similar manner chromatograph the *Resolution solution:* the resolution, *R*, between the sorbitol and mannitol peaks is not less than 2.0.

Procedure—Separately inject equal volumes (about 20 µL) of the *Assay preparation* and the *Standard preparation* into the chromatograph, record the chromatograms, and measure the responses for the major peaks. Calculate the quantity, in mg, of $C_6H_{14}O_6$, in the Sorbitol taken by the formula:

$$50C(r_U / r_S),$$

in which *C* is the concentration, in mg per mL, of USP Sorbitol RS in the *Standard preparation*, and r_U and r_S are the peak responses obtained from the *Assay preparation* and the *Standard preparation*, respectively.

Sorbitol Solution—*see* Sorbitol Solution USP

Soybean Oil—*see* Soybean Oil USP

Squalane

```
             CH3          CH3          CH3          CH3
             |            |            |            |
(CH3)2CH(CH2)3CH(CH2)3CH(CH2)4CH(CH2)3CH(CH2)3CH(CH3)2
```

$C_{30}H_{62}$ 422.82

Tetracosane, 2,6,10,15,19,23-hexamethyl-.
2,6,10,15,19,23-Hexamethyltetracosane [*111-01-3*].

» Squalane is a saturated hydrocarbon obtained by hydrogenation of squalene, an aliphatic triterpene occurring in some fish oils.

Packaging and storage—Preserve in tight containers.

Reference standard—*USP Squalane Reference Standard*—Do not dry before using. After opening ampul transfer the contents to a tightly closed container.

Identification—The infrared absorption spectrum of a film of it, spread between sodium chloride plates, exhibits maxima only at the same wavelengths as that of a similar preparation of USP Squalane RS.

Specific gravity ⟨841⟩: between 0.807 and 0.810 at 20°.

Refractive index ⟨831⟩: between 1.4510 and 1.4525 at 20°.

Residue on ignition ⟨281⟩: not more than 0.5%.

Acid value ⟨401⟩: not more than 0.2.

Iodine value ⟨401⟩: not more than 4.

Saponification value ⟨401⟩: not more than 2.

Chromatographic purity—Inject about 2 µL of a 2 in 100 solution of Squalane in *n*-hexane into a suitable gas chromatograph, equipped with a flame-ionization detector. Under typical conditions the instrument contains a 1.8-m × 3-mm column packed with 3 percent phase G1 on packing S1A, the column is temperature-programmed at about 6° per minute from 130° to 270°, the injection port and the detector block are maintained at 280°, and nitrogen is used as the carrier gas at a flow rate of 25 mL per minute. Identify the squalane peak by comparison of the retention time with that of the main peak obtained by chromatographing a similar solution of USP Squalane RS under the same conditions: the area of the squalane peak is not less than 97% of the sum of the areas of all of the peaks in the chromatogram.

Starch

Starch.
Starch [*9005-25-8*].

» Starch consists of the granules separated from the mature grain of corn [*Zea mays* Linné (Fam. Gramineae)] or of wheat [*Triticum aestivum* Linné (Fam. Gramineae)], or from tubers of the potato [*Solanum tuberosum* Linné (Fam. Solanaceae)].

NOTE—Starches obtained from different botanical sources may not have identical properties with respect to their use for specific pharmaceutical purposes, e.g., as a tablet-disintegrating agent. Therefore, types of starch should not be interchanged unless performance equivalency has been ascertained.

Packaging and storage—Preserve in well-closed containers.

Labeling—Label it to indicate the botanical source from which it was derived.

Botanic characteristics—

Corn starch—Polygonal, rounded or spheroidal granules up to about 35 µm in diameter and usually having a circular or several-rayed central cleft.

Wheat starch—Two distinct types of granules are simple lenticular large granules 20 µm to 25 µm or up to 50 µm in diameter, and small spherical granules 5 µm to 10 µm in diameter. Striations are faintly marked and concentric.

Potato starch—Simple granules, irregularly ovoid or spherical, 30 µm to 100 µm in diameter, and subspherical granules 10 µm to 35 µm in diameter. Striations are well marked and concentric.

Identification—

A: Prepare a smooth mixture of 1 g of it with 2 mL of cold water, stir it into 15 mL of boiling water, boil gently for 2 minutes, and cool: the product is a translucent, whitish jelly.

B: A water slurry of it is colored reddish violet to deep blue by iodine TS.

Microbial limits—It meets the requirements of the tests for absence of *Salmonella* species and *Escherichia coli* under *Microbial Limit Tests* ⟨61⟩.

pH ⟨791⟩—Prepare a slurry by weighing 20.0 g ± 100 mg of Starch, transferring to a suitable non-metallic container, and adding 100 mL of water. Agitate continuously at a moderate rate for 5 minutes, then stop agitation, and immediately determine the pH to the nearest 0.1 unit: the pH, determined potentiometrically, is between 4.5 and 7.0 for *Corn starch* and *Wheat starch*, and is between 5.0 and 8.0 for *Potato starch*.

Loss on drying ⟨731⟩—Dry it at 120° for 4 hours: it loses not more than 14.0% of its weight.

Residue on ignition ⟨281⟩: not more than 0.5%, determined on a 2.0-g test specimen.

Iron ⟨241⟩—Dissolve the residue obtained in the test for *Residue on ignition* in 8 mL of hydrochloric acid with the aid of gentle heating, dilute with water to 100 mL, and mix. Dilute 25 mL of this solution with water to 47 mL: the limit is 0.002%.

Oxidizing substances—Transfer 4.0 g to a glass-stoppered, 125-mL conical flask, and add 50.0 mL of water. Insert the stopper, and swirl for 5 minutes. Decant into a glass-stoppered, 50-mL centrifuge tube, and centifuge to clarify. Transfer 30.0 mL of clear supernatant liquid to a glass-stoppered, 125-mL conical flask. Add 1 mL of glacial acetic acid and 0.5 g to 1.0 g of potassium iodide. Insert the stopper, swirl, and allow to stand for 25 to 30 minutes in the dark. Add 1 mL of starch TS, and titrate with 0.002 *N* sodium thiosulfate VS to the disappearance of the starch-iodine color. Perform a blank determination, and make any necessary correction. Each mL of 0.002 *N* sodium thiosulfate is equivalent to 34 µg of oxidant, calculated as hydrogen peroxide. Not more than 1.4 mL of 0.002 *N* sodium thiosulfate is required (0.002%).

Sulfur dioxide—Mix 20 g with 200 mL of water until a smooth suspension is obtained, and filter. To 100 mL of the clear filtrate add 3 mL of starch TS, and titrate with 0.010 *N* iodine to the first permanent blue color: not more than 2.7 mL is consumed (0.008%).

Pregelatinized Starch

» Pregelatinized Starch is Starch that has been chemically and/or mechanically processed to rupture all or part of the granules in the presence of water and subsequently dried. Some types of Pregelatinized Starch may be modified to render them compressible and flowable in character.

pH ⟨791⟩—Prepare a slurry by weighing 10.0 ± 0.1 g in 10 mL of alcohol and diluting with water to 100 mL. Agitate continuously at a moderate rate for 5 minutes, then cease agitation, and immediately determine the pH to the nearest 0.1 unit: the pH is between 4.5 and 7.0, determined potentiometrically.

Iron ⟨241⟩—Dissolve the residue obtained in the test for *Residue on ignition* in 8 mL of hydrochloric acid with the aid of gentle heating, dilute with water to 100 mL, and mix. Dilute 25 mL of this solution with water to 47 mL: the limit is 0.002%.

Oxidizing substances—To 5 g add 20 mL of a mixture of equal volumes of methanol and water, then add 1 mL of 6 *N* acetic acid, and stir until a homogeneous suspension is obtained. Add 0.5 mL of a freshly prepared, saturated solution of potassium iodide, mix, and allow to stand for 5 minutes: no distinct blue, brown, or purple color is observed.

Sulfur dioxide—Mix 20 g with 200 mL of sodium sulfate solution (1 in 5), and filter. To 100 mL of the clear filtrate add 3 mL of starch TS, and titrate with 0.010 *N* iodine to the first permanent blue color: not more than 2.7 mL is consumed (0.008%).

Other requirements—It responds to *Identification test B* and meets the requirements for *Packaging and storage*, *Microbial limits*, *Loss on drying*, and *Residue on ignition* under *Starch*.

Stearic Acid

Octadecanoic acid.
Stearic acid [*57-11-4*].

» Stearic Acid is manufactured from fats and oils derived from edible sources and is a mixture of Stearic Acid ($C_{18}H_{36}O_2$) and palmitic acid ($C_{16}H_{32}O_2$). The content of $C_{18}H_{36}O_2$ is not less than 40.0 percent, the content of $C_{16}H_{32}O_2$ is not less than 40.0 percent, and the sum of the two is not less than 90.0 percent.

NOTE—Stearic Acid labeled solely for external use is exempt from the requirement that it be prepared from edible sources.

Packaging and storage—Preserve in well-closed containers.

Labeling—If it is for external use only, the labeling so indicates.

Reference standards—*USP Stearic Acid Reference Standard*—Dry over silica gel for 4 hours before using. *USP Palmitic Acid Reference Standard*—Dry over silica gel for 4 hours before using.

Congealing temperature ⟨651⟩: not lower than 54°.

Residue on ignition ⟨281⟩: not more than 4 mg, determined on a 4-g portion (0.1%).

Heavy metals, *Method II* ⟨231⟩: 0.001%.

Mineral acid—Shake 5 g of melted Stearic Acid with an equal volume of hot water for 2 minutes, cool, and filter: the filtrate is not reddened by the addition of 1 drop of methyl orange TS.

Neutral fat or paraffin—Add 1 g of Stearic Acid to 30 mL of anhydrous sodium carbonate solution (1 in 60) in a flask, and boil the mixture: the resulting solution, while hot, shows not more than a faint opalescence.

Iodine value ⟨401⟩: not more than 4.

Assay—Place about 100 mg of Stearic Acid in a small conical flask fitted with a suitable reflux attachment. Place about 50 mg of USP Stearic Acid RS and about 50 mg of USP Palmitic Acid RS in a similar flask. Treat each flask as follows. Add 5.0 mL of a solution prepared by dissolving 14 g of boron trifluoride in methanol to make 100 mL, swirl to mix, and reflux for 15 minutes or until the solid is dissolved. Cool, transfer the reaction mixture with the aid of 10 mL of chromatographic solvent hexane to a 60-mL separator, and add 10 mL of water and 10 mL of saturated sodium chloride solution. Shake, allow to separate, then drain and discard the lower, aqueous layer. Pass the hexane layer through 6 g of anhydrous sodium sulfate (previously washed with chromatographic solvent hexane) into a suitable flask. Using a syringe fitted with a suitable needle, introduce a 1-µL to 2-µL portion of the assay preparation (which contains the Stearic Acid) into a suitable gas chromatograph equipped with a flame-ionization detector. The column preferably is of glass, 1.5 m in length and 3 mm in inside diameter, and it is packed with 15 percent G4 on support S1A. The carrier gas is helium, passed through a bed of molecular sieve for drying, if necessary. The temperatures of the port and the detector are maintained at 210°, and the column temperature is maintained at 165°.

System suitability—In a suitable chromatogram, the resolution factor, *R* (see *Chromatography* ⟨621⟩), is not less than 2.0 between the peaks from methyl palmitate and methyl stearate (located by comparison with the chromatogram of the standard preparation), and five replicate injections of a single sample show a coefficient of variation of not more than 1.5% in the percentage of methyl stearate and methyl palmitate, respectively. Measure the peak areas of the fatty acid esters in the chromatogram, and determine the percentage of $C_{18}H_{36}O_2$ in the portion of Stearic Acid taken by the formula:

$$100(A/B),$$

in which *A* is the area due to the methyl stearate peak, and *B* is the sum of the areas of all of the fatty acid ester peaks in the chromatogram. Similarly, determine the percentage of $C_{16}H_{32}O_2$.

Purified Stearic Acid

» Purified Stearic Acid is manufactured from fats and oils derived from edible sources and is a mixture of Stearic Acid ($C_{18}H_{36}O_2$) and palmitic acid ($C_{16}H_{32}O_2$), which together constitute not less than 96.0 percent of the total content. The content of $C_{18}H_{36}O_2$ is not less than 90.0 percent of the total.

NOTE—Purified Stearic Acid labeled solely for external use is exempt from the requirement that it be prepared from edible sources.

Packaging and storage—Preserve in well-closed containers.

Labeling—If it is for external use only, the labeling so indicates.

Reference standards—*USP Stearic Acid Reference Standard*—Dry over silica gel for 4 hours before using. *USP Palmitic Acid Reference Standard*—Dry over silica gel for 4 hours before using.

Congealing temperature ⟨651⟩: between 66° and 69°.

Acid value ⟨401⟩: between 195 and 200, about 1 g, accurately weighed, being used.

Iodine value ⟨401⟩: not more than 1.5.

Other requirements—It meets the requirements for *Residue on ignition*, *Heavy metals*, *Mineral acid*, *Neutral fat or paraffin*, and *Assay* under *Stearic Acid*.

Stearyl Alcohol

1-Octadecanol.
1-Octadecanol [*112-92-5*].

» Stearyl Alcohol contains not less than 90.0 percent of stearyl alcohol ($C_{18}H_{38}O$), the remainder consisting chiefly of related alcohols.

Packaging and storage—Preserve in well-closed containers.

Reference standards—*USP Stearyl Alcohol Reference Standard*—Do not dry before using. *USP Cetyl Alcohol Reference Standard*—Do not dry before using.

Identification—The retention time of the major peak obtained in the *Assay* is the same as that obtained from the standard solution.

Melting range ⟨741⟩: between 55° and 60°.

Acid value ⟨401⟩: not more than 2.

Iodine value ⟨401⟩: not more than 2.

Hydroxyl value—Place about 2 g, accurately weighed, in a dry, glass-stoppered, 250-mL flask, and add 2 mL of pyridine, followed by 10 mL of toluene. To the mixture add 10.0 mL of a solution of acetyl chloride, prepared by mixing 10 mL of acetyl chloride with 90 mL of toluene. Insert the stopper in the flask, and immerse in a water bath heated at 60° to 65° for 20 minutes. Add 25 mL of water, again insert the stopper in the flask, and shake vigorously for several minutes to decompose the excess acetyl chloride. Add 0.5 mL of phenolphthalein TS, and titrate to a permanent pink end-point with 1 *N* sodium hydroxide VS, shaking the flask vigorously toward the end of the titration to maintain the contents in an emulsified condition. Perform a blank test with the same quantities of the same reagents and in the same manner. The difference between the number of mL of 1 *N* sodium hydroxide consumed in the test with the sample and that consumed in the blank test, multiplied by 56.1, and the result divided by the weight, in g, of the Stearyl Alcohol used, represents the hydroxyl value of the Stearyl Alcohol, which is between 195 and 220.

Assay—

Chromatographic system (see *Chromatography* ⟨621⟩)—The gas chromatograph is equipped with a flame-ionization detector and a 2-m × 3-mm column packed with 10 percent phase G2 on support S1A. Dry helium is used as the carrier gas. The column is maintained at a temperature of about 205°, and the injection port and the detector are maintained at temperatures of about 275° and 250°, respectively.

System suitability solution—Prepare a system suitability solution by dissolving about 90 mg of USP Stearyl Alcohol RS and 10 mg of USP Cetyl Alcohol RS in 10.0 mL of alcohol. Inject 2 µL of this solution and calculate the resolution factor, *R* (see *Chromatography* ⟨621⟩), for stearyl alcohol and cetyl alcohol. A suitable resolution is not less than 4.0. Inject 2-µL portions of the suitability standard solution until the area ratio of stearyl alcohol to cetyl alcohol for each of five sequential injections is within 1.5% of the average of the area ratios for the five injections.

Procedure—Inject a suitable volume (about 2 µL) of a solution of Stearyl Alcohol in dehydrated alcohol (1 in 100) into the gas chromatograph. Measure the peak areas of the long-chain alcohol components in the chromatogram, and determine the percentage of $C_{18}H_{38}O$ in the portion of Stearyl Alcohol taken by the formula:

$$100A/B,$$

in which *A* is the area due to the stearyl alcohol peak and *B* is the sum of the areas of all of the long-chain alcohol peaks in the chromatogram.

Strong Ammonia Solution—*see* Ammonia Solution, Strong

Stronger Rose Water—*see* Rose Water, Stronger

Sucrose

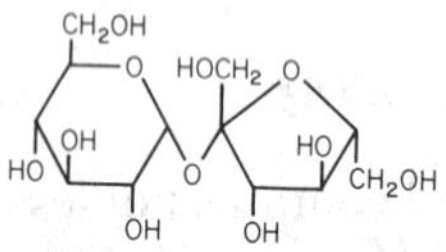

$C_{12}H_{22}O_{11}$ 342.30

α-D-Glucopyranoside, β-D-fructofuranosyl-.

Sucrose [*57-50-1*].

» Sucrose is a sugar obtained from *Saccharum officinarum* Linné (Fam. Gramineae), *Beta vulgaris* Linné (Fam. Chenopodiaceae), and other sources. It contains no added substances.

Packaging and storage—Preserve in well-closed containers.

Specific rotation ⟨781⟩: not less than +65.9°, determined in a solution containing 2.6 g in each 10 mL, the sample having been previously dried at 105° for 2 hours.

Residue on ignition ⟨281⟩: not more than 0.05%, a 5-g specimen being used.

Chloride ⟨221⟩—A 2.0-g portion shows no more chloride than corresponds to 0.10 mL of 0.020 *N* hydrochloric acid (0.0035%).

Sulfate ⟨221⟩—A 5.0-g portion shows no more sulfate than corresponds to 0.30 mL of 0.020 *N* sulfuric acid (0.006%).

Calcium—To 10 mL of a solution (1 in 10) add 1 mL of ammonium oxalate TS: the solution remains clear for at least 1 minute.

Heavy metals ⟨231⟩—Dissolve 4.0 g in 15 mL of water, add 1 mL of 0.12 *N* hydrochloric acid, and dilute with water to 25 mL: the limit is 5 ppm.

Invert sugar—Dissolve 20 g in water to make 100 mL, and filter if necessary. Place 50 mL of the clear liquid in a 250-mL beaker, add 50 mL of alkaline cupric tartrate TS, cover the beaker with a watch glass, and heat the mixture at such a rate that it comes to a boil in approximately 4 minutes, and boil for 2 minutes, accurately timed. Add at once 100 mL of cold, recently boiled water, and immediately collect the precipitated cuprous oxide on a tared filtering crucible containing a sintered-glass disk of medium pore size, or suitable equivalent. Thoroughly wash the residue on the filter with hot water, then with 10 mL of alcohol, and finally with 10 mL of ether, and dry at 105° for 1 hour: the weight of the cuprous oxide does not exceed 112 mg.

Sucrose Octaacetate

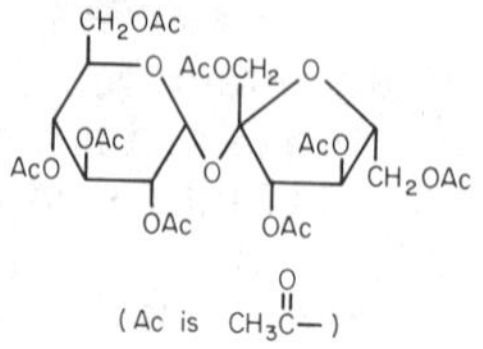

(Ac is $CH_3C(=O)-$)

$C_{28}H_{38}O_{19}$ 678.60

α-D-Glucopyranoside, 1,3,4,6-tetra-*O*-acetyl-β-D-fructofuranosyl, tetraacetate.

Sucrose octaacetate [*126-14-7*].

» Sucrose Octaacetate contains not less than 98.0 percent and not more than 100.5 percent of $C_{28}H_{38}O_{19}$, calculated on the anhydrous basis.

Packaging and storage—Preserve in tight containers.

Melting temperature ⟨741⟩: not lower than 78°.

Acidity—Dissolve 1 g in 20 mL of neutralized alcohol, and add 2 drops of phenolphthalein TS: not more than 2 drops of 0.1 *N* sodium hydroxide are required to produce a red color.

Water, *Method I* ⟨921⟩: not more than 1.0%.

Residue on ignition ⟨281⟩: not more than 0.1%.

Assay—Transfer about 100 mg of Sucrose Octaacetate, accurately weighed, to a 500-mL conical flask, and dissolve in about 50 mL of 70 percent alcohol. Neutralize the solution with 0.1 *N* sodium hydroxide VS, using phenolphthalein TS as the indicator. Add 25.0 mL of 0.1 *N* sodium hydroxide VS, attach an air condenser to the flask, protect from absorption of carbon dioxide, and reflux on a steam bath for 1 hour. Remove from the steam bath, cool quickly, and titrate the excess alkali with 0.1 *N* sulfuric acid VS, using phenolphthalein TS as the indicator. Perform a blank determination (see *Residual Titrations* under *Titrimetry* ⟨541⟩). Each mL of 0.1 *N* sodium hydroxide is equivalent to 8.483 mg of $C_{28}H_{38}O_{19}$.

Compressible Sugar

» Compressible Sugar, previously dried at 105° for 4 hours, contains not less than 95.0 percent and not more than 98.0 percent of sucrose ($C_{12}H_{22}O_{11}$). It may contain starch, malto-dextrin, or invert sugar, and may contain a suitable lubricant.

Packaging and storage—Preserve in well-closed containers.

Identification—The specific rotation of the uninverted solution obtained in the *Assay* is not less than 62.6°, and the acid-inverted solution obtained in the *Assay* is levorotatory.

Microbial limits—It meets the requirements of the tests for the absence of *Salmonella* species and *Escherichia coli* under *Microbial Limit Tests* ⟨61⟩.

Loss on drying ⟨731⟩—Dry it at 105° for 4 hours: it loses between 0.25% and 1.0% of its weight.

Residue on ignition ⟨281⟩: not more than 0.1%.

Chloride, Sulfate, Calcium, and Heavy metals—Transfer about 20 g, accurately weighed, to a 100-mL volumetric flask, add 80 mL of water, shake to dissolve the sucrose, then add water to volume, and mix. Separate the solubilized sucrose from any insoluble matter by filtration until the filtrate is sparkling clear, and use the freshly prepared, clear filtrate for the following tests.

Chloride ⟨221⟩—A 10-mL portion shows no more chloride than corresponds to 0.40 mL of 0.020 *N* hydrochloric acid (0.014%).

Sulfate ⟨221⟩—A 25-mL portion shows no more sulfate than corresponds to 0.50 mL of 0.020 *N* sulfuric acid (0.010%).

Calcium ⟨191⟩—To a 5-mL portion add 1 mL of ammonium oxalate TS: the solution remains clear for not less than 1 minute.

Heavy metals ⟨231⟩—To a 20-mL portion add 4 mL of water and 1 mL of 0.1 *N* hydrochloric acid: the limit is 5 ppm.

Assay—Transfer 26.0 g of Compressible Sugar, previously dried and accurately weighed, to a 100-mL volumetric flask, add about 0.3 mL of a saturated aqueous solution of lead acetate, shake with about 90 mL of water, dilute with water to volume, and mix. Distribute evenly on the surface of a sheet of medium-fast filter paper about 8 g of chromatographic siliceous earth suitable for column partition chromatography (see under *Reagents* in the section, *Reagents, Indicators, and Solutions*), and filter the solution, with the aid of vacuum, discarding the first 20 mL of the filtrate. Pipet 25 mL of the subsequent clear filtrate into each of two 50-mL volumetric flasks. Slowly add 6 mL of dilute hydrochloric acid (1 in 2) to one flask while rotating it, dilute with water nearly to volume, and mix. Place the flask in a water bath maintained at a temperature of 60°, continuously shake the flask in the bath for about 3 minutes, and allow the flask to remain in the bath for a total of 10 minutes. Immediately cool to 20° by plunging the flask into a cold bath, dilute with water to volume at 20°, and mix. Cool the contents of the second flask to 20°, dilute with water to volume at 20°, and mix. Maintain both flasks at a temperature of 20° for 30 minutes. Determine the specific rotation of each solution at 20° as directed under *Optical Rotation* ⟨781⟩. Calculate the percentage of $C_{12}H_{22}O_{11}$ taken by the formula:

$$100(\alpha_o - \alpha_i)/88.3,$$

in which α_o and α_i are the specific rotations of the uninverted and acid-inverted solutions, respectively.

Confectioner's Sugar

» Confectioner's Sugar is Sucrose ground together with corn starch to a fine powder. It contains not less than 95.0 percent of sucrose ($C_{12}H_{22}O_{11}$), calculated on the dried basis.

Packaging and storage—Preserve in well-closed containers.

Identification, Specific rotation, Chloride, Calcium, Sulfate, and Heavy metals—Transfer about 20 g, accurately weighed, to a 100-mL volumetric flask, add 80 mL of water, shake to dissolve the sucrose, then add water to volume, and mix. Separate the solubilized sucrose from the insoluble starch component by filtration until the filtrate is sparkling clear. Use the insoluble portion for the *Identification test*, and use the freshly prepared, clear filtrate for the other tests that follow.

Identification—A water slurry of the insoluble portion responds to *Identification test B* under *Starch*.

Specific rotation ⟨781⟩: not less than +62.6°, determined on a portion of the filtrate, corresponding to not less than 95.0% of $C_{12}H_{22}O_{11}$, calculated on the dried basis.

Chloride ⟨221⟩—A 10-mL portion of the filtrate shows no more chloride than corresponds to 0.40 mL of 0.020 *N* hydrochloric acid (0.014%).

Calcium ⟨191⟩—To 5 mL of the filtrate add 5 mL of water and 1 mL of ammonium oxalate TS: the solution remains clear for not less than 1 minute.

Sulfate ⟨221⟩—A 25-mL portion of the filtrate shows no more sulfate than corresponds to 0.30 mL of 0.020 *N* sulfuric acid (0.006%).

Heavy metals ⟨231⟩—To 20 mL of the filtrate add 4 mL of water and 1 mL of 0.1 *N* hydrochloric acid: the limit is 5 ppm.

Microbial limits—It meets the requirements of the tests for absence of *Salmonella* species and *Escherichia coli* under *Microbial Limit Tests* ⟨61⟩.

Loss on drying ⟨731⟩—Dry it at 105° for 4 hours: it loses not more than 1.0% of its weight.

Residue on ignition ⟨281⟩: not more than 0.08%.

Sugar Spheres

» Sugar Spheres contain not less than 62.5 percent and not more than 91.5 percent of sucrose ($C_{12}H_{22}O_{11}$), calculated on the dried basis, the remainder consisting chiefly of starch. They consist of approximately spherical particles of a labeled nominal size range. They may contain color additives permitted by the federal Food and Drug Administration for use in drugs.

Packaging and storage—Preserve in well-closed containers.

Labeling—The label states the nominal particle size range.

Identification and Specific rotation—Transfer about 20 g, accurately weighed, to a 200-mL volumetric flask, add 160 mL of water, shake to dissolve the sucrose, add water to volume, and mix. Separate the solubilized sucrose from the insoluble starch component by vacuum filtration through fine filter paper until the filtrate is clear. Use the insoluble portion for the *Identification test*, and use the freshly prepared, clear filtrate for the *Specific rotation* test.

Identification—A water slurry of the insoluble portion responds to *Identification test B* under *Starch*.

Specific rotation ⟨781⟩: not less than +41° and not more than +61°, determined on a portion of the filtrate, corresponding to not less than 62.5% and not more than 91.5% of sucrose ($C_{12}H_{22}O_{11}$), calculated on the dried basis.

Microbial limits ⟨61⟩—The Spheres meet the requirements of the tests for absence of *Salmonella* species, *Escherichia coli*, *Staphylococcus aureus*, and *Pseudomonas aeruginosa*, and the total aerobic microbial count does not exceed 100 per g.

Loss on drying ⟨731⟩—Dry the Spheres at 105° for 4 hours: the material loses not more than 4.0% of its weight.

Residue on ignition ⟨281⟩: not more than 0.25%, determined on a 2.0-g specimen ignited at a temperature of 700 ± 25°.

Particle size (see *Powder Fineness* ⟨811⟩)—Test a portion of the Spheres in accordance with the procedure for coarse powders. Not less than 90.0% of it passes the larger nominal standard sieve; 100.0% passes the next larger standard sieve; not more than 10.0% passes the smaller nominal sieve. [NOTE—Use a mechanical sieve-shaking unit that employs both rotary horizontal motion and tapping, in order to ensure reliability of this test.]

Heavy metals, *Method II* ⟨231⟩: 5 ppm.

Sulfur Dioxide

SO_2 64.06
Sulfur dioxide.
Sulfur dioxide [*7446-09-5*].

» Sulfur Dioxide contains not less than 97.0 percent, by volume, of SO_2.

Caution—Sulfur Dioxide is poisonous.

Packaging and storage—Preserve in cylinders.

NOTE—Sulfur Dioxide is used most in the form of a gas in pharmaceutical applications, and is described herein for such purposes. However, it is usually packaged under pressure, hence the following specifications are designed for testing it in liquid form.

Water, *Method I* ⟨921⟩—Taking precautions to avoid absorption of moisture, transfer 3 g (about 2.1 mL) to a suitable flask, and add 20 mL of anhydrous pyridine: not more than 2.0% is found.

Nonvolatile residue—Transfer 300 g (about 209 mL) to a tared, 250-mL conical flask, and allow the liquid to evaporate spontaneously in a well-ventilated hood. When evaporation appears complete, blow a current of dry, filtered air through the flask until the odor of sulfur dioxide is no longer apparent: the weight of the residue does not exceed 7.5 mg (0.0025%).

Sulfuric acid—To the flask containing the residue obtained in the test for *Nonvolatile residue* add 25 mL of water previously neutralized to methyl red TS. Swirl the flask, and titrate with 0.10 *N* sodium hydroxide: not more than 1.3 mL is required (about 0.002%).

Assay—Collect 100.0 mL of gaseous Sulfur Dioxide over mercury, and note the temperature of the sample and the pressure upon it. Slowly introduce 50.0 mL of 0.1 *N* sodium hydroxide into the air space over the mercury, and absorb the sample in the solution by shaking. When absorption is complete, transfer the solution to a 250-mL conical flask, add 3 mL of starch TS, and titrate with 0.1 *N* iodine VS until the solution is pale blue in color. Each mL of 0.1 *N* iodine is equivalent to 1.094 mL of SO_2 at a temperature of 0° and a pressure of 760 mm of mercury.

Sulfuric Acid

H_2SO_4 98.07
Sulfuric acid.
Sulfuric acid [*7664-93-9*].

» Sulfuric Acid contains not less than 95.0 percent and not more than 98.0 percent, by weight, of H_2SO_4.

Caution—When Sulfuric Acid is to be mixed with other liquids, always add it to the diluent, and exercise great caution.

Packaging and storage—Preserve in tight containers.

Identification—It responds to the tests for *Sulfate* ⟨191⟩.

Residue on ignition ⟨281⟩—Evaporate 22 mL (40 g) to dryness, and ignite: not more than 2 mg of residue remains (0.005%).

Chloride ⟨221⟩—A dilution of 1.1 mL (2.0 g) in water shows no more chloride than corresponds to 0.15 mL of 0.020 *N* hydrochloric acid (0.005%).

Arsenic ⟨211⟩—Add 1.6 mL (3.0 g) to 3 mL of nitric acid and 20 mL of water, and evaporate until dense fumes of sulfur trioxide form. Cool, and cautiously wash the solution into an arsine generating flask with 50 mL of water; the resulting solution meets the requirements of the test, the addition of 20 mL of dilute sulfuric acid (1 in 5) specified under *Procedure* being omitted (1 ppm).

Heavy metals ⟨231⟩—Add 2.2 mL (4.0 g) to about 10 mg of sodium carbonate dissolved in 10 mL of water. Heat until almost dry, add 1 mL of nitric acid, evaporate to dryness, add 2 mL of 1 *N* acetic acid to the residue, and dilute with water to 25 mL: the limit is 5 ppm.

Reducing substances—Carefully dilute 4.4 mL (8.0 g) with about 50 mL of ice cold water, keeping the solution cold during the addition. Add 0.10 mL of 0.10 *N* potassium permanganate: the solution remains pink for 5 minutes.

Assay—Place about 1 mL of Sulfuric Acid in an accurately weighed, glass-stoppered flask containing about 20 mL of water, and weigh again to obtain the weight of the test specimen. Dilute with about 25 mL of water, cool, add methyl orange TS, and titrate with 1 *N* sodium hydroxide VS. Each mL of 1 *N* sodium hydroxide is equivalent to 49.04 mg of H_2SO_4.

Syrup

» Syrup is a solution of Sucrose in Purified Water. It contains a preservative unless it is used when freshly prepared.

Sucrose	850 g
Purified Water, a sufficient quantity, to make	1000 mL

Syrup may be prepared by the use of boiling water or, preferably, without heat, by the following process. Place the Sucrose in a suitable percolator the neck of which is nearly filled with loosely packed cotton moistened, after packing, with a few drops of water. Pour carefully about 450 mL of Purified Water upon the Sucrose, and regulate the outflow to a steady drip of percolate. Return the percolate, if necessary, until all of the Sucrose has dissolved. Then wash the inside of the percolator and the cotton with sufficient Purified Water to bring the volume of the percolate to 1000 mL, and mix.

Packaging and storage—Preserve in tight containers, preferably in a cool place.

Specific gravity ⟨841⟩: not less than 1.30.

Syrups—*see complete list in index*
Talc—*see* Talc USP

Tartaric Acid

```
      COOH
       |
   H—C—OH
       |
  HO—C—H
       |
      COOH
```

$C_4H_6O_6$ 150.09
Butanedioic acid, 2,3-dihydroxy-; Butanedioic acid, 2,3-dihydroxy-, [*R*-(*R**,*R**)]-.
Tartaric acid; L-(+)-Tartaric acid [*526-83-0*] and [*87-69-4*].

» Tartaric Acid, dried over phosphorus pentoxide for 3 hours, contains not less than 99.7 percent and not more than 100.5 percent of $C_4H_6O_6$.

Packaging and storage—Preserve in well-closed containers.

Identification—

A: It responds to the tests for *Tartrate* ⟨191⟩.

B: When ignited, it gradually decomposes, emitting an odor resembling that of burning sugar (*distinction from citric acid*).

Specific rotation ⟨781⟩: between +12.0° and +13.0°, calculated on the dried basis, determined in a solution containing 2 g in each 10 mL.

Loss on drying ⟨731⟩—Dry it over phosphorus pentoxide for 3 hours: it loses not more than 0.5% of its weight.

Residue on ignition ⟨281⟩: not more than 0.1%.

Oxalate—Nearly neutralize 10 mL of a solution of it (1 in 10) with 6 *N* ammonium hydroxide, and add 10 mL of calcium sulfate TS: no turbidity is produced.

Sulfate ⟨221⟩—To 10 mL of a solution (1 in 100) add 3 drops of hydrochloric acid and 1 mL of barium chloride TS: no turbidity is produced.

Heavy metals, *Method II* ⟨231⟩: 0.001%.

Assay—Place about 2 g of Tartaric Acid, previously dried and accurately weighed, in a conical flask. Dissolve it in 40 mL of water, add phenolphthalein TS, and titrate with 1 *N* sodium hydroxide VS. Each mL of 1 *N* sodium hydroxide is equivalent to 75.04 mg of $C_4H_6O_6$.

Thimerosal—*see* Thimerosal USP

Thymol

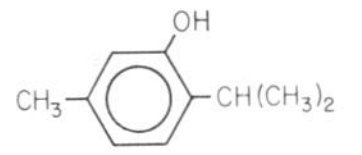

$C_{10}H_{14}O$ 150.22
Phenol, 5-methyl-2-(1-methylethyl)-.
Thymol.
p-Cymen-3-ol [*89-83-8*].

» Thymol contains not less than 99.0 percent and not more than 101.0 percent of $C_{10}H_{14}O$.

Packaging and storage—Preserve in tight, light-resistant containers.

Identification—

A: Triturate it with about an equal weight of camphor or menthol: the mixture liquefies.

B: Dissolve a very small crystal of it in 1 mL of glacial acetic acid, and add 6 drops of sulfuric acid and 1 drop of nitric acid: the liquid shows a deep bluish green color when viewed by reflected light.

C: Place about 1 g in a test tube, add 5 mL of sodium hydroxide solution (1 in 10), and heat in a water bath: a clear, colorless or pale red, solution is produced, and it becomes darker on standing, without the separation of oily drops. Add to this solution a few drops of chloroform, and agitate the mixture: a violet color is produced.

Melting range ⟨741⟩: between 48° and 51°; but when melted, Thymol remains liquid at a considerably lower temperature.

Nonvolatile residue—Volatilize about 2 g, accurately weighed, on a steam bath, and dry at 105° to constant weight: not more than 0.05% of residue remains.

Assay—Transfer about 100 mg of Thymol, accurately weighed, to a 250-mL iodine flask, and dissolve in 25 mL of 1 *N* sodium hydroxide. Add 20 mL of hot dilute hydrochloric acid (1 in 2), and immediately titrate with 0.1 *N* bromine VS to within 1 to 2 mL of the calculated end-point. Warm the solution to between 70° and 80°, add 2 drops of methyl orange TS, and continue the titration slowly, swirling vigorously after each addition. When the color of the methyl orange is bleached, add 2 drops of 0.1 *N* bromine VS, shake for 10 seconds, add 1 drop of methyl orange TS, and shake vigorously. If the solution is red, continue the titration, dropwise and with shaking, until the color is discharged. Repeat the alternate addition of the titrant and the methyl orange TS until the red color is discharged after the addition of the TS. Each mL of 0.1 *N* bromine is equivalent to 3.755 mg of $C_{10}H_{14}O$.

Tinctures—*see complete list in index*
Titanium Dioxide—*see* Titanium Dioxide USP

Tocopherols Excipient

» Tocopherols Excipient is a vegetable oil solution containing not less than 50.0 percent of total tocopherols, of which not less than 80.0 percent consists of varying amounts of beta, gamma, and delta tocopherols.

Packaging and storage—Preserve in tight containers, protected from light. Protect with a blanket of an inert gas.

Labeling—Label it to indicate the content, in mg per g, of total tocopherols and of the sum of beta, gamma, and delta tocopherols.

Reference standard—*USP Alpha Tocopherol Reference Standard*—Keep container tightly closed and protected from light. Do not dry before using.

Identification—

A: Dissolve about 50 mg in 10 mL of dehydrated alcohol, add, with swirling, 2 mL of nitric acid, and heat at about 75° for 15 minutes: a bright red or orange color develops.

B: The retention time of the third major peak (i.e., the peak occurring just before that of the internal standard) in the chromatogram of the *Assay preparation* corresponds to that of the *Standard preparation*, both relative to that of the internal standard, as obtained in the *Assay*.

Acidity—Dissolve 1.0 g in 25 mL of a mixture of equal volumes of alcohol and ether (which has been neutralized to phenolphthalein with 0.1 *N* sodium hydroxide), add 0.5 mL of phenolphthalein TS, and titrate with 0.10 *N* sodium hydroxide until the solution remains faintly pink after being shaken for 30 seconds: not more than 1.0 mL of 0.10 *N* sodium hydroxide is required.

Assay—

Internal standard solution—Transfer about 600 mg of hexadecyl hexadecanoate, accurately weighed, to a 200-mL volumetric flask, dissolve in a diluting solution containing 2 volumes of pyridine and 1 volume of propionic anhydride, dilute with the diluting solution to volume, and mix.

Standard preparations—[NOTE—Use low-actinic glassware.] Transfer 12-, 25-, 37-, and 50-mg portions of USP Alpha Tocopherol RS, accurately weighed, to separate 50-mL conical flasks having 19/38 standard-taper ground-glass necks. Pipet 25 mL of the *Internal standard solution* into each flask, mix, and reflux for 10 minutes under water-cooled condensers.

Assay preparation—[NOTE—Use low-actinic glassware.] Transfer about 60 mg of Tocopherols Excipient, accurately weighed, to a 50-mL conical flask similar to the flasks used in preparing the *Standard preparations*, add 10.0 mL of *Internal standard solution*, mix, and reflux for 10 minutes under a water-cooled condenser.

Chromatographic system (see *Chromatography* ⟨621⟩)—The gas chromatograph is equipped with a flame-ionization detector, and contains a 2-m × 4-mm borosilicate glass column packed with 2 percent to 5 percent liquid phase G2 on 80- to 100-mesh support S1AB utilizing either a glass-lined sample introduction system or on-column injection. The column is maintained isothermally at a temperature between 245° and 265°, and the injection port and detector block are maintained at about 10° higher than the column temperature. The flow rate of dry carrier gas is adjusted to obtain a hexadecyl hexadecanoate peak 30 to 32 minutes after sample introduction. [NOTE—Cure and condition the column as necessary.]

System suitability—Chromatograph a sufficient number of injections of the *Assay preparation*, as directed under *Calibration*, to ensure that the resolution, *R*, between the major peaks occurring at retention times of approximately 0.50 (delta tocopheryl propionate) and 0.63 (beta plus gamma tocopheryl propionates), relative to hexadecyl hexadecanoate at 1.00, is not less than 2.5.

Calibration—Chromatograph 2- to 5-µL portions of each *Standard preparation*, and record the peak areas as directed under *Procedure*. Calculate the relative response factor, *F*, for each concentration of the *Standard preparation* by the formula:

$$(A_S/A_D)(C_D/C_S),$$

in which C_D and C_S are the concentrations, in mg per mL, of hexadecyl hexadecanoate and of USP Alpha Tocopherol RS, respectively, in the *Standard preparation*. Successively chromatograph a sufficient number of portions of each *Standard preparation* to ensure that the factor, *F*, is constant within a range of 2.0%. Prepare a relative response factor curve by plotting *F* versus the alpha tocopheryl propionate peak area.

Procedure—Inject a suitable portion (2 µL to 5 µL) of the *Assay preparation* into the chromatograph, and record the chromatogram. Measure the areas under the 4 major peaks occurring at relative retention times of approximately 0.50, 0.63, 0.76, and 1.00, and record the values as a_δ, $a_{\beta\gamma}$, a_α, and a_D, corresponding to delta tocopheryl propionate, beta plus gamma tocopheryl propionates, alpha tocopheryl propionate, and hexadecyl hexadecanoate, respectively. Calculate the quantity, in mg, of each tocopherol form in the Tocopherols Excipient taken by the formulas:

$$\text{delta tocopherol} = (10C_D/F)(a_\delta/a_D);$$
$$\text{beta plus gamma tocopherols} = (10C_D/F)(a_{\beta\gamma}/a_D);$$
$$\text{alpha tocopherol} = (10C_D/F)(a_\alpha/a_D),$$

in which F is obtained from the relative response factor curve (see *Calibration*) for each of the corresponding areas under the delta, beta plus gamma, and alpha tocopheryl propionate peaks produced by the *Assay preparation*. [NOTE—The relative response factor for delta tocopheryl propionate and for beta plus gamma tocopheryl propionates has been determined empirically to be the same as for alpha tocopheryl propionate.]

Tolu Balsam Syrup

» Prepare Tolu Balsam Syrup as follows:

Tolu Balsam Tincture	50 mL
Magnesium Carbonate	10 g
Sucrose	820 g
Purified Water, a sufficient quantity, to make	1000 mL

Add the tincture all at once to the Magnesium Carbonate and 60 g of the Sucrose in a mortar, and mix. Gradually add 430 mL of Purified Water with trituration, and filter. Dissolve the remainder of the Sucrose in the clear filtrate with gentle heating, strain the syrup while warm, and add sufficient Purified Water through the strainer to make the product measure 1000 mL. Mix.

NOTE—Tolu Balsam Syrup may be prepared also as follows. Place 760 g of the Sucrose in a suitable percolator, the neck of which is nearly filled with loosely packed cotton, moistened after packing with a few drops of water. Pour the filtrate, obtained as directed in the preceding instructions, upon the Sucrose, and regulate the outflow to a steady drip of percolate. When all of the liquid has run through, return portions of the percolate, if necessary, to dissolve all the Sucrose. Then pass enough Purified Water through the cotton to make the product measure 1000 mL. Mix.

Packaging and storage—Preserve in tight containers, at a temperature not above 25°.

Alcohol content, *Method II* ⟨611⟩: between 3.0% and 5.0% of C_2H_5OH.

Tolu Balsam Tincture

» Prepare Tolu Balsam Tincture as follows:

Tolu Balsam	200 g
To make	1000 mL

Prepare a tincture by Process M (see *Pharmaceutical Dosage Forms* ⟨1151⟩), using alcohol as the menstruum.

Packaging and storage—Preserve in tight, light-resistant containers, and avoid exposure to direct sunlight and to excessive heat.

Alcohol content, *Method I* ⟨611⟩: between 77.0% and 83.0% of C_2H_5OH.

Tragacanth

» Tragacanth is the dried gummy exudation from *Astragalus gummifer* Labillardière, or other Asiatic species of *Astragalus* (Fam. Leguminosae).

Packaging and storage—Preserve in well-closed containers.

Botanic characteristics—

Tragacanth—Flattened, lamellated, frequently curved fragments or straight or spirally twisted linear pieces from 0.5 mm to 2.5 mm in thickness. Is white to weak yellow in color, translucent, and horny in texture. Its fracture is short. Is rendered more easily pulverizable by heating to 50°. Is odorless and has an insipid, mucilaginous taste.

Histology—Pieces of Tragacanth softened in water and mounted in water or glycerin show numerous lamellae and a few starch grains.

Powdered Tragacanth—White to yellowish white. When examined in water mounts, it shows numerous angular fragments of mucilage with circular or irregular lamellae, and occasional starch grains up to 25 µm in diameter, mostly simple, spherical to elliptical, with occasional 2- to 4-compound grains, a few of the grains being swollen and more or less altered. The powder shows few or no fragments of lignified vegetable tissue (*Indian gum*).

Identification—Add 1 g to 50 mL of water: it swells and forms a smooth, nearly uniform, stiff, opalescent mucilage free from cellular fragments.

Microbial limits—It meets the requirements of the tests for absence of *Salmonella* species and *Escherichia coli* under *Microbial Limit Tests* ⟨61⟩.

Arsenic, *Method II* ⟨211⟩: 3 ppm.

Lead ⟨251⟩: 0.001%.

Heavy metals, *Method II* ⟨231⟩: 0.004%.

Karaya gum—Boil 1 g with 20 mL of water until a mucilage is formed, add 5 mL of hydrochloric acid, and again boil the mixture for 5 minutes: no pink or red color develops.

Triacetin—*see* Triacetin USP

Tribasic Calcium Phosphate—*see* Calcium Phosphate, Tribasic

Trichloromonofluoromethane

CCl_3F

CCl_3F 137.37

Methane, trichlorofluoro-.

Trichlorofluoromethane [*75-69-4*].

Packaging and storage—Preserve in tight cylinders, and avoid exposure to excessive heat.

Identification—The infrared absorption spectrum of it, determined in a 10-cm cell with sodium chloride windows, at atmospheric pressure, exhibits maxima, among others, at the following wavelengths, in µm: 4.67 (*m*), 5.95 (*m*), 7.28 (*s*), 8.06 (*m*), 9.2 (*vs*), 10.7 (*vs*), 11.8 (*vs*), and 13.4 (*m*). The stronger maxima are best obtained at pressures less than 10 mm of mercury.

Boiling temperature: approximately 24°, determined as directed under *Approximate Boiling Temperature* (see *Aerosols* ⟨601⟩).

Water: not more than 0.001%, determined as directed under *Water Content* (see *Aerosols* ⟨601⟩).

High-boiling residues: not more than 0.01%, determined as directed for *High-boiling Residues, Method I*, under *Aerosols* ⟨601⟩.

Inorganic chlorides—Place 5 mL of anhydrous methanol in a test tube, add 3 drops of a saturated solution of silver nitrate in anhydrous methanol, shake, and add 7 g of Trichloromonofluoromethane: no opalescence or turbidity is produced.

Triethyl Citrate

$C_{12}H_{20}O_7$ 276.29

» Triethyl Citrate contains not less than 99.0 percent and not more than 100.5 percent of $C_{12}H_{20}O_7$, calculated on the anhydrous basis.

Packaging and storage—Preserve in tight containers.

Specific gravity ⟨841⟩: between 1.135 and 1.139.

Refractive index ⟨831⟩: between 1.439 and 1.441.

Acidity—Dissolve 32.0 g in 30 mL of alcohol, previously neutralized to phenolphthalein, add phenolphthalein TS, and titrate with 0.10 *N* sodium hydroxide. Not more than 1.0 mL is required.

Water, *Method I* ⟨921⟩: not more than 0.25%.

Assay—Weigh accurately about 1.5 g of Triethyl Citrate into a 500-mL flask equipped with a standard-taper ground joint. Add 25 mL of isopropyl alcohol and 25 mL of water. Pipet 50 mL of 0.5 *N* potassium hydroxide VS into the mixture, add a few boiling chips, and attach a water-cooled condenser. Reflux for 1.5 hours, cool, wash down the condenser with about 20 mL of water, add 5 drops of bromothymol blue TS, and titrate the excess alkali with 0.5 *N* sulfuric acid VS. Perform a blank determination, and make any necessary correction. Each mL of 0.5 *N* potassium hydroxide is equivalent to 46.05 mg of $C_{12}H_{20}O_7$.

Trolamine

$C_6H_{15}NO_3$ 149.19
Ethanol, 2,2′,2″-nitrilotris-.
2,2′,2″-Nitrilotriethanol [*102-71-6*].

» Trolamine is a mixture of alkanolamines consisting largely of triethanolamine [$N(C_2H_4OH)_3$] containing some diethanolamine [$NH(C_2H_4OH)_2$] and monoethanolamine [$NH_2(C_2H_4OH)$]. It contains not less than 99.0 percent and not more than 107.4 percent of alkanolamines, calculated on the anhydrous basis as $N(C_2H_4OH)_3$.

Packaging and storage—Preserve in tight, light-resistant containers.

Identification—

A: To 1 mL add 0.1 mL of cupric sulfate TS: a deep blue color is produced. Add 5 mL of 1 *N* sodium hydroxide, and concentrate to one-third of the original volume by boiling: the blue color remains.

B: To 1 mL add 0.3 mL of cobaltous chloride TS: a carmine-red color is produced.

C: Heat 1 mL gently in a test tube: the vapors turn moistened red litmus paper blue.

Specific gravity ⟨841⟩: between 1.120 and 1.128.

Refractive index ⟨831⟩: between 1.481 and 1.486 at 20°.

Water, *Method I* ⟨921⟩: not more than 0.5%, a mixture of 5.0 mL of glacial acetic acid and 20 mL of methanol being used as the solvent.

Residue on ignition ⟨281⟩: not more than 0.05%.

Assay—Transfer about 2 g of Trolamine, accurately weighed, to a 300-mL conical flask. Add 75 mL of water and 2 drops of methyl red TS, and titrate with 1 *N* hydrochloric acid VS: each mL of 1 *N* hydrochloric acid is equivalent to 149.2 mg of triethanolamine, expressed as $N(C_2H_4OH)_3$.

Tyloxapol—*see* Tyloxapol USP

Vanilla

» Vanilla is the cured, full-grown, unripe fruit of *Vanilla planifolia* Andrews, often known in commerce as Mexican or Bourbon Vanilla, or of *Vanilla tahitensis* J. W. Moore, known in commerce as Tahiti Vanilla (Fam. Orchidaceae).

Vanilla yields not less than 12.0 percent of anhydrous diluted alcohol-soluble extractive.

Packaging and storage—Preserve in tight containers, and store in a cold place.

Labeling—The commercial variety of Vanilla, whether Mexican, Bourbon, or Tahiti, is stated on the label.

NOTE—Do not use Vanilla that has become brittle.

Botanic characteristics—

Unground Vanilla—Linear, flattened capsule, from 12 cm to 35 cm in length and from 5 mm to 9 mm in width; having an apex terminating in a flat, circular scar, and a base gradually tapering, more or less curved or hooked, or in Tahiti Vanilla, broad in the middle and tapering toward either end, the base closely resembling the summit. It is flexible and tough; externally nearly black, dusky brown to moderate brown, longitudinally wrinkled, moist, glossy, and occasionally with an efflorescence of acicular or prismatic crystals of vanillin. Internally it is unilocular, with a brownish black pulp and numerous minute seeds. Occasional capsules are split near the summit into three parts.

Histology—Epidermis with a distinct cuticle and occasional stomata, the epidermal cells containing red to brown bodies, occasional prisms of calcium oxalate or crystals of vanillin; a collenchyma layer of one or two rows of cells; a thick sarcocarp, composed of parenchyma and an interrupted circle of fibrovascular bundles, the latter leptocentric with a few vessels, and an outer circle of fibers with thin, strongly lignified walls and numerous transverse simple pits; the vessels with walls having slit-like pits or spiral thickenings; the parenchyma cells usually thin-walled and deeply undulate, some thick-walled with oblique, slit-like pits or broad spiral bands, and containing occasional bundles of acicular crystals of calcium oxalate up to 400 µm in length, or a thin protoplasmic layer enclosing numerous oil globules; an endocarp composed of placental and interplacental regions; the placental region consisting of six bifid placentas extending into the cavity of the fruit and bearing irregularly trianguloid, black to reddish, flattened seeds having a deeply reticulate seed-coat and up to about 250 µm in diameter; the interplacental regions showing long, nearly straight hairs more or less matted together by their gummy resinous secretion.

Powdered Vanilla—Dusky brown to nearly black. The principal elements of identification are fragments of parenchyma of the sarcocarp with long, oblique, slit-like walls or broad spiral bands, calcium oxalate crystals of acicular outline and up to 400 µm in length, and monoclinic prisms up to 35 µm in length, numerous unicellular, nearly straight glandular hairs, fragments of the seed-coat with polygonal stone cells, and the slender crystals of vanillin.

Test for vanillin—Place a few of the crystals, occurring as an efflorescence on the fruit, on a microslide or watch glass, and add 1 drop of phloroglucinol TS and 1 drop of hydrochloric acid: the solution immediately acquires a red color.

Assay—Place 2 g of Vanilla, finely cut or in coarse powder and accurately weighed, in a suitable flask. Add 70 mL of diluted alcohol, shake the mixture for 2 hours in a mechanical shaker or during 8 hours at 30-minute intervals, and allow to stand overnight. Decant the liquid onto a filter, and wash the flask and residue with small portions of diluted alcohol, passing the washings through the filter until the filtrate measures 100.0 mL. Mix the filtrate well, evaporate a 50.0-mL portion in a suitable tared container on a steam bath to dryness, and dry the residue at 105° for 4 hours. The weight obtained represents the yield of anhydrous diluted alcohol-soluble extractive from one-half of the Vanilla taken.

Vanilla Tincture

» Prepare Vanilla Tincture as follows:

Vanilla, cut into small pieces	100 g
Sucrose, in coarse granules	200 g
Alcohol,	
Diluted Alcohol,	
Purified Water, each a sufficient quantity, to make	1000 mL

Add 200 mL of water to the comminuted Vanilla in a suitable covered container, and macerate for 12 hours, preferably in a warm place. Add 207 mL of Alcohol to the mixture, mix, and macerate for about 3 days. Transfer the mixture to a percolator containing the Sucrose, and drain. Then pack the drug firmly, and percolate slowly, using Diluted Alcohol as the menstruum.

Packaging and storage—Preserve in tight, light-resistant containers, and avoid exposure to direct sunlight and to excessive heat.

Alcohol content, *Method I* ⟨611⟩: between 38.0% and 42.0% of C_2H_5OH.

Vanillin

CHO / OCH3 / OH

$C_8H_8O_3$ 152.15
Benzaldehyde, 4-hydroxy-3-methoxy-.
Vanillin [*121-33-5*].

» Vanillin contains not less than 97.0 percent and not more than 103.0 percent of $C_8H_8O_3$, calculated on the dried basis.

Packaging and storage—Preserve in tight, light-resistant containers.

Reference standard—*USP Vanillin Reference Standard*—Dry over silica gel for 4 hours before using.

Identification—

A: The infrared absorption spectrum of a potassium bromide dispersion of it exhibits maxima only at the same wavelengths as that of a similar preparation of USP Vanillin RS.

B: The ultraviolet absorption spectrum of a 1 in 125,000 solution in methanol exhibits maxima and minima at the same wavelengths as that of a similar solution of USP Vanillin RS, concomitantly measured.

Melting range ⟨741⟩: between 81° and 83°.

Loss on drying ⟨731⟩—Dry it over silica gel for 4 hours: it loses not more than 1.0% of its weight.

Residue on ignition ⟨281⟩: not more than 0.05%.

Assay—Transfer about 100 mg of Vanillin, accurately weighed, to a 250-mL volumetric flask, add methanol to volume, and mix. Pipet 2 mL of this solution into a 100-mL volumetric flask, add methanol to volume, and mix. Dissolve an accurately weighed quantity of USP Vanillin RS in methanol, and dilute quantitatively and stepwise with methanol to obtain a Standard solution having a known concentration of about 8 µg per mL. Concomitantly determine the absorbances of both solutions in 1-cm cells at the wavelength of maximum absorbance at about 308 nm, with a suitable spectrophotometer, using methanol as the blank. Calculate the quantity, in mg, of $C_8H_8O_3$ in the portion of Vanillin taken by the formula:

$$12.5C(A_U / A_S),$$

in which C is the concentration, in µg per mL, of USP Vanillin RS in the Standard solution, and A_U and A_S are the absorbances of the solution of Vanillin and the Standard solution, respectively.

Hydrogenated Vegetable Oil

» Hydrogenated Vegetable Oil is refined, bleached, hydrogenated, and deodorized vegetable oil stearins consisting mainly of the triglycerides of stearic and palmitic acids.

Packaging and storage—Preserve in tight containers, in a cool place.

Melting range, *Class II* ⟨741⟩: between 61° and 66°.

Loss on drying ⟨731⟩—Dry it at 105° for 4 hours: it loses not more than 0.1% of its weight.

Heavy metals, *Method II* ⟨231⟩: 0.001%.

Acid value ⟨401⟩—Weigh accurately 20 g into a conical flask, melt on a steam bath, add 100 mL of hot alcohol that previously has been neutralized with 0.1 *N* sodium hydroxide to phenolphthalein TS, swirl, and add 1 mL of phenolphthalein TS. Titrate with 0.10 *N* sodium hydroxide until the solution remains faintly pink after being shaken for 15 seconds: the acid value is not more than 2.0.

Iodine value ⟨401⟩: not more than 5.

Saponification value ⟨401⟩: between 188 and 198.

Unsaponifiable matter ⟨401⟩: not more than 0.8%.

Water for Injection—*see* Water for Injection USP
Water for Injection, Sterile—*see* Water for Injection, Sterile USP
Water for Irrigation, Sterile—*see* Water for Irrigation, Sterile USP
Water, Peppermint—*see* Peppermint Water
Water, Purified—*see* Water, Purified USP
Water, Stronger Rose—*see* Rose Water, Stronger

Carnauba Wax

» Carnauba Wax is obtained from the leaves of *Copernicia cerifera* Mart. (Fam. Palmae).

Packaging and storage—Preserve in well-closed containers.

Melting range, *Class II* ⟨741⟩: between 81° and 86°.

Residue on ignition ⟨281⟩—Heat 2 g in an open porcelain or platinum dish over a flame: it volatilizes without emitting an acrid odor. Ignite: the weight of the residue does not exceed 5 mg (0.25%).

Heavy metals, *Method II* ⟨231⟩: 0.004%.

Acid value—Weigh accurately 3 g into a 250-mL flask attached to a reflux condenser, add 50 mL of a mixture of isopropyl alcohol and toluene (5:4), and boil gently until the wax is completely dissolved. Remove the flask from the condenser, add about 1 mL of phenolphthalein TS, and titrate with 0.5 *N* alcoholic potassium hydroxide VS to a faint, reddish yellow color. Calculate the acid value as the number of mg of potassium hydroxide required to neutralize the free acids in 1 g of Carnauba Wax. The acid value so obtained is between 2 and 7.

Saponification value ⟨401⟩—To the solution from the test for *Acid value* add 15.0 mL of 0.5 *N* alcoholic potassium hydroxide VS, reflux for 3 hours, and titrate the excess alkali with 0.5 *N* hydrochloric acid VS to a yellow-amber color. Perform a blank determination (see *Residual Titrations* under *Titrimetry* ⟨541⟩). The saponification value is the summation of the ester value so obtained and the *Acid value*, and it is between 78 and 95.

Wax, Cetyl Esters—*see* Cetyl Esters Wax

Emulsifying Wax

» Emulsifying Wax is a waxy solid prepared from Cetostearyl Alcohol containing a polyoxyethylene derivative of a fatty acid ester of sorbitan.

Packaging and storage—Preserve in well-closed containers.

Melting range, *Class III* ⟨741⟩: between 48° and 52°.

pH ⟨791⟩: between 5.5 and 7.0, in a dispersion (3 in 100).

Hydroxyl value ⟨401⟩: between 178 and 192.

Iodine value ⟨401⟩: not more than 3.5.

Saponification value ⟨401⟩: not more than 14.

Microcrystalline Wax

» Microcrystalline Wax is a mixture of straight-chain, branched-chain, and cyclic hydrocarbons, obtained by solvent fractionation of the still bottom fraction of petroleum by suitable dewaxing or deoiling means.

Packaging and storage—Preserve in tight containers.

Labeling—Label it to indicate the name and proportion of any added stabilizer.

Color—Melt about 10 g on a steam bath, and pour 5 mL of the liquid into a clear-glass, 16- × 150-mm bacteriological test tube: the warm, melted liquid is not darker than a solution made by mixing 3.8 mL of ferric chloride CS and 1.2 mL of cobaltous chloride CS in a similar tube, the comparison of the two being made in reflected light against a white background, the tubes being held directly against the background at such an angle that there is no fluorescence.

Melting range, *Class III* ⟨741⟩: between 54° and 102°.

Consistency—

Apparatus—Determine the consistency of Microcrystalline Wax by means of a penetrometer fitted with a polished metal needle weighing 2.5 ± 0.05 g and having a truncated symmetric tapered angle of 9°10′ ± 15′. The needle is tapered, with a length of 25.4 mm, and the shaft attached to the needle is 58 mm in length and 3.17 mm in diameter. The plunger that fits into the penetrometer and guides the path of the needle weighs 47.5 ± 0.05 g. An additional weight of 50 ± 0.05 g is added to the top of the plunger to give a total load of 100 g.

Procedure—The wax specimen is cast in a brass cylinder open at both ends. The cylinder has an inside diameter of 25.4 mm and is 31.8 mm in height. Place the cylinder on a brass plate wetted with an equal volume mixture of glycerin and water, and place the plate on two corks. Pour the wax, melted at approximately 17° above its congealing point, into the cylinder. Continue pouring the wax until a convex meniscus is formed above the cylinder. Allow the specimen to cool for 1 hour at approximately 24°. Shave excess wax from the top of the cylinder and remove the plate. With the smooth wax surface in the up position, condition the specimen in a water bath at 25° for 1 hour.

Arrange the penetrometer so that the wax specimen is completely immersed in the water bath while penetration is run. Lower the needle until the tip just touches the top surface of the specimen. Release the needle for 5 seconds and read the depth of penetration in tenths of millimeters. Perform four determinations and calculate the average value of the four readings. The consistency value of Microcrystalline Wax is between 3 and 100 (not less than 0.3 mm and not more than 10.0 mm).

Acidity—If the addition of phenolphthalein TS in the test for *Alkalinity* produces no pink color, add 0.1 mL of methyl orange TS: no red or pink color is produced.

Alkalinity—Introduce 35 g into a 250-mL separator, add 100 mL of boiling water, and shake vigorously for 5 minutes. Draw off the separated water into a casserole, wash further with two 50-mL portions of boiling water, and add the washings to the casserole. To the pooled washings add 1 drop of phenolphthalein TS, and boil: the solution does not acquire a pink color.

Residue on ignition ⟨281⟩—Heat 2 g in an open porcelain or platinum dish over a flame: it volatilizes without emitting an acrid odor, and on ignition yields not more than 0.1% of residue.

Organic acids—To 20 g add 100 mL of a mixture of neutralized alcohol and water (1 in 2), agitate thoroughly, and heat to boiling. Add 1 mL of phenolphthalein TS, and titrate rapidly with 0.1 *N* sodium hydroxide VS, with vigorous agitation, to a sharp pink end-point in the alcohol-water layer: not more than 0.4 mL of 0.1 *N* sodium hydroxide is required.

Fixed oils, fats, and rosin—Digest 10 g with 50 mL of sodium hydroxide solution (1 in 5) at 100° for 30 minutes. Separate the water layer, and acidify it with 2 *N* sulfuric acid: no oily or solid matter separates.

White Wax

» White Wax is the product of bleaching and purifying Yellow Wax that is obtained from the honeycomb of the bee [*Apis mellifera* Linné (Fam. Apidae)] and that meets the requirements of the *Saponification cloud test.*

Packaging and storage—Preserve in well-closed containers.

Melting range, *Class II* ⟨741⟩: between 62° and 65°.

Saponification cloud test—Place 3.00 g in a round-bottom, 100-mL boiling flask fitted with a ground-glass joint. Add 30 mL of a solution prepared by dissolving 40 g of potassium hydroxide in about 900 mL of aldehyde-free alcohol maintained at a temperature not exceeding 15°, and then when solution is complete, warming to room temperature and adding aldehyde-free alcohol to make 1000 mL. Reflux the mixture gently for 2 hours. At the end of this period, open the flask, insert a thermometer into the solution, and place the flask in a container of water at a temperature of 80°. Rotate the flask in the bath while both the bath and the solution cool: the solution shows no cloudiness or globule formation before the temperature reaches 65°.

Fats or fatty acids, Japan wax, rosin, and soap—Boil 1 g for 30 minutes with 35 mL of 3.5 *N* sodium hydroxide contained in a 100-mL beaker, maintaining the volume by the occasional addition of water, and allow the mixture to cool at room temperature for about 2 hours: the wax separates, leaving the liquid clear, turbid, or translucent, but not opaque. Filter the cool mixture, and acidify the clear filtrate with hydrochloric acid: the liquid remains clear or shows not more than a slight amount of turbidity or precipitate.

Acid value ⟨401⟩—Warm about 3 g, accurately weighed, in a 200-mL flask with 25 mL of neutralized dehydrated alcohol until melted, shake the mixture, add 1 mL of phenolphthalein TS, and titrate the warm liquid with 0.5 *N* alcoholic potassium hydroxide

VS to produce a permanent, faint pink color: the acid value so obtained is between 17 and 24.

Ester value ⟨401⟩—To the solution resulting from the determination of *Acid value* add 25.0 mL of 0.5 *N* alcoholic potassium hydroxide VS and 50 mL of aldehyde-free alcohol, reflux the mixture for 4 hours, and titrate the excess alkali with 0.5 *N* hydrochloric acid VS. Perform a blank determination (see *Residual Titrations* under *Titrimetry* ⟨541⟩). The ester value so obtained is between 72 and 79.

Yellow Wax

» Yellow Wax is the purified wax from the honeycomb of the bee [*Apis mellifera* Linné (Fam. Apidae)].

NOTE—To meet the specifications of this monograph, the crude beeswax used to prepare Yellow Wax conforms to the *Saponification cloud test.*

Packaging and storage—Preserve in well-closed containers.

Other requirements—It meets the requirements for *Melting range, Saponification cloud test, Fats or fatty acids, Japan wax, rosin, and soap, Acid value,* and *Ester value* under *White Wax.*

White Ointment—*see* Ointment, White USP
White Petrolatum—*see* Petrolatum, White USP
White Wax—*see* Wax, White

Xanthan Gum

» Xanthan Gum is a high molecular weight polysaccharide gum produced by a pure-culture fermentation of a carbohydrate with *Xanthomonas campestris*, then purified by recovery with Isopropyl Alcohol, dried, and milled. It contains D-glucose and D-mannose as the dominant hexose units, along with D-glucuronic acid, and is prepared as the sodium, potassium, or calcium salt. It yields not less than 4.2 percent and not more than 5.0 percent of carbon dioxide, calculated on the dried basis, corresponding to not less than 91.0 percent and not more than 108.0 percent of Xanthan Gum.

Packaging and storage—Preserve in well-closed containers.

Identification—To 300 mL of water in a 400-mL beaker, previously heated to 80° and stirred rapidly with a mechanical stirrer, add, at the point of maximum agitation, a dry blend of 1.5 g of Xanthan Gum and 1.5 g of locust bean gum. Stir until the mixture dissolves, and then continue stirring for 30 minutes longer. Do not allow the temperature of the mixture to drop below 60° during the stirring. Discontinue stirring, and allow the mixture to cool at room temperature for not less than 2 hours: a firm, rubbery gel forms after the temperature drops below 40°, but no such gel forms in a control solution prepared in the same manner with 3.0 g of Xanthan Gum and without locust bean gum.

Viscosity ⟨911⟩—Place 250 mL of water in a 400-mL beaker, and add a dry blend of 3.0 g of Xanthan Gum and 3.0 g of potassium chloride slowly while stirring at 800 rpm, using a low-pitched propeller-type stirrer. Add an additional quantity of 44 mL of water, rinsing the walls of the beaker. Continue stirring for 2 hours, adjust the temperature to 24 ± 1°, vigorously stirring by hand in a vertical motion to eliminate any possibility of thixotropic effects or layering. Equip a suitable rotational viscosimeter with a spindle having a cylinder 1.27 cm in diameter and 0.16 cm high attached to a shaft 0.32 cm in diameter, the distance from the top of the cylinder to the lower tip of the shaft being 2.54 cm, and the immersion depth being 5.00 cm (No. 3 spindle). With the spindle rotating at 60 rpm, immediately observe and record the scale reading. While stirring, raise the temperature of the solution to 66 ± 1°, and measure the viscosity similarly. Convert the scale readings to centipoises by multiplying the readings by the constant for the viscosimeter spindle and speed employed. The viscosity at 24° is not less than 600 centipoises, and the ratio of the viscosity at 24° to that at 66° is between 1.02 and 1.45.

Microbial limits—It meets the requirements of the tests for *Salmonella* species and *Escherichia coli* under *Microbial Limit Tests* ⟨61⟩.

Loss on drying ⟨731⟩—Dry it at 105° for 2.5 hours: it loses not more than 15.0% of its weight.

Ash—Weigh accurately about 3 g in a tared crucible, and incinerate at about 650° until free from carbon. Cool the crucible and its contents in a desiccator, and weigh: the weight of the ash is between 6.5% and 16.0%, calculated on the dried basis.

Arsenic, *Method II* ⟨211⟩: 3 ppm.

Heavy metals, *Method II* ⟨231⟩—[NOTE—Use a platinum crucible for the ignition.] The limit is 0.003%.

Lead ⟨251⟩—Prepare a *Test Preparation* as directed for organic compounds, and use 5 mL of *Diluted Standard Lead Solution* (5 µg of Pb) for the test: the limit is 5 ppm.

Isopropyl alcohol—

Internal standard solution—Dissolve about 500 mg of tertiary butyl alcohol in about 500 mL of water, and mix.

Standard solution—Dissolve a suitable quantity of isopropyl alcohol, accurately weighed, in water to obtain a solution having a known concentration of about 1 mg of isopropyl alcohol per mL.

Standard preparation—Pipet 4 mL of the *Standard solution* and 4 mL of the *Internal standard solution* into a 100-mL volumetric flask, dilute with water to volume, and mix.

Test preparation—Disperse 1 mL of a suitable antifoam emulsion in 200 mL of water contained in a 1000-mL, round-bottom distilling flask having a 24/40 standard taper ground joint. Add about 5 g of Xanthan Gum, accurately weighed, and shake for 1 hour on a wrist-action mechanical shaker. Connect the flask to a fractionating column, and distil about 100 mL, adjusting the heat so that foam does not enter the column. Add by pipet 4 mL of *Internal standard solution*, and mix.

Chromatographic system (see *Chromatography* ⟨621⟩)—Under typical conditions, the gas chromatograph is equipped with a flame-ionization detector and contains a 1.8-m × 3.2-mm stainless steel column packed with 80- to 100-mesh surface silanized packing S3, or equivalent, the column is maintained at 165°, the injection port and detector block are maintained at 200°, and helium is used as the carrier gas.

Procedure—Inject separately equal volumes (about 4 to 5 µL) of the *Standard preparation* and the *Test preparation* into a suitable gas chromatograph. Record the chromatograms, and determine the peak responses of isopropyl alcohol and tertiary butyl alcohol in each chromatogram. The retention time of tertiary butyl alcohol is about 1.5 relative to that of isopropyl alcohol. Calculate the weight, in mg, of isopropyl alcohol in the quantity of Xanthan Gum taken by the formula:

$$4C(R_U / R_S),$$

in which C is the concentration, in mg per mL, of isopropyl alcohol in the *Standard solution*, and R_U and R_S are the ratios of the peak responses of the isopropyl alcohol peak to the tertiary butyl alcohol peak obtained from the *Test preparation* and the *Standard preparation*, respectively: not more than 0.075% is found.

Pyruvic acid—

Standard preparation—Transfer 45 mg of pyruvic acid, accurately weighed, to a 500-mL volumetric flask, dissolved in water, dilute with water to volume, and mix. Transfer 10.0 mL of this solution to a glass-stoppered, 50-mL flask, and proceed as directed under *Test preparation*, beginning with "Add 20.0 mL of 1 *N* hydrochloric acid."

Test preparation—Dissolve 600 mg of Xanthan Gum, accurately weighed, in water to make 100.0 mL, and transfer 10.0 mL of the solution to a glass-stoppered, 50-mL flask. Add 20.0

mL of 1 *N* hydrochloric acid, weigh the flask, and reflux for 3 hours, taking precautions to prevent loss of vapors. Cool, and add water to make up for any weight loss during refluxing. Transfer 2.0 mL of this solution to a 30-mL separator containing 1.0 mL of a 1 in 200 solution of 2,4-dinitrophenylhydrazine in 2 *N* hydrochloric acid, mix, and allow to stand for 5 minutes. Extract the mixture with 5 mL of ethyl acetate, and discard the aqueous layer. Extract the hydrazone from the ethyl acetate with three 5-mL portions of sodium carbonate TS, collect the extracts in a 50-mL volumetric flask, dilute with sodium carbonate TS to volume, and mix.

Procedure—Determine the absorbances of the solutions in 1-cm cells at the wavelength of maximum absorbance at about 375 nm, with a suitable spectrophotometer, using sodium carbonate TS as the blank. The absorbance of the *Test preparation* is not less than that of the *Standard preparation*, corresponding to not less than 1.5% of pyruvic acid.

Assay—Proceed with Xanthan Gum as directed for *Procedure* under *Alginates Assay* ⟨311⟩, using about 1.2 g of Xanthan Gum, accurately weighed.

Yellow Ointment—*see* Ointment, Yellow USP
Yellow Wax—*see* Wax, Yellow

Zein

» Zein is a prolamine derived from corn (*Zea mays* Linné [Fam. Gramineae]).

Packaging and storage—Preserve in tight containers.

Identification—

A: Dissolve about 0.1 g in 10 mL of 0.1 *N* sodium hydroxide, and add a few drops of cupric sulfate TS. Warm in a water bath: a purple color develops.

B: To a test tube containing 25 mg of Zein add 1 mL of nitric acid. Agitate vigorously: the solution becomes light yellow. Further addition of about 10 mL of 6 *N* ammonium hydroxide produces an orange color.

C: It is insoluble in water, but 1 g of it in 10 mL of a mixture of isopropyl alcohol and water (85:15), at a temperature of about 37°, yields a clear to cloudy solution.

Microbial limits ⟨61⟩—The total bacterial count does not exceed 1000 per g, and the tests for *Salmonella* species and *Escherichia coli* are negative.

Loss on drying ⟨731⟩—Dry it at 105° for 2 hours: it loses not more than 8.0% of its weight.

Residue on ignition ⟨281⟩: not more than 2.0%.

Arsenic, *Method II* ⟨211⟩: 3 ppm.

Heavy metals, *Method II* ⟨231⟩: 0.002%.

Nitrogen content—Proceed as directed under *Nitrogen Determination, Method I* ⟨461⟩: the nitrogen content is not less than 13.1% and not more than 17.0%, on the dried basis.

Zinc Stearate—*see* Zinc Stearate USP

Combined Index to USP XXII *and* NF XVII

Where an entry includes the cross-reference, "*see display listing*," this and related items are to be found in a group listing under that heading on the same page or a closely following page of this index. In the "General chapters" display listing, numbers in angle brackets, as ⟨421⟩, refer to chapter numbers in the General Chapters section.

A

Acid

Aerosol

Albumin

Alcohol

Anhydrous

B

Blood

Buffer

C

Calcium

Capsules

Capsules *continued*

Chromatography

Cream

D

Diluted

Dried

E

Elixir

Ether

Ethyl

Extract

F

G

Gel

General chapters

H

I

Inhalation

Injection

Injection *continued*

J

K

L

Lotion

M

N

Nasal solution

O

Oil

Ointment

Ophthalmic ointment

Ophthalmic ointment *continued*

Ophthalmic solution

Ophthalmic suspension

Oral solution

Oral suspension

P

Potassium

Potassium *continued*

Powder

Q

R

Radioactive pharmaceuticals

Reagents

Resin

S

Sodium

Sodium *continued*

Solution

Solution *continued*

Starch

Sterile

Sterile *continued*

Suppositories

Suspension

Suspension *continued*

Syrup

T

Tables

Tablets

Tablets *continued*

Tincture

Topical solution

U

V

Vaccine

Water

X

Y

Z

NOTES

NOTES

NOTES

NOTES

NOTES

NOTES

ATTACH
STAMP
HERE

U. S. Pharmacopeial Convention, Inc.
Order Processing Dept.
12601 Twinbrook Parkway
Rockville, MD 20852